Will Help You Succeed

INVESTIGATING**LIFE**

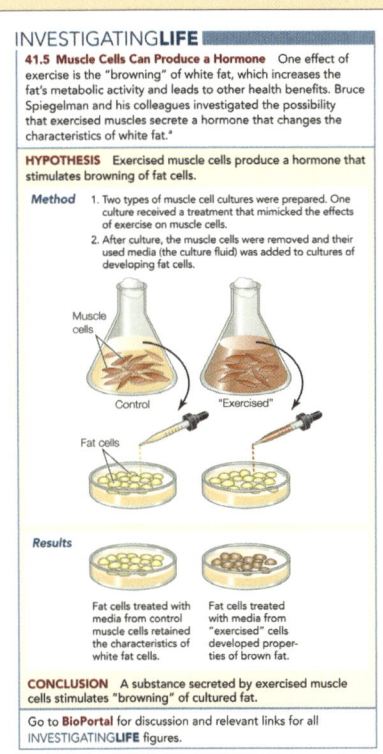

41.5 Muscle Cells Can Produce a Hormone One effect of exercise is the "browning" of white fat, which increases the fat's metabolic activity and leads to other health benefits. Bruce Spiegelman and his colleagues investigated the possibility that exercised muscles secrete a hormone that changes the characteristics of white fat.*

HYPOTHESIS Exercised muscle cells produce a hormone that stimulates browning of fat cells.

Method
1. Two types of muscle cell cultures were prepared. One culture received a treatment that mimicked the effects of exercise on muscle cells.
2. After culture, the muscle cells were removed and their used media (the culture fluid) was added to cultures of developing fat cells.

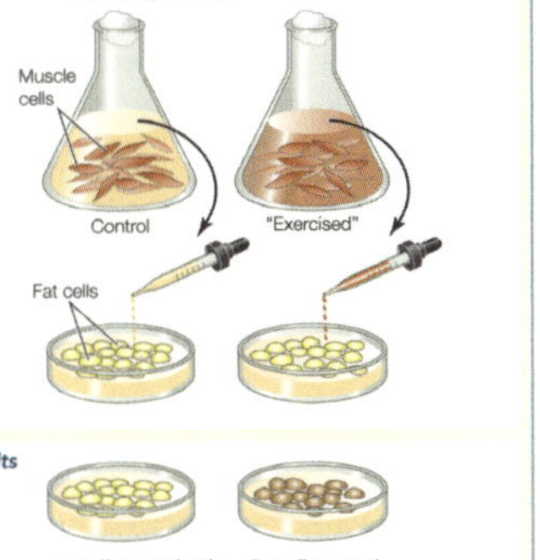

Muscle cells
Control
"Exercised"
Fat cells
Results

Fat cells treated with media from control muscle cells retained the characteristics of white fat cells.

Fat cells treated with media from "exercised" cells developed properties of brown fat.

CONCLUSION A substance secreted by exercised muscle cells stimulates "browning" of cultured fat.

Go to **BioPortal** for discussion and relevant links for all INVESTIGATING**LIFE** figures.

*Boström, P. et al. 2012. *Nature* 481: 463–469.

INVESTIGATING LIFE FIGURES emphasize the process of scientific inquiry and give a realistic sense of how science is done. Each Investigating Life figure is organized in order of hypothesis, method, results, and conclusion, and includes the original citation(s). BioPortal provides further discussion and links to the original papers, related research, and additional material.

RESEARCH TOOLS FIGURES introduce the important techniques and quantitative methods scientists use in the practice of biology.

WORKING WITH DATA EXERCISES involve you in answering questions based on a subset of data from some of the experiments presented in the textbook, helping you build quantitative and analytical skills. These exercises are available both in the textbook and online, in BioPortal.

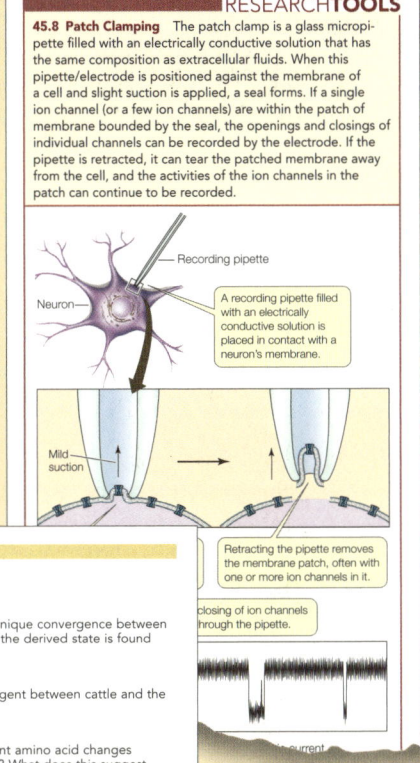

RESEARCH**TOOLS**

45.8 Patch Clamping The patch clamp is a glass micropipette filled with an electrically conductive solution that has the same composition as extracellular fluids. When this pipette/electrode is positioned against the membrane of a cell and slight suction is applied, a seal forms. If a single ion channel (or a few ion channels) are within the patch of membrane bounded by the seal, the openings and closings of individual channels can be recorded by the electrode. If the pipette is retracted, it can tear the patched membrane away from the cell, and the activities of the ion channels in the patch can continue to be recorded.

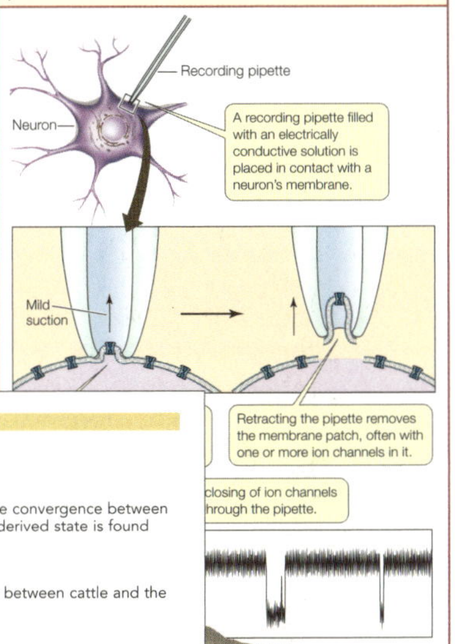

Recording pipette
Neuron
A recording pipette filled with an electrically conductive solution is placed in contact with a neuron's membrane.
Mild suction
Retracting the pipette removes the membrane patch, often with one or more ion channels in it.
closing of ion channels through the pipette.

WORKING WITH**DATA:**

Detecting Convergence in Lysozyme Sequences

Original Paper

Stewart, C.-B., J. W. Schilling, and A. C. Wilson. 1987. Adaptive evolution in the stomach lysozymes of foregut fermenters. *Nature* 330: 401–404.

Analyze the Data

Caro-Beth Stewart and her colleagues collected lysozyme sequences from six species of mammals. A small sample of their data is shown in the table. The phylogeny of these six species is well supported from analysis of many genes and much morphological data. Using the phylogenetic tree (Figure 24.7), plot the amino acid changes across th…

QUESTION 1

Which amino acid positions show unique convergence between the langur and cattle lineages (i.e., the derived state is found *only* in cattle and langurs)?

QUESTION 2

Which additional position is convergent between cattle and the ancestor of langurs and baboons?

QUESTION 3

Did you detect any other convergent amino acid changes between any other pair of lineages? What does this suggest … the convergent changes you observed between cattle …

BALLOON CAPTIONS guide you step-by-step through experiments and biological processes.

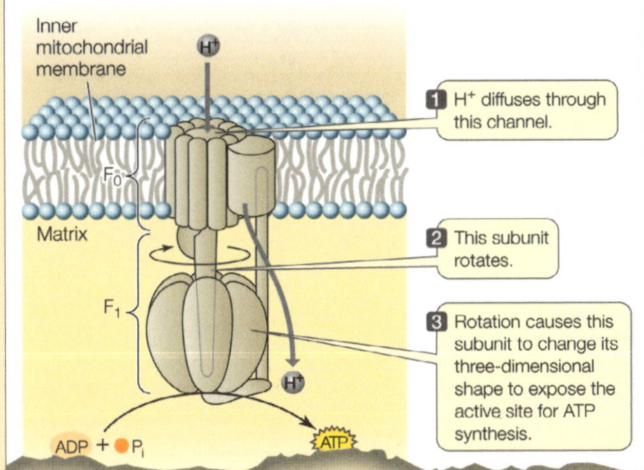

Inner mitochondrial membrane
H^+
F_0
Matrix
F_1
1 H^+ diffuses through this channel.
2 This subunit rotates.
3 Rotation causes this subunit to change its three-dimensional shape to expose the active site for ATP synthesis.
H^+
ADP + P_i
ATP

The real world

Our conclusion	Null hypothesis true (*not more females*)	Null hypothesis false (*more females*)
Null hypothesis true (*not more females*)	√	Type 2 error (*false negative*)
Null hypothesis false (*more females*)	Type 1 error (*false positive*)	√

FIGURE B10 Two Types of Error Possible outcomes of a statistical test. Statistical inference can result in correct and incorrect conclusions about the population of interest.

JUMPING TO THE WRONG CONCLUSIONS There are two ways that a statistical test can go wrong (**FIGURE B10**). We can reject the …

STATISTICS PRIMER APPENDIX introduces some of the basic statistical techniques used in biological research, including common techniques and methods of analyzing data (such as those used in some of the Working with Data exercises).

INSTANT ACCESS (QR) CODES AND DIRECT WEB LINKS link you directly to dynamic Animated Tutorials, Media Clips, and Interactive Summaries.

Go to Media Clip 13.1
Discovery of the Double Helix
Life10e.com/mc13.1

Wherever you see a QR code like this, scan it with your smartphone or tablet to jump directly to that media resource.

Never used QR codes before? To scan QR codes, all you need is a mobile device with a camera and a free QR scanning app. If you don't have a QR scanning app installed, simply go to your device's app store and search for "QR scanner."

Not using a smartphone or tablet? Just type the short Web address shown into your computer's browser.

NOTE: All of the resources referenced in the book are accessible directly via their QR codes and Web addresses, and they are all also available within BioPortal. (See the following page for more about BioPortal.)

BioPortal FOR LIFE TENTH EDITION

ONE-SITE ACCESS AT yourBioPortal.com

In one convenient location, BioPortal combines the complete eBook with all of the media resources, self-quizzes, and many other study tools that can help you succeed in the course. Your instructor may also use BioPortal to post online assignments, course documents, and other resources.

BioPortal Features...

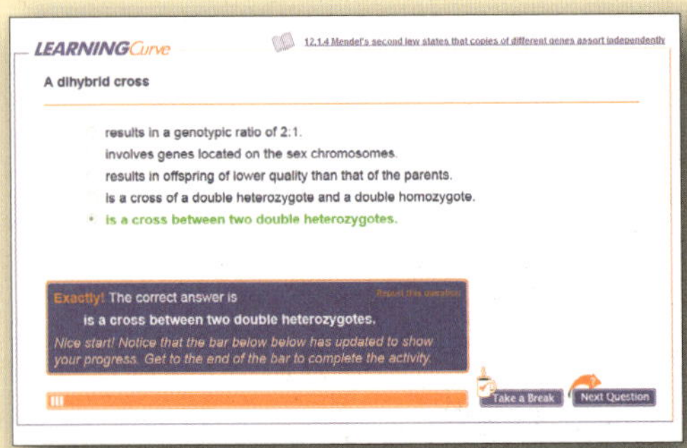

LEARNINGCurve
A dihybrid cross
results in a genotypic ratio of 2:1.
involves genes located on the sex chromosomes.
results in offspring of lower quality than that of the parents.
is a cross of a double heterozygote and a double homozygote.
is a cross between two double heterozygotes.
Exactly! The correct answer is
is a cross between two double heterozygotes.
Nice start! Notice that the bar below below has updated to show your progress. Get to the end of the bar to complete the activity.
Take a Break
Next Question

LEARNINGCURVE ▲

In an engaging, game-like format, LearningCurve activities offer a series of questions that adapt to your level of understanding for each topic. The more questions you answer correctly, the more advanced the questions become. All questions come with links to relevant eBook sections, to help you learn the material faster. The LearningCurve report helps you focus your study time where it's needed most.

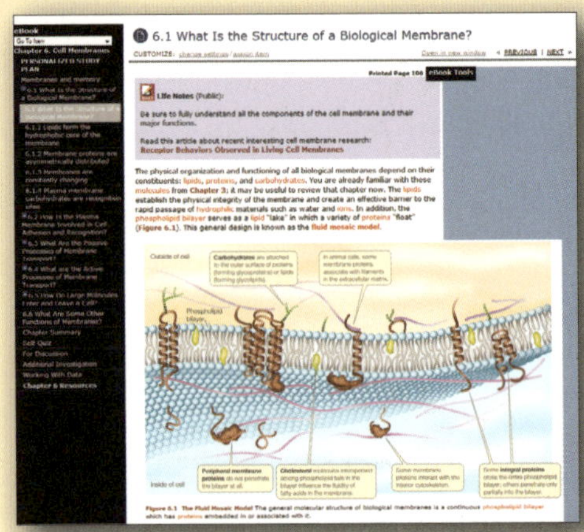

6.1 What Is the Structure of a Biological Membrane?

◀ INTERACTIVE eBOOK

The fully interactive version of *LIFE*, Tenth Edition integrates all student media resources into the text right where you need them. The eBook also includes helpful study tools such as highlighting, bookmarking, and note-taking.

DIAGNOSTIC QUIZZING

Take these quizzes to see how well you understand the material in each chapter. Feedback comes in the form of a "Personalized Study Plan," showing which answers you got right, and linking you to eBook sections and other media resources to help you with the topics you need to study further.

INTERACTIVE SUMMARIES

These reviews revisit the main points of each chapter, with links to relevant textbook figures, animated tutorials and activities.

ANIMATED TUTORIALS ▶

Animations present complex topics in a clear, easy-to-follow format. In BioPortal, these are followed by a brief quiz.

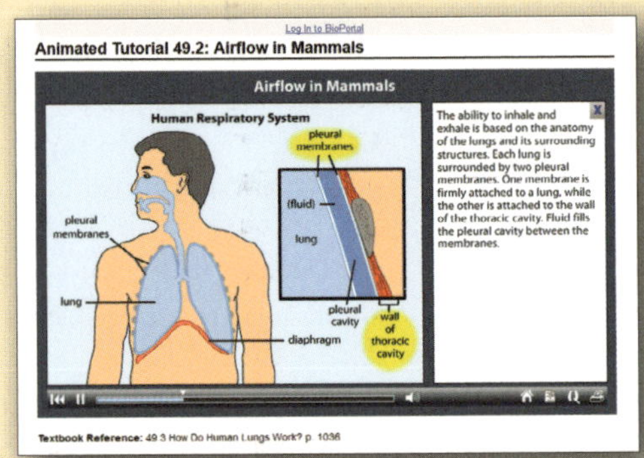

Animated Tutorial 49.2: Airflow in Mammals
Airflow in Mammals
Human Respiratory System
pleural membranes
lung
diaphragm
pleural cavity
wall of thoracic cavity
Textbook Reference: 49.5 How Do Human Lungs Work? p. 1036

MEDIA CLIPS

Compelling short videos bring concepts to life and show fascinating examples.

ACTIVITIES

Solidify your grasp of important facts and concepts through activities that have you label steps in processes, identify parts of structures, build diagrams, and distinguish different types of organisms.

BIONEWS FROM *SCIENTIFIC AMERICAN*

Keep up with the latest in science commentary, and other biology-related stories and images with this newsfeed. Content includes news updates, podcasts, magazine articles, science blog entries, "strange but true" stories, and more.

ALSO INCLUDED:

- **Working with Data Exercises:** Online versions of the exercises included in the textbook, which can be assigned by your instructor.
- **Investigating Life Links:** Links and additional information for all of the Investigating Life figures.
- **Flashcards & Key Terms:** Online flashcard sets ideal for learning and reviewing all the important terminology from each chapter.
- **Lecture Notebook:** PDFs of all the figures and tables from each chapter—great for taking notes in class.
- **Full Glossary:** Includes audio pronunciations for all terms.
- **Math for Life:** A primer on the math skills needed in the biology laboratory.
- **Survival Skills:** A guide to effective study habits and time-management strategies.
- **Tree of Life:** An online interactive version of the tree of life from the Appendix.

INVESTIGATING**LIFE**

INVESTIGATING**LIFE**

WORKING WITH**DATA**

WORKING WITH**DATA:**

RESEARCH**TOOLS**

RESEARCH**TOOLS**

LIFE

The Science of Biology

TENTH EDITION

DAVID
SADAVA
The Claremont Colleges

DAVID M.
HILLIS
University of Texas

H. CRAIG
HELLER
Stanford University

MAY R.
BERENBAUM
University of Illinois

SINAUER MACMILLAN

LIFE
The Science of Biology
TENTH EDITION
SADAVA
HILLIS
HELLER
BERENBAUM

THE COVER
The sea slug *Elysia crispata*. This animal is able to carry out photosynthesis using chloroplasts incorporated from the algae it feeds on (see back cover). Photograph © Alex Mustard/Naturepl.com.

THE FRONTISPIECE
Red-crowned cranes, *Grus japonensis*, gather on a river in Hokkaido, Japan. ©Steve Bloom Images/Alamy.

LIFE: The Science of Biology, Tenth Edition

ADDRESS EDITORIAL CORRESPONDENCE TO:
Sinauer Associates, Inc., 23 Plumtree Road, Sunderland, MA 01375 U.S.A.

www.sinauer.com
publish@sinauer.com

ADDRESS ORDERS TO:
MPS / W. H. Freeman & Co., Order Dept., 16365 James Madison Highway, U.S. Route 15, Gordonsville, VA 22942 U.S.A.

EXAMINATION COPY INFORMATION: 1-800-446-8923

SUSTAINABLE FORESTRY INITIATIVE
Certified Chain of Custody
Promoting Sustainable Forestry
www.sfiprogram.org
SFI-00712

GREEN EDITION
Planet Friendly Publishing
✔ Made in the United States
✔ Printed on Recycled Paper
Text: 10% Cover: 10%
Learn more: www.greenedition.org

Courier Corporation, the manufacturer
of this book, owns the *Green Edition* Trademark

Library of Congress Cataloging-in-Publication Data

Life : the science of biology / David Sadava ... [et al.]. -- 10th ed.
p. cm.
Includes bibliographical references and index.
ISBN 978-1-4292-9864-3 (casebound) — 978-1-4641-4122-5 (pbk. : v. 1) —
ISBN 978-1-4641-4123-2 (pbk. : v. 2) — ISBN 978-1-4641-4124-9 (pbk. : v. 3)
1. Biology--Textbooks. I. Sadava, David E.
QH308.2.L565 2013
570--dc23 2012039164

Printed in U.S.A.
First Printing December 2012
The Courier Companies, Inc.

To all the educators who have worked tirelessly
for quality biology education

The Authors

DAVID HILLIS MAY BERENBAUM CRAIG HELLER DAVID SADAVA

DAVID SADAVA is the Pritzker Family Foundation Professor of Biology, Emeritus at the Keck Science Center of Claremont McKenna, Pitzer, and Scripps, three of The Claremont Colleges. In addition, he is Adjunct Professor of Cancer Cell Biology at the City of Hope Medical Center in Duarte, California. Twice winner of the Huntoon Award for superior teaching, Dr. Sadava has taught courses on introductory biology, biotechnology, biochemistry, cell biology, molecular biology, plant biology, and cancer biology. In addition to *Life: The Science of Biology and Principles of Life*, he is the author or coauthor of books on cell biology and on plants, genes, and crop biotechnology. His research has resulted in many papers coauthored with his students, on topics ranging from plant biochemistry to pharmacology of narcotic analgesics to human genetic diseases. For the past 15 years, he has investigated multidrug resistance in human small-cell lung carcinoma cells with a view to understanding and overcoming this clinical challenge. At the City of Hope, his current work focuses on new anti-cancer agents from plants. He is the featured lecturer in "Understanding Genetics: DNA, Genes and their Real-World Applications," a video course for The Great Courses series.

DAVID M. HILLIS is the Alfred W. Roark Centennial Professor in Integrative Biology and the Director of the Dean's Scholars Program at the University of Texas at Austin, where he also has directed the School of Biological Sciences and the Center for Computational Biology and Bioinformatics. Dr. Hillis has taught courses in introductory biology, genetics, evolution, systematics, and biodiversity. He has been elected to the National Academy of Sciences and the American Academy of Arts and Sciences, awarded a John D. and Catherine T. MacArthur fellowship, and has served as President of the Society for the Study of Evolution and of the Society of Systematic Biologists. He served on the National Research Council committee that wrote the report *BIO 2010: Transforming Undergraduate Biology Education for Research Biologists*. His research interests span much of evolutionary biology, including experimental studies of viral evolution, empirical studies of natural molecular evolution, applications of phylogenetics, analyses of biodiversity, and evolutionary modeling. He is particularly interested in teaching and research about the practical applications of evolutionary biology.

H. CRAIG HELLER is the Lorry I. Lokey/Business Wire Professor in Biological Sciences and Human Biology at Stanford University. He has taught in the core biology courses at Stanford since 1972 and served as Director of the Program in Human Biology, Chairman of the Biological Sciences Department, and Associate Dean of Research. Dr. Heller is a fellow of the American Association for the Advancement of Science and a recipient of the Walter J. Gores Award for excellence in teaching and the Kenneth Cuthberson Award for Exceptional Service to Stanford University. His research is on the neurobiology of sleep and circadian rhythms, mammalian hibernation, the regulation of body temperature, the physiology of human performance, and the neurobiology of learning. He has done research on a huge variety of animals and physiological problems, including from sleeping kangaroo rats, diving seals, hibernating bears, photoperiodic hamsters, and exercising athletes. Dr. Heller has extended his enthusiasm for promoting active learning via the development of a two-year curriculum in human biology for the middle grades, through the production of Virtual Labs—interactive computer-based modules to teach physiology.

MAY BERENBAUM is the Swanlund Professor and Head of the Department of Entomology at the University of Illinois at Urbana-Champaign. She has taught courses in introductory animal biology, entomology, insect ecology, and chemical ecology and has received teaching awards at the regional and national levels from the Entomological Society of America. A fellow of the National Academy of Sciences, the American Academy of Arts and Sciences, and the American Philosophical Society, she served as President of the American Institute for Biological Sciences in 2009 and currently serves on the Board of Directors of AAAS. Her research addresses insect–plant coevolution and ranges from molecular mechanisms of detoxification to impacts of herbivory on community structure. Concerned with the practical application of ecological and evolutionary principles, she has examined impacts of genetic engineering, global climate change, and invasive species on natural and agricultural ecosystems. In recognition of her work, she received the 2011 Tyler Prize for Environmental Achievement. Devoted to fostering science literacy, she has published numerous articles and five books on insects for the general public.

Contents in Brief

Preface

Biology is a constantly changing scientific field. New discoveries about the living world are being made every day, and more than 1 million new research articles in biology are published each year. Beyond the constant need to update the concepts and facts presented in any science textbook, in recent years ideas about how best to educate the upcoming generation of biologists have undergone dynamic and exciting change.

Although we and many of our colleagues had thought about the nature of biological education as individuals, it is only recently that biologists have come together to discuss these issues. Reports from the National Academy of Sciences, Howard Hughes Medical Institute, and College Board AP Biology Program not only express concern about how best to instruct undergraduates in biology, but offer concrete suggestions about how to design the introductory biology course—and by extension, our book. We have followed these discussions closely and have been especially impressed with the report "Vision and Change in Undergraduate Biology Education" (visionandchange.org). As participants in the educational enterprise, we have answered the report's call to action with this textbook and its associated ancillary materials.

The "Vision and Change" report proposes five core concepts for biological literacy:

1. Evolution
2. Structure and function
3. Information flow, exchange, and storage
4. Pathways and transformations of energy and matter
5. Systems

These five concepts have always been recurring themes in Life, but in this Tenth Edition we have brought them even more "front and center."

"Vision and Change" also advocates that students learn and demonstrate core competencies, including the ability to apply the process of science using quantitative reasoning. Life has always emphasized the experimental nature of biology. This edition responds further to these core competency issues with a new working with data feature and the addition of a statistics primer (Appendix B). The authors' multiple educational perspectives and areas of expertise, as well as input from many colleagues and students who used previous editions, have informed the approach to this new edition.

Enduring Features

We remain committed to blending the presentation of core ideas with an emphasis on introducing students to the process of scientific inquiry. Having pioneered the idea of depicting important experiments in unique figures designed to help students understand and appreciate the way scientific investigations work, we continue to develop this approach in the book's 70 **Investigating Life** figures. Each of these figures sets the experiment in perspective and relates it to the accompanying text. As in previous editions, these figures employ the structure Hypothesis, Method, Results, and Conclusion. We have added new information focusing on the individuals who performed these experiments so students can appreciate more fully that science is a human and very personal activity. Each Investigating Life figure has a reference to BioPortal (*yourBioPortal.com*), where discussion and references to follow-up research can be found. A related feature is the **Research Tools** figures, which depict laboratory and field methods used in biology. These, too, have been expanded to provide more useful context for their importance.

Some 15 years ago, *Life's* authors and publishers pioneered the use of **balloon captions** in our figures. We recognized then that many students are visual learners, and this fact is even truer today. *Life*'s balloon captions bring the crucial explanations of intricate, complex processes directly into the illustration, allowing students to integrate information without repeatedly going back and forth between the figure, its legend, and the text.

We continue to refine our chapter organization. Our **opening stories** have always provide historical, medical, or social context to intrigue students and show how the subject of each chapter relates to the world around them. In the Tenth Edition, the opening stories all end with a question that is revisited throughout the chapter. At the end of each chapter the answer is presented in the light of material the student encountered in the body of the chapter.

A **chapter outline** asks questions to emphasize scientific inquiry, each of which is answered in a major section of the chapter. A **Recap** summarizes each section's key concepts and poses questions that help the student review and test their mastery of these concepts. The recap questions are similar in form to the learning objectives used in many introductory biology courses. The **Chapter Summaries** highlight each chapter's key figures and defined terms, while restating the major concepts presented in the chapter in a concise and student-friendly manner, with references to specific figures and to the activities and animated tutorials available in BioPortal.

At the end of the book, students will find a much-expanded glossary that continues *Life's* practice of providing Latin or Greek derivations for many of the defined terms. As students become gradually (and painlessly) more familiar with such root words, the mastery of vocabulary as they continue in their biological or medical studies will be easier. In addition, the popular **Tree of Life appendix** (Appendix A) presents the phylogenetic tree of life as a reference tool that allows students to place any group of organisms mentioned in the text into the context of the rest of life. The web-based version of Appendix A provides links to photos, keys, species lists, distribution maps, and other information (via the online database at DiscoverLife.org) to help students explore biodiversity in greater detail.

New Features

The Tenth Edition of *Life* has a different look and feel from its predecessors. The new color palette and more open design will, we hope, be more accessible to students. And, in keeping with our heightened emphasis on scientific inquiry and quantitative analysis, we have added **Working with Data** exercises to almost all chapters. In these innovative exercises, we describe the context and approach of a research paper that provides the basis of the analysis. We then ask questions that require students to analyze data, make calculations, and draw conclusions. Answers (or suggested possible answers) to these questions are included in BioPortal and can be made available to students at the instructor's discretion.

Because many of the questions in the Working with Data exercises require the use of basic statistical methods, we have included a **Statistics Primer** as the book's Appendix B, describing the concepts and some methods of statistical analysis. We hope that the Working with Data exercises and statistics primer will reinforce students' skills and their ability to apply quantitative analysis to biology.

We have added links to **Media Clips** in the body of the text, with at least one per chapter. These brief clips are intended to enlighten and entertain. Recognizing the widespread use of "smart phones" by students, the textbook includes **instant access (QR) codes** that bring the Media Clips, Animated Tutorials, and Interactive Summaries directly to the screen in your hand. If you do not have a smart phone, never fear, we also provide direct web addresses to these features.

As educators, we follow current discussions of pedagogy in biological education. The chapter-ending **Chapter Reviews** now contain multiple levels of questions based on Bloom's taxonomy: Remembering, Understanding and Applying, and Analyzing and Evaluating. Answers to these questions appear at the end of the book.

For a detailed description of the media and supplements available for the Tenth Edition, please turn to "*Life's* Media and Supplements" on page xvii.

The Ten Parts

PART ONE, THE SCIENCE OF LIFE AND ITS CHEMICAL BASIS Chapter 1 introduces the core concepts set forth in the "Vision and Change" report and continues the much-praised approach of focusing on a specific series of experiments that introduces students to biology as an experimentally based and constantly expanding science. Chapter 1 emphasizes the principles of biology that are the foundation for the rest of the book, including the unity of life at the cellular level and how evolution unites the living world. Chapters 2–4 cover the chemical principles and building blocks that underlie life. Chapter 4 also includes a discussion of how life could have evolved from inanimate chemicals.

PART TWO, CELLS The nature of cells and their role as the structural and functional basis of life is foundational to biology. These revised chapters include expanded explanations of how experimental manipulations of living systems have been used to discover cause and effect in biology. Students who are intrigued by the question "Where did the first cells come from?" will appreciate the updated discussion of ideas on the origin of cells and organelles, as well as expanded discussion of the evolution of multicellularity and cell interactions. In response to reviewer comments, the discussion of membrane potential has been moved to Chapter 45, where students may find it to be more relevant.

PART THREE, CELLS AND ENERGY The biochemistry of life and energy transformations are among the most challenging topics for many students. We have worked to clarify such concepts as enzyme inhibition, allosteric enzymes, and the integration of biochemical systems. Revised presentations of glycolysis and the citric acid cycle now focus, in both text and figures, on key concepts and attempt to limit excessive detail. There are also revised discussions of the ecological roles of alternate pathways of photosynthetic carbon fixation, as well as the roles of accessory pigments and reaction center in photosynthesis.

PART FOUR, GENES AND HEREDITY This crucial section of the book is revised to improve clarity, link related concepts, and provide updates from recent research results. Rather than being segregated into separate chapters, material on prokaryotic genetics and molecular medicine are now interwoven into relevant chapters. Chapter 11 on the cell cycle includes a new discussion of how the mechanisms of cell division are altered in cancer cells. Chapter 12 on transmission genetics now includes coverage of this phenomenon in prokaryotes. Chapters 13 and 14 cover gene expression and gene regulation, including new discoveries about the roles of RNA and an expanded discussion of epigenetics. Chapter 15 covers the subject of gene mutations and describes updated applications of medical genetics.

PART FIVE, GENOMES This extensive and up-to-date coverage of genomes expands and reinforces the concepts covered in Part Four. The first chapter of Part Five describes how genomes are analyzed and what they tell us about the biology of prokaryotes and eukaryotes, including humans. Methods of DNA sequencing and genome analysis, familiar to many students in a general way, are rapidly improving, and we discuss these advances as well as how bioinformatics is used. This leads to a chapter describing how our knowledge of molecular biology and genetics underpins biotechnology—the application of this knowledge to practical problems and issues such as stem cell research. Part Five closes with a unique sequence of two chapters that explore the interface of developmental processes with molecular biology (Chapter 19) and with evolution (Chapter 20), providing students with a link between these two crucial topics and a bridge to Part Six.

PART SIX, THE PATTERNS AND PROCESSES OF EVOLUTION Many students come to the introductory biology course with ideas about evolution already firmly in place. One common view, that evolution is only about Darwin, is firmly put to rest at the start of Chapter 21, which not only illustrates the practical value of fully understanding modern evolutionary biology, but briefly and succinctly traces the history of "Darwin's dangerous idea" through the twentieth century and up to the present syntheses of molecular evolutionary genetics and evolutionary developmental biology—fields of study that uphold and support the principles of evolutionary biology as the basis for comparing and comprehending all other aspects of biology. The remaining sections of Chapter 21 describe the mechanisms of evolution in clear, matter-of-fact terms. Chapter 22 describes phylogenetic trees as a tool not only of classification but also of evolutionary inquiry. The remaining chapters cover speciation and molecular evolution, concluding with an overview of the evolutionary history of life on Earth.

PART SEVEN, THE EVOLUTION OF DIVERSITY Continuing the theme of how evolution has shaped our world, Part Seven introduces the latest views on biodiversity and the evolutionary relationships among organisms. The chapters have been revised with the aim of making it easier for students to appreciate the major evolutionary changes that have taken place within the different groups of organisms. These chapters emphasize understanding the big picture of organismal diversity—the tree of life—as opposed to memorizing a taxonomic hierarchy and names. Throughout the book, the tree of life is emphasized as a way of understanding and organizing biological information.

PART EIGHT, FLOWERING PLANTS: FORM AND FUNCTION The emphasis of this modern approach to plant form and function is not only on the basic findings that led to the elucidation of mechanisms for plant growth and reproduction, but also on the use of genetics of model organisms. In response to users of earlier editions, material covering recent discoveries in plant molecular biology and signaling has been reorganized and streamlined to make it more accessible to students. There are also expanded and clearer explanations of such topics as water relations, the plant body plan, and gamete formation and double fertilization.

PART NINE, ANIMALS: FORM AND FUNCTION This overview of animal physiology begins with a sequence of chapters covering the systems of information—endocrine, immune, and neural. Learning about these information systems provides important groundwork and explains the processes of control and regulation that affect and integrate the individual physiological systems covered in the remaining chapters of the Part. Chapter 45, "Neurons and Nervous Systems," has been rearranged and contains descriptions of exciting new discoveries about glial cells and their role in the vertebrate nervous system. The organization of several other chapters has been revised to reflect recent findings and to allow the student to more readily identify the most important concepts to be mastered.

PART TEN, ECOLOGY Part Ten continues *Life*'s commitment to presenting the experimental and quantitative aspects of biology, with increased emphasis on how ecologists design and conduct experiments. New exercises provide opportunities for students to see how ecological data are acquired in the laboratory and in the field, how these data are analyzed, and how the results are applied to answer questions. There is also an expanded discussion of aquatic biomes and a more synthetic explanation of how aquatic, terrestrial, and atmospheric components integrate to influence the distribution and abundance of life on Earth. In addition there is an expanded emphasis on examples of successful strategies proposed by ecologists to mitigate human impacts on the environment; rather than an inventory of ways human activity adversely affects natural systems, this revised Tenth Edition provides more examples of ways that ecological principles can be applied to increase the sustainability of these systems.

Exceptional Value Formats

We again provide *Life* both as the full book and as a set of paperback volumes. Thus, instructors who want to use less than the whole book can choose from these split volumes, each of which contains the book's front matter, appendices, glossary, and index.

- Volume I, *The Cell and Heredity*, includes: Part One, The Science of Life and Its Chemical Basis (Chapters 1–4); Part Two, Cells (Chapters 5–7); Part Three, Cells and Energy (Chapters 8–10); Part Four, Genes and Heredity (Chapters 11–16); and Part Five, Genomes (Chapters 17–20).
- Volume II, *Evolution, Diversity, and Ecology*, includes: Chapter 1, Studying Life; Part Six, The Patterns and Processes of Evolution (Chapters 21–25); Part Seven, The Evolution of Diversity (Chapters 26–33); and Part Ten, Ecology (Chapters 54–59).
- Volume III, *Plants and Animals*, includes: Chapter 1, Studying Life; Part Eight, Flowering Plants: Form and Function (Chapters 34–39); and Part Nine, Animals: Form and Function (Chapters 40–53).

Responding to student concerns, there also are two ways to obtain the entire book at a significantly reduced cost. The looseleaf edition of *Life* is a shrink-wrapped, unbound, three-hole-punched version that fits into a three-ring binder. Students take only what they need to class and can easily integrate instructor handouts and other resources.

Life was the first comprehensive biology text to offer the entire book as a truly robust eBook, and we offer the Tenth Edition in this flexible, interactive format that gives students a different way to read the text and learn the material. The eBook integrates student media resources (animations, activities, interactive summaries, and quizzes) and offers instructors a powerful way to customize the textbook with their own text, images, web links, and, in BioPortal, quizzes, and other materials.

We are proud that our print edition is a greener *Life* that minimizes environmental impact. *Life* was the first introductory biology text to be printed on paper earning the Forest Stewardship Council label, the "gold standard in green paper," and it continues to be manufactured from wood harvested from sustainable forests.

Many People to Thank

One of the wisest pieces of advice ever given to a textbook author is to "be passionate about your subject, but don't put your ego on the page." Considering all the people who looked over our shoulders throughout the process of creating this book, this advice could not be more apt. We are indebted to the many people who help to make this book what it is. First and foremost among these are our colleagues, biologists from over 100 institutions. Before we set pen to paper, we solicited the advice of users of *Life*'s Ninth Edition, as well as users of other books. These reviewers gave detailed suggestions for improvements. Other colleagues acted as reviewers when the book was almost completed, pointing out inaccuracies or lack of clarity. All of these biologists are listed in the reviewer credits, along with the dozens who reviewed all of the revised assessment resources.

Once we began writing, we had the superb advice of a team of experienced, knowledgeable, and patient biologists working as development and line editors. Laura Green of Sinauer Associates headed the team and coordinated her own fine work with that of Jane Murfett, Norma Roche, and Liz Pierson to produce a polished and professional text. We are especially indebted to Laura for her work on the important Investigating Life and new Working with Data elements. For the tenth time in ten editions, Carol Wigg oversaw the editorial process. Her positive influence pervades the entire book. Artist Elizabeth Morales again translated our crude sketches into beautiful new illustrations. We hope you agree that our art program remains superbly clear and elegant. Johannah Walkowicz effectively coordinated the hundreds of reviews described above. David McIntyre, photo editor extraordinaire, researched and provided us with new photographs, including many of his own, to enrich the book's content and visual statement. Joanne Delphia is responsible for the crisp new design and layout that make this edition of *Life* not just clear and readable but beautiful as well. Christopher Small headed Sinauer's production team and contributed in innumerable ways to bringing *Life* to its final form. Jason Dirks coordinated the creation of our array of media and instructor resources, with Mary Tyler, Mitch Walkowicz, and Carolyn Wetzel serving as editors for our expanded assessment supplements.

W. H. Freeman continues to bring *Life* to a wider audience. Associate Director of Marketing Debbie Clare, the regional specialists, regional managers, and experienced sales force are effective ambassadors and skillful transmitters of the features and unique strengths of our book. We depend on their expertise and energy to keep us in touch with how *Life* is perceived by its users. Thanks also to the Freeman media group for eBook and BioPortal production.

Finally, we thank our friend Andy Sinauer. Like ours, his name is on the cover of the book, and he truly cares deeply about what goes into it.

DAVID SADAVA

DAVID HILLIS

CRAIG HELLER

MAY BERENBAUM

Reviewers for the Tenth Edition

Between Edition Reviewers

Shivanthi Anandan, Drexel University
Brian Bagatto, The University of Akron
Mary Bisson, University at Buffalo, The State University of New York
Meredith Blackwell, Louisiana State University
Randy Brooks, Florida Atlantic University
Heather Caldwell, Kent State University
Jeffrey Carrier, Albion College
David Champlin, University of Southern Maine
Wesley Colgan, Pikes Peak Community College
Emma Creaser, Unity College
Karen Curto, University of Pittsburgh
John Dennehy, Queens College, The City University of New York
Rajinder Dhindsa, McGill University
James A. Doyle, University of California, Davis
Scott Edwards, Harvard University
David Eldridge, Baylor University
Joanne Ellzey, The University of Texas at El Paso
Douglas Gayou, University of Missouri
Stephen Gehnrich, Salisbury University
Arundhati Ghosh, University of Pittsburgh
Nathalia Glickman Holtzman, Queens College, The City University of New York
Elizabeth Good, University of Illinois at Urbana-Champaign
Harry Greene, Cornell University
Alice Heicklen, Columbia University
Albert Herrera, University of Southern California
David Hibbett, Clark University
Mark Holbrook, University of Iowa
Craig Jordan, The University of Texas at San Antonio
Walter Judd, University of Florida
John M. Labavitch, University of California, Davis
Nathan H. Lents, John Jay College of Criminal Justice, The City University of New York
Barry Logan, Bowdoin College
Barbara Lom, Davidson College
David Low, University of California, Davis
Janet Loxterman, Idaho State University
Sharon Lynn, The College of Wooster
Julin Maloof, University of California, Davis
Richard McCarty, Johns Hopkins University
Sheila McCormick, University of California, Berkeley
Marcie Moehnke, Baylor University
Roberta Moldow, Seton Hall University
Tsafrir Mor, Arizona State University
Alexander Motten, Duke University
Barbara Musolf, Clayton State University
Stuart Newfeld, Arizona State University
Bruce Ostrow, Grand Valley State University
Laura K. Palmer, The Pennsylvania State University, Altoona
Robert Pennock, Michigan State University
Kamini Persaud, University of Toronto, Scarborough
Roger Persell, Hunter College, The City University of New York
Matthew Rand, Carleton College
Susan Richardson, Florida Atlantic University
Brian C. Ring, Valdosta State University
Jay Rosenheim, University of California, Davis
Ben Rowley, University of Central Arkansas
Ann Rushing, Baylor University
Mikal Saltveit, University of California, Davis
Joel Schildbach, Johns Hopkins University
Christopher J. Schneider, Boston University
Paul Schulte, University of Nevada, Las Vegas
Leah Sheridan, University of Northern Colorado
Gary Shin, University of California, Los Angeles
Mitchell Singer, University of California, Davis
William Taylor, The University of Toledo
Sharon Thoma, University of Wisconsin, Madison
James F. A. Traniello, Boston University
Terry Trier, Grand Valley State University
Sara Via, University of Maryland
Curt Walker, Dixie State College
Fred Wasserman, Boston University
Alexander J. Werth, Hampden-Sydney College
Elizabeth Willott, University of Arizona

Accuracy Reviewers

Rebecca Rashid Achterman, Western Washington University
Maria Ambrosetti, Emory University
Miriam Ashley-Ross, Wake Forest University
Felicitas Avendaño, Grand View University
David Bailey, St. Norbert College
Chhandak Basu, California State University, Northridge
Jim Bednarz, Arkansas State University
Charlie Garnett Benson, Georgia State University
Katherine Boss-Williams, Emory University
Ben Brammell, Asbury University

Christopher I. Brandon, Jr., Georgia Gwinnett College
Carolyn J. W. Bunde, Idaho State University
Darlene Campbell, Cornell University
Jeffrey Carmichael, University of North Dakota
David J. Carroll, Florida Institute of Technology
Ethan Carver, The University of Tennessee at Chattanooga
Peter Chabora, Queens College, The City University of New York
Heather Cook, Wagner College
Hsini Lin Cox, The University of Texas at El Paso
Douglas Darnowski, Indiana University Southeast
Stephen Devoto, Wesleyan University
Rajinder Dhindsa, McGill University
Jesse Dillon, California State University, Long Beach
James A. Doyle, University of California, Davis
Devin Drown, Indiana University
Richard E. Duhrkopf, Baylor University
Weston Dulaney, Nashville State Community College
David Eldridge, Baylor University
Kenneth Filchak, University of Notre Dame
Kerry Finlay, University of Regina
Kevin Folta, University of Florida
Douglas Gayou, University of Missouri
David T. Glover, Food and Drug Administration
Russ Goddard, Valdosta State University
Elizabeth Godrick, Boston University
Leslie Goertzen, Auburn University
Elizabeth Good, University of Illinois at Urbana-Champaign
Ethan Graf, Amherst College
Eileen Gregory, Rollins College
Julie C. Hagelin, University of Alaska, Fairbanks
Nathalia Glickman Holtzman, Queens College, The City University of New York
Dianne Jennings, Virginia Commonwealth University
Jamie Jensen, Bringham Young University
Glennis E. Julian
Erin Keen-Rhinehart, Susquehanna University
Henrik Kibak, California State University, Monterey Bay
Brandi Brandon Knight, Emory University
Daniel Kueh, Emory University
John G. Latto, University of California, Santa Barbara
Kristen Lennon, Frostburg State University
David Low, University of California, Santa Barbara
Jose-Luis Machado, Swarthmore College
Jay Mager, Ohio Northern University
Stevan Marcus, University of Alabama
Nilo Marin, Broward College
Marlee Marsh, Columbia College South Carolina
Erin Martin, University of South Florida, Sarasota-Manatee
Brad Mehrtens, University of Illinois at Urbana-Champaign
Michael Meighan, University of California, Berkeley
Tsafrir Mor, Arizona State University
Roderick Morgan, Grand Valley State University
Jacalyn Newman, University of Pittsburgh
Alexey Nikitin, Grand Valley State University
Zia Nisani, Antelope Valley College
Laura K. Palmer, The Pennsylvania State University, Altoona
Nancy Pencoe, State University of West Georgia
David P. Puthoff, Frostburg State University
Brett Riddle, University of Nevada, Las Vegas
Leslie Riley, Ohio Northern University
Brian C. Ring, Valdosta State University
Heather Roffey, McGill University
Lori Rose, Hill College
Naomi Rowland, Western Kentucky University
Beth Rueschhoff, Indiana University Southeast
Ann Rushing, Baylor University
Illya Ruvinsky, University of Chicago
Paul Schulte, University of Nevada, Las Vegas
Susan Sharbaugh, University of Alaska, Fairbanks
Jonathan Shenker, Florida Institute of Technology
Gary Shin, California State University, Long Beach
Ken Spitze, University of West Georgia
Bruce Stallsmith, The University of Alabama in Huntsville
Robert M. Steven, The University of Toledo
Zuzana Swigonova, University of Pittsburgh
Rebecca Symula, The University of Mississippi
Mark Taylor, Baylor University
Mark Thogerson, Grand Valley State University
Elethia Tillman, Spelman College
Terry Trier, Grand Valley State University
Michael Troyan, The Pennsylvania State University, University Park
Sebastian Velez, Worcester State University
Sheela Vemu, Northern Illinois University
Andrea Ward, Adelphi University
Katherine Warpeha, University of Illinois at Chicago
Fred Wasserman, Boston University
Michelle Wien, Bryn Mawr College
Robert Wisotzkey, California State University, East Bay
Greg Wray, Duke University
Joanna Wysocka-Diller, Auburn University
Catherine Young, Ohio Northern University
Heping Zhou, Seton Hall University

Assessment Reviewers

Maria Ambrosetti, Georgia State University
Cecile Andraos-Selim, Hampton University
Felicitas Avendaño, Grand View University
David Bailey, St. Norbert College
Jim Bednarz, Arkansas State University
Charlie Garnett Benson, Georgia State University
Katherine Boss-Williams, Emory University
Ben Brammell, Asbury University
Christopher I. Brandon, Jr., Georgia Gwinnett College
Brandi Brandon Knight, Emory University

Ethan Carver, The University of Tennessee, Chattanooga
Heather Cook, Wagner College
Hsini Lin Cox, The University of Texas at El Paso
Douglas Darnowski, Indiana University Southeast
Jesse Dillon, California State University, Long Beach
Devin Drown, Indiana University
Richard E. Duhrkopf, Baylor University
Weston Dulaney, Nashville State Community College
Kenneth Filchak, University of Notre Dame
Elizabeth Godrick, Boston University
Elizabeth Good, University of Illinois at Urbana-Champaign
Susan Hengeveld, Indiana University Bloomington
Nathalia Glickman Holtzman, Queens College, The City College of New York
Glennis E. Julian
Erin Keen-Rhinehart, Susquehanna University
Stephen Kilpatrick, University of Pittsburgh
Daniel Kueh, Emory University
Stevan Marcus, University of Alabama
Nilo Marin, Broward College
Marlee Marsh, Columbia College
Erin Martin, University of South Florida, Sarasota-Manatee
Brad Mehrtens, University of Illinois at Urbana-Champaign
Darlene Mitrano, Christopher Newport University
Anthony Moss, Auburn University
Jacalyn Newman, University of Pittsburgh
Alexey Nikitin, Grand Valley State University
Zia Nisani, Antelope Valley College
Sabiha Rahman, University of Ottawa
Nancy Rice, Western Kentucky University
Brian C. Ring, Valdosta State University
Naomi Rowland, Western Kentucky University
Jonathan Shenker, Florida Institute of Technology
Gary Shin, California State University, Long Beach
Jacob Shreckengost, Emory University
Michael Smith, Western Kentucky University
Ken Spitze, University of West Georgia
Bruce Stallsmith, The University of Alabama in Huntsville
Zuzana Swigonova, University of Pittsburgh
William Taylor, The University of Toledo
Mark Thogerson, Grand Valley State University
Elethia Tillman, Spelman College
Michael Troyan, The Pennsylvania State University
Ximena Valderrama, Ramapo College of New Jersey
Sheela Vemu, Northern Illinois University
Suzanne Wakim, Butte College
Katherine Warpeha, University of Illinois at Chicago
Fred Wasserman, Boston University
Michelle Wien, Bryn Mawr College
Robert Wisotzkey, California State University, East Bay
Heping Zhou, Seton Hall University

LIFE's Media and Supplements

BIOPORTAL

yourBioPortal.com

BioPortal is the online gateway to all of *Life*'s digital resources, including the fully interactive eBook, a wide range of student and instructor media resources, and powerful assessment tools. BioPortal includes the following features and resources:

Life, Tenth Edition eBook (eBook also available stand-alone)

- Complete online version of the textbook
- Integration of all Media Clips, Activities, Animated Tutorials, and other media resources
- In-text links to all glossary entries, with audio pronunciations
- A flexible notes feature and easy text highlighting
- Searchable glossary and index
- Full-text search

Additional eBook features for instructors:

- *Content Customization*: Instructors can easily hide chapters or sections that they don't cover in their course, re-arrange the order of chapters and sections, and add their own content directly into the eBook.
- *Instructor Notes*: Instructors can annotate the eBook with their own notes and content on any page. Instructor notes can include text, Web links, images, links to BioPortal resources, uploaded documents, and more.

LearningCurve

New for the Tenth Edition, LearningCurve is a powerful adaptive quizzing system with a game-like format that engages students. Rather than simply answering a fixed set of questions, students answer dynamically-selected questions to progress toward a target level of understanding. At any point, students can view a report of how well they are performing in each topic area (with links to eBook sections and media resources), to help them focus on problem areas.

Student BioPortal Resources

DIAGNOSTIC QUIZZING. The pre-built diagnostic quizzes assesses student understanding of each section of each chapter, and generates a Personalized Study Plan to effectively focus student study time. The plan includes links to specific textbook sections, animated tutorials, and activities.

INTERACTIVE SUMMARIES. For each chapter, these dynamic summaries combine a review of important concepts with links to all of the key figures, Activities, and Animated Tutorials.

ANIMATED TUTORIALS. In-depth tutorials that present complex topics in a clear, easy-to-follow format that combines a detailed animation or simulation with an introduction, conclusion, and brief quiz.

MEDIA CLIPS. New for the Tenth Edition, these short, engaging video clips depict fascinating examples of some of the many organisms, processes, and phenomena discussed in the textbook.

ACTIVITIES. A range of interactive activities that help students learn and review key facts and concepts through labeling diagrams, identifying steps in processes, and matching concepts.

LECTURE NOTEBOOK. New for the Tenth Edition, the Lecture Notebook is included online in BioPortal. The Notebook includes all of the textbook's figures and tables, with space for note-taking, and is available as downloadable PDF files.

BIONEWS FROM SCIENTIFIC AMERICAN. BioNews makes it easy for instructors to bring the dynamic nature of the biological sciences and up-to-the minute currency into their course, via an automatically updated news feed.

BIONAVIGATOR. A unique visual way to explore all of the Animated Tutorials and Activities across the various levels of biological inquiry—from the global scale down to the molecular scale.

WORKING WITH DATA. Online versions of the Working with Data exercises that are included in the textbook.

FLASHCARDS AND KEY TERMS. The Flashcards and Key Terms provide an ideal way for students to learn and review the extensive terminology of introductory biology, featuring a review mode and a quiz mode.

INVESTIGATING LIFE LINKS. For each Investigating Life figure in the textbook, BioPortal includes an overview of the experiment featured in the figure with links to the original paper(s), related research or applications that followed, and additional information related to the experiment.

GLOSSARY. The full glossary, with audio pronunciations for all terms.

TREE OF LIFE. An interactive version of the Tree of Life from Appendix A. The online Tree links to a wealth of information on each group listed.

MATH FOR LIFE. A collection of mathematical shortcuts and references to help students with the quantitative skills they need in the biology laboratory.

SURVIVAL SKILLS. A guide to more effective study habits, including time management, note-taking, effective highlighting, and exam preparation.

Instructor BioPortal Resources

Assessment

- LearningCurve and Diagnostic Quizzing reports provide instructors with a wealth of information on student comprehension, by textbook section, along with targeted lecture resources for those areas requiring the most attention.
- Comprehensive question banks include questions from the Test Bank, LearningCurve, Diagnostic Quizzes, Study Guide, and textbook Chapter Review.
- Question filtering allows instructors to select questions based on Bloom's category and/or textbook section, in order to easily select the desired mix of question types.
- Easy-to-use assessment tools allow instructors to create quizzes and many other types of assignments using any combination of publisher-provided questions and those created by the instructor.

Media Resources

(see Instructor's Media Library below for details)

- Videos
- PowerPoint Presentations (Figures & Tables, Lecture, Editable Labels, Layered Art)
- Supplemental Photos
- Active Learning Exercises
- Instructor's Manual
- Lecture Notes
- Answers to Working with Data Exercises
- Course management features
- Complete course customization capabilities
- Custom resources/document posting
- Robust gradebook
- Communication Tools: Announcements, Calendar, Course Email, Discussion Boards

Student Supplements

Life, Tenth Edition Study Guide

(Paper, ISBN 978-1-4641-2365-8)

The *Life* Study Guide offers a variety of study and review resources to accompany each chapter of the textbook. The opening Big Picture section gives students a concise overview of the main concepts covered in the chapter. The Study Strategies section points out common problem areas that students may find more challenging, and suggests strategies for learning the material most effectively. The Key Concept Review section combines a detailed review of each section with questions that help students synthesize and apply what they have learned, including diagram questions, short-answer questions, and more open-ended questions. Each chapter concludes with a Test Yourself section that allows students to test their comprehension. All questions include answers, explanations, and references to textbook sections.

Life Flashcards App

Available for iPhone/iPad and Android, the *Life* Flashcards App is a great way for students to learn and review all the key terminology from the textbook, whenever and wherever they want to study, in an intuitive flashcard interface. Available in the iTunes App Store and Google Play.

CatchUp Math & Stats

Michael Harris, Gordon Taylor, and Jacquelyn Taylor
(ISBN 978-1-4292-0557-3)

Presented in brief, accessible units, this primer will help students quickly brush up on the quantitative skills they need to succeed in biology.

Student Handbook for Writing in Biology, Third Edition

Karen Knisely (ISBN 978-1-4292-3491-7)

This book provides practical advice to students who are learning to write according to the conventions in biology, using the standards of journal publication as a model.

Bioethics and the New Embryology: Springboards for Debate

Scott F. Gilbert, Anna Tyler, and Emily Zackin
(ISBN 978-0-7167-7345-0)

Our ability to alter the course of human development ranks among the most significant changes in modern science and has brought embryology into the public domain. The question that must be asked is: Even if we can do such things, should we?

BioStats Basics: A Student Handbook

James L. Gould and Grant F. Gould (ISBN 978-0-7167-3416-1)

Engaging and informal, *BioStats Basics* provides introductory-level biology students with a practical, accessible introduction to statistical research.

Inquiry Biology: A Laboratory Manual, Volumes 1 and 2

Mary Tyler, Ryan W. Cowan, and Jennifer L. Lockhart (Volume 1 ISBN 978-1-4292-9288-7; Volume 2 ISBN 978-1-4292-9289-4)

This introductory biology laboratory manual is inquiry-based—instructing in the process of science by allowing students to ask their own questions, gather background information, formulate hypotheses, design and carry out experiments, collect and analyze data, and formulate conclusions.

Hayden-McNeil Life Sciences Lab Notebook

(ISBN 978-1-4292-3055-1)

This carbonless laboratory notebook is of the highest quality and durability, allowing students to hand in originals or copies, not entire composition books. Contains Hayden-McNeil's unique white paper carbonless copies and biology-specific reference materials.

Instructor Media & Supplements

Instructor's Media Library

(Available both online via BioPortal and on disc; disc version ISBN 978-1-4641-2364-1)

The *Life,* Tenth Edition Instructor's Media Library includes a wide range of electronic resources to help instructors plan their course, present engaging lectures, and effectively assess their students. The Media Library includes the following resources:

TEXTBOOK FIGURES AND TABLES. Every figure and table from the textbook (including all photos and all un-numbered figures) is provided in both JPEG (high- and low-resolution) and PDF formats, in multiple versions.

UNLABELED FIGURES. Every figure is provided in an unlabeled format, useful for student quizzing and custom presentations.

SUPPLEMENTAL PHOTOS. The supplemental photograph collection contains over 1,500 photographs, giving instructors a wealth of additional imagery to draw upon.

ANIMATIONS. An extensive collection of detailed animations, all built specifically for Life, and viewable in either narrated or step-through mode.

VIDEOS. Featuring many new segments for the Tenth Edition, the wide-ranging collection of video segments help demonstrate the complexity and beauty of life.

POWERPOINT RESOURCES. For each chapter of the textbook, many different PowerPoint presentations are available, providing instructors the flexibility to build presentations in the manner that best suits their needs, including the following:

- Textbook Figures and Tables
- Lecture Presentation
- Figures with Editable Labels
- Layered Art Figures
- Supplemental Photos
- Videos
- Animations
- Active Learning Exercises

INSTRUCTOR'S MANUAL, LECTURE NOTES, and **TEST BANK** are available in Microsoft Word format for easy use in lecture and exam preparation.

MEDIA GUIDE. A PDF version of the Media Guide from the Instructor's Resource Kit, convenient for searching.

ACTIVE LEARNING EXERCISES. Set up for easy integration into lectures, each exercise poses a question or problem for the class to discuss or solve during lecture. Each also includes a multiple-choice element, for easy use with clicker systems.

ANSWERS TO WORKING WITH DATA EXERCISES. Complete answers to all of the Working with Data exercises.

Instructor's Resource Kit

(Binder, ISBN 978-1-4641-4131-7)

The *Life,* Tenth Edition Instructor's Resource Kit includes a wealth of information to help instructors in the planning and teaching of their course. The Kit includes:

INSTRUCTOR'S MANUAL

- *Chapter Overview*: A brief, high-level synopsis of the chapter.
- *What's New*: A guide to the revisions, updates, and new content added to the Tenth Edition.
- *Key Concepts & Learning Objectives*: New for the Tenth Edition, this section includes the major learning goals for the chapter, a detailed set of key concepts, and specific learning objectives for each key concept.
- *Chapter Outline*: All of the chapter's section headings and sub-headings.
- *Key Terms*: All of the important terms introduced in the chapter.

LECTURE NOTES. Detailed lecture outlines for each chapter, including references to relevant figures and media resources.

MEDIA GUIDE. A visual guide to the extensive media resources available with Life, including all animations, activities, videos, and supplemental photos.

Overhead Transparencies

(ISBN 978-1-4641-4127-0)

The set of overheads includes over 1,000 transparencies—including all of the four-color line art and all of the tables from the text—in two convenient binders. All figures have been formatted and color-enhanced for clear projection in a wide range of conditions. Labels and images have been resized for improved readability.

Test File

(Paper, ISBN 978-1-4292-5579-0)

The *Life*, Tenth Edition Test File includes over 5,000 questions and has been revised and reviewed for both accuracy and effectiveness. All questions are referenced to specific textbook headings and categorized according to Bloom's taxonomy. This allows instructors to easily build quizzes and exams with the desired mix of content, coverage, and question types (factual, conceptual, analyzing/applying, etc.). Each chapter includes a wide range of multiple choice and fill-in-the-blank questions, in addition to diagram questions that involve the student in working with illustrations of structures, graphs, steps in processes, and more.

Computerized Test Bank

(CD, ISBN 978-1-4641-4128-7)

The entire Test File, plus the Diagnostic Quizzes, LearningCurve questions, Study Guide questions, and Textbook End-of Chapter Review questions are all included in Wimba's easy-to-use Diploma program (software included). Designed for both novice and advanced users, Diploma allows instructors to quickly and easily create or edit questions, create quizzes or exams with a "drag-and-drop" feature (using any combination of publisher-provided and instructor-added questions), publish to online courses, and print paper-based assessments.

Figure Correlation Tool

An invaluable resource for instructors switching to *Life*, Tenth Edition from another textbook or from *Life*, Ninth Edition, this online tool provides correlations between all of the figures in *Life*, Tenth Edition and figures in other majors biology textbooks and *Life*, Ninth Edition.

Course Management System Support

As a service for *Life* adopters using Blackboard, WebCT, ANGEL, or other course management systems, full electronic course packs are available.

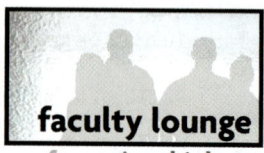

faculty lounge for majors biology

Faculty Lounge for Majors Biology is the first publisher-provided website for the majors biology community that lets instructors freely communicate and share peer-reviewed lecture and teaching resources. The Faculty Lounge offers convenient access to peer-recommended and vetted resources, including the following categories: Images, News, Videos, Labs, Lecture Resources, and Educational Research.
majorsbio.facultylounge.whfreeman.com

iclicker

Developed for educators by educators, iclicker is a hassle-free radio-frequency classroom response system that makes it easy for instructors to ask questions, record responses, take attendance, and direct students through lectures as active participants. For more information, visit **www.iclicker.com**.

LabPartner

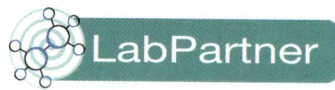

LabPartner is a site designed to facilitate the creation of customized lab manuals. Its database contains a wide selection of experiments published by W. H. Freeman and Hayden-McNeil Publishing. Instructors can preview, choose, and re-order labs, interleave their own original experiments, add carbonless graph paper and a pocket folder, customize the cover both inside and out, and select a binding type. Manuals are printed on-demand.
www.whfreeman.com/labpartner

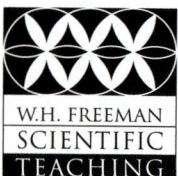

W.H. FREEMAN SCIENTIFIC TEACHING

The Scientific Teaching Book Series is a collection of practical guides, intended for all science, technology, engineering and mathematics (STEM) faculty who teach undergraduate and graduate students in these disciplines. The purpose of these books is to help faculty become more successful in all aspects of teaching and learning science, including classroom instruction, mentoring students, and professional development. Authored by well-known science educators, the Series provides concise descriptions of best practices and how to implement them in the classroom, the laboratory, or the department. For readers interested in the research results on which these best practices are based, the books also provide a gateway to the key educational literature.

Scientific Teaching

Jo Handelsman, Sarah Miller, and Christine Pfund (ISBN 978-1-4292-0188-9)

Transformations: Approaches to College Science Teaching

Deborah Allen and Kimberly Tanner (ISBN 978-1-4292-5335-2)

Entering Research: A Facilitator's Manual

Workshops for Students Beginning Research in Science

Janet L. Branchaw, Christine Pfund, and Raelyn Rediske (ISBN 978-1-429-25857-9)

Discipline-Based Science Education Research: A Scientist's Guide

Stephanie Slater, Tim Slater, and Janelle M. Bailey (ISBN 978-1-4292-6586-7)

Assessment in the College Classroom

Clarissa Dirks, Mary Pat Wenderoth, Michelle Withers (ISBN 978-1-4292-8197-3)

Contents

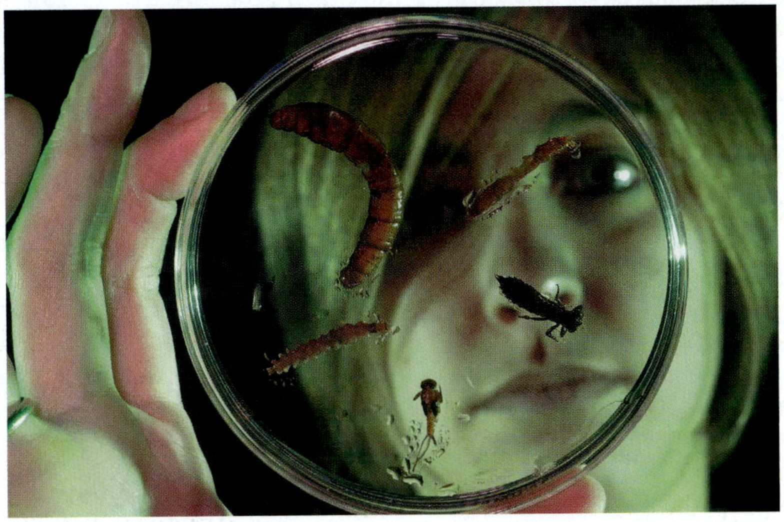

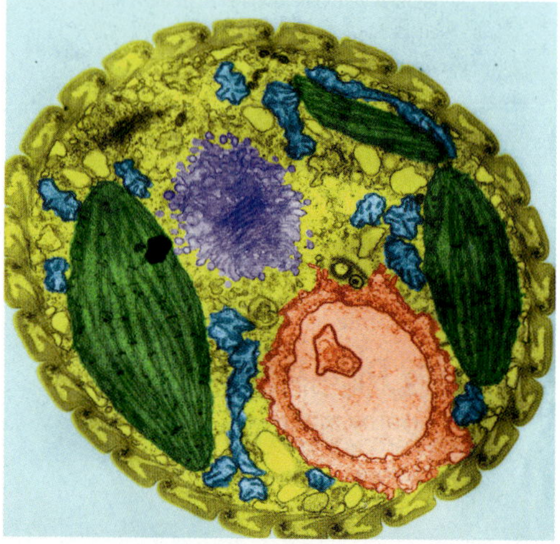

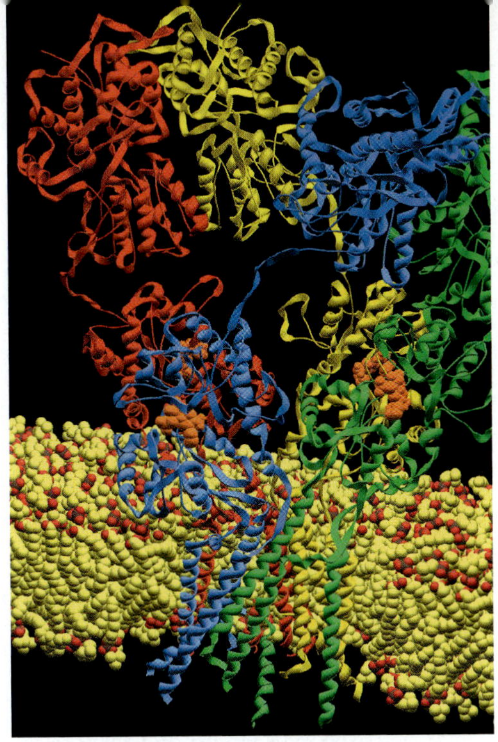

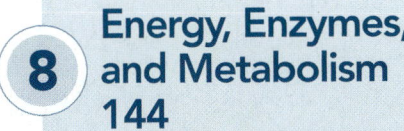

PART FOUR
Genes and Heredity

11 The Cell Cycle and Cell Division 205

12 Inheritance, Genes, and Chromosomes 232

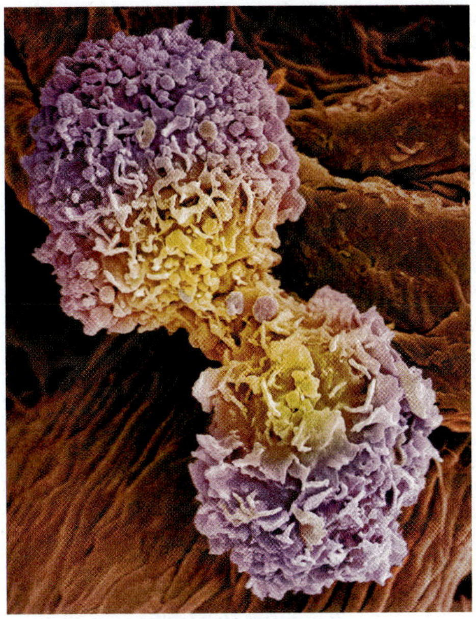

13 DNA and Its Role in Heredity 259

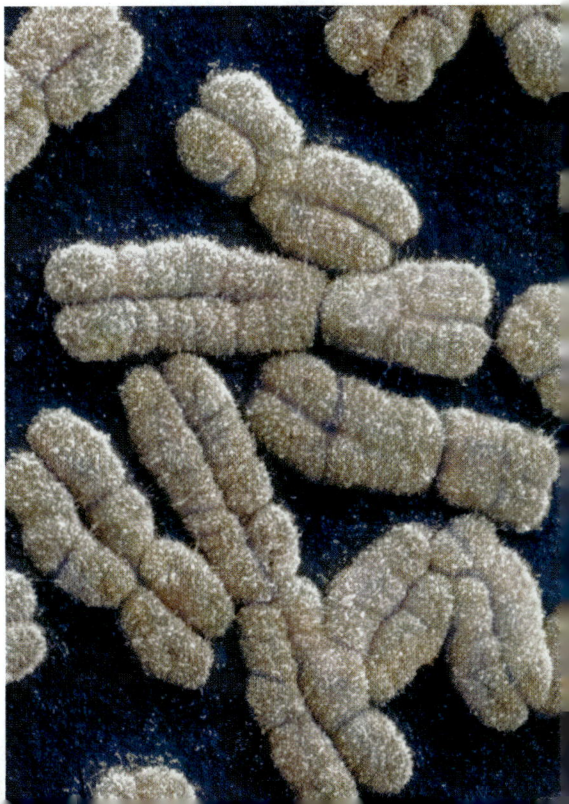

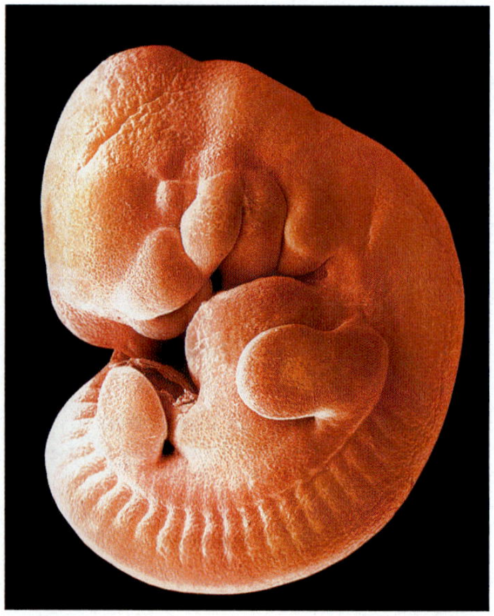

1
2
3
4

PART SEVEN
The Evolution of Diversity

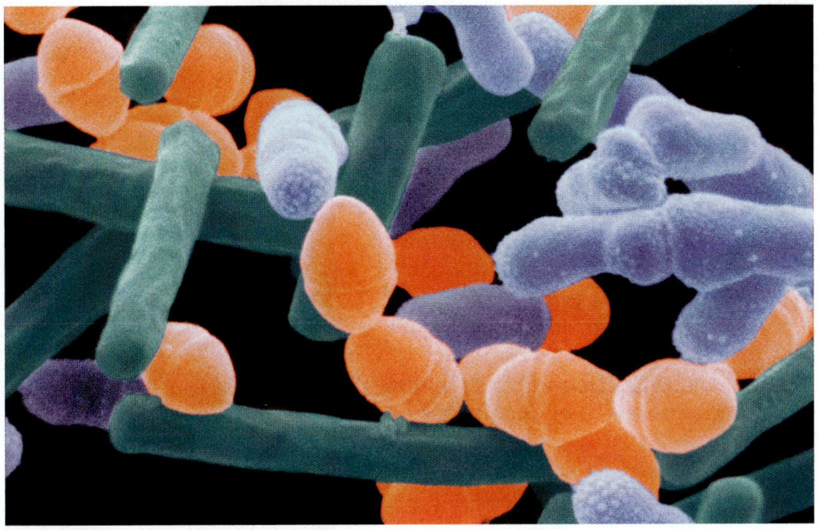

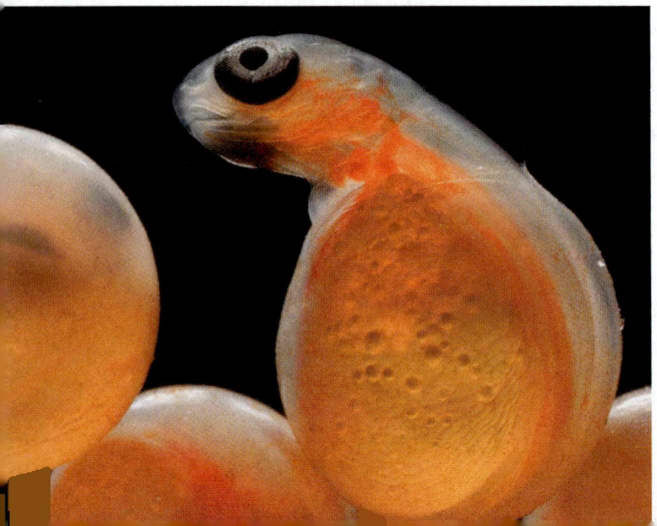

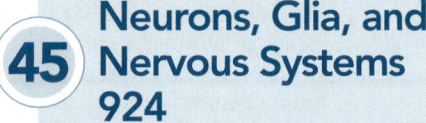

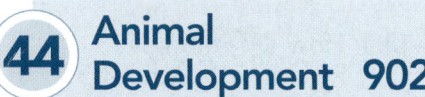

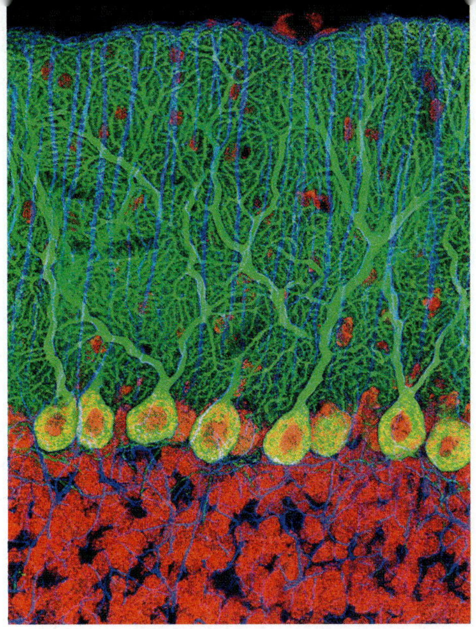

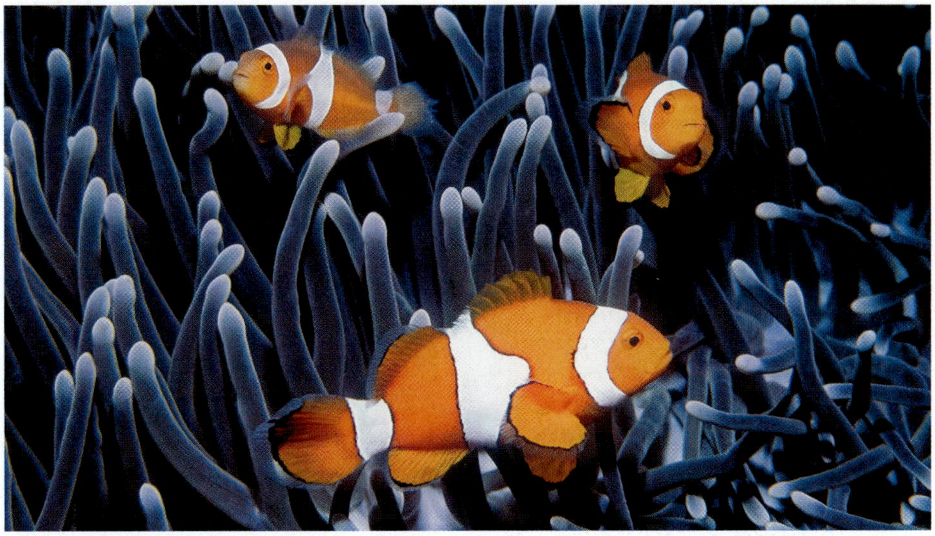

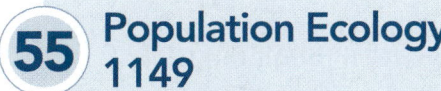

PART ONE
The Science of Life and Its Chemical Basis

1 Studying Life

CHAPTER OUTLINE

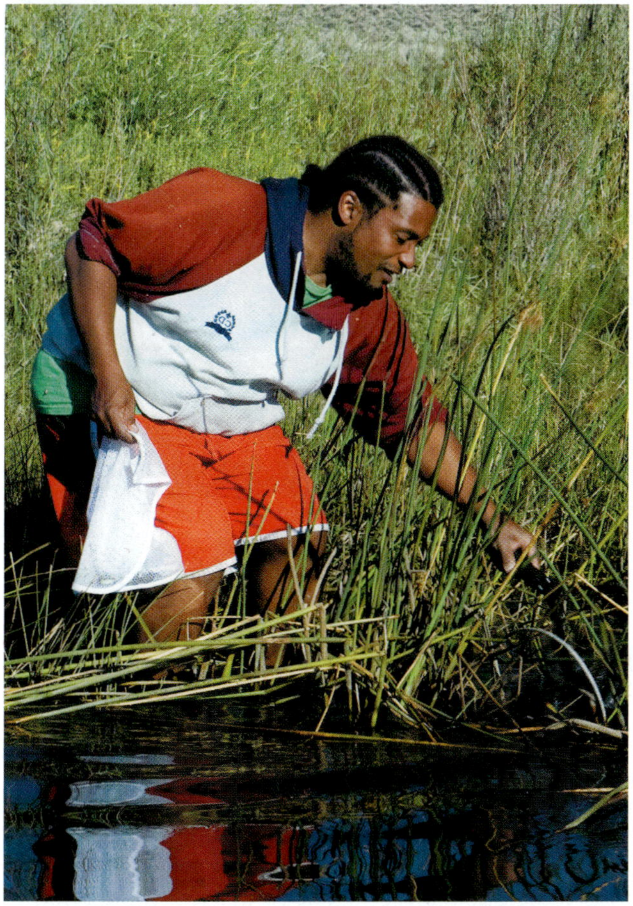

What's Happening to the Frogs? Tyrone Hayes grew up near the great Congaree Swamp in South Carolina collecting turtles, snakes, frogs, and toads. He is now a professor of biology at the University of California at Berkeley. In the laboratory and in the field, he is studying how and why populations of frogs are endangered by agricultural pesticides.

AMPHIBIANS—frogs, salamanders, and wormlike caecilians—have been around so long they watched the dinosaurs come and go. But for the last three decades, amphibian populations around the world have been declining dramatically. Today more than a third of the world's amphibian species are threatened with extinction. Why are these animals disappearing?

Tyrone Hayes, a biologist at the University of California at Berkeley, probed the effects of certain chemicals that are applied to croplands in large quantities and that accumulate in the runoff water from the fields. Hayes focused on the effects on amphibians of atrazine, a weed killer (herbicide) widely used in the United States and some other countries, where it is a common contaminant in fresh water (its use has been banned in the European Union). In the U.S., atrazine is usually applied in the spring, when many amphibians are breeding and thousands of tadpoles swim in the ditches, ponds, and streams that receive runoff from farms.

In his laboratory, Hayes and his associates raised frog tadpoles in water containing no atrazine and also in water with concentrations ranging from 0.01 parts per billion (ppb) up to 25 ppb. Concentrations as low as 0.1 ppb had a dramatic effect on tadpole development: it feminized the males. When these males became adults, their vocal structures—which are used in mating calls and thus are crucial for successful reproduction—were smaller than normal; in some, eggs were growing in the testes; some developed female sex organs. In other studies, normal adult male frogs exposed to 25 ppb had a tenfold reduction in testosterone levels and did not produce sperm. You can imagine the disastrous effects of such developmental and hormonal changes on the capacity of frogs to breed and reproduce.

But these experiments were performed in the laboratory, with a species of frog bred for laboratory use. Would the results be the same in nature? To find out, Hayes and his students traveled from Utah to Iowa, sampling water and collecting frogs. They analyzed the water for atrazine and examined the frogs. The only site where the frogs were normal was one where atrazine was undetectable. At all other sites, male frogs had abnormalities of the sex organs.

Like other biologists, Hayes made observations. He then made predictions based on those observations, and designed and carried out experiments to test his predictions.

?

Could atrazine in the environment affect species other than amphibians?

See answer on p. 18.

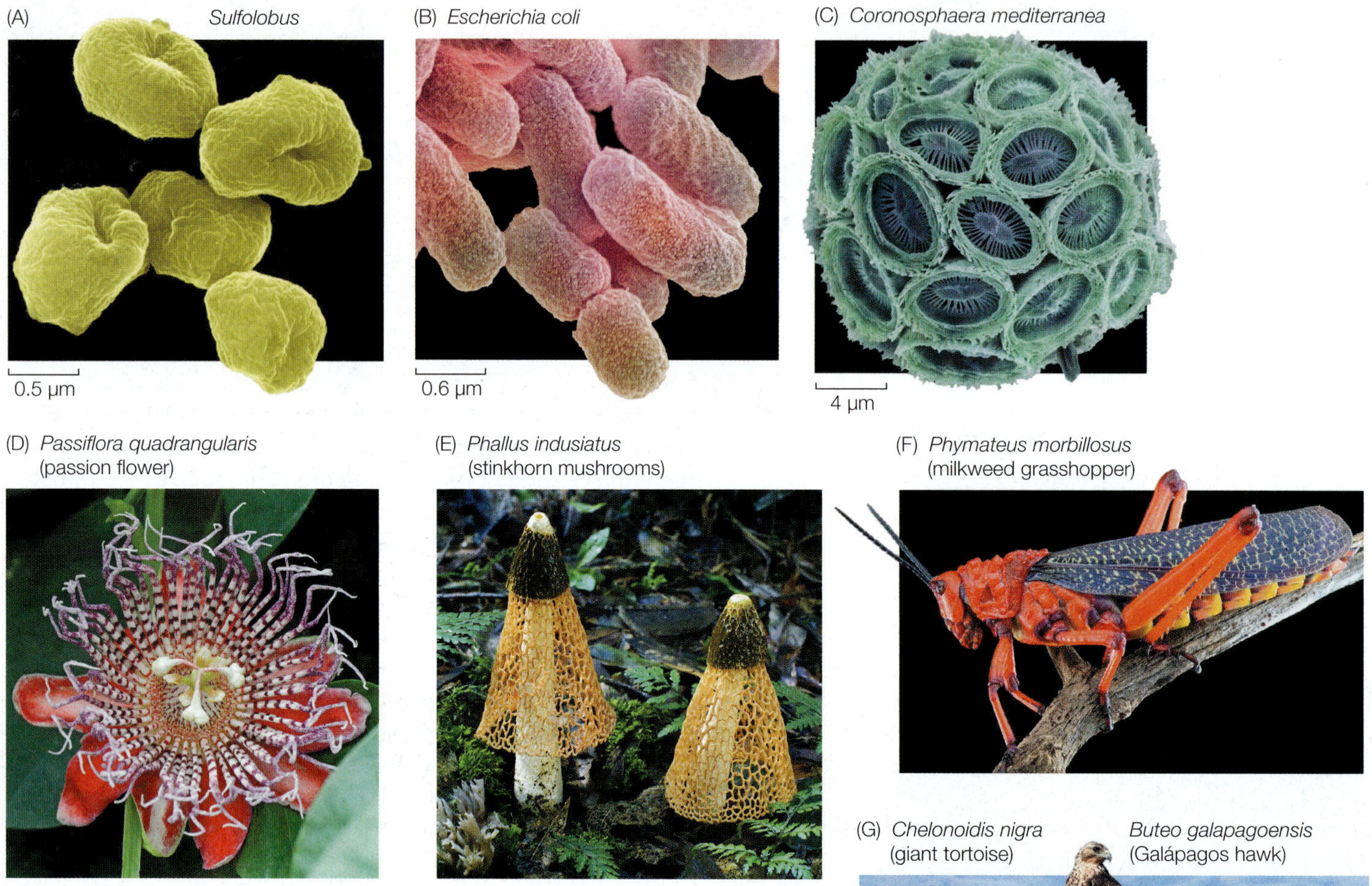
(A) Sulfolobus
0.5 μm
(B) Escherichia coli
0.6 μm
(C) Coronosphaera mediterranea
4 μm
(D) Passiflora quadrangularis
(passion flower)
(E) Phallus indusiatus
(stinkhorn mushrooms)
(F) Phymateus morbillosus
(milkweed grasshopper)

(G) Chelonoidis nigra
(giant tortoise)
Buteo galapagoensis
(Galápagos hawk)

1.1 The Many Faces of Life The processes of evolution have led to the millions of diverse organisms living on Earth today. Archaea (A) and bacteria (B) are all single-celled, prokaryotic organisms, as described in Chapter 26. (C) Many protists are unicellular but, as discussed in Chapter 27, their cell structures are more complex than those of the prokaryotes. This protist has manufactured "plates" of calcium carbonate that surround and protect its single cell. (D–G) Most of the visible life on Earth is multicellular. Chapters 28 and 29 cover the green plants (D). The other broad groups of multicellular organisms are the fungi (E), discussed in Chapter 30, and the animals (F, G), covered in Chapters 31–33.

1.1 What Is Biology?

Biology is the scientific study of living things, which we call organisms (**Figure 1.1**). The living organisms we know about are all descended from a common origin of life on Earth that occurred almost 4 billion years ago. Living organisms share many characteristics that allow us to distinguish them from the nonliving world:

- Organisms are made up of a common set of chemical components, including particular carbohydrates, fatty acids, nucleic acids, and amino acids, among others.
- The building blocks of most organisms are **cells**—individual structures enclosed by plasma membranes.
- The cells of living organisms convert molecules obtained from their environment into new biological molecules.
- Cells extract energy from the environment and use it to do biological work.
- Organisms contain genetic information that uses a nearly universal code to specify the assembly of proteins.
- Organisms share similarities among a fundamental set of genes and replicate this genetic information when reproducing themselves.
- Organisms exist in populations that evolve through changes in the frequencies of genetic variants within the populations over time.
- Living organisms self-regulate their internal environments, thus maintaining the conditions that allow them to survive.

Taken together, these characteristics logically lead to the conclusion that all life has a common ancestry, and that the diverse organisms alive today all originated from one life form. If life had multiple origins, we would not expect to see the striking similarities across gene sequences, the nearly universal genetic code, or the common set of amino acids that characterizes every known living organism. Organisms from a separate origin of life—say, on another planet—might be similar in some ways to life on Earth. For example, such life forms would probably possess heritable genetic information that they could pass on to offspring. But we would not expect the details of their genetic code or the fundamental sequences of their genomes to be the same as or even similar to ours.

The list is necessarily simplified, and some forms of life may not display all of the listed characteristics all of the time. For example, the seed of a desert plant may go for many years without extracting energy from the environment, converting molecules, regulating its internal environment, or reproducing; yet the seed is alive. And there are viruses, which are not composed of cells and cannot carry out physiological functions on their own (they parasitize host cells to function for them). Yet viruses contain genetic information, and they mutate and evolve. So even though viruses are not independent cellular organisms, their existence depends on cells. In addition, it is highly probable that viruses evolved from cellular life forms. Thus most biologists consider viruses to be a part of life.

This book will explore the details of the common characteristics of life, how these characteristics arose, and how they work together to enable organisms to survive and reproduce. Not all organisms survive and reproduce with equal success, and it is through differential survival and reproduction that living systems evolve and become adapted to Earth's many environments. The processes of evolution have generated the enormous diversity of life on Earth, and evolution is a central theme of biology.

Life arose from non-life via chemical evolution

Geologists estimate that Earth formed between 4.6 and 4.5 billion years ago. At first the planet was not a very hospitable place. It was some 600 million years or more before the earliest life evolved. If we picture the 4.6-billion-year history of Earth as a 30-day month, life first appeared some time around the end of the first week (**Figure 1.2**).

When we consider how life might have arisen from nonliving matter, we must take into account the properties of the young Earth's atmosphere, oceans, and climate, all of which were very different than they are today. Biologists postulate that complex biological molecules first arose through the random physical association of chemicals in that environment. Experiments simulating the conditions on early Earth have confirmed that the generation of complex molecules under such conditions is possible, even probable. The critical step for the evolution of life, however, was the appearance of **nucleic acids**—molecules that could reproduce themselves and also serve as templates for the synthesis of **proteins**, large molecules with complex but stable shapes. The variation in the shapes of these proteins enabled them to participate in increasing numbers and kinds of chemical reactions with other molecules. These subjects are covered in Part One of this book.

Each "day" represents about 150 million years.
Life appeared some time around day 5, a little less than 4 billion years ago.
1
2
3
4
5
Origin of life
6
7
8
9
10
11
12
Oldest fossils
13
Origin of photo-synthesis
14
15
16
17
18
19
20
Origin of eukaryotic cells
21
22
23
24
Multi-cellular organisms
25
26
27
28
29
30
27
Abundant fossils of aquatic life
28
First land plants
First land animals
29
Coal-forming forests
Insects abundant
First mammals
Dinosaurs dominant
30
First birds
First flowering plants
Rise of mammals
First hominids
Homo sapiens (modern humans) arose in the last 5 minutes of day 30 (around 500,000 years ago).
Recorded history covers the last few seconds of day 30.
12
3
6
9

1.2 Life's Timeline Depicting the 4.6 billion years of Earth's history on the scale of a 30-day month provides a sense of the immensity of evolutionary time.

Cellular structure evolved in the common ancestor of life

Another important step in the history of life was the enclosure of complex proteins and other biological molecules by membranes that contained them in a compact internal environment separate from the surrounding (external) environment. Molecules called fatty acids played a critical role because these molecules do not dissolve in water; rather they form membranous

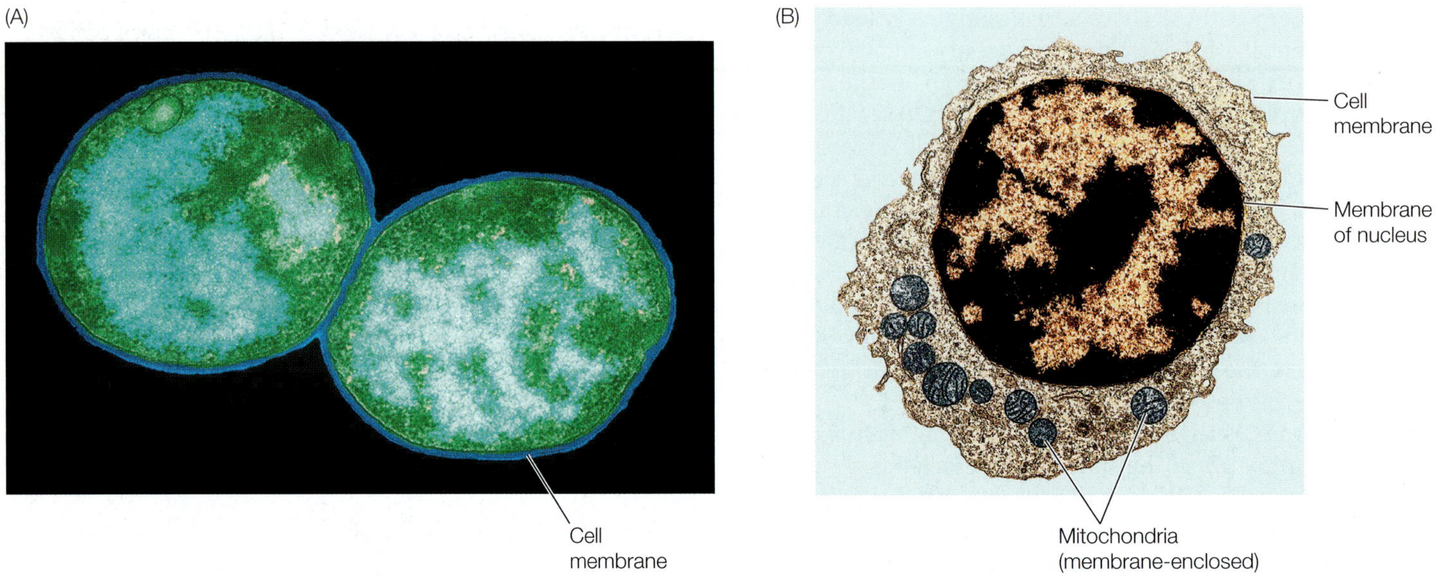
(A)
Cell
membrane
(B)
Cell
membrane
Membrane
of nucleus
Mitochondria
(membrane-enclosed)

1.3 Cells Are Building Blocks for Life These photographs of cells were taken with a transmission electron microscope (see Figure 5.3) and enhanced with added color to highlight details. (A) Two prokaryotic cells of an *Enterococcus* bacterium that lives in the human digestive system. Prokaryotes are unicellular organisms with genetic and biochemical material enclosed inside a single membrane. (B) A human white blood cell (lymphocyte) represents one of the many specialized cell types that make up a multicellular eukaryote. Multiple membranes within the cell-enclosing outer membrane segregate the different biochemical processes of eukaryotic cells.

films that, when agitated, can form spherical structures. These membranous structures could have enveloped assemblages of biological molecules. The creation of an internal environment that concentrated the reactants and products of chemical reactions opened up the possibility that those reactions could be integrated and controlled within a tiny cell (**Figure 1.3**). Scientists postulate that this natural process of membrane formation resulted in the first cells with the ability to reproduce—that is, the evolution of the first cellular organisms.

For the first few billion years of cellular life, all the organisms that existed were unicellular and were enclosed by a single outer membrane. Such organisms, like the bacteria that are still abundant on Earth today, are called **prokaryotes**. Two main groups of prokaryotes emerged early in life's history: the **bacteria** and **archaea**. Some representatives of each of these groups began to live in a close, interdependent relationship with one another, and eventually merged to form a third major lineage of life, the **eukaryotes**. In addition to their outer membranes, the cells of eukaryotes have internal membranes that enclose specialized organelles within their cells. Eukaryote organelles include the nucleus that contains the genetic material and the mitochondria that power the cell. The structure of prokaryote and eukaryote cells and their membranes are the subjects of Part Two.

At some point, the cells of some eukaryotes failed to separate after cell division, remaining attached to each other. Such permanent colonial aggregations of cells made it possible for some of the associated cells to specialize in certain functions, such as reproduction, while other cells specialized in other functions, such as absorbing nutrients. This **cellular specialization** enabled multicellular eukaryotes to increase in size and become more efficient at gathering resources and adapting to specific environments.

Photosynthesis allows some organisms to capture energy from the sun

Living cells require energy in order to function, and the biochemistry of the fundamental processes of energy conversion that drive life is covered in Part Three.

To fuel their cellular **metabolism** (energy transformations), the earliest prokaryotes took in small molecules directly from their environment and broke them down to their component atoms, thus releasing and using the energy contained in the chemical bonds. Many modern prokaryotes still function this way, and they function very successfully. But about 2.5 billion years ago, the emergence of **photosynthesis** changed the nature of life on Earth.

The chemical reactions of photosynthesis transform the energy of sunlight into a form of biological energy that powers the synthesis of large molecules. These large molecules can then be broken down to provide metabolic energy. Photosynthesis is the basis of much of life on Earth today because its energy-capturing processes provide food for other organisms. Early photosynthetic cells were probably similar to present-day prokaryotes called cyanobacteria (**Figure 1.4**). Over time, photosynthetic prokaryotes became so abundant that vast quantities of oxygen gas (O_2), which is a by-product of photosynthesis, began to accumulate in the atmosphere.

During the early eons of life, there was no O_2 in Earth's atmosphere. In fact, O_2 was poisonous to many of the prokaryotes living at that time. As O_2 levels increased, however, those

(A)
0.5 cm
(B)
Stromatolites form as small grains of sediment are cemented together by communities of micro-organisms, especially cyanobacteria.
10 cm

1.4 Photosynthetic Organisms Changed Earth's Atmosphere (A) Colonies of photosynthetic cyanobacteria and other microorganisms produced structures called stromatolites that were preserved in the ancient fossil record. This section of fossilized stromatolite reveals layers representing centuries of growth. (B) Living stromatolites can still be found in appropriate environments.

organisms that *did* tolerate O_2 were able to proliferate. The abundance of O_2 opened up vast new avenues of evolution because **aerobic metabolism**—a biochemical process that uses O_2 to extract energy from nutrient molecules—is far more efficient than **anaerobic metabolism** (which does not use O_2). Aerobic metabolism allows organisms to grow larger and is used by the majority of organisms today.

Oxygen in the atmosphere also made it possible for life to move onto land. For most of life's history, UV radiation falling on Earth's surface was so intense that it destroyed any organism that was not well shielded by water. But the accumulation of photosynthetically generated O_2 in the atmosphere for more than 2 billion years gradually produced a thick layer of ozone (O_3) in the upper atmosphere. By about 500 million years ago, the ozone layer was sufficiently dense and absorbed enough of the sun's UV radiation to make it possible for organisms to leave the protection of the water and live on land.

Biological information is contained in a genetic language common to all organisms

The information that specifies what an organism will look like and how it will function—its "blueprint" for existence—is contained in the organism's **genome**: the sum total of all the DNA molecules contained in each of its cells. **DNA** (deoxyribonucleic acid) molecules are long sequences of four different subunits called **nucleotides**. The sequence of these four nucleotides contains genetic information. **Genes** are specific segments of DNA that encode the information the cell uses to create amino acids and form them into proteins (**Figure 1.5**). Protein molecules govern the chemical reactions within cells and form much of an organism's structure.

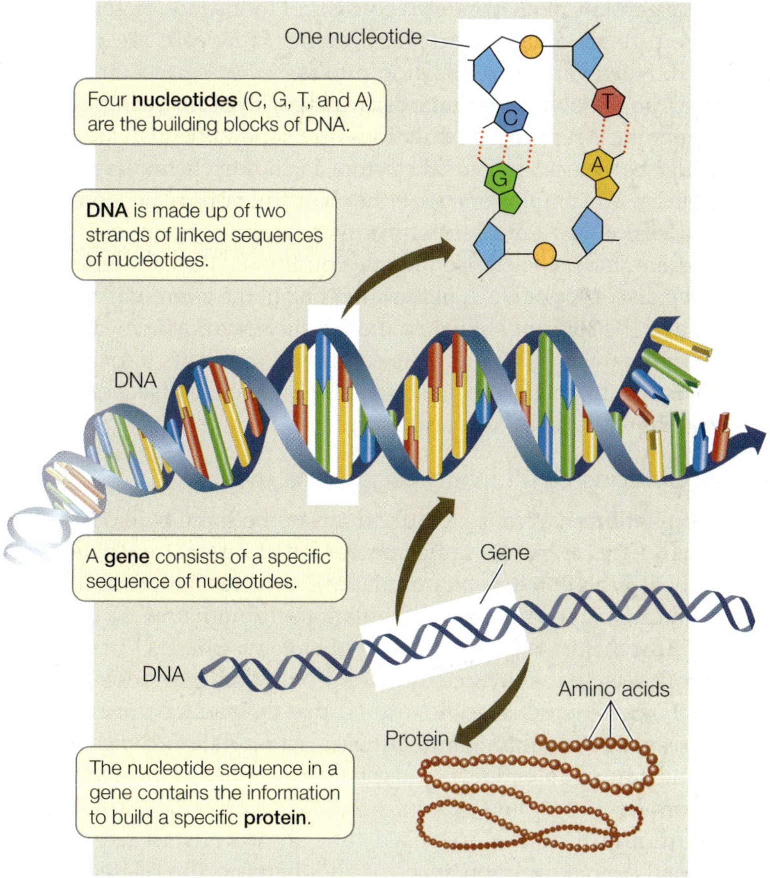
One nucleotide
Four **nucleotides** (C, G, T, and A) are the building blocks of DNA.
C
T
G
A
DNA is made up of two strands of linked sequences of nucleotides.
DNA
A **gene** consists of a specific sequence of nucleotides.
Gene
DNA
Amino acids
Protein
The nucleotide sequence in a gene contains the information to build a specific **protein**.

1.5 DNA Is Life's Blueprint The instructions for life are contained in the sequences of nucleotides in DNA molecules. Specific DNA nucleotide sequences comprise genes. The average length of a single human gene is 16,000 nucleotides. The information in each gene provides the cell with the information it needs to manufacture molecules of a specific protein.

By analogy with a book, the nucleotides of DNA are like the letters of an alphabet, and protein molecules are sentences. Combinations of proteins that form structures and control biochemical processes are the paragraphs. The structures and processes that are organized into different systems with specific tasks (such as digestion or transport) are the chapters of the book, and the complete book is the organism. If you were to write out your own genome using four letters to represent the four nucleotides, you would write more than 3 billion letters. Using the size type you are reading now, your genome would fill about 1,000 books the size of this one. The mechanisms of evolution are the authors and editors of all the books in the library of life.

All the cells of a multicellular organism contain essentially the same genome, yet different cells have different functions and form different structures—contractile proteins form in muscle cells, hemoglobin in red blood cells, digestive enzymes in gut cells, and so on. Therefore different types of cells in an organism must express different parts of the genome. How cells control gene expression in ways that enable a complex organism to develop and function is a major focus of current biological research.

The genome of an organism consists of thousands of genes. This entire genome must be replicated as new cells are produced. However, the replication process is not perfect, and a few errors, known as mutations, are likely to occur each time the genome is replicated. Mutations occur spontaneously; they can also be induced by outside factors, including chemicals and radiation. Most mutations are either harmful or have no effect, but occasionally a mutation improves the functioning of the organism under the environmental conditions it encounters.

The discovery of DNA in the latter half of the twentieth century and the subsequent elucidation of the remarkable mechanisms by which this material encodes and transmits information transformed biological science. These crucial discoveries are detailed in Parts Four and Five.

Populations of all living organisms evolve

A **population** is a group of individuals of the same type of organism—that is, of the same **species**—that interact with one another. **Evolution** acts on populations; it is the change in the genetic makeup of biological populations through time. Evolution is the major unifying principle of biology. Charles Darwin compiled factual evidence for evolution in his 1859 book *On the Origin of Species*. Darwin argued that differential survival and reproduction among individuals in a population, which he termed **natural selection**, could account for much of the evolution of life.

Although Darwin proposed that all organisms are descended from a common ancestor and therefore are related to one another, he did not have the advantage of understanding the mechanisms of genetic inheritance and mutation. Even so, he observed that offspring resembled their parents; therefore, he surmised, such mechanisms had to exist. Part Six will describe how Darwin's theory of natural selection is both supported and explained by the massive body of molecular genetic data elucidated during the twentieth century, and how these elements coincide and mesh in the modern field of evolutionary biology.

If all the organisms on Earth today are the descendants of a single kind of unicellular organism that lived almost 4 billion years ago, how have they become so different? As mentioned earlier, organisms reproduce by replicating their genomes, and mutations are introduced almost every time a genome is replicated. Some of these mutations give rise to structural and functional changes in organisms. As individuals mate with one another, the genetic variants stemming from mutation can change in frequency within a population, and the population is said to evolve.

Any population of a plant or animal species displays variation, and if you select breeding pairs on the basis of some particular trait, that trait is more likely to be present in their offspring than in the general population. Darwin himself bred pigeons, and was well aware of how pigeon fanciers selected breeding pairs to produce offspring with unusual feather patterns, beak shapes, or body sizes (see Figure 21.5). He realized that if humans could select for specific traits in domesticated animals, the same process could operate in nature; hence the term "natural selection" as opposed to artificial (human-imposed) selection.

How does natural selection function? Darwin postulated that different probabilities of survival and reproductive success would do the job. He reasoned that the reproductive capacity of plants and animals, if unchecked, would result in unlimited growth of populations, but we do not observe such growth in nature; in most species, only a small percentage of an individual's offspring will survive to reproduce. Thus any trait that confers even a small increase in the probability that its possessor will survive and reproduce would spread in the population.

Because organisms with certain traits survive and reproduce best under specific sets of conditions, natural selection leads to **adaptations**: structural, physiological, or behavioral traits that enhance an organism's chances of survival and reproduction in its environment (**Figure 1.6**). In addition to natural selection, evolutionary processes such as sexual selection (for example, selection due to mate choice) and genetic drift (the random fluctuation of gene frequencies in a population due to chance events) contribute to the rise of biodiversity. These processes operating over evolutionary history have led to the remarkable diversity of life on Earth.

Biologists can trace the evolutionary tree of life

As populations become geographically isolated from one another, they evolve differences. As populations diverge from one another, individuals in each population become less likely to reproduce with individuals of the other population. Eventually these differences between populations become so great that the two populations are considered different species. Thus species that share a fairly recent evolutionary history are generally more similar to each other than species

(A) *Dyscophus guineti*
(B) *Xenopus laevis*
(C) *Agalychnis callidryas*
(D) *Rhacophorus nigropalmatus*

1.6 Adaptations to the Environment The limbs of frogs show adaptations to the different environments of each species. (A) This terrestrial frog walks across the ground using its short legs and peglike digits (toes). (B) Webbed rear feet are evident in this highly aquatic species of frog. (C) This arboreal species has toe pads, which are adaptations for climbing. (D) A different arboreal species has extended webbing between the toes, which increases surface area and allows the frog to glide from tree to tree.

that share an ancestor in the more distant past. By identifying, analyzing, and quantifying similarities and differences between species, biologists can construct **phylogenetic trees** that portray the evolutionary histories of the different groups of organisms.

Tens of millions of species exist on Earth today; many times that number lived in the past but are now extinct. Biologists give each of these species a distinctive scientific name formed from two Latinized names—a **binomial**. The first name identifies the species' **genus** (plural *genera*)—a group of species that share a recent common ancestor. The second is the name of the species. For example, the scientific name for the human species is *Homo sapiens*: *Homo* is our genus, *sapiens* our species. *Homo* is Latin for "man," and *sapiens* is from the Latin word for "wise" or "rational." Our closest relatives in the genus *Homo* are the Neanderthals, *Homo neanderthalensis*. Neanderthals are now extinct and are known only from their fossil remains.

Much of biology is based on comparisons among species, and these comparisons are useful precisely because we can place species in an evolutionary context relative to one another. Our ability to do this has been greatly enhanced in recent decades by our ability to sequence and compare the genomes of different species. Genome sequencing and other molecular techniques have allowed biologists to augment evolutionary knowledge based on the fossil record with a vast array of molecular evidence. The result is the ongoing compilation of phylogenetic trees that document and diagram evolutionary relationships as part of an overarching tree of life, the broadest categories of which are shown in **Figure 1.7** and will be surveyed in more detail in Part Seven. (The tree is expanded in Appendix A, and you can also explore the tree interactively.)

Although many details remain to be clarified, the broad outlines of the tree of life have been determined. Its branching patterns are based on a rich array of evidence from fossils, structures, metabolic processes, behavior, and molecular

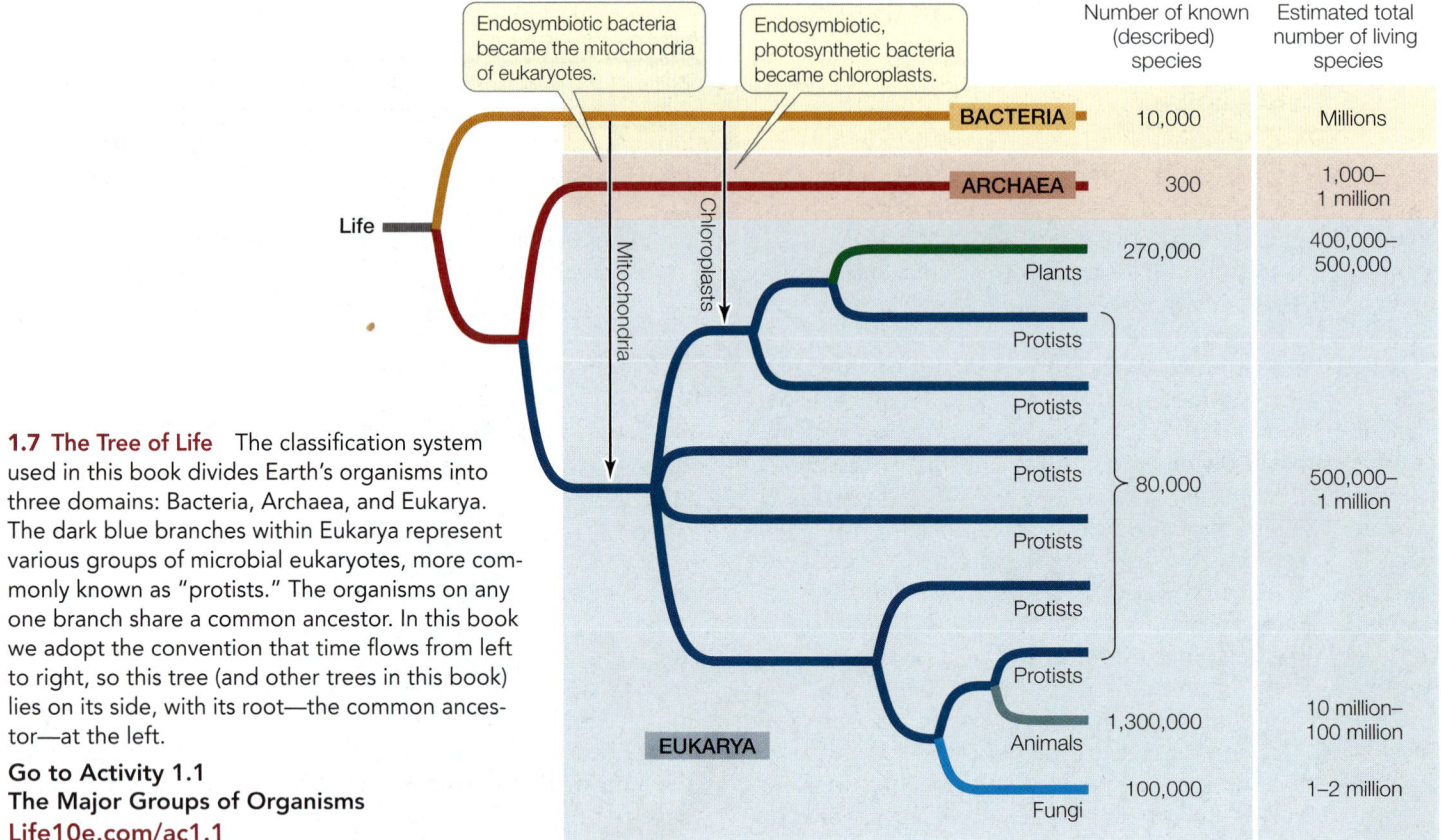

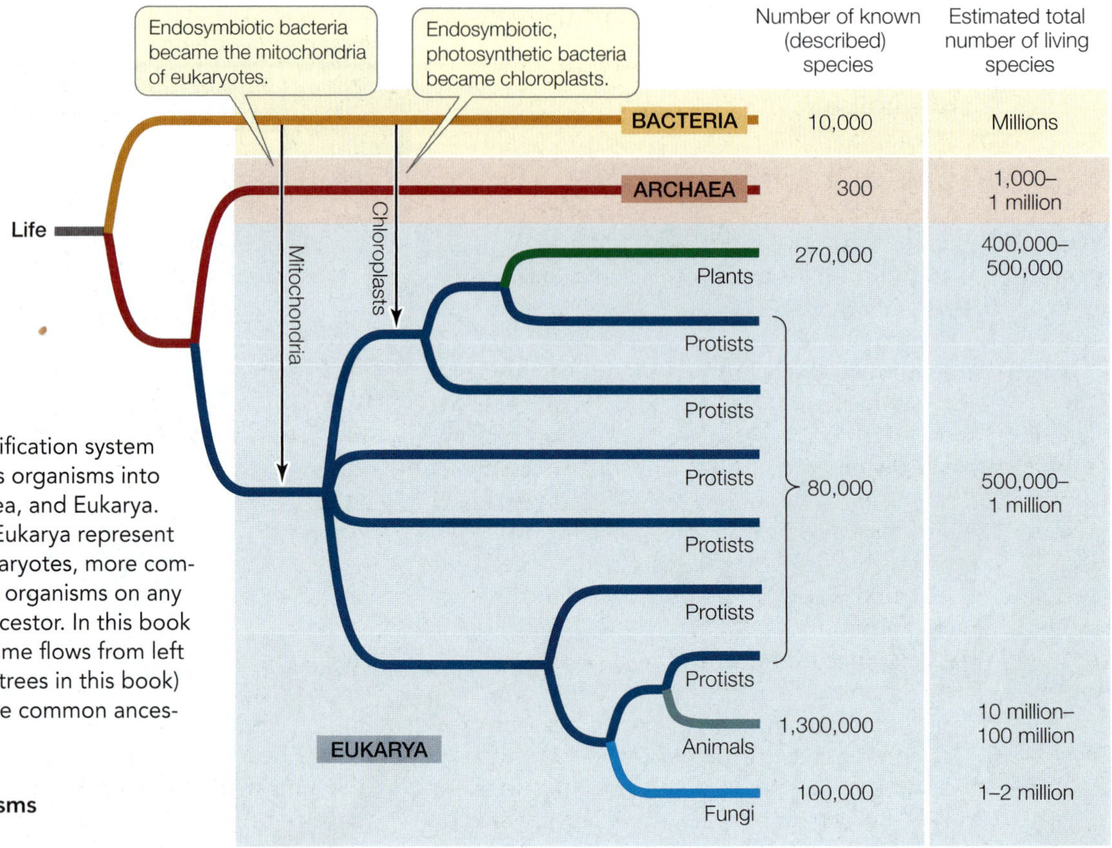
Endosymbiotic bacteria became the mitochondria of eukaryotes.
Endosymbiotic, photosynthetic bacteria became chloroplasts.
Number of known (described) species
Estimated total number of living species
Life
BACTERIA
10,000
Millions
ARCHAEA
300
1,000–1 million
Mitochondria
Chloroplasts
Plants
270,000
400,000–500,000
Protists
Protists
Protists
80,000
500,000–1 million
Protists
Protists
Protists
Animals
1,300,000
10 million–100 million
Fungi
100,000
1–2 million
EUKARYA

1.7 The Tree of Life The classification system used in this book divides Earth's organisms into three domains: Bacteria, Archaea, and Eukarya. The dark blue branches within Eukarya represent various groups of microbial eukaryotes, more commonly known as "protists." The organisms on any one branch share a common ancestor. In this book we adopt the convention that time flows from left to right, so this tree (and other trees in this book) lies on its side, with its root—the common ancestor—at the left.

Go to Activity 1.1 The Major Groups of Organisms
Life10e.com/ac1.1

analyses of genomes. Two of the three main domains of life—Archaea and Bacteria—are single-celled prokaryotes, as mentioned earlier in this chapter. However, members of these two groups differ so fundamentally in their metabolic processes that they are believed to have separated into distinct evolutionary lineages very early. Species belonging to the third domain—Eukarya—have eukaryotic cells whose mitochondria and chloroplasts originated from endosymbioses of bacteria.

Plants, fungi, and animals are examples of familiar multicellular eukaryotes that evolved independently, from different groups of the unicellular eukaryotes informally known as protists. We know that plants, fungi, and animals had independent origins of multicellularity because each of these three groups is most closely

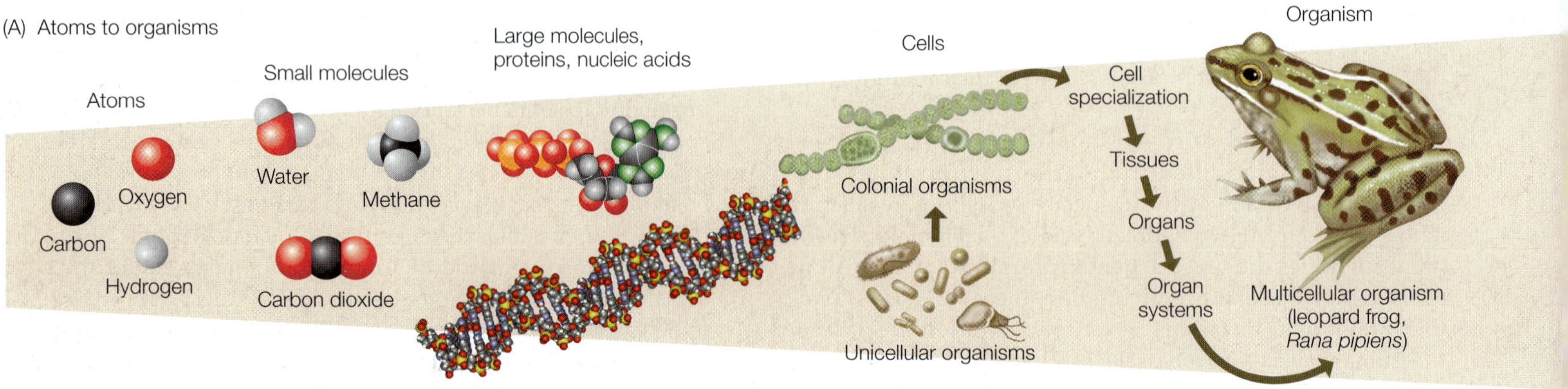
(A) Atoms to organisms
Atoms
Carbon
Oxygen
Hydrogen
Small molecules
Water
Methane
Carbon dioxide
Large molecules, proteins, nucleic acids
Cells
Colonial organisms
Unicellular organisms
Cell specialization
Tissues
Organs
Organ systems
Organism
Multicellular organism (leopard frog, *Rana pipiens*)

1.8 Biology Is Studied at Many Levels of Organization (A) Life's properties emerge when DNA and other molecules are organized in cells, which form building blocks for organisms. (B) Organisms exist in populations and interact with other populations to form communities, which interact with the physical environment to make up the many ecosystems of the biosphere.

Go to Activity 1.2 The Hierarchy of Life
Life10e.com/ac1.2

related to different groups of unicellular protists, as can be seen from the branching pattern of Figure 1.7.

Cellular specialization and differentiation underlie multicellular life

Looking back at Figure 1.2, you can see that for more than half of Earth's history, all life was unicellular. Unicellular species remain ubiquitous and highly successful in the present, even though the diverse multicellular organisms, owing to their much larger size, may seem to us to dominate the planet.

With the evolution of cells specialized for different functions within the same organism, these differentiated cells lost many of the functions carried out by single-celled organisms, and a **biological hierarchy** emerged (**Figure 1.8A**). To accomplish their specialized tasks, assemblages of differentiated cells are organized into **tissues**. For example, a single muscle cell cannot generate much force, but when many cells combine to form the tissue of a working muscle, considerable force and movement can be generated. Different tissue types are organized to form **organs** that accomplish specific functions. The heart, brain, and stomach are each constructed of several types of tissues, as are the roots, stems, and leaves of plants. Organs whose functions are interrelated can be grouped into **organ systems**; the esophagus, stomach, and intestines, for example, are all part of the digestive system. The physiology of two major groups of multicellular organisms (land plants and animals) is discussed in detail in Parts Eight and Nine, respectively.

Living organisms interact with one another

Organisms do not live in isolation, and the internal hierarchy of the individual organism is matched by the external hierarchy of the biological world (**Figure 1.8B**). As mentioned earlier in this section, a group of individuals of the same species that interact with one another is a population. The populations of all the species that live and interact in a defined area (areas are defined in different ways and can be small or large) are called a **community**. Communities together with their abiotic (nonliving) environment constitute an **ecosystem**.

Individuals in a population interact in many different ways. Animals eat plants and other animals (usually members of another species) and compete with other species for food and other resources. Some animals prevent other individuals of their own species from exploiting a resource, be it food, nesting sites, or mates. Animals may also cooperate with members of their own species, forming social units such as a termite colony or a flock of birds. Such interactions have resulted in the evolution of social behaviors such as communication and courtship displays.

Plants also interact with their external environment, which includes other plants, fungi, animals, and microorganisms. All terrestrial plants depend on partnerships with fungi, bacteria, and animals. Some of these partnerships are necessary to obtain nutrients, some to produce fertile seeds, and still others to disperse seeds. Plants compete with each other for light and water and have ongoing evolutionary interactions with the animals that eat them. Through time, many adaptations have evolved in plants that protect them from predation (such as thorns) or that help then attract the animals that assist in their reproduction (such as sweet nectar or colorful flowers). The interactions of populations of plant and animal species in a community are major evolutionary forces that produce specialized adaptations.

Communities interacting over a broad geographic area with distinguishing physical features form ecosystems; examples

(B) Organisms to ecosystems

Population
Community
Ecosystem
Biosphere

(B) *Spermophilus parryii*

(A) *Propithecus verreauxi*

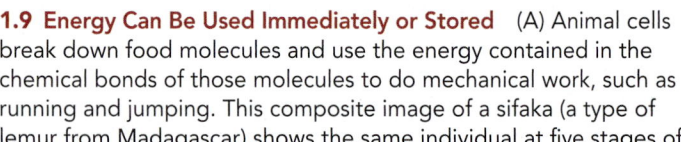

Go to Media Clip 1.1
Leaping Lemurs
Life10e.com/mc1.1

1.9 Energy Can Be Used Immediately or Stored (A) Animal cells break down food molecules and use the energy contained in the chemical bonds of those molecules to do mechanical work, such as running and jumping. This composite image of a sifaka (a type of lemur from Madagascar) shows the same individual at five stages of a single jump. (B) The cells of this Arctic ground squirrel have broken down the complex carbohydrates in the plants it consumed and converted those molecules into fats. The fats are stored in the animal's body to provide an energy supply for the cold months.

include Arctic tundra, coral reef, and tropical rainforest. The ways in which species interact with one another and with their environment in populations, communities, and ecosystems is the subject of ecology, covered in Part Ten of this book.

Nutrients supply energy and are the basis of biosynthesis

Living organisms acquire nutrients from the environment. Nutrients supply the organism with energy and raw materials for carrying out biochemical reactions. Life depends on thousands of biochemical reactions that occur inside cells. Some of these reactions break down nutrient molecules into smaller chemical units, and in the process some of the energy contained in the chemical bonds of the nutrients is captured by high-energy molecules that can be used to do different kinds of cellular work.

One obvious kind of work cells do is mechanical—moving molecules from one cellular location to another, moving whole cells or tissues, or even moving the organism itself, as muscles do (**Figure 1.9A**). The most basic cellular work is the building, or synthesis, of new complex molecules and structures from smaller chemical units. For example, we are all familiar with the fact that carbohydrates eaten today may be deposited in the body as fat tomorrow (**Figure 1.9B**). Still another kind of work is the electrical work that is the essence of information processing in nervous systems.

The myriad biochemical reactions that take place in cells are integrally linked in that the products of one reaction are the raw materials of the next. These complex networks of reactions must be integrated and precisely controlled; when they are not, the result is malfunction and disease.

Living organisms must regulate their internal environment

The specialized cells, tissues, and organ systems of multicellular organisms exist in and depend on an **internal environment** that is made up of extracellular fluids. Because this environment serves the needs of the cells, its physical and chemical composition must be maintained within a narrow range of physiological conditions that support survival and function. The maintenance of this narrow range of conditions is known as **homeostasis**. A relatively stable internal (but extracellular) environment means that cells can function efficiently even when conditions outside the organism's body become unfavorable for cellular processes.

The organism's regulatory systems obtain information from sensory cells that provide information about both the internal and external conditions the organism is subject to at a given time. The cells of regulatory systems process and integrate this information and send signals to components of physiological systems, which can change in response to these signals so that the organism's internal environment remains reasonably constant.

The concept of homeostasis extends beyond the internal environment of multicellular organisms, however. In both unicellular and multicellular organisms, individual cells must regulate physiological parameters (such as acidity and salinity), maintaining them within a range that allows those cells to survive and function. Individual cells regulate these properties through actions of the plasma membrane that encloses them and are the cell's interface with its environment (either internal or external). Thus self-regulation to maintain a more or less constant internal environment is a general attribute of all living organisms.

RECAP 1.1

All organisms are related by common descent from a single ancestral form. They contain genetic information that encodes how they look and how they function. They also reproduce, extract energy from their environment, and use energy to do biological work, synthesize complex molecules to construct biological structures, regulate their internal environment, and interact with one another.

- Why did the evolution of photosynthesis so radically affect the course of life on Earth? See pp. 4–5
- Describe the relationship between evolution by natural selection and the genetic code. See p. 6
- What information have biologists used to construct a tree of life? See pp. 6–8 and Figure 1.7
- What do we mean by "homeostasis," and why is it crucial to living organisms? See p. 10

The preceding section briefly outlined the major features of life—features that will be covered in depth in subsequent chapters of this book. Before going into the details of what we know about life, however, it is important to understand how scientists obtain information and how they use that information in broadening our understanding of Earth's diverse living organisms and putting this understanding to practical use.

1.2 How Do Biologists Investigate Life?

Scientific investigations are based on observation, data, experimentation, and logic. Scientists use many different tools and methods in making observations, collecting data, designing experiments, and applying logic, but they are always guided by established principles that allow us to discover new aspects about the structure, function, evolution, and interactions of organisms.

Observing and quantifying are important skills

Biologists have always observed the world around them, but today our ability to observe is greatly enhanced by technologies such as electron microscopes, rapid genome sequencing, magnetic resonance imaging, and global positioning satellites. These technologies allow us to observe everything from the distribution of molecules in the body to the movement of animals across continents and oceans.

Observation is a basic tool of biology, but as scientists we must also be able to quantify the information, or **data**, we collect as we observe. Whether we are testing a new drug or mapping the migrations of the great whales, applying mathematical and statistical calculations to the data we collect is essential. For example, biologists once classified organisms based entirely on qualitative descriptions of the physical differences among them. There was no way of objectively determining evolutionary relationships of organisms, and biologists had to depend on the fossil record for insight. Today our ability to quantify the molecular and physical differences among species, combined with explicit mathematical models of the evolutionary process, enables quantitative analyses of evolutionary history. These mathematical calculations, in turn, facilitate comparative investigations of all other aspects of an organism's biology.

Scientific methods combine observation, experimentation, and logic

Textbooks often describe "*the* scientific method," as if there is a single, simple flow chart that all scientists follow. This is an oversimplification. Although flow charts such as the one shown in Figure 1.10 incorporate much of what scientists do, you should not conclude that scientists necessarily progress through the steps of the process in one prescribed, linear order.

Observations lead to questions, and scientists make additional observations and often do experiments to answer those

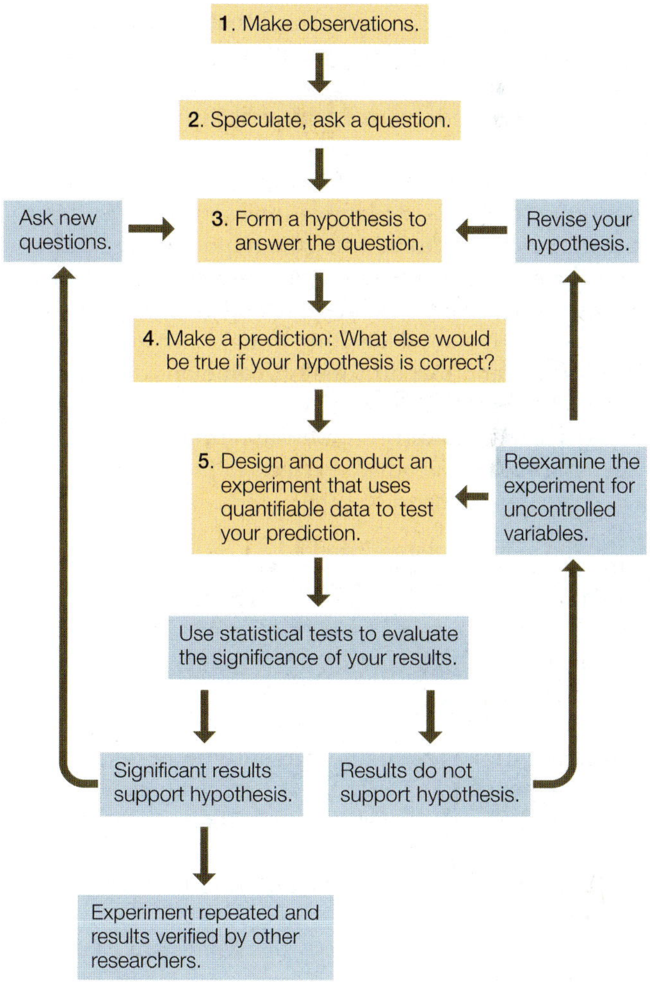
1. Make observations.
2. Speculate, ask a question.
Ask new questions.
3. Form a hypothesis to answer the question.
Revise your hypothesis.
4. Make a prediction: What else would be true if your hypothesis is correct?
5. Design and conduct an experiment that uses quantifiable data to test your prediction.
Reexamine the experiment for uncontrolled variables.
Use statistical tests to evaluate the significance of your results.
Significant results support hypothesis.
Results do not support hypothesis.
Experiment repeated and results verified by other researchers.

1.10 Scientific Methodology The process of observation, speculation, hypothesis, prediction, and experimentation is a cornerstone of modern science, although scientists may initiate their research at several different points. Answers gleaned through experimentation lead to new questions, more hypotheses, further experiments, and expanding knowledge.

questions. This hypothesis–prediction approach traditionally has five steps: (1) making observations; (2) asking questions; (3) forming hypotheses, which are tentative answers to the questions; (4) making predictions based on the hypotheses; and (5) testing the predictions by making additional observations or conducting experiments.

After posing a question, a scientist often uses **inductive logic** to propose a tentative answer. Inductive logic involves taking observations or facts and creating a new proposition that is compatible with those observations or facts. Such a tentative proposition is a **hypothesis** (plural *hypotheses*). In formulating a hypothesis, scientists put together the facts and data at their disposal to formulate one or more possible answers to the question. For example, at the opening of this chapter you learned that scientists have observed the rapid decline of amphibian populations worldwide and are asking why. Some scientists have hypothesized that a fungal disease is a cause; other scientists have hypothesized that increased exposure to ultraviolet radiation is a cause. Tyrone Hayes hypothesized that exposure to agricultural chemicals, specifically the widely used herbicide atrazine, could be a cause.

The next step in the scientific method is to apply a different form of logic—**deductive logic**—that starts with a statement believed to be true (the hypothesis) and then goes on to predict what facts would also have to be true to be compatible with that statement. Hayes knew that atrazine is commonly applied in the spring, when amphibians are breeding, and that atrazine is a common contaminant in the waters in which amphibians live as they develop into adults. Thus he predicted that frog tadpoles exposed to atrazine would show adverse effects of the chemical once they reached adulthood.

Go to Animated Tutorial 1.1
Using Scientific Methodology
Life10e.com/at1.1

Good experiments have the potential to falsify hypotheses

Once predictions are made from a hypothesis, experiments can be designed to test those predictions. The most informative experiments are those that have the ability to show that the prediction is wrong. If the prediction is wrong, the hypothesis must be questioned, modified, or rejected.

There are two general types of experiments, both of which compare data from different groups or samples. A **controlled experiment** manipulates one or more of the factors being tested; **comparative experiments** compare unmanipulated data gathered from different sources. As described at the opening of this chapter, Tyrone Hayes and his colleagues conducted both types of experiments to test the prediction that the herbicide atrazine, a contaminant in freshwater ponds and streams throughout the world, affects the development of frogs.

INVESTIGATING**LIFE**

1.11 Controlled Experiments Manipulate a Variable The Hayes laboratory created controlled environments that differed only in the concentrations of atrazine in the water. Eggs from leopard frogs (*Rana pipiens*) raised specifically for laboratory use were allowed to hatch and the tadpoles were separated into experimental tanks containing water with different concentrations of atrazine.[a]

HYPOTHESIS Exposure to atrazine during larval development causes abnormalities in the reproductive tissues of male frogs.

Method

1. Establish 9 tanks in which all attributes are held constant except the water's atrazine concentration. Establish 3 atrazine conditions (3 replicate tanks per condition): 0 ppb (control condition), 0.1 ppb, and 25 ppb.
2. Place *Rana pipiens* tadpoles from laboratory-reared eggs in the 9 tanks (30 tadpoles per replicate).
3. When tadpoles have transitioned into adults, sacrifice the animals and evaluate their reproductive tissues.
4. Test for correlation of degree of atrazine exposure with the presence of abnormalities in the gonads (testes) of male frogs.

Results

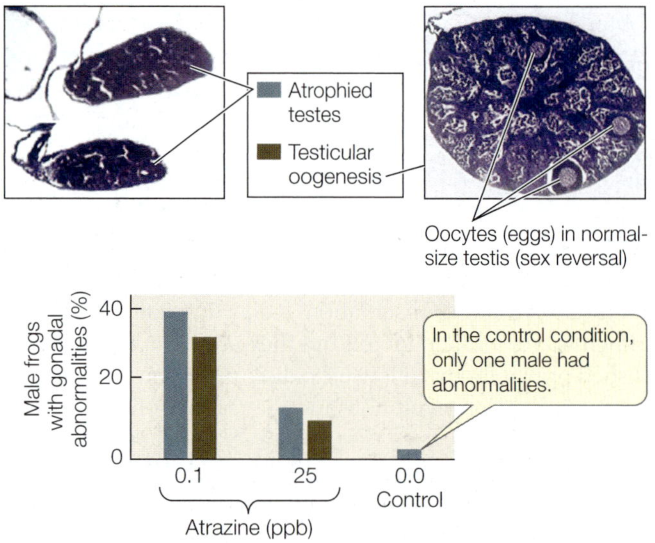
Atrophied testes
Testicular oogenesis
Oocytes (eggs) in normal-size testis (sex reversal)
Male frogs with gonadal abnormalities (%)
40
20
0
0.1
25
0.0
Control
Atrazine (ppb)
In the control condition, only one male had abnormalities.

CONCLUSION Exposure to atrazine at concentrations as low as 0.1 ppb induces abnormalities in the gonads of male frogs. The effect is not proportional to the level of exposure.

Go to **BioPortal** for discussion and relevant links for all INVESTIGATING**LIFE** figures.

[a]Hayes, T. et al. 2003. *Environmental Health Perspectives III*: 568–575.

In a controlled experiment, we start with groups or samples that are as similar as possible. We predict on the basis of our hypothesis that some critical factor, or variable, has an effect on the phenomenon we are investigating. We devise some method to manipulate *only that variable* in an "experimental" group and compare the resulting data with data from an unmanipulated "control" group. If the predicted difference occurs, we then apply statistical tests to ascertain the probability that the manipulation created the difference (as opposed to the difference being the result of random

INVESTIGATING**LIFE**

1.12 Comparative Experiments Look for Differences among Groups To see whether the presence of atrazine correlates with testicular abnormalities in male frogs, the Hayes lab collected frogs and water samples from different locations around the U.S. The analysis that followed was "blind," meaning that the frogs and water samples were coded so that experimenters working with each specimen did not know which site the specimen came from.[a]

HYPOTHESIS Presence of the herbicide atrazine in environmental water correlates with gonadal abnormalities in frog populations.

Method

1. Based on commercial sales of atrazine, select 4 sites (sites 1–4) less likely and 4 sites (sites 5–8) more likely to be contaminated with atrazine.
2. Visit all sites in the spring (i.e., when frogs have transitioned from tadpoles into adults); collect frogs and water samples.
3. In the laboratory, sacrifice frogs and examine their reproductive tissues, documenting abnormalities.
4. Analyze the water samples for atrazine concentration (the sample for site 7 was not tested).
5. Quantify and correlate the incidence of reproductive abnormalities with environmental atrazine concentrations.

Results

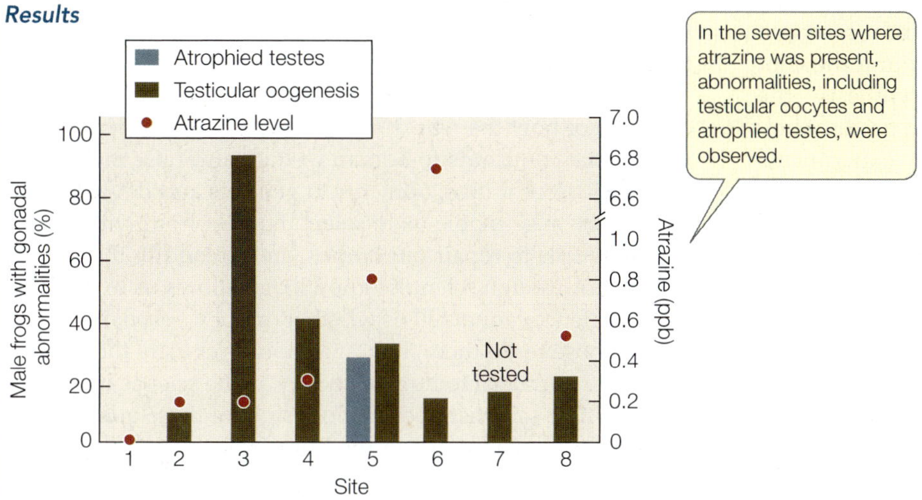
Atrophied testes
Testicular oogenesis
Atrazine level
Male frogs with gonadal abnormalities (%)
100
80
60
40
20
0
1
2
3
4
5
6
7
8
Site
Not tested
7.0
6.8
6.6
1.0
0.8
0.6
0.4
0.2
0
Atrazine (ppb)
In the seven sites where atrazine was present, abnormalities, including testicular oocytes and atrophied testes, were observed.

CONCLUSION Reproductive abnormalities exist in frogs from environments in which aqueous atrazine concentration is 0.2 ppb or above. The incidence of abnormalities does not appear to be proportional to atrazine concentration at the time of transition to adulthood.

Go to **BioPortal** for discussion and relevant links for all INVESTIGATING**LIFE** figures.

[a]Hayes, T. et al. 2003. *Nature* 419: 895–896.

chance). **Figure 1.11** describes one of the many controlled experiments performed by the Hayes laboratory to quantify the effects of atrazine on male frogs.

The basis of controlled experiments is that one variable is manipulated while all others are held constant. The variable that is manipulated is called the **independent variable**, and the response that is measured is the **dependent variable**. A good controlled experiment is not easy to design because biological variables are so interrelated that it is difficult to alter just one.

A comparative experiment starts with the prediction that there will be a difference between samples or groups based on the hypothesis. In comparative experiments, however, we cannot control the variables; often we cannot even identify all the variables that are present. We are simply gathering and comparing data from different sample groups.

When his controlled experiments indicated that atrazine indeed affects reproductive development in frogs, Hayes and his colleagues performed a comparative experiment. They collected frogs and water samples from eight widely separated sites across the United States and compared the incidence of abnormal frogs from environments with very different levels of atrazine (**Figure 1.12**). Of course, the sample sites differed in many ways besides the level of atrazine present.

The results of experiments frequently reveal that the situation is more complex than the hypothesis anticipated, thus raising new questions. In the Hayes experiments, for example, there was no clear direct relationship between the *amount* of atrazine present and the percentage of abnormal frogs: there were fewer abnormal frogs at the highest concentrations of atrazine than at lower concentrations. There are no "final answers" in science. Investigations consistently reveal more complexity than we expect, so scientists must design systematic approaches to identify, assess, and understand that complexity.

Statistical methods are essential scientific tools

Whether we do comparative or controlled experiments, at the end we have to decide whether there is a difference between the samples, individuals, groups, or populations in the study. How do we decide whether a measured difference is enough to support or falsify a hypothesis? In other words, how do we decide in an unbiased, objective way that the measured difference is significant?

Significance can be measured with statistical methods. Scientists use statistics because they recognize that variation is always present in any set of measurements. Statistical tests calculate the probability that the differences observed in an experiment could be due to random variation. The results of statistical tests are therefore probabilities. A statistical test starts with a **null hypothesis**—the premise that any observed differences are simply the result of random differences that arise from drawing two finite samples from the same population. When quantified observations, or data, are collected, statistical methods are applied to those data to calculate the likelihood that the null hypothesis is correct.

More specifically, statistical methods tell us the probability of obtaining the same results by chance even if the null hypothesis were true. *We need to eliminate, insofar as possible, the chance that any differences showing up in the data are merely the result of random variation in the samples tested.* Scientists generally conclude that the differences they measure are significant if statistical tests show that the probability of error (that is, the probability that a difference as large as the one observed could be obtained by mere chance) is 5 percent or lower, although more stringent levels of significance may be set for some problems. Appendix B of this book is a short primer on statistical methods that you can refer to as you analyze data that will be presented throughout the text.

Discoveries in biology can be generalized

Because all life is related by descent from a common ancestor, shares a genetic code, and consists of similar biochemical building blocks, knowledge gained from investigations of one type of organism can, with thought and care, be generalized to other organisms. Biologists use **model systems** for research, knowing that they can extend their findings from such systems to other organisms. For example, our basic understanding of the chemical reactions in cells came from research on bacteria but is applicable to all cells, including those of humans. Similarly, the biochemistry of photosynthesis—the process by which all green plants use sunlight to produce biological molecules—was largely worked out from experiments on *Chlorella*, a unicellular green alga. Much of what we know about the genes that control plant development is the result of work on *Arabidopsis thaliana*, a relative of the mustard plant. Knowledge about how animals, including humans, develop has come from work on sea urchins, frogs, chickens, roundworms, mice, and fruit flies. Being able to generalize from model systems is a powerful tool in biology.

Not all forms of inquiry are scientific

Science is a unique human endeavor that has certain standards of practice. Other areas of scholarship share with science the practice of making observations and asking questions, but scientists are distinguished by *what they do with their observations* and *how they frame the answers*. Quantifiable data, subjected to appropriate statistical analysis, are critical in evaluating hypotheses (the Working with Data exercises you will find throughout this book are intended to reinforce this way of thinking). In short, scientific observation and evaluation is the most powerful approach humans have devised for learning about the world and how it works.

Scientific explanations for natural processes are objective and reliable because *a hypothesis must be testable* and *a hypothesis must have the potential of being rejected* by direct observations and experiments. Scientists must clearly describe the methods they use to test hypotheses so that other scientists can repeat their results. Not all experiments are repeated, but surprising or controversial results are always subjected to independent verification. Scientists worldwide share this process of testing and rejecting hypotheses, contributing to a common body of scientific knowledge.

If you understand the methods of science, you can distinguish science from non-science. Art, music, and literature all contribute to the quality of human life, but they are not science. They do not use scientific methods to establish what is fact. Religion is not science, although religions have historically attempted to explain natural events ranging from unusual weather patterns to crop failures to human diseases. Most such phenomena that at one time were mysterious can now be explained in terms of scientific principles. Fundamental tenets of religious faith, such as the existence of a supreme deity or deities, cannot be confirmed or refuted by experimentation and are thus outside the realm of science.

The power of science derives from strict objectivity and absolute dependence on evidence based on *reproducible and quantifiable observations*. A religious or spiritual explanation of a natural phenomenon may be coherent and satisfying for the person holding that view, but it is not testable and therefore it is not science. To invoke a supernatural explanation (such as a "creator" or "intelligent designer" with no known bounds) is to depart from the world of science. Science does not necessarily say that religious beliefs are wrong; they are simply not part of the world of science, and many religious beliefs are untestable using scientific methods.

Science describes how the world works. It is silent on the question of how the world "ought to be." Many scientific advances that contribute to human welfare also raise major ethical issues. Recent developments in genetics and developmental biology may enable us to select the sex of our children, to use stem cells to repair our bodies, and to modify the human genome. Although scientific knowledge allows us to do these things, science cannot tell us whether or not we *should* do so or, if we choose to do them, how we should regulate them. Such issues are as crucial to human society as the science itself, and a responsible scientist does not lose sight of these questions or neglect the contributions of the humanities or social sciences in attempting to come to grips with them.

RECAP 1.2

Scientific methods of inquiry start with the formulation of hypotheses based on observations and data. Comparative and controlled experiments are carried out to test hypotheses.

- Explain the relationship between a hypothesis and an experiment. **See pp. 11–12 and Figure 1.10**
- What is controlled in a controlled experiment? **See pp. 11–12 and Figure 1.11**
- What features characterize questions that can be answered only by using a comparative approach? **See p. 13 and Figure 1.12**
- Explain why arguments must be supported by quantifiable and reproducible data in order to be considered scientific. **See pp. 13–14**
- Why can the results of biological research on one species often be generalized to very different species? **See p. 14**

The vast body of scientific knowledge accumulated over centuries of human civilization allows us to understand and manipulate aspects of the natural world in ways that no other species can. These abilities present us with challenges, opportunities, and above all, responsibilities.

1.3 Why Does Biology Matter?

Human beings exist in and depend on a world of living organisms. The oxygen in the air we breathe is produced by photosynthesis conducted by countless billions of individual organisms. The food that fuels our bodies comes from the tissues of other living organism. The fuels that drive our cars and power our electric plants are, for the most part, various forms of carbon molecules produced by living organisms—mostly millions of years ago. Inside and out, our bodies are covered in complex communities of living unicellular organisms, most of which help us maintain our health. There are also harmful species that invade our bodies and can cause mild to serious diseases, or even death. These interactions with other species are not limited to humans. Ecosystem function depends on thousands of complex interactions among the millions of species that inhabit Earth. In other words, understanding biological principles is essential to our lives and for maintaining the functioning of Earth as we know it and depend on it.

Modern agriculture depends on biology

Agriculture represents some of the earliest human applications of biological principles. Even in prehistoric times, farmers selected the most productive or otherwise favorable plants and animals to use as seed stock for propagation, and over generations farmers continued and refined these practices. His knowledge of this kind of artificial selection helped Charles Darwin understand the importance of natural selection in evolution across all of life.

In modern times, increasing knowledge of plant biology has transformed agriculture in many ways and has resulted in huge boosts in food production (Figure 1.13), which in turn has allowed the planet to support a far larger human population than it once could have. Over the past few decades, detailed knowledge of the genomes of many domestic species and the development of technology for directly recombining genes have allowed biologists to develop new breeds and strains of animals, plants, and fungi of agricultural interest. For example, new strains of crop plants are being developed that are resistant to pests or can tolerate drought. Moreover, understanding evolutionary theory allows biologists to devise strategies for the application of pesticides that minimize the evolution of pest resistance. And better understanding of plant–fungus relationships results in better plant health and higher productivity. These are just a few of the many ways that biology continues to inform and improve agricultural practice.

1.13 A Green Revolution The agricultural advancements of the last 100 years have vastly increased yields and nutritional value of crops such as grains that sustain the expanding human population. In the last 30 years, these advancements have included genetic recombination techniques. Here a researcher with the U.S. Department of Agriculture works with a strain of "supernutritious" rice that provides high levels of the amino acid lysine.

Biology is the basis of medical practice

People have speculated about the causes of diseases and searched for methods to combat them since ancient times. Long before the microbial causes of many diseases were known, people recognized that infections could be passed from one person to another, and the isolation of infected persons has been practiced as long as written records have been available.

Modern biological research informs us about how living organisms work, and about why they develop the problems and infections that we call disease. In addition to diseases caused by infection of other organisms, we now know that many diseases are genetic—meaning that variants of genes in our genomes cause particular problems in the way we function. Developing appropriate treatments or cures for diseases depends on understanding the origin, basis, and effects of these diseases, as well understanding the consequences of any changes that we make. For example, the recent resurgence of tuberculosis is the result of the evolution of bacteria that are resistant to antibiotics. Dealing with future tuberculosis epidemics requires understanding aspects of molecular biology, physiology, microbial ecology, and evolution—in other words, many of the general principles of modern biology.

Many of the microbial organisms that are periodically epidemic in human populations have short generation times and high mutation rates. For example, we need yearly vaccines for flu because of the high rate of evolution of influenza viruses, the causative agent of flu. Evolutionary principles help us understand how influenza viruses are changing, and can even help us predict which strains of influenza virus are likely to lead to future flu epidemics. This medical understanding—which combines an application of molecular biology, evolutionary

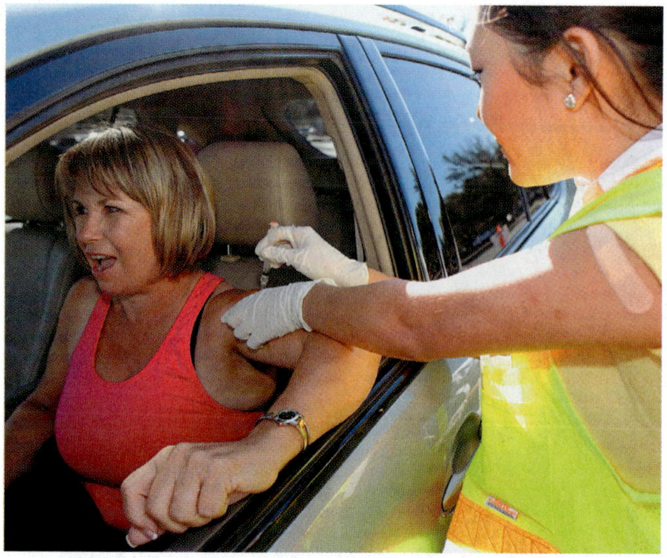

1.14 Medical Applications of Biology Improve Human Health Vaccination to prevent disease is a biologically based medical practice that began in the eighteenth century. Today evolutionary biology and genomics provide the basis for constant updates to vaccines that protect humans from virus-borne diseases such as flu. In the developed world, vaccinations have become so commonplace that some are offered on a "drive-through" basis.

theory, and basic principles of ecology—allows medical researchers to develop effective vaccines and other strategies for the control of major epidemics (Figure 1.14).

Biology can inform public policy

Thanks to the deciphering of genomes and our newfound ability to manipulate them, vast new possibilities now exist for controlling human diseases and increasing agricultural productivity—but these capabilities raise ethical and policy issues. How much and in what ways should we tinker with the genes of humans and other species? Does it matter whether the genomes of our crop plants and domesticated animals are changed by traditional methods of controlled breeding and crossbreeding or by the biotechnology of gene transfer? What rules should govern the release of genetically modified organisms into the environment? Science alone cannot provide all the answers, but wise policy decisions must be based on accurate scientific information.

Biologists are increasingly called on to advise government agencies concerning the laws, rules, and regulations by which society deals with the increasing number of challenges that have a biological basis. As an example of the value of scientific knowledge for the assessment and formulation of public policy, consider a management problem. Scientists and fishermen have long known that Atlantic bluefin tuna (*Thunnus thynnus*) have a western breeding ground in the Gulf of Mexico and an eastern breeding ground in the Mediterranean Sea (Figure 1.15). Overfishing led to declining numbers of bluefin tuna,

(A)

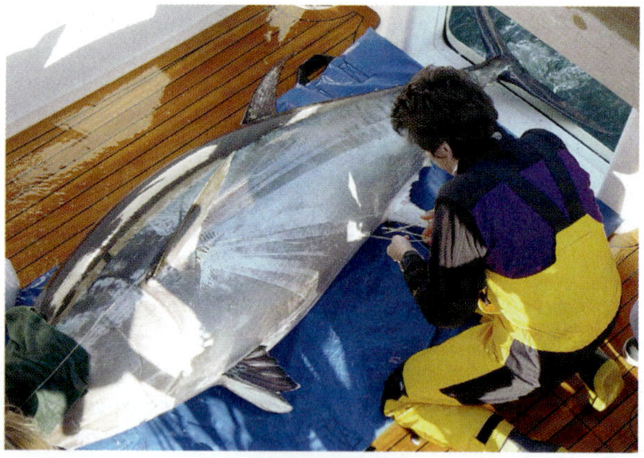

(B)

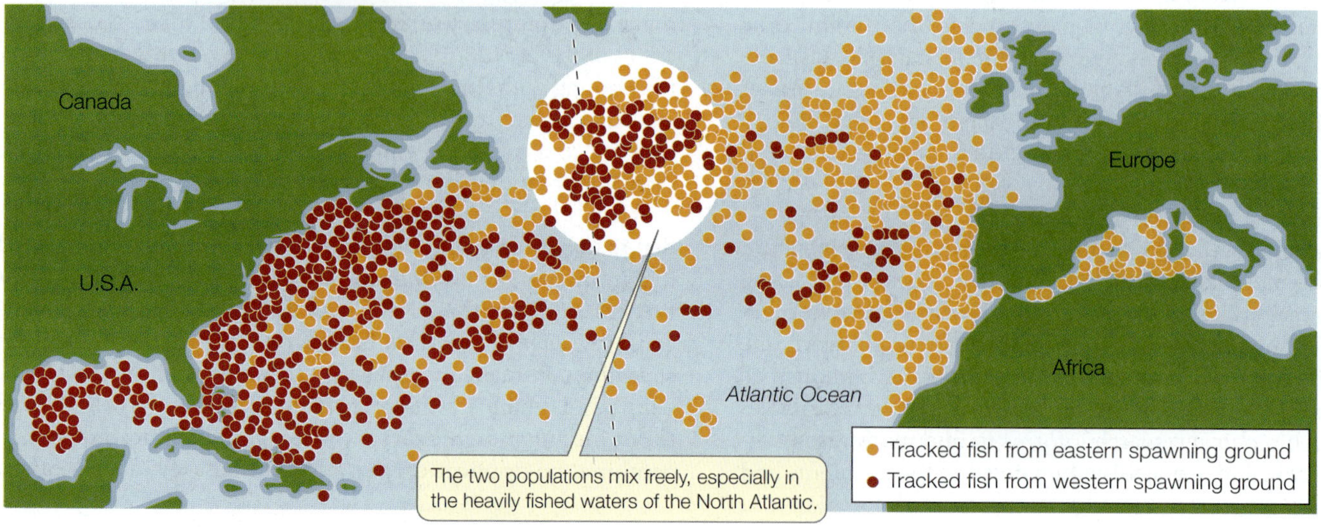
Canada
U.S.A.
Europe
Africa
Atlantic Ocean
The two populations mix freely, especially in the heavily fished waters of the North Atlantic.
Tracked fish from eastern spawning ground
Tracked fish from western spawning ground

1.15 Bluefin Tuna Do Not Recognize Boundaries (A) Marine biologist Barbara Block attaches computerized data-recording tracking tags to a live bluefin tuna before returning it to the Atlantic Ocean, where its travels will be monitored. (B) At one time we assumed that bluefins from western- and eastern-breeding populations also fed on their respective sides of the Atlantic, so separate fishing quotas for each side (dashed line) in an attempt to speed recovery of the endangered western population. Now, however, tracking data have shown that the two populations *do not* remain separate after spawning, so in fact the arbitrary boundary and quotas do not protect the endangered population.

(A) 1941
Riggs Glacier
Muir Glacier
(B) 2004
Riggs Glacier

1.16 A Warmer World Earth's climate has been steadily warming for the last 150 years. The rate of this warming trend has also steadily increased, resulting in the rapid melting of polar ice caps, glaciers, and alpine (mountaintop) snow and ice. This photograph shows the effects of 63 years of climate change on two ancient, longstanding glaciers in Alaska. Over that time, Muir Glacier retreated some 7 kilometers and can no longer be seen from the original vantage point. Understanding how biological populations respond to such change requires integration of biological principles from molecular biology to ecosystem ecology.

especially in the western-breeding populations, to the point of these populations being endangered.

Initially it was assumed by scientists, fishermen, and policy makers alike that the eastern and western populations had geographically separate feeding grounds as well as separate breeding grounds. Acting on this assumption, an international commission drew a line down the middle of the Atlantic Ocean and established strict fishing quotas on the western side of the line, with the intent of allowing the western population to recover. Modern tracking data, however, revealed that in fact the eastern and western bluefin populations mix freely on their feeding grounds across the entire North Atlantic—a swath of ocean that includes the most heavily fished waters in the world. Tuna caught on the eastern side of the line could just as likely be from the western breeding population as the eastern; thus the established policy could not achieve its intended goal.

Policy makers take more things into consideration than scientific knowledge and recommendations. For example, studies on the effects of atrazine on amphibians have led one U.S. group, the Natural Resources Defense Council, to take legal action to have atrazine banned on the basis of the Endangered Species Act. The U.S. Environmental Protection Agency, however, must also consider the potential loss to agriculture that such a ban would create and thus has continued to approve atrazine's use as long as environmental levels do not exceed 30 to 40 ppb—which is 300 to 400 times the levels shown to induce abnormalities in the Hayes studies. Scientific conclusions do not always prevail in the political world. Some scientific conclusions may have more influence than others, however, especially when they indicate a strong possibility of negative effects on humans.

Biology is crucial for understanding ecosystems

The world has been changing since its formation and continues to change with every passing day. Human activity, however, is resulting in an unprecedented *rate* of change in the world's ecosystems. For example, the mining and consumption of fossil fuels is releasing massive quantities of carbon dioxide into Earth's atmosphere. This anthropogenic (human-generated) increase in atmospheric carbon dioxide is largely responsible for the rapid rate of climate warming recorded over the last 50 years (**Figure 1.16**).

Our use of natural resources is putting stress on the ability of Earth's ecosystems to continue to produce the goods and services on which our society depends. Human activities are changing global climates at an unprecedented rate and are leading to the extinctions of large numbers of species (such as the amphibians featured in this chapter). The modern, warmer world is also experiencing the spread of new diseases and the resurgence of old ones. Biological knowledge is vital for determining the causes of these changes and for devising policies to deal with them.

Biology helps us understand and appreciate biodiversity

Beyond issues of policy and pragmatism lies the human "need to know." Humans are fascinated by the richness and diversity of life, and most people want to know more about organisms and how they interact. Human curiosity might even be seen as an adaptive trait—it is possible that such a trait could have been selected for if individuals who were motivated to learn about their surroundings were likely to have survived and reproduced better, on average, than their less curious relatives.

1.17 Discovering Life on Earth These biologists are collecting insects in the top boughs of a spruce tree in the Carmanah Valley of Vancouver, Canada. Biologists estimate that the number of species discovered to date is only a small percentage of the number of species that inhabit Earth. To fill this gap in our knowledge, biologists around the world are applying thorough sampling techniques and new genetic tools to document and understand the Earth's biodiversity.

Far from ending the process, new discoveries and greater knowledge typically engender questions no one thought to ask before. There are vast numbers of questions for which we do not yet have answers, and the most important motivator of most scientists is curiosity.

Observing the living world motivates many biologists to learn more and to constantly collect new information (Figure 1.17). An intimate understanding of the **natural history** of a group of organisms—that is, how those organisms get their food, reproduce, behave, regulate their internal environments, and interact with other organisms—facilitates observations and provides a stronger basis for framing hypotheses about about those observations. The more information biologists have and the more the observer knows about general principles, the more he or she is likely to gain new insights from observing nature.

Most humans engage in activities that depend on biodiversity. You may be an avid birdwatcher, or enjoy gardening, or seek out particular species if you hunt or fish. Some people like to observe or collect butterflies, or mushrooms, or other groups of plants, animals, and fungi. Displays of spring wildflowers bring out throngs of human viewers in many areas of the world. Hiking and camping in natural areas full of diverse species are activities enjoyed by millions. All of these interests support the growing industry of eco-tourism, which depends on the observation of rare or unusual species. Learning about biology greatly increases our enjoyment of these activities.

RECAP 1.3

Biology informs us about the structure, processes, and interactions of the living organisms that make up our world. Informed decisions about food and energy production, health, and our environment depend on biological knowledge. Biology also addresses the human need to understand the world around us, and helps us appreciate the diverse planet we call home.

- Describe an example of how modern biology is applied to agriculture. **See p. 15**
- Why are some antibiotics not as effective for treating bacterial diseases as they were when the drugs were originally introduced? **See p. 15**
- What is an example of a biological problem that is directly related to global climate change? **See p. 17**

This chapter has provided a brief roadmap of the rest of the book. Thinking about the principles outlined here may help you to clarify and make sense of the pages of detailed description to come. At the end of the course you may wish to revisit Chapter 1 and see if you have a different perspective on the world of biology.

?

Could atrazine in the environment affect species other than amphibians?

ANSWER

An important aspect of the scientific process is the replication of experimental results. In some cases the exact same experiment is repeated in another laboratory by other investigators and the results are compared. In other cases the experiment is repeated on other species to test the generality of the findings.

Following the publications by Hayes and his students, other investigators tested the effects of atrazine on other species of amphibians as well as on vertebrates other than amphibians. Feminizing effects of atrazine have now been demonstrated in fish, reptiles, and mammals. These results are not surprising, because as you will learn in Chapters 41 and 43, the hormonal controls of sex development and function are the same, and therefore the effects of atrazine should generalize to other vertebrate species.

Biologists have now studied the molecular mechanisms of the effects of atrazine on the hormonal control of sex and found that very similar responses to atrazine are seen in fish and in cultures of human cells. So atrazine in the environment is increasingly a concern for the health of many other species—and that includes humans.

CHAPTERSUMMARY 1

1.1 What Is Biology?

- **Biology** is the scientific study of living organisms, including their characteristics, functions, and interactions.
- All living organisms are related to one another through common descent. Shared features of all living organisms, such as specific chemical building blocks, a nearly universal genetic code, and sequence similarities across fundamental genes, support the common ancestry of life.
- Cells evolved early in the history of life. **Cellular specialization** allowed multicellular organisms to increase in size and diversity. **Review Figure 1.2**
- The instructions for a cell are contained in its **genome**, which consists of **DNA** molecules made up of sequences of **nucleotides**. Specific segments of DNA called **genes** contain the information the cell uses to make **proteins**. **Review Figure 1.5**
- **Photosynthesis** provided a means of capturing energy directly from sunlight and over time changed Earth's atmosphere.
- **Evolution**—change in the genetic makeup of biological **populations** through time—is a fundamental principle of life. Populations evolve through several different processes, including **natural selection**, which is responsible for the diversity of **adaptations** found in living organisms.
- Biologists use fossils, anatomical similarities and differences, and molecular comparisons of genomes to reconstruct the history of life. Three domains—**Bacteria**, **Archea**, and **Eukarya**—represent the major divisions, which were established very early in life's history. **Review Figure 1.7, ACTIVITY 1.1**
- Life can be studied at different levels of organization within a **biological hierarchy**. The specialized cells of multicellular organisms are organized into **tissues**, **organs**, and **organ systems**. Individual organisms form populations and interact with other organisms of their own and other species. The populations that live and interact in a defined area form a **community**, and communities together with their abiotic (nonliving) environment constitute an **ecosystem**. **Review Figure 1.9, ACTIVITY 1.2**
- Living organisms, whether unicellular or multicellular, must regulate their internal environment to maintain **homeostasis**, the range of physical conditions necessary for their survival and function.

1.2 How Do Biologists Investigate Life?

- Scientific methods combine observation, gathering information (**data**), experimentation, and logic to study the natural world. Many scientific investigations involve five steps: making observations, asking questions, forming hypotheses, making predictions, and testing those predictions. **Review Figure 1.10**
- **Hypotheses** are tentative answers to questions. Predictions made on the basis of a hypothesis are tested with additional observations and two kinds of experiments, **comparative** and **controlled experiments**. **Review Figures 1.11, 1.12, ANIMATED TUTORIAL 1.1**
- Quantifiable data are critical in evaluating hypotheses. Statistical methods are applied to quantitative data to establish whether or not the differences observed could be the result of chance. These methods start with the **null hypothesis** that there are no differences. **See Appendix B**
- Biological knowledge obtained from a **model system** may be generalized to other species.

1.3 Why Does Biology Matter?

- Application of biological knowledge is responsible for vastly increased agricultural production.
- Understanding and treatment of human disease requires an integration of a wide range of biological principles, from molecular biology through cell biology, physiology, evolution, and ecology.
- Biologists are often called on to advise government agencies on the solution of important problems that have a biological component.
- Biology is increasing important for understanding how organisms interact in a rapidly changing world.
- Biology helps us understand and appreciate the diverse living world.

Go to the Interactive Summary to review key figures, Animated Tutorials, and Activities Life10e.com/is1

CHAPTERREVIEW

REMEMBERING

1. Which of the following is *not* an attribute common to all living organisms?
   a. They are made up of a common set of chemical components, including particular nucleic and amino acids.
   b. They contain genetic information that uses a nearly universal code to specify the assembly of proteins.
   c. They share sequence similarities among their genes.
   d. They exist in populations that evolve over time.
   e. They extract energy from the sun in a process called photosynthesis.

2. In describing the hierarchy of life, which of the following descriptions of relationships is *not* accurate?
   a. An organ is a structure consisting of different types of cells and tissues.
   b. A population consists of all of the different animals in a particular type of environment.
   c. An ecosystem includes different communities.
   d. A tissue consists of a particular type of cells.
   e. A community consists of populations of different species.

3. Which of the following is a property of a good hypothesis?
   a. It is a statement of facts.
   b. It is general enough to explain a variety of possible experimental outcomes.
   c. It is independent of any observations.
   d. It explains things that are not addressable by experimentation.
   e. It can be falsified by experiments.

4. Which of the following events was most directly responsible for increasing oxygen in Earth's atmosphere?
   a. The cooling of the planet
   b. The origin of eukaryotes
   c. The origin of multicellularity
   d. The origin of photosynthesis
   e. The origin of prokaryotes

5. Which of the following is a reason to use statistics to evaluate data?
   a. It enables you to prove that your hypothesis is correct.
   b. It enables you to exclude data that do not fit your hypothesis.
   c. It makes it possible to exclude the null hypothesis.
   d. It enables you to predict experimental results.
   e. It accounts for variation in scientific measurements.

UNDERSTANDING & APPLYING

6. Why is it important in science to design and perform experiments that are capable of falsifying a hypothesis?

7. What is the significance of the fact that mitochondria and chloroplasts contain the DNA that instructs their form and function?

8. The results in Dr. Hayes's comparative experiments were more variable than the results from his controlled experiments. How would you explain this?

ANALYZING & EVALUATING

9. Biologists can now isolate genes from organisms and decode their DNA. When the nucleotide sequences from the same gene in different species are compared, differences are discovered. How could you use those data to deduce the evolutionary relationships among the organisms in your comparison?

10. Mitochondria are cell organelles that have their own DNA and replicate independently of the cell itself. In most organisms, mitochondria are inherited only from the mother. Based on this observation, when might it be advantageous or disadvantageous to use mitochondrial DNA rather than nuclear DNA for studying evolutionary relationships among populations?

Go to BioPortal at **yourBioPortal.com** for Animated Tutorials, Activities, LearningCurve Quizzes, Flashcards, and many other study and review resources.

2 Small Molecules and the Chemistry of Life

CHAPTER**OUTLINE**

Big Teeth Isotopes in *Camarasaurus* teeth yield clues about the behavior of these huge dinosaurs—150 million years after the last of them disappeared.

"YOU ARE WHAT YOU EAT—and that applies to teeth" is a modification of a famous saying about body chemistry. As we pointed out in Chapter 1, living things are made up of the same kinds of atoms that make up the inanimate universe. One of these atoms is oxygen (O), which is part of water (H_2O). Oxygen has two naturally occurring variants called isotopes; they have the same chemical properties but different weights because their nuclei have different numbers of neutrons. Both isotopes of O are incorporated into the bodies of animals that consume the isotopes in water and food.

The hard surface of teeth, called enamel, is made up largely of calcium phosphate, which has the chemical formula $Ca_3(PO_4)_2$. Calcium phosphate has a lot of oxygen, and the isotopic composition of the oxygen in enamel varies depending on where an animal was living when the enamel was made. When water evaporates from the ocean, it forms clouds that move inland and release rain. Water made up of the heavier isotope of O is heavier, and tends to fall more readily than water containing the lighter isotope. Regions of the world that are closer to the ocean receive rain containing more heavy water than regions further away, and these differences are reflected in the bodies of animals that dwell in these regions.

This property has been used to reveal an astounding fact about dinosaurs that lived in the great basins of southwestern North America about 150 million years ago. *Camarasaurus* was big, really big—up to 75 feet long and weighing up to 50 tons. Henry Ficke from Colorado College analyzed the oxygen isotopes in the enamel of *Camarasaurus* fossils and found two kinds of teeth: Some had the heavy oxygen content typical of rains and rocks in the basin region. But others, surprisingly, had a lower proportion of heavy oxygen, indicating that the animals had lived at higher elevations 300 km to the west. This indicates for the first time that dinosaurs migrated a long way from west to east. The reason for this migration is not clear. *Camarasaurus* ate a plant-based diet, and perhaps the migration was directed at finding food.

Life millions of years ago, as today, was based on chemistry. Just like the dinosaurs, we are what we eat—including our teeth. Indeed, biologists accept that life is based on chemistry and obeys universal laws of chemistry and physics. This physical–chemical view of life forms much of the basis of this book, and has led to great advances in biological science.

?

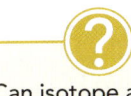

Can isotope analysis of water be used to detect climate change?

See answer on p. 36.

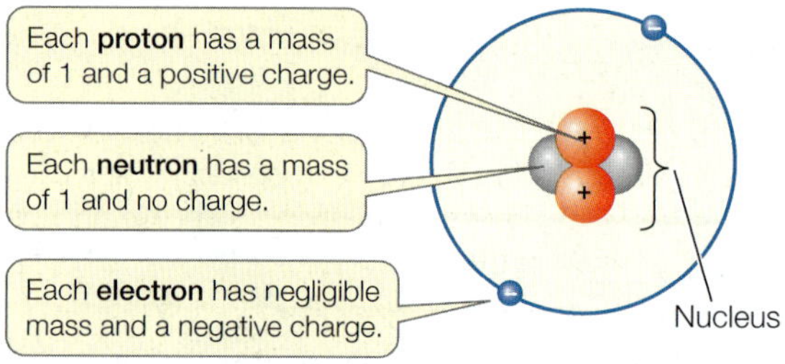
Each **proton** has a mass of 1 and a positive charge.
Each **neutron** has a mass of 1 and no charge.
Each **electron** has negligible mass and a negative charge.
Nucleus

2.1 The Helium Atom This representation of a helium atom is called a Bohr model. Although the nucleus accounts for virtually all of the atomic weight, it occupies only 1/10,000 of the atom's volume.

2.1 How Does Atomic Structure Explain the Properties of Matter?

All matter is composed of **atoms**. Atoms are tiny—more than a trillion (10^{12}) of them could fit on top of the period at the end of this sentence. Each atom consists of a dense, positively charged **nucleus**, around which one or more negatively charged **electrons** move (**Figure 2.1**). The nucleus contains one or more positively charged **protons** and may contain one or more **neutrons** with no electric charge. Atoms and their component particles have volume and mass, which are characteristics of all matter. **Mass** is a measure of the quantity of matter present; the greater the mass, the greater the quantity of matter.

The mass of a proton serves as a standard unit of measure called the **dalton** (named after the English chemist John Dalton). A single proton or neutron has a mass of about 1 dalton (Da), which is 1.7×10^{-24} grams (0.0000000000000000000000017 g), but an electron is even tinier at 9×10^{-28} g (0.0005 Da). Because the mass of an electron is negligible compared with the mass of a proton or a neutron, the contribution of electrons to the mass of an atom can usually be ignored when measurements and calculations are made. It is electrons, however, that determine how atoms will combine with other atoms to form stable associations.

Each proton has a positive electric charge, defined as +1 unit of charge. An electron has a negative charge equal and opposite to that of a proton (–1). The neutron, as its name suggests, is electrically neutral, so its charge is 0. Charges that are different (+/–) attract each other, whereas charges that are alike (+/+, –/–) repel each other. Generally, atoms are electrically neutral because the number of electrons in an atom equals the number of protons.

An element consists of only one kind of atom

An **element** is a pure substance that contains only one kind of atom. The element hydrogen consists only of hydrogen atoms; the element iron consists only of iron atoms. The atoms of each element have certain characteristics or properties that distinguish them from the atoms of other elements. These physical and chemical (reactive) properties depend on the numbers of subatomic particles the atoms contain. Such properties include mass and how the atoms interact and associate with other atoms.

There are 94 elements in nature and at least another 24 have been made in physics laboratories. About 98 percent of the mass of every living organism is composed of just six elements:

Carbon (symbol C)	Hydrogen (H)	Nitrogen (N)
Oxygen (O)	Phosphorus (P)	Sulfur (S)

The biological roles of these elements will be our major concern in this book, but other elements are found in living organisms as well. Sodium and potassium, for example, are essential for nerve function; calcium can act as a biological signal; iodine is a component of a vital hormone; and magnesium is bound to chlorophyll in plants.

Each element has a unique number of protons

An element differs from other elements by the number of protons in the nucleus of each of its atoms; the number of protons is designated the **atomic number**. This atomic number is unique to each element and does not change. The atomic number of helium is 2, and an atom of helium always has two protons; the atomic number of oxygen is 8, and an atom of oxygen always has eight protons. Since the number of protons (and electrons) determines how an element behaves in chemical reactions, it is possible to arrange the elements in a table such that those with similar chemical properties are grouped together. This is the familiar **periodic table** that is shown in **Figure 2.2**.

Go to Media Clip 2.1
The Elements Song
Life10e.com/mc2.1

Along with a definitive number of protons, every element except hydrogen has one or more neutrons in its nucleus. The **mass number** of an atom is the total number of protons and neutrons in its nucleus. The nucleus of a carbon atom contains six protons and six neutrons and has a mass number of 12. Oxygen has eight protons and eight neutrons and has a mass number of 16. Since the mass of an electron is negligible, the mass number is essentially the mass of the atom (see below) in daltons.

By convention, we often print the symbol for an element with the atomic number at the lower left and the mass number at the upper left, both immediately preceding the symbol. Thus hydrogen, carbon, and oxygen can be written as ${}^{1}_{1}H$, ${}^{12}_{6}C$, and ${}^{16}_{8}O$, respectively.

The number of neutrons differs among isotopes

In some elements, the number of neutrons in the atomic nucleus is not always the same. Different **isotopes** of the same element have the same number of protons but different numbers of neutrons, as you saw in the opening story of this chapter. Many elements have several isotopes. Generally, isotopes are formed when atoms combine and/or release particles (decay). The isotopes of hydrogen shown below each have special names, but the isotopes of most elements do not have distinct names.

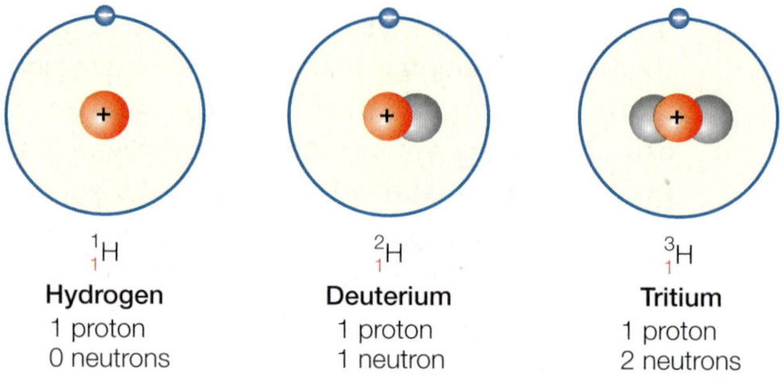
${}^{1}_{1}H$
Hydrogen
1 proton
0 neutrons
${}^{2}_{1}H$
Deuterium
1 proton
1 neutron
${}^{3}_{1}H$
Tritium
1 proton
2 neutrons

2.2 The Periodic Table The periodic table groups the elements according to their physical and chemical properties. Elements 1–94 occur in nature; elements with atomic numbers above 94 were created in the laboratory.

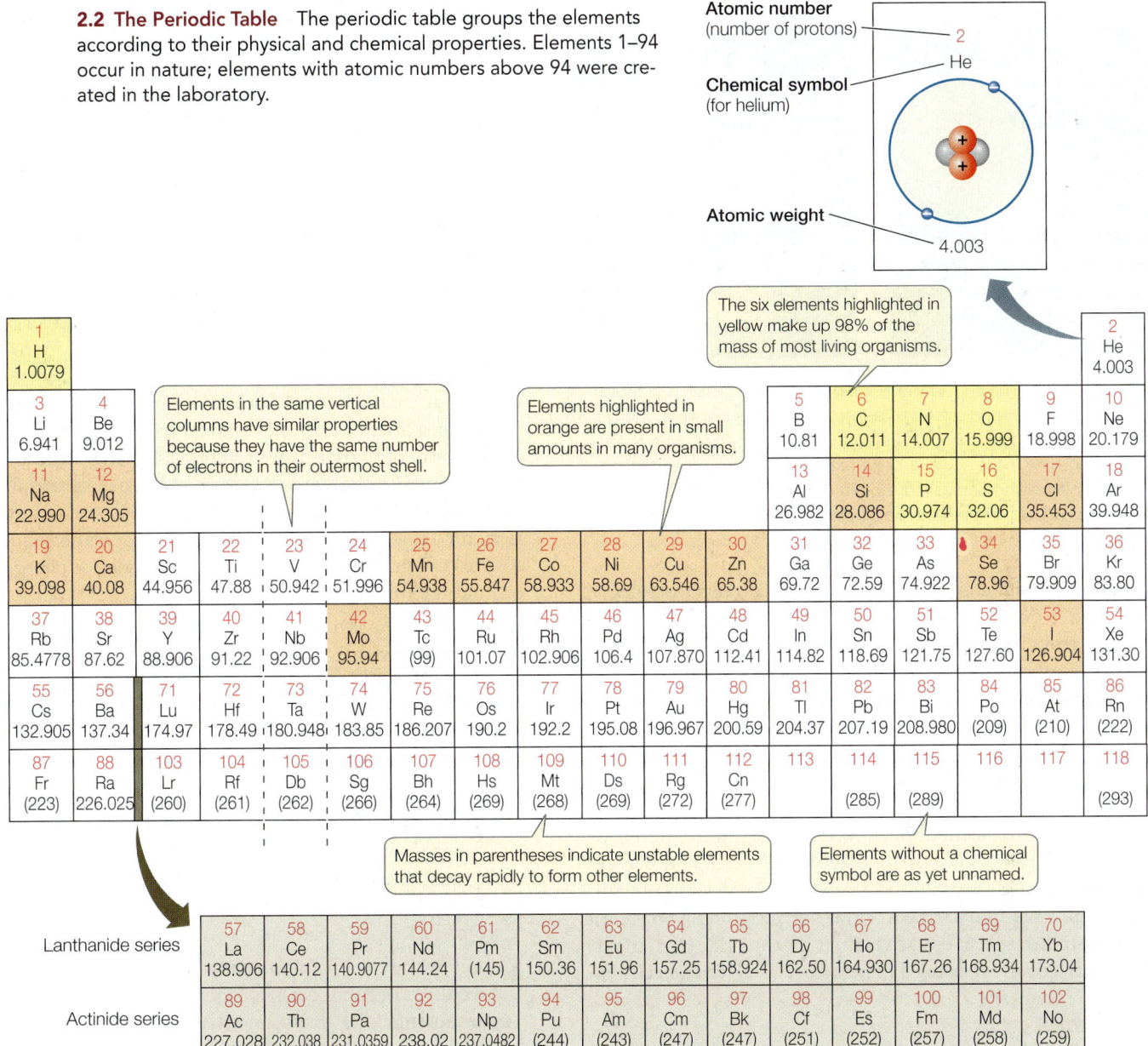

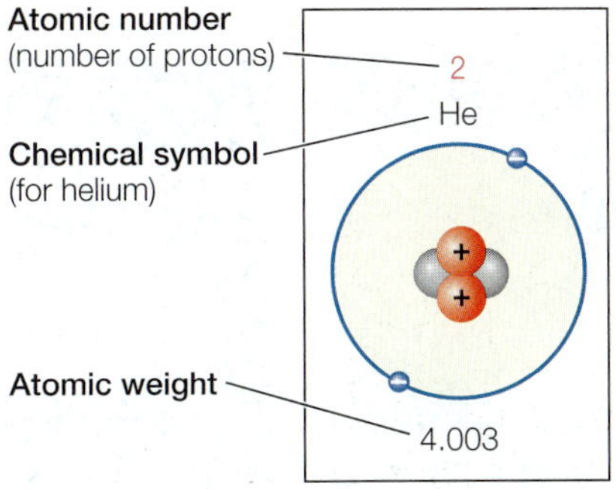
Atomic number
(number of protons)
2
He
Chemical symbol
(for helium)
Atomic weight
4.003

The six elements highlighted in yellow make up 98% of the mass of most living organisms.

Elements in the same vertical columns have similar properties because they have the same number of electrons in their outermost shell.

Elements highlighted in orange are present in small amounts in many organisms.

1 H 1.0079																	2 He 4.003
3 Li 6.941	4 Be 9.012											5 B 10.81	6 C 12.011	7 N 14.007	8 O 15.999	9 F 18.998	10 Ne 20.179
11 Na 22.990	12 Mg 24.305											13 Al 26.982	14 Si 28.086	15 P 30.974	16 S 32.06	17 Cl 35.453	18 Ar 39.948
19 K 39.098	20 Ca 40.08	21 Sc 44.956	22 Ti 47.88	23 V 50.942	24 Cr 51.996	25 Mn 54.938	26 Fe 55.847	27 Co 58.933	28 Ni 58.69	29 Cu 63.546	30 Zn 65.38	31 Ga 69.72	32 Ge 72.59	33 As 74.922	34 Se 78.96	35 Br 79.909	36 Kr 83.80
37 Rb 85.4778	38 Sr 87.62	39 Y 88.906	40 Zr 91.22	41 Nb 92.906	42 Mo 95.94	43 Tc (99)	44 Ru 101.07	45 Rh 102.906	46 Pd 106.4	47 Ag 107.870	48 Cd 112.41	49 In 114.82	50 Sn 118.69	51 Sb 121.75	52 Te 127.60	53 I 126.904	54 Xe 131.30
55 Cs 132.905	56 Ba 137.34	71 Lu 174.97	72 Hf 178.49	73 Ta 180.948	74 W 183.85	75 Re 186.207	76 Os 190.2	77 Ir 192.2	78 Pt 195.08	79 Au 196.967	80 Hg 200.59	81 Tl 204.37	82 Pb 207.19	83 Bi 208.980	84 Po (209)	85 At (210)	86 Rn (222)
87 Fr (223)	88 Ra 226.025	103 Lr (260)	104 Rf (261)	105 Db (262)	106 Sg (266)	107 Bh (264)	108 Hs (269)	109 Mt (268)	110 Ds (269)	111 Rg (272)	112 Cn (277)	113	114 (285)	115 (289)	116	117	118 (293)

Masses in parentheses indicate unstable elements that decay rapidly to form other elements.

Elements without a chemical symbol are as yet unnamed.

Lanthanide series	57 La 138.906	58 Ce 140.12	59 Pr 140.9077	60 Nd 144.24	61 Pm (145)	62 Sm 150.36	63 Eu 151.96	64 Gd 157.25	65 Tb 158.924	66 Dy 162.50	67 Ho 164.930	68 Er 167.26	69 Tm 168.934	70 Yb 173.04
Actinide series	89 Ac 227.028	90 Th 232.038	91 Pa 231.0359	92 U 238.02	93 Np 237.0482	94 Pu (244)	95 Am (243)	96 Cm (247)	97 Bk (247)	98 Cf (251)	99 Es (252)	100 Fm (257)	101 Md (258)	102 No (259)

The natural isotopes of carbon, for example, are ^{12}C (six neutrons in the nucleus), ^{13}C (seven neutrons), and ^{14}C (eight neutrons). Note that all three (called "carbon-12," "carbon-13," and "carbon-14") have six protons, so they are all carbon. Most carbon atoms are ^{12}C, about 1.1 percent are ^{13}C, and a tiny fraction are ^{14}C. All have virtually the same chemical reactivity, which is an important property for their use in experimental biology and medicine.

An element's **atomic weight** (or relative atomic mass) is equivalent to the average of the mass numbers of a representative sample of atoms of that element, with all the isotopes in their normally occurring proportions. More precisely, an element's atomic weight is defined as the ratio of the average mass per atom of the element to 1/12 of the mass of an atom of ^{12}C. Because it is a ratio, atomic weight is a dimensionless physical quantity—it is not expressed in units. The atomic weight of hydrogen, taking into account all of its isotopes and their typical abundances, is 1.00794. This number is fractional because it is the average of the contributing masses of all of the isotopes. This definition implies that in any given sample of hydrogen atoms of a particular element found on Earth, the average composition of isotopes will be constant. But as you saw in the opening to this chapter, that is not necessarily so. Some water has more of the heavy isotopes. So chemists are now listing atomic weights as ranges, for example, H: 1.00784–1.00811.

Most isotopes are stable. But some, called **radioisotopes**, are unstable and spontaneously give off energy in the form of α (alpha), β (beta), or γ (gamma) radiation from the atomic nucleus. Known as **radioactive decay**, this release of energy transforms the original atom. The type of transformation varies depending on the radioisotope, but some result in a different number of protons, so that the original atom becomes a different element.

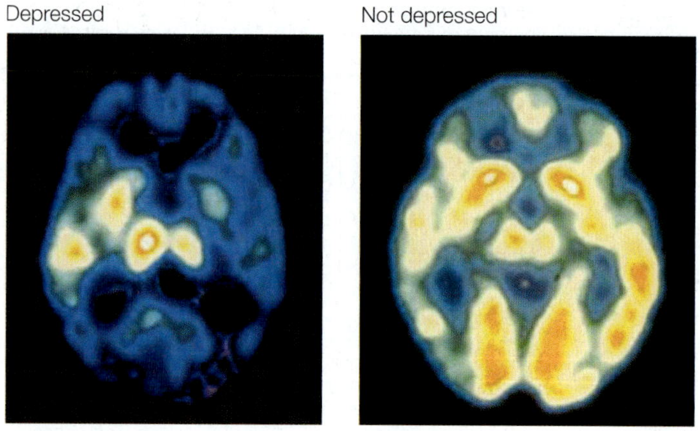
Depressed
Not depressed

2.3 Tagging the Brain In these images from live persons, a radioactively labeled sugar is used to detect differences between the brain activity of a depressed person (left) and that of a person who is not depressed. The more active a brain region, the more sugar it takes up (shown as orange areas). The brain of the depressed person (left) shows less activity than the brain of the person who is not depressed.

With sensitive instruments, scientists can use the released radiation to detect the presence of radioisotopes. For instance, if an earthworm is given food containing a radioisotope, its path through the soil can be followed using a simple detector called a Geiger counter. Most atoms in living organisms are organized into stable associations called **molecules**. If a radioisotope is incorporated into a molecule, it acts as a tag or label, allowing a researcher or physician to track the molecule in an experiment or in the body (**Figure 2.3**). Radioisotopes are also used to date fossils, an application described in Section 25.1.

Although radioisotopes are useful in research and in medicine, even a low dose of the radiation they emit has the potential to damage molecules and cells. However, these damaging effects are sometimes used to our advantage; for example, the radiation from ^{60}Co (cobalt-60) is used in medicine to kill cancer cells.

The behavior of electrons determines chemical bonding and geometry

The number of electrons in an atom determines how it will combine with other atoms. Biologists are interested in how chemical changes take place in living cells. When considering atoms, they are concerned primarily with electrons because the behavior of electrons explains how chemical reactions occur. Chemical reactions alter the atomic compositions of substances and thus alter their properties. Reactions usually involve changes in the distribution of electrons between atoms.

The location of a given electron in an atom at any given time is impossible to determine. We can only describe a volume of space within the atom where the electron is likely to be. The region of space where the electron is found at least 90 percent of the time is the electron's **orbital**. Orbitals have characteristic shapes and orientations, and a given orbital can be occupied by a maximum of two electrons (**Figure 2.4**). Thus any atom larger than helium (atomic number 2) must have electrons in two or more orbitals. As we move from lighter to heavier atoms in the periodic table, the orbitals are filled in a specific sequence, in a series of what are known as **electron shells**, or energy levels, around the nucleus.

- *First shell*: The innermost electron shell consists of just one orbital, called an *s* orbital. A hydrogen atom ($_1H$) has one electron in its first shell; helium ($_2He$) has two. Atoms of all other elements have two or more shells to accommodate orbitals for additional electrons.
- *Second shell*: The second shell contains four orbitals (an *s* orbital and three *p* orbitals) and hence holds up to eight electrons. As depicted in Figure 2.4, *s* orbitals have the shape of a sphere, whereas *p* orbitals are oriented at right angles to one another. The orientations of these orbitals in space contribute to the three-dimensional shapes of molecules when atoms link to other atoms.
- *Additional shells*: Elements with more than ten electrons have three or more electron shells. The farther a shell is from the nucleus, the higher the energy level is for an electron occupying that shell.

Go to Activity 2.1 Electron Orbitals Life10e.com/ac2.1

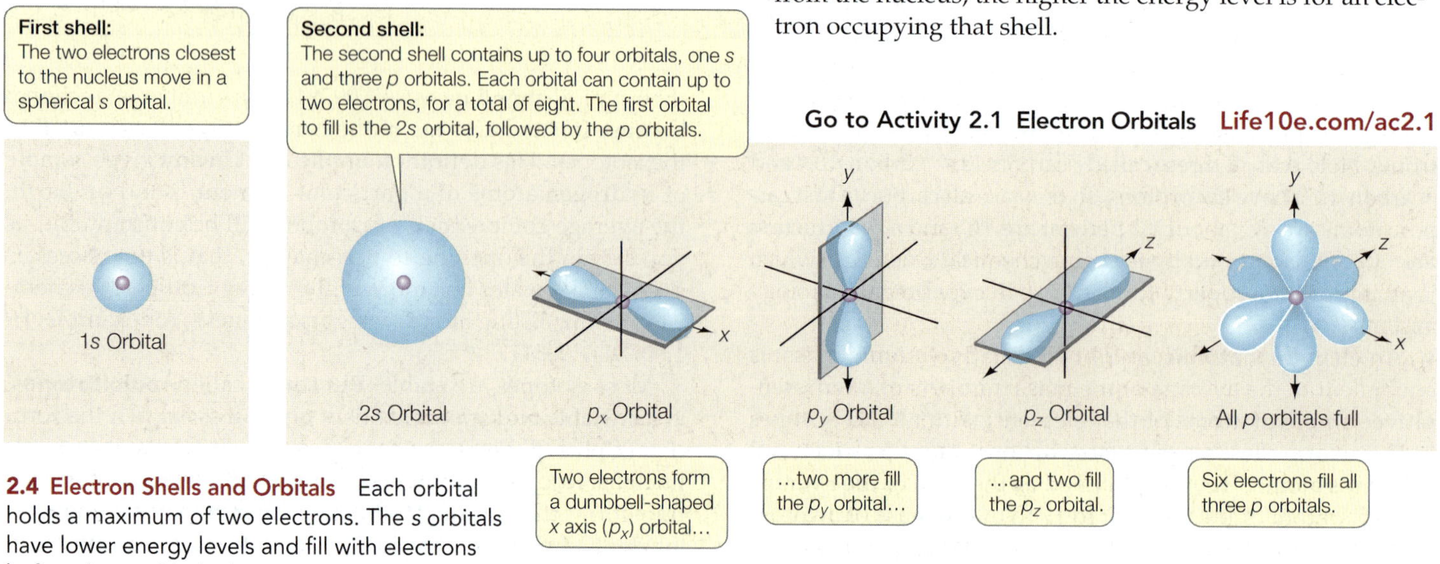
First shell:
The two electrons closest to the nucleus move in a spherical *s* orbital.
Second shell:
The second shell contains up to four orbitals, one *s* and three *p* orbitals. Each orbital can contain up to two electrons, for a total of eight. The first orbital to fill is the 2*s* orbital, followed by the *p* orbitals.
1s Orbital
2s Orbital
p_x Orbital
p_y Orbital
p_z Orbital
All *p* orbitals full
x
y
z
Two electrons form a dumbbell-shaped *x* axis (p_x) orbital...
...two more fill the p_y orbital...
...and two fill the p_z orbital.
Six electrons fill all three *p* orbitals.

2.4 Electron Shells and Orbitals Each orbital holds a maximum of two electrons. The *s* orbitals have lower energy levels and fill with electrons before the *p* orbitals do.

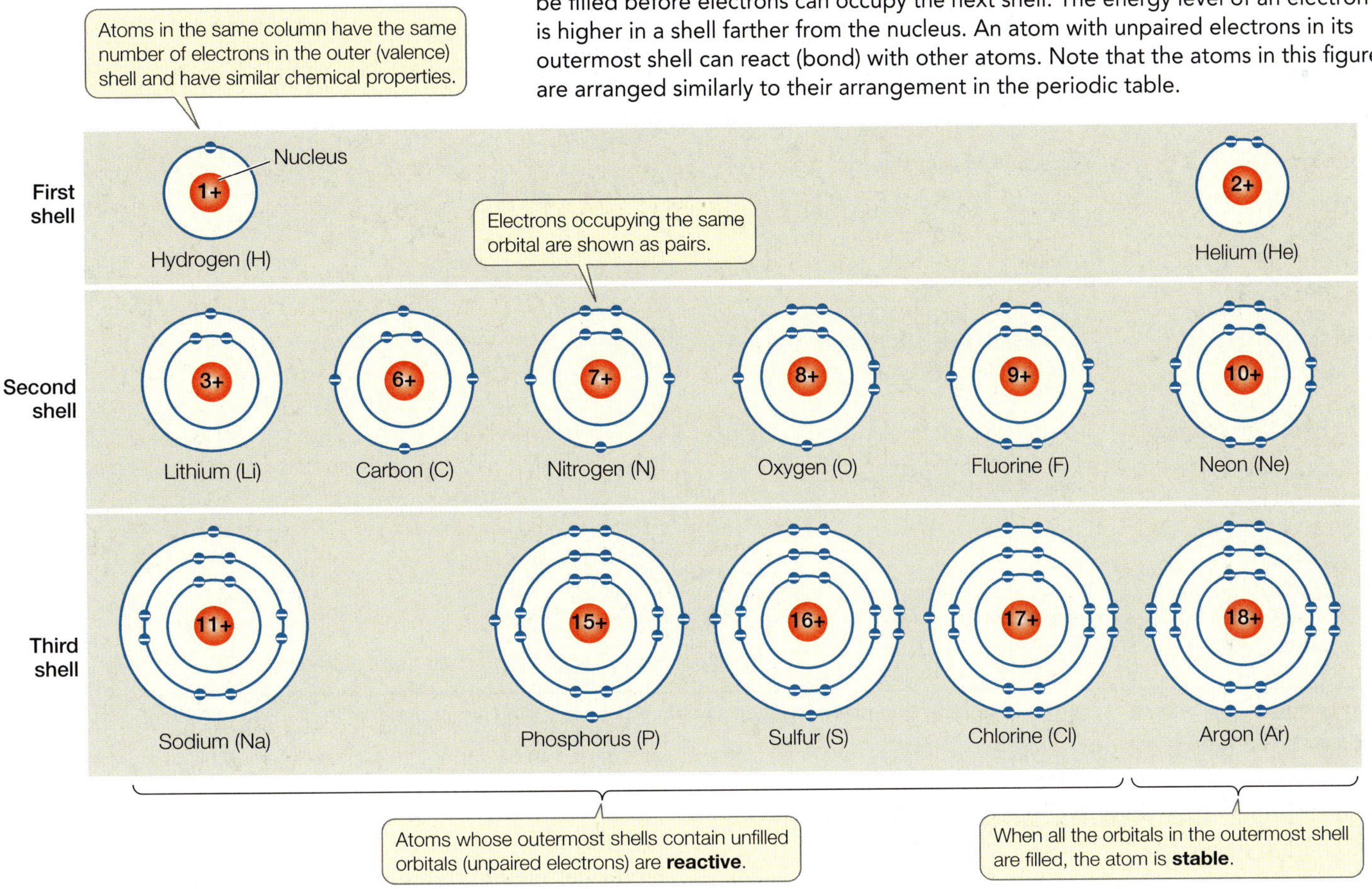
Atoms in the same column have the same number of electrons in the outer (valence) shell and have similar chemical properties.
Nucleus
First shell
1+
Hydrogen (H)
Electrons occupying the same orbital are shown as pairs.
2+
Helium (He)
Second shell
3+
Lithium (Li)
6+
Carbon (C)
7+
Nitrogen (N)
8+
Oxygen (O)
9+
Fluorine (F)
10+
Neon (Ne)
Third shell
11+
Sodium (Na)
15+
Phosphorus (P)
16+
Sulfur (S)
17+
Chlorine (Cl)
18+
Argon (Ar)
Atoms whose outermost shells contain unfilled orbitals (unpaired electrons) are reactive.
When all the orbitals in the outermost shell are filled, the atom is stable.

2.5 Electron Shells Determine the Reactivity of Atoms Each shell can hold a specific maximum number of electrons. Going out from the nucleus, each shell must be filled before electrons can occupy the next shell. The energy level of an electron is higher in a shell farther from the nucleus. An atom with unpaired electrons in its outermost shell can react (bond) with other atoms. Note that the atoms in this figure are arranged similarly to their arrangement in the periodic table.

The *s* orbitals fill with electrons first, and their electrons have the lowest energy level. Subsequent shells have different numbers of orbitals, but the outermost shells usually hold only eight electrons. In any atom, the outermost electron shell (the **valence shell**) determines how the atom combines with other atoms—that is, how the atom behaves chemically. When a valence shell with four orbitals contains eight electrons, there are no unpaired electrons and the atom is stable—it is least likely to react with other atoms (Figure 2.5). Examples of chemically stable elements are helium, neon, and argon. By contrast, atoms that have one or more unpaired electrons in their outer shells are capable of reacting with other atoms.

Atoms with unpaired electrons (i.e., partially filled orbitals) in their outermost electron shells are unstable and will undergo reactions in order to fill their outermost shells. *Reactive atoms can attain stability either by sharing electrons with other atoms or by losing or gaining one or more electrons.* In either case, the atoms involved are bonded together into stable associations called molecules. The tendency of atoms to form stable molecules so that they have eight electrons in their outermost shells is known as the octet rule. Many atoms in biologically important molecules—for example, carbon (C) and nitrogen (N)—follow this rule. An important exception is hydrogen (H), which attains stability when two electrons occupy its single shell (consisting of just one *s* orbital).

RECAP 2.1

Living organisms are composed of the same set of chemical elements as the rest of the universe. An atom consists of a nucleus of protons and neutrons, and a characteristic configuration of electrons in orbitals around the nucleus. This structure determines the atom's chemical properties.

- Describe the arrangement of protons, neutrons, and electrons in an atom. See Figure 2.1
- Use the periodic table to identify some of the similarities and differences in atomic structure among oxygen, carbon, and helium. How does the configuration of the valence shell influence the placement of an element in the periodic table? See p. 23 and Figures 2.2, 2.5
- How does bonding help a reactive atom achieve stability? See p. 25 and Figure 2.5

We have introduced the individual players on the biochemical stage—the atoms. We have shown how the number of unpaired electrons in an atom's valence shell drives its "quest for stability." Next we will describe the different types of chemical bonds that can lead to stability—joining atoms together into molecular structures with hosts of different properties.

TABLE 2.1 Chemical Bonds and Interactions

Name	Basis of Interaction	Structure	Bond Energy[a]
Covalent bond	Sharing of electron pairs	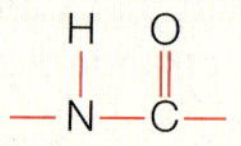	50–110
Ionic attraction	Attraction of opposite charges	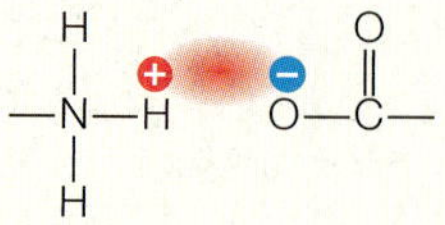	3–7
Hydrogen bond	Electrical attraction between a covalently bonded H atom and an electronegative atom	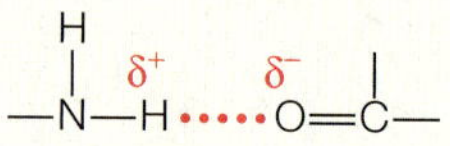	3–7
Hydrophobic interaction	Interaction of nonpolar substances in the presence of polar substances (especially water)	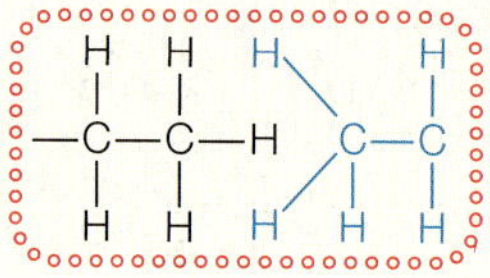	1–2
van der Waals interaction	Interaction of electrons of nonpolar substances	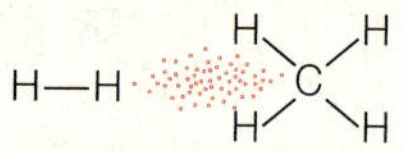	1

[a]Bond energy is the amount of energy in kcal/mol needed to separate two bonded or interacting atoms under physiological conditions.

2.2 How Do Atoms Bond to Form Molecules?

A **chemical bond** is an attractive force that links two atoms together in a molecule. There are several kinds of chemical bonds (Table 2.1). In this section we will begin with covalent bonds, the strong bonds that result from the sharing of electrons. Next we will examine ionic attractions, which form when an atom gains or loses one or more electrons to achieve stability. We will then consider other, weaker kinds of interactions, including hydrogen bonds.

Go to Animated Tutorial 2.1
Chemical Bond Formation
Life10e.com/at2.1

Covalent bonds consist of shared pairs of electrons

A **covalent bond** forms when two atoms attain stable electron numbers in their outermost shells by sharing one or more pairs of electrons. Consider two hydrogen atoms coming into close proximity, each with an unpaired electron in its single shell (Figure 2.6). When the electrons pair up, a stable association is formed, and this links the two hydrogen atoms in a covalent bond, forming the molecule H_2.

A **compound** is a pure substance made up of two or more different elements bonded together in a fixed ratio. Chemical symbols identify the different elements in a compound, and subscript numbers indicate how many atoms of each element are present (e.g., H_2O has two atoms of hydrogen bonded to a single oxygen atom). Every compound has a **molecular weight** (relative molecular mass) that is the sum of the atomic weights of all atoms in the molecule. Looking at

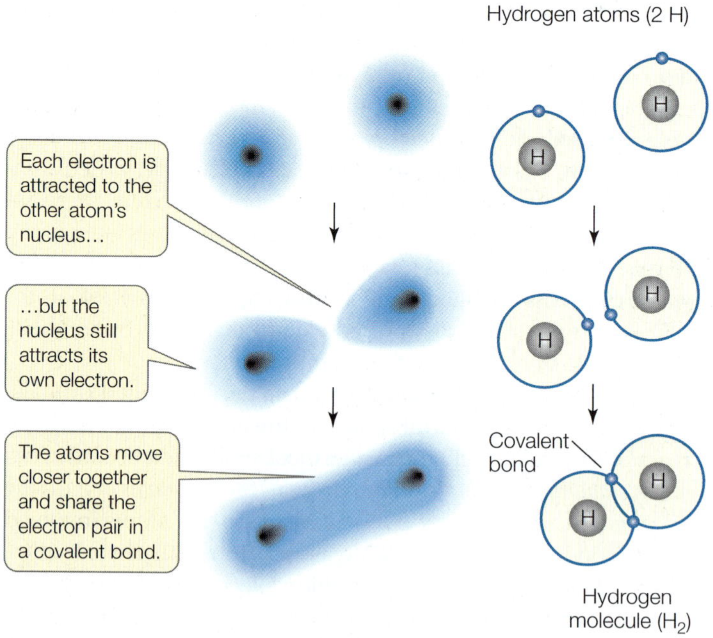
Hydrogen atoms (2 H)
Each electron is attracted to the other atom's nucleus...
...but the nucleus still attracts its own electron.
The atoms move closer together and share the electron pair in a covalent bond.
Covalent bond
Hydrogen molecule (H_2)

2.6 Electrons Are Shared in Covalent Bonds Two hydrogen atoms can combine to form a hydrogen molecule. A covalent bond forms when the electron orbitals of the two atoms overlap in an energetically stable manner.

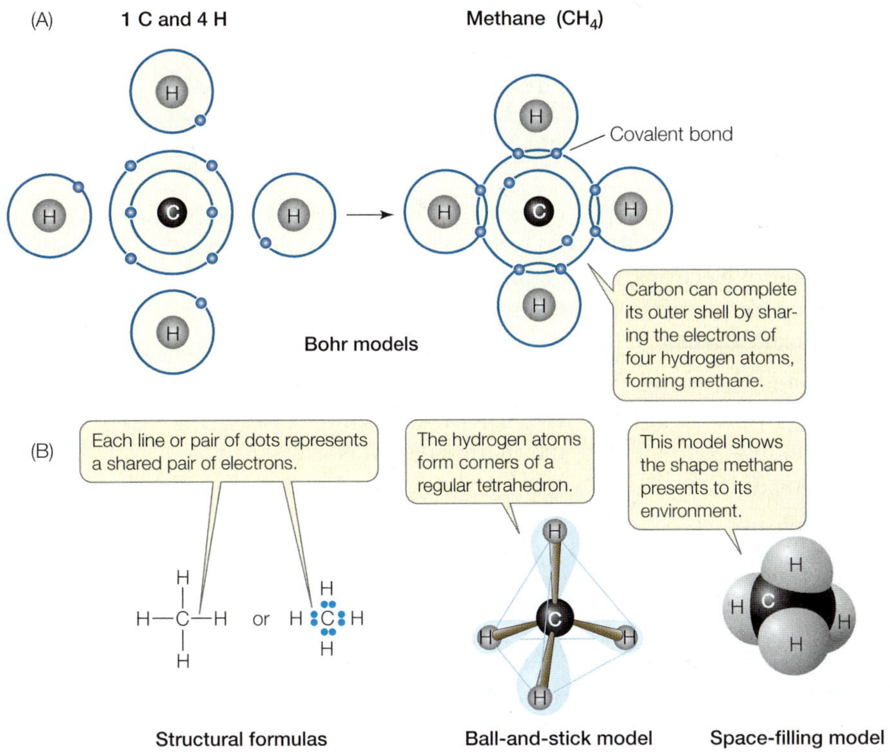

(A)
1 C and 4 H
Methane (CH_4)
Covalent bond
Carbon can complete its outer shell by sharing the electrons of four hydrogen atoms, forming methane.
Bohr models
(B)
Each line or pair of dots represents a shared pair of electrons.
The hydrogen atoms form corners of a regular tetrahedron.
This model shows the shape methane presents to its environment.
or
Structural formulas
Ball-and-stick model
Space-filling model

2.7 Covalent Bonding Can Form Compounds (A) Bohr models showing the formation of covalent bonds in methane, whose molecular formula is CH_4. Electrons are shown in shells around the nucleus. (B) Three additional ways of representing the structure of methane. The ball-and-stick model and the space-filling model show the spatial orientations of the bonds. The space-filling model indicates the overall shape and surface of the molecule. In the chapters that follow, different conventions will be used to depict molecules. Bear in mind that these are models to illustrate certain properties, not accurate portrayals of how atoms would actually appear.

the periodic table in Figure 2.2, you can calculate the molecular weight of water to be 18.01. (But remember that this value comes from the average atomic weights of hydrogen and oxygen; the molecular weight of the heavy water in our opening story is higher because it is formed from heavier isotopes.) Molecules that make up living organisms can have molecular weights of up to half a billion, and covalent bonds are found in all.

How are the covalent bonds formed in a molecule of methane gas (CH_4)? The carbon atom has six electrons: two electrons fill its inner shell, and four unpaired electrons travel in its outer shell. Because its outer shell can hold up to eight electrons, carbon can share electrons with up to four other atoms—*it can form four covalent bonds* (**Figure 2.7A**). When an atom of carbon reacts with four hydrogen atoms, methane forms. Thanks to electron sharing, the outer shell of methane's carbon atom is now filled with eight electrons, a stable configuration. The outer shell of each of the four hydrogen atoms is also filled. Four covalent bonds—four shared electron pairs—hold methane together. **Figure 2.7B** shows several different ways to represent the molecular structure of methane. **Table 2.2** shows the covalent bonding capacities of some biologically significant elements.

STRENGTH AND STABILITY Covalent bonds are very strong, meaning that it takes a lot of energy to break them. At temperatures where life exists, the covalent bonds of biological molecules are quite stable, as are their three-dimensional structures. However, this stability does not preclude change, as we will discover.

TABLE 2.2 Covalent Bonding Capabilities of Some Biologically Important Elements

Element	Usual Number of Covalent Bonds
Hydrogen (H)	1
Oxygen (O)	2
Sulfur (S)	2
Nitrogen (N)	3
Carbon (C)	4
Phosphorus (P)	5

ORIENTATION For a given pair of elements—for example, carbon bonded to hydrogen—the length of the covalent bond is always the same. And for a given atom within a molecule, the angle of each of its covalent bonds, with respect to the other bonds, is generally the same. This is true regardless of the type of larger molecule that contains the atom. For example, the four filled orbitals around the carbon atom in methane are always distributed in space so that the bonded hydrogen atoms point to the corners of a regular tetrahedron, with carbon in the center (see Figure 2.7B). Even when carbon is bonded to four atoms other than hydrogen, this three-dimensional orientation is more or less maintained. The orientation of covalent bonds in space gives the molecules their three-dimensional geometry, and the shapes of molecules contribute to their biological functions, as we will see in Section 3.1.

Even though the orientations of bonds around each atom are fairly stable, the shapes of molecules can change. Think of a single covalent bond as an axle around which the two atoms, along with their other bonded atoms, can rotate.

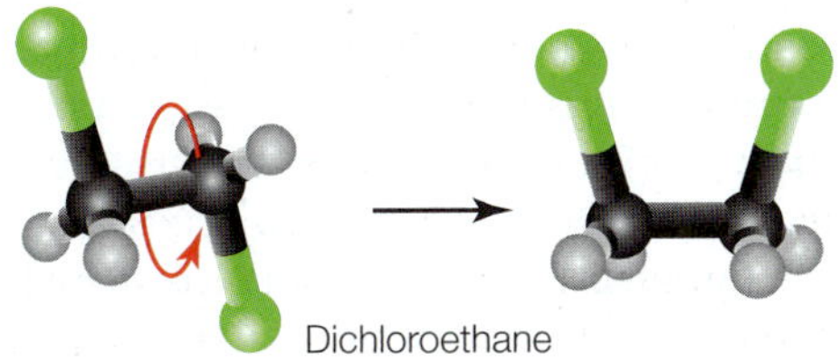

Dichloroethane

This phenomenon has enormous implications for the large molecules that make up living tissues. In long chains of atoms (especially carbons) that can rotate freely, there are many possibilities for the arrangement of atoms within the chain. This allows molecules to alter their structures, for example, to fit other molecules.

MULTIPLE COVALENT BONDS Two atoms can share more than one pair of electrons, forming multiple covalent bonds. These can be represented by lines between the chemical symbols for the linked atoms:

- A single bond involves the sharing of a single pair of electrons (for example, H—H or C—H).
- A double bond involves the sharing of four electrons (two pairs) (C=C).
- Triple bonds—six shared electrons—are rare, but there is one in nitrogen gas (N≡N), which is the major component of the air we breathe.

UNEQUAL SHARING OF ELECTRONS If two atoms of the same element are covalently bonded, there is an equal sharing of the pair(s) of electrons in their outermost shells. However, when the two atoms are of different elements, the sharing is not necessarily equal. One nucleus may exert a greater attractive force on the electron pair than the other nucleus, so that the pair tends to be closer to that atom.

The attractive force that an atomic nucleus exerts on electrons in a covalent bond is called its **electronegativity**. The electronegativity of an atom depends on how many positive charges it has (atoms with more protons are more positive and thus more attractive to electrons) and on the distance between the nucleus and the electrons in the outer (valence) shell (the closer the electrons, the greater the electronegative pull). Table 2.3 shows the electronegativities (which are calculated to produce dimensionless quantities) of some elements important in biological systems.

If two atoms are close to each other in electronegativity, they will share electrons equally in what is called a **nonpolar covalent bond**. Two oxygen atoms, for example, each with an electronegativity of 3.5, will share electrons equally. So will two hydrogen atoms (each with an electronegativity of 2.1). But when hydrogen bonds with oxygen to form water, the electrons involved are unequally shared; they tend to be nearer to the oxygen nucleus because it is the more electronegative of the two. When electrons are drawn to one nucleus more than to the other, the result is a **polar covalent bond** (Figure 2.8).

Because of this unequal sharing of electrons, the oxygen end of the hydrogen–oxygen bond has a slightly negative charge (symbolized by δ^- and spoken of as "delta negative," meaning a partial unit of charge), and the hydrogen end has a slightly positive charge (δ^+). The bond is **polar** because these opposite charges are separated at the two ends, or poles, of the bond. The partial charges that result from polar covalent bonds produce polar molecules or polar regions of large molecules. Polar bonds within molecules greatly influence the interactions they have with other polar molecules. Water (H_2O) is a polar compound, and this polarity has significant effects on its physical properties and chemical reactivity, as we will see in later chapters.

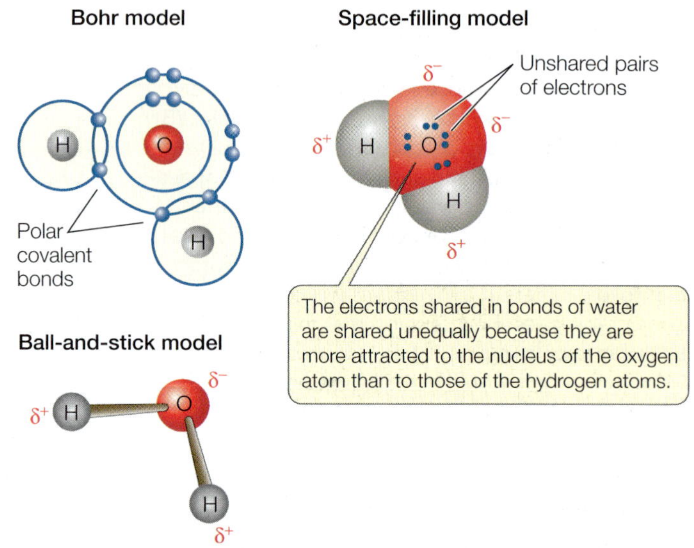
Bohr model
Space-filling model
H
O
H
Polar covalent bonds
δ−
Unshared pairs of electrons
δ+
δ−
δ+
The electrons shared in bonds of water are shared unequally because they are more attracted to the nucleus of the oxygen atom than to those of the hydrogen atoms.
Ball-and-stick model
δ+
δ−
δ+

2.8 Water's Covalent Bonds Are Polar These three representations all illustrate polar covalent bonding in water (H_2O). When atoms with different electronegativities, such as oxygen and hydrogen, form a covalent bond, the electrons are drawn to one nucleus more than to the other. A molecule held together by such a polar covalent bond has partial (δ^+ and δ^-) charges at different surfaces. In water, the shared electrons are displaced toward the oxygen atom's nucleus.

Ionic attractions form by electrical attraction

When one interacting atom is much more electronegative than the other, a complete transfer of one or more electrons may take place. Consider sodium (electronegativity 0.9) and chlorine (3.1). A sodium atom has only one electron in its outermost shell; this condition is unstable. A chlorine atom has seven electrons in its outermost shell—another unstable condition. Since the electronegativity of chlorine is so much greater than that of sodium, any electrons involved in bonding will tend to transfer completely

TABLE 2.3 Some Electronegativities

Element	Electronegativity
Oxygen (O)	3.5
Chlorine (Cl)	3.1
Nitrogen (N)	3.0
Carbon (C)	2.5
Phosphorus (P)	2.1
Hydrogen (H)	2.1
Sodium (Na)	0.9
Potassium (K)	0.8

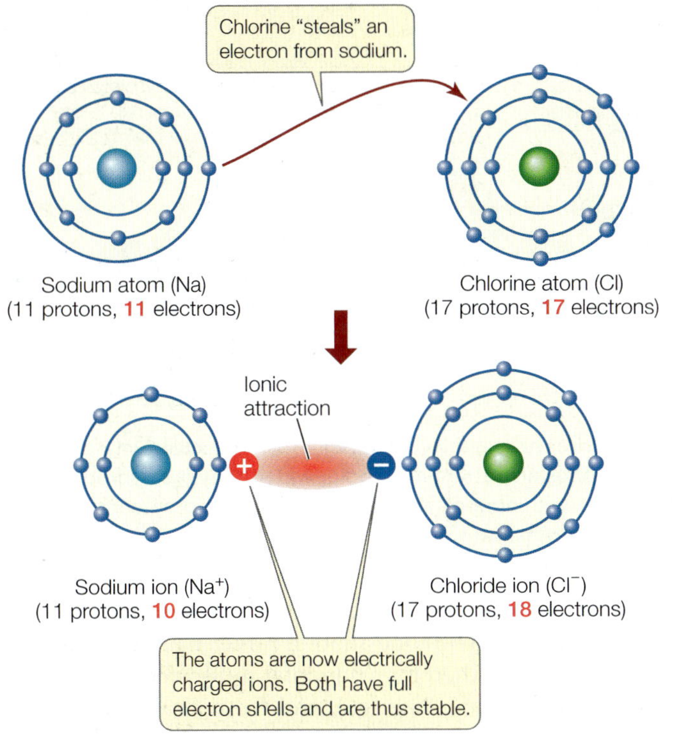
Chlorine "steals" an electron from sodium.
Sodium atom (Na)
(11 protons, 11 electrons)
Chlorine atom (Cl)
(17 protons, 17 electrons)
Ionic attraction
Sodium ion (Na+)
(11 protons, 10 electrons)
Chloride ion (Cl−)
(17 protons, 18 electrons)
The atoms are now electrically charged ions. Both have full electron shells and are thus stable.

2.9 Formation of Sodium and Chloride Ions When a sodium atom reacts with a chlorine atom, the more electronegative chlorine fills its outermost shell by "stealing" an electron from the sodium. In so doing, the chlorine atom becomes a negatively charged chloride ion (Cl^-). With one less electron, the sodium atom becomes a positively charged sodium ion (Na^+).

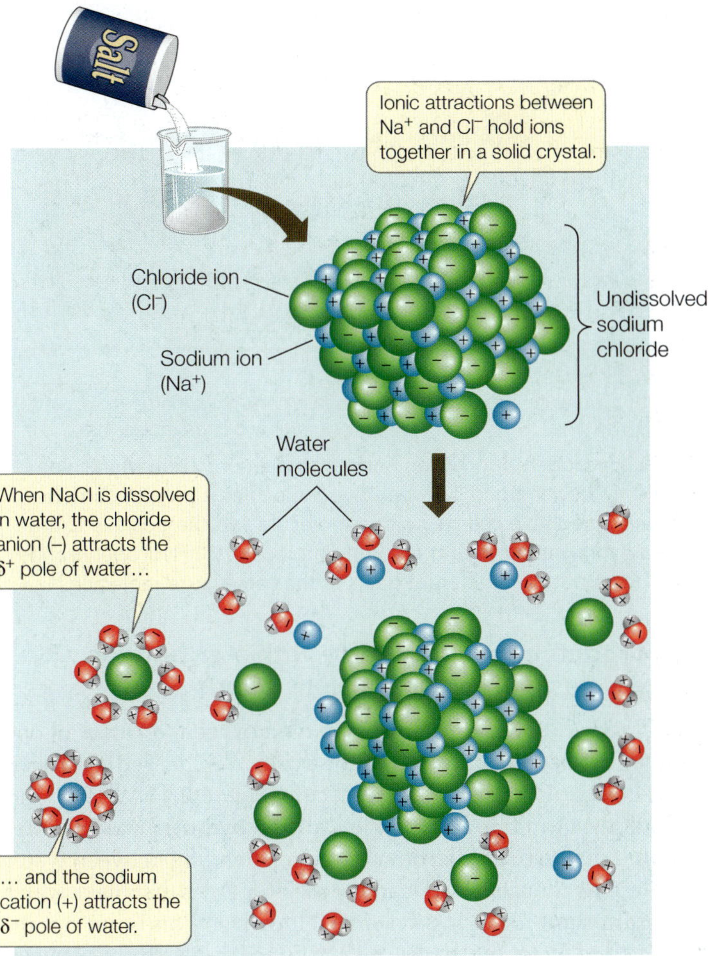
Salt
Ionic attractions between Na+ and Cl− hold ions together in a solid crystal.
Chloride ion (Cl−)
Sodium ion (Na+)
Undissolved sodium chloride
Water molecules
When NaCl is dissolved in water, the chloride anion (–) attracts the δ+ pole of water…
… and the sodium cation (+) attracts the δ− pole of water.

2.10 Water Molecules Surround Ions When an ionic solid dissolves in water, polar water molecules cluster around the cations and anions, preventing them from reassociating.

from sodium's outermost shell to that of chlorine (**Figure 2.9**). This reaction between sodium and chlorine makes the resulting atoms more stable because they both have eight fully paired electrons in their outer shells. The result is two **ions**.

Ions are electrically charged particles that form when atoms gain or lose one or more electrons:

- The sodium ion (Na^+) in our example has a +1 unit of charge because it has one less electron than it has protons. The outermost electron shell of the sodium ion is full, with eight electrons, so the ion is stable. Positively charged ions are called **cations**.
- The chloride ion (Cl^-) has a –1 unit of charge because it has one more electron than it has protons. This additional electron gives Cl^- a stable outermost shell with eight electrons. Negatively charged ions are called **anions**.

Some elements can form ions with multiple charges by losing or gaining more than one electron. Examples are Ca^{2+} (the calcium ion, a calcium atom that has lost two electrons) and Mg^{2+} (the magnesium ion). Two biologically important elements can each yield more than one stable ion. Iron yields Fe^{2+} (the ferrous ion) and Fe^{3+} (the ferric ion), and copper yields Cu^+ (the cuprous ion) and Cu^{2+} (the cupric ion). Groups of covalently bonded atoms that carry an electric charge are called **complex ions**; examples include NH_4^+ (the ammonium ion), SO_4^{2-} (the sulfate ion), and PO_4^{3-} (the phosphate ion). Once formed, ions are usually stable and no more electrons are lost or gained.

Ionic attractions are bonds formed as a result of the electrical attraction between ions bearing opposite charges. Ions can form bonds that result in stable solid compounds, which are referred to by the general term salts. Examples are sodium chloride (NaCl) and potassium phosphate (K_3PO_4). In sodium chloride—familiar to us as table salt—cations and anions are held together by ionic attractions. In solids, the attractions are strong because the ions are close together. However, when ions are dispersed in water, the distances between them can be large; the strength of the attraction is thus greatly reduced. Under the conditions in living cells, an ionic attraction is less strong than a nonpolar covalent bond (see Table 2.1).

Not surprisingly, ions can interact with polar molecules, since both are charged. This interaction results when a solid salt such as NaCl dissolves in water. Water molecules surround the individual ions, separating them (**Figure 2.10**). The negatively charged chloride ions attract the positive poles of the water molecules, while the positively charged sodium ions attract the negative poles of the water molecules. This special property of water (its polarity) is one reason it is such a good biological solvent (see Section 2.4).

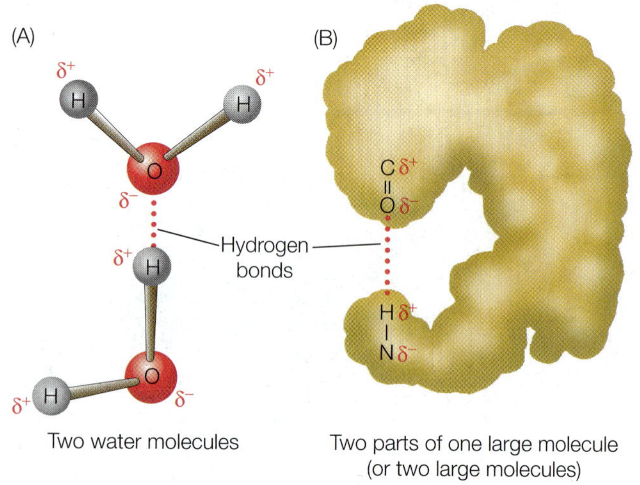
(A)
(B)
Hydrogen bonds
Two water molecules
Two parts of one large molecule (or two large molecules)

2.11 Hydrogen Bonds Can Form between or within Molecules (A) A hydrogen bond between two molecules is an attraction between a negative charge on one molecule and the positive charge on a hydrogen atom of the second molecule. (B) Hydrogen bonds can form between different parts of the same large molecule.

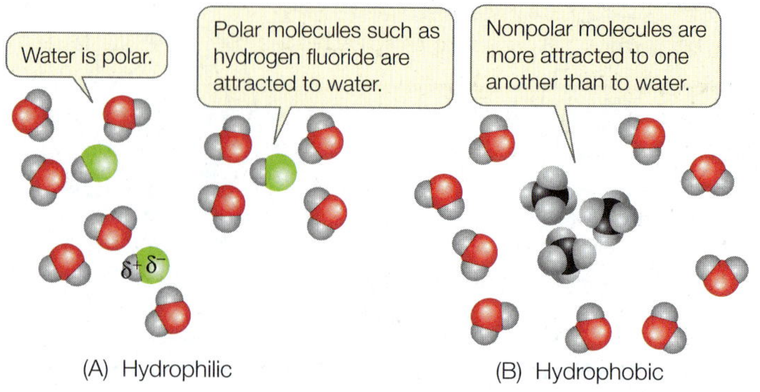
Water is polar.
Polar molecules such as hydrogen fluoride are attracted to water.
Nonpolar molecules are more attracted to one another than to water.
(A) Hydrophilic
(B) Hydrophobic

2.12 Hydrophilic and Hydrophobic (A) Molecules with polar covalent bonds are attracted to polar water (they are hydrophilic). (B) Molecules with nonpolar covalent bonds show greater attraction to one another than to water (they are hydrophobic).

Hydrogen bonds may form within or between molecules with polar covalent bonds

In liquid water, the negatively charged oxygen (δ^-) atom of one water molecule is attracted to the positively charged hydrogen (δ^+) atoms of another water molecule (**Figure 2.11A**). The bond resulting from this attraction is called a **hydrogen bond**. Later in this chapter we'll see how hydrogen bonding between water molecules contributes to many of the properties that make water so important for living systems. Hydrogen bonds are not restricted to water molecules. Such a bond can also form between a strongly electronegative atom in one molecule and a hydrogen atom that is involved in a polar covalent bond in another molecule, or another part of the same molecule (**Figure 2.11B**).

A hydrogen bond is weaker than most ionic attractions because its formation is due to partial charges (δ^+ and δ^-). It is much weaker than a covalent bond between a hydrogen atom and an oxygen atom (see Table 2.1). Although individual hydrogen bonds are weak, there can be many of them within a single molecule or between two molecules. In these cases, the hydrogen bonds together have considerable strength and can greatly influence the structure and properties of substances. For example, hydrogen bonds play important roles in determining and maintaining the three-dimensional shapes of giant molecules such as DNA and proteins (see Section 3.2).

Hydrophobic interactions bring together nonpolar molecules

Just as water molecules can interact with one another through hydrogen bonds, any molecule that is polar can interact with other polar molecules through the weak (δ^+ to δ^-) attractions of hydrogen bonds. If a polar molecule interacts with water in this way, it is called **hydrophilic** ("water-loving") (**Figure 2.12A**).

Nonpolar molecules, in contrast, tend to interact with other nonpolar molecules. For example, carbon (electronegativity 2.5) forms nonpolar bonds with hydrogen (electronegativity 2.1), and molecules containing only hydrogen and carbon atoms—called **hydrocarbon molecules**—are nonpolar. In water these molecules tend to aggregate with one another rather than with the polar water molecules. Therefore, nonpolar molecules are known as **hydrophobic** ("water-hating"), and the interactions between them are called **hydrophobic interactions** (**Figure 2.12B**). Of course, hydrophobic substances do not really "hate" water; they can form weak interactions with it, since the electronegativities of carbon and hydrogen are not exactly the same. But these interactions are far weaker than the hydrogen bonds between the water molecules (see Table 2.1), so the nonpolar substances tend to aggregate.

van der Waals forces involve contacts between atoms

The interactions between nonpolar substances are enhanced by **van der Waals forces**, which occur when the atoms of two molecules are in close proximity. These brief interactions result from random variations in the electron distribution in one molecule, which create opposite charge distributions in the adjacent molecule. So there will be a weak, temporary δ^+ to δ^- attraction. Although a single van der Waals interaction is brief and weak, the sum of many such interactions over the entire span of a large nonpolar molecule can result in substantial attraction. This is important when hydrophobic regions of different molecules such as an enzyme and a substrate come together (see Chapter 8).

RECAP 2.2

Some atoms form strong covalent bonds with other atoms by sharing one or more pairs of electrons. Unequal sharing of electrons produces polarity. Other atoms become ions by losing or gaining electrons, and they interact with other ions or polar molecules.

- Why is a covalent bond stronger than an ionic attraction? **See pp. 26–29 and Table 2.1**
- How do variations in electronegativity result in the unequal sharing of electrons in polar molecules? **See p. 28, Table 2.3, and Figure 2.8**
- What is a hydrogen bond and how is it important in biological systems? **See p. 30 and Figure 2.11**

The bonding of atoms into molecules is not necessarily a permanent affair. The dynamic of life involves constant change, even at the molecular level. In the next section we will examine how molecules interact with one another—how they break up, how they find new partners, and what the consequences of those changes can be.

2.3 How Do Atoms Change Partners in Chemical Reactions?

A **chemical reaction** occurs when moving atoms collide with sufficient energy to combine or to change their bonding partners. Consider the combustion reaction that takes place in the flame of a propane stove. When propane (C_3H_8) reacts with oxygen gas (O_2), the carbon atoms become bonded to oxygen atoms instead of hydrogen atoms, and the hydrogen atoms become bonded to oxygen instead of carbon (Figure 2.13). As the covalently bonded atoms change partners, the composition of the matter changes; propane and oxygen gas become carbon dioxide and water. This chemical reaction can be represented by the equation

$$C_3H_8 + 5\,O_2 \rightarrow 3\,CO_2 + 4\,H_2O + \text{Energy}$$
$$\text{Reactants} \rightarrow \text{Products}$$

In this equation, the propane and oxygen are the **reactants**, and the carbon dioxide and water are the **products**. In fact, this is a special type of reaction called an oxidation–reduction reaction. Electrons and protons (i.e., hydrogen atoms) are transferred from propane (the reducing agent) to oxygen (the oxidizing agent) to form water. You will see this kind of reaction involving electron/proton transfer many times in later chapters.

The products of a chemical reaction can have very different properties from the reactants. In the case shown in Figure 2.13, the reaction is *complete*: all the propane and oxygen are used up in forming the two products. The arrow symbolizes the direction of the chemical reaction. The numbers preceding the molecular formulas indicate how many molecules are used or produced.

Note that in this and all other chemical reactions, *matter is neither created nor destroyed*. The total number of carbon atoms on the left side of the equation (3) equals the total number of carbon atoms on the right (3). In other words, the equation is *balanced*. However, there is another aspect of this reaction: the heat and light of the stove's flame reveal that the reaction between propane and oxygen releases a great deal of energy.

Energy is defined as the capacity to do work, but in the context of chemical reactions, it can be thought of as the capacity for change. Chemical reactions do not create or destroy energy, but *changes in the form of energy* usually accompany chemical reactions.

In the reaction between propane and oxygen, a large amount of heat energy is released. This energy was present in another form, called potential chemical energy, in the covalent bonds within the propane and oxygen gas molecules. Not all reactions release energy; indeed, many chemical reactions require that energy be supplied from the environment. Some of this energy is then stored as potential chemical energy in the bonds formed in the products. We will see in future chapters how reactions that release energy and reactions that require energy can be linked together.

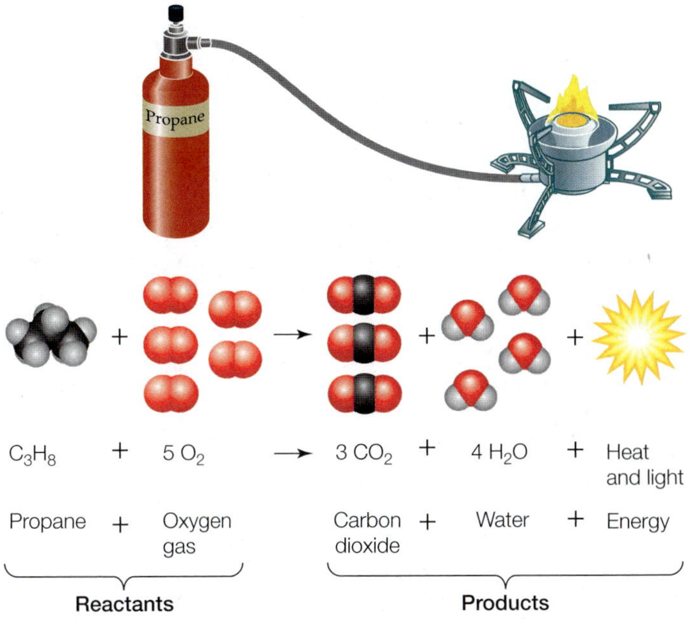
Propane
C_3H_8 + $5\,O_2$ → $3\,CO_2$ + $4\,H_2O$ + Heat and light
Propane + Oxygen gas
Carbon dioxide + Water + Energy
Reactants
Products

2.13 Bonding Partners and Energy May Change in a Chemical Reaction One molecule of propane (a gas used for cooking) from this burner reacts with five molecules of oxygen gas to give three molecules of carbon dioxide and four molecules of water. This reaction releases energy in the form of heat and light.

Many chemical reactions take place in living cells, and some of these have a lot in common with the oxidation–reduction reaction that happens in the combustion of propane. In cells, the reactants are different (they may be sugars or fats), and the reactions proceed by many intermediate steps that permit the released energy to be harvested and put to use by the cells. But the products are the same: carbon dioxide and water. We will discuss energy changes, oxidation–reduction reactions, and several other types of chemical reactions that are prevalent in living systems in Part Three of this book.

RECAP 2.3

In a chemical reaction, a set of reactants is converted to a set of products with different chemical compositions. This is accomplished by breaking old bonds and making new ones. A reaction may release energy or require its input.

- Explain how a chemical equation is balanced. **See p. 31 and Figure 2.13**
- How can the form of energy change during a chemical reaction? **See p.31**

We will return to chemical reactions and how they occur in living systems in Part Three of this book. First, however, we will examine the unique properties of the substance in which most biochemical reactions take place: water.

2.4 What Makes Water So Important for Life?

A human body is more than 70 percent water by weight, excluding the minerals contained in bones. Water is the dominant component of virtually all living organisms, and most biochemical reactions take place in this watery, or aqueous, environment. What makes water so important?

Water is an unusual substance with unusual properties. Under conditions on Earth, water exists in solid, liquid, and gas forms, all of which have relevance to living systems. Water allows chemical reactions to occur inside living organisms, and it is necessary for the formation of certain biological structures. In this section we will explore how the structure and interactions of water molecules make water essential to life.

Water has a unique structure and special properties

The molecule H_2O has unique chemical features. As we have already learned, water is a polar molecule that can form hydrogen bonds. The four pairs of electrons in the outer shell of the oxygen atom repel one another, giving the water molecule a tetrahedral shape:

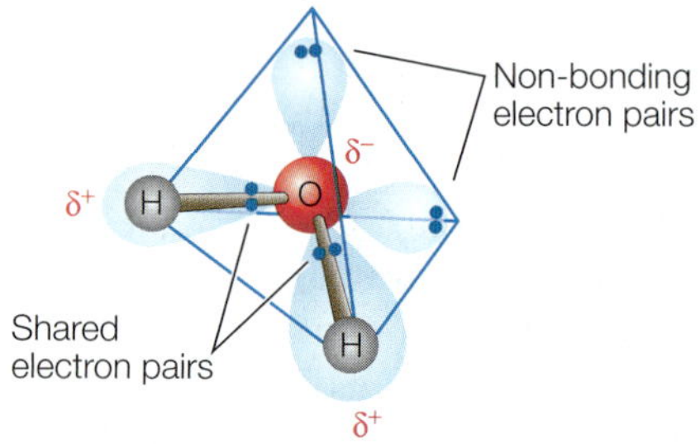
Non-bonding electron pairs
δ−
δ+
H
O
Shared electron pairs
H
δ+

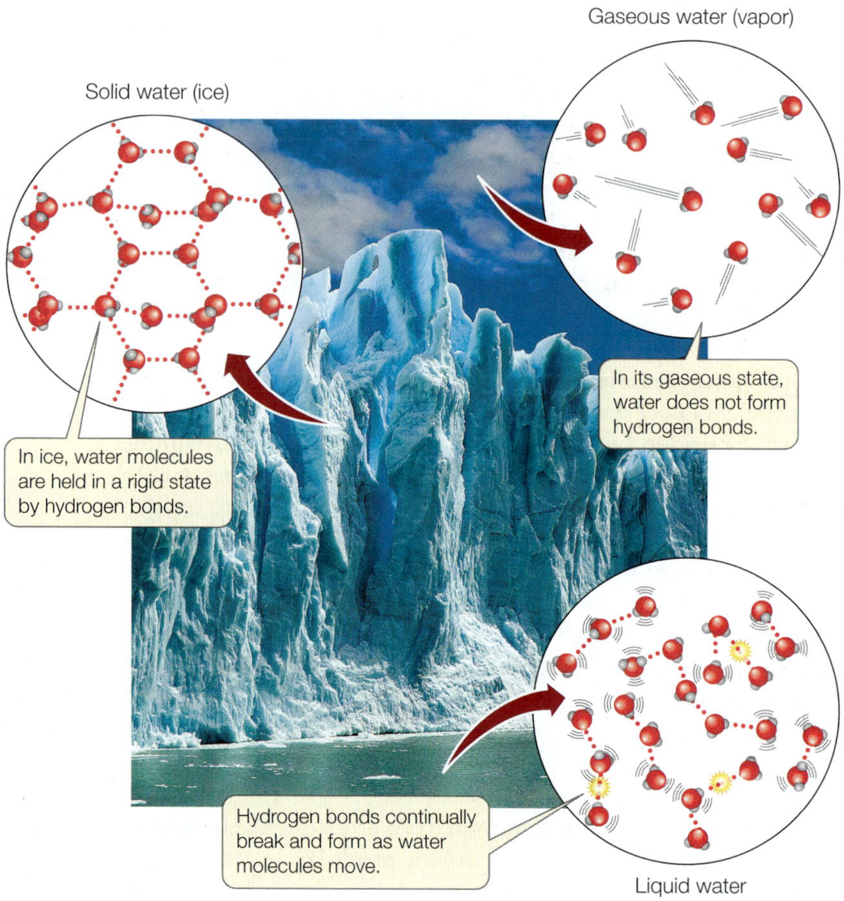
Gaseous water (vapor)
Solid water (ice)
In its gaseous state, water does not form hydrogen bonds.
In ice, water molecules are held in a rigid state by hydrogen bonds.
Hydrogen bonds continually break and form as water molecules move.
Liquid water

2.14 Hydrogen Bonding and the Properties of Water Hydrogen bonding occurs between the molecules of water in both its liquid and solid states. Ice is more structured but less dense than liquid water, which is why ice floats. Water forms a gas when its hydrogen bonds are broken and the molecules move farther apart.

These chemical features explain some of the interesting properties of water, such as the ability of ice to float, the melting and freezing temperatures of water, the ability of water to store heat, the formation of water droplets, water's ability to dissolve many substances, and its inability to dissolve many others.

ICE FLOATS In water's solid state (ice), individual water molecules are held in place by hydrogen bonds. Each molecule is bonded to four other molecules in a rigid, crystalline structure (**Figure 2.14**). Although the molecules are held firmly in place, they are farther apart from one another than they are in liquid water, where the molecules are moving about. In other words, solid water is less dense than liquid water, which is why ice floats.

Think of the biological consequences if ice were to sink in water. A pond would freeze from the bottom up, becoming a solid block of ice in winter and killing most of the organisms living there. Once the whole pond was frozen, its temperature could drop well below the freezing point of water. But in fact ice floats, forming an insulating layer on the top of the pond, and reducing heat flow to the cold air above. Thus fish, plants, and other organisms in the pond are not subjected to temperatures lower than 0°C, which is the freezing point of pure water.

MELTING, FREEZING, AND HEAT CAPACITY Compared with many other substances that have molecules of similar size, ice requires a great deal of heat energy to melt. This is because so many hydrogen bonds must be broken in order for water to change from solid to liquid. In the opposite process—freezing—a great deal of energy is released to the environment.

This property of water contributes to the surprising constancy of the temperatures found in oceans and other large bodies of water throughout the year. The temperature changes of coastal land masses are also moderated by large bodies of water. Indeed, water helps minimize variations in atmospheric temperature across the planet. This moderating ability is a result of the high heat capacity of liquid water, which is in turn a result of its high specific heat.

The **specific heat** of a substance is the amount of heat energy required to raise the temperature of 1 gram of that substance by 1°C. Raising the temperature of liquid water takes a relatively large amount of heat because much of the heat energy is used to break the hydrogen bonds that hold the liquid together. Compared with other small molecules that are liquids, water has a high specific heat. For example, water has twice the specific heat of ethyl alcohol.

Water also has a high **heat of vaporization**, which means that a lot of heat is required to change water from its liquid to its gaseous state (the process of evaporation). Once again, much of the heat energy is used to break the many hydrogen bonds between the water molecules. This heat must be absorbed from the environment in contact with the water. Evaporation thus has a cooling effect on the environment—whether a leaf, a forest, or an entire land mass. This effect explains why sweating cools the human body: as sweat evaporates from the skin, it uses up some of the adjacent body heat (Figure 2.15A).

COHESION AND SURFACE TENSION In liquid water, individual molecules are able to move about. The hydrogen bonds between the molecules continually form and break (see Figure 2.14). Chemists estimate that this occurs about a trillion times a minute for a single water molecule, making it a truly dynamic structure.

At any given time, a water molecule will form on average 3.4 hydrogen bonds with other water molecules. These hydrogen bonds explain the cohesive strength of liquid water. This cohesive strength, or **cohesion**, is defined as the capacity of water molecules to resist coming apart from one another when placed under tension. Water's cohesive strength permits narrow columns of liquid water to move from the roots to the leaves of tall trees. When water evaporates from the leaves, the entire column moves upward in response to the pull of the molecules at the top (Figure 2.15B).

The surface of liquid water exposed to the air is difficult to puncture because the water molecules at the surface are hydrogen-bonded to other water molecules below them. This surface tension of water permits a container to be filled slightly above its rim without overflowing, and it permits spiders to walk on the surface of a pond (Figure 2.15C).

High heat of vaporization: Sweating uses evaporation of water to cool the body.
Cohesion: Water's cohesive strength helps it to flow from the roots to the leaves in a tree.
Surface tension: Water molecules stick to one another and help prevent this wolf spider from sinking.

2.15 Water in Biology These three properties of water make it beneficial to organisms.

The reactions of life take place in aqueous solutions

A **solution** is produced when a substance (the **solute**) is dissolved in a liquid (the **solvent**). If the solvent is water, then the solution is an aqueous solution. Many of the important molecules in biological systems are polar, and therefore soluble in water. Many important biochemical reactions occur in aqueous solutions within cells.

Biologists who are interested in the biochemical reactions within cells need to identify the reactants and products and to determine their amounts:

- Qualitative analyses deal with the identification of substances involved in chemical reactions. For example, a qualitative analysis would be used to investigate the steps involved and the products formed during respiration, when carbon-containing compounds are broken down to release energy in living tissues.
- Quantitative analyses measure concentrations or amounts of substances. For example, a biochemist would use a quantitative analysis to measure how much of a certain product is formed in a chemical reaction. What follows is a brief introduction to some of the quantitative chemical terms you will see in this book.

Fundamental to quantitative thinking in chemistry and biology is the concept of the mole. A **mole** is the amount of a substance (in grams) that is numerically equal to its molecular weight. So a mole of table sugar ($C_{12}H_{22}O_{11}$) weighs about 342 grams; a mole of sodium ion (Na^+) weighs 23 grams; and a mole of hydrogen gas (H_2) weighs 2 grams.

Quantitative analyses do not yield direct counts of molecules. Because the amount of a substance in 1 mole is directly related to its molecular weight, it follows that the number of molecules in 1 mole is constant for all substances. So 1 mole of salt contains the same number of molecules as 1 mole of table sugar. This constant number of molecules in a mole is called **Avogadro's number**, and it is 6.02×10^{23} molecules per mole. Chemists work with moles of substances (which can be weighed in the laboratory) instead of actual molecules, which are too numerous to be counted. Consider 34.2 grams (just over 1 ounce) of table sugar, $C_{12}H_{22}O_{11}$. This is one-tenth of a mole, or as Avogadro puts it, 6.02×10^{23} molecules.

A chemist can dissolve a mole of sugar (342 g) in water to make 1 liter of solution, knowing that the mole contains 6.02×10^{23} individual sugar molecules. This solution—1 mole of a substance dissolved in water to make 1 liter—is called a 1 molar (1M) solution. When a physician injects a certain volume and molar concentration of a drug into the bloodstream of a patient, a rough calculation can be made of the actual number of drug molecules that will interact with the patient's cells.

The many molecules dissolved in the water of living tissues are not present at concentrations anywhere near 1 molar. Most are in the micromolar (millionths of a mole per liter of solution; µM) to millimolar (thousandths of a mole per liter; mM) range. Some, such as hormone molecules, are even less concentrated than that. While these molarities seem to indicate very low concentrations, remember that even a 1 µM solution has 6.02×10^{17} molecules of the solute per liter.

Aqueous solutions may be acidic or basic

When some substances dissolve in water, they release hydrogen ions (H^+), which are actually single, positively charged protons. Hydrogen ions can interact with other molecules and change their properties. For example, the protons in "acid rain" can damage plants, and you probably have experienced the excess of hydrogen ions that we call "acid indigestion."

Here we will examine the properties of **acids** (defined as substances that release H^+) and **bases** (defined as substances that accept H^+). We will distinguish between strong and weak acids and bases, and provide a quantitative means for stating the concentration of H^+ in solutions: the pH scale.

ACIDS RELEASE H^+ When hydrochloric acid (HCl) is added to water, it dissolves, releasing the ions H^+ and Cl^-:

$$HCl \rightarrow H^+ + Cl^-$$

Because its H^+ concentration has increased, the solution is acidic.

Acids are substances that *release* H^+ ions in solution. HCl is an acid, as is H_2SO_4 (sulfuric acid). One molecule of sulfuric acid will ionize to yield two H^+ and one SO_4^{2-}. Biological compounds that contain —COOH (the carboxyl group) are also acids because the carboxyl group ionizes to $—COO^-$, releasing H^+:

$$—COOH \rightarrow —COO^- + H^+$$

Acids that fully ionize in solution, such as HCl and H_2SO_4 are called strong acids. However, not all acids ionize fully in water. For example, if acetic acid (CH_3COOH) is added to water, some of it will dissociate into two ions (CH_3COO^- and H^+), but some of the original acetic acid will remain as well. Because the reaction is not complete, acetic acid is a weak acid.

BASES ACCEPT H^+ Bases are substances that *accept* H^+ in solution. As with acids, there are strong and weak bases. If NaOH (sodium hydroxide) is added to water, it dissolves and ionizes, releasing OH^- and Na^+ ions:

$$NaOH \rightarrow Na^+ + OH^-$$

Because OH^- absorbs H^+ to form water, such a solution is basic. This reaction is complete, and so NaOH is a strong base.

Weak bases include the bicarbonate ion (HCO_3^-), which can accept an H^+ ion and become carbonic acid (H_2CO_3), and ammonia (NH_3), which can accept H^+ and become an ammonium ion (NH_4^+). Biological compounds that contain $—NH_2$ (the amino group) are also bases because $—NH_2$ accepts H^+:

$$—NH_2 + H^+ \rightarrow —NH_3^+$$

ACID–BASE REACTIONS MAY BE REVERSIBLE When acetic acid is dissolved in water, two reactions happen. First, the acetic acid forms its ions:

$$CH_3COOH \rightarrow CH_3COO^- + H^+$$

Then, once the ions are formed, some of them re-form acetic acid:

$$CH_3COO^- + H^+ \rightarrow CH_3COOH$$

This pair of reactions is reversible. A **reversible reaction** can proceed in either direction—left to right or right to left—depending on the relative starting concentrations of the reactants and products. The formula for a reversible reaction can be written using a double arrow:

$$CH_3COOH \rightleftharpoons CH_3COO^- + H^+$$

In terms of acids and bases, there are two types of reactions, depending on the extent of the reversibility:

- The ionization of strong acids and bases in water is virtually irreversible.
- The ionization of weak acids and bases in water is somewhat reversible.

WATER IS A WEAK ACID AND A WEAK BASE The water molecule has a slight but significant tendency to ionize into a hydroxide ion (OH^-) and a hydrogen ion (H^+). Actually, two water molecules participate in this reaction. One of the two molecules "captures" a hydrogen ion from the other, forming a hydroxide ion and a hydronium ion:

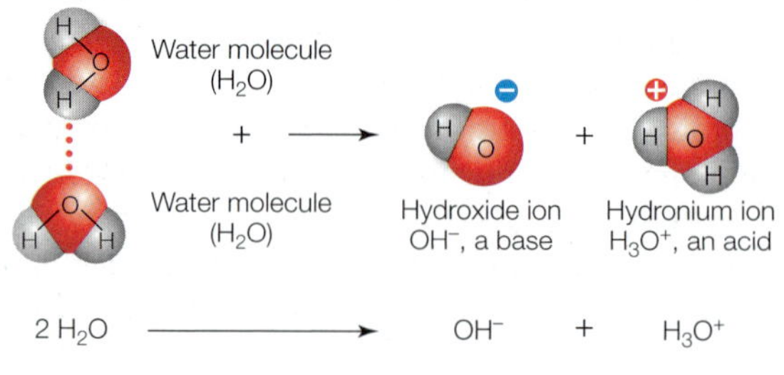

Water molecule
(H_2O)
+
Water molecule
(H_2O)
Hydroxide ion
OH^-, a base
+
Hydronium ion
H_3O^+, an acid
2 H_2O
OH^-
+
H_3O^+

The hydronium ion is, in effect, a hydrogen ion bound to a water molecule. For simplicity, biochemists tend to use a modified representation of the ionization of water:

$$H_2O \rightarrow H^+ + OH^-$$

The ionization of water is important to all living creatures. This fact may seem surprising, since only about 1 water molecule in 500 million is ionized at any given time. But this is less surprising if we focus on the abundance of water in living systems, and the reactive nature of the H^+ ions produced by ionization.

pH: HYDROGEN ION CONCENTRATION As we have seen, compounds can be either acids or bases, and thus solutions can be either acidic or basic. We can measure how acidic or basic a solution is by measuring its concentration of H^+ in moles per liter (its molarity; see p. 34). Here are some examples:

- Pure water has a H^+ concentration of 10^{-7} M.
- A 1 M HCl solution has a H^+ concentration of 1 M (recall that all the HCl dissociates into its ions).
- A 1 M NaOH solution has a H^+ concentration of 10^{-14} M.

This is a very wide range of numbers to work with—think about the decimals! It is easier to work with the logarithm of the H^+ concentration, because logarithms compress this range: the $\log_{10}$ of 100, for example, is 2; and the $\log_{10}$ of 0.01 is –2. Because most H^+ concentrations in living systems are less than 1 M, their $\log_{10}$ values are negative. For convenience, we convert these negative numbers into positive ones by using the *negative* of the logarithm of the H^+ molar concentration. This number is called the **pH** of the solution.

Since the H^+ concentration of pure water is 10^{-7} M, its pH is $-\log(10^{-7}) = -(-7)$, or 7. A smaller negative logarithm means a larger number. In practical terms, a lower pH means a higher H^+ concentration, or greater acidity. In 1 M HCl, the H^+ concentration is 1 M, so the pH is the negative logarithm of 1 ($-\log 10^0$), or 0. The pH of 1 M NaOH is the negative logarithm of 10^{-14}, or 14.

A solution with a pH of less than 7 is acidic—it contains more H^+ ions than OH^- ions. A solution with a pH of 7 is referred to as neutral, and a solution with a pH value greater than 7 is basic. Figure 2.16 shows the pH values of some common substances.

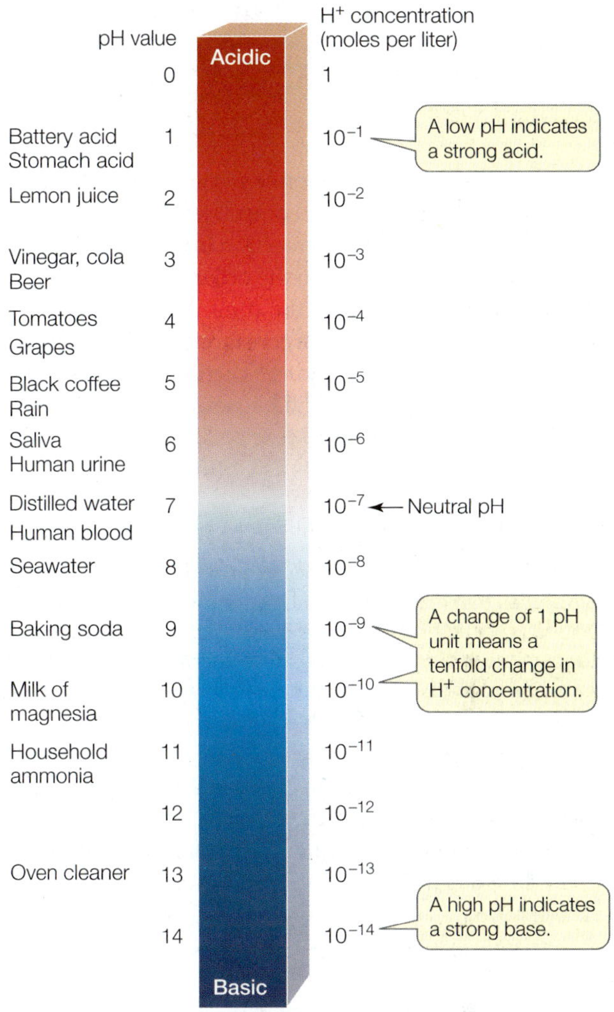
pH value
Acidic
Basic
H+ concentration (moles per liter)
0
1
Battery acid
Stomach acid
Lemon juice
Vinegar, cola
Beer
Tomatoes
Grapes
Black coffee
Rain
Saliva
Human urine
Distilled water
Human blood
Seawater
Baking soda
Milk of magnesia
Household ammonia
Oven cleaner
1
2
3
4
5
6
7
8
9
10
11
12
13
14
10−1
10−2
10−3
10−4
10−5
10−6
10−7
10−8
10−9
10−10
10−11
10−12
10−13
10−14
Neutral pH
A low pH indicates a strong acid.
A change of 1 pH unit means a tenfold change in H+ concentration.
A high pH indicates a strong base.

2.16 pH Values of Some Familiar Substances

Why is this discussion of pH so relevant to biology? Many reactions involve the transfer of an ion or charged group from one molecule to another, and the presence of positive or negative ions in the environment can greatly influence the rates of such reactions. Furthermore, pH can influence the shapes of molecules. Many biologically important molecules contain charged groups (e.g., $—COO^-$) that can interact with the polar regions of water, and these interactions influence the way such molecules fold up into three-dimensional shapes. If these charged groups combine with H^+ or other ions in their environment to form uncharged groups (e.g., —COOH, see above), they will have a reduced tendency to interact with water. These uncharged (hydrophobic) groups might induce the molecule to fold up differently so that they are no longer in contact with the watery environment. Since the three-dimensional structures of biological molecules greatly affect the way they function, organisms do all they can to minimize changes in the pH of their cells and tissues. An important way to do this is with buffers.

BUFFERS The maintenance of internal constancy—homeostasis—is a hallmark of all living things and extends to pH. If biological molecules lose or gain H^+ ions, their properties can change, thus upsetting homeostasis. Internal constancy is achieved with buffers: solutions that maintain a relatively constant pH even when substantial amounts of acid or base are added. How does this work?

A **buffer** is a mixture of a weak acid and its corresponding base, or a weak base and its corresponding acid. For example, a weak acid is carbonic acid (H_2CO_3), and its corresponding base is the bicarbonate ion (HCO_3^-). If another acid is added to a solution containing this mixture (a buffered solution), not all the H^+ ions from the acid remain in solution. Instead, many of them combine with the bicarbonate ions to produce more carbonic acid:

$$HCO_3^- + H^+ \rightleftharpoons H_2CO_3$$

This reaction uses up some of the H^+ ions in the solution and decreases the acidifying effect of the added acid. If a base is added, the reaction essentially reverses. Some of the carbonic acid ionizes to produce bicarbonate ions and more H^+, which counteracts some of the added base. In this way, the buffer minimizes the effect that an added acid or base has on pH. The carbonic acid/bicarbonate buffering system is present in the blood, where it is important for preventing significant changes in pH that could disrupt the ability of the blood to carry vital oxygen to tissues. A given amount of acid or base causes a smaller pH change in a buffered solution than in a non-buffered one (Figure 2.17).

Buffers illustrate an important chemical principle of reversible reactions, called the **law of mass action**. Addition of a

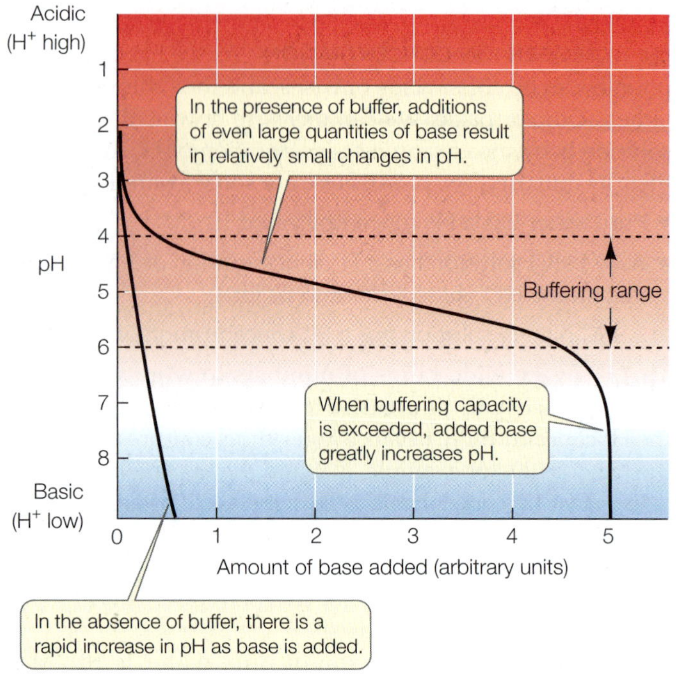
Acidic
(H+ high)
pH
Basic
(H+ low)
1
2
3
4
5
6
7
8
0
1
2
3
4
5
Amount of base added (arbitrary units)
In the presence of buffer, additions of even large quantities of base result in relatively small changes in pH.
Buffering range
When buffering capacity is exceeded, added base greatly increases pH.
In the absence of buffer, there is a rapid increase in pH as base is added.

2.17 Buffers Minimize Changes in pH When a base is added to a solution, the pH of the solution increases. Without a buffer, the change is large and the slope of the pH graph is steep. In the presence of a buffer, however, the slope within the buffering range is shallow.

reactant on one side of a reversible system drives the reaction in the direction that uses up that compound. In the case of buffers, addition of an acid drives the reaction in one direction; addition of a base drives the reaction in the other direction.

We use a buffer to relieve the common problem of indigestion. The lining of the stomach constantly secretes hydrochloric acid, making the stomach contents acidic. But excessive stomach acid inhibits digestion and causes discomfort. We can relieve this discomfort by ingesting a salt such as $NaHCO_3$ (sodium bicarbonate), which acts as a buffer.

RECAP 2.4

Most of the chemistry of life occurs in water, which has unique properties that make it an ideal medium for supporting life. Aqueous solutions can be acidic or basic, depending on the concentration of hydrogen ions. The cells and tissues of organisms are buffered, however, because changes in pH can change the properties of biological molecules.

- What are some biologically important properties of water that arise from its molecular structure? **See pp. 32–33 and Figure 2.14**
- What is a solution, and why do we call water "the medium of life"? **See p. 33**
- What is the relationship among hydrogen ions, acids, and bases? Explain what the pH scale measures. **See p.35 and Figure 2.16**
- How does a buffer work, and why is buffering important to living systems? **See pp. 35–36 and Figure 2.17**

An Overview and a Preview

Now that we have covered the major properties of atoms and molecules, let's review them and see how these properties relate to the major molecules of biological systems.

- *Molecules vary in size.* Some are small, such as those of hydrogen gas (H_2) and methane (CH_4). Others are larger, such as a molecule of table sugar ($C_{12}H_{22}O_{11}$), which has 45 atoms. Still others, especially proteins and nucleic acids, are gigantic, containing tens of thousands or even millions of atoms.
- *Each molecule can have a specific three-dimensional shape.* For example, the orientations of the bonding orbitals around the carbon atom give the methane molecule (CH_4) the shape of a regular tetrahedron (see Figure 2.7B). Larger molecules have complex shapes that result from the numbers and kinds of atoms present, and the ways in which they are linked together. Some large molecules, such as the protein hemoglobin (the oxygen carrier in red blood cells), have compact, ball-like shapes. Others, such as the protein keratin that makes up hair, have long, thin, ropelike structures. Their shapes relate to the roles these molecules play in living cells.
- *Molecules are characterized by certain chemical properties* that determine their biological roles. Chemists use atomic composition, structure (three-dimensional shape), reactivity, and solubility to distinguish a pure sample of one molecule from a sample of a different molecule. The presence of certain groups of atoms can impart distinctive chemical properties to a molecule.

Between the small molecules discussed in this chapter and the world of the living cell are the macromolecules. We will discuss these larger molecules—proteins, lipids, carbohydrates, and nucleic acids—in the next two chapters.

?

Can isotope analysis of water be used to detect climate change?

ANSWER

Water evaporates in warmer regions at the tropical latitudes on Earth and moves toward the cooler poles. As an air mass moves from a warmer to a cooler region, water vapor condenses and is removed as precipitation. The heavy isotopes of H and O tend to fall as precipitation more readily than the lighter isotopes, so as the water vapor moves toward the poles, it becomes enriched in the lighter isotopes. The ratio of heavy to light isotopes that reach the poles depends on the climate—the cooler the climate, the lower the ratio, because more water precipitates as it moves toward the poles, depleting more of the heavier isotopes. Analyses of polar ice cores show that heavy-to-light isotope ratios vary over geological time scales. This has allowed scientists to reconstruct climate change in the past, and to relate it to fossil organisms that lived at those times.

CHAPTER SUMMARY 2

2.1 How Does Atomic Structure Explain the Properties of Matter?

- Matter is composed of atoms. Each **atom** consists of a positively charged **nucleus** made up of **protons** and **neutrons**, surrounded by **electrons** bearing negative charges. **Review Figure 2.1**
- The number of protons in the nucleus defines an **element**. There are many elements in the universe, but only a few of them make up the bulk of living organisms: C, H, O, P, N, and S. **Review Figure 2.2**
- **Isotopes** of an element differ in their numbers of neutrons. **Radioisotopes** are radioactive, emitting radiation as they break down.
- Electrons are distributed in **electron shells**, which are volumes of space defined by specific numbers of orbitals. Each **orbital** contains a maximum of two electrons. **Review Figures 2.4, 2.5, ACTIVITY 2.1**
- In losing, gaining, or sharing electrons to become more stable, an atom can combine with other atoms to form a **molecule**.

2.2 How Do Atoms Bond to Form Molecules?
See ANIMATED TUTORIAL 2.1

- A **chemical bond** is an attractive force that links two atoms together in a molecule. **Review Table 2.1**
- A **compound** is a substance made up of molecules with two or more different atoms bonded together in a fixed ratio, such as water (H_2O).
- **Covalent bonds** are strong bonds formed when two atoms share one or more pairs of electrons. **Review Figure 2.6**
- When two atoms of unequal electronegativity bond with each other, a **polar covalent bond** is formed. The two ends, or poles, of the bond have partial charges (δ^+ or δ^-). **Review Figure 2.8**
- An **ion** is an electrically charged body that forms when an atom gains or loses one or more electrons in order to form a more stable electron configuration. **Anions** and **cations** are negatively and positively charged ions, respectively. Different charges attract, and like charges repel each other.
- **Ionic attractions** occur between oppositely charged ions. Ionic attractions are strong in solids (salts) but weaken when the ions are separated from one another in solution. **Review Figure 2.9**
- A **hydrogen bond** is a weak electrical attraction that forms between a δ^+ hydrogen atom in one molecule and a δ^- atom in another molecule (or in another part of the same, large molecule). Hydrogen bonds are abundant in water. **Review Figure 2.11**
- Nonpolar molecules interact very little with polar molecules, including water. Nonpolar molecules are attracted to one another by very weak bonds called **van der Waals forces**.

2.3 How Do Atoms Change Partners in Chemical Reactions?

- In **chemical reactions**, atoms combine or change their bonding partners. **Reactants** are converted into **products**.
- Some chemical reactions release **energy** as one of their products; other reactions can occur only if energy is provided to the reactants.
- Neither matter nor energy is created or destroyed in a chemical reaction, but both change form. **Review Figure 2.13**
- Some chemical reactions, especially in biology, are reversible. That is, the products formed may be converted back to the reactants.
- In organisms, chemical reactions take place in multiple steps so that released energy can be harvested for cellular activities.

2.4 What Makes Water So Important for Life?

- Water's molecular structure and its capacity to form hydrogen bonds give it unique properties that are significant for life. **Review Figure 2.14**
- The high **specific heat** of water means that water gains or loses a great deal of heat when it changes state. Water's high **heat of vaporization** ensures effective cooling when water evaporates.
- The **cohesion** of water molecules refers to their capacity to resist coming apart from one another. Hydrogen bonding between the water molecules plays an essential role in this property.
- A **solution** is produced when a solid substance (the **solute**) dissolves in a liquid (the **solvent**). Water is the critically important solvent for life.

Go to the Interactive Summary to review key figures, Animated Tutorials, and Activities
Life10e.com/is2

CHAPTER**REVIEW**

REMEMBERING

1. The atomic number of an element
   a. equals the number of neutrons in an atom.
   b. equals the number of protons in an atom.
   c. equals the number of protons minus the number of neutrons.
   d. equals the number of neutrons plus the number of protons.
   e. depends on the isotope.

2. The mass number of an element
   a. equals the number of neutrons in an atom.
   b. equals the number of protons in an atom.
   c. equals the number of electrons in an atom.
   d. equals the number of neutrons plus the number of protons.
   e. depends on the relative abundances of its electrons and neutrons.

3. Which of the following statements about the isotopes of an element is *not* true?
   a. They all have the same atomic number.
   b. They all have the same number of protons.
   c. They all have the same number of neutrons.
   d. They all have the same number of electrons.
   e. They all have identical chemical properties.

4. Which of the following statements about covalent bonds is *not* true?
   a. A covalent bond is stronger than a hydrogen bond.
   b. A covalent bond can form between atoms of the same element.
   c. Only a single covalent bond can form between two atoms.
   d. A covalent bond results from the sharing of electrons by two atoms.
   e. A covalent bond can form between atoms of different elements.

5. Which of the following statements about water is *not* true?
   a. It releases a large amount of heat when changing from liquid into vapor.
   b. Its solid form is less dense than its liquid form.
   c. It is the most effective solvent for polar molecules.
   d. It is typically the most abundant substance in a living organism.
   e. It takes part in some important chemical reactions.

6. The reaction $HCl \rightarrow H^+ + Cl^-$ in the human stomach is an example of the
   a. cleavage of a hydrophobic bond.
   b. formation of a hydrogen bond.
   c. elevation of the pH of the stomach.
   d. formation of ions by dissociation of an acid.
   e. formation of polar covalent bonds.

UNDERSTANDING & APPLYING

7. Using the information in the periodic table (Figure 2.2), draw a Bohr model (see Figures 2.5 and 2.7) of silicon dioxide, showing electrons shared in covalent bonds.

8. Compare a covalent bond between two hydrogen atoms with a hydrogen bond between a hydrogen and an oxygen atom, with regard to the electrons involved, the role of polarity, and the strength of the bond.

9. Use Tables 2.2 and 2.3 to determine for each of the pairs of bonded atoms below:
   a. whether the bond is polar or nonpolar;
   b. if polar, which end is δ^-; and
   c. whether the bond is hydrophilic or hydrophobic.

   C–H C=O O–P C–C

ANALYZING & EVALUATING

10. Geckos are lizards that are amazing climbers. A gecko can climb up a glass surface and stick to it with a single toe. Professor Kellar Autumn at Lewis and Clark College and his students and collaborators have shown that each toe of a gecko has millions of micrometer-sized hairs, and that each hair splits into hundreds of 200-nanometer tips that provide intimate contact with a surface. Careful measurements show that a million of these tips could easily support the animal, but it has far more. The toes stick well on hydrophilic and hydrophobic surfaces. Bending the hairs allows the gecko to detach. What kind of noncovalent force is involved in gecko sticking?

11. Would you expect the elemental composition of Earth's crust to be the same as that of the human body?

Go to BioPortal at **yourBioPortal.com** for Animated Tutorials, Activities, LearningCurve Quizzes, Flashcards, and many other study and review resources.

3 Proteins, Carbohydrates, and Lipids

CHAPTER**OUTLINE**

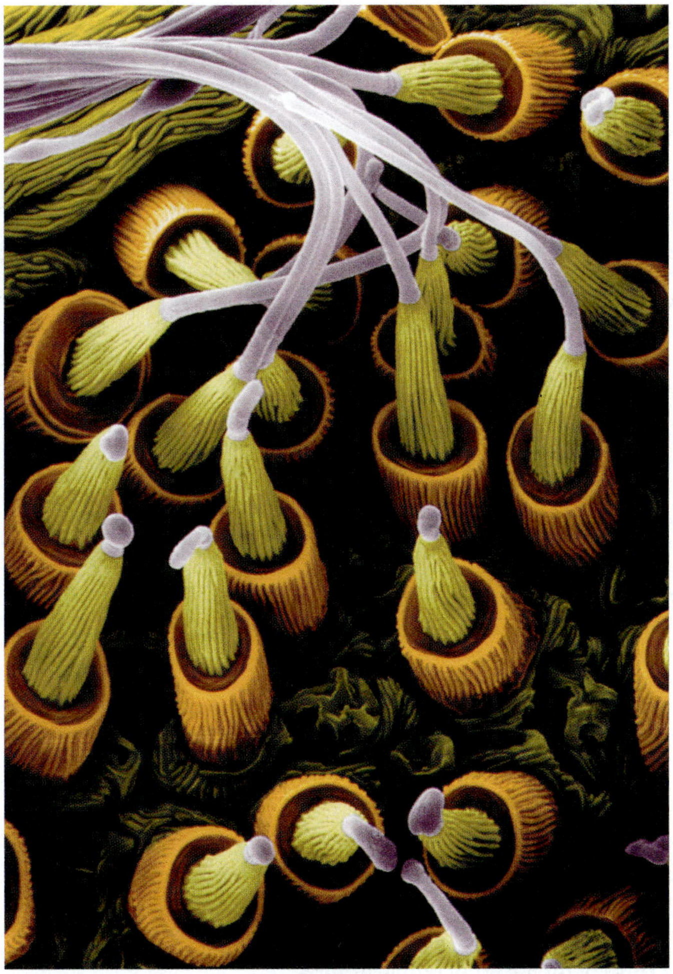

A Complex Macromolecule Spider silk (purple) being spun from a gland by the shiny black spider, *Castercantha*.

A SPIDER WEB is an amazing structure. It is not only beautiful to look at, but it is an architectural wonder that is the spider's home, its mating place, and its way to capture food. Think of a fly that chances to interact with a spider web. The fibers of the web must slow down the fly, but they cannot break, so they need to stretch to dissipate the energy of the fly's movement. The fibers holding the fly cannot stretch too much, however. They must be strong enough to hold the web in place and not let it wobble out of control. Web fibers are far thinner than a human hair, yet they are five times tougher than steel and in some cases more elastic than nylon. The fibers can also be long; for example, the Darwin's bark spider makes strands up to 25 meters long.

Spider silk is composed of variations on a single type of large molecule—a macromolecule called protein. Proteins are polymers: long chains of individual smaller units called amino acids. The proteins in spider silks have characteristic structures and amino acid compositions depending on their particular functions. Proteins in the stretchy web fibers have amino acids that allow them to curl into spirals, and these spirals can slip along one another to change the fiber's length. Another kind of spider silk is the dragline silk, which is less stretchy and used to construct the outline of the web, its spokes, and the lifeline of the spider. The proteins in these strong fibers are made up of amino acids that cause the proteins to fold into flat sheets with ratchets, so that parallel sheets can fit together like Lego blocks. This arrangement makes these fibers hard to pull apart. The relationship between chemical structure and biological function is a recurring theme in biochemistry, as you will see in this and the succeeding chapters.

Proteins are one of the four major kinds of large molecules that characterize living systems. These macromolecules, which also include carbohydrates, lipids, and nucleic acids, differ in several significant ways from the small molecules and ions described in Chapter 2. First—no surprise—they are larger; the molecular masses of some nucleic acids reach billions of daltons. Second, these molecules all contain carbon atoms, and so belong to a group known as organic compounds. Third, the atoms of individual macromolecules are held together mostly by covalent bonds, which gives them structural stability and distinctive three-dimensional geometries. These distinctive shapes are the basis of many of the functions of macromolecules, particularly the proteins.

?

Can knowledge of spider web protein structure be put to practical use?

See answer on p. 59.

3.1 What Kinds of Molecules Characterize Living Things?

Four kinds of molecules are characteristic of living things: proteins, carbohydrates, lipids, and nucleic acids. With the exception of the lipids, these biological molecules are **polymers** (*poly*, "many"; *mer*, "unit") constructed by the covalent bonding of smaller molecules called **monomers**. Each kind of biological molecule is made up of monomers with similar chemical structures:

- Proteins are formed from different combinations of 20 amino acids, all of which share chemical similarities.
- Carbohydrates can form giant molecules by linking together chemically similar sugar monomers (monosaccharides) to form polysaccharides.
- Nucleic acids are formed from four kinds of nucleotide monomers linked together in long chains.
- Lipids also form large structures from a limited set of smaller molecules, but in this case noncovalent forces maintain the interactions between the lipid monomers.

Polymers with molecular weights exceeding 1,000 are considered to be **macromolecules**. The proteins, carbohydrates, and nucleic acids of living systems certainly fall into this category. Although large lipid structures are not polymers in the strictest sense, it is convenient to treat them as a special type of macromolecule (see Section 3.4).

How the macromolecules function and interact with other molecules depends on the properties of certain chemical groups in their monomers, the functional groups.

Go to Animated Tutorial 3.1 Macromolecules
Life10e.com/at3.1

Functional groups give specific properties to biological molecules

Certain small groups of atoms, called **functional groups**, occur frequently in biological molecules (Figure 3.1). Each functional group has specific chemical properties, and when it is attached to a larger molecule, it confers those properties on the larger molecule. One of these properties is polarity. Looking at the structures in Figure 3.1, can you determine which functional groups are the most polar? (Hint: look for C—O, N—H, and P—O bonds.) The consistent chemical behavior of functional groups helps us understand the properties of the molecules that contain them.

Because macromolecules are so large, they contain many different functional groups. A single large protein may contain hydrophobic, polar, and charged functional groups, each of which gives different specific properties to local sites on the macromolecule. As we will see, sometimes these different groups interact within the same macromolecule. They help determine the shape of the macromolecule as well as how it interacts with other macromolecules and with smaller molecules.

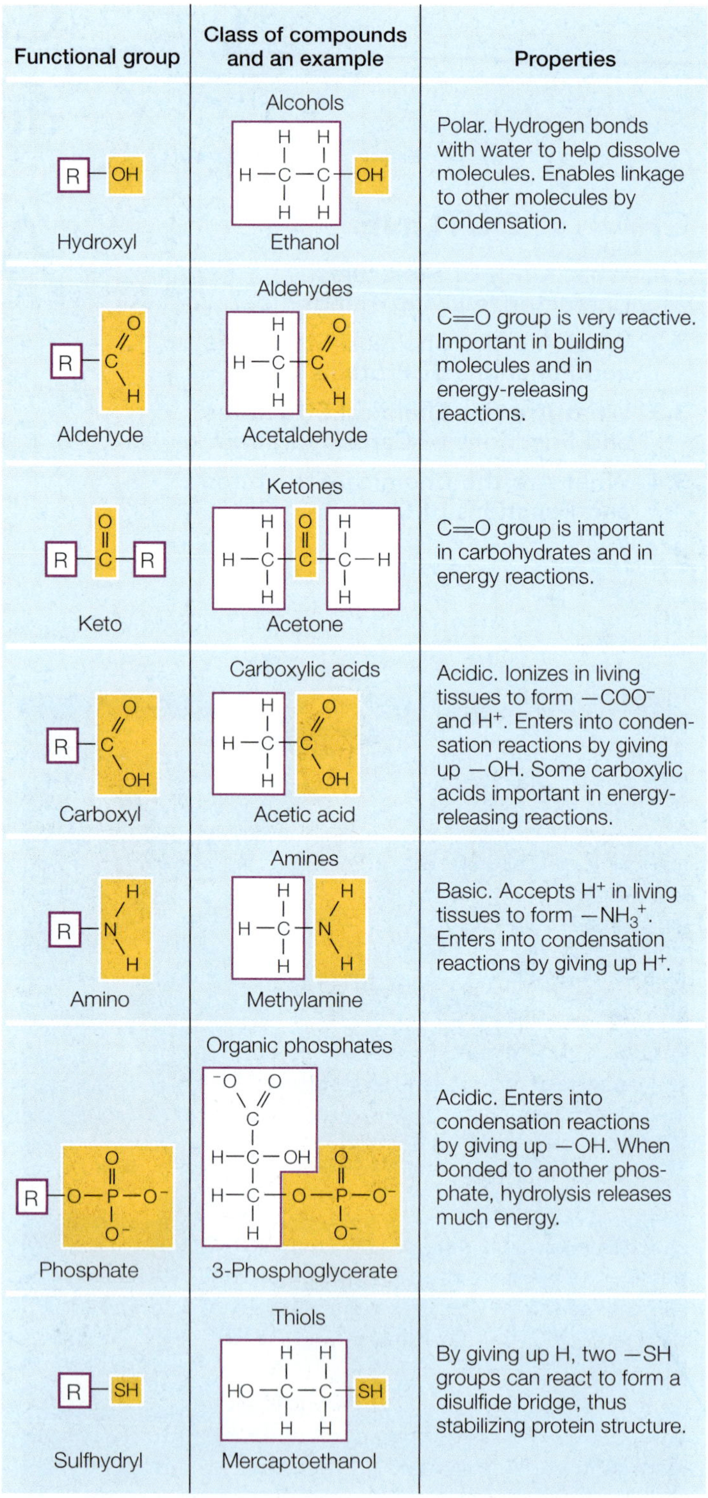
Functional group
Class of compounds and an example
Properties
Alcohols
Hydroxyl
Ethanol
Polar. Hydrogen bonds with water to help dissolve molecules. Enables linkage to other molecules by condensation.
Aldehydes
Aldehyde
Acetaldehyde
C=O group is very reactive. Important in building molecules and in energy-releasing reactions.
Ketones
Keto
Acetone
C=O group is important in carbohydrates and in energy reactions.
Carboxylic acids
Carboxyl
Acetic acid
Acidic. Ionizes in living tissues to form —COO⁻ and H⁺. Enters into condensation reactions by giving up —OH. Some carboxylic acids important in energy-releasing reactions.
Amines
Amino
Methylamine
Basic. Accepts H⁺ in living tissues to form —NH3⁺. Enters into condensation reactions by giving up H⁺.
Organic phosphates
Phosphate
3-Phosphoglycerate
Acidic. Enters into condensation reactions by giving up —OH. When bonded to another phosphate, hydrolysis releases much energy.
Thiols
Sulfhydryl
Mercaptoethanol
By giving up H, two —SH groups can react to form a disulfide bridge, thus stabilizing protein structure.

3.1 Some Functional Groups Important to Living Systems Highlighted here are the seven functional groups most commonly found in biologically important molecules. "R" is a variable chemical grouping.

Go to Activity 3.1 Functional Groups
Life10e.com/ac3.1

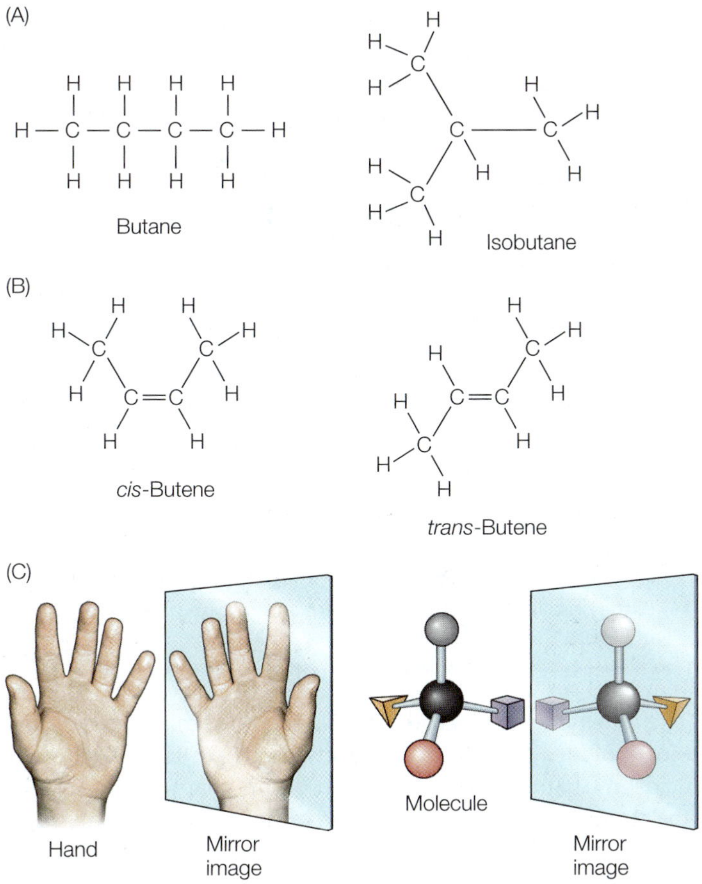
(A)
Butane
Isobutane
(B)
cis-Butene
trans-Butene
(C)
Hand
Mirror image
Molecule
Mirror image

3.2 Isomers Isomers have the same chemical formula, but the atoms are arranged differently. Pairs of isomers often have different chemical properties.

Isomers have different arrangements of the same atoms

Isomers are molecules that have the same chemical formula—the same kinds and numbers of atoms—but with the atoms arranged differently. (The prefix *iso-*, meaning "same," is encountered in many biological terms.) Of the different kinds of isomers, we will consider three: structural isomers, *cis-trans* isomers, and optical isomers.

Structural isomers differ in how their atoms are joined together. Consider two simple molecules, each composed of four carbon and ten hydrogen atoms bonded covalently, both with the formula C_4H_{10}. These atoms can be linked in two different ways, resulting in different molecules (Figure 3.2A).

In biological molecules, ***cis-trans* isomers** typically involve a double bond between two carbon atoms, where the carbons share two pairs of electrons. When the remaining two bonds of each of these carbons are to two different atoms or groups of atoms (e.g., a hydrogen and a methyl group; Figure 3.2B), these can be oriented on the same side or different sides of the double-bonded molecule. If the different atoms or groups of atoms are on the same side, the double bond is called *cis*; if they are on opposite sides, the bond is *trans*. These molecules can have very different properties.

Optical isomers occur when a carbon atom has four different atoms or groups of atoms attached to it. This pattern allows for two different ways of making the attachments, each the mirror image of the other (Figure 3.2C). Such a carbon atom is called an asymmetric carbon, and the two resulting molecules are optical isomers of one another. You can envision your right and left hands as optical isomers. Just as a glove is specific for a particular hand, some biochemical molecules that can interact with one optical isomer of a carbon compound are unable to "fit" the other.

The structures of macromolecules reflect their functions

The four kinds of biological macromolecules are present in roughly the same proportions in all living organisms (Figure 3.3). Furthermore, a protein that has a certain function in an apple tree probably has a similar function in a human being, because the protein's chemistry is the same wherever it is found. Such biochemical unity reflects the evolution of all life from a common ancestor, by descent with modification. An important advantage of biochemical unity is that some organisms can acquire needed raw materials by eating other organisms. When you eat an apple, the molecules you take in include carbohydrates, lipids, and proteins that can be broken down and rebuilt into the varieties of those molecules needed by humans.

Each type of macromolecule performs one or more functions such as energy storage, structural support, catalysis (speeding up of chemical reactions), transport of other molecules, regulation of other molecules, defense, movement, or information storage. These roles are not necessarily exclusive; for example, both carbohydrates and proteins can play structural roles, supporting and protecting tissues and organs. However, only the nucleic acids specialize in information storage and transmission. These macromolecules function as hereditary material, carrying the traits of both species and individuals from generation to generation.

The functions of macromolecules are directly related to their three-dimensional shapes and to the sequences and chemical properties of their monomers. Some macromolecules fold into

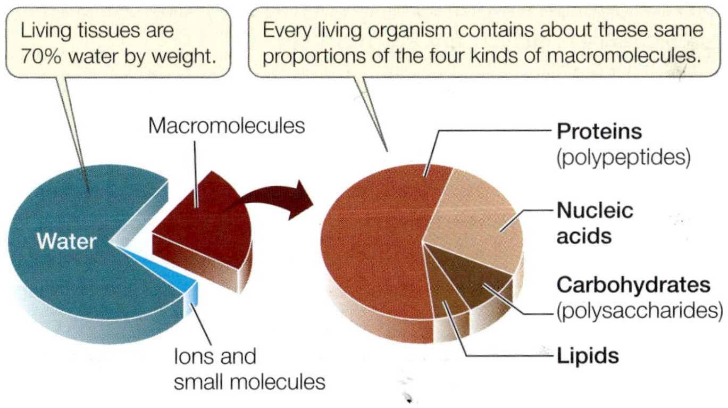
Living tissues are 70% water by weight.
Every living organism contains about these same proportions of the four kinds of macromolecules.
Macromolecules
Water
Ions and small molecules
Proteins (polypeptides)
Nucleic acids
Carbohydrates (polysaccharides)
Lipids

3.3 Substances Found in Living Tissues The substances shown here make up the nonmineral components of living tissues (bone would be an example of a mineral component).

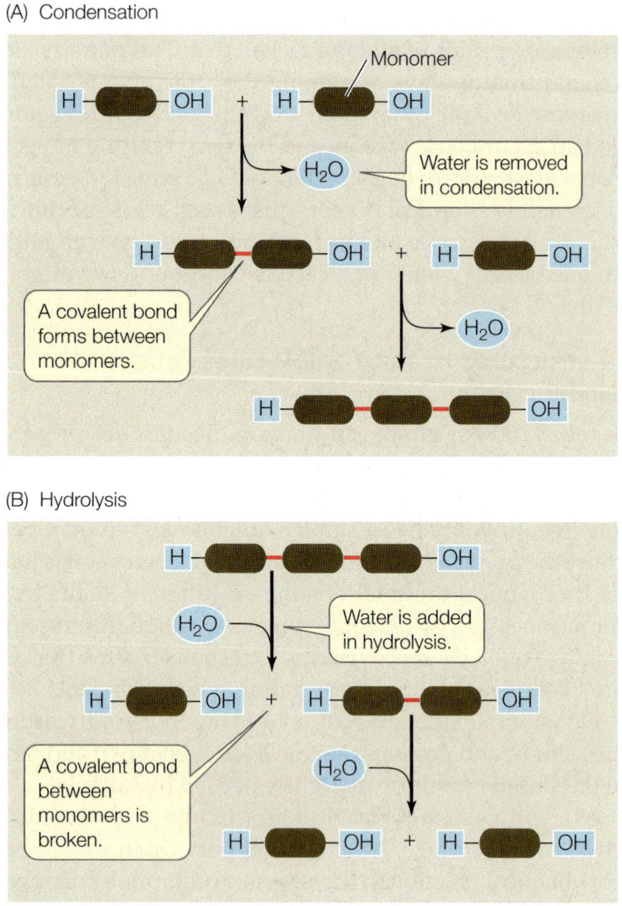
(A) Condensation
Monomer
H OH + H OH
H₂O
Water is removed in condensation.
H OH + H OH
A covalent bond forms between monomers.
H₂O
H OH
(B) Hydrolysis
H OH
H₂O
Water is added in hydrolysis.
H OH + H OH
A covalent bond between monomers is broken.
H₂O
H OH + H OH

3.4 Condensation and Hydrolysis of Polymers (A) Condensation reactions link monomers into polymers and produce water. (B) Hydrolysis reactions break polymers into individual monomers and consume water.

compact forms with surface features that make them water-soluble and capable of intimate interactions with other molecules. Some proteins and carbohydrates form long, fibrous structures (such as those found in hair or spider silk) that provide strength and rigidity to cells and tissues. The long, thin assemblies of proteins in muscles can contract, resulting in movement.

Most macromolecules are formed by condensation and broken down by hydrolysis

Polymers are formed from monomers by a series of **condensation reactions** (sometimes called dehydration reactions; both terms refer to the loss of water). Condensation reactions result in the formation of covalent bonds between monomers. A molecule of water is released with each covalent bond formed (**Figure 3.4A**). The condensation reactions that produce the different kinds of polymers differ in detail, but in all cases polymers form only if water molecules are removed and energy is added to the system. In living systems, specific energy-rich molecules supply the necessary energy.

The reverse of a condensation reaction is a **hydrolysis reaction** (*hydro*, "water"; *lysis*, "break"). Hydrolysis reactions result in the breakdown of polymers into their component monomers. Water reacts with the covalent bonds that link the polymer together. For each covalent bond that is broken, a water molecule splits into two ions (H^+ and OH^-), which each become part of one of the products (**Figure 3.4B**). Hydrolysis releases energy.

RECAP 3.1

The four kinds of large molecules that distinguish living tissues are proteins, lipids, carbohydrates, and nucleic acids. Most are polymers: chains of linked monomers. Very large polymers are called macromolecules. Biological molecules carry out a variety of life-sustaining functions.

- How do functional groups affect the structures and functions of macromolecules? (Keep this question in mind as you read the rest of this chapter.) **See p. 40 and Figure 3.1**
- What are the differences between structural, *cis-trans*, and optical isomers? **See p. 41 and Figure 3.2**
- How do monomers link up to form polymers, and how do they break down into monomers again? **See p. 42 and Figure 3.4**

The four types of macromolecules can be seen as the building blocks of life. We will cover the unique properties of the nucleic acids in Chapter 4. The remainder of this chapter will describe the structures and functions of the proteins, carbohydrates, and lipids.

3.2 What Are the Chemical Structures and Functions of Proteins?

Proteins have very diverse roles. In virtually every chapter of this book you will study examples of their extensive functions (**Table 3.1**). Among the functions of macromolecules listed in Section

TABLE 3.1 Proteins and Their Functions

Category	Function
Enzymes	Catalyze (speed up) biochemical reactions
Structural proteins	Provide physical stability and movement
Defensive proteins	Recognize and respond to nonself substances (e.g., antibodies)
Signaling proteins	Control physiological processes (e.g., hormones)
Receptor proteins	Receive and respond to chemical signals
Membrane transporters	Regulate passage of substances across cellular membranes
Storage proteins	Store amino acids for later use
Transport proteins	Bind and carry substances within the organism
Gene regulatory proteins	Determine the rate of expression of a gene

3.1, only two—energy storage and information storage—are not usually performed by proteins.

All **proteins** are polymers made up of 20 amino acids in different proportions and sequences. Proteins range in size from small ones such as insulin, which has 51 amino acids and a molecular weight of 5,733, to huge molecules such as the muscle protein titin, with 26,926 amino acids and a molecular weight of 2,993,451. Proteins consist of one or more **polypeptide chains**—unbranched (linear) polymers of covalently linked amino acids. Variation in the sequences of amino acids in the polypeptide chains allows for the vast diversity in protein structure and function. Each chain folds into a particular three-dimensional shape that is specified by the sequence of amino acids present in the chain.

Amino acids are the building blocks of proteins

Each **amino acid** has both a carboxyl functional group and an amino functional group (see Figure 3.1) attached to the same carbon atom, called the α (alpha) carbon. Also attached to the α carbon atom are a hydrogen atom and a **side chain**, or **R group**, designated by the letter R.

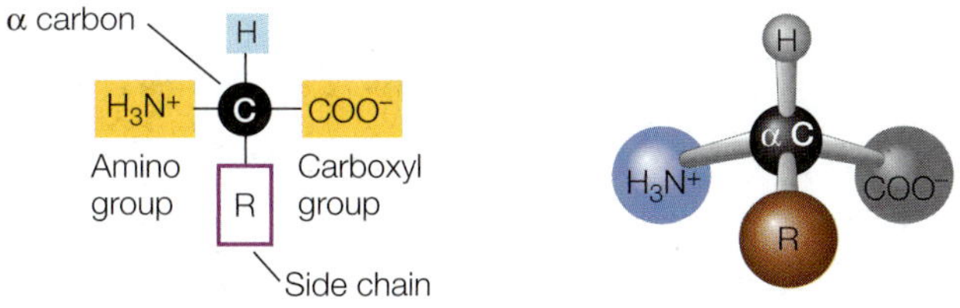
α carbon
H
H_3N^+
C
COO^-
Amino group
Carboxyl group
R
Side chain
H
α C
H_3N^+
COO^-
R

The α carbon is asymmetrical because it is bonded to four different atoms or groups of atoms. Therefore, amino acids can exist as optical isomers called D-amino acids and L-amino acids. D and L are abbreviations of the Latin terms for right (*dextro*) and left (*levo*). Only L-amino acids (with the configuration shown above) are commonly found in the proteins of most organisms, and their presence is an important chemical "signature" of life.

At the pH levels typically found in cells (usually about pH 7), both the carboxyl and amino groups of amino acids are ionized: the carboxyl group has lost a hydrogen ion:

$$\text{—COOH} \rightarrow \text{—COO}^- + \text{H}^+$$

and the amino group has gained a hydrogen ion:

$$\text{—NH}_2 + \text{H}^+ \rightarrow \text{—NH}_3^+$$

Thus amino acids are simultaneously acids and bases.

The side chains (or R groups) of amino acids contain functional groups that are important in determining the three-dimensional structure and thus the function of the protein. As Table 3.2 shows, the 20 amino acids found in living organisms are grouped and distinguished by their side chains:

- Five amino acids have electrically charged (ionized) side chains at pH levels typical of living cells. These side chains attract water (are hydrophilic) and attract oppositely charged ions of all sorts.
- Five amino acids have polar side chains. They are also hydrophilic and attract other polar or charged molecules.

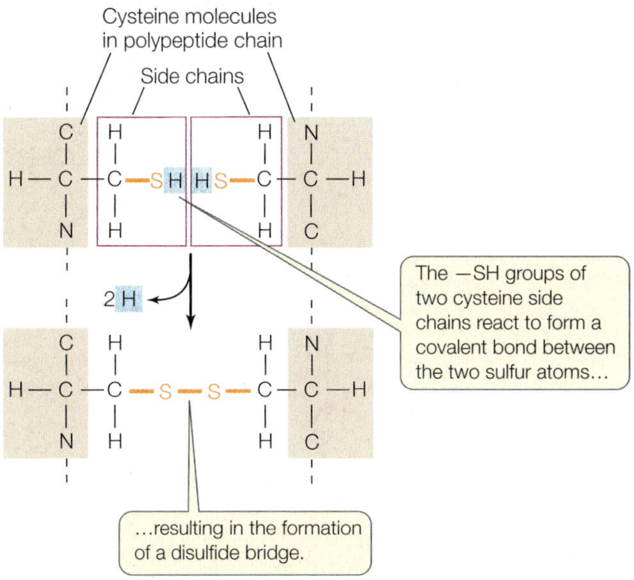
Cysteine molecules in polypeptide chain
Side chains
The —SH groups of two cysteine side chains react to form a covalent bond between the two sulfur atoms...
2 H
...resulting in the formation of a disulfide bridge.

3.5 A Disulfide Bridge Two cysteine molecules in a polypeptide chain can form a disulfide bridge (—S—S—) by oxidation (removal of H atoms).

- Seven amino acids have side chains that are nonpolar and thus hydrophobic. In the watery environment of the cell, these hydrophobic groups may cluster together in the interior of the protein.

Three amino acids—cysteine, glycine, and proline—are special cases, although the side chains of the latter two are generally hydrophobic.

- The cysteine side chain, which has a terminal —SH group, can react with another cysteine side chain in an oxidation reaction to form a covalent bond (Figure 3.5). Such a bond, called a **disulfide bridge** or disulfide bond (—S—S—), helps determine how a polypeptide chain folds.
- The glycine side chain consists of a single hydrogen atom. It is small enough to fit into tight corners in the interiors of protein molecules where larger side chains could not fit.
- Proline possesses a modified amino group that lacks a hydrogen and instead forms a covalent bond with the hydrocarbon side chain, resulting in a ring structure. This limits both its hydrogen-bonding ability and its ability to rotate about the α carbon. Thus proline is often found where a protein bends or loops.

Go to Activity 3.2 Features of Amino Acids Life10e.com/ac3.2

Peptide linkages form the backbone of a protein

When amino acids polymerize, the carboxyl and amino groups attached to the α carbon are the reactive groups. The carboxyl group of one amino acid reacts with the amino group of another, undergoing a condensation reaction that forms a **peptide linkage** (also called a peptide bond). Figure 3.6 gives a simplified description of this reaction.

TABLE 3.2
The Twenty Amino Acids

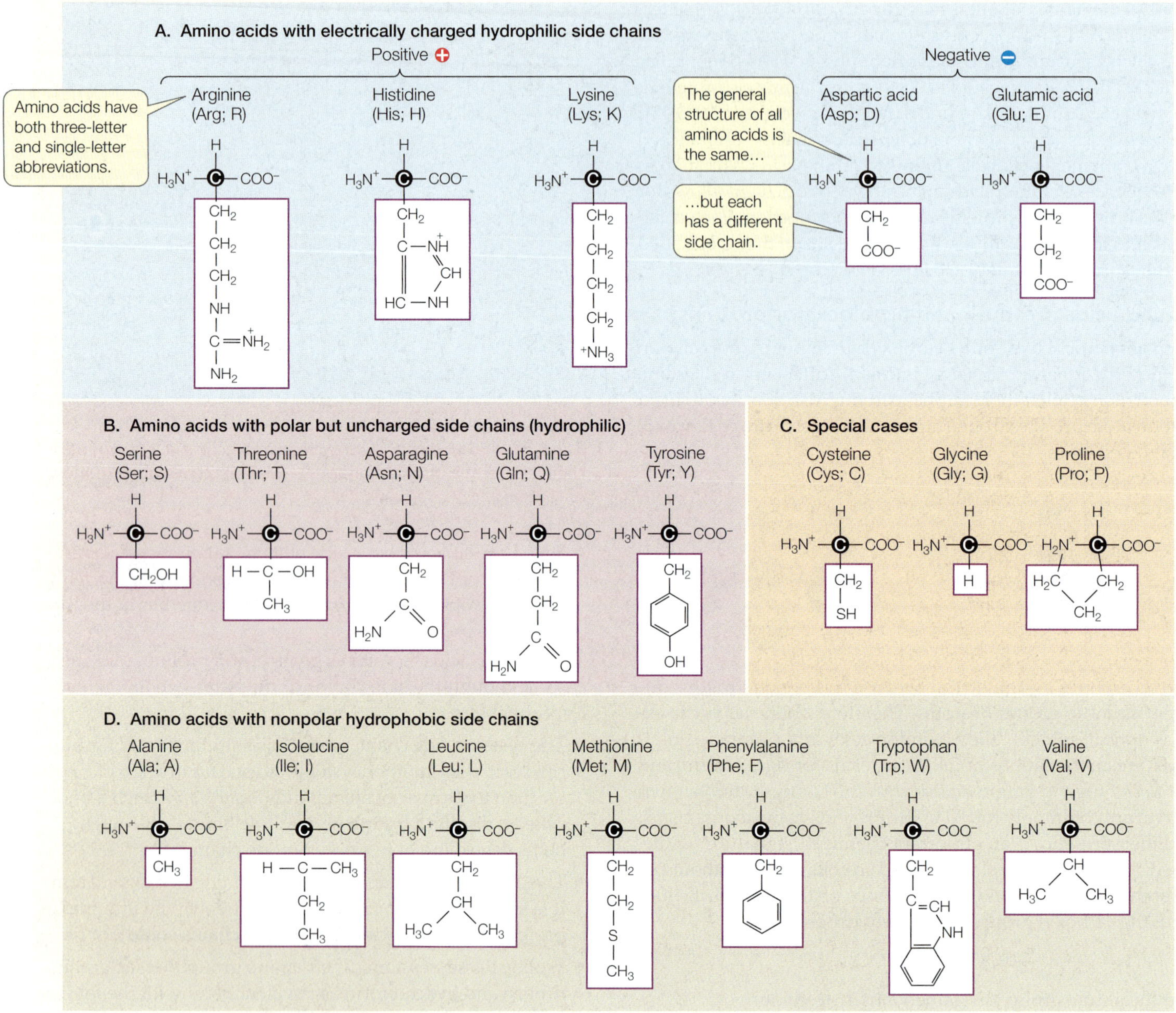
A. Amino acids with electrically charged hydrophilic side chains
Positive
Negative
Amino acids have both three-letter and single-letter abbreviations.
Arginine (Arg; R)
Histidine (His; H)
Lysine (Lys; K)
The general structure of all amino acids is the same...
...but each has a different side chain.
Aspartic acid (Asp; D)
Glutamic acid (Glu; E)
B. Amino acids with polar but uncharged side chains (hydrophilic)
Serine (Ser; S)
Threonine (Thr; T)
Asparagine (Asn; N)
Glutamine (Gln; Q)
Tyrosine (Tyr; Y)
C. Special cases
Cysteine (Cys; C)
Glycine (Gly; G)
Proline (Pro; P)
D. Amino acids with nonpolar hydrophobic side chains
Alanine (Ala; A)
Isoleucine (Ile; I)
Leucine (Leu; L)
Methionine (Met; M)
Phenylalanine (Phe; F)
Tryptophan (Trp; W)
Valine (Val; V)

Just as a sentence begins with a capital letter and ends with a period, polypeptide chains have a beginning and an end. The "capital letter" marking the beginning of a polypeptide is the amino group of the first amino acid added to the chain and is known as the N terminus. The "period" is the carboxyl group of the last amino acid added; this is the C terminus.

Two characteristics of the peptide bond are especially important in the three-dimensional structures of proteins:

- In the C—N linkage, the adjacent α carbons (α-C—C—N—α-C) are not free to rotate fully, which limits the folding of the polypeptide chain.
- The oxygen bound to the carbon (C=O) in the carboxyl group carries a slight negative charge (δ^-), whereas the hydrogen bound to the nitrogen (N—H) in the amino group is slightly positive (δ^+). This asymmetry of charge favors hydrogen bonding within the protein molecule itself and between molecules. These bonds contribute to the structures and functions of many proteins.

In addition to these characteristics of the peptide linkage, the particular sequence of amino acids—with their various R groups—in the polypeptide chain also plays a vital role in determining a protein's structure and function.

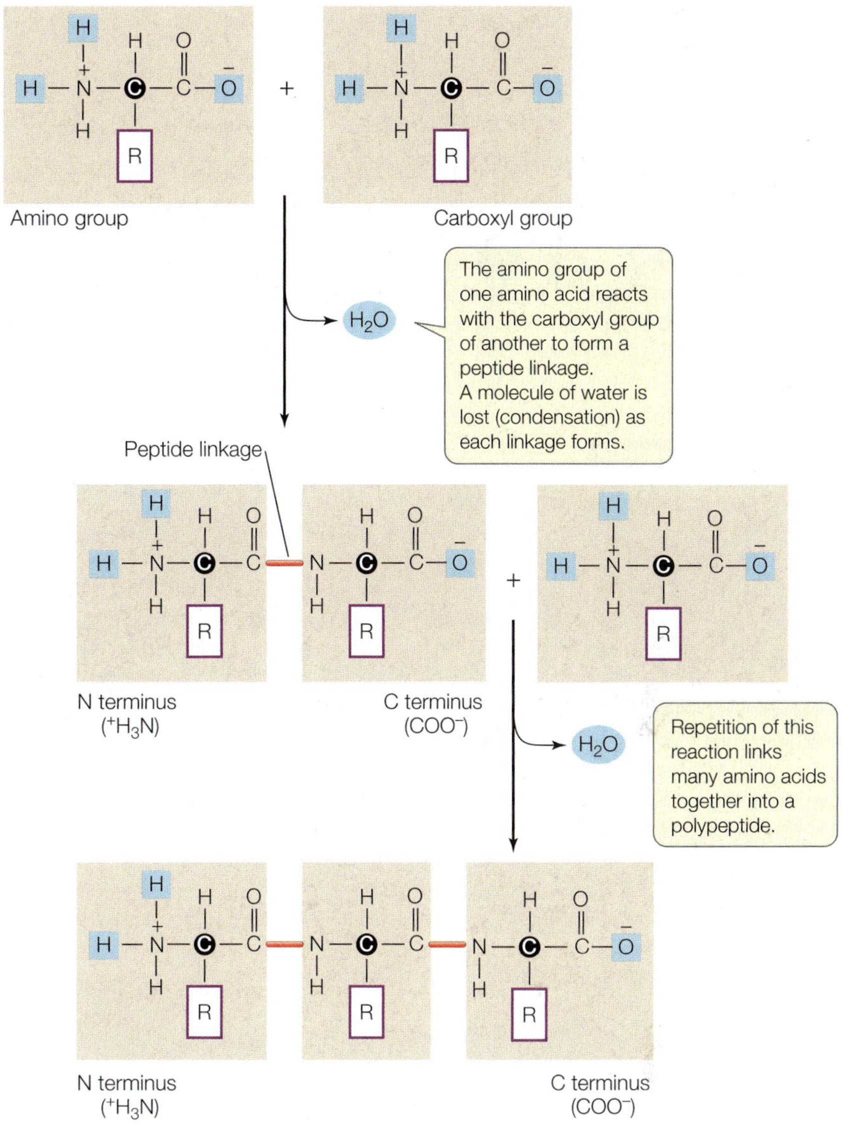
Amino group
Carboxyl group
H_2O
The amino group of one amino acid reacts with the carboxyl group of another to form a peptide linkage. A molecule of water is lost (condensation) as each linkage forms.
Peptide linkage
N terminus (^+H_3N)
C terminus (COO^-)
H_2O
Repetition of this reaction links many amino acids together into a polypeptide.
N terminus (^+H_3N)
C terminus (COO^-)

3.6 Formation of Peptide Linkages In living things, the reaction leading to a peptide linkage (also called a peptide bond) has many intermediate steps, but the reactants and products are the same as those shown in this simplified diagram.

The primary structure of a protein is its amino acid sequence

The precise sequence of amino acids in a polypeptide chain held together by peptide bonds constitutes the **primary structure** of a protein (Figure 3.7A). The backbone of the polypeptide chain consists of the repeating sequence —N—C—C— made up of the N atom from the amino group, the α C atom, and the C atom from the carboxyl group in each amino acid.

The single-letter abbreviations for amino acids (see Table 3.2) are used to record the amino acid sequence of a protein. Here, for example, are the first 20 amino acids (out of a total of 124) in the protein ribonuclease from a cow:

KETAAAKFERQHMDSSTSAA

The theoretical number of different proteins is enormous. Since there are 20 different amino acids, there could be 20 × 20 = 400 distinct dipeptides (two linked amino acids) and 20 × 20 × 20 = 8,000 different tripeptides (three linked amino acids). Imagine this process of multiplying by 20 extended to a protein made up of 100 amino acids (which would be considered a small protein). There could be 20^{100} (that's approximately 10^{130}) such small proteins, each with its own distinctive primary structure. How large is the number 20^{100}? Physicists tell us that there aren't that many electrons in the entire universe.

The sequence of amino acids in the polypeptide chain(s) determines its final shape. The properties associated with each functional group in the side chains of the amino acids (see Table 3.2) determine how the protein can twist and fold, thus adopting a specific stable structure that distinguishes it from every other protein.

The secondary structure of a protein requires hydrogen bonding

A protein's **secondary structure** consists of regular, repeated spatial patterns in different regions of a polypeptide chain. There are two basic types of secondary structure, both determined by hydrogen bonding between the amino acids that make up the primary structure: the α helix and the β pleated sheet.

THE ALPHA HELIX The α (**alpha**) **helix** is a right-handed coil that turns in the same direction as a standard wood screw (Figure 3.7B and Figure 3.8). The R groups extend outward from the peptide backbone of the helix. The coiling results from hydrogen bonds that form between the δ^+ hydrogen of the N—H of one amino acid and the δ^- oxygen of the C═O of another. When this pattern of hydrogen bonding is established repeatedly over a segment of the protein, it stabilizes the coil.

THE BETA PLEATED SHEET A β (**beta**) **pleated sheet** is formed from two or more polypeptide chains that are almost completely extended and aligned. The sheet is stabilized by hydrogen bonds between the N—H groups on one chain and the C═O groups on the other (Figure 3.7C). A β pleated sheet may form between separate polypeptide chains or between different regions of a single polypeptide chain that is bent back on itself. The ratcheted, stacked sheets in dragline spider silks (see the opening story at the beginning of the chapter) are made up of β pleated sheets. Many proteins contain regions of both α helix and β pleated sheet in the same polypeptide chain.

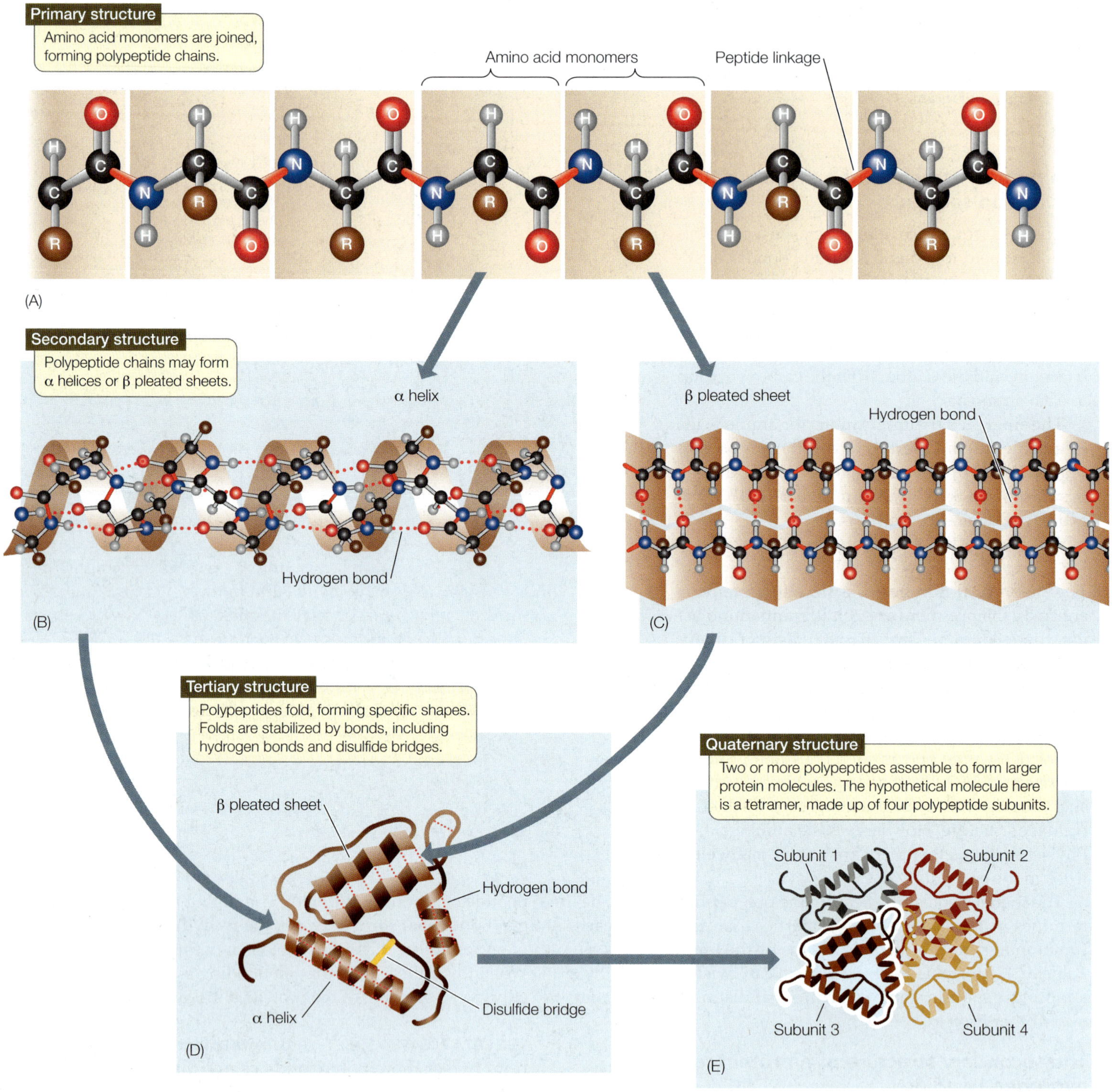

Primary structure
Amino acid monomers are joined, forming polypeptide chains.
Amino acid monomers
Peptide linkage
(A)
Secondary structure
Polypeptide chains may form α helices or β pleated sheets.
α helix
Hydrogen bond
(B)
β pleated sheet
Hydrogen bond
(C)
Tertiary structure
Polypeptides fold, forming specific shapes. Folds are stabilized by bonds, including hydrogen bonds and disulfide bridges.
β pleated sheet
Hydrogen bond
α helix
Disulfide bridge
(D)
Quaternary structure
Two or more polypeptides assemble to form larger protein molecules. The hypothetical molecule here is a tetramer, made up of four polypeptide subunits.
Subunit 1
Subunit 2
Subunit 3
Subunit 4
(E)

3.7 The Four Levels of Protein Structure Secondary, tertiary, and quaternary structure all arise from the primary structure of the protein.

The tertiary structure of a protein is formed by bending and folding

In many proteins, the polypeptide chain is bent at specific sites and then folded back and forth, resulting in the **tertiary structure** of the protein (Figure 3.7D). Although α helices and β pleated sheets contribute to the tertiary structure, usually only portions of the macromolecule have these secondary structures, and large regions consist of tertiary structure unique to a particular protein. For example, the proteins found in stretchy spider silks have repeated amino acid sequences that cause the proteins to fold into structures called right-handed β-spirals. Tertiary structure is a macromolecule's definitive three-dimensional shape, often including a buried interior as well as a surface that is exposed to the environment.

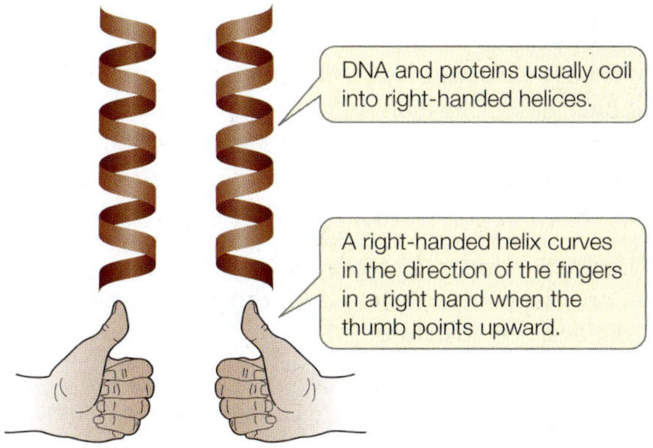
DNA and proteins usually coil into right-handed helices.
A right-handed helix curves in the direction of the fingers in a right hand when the thumb points upward.

3.8 Left- and Right-Handed Helices A protein will often have one or more right-handed helices as part of its secondary structure.

The protein's exposed outer surfaces present functional groups capable of interacting with other molecules in the cell. These molecules might be other macromolecules, including proteins, nucleic acids, carbohydrates, and lipid structures, or smaller chemical substances.

Whereas hydrogen bonding between the N—H and C═O groups within and between chains is responsible for secondary structure, the interactions between R groups—the amino acid side chains—and between R groups and the environment determine tertiary structure. We described the various strong and weak interactions between atoms in Section 2.2. Many of these interactions are involved in determining and maintaining tertiary structure.

- Covalent disulfide bridges can form between specific cysteine side chains (see Figure 3.5), holding a folded polypeptide in place.
- Hydrogen bonds between side chains also stabilize folds in proteins.
- Hydrophobic side chains can aggregate together in the interior of the protein, away from water, folding the polypeptide in the process. Close interactions between the hydrophobic side chains are stabilized by van der Waals forces.
- Ionic attractions can form between positively and negatively charged side chains, forming salt bridges between amino acids. Salt bridges can be near the surfaces of polypeptides or buried deep within a protein, away from water. These interactions occur between positively and negatively charged amino acids, for example glutamic acid (which has a negatively charged R group) and arginine (which is positively charged) (see Table 3.2):

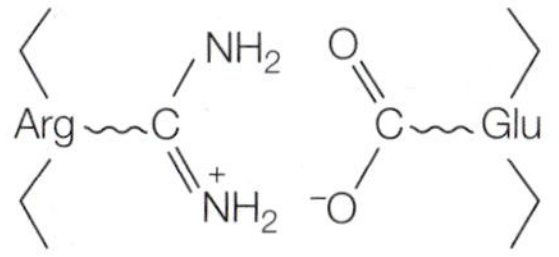
NH_2
Arg
C
$\overset{+}{N}H_2$
O
C
Glu
^{-}O

A complete description of a protein's tertiary structure would specify the location of every atom in the molecule in three-dimensional space relative to all the other atoms. **Figure 3.9** shows three models of the structure of the protein lysozyme. The space-filling model might be used to study how other molecules interact with specific sites and R groups on the protein's surface. The stick model emphasizes the sites where bends occur, resulting in

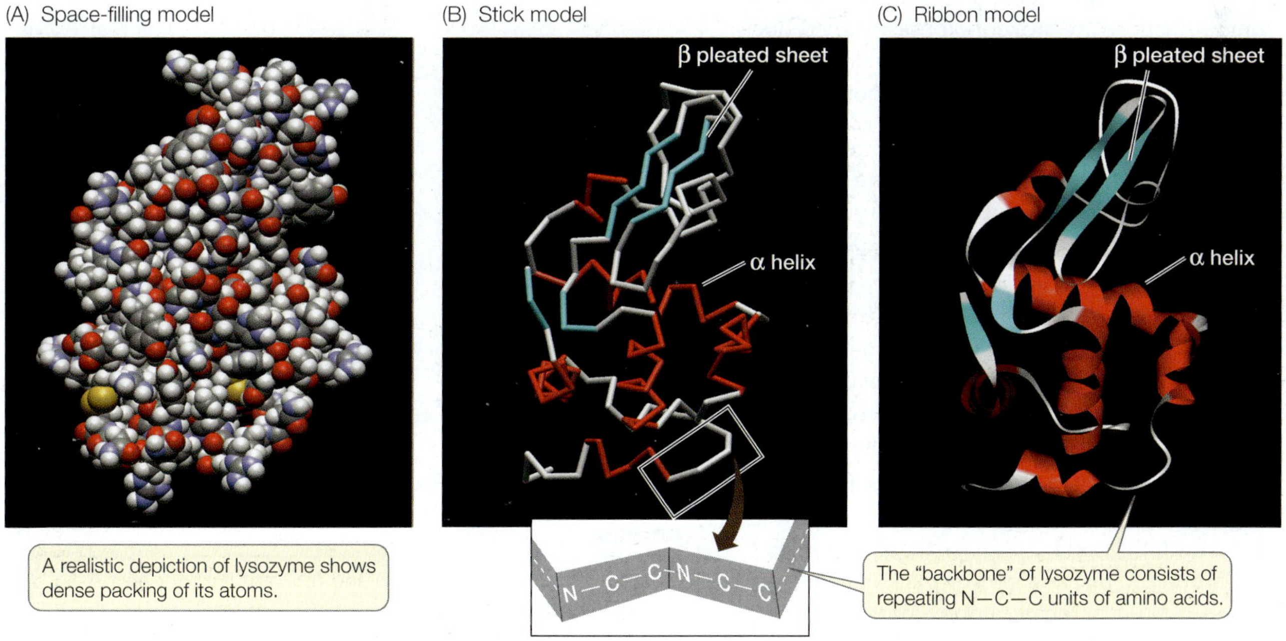
(A) Space-filling model
(B) Stick model
(C) Ribbon model
β pleated sheet
α helix
β pleated sheet
α helix
A realistic depiction of lysozyme shows dense packing of its atoms.
N—C—C—N—C—C
The "backbone" of lysozyme consists of repeating N—C—C units of amino acids.

3.9 Three Representations of Lysozyme Different molecular representations of a protein emphasize different aspects of its tertiary structure: surface features, sites of bends and folds, or sites where alpha or beta structures predominate. These three representations of lysozyme are similarly oriented.

Go to Media Clip 3.1
Protein Structures in 3D
Life10e.com/mc3.1

folds in the polypeptide chain. The ribbon model, perhaps the most widely used, shows the different types of secondary structure and how they fold into the tertiary structure.

Remember that both secondary and tertiary structure derive from primary structure. If a protein is heated slowly and moderately, the heat energy will disrupt only the weak interactions, causing the secondary and tertiary structure to break down. The protein is then said to be **denatured**. But in some cases the protein can return to its normal tertiary structure when it cools, demonstrating that all the information needed to specify the unique shape of a protein is contained in its primary structure. This was first shown (using chemicals instead of heat to denature the protein) by biochemist Christian Anfinsen for the protein ribonuclease (**Figure 3.10**).

The quaternary structure of a protein consists of subunits

Many functional proteins contain two or more polypeptide chains, called subunits, each of them folded into its own unique tertiary structure. The protein's **quaternary structure** results from the ways in which these subunits bind together and interact (**Figure 3.7E**).

The models of hemoglobin in **Figure 3.11** illustrate quaternary structure. Hydrophobic interactions, van der Waals forces, hydrogen bonds, and ionic attractions all help hold the four subunits together to form a hemoglobin molecule. However, the weak nature of these forces permits small changes in the quaternary structure to aid the protein's function—which is to carry oxygen in red blood cells. As hemoglobin binds one O_2 molecule, the four subunits shift their relative positions slightly, changing the quaternary structure. Ionic attractions are broken, exposing buried side chains that enhance the binding of additional O_2 molecules. The quaternary structure changes back when hemoglobin releases its O_2 molecules to the cells of the body.

Shape and surface chemistry contribute to protein function

The shapes and structures of proteins allow specific sites on their exposed surfaces to bind noncovalently to other molecules, which may be large or small. The binding is usually very specific because only certain compatible chemical groups will bind to one another. The specificity of protein binding depends on two general properties of the protein: its shape, and the chemistry of its exposed surface groups.

- *Shape.* When a small molecule collides with and binds to a much larger protein, it is like a baseball being caught by a catcher's mitt: the mitt has a shape that binds to the ball and fits around it. Just as a hockey puck or a Ping-Pong ball does not fit a baseball catcher's mitt, a given molecule will not bind to a protein unless there is a general "fit" between their three-dimensional shapes.

INVESTIGATING**LIFE**

3.10 Primary Structure Specifies Tertiary Structure Using the protein ribonuclease, Christian Anfinsen showed that proteins spontaneously fold into functionally correct three-dimensional configurations.[a] As long as the primary structure is not disrupted, the information for correct folding (under the right conditions) is retained.

HYPOTHESIS Under controlled conditions that simulate the normal cellular environment, a denatured protein can refold into a functional three-dimensional structure.

Method Chemically denature a functional ribonuclease so that only its primary structure (i.e., an unfolded polypeptide chain) remains. Once denaturation is complete, remove the disruptive chemicals.

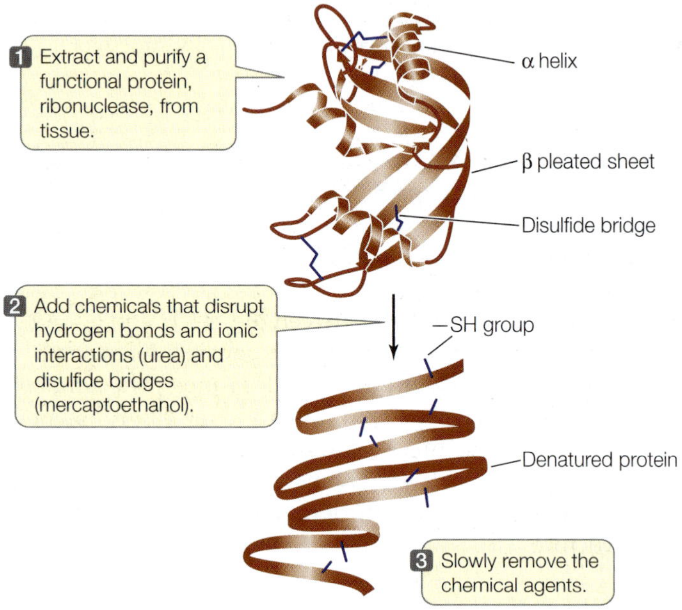
1 Extract and purify a functional protein, ribonuclease, from tissue.
α helix
β pleated sheet
Disulfide bridge
2 Add chemicals that disrupt hydrogen bonds and ionic interactions (urea) and disulfide bridges (mercaptoethanol).
—SH group
Denatured protein
3 Slowly remove the chemical agents.

Results When the disruptive agents are removed, three-dimensional structure is restored and the protein once again is functional.

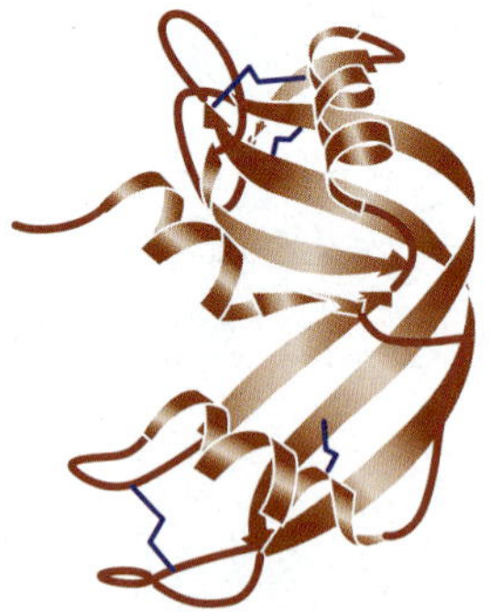

CONCLUSION In normal cellular conditions, the primary structure of a protein specifies how it folds into a functional, three-dimensional structure.

Go to **BioPortal** for discussion and relevant links for all INVESTIGATING**LIFE** figures.

[a]Anfinsen, C. B. et al. 1961. *Proceedings of the National Academy of Sciences USA* 47: 1309–1314.

WORKING WITH **DATA**:

Primary Structure Specifies Tertiary Structure

Original Papers

Anfinsen, C. B., E. Haber, M. Sela, and F. White, Jr. 1961. The kinetics of formation of native ribonuclease during oxidation of the reduced polypeptide chain. *Proceedings of the National Academy of Sciences USA* 47: 1309–1314.

White, Jr., F. 1961. Regeneration of native secondary and tertiary structures by air oxidation of reduced ribonuclease. *Journal of Biological Chemistry* 236: 1353–1360.

Analyze the Data

After the tertiary structures of proteins were shown to be highly specific, the question arose as to how the order of amino acids determined the three-dimensional structure. The second protein whose structure was determined was ribonuclease A (RNase A). This enzyme was readily available from cow pancreases at slaughterhouses and, because it works in the highly acidic environment of the cow stomach, was stable compared with most proteins and easy to purify. RNase A has 124 amino acids. Among these are eight cysteine residues, which form four disulfide bridges. Were these covalent links between cysteines essential for the three-dimensional structure of RNase A? Christian Anfinsen and his colleagues set out to answer this question. They first destroyed these links by reducing the S—S bonds to —SH and —SH. With the links destroyed, they looked at the three-dimensional structure of the protein (the extent of denaturation) and assessed protein function by measuring the loss of enzyme activity. They then removed the reducing agent (mercaptoethanol) and allowed the S—S bonds to re-form. They found that links between amino acids were indeed essential for tertiary structure and function. Anfinsen was awarded the Nobel Prize in Chemistry in 1973.

QUESTION 1

Initially, the disulfide bonds (S—S) in RNase A were eliminated because the sulfur atoms in cysteine residues were all reduced (—SH). At time zero, reoxidation began; and at various times, the amount of S—S bond re-formation and the activity of the enzyme were measured by chemical methods. The data are shown in **FIGURE A**.

At what time did disulfide bonds begin to form? At what time did enzyme activity begin to appear? Explain the difference between these times.

QUESTION 2

The three-dimensional structure of RNase A was examined by ultraviolet spectroscopy. In this technique, the protein was exposed to different wavelengths of ultraviolet light (measured in nanometers) and the amount of light absorbed by the protein at each wavelength was measured (E). The results are plotted in **FIGURE B**.

Look carefully at the plots. What are the differences between the peak absorbances of native (untreated) and reduced (denatured) RNase A? What happened when reduced RNase A was reoxidized (renatured)? What can you conclude about the structure of RNase A from these experiments?

FIGURE A

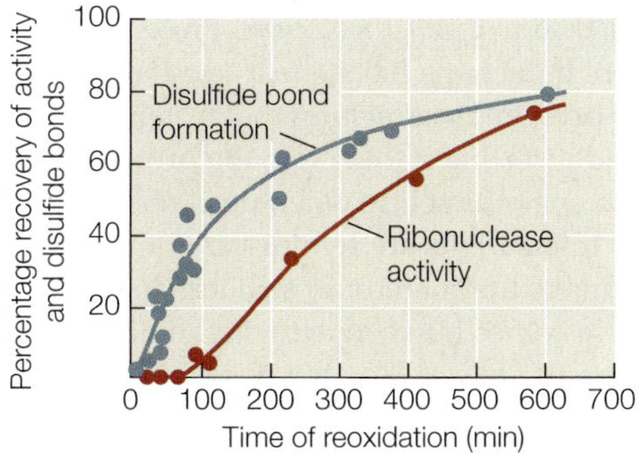
Percentage recovery of activity and disulfide bonds
100
80
60
40
20
Disulfide bond formation
Ribonuclease activity
0
100
200
300
400
500
600
700
Time of reoxidation (min)

FIGURE B

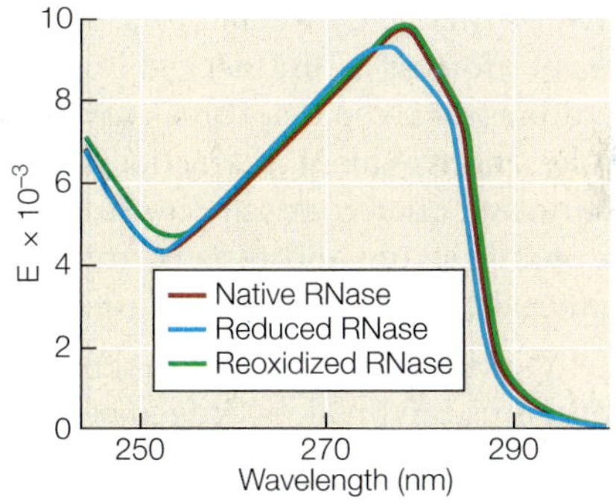
10
8
6
4
2
0
E × 10⁻³
Native RNase
Reduced RNase
Reoxidized RNase
250
270
290
Wavelength (nm)

Go to **BioPortal** for all WORKING WITH**DATA** exercises

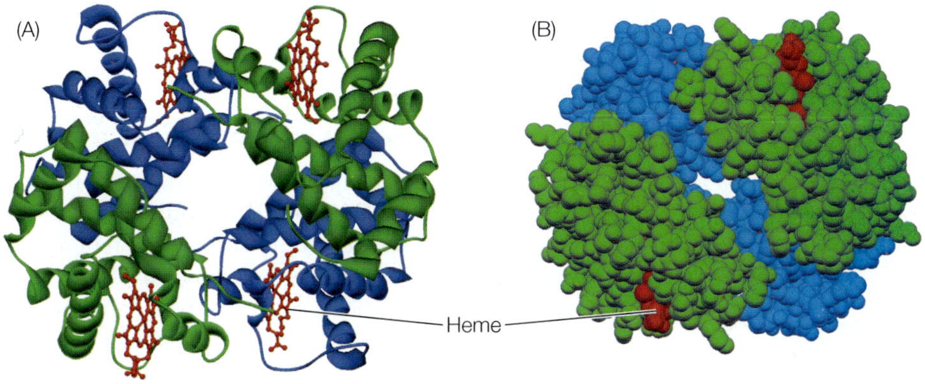
(A)
(B)
Heme

3.11 Quaternary Structure of a Protein Hemoglobin consists of four folded polypeptide subunits that assemble themselves into the quaternary structure represented by the ribbon model (A) and space-filling model (B). In both graphic representations, each type of subunit is a different color (α subunits are blue and β subunits are green). The heme groups (red) contain iron and are the oxygen-carrying sites.

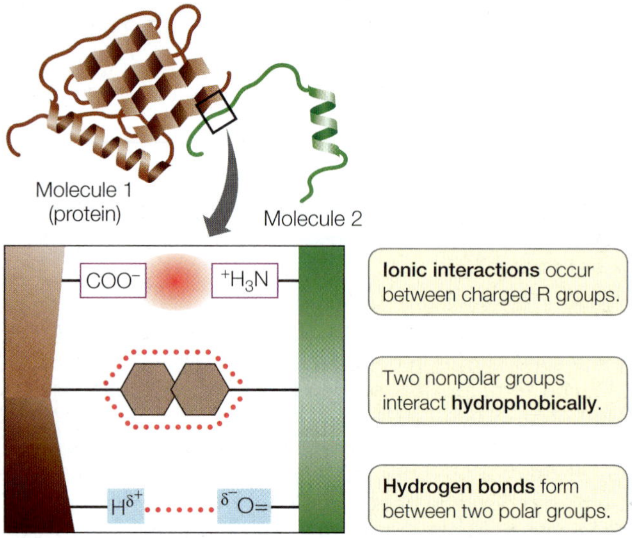
Molecule 1
(protein)
Molecule 2
COO⁻
⁺H₃N
Ionic interactions occur between charged R groups.
Two nonpolar groups interact hydrophobically.
Hydrogen bonds form between two polar groups.

3.12 Noncovalent Interactions between Proteins and Other Molecules Noncovalent interactions (see p. 26) allow a protein (brown) to bind tightly to another molecule (green) with specific properties. Noncovalent interactions also allow regions within the same protein to interact with one another.

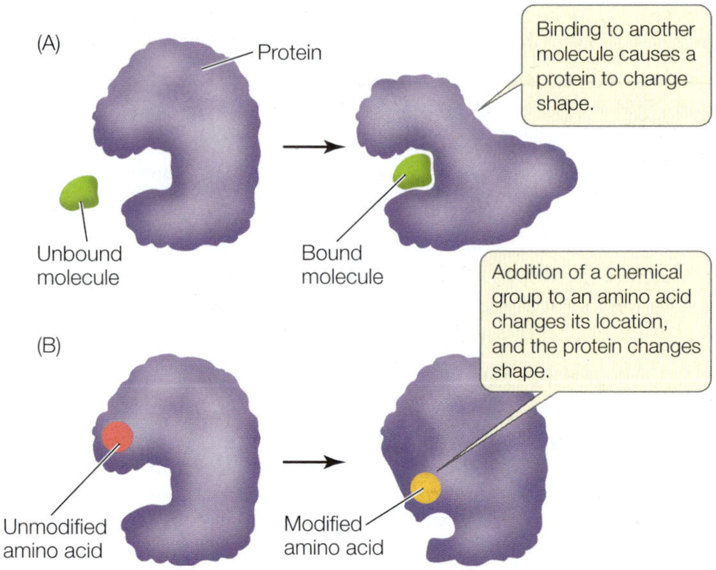
(A)
Protein
Binding to another molecule causes a protein to change shape.
Unbound molecule
Bound molecule
(B)
Addition of a chemical group to an amino acid changes its location, and the protein changes shape.
Unmodified amino acid
Modified amino acid

3.13 Protein Structure Can Change Proteins can change their tertiary structure when they bind to other molecules (A) or are modified chemically (B).

- *Chemistry*. The exposed R groups on the surface of a protein permit chemical interactions with other substances (**Figure 3.12**). Three types of interactions may be involved: ionic, hydrophobic, or hydrogen bonding. Many important functions of proteins involve interactions between surface R groups and other molecules.

Environmental conditions affect protein structure

Because they are determined by weak forces, the three-dimensional structures of proteins are influenced by environmental conditions. Conditions that would not break covalent bonds can disrupt the weaker, noncovalent interactions that determine secondary, tertiary, and quaternary structure. Such alterations may affect a protein's shape and thus its function. Various conditions can alter the weak, noncovalent interactions:

- Increases in temperature cause more rapid molecular movements and thus can break hydrogen bonds and hydrophobic interactions.
- Alterations in pH can change the pattern of ionization of exposed carboxyl and amino groups in the R groups of amino acids, thus disrupting the pattern of ionic attractions and repulsions.
- High concentrations of polar substances such as urea can disrupt the hydrogen bonding that is crucial to protein structure. Urea was used in the experiment on reversible protein denaturation shown in Figure 3.10.
- Nonpolar substances may also disrupt normal protein structure in cases where hydrophobic interactions are essential to maintain the structure.

Although denaturation is reversible in many cases (see Figure 3.10), in other cases it can be irreversible, such as when amino acids that were buried in the interior of the protein become exposed at the surface, or vice versa. This can result in the formation of new structures with different properties. Boiling an egg denatures its proteins and is, as you know, not reversible.

Protein shapes can change

As we saw in the case of hemoglobin, which undergoes subtle shape changes when it binds oxygen, the shapes of proteins can change as a result of their interactions with other molecules. Proteins can also change shape if they undergo covalent modifications.

- *Proteins interact with other molecules.* Proteins do not exist in isolation. In fact, if a biochemist "goes fishing" with a particular protein, by attaching the protein to a chemical "hook" and inserting it into cells, the protein will often be attached to something else when it is "reeled in." These molecular interactions are reminiscent of the interactions that make up quaternary structure (see above). If a polypeptide comes into contact with another molecule, R groups on its surface may form weak interactions (e.g., hydrophobic, van der Waals) with groups on the surface of the other molecule. This may disrupt some of the interactions between R groups within the polypeptide, causing it to undergo a change in shape (**Figure 3.13A**). You will see many instances of this in the coming chapters. An important example is an enzyme, which changes shape when it comes into contact with a reactant in a biochemical reaction (see Section 8.4).
- *Proteins undergo covalent modifications.* After it is made, the structure of a protein can be modified by the covalent bonding of a chemical group to the side chain of one or more of its amino acids. The chemical modification of just one amino acid can alter the shape and function of a protein. An example is the addition of a charged phosphate group to a relatively nonpolar R group. This can cause the amino acid to become more hydrophilic and to move to the outer surface of the protein, altering the shape of the protein in the region near the amino acid (**Figure 3.13B**).

Molecular chaperones help shape proteins

Within a living cell, a polypeptide chain is sometimes in danger of binding the wrong substance. There are two major situations when this can occur:

- *Just after a protein is made.* When a protein has not yet folded completely, it can present a surface that binds the wrong molecule.
- *Following denaturation.* Certain conditions, such as moderate heat, can cause some proteins in a living cell to denature without killing the organism. Before the protein can re-fold, it may present a surface that binds the wrong molecule. In these cases, the inappropriate binding may be irreversible. Many cells have a special class of proteins, called **chaperones**, that protect the three-dimensional structures of other proteins. Like the chaperones at a high school dance, they prevent inappropriate interactions and enhance appropriate ones. Typically, a chaperone protein has a cagelike structure that pulls in a polypeptide, causes it to fold into the correct shape, and then releases it (**Figure 3.14**). Tumors make chaperone proteins, possibly to stabilize proteins important in the cancer process, and so chaperone-inhibiting drugs are being designed for use in chemotherapy. In some clinical situations, treatment with these inhibitors results in the inappropriate folding of proteins in tumor cells, causing the tumors to stop growing.

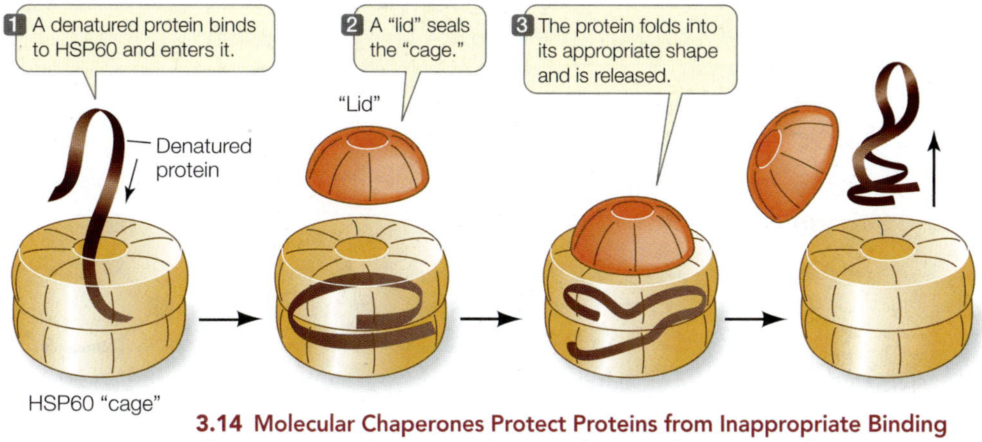
1 A denatured protein binds to HSP60 and enters it.
Denatured protein
HSP60 "cage"
2 A "lid" seals the "cage."
"Lid"
3 The protein folds into its appropriate shape and is released.

3.14 Molecular Chaperones Protect Proteins from Inappropriate Binding Chaperone proteins surround new or denatured proteins and prevent them from binding to the wrong substances. Heat shock proteins such as HSP60, shown here, make up one class of chaperone proteins.

RECAP 3.2

Proteins are polymers of amino acids. The sequence of amino acids in a protein determines its primary structure. Secondary, tertiary, and quaternary structures arise through interactions among the amino acids. A protein's three-dimensional shape and exposed chemical groups establish its binding specificity for other substances.

- What are the attributes of an amino acid's R group that would make it hydrophobic? Hydrophilic? See p. 43 and Table 3.2
- Sketch and explain how two amino acids link together to form a peptide linkage. See pp. 43–45 and Figure 3.6
- What are the four levels of protein structure, and how are they all ultimately determined by the protein's primary structure (i.e., its amino acid sequence)? See pp. 45–48 and Figure 3.7
- How do environmental factors such as temperature and pH affect the weak interactions that give a protein its specific shape and function? See p. 50

The seemingly infinite number of protein configurations made possible by the biochemical properties of the 20 amino acids has driven the evolution of life's diversity. The linkage configurations of sugar monomers (monosaccharides) determine the structures of the next group of macromolecules, the carbohydrates, which provide energy for life.

3.3 What Are the Chemical Structures and Functions of Carbohydrates?

Carbohydrates make up a large group of molecules that all have similar atomic compositions but differ greatly in size, chemical properties, and biological functions. Carbohydrates usually have the general formula $C_mH_{2n}O_n$ (where *m* and *n* stand for numbers), which makes them appear as hydrates of carbon [associations between water molecules and carbon in the ratio $C_m(H_2O)_n$], hence their name. However, carbohydrates are not really "hydrates" because the water molecules are not intact. Rather, the linked carbon atoms are bonded with hydrogen atoms (—H) and hydroxyl groups (—OH), the components of water. Carbohydrates have three major biochemical roles:

- They are a source of stored energy that can be released in a form usable by organisms.
- They are used to transport stored energy within complex organisms.
- They serve as carbon skeletons that can be rearranged to form new molecules.

Some carbohydrates are relatively small, with molecular weights of less than 100. Others are true macromolecules, with molecular weights in the hundreds of thousands.

There are four categories of biologically important carbohydrate defined by the number of monomers:

- **Monosaccharides** (*mono*, "one"; *saccharide*, "sugar"), such as glucose, are simple sugars. They are the monomers from which the larger carbohydrates are constructed.
- **Disaccharides** (*di*, "two") consist of two monosaccharides linked together by covalent bonds. The most familiar is sucrose, which is made up of covalently bonded glucose and fructose molecules.
- **Oligosaccharides** (*oligo*, "several") are made up of several (3–20) monosaccharides.
- **Polysaccharides** (*poly*, "many"), such as starch, glycogen, and cellulose, are polymers made up of hundreds or thousands of monosaccharides.

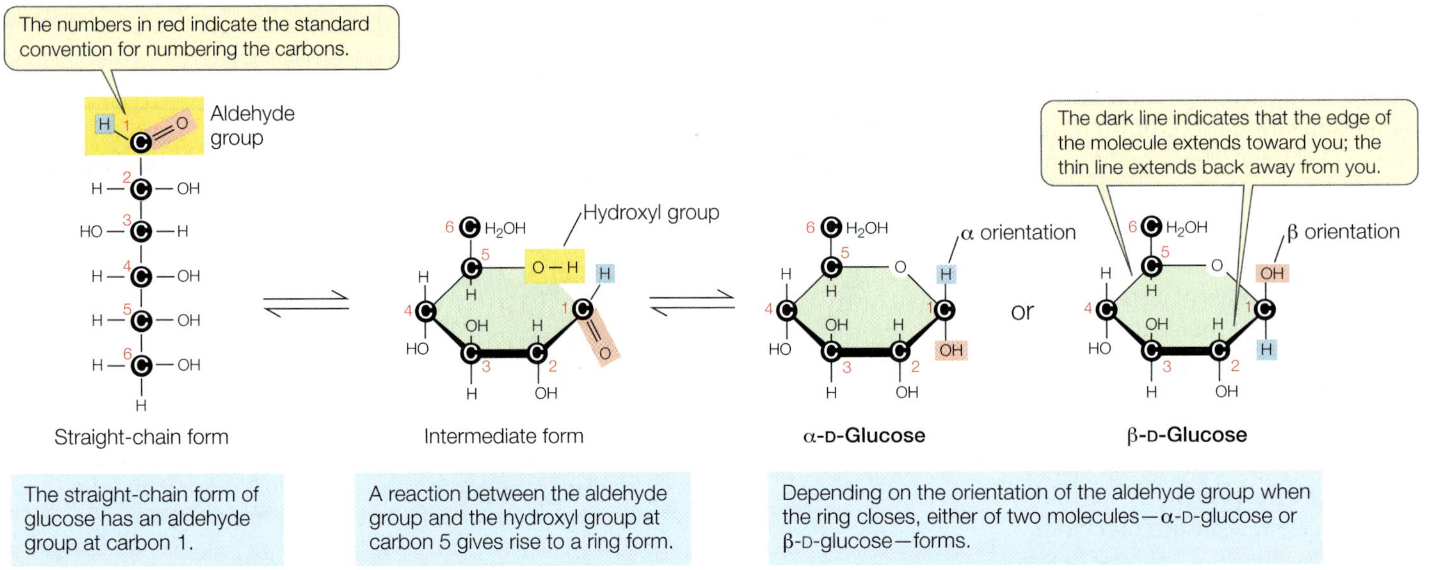
The numbers in red indicate the standard convention for numbering the carbons.
Aldehyde group
Hydroxyl group
α orientation
or
The dark line indicates that the edge of the molecule extends toward you; the thin line extends back away from you.
β orientation
Straight-chain form
Intermediate form
α-D-Glucose
β-D-Glucose
The straight-chain form of glucose has an aldehyde group at carbon 1.
A reaction between the aldehyde group and the hydroxyl group at carbon 5 gives rise to a ring form.
Depending on the orientation of the aldehyde group when the ring closes, either of two molecules—α-D-glucose or β-D-glucose—forms.

3.15 From One Form of Glucose to the Other All glucose molecules have the formula $C_6H_{12}O_6$, but their structures vary. When dissolved in water, the α and β "ring" forms of glucose interconvert. The convention used here for numbering the carbon atoms is standard in biochemistry.

Go to Activity 3.3 Forms of Glucose Life10e.com/ac3.3

Monosaccharides are simple sugars

All living cells contain the monosaccharide **glucose**; it is the familiar "blood sugar," used to transport energy in humans. Cells use glucose as an energy source, breaking it down through a series of reactions that release stored energy and produce water and carbon dioxide; this is a cellular form of the combustion reaction described in Section 2.3.

Glucose exists in straight chains and in ring forms. The ring forms predominate in virtually all biological circumstances because they are more stable under physiological conditions. There are two versions of the glucose ring, called α- and β-glucose, which differ only in the orientation of the —H and —OH groups attached to carbon 1 (**Figure 3.15**). The α and β forms interconvert and exist in equilibrium when dissolved in water.

Different monosaccharides contain different numbers of carbons. Some monosaccharides are structural isomers, with the same kinds and numbers of atoms but in different arrangements (**Figure 3.16**). Such seemingly small structural changes can significantly alter their properties. Most of the monosaccharides in living systems belong to the D (right-handed) series of optical isomers.

Pentoses (*pente*, "five") are five-carbon sugars. Two pentoses are of particular biological importance: the backbones of the nucleic acids RNA and DNA contain ribose and deoxyribose, respectively (see Section 4.1). These two pentoses are not isomers of each other; rather, one oxygen atom is missing from carbon 2 in deoxyribose (*de-*, "absent"). The absence of this oxygen atom is an important distinction between RNA and DNA.

The **hexoses** (*hex*, "six") shown in Figures 3.15 and 3.16 are a group of structural isomers with the formula $C_6H_{12}O_6$. Common hexoses are glucose, fructose (so named because it was first found in fruits), mannose, and galactose.

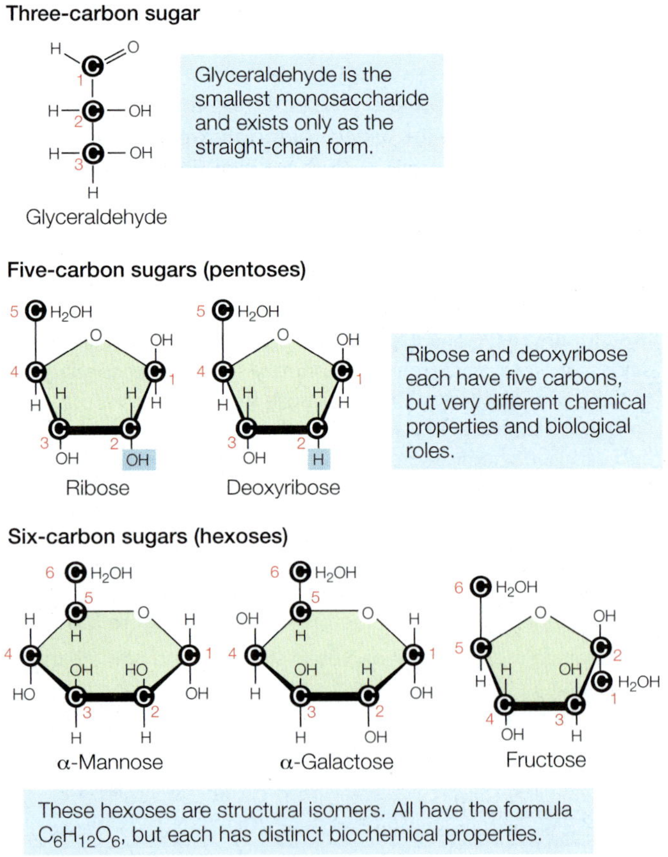
Three-carbon sugar
Glyceraldehyde is the smallest monosaccharide and exists only as the straight-chain form.
Glyceraldehyde
Five-carbon sugars (pentoses)
Ribose
Deoxyribose
Ribose and deoxyribose each have five carbons, but very different chemical properties and biological roles.
Six-carbon sugars (hexoses)
α-Mannose
α-Galactose
Fructose
These hexoses are structural isomers. All have the formula C6H12O6, but each has distinct biochemical properties.

3.16 Monosaccharides Are Simple Sugars Monosaccharides are made up of varying numbers of carbons. Some hexoses are structural isomers that have the same kind and number of atoms, but the atoms are arranged differently. Fructose, for example, is a hexose but forms a five-membered ring like the pentoses.

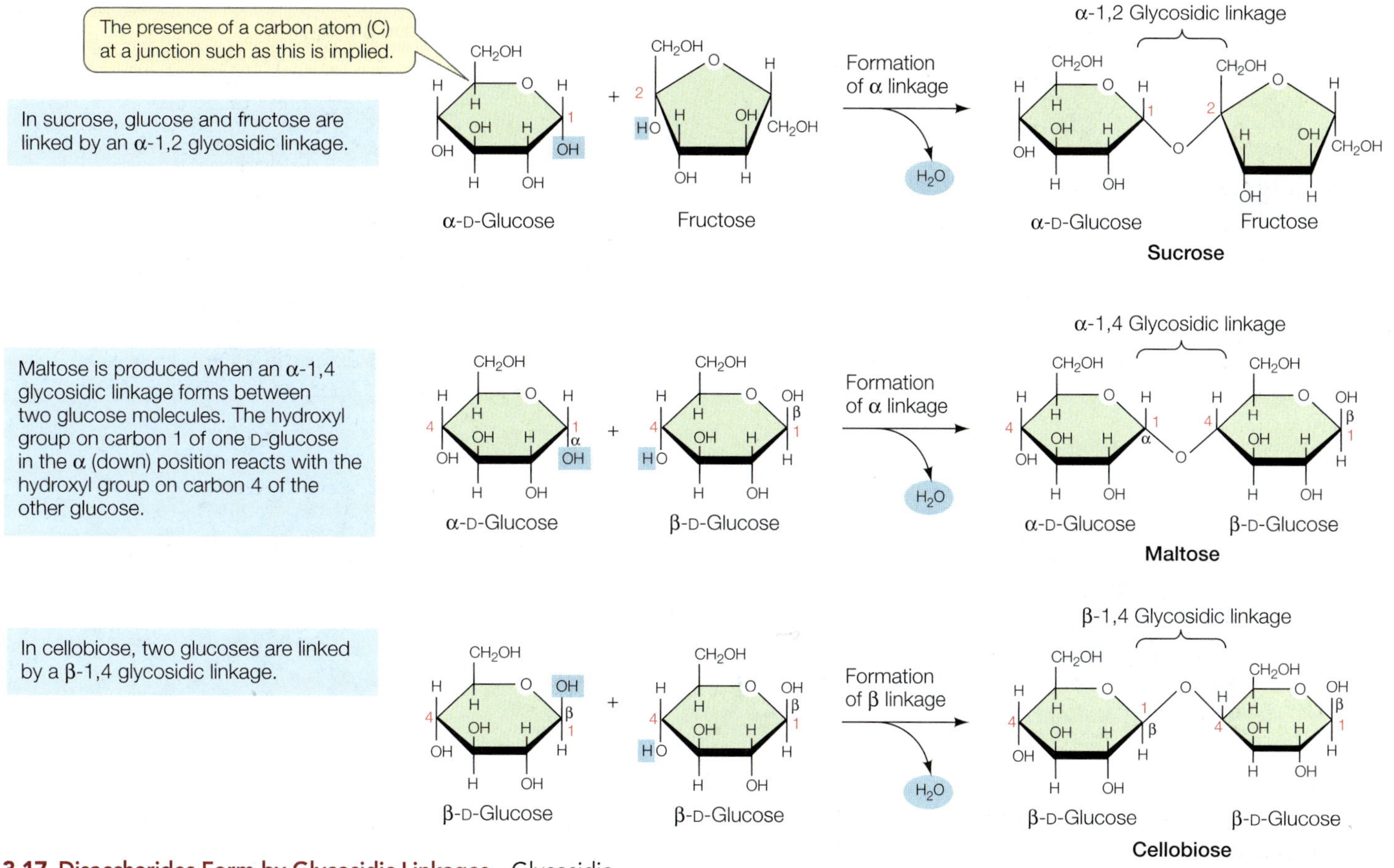
The presence of a carbon atom (C) at a junction such as this is implied.
In sucrose, glucose and fructose are linked by an α-1,2 glycosidic linkage.
α-D-Glucose
Fructose
Formation of α linkage
H_2O
α-1,2 Glycosidic linkage
α-D-Glucose
Fructose
Sucrose
Maltose is produced when an α-1,4 glycosidic linkage forms between two glucose molecules. The hydroxyl group on carbon 1 of one D-glucose in the α (down) position reacts with the hydroxyl group on carbon 4 of the other glucose.
α-D-Glucose
β-D-Glucose
Formation of α linkage
H_2O
α-1,4 Glycosidic linkage
α-D-Glucose
β-D-Glucose
Maltose
In cellobiose, two glucoses are linked by a β-1,4 glycosidic linkage.
β-D-Glucose
β-D-Glucose
Formation of β linkage
H_2O
β-1,4 Glycosidic linkage
β-D-Glucose
β-D-Glucose
Cellobiose

3.17 Disaccharides Form by Glycosidic Linkages Glycosidic linkages between two monosaccharides can create many different disaccharides. The particular disaccharide formed depends on which monosaccharides are linked, on the site of linkage (i.e., which carbon atoms are involved), and on the form (α or β) of the linkage.

Glycosidic linkages bond monosaccharides

The disaccharides, oligosaccharides, and polysaccharides are all constructed from monosaccharides that are covalently bonded together by condensation reactions that form **glycosidic linkages** (**Figure 3.17**). A single glycosidic linkage between two monosaccharides forms a disaccharide. For example, sucrose—common table sugar in the human diet and a major disaccharide in plants—is formed from a glucose and a fructose molecule. The disaccharides maltose and cellobiose are made from two glucose molecules (see Figure 3.17). Maltose and cellobiose are structural isomers, both having the formula $C_{12}H_{22}O_{11}$. However, they have different chemical properties and are recognized by different enzymes in biological tissues. For example, maltose can be hydrolyzed into its monosaccharides in the human body, whereas cellobiose cannot.

Oligosaccharides contain several monosaccharides bound by glycosidic linkages at various sites. Many oligosaccharides have additional functional groups, which give them special properties. Oligosaccharides are often covalently bonded to proteins and lipids on the outer cell surface, where they serve as recognition signals. The different human blood groups (for example, the ABO blood types) get their specificities from oligosaccharide chains.

Polysaccharides store energy and provide structural materials

Polysaccharides are large (sometimes gigantic) polymers of monosaccharides connected by glycosidic linkages (**Figure 3.18**). In contrast to proteins, polysaccharides are not necessarily linear chains of monomers. Each monomer unit has several sites that are capable of forming glycosidic linkages, and thus branched molecules are possible.

STARCH **Starches** comprise a family of giant molecules of broadly similar structure. While all starches are polysaccharides of glucose with α-glycosidic linkages (α–1,4 and α–1,6 glycosidic bonds; see Figure 3.18A), the different starches can be distinguished by the amount of branching that occurs at carbons 1 and 6 (see Figure 3.18B). Starch is the principal energy storage compound of plants. Some plant starches, such as amylose, are unbranched; others are moderately branched (for example, amylopectin). Starch readily binds water. When the water is removed, however, hydrogen bonds tend to form between the unbranched polysaccharide chains, which then aggregate. Large starch aggregates called starch grains can be observed in the storage tissues of plant seeds (see Figure 3.18C).

GLYCOGEN **Glycogen** is a water-insoluble, highly branched polymer of glucose. It is used to store glucose in the liver and muscles and is thus an energy storage compound for animals, as starch is for plants. Both glycogen and starch are readily

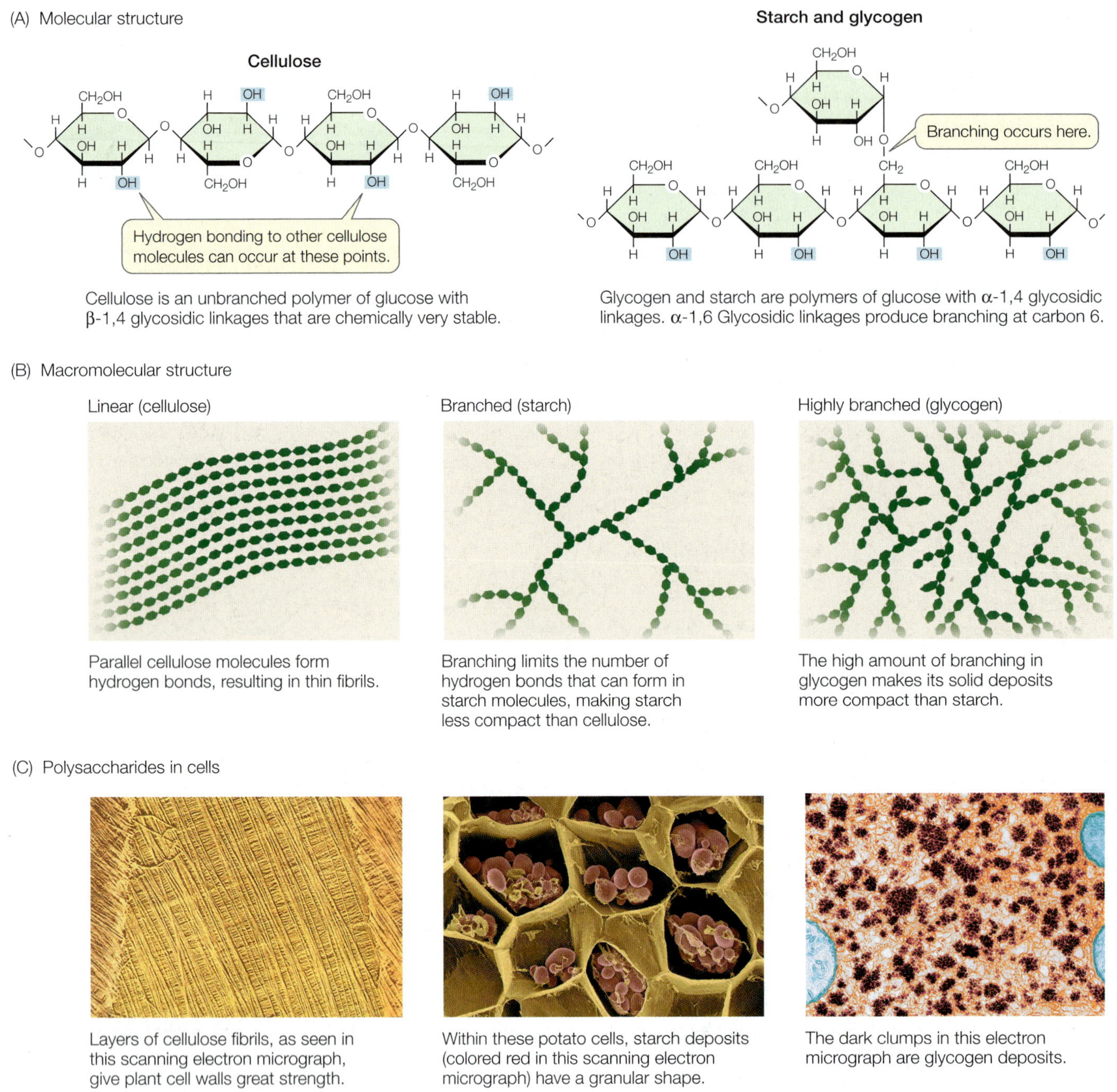
(A) Molecular structure
Cellulose
CH_2OH
OH
H
Hydrogen bonding to other cellulose molecules can occur at these points.
Cellulose is an unbranched polymer of glucose with β-1,4 glycosidic linkages that are chemically very stable.
Starch and glycogen
Branching occurs here.
CH_2
Glycogen and starch are polymers of glucose with α-1,4 glycosidic linkages. α-1,6 Glycosidic linkages produce branching at carbon 6.
(B) Macromolecular structure
Linear (cellulose)
Branched (starch)
Highly branched (glycogen)
Parallel cellulose molecules form hydrogen bonds, resulting in thin fibrils.
Branching limits the number of hydrogen bonds that can form in starch molecules, making starch less compact than cellulose.
The high amount of branching in glycogen makes its solid deposits more compact than starch.
(C) Polysaccharides in cells
Layers of cellulose fibrils, as seen in this scanning electron micrograph, give plant cell walls great strength.
Within these potato cells, starch deposits (colored red in this scanning electron micrograph) have a granular shape.
The dark clumps in this electron micrograph are glycogen deposits.

3.18 Representative Polysaccharides Cellulose, starch, and glycogen have different levels of branching and compaction of the polysaccharides.

hydrolyzed into glucose monomers, which in turn can be broken down to liberate their stored energy.

But if it is glucose that is needed for fuel, why store it in the form of glycogen? The reason is that 1,000 glucose molecules would exert 1,000 times the osmotic pressure of a single glycogen molecule, causing water to enter cells where glucose is stored (see Section 6.3). If it were not for polysaccharides, many organisms would expend a lot of energy expelling excess water from their cells.

CELLULOSE As the predominant component of plant cell walls, **cellulose** is by far the most abundant organic compound on Earth. Like starch and glycogen, cellulose is a polysaccharide of glucose, but its individual monosaccharides are connected by β- rather than by α-glycosidic linkages. Starch is easily degraded by the actions of chemicals or enzymes. Cellulose, however, is chemically more stable because of its β-glycosidic linkages. Thus whereas starch is easily broken down to supply glucose for energy-producing reactions, cellulose is an excellent structural material that can withstand harsh environmental conditions without substantial change.

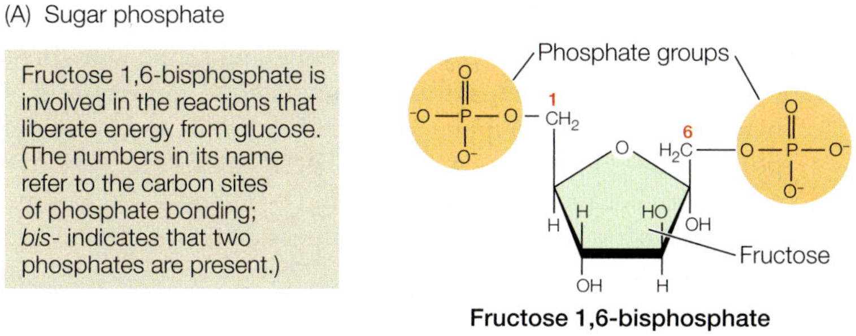
(A) Sugar phosphate
Fructose 1,6-bisphosphate is involved in the reactions that liberate energy from glucose. (The numbers in its name refer to the carbon sites of phosphate bonding; *bis*- indicates that two phosphates are present.)
Phosphate groups
Fructose
Fructose 1,6-bisphosphate

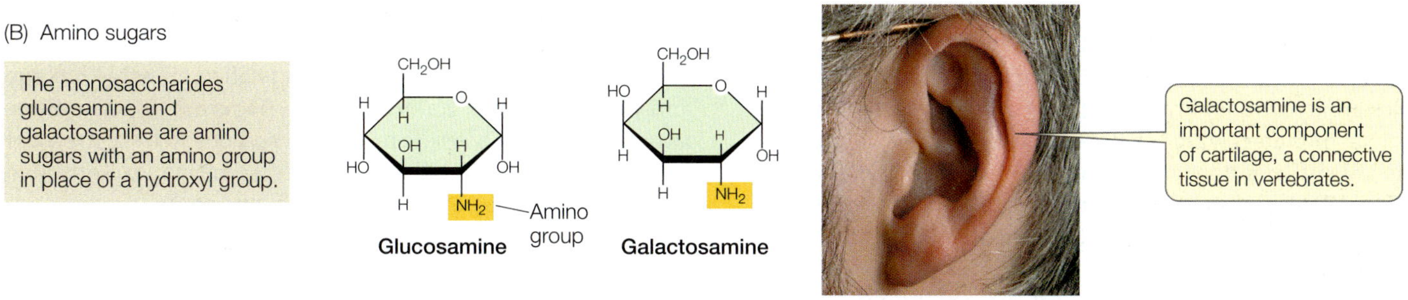
(B) Amino sugars
The monosaccharides glucosamine and galactosamine are amino sugars with an amino group in place of a hydroxyl group.
Amino group
Glucosamine
Galactosamine
Galactosamine is an important component of cartilage, a connective tissue in vertebrates.

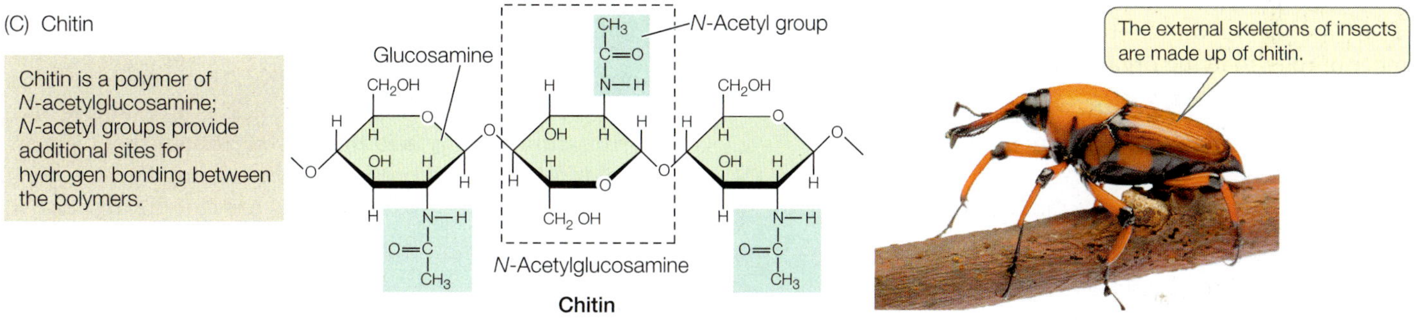
(C) Chitin
Chitin is a polymer of *N*-acetylglucosamine; *N*-acetyl groups provide additional sites for hydrogen bonding between the polymers.
Glucosamine
N-Acetyl group
N-Acetylglucosamine
Chitin
The external skeletons of insects are made up of chitin.

3.19 Chemically Modified Carbohydrates Added functional groups can modify the form and properties of a carbohydrate.

Chemically modified carbohydrates contain additional functional groups

Some carbohydrates are chemically modified by oxidation–reduction reactions, or by the addition of functional groups such as phosphate, amino, or *N*-acetyl groups (**Figure 3.19**). For example, carbon 6 in glucose may be oxidized from —CH_2OH to a carboxyl group (—COOH), producing glucuronic acid. Or a phosphate group may be added to one or more of the —OH sites. Some of the resulting sugar phosphates, such as fructose 1,6-bisphosphate, are important intermediates in cellular energy reactions, which we will discuss in Chapter 9.

When an amino group is substituted for an —OH group, amino sugars, such as glucosamine and galactosamine, are produced. These compounds are important in the extracellular matrix (see Section 5.4), where they form parts of glycoproteins, which are molecules involved in keeping tissues together. Galactosamine is a major component of cartilage, the material that forms caps on the ends of bones and stiffens the ears and nose. A derivative of glucosamine is present in the polymer chitin, the principal structural polysaccharide in the external skeletons of insects and many crustaceans (such as crabs and lobsters), and a component of the cell walls of fungi. Because these are among the most abundant complex organisms on Earth, chitin rivals cellulose as one of the most abundant substances in the living world.

RECAP 3.3

Carbohydrates are composed of carbon, hydrogen, and oxygen and have the general formula $C_mH_{2n}O_n$. They provide energy and structure to cells and are precursors of numerous important biological molecules. Monosaccharide monomers can be connected by glycosidic linkages to form disaccharides, oligosaccharides, and polysaccharides.

- Draw the chemical structure of a disaccharide formed from two monosaccharides. **See Figure 3.17**
- What qualities of the polysaccharides starch and glycogen make them useful for energy storage? **See pp. 53–54 and Figure 3.18**
- After looking at the cellulose molecule in Figure 3.18A, can you see why a large number of hydrogen bonds are present in the linear structure of cellulose shown in Figure 3.18B? Why is this structure so strong? **See p. 54**

We have seen how amino acid monomers form protein polymers and how sugar monomers form the polymers of carbohydrates. Now we will look at the lipids, which are unique among the four classes of large biological molecules in that they are not, strictly speaking, polymers.

3.4 What Are the Chemical Structures and Functions of Lipids?

Lipids—colloquially called fats—are hydrocarbons that are insoluble in water because of their many nonpolar covalent bonds. As we saw in Section 2.2, nonpolar hydrocarbon molecules are hydrophobic and preferentially aggregate together, away from water, which is polar. When nonpolar hydrocarbons are sufficiently close to one another, weak but additive van der Waals forces help hold them together. The huge macromolecular aggregations that can form are not polymers in a strict chemical sense, because the individual lipid molecules are not covalently bonded. With this understanding, it is still useful to consider aggregations of individual lipids as a different sort of polymer.

There are several different types of lipids, and they play a number of roles in living organisms:

- Fats and oils store energy.
- Phospholipids play important structural roles in cell membranes.
- Carotenoids and chlorophylls help plants capture light energy.
- Steroids and modified fatty acids play regulatory roles as hormones and vitamins.
- Fat in animal bodies serves as thermal insulation.
- A lipid coating around nerves provides electrical insulation.
- Oil or wax on the surfaces of skin, fur, feathers, and leaves repels water and prevents excessive evaporation of water from terrestrial animals and plants.

Fats and oils are triglycerides

Chemically, fats and oils are **triglycerides**, also known as simple lipids. Triglycerides that are solid at room temperature (around 20°C) are called **fats**; those that are liquid at room temperature are called **oils**.

Triglycerides are composed of two types of building blocks: fatty acids and glycerol. **Glycerol** is a small molecule with three hydroxyl (—OH) groups (thus it is an alcohol). A **fatty acid** is made up of a long nonpolar hydrocarbon chain and an acidic polar carboxyl group (—COOH). These chains are very hydrophobic because of their abundant C—H and C—C bonds, which have low electronegativity values and are nonpolar (see Section 2.2).

A triglyceride contains three fatty acid molecules and one molecule of glycerol. Synthesis of a triglyceride involves three condensation (dehydration) reactions. In each reaction, the carboxyl group of a fatty acid bonds with a hydroxyl group of glycerol, resulting in a covalent bond called an **ester linkage** and the release of a water molecule (**Figure 3.20**). The three fatty acids in a triglyceride molecule need not all have the same hydrocarbon chain length or structure; some may be saturated fatty acids, whereas others may be unsaturated:

- In **saturated fatty acids**, all the bonds between the carbon atoms in the hydrocarbon chain are single bonds—there are no double bonds. That is, all the bonds are saturated with hydrogen atoms (**Figure 3.21A**). These fatty acid molecules are relatively straight, and they pack together tightly, like pencils in a box.
- In **unsaturated fatty acids**, the hydrocarbon chain contains one or more double bonds. Linoleic acid is an example of a polyunsaturated fatty acid that has two double bonds near the middle of the hydrocarbon chain, causing kinks in the molecule (**Figure 3.21B**). Such kinks prevent the unsaturated fat molecules from packing together tightly.

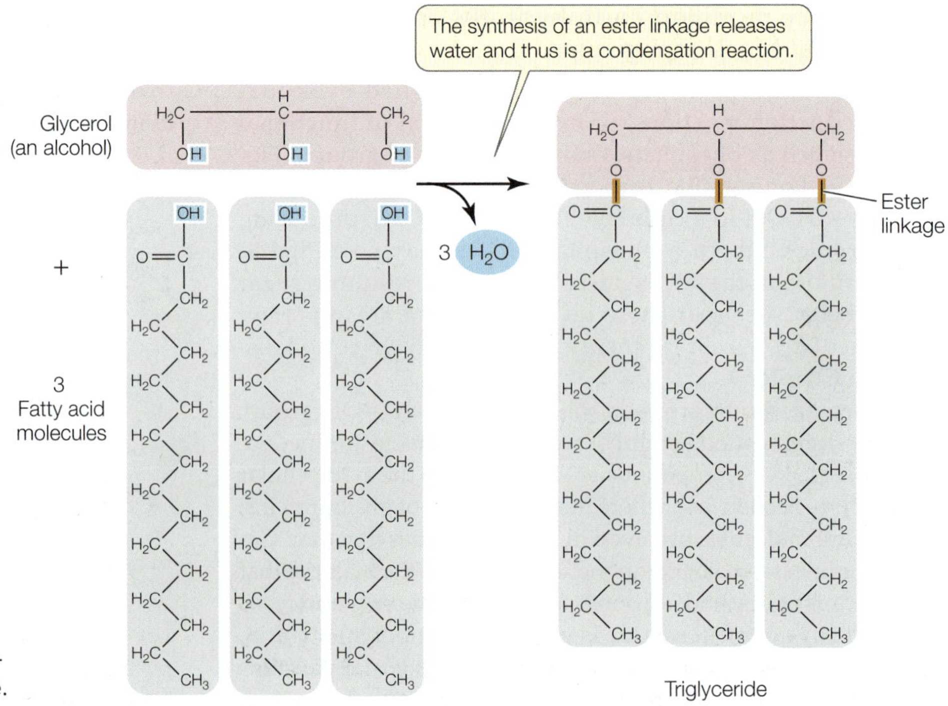
The synthesis of an ester linkage releases water and thus is a condensation reaction.
Glycerol (an alcohol)
+
3 Fatty acid molecules
3 H_2O
Ester linkage
Triglyceride

3.20 Synthesis of a Triglyceride In living things, the reaction that forms a triglyceride is more complex, but the end result is the same as shown here.

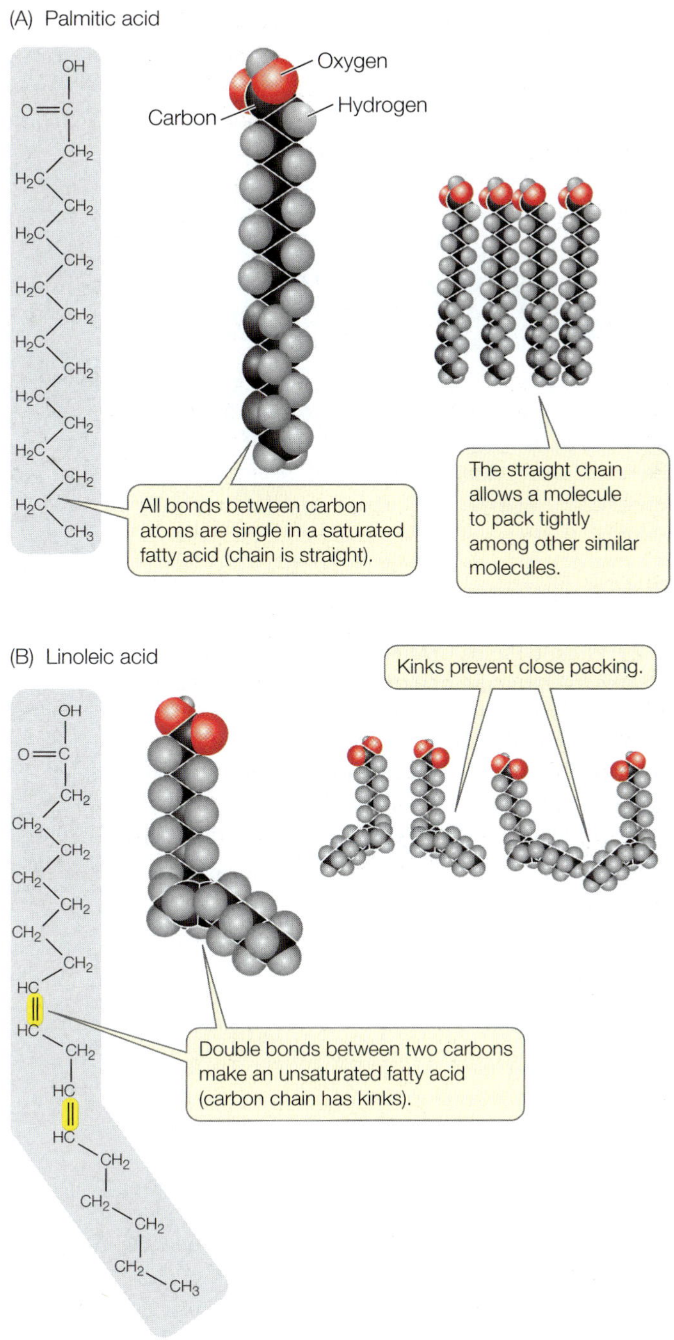
(A) Palmitic acid
Oxygen
Hydrogen
Carbon
All bonds between carbon atoms are single in a saturated fatty acid (chain is straight).
The straight chain allows a molecule to pack tightly among other similar molecules.
(B) Linoleic acid
Kinks prevent close packing.
Double bonds between two carbons make an unsaturated fatty acid (carbon chain has kinks).

3.21 Saturated and Unsaturated Fatty Acids (A) The straight hydrocarbon chain of a saturated fatty acid allows the molecule to pack tightly with other, similar molecules. (B) In unsaturated fatty acids, kinks in the chain prevent close packing. The color convention in the models shown here (gray, H; red, O; black, C) is commonly used.

The kinks in fatty acid molecules are important in determining the fluidity and melting points of lipids. The triglycerides of animal fats tend to have many long-chain saturated fatty acids packed tightly together; these fats are usually solids at room temperature and have high melting points. The triglycerides of plants, such as corn oil, tend to have short or unsaturated fatty acids. Because of their kinks, these fatty acids pack together poorly and have low melting points, and these triglycerides are usually liquids at room temperature.

EXTRA VIRGIN OLIVE OIL
BUTTER AA
BUTTER AA
BUTTER AA
BUTTER AA

Fatty acids are excellent storehouses for chemical energy. As you will see in Chapter 9, when the C—H bond is broken, it releases significant energy that an organism can use for its own purposes, such as movement or building up other complex molecules.

Phospholipids form biological membranes

We have mentioned the hydrophobic nature of the many C—C and C—H bonds in fatty acids. But what about the carboxyl functional group at the end of the molecule? When it ionizes and forms COO^-, it is strongly hydrophilic. So a fatty acid is a molecule with a hydrophilic end and a long hydrophobic tail. It has two opposing chemical properties; the technical term for this is **amphipathic**. When fatty acids are bonded to glycerol, their carboxyl groups are incorporated into the ester bonds, and the resulting triglyceride is hydrophobic.

Like triglycerides, **phospholipids** contain fatty acids bound to glycerol by ester linkages. In phospholipids, however, any one of several phosphate-containing compounds replaces one of the fatty acids, giving phospholipids amphipathic properties (Figure 3.22A). The phosphate functional group has a negative electric charge, so this portion of the molecule is hydrophilic, attracting polar water molecules. But the two fatty acids are hydrophobic, so they tend to avoid water and aggregate together or with other hydrophobic substances.

In an aqueous environment, phospholipids line up in such a way that the nonpolar, hydrophobic "tails" pack tightly together and the phosphate-containing "heads" face outward, where they interact with water. The phospholipids thus form a **bilayer**: a sheet two molecules thick, with water excluded from the core (Figure 3.22B). Biological membranes have this kind of **phospholipid bilayer** structure, and we will devote Chapter 6 to their biological functions.

Some lipids have roles in energy conversion, regulation, and protection

In the paragraphs above we focused on triglycerides and phospholipids—lipids that are involved in energy storage and cell structure. However, there are other nonpolar and amphipathic lipids that have different structures and roles.

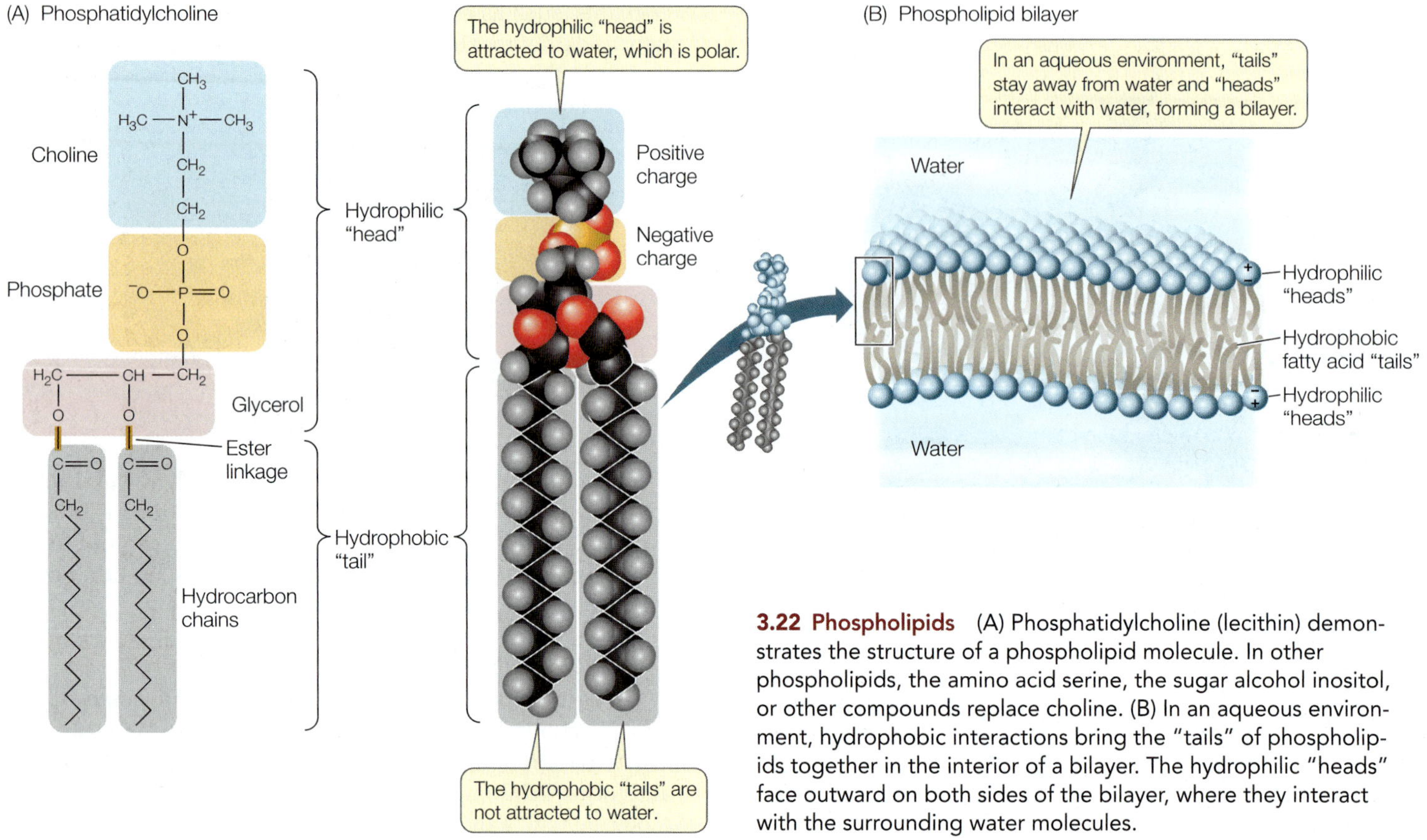
(A) Phosphatidylcholine
Choline
Phosphate
Glycerol
Ester linkage
Hydrocarbon chains
Hydrophilic "head"
Hydrophobic "tail"
The hydrophilic "head" is attracted to water, which is polar.
Positive charge
Negative charge
The hydrophobic "tails" are not attracted to water.
(B) Phospholipid bilayer
In an aqueous environment, "tails" stay away from water and "heads" interact with water, forming a bilayer.
Water
Hydrophilic "heads"
Hydrophobic fatty acid "tails"
Hydrophilic "heads"
Water

3.22 Phospholipids (A) Phosphatidylcholine (lecithin) demonstrates the structure of a phospholipid molecule. In other phospholipids, the amino acid serine, the sugar alcohol inositol, or other compounds replace choline. (B) In an aqueous environment, hydrophobic interactions bring the "tails" of phospholipids together in the interior of a bilayer. The hydrophilic "heads" face outward on both sides of the bilayer, where they interact with the surrounding water molecules.

CAROTENOIDS The carotenoids are a family of light-absorbing pigments found in plants and animals. Beta-carotene (β-carotene) is one of the pigments that traps light energy in leaves during photosynthesis. In humans, a molecule of β-carotene can be broken down into two vitamin A molecules. Vitamin A is used to make the pigment *cis*-retinal, which is required for vision.

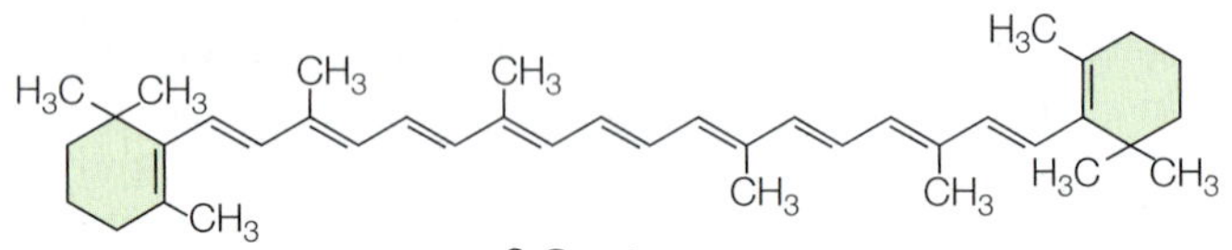

β-Carotene

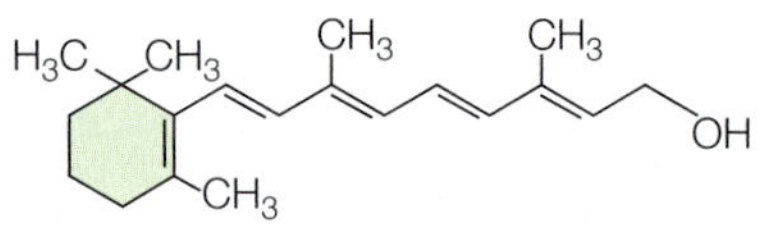

Vitamin A

Carotenoids are responsible for the colors of carrots, tomatoes, pumpkins, egg yolks, and butter. The brilliant yellows and oranges of autumn leaves are also from carotenoids.

STEROIDS The steroids are a family of organic compounds whose multiple rings are linked through shared carbons. The steroid cholesterol is an important constituent of membranes, helping maintain membrane integrity (see Section 6.1).

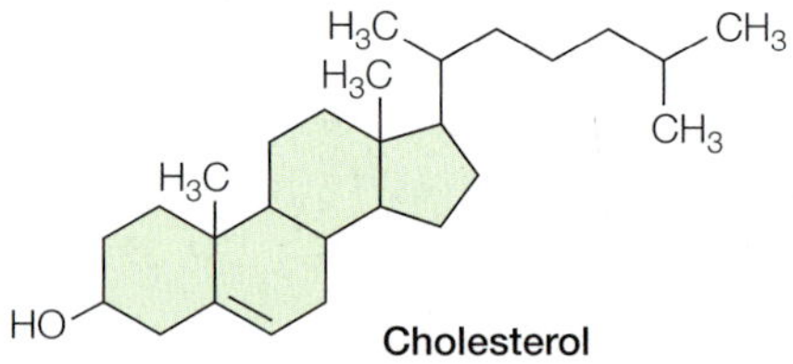

Cholesterol

Other steroids function as hormones: chemical signals that carry messages from one part of the body or in some cases are synthesized in inadequate amounts to another (see Chapter 41). Cholesterol is synthesized in the liver and is the starting material for making steroid hormones such as testosterone and estrogen.

VITAMINS Vitamins are small molecules that are not synthesized by the human body or in some cases are synthesized in inadequate amounts and so must be acquired from the diet (see Chapter 51). For example, vitamin A is formed from the β-carotene found in green and yellow vegetables (see above). In humans, a deficiency of vitamin A leads to dry skin, eyes, and internal body surfaces, retarded growth and development, and night blindness, which is a diagnostic symptom for the deficiency. Vitamins D, E, and K are also lipids.

WAXES Birds and mammals have glands in their skins that secrete waxy coatings onto their hair or feathers. These coatings repel water and help keep the hair and feathers pliable. The shiny leaves of plants such as holly, familiar during winter holidays, also have waxy coatings. Waxy coatings on plants can help them retain water and exclude pathogens. Bees make their honeycombs out of wax. Waxes are substances that are hydrophobic and plastic, or malleable, at room temperature. Each wax molecule consists of a saturated, long-chain fatty acid and a saturated, long-chain alcohol joined by an ester linkage. The result is a very long molecule with 40–60 CH_2 groups.

RECAP 3.4

Lipids include both hydrophobic and amphipathic molecules that are largely composed of carbon and hydrogen. They are important in energy storage, light absorption, regulation, and biological structures. A phospholipid is composed of two hydrophobic fatty acids linked to glycerol and a hydrophilic phosphate group. Cell membranes contain phospholipid bilayers.

- Draw the molecular structures of fatty acids and glycerol and show how they are linked to form a triglyceride. **See p. 56 and Figure 3.20**
- What is the difference between fats and oils? **See pp. 56–57 and Figure 3.21**
- How does the polar nature of phospholipids result in their forming a bilayer? **See p. 57 and Figure 3.22**
- Why are steroids and some vitamins classified as lipids? **See p. 58**

In this chapter we discussed three of the classes of macromolecules that are characteristic of living organisms, but a final class of biological macromolecules has special importance to the living world. Nucleic acids transmit life's "blueprint" to each new organism. This chapter illustrated the wonderful biochemical unity of life, implying that all life has a common origin (see Section 1.1). Essential to this origin were the monomeric nucleotides and their polymers, nucleic acids. In the next chapter we will turn to the related topics of nucleic acids and the origin of life.

?

Can knowledge of spider web protein structure be put to practical use?

ANSWER

Because of its strength, spider silk is much desired for human uses, ranging from surgical sutures in medicine to bulletproof vests in the military. "Farming" live spiders is tedious, costly, and gives a low yield of usable silk for industry. Unlike your hair, which grows continuously, spider silk is synthesized and stored as a liquid precursor solution in silk glands, and then "spun" out into fibers as needed. Recently, biotechnology has been used to genetically engineer silkworms, which produce their own form of silk, to make spider silk instead; even bacteria have been coaxed into making massive amounts of the protein. Moreover, by carefully studying how spiders do it, scientists have successfully spun out usable fibers from these artificial systems.

CHAPTER**SUMMARY** 3

3.1 What Kinds of Molecules Characterize Living Things?

See ANIMATED TUTORIAL 3.1

- **Macromolecules** are **polymers** constructed by the formation of covalent bonds between smaller molecules called **monomers**. Macromolecules in living organisms include polysaccharides, proteins, and nucleic acids. Large lipid structures may also be considered macromolecules.
- **Functional groups** are small groups of atoms that are consistently found together in a variety of different macromolecules. Functional groups have particular chemical properties that they confer on any larger molecule of which they are a part. **Review Figure 3.1, ACTIVITY 3.1**
- **Structural**, ***cis-trans***, and **optical isomers** have the same kinds and numbers of atoms but differ in their structures and properties. **Review Figure 3.2**
- The many functions of macromolecules are directly related to their three-dimensional shapes, which in turn result from the sequences and chemical properties of their monomers.
- Monomers are joined by **condensation reactions**, which release a molecule of water for each bond formed. **Hydrolysis reactions** use water to break polymers into monomers. **Review Figure 3.4**

3.2 What Are the Chemical Structures and Functions of Proteins?

- The functions of **proteins** include support, protection, catalysis, transport, defense, regulation, and movement. **Review Table 3.1**
- Proteins consist of one or more **polypeptide chains**, which are polymers of **amino acids**. Four atoms or groups are attached to a central carbon atom: a hydrogen atom, an amino group, a carboxyl group, and a variable R group. The particular properties of each amino acid depend on its **side chain**, or **R group**, which may be charged, polar, or hydrophobic. **Review Table 3.2, ACTIVITY 3.2**
- **Peptide linkages**, also called peptide bonds, covalently link amino acids into polypeptide chains. These bonds form by condensation reactions between the carboxyl and amino groups. **Review Figure 3.6**
- The **primary structure** of a protein is the sequence of amino acids in the chain. This chain is folded into a **secondary structure**, which in different parts of the protein may form an **α helix** or a **β pleated sheet**. **Review Figure 3.7A–C**
- **Disulfide bridges** and noncovalent interactions between amino acids cause polypeptide chains to fold into three-dimensional **tertiary structures**. Weak, noncovalent interactions allow multiple poly-peptide chains to form **quaternary structures**. **Review Figure 3.7D, 3.7E**

continued

- Heat, alterations in pH, or certain chemicals can all result in a protein becoming **denatured**. This involves the loss of tertiary and/or secondary structure as well as biological function. **Review Figure 3.10**
- The specific shape and structure of a protein allows it to bind noncovalently to other molecules. In addition, amino acids may be modified by the covalent bonding of chemical groups to their side chains. Such binding may result in the protein changing its shape. **Review Figures 3.12, 3.13**
- **Chaperone proteins** enhance correct protein folding and prevent inappropriate binding to other molecules. **Review Figure 3.14**

3.3 What Are the Chemical Structures and Functions of Carbohydrates?

- **Carbohydrates** contain carbon bonded to hydrogen and oxygen atoms and have the general formula $C_mH_{2n}O_n$.
- **Monosaccharides** are the monomers that make up carbohydrates. **Hexoses** such as **glucose** are six-carbon monosaccharides; **pentoses** have five carbons. **Review Figure 3.16, ACTIVITY 3.3**
- **Glycosidic linkages**, which have either an α or a β orientation in space, are covalent bonds between monosaccharides. Two linked monosaccharides are called **disaccharides**; larger units are **oligosaccharides** and **polysaccharides**. **Review Figure 3.17**
- **Starch** is a polymer of glucose that stores energy in plants, and **glycogen** is an analogous polymer in animals. They can be easily broken down to release stored energy. **Review Figure 3.18**
- **Cellulose** is a very stable glucose polymer and is the principal structural component of plant cell walls.

3.4 What Are the Chemical Structures and Functions of Lipids?

- **Lipids** are hydrocarbons that are insoluble in water because of their many nonpolar covalent bonds. They play roles in energy storage, membrane structure, light harvesting, regulation, and protection.
- **Fats** and **oils** are **triglycerides**. A triglyceride is composed of three **fatty acids** covalently bonded to a molecule of glycerol by **ester linkages**. **Review Figure 3.20**
- A **saturated fatty acid** has a hydrocarbon chain with no double bonds. These molecules can pack together tightly. The hydrocarbon chain of an **unsaturated fatty acid** has one or more double bonds that bend the chain, preventing close packing. **Review Figure 3.21**
- A **phospholipid** has a hydrophobic hydrocarbon "tail" and a hydrophilic phosphate "head"; that is, it is **amphipathic**. In water, the interactions of the tails and heads of phospholipids generate a **phospholipid bilayer**. The heads are directed outward, where they interact with the surrounding water. The tails are packed together in the interior of the bilayer, away from water. **Review Figure 3.22**
- Other lipids include vitamins A, D, E, and K, steroids, and plant pigments such as carotenoids.

Go to the Interactive Summary to review key figures, Animated Tutorials, and Activities Life10e.com/is3

CHAPTER**REVIEW**

REMEMBERING

1. The most abundant molecule in the cell is
   a. a carbohydrate.
   b. a lipid.
   c. a nucleic acid.
   d. a protein.
   e. water.

2. All lipids are
   a. triglycerides.
   b. polar.
   c. hydrophilic.
   d. polymers of fatty acids.
   e. more soluble in nonpolar solvents than in water.

3. All carbohydrates
   a. are polymers.
   b. are simple sugars.
   c. consist of one or more simple sugars.
   d. are found in biological membranes.
   e. are more soluble in nonpolar solvents than in water.

4. Which of the following statements about the primary structure of a protein is *not* true?
   a. It may be branched.
   b. It is held together by covalent bonds.
   c. It is unique to that protein.
   d. It determines the tertiary structure of the protein.
   e. It is the sequence of amino acids in the protein.

5. The amino acid leucine
   a. is found in all proteins.
   b. cannot form peptide linkages.
   c. has a hydrophobic side chain.
   d. has a hydrophilic side chain.
   e. is identical to the amino acid lysine.

6. The amphipathic nature of phospholipids is
   a. determined by the fatty acid composition.
   b. important in membrane structure.
   c. polar but not nonpolar.
   d. shown only if the lipid is in a nonpolar solvent.
   e. important in energy storage by lipids.

UNDERSTANDING & APPLYING

7. A single amino acid change in a protein can change its shape. Normally, at a certain position in a protein is the amino acid glycine (see Table 3.2). If glycine is replaced with either glutamic acid or arginine, the protein shape near that amino acid changes significantly. There are two possible explanations for this:

   a. A small amino acid at that position in the polypeptide is necessary for normal shape.
   b. An uncharged amino acid is necessary for normal shape.

   Further amino acid substitutions are done to distinguish between these possibilities. Replacing glycine with serine or alanine results in normal shape; but replacing glycine with valine changes the shape. Which of the two possible explanations is supported by the observations? Explain your answer.

8. Examine the hexose isomers mannose and galactose below. What makes them structural isomers of one another? Which functional groups do these carbohydrates contain, and what properties do these functional groups give to the molecules?

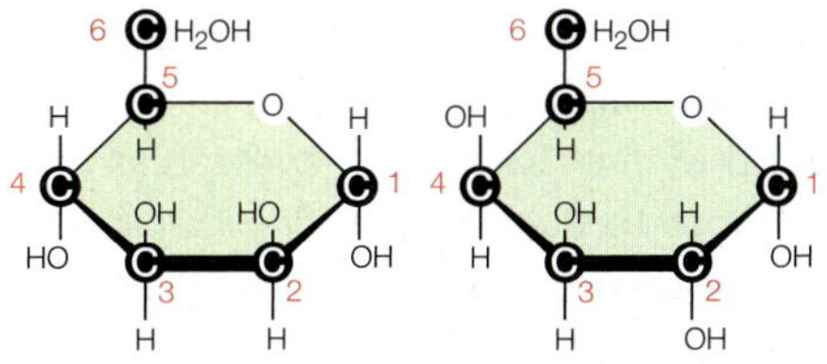
6
H_2OH
5
O
H
H
4
1
OH
HO
HO
OH
3
2
H
H
α-Mannose
6
H_2OH
5
O
OH
H
H
4
1
OH
H
H
OH
3
2
H
OH
α-Galactose

9. How does high temperature affect protein structure? When an organism is exposed to high temperature, it often makes a special class of molecular chaperones called heat shock proteins. How do you think these proteins work?

ANALYZING & EVALUATING

10. Suppose that, in a given protein, one lysine is replaced by aspartic acid (see Table 3.2). Does this change occur in the primary structure or in the secondary structure? How might it result in a change in tertiary structure? In quaternary structure?

11. Human hair is composed of the protein keratin. At the hair salon, two techniques are used to modify the three-dimensional shape of hair. Styling involves heat, and a perm involves cleaving and re-forming disulfide bonds. How would you investigate these phenomena in terms of protein structure?

Go to BioPortal at **yourBioPortal.com** for Animated Tutorials, Activities, LearningCurve Quizzes, Flashcards, and many other study and review resources.

4 Nucleic Acids and the Origin of Life

CHAPTER OUTLINE

Fast Cat Cheetahs are among the swiftest of the world's land animals. The 7,000 cheetahs living today have almost identical DNA sequences in their genomes, resulting from an evolutionary event about 10,000 years ago that wiped out all but a few individuals. The history of life is largely written in its DNA.

THE NAMES OF THE CHEETAH, *Acinonyx jubatus*, describe it well. "Cheetah" comes from the Hindi word *chiita*, meaning "spotted," for the small black spots on the animal's yellow fur. *Acinonyx* means "no-move claw" in Greek. Cheetahs cannot fully retract their claws—an advantage for running fast and hunting. In Latin, *jubatus* means "maned"—a characteristic of cheetah cubs.

This sleek, muscular cat is a solitary hunter of mammals such as gazelles and hares. It stalks its prey until it is 10–30 meters away and then chases it at speeds of up to 110 km/h (70 mph). Usually, the chase is over within a minute.

There are only about 7,000 cheetahs in the world today, most of them in Africa. The recent decline in their numbers is mostly due to humans: loss of habitat, and killing by farmers trying to protect their livestock. But—written in the cheetah's DNA—is more to the story of their decline.

Like proteins and polysaccharides, DNA is a macromolecule, in this case composed of a set of four different monomers called nucleotides. The nucleotide sequence of DNA is essential to its function, which is to carry information that determines an organism's characteristics. If you compare the sequence of the billions of nucleotides in your own DNA with that of an unrelated person in your class, the sequences will be about 0.5 percent different. This variation is reflected in the many differences among individual humans.

The genomes of cheetahs have a remarkable degree of similarity, almost as if all cheetahs descended from a single set of parents. The modern cheetah probably evolved about 15 million years ago and was widespread until the last ice age, which ended about 10,000 years ago. At that point many other large mammals (e.g.,the sabre-toothed tiger)died out, but a few cheetahs apparently survived and were the ancestors of the modern animals. So it is presumed that all current cheetahs—and their DNA—derive from the few individuals that survived an event that almost wiped out the species.

DNA belongs to a class of large molecules called nucleic acids. In Chapters 2 and 3 we described molecules that are important for biological structure and function. Here we turn to the nucleic acids, which are involved in perpetuating of life.

? Can DNA analysis be used in the conservation and expansion of the cheetah population?

See answer on p. 75.

4.1 What Are the Chemical Structures and Functions of Nucleic Acids?

From medicine to evolution, from agriculture to forensics, the properties of nucleic acids affect our lives every day. It is with nucleic acids that the concept of "information" entered the biological vocabulary. Nucleic acids are uniquely capable of coding for and transmitting biological information.

Nucleic acids are polymers specialized for the storage, transmission, and use of genetic information. There are two types of nucleic acids: **DNA** (**deoxyribonucleic acid**) and **RNA** (**ribonucleic acid**). DNA is a macromolecule that encodes hereditary information and passes it from generation to generation. Through RNA intermediates, the information encoded in DNA is used to specify the amino acid sequences of proteins and control the expression synthesis of other RNAs. During cell division and reproduction, information flows from existing DNA to the newly formed DNA in a new cell or organism. In the nonreproductive activities of the cell, information flows from DNA to RNA to proteins. It is the proteins that ultimately carry out many of life's functions.

Nucleotides are the building blocks of nucleic acids

Nucleic acids are polymers composed of monomers called nucleotides. A **nucleotide** consists of three components: a nitrogen-containing **base**, a pentose sugar, and one to three phosphate groups (Figure 4.1). Molecules consisting of a pentose sugar and a nitrogenous base—but no phosphate group—are called **nucleosides**. The nucleotides that make up nucleic acids contain just one phosphate group—they are nucleoside *mono*phosphates.

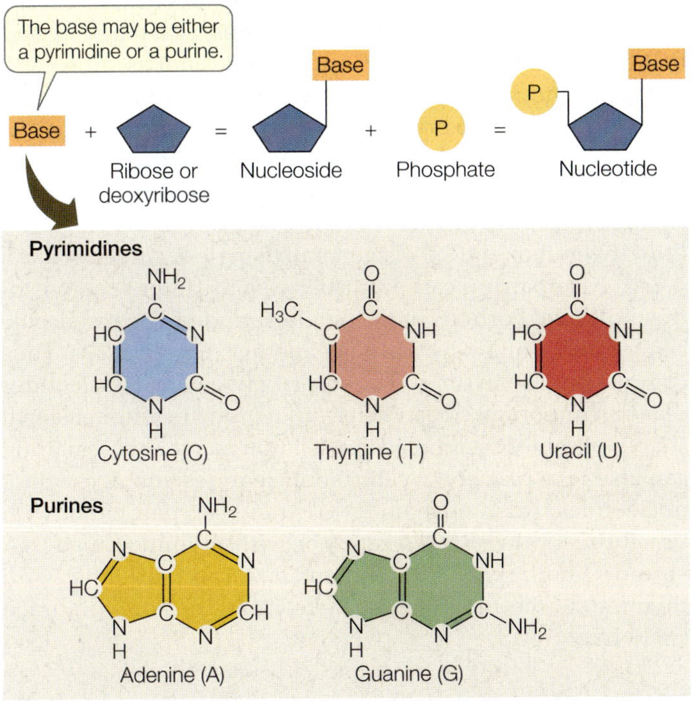

The base may be either a pyrimidine or a purine.
Base
+
Ribose or deoxyribose
=
Base
Nucleoside
+
P
Phosphate
=
P
Base
Nucleotide
Pyrimidines
Cytosine (C)
Thymine (T)
Uracil (U)
Purines
Adenine (A)
Guanine (G)

4.1 Nucleotides Have Three Components Nucleotide monomers are the building blocks of DNA and RNA polymers.

Go to Activity 4.1 Nucleic Acid Building Blocks
Life10e.com/ac4.1

TABLE 4.1 Distinguishing RNA from DNA

Nucleic Acid	Sugar	Bases	Name of Nucleoside	Strands
RNA	Ribose	Adenine	Adenosine	Single
		Cytosine	Cytidine	
		Guanine	Guanosine	
		Uracil	Uridine	
DNA	Deoxyribose	Adenine	Deoxyadenosine	Double
		Cytosine	Deoxycytidine	
		Guanine	Deoxyguanosine	
		Thymine	Deoxythymidine	

The bases of the nucleic acids take one of two chemical forms: a six-membered single-ring structure called a **pyrimidine**, or a fused double-ring structure called a **purine** (see Figure 4.1). In DNA, the pentose sugar is **deoxyribose**, which differs from the **ribose** found in RNA by the absence of one oxygen atom (see Figure 3.16).

During the formation of a nucleic acid, new nucleotides are added to an existing chain one at a time. The pentose sugar in the last nucleotide of the existing chain and the phosphate on the new nucleotide undergo a condensation reaction (see Figure 3.4), and the resulting bond is called a **phosphodiester linkage** (Figure 4.2). The phosphate on the new nucleotide is attached to the 5′-carbon atom of its sugar, and the linkage occurs between it and the 3′-carbon on the last sugar of the existing chain. Because each nucleotide is added to the 3′-carbon of the last sugar, nucleic acids are said to *grow in the 5′-to-3′ direction*.

As with carbohydrates (see Section 3.3), nucleic acids can range in size. Oligonucleotides are relatively short, with about 20 nucleotide monomers, whereas polynucleotides can be much longer.

- Oligonucleotides include RNA molecules that function as "primers" to begin the duplication of DNA; RNA molecules that regulate the expression of genes; and synthetic DNA molecules used for amplifying and analyzing other, longer nucleotide sequences.
- Polynucleotides, more commonly referred to as nucleic acids, include DNA and most RNA. Polynucleotides can be very long, and indeed are the longest polymers in the living world. Some DNA molecules in humans contain hundreds of millions of nucleotides.

Base pairing occurs in both DNA and RNA

DNA and RNA differ somewhat in their sugar groups, bases, and strand structure (Table 4.1). Four bases are found in DNA: **adenine** (**A**), **cytosine** (**C**), **guanine** (**G**), and **thymine** (**T**). RNA is also made up of four different monomers, but its nucleotides include **uracil** (**U**) instead of thymine. The sugar in DNA is deoxyribose, whereas the sugar in RNA is ribose.

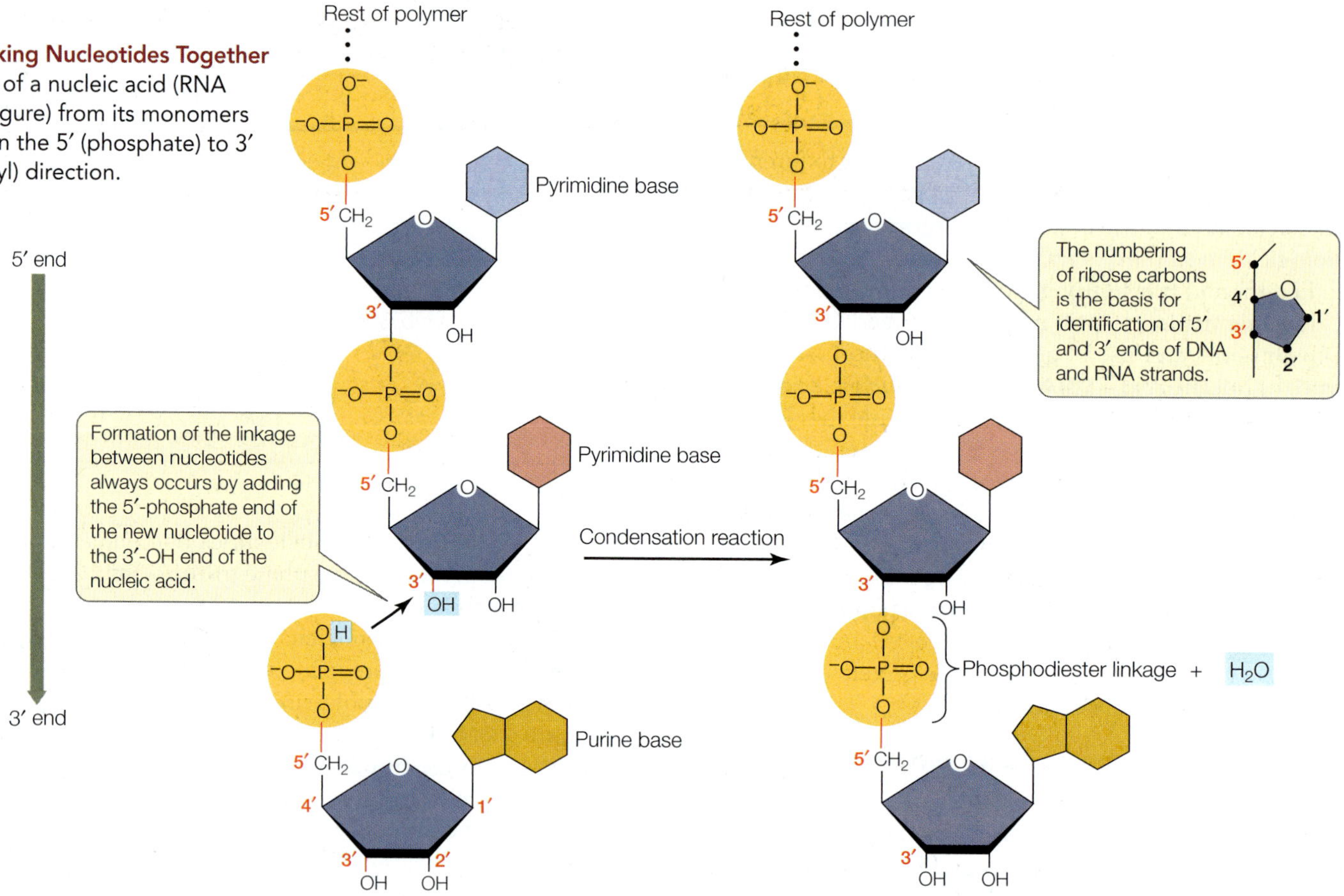

Rest of polymer
5′ end
3′ end
Pyrimidine base
Pyrimidine base
Purine base
Formation of the linkage between nucleotides always occurs by adding the 5′-phosphate end of the new nucleotide to the 3′-OH end of the nucleic acid.
Condensation reaction
The numbering of ribose carbons is the basis for identification of 5′ and 3′ ends of DNA and RNA strands.
Phosphodiester linkage + H2O

4.2 Linking Nucleotides Together Growth of a nucleic acid (RNA in this figure) from its monomers occurs in the 5′ (phosphate) to 3′ (hydroxyl) direction.

The key to understanding the structure and function of nucleic acids is the principle of **complementary base pairing**. In DNA, thymine and adenine always pair (T-A), and cytosine and guanine always pair (C-G). In RNA, the base pairs are A-U and C-G.

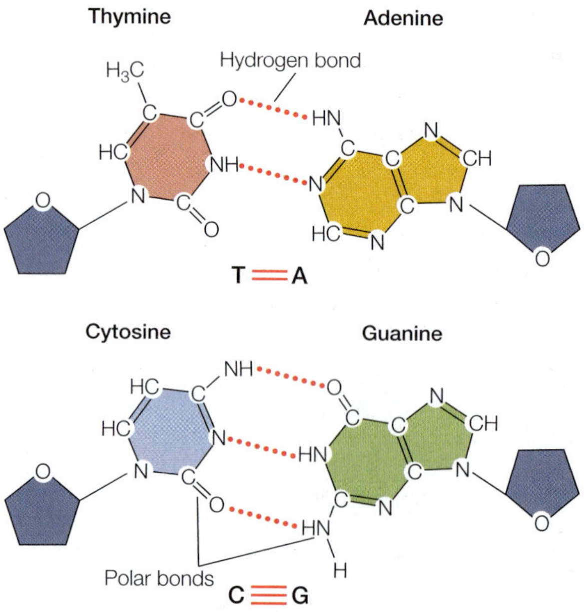

Thymine
Adenine
Hydrogen bond
T=A
Cytosine
Guanine
Polar bonds
C≡G

Base pairs are held together primarily by hydrogen bonds. As you can see, there are polar C=O and N—H covalent bonds in the bases; these can form hydrogen bonds between the δ^- on an oxygen or nitrogen of one base and the δ^+ on a hydrogen of another base.

Individual hydrogen bonds are relatively weak, but there are so many of them in a DNA or RNA molecule that collectively they provide a considerable force of attraction, which can bind together two polynucleotide strands, or a single strand that folds back onto itself. This attraction is not as strong as a covalent bond, however. This means that individual base pairs are relatively easy to break with a modest input of energy. As you will see, the breaking and making of hydrogen bonds in nucleic acids is vital to their role in living systems.

RNA Even though RNA is generally single-stranded (**Figure 4.3A**), base pairing can occur between different regions of the molecule. Portions of the single-stranded RNA molecule can fold back and pair with one another (**Figure 4.3B**). Thus complementary hydrogen bonding between ribonucleotides plays an important role in determining the three-dimensional shapes of some RNA molecules. Complementary base pairing can also take place between ribonucleotides and deoxyribonucleotides. Adenine in an RNA strand can pair either with uracil (in another RNA strand) or with thymine (in a DNA strand). Similarly, an adenine in DNA can pair either with thymine (in the complementary DNA strand) or with uracil (in RNA).

DNA Usually, DNA is double-stranded; that is, it consists of two separate polynucleotide strands of the same length that are held together by hydrogen bonds between base pairs (**Figure 4.4A**). In contrast to RNA's diversity in three-dimensional structure, DNA is remarkably uniform. The A-T and G-C base pairs are about

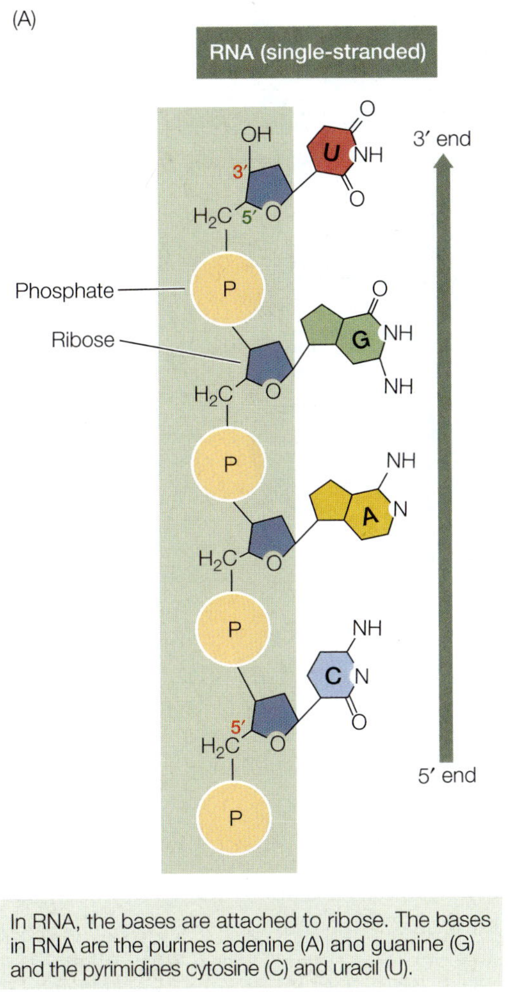
(A)
RNA (single-stranded)
OH
3′
O
U
NH
O
H₂C 5′ O
3′ end
Phosphate
P
Ribose
O
G
NH
NH
H₂C
O
P
NH
A
N
H₂C
O
P
NH
C
N
O
5′
H₂C
O
5′ end
P
In RNA, the bases are attached to ribose. The bases in RNA are the purines adenine (A) and guanine (G) and the pyrimidines cytosine (C) and uracil (U).

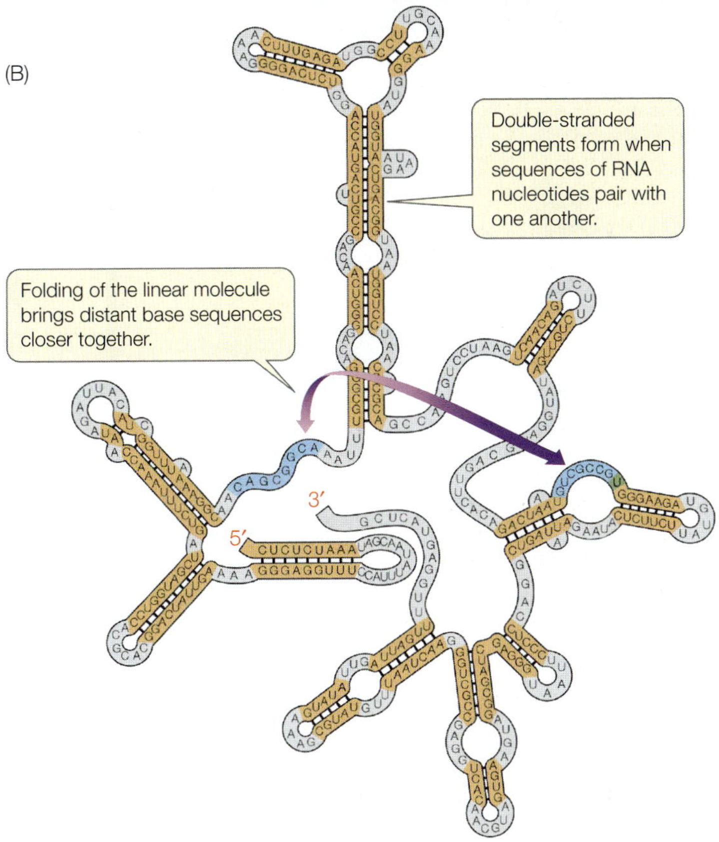
(B)
Double-stranded segments form when sequences of RNA nucleotides pair with one another.
Folding of the linear molecule brings distant base sequences closer together.
3′
5′
CUCUCUAAA
GGGAGGUUU

4.3 RNA (A) RNA is usually a single strand. (B) When a single-stranded RNA folds back on itself, hydrogen bonds between complementary sequences can stabilize it into a three-dimensional shape with complex surface characteristics.

the same size (each is a purine paired with a pyrimidine), and the two polynucleotide strands form a "ladder" that twists into a **double helix** (Figure 4.4B). The sugar–phosphate groups form the sides of the ladder, and the bases with their hydrogen bonds form the "rungs" on the inside. DNA carries genetic information in its sequence of base pairs rather than in its three-dimensional structure. The key differences among DNA molecules are their different nucleotide base sequences.

Go to Activity 4.2 DNA Structure
Life10e.com/ac4.2

DNA carries information and is expressed through RNA

DNA is a purely informational molecule. The information is encoded in the sequence of bases carried in its strands. For example, the information encoded in the sequence TCAGCA is different from the information in the sequence CCAGCA. DNA transmits information in two ways:

- DNA can be reproduced exactly. This is called **DNA replication**. It is done by polymerization using an existing strand as a base-pairing template.
- Certain DNA sequences can be copied into RNA, in a process called **transcription**. The nucleotide sequence in the RNA can then be used to specify a sequence of amino acids in a polypeptide chain. This process is called **translation**. The overall process of transcription and translation is called **gene expression**.

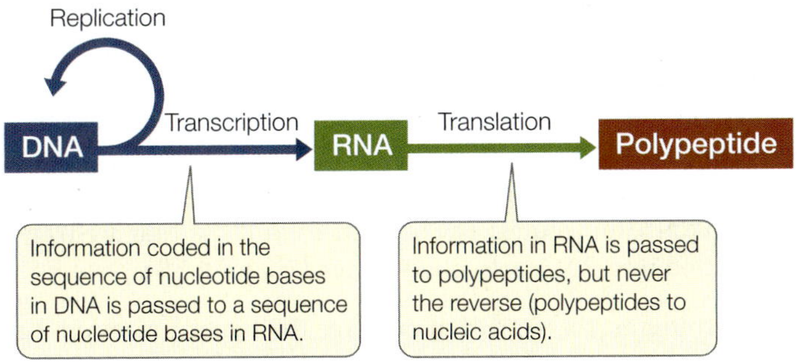
Replication
DNA
Transcription
RNA
Translation
Polypeptide
Information coded in the sequence of nucleotide bases in DNA is passed to a sequence of nucleotide bases in RNA.
Information in RNA is passed to polypeptides, but never the reverse (polypeptides to nucleic acids).

The details of these important processes are described in later chapters, but it is important to realize two things at this point:

1. *DNA replication and transcription depend on the base-pairing properties of nucleic acids.* Recall that the hydrogen-bonded base pairs are A-T and G-C in DNA and A-U and G-C in RNA. Consider, for example, this double-stranded DNA region:

5′-TCAGCA-3′
3′-AGTCGT-5′

Transcription of the lower strand will result in a single strand of RNA with the sequence 5′-UCAGCA-3′. Can you figure out the sequence that the top strand would produce?

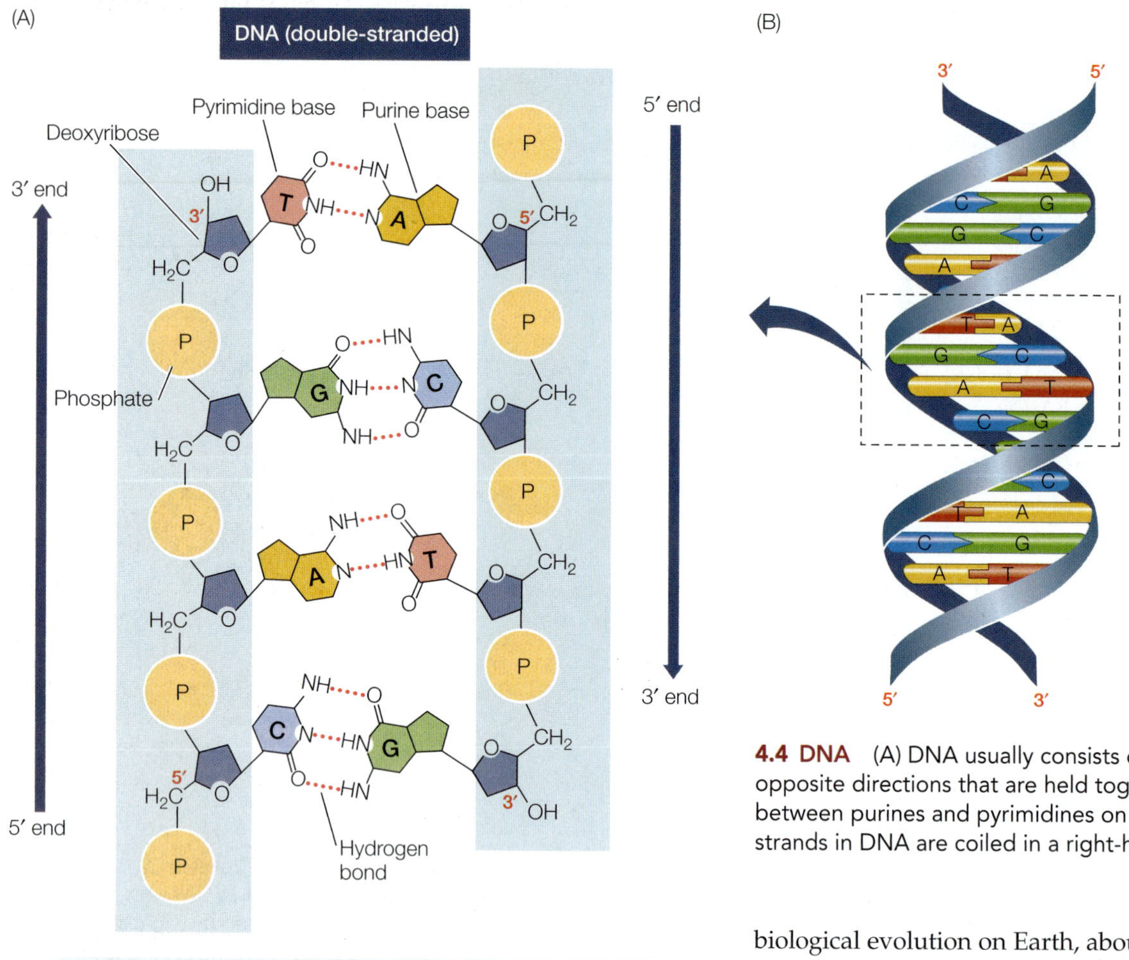
(A)
DNA (double-stranded)
Pyrimidine base
Purine base
Deoxyribose
3′ end
5′ end
OH
Phosphate
P
T
A
G
C
Hydrogen bond
5′ end
3′ end
(B)
3′
5′
5′
3′

4.4 DNA (A) DNA usually consists of two strands running in opposite directions that are held together by hydrogen bonds between purines and pyrimidines on the two strands. (B) The two strands in DNA are coiled in a right-handed double helix.

In DNA, the bases are attached to deoxyribose, and the base thymine (T) is found instead of uracil. Hydrogen bonds between purines and pyrimidines hold the two strands of DNA together.

2. *DNA replication usually involves the entire DNA molecule.* Since DNA holds essential information, it must be replicated completely and accurately so that each new cell or new organism receives a complete set of DNA from its parent (**Figure 4.5A**). The complete set of DNA in a living organism is called its **genome**. However, not all of the information in the genome is needed at all times and in all tissues, and only small sections of the DNA are transcribed into RNA molecules. The sequences of DNA that are transcribed into RNA are called **genes** (**Figure 4.5B**).

In humans, the gene that encodes the major protein in hair (keratin) is expressed only in skin cells that produce hair. The genetic information in the keratin-encoding gene is transcribed into RNA and then translated into a keratin polypeptide. In other tissues such as the muscles, the keratin gene is not transcribed, but other genes are—for example, the genes that encode proteins present in muscles but not in skin or hair.

The DNA base sequence reveals evolutionary relationships

DNA carries hereditary information from one generation to the next, gradually accumulating changes in its base sequences over long periods of time. A series of DNA molecules stretches back through the lineage of every organism to the beginning of biological evolution on Earth, about 4 billion years ago. Therefore closely related living species have more similar base sequences than species that are more distantly related. The same is true for closely related versus distantly related individuals within a species. The details of how scientists use this information are covered in Chapter 24. We described one such analysis, of the cheetah, in the opening story of this chapter.

Remarkable developments in DNA sequencing and computer technology have enabled scientists to determine the entire DNA base sequences—the genome—of many organisms, including humans, whose genome contains about 3 billion base pairs. These studies have confirmed many of the evolutionary relationships that had been inferred previously from more traditional comparisons of body structure, biochemistry, and physiology. For example, traditional comparisons had indicated that the closest living relative of humans (*Homo sapiens*) is the chimpanzee (genus *Pan*). In fact, the chimpanzee genome shares more than 98 percent of its DNA base sequence with the human genome. Increasingly, scientists turn to DNA analyses to elucidate evolutionary relationships when other comparisons are not possible or are not conclusive. For example, DNA studies revealed a close relationship between starlings and mockingbirds that was not expected on the basis of their anatomy or behavior.

Nucleotides have other important roles

Nucleotides are more than just the building blocks of nucleic acids. As we will describe in later chapters, there are several nucleotides (or modified nucleotides) with other functions:

- ATP (adenosine triphosphate) acts as an energy transducer in many biochemical reactions (see Section 8.2).

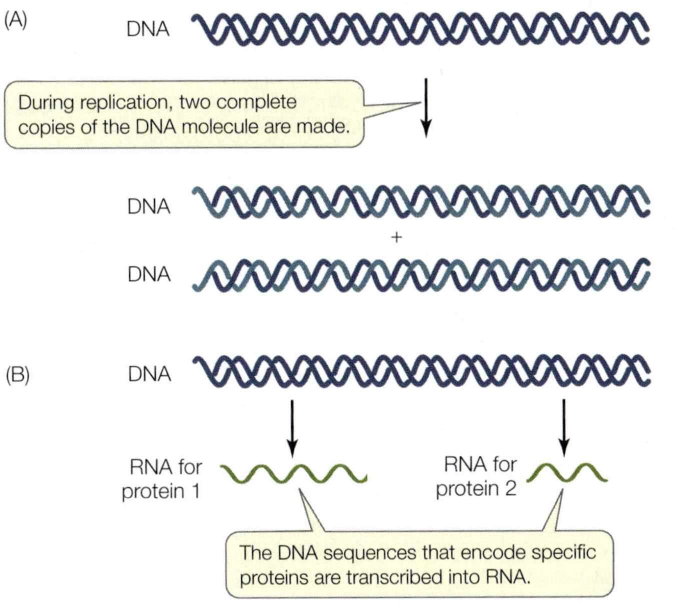

(A)
DNA
During replication, two complete copies of the DNA molecule are made.
DNA
+
DNA
(B)
DNA
RNA for protein 1
RNA for protein 2
The DNA sequences that encode specific proteins are transcribed into RNA.

4.5 DNA Replication and Transcription DNA is usually completely replicated (A) but only partially transcribed (B). RNA transcripts are produced from genes that code for specific proteins. Transcription of different genes occurs at different times and, in multicellular organisms, in different cells of the body.

- GTP (guanosine triphosphate) serves as an energy source, especially in protein synthesis. It also plays a role in the transfer of information from the environment to cells (see Section 7.2).
- cAMP (cyclic adenosine monophosphate) is a special nucleotide with an additional bond between the sugar and the phosphate group. It is essential in many processes, including the actions of hormones and the transmission of information by the nervous system (see Section 7.3).
- Nucleotides play roles as carriers in the synthesis and breakdown of carbohydrates and lipids.

RECAP 4.1

The nucleic acids DNA and RNA are polymers made up of nucleotide monomers. The sequence of nucleotides in DNA carries the information that is used by RNA to specify primary protein structure. The genetic information in DNA is passed from generation to generation and can be used to understand evolutionary relationships.

- List the key differences between DNA and RNA, and between purines and pyrimidines. See pp. 63–65, Table 4.1, and Figure 4.1
- How do purines and pyrimidines pair up in complementary base pairing? See p. 64
- What are the differences between DNA replication and transcription? See pp. 65–66 and Figure 4.5
- How can DNA molecules be very diverse, even though they appear to be structurally similar? See p. 65

We have seen that the nucleic acids RNA and DNA carry the blueprint of life, and that the inheritance of these macromolecules reaches back to the beginning of evolutionary time. But when, where, and how did nucleic acids arise on Earth? How did the building blocks of life, such as amino acids and sugars, originally arise?

4.2 How and Where Did the Small Molecules of Life Originate?

Chapter 2 pointed out that living things are composed of the same atomic elements as the inanimate universe. But the arrangements of these atoms into molecules are unique in biological systems. You will not find biological molecules in inanimate matter unless they came from a once-living organism.

It is impossible to know for certain how life on Earth began. But one thing is sure: life (or at least life as we know it) is not constantly being restarted. That is, **spontaneous generation** of life from inanimate nature is not happening repeatedly before our eyes. Now and for many millenia in the past, all life has come from life that existed before. But people, including scientists, did not always believe this.

Experiments disproved the spontaneous generation of life

The idea that life can originate repeatedly from nonliving matter has been common in many cultures and religions. During the European Renaissance (from the fourteenth to seventeenth centuries, a period that witnessed the birth of modern science), most people thought that at least some forms of life arose repeatedly and directly from inanimate or decaying matter by spontaneous generation. Many thought that mice arose from sweaty clothes placed in dim light; that frogs sprang directly from moist soil; and that rotting meat produced flies. Scientists such as the Italian physician and poet Francesco Redi, however, doubted these assumptions. Redi proposed that flies arose not by some mysterious transformation of decaying meat, but from other flies that laid their eggs on the meat. In 1668, Redi performed a scientific experiment—a relatively new concept at the time—to test his hypothesis. He set out several jars containing chunks of meat.

- One jar contained meat exposed to both air and flies.
- A second jar was covered with a fine cloth so that the meat was exposed to air but not to flies.
- The third jar was sealed with a lid so the meat was exposed to neither air nor flies.

No lid
Fine cloth cover
Lid

As he had hypothesized, Redi found maggots, which then hatched into flies, only in the first jar. This finding demonstrated that maggots could occur only where flies were present before. The idea that a complex organism like a fly could appear spontaneously from a nonliving substance in the meat, or from "something in the air," was laid to rest. Well, perhaps not quite to rest.

In the 1660s, newly developed microscopes revealed a vast new biological world. Under microscopic observation, virtually every environment on Earth was found to be teeming with tiny organisms. Some scientists believed these organisms arose spontaneously from their rich chemical environment, by the action of a "life force." But experiments in the nineteenth century by the great French scientist Louis Pasteur showed that microorganisms can arise only from other microorganisms, and that an environment without life remains lifeless (**Figure 4.6**).

Go to Animated Tutorial 4.1
Pasteur's Experiment
Life10e.com/at4.1

Pasteur's and Redi's experiments indicated that living organisms cannot arise from nonliving materials *under the conditions that exist on Earth now*. But their experiments did not prove that spontaneous generation never occurred. Eons ago, conditions on Earth and in the atmosphere above it were vastly different than they are today. Indeed, conditions similar to those found on primitive Earth may have existed, or may exist now, on other bodies in our solar system and elsewhere. This has led scientists to ask whether life has originated on other bodies in space, as it did on Earth.

Life began in water

As we emphasized in Chapter 2, water is an essential component of life as we know it. This is why there was great excitement when remote laboratories sent from Earth detected water ice on Mars. Astronomers believe our solar system began forming about 4.6 billion years ago, when a star exploded and collapsed to form the sun and about 500 bodies called planetesimals. These planetesimals collided with one another to form the inner planets, including Earth and Mars. The first chemical signatures indicating the presence of life on Earth are about 4 billion years old. So it took 600 million years for the chemical conditions on Earth to become just right for life. Key among those conditions was the presence of water.

Ancient Earth probably had a lot of water high in its atmosphere. But the new planet was hot, and the water remained in vapor form and dissipated into space. As Earth cooled, it became possible for water to condense on the planet's surface—but where did that water come from? One current view is that comets (loose agglomerations of dust and ice that have orbited the sun since the planets formed) struck Earth

INVESTIGATING**LIFE**

4.6 Disproving the Spontaneous Generation of Life Previous experiments disproving the spontaneous generation of larger organisms were called into question when microorganisms were discovered. Louis Pasteur's classic experiments disproved the spontaneous generation of microorganisms.[a]

HYPOTHESIS Microorganisms come only from other microorganisms and cannot arise by spontaneous generation.

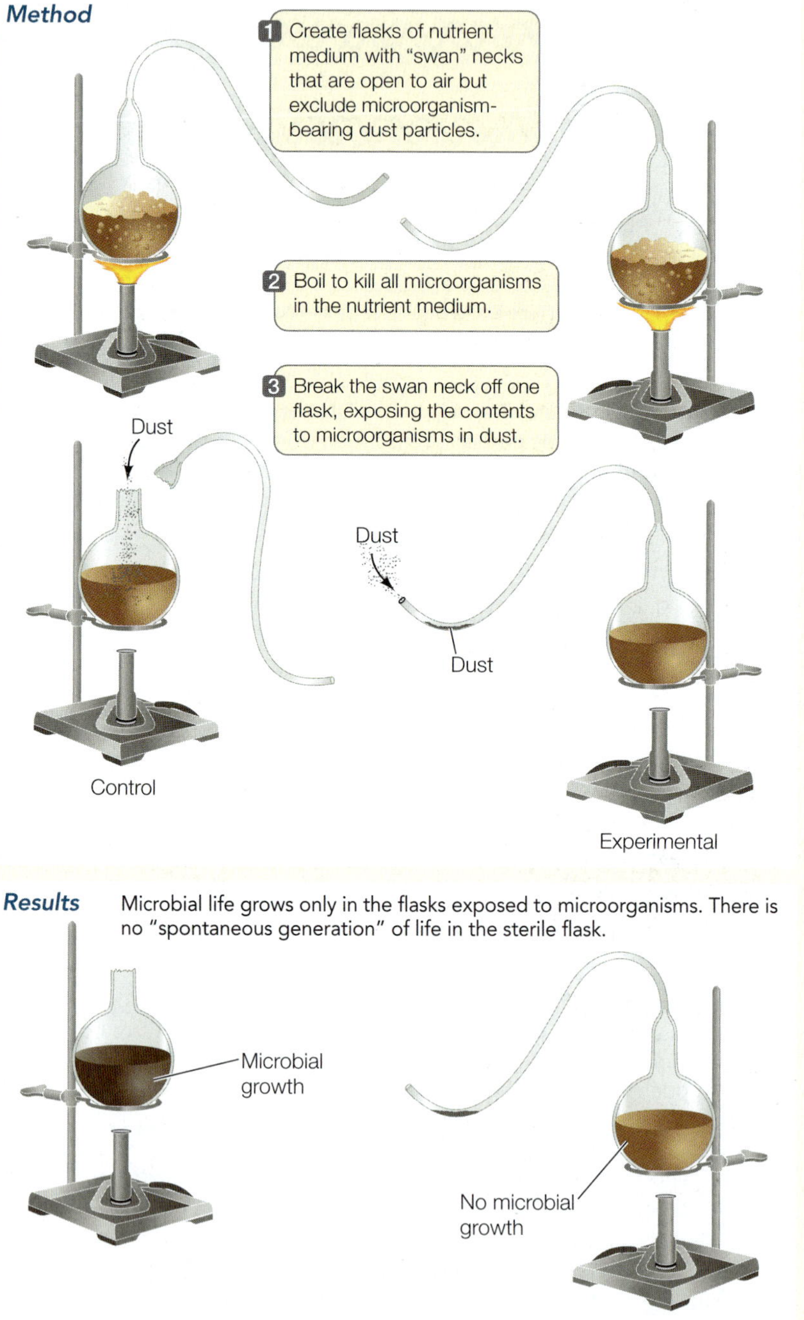
Method
1 Create flasks of nutrient medium with "swan" necks that are open to air but exclude microorganism-bearing dust particles.
2 Boil to kill all microorganisms in the nutrient medium.
3 Break the swan neck off one flask, exposing the contents to microorganisms in dust.
Dust
Dust
Dust
Control
Experimental
Results
Microbial life grows only in the flasks exposed to microorganisms. There is no "spontaneous generation" of life in the sterile flask.
Microbial growth
No microbial growth

CONCLUSION All life comes from pre-existing life. An environment without life remains lifeless.

Go to **BioPortal** for discussion and relevant links for all INVESTIGATING**LIFE** figures.

[a]Pasteur gave a talk on his research at the "Sorbonne Scientific Soirée" on April 7, 1864. This talk has been translated into English: http://rc.usf.edu/~levineat/pasteur.pdf

and Mars repeatedly, bringing to those planets not only water but also other chemical components of life, such as nitrogen.

As the planets cooled and chemicals from their crusts dissolved in the water, simple chemical reactions would have taken place. Some of these reactions might have led to life, but impacts by large comets and rocky meteorites released enough energy to heat the developing oceans almost to boiling, thus destroying any early life that might have existed. On Earth, these large impacts eventually subsided, and some time around 3.8 to 4 billion years ago, life gained a foothold. There has been life on Earth ever since.

Several models have been proposed to explain the origin of life on Earth. The next sections will discuss two alternative theories: that life came from outside Earth, or that life arose on Earth through chemical evolution.

Life may have come from outside Earth

In 1969 a remarkable event led to the discovery that a meteorite from space carried molecules that were characteristic of life on Earth. On September 28 of that year, fragments of a meteorite fell around the town of Murchison, Australia. Using gloves to avoid Earth-derived contamination, scientists immediately shaved off tiny pieces of the rock, put them in test tubes, and extracted them in water (Figure 4.7). They found several of the molecules that are unique to life, including purines, pyrimidines, sugars, and ten amino acids.

4.7 The Murchison Meteorite Pieces from a fragment of the meteorite that landed in Australia in 1969 were put into test tubes with water. Soluble molecules present in the rock—including amino acids, nucleotide bases, and sugars—dissolved in the water. Plastic gloves and sterile instruments were used to reduce the possibility of contamination with substances from Earth.

Go to Media Clip 4.1
DNA Building Blocks from Space
Life10e.com/mc4.1

Were these molecules truly brought from space as part of the meteorite, or did they get there after the rock landed on Earth? There are a number of reasons to believe the molecules were not Earthly contaminants:

- The scientists took great care to avoid contamination. They used gloves and sterile instruments, took pieces from below the rock's surface, and did their work very soon after it landed (they hoped before organisms from Earth could contaminate the samples).
- Amino acids in living organisms on Earth are L-amino acids: they are found in only one of the two possible optical isomeric forms (see Figure 3.2). The amino acids in the meteorite were a mixture of L- and D-isomers, with a slight preponderance of the L form. Thus the amino acids in the meteorite were not likely to have come from a living organism on Earth.
- In the story that opened Chapter 2, we described how the ratio of isotopes in a living organism reflects the ratio of the same isotopes in the environment where the organism lives. The isotope ratios for carbon and hydrogen in the sugars from the meteorite were different from the ratios of those elements found on Earth.

More than 90 meteorites from Mars have been recovered on Earth. Many show signs of water, for example minerals such as carbonates that are precipitated from aqueous solution. Some also contain organic molecules that are the chemical signatures of life. While the presence of such molecules suggests that these rocks once harbored life, it does not prove that there were living organisms in the rocks when they landed on Earth. Many scientists find it hard to believe that an organism could survive thousands of years of traveling through space in a meteorite, followed by intense heat as the meteorite passed through Earth's atmosphere. But there is evidence that the heat at the centers of some meteorites may not have been severe. If this was the case, then a long interplanetary trip by living organisms might have been possible.

Prebiotic synthesis experiments model early Earth

It is clear that other bodies in the solar system have, or once had, water and other simple organic molecules. Possibly, a meteorite was the source of the simple molecules that were the original building blocks for life on Earth. But a second theory for the origin of life on Earth, **chemical evolution**, holds that conditions on primitive Earth led to the formation of these simple molecules (prebiotic synthesis), and that these molecules led to the formation of life forms. Scientists have sought to reconstruct those primitive conditions, both physically (by varying temperature) and chemically (by re-creating the mixes of elements that may have been present).

HOT CHEMISTRY In oxygenated water, some trace metals such as molybdenum and rhenium are soluble, and their presence in sediments under oceans and lakes is directly proportional to the amount of oxygen gas (O_2) that was present in and above the water at the times the rocks were formed. Measurements of dated sedimentary cores indicate that none of these rare metals was present prior to 2.5 billion years ago. This and other lines of evidence suggest that there was little O_2 in Earth's early atmosphere. Oxygen gas is thought to have accumulated about 2.5 billion years ago as the by-product of photosynthesis by single-celled life forms; today 21 percent of our atmosphere is O_2.

In the 1950s Stanley Miller and Harold Urey at the University of Chicago set up an experimental "atmosphere" containing the gases they thought were present in Earth's early atmosphere: hydrogen gas, ammonia, methane gas, and water vapor. They passed an electrical spark through these gases to simulate lightning, a source of energy to drive chemical reactions. Then they cooled the system so the gases would condense and collect in a watery solution, or "ocean" (**Figure 4.8**). After a week of continuous operation, the system contained numerous organic molecules, including a variety of amino acids—the building blocks of proteins.

Go to Animated Tutorial 4.2
Synthesis of Prebiotic Molecules
Life10e.com/at4.2

COLD CHEMISTRY Stanley Miller also performed a long-term experiment in which the electrical spark was not used. In 1972 he filled test tubes with ammonia gas, water vapor, and cyanide (HCN), another molecule that is thought to have formed on primitive Earth. After checking that there were no contaminating substances or organisms that might confound the results, he sealed the tubes and cooled them to –78°C, the temperature of the ice that covers Europa, one of Jupiter's moons. Opening the tubes 27 years later, Miller found amino acids and nucleotide bases. Apparently, pockets of liquid water within the ice had allowed high concentrations of the starting materials to accumulate, thereby speeding up chemical reactions. The important conclusion is that the cold water within ice on ancient Earth, and other celestial bodies such as Mars, Europa, and Enceladus (one of Saturn's moons; satellite photos have revealed geysers of liquid water coming from its interior), may have provided environments for the prebiotic synthesis of molecules required for the subsequent formation of simple living systems.

The results of these experiments were profoundly important in giving weight to speculations about the chemical origin of life on Earth and elsewhere in the universe. Decades of experimental work and critical evaluation followed Miller and Urey's original experiments. In science, an experiment and its results must be repeatable and be reinterpreted and refined as more knowledge accumulates. For example, ideas about Earth's original atmosphere have changed. There is abundant evidence indicating that major volcanic eruptions occurred 4 billion years ago; these would have released carbon dioxide (CO_2), nitrogen (N_2), hydrogen sulfide (H_2S), and sulfur dioxide (SO_2) into the atmosphere. Experiments using these gases in addition to the ones in the original Miller–Urey experiment have produced a more diverse list of organic products:

- All five bases that are present in DNA and RNA (i.e., A, T, C, G, and U)
- All of the 20 amino acids used in protein synthesis
- Many 3- to 6-carbon sugars
- Certain fatty acids
- Vitamin B_6 (pantothenic acid, a component of coenzyme A)
- Nicotinamide (part of NAD, which is involved in energy metabolism)
- Carboxylic acids such as succinic and lactic acids (also involved in energy metabolism)

INVESTIGATINGLIFE

4.8 Miller and Urey Synthesized Prebiotic Molecules in an Experimental Atmosphere With an increased understanding of the atmospheric conditions that existed on primitive Earth, the researchers devised an experiment to see if these conditions could lead to the formation of organic molecules.[a,b]

HYPOTHESIS Organic chemical compounds can be generated under conditions similar to those that existed in the atmosphere of primitive Earth.

Method

Results

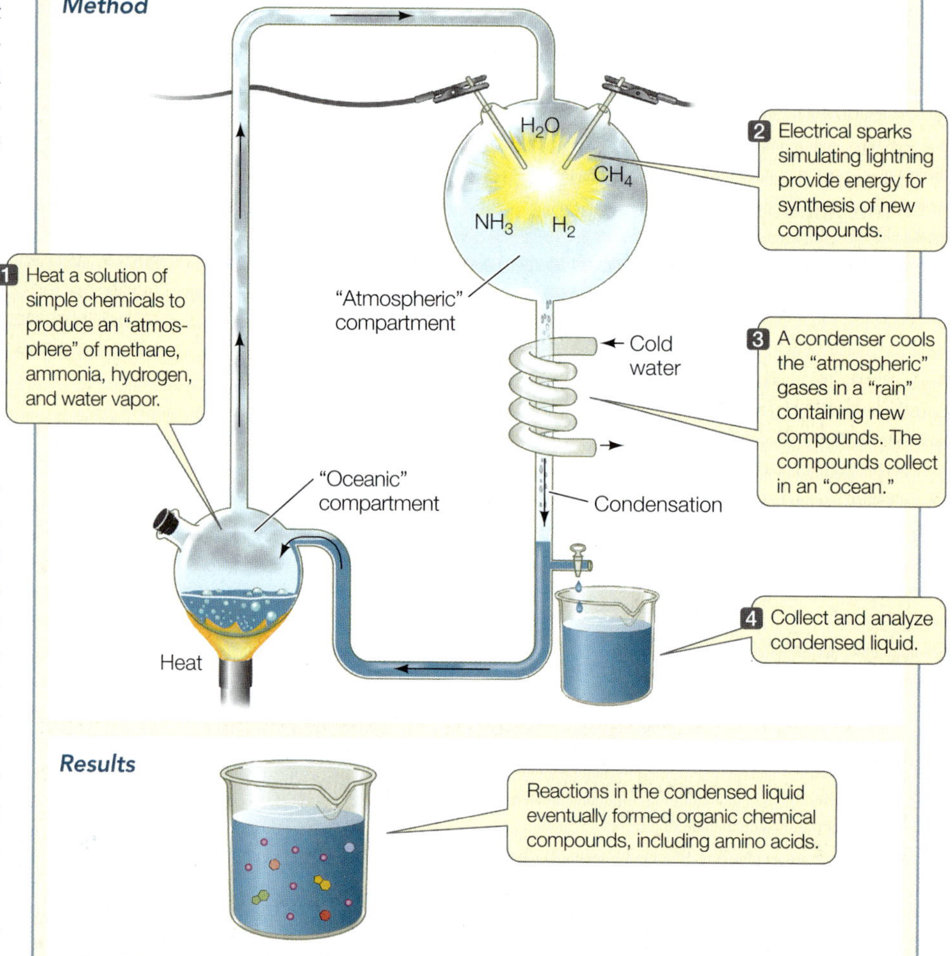
H2O
CH4
NH3
H2
2 Electrical sparks simulating lightning provide energy for synthesis of new compounds.
1 Heat a solution of simple chemicals to produce an "atmosphere" of methane, ammonia, hydrogen, and water vapor.
"Atmospheric" compartment
Cold water
3 A condenser cools the "atmospheric" gases in a "rain" containing new compounds. The compounds collect in an "ocean."
"Oceanic" compartment
Condensation
4 Collect and analyze condensed liquid.
Heat
Reactions in the condensed liquid eventually formed organic chemical compounds, including amino acids.

CONCLUSION The chemical building blocks of life could have been generated in the probable atmosphere of early Earth.

Go to **BioPortal** for discussion and relevant links for all INVESTIGATING**LIFE** figures.

[a]Miller, S. L. 1953. *Science* 117: 528–519.
[b]Miller, S. L. and H. C. Urey. 1959. *Science* 130: 245–251.

WORKING WITH**DATA**:

Could Biological Molecules Have Been Formed from Chemicals Present in Earth's Early Atmosphere?

Original Papers

Miller, S. L. 1953. A production of amino acids under possible primitive earth conditions. *Science* 117: 528–519.

Miller, S. L. and H. C. Urey. 1959. Organic compound synthesis on the primitive earth. *Science* 130: 245–251.

Analyze the Data

In the 1950s the Nobel Prize–winning chemist Harold Urey proposed that the molecules present in primitive Earth's atmosphere were methane (CH_4), ammonia (NH_3), hydrogen (H_2), and water (H_2O). He suggested that it might be possible to generate the building blocks of life, such as amino acids, in a laboratory simulation of these early conditions. Urey's graduate student Stanley Miller ran the simulation, which consisted of the four gases in an enclosed apparatus, an electric discharge to provide energy, and a cooling condenser to allow any substances that formed to dissolve in a watery "ocean" (see Figure 4.8). After a week, Miller analyzed the water using paper chromatography and found amino acids. This hallmark experiment was the first to demonstrate that organic molecules may have formed on Earth before life appeared.

The data Miller and Urey gave for sources of energy impinging on Earth are shown in the table.

QUESTION 1
Of the total energy from the sun, only a small fraction is in the ultraviolet range, less than 250 nm. What proportion of total solar energy is the energy with wavelengths below 250 nm?

QUESTION 2
The molecules CH_4, H_2O, NH_3, and CO_2 absorb light at wavelengths of less than 200 nm. What fraction of total solar radiation is in this range?

QUESTION 3
Miller and Urey used electric discharges as their energy source. What other sources of energy could be used in similar experiments?

Source	Energy (cal cm^{-2} yr^{-1})
Total radiation from sun	260,000
Ultraviolet light	
Wavelength <250 nm[a]	570
Wavelength <200 nm	85
Wavelength <150 nm	3.5
Electric discharges	4
Cosmic rays	0.0015
Radioactivity	0.8
Volcanoes	0.13

[a]Nanometer, 10^{-9} meters.

Go to **BioPortal** for all WORKING WITH**DATA** exercises

RECAP 4.2

Life does not arise repeatedly through spontaneous generation, but comes from pre-existing life. Water is an essential ingredient for the emergence of life. Meteorites that have landed on Earth provide some evidence for an extraterrestrial origin of life. Chemical synthesis experiments provide support for the idea that life's simple molecules formed in the prebiotic environment on Earth.

- Explain how Redi's and Pasteur's experiments disproved spontaneous generation. **See pp. 67–68 and Figure 4.6**
- What is the evidence that life on Earth came from other bodies in the solar system? **See p. 69**
- What is the significance of the Miller–Urey experiment, what did it find, and what were its limitations? **See pp. 70–71 and Figure 4.8**

Chemistry experiments modeling the conditions of ancient Earth provide clues about the origins of the monomers (such as amino acids) that make up the polymers (such as proteins) that characterize life. How did these polymers develop?

4.3 How Did the Large Molecules of Life Originate?

The Miller–Urey experiment and others that followed provide a plausible scenario for the formation of the building blocks of life under conditions that prevailed on primitive Earth. The next step in forming and supporting a general theory on the origin of life would be an explanation for how polymers formed from these monomers.

Chemical evolution may have led to polymerization

Scientists have used a number of model systems to try to simulate conditions under which polymers might have been made. Each of these systems is based on several observations and speculations:

- Solid mineral surfaces, such as powderlike clays, have large surface areas. Scientists speculate that the silicates in clay may have catalyzed (speeded up) the condensation reactions that resulted in organic polymers.
- Hydrothermal vents deep in the ocean, where hot water emerges from beneath Earth's crust, lack oxygen gas and contain metals such as iron and nickel. In laboratory experiments, these metals have been shown to catalyze the polymerization of amino acids in the absence of oxygen.
- In hot pools at the edges of oceans, evaporation may have concentrated monomers to the point where polymerization was favored (the "primordial soup" hypothesis).

In whatever ways the earliest stages of chemical evolution occurred, they resulted in the emergence of monomers and polymers that have probably remained unchanged in their general structures and functions for several billion years.

RNA may have been the first biological catalyst

A hallmark of living organisms is chemical change, for example, DNA replication and RNA synthesis, which we described earlier in this chapter (see Figure 4.5). There are many other chemical changes that occur in living systems, often involving the hydrolysis or synthesis of macromolecules or conversions among small molecules. As you will see in Chapter 8, these chemical changes can occur spontaneously in aqueous solutions like those that exist in biological systems, but most would occur extremely slowly. The evolution of catalysts—molecules that speed up biochemical conversions—solved this problem. So a key to the origin of life is the appearance of catalysts.

In life today, the main catalysts are proteins called enzymes. The myriad shapes of proteins allow them to bind to diverse substances in solution and speed up chemical reactions. But we know that proteins are made from information in nucleic acids (see p. 65). So we have a problem: If proteins are needed for life, nucleic acids must have appeared first, so that the proteins could be made. But if nucleic acids appeared before proteins, the proteins could not have been the first catalysts. Could nucleic acids be catalysts, in addition to their role as blueprints for protein synthesis? The answer is yes.

Like a protein, the three-dimensional structure of a folded RNA molecule presents a unique surface to the external environment (see Figure 4.3). The surfaces of RNA molecules can be every bit as specific as those of proteins. The three-dimensional shapes and other chemical properties of certain RNA molecules allow them to function as catalysts. Catalytic RNAs, called **ribozymes**, can speed up reactions involving their own nucleotides as well as other cellular substances. Although in retrospect it is not too surprising, the discovery of catalytic RNAs was a major shock to a community of biologists who were convinced that all biological catalysts were proteins (enzymes). It took almost a decade for the work of the scientists involved, Thomas Cech and Sidney Altman, to be fully accepted by other scientists. Later, they were awarded the Nobel Prize.

Given that RNA can be both informational (in its nucleotide sequence) and catalytic (because of its ability to form unique three-dimensional shapes), it has been hypothesized that early life consisted of an "RNA world"—a world before DNA. It is thought that when RNA was first made, it could have acted as a catalyst for its own replication as well as for the synthesis of proteins. DNA could eventually have evolved from RNA (**Figure 4.9**). Several lines of evidence support this scenario:

- In living organisms today, the formation of peptide linkages (see Figure 3.6) is catalyzed by ribozymes.
- In certain viruses called retroviruses, there is an enzyme called reverse transcriptase that catalyzes the synthesis of DNA from RNA.
- When a short, naturally occuring RNA molecule is added to a mixture of nucleotides, RNA polymers are formed at a rate 7 million times greater than the formation of polymers without the added RNA. This indicates that the added RNA is a catalyst, not just a template.
- An artificial ribozyme has been developed that can catalyze the assembly of short RNAs into a longer molecule that is an exact copy of itself (**Figure 4.10**). This may be how nucleic acid replication evolved.

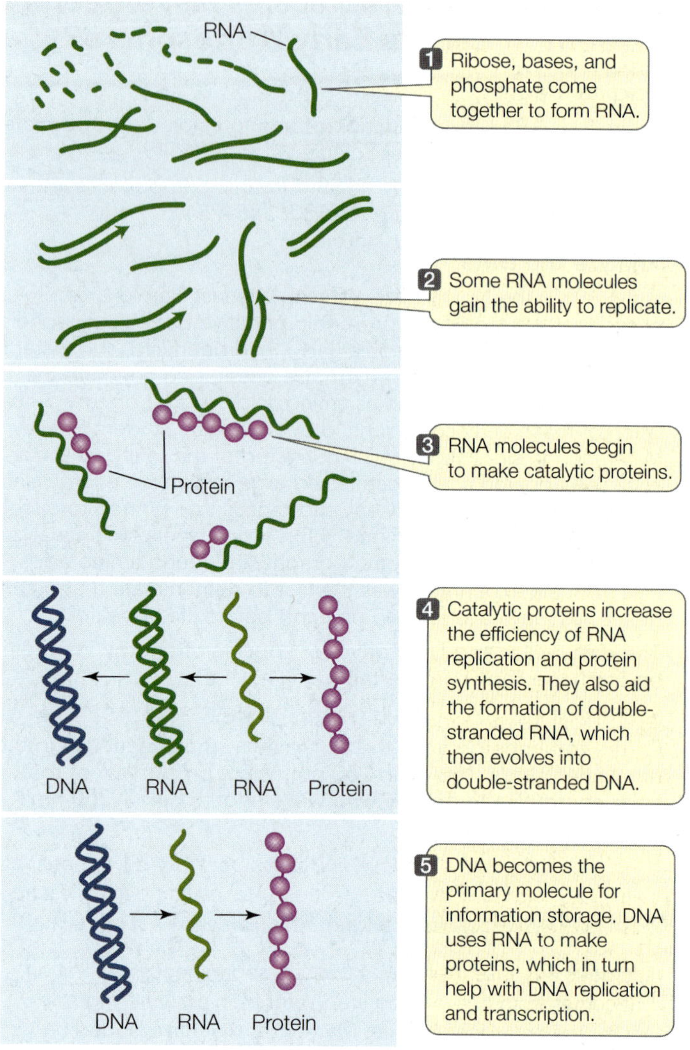
RNA
1 Ribose, bases, and phosphate come together to form RNA.
2 Some RNA molecules gain the ability to replicate.
Protein
3 RNA molecules begin to make catalytic proteins.
DNA
RNA
RNA
Protein
4 Catalytic proteins increase the efficiency of RNA replication and protein synthesis. They also aid the formation of double-stranded RNA, which then evolves into double-stranded DNA.
DNA
RNA
Protein
5 DNA becomes the primary molecule for information storage. DNA uses RNA to make proteins, which in turn help with DNA replication and transcription.

4.9 The "RNA World" Hypothesis This view postulates that in a world before DNA, RNA alone was both the blueprint for protein synthesis and a catalyst for its own replication. Eventually, the information storage molecules of DNA could have evolved from RNA.

RECAP 4.3

The formation of the large polymers that are characteristic of life may have occurred on the surfaces of clay particles, near hydrothermal vents, or in hot pools at the edges of oceans. RNA may have been the first genetic material and catalyst.

- Why was the discovery of ribozymes important for the development of the "RNA world" hypothesis? **See p. 72**
- How can RNA self-replicate? **See p. 72 and Figure 4.10**

The discovery of mechanisms for the formation of small and large molecules is essential to answering questions about the origin of life on Earth. But we also need to understand how organized living systems formed. Such systems display the characteristic properties of life, including reproduction, energy processing, and responsiveness to the environment. These are properties of cells, whose origin we will explore in the next section.

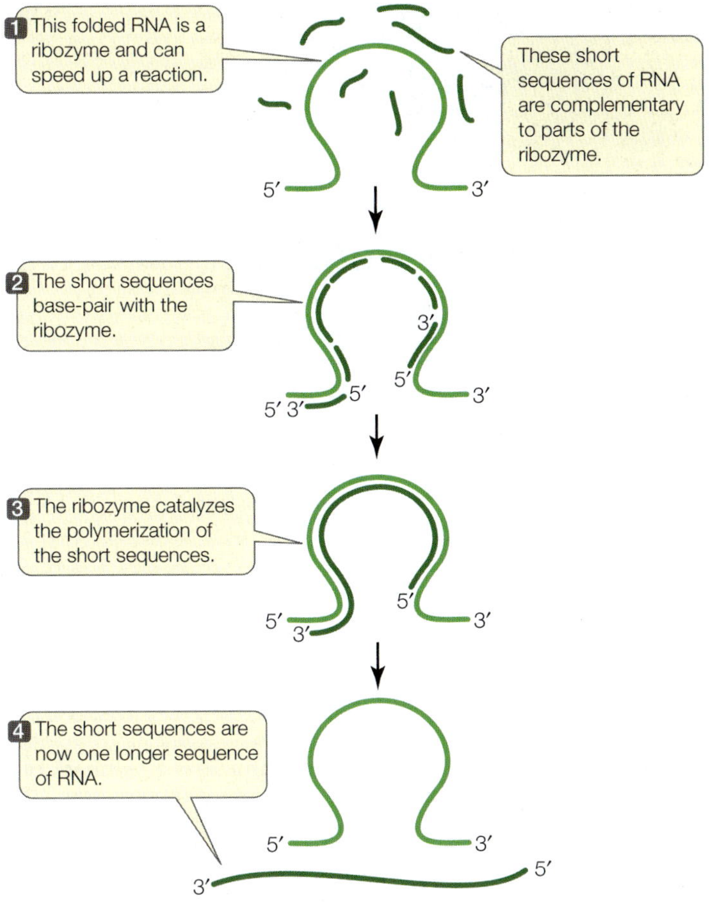

1 This folded RNA is a ribozyme and can speed up a reaction.
These short sequences of RNA are complementary to parts of the ribozyme.
5′
3′
2 The short sequences base-pair with the ribozyme.
3′
5′
5′ 3′
5′
3′
3 The ribozyme catalyzes the polymerization of the short sequences.
5′
3′
5′
3′
4 The short sequences are now one longer sequence of RNA.
5′
3′
3′
5′

4.10 An Early Catalyst for Life? In the laboratory, a synthetic ribozyme (a folded RNA molecule) can catalyze the polymerization of shorter RNA strands into a longer molecule that is identical to itself. This may be how the earliest nucleic acids replicated.

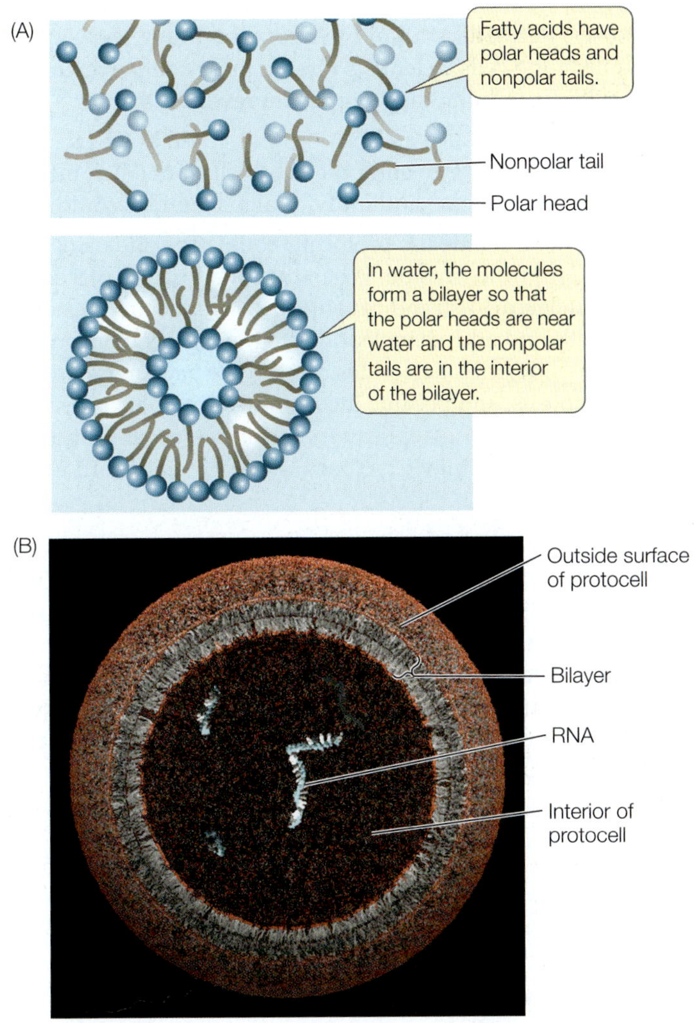

(A)
Fatty acids have polar heads and nonpolar tails.
Nonpolar tail
Polar head
In water, the molecules form a bilayer so that the polar heads are near water and the nonpolar tails are in the interior of the bilayer.
(B)
Outside surface of protocell
Bilayer
RNA
Interior of protocell

4.11 Protocells (A) In a series of experiments, Jack Szostak and his colleagues mixed fatty acid molecules in water. The molecules formed spherical structures called protocells, with water surrounded by bilayers of fatty acids. (B) A model of a protocell. A portion of the "membrane" has been cut away to reveal the inside of the protocell and the membrane's bilayer structure. Nutrients and nucleotides pass through the "membrane" and enter the protocell, where they copy an already present RNA template. The new copies of RNA remain in the protocell.

4.4 How Did the First Cells Originate?

As you have seen from many of the theories for the origin of life, the evolution of biochemistry occurred under localized conditions. That is, the chemical reactions of life could not occur in a dilute aqueous environment. There had to be a compartment of some sort that brought together and concentrated the compounds involved in these events. Biologists have proposed that initially this compartment may have simply been a tiny droplet of water on the surface of a rock. But another major event in the origin of life was necessary.

Life as we know it is separated from the environment within structurally defined units called **cells**. The internal contents of a cell are separated from the nonbiological environment by a special barrier—a **membrane**. The membrane is not just a barrier; it regulates what goes into and out of the cell, as we will describe in Chapter 6. This role of the surface membrane is very important because it permits the interior of the cell to maintain a chemical composition that is different from its external environment. How did the first cells with membranes come into existence?

Experiments explore the origin of cells

Jack Szostak and his colleagues at Harvard University built a laboratory model that gives insights into the origin of cells. To do this, they first put fatty acids (which can be made in prebiotic experiments) into water. Recall from Chapter 3 that fatty acids are amphipathic: they have a hydrophilic polar head and a long, nonpolar tail that is hydrophobic (see Figure 3.22). When placed in water, fatty acids will arrange themselves in a round "huddle" much like a football team: the hydrophilic heads point outward to interact with the aqueous environment, and the fatty acid tails point inward, away from the water molecules.

What if some water becomes trapped in the interior of this "huddle"? Now the layer of hydrophobic fatty acid tails is in water, which is an unstable situation. To stabilize this structure, a second layer of fatty acids forms. This **lipid bilayer** has the polar heads of the fatty acids facing both outward and inward, because they are attracted to the polar water molecules present on each side of the double layer. The nonpolar tails form the interior of the bilayer (**Figure 4.11**). These prebiotic, water-filled structures, defined by a lipid bilayer membrane, very much resemble living cells. Scientists refer to these compartments as **protocells**. Examining their properties revealed that

- Large molecules such as DNA or RNA could not pass through the bilayer to enter the protocells, but small molecules such as sugars and individual nucleotides could.
- Nucleic acids inside the protocells could replicate using the nucleotides from outside. When the investigators placed a short nucleic acid strand capable of self-replication inside protocells and added nucleotides to the watery environment outside, the nucleotides crossed the barrier, entered the protocells, and became incorporated into new polynucleotide chains. This replication, which can occur without protein catalysis, may have been the first step toward cell reproduction.

Were these protocells truly cells, and was the lipid bilayer produced in these experiments a true cell membrane? Certainly not. The protocells could not fully reproduce, nor could they carry out all the metabolic reactions that take place in modern cells. The simple lipid bilayer had few of the sophisticated functions of modern cell membranes. Nevertheless, the protocell may be a reasonable facsimile of a cell as it evolved billions of years ago:

- It can act as an organized system of parts, with substances interacting and reacting, in some cases catalytically.
- It includes an interior that is distinct from the exterior environment.
- It is capable of self-replication.

These are all fundamental characteristics of living cells.

Some ancient cells left a fossil imprint

In the 1990s scientists made a rare find: a formation of ancient rocks in Australia that had remained relatively unchanged since the rocks first formed 3.5 billion years ago. In one of these rock samples, geologist J. William Schopf of the University of California, Los Angeles, saw chains and clumps of what looked tantalizingly like contemporary cyanobacteria, or "blue-green" bacteria (**Figure 4.12**). Cyanobacteria are believed to have been among the first organisms because they can perform photosynthesis, converting CO_2 and water into carbohydrates. Schopf needed to prove that the chains were once alive, not just the results of simple chemical reactions. He and his colleagues looked for chemical evidence of photosynthesis in the rock samples.

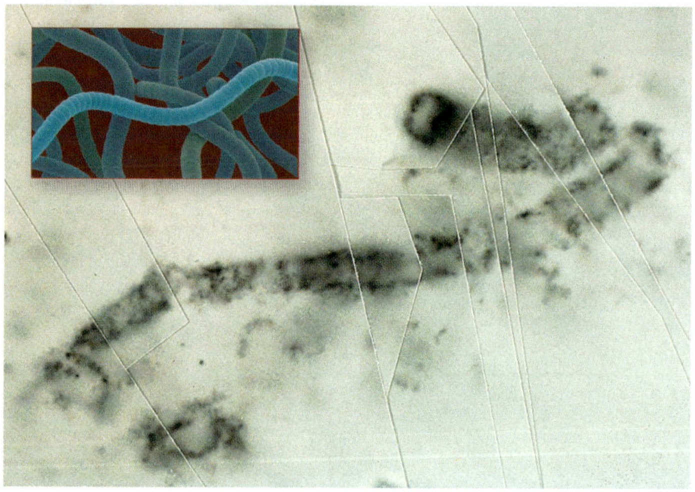

4.12 The Earliest Cells? This fossil from Western Australia is 3.5 billion years old. Its form is similar to that of modern filamentous cyanobacteria (inset).

The use of carbon dioxide in photosynthesis is a hallmark of life and leaves a unique chemical signature—a specific ratio of carbon isotopes (^{13}C:^{12}C) in the resulting carbohydrates. Schopf showed that the Australian material had this isotope signature. Furthermore, microscopic examination of the chains revealed *internal* substructures that are characteristic of living systems and were not likely to be the result of simple chemical reactions. Schopf's evidence suggests that the Australian sample is indeed the remains of a truly ancient living organism.

In 2011 a different team of scientists, working about 20 miles from Schopf's discovery, found similar-looking microfossil structures in sandstone rocks that were about 3.4 billion years old. In this case, a chemical analysis of the rocks indicated that these cells used sulfur instead of oxygen in the series of cellular reactions that release chemical energy.

Taking geological, chemical, and biological evidence into account, it is plausible that it took about 500 million to a billion years from the formation of Earth until the appearance of the first cells (**Figure 4.13**). Life has been cellular ever since. In the next chapter we will begin our study of cell structure and function.

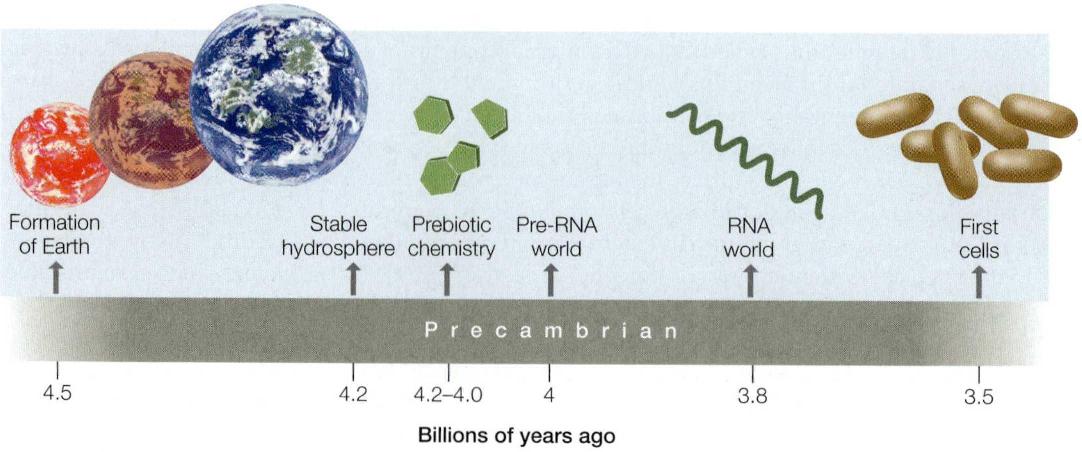
Formation of Earth
Stable hydrosphere
Prebiotic chemistry
Pre-RNA world
RNA world
First cells
Precambrian
4.5
4.2
4.2–4.0
4
3.8
3.5
Billions of years ago

4.13 The Origin of Life This highly simplified timeline gives a sense of the major events that culminated in the origin of life more than 3.5 billion years ago.

RECAP 4.4

The chemical reactions that preceded living organisms probably occurred in specialized compartments, such as water droplets on the surfaces of minerals. Life as we know it did not begin until the emergence of cells. Protocells made in the laboratory have some of the properties of modern cells. Cell-like structures fossilized in ancient rocks date the first cells to about 3.5 billion years ago.

- Explain the importance of the cell membrane to the evolution of living organisms. **See p. 73**
- What is the evidence that ancient rocks contain the fossils of cells? **See p. 74**

?

Can DNA analysis be used in the conservation and expansion of the cheetah population?

ANSWER

At the Cheetah Conservation Fund in Otjiwarongo, Namibia, a DNA sequencing laboratory analyzes populations of wild cheetahs that live in that area. The aim is to mate pairs of cheetahs that differ the most in their DNA sequences so that their offspring will have the greatest possible genetic diversity. Because of their genetic homogeneity, cheetah males produce a low amount of sperm, and the scientists in Namibia have developed artificial insemination methods to overcome this.

CHAPTER SUMMARY 4

4.1 What Are the Chemical Structures and Functions of Nucleic Acids?

- The unique functions of the **nucleic acids—DNA** and **RNA**—are information storage and transfer. DNA is the hereditary material that passes genetic information from one generation to the next, and RNA uses that information to specify the structures of proteins.
- Nucleic acids are polymers of nucleotides. A **nucleotide** consists of a phosphate group, a pentose sugar (**ribose** in RNA and **deoxyribose** in DNA), and a nitrogen-containing **base**. **Review Figure 4.1**
- In DNA, the nucleotide bases are **adenine** (**A**), **guanine** (**G**), **cytosine** (**C**), and **thymine** (**T**). **Uracil** (**U**) replaces thymine in RNA. C, T, and U have single-ring structures and are **pyrimidines**. A and G have double-ring structures and are **purines**.
- The nucleotides in DNA and RNA are joined by **phosphodiester linkages** involving the sugar of one nucleotide and the phosphate of the next, forming a nucleic acid polymer. **Review Figure 4.2, ACTIVITY 4.1**
- **Complementary base pairing** due to hydrogen bonds between A and T, A and U, and G and C occurs in nucleic acids. In RNA, the hydrogen bonds result in a folded molecule. In DNA, the hydrogen bonds connect two strands into a double helix. **Review Figures 4.3, 4.4, ACTIVITY 4.2**
- The information content of DNA and RNA resides in their base sequences.
- DNA is expressed as RNA in **transcription**. RNA can then specify the amino acid sequence of a protein in **translation**. **Review Figure 4.5**

4.2 How and Where Did the Small Molecules of Life Originate?

- Historically, many cultures believed that life originated repeatedly by **spontaneous generation**. This was disproven experimentally. **Review Figure 4.6, ANIMATED TUTORIAL 4.1**
- A prerequisite for life is the presence of water.
- Some meteorites that have landed on Earth contain organic molecules, suggesting that life might have originated extraterrestrially.
- An alternative hypothesis is **chemical evolution**: the idea that organic molecules were formed on Earth before life began.
- Chemical experiments modeling the prebiotic conditions on Earth support the idea of chemical evolution. **Review Figure 4.8, ANIMATED TUTORIAL 4.2**

4.3 How Did the Large Molecules of Life Originate?

- Chemical evolution may have led to the polymerization of small molecules into polymers. This may have occurred on the surfaces of clay particles, in hydrothermal vents, or in hot pools at the edges of oceans.
- A catalyst speeds up a chemical reaction. Today most catalysts are proteins, but some RNA molecules can function as both catalysts and information molecules. A catalytic RNA is called a **ribozyme**.
- The existence of ribozymes supports the idea of an "RNA world"—a world before DNA. On early Earth, RNA may have acted as a catalyst for its own replication as well as for the synthesis of proteins. DNA could eventually have evolved from RNA. **Review Figure 4.9**
- In support of the "RNA world" hypothesis, an artificial self-replicating ribozyme was developed in the laboratory. **Review Figure 4.10**

4.4 How Did the First Cells Originate?

- A key to the emergence of living cells was the prebiotic generation of compartments enclosed by **membranes**. Such enclosed compartments permitted the generation and maintenance of internal chemical conditions that were different from those in the exterior environment.
- In the laboratory, fatty acids assemble into **protocells** that have some of the characteristics of cells. **Review Figure 4.11**
- Ancient rocks (3.5 billion years old) have been found with imprints that are probably fossils of early cells.

Go to the Interactive Summary to review key figures, Animated Tutorials, and Activities Life10e.com/is4

CHAPTER**REVIEW**

REMEMBERING

1. A nucleotide in DNA is made up of
   a. four bases.
   b. a base plus a ribose sugar.
   c. a base plus a deoxyribose sugar plus a phosphate.
   d. a sugar plus a phosphate.
   e. a sugar and a base.

2. Nucleotides in RNA are connected to one another in the polynucleotide chain by
   a. covalent bonds between bases.
   b. covalent bonds between sugars.
   c. covalent bonds between sugar and phosphate.
   d. hydrogen bonds between purines.
   e. hydrogen bonds between any bases.

3. Which is a difference between DNA and RNA?
   a. DNA is single-stranded and RNA is double-stranded.
   b. DNA is only informational and RNA is only catalytic.
   c. DNA contains deoxyribose and RNA contains ribose.
   d. DNA is transcribed and RNA is replicated.
   e. DNA contains uracil (U) and RNA contains thymine (T).

4. The components in the atmosphere for the Miller–Urey experiment on prebiotic synthesis did not include
   a. H_2.
   b. H_2O.
   c. O_2.
   d. NH_3.
   e. CH_4.

5. The "RNA world" hypothesis proposes that
   a. RNA formed from DNA.
   b. RNA was both a catalyst and genetic material.
   c. RNA was a catalyst only.
   d. RNA formed after proteins.
   e. DNA formed after RNA was broken down.

6. Findings in ancient rocks indicate cells first appeared
   a. about 4.5 billion years ago.
   b. about 3.5 billion years ago.
   c. about 2 billion years ago.
   d. before rocks were formed.
   e. before water arrived on Earth.

UNDERSTANDING & APPLYING

7. What conditions existing on Earth today might preclude the origin of life from the prebiotic molecules Miller and Urey used?

8. Applied biologists are trying to develop reagents with specific three-dimensional shapes to bind to target molecules. An active research area in this regard is the development of oligonucleotides of RNA. What properties of oligonucleotide chains cause them to fold into a precise shape? How long do you think a chain would have to be to take a unique shape?

9. Why was the evolution of a self-contained cell essential for life as we know it?

ANALYZING & EVALUATING

10. The interpretation of Pasteur's experiment (see Figure 4.6) depended on the inactivation of microorganisms by heat. We now know of microorganisms that can survive extremely high temperatures (see Chapter 26). Does this change the interpretation of Pasteur's experiment? What experiments would you do to inactivate such microbes?

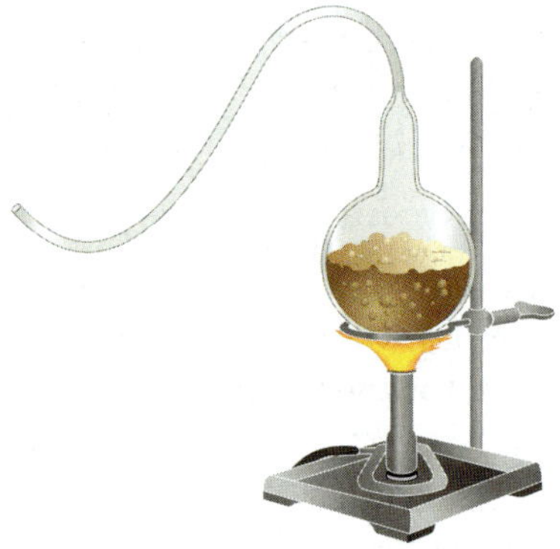

11. The Miller–Urey experiment (see Figure 4.8) showed that it was possible for amino acids to be formed from gases that were hypothesized to have been in Earth's early atmosphere. These amino acids were dissolved in water. Knowing what you do about the polymerization of amino acids into proteins (see Figure 3.6), how would you set up experiments to show that proteins can form under the conditions of early Earth?

Go to BioPortal at **yourBioPortal.com** for Animated Tutorials, Activities, LearningCurve Quizzes, Flashcards, and many other study and review resources.

5 Cells: The Working Units of Life

CHAPTER**OUTLINE**

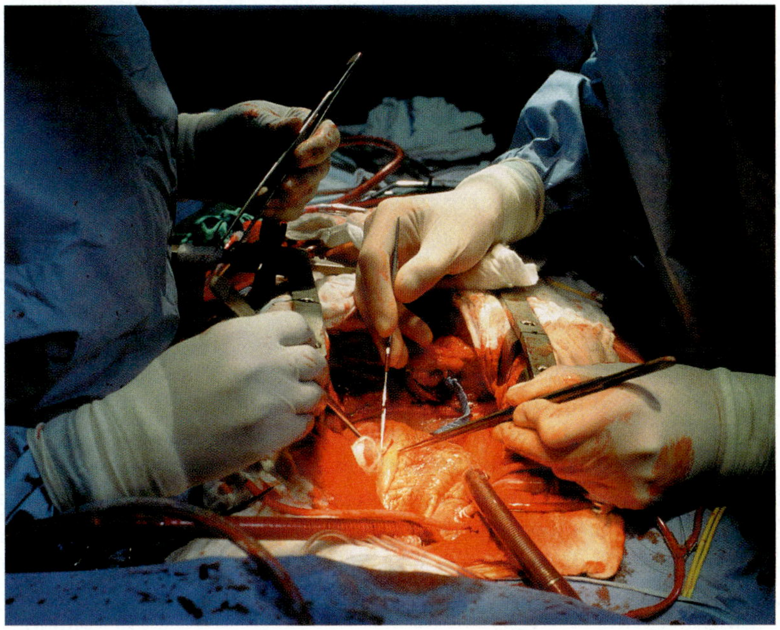

New Heart Tissue Cardiac problems are often treated with surgery. Stem cells that are coaxed to differentiate into heart cells may be used to repair a damaged heart.

IT IS A DAY in the not-too-distant future. Decades of eating fatty foods, combined with an inherited tendency to deposit cholesterol in his arteries, have finally caught up with 70-year-old Don. A blood clot has closed off blood flow to part of his heart, leading to a heart attack and severe damage to that vital organ.

If this had happened today, Don would have been faced with a long period of rehabilitation, taking medications to manage his weakened heart. Instead, his physicians take a pinch of skin tissue from his arm and bring it to a laboratory. After certain DNA sequences are added, Don's skin cells no longer look and act like skin cells: they are undifferentiated (unspecialized) and reproduce continuously in the laboratory dish. These cells are also multipotent stem cells, able to differentiate into almost any type of cell in the body if given the right environment. When they are injected directly into Don's heart, his stem cells soon become heart muscle cells, repairing the damage caused by the heart attack. Don leaves the hospital with full cardiac function and recommendations for a healthy diet.

The potential uses of stem cells in medicine have generated a lot of excitement in recent years. Such widely read periodicals as *Time* have hailed advances in stem cell research as "breakthroughs of the year." Patients with the neurological disorder Parkinson's disease dream of the day when their skin cells can be turned into brain cells to fix their damaged nervous systems. People with diabetes hope for stem cells to repair their pancreas. The list is long.

Behind all of this hope and the research it has inspired is a cornerstone of biological science: the cell theory. As you saw in the last chapter, a key event in the emergence of life was the enclosure of biochemical reactions inside cells, thus concentrating them and separating them from the external environment. The emergence of cells was essential for the evolution of life as we know it—and this is reflected in the first two basic tenets of cell theory: cells are the fundamental units of life; and all living organisms are composed of cells. Don's stem cells contain not just the activities of a living entity, but also the potential to change those activities in new directions. The third (and equally important) tenet of cell theory states that the cell is the unit of reproduction: all cells come from pre-existing cells. Stem cell therapy does not create new cells out of thin air; it coaxes existing ones to differentiate and reproduce along the desired path.

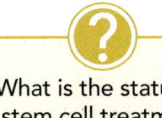

?

What is the status of stem cell treatment for heart disease?

See answer on p. 102.

5.1 What Features Make Cells the Fundamental Units of Life?

In Chapter 1 we introduced some of the characteristics of life: chemical complexity, growth and reproduction, the ability to refashion substances from the environment, and the ability to move specific substances into and out of the organism. These characteristics are all demonstrated by cells. Just as atoms are the building blocks of chemistry, cells are the building blocks of life.

The **cell theory** is an important unifying principle of biology. There are three critical components of the cell theory:

- Cells are the fundamental units of life.
- All living organisms are composed of cells.
- All cells come from preexisting cells.

To the original cell theory, first stated in 1838, should be added:

- Evolution through natural selection explains the diversity of modern cells.

Cells contain water and the other small and large molecules, which we examined in Chapters 2–4. Each cell contains at least 10,000 different types of molecules, most of them present in many copies. Cells use these molecules to transform matter and energy, to respond to their environments, and to reproduce themselves.

The cell theory has three important implications:

- Studying cell biology is in some sense the same as studying life. The principles that underlie the functions of the single cell of a bacterium are similar to those governing the approximately 60 trillion cells of your body.
- Life is continuous. All those cells in your body came from a single cell, a fertilized egg (zygote). That zygote came from the fusion of two cells, a sperm and an egg, from your parents. The cells of your parents' bodies were all derived from their parents, and so on back through generations and evolution to the first living cell.
- The origin of life on Earth was marked by the origin of the first cells (see Chapter 4).

Even the largest creatures on Earth are composed of cells, but the cells themselves are usually too small for the naked eye to see. Why are cells so small?

Cell size is limited by the surface area-to-volume ratio

Most cells are tiny. In 1665 Robert Hooke estimated that in one square inch of cork, which he examined under his magnifying lens, there were 1,259,712,000 cells! The diameters of cells range from about 1 to 100 micrometers (μm). There are some exceptions: the eggs of birds are single cells that are, relatively speaking, enormous, and individual cells of several types of algae and bacteria are large enough to be viewed with the unaided eye (**Figure 5.1**).

Small cell size is a practical necessity arising from the change in the **surface area-to-volume ratio** of any object as it increases in size. As an object increases in volume, its surface area also increases, but not at the same rate (**Figure 5.2**). This phenomenon has great biological significance. To appreciate this point,

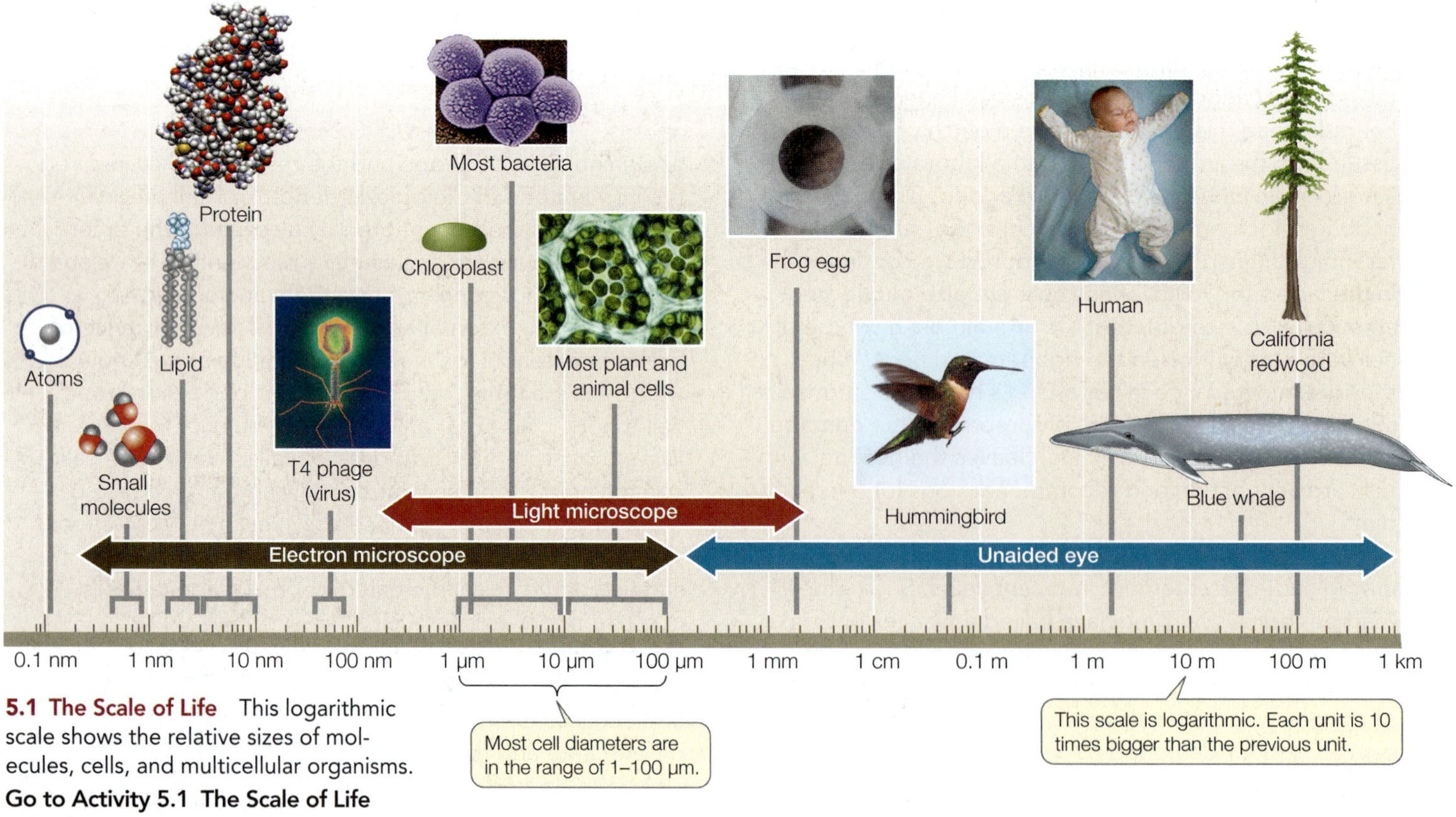
Protein
Most bacteria
Chloroplast
Lipid
Atoms
Small molecules
T4 phage (virus)
Most plant and animal cells
Frog egg
Human
California redwood
Hummingbird
Blue whale
Light microscope
Electron microscope
Unaided eye
0.1 nm
1 nm
10 nm
100 nm
1 μm
10 μm
100 μm
1 mm
1 cm
0.1 m
1 m
10 m
100 m
1 km
Most cell diameters are in the range of 1–100 μm.
This scale is logarithmic. Each unit is 10 times bigger than the previous unit.

5.1 The Scale of Life This logarithmic scale shows the relative sizes of molecules, cells, and multicellular organisms.

Go to Activity 5.1 The Scale of Life
Life10e.com/ac5.1

Diameter	1 μm	2 μm	3 μm
Surface area $4 \pi r^2$	3.14 μm^2	12.56 μm^2	28.26 μm^2
Volume $4/3 \pi r^3$	0.52 μm^3	4.19 μm^3	14.18 μm^3
Surface area-to-volume ratio	6:1	3:1	2:1

5.2 Why Cells Are Small As an object grows larger, its volume increases more rapidly than its surface area. Cells must maintain a large surface area-to-volume ratio in order to function. This explains why large organisms are composed of many small cells rather than a few huge ones.

let's assume that the amount of chemical activity carried out by a cell is proportional to its volume. The surface area of the cell determines the amount of substances that can enter it from the outside environment, and the amount of waste products that can exit to the environment.

As a living cell grows larger, its chemical activity, and thus its need for resources and its rate of waste production, increases faster than its surface area. (The surface area being two-dimensional, increases in proportion to the square of the radius, whereas the volume being three-dimensional, increases much more—in proportion to the cube of the radius.) In addition, substances must move from one site to another within the cell; the smaller the cell, the more easily this is accomplished. This explains why large organisms must consist of many small cells: cells must be small in volume in order to maintain a large enough surface area-to-volume ratio and an ideal internal volume. The large surface area represented by the many small cells of a multicellular organism enables it to carry out the many different functions required for survival.

Microscopes reveal the features of cells

Microscopes do two different things that allow cells and details within them to be seen by the human eye. They provide the ability to see great detail, which then allows the viewer to magnify the image of interest. The property that allows detail to be seen is called resolution. Formally defined, resolution is the minimum distance two objects can be apart and still be seen as two objects. Resolution for the human eye is about 0.2 mm (200 μm). Most cells are much smaller than 200 μm and thus are invisible to the human eye. Microscopes magnify and increase resolution so that cells and their internal structures can be seen clearly (**Figure 5.3**).

There are two basic types of microscopes—light microscopes and electron microscopes—that use different forms of radiation (see Figure 5.3). While the resolution is better in electron microscopy, only dead cells are visualized because they must be prepared in a vacuum. Light microscopes, by contrast, can be used to visualize living cells (for example, by phase-contrast microscopy; see Figure 5.3).

Before we delve into the details of cell structure, it is useful to consider the many uses of microscopy. An entire branch of medicine, pathology, makes use of many different methods of microscopy to aid in the analysis of cells and the diagnosis of diseases. For instance, a surgeon might remove from a body some tissue suspected of being cancerous. The pathologist might:

- examine the tissue quickly by phase-contrast microscopy or interference-contrast microscopy to determine the size, shape, and spread of the cells;
- stain the tissue with a general dye and examine it by bright-field microscopy to bring out features such as the shapes of the nuclei, or cell division characteristics;
- examine the tissue under the electron microscope to observe internal structures such as the mitochondria or the chromatin (these are described in Section 5.3);
- stain the tissue with a specific dye and examine it by microscopy for the presence of proteins that are diagnostic of particular cancers. The results can influence the choice of therapy.

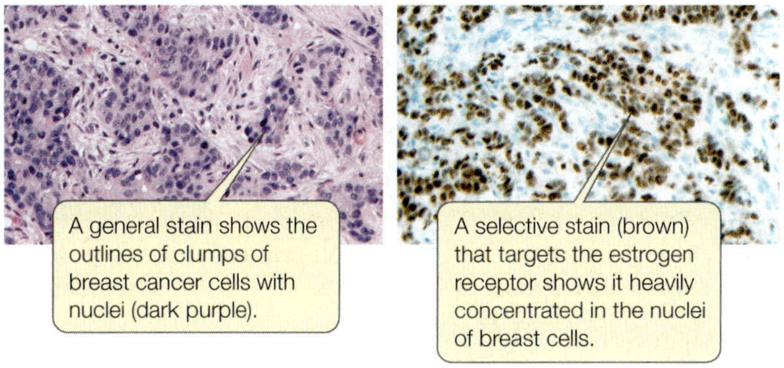
A general stain shows the outlines of clumps of breast cancer cells with nuclei (dark purple).
A selective stain (brown) that targets the estrogen receptor shows it heavily concentrated in the nuclei of breast cells.

The plasma membrane forms the outer surface of every cell

While the structural diversity of cells can often be observed using light microscopy, the **plasma membrane** is best observed with an electron microscope. This very thin structure forms the outer surface of every cell, and it has more or less the same thickness and molecular structure in all cells. We will describe the membrane in more detail in Chapter 6. For now we should keep in mind that it consists of a phospholipid bilayer (see Section 3.4), and that a variety of proteins are embedded within the bilayer. There is much compositional and functional diversity in the proteins associated with the plasma membrane. The membrane has several important roles:

- The plasma membrane acts as a selectively permeable barrier, preventing some substances from crossing it while permitting other substances to enter and leave the cell. For example, macromolecules such as DNA and proteins cannot normally cross the plasma membrane, but some smaller molecules such as oxygen can. In addition to size, other factors (particularly polarity) determine a molecule's ability to cross the plasma membrane. Because the membrane is composed mostly of hydrophobic fatty acids,

RESEARCH**TOOLS**

5.3 Looking at Cells The six images on this page show some techniques used in light microscopy. The three images on the following page were created using electron microscopes. All of these images are of a particular type of cultured cell known as HeLa cells. Note that the images in most cases are flat, two-dimensional views. As you look at images of cells, keep in mind that they are three-dimensional structures.

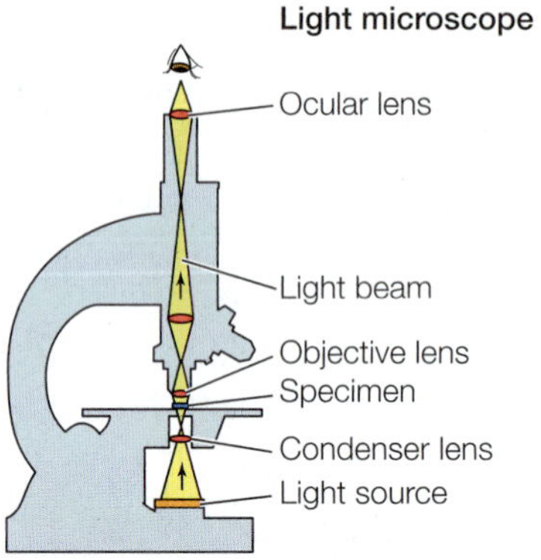
Light microscope
Ocular lens
Light beam
Objective lens
Specimen
Condenser lens
Light source

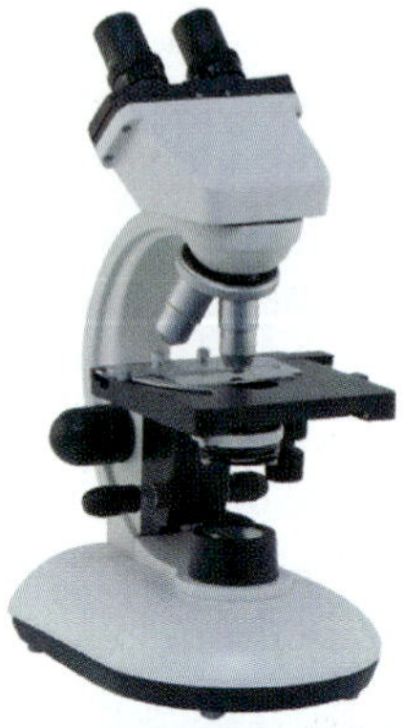

In a *light microscope*, glass lenses and visible light are used to form an image. The resolution is about 0.2 µm, which is 1,000 times greater than that of the human eye. Light microscopy allows visualization of cell sizes and shapes and some internal cell structures. Internal structures are hard to see under visible light, so cells are often chemically treated and stained with various dyes to make certain structures stand out by increasing contrast.

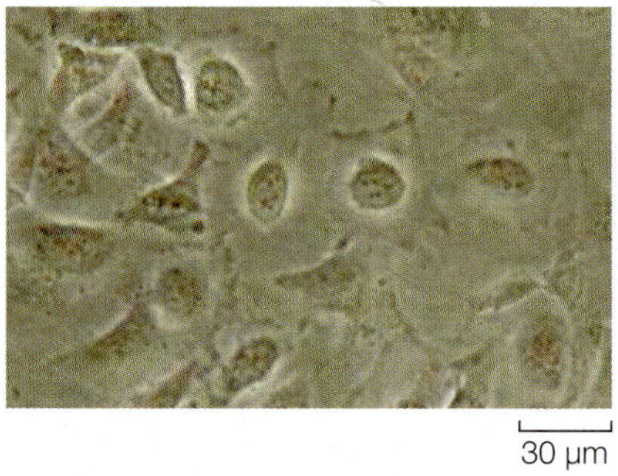
30 µm

In **bright-field microscopy**, light passes directly through these human cells. Unless natural pigments are present, there is little contrast and details are not distinguished.

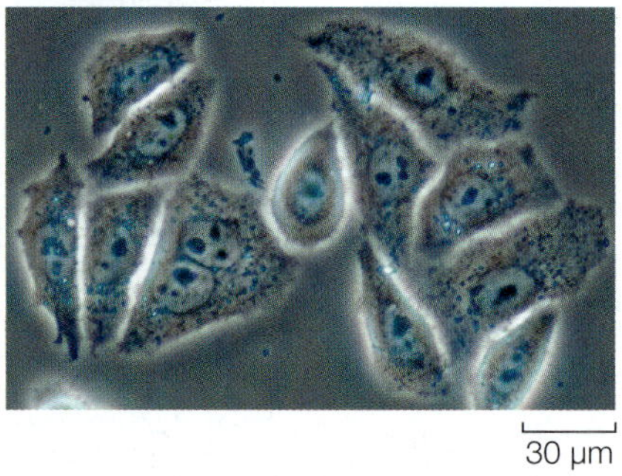
30 µm

In **phase-contrast microscopy**, contrast in the image is increased by emphasizing differences in refractive index (the capacity to bend light), thereby enhancing light and dark regions in the cell.

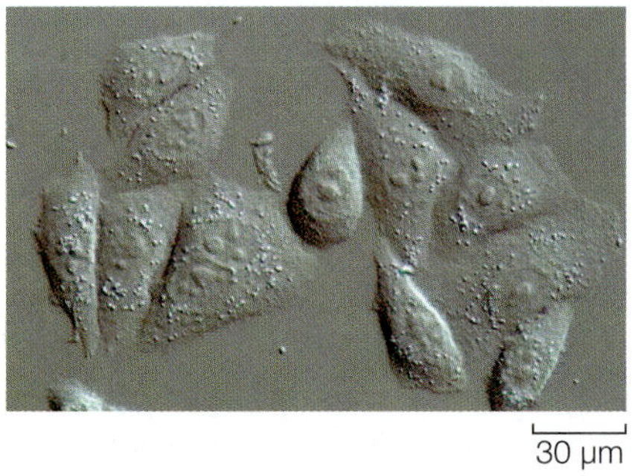
30 µm

Differential interference-contrast microscopy uses two beams of polarized light. The combined images look as if the cell is casting a shadow on one side.

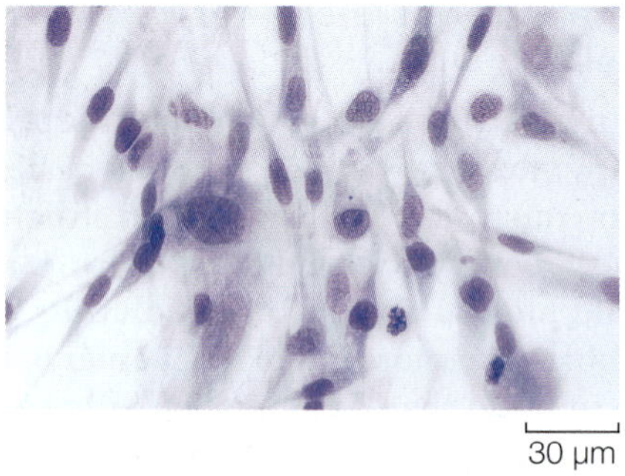
30 µm

In **stained bright-field microscopy**, a stain enhances contrast and reveals details not otherwise visible. Stains differ greatly in their chemistry and their capacity to bind to cell materials, so many choices are available.

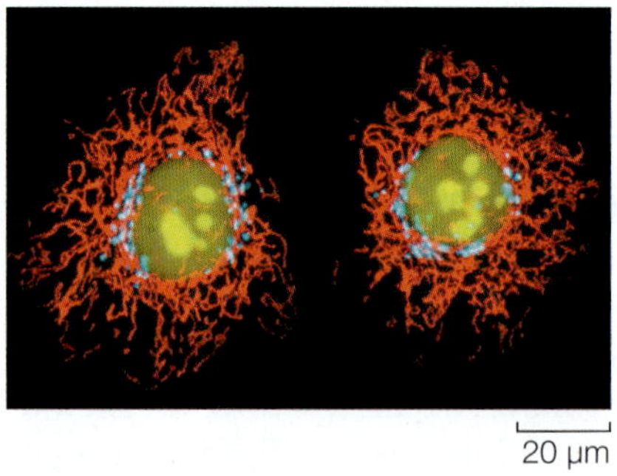
20 µm

In **fluorescence microscopy**, a natural substance in the cell or a fluorescent dye that binds to a specific cell material is stimulated by a beam of light, and the longer-wavelength fluorescent light is observed coming directly from the dye.

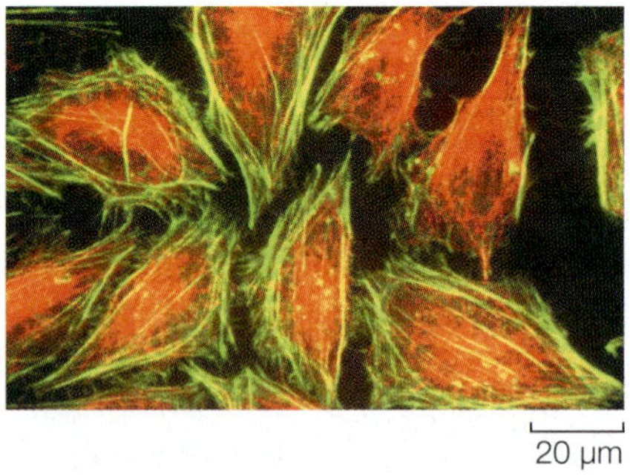
20 µm

Confocal microscopy uses fluorescent materials but adds a system of focusing both the stimulating and emitted light so that a single plane through the cell is seen. The result is a sharper two-dimensional image than with standard fluorescence microscopy.

nonpolar molecules cross it more easily than polar or charged molecules.

- The plasma membrane allows the cell to maintain a more or less constant internal environment. The maintenance of a constant internal environment (known as homeostasis) is a key characteristic of life and will be discussed in detail in Chapter 40. The membrane contributes to homeostasis by actively regulating the transport of substances across it. This dynamic process is distinct from the more passive process of diffusion, which is dependent on only the size of a molecule.

RESEARCH**TOOLS**

5.3 Looking at Cells *(continued)*

Go to Activity 5.2 Know Your Techniques Life10e.com/ac5.2

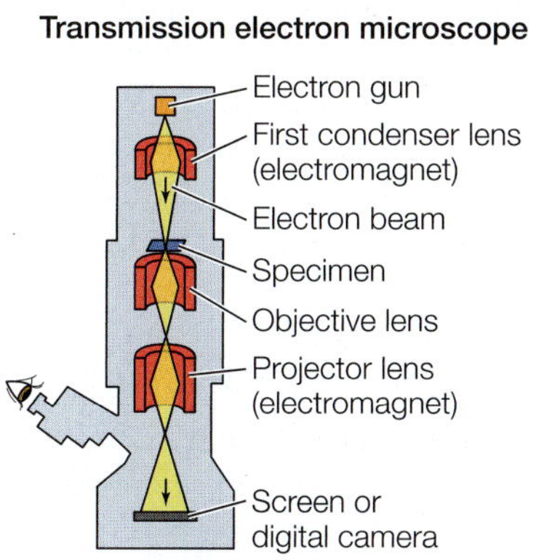
Transmission electron microscope
Electron gun
First condenser lens (electromagnet)
Electron beam
Specimen
Objective lens
Projector lens (electromagnet)
Screen or digital camera

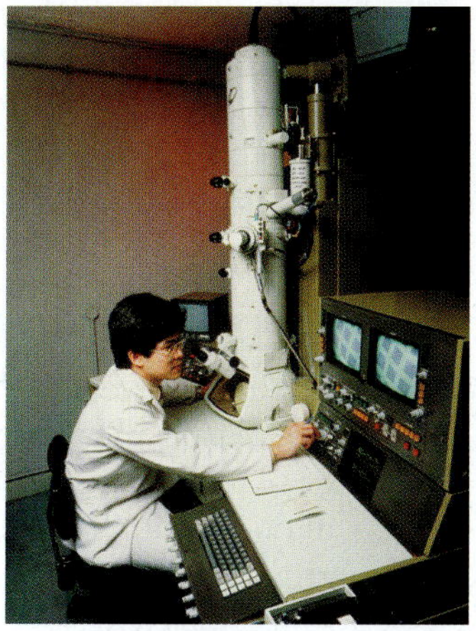

In an *electron microscope*, electromagnets are used to focus an electron beam, much as a light microscope uses glass lenses to focus a beam of light. Since we cannot see electrons, the electron microscope directs them through a vacuum at a fluorescent screen or digital camera to create a visible image. The resolution of electron microscopes is about 2 nm, which is about 100,000 times greater than that of the human eye. This resolution permits the details of many subcellular structures to be distinguished.

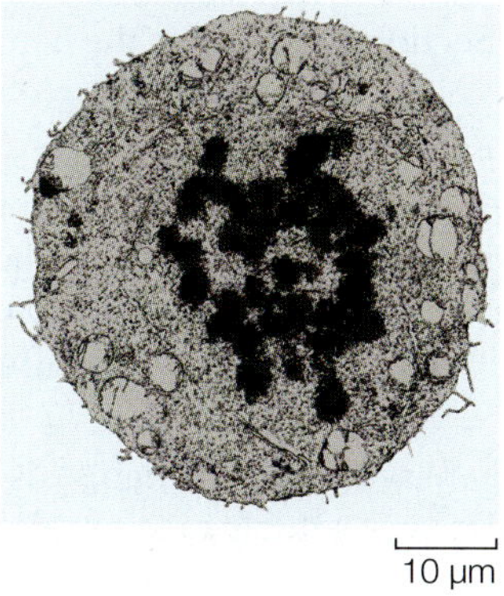
10 μm

In **transmission electron microscopy** (TEM), a beam of electrons is focused on the object by magnets. Objects appear darker if they absorb the electrons. If the electrons pass through they are detected on a fluorescent screen.

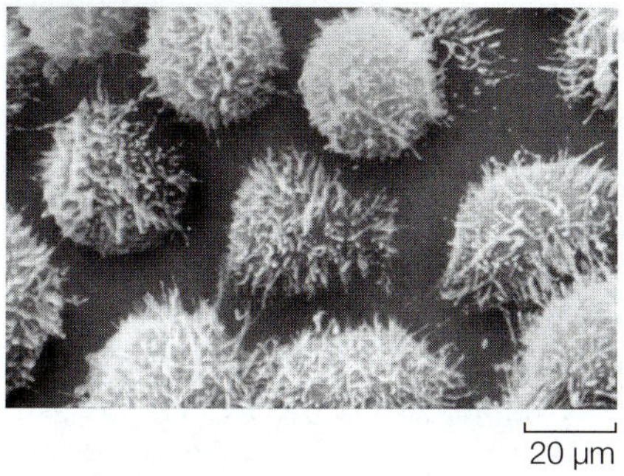
20 μm

Scanning electron microscopy (SEM) directs electrons to the surface of the sample, where they cause other electrons to be emitted. These electrons are viewed on a screen. The three-dimensional surface of the sample can be visualized.

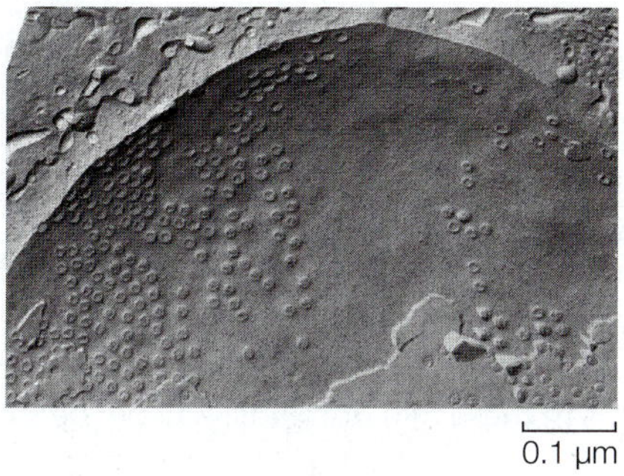
0.1 μm

In **freeze-fracture microscopy**, cells are frozen and then a knife is used to crack them open. The crack often passes through the interior of plasma and internal membranes. The "bumps" that appear are usually large proteins or aggregates embedded in the interior of the membrane.

- As the cell's boundary with the outside environment, the plasma membrane is important in communicating with adjacent cells and receiving signals from the environment. We will describe this function in Chapter 7.
- The plasma membrane often has proteins protruding from it that are responsible for binding and adhering to adjacent cells. Thus the plasma membrane plays an important structural role and contributes to cell shape.

Cells are classified as either prokaryotic or eukaryotic

As we learned in Section 1.1, biologists classify all living things into three domains: Archaea, Bacteria, and Eukarya. The organisms in Archaea and Bacteria are collectively called **prokaryotes**, and they have in common a prokaryotic cell organization. A prokaryotic cell does not typically have membrane-enclosed internal compartments; in particular, it does not have a nucleus. The first cells were probably similar in organization to those of modern prokaryotes.

Eukaryotic cell organization is found in members of the domain Eukarya (**eukaryotes**), which includes the protists, plants, fungi, and animals. In contrast to prokaryotic cells, eukaryotic cells contain membrane-enclosed compartments called **organelles**. The most notable organelle is the cell **nucleus**, where most of the cell's DNA is located and where gene expression begins:

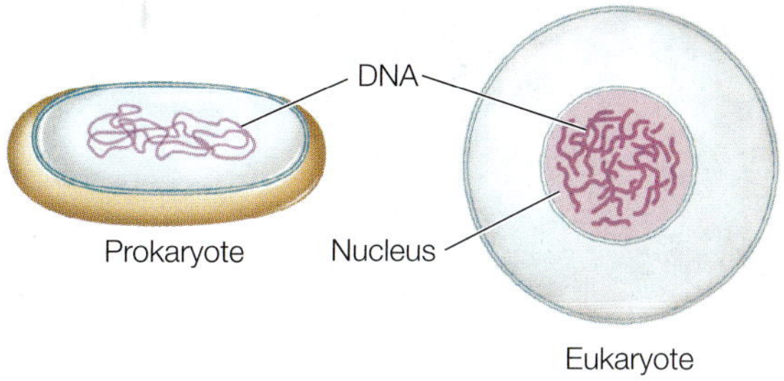
DNA
Prokaryote
Nucleus
Eukaryote

Just as a cell is an enclosed compartment, separating its contents from the surrounding environment, so each organelle provides a compartment that separates molecules and biochemical reactions from the rest of the cell. This "division of labor" provides possibilities for regulation and efficiency that were important in the evolution of complex organisms and helps to explain the complexity of eukaryotic cells relative to prokaryotic cells.

RECAP 5.1

The cell theory is a unifying principle of biology. Cell size is limited in order to maintain a high surface area-to-volume ratio. Both prokaryotic and eukaryotic cells are enclosed within a plasma membrane, but prokaryotic cells lack the membrane-enclosed organelles that are found in eukaryotic cells.

- How does cell biology embody all the principles of life? See p. 78
- Why are cells small? See pp. 78–79 and Figure 5.2
- Explain the importance of the plasma membrane and the membranes that surround organelles. See pp. 79–81

As we mentioned in this section, there are two structural themes in cell architecture: prokaryotic and eukaryotic. We will now turn to the organization of prokaryotic cells.

5.2 What Features Characterize Prokaryotic Cells?

In terms of sheer numbers, prokaryotes are the most successful organisms on Earth. As we examine prokaryotic cells in this section, bear in mind that there are vast numbers of prokaryotic species, and that the Bacteria and Archaea are distinguished in numerous ways. These differences, and the vast diversity of organisms in these two domains, will be the subject of Chapter 26.

Prokaryotic cells, with diameters or lengths in the range 1–10 μm (micrometers), are generally smaller than eukaryotic cells, whose diameters are usually in the range of 10–100 μm. Each individual prokaryote is a single cell, but many types of prokaryotes form chains or small clusters of cells, and some occur in large clusters containing hundreds of cells. In this section we will first consider the features shared by cells in the domains Bacteria and Archaea. Then we will examine structural features that are found in some, but not all, prokaryotes.

Prokaryotic cells share certain features

All prokaryotic cells have the same basic structure (Figure 5.4):

- The plasma membrane encloses the cell, regulating the traffic of materials into and out of the cell, and separating its interior from the external environment.
- The **nucleoid** is a region in the cell where the DNA is located. As we described in Section 4.1, DNA is the hereditary material that controls cell growth, maintenance, and reproduction.
- The rest of the material enclosed in the plasma membrane is called the **cytoplasm**. The cytoplasm consists of a liquid component, the cytosol, and a variety of insoluble filaments and particles, the most abundant of which are ribosomes (see below).
- The **cytosol** consists mostly of water that contains dissolved ions, small molecules, and soluble macromolecules such as proteins.

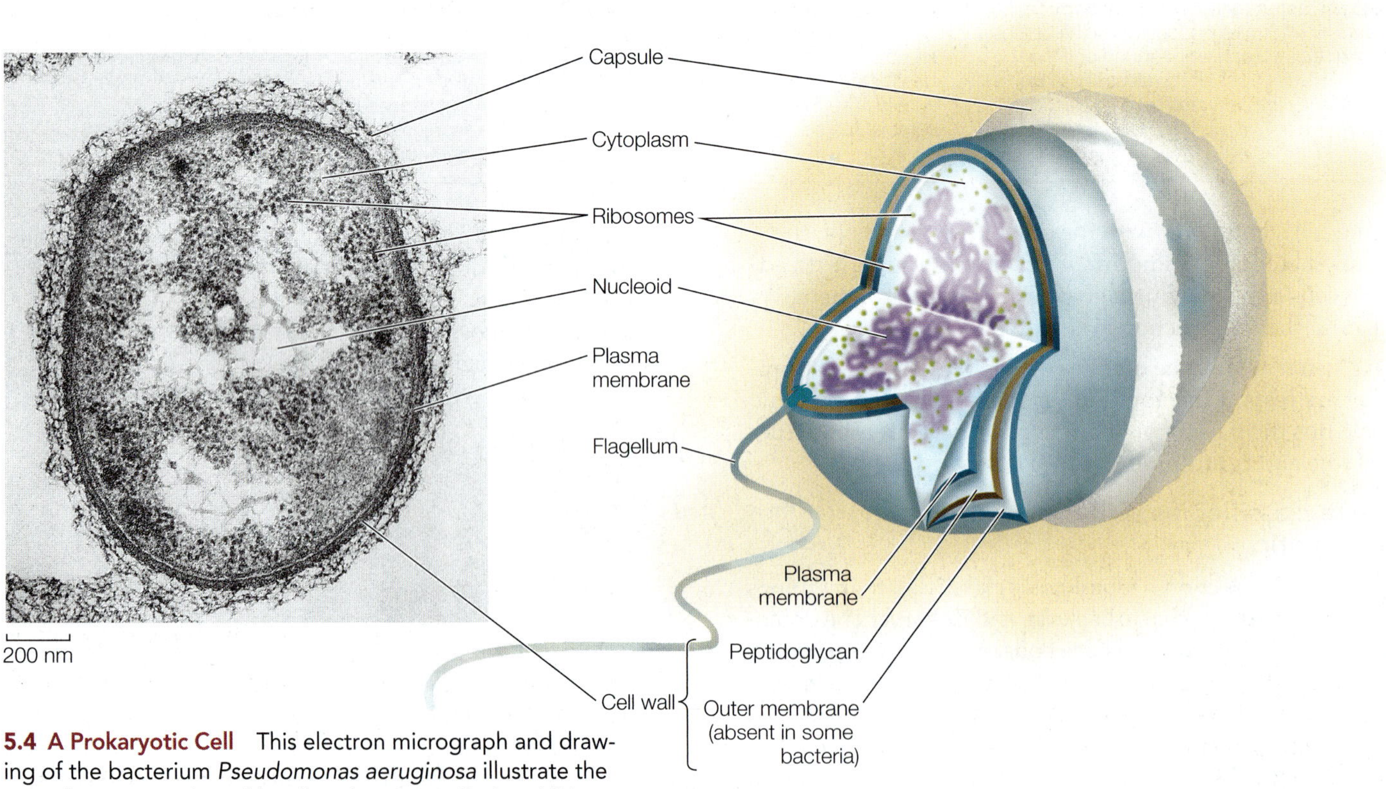

Capsule
Cytoplasm
Ribosomes
Nucleoid
Plasma membrane
Flagellum
Plasma membrane
Peptidoglycan
Cell wall
Outer membrane (absent in some bacteria)
200 nm

5.4 A Prokaryotic Cell This electron micrograph and drawing of the bacterium *Pseudomonas aeruginosa* illustrate the typical structures shared by all prokaryotic cells. In addition, this bacterium has a protective outer membrane that is not present in all prokaryotes. The flagellum and capsule are also structures found in some, but not all, prokaryotic cells.

- **Ribosomes** are complexes of RNA and proteins that are about 25 nm (nanometers) in diameter. They can only be visualized with the electron microscope. They are the sites of protein synthesis, where information coded for in nucleic acids directs the sequential linking of amino acids to form proteins.

The cytoplasm is not a static region. Rather, the substances in this environment are in constant motion. For example, a typical protein moves around the entire cell within a minute, and it collides with many other molecules along the way. This motion helps ensure that biochemical reactions proceed at rates sufficient to meet the needs of the cell. Although they are structurally less complex than eukaryotic cells, prokaryotic cells are functionally complex, carrying out thousands of biochemical reactions.

Specialized features are found in some prokaryotes

As they evolved, some prokaryotes developed specialized structures that gave a selective advantage to those that had them: cells with these structures were better able to survive and reproduce in particular environments than cells lacking them.

CELL WALLS Most prokaryotes have a **cell wall** located outside the plasma membrane. The rigidity of the cell wall supports the cell and determines its shape. The cell walls of most bacteria, but not archaea, contain peptidoglycan, a polymer of amino sugars that is linked at regular intervals to short peptides. Cross-linking among these peptides results in a single giant molecule around the entire cell. In some bacteria, another layer, the **outer membrane** (a polysaccharide-rich phospholipid membrane), encloses the peptidoglycan layer (see Figure 5.4). Unlike the plasma membrane, this outer membrane is not a major barrier to the movement of molecules across it.

Enclosing the cell wall in some bacteria is a slimy layer composed mostly of polysaccharides and referred to as a **capsule**. In some cases these capsules protect the bacteria from attack by white blood cells in the animals they infect. Capsules also help keep the cells from drying out, and sometimes they help bacteria attach to other cells. Many prokaryotes produce no capsule, and those that do have capsules can survive even if they lose them, so the capsule is not essential to prokaryotic life.

INTERNAL MEMBRANES Some groups of bacteria—including the cyanobacteria—carry out photosynthesis: they use energy from the sun to convert carbon dioxide and water into carbohydrates. These bacteria have an **internal membrane** system that contains molecules needed for photosynthesis. The development of photosynthesis, which requires membranes, was an important event in the early evolution of life on Earth. Other prokaryotes have internal membrane folds that are attached to the plasma membrane. These folds may function in cell division or in various energy-releasing reactions.

FLAGELLA AND PILI Some prokaryotes swim by using appendages called **flagella**, which sometimes look like tiny corkscrews (**Figure 5.5A**). In bacteria, the filament of the flagellum is made of a protein called flagellin. A complex motor protein spins the flagellum on its axis like a propeller, driving the cell along. The motor protein is anchored to the plasma membrane and,

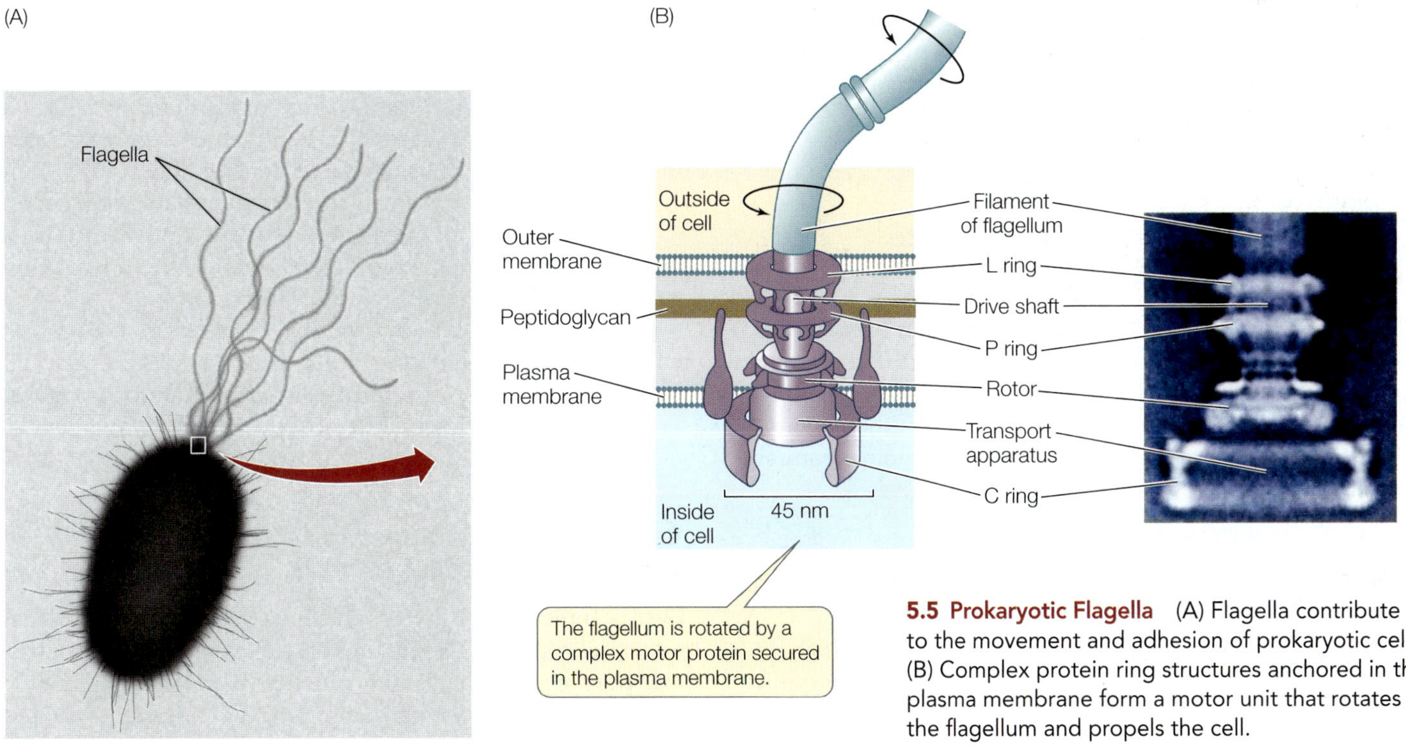

(A)
Flagella
(B)
Outside of cell
Outer membrane
Peptidoglycan
Plasma membrane
Inside of cell
45 nm
Filament of flagellum
L ring
Drive shaft
P ring
Rotor
Transport apparatus
C ring
The flagellum is rotated by a complex motor protein secured in the plasma membrane.

5.5 Prokaryotic Flagella (A) Flagella contribute to the movement and adhesion of prokaryotic cells. (B) Complex protein ring structures anchored in the plasma membrane form a motor unit that rotates the flagellum and propels the cell.

in some bacteria, to the outer membrane of the cell wall (**Figure 5.5B**). We know that the flagella cause the motion of cells because if they are removed, the cells do not move.

Pili are structures made of protein that project from the surfaces of some types of bacterial cells. These hairlike structures are shorter than flagella and are used for adherence. Conjugative pili (sex pili) help bacteria join to one another to exchange genetic material. Fimbriae are composed of the same proteins as pili but are shorter, and help cells adhere to surfaces such as animal cells, for food and protection.

CYTOSKELETON The **cytoskeleton** is the collective name for protein filaments that play roles in cell division or in maintaining the shapes of cells. One such protein forms a ring structure that constricts during cell division, whereas another forms helical structures that extend down the lengths of rod-shaped cells, helping maintain their shapes. In the past it was thought that only eukaryotic cells had cytoskeletons (see Section 5.3), but more recently, biologists have recognized that cytoskeletal components are also widely distributed among prokaryotes.

RECAP 5.2

Prokaryotic cells share basic features, including the plasma membrane, the nucleoid, and the cytoplasm, which consists of the liquid cytosol and insoluble filaments and particles, including ribosomes. Other features, such as cell walls, internal membranes, and flagella, are present in some but not all prokaryotes.

- What structures are present in all prokaryotic cells? **See pp. 82–83 and Figure 5.4**
- Describe the structure and function of a specialized prokaryotic cell feature, such as the cell wall, capsule, flagellum, or pilus. **See pp. 83–84 and Figure 5.5**

As we mentioned earlier, the prokaryotic cell is one of two types of cell recognized in cell biology. The other is the eukaryotic cell. Eukaryotic cells are more structurally and functionally complex than prokaryotic cells.

5.3 What Features Characterize Eukaryotic Cells?

In the opening story of this chapter we saw that human cells arise by the differentiation of stem cells. This differentiation results in hundreds of different cell types, all with specialized functions in the human body. But these and all other eukaryotic cells share many features, and it is these common features that we will discuss in this section.

Go to Animated Tutorial 5.1
Eukaryotic Cell Tour
Life10e.com/at5.1

Eukaryotic cells generally have lengths or diameters about ten times greater than those of prokaryotes. Like prokaryotic cells, eukaryotic cells have a plasma membrane, cytoplasm, and ribosomes. But as you learned earlier in this chapter, eukaryotic cells also have compartments within the cytoplasm whose interiors are separated from the cytosol by membranes.

Compartmentalization is the key to eukaryotic cell function

The membranous compartments of eukaryotic cells are called organelles. Each type of organelle has a specific role: some organelles have been characterized as factories that make specific products, whereas others are like power plants that take in energy in one form and convert it to a more useful form. These functional roles are defined by the chemical reactions that occur within the organelles. Eukaryotic cells also have some structures that are analogous to those in prokaryotes. For example, they have a cytoskeleton composed of protein fibers and, outside the cell membrane, an extracellular matrix.

Animal and plant cells have many organelles and structures in common—the most obvious is the cell nucleus. But they also have some differences. For example, many plant cells have chloroplasts that perform photosynthesis.

Organelles can be studied by microscopy or isolated for chemical analysis

Cell organelles and structures were first detected by light and then by electron microscopy. The functions of the organelles could sometimes be inferred by observations and experiments, leading, for example, to the hypothesis (later confirmed) that the nucleus contained the genetic material. Later, the use of stains targeted to specific macromolecules allowed cell biologists to determine the chemical compositions of organelles (see Figure 5.14, which shows three different cytoskeletal proteins in a single cell).

Another way to analyze cells is to take them apart in a process called cell fractionation. This process permits cell organelles and other cytoplasmic structures to be separated from each other and examined using chemical methods. Cell fractionation begins with the destruction of the plasma membrane, which allows the cytoplasmic components to flow out into a test tube. The various organelles can then be separated from one another on the basis of size or density (**Figure 5.6**). Biochemical analyses can then be done on the isolated organelles.

Microscopy and cell fractionation have complemented each other, giving us a more complete picture of the composition and function of each organelle and structure.

Microscopy of plant and animal cells has revealed that many of the organelles are similar in appearance in each cell type (**Figure 5.7**). By comparing Figures 5.7 and 5.4 you can see some of the prominent differences between eukaryotic cells and prokaryotic cells.

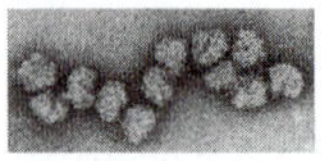

Ribosomes are factories for protein synthesis

The ribosomes of prokaryotes and eukaryotes are similar in that both types consist of two different-sized subunits. Eukaryotic ribosomes are somewhat larger than those of prokaryotes, but the structure of prokaryotic ribosomes is better understood. Chemically, ribosomes consist of a special type of RNA called ribosomal

RESEARCH**TOOLS**

5.6 Cell Fractionation Organelles can be separated from one another after cells are broken open and their contents suspended in an aqueous medium. The medium is placed in a tube and spun in a centrifuge, which rotates about an axis at high speed. Centrifugal forces (measured in multiples of gravity, × *g*) cause particles to sediment (form a pellet) at the bottom of the tube, which may be collected for biochemical study. Heavier particles sediment at lower speeds (lower centrifugal forces) than lighter particles. By adjusting the speed of centrifugation, researchers can separate and partially purify cellular organelles and large particles such as ribosomes.

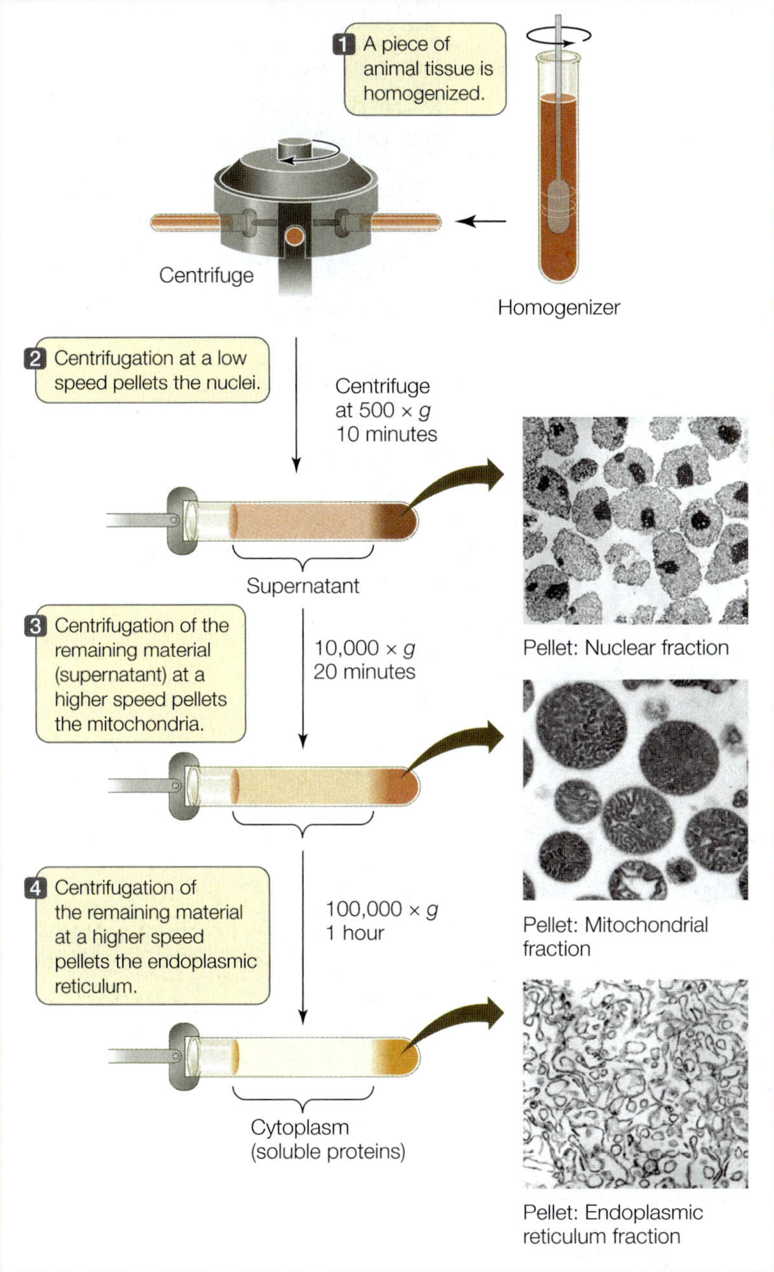

1 A piece of animal tissue is homogenized.
Centrifuge
Homogenizer
2 Centrifugation at a low speed pellets the nuclei.
Centrifuge at 500 × g 10 minutes
Supernatant
Pellet: Nuclear fraction
3 Centrifugation of the remaining material (supernatant) at a higher speed pellets the mitochondria.
10,000 × g 20 minutes
Pellet: Mitochondrial fraction
4 Centrifugation of the remaining material at a higher speed pellets the endoplasmic reticulum.
100,000 × g 1 hour
Cytoplasm (soluble proteins)
Pellet: Endoplasmic reticulum fraction

RNA (rRNA). Ribosomes also contain more than 50 different protein molecules, which are noncovalently bound to the rRNA.

In prokaryotic cells, ribosomes generally float freely in the cytoplasm. In eukaryotic cells they are found in multiple places: in the cytoplasm, where they may be free or attached to the surface of the endoplasmic reticulum (a membrane-bound organelle; see below), inside the mitochondria, and inside the chloroplasts in plant cells. In each of these locations, the ribosomes are molecular factories where proteins are synthesized. Although they seem small in comparison with the cells that contain them, by molecular standards ribosomes are huge complexes (about 25 nm in diameter), made up of several dozen different molecules.

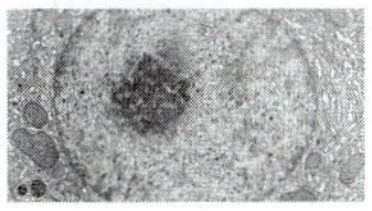

The nucleus contains most of the genetic information

As we discussed in Chapter 4, hereditary information is stored in the sequence of nucleotides in DNA molecules. Most of the DNA in eukaryotic cells resides in the nucleus (see Figure 5.7). Information encoded in the DNA is translated into proteins at the ribosomes. This process is described in detail in Chapter 14.

Most cells have a single nucleus, which is usually the largest organelle. The nucleus of a typical animal cell is approximately 5 μm (micrometers) in diameter—substantially larger than many prokaryotic cells (**Figure 5.8A**). The nucleus has several functions in the cell:

- It is the location of most of the cell's DNA and the site of DNA replication.
- It is the site where gene transcription is turned on or off.
- A region within the nucleus, the **nucleolus**, is where ribosomes begin to be assembled from RNA and proteins.

The contents of the nucleus, aside from the nucleolus, are referred to as the nucleoplasm. Similar to the cytoplasm, the nucleoplasm consists of the liquid content of the nucleus and the insoluble molecules suspended within it.

The nucleus is surrounded by an integrated structure comprised of two membranes, called the **nuclear envelope**. This structure separates the genetic material from the cytoplasm. Functionally, it separates DNA transcription (which occurs in the nucleus) from translation (which occurs in the cytoplasm) (see Section 4.1). The two membranes of the nuclear envelope are perforated by thousands of nuclear pores, each measuring approximately 9 nm in diameter, which connect the nucleoplasm with the cytoplasm.

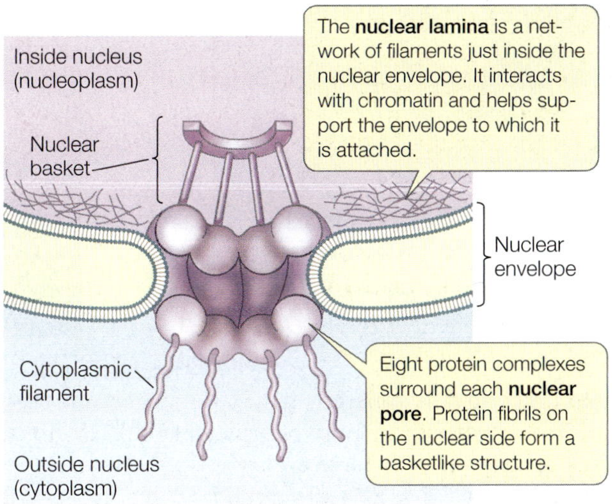

Inside nucleus (nucleoplasm)
The nuclear lamina is a network of filaments just inside the nuclear envelope. It interacts with chromatin and helps support the envelope to which it is attached.
Nuclear basket
Nuclear envelope
Cytoplasmic filament
Eight protein complexes surround each nuclear pore. Protein fibrils on the nuclear side form a basketlike structure.
Outside nucleus (cytoplasm)

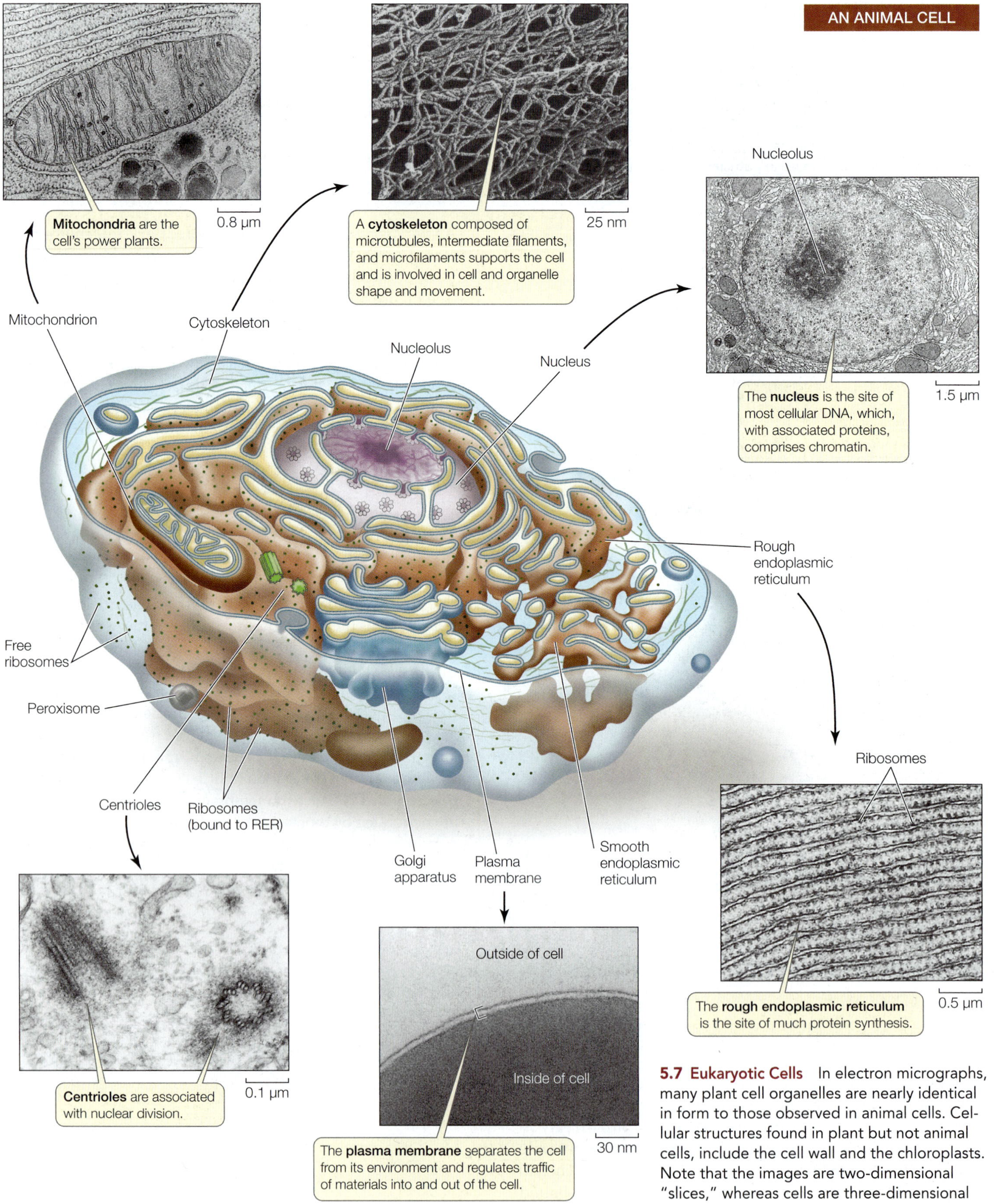
AN ANIMAL CELL
Mitochondria are the cell's power plants.
0.8 μm
A cytoskeleton composed of microtubules, intermediate filaments, and microfilaments supports the cell and is involved in cell and organelle shape and movement.
25 nm
Nucleolus
The nucleus is the site of most cellular DNA, which, with associated proteins, comprises chromatin.
1.5 μm
Mitochondrion
Cytoskeleton
Nucleolus
Nucleus
Rough endoplasmic reticulum
Free ribosomes
Peroxisome
Centrioles
Ribosomes (bound to RER)
Golgi apparatus
Plasma membrane
Smooth endoplasmic reticulum
Ribosomes
Centrioles are associated with nuclear division.
0.1 μm
Outside of cell
Inside of cell
The plasma membrane separates the cell from its environment and regulates traffic of materials into and out of the cell.
30 nm
The rough endoplasmic reticulum is the site of much protein synthesis.
0.5 μm

5.7 Eukaryotic Cells In electron micrographs, many plant cell organelles are nearly identical in form to those observed in animal cells. Cellular structures found in plant but not animal cells, include the cell wall and the chloroplasts. Note that the images are two-dimensional "slices," whereas cells are three-dimensional structures.

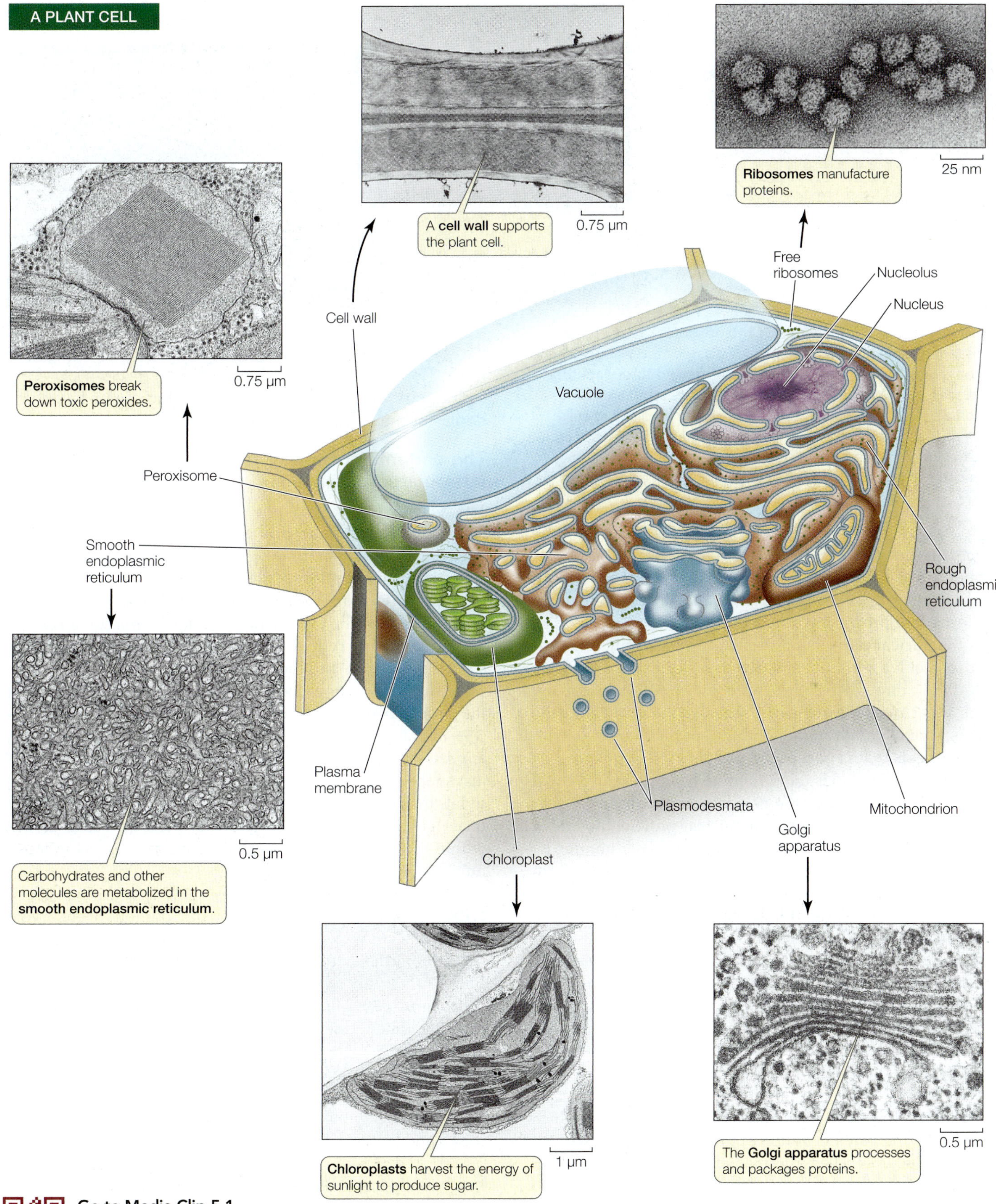
A PLANT CELL
A cell wall supports the plant cell.
0.75 µm
Ribosomes manufacture proteins.
25 nm
Peroxisomes break down toxic peroxides.
0.75 µm
Cell wall
Free ribosomes
Nucleolus
Nucleus
Vacuole
Peroxisome
Smooth endoplasmic reticulum
Rough endoplasmic reticulum
Plasma membrane
Plasmodesmata
Mitochondrion
Golgi apparatus
Chloroplast
0.5 µm
Carbohydrates and other molecules are metabolized in the smooth endoplasmic reticulum.
1 µm
Chloroplasts harvest the energy of sunlight to produce sugar.
0.5 µm
The Golgi apparatus processes and packages proteins.

Go to Media Clip 5.1
The Inner Life of a Cell
Life10e.com/mc5.1

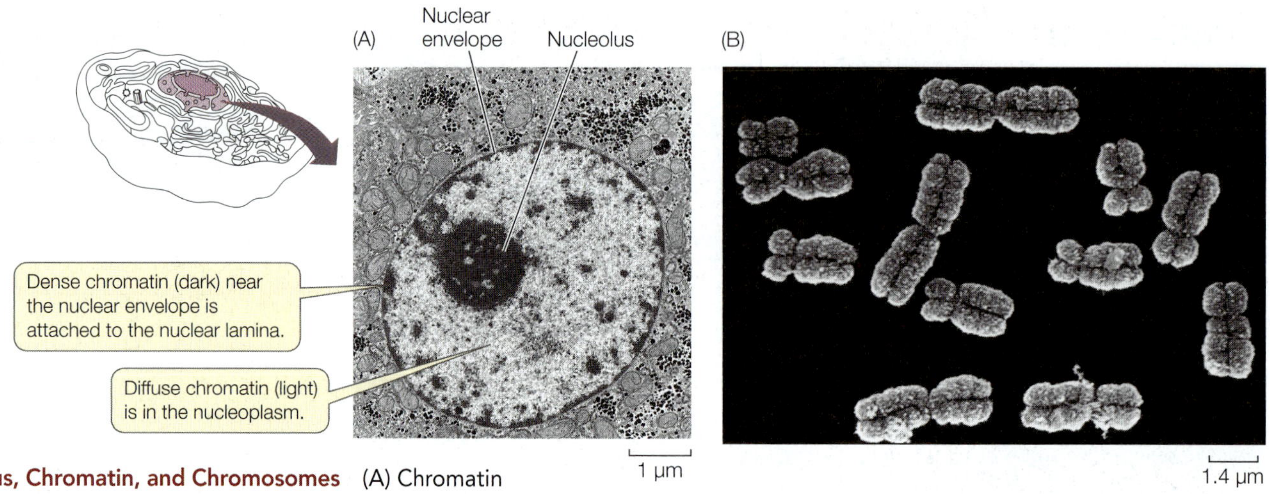

(A)
Nuclear envelope
Nucleolus
Dense chromatin (dark) near the nuclear envelope is attached to the nuclear lamina.
Diffuse chromatin (light) is in the nucleoplasm.
1 μm
(B)
1.4 μm

5.8 The Nucleus, Chromatin, and Chromosomes (A) Chromatin consists of nuclear DNA and the proteins associated with it. When the cell is not dividing, the chromatin is dispersed throughout the nucleus. This two-dimensional image was made using a transmission electron microscope. (B) The chromatin in dividing cells becomes highly condensed so that the individual chromosomes can be seen. This three-dimensional image of isolated metaphase chromosomes was produced using a scanning electron microscope.

The pores regulate the traffic between these two cellular compartments by allowing some molecules to enter or exit the nucleus and blocking others. This allows the nucleus to regulate its information-processing functions.

At the nuclear pore, small substances, including ions and other molecules with molecular weights of less than 10,000, freely diffuse through the pore. Larger molecules, such as many proteins that are made in the cytoplasm and imported into the nucleus, cannot get through without a specific short sequence of amino acids that is part of the protein. This sequence acts as a signal that identifies the protein to be imported. We will describe this sequence and the evidence for its role in Chapter 14 (see Figure 14.19). For a typical nuclear protein, the rate of import into the nucleus is about 100 molecules per minute.

Inside the nucleus, DNA is combined with proteins to form a fibrous complex called **chromatin**. Chromatin occurs in the form of exceedingly long, thin threads called **chromosomes**. Different eukaryotic organisms have different numbers of chromosomes (ranging from two in one kind of Australian ant to hundreds in some plants). Prior to cell division, the chromatin becomes tightly compacted and condensed so that the individual chromosomes are visible under a light microscope. This facilitates distribution of the DNA during cell division (Figure 5.8B).

At the interior periphery of the nucleus, the chromatin is attached to a protein meshwork, called the nuclear lamina, which is formed by the polymerization of proteins called lamins into long thin structures called intermediate filaments. The nuclear lamina maintains the shape of the nucleus by its attachment to both the chromatin and the nuclear envelope.

At the exterior of the nucleus, the outer membrane of the nuclear envelope folds outward into the cytoplasm and is continuous with the membrane of another organelle, the endoplasmic reticulum, which we will discuss next.

The endomembrane system is a group of interrelated organelles

Much of the volume of some eukaryotic cells is taken up by an extensive **endomembrane system**. This is an interconnected system of membrane-enclosed compartments that are sometimes flattened into sheets and sometimes have other characteristic shapes (see Figure 5.7). The endomembrane system includes the plasma membrane, the nuclear envelope, the endoplasmic reticulum, the Golgi apparatus, and lysosomes, which are derived from the Golgi. Tiny, membrane-surrounded droplets called vesicles shuttle substances between the various components of the endomembrane system (Figure 5.9). In drawings and electron microscope pictures, this system appears static, fixed in space and time. But these depictions are just snapshots; in the living cell, the membranes and the materials they contain are in constant motion. Membrane components have been observed to shift from one organelle to another within the endomembrane system. Thus all these membranes must be functionally related.

ENDOPLASMIC RETICULUM Electron micrographs of eukaryotic cells reveal networks of interconnected membranes branching throughout the cytoplasm, forming tubes and flattened sacs. These membranes are collectively called the **endoplasmic reticulum**, or **ER**. The interior compartment of the ER, referred to as the lumen, is separate and distinct from the surrounding cytoplasm (see Figure 5.9). The ER can enclose up to 10 percent of the interior volume of the cell, and its folds result in a surface area many times greater than that of the plasma membrane. There are two types of endoplasmic reticulum, the so-called rough and smooth.

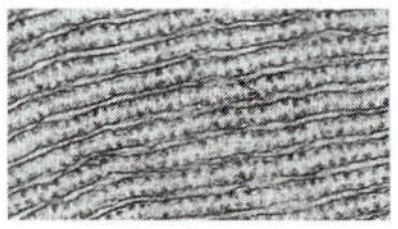

Rough endoplasmic reticulum (**RER**) is called "rough" because of the many ribosomes attached to the outer surface of the membrane, giving it a "rough" appearance

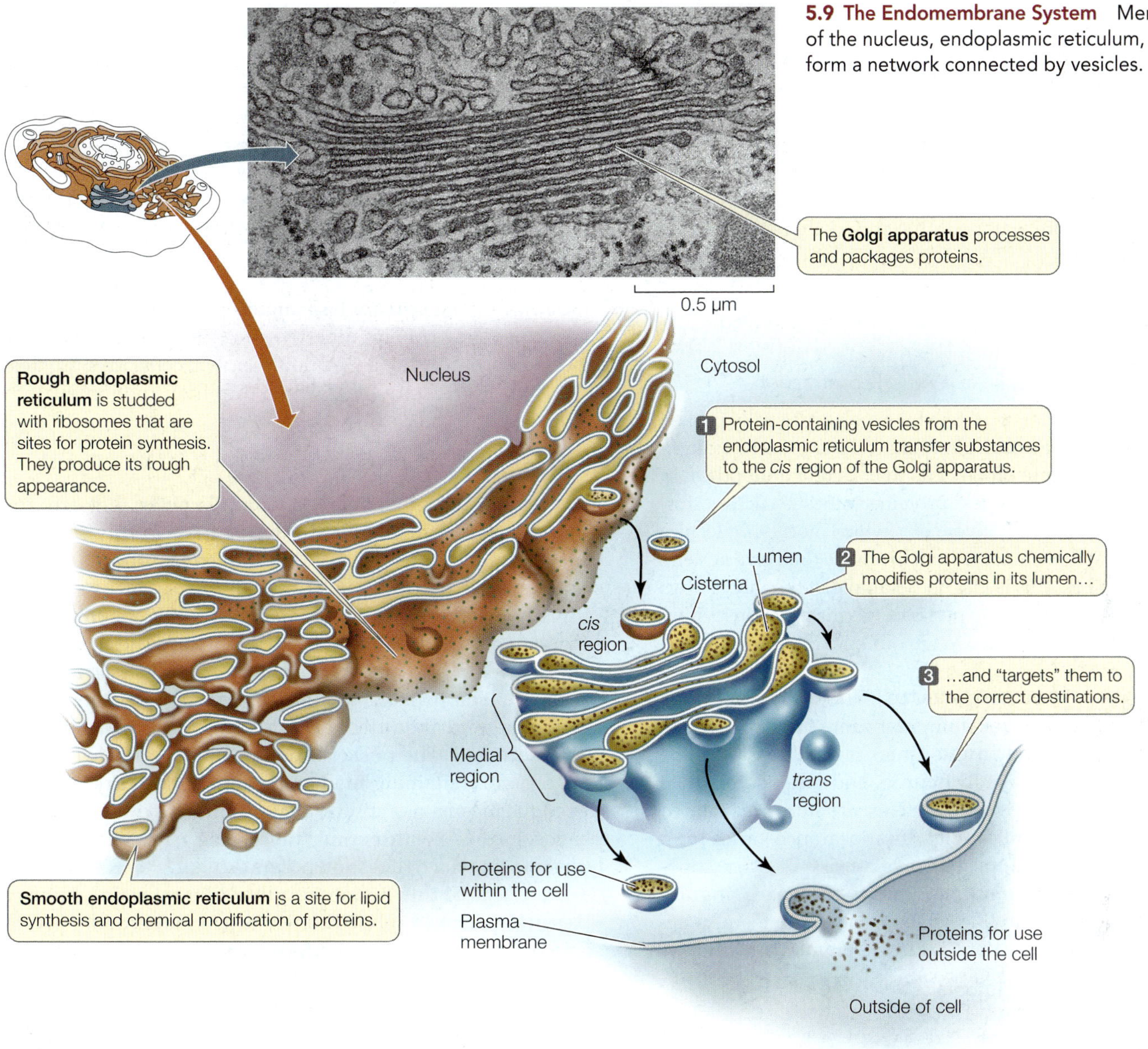

The Golgi apparatus processes and packages proteins.
0.5 µm
Rough endoplasmic reticulum is studded with ribosomes that are sites for protein synthesis. They produce its rough appearance.
Nucleus
Cytosol
1 Protein-containing vesicles from the endoplasmic reticulum transfer substances to the cis region of the Golgi apparatus.
Lumen
Cisterna
2 The Golgi apparatus chemically modifies proteins in its lumen…
cis region
3 …and "targets" them to the correct destinations.
Medial region
trans region
Proteins for use within the cell
Plasma membrane
Proteins for use outside the cell
Outside of cell
Smooth endoplasmic reticulum is a site for lipid synthesis and chemical modification of proteins.

5.9 The Endomembrane System Membranes of the nucleus, endoplasmic reticulum, and Golgi form a network connected by vesicles.

in electron microscopy (see Figure 5.7). The bound ribosomes are actively involved in protein synthesis, but that is not the entire story:

- The RER receives into its lumen certain newly synthesized proteins (including exported proteins and those destined for lysosomes and the plasma membrane), segregating them away from the cytoplasm. The RER also participates in transporting these proteins to other locations in the cell.
- While inside the RER, proteins can be chemically modified to alter their functions and to "tag" them for delivery to specific cellular destinations.
- Proteins are shipped to destinations elsewhere in the cell enclosed within vesicles that pinch off from the RER.
- Most membrane-bound proteins are made in the RER.

A protein enters the lumen of the RER through a pore as it is synthesized. As with a protein passing through a nuclear pore, this is accomplished via a sequence of amino acids on the protein, which acts as a RER localization signal ("address"; see Section 14.6). Once in the lumen of the RER, proteins undergo several changes, including the formation of disulfide bridges and folding into their tertiary structures (see Figure 3.7).

Some proteins are covalently linked to carbohydrate groups in the RER, thus becoming glycoproteins. In the case of proteins directed to the lysosomes, the carbohydrate groups are part of an "addressing" system that ensures that the right proteins are directed to those organelles. This addressing system is very important because the enzymes within the lysosomes are some of the most destructive the cell makes. Were they not properly addressed and contained, they could destroy the cell.

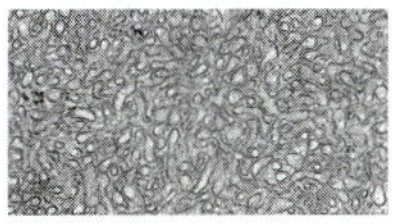

The **smooth endoplasmic reticulum (SER)** lacks ribosomes and is more tubular (and less like flattened sacs) than the RER, but it shows continuity with portions of the RER (see Figure 5.9). Certain proteins that are synthesized in the RER are chemically modified within the lumen of the SER. The SER has four other important roles:

- It is responsible for the chemical modification of small molecules taken in by the cell that may be toxic to the cell. These modifications make the targeted molecules more polar, so they are more water-soluble and easily removed.
- It is the site for glycogen degradation in animal cells. We discuss this important process in Chapter 9.
- It is the site where lipids and steroids are synthesized.
- It stores calcium ions, which when released trigger a number of cell responses, such as a muscle contraction.

Cells that synthesize a lot of protein for export are usually packed with RER. Examples include glandular cells that secrete digestive enzymes and white blood cells that secrete antibodies. In contrast, cells that carry out less protein synthesis (such as storage cells) contain less RER. Liver cells, which modify molecules (including toxins) that enter the body from the digestive system, have abundant SER. This is just one example of the many ways individual cells differentiate to perform specific functions within a multicellular organism (like Don in our opening story).

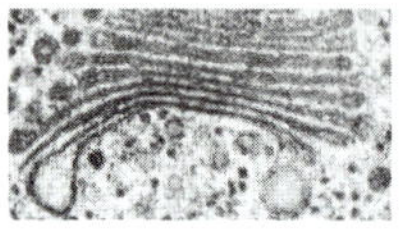

GOLGI APPARATUS The **Golgi apparatus** (or Golgi complex), more often referred to merely as the Golgi, is another part of the diverse, dynamic, and extensive endomembrane system (see Figure 5.9). This structure was named after its discoverer, Camillo Golgi. Its appearance varies from species to species, but it almost always consists of two components: flattened membranous sacs called cisternae (singular cisterna) that are piled up like saucers, and small membrane-enclosed vesicles. The entire apparatus is about 1 μm long.

Go to Animated Tutorial 5.2
The Golgi Apparatus
Life10e.com/at5.2

The Golgi has several roles:

- It receives protein-containing vesicles from the RER.
- It modifies, concentrates, packages, and sorts proteins before they are sent to their cellular or extracellular destinations.
- It adds carbohydrates to proteins and modifies other carbohydrates that were attached to proteins in the RER.
- It is where some polysaccharides for the plant cell wall are synthesized.

The cisternae of the Golgi apparatus have three functionally distinct regions: the *cis* region lies nearest to the nucleus or a patch of RER, the *trans* region lies closest to the plasma membrane, and the medial region lies in between (see Figure 5.9). (The terms *cis*, *trans*, and medial derive from Latin words meaning, respectively, "on the same side," "on the opposite side," and "in the middle.") These three parts of the Golgi apparatus contain different enzymes and perform different functions.

Protein-containing vesicles from the RER fuse with the *cis* membrane of the Golgi apparatus, releasing its cargo into the lumen of the Golgi cisterna. Other vesicles may move between the cisternae, transporting proteins, and it appears that some proteins move from one cisterna to the next through tiny channels. Vesicles budding off from the *trans* region carry their contents away from the Golgi apparatus. These vesicles go to the plasma membrane, or to another organelle in the endomembrane system called the lysosome.

LYSOSOMES The **primary lysosomes** originate from the Golgi apparatus. They contain digestive enzymes, and are the sites where macromolecules—proteins, polysaccharides, nucleic acids, and lipids—are hydrolyzed into their monomers (see Figure 3.4).

$$R_1\text{—}R_2 \text{ (linked monomers)} + H_2O \rightarrow R_1\text{—OH} + R_2\text{—H}$$

Lysosomes are about 1 micrometer in diameter; they are surrounded by a single membrane and have a densely staining, featureless interior (**Figure 5.10**). There may be dozens of lysosomes in a cell, depending on its needs.

Lysosomes are sites for the breakdown of food, other cells, or foreign objects that are taken up by the cell. These materials enter the cell by a process called **phagocytosis** (*phago*, "eat"; *cytosis*, "cellular"). In this process, a pocket forms in the plasma membrane and then deepens and encloses material from outside the cell. The pocket becomes a small vesicle called a phagosome, containing food or other material, which breaks free of the plasma membrane to move into the cytoplasm. The phagosome fuses with a primary lysosome to form a **secondary lysosome**, in which digestion occurs.

The effect of this fusion is rather like releasing hungry foxes into a chicken coop: the enzymes in the secondary lysosome quickly hydrolyze the food particles. These reactions are enhanced by the mild acidity of the lysosome's interior, where the pH is lower than in the surrounding cytoplasm. The products of digestion pass through the membrane of the lysosome, providing energy and raw materials for other cellular processes. The "used" secondary lysosome, now containing undigested particles, then moves to the plasma membrane, fuses with it, and releases the undigested contents to the environment.

Phagocytes (see Section 42.1) are specialized cells that have an essential role in taking up and breaking down materials; they are found in nearly all animals and many protists. You will encounter them and their activities again at many places in this book, but at this point one example suffices: in the human liver and spleen, phagocytes digest approximately 10 billion aged or damaged blood cells each day! The digestion products are then used to make new cells to replace those that are digested.

Lysosomes are active even in cells that do not perform phagocytosis. Cells are dynamic systems; some cell components are continually being broken down and replaced by new ones. The programmed destruction of cell components is called **autophagy**, and lysosomes are where the cell breaks down its

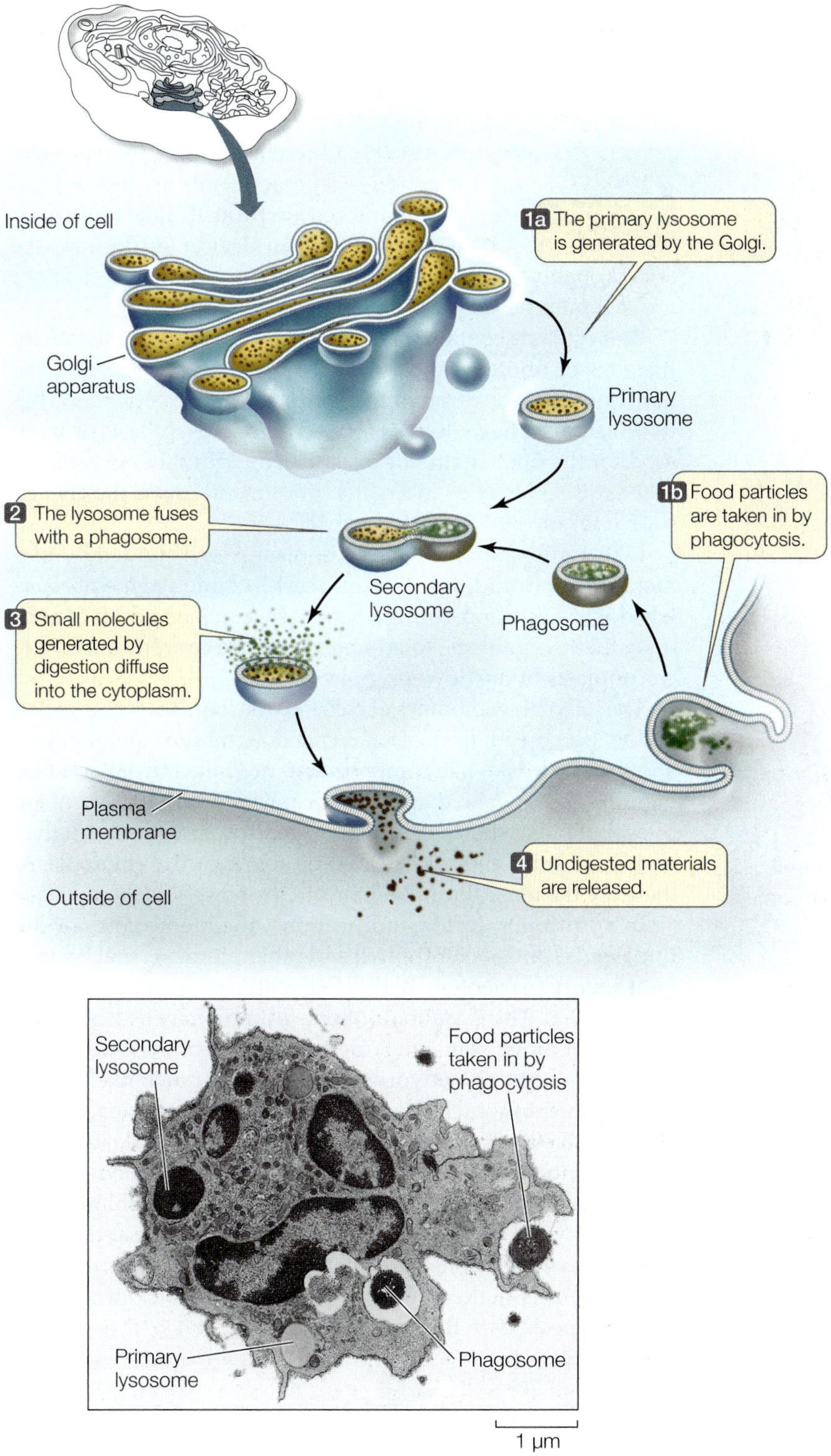

Inside of cell
1a The primary lysosome is generated by the Golgi.
Golgi apparatus
Primary lysosome
2 The lysosome fuses with a phagosome.
1b Food particles are taken in by phagocytosis.
Secondary lysosome
Phagosome
3 Small molecules generated by digestion diffuse into the cytoplasm.
Plasma membrane
4 Undigested materials are released.
Outside of cell
Secondary lysosome
Food particles taken in by phagocytosis
Primary lysosome
Phagosome
1 μm

5.10 Lysosomes Isolate Digestive Enzymes from the Cytoplasm Lysosomes are sites for the hydrolysis of material taken into the cell by phagocytosis.

Go to Activity 5.3 Lysosomal Digestion Life10e.com/ac5.3

own materials. With the proper signal, lysosomes can engulf entire organelles, hydrolyzing their constituents.

How important is autophagy? An entire class of human diseases called lysosomal storage diseases are caused by the failure of lysosomes to digest specific cellular components; these diseases are invariably very harmful or fatal. An example is Tay-Sachs disease, in which a particular lipid called a ganglioside is not broken down in the lysosomes and instead accumulates in brain cells. In the most common form of this disease, a baby starts exhibiting neurological symptoms and becomes blind, deaf, and unable to swallow after six months of age. Death occurs before age 4.

Plant cells do not appear to contain lysosomes, but the vacuole of a plant cell (which we will describe below) may function in an equivalent capacity because it, like lysosomes, contains many digestive enzymes.

Some organelles transform energy

A cell requires energy to make the molecules it needs for activities such as growth, reproduction, responsiveness, and movement. Energy is harvested from fuel molecules in the mitochondria (found in all eukaryotic cells) and from sunlight in the chloroplasts of plant cells. In contrast, energy transformations in prokaryotic cells are associated with enzymes attached to the inner surface of the plasma membrane or to extensions of the plasma membrane that protrude into the cytoplasm.

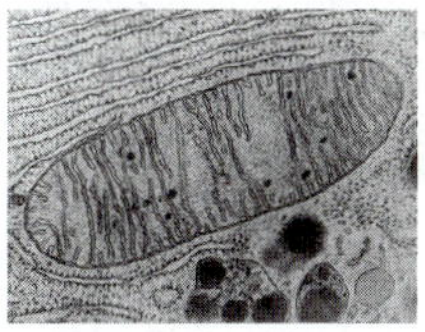

MITOCHONDRIA In eukaryotic cells, the breakdown of fuel molecules such as glucose begins in the cytosol. The molecules that result from this partial degradation enter the **mitochondria** (singular *mitochondrion*), whose primary function is to harvest the chemical energy of those fuel molecules in a form that the cell can use, namely the energy-rich molecule ATP (adenosine triphosphate) (see Section 8.2). The production of ATP in the mitochondria, using fuel molecules and molecular oxygen (O_2), is called **cellular respiration**.

Typical mitochondria are somewhat less than 1.5 μm in diameter and are 2–8 μm in length—about the size of many bacteria. They can divide independently of the central nucleus. The number of mitochondria per cell ranges from one gigantic organelle in some unicellular protists to a few hundred thousand in large egg cells. An average human liver cell contains more than 1,000 mitochondria. Cells that are active in movement and growth require the most chemical energy, and these tend to have the most mitochondria per unit of volume.

Mitochondria have two membranes. The outer membrane is smooth and protective, and it offers little resistance to the movement of substances into and out of the organelle. Immediately inside the outer membrane is an inner membrane, which folds inward in many places and thus has a surface area much

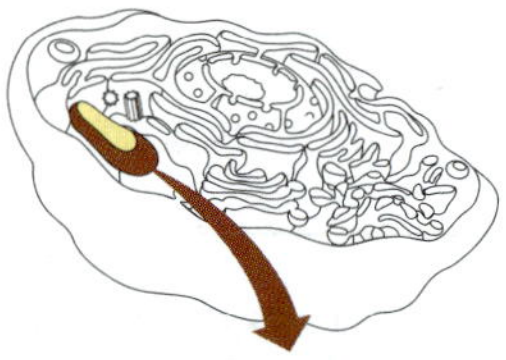

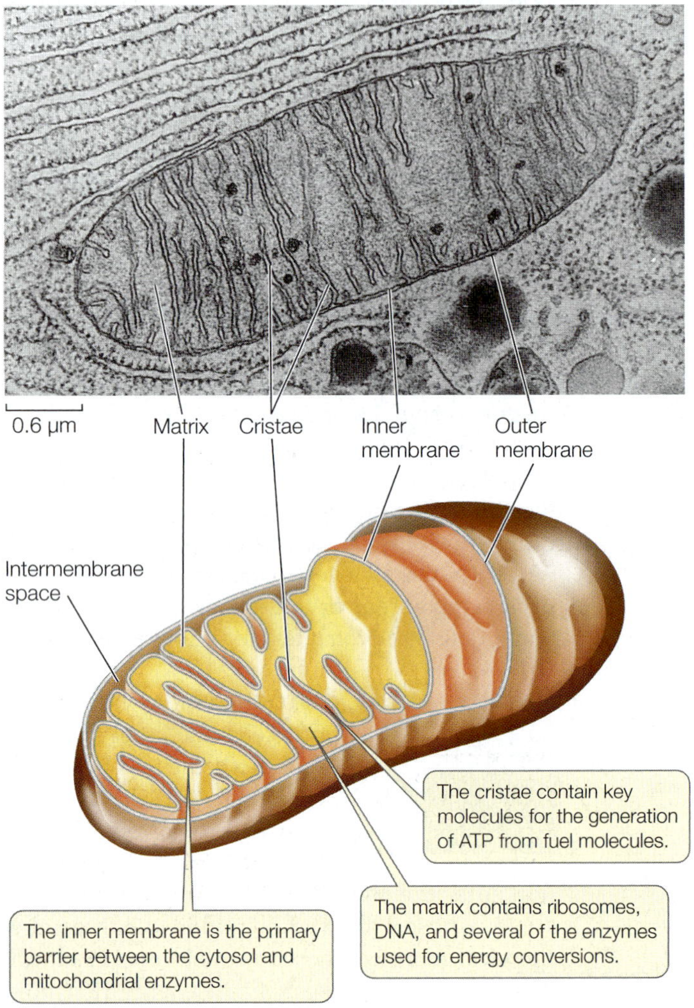
0.6 μm
Matrix
Cristae
Inner membrane
Outer membrane
Intermembrane space
The cristae contain key molecules for the generation of ATP from fuel molecules.
The matrix contains ribosomes, DNA, and several of the enzymes used for energy conversions.
The inner membrane is the primary barrier between the cytosol and mitochondrial enzymes.

5.11 A Mitochondrion Converts Energy from Fuel Molecules into ATP The electron micrograph is a two-dimensional slice through a three-dimensional organelle. As the drawing emphasizes, the cristae are extensions of the inner mitochondrial membrane.

greater than that of the outer membrane (**Figure 5.11**). The folds tend to be quite regular, giving rise to shelflike structures called cristae. The inner membrane exerts much more control over what enters and leaves the space it encloses than does the outer membrane. Embedded in the inner mitochondrial membrane are many large protein complexes that participate in cellular respiration.

The space enclosed by the inner membrane is referred to as the mitochondrial matrix. In addition to many enzymes, the matrix contains ribosomes and DNA that are used to make some of the proteins needed for cellular respiration. As we will discuss later in this chapter, it is likely that this DNA is the remnant of a larger, complete chromosome from a prokaryote that may have been the mitochondrion's progenitor. In Chapter 9 we will discuss how the different parts of the mitochondrion work together in cellular respiration.

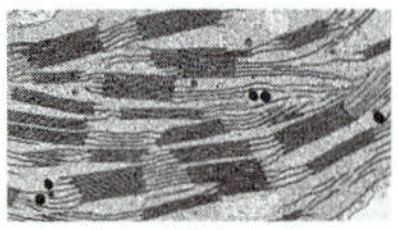

PLASTIDS One class of organelles—the plastids—is present only in the cells of plants and certain protists. Like mitochondria, plastids can divide autonomously and probably evolved from independent prokaryotes. There are several types of plastids, with different functions.

Chloroplasts contain the green pigment chlorophyll and are the sites of photosynthesis (**Figure 5.12**). In photosynthesis, light energy is converted into the chemical energy of bonds between atoms. The molecules formed by photosynthesis provide food for the photosynthetic organism and for other organisms that eat it. Directly or indirectly, photosynthesis is the energy source for most of the living world.

Like a mitochondrion, a chloroplast is surrounded by two membranes. In addition, there is a series of internal membranes whose structure and arrangement vary from one group of photosynthetic organisms to another. Here we concentrate on the chloroplasts of the flowering plants.

The internal membranes of chloroplasts look like stacks of flat, hollow pita bread. Each stack is called a granum (plural grana) and the pita bread–like compartments are called **thylakoids** (see Figure 5.12). Thylakoid lipids are distinctive: only 10 percent are phospholipids, whereas the rest are galactose-substituted diglycerides and sulfolipids. Because of the abundance of chloroplasts, these are the most abundant lipids in the biosphere.

In addition to lipids and proteins, the membranes of the thylakoids contain chlorophyll and other pigments that harvest light energy for photosynthesis (we will see how they do this in Section 10.2). The thylakoids of one granum may be connected to those of other grana, making the interior of the chloroplast a highly developed network of membranes, much like the ER.

The fluid in which the grana are suspended is called the stroma. Like the mitochondrial matrix, the chloroplast stroma contains ribosomes and DNA, which are used to synthesize some, but not all, of the proteins that make up the chloroplast.

Other types of plastids, such as chromoplasts and leucoplasts, have functions different from those of chloroplasts. Chromoplasts make and store red, yellow, and orange pigments, especially in flowers and fruits.

Chromoplast
5 μm

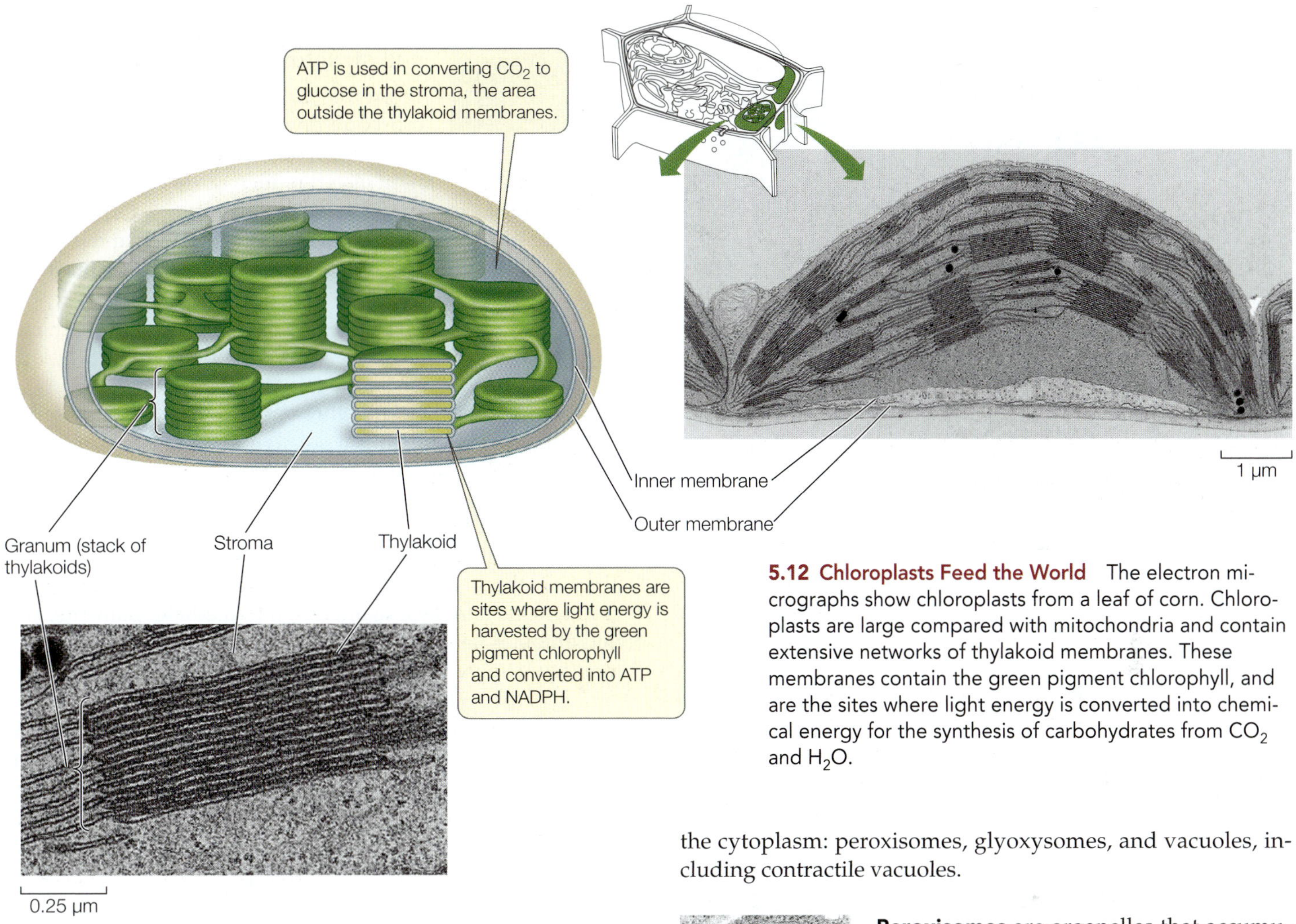

ATP is used in converting CO_2 to glucose in the stroma, the area outside the thylakoid membranes.
Granum (stack of thylakoids)
Stroma
Thylakoid
Thylakoid membranes are sites where light energy is harvested by the green pigment chlorophyll and converted into ATP and NADPH.
Inner membrane
Outer membrane
0.25 μm
1 μm

5.12 Chloroplasts Feed the World The electron micrographs show chloroplasts from a leaf of corn. Chloroplasts are large compared with mitochondria and contain extensive networks of thylakoid membranes. These membranes contain the green pigment chlorophyll, and are the sites where light energy is converted into chemical energy for the synthesis of carbohydrates from CO_2 and H_2O.

Leucoplasts are storage organelles that do not contain pigments. An amyloplast is a leucoplast that stores starch.

Leucoplast
Starch grains
1 μm

There are several other membrane-enclosed organelles

There are several other organelles whose boundary membranes separate their specialized chemical reactions and contents from the cytoplasm: peroxisomes, glyoxysomes, and vacuoles, including contractile vacuoles.

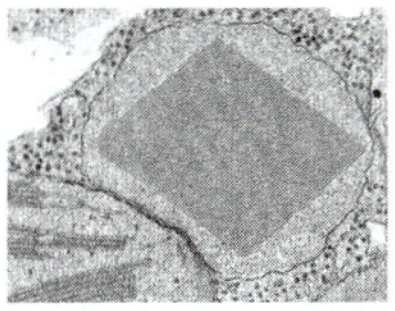

Peroxisomes are organelles that accumulate toxic peroxides, such as hydrogen peroxide (H_2O_2), that occur as by-products of some biochemical reactions. These peroxides are safely broken down inside the peroxisomes without mixing with other parts of the cell.

$$RH_2 + O_2 \rightarrow R + H_2O_2 \text{ (cellular reactions)}$$

$$2\, H_2O_2 \rightarrow 2\, H_2O + O_2 \text{ (inside peroxisome)}$$

Peroxisomes are small organelles, about 0.2–1.7 micrometers in diameter. They have a single membrane and a granular interior containing specialized enzymes. Peroxisomes are found in at least some of the cells of almost every eukaryotic species.

As with lysosomes, there are rare inherited diseases in humans that involve peroxisomes. In Zellweger syndrome there is a defect in peroxisome assembly, and affected infants are born without peroxisomes. As you can imagine, a consequence of this is the accumulation of toxic peroxides, and the infants seldom live beyond one year of age.

Glyoxysomes are similar to peroxisomes and are found only in plants. They are most abundant in young plants and are the locations where stored lipids are converted into carbohydrates for transport to growing cells.

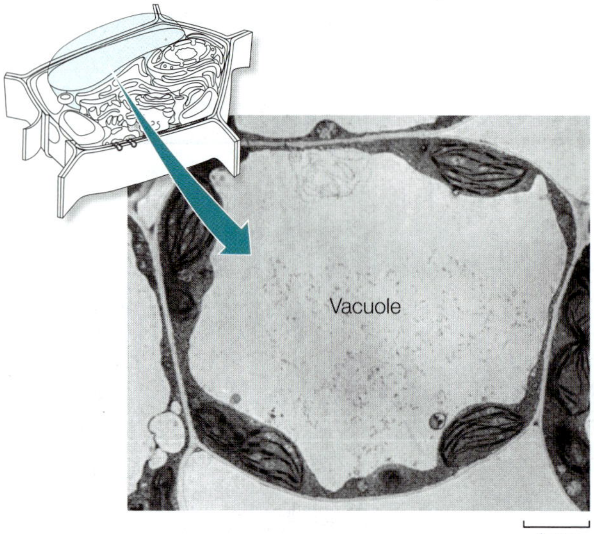
Vacuole
4 μm

5.13 Vacuoles in Plant Cells Are Usually Large The large central vacuole in this cell is typical of mature plant cells.

Vacuoles occur in many eukaryotic cells but particularly those of plants, fungi, and protists. Plant vacuoles (**Figure 5.13**) have several functions:

- *Storage*: Plant cells produce a number of toxic by-products and waste products, many of which are simply stored within vacuoles. Because they are poisonous or distasteful, these stored materials deter some animals from eating the plants and may thus contribute to plant defense and survival.
- *Structure*: In many plant cells, vacuoles take up more than 90 percent of the cell volume and grow as the cell grows. The presence of dissolved substances in the vacuole causes water to enter it from the cytoplasm, making the vacuole swell like a balloon. A mature plant cell does not swell when the vacuole fills with water, since it has a rigid cell wall. Instead, it stiffens from the increase in water pressure (called turgor), and this supports the plant (see Figure 6.9).
- *Reproduction*: Vacuoles contain some of the pigments (especially blue and pink ones) in the petals and fruits of flowering plants. These pigments—the anthocyanins—are visual cues, which help attract animals that assist in pollination or seed dispersal.
- *Digestion*: In some plants, the vacuoles in seeds contain enzymes that hydrolyze stored proteins into monomers. During seed germination, the monomers are used as food by the developing plant seedlings.

Contractile vacuoles are found in many freshwater protists. Their function is to get rid of the excess water that rushes into the cell because of the imbalance in solute concentration between the interior of the cell and its freshwater environment. The contractile vacuole enlarges as water enters, then abruptly contracts, forcing the water out of the cell through a special pore structure.

So far we have discussed numerous membrane-enclosed organelles. Now we will turn to a group of cytoplasmic structures without membranes.

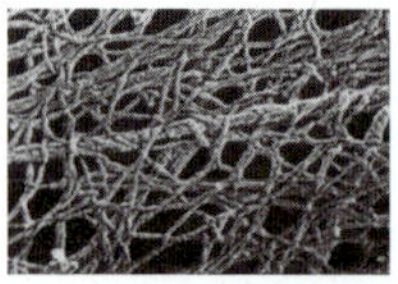

The cytoskeleton is important in cell structure and movement

From the earliest observations, light microscopy revealed distinctive cell shapes that would sometimes change, and rapid movements within cells. With the advent of electron microscopy, a new world of cellular substructure was revealed, including a meshwork of filaments inside cells. Experimentation showed that this meshwork—called the cytoskeleton—fills several important roles:

- It supports the cell and maintains its shape.
- It holds cell organelles and other particles in position within the cell.
- It moves organelles and other particles around in the cell.
- It is involved with movements of the cytoplasm, called cytoplasmic streaming.
- It interacts with extracellular structures, helping anchor the cell in place.

There are three components of the eukaryotic cytoskeleton: microfilaments (smallest diameter), intermediate filaments, and microtubules (largest diameter). These filaments have very different functions.

MICROFILAMENTS **Microfilaments** can exist as single filaments, in bundles, or in networks. They are about 7 nanometers in diameter and up to several micrometers long. Microfilaments have two major roles:

- They help the entire cell or parts of the cell move.
- They determine and stabilize cell shape.

Microfilaments are assembled from monomers of **actin**, a protein that exists in several forms and has many functions, especially in animals. The actin found in microfilaments (which are also known as actin filaments) has distinct ends, designated "plus" and "minus." These ends permit actin monomers to interact with one another to form long, double helical chains (**Figure 5.14A**). Within cells, the polymerization of actin into microfilaments is reversible, and the microfilaments can disappear from cells by breaking down into monomers of free actin. Special actin-binding proteins mediate these processes.

In the muscle cells of animals, actin filaments are associated with another protein, the "motor protein" **myosin**, and the interactions of these two proteins account for the contraction of muscles (described in Section 48.1). In non-muscle cells, actin filaments are associated with localized changes in cell shape. For example, microfilaments are involved in the flowing movement of the cytoplasm called cytoplasmic streaming, in amoeboid movement, and in the "pinching" contractions that divide an animal cell into two daughter cells. Microfilaments are also involved in the formation of cellular extensions called pseudopodia (*pseudo*, "false"; *podia*, "feet") that enable some cells to move (**Figure 5.15**). As you will see in Chapter 42, cells of the immune system must move toward other cells during the immune response.

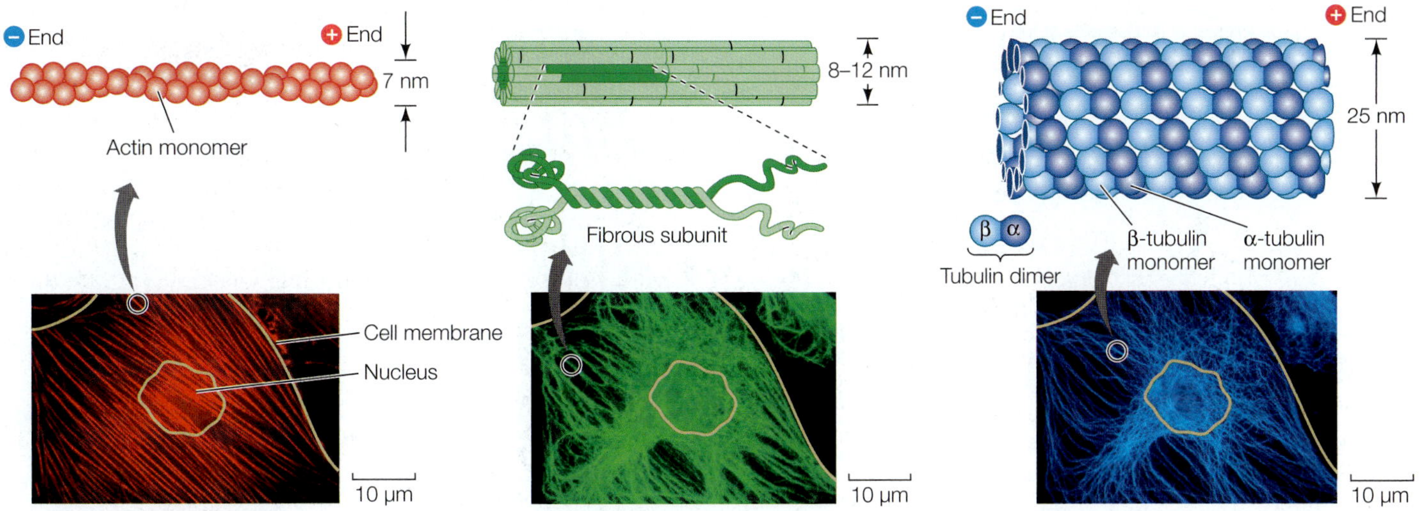
End
End
7 nm
Actin monomer
Cell membrane
Nucleus
10 μm
8–12 nm
Fibrous subunit
10 μm
End
End
25 nm
β α
Tubulin dimer
β-tubulin monomer
α-tubulin monomer
10 μm

(A) **Microfilaments**
Made up of strands of the protein actin; often interact with strands of other proteins.

(B) **Intermediate filaments**
Made up of fibrous proteins organized into tough, ropelike assemblages that stabilize a cell's structure and help maintain its shape.

(C) **Microtubules**
Long, hollow cylinders made up of many molecules of the protein tubulin. Tubulin consists of two subunits, α-tubulin and β-tubulin.

5.14 The Cytoskeleton Three highly visible and important structural components of the cytoskeleton are shown here in detail. The photographs are all of the same cell, treated with different fluorescent antibodies that detect microfilaments (A), intermediate filaments (B), or microtubules (C). These structures maintain and reinforce cell shape and contribute to cell movement. The position of the cell's nucleus is near the center of the photos.

In some cell types, microfilaments form a meshwork just inside the plasma membrane. Actin-binding proteins then cross-link the microfilaments to form a rigid netlike structure that supports the cell. For example, microfilaments support the tiny microvilli that line the human intestine, giving it a larger surface area through which to absorb nutrients (**Figure 5.16**).

INTERMEDIATE FILAMENTS There are at least 50 different kinds of **intermediate filaments**, many of them specific to a few cell types. They generally fall into six molecular classes (based on amino acid sequence) that share the same general structure. One of these classes consists of fibrous proteins of the keratin family, which also includes the proteins that make up hair and fingernails. Intermediate filaments are tough, ropelike protein assemblages 8–12 nanometers in diameter (**Figure 5.14B**). They are more permanent than the other two types in that they do not continually form and re-form, as the microtubules and microfilaments do.

Intermediate filaments have two major structural functions:

- They anchor cell structures in place. In some cells, intermediate filaments radiate from the nuclear envelope and help maintain the positions of the nucleus and other organelles in the cell. The lamins of the nuclear lamina are intermediate filaments. Other kinds of intermediate filaments help hold in place the complex apparatus of microfilaments in the microvilli of intestinal cells (see Figure 5.16).
- They resist tension. For example, they maintain rigidity in body surface tissues by stretching through the cytoplasm and connecting specialized membrane structures called desmosomes (see Figure 6.7).

MICROTUBULES The largest-diameter components of the cytoskeletal system, **microtubules**, are long, hollow, unbranched cylinders about 25 nanometers in diameter and up to several micrometers long. Microtubules have two roles in the cell:

- They form a rigid internal skeleton for some cells.
- They act as a framework along which motor proteins can move structures within the cell.

Microtubules are assembled from dimers of the protein **tubulin**. A dimer is a molecule made up of two monomers. The polypeptide monomers that make up the tubulin dimer are known as α-tubulin and β-tubulin. Thirteen chains of tubulin dimers surround the central cavity of the microtubule (**Figure 5.14C**; see also Figure 5.17B). As in microfilaments, the two ends of a microtubule are different: one is designated the "plus" end and the other the "minus" end. Tubulin dimers can be rapidly added or subtracted, mainly at the plus end, lengthening or shortening the microtubule.

Many microtubules radiate from a region of the cell called the microtubule organizing center. Tubulin polymerization results in a rigid structure (sometimes called an endoskeleton), and tubulin depolymerization leads to its collapse. The capacity to change length rapidly makes microtubules dynamic structures: they are readily adapted for new purposes in the cell. For example, by disassembly and reassembly,

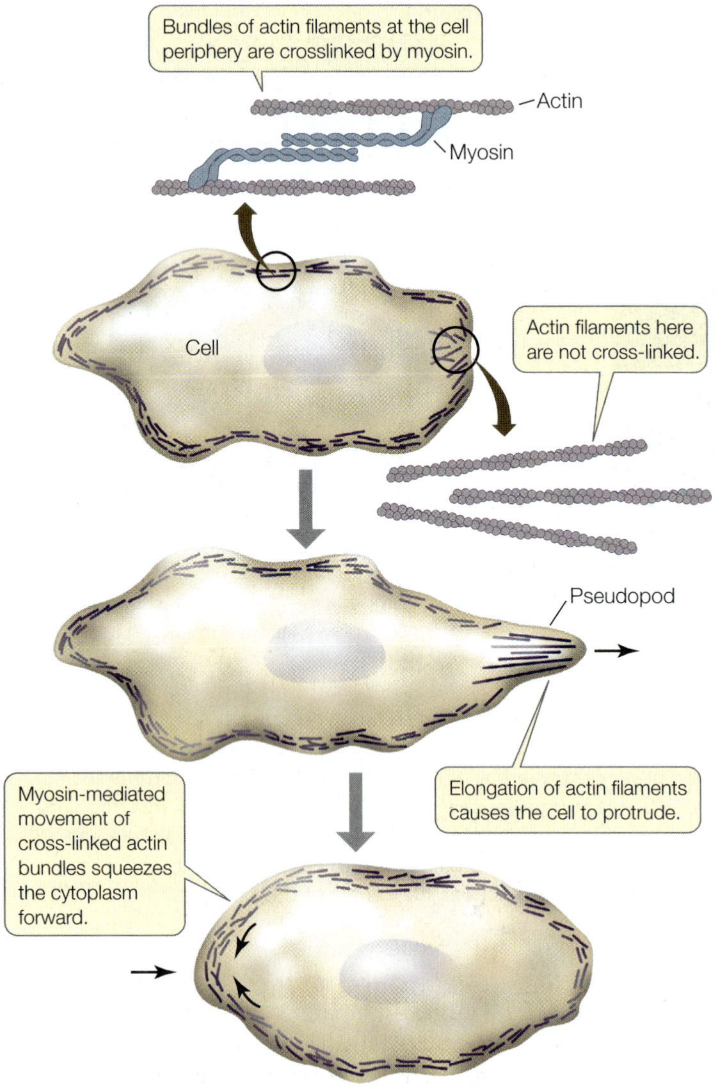
Bundles of actin filaments at the cell periphery are crosslinked by myosin.
Actin
Myosin
Cell
Actin filaments here are not cross-linked.
Pseudopod
Elongation of actin filaments causes the cell to protrude.
Myosin-mediated movement of cross-linked actin bundles squeezes the cytoplasm forward.

5.15 Microfilaments and Cell Movements Microfilaments mediate the movement of whole cells (as illustrated here for amoeboid movement), as well as the movement of cytoplasm within a cell.

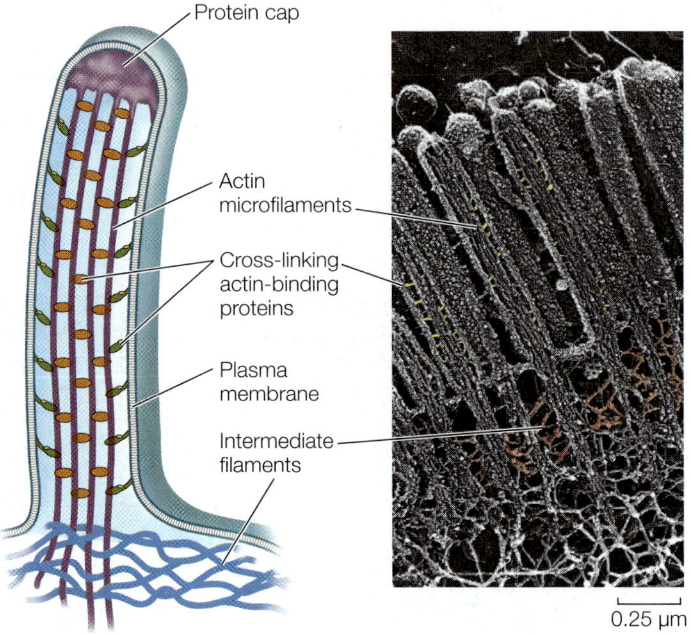
Protein cap
Actin microfilaments
Cross-linking actin-binding proteins
Plasma membrane
Intermediate filaments
0.25 μm

5.16 Microfilaments for Support Cells that line the intestine are folded into tiny projections called microvilli, which are supported by microfilaments. The microfilaments interact with intermediate filaments at the base of each microvillus. Microvilli increase the surface area of a cell, facilitating its absorption of small molecules.

microtubules can move to new parts of the cell and assemble new structures needed for cell division. Microtubules from all eukaryotes have this dynamic property, indicating that it is evolutionarily advantageous over a static, unchanging structure.

In plants, microtubules help control the arrangement of the cellulose fibers of the cell wall. Electron micrographs of plants frequently show microtubules lying just inside the plasma membranes of cells that are forming or extending their cell walls. If the orientation of these microtubules is altered experimentally, it leads to a similar change in the cell wall and a new shape for the cell.

Microtubules serve as tracks for **motor proteins**, specialized molecules that use cellular energy to change their shapes and move. Motor proteins bind to and move along the microtubules, carrying materials from one part of the cell to another. Microtubules are also essential in distributing chromosomes to daughter cells during cell division. Because of this, drugs such as vincristine and taxol, which disrupt microtubule dynamics, also disrupt cell division. These drugs are useful for treating cancer, where cell division is excessive.

CILIA AND FLAGELLA Microtubules and their associated proteins line the interior of certain movable appendages on eukaryotic cells: the **cilia** (Figure 5.17A) and flagella. Many cells have one or the other of these appendages, which form from projections of the plasma membrane:

- Cilia are only 0.25 μm in length. They occur by the hundreds on individual cells and move stiffly to either propel the cell (for example, in protists) or to move fluid over a stationary cell (as in the human respiratory system).
- Flagella are longer—100 to 200 μm—and occur singly or in pairs. They can push or pull a cell through its aqueous environment.

In cross section, a typical cilium or eukaryotic flagellum is surrounded by the plasma membrane and contains a "9 + 2" array of microtubules. As Figure 5.17B shows, nine fused pairs of microtubules—called doublets—form an outer cylinder, and one pair of unfused microtubules runs up the center. Each doublet is connected to the center of the structure by a radial spoke. This structure is essential to the bending motion of both cilia and flagella. How does this bending occur?

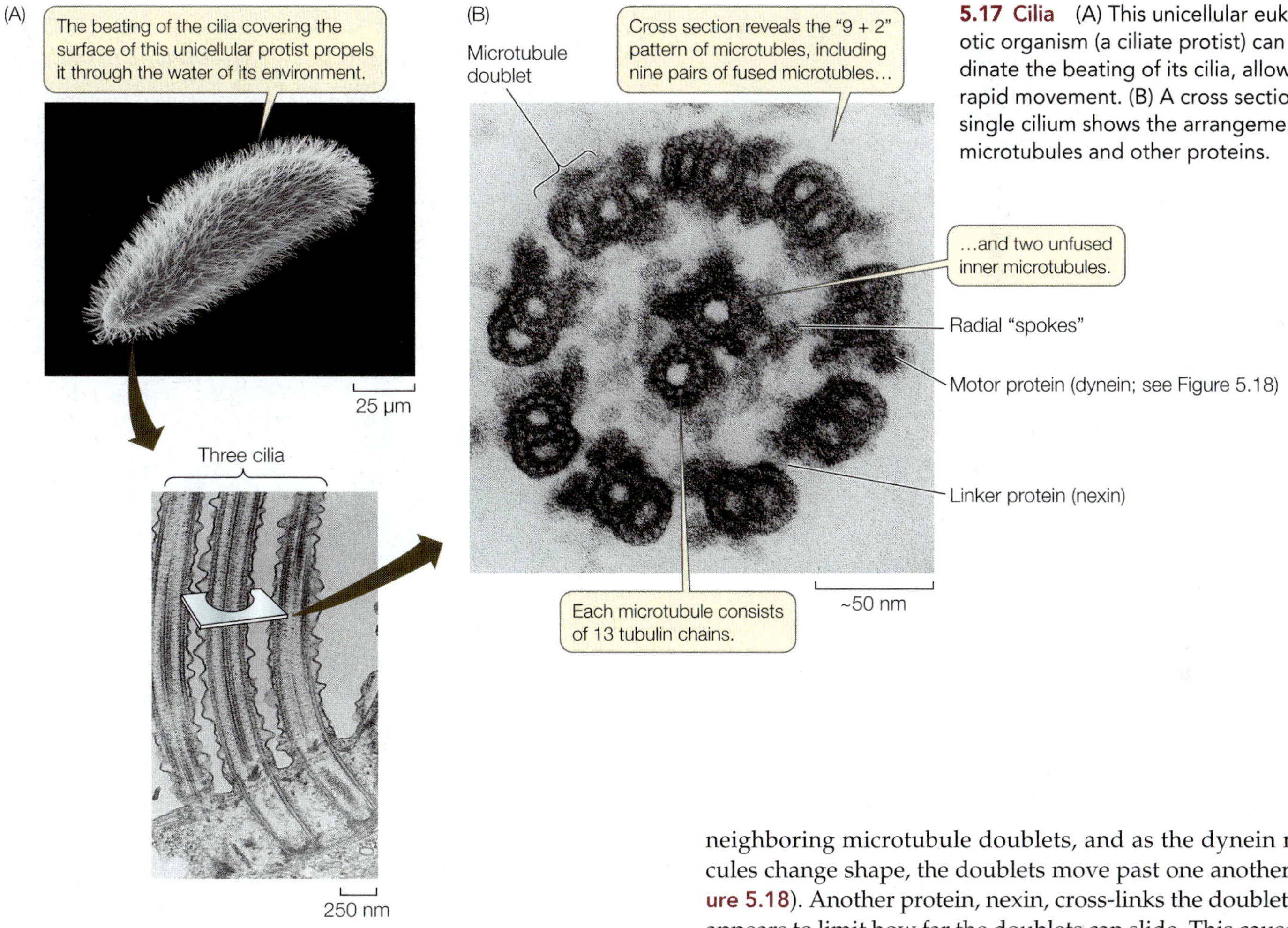
(A)
The beating of the cilia covering the surface of this unicellular protist propels it through the water of its environment.
25 µm
Three cilia
250 nm
(B)
Microtubule doublet
Cross section reveals the "9 + 2" pattern of microtubules, including nine pairs of fused microtubules…
…and two unfused inner microtubules.
Radial "spokes"
Motor protein (dynein; see Figure 5.18)
Linker protein (nexin)
Each microtubule consists of 13 tubulin chains.
~50 nm

5.17 Cilia (A) This unicellular eukaryotic organism (a ciliate protist) can coordinate the beating of its cilia, allowing rapid movement. (B) A cross section of a single cilium shows the arrangement of microtubules and other proteins.

The motion of cilia and flagella results from the sliding of the microtubule doublets past one another. This sliding is driven by the motor protein dynein, which, like other motor proteins, works by undergoing reversible shape changes that require chemical energy. Dynein molecules bind between two neighboring microtubule doublets, and as the dynein molecules change shape, the doublets move past one another (Figure 5.18). Another protein, nexin, cross-links the doublets and appears to limit how far the doublets can slide. This causes the cilium or flagellum to bend.

Other motor proteins, including kinesin, carry protein-laden vesicles or other organelles from one part of the cell to another (Figure 5.19). These proteins bind to the organelle and "walk" it along a microtubule by a repeated series of shape changes. Recall that microtubules are directional, with a plus

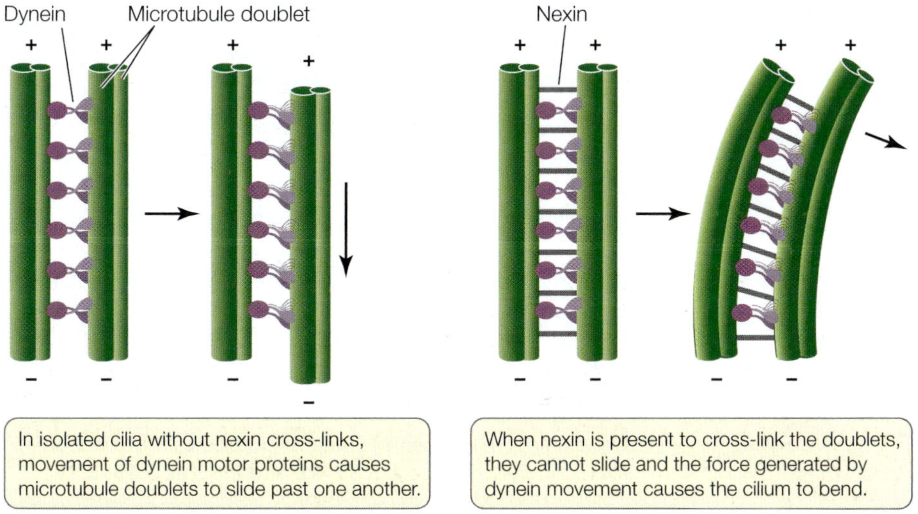
Dynein
Microtubule doublet
Nexin
In isolated cilia without nexin cross-links, movement of dynein motor proteins causes microtubule doublets to slide past one another.
When nexin is present to cross-link the doublets, they cannot slide and the force generated by dynein movement causes the cilium to bend.

5.18 A Motor Protein Moves Microtubules in Cilia and Flagella A motor protein, dynein, causes microtubule doublets to slide past one another. In a flagellum or cilium, anchorage of the microtubule doublets to one another results in bending.

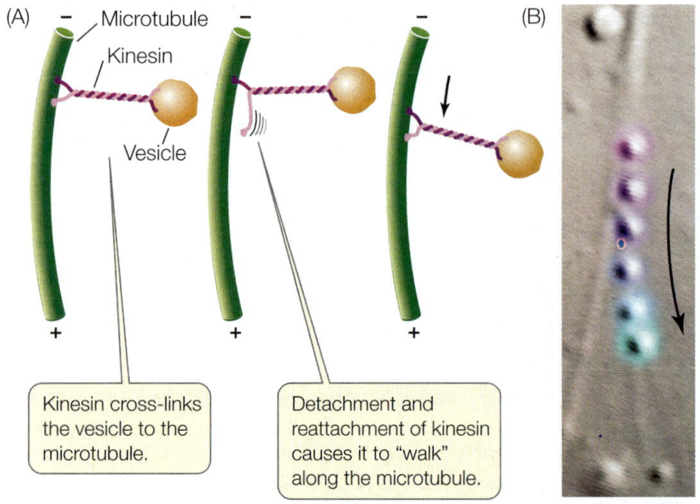
(A)
Microtubule
Kinesin
Vesicle
Kinesin cross-links the vesicle to the microtubule.
Detachment and reattachment of kinesin causes it to "walk" along the microtubule.
(B)

5.19 A Motor Protein Pulls Vesicles along Microtubules
(A) Kinesin delivers vesicles to various parts of the cell by moving along microtubule "railroad tracks." (B) A vesicle is pulled by kinesin along a microtubule in the protist *Dictyostelium*. The time sequence, with half-second intervals, is shown by the color changes from purple to blue.

and a minus end. Cytoplasmic dynein (which has a different role than the one found in cilia and flagella) moves attached organelles toward the minus end, whereas kinesin moves them toward the plus end (see Figure 5.14).

Biologists can manipulate living systems to establish cause and effect

How do we know that the structural fibers of the cytoskeleton can achieve all these dynamic functions? We can observe an individual structure under the microscope, and we can observe the functions of living cells that contain that structure. These observations may suggest that the structure carries out a particular function, but mere correlation does not show cause and effect. For example, light microscopy of living cells reveals that the cytoplasm is actively streaming around the cell, and that cytoplasm flows into an extended portion of an amoeboid cell during movement. The observed presence of cytoskeletal components *suggests, but does not prove*, their role in this process. Science seeks to show the specific links that relate one process, A, to a function, B. In cell biology, two approaches are often used to show that a structure or process A causes function B:

- *Inhibition*: use a drug that inhibits A and see if B still occurs. If it does not, then A is probably a causative factor for B. **Figure 5.20** shows an experiment with such a drug (an inhibitor) that demonstrates cause and effect in the case of the cytoskeleton and cell movement.
- *Mutation*: examine a cell that lacks the gene (or genes) for A and see if B still occurs. If it does not, then A is probably a causative factor for B. Part Four of this book describes many experiments using this genetic approach.

INVESTIGATING**LIFE**

5.20 The Role of Microfilaments in Cell Movement—Showing Cause and Effect in Biology After a test tube demonstration that the drug cytochalasin B prevented microfilament formation from monomeric precursors, the question was asked: Will the drug work like this in living cells and inhibit cell movement in *Amoeba*? Complementary experiments showed that the drug did not poison other cellular processes.

HYPOTHESIS Amoeboid cell movements are caused by the cytoskeleton.

Method

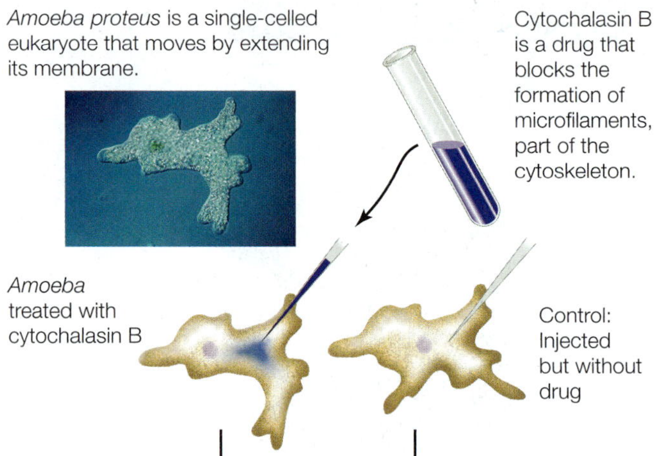
Amoeba proteus is a single-celled eukaryote that moves by extending its membrane.
Cytochalasin B is a drug that blocks the formation of microfilaments, part of the cytoskeleton.
Amoeba treated with cytochalasin B
Control: Injected but without drug

Results

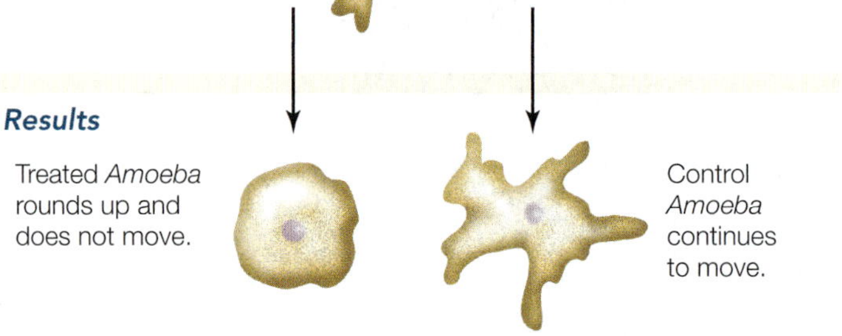
Treated *Amoeba* rounds up and does not move.
Control *Amoeba* continues to move.

CONCLUSION Microfilaments of the cytoskeleton are essential for amoeboid cell movement.

Go to **BioPortal** for discussion and relevant links for all INVESTIGATING**LIFE** figures.

Pollard, T. D. and R. R. Weihing. 1974. *CRC Critical Reviews of Biochemistry* 2: 1–65.

RECAP 5.3

The hallmark of eukaryotic cells is compartmentalization. Membrane-enclosed organelles process information, transform energy, form internal compartments for transporting proteins, and carry out intracellular digestion. An internal cytoskeleton plays several structural roles.

- What are some advantages of organelle compartmentalization? **See p. 84**
- Describe the structural and functional differences between rough and smooth endoplasmic reticulum. **See pp. 88–90 and Figure 5.9**
- Explain how motor proteins and microtubules move materials within the cell. **See pp. 97–98 and Figures 5.18, 5.19**

WORKING WITH DATA:

The Role of Microfilaments in Cell Movement

Original Paper

Pollard, T. D. and R. R. Weihing. 1974. Actin and myosin in cell movement. *CRC Critical Reviews of Biochemistry* 2: 1–65.

Analyze the Data

In a search for natural molecules that have effects on cells—particularly cancer cells—a team of chemists and biologists at Imperial Chemical Industries examined extracts of the fungus *Helminthosporium dematiodeium*. When the extracts appeared to inhibit cell division, the scientists purified the active ingredient and called it cytochalasin B (from the Greek *cyto*, "cell" and *chalasis*, "dislocation"). Remarkably, application of cytochalasin B to dividing cells blocked the division of the cytoplasm but not division of the nucleus, so the result was a binucleate cell. In addition, the drug inhibited cell movement and phagocytosis. These dynamic processes were both hypothesized to involve cytoplasmic microfilaments (actin filaments). In the test tube, cytochalasin B blocked the polymerization of actin monomers into actin filaments. This prompted the use of cytochalasin B in cause-and-effect experiments: if a cellular process was inhibited by the drug, that process must involve microfilaments. This is the basis of the experiment shown in Figure 5.20.

Several important controls were done to validate the conclusions of the experiment. The experiment was repeated in the presence of the following drugs: cycloheximide, which inhibits new protein synthesis; dinitrophenol, which inhibits new ATP formation (energy); and colchicine, which inhibits the polymerization of microtubules. The results are shown in the table.

Condition	Rounded cells (%)
No drug	3
Cytochalasin B	95
Colchicine	4
Cycloheximide	3
Cycloheximide + cytochalasin B	94
Dinitrophenol	5
Dinitrophenol + cytochalasin B	85

QUESTION 1

Explain the reasoning behind each experiment. Why were these controls important?

QUESTION 2

Interpret the results of each experiment. What can you conclude about movements in *Amoeba* and the cytoskeleton?

Go to **BioPortal** for all WORKING WITH DATA exercises

All cells interact with their environments. Many eukaryotic cells are parts of multicellular organisms and must closely coordinate their activities with other cells. The plasma membrane plays a crucial role in these interactions, but other structures outside that membrane are involved as well.

5.4 What Are the Roles of Extracellular Structures?

Although the plasma membrane is the functional barrier between the inside and the outside of a cell, many structures are produced by cells and secreted to the outside of the plasma membrane, where they play essential roles in protecting, supporting, or attaching cells to each other. Because they are outside the plasma membrane, these structures are said to be extracellular. The peptidoglycan cell wall of bacteria is an example of an extracellular structure (see Figure 5.4). In eukaryotes, other extracellular structures—the cell walls of plants and the extracellular matrices found between the cells of animals—play similar roles. Each of these structures is made up of two main components: a prominent fibrous macromolecule and a gel-like medium in which the fibers are embedded.

The plant cell wall is an extracellular structure

The plant cell wall performs the same role as skeletal structures in animals. It is a semirigid structure outside the plasma membrane (**Figure 5.21**). We will consider the structure and role of

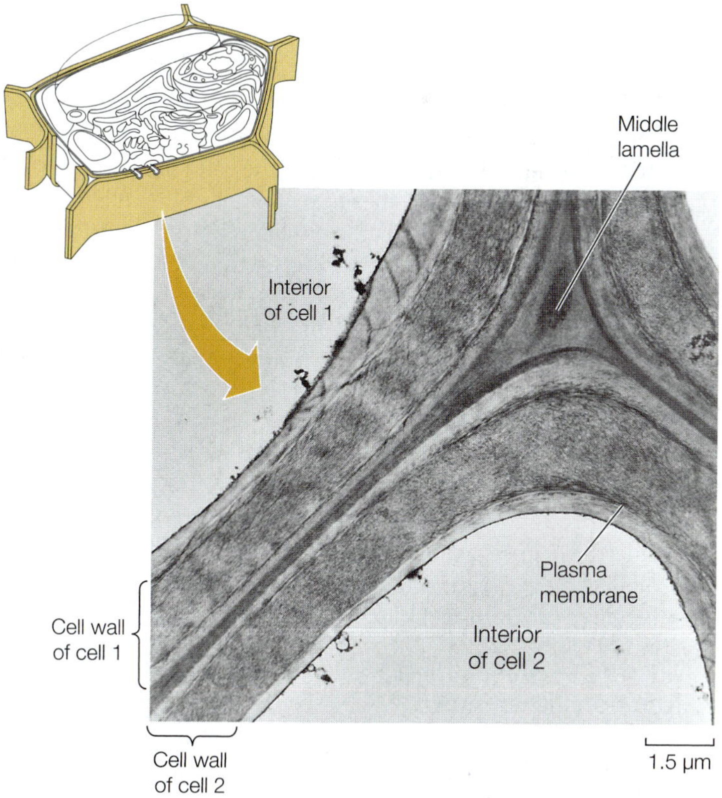

Middle lamella
Interior of cell 1
Plasma membrane
Interior of cell 2
Cell wall of cell 1
Cell wall of cell 2
1.5 μm

5.21 The Plant Cell Wall The semirigid cell wall provides support for plant cells. It is composed of cellulose fibrils embedded in a matrix of polysaccharides and proteins.

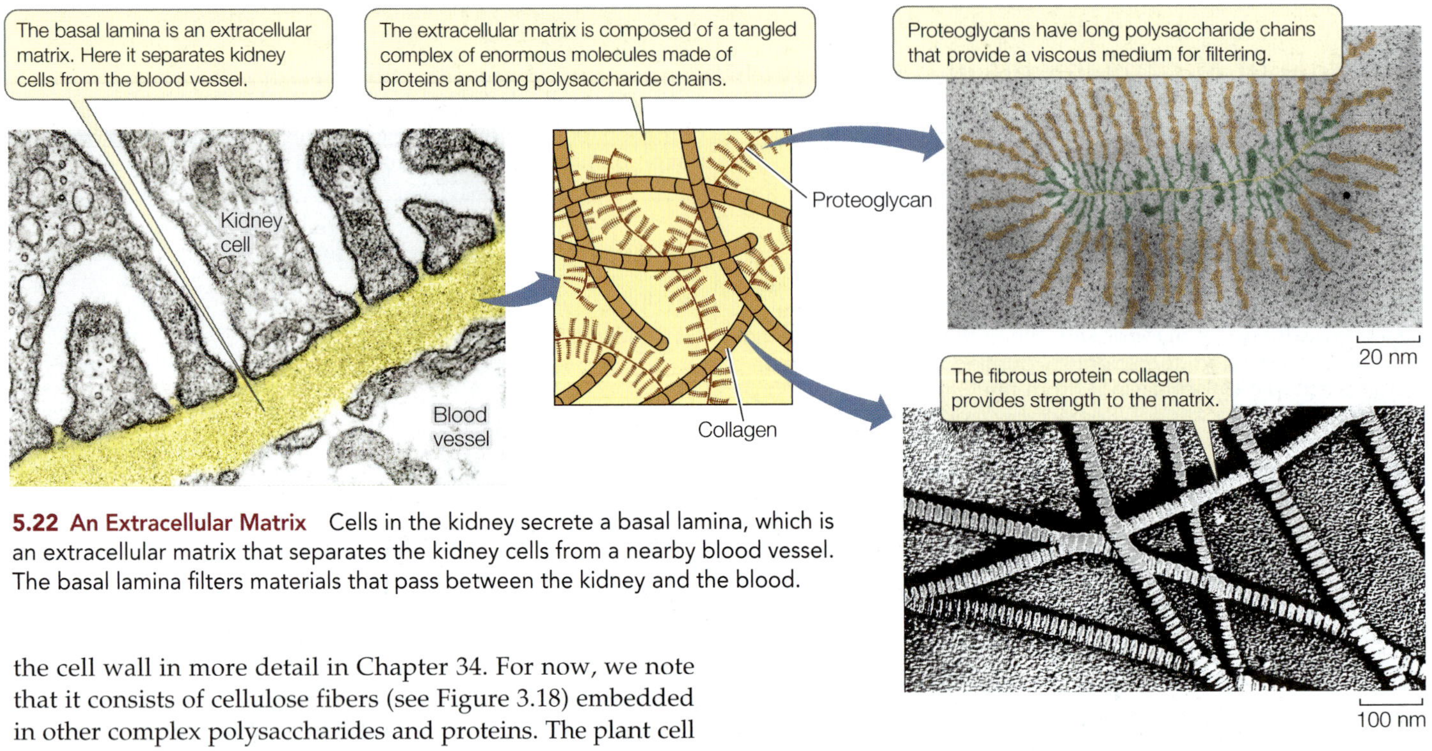
The basal lamina is an extracellular matrix. Here it separates kidney cells from the blood vessel.
The extracellular matrix is composed of a tangled complex of enormous molecules made of proteins and long polysaccharide chains.
Proteoglycans have long polysaccharide chains that provide a viscous medium for filtering.
Kidney cell
Blood vessel
Proteoglycan
Collagen
20 nm
The fibrous protein collagen provides strength to the matrix.
100 nm

5.22 An Extracellular Matrix Cells in the kidney secrete a basal lamina, which is an extracellular matrix that separates the kidney cells from a nearby blood vessel. The basal lamina filters materials that pass between the kidney and the blood.

the cell wall in more detail in Chapter 34. For now, we note that it consists of cellulose fibers (see Figure 3.18) embedded in other complex polysaccharides and proteins. The plant cell wall has three major roles:

- It provides support for the cell and plant by remaining rigid. Yet it is flexible enough that it can allow the plant to bend in the wind, for example.
- It acts as a barrier to infection by fungi and other organisms that can cause plant diseases.
- It contributes to plant form by growing as the plant cells expand.

In some cells, such as those in a leaf, the cell wall is porous to allow the passage of molecules into and out of the cell. In other cells, such as those of the plant's vascular system (which transports water and small molecules between organs), the wall is not porous.

Because of their thick cell walls, plant cells viewed under a light microscope appear to be entirely isolated from one another. But electron microscopy reveals that this is not the case. The cytoplasms of adjacent plant cells are connected by numerous plasma membrane–lined channels called **plasmodesmata**, which are about 20–40 nanometers in diameter and extend through the cell walls (see Figure 5.7). Plasmodesmata permit the diffusion of water, ions, small molecules, RNA, and proteins between connected cells, allowing for the use of these substances far from their sites of synthesis.

The extracellular matrix supports tissue functions in animals

Animal cells lack the semirigid wall that is characteristic of plant cells, but many animal cells are surrounded by, or in contact with, an **extracellular matrix**. This matrix is composed of three types of molecules: fibrous proteins such as **collagen** (the most abundant protein in mammals, constituting over 25 percent of the protein in the human body); a matrix of glycoproteins termed **proteoglycans**, consisting primarily of sugars; and a third group of proteins that link the fibrous proteins and the gel-like proteoglycan matrix together (**Figure 5.22**). These proteins and proteoglycans are secreted, along with other substances that are specific to certain body tissues, by cells that are present in or near the matrix.

The functions of the extracellular matrix are many:

- It holds cells together in tissues. In Chapter 6 we will see how there is an intercellular "glue" that is involved in both cell recognition and adhesion.
- It contributes to the physical properties of cartilage, skin, and other tissues. For example, the mineral component of bone is laid down on an organized extracellular matrix.
- It helps filter materials passing between different tissues. This is especially important in the kidney.
- It helps orient cell movements during embryonic development and during tissue repair.
- It plays a role in chemical signaling from one cell to another. Proteins connect the cell's plasma membrane to the extracellular matrix. These proteins (for example, integrin) span the plasma membrane and are involved with transmitting signals to the interior of the cell. This allows communication between the extracellular matrix and the cytoplasm of the cell.

RECAP 5.4

Extracellular structures are produced by cells and secreted outside the plasma membrane. Most consist of a fibrous component in a gel-like medium.

- What are the functions of the cell wall in plants and the extracellular matrix in animals? **See p. 100**

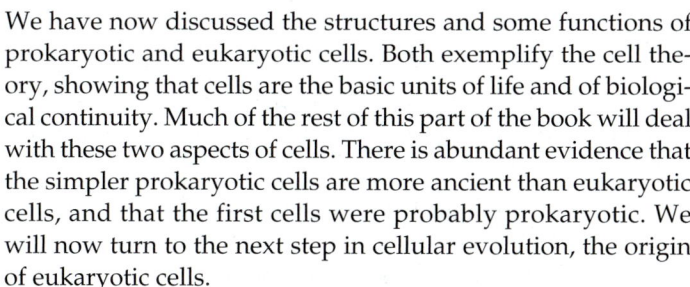

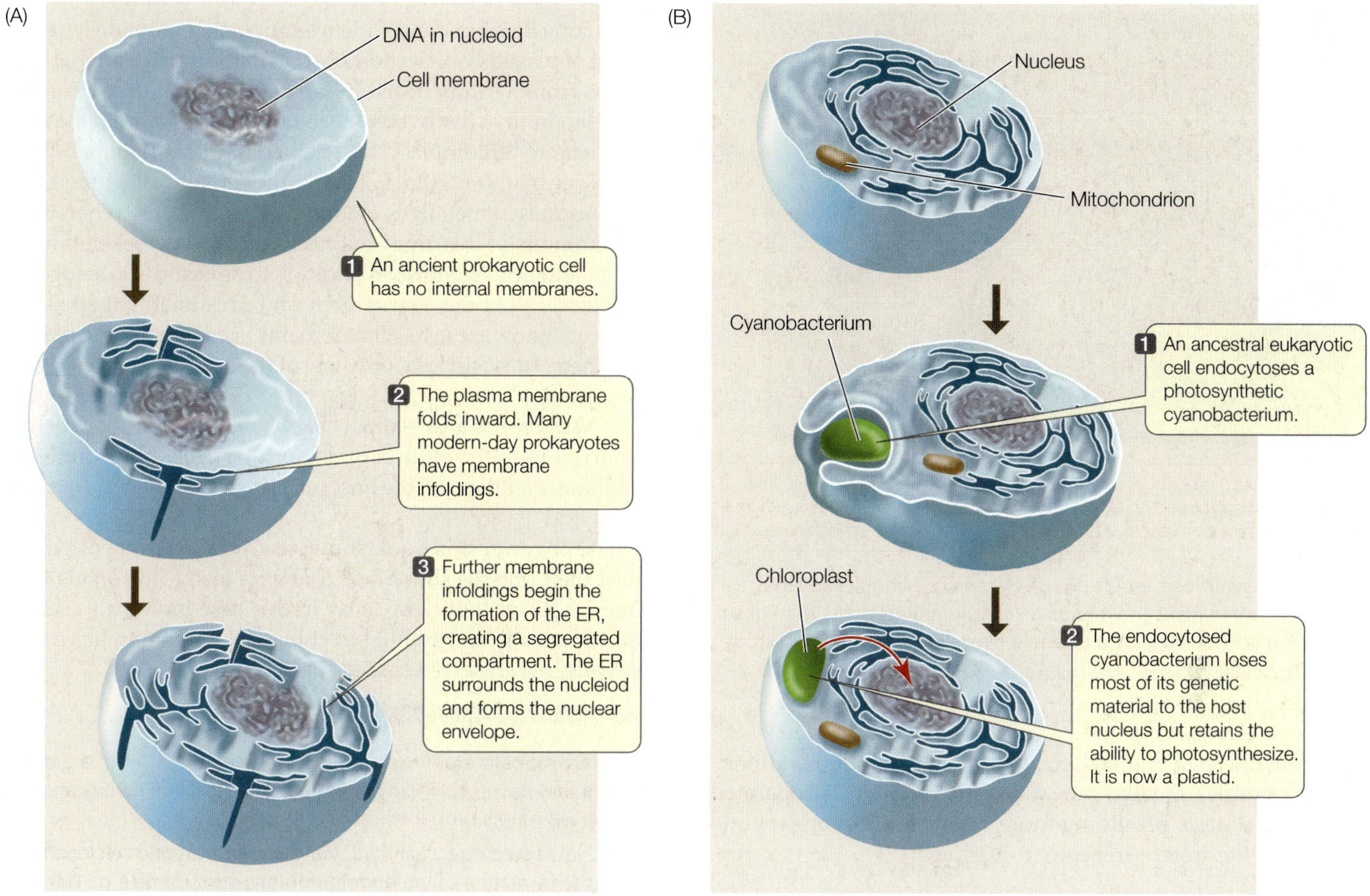
(A)
DNA in nucleoid
Cell membrane
1 An ancient prokaryotic cell has no internal membranes.
2 The plasma membrane folds inward. Many modern-day prokaryotes have membrane infoldings.
3 Further membrane infoldings begin the formation of the ER, creating a segregated compartment. The ER surrounds the nucleiod and forms the nuclear envelope.
(B)
Nucleus
Mitochondrion
Cyanobacterium
1 An ancestral eukaryotic cell endocytoses a photosynthetic cyanobacterium.
Chloroplast
2 The endocytosed cyanobacterium loses most of its genetic material to the host nucleus but retains the ability to photosynthesize. It is now a plastid.

5.23 The Origins of Organelles (A) The endomembrane system and nuclear envelope may have been formed by infolding and then fusion of the plasma membrane. (B) The endosymbiosis theory proposes that some organelles may be descended from prokaryotes that were engulfed by other, larger cells.

We have now discussed the structures and some functions of prokaryotic and eukaryotic cells. Both exemplify the cell theory, showing that cells are the basic units of life and of biological continuity. Much of the rest of this part of the book will deal with these two aspects of cells. There is abundant evidence that the simpler prokaryotic cells are more ancient than eukaryotic cells, and that the first cells were probably prokaryotic. We will now turn to the next step in cellular evolution, the origin of eukaryotic cells.

5.5 How Did Eukaryotic Cells Originate?

Life on Earth was entirely prokaryotic for about 2 billion years—from the time when prokaryotic cells first appeared until about 1.5 billion years ago, when eukaryotic cells arrived on the scene. The advent of compartmentalization—the hallmark of eukaryotes—was a major event in the history of life. It permitted many more biochemical functions to exist in the same cell than had previously been possible. Compared with a typical eukaryote, a single prokaryotic cell is often biochemically specialized, limited in the resources it can use and the functions it can perform.

What is the origin of compartmentalization? We will describe the evolution of eukaryotic organelles in more detail in Section 27.1. Here we outline two major themes in this process.

Internal membranes and the nuclear envelope probably came from the plasma membrane

We noted earlier that some bacteria contain internal membranes. How could these arise? In electron micrographs, the internal membranes of prokaryotes often appear to be inward folds of the plasma membrane. This has led to a theory that the endomembrane system and the cell nucleus originated by related processes (**Figure 5.23A**). The close relationship between the endoplasmic reticulum (ER) and the nuclear envelope in today's eukaryotes is consistent with this theory.

A bacterium with enclosed compartments would have several evolutionary advantages. Chemicals could be concentrated in particular regions of the cell, allowing chemical reactions to proceed more efficiently. A biochemical process could be segregated within an organelle with, for example, a different

Daughter cells
This cell has a green, photosynthetic plastid acquired from an algal cell.
After cell division, only one of the daughter cells inherits the plastid; the other cell must ingest a new one from the environment.

5.24 Endosymbiosis in Action A *Hatena* cell engulfs an algal cell, which then loses most of its cellular functions other than photosynthesis. This reenacts a possible event in the origin of plastids in eukaryotic cells.

pH from the rest of the cell, creating more favorable conditions for that process. Finally, gene transcription could be separated from translation, providing more opportunities for separate control of these steps in gene expression.

Some organelles arose by endosymbiosis

Symbiosis means "living together," and often refers to two organisms that coexist, each one supplying something that the other needs. Biologists have proposed that some organelles—the mitochondria and the plastids—arose not by an infolding of the plasma membrane but by one cell ingesting (but not digesting) another cell, giving rise to a symbiotic relationship. Eventually, the ingested cell lost its autonomy and some of its functions. In addition, many of the ingested cell's genes were transferred to the host's DNA. Mitochondria and plastids in today's eukaryotic cells are the remnants of these symbionts, retaining some specialized functions that benefit their host cells. This is the essence of the **endosymbiosis theory** for the origin of organelles.

Consider the case of the plastid. About 2.5 billion years ago some prokaryotes (the cyanobacteria) developed photosynthesis (see Figure 1.4). The emergence of these prokaryotes was a key event in the evolution of complex organisms because they increased the O_2 concentration in Earth's atmosphere (see Section 1.1)

According to the endosymbiosis theory, photosynthetic prokaryotes also provided the precursor of the modern-day plastid. Cells without cell walls can engulf relatively large particles by phagocytosis (see Figure 5.10). In some cases, such as that of phagocytes in the human immune system, the engulfed particle can be an entire cell, such as a bacterium. Plastids may have arisen by a similar event involving an ancestral eukaryote and a cyanobacterium (**Figure 5.23B**).

Among the abundant evidence supporting the endosymbiotic origin of plastids (see Section 27.1), perhaps the most remarkable comes from a sandy beach in Japan. In 2006 Noriko Okamoto and Isao Inouye discovered a single-celled eukaryote that contains a large "chloroplast" and named it *Hatena* (**Figure 5.24**). It turns out that the "chloroplast" is the remains of a green alga, *Nephroselmis*, which lives among the *Hatena* cells. When living autonomously, this algal cell has flagella, a cytoskeleton, ER, Golgi, and mitochondria in addition to a plastid. Once ingested by *Hatena*, all of these structures, and presumably their associated functions, are lost. What remains is essentially the plastid.

When *Hatena* divides, only one of the two daughter cells ends up with the "chloroplast." The other cell finds and ingests its own *Nephroselmis* alga—almost like a "replay" of what may have occurred in the evolution of eukaryotic cells. No wonder the Japanese scientists call the host cell *Hatena*: in Japanese, it means "how odd"!

The cover of this book shows another example of chloroplast theft: the sea slug *Elysia chlorotica* also feeds on algae for their photosynthetic capacity. In this case biologists recently found that a gene essential to chloroplast function has apparently been transferred to the slug nucleus.

RECAP 5.5

Eukaryotic cells arose long after prokaryotic cells. Some organelles may have evolved by infolding of the plasma membrane, whereas others probably evolved by endosymbiosis.

- How could membrane infolding in a prokaryotic cell lead to the formation of the endomembrane system? **See p. 101 and Figure 5.23A**
- Explain the endosymbiosis theory for the origin of chloroplasts. **See p. 102 and Figure 5.23B**

In this chapter we presented an overview of the components of cells, with some ideas about their structures, functions, and origins. As you now embark on the study of major cellular processes, keep in mind that cellular components do not exist in isolation: they are part of a dynamic, interacting system. In Chapter 6 we will show that the plasma membrane is far from a passive barrier, but instead is a multifunctional system that connects the inside of the cell with its extracellular environment.

?

What is the status of stem cell treatment for heart disease?

ANSWER

Although there are active clinical trials for using stem cell therapy in patients with heart disease in the United States and western Europe, by far the most active use of this treatment occurs in China. Stem cells are typically collected from umbilical cords at childbirth and stored for later use. There have been anecdotal reports of great successes in using such treatments to repair damaged hearts. An examination of treated and functioning hearts indicated that the stem cells were involved in repair of the heart muscle and the blood vessels that supply it.

CHAPTERSUMMARY 5

5.1 What Features Make Cells the Fundamental Units of Life?

- The **cell theory** is the unifying theory of cell biology. All living things are composed of cells, and all cells come from preexisting cells.
- A cell is small in order to maintain a large **surface area-to-volume ratio**. This allows it to exchange adequate quantities of materials with its environment. Review Figures 5.1, 5.2, ACTIVITY 5.1
- Cell structures can be studied with light and electron microscopes. Review Figure 5.3, ACTIVITY 5.2
- All cells are enclosed by a selectively permeable **plasma membrane** that separates their contents from the external environment.
- Whereas certain biochemical processes, molecules, and structures are shared by all kinds of cells, there are two categories of organisms—**prokaryotes** and **eukaryotes**—that can be distinguished by characteristic cell structures.
- Eukaryotic cells are generally larger and more complex than prokaryotic cells. They contain membrane-bound **organelles**, including the **nucleus**.

5.2 What Features Characterize Prokaryotic Cells?

- Prokaryotic cells have no internal compartments but have a **nucleoid** region containing DNA, and a **cytoplasm** containing **cytosol**, **ribosomes**, proteins, and small molecules. Some prokaryotes have additional protective structures, including a **cell wall**, an **outer membrane**, and a **capsule**. Review Figure 5.4
- Some prokaryotes have folded **internal membranes** such as those used in photosynthesis, and some have **flagella** or **pili** for motility or attachment. Review Figure 5.5
- Filamentous proteins in the cytoplasm make up the **cytoskeleton**, which assists in cell division and the maintenance of cell shape.

5.3 What Features Characterize Eukaryotic Cells?

See ANIMATED TUTORIAL 5.1

- Eukaryotic cells are larger than prokaryotic cells and contain many membrane-enclosed **organelles**. The membranes that envelop organelles ensure compartmentalization of their functions. Review Figure 5.7
- The **nucleus** contains most of the cell's DNA and participates in the control of protein synthesis. The DNA and the proteins associated with it form a material called **chromatin**. Each long, thin DNA molecule occurs in a discrete chromatin structure called a **chromosome**. Review Figure 5.8
- Within the nucleus is the **nucleolus**, where ribosome assembly begins. After partial assembly, the **ribosomes** are transported to the cytoplasm, where they are completed and function as sites of protein synthesis.
- The **endomembrane system**—consisting of the **endoplasmic reticulum** and the **Golgi apparatus**—is a series of interrelated compartments enclosed by membranes. It segregates proteins and modifies them. **Lysosomes** contain many digestive enzymes. Review Figures 5.9, 5.10, ACTIVITY 5.3, ANIMATED TUTORIAL 5.2
- **Mitochondria** and **chloroplasts** are semiautonomous organelles that process energy. Mitochondria are present in most eukaryotic organisms and contain the enzymes needed for **cellular respiration**. The cells of photosynthetic eukaryotes contain chloroplasts that harvest light energy for photosynthesis. Review Figures 5.11, 5.12
- Large **vacuoles** are present in many plant cells. A vacuole consists of a membrane-enclosed compartment full of water and dissolved substances.
- The **microfilaments**, **intermediate filaments**, and **microtubules** of the cytoskeleton provide the cell with shape, strength, and movement. Review Figure 5.14
- **Motor proteins** use cellular energy to change shape and move. They drive the bending movements of **cilia** and **flagella**, and transport organelles along microtubules within the cell. Review Figures 5.18, 5.19

5.4 What Are the Roles of Extracellular Structures?

- The plant **cell wall** consists principally of cellulose. Cell walls are pierced by **plasmodesmata** that join the cytoplasms of adjacent cells.
- In animals, the **extracellular matrix** consists of different kinds of proteins, including **collagen** and **proteoglycans**. Review Figure 5.22

5.5 How Did Eukaryotic Cells Originate?

- Infoldings of the plasma membrane could have led to the formation of some membrane-enclosed organelles, such as the endomembrane system and the nucleus. Review Figure 5.23A
- **Symbiosis** means "living together." The **endosymbiosis theory** states that mitochondria and chloroplasts originated when larger cells engulfed, but did not digest, smaller cells. Mutual benefits permitted this symbiotic relationship to be maintained, allowing the smaller cells to evolve into the eukaryotic organelles observed today. Review Figure 5.23B

Go to the Interactive Summary to review key figures, Animated Tutorials, and Activities
Life10e.com/is5

CHAPTER**REVIEW**

REMEMBERING

1. Which structure is generally present in both prokaryotic cells and eukaryotic plant cells?
   a. Chloroplasts
   b. Cell wall
   c. Nucleus
   d. Mitochondria
   e. Microtubules

2. The major factor limiting cell size is the
   a. concentration of water in the cytoplasm.
   b. need for energy.
   c. presence of membrane-enclosed organelles.
   d. ratio of surface area to volume.
   e. composition of the plasma membrane.

3. Which statement about plastids is true?
   a. They are found in prokaryotes.
   b. They are surrounded by a single membrane.
   c. They are the sites of cellular respiration.
   d. They are found only in fungi.
   e. They may contain various pigments or polysaccharides.

4. Which structure is *not* surrounded by one or more membranes?
   a. Ribosome
   b. Chloroplast
   c. Mitochondrion
   d. Peroxisome
   e. Vacuole

5. The cytoskeleton consists of
   a. cilia, flagella, and microfilaments.
   b. cilia, microtubules, and microfilaments.
   c. internal cell walls.
   d. microtubules, intermediate filaments, and microfilaments.
   e. calcified microtubules.

6. Microfilaments
   a. are composed of polysaccharides.
   b. are composed of actin.
   c. allow cilia and flagella to move.
   d. make up the spindle that aids the movement of chromosomes.
   e. maintain the position of the chloroplast in the cell.

UNDERSTANDING & APPLYING

7. If all the lysosomes within a cell suddenly ruptured, what would be the most likely result?
   a. The macromolecules in the cytosol would break down.
   b. More proteins would be made.
   c. The DNA in mitochondria would break down.
   d. The mitochondria and chloroplasts would divide.
   e. There would be no change in cell function.

8. Through how many membranes would a molecule have to pass in moving from the interior (stroma) of a chloroplast to the interior (matrix) of a mitochondrion? From the interior of a lysosome to the outside of a cell? From one ribosome to another?

9. Compare the extracellular matrix of the animal cell with the plant cell wall, with respect to composition of the fibrous and nonfibrous components, rigidity, and connectivity of cells.

ANALYZING & EVALUATING

10. The drug vincristine is used to treat many cancers. It apparently works by causing microtubules to depolymerize. Vincristine use has many side effects, including loss of dividing cells and nerve problems. Explain why this might be so.

11. The movements of newly synthesized proteins can be followed through cells using a "pulse–chase" experiment. During synthesis, proteins are tagged with a radioactive isotope (the "pulse"), and then the cells are allowed to process the proteins for varying periods of time. The locations of the radioactive proteins are then determined by isolating cell organelles and quantifying their levels of radioactivity. What results would you expect for (a) a lysosomal enzyme and (b) a protein that is released from the cell?

Go to BioPortal at **yourBioPortal.com** for Animated Tutorials, Activities, LearningCurve Quizzes, Flashcards, and many other study and review resources.

6 Cell Membranes

CHAPTER**OUTLINE**

Fender
Marshall
Marshall

Sweating: A Membrane Activity During physical activity, water is transported across the cell membranes of sweat glands and out of the skin by exocytosis.

A FEW DAYS AFTER HE BECAME PRIME MINISTER, as World War II spread across Europe, Winston Churchill told the British Parliament, "I have nothing to offer but blood, toil, tears and sweat." He may not have known that the last two, tears and sweat, are transported across cell membranes inside vesicles. The harder we work, the hotter we get and the more we sweat. As you saw in Chapter 2, sweating is a way to reduce body heat by using excess heat to evaporate water. At peak activity, we may lose as much as 2 liters of water an hour, and if you know anything about the German air attacks on London during the war, you know that the people indeed toiled hard and must have sweated a lot.

The sweat glands lie just below the surface of the skin. They are essentially cell-lined tubes surrounded by extracellular fluid. When sweating is triggered, these tubes fill with water and dissolved substances. To get from the extracellular fluid into the tube, water must go through the cells that line the tube.

A hallmark of living cells is the ability to regulate what enters and leaves their cytoplasms. This is a function of the plasma (or cell) membrane, a hydrophobic lipid bilayer with associated proteins. Because it is insoluble in the aqueous environment both inside and outside cells, the membrane is a physical barrier. But it is also a functional barrier. Whereas water is polar, the interior of the membrane is nonpolar—so water has a natural tendency to avoid the membrane. The rate of movement of water across a lipid bilayer is modest. When you engage in normal activities such as reading this book, the cell membranes enclosing the cells lining the sweat glands do not allow much water to enter or leave. But when you exercise vigorously, tiny membrane-enclosed vesicles inside the cells fill with water and some dissolved salts. In a process called exocytosis, these vesicles fuse with the cell membrane and release their watery contents (sweat) into the tubes. From there the sweat flows to the surface of the skin and evaporates.

Vesicles are not the only way to get polar water across a nonpolar membrane. In the mammalian kidney, and in plant roots, stems, and leaves, special pores called aquaporins occur in the cell membrane. Water can flow through them readily, as the proteins lining the channel have a hydrophilic inner surface. The water-carrying function of aquaporins is so well known that there is a cosmetic moisturizing cream called Aquaporin Active!

> ?
>
> Water purity is a worldwide problem. Can aquaporin membrane channels be used in water purification?
>
> See answer on p. 122.

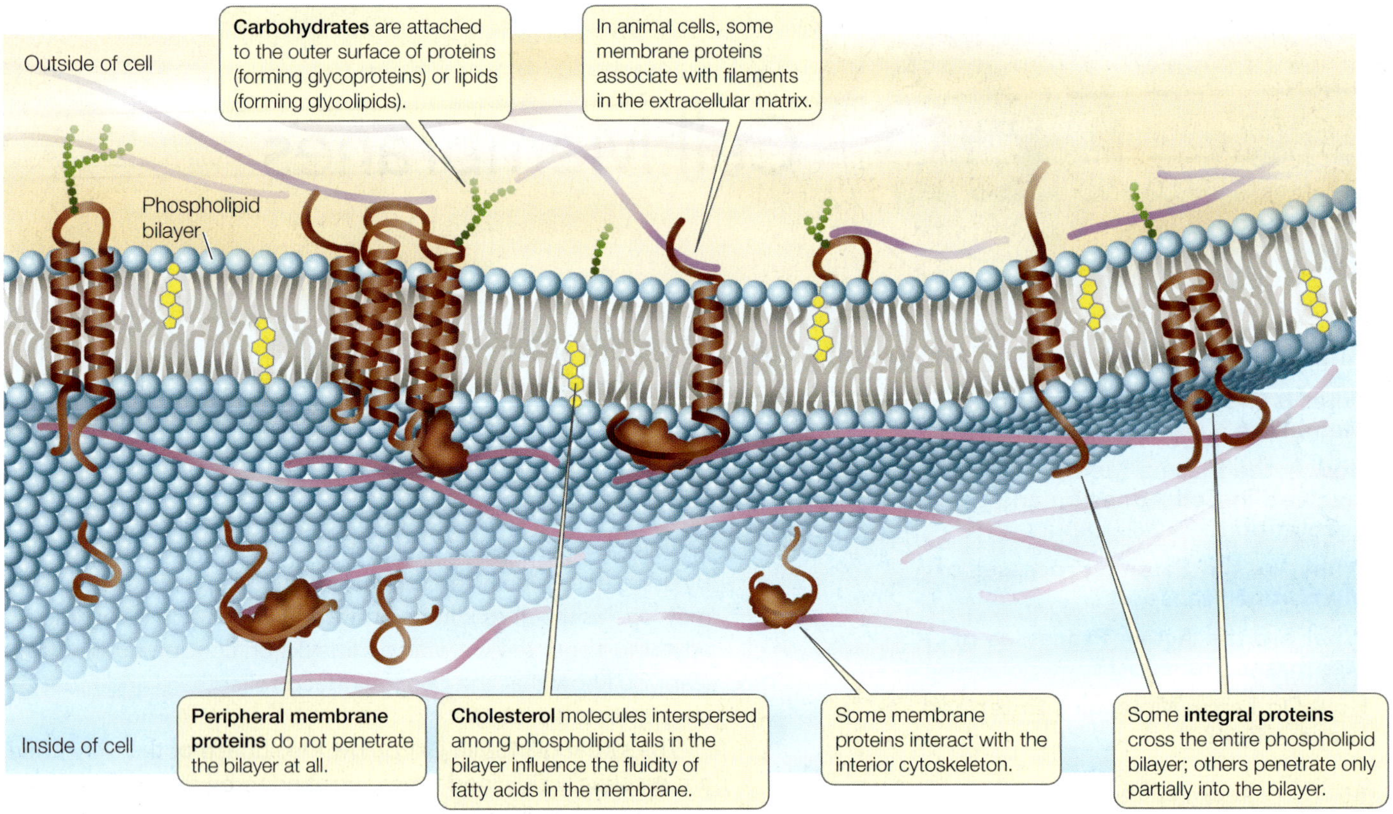

Outside of cell
Carbohydrates are attached to the outer surface of proteins (forming glycoproteins) or lipids (forming glycolipids).
In animal cells, some membrane proteins associate with filaments in the extracellular matrix.
Phospholipid bilayer
Inside of cell
Peripheral membrane proteins do not penetrate the bilayer at all.
Cholesterol molecules interspersed among phospholipid tails in the bilayer influence the fluidity of fatty acids in the membrane.
Some membrane proteins interact with the interior cytoskeleton.
Some integral proteins cross the entire phospholipid bilayer; others penetrate only partially into the bilayer.

6.1 The Fluid Mosaic Model The general molecular structure of a biological membrane is a continuous phospholipid bilayer which has proteins embedded in it or associated with it.

Go to Activity 6.1 The Fluid Mosaic Model Life10e.com/ac6.1

6.1 What Is the Structure of a Biological Membrane?

The physical organization and functioning of all biological membranes depend on their constituents: lipids, proteins, and carbohydrates. You are already familiar with these molecules from Chapter 3; it may be useful to review that chapter now. The lipids establish the physical integrity of the membrane and create an effective barrier to the rapid passage of hydrophilic materials such as water and ions. In addition, the phospholipid bilayer serves as a lipid "lake" in which a variety of proteins "float" (**Figure 6.1**). This general design is known as the **fluid mosaic model**. It is *mosaic* because it is made up of many discrete components, and *fluid* because they can move freely.

In the fluid mosaic model for biological membranes, the proteins are noncovalently embedded in the phospholipid bilayer by their hydrophobic regions (or domains), but their hydrophilic domains are exposed to the watery conditions on either side of the bilayer. These membrane proteins have several functions, including moving materials through the membrane and receiving chemical signals from the cell's external environment. Each membrane has a set of proteins suitable for the specialized functions of the cell or organelle it surrounds.

The carbohydrates associated with membranes are attached either to the lipids or to protein molecules. In plasma membranes, carbohydrates are located on the outside of the cell, where they may interact with substances in the external environment. Like some of the membrane proteins, carbohydrates are crucial in recognizing specific molecules, such as those on the surfaces of adjacent cells.

Although the fluid mosaic model is largely valid for membrane structure, it does not say much about membrane composition. As you read about the various molecules in membranes in the next sections, keep in mind that some membranes have more protein than lipids, others are lipid-rich, others have significant amounts of cholesterol or other sterols, and still others are rich in carbohydrates.

Lipids form the hydrophobic core of the membrane

The lipids in biological membranes are usually phospholipids. Recall from Section 2.2 that some compounds are hydrophilic ("water-loving") and others are hydrophobic ("water-hating"), and from Section 3.4 that a phospholipid molecule has regions of both kinds:

- *Hydrophilic regions*: The phosphorus-containing "head" of the phospholipid is electrically charged and therefore associates with polar water molecules.
- *Hydrophobic regions*: The long, nonpolar fatty acid "tails" of the phospholipid associate with other nonpolar materials; they do not dissolve in water or associate with hydrophilic substances.

Because of these properties, one way in which phospholipids can coexist with water is to form a bilayer, with the fatty acid "tails" of the two layers interacting with each other and the polar

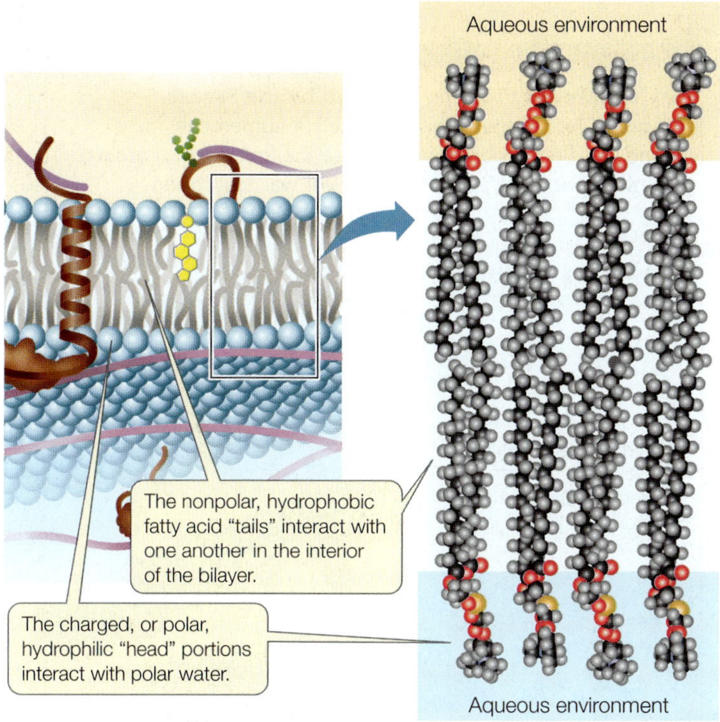
Aqueous environment
The nonpolar, hydrophobic fatty acid "tails" interact with one another in the interior of the bilayer.
The charged, or polar, hydrophilic "head" portions interact with polar water.
Aqueous environment

6.2 A Phospholipid Bilayer The phospholipid bilayer separates two aqueous regions. The eight phospholipid molecules shown on the right represent a small cross section of a membrane bilayer.

"heads" facing the outside aqueous environment (**Figure 6.2**). The thickness of a biological membrane is about 8 nanometers (0.008 μm), which is twice the length of a typical phospholipid—another indication that the membrane consists of a lipid bilayer. A typical sheet of paper is about 8,000 times thicker than this.

In the laboratory, it is easy to make artificial bilayers with the same organization as natural membranes. Small holes in such bilayers seal themselves spontaneously. This capacity of lipids to associate with one another and maintain a bilayer organization helps biological membranes fuse during vesicle formation, phagocytosis, and related processes.

All biological membranes have a similar structure, but they differ in the kinds of proteins and lipids they contain. Membranes from different cells or organelles may differ greatly in their lipid composition. Phospholipids can differ in terms of fatty acid chain length (number of carbon atoms), degree of unsaturation (number of double bonds) in the fatty acids, and the polar groups present (see Chapter 3). The saturated chains allow close packing of fatty acids in the bilayer, whereas the "kinks" in unsaturated fatty acids (see Figure 3.21) make for a less dense, more fluid packing.

Up to 25 percent of the lipid content of an animal cell plasma membrane may be the steroid cholesterol (see Section 3.4). Cholesterol preferentially associates with saturated fatty acids. When present, cholesterol is important for membrane integrity; the cholesterol in your membranes is not hazardous to your health.

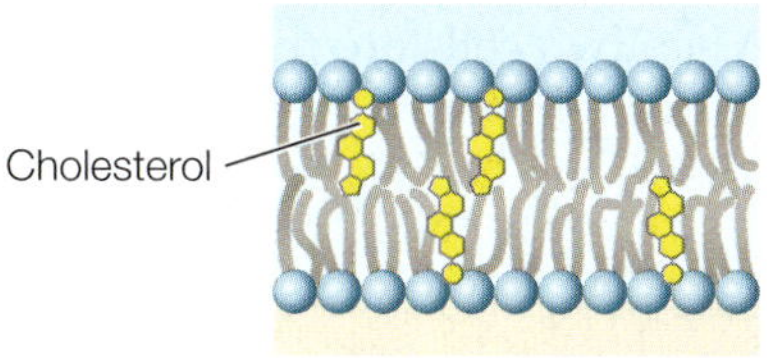
Cholesterol

The fatty acids of the phospholipids make the hydrophobic interior of the membrane somewhat fluid—about as fluid as lightweight olive oil. This fluidity permits some molecules to move laterally within the plane of the membrane. A given phospholipid molecule in the plasma membrane can travel from one end of the cell to the other in a little more than a second! However, a phospholipid molecule in one half of the bilayer is unlikely to spontaneously flip over to the other side. For that to happen, the polar part of the molecule would have to move through the hydrophobic interior of the membrane. Since spontaneous phospholipid flip-flops are rare, the inner and outer halves of the bilayer may be quite different in the kinds of phospholipids they contain.

Membrane fluidity is affected by several factors, two of which are particularly important:

- *Lipid composition*: Cholesterol and long-chain, saturated fatty acids pack tightly beside one another, with little room for movement. This close packing results in less-fluid membranes. A membrane with shorter-chain fatty acids, unsaturated fatty acids, or less cholesterol is more fluid.
- *Temperature*: Because molecules move more slowly and fluidity decreases at reduced temperatures, cellular processes that take place within the membrane may slow down or stop under cold conditions in organisms that cannot keep their bodies warm. To address this problem, some organisms simply change the lipid composition of their membranes when they get cold, replacing saturated with unsaturated fatty acids and using fatty acids with shorter tails. These changes play a role in the survival of plants, bacteria, and hibernating animals during the winter.

Go to Animated Tutorial 6.1
Lipid Bilayer Composition
Life10e.com/at6.1

Membrane proteins are asymmetrically distributed

All biological membranes contain proteins. Typically, plasma membranes have 1 protein molecule for every 25 lipid molecules. This ratio varies depending on membrane function. In the inner membrane of the mitochondrion, which is specialized for energy processing, there is 1 protein for every 15 lipids. However, myelin—a membrane that encloses portions of some neurons (nerve cells) and acts as an electrical insulator—has only 1 protein for every 70 lipids.

Membrane proteins are very diverse. In fact, about one-fourth of the protein-coding genes in the eukaryotic genome encode membrane proteins. There are two general types of membrane proteins: peripheral proteins and integral proteins.

Peripheral membrane proteins lack exposed hydrophobic groups and are not embedded in the bilayer. Instead, they have polar or charged regions that interact with exposed parts of integral membrane proteins, or with the polar heads of phospholipid molecules (see Figure 6.1).

Integral membrane proteins are at least partly embedded in the phospholipid bilayer (see Figure 6.1). Like phospholipids,

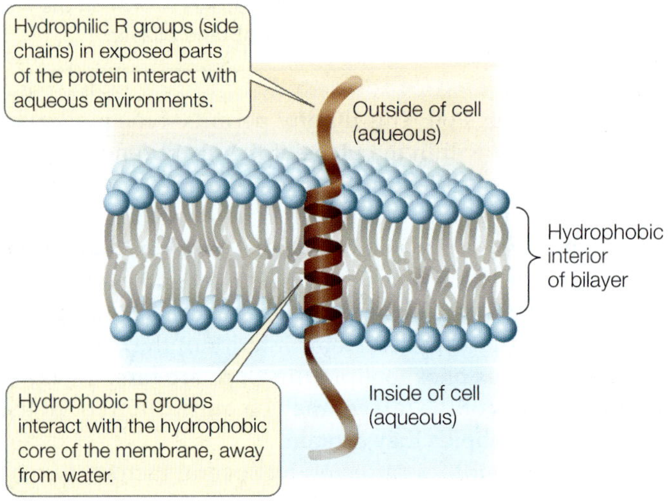
Hydrophilic R groups (side chains) in exposed parts of the protein interact with aqueous environments.
Outside of cell (aqueous)
Hydrophobic interior of bilayer
Hydrophobic R groups interact with the hydrophobic core of the membrane, away from water.
Inside of cell (aqueous)

6.3 Interactions of Integral Membrane Proteins An integral membrane protein is held in the membrane by the distribution of the hydrophilic and hydrophobic side chains on its amino acids. The hydrophilic parts of the protein extend into the aqueous cell exterior and the internal cytoplasm. The hydrophobic side chains interact with the hydrophobic lipid core of the membrane.

RESEARCH**TOOLS**

6.4 Membrane Proteins Revealed by the Freeze-Fracture Technique This HeLa cell (a human cell) membrane was first frozen to immobilize the lipids and proteins, and then fractured so that the bilayer was split open.

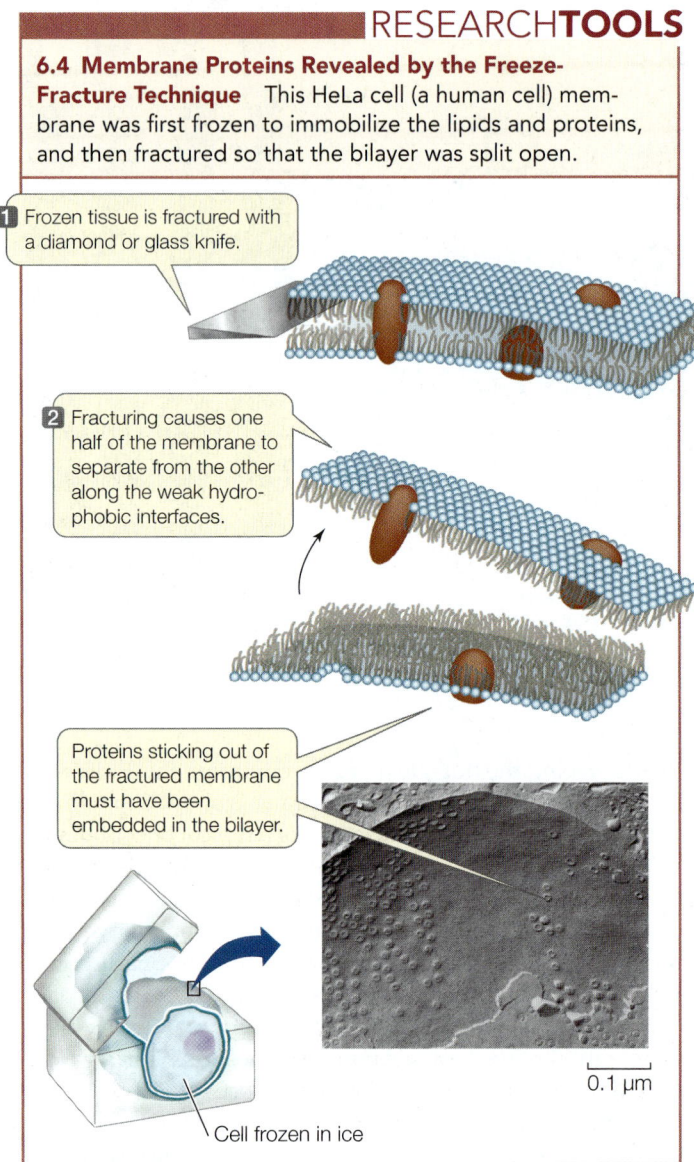

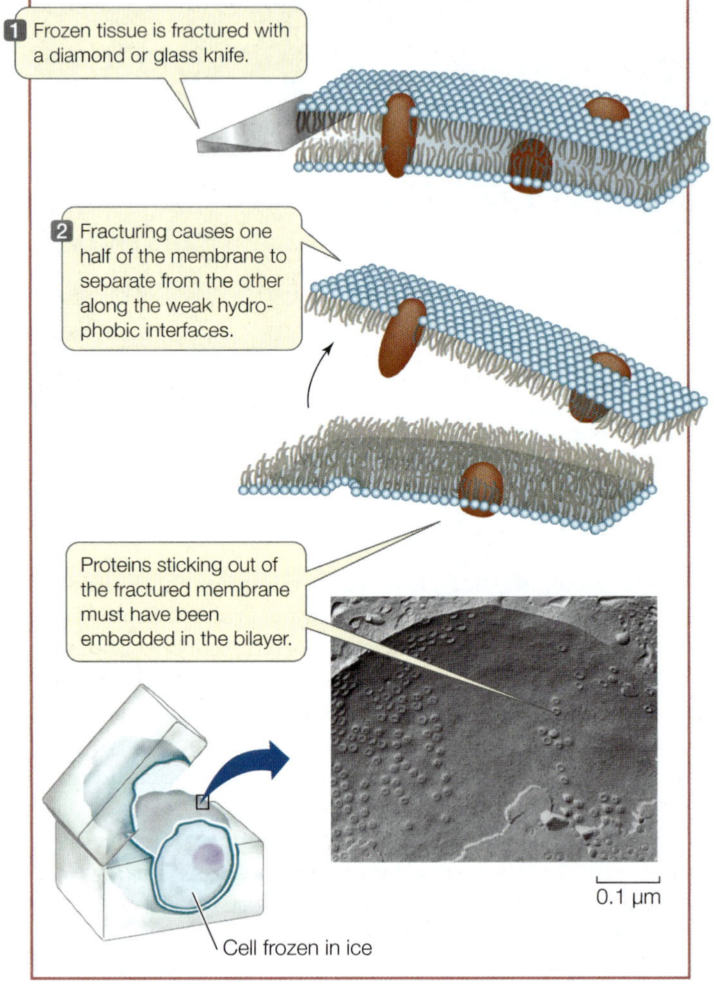
1 Frozen tissue is fractured with a diamond or glass knife.
2 Fracturing causes one half of the membrane to separate from the other along the weak hydrophobic interfaces.
Proteins sticking out of the fractured membrane must have been embedded in the bilayer.
0.1 μm
Cell frozen in ice

these proteins have both hydrophilic and hydrophobic regions (domains) (**Figure 6.3**).

- *Hydrophilic domains*: Stretches of amino acids with hydrophilic side chains (R groups; see Table 3.2) give certain regions of the protein a polar character. These hydrophilic domains interact with water and stick out into the aqueous environment inside or outside the cell.
- *Hydrophobic domains*: Stretches of amino acids with hydrophobic side chains give other regions of the protein a nonpolar character. These domains interact with the fatty acids in the interior of the phospholipid bilayer, away from water.

A special preparation method for electron microscopy, called freeze-fracturing, reveals proteins that are embedded in the phospholipid bilayers of cellular membranes (**Figure 6.4**). When the two lipid leaflets (or layers) that make up the bilayer are separated, the proteins can be seen as bumps that protrude from the interior of each membrane. The bumps are not observed when artificial bilayers of pure lipid are freeze-fractured.

Membrane proteins and lipids generally interact only noncovalently. The polar ends of proteins can interact with the polar ends of lipids, and the nonpolar regions of both molecules can interact hydrophobically. However, some membrane proteins have fatty acids or other lipid groups covalently attached to them. Their hydrophobic lipid components allow these proteins to tether themselves to the phospholipid bilayer.

Proteins are asymmetrically distributed on the inner and outer surfaces of membranes. An integral protein that extends all the way through the phospholipid bilayer and protrudes on both sides is known as a **transmembrane protein**. In addition to one or more **transmembrane domains** that extend through the bilayer, such a protein may have domains with other specific functions on the inner and outer sides of the membrane. Peripheral membrane proteins are located on one side of the membrane or the other. This asymmetrical arrangement of membrane proteins gives the two surfaces of the membrane different properties. As we will soon see, these differences have great functional significance.

Like lipids, some membrane proteins move around relatively freely within the phospholipid bilayer. Experiments that involve the technique of cell fusion illustrate this migration dramatically. When two cells are fused, a single continuous membrane forms and surrounds both cells, and some proteins from each cell distribute themselves uniformly around this membrane (**Figure 6.5**).

Although some proteins are free to migrate in the membrane, others are not, but rather appear to be "anchored" to a specific region of the membrane. These membrane regions are

INVESTIGATING**LIFE**

6.5 Rapid Diffusion of Membrane Proteins Two animal cells can be fused together in the laboratory, forming a single large cell (heterokaryon). This phenomenon was used to test whether membrane proteins can diffuse independently in the plane of the plasma membrane.[a]

HYPOTHESIS Proteins embedded in a membrane can diffuse freely within the membrane.

Method

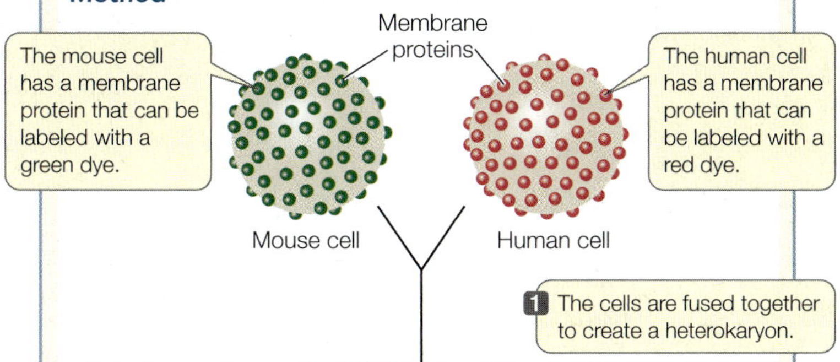

Membrane proteins
The mouse cell has a membrane protein that can be labeled with a green dye.
The human cell has a membrane protein that can be labeled with a red dye.
Mouse cell
Human cell
1 The cells are fused together to create a heterokaryon.

Results

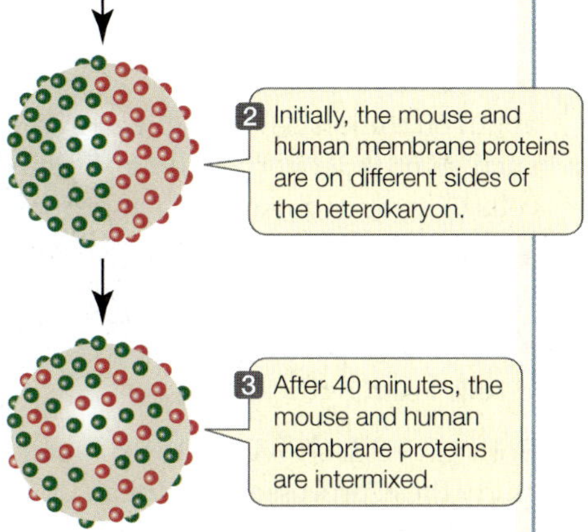

2 Initially, the mouse and human membrane proteins are on different sides of the heterokaryon.
3 After 40 minutes, the mouse and human membrane proteins are intermixed.

CONCLUSION Membrane proteins can diffuse rapidly in the plane of the membrane.

Go to **BioPortal** for discussion and relevant links for all INVESTIGATING**LIFE** figures.

[a]Frye, L. D. and M. Edidin. 1970. *Journal of Cell Science* 7: 319–335.

like a corral of horses on a farm: the horses are free to move around within the fenced area but not outside it. An example is the protein in the plasma membrane of a muscle cell that recognizes a chemical signal from a neuron. This protein is normally found only at the specific region where the neuron meets the muscle cell. How does this happen?

Proteins inside the cell can restrict the movement of proteins within a membrane. The cytoskeleton may have components just below the inner face of the membrane that are attached to membrane proteins protruding into the cytoplasm. The stability of the cytoskeletal components may thus restrict movement of attached membrane proteins.

Membranes are constantly changing

Membranes in eukaryotic cells are constantly forming, transforming from one type to another, fusing with one another, and breaking down. As we discussed in Chapter 5, fragments of membrane move, in the form of vesicles, from the endoplasmic reticulum (ER) to the Golgi, and from the Golgi to the plasma membrane (see Figure 5.9). Secondary lysosomes form when primary lysosomes from the Golgi fuse with phagosomes from the plasma membrane (see Figure 5.10).

Because all membranes appear similar under the electron microscope, and because they interconvert readily, we might expect all subcellular membranes to be chemically identical. However, that is not the case: there are major chemical differences among the membranes of even a single cell. Membranes are changed chemically when they form parts of certain organelles. In the Golgi apparatus, for example, the membranes of the *cis* face closely resemble those of the ER in chemical composition, but those of the *trans* face are more similar to the plasma membrane.

Plasma membrane carbohydrates are recognition sites

In addition to lipids and proteins, the plasma membrane contains carbohydrates (see Figure 6.1). The carbohydrates are located on the outer surface of the plasma membrane and serve as recognition sites for other cells and molecules, as you will see in Section 6.2.

Membrane-associated carbohydrates may be covalently bonded to lipids or to proteins:

- A **glycolipid** consists of a carbohydrate covalently bonded to a lipid. Extending out from the cell surface, the carbohydrate may serve as a recognition signal for interactions between cells. For example, the carbohydrates on some glycolipids change when cells become cancerous. This change may allow white blood cells to target cancer cells for destruction.
- A **glycoprotein** consists of one or more short carbohydrate chains covalently bonded to a protein. The bound carbohydrates are oligosaccharides, usually not exceeding 15 monosaccharide units in length (see Section 3.3). A proteoglycan (see Section 5.4) is a more heavily glycosylated protein: it has more carbohydrate molecules attached to it, and the carbohydrate chains are often longer than they are in glycoproteins. The carbohydrates of glycoproteins and proteoglycans often function in cell recognition and adhesion.

The "alphabet" of monosaccharides on the outer surfaces of membranes can generate a large diversity of messages. Recall from Section 3.3 that monosaccharides are simple carbohydrates, often containing five or six carbons in a ring structure, which can bond with one another in various configurations. They may form linear or branched oligosaccharides with many different three-dimensional shapes. An oligosaccharide of a specific shape on one cell can bind to a complementary shape on an adjacent cell. This binding is the basis of cell–cell adhesion.

WORKING WITH**DATA:**

Rapid Diffusion of Membrane Proteins

Original paper

Frye, L. D. and M. Edidin. 1970. The rapid intermixing of cell surface antigens after formation of mouse-human heterokaryons. *Journal of Cell Science* 7: 319–335.

Analyze the Data

One of the key experiments providing evidence for the fluid mosaic model was performed by Louis Frye and Michael Edidin at Johns Hopkins University. The scientists took advantage of the recently developed technique of cell fusion and used it to show that membrane proteins rapidly diffuse within the plane of the membrane (see Figure 6.5). Under the right conditions, two cells could fuse together, forming a binucleate cell with one continuous membrane. In this case, mouse and human cells were fused, and membrane proteins specific to each cell type were visualized using antibodies—the mouse one coupled to a green dye and the human to a red dye. Immediately after fusion, half of the cell membrane stained red, and half green. Then over time, the colors intermixed, demonstrating that the mouse and human proteins were diffusing within the membrane. The percentage of cells that had red and green colors fully intermixed was calculated over time after cell fusion. The results are shown in the table.

QUESTION 1

Plot the percentage of fully mixed cells over time. How long did it take for complete mixing?

QUESTION 2

What does your answer to Question 1 indicate about the rate of diffusion of the mouse and human proteins?

Time (min)	Cells with fully mixed proteins (%)
5	0
10	3
25	40
40	94
120	100

Go to **BioPortal** for all WORKING WITH**DATA** exercises

RECAP 6.1

The fluid mosaic model applies to the plasma membrane and the membranes of organelles. An integral membrane protein has both hydrophilic and hydrophobic domains, which affect its position and function in the membrane. Carbohydrates that attach to lipids and proteins on the outside of the membrane serve as recognition sites.

- What are some of the features of the fluid mosaic model of biological membranes? **See p. 106**
- Explain how the hydrophobic and hydrophilic regions of phospholipids cause a membrane bilayer to form. **See Figures 6.1 and 6.2**
- What differentiates an integral protein from a peripheral protein? **See pp. 107–108 and Figure 6.1**
- What is the experimental evidence that membrane proteins can diffuse in the plane of the membrane? **See p. 108 and Figure 6.5**

Now that you understand the structure of biological membranes, let's see how their components function. In the next section we'll focus on the membrane that surrounds individual cells: the plasma membrane. We'll then look at how the plasma membrane allows individual cells to be grouped together into multicellular systems of tissues.

6.2 How Is the Plasma Membrane Involved in Cell Adhesion and Recognition?

Often the cells of multicellular organisms exist in specialized groups with similar functions, called tissues. Your body has about 60 trillion cells organized into various kinds of tissues—such as muscle, nerve, and epithelium. Two processes allow cells to arrange themselves in groups:

- **Cell recognition**, in which one cell specifically binds to another cell of a certain type
- **Cell adhesion**, in which the connection between the two cells is strengthened

Both processes involve the plasma membrane. One way to study these processes is to break down a tissue into its individual cells and then allow them to adhere to one another again. This type of experiment is most easily done in relatively simple organisms, such as sponges, which provide good models for studying processes that also occur in the complex tissues of larger species.

Sponges are multicellular marine animals that have only a few distinct cell layers (see Section 31.5). The cells of a sponge adhere to one another but can be separated mechanically by passing the animal several times through a fine wire screen (**Figure 6.6**). Through this process, what was a single animal becomes hundreds of individual cells suspended in seawater. If such cells are stirred gently for a few hours, cell recognition occurs: The cells bump into and recognize one another, sticking together in the same shape and tissue organization as the original sponge. This recognition is species-specific; if disaggregated sponge cells from two different species are placed in the same container and shaken, individual cells will stick only to other cells of the same species. Two different sponges form, just like the ones at the start of the experiment.

Such tissue-specific and species-specific cell recognition and cell adhesion are essential to the formation and maintenance of tissues in multicellular organisms. Think of your own body.

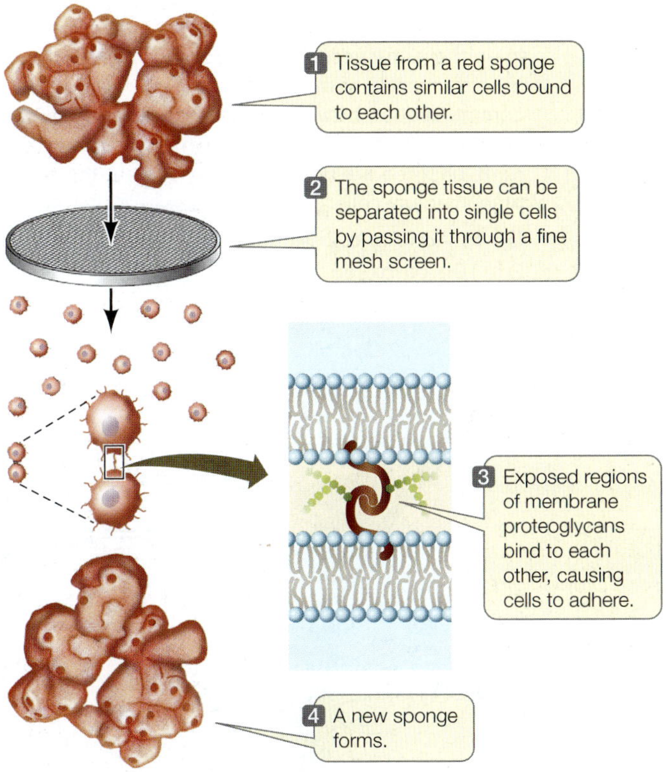
1 Tissue from a red sponge contains similar cells bound to each other.
2 The sponge tissue can be separated into single cells by passing it through a fine mesh screen.
3 Exposed regions of membrane proteoglycans bind to each other, causing cells to adhere.
4 A new sponge forms.

6.6 Cell Recognition and Adhesion In most cases (including the aggregation of animal cells into tissues), the binding between molecules is homotypic (same to same).

What keeps muscle cells bound to muscle cells and skin to skin? Specific cell adhesion is so obvious a characteristic of complex organisms that it is easy to overlook. You will see many examples of specific cell adhesion throughout this book; here we describe its general principles. As you will see, cell recognition and cell adhesion depend on plasma membrane proteins.

Cell recognition and adhesion involve proteins and carbohydrates at the cell surface

The molecules responsible for cell recognition and adhesion in sponges are proteoglycans (often 80% carbohydrate by molecular weight) that carry two kinds of carbohydrates. One kind is relatively small and binds to membrane components, keeping the proteoglycan attached to the cell. The other kind of carbohydrate is a larger, sulfated polysaccharide. If the sulfated polysaccharide from a particular species of sponge is purified and attached to cellulose beads, the beads will aggregate together or with sponge cells—but only with cells of the same species from which the polysaccharide was purified. This demonstrates that the sulfated polysaccharide is responsible for both the specific recognition and adhesion of the sponge cells.

Cell adhesion can result from interactions between the carbohydrates that are parts of glycolipids, glycoproteins, or proteoglycans—as is the case in sponge cells. In other cases, a carbohydrate on one cell interacts with a membrane protein on another cell. Or two proteins can interact directly. As we described in Section 3.2, a protein not only has a specific shape, it also has specific chemical groups exposed on its surface where they can interact with other substances, including other proteins. Both of these features allow binding to other specific molecules. Cell adhesion occurs in all kinds of multicellular organisms. In plants, cell adhesion may be mediated by both integral membrane proteins and specific carbohydrates in the cell walls.

In most cases, the binding of cells in a tissue is **homotypic**; that is, the same molecule sticks out of both cells, and the exposed surfaces bind to each other. But **heterotypic** binding (between different molecules on different cells) also occurs. In this case, different chemical groups on different surface molecules have an affinity for one another. For example, when the mammalian sperm meets the egg, different proteins on the two types of cells have complementary binding surfaces. Similarly, some algae form male and female reproductive cells (analogous to sperm and eggs) that have flagella to propel them toward each other. Male and female cells can recognize each other by heterotypic glycoproteins on their flagella.

Three types of cell junctions connect adjacent cells

In a complex multicellular organism, cell recognition molecules allow specific types of cells to bind to one another. Often, after the initial binding, both cells contribute material to form additional membrane structures that connect them to one another. These specialized structures, called **cell junctions**, are most evident in electron micrographs of epithelial tissues, which are layers of cells that line body cavities or cover body surfaces. These surfaces often receive stresses or must retain their contents under pressure, so it is particularly important that their cells adhere tightly. We will examine three types of cell junctions that enable animal cells to seal intercellular spaces, reinforce attachments to one another, and communicate with each other. Tight junctions, desmosomes, and gap junctions, respectively, perform these three functions (Figure 6.7).

- **Tight junctions** prevent substances from moving through the spaces between cells. For example, cells lining the bladder have tight junctions so urine cannot leak out into the body cavity. Another important function of tight junctions is to maintain distinct faces of a cell within a tissue by restricting the migration of membrane proteins over the cell surface from one face to the other.
- **Desmosomes** hold neighboring cells firmly together, acting like spot welds or rivets. Materials can still move around in the extracellular matrix. This provides mechanical stability for tissues such as skin that receive physical stress.
- **Gap junctions** are channels that run between membrane pores in adjacent cells, allowing substances to pass between cells. In the heart, for example, gap junctions allow the rapid spread of electric current (mediated by ions) so the heart muscle cells beat in unison.

Cell membranes adhere to the extracellular matrix

In Section 5.4 we described the extracellular matrices of animal cells, which are composed of collagen protein arranged in

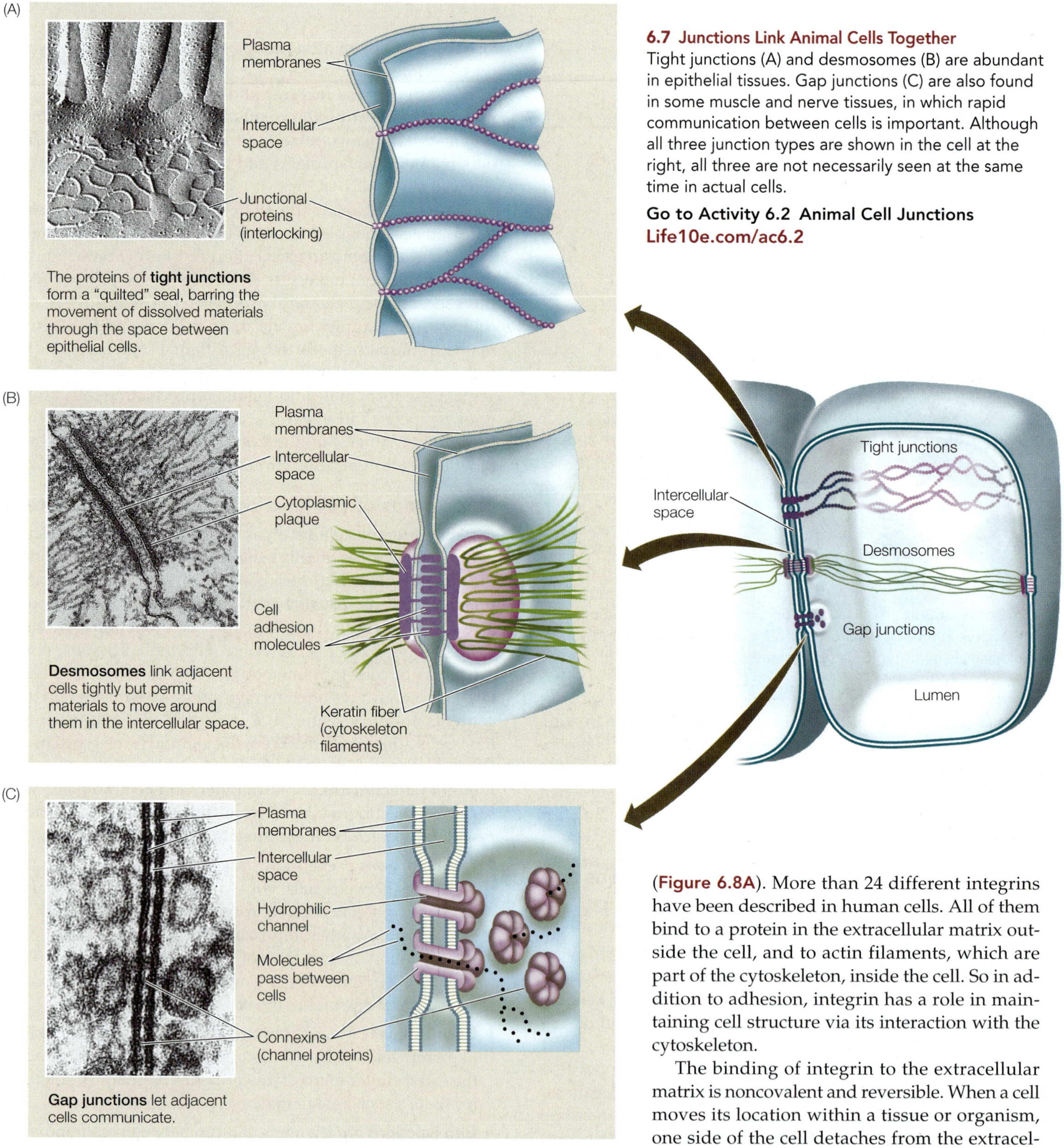
(A)
Plasma membranes
Intercellular space
Junctional proteins (interlocking)
The proteins of **tight junctions** form a "quilted" seal, barring the movement of dissolved materials through the space between epithelial cells.
(B)
Plasma membranes
Intercellular space
Cytoplasmic plaque
Cell adhesion molecules
Keratin fiber (cytoskeleton filaments)
Desmosomes link adjacent cells tightly but permit materials to move around them in the intercellular space.
(C)
Plasma membranes
Intercellular space
Hydrophilic channel
Molecules pass between cells
Connexins (channel proteins)
Gap junctions let adjacent cells communicate.
Intercellular space
Tight junctions
Desmosomes
Gap junctions
Lumen

6.7 Junctions Link Animal Cells Together Tight junctions (A) and desmosomes (B) are abundant in epithelial tissues. Gap junctions (C) are also found in some muscle and nerve tissues, in which rapid communication between cells is important. Although all three junction types are shown in the cell at the right, all three are not necessarily seen at the same time in actual cells.

Go to Activity 6.2 Animal Cell Junctions
Life10e.com/ac6.2

fibers in a gelatinous matrix of proteoglycans. The attachment of a cell to the extracellular matrix is important in maintaining the integrity of a tissue. In addition, some cells can detach from their neighbors, move, and attach to other cells; this is often mediated by interactions with the extracellular matrix.

A transmembrane protein called **integrin** often mediates the attachment of epithelial cells to the extracellular matrix (**Figure 6.8A**). More than 24 different integrins have been described in human cells. All of them bind to a protein in the extracellular matrix outside the cell, and to actin filaments, which are part of the cytoskeleton, inside the cell. So in addition to adhesion, integrin has a role in maintaining cell structure via its interaction with the cytoskeleton.

The binding of integrin to the extracellular matrix is noncovalent and reversible. When a cell moves its location within a tissue or organism, one side of the cell detaches from the extracellular matrix while the other side extends in the direction of movement, forming new attachments in that direction (**Figure 6.8B**). The integrin at the "back" of the cell (away from the direction of movement) is brought into the cytoplasm by endocytosis (see Section 6.5) so that it can be recycled and used for new attachments at the "front" of the cell. These events are important for cell movement within the developing embryo, and for the spread of cancer cells.

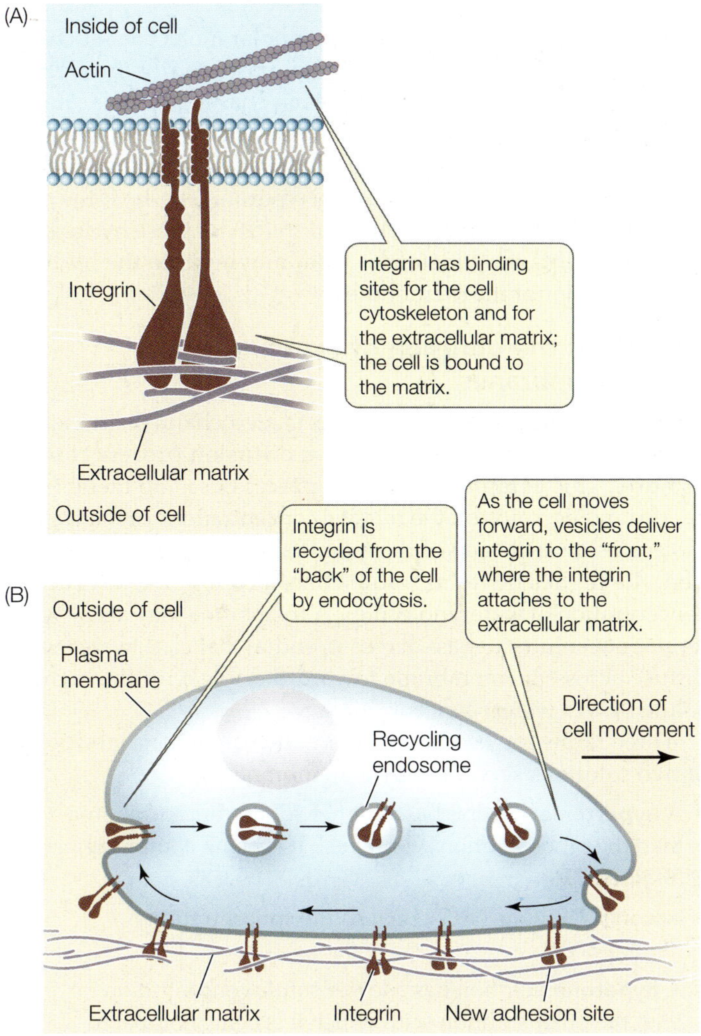
(A)
Inside of cell
Actin
Integrin
Integrin has binding sites for the cell cytoskeleton and for the extracellular matrix; the cell is bound to the matrix.
Extracellular matrix
Outside of cell
(B)
Outside of cell
Integrin is recycled from the "back" of the cell by endocytosis.
As the cell moves forward, vesicles deliver integrin to the "front," where the integrin attaches to the extracellular matrix.
Plasma membrane
Recycling endosome
Direction of cell movement
Extracellular matrix
Integrin
New adhesion site

6.8 Integrins and the Extracellular Matrix (A) Integrins mediate the attachment of cells to the extracellular matrix. (B) Cell movements are mediated by integrin attachment.

RECAP 6.2

In multicellular organisms, cells arrange themselves into tissues via the processes of cell recognition and cell adhesion. These processes are mediated by membrane-associated proteins and carbohydrates. Cell membrane proteins also interact with the extracellular matrix. Cell junctions assist in strengthening tissues and allow cells to communicate with one another.

- Describe the difference between cell recognition and cell adhesion. **See p. 110**
- How do the three types of cell junctions regulate the passage of materials between cells and through the intercellular space? **See p. 111 and Figure 6.7**

We have just examined how the plasma membrane and molecules associated with it accommodate binding between cells and the maintenance of cell adhesion. We'll turn now to another major function of membranes: regulating the substances that enter or leave a cell or organelle.

6.3 What Are the Passive Processes of Membrane Transport?

As you have already learned, biological membranes have many functions, and control of the cell's internal composition is one of the most important. Biological membranes allow some substances, but not others, to pass through them. This characteristic of membranes is called **selective permeability**. Selective permeability allows the membrane to determine what substances enter or leave a cell or organelle.

There are two fundamentally different processes by which substances cross biological membranes:

- The processes of **passive transport** do not require the input of chemical energy to drive them.
- The processes of **active transport** require the input of chemical energy (metabolic energy).

This section focuses on the passive processes by which substances cross membranes. The energy for the passive transport of a substance comes from the difference between its concentration on one side of the membrane and its concentration on the other—its **concentration gradient**. Passive transport can involve either of two types of diffusion: simple diffusion through the phospholipid bilayer, or facilitated diffusion via channel proteins or carrier proteins.

Go to Animated Tutorial 6.2
Passive Transport
Life10e.com/at6.2

Diffusion is the process of random movement toward a state of equilibrium

In a solution, there is a tendency for all of the components to be evenly distributed. You can see this when a drop of ink is allowed to fall into a gelatin suspension. Initially the pigment molecules are very concentrated, but they will move about at random, slowly spreading until the intensity of the color is exactly the same throughout the gelatin.

A solution in which the solute particles are uniformly distributed is said to be at equilibrium because there will be no future net change in their concentration. Being at equilibrium does not mean that the particles have stopped moving; it just means that they are moving in such a way that their overall distribution does not change.

Diffusion is the process of random movement toward a state of equilibrium. Although the motion of each individual particle is absolutely random, the net movement of particles is directional until equilibrium is reached. Diffusion is thus a net movement from regions of greater concentration to regions of lesser concentration.

In a complex solution (one with many different solutes), the diffusion of each solute is independent of those of the others. How fast a substance diffuses depends on three factors:

- The *diameter* of the molecules or ions: smaller molecules diffuse faster.
- The *temperature* of the solution: higher temperatures lead to faster diffusion because ions or molecules have more energy, and thus move more rapidly, at higher temperatures.
- The *concentration gradient* in the system—that is, the change in solute concentration with distance in a given direction: the greater the concentration gradient, the more rapidly a substance diffuses.

DIFFUSION WITHIN CELLS AND TISSUES In a small volume such as that inside a cell, solutes distribute themselves rapidly by diffusion. Small molecules and ions may move from one end of an organelle to another in a millisecond (10^{-3} s, or one-thousandth of a second). However, the usefulness of diffusion as a transport mechanism declines drastically as distances become greater. In the absence of mechanical stirring, diffusion across more than a centimeter may take an hour or more, and diffusion across meters may take years! Diffusion would not be adequate to distribute materials over the length of a human body, much less that of a larger organism. But within our cells or across layers of one or two cells, diffusion is rapid enough to distribute small molecules and ions almost instantaneously.

DIFFUSION ACROSS MEMBRANES In a solution without barriers, all the solutes diffuse at rates determined by temperature, their physical properties, and their concentration gradients. If a biological membrane divides the solution into separate compartments, then the movement of the different solutes can be affected by the properties of the membrane. The membrane is said to be permeable to solutes that can cross it more or less easily, but impermeable to substances that cannot move across it.

Molecules to which the membrane is impermeable remain in separate compartments, and their concentrations may be different on the two sides of the membrane. Molecules to which the membrane is permeable diffuse from one compartment to the other until their concentrations are equal on both sides of the membrane, and equilibrium is reached. After that point, individual molecules will continue to pass through the membrane, but *there will be no net change in concentration.*

Simple diffusion takes place through the phospholipid bilayer

In **simple diffusion**, small molecules pass through the phospholipid bilayer of the membrane. A molecule that is itself hydrophobic, and is therefore soluble in lipids, enters the membrane readily and is able to pass through it. The more lipid-soluble the molecule is, the more rapidly it diffuses through the membrane bilayer. This statement holds true over a wide range of molecular weights.

By contrast, electrically charged or polar molecules, such as amino acids, sugars, and ions, do not pass readily through a membrane for two reasons. First, such charged or polar molecules are not very soluble in the hydrophobic interior of the bilayer. Second, such substances form many hydrogen bonds with water and ions in the aqueous environment, be it the cytoplasm or the cell exterior. The multiplicity of these hydrogen bonds prevents the substances from moving into the hydrophobic interior of the membrane.

Osmosis is the diffusion of water across membranes

Water molecules pass through specialized channels in membranes (see the opening story) by a diffusion process called **osmosis**. This completely passive process uses no metabolic energy and depends on the relative concentrations of the water molecules on each side of the membrane. In a particular solution, the higher the total solute concentration, the lower the concentration of water molecules. A membrane may allow water but not solutes to pass across it, and in that case, water will diffuse across the membrane toward the side with the higher solute (lower water) concentration.

Three terms are used to compare the solute concentrations of two solutions separated by a membrane:

- A **hypertonic** solution has a higher solute concentration than the other solution with which it is being compared (**Figure 6.9A**).
- **Isotonic** solutions have equal solute concentrations (**Figure 6.9B**).
- A **hypotonic** solution has a lower solute concentration than the other solution with which it is being compared (**Figure 6.9C**).

Water moves from a hypotonic solution across a membrane to a hypertonic solution.

When we say that "water moves," bear in mind that we are referring to the net movement of water. Since it is so abundant, water is constantly moving through protein channels across the plasma membrane into and out of cells. What concerns us here is whether the overall movement is greater in one direction or the other.

The concentration of solutes in the environment determines the direction of osmosis in all animal cells. A red blood cell takes up water from a solution that is hypotonic to the cell's contents. The cell bursts because its plasma membrane cannot withstand the pressure created by the water entry and the resultant swelling. Conversely, the cell shrinks if the solution surrounding it is hypertonic to its contents. The integrity of red and white blood cells is absolutely dependent on the maintenance of a constant solute concentration in the blood plasma: the plasma must be isotonic to the blood cells if the cells are not to burst or shrink. Regulation of the solute concentration of body fluids is thus an important process for organisms without cell walls.

In contrast to animal cells, the cells of plants, archaea, bacteria, fungi, and some protists have cell walls that limit their volumes and keep them from bursting. Cells with sturdy walls

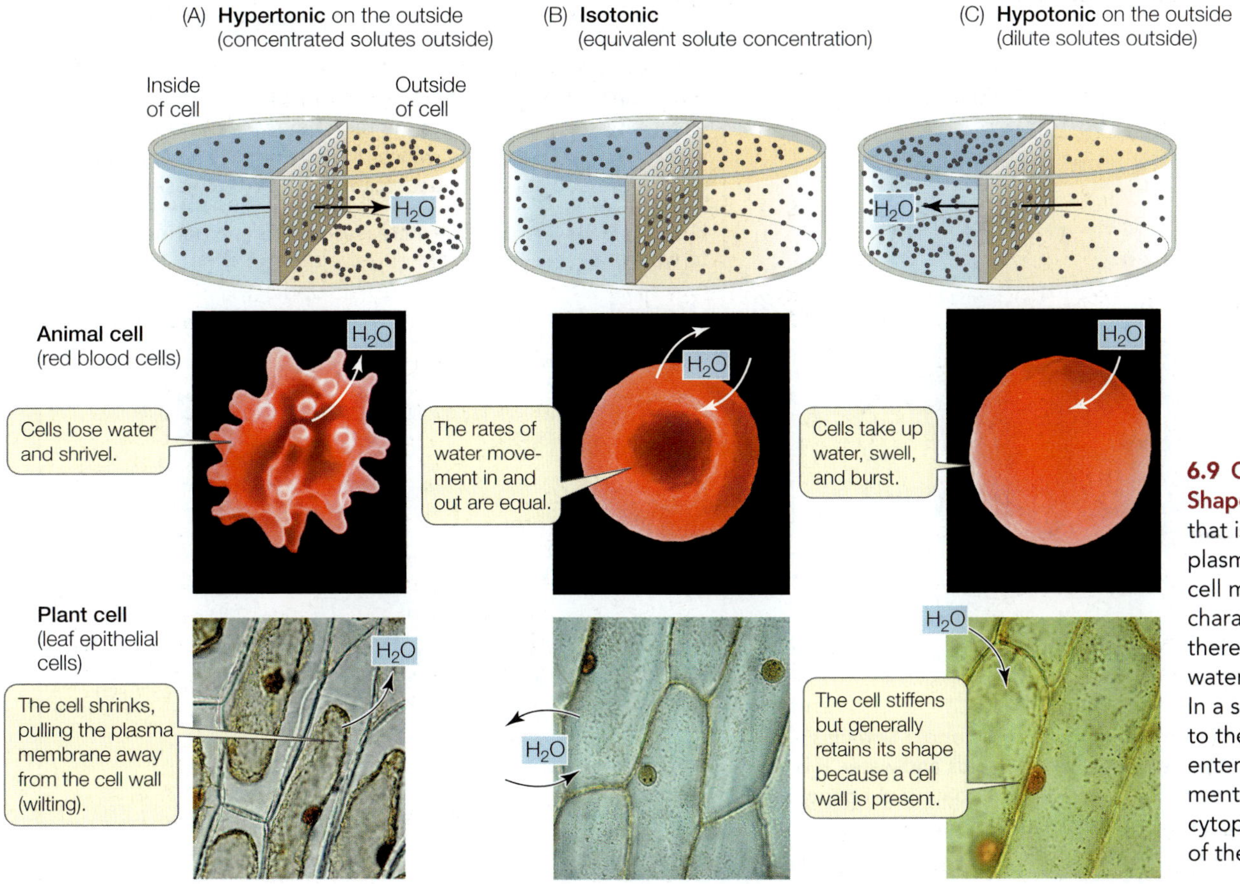
(A) Hypertonic on the outside (concentrated solutes outside)
(B) Isotonic (equivalent solute concentration)
(C) Hypotonic on the outside (dilute solutes outside)
Inside of cell
Outside of cell
H_2O
Animal cell (red blood cells)
Cells lose water and shrivel.
The rates of water movement in and out are equal.
Cells take up water, swell, and burst.
Plant cell (leaf epithelial cells)
The cell shrinks, pulling the plasma membrane away from the cell wall (wilting).
The cell stiffens but generally retains its shape because a cell wall is present.

6.9 Osmosis Can Modify the Shapes of Cells In a solution that is isotonic with the cytoplasm (B), a plant or animal cell maintains a consistent, characteristic shape because there is no net movement of water into or out of the cell. In a solution that is hypotonic to the cytoplasm (C), water enters the cell. An environment that is hypertonic to the cytoplasm (A) draws water out of the cell.

take up a limited amount of water, and in so doing they build up internal pressure against the cell wall, which prevents further water from entering. This pressure within the cell is called **turgor pressure**. Turgor pressure keeps plants upright (and lettuce crisp) and is the driving force for the enlargement of plant cells. It is a normal and essential component of plant growth. If enough water leaves the cells, turgor pressure drops and the plant wilts. Turgor pressure reaches about 100 pounds per square inch (0.7 kg/cm^2)—several times greater than the pressure in automobile tires. This pressure is so great that the cells would change shape and detach from one another were it not for adhesive molecules in the plant cell walls.

Diffusion may be aided by channel proteins

As we saw earlier, polar or charged substances such as water, amino acids, sugars, and ions do not readily diffuse across membranes. But they can cross the hydrophobic phospholipid bilayer passively (that is, without the input of energy) in one of two ways, depending on the substance:

- **Channel proteins** are integral membrane proteins that form channels across the membrane through which certain substances can pass.
- **Carrier proteins** bind substances and speed up their diffusion through the phospholipid bilayer.

Both of these processes are forms of **facilitated diffusion**. That is, the substances diffuse according to their concentration gradients, but their diffusion is facilitated by protein channels or carriers.

ION CHANNELS The best-studied channel proteins are the **ion channels**. As you will see in later chapters, the movement of ions across membranes is important in many biological processes such as respiration within the mitochondria, the electrical activity of the nervous system, and the opening of pores in leaves to allow gas exchange with the environment. Several types of ion channel have been identified—each specific for a particular ion. All show the same basic structure of a hydrophilic pore that allows a particular ion to move through its center (**Figure 6.10**).

Just as a fence may have a gate that can be opened or closed, most ion channels are gated: they can be opened or closed to ion passage. A **gated channel** opens when a stimulus causes a change in the three-dimensional shape of the channel. In some cases, this stimulus is the binding of a chemical signal, or **ligand** (see Figure 6.10). Channels controlled in this way are called ligand-gated channels. In contrast, a voltage-gated channel is stimulated to open or close by a change in the voltage (electric charge difference) across the membrane. As you will see in Section 45.2, voltage-gated channels are important for nerve cell function.

THE SPECIFICITY OF ION CHANNELS How does an ion channel allow one ion, but not another, to pass through? It is not simply a matter of charge and size of the ion. For example, a sodium ion (Na^+), with a radius of 0.095 nanometers (nm), is smaller than K^+ (0.130 nm), and both carry the same positive charge. Yet the potassium channel lets only K^+ pass through the membrane, and not the smaller Na^+. Nobel laureate Roderick MacKinnon at The Rockefeller University found an elegant explanation for this when he deciphered the structure of a potassium channel from a bacterium.

Being charged, both Na^+ and K^+ are attracted to water molecules. They are surrounded by water "shells" in solution, held by the attraction of their positive charges to the negatively

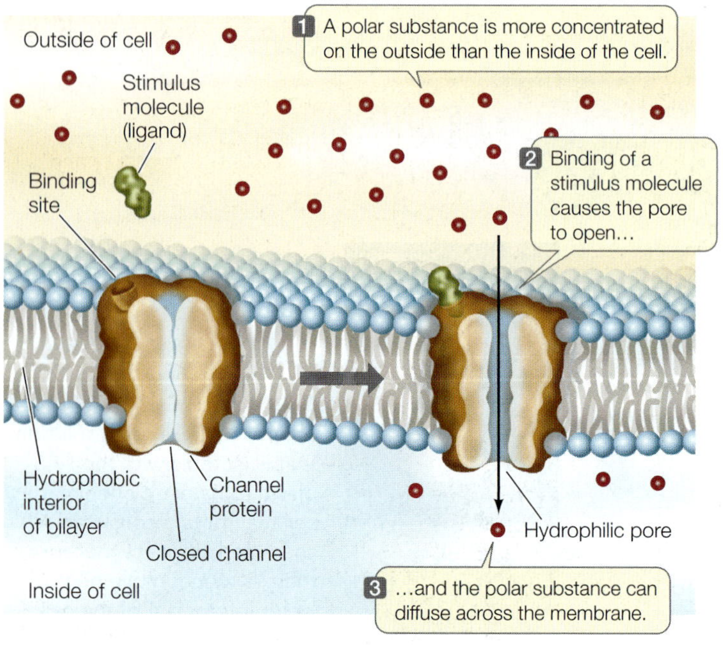
Outside of cell
1 A polar substance is more concentrated on the outside than the inside of the cell.
Stimulus molecule (ligand)
Binding site
2 Binding of a stimulus molecule causes the pore to open...
Hydrophobic interior of bilayer
Channel protein
Closed channel
Hydrophilic pore
Inside of cell
3 ...and the polar substance can diffuse across the membrane.

6.10 A Gated Channel Protein Opens in Response to a Stimulus The channel protein has a pore of polar amino acids and water. It is anchored in the hydrophobic bilayer by its outer coating of nonpolar R groups. The protein changes its three-dimensional shape when a stimulus molecule (ligand) binds to it, opening the pore so that specific hydrophilic substances can pass through. Other (voltage) gated channels open in response to an electrical potential (voltage).

charged oxygen atoms on the water molecules (see Figure 2.10). The potassium channel contains highly polar oxygen atoms at its opening. The gap enclosed by these atoms is exactly the right size so that when a K^+ ion approaches the opening, it is more strongly attracted to the oxygen atoms there than to those of the water molecules in its shell. It sheds its water shell and passes through the channel. The smaller Na^+ ion, however, is kept a bit more distant from the oxygen atoms at the opening of the channel because extra water molecules can fit between the ion (with its shell) and the oxygen atoms at the opening. This hydration prevents entry of Na^+ into the potassium channel.

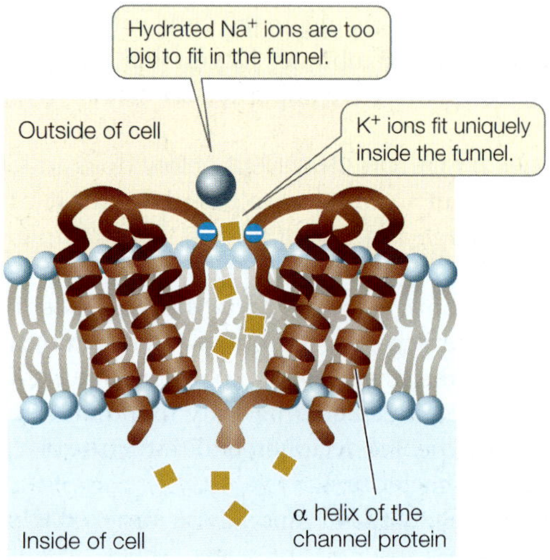
Hydrated Na+ ions are too big to fit in the funnel.
Outside of cell
K+ ions fit uniquely inside the funnel.
Inside of cell
α helix of the channel protein

INVESTIGATINGLIFE

6.11 Aquaporins Increase Membrane Permeability to Water A protein was isolated from the membranes of cells in which water diffuses rapidly across the membranes. When the protein was inserted into oocytes, which do not normally have it, the water permeability of the oocytes was greatly increased.[a]

HYPOTHESIS Aquaporin increases membrane permeability to water.

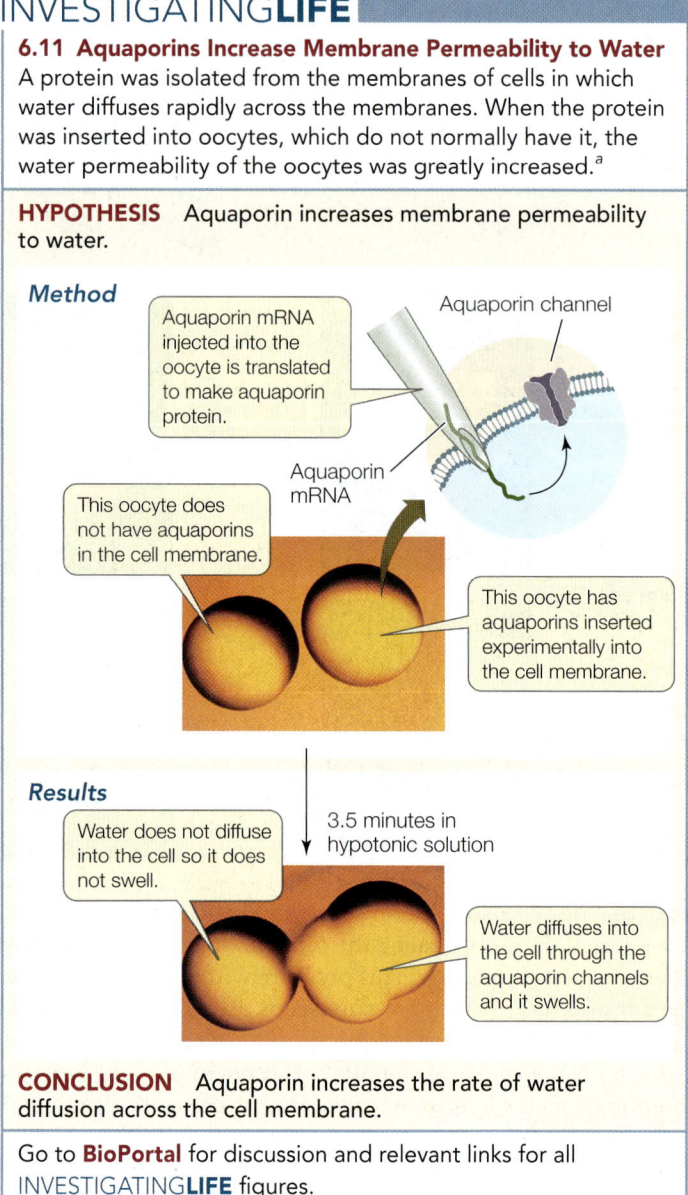

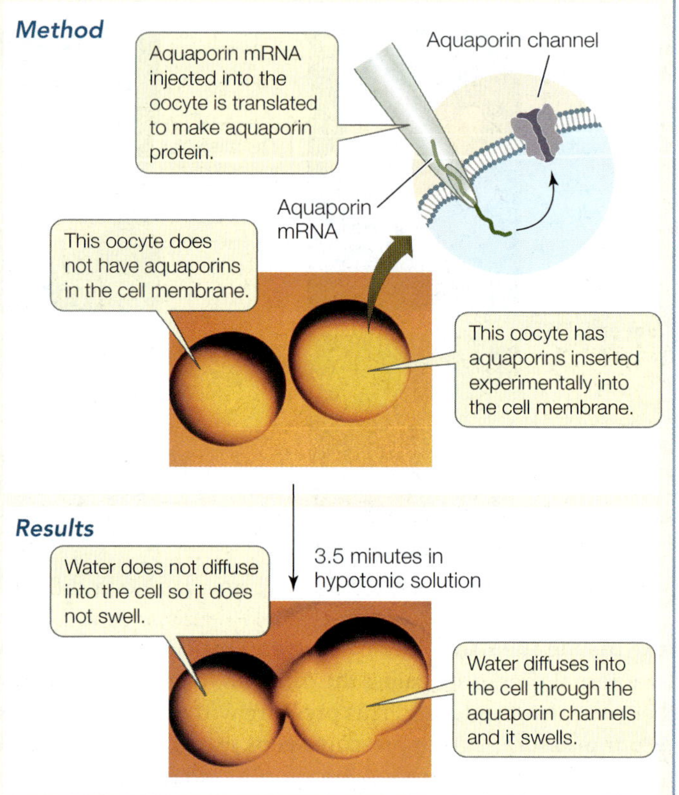
Method
Aquaporin mRNA injected into the oocyte is translated to make aquaporin protein.
Aquaporin channel
Aquaporin mRNA
This oocyte does not have aquaporins in the cell membrane.
This oocyte has aquaporins inserted experimentally into the cell membrane.
Results
Water does not diffuse into the cell so it does not swell.
3.5 minutes in hypotonic solution
Water diffuses into the cell through the aquaporin channels and it swells.

CONCLUSION Aquaporin increases the rate of water diffusion across the cell membrane.

Go to **BioPortal** for discussion and relevant links for all INVESTIGATING**LIFE** figures.

[a]Preston, G. M. et al. 1992. *Science* 256: 385–387.

Such a potassium channel is voltage-gated. The mechanism for opening and closing it depends on interactions between positively charged arginine residues on the protein and negative charges on membrane phospholipids.

AQUAPORINS FOR WATER As you saw in the opening story of this chapter, water can cross membranes through protein channels called **aquaporins**. These channels function as a cellular plumbing system for moving water. Like the K^+ channel, the aquaporin channel is highly specific. Water molecules move in single file through the channel, which excludes ions so that the electrical properties of the cell are maintained. Aquaporins were first identified when a protein from red blood cell membranes was inserted into frog oocytes (immature egg cells). The membranes of these cells are normally impermeable to water, but the membranes of the cells treated with aquaporins became much more permeable (Figure 6.11).

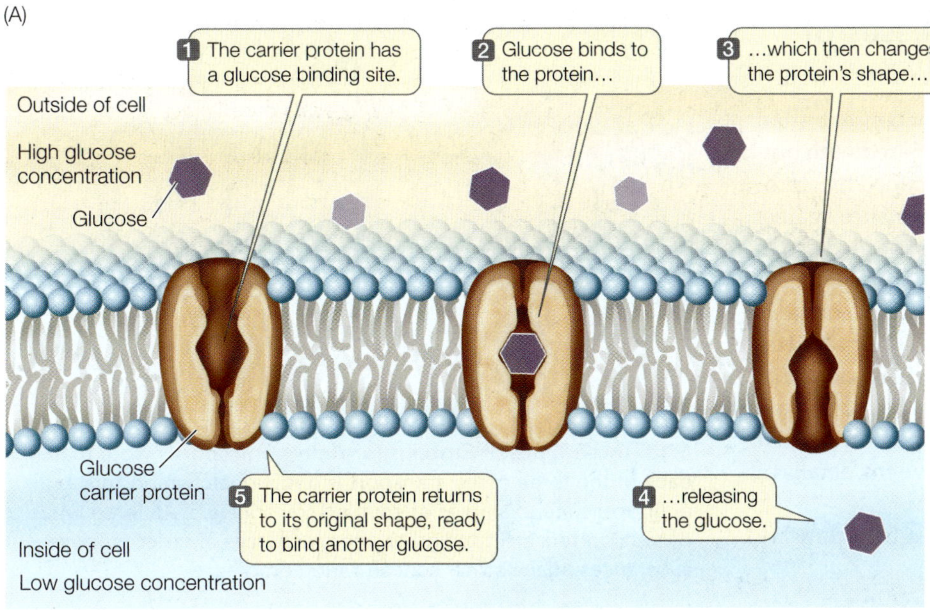
(A)
1 The carrier protein has a glucose binding site.
2 Glucose binds to the protein...
3 ...which then changes the protein's shape...
Outside of cell
High glucose concentration
Glucose
Glucose carrier protein
5 The carrier protein returns to its original shape, ready to bind another glucose.
4 ...releasing the glucose.
Inside of cell
Low glucose concentration

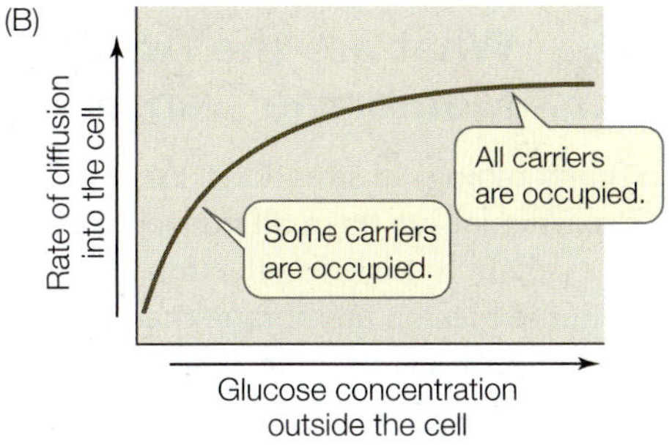
(B)
Rate of diffusion into the cell
All carriers are occupied.
Some carriers are occupied.
Glucose concentration outside the cell

6.12 A Carrier Protein Facilitates Diffusion The glucose transporter is a carrier protein that allows glucose to enter the cell at a faster rate than would be possible by simple diffusion. (A) The transporter binds to glucose, brings it into the membrane interior, then changes shape, releasing glucose into the cell cytoplasm. (B) The graph shows the rate of glucose entry via a carrier versus the concentration of glucose outside the cell. As the glucose concentration increases, the rate of diffusion increases until the point at which all the available transporters are being used (the system is saturated).

Carrier proteins aid diffusion by binding substances

As we mentioned earlier, another type of facilitated diffusion involves the binding of the transported substance to a membrane protein called a carrier protein. Like channel proteins, carrier proteins facilitate the passive diffusion of substances into or out of cells or organelles. Carrier proteins transport polar molecules such as sugars and amino acids.

Glucose is the major energy source for most cells, and living systems require a great deal of it. Glucose is polar and cannot readily diffuse across membranes. Eukaryotic cell membranes contain a carrier protein—the glucose transporter—that facilitates glucose uptake into the cell. Binding of glucose to a specific three-dimensional site on one side of the transporter protein causes the protein to change its shape and release glucose on the other side of the membrane (**Figure 6.12A**). Since glucose is either broken down or otherwise removed almost as soon as it enters a cell, there is almost always a strong concentration gradient favoring glucose entry (that is, a higher concentration outside the cell than inside).

Transport by carrier proteins is different from simple diffusion. In simple diffusion, the rate of movement depends on the concentration gradient across the membrane. This is also true for carrier-mediated transport, up to a point. In carrier-mediated transport, as the concentration gradient increases, the diffusion rate also increases, but its *rate* of increase slows, and a point is reached at which the diffusion rate becomes constant. At this point, the facilitated diffusion system is said to be *saturated* (**Figure 6.12B**). A particular cell has a specific number of carrier protein molecules in its plasma membrane. The rate of diffusion reaches a maximum when all the carrier molecules are fully loaded with solute molecules. Think of waiting for the elevator on the ground floor of a hotel with 50 other people. You can't all get in the elevator (carrier) at once, so the rate of transport (say, ten people at a time) is at its maximum, and the transport system is "saturated." As a consequence, cells that require large amounts of energy, such as muscle cells, have high concentrations of glucose transporters in their membranes so that the maximum rate of facilitated diffusion is greater. Likewise, the human brain has high glucose needs, and the blood vessels that nourish it have high concentrations of glucose transporters.

RECAP 6.3

Diffusion is the movement of ions or molecules from a region of greater concentration to a region of lesser concentration. Osmosis is the diffusion of water through a selectively permeable cell membrane. Channel proteins and carrier proteins can facilitate the diffusion of charged and polar substances, including water, across cell membranes.

- What properties of a substance determine whether, and how fast, it will diffuse across a membrane? **See p. 114**
- Describe osmosis and explain the terms hypertonic, hypotonic, and isotonic. **See p. 114 and Figure 6.9**
- How does a channel protein facilitate diffusion? **See pp. 115–116 and Figure 6.10**

The process of diffusion tends to equalize the concentrations of substances outside and inside cells. However, one hallmark of a living cell is that it can have an internal composition quite different from that of its environment. To achieve this, a cell must sometimes move substances against their concentration gradients. This process requires an input of energy and is known as active transport.

6.4 What are the Active Processes of Membrane Transport?

In many biological situations, there is a different concentration of a particular ion or small molecule inside compared with outside a cell. In these cases, the imbalance is maintained by a protein in the plasma membrane that moves the substance against its concentration and/or electrical gradient. This is called active transport, and because it is acting "against the normal flow," it requires the expenditure of energy. Often the energy source is adenosine triphosphate (ATP). In eukaryotes, ATP is produced in the mitochondria. It has chemical energy stored in its terminal phosphate bond. This energy is released when ATP is converted to adenosine diphosphate (ADP) in a hydrolysis reaction that breaks the terminal phosphate bond. We will give the details of how ATP provides energy to cells in Section 8.2.

The differences between diffusion and active transport are summarized in Table 6.1.

Go to Animated Tutorial 6.3
Active Transport
Life10e.com/at6.3

Active transport is directional

In many cases of simple and facilitated diffusion, ions or molecules can move down their concentration gradients in either direction across the cell membrane. In contrast, active transport is directional, and moves a substance either into or out of the cell or organelle, depending on need. There are three kinds of membrane proteins that carry out active transport (Figure 6.13):

- A **uniporter** moves a single substance in one direction. For example, a calcium-binding protein found in the plasma membrane and the ER membranes of many cells actively transports Ca^{2+} to locations where it is more highly concentrated, either outside the cell or inside the lumen of the ER.
- A **symporter** moves two substances in the same direction. For example, a symporter in the cells that line the intestine must bind Na^+ in addition to an amino acid in order to absorb amino acids from the intestine.
- An **antiporter** moves two substances in opposite directions, one into the cell (or organelle) and the other out of the cell (or organelle). For example, many cells have a sodium–potassium pump that moves Na^+ out of the cell and K^+ into it.

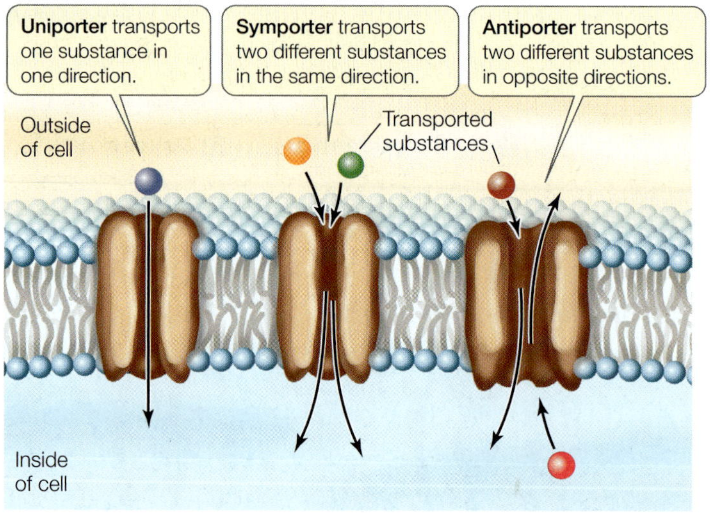

Uniporter transports one substance in one direction.
Symporter transports two different substances in the same direction.
Antiporter transports two different substances in opposite directions.
Outside of cell
Transported substances
Inside of cell

6.13 Three Types of Proteins for Active Transport Note that in each of the three cases, transport is directional. Symporters and antiporters are examples of coupled transporters. All three types of transporters are coupled to energy sources in order to move substances against their concentration gradients.

Symporters and antiporters are also known as **coupled transporters** because they move two substances at once.

Different energy sources distinguish different active transport systems

There are two basic types of active transport:

- **Primary active transport** involves the direct hydrolysis of ATP, which provides the energy required for transport.
- **Secondary active transport** does not use ATP directly. Instead, its energy is supplied by an ion concentration gradient established by primary (ATP-driven) active transport. Secondary active transport uses the energy of ATP indirectly in the form of the gradient.

In primary active transport, energy released by the hydrolysis of ATP drives the movement of specific ions against their concentration gradients. For example, we mentioned earlier that concentrations of potassium ions (K^+) inside a cell are often much higher than in the fluid bathing the cell. However, the concentration of sodium ions (Na^+) is often much higher outside the cell. A protein in the plasma membrane pumps Na^+ out of the cell and K^+ into the cell against these concentration gradients, ensuring that the gradients are maintained (Figure 6.14). This **sodium–potassium (Na^+–K^+) pump** is found in all animal cells. The pump is an integral membrane glycoprotein.

TABLE 6.1 Membrane Transport Mechanisms

	Simple Diffusion	Facilitated Diffusion (through Channel or Carrier)	Active Transport
Cellular energy required?	No	No	Yes
Driving force	Concentration gradient	Concentration gradient	ATP hydrolysis (against concentration gradient)
Membrane protein required?	No	Yes	Yes
Specificity	No	Yes	Yes

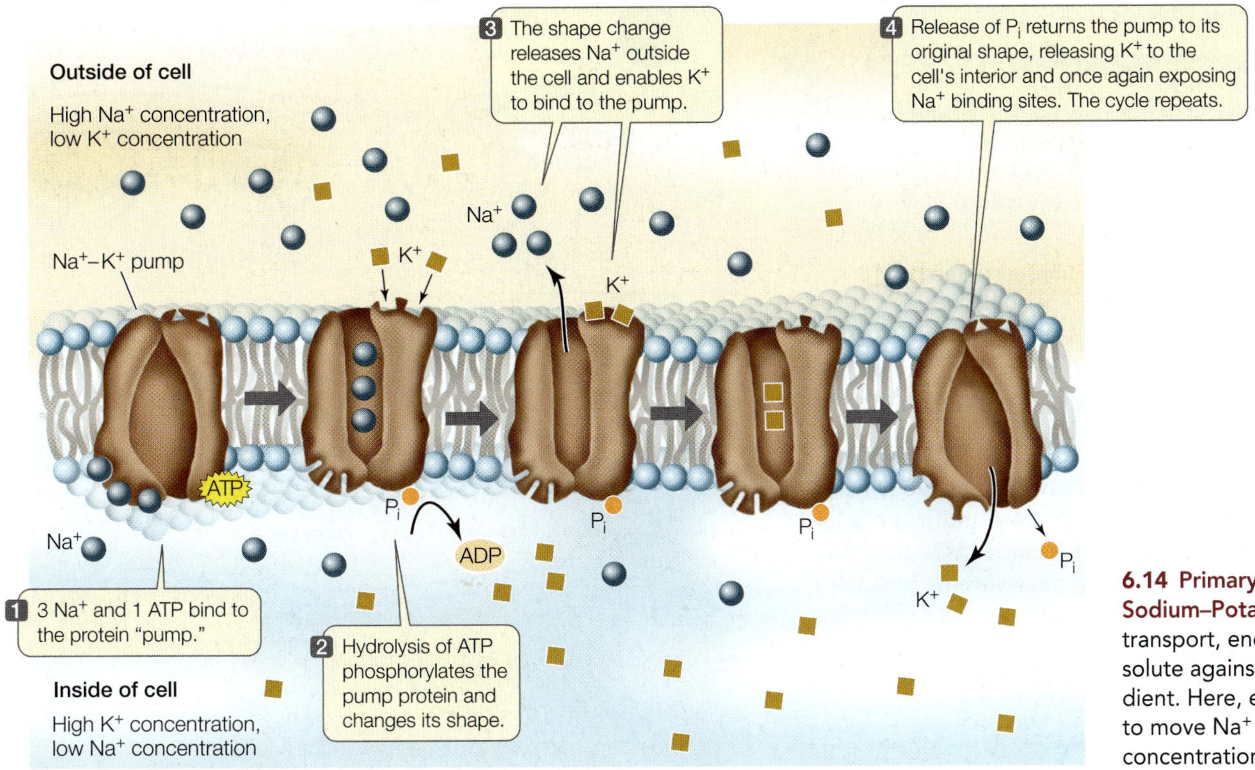

Outside of cell
High Na^+ concentration, low K^+ concentration
Na^+–K^+ pump
1 3 Na^+ and 1 ATP bind to the protein "pump."
ATP
Na^+
2 Hydrolysis of ATP phosphorylates the pump protein and changes its shape.
P_i
ADP
K^+
3 The shape change releases Na^+ outside the cell and enables K^+ to bind to the pump.
Na^+
4 Release of P_i returns the pump to its original shape, releasing K^+ to the cell's interior and once again exposing Na^+ binding sites. The cycle repeats.
Inside of cell
High K^+ concentration, low Na^+ concentration

6.14 Primary Active Transport: The Sodium–Potassium Pump In active transport, energy is used to move a solute against its concentration gradient. Here, energy from ATP is used to move Na^+ and K^+ against their concentration gradients.

It breaks down a molecule of ATP to ADP and a free phosphate ion (P_i) and uses the energy released to bring two K^+ ions into the cell and export three Na^+ ions. The Na^+–K^+ pump is thus an antiporter because it moves two substances in different directions.

In secondary active transport, the movement of a substance against its concentration gradient is accomplished using energy "regained" by letting ions move across the membrane with their concentration gradients. For example, once the sodium–potassium pump establishes a concentration gradient of sodium ions, the passive diffusion of some Na^+ back into the cell can provide energy for the secondary active transport of glucose into the cell (**Figure 6.15**). This occurs when glucose is absorbed into the bloodstream from the digestive tract. Secondary active transport aids in the uptake of amino acids and sugars, which are essential raw materials for cell maintenance and growth. Both types of coupled transport proteins—symporters and antiporters—are used for secondary active transport.

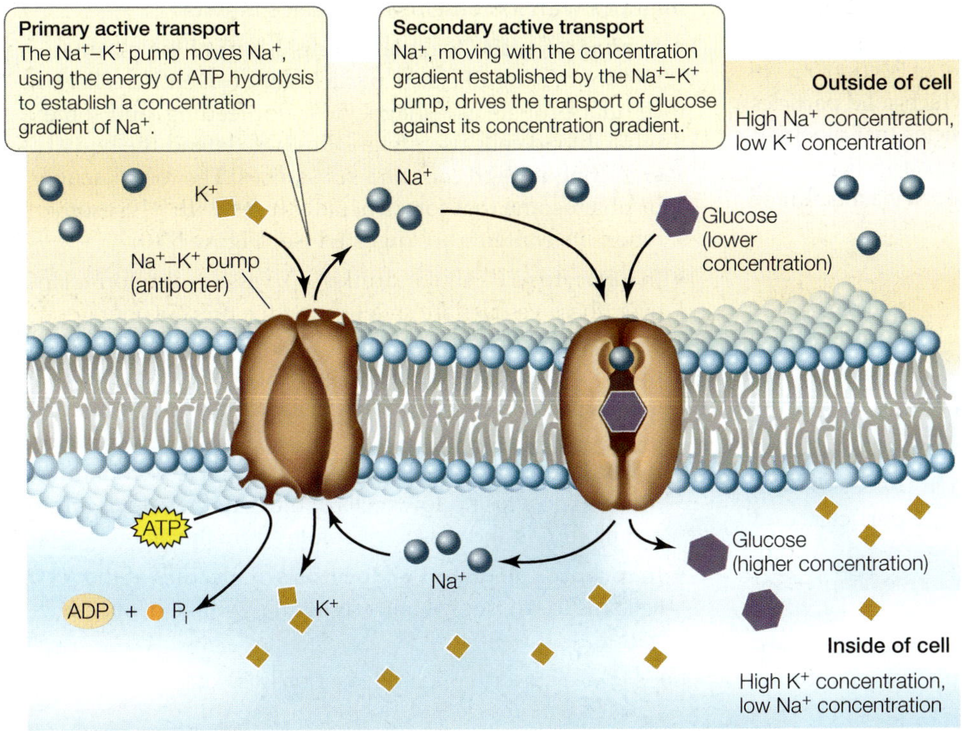

Primary active transport
The Na^+–K^+ pump moves Na^+, using the energy of ATP hydrolysis to establish a concentration gradient of Na^+.
Secondary active transport
Na^+, moving with the concentration gradient established by the Na^+–K^+ pump, drives the transport of glucose against its concentration gradient.
Outside of cell
High Na^+ concentration, low K^+ concentration
K^+
Na^+
Na^+–K^+ pump (antiporter)
Glucose (lower concentration)
ATP
ADP + P_i
K^+
Na^+
Glucose (higher concentration)
Inside of cell
High K^+ concentration, low Na^+ concentration

6.15 Secondary Active Transport The Na^+ concentration gradient established by primary active transport (left) powers the secondary active transport of glucose (right). A symporter protein couples the movement of glucose across the membrane against its concentration gradient to the passive movement of Na^+ into the cell.

RECAP 6.4

Active transport across a membrane is directional and requires an input of energy to move substances against their concentration gradients. Active transport allows a cell to maintain small molecules and ions at concentrations very different from those in the surrounding environment.

- Why is energy required for active transport? **See p. 118**
- Why is the sodium–potassium (Na^+–K^+) pump classified as an antiporter? **See p. 118–119 and Figure 6.14**
- Explain the difference between primary active transport and secondary active transport. **See p. 118 and Figure 6.15**

We have examined a number of passive and active ways in which ions and small molecules can enter and leave cells. But what about large molecules such as proteins? Many proteins are so large that they diffuse very slowly, and their bulk makes it difficult for them to pass through the phospholipid bilayer. A completely different mechanism is needed to move intact large molecules across membranes.

6.5 How Do Large Molecules Enter and Leave a Cell?

Macromolecules such as proteins, polysaccharides, and nucleic acids are simply too large and too charged or polar to pass through biological membranes. This is actually a fortunate property—think of the consequences if such molecules diffused out of cells. A red blood cell would not retain its hemoglobin! As we discussed in Chapter 5, the development of a selectively permeable membrane was essential for the functioning of the first cells when life on Earth began. The interior of a cell can be maintained as a separate compartment with a different composition from that of the exterior environment, which is subject to abrupt changes. However, cells must sometimes take up or secrete (release to the external environment) intact large molecules. In Section 5.3 we described phagocytosis, the mechanism by which solid particles can be brought into the cell by means of vesicles that pinch off from the plasma membrane. The general terms for the mechanisms by which substances enter and leave the cell via membrane vesicles are endocytosis and exocytosis.

Go to Animated Tutorial 6.4
Endocytosis and Exocytosis
Life10e.com/at6.4

Macromolecules and particles enter the cell by endocytosis

Endocytosis is a general term for a group of processes that bring small molecules, macromolecules, large particles, and even small cells into the eukaryotic cell (Figure 6.16A). We described an example of endocytosis earlier in the chapter, involving integrins (see Figure 6.8). Generally, there are three types of endocytosis: phagocytosis, pinocytosis, and receptor-mediated endocytosis. In all three, the plasma membrane

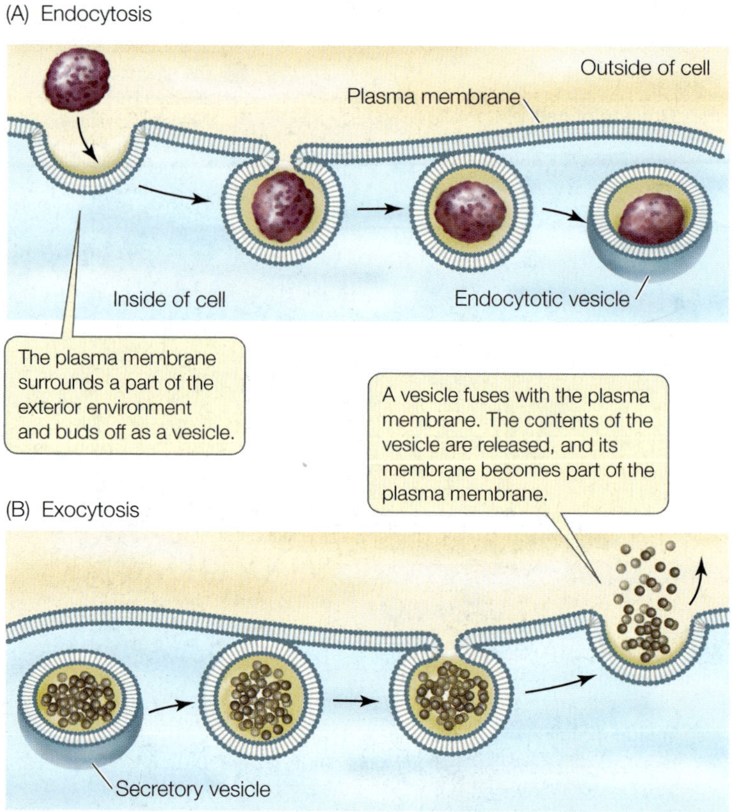
(A) Endocytosis
Outside of cell
Plasma membrane
Inside of cell
Endocytotic vesicle
The plasma membrane surrounds a part of the exterior environment and buds off as a vesicle.
A vesicle fuses with the plasma membrane. The contents of the vesicle are released, and its membrane becomes part of the plasma membrane.
(B) Exocytosis
Secretory vesicle

6.16 Endocytosis and Exocytosis Endocytosis (A) and exocytosis (B) are used by eukaryotic cells to take up and release fluids, large molecules, and particles. Smaller cells, such as invading bacteria, can be taken up by endocytosis.

Go to Media Clip 6.1
An Amoeba Eats by Phagocytosis
Life10e.com/mc6.1

invaginates (folds inward), forming a small pocket around materials from the environment. The pocket deepens, forming a vesicle. This vesicle separates from the plasma membrane and migrates with its contents to the cell's interior.

- In **phagocytosis** ("cellular eating"), part of the plasma membrane engulfs large particles or even entire cells. Unicellular protists use phagocytosis for feeding, and some white blood cells use phagocytosis to defend the body by engulfing foreign cells and substances. The food vacuole or phagosome that forms usually fuses with a lysosome, where its contents are digested (see Figure 5.10).
- In **pinocytosis** ("cellular drinking"), vesicles also form. However, these vesicles are smaller, and the process operates to bring fluids and dissolved substances into the cell. Like phagocytosis, pinocytosis can be relatively nonspecific regarding what it brings into the cell. For example, pinocytosis occurs constantly in the endothelium, the single layer of cells that separates a tiny blood capillary from the surrounding tissue. Pinocytosis allows cells of the endothelium to rapidly acquire fluids and dissolved solutes from the blood.
- In **receptor-mediated endocytosis**, molecules at the cell surface recognize and trigger the uptake of specific materials.

Let's take a closer look at this last process.

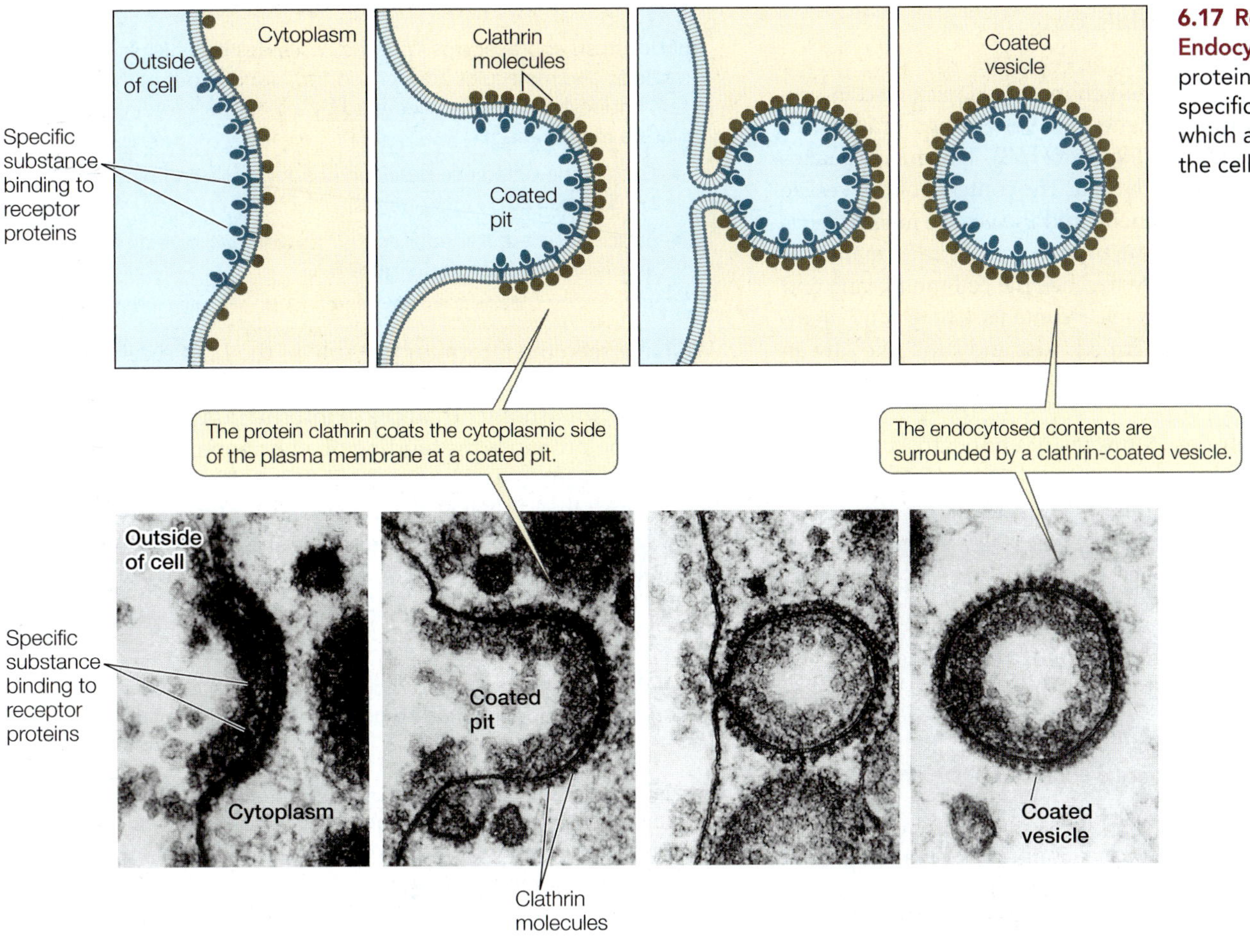
Outside of cell
Cytoplasm
Specific substance binding to receptor proteins
Clathrin molecules
Coated pit
Coated vesicle
The protein clathrin coats the cytoplasmic side of the plasma membrane at a coated pit.
The endocytosed contents are surrounded by a clathrin-coated vesicle.
Outside of cell
Specific substance binding to receptor proteins
Cytoplasm
Coated pit
Clathrin molecules
Coated vesicle

6.17 Receptor-Mediated Endocytosis The receptor proteins in a coated pit bind specific macromolecules, which are then carried into the cell by a coated vesicle.

Receptor-mediated endocytosis is highly specific

Receptor-mediated endocytosis is used by animal cells to capture specific macromolecules from the cell's environment. This process depends on **receptor proteins**, which are proteins that can bind to specific molecules within the cell or in the cell's external environment. In receptor-mediated endocytosis, the receptors are integral membrane proteins located at particular regions on the extracellular surface of the plasma membrane. These membrane regions are called coated pits because they form slight depressions in the plasma membrane and their cytoplasmic surfaces are coated by other proteins, such as clathrin. The uptake process is similar to that in phagocytosis.

When a receptor protein binds to its specific ligand (in this case, the macromolecule to be taken into the cell), its coated pit invaginates and forms a coated vesicle around the bound macromolecule. The clathrin molecules strengthen and stabilize the vesicle, which carries the macromolecule away from the plasma membrane and into the cytoplasm (Figure 6.17). Once inside, the vesicle loses its clathrin coat and may fuse with a lysosome, where the engulfed material is digested (by the hydrolysis of polymers to monomers) and the products released into the cytoplasm. Because of its specificity for particular macromolecules, receptor-mediated endocytosis is an efficient method of taking up substances that may exist at low concentrations in the cell's environment.

Receptor-mediated endocytosis is the method by which cholesterol is taken up by most mammalian cells. Water-insoluble cholesterol and triglycerides are packaged by liver cells into lipoprotein particles. Most of the cholesterol is packaged into a type of lipoprotein particle called low-density lipoprotein, or LDL, which is circulated via the bloodstream. When a particular cell requires cholesterol, it produces specific LDL receptors, which are inserted into the plasma membrane in clathrin-coated pits. Binding of LDLs to the receptor proteins triggers the uptake of the LDLs via receptor-mediated endocytosis. Within the resulting vesicle, the LDL particles are freed from the receptors. The receptors segregate to a region that buds off and forms a new vesicle, which is recycled to the plasma membrane. The freed LDL particles remain in the original vesicle, which fuses with a lysosome. There, the LDLs are digested and the cholesterol made available for cell use.

In healthy individuals, the liver takes up unused LDLs for recycling. People with the inherited disease familial hypercholesterolemia have a deficient LDL receptor in their livers. This prevents receptor-mediated endocytosis of LDLs, resulting in dangerously high levels of cholesterol in the blood. The cholesterol builds up in the arteries that nourish the heart and causes heart attacks. In extreme cases where only the deficient receptor is present, children and teenagers can have severe cardiovascular disease.

Exocytosis moves materials out of the cell

Exocytosis is the process by which materials packaged in vesicles are secreted from a cell when the vesicle membrane fuses with the plasma membrane (Figure 6.16B). This fusing makes an opening to the outside of the cell. The contents of the vesicle are released into the environment, and the vesicle membrane is smoothly incorporated into the plasma membrane. In another form of exocytosis, the vesicle touches the cell membrane and a pore forms, releasing the vesicle's contents. There is no membrane fusion in this process, termed "kiss and run." We saw an example of exocytosis in describing sweat glands at the start of the chapter.

Table 6.2 summarizes examples of endocytosis and exocytosis.

TABLE 6.2 Endocytosis and Exocytosis

Type of Process	Example
Endocytosis	
Receptor-mediated endocytosis	Specific uptake of large molecules, e.g., LDL
Pinocytosis	Nonspecific uptake of extracellular fluid, e.g., fluids and dissolved substances from blood
Phagocytosis	Nonspecific uptake of large undissolved particles, e.g., invading bacteria by cells of the immune system
Exocytosis	
Release of large molecules	Vesicle fusion with cell membrane, e.g., digestive enzymes in the pancreas
Release of small molecules	Vesicle fusion with cell membrane, e.g., neurotransmitters at the synapse

RECAP 6.5

Endocytosis and exocytosis are the processes by which large particles and molecules are transported into and out of the cell. Endocytosis may be mediated by a receptor protein in the plasma membrane.

- Explain the difference between phagocytosis and pinocytosis. See p. 120
- Describe receptor-mediated endocytosis and give an example. See p. 121 and Figure 6.17

We have seen the informational role of the LDL receptor protein in the recognition and endocytosis of LDL, with its cargo of cholesterol. Another example of information processing by a membrane protein is the binding of a hormone such as insulin to specific receptors on a target cell. When insulin binds to receptors on a liver cell, it elicits the uptake of glucose. In Chapter 7 we will see many other examples of the roles of membrane proteins in information processing.

?

Water purity is a worldwide problem. Can aquaporin membrane channels be used in water purification?

ANSWER

There are more than ten genes encoding aquaporins in humans, each channel having its particular location in the body (for example, salivary glands and the kidney). All of these aquaporins, and those of other organisms, including plants, share a common structure that spans the plasma membrane and has a channel through which water molecules pass in single file. Efforts are underway to insert aquaporins into synthetic membranes used for industrial applications. Because aquaporins allow only water to pass through (no solutes), such membranes could be used to purify contaminated fresh water or to desalinate seawater for drinking.

CHAPTER SUMMARY 6

6.1 What Is the Structure of a Biological Membrane?

- Biological membranes consist of lipids, proteins, and carbohydrates. The **fluid mosaic model** of membrane structure describes a phospholipid bilayer in which proteins can move about within the plane of the membrane. Review ACTIVITY 6.1
- The two layers of a membrane may have different properties because of their different lipid compositions. Animal cell membranes may contain high concentrations (up to 25%) of cholesterol. Review ANIMATED TUTORIAL 6.1
- The properties of membranes also depend on the **integral membrane proteins** and **peripheral membrane proteins** associated with them. Some proteins, called **transmembrane proteins**, span the membrane. Review Figure 6.1
- Carbohydrates, attached to proteins in **glycoproteins** or to phospholipids in **glycolipids**, project from the external surface of the plasma membrane and function as recognition signals.
- Membranes are not static structures, but are constantly forming, exchanging components, and breaking down.

6.2 How Is the Plasma Membrane Involved in Cell Adhesion and Recognition?

- In order for cells to assemble into tissues, they must recognize and adhere to one another. **Cell recognition** and **cell adhesion** depend on membrane-associated proteins and carbohydrates. Review Figure 6.6
- Adhesion can involve binding between identical (**homotypic**) or different (**heterotypic**) molecules on adjacent cells.
- **Cell junctions** connect adjacent cells. **Tight junctions** prevent the passage of molecules through the intercellular spaces between cells, and they restrict the migration of membrane proteins over the cell surface. **Desmosomes** cause cells to adhere firmly to one another. **Gap junctions** provide channels for communication between adjacent cells. Review Figure 6.7, ACTIVITY 6.2
- **Integrins** mediate the attachment of animal cells to the extracellular matrix. Detachment and recycling of integrins allow cells to move. Review Figure 6.8

continued

6.3 What Are the Passive Processes of Membrane Transport?
See ANIMATED TUTORIAL 6.2

- Membranes exhibit **selective permeability**, regulating which substances pass through them. Substances can cross the membrane by either **passive transport**, which requires no input of chemical energy, or **active transport**, which uses chemical energy.
- **Diffusion** is the movement of a solute from a region of higher concentration to a region of lower concentration. Equilibrium is reached when there is no further net change in concentration.
- In **osmosis**, water diffuses across a membrane from a region of higher water concentration to a region of lower water concentration.
- Most cells are in an **isotonic** environment, where total solute concentrations on both sides of the plasma membrane are equal. If the solution surrounding a cell is **hypotonic** to the cell interior, more water enters the cell than leaves it, causing it to swell. In plant cells, this contributes to **turgor pressure**. In a **hypertonic** solution, more water leaves the cell than enters it, causing it to shrivel. **Review Figure 6.9**
- A substance can diffuse passively across a membrane by either **simple diffusion** or **facilitated diffusion**, via a **channel protein** or a carrier protein.
- **Ion channels** are membrane proteins that allow the rapid facilitated diffusion of ions through membranes. **Gated channels** can be opened or closed by either chemical **ligands** or changes in membrane voltage. **Review Figure 6.10**
- **Aquaporins** are water channels. **Review Figure 6.11**
- **Carrier proteins** bind to polar molecules such as sugars and amino acids and transport them across the membrane. The maximum rate of this type of facilitated diffusion is limited by the number of carrier (transporter) proteins in the membrane. **Review Figure 6.12**

6.4 What Are the Active Processes of Membrane Transport?
See ANIMATED TUTORIAL 6.3

- **Active transport** requires the use of chemical energy to move substances across membranes against their concentration or electrical gradients. Active transport proteins may be **uniporters**, **symporters**, or **antiporters**. **Review Figure 6.13**
- In **primary active transport**, energy from the hydrolysis of ATP is used to move ions into or out of cells. The **sodium–potassium pump** is an important example. **Review Figure 6.14**
- **Secondary active transport** couples the passive movement of one substance down its concentration gradient to the movement of another substance against its concentration gradient. Energy from ATP is used indirectly to establish the concentration gradient that results in the movement of the first substance. **Review Figure 6.15**

6.5 How Do Large Molecules Enter and Leave a Cell?
See ANIMATED TUTORIAL 6.4

- **Endocytosis** is the transport of macromolecules, large particles, and small cells into eukaryotic cells via the invagination of the plasma membrane and the formation of vesicles. **Phagocytosis** and **pinocytosis** are types of endocytosis. **Review Figure 6.16A**
- In **exocytosis**, materials in vesicles are secreted from the cell when the vesicles fuse with the plasma membrane. **Review Figure 6.16B**
- In **receptor-mediated endocytosis**, a specific **receptor protein** on the plasma membrane binds to a particular macromolecule. **Review Figure 6.17**

Go to the Interactive Summary to review key figures, Animated Tutorials, and Activities
Life10e.com/is6

CHAPTER**REVIEW**

REMEMBERING

1. When a hormone molecule binds to a specific protein on the plasma membrane, the protein it binds to is called a
   a. ligand.
   b. clathrin.
   c. receptor protein.
   d. hydrophobic protein.
   e. cell adhesion molecule.

2. Which statement about membrane proteins is *not* true?
   a. They all extend from one side of the membrane to the other.
   b. Some serve as channels for ions to cross the membrane.
   c. Many are free to migrate laterally within the membrane.
   d. Their position in the membrane is determined by their structure.
   e. Some have both hydrophobic and hydrophilic regions.

3. Which statement about membrane carbohydrates is *not* true?
   a. Some are bound to proteins.
   b. Some are bound to lipids.
   c. They are added to proteins in the Golgi apparatus.
   d. They show little diversity.
   e. They are important in recognition reactions at the cell surface.

4. Which statement about ion channels is *not* true?
   a. They form pores in the membrane.
   b. They are proteins.
   c. All ions pass through the same type of channel.
   d. Movement through them is from regions of high concentration to regions of low concentration.
   e. Movement through them is by diffusion.

5. Facilitated diffusion and active transport both
   a. require ATP.
   b. require the use of proteins as carriers or channels.
   c. carry ions and not small molecules.
   d. increase without limit as the concentration gradient increases.
   e. depend on the solubility of the solute in lipids.

6. Primary and secondary active transport both
   a. generate ATP.
   b. are based on passive movement of Na^+ ions.
   c. include the passive movement of glucose molecules.
   d. use ATP directly.
   e. can move solutes against their concentration gradients.

UNDERSTANDING & APPLYING

7. You are studying how the protein transferrin enters cells. When you examine cells that have taken up transferrin, you find it inside clathrin-coated vesicles. Therefore the most likely mechanism for uptake of transferrin is
   a. facilitated diffusion.
   b. an antiporter.
   c. receptor-mediated endocytosis.
   d. gap junctions.
   e. ion channels.

8. Muscle function requires calcium ions (Ca^{2+}) to be pumped into a subcellular compartment against a concentration gradient. What types of molecules are required for this to happen?

9. Section 27.2 will describe the diatoms, which are protists that have complex glassy structures in their cell walls (see Figure 27.8). These structures form within the Golgi apparatus. How do these structures reach the cell wall without having to pass through a membrane?

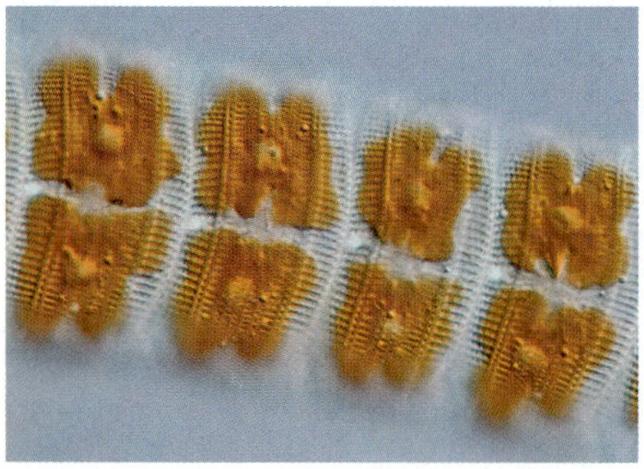

Eight diatom cells inside their ornate cell walls

ANALYZING & EVALUATING

10. Organisms that live in fresh water are almost always hypertonic to their environment. In what way is this a serious problem? How could some organisms cope with this problem?

11. When a normal lung cell becomes a lung cancer cell, there are several important changes in plasma membrane properties. How would you investigate the following phenomena? (a) The cancer cell membrane is more fluid, with more rapid diffusion in the plane of the membrane of both lipids and proteins. (b) The cancer cell has altered cell adhesion properties, binding to other tissues in addition to lung cells.

Go to BioPortal at **yourBioPortal.com** for Animated Tutorials, Activities, LearningCurve Quizzes, Flashcards, and many other study and review resources.

7 Cell Communication and Multicellularity

CHAPTER OUTLINE

Voles Prairie voles display extensive bonding behaviors after mating. These behaviors are mediated by peptides acting as intercellular signals.

PRAIRIE VOLES (*Microtus ochrogaster*) are small rodents that live in temperate climates, where they dig tunnels in fields. When a male prairie vole encounters a female, mating often ensues. After mating (which can take as long as a day) the pair stays together, building a nest and raising their pups together. The bond between the two voles is so strong that they stay together for life. Contrast this behavior with that of the montane vole (*M. montanus*), which is closely related to the prairie vole and lives in the hills not far away. In this species, mating is quick, and afterward the pair separates. The male looks for new mates and the female abandons her young soon after they are born.

The explanation for these dramatic behavioral differences lies in the brains of these two species. When prairie voles mate, the brains of both males and females release specific peptides consisting of nine amino acids. In females, the peptide is oxytocin; in males, it is vasopressin. The peptides are circulated in the bloodstream and reach all tissues in the body, but they bind to only a few cell types. These cells have surface proteins called receptors, to which the peptides specifically bind, like a key inserting into a lock.

The interaction of peptide and receptor causes the receptor, which extends across the plasma membrane, to change shape. Within the cytoplasm, this change sets off a series of events that ultimately result in changes in behavior. The receptors for oxytocin and vasopressin in prairie voles are most concentrated in the regions of the brain that are responsible for behaviors such as bonding and caring for the young. In montane voles, there are far fewer receptors for these peptides, and as a result, fewer bonding and caring behaviors. Clearly, oxytocin and vasopressin are signals that induce these behaviors.

Intercellular signaling is a hallmark of multicellular organisms. Even in the simplest such organisms, the differentiation of a group of cells into a specialized tissue (e.g., reproductive cells) must be integrated into the organism as a whole. A cell's response to a signal molecule takes place in three sequential steps. First, the signal binds to a receptor in the cell, often embedded in the outside surface of the plasma membrane. Second, signal binding conveys a message to the cell. Third, the cell changes its activity in response to the signal. In a multicellular organism, these steps lead to changes in that organism's functioning.

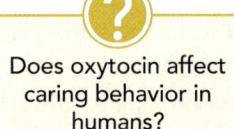

? Does oxytocin affect caring behavior in humans?

See answer on p. 141.

7.1 What Are Signals, and How Do Cells Respond to Them?

Both prokaryotic and eukaryotic cells process information from their environments. This information can be in the form of a physical stimulus, such as the light reaching your eyes as you read this book, or chemicals that bathe a cell, such as lactose in a bacterial growth medium. It may come from outside the organism, such as the scent of a female moth seeking a mate in the dark, or from a neighboring cell within the organism, such as in the heart, where thousands of muscle cells contract in unison by transmitting signals to one another.

Of course, the mere presence of a signal does not mean that a cell will respond to it, just as you do not pay close attention to every stimulus in your environment as you study. To respond to a signal, a cell must have a specific receptor that can detect it and a way to use that information to influence cellular processes. A **signal transduction pathway** is a sequence of molecular events and chemical reactions that lead to a cell's response to a signal. Signal transduction pathways vary greatly in their details, but every such pathway involves a signal, a receptor, and a response. In this section we will discuss signals and provide a brief overview of signal transduction. We will consider receptors in Section 7.2, and other aspects of signal transduction in Sections 7.3 and 7.4.

Cells receive signals from the physical environment and from other cells

The physical environment is full of signals. For example, our sense organs allow us to respond to light (a physical signal), or odors and tastes (chemical signals) from our environment. Bacteria and protists can respond to minute chemical changes in their surroundings. Plants respond to light as a signal as well as an energy source, for example, by growing toward the source of light. However, a cell deep inside a large multicellular organism is far away from the exterior environment—its signals come from neighboring cells and the surrounding extracellular fluids. In such organisms, chemical signals are often made in one part of the body and arrive at target cells by local diffusion or by circulation in the blood or the plant vascular system. These signals are usually present in tiny concentrations (as low as 10^{-10} M) (see Chapter 2 for an explanation of molar concentrations) and differ in their sources and mode of delivery (Figure 7.1):

- **Autocrine** signals diffuse to and affect the cells that make them. For example, many tumor cells reproduce uncontrollably because they both make, and respond to, signals that stimulate cell division.
- **Juxtacrine** signals affect only cells adjacent to the cell producing the signal. This is especially common during development.
- **Paracrine** signals diffuse to and affect nearby cells. An example is a neurotransmitter made by one nerve cell that diffuses to a nearby cell and stimulates it (see Section 45.3).

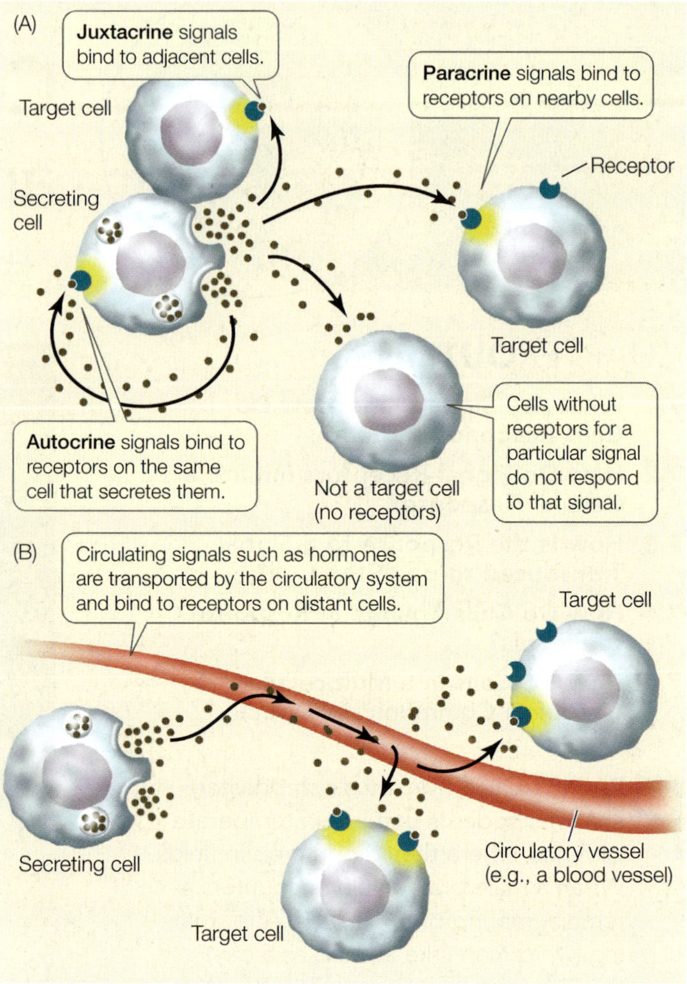
(A)
Juxtacrine signals bind to adjacent cells.
Paracrine signals bind to receptors on nearby cells.
Target cell
Receptor
Secreting cell
Target cell
Autocrine signals bind to receptors on the same cell that secretes them.
Not a target cell (no receptors)
Cells without receptors for a particular signal do not respond to that signal.
(B)
Circulating signals such as hormones are transported by the circulatory system and bind to receptors on distant cells.
Target cell
Secreting cell
Circulatory vessel (e.g., a blood vessel)
Target cell

7.1 Chemical Signaling Systems (A) A signal molecule can diffuse to and act on the cell that produces it, an adjacent cell, or a nearby cell. (B) Many signals act on distant cells and must be transported by the organism's circulatory system.

Go to Activity 7.1 Chemical Signaling Systems
Life10e.com/ac7.1

- Signals that travel through the circulatory systems of animals or the vascular systems of plants are generally called **hormones**.

A signal transduction pathway involves a signal, a receptor, and responses

As we said earlier, the basic elements of any signal transduction pathway are a signal, a receptor, and a response (Figure 7.2). For the information from a signal to be transmitted to a cell, the target cell must be able to receive the signal and respond to it. This is the job of receptors. In a mammal, all cells may be exposed to a chemical signal that is circulated in the blood, such as oxytocin or vasopressin (see the opening of this chapter). However, most body cells are not capable of responding to these signals. *Only cells with the appropriate receptors can respond.*

The "response" can involve enzymes, which catalyze biochemical reactions, and transcription factors, which are proteins that turn the expression of particular genes on and off. An important feature of signal transduction is that the activities of

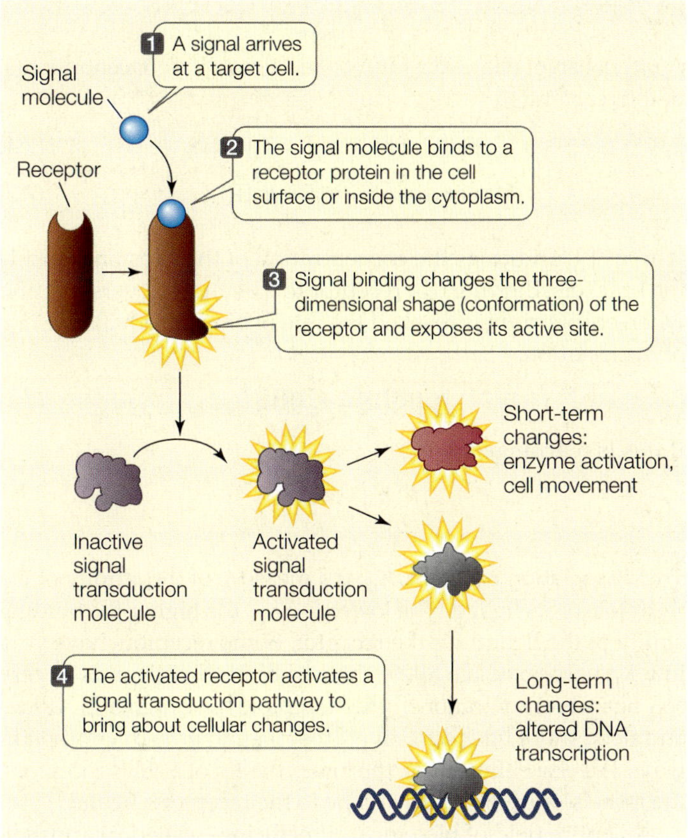

1 A signal arrives at a target cell.
Signal molecule
2 The signal molecule binds to a receptor protein in the cell surface or inside the cytoplasm.
Receptor
3 Signal binding changes the three-dimensional shape (conformation) of the receptor and exposes its active site.
Short-term changes: enzyme activation, cell movement
Inactive signal transduction molecule
Activated signal transduction molecule
4 The activated receptor activates a signal transduction pathway to bring about cellular changes.
Long-term changes: altered DNA transcription

7.2 A Signal Transduction Pathway This general pathway is common to many cells and situations. The ultimate effects on the cell are either short-term or long-term molecular changes, or both.

specific enzymes and transcription factors are regulated: they are either activated or inactivated to bring about cellular changes (see Figure 7.2). For example, an enzyme may be activated by the addition of a phosphate group (phosphorylation) to a particular site on the protein, thereby changing the enzyme's shape (see Figure 3.13B) and exposing its active site. The activity of a protein can also be regulated by mechanisms that control its location in the cell. For example, a transcription factor located in the cytoplasm is inactive because it is separated from the genetic material in the nucleus; a signal transduction pathway may result in the transport of the factor to the nucleus, where it can affect gene expression. There are many other ways in which enzymes and transcription factors are regulated. We will encounter some of these mechanisms in Sections 7.3 and 7.4.

In this chapter we consider signal transduction pathways in isolation from one another. But life is not that simple. In fact, there is a great deal of **crosstalk**: interactions between different signal transduction pathways. For example, signal transduction pathways often branch: a single activated protein (receptor or enzyme) might activate enzymes or transcription factors in multiple pathways, leading to multiple responses to a single stimulus. Multiple signal transduction pathways might converge on a single transcription factor, allowing the transcription of a single gene to be adjusted in response to several different signals. Crosstalk can also result in the activation of one pathway and the inhibition of another. This phenomenon inside the cell is analogous to the "crosstalk" that occurs at the level of the whole body. For example, in your limbs you have opposing muscles. When you bend your elbow, you contract one set of muscles and relax the opposing muscles, so that the arm will bend. Because of crosstalk, biologists often refer to "signaling networks" rather than signal transduction pathways, reflecting the high degree of complexity in cellular signaling.

RECAP 7.1

Cells are constantly exposed to molecular signals that can come from the external environment or from within the body of a multicellular organism. To respond to a signal, the cell must have a specific receptor that detects the signal and activates some type of cellular response. Signal transduction pathways involve regulation of enzymes and transcription factors, and crosstalk occurs between pathways.

- What are the differences between an autocrine signal, a juxtacrine signal, a paracrine signal, and a hormone? See p. 126 and Figure 7.1
- Describe three components in a cell's response to a signal. See pp. 126–127 and Figure 7.2

The general features of signal transduction pathways described in this section will recur in more detail throughout the chapter. First let's consider more closely the nature of the receptors that bind signal molecules.

7.2 How Do Signal Receptors Initiate a Cellular Response?

Any given cell in a multicellular organism is bombarded with many signals. However, it responds to only some of them, because no cell makes receptors for all signals. A **receptor** protein recognizes its signal very specifically, in much the same way that a membrane transport protein recognizes and binds to the substance it transports. This specificity ensures that only those cells that make a specific receptor will respond to a given signal.

Receptors that recognize chemical signals have specific binding sites

A specific chemical signal molecule fits into a three-dimensional site on its protein receptor (**Figure 7.3A**). A molecule that binds to a receptor site on another molecule in this way is called a **ligand**. Binding of the signaling ligand causes the receptor protein to change its three-dimensional shape, and that conformational change initiates a cellular response. The ligand does not contribute further to this response. In fact, the ligand is usually not metabolized into a useful product; its role is purely to "knock on the door." (This is in sharp contrast to the enzyme–substrate interaction, which we will describe in Chapter 8. The entire purpose of that interaction is to convert the substrate into a useful product.)

The sensitivity of a cell to a signal is determined in part by the affinity of the cell's receptors for the signal ligand—the likelihood that the receptor will bind to the ligand at any given

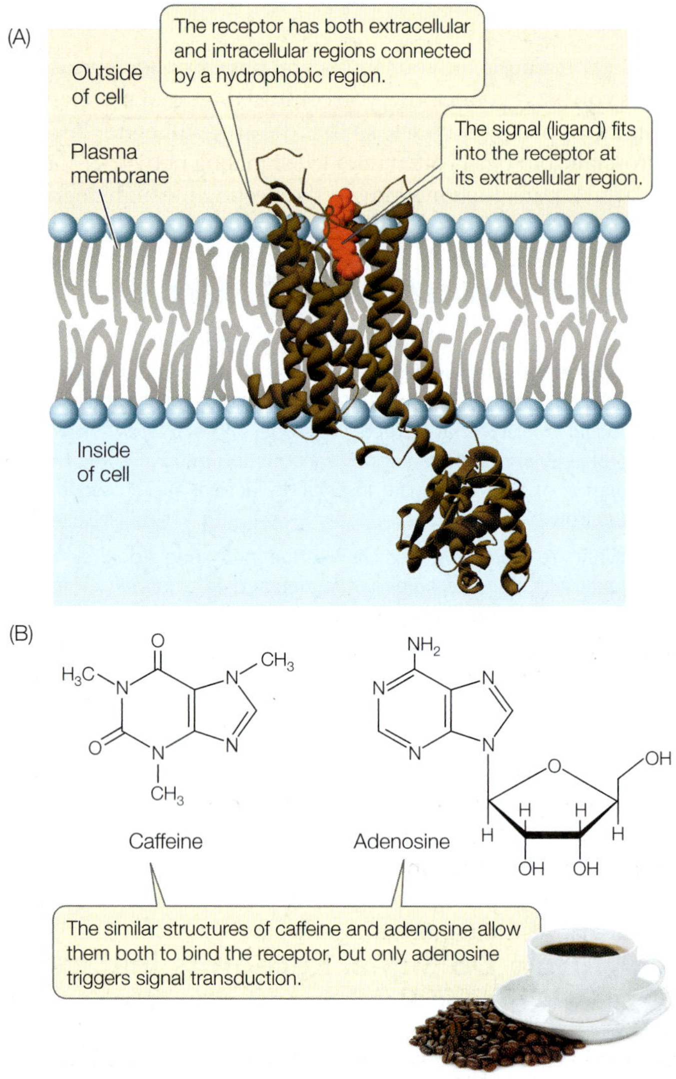
(A)
The receptor has both extracellular and intracellular regions connected by a hydrophobic region.
Outside of cell
Plasma membrane
The signal (ligand) fits into the receptor at its extracellular region.
Inside of cell
(B)
Caffeine
Adenosine
The similar structures of caffeine and adenosine allow them both to bind the receptor, but only adenosine triggers signal transduction.

7.3 A Signal and Its Receptor (A) The adenosine 2A receptor occurs in the human brain, where it is involved in inhibiting arousal. (B) Adenosine is the normal ligand for the receptor. Caffeine has a similar structure to that of adenosine and can act as an antagonist that binds the receptor and prevents its normal functioning.

ligand concentration. Receptors (R) bind to their ligands (L) according to chemistry's law of mass action. This means that the binding is reversible:

$$R + L \rightleftharpoons RL \quad (7.1)$$

For most ligand–receptor complexes (RL), the equilibrium point is far to the right—that is, binding is favored. Reversibility is important, however, because if the ligand were never released, the receptor would be continuously stimulated and the cell would never stop responding.

As with any reversible chemical reaction, the binding and dissociation processes each have a rate constant, here designated K_1 and K_2:

$$\text{Binding:} \quad R + L \xrightarrow{K_1} RL \quad (7.2)$$

$$\text{Dissociation:} \quad RL \xrightarrow{K_2} R + L \quad (7.3)$$

A rate constant relates the rate of a reaction to the concentration(s) of the reactant(s):

$$\text{Rate of binding} = K_1[R][L] \quad (7.4)$$

$$\text{Rate of dissociation} = K_2[RL] \quad (7.5)$$

where "[ ]" indicates the concentration of the substance inside the brackets. Binding of a receptor to a ligand is reversible, and when equilibrium is reached the rate of binding equals the rate of dissociation:

$$K_1[R][L] = K_2[RL] \quad (7.6)$$

If this is rearranged, we get:

$$\frac{[R]\,[L]}{[RL]} = \frac{K_2}{K_1} = K_D \quad (7.7)$$

The **dissociation constant, K_D**, is a measure of the affinity of the receptor for its ligand. The lower the K_D, the higher the binding ability of the ligand for the receptor. Some receptors have very low K_D values, which allow them to bind their ligands at very low ligand concentrations; other receptors have higher K_D values and need more ligand to set off their signal transduction pathways. All else being equal, the lower the K_D of a cell's receptors, the more sensitive the cell will be to the receptor's ligand.

An entire field of biology and medicine—called pharmacology—is devoted to the study of drugs. Drugs function as ligands that bind specific receptors. In the discovery and design of new drugs, it is helpful to know the specific receptor that the drug will bind, because then it is possible to determine the K_D value of its binding. This is one factor that can be taken into consideration when determining dosage levels. Of course, many drugs have side effects, and these are also dosage-dependent!

An inhibitor (or antagonist) can also bind to a receptor protein, instead of the normal ligand. There are both natural and artificial antagonists of receptor binding. Many substances that alter human behavior bind to specific receptors in the brain, and prevent the binding of the receptors' specific ligands. One example is caffeine, which is probably the world's most widely consumed stimulant. In the brain, the nucleoside adenosine acts as a ligand that binds to a receptor on nerve cells, initiating a signal transduction pathway that reduces brain activity, especially arousal. Because caffeine has a similar molecular structure to that of adenosine, it also binds to the adenosine receptor (Figure 7.3B). But in this case binding does not initiate a signal transduction pathway. Rather, it "ties up" the receptor, preventing adenosine binding and thereby allowing continued nerve cell activity and arousal.

Receptors can be classified by location and function

The chemistry of ligand signals is quite variable: some ligands can diffuse through membranes whereas others cannot. Physical signals such as light also vary in their ability to penetrate cells and tissues. Correspondingly, a receptor can be classified by its location in the cell, which largely depends on the nature of its signal (Figure 7.4):

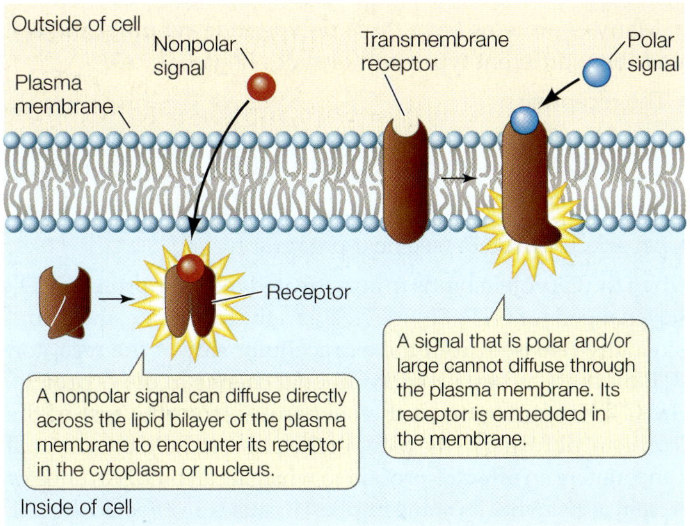
Outside of cell
Nonpolar signal
Transmembrane receptor
Polar signal
Plasma membrane
Receptor
A nonpolar signal can diffuse directly across the lipid bilayer of the plasma membrane to encounter its receptor in the cytoplasm or nucleus.
A signal that is polar and/or large cannot diffuse through the plasma membrane. Its receptor is embedded in the membrane.
Inside of cell

7.4 Two Locations for Receptors Receptors can be located inside the cell (in the cytoplasm or nucleus) or in the plasma membrane.

- *Membrane receptors*: Large or polar ligands cannot cross the lipid bilayer. Insulin, for example, is a protein hormone that cannot diffuse through the plasma membrane; instead, it binds to a transmembrane receptor with an extracellular binding domain.
- *Intracellular receptors*: Small or nonpolar ligands can diffuse across the nonpolar phospholipid bilayer of the plasma membrane and enter the cell. Estrogen, for example, is a lipid-soluble steroid hormone that can easily diffuse across the plasma membrane; it binds to a receptor inside the cell. Light of certain wavelengths can penetrate the cells in a plant leaf quite easily, and many plant light receptors (photoreceptors) are also intracellular.

In complex eukaryotes such as mammals and higher plants, there are three well-studied categories of plasma membrane receptors that are grouped according to their functions: ion channels, protein kinase receptors, and G protein-linked receptors.

ION CHANNEL RECEPTORS As described in Section 6.3, the plasma membranes of many types of cells contain gated **ion channels** that allow ions such as Na^+, K^+, Ca^{2+}, or Cl^- to enter or leave the cell. The gate-opening mechanism is an alteration in the three-dimensional shape of the channel protein upon interaction with a signal; thus these proteins function as receptors. Each type of ion channel responds to a specific signal, including sensory stimuli such as light, sound, and electric charge differences across the plasma membrane, as well as chemical ligands such as hormones and neurotransmitters.

The acetylcholine receptor, which is located in the plasma membrane of skeletal muscle cells, is an example of a gated ion channel. This receptor protein is a sodium channel that binds the ligand acetylcholine, which is a neurotransmitter—a chemical signal released from neurons (nerve cells) (Figure 7.5). When two molecules of acetylcholine bind to the receptor, it opens for about a thousandth of a second. That

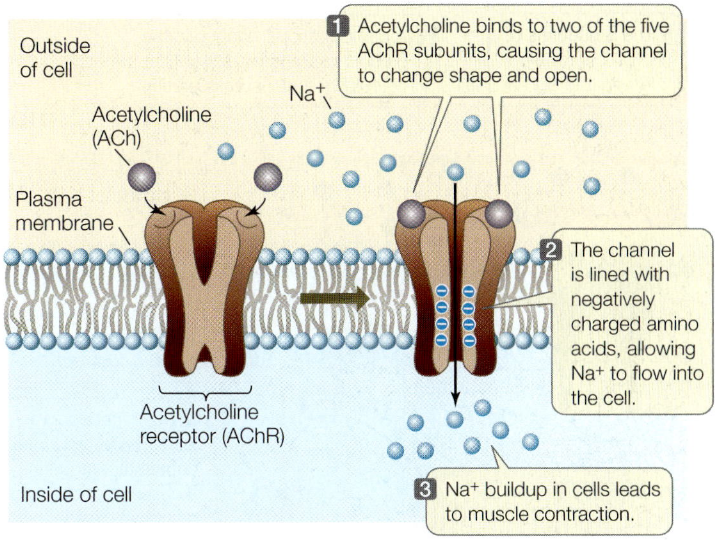
Outside of cell
1 Acetylcholine binds to two of the five AChR subunits, causing the channel to change shape and open.
Na+
Acetylcholine (ACh)
Plasma membrane
2 The channel is lined with negatively charged amino acids, allowing Na+ to flow into the cell.
Acetylcholine receptor (AChR)
Inside of cell
3 Na+ buildup in cells leads to muscle contraction.

7.5 A Gated Ion Channel The acetylcholine receptor (AChR) is a ligand-gated ion channel for sodium ions. It is made up of five polypeptide subunits. When acetylcholine molecules (ACh) bind to two of the subunits, the gate opens and Na^+ flows into the cell. This channel helps regulate membrane polarity (see Chapter 45).

is enough time for Na^+, which is more concentrated outside the cell than inside, to rush into the cell, moving in response to both concentration and electrical potential gradients. The change in Na^+ concentration in the cell initiates a series of events that result in muscle contraction.

PROTEIN KINASE RECEPTORS Some eukaryotic receptor proteins are **protein kinases**. When they are activated, they catalyze the phosphorylation of themselves and/or other proteins, thus changing their shapes and therefore their functions.

$$\text{Target protein} + \text{ATP} \xrightarrow{\text{Protein kinase}} \text{protein-}Ⓟ + \text{ADP}$$
(altered shape and function)

The receptor for insulin is an example of a protein kinase receptor. Insulin is a protein hormone made by the mammalian pancreas. Its receptor has two copies each of two different polypeptide subunits called α and β (Figure 7.6). When insulin binds to the receptor, the receptor becomes activated and is able to phosphorylate itself and certain cytoplasmic proteins that are appropriately called insulin response substrates. These proteins then initiate many cellular responses, including the insertion of glucose transporters (see Figure 6.12) into the plasma membrane.

G PROTEIN-LINKED RECEPTORS A third category of eukaryotic plasma membrane receptors is the **G protein-linked receptors**, also referred to as the seven-transmembrane domain receptors. These receptors have many physiological roles, including light detection in the mammalian retina (photoreceptors), detection of odors (olfactory receptors), and the regulation of mood and behavior. For example, the receptors that bind the hormones oxytocin and vasopressin, which affect mating behavior in voles (see the opening story), are G protein-linked receptors.

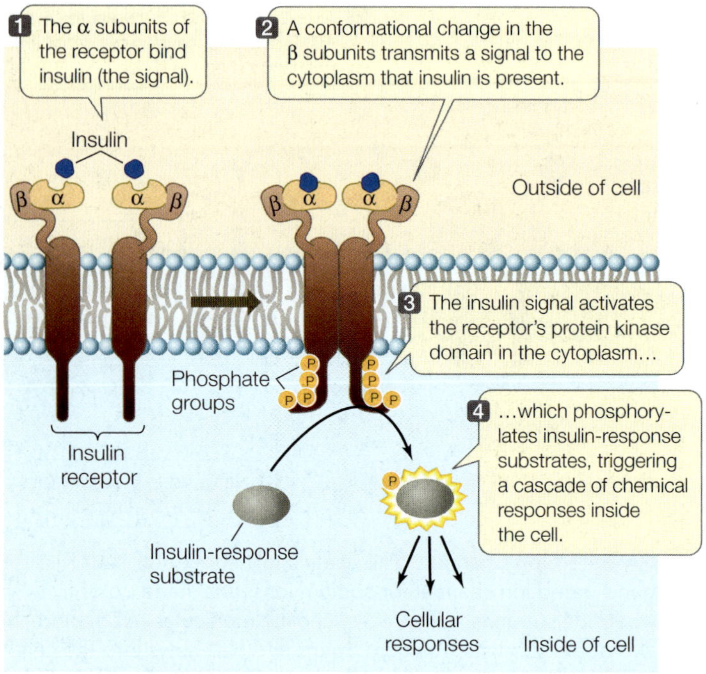
1 The α subunits of the receptor bind insulin (the signal).
2 A conformational change in the β subunits transmits a signal to the cytoplasm that insulin is present.
Insulin
β α α β
Outside of cell
3 The insulin signal activates the receptor's protein kinase domain in the cytoplasm…
Phosphate groups
P
4 …which phosphorylates insulin-response substrates, triggering a cascade of chemical responses inside the cell.
Insulin receptor
Insulin-response substrate
Cellular responses
Inside of cell

7.6 A Protein Kinase Receptor The mammalian hormone insulin binds to a receptor on the outside surface of the cell and initiates a response.

The descriptive name identifies a fascinating group of receptors, each of which is composed of a single protein with seven transmembrane domains. These seven domains pass through the phospholipid bilayer and are separated by short loops that extend either outside or inside the cell. Ligand binding on the extracellular side of the receptor changes the shape of its cytoplasmic region, exposing a site that binds to a mobile membrane protein called a **G protein**. The G protein is partially inserted into the lipid bilayer and partially exposed on the cytoplasmic surface of the membrane.

Many G proteins have three polypeptide subunits and can bind three different types of molecules (**Figure 7.7A**):

- The receptor
- GDP and GTP (guanosine diphosphate and triphosphate, respectively; these are nucleoside phosphates like ADP and ATP)
- An effector protein (see next paragraph)

When the G protein binds to an activated receptor protein, GDP is exchanged for GTP (**Figure 7.7B**). At the same time, the ligand is usually released from the extracellular side of the receptor. GTP binding causes a conformational change in the G protein. The GTP-bound subunit then separates from the rest of the G protein, diffusing in the plane of the phospholipid bilayer until it encounters an **effector protein** to which it can bind. An effector protein is just what its name implies: it causes an effect in the cell. The binding of the GTP-bearing G protein subunit activates the effector—which may be an enzyme or an ion channel—thereby causing changes in cell function (**Figure 7.7C**).

After activation of the effector protein, the GTP bound to the G protein is hydrolyzed to GDP. The now inactive G protein subunit separates from the effector protein and diffuses in the membrane to collide with and bind to the other two G protein subunits. When the three components of the G protein are reassembled, the protein is capable of binding again to an activated receptor. After binding, the activated receptor exchanges the GDP on the G protein for a GTP, and the cycle begins again.

Intracellular receptors are located in the cytoplasm or the nucleus

Intracellular receptors are located inside the cell and respond to physical signals such as light (for example, some photoreceptors in plants) or chemical signals that can diffuse across the plasma membrane (for example, steroid hormones in animals). Many intracellular receptors are transcription factors. Some are

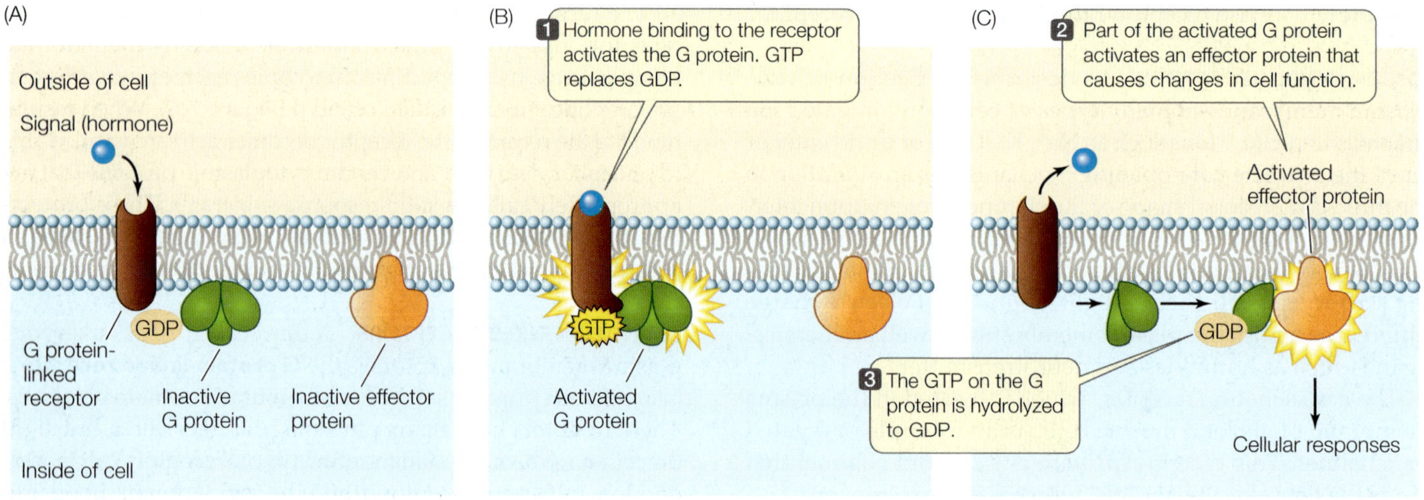
(A)
Outside of cell
Signal (hormone)
GDP
G protein-linked receptor
Inactive G protein
Inactive effector protein
Inside of cell
(B)
1 Hormone binding to the receptor activates the G protein. GTP replaces GDP.
GTP
Activated G protein
(C)
2 Part of the activated G protein activates an effector protein that causes changes in cell function.
Activated effector protein
GDP
3 The GTP on the G protein is hydrolyzed to GDP.
Cellular responses

7.7 A G Protein-Linked Receptor
The G protein is an intermediary between the receptor and its effector.

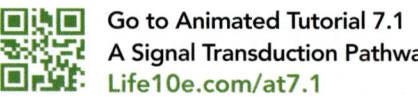

Go to Animated Tutorial 7.1
A Signal Transduction Pathway
Life10e.com/at7.1

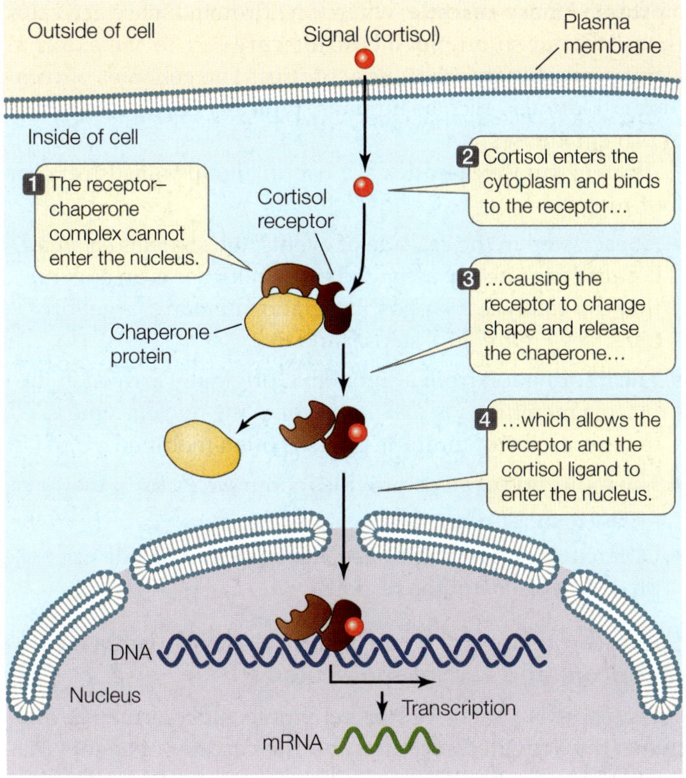
Outside of cell
Signal (cortisol)
Plasma membrane
Inside of cell
1 The receptor–chaperone complex cannot enter the nucleus.
Cortisol receptor
2 Cortisol enters the cytoplasm and binds to the receptor…
3 …causing the receptor to change shape and release the chaperone…
Chaperone protein
4 …which allows the receptor and the cortisol ligand to enter the nucleus.
DNA
Nucleus
Transcription
mRNA

7.8 An Intracellular Receptor The receptor for cortisol is in the cytoplasm bound to a chaperone protein that is released when cortisol binds to the receptor.

located in the cytoplasm until they are activated; after binding their ligands, these transcription factors move to the nucleus where they bind to DNA and alter the expression of specific genes. A typical example is the receptor for the steroid hormone cortisol. This receptor is normally bound to a chaperone protein that blocks it from entering the nucleus. Binding of the hormone causes the receptor to change its shape so that the chaperone is released (**Figure 7.8**). This release allows the receptor to enter the nucleus, where it affects DNA transcription. Another group of intracellular receptors are always located in the nucleus, and their ligands must enter the nucleus before binding.

RECAP 7.2

Receptors are proteins that bind, or are changed by, specific ligands or physical signals. The changed receptor initiates a response in the cell. These receptors are located in the plasma membrane or inside the cell.

- What is the nature and importance of specificity in the binding of a receptor to its particular ligand? **See p. 127**
- How is a dissociation constant calculated, and what is its relevance to ligand–receptor binding? **See p. 128**
- Describe three important categories of plasma membrane receptors that are seen in complex eukaryotes. **See pp. 129–130 and Figures 7.5, 7.6, 7.7**

Now that we have discussed signals and receptors, let's examine the characteristics of the molecules (transducers) that mediate the cellular response.

7.3 How Is the Response to a Signal Transduced through the Cell?

As we have seen, there are many different kinds of signals and receptors. Not surprisingly, the ways that signals are transduced, and the resulting cellular responses, are also highly varied. Some signal transduction pathways are quite simple and direct, whereas others involve multiple steps. As we mentioned in Section 7.1, signal transduction pathways can involve enzymes and transcription factors. In addition, small nonprotein molecules called **second messengers** can diffuse throughout the cytoplasm and mediate further steps in pathways.

In many cases, the signal can initiate a chain (cascade) of events, in which proteins interact with other proteins, which interact with still other proteins until the final responses are achieved. Through such a cascade, an initial signal can be both amplified and distributed to cause several different responses in the target cell. In this section we will examine the kinds of molecules that transduce signals and look at several different signal transduction pathways.

A protein kinase cascade amplifies a response to ligand binding

We have seen that when a signal binds to a protein kinase receptor, the receptor's conformation changes, exposing a protein kinase active site on the receptor's cytoplasmic domain. The protein kinase then catalyzes the phosphorylation of target proteins. Protein kinase receptors are important in binding signals called growth factors that stimulate cell division in both plants and animals.

Scientists worked out the signal transduction pathway for one growth factor by studying a cell that went wrong. Many human bladder cancers contain an abnormal form of a protein called Ras (so named because a similar protein was previously isolated from a *rat* sarcoma tumor). Investigations of these bladder cancers showed that Ras was a G protein and that the abnormal form was always active because it was permanently bound to GTP and thus caused continuous cell division (**Figure 7.9**). If this abnormal form of Ras was inhibited, the cells stopped dividing. This discovery has led to a major effort to develop specific Ras inhibitors for cancer treatment.

Go to Animated Tutorial 7.2
Signal Transduction and Cancer
Life10e.com/at7.2

Other cancer cells have abnormalities in different parts of the same signal transduction pathway. Biologists compared the defects in these cells with the normal signaling process in non-cancer cells and thus worked out the entire signaling pathway (**Figure 7.10**). In Section 7.2 we discussed G protein-linked receptors, but other kinds of receptors can

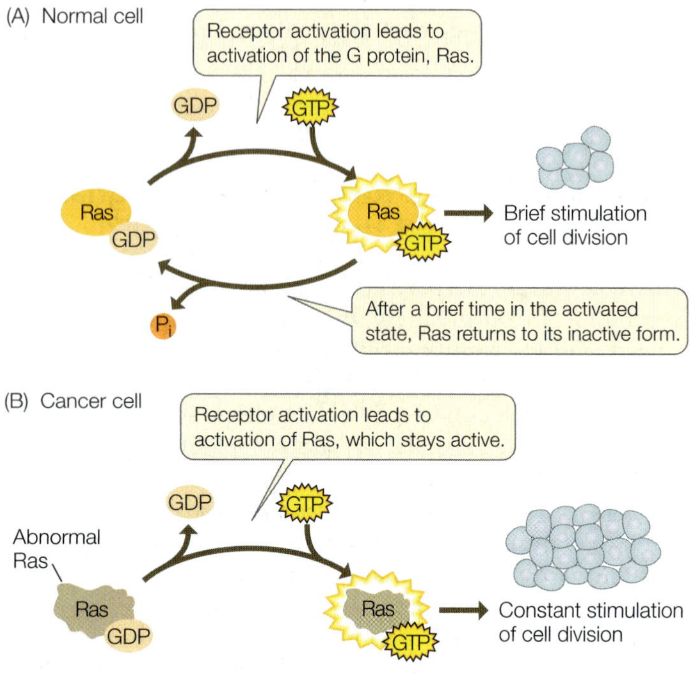

(A) Normal cell
Receptor activation leads to activation of the G protein, Ras.
GDP
GTP
Ras
GDP
Ras
GTP
Brief stimulation of cell division
Pi
After a brief time in the activated state, Ras returns to its inactive form.
(B) Cancer cell
Receptor activation leads to activation of Ras, which stays active.
GDP
GTP
Abnormal Ras
Ras
GDP
Ras
GTP
Constant stimulation of cell division

7.9 Signal Transduction and Cancer (A) Ras is a G protein that regulates cell division. (B) In some tumors, the Ras protein is permanently active, resulting in uncontrolled cell division.

interact with G proteins as well. The G protein Ras mediates a response after activation by the protein kinase receptor. The resulting signal transduction pathway is an example of a **protein kinase cascade**, where one protein kinase activates the next, and so on. Such cascades are key to the external regulation of many cellular activities. The genomes of complex eukaryotes, such as humans, typically encode hundreds of protein kinases.

Protein kinase cascades are useful signal transducers for four reasons:

- At each step in the cascade of events, the signal is *amplified*, because each newly activated protein kinase is an enzyme that can catalyze the phosphorylation of many target proteins (see Figure 7.10, steps 5 and 6).
- The information from a signal that originally arrived at the plasma membrane is *communicated* to the nucleus where the expression of multiple genes is often modified.
- The multitude of steps provides some *specificity* to the process.
- Different target proteins at each step in the cascade can provide *variation* in the response.

Second messengers can amplify signals between receptors and target molecules

Often, there is a small, nonprotein molecule intermediary between the activated receptor and the cascade of events that ensues. Earl Sutherland and his colleagues at Case Western Reserve University discovered one such molecule when they were investigating the activation of the liver enzyme glycogen phosphorylase by the hormone epinephrine. The

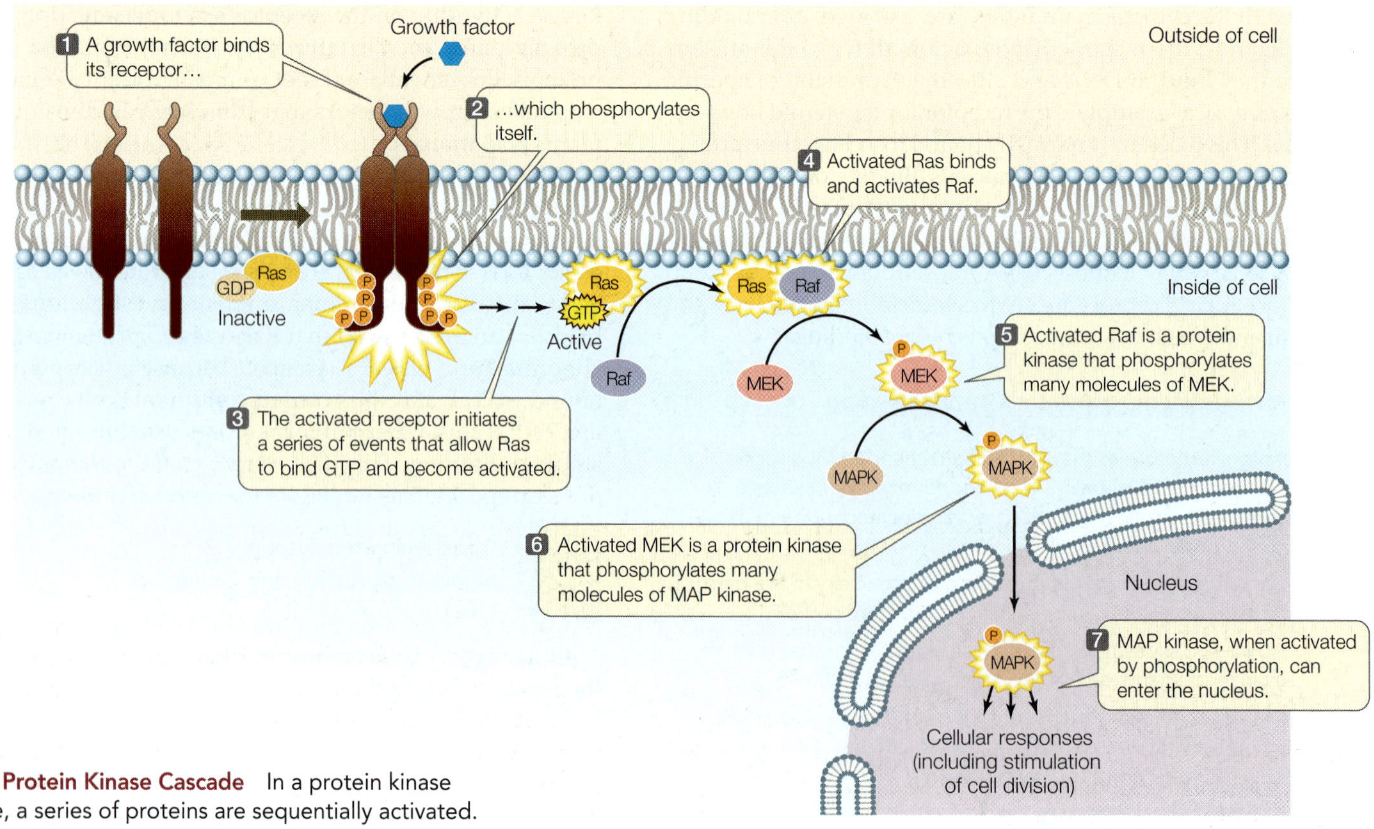

1 A growth factor binds its receptor...
Growth factor
Outside of cell
2 ...which phosphorylates itself.
4 Activated Ras binds and activates Raf.
Ras
GDP
Inactive
P
Ras
GTP
Active
Ras
Raf
Inside of cell
Raf
MEK
MEK
5 Activated Raf is a protein kinase that phosphorylates many molecules of MEK.
3 The activated receptor initiates a series of events that allow Ras to bind GTP and become activated.
MAPK
MAPK
6 Activated MEK is a protein kinase that phosphorylates many molecules of MAP kinase.
Nucleus
MAPK
7 MAP kinase, when activated by phosphorylation, can enter the nucleus.
Cellular responses (including stimulation of cell division)

7.10 A Protein Kinase Cascade In a protein kinase cascade, a series of proteins are sequentially activated.

INVESTIGATING**LIFE**

7.11 The Discovery of a Second Messenger Glycogen phosphorylase is activated in liver cells after epinephrine binds to a membrane receptor. Sutherland and his colleagues observed that this activation could occur in vitro only if fragments of the plasma membrane were present. They designed experiments to show that a second messenger caused the activation of glycogen phosphorylase.[a]

HYPOTHESIS A second messenger mediates between receptor activation at the plasma membrane and enzyme activation in the cytoplasm.

Method

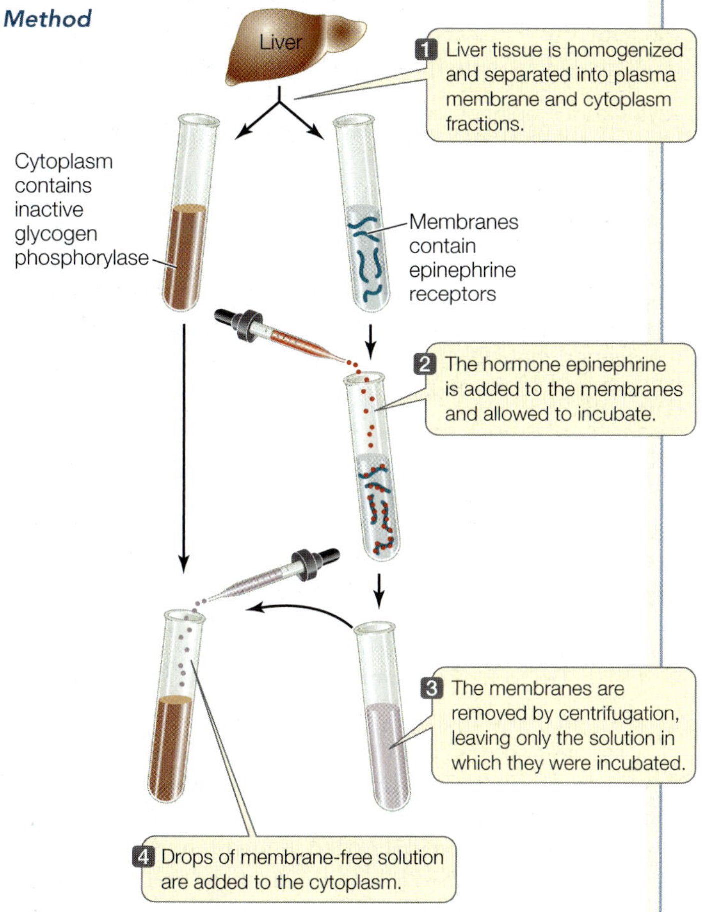
Liver
1 Liver tissue is homogenized and separated into plasma membrane and cytoplasm fractions.
Cytoplasm contains inactive glycogen phosphorylase
Membranes contain epinephrine receptors
2 The hormone epinephrine is added to the membranes and allowed to incubate.
3 The membranes are removed by centrifugation, leaving only the solution in which they were incubated.
4 Drops of membrane-free solution are added to the cytoplasm.

Results Glycogen phosphorylase in the cytoplasm is activated.

CONCLUSION A soluble second messenger, produced by hormone-activated membranes, is present in the solution and activates enzymes in the cytoplasm.

Go to **BioPortal** for discussion and relevant links for all INVESTIGATING**LIFE** figures.

[a]Rall, T. W. et al. 1957. *Journal of Biological Chemistry* 224: 463.

hormone is released when an animal faces life-threatening conditions and needs energy fast for the fight-or-flight response (see Figure 41.3). Glycogen phosphorylase catalyzes the breakdown of glycogen stored in the liver so that the resulting glucose molecules can be released to the blood. The

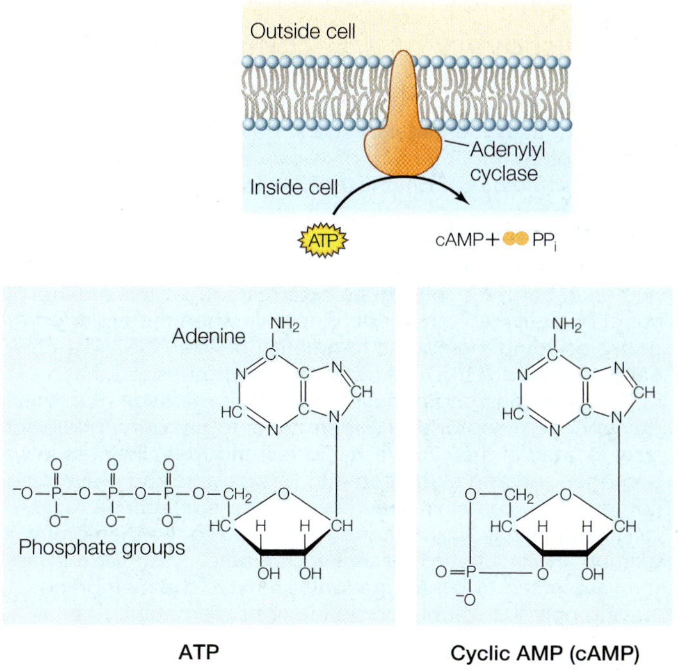
Outside cell
Adenylyl cyclase
Inside cell
ATP
cAMP + PP$_i$
Adenine
Phosphate groups
ATP
Cyclic AMP (cAMP)

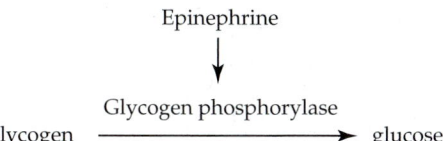

7.12 The Formation of Cyclic AMP The formation of cAMP from ATP is catalyzed by adenylyl cyclase, an enzyme that is activated by G proteins.

enzyme is present in the liver cell cytoplasm but is inactive unless the liver cells are exposed to epinephrine:

$$\text{Glycogen} \xrightarrow[\quad]{\text{Glycogen phosphorylase} \;\uparrow\; \text{Epinephrine}} \text{glucose}$$

The researchers found that epinephrine could activate glycogen phosphorylase in liver cells that had been broken open, but only if the entire cell contents, including plasma membrane fragments, were present. Under these conditions epinephrine was bound to the plasma membrane fragments (the location of its receptor), but the active phosphorylase was present in the solution. Adding epinephrine to just cytoplasm with inactive phosphorylase did not result in activation. They hypothesized that there must be a "second messenger" that transmits the epinephrine signal (epinephrine being the "first messenger"). They investigated this by separating plasma membrane fragments from the cytoplasmic fractions of broken liver cells and following the sequence of steps described in **Figure 7.11**. The experiments confirmed the existence of a second messenger, later identified as **cyclic AMP (cAMP)**, which is produced from ATP by the enzyme adenylyl cyclase (**Figure 7.12**). Adenylyl cyclase is activated via a G protein-linked epinephrine receptor (see the first steps in Figure 7.18).

In contrast to the specificity of receptor binding, second messengers such as cAMP allow a cell to respond to a single event at the plasma membrane with many events inside the cell. Thus second messengers serve to rapidly amplify and distribute the signal—for example, binding of a single epinephrine molecule leads to the production of many molecules of cAMP, which

WORKING WITH**DATA:**

The Discovery of a Second Messenger

Original Paper

Rall, T. W., E. W. Sutherland, and J. Berthet. 1957. The relationship of epinephrine and glucagon to liver phosphorylase. *Journal of Biological Chemistry* 224: 463.

Analyze the Data

While studying the action of glycogen phosphorylase, Earl Sutherland and his colleagues determined that this enzyme could be activated by epinephrine only when the entire contents, including membrane fragments, of liver cells were present (see Figure 7.11). The researchers hypothesized that a cytoplasmic messenger must transmit the message from the epinephrine receptor at the membrane to glycogen phosphorylase, located in the cytoplasm. To test this idea, liver tissue was homogenized and separated into cytoplasmic and membrane components, containing the enzyme and epinephrine receptors, respectively. Epinephrine was added to the membrane fraction and incubated for a period of time. This fraction was then subjected to centrifugation to remove the membranes, leaving only the soluble portion in the supernatant. A small sample of the membrane-free solution was added to the cytoplasmic fraction, which was then assayed for the presence of glycogen phosphorylase activity. The assay showed that active glycogen phosphorylase was indeed present in the cytoplasmic fraction. These results confirmed the hypothesis that a soluble second messenger was produced in response to epinephrine binding to its receptor in the membrane, and then diffused into the cytoplasm to activate the enzyme. Later research by Sutherland identified cAMP as the second messenger involved in the mechanism of action of epinephrine, as well as many other hormones. Sutherland's research was highly regarded in the scientific community, and in 1971 he won the Nobel Prize in Physiology or Medicine for his discoveries concerning "the mechanisms of the action of hormones."

QUESTION 1

As part of Sutherland's research, the activity of glycogen phosphorylase was measured in various liver cell fractions, with or without incubation with epinephrine. The table shows the results. Explain how these data support the hypothesis that there is a soluble second messenger that activates the enzyme.

Condition	Enzyme activity (units)
Homogenate	0.4
Homogenate + epinephrine	2.5
Cytoplasm	0.2
Cytoplasm + epinephrine	0.4
Membranes + epinephrine	0.4
Cytoplasm + membranes + epinephrine	2.0

QUESTION 2

Propose an experiment to test whether the factor (second messenger) that activates the enzyme is stable on heating (and therefore probably not a protein), and give predicted data.

QUESTION 3

The second messenger, cAMP, was purified from the hormone-treated membrane fraction. Propose experiments to show that cAMP could replace the membrane fraction and hormone treatment in the activation of glycogen phosphorylase, and create a table to show possible results.

Go to **BioPortal** for all WORKING WITH**DATA** exercises

then activate many enzyme targets by binding to them noncovalently. In the case of epinephrine and the liver cell, glycogen phosphorylase is just one of several enzymes that are activated.

In addition, second messengers are often involved in cross-talk between different signaling pathways. Activation of the epinephrine receptor is not the only way for a cell to produce cAMP; and as noted, there are multiple targets of cAMP in the cell, and these targets are parts of other pathways.

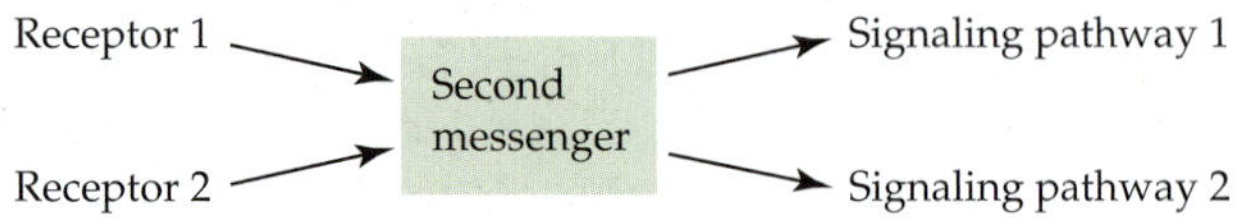
Receptor 1
Receptor 2
Second messenger
Signaling pathway 1
Signaling pathway 2

Several other classes of second messengers have since been identified, including lipid-derived second messengers, calcium ions, and nitric oxide.

LIPID-DERIVED SECOND MESSENGERS In addition to their role as structural components of the plasma membrane, phospholipids are also involved in signal transduction. When certain phospholipids are hydrolyzed into their component parts by enzymes called phospholipases, second messengers are formed.

The best-studied examples of lipid-derived second messengers come from the hydrolysis of the phospholipid **phosphatidyl inositol-bisphosphate (PIP$_2$)**. Like all phospholipids, PIP$_2$ has a hydrophobic portion embedded in the plasma membrane: two fatty acid tails attached to a molecule of glycerol, which together form **diacylglycerol**, or **DAG**. The hydrophilic portion of PIP$_2$ is **inositol trisphosphate**, or **IP$_3$**, which projects into the cytoplasm.

As with cAMP, the receptors involved in this second-messenger system are often G protein-linked receptors. A G protein subunit is activated by the receptor, then diffuses within the plasma membrane and activates phospholipase C, an enzyme that is also located in the membrane. This enzyme cleaves off the IP$_3$ from PIP$_2$, leaving the diacylglycerol (DAG) in the phospholipid bilayer:

$$\underset{\text{In membrane}}{\text{PIP}_2} \xrightarrow{\text{Phospholipase C}} \underset{\text{Released to cytoplasm}}{\text{IP}_3} + \underset{\text{In membrane}}{\text{DAG}}$$

IP$_3$ and DAG are both second messengers; they have different modes of action that build on each other to activate protein kinase C (PKC) (**Figure 7.13**). PKC refers to a family of protein

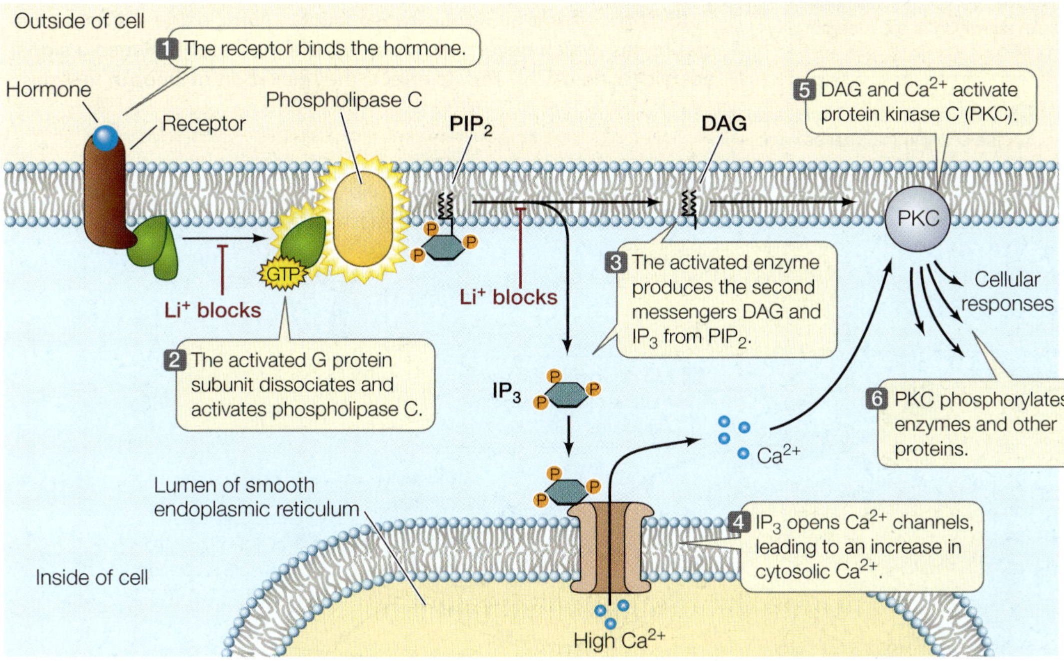
Outside of cell
1 The receptor binds the hormone.
Hormone
Receptor
Phospholipase C
PIP2
DAG
5 DAG and Ca2+ activate protein kinase C (PKC).
PKC
GTP
Li+ blocks
Li+ blocks
3 The activated enzyme produces the second messengers DAG and IP3 from PIP2.
Cellular responses
2 The activated G protein subunit dissociates and activates phospholipase C.
IP3
6 PKC phosphorylates enzymes and other proteins.
Ca2+
Lumen of smooth endoplasmic reticulum
4 IP3 opens Ca2+ channels, leading to an increase in cytosolic Ca2+.
Inside of cell
High Ca2+

7.13 The IP_3/DAG Second-Messenger System Phospholipase C hydrolyzes the phospholipid PIP_2 into its components, IP_3 and DAG, both of which are second messengers. Lithium ions (Li^+) block this pathway and are used to treat bipolar disorder (red type).

kinases that can phosphorylate a wide variety of target proteins, leading to a multiplicity of cellular responses that vary depending on the tissue or cell type.

The IP_3/DAG pathway is apparently a target for the ion lithium (Li^+), which has been used for many years as a psychoactive drug to treat bipolar (manic-depressive) disorder. This serious illness occurs in about 1 in every 100 people. In these patients, an overactive IP_3/DAG signal transduction pathway in the brain leads to excessive brain activity in certain regions. Lithium "tones down" this pathway in two ways, as indicated by the red notations in Figure 7.13: it inhibits G protein activation of phospholipase C, as well as the synthesis of IP_3. The overall result is that brain activity returns to normal.

CALCIUM IONS Calcium ions (Ca^{2+}) are scarce inside most cells, which have cytosolic Ca^{2+} concentrations of only about 0.1 μM. Ca^{2+} concentrations outside cells and within the endoplasmic reticulum are usually much higher. Active transport proteins in the plasma and ER membranes maintain this concentration difference by pumping Ca^{2+} out of the cytosol. In contrast to cAMP and the lipid-derived second messengers, Ca^{2+} cannot be synthesized to increase the intracellular Ca^{2+} concentration. Instead, Ca^{2+} ion levels are regulated via the opening and closing of ion channels, and the action of membrane pumps.

There are many signals that can cause calcium channels to open, including IP_3 (see Figure 7.13). The entry of a sperm into an egg is a very important signal that causes a massive opening of calcium channels, resulting in numerous and dramatic changes that prepare the now fertilized egg for cell division and development (**Figure 7.14**). Whatever the initial signal that causes calcium channels to open, their opening results in a dramatic increase in cytosolic Ca^{2+} concentration, which can increase up to 100-fold within a fraction of a second. As we saw earlier, this increase activates protein kinase C. In addition, Ca^{2+} controls other ion channels and stimulates secretion by exocytosis in many cell types.

NITRIC OXIDE Most signaling molecules and second messengers are solutes that remain dissolved in either the aqueous or hydrophobic components of cells. It was a great surprise to find that a gas could also be active in signal transduction. Nitric oxide (NO) is a second messenger in the signal transduction pathway between the neurotransmitter acetylcholine

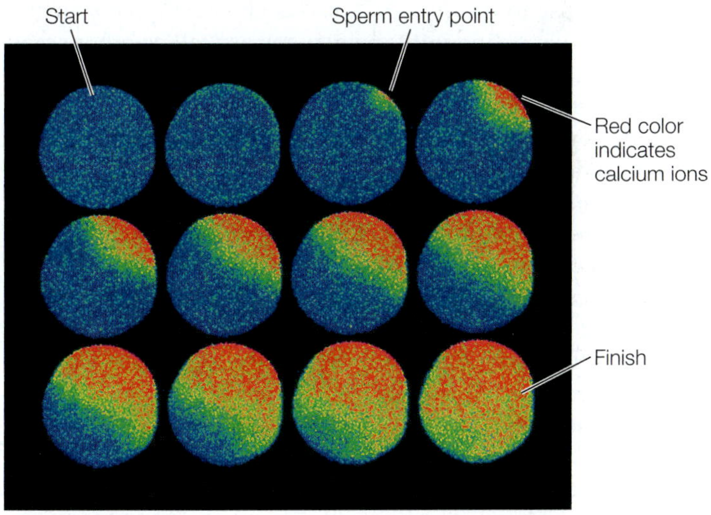
Start
Sperm entry point
Red color indicates calcium ions
Finish

7.14 Calcium Ions as Second Messengers The concentration of Ca^{2+} can be measured using a dye that fluoresces when it binds the ion. Here, fertilization in a starfish egg causes a rush of Ca^{2+} from the environment into the cytoplasm. Areas of high Ca^{2+} concentration are indicated by the red color, and the events are photographed at 5-second intervals. Calcium signaling occurs in virtually all animal groups and triggers cell division in fertilized eggs, initiating the development of new individuals.

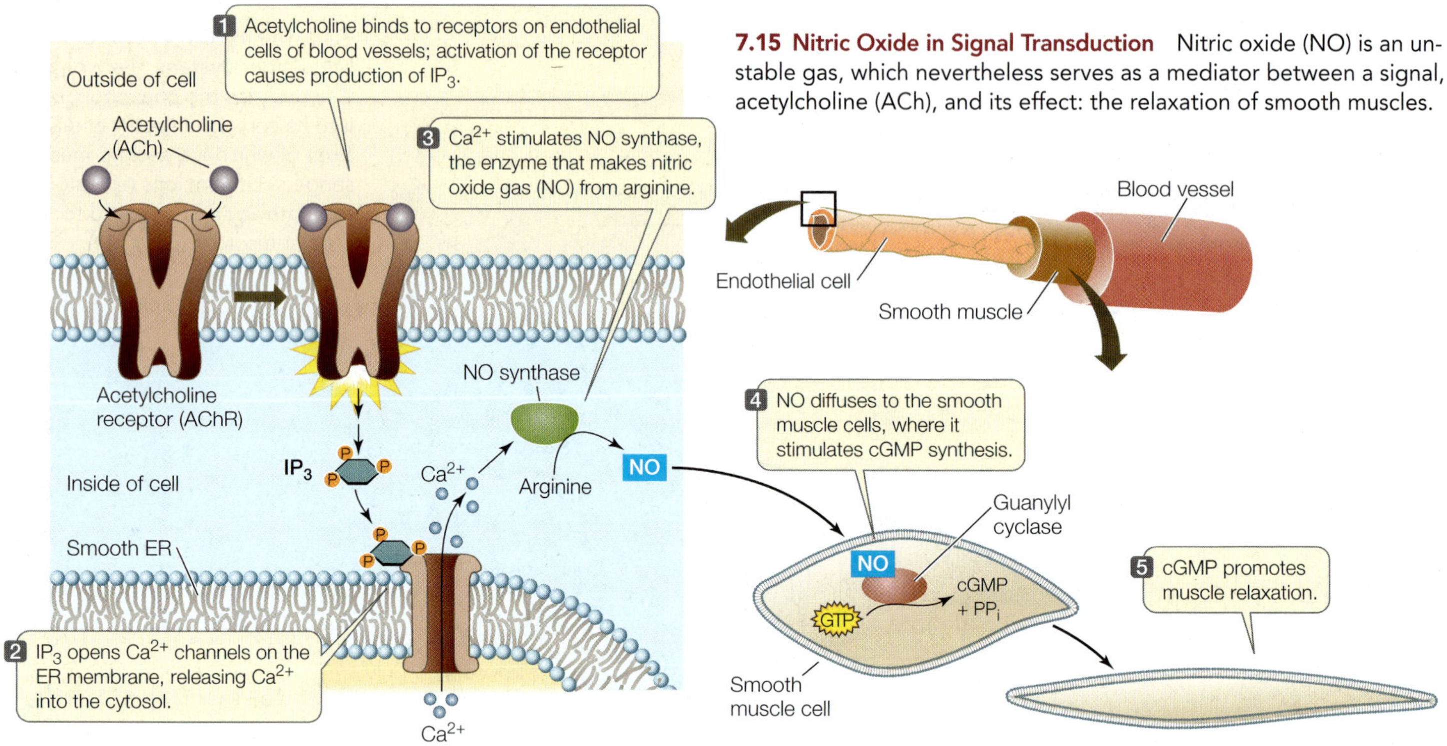

1 Acetylcholine binds to receptors on endothelial cells of blood vessels; activation of the receptor causes production of IP3.
Outside of cell
Acetylcholine (ACh)
3 Ca2+ stimulates NO synthase, the enzyme that makes nitric oxide gas (NO) from arginine.
Blood vessel
Endothelial cell
Smooth muscle
Acetylcholine receptor (AChR)
NO synthase
4 NO diffuses to the smooth muscle cells, where it stimulates cGMP synthesis.
IP3
Ca2+
Inside of cell
Arginine
NO
Guanylyl cyclase
Smooth ER
NO
cGMP + PPi
GTP
5 cGMP promotes muscle relaxation.
2 IP3 opens Ca2+ channels on the ER membrane, releasing Ca2+ into the cytosol.
Smooth muscle cell
Ca2+

7.15 Nitric Oxide in Signal Transduction Nitric oxide (NO) is an unstable gas, which nevertheless serves as a mediator between a signal, acetylcholine (ACh), and its effect: the relaxation of smooth muscles.

(see Section 7.2) and the relaxation of smooth muscles lining blood vessels, which allows more blood flow (**Figure 7.15**). In the body, NO is made from the amino acid arginine by the enzyme NO synthase. When the acetylcholine receptor on the surface of an endothelial cell is activated, IP_3 is released from the membrane (via the pathway shown in Figure 7.13), causing a calcium channel in the ER membrane to open and a subsequent increase in cytosolic Ca^{2+}. The Ca^{2+} then activates NO synthase to produce NO. NO is chemically very unstable, readily reacting with oxygen gas as well as other small molecules. Although NO diffuses readily, it does not get far. Conveniently, the endothelial cells are close to the underlying smooth muscle cells, where NO activates an enzyme called guanylyl cyclase (a close relative of adenylyl cyclase). This enzyme catalyzes the formation of cyclic GMP (cGMP): yet another second messenger that contributes to the relaxation of muscle cells.

The discovery of NO as a participant in signal transduction explained the action of nitroglycerin, a drug that has been used for more than a century to treat angina, the chest pain caused by insufficient blood flow to the heart. Nitroglycerin releases NO, which results in relaxation of the blood vessels and increased blood flow. The drug sildenafil (Viagra) was developed to treat angina via the NO signal transduction pathway but was only modestly useful for that purpose. However, men taking it reported more pronounced penile erections. During sexual stimulation, NO acts as a signal, causing an increase in cGMP and a subsequent relaxation of the smooth muscles surrounding the arteries in the corpus cavernosum of the penis. As a result of this signal, the penis fills with blood, producing an erection. Sildenafil acts by inhibiting an enzyme (a phosphodiesterase) that breaks down cGMP—resulting in more cGMP and better erections.

Signal transduction is highly regulated

There are several ways in which cells can regulate the activity of a transducer molecule. The concentration of NO, which breaks down quickly, can be regulated only by how much of it is made. By contrast, membrane pumps and ion channels regulate the cytosolic concentration of Ca^{2+}, as we have seen. To regulate protein kinase cascades, G proteins, and cAMP, there are enzymes that inactivate the activated transducer (**Figure 7.16**).

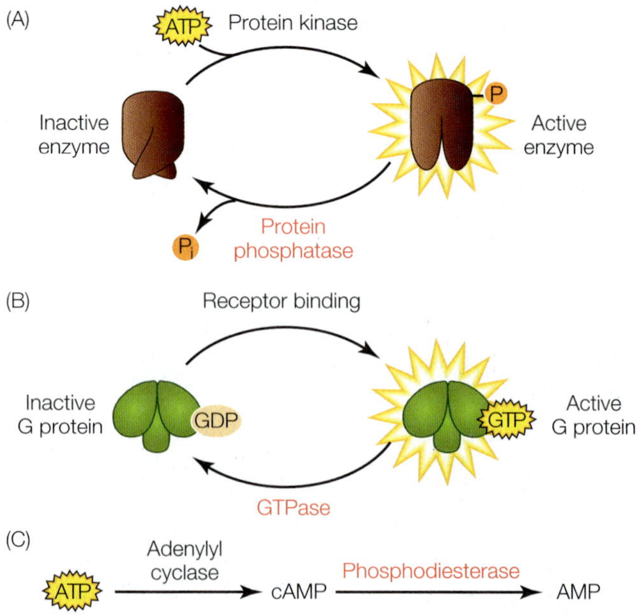

(A)
ATP
Protein kinase
Inactive enzyme
P
Active enzyme
Pi
Protein phosphatase
(B)
Receptor binding
Inactive G protein
GDP
GTP
Active G protein
GTPase
(C)
ATP
Adenylyl cyclase
cAMP
Phosphodiesterase
AMP

7.16 Regulation of Signal Transduction Some signals lead to the production of active transducers such as (A) protein kinases, (B) G proteins, and (C) cAMP. Other enzymes (shown in red) inactivate or remove these transducers.

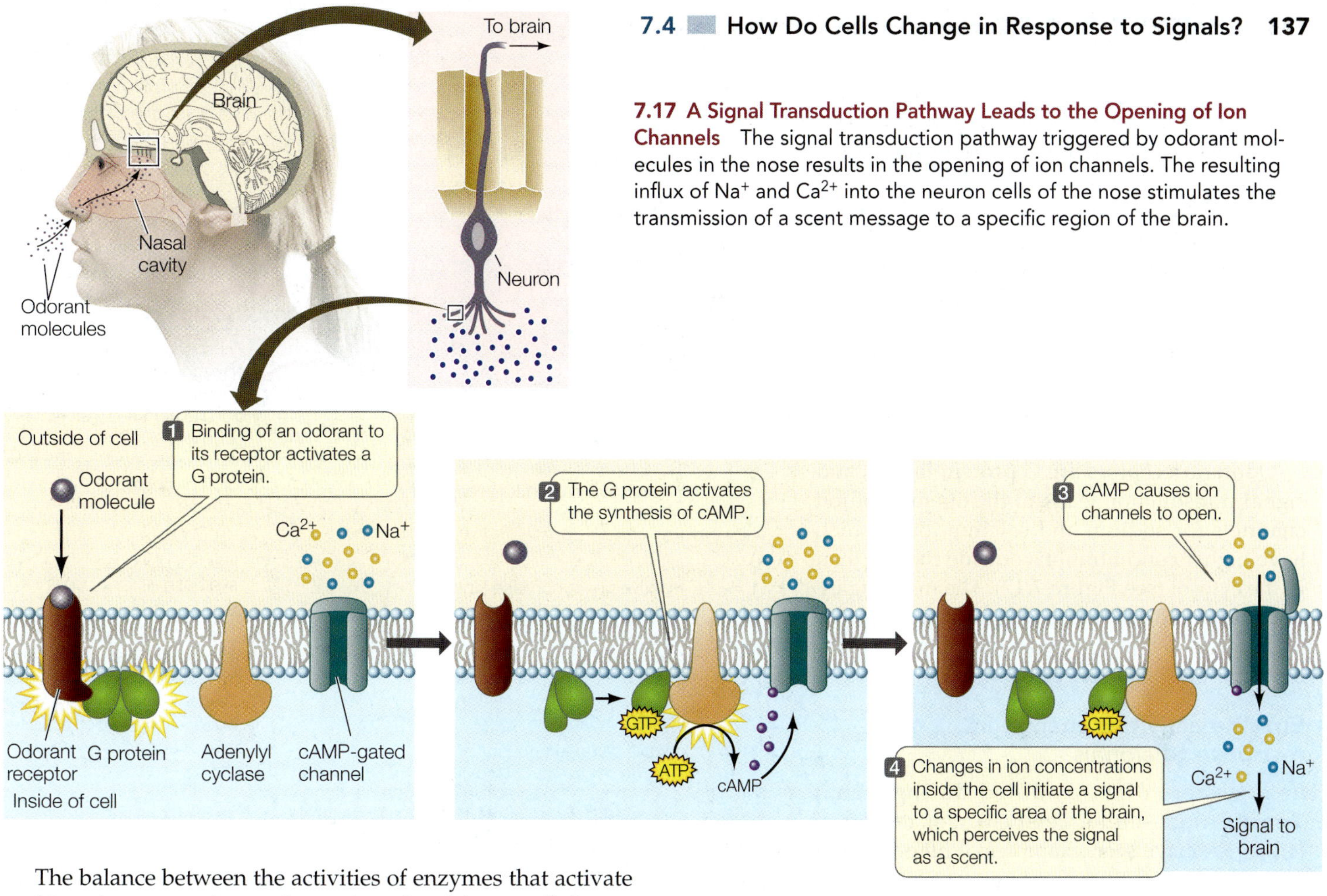

To brain
Brain
Nasal cavity
Odorant molecules
Neuron
Outside of cell
1 Binding of an odorant to its receptor activates a G protein.
Odorant molecule
Ca^{2+}
Na^+
Odorant receptor
G protein
Adenylyl cyclase
cAMP-gated channel
Inside of cell
2 The G protein activates the synthesis of cAMP.
GTP
ATP
cAMP
3 cAMP causes ion channels to open.
4 Changes in ion concentrations inside the cell initiate a signal to a specific area of the brain, which perceives the signal as a scent.
Signal to brain

7.17 A Signal Transduction Pathway Leads to the Opening of Ion Channels The signal transduction pathway triggered by odorant molecules in the nose results in the opening of ion channels. The resulting influx of Na^+ and Ca^{2+} into the neuron cells of the nose stimulates the transmission of a scent message to a specific region of the brain.

The balance between the activities of enzymes that activate and inactivate transducers determines the ultimate cellular response to a signal. Cells can alter this balance in several ways:

- *Synthesis or breakdown of the enzymes.* For example, synthesis of adenylyl cyclase (which synthesizes cAMP) and breakdown of phosphodiesterase (which breaks down cAMP) would tilt the balance in favor of more cAMP in the cell.
- *Activation or inhibition of the enzymes by other molecules.* An example is the inhibition of phosphodiesterase by sildenafil.

Because cell signaling is so important in diseases such as cancer, a search is under way for new drugs that can modulate the activities of enzymes that participate in signal transduction pathways.

RECAP 7.3

Signal transduction is the series of steps between the binding of a signal to a receptor and the ultimate cellular response. A protein kinase cascade amplifies a signal through a series of protein phosphorylation reactions. In many cases, a second messenger serves to amplify and distribute the downstream effects of the signal.

- How does a protein kinase cascade amplify a signal's message inside the cell? **See p. 132 and Figure 7.10**
- What is the role of cAMP as a second messenger? **See p. 133 and Figure 7.11**
- How are signal transduction cascades regulated? **See pp. 136–137 and Figure 7.16**

We have seen how the binding of a signal to its receptor initiates the response of a cell to the signal, and how signal transduction pathways amplify the signal and distribute its effects to numerous targets in the cell. In the next section we will consider the third step in the signal transduction process, the actual effects of the signal on cell function.

7.4 How Do Cells Change in Response to Signals?

The effects of a signal on cell function take three primary forms: the opening of ion channels, changes in the activities of enzymes, or differential gene expression. These events set the cell on a path for further and sometimes dramatic changes in form and function.

Ion channels open in response to signals

We have seen that ion channels can function as receptors in cell signaling, and as components of more complex signal transduction pathways: for example, the calcium ion channel in the pathway shown in Figure 7.13. In some cases, the opening of an ion channel is itself the cellular response to a signal. For example, the opening of ion channels is a key response in the nervous system. In sense organs, specialized cells have receptors that respond to external stimuli such as light, sound, taste, odor, or pressure. The alteration of the receptor results in the opening of ion channels. We will focus here on one such signal transduction pathway, that for the sense of smell, which responds to gaseous molecules in the environment (**Figure 7.17**).

The sense of smell is well developed in mammals. Each of the thousands of neurons in the nose expresses one of many different odorant receptors. The identification of which chemical signal, or odorant, activates which receptor is just getting under way. Humans have the genetic capacity to make about 950 different odorant receptor proteins, but very few people express more than 400 of them. Some express far fewer, which may explain why you are able to smell certain things that your roommate cannot, or vice versa.

Odorant receptors are G protein-linked, and signal transduction leads to the opening of ion channels for sodium and calcium ions, which have higher concentrations outside the cell than in the cytosol (see Figure 7.17). The resulting influx of Na^+ and Ca^{2+} causes the neuron to become stimulated so that it sends a signal to the brain that a particular odor is present.

Enzyme activities change in response to signals

Enzymes are often modified during signal transduction—either covalently or noncovalently. We have seen examples of both types of protein modification in earlier sections of this chapter. For example, addition of a phosphate group to an enzyme by a protein kinase is a covalent change; cAMP binding is noncovalent. Both types of modification change the enzyme's shape, activating or inhibiting its function. In the case of activation, the shape change exposes a previously inaccessible active site, and the target enzyme goes on to perform a new cellular role.

The G protein-mediated protein kinase cascade that is stimulated by epinephrine in liver cells results in the activation by cAMP of a key signaling molecule, protein kinase A. In turn, protein kinase A phosphorylates two other enzymes, with opposite effects:

- *Inhibition*: Glycogen synthase, which catalyzes the joining of glucose molecules to synthesize the energy-storing molecule glycogen, is inactivated when a phosphate group is added to it by protein kinase A. Thus the epinephrine signal *prevents glucose from being stored* in the form of glycogen (**Figure 7.18, step 1**).
- *Activation*: Phosphorylase kinase is activated when a phosphate group is added to it. It is part of a protein kinase cascade that ultimately leads to the activation of glycogen phosphorylase, another key enzyme in glucose metabolism. This enzyme results in the *liberation of glucose molecules* from glycogen (**Figure 7.18, steps 2 and 3**).

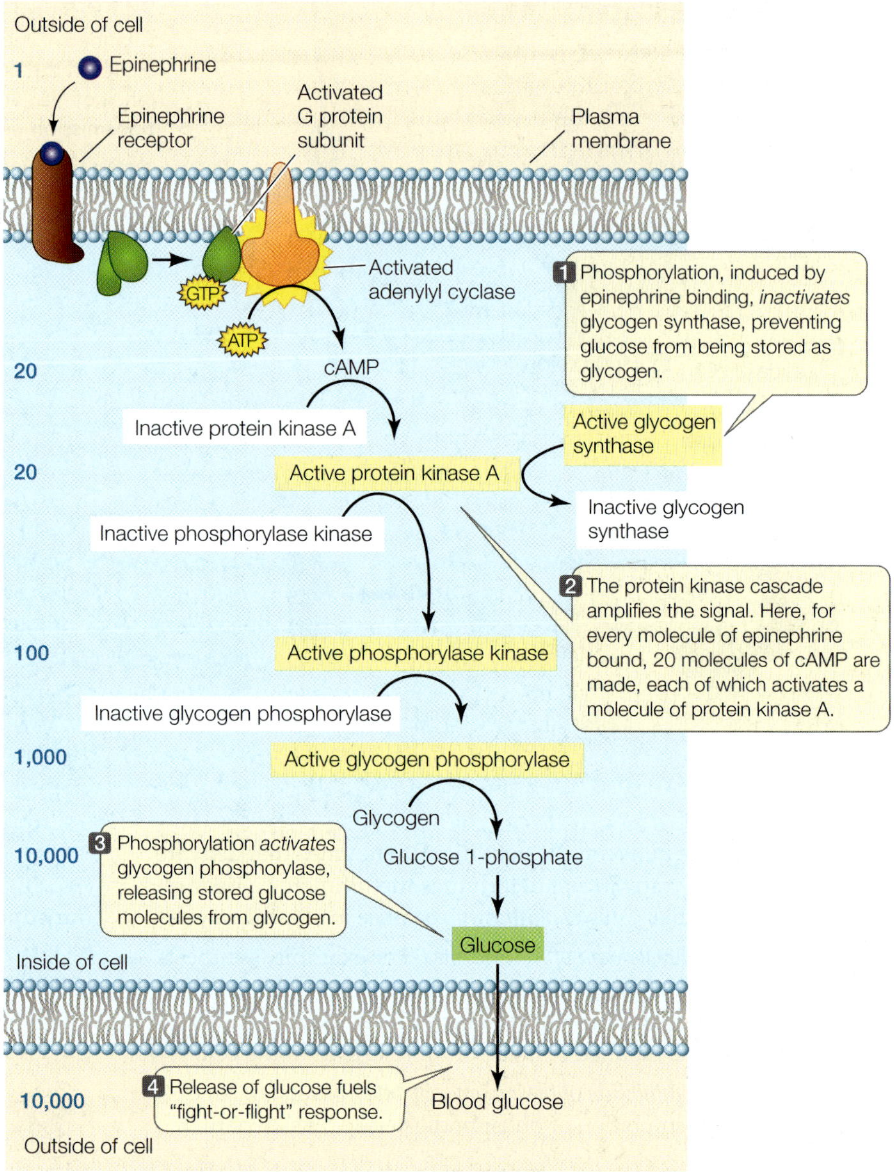
Outside of cell
1
Epinephrine
Epinephrine receptor
Activated G protein subunit
Plasma membrane
GTP
ATP
Activated adenylyl cyclase
1 Phosphorylation, induced by epinephrine binding, *inactivates* glycogen synthase, preventing glucose from being stored as glycogen.
20
cAMP
Inactive protein kinase A
Active glycogen synthase
20
Active protein kinase A
Inactive glycogen synthase
Inactive phosphorylase kinase
2 The protein kinase cascade amplifies the signal. Here, for every molecule of epinephrine bound, 20 molecules of cAMP are made, each of which activates a molecule of protein kinase A.
100
Active phosphorylase kinase
Inactive glycogen phosphorylase
1,000
Active glycogen phosphorylase
Glycogen
10,000
3 Phosphorylation *activates* glycogen phosphorylase, releasing stored glucose molecules from glycogen.
Glucose 1-phosphate
Glucose
Inside of cell
10,000
4 Release of glucose fuels "fight-or-flight" response.
Blood glucose
Outside of cell

7.18 A Cascade of Reactions Leads to Altered Enzyme Activity Liver cells respond to epinephrine by activating G proteins, which in turn activate the synthesis of the second messenger cAMP. Cyclic AMP initiates a protein kinase cascade, greatly amplifying the epinephrine signal, as indicated by the blue numbers. The cascade both inhibits the conversion of glucose to glycogen and stimulates the release of previously stored glucose.

The amplification of the signal in this pathway is impressive; as detailed in Figure 7.18, each molecule of epinephrine that arrives at the plasma membrane ultimately results in the release of 10,000 molecules of glucose into the bloodstream:

1	molecule of epinephrine bound to the membrane leads to
20	molecules of cAMP, which activate
20	molecules of protein kinase A, which activate
100	molecules of phosphorylase kinase, which activate
1,000	molecules of glycogen phosphorylase, which produce
10,000	molecules of glucose 1-phosphate, which produce
10,000	molecules of blood glucose

Signals can initiate DNA transcription

As we introduced in Section 4.1, the genetic material, DNA, is expressed by transcription as RNA, which is then translated into a protein whose amino acid sequence is specified by the original DNA sequence. Proteins are important in all cellular functions, so a key way to regulate specific functions in a cell is to regulate which proteins are made, and therefore which DNA sequences are transcribed.

Signal transduction plays an important role in determining which DNA sequences are transcribed. Common targets of signal transduction are proteins called transcription factors, which bind to specific DNA sequences in the cell nucleus and activate or inactivate transcription of the adjacent DNA regions. For example, the Ras signaling pathway (see Figure 7.10) ends in the nucleus. The final protein kinase in the Ras signaling cascade, MAPK (mitogen-activated protein kinase; a mitogen is a type of signal that stimulates cell division), enters the nucleus and phosphorylates a protein that stimulates the expression of several genes involved in cell proliferation.

In this chapter we have concentrated on signaling pathways that occur in animal cells. However, signal transduction pathways play equally important roles in other organisms, including plants, as you will see in Part Eight of this book.

RECAP 7.4

Cells respond to signal transduction by opening membrane channels, activating or inactivating enzymes, and stimulating or inhibiting gene transcription.

- What role does cAMP play in the sense of smell? **See pp. 137–138 and Figure 7.17**
- How does amplification of a signal occur, and why is it important in a cell's response to changes in its environment? **See p. 138 and Figure 7.18**

We have described how signals from a cell's environment can influence the cell. But the environment of a cell in a multicellular organism is more than the extracellular medium—it includes neighboring cells as well. In the next section we'll see how specialized junctions between cells allow them to pass signals from one to another.

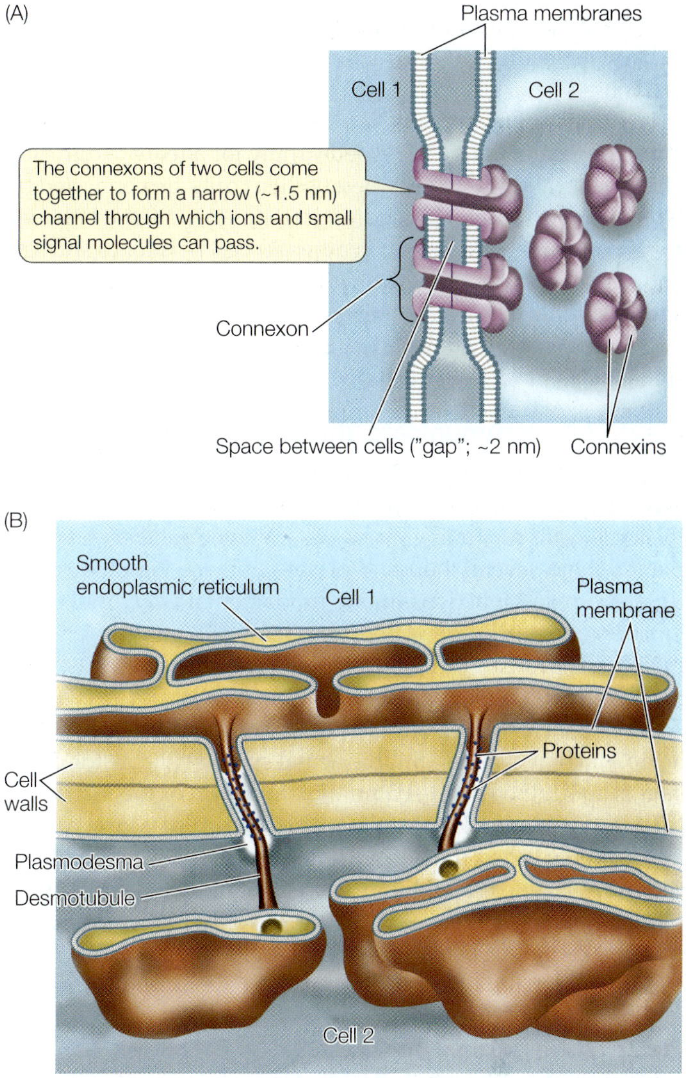
(A)
Plasma membranes
Cell 1
Cell 2
The connexons of two cells come together to form a narrow (~1.5 nm) channel through which ions and small signal molecules can pass.
Connexon
Space between cells ("gap"; ~2 nm)
Connexins
(B)
Smooth endoplasmic reticulum
Cell 1
Plasma membrane
Proteins
Cell walls
Plasmodesma
Desmotubule
Cell 2

7.19 Communicating Junctions (A) An animal cell may contain hundreds of gap junctions connecting it to neighboring cells. The pores of gap junctions allow small molecules to pass from cell to cell, ensuring similar concentrations of important signaling molecules in adjacent cells so that the cells can coordinate their activities. (B) Plasmodesmata connect plant cells. The desmotubule, derived from the smooth endoplasmic reticulum, fills up most of the space inside a plasmodesma, leaving a tiny gap through which small metabolites and ions can pass.

7.5 How Do Cells in a Multicellular Organism Communicate Directly?

The hallmark of multicellular organisms is their ability to have specialized functions in subsets of cells within their bodies. How do these cells communicate with one another so that they can work together for the good of the entire organism? As we learned in Section 7.1, some intercellular signals travel through the circulatory system to reach their target cells. But cells also have more direct ways of communicating. Cells that are packed together within a tissue can communicate directly with their neighbors via specialized intercellular junctions: gap junctions in animals (see Figure 6.7) and plasmodesmata in plants (see Figure 5.7).

Animal cells communicate through gap junctions

Gap junctions are channels between adjacent cells that occur in many animals, occupying up to 25 percent of the area of the plasma membrane (**Figure 7.19A**). Gap junctions traverse the narrow space between the plasma membranes of two adjacent cells (the "gap") by means of channel structures called **connexons**. The walls of a connexon are composed of six subunits of the integral membrane protein connexin. In adjacent cells, two connexons come together to form a gap junction that links the cytoplasms of the two cells. There may be hundreds of these channels between a cell and its neighbors. The channel pores are about

1.5 nanometers (nm) in diameter—too narrow for the passage of large molecules but adequate for passage of small molecules and ions.

In the lens of the mammalian eye, only the cells at the periphery are close enough to the blood supply for adequate diffusion of nutrients and wastes. But because lens cells are connected by large numbers of gap junctions, material can diffuse between them rapidly and efficiently. In other tissues, hormones and second messengers can move through gap junctions. Sometimes just a few cells in a tissue have the receptor for a particular signal; in such cases, gap junctions allow a coordinated response to the signal by all the cells in the tissue.

Plant cells communicate through plasmodesmata

Instead of gap junctions, plants have **plasmodesmata** (singular *plasmodesma*), which are membrane-lined tunnels that traverse the thick cell walls separating plant cells from one another. A typical plant cell has several thousand plasmodesmata. Plasmodesmata differ from gap junctions in one fundamental way: unlike gap junctions, in which the wall of the channel is made of integral membrane proteins from the adjacent plasma membranes, plasmodesmata are lined by the fused plasma membranes themselves.

The diameter of a plasmodesma is about 6 nm, far larger than a gap junction channel. But the actual space available for diffusion is about the same—1.5 nm. Examination of the interior of the plasmodesma by transmission electron microscopy reveals that a tubule called the **desmotubule**, apparently derived from the endoplasmic reticulum, fills up most of the opening of the plasmodesma (**Figure 7.19B**). Typically, only small metabolites and ions can move between plant cells.

Plasmadesmata play an important role in plant physiology because the bulk transport system in plants, the vascular system, lacks the tiny circulatory vessels (capillaries) that many animals have for bringing gases and nutrients to every cell. For example, diffusion from cell to cell across plasma membranes is probably inadequate to account for the movement of a plant hormone from the site of production to the site of action. Instead, plants rely on more rapid diffusion through plasmodesmata to ensure that all cells of a tissue respond to a signal at the same time. There are cases in which larger molecules or particles can pass between cells via plasmodesmata. For example, some viruses can move through plasmodesmata by using "movement proteins" to assist their passage.

Modern organisms provide clues about the evolution of cell–cell interactions and multicellularity

Even though single-celled organisms continue to be highly successful on Earth, over time complex multicellular organisms evolved, along with their division of biological labor among specialized cells. The transition from single-celled to multicellular life took a long time. Indeed, while there is evidence that single-celled organisms arose about 500 million to a billion years after the formation of Earth (see Chapter 4), the first evidence of true multicellular organisms dates from more than a billion years later.

Studying the evolutionary origin of multicellularity is a challenge because it happened so long ago. The closest unicellular relatives of most modern animals and plants probably existed hundreds of millions of years ago. The evolution from single-celled to multicellular organisms may have occurred in several steps:

- Aggregation of cells into a cluster
- Intercellular communication within the cluster
- Specialization of some cells within the cluster
- Organization of specialized cells into groups (tissues)

A key event would have been the evolution of intercellular communication, which is necessary to coordinate the activities of different cells within a multicellular organism.

We can visualize how the evolution of multicellularity might have occurred by looking at the "Volvocine line" of aquatic green algae (Chlorophyta). These plants range from single cells to complex multicellular organisms with differentiated cell clusters (**Figure 7.20**). Included in this range are a single-celled organism (*Chlamydomonas*); an organism that occurs

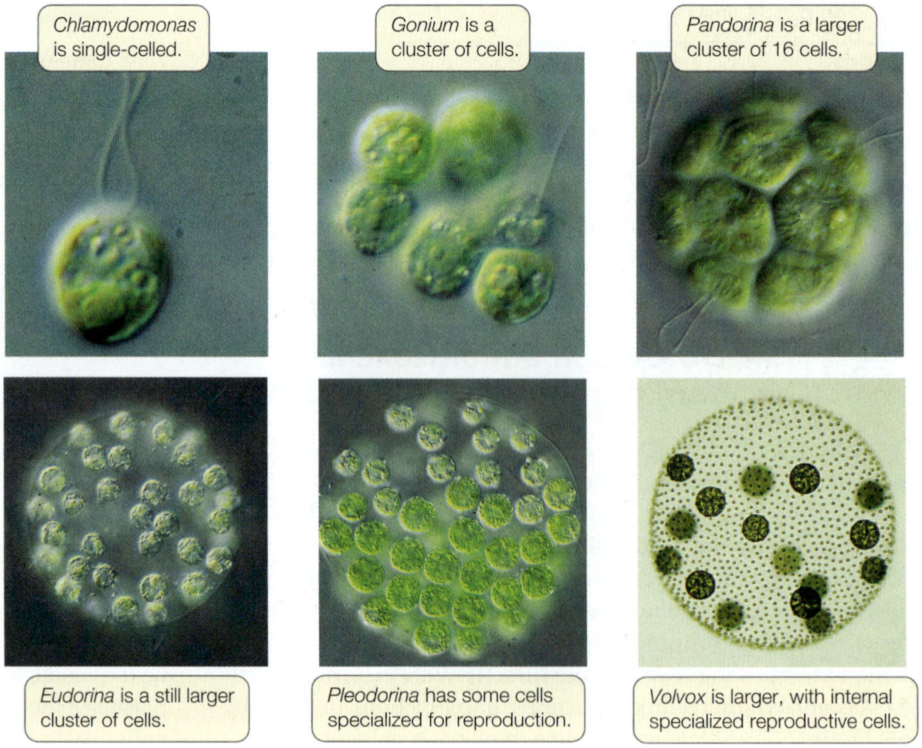
Chlamydomonas is single-celled.
Gonium is a cluster of cells.
Pandorina is a larger cluster of 16 cells.
Eudorina is a still larger cluster of cells.
Pleodorina has some cells specialized for reproduction.
Volvox is larger, with internal specialized reproductive cells.

7.20 Multicellularity The evolution of intercellular interactions in a multicellular organism can be inferred from these green algae.

in small cell clusters (*Gonium*); species with larger cell clusters (*Pandorina* and *Eudorina*); a colony of somatic and reproductive cells (*Pleodorina*); and a larger, 1,000-celled alga with somatic and reproductive cells organized into separate tissues (*Volvox*).

Chlamydomonas is the single-celled member of this group. It has two cellular phases: a swimming phase, when the cells have flagella and move about, and a non-swimming phase, when the flagella are reabsorbed (disaggregated) and the cell undergoes cell division (reproduction). Compare this with *Volvox*: most of the cells of this multicellular, spherical organism are on the surface; the beating of their flagella gives the organism a rolling motion as it swims toward light, where it can perform photosynthesis. But some *Volvox* cells are larger and located inside the sphere. These cells are specialized for reproduction: they lose their flagella and then divide to form offspring.

The separation of somatic and reproductive functions in *Volvox* is possible because of a key intercellular signaling mechanism that coordinates the activities of the separate tissues within the organism. *Volvox* has a gene whose protein product is produced by the outer, motile cells and travels to the reproductive cells, causing them to lose their flagella and divide. This gene is not active in species such as *Gonium* and *Pandorina*, which show cell aggregation but no cell specialization.

RECAP 7.5

Cells can communicate with their neighbors through specialized cell junctions. In animals, these structures are gap junctions; in plants, they are plasmodesmata. The evolution of intercellular communication and tissue formation can be inferred from existing organisms, such as certain related green algae.

- What are the roles that gap junctions and plasmodesmata play in cell signaling? See pp. 139–140 and Figure 7.19
- How does the Volvocine line of green algae show possible steps in the evolution of cell communication and tissue formation? See pp. 140–141 and Figure 7.20

Go to Media Clip 7.1
Social Amoebas Aggregate on Cue
Life10e.com/mc7.1

Does oxytocin affect caring behavior in humans?

ANSWER

Oxytocin is released during sexual activity in humans, and the release results in bonding behaviors just as it does in voles. Other behaviors are affected by oxytocin as well. Human volunteers given a nasal spray of oxytocin show more trust when investing money with a stranger than do volunteers given an inert spray. This kind of experiment has opened up a new field called neuroeconomics.

CHAPTER**SUMMARY** 7

7.1 What Are Signals, and How Do Cells Respond to Them?

- Cells receive many signals from the physical environment and from other cells. Chemical signals are often at very low concentrations. **Autocrine** signals affect the cells that make them; **juxtacrine** signals affect adjacent cells; **paracrine** signals diffuse to and affect nearby cells; and **hormones** are carried through the circulatory systems of animals or the vascular systems of plants. Review Figure 7.1, ACTIVITY 7.1
- A **signal transduction pathway** involves the interaction of a signal molecule with a receptor; the transduction of the signal via a series of steps within the cell; and effects on the function of the cell. Review Figure 7.2
- Signal transduction pathways involve regulation of enzymes and transcription factors. A great deal of **crosstalk** occurs between pathways.

7.2 How Do Signal Receptors Initiate a Cellular Response?

- Cells respond to signals only if they have specific **receptor** proteins that can recognize those signals.
- Binding of a signal **ligand** to its receptor obeys the chemical law of mass action. A key measurement of the strength of binding is the **dissociation constant** (K_D).
- Depending on the nature of its signal or ligand, a receptor may be located in the plasma membrane or inside the target cell. Review Figure 7.4
- Receptors located in the plasma membrane include **ion channels**, **protein kinases**, and **G protein-linked receptors**.
- Ion channel receptors are "gated." The gate "opens" when the three-dimensional structure of the channel protein is altered by ligand binding. Review Figure 7.5
- Protein kinase receptors catalyze the phosphorylation of themselves and/or other proteins. Review Figure 7.6
- A **G protein** has three important binding sites, which bind a G protein-linked receptor, GDP or GTP, and an **effector protein**. A G protein can either activate or inhibit an effector protein. Review Figure 7.7, ANIMATED TUTORIAL 7.1
- **Intracellular receptors** include certain photoreceptors in plants and steroid hormone receptors in animals. A lipid-soluble ligand such as a steroid hormone may enter the cytoplasm or the nucleus before binding. Many intracellular receptors are transcription factors. Review Figure 7.8

7.3 How Is the Response to a Signal Transduced through the Cell?

- A **protein kinase cascade** amplifies the response to receptor binding. Review Figure 7.10, ANIMATED TUTORIAL 7.2
- Second messengers include **cyclic AMP (cAMP)**, **inositol trisphosphate (IP_3)**, **diacylglycerol (DAG)**, and **calcium** ions. IP_3 and DAG are derived from the phospholipid **phosphatidyl inositol-bisphosphate (PIP_2)**.
- The gas nitric oxide (NO) is involved in signal transduction in human smooth muscle cells. Review Figure 7.15

continued

- Signal transduction can be regulated in several ways. The balance between activating and inactivating the molecules involved determines the ultimate cellular response to a signal. **Review Figure 7.16**

7.4 How Do Cells Change in Response to Signals?

- The cellular responses to signals may include the opening of ion channels, the alteration of enzyme activities, or changes in gene expression. **Review Figure 7.17**
- Activated enzymes may activate other enzymes in a signal transduction pathway, leading to impressive amplification of a signal. **Review Figure 7.18**
- Protein kinases covalently add phosphate groups to target proteins; cAMP binds target proteins noncovalently. Both kinds of binding change the target protein's conformation to expose or hide its active site.

7.5 How Do Cells in a Multicellular Organism Communicate Directly?

- Many adjacent animal cells can communicate with one another directly through small pores in their plasma membranes called **gap junctions**. Protein structures called **connexons** form thin channels between two adjacent cells through which small signal molecules and ions can pass. **Review Figure 7.19A**
- Plant cells are connected by somewhat larger pores called **plasmodesmata**, which traverse both plasma membranes and cell walls. The **desmotubule** narrows the opening of the plasmodesma. **Review Figure 7.19B**
- The evolution of cell communication and tissue formation can be inferred from existing organisms, such as certain green algae. **Review Figure 7.20**

See ACTIVITY 7.2 for a concept review of this chapter.

Go to the Interactive Summary to review key figures, Animated Tutorials, and Activities Life10e.com/is7

CHAPTER**REVIEW**

REMEMBERING

1. What is the correct order for the following events in the interaction of a cell with a signal? (1) Alteration of cell function; (2) signal binds to receptor; (3) signal released from source; (4) signal transduction.
   a. 1234
   b. 2314
   c. 3214
   d. 3241
   e. 3421

2. Steroid hormones such as estrogen act on target cells by
   a. initiating second-messenger activity.
   b. binding to membrane proteins.
   c. initiating gene expression.
   d. activating enzymes.
   e. binding to membrane lipids.

3. Which of the following is *not* a consequence of a signal binding to a receptor?
   a. Activation of receptor enzyme activity
   b. Diffusion of the receptor in the plasma membrane
   c. Change in conformation of the receptor protein
   d. Breakdown of the receptor to amino acids
   e. Release of the signal from the receptor

4. A nonpolar ligand such as a steroid hormone usually binds to a/an
   a. intracellular receptor.
   b. protein kinase.
   c. ion channel.
   d. phospholipid.
   e. second messenger.

5. Which of the following is *not* true of a protein kinase cascade?
   a. The signal is amplified.
   b. A second messenger is formed.
   c. Target proteins are phosphorylated.
   d. The cascade ends up at the mitochondrion.
   e. The cascade begins at the plasma membrane.

6. Plasmodesmata and gap junctions
   a. allow small molecules and ions to pass rapidly between cells.
   b. are both membrane-lined channels.
   c. are channels about 1 millimeter in diameter.
   d. are present only once per cell.
   e. are involved in cell recognition.

UNDERSTANDING & APPLYING

7. Why do some signals ("first messengers") trigger "second messengers" to activate target cells?
   a. The first messenger requires activation by ATP.
   b. The first messenger is not water-soluble.
   c. The first messenger binds to many types of cells.
   d. The first messenger cannot cross the plasma membrane.
   e. There are no receptors for the first messenger.

8. The major difference between a cell that responds to a signal and one that does not is the presence of a
   a. DNA sequence that binds to the signal.
   b. nearby blood vessel.
   c. receptor.
   d. second messenger.
   e. transduction pathway.

9. Cyclic AMP is a second messenger in many different responses. How can the same messenger act in different ways in different cells?

10. Compare direct communication via plasmodesmata or gap junctions with receptor-mediated communication between cells. What are the advantages of one method over the other?

ANALYZING & EVALUATING

11. Like the Ras protein itself, the various components of the Ras signaling pathway are changed in cancer cells. What might be the biochemical consequences of mutations in the genes coding for (*a*) Raf and (*b*) MAP kinase that result in rapid cell division?

12. The tiny invertebrate *Hydra* has an apical region with tentacles and a long, slender body. *Hydra* can reproduce asexually when cells on the body wall differentiate and form a bud, which then breaks off as a new organism. Buds form only at certain distances from the apex, leading to the idea that the apex releases a signal molecule that diffuses down the body and, at high concentrations (i.e., near the apex), inhibits bud formation. *Hydra* lacks a circulatory system, so this inhibitor must diffuse from cell to cell. If you had an antibody that binds to connexons and plugs up the gap junctions, how would you test the hypothesis that *Hydra*'s inhibitory factor passes through these junctions?

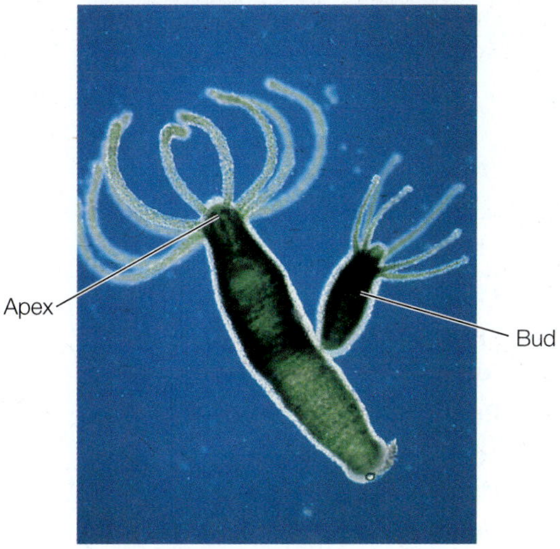
Apex
Bud

Go to BioPortal at **yourBioPortal.com** for Animated Tutorials, Activities, LearningCurve Quizzes, Flashcards, and many other study and review resources.

8 Energy, Enzymes, and Metabolism

CHAPTER OUTLINE

Cleaning Aids Enzymes are an important component of detergents. By hydrolyzing the macromolecules that make up stains, enzymes help the stains dissolve in wash water.

"CLEAN THE DISHES!" probably gives you the same reaction as "Clean your clothes!" For most people, washing things has never been a favorite activity. In chemical terms, the problem is that nonpolar substances, such as food and dirt, stick to nonpolar surfaces on dishes and clothes. Water alone doesn't do a very good job of unsticking them. Perhaps the best-known cleaning aid is soap, which has polar and nonpolar regions that help separate nonpolar substances from one another. The polar regions allow the soap to dissolve in water, and the nonpolar regions solubilize the targeted stuff on your dishes and clothes. Even though synthetic detergents have now improved on traditional soaps, the TV ads will tell you that's still not enough. The "really tough stains" need something more.

A century ago, German chemist Otto Rohm came up with the idea of using enzymes to make the stains on clothes and dishes more water-soluble. Stains from meat or blood contain insoluble polymers such as protein. Rohm's idea was to mimic the mammalian digestive system, where insoluble polymers are hydrolyzed to soluble monomers by enzymes—proteins that speed up chemical reactions. Rohm isolated the digestive enzyme trypsin, which speeds up the hydrolysis of proteins, and added it to a detergent. The result was a dramatic improvement in the removal of stains. Nevertheless, it was many decades before this approach was used widely, because there were several challenges that had to be surmounted. First, the supply of enzymes from Rohm's original source (the pancreas) was limited. To solve this, biologists used bacteria, yeast, and fungi to make enzymes in huge amounts. Second, enzymes are proteins, and their three-dimensional structure is vital to their activity. The detergents and ions used for washing tend to destroy enzyme structure. To solve this, scientists screened many organisms for enzymes that would work well under these conditions and in a wide range of temperatures. They also performed genetic manipulations to improve on nature. Finally, not all stains are proteins. The spaghetti sauce you might eat stains your shirt with fats and starch. To solve this, scientists added lipases that hydrolyze lipids, and amylases that hydrolyze starch.

The result of all this chemistry is a range of modern cleaning products that attack dirt and grime on many fronts. The enzymes in these products perform specific chemical transformations and are active in washing conditions.

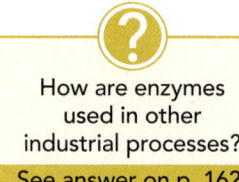

> ? How are enzymes used in other industrial processes?
>
> See answer on p. 162.

8.1 What Physical Principles Underlie Biological Energy Transformations?

A **chemical reaction** occurs when atoms have sufficient energy to combine or change their bonding partners. Consider the hydrolysis of the disaccharide sucrose to its component monomers, glucose and fructose (see p. 53 for the chemical structures of these sugars). We can express this reaction by a chemical equation:

$$\underset{(C_{12}H_{22}O_{11})}{\text{Sucrose}} + H_2O \rightarrow \underset{(C_6H_{12}O_6)}{\text{glucose}} + \underset{(C_6H_{12}O_6)}{\text{fructose}}$$

In this equation, sucrose and water are the **reactants**, and glucose and fructose are the **products**. During the reaction, some of the bonds in sucrose and water are broken and new bonds are formed, resulting in products with chemical properties that are very different from those of the reactants. The sum total of all the chemical reactions occurring in a biological system at a given time is called **metabolism**. Metabolic reactions involve energy changes; for example, the energy contained in the chemical bonds of sucrose (reactants) is greater than the energy in the bonds of the two products, glucose and fructose.

What is energy? Physicists define it as the capacity to do work, which occurs when a force operates on an object over a distance. In biochemistry, it is more useful to consider energy as *the capacity for change*. In biochemical reactions, energy changes are usually associated with changes in the chemical compositions and properties of molecules.

There are two basic types of energy

Energy comes in many forms: chemical, electrical, heat, light, and mechanical (Table 8.1). But all forms of energy can be considered as one of two basic types (Figure 8.1):

- **Potential energy** is the energy of state or position—that is, stored energy. It can be stored in many forms: in chemical bonds, as a concentration gradient, or even as an electric charge imbalance.

TABLE **8.1**
Energy in Biology

Form of Energy	Example in Biology
Chemical: Stored in bonds	Chemical energy is released during the hydrolysis of polymers
Electrical: Separation of charges	Electrical gradients across cell membranes help drive the movement of ions through channels
Heat: Transfer due to temperature difference	Heat can be released by chemical reactions
Light: Electromagnetic radiation stored as photons	Light energy is captured by pigments in the eye
Mechanical: Energy of motion	Mechanical energy is used in muscle movements

Potential chemical energy is converted into kinetic mechanical energy when the cat leaps.
Potential chemical energy is stored in the muscles of the cat.

8.1 Energy Conversions and Work A leaping cat illustrates both the conversion between potential and kinetic energy and the conversion of energy from one form (chemical) to another (mechanical).

- **Kinetic energy** is the energy of movement—that is, the type of energy that does work, that makes things change. For example, heat causes molecular motions and can even break chemical bonds.

Potential energy can be converted into kinetic energy and vice versa, and the form that the energy takes can also be converted. Think of reading this book: light energy is converted to chemical energy in your eyes, and then is converted to electrical energy in the nerve cells that carry messages to your brain. When you decide to turn a page, the electrical and chemical energy of nerves and muscles are converted to kinetic energy for movement of your hand and arm.

There are two basic types of metabolism

Energy changes in living systems usually occur as chemical changes, in which energy is stored in, or released from, chemical bonds.

Anabolic reactions (collectively anabolism) link simple molecules to form more complex molecules (for example, the synthesis of sucrose from glucose and fructose). Anabolic reactions require an input of energy. Energy is captured in the chemical bonds that are formed (for example, the glycosidic bond between the two monosaccharides). This captured energy is stored in the chemical bonds as potential energy.

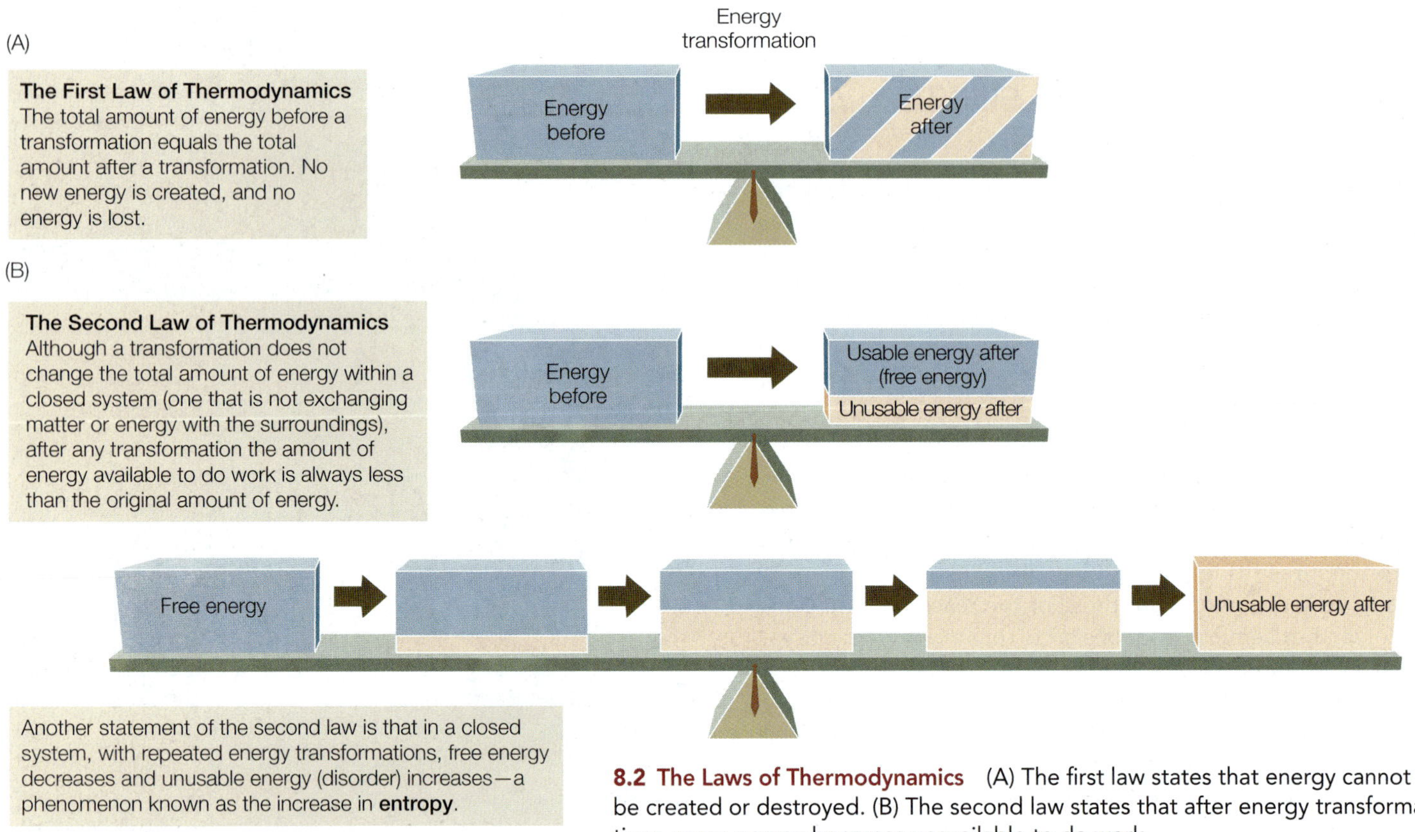
(A)
Energy transformation
The First Law of Thermodynamics
The total amount of energy before a transformation equals the total amount after a transformation. No new energy is created, and no energy is lost.
Energy before
Energy after
(B)
The Second Law of Thermodynamics
Although a transformation does not change the total amount of energy within a closed system (one that is not exchanging matter or energy with the surroundings), after any transformation the amount of energy available to do work is always less than the original amount of energy.
Energy before
Usable energy after (free energy)
Unusable energy after
Free energy
Unusable energy after
Another statement of the second law is that in a closed system, with repeated energy transformations, free energy decreases and unusable energy (disorder) increases—a phenomenon known as the increase in entropy.

8.2 The Laws of Thermodynamics (A) The first law states that energy cannot be created or destroyed. (B) The second law states that after energy transformations, some energy becomes unavailable to do work.

Catabolic reactions (collectively catabolism) break down complex molecules into simpler ones and release the energy stored in the chemical bonds. For example, when sucrose is hydrolyzed, energy is released. In a biological system the released energy may be recaptured in new chemical bonds, or it may be used as kinetic energy—moving atoms, molecules, cells, or the whole organism.

Catabolic and anabolic reactions are often linked. The energy released in catabolic reactions is often used to drive anabolic reactions—that is, to do biological work. For example, the energy released by the breakdown of glucose (catabolism) is used to drive anabolic reactions such as the synthesis of triglycerides. This is why you accumulate fat if you eat food in excess of your energy requirements.

The **laws of thermodynamics** (*thermo*, "energy"; *dynamics*, "change") were derived from studies of the fundamental physical properties of energy, and the ways it interacts with matter. The laws apply to all matter and all energy transformations in the universe. Their application to living systems helps us understand how organisms and cells harvest and transform energy to sustain life.

The first law of thermodynamics: Energy is neither created nor destroyed

The first law of thermodynamics states that in any energy conversion, energy is neither created nor destroyed. In other words, during any conversion of energy, the total energy before and after the conversion is the same (Figure 8.2A). As you will see in the next two chapters, the potential energy present in the chemical bonds of carbohydrates and lipids can be converted to potential energy in the form of adenosine triphosphate (ATP). This can then be converted into kinetic energy to do mechanical work (such as in muscle contractions) or biochemical work (such as protein synthesis).

The second law of thermodynamics: Disorder tends to increase

Although energy cannot be created or destroyed, the second law of thermodynamics states that when energy is converted from one form to another, some of that energy becomes unavailable for doing work (Figure 8.2B). In other words, no physical process or chemical reaction is 100 percent efficient; some of the released energy is lost to a form associated with disorder. Think of disorder as a kind of randomness that is due to the thermal motion of particles; this energy is of such a low value and so dispersed that it is unusable. **Entropy** is a measure of the disorder in a system.

It takes energy to impose order on a system. Unless energy is applied to a system, it will be randomly arranged or disordered. The second law applies to all energy transformations, but we will focus here on chemical reactions in living systems.

NOT ALL ENERGY CAN BE USED In any system, the total energy includes the usable energy that can do work and the unusable energy that is lost to disorder:

$$\text{Total energy} = \text{usable energy} + \text{unusable energy}$$

In biological systems, the total energy is called **enthalpy** (***H***). The usable energy that can do work is called **free energy** (***G***). Free energy is what cells require for all the chemical reactions involved in growth, cell division, and maintenance. The unusable energy is represented by entropy (S) multiplied by the absolute temperature (T). Thus we can rewrite the word equation above more precisely as:

$$H = G + TS \quad (8.1)$$

Because we are interested in usable energy, we rearrange Equation 8.1:

$$G = H - TS \quad (8.2)$$

Although we cannot measure G, H, or S absolutely, we can determine the change in each at a constant temperature. Such energy changes are measured in calories (cal) or joules (J).* A change in energy is represented by the Greek letter delta (Δ). The change in free energy (ΔG) of any chemical reaction is equal to the difference in free energy between the products and the reactants:

$$\Delta G_{reaction} = G_{products} - G_{reactants} \quad (8.3)$$

Such a change can be either positive or negative; that is, the free energy of the products can be more or less than the free energy of the reactants. If the products have more free energy than the reactants, then there must have been some input of energy into the reaction. (Remember that energy cannot be created, so some energy must have been added from an external source.)

At a constant temperature, ΔG is defined in terms of the change in total energy (ΔH) and the change in entropy (ΔS):

$$\Delta G = \Delta H - T\Delta S \quad (8.4)$$

Equation 8.4 tells us whether free energy is released or consumed by a chemical reaction:

- If ΔG is negative ($\Delta G < 0$), free energy is released.
- If ΔG is positive ($\Delta G > 0$), free energy is required (consumed).

If the necessary free energy is not available, the reaction does not occur. The sign and magnitude of ΔG depend on the two factors on the right side of the equation:

- ΔH: In a chemical reaction, ΔH is the total amount of energy added to the system ($\Delta H > 0$) or released ($\Delta H < 0$).
- ΔS: Depending on the sign and magnitude of ΔS, the entire term, $T\Delta S$, may be negative or positive, large or small. In other words, in living systems at a constant temperature (no change in T), the magnitude and sign of ΔG can depend a lot on changes in entropy.

If a chemical reaction increases entropy, its products are more disordered or random than its reactants. If there are more products than reactants, as in the hydrolysis of a protein to its amino acids, the products have considerable freedom to move around. The disorder in a solution of amino acids will be large compared with that in the protein, in which peptide bonds and other forces prevent free movement. So in hydrolysis, the change in entropy (ΔS) will be positive. Conversely, if there are fewer products and they are more restrained in their movements than the reactants (as for amino acids being joined in a protein), ΔS will be negative.

*A calorie is the amount of heat energy needed to raise the temperature of 1 gram of pure water from 14.5°C to 15.5°C. In the SI system, energy is measured in joules. 1 J = 0.239 cal; conversely, 1 cal = 4.184 J. Thus, for example, 486 cal = 2,033 J, or 2.033 kJ. Although they are defined here in terms of heat, the calorie and the joule are measures of mechanical, electrical, or chemical energy. When you compare data on energy, always compare joules with joules and calories with calories.

DISORDER TENDS TO INCREASE The second law of thermodynamics also predicts that, as a result of energy transformations, disorder tends to increase; some energy is always lost to random thermal motion (entropy). Chemical changes, physical changes, and biological processes all tend to increase entropy (see Figure 8.2B), and this tendency gives direction to these processes. It explains why some reactions proceed in one direction rather than another.

How does the second law apply to organisms? Consider the human body, with its highly organized tissues and organs composed of large, complex molecules. This level of complexity appears to be in conflict with the second law but is not for two reasons. First, the construction of complexity is coupled to the generation of disorder. Constructing 1 kg of a human body requires the catabolism of about 10 kg of highly ordered biological materials (our food), which are converted into CO_2, H_2O, and other simple molecules that move independently and randomly. So metabolism creates far more disorder (more energy is lost to entropy) than the amount of order (total energy; enthalpy) stored in 1 kg of flesh. Second, life requires a constant input of energy to maintain order. Without this energy, the complex structures of living systems would break down. Because energy is used to generate and maintain order, there is no conflict with the second law of thermodynamics.

Having seen that the laws of thermodynamics apply to living things, we will now turn to a consideration of how these laws apply to biochemical reactions.

Chemical reactions release or consume energy

As we saw earlier, anabolic reactions link simple molecules to form more complex molecules, so they tend to increase complexity (order) in the cell. By contrast, catabolic reactions break down complex molecules into simpler ones, so they tend to decrease complexity (generate disorder).

- Catabolic reactions may break down an ordered reactant into smaller, more randomly distributed products. Reactions that release free energy ($-\Delta G$) are called **exergonic** (or exothermic) reactions (**Figure 8.3A**). For example:

  Complex molecules $\rightarrow$ free energy + small molecules

- Anabolic reactions may make a single product (a highly ordered substance) out of many smaller reactants (less ordered). Reactions that require or consume free energy ($+\Delta G$) are called **endergonic** (or endothermic) reactions (**Figure 8.3B**). For example:

  Free energy + small molecules $\rightarrow$ complex molecules

In principle, chemical reactions are reversible and can run both forward and backward. For example, if compound A can be converted into compound B (A $\rightarrow$ B), then B, in principle,

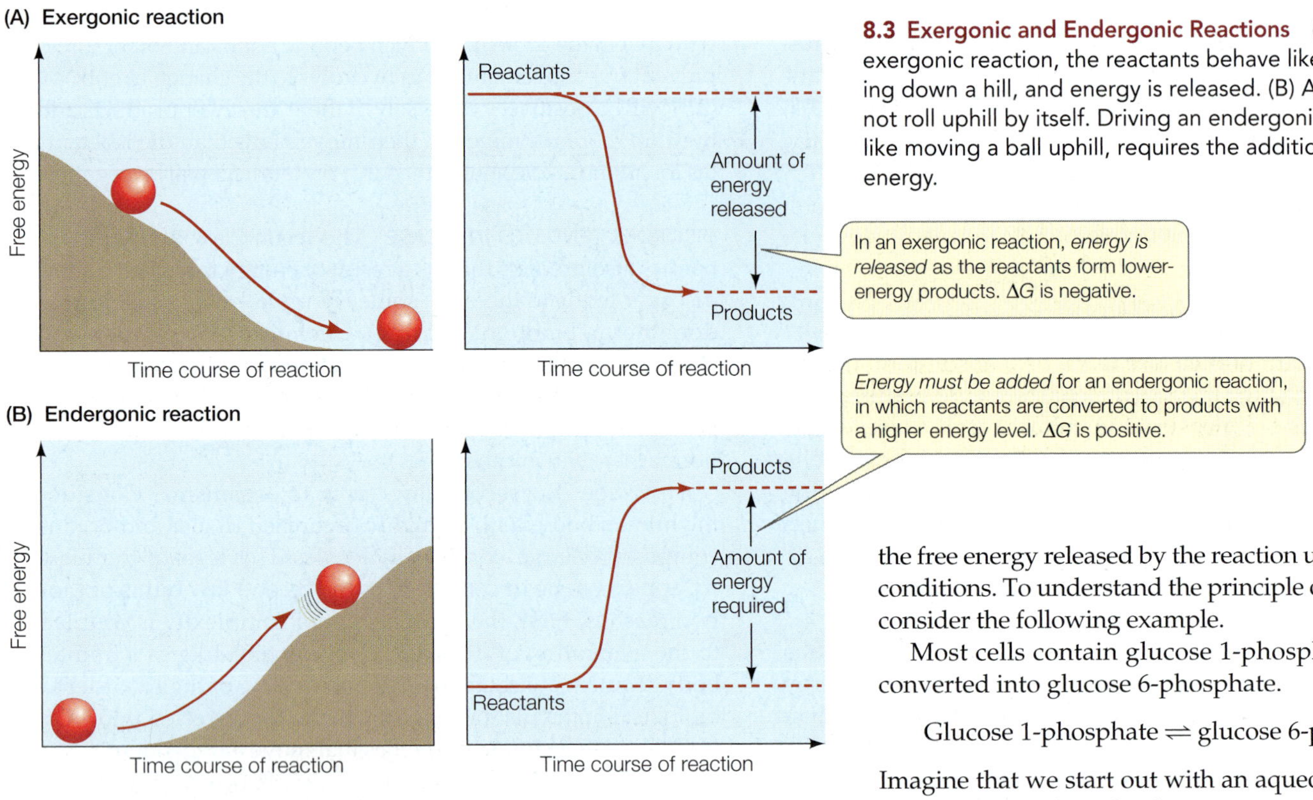
(A) Exergonic reaction
Free energy
Time course of reaction
Reactants
Amount of energy released
Products
Time course of reaction
In an exergonic reaction, *energy is released* as the reactants form lower-energy products. ΔG is negative.
(B) Endergonic reaction
Free energy
Time course of reaction
Products
Amount of energy required
Reactants
Time course of reaction
Energy must be added for an endergonic reaction, in which reactants are converted to products with a higher energy level. ΔG is positive.

8.3 Exergonic and Endergonic Reactions (A) In an exergonic reaction, the reactants behave like a ball rolling down a hill, and energy is released. (B) A ball will not roll uphill by itself. Driving an endergonic reaction, like moving a ball uphill, requires the addition of free energy.

can be converted into A (B → A), although *the concentrations of A and B determine which of these directions will be favored*. Think of the overall reaction as resulting from competition between the forward and reverse reactions (A ⇌ B). According to the law of mass action, increasing the concentration of A makes the forward reaction happen more often relative to the reverse reaction, just as B favors the reverse reaction.

There are concentrations of A and B at which the forward and reverse reactions take place at the same rate. At these concentrations, no further net change in the system is observable, although individual molecules are still forming and breaking apart. This balance between forward and reverse reactions is known as **chemical equilibrium**. Chemical equilibrium is a state of no net change, and a state in which $\Delta G = 0$.

Chemical equilibrium and free energy are related

Every chemical reaction proceeds to a certain extent, but not necessarily to completion (all reactants converted into products). Each reaction has a specific equilibrium point, which is related to the free energy released by the reaction under specified conditions. To understand the principle of equilibrium, consider the following example.

Most cells contain glucose 1-phosphate, which is converted into glucose 6-phosphate.

$$\text{Glucose 1-phosphate} \rightleftharpoons \text{glucose 6-phosphate}$$

Imagine that we start out with an aqueous solution of glucose 1-phosphate that has a concentration of 0.02 M. (M stands for molar concentration; see Section 2.4.) The solution is maintained under constant environmental conditions (25°C and pH 7). As the reaction proceeds to equilibrium, the concentration of the product, glucose 6-phosphate, rises from 0 to 0.019 M, while the concentration of the reactant, glucose 1-phosphate, falls to 0.001 M. At this point, equilibrium is reached (**Figure 8.4**). At equilibrium, the reverse reaction, from glucose 6-phosphate to glucose 1-phosphate, progresses at the same rate as the forward reaction.

At equilibrium, then, this reaction has a product-to-reactant ratio of 19:1 (0.019/0.001), so the forward reaction has gone 95 percent of the way to completion ("to the right," as written above). This result is obtained every time the experiment is run under the same conditions.

The change in free energy (ΔG) for any reaction is related directly to its point of equilibrium. The further toward completion the point of equilibrium lies, the more free energy is released. In an exergonic reaction, ΔG is a negative number. The value of ΔG also depends on the beginning concentrations of the reactants and products and other conditions such as temperature,

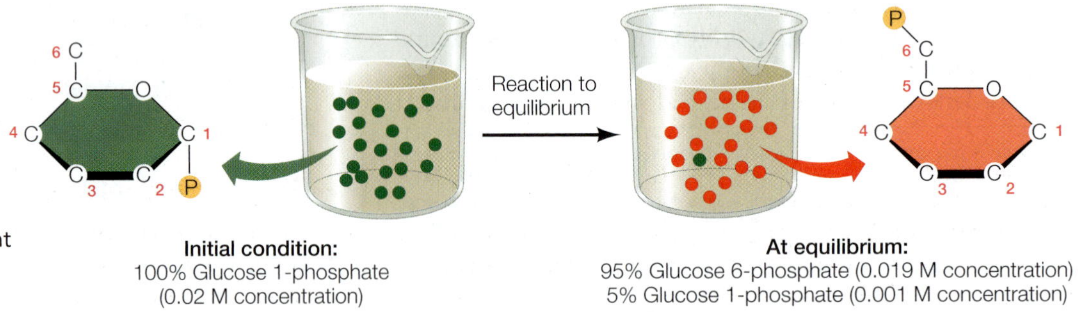
6 C
5 C
O
4 C
C 1
C
C
3
2
P
Reaction to equilibrium
P
6 C
5 C
O
4 C
C 1
C
C
3
2
Initial condition:
100% Glucose 1-phosphate
(0.02 M concentration)
At equilibrium:
95% Glucose 6-phosphate (0.019 M concentration)
5% Glucose 1-phosphate (0.001 M concentration)

8.4 Chemical Reactions Run to Equilibrium No matter what quantities of glucose 1-phosphate and glucose 6-phosphate are dissolved in water, when equilibrium is attained, there will always be 95 percent glucose 6-phosphate and 5 percent glucose 1-phosphate.

pressure, and pH of the solution. Biochemists often calculate ΔG using standard laboratory conditions: 25°C, one atmosphere pressure, one molar (1M) concentrations of the solutes, and pH 7. The standard free energy change calculated using these conditions is designated $\Delta G^{0\prime}$. In our example of the conversion of glucose 1-phosphate to glucose 6-phosphate, $\Delta G^{0\prime} = -1.7$ kcal/mol, or -7.1 kJ/mol.

A large, positive ΔG for a reaction means that it proceeds hardly at all to the right ($A \rightarrow B$). If the concentration of B is initially high relative to that of A, such a reaction runs "to the left" ($A \leftarrow B$), and at equilibrium nearly all of B is converted into A. A ΔG value near zero is characteristic of a readily reversible reaction: reactants and products have almost the same free energies.

In Chapters 9 and 10 we will examine the metabolic reactions that harvest energy from food and light. In turn, this energy is used to synthesize carbohydrates, lipids, and proteins. All of the chemical reactions carried out by living organisms are governed by the principles of thermodynamics and equilibrium.

RECAP 8.1

Two laws of thermodynamics govern energy transformations in biological systems. A biochemical reaction can release or consume energy, and it may not run to completion, but instead end up at a point of equilibrium.

- What is the difference between potential energy and kinetic energy? Between anabolism and catabolism? **See pp. 145–146**
- What are the laws of thermodynamics? How do they relate to biology? **See pp. 146–147 and Figure 8.2**
- What is the difference between endergonic and exergonic reactions, and what is the importance of ΔG? **See p. 147 and Figure 8.3**

The principles of thermodynamics that we have been discussing apply to all energy transformations in the universe, so they are very powerful and useful. Next we'll apply them to reactions in cells that involve the currency of biological energy, ATP.

8.2 What Is the Role of ATP in Biochemical Energetics?

Cells rely on adenosine triphosphate (ATP) for the capture and transfer of the free energy they require to do chemical work. ATP operates as a kind of "energy currency." Just as it is more effective, efficient, and convenient for you to trade money for a lunch than to trade your actual labor, it is useful for cells to have a single currency for transferring energy between different reactions and cell processes. So some of the free energy that is released by exergonic reactions is captured in the formation of ATP from adenosine diphosphate (ADP) and inorganic phosphate (HPO_4^{2-}, which is commonly abbreviated to P_i). The ATP can then be hydrolyzed at other sites in the cell to release free energy to drive endergonic reactions. [In some reactions, guanosine triphosphate (GTP) is used as the energy transfer molecule instead of ATP, but we will focus on ATP here.]

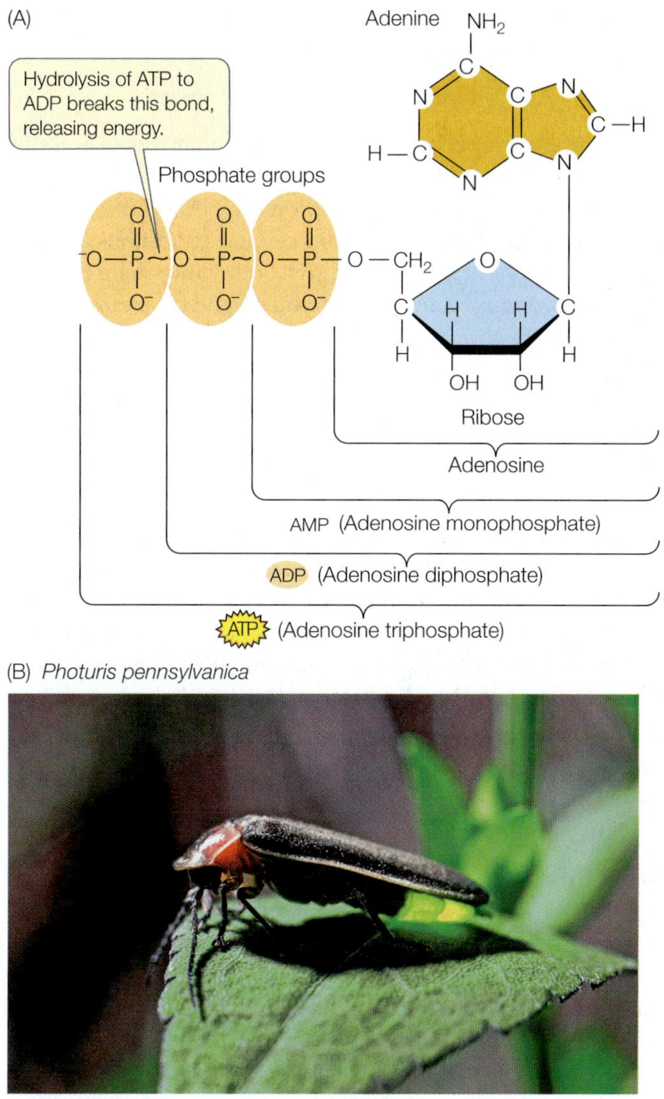
(A)
Adenine
NH_2
Hydrolysis of ATP to ADP breaks this bond, releasing energy.
Phosphate groups
Ribose
Adenosine
AMP (Adenosine monophosphate)
ADP (Adenosine diphosphate)
ATP (Adenosine triphosphate)
(B) *Photuris pennsylvanica*

8.5 ATP (A) ATP is richer in energy than its relatives ADP and AMP. (B) Fireflies use ATP to initiate the oxidation of luciferin. This process converts chemical energy into light energy, emitting rhythmic flashes that signal the insect's readiness to mate.

ATP has another important role in the cell beyond its use as an energy currency: it can be converted into a building block for nucleic acids (see Chapter 4). The structure of ATP is similar to those of other nucleoside triphosphates, but two things about ATP make it especially useful to cells.

- ATP releases a relatively large amount of energy when hydrolyzed to ADP and P_i.
- ATP can phosphorylate (donate a phosphate group to) many different molecules, which gain some of the energy that was stored in the ATP.

ATP hydrolysis releases energy

An ATP molecule consists of the nitrogenous base adenine bonded to ribose (a sugar), which is attached to a sequence of three phosphate groups (**Figure 8.5A**). The hydrolysis of a molecule of ATP yields free energy, as well as ADP and an inorganic phosphate ion (P_i). Thus:

$$ATP + H_2O \rightarrow ADP + P_i + \text{free energy}$$

The important property of this reaction is that it is exergonic, releasing free energy. Under standard laboratory conditions, the change in free energy for this reaction (ΔG) is about –7.3 kcal/mol (–30 kJ/mol). However, under cellular conditions, ΔG can be as much as –14 kcal/mol. We give both values here because you will encounter them both and you should be aware of their origins. Both are correct, but in different conditions.

A molecule of ATP can be hydrolyzed either to ADP and P_i, or to adenosine monophosphate (AMP) and a pyrophosphate ion ($P_2O_7^{4-}$; commonly abbreviated as PP_i). Two characteristics of ATP account for the free energy released by the loss of one or two of its phosphate groups:

- Because phosphate groups are negatively charged and so repel each other, it takes energy to get two phosphates near enough to each other to make the covalent bond that links them together. Some of this energy is stored as potential energy in the P~O bonds between the phosphates in ATP (the wavy line indicates a high energy bond).
- The free energy of this P~O bond (called a phosphoric acid anhydride bond) is much higher than the energy of the O—H bond that forms as a result of hydrolysis. So some usable energy is released by hydrolysis.

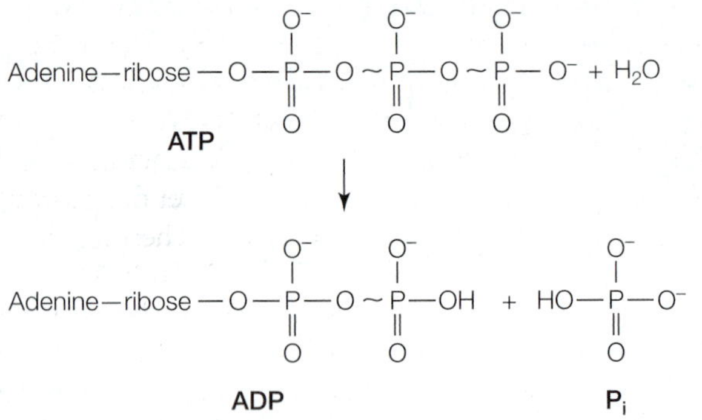
Adenine—ribose
+ H2O
ATP
ADP
Pi

Cells use the energy released by ATP hydrolysis to fuel endergonic reactions (such as the biosynthesis of complex molecules), for active transport, and for movement. Another interesting example of the use of ATP involves converting its chemical energy into light energy.

BIOLUMINESCENCE The production of light by living organisms is referred to as **bioluminesence** (Figure 8.5B). It is an example of an endergonic reaction driven by ATP hydrolysis that involves an interconversion of energy forms (chemical to light). The chemical that becomes luminescent is called luciferin (after the light-bearing fallen angel, Lucifer):

$$\text{Luciferin} + O_2 + ATP \xrightarrow{\text{Luciferase}} \text{oxyluciferin} + AMP + PP_i + \text{light}$$

This reaction and the enzyme that catalyzes it (luciferase) occur in a wide variety of organisms in addition to the familiar firefly. These include a variety of marine organisms, microorganisms, worms, and mushrooms. The light is generally used to avoid predators or to attract potential mates.

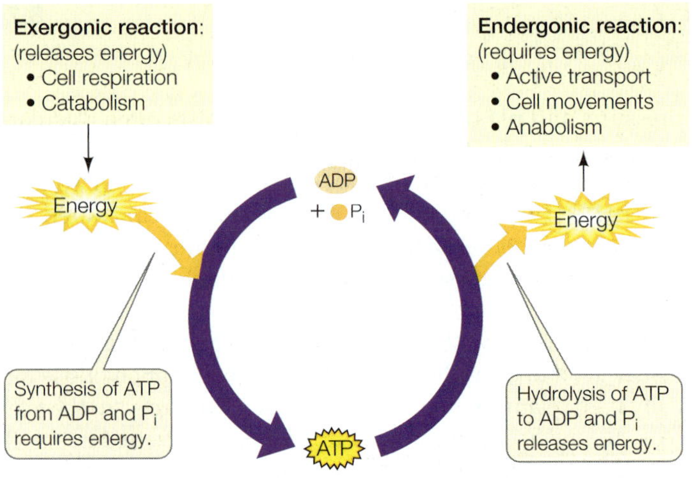
Exergonic reaction:
(releases energy)
• Cell respiration
• Catabolism
Endergonic reaction:
(requires energy)
• Active transport
• Cell movements
• Anabolism
Energy
ADP
+ Pi
Energy
Synthesis of ATP from ADP and Pi requires energy.
ATP
Hydrolysis of ATP to ADP and Pi releases energy.

8.6 Coupling of Reactions Exergonic cellular reactions release the energy needed to make ATP from ADP. The energy released from the conversion of ATP back to ADP can be used to fuel endergonic reactions.

Go to Activity 8.1 ATP and Coupled Reactions
Life10e.com/ac8.1

Go to Media Clip 8.1
Bioluminescence in the Deep Sea
Life10e.com/mc8.1

ATP couples exergonic and endergonic reactions

As we have just seen, the hydrolysis of ATP is exergonic and yields ADP, P_i, and free energy (or AMP, PP_i, and free energy). The reverse reaction, the formation of ATP from ADP and P_i, is endergonic and consumes as much free energy as is released by the hydrolysis of ATP:

$$ADP + P_i + \text{free energy} \rightarrow ATP + H_2O$$

Many different exergonic reactions in the cell can provide the energy to convert ADP into ATP. For eukaryotes and many prokaryotes, the most important of these reactions is cellular respiration, in which some of the energy released from fuel molecules is captured in ATP. The formation and hydrolysis of ATP constitute what might be called an "energy-coupling cycle," in which ADP picks up energy from exergonic reactions to become ATP, which then donates energy to endergonic reactions. ATP is the common component of these reactions and is the agent of coupling, as illustrated in Figure 8.6.

Coupling of exergonic and endergonic reactions is very common in metabolism. Free energy is captured and retained in the P~O bonds of ATP. ATP then diffuses to another site in the cell, where its hydrolysis releases the free energy to drive an endergonic reaction. For example, the formation of glucose 6-phosphate from glucose (Figure 8.7), which has a positive ΔG (is endergonic), will not proceed without the input of free energy from ATP hydrolysis, which has a negative ΔG (is exergonic). The overall ΔG for the coupled reactions (when the two ΔGs are added together) is negative. Hence the reactions proceed exergonically when they are coupled, and glucose 6-phosphate is synthesized. As you will see in Chapter 9, this is the initial reaction in the catabolism of glucose.

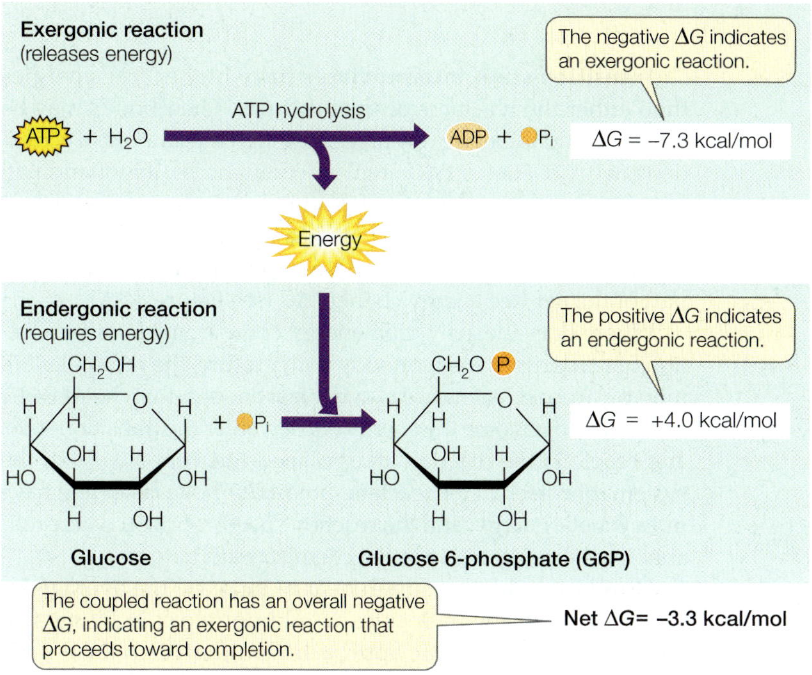

Exergonic reaction
(releases energy)
The negative ΔG indicates an exergonic reaction.
ATP hydrolysis
ATP + H2O
ADP + Pi
ΔG = –7.3 kcal/mol
Energy
Endergonic reaction
(requires energy)
The positive ΔG indicates an endergonic reaction.
CH2OH
CH2O P
+ Pi
ΔG = +4.0 kcal/mol
Glucose
Glucose 6-phosphate (G6P)
The coupled reaction has an overall negative ΔG, indicating an exergonic reaction that proceeds toward completion.
Net ΔG= –3.3 kcal/mol

8.7 Coupling of ATP Hydrolysis to an Endergonic Reaction The addition of phosphate derived from the hydrolysis of ATP to glucose forms the molecule glucose 6-phosphate (in a reaction catalyzed by hexokinase). ATP hydrolysis is exergonic and the energy released drives the second reaction, which is endergonic.

An active cell requires the production of millions of molecules of ATP per second to drive its biochemical machinery. You are already familiar with some of the activities in the cell that require energy from the hydrolysis of ATP:

- Active transport across a membrane (see Figure 6.14)
- Condensation reactions that use enzymes to form polymers (see Figure 3.4A)
- Modifications of cell signaling proteins by protein kinases (see Figure 7.16)
- Motor proteins that move vesicles along microtubules (see Figure 5.19)

An ATP molecule is typically consumed within a second of its formation. At rest, an average person produces and hydrolyzes about 40 kg of ATP per day—as much as some people weigh. This means that each ATP molecule undergoes about 10,000 cycles of synthesis and hydrolysis every day!

RECAP 8.2

ATP is the "energy currency" of cells. Some of the free energy released by exergonic reactions can be captured in the form of ATP. This energy can then be released by ATP hydrolysis and used to drive endergonic reactions.

- How does ATP store energy? **See pp. 149–150**
- What are coupled reactions? **See pp. 150–151 and Figure 8.7**

ATP is synthesized and used up very rapidly. But these biochemical reactions—and nearly all the others that take place inside a cell—could not proceed so rapidly without the help of enzymes.

8.3 What Are Enzymes?

When we know the change in free energy (ΔG) of a reaction, we know where the equilibrium point of the reaction lies: the more negative the ΔG value is, the further the reaction proceeds toward completion. However, ΔG tells us nothing about the *rate* of a reaction—the speed at which it moves toward equilibrium. The reactions that cells depend on have spontaneous rates that are so slow that the cells would not survive without a way to speed up the reactions. That is the role of catalysts: substances that speed up reactions without themselves being permanently altered. A catalyst does not cause a reaction to occur that would not proceed without it, *but merely increases the rate of the reaction*, allowing equilibrium to be approached more rapidly. This is an important point: *no catalyst makes a reaction occur that cannot otherwise occur.*

Most biological catalysts are proteins called enzymes. Although we will focus here on proteins, some catalysts are RNA molecules called ribozymes (see Section 4.3). A biological catalyst, whether protein or RNA, is a framework or scaffold within which chemical catalysis takes place. This molecular framework binds the reactants and sometimes participates in the reaction itself; however, such participation does not permanently change the enzyme. The catalyst ends up in exactly the same chemical condition after a reaction as before it. Although there is considerable evidence that the first enzymes to evolve were ribozymes, cells now use proteins rather than RNA to catalyze most biochemical reactions. Compared with RNA, proteins show greater diversity in their three-dimensional structures, and in the chemical functions provided by their functional groups (see Table 3.2).

In this section we will discuss the energy barrier that controls the rate of a chemical reaction. Then we will focus on the roles of enzymes: how they interact with specific reactants, how they lower the energy barrier, and how they permit reactions to proceed more quickly.

To speed up a reaction, an energy barrier must be overcome

An exergonic reaction may release free energy, but without a catalyst it will take place very slowly. This is because there is an energy barrier between reactants and products. Think about the hydrolysis of sucrose, which we described in Section 8.1:

$$\text{Sucrose} + H_2O \rightarrow \text{glucose} + \text{fructose}$$

In humans, this reaction is part of the process of digestion. Even if water is abundant, the sucrose molecule will only very rarely bind the H atom and the —OH group of water at the appropriate locations to break the covalent bond between glucose and fructose *unless there is an input of energy to initiate the reaction.* Such an input of energy will place the sucrose into a reactive mode called the **transition state**. The energy input required for sucrose to reach this state is called the **activation energy** (E_a).

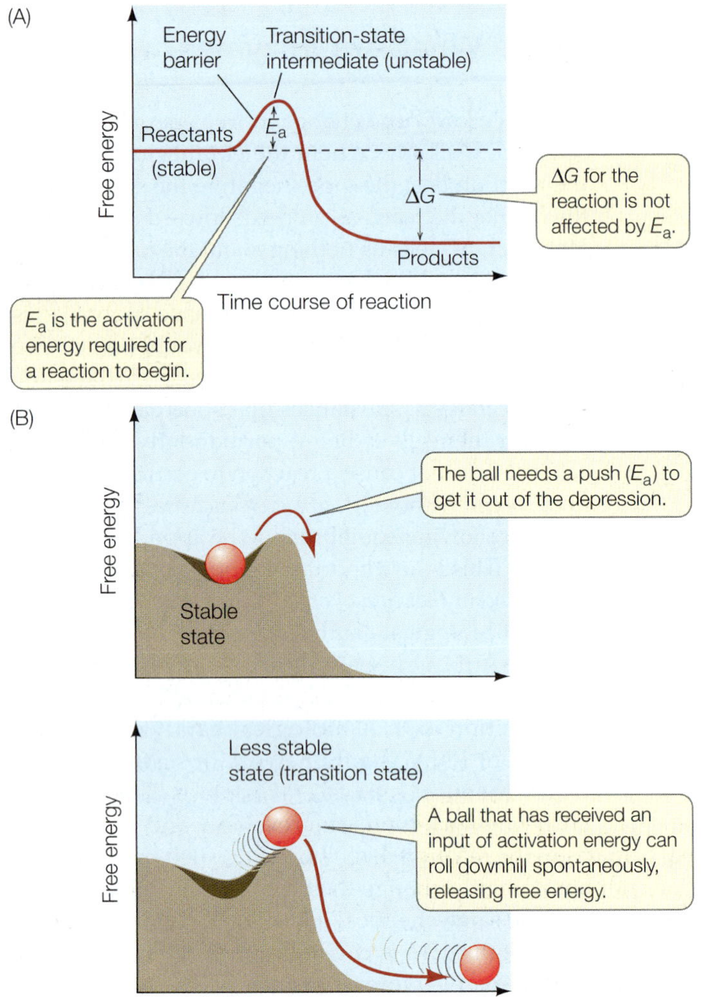
(A)
Energy barrier
Transition-state intermediate (unstable)
Reactants (stable)
E_a
ΔG
ΔG for the reaction is not affected by E_a.
Products
Free energy
Time course of reaction
E_a is the activation energy required for a reaction to begin.
(B)
Free energy
The ball needs a push (E_a) to get it out of the depression.
Stable state
Less stable state (transition state)
A ball that has received an input of activation energy can roll downhill spontaneously, releasing free energy.

8.8 Activation Energy Initiates Reactions (A) In any chemical reaction, an initial stable state must become less stable before change is possible. (B) A ball on a hillside provides a physical analogy to the biochemical principle graphed in (A).

The following example will help illustrate the ideas of activation energy and transition state:

$$\text{Fireworks} + O_2 \rightarrow CO_2 + H_2O + \text{energy (heat and light)}$$

Activation energy in the form of a spark is needed to excite the molecules in the fireworks so they will react with oxygen in the air. Once the transition state is reached, the reaction occurs.

In general, exergonic reactions proceed only after the reactants are pushed over the energy barrier by some added energy. The energy barrier thus represents the amount of energy needed to start the reaction, the activation energy (**Figure 8.8A**). Recall the ball rolling down the hill in Figure 8.3A. The ball has a lot of potential energy at the top of the hill. However, if it is stuck in a small depression, it will not roll down the hill, even though that action is exergonic. To start the ball rolling, a small amount of energy (activation energy) is needed to push it out of the depression (**Figure 8.8B**). In a chemical reaction, the activation energy is the energy needed to change the reactants into unstable molecular forms called transition-state intermediates.

Transition-state intermediates have higher free energies than either the reactants or the products. Their bonds may be stretched and therefore unstable. Although the amount of activation energy needed for different reactions varies, it is often small compared with the change in free energy of the overall reaction. The activation energy put in to start a reaction is recovered during the ensuing "downhill" phase of the reaction, so it is not a part of the net free energy change, ΔG (see Figure 8.8A).

Where does the activation energy come from? In any collection of reactants at room or body temperature, the molecules are moving around. A few are moving fast enough that their kinetic energy can overcome the energy barrier, enter the transition state, and react. So the reaction takes place—but very slowly. If the system is heated, all the reactant molecules move faster and have more kinetic energy, and the reaction speeds up. You have probably used this technique in the chemistry laboratory.

However, adding enough heat to increase the average kinetic energy of the molecules would not work in living systems. Such a nonspecific approach would accelerate all reactions, including destructive ones such as the denaturation of proteins (see Chapter 3). A more effective way to speed up a reaction in a living system is to lower the energy barrier by bringing the reactants close together. In living cells, enzymes and ribozymes accomplish this task.

Enzymes bind specific reactants at their active sites

Catalysts increase the rates of chemical reactions. Most nonbiological catalysts are nonspecific. For example, powdered platinum catalyzes virtually any reaction in which molecular hydrogen (H_2) is a reactant. In contrast, most biological catalysts are highly specific. An enzyme or ribozyme usually recognizes and binds to only one or a few closely related reactants, and it catalyzes only a single chemical reaction. In the discussion that follows, we focus on enzymes, but remember that similar rules of chemical behavior apply to ribozymes as well.

In an enzyme-catalyzed reaction, the reactants are called **substrates**. Substrate molecules bind to a particular site on the enzyme, called the **active site**, where catalysis takes place (**Figure 8.9**). The specificity of an enzyme results from the exact three-dimensional shape and structure of its active site, into which only a narrow range of substrates can fit. Other molecules—with different shapes, different functional groups, and different properties—cannot fit properly and bind to the active site. This specificity is comparable to the specific binding of a membrane transport protein or receptor protein to its specific ligand, as described in Chapters 6 and 7.

The names of enzymes often reflect their functions and end with the suffix "ase." For example the enzyme sucrase catalyzes the hydrolysis of sucrose, which we have referred to above:

$$\text{Sucrose} + H_2O \xrightarrow{\text{Sucrase}} \text{glucose} + \text{fructose}$$

And as we saw in the opening story, lipases and amylases catalyze the hydrolysis of lipids and starch, respectively (starch is sometimes referred to as amylum).

The binding of a substrate to the active site of an enzyme produces an **enzyme–substrate complex (ES)** that is held

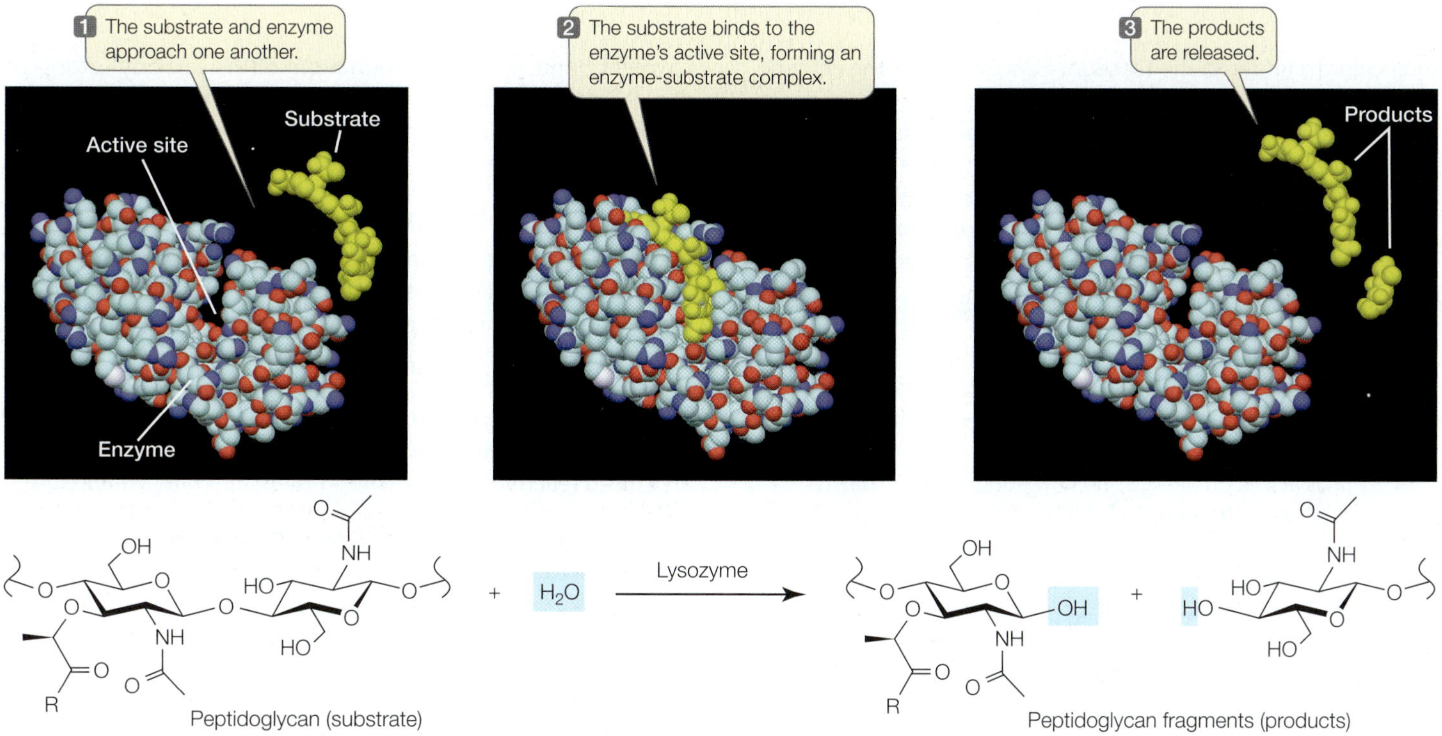
1 The substrate and enzyme approach one another.
2 The substrate binds to the enzyme's active site, forming an enzyme-substrate complex.
3 The products are released.
Active site
Substrate
Enzyme
Products
Lysozyme
Peptidoglycan (substrate)
Peptidoglycan fragments (products)

8.9 Enzyme and Substrate A reaction involving an enzyme is illustrated by lysozyme. Lysozyme catalyzes breakage of bonds in the peptidoglycans of bacterial cell walls. (See Section 5.2 for a description of peptidoglycans.)

together by several interactions, such as hydrogen bonding, electrical attraction, or temporary covalent bonding. The enzyme–substrate complex gives rise to product and free enzyme:

$$E + S \rightarrow ES \rightarrow E + P$$

where E is the enzyme, S is the substrate, P is the product, and ES is the enzyme–substrate complex. The free enzyme (E) is in the same chemical form at the end of the reaction as at the beginning. While bound to the substrate, it may change chemically, but by the end of the reaction it has been restored to its initial form and is ready to bind more substrate.

The equation that forms ES may be familiar to you from Chapter 7: it is the same as the binding of a receptor and ligand (RL) in cell signaling. In that chapter, we defined the dissociation constant (K_D) as a measure of the affinity of the receptor for its ligand. The lower the K_D, the tighter the binding. For enzymes and their substrates, K_D values are often in the range 10^{-5} to 10^{-6} M. This favors the formation of ES. In practical terms, it means that the binding between enzyme and substrate is somewhat reversible; the substrate can be released before the reaction. However, this is counteracted by the fact that ES is short-lived and the product(s) form quickly.

Enzymes lower the energy barrier but do not affect equilibrium

When reactants are bound to the enzyme, forming an enzyme–substrate complex, they require less activation energy than the transition-state intermediates in the corresponding uncatalyzed reaction (**Figure 8.10**). Thus the enzyme lowers the energy barrier for the reaction—it offers the reaction an easier path, speeding it up. When an enzyme lowers the energy barrier, both the forward and the reverse reactions speed up, so the enzyme-catalyzed reaction proceeds toward equilibrium more rapidly than the uncatalyzed reaction. *The final equilibrium is the same with or without the enzyme.* Similarly, adding an enzyme to a reaction does not change the difference in free energy (ΔG) between the reactants and the products (see Figure 8.10).

Enzymes can change the rate of a reaction substantially. For example, if a particular protein that has arginine as its terminal amino acid just sits in solution, the protein molecules tend toward disorder and the terminal peptide bonds break, releasing the arginine residues (ΔS increases). Without an enzyme this is a very slow reaction—it takes about 7 years for half of the protein

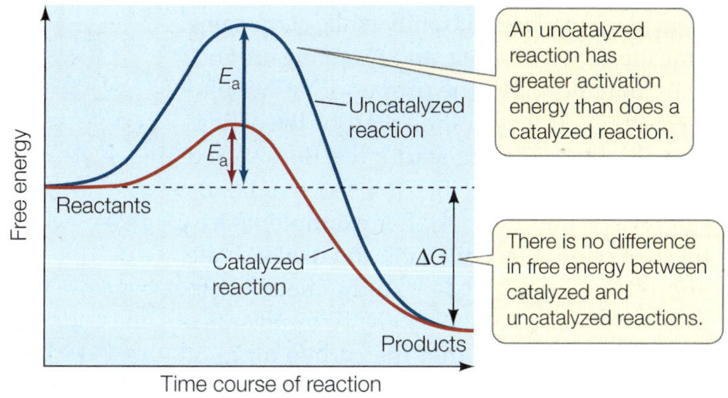
Free energy
E_a
E_a
Uncatalyzed reaction
Reactants
Catalyzed reaction
ΔG
Products
Time course of reaction
An uncatalyzed reaction has greater activation energy than does a catalyzed reaction.
There is no difference in free energy between catalyzed and uncatalyzed reactions.

8.10 Enzymes Lower the Energy Barrier Although the activation energy is lower in an enzyme-catalyzed reaction than in an uncatalyzed reaction, the energy released is the same with or without catalysis. In other words, E_a is lower, but ΔG is unchanged. Lower activation energy means the reaction will take place at a faster rate.

Go to Activity 8.2 Free Energy Changes Life10e.com/ac8.2

molecules to undergo the reaction. However, with the enzyme carboxypeptidase A catalyzing the reaction, half the arginines are released in less than a second! Rate enhancement by enzymes varies from 1 million times to an amazing 10^{17} times for the champion enzyme orotidine monophosphate decarboxylase! The consequence of catalysis for living cells is not difficult to imagine. Such increased reaction rates make new realities possible.

RECAP 8.3

A chemical reaction requires a "push" over the energy barrier to get started. An enzyme reduces the activation energy needed to start a reaction by binding the reactants (substrates). This speeds up the reaction.

- What is activation energy, and how is it supplied in the chemistry lab? **See pp. 151–152 and Figure 8.8**
- Explain how the structure of an enzyme makes that enzyme specific. **See p.152 and Figure 8.9**
- How does an enzyme speed up a reaction? **See Figure 8.10**
- What is the relationship between an enzyme and the equilibrium point of a reaction? **See p. 153**

Now that you have a general understanding of the structures, functions, and specificities of enzymes, let's look more closely at how they work.

8.4 How Do Enzymes Work?

During and after the formation of the enzyme–substrate complex, chemical interactions occur. These interactions contribute directly to the breaking of old bonds and the formation of new ones. In catalyzing a reaction, an enzyme may use one or more mechanisms.

Enzymes can orient substrates

When free in solution, substrates are moving from place to place randomly while at the same time vibrating, rotating, and tumbling around. They may not have the proper orientation to interact when they collide. Part of the activation energy needed to start a reaction is used to bring together specific atoms so that bonds can form (**Figure 8.11A**). For example, if acetyl coenzyme A (acetyl CoA) and oxaloacetate are to form citrate (a step in the metabolism of glucose; see Section 9.2), the two substrates must be oriented so that the carbon atom of the methyl group of acetyl CoA can form a covalent bond with the carbon atom of the carbonyl group of oxaloacetate. The active site of the enzyme citrate synthase has just the right shape to bind these two molecules so that these atoms are adjacent.

Enzymes can induce strain in the substrate

Once a substrate has bound to its active site, an enzyme can cause bonds in the substrate to stretch, putting it in an unstable transition state (**Figure 8.11B**). For example, lysozyme is a protective enzyme abundant in tears and saliva that destroys invading bacteria by cleaving peptidoglycans in their cell walls (see Figure 8.9). Lysozyme's active site "stretches" the bonds between the glycan monomers, rendering the bonds unstable and more reactive to lysozyme's other substrate, water.

Enzymes can temporarily add chemical groups to substrates

The side chains (R groups) of an enzyme's amino acids may be direct participants in making its substrates more chemically reactive (**Figure 8.11C**).

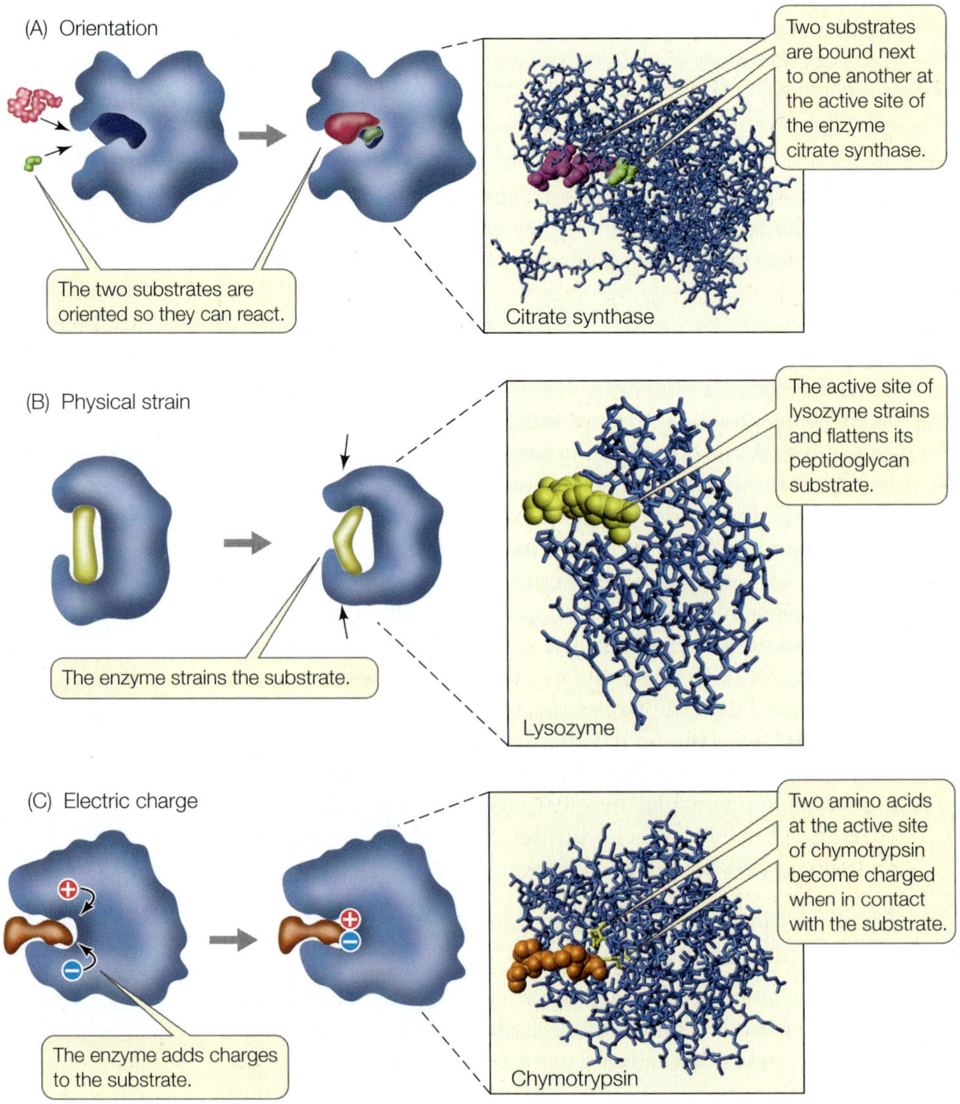

(A) Orientation
Two substrates are bound next to one another at the active site of the enzyme citrate synthase.
The two substrates are oriented so they can react.
Citrate synthase
(B) Physical strain
The active site of lysozyme strains and flattens its peptidoglycan substrate.
The enzyme strains the substrate.
Lysozyme
(C) Electric charge
Two amino acids at the active site of chymotrypsin become charged when in contact with the substrate.
The enzyme adds charges to the substrate.
Chymotrypsin

8.11 Life at the Active Site Enzymes have several ways of causing their substrates to enter the transition state: (A) orientation, (B) physical strain, and (C) chemical charge.

- In *acid–base catalysis*, the acidic or basic side chains of the amino acids in the active site transfer H^+ to or from the substrate, destabilizing a covalent bond in the substrate, and permitting it to break.
- In *covalent catalysis*, a functional group in a side chain forms a temporary covalent bond with a portion of the substrate.
- In *metal ion catalysis*, metal ions such as copper, iron, and manganese, which are often firmly bound to side chains of enzymes, can lose or gain electrons without detaching from the enzymes. This ability makes them important participants in oxidation–reduction reactions, which involve the loss or gain of electrons.

Molecular structure determines enzyme function

Most enzymes are much larger than their substrates. An enzyme is typically a protein containing hundreds of amino acids. It may consist of a single folded polypeptide chain or of several subunits (see Section 3.2). Its substrate is generally a small molecule or a small part of a large molecule. The active site of the enzyme is usually quite small, not more than 6 to 12 amino acids. Two questions arise from these observations:

- What features of the active site allow it to recognize and bind the substrate?
- What is the role of the rest of the huge protein?

THE ACTIVE SITE IS SPECIFIC TO THE SUBSTRATE(S) The remarkable ability of an enzyme to select exactly the right substrate(s) depends on a precise interlocking of molecular shapes and interactions of chemical groups at the active site. The binding of a substrate to the active site depends on the same kinds of forces that maintain the tertiary structure of the enzyme: hydrogen bonds, the attraction and repulsion of electrically charged groups, and hydrophobic interactions. The specific fit of the substrate in the active site of lysozyme is illustrated in Figures 8.9 and 8.11B.

AN ENZYME CHANGES SHAPE WHEN IT BINDS A SUBSTRATE Just as a membrane receptor protein may undergo precise changes in conformation upon binding to its ligand (see Chapter 7), some enzymes change their shapes when they bind their substrate(s). These shape changes, which are called **induced fit**, alter the shape of the active site(s) of the enzyme.

An example of induced fit can be seen in the enzyme hexokinase (see Figure 8.7), which catalyzes the reaction

$$\text{Glucose} + \text{ATP} \rightarrow \text{glucose 6-phosphate} + \text{ADP}$$

Induced fit brings reactive side chains from the hexokinase active site into alignment with the substrates (**Figure 8.12**), facilitating its catalytic mechanisms. Equally important, the folding of hexokinase to fit around the substrates (glucose and ATP) excludes water from the active site. This is essential, because if water were present, the ATP could be hydrolyzed to ADP and P_i. But since water is absent, the transfer of a phosphate from ATP to glucose is favored.

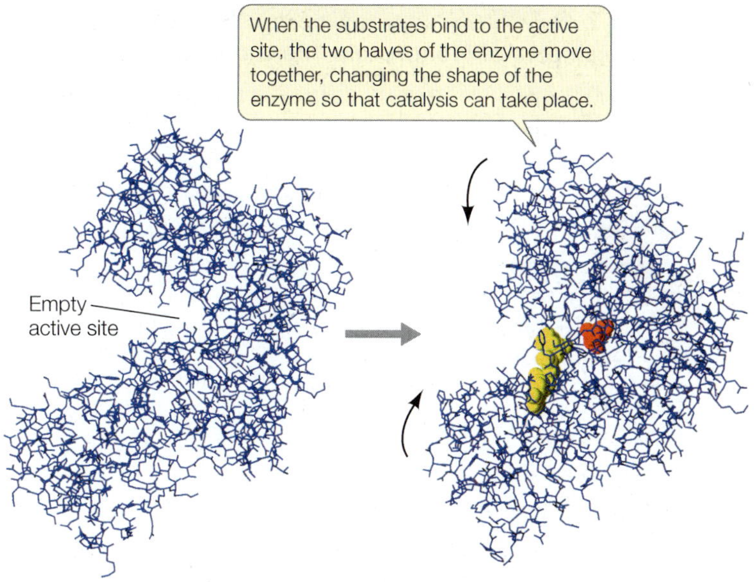
When the substrates bind to the active site, the two halves of the enzyme move together, changing the shape of the enzyme so that catalysis can take place.
Empty active site

8.12 Some Enzymes Change Shape When Substrate Binds to Them Shape changes result in an induced fit between enzyme and substrate, improving the catalytic ability of the enzyme. Induced fit can be observed in the enzyme hexokinase, seen here with and without its substrates, glucose (red) and ATP (yellow).

Induced fit at least partly explains why enzymes are so large. The rest of the macromolecule may have one or more of the following three roles:

- It provides a framework so that the amino acids of the active site are properly positioned in relation to the substrate(s).
- It participates in significant changes in protein shape and structure that result in induced fit.
- It provides binding sites for regulatory molecules (see Section 8.5).

Some enzymes require other molecules in order to function

As large and complex as enzymes are, many of them require the presence of nonprotein chemical "partners" in order to function (**Table 8.2**):

- *Prosthetic groups* are distinct, non–amino acid atoms or molecular groupings that are permanently bound to their enzymes. An example is flavin adenine dinucleotide (FAD), which is bound to succinate dehydrogenase, an important enzyme in cellular respiration (see Section 9.3).
- *Inorganic cofactors* include ions such as copper, zinc, and iron that are permanently bound to certain enzymes. For example, the enzyme alcohol dehydrogenase contains the cofactor zinc.
- A *coenzyme* is a nonprotein carbon-containing molecule that is required for the action of one or more enzymes. It is usually relatively small compared with the enzyme to which it temporarily binds.

A coenzyme moves from enzyme to enzyme, adding or removing chemical groups from the substrate. A coenzyme is like a

TABLE **8.2** Some Examples of Nonprotein "Partners" of Enzymes

Type of Molecule	Role in Catalyzed Reactions
Prosthetic groups	
Heme	Binds ions, O_2, and electrons
FAD	Carries electrons/protons
Retinal	Converts light energy
Inorganic cofactors	
Iron (Fe^{2+} or Fe^{3+})	Oxidation/reduction
Copper (Cu^{+} or Cu^{2+})	Oxidation/reduction
Zinc (Zn^{2+})	Stabilizes DNA binding structure
Coenzymes	
Biotin	Carries —COO^-
Coenzyme A	Carries —CO—CH_3
NAD	Carries electrons/protons
ATP	Provides/extracts energy

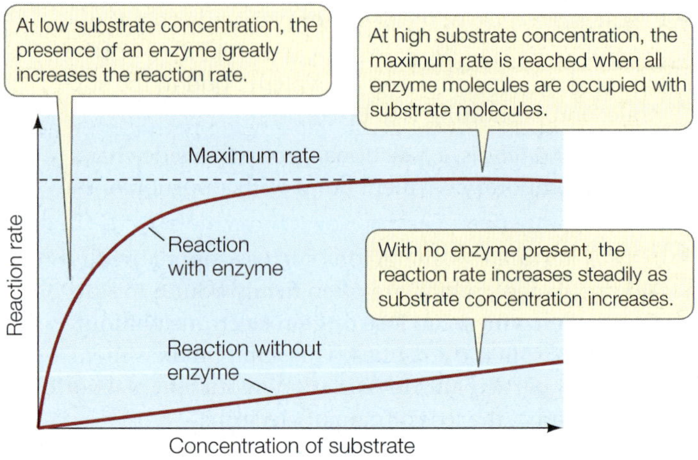
At low substrate concentration, the presence of an enzyme greatly increases the reaction rate.
At high substrate concentration, the maximum rate is reached when all enzyme molecules are occupied with substrate molecules.
Maximum rate
Reaction with enzyme
With no enzyme present, the reaction rate increases steadily as substrate concentration increases.
Reaction without enzyme
Reaction rate
Concentration of substrate

8.13 Catalyzed Reactions Reach a Maximum Rate Because there is usually less enzyme than substrate present, the reaction rate levels off when the enzyme becomes saturated.

substrate in that it does not permanently bind to the enzyme: it binds to the active site, changes chemically during the reaction, and then separates from the enzyme to participate in other reactions. There is actually no clear distinction between the functions of coenzymes and some substrates. For example, ATP and ADP have been described as coenzymes, even though they are really substrates that gain or lose phosphate groups during chemical reactions. The term coenzyme was coined before the functions of these molecules were fully understood. Biochemists continue to use the term, and for consistency with the field, we use it in this book.

In the next chapter we will encounter other coenzymes that function in energy-harvesting reactions by accepting or donating electrons or hydrogen atoms. In animals, some coenzymes are produced from vitamins—substances that must be obtained from food because they cannot be synthesized by the body. For example, the B vitamin niacin is used to make the coenzyme nicotinamide adenine dinucleotide (NAD).

The substrate concentration affects the reaction rate

For a reaction of the type A → B, the rate of the uncatalyzed reaction is directly proportional to the concentration of A. The higher the concentration of substrate, the faster the rate of the reaction. The appropriate enzyme not only speeds up the reaction; it also changes the shape of a plot of rate versus substrate concentration (**Figure 8.13**). For a given concentration of enzyme, the rate of the enzyme-catalyzed reaction initially increases as the substrate concentration increases from zero, but then it levels off. At some point, further increases in the substrate concentration do not significantly increase the reaction rate—the maximum rate has been reached.

Since the concentration of an enzyme is usually much lower than that of its substrate and does not change as substrate concentration changes, what we see is a saturation phenomenon like the one that occurs in facilitated diffusion (see Figure 6.12). When all the enzyme molecules are bound to substrate molecules, the enzyme is working as fast as it can—at its maximum rate. Nothing is gained by adding more substrate, because no free enzyme molecules are left to act as catalysts. Under these conditions the active sites are said to be saturated.

The maximum rate of a catalyzed reaction can be used to measure how efficient the enzyme is. The turnover number is the maximum number of substrate molecules that one enzyme molecule can convert to product per unit of time. This number ranges from 1 molecule every 2 seconds for lysozyme to an amazing 40 million molecules per second for the liver enzyme catalase.

RECAP 8.4

Enzymes orient their substrates to bring together specific atoms so that bonds can form. An enzyme can participate in the reaction it catalyzes by temporarily changing shape or destabilizing the enzyme–substrate complex. Some enzymes require prosthetic groups, inorganic cofactors, or coenzymes in order to function.

- What are three mechanisms of enzyme catalysis? **See p.154 and Figure 8.11**
- What are the chemical roles of coenzymes in enzymatic reactions? **See pp. 155–156 and Table 8.2**

We've seen in this section how individual enzymes work on their substrates. However, enzymes inside organisms don't operate in isolation—there may be thousands of different enzymes within a given cell. Let's see how all these different enzymes work together in a complex organism.

8.5 How Are Enzyme Activities Regulated?

A major characteristic of life is homeostasis—the maintenance of stable internal conditions (see Chapter 40). How does a cell maintain a relatively constant internal environment while thousands of chemical reactions are going on? These chemical reactions operate within metabolic pathways in which the product of one reaction is a reactant for the next. These pathways do not exist in isolation, but interact extensively, and each reaction in each pathway is catalyzed by a specific enzyme.

Within a cell or organism, the presence and activity of enzymes determine the "flow" of chemicals through different metabolic pathways. The amount of enzyme activity, in turn, is controlled in part via the regulation of gene expression. Many signal transduction pathways (described in Chapter 7) end with changes in gene expression, and often the genes that are switched on or off encode enzymes. But the simple presence of an enzyme does not ensure that it is functioning. Another way cells can control which pathways are active at a particular time is by the activation or inactivation of existing enzymes. If one enzyme in the pathway is inactive, that step and all subsequent steps shut down. Thus some enzymes are target points for the regulation of entire sequences of chemical reactions.

Regulation of the rates at which thousands of different enzymes operate contributes to homeostasis within an organism. Such control permits cells to make orderly changes in their functions in response to changes in the external environment. In Chapter 7 we described a number of enzymes that become activated in signal transduction pathways, illustrating how enzyme activation can dramatically alter cell functions. (For example, see the activation of glycogen phosphorylase in Figure 7.18.)

The flow of chemicals such as carbon atoms through interacting metabolic pathways can be studied, but this process becomes complicated quickly, because each pathway influences the others. Computer algorithms are used to model these pathways and show how they mesh in an interdependent system (**Figure 8.14**). Such models can help predict what will happen if the concentration of one molecule or another is altered. This new field of biology is called **systems biology**, and it has numerous applications.

In this section we will investigate the roles of enzymes in organizing and regulating metabolic pathways. We will also examine how the environment—particularly temperature and pH—affects enzyme activity.

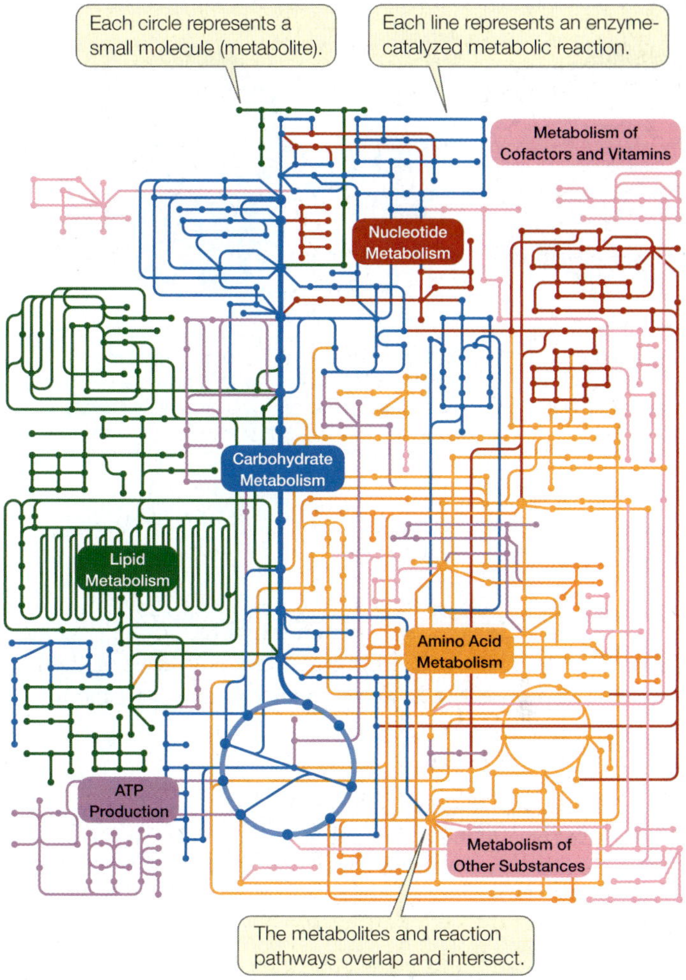
Each circle represents a small molecule (metabolite).
Each line represents an enzyme-catalyzed metabolic reaction.
Metabolism of Cofactors and Vitamins
Nucleotide Metabolism
Carbohydrate Metabolism
Lipid Metabolism
Amino Acid Metabolism
ATP Production
Metabolism of Other Substances
The metabolites and reaction pathways overlap and intersect.

8.14 Metabolic Pathways The complex interactions of metabolic pathways can be modeled by the tools of systems biology. In cells, the main elements controlling these pathways are enzymes.

Enzymes can be regulated by inhibitors

Various chemical inhibitors can bind to enzymes, slowing down the rates of the reactions they catalyze. Some inhibitors occur naturally in cells; others are artificial. Naturally occurring inhibitors regulate metabolism; artificial ones can be used to treat disease, to kill pests, or to study how enzymes work. In some cases the inhibitor binds the enzyme irreversibly, and the enzyme becomes permanently inactivated. In other cases the inhibitor has reversible effects; it can separate from the enzyme, allowing the enzyme to function fully as before. The removal of a natural reversible inhibitor increases an enzyme's rate of catalysis.

IRREVERSIBLE INHIBITION If an inhibitor covalently binds to certain side chains at the active site of an enzyme, it will permanently inactivate the enzyme by destroying its capacity to interact with its normal substrate. An example of an irreversible inhibitor is DIPF (diisopropyl phosphorofluoridate), which reacts with serine (**Figure 8.15**). DIPF is an irreversible inhibitor of acetylcholinesterase, whose operation is essential for the normal functioning of the nervous system. Because of their effect on acetylcholinesterase, DIPF and other similar compounds are classified as nerve gases, and were developed for biological warfare. One of these compounds, Sarin, was used in an attack on the Tokyo subway in 1995, resulting in a dozen deaths and the hospitalization of hundreds more. The widely used insecticide malathion is a derivative of DIPF that inhibits only insect acetylcholinesterase, not the mammalian enzyme. The irreversible inhibition of enzymes is of practical use to humans, but this form of regulation is not common in the cell, because the enzyme is permanently inactivated and cannot be recycled. Instead, cells use reversible inhibition.

REVERSIBLE INHIBITION In some cases an inhibitor is similar enough to a particular enzyme's natural substrate to bind noncovalently to its active site, yet different enough that the

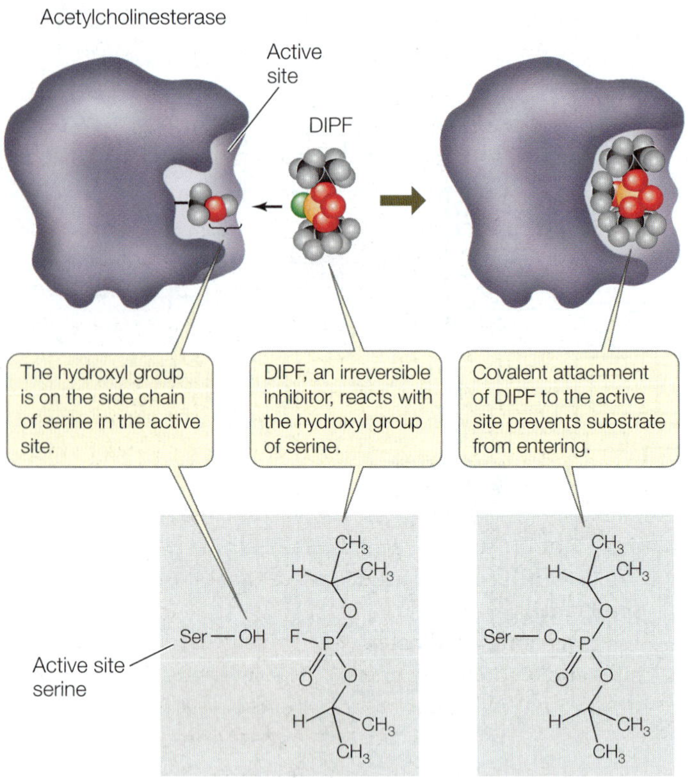

Acetylcholinesterase
Active site
DIPF
The hydroxyl group is on the side chain of serine in the active site.
DIPF, an irreversible inhibitor, reacts with the hydroxyl group of serine.
Covalent attachment of DIPF to the active site prevents substrate from entering.
Active site serine
Ser—OH
Ser—O

8.15 Irreversible Inhibition DIPF forms a stable covalent bond with the side chain of the amino acid serine at the active site of the enzyme acetylcholinesterase, thus irreversibly disabling the enzyme.

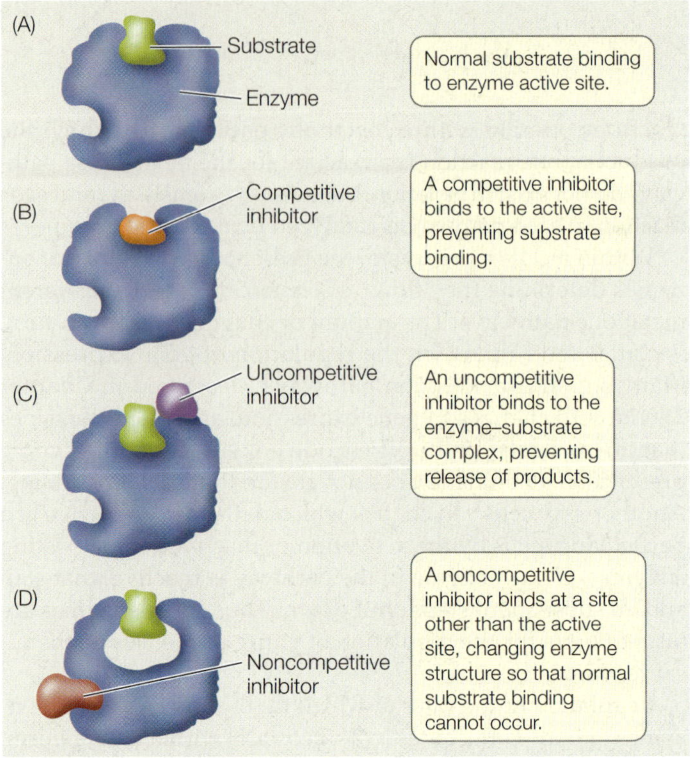

(A)
Substrate
Enzyme
Normal substrate binding to enzyme active site.
(B)
Competitive inhibitor
A competitive inhibitor binds to the active site, preventing substrate binding.
(C)
Uncompetitive inhibitor
An uncompetitive inhibitor binds to the enzyme–substrate complex, preventing release of products.
(D)
Noncompetitive inhibitor
A noncompetitive inhibitor binds at a site other than the active site, changing enzyme structure so that normal substrate binding cannot occur.

8.16 Reversible Inhibition (A) Normal enzyme–substrate binding. (B) Competitive inhibition. (C) Uncompetitive inhibition. (D) Noncompetitive inhibition.

Go to Animated Tutorial 8.1
Enzyme Catalysis
Life10e.com/at8.1

enzyme catalyzes no chemical reaction. While such a molecule is bound to the enzyme, the natural substrate cannot enter the active site and the enzyme is unable to function. Such a molecule is called a **competitive inhibitor** because it competes with the natural substrate for the active site (Figure 8.16A, B). In this case, the degree of inhibition depends on the relative concentrations of the substrate and the inhibitor: if the inhibitor concentration is higher, it is more likely to bind the active site of the enzyme than the substrate, and vice versa. The inhibition is reversible because if the concentration of substrate is increased or if the concentration of inhibitor is reduced, the substrate is more likely to bind, and the enzyme is active again.

An example of a competitive inhibitor is the drug methotrexate. An important coenzyme in the formation of purines (components of nucleic acids) is tetrahydrofolate, which is formed from dihydrofolate in a reaction catalyzed by dihydrofolate reductase (DHFR):

$$\text{Dihydrofolate} \xrightarrow{\text{DHFR}} \text{tetrahydrofolate}$$

When cancer cells reproduce, they need to replicate their DNA, and so they need to produce purines. This makes DHFR an ideal target for an anticancer drug. A team led by Sidney Farber at Harvard Medical School first showed that an analog of dihydrofolate could treat leukemia. The drug used by Farber (aminopterin) has since been replaced by a similar analog, methotrexate:

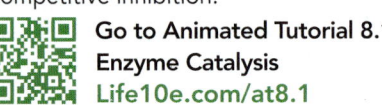

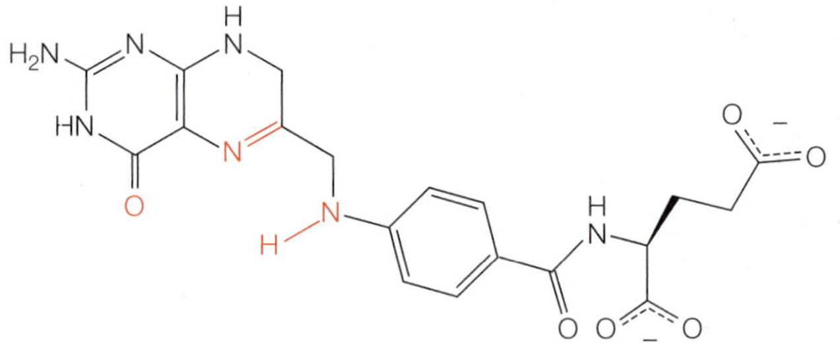

Dihydrofolate

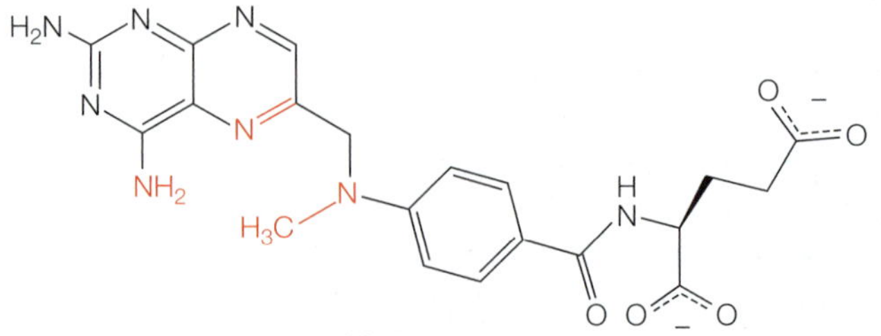

Methotrexate

This successful drug is used to treat inflammatory diseases such as psoriasis and rheumatoid arthritis, as well as cancer.

An **uncompetitive inhibitor** (Figure 8.16C) binds to the enzyme–substrate complex, preventing the complex from releasing products. Unlike competitive inhibition, it cannot be overcome by adding more substrate.

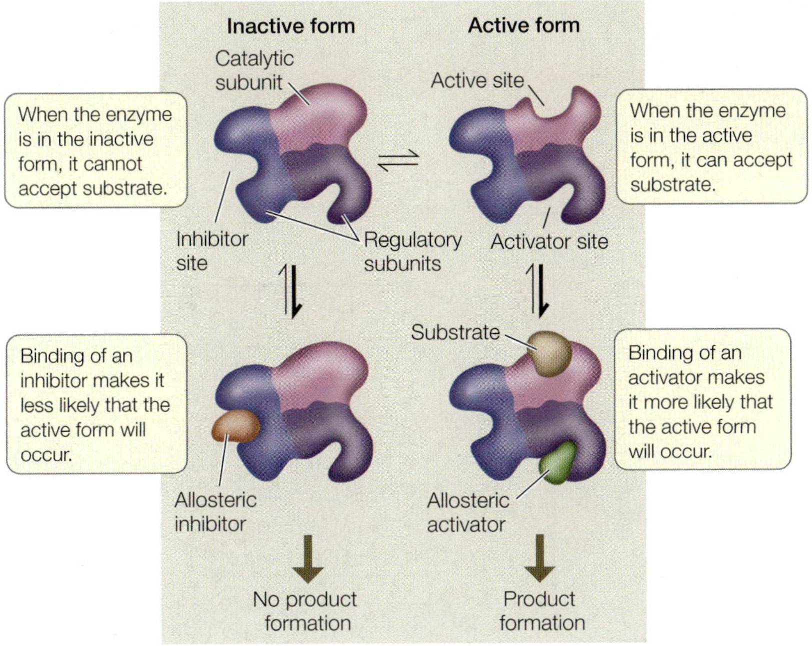

Inactive form
Active form
Catalytic subunit
Active site
When the enzyme is in the inactive form, it cannot accept substrate.
When the enzyme is in the active form, it can accept substrate.
Inhibitor site
Regulatory subunits
Activator site
Substrate
Binding of an inhibitor makes it less likely that the active form will occur.
Binding of an activator makes it more likely that the active form will occur.
Allosteric inhibitor
Allosteric activator
No product formation
Product formation

8.17 Allosteric Regulation of Enzymes Active and inactive forms of an enzyme can be interconverted, depending on the binding of effector molecules at sites other than the active site. Binding an inhibitor stabilizes the inactive form, and binding an activator stabilizes the active form.

Go to Animated Tutorial 8.2
Allosteric Regulation of Enzymes
Life10e.com/at8.2

A **noncompetitive inhibitor** binds to an enzyme at a site distinct from the active site. This binding causes a change in the shape of the enzyme that alters its activity (Figure 8.16D). The active site may no longer bind the substrate, or if it does, the rate of product formation may be reduced. Like competitive inhibitors, noncompetitive inhibitors can become unbound, so their effects are reversible.

Allosteric enzymes are controlled via changes in shape

The change in enzyme shape that is due to noncompetitive inhibitor binding is an example of allostery (*allo*, "different"; *stereos*, "shape"). **Allosteric regulation** occurs when an effector molecule binds to a site other than the active site of an enzyme, *inducing the enzyme to change its shape*. The change in shape alters the affinity of the active site for the substrate, and so the rate of the reaction is changed.

Often, an enzyme will exist in the cell in more than one possible shape (Figure 8.17):

- The *active form* of the enzyme has the proper shape for substrate binding.
- The *inactive form* of the enzyme has a shape that cannot bind the substrate.

Other molecules, collectively referred to as effectors, can influence which form the enzyme takes:

- Binding of an inhibitor to a site other than the active site can stabilize the inactive form of the enzyme, making it less likely to convert to the active form.
- The active form can be stabilized by the binding of an activator to another site on the enzyme.

Like substrate binding, the binding of inhibitors and activators to their regulatory sites (also called allosteric sites) is highly specific. Most (but not all) enzymes that are allosterically regulated are proteins with quaternary structure; that is, they are made up of multiple polypeptide subunits. The polypeptide that has the active site is called the catalytic subunit. The allosteric sites are often located on different polypeptides, called the regulatory subunits (see Figure 8.17).

Some enzymes have multiple subunits containing active sites, and the binding of substrate to one of the active sites causes allosteric effects. When substrate binds to one subunit, there is a slight change in protein structure that influences the adjacent subunit. The slight change to the second subunit makes its active site more likely to bind to the substrate. So the reaction speeds up as the sites become sequentially activated.

As a result, an allosteric enzyme with multiple active sites and a nonallosteric enzyme with a single active site differ greatly in their reaction rates when the substrate concentration is low. Graphs of reaction rates plotted against substrate concentrations show this relationship. For a nonallosteric enzyme, the plot is hyperbolic:

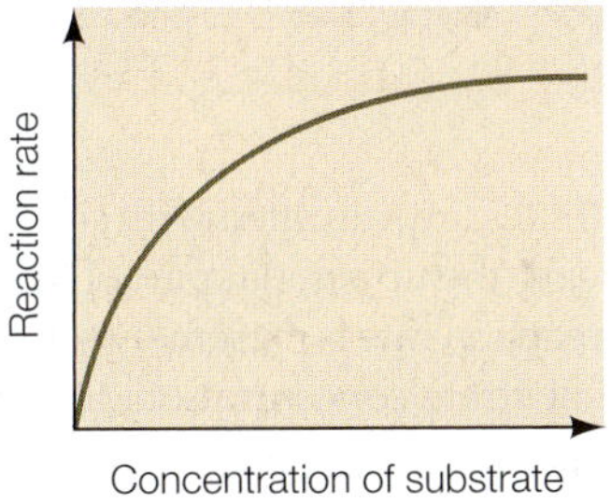

Reaction rate
Concentration of substrate

The reaction rate first increases sharply with increasing substrate concentration, then tapers off to a constant maximum rate as the supply of enzyme becomes saturated.

For a multisubunit allosteric enzyme, the graph looks different, having a sigmoid (S-shaped) appearance:

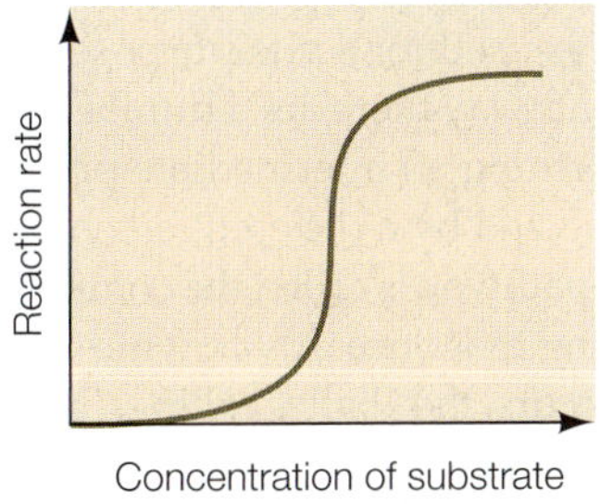

Reaction rate
Concentration of substrate

At low substrate concentrations, the reaction rate increases only gradually as substrate concentration increases. After the substrate binds to the first active site of the enzyme (the slowly increasing part of the curve), there is a change in the quaternary structure such that the other sites become more likely to bind substrate, so the reaction speeds up (the rapidly increasing part of

WORKING WITHDATA:

How Does an Herbicide Work?

Original Paper

Boocock, M. R. and J. R. Coggins. 1983. Kinetics of 5-enolpyruvylshikimate-3-phosphate synthase inhibition by glyphosate. *FEBS Letters* 154: 127–133.

Analyze the Data

Glyphosate (also called Roundup) is used to kill weeds growing in farmer's fields or backyard gardens. Glyphosate kills plants by inhibiting an enzyme (5-enolpyruvylshikimate-3-phosphate synthase; EPSP synthase) in the metabolic pathway used to synthesize the amino acids phenylalanine, tyrosine, and tryptophan. Plants treated with glyphosate can't make these amino acids and die.

To investigate how glyphosate inhibits EPSP synthase, Boocock and Coggins isolated the enzyme from *Neurospora crassa* (a fungus, which was a more convenient source than plants). They measured the rate of the EPSP synthase reaction in the presence of different concentrations of glyphosate and of one of EPSP synthase's substrates, phosphoenolpyruvate (PEP). Their results are shown in the graph (presented in a different format than in the original paper).

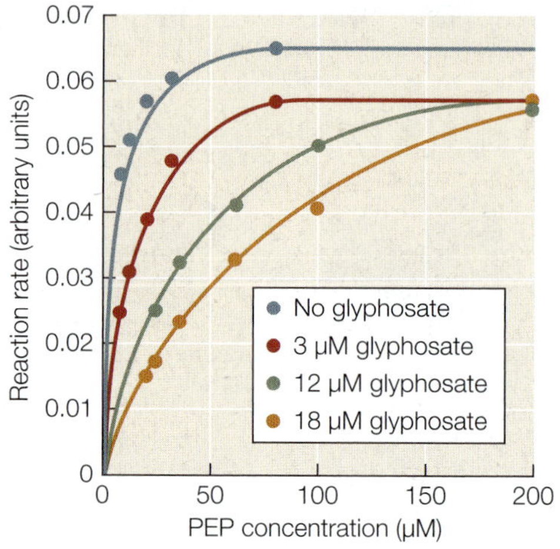
Reaction rate (arbitrary units)
0.07
0.06
0.05
0.04
0.03
0.02
0.01
0
0
50
100
150
200
PEP concentration (μM)
No glyphosate
3 μM glyphosate
12 μM glyphosate
18 μM glyphosate

QUESTION 1

At about what substrate concentration does EPSP synthase become saturated when no glyphosate is present? How much substrate is needed to saturate EPSP synthase in the presence of 18 μM glyphosate? In each case, what is the reaction rate at saturation?

QUESTION 2

Looking at the curve for the reaction rate without inhibitor, is EPSP synthase a multi-subunit allosteric enzyme? Explain your answer.

QUESTION 3

Based on these data, what is the most likely mechanism for glyphosate inhibition of EPSP synthase: competitive, noncompetitive, or uncompetitive? Why?

Go to **BioPortal** for all WORKING WITH**DATA** exercises

the curve). Once all sites are saturated with substrate, the reaction rate reaches a plateau (the upper, flat part of the curve). Within a certain range, the reaction rate is extremely sensitive to relatively small changes in substrate concentration. In addition, allosteric enzymes are very sensitive to low concentrations of inhibitors. Because of this sensitivity, allosteric enzymes are important in regulating entire metabolic pathways.

Allosteric effects regulate many metabolic pathways

Metabolic pathways typically involve a starting material, various intermediate products, and an end product that is used for some purpose by the cell. In each pathway there are a number of reactions, each forming an intermediate product and each catalyzed by a different enzyme. The first step in a pathway is called the commitment step, meaning that once this enzyme-catalyzed reaction occurs, the "ball is rolling," and the other reactions happen in sequence, leading to the end product. But what if the cell has no requirement for that product—for example, if that product is available from its environment in adequate amounts? It would be energetically wasteful for the cell to continue making something it does not need.

One way to avoid this problem is to shut down the metabolic pathway by having the final product inhibit the enzyme that catalyzes the commitment step (**Figure 8.18**). Often this inhibition occurs allosterically. When the end product is present at a high concentration, some of it binds to an allosteric site on the commitment step enzyme, thereby causing it to become inactive. Thus the final product acts as a noncompetitive inhibitor (described earlier in this section)

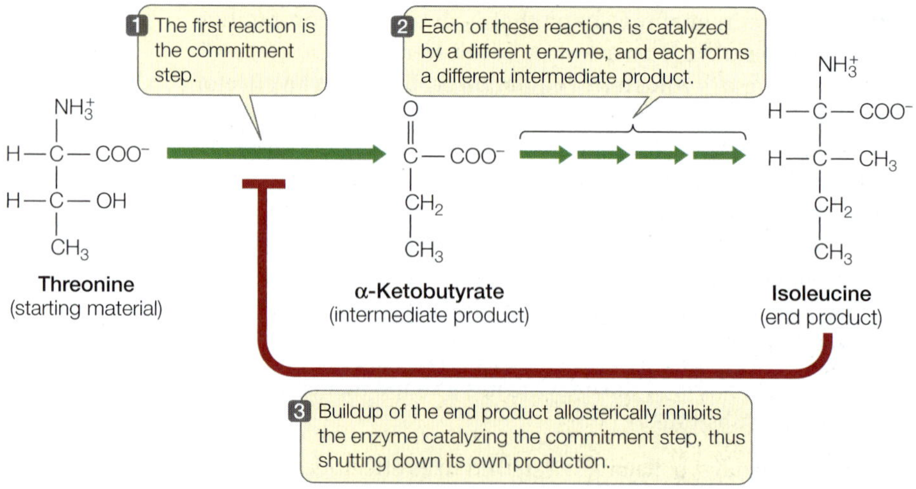
1 The first reaction is the commitment step.
2 Each of these reactions is catalyzed by a different enzyme, and each forms a different intermediate product.
Threonine
(starting material)
α-Ketobutyrate
(intermediate product)
Isoleucine
(end product)
3 Buildup of the end product allosterically inhibits the enzyme catalyzing the commitment step, thus shutting down its own production.

8.18 Feedback Inhibition of Metabolic Pathways The first reaction in a metabolic pathway is referred to as the commitment step. It is often catalyzed by an enzyme that can be allosterically inhibited by the end product of the pathway. The specific pathway shown here is the synthesis of isoleucine from threonine in bacteria. It is typical of many enzyme-catalyzed biosynthetic pathways.

of the first enzyme in the pathway. This mechanism is known as feedback inhibition or end-product inhibition. We will describe many other examples of such inhibition in later chapters.

Many enzymes are regulated through reversible phosphorylation

As we saw in Chapter 7, many enzymes involved in signal transduction are regulated via reversible phosphorylation (see Figure 7.16A). An enzyme can be activated by a protein kinase, which adds a phosphate to one or more specific amino acids. This results in a change in the shape of the enzyme, making it active. Such activation is reversible because another enzyme called a protein phosphatase can remove the phosphate groups, so that the enzyme becomes inactive again. In addition to the enzymes involved in signal transduction, many other enzymes and proteins in the cell (such as ion channels) are regulated via reversible phosphorylation. Reflecting the important role of protein phosphorylation in cell functions, the human genome contains about 500 protein kinase genes: about 2 percent of all the protein-coding genes we have.

Enzymes are affected by their environment

Enzymes enable cells to perform chemical reactions and carry out complex processes rapidly without using the extremes of temperature and pH employed by chemists in the laboratory. However, because of their three-dimensional structures and the chemistry of the side chains in their active sites, enzymes (and their substrates) are highly sensitive to changes in temperature and pH. This was one of the obstacles to the use of enzymes in laundry detergent (see the opening story). In Section 3.2 we described the general effects of these environmental factors on proteins. Here we will examine their effects on enzyme function (which, of course, depends on enzyme structure and chemistry).

pH AFFECTS ENZYME ACTIVITY The rates of most enzyme-catalyzed reactions depend on the pH of the solution in which they occur. While the water inside cells is generally at a neutral pH of 7, the presence of acids, bases, and buffers can alter this. Each enzyme is most active at a particular pH; its activity decreases as the solution is made more acidic or more basic than the ideal (optimal) pH (**Figure 8.19**). As an example, consider the human digestive system (see Section 51.3). The pH inside the human stomach is highly acidic, around pH 1.5. However, many enzymes that hydrolyze macromolecules in the intestines, such as proteases, have pH optima in the neutral range. So when food enters the small intestine, a buffer (bicarbonate) is secreted into the intestine to raise the pH to 6.5. This allows the hydrolytic enzymes to be active and digest the food.

An important factor in the effect of pH on enzyme function is ionization of the carboxyl, amino, and other groups on either the substrate or the enzyme. In neutral or basic solutions, carboxyl groups (—COOH) release H^+ to become negatively charged carboxylate groups (—COO^-). However, in neutral or acidic solutions, amino groups (—NH_2) accept H^+ to become positively charged —NH_3^+ groups (see the discussion of acids and bases in Section 2.4). Thus in a neutral solution, an amino group is electrically attracted to a carboxyl group on another molecule or another part of the same molecule, because both groups are ionized and have opposite charges. If the pH changes, however, the ionization of these groups may change. For example, at a low pH (high H^+ concentration, such as the stomach contents where the enzyme pepsin is active), the excess H^+ may react with —COO^- to form —COOH. If this happens, the group is no longer negatively charged and can no longer interact with positively charged groups in the protein, so the folding of the protein may be altered. If such a change occurs at the active site of an enzyme, the enzyme may no longer be able to bind to its substrate.

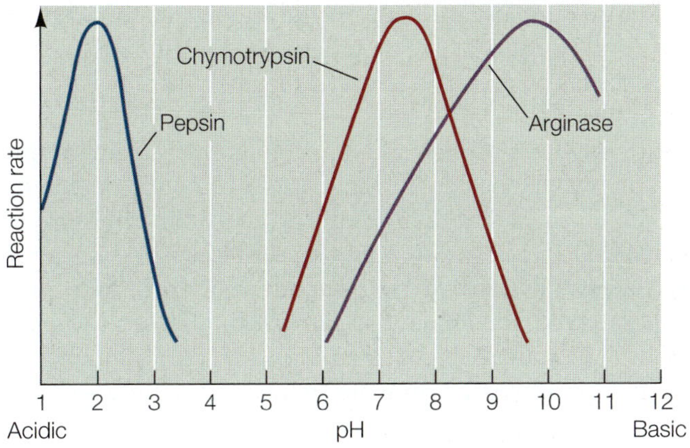
Chymotrypsin
Pepsin
Arginase
Reaction rate
1
2
3
4
5
6
7
8
9
10
11
12
Acidic
pH
Basic

8.19 pH Affects Enzyme Activity An enzyme catalyzes its reaction at a maximum rate. The activity curve for each enzyme peaks at its optimal pH. For example, pepsin is active in the acidic environment of the stomach, whereas chymotrypsin is active in the small intestine.

TEMPERATURE AFFECTS ENZYME ACTIVITY In general, warming increases the rate of a chemical reaction because a greater proportion of the reactant molecules have enough kinetic energy to provide the activation energy for the reaction. Enzyme-catalyzed reactions are no different (**Figure 8.20**). However, temperatures that are too high inactivate enzymes, because at high temperatures enzyme molecules vibrate and twist so rapidly that some of their noncovalent bonds break. When an enzyme's tertiary structure is changed by heat it loses its function. Some enzymes

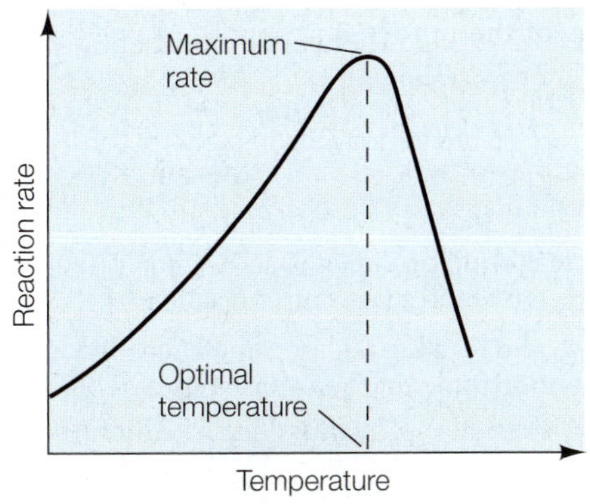
Maximum
rate
Reaction rate
Optimal
temperature
Temperature

8.20 Temperature Affects Enzyme Activity Each enzyme is most active at a particular optimal temperature. At higher temperatures the enzyme becomes denatured and inactive; this explains why the activity curve falls off abruptly at temperatures above the optimal.

denature at temperatures only slightly above that of the human body, but a few are stable even at the boiling point (or freezing point) of water. All enzymes, however, have an optimal temperature for activity.

Individual organisms adapt to changes in the environment in many ways, one of which is based on groups of enzymes, called **isozymes**, that catalyze the same reaction but have different amino acid compositions and physical properties. Different isozymes within a given group may have different optimal temperatures. The rainbow trout, for example, has several isozymes of the enzyme acetylcholinesterase. If a rainbow trout is transferred from warm water to near-freezing water (2°C), the fish produces an isozyme of acetylcholinesterase that is different from the one it produces at the higher temperature. The new isozyme has a lower optimal temperature, allowing the fish's nervous system to perform normally in the colder water.

In general, enzymes adapted to warm temperatures do not denature at those temperatures, because their tertiary structures are held together largely by covalent bonds, such as charge interactions or disulfide bridges, instead of the more heat-sensitive weak chemical interactions. Most enzymes in humans are more stable at high temperatures than those of the bacteria that infect us, so that a moderate fever tends to denature bacterial enzymes, but not our own.

RECAP 8.5

The rates of most enzyme-catalyzed reactions are affected by interacting molecules (such as inhibitors and activators) and by environmental factors (such as temperature and pH). Reversible phosphorylation is another important mechanism for regulating enzyme activity.

- What is the difference between reversible and irreversible enzyme inhibition? **See p. 157**
- How are allosteric enzymes regulated? **See pp. 159–160 and Figure 8.17**
- Explain the concept of feedback inhibition. How might the reactions shown in Figure 8.18 fit into a systems diagram such as the one shown in Figure 8.14? **See p. 160**

?

How are enzymes used in other industrial processes?

ANSWER

The commercial application of purified enzymes is a multibillion-dollar industry. Examples in the food industry include pectinase, an enzyme that hydrolyzes plant cell wall components and is used in clarifying fruit juices; and glucoamylase, which hydrolyzes starch and is used to make the widely used sweetener, high-fructose corn syrup. Many enzymes used in industry come from microbes such as bacteria and fungi. One reason is that microbes are relatively easy to grow in large quantities in a controlled industrial setting. Another reason is that microbes use enzymes to hydrolyze molecules in their environment, so they often either secrete the enzymes or carry them exposed on the cell surface. This makes extracting and purifying these enzymes relatively straightforward.

CHAPTER SUMMARY 8

8.1 What Physical Principles Underlie Biological Energy Transformations?

- Energy is the capacity to do work. In a biological system, the usable energy is called **free energy** (***G***). The unusable energy is **entropy** (***S***), a measure of the disorder in the system.
- **Potential energy** is the energy of state or position; it includes the energy stored in chemical bonds. **Kinetic energy** is the energy of motion; it is the type of energy that can do work.
- The **laws of thermodynamics** apply to living organisms. The first law states that energy cannot be created or destroyed. The second law states that energy transformations decrease the amount of energy available to do work (free energy) and increase disorder. **Review Figure 8.2**
- The change in free energy (ΔG) of a reaction determines its point of **chemical equilibrium**, at which the forward and reverse reactions proceed at the same rate.
- An **exergonic** reaction releases free energy and has a negative ΔG. An **endergonic** reaction consumes or requires free energy and has a positive ΔG. Endergonic reactions proceed only if free energy is provided. **Review Figure 8.3**
- **Metabolism** is the sum of all the biochemical (metabolic) reactions in an organism. **Catabolic reactions** are associated with the breakdown of complex molecules and release energy (are exergonic). **Anabolic reactions** build complexity in the cell and are endergonic.

8.2 What Is the Role of ATP in Biochemical Energetics?

- Adenosine triphosphate (ATP) serves as an energy currency in cells. Hydrolysis of ATP releases a relatively large amount of free energy.
- The ATP cycle couples exergonic and endergonic reactions, harvesting free energy from exergonic reactions, and providing free energy for endergonic reactions. **Review Figure 8.6, ACTIVITY 8.1**

8.3 What Are Enzymes?

- The rate of a chemical reaction is independent of ΔG but is determined by the energy barrier. **Review Figure 8.8**
- Enzymes are protein catalysts that affect the rates of biological reactions by lowering the energy barrier, supplying the **activation energy** (E_a) needed to initiate reactions. **Review Figure 8.10, ACTIVITY 8.2**
- A **substrate** binds to the enzyme's active site—the site of catalysis—forming an **enzyme–substrate** (**ES**) **complex**. Enzymes are highly specific for their substrates. **Review Figure 8.9**

continued

8.4 How Do Enzymes Work?

- At the active site, a substrate can be oriented correctly, chemically modified, or strained. As a result, the substrate readily forms its **transition state**, and the reaction proceeds. **Review Figure 8.11**
- Binding substrate causes many enzymes to change shape, exposing their active site(s) and allowing catalysis. The change in enzyme shape caused by substrate binding is known as **induced fit**. **Review Figure 8.12**
- Some enzymes require other substances, known as cofactors, to carry out catalysis. Prosthetic groups are permanently bound to enzymes; coenzymes are not. A coenzyme can be considered a substrate, as it is changed by the reaction and then released from the enzyme.
- Substrate concentration affects the rate of an enzyme-catalyzed reaction.

8.5 How Are Enzyme Activities Regulated?

- Metabolism is organized into pathways in which the product of one reaction is a reactant for the next reaction. Each reaction in the pathway is catalyzed by a different enzyme.
- Enzyme activity is subject to regulation. Some inhibitors bind irreversibly to enzymes. Others bind reversibly. **Review Figures 8.15, 8.16, ANIMATED TUTORIAL 8.1**
- An allosteric effector binds to a site other than the active site and stabilizes the active or inactive form of an enzyme. **Review Figure 8.17, ANIMATED TUTORIAL 8.2**
- The end product of a metabolic pathway may inhibit an enzyme that catalyzes the commitment step of that pathway. **Review Figure 8.18**
- Reversible phosphorylation is another important mechanism for regulating enzyme activity.
- Enzymes are sensitive to their environments. Both pH and temperature affect enzyme activity. **Review Figures 8.19, 8.20**

Go to the Interactive Summary to review key figures, Animated Tutorials, and Activities Life10e.com/is8

CHAPTER**REVIEW**

REMEMBERING

1. Coenzymes differ from enzymes in that coenzymes are
   a. only active outside the cell.
   b. polymers of amino acids.
   c. smaller molecules, such as vitamins.
   d. specific for one reaction.
   e. always carriers of high-energy phosphate.

2. Which statement about thermodynamics is true?
   a. Free energy is used up in an exergonic reaction.
   b. Free energy cannot be used to do work.
   c. The total amount of energy can change after a chemical transformation.
   d. Free energy can be kinetic but not potential energy.
   e. Entropy has a tendency to increase.

3. The active site of an enzyme
   a. never changes shape.
   b. forms no chemical bonds with substrates.
   c. determines, by its structure, the specificity of the enzyme.
   d. looks like a lump projecting from the surface of the enzyme.
   e. changes the ΔG of the reaction.

4. The molecule ATP is
   a. a component of most proteins.
   b. high in energy because of the presence of adenine.
   c. required for many energy-transforming biochemical reactions.
   d. a catalyst.
   e. used in some exergonic reactions to provide energy.

5. In an enzyme-catalyzed reaction,
   a. a substrate does not change.
   b. the rate decreases as substrate concentration increases.
   c. the enzyme can be permanently changed.
   d. strain may be added to a substrate.
   e. the rate is not affected by substrate concentration.

6. Which statement about enzyme inhibitors is *not* true?
   a. A competitive inhibitor binds to the active site of the enzyme.
   b. An allosteric inhibitor binds to a site on the active form of the enzyme.
   c. A noncompetitive inhibitor binds to 2-3,10-1118-19,20-21
   d. a site other than the active site.
   e. Noncompetitive inhibition cannot be completely overcome by the addition of more substrate.
   f. Competitive inhibition can be completely overcome by the addition of more substrate.

UNDERSTANDING & APPLYING

7. What makes it possible for endergonic reactions to proceed in organisms?

8. Consider two proteins: one is an enzyme dissolved in the cytosol of a cell, the other is an ion channel in its plasma membrane. Contrast the structures of the two proteins, indicating at least two important differences.

9. Plot free energy versus the time course of an endergonic reaction, and the same for an exergonic reaction. Include the activation energy on both plots. Label E_a and ΔG on both graphs.

10. When potatoes are peeled, the enzyme polyphenol oxidase causes discoloration by catalyzing the oxidation of certain molecules, using O_2 as a substrate. Explain the following observations:
    a. If potatoes are peeled under water and kept there, browning is reduced.
    b. Potatoes that have been boiled at 100°C and then sliced do not turn brown.
    c. If lemon juice (pH 3) is applied to newly peeled potatoes, they do not brown.

ANALYZING & EVALUATING

11. Consider an enzyme that is subject to allosteric regulation. If a competitive inhibitor (not an allosteric inhibitor) is added to a solution containing such an enzyme, the ratio of enzyme molecules in the active form to those in the inactive form increases. Explain this observation.

12. In humans, hydrogen peroxide (H_2O_2) is a dangerous toxin produced as a by-product of several metabolic pathways. The accumulation of H_2O_2 is prevented by its conversion to harmless H_2O, a reaction catalyzed by the appropriately named enzyme catalase. Air pollutants can inhibit this enzyme and leave individuals susceptible to tissue damage by H_2O_2. How would you investigate whether catalase has an allosteric or a nonallosteric mechanism, and whether the pollutants are acting as competitive or noncompetitive inhibitors?

$$2\,H_2O_2 \xrightarrow{\text{Catalase}} O_2 + 2\,H_2O$$

Go to BioPortal at **yourBioPortal.com** for Animated Tutorials, Activities, LearningCurve Quizzes, Flashcards, and many other study and review resources.

9 Pathways That Harvest Chemical Energy

CHAPTER OUTLINE

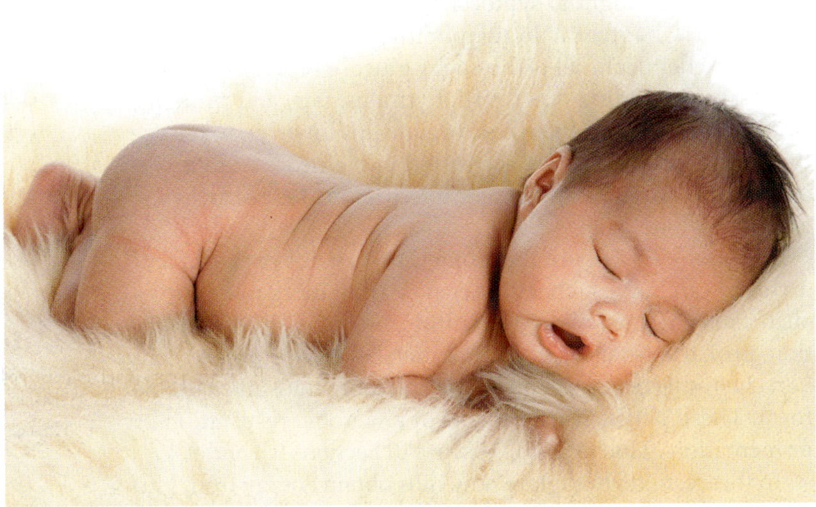

Built-in Heater A newborn baby has abundant brown fat stores, which can be oxidized to release heat instead of chemical energy.

THE "OBESITY EPIDEMIC" is probably the biggest recent story about the health of people in the developed world. There have been so many declarations by physicians, public health experts, and politicians that it might be easy to think it is all exaggerated. However, with 15 percent of children and 30 percent of adults in the United States described as obese, we face dramatic increases in diseases associated with obesity such as diabetes, heart disease, and cancer. The human and financial toll of this will be huge. What can be done?

Most people agree that in the majority of cases, obesity can be prevented or reduced if we eat less and exercise more. It is all a matter of energy: if we eat more energy-yielding molecules than we need to build up our bodies and to fuel activities such as brain functions and physical activity, we will store the unneeded energy as fat. Evolutionarily, this is usually advantageous, because fat stores energy in its C—C and C—H bonds for later use when food is scarce. But excess fat, as noted above, has adverse consequences.

Not all fat tissues, or adipose tissues, are the same. White adipose tissue (sometimes referred to simply as "white fat") is used primarily to store energy. But some adipose tissue is brown because of its high concentration of mitochondria, which have iron-containing pigments. When energy-rich molecules in brown fat are catabolized, the stored energy is released not as chemical energy but as heat. The cells in brown fat make a protein called UCP1 (uncoupling protein 1) that inserts into the inner membranes of mitochondria, making them permeable to protons (H^+). As you will see, the general impermeability of these membranes to H^+ is key to coupling the catabolism of molecules such as fats to the release of their stored energy in chemical form (to make ATP). If the membranes become permeable to H^+, this coupling is lost and the stored energy is released as heat.

Human infants are born with a lot of brown fat in their back and shoulder regions—it comprises about 5 percent of their body weight. Because infants have a high surface area-to-volume ratio, they tend to lose a lot of heat. One way that they keep warm is to produce heat in their brown fat tissues. As a child grows up, the brown fat content of the body is reduced. Adults have mostly white fat, which has less UCP1 and generates less heat when the fat is catabolized.

It used to be thought that brown fat was not present in adult humans. But recently it was confirmed that adults do have some brown fat, in the upper chest and neck area. Interestingly, obese people have less brown fat than lean people, leading to the suggestion that brown fat may be somehow associated with a tendency to remain lean. A tantalizing possibility is that recent discoveries about brown fat and UCP1 may lead to new methods for weight loss.

> ?
>
> Can brown fat in adults be a target for weight loss?
>
> See answer on p. 182.

9.1 How Does Glucose Oxidation Release Chemical Energy?

Energy is stored in the covalent bonds of fuels, and it can be released and transformed. Wood burning in a campfire releases energy as heat and light. In cells, fuel molecules release chemical energy that is used to make ATP, which in turn drives endergonic reactions. ATP is central to the energy transformations of all living organisms. Photosynthetic cells and organisms use energy from sunlight to synthesize their own fuels, as we will describe in Chapter 10. In nonphotosynthetic cells, the most common chemical fuel is the sugar glucose ($C_6H_{12}O_6$). Other molecules, including other carbohydrates, fats, and proteins, can supply energy to the whole organism. However, to release their energy they must be converted into glucose or intermediate compounds that can enter into the various pathways of glucose metabolism.

We should note here that some prokaryotes (bacteria and archaea) can harvest chemical energy from inorganic sources such as metal ions, hydrogen sulfide, and ammonia. In addition, prokaryotes use a variety of metabolic pathways to convert chemical energy (both organic and inorganic) into a usable form. These pathways are forms of anaerobic respiration or fermentation, which we will discuss in Sections 9.3 and 9.4.

In this section we explore how cells obtain energy from glucose by the chemical process of oxidation, which is carried out through a series of metabolic pathways. Five principles govern metabolic pathways:

- A complex chemical transformation occurs in a series of separate reactions that form a metabolic pathway.
- Each reaction is catalyzed by a specific enzyme.
- Many metabolic pathways are similar in all organisms, from bacteria to humans.
- In eukaryotes, many metabolic pathways are compartmentalized, with certain reactions occurring inside specific organelles.
- Some key enzymes in each metabolic pathway can be inhibited or activated to alter the rate of the pathway.

Cells trap free energy while metabolizing glucose

As we saw in Section 2.3, the familiar process of combustion (burning) is very similar to the chemical processes that release energy in cells. If glucose is burned in a flame, it reacts with oxygen gas (O_2), forming carbon dioxide and water and releasing energy in the form of heat. The balanced equation for the complete reaction is

$$C_6H_{12}O_6 + 6\,O_2 \rightarrow 6\,CO_2 + 6\,H_2O + \text{free energy} \quad (\Delta G = -686 \text{ kcal/mol})$$

This is an oxidation–reduction reaction (see next page), in which glucose loses electrons (becomes oxidized) and oxygen gains them (becomes reduced). The energy that is released can be used to do work. The same equation applies to the overall catabolism of glucose in cells. However, in contrast to combustion, the catabolism of glucose is a multistep pathway. Each step is catalyzed by an enzyme, and the process is compartmentalized.

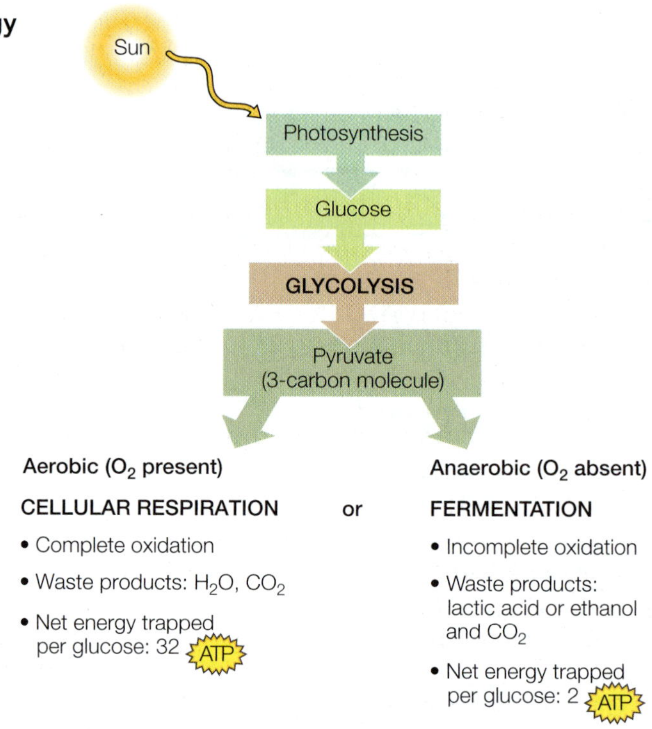
Sun
Photosynthesis
Glucose
GLYCOLYSIS
Pyruvate
(3-carbon molecule)
Aerobic (O_2 present)
CELLULAR RESPIRATION
• Complete oxidation
• Waste products: H_2O, CO_2
• Net energy trapped per glucose: 32 ATP
or
Anaerobic (O_2 absent)
FERMENTATION
• Incomplete oxidation
• Waste products: lactic acid or ethanol and CO_2
• Net energy trapped per glucose: 2 ATP

9.1 Energy for Life Many prokaryotes and all eukaryotes obtain their energy from the food compounds produced by photosynthesis. They convert these compounds into glucose, which they metabolize to trap energy in ATP.

Unlike combustion, glucose catabolism is tightly regulated and occurs at temperatures compatible with life.

The glucose catabolism pathway "extracts" the energy stored in the covalent bonds of glucose and stores it instead in ATP molecules via the phosphorylation reaction:

$$ADP + P_i + \text{free energy} \rightarrow ATP$$

ATP is the energy currency of cells (see Chapter 8). The energy trapped in ATP can be used to do cellular work—such as movement of muscles or active transport across membranes—just as the energy captured from combustion can be used to do mechanical work.

The standard free energy change resulting from the complete conversion of glucose and O_2 to CO_2 and water, whether by combustion or by metabolism, is –686 kcal/mol (–2,870 kJ/mol). Thus the overall reaction is highly exergonic and can drive the endergonic formation of a great deal of ATP from ADP and phosphate. Note that in the discussion that follows, "energy" means free energy.

Three catabolic processes harvest the energy in the chemical bonds of glucose: glycolysis, cellular respiration, and fermentation (Figure 9.1). All three processes involve pathways made up of many distinct chemical reactions.

- **Glycolysis** begins glucose catabolism. Through a series of chemical rearrangements, glucose is converted to two molecules of the three-carbon product **pyruvate**, and a small amount of energy is captured in usable forms. Glycolysis is an **anaerobic** process because it does not require O_2.
- **Cellular respiration** uses O_2 from the environment, and thus it is **aerobic**. Each pyruvate molecule is completely converted into three molecules of CO_2 through a set of catabolic pathways including pyruvate oxidation, the citric acid cycle,

and an electron transport system (the respiratory chain). In the process, a great deal of the energy stored in the covalent bonds of pyruvate is captured to form ATP.

- **Fermentation** does not involve O_2 (it is anaerobic). Fermentation converts pyruvate into lactic acid or ethyl alcohol (ethanol), which are still relatively energy-rich molecules. Because the breakdown of glucose is incomplete, much less energy is released when glycolysis is coupled to fermentation than when it is coupled to cellular respiration.

Redox reactions transfer electrons and energy

As we discussed in Section 8.2, the addition of a phosphate group to ADP to make ATP is an endergonic reaction (see Figure 8.6). It is achieved by coupling an exergonic reaction to ATP production: the energy released in the exergonic reaction is used to drive ATP synthesis. Electrons are transferred in the exergonic reaction. A reaction in which one substance transfers one or more electrons to another substance is called an **oxidation–reduction**, or **redox**, **reaction**.

- **Reduction** is the gain of one or more electrons by an atom, ion, or molecule.
- **Oxidation** is the loss of one or more electrons.

Oxidation and reduction *always occur together*: as one chemical is oxidized, the electrons it loses are transferred to another chemical, reducing it. In a redox reaction, we call the reactant that becomes reduced an oxidizing agent and the one that becomes oxidized a reducing agent:

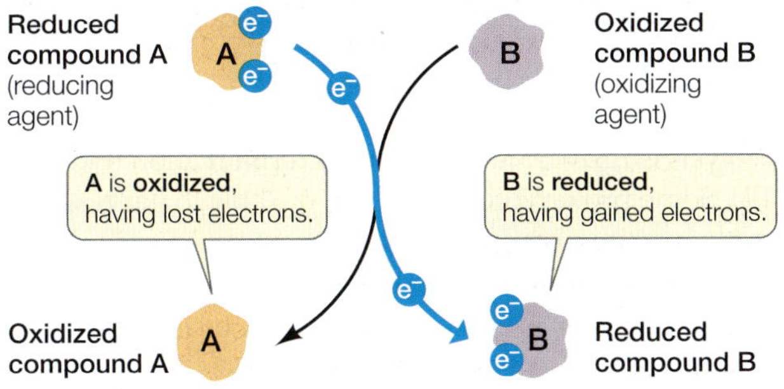

Reduced compound A (reducing agent)
Oxidized compound B (oxidizing agent)
A is oxidized, having lost electrons.
B is reduced, having gained electrons.
Oxidized compound A
Reduced compound B

In the metabolism of glucose, glucose is the reducing agent (electron donor) and O_2 is the oxidizing agent (electron acceptor).

Although oxidation and reduction are always defined in terms of electron traffic, it is often simpler to think in terms of the gain or loss of hydrogen atoms. The transfer of electrons is often associated with the transfer of hydrogen ions ($H = H^+ + e^-$). So when a molecule loses hydrogen atoms, it becomes oxidized.

In general, the more reduced a molecule is, the more energy is stored in its covalent bonds (**Figure 9.2**). In a redox reaction, some energy is transferred from the reducing agent to the reduced product. The rest remains in the reducing agent or is lost to entropy. As we will see, some of the key reactions of glycolysis and cellular respiration are highly exergonic redox reactions.

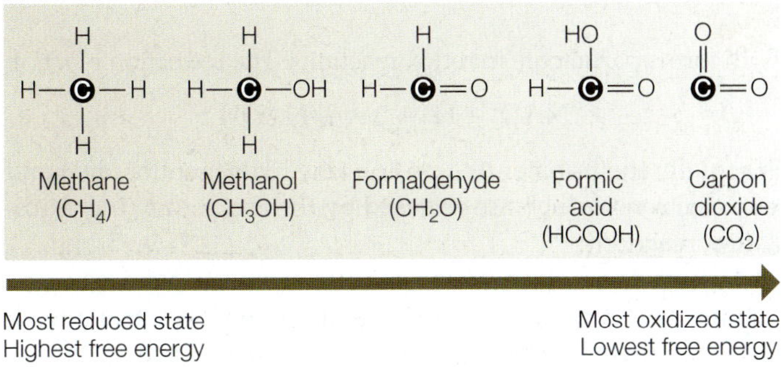

Methane (CH_4)
Methanol (CH_3OH)
Formaldehyde (CH_2O)
Formic acid (HCOOH)
Carbon dioxide (CO_2)
Most reduced state
Highest free energy
Most oxidized state
Lowest free energy

9.2 Oxidation, Reduction, and Energy The more oxidized a carbon atom in a molecule is, the less its stored free energy.

The coenzyme NAD^+ is a key electron carrier in redox reactions

Section 8.4 describes the role of coenzymes, small molecules that assist in enzyme-catalyzed reactions. ADP acts as a coenzyme when it picks up energy released in an exergonic reaction and packages it to form ATP. The coenzyme nicotinamide adenine dinucleotide (NAD^+) acts as an electron carrier in redox reactions:

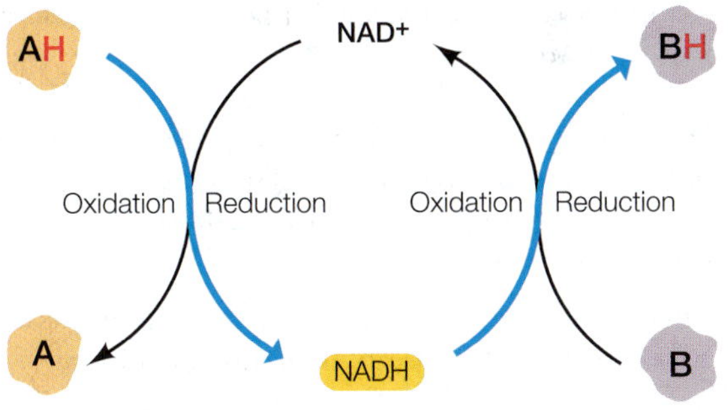

AH
NAD+
BH
Oxidation
Reduction
Oxidation
Reduction
A
NADH
B

As you can see, NAD^+ exists in two chemically distinct forms, one oxidized (NAD^+) and the other reduced (NADH) (**Figure 9.3**).

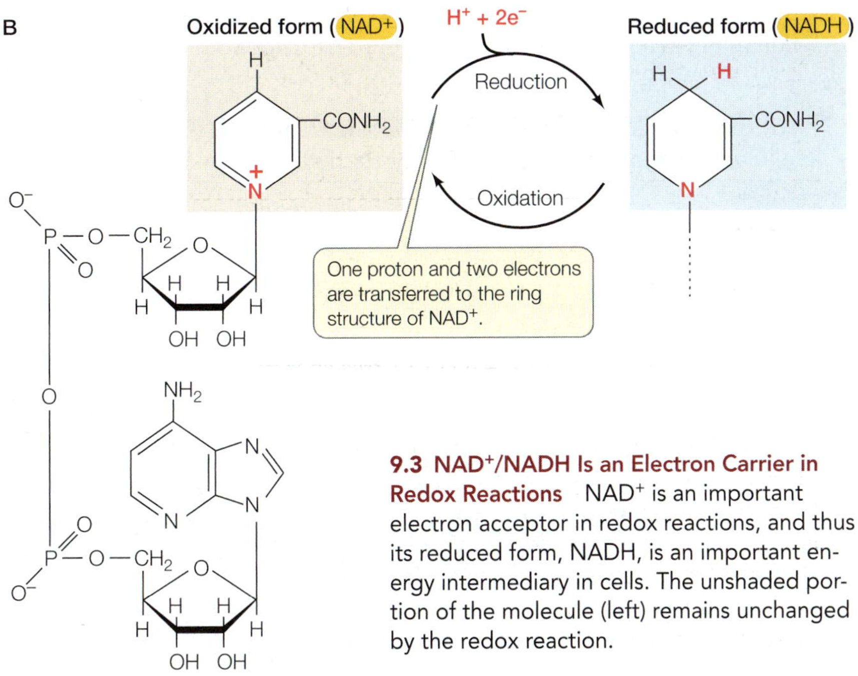

Oxidized form (NAD^+)
$H^+ + 2e^-$
Reduced form (NADH)
Reduction
Oxidation
$CONH_2$
One proton and two electrons are transferred to the ring structure of NAD^+.

9.3 NAD^+/NADH Is an Electron Carrier in Redox Reactions NAD^+ is an important electron acceptor in redox reactions, and thus its reduced form, NADH, is an important energy intermediary in cells. The unshaded portion of the molecule (left) remains unchanged by the redox reaction.

Both forms participate in redox reactions. The reduction reaction

$$NAD^+ + H^+ + 2\,e^- \rightarrow NADH$$

is actually the transfer of a proton (the hydrogen ion, H^+) and two electrons, which are released by the accompanying oxidization reaction.

The electrons do not remain with the coenzyme. Oxygen is highly electronegative and readily accepts electrons from NADH. The oxidation of NADH by O_2 (which occurs in several steps)

$$NADH + H^+ + \tfrac{1}{2}\,O_2 \rightarrow NAD^+ + H_2O$$

is highly exergonic, with a standard free energy change at pH 7 (ΔG) of –52.4 kcal/mol (–219 kJ/mol). Note that the oxidizing agent appears here as "½ O_2" instead of "O." This notation emphasizes that it is molecular oxygen, O_2, that acts as the oxidizing agent.

Just as ATP can be thought of as a package of 7.3 kcal/mol (30.5 kJ/mol) of free energy, NADH can be thought of as a larger package of free energy (52.4 kcal/mol, see above). NAD^+ is a common electron carrier in cells, but not the only one. Another carrier, flavin adenine dinucleotide (FAD), also transfers electrons during glucose metabolism.

An overview: Harvesting energy from glucose

Both eukaryotic and prokaryotic cells can harvest energy from glucose using different combinations of the following metabolic pathways:

- Under aerobic conditions, when O_2 is available as the final electron acceptor, four pathways operate (**Figure 9.4A**). Glycolysis is followed by the three pathways of cellular respiration: pyruvate oxidation, the citric acid cycle (also called the Krebs cycle or the tricarboxylic acid cycle), and electron transport/ATP synthesis (also called the respiratory chain).
- In eukaryotes and many prokaryotes, pyruvate oxidation, the citric acid cycle, and the respiratory chain do not function under anaerobic conditions. The pyruvate produced by glycolysis is further metabolized by fermentation (**Figure 9.4B**). Some prokaryotes, however, are able to harvest energy in pathways involving respiratory chains even in the absence of oxygen (anaerobic respiration; see Section 9.3).

The five pathways shown in Figure 9.4 occur in different locations in the cell (**Table 9.1**).

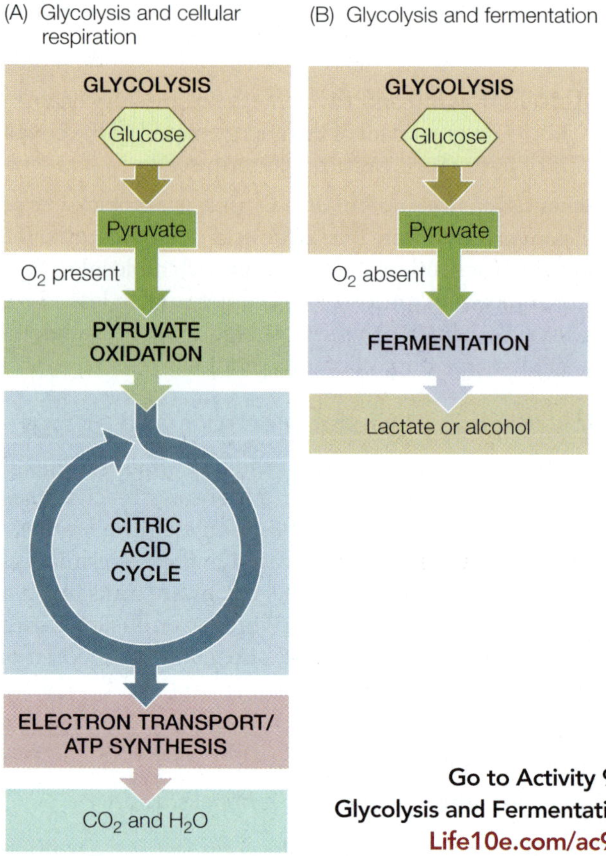
(A) Glycolysis and cellular respiration
(B) Glycolysis and fermentation
GLYCOLYSIS
Glucose
Pyruvate
O_2 present
PYRUVATE OXIDATION
CITRIC ACID CYCLE
ELECTRON TRANSPORT/ ATP SYNTHESIS
CO_2 and H_2O
GLYCOLYSIS
Glucose
Pyruvate
O_2 absent
FERMENTATION
Lactate or alcohol

Go to Activity 9.1 Glycolysis and Fermentation Life10e.com/ac9.1

9.4 Energy-Yielding Metabolic Pathways Energy-yielding reactions can be grouped into five metabolic pathways: glycolysis, pyruvate oxidation, the citric acid cycle, the respiratory chain/ATP synthesis, and fermentation. (A) The three lower pathways occur only in the presence of O_2 and are collectively referred to as cellular respiration. (B) When O_2 is unavailable, glycolysis is followed by fermentation.

TABLE **9.1** Cellular Locations for Major Energy Pathways in Eukaryotes and Prokaryotes

Eukaryotes	Prokaryotes
In cytoplasm	In cytoplasm
Glycolysis	Glycolysis
Fermentation	Fermentation
	Citric acid cycle
Inside mitochondrion	On plasma membrane
Matrix	Pyruvate oxidation
Citric acid cycle	Respiratory chain
Pyruvate oxidation	
Inner membrane	
Respiratory chain	

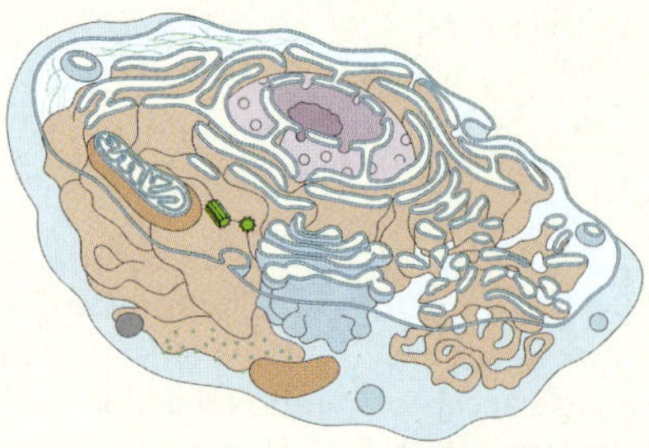

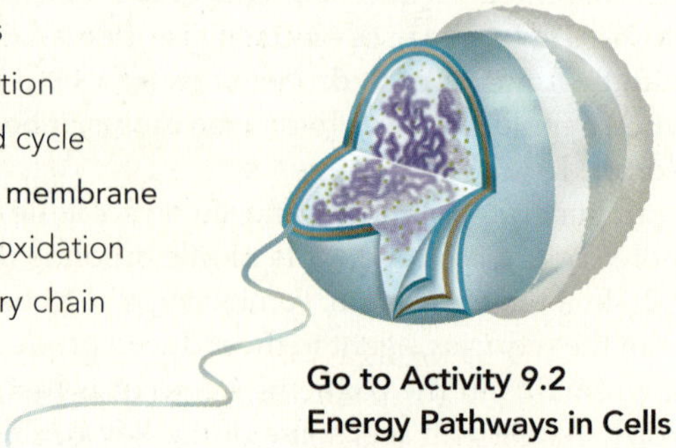

Go to Activity 9.2 Energy Pathways in Cells Life10e.com/ac9.2

RECAP 9.1

The free energy released from the oxidation of glucose is trapped in the form of ATP. In many prokaryotes and all eukaryotes, five major pathways combine in different ways to produce ATP, which supplies the energy for myriad other reactions in living cells.

- What principles govern metabolic pathways in cells? See p. 166
- Describe how the coupling of oxidation and reduction transfers energy from one molecule to another. See p. 167
- Explain the roles of NAD^+ and O_2 with respect to electrons in a redox reaction. See pp. 167–168 and Figures 9.2 and 9.3

Now that you have an overview of the metabolic pathways that harvest energy from glucose, let's take a closer look at the three pathways involved in aerobic glucose catabolism: glycolysis, pyruvate oxidation, and the citric acid cycle.

9.2 What Are the Aerobic Pathways of Glucose Catabolism?

The aerobic pathways of glucose catabolism oxidize glucose completely to CO_2 and H_2O. Initially, the glycolysis reactions convert the six-carbon glucose molecule to two three-carbon pyruvate molecules (**Figure 9.5**). Pyruvate is then converted to CO_2 in a second series of reactions beginning with pyruvate oxidation and followed by the citric acid cycle. In addition to generating CO_2, the oxidation events are coupled with the reduction of electron carriers, mostly NAD^+. Thus much of the chemical energy in the C—C and C—H bonds of glucose is transferred to NAD^+ to form NADH. Ultimately this energy will be transferred to ATP, but this comes in a separate series of reactions involving electron transport, called the respiratory chain. In the respiratory chain, redox reactions result in the oxidative phosphorylation of ADP by ATP synthase. We will begin our consideration of the catabolism of glucose with a closer look at glycolysis.

In glycolysis, glucose is partially oxidized and some energy is released

Glycolysis takes place in the cytosol and involves ten enzyme-catalyzed reactions. During glycolysis, some of the covalent bonds between carbon and hydrogen atoms in the glucose molecule are oxidized, releasing some of the stored energy. The products are two molecules of pyruvate (pyruvic acid), two molecules of ATP, and two molecules of NADH. Glycolysis can be divided into two stages: the initial energy-investing reactions that consume ATP, and the energy-harvesting reactions that produce ATP and NADH (see Figure 9.5).

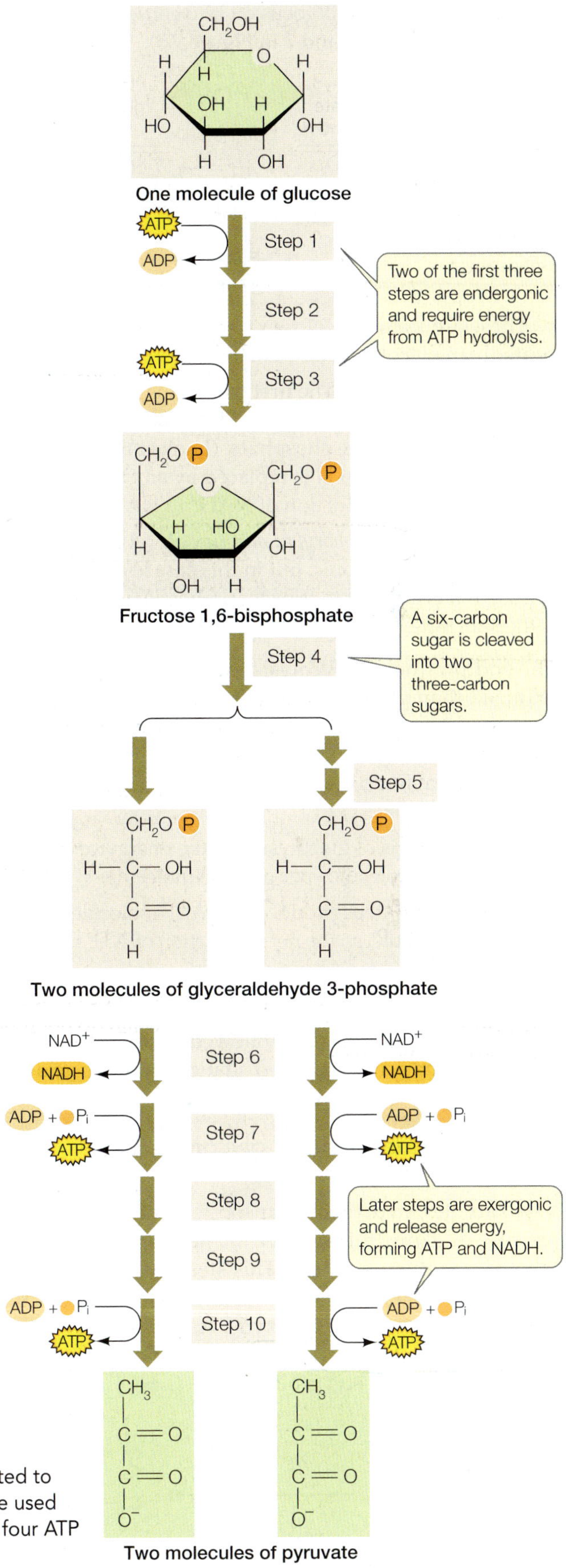

One molecule of glucose
ATP
ADP
Step 1
Step 2
Two of the first three steps are endergonic and require energy from ATP hydrolysis.
ATP
ADP
Step 3
Fructose 1,6-bisphosphate
Step 4
A six-carbon sugar is cleaved into two three-carbon sugars.
Step 5
Two molecules of glyceraldehyde 3-phosphate
NAD+
NADH
Step 6
ADP + Pi
ATP
Step 7
Step 8
Later steps are exergonic and release energy, forming ATP and NADH.
Step 9
ADP + Pi
ATP
Step 10
Two molecules of pyruvate

9.5 Glycolysis Converts Glucose into Pyruvate Glucose is converted to pyruvate in ten enzyme-catalyzed steps. Along the way, two ATP are used (Steps 1 and 3), two NAD^+ are reduced to two NADH (Step 6), and four ATP are produced (Steps 7 and 10).

To help you understand the process without getting into extensive detail, we will focus on two consecutive reactions in this pathway (Steps 6 and 7 in Figure 9.5).

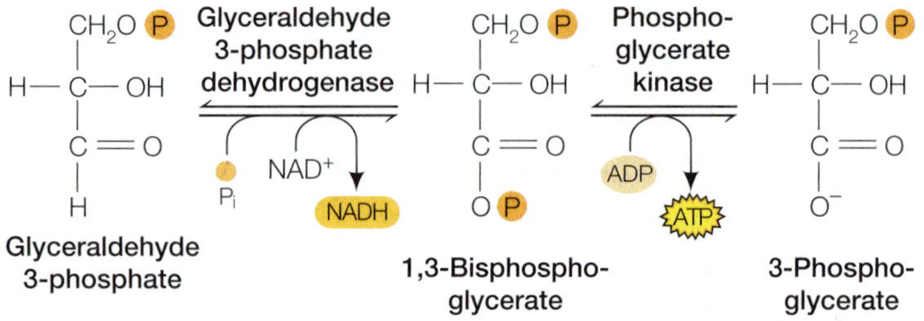
Glyceraldehyde 3-phosphate dehydrogenase
Phospho-glycerate kinase
NAD+
NADH
Pi
ADP
ATP
Glyceraldehyde 3-phosphate
1,3-Bisphospho-glycerate
3-Phospho-glycerate

These are examples of two types of reaction that occur repeatedly in glycolysis and in many other metabolic pathways:

1. *Oxidation–reduction*: The first reaction is exergonic—more than 50 kcal/mol of energy are released in the oxidation of glyceraldehyde 3-phosphate. (Look at the bottom carbon atom, where an H is replaced by an O.) The energy is trapped via the reduction of NAD^+ to NADH.
2. *Substrate-level phosphorylation*: The second reaction in this series is also exergonic, but in this case less energy is released. It is enough to transfer a phosphate directly from the substrate to ADP, forming ATP.

The end product of glycolysis, pyruvate, is somewhat more oxidized than glucose. In the presence of O_2, further oxidation can occur. In prokaryotes these subsequent reactions take place in the cytosol, but in eukaryotes they take place in the mitochondrial matrix.

To summarize:

- The initial steps of glycolysis use the energy of hydrolysis of two ATP molecules per glucose molecule.
- The remaining steps produce four ATP molecules per glucose molecule, so the net production of ATP is two molecules.
- Glycolysis produces two molecules of NADH.

If O_2 is present, glycolysis is followed by the three stages of cellular respiration: pyruvate oxidation, the citric acid cycle, and the respiratory chain/ATP synthesis.

Pyruvate oxidation links glycolysis and the citric acid cycle

In eukaryotes, pyruvate is transported into the mitochondrial matrix (see Figure 5.11), where the next step in the aerobic catabolism of glucose occurs. This step involves the oxidation of pyruvate to a two-carbon acetate molecule and CO_2. The acetate is then bound to coenzyme A to form **acetyl coenzyme A** (**acetyl CoA**); CoA is used in various biochemical reactions as a carrier of the acetyl group ($H_3C—C{=}O$).

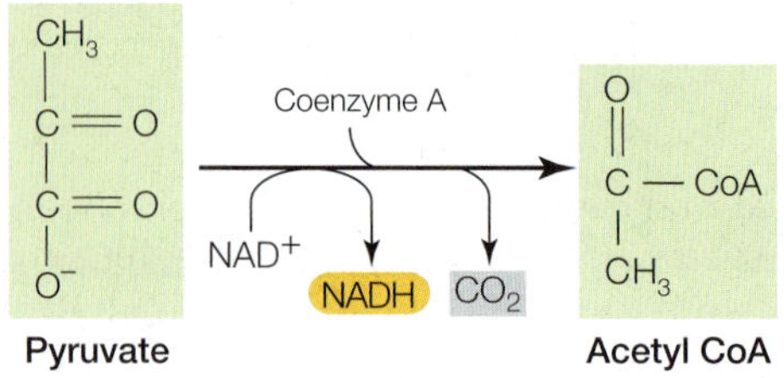
Coenzyme A
NAD+
NADH
CO2
Pyruvate
Acetyl CoA

This is the link between glycolysis and further oxidative reactions (see Figure 9.4).

The formation of acetyl CoA is a multistep reaction catalyzed by the pyruvate dehydrogenase complex, which contains 60 individual proteins and 5 different coenzymes. The overall reaction is exergonic, and one molecule of NAD^+ is reduced to NADH. But the main role of acetyl CoA is to donate its acetyl group to the four-carbon compound oxaloacetate, forming the six-carbon molecule citrate. This initiates the citric acid cycle, one of life's most important energy-harvesting pathways.

The citric acid cycle completes the oxidation of glucose to CO_2

Acetyl CoA is the starting point for the citric acid cycle. This pathway of eight reactions completely oxidizes the two-carbon acetyl group to two molecules of CO_2. The free energy released from these reactions is captured by GDP and the electron carriers NAD^+ and FAD (**Figure 9.6**). [Remember from Section 7.2 that GDP (guanosine diphosphate) is a nucleoside diphosphate

Go to Activity 9.3 The Citric Acid Cycle Life10e.com/ac9.3

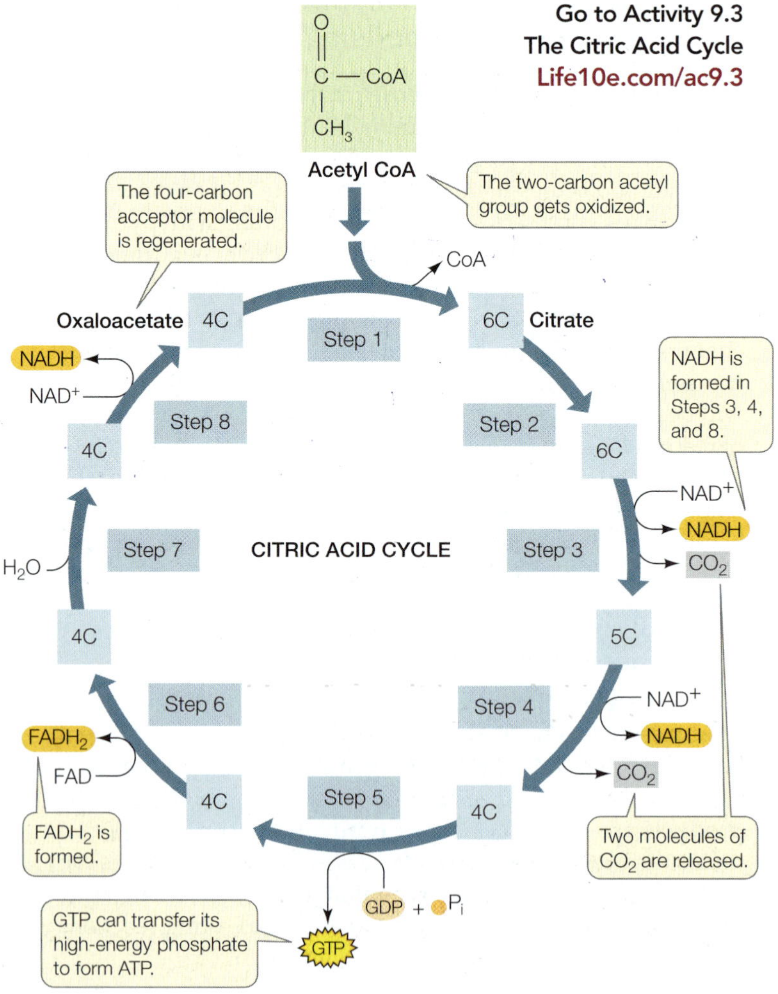
Acetyl CoA
The four-carbon acceptor molecule is regenerated.
The two-carbon acetyl group gets oxidized.
CoA
Oxaloacetate 4C
6C Citrate
Step 1
NADH
NAD+
Step 8
Step 2
4C
6C
NADH is formed in Steps 3, 4, and 8.
NAD+
NADH
CO2
H2O
Step 7
CITRIC ACID CYCLE
Step 3
4C
5C
Step 6
Step 4
NAD+
NADH
FADH2
FAD
CO2
4C
Step 5
4C
FADH2 is formed.
Two molecules of CO2 are released.
GDP + Pi
GTP can transfer its high-energy phosphate to form ATP.
GTP

9.6 The Citric Acid Cycle The citric acid cycle involves eight steps; in the last step, the starting material acceptor, oxaloacetate, is regenerated. Energy is released and captured by reducing NAD^+ or FAD, or by producing GTP. "6C," "5C," and so on indicate the number of carbon atoms in each intermediate in the cycle.

like ADP.] This is a cycle because the starting material, oxaloacetate, is regenerated in the last step and is ready to accept another acetate group from acetyl CoA. The citric acid cycle operates twice for each glucose molecule that enters glycolysis (once for each pyruvate that enters the mitochondrion).

Let's focus on the final reaction of the cycle (Step 8 in Figure 9.6), as an example of the kind of reaction that occurs:

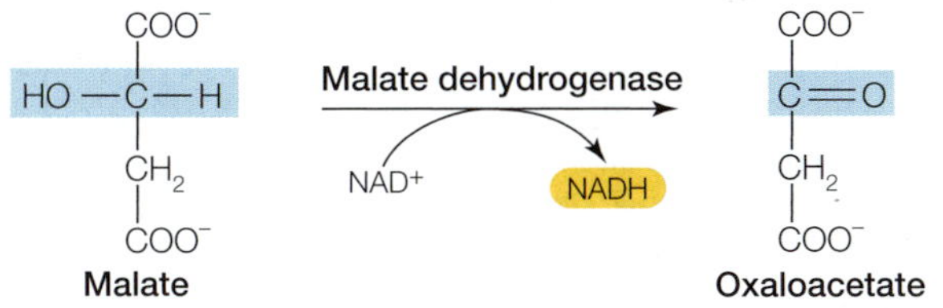
Malate dehydrogenase
NAD+
NADH
Malate
Oxaloacetate

This oxidation reaction (see the carbon atom highlighted in blue) is exergonic, and the released energy is trapped by NAD^+, forming NADH. With four such reactions (the $FADH_2$ produced in Step 6 is a reduced coenzyme similar to NADH), the citric acid cycle harvests a great deal of chemical energy from the oxidation of acetyl CoA.

To summarize:

- The inputs to the citric acid cycle are acetate (in the form of acetyl CoA), water, GDP, and the oxidized electron carriers NAD^+ and FAD.
- The outputs are carbon dioxide, reduced electron carriers (NADH and $FADH_2$), and a small amount of GTP. The energy in the terminal phosphate of GTP is transferred to ATP:

$$GTP + ADP \rightarrow ATP + GDP$$

Thus the citric acid cycle releases two carbons as CO_2 and produces four reduced electron carrier molecules.

Overall, for each molecule of glucose that is oxidized, two molecules of pyruvate are produced during glycolysis, and after oxidation these feed two turns of the citric acid cycle. So the oxidation of one glucose molecule yields:

- Six CO_2
- Ten NADH (two in glycolysis, two in pyruvate oxidation, and six in the citric acid cycle)
- Two $FADH_2$
- Four ATP

Pyruvate oxidation and the citric acid cycle are regulated by the concentrations of starting materials

We have seen how pyruvate, a three-carbon molecule, is completely oxidized to CO_2 by pyruvate dehydrogenase and the citric acid cycle. For the cycle to continue, the starting molecules—acetyl CoA and oxidized electron carriers—must all be replenished. The electron carriers are reduced during the citric acid cycle and in Step 6 of glycolysis (see Figure 9.5) and they must be reoxidized:

$$NADH \rightarrow NAD^+ + H^+ + 2\,e^-$$
$$FADH_2 \rightarrow FAD + 2\,H^+ + 2\,e^-$$

These oxidation reactions occur as coupled redox reactions in which other molecules get reduced. When it is present, O_2 is the molecule that eventually accepts these electrons, and it is reduced to form H_2O.

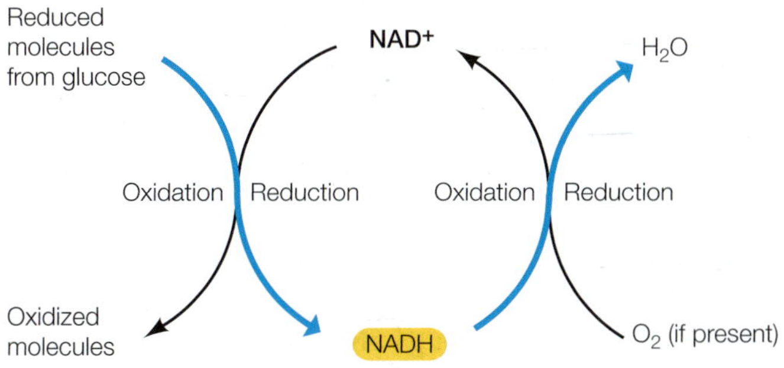
Reduced molecules from glucose
NAD+
H2O
Oxidation
Reduction
Oxidation
Reduction
Oxidized molecules
NADH
O2 (if present)

RECAP 9.2

The oxidation of glucose in the presence of O_2 involves glycolysis, pyruvate oxidation, and the citric acid cycle. In glycolysis, glucose is converted to pyruvate with some energy capture. Pyruvate is oxidized first to acetyl CoA by pyruvate dehydrogenase, then all the way to CO_2 by the citric acid cycle, releasing energy that is captured in the form of reduced electron carriers.

- What is the net energy yield of glycolysis in terms of energy invested and energy harvested? See p. 169 and Figure 9.5
- What role does pyruvate oxidation play in relation to the citric acid cycle? See p. 170
- Explain why reoxidation of NADH is crucial for the continuation of the citric acid cycle. See p. 171 and Figure 9.6

Pyruvate oxidation and the citric acid cycle cannot continue operating unless O_2 is available to receive electrons during the reoxidation of reduced electron carriers. However, these electrons are not passed directly to O_2, as you will learn next.

9.3 How Does Oxidative Phosphorylation Form ATP?

The overall process of ATP synthesis resulting from the reoxidation of electron carriers in the presence of O_2 is called **oxidative phosphorylation**. In this section we describe oxidative phosphorylation as it occurs in mitochondria, but the same process occurs in prokaryotes on the plasma membrane (see Table 9.1). Two components of the process can be distinguished:

1. *Electron transport.* The electrons from NADH and $FADH_2$ pass through the **respiratory chain**, a series of membrane-associated electron carriers. The flow of electrons along this pathway results in the active transport of protons out of the mitochondrial matrix and across the inner mitochondrial membrane, creating a proton concentration gradient.
2. *Chemiosmosis.* The protons diffuse back into the mitochondrial matrix through a channel protein, **ATP synthase**, which couples this diffusion to the synthesis of ATP. As we mentioned in the opening story, the inner mitochondrial membrane is normally impermeable to protons, so the only way for them to follow their concentration gradient is through the channel.

Before we proceed with the details of these pathways, let's consider an important question: Why should the respiratory chain be such a complex process? Why don't cells use the following single step?

$$2\ NADH + 2\ H^+ + O_2 \rightarrow 2\ NAD^+ + 2\ H_2O$$

The answer is that this reaction would be untamable. It is extremely exergonic—and oxidizing NADH this way would be rather like setting off a stick of dynamite in the cell. There is no biochemical way to harvest that burst of energy efficiently and put it to physiological use (that is, no single metabolic reaction is so endergonic as to consume a significant fraction of that energy in a single step). To control the release of energy during the oxidation of glucose, cells have evolved a lengthy respiratory chain: a series of reactions, each of which releases a small amount of energy, one step at a time.

The respiratory chain transfers electrons and protons, and releases energy

The respiratory chain is located in the inner mitochondrial membrane. Because of the extensive folding of the membrane, there is much more space for the proteins involved in the chain than there would be in a membrane with less surface area. There are several interacting components, including large integral proteins, a small peripheral protein, and a small lipid molecule. **Figure 9.7** shows a plot of the free energy released as electrons are passed between the carriers.

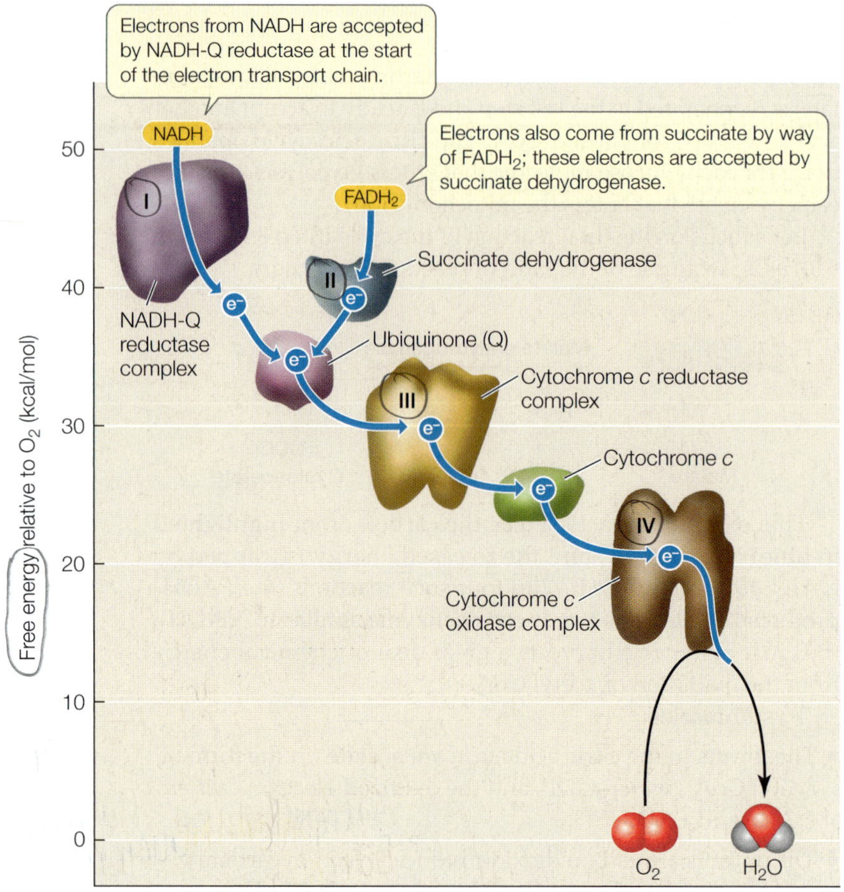
Electrons from NADH are accepted by NADH-Q reductase at the start of the electron transport chain.
Electrons also come from succinate by way of $FADH_2$; these electrons are accepted by succinate dehydrogenase.
NADH
$FADH_2$
I
II
III
IV
e^-
Succinate dehydrogenase
NADH-Q reductase complex
Ubiquinone (Q)
Cytochrome c reductase complex
Cytochrome c
Cytochrome c oxidase complex
Free energy relative to O_2 (kcal/mol)
50
40
30
20
10
0
O_2
H_2O

9.7 The Oxidation of NADH and $FADH_2$ in the Respiratory Chain Electrons from NADH and $FADH_2$ are passed along the respiratory chain, a series of protein complexes in the inner mitochondrial membrane containing electron carriers and enzymes. The carriers gain free energy when they become reduced and release free energy when they are oxidized. This illustration shows the standard free energy changes along the respiratory chain.

Go to Activity 9.4 Respiratory Chain Life10e.com/ac9.4

- Four large protein complexes (I, II, III, and IV) contain electron carriers and associated enzymes. In eukaryotes they are integral proteins of the inner mitochondrial membrane (see Figure 5.11), and three are transmembrane proteins.
- Cytochrome *c* is a small peripheral protein that lies in the intermembrane space. It is loosely attached to the outer surface of the inner mitochondrial membrane.
- Ubiquinone (often referred to as coenzyme Q10; abbreviated Q) is a small, nonpolar, lipid molecule that moves freely within the hydrophobic interior of the phospholipid bilayer of the inner mitochondrial membrane.

As illustrated in Figure 9.7, NADH passes electrons to protein complex I (called NADH-Q reductase), which in turn passes the electrons to Q. This electron transfer is accompanied by a large drop in free energy. Complex II (succinate dehydrogenase) passes electrons to Q from $FADH_2$, which was generated in Step 6 of the citric acid cycle (see Figure 9.6). These electrons enter the chain later than those from NADH and will ultimately produce less ATP.

Complex III (cytochrome *c* reductase) receives electrons from Q and passes them to cytochrome *c*. Complex IV (cytochrome *c* oxidase) receives electrons from cytochrome *c* and passes them to oxygen. Finally the reduction of oxygen to H_2O occurs:

$$O_2 + 4\ H^+ + 4\ e^- \rightarrow 2\ H_2O$$

Notice that four protons (H^+) are also consumed in this reaction. This contributes to the proton concentration gradient across the inner mitochondrial membrane by reducing the proton concentration in the mitochondrial matrix.

During electron transport, protons are also actively transported across the membrane: electron transport within each of the three transmembrane complexes (I, III, and IV) results in the transfer of protons from the matrix to the intermembrane space (**Figure 9.8**). The mechanisms by which the three complexes transfer protons across the membrane are still not well understood but are likely to involve Q. The transfer sets up an imbalance of protons, with the impermeable inner mitochondrial membrane as a barrier. The concentration of H^+ in the intermembrane space is higher than in the matrix, and this gradient represents a source of potential energy. The diffusion

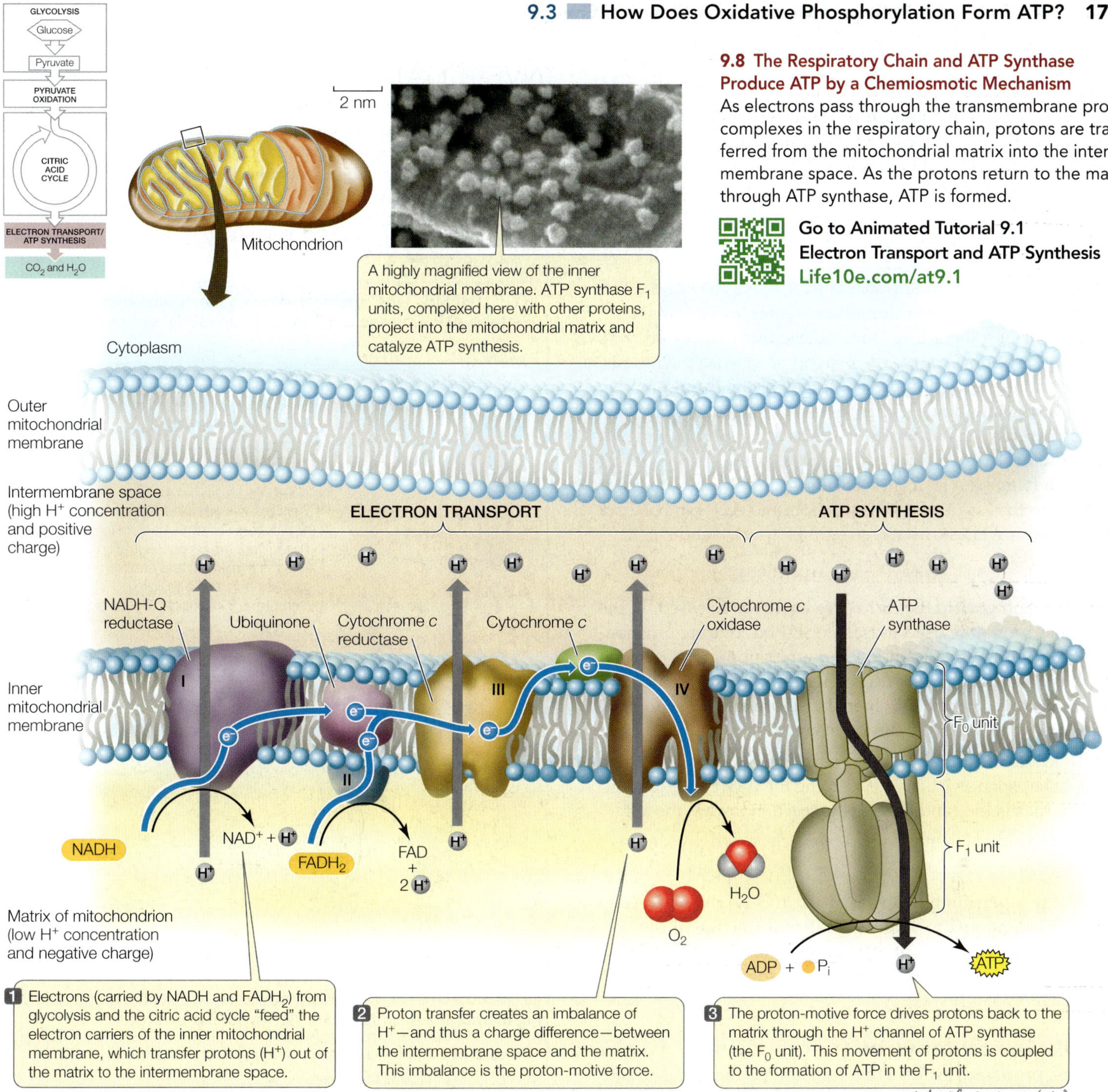
GLYCOLYSIS
Glucose
Pyruvate
PYRUVATE OXIDATION
CITRIC ACID CYCLE
ELECTRON TRANSPORT/ ATP SYNTHESIS
CO_2 and H_2O
2 nm
Mitochondrion
A highly magnified view of the inner mitochondrial membrane. ATP synthase F_1 units, complexed here with other proteins, project into the mitochondrial matrix and catalyze ATP synthesis.
Cytoplasm
Outer mitochondrial membrane
Intermembrane space (high H^+ concentration and positive charge)
ELECTRON TRANSPORT
ATP SYNTHESIS
NADH-Q reductase
Ubiquinone
Cytochrome *c* reductase
Cytochrome *c*
Cytochrome *c* oxidase
ATP synthase
Inner mitochondrial membrane
I
II
III
IV
F_0 unit
F_1 unit
NADH
$NAD^+ + H^+$
$FADH_2$
FAD + 2 H^+
O_2
H_2O
ADP + P_i
ATP
Matrix of mitochondrion (low H^+ concentration and negative charge)
1 Electrons (carried by NADH and $FADH_2$) from glycolysis and the citric acid cycle "feed" the electron carriers of the inner mitochondrial membrane, which transfer protons (H^+) out of the matrix to the intermembrane space.
2 Proton transfer creates an imbalance of H^+—and thus a charge difference—between the intermembrane space and the matrix. This imbalance is the proton-motive force.
3 The proton-motive force drives protons back to the matrix through the H^+ channel of ATP synthase (the F_0 unit). This movement of protons is coupled to the formation of ATP in the F_1 unit.

9.8 The Respiratory Chain and ATP Synthase Produce ATP by a Chemiosmotic Mechanism As electrons pass through the transmembrane protein complexes in the respiratory chain, protons are transferred from the mitochondrial matrix into the intermembrane space. As the protons return to the matrix through ATP synthase, ATP is formed.

Go to Animated Tutorial 9.1
Electron Transport and ATP Synthesis
Life10e.com/at9.1

of those protons across the membrane is coupled with the formation of ATP. Thus the energy originally contained in glucose and other fuel molecules is finally captured in the cellular energy currency, ATP. For each pair of electrons passed along the chain from NADH to oxygen, about 2.5 molecules of ATP are formed. $FADH_2$ oxidation produces about 1.5 ATP molecules.

Proton diffusion is coupled to ATP synthesis

All the electron carriers and enzymes of the respiratory chain, except cytochrome *c*, are embedded in the inner mitochondrial membrane. As we have just seen, the operation of the respiratory chain results in the active transport of protons from the mitochondrial matrix to the intermembrane space. The transmembrane protein complexes (I, III, and IV) act as proton carriers, and as a result, the intermembrane space is more acidic than the matrix.

Because of the positive charge carried by a proton (H^+), the respiratory chain creates not only a concentration gradient but also a difference in electric charge across the inner mitochondrial membrane, making the mitochondrial matrix more negative than the intermembrane space. Together, the proton concentration gradient and the electric charge difference constitute a source of potential energy called the **proton-motive force**. This force tends to drive the protons back across the membrane, just as the charge on a battery drives the flow of electrons to discharge the battery.

The hydrophobic interior of the lipid bilayer is essentially impermeable to protons, so the potential energy of the proton-motive force cannot be discharged by simple diffusion of protons across the membrane. However, protons can diffuse across the membrane by passing through a specific proton channel, called ATP synthase, which couples proton movement to the synthesis of ATP. This coupling of the proton-motive force and ATP synthesis is called the chemiosmotic mechanism—or **chemiosmosis**—and is found in all respiring cells.

THE CHEMIOSMOTIC MECHANISM FOR ATP SYNTHESIS The chemiosmotic mechanism involves a complex of transmembrane proteins—including a proton channel and the enzyme ATP synthase—that couples proton diffusion to ATP synthesis. The potential energy of the H^+ gradient, or the proton-motive force (described above), is harnessed by ATP synthase. This protein complex has two roles: it acts as a channel allowing H^+ to diffuse back into the matrix, and it uses the energy of that diffusion to make ATP from ADP and P_i.

ATP synthesis is a reversible reaction, and ATP synthase can also act as an ATPase, hydrolyzing ATP to ADP and P_i:

$$\text{ATP} \rightleftharpoons \text{ADP} + \text{P}_i + \text{free energy}$$

If the reaction goes to the right, free energy is released. In the mitochondrion, it is used to transfer H^+ out of the mitochondrial matrix—not the usual mode of operation. If the reaction goes to the left, it uses the free energy from H^+ diffusion into the matrix to make ATP. What makes it prefer ATP synthesis? There are two answers to this question:

- ATP leaves the mitochondrial matrix for use elsewhere in the cell as soon as it is made, keeping the ATP concentration in the matrix low, and driving the reaction toward the left.
- The H^+ gradient is maintained by electron and proton transport.

Every day a person hydrolyzes about 10^{25} ATP molecules to ADP. This amounts to 9 kg, a significant fraction of the persons's entire body weight! The vast majority of this ADP is "recycled"—converted back to ATP—using free energy from the oxidation of glucose.

EXPERIMENTAL DEMONSTRATION OF CHEMIOSMOSIS When it was first proposed, the idea that a proton gradient was the energy intermediate linking electron transport to ATP synthesis was a departure from the conventional thinking of the time. Scientists assumed that the energy for ATP synthesis came directly from the substrate donating the phosphate group to form ATP (i.e., from substrate-level phosphorylation; see Section 9.2). However, no intermediate could be found for this reaction in the mitochondrion, leading to the new hypothesis that chemiosmosis was the mechanism of oxidative phosphorylation. Experimental evidence was needed to support this hypothesis. The key experiment that demonstrated that a proton (H^+) gradient across a membrane could drive ATP synthesis was first performed using chloroplasts, the organelles in plants that convert the energy in sunlight into chemical energy (photosynthesis; see Chapter 10) (**Figure 9.9**). Soon thereafter, the same mechanism was shown to work in mitochondria.

INVESTIGATING**LIFE**

9.9 An Experiment Demonstrates the Chemiosmotic Mechanism The chemiosmosis hypothesis was a bold departure from the conventional scientific thinking of the time. It required an intact compartment enclosed by a membrane. Could a proton gradient drive the synthesis of ATP? The first experiments to answer this question used chloroplasts, plant organelles that use the same mechanism as mitochondria to synthesize ATP.[a]

HYPOTHESIS A H^+ gradient across a membrane that contains ATP synthase is sufficient to drive ATP synthesis.

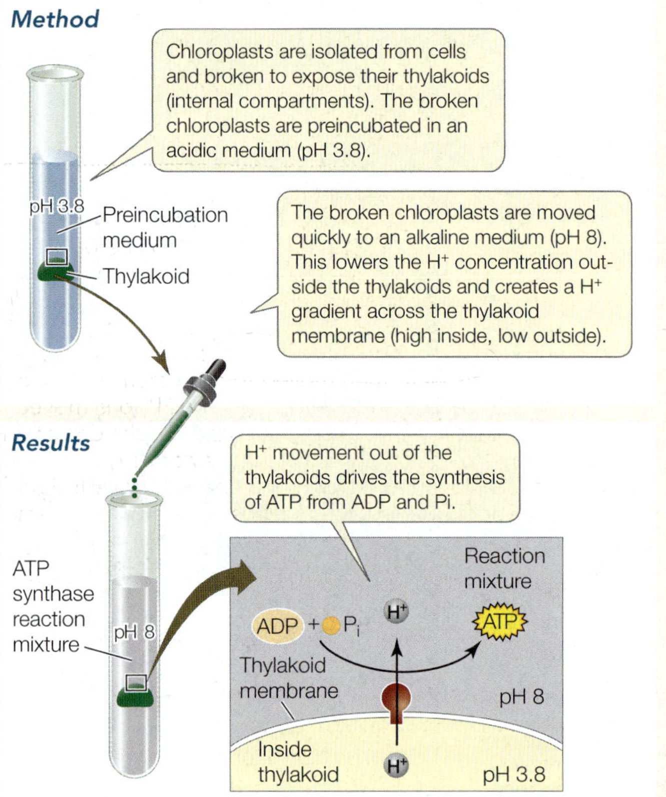
Method
Chloroplasts are isolated from cells and broken to expose their thylakoids (internal compartments). The broken chloroplasts are preincubated in an acidic medium (pH 3.8).
pH 3.8
Preincubation medium
Thylakoid
The broken chloroplasts are moved quickly to an alkaline medium (pH 8). This lowers the H+ concentration outside the thylakoids and creates a H+ gradient across the thylakoid membrane (high inside, low outside).
Results
H+ movement out of the thylakoids drives the synthesis of ATP from ADP and Pi.
ATP synthase reaction mixture
pH 8
Reaction mixture
ADP + Pi
H+
ATP
Thylakoid membrane
pH 8
Inside thylakoid
H+
pH 3.8

CONCLUSION A H^+ gradient across an ATP synthase–containing membrane is sufficient for ATP synthesis by organelles.

Go to **BioPortal** for discussion and relevant links for all INVESTIGATING**LIFE** figures.

[a]Jagendorf, A. T. and E. Uribe. 1966. *Proceedings of the National Academy of Sciences USA* 55: 170–177.

Go to Animated Tutorial 9.2
Two Experiments Demonstrate the Chemiosmotic Mechanism Life10e.com/at9.2

As we have seen, the coupling of electron transport (which generates the proton gradient) with chemiosmosis is vital for the capture of free energy in the form of ATP. Uncoupling protein 1 (UCP1), which is found in the mitochondria of brown fat cells (see the opening story), demonstrates the importance of this coupling. By disrupting the gradient, UCP1 allows the energy released during electron transport to be in the form of heat, rather than chemical energy trapped in ATP.

WORKING WITH**DATA:**

Experimental Demonstration of the Chemiosmotic Mechanism

Original Paper

Jagendorf, A. T. and E. Uribe. 1966. ATP formation caused by acid-base transition of spinach chloroplasts. *Proceedings of the National Academy of Sciences USA* 55: 170–177.

Analyze the Data

Previous research into the force driving ATP synthesis in mitochondria led to the postulation of the chemiosmotic mechanism by Peter Mitchell. A key prediction of his hypothesis was that a chemical potential across a membrane, in this case driven by an imbalance of protons (H^+), could provide energy for the synthesis of ATP from ADP and P_i. Working with chloroplasts—which as you will see in Chapter 10 also use the chemiosmotic mechanism to make ATP—André Jagendorf and Ernest Uribe at Johns Hopkins University performed a clever experiment, using rapid changes in pH to create a proton gradient across a membrane and to drive ATP synthesis (see Figure 9.9). They isolated chloroplasts from plant cells, then broke the chloroplasts' outer membranes to expose the thylakoids, internal membrane-enclosed compartments that contain the chloroplast ATP synthase (see Figure 10.10). The broken chloroplasts were placed in a medium at a relatively high H^+ concentration (acid pH) until equilibrium was reached (the interior of the thylakoids became acidic). The broken chloroplasts were then moved to a medium with a relatively low H^+ concentration (alkaline pH). The H^+ ions moved down the concentration gradient (out of the thylakoids), driving the synthesis of ATP. Although chloroplasts normally make ATP during photosynthesis in the light, in these experiments light was not required—just the H^+ gradient across the thylakoid membrane. Mitchell was a brilliant theoretician; Jagendorf, Uribe, and others provided experimental confirmation for his ideas, which won Mitchell the Nobel Prize in 1978.

In Jagendorf and Uribe's experiment, thylakoids were preincubated in a solution at pH 3.8 (or in one experiment, pH 7) and then quickly transferred to an ATP synthase reaction mixture containing ADP and inorganic phosphate (P_i) at pH 8. ATP formation was measured in two ways. The first used the enzyme luciferase, which catalyzes the formation of a luminescent molecule if ATP is present. The second used molybdate to measure phosphorylation directly. The table shows the data from the paper.

		ATP formation[a]	
Preincubation pH	ATP synthase reaction mixture (pH 8)	Luciferase assay	Molybdate assay
3.8	Complete	141	166
7.0	Complete	12	3
3.8	– P_i[b]	12	3
3.8	– ADP	4	3
3.8	– thylakoids	7	3

[a]In nmoles ATP per mg chlorophyll (a pigment in thylakoids).
[b]A minus sign indicates a component that was omitted from the reaction mixture.

QUESTION 1
Which two experiments show that a proton gradient is necessary and sufficient to stimulate ATP formation? Explain your reasoning.

QUESTION 2
Why was there less ATP production in the absence of Pi?

Go to **BioPortal** for all WORKING WITH**DATA** exercises

HOW ATP SYNTHASE WORKS: A MOLECULAR MOTOR Now that we have established that the H^+ gradient is needed for ATP synthesis, a question remains: How does the enzyme actually make ATP from ADP and P_i? This is certainly a fundamental question in biology, as it underlies energy harvesting in most cells. The structure and mechanism of ATP synthase, illustrated in **Figure 9.10A**, are shared by living organisms as diverse as bacteria and humans. ATP synthase is a molecular motor composed of two parts: the F_0 unit, a transmembrane region that is the H^+ channel; and the F_1 unit, which contains the active sites for ATP synthesis. F_1 consists of six subunits (three each of two polypeptide chains), arranged like the segments of an orange around a central polypeptide that interacts with the membrane-embedded F_0. Electron transport sets up an H^+ gradient across the membrane. This gradient has potential energy, and when H^+ diffuses through the channel, the potential energy is converted to kinetic energy, causing the central polypeptide to rotate. The energy is transmitted to the catalytic subunits of F_1, resulting in ATP synthesis. These molecular motors make ATP at rates up to 100 molecules per second.

An ingenious experiment confirmed this rotary motor mechanism (**Figure 9.10B**). Masasuke Yoshida and his colleagues at the Tokyo Institute of Technology isolated the F_1 portion of the ATP synthase and attached it to a glass slide. Fluorescently labeled microfilaments were attached to the central peptide, and the slide was incubated in a solution containing ATP. In this case there was no proton gradient to drive the molecular motor in the direction of ATP synthesis. Instead, ATP was hydrolyzed to ADP and P_i. This caused the motor to spin, and the rotation of the labeled microfilament was visible under a microscope. The result was amazing, with the labeled filaments clearly rotating like propellers.

ELECTRON TRANSPORT CAN BE TOXIC Oxygen is an excellent electron acceptor, forming water as a result:

$$O_2 + 4\,H^+ + 4\,e^- \rightarrow 2\,H_2O \quad (9.1)$$

Unfortunately, the consequences of an incomplete transfer of electrons are far from harmless because the intermediates are toxic:

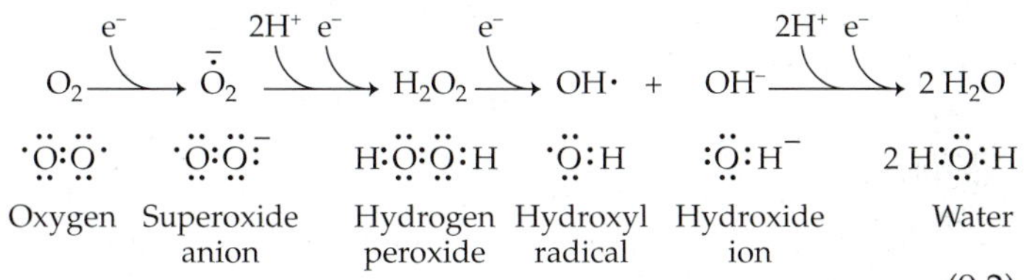
e⁻
2H⁺ e⁻
e⁻
2H⁺ e⁻
O2
O2
H2O2
OH·
+
OH⁻
2 H2O
Oxygen
Superoxide anion
Hydrogen peroxide
Hydroxyl radical
Hydroxide ion
Water

(9.2)

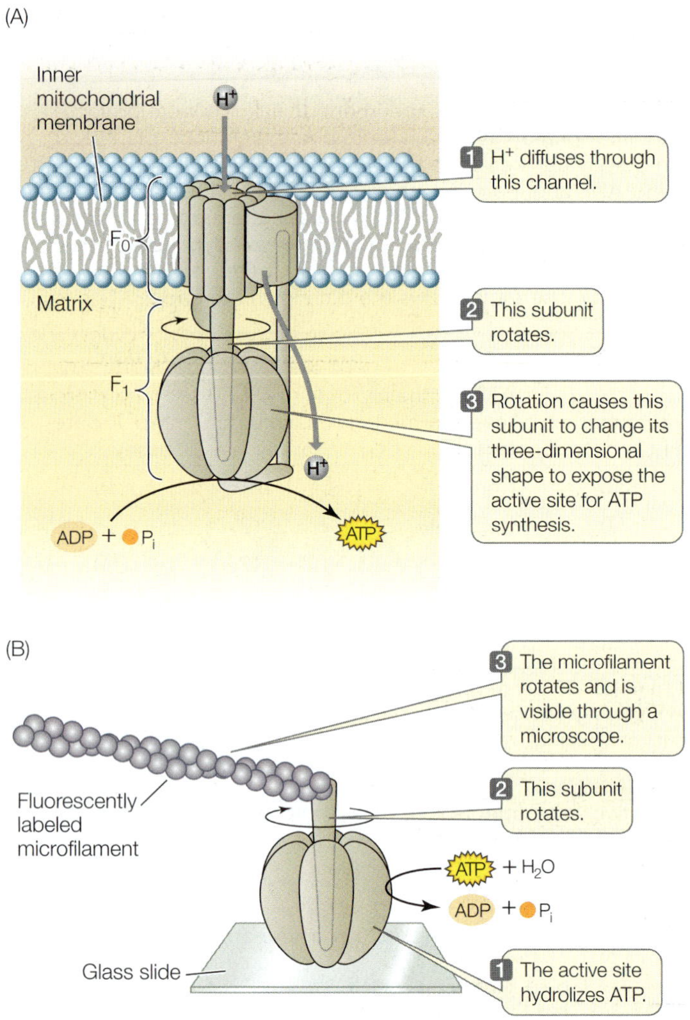

(A)
Inner mitochondrial membrane
H+
1 H+ diffuses through this channel.
F0
Matrix
2 This subunit rotates.
F1
3 Rotation causes this subunit to change its three-dimensional shape to expose the active site for ATP synthesis.
H+
ADP + Pi
ATP
(B)
3 The microfilament rotates and is visible through a microscope.
Fluorescently labeled microfilament
2 This subunit rotates.
ATP + H2O
ADP + Pi
Glass slide
1 The active site hydrolizes ATP.

9.10 How ATP Is Made (A) Mitochondrial ATP synthase is a rotary motor. (B) A clever experiment was used to visualize the rotary motor.

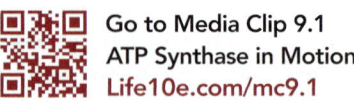

Go to Media Clip 9.1
ATP Synthase in Motion
Life10e.com/mc9.1

Because they have unpaired electrons, superoxide and the hydroxyl radical are highly reactive and "seek out" other molecules to react with to gain or lose electrons and achieve a more stable state. These molecules can include lipids and nucleic acids, whose properties are changed dramatically by these reactions. Superoxide damage has been implicated in many human conditions, ranging from lung damage in emphysema to renal failure, stroke, and even aging. Cells have enzymes that "scavenge" these oxidizing substances before they can do harm, reducing them to harmless substances. Superoxide dismutase converts superoxide to peroxide, and catalase (in peroxisomes; see Section 5.3) converts peroxide to water:

$$O_2^- + 2\,H^+ \xrightarrow{\text{Superoxide dismutase}} O_2 + H_2O_2 \qquad (9.3)$$

$$2\,H_2O_2 \xrightarrow{\text{Catalase}} O_2 + 2\,H_2O \qquad (9.4)$$

Antioxidant vitamins such as vitamin E act in a similar way by reducing these harmful molecules.

TABLE 9.2 Electron Acceptors Used in the Respiratory Chain of Anaerobic Microbes

Terminal Electron Acceptor	Product Formed	Organism
SO_4^{-2}	H_2S	*Desulfovibrio desulfuricans*
Fe^{3+}	Fe^{2+}	*Geobacter metallireducens*
NO_3^-	NO_2^-	*Escherichia coli*
CO_2	CH_4	*Methanosarcina barkeri*
CO_2	CH_3COO^-	*Clostridium aceticum*
Fumarate	Succinate	*Wolinella succinogenes*

Some microorganisms use non-O_2 electron acceptors

A more general way to describe the last reaction in electron transport (Equation 9.1, above) is:

$$X_{\text{oxidized}} + e^- \rightarrow X_{\text{reduced}} \qquad (9.5)$$

Part of the amazing success of bacteria and archaea is that they have evolved biochemical pathways that allow them to exist in environments where O_2 is scarce or absent. As you will see in the next section, for most animals and plants, the anaerobic (no O_2) catabolism of glucose generally yields much less energy than aerobic catabolism. However, many bacteria and archaea exploit their environments to use alternative electron acceptors. This allows them to complete the electron transport chain and produce ATP even in the absence of O_2. Table 9.2 summarizes some of these pathways, which are referred to as **anaerobic respiration**. Note that some of these microbes use ions as electron acceptors while others use small molecules.

RECAP 9.3

The oxidation of reduced electron carriers in the respiratory chain drives the active transport of protons across the inner mitochondrial membrane, generating a proton-motive force. Diffusion of protons down their electrochemical gradient through ATP synthase is coupled to the synthesis of ATP. Electron transport can form toxic intermediates. Some bacteria and archaea can respire using alternative electron acceptors instead of O_2.

- What are the roles of oxidation and reduction in the respiratory chain? See Figures 9.7 and 9.8
- What is the proton-motive force, and how does it drive chemiosmosis? See pp. 173–174
- Explain how the experiment described in Figure 9.9 demonstrates the chemiosmotic mechanism. See pp. 174–175 and Figure 9.9

Oxidative phosphorylation captures a great deal of energy in ATP. But it does not occur if O_2 is absent. We will turn now to the metabolism of glucose in anaerobic conditions.

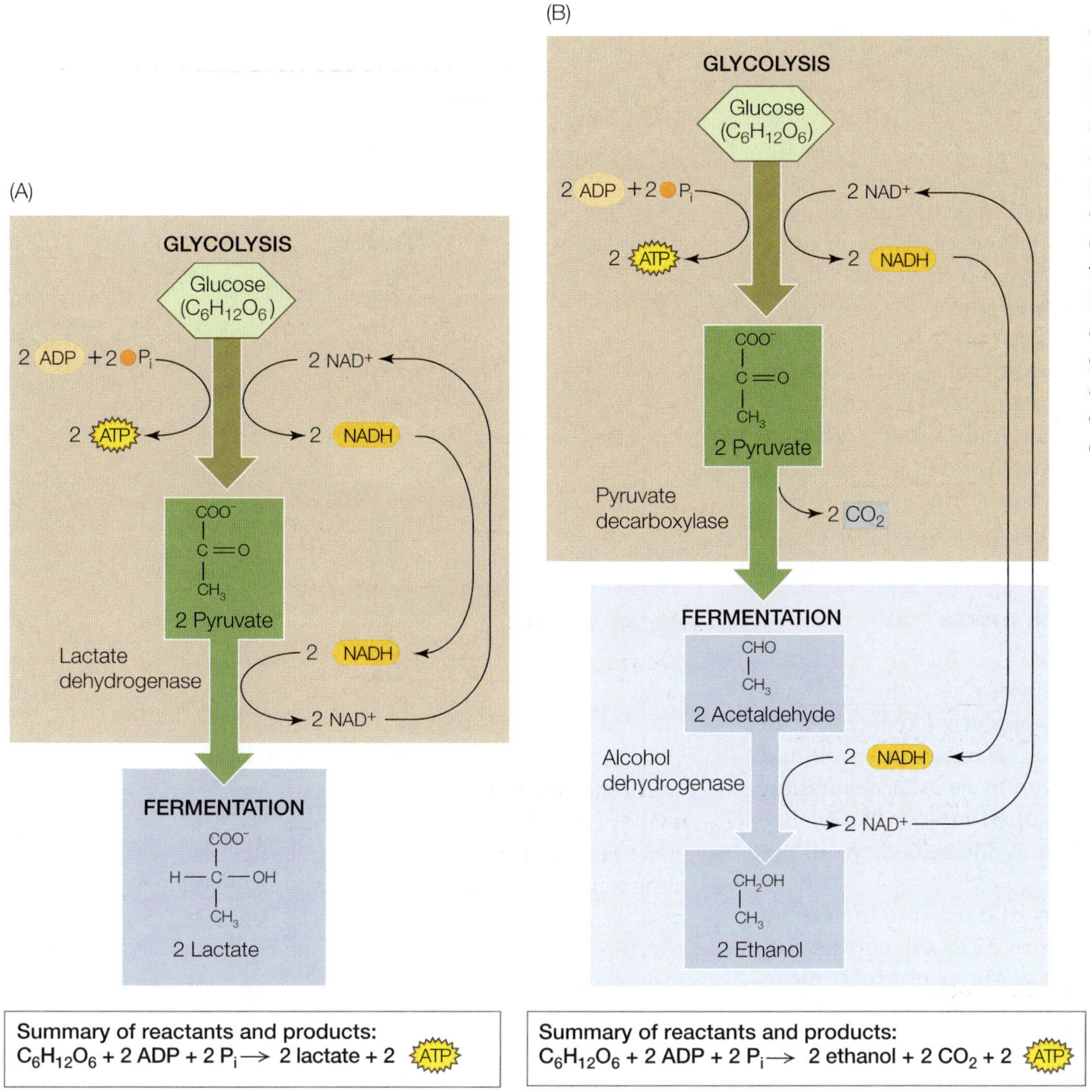
(A)
GLYCOLYSIS
Glucose ($C_6H_{12}O_6$)
2 ADP + 2 P_i
2 ATP
2 NAD+
2 NADH
COO⁻ C=O CH₃
2 Pyruvate
Lactate dehydrogenase
2 NADH
2 NAD+
FERMENTATION
COO⁻ H—C—OH CH₃
2 Lactate
Summary of reactants and products:
$C_6H_{12}O_6$ + 2 ADP + 2 P_i ⟶ 2 lactate + 2 ATP
(B)
GLYCOLYSIS
Glucose ($C_6H_{12}O_6$)
2 ADP + 2 P_i
2 ATP
2 NAD+
2 NADH
COO⁻ C=O CH₃
2 Pyruvate
Pyruvate decarboxylase
2 CO_2
FERMENTATION
CHO CH₃
2 Acetaldehyde
Alcohol dehydrogenase
2 NADH
2 NAD+
CH₂OH CH₃
2 Ethanol
Summary of reactants and products:
$C_6H_{12}O_6$ + 2 ADP + 2 P_i ⟶ 2 ethanol + 2 CO_2 + 2 ATP

9.11 Fermentation Glycolysis produces pyruvate, ATP, and NADH from glucose. (A) Lactic acid fermentation uses NADH as a reducing agent to reduce pyruvate to lactate, thus regenerating NAD^+ to keep glycolysis operating. (B) In alcoholic fermentation, pyruvate from glycolysis is converted into acetaldehyde, and CO_2 is released. NADH from glycolysis is used to reduce acetaldehyde to ethanol, thus regenerating NAD^+ to keep glycolysis operating.

9.4 How Is Energy Harvested from Glucose in the Absence of Oxygen?

In eukaryotes, in the absence of O_2 (anaerobic conditions), a small amount of ATP can be produced by glycolysis and fermentation. Like glycolysis, fermentation pathways occur in the cytoplasm. There are many different types of fermentation, but they all operate to regenerate NAD^+ so that the NAD^+-requiring reactions of glycolysis can continue. The two best-understood fermentation pathways are found in a wide variety of organisms, including eukaryotes:

- Lactic acid fermentation, the end product of which is lactic acid (lactate)
- Alcoholic fermentation, the end product of which is ethyl alcohol (ethanol)

In **lactic acid fermentation**, pyruvate serves as the electron acceptor and lactate is the product (**Figure 9.11A**). This process takes place in many microorganisms and complex organisms, including higher plants and vertebrates. A notable example of lactic acid fermentation occurs in vertebrate muscle tissue. Usually, vertebrates get their energy for muscle contraction aerobically, with the circulatory system supplying O_2 to muscles. In small vertebrates, this is almost always adequate: for example, birds can fly long distances without resting. But in larger vertebrates such as humans, the circulatory system is not up to the task of delivering enough O_2 when the need is great, such as during intense activity. At this point, the muscle cells break down glycogen (a stored polysaccharide; see Figure 3.18) and undergo lactic acid fermentation.

Lactate buildup becomes a problem after prolonged periods of intense exercise because it is associated with an increase in the H^+ concentration in the cell, lowering the pH. This affects cellular activities and causes muscle cramps, resulting in muscle pain, which abates upon resting. Lactate dehydrogenase, the enzyme that catalyzes the fermentation reaction, works in both directions. That is, when O_2 is available it can catalyze the oxidation of lactate to form pyruvate, which is then catabolized to CO_2 with concomitant energy release to form ATP. When lactate levels are decreased, muscle activity can resume.

Alcoholic fermentation takes place in certain yeasts (eukaryotic microbes) and some plant cells under anaerobic conditions. This process requires two enzymes, pyruvate decarboxylase and alcohol dehydrogenase, which metabolize pyruvate to ethanol (Figure 9.11B). As with lactic acid fermentation, the reactions are essentially reversible. For thousands of years, humans have used anaerobic fermentation by yeast cells to produce alcoholic beverages. The cells use sugars from plant sources (glucose from grapes or maltose from barley) to produce the end product, ethanol, in wine and beer.

By recycling NAD^+, fermentation allows glycolysis to continue, thus producing small amounts of ATP through substrate-level phosphorylation. The net yield of two ATPs per glucose molecule is much lower than the energy yield from cellular respiration. For this reason, most organisms existing in anaerobic environments are small microbes that grow relatively slowly.

Cellular respiration yields much more energy than fermentation

The total net energy yield from glycolysis plus fermentation is two molecules of ATP per molecule of glucose oxidized. The maximum yield of ATP that can be harvested from a molecule of glucose through glycolysis followed by cellular respiration is much greater—about 32 molecules of ATP (Figure 9.12). (Review Figures 9.5, 9.6, and 9.8, and p. 173 to see where all the ATP molecules come from.)

Why do the metabolic pathways that operate in aerobic environments produce so much more ATP? Glycolysis and fermentation only partially oxidize glucose. Much more energy remains in the end products of fermentation (lactic acid and ethanol) than in CO_2, the end product of cellular respiration. In cellular respiration, carriers (mostly NAD^+) are reduced during pyruvate oxidation and the citric acid cycle. Then the reduced carriers are oxidized by the respiratory chain, with the accompanying production of ATP by chemiosmosis (about 2.5 ATP for each NADH and 1.5 ATP for each $FADH_2$). In an aerobic environment, a cell or organism capable of aerobic metabolism will have the advantage over one that is limited to fermentation, in terms of its ability to harvest chemical energy. Two key events in the evolution of multicellular organisms were the rise in atmospheric O_2 levels (see Chapter 1) and the development of metabolic pathways to use that O_2.

The yield of ATP is reduced by the impermeability of mitochondria to NADH

The total gross yield of ATP from the oxidation of one molecule of glucose to CO_2 is about 32. However, in many eukaryotes the inner mitochondrial membrane is impermeable to NADH, and a "toll" of one ATP must be paid for each NADH molecule that is produced in glycolysis and must be "shuttled" into the mitochondrial matrix. So in these organisms, the net yield of ATP is 30.

NADH shuttle systems transfer the electrons captured by glycolysis onto substrates that are capable of movement across the mitochondrial membranes. In muscle and liver tissues (and

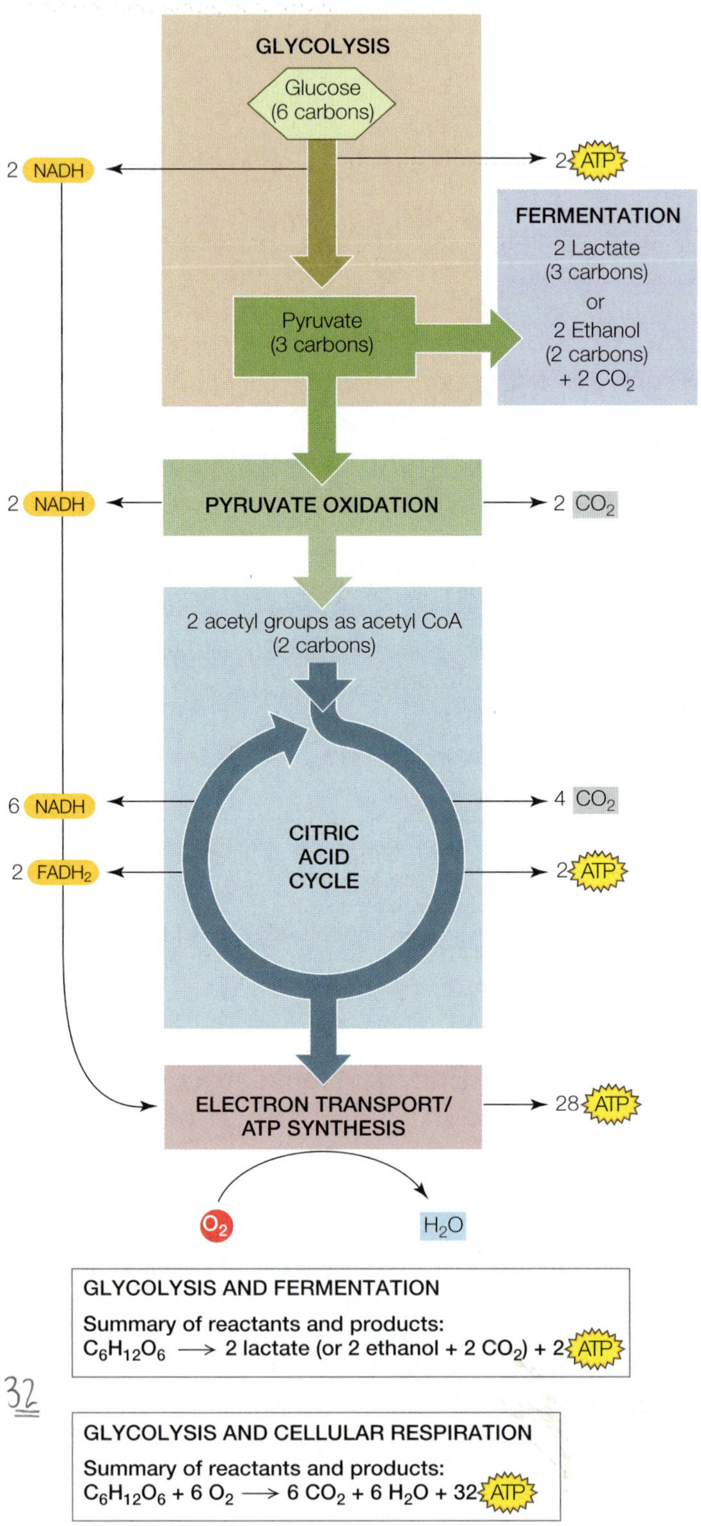
GLYCOLYSIS
Glucose (6 carbons)
2 NADH
2 ATP
FERMENTATION
2 Lactate (3 carbons)
or
2 Ethanol (2 carbons) + 2 CO_2
Pyruvate (3 carbons)
2 NADH
PYRUVATE OXIDATION
2 CO_2
2 acetyl groups as acetyl CoA (2 carbons)
6 NADH
2 $FADH_2$
CITRIC ACID CYCLE
4 CO_2
2 ATP
ELECTRON TRANSPORT/ ATP SYNTHESIS
28 ATP
O_2
H_2O
GLYCOLYSIS AND FERMENTATION
Summary of reactants and products:
$C_6H_{12}O_6 \longrightarrow$ 2 lactate (or 2 ethanol + 2 CO_2) + 2 ATP
GLYCOLYSIS AND CELLULAR RESPIRATION
Summary of reactants and products:
$C_6H_{12}O_6 + 6\ O_2 \longrightarrow 6\ CO_2 + 6\ H_2O$ + 32 ATP

9.12 Cellular Respiration Yields More Energy Than Fermentation Electron carriers are reduced in pyruvate oxidation and the citric acid cycle, then oxidized by the respiratory chain. These reactions produce ATP via chemiosmosis.

Go to Activity 9.5 Energy Levels
Life10e.com/ac9.5

the brown fat in the opening story), an important shuttle involves glycerol 3-phosphate. In the cytosol,

$$\text{NADH (from glycolysis)} + \text{dihydroxyacetone phosphate (DHAP)} \rightarrow NAD^+ + \text{glycerol 3-phosphate}$$

Glycerol 3-phosphate is transferred to the outer surface of the inner mitochondrial membrane. At that surface,

$$\text{FAD} + \text{glycerol 3-phosphate} \rightarrow FADH_2 + \text{DHAP}$$

The electrons flow from $FADH_2$ into the electron transport chain via ubiquinone (Q) (see Section 9.3). DHAP is able to move back to the cytosol, where it is available to repeat the process. Note that the reducing electrons are transferred from NADH to $FADH_2$. As you know from Figure 9.8, the energy yield in terms of ATP from $FADH_2$ is lower than that from NADH. This lowers the overall energy yield.

RECAP 9.4

In the absence of O_2, fermentation pathways use NADH formed by glycolysis to reduce pyruvate and regenerate NAD^+. The energy yield of glycolysis coupled to fermentation is low because glucose is only partially oxidized. When O_2 is present, the electron carriers of cellular respiration allow for the full oxidation of glucose, so the energy yield from glucose is much higher.

- Why is replenishing NAD^+ crucial to cellular metabolism? **See pp. 177–178**
- How does fermentation replenish NAD^+? **Review Figure 9.11**
- What is the total energy yield from glucose in human cells in the presence versus the absence of O_2? **See p. 178 and Figure 9.12**

Now that you've seen how cells harvest energy, let's see how that energy moves through other metabolic pathways in the cell.

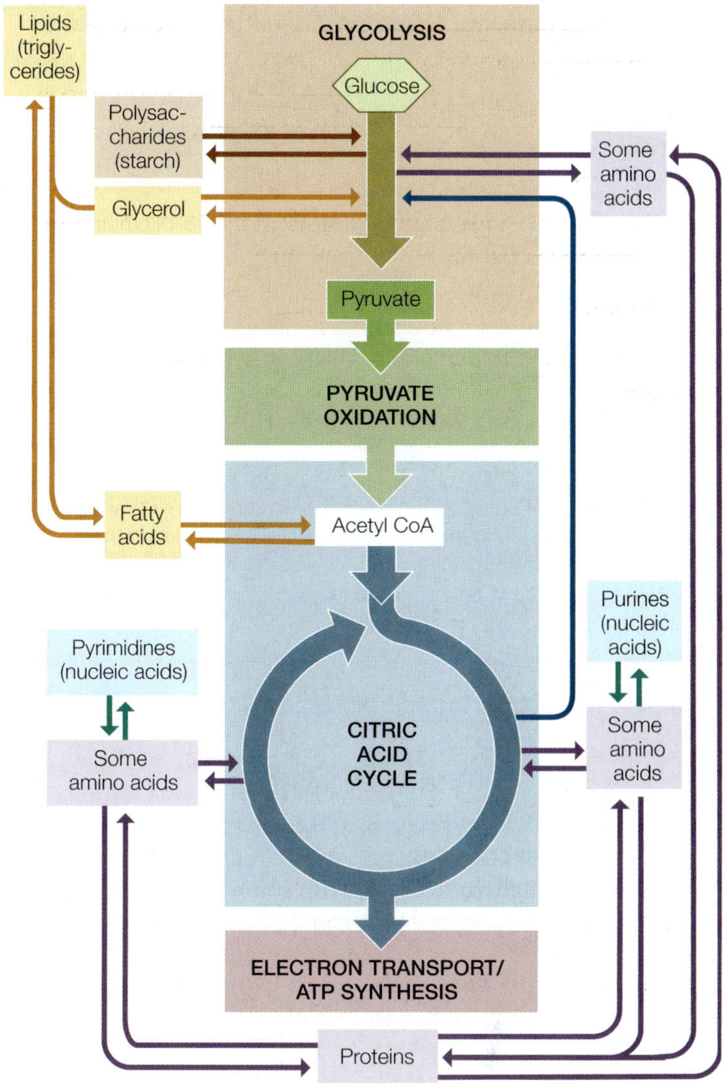

Lipids (triglycerides)
GLYCOLYSIS
Glucose
Polysaccharides (starch)
Some amino acids
Glycerol
Pyruvate
PYRUVATE OXIDATION
Fatty acids
Acetyl CoA
Purines (nucleic acids)
Pyrimidines (nucleic acids)
CITRIC ACID CYCLE
Some amino acids
Some amino acids
ELECTRON TRANSPORT/ ATP SYNTHESIS
Proteins

9.13 Relationships among the Major Metabolic Pathways of the Cell Note the central positions of glycolysis and the citric acid cycle in this network of metabolic pathways. Also note that many of the pathways can operate essentially in reverse.

9.5 How Are Metabolic Pathways Interrelated and Regulated?

Glycolysis and the pathways of cellular respiration do not operate in isolation. Rather, there is an interchange of molecules into and out of these pathways, to and from the metabolic pathways for the synthesis and breakdown of amino acids, nucleotides, fatty acids, and other building blocks of life (see Figure 8.14). Carbon skeletons (i.e., the carbon backbones of organic molecules) can enter the catabolic pathways and be broken down to release their energy, or they can enter anabolic pathways to be used in the formation of the macromolecules that are the major constituents of the cell. These relationships are summarized in Figure 9.13. In this section we will explore how pathways are interrelated by the sharing of intermediate molecules, and we will see how pathways are regulated by the inhibitors of key enzymes.

Catabolism and anabolism are linked

A hamburger or veggie burger on a bun contains three major sources of carbon skeletons: carbohydrates, mostly in the form of starch (a polysaccharide); lipids, mostly as triglycerides (three fatty acids attached to glycerol); and proteins (polymers of amino acids). Look at Figure 9.13 to see how each of these three types of macromolecules can be hydrolyzed and used in catabolism or anabolism.

CATABOLIC INTERCONVERSIONS Polysaccharides, lipids, and proteins can all be broken down to provide energy:

- *Polysaccharides* are hydrolyzed to glucose. Glucose then passes through glycolysis and cellular respiration, where its energy is captured in ATP.
- *Lipids* are broken down into their constituents, glycerol and fatty acids. Glycerol is converted into dihydroxyacetone phosphate (DHAP), an intermediate in glycolysis. Fatty acids are highly reduced molecules that are converted to acetyl CoA inside the mitochondrion by a series of oxidation enzymes, in a process known as β-oxidation. For example,

glycerol → DHAP
fatty acids → acetyl CoA

the β-oxidation of a 16-carbon (C_{16}) fatty acid occurs in several steps:

C_{16} fatty acid + CoA → C_{16} fatty acyl CoA
C_{16} fatty acyl CoA + CoA → C_{14} fatty acyl CoA + acetyl CoA
Repeat 6 times → 8 acetyl CoA

The acetyl CoA can then enter the citric acid cycle and be catabolized to CO_2.

- *Proteins* are hydrolyzed to their amino acid building blocks. The 20 different amino acids feed into glycolysis or the citric acid cycle at different points determined by their structures. For example, the amino acid glutamate is converted into α-ketoglutarate, an intermediate in the citric acid cycle (the five-carbon molecule in Figure 9.6).

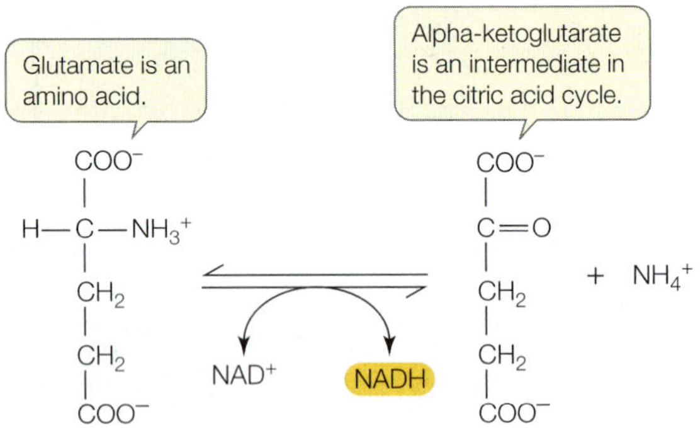
Glutamate is an amino acid.
Alpha-ketoglutarate is an intermediate in the citric acid cycle.
COO⁻
H—C—NH3+
CH2
CH2
COO⁻
NAD+
NADH
C=O
+ NH4+

ANABOLIC INTERCONVERSIONS Many catabolic pathways can operate essentially in reverse, with some modifications. Glycolytic and citric acid cycle intermediates, instead of being oxidized to form CO_2, can be reduced and used to form glucose in a process called **gluconeogenesis** (which means "new formation of glucose"). Likewise, acetyl CoA can be used to form fatty acids. The most common fatty acids have even numbers of carbons: 14, 16, or 18. These are formed by the addition of two-carbon acetyl CoA "units" one at a time until the appropriate chain length is reached. Acetyl CoA is also a building block for various pigments, plant growth substances, rubber, steroid hormones, and other molecules.

Some intermediates in the citric acid cycle are reactants in pathways that synthesize important components of nucleic acids. For example, α-ketoglutarate is a starting point for purines, and oxaloacetate for pyrimidines. In addition, α-ketoglutarate is a starting point for the synthesis of chlorophyll (used in photosynthesis; see Chapter 10) and the amino acid glutamate (used in protein synthesis).

Catabolism and anabolism are integrated

A carbon atom from a protein in your burger can end up in DNA, fat, or CO_2, among other fates. How does the organism "decide" which metabolic pathways to follow, in which cells? With all of the possible interconversions, you might expect that cellular concentrations of various biochemical molecules would vary widely. Remarkably, the levels of these substances in what is called the metabolic pool—the sum total of all the biochemical molecules in a cell—are quite constant. Organisms regulate the enzymes in various cells in order to maintain a balance between catabolism and anabolism. For example, let's look at how glucose levels are maintained in the body during exercise (**Figure 9.14**).

When a person is walking or jogging, there are two major organs where energy is most needed: the leg muscles to power movement and the heart muscles to circulate blood. The energy comes from the catabolism of glucose by glycolysis, the citric acid cycle, and oxidative phosphorylation. These muscles therefore need a lot of glucose. In the leg muscles, there is some

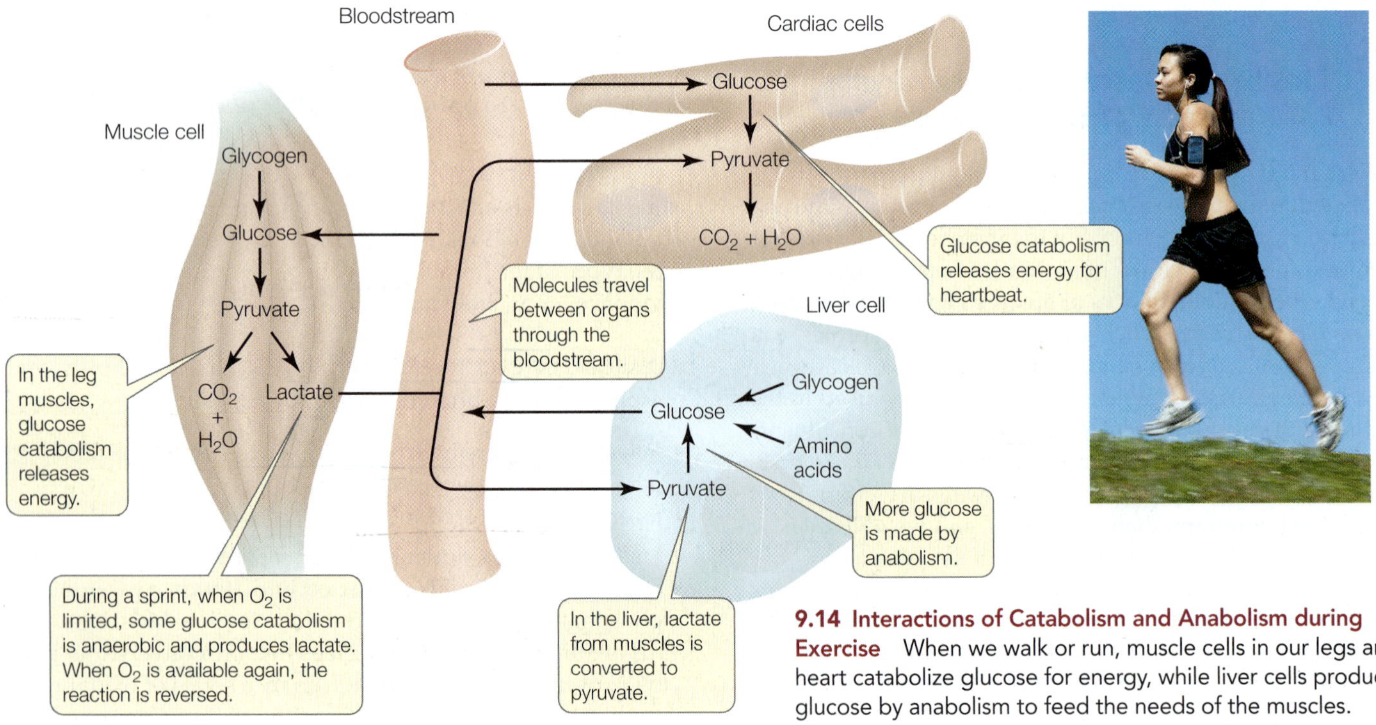
Bloodstream
Cardiac cells
Glucose
Pyruvate
CO2 + H2O
Glucose catabolism releases energy for heartbeat.
Muscle cell
Glycogen
Glucose
Pyruvate
CO2 + H2O
Lactate
In the leg muscles, glucose catabolism releases energy.
Molecules travel between organs through the bloodstream.
Liver cell
Glycogen
Glucose
Amino acids
Pyruvate
More glucose is made by anabolism.
During a sprint, when O2 is limited, some glucose catabolism is anaerobic and produces lactate. When O2 is available again, the reaction is reversed.
In the liver, lactate from muscles is converted to pyruvate.

9.14 Interactions of Catabolism and Anabolism during Exercise When we walk or run, muscle cells in our legs and heart catabolize glucose for energy, while liver cells produce glucose by anabolism to feed the needs of the muscles.

stored glycogen, which is hydrolyzed to glucose monomers. In addition, glycogen is hydrolyzed in the liver, which releases the glucose into the blood. Both the heart and leg muscles receive glucose this way. As glucose is used up by the working muscles, more glucose is made in the liver by anabolism from amino acids and pyruvate. Some of the pyruvate used in this anabolism comes from lactate that was made by fermentation in the leg muscles and then transported back to the liver in the blood. Of course, all glucose comes originally from food, and is stored in the liver and the muscles in the form of glycogen. Glucose is catabolized in all cells at all times to provide them with energy for basic cellular activities.

The integration of catabolism and anabolism shown here implies that there are control points in the biochemical pathways. For example, how does the liver "know" that it should be making glucose rather than catabolizing it or storing it?

Metabolic pathways are regulated systems

The regulation of interconnecting biochemical pathways is a problem of systems biology, which seeks to understand how biochemical pathways interact (see Section 8.5). It is a bit like trying to predict traffic patterns in a city: if an accident blocks traffic on a major road, drivers take alternate routes, where the traffic volume consequently changes.

Consider what happens to the starch in your burger bun. In the digestive system, starch is hydrolyzed to glucose, which enters the blood for distribution to the rest of the body. But before the glucose is distributed, a regulatory check must be made: If there is already enough glucose in the blood to supply the body's needs, the excess glucose is converted into glycogen and stored in the liver and muscles. If not enough glucose is supplied by food, glycogen is broken down, or other molecules are used to make glucose by gluconeogenesis. The end result is that the level of glucose in the blood is remarkably constant. How does the body accomplish this?

Glycolysis, the citric acid cycle, and the respiratory chain are subject to allosteric regulation (see Section 8.5) of key enzymes involved. In a metabolic pathway, a high concentration of the final product can inhibit the action of an enzyme that catalyzes an earlier reaction (see Figure 8.18). Furthermore, an excess of the product of one pathway can activate an enzyme in another pathway, speeding up its reactions and diverting raw materials away from synthesis of the first product (**Figure 9.15**). These negative and positive feedback mechanisms are used at many points in the energy-harvesting pathways and are summarized in **Figure 9.16**.

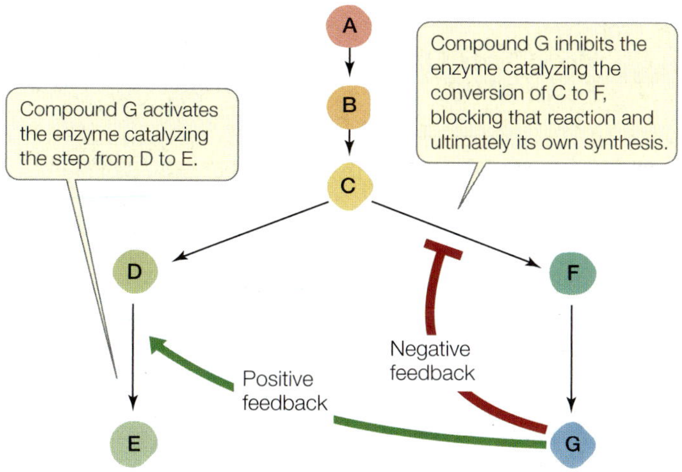

A
B
C
D
E
F
G
Compound G activates the enzyme catalyzing the step from D to E.
Compound G inhibits the enzyme catalyzing the conversion of C to F, blocking that reaction and ultimately its own synthesis.
Positive feedback
Negative feedback

9.15 Regulation by Negative and Positive Feedback
Allosteric feedback regulation plays an important role in metabolic pathways. The accumulation of some products can shut down their synthesis, or can stimulate other pathways that require the same raw materials.

Go to Activity 9.6 Regulation of Energy Pathways
Life10e.com/ac9.6

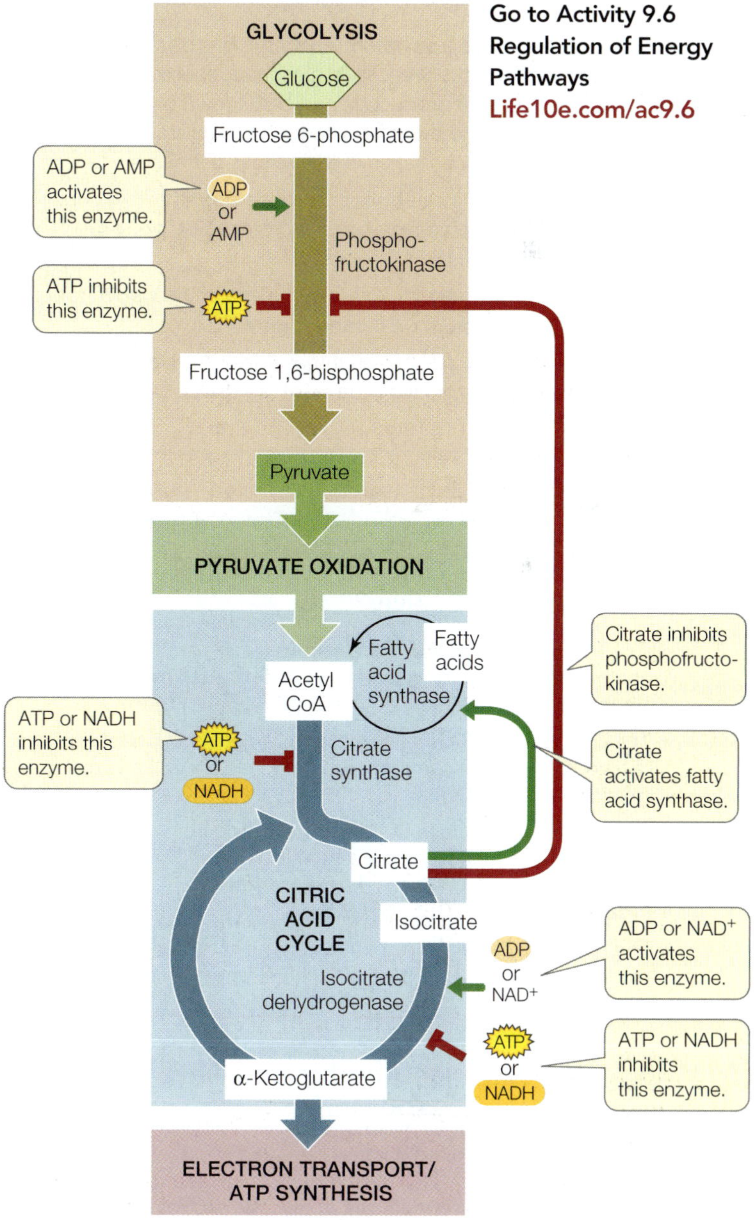

GLYCOLYSIS
Glucose
Fructose 6-phosphate
ADP or AMP activates this enzyme.
ADP or AMP
Phospho-fructokinase
ATP inhibits this enzyme.
ATP
Fructose 1,6-bisphosphate
Pyruvate
PYRUVATE OXIDATION
Acetyl CoA
Fatty acid synthase
Fatty acids
Citrate inhibits phosphofructo-kinase.
Citrate activates fatty acid synthase.
ATP or NADH inhibits this enzyme.
ATP or NADH
Citrate synthase
Citrate
CITRIC ACID CYCLE
Isocitrate
Isocitrate dehydrogenase
ADP or NAD+
ADP or NAD+ activates this enzyme.
ATP or NADH
ATP or NADH inhibits this enzyme.
α-Ketoglutarate
ELECTRON TRANSPORT/ ATP SYNTHESIS

9.16 Allosteric Regulation of Glycolysis and the Citric Acid Cycle
Allosteric regulation controls glycolysis and the citric acid cycle at crucial early steps, increasing their efficiency and preventing the excessive buildup of intermediates.

- The main control point in glycolysis is the glycolytic enzyme phosphofructokinase, which catalyzes Step 3 in Figure 9.5. This enzyme is allosterically inhibited by ATP or citrate, and activated by ADP or AMP. Under anaerobic conditions, fermentation yields a relatively small amount of ATP, and phosphofructokinase operates at a high rate. However, when conditions are aerobic, respiration makes 16 times more ATP than fermentation does, and the abundant ATP allosterically inhibits phosphofructokinase. So glycolysis slows down.
- The main control point in the citric acid cycle is the enzyme *isocitrate dehydrogenase* (catalyzes Step 3 in Figure 9.6). This enzyme is activated by increases in substrate concentrations (ADP, NAD^+, and isocitrate) and is inhibited by products of the citric acid cycle: ATP and NADH. If too much ATP or NADH accumulates, the reaction is slowed, and the citric acid cycle shuts down.
- Another control point involves *acetyl CoA*. If the level of ATP is high and the citric acid cycle shuts down, the accumulation of citrate activates fatty acid synthase, diverting acetyl CoA to the synthesis of fatty acids for storage. That is one reason why people who eat too much accumulate fat. These fatty acids may be metabolized later to produce more acetyl CoA.

RECAP 9.5

Glucose can be made from intermediates in glycolysis and the citric acid cycle by a process called gluconeogenesis. The metabolic pathways for the production and breakdown of lipids and amino acids are tied to those of glucose metabolism. Reaction products regulate key enzymes in the various pathways.

- Give examples of a catabolic interconversion of a lipid and of an anabolic interconversion of a protein. **See pp. 179–180 and Figure 9.13**
- How does phosphofructokinase serve as a control point for glycolysis? **See pp.181–182 and Figure 9.16**
- Describe what would happen if there were no allosteric mechanism for modulating the level of acetyl CoA. **See p. 182**

?

Can brown fat in adults be a target for weight loss?

ANSWER

Yuxiang Sun and her colleagues at the Baylor College of Medicine in Houston, Texas, have bred mice that make more UCP1 as adults. Like normal mice, they have brown fat for the generation of heat as newborns. But these genetically altered mice continue to use UCP1 for heat generation as adults. They eat as much and exercise as much as the normal mice, but they stay thinner by breaking down fat to release heat instead of chemical energy. By targeting the mitochondrial protein UCP1, scientists may be able to turn up brown fat's furnace in humans.

CHAPTER SUMMARY 9

9.1 How Does Glucose Oxidation Release Chemical Energy?

- As a material is **oxidized**, the electrons it loses are transferred to another material, which is thereby **reduced**. Such **redox reactions** transfer large amounts of energy. **Review Figure 9.2**
- The coenzyme NAD^+ is a key electron carrier in biological redox reactions. It exists in two forms, one oxidized (NAD^+) and the other reduced (NADH).
- **Glycolysis** does not use O_2. Under **aerobic** conditions, **cellular respiration** continues the process of breaking down glucose. Under **anaerobic** conditions, **fermentation** occurs. **Review Figure 9.4, ACTIVITIES 9.1, 9.2**
- The pathways of cellular respiration after glycolysis are pyruvate oxidation, the citric acid cycle, and the electron transport/**ATP synthesis**.

9.2 What Are the Aerobic Pathways of Glucose Catabolism?

- Glycolysis consists of ten enzyme-catalyzed reactions that occur in the cell cytoplasm. Two **pyruvate** molecules are produced for each partially oxidized molecule of glucose, providing the starting material for both cellular respiration and fermentation. **Review Figure 9.5**
- Pyruvate oxidation follows glycolysis and links glycolysis to the citric acid cycle. This pathway converts pyruvate into **acetyl CoA**.
- Acetyl CoA is the starting point of the citric acid cycle. It reacts with oxaloacetate to produce citrate. A series of eight enzyme-catalyzed reactions oxidize citrate and regenerate oxaloacetate, continuing the cycle. **Review Figure 9.6, ACTIVITY 9.3**

9.3 How Does Oxidative Phosphorylation Form ATP?

- Oxidation of electron carriers in the presence of O_2 releases energy that can be used to form ATP in a process called **oxidative phosphorylation**.
- The NADH and $FADH_2$ produced in glycolysis, pyruvate oxidation, and the citric acid cycle are oxidized by the **respiratory chain**, regenerating NAD^+ and FAD. Oxygen (O_2) is the final acceptor of electrons and protons, forming water (H_2O). **Review Figure 9.7, ACTIVITY 9.4**
- The respiratory chain not only transports electrons, but also transfers protons across the inner mitochondrial membrane, creating the **proton-motive force**.
- Protons driven by the proton-motive force can return to the mitochondrial matrix via **ATP synthase**, a molecular motor that couples this movement of protons to the synthesis of ATP. This process is called **chemiosmosis**. **Review Figure 9.8, ANIMATED TUTORIALS 9.1, 9.2**

continued

- Incomplete transfer of electrons can result in the formation of toxic superoxides and hydroxyl radicals. Special enzymes remove them to protect the cell from damage.

9.4 How Is Energy Harvested from Glucose in the Absence of Oxygen?

- In the absence of O_2, glycolysis is followed by fermentation. Together, these pathways partially oxidize pyruvate and generate end products such as lactic acid or ethanol. In the process, NAD^+ is regenerated from NADH so that glycolysis can continue, thus generating a small amount of ATP. **Review Figure 9.11**
- For each molecule of glucose used, glycolysis plus fermentation yields two molecules of ATP. In contrast, glycolysis operating with pyruvate oxidation, the citric acid cycle, and the respiratory chain/ATP synthase yields up to 32 molecules of ATP per molecule of glucose. **Review Figure 9.12, ACTIVITY 9.5**

9.5 How Are Metabolic Pathways Interrelated and Regulated?

- The catabolic pathways for the breakdown of carbohydrates, fats, and proteins feed into the energy-harvesting metabolic pathways. **Review Figure 9.13**
- Anabolic pathways use intermediate components of the energy-harvesting pathways to synthesize fats, amino acids, and other essential building blocks.
- The formation of glucose from intermediates of glycolysis and the citric acid cycle is called **gluconeogenesis**.
- The rates of glycolysis and the citric acid cycle are controlled by allosteric regulation and by the diversion of excess acetyl CoA into fatty acid synthesis. Key regulated enzymes include phosphofructokinase, citrate synthase, isocitrate dehydrogenase, and fatty acid synthase. **See Figure 9.16, ACTIVITY 9.6**

Go to the Interactive Summary to review key figures, Animated Tutorials, and Activities
Life10e.com/is9

CHAPTER**REVIEW**

REMEMBERING

1. The role of oxygen gas in our cells is to
   a. catalyze reactions in glycolysis.
   b. produce CO_2.
   c. form ATP.
   d. accept electrons from the respiratory chain.
   e. react with glucose to split water.

2. Oxidation and reduction
   a. entail the gain or loss of proteins.
   b. are defined as the loss of electrons.
   c. are both endergonic reactions.
   d. always occur together.
   e. proceed only under aerobic conditions.

3. Glycolysis
   a. takes place in the mitochondrion.
   b. produces no ATP.
   c. is part of the respiratory chain.
   d. is the same thing as fermentation.
   e. reduces two molecules of NAD^+ for every glucose molecule processed.

4. Fermentation
   a. takes place in the mitochondrion.
   b. takes place in all animal cells.
   c. does not require O_2.
   d. requires lactic acid.
   e. prevents glycolysis.

5. The respiratory chain
   a. is located in the mitochondrial matrix.
   b. includes only peripheral membrane proteins.
   c. always produces ATP.
   d. reoxidizes reduced coenzymes.
   e. operates simultaneously with fermentation.

6. Compared with glycolysis coupled with fermentation, the aerobic pathways of glucose metabolism produce
   a. more ATP.
   b. pyruvate.
   c. fewer protons for transfer in the mitochondria.
   d. less CO_2.
   e. more oxidized coenzymes.

UNDERSTANDING & APPLYING

7. Trace the sequence of chemical changes that occurs in mammalian tissue when the oxygen supply is cut off. The first change is that all of the cytochrome *c* becomes reduced, because electrons can still flow from cytochrome *c*, but there is no oxygen to accept electrons from cytochrome *c* oxidase. What are the remaining changes?

8. You eat a burger that contains polysaccharides, proteins, and lipids. Using what you know of the integration of biochemical pathways, explain how the amino acids in the proteins and the glucose in the polysaccharides can end up as fats.

9. The following reaction occurs in the citric acid cycle:

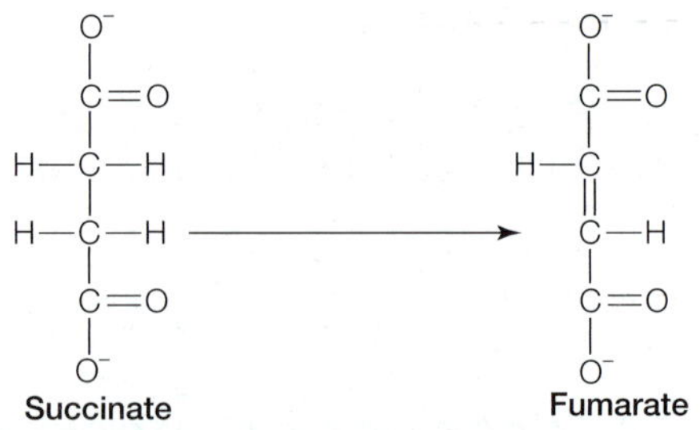
Succinate
Fumarate

Answer each of the following questions, and explain your answers:

a. Is this reaction an oxidation or reduction?
b. Is the reaction exergonic or endergonic?
c. This reaction requires a coenzyme. What kind of coenzyme?
d. What happens to the fumarate after the reaction is completed?
e. What happens to the coenzyme after the reaction is completed?

ANALYZING & EVALUATING

10. Some cells that use the aerobic pathways of glucose metabolism can also thrive by using fermentation under anaerobic conditions. Given the lower yield of ATP (per molecule of glucose) in fermentation, how can these cells function so efficiently under anaerobic conditions?

11. The drug antimycin A blocks electron transport in mitochondria. Explain what would happen if the experiment in Figure 9.9 were repeated in the presence of this drug.

Go to BioPortal at **yourBioPortal.com** for Animated Tutorials, Activities, LearningCurve Quizzes, Flashcards, and many other study and review resources.

10 Photosynthesis: Energy from Sunlight

CHAPTER OUTLINE

FACE Free-air concentration enrichment uses pipes to release CO_2 around plants in a field, to estimate the effects of rising atmospheric CO_2 on photosynthesis and plant growth.

IF ALL THE CARBOHYDRATES produced by photosynthesis in a year were in the form of sugar cubes, there would be 300 quadrillion of them. As you may have learned from previous courses, photosynthetic organisms use atmospheric carbon dioxide (CO_2) to produce carbohydrates. The simplified equation says it all:

$$CO_2 + H_2O \xrightarrow{\text{Sunlight}} \text{carbohydrates} + O_2$$

Given the role of CO_2, will the photosynthesis rate change with increasing levels of atmospheric CO_2? Over the past 200 years, the concentration of atmospheric CO_2 has increased—from 280 parts per million (ppm) in 1800 to 395 ppm in 2011—and the increase will probably continue for some time. Carbon dioxide is a "greenhouse gas" that traps heat in the atmosphere, and the rising CO_2 level is predicted by climate scientists to result in global climate change. Plant biologists are being asked two questions about the rise in CO_2: will it lead to an increased rate of photosynthesis, and if so, will it lead to increased plant growth?

To answer these questions under realistic conditions, scientists developed a way to expose plants to high levels of CO_2 in the field. *Free-air concentration enrichment* (FACE) uses rings of pipes to release CO_2 into the air surrounding plants in fields or forests. Wind speed and direction are monitored by a computer, which constantly controls which pipes release CO_2. Data from these experiments confirm that photosynthetic rates increase as the concentration of CO_2 rises, and these measurements indicate that as atmospheric CO_2 rises globally, there will be an increase in photosynthesis.

Does this increase in photosynthesis result in an increase in plant growth? Keep in mind that plants, like all organisms, use carbohydrates as an energy source. They perform cellular respiration with the general equation:

$$\text{Carbohydrates} + O_2 \rightarrow CO_2 + H_2O + \text{energy}$$

The challenge facing plant biologists is to determine the balance between photosynthesis and respiration and how this affects the rate of plant growth. The FACE experiments indicate that plant growth and crop yields increase under higher CO_2 concentrations, suggesting that the overall increase in photosynthesis is greater than the increase in respiration.

> ? What possible effects will increased atmospheric CO_2 have on global food production?
>
> See answer on p. 202.

10.1 What Is Photosynthesis?

The energy released by catabolic pathways in almost all organisms (with the exception of those living near deep sea vents) ultimately comes from the sun. **Photosynthesis** (literally, "synthesis from light") is an anabolic process by which the energy of sunlight is captured and used to convert carbon dioxide (CO_2) into more complex carbon-containing compounds. Plants, algae, and cyanobacteria live in aerobic environments and carry out oxygenic photosynthesis: the conversion of CO_2 and water (H_2O) into carbohydrates (which we will represent as a six-carbon sugar, $C_6H_{12}O_6$) and oxygen gas (O_2) (**Figure 10.1**).

oxygenic photosynthesis

$$6\ CO_2 + 6\ H_2O \rightarrow C_6H_{12}O_6 + 6\ O_2 \qquad (10.1)$$

Some kinds of bacteria live in anaerobic environments and carry out anoxygenic photosynthesis, in which energy from the sun is used to convert CO_2 to more complex molecules without the production of O_2. We will describe this process in a little more detail below, but first let's look at oxygenic photosynthesis, summarized in Equation 10.1 above.

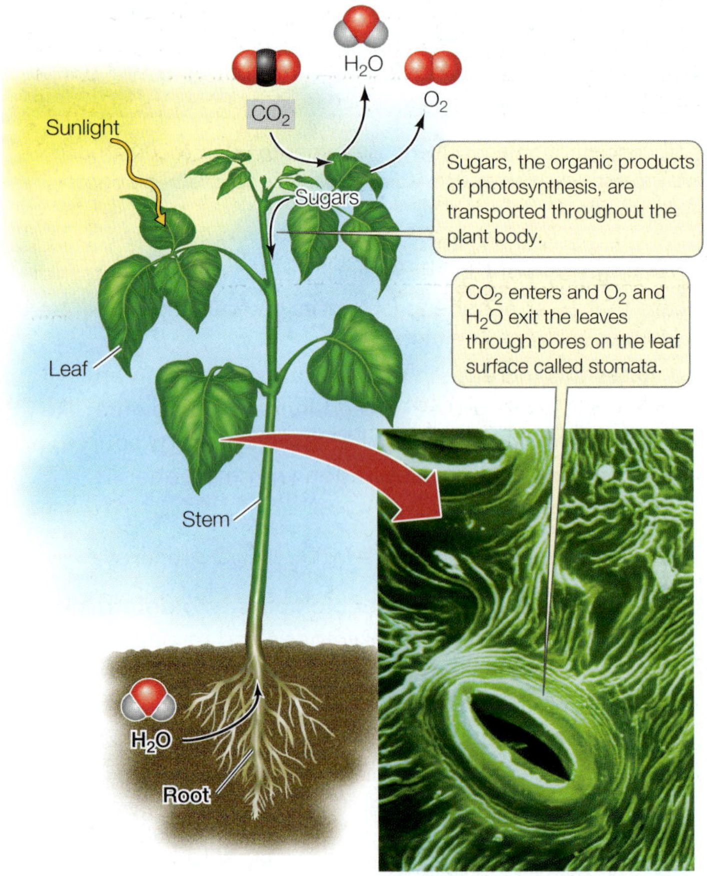

H_2O
CO_2
O_2
Sunlight
Sugars
Sugars, the organic products of photosynthesis, are transported throughout the plant body.
CO_2 enters and O_2 and H_2O exit the leaves through pores on the leaf surface called stomata.
Leaf
Stem
H_2O
Root

10.1 The Ingredients for Photosynthesis A typical terrestrial plant uses light from the sun, water from the soil, and carbon dioxide from the atmosphere to form organic compounds by photosynthesis.

Equation 10.1 shows a very endergonic reaction. And while it is essentially correct, it is too general for a real understanding of the processes involved. Several questions arise: What are the precise chemical reactions of photosynthesis? What role does light play in these reactions? How do carbons become linked to form carbohydrates? What carbohydrates are formed? And where does the oxygen gas come from: CO_2 or H_2O?

Experiments with isotopes show that O_2 comes from H_2O in oxygenic photosynthesis

In 1941 Samuel Ruben and Martin Kamen at the University of California, Berkeley performed experiments using the isotopes ^{18}O and ^{16}O to identify the source of the O_2 produced during photosynthesis (**Figure 10.2**). Their results showed that all the oxygen gas produced during photosynthesis comes from water, as is reflected in the revised balanced equation:

$$6\ CO_2 + 12\ H_2O \rightarrow C_6H_{12}O_6 + 6\ O_2 + 6\ H_2O \qquad (10.2)$$

INVESTIGATING**LIFE**

10.2 The Source of the Oxygen Produced by Photosynthesis Although it was clear that O_2 was made during photosynthesis, its molecular source was not known. Two possibilities were the reactants, CO_2 and H_2O. In two separate experiments, Samuel Ruben and Martin Kamen labeled the oxygen in these molecules with the isotope ^{18}O, then tested the O_2 produced by a green plant to find out which molecules contributed the oxygen.[a]

HYPOTHESIS The oxygen released by photosynthesis comes from water rather than CO_2.

Method

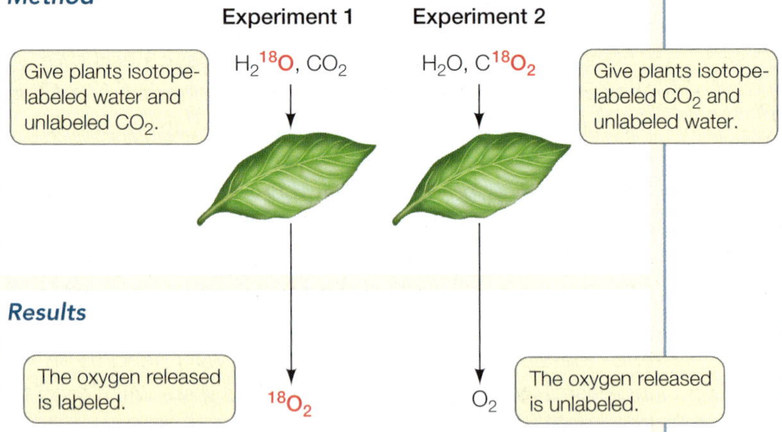

Experiment 1
Experiment 2
Give plants isotope-labeled water and unlabeled CO_2.
$H_2{}^{18}O$, CO_2
H_2O, $C^{18}O_2$
Give plants isotope-labeled CO_2 and unlabeled water.
Results
The oxygen released is labeled.
$^{18}O_2$
O_2
The oxygen released is unlabeled.

CONCLUSION Water is the source of the oxygen atoms in the O_2 produced by photosynthesis.

Go to **BioPortal** for discussion and relevant links for all INVESTIGATING**LIFE** figures.

[a]Ruben, S., et al. 1941. *Journal of the American Chemical Society* 63: 877–879.

Go to Animated Tutorial 10.1
The Source of the Oxygen Produced by Photosynthesis
Life10e.com/at10.1

Water appears on both sides of the equation because it is both used as a reactant (the 12 molecules on the left) and released as a product (the 6 new ones on the right). This revised equation accounts for all the water molecules needed for all the oxygen gas produced.

The realization that water was the source of photosynthetic O_2 led to an understanding of photosynthesis in terms of oxidation and reduction. As we described in Chapter 9, oxidation–reduction (redox) reactions are coupled: when one molecule becomes oxidized in a reaction, another gets reduced. In this case, oxygen atoms in the reduced state in H_2O get oxidized to O_2:

$$12\ H_2O \rightarrow 24\ H^+ + 24\ e^- + 6\ O_2 \qquad (10.3)$$

while carbon atoms in the oxidized state in CO_2 get reduced to carbohydrate, with the simultaneous production of water:

$$6\ CO_2 + 24\ H^+ + 24\ e^- \rightarrow C_6H_{12}O_6 + 6\ H_2O \qquad (10.4)$$

Adding Equations 10.3 and 10.4 (chemistry students will recognize them as half-cell reactions) gives the overall Equation 10.2 shown above.

As we have just seen, water is the donor of protons and electrons in oxygenic photosynthesis. In anoxygenic photosynthesis, other molecules are used as electron donors in the reduction of CO_2 to carbohydrates. For example, purple sulfur bacteria use hydrogen sulfide (H_2S) as the electron donor:

$$12\ H_2S + 6\ CO_2 + \text{light} \rightarrow C_6H_{12}O_6 + 6\ H_2O + 12\ S \qquad (10.5)$$

Green sulfur bacteria use sulfide ions, hydrogen, or ferrous iron as electron donors, whereas another group of bacteria use compounds derived from arsenic. The remainder of this chapter will focus on oxygenic photosynthesis, which produces the vast majority of the organic carbon used by life on Earth today and replenishes the O_2 in our atmosphere.

WORKING WITH**DATA:**

Water Is the Source of the Oxygen Produced by Photosynthesis

Original Paper

Ruben, S., M. Randall, M. D. Kamen, and J. L. Hyde. 1941. Heavy oxygen (^{18}O) as a tracer in the study of photosynthesis. *Journal of the American Chemical Society* 63(3): 877–879.

Analyze the Data

In the 1930s, Cornelius van Niel (1897–1985), then a graduate student at Stanford University, was the first to propose that the oxygen released during photosynthesis is not actually derived from carbon dioxide, but rather from the water molecules consumed in the reaction. This hypothesis was formed on the basis of the discovery that the anaerobic purple sulfur bacteria do not release oxygen during photosynthesis. Instead, these organisms convert hydrogen sulfide (H_2S) into elemental sulfur in their photosynthetic pathway (see Equation 10.5).

Given the similarities between this photosynthetic process and that performed by green plants, van Niel proposed that during aerobic photosynthesis, hydrogen is extracted from water and incorporated into glucose, leaving the oxygen to be released as elemental oxygen. This hypothesis was later confirmed by Samuel Ruben and Martin Kamen, who used the "heavy" isotope of oxygen, ^{18}O, to trace the flow of oxygen in plants. Their results showed that water is the source of the oxygen produced by photosynthesis (see Figure 10.2).

As part of the expanding research on radioisotopes during World War II, the U.S. government set up a radiation laboratory at the University of California, Berkeley. Out of this lab came key experiments that described the light-dependent and light-independent pathways of photosynthesis. In this set of experiments, *Chlorella* algal cells were exposed to water and CO_2, the latter coming from potassium carbonate (K_2CO_3) and potassium bicarbonate ($KHCO_3^-$), which dissolve in water to form CO_2. In Experiment 1 the water contained more ^{18}O than ^{16}O, whereas in Experiment 2 the CO_2 contained more ^{18}O than ^{16}O. A mass spectrometer was used to measure the isotopic contents of the reactants and the O_2 produced, and the data were presented as the isotopic ratio ($^{18}O/^{16}O$). These data are shown in the table.

QUESTION 1
In Experiment 1, was the isotopic ratio of O_2 similar to that of H_2O or to that of CO_2? What about in Experiment 2?

QUESTION 2
What can you conclude from these data?

	Time before start of O_2 collection (min)	Time at end of O_2 collection (min)	$^{18}O/^{16}O$ (percent ^{18}O in compound)		
			H_2O	$HCO_3^- + CO_3^{2-}$ (CO_2 sources)	O_2
Experiment 1: 0.09 M $KHCO_3$ + 0.09 M K_2CO_3 (^{18}O in H_2O)	0		0.85	0.20	
	45	110	0.85	0.41	0.84
	110	223	0.85	0.55	0.85
	225	350	0.85	0.61	0.86
Experiment 2: 0.14 M $KHCO_3$ + 0.06 M K_2CO_3 (^{18}O in CO_2)	0		0.20		
	40	110	0.20	0.50	0.20
	110	185	0.20	0.40	0.20

Go to **BioPortal** for all WORKING WITH**DATA** exercises

Photosynthesis involves two pathways

Equation 10.2 above summarizes the overall process of oxygenic photosynthesis, but not the stages by which it is completed. Water serves as the electron donor, but there is an intermediary carrier of the H^+ and electrons between the oxidation and reduction reactions. The carrier is the coenzyme nicotinamide adenine dinucleotide phosphate ($NADP^+$).

Like glycolysis and the other metabolic pathways that harvest energy in cells, photosynthesis is a process consisting of many reactions. These reactions are commonly divided into two main pathways:

- The **light reactions** convert light energy into chemical energy in the form of ATP and the reduced electron carrier NADPH. This molecule is similar to NADH (see Section 9.1) but with an additional phosphate group attached to the sugar of its adenosine. In general, NADPH acts as a reducing agent in photosynthesis and other anabolic reactions.
- The **light-independent reactions** (carbon-fixation reactions) do not use light directly, but instead use ATP, NADPH (made by the light reactions), and CO_2 to produce carbohydrate.

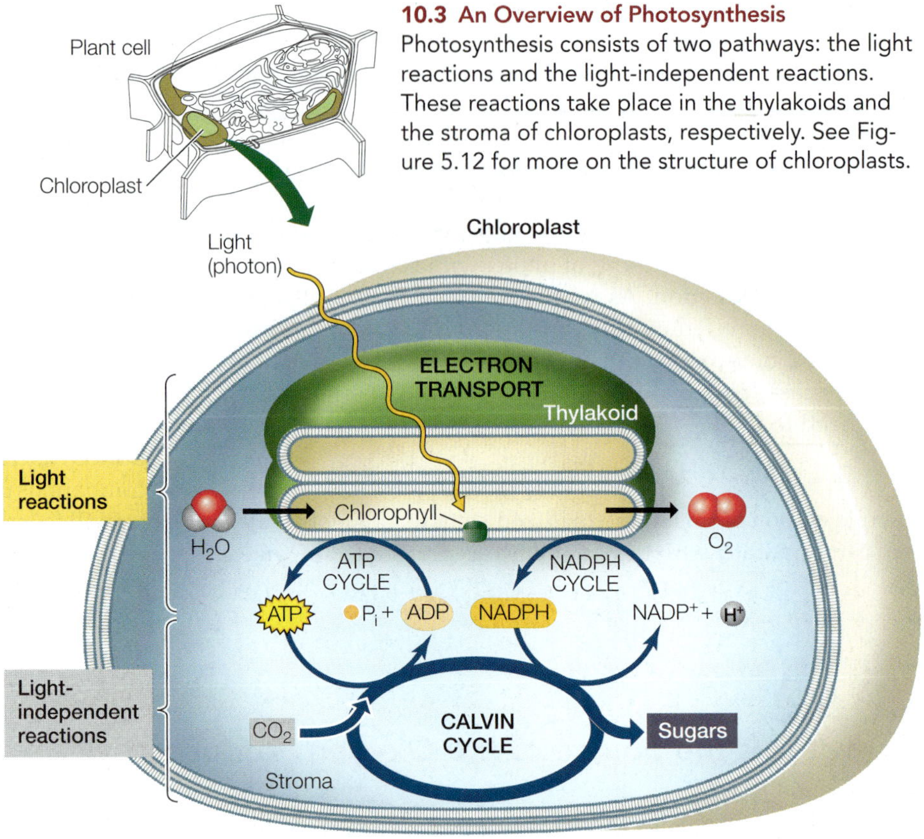
Plant cell
Chloroplast
Light (photon)
Chloroplast
ELECTRON TRANSPORT
Thylakoid
Light reactions
Chlorophyll
H_2O
O_2
ATP CYCLE
NADPH CYCLE
ATP
P_i + ADP
NADPH
$NADP^+$ + H^+
Light-independent reactions
CO_2
CALVIN CYCLE
Sugars
Stroma

10.3 An Overview of Photosynthesis Photosynthesis consists of two pathways: the light reactions and the light-independent reactions. These reactions take place in the thylakoids and the stroma of chloroplasts, respectively. See Figure 5.12 for more on the structure of chloroplasts.

The light-independent reactions are sometimes called the dark reactions because they do not directly require light energy. They are also called the carbon-fixation reactions. However, in most plants the light reactions and the light-independent reactions *stop in the dark* because ATP synthesis and $NADP^+$ reduction require light. The reactions of both pathways proceed within the chloroplast, but they occur in different parts of that organelle (**Figure 10.3**).

As we describe these two series of reactions in more detail, you will see that they conform to the principles of biochemistry that we discussed in Chapters 8 and 9: energy transformations, oxidation–reduction, and the stepwise nature of biochemical pathways.

RECAP 10.1

The light reactions of photosynthesis convert light energy into chemical energy. The light-independent reactions use that chemical energy to reduce CO_2 to carbohydrates. While most photosynthetic organisms use water as the electron donor for reduction of CO_2, some use other molecules, such as hydrogen sulfide (H_2S).

- What is the experimental evidence that water is the source of the O_2 produced during photosynthesis? **See p. 186 and Figure 10.2**
- What is the relationship between the light reactions and the light-independent reactions of photosynthesis? **See p. 188 and Figure 10.3**

We will describe the light reactions and the light-independent reactions separately. In the next section we will begin by discussing the physical nature of light and the specific photosynthetic molecules that capture its energy.

10.2 How Does Photosynthesis Convert Light Energy into Chemical Energy?

Light is a form of energy, and it can be converted to other forms of energy such as heat or chemical energy. Our focus here will be on light as the source of energy to drive the formation of ATP (from ADP and P_i) and NADPH (from $NADP^+$ and H^+).

Light energy is absorbed by chlorophyll and other pigments

It is helpful here to discuss light in terms of its photochemistry and photobiology.

PHOTOCHEMISTRY Light is a form of **electromagnetic radiation**. Electromagnetic radiation is propagated in waves, and the amount of energy in the radiation is inversely proportional to its **wavelength**—the shorter the wavelength, the greater the energy. The visible portion of the electromagnetic spectrum (**Figure 10.4**) encompasses a wide range of wavelengths and energy levels. In addition to traveling in waves, light also behaves as particles, called **photons**, which have no mass. In plants and other photosynthetic organisms, receptive molecules absorb photons in order to harvest their energy for biological processes. These receptive molecules absorb only

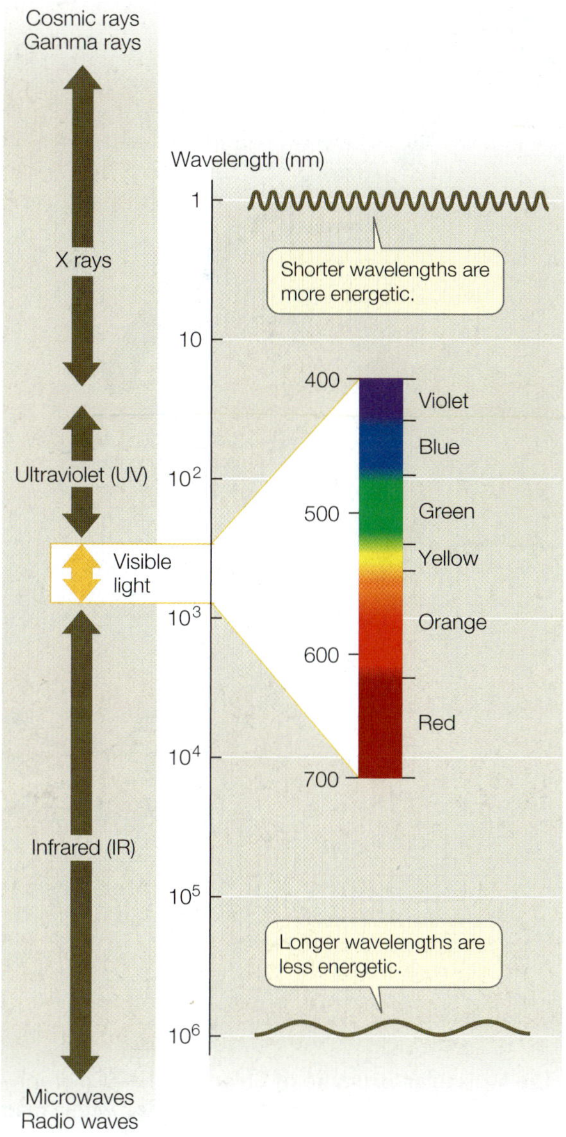
Cosmic rays
Gamma rays
X rays
Ultraviolet (UV)
Visible light
Infrared (IR)
Microwaves
Radio waves
Wavelength (nm)
1
10
10^2
10^3
10^4
10^5
10^6
Shorter wavelengths are more energetic.
Longer wavelengths are less energetic.
400
500
600
700
Violet
Blue
Green
Yellow
Orange
Red

10.4 The Electromagnetic Spectrum The portion of the electromagnetic spectrum that is visible to humans as light is shown in detail at the right.

specific wavelengths of light—photons with specific amounts of energy.

When a photon meets a molecule, one of three things can happen:

- The photon may bounce off the molecule—it may be scattered or reflected.
- The photon may pass through the molecule—it may be transmitted.
- The photon may be absorbed by the molecule, adding energy to the molecule.

Neither of the first two outcomes causes any change in the molecule. However, in the case of absorption, the photon disappears and its energy is absorbed by the molecule. The photon's energy cannot disappear, because according to the first law of thermodynamics, energy is neither created nor destroyed. When the molecule acquires the energy of the photon, it is raised from a ground state (with lower energy) to an excited state (with higher energy):

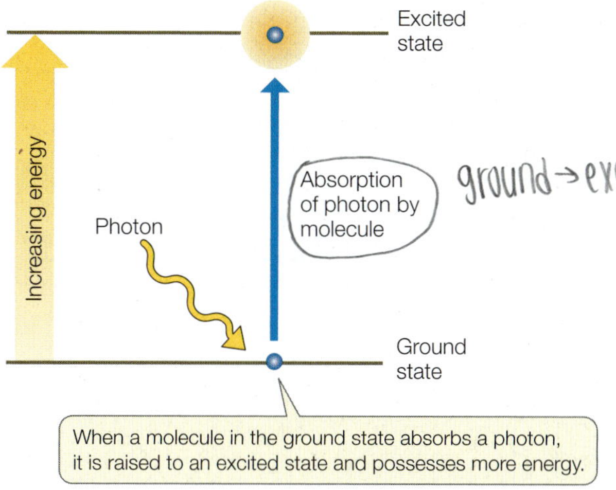
Excited state
Increasing energy
Absorption of photon by molecule
Photon
Ground state
When a molecule in the ground state absorbs a photon, it is raised to an excited state and possesses more energy.

The difference in free energy between the molecule's excited state and its ground state is approximately equal to the free energy of the absorbed photon (a small amount of energy is lost to entropy, according to the second law of thermodynamics). The increase in energy boosts one of the electrons in the molecule into a shell farther from its nucleus; this electron is now held less firmly, making the molecule unstable and more chemically reactive.

PHOTOBIOLOGY The specific wavelengths absorbed by a particular molecule are characteristic of that type of molecule. Molecules that absorb wavelengths in the visible spectrum are called **pigments**.

When a beam of white light (containing all the wavelengths of visible light) falls on a pigment, certain wavelengths are absorbed. The remaining wavelengths are scattered or transmitted and make the pigment appear to us as colored. For example, the pigment **chlorophyll** absorbs both blue and red light, and we see the remaining light, which is primarily green. If we plot light absorbed by a purified pigment against wavelength, the result is an **absorption spectrum** for that pigment.

In contrast to the absorption spectrum, an **action spectrum** is a plot of the rate of photosynthesis carried out by an organism against the wavelengths of light to which it is exposed. An action spectrum for photosynthesis can be determined as follows:

1. Place the organism in a closed container.
2. Expose it to light of a certain wavelength for a period of time.
3. Measure the rate of photosynthesis by the amount of O_2 released.
4. Repeat with light of other wavelengths.

Figure 10.5 shows the absorption spectra of pigments in *Anacharis*, a common aquarium plant, and the action spectrum for photosynthesis by that plant. The two spectra can be compared to see which pigments in *Anacharis* contribute most to light harvesting for photosynthesis.

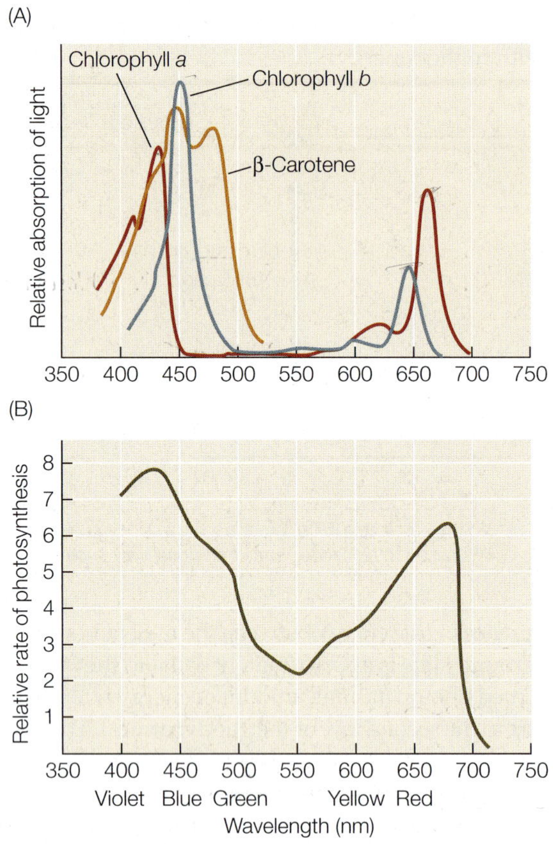

(A)
Chlorophyll a
Chlorophyll b
β-Carotene
Relative absorption of light
350 400 450 500 550 600 650 700 750
(B)
Relative rate of photosynthesis
8
7
6
5
4
3
2
1
350 400 450 500 550 600 650 700 750
Violet Blue Green Yellow Red
Wavelength (nm)

10.5 Absorption and Action Spectra (A) The absorption spectra of purified pigments from the common aquarium plant *Anacharis*. (B) The action spectrum for photosynthesis of that plant.

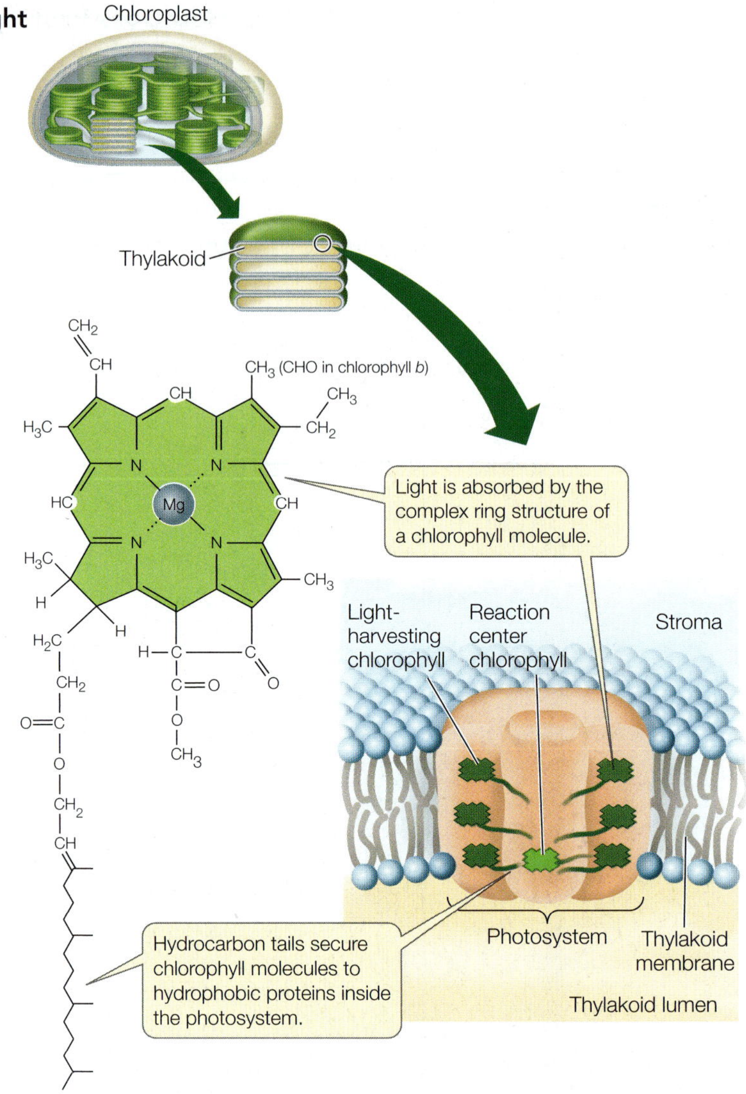

Chloroplast
Thylakoid
CH_3 (CHO in chlorophyll b)
Mg
Light is absorbed by the complex ring structure of a chlorophyll molecule.
Light-harvesting chlorophyll
Reaction center chlorophyll
Stroma
Photosystem
Thylakoid membrane
Thylakoid lumen
Hydrocarbon tails secure chlorophyll molecules to hydrophobic proteins inside the photosystem.

10.6 The Molecular Structure of Chlorophyll *a* Chlorophyll *a* consists of a complex ring structure (green area) with a magnesium atom at the center, plus a hydrocarbon "tail." The tail anchors the molecule to integral membrane proteins in the thylakoid membrane. Chlorophyll *b* is identical except for the replacement of a methyl group (—CH_3) with an aldehyde group (—CHO) at the upper right. These pigments are components of multi-protein complexes called photosystems, which span the thylakoid membrane.

The major pigment used to drive the light reactions of oxygenic photosynthesis is chlorophyll *a* (**Figure 10.6**). (In Figure 10.5 you can see that the wavelengths at which photosynthesis is highest in *Anacharis* are the same wavelengths at which chlorophyll *a* absorbs the most light.) Chlorophyll *a* has a complex ring structure, similar to that of the heme group of hemoglobin, with a magnesium ion at the center. A long hydrocarbon "tail" anchors the molecule to proteins within a large multi-protein complex called a **photosystem**, which spans the thylakoid membrane. Molecules of chlorophyll *a* and various accessory pigments (such as chlorophylls *b* and *c*, carotenoids, and phycobilins; see below) are arranged into **light-harvesting complexes**, also called antenna systems. Multiple light-harvesting complexes surround a single **reaction center** within the photosystem. Light energy is captured by the light-harvesting complexes and transferred to the reaction center, where chlorophyll *a* molecules participate in redox reactions that result in the conversion of the light energy to chemical energy.

As mentioned above, chlorophyll absorbs blue and red light, which are near the two ends of the visible spectrum (see Figure 10.4). The various accessory pigments absorb light in other parts of the spectrum and thus function to broaden the range of wavelengths that can be used for photosynthesis. You can see this in the action spectrum in Figure 10.5B: *Anacharis* is able to photosynthesize at wavelengths of light that chlorophyll *a* doesn't absorb but that other pigments absorb (e. g., 500 nm). Different photosynthetic organisms have different combinations of accessory pigments. Higher plants and green algae have chlorophyll *b* (with a structure and absorption spectrum very similar to that of chlorophyll *a*) and carotenoids such as β-carotene, which absorb photons in the blue and blue-green wavelengths. Phycobilins, which are found in red algae and cyanobacteria, absorb various yellow-green, yellow, and orange wavelengths.

Light absorption results in photochemical change

When a pigment molecule absorbs light, it enters an excited state. This is an unstable situation, and the molecule rapidly returns to its ground state, releasing most of the absorbed energy. This is an extremely rapid process—measured in picoseconds (trillionths of a second). Within a light-harvesting complex

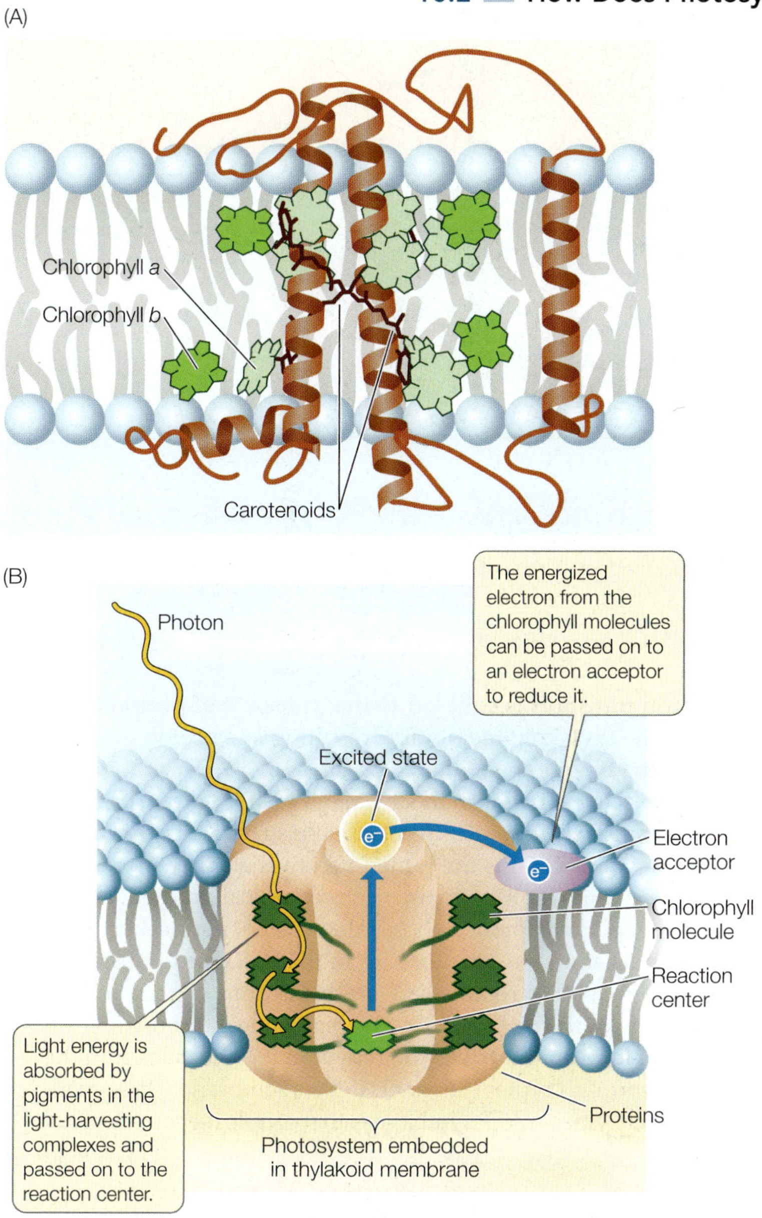
(A)
Chlorophyll a
Chlorophyll b
Carotenoids
(B)
Photon
The energized electron from the chlorophyll molecules can be passed on to an electron acceptor to reduce it.
Excited state
Electron acceptor
Chlorophyll molecule
Reaction center
Proteins
Light energy is absorbed by pigments in the light-harvesting complexes and passed on to the reaction center.
Photosystem embedded in thylakoid membrane

10.7 Energy Transfer and Electron Transport (A) The molecular structure of a single light-harvesting complex shows the polypeptide in brown with three helices that span the thylakoid membrane. Pigment molecules (carotenoids and chlorophylls *a* and *b*) are bound to the polypeptide. (B) This simplified illustration of the entire photosystem uses chlorophyll molecules to represent the light-harvesting complexes. Energy from a photon is transferred from one pigment molecule to another, until it reaches a chlorophyll *a* molecule in the reaction center. The chlorophyll *a* molecule can give up its excited electron to an electron acceptor.

(**Figure 10.7A**), the energy released by a pigment molecule (for example, chlorophyll *b*) is absorbed by other, adjacent pigment molecules. The energy (not as electrons but in the form of chemical energy called resonance) is passed from molecule to molecule until it reaches a chlorophyll *a* molecule at the reaction center of the photosystem (**Figure 10.7B**).

A ground-state chlorophyll *a* molecule at the reaction center (symbolized by Chl) absorbs the energy from the adjacent chlorophylls and becomes excited (Chl*), but to return to the ground state this chlorophyll does not pass the energy to another pigment molecule—something very different occurs. *The reaction center converts the absorbed light energy into chemical energy*. The chlorophyll molecule in the reaction center absorbs sufficient energy that it actually *gives up its excited electron to a chemical acceptor*:

$$\text{Chl}^* + \text{acceptor} \rightarrow \text{Chl}^+ + \text{acceptor}^- \quad (10.6)$$

This, then, is the first consequence of light absorption by chlorophyll: *the reaction center chlorophyll (Chl*) loses its excited electron in a redox reaction and becomes Chl*$^+$. As a result of this transfer of an electron, the chlorophyll gets oxidized, while the acceptor molecule is reduced.

Reduction leads to ATP and NADPH formation

The electron acceptor that is reduced by Chl* is the first in a chain of electron carriers in the thylakoid membrane. Electrons are passed from one carrier to another in an energetically "downhill" series of reductions and oxidations. Thus the thylakoid membrane has an electron transport system similar to the respiratory chain of mitochondria (see Section 9.3). The final electron acceptor is $NADP^+$, which gets reduced:

$$\text{NADP}^+ + \text{H}^+ + 2\,e^- \rightarrow \text{NADPH} \quad (10.7)$$

As in mitochondria, ATP is produced chemiosmotically during the process of electron transport. **Figure 10.8** shows the series of **noncyclic electron transport** reactions that use the energy from light to generate NADPH and ATP. There are two photosystems, each with its own reaction center:

- **Photosystem I** (containing the "P_{700}" chlorophylls at its reaction center) absorbs light energy best at 700 nanometers (nm) and passes its excited electrons (via intermediate molecules) to $NADP^+$, reducing it to NADPH.
- **Photosystem II** (with "P_{680}" chlorophylls at its reaction center) absorbs light energy best at 680 nm, oxidizes water molecules, and passes its energized electrons through a series of carriers to produce ATP.

Let's look in more detail at these photosystems, beginning with photosystem II.

PHOTOSYSTEM II After an excited chlorophyll in the reaction center (Chl*) gives up its energetic electron to reduce a chemical acceptor molecule, the chlorophyll lacks an electron and is very unstable. It has a strong tendency to "grab" an electron from another molecule to replace the one it lost—in chemical terms, it is a strong oxidizing agent. The replenishing electrons come from water, splitting the H—O—H bonds:

$$H_2O \rightarrow \tfrac{1}{2}\,O_2 + 2\,H^+ + 2\,e^- \quad (10.8)$$

$$2\,e^- + 2\,\text{Chl}^+ \rightarrow 2\,\text{Chl} \quad (10.9)$$

$$\text{Overall: } 2\,\text{Chl}^+ + H_2O \rightarrow 2\,\text{Chl} + 2\,H^+ + \tfrac{1}{2}\,O_2 \quad (10.10)$$

Notice that the source of O_2 in photosynthesis is H_2O.

Back to the electron acceptor in the electron transport system: the energetic electrons are passed through a series of membrane-bound carriers to a final acceptor at a lower energy

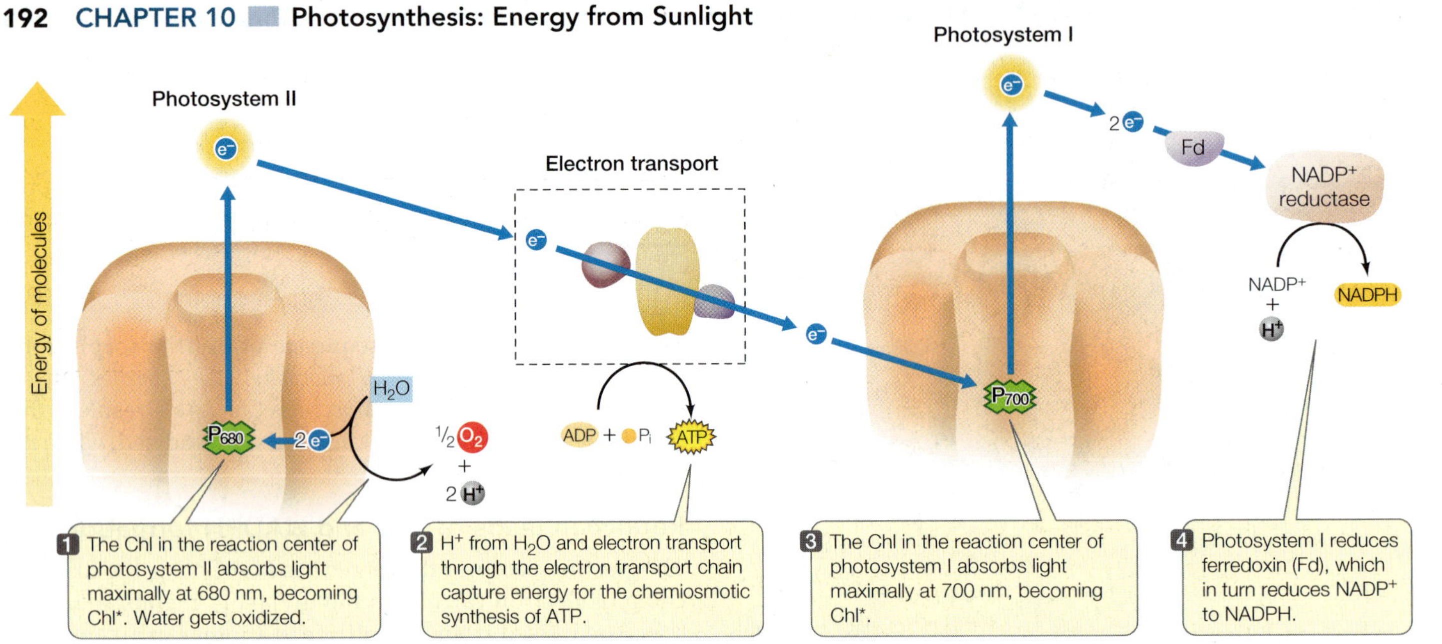
Photosystem II
Electron transport
Photosystem I
Energy of molecules
H_2O
P_{680}
2 e⁻
½ O_2 + 2 H^+
ADP + P_i
ATP
P_{700}
2 e⁻
Fd
NADP⁺ reductase
NADP⁺ + H⁺
NADPH
1 The Chl in the reaction center of photosystem II absorbs light maximally at 680 nm, becoming Chl*. Water gets oxidized.
2 H⁺ from H_2O and electron transport through the electron transport chain capture energy for the chemiosmotic synthesis of ATP.
3 The Chl in the reaction center of photosystem I absorbs light maximally at 700 nm, becoming Chl*.
4 Photosystem I reduces ferredoxin (Fd), which in turn reduces NADP⁺ to NADPH.

10.8 Noncyclic Electron Transport Uses Two Photosystems
Absorption of light energy by chlorophyll molecules in the reaction centers of photosystems I and II allows them to pass electrons into a series of redox reactions.

level. As in the mitochondrion, a proton gradient is generated and is used by ATP synthase to make ATP (see below).

PHOTOSYSTEM I In photosystem I, an excited electron from the Chl* at the reaction center reduces an acceptor. The oxidized chlorophyll (Chl^+) now "grabs" an electron, but in this case the electron comes from the last carrier in the electron transport system. This links the two photosystems chemically. They are also linked spatially, with the two photosystems adjacent to one another in the thylakoid membrane. The energetic electrons from photosystem I pass through several molecules and end up reducing $NADP^+$ to NADPH.

Next in the process of harvesting light energy to produce carbohydrates is the series of carbon-fixation reactions. These reactions require more ATP than NADPH. If the pathway we just described—the linear or noncyclic pathway—were the only set of light reactions operating, there might not be sufficient ATP for carbon fixation. **Cyclic electron transport** makes up for this imbalance. This pathway uses photosystem I and the electron transport system to produce ATP but not NADPH; it is cyclic because an electron is passed from an excited chlorophyll and recycles back to the same chlorophyll (**Figure 10.9**).

Chemiosmosis is the source of the ATP produced in photophosphorylation

In Chapter 9 we described the chemiosmotic mechanism for ATP formation in the mitochondrion. A similar mechanism, called **photophosphorylation**, operates in the chloroplast, where electron transport is coupled to the transport of protons (H^+) across the thylakoid membrane, resulting in a proton gradient across the membrane (**Figure 10.10**).

The electron carriers in the thylakoid membrane are oriented so that protons are transferred from the stroma into the lumen of the thylakoid. Thus the lumen becomes acidic with respect to the stroma, resulting in an electrochemical gradient across the thylakoid membrane, whose bilayer is not permeable to H^+. Water oxidation creates more H^+ in the thylakoid lumen and $NADP^+$ reduction removes H^+ in the stroma. Both reactions contribute to the H^+ gradient. The high concentration of H^+ in the thylakoid space drives the movement of H^+ back into the stroma through protein channels in the membrane. These channels are enzymes—ATP synthases—that couple the movement of

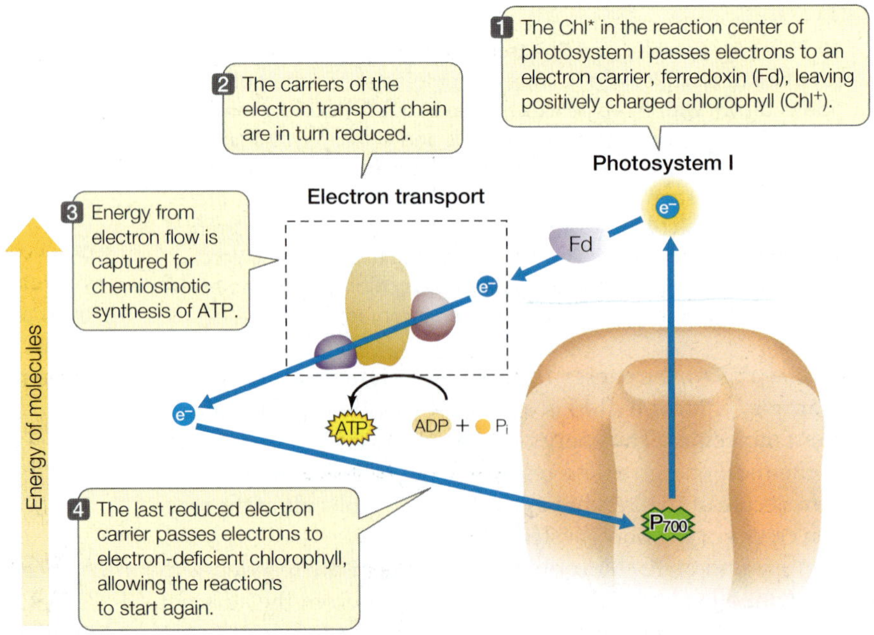
1 The Chl* in the reaction center of photosystem I passes electrons to an electron carrier, ferredoxin (Fd), leaving positively charged chlorophyll (Chl⁺).
2 The carriers of the electron transport chain are in turn reduced.
3 Energy from electron flow is captured for chemiosmotic synthesis of ATP.
Electron transport
Photosystem I
Fd
Energy of molecules
ATP
ADP + P_i
P_{700}
4 The last reduced electron carrier passes electrons to electron-deficient chlorophyll, allowing the reactions to start again.

10.9 Cyclic Electron Transport Traps Light Energy as ATP
Cyclic electron transport produces ATP but no NADPH.

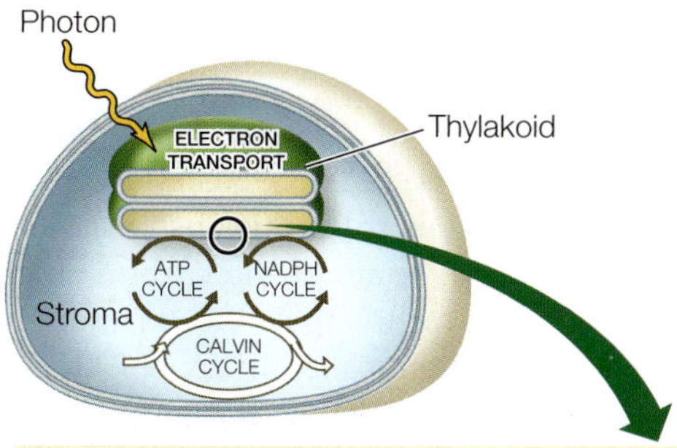
Photon
ELECTRON TRANSPORT
Thylakoid
ATP CYCLE
NADPH CYCLE
Stroma
CALVIN CYCLE

10.10 Photophosphorylation In the thylakoid membrane, electrons are passed from photosystem II to photosystem I via a series of electron carriers, including plastoquinone (PQ), cytochrome (Cyt), and plastocyanin (PC). From photosystem I the electrons are passed to ferredoxin (Fd) and then to NADP reductase. This process results in a proton gradient across the membrane, which drives ATP synthesis. Compare this illustration with Figure 9.8, where a similar process is depicted in mitochondria.

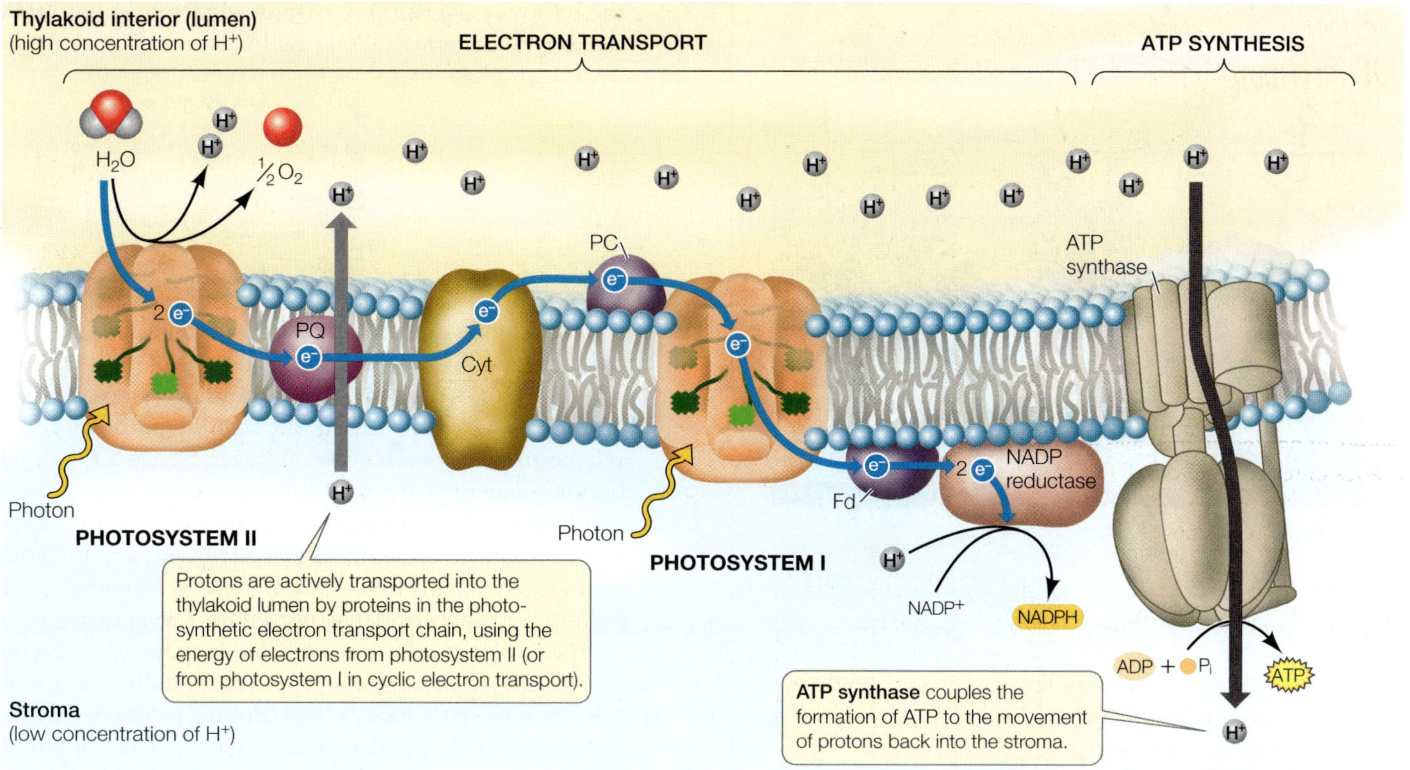
Thylakoid interior (lumen)
(high concentration of H+)
ELECTRON TRANSPORT
ATP SYNTHESIS
H2O
½O2
H+
2 e−
PQ
Cyt
PC
e−
ATP synthase
Photon
PHOTOSYSTEM II
Photon
PHOTOSYSTEM I
Fd
NADP reductase
NADP+
NADPH
ADP + Pi
ATP
Protons are actively transported into the thylakoid lumen by proteins in the photosynthetic electron transport chain, using the energy of electrons from photosystem II (or from photosystem I in cyclic electron transport).
ATP synthase couples the formation of ATP to the movement of protons back into the stroma.
Stroma
(low concentration of H+)

Go to Animated Tutorial 10.2
Photophosphorylation
Life10e.com/at10.2

protons to the formation of ATP, as they do in mitochondria (see Figure 9.8). Indeed, chloroplast ATP synthase is about 60 percent identical to human mitochondrial ATP synthase—a remarkable similarity, given that plants and animals had their most recent common ancestor more than a billion years ago. This is testimony to the evolutionary unity of life.

The mechanisms of the two enzymes are similar, but their orientations differ. In chloroplasts, protons flow through the ATP synthase out of the thylakoid lumen into the stroma (where the ATP is synthesized). In mitochondria the protons flow out of the intermembrane space into the mitochondrial matrix.

RECAP 10.2

Conversion of light energy into chemical energy occurs when pigments absorb photons. Light energy is used to drive a series of protein-associated redox reactions in the thylakoid membranes of the chloroplast.

- How does chlorophyll absorb and transfer light energy? **See pp. 190–191 and Figure 10.7**
- How are electrons produced in photosystem II, and how do they flow to photosystem I? **See pp. 191–192 and Figure 10.8**
- How does cyclic electron transport in photosystem I result in the production of ATP? **See p. 192 and Figure 10.9**

We have seen how light energy drives the synthesis of ATP and NADPH in the stroma of chloroplasts. We will now turn to the light-independent reactions of photosynthesis, which use energy-rich ATP and NADPH to reduce CO_2 and form carbohydrates.

10.3 How Is Chemical Energy Used to Synthesize Carbohydrates?

Most of the enzymes that catalyze the reactions of CO_2 fixation are dissolved in the stroma of the chloroplast, where those reactions take place. These enzymes use the energy in ATP and NADPH to reduce CO_2 to carbohydrates. Therefore, with some exceptions, CO_2 fixation occurs only in the light, when ATP and NADPH are being generated.

Radioisotope labeling experiments revealed the steps of the Calvin cycle

To identify the reactions by which the carbon from CO_2 ends up in carbohydrates, scientists found a way to label CO_2 so that they could isolate and identify the compounds formed

INVESTIGATINGLIFE

10.11 Tracing the Pathway of CO_2 How is CO_2 incorporated into carbohydrate during photosynthesis? What is the first stable covalent linkage that forms with the carbon of CO_2? Melvin Calvin and his colleagues used short exposures to $^{14}CO_2$ to identify the first compound formed from CO_2.[a]

HYPOTHESIS The first product of CO_2 fixation is a 3-carbon molecule.

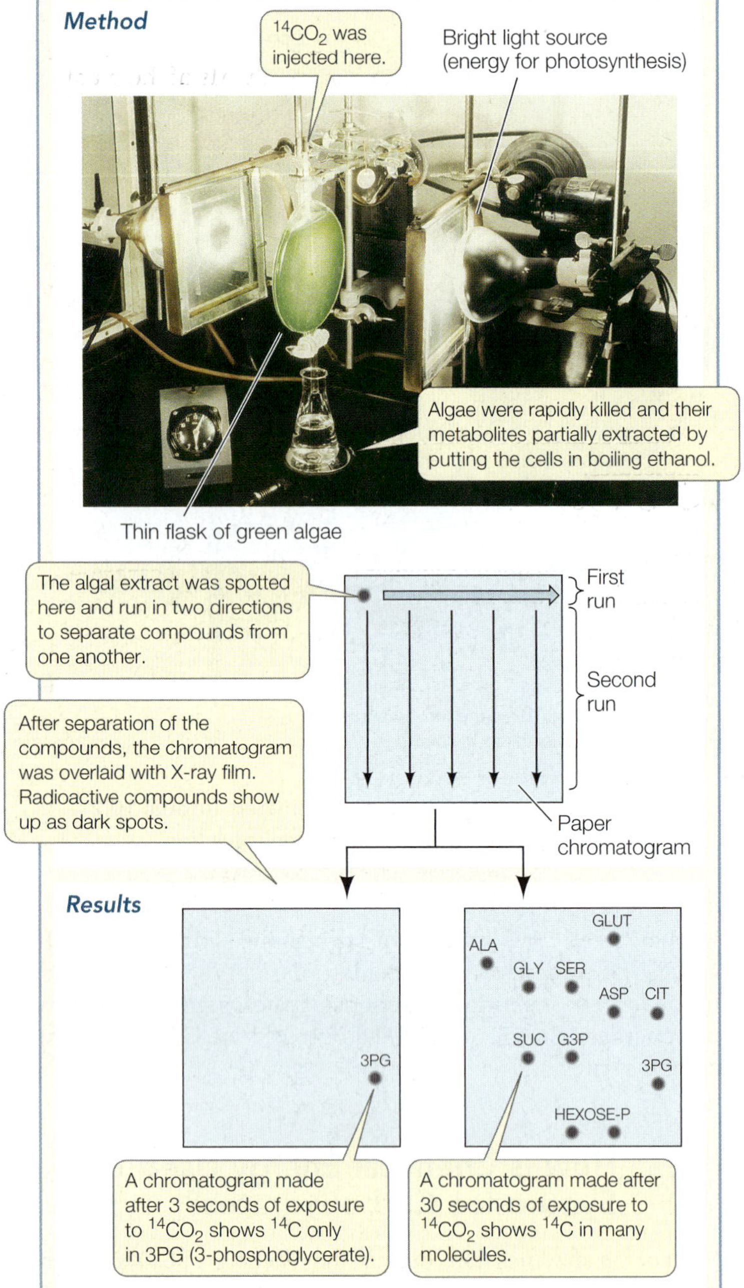

Method
$^{14}CO_2$ was injected here.
Bright light source (energy for photosynthesis)
Algae were rapidly killed and their metabolites partially extracted by putting the cells in boiling ethanol.
Thin flask of green algae
The algal extract was spotted here and run in two directions to separate compounds from one another.
First run
Second run
After separation of the compounds, the chromatogram was overlaid with X-ray film. Radioactive compounds show up as dark spots.
Paper chromatogram
Results
3PG
GLUT
ALA
GLY SER
ASP CIT
SUC G3P
3PG
HEXOSE-P
A chromatogram made after 3 seconds of exposure to $^{14}CO_2$ shows ^{14}C only in 3PG (3-phosphoglycerate).
A chromatogram made after 30 seconds of exposure to $^{14}CO_2$ shows ^{14}C in many molecules.

CONCLUSION The initial product of CO_2 fixation is 3PG. Later, the carbon from CO_2 ends up in many molecules.

Go to **BioPortal** for discussion and relevant links for all INVESTIGATINGLIFE figures.

[a]Benson, A. A., et al. 1950. *Journal of the American Chemical Society* 72: 1710–1718.

Go to Animated Tutorial 10.3
Tracing the Pathway of CO_2
Life10e.com/at10.3

from it during photosynthesis. In the 1950s, Melvin Calvin, Andrew Benson, and their colleagues used radioactively labeled CO_2 in which some of the carbon atoms were the radioisotope ^{14}C rather than the normal ^{12}C. They were able to trace the chemical pathway of CO_2 fixation (**Figure 10.11**). The first molecule that appeared in the pathway was a three-carbon sugar phosphate called 3-phosphoglycerate (3PG) (the ^{14}C is shown in red):

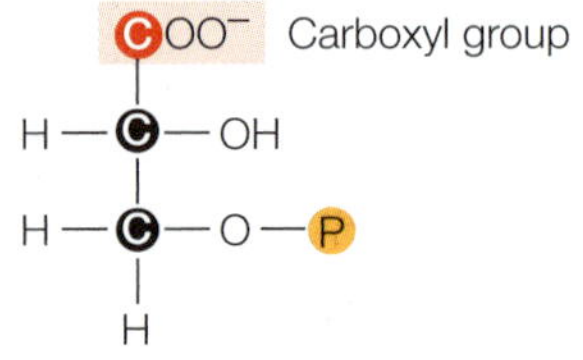

Carboxyl group

3-Phosphoglycerate (3PG)

Using successively longer exposures to $^{14}CO_2$, Calvin and his colleagues were able to trace the route of ^{14}C as it moved through a series of compounds, including monosaccharides and amino acids. It turned out that the pathway the ^{14}C moved through was a cycle. In this cycle, the CO_2 initially bonds covalently to a five-carbon acceptor molecule. The resulting six-carbon intermediate quickly breaks into two three-carbon molecules. As the cycle repeats, a carbohydrate is produced and the initial CO_2 acceptor is regenerated. This pathway was appropriately named the **Calvin cycle**.

The initial reaction in the Calvin cycle adds the one-carbon CO_2 to the five-carbon acceptor molecule ribulose 1,5-bisphosphate (RuBP). The product is an intermediate six-carbon compound, which quickly breaks down and forms two molecules of 3PG (**Figure 10.12**). The intermediate compound is broken down so rapidly that Calvin did not observe radioactive label appearing in it first. But the enzyme that catalyzes its formation, **ribulose bisphosphate carboxylase/oxygenase** (**rubisco**), is the most abundant protein in the world! It constitutes up to 50 percent of all the protein in every plant leaf.

The Calvin cycle is made up of three processes

The Calvin cycle uses the ATP and NADPH made in the light to reduce CO_2 in the stroma to a carbohydrate. Like all biochemical pathways, each reaction is catalyzed by a specific enzyme. The cycle is composed of three distinct processes (**Figure 10.13**):

- *Fixation* of CO_2. As we have seen, this reaction is catalyzed by rubisco, and its stable product is 3PG.
- *Reduction* of 3PG to form glyceraldehyde 3-phosphate (G3P). This series of reactions involves a phosphorylation (using the ATP made in the light reactions) and a reduction (coupled to the oxidation of NADPH made in the light reactions).
- *Regeneration* of the CO_2 acceptor, RuBP. Most of the G3P ends up as ribulose monophosphate (RuMP), and ATP is used to convert this compound into RuBP. So for every "turn" of the cycle, one CO_2 is fixed and one CO_2 acceptor is regenerated.

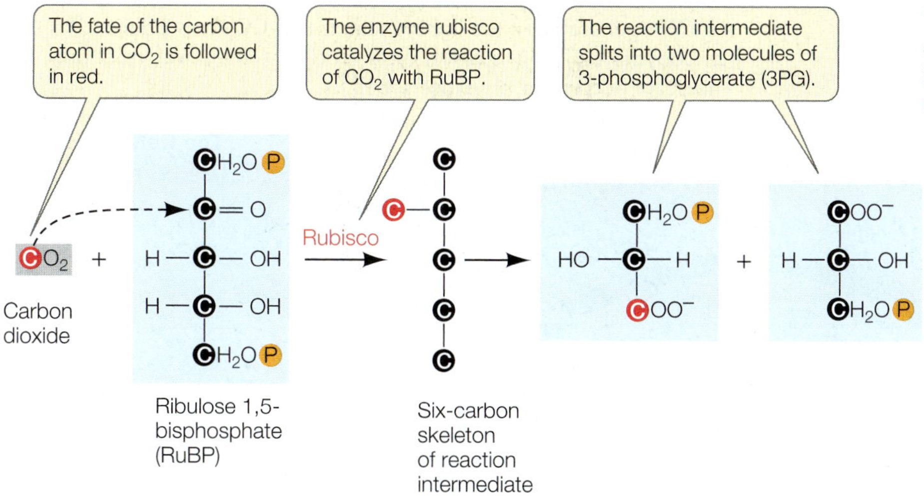
The fate of the carbon atom in CO_2 is followed in red.
The enzyme rubisco catalyzes the reaction of CO_2 with RuBP.
The reaction intermediate splits into two molecules of 3-phosphoglycerate (3PG).
CO_2
Carbon dioxide
Rubisco
Ribulose 1,5-bisphosphate (RuBP)
Six-carbon skeleton of reaction intermediate

10.12 RuBP Is the Carbon Dioxide Acceptor CO_2 is added to a five-carbon compound, RuBP. The resulting six-carbon compound immediately splits into two molecules of 3PG.

The product of this cycle is **glyceraldehyde 3-phosphate (G3P)**, which is a three-carbon sugar phosphate, also called triose phosphate:

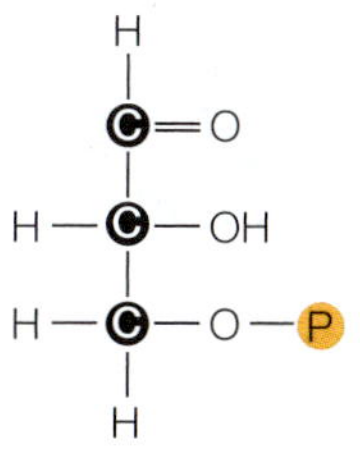

Glyceraldehyde 3-phosphate (G3P)

In a typical leaf, five-sixths of the G3P is recycled into RuBP. There are two fates for the remaining G3P, depending on the time of day and the needs of different parts of the plant:

- Some of the G3P is exported out of the chloroplast to the cytosol, where it is converted to hexoses (glucose and fructose). These molecules may be used in glycolysis and mitochondrial respiration to power the activities of photosynthetic cells (see Chapter 9) or they may be converted into the disaccharide sucrose, which is transported out of the leaf to other organs in the plant. There the sucrose is hydrolyzed to its constituent monosaccharides, which can be used as sources of energy or as building blocks for other molecules.
- Some of the G3P is used to synthesize glucose inside the chloroplast. As the day wears on, glucose molecules accumulate and are linked together to form the polysaccharide starch. This stored carbohydrate can then be drawn upon during the night so that the photosynthetic tissues can continue to export sucrose to the rest of the plant, even when photosynthesis is not taking place. In addition, starch is

WORKING WITH**DATA:**

Tracing the Pathway of CO_2

Original Papers

Calvin and his colleagues described their experiments in a series of 26 papers titled "The Path of Carbon in Photosynthesis." Perhaps the most important was one that showed how paper chromatography and labeled CO_2 could be used as tracers:

Benson, A. A., J. A. Bassham, M. Calvin, T. C. Goodale, V. A. Haas, and W. Stepka. 1950. The path of carbon in photosynthesis. V. Paper chromatography and radioautography of the products. *Journal of the American Chemical Society* 72: 1710–1718.

Analyze the Data

To elucidate the sequence of reactions that allow carbon fixation, Melvin Calvin and colleagues exposed suspensions of the green alga *Chlorella* to $^{14}CO_2$ (see Figure 10.11). After 3 seconds of photosynthesis, the ^{14}C from $^{14}CO_2$ was found only in 3-phosphoglycerate (3PG), but after 30 seconds many compounds were radioactive. Calvin and his colleagues then expanded on these results and were able to determine the exact sequence of reactions and reaction intermediates in the Calvin cycle by exposing the cells to $^{14}CO_2$ for various periods of time.

The first reaction in CO_2 fixation can occur in the dark. To show this, Calvin and his colleagues exposed *Chlorella* cells to $^{14}CO_2$ under bright lights for 20 minutes and harvested the cells. Then they repeated the experiment, but with the addition of various periods of darkness (30 sec, 2 min, and 5 min) following the 20 minutes of light. They used the harvested cells to make chromatograms and identified the labeled compounds, using a radioactivity detector to quantify the amount of ^{14}C in each compound. The data are shown in the table.

	Relative amount of radioactivity after:			
Compound	20 min light	20 min light + 30 sec dark	20 min light + 2 min dark	20 min light + 5 min dark
3PG	5,500	10,100	10,000	5,200
RuBP	4,900	680	1,850	1,800
Sucrose	13,000	13,500	15,000	14,750

QUESTION 1

Using the data in the table, plot radioactivity in 3PG versus time. What do the data show?

QUESTION 2

Why did the amount of radioactively labeled RuBP go down after 30 seconds in the dark?

Go to **BioPortal** for all WORKING WITH**DATA** exercises

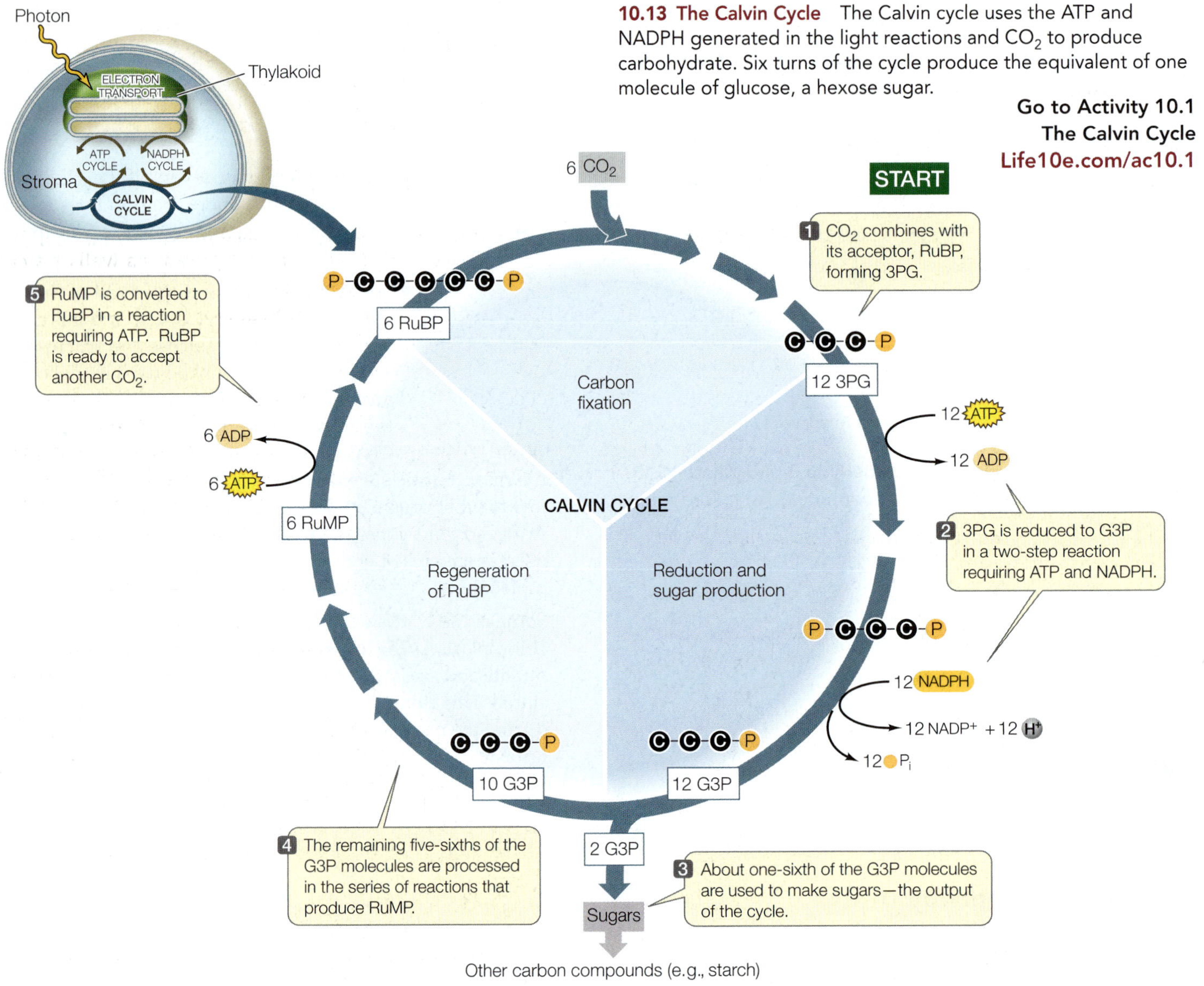
Photon
ELECTRON TRANSPORT
Thylakoid
ATP CYCLE
NADPH CYCLE
Stroma
CALVIN CYCLE
6 CO_2
START
1 CO_2 combines with its acceptor, RuBP, forming 3PG.
12 3PG
12 ATP
12 ADP
2 3PG is reduced to G3P in a two-step reaction requiring ATP and NADPH.
12 NADPH
12 $NADP^+$ + 12 H^+
12 P_i
12 G3P
2 G3P
Sugars
Other carbon compounds (e.g., starch)
3 About one-sixth of the G3P molecules are used to make sugars—the output of the cycle.
10 G3P
4 The remaining five-sixths of the G3P molecules are processed in the series of reactions that produce RuMP.
6 RuMP
6 ATP
6 ADP
5 RuMP is converted to RuBP in a reaction requiring ATP. RuBP is ready to accept another CO_2.
6 RuBP
Carbon fixation
CALVIN CYCLE
Regeneration of RuBP
Reduction and sugar production

10.13 The Calvin Cycle The Calvin cycle uses the ATP and NADPH generated in the light reactions and CO_2 to produce carbohydrate. Six turns of the cycle produce the equivalent of one molecule of glucose, a hexose sugar.

Go to Activity 10.1 The Calvin Cycle
Life10e.com/ac10.1

abundant in nonphotosynthetic organs such as roots, underground stems, and seeds, where it provides a ready supply of glucose to fuel cellular activities, including plant growth.

The plant uses the carbohydrates produced in photosynthesis to make other compounds. The carbon molecules are incorporated into amino acids, lipids, and the building blocks of nucleic acids—in fact all the organic molecules in the plant.

The products of the Calvin cycle are of crucial importance to Earth's entire biosphere. For the majority of living organisms on Earth, the C—C and C—H covalent bonds generated by the cycle provide almost all of the energy for life. Photosynthetic organisms, which are also called **autotrophs** ("self-feeders"), release most of this energy by glycolysis and cellular respiration, and use it to support their own growth, development, and reproduction. But plants are also the source of energy for other organisms. Much plant matter ends up being consumed by **heterotrophs** ("other-feeders"), such as animals, which cannot photosynthesize. Heterotrophs depend on autotrophs for both raw materials and energy. Free energy is released from food by glycolysis and cellular respiration in the cells of heterotrophs.

Light stimulates the Calvin cycle

As we have seen, the Calvin cycle uses NADPH and ATP, which are generated using energy from light. Two other processes connect the light reactions with this CO_2 fixation pathway. Both connections are indirect but significant:

- Light-induced pH changes in the stroma activate some Calvin cycle enzymes. Proton transfer from the stroma into the thylakoid lumen causes an increase in the pH of the stroma from 7 to 8 (a tenfold decrease in H^+ concentration). This favors the activation of rubisco.
- Light-induced electron transport reduces disulfide bridges in four of the Calvin cycle enzymes, thereby activating them (Figure 10.14). When ferredoxin is reduced in photosystem I (see Figure 10.8), it passes some electrons to a small, soluble protein called thioredoxin, and this protein

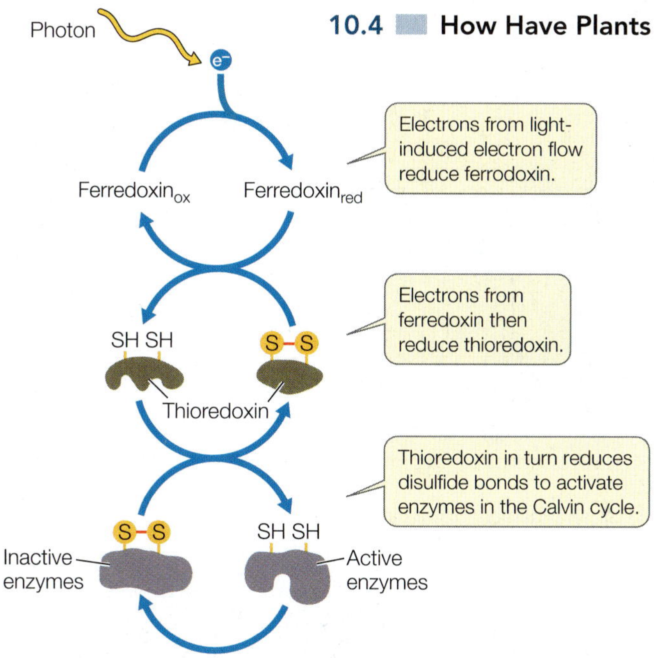
Photon
e⁻
Ferredoxin ox
Ferredoxin red
Electrons from light-induced electron flow reduce ferrodoxin.
SH SH
S–S
Electrons from ferredoxin then reduce thioredoxin.
Thioredoxin
Thioredoxin in turn reduces disulfide bonds to activate enzymes in the Calvin cycle.
S–S
SH SH
Inactive enzymes
Active enzymes

10.14 The Photochemical Reactions Stimulate the Calvin Cycle By reducing (breaking) disulfide bridges, electrons from the light reactions activate enzymes in CO_2 fixation.

passes electrons to four enzymes in the CO_2 fixation pathway. Reduction of the sulfurs in the disulfide bridges of these enzymes (see Figure 3.5) forms SH groups and breaks the bridges. The resulting changes in their three-dimensional shapes activate the enzymes and increase the rate at which the Calvin cycle operates.

RECAP 10.3

ATP and NADPH produced in the light reactions power the synthesis of carbohydrates by the Calvin cycle. This cycle fixes CO_2, reduces it, and regenerates the acceptor, RuBP, for further fixation.

- Describe the experiments that led to the identification of 3PG as the initial product of carbon fixation. **See p. 194 and Figure 10.11**
- What are the three processes of the Calvin cycle? **See p. 194 and Figure 10.13**
- In what ways does light stimulate the Calvin cycle? **See pp. 196–197 and Figure 10.14**

Although all green plants carry out the Calvin cycle, some plants have evolved variations on, or additional steps in, the light-independent reactions. These variations and additions have permitted plants to adapt to and thrive in certain environmental conditions. Let's look at these environmental limitations and the metabolic bypasses that have evolved to circumvent them.

10.4 How Have Plants Adapted Photosynthesis to Environmental Conditions?

In addition to fixing CO_2 during photosynthesis, rubisco can react with O_2. This reaction, which leads to a process called photorespiration, lowers the overall rate of CO_2 fixation in some plants. After examining this problem, we'll look at some biochemical pathways and features of plant anatomy that compensate for the limitations of rubisco.

Rubisco catalyzes the reaction of RuBP with O_2 or CO_2

As its full name indicates, rubisco (ribulose bisphosphate carboxylase/oxygenase) is an oxygenase as well as a carboxylase—it can add O_2 to the acceptor molecule RuBP instead of CO_2. The affinity of rubisco for CO_2 is about ten times stronger than its affinity for O_2. This means that inside a leaf with a normal exchange of air with the outside, CO_2 fixation is favored even though the concentration of CO_2 in the air is far less than that of O_2. But if there is an even higher concentration of O_2 in the leaf, then the O_2 competes with the CO_2, and rubisco combines RuBP with O_2 rather than CO_2. This reduces the overall amount of CO_2 that is converted into carbohydrates, and may play a role in limiting plant growth.

When O_2 is added to RuBP, one of the products is a two-carbon compound, phosphoglycolate:

$$\text{RuBP} + O_2 \rightarrow \text{phosphoglycolate} + \text{3-phosphoglycerate (3PG)} \quad (10.11)$$

The 3PG formed by rubisco's oxygenase activity enters the Calvin cycle, but the phosphoglycolate does not. Plants have evolved a metabolic pathway that can partially recover the carbon in phosphoglycolate. The phosphoglycolate is hydrolyzed to glycolate, which diffuses into peroxisomes (**Figure 10.15**). There, a series of reactions converts it into the amino acid glycine:

$$\text{Glycolate} + O_2 \rightarrow \text{glycine} \quad (10.12)$$

The glycine then diffuses into a mitochondrion, where two glycine molecules are converted in a series of reactions into the amino acid serine, releasing CO_2:

$$\text{2 Glycine} \rightarrow \text{serine} + CO_2 \quad (10.13)$$

The serine moves into the peroxisome, where it is converted to glycerate. The glycerate then moves into the chloroplast, where it is phosphorylated to make 3PG, which enters the Calvin cycle. Note that it takes two phosphoglycolate molecules from Equation 10.11 to produce the two glycines used in Equation 10.13. So overall:

$$\text{2 Phosphoglycolate (4 carbons)} + O_2 \rightarrow \text{3PG (3 carbons)} + CO_2 \quad (10.14)$$

This pathway thus reclaims 75 percent of the carbons from phosphoglycolate for the Calvin cycle. In other words, the reaction of RuBP with O_2 instead of CO_2 reduces the net carbon fixed by the Calvin cycle by 25 percent. The pathway is called **photorespiration** because it consumes O_2 and releases CO_2 and because it occurs only in the light (because of the same enzyme activation processes that were mentioned above with regard to the Calvin cycle).

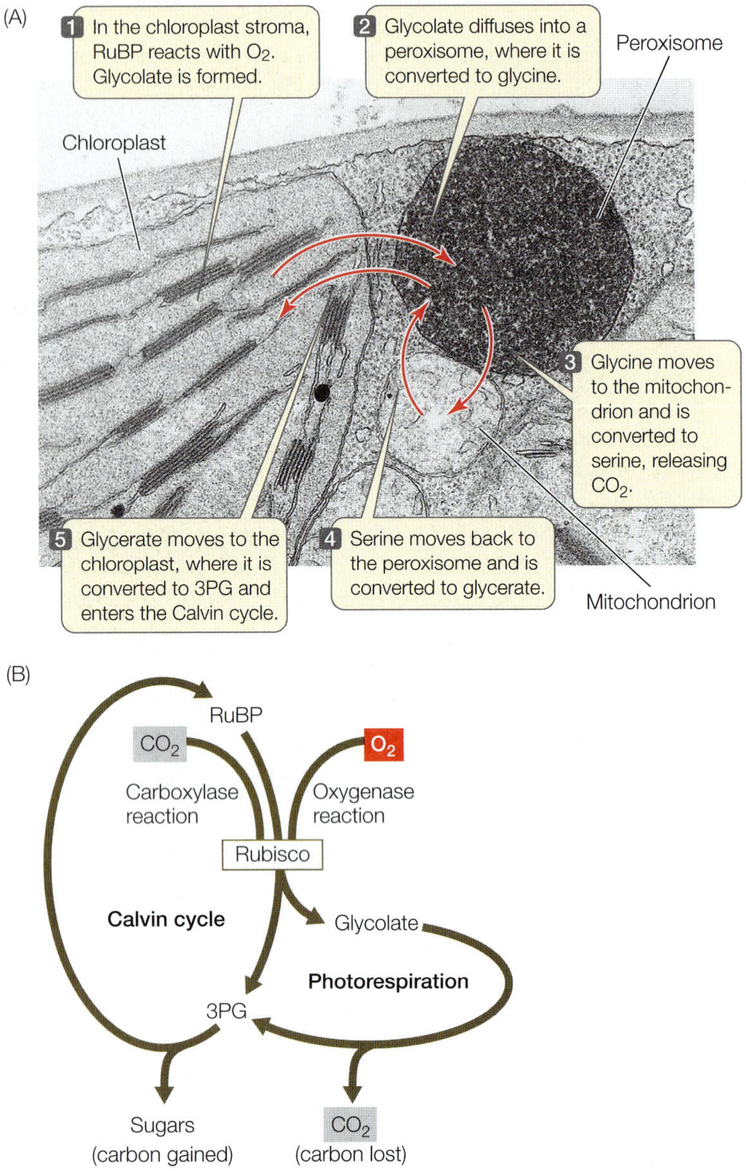
(A)
1 In the chloroplast stroma, RuBP reacts with O_2. Glycolate is formed.
2 Glycolate diffuses into a peroxisome, where it is converted to glycine.
Peroxisome
Chloroplast
3 Glycine moves to the mitochondrion and is converted to serine, releasing CO_2.
5 Glycerate moves to the chloroplast, where it is converted to 3PG and enters the Calvin cycle.
4 Serine moves back to the peroxisome and is converted to glycerate.
Mitochondrion
(B)
RuBP
CO_2
O_2
Carboxylase reaction
Oxygenase reaction
Rubisco
Calvin cycle
Glycolate
Photorespiration
3PG
Sugars (carbon gained)
CO_2 (carbon lost)

10.15 Organelles of Photorespiration (A) The reactions of photorespiration take place in the chloroplasts, peroxisomes, and mitochondria. (B) Overall, photorespiration consumes O_2 and releases CO_2.

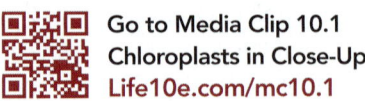

Go to Media Clip 10.1 Chloroplasts in Close-Up
Life10e.com/mc10.1

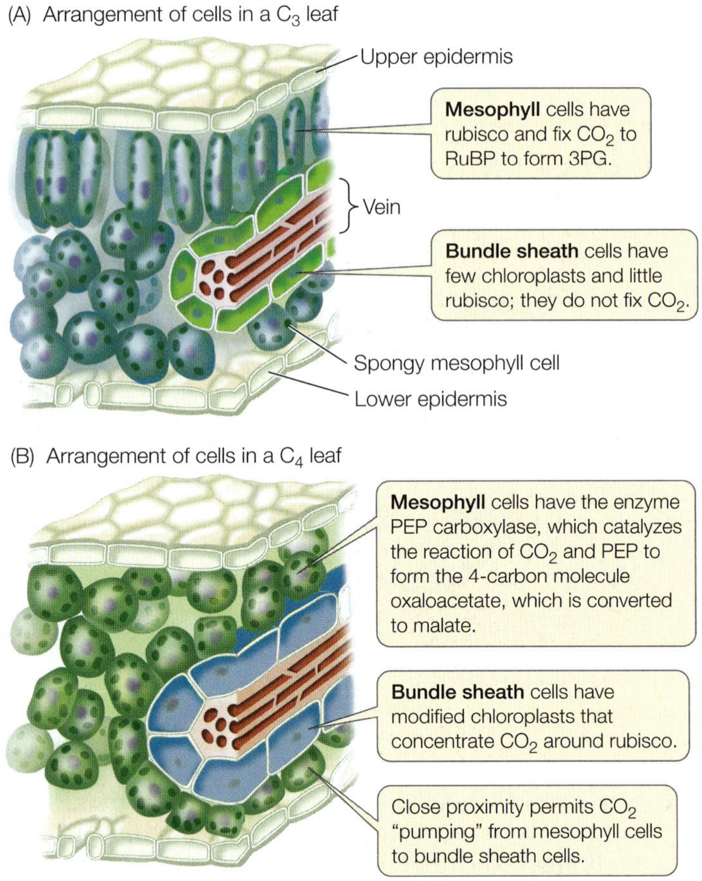
(A) Arrangement of cells in a C_3 leaf
Upper epidermis
Mesophyll cells have rubisco and fix CO_2 to RuBP to form 3PG.
Vein
Bundle sheath cells have few chloroplasts and little rubisco; they do not fix CO_2.
Spongy mesophyll cell
Lower epidermis
(B) Arrangement of cells in a C_4 leaf
Mesophyll cells have the enzyme PEP carboxylase, which catalyzes the reaction of CO_2 and PEP to form the 4-carbon molecule oxaloacetate, which is converted to malate.
Bundle sheath cells have modified chloroplasts that concentrate CO_2 around rubisco.
Close proximity permits CO_2 "pumping" from mesophyll cells to bundle sheath cells.

10.16 Leaf Anatomy of C_3 and C_4 Plants Carbon dioxide fixation occurs in different organelles and cells of the leaves in (A) C_3 plants and (B) C_4 plants. Cells that are tinted blue have rubisco.

Go to Activity 10.2 C_3 and C_4 Leaf Anatomy
Life10e.com/ac10.2

Why does rubisco act as an oxygenase as well as a carboxylase? Several factors are involved: active site affinities, concentrations of CO_2 and O_2, and temperature:

- As noted above, rubisco has a ten times higher affinity for CO_2 than for O_2, and this favors CO_2 fixation.
- In the leaf, the relative concentrations of CO_2 and O_2 vary. If O_2 is relatively abundant, rubisco acts as an oxygenase and photorespiration ensues. If CO_2 predominates, rubisco fixes it for the Calvin cycle.
- Photorespiration is more likely at high temperatures. On a hot, dry day, small pores in the leaf surface called **stomata** close to prevent water from evaporating from the leaf (see Figure 10.1). But this also prevents gases from entering and leaving the leaf. The CO_2 concentration in the leaf falls as CO_2 is used up in photosynthetic reactions, and the O_2 concentration rises because of these same reactions. As the ratio of CO_2 to O_2 falls, the oxygenase activity of rubisco is favored, and photorespiration proceeds.

C_3 plants undergo photorespiration but C_4 plants do not

Plants differ in how they fix CO_2, and can be distinguished as C_3 or C_4 plants, based on whether the first product of CO_2 fixation is a three- or four-carbon molecule. In **C_3 plants** such as roses, wheat, and rice, the first product is the three-carbon molecule 3PG—as we have just described for the Calvin cycle. In these plants the cells of the mesophyll, which makes up the main body of the leaf, are full of chloroplasts containing rubisco (Figure 10.16A). On a hot day, these leaves close their stomata to conserve water, and as a result, rubisco acts as an oxygenase as well as a carboxylase, and photorespiration occurs.

C_4 plants, which include corn, sugarcane, and tropical grasses, make the four-carbon molecule **oxaloacetate** as the first product of CO_2 fixation (Figure 10.16B). On a hot day, they partially close their stomata to conserve water, but their rate of photosynthesis does not fall. What do they do differently?

C_4 plants have evolved a mechanism that increases the concentration of CO_2 around the rubisco enzyme while at the same time isolating the rubisco from atmospheric O_2. Thus in these plants the carboxylase reaction is favored over the oxygenase reaction; the Calvin cycle operates, but photorespiration does not occur. This mechanism involves the initial fixation of CO_2 in the mesophyll cells and then the transfer of the fixed carbon (as a four-carbon molecule) to the **bundle sheath cells**, where the fixed CO_2 is released for use in the Calvin cycle (**Figure 10.17**). The bundle sheath cells are located in the interior of the leaf where less atmospheric O_2 can reach them than reaches cells near the surface of the leaf.

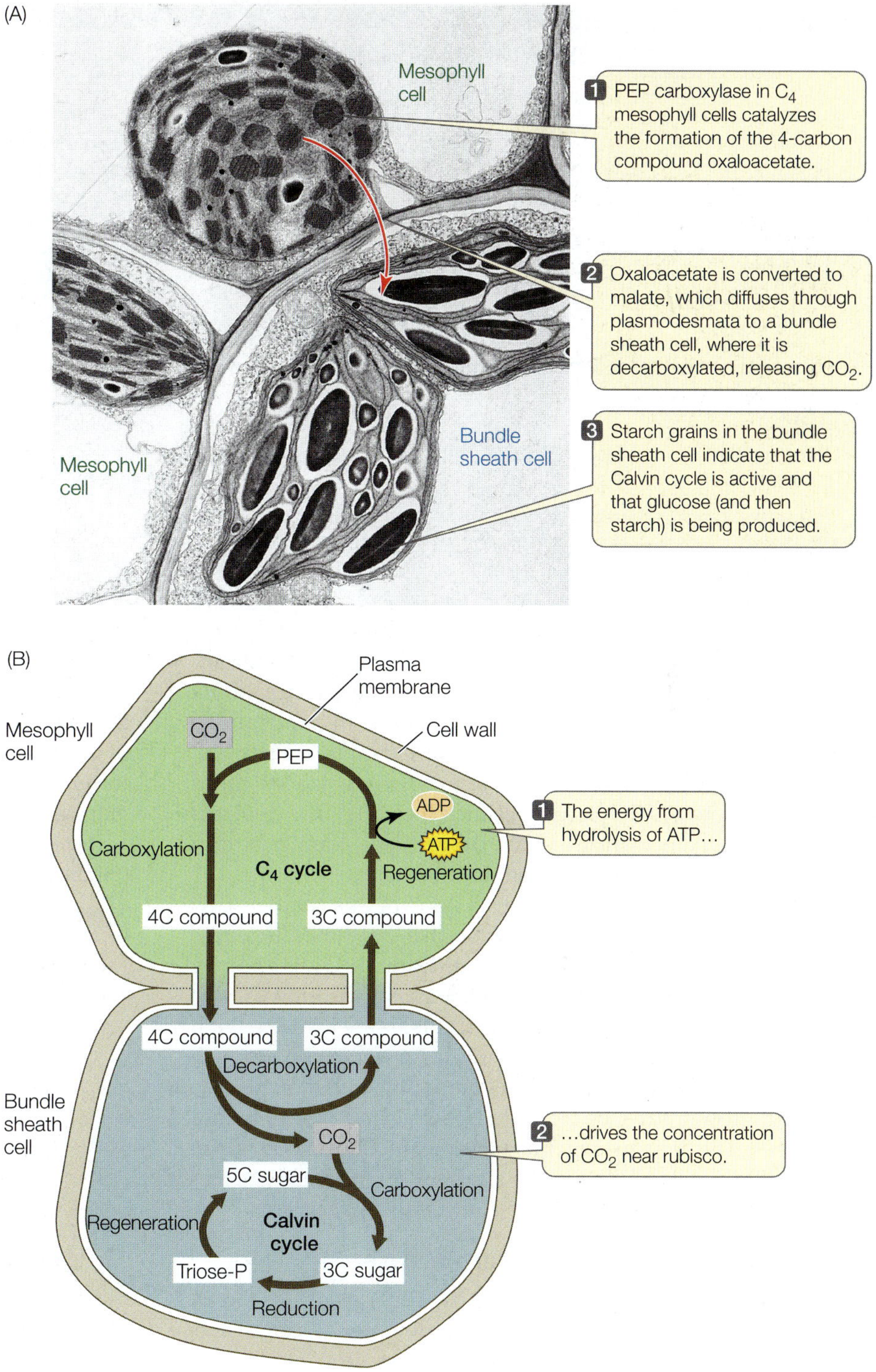
(A)
Mesophyll cell
1 PEP carboxylase in C_4 mesophyll cells catalyzes the formation of the 4-carbon compound oxaloacetate.
2 Oxaloacetate is converted to malate, which diffuses through plasmodesmata to a bundle sheath cell, where it is decarboxylated, releasing CO_2.
Bundle sheath cell
3 Starch grains in the bundle sheath cell indicate that the Calvin cycle is active and that glucose (and then starch) is being produced.
Mesophyll cell
(B)
Plasma membrane
Mesophyll cell
Cell wall
CO_2
PEP
ADP
ATP
1 The energy from hydrolysis of ATP…
Carboxylation
C_4 cycle
Regeneration
4C compound
3C compound
4C compound
3C compound
Decarboxylation
Bundle sheath cell
CO_2
2 …drives the concentration of CO_2 near rubisco.
5C sugar
Carboxylation
Regeneration
Calvin cycle
Triose-P
3C sugar
Reduction

10.17 The Anatomy and Biochemistry of C_4 Carbon Fixation (A) Carbon dioxide is fixed initially in the mesophyll cells but enters the Calvin cycle in the bundle sheath cells. (B) The two cell types share an interconnected biochemical pathway for CO_2 assimilation.

The first enzyme in this C_4 carbon fixation process, called **PEP carboxylase**, is present in the cytosols of mesophyll cells near the leaf's surface. This enzyme fixes CO_2 to a three-carbon acceptor compound, **phosphoenolpyruvate** (**PEP**), to produce the four-carbon fixation product, oxaloacetate. PEP carboxylase has two advantages over rubisco:

- It does not have oxygenase activity.
- It fixes CO_2 even at very low CO_2 levels.

So even on a hot day when the stomata are partially closed and the ratio of O_2 to CO_2 rises, PEP carboxylase just keeps on fixing CO_2.

Oxaloacetate is converted to malate, which diffuses out of the mesophyll cells and into the bundle sheath cells (see Figure 10.16B). (Some C_4 plants convert the oxaloacetate to aspartate instead of malate, but we will only discuss the malate pathway here.) The bundle sheath cells contain modified chloroplasts that are designed to concentrate CO_2 around the rubisco. There, the four-carbon malate loses one carbon (is decarboxylated), forming CO_2 and pyruvate. The latter moves back to the mesophyll cells where the three-carbon acceptor compound, PEP, is regenerated at the expense of ATP. So the "expenditure" of ATP in the mesophyll cell "pumps up" the CO_2 concentration around rubisco in the bundle sheath cell, so that it acts as a carboxylase and begins the Calvin cycle.

Under relatively cool or cloudy conditions, C_3 plants have an advantage over C_4 plants in that they don't expend energy to "pump up" the concentration of CO_2 near rubisco. But this advantage begins to be outweighed under conditions that favor photorespiration, such as warmer seasons and climates. Under these conditions C_4 plants have the advantage, especially if there is ample light to supply the extra ATP required for C_4 photosynthesis. For example, Kentucky bluegrass is a C_3 plant that thrives on lawns in April and May. But in the heat of summer it does not do as well, and Bermuda grass, a C_4 plant, takes over the lawn. The same is true on a global scale for crops: C_3 plants such as soybean, wheat, and barley have been adapted for human food production in temperate climates, whereas C_4 plants such as corn and sugarcane originated and are grown mainly in the tropics.

TABLE 10.1
Comparison of Photosynthesis in C_3, C_4, and CAM Plants

	C_3 plants	C_4 plants	CAM plants
Calvin cycle used?	Yes	Yes	Yes
Primary CO_2 acceptor	RuBP	PEP	PEP
CO_2-fixing enzyme	Rubisco	PEP carboxylase	PEP carboxylase
First product of CO_2 fixation	3PG (3-carbon)	Oxaloacetate (4-carbon)	Oxaloacetate (4-carbon)
Affinity of carboxylase for CO_2	Moderate	High	High
Photosynthetic cells of leaf	Mesophyll	Mesophyll and bundle sheath	Mesophyll with large vacuoles
Photorespiration	Extensive	Minimal	Minimal

THE EVOLUTION OF CO_2 FIXATION PATHWAYS C_3 plants are more ancient than C_4 plants. Whereas C_3 photosynthesis appears to have begun about 2.5 billion years ago, C_4 plants appeared about 12 million years ago. A possible factor in the emergence of the C_4 pathway is the decline in atmospheric CO_2. When dinosaurs dominated Earth 100 million years ago, the concentration of CO_2 in the atmosphere was four times what it is now. As CO_2 levels declined thereafter, the C_4 plants would have gained an advantage over their C_3 counterparts in high-temperature, high-light environments.

As we described in the opening essay of this chapter, CO_2 levels have been increasing over the past 200 years. Currently, the level of CO_2 is not enough for maximal CO_2 fixation by rubisco, so photorespiration occurs, reducing the growth rates of C_3 plants. Under hot conditions, C_4 plants are favored. But if CO_2 levels in the atmosphere continue to rise, the reverse will occur and C_3 plants will have a comparative advantage. The overall growth rates of crops such as rice and wheat should increase. This may or may not translate into more food, given that other effects of the human-spurred CO_2 increase (such as global climate change) will also alter Earth's ecosystems.

CAM plants also use PEP carboxylase

Other plants besides the C_4 plants use PEP carboxylase to fix and accumulate CO_2. They include some water-storing plants (succulents) of the family Crassulaceae, many cacti, pineapples, and several other kinds of flowering plants. The CO_2 metabolism of these plants is called **crassulacean acid metabolism**, or **CAM**, after the family of succulents in which it was discovered. CAM is much like the metabolism of C_4 plants in that CO_2 is initially fixed into a four-carbon compound. But in CAM plants the initial CO_2 fixation and the Calvin cycle are *separated in time rather than space*.

- At night, when it is cooler and water loss is minimized, the stomata open. CO_2 is fixed in mesophyll cells to form the four-carbon compound oxaloacetate, which is converted into malate and stored in the vacuole.
- During the day, when the stomata close to reduce water loss, the accumulated malate is shipped from the vacuole to the chloroplasts, where its decarboxylation supplies the CO_2 for the Calvin cycle and the light reactions supply the necessary ATP and NADPH.

CAM benefits the plant by allowing it to close its stomata during the day. As you will learn in Chapter 35, plants lose most of the water that they take up in their roots by evaporation through the leaves (transpiration). In dry climates, closing stomata is a key to water conservation and survival.

Table 10.1 compares photosynthesis in C_3, C_4, and CAM plants.

RECAP 10.4

Rubisco catalyzes the carboxylation of RuBP to form two 3PG, and the oxygenation of RuBP to form one 3PG and one phosphoglycolate. The diversion of rubisco to its oxygenase function decreases net CO_2 fixation. C_4 photosynthesis and CAM allow plants to fix CO_2 under warm, dry conditions when stomata are closed and CO_2 entry into the leaf is limited.

- Explain how photorespiration recovers some of the carbon that is channeled away from the Calvin cycle. **See p. 197 and Figure 10.15**
- How do C_4 plants keep the concentration of CO_2 around rubisco high, and why? **See pp. 198–199 and Figure 10.17**
- What is the pathway for CO_2 fixation in CAM plants? **See p. 200**

Now that we understand how photosynthesis produces carbohydrates, let's see how the pathways of photosynthesis are connected to other metabolic pathways.

10.5 How Does Photosynthesis Interact with Other Pathways?

Green plants are autotrophs and can synthesize all the molecules they need from simple starting materials: CO_2, H_2O, phosphate, sulfate, ammonium ions (NH_4^+), and small quantities of other mineral nutrients. The NH_4^+ is needed to synthesize amino acids and nucleotides, and it comes from

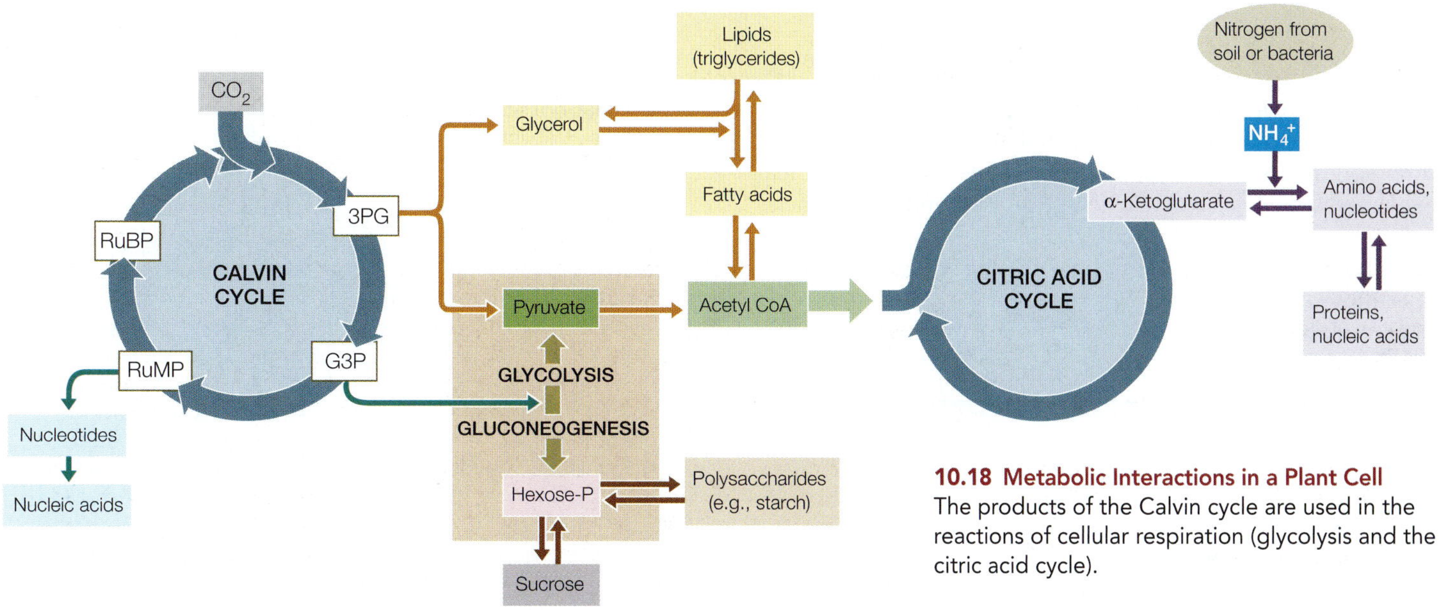

Lipids (triglycerides)
Nitrogen from soil or bacteria
CO_2
Glycerol
NH_4^+
3PG
Fatty acids
α-Ketoglutarate
Amino acids, nucleotides
RuBP
CALVIN CYCLE
Pyruvate
Acetyl CoA
CITRIC ACID CYCLE
Proteins, nucleic acids
RuMP
G3P
GLYCOLYSIS
GLUCONEOGENESIS
Nucleotides
Polysaccharides (e.g., starch)
Hexose-P
Nucleic acids
Sucrose

10.18 Metabolic Interactions in a Plant Cell The products of the Calvin cycle are used in the reactions of cellular respiration (glycolysis and the citric acid cycle).

nitrogen-containing molecules in soil water or from N_2 gas fixed from the atmosphere by bacteria, as we will see in Chapter 36.

Plants use the carbohydrates generated in photosynthesis to provide energy for processes such as active transport and anabolism. Both cellular respiration and fermentation can occur in plants, although the former is far more common. Unlike photosynthesis, plant cellular respiration occurs all the time in both the light and the dark.

Photosynthesis and respiration are closely linked through the Calvin cycle (**Figure 10.18**). The partitioning of G3P is particularly important:

- Some G3P from the Calvin cycle enters glycolysis and is converted into pyruvate in the cytosol. This pyruvate can be used in cellular respiration for energy, or its carbon skeletons can be used in anabolic reactions to make lipids, proteins, and other carbohydrates (see Figure 9.13).
- Some G3P can enter a pathway that is the reverse of glycolysis (gluconeogenesis; see Section 9.5). In this case, hexose phosphates (hexose P) and then sucrose are formed and transported to the nonphotosynthetic tissues of the plant (such as the root).

Energy flows from sunlight to reduced carbon in photosynthesis, then to ATP in respiration. Energy can also be stored in the bonds of macromolecules such as polysaccharides, lipids, and proteins. For a plant to grow, energy storage (as body structures) must exceed energy release; that is, overall carbon fixation by photosynthesis must exceed respiration. This principle is the basis of the ecological food chain, as we will see in later chapters.

Photosynthesis provides most of the energy that we need for life. Given the uncertainties about the future of photosynthesis (because of changes in CO_2 levels and climate change), it would be wise to seek ways to improve photosynthetic efficiency. **Figure 10.19** shows the various ways in which solar energy is used by plants or lost. In essence, only about 5 percent of

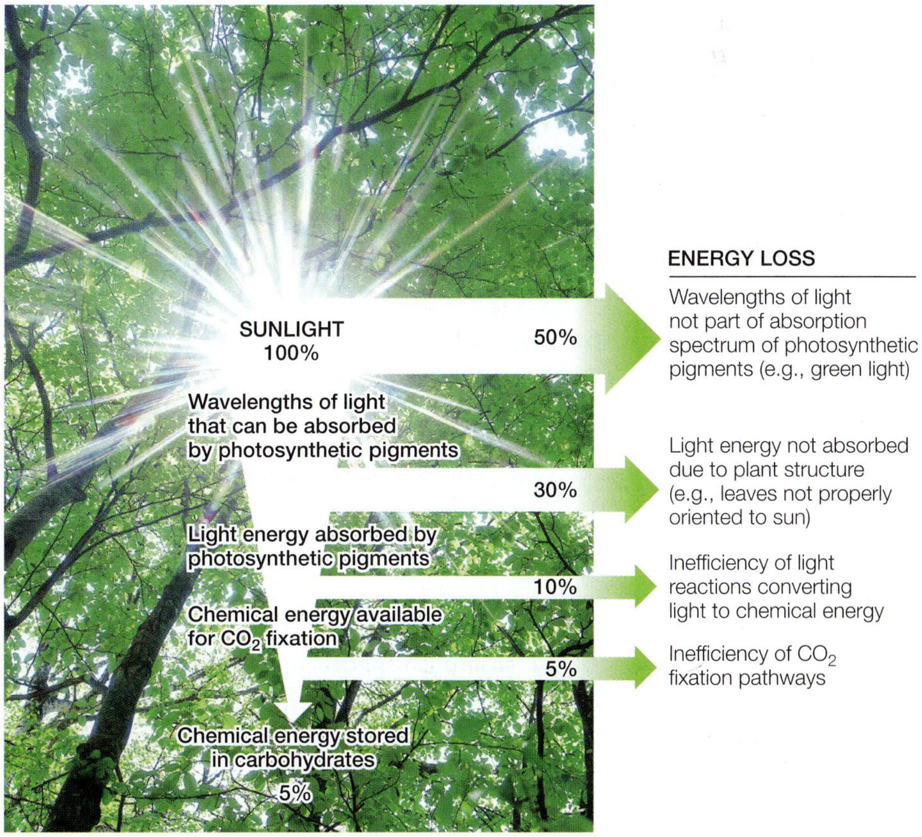

SUNLIGHT 100%
50%
Wavelengths of light that can be absorbed by photosynthetic pigments
30%
Light energy absorbed by photosynthetic pigments
10%
Chemical energy available for CO_2 fixation
5%
Chemical energy stored in carbohydrates
5%
ENERGY LOSS
Wavelengths of light not part of absorption spectrum of photosynthetic pigments (e.g., green light)
Light energy not absorbed due to plant structure (e.g., leaves not properly oriented to sun)
Inefficiency of light reactions converting light to chemical energy
Inefficiency of CO_2 fixation pathways

10.19 Energy Losses in Photosynthesis Photosynthetic pathways preserve at most about 5 percent of the sun's energy input as chemical energy in carbohydrates.

the sunlight that reaches Earth is converted into plant growth. The inefficiencies of photosynthesis involve basic chemistry and physics (some light energy is not absorbed by photosynthetic pigments) as well as biology (plant anatomy and leaf exposure, the oxygenase reaction of rubisco, and inefficiencies in metabolic pathways). While it is hard to change chemistry and physics, biologists might be able to use their knowledge of plants to improve on the basic biology of photosynthesis. This could result in a more efficient use of resources and better food production.

RECAP 10.5

The products of photosynthesis are used in glycolysis and the citric acid cycle, as well as in the synthesis of lipids, proteins, and other large molecules.

- How do common intermediates link the pathways of glycolysis, the citric acid cycle, and photosynthesis? **See p. 201 and Figure 10.18**
- Why is only 5 percent of the solar radiation that reaches Earth captured by plants? **See pp. 201–202 and Figure 10.19**

?

What possible effects will increased atmospheric CO_2 have on global food production?

ANSWER

Crops could be affected by increased atmospheric CO_2 in multiple ways. Higher CO_2 levels generally lead to increased photosynthesis. This is especially true for C_3 plants, which are more sensitive than C_4 plants to CO_2 levels. Because increased photosynthesis leads to greater plant growth, C_3 crops such as wheat and rice will tend to grow more. However, it is unclear whether this growth will be in the vegetative parts of the plant (stems and leaves) or in the part we eat (grain). To further complicate matters, such increases in plant growth may be counteracted by the effects of increased CO_2 on climate. For example, increased temperatures would increase the rate of photosynthesis and extend the growing season, but might alter rainfall patterns. In some areas of the world there might be less rain, and this could limit plant growth. Overall, while there will be significant regional variations, biologists estimate that increased CO_2 will result in moderately increased food production.

CHAPTER**SUMMARY** 10

10.1 What Is Photosynthesis?

- In the process of **photosynthesis**, the energy of sunlight is captured and used to convert CO_2 into more complex carbon-containing compounds. **See ANIMATED TUTORIAL 10.1**
- Plants, algae, and cyanobacteria live in aerobic environments and carry out oxygenic photosynthesis: the conversion of CO_2 and H_2O into carbohydrates.
- Some bacteria that live in anaerobic environments carry out anoxygenic photosynthesis, in which energy from the sun is used to fix CO_2 without the use of H_2O and the production of O_2.
- The **light reactions** of photosynthesis convert light energy into chemical energy. They produce ATP and reduce $NADP^+$ to NADPH. **Review Figure 10.3**
- The **light-independent reactions** do not use light directly but instead use ATP and NADPH to reduce CO_2, forming carbohydrates.

10.2 How Does Photosynthesis Convert Light Energy into Chemical Energy?

- Light is a form of electromagnetic radiation. It is emitted in particle-like packets called **photons** but has wavelike properties.
- Molecules that absorb light in the visible spectrum are called **pigments**. Photosynthetic organisms have several pigments, most notably **chlorophylls**, but also accessory pigments such as carotenoids and phycobilins.
- Absorption of a photon 5puts an electron of a pigment molecule in an excited state that has more energy than its ground state.
- Each pigment has a characteristic **absorption spectrum**. An **action spectrum** reflects the rate of photosynthesis carried out by a photosynthetic organism at a given wavelength of light. **Review Figure 10.5**
- The pigments in photosynthetic organisms are arranged into **light-harvesting complexes** that absorb energy from light and funnel this energy to chlorophyll *a* molecules in the reaction center of the **photosystem**. Chlorophyll can act as a reducing agent, transferring excited electrons to other molecules. **Review Figure 10.7**
- **Noncyclic electron transport** uses photosystems I and II to produce ATP, NADPH, and O_2. **Cyclic electron transport** uses only photosystem I and produces only ATP. Both systems generate ATP via the electron transport chain. **Review Figures 10.8, 10.9**
- Chemiosmosis is the mechanism of ATP production in **photophosphorylation**. **Review Figure 10.10, ANIMATED TUTORIAL 10.2**

10.3 How Is Chemical Energy Used to Synthesize Carbohydrates?

- The **Calvin cycle** makes carbohydrates from CO_2. The cycle consists of three processes: fixation of CO_2, reduction and carbohydrate production, and regeneration of RuBP. **See ANIMATED TUTORIAL 10.3**
- RuBP is the initial CO_2 acceptor, and 3PG is the first stable product of CO_2 fixation. The enzyme **rubisco** catalyzes the reaction of CO_2 and RuBP to form 3PG. **Review Figure 10.12**
- ATP and NADPH formed by the light reactions are used in the reduction of 3PG to form **G3P**. **Review Figure 10.13, ACTIVITY 10.1**
- Light stimulates enzymes in the Calvin cycle, further integrating the light-dependent and light-independent pathways. **Review Figure 10.14**

continued

10.4 How Have Plants Adapted Photosynthesis to Environmental Conditions?

- Rubisco can catalyze a reaction between O_2 and RuBP in addition to the reaction between CO_2 and RuBP. At high temperatures and low CO_2 concentrations, the oxygenase function of rubisco is favored over its carboxylase function.
- The oxygenase reaction catalyzed by rubisco significantly reduces the efficiency of photosynthesis. The subsequent reactions of **photorespiration** recover some of the fixed carbon that otherwise would be lost. **Review Figure 10.15**
- In **C_4 plants**, CO_2 reacts with **phosphoenolpyruvate (PEP)** to form a four-carbon intermediate in mesophyll cells. The four-carbon product releases its CO_2 to rubisco in the **bundle sheath cells** in the interior of the leaf. **Review Figures 10.16, 10.17, ACTIVITY 10.2**
- **CAM** plants operate much like C_4 plants, but their initial CO_2 fixation by PEP carboxylase is temporally separated from the Calvin cycle, rather than spatially separated as in C_4 plants.

10.5 How Does Photosynthesis Interact with Other Pathways?

- Photosynthesis and cellular respiration are linked through the **Calvin cycle**, the citric acid cycle, and glycolysis. **Review Figure 10.18**
- To survive, a plant must photosynthesize more than it respires.
- Photosynthesis uses only a small portion of the energy of sunlight. **Review Figure 10.19**

Go to the Interactive Summary to review key figures, Animated Tutorials, and Activities Life10e.com/is10

CHAPTER**REVIEW**

REMEMBERING

1. In noncyclic photosynthetic electron transport, water is used to
   a. excite chlorophyll.
   b. hydrolyze ATP.
   c. reduce P_i.
   d. oxidize NADPH.
   e. reduce chlorophyll.

2. In cyclic electron transport,
   a. oxygen gas is released.
   b. ATP is formed.
   c. water donates electrons and protons.
   d. NADPH forms.
   e. CO_2 reacts with RuBP.

3. In chloroplasts,
   a. light leads to the flow of protons out of the thylakoids.
   b. ATP is formed when protons flow into the thylakoid lumen.
   c. light causes the thylakoid lumen to become less acidic than the stroma.
   d. protons return actively to the stroma through protein channels.
   e. proton transfer requires ATP.

4. Which statement about chlorophylls is *not* true?
   a. Chlorophylls absorb light near both ends of the visible spectrum.
   b. Chlorophylls can accept energy from other pigments, such as carotenoids.
   c. Excited chlorophyll can either reduce another substance or release light energy.
   d. Excited chlorophyll cannot be an oxidizing agent.
   e. Chlorophylls contain magnesium.

5. Which statement about the Calvin cycle is *not* true?
   a. CO_2 reacts with RuBP to form 3PG.
   b. RuBP forms by the metabolism of 3PG.
   c. ATP and NADPH form when 3PG is reduced.
   d. The concentration of 3PG rises if the light is switched off.
   e. Rubisco catalyzes the reaction of CO_2 and RuBP.

6. Photosynthesis in green plants occurs only during the day. Respiration in plants occurs
   a. only at night.
   b. only when there is enough ATP.
   c. only during the day.
   d. all the time.
   e. in the chloroplast after photosynthesis.

UNDERSTANDING & APPLYING

7. Both photosynthetic electron transport and the Calvin cycle stop in the dark. Which specific reaction stops first? Which stops next? Continue answering the question "Which stops next?" until you have explained why both pathways have stopped.

8. Differentiate between cyclic and noncyclic electron transport in terms of (*a*) the products and (*b*) the source of electrons for the reduction of oxidized chlorophyll.

9. Trace the pathway of carbon fixed by CO_2 in photosynthesis to a carbon atom in a protein.

ANALYZING & EVALUATING

10. If water labeled with ^{18}O is added to a suspension of photosynthesizing chloroplasts, which of the following compounds will first become labeled with ^{18}O: ATP, NADPH, O_2, or 3PG? If water labeled with ^3H is added, which of the same compounds will first become radioactive? Which will be first if CO_2 labeled with ^{14}C is added?

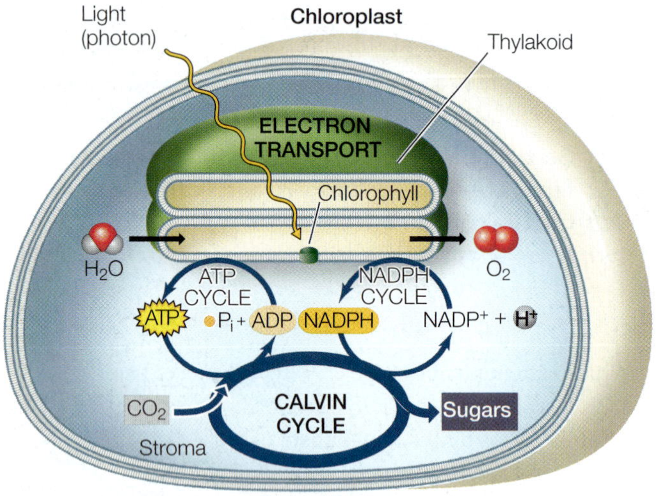
Light (photon)
Chloroplast
Thylakoid
ELECTRON TRANSPORT
Chlorophyll
H2O
O2
ATP CYCLE
NADPH CYCLE
ATP
Pi + ADP
NADPH
NADP+ + H+
CO2
CALVIN CYCLE
Sugars
Stroma

11. The Viking I Lander arrived on Mars in 1976 to detect signs of life. Explain the rationale behind the following experiments this unmanned probe performed:
    a. A scoop of dirt was inserted into a container and $^{14}CO_2$ was added. After a while during the Martian day, the $^{14}CO_2$ was removed and the dirt was heated to a high temperature. Scientists monitoring the experiment back on Earth looked for the release of $^{14}CO_2$ as a sign of life.
    b. The same experiment was performed, except that the dirt was heated to a high temperature for 30 minutes and then allowed to cool to Martian temperature right after scooping, and before the $^{14}CO_2$ was added.

Go to BioPortal at **yourBioPortal.com** for Animated Tutorials, Activities, LearningCurve Quizzes, Flashcards, and many other study and review resources.

PART FOUR Genes and Heredity

11 The Cell Cycle and Cell Division

CHAPTER OUTLINE

11.1 How Do Prokaryotic and Eukaryotic Cells Divide?

11.2 How Is Eukaryotic Cell Division Controlled?

11.3 What Happens during Mitosis?

11.4 What Role Does Cell Division Play in a Sexual Life Cycle?

11.5 What Happens during Meiosis?

11.6 In a Living Organism, How Do Cells Die?

11.7 How Does Unregulated Cell Division Lead to Cancer?

Henrietta Lacks and HeLa Cells
Lacks' tumor cells have far outlived her sad demise from cancer. They reproduce rapidly on the surface of a solid medium. They have been cultured in a laboratory since her death in 1951.

ON JANUARY 29, 1951, 30-year-old Henrietta Lacks visited the nearby Johns Hopkins Hospital in Baltimore, Maryland, because she had been bleeding abnormally after the birth of her last child. The physician found the reason for the bleeding: a tumor the size of a quarter on her cervix. A piece of the tumor was sent to a pathologist in a clinical laboratory, who reported that the tumor was malignant.

A week later Lacks was back at the hospital, where physicians treated her tumor with radiation to try to kill it. But before the treatment began, they took a small sample of cells and sent them to the research lab of George and Margaret Gey, two scientists at the hospital who had been trying for 20 years to coax human cells to live and multiply outside the body. If they could do so, they thought, they might find a cure for cancer. The Geys hit pay dirt with Lacks' cells; they grew more vigorously than any cells they had ever seen. Unfortunately, they also grew fast in Lacks' body, and in a few months they had spread to almost all of her organs. On the day she died, October 4, 1951, Dr. George Gey appeared on national television with a test tube of her cells, which he called HeLa cells. "It is possible that, from a fundamental study such as this, we will ... learn a way by which cancer can be completely wiped out," he said.

Because of their robust ability to reproduce, HeLa cells quickly became a staple of cell biology research. In controlled settings they could be infected with viruses, and they were instrumental in developing the supply of polioviruses that led to the first vaccine against that dread disease. HeLa cells have been used for important basic and applied research ever since. Although Lacks had never been outside Virginia and Maryland, her cells have traveled all over the world and even into space on the space shuttle. Over the past 60 years, tens of thousands of research articles have been published using information obtained from Lacks' cells. Her cells grow so well in the lab that they have sometimes contaminated and taken over cell cultures of other cell types. If they aren't careful, researchers who think they are studying, say, kidney cells may be studying HeLa cells instead.

Understanding the cell division cycle and its control is clearly an important subject for understanding cancer. But cell division is not just important in medicine. It underlies the growth, development, and reproduction of all organisms.

?

What makes HeLa cells reproduce so well in the laboratory??

See answer on p. 229.

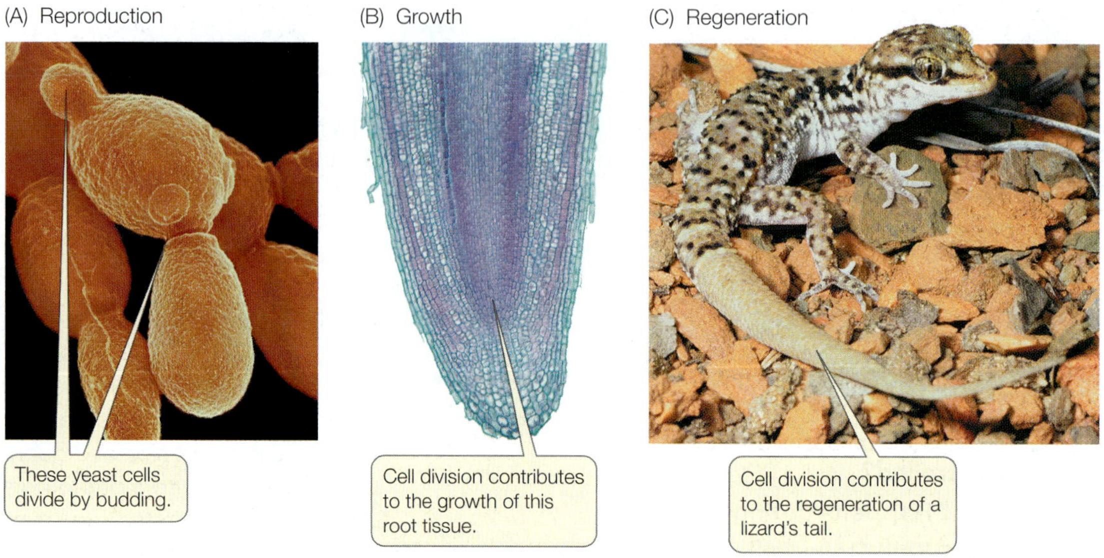
(A) Reproduction
These yeast cells divide by budding.
(B) Growth
Cell division contributes to the growth of this root tissue.
(C) Regeneration
Cell division contributes to the regeneration of a lizard's tail.

11.1 Important Consequences of Cell Division Cell division is the basis for (A) reproduction, (B) growth, and (C) repair and regeneration of tissues.

11.1 How Do Prokaryotic and Eukaryotic Cells Divide?

The life cycle of an organism, from birth to death, is intimately linked to cell division. Cell division plays important roles in the growth and repair of tissues in multicellular organisms, as well as in the reproduction of all organisms (**Figure 11.1**).

In order for any cell to divide, four events must occur:

- There must be a **reproductive signal**. This signal initiates cell division and may originate from either inside or outside the cell.
- **Replication** of DNA (the genetic material) must occur so that each of the two new cells will have a complete, identical set of genes.
- The cell must distribute the replicated DNA to each of the two new cells. This process is called **segregation**.
- Enzymes and organelles for the new cells must be synthesized, and new material must be added to the plasma membrane (and the cell wall, in organisms that have one), in order to separate the two new cells by a process called **cytokinesis**.

These four events proceed somewhat differently in prokaryotes and eukaryotes.

Prokaryotes divide by binary fission

In prokaryotes, cell division results in the reproduction of the entire single-celled organism. The cell grows in size, replicates its DNA, and then separates the cytoplasm and DNA into two new cells by a process called **binary fission**.

REPRODUCTIVE SIGNALS External factors such as environmental conditions and nutrient concentrations are common signals for the initiation of cell division in prokaryotes. The bacterium *Escherichia coli*, which is widely used in genetic studies, is a "cell division machine." If abundant sources of carbohydrates and mineral nutrients are available, it can divide as often as every 20 minutes. Another bacterium, *Bacillus subtilis*, not only slows its growth when nutrient levels are low but also stops dividing altogether, and then resumes dividing when conditions improve.

REPLICATION OF DNA As we saw in Section 5.3, a **chromosome** consists of a long, thin DNA molecule with proteins attached to it. When a cell divides, all of its chromosomes, which contain the genetic information for the organism, must be replicated, and one copy of each chromosome must find its way into each of the two new cells.

Most prokaryotes have just one main chromosome—a single long DNA molecule with its associated proteins. In *E. coli*, the ends of the DNA molecule are joined to create a circular chromosome. If the *E. coli* DNA were spread out into an actual circle, it would be about 500 micrometers (μm) in diameter. The bacterium itself is only about 2 μm long and 1μm in diameter. Thus if the bacterial DNA were fully extended, it would form a circle more than 200 times larger than the cell! To fit into the cell, bacterial DNA must be compacted. The DNA folds in on itself, and positively charged (basic) proteins bound to the negatively charged (acidic) DNA contribute to this folding.

Two regions of the prokaryotic chromosome play functional roles in cell reproduction:

- *ori*: the site where replication of the circular chromosome starts (the *ori*gin of replication)
- *ter*: the site where replication ends (the *ter*minus of replication)

Chromosome replication takes place as the DNA is threaded through a "replication complex" of proteins near the center of the cell. Replication begins at the *ori* site and moves toward the *ter* site. While the DNA replicates, anabolic metabolism is active and the cell grows. When replication is complete, the two daughter DNA molecules separate and segregate from one another at opposite ends of the cell. In rapidly dividing prokaryotes, DNA replication occupies the entire time between cell divisions.

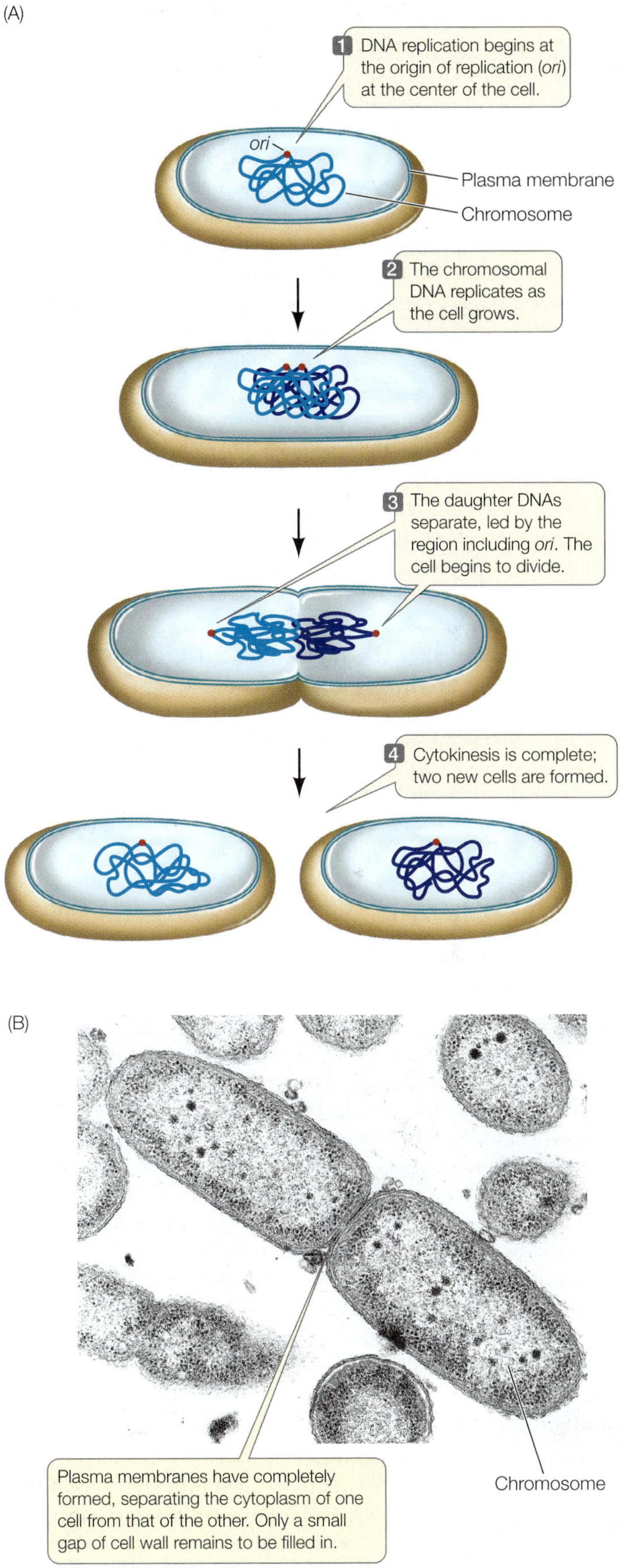
(A)
1 DNA replication begins at the origin of replication (*ori*) at the center of the cell.
ori
Plasma membrane
Chromosome
2 The chromosomal DNA replicates as the cell grows.
3 The daughter DNAs separate, led by the region including *ori*. The cell begins to divide.
4 Cytokinesis is complete; two new cells are formed.
(B)
Plasma membranes have completely formed, separating the cytoplasm of one cell from that of the other. Only a small gap of cell wall remains to be filled in.
Chromosome

11.2 Prokaryotic Cell Division (A) The process of cell division in a bacterium. (B) These two cells of the bacterium *Pseudomonas aeruginosa* have almost completed cytokinesis.

SEGREGATION OF DNA MOLECULES Replication begins near the center of the cell, and as it proceeds, the *ori* regions move toward opposite ends of the cell (**Figure 11.2A**). DNA sequences adjacent to the *ori* region bind proteins that are essential for this segregation. This is an active process, since the binding proteins hydrolyze ATP.

CYTOKINESIS Immediately after chromosome replication is finished, cytokinesis begins. At first, the plasma membrane pinches in to form a ring of fibers similar to a purse string. The major component of these fibers is a protein that is related to eukaryotic tubulin (which makes up microtubules; see Figure 5.14). As the membrane pinches in, new cell wall materials are deposited, which finally separate the two cells (**Figure 11.2B**).

Eukaryotic cells divide by mitosis or meiosis followed by cytokinesis

As in prokaryotes, cell reproduction in eukaryotes entails reproductive signals, DNA replication, segregation, and cytokinesis. The details, however, are quite different:

- *Reproductive signal*: Unlike prokaryotes, eukaryotic cells do not constantly divide whenever environmental conditions are adequate. In fact, most eukaryotic cells that are part of a multicellular organism and have become specialized seldom divide. In a eukaryotic organism, the signals for cell division are usually not related to the environment of a single cell, but to the function of the entire organism.
- *Replication*: Whereas most prokaryotes have a single main chromosome, eukaryotes usually have many (humans have 46). Consequently the processes of replication and segregation are more intricate in eukaryotes than in prokaryotes. In eukaryotes, DNA replication is usually limited to a portion of the period between cell divisions.
- *Segregation*: In eukaryotes, the newly replicated chromosomes are closely associated with each other (thus they are known as **sister chromatids**), and a mechanism called **mitosis** segregates them into two new nuclei.
- *Cytokinesis*: Cytokinesis proceeds differently in plant cells (which have a cell wall) than in animal cells (which do not).

The cells resulting from mitosis are identical to the parent cell in the amount and kind of DNA they contain. This contrasts with the second mechanism of nuclear division, meiosis.

Meiosis is the process of nuclear division that occurs in cells involved with sexual reproduction. Whereas the two products of mitosis are genetically identical to the cell that produced them—they both have the same DNA—the products of meiosis are not. As we will see in Section 11.5, meiosis generates diversity by shuffling the genetic material, resulting in new gene combinations. Meiosis plays a key role in the sexual life cycle, which we will discuss in Section 11.4.

RECAP 11.1

Four events are required for cell division: a reproductive signal, replication of the genetic material (DNA), segregation of replicated DNA, and separation of the two daughter cells (cytokinesis). Prokaryotes often have just one chromosome, and cell division can be rapid. Eukaryotes usually have multiple chromosomes, and the process of cell division is more intricate, involving either mitosis or meiosis.

- What is the reproductive signal that leads the bacterium *Bacillus subtilis* to divide? See p. 206
- Explain why DNA must be replicated and segregated *before* a cell can divide. See p. 206
- Describe the major steps in binary fission. See pp. 206–207 and Figure 11.2

What determines whether a eukaryotic cell will divide? How does mitosis lead to identical cells, and meiosis to diversity? Why do most eukaryotic organisms reproduce sexually? In the sections that follow, we will describe the details of mitosis and meiosis and discuss their roles in development and evolution.

11.2 How Is Eukaryotic Cell Division Controlled?

As you will see throughout this book, different cells have different rates of cell division. Some cells, such as those in an early embryo, divide rapidly and continuously. Others, such as neurons in the brain, don't divide at all. Clearly, the signaling pathways for cells to divide are highly controlled.

The period from one cell division to the next is referred to as the **cell cycle**. The cell cycle can be divided into mitosis/cytokinesis and interphase. During **interphase**, the cell nucleus is visible and typical cell functions occur, including DNA replication. This phase of the cell cycle begins when cytokinesis is completed and ends when mitosis (M) begins (Figure 11.3). In this section we will describe the events of interphase, especially those that trigger mitosis.

The duration of the cell cycle varies considerably in different cell types. In the early embryo the cell cycle may be as short as 30 minutes, whereas rapidly dividing cells in an adult human typically complete the cycle in about 24 hours. And as mentioned above, many cell types in a mature organism do not divide at all. In general, cells spend most of their time in interphase. So if we take a snapshot through the microscope of a cell population, only a few will be in mitosis or cytokinesis at any given moment. Interphase has three subphases called G1, S, and G2. In a cell cycle of 24 hours, these subphases would typically last for 11 hours (G1), 8 hours (S), and 4 hours (G2), with the remaining 1 hour spent in mitosis.

- *G1 phase.* During G1, each chromosome is a single, unreplicated DNA molecule with associated proteins. Variations in the duration of G1 account for most of the variability in the length of the cell cycle in different cell types. Some rapidly dividing embryonic cells dispense with it entirely, whereas other cells may remain in G1 for weeks or even years. In many cases these cells enter a resting phase called G0. Special internal and external signals are needed to prompt a cell to leave G0 and reenter the cell cycle at G1.
- *The G1-to-S transition.* At the **G1-to-S transition** the commitment is made to DNA replication and subsequent cell division.
- *S phase.* DNA replication occurs during **S phase** (see Section 13.3 for a detailed description of DNA replication). Each chromosome is duplicated and thereafter consists of two sister chromatids (the products of DNA replication). The sister chromatids remain joined together until mitosis, when they segregate into two daughter cells.
- *G2 phase.* During **G2 phase**, the cell makes preparations for mitosis—for example, by synthesizing and assembling the structures that move the chromatids to opposite ends of the dividing cell.

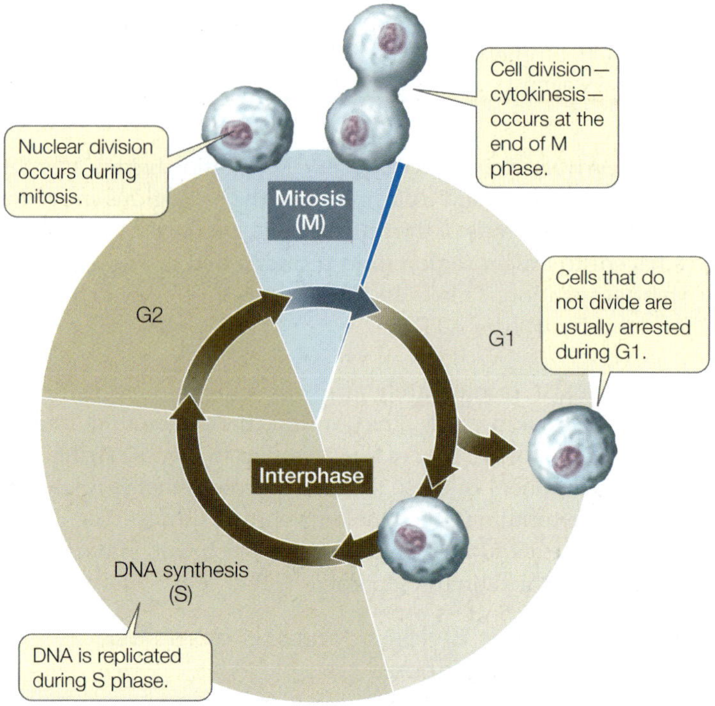
Nuclear division occurs during mitosis.
Cell division—cytokinesis—occurs at the end of M phase.
Mitosis (M)
G2
G1
Cells that do not divide are usually arrested during G1.
Interphase
DNA synthesis (S)
DNA is replicated during S phase.

11.3 The Eukaryotic Cell Cycle The cell cycle consists of a mitotic (M) phase, during which mitosis and cytokinesis take place, and a long period of growth known as interphase. Interphase has three subphases (G1, S, and G2) in cells that divide.

The initiation, termination, and operations of these phases are regulated by specific signals.

Specific internal signals trigger events in the cell cycle

Cell fusion experiments were used to reveal the existence of internal signals that control the transitions between stages of the cell cycle. For example, an experiment involving the fusion of HeLa cells (the cells described in the opening story) at different phases of the cell cycle showed that a cell in S phase produces

INVESTIGATING**LIFE**

11.4 Regulation of the Cell Cycle Nuclei in G1 do not undergo DNA replication, but nuclei in S phase do. To determine if there is some signal in the S cells that stimulates G1 cells to replicate their DNA, Rao and Johnson fused together cells in G1 and S phases, creating cells with both G1 and S properties.[a]

HYPOTHESIS A cell in S phase contains an activator of DNA replication.

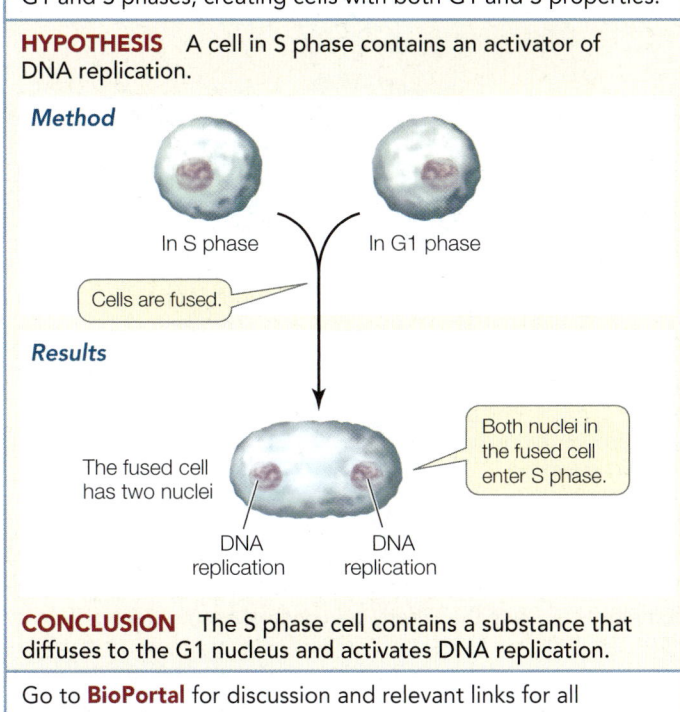

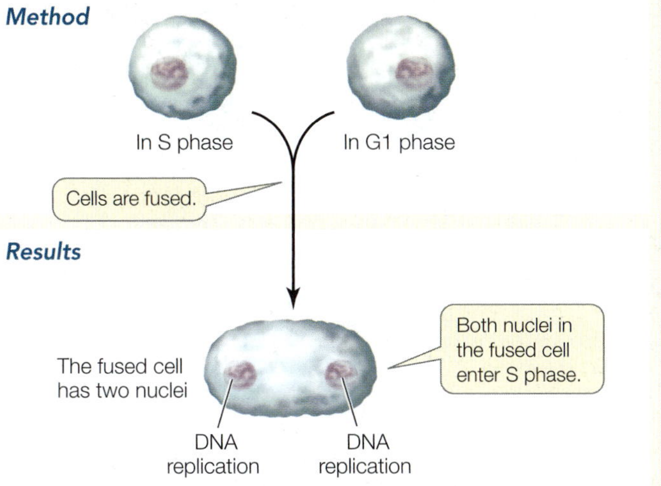
Method
In S phase
In G1 phase
Cells are fused.
Results
The fused cell has two nuclei
Both nuclei in the fused cell enter S phase.
DNA replication
DNA replication

CONCLUSION The S phase cell contains a substance that diffuses to the G1 nucleus and activates DNA replication.

Go to **BioPortal** for discussion and relevant links for all INVESTIGATING**LIFE** figures.

[a]Rao, P. N. and R. T. Johnson. 1970. *Nature* 225: 159–164.

a substance that activates DNA replication (**Figure 11.4**). Similar experiments pointed to the existence of signals controlling entry into M phase. As you will see, the signals that control progress through the cell cycle act through protein kinases.

Progress through the cell cycle depends on the activities of **cyclin-dependent kinases**, or **Cdk's**. Recall from Section 7.2 that a protein kinase is an enzyme that catalyzes the transfer of a phosphate group from ATP to a target protein; this phosphate transfer is called phosphorylation.

$$\underset{\text{(Inactive)}}{\text{Protein + ATP}} \xrightarrow{\text{Protein kinase}} \underset{\text{(Active)}}{\text{protein-P + ADP}}$$

By catalyzing the phosphorylation of certain target proteins, Cdk's play important roles at various points in the cell cycle. The discovery that Cdk's induce cell division is a beautiful example of how research on different organisms and different cell types can converge on a single mechanism. One group of scientists, led by James Maller at the University of Colorado, was studying immature sea urchin eggs, trying to find out how they are stimulated to divide and eventually form mature eggs. A protein called maturation promoting factor was purified from maturing eggs, which by itself prodded immature egg cells to divide.

WORKING WITH**DATA:**

Regulation of the Cell Cycle

Original Paper

Rao, P. N. and R. T. Johnson. 1970. Mammalian cell fusion: Studies on the regulation of DNA synthesis and mitosis. *Nature* 225: 159–164.

Analyze the Data

The fusion of cellular membranes is a natural process; it occurs during endocytosis and exocytosis, and in fertilization (the fusion of gametes). Membrane fusion also occurs when membrane-enclosed viruses infect their host cells. Occasionally, these viruses also induce the fusion of adjacent host cells, creating a multinucleate cell. This observation led to the use of Sendai virus, a membrane-enclosed mouse respiratory virus, as a tool in the laboratory to fuse cells experimentally. Rao and Johnson used this strategy to study the regulation of the cell cycle (see Figure 11.4).

In their experiment, Rao and Johnson used HeLa cells, which divide continuously (see the opening story of this chapter). They first synchronized separate populations of the cells in G1 or S phase. Before fusion, the cells in S phase were exposed to a radioactively labeled component of DNA (thymidine). The radioactivity was incorporated into these cells' newly replicated DNA, labeling their nuclei. The S and G1 cells were then fused using Sendai virus (resulting in G1/S fusions) and again exposed to labeled thymidine. At various times after fusion, the scientists calculated the percent of previously unlabeled (G1) nuclei that had incorporated new label (i.e., had replicated their DNA) (**FIGURE A**). In a second series of experiments, S and G2 cells were fused in various combinations and then the numbers of cells in mitosis were counted and expressed as a percent of all cells in the population (**FIGURE B**).

QUESTION 1

According to Figure A, how long did it take for all the G1 nuclei in the G1/S cells to become labeled?

QUESTION 2

Examine the data for fused G1/G1 cells and unfused G1 cells in Figure A. Explain why these were appropriate controls for the experiments. When did these nuclei become labeled? Compare these times with each other and with the G1/S nuclei and discuss.

QUESTION 3

Examine the data in Figure B. Why did it take longer for the cells in S phase to begin mitosis than it did for the cells in G2?

QUESTION 4

According to Figure B, did fusion with G2 cells alter the timing of mitosis in the S cell nuclei? Explain what this means in terms of control of the cell cycle.

FIGURE A

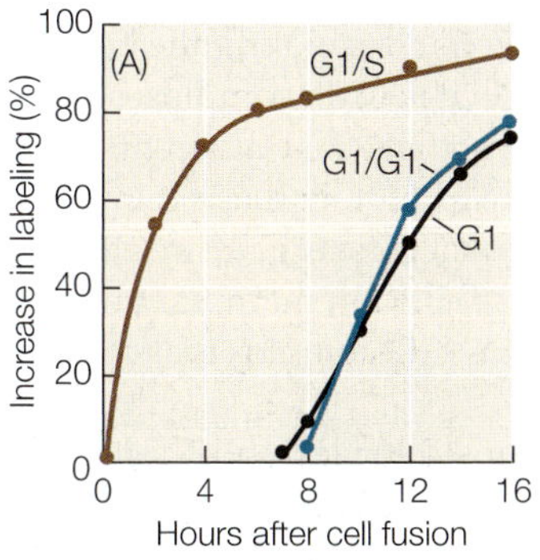
(A)
G1/S
G1/G1
G1
Increase in labeling (%)
100
80
60
40
20
0
0
4
8
12
16
Hours after cell fusion

FIGURE B

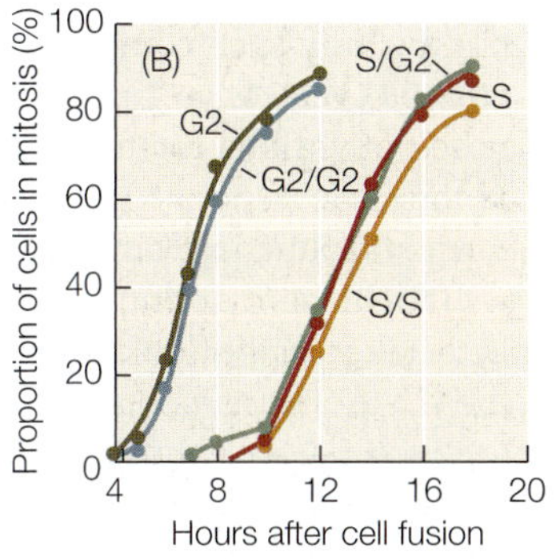
(B)
S/G2
S
G2
G2/G2
S/S
Proportion of cells in mitosis (%)
100
80
60
40
20
0
4
8
12
16
20
Hours after cell fusion

Go to **BioPortal** for all WORKING WITH**DATA** exercises

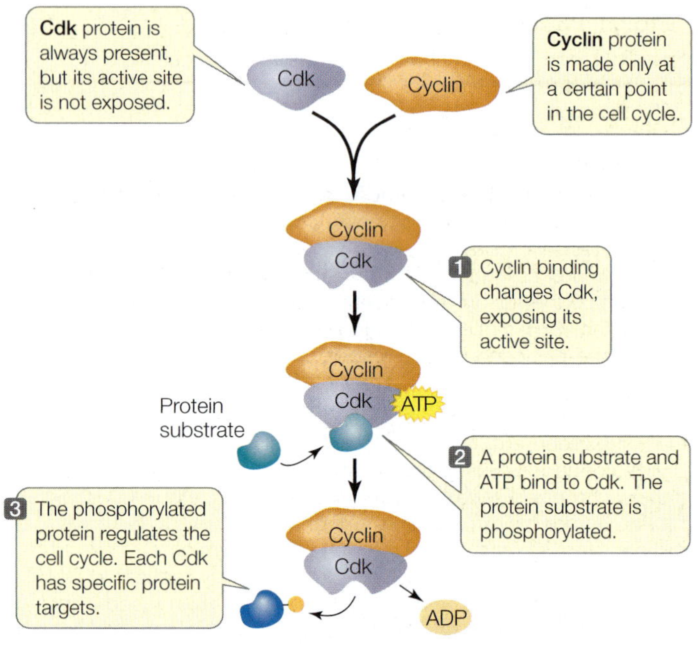
Cdk protein is always present, but its active site is not exposed.
Cdk
Cyclin
Cyclin protein is made only at a certain point in the cell cycle.
Cyclin
Cdk
1 Cyclin binding changes Cdk, exposing its active site.
Protein substrate
Cyclin
Cdk
ATP
2 A protein substrate and ATP bind to Cdk. The protein substrate is phosphorylated.
3 The phosphorylated protein regulates the cell cycle. Each Cdk has specific protein targets.
Cyclin
Cdk
ADP

11.5 Cyclin Binding Activates Cdk Binding of a cyclin changes the three-dimensional structure of an inactive Cdk, making it an active protein kinase. Each cyclin–Cdk complex phosphorylates a specific target protein in the cell cycle.

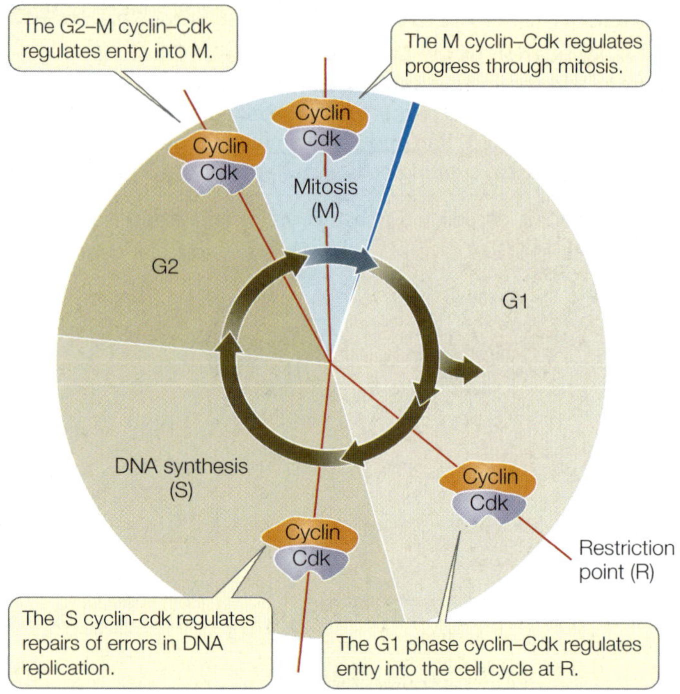
The G2–M cyclin–Cdk regulates entry into M.
The M cyclin–Cdk regulates progress through mitosis.
Cyclin
Cdk
Cyclin
Cdk
Mitosis (M)
G2
G1
DNA synthesis (S)
Cyclin
Cdk
Cyclin
Cdk
Restriction point (R)
The S cyclin-cdk regulates repairs of errors in DNA replication.
The G1 phase cyclin–Cdk regulates entry into the cell cycle at R.

11.6 Cyclin-Dependent Kinases Regulate Progress through the Cell Cycle By acting at checkpoints (red lines), different cyclin–Cdk complexes regulate the orderly sequence of events in the cell cycle.

Meanwhile, Leland Hartwell at the University of Washington was studying the cell cycle in yeast (a single-celled eukaryote; see Figure 11.1A) and found a strain that was stalled at the G1–S boundary because it lacked a Cdk. It turned out that this yeast Cdk and the sea urchin maturation promoting factor had similar properties, and further work confirmed that the sea urchin protein was indeed a Cdk. Similar Cdk's were soon found to control the G1-to-S transition in many other organisms, including humans. This control point in the cell cycle is now called the **restriction (R) point**. Other Cdk's were found to control other parts of the cell cycle.

Cdk's are not active on their own. As their name implies, cyclin-dependent kinases need to be activated by binding to a second type of protein, called **cyclin**. This binding—an example of allosteric regulation (see Section 8.5)—activates the Cdk by altering the shape of its active site (**Figure 11.5**).

The cyclin–Cdk that controls passage from G1 to S phase is not the only such complex involved in regulating the eukaryotic cell cycle. There are different cyclin–Cdk complexes, composed of various cyclins and Cdk's, that act at different stages of the cycle (**Figure 11.6**). The details of how these complexes form and function vary among eukaryotic organisms, but we will focus here on the complexes found in mammalian cells. Let's take a closer look at the cyclin–Cdk complex that controls the G1-to-S transition.

Cyclin–Cdk catalyzes the phosphorylation of a protein called retinoblastoma protein (RB, named because of its role in cancer; see Section 11.7). In many cells, RB or a protein like it acts as an inhibitor of the cell cycle at the R point. To begin S phase, a cell must get by the RB block. Here is where cyclin–Cdk comes in: it catalyzes the phosphorylation at multiple sites on the RB molecule. This causes a change in the three-dimensional structure of RB, thereby inactivating it. With RB out of the way, the cell cycle can proceed. To summarize:

$$\text{RB} \xrightarrow{\text{Cyclin–Cdk}} \text{RB-P}$$

(Active—blocks cell cycle) (Inactive—allows cell cycle)

Progress through the cell cycle is regulated by the activities of Cdk's, and so regulating *them* is a key to regulating cell division. An effective way to regulate Cdk's is to regulate the presence or absence of cyclins (**Figure 11.7**). Simply put, if a cyclin is not present, its partner Cdk is not active. As their name suggests, cyclins are present cyclically: they are made only at certain times in the cell cycle.

The different cyclin–Cdk's act at **cell cycle checkpoints**, signaling pathways that regulate the cell cycle's progress. For example, if a cell's DNA is substantially damaged by radiation or toxic chemicals, the cell may be prevented from successfully completing a cell cycle. During interphase, there are three checkpoints, with a fourth during mitosis (see Figure 11.6; **Table 11.1**). The table lists the triggers that will cause the cell cycle to pause at each point.

Let's consider the G1 checkpoint (R). If DNA is damaged by radiation during G1, a signaling pathway results in the production of a protein called p21. (The *p* stands for "protein" and the *21* stands for its molecular weight—about 21,000.) The p21

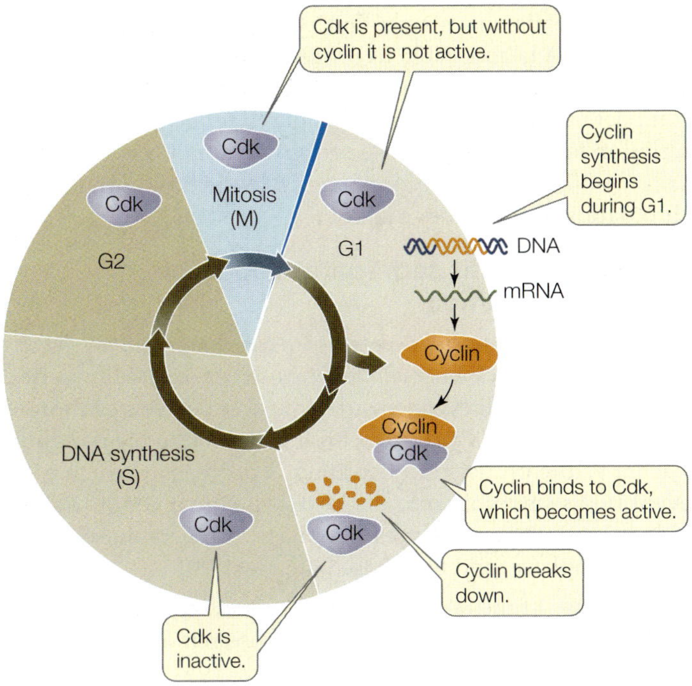
Cdk is present, but without cyclin it is not active.
Cyclin synthesis begins during G1.
Cdk
Mitosis (M)
Cdk
Cdk
G2
G1
DNA
mRNA
Cyclin
Cyclin
Cdk
DNA synthesis (S)
Cyclin binds to Cdk, which becomes active.
Cdk
Cdk
Cyclin breaks down.
Cdk is inactive.

11.7 Cyclins Are Transient in the Cell Cycle Cyclins are made at a particular time and then break down. In this case, the cyclin is present during G1 and activates a Cdk at that time.

protein can bind to the G1–S Cdk, preventing its activation by cyclin. So the cell cycle stops while repairs are made to the DNA (you will learn more about DNA repair in Section 13.4). When the DNA damage pathway is no longer operating, p21 breaks down, allowing the cyclin–Cdk's to function, and the cell cycle proceeds. If DNA damage is severe and it cannot be repaired, the cell will undergo programmed cell death (apoptosis, which we will discuss later in this chapter). Such controls prevent defective cells from proliferating and potentially harming an organism.

Growth factors can stimulate cells to divide

Cyclin–Cdk's provide cells with internal controls of their progress through the cell cycle, but the cell cycle is also influenced by external signals. Not all cells in an organism go through the cell cycle on a regular basis. Some cells either no longer go through the cell cycle and enter G0, or go through it slowly and divide infrequently. If such cells are to divide, they must be stimulated by external chemical signals called **growth factors**:

- If you cut yourself and bleed, specialized cell fragments called platelets gather at the wound to initiate blood clotting. The platelets produce and release a protein called platelet-derived growth factor that diffuses to the adjacent cells in the skin and stimulates them to divide and heal the wound.
- Red and white blood cells have limited lifetimes and must be replaced through the division of immature, unspecialized blood cell precursors in the bone marrow. Two types of growth factors, interleukins and erythropoietin, stimulate the division and specialization, respectively, of precursor cells of white blood cells and red blood cells.

In these and other examples, growth factors bind to specific receptors on target cells and activate signal transduction pathways that end with cyclin synthesis, thereby activating Cdk's and the cell cycle.

TABLE 11.1 Cell Cycle Checkpoints

Cell Cycle Phase	Checkpoint Trigger
G1	DNA damage
S	Incomplete replication or DNA damage
G2	DNA damage
M	Chromosome unattached to spindle

RECAP 11.2

The eukaryotic cell cycle is under both external and internal control. Cdk's control the eukaryotic cell cycle and are themselves controlled by cyclins. External signals such as growth factors can initiate the cell cycle.

- Draw a cell cycle diagram showing the various stages of interphase. See p. 208 and Figure 11.3
- How do cyclin–Cdk's control the progress of the cell cycle? See pp. 209–210 and Figure 11.6
- What are growth factors, and how do they act to control the cell cycle? See p. 211

11.3 What Happens during Mitosis?

Segregation of the replicated DNA occurs during mitosis. Prior to segregation, the DNA molecules and their associated proteins in each chromosome become condensed into more compact structures. After segregation by mitosis, cytokinesis separates the two cells. Let's now look at these steps more closely.

Go to Animated Tutorial 11.1
Mitosis
Life10e.com/at11.1

Prior to mitosis, eukaryotic DNA is packed into very compact chromosomes

A eukaryotic chromosome consists of one or two long, linear, double-stranded DNA molecules bound with many proteins (the complex of DNA and proteins is referred to as **chromatin**). Before S phase, each chromosome contains only one double-stranded DNA molecule. After it replicates during S phase, however, there are two double-stranded DNA molecules: the sister chromatids (Figure 11.8). Throughout G2 the sister chromatids are held together along most of their length by a protein complex called

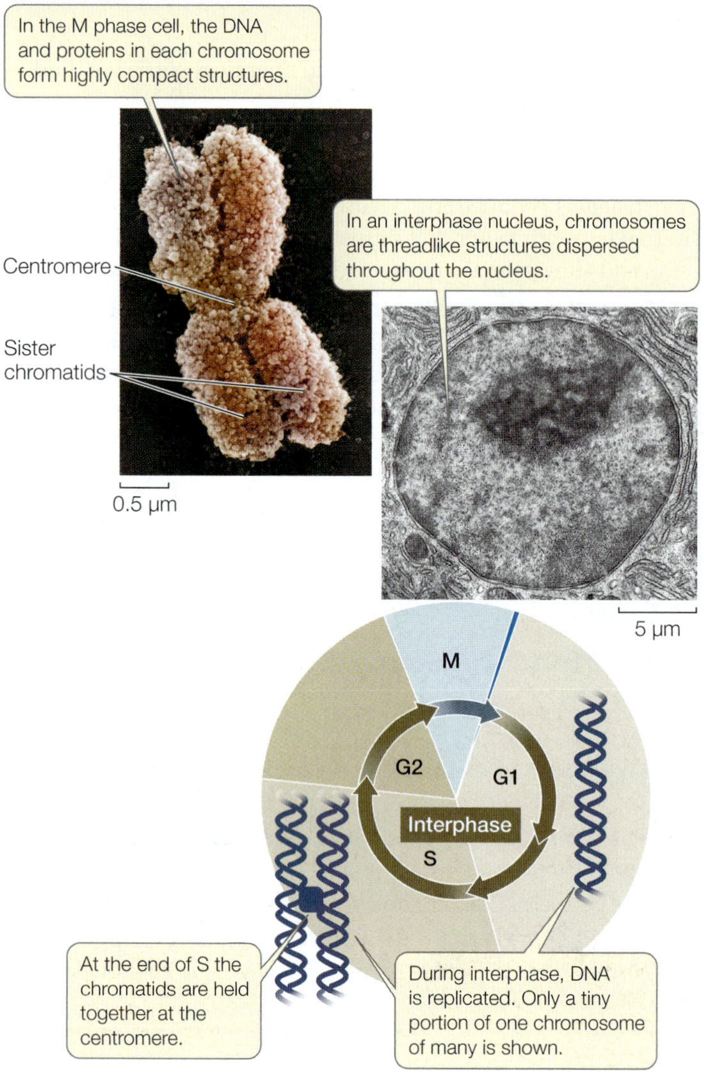
In the M phase cell, the DNA and proteins in each chromosome form highly compact structures.
Centromere
Sister chromatids
0.5 µm
In an interphase nucleus, chromosomes are threadlike structures dispersed throughout the nucleus.
5 µm
M
G2
G1
Interphase
S
At the end of S the chromatids are held together at the centromere.
During interphase, DNA is replicated. Only a tiny portion of one chromosome of many is shown.

11.8 Chromosomes, Chromatids, and Chromatin DNA in the interphase nucleus is diffuse and becomes compacted as mitosis begins.

cohesin. At mitosis most of the cohesin is removed, except in a region called the **centromere**, where the chromatids remain held together. At the end of G2 and the beginning of mitosis, a second group of proteins called condensins coat the DNA molecules and make them more compact.

If all of the DNA in a typical human cell were put end to end, it would be nearly 2 meters long. Yet the nucleus is only 5 µm (0.000005 meters) in diameter. So eukaryotic DNA is extensively packaged in a highly organized way (**Figure 11.9**). This packing is achieved largely by proteins call histones (*histos*, "web" or "loom"), which are positively charged at cellular pH because of their high content of the basic amino acids lysine and arginine. The charged R groups on these amino acids bind to the negatively charged phosphate groups on DNA by ionic attractions. These DNA–histone interactions, as well as histone–histone interactions, result in the formation of beadlike units called **nucleosomes** (see Figure 11.9).

During interphase, the chromatin that makes up each chromosome is less densely packaged and consists of single DNA molecules running around vast numbers of nucleosomes that resemble beads on a string. During this phase of the cell cycle, the DNA is accessible to proteins involved in replication and transcription. Once a mitotic chromosome is formed, its compact nature makes it inaccessible to replication and transcription factors, and so these processes cannot occur. Further coiling of the chromatin continues up to the time at which the chromatids begin to move apart.

Overview: Mitosis segregates copies of genetic information

In mitosis, a single nucleus gives rise to two nuclei that are genetically identical to each other and to the parent nucleus. Mitosis (the M phase of the cell cycle) ensures the accurate segregation of the eukaryotic cell's multiple chromosomes into the daughter nuclei. While mitosis is a continuous process in which each event flows smoothly into the next, it is convenient to subdivide it into a series of stages: prophase, prometaphase, metaphase, anaphase, and telophase (**Figure 11.10**).

- **Prophase**. At the beginning of mitosis the chromatin becomes condensed and the separate chromatids become visible through a light microscope.
- **Prometaphase**. The nuclear envelope breaks down and the compacted chromosomes, each consisting of two chromatids, attach to the spindle apparatus (see below).
- **Metaphase**. The chromosomes line up at the midline of the cell (equatorial position).
- **Anaphase**. The chromatids separate and move away from each other toward opposite poles.
- **Telophase**. A nuclear envelope forms around each set of chromosomes, nucleoli appear, and the chromosomes become less compact. The spindle disappears. As a result, there are two new nuclei in a single cell.

Let's take a closer look at two cellular structures that contribute to the orderly segregation of the chromosomes during mitosis—the centrosomes and the spindle.

The centrosomes determine the plane of cell division

The **spindle apparatus** (also called the *mitotic spindle* or simply the *spindle*) is the dynamic structure that moves sister chromatids apart during mitosis. It is made up of microtubules. Before the spindle can form, its orientation must be determined. This is accomplished by the **centrosome** ("central body"), an organelle in the cytoplasm near the nucleus. In many organisms the centrosome consists of a pair of **centrioles**, each one a hollow tube formed by nine microtubule triplets. The two tubes are at right angles to each other. During S phase the centrosome doubles, and at the beginning of prophase the two centrosomes separate from one another, moving to opposite ends of the nuclear envelope. These identify the "poles" toward which chromosomes move during anaphase. The cells of plants and fungi lack centrosomes, but distinct microtubule organizing centers at each end of the cell play the same role.

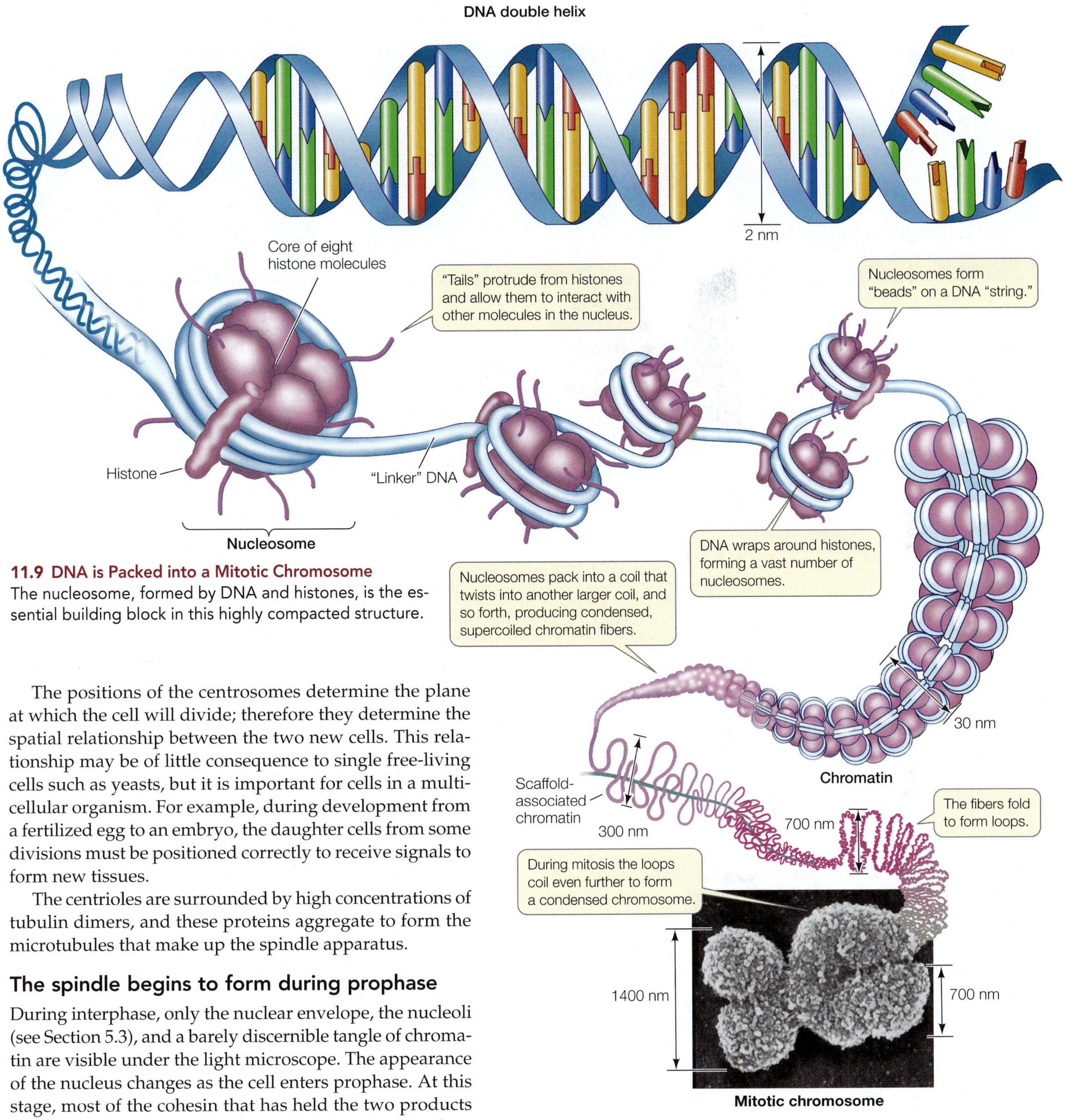

DNA double helix
2 nm
Core of eight histone molecules
"Tails" protrude from histones and allow them to interact with other molecules in the nucleus.
Nucleosomes form "beads" on a DNA "string."
Histone
"Linker" DNA
Nucleosome
DNA wraps around histones, forming a vast number of nucleosomes.
Nucleosomes pack into a coil that twists into another larger coil, and so forth, producing condensed, supercoiled chromatin fibers.
30 nm
Chromatin
Scaffold-associated chromatin
300 nm
700 nm
The fibers fold to form loops.
During mitosis the loops coil even further to form a condensed chromosome.
1400 nm
700 nm
Mitotic chromosome

11.9 DNA is Packed into a Mitotic Chromosome The nucleosome, formed by DNA and histones, is the essential building block in this highly compacted structure.

The positions of the centrosomes determine the plane at which the cell will divide; therefore they determine the spatial relationship between the two new cells. This relationship may be of little consequence to single free-living cells such as yeasts, but it is important for cells in a multicellular organism. For example, during development from a fertilized egg to an embryo, the daughter cells from some divisions must be positioned correctly to receive signals to form new tissues.

The centrioles are surrounded by high concentrations of tubulin dimers, and these proteins aggregate to form the microtubules that make up the spindle apparatus.

The spindle begins to form during prophase

During interphase, only the nuclear envelope, the nucleoli (see Section 5.3), and a barely discernible tangle of chromatin are visible under the light microscope. The appearance of the nucleus changes as the cell enters prophase. At this stage, most of the cohesin that has held the two products of DNA replication together since S phase is removed, so the individual chromatids become visible. They are still held together by a small amount of cohesin at the centromere. Late in prophase, specialized three-layered structures called **kinetochores** develop in the centromere region, one on each chromatid. These structures will be important for chromosome movement.

Each of the two centrosomes, now on opposite sides of the nucleus, serves as a mitotic center, or pole, toward which the chromosomes will move (**Figure 11.11A**). During prophase and prometaphase, microtubules form between the poles and the chromosomes to make up the spindle. The spindle serves as a structure to which the chromosomes attach and as a framework keeping the two poles apart. Each half of the spindle develops as tubulin dimers aggregate from around the centrioles and form long fibers that extend into the middle region of the cell. The

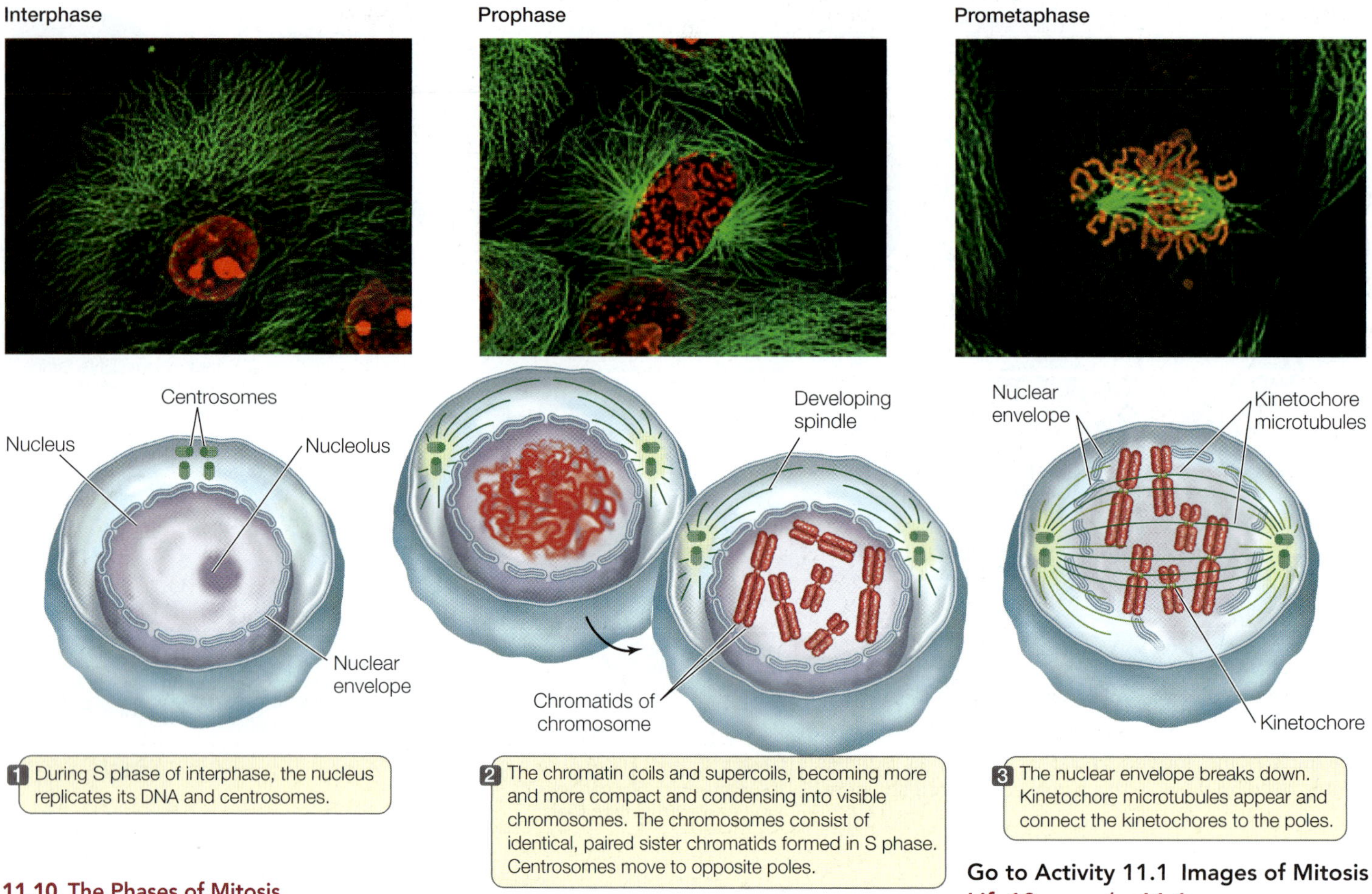

Interphase
Prophase
Prometaphase
Centrosomes
Nucleus
Nucleolus
Nuclear envelope
Developing spindle
Chromatids of chromosome
Nuclear envelope
Kinetochore microtubules
Kinetochore
1 During S phase of interphase, the nucleus replicates its DNA and centrosomes.
2 The chromatin coils and supercoils, becoming more and more compact and condensing into visible chromosomes. The chromosomes consist of identical, paired sister chromatids formed in S phase. Centrosomes move to opposite poles.
3 The nuclear envelope breaks down. Kinetochore microtubules appear and connect the kinetochores to the poles.

11.10 The Phases of Mitosis
Mitosis results in two new nuclei that are genetically identical to each other and to the nucleus from which they were formed. In the micrographs, the green dye stains microtubules (and thus the spindle); the red dye stains the chromosomes. The chromosomes in the diagrams are stylized to emphasize the fates of the individual chromatids.

Go to Activity 11.1 Images of Mitosis
Life10e.com/ac11.1

Go to Media Clip 11.1 Mitosis: Live and Up Close
Life10e.com/mc11.1

microtubules are initially unstable, constantly forming and falling apart, until they contact kinetochores or microtubules from the other half-spindle and become more stable.

There are two types of microtubule in the spindle:

- Polar microtubules form the framework of the spindle and run from one pole to the other.
- Kinetochore microtubules, which form later, attach to the kinetochores on the chromosomes. The two sister chromatids in each chromosome pair become attached to kinetochore microtubules in opposite halves of the spindle (**Figure 11.11B**). This ensures that the two chromatids will eventually move to opposite poles.

Movement of the chromatids is the central feature of mitosis. It accomplishes the segregation that is needed for cell division and completion of the cell cycle. Prophase prepares for this movement, and the actual segregation takes place in the next three phases of mitosis.

Chromosome separation and movement are highly organized

During the next three phases of mitosis—prometaphase, metaphase, and anaphase—dramatic changes take place in the cell and the chromosomes (see above and Figure 11.10). During prometaphase, the nuclear envelope breaks down and spindle formation is completed. During metaphase, the chromosomes line up at the equatorial position of the cell. Now let's consider two key processes of anaphase: separation of the chromatids, and the mechanism of their actual movement toward the poles.

CHROMATID SEPARATION The separation of chromatids occurs at the beginning of anaphase. It is controlled by an M phase cyclin–Cdk (see Figure 11.6), which activates another protein complex called the anaphase-promoting complex (APC). Separation occurs because one subunit of the cohesin protein holding the sister chromatids together is hydrolyzed

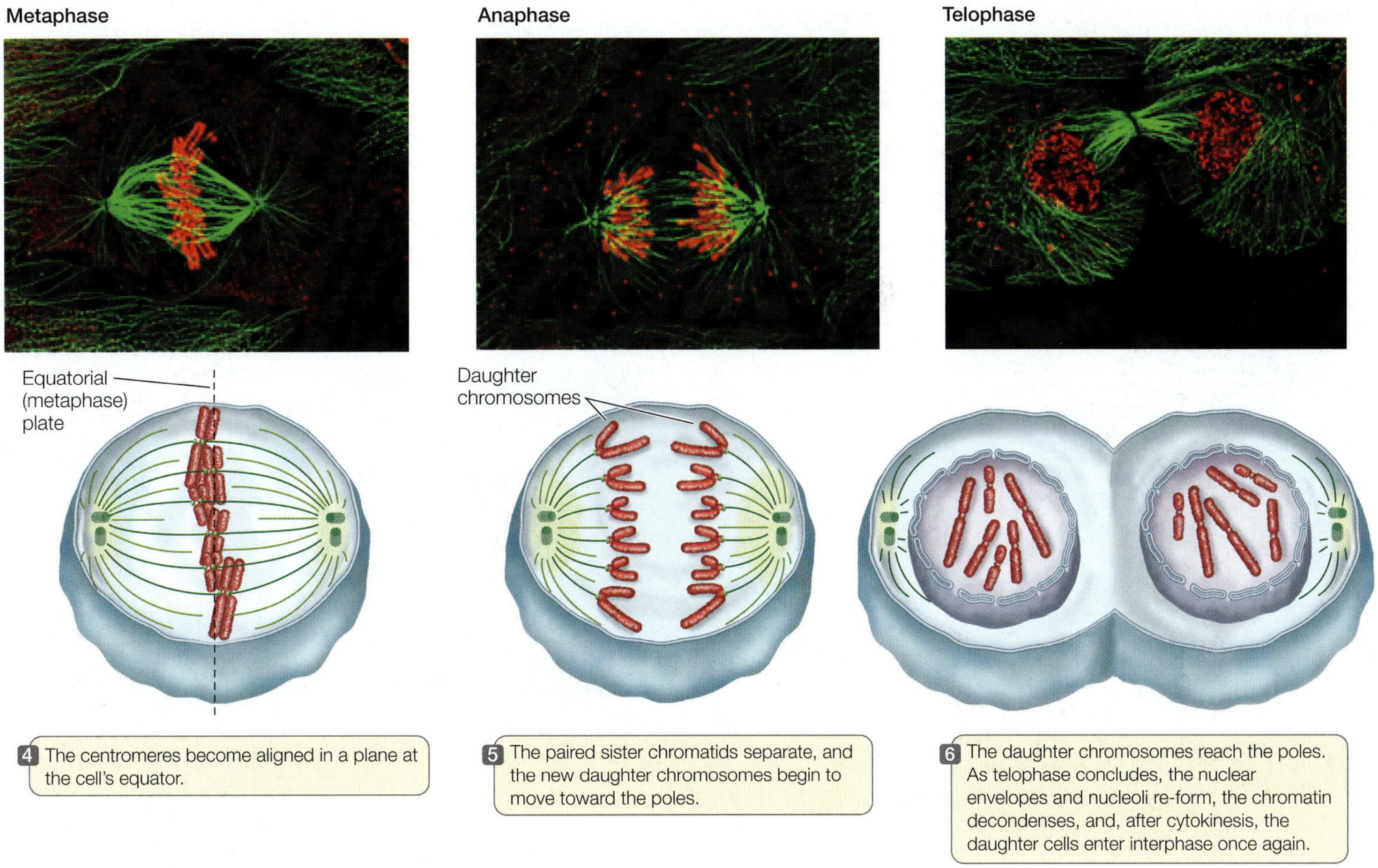
Metaphase
Anaphase
Telophase
Equatorial (metaphase) plate
Daughter chromosomes
4 The centromeres become aligned in a plane at the cell's equator.
5 The paired sister chromatids separate, and the new daughter chromosomes begin to move toward the poles.
6 The daughter chromosomes reach the poles. As telophase concludes, the nuclear envelopes and nucleoli re-form, the chromatin decondenses, and, after cytokinesis, the daughter cells enter interphase once again.

by a specific protease, appropriately called separase (**Figure 11.12**). A cell cycle checkpoint, often called the spindle assembly checkpoint, occurs at the end of metaphase to inhibit the APC if one of chromosomes is not attached properly to the spindle. When all chromosomes are attached, the APC is activated and the chromatids separate. After separation the chromatids are called **daughter chromosomes**. Note the *difference between chromatids and chromosomes*:

- Chromatids share a centromere.
- Chromosomes have their own centromere.

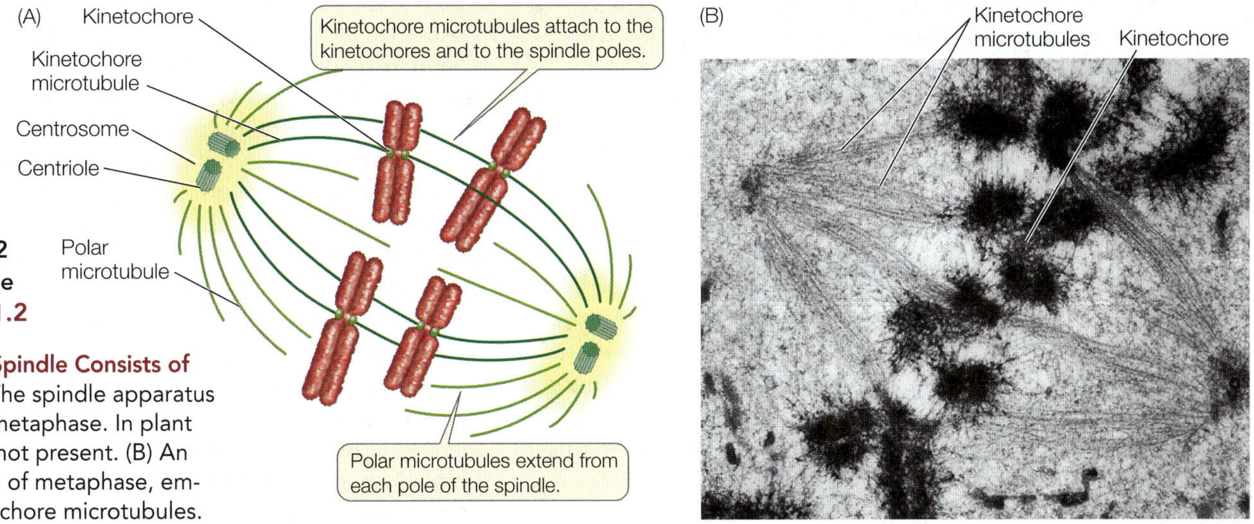
(A)
Kinetochore
Kinetochore microtubule
Centrosome
Centriole
Polar microtubule
Kinetochore microtubules attach to the kinetochores and to the spindle poles.
Polar microtubules extend from each pole of the spindle.
(B)
Kinetochore microtubules
Kinetochore

Go to Activity 11.2 The Mitotic Spindle
Life10e.com/ac11.2

11.11 The Mitotic Spindle Consists of Microtubules (A) The spindle apparatus in an animal cell at metaphase. In plant cells, centrioles are not present. (B) An electron micrograph of metaphase, emphasizing the kinetochore microtubules.

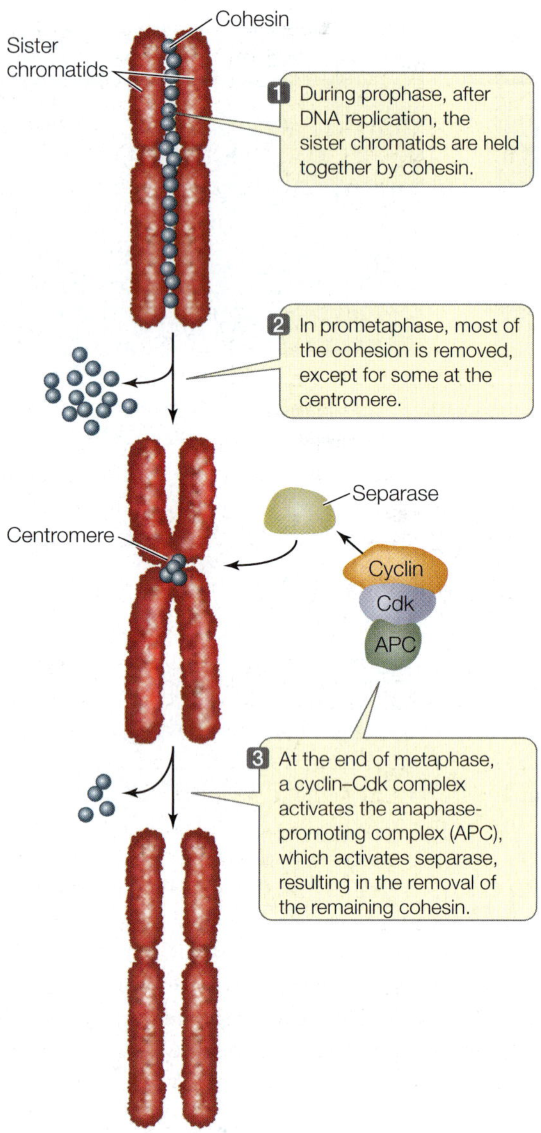
Cohesin
Sister chromatids
1 During prophase, after DNA replication, the sister chromatids are held together by cohesin.
2 In prometaphase, most of the cohesion is removed, except for some at the centromere.
Separase
Centromere
Cyclin
Cdk
APC
3 At the end of metaphase, a cyclin–Cdk complex activates the anaphase-promoting complex (APC), which activates separase, resulting in the removal of the remaining cohesin.

11.12 Chromatid Attachment and Separation The cohesin protein complex holds sister chromatids together at the centromere. The enzyme separase hydrolyzes cohesin at the end of metaphase, allowing the chromatids to separate into daughter chromosomes.

CHROMOSOME MOVEMENT The migration of the two sets of daughter chromosomes to the poles of the cell is a highly organized, active process. Two mechanisms operate to move the chromosomes along. First, the kinetochores contain molecular motor proteins, including kinesins and cytoplasmic dynein (see Figures 5.18 and 5.19), which use energy from ATP hydrolysis to do the work of moving the chromosomes along the microtubules. Second, the kinetochore microtubules shorten from the poles, drawing the chromosomes toward them.

The last stage of mitosis is telophase, when the spindle disappears and a nuclear envelope forms around each set of chromosomes (see Figure 11.10). Finally, the cytoplasms of the two daughter cells separate during cytokinesis: the last stage of cell division.

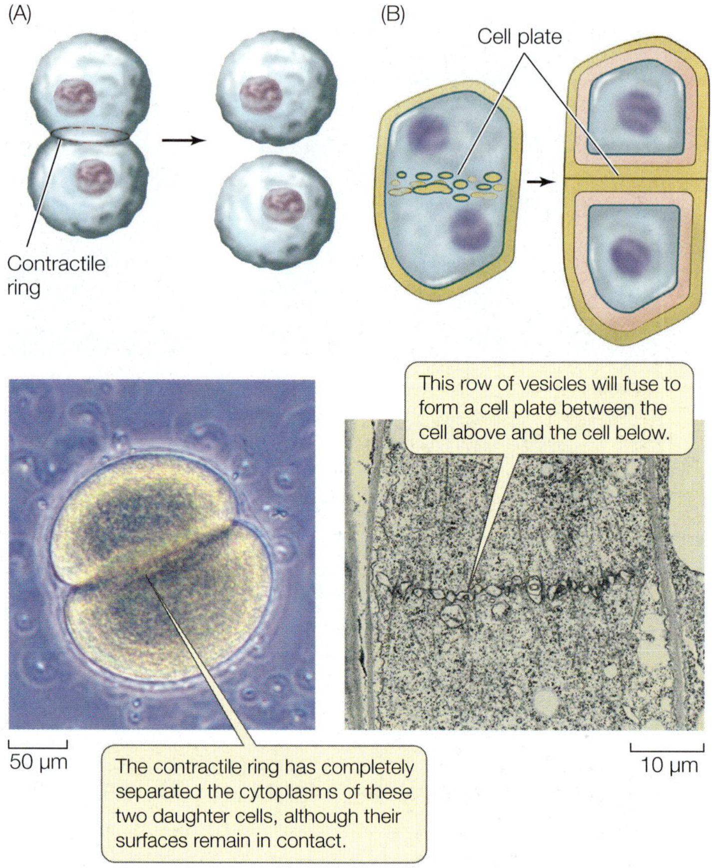
(A)
Contractile ring
(B)
Cell plate
This row of vesicles will fuse to form a cell plate between the cell above and the cell below.
50 μm
The contractile ring has completely separated the cytoplasms of these two daughter cells, although their surfaces remain in contact.
10 μm

11.13 Cytokinesis Differs in Animal and Plant Cells (A) A sea urchin zygote (fertilized egg) that has just completed cytokinesis at the end of the first cell division of its development into an embryo. (B) A dividing plant cell in late telophase. Plant cells divide differently from animal cells because plant cells have cell walls.

Cytokinesis is the division of the cytoplasm

Cytokinesis divides the cell's cytoplasm after nuclear division in mitosis. There are substantial differences between the process in animal and plant cells. In animal cells, cytokinesis begins with a furrowing of the plasma membrane, as if an invisible thread were cinching the cytoplasm between the two nuclei (**Figure 11.13A**). This contractile ring is composed of microfilaments of actin and myosin (see Figure 5.15), which form a ring on the cytoplasmic surface of the plasma membrane. These two proteins interact to produce a contraction, pinching the cell in two. The microfilaments assemble rapidly from actin monomers that are present in the interphase cytoskeleton. Their assembly is under the control of calcium ions that are released from storage sites in the center of the cell.

The plant cell cytoplasm divides differently because plants have rigid cell walls. In plant cells, as the spindle breaks down after mitosis, membranous vesicles derived from the Golgi apparatus appear along the plane of cell division, roughly midway between the two daughter nuclei. The vesicles are propelled along microtubules by the motor protein kinesin, and fuse to form a new plasma membrane. At the same time they

TABLE 11.2
Summary of Cell Cycle Events

Phase	Events
Interphase:	
G1	Growth; restriction point at end
S	DNA replication
G2	Spindle synthesis begins; preparation for mitosis
Mitosis:	
Prophase	Condensation of chromosomes; spindle assembly
Prometaphase	Nuclear envelope breakdown; chromosome attachment to spindle
Metaphase	Alignment of chromosomes at equatorial plate
Anaphase	Separation of chromatids; migration to poles
Telophase	Chromosomes decondense; nuclear envelope re-forms
Cytokinesis	Cell separation; cell membrane and/or wall formation

contribute their contents to a cell plate, which is the beginning of a new cell wall (Figure 11.13B).

Following cytokinesis, each daughter cell contains all the components of a complete cell. A precise distribution of chromosomes is ensured by mitosis. In contrast, organelles such as ribosomes, mitochondria, and chloroplasts do not need to be distributed equally between daughter cells, as long as some of each are present in each cell. Accordingly, there is no mechanism with a precision comparable to that of mitosis to provide for their equal allocation to daughter cells. As we will see in Chapter 19, the unequal distribution of cytoplasmic components during development can have functional significance for the two new cells.

Table 11.2 summarizes the major events of the cell cycle.

RECAP 11.3

Mitosis is the ordered division of a eukaryotic cell nucleus into two nuclei with identical sets of chromosomes. The process of mitosis, while continuous, can be viewed as a series of events (prophase, prometaphase, metaphase, anaphase, and telophase). Once two nuclei have formed, the cell divides into two cells by cytokinesis.

- What is the difference between a chromosome, a chromatid, and a daughter chromosome? See Figures 11.8, 11.10
- What are the various levels of "packing" by which the genetic information contained in linear DNA is condensed during prophase? See p. 212 and Figure 11.9
- Describe how chromosomes move during mitosis. See p. 216 and Figure 11.10
- What are the differences in cytokinesis between plant and animal cells? See pp. 216–217 and Figure 11.13

The intricate process of mitosis results in two cells that are genetically identical. But as we mentioned earlier, there is another eukaryotic cell division process, called meiosis, that results in genetic diversity. In the next section we will discuss the roles of mitosis and meiosis in sexual reproduction, and then turn to the details of meiosis in Section 11.5.

11.4 What Role Does Cell Division Play in a Sexual Life Cycle?

The mitotic cell cycle repeats itself, and by this process a single cell can give rise to many cells with identical nuclear DNA. Meiosis, by contrast, produces just four daughter cells. Mitosis and meiosis are both involved in reproduction but in different ways: asexual reproduction involves only mitosis, whereas the sexual reproduction cycle involves both mitosis and meiosis.

Asexual reproduction by mitosis results in genetic constancy

Asexual reproduction, sometimes called vegetative reproduction, is based on the mitotic division of the nucleus. An organism that reproduces asexually may be single-celled like yeast, reproducing itself with each cell cycle, or it may be multicellular like an aspen in a forest in the Wasatch Mountains of Utah (Figure 11.14). Aspen can reproduce sexually, with male and female plants, but in many aspen stands all the trees are the same sex, and DNA analyses have shown that they are **clones** of a single parent organism; the offspring are genetically identical to the parent. In such stands, an extensive root system spreads through the soil and at intervals new stems sprout and grow into trees. Any genetic variation among the trees is most likely due to small environmentally caused changes in the DNA. As you will see, this small amount of variation contrasts with the extensive variation possible in sexually reproducing organisms.

11.14 Asexual Reproduction on a Large Scale In this forest, the aspen trees arose from a single tree by asexual reproduction. They are virtually identical genetically.

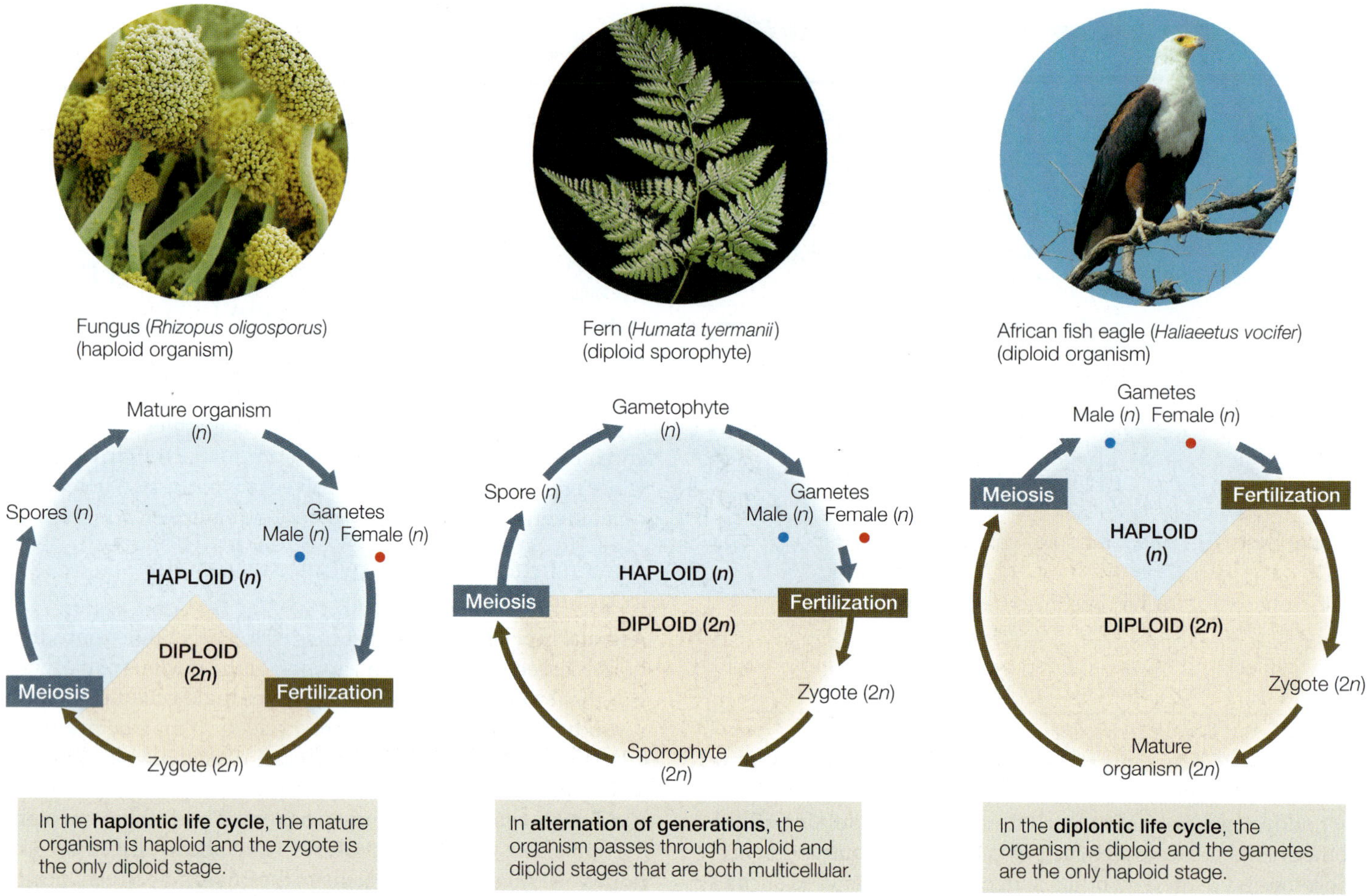

Fungus (*Rhizopus oligosporus*)
(haploid organism)
Fern (*Humata tyermanii*)
(diploid sporophyte)
African fish eagle (*Haliaeetus vocifer*)
(diploid organism)
Mature organism (n)
Spores (n)
Gametes
Male (n) Female (n)
HAPLOID (n)
DIPLOID (2n)
Meiosis
Fertilization
Zygote (2n)
In the **haplontic life cycle**, the mature organism is haploid and the zygote is the only diploid stage.
Gametophyte (n)
Spore (n)
Gametes
Male (n) Female (n)
HAPLOID (n)
Meiosis
Fertilization
DIPLOID (2n)
Zygote (2n)
Sporophyte (2n)
In **alternation of generations**, the organism passes through haploid and diploid stages that are both multicellular.
Gametes
Male (n) Female (n)
Meiosis
Fertilization
HAPLOID (n)
DIPLOID (2n)
Zygote (2n)
Mature organism (2n)
In the **diplontic life cycle**, the organism is diploid and the gametes are the only haploid stage.

11.15 Fertilization and Meiosis Alternate in Sexual Reproduction In sexual reproduction, haploid (n) cells or organisms alternate with diploid ($2n$) cells or organisms.

Go to Activity 11.3 Sexual Life Cycle Life10e.com/ac11.3

Sexual reproduction by meiosis results in genetic diversity

Unlike asexual reproduction, **sexual reproduction** results in an organism that is not identical to its parents. Sexual reproduction requires **gametes** created by meiosis; two parents each contribute one gamete to each of their offspring. Meiosis can produce gametes—and thus offspring—that differ genetically from each other and from the parents. Because of this genetic variation, some offspring may be better adapted than others to survive and reproduce in a particular environment. Meiosis thus generates the genetic diversity that is the raw material for natural selection and evolution.

In most multicellular organisms, the body cells that are *not* specialized for reproduction, called **somatic cells**, each contain two sets of chromosomes, which are found in pairs. One chromosome of each pair comes from each of the organism's two parents; for example, in humans with 46 chromosomes, 23 come from the mother and 23 from the father. The members of such a **homologous pair** are similar in size and appearance. The exception is the sex chromosomes, which are found in some species. The two chromosomes in a homologous pair (called **homologs**) bear corresponding, though not identical, genetic information. For example, a homologous pair of chromosomes in a plant may carry different versions of a gene that controls seed shape. One homolog may carry the version for wrinkled seeds while the other may carry the version for smooth seeds.

There is no simple relationship between the size of an organism and its chromosome number. A housefly has 5 chromosome pairs and a horse has 32, but the smaller carp (a fish) has 52 pairs. Probably the highest number of chromosomes in any organism is in the fern *Ophioglossum reticulatum,* which has 1,260 (630 pairs)!

In contrast to somatic cells, gametes contain only a single set of chromosomes—that is, one homolog from each pair. The number of chromosomes in a gamete is denoted by n, and the cell is said to be **haploid**. During reproduction, two haploid gametes fuse to form a **zygote** in a process called **fertilization**. The zygote thus has two sets of chromosomes, just as the somatic cells do. Its chromosome number is denoted by $2n$, and the zygote is said to be **diploid**. Depending on the organism, the zygote may divide by either meiosis or mitosis. Either way, a new mature organism develops that is capable of sexual reproduction.

All sexual life cycles involve meiosis to produce haploid cells. Figure 11.15 presents three types. In some life cycles the products of meiosis undergo cell division, resulting in a mature organism with haploid cells. Specialized cells in these organisms become gametes. In other life cycles the gametes form directly from the products of meiosis. In all cases, the gametes fuse to produce a zygote, beginning the diploid stage of the life cycle. Since the origin of sexual reproduction, evolution has generated many different versions of the sexual life cycle.

The essence of sexual reproduction is the *random selection of half of the diploid chromosome set* to make a haploid gamete, followed by fusion of two haploid gametes to produce a diploid cell. Both of these steps contribute to a shuffling of genetic information in the population, so that no two individuals have exactly the same genetic makeup (unless they are identical twins). The diversity provided by sexual reproduction opens up enormous opportunities for evolution.

RECAP 11.4

Meiosis is necessary for sexual reproduction, in which haploid gametes fuse to produce a diploid zygote. Sexual reproduction increases genetic diversity, the raw material of evolution.

- What is the difference, in terms of genetics, between asexual and sexual reproduction? **See pp. 217–218**
- How does fertilization produce a diploid organism? **See p. 218**
- What general features do all sexual life cycles have in common? **See pp. 218–219 and Figure 11.15**

Meiosis, unlike mitosis, results in daughter cells that differ genetically from the parent cell. We will now look at the details of meiosis to see how this genetic shuffling occurs.

11.5 What Happens during Meiosis?

In the last section we described the role and importance of meiosis in sexual reproduction. Now we will see how meiosis accomplishes the orderly and precise generation of haploid cells.

Meiosis consists of *two* nuclear divisions that together reduce the number of chromosomes to the haploid number, in preparation for sexual reproduction. *Although the nucleus divides twice during meiosis, the DNA is replicated only once.* Unlike the products of mitosis, the products of meiosis are genetically different from one another and from the parent cell. To understand the process of meiosis and its specific details, it is useful to keep in mind the overall functions of meiosis:

- To reduce the chromosome number from diploid to haploid
- To ensure that each of the haploid products has a complete set of chromosomes
- To generate genetic diversity among the products

The events of meiosis are illustrated in **Figure 11.16**. In this section we will discuss some of the key features that distinguish meiosis from mitosis.

Go to Animated Tutorial 11.2
Meiosis
Life10e.com/at11.2

Meiotic division reduces the chromosome number

As noted above, meiosis consists of two nuclear divisions, meiosis I and meiosis II. Two unique features characterize **meiosis I**.

- *Homologous chromosomes come together to pair* along their entire lengths. No such pairing occurs in mitosis.
- *The homologous chromosome pairs separate*, but the individual chromosomes, each consisting of two sister chromatids, remain intact. (The chromatids will separate during meiosis II.)

Like mitosis, meiosis I is preceded by an interphase with an S phase, during which each chromosome is replicated. As a result, each chromosome consists of two sister chromatids, held together by cohesin proteins. At the end of meiosis I, two nuclei form, each with half of the original chromosomes (one member of each homologous pair). Since the centromeres did not separate, these chromosomes are still composed of two sister chromatids. The sister chromatids are separated during **meiosis II**, which is *not* preceded by DNA replication. As a result, the products of meiosis I and II are four cells, each containing the haploid number of chromosomes. But these four cells are not genetically identical.

Chromatid exchanges during meiosis I generate genetic diversity

Meiosis I begins with a long prophase I (the first three panels of Figure 11.16), during which the chromosomes change markedly. The homologous chromosomes pair by adhering along their lengths in a process called **synapsis**. (This does not usually happen in mitosis.) This pairing process lasts from prophase I to the end of metaphase I. The four chromatids of each pair of homologous chromosomes form a **tetrad**, or bivalent. For example, in a human cell at the end of prophase I there are 23 tetrads, each consisting of four chromatids. The four chromatids come from the two partners in each homologous pair of chromosomes.

Throughout prophase I and metaphase I, the chromatin continues to coil and compact, so that the chromosomes appear ever thicker. At a certain point, the homologous chromosomes appear to repel each other, especially near the centromeres, but they remain held together by physical attachments mediated by cohesins. Later in prophase, regions having these attachments take on an X-shaped appearance (**Figure 11.17**) and are called **chiasmata** (singular *chiasma*, "cross").

A chiasma reflects an *exchange of genetic material* between nonsister chromatids on homologous chromosomes—what geneticists call **crossing over** (**Figure 11.18**). The chromosomes usually begin exchanging material shortly after synapsis begins, but chiasmata do not become visible until later, when the homologs are repelling each other. Crossing over results in **recombinant chromatids**, and it increases genetic variation among the products of meiosis by shuffling genetic information among the homologous pairs. In Chapter 12 we will explore further the genetic consequences of crossing over. Mitosis seldom takes more than an hour or two, but meiosis can take much longer. In human males, the cells in

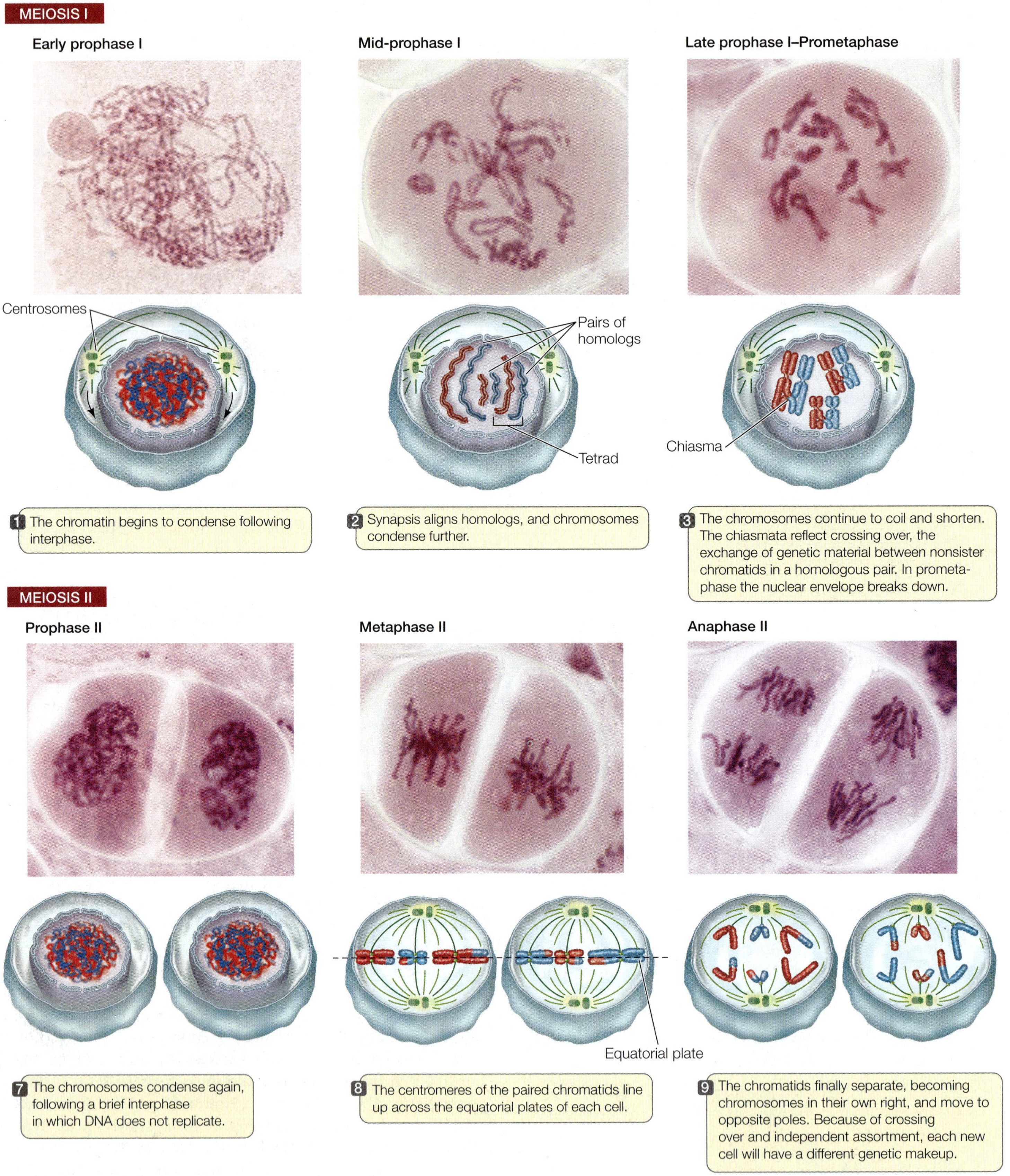
MEIOSIS I
Early prophase I
Mid-prophase I
Late prophase I–Prometaphase
Centrosomes
Pairs of homologs
Tetrad
Chiasma
1 The chromatin begins to condense following interphase.
2 Synapsis aligns homologs, and chromosomes condense further.
3 The chromosomes continue to coil and shorten. The chiasmata reflect crossing over, the exchange of genetic material between nonsister chromatids in a homologous pair. In prometaphase the nuclear envelope breaks down.
MEIOSIS II
Prophase II
Metaphase II
Anaphase II
Equatorial plate
7 The chromosomes condense again, following a brief interphase in which DNA does not replicate.
8 The centromeres of the paired chromatids line up across the equatorial plates of each cell.
9 The chromatids finally separate, becoming chromosomes in their own right, and move to opposite poles. Because of crossing over and independent assortment, each new cell will have a different genetic makeup.

the testis that undergo meiosis take about a week for prophase I and about a month for the entire meiotic cycle. In females, prophase I begins long before a woman's birth, during her early fetal development, and ends as much as decades later, during the monthly ovarian cycle (see Chapter 43).

During meiosis homologous chromosomes separate by independent assortment

A diploid organism has two sets of chromosomes ($2n$): one set derived from its male parent, and the other from its female parent. As the organism grows and develops, its cells undergo

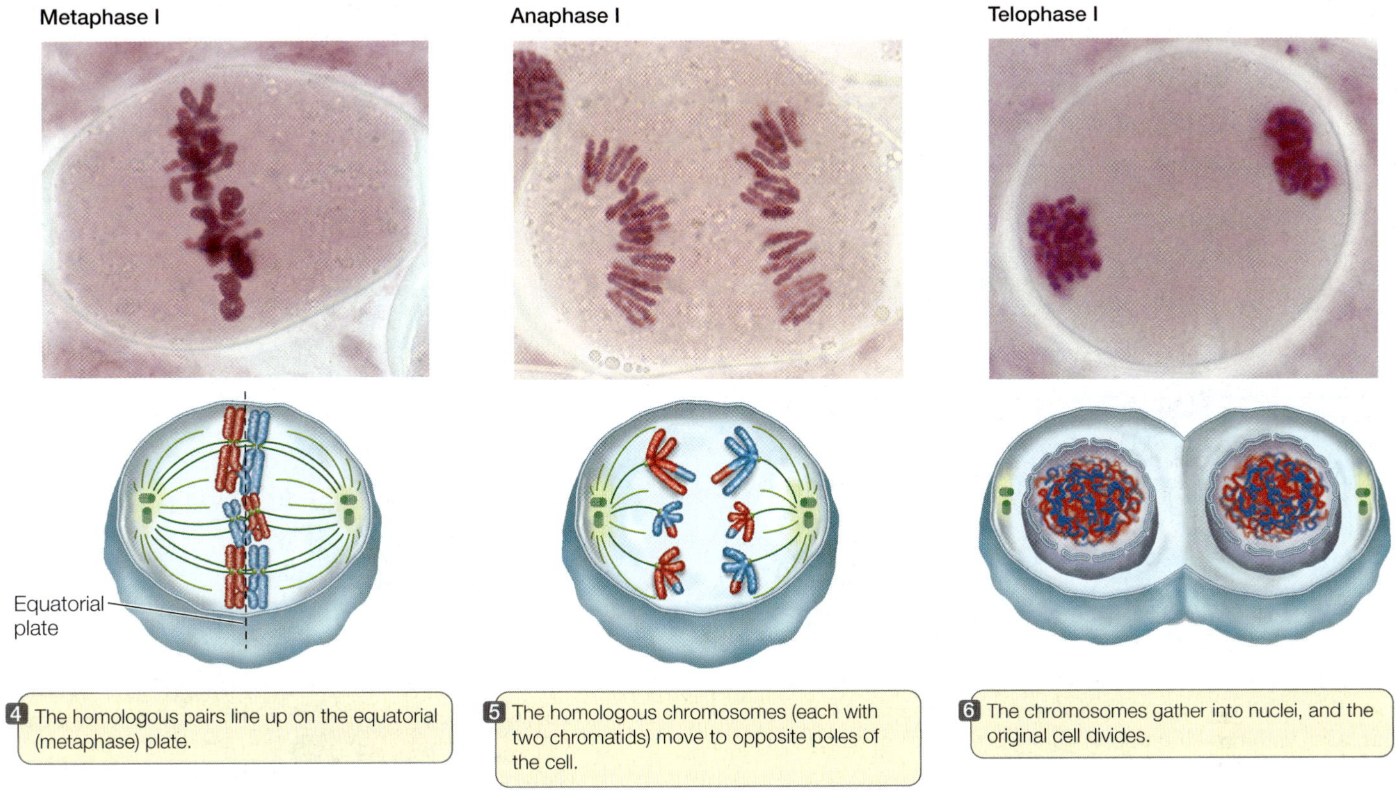

Metaphase I
Anaphase I
Telophase I
Equatorial plate
4 The homologous pairs line up on the equatorial (metaphase) plate.
5 The homologous chromosomes (each with two chromatids) move to opposite poles of the cell.
6 The chromosomes gather into nuclei, and the original cell divides.

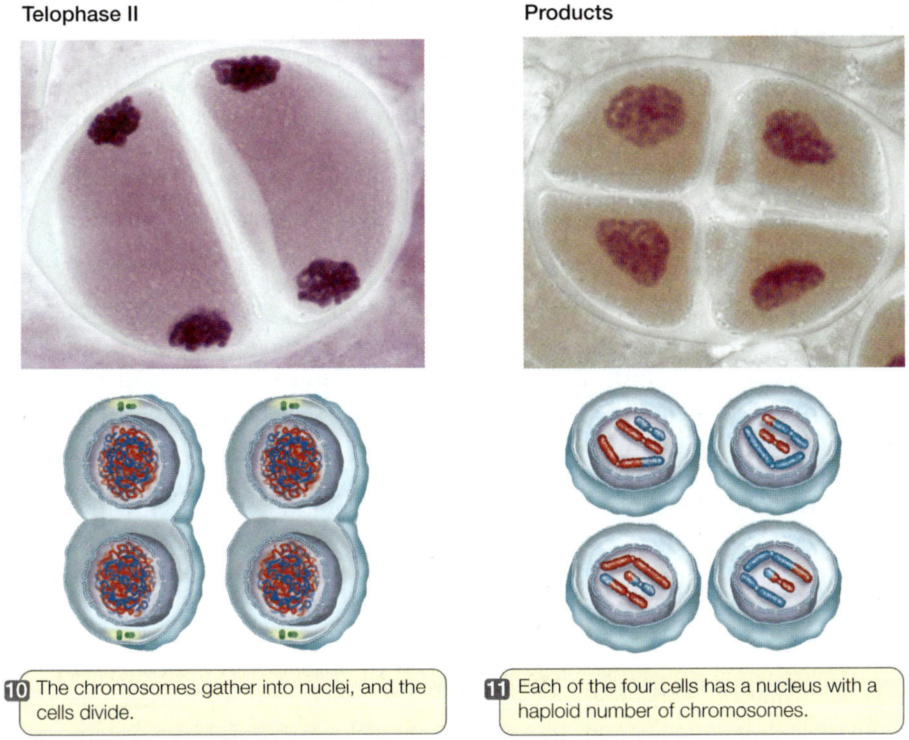

Telophase II
Products
10 The chromosomes gather into nuclei, and the cells divide.
11 Each of the four cells has a nucleus with a haploid number of chromosomes.

11.16 Meiosis: Generating Haploid Cells In meiosis, four daughter nuclei are produced, each of which has half as many chromosomes as the original cell. Four haploid cells are the result of two successive nuclear divisions. The micrographs show meiosis in the male reproductive organ of a lily; the diagrams show the corresponding phases in an animal cell. (For instructional purposes, the chromosomes from one parent are colored blue and those from the other parent are red.)

Go to Activity 11.4 Images of Meiosis
Life10e.com/ac11.4

mitotic divisions. In mitosis, each chromosome behaves independently of its homolog, and its two chromatids are sent to opposite poles during anaphase. Each daughter nucleus ends up with $2n$ chromosomes. In meiosis, things are very different. **Figure 11.19** compares the two processes.

In meiosis I, chromosomes of maternal origin pair with their paternal homologs during synapsis. *This pairing does not occur in mitosis.* Segregation of the homologs during meiotic anaphase I ensures that each newly formed cell receives one member of each homologous pair (see steps 4–6 of Figure 11.16). For example, at the end of meiosis I in humans, each daughter nucleus contains 23 of the original 46 chromosomes. In this way, the chromosome number is decreased from diploid to haploid. Furthermore, meiosis I guarantees that each daughter nucleus gets one full set of chromosomes.

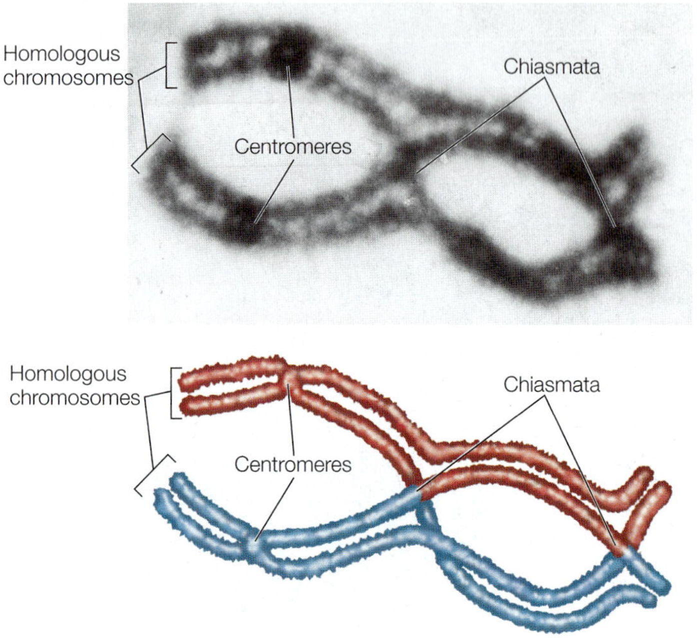
Homologous chromosomes
Chiasmata
Centromeres
Homologous chromosomes
Chiasmata
Centromeres

11.17 Chiasmata: Evidence of Genetic Exchange between Chromatids This micrograph shows a pair of homologous chromosomes, each with two chromatids, during prophase I of meiosis in a salamander. Two chiasmata are visible.

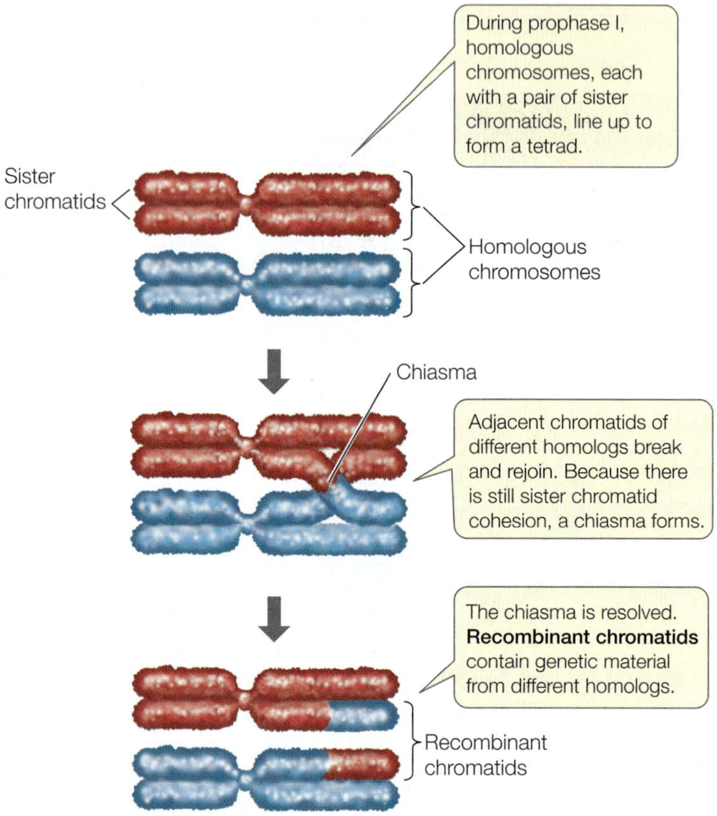
During prophase I, homologous chromosomes, each with a pair of sister chromatids, line up to form a tetrad.
Sister chromatids
Homologous chromosomes
Chiasma
Adjacent chromatids of different homologs break and rejoin. Because there is still sister chromatid cohesion, a chiasma forms.
The chiasma is resolved. Recombinant chromatids contain genetic material from different homologs.
Recombinant chromatids

11.18 Crossing Over Forms Genetically Diverse Chromosomes The exchange of genetic material by crossing over results in new combinations of genetic information on the recombinant chromosomes. The two different colors distinguish the chromosomes contributed by the male and female parents.

Crossing over is one reason for the genetic diversity among the products of meiosis; another is **independent assortment**. It is a matter of chance which member of a homologous pair goes to which daughter cell at anaphase I. For example, imagine there are two homologous pairs of chromosomes in the diploid parent nucleus. A particular daughter nucleus could receive the paternal chromosome 1 and the maternal chromosome 2. Or it could get paternal 2 and maternal 1, or both maternal, or both paternal. It all depends on the way in which the homologous pairs line up at metaphase I. This phenomenon is termed independent assortment.

Note that of the four possible chromosome combinations just described, only two produce daughter nuclei with full complements of either maternal or paternal chromosome sets (apart from the material exchanged by crossing over). *The greater the number of chromosomes, the less probable it is that the original parental combinations will be reestablished, and the greater the potential for genetic diversity.* Most species of diploid organisms have more than two pairs of chromosomes. In humans, with 23 chromosome pairs, 2^{23} (8,388,608) different combinations can be produced just by the mechanism of independent assortment. Taking the extra genetic shuffling afforded by crossing over into account, the number of possible combinations is very large indeed. Crossing over and independent assortment, along with the processes that result in mutations, provide the genetic diversity needed for the differential survival and reproduction of diverse individuals—the basis of evolution by natural selection.

We have seen how meiosis I is fundamentally different from mitosis. In contrast, meiosis II is similar to mitosis in that it involves the separation of sister chromatids into daughter nuclei (see steps 7–11 in Figure 11.16). However, because of crossing over during meiosis I, the sister chromatids are not necessarily identical to one another as they would be in mitosis. Chance assortment of the chromatids during meiosis II contributes further to the genetic diversity of the meiotic products. The final products of meiosis I and meiosis II are four haploid daughter cells, each with one set (n) of chromosomes.

Meiotic errors lead to abnormal chromosome structures and numbers

In the complex processes of mitosis and meiosis, things occasionally go wrong. In meiosis I, a pair of homologous chromosomes may fail to separate, and in mitosis or meiosis II, sister chromatids may fail to separate. Conversely, homologous chromosomes may fail to remain together during metaphase I of meiosis, and then both may migrate to the same pole in anaphase I. These are all examples of **nondisjunction**, which results in the production of aneuploid cells. **Aneuploidy** is a condition in which one or more chromosomes are either lacking or present in excess. If nondisjunction occurs

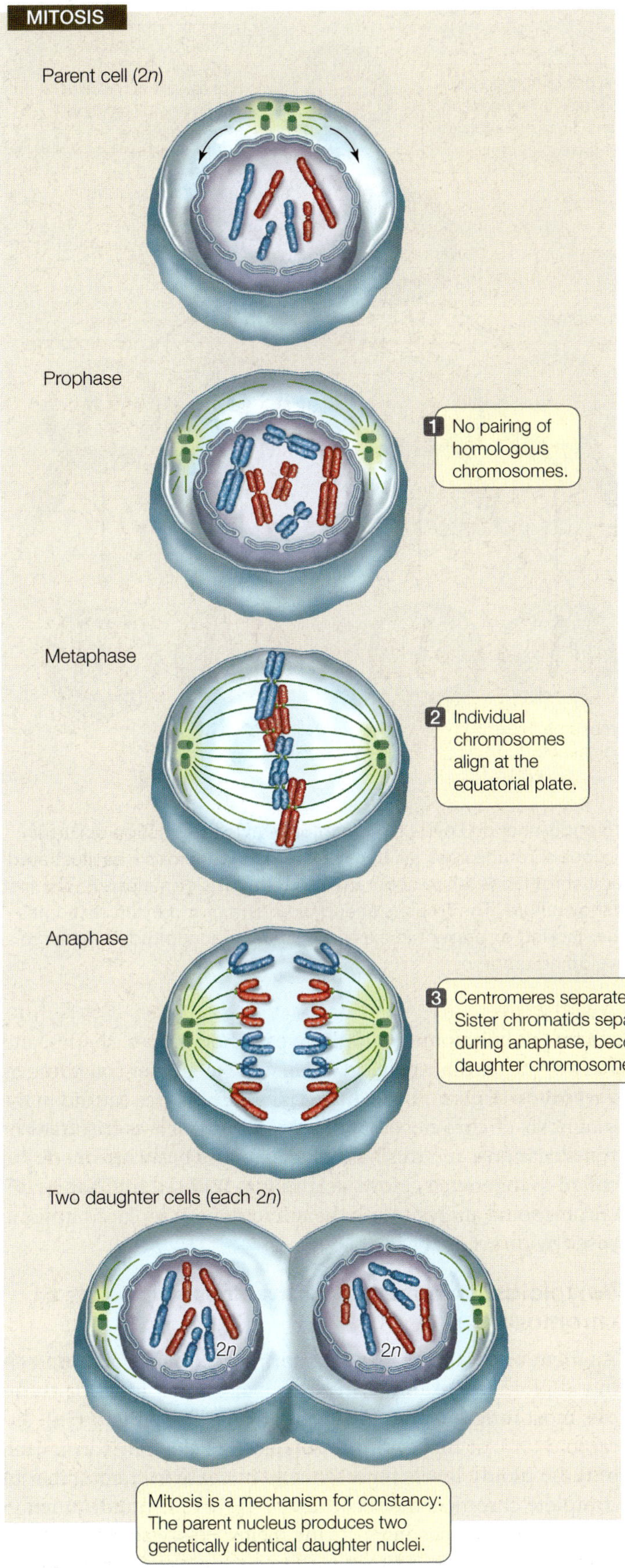
MITOSIS
Parent cell (2n)
Prophase
1 No pairing of homologous chromosomes.
Metaphase
2 Individual chromosomes align at the equatorial plate.
Anaphase
3 Centromeres separate. Sister chromatids separate during anaphase, becoming daughter chromosomes.
Two daughter cells (each 2n)
2n
2n
Mitosis is a mechanism for constancy: The parent nucleus produces two genetically identical daughter nuclei.

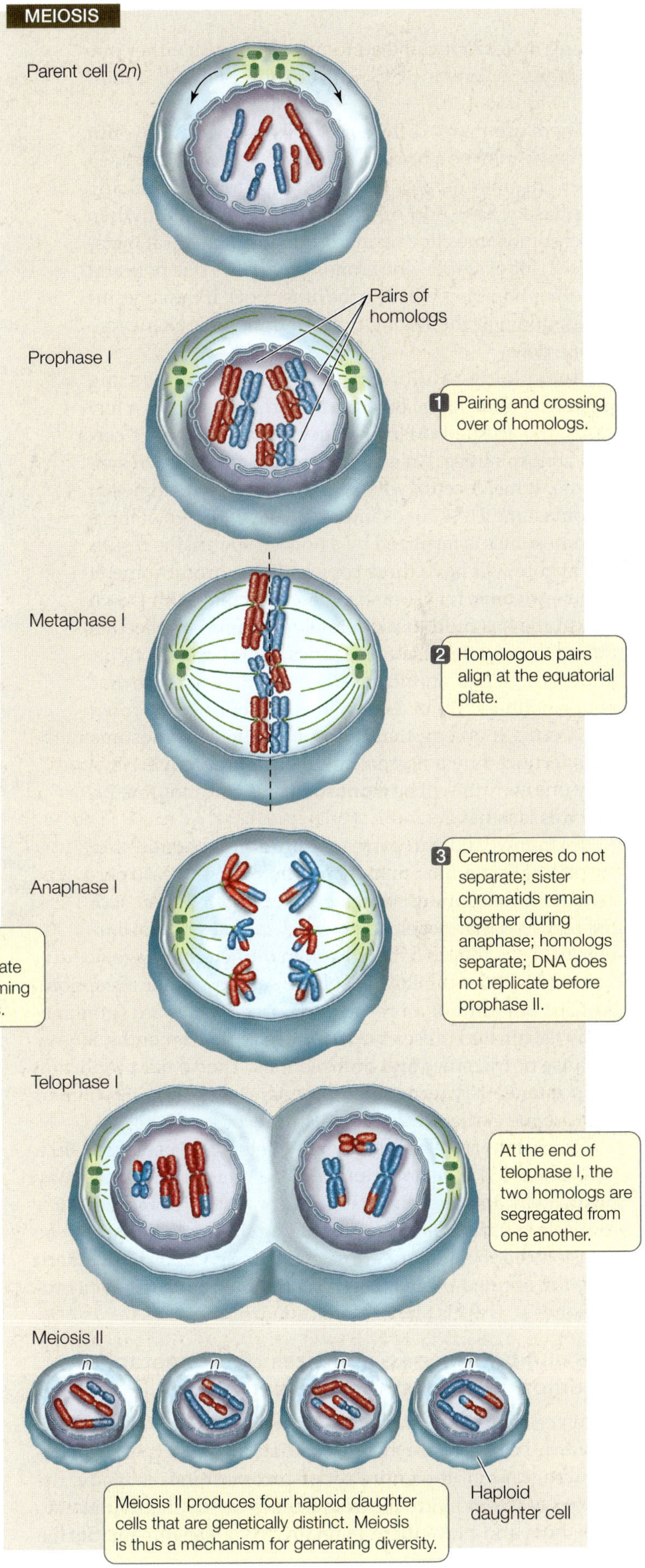
MEIOSIS
Parent cell (2n)
Pairs of homologs
Prophase I
1 Pairing and crossing over of homologs.
Metaphase I
2 Homologous pairs align at the equatorial plate.
Anaphase I
3 Centromeres do not separate; sister chromatids remain together during anaphase; homologs separate; DNA does not replicate before prophase II.
Telophase I
At the end of telophase I, the two homologs are segregated from one another.
Meiosis II
n
n
n
n
Haploid daughter cell
Meiosis II produces four haploid daughter cells that are genetically distinct. Meiosis is thus a mechanism for generating diversity.

11.19 Mitosis and Meiosis: A Comparison Meiosis differs from mitosis chiefly by the pairing of homologs and by the failure of the centromeres to separate at the end of metaphase I.

during meiosis, it can lead to offspring with either one too many, or one too few, chromosomes in all of their cells (Figure 11.20).

There are many different causes of aneuploidy, but one cause may be a breakdown in the cohesins that keep sister chromatids and tetrads joined together during prophase. These and other proteins ensure that when the chromosomes line up at the equatorial plate at metaphase I, for example, one homolog will face one pole and the other homolog will face the other pole. If the cohesins break down at the wrong time, both homologs may go to one pole.

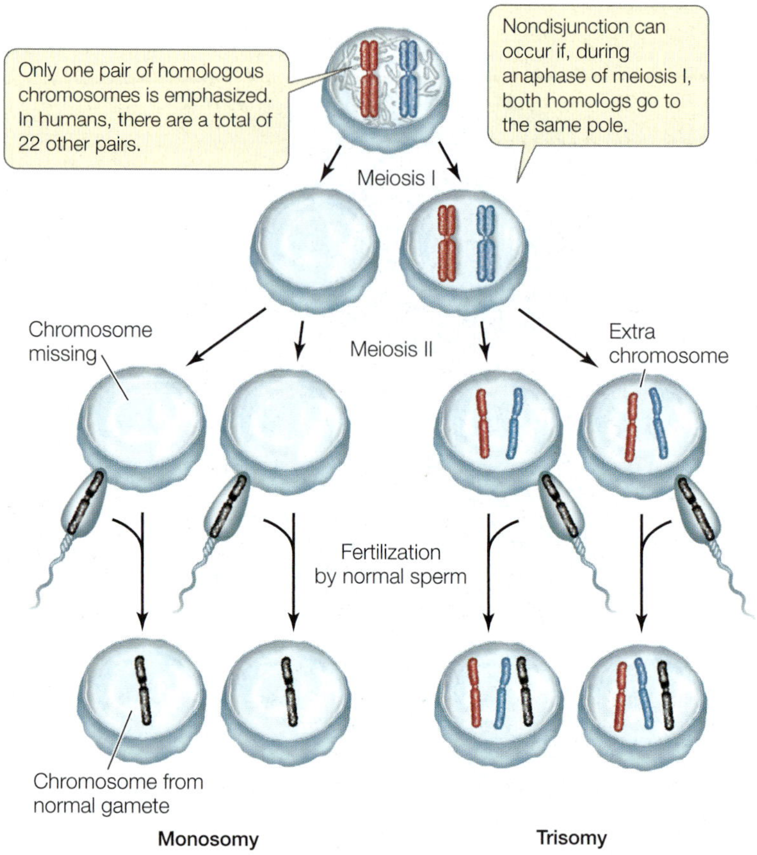
Only one pair of homologous chromosomes is emphasized. In humans, there are a total of 22 other pairs.
Nondisjunction can occur if, during anaphase of meiosis I, both homologs go to the same pole.
Meiosis I
Chromosome missing
Meiosis II
Extra chromosome
Fertilization by normal sperm
Chromosome from normal gamete
Monosomy
Trisomy

11.20 Nondisjunction Leads to Aneuploidy Nondisjunction occurs if homologous chromosomes fail to separate during meiosis I, as illustrated here, or if chromatids fail to separate during mitosis or meiosis II. The first case is shown here. The result is aneuploidy: one or more chromosomes are either lacking or present in excess. Generally, aneuploidy is lethal to the developing embryo.

Aneuploidy resulting from nondisjunction during meiosis is often lethal for the affected offspring. In a few cases, the affected offspring survive but may have certain abnormalities. An example in humans is Down syndrome, which occurs when a gamete has two copies of chromosome 21. If, for example, an egg with two of these chromosomes is fertilized by a normal sperm, the resulting zygote will have three copies of the chromosome: it will be **trisomic** for chromosome 21. A child with Down syndrome has mild to moderately impaired intellectual ability; characteristic abnormalities of the hands, tongue, and eyelids; and an increased susceptibility to cardiac abnormalities. About 1 child in 800 is born with Down syndrome. If an egg that did not receive chromosome 21 is fertilized by a normal sperm, the zygote will have only one copy: it will be **monosomic** for chromosome 21, and this is lethal.

Trisomies and the corresponding monosomies are surprisingly common in human zygotes, with 10–30 percent of all conceptions showing aneuploidy. But most of the embryos that develop from such zygotes do not survive to birth, and those that do often die before the age of 1 year (trisomies for chromosome 21 are the viable exception). At least one-fifth of all recognized pregnancies are spontaneously terminated (miscarried) during the first 2 months, largely because of trisomies and monosomies. The actual proportion of spontaneously terminated pregnancies is certainly higher, because the earliest ones often go unrecognized.

Other abnormal chromosomal events can also occur. In a process called **translocation**, a piece of a chromosome may break away and become attached to another chromosome. For example, a particular large part of one chromosome 21 may be translocated to another chromosome. Individuals who inherit this translocated piece along with two normal copies of chromosome 21 will also have Down syndrome.

The number, shapes, and sizes of the metaphase chromosomes constitute the karyotype

When cells are in metaphase of mitosis, it is often possible to count and characterize their individual chromosomes. If a photomicrograph of the entire set of chromosomes is made, the images of the individual chromosomes can be manipulated to pair them and place them in an orderly arrangement. Such a rearranged photomicrograph reveals the number, shapes, and sizes of the chromosomes in a cell, which together constitute its **karyotype** (Figure 11.21). In humans, karyotypes can aid in the diagnosis of chromosomal abnormalities such as trisomies or translocations, and this has led to an entire branch of medicine called cytogenetics. However, as you will see in Chapter 15, chromosome analysis with the microscope is replaced in some cases by direct analysis of DNA.

Polyploids have more than two complete sets of chromosomes

As mentioned in Section 11.4, mature organisms are often either diploid (for example, most animals) or haploid (for example, most fungi). Under some circumstances, triploid ($3n$), tetraploid ($4n$), or higher-order **polyploid** nuclei may form. Each of these ploidy levels represents an increase in the number of complete chromosome sets present. If there is nondisjunction of all of the chromosomes during meiosis I (see above), diploid gametes will form. This can lead to *auto*polyploidy after fertilization. Autotriploids and autotetraploids have been important in some cases in species formation. A diploid nucleus can

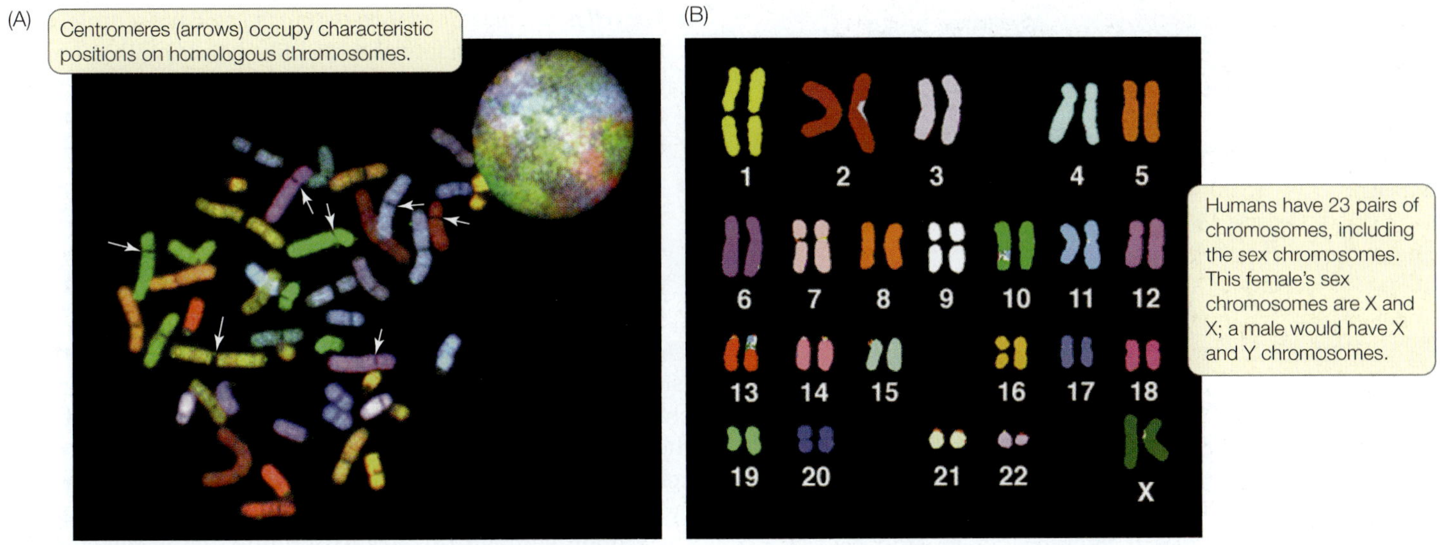
(A)
Centromeres (arrows) occupy characteristic positions on homologous chromosomes.
(B)
1
2
3
4
5
6
7
8
9
10
11
12
13
14
15
16
17
18
19
20
21
22
X
Humans have 23 pairs of chromosomes, including the sex chromosomes. This female's sex chromosomes are X and X; a male would have X and Y chromosomes.

11.21 The Human Karyotype (A) Chromosomes from a human cell in metaphase. The DNA of each chromosome pair has a specific nucleotide sequence that is stained by a particular colored dye, so that the chromosomes in a homologous pair share a distinctive color. Each chromosome at this stage is composed of two chromatids, but they cannot be distinguished. At the upper right is an interphase nucleus. (B) This karyogram, produced by computerized analysis of the image on the left, shows homologous pairs lined up together and numbered, clearly revealing the individual's karyotype.

undergo normal meiosis because there are two sets of chromosomes to make up homologous pairs, which separate during anaphase I. Similarly, a tetraploid nucleus has an even number of each kind of chromosome, so each chromosome can pair with its homolog. However, a triploid nucleus cannot undergo normal meiosis because one-third of the chromosomes would lack partners. Triploid individuals are usually sterile.

Because polyploid nuclei have more chromosome sets, their cells tend to be larger. This has led to the use of polyploid plants in agriculture. Diploid bananas are smaller and produce inedible seeds; triploid bananas are larger and seedless. A similar phenomenon is seen in triploid seedless watermelon. Perhaps the best known, and certainly the most important, polyploid crop plant is wheat. In this case, hybridization occurred between species, forming new *allo*polyploid species:

- Haploid gametes from two species (A and B) mated to form a diploid zygote (chromosomes AB).
- Nondisjunction of all chromosomes occurred during mitosis in the fertilized egg, resulting in a tetraploid (AABB).
- The tetraploid grew up to be a fertile adult.

Modern bread wheat is the result of two such events, which occurred around 8,000 to 10,000 years ago, resulting in a hexaploid (Figure 11.22). Wheat's properties of grain formation and environmental adaptation thus come from three different ancestral species. Other allopolyploid crops include cotton, oats, and sugarcane.

RECAP 11.5

Meiosis produces four daughter cells in which the chromosome number is reduced from diploid to haploid. Because of the independent assortment of chromosomes and the crossing over of homologous chromatids, the four products of meiosis are not genetically identical. Meiotic errors, such as the failure of a homologous chromosome pair to separate, can lead to abnormal numbers of chromosomes. Several important crop plants, such as wheat, are polyploid.

- How do crossing over and independent assortment result in unique daughter nuclei? **See pp. 219–222 and Figures 11.16, 11.18**
- What are the differences between meiosis and mitosis? **See pp. 221–222 and Figure 11.19**
- What is aneuploidy, and how can it arise from nondisjunction during meiosis? **See pp. 222–224 and Figure 11.20**

An essential role of cell division in complex eukaryotes is to replace dead cells. What causes cells to die?

11.6 In a Living Organism, How Do Cells Die?

Cells within a living organism can die in one of two ways. The first type of cell death, **necrosis**, occurs when cells and tissues are damaged by mechanical means or toxins, or are starved of oxygen or nutrients. These cells often swell up and burst, releasing their contents into the extracellular environment. This process often results in inflammation (see Section 42.2).

Another kind of cell death occurs during normal developmental processes. **Apoptosis** (Greek, "falling apart") is a programmed series of events that results in cell death. Why would a cell initiate apoptosis, which is essentially cell suicide? In animals, there are two possible reasons:

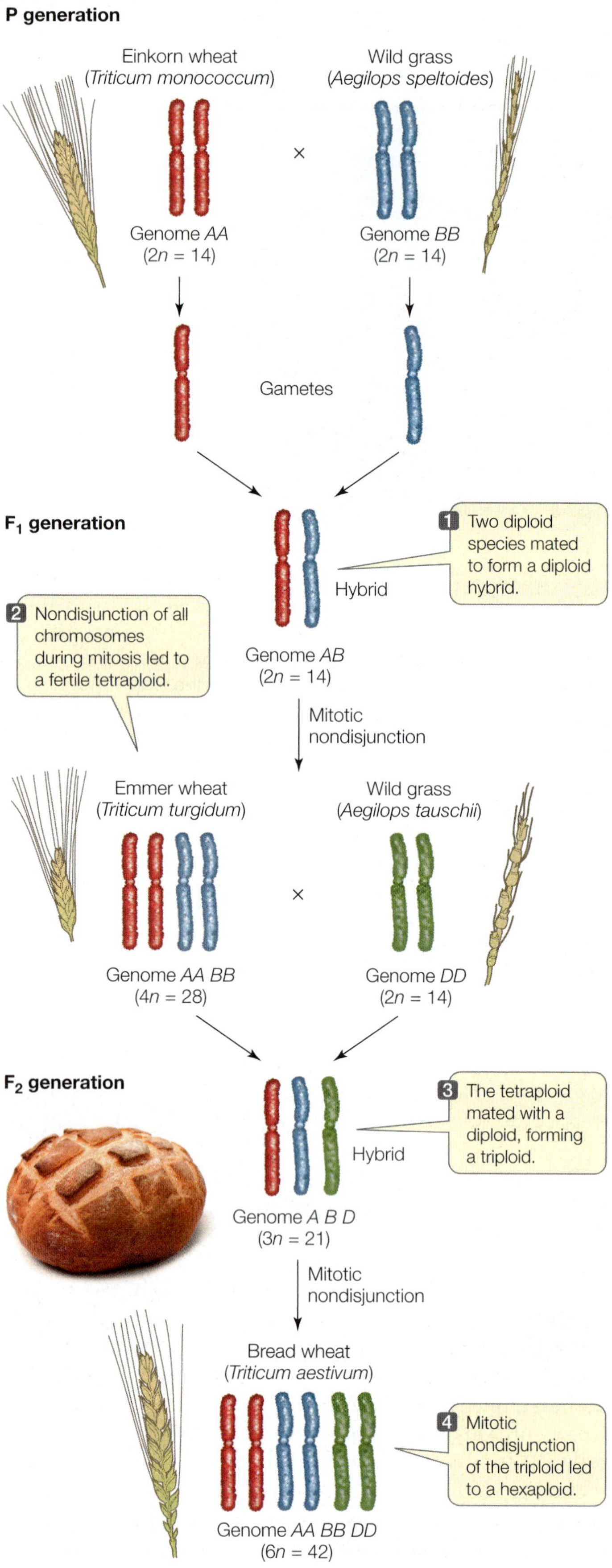
P generation
Einkorn wheat
(Triticum monococcum)
Wild grass
(Aegilops speltoides)
×
Genome AA
(2n = 14)
Genome BB
(2n = 14)
Gametes
F1 generation
1 Two diploid species mated to form a diploid hybrid.
Hybrid
2 Nondisjunction of all chromosomes during mitosis led to a fertile tetraploid.
Genome AB
(2n = 14)
Mitotic nondisjunction
Emmer wheat
(Triticum turgidum)
Wild grass
(Aegilops tauschii)
×
Genome AA BB
(4n = 28)
Genome DD
(2n = 14)
F2 generation
3 The tetraploid mated with a diploid, forming a triploid.
Hybrid
Genome A B D
(3n = 21)
Mitotic nondisjunction
Bread wheat
(Triticum aestivum)
4 Mitotic nondisjunction of the triploid led to a hexaploid.
Genome AA BB DD
(6n = 42)

11.22 Polyploidy and the Origin of Bread Wheat Three species and two events of mitotic nondisjunction led to modern wheat.

- *The cell is no longer needed by the organism.* For example, before birth, a human fetus has weblike hands, with connective tissue between the fingers. As development proceeds, this unneeded tissue disappears as the cells undergo apoptosis in response to specific signals.
- *The longer cells live, the more prone they are to genetic damage that could lead to cancer.* This is especially true of epithelial cells on the surface of an organism, which may be exposed to radiation or toxic substances. Such cells normally die after only days or weeks and are replaced by new cells.

The outward events of apoptosis are similar in many organisms. The cell becomes detached from its neighbors and its chromatin is digested by enzymes that cut the DNA (between the nucleosomes) into fragments of about 180 base pairs. The cell forms membranous lobes, or "blebs," that break up into cell fragments (**Figure 11.23A**). In a remarkable example of the economy of nature, the surrounding, living cells usually ingest the remains of the dead cell by phagocytosis. Neighboring cells digest the apoptotic cell contents in their lysosomes, and the digested components are recycled.

Apoptosis is also used by plant cells, in an important defense mechanism called the hypersensitive response. Plants can protect themselves from disease by undergoing apoptosis at the site of infection by a fungus or bacterium. With no living tissue to grow in, the invading organism is not able to spread to other parts of the plant. Because of their rigid cell walls, plant cells do not form blebs the way animal cells do. Instead, they digest their own cell contents in the vacuole and then release the digested components into the vascular system.

Despite these differences between plant and animal cells, they share many of the signal transduction pathways that lead to apoptosis. A variety of signals, either external or from inside the cell, can lead to programmed cell death (**Figure 11.23B**). Such signals include hormones, growth factors, viral infection, certain toxins, or extensive DNA damage. These signals activate specific receptors, which in turn activate signal transduction pathways leading to apoptosis. Some apoptotic pathways target the mitochondria, for example, by increasing the permeability of mitochondrial membranes. The cell quickly dies if its mitochondria can't carry out cellular respiration. An important class of enzymes called **caspases** are activated during apoptosis. These enzymes are proteases that hydrolyze target molecules in a cascade of events. The cell dies as the caspases hydrolyze proteins of the nuclear envelope, nucleosomes, and plasma membrane.

RECAP 11.6

Cell death can occur either by necrosis or by apoptosis. Apoptosis is governed by precise molecular controls.

- What are some differences between apoptosis and necrosis? See p. 225
- In what situations is apoptosis necessary? See p. 226
- How is apoptosis regulated? See Figure 11.23

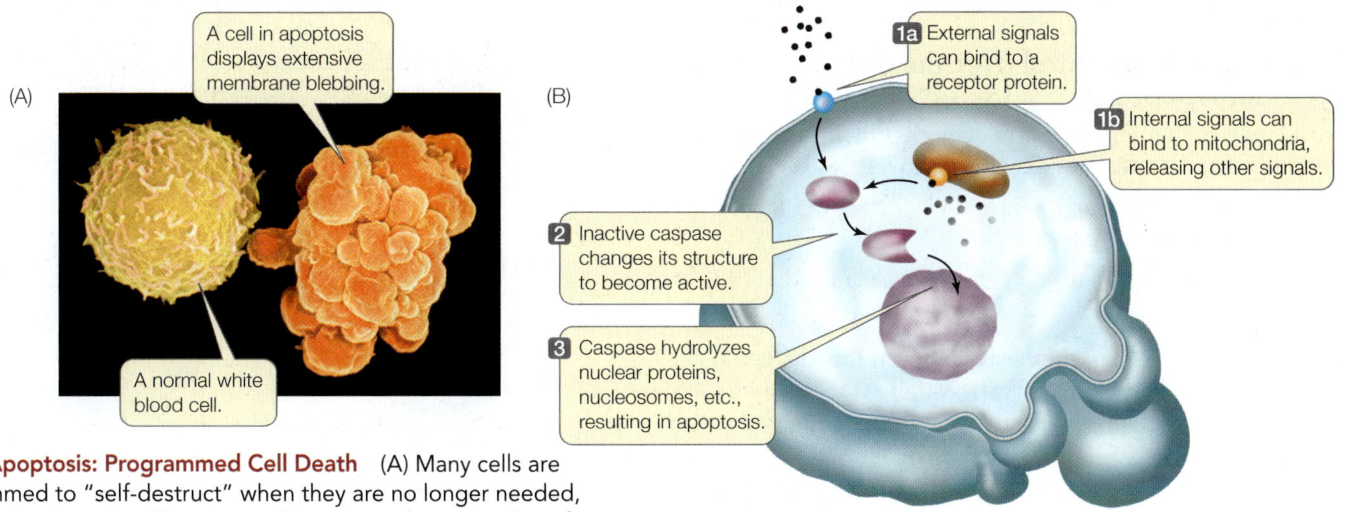
(A)
A cell in apoptosis displays extensive membrane blebbing.
A normal white blood cell.
(B)
1a External signals can bind to a receptor protein.
1b Internal signals can bind to mitochondria, releasing other signals.
2 Inactive caspase changes its structure to become active.
3 Caspase hydrolyzes nuclear proteins, nucleosomes, etc., resulting in apoptosis.

11.23 Apoptosis: Programmed Cell Death (A) Many cells are programmed to "self-destruct" when they are no longer needed, or when they have lived long enough to accumulate a burden of DNA damage that might harm the organism. (B) Both external and internal signals stimulate caspases, the enzymes that break down specific cell constituents, resulting in apoptosis.

Mitosis adds cells to organisms, and apoptosis removes them. Under normal circumstances these processes are balanced so as to benefit the organism as a whole. In the next section we will examine what happens when this balance is disturbed and cell production runs out of control.

11.7 How Does Unregulated Cell Division Lead to Cancer?

Perhaps no malady affecting people in the industrialized world instills more fear than cancer, and most people realize that it involves an inappropriate increase in cell numbers. One in three Americans will have some form of cancer in his or her lifetime, and at present, one in four will die of it. With 1.5 million new cases and half a million deaths in the United States annually, cancer ranks second only to heart disease as a killer.

Cancer cells differ from normal cells

Cancer cells differ from the normal cells from which they originate in two ways:

- Cancer cells lose control over cell division.
- Cancer cells can migrate to other locations in the body.

Most cells in the body divide only if they are exposed to extracellular signals such as growth factors. Cancer cells do not respond to these controls, and instead divide more or less continuously, ultimately forming **tumors** (large masses of cells). By the time a physician can feel a tumor or see one on an X-ray film or CAT scan, it already contains millions of cells. Tumors can be benign or malignant.

Benign tumors resemble the tissue they came from, grow slowly, and remain localized where they develop. For example, a lipoma is a benign tumor of fat cells that may arise in the armpit and remain there. Benign tumors are not cancers, but they must be removed if they impinge on an organ, obstructing its function.

Malignant tumors do not look like their parent tissue at all. A flat, specialized epithelial cell in the lung wall may turn into a relatively featureless, round, malignant lung cancer cell (Figure 11.24). Malignant cells often have irregular structures, such as variable nucleus sizes and shapes. This characteristic was used to identify the cells in Henrietta Lacks' tumor as malignant (see the opening story of this chapter).

The second and most fearsome characteristic of cancer cells is their ability to invade surrounding tissues and spread to other parts of the body by traveling through the bloodstream or lymphatic ducts. When malignant cells become lodged in some distant part of the body, they go on dividing and growing, establishing a tumor at that new site. This spreading, called **metastasis**, results in organ failures and makes the cancer very hard to treat.

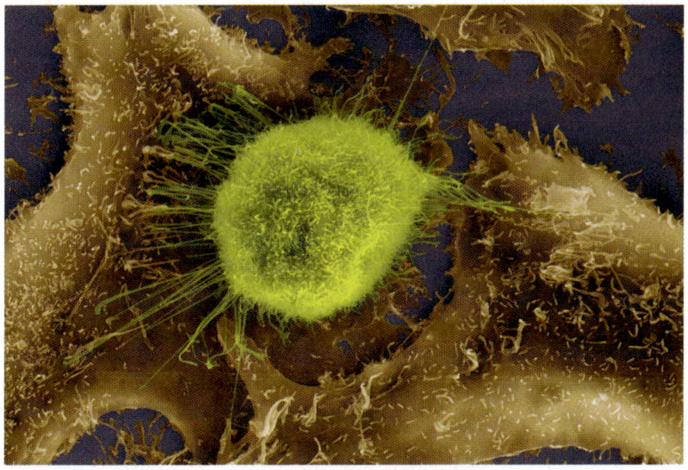

11.24 A Cancer Cell with its Normal Neighbors This lung cancer cell (yellow-green) is quite different from the normal lung cells surrounding it. The cancer cell can divide more rapidly than its normal counterparts, and it can spread to other organs. This form of small-cell cancer is lethal, with a 5-year survival rate of only 10 percent. Most cases are caused by tobacco smoking.

Cancer cells lose control over the cell cycle and apoptosis

Earlier in this chapter you learned about proteins that regulate the progress of a eukaryotic cell through the cell cycle:

- Positive regulators such as growth factors stimulate the cell cycle: they are like "gas pedals."
- Negative regulators such as retinoblastoma protein (RB) inhibit the cell cycle: they are like "brakes."

Just as driving a car requires stepping on the gas pedal *and* releasing the brakes, a cell will go through a division cycle only if the positive regulators are active and the negative regulators are inactive.

In most cells, the two regulatory systems ensure that the cells divide only when needed. In cancer cells, these two processes are abnormal.

- **Oncogene** proteins are positive regulators in cancer cells. They are derived from normal positive regulators that have become mutated to be overly active or that are present in excess, and they stimulate the cancer cells to divide more often. Oncogene products could be growth factors, their receptors, or other components in the signal transduction pathways (see Chapter 7) that stimulate cell division. An example of an oncogene protein is a growth factor receptor in a breast cancer cell (**Figure 11.25A**). Normal breast cells have relatively low numbers of the human epidermal growth factor receptor HER2. So breast cells are not usually stimulated to multiply in the presence of this growth factor. In about 25 percent of breast cancers, a DNA change results in the increased production of the HER2 receptor. This results in positive stimulation of the cell cycle, and a rapid proliferation of cells with the altered DNA.
- **Tumor suppressors** are negative regulators in both cancer and normal cells, but in cancer cells they are inactive. An example is the RB protein, which when active acts at R (the restriction point) in G1 to block the cell cycle (see Figure 11.6). In some cancer cells RB is inactive, allowing the cell cycle to proceed. Some viral proteins can inactivate tumor suppressors. For example, the human papillomavirus infects cells of the cervix and produces a protein called E7. E7 binds to the RB protein and prevents it from inhibiting the cell cycle (**Figure 11.25B**). Another important tumor suppressor is p53, a transcription factor that is involved in cell cycle checkpoint pathways and apoptosis. The importance of p53 as a tumor suppressor is illustrated by the fact that more than 50 percent of human tumors have mutations in the gene that encodes p53.

The discovery of apoptosis (see Section 11.6) changed the way biologists think about cancer. In a population of cells, the net increase in cell numbers over time (the growth rate) is a function of cells added (the rate of cell division) and cells lost (the rate of apoptosis):

$$\text{Growth rate of cell population} = \text{rate of cell division} - \text{rate of apoptosis}$$

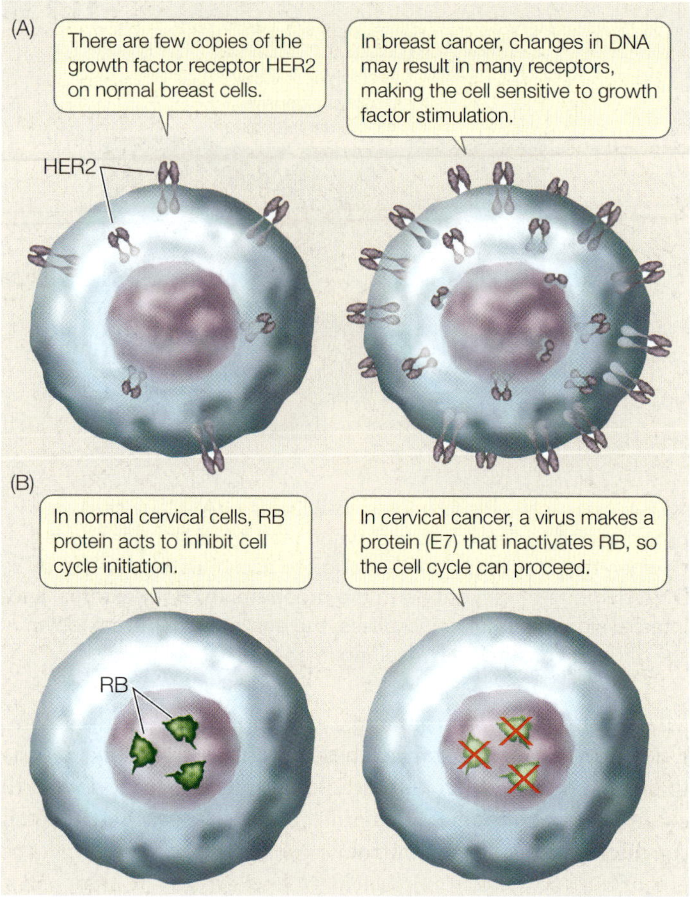
(A)
There are few copies of the growth factor receptor HER2 on normal breast cells.
In breast cancer, changes in DNA may result in many receptors, making the cell sensitive to growth factor stimulation.
HER2
(B)
In normal cervical cells, RB protein acts to inhibit cell cycle initiation.
In cervical cancer, a virus makes a protein (E7) that inactivates RB, so the cell cycle can proceed.
RB

11.25 Molecular Changes in Cancer Cells In cancer, oncogene proteins become active (A) and tumor suppressor proteins become inactive (B).

In normal nongrowing tissues, the rate of cell division equals the rate of apoptosis, so the cell population as a whole does not grow. Cancer cells are defective in their regulation of the cell cycle, resulting in increased rates of cell division. In addition, cancer cells can lose the ability to respond to positive regulators of apoptosis (see Figure 11.23), and this results in lowered rates of cell death. Both of these defects favor an increased growth rate of the cancer cell population.

Cancer treatments target the cell cycle

The most successful and widely used treatment for cancer is surgery. While physically removing a tumor is optimal, it is often difficult for a surgeon to get all of the tumor cells. (A tumor about 1 centimeter in diameter already has a billion cells!) Tumors are generally embedded in normal tissues. Added to this is the probability that cells of the tumor may have broken off and spread to other organs. This makes it unlikely that localized surgery will be curative. So other approaches are taken to treat or cure cancer, and these generally target the cell cycle (**Figure 11.26**). The goal is to decrease the rate of cell division and/or increase the rate of apoptosis so that the cancer cell population decreases.

An example of a cancer drug that targets the cell cycle is 5-fluorouracil, which blocks the synthesis of thymine, one of the four bases in DNA. The drug paclitaxel prevents the functioning of microtubules in the mitotic spindle. Both drugs inhibit the cell cycle, and apoptosis causes tumor shrinkage. More dramatic is

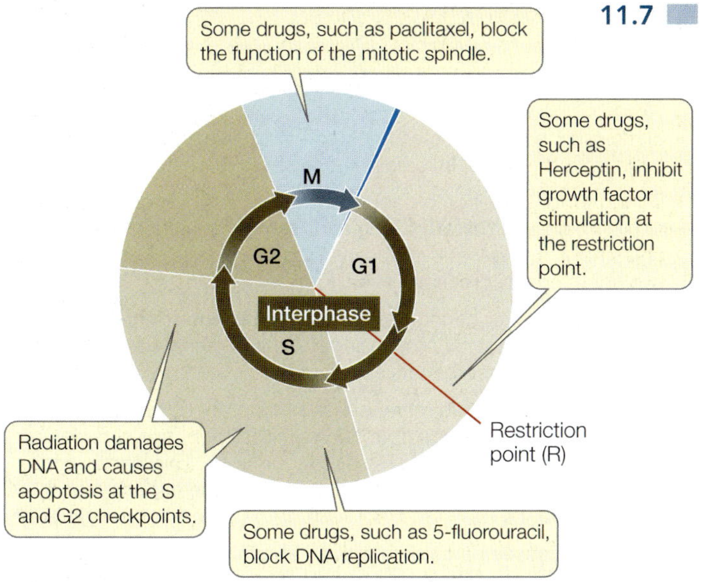
Some drugs, such as paclitaxel, block the function of the mitotic spindle.
Some drugs, such as Herceptin, inhibit growth factor stimulation at the restriction point.
M
G2
G1
Interphase
S
Restriction point (R)
Radiation damages DNA and causes apoptosis at the S and G2 checkpoints.
Some drugs, such as 5-fluorouracil, block DNA replication.

11.26 Cancer Treatment and the Cell Cycle To prevent cancer cells from dividing, physicians use combinations of therapies that attack the cell cycle at different points.

radiation treatment, in which a beam of high-energy radiation is focused on the tumor. DNA damage is extensive, and the cell cycle checkpoint for DNA repair is overwhelmed. As a result, the cell undergoes apoptosis. A major problem with these treatments is that they target normal cells as well as the tumor cells. These treatments are toxic to tissues with large populations of normal dividing cells such as those in the intestine, skin, and bone marrow (where blood cells are produced).

A major effort in cancer research is to find treatments that target only cancer cells. A promising recent example is Herceptin, which targets the HER2 growth factor receptor that occurs at high levels on the surfaces of some breast cancer cells (see Figure 11.25A). Herceptin binds specifically to the HER2 receptor but does not stimulate it. This prevents the natural growth factor from binding, and so the cells are not stimulated to divide. As a result, the tumor shrinks because the apoptosis rate remains the same. More such treatments are on the way.

RECAP 11.7

Cancer cells differ from normal cells in terms of their rapid cell division and their ability to spread (metastasis). Many proteins regulate the cell cycle, either positively (oncogenes) or negatively (tumor suppressors). In cancer, one or another of these proteins is altered in some way, making its activity abnormal. Radiation and many cancer drugs target proteins involved in the cell cycle.

- How are oncogene proteins and tumor suppressor proteins involved in cell cycle control in normal and cancer cells? **Review p. 228 and Figure 11.25**
- How does cancer treatment target the cell cycle? **Review pp. 228–229 and Figure 11.26**

We have now looked at the cell cycle and at cell division by binary fission, mitosis, and meiosis. We have described the normal cell cycle and how its regulation is disrupted in cancer. We have seen how meiosis produces haploid cells in sexual life cycles. In the coming chapters we will examine heredity, genes, and DNA.

?

What makes HeLa cells reproduce so well in the laboratory?

ANSWER

In normal tissues, the rate of cell division is offset by the rate of cell death. Unlike most normal cells, HeLa cells keep growing because they have a genetic imbalance that heavily favors cell reproduction over cell death. Henrietta Lacks was infected with the human papillomavirus, which stimulates cell division in cervical cells. In addition, an enzyme called telomerase, which keeps DNA intact and prevents cell death, is overexpressed in HeLa cells. This combination of traits—increased cell division plus decreased apoptosis—leads to the extraordinary growth rate of HeLa cells.

CHAPTER SUMMARY 11

11.1 How Do Prokaryotic and Eukaryotic Cells Divide?

- Cell division is necessary for the reproduction, growth, and repair of organisms.
- Cell division must be initiated by a **reproductive signal**. Before a cell can divide, the genetic material (DNA) must be **replicated** and **segregated** to separate portions of the cell. **Cytokinesis** then divides the cytoplasm into two cells.
- In prokaryotes, most cellular DNA is a single molecule, usually in the form of a circular **chromosome**. Prokaryotes reproduce by **binary fission**. **Review Figure 11.2**
- In eukaryotes, cells divide by either **mitosis** or **meiosis**. Eukaryotic cell division follows the same general pattern as binary fission, but with significant differences. For example, a eukaryotic cell has a distinct nucleus whose chromosomes must be replicated prior to separating the two daughter cells.

11.2 How Is Eukaryotic Cell Division Controlled?

- The eukaryotic cell cycle has two main phases: **interphase**, during which cells are not dividing and the DNA is replicating, and mitosis or M phase, when the cells are dividing.
- During most of the eukaryotic cell cycle, the cell is in interphase, which is divided into three subphases: S, G1, and G2. DNA is replicated during **S phase**. Mitosis (M phase) and cytokinesis follow. **Review Figure 11.3**
- **Cyclin–Cdk** complexes regulate the passage of cells through checkpoints in the cell cycle. Retinoblastoma protein (RB) inhibits the cell cycle at the **restriction point**. The cyclin–Cdk functions by inactivating RB and allows the cell cycle to progress. **Review Figures 11.5, 11.6**
- External controls such as **growth factors** can stimulate the cell to begin a division cycle.

continued

11.3 What Happens during Mitosis?
See ANIMATED TUTORIAL 11.1

- In mitosis, a single nucleus gives rise to two nuclei that are genetically identical to each other and to the parent nucleus.
- DNA is wrapped around proteins called histones, forming bead-like units called **nucleosomes**. A eukaryotic chromosome contains strings of nucleosomes bound to proteins in a complex called **chromatin**. **Review Figure 11.9**
- At mitosis, the replicated chromosomes (**sister chromatids**) are held together at the **centromere**. Each chromatid consists of one double-stranded DNA molecule. During mitosis sister chromatids, attached by cohesin, line up at the equatorial plate and attach to the **spindle apparatus**. The chromatids separate (becoming **daughter chromosomes**) and migrate to opposite ends of the cell. **Review Figures 11.10, 11.11, ACTIVITIES 11.1, 11.2**
- Mitosis can be divided into several phases called **prophase**, **prometaphase**, **metaphase**, **anaphase**, and **telophase**.
- Nuclear division is usually followed by cytokinesis. Animal cell cytoplasms divide via a contractile ring made up of actin microfilaments and myosin. In plant cells, cytokinesis is accomplished by vesicles that fuse to form a cell plate. **Review Figure 11.13**

11.4 What Role Does Cell Division Play in a Sexual Life Cycle?

- **Asexual reproduction** produces **clones**, new organisms that are genetically identical to the parent. Any genetic variation is the result of changes in genes.
- In **sexual reproduction**, two **haploid** gametes—one from each parent—unite in **fertilization** to form a genetically unique, **diploid zygote**. Sexual life cycles can be haplontic, diplontic, or involve alternation of generations. **Review Figure 11.15, ACTIVITY 11.3**
- In non-haplontic sexually reproducing organisms, certain cells in the adult undergo meiosis, a process by which a diploid cell produces haploid gametes.
- Each gamete contains one of each **homologous pair** of chromosomes from the parent.

11.5 What Happens during Meiosis?
See ANIMATED TUTORIAL 11.2

- Meiosis consists of two nuclear divisions, **meiosis I** and **meiosis II**, that together reduce the chromosome number from diploid to haploid. Meiosis ensures that each haploid cell contains one member of each chromosome pair, and results in four genetically diverse haploid cells, usually gametes. **Review Figure 11.16, ACTIVITY 11.4**
- In meiosis I, entire chromosomes, each with two chromatids, migrate to the poles. In meiosis II, the sister chromatids separate.
- During prophase I, homologous chromosomes undergo **synapsis** to form pairs in a **tetrad**. Chromatids can form junctions called **chiasmata**, and genetic material may be exchanged between the two homologs by **crossing over**. **Review Figures 11.17, 11.18**
- Both crossing over during prophase I and **independent assortment** of the homologs as they separate during anaphase I ensure that the gametes are genetically diverse.
- In **nondisjunction**, two members of a homologous pair of chromosomes go to the same pole during meiosis I, or two chromatids go to the same pole during meiosis II or mitosis. This leads to one gamete having an extra chromosome and another lacking that chromosome. **Review Figure 11.20**
- The union between a gamete with an abnormal chromosome number and a normal haploid gamete results in **aneuploidy**. Such genetic abnormalities can be harmful or lethal to the organism.
- The numbers, shapes, and sizes of the metaphase chromosomes constitute the **karyotype** of an organism.
- **Polyploids** have more than two sets of haploid chromosomes. Sometimes these sets come from different species. **Review Figure 11.22**

11.6 In a Living Organism, How Do Cells Die?

- A cell may die by **necrosis**, or it may self-destruct by **apoptosis**, a genetically programmed series of events that includes the fragmentation of its nuclear DNA.
- Apoptosis is regulated by external and internal signals. These signals result in activation of a class of enzymes called **caspases** that hydrolyze proteins in the cell. **Review Figure 11.23**

11.7 How Does Unregulated Cell Division Lead to Cancer?

- Cancer cells divide more rapidly than normal cells and can be **metastatic**, spreading to distant organs in the body.
- Cancer can result from changes in either of two types of proteins that regulate the cell cycle. **Oncogene** proteins stimulate cell division and are activated in cancer. **Tumor suppressor** proteins normally inhibit the cell cycle, but in cancer they are inactive. **Review Figure 11.25**
- Cancer treatment often targets the cell cycle in tumor cells. **Review Figure 11.26**

Go to the Interactive Summary to review key figures, Animated Tutorials, and Activities.
Life10e.com/is11

CHAPTERREVIEW

REMEMBERING

1. Which statement about eukaryotic chromosomes is *not* true?
   a. They sometimes consist of two chromatids.
   b. They sometimes consist only of a single chromatid.
   c. They normally possess a single centromere.
   d. They consist only of proteins.
   e. During metaphase they are visible under the light microscope.

2. Which statement about the cell cycle is *not* true?
   a. It consists of interphase, mitosis, and cytokinesis.
   b. The cell's DNA replicates during G1.
   c. A cell can remain in G1 for weeks or much longer.
   d. DNA is not replicated during G2.
   e. Cells enter the cell cycle as a result of internal or external signals.

3. Which statement about mitosis is *not* true?
   a. A single nucleus gives rise to two identical daughter nuclei.
   b. The daughter nuclei are genetically identical to the parent nucleus.
   c. The centromeres separate at the onset of anaphase.
   d. Homologous chromosomes synapse in prophase.
   e. The centrosomes organize the microtubules of the spindle fibers.

4. Apoptosis
   a. occurs in all cells.
   b. involves the formation of the plasma membrane.
   c. does not occur in an embryo.
   d. is a series of programmed events resulting in cell death.
   e. is the same as necrosis.

5. In meiosis,
   a. meiosis II reduces the chromosome number from diploid to haploid.
   b. DNA replicates between meiosis I and meiosis II.
   c. the chromatids that make up a chromosome in meiosis II are identical.
   d. each chromosome in prophase I consists of four chromatids.
   e. homologous chromosomes separate from one another in anaphase I.

6. Which statement about cytokinesis is true?
   a. In animals, a cell plate forms.
   b. In plants, it is initiated by furrowing at the membrane.
   c. It immediately follows the G1 phase of the cell cycle.
   d. In animal cells, actin and myosin play an important role.
   e. It is the division of the nucleus.

UNDERSTANDING & APPLYING

7. An animal has a diploid chromosome number of 12. An egg cell of that animal has 5 chromosomes. The most probable explanation is
   a. normal mitosis.
   b. normal meiosis.
   c. nondisjunction in meiosis I.
   d. nondisjunction in meiosis I or II.
   e. nondisjunction in mitosis.

8. The number of daughter chromosomes in a human cell (diploid number 46) in anaphase II of meiosis is
   a. 2.
   b. 23.
   c. 46.
   d. 69.
   e. 92.

9. Contrast mitotic prophase and prophase I of meiosis. Contrast mitotic anaphase and anaphase I of meiosis.

10. The tumor suppressor p53 normally acts by stimulating the expression of p21. Explain why loss of p53 function would lead to uncontrolled cell division.

ANALYZING & EVALUATING

11. Cancer-fighting drugs are usually given in combination to target several different stages of the cell cycle. Why might this be a better approach than a single drug?

12. Studying the cell cycle is much easier if the cells under investigation are synchronous—all in the same stage of the cell cycle. Some populations of cells are naturally synchronous. The anther (male sex organ) of a lily plant contains cells that become pollen grains (male gametes). As anthers develop in the flower, their lengths correlate precisely with the stage of the meiotic cycle in those cells. An anther that is 1.5 millimeters long, for example, contains cells in early prophase I. How would you use lily anthers to investigate the roles of cyclins and Cdk's in the meiotic cell cycle?

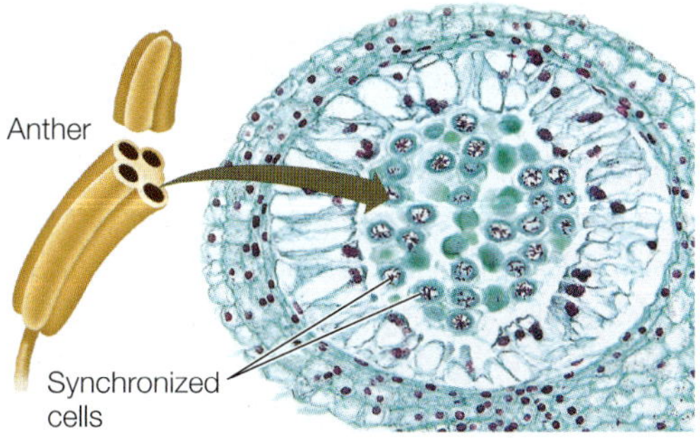
Anther
Synchronized cells

Go to BioPortal at **yourBioPortal.com** for Animated Tutorials, Activities, LearningCurve Quizzes, Flashcards, and many other study and review resources.

12 Inheritance, Genes, and Chromosomes

CHAPTER**OUTLINE**

Devil Genetics The Tasmanian devil (*Sarcophilus harrisii*) is in danger of extinction because its genetic makeup has made it susceptible to cancer.

THE TASMANIAN DEVIL (*Sarcophilus harrisii*) does not deserve the reputation that comes with its nickname and certainly does not deserve what has happened to its gene pool. A marsupial (a type of mammal that gives birth to immature young) native to the Australian continent, the devil resembles a small, black dog with patches of white fur. Generally kind and playful, it eats a diet of various dead animals. It is harmless to humans and, cartoon depictions to the contrary, does not eat live rabbits. When startled, a devil often bares its teeth and shrieks to scare off a possible predator; this is how European settlers gave the creature its common name.

Over the years, the devil population was reduced because of hunting and disease; more recently, it has increased again from the few individuals that survived. The close relatedness of individuals within the current population was confirmed by sequencing the genomes of two devils from opposite ends of the island of Tasmania. Of the 3 billion nucleotides in the devil genome, there were fewer than a million differences; humans show about four times as much variation.

Thanks to conservation efforts in the last century, the population of wild Tasmanian devils had risen to about 100,000 by 1996. But lurking within their chromosomes was a "genetic time bomb." That year, biologists noticed a particularly gruesome tumor affecting the faces and necks of devils. The cancer prevents them from feeding and is invariably lethal. It has led to a drastic reduction in the devil population, and at the current rate of spread, the species could be extinct in a few years.

At first, biologists suspected a cancer-causing infectious virus, but they could not find one. Instead, they found that the cancer cells themselves pass from an affected devil to an unaffected one when they bite each other during mating or feeding; the tumor then grows in the second animal. This is highly unusual. If you accidentally got cancer cells from an unrelated person, the genetic differences between the two of you would result in your immune system recognizing the incoming cells as "nonself," and they would be rejected and destroyed. But devils are so closely related genetically that this rejection does not occur.

A massive effort to preserve the remaining Tasmanian devil population is underway. But the lessons from genetics are clear: Related individuals share most of their genes, and genetic diversity is important in disease resistance.

?

How can knowledge of genetics be used to save the Tasmanian devil?

See answer on p. 255.

12.1 What Are the Mendelian Laws of Inheritance?

Genetics, the field of biology concerned with inheritance, has a long history. There is good evidence that people were deliberately breeding animals (horses) and plants (the date palm tree) for desirable characteristics as long as 5,000 years ago. The general idea was to examine the natural variation among the individuals of a species and "breed the best to the best and hope for the best." This was a hit-or-miss method—sometimes the resulting offspring had all the good characteristics of the parents, but often they did not.

By the mid-nineteenth century, two theories had emerged to explain the results of breeding experiments:

- The theory of *blending inheritance* proposed that gametes contained hereditary determinants (what we now call genes) that blended when the gametes fused during fertilization. Like inks of different colors, the two different determinants lost their individuality after blending and could never be separated. For example, if a plant that made smooth, round seeds was crossed (bred) with a plant that made wrinkled seeds, the offspring would be intermediate between the two and the determinants for the two parental characteristics would be lost.
- The theory of *particulate inheritance* proposed that each determinant had a physically distinct nature; when gametes fused in fertilization, the determinants remained intact. According to this theory, if a plant that made round seeds was crossed with a plant that made wrinkled seeds, the offspring (no matter the shape of their seeds) would still contain the determinants for the two characteristics.

The story of how these competing theories were tested provides a great example of how the scientific method can be used to support one theory and reject another. In the following sections we will look in detail at experiments performed in the 1860s by an Austrian monk and scientist, Gregor Mendel, whose work clearly supported the particulate theory.

Mendel used the scientific method to test his hypotheses

After entering the priesthood at a monastery in Brno, in what is now the Czech Republic, Gregor Mendel was sent to the University of Vienna, where he studied biology, physics, and mathematics. He returned to the monastery in 1853 to teach. The abbot in charge had set up a small plot of land to do experiments with plants and encouraged Mendel to continue with them (**Figure 12.1**). Over seven years, Mendel made crosses with many thousands of plants. Analysis of his meticulously gathered data suggested to him that inheritance was due to particulate factors.

Mendel presented his theories in two public lectures in 1865 and a detailed written publication in 1866, but his work was ignored by mainstream scientists until 1900. By that time, the discovery of chromosomes had suggested to biologists that genes might be carried on chromosomes. When they read Mendel's work on particulate inheritance, the biologists connected the dots between genes and chromosomes.

Mendel chose to study the common garden pea because of its ease of cultivation and the feasibility of making controlled crosses. Pea flowers have both male and female sex organs: stamens and pistils, which produce gametes that are contained within the pollen and ovules, respectively.

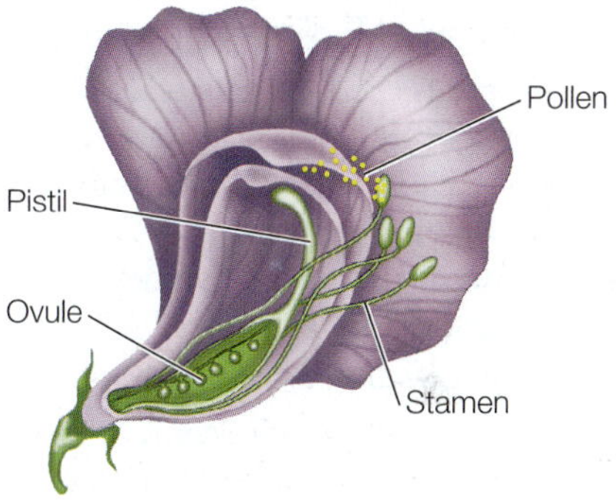
Pollen
Pistil
Ovule
Stamen

Pea flowers normally self-fertilize. However, the male organs can be removed from a flower so that it can be manually fertilized with pollen from a different flower.

Go to Media Clip 12.1
Mendel's Discoveries
Life10e.com/mc12.1

12.1 Gregor Mendel and His Garden The Austrian monk Gregor Mendel (left) did his groundbreaking genetics experiments in a garden at the monastery at Brno, in what is now the Czech Republic.

There are many varieties of pea plants with easily recognizable characteristics. A **character** is an observable physical feature, such as seed shape. A **trait** is a particular form of a character, such as round or wrinkled seeds. Mendel worked with seven pairs of varieties with contrasting traits for characters such as seed shape, seed color, and flower color. These varieties were true-breeding: that is, when he crossed a plant that produced wrinkled seeds with another of the same variety, all of the offspring plants produced wrinkled seeds.

As we will see, Mendel developed a set of hypotheses to explain the inheritance of particular pea traits, and then designed crossing experiments to test his hypotheses. He performed his crosses in the following manner:

- He removed the stamens from (emasculated) flowers of one parental variety so that it couldn't self-fertilize. Then he collected pollen from another parental variety and placed it on the pistils of the emasculated flowers. The plants providing and receiving the pollen were the **parental generation**, designated **P**.
- In due course, seeds formed and were planted. The seeds and the resulting new plants constituted the **first filial generation**, or $\mathbf{F_1}$. (The word "filial" refers to the relationship between offspring and parents, from the Latin *filius*, "son.") Mendel examined each F_1 plant to see which traits it bore and then recorded the number of F_1 plants expressing each trait.
- In some experiments the F_1 plants were allowed to self-pollinate and produce a **second filial generation**, or $\mathbf{F_2}$. Again, each F_2 plant was characterized and counted.

Mendel's first experiments involved monohybrid crosses

The term "hybrid" refers to the offspring of crosses between organisms differing in one or more characters. In Mendel's first experiments, he crossed parental (P) varieties with contrasting traits for a single character, producing monohybrids (from the Greek *monos*, "single") in the F_1 generation. He subsequently planted the F_1 seeds and allowed the resulting plants to self-pollinate to produce the F_2 generation. This technique is referred to as a **monohybrid cross**.

Mendel performed the same experiment for seven pea characters. His method is illustrated in **Figure 12.2**, using seed shape as an example. When he crossed a strain that made round seeds with one that made wrinkled seeds, all of the F_1 seeds were round—it was as if the wrinkled seed trait had disappeared completely. However, when F_1 plants were allowed to self-pollinate to produce F_2 seeds, about one-fourth of the seeds were wrinkled. These observations were key to distinguishing the two theories noted above:

- The F_1 offspring were not a blend of the two traits of the parents. Only one of the traits was present (in this case, round seeds).
- Some F_2 offspring had wrinkled seeds. The trait had not disappeared because of blending.

These observations led to a rejection of the blending theory of inheritance and provided support for the particulate theory.

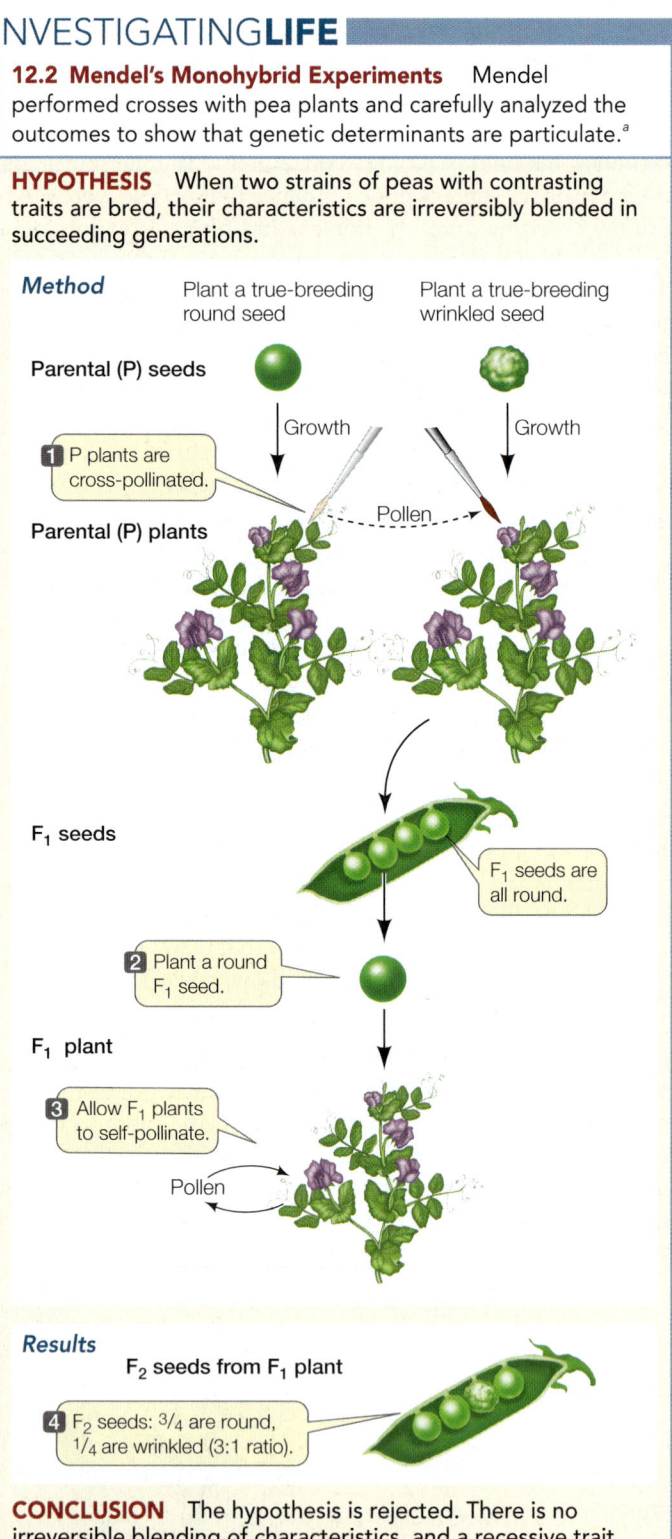

INVESTIGATING**LIFE**

12.2 Mendel's Monohybrid Experiments Mendel performed crosses with pea plants and carefully analyzed the outcomes to show that genetic determinants are particulate.[a]

HYPOTHESIS When two strains of peas with contrasting traits are bred, their characteristics are irreversibly blended in succeeding generations.

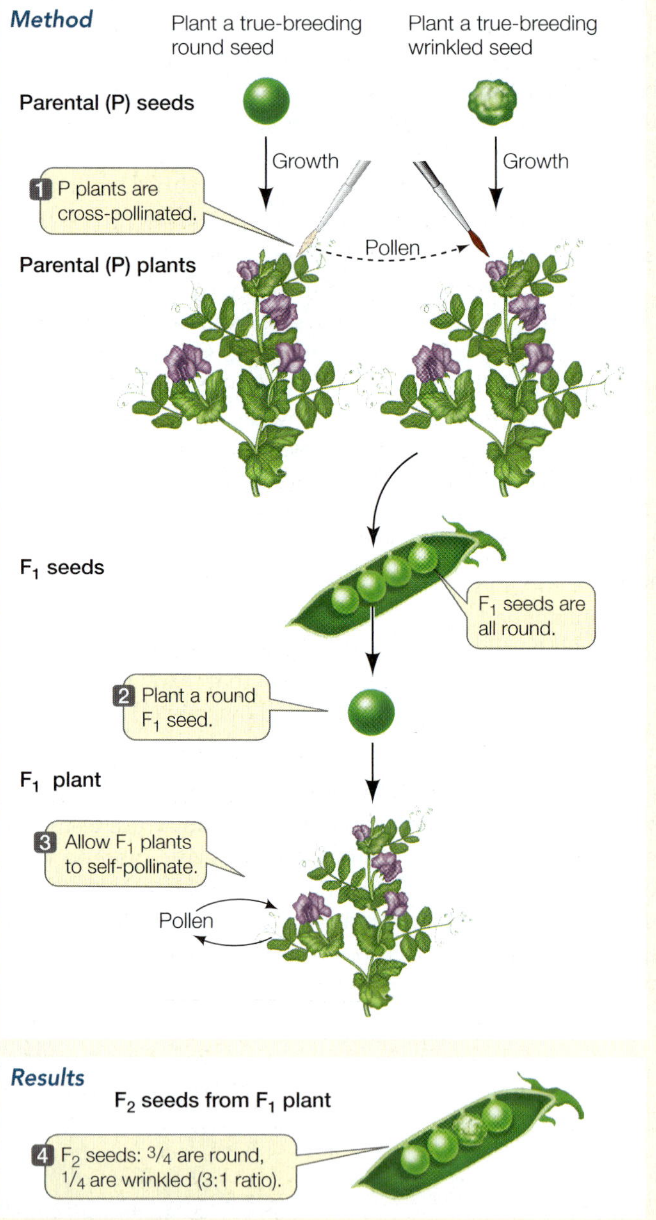
Method
Plant a true-breeding round seed
Plant a true-breeding wrinkled seed
Parental (P) seeds
Growth
Growth
1 P plants are cross-pollinated.
Pollen
Parental (P) plants
F_1 seeds
F_1 seeds are all round.
2 Plant a round F_1 seed.
F_1 plant
3 Allow F_1 plants to self-pollinate.
Pollen
Results
F_2 seeds from F_1 plant
4 F_2 seeds: 3/4 are round, 1/4 are wrinkled (3:1 ratio).

CONCLUSION The hypothesis is rejected. There is no irreversible blending of characteristics, and a recessive trait can reappear in succeeding generations.

Go to **BioPortal** for discussion and relevant links for all INVESTIGATING**LIFE** figures.

[a]See www.mendelweb.org/Mendel.plain.html

We now know that hereditary determinants are not actually "particulate," but they *are* physically distinct entities: sequences of DNA carried on chromosomes.

WORKING WITH**DATA**:

Mendel's Monohybrid Experiments

Original Paper

The original German version of Mendel's paper, *Versuche uber Pflanzen-Hybriden*, with an English translation and extensive explanatory notes, is available online: www.mendelweb.org/Mendel.plain.html.

Analyze the Data

Mendel's monohybrid crosses were key to his rejection of the theory of blending inheritance (see Figure 12.2). One of his monohybrid crosses was between true-breeding green-seeded and yellow-seeded pea plants. He observed that all of the pea plants' F_1 generation had yellow seeds. Mendel then allowed the F_1 plants to self-pollinate, and the seed colors of the resulting F_2 generation were analyzed. The table shows actual data from individual plants in the F_2 generation as reported in Mendel's paper. Mendel made mathematical calculations, and in a later part of the paper he showed the overall ratios for these two traits. However, he did not perform a statistical analysis to determine whether the variations in the data reflected a general pattern of inheritance or were simply due to chance.

QUESTION 1

Use the hypothesis that the ratio of yellow to green seeds in the F_2 generation would be 3:1 and perform a chi-square test to analyze the results for each plant in the table (refer to Appendix B for information about the chi-square test). What can you conclude about this hypothesis from the individual plants? How many crosses have *P*-values > 0.05?

QUESTION 2

Now total the data from all the plants and rerun the chi-square analysis. What can you conclude? What does your analysis indicate about the need for using a large number of organisms in studies of genetics?

	Seed color	
Plant	Yellow	Green
1	25	11
2	32	7
3	14	5
4	70	27
5	24	13
6	20	6
7	32	13
8	44	9
9	50	14
10	44	18

Go to **BioPortal** for all WORKING WITH**DATA** exercises

All seven crosses between varieties with contrasting traits gave the same kind of data (**Table 12.1**). In the F_1 generation only one of the two traits was seen, but the other trait reappeared in about one-fourth of the offspring in the F_2 generation. Mendel called the trait that appeared in the F_1 and was more abundant in the F_2 the **dominant** trait, and the other trait **recessive**. In the F_2 generation, the *ratio* of dominant to recessive traits was about 3:1. (To calculate the ratios shown in Table 12.1, divide the number of F_2 plants with the dominant trait by the number with the recessive trait.)

You can see in Table 12.1 that for each character, Mendel counted hundreds or even thousands of F_2 seeds or plants to

TABLE 12.1

Mendel's Results from Monohybrid Crosses

	Parental generation phenotypes			F_2 Generation phenotypes			
	Dominant	Recessive		Dominant	Recessive	Total	Ratio
	Round seeds ×	Wrinkled seeds		5,474	1,850	7,324	2.96:1
	Yellow seeds ×	Green seeds		6,022	2,001	8,023	3.01:1
	Purple flowers ×	White flowers		705	224	929	3.15:1
	Inflated pods ×	Constricted pods		882	299	1,181	2.95:1
	Green pods ×	Yellow pods		428	152	580	2.82:1
	Axial flowers ×	Terminal flowers		651	207	858	3.14:1
	Tall stems (1 m) ×	Dwarf stems (0.3 m)		787	277	1,064	2.84:1

see how many carried each trait. As we will discuss in more detail below, the probability of a plant inheriting a particular trait is independent of the probability of another plant inheriting the same trait. If Mendel had looked at only a few F_2 progeny from the "round × wrinkled" cross, he might, by chance, have found only round seeds. Or he might have found a higher proportion of wrinkled seeds than he did. In order to discover recurring patterns and to develop his laws of inheritance, Mendel used very large numbers of plants.

Mendel went on to expand on the particulate theory. He proposed that hereditary determinants—we will call them **genes** here, though Mendel did not use that term—occur in pairs and segregate (separate) from one another during the formation of gametes. He concluded that each pea plant has two genes for each character (such as seed shape), one inherited from each parent. We now use the term **diploid** to describe the state of having two copies of each gene; **haploids** have just a single copy.

Mendel concluded that while each gamete contains one copy of each gene, the resulting zygote contains two copies, because it is produced by the fusion of two gametes. Furthermore, different traits arise from different forms of a gene (now called **alleles**) for a particular character. For example, Mendel studied two alleles for seed shape: one that caused round seeds and the other causing wrinkled seeds.

An organism that is **homozygous** for a gene has two alleles that are the same (for example, two copies of the allele for round seeds). An organism that is **heterozygous** for a gene has two different alleles (for example, one allele for round seeds and one allele for wrinkled seeds). In a heterozygote, one of the two alleles may be dominant (such as round, *R*) and the other recessive (wrinkled, *r*). By convention, dominant alleles are designated with uppercase letters and recessive alleles with lowercase letters.

The physical appearance of an organism is its **phenotype**. Mendel proposed that the phenotype is the result of the **genotype**, or genetic constitution, of the organism showing the phenotype. Round seeds and wrinkled seeds are two phenotypes resulting from three possible genotypes: the wrinkled seed phenotype is produced by the genotype *rr*, whereas the round seed phenotype is produced by either of the genotypes *RR* or *Rr* (because the *R* allele is dominant to the *r* allele).

Mendel's first law states that the two copies of a gene segregate

How do Mendel's theories explain the proportions of traits seen in the F_1 and F_2 generations of his monohybrid crosses? Mendel's first law—the **law of segregation**—states that *when any individual produces gametes, the two copies of a gene separate, so that each gamete receives only one copy*. Thus gametes from a parent with the *RR* genotype will all carry *R*; gametes from an *rr* parent will all carry *r*; and the progeny derived from a cross between these parents will all be *Rr*, producing seeds with a round phenotype (Figure 12.3).

Now let's consider the composition of the F_2 generation. Because the alleles segregate, half of the gametes produced by the F_1 generation will have the *R* allele and the other half will have the *r* allele. What genotypes are produced when these gametes fuse to form the next (F_2) generation?

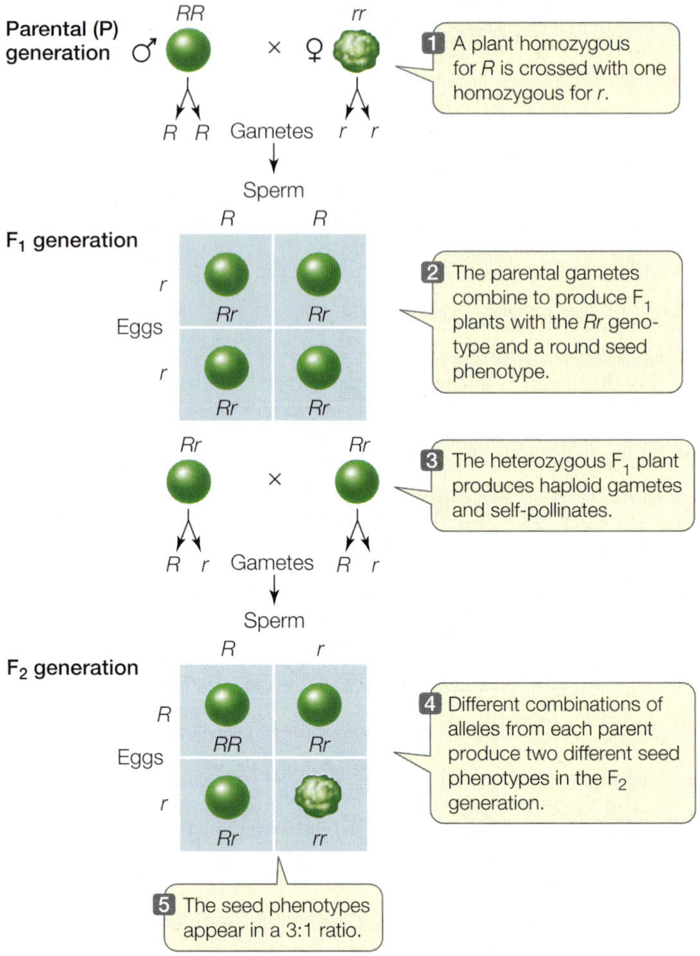

Parental (P) generation
RR
rr
Gametes
R R
r r
Sperm
F1 generation
Eggs
Rr
F2 generation
RR
1 A plant homozygous for R is crossed with one homozygous for r.
2 The parental gametes combine to produce F1 plants with the Rr genotype and a round seed phenotype.
3 The heterozygous F1 plant produces haploid gametes and self-pollinates.
4 Different combinations of alleles from each parent produce two different seed phenotypes in the F2 generation.
5 The seed phenotypes appear in a 3:1 ratio.

12.3 Mendel's Explanation of Inheritance Mendel concluded that inheritance depends on discrete factors from each parent that do not blend in the offspring.

The allele combinations that will result from a cross can be predicted using a **Punnett square**, a method devised in 1905 by the British geneticist Reginald Punnett. This device ensures that we consider all possible combinations of gametes when calculating expected genotype frequencies of the resulting offspring. A Punnett square looks like this:

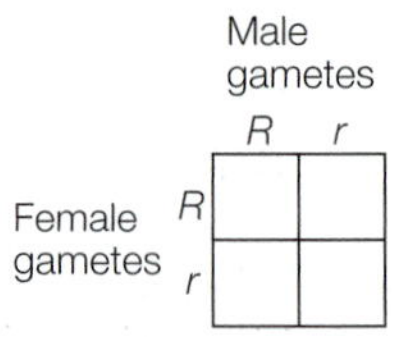

Male gametes
R r
Female gametes
R
r

It is a simple grid with all possible male gamete (haploid sperm) genotypes shown along the top and all possible female gamete (haploid egg) genotypes along the left side. The grid is completed by filling in each square with the diploid genotype that can be generated from each combination of gametes. In this example, to fill in the top right square, we put in the *R* from the female gamete (the egg cell) and the *r* from the male gamete (a sperm cell in the pollen tube), yielding *Rr*.

Once the Punnett square is filled in, we can readily see that there are four possible combinations of alleles in the F_2 generation: *RR*, *Rr*, *rR*, and *rr* (see Figure 12.3). Since *R* is dominant, there are three ways to get round seeds in the F_2 generation

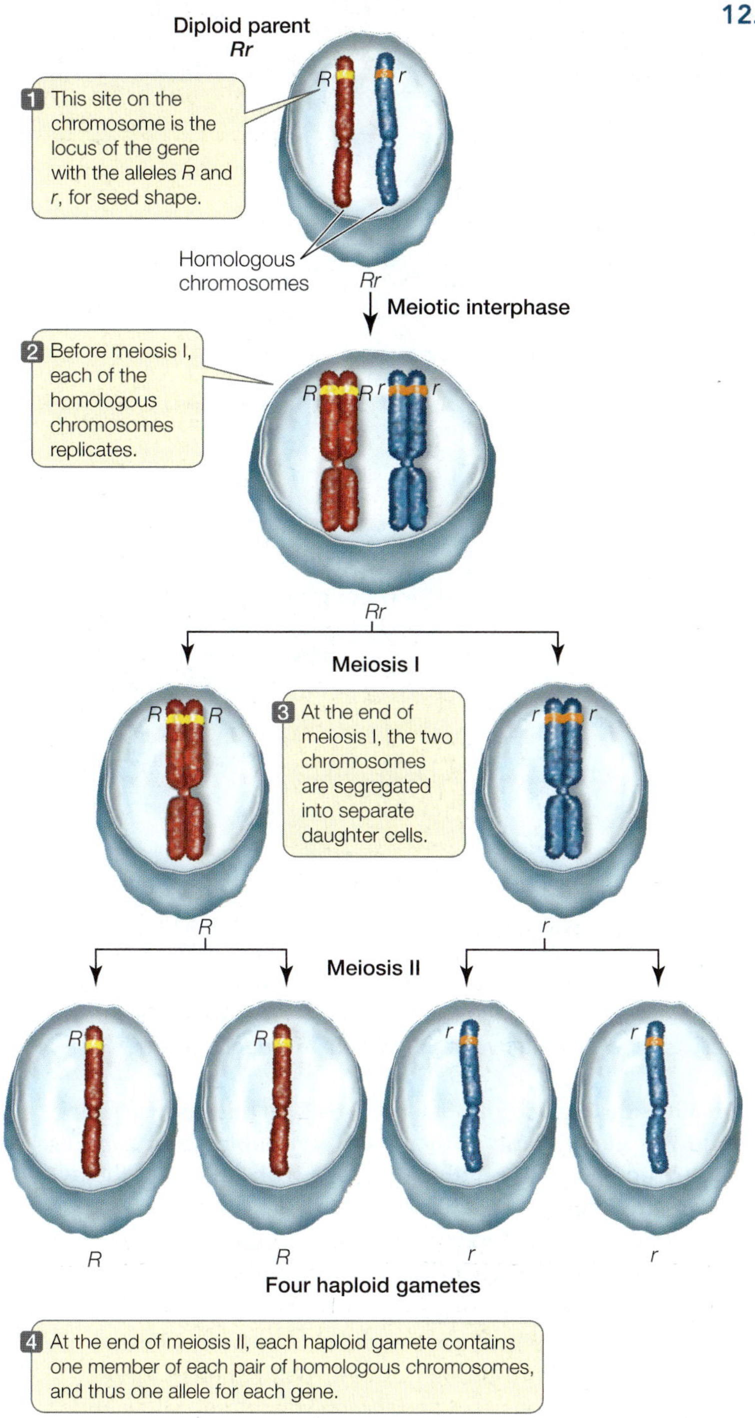
Diploid parent
Rr
1 This site on the chromosome is the locus of the gene with the alleles R and r, for seed shape.
R
r
Homologous chromosomes
Rr
Meiotic interphase
2 Before meiosis I, each of the homologous chromosomes replicates.
R R r r
Rr
Meiosis I
R R
3 At the end of meiosis I, the two chromosomes are segregated into separate daughter cells.
r r
R
r
Meiosis II
R R r r
R R r r
Four haploid gametes
4 At the end of meiosis II, each haploid gamete contains one member of each pair of homologous chromosomes, and thus one allele for each gene.

12.4 Meiosis Accounts for the Segregation of Alleles Although Mendel had no knowledge of chromosomes or meiosis, we now know that a pair of alleles resides on homologous chromosomes, and that those alleles segregate during meiosis.

(genotype *RR*, *Rr*, or *rR*), but only one way to get wrinkled seeds (genotype *rr*). Therefore we predict a 3:1 ratio of these phenotypes in the F_2 generation, remarkably close to the ratios Mendel found experimentally for all the traits he compared (see Table 12.1).

Mendel did not live to see his theories placed on a sound physical footing with the discoveries of chromosomes and DNA. Genes are now known to be relatively short sequences of DNA (a few thousand base pairs in length) found on the much longer DNA molecules that make up chromosomes (which are often millions of base pairs long). Today we can picture the different alleles of a gene segregating as chromosomes separate during meiosis I (**Figure 12.4**).

We also know now that genes determine phenotypes mostly by producing proteins with particular functions, such as enzymes. In many cases a dominant gene is expressed (transcribed and translated) to produce a functional protein, while a recessive gene is mutated so that it is no longer expressed, or it encodes a mutant protein that is nonfunctional. For example, the wrinkled seed phenotype of *rr* peas is caused by the absence of an enzyme called starch branching enzyme 1 (SBE1), which is essential for starch synthesis. With less starch, the developing seed has more sucrose and this causes an inflow of water by osmosis. When the seed matures and dries out, the water is lost, leaving a shrunken seed. A single copy of the *R* allele produces enough functional SBE1 to prevent the wrinkled phenotype, which accounts for the dominance of *R* over *r*.

Mendel verified his hypotheses by performing test crosses

As mentioned above, Mendel arrived at his laws of inheritance by developing a series of hypotheses and then designing experiments to test them. One such hypothesis was that there are two possible allele combinations (*RR* or *Rr*) for seeds with the round phenotype. Mendel verified this hypothesis by performing test crosses with F_1 seeds derived from a variety of other crosses. A **test cross** is used to determine whether an individual showing a dominant trait is homozygous or heterozygous. The individual in question is crossed with an individual that is homozygous for the recessive trait—easy to identify, because all individuals with the recessive phenotype are homozygous for that trait.

The recessive homozygote for the seed shape gene has wrinkled seeds and the genotype *rr*. The individual being tested may be described initially as *R*_ because we do not yet know the identity of the second allele. We can predict two possible results:

- If the individual being tested is homozygous dominant (*RR*), all offspring of the test cross will be *Rr* and show the dominant trait (round seeds) (**Figure 12.5, left**).
- If the individual being tested is heterozygous (*Rr*), then approximately half the offspring of the test cross will be heterozygous and show the dominant trait (*Rr*), and the other half will be homozygous for the recessive trait (*rr*) (**Figure 12.5, right**).

Mendel obtained results consistent with both of these predictions; thus his hypothesis accurately predicted the results of his test crosses.

Mendel's second law states that copies of different genes assort independently

Consider an organism that is heterozygous for two genes (*RrYy*). In this example, the dominant *R* and *Y* alleles came from one parent, and the recessive *r* and *y* alleles came from the other parent. When this organism produces gametes, do the *R* and *Y* alleles always go together in one gamete, and *r* and *y*

INVESTIGATING**LIFE**

12.5 Homozygous or Heterozygous? An individual with a dominant phenotype may have either a homozygous or a heterozygous genotype. The test cross determines which.[a]

HYPOTHESIS The progeny of a test cross can reveal whether an organism is homozygous or heterozygous.

Method

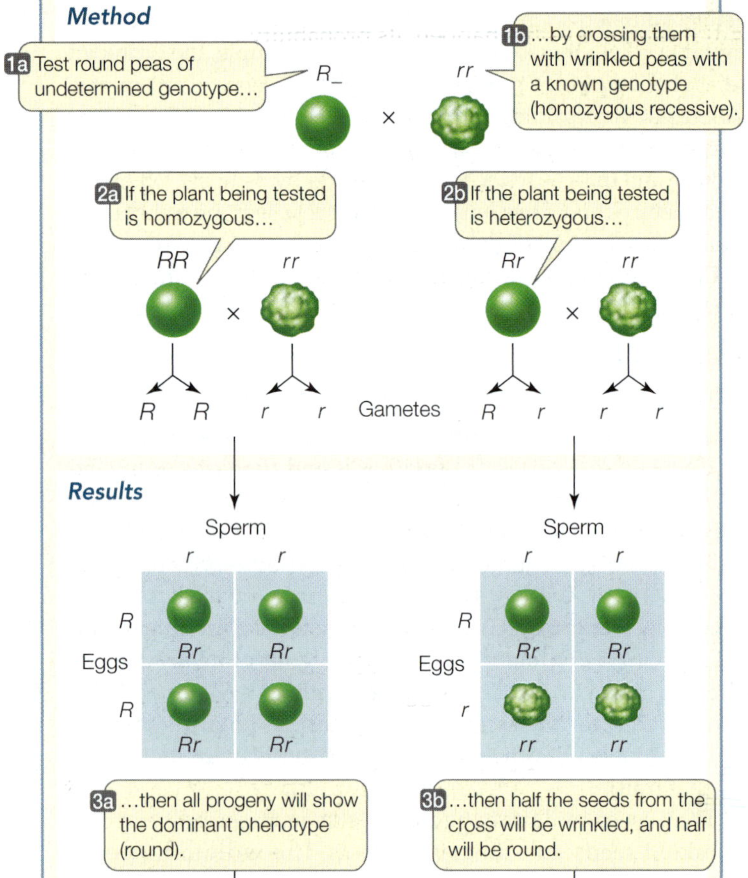

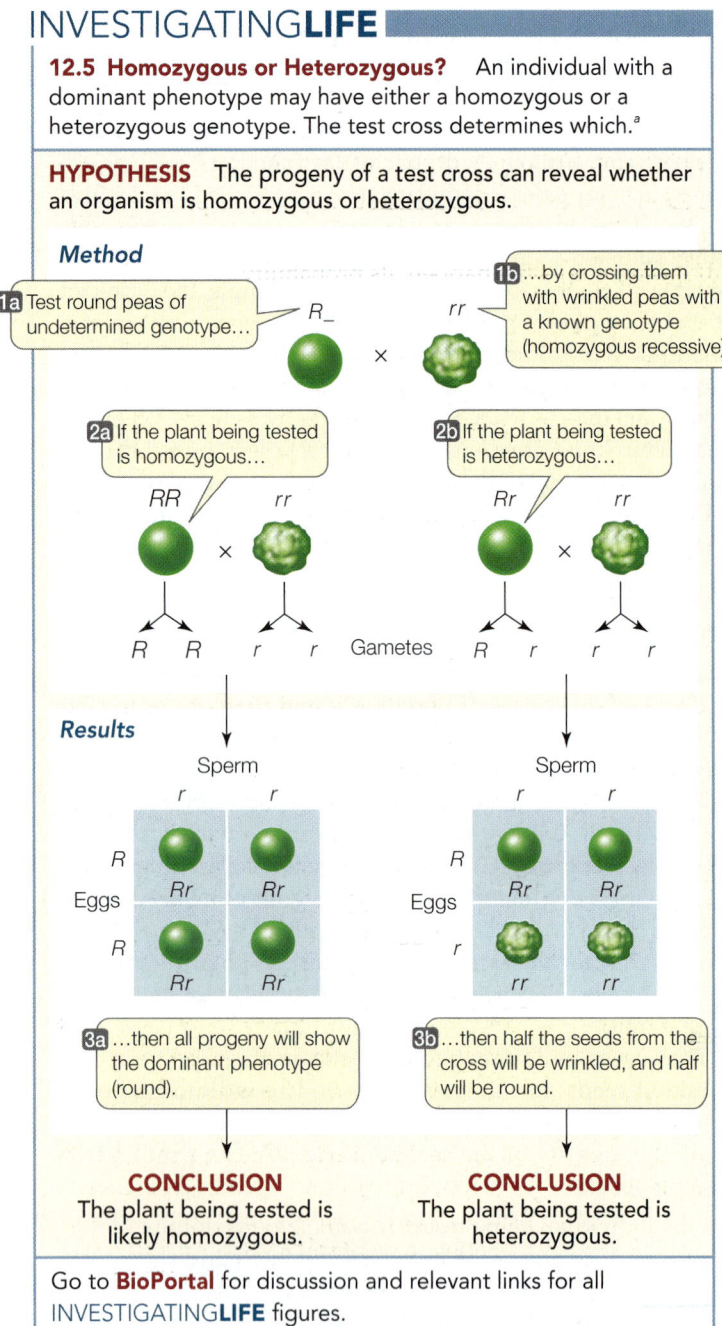

1a Test round peas of undetermined genotype...
R_
×
rr
1b ...by crossing them with wrinkled peas with a known genotype (homozygous recessive).
2a If the plant being tested is homozygous...
RR
×
rr
R R
r r
Gametes
2b If the plant being tested is heterozygous...
Rr
×
rr
R r
r r
Results
Sperm
r r
R
Eggs
R
Rr Rr
Rr Rr
Sperm
r r
R
Eggs
r
Rr Rr
rr rr
3a ...then all progeny will show the dominant phenotype (round).
3b ...then half the seeds from the cross will be wrinkled, and half will be round.

CONCLUSION
The plant being tested is likely homozygous.

CONCLUSION
The plant being tested is heterozygous.

Go to **BioPortal** for discussion and relevant links for all INVESTIGATING**LIFE** figures.

[a]See www.mendelweb.org/Mendel.plain.html

Go to Activity 12.1 Homozygous or Heterozygous?
Life10e.com/ac12.1

alleles in another? Or can a single gamete receive one recessive and one dominant allele (*R* and *y* or *r* and *Y*)?

Mendel performed another series of experiments to answer these questions. He began with peas that differed in *two* characters: seed shape and seed color. One parental variety produced only round, yellow seeds (*RRYY*), and the other produced only wrinkled, green ones (*rryy*). A cross between these two varieties produced an F_1 generation in which all the plants were *RrYy*. Because the *R* and *Y* alleles were dominant, the F_1 seeds were all round and yellow.

Mendel continued this experiment into the F_2 generation by performing a **dihybrid cross**—a cross between individuals

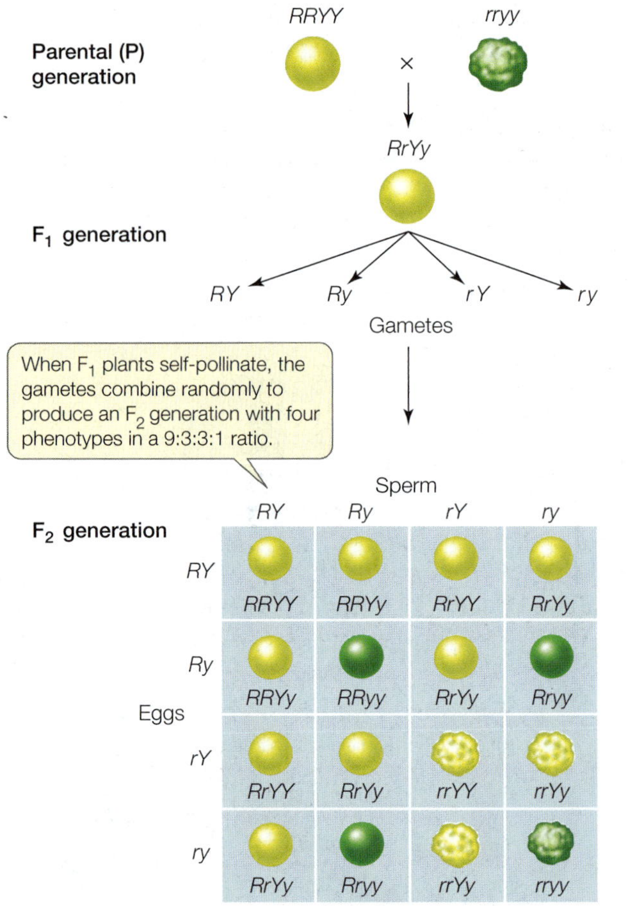

RRYY
rryy
Parental (P) generation
×
RrYy
F1 generation
RY
Ry
rY
ry
Gametes
When F1 plants self-pollinate, the gametes combine randomly to produce an F2 generation with four phenotypes in a 9:3:3:1 ratio.
F2 generation
Sperm
RY Ry rY ry
Eggs
RY
RRYY RRYy RrYY RrYy
Ry
RRYy RRyy RrYy Rryy
rY
RrYY RrYy rrYY rrYy
ry
RrYy Rryy rrYy rryy

12.6 Independent Assortment The 16 possible combinations of gametes in this dihybrid cross result in nine different genotypes. Because *R* and *Y* are dominant over *r* and *y*, respectively, the nine genotypes result in four phenotypes in a ratio of 9:3:3:1. These results show that the two genes segregate independently.

that are identical double heterozygotes. In this case, he simply allowed the F_1 plants, which were all double heterozygotes, to self-pollinate. Depending on whether the alleles of the two genes are inherited together or separately, there are two possible outcomes, as Mendel saw:

1. *The alleles could maintain the associations they had in the parental generation—they could be linked.* If this were the case, the F_1 plants would produce two types of gametes (*RY* and *ry*). The F_2 progeny resulting from self-pollination of these F_1 plants would consist of *two phenotypes*: round yellow and wrinkled green in the ratio of 3:1, just as in the monohybrid cross.
2. *The segregation of R from r could be independent of the segregation of Y from y—the two genes could be unlinked.* In this case, four kinds of gametes would be produced in equal numbers: *RY*, *Ry*, *rY*, and *ry*. When these gametes combine at random, they should produce an F_2 generation with *four phenotypes* (round yellow, round green, wrinkled yellow, wrinkled green). Putting these possibilities into a Punnett square, we can predict that these four phenotypes would occur in a ratio of 9:3:3:1.

Mendel's dihybrid crosses supported the second prediction: four different phenotypes appeared in the F_2 generation in a ratio of about 9:3:3:1 (**Figure 12.6**). On the basis of such experiments, Mendel proposed his second law—the **law of**

independent assortment: *alleles of different genes assort independently of one another during gamete formation*. In the example above, the segregation of the *R* and *r* alleles is independent of the segregation of the *Y* and *y* alleles. As you will see later in this chapter, this is not as universal as the law of segregation because it does not apply to genes located near one another on the same chromosome. However, it is correct to say that *chromosomes segregate independently* during the formation of gametes, and so do any two genes located on separate chromosome pairs (**Figure 12.7**).

Probability can be used to predict inheritance

One key to Mendel's success was his use of large sample sizes. By counting many progeny from each cross, he observed clear patterns that allowed him to formulate his theories. After his work became widely recognized, geneticists began using simple probability calculations to predict the ratios of genotypes and phenotypes in the progeny of a given cross or mating. They use statistics to determine whether the actual results match the prediction (see Working with Data, p. 235).

You can think of probabilities by considering a coin toss. The basic conventions of probability are simple:

- If an event is absolutely certain to happen, its probability is 1.
- If it cannot possibly happen, its probability is 0.
- All other events have a probability between 0 and 1.

There are two possible outcomes of a coin toss, and both are equally likely, so the probability of heads is $1/2$—as is the probability of tails.

If two coins (say a penny and a dime) are tossed, each acts independently of the other (**Figure 12.8**). What is the probability of both coins coming up heads? In half of the tosses, the penny comes up heads, and in half of that fraction, the dime comes up heads. The probability of both coins coming up heads is $1/2 \times 1/2 = 1/4$. In general, *the probability of two independent outcomes occurring together is found by multiplying the two individual probabilities*. This can be applied to a monohybrid cross (see Figure 12.3). After the self-pollination of an *Rr* F_1 plant, the probability that an F_2 plant will have the genotype *RR* is $1/2 \times 1/2 = 1/4$, because the chance that the sperm will have the genotype *R* is $1/2$, and the chance that the egg will have the genotype *R* is also $1/2$. Similarly, the probability of *rr* offspring is also $1/4$.

Probability can also be used to predict the proportions of phenotypes in a dihybrid cross. Let's see how this works for the experiment shown in Figure 12.6. Using the principles described above, we can calculate the probability of an F_2 seed being round. This is found by adding the probability of an *Rr* heterozygote ($1/2$) to the probability of an *RR* homozygote ($1/4$): a total of $3/4$. By the same reasoning, the probability that a seed will be yellow is also $3/4$. The two characters are determined by separate genes and are independent of each other, so:

- The joint probability for both round and yellow is $3/4 \times 3/4 = 9/16$.

What is the probability of F_2 seeds being both wrinkled and yellow? The probability

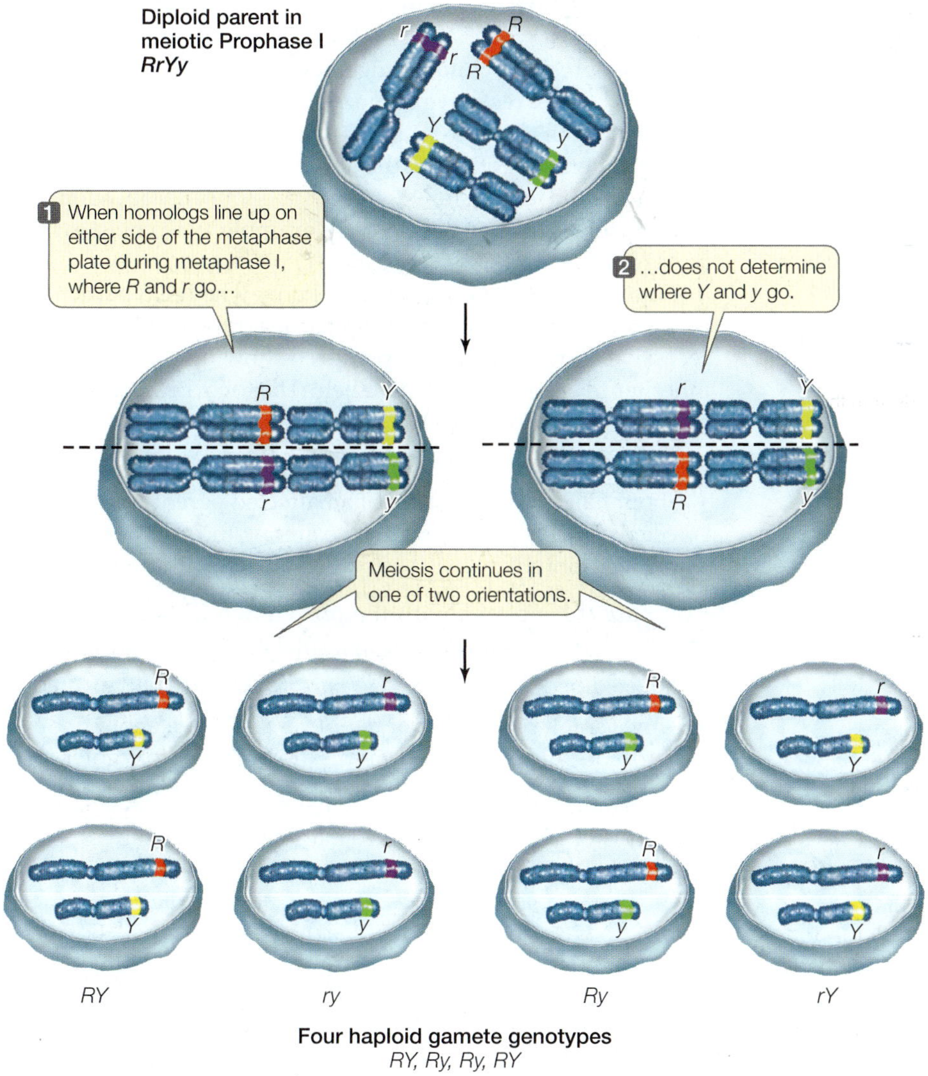

Diploid parent in meiotic Prophase I
RrYy
r
r
R
R
Y
Y
y
y
1 When homologs line up on either side of the metaphase plate during metaphase I, where *R* and *r* go...
2 ...does not determine where *Y* and *y* go.
R
r
Y
y
r
R
Y
y
Meiosis continues in one of two orientations.
RY
ry
Ry
rY
Four haploid gamete genotypes
RY, Ry, Ry, RY

12.7 Meiosis Accounts for Independent Assortment of Alleles We now know that copies of genes on different chromosomes are segregated independently during metaphase I of meiosis. Thus a parent of genotype *RrYy* can form gametes with four different genotypes.

Go to Animated Tutorial 12.1
Independent Assortment of Alleles
Life10e.com/at12.1

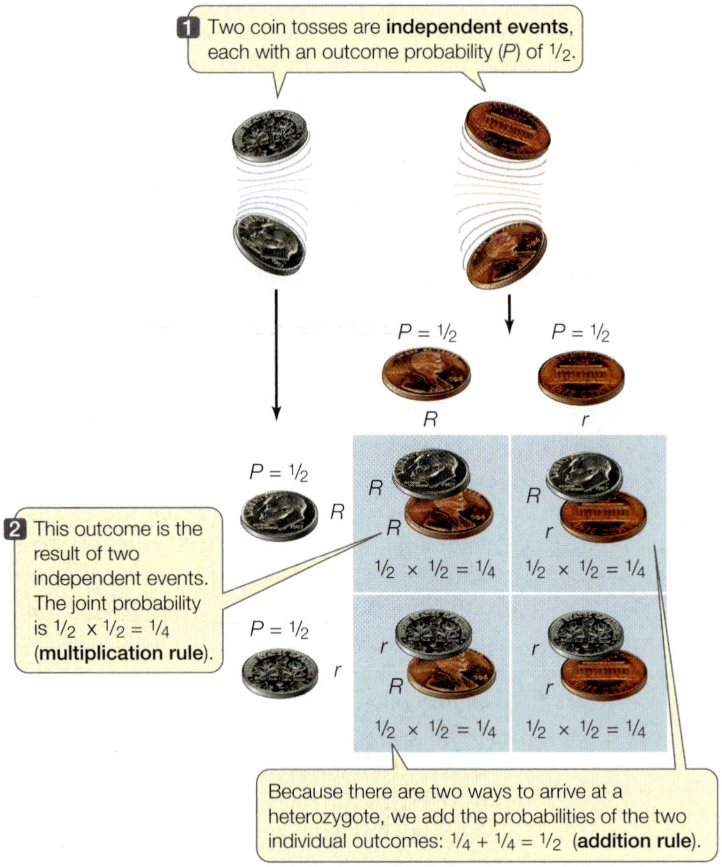
1 Two coin tosses are independent events, each with an outcome probability (P) of 1/2.
P = 1/2
P = 1/2
R
r
P = 1/2
R
P = 1/2
r
R R
R r
r R
r r
1/2 × 1/2 = 1/4
1/2 × 1/2 = 1/4
1/2 × 1/2 = 1/4
1/2 × 1/2 = 1/4
2 This outcome is the result of two independent events. The joint probability is 1/2 x 1/2 = 1/4 (multiplication rule).
Because there are two ways to arrive at a heterozygote, we add the probabilities of the two individual outcomes: 1/4 + 1/4 = 1/2 (addition rule).

12.8 Using Probability Calculations in Genetics Like the results of a coin toss, the probability of any given combination of alleles appearing in the offspring of a cross can be obtained by multiplying the probabilities of each event. Since a heterozygote can be formed in two ways, these two probabilities are added together.

of being yellow is again $3/4$; the probability of being wrinkled is $1/2 \times 1/2 = 1/4$, so:

- The joint probability for both wrinkled and yellow is $1/4 \times 3/4 = 3/16$.

By the same reasoning:

- The joint probability for both round ($3/4$) and green ($1/4$) is also $3/16$.

Finally:

- The joint probability for both wrinkled and green is $1/4 \times 1/4 = 1/16$.

Looking at all four phenotypes, we see that they are expected to occur in the ratio of 9:3:3:1.

A Punnett square or these simple probability calculations can be used to determine the *expected* proportions of offspring with particular phenotypes. In the dihybrid cross discussed above, about one-sixteenth of the F_2 seeds are expected to be wrinkled and green. But this does not mean that among 16 F_2 seeds there will always be exactly one wrinkled, green seed. For any toss of a coin, the probability of heads is independent of what happened in all the previous tosses. Even if you get three heads in a row, the chance of a head in the next toss is still $1/2$, and it is quite possible to toss a coin four times and get four heads. But if you toss the coin many times, you are highly likely to get heads in about half of the tosses. If Mendel had examined only a few progeny in each of his crosses, it is unlikely that he would have observed the phenotypic ratios that he did observe. It was his large sample sizes that allowed him to identify the underlying patterns of inheritance.

Mendel's laws can be observed in human pedigrees

How are Mendel's laws of inheritance applied to humans? Mendel worked out his laws by performing many planned crosses and counting many offspring. Neither of these approaches is possible with humans, so human geneticists rely on **pedigrees**: family trees that show the occurrence of phenotypes (and alleles) in several generations of related individuals.

Because humans have such small numbers of offspring, human pedigrees do not show the clear proportions of phenotypes that Mendel saw in his pea plants. For example, when a man and a woman who are both heterozygous for a recessive allele (say, *Aa*) have children together, each child has a $1/4$ probability of being a recessive homozygote (*aa*). If this couple were to have dozens of children, about one-fourth of them would be recessive homozygotes. But the offspring of a single couple are likely to be too few to show the exact one-fourth proportion. In a family with only two children, for example, both could easily be *aa* (or *Aa*, or *AA*).

Human geneticists may wish to know whether a particular rare allele that causes an abnormal phenotype is dominant or recessive. **Figure 12.9A** is a pedigree showing the pattern of inheritance of a rare dominant allele. The following are the key features to look for in such a pedigree:

- Every affected person has an affected parent.
- About half of the offspring of an affected parent are also affected. (This is easiest to observe among the twelve cousins in generation III.)

Compare this pattern with the one shown in **Figure 12.9B**, which is typical for the inheritance of a rare recessive allele:

- Affected people can have two parents who are not affected.
- Only a small proportion of people are affected: about one-fourth of children whose parents are both heterozygotes.

In the families of individuals who have a rare recessive phenotype, it is not uncommon to find a marriage of two relatives. This observation is a result of the rarity of recessive alleles that give rise to abnormal phenotypes. For two phenotypically normal parents to have an affected child (*aa*), the parents must both be heterozygous (*Aa*). If a particular recessive allele is rare in the general population, the chance of two people marrying who are both carrying that allele is quite low. However, if that allele is present in a family, two cousins might share it (see Figure 12.9B). For this reason, studies on populations that are isolated either culturally (by religion, as with the Amish in the United States) or geographically (as on islands) have been extremely valuable

12.9 Pedigree Analysis and Inheritance (A) This pedigree represents a family affected by Huntington's disease, which results from a rare dominant allele. Everyone who inherits this allele is affected. (B) The family in this pedigree carries the allele for albinism, a recessive trait. Because the trait is recessive, heterozygotes do not have the albino phenotype, but they can pass the allele on to their offspring. In this family, in generation III the heterozygous parents are cousins; however, the same result could occur if the parents were unrelated but heterozygous.

Go to Animated Tutorial 12.2
Pedigree Analysis
Life10e.com/at12.2

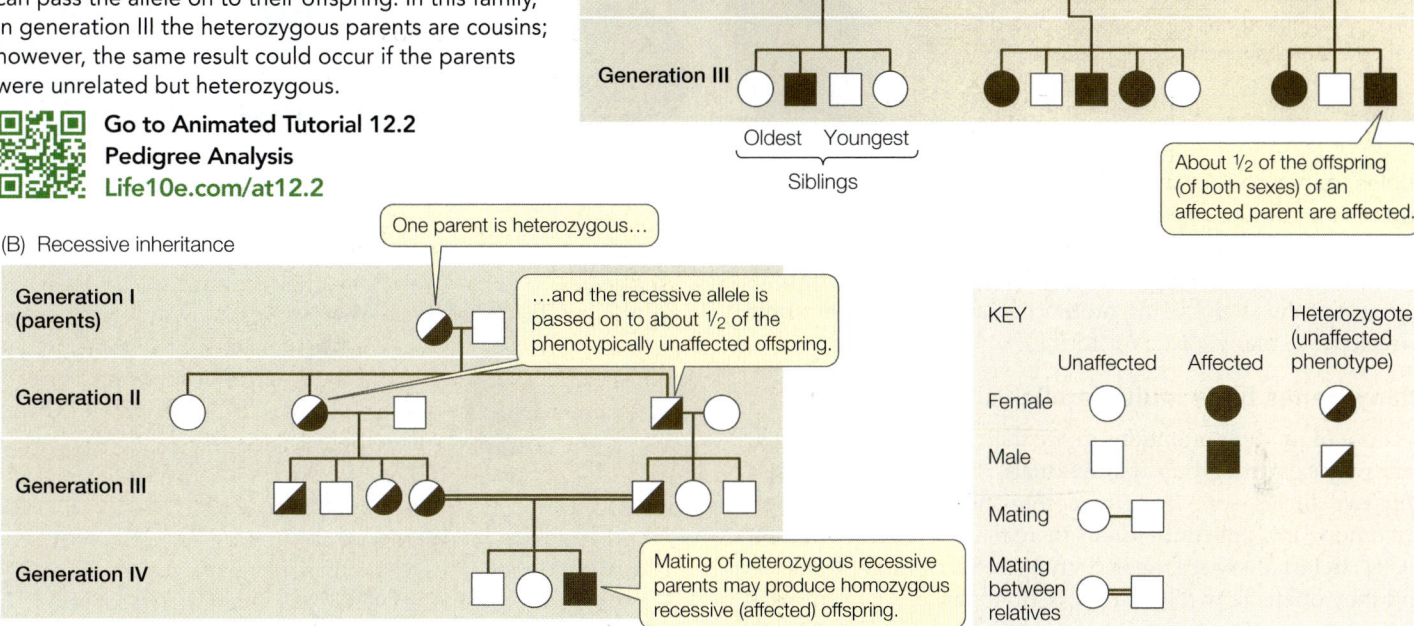

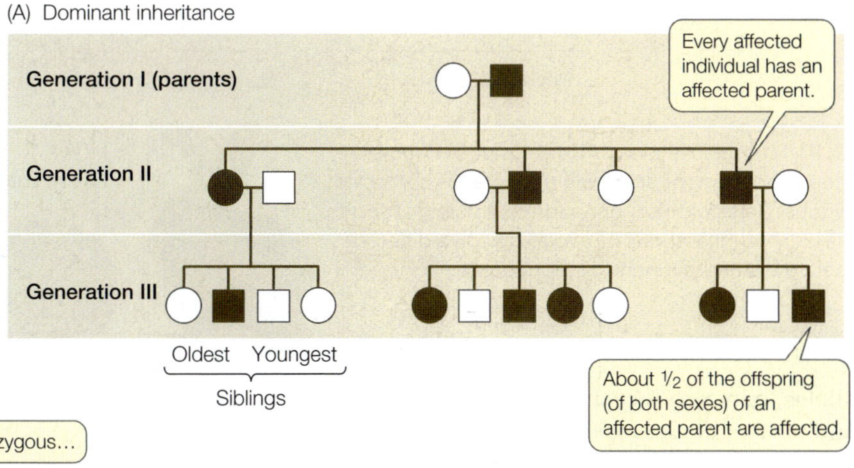
(A) Dominant inheritance
Generation I (parents)
Generation II
Generation III
Every affected individual has an affected parent.
Oldest Youngest
Siblings
About 1/2 of the offspring (of both sexes) of an affected parent are affected.

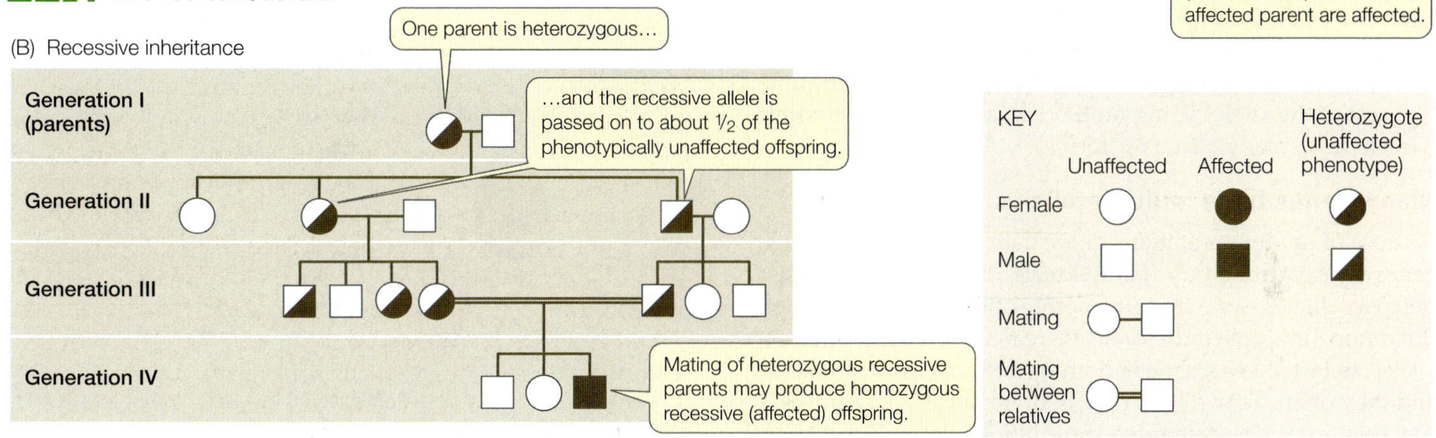
(B) Recessive inheritance
One parent is heterozygous...
Generation I (parents)
...and the recessive allele is passed on to about 1/2 of the phenotypically unaffected offspring.
Generation II
Generation III
Generation IV
Mating of heterozygous recessive parents may produce homozygous recessive (affected) offspring.
KEY
Unaffected
Affected
Heterozygote (unaffected phenotype)
Female
Male
Mating
Mating between relatives

to human geneticists. People in these groups are more likely to marry relatives who may carry the same rare recessive alleles.

RECAP 12.1

Mendel showed that genetic determinants are particulate and do not "blend" when the genes from two gametes combine. Mendel's first law of inheritance states that two copies of a gene segregate during gamete formation. His second law states that genes assort independently during gamete formation. The frequencies with which different allele combinations will be expressed in offspring can be calculated with a Punnett square or using probability theory.

- What results seen in the F_1 and F_2 generations of Mendel's monohybrid cross experiments refuted the blending theory of inheritance? **See p. 234, Figures 12.2, 12.3, and Table 12.1**
- How do events in meiosis explain Mendel's monohybrid cross results? **See p. 237 and Figure 12.4**
- How do events in meiosis explain the independent assortment of alleles in Mendel's dihybrid cross experiments? **See p. 239 and Figures 12.6, 12.7**
- Draw human pedigrees for dominant and recessive inheritance. **See p. 240 and Figure 12.9**

Mendel's laws of inheritance remain valid today; his discoveries laid the groundwork for all future studies of genetics. Inevitably, however, we have learned that things are more complicated than they seemed at first. Let's take a look at some of these complications, beginning with the interactions between alleles.

12.2 How Do Alleles Interact?

Over time genes accumulate changes, giving rise to new alleles. Thus there can be many alleles for a single character. In addition, alleles do not always show simple dominant–recessive relationships. Furthermore, a single allele may have multiple phenotypic effects.

New alleles arise by mutation

Genes are subject to **mutations**, which are stable, inherited changes in the genetic material. In other words, an allele can mutate to become a different allele. For example, you can imagine that at one time all pea plants were tall and had the height allele *T*. A mutation occurred in that allele that resulted in a new allele, *t* (conferring a short phenotype). If this mutation was in a cell that underwent meiosis to form gametes, some of the resulting gametes would carry the *t* allele, and some offspring of this pea plant would carry the *t* allele. Mutation will be discussed in more detail in Chapter 15. By creating variety, mutations provide the raw material for evolution.

Geneticists usually define one particular allele of a gene as the **wild type**; this allele is the one that is present in most individuals in nature ("the wild") and gives rise to an expected trait or phenotype. Other alleles of that gene, often called mutant

Possible genotypes	CC, Cc^{chd}, Cc^{h}, Cc	$c^{chd}c^{chd}$, $c^{chd}c$	$c^{h}c^{h}$, $c^{h}c$	cc
Phenotype	Dark gray	Chinchilla	Point restricted	Albino

12.10 Multiple Alleles for Coat Color in Rabbits These photographs show the phenotypes conferred by four alleles of the *C* gene for coat color in rabbits. Different combinations of two alleles give different coat colors and pigment distributions.

alleles, may produce a different phenotype. The wild-type and mutant alleles reside at the same genetic **locus**, which is their specific position on a chromosome. A genetic locus with a wild-type allele that is present less than 99 percent of the time (the rest of the alleles being mutant) is said to be **polymorphic** (Greek *poly*, "many"; *morph*, "form").

Many genes have multiple alleles

Because of random mutations, more than two alleles of a given gene may exist in a group of individuals. Any one individual has only two alleles—one from its mother and one from its father. But among multiple individuals there may be several different alleles. In fact, there are many examples of such multiple alleles, and they often show a hierarchy of dominance. An example is coat color in rabbits, determined by multiple alleles of the *C* gene:

- C determines dark gray
- c^{chd} determines chinchilla, a lighter gray
- c^{h} determines Himalayan, where pigment is restricted to the extremities (point restricted)
- c determines albino

The hierarchy of dominance for these alleles is $C > C^{chd}, C^{h} > c$. Any rabbit with the *C* allele (paired with itself or any other allele) is dark gray, and a *cc* rabbit is albino. Intermediate colors result from different allele combinations, as shown in **Figure 12.10**. As this example illustrates, multiple alleles can increase the number of possible phenotypes.

Dominance is not always complete

In the pairs of alleles studied by Mendel, dominance is complete in heterozygous individuals. That is, an *Rr* individual always expresses the *R* phenotype. However, many genes have alleles that are not dominant or recessive to one another. Instead, the heterozygotes show an intermediate phenotype—at first glance, like that predicted by the old blending theory of inheritance. For example, if a true-breeding eggplant that produces the familiar purple fruit is crossed with a true-breeding white eggplant, all the F_1 plants produce violet fruit, an intermediate between the two parents. However, further crosses indicate that this apparent blending phenomenon can still be explained in terms of Mendelian genetics (**Figure 12.11**). The purple and white alleles have not disappeared, as those colors reappear when the F_1 plants are interbred.

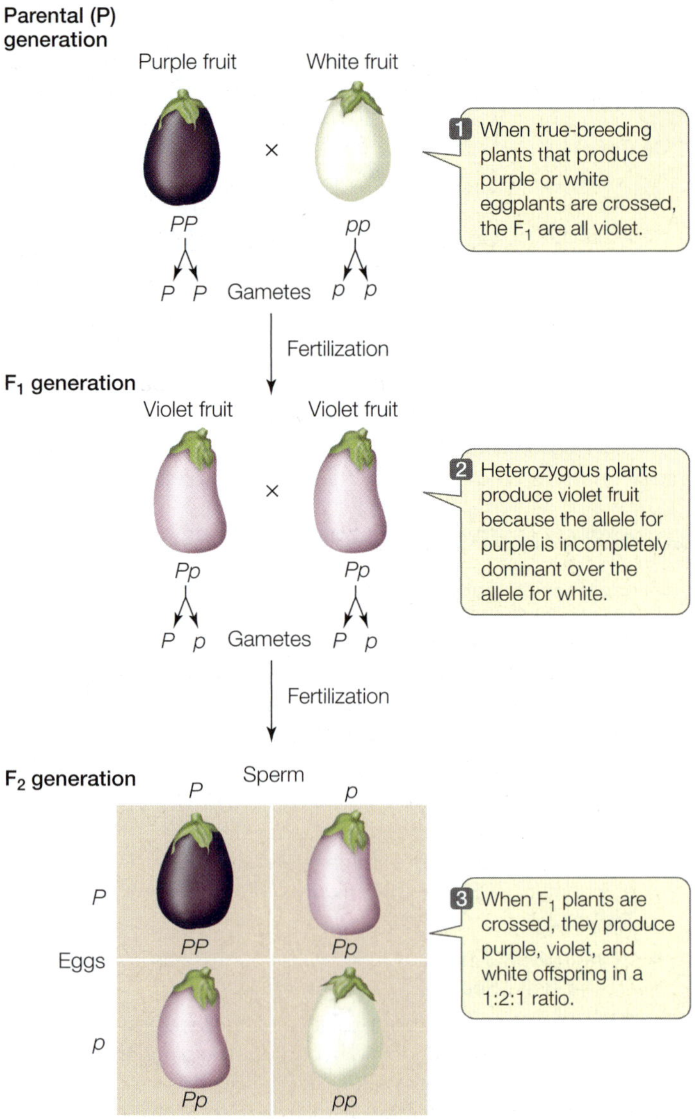
Parental (P) generation
Purple fruit
White fruit
×
PP
pp
P P Gametes p p
1 When true-breeding plants that produce purple or white eggplants are crossed, the F1 are all violet.
Fertilization
F1 generation
Violet fruit
Violet fruit
×
Pp
Pp
P p Gametes P p
2 Heterozygous plants produce violet fruit because the allele for purple is incompletely dominant over the allele for white.
Fertilization
F2 generation
Sperm
P
p
Eggs
P
p
PP
Pp
Pp
pp
3 When F1 plants are crossed, they produce purple, violet, and white offspring in a 1:2:1 ratio.

12.11 Incomplete Dominance Follows Mendel's Laws An intermediate phenotype can occur in heterozygotes when neither allele is dominant. The heterozygous phenotype (here, violet fruit) may give the appearance of a blended trait, but the traits of the parental generation reappear in their original forms in succeeding generations, as predicted by Mendel's laws of inheritance.

12.12 ABO Blood Reactions Are Important in Transfusions This table shows the results of mixing red blood cells of types A, B, AB, and O with serum containing anti-A or anti-B antibodies. As you look down the columns, note that each of the types, when mixed separately with anti-A and with anti-B, gives a unique pair of results; this is the basic method by which blood is typed. People with type O blood are good blood donors because O cells do not react with either anti-A or anti-B antibodies. People with type AB blood are good recipients, since they make neither type of antibody. When blood transfusions are incompatible, the reaction (clumping of red blood cells) can have severely adverse consequences for the recipient.

Blood type of cells	Genotype	Blood cell types that body rejects	Reaction to added antibodies	
			Anti-A	Anti-B
A	I^AI^A or I^AI^O	B		
B	I^BI^B or I^BI^O	A		
AB	I^AI^B	Neither A nor B		
O	I^OI^O	A, B, and AB		

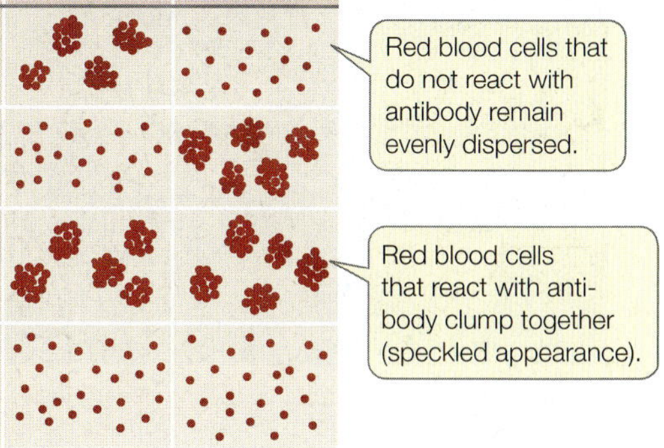
Red blood cells that do not react with antibody remain evenly dispersed.
Red blood cells that react with antibody clump together (speckled appearance).

When heterozygotes show a phenotype that is intermediate between those of the two homozygotes, the gene is said to be governed by **incomplete dominance**. In other words, neither of the two alleles is dominant. Incomplete dominance is common in nature; in fact, Mendel's study of pea-plant traits is unusual in that all the traits happened to be characterized by complete dominance.

In codominance, both alleles at a locus are expressed

Sometimes the two alleles at a locus produce two different phenotypes that *both* appear in heterozygotes, a phenomenon called **codominance**. Note that this is different from incomplete dominance, where the phenotype of a heterozygote is a blend of the phenotypes of the parents. A good example of codominance is seen in the ABO blood group system in humans (this is also an example of multiple alleles).

Early attempts at blood transfusion frequently killed the patient. Around 1900, the Austrian scientist Karl Landsteiner mixed blood cells and serum (blood from which cells have been removed) from different individuals. He found that only certain combinations of blood and serum are compatible. In other combinations, the red blood cells from one individual form clumps in the presence of serum from the other individual. This discovery led to our ability to administer compatible blood transfusions that do not kill the recipient.

The formation of clumps occurs because of the immune system, which protects the body from invasion by "nonself" molecules or organisms. (This ability to recognize nonself cells is why people do not "catch" cancer like the Tasmanian devils in the opening story.) People make specific proteins in the serum, called antibodies, which react with foreign molecules or particles. The specific part of a molecule that is recognized by an antibody is called an antigen. Oligosaccharides on the surfaces of red blood cells can function as antigens. For example, people with the A blood group make the A antigen, and those with the B group make the B antigen: these antigens are specific oligosaccharides on the surfaces of their red blood cells. If a person with the A blood group is given a transfusion of blood from a person with the B group, their immune system will recognize the B antigen as nonself and make antibodies against it (**Figure 12.12**). Likewise, a person with the B blood group will make antibodies against the A antigen. However, someone with the codominant AB blood group makes both the A and the B antigens and will not make antibodies against either antigen. A person with the O blood group makes neither the A nor the B antigen (see Figure 12.12).

The oligosaccharides on the surfaces of red blood cells are made by enzymes that catalyze the formation of bonds between specific sugars. The ABO genetic locus encodes one such enzyme and has three alleles, I^A, I^B, and I^O, each producing a different version of the enzyme. The product of the I^A allele adds N-acetylgalactosamine to the end of a pre-existing oligosaccharide chain, resulting in the A antigen. The product of the I^B allele adds galactose to the same chain, making the B antigen. The I^O allele encodes a protein that has no enzymatic activity:

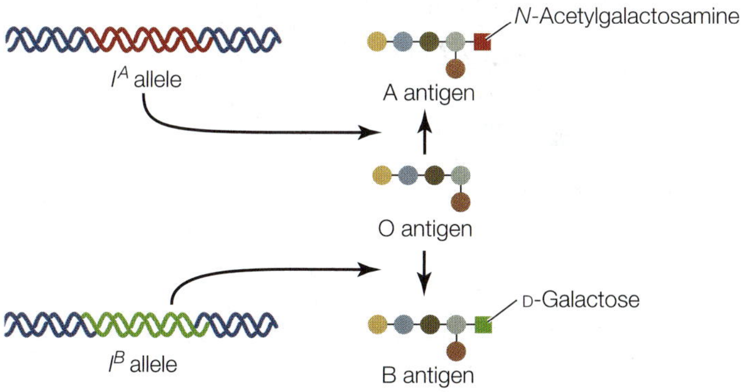
N-Acetylgalactosamine
I^A allele
A antigen
O antigen
D-Galactose
I^B allele
B antigen

Since people inherit one allele from each parent, they may have any combination of these alleles: I^AI^B, I^AI^O, I^AI^A, and so on. The I^A and I^B alleles are codominant because a person with both alleles makes both the A and the B antigens, and both kinds of oligosaccharide occur on their red blood cells. We will learn much more about the functions of antibodies and antigens in Chapter 42.

Some alleles have multiple phenotypic effects

Mendel's principles were further extended when it was discovered that a single allele can influence more than one phenotype. In such a case, we say that the allele is **pleiotropic**. A classic example is the heritable human disease phenylketonuria, which causes mental retardation and reduced hair and skin pigmentation. The disease occurs in people who have a mutation in the gene for a liver enzyme that converts the amino acid

A dog with alleles *B* and *E* is black.
A dog with alleles *bb* and *E* is brown.
A dog with *ee* is yellow, regardless of its *Bb* alleles.
Black (*B_E_*)
Chocolate (*bbE_*)
Yellow (*_ _ee*)

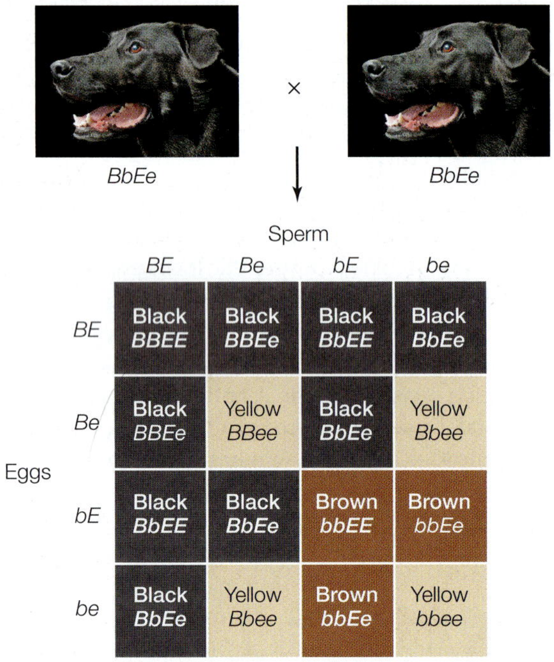
BbEe × *BbEe*
Sperm
BE *Be* *bE* *be*
Eggs
BE: Black *BBEE*, Black *BBEe*, Black *BbEE*, Black *BbEe*
Be: Black *BBEe*, Yellow *BBee*, Black *BbEe*, Yellow *Bbee*
bE: Black *BbEE*, Black *BbEe*, Brown *bbEE*, Brown *bbEe*
be: Black *BbEe*, Yellow *Bbee*, Brown *bbEe*, Yellow *bbee*

12.13 Genes May Interact Epistatically Epistasis occurs when one gene alters the phenotypic effect of another gene. In Labrador retrievers, the *Ee* gene determines the expression of the *Bb* gene.

phenylalanine to tyrosine. Without a functional form of this enzyme, phenylalanine builds up in the body to toxic levels, and this affects development in a variety of ways. Given what we now know about genes and their functions, it is not surprising that a gene with such an important metabolic role should have pleiotropic effects. Other examples of pleiotropy include plant and animal genes whose products affect hormone levels, since many hormones play multiple roles in the body.

RECAP 12.2

Genes are subject to random mutations that give rise to new alleles; thus many genes have more than two alleles within a population. Dominance is not necessarily an all-or-nothing phenomenon. Some genes have multiple effects on phenotype.

- How do the crosses in Figure 12.11 demonstrate incomplete dominance? **See pp. 242–243**
- Explain how blood type AB results from codominance. **See p. 243 and Figure 12.12**

Thus far we have discussed phenotypic characters that are affected by single genes. In many cases, however, several genes interact to determine a phenotype. To complicate things further, the physical environment may interact with the genetic constitution of an individual in determining the phenotype.

12.3 How Do Genes Interact?

We have just seen how two alleles of the same gene can interact to produce a phenotype. If you consider most complex phenotypes, such as human height, you will realize that they are influenced by the products of many genes. We now turn to the genetics of such gene interactions.

Epistasis ("to stand upon") occurs when the phenotypic expression of one gene is affected by another gene. For example, two genes (*B* and *E*) encode proteins that determine coat color in Labrador retrievers:

- Allele *B* (black pigment) is dominant to *b* (brown).
- Allele *E* (pigment deposition in hair) is dominant to *e* (no deposition, so hair is yellow).

An *EE* or *Ee* dog with *BB* or *Bb* is black, and one with *bb* is brown. An *ee* dog is yellow regardless of the *B* gene alleles present (**Figure 12.13**). So the product of the *E* allele is needed for the expression of both the *B* and the *b* alleles, and *E* is said to be epistatic to *B*.

Hybrid vigor results from new gene combinations and interactions

In 1876, Charles Darwin reported that when he crossed two different true-breeding, homozygous genetic strains of corn, the offspring were 25 percent taller than either of the parent strains. Darwin's observation was largely ignored for the next 30 years. In 1908, George Shull "rediscovered" this idea, reporting that not just plant height but the weight of the corn grain produced was dramatically higher in the offspring (**Figure 12.14**). Agricultural scientists took note, and Shull's paper had a lasting impact on the field of applied genetics.

Farmers have known for centuries that matings among close relatives (known as **inbreeding**) can result in offspring

B73
Hybrid
Mo17

12.14 Hybrid Vigor in Corn Two homozygous parent lines of corn, B73 and Mo17, were crossed to produce the more vigorous hybrid line.

The temperature of the extremities is lower and allows expression of the black coat color gene.
The temperature of most of the body is too high for the expression of the black coat color gene.

12.15 The Environment Influences Gene Expression The rabbit and cat express a coat pattern called "point restriction." Their genotypes specify dark hair/fur, but the enzyme for dark color is inactive at normal body temperature, so only the extremities—the coolest regions of the body—express the phenotype.

of lower quality than matings between unrelated individuals. Agricultural scientists call this inbreeding depression. (This is another of the concerns of conservationists working with the Tasmanian devils in the opening story.) The problems with inbreeding arise because close relatives tend to have the same recessive alleles, some of which may be harmful. The "hybrid vigor" after crossing inbred lines is called **heterosis** (short for heterozygosis). The cultivation of hybrid corn spread rapidly in the United States and all over the world, quadrupling grain production. The practice of hybridization has spread to many other crops and animals used in agriculture. For example, beef cattle that are crossbred are larger and live longer than cattle bred within their own genetic strain.

There has been much controversy over the mechanism by which heterosis works. The dominance hypothesis assumes that all of the extra growth in hybrids can be explained by the lack of inbreeding depression, since hybrids are unlikely to be homozygous for deleterious, recessive alleles. The overdominance hypothesis postulates that in hybrids, new combinations of alleles from the parental strains interact with one another, resulting in superior traits that cannot occur in the parental lines. Many of the characters in question are controlled by multiple genes (see below), and from recent studies it appears that both dominance and overdominance can contribute to heterosis in specific characters.

The environment affects gene action

The phenotype of an individual does not result from its genotype alone. *Genotype and environment interact to determine the phenotype of an organism.* This is especially important to remember in the era of genome sequencing (see Chapter 17). When the sequence of the human genome was completed in 2003, it was hailed as the "book of life," and public expectations of the benefits gained from this knowledge were (and are) high.

But this kind of "genetic determinism" is wrong. Common knowledge tells us that environmental variables such as light, temperature, and nutrition can affect the phenotypic expression of a genotype.

A familiar example of this phenomenon involves "point restriction" coat patterns found in Siamese cats and certain rabbit breeds (**Figure 12.15**). These animals carry a mutant allele of a gene that controls the growth of dark fur all over the body. As a result of this mutation, the enzyme encoded by the gene is inactive at temperatures above a certain point (usually around 35°C). The animals maintain a body temperature above this point, and so their fur is mostly light. However, the extremities—feet, ears, nose, and tail—are cooler, about 25°C, so the fur on these regions is dark. These animals are all white when they are born, because the extremities were kept warm in the mother's womb.

A simple experiment shows that the dark fur is temperature-dependent. If a patch of white fur on a point-restricted rabbit's back is removed and an ice pack is placed on the skin where the patch was, the fur that grows back will be dark. This indicates that although the gene for dark fur was expressed all along, the environment inhibited the activity of the mutant enzyme.

Two parameters describe the effects of genes and environment on the phenotype:

- **Penetrance** is the proportion of individuals in a group with a given genotype that actually show the expected phenotype. For example, many people who inherit a mutant allele of the gene *BRCA1* develop breast cancer. But for some reason, some people with the mutation do not. So the *BRCA1* allele is said to be incompletely penetrant.

4:10
5:0
5:1
5:2
5:3
5:4
5:5
5:6
5:7
5:8
5:9
5:10
5:11
6:0
6:1
6:2
6:3
6:4
6:5

12.16 Quantitative Variation Quantitative variation is produced by the interaction of genes at multiple loci and the environment. These students (women [in white] are shorter; men [in blue] are taller) show continuous variation in height that is the result of interactions between many genes and the environment.

- **Expressivity** is the degree to which a genotype is expressed in an individual. For example, a woman with the mutant *BRCA1* allele may develop both breast and ovarian cancer as part of the phenotype, but another woman with the same mutation may only develop breast cancer. So the mutant allele is said to have variable expressivity.

Most complex phenotypes are determined by multiple genes and the environment

Certain simple characters, such as those that Mendel studied in peas, differ in discrete, **qualitative** ways. Mendel used true-breeding parental pea plants that were either short or tall, had purple or white flowers, or had round or wrinkled seeds. But for most complex characters, such as height in humans, the phenotype varies more or less continuously over a range. Some people are short, others are tall, and many are in between the two extremes. Such variation within a population is called **quantitative**, or continuous, variation (**Figure 12.16**).

Sometimes this variation results largely from the alleles that an individual possesses. For instance, much of human eye color is the result of a number of genes controlling the synthesis and distribution of dark melanin pigment. Dark eyes have a lot of it, brown eyes less, and green, gray, and blue eyes even less. In the latter cases, the distribution of other pigments in the eye is what determines light reflection and color.

In most cases, however, *quantitative variation is due to both genes and environment*. Height in humans certainly falls into this category. If you look at families, you often see that parents and their offspring all tend to be tall or short. However, nutrition also plays a role in height: American 18-year-olds today are about 20 percent taller than their great-grandparents were at the same age. Three generations are not enough time for mutations that would exert such a dramatic effect to spread throughout the general population, so the height difference must be due to environmental factors.

Geneticists call the genes that together determine such complex characters **quantitative trait loci**. Identifying these loci is a major challenge, and an important one. For example, the amount of grain that a variety of rice produces in a growing season is determined by many interacting genetic factors. Crop plant breeders have worked hard to decipher these factors in order to breed higher-yielding rice strains. In a similar way, human characteristics such as disease susceptibility and behavior are caused in part by quantitative trait loci. Recently, one of the many genes involved with human height was identified. The gene, *HMGA2*, has an allele that apparently has the potential to add 4 millimeters to human height.

RECAP 12.3

In epistasis, one gene affects the expression of another. Perhaps the most challenging problem for genetics is the explanation of complex phenotypes that are caused by many interacting genes and the environment.

- Explain the difference between penetrance and expressivity. **See pp. 245–246**
- How is quantitative variation different from qualitative variation? **See p. 246**

In the next section we'll see how the discovery that genes occupy specific positions on chromosomes enabled Mendel's successors to provide a physical explanation for his model of inheritance, and to provide an explanation for those cases where Mendel's second law does not apply.

12.4 What Is the Relationship between Genes and Chromosomes?

There are far more genes than chromosomes. Studies of different genes that are physically linked on the same chromosome reveal inheritance patterns that are not Mendelian. These patterns have been useful for identifying genes that are linked to one another, and for determining how far apart they are on the chromosome.

Genetic linkage was first discovered in the fruit fly *Drosophila melanogaster*. Its small size, the ease with which it can be bred, its few chromosomes, and its short generation time make this animal an attractive experimental subject. Beginning in 1909, Thomas Hunt Morgan and his students at Columbia University pioneered the study of *Drosophila*, and today it remains a very important model organism for studies of genetics.

Go to Animated Tutorial 12.3
Alleles That Do Not Assort Independently
Life10e.com/at12.3

Genes on the same chromosome are linked

Some of the crosses Morgan performed with fruit flies yielded phenotypic ratios that were not in accordance with those predicted by Mendel's law of independent assortment. Morgan crossed *Drosophila* with known genotypes at two loci, *B* and *Vg*:

- *B* (wild-type gray body) is dominant over *b* (black body).
- *Vg* (wild-type wing) is dominant over *vg* (vestigial, a very small wing).

Morgan first made an F_1 generation by crossing homozygous dominant *BBVgVg* flies with homozygous recessives (*bbvgvg*). He then performed a test cross with the F_1 flies: *BbVgvg* × *bbvgvg*.* Morgan expected to see four phenotypes in a ratio of 1:1:1:1, but that is not what he observed. The body color gene and the wing size gene were not assorting independently; rather, they were usually inherited together (**Figure 12.17**).

These results became understandable to Morgan when he considered the possibility that the two loci are on the same chromosome—that is, that they might be linked. Suppose that the *B* and *Vg* loci are indeed located on the same chromosome. Why didn't all of Morgan's F_1 flies have the parental phenotypes—that is, why didn't his cross result in gray flies with normal wings and black flies with vestigial wings, in a 1:1 ratio? If linkage were absolute—that is, if chromosomes always remained intact and unchanged—we would expect to see just those two types of progeny. However, this does not always happen.

* Do you recognize this type of cross? It is a test cross for the two gene pairs; see Figure 12.5.

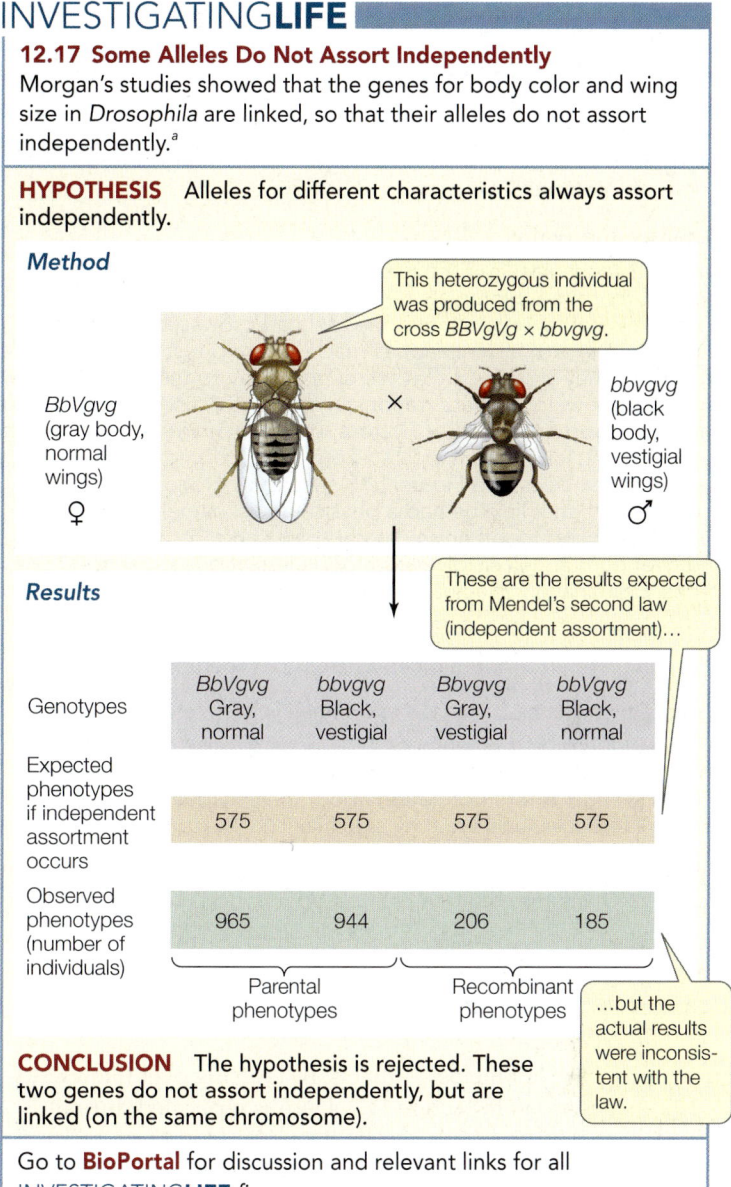

INVESTIGATING**LIFE**

12.17 Some Alleles Do Not Assort Independently
Morgan's studies showed that the genes for body color and wing size in *Drosophila* are linked, so that their alleles do not assort independently.[a]

HYPOTHESIS Alleles for different characteristics always assort independently.

Method

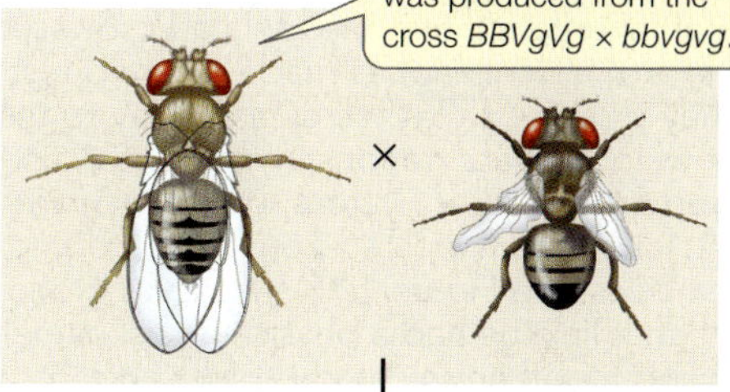
This heterozygous individual was produced from the cross *BBVgVg* × *bbvgvg*.
BbVgvg (gray body, normal wings) ♀
×
bbvgvg (black body, vestigial wings) ♂

Results

These are the results expected from Mendel's second law (independent assortment)...

Genotypes	*BbVgvg* Gray, normal	*bbvgvg* Black, vestigial	*Bbvgvg* Gray, vestigial	*bbVgvg* Black, normal
Expected phenotypes if independent assortment occurs	575	575	575	575
Observed phenotypes (number of individuals)	965	944	206	185
	Parental phenotypes		Recombinant phenotypes	

...but the actual results were inconsistent with the law.

CONCLUSION The hypothesis is rejected. These two genes do not assort independently, but are linked (on the same chromosome).

Go to **BioPortal** for discussion and relevant links for all INVESTIGATING**LIFE** figures.

[a]Morgan, T. H. 1912. *Science* 36: 719–720.

Genes can be exchanged between chromatids and mapped

If linkage were absolute, Mendel's law of independent assortment would apply only to loci on different chromosomes. Instead, genes at different loci on the same chromosome *do* sometimes separate from one another during meiosis. Genes may recombine when two homologous chromosomes physically exchange corresponding segments

WORKING WITH**DATA**:

Some Alleles Do Not Assort Independently

Original Paper

Morgan, T. H. 1912. Complete linkage in the second chromosome of *Drosophila*. *Science* 36: 719–720.

Analyze the Data

Mendel's work was "rediscovered" 40 years after its publication. At that time, biologists began to find some exceptions to the rules of inheritance that Mendel had proposed. Thomas Hunt Morgan and his colleagues made dihybrid test crosses in fruit flies. They proposed that the clearest way to test for linkage was not to look at aberrations in the 9:3:3:1 phenotypic ratio expected from an $F_1 \times F_1$ cross, but to examine aberrations in the 1:1:1:1 ratio expected from an $F_1 \times$ homozygous recessive test cross (see Figure 12.5). Morgan's group then hypothesized that linkage had a physical basis, namely that genes are linked together on chromosomes and that rare crossing over during meiosis gives rise to the less frequent phenotypes. Examination of actual chromosomal events confirmed this.

QUESTION 1

Morgan first performed a dihybrid cross between black, normal-winged flies (*bbVgVg*) and gray, vestigial-winged flies (*BBvgvg*). The F_1 flies were interbred, yielding the F_2 phenotypes shown in the table (Experiment 1). Compare these data with the expected data in a 9:3:3:1 ratio by using the chi-square test (see Appendix B for information about the chi-square test). Are there differences, and are they significant?

QUESTION 2

To quantify linkage, Morgan crossed homozygous black, normal-winged females with homozygous gray, vestigial-winged males. He then crossed the F_1 females with black, vestigial-winged males. (You should note that this is not the same test cross as the one shown in Figure 12.17. In that case, the original parents were *BBVgVg* and *bbvgvg*.) The results of this test cross are shown in the table (Experiment 2). Are these genes linked? If they are linked, what is the map distance between the genes? Explain why these data are so different from the data shown in Figure 12.17.

QUESTION 3

In a third experiment, Morgan crossed two genetic strains of flies that were homozygous for the body color and wing genes. The F_1 flies were all gray and normal-winged, and these were interbred. The results are shown in the table (Experiment 3). What were the genotypes and phenotypes of the original parents that produced the F_1?

	Number of progeny showing each phenotype			
Experiment	Gray, normal	Black, normal	Gray, vestigial	Black, vestigial
1	2,316	1,146	737	0
2	578	1,413	1,117	307
3	246	9	65	18

Go to **BioPortal** for all WORKING WITH**DATA** exercises

during prophase I of meiosis—that is, by crossing over (**Figure 12.18**). As described in Section 11.2, DNA is replicated during S phase, so that by prophase I, when homologous chromosome pairs come together to form tetrads, each chromosome consists of two chromatids.

Note that the exchange event involves *only two of the four chromatids* in a tetrad, one from each member of the homologous pair, and can occur at any point along the length of the chromosome. The chromosome segments involved are exchanged reciprocally, so both chromatids involved in crossing over become recombinant (that is, each chromatid ends up with genes from both of the organism's parents). Usually several exchange events occur along the length of each homologous pair.

When crossing over takes place between two linked genes, not all the progeny of a cross have the parental phenotypes. Instead, recombinant offspring appear as well, as they did in Morgan's test cross (see Figure 12.17). They appear in proportions called **recombinant frequencies**, which are calculated by dividing the number of recombinant progeny by the total number of progeny (**Figure 12.19**). Recombinant frequencies will be *greater for loci that are farther apart* on the chromosome than for loci that are closer together because an exchange event is more likely to occur between genes that are far apart.

By calculating recombination frequencies, geneticists can infer the locations of genes along a chromosome and generate a genetic map. Below is a map showing five genes on a fruit fly chromosome constructed using the recombination frequencies generated by test crosses involving various pairs of genes:

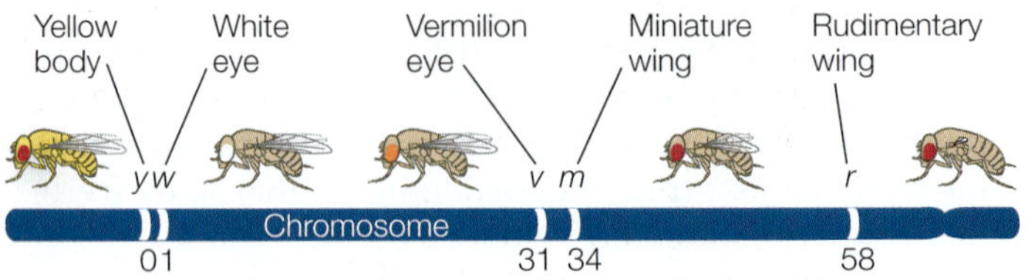
Yellow body
White eye
Vermilion eye
Miniature wing
Rudimentary wing
y w
v m
r
Chromosome
01
31 34
58

In the chromosome shown above, the recombination frequency between *y* and *w* is low, so they are close together on the map. Recombination between *y* and *v* is more frequent, so they are farther apart. The recombination frequencies are converted to map units (also called centimorgans, cM); one map unit is equivalent to an average recombination frequency of 0.01 (1 percent).

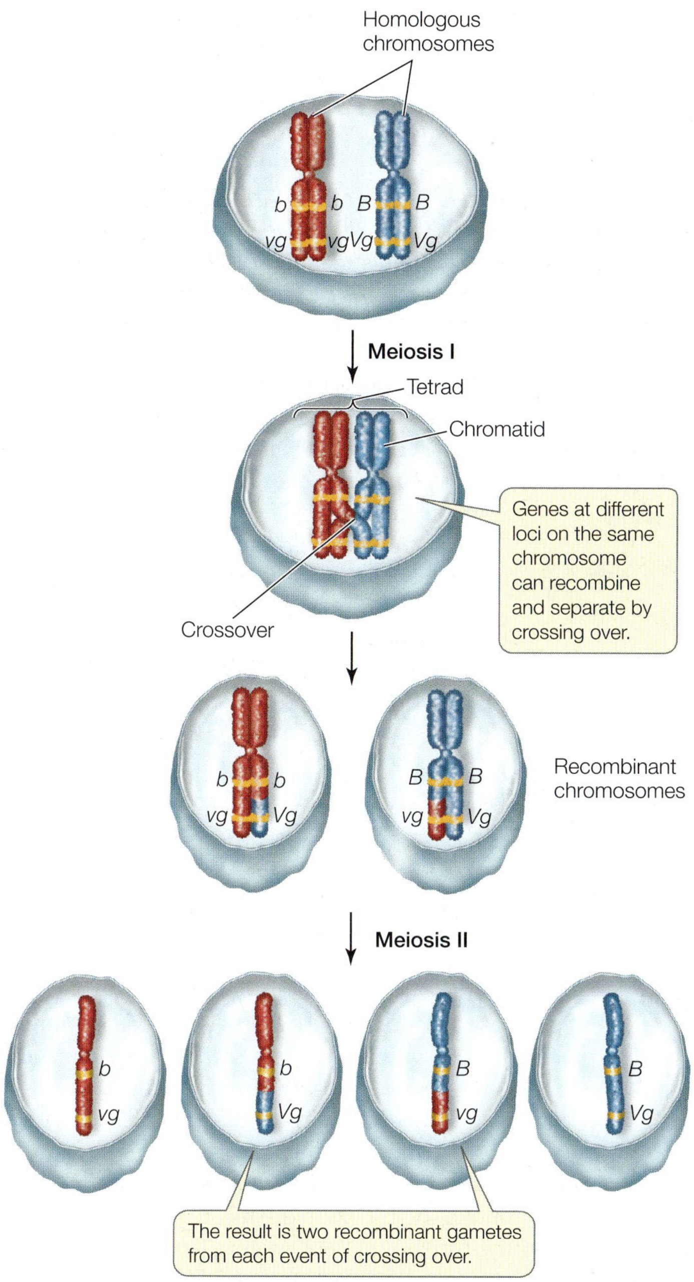

Homologous chromosomes
b
b
B
B
vg
vg
Vg
Vg
Meiosis I
Tetrad
Chromatid
Genes at different loci on the same chromosome can recombine and separate by crossing over.
Crossover
b
b
vg
Vg
B
B
vg
Vg
Recombinant chromosomes
Meiosis II
b
vg
b
Vg
B
vg
B
Vg
The result is two recombinant gametes from each event of crossing over.

12.18 Crossing Over Results in Genetic Recombination Recombination accounts for why linked alleles are not always inherited together. Alleles at different loci on the same chromosome can be recombined by crossing over, and separated from one another. Such recombination occurs during prophase I of meiosis.

The era of gene sequencing has made mapping less important in some areas of genetics research. However, mapping it still a way to verify that a particular DNA sequence corresponds with a particular phenotype. Linkage has allowed biologists to isolate genes and to identify genetic markers linked to important genes. This is important in breeding new crops and animals for agriculture, and for identifying people carrying medically significant mutations.

Linkage is revealed by studies of the sex chromosomes

In Mendel's work, reciprocal crosses always gave similar results; it did not matter whether a dominant allele was contributed by the female parent or the male parent. But in some cases, the parental origin of a chromosome does matter. For example, human males inherit a bleeding disorder called hemophilia from their mothers, not from their fathers. To understand the types of inheritance in which the parental origin of an allele is important, we must consider the ways in which sex is determined in different species.

SEX DETERMINATION BY CHROMOSOMES In corn, every diploid adult has both male and female reproductive structures. The tissues in these two types of structure are genetically identical, just as roots and leaves are genetically identical. Organisms such as corn, in which the same individual produces both male and female gametes, are said to be **monoecious** (Greek, "one house"). Other organisms, such as date palms, oak trees, and most animals, are **dioecious** ("two houses"), meaning that some individuals can produce only male gametes and others can produce only female gametes. In other words, in dioecious organisms the different sexes are different individuals.

In mammals and birds, sex is determined by differences in the chromosomes, but such determination operates in different ways in different groups of organisms. For example, in many animals, including mammals, sex is determined by a single pair of **sex chromosomes**, which differ from one another. The remaining chromosomes, called **autosomes**, occur in pairs in males and females. For example, in humans there are 22 pairs of autosomes in males and females, and 1 pair of sex chromosomes. The chromosomal bases for sex determination in various groups of animals are summarized in Table 12.2.

The sex chromosomes of female mammals consist of a pair of X chromosomes. Male mammals, by contrast, have one X chromosome and a sex chromosome that is not found in females, the Y chromosome. Females may be represented as XX and males as XY.

MALE MAMMALS PRODUCE TWO KINDS OF GAMETES Each gamete produced by a male mammal has a complete set of autosomes, but half the gametes carry an X chromosome and the

TABLE **12.2**

Sex Determination in Animals

Animal Group	Mechanism
Bees	Males are haploid, females are diploid
Fruit flies	Fly is female if ratio of X chromosomes to sets of autosomes is 1 or more
Birds	Males ZZ (homogametic), females WZ (heterogametic)
Mammals	Males XY (heterogametic), females XX (homogametic)

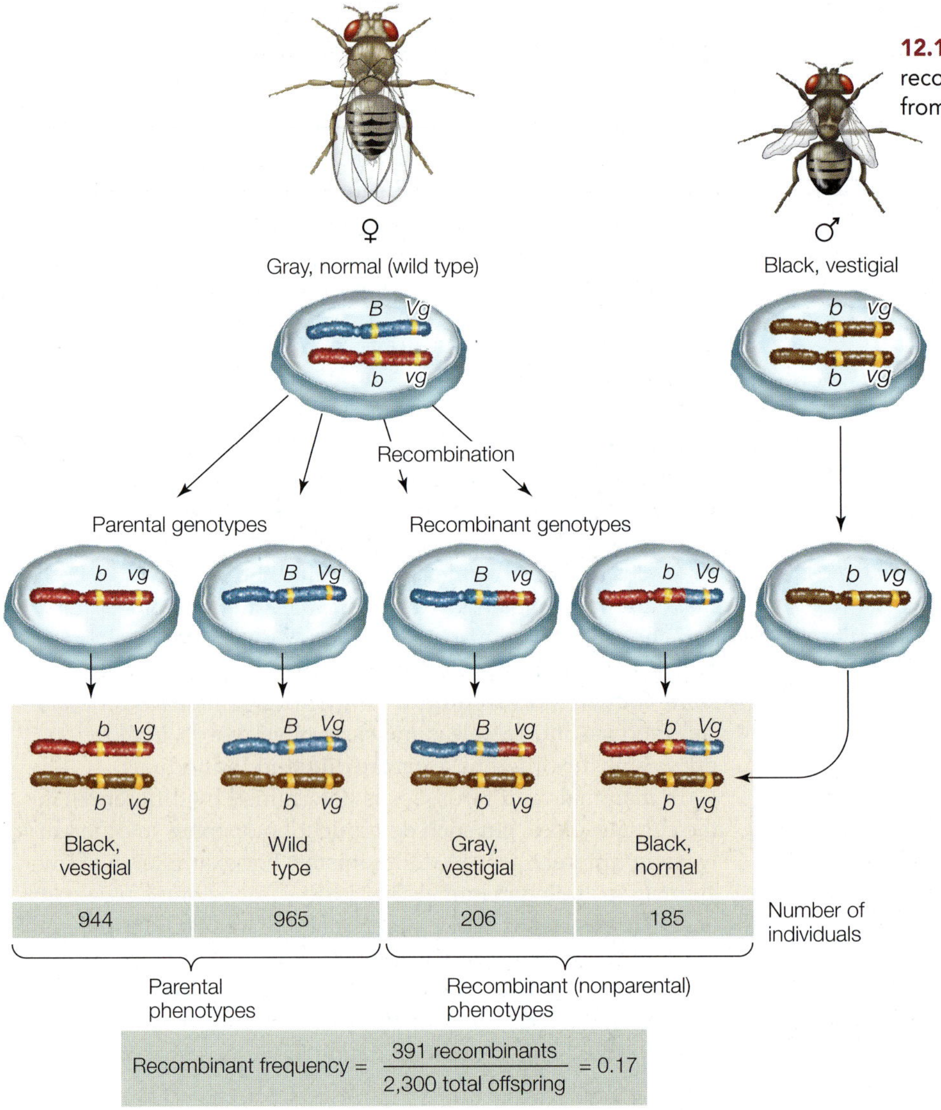

♀
Gray, normal (wild type)
♂
Black, vestigial
Recombination
Parental genotypes
Recombinant genotypes
Black, vestigial
Wild type
Gray, vestigial
Black, normal
944
965
206
185
Number of individuals
Parental phenotypes
Recombinant (nonparental) phenotypes
Recombinant frequency = 391 recombinants / 2,300 total offspring = 0.17

12.19 Recombinant Frequencies The frequency of recombinant offspring (those with a phenotype different from either parent) can be calculated.

other half carry a Y. When an X-bearing sperm fertilizes an egg, the resulting XX zygote is female; when a Y-bearing sperm fertilizes an egg, the resulting XY zygote is male.

SEX CHROMOSOME ABNORMALITIES REVEALED THE GENE THAT DETERMINES SEX Can we determine which chromosome, X or Y, carries the sex-determining gene, and can the gene be identified? One way to determine cause (e.g., the presence of a gene on the Y chromosome) and effect (e.g., maleness) is to look at cases of biological error, in which the expected outcome does not happen.

We can learn something about the functions of X and Y chromosomes from abnormal sex chromosome arrangements resulting from nondisjunction during meiosis or mitosis (see Section 11.5). As you will recall, nondisjunction occurs when a pair of homologous chromosomes (in meiosis I) or sister chromatids (in mitosis or meiosis II) fail to separate. As a result, a gamete may have one too few or one too many chromosomes. If this gamete fuses with another gamete that has the full haploid chromosome set, the resulting offspring will be aneuploid, with fewer or more chromosomes than normal.

In humans, XO individuals sometimes appear. The O indicates that a chromosome is missing—that is, individuals that are XO have only one sex chromosome (an X). Human XO individuals are females who are moderately abnormal physically but normal mentally; usually they are also sterile. The XO condition in humans is called Turner syndrome. It is the only known case in which a person can survive with only one member of a chromosome pair (here, the XY pair), although most XO conceptions are spontaneously terminated early in development. XXY individuals also occur; this condition, which affects males, is called Klinefelter syndrome and results in overlong limbs and sterility.

These observations suggest that the gene controlling maleness is located on the Y chromosome. Observations of people with other types of chromosomal abnormalities helped researchers pinpoint the location of that gene:

- Some women are genetically XY but lack a small portion of the Y chromosome.
- Some men are genetically XX but have a small piece of the Y chromosome attached to another chromosome.

The Y fragments that are respectively missing and present in these two cases are the same and contain the maleness-determining gene, which was named *SRY* (*s*ex-determining *r*egion on the *Y* chromosome).

The *SRY* gene encodes a protein involved in **primary sex determination**—that is, the determination of the kinds of gametes that an individual will produce and the organs that will make them (the male and female gonads). In the presence of the functional SRY protein, an embryo develops sperm-producing testes. (Notice that *italic type* is used for the name of a gene, but roman type is used for the name of a protein.) If the embryo has no Y chromosome, the *SRY* gene is absent, and thus the SRY protein is not made. In the absence of the SRY protein, the embryo develops egg-producing ovaries. In this case, a gene on the X chromosome called *DAX1* produces an anti-testis factor. So the role of SRY in a male is to inhibit the maleness inhibitor encoded by *DAX1*. The SRY protein does this in male cells, but since it is not present in females, DAX1 can act to inhibit maleness.

One function of the gonads is to produce hormones (such as testosterone and estrogen) that send signals to the rest of the body and control the development of **secondary sex characteristics**. These are outward manifestations of maleness and femaleness, such as differences in body type, breast development, body hair, and voice. Secondary sex characteristics distinguish males and females but are not directly part of the reproductive system.

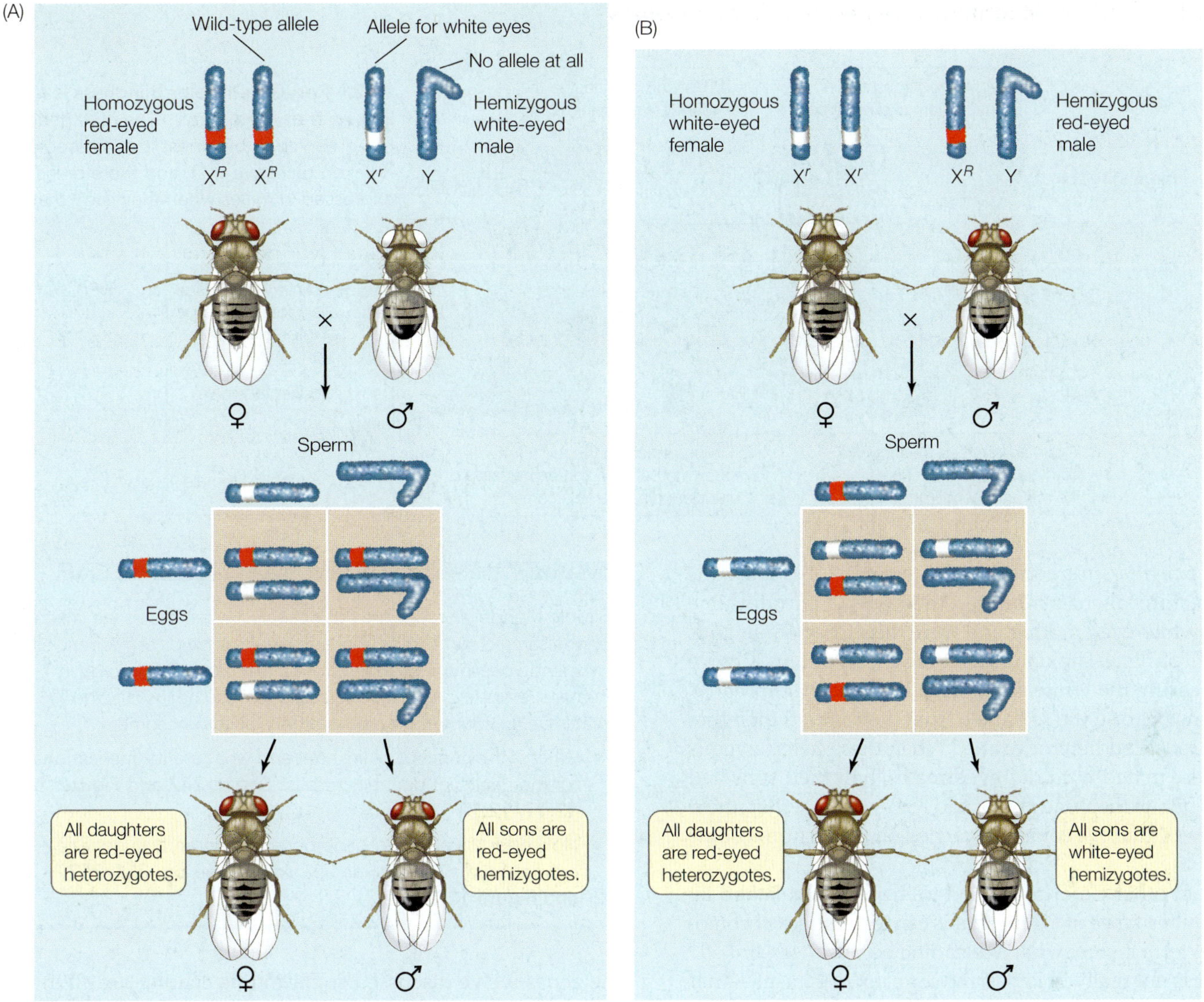
(A)
Wild-type allele
Allele for white eyes
No allele at all
Homozygous red-eyed female
Hemizygous white-eyed male
X^R X^R
X^r Y
Sperm
Eggs
All daughters are red-eyed heterozygotes.
All sons are red-eyed hemizygotes.
(B)
Homozygous white-eyed female
Hemizygous red-eyed male
X^r X^r
X^R Y
Sperm
Eggs
All daughters are red-eyed heterozygotes.
All sons are white-eyed hemizygotes.

12.20 Eye Color Is a Sex-Linked Trait in *Drosophila* Morgan demonstrated that a mutant allele that causes white eyes in *Drosophila* is carried on the X chromosome. Note that in this case, the reciprocal crosses do not have the same results.

SEX-LINKED INHERITANCE IN FRUIT FLIES As noted in Table 12.2, sex determination in fruit flies is based on the *proportions* of sex chromosomes, because the numbers of these chromosomes can vary. But most commonly, the fruit fly genome has four pairs of chromosomes: three pairs of autosomes (shown in blue) and (as in humans) a pair of sex chromosomes (green) that differ in size:

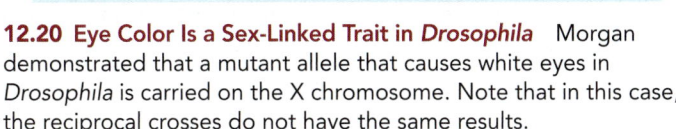

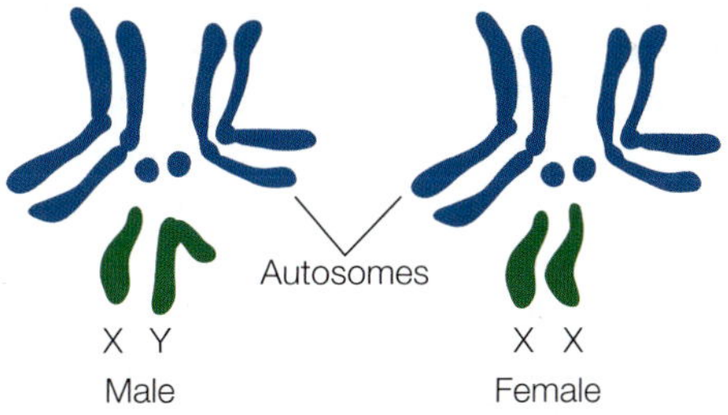
Autosomes
X Y
Male
X X
Female

In this case, the female fly has two X chromosomes and the male has only one, the other being the Y chromosome—so the female is XX and the male is XY. As in other organisms, the X and Y chromosomes are not true homologs of one another: *many genes on the X chromosome are not present on the Y*. The X chromosome of *Drosophila* was one of the first to have specific genes assigned to it.

Thomas Morgan identified a **sex-linked** gene that controls eye color in *Drosophila*. The wild-type allele of the gene confers red eyes, whereas a recessive mutant allele confers white eyes. Morgan's experimental crosses demonstrated that this eye color locus is on the X chromosome. If we abbreviate the eye color alleles as *R* (red eyes) and *r* (white eyes), the presence of the alleles on the X chromosome is designated by X^R and X^r.

Morgan crossed a homozygous red-eyed female (X^RX^R) with a white-eyed male. The male is designated X^rY because the Y chromosome does not carry any allele for this gene. (Any gene that is present as a single copy in a diploid organism is called **hemizygous**.) All the sons and daughters from this cross had red eyes, because the red phenotype is dominant over white and all the progeny had inherited a wild-type X chromosome (X^R) from their mother (**Figure 12.20A**). This phenotypic outcome would have occurred even if the *R* gene had been present on an autosome rather than a sex chromosome. In that case, the male would have been homozygous recessive—*rr*.

When Morgan performed the reciprocal cross, in which a white-eyed female (X^rX^r) was mated with a red-eyed male (X^RY), the results were unexpected: *all the sons were white-eyed*

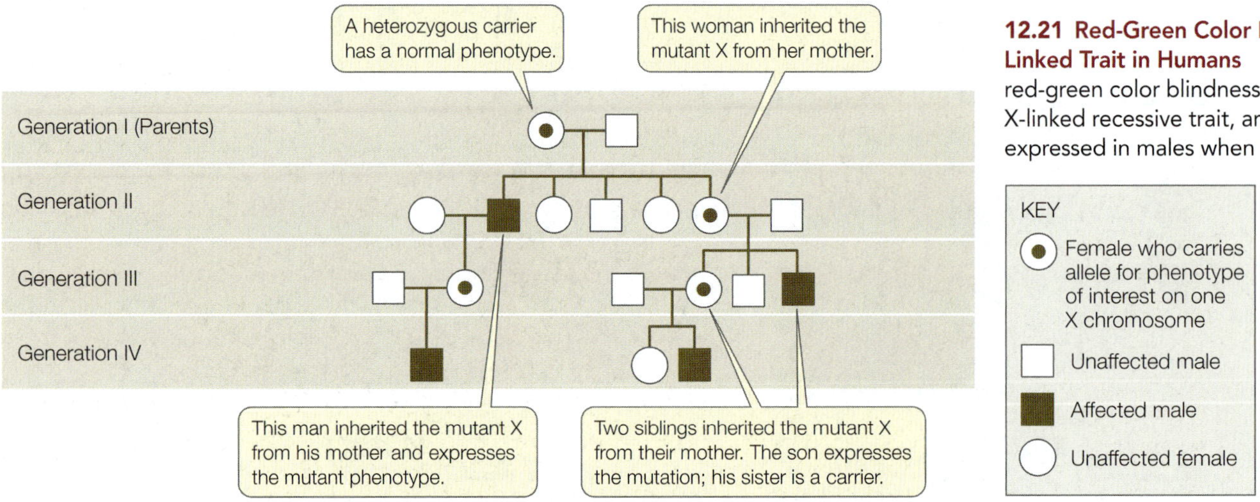
A heterozygous carrier has a normal phenotype.
This woman inherited the mutant X from her mother.
Generation I (Parents)
Generation II
Generation III
Generation IV
This man inherited the mutant X from his mother and expresses the mutant phenotype.
Two siblings inherited the mutant X from their mother. The son expresses the mutation; his sister is a carrier.
KEY
Female who carries allele for phenotype of interest on one X chromosome
Unaffected male
Affected male
Unaffected female

12.21 Red-Green Color Blindness Is a Sex-Linked Trait in Humans The mutant allele for red-green color blindness is expressed as an X-linked recessive trait, and therefore is always expressed in males when they carry that allele.

and all the daughters were red-eyed (**Figure 12.20B**). The sons from the reciprocal cross inherited their only X chromosome from their white-eyed mother and were therefore hemizygous for the white allele. The daughters, however, got an X chromosome bearing the white allele from their mother and an X chromosome bearing the red allele from their father; therefore they were red-eyed heterozygotes. When these heterozygous females were mated with red-eyed males, half their sons had white eyes but all their daughters had red eyes. Together, these results showed that eye color was carried on the X chromosome and not on the Y.

These and other experiments led to the term **sex-linked inheritance**: inheritance of a gene that is carried on a sex chromosome. (This term is somewhat misleading because "sex-linked" inheritance is not really linked to the sex of an organism—after all, both males and females carry X chromosomes.) In mammals, the X chromosome is larger and carries more genes than the Y. For this reason, most examples of sex-linked inheritance involve genes that are carried on the X chromosome.

Many sexually reproducing species, including humans, have sex chromosomes. As in most fruit flies, human males are XY, females are XX, and relatively few of the genes that are present on the X chromosome are present on the Y. Pedigree analyses of X-linked recessive phenotypes like the one in **Figure 12.21** reveal the following patterns (compare with the pedigrees of non–X-linked phenotypes in Figure 12.9):

- The phenotype appears much more often in males than in females, because only one copy of the rare allele is needed for its expression in males, whereas two copies must be present in females.
- A male with the mutation can pass it on only to his daughters; all his sons get his Y chromosome.
- Daughters who receive one X-linked mutation are heterozygous carriers. They are phenotypically normal, but they can pass the mutant allele to their sons or daughters. On average, half their children will inherit the mutant allele since half of their X chromosomes carry the normal allele.
- The mutant phenotype can skip a generation if the mutation passes from a male to his daughter (who will be phenotypically normal) and then to her son.

RECAP 12.4

Simple Mendelian ratios are not observed when genes are linked on the same chromosome. Linkage results in atypical frequencies of phenotypes in the offspring from a test cross. Sex linkage in humans refers to genes on one sex chromosome (usually the X) that have no counterpart on the other sex chromosome.

- What is the concept of linkage, and what are its implications for the results of genetic crosses? **See p. 247 and Figures 12.17, 12.19**
- How does a sex-linked gene behave differently in genetic crosses from a gene on an autosome? **See pp. 251–252 and Figure 12.20**

The genes we've discussed so far in this chapter are all in the cell nucleus. But other organelles, including mitochondria and plastids, also carry genes. What are these genes, and how are they inherited?

12.5 What Are the Effects of Genes Outside the Nucleus?

The nucleus is not the only organelle in a eukaryotic cell that carries genetic material. As described in Section 5.5, mitochondria and plastids contain small numbers of genes, which are remnants of the entire genomes of endosymbiotic prokaryotes that eventually gave rise to these organelles. For example, in humans there are about 21,000 genes coding for proteins in the nuclear genome and 37 in the mitochondrial genome. Plastid genomes are about five times larger than those of mitochondria.

The inheritance of organelle genes differs from that of nuclear genes for several reasons:

- In most organisms, mitochondria and plastids are inherited only from the mother. As you will learn in Chapter 43, eggs contain abundant cytoplasm and organelles, but the only part of the sperm that survives to take part in the union of haploid gametes is the nucleus. So you have inherited your mother's mitochondria (with their genes), but not your father's.

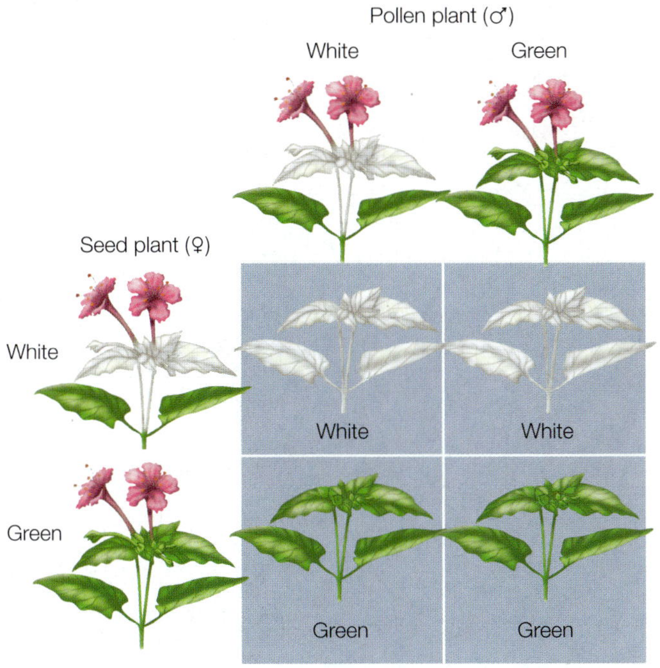
Pollen plant (♂)
White
Green
Seed plant (♀)
White
White
White
Green
Green
Green

12.22 Cytoplasmic Inheritance In four o'clock plants, leaf color is inherited through the female plant only. The white leaf color is caused by a chloroplast mutation that occurs during the life of the parent plant; the leaves that form before the mutation occurs are green. The mutation is passed on to the germ cells, and the offspring that inherit the mutation are entirely white.

- There may be hundreds of mitochondria or plastids in a cell. So a cell is not diploid for organelle genes.
- Organelle genes tend to mutate at much faster rates than nuclear genes, so organelle genes often have multiple alleles.

Several of the genes carried by cytoplasmic organelles are important for organelle assembly and function, and mutations of these genes can have profound effects on the organism. The phenotypes resulting from such mutations reflect the organelles' roles. For example, in plants and some photosynthetic protists, certain plastid gene mutations affect the proteins that assemble chlorophyll molecules into photosystems. These mutations result in a phenotype that is essentially white instead of green. The inheritance of this phenotype follows a non-Mendelian, maternal pattern (**Figure 12.22**). Mitochondrial gene mutations that affect the respiratory chain result in less ATP production. In animals, these mutations have particularly noticeable effects in tissues with high energy requirements, such as the nervous system, muscles, and kidneys.

RECAP 12.5

Genes in the genomes of organelles, specifically plastids and mitochondria, do not behave in a Mendelian fashion.

- Why are genes carried in organelles usually inherited only from the mother? See p. 252

Mendel and those who followed him focused on eukaryotes, with diploid organisms and haploid gametes. A half-century after the rediscovery of Mendel's work, a process that allows genetic recombination was discovered in prokaryotes as well. We will now turn to that process.

12.6 How Do Prokaryotes Transmit Genes?

As described in Chapter 5, prokaryotic cells lack nuclei; they contain their genetic material mostly as single chromosomes in central regions of their cells. Prokaryotes reproduce asexually by binary fission, a process that gives rise to progeny that are virtually identical genetically. That is, the offspring of cell reproduction in prokaryotes constitute a clone (see Chapter 11). You might expect, therefore, that there is no way for individuals of these organisms to exchange genes, as in sexual reproduction.

How then do prokaryotes evolve? Mutations occur in prokaryotes just as they do in eukaryotes, and the resulting new alleles increase genetic diversity. In addition, it turns out that prokaryotes do have a sexual process for transferring genes between cells. Along with mutation, this process provides for genetic diversity among prokaryotes.

Bacteria exchange genes by conjugation

To illustrate the kind of experiment that led to the discovery of bacterial DNA transfer, let's consider two strains of the bacterium *E. coli* with different alleles for each of six genes. One stain carries the dominant (wild-type) alleles for three of the genes and the recessive (mutant) alleles for the other genes. This situation is reversed in the other strain. Simply put, the two strains have the following genotypes (remember that bacteria are haploid):

ABCdef and *abcDEF*

where capital letters indicate wild-type alleles and lowercase letters indicate mutant alleles.

When the two strains are grown together in the laboratory, most of the cells produce clones. That is, almost all of the cells that grow have the original genotypes. However, out of millions of bacteria, a few occur that have the genotype

ABCDEF

How could these completely wild-type bacteria arise? One possibility is mutation: in the *abcDEF* bacteria, the *a* allele could have mutated to *A*, the *b* allele to *B*, and the *c* allele to *C*. The problem with this explanation is that a mutation at any particular point in an organism's DNA sequence is a very rare event. The probability of all three events occurring in the same cell is extremely low—much lower than the actual rate of appearance of cells with the *ABCDEF* genotype. So the mutant cells must have acquired wild-type genes some other way—and this turns out to be the transfer of DNA between cells.

Electron microscopy shows that genetic transfers between bacteria can happen via physical contact between the cells (**Figure 12.23A**). Contact is initiated by a thin projection called a **sex pilus** (plural *pili*), which extends from one cell (the donor), attaches to another (the recipient), and draws the two cells together. Genetic material can then pass from the donor cell to the recipient through a thin cytoplasmic bridge called a conjugation tube. There is no reciprocal transfer of DNA from the recipient to the donor. This process is referred to as **bacterial conjugation**.

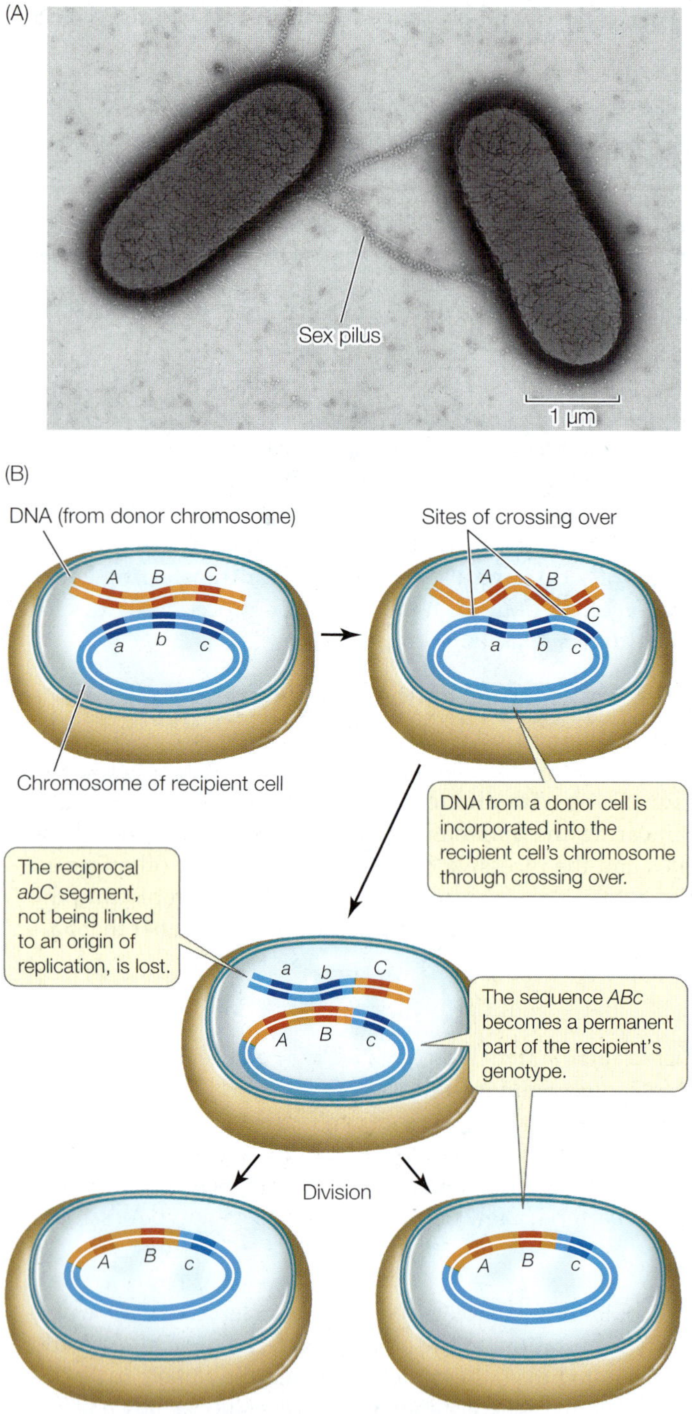

(A)
Sex pilus
1 μm
(B)
DNA (from donor chromosome)
A B C
a b c
Chromosome of recipient cell
Sites of crossing over
A B
C
a b c
DNA from a donor cell is incorporated into the recipient cell's chromosome through crossing over.
The reciprocal abC segment, not being linked to an origin of replication, is lost.
a b C
A B c
The sequence ABc becomes a permanent part of the recipient's genotype.
Division
A B c
A B c

12.23 Bacterial Conjugation and Recombination (A) Sex pili draw two bacteria into close contact, so that a cytoplasmic conjugation tube can form. DNA is transferred from one cell to the other via the conjugation tube. (B) DNA from a donor cell can become incorporated into a recipient cell's chromosome through crossing over.

Once the donor DNA is inside the recipient cell, it can recombine with the recipient cell's genome. In much the same way that chromosomes pair up, gene for gene, in prophase I of meiosis, the donor DNA can line up beside its homologous genes in the recipient, and crossing over can occur. Some of the genes from the donor can become integrated into the genome of the recipient, thus changing the recipient's genetic constitution (**Figure 12.23B**). When the recipient cells proliferate, the integrated donor genes are passed on to all progeny cells.

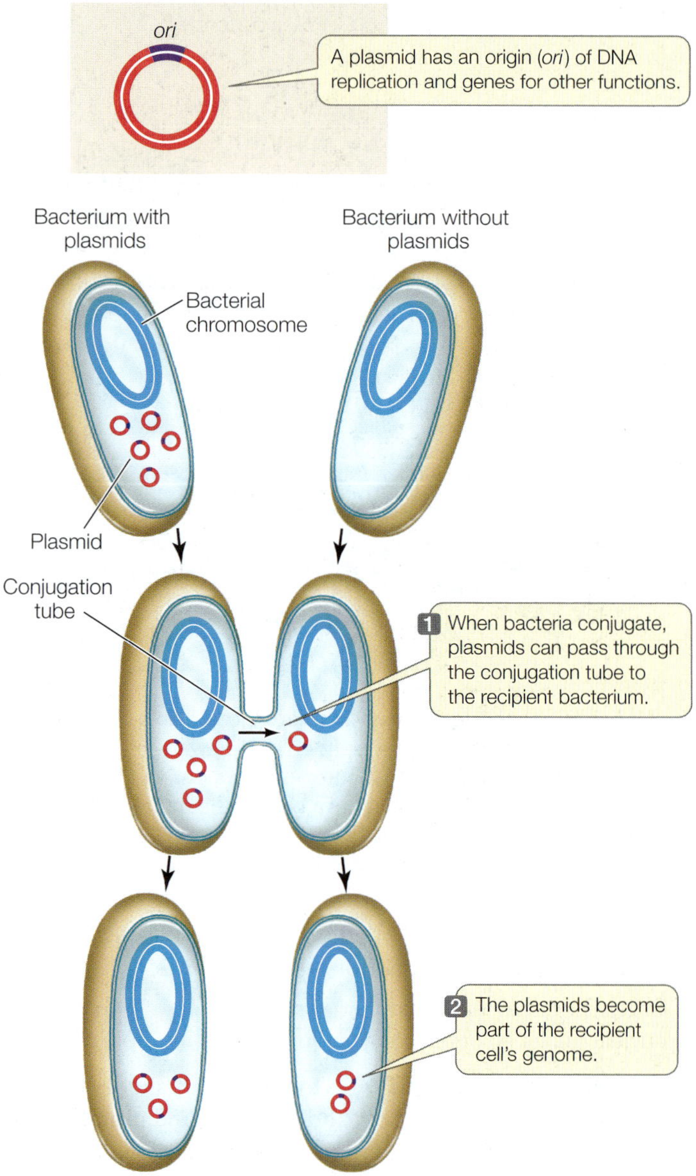

ori
A plasmid has an origin (ori) of DNA replication and genes for other functions.
Bacterium with plasmids
Bacterium without plasmids
Bacterial chromosome
Plasmid
Conjugation tube
1 When bacteria conjugate, plasmids can pass through the conjugation tube to the recipient bacterium.
2 The plasmids become part of the recipient cell's genome.

12.24 Gene Transfer by Plasmids When plasmids enter a cell via conjugation, their genes can be expressed in the recipient cell.

Bacterial conjugation is controlled by plasmids

In addition to their main chromosome, many bacteria harbor additional smaller, circular DNA molecules called **plasmids**, which replicate independently of the main chromosome. Plasmids typically contain at most a few dozen genes, which may fall into one of several categories:

- *Genes for unusual metabolic capacities*. For example, bacteria carrying plasmids that confer the ability to break down hydrocarbons can be used to clean up oil spills.
- *Genes for antibiotic resistance*. Plasmids carrying such genes are called R factors, and since they can be transferred between bacteria via conjugation, they are a major threat to human health.
- *Genes that confer the ability to make a sex pilus.*

During bacterial conjugation it is usually plasmids that are transferred from one bacterium to another (**Figure 12.24**). A single strand of the donor plasmid is transferred to the recipient, and then synthesis of a complementary DNA strand results in two complete copies of the plasmid, one in the donor and one in the recipient. Plasmids can replicate independently of the main chromosome, but sometimes they become integrated into the chromosome. When this happens, the genes for conjugation are still expressed, and the plasmid DNA can be transferred between cells. But because the plasmid is part of the main chromosome, DNA from the chromosome is transferred along with the plasmid DNA. The amount of chromosomal DNA that can be transferred this way depends on how long the bacterial cells are in contact; for example, it takes about 100 minutes for the entire *E. coli* chromosome to be transferred by conjugation.

RECAP 12.6

Although they are haploid and reproduce asexually, prokaryotes have the ability to transfer genes from one cell to another. These genes are usually carried on small, circular DNA molecules called plasmids, but chromosomal DNA is sometimes transferred as well.

- How were prokaryotic gene transfer and recombination discovered? See p. 253
- What are the differences between recombination after conjugation in prokaryotes and recombination during meiosis in eukaryotes?

How can knowledge of genetics be used to save the Tasmanian devil?

ANSWER

The Tasmanian Devil Genome Project, based in the United States but involving biologists from Australia and other countries, is dedicated to preserving the species by determining the genetic diversity of as many devils as possible. Matings are planned between the most genetically diverse individuals in the wild to maximize heterozygosity. (This is not as easy to do as it is to write!) In addition, a captive "insurance population" from an area where devils are cancer-free is being developed. These animals, too, are being genotyped, in the hope of generating a larger uninfected population.

CHAPTER SUMMARY 12

12.1 What Are the Mendelian Laws of Inheritance?

- Physical features of organisms, or **characters**, can exist in different forms, or **traits**. A heritable trait is one that can be passed from parent to offspring. A **phenotype** is the physical appearance of an organism; a **genotype** is the genetic constitution of the organism.
- The different forms of a **gene** are called **alleles**. Organisms that have two identical alleles for a trait are called **homozygous**; organisms that have two different alleles for a trait are called **heterozygous**. A gene resides at a particular site on a chromosome called a **locus**.
- Mendel's experiments included reciprocal crosses and **monohybrid crosses** between true-breeding pea plants. Analysis of his meticulously tabulated data led Mendel to propose a particulate theory of inheritance stating that discrete units (now called genes) are responsible for the inheritance of specific traits, to which both parents contribute equally.
- Mendel's first law, the **law of segregation**, states that when any individual produces gametes, the two copies of a gene separate, so that each gamete receives only one member of the pair. Thus every individual in the F_1 inherits one copy from each parent. **Review Figures 12.3, 12.4**
- Mendel used a **test cross** to find out whether an individual showing a dominant phenotype was homozygous or heterozygous. **Review Figure 12.5, ACTIVITY 12.1**
- Mendel's use of **dihybrid crosses** to study the inheritance of two characters led to his second law: the **law of independent assortment**. The independent assortment of genes in meiosis leads to nonparental combinations of phenotypes in the offspring of a dihybrid cross. **Review Figures 12.6, 12.7, ANIMATED TUTORIAL 12.1**
- Probability calculations and **pedigrees** help geneticists trace Mendelian inheritance patterns. **Review Figures 12.8, 12.9, ANIMATED TUTORIAL 12.2**

12.2 How Do Alleles Interact?

- New alleles arise by random **mutation**. Many genes have multiple alleles. A wild-type allele gives rise to the predominant form of a trait. When the wild-type allele is present at a locus less than 99 percent of the time, the locus is said to be **polymorphic**. **Review Figure 12.10**
- In **incomplete dominance**, neither of two alleles is dominant. The heterozygous phenotype is intermediate between the homozygous phenotypes. **Review Figure 12.11**
- **Codominance** exists when two alleles at a locus produce two different phenotypes that both appear in heterozygotes.
- An allele that affects more than one trait is said to be **pleiotropic**.

12.3 How Do Genes Interact?

- In **epistasis**, one gene affects the expression of another. **Review Figure 12.13**
- Environmental conditions can affect the expression of a genotype.
- **Penetrance** is the proportion of individuals in a group with a given genotype that show the expected phenotype. **Expressivity** is the degree to which a genotype is expressed in an individual.
- Variations in phenotypes can be **qualitative** (discrete) or **quantitative** (graduated, continuous). Most quantitative traits result from the effects of several genes and the environment. Genes that together determine quantitative characters are called **quantitative trait loci**.

continued

12.4 What Is the Relationship between Genes and Chromosomes?

See ANIMATED TUTORIAL 12.3

- Each chromosome carries many genes. Genes on the same chromosome are referred to as a linkage group.
- Genes on the same chromosome can recombine by crossing over. The resulting recombinant chromosomes have new combinations of alleles. Review Figures 12.18, 12.19
- **Sex chromosomes** carry genes that determine whether the organism will produce male or female gametes. All other chromosomes are called **autosomes**. The specific functions of sex chromosomes differ among different groups of organisms.
- **Primary sex determination** in mammals is usually a function of the presence or absence of the *SRY* gene. **Secondary sex characteristics** are the outward manifestations of maleness and femaleness.
- In fruit flies and mammals, the X chromosome carries many genes, but the Y chromosome has only a few. Males have only one allele (are **hemizygous**) for X-linked genes, so recessive **sex-linked** mutations are expressed phenotypically more often in males than in females. Females may be unaffected carriers of such alleles. Review Figure 12.21

12.5 What Are the Effects of Genes Outside the Nucleus?

- Cytoplasmic organelles such as plastids and mitochondria contain small numbers of genes. In many organisms, cytoplasmic genes are inherited only from the mother because the male gamete contributes only its nucleus (i.e., no cytoplasm) to the zygote at fertilization. Review Figure 12.22

12.6 How Do Prokaryotes Transmit Genes?

- Prokaryotes reproduce primarily asexually but can exchange genes in a sexual process called **conjugation**. Review Figure 12.23
- **Plasmids** are small, extra chromosomes in bacteria that carry genes involved in important metabolic processes and that can be transmitted from one cell to another. Review Figure 12.24

See ACTIVITIES 12.2, 12.3 for a concept review of this chapter

Go to the Interactive Summary to review key figures, Animated Tutorials, and Activities
Life10e.com/is12

CHAPTER**REVIEW**

REMEMBERING

1. In a simple Mendelian monohybrid cross, true-breeding tall and short plants are crossed, and the F_1 plants, which are all tall, are allowed to self-pollinate. What fraction of the F_2 generation are both tall and heterozygous?
   a. $1/8$
   b. $1/4$
   c. $1/3$
   d. $2/3$
   e. $1/2$

2. The phenotype of an individual
   a. depends at least in part on the genotype.
   b. is either homozygous or heterozygous.
   c. determines the individual's genotype.
   d. is the genetic constitution of the organism.
   e. is either monohybrid or dihybrid.

3. Which statement about an individual that is homozygous for an allele is *not* true?
   a. Each of its cells possesses two copies of that allele.
   b. Each of its gametes contains one copy of that allele.
   c. It is true-breeding with respect to that allele.
   d. Its parents were necessarily homozygous for that allele.
   e. It can pass that allele to its offspring.

4. Which statement about a monohybrid test cross is *not* true?
   a. It tests whether an individual of unknown genotype is homozygous or heterozygous.
   b. The test individual is crossed with a homozygous recessive individual.
   c. If the test individual is heterozygous, the progeny will have a 1:1 phenotypic ratio.
   d. If the test individual is homozygous, the progeny will have a 3:1 phenotypic ratio.
   e. Test cross results are consistent with Mendel's model of inheritance.

5. Linked genes
   a. must be immediately adjacent to one another on a chromosome.
   b. have alleles that assort independently of one another.
   c. never show crossing over.
   d. are on the same chromosome.
   e. always have multiple alleles.

6. The genetic sex of a human is determined by
   a. ploidy, with the male being haploid.
   b. the Y chromosome.
   c. X and Y chromosomes, the male being XX.
   d. the number of X chromosomes, the male being XO.
   e. Z and W chromosomes, the male being ZZ.

7. In epistasis
   a. the phenotype does not change from generation to generation.
   b. one gene affects the expression of another.
   c. a portion of a chromosome is deleted.
   d. a portion of a chromosome is inverted.
   e. the behavior of two genes is entirely independent.

UNDERSTANDING & APPLYING

8. The ABO blood groups in humans are determined by a multiple-allele system in which I^A and I^B are codominant and are both dominant to I^O. A newborn infant is type A. The mother is type O. Possible phenotypes of the father are
   a. A, B, or AB.
   b. A, B, or O.
   c. O only.
   d. A or AB.
   e. A or O.

9. In humans, spotted teeth are caused by a dominant sex-linked gene. A man with spotted teeth whose father had normal teeth marries a woman with normal teeth. Therefore
   a. all of their daughters will have normal teeth.
   b. all of their daughters will have spotted teeth.
   c. all of their children will have spotted teeth.
   d. half of their sons will have spotted teeth.
   e. all of their sons will have spotted teeth.

10. In guinea pigs, black body color (*B*) is completely dominant over albino (*b*). For the crosses below, give the genotypes of the parents:

Parental phenotypes	Black offspring	Albino offspring	Parental genotypes?
Black × albino	12	0	
Albino × albino	0	12	
Black × albino	5	7	
Black × black	9	3	

11. In the genetic cross *AaBbCcDdEE* × *AaBBCcDdEe*, what fraction of the offspring will be heterozygous for all of these genes (*AaBbCcDdEe*)? Assume all genes are unlinked.

12. The pedigree below shows the inheritance of a rare mutant phenotype in humans, congenital cataracts (black symbols).

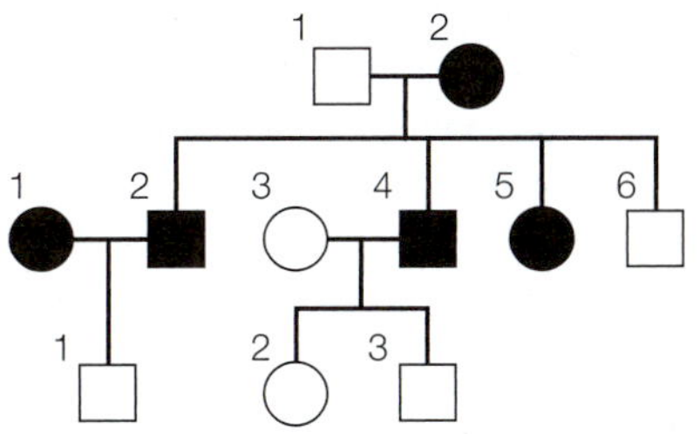
1
2
1
2
3
4
5
6
1
2
3

   a. Are cataracts inherited as an autosomal dominant trait? Autosomal recessive? X-linked dominant? X-linked recessive?
   b. Person #5 in the second generation marries a man who does not have cataracts. Two of their four children, a boy and a girl, develop cataracts. What is the chance that their next child will be a girl with cataracts?

13. In cats, black coat (*B*) is codominant with yellow (*b*). The coat color gene is on the X chromosome. Calico cats, which have coats with black and yellow patches, are heterozygous for the coat color alleles.
   a. Why are most calico cats females?
   b. A calico female, Pickle, had a litter with one yellow male, two black males, two yellow females, and three calico females. What were the genotype and phenotype of the father?

14. In *Drosophila,* three autosomal genes have alleles as follows:

   Gray body color (*G*) is dominant over black (*g*).
   Normal wings (*A*) is dominant over vestigial (*a*).
   Red eye (*R*) is dominant over sepia (*r*).

   Two crosses were performed, with the following results:

   Cross I: Parents: heterozygous red, normal × sepia, vestigial
   Offspring: 131 red, normal
   120 sepia, vestigial
   122 red, vestigial
   127 sepia, normal

   Cross II: Parents: heterozygous gray, normal × black, vestigial
   Offspring: 236 gray, normal
   253 black, vestigial
   50 gray, vestigial
   61 black, normal

   Are any of the three genes linked on the same chromosome? If so, what is the distance between the linked genes (in map units)?

15. In a particular plant species, two alleles control flower color, which can be yellow, blue, or white. Crosses of these plants produce the following offspring:

Parental phenotypes	Offspring phenotypes (ratio)
Yellow × yellow	All yellow
Blue × yellow	Blue or yellow (1:1)
Blue × white	Blue or white (1:1)
White × white	All white

   What will be the phenotype, and ratio, of the offspring of a cross of blue × blue?

16. In *Drosophila* the recessive allele *p*, when homozygous, determines pink eyes. *Pp* or *PP* results in wild-type eye color. Another gene on a different chromosome has a recessive allele, *sw*, that produces short wings when homozygous. Consider a cross between females of genotype *PPSwSw* and males of genotype *ppswsw*. Describe the phenotypes and genotypes of the F_1 generation and of the F_2 generation, produced by allowing the F_1 progeny to mate with one another.

17. On the same chromosome of *Drosophila* that carries the *p* (pink eyes) locus, there is another locus that affects the wings. Homozygous recessives, *byby*, have blistery wings, while the dominant allele *By* produces wild-type wings. The *P* and *By* loci are very close together on the chromosome; that is, the two loci are tightly linked. In answering Questions 17*a* and 17*b*, assume that no crossing over occurs, and that the F_2 generation is produced by interbreeding the F_1 progeny.
   a. For the cross *PPByBy* × *ppbyby*, give the phenotypes and genotypes of the F_1 and F_2 generations.
   b. For the cross *PPbyby* × *ppByBy*, give the phenotypes and genotypes of the F_1 and F_2 generations.
   c. For the cross of Question 17*b*, what further phenotype(s) would appear in the F_2 generation if crossing over occurred?
   d. Draw a nucleus undergoing meiosis at the stage in which the crossing over (Question 17*c*) occurred. In which generation (P, F_1, or F_2) did this crossing over take place?

18. In chickens, when the dominant alleles of the genes for rose comb (*R*) and pea comb (*A*) are present together (*R_A_*), the result is a bird with a walnut comb. Chickens that are homozygous recessive for both genes produce a single comb. A rose-combed bird mated with a walnut-combed bird and produced offspring in the proportion: $\frac{3}{8}$ walnut : $\frac{3}{8}$ rose : $\frac{1}{8}$ pea : $\frac{1}{8}$ single.

    What were the genotypes of the parents?

Rose comb

Pea comb

Walnut comb

Single comb

19. In *Drosophila*, white (*w*), eosin (w^e), and wild-type red (w^+) are multiple alleles at a single locus for eye color. This locus is on the X chromosome. A female that has eosin (pale orange) eyes is crossed with a male that has wild-type eyes. All the female progeny are red-eyed; half the male progeny have eosin eyes, and half have white eyes. Assume the female has two X chromosomes and the male has one X and one Y.
   a. What is the order of dominance of these alleles?
   b. What are the genotypes of the parents and progeny?

20. In humans, red-green color blindness is determined by an X-linked recessive allele (*a*), whereas eye color is determined by an autosomal gene, where brown (*B*) is dominant over blue (*b*).
   a. What gametes can be formed with respect to these genes by a heterozygous, brown-eyed, color-blind male?
   b. If a blue-eyed mother with normal vision has a brown-eyed, color-blind son and a blue-eyed, color-blind daughter, what are the genotypes of both parents and children?

21. If the dominant allele *A* is necessary for hearing in humans, and another allele, *B*, located on a different chromosome, results in deafness no matter what other genes are present, what percentage of the offspring of the marriage of *aaBb* × *Aabb* will be deaf?

22. The disease Leber's optic neuropathy is caused by a mutation in a gene carried on mitochondrial DNA. What would be the phenotype of their first child if a man with this disease married a woman who did not have the disease? What would be the result if the wife had the disease and the husband did not?

ANALYZING & EVALUATING

23. Sometimes scientists get lucky. Consider Mendel's dihybrid cross shown in Figure 12.6. Peas have a haploid number of seven chromosomes, so many of their genes are linked. What would Mendel's results have been if the genes for seed color and seed shape were linked with a map distance of 10 units? Now, consider Morgan's fruit flies (see Figure 12.19). Suppose that the genes for body color and wing shape were not linked? What results would Morgan have obtained?

13 DNA and Its Role in Heredity

CHAPTER OUTLINE

Observation, Insight, and Discovery Louis Pasteur's observations and "prepared mind" led to unique insights and scientific breakthroughs in the late nineteenth century. This pattern of research continues to advance the scientific enterprise, resulting in discoveries such as the drug cisplatin.

OVER A CENTURY AGO, the great French microbiologist Louis Pasteur said "chance favors only the prepared mind." He meant that great discoveries come not just from a flash of insight, but from careful observations made by people whose background has made them ready to interpret their observations in a new way. Such "prepared minds" have sparked the development of many new cancer drugs.

Testicular cancer occurs without warning with rapid cell divisions of germ cells. It often spreads to other organs, such as the lungs and brain. With a lifetime risk of about 1 in 250 males, it typically strikes men in their twenties and is the most common cancer in young men. Despite its potential lethality, testicular cancer is one of the few tumors of adults that is highly curable. The cure is primarily due to a drug called cisplatin.

Dr. Barnett Rosenberg, a scientist at Michigan State University, was curious about how electric fields might affect cells. He put bacteria into a growth medium with platinum electrodes connected to a battery. The result was striking: the bacteria stopped dividing. Thinking he was on the road to a major discovery about electromagnetism and cells, Rosenberg tried the experiment again, this time using copper and zinc electrodes. (You are probably familiar with the Cu/Zn system if you've taken a chemistry course.) This time the bacteria kept dividing, with no adverse effects. Only platinum electrodes inhibited cell division.

In light of the data, Rosenberg revised his hypothesis to propose that something leaked out of the platinum electrodes into the medium, and that this "something" blocked cell division. He confirmed his hypothesis by treating bacteria with the medium in which the platinum electrodes had been inserted; the bacteria did not divide. Realizing that cancer cells have uncontrolled cell division, he duplicated his experiments with tumor cells in a laboratory dish. This led to the isolation and development of cisplatin. The drug was so successful with testicular cancer that it has also been used with some success on other tumors.

An essential event for cell division is the complete and precise duplication of the genetic material, DNA. The two strands of DNA unwind and separate, each strand acting as a template for the building of a new strand. Strand separation is possible because the two strands are held together by weak forces, including hydrogen bonds. Cisplatin forms covalent bonds with nucleotides on opposite strands of the DNA, irreversibly cross-linking the two strands together. As a result, the DNA strands cannot separate for replication or expression. With such severe damage to its DNA, the cell then undergoes programmed cell death.

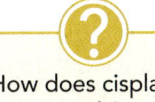

How does cisplatin work?

See answer on p. 278.

INVESTIGATING**LIFE**

13.1 Genetic Transformation Griffith's experiments demonstrated that something in the virulent S strain of pneumococcus could transform nonvirulent R strain bacteria into a lethal form, even when the S strain bacteria had been killed by high temperatures.[a]

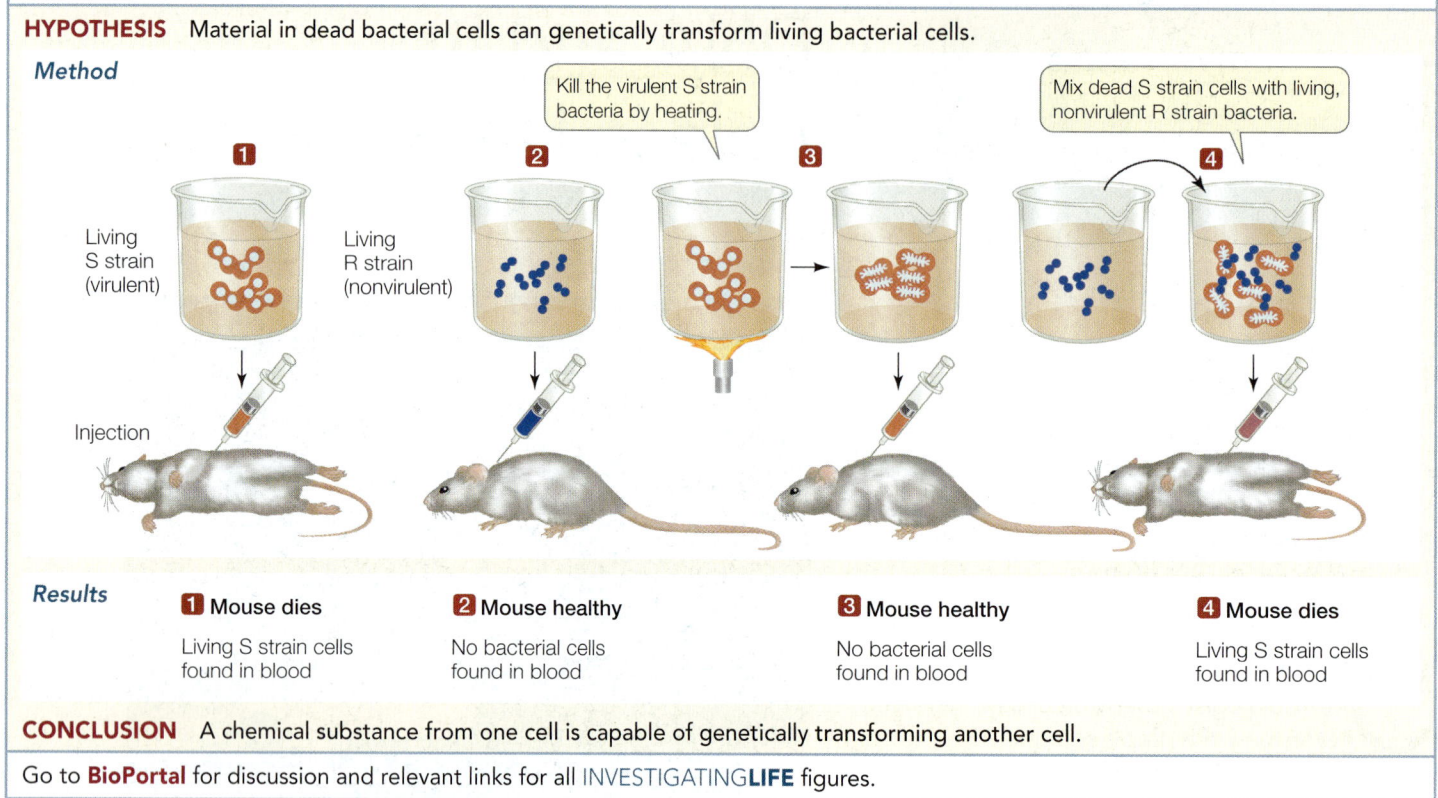

HYPOTHESIS Material in dead bacterial cells can genetically transform living bacterial cells.

Method

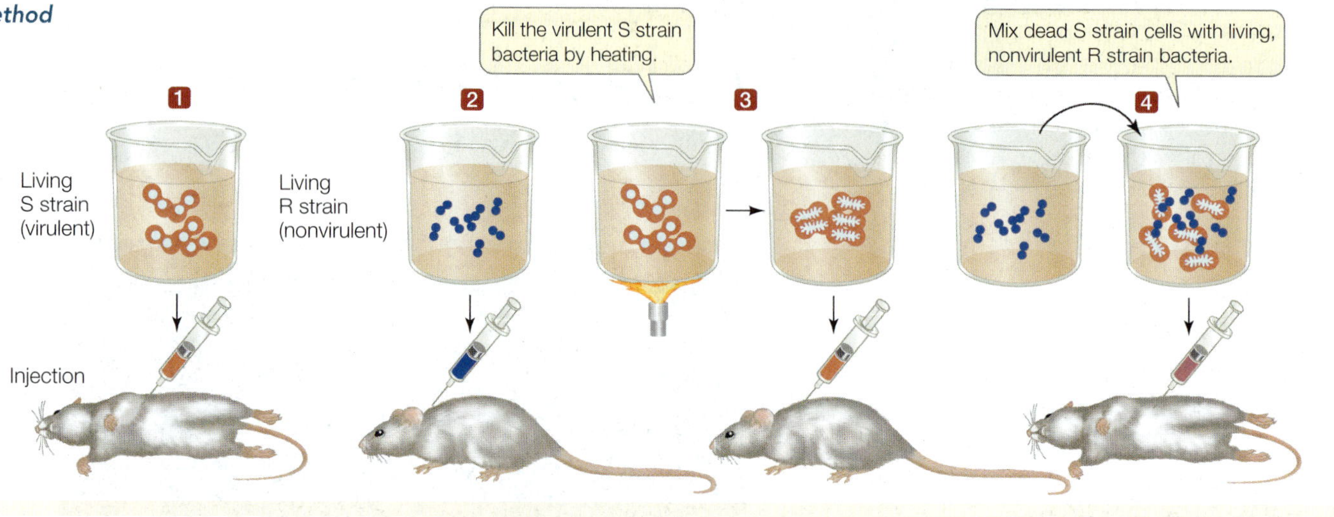

Kill the virulent S strain bacteria by heating.
Mix dead S strain cells with living, nonvirulent R strain bacteria.
1
2
3
4
Living S strain (virulent)
Living R strain (nonvirulent)
Injection

Results

1 Mouse dies — Living S strain cells found in blood

2 Mouse healthy — No bacterial cells found in blood

3 Mouse healthy — No bacterial cells found in blood

4 Mouse dies — Living S strain cells found in blood

CONCLUSION A chemical substance from one cell is capable of genetically transforming another cell.

Go to **BioPortal** for discussion and relevant links for all INVESTIGATING**LIFE** figures.

[a]Griffith, F. 1928. *Journal of Hygiene* 27: 113–159.

13.1 What Is the Evidence that the Gene Is DNA?

By the early twentieth century, geneticists had associated the presence of genes with chromosomes. Research began to focus on exactly which chemical component of chromosomes comprised this genetic material.

By the 1920s, scientists knew that chromosomes were made up of DNA and proteins. At this time a new dye was developed by Robert Feulgen that could bind specifically to DNA and that stained cell nuclei red in direct proportion to the amount of DNA present in the cell. This technique provided circumstantial evidence that DNA was the genetic material:

- *DNA was in the right place.* DNA was confirmed to be an important component of the nucleus and the chromosomes, which were known to carry genes.
- *DNA was present in the right amounts.* The amount of DNA in somatic cells (body cells not specialized for reproduction) was twice that in reproductive cells (eggs or sperm)—as might be expected for diploid and haploid cells, respectively.
- *DNA varied among species.* When cells from different species were stained with the dye and their color intensity measured, each species appeared to have its own specific amount of nuclear DNA.

But circumstantial evidence is *not* a scientific demonstration of cause and effect. After all, proteins are also present in cell nuclei. Science relies on experiments to test hypotheses. The convincing demonstration that DNA is the genetic material came from two sets of experiments, one with bacteria and the other with viruses.

DNA from one type of bacterium genetically transforms another type

In science, research on one specific topic often contributes to another, apparently unrelated area. Such a case of serendipity is seen in the work of Frederick Griffith, an English physician. In the 1920s Griffith was studying the bacterium *Streptococcus pneumoniae*, or pneumococcus, one of the agents that cause pneumonia in humans. He was trying to develop a vaccine against this devastating illness (antibiotics had not yet been discovered). Griffith was working with two strains of pneumococcus:

- Cells of the S strain produced colonies that looked smooth (S). Covered by a polysaccharide capsule, these cells were protected from attack by a host's immune system. When S cells were injected into mice, they reproduced and caused pneumonia (the strain was virulent).
- Cells of the R strain produced colonies that looked rough (R), lacked the protective capsule, and were *not* virulent.

When Griffith inoculated mice with heat-killed S-type pneumococcus cells, the cells did not produce infection. However, when he inoculated other mice with a mixture of living R-type cells and heat-killed S-type cells, to his astonishment, the mice died of pneumonia (**Figure 13.1**). When he examined blood from these mice, he found it full of living bacteria—many of them with characteristics of the virulent S strain! Griffith concluded that in the presence of the dead S-type pneumococcus cells, some of the living R-type cells had been transformed into virulent S cells. These cells were able to grow in the bodies of the mice, causing pneumonia and multiplying in the blood. The fact that these S-type cells reproduced to make more S-type cells showed that the change from R-type to S-type was genetic.

Did this transformation of the bacteria depend on something that happened in the mouse's body? No. The same transformation could be achieved in a test tube by mixing living R cells with heat-killed S cells, or even with a cell-free extract of the heat-killed S cells. (A cell-free extract contains all the contents of ruptured cells, but no intact cells.) These results demonstrated that some substance from the dead S pneumococcus cells could cause a heritable change in the affected R cells.

Oswald Avery and his colleagues at what is now The Rockefeller University identified the substance causing bacterial transformation in two ways:

- *Eliminating other possibilities.* Cell-free extracts containing the transforming substance were treated with enzymes that destroyed candidates for the genetic material, such as proteins, RNA, and DNA. When the treated samples were tested, the ones treated with RNase and protease (which destroy RNA and proteins, respectively) were still able to transform R-type bacteria into the S-type. But the transforming activity was lost in the extract treated with DNase (which destroys DNA) (**Figure 13.2**).
- *Positive experiment.* They isolated virtually pure DNA from a cell-free extract containing the transforming substance. The DNA alone caused bacterial transformation.

We now know that the gene for the enzyme that catalyzes the synthesis of the polysaccharide capsule, which makes the bacterial colony look "smooth," was transferred into the R cells during transformation.

INVESTIGATING**LIFE**

13.2 Genetic Transformation by DNA Experiments by Avery and his colleagues showed that DNA from the virulent S strain of pneumococcus was responsible for the transformation in Griffith's experiments (see Figure 13.1).[a]

HYPOTHESIS The chemical nature of the transforming substance from pneumococcus is DNA.

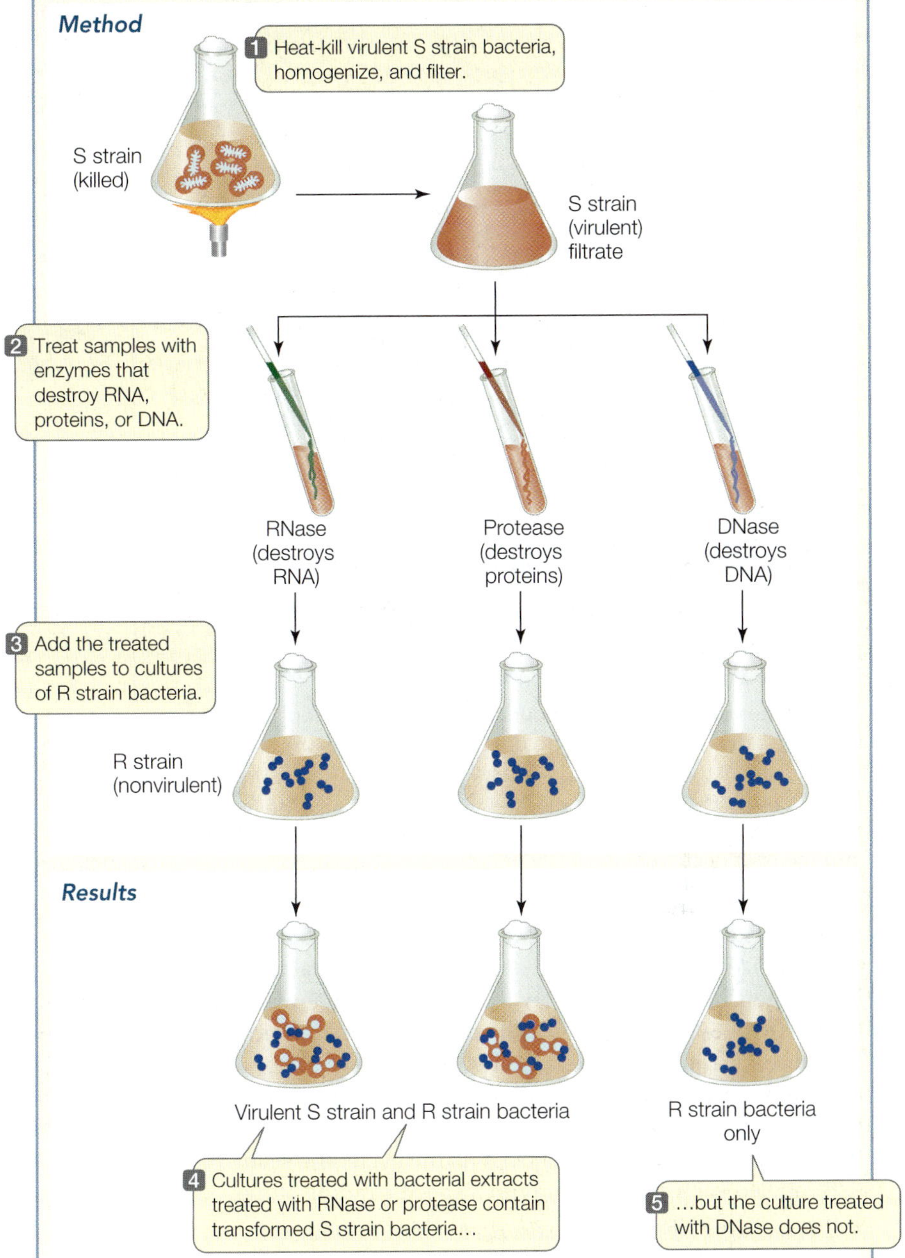

Method
1 Heat-kill virulent S strain bacteria, homogenize, and filter.
S strain (killed)
S strain (virulent) filtrate
2 Treat samples with enzymes that destroy RNA, proteins, or DNA.
RNase (destroys RNA)
Protease (destroys proteins)
DNase (destroys DNA)
3 Add the treated samples to cultures of R strain bacteria.
R strain (nonvirulent)
Results
Virulent S strain and R strain bacteria
R strain bacteria only
4 Cultures treated with bacterial extracts treated with RNase or protease contain transformed S strain bacteria...
5 ...but the culture treated with DNase does not.

CONCLUSION Because only DNase destroyed the transforming substance, the transforming substance is DNA.

Go to **BioPortal** for discussion and relevant links for all INVESTIGATING**LIFE** figures.

[a]Avery, O. T. et al. 1944. *Journal of Experimental Medicine* 79: 137–158.

Viral infection experiments confirmed that DNA is the genetic material

Even with the bacterial transformation experiments, many biologists were still not convinced that DNA is the genetic material. One problem was that DNA, being made up of only four simple nucleotides (see Section 4.1), seemed too uniform a substance to be able to confer all the functions and variety of life. The possibility remained that proteins, with all their chemical and structural diversity, fulfilled that role. Experiments with a virus were designed to distinguish between these alternatives.

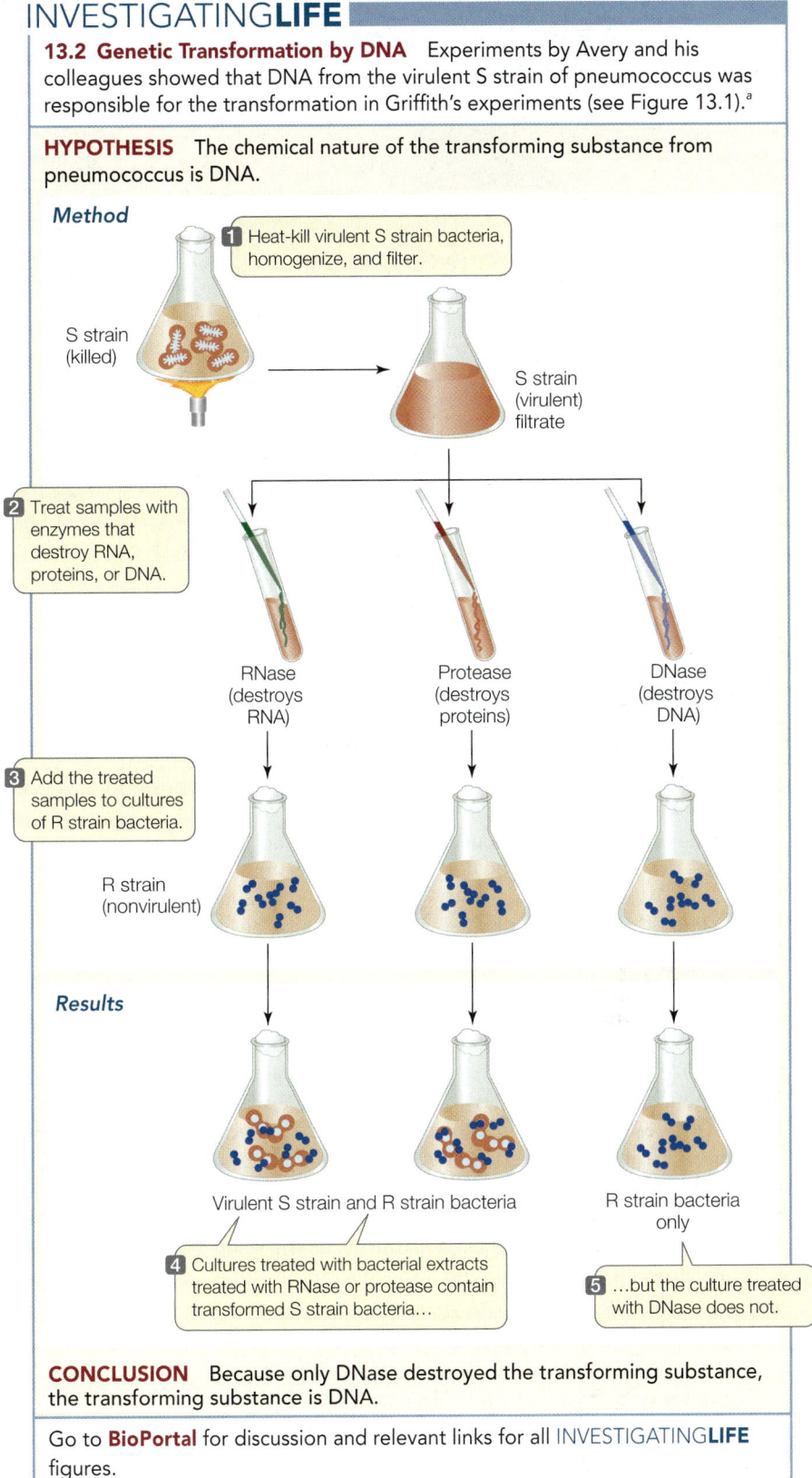

Alfred Hershey and Martha Chase of the Cold Spring Harbor Laboratory in New York studied bacteriophage T2 (phage T2), which infects the bacterium *Escherichia coli.* T2 phage consists of a DNA core packed inside a protein coat (**Figure 13.3**). When it attacks a bacterium, part (but not all) of the virus enters

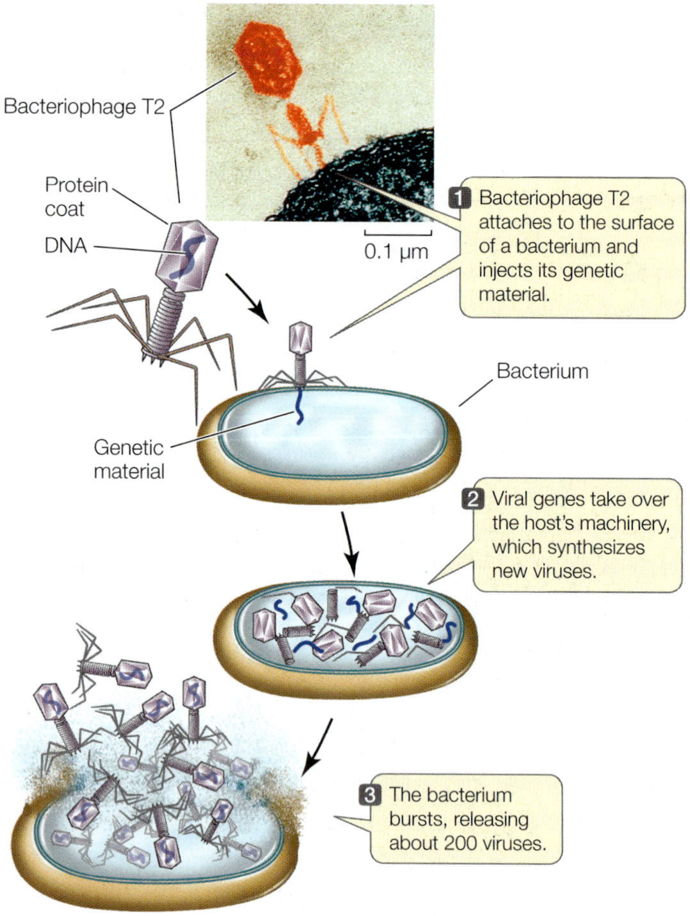

Bacteriophage T2
Protein coat
DNA
0.1 µm
1 Bacteriophage T2 attaches to the surface of a bacterium and injects its genetic material.
Bacterium
Genetic material
2 Viral genes take over the host's machinery, which synthesizes new viruses.
3 The bacterium bursts, releasing about 200 viruses.

13.3 Bacteriophage T2: Reproduction Cycle Bacteriophage T2 is parasitic on *E. coli*, depending on the bacterium to produce new viruses. The external structures of bacteriophage T2 consist entirely of protein, and the DNA is contained within the protein coat. When the virus infects an *E. coli* cell, its genetic material is injected into the host bacterium.

the bacterial cell. About 20 minutes later, the cell bursts, releasing dozens of particles that are virtually identical to the infecting virus particle. Clearly the virus is somehow able to convert the host cell from its own genetic program into a viral replication machine. Hershey and Chase set out to determine what part of the virus—DNA or protein—enters the host cell to bring about this genetic change. To trace the two components of the virus over its life cycle, the scientists labeled each component with a specific radioisotope:

- *Proteins were labeled with radioactive sulfur*. Proteins contain some sulfur (in the amino acids cysteine and methionine), but DNA does not. Sulfur has a radioactive isotope, ^{35}S. Hershey and Chase grew bacteriophage T2 in a bacterial culture in the presence of ^{35}S, so the proteins of the resulting viruses were labeled with (contained) the radioisotope.
- *DNA was labeled with radioactive phosphorus*. DNA contains a lot of phosphorus (in the deoxyribose–phosphate backbone—see Figure 4.4), whereas proteins contain little or none. Phosphorus also has a radioisotope, ^{32}P. The researchers grew another batch of T2 in a bacterial culture in the presence of ^{32}P, thus labeling the viral DNA with ^{32}P.

INVESTIGATING**LIFE**

13.4 The Hershey–Chase Experiment When Hershey and Chase infected bacterial cells with radioactively labeled T2 bacteriophage, only labeled DNA was found in the bacteria.[a] The infected cells were agitated to remove the viral coats from the bacteria and were then centrifuged to pellet the bacteria. The labeled protein remained in the supernatant. This showed that DNA, not protein, is the genetic material.

HYPOTHESIS Either component of a bacteriophage—DNA or protein—might be the hereditary material that enters a bacterial cell to direct the assembly of new viruses.

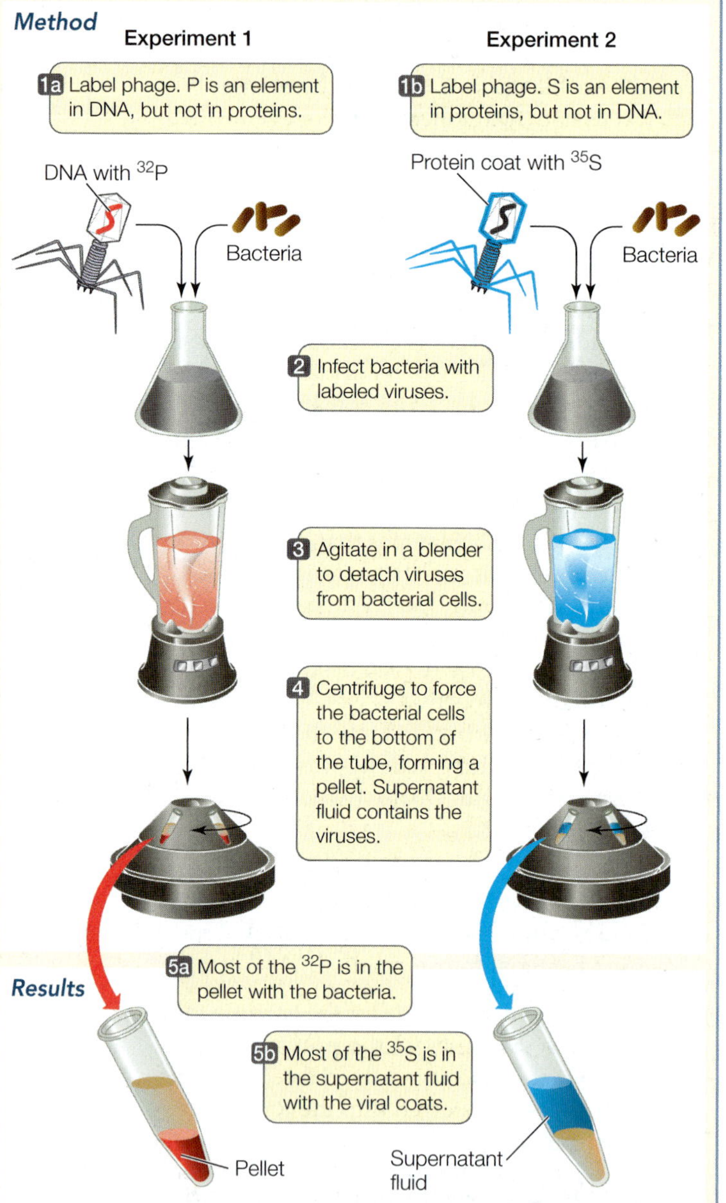

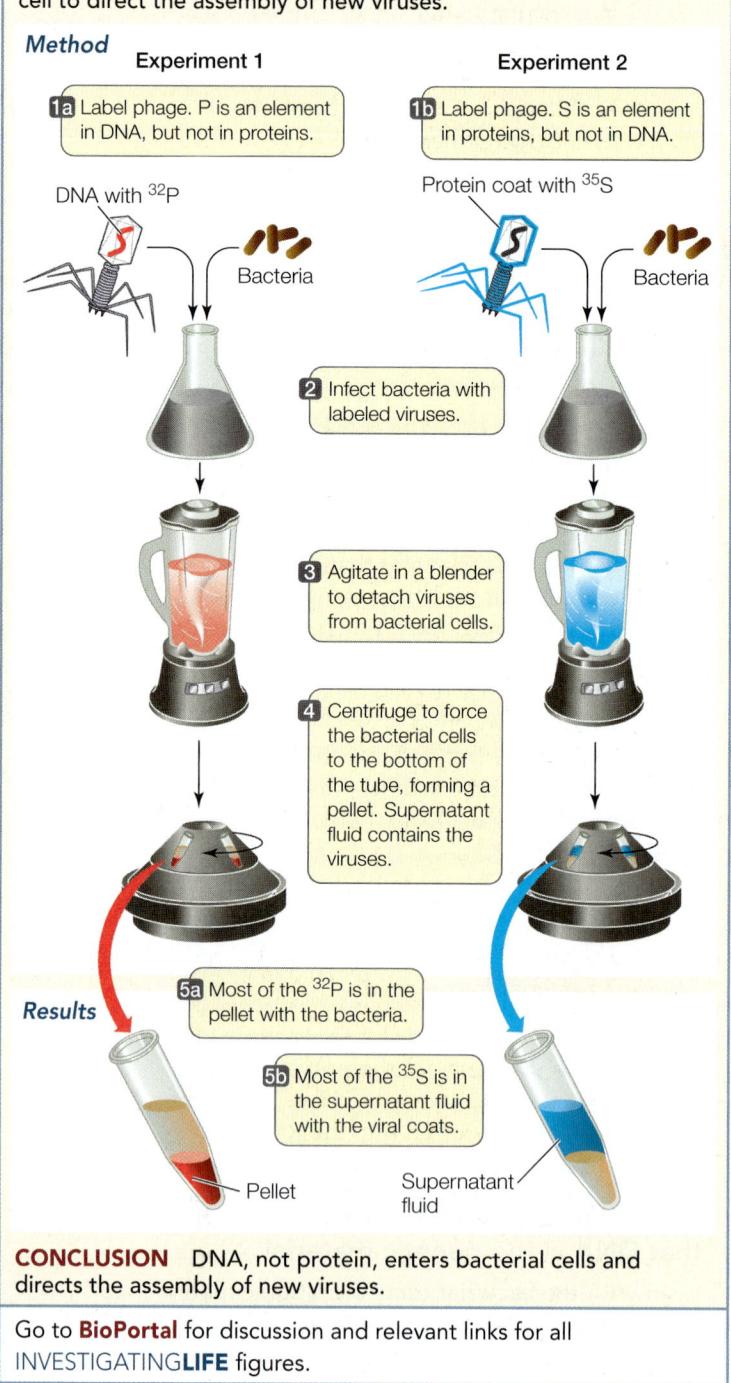

Method
Experiment 1
Experiment 2
1a Label phage. P is an element in DNA, but not in proteins.
1b Label phage. S is an element in proteins, but not in DNA.
DNA with 32P
Protein coat with 35S
Bacteria
Bacteria
2 Infect bacteria with labeled viruses.
3 Agitate in a blender to detach viruses from bacterial cells.
4 Centrifuge to force the bacterial cells to the bottom of the tube, forming a pellet. Supernatant fluid contains the viruses.
Results
5a Most of the 32P is in the pellet with the bacteria.
5b Most of the 35S is in the supernatant fluid with the viral coats.
Pellet
Supernatant fluid

CONCLUSION DNA, not protein, enters bacterial cells and directs the assembly of new viruses.

Go to **BioPortal** for discussion and relevant links for all INVESTIGATING**LIFE** figures.

[a]Hershey, A. D. and M. Chase. 1952. *The Journal of General Physiology* 36: 39–56.

Hershey and Chase used these radioactively labeled viruses in their experiments (**Figure 13.4**). In one experiment, they allowed ^{32}P-labeled bacteriophage to infect bacteria; in the other, the bacteria were infected with ^{35}S-labeled bacteriophage. After a few minutes they agitated each mixture vigorously in a kitchen blender, stripping away the parts of the viruses that had not penetrated the bacteria, without bursting the bacteria. Then they separated the bacteria from the rest of the material (the remains of the viruses) in a centrifuge. The result was that the bacterial cells in the centrifuge pellet contained most of the ^{32}P (and thus the viral DNA), and the supernatant fluid with the viral remains contained most of the ^{35}S (and thus the viral protein). These results indicated that it was the DNA that had been transferred into the bacteria, and that DNA was the molecule responsible for redirecting the genetic program of the bacterial cell.

Go to Animated Tutorial 13.1
The Hershey–Chase Experiment
Life10e.com/at13.1

Eukaryotic cells can also be genetically transformed by DNA

The transformation of eukaryotic cells by DNA is often called **transfection**. This can be demonstrated using a **genetic marker**, a gene whose presence in the recipient cells confers an observable phenotype. When transforming both prokaryotes and eukaryotes, researchers often use an antibiotic resistance or nutritional marker gene, which permits the growth of transformed cells but not of nontransformed cells. A common marker in mammalian transfection experiments is a gene that confers resistance to the antibiotic neomycin (**Figure 13.5**). There are many methods for transfection, including chemical treatments that allow the DNA to be taken up by the cells. Any cell can be transfected, even an egg cell. In this case, a whole new genetically transformed organism can result; such an organism is referred to as transgenic. Transformation in eukaryotes is the final line of evidence for DNA as the genetic material.

RECAP 13.1

Experiments on bacteria and on viruses demonstrated that DNA is the genetic material.

- At the time of Griffith's experiments in the 1920s, what circumstantial evidence suggested to scientists that DNA might be the genetic material? **See p. 260**
- How did the experiments of Avery and his colleagues provide further evidence that DNA was the genetic material? **See p. 261 and Figure 13.2**
- What attributes of bacteriophage T2 were key to the Hershey–Chase experiments demonstrating that DNA, rather than protein, is the genetic material? **See pp. 261–262 and Figure 13.4**

The transformation and viral infection experiments convinced biologists that the genetic material is DNA. It had been known for several decades that chemically, DNA is a polymer of nucleotides. Next, scientists turned to DNA's precise three-dimensional structure.

INVESTIGATING**LIFE**

13.5 Transfection in Eukaryotic Cells A DNA molecule can be treated chemically so that it is taken up from a solution by mammalian cells.[a] The inclusion of a marker gene shows that the cells have been genetically transformed by the DNA.

HYPOTHESIS DNA can transform eukaryotic cells.

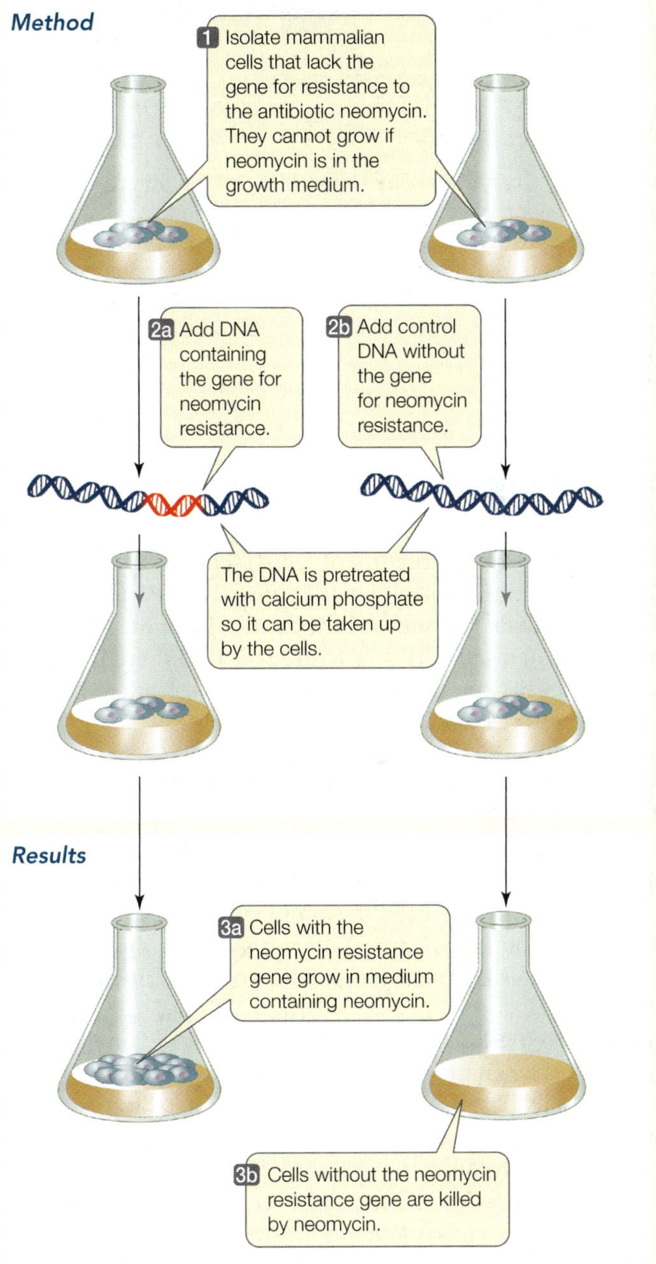
Method
1 Isolate mammalian cells that lack the gene for resistance to the antibiotic neomycin. They cannot grow if neomycin is in the growth medium.
2a Add DNA containing the gene for neomycin resistance.
2b Add control DNA without the gene for neomycin resistance.
The DNA is pretreated with calcium phosphate so it can be taken up by the cells.
Results
3a Cells with the neomycin resistance gene grow in medium containing neomycin.
3b Cells without the neomycin resistance gene are killed by neomycin.

CONCLUSION The cells were transformed by DNA.

Go to **BioPortal** for discussion and relevant links for all INVESTIGATING**LIFE** figures.

[a]Bacchetti, S. and F. L. Graham. 1977. *Proceedings of the National Academy of Sciences USA* 74: 1590–1594.

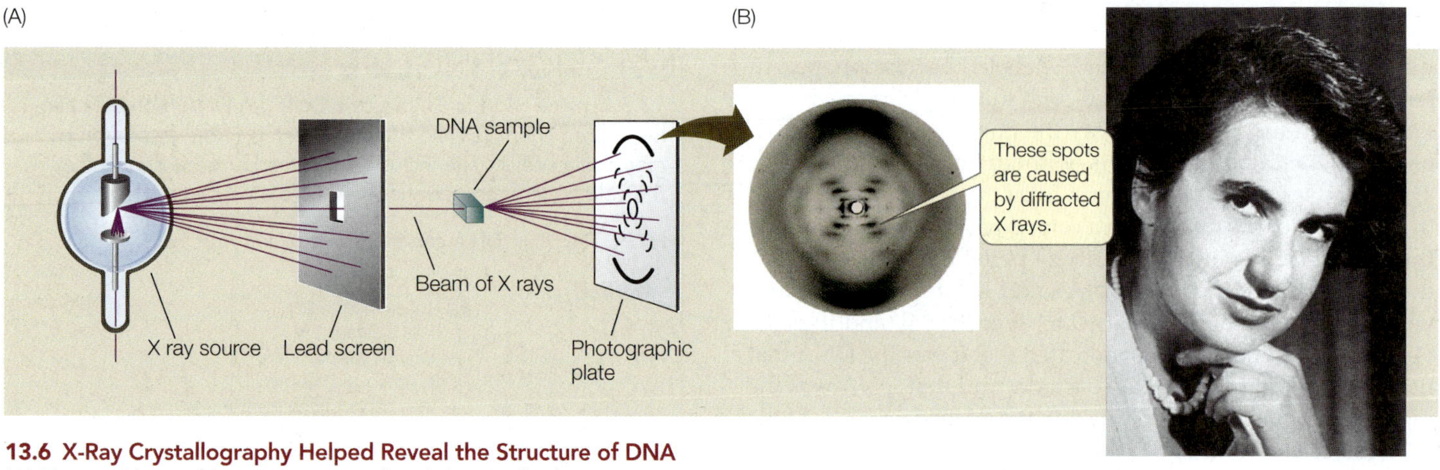
(A)
(B)
DNA sample
Beam of X rays
X ray source
Lead screen
Photographic plate
These spots are caused by diffracted X rays.

13.6 X-Ray Crystallography Helped Reveal the Structure of DNA (A) The positions of atoms in a crystallized chemical substance can be inferred by the pattern of diffraction of X rays passed through it. The pattern of DNA is both highly regular and repetitive. (B) Rosalind Franklin's crystallography and her "photograph 51" (shown) helped scientists visualize the helical structure of the DNA molecule.

13.2 What Is the Structure of DNA?

In determining the structure of DNA, scientists hoped to find the answers to two questions: (1) how is DNA replicated between cell divisions, and (2) how does it direct the synthesis of specific proteins? DNA's structure was deciphered only after many types of experimental evidence were considered together in a theoretical framework.

Watson and Crick used modeling to deduce the structure of DNA

Once pure DNA fibers could be isolated, biophysicists and biochemists examined the DNA for hints about its structure. The evidence eventually used to solve DNA's structure included crucial data obtained using X-ray crystallography and a thorough characterization of the chemical composition of DNA.

PHYSICAL EVIDENCE FROM X-RAY DIFFRACTION Some chemical substances, when they are isolated and purified, can be made to form crystals. The positions of atoms in a crystallized substance can be inferred from the diffraction pattern of X rays passing through the substance. In the early 1950s the New Zealand-born biophysicist Maurice Wilkins discovered a way to make highly ordered fibers of DNA that were suitable for X-ray diffraction studies. His samples were analyzed by X-ray crystallographer Rosalind Franklin of Kings College, London (**Figure 13.6**). Franklin's data suggested that DNA was a double (two-stranded) helix with ten nucleotides in each full turn, and that each full turn was 3.4 nanometers (nm) in length. The molecule's diameter of 2 nm suggested that the sugar–phosphate backbone of each DNA strand must be on the outside of the helix.

CHEMICAL EVIDENCE FROM BASE COMPOSITION Biochemists knew that DNA was a polymer of nucleotides. Each nucleotide consists of a molecule of the sugar deoxyribose, a phosphate group, and a nitrogen-containing base (see Figure 4.1). The only differences among the four nucleotides of DNA are their nitrogenous bases: the purines **adenine** (**A**) and **guanine** (**G**), and the pyrimidines **cytosine** (**C**) and **thymine** (**T**).

In the early 1950s, biochemist Erwin Chargaff and his colleagues at Columbia University reported that DNA from many different species—and from different sources within a single organism—exhibits certain regularities. This led to the following rule: In any DNA sample, the amount of adenine equals the amount of thymine (A = T), and the amount of guanine equals the amount of cytosine (G = C). As a result, the total abundance of purines (A + G) equals the total abundance of pyrimidines (T + C):

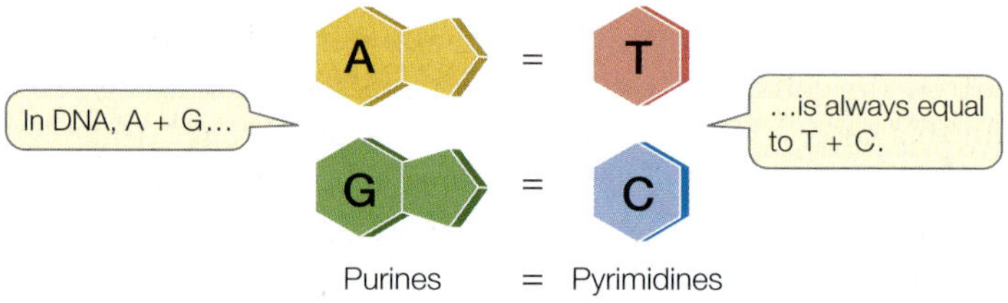
In DNA, A + G...
A
=
T
G
=
C
...is always equal to T + C.
Purines
=
Pyrimidines

This provided an important clue about the way the bases are arranged in a DNA double helix. Chargaff and colleagues found that this rule held for every organism they examined. However, the relative abundances of A + T versus G + C vary slightly among organisms; in humans the percentages are A= T = 30 percent and G = C = 20 percent.

WATSON AND CRICK'S MODEL If you have taken chemistry courses, you may be familiar with model building, where balls (atoms) and sticks (bonds) are used to put together molecules based on known physical and chemical properties and bond angles. A physicist, Francis Crick, and a geneticist, James D. Watson (**Figure 13.7A**), who were then at the Cavendish Laboratory of Cambridge University, used model building to solve the structure of DNA. They used the physical and chemical evidence we just described:

- To be consistent with Franklin's X-ray diffraction images, Watson and Crick's model had the nucleotide bases on the interior of the two strands, with a sugar–phosphate

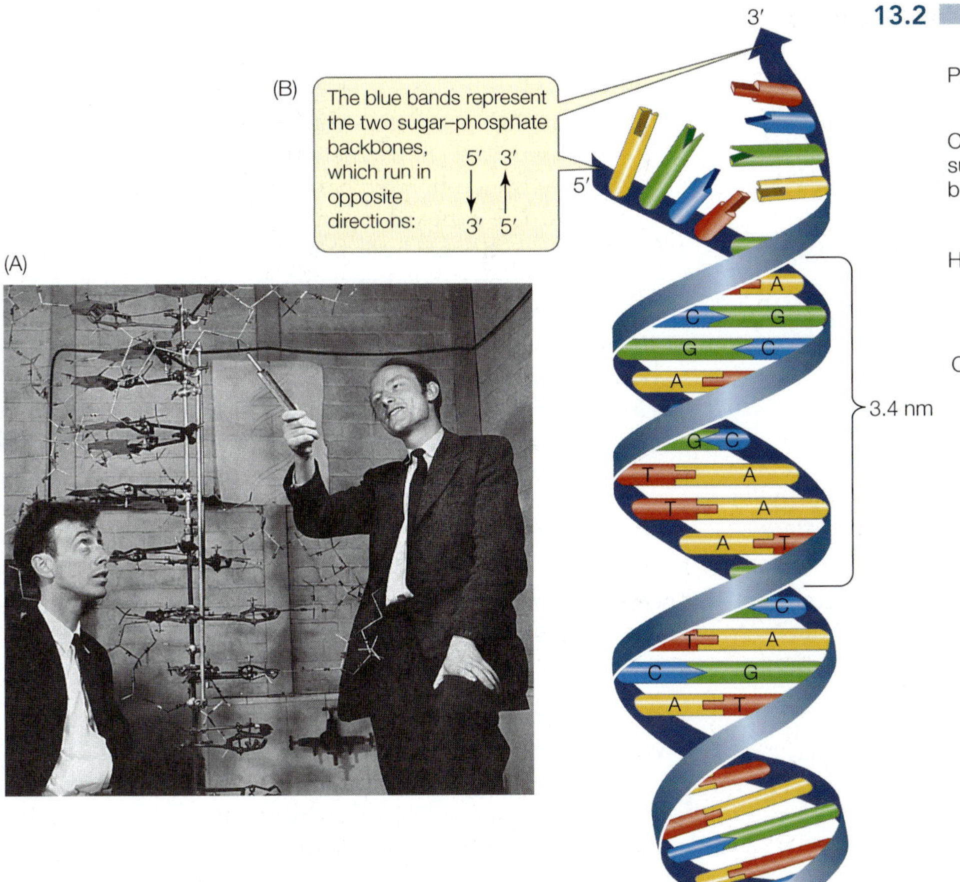
(A)
(B)
The blue bands represent the two sugar–phosphate backbones, which run in opposite directions:
5′ 3′
3′ 5′
3′
5′
3.4 nm
A
C G
G C
A
G C
T A
T A
A T
C
T A
C G
A T

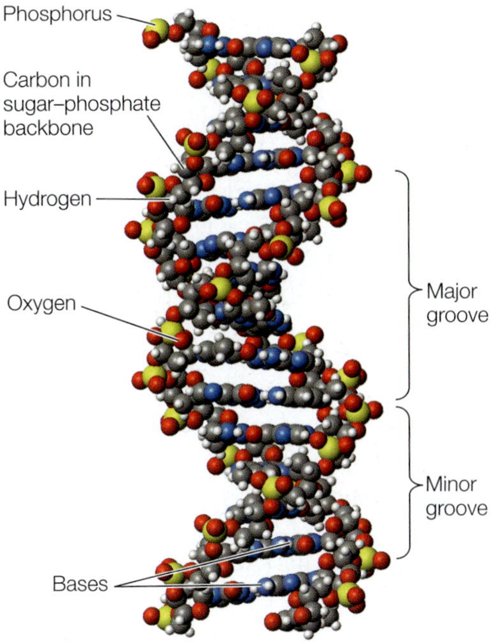
Phosphorus
Carbon in sugar–phosphate backbone
Hydrogen
Oxygen
Major groove
Minor groove
Bases

13.7 DNA Is a Double Helix (A) James Watson (left) and Francis Crick (right) proposed that the DNA molecule has a double-helical structure. (B) Biochemists can now pinpoint the position of every atom in a DNA molecule. To see that the essential features of the original Watson–Crick model have been verified, follow with your eyes the double-helical chains of sugar–phosphate groups and note the horizontal rungs of the bases.

Go to Media Clip 13.1
Discovery of the Double Helix
Life10e.com/mc13.1

"backbone" on the outside. In addition, the two DNA strands ran in opposite directions, that is, they were **antiparallel**. The two strands would not fit together otherwise:

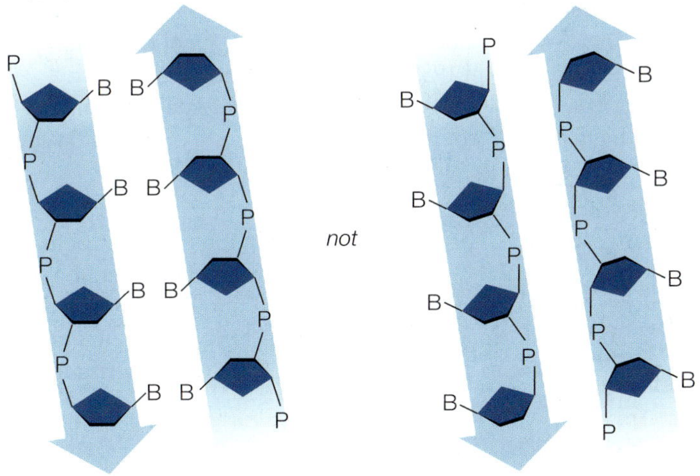
P
B B
P
not
B

- To satisfy Chargaff's rule (A = T and G = C), Watson and Crick's model always paired a purine on one strand with a pyrimidine on the opposite strand. These **base pairs** (A-T and G-C) have the same width down the double helix, a uniformity that was also confirmed by X-ray diffraction:

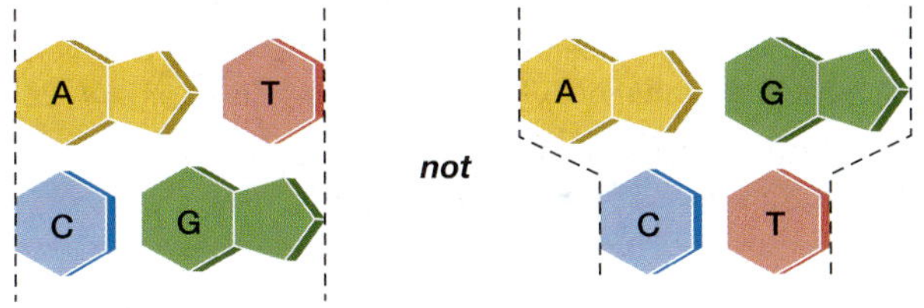
A
T
C
G
not
A
G
C
T

Crick and Watson built their tin model of DNA in late February 1953. This structure explained all the known chemical properties of DNA, and it opened the door to understanding its biological functions. There have been minor amendments to that first published structure, but its principal features remain unchanged.

Four key features define DNA structure

Four features summarize the molecular architecture of the DNA molecule (**Figure 13.7B**):

- DNA is a *double-stranded helix*, with a sugar–phosphate backbone on the outside and base pairs lined up on the inside.
- DNA is usually a *right-handed helix*. If you curl the fingers of your right hand and point your thumb upward, the curve of the helix follows the direction of your fingers, and it winds upward in the direction of your thumb.
- DNA is *antiparallel* (the two strands run in opposite directions).
- DNA has *major and minor grooves* in which the outer edges of the nitrogenous bases are exposed.

THE HELIX The sugar–phosphate backbones of the polynucleotide chains form a coil around the outside of the helix, and the nitrogenous bases point toward the center. The chains are held together by two chemical forces:

1. *Hydrogen bonding* between specifically paired bases. Consistent with Chargaff's rule, adenine (A) pairs with thymine (T) by forming two hydrogen bonds, and guanine (G) pairs with cytosine (C) by forming three hydrogen bonds:

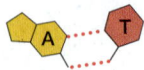

A
T
C
G

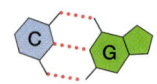

   Every base pair consists of one purine (A or G) and one pyrimidine (T or C). This pattern is known as **complementary base pairing**.

2. *Van der Waals* forces between adjacent bases on the same strand. When the base rings come near one another, they tend to stack like poker chips because of these weak attractions.

ANTIPARALLEL STRANDS The backbone of each DNA strand contains repeating units of the five-carbon monosaccharide deoxyribose:

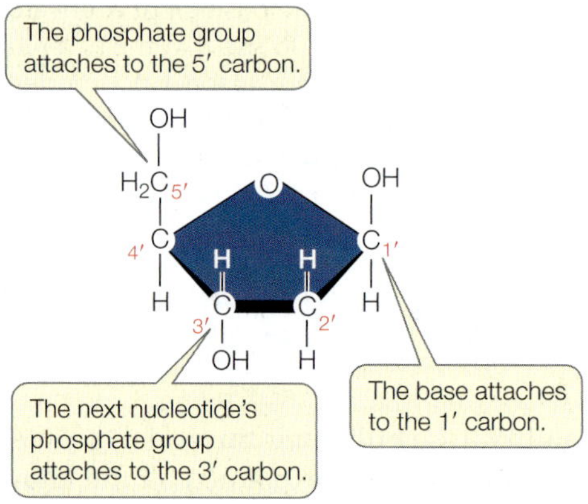

The phosphate group attaches to the 5′ carbon.
The next nucleotide's phosphate group attaches to the 3′ carbon.
The base attaches to the 1′ carbon.

The number followed by a prime (′) designates the position of a carbon atom in this sugar molecule. In the sugar–phosphate backbone of DNA, the phosphate groups are connected to the 3′ carbon of one deoxyribose molecule and the 5′ carbon of the next, linking successive sugars together.

Thus the two ends of a polynucleotide chain differ. At one end of a chain is a free (not connected to another nucleotide) 5′ phosphate group (—OPO_3^-); this is called the **5′ end**. At the other end is a free 3′ hydroxyl group (—OH); this is called the **3′ end**. In a DNA double helix, the 5′ end of one strand is paired with the 3′ end of the other strand, and vice versa. In other words, if you drew an arrow for each strand running from 5′ to 3′, the arrows would point in opposite directions (see also Figure 4.4A).

BASE EXPOSURE IN THE GROOVES Look back at Figure 13.7B and note the major and minor grooves in the helix. These grooves exist because the backbones of the two strands are closer together on one side of the double helix (forming the minor groove) than on the other side (forming the major groove). **Figure 13.8** shows the four possible configurations of the flat, hydrogen-bonded base pairs in the major and minor grooves. The exposed outer edges of the base pairs are accessible for additional hydrogen bonding. Note that the arrangements of unpaired atoms and groups differ in A-T and G-C pairs. Thus, the *surfaces of the A-T and C-G base pairs are chemically distinct*, allowing other molecules such as proteins to recognize specific base pair sequences and bind to them. *The binding of proteins to specific base pair sequences is the key to protein–DNA interactions*, which are necessary for the replication and expression of the genetic information in DNA.

The double-helical structure of DNA is essential to its function

The genetic material performs four important functions, and the DNA structure proposed by Watson and Crick was elegantly suited to three of them.

- *The genetic material stores an organism's genetic information.* With its millions of nucleotides, the base sequence of a DNA molecule can encode and store an enormous amount of information. Variations in DNA sequences can account for species and individual differences. DNA fits this role nicely.
- *The genetic material is susceptible to mutations* (permanent changes) *in the information it encodes*. For DNA, mutations might be simple changes in the linear sequence of base pairs.
- *The genetic material is precisely replicated in the cell division cycle.* Replication could be accomplished by complementary base pairing, A with T and G with C. In the original publication of their findings in 1953, Watson and Crick coyly pointed out, "It has not escaped our notice that the specific pairing we have postulated immediately suggests a possible copying mechanism for the genetic material."
- *The genetic material (the coded information in DNA) is expressed as the phenotype.* This function is not obvious in the structure of DNA. However, as we will see in the next chapter, the nucleotide sequence of DNA is copied into RNA, which uses the coded information to specify a linear sequence of amino acids—a protein. The folded forms of proteins determine many of the phenotypes of an organism.

RECAP 13.2

DNA is a double helix made up of two antiparallel polynucleotide chains. The two chains are joined by hydrogen bonds between the nucleotide bases, which pair specifically: A with T, and G with C. Chemical groups on the bases that are exposed in the grooves of the helix are available for hydrogen bonding with other molecules, such as proteins. These molecules can recognize specific sequences of nucleotide bases.

- Describe the evidence that Watson and Crick used to come up with the double helix model for DNA. **See pp. 264–265**
- How does the double-helical structure of DNA relate to its function? **See p. 266 and Figures 13.7, 13.8**

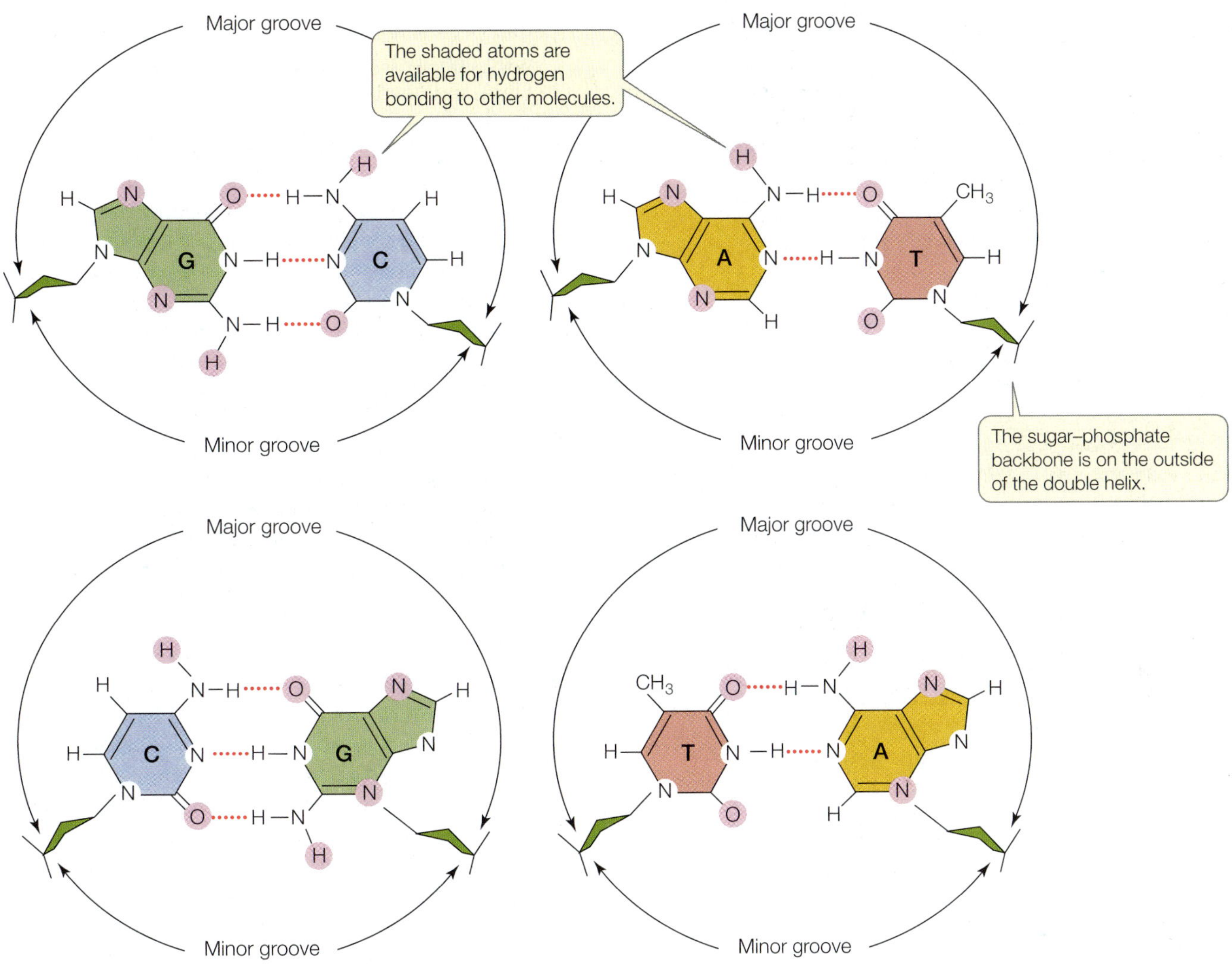

Major groove
Minor groove
The shaded atoms are available for hydrogen bonding to other molecules.
The sugar–phosphate backbone is on the outside of the double helix.
G
C
A
T

13.8 Base pairs in DNA Can Interact with Other Molecules
These diagrams show the four possible configurations of base pairs within the double helix. Atoms shaded in purple are available for hydrogen bonding with other molecules, such as proteins.

Once the structure of DNA was understood, it was possible to investigate how DNA replicates itself. Next we will examine the experiments that taught us how this elegant process works.

13.3 How Is DNA Replicated?

The mechanism of DNA replication that suggested itself to Watson and Crick was soon confirmed. First, researchers showed that DNA could be replicated in a test tube containing simple substrates and an enzyme. A subsequent study showed that each of the two strands of the double helix can serve as a template for a new strand of DNA.

Go to Animated Tutorial 13.2
Replication and DNA Polymerization
Life10e.com/at13.2

Three modes of DNA replication appeared possible

New DNA molecules with the same base sequence as an original molecule can be synthesized in a test tube containing the following substances:

- The deoxyribonucleoside triphosphates dATP, dCTP, dGTP, and dTTP. These are the monomers from which the DNA polymers are formed.
- DNA molecules of a particular sequence that serve as **templates** to direct the sequence of nucleotides in the new molecules.
- A **DNA polymerase** enzyme to catalyze the polymerization reaction.
- Salts and a pH buffer, to create an appropriate chemical environment for the DNA polymerase.

The fact that DNA could be synthesized in a test tube confirmed that a DNA molecule contains the information needed

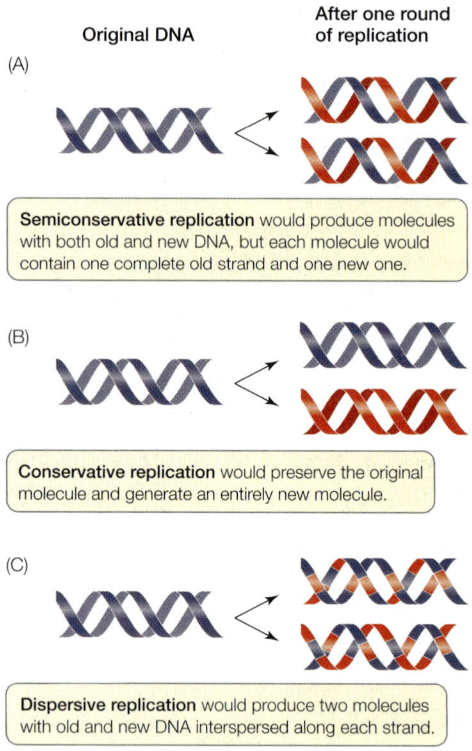
Original DNA
After one round of replication
(A)
Semiconservative replication would produce molecules with both old and new DNA, but each molecule would contain one complete old strand and one new one.
(B)
Conservative replication would preserve the original molecule and generate an entirely new molecule.
(C)
Dispersive replication would produce two molecules with old and new DNA interspersed along each strand.

13.9 Three Models for DNA Replication In each model, the original DNA is shown in blue and the newly synthesized DNA is in red.

for its own replication. The next challenge was to determine which of three possible replication patterns occurs during DNA replication:

- *Semiconservative replication*, in which each parent strand serves as a template for a new strand, and the two new DNA molecules each have one old and one new strand (**Figure 13.9A**)
- *Conservative replication*, in which the original double helix serves as a template for, but does not contribute to, a new double helix (**Figure 13.9B**)
- *Dispersive replication*, in which fragments of the original DNA molecule serve as templates for assembling two new molecules, each containing old and new parts, perhaps at random (**Figure 13.9C**)

Watson and Crick's original paper suggested that DNA replication was semiconservative, but the test tube demonstration described above did not provide a basis for choosing among these three models.

An elegant experiment demonstrated that DNA replication is semiconservative

In 1958 Matthew Meselson and Franklin Stahl at the California Institute of Technology convinced the scientific community that DNA is reproduced by **semiconservative replication**. They used density labeling to distinguish between parent strands of DNA and newly copied ones.

The key to their experiment was the use of a "heavy" isotope of nitrogen. Heavy nitrogen (^{15}N) is a rare, nonradioactive isotope that makes molecules containing it denser than chemically identical molecules containing the common isotope ^{14}N. Two cultures of the bacterium *E. coli* were grown for many generations, one in a medium containing ^{15}N and the other in a medium containing ^{14}N. When DNA extracts from the two cultures were combined and centrifuged in a solution of cesium chloride, which forms a density gradient under centrifugation, two separate bands of DNA formed in the centrifugation tube. The DNA from the ^{15}N culture was heavier than the DNA from the ^{14}N culture, so it formed a band at a different position in the density gradient.

Next, Meselson and Stahl grew another *E. coli* culture in ^{15}N medium, then transferred the bacteria to normal ^{14}N medium and allowed them to continue growing (**Figure 13.10**). The cells replicated their DNA and divided every 20 minutes. Meselson and Stahl collected some of the bacteria at time intervals and extracted DNA from the samples. You can follow their results for the first two generations in Figure 13.10. The results can be explained only by the semiconservative model of DNA replication. The crucial observations demonstrating this model were that all the DNA at the end of the first generation was of intermediate density, while at the end of the second generation there were two discrete bands: one of intermediate and one of light DNA. If the conservative model had been true, there would have been no intermediate density DNA. If the dispersive model were correct, then the DNA would all have been intermediate for the first few generations, with the single intermediate band becoming progressively lighter.

There are two steps in DNA replication

Semiconservative DNA replication in the cell involves several different enzymes and other proteins. It takes place in two general steps:

- The DNA double helix is unwound to separate the two template strands and make them available for new base pairing (this is the step that cisplatin prevents; see the opening story).
- As new nucleotides form complementary base pairs with template DNA, they are covalently linked together by phosphodiester bonds, forming a polymer whose base sequence is complementary to the bases in the template strand.

The nucleotides that make up DNA are deoxyribonucleoside monophosphates because they each contain deoxyribose and one phosphate group (see Figure 4.1). The free monomers that are brought together to form DNA have three phosphate groups. They are the deoxyribonucleoside triphosphates dATP, dTTP, dCTP, and dGTP, collectively referred to as dNTPs. The

INVESTIGATING**LIFE**

13.10 The Meselson–Stahl Experiment A centrifuge was used to separate DNA molecules labeled with isotopes of different densities. This experiment revealed a pattern that supports the semiconservative model of DNA replication.[a]

HYPOTHESIS DNA replicates semiconservatively.

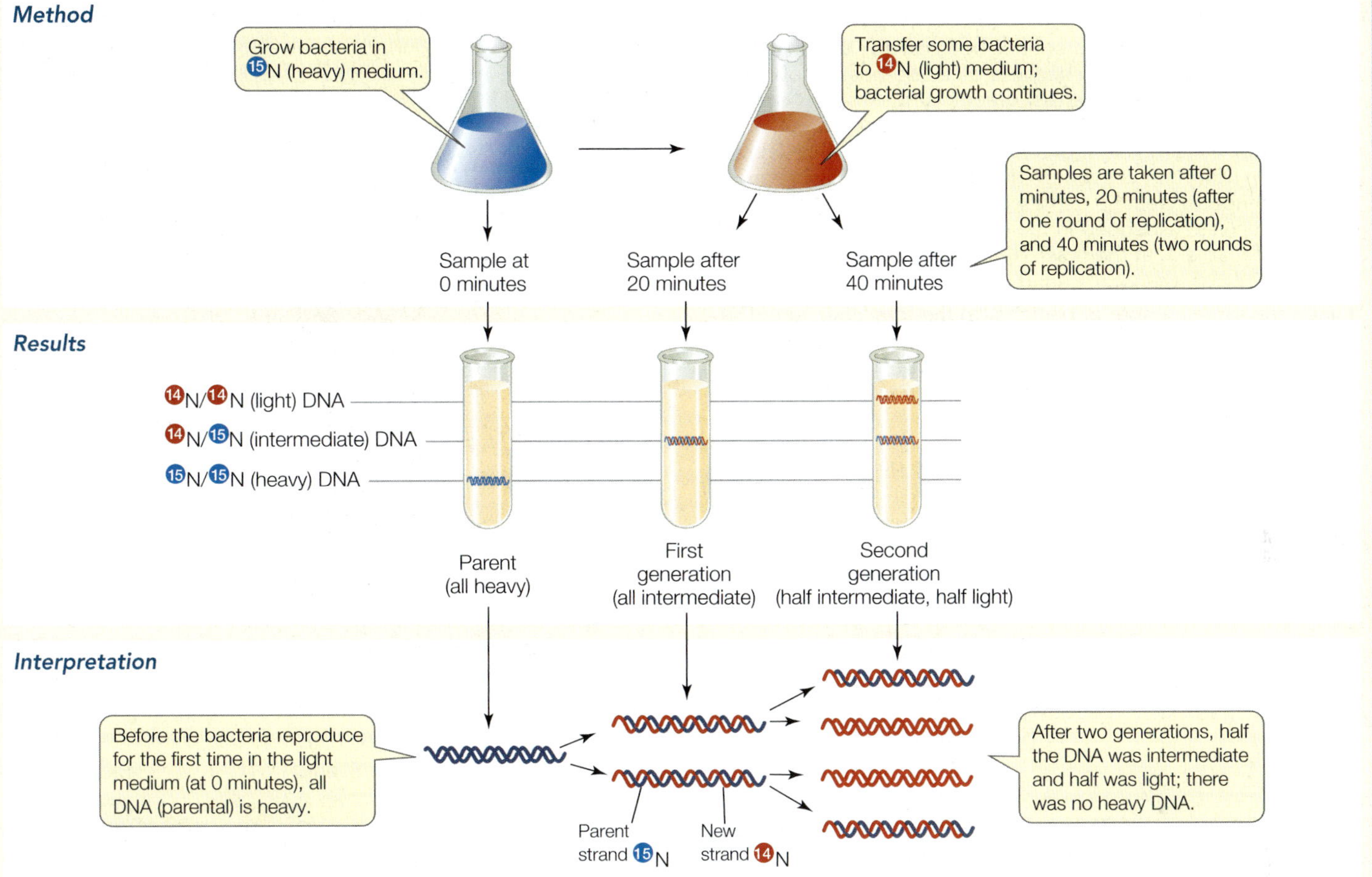
Method
Grow bacteria in 15N (heavy) medium.
Transfer some bacteria to 14N (light) medium; bacterial growth continues.
Samples are taken after 0 minutes, 20 minutes (after one round of replication), and 40 minutes (two rounds of replication).
Sample at 0 minutes
Sample after 20 minutes
Sample after 40 minutes
Results
14N/14N (light) DNA
14N/15N (intermediate) DNA
15N/15N (heavy) DNA
Parent (all heavy)
First generation (all intermediate)
Second generation (half intermediate, half light)
Interpretation
Before the bacteria reproduce for the first time in the light medium (at 0 minutes), all DNA (parental) is heavy.
Parent strand 15N
New strand 14N
After two generations, half the DNA was intermediate and half was light; there was no heavy DNA.

CONCLUSION This pattern could only have been observed if each DNA molecule contains a template strand from the parental DNA; thus DNA replication is semiconservative.

Go to **BioPortal** for discussion and relevant links for all INVESTIGATING**LIFE** figures.

[a]Meselson, M. and F. Stahl. 1958. *Proceedings of the National Academy of Sciences USA* 44: 671–682.

Go to Animated Tutorial 13.3
The Meselson–Stahl Experiment
Life10e.com/at13.3

three phosphate groups are attached to the 5′ carbon on the deoxyribose sugar (see p. 266).

A key observation regarding DNA replication is that *nucleotides are added to the growing new strand at the 3′ end*—the end at which the DNA strand has a free hydroxyl (—OH) group on the 3′ carbon of its terminal deoxyribose. In the formation of the phosphodiester linkage (a condensation reaction), two of the phosphate groups on an incoming dNTP are removed [as pyrophosphate (PP_i)], and the remaining phosphate is bonded to the 3′ carbon on the terminal deoxyribose (**Figure 13.11**; see also Figure 4.2). Just as energy is released when ATP is hydrolyzed to AMP (with subsequent hydrolysis of PP_i to two phosphates), energy is released by the hydrolysis of the dNTP, and this energy is used to drive the condensation reaction.

DNA polymerases add nucleotides to the growing chain

DNA replication begins with the binding of a large protein complex (the pre-replication complex) to a specific site on the DNA molecule. This complex contains several different proteins, including the enzyme DNA polymerase, which catalyzes the addition of nucleotides as the new DNA chain grows. All chromosomes have at least one region called the **origin of replication** (***ori***), to which the pre-replication complex binds. Binding occurs when proteins in the complex recognize specific DNA sequences within the *ori*.

ORIGINS OF REPLICATION The single circular chromosome of the bacterium *E. coli* has 4×10^6 base pairs (bp) of DNA.

WORKING WITH**DATA:**

The Meselson–Stahl Experiment

Original Paper

Meselson, M. and F. Stahl. 1958. The replication of DNA in *Escherichia coli. Proceedings of the National Academy of Sciences USA* 44: 671–682.

Analyze the Data

The Meselson–Stahl experiment has been called "the most beautiful experiment in biology" because of its essential simplicity. Meselson and Stahl used density gradients to examine how DNA molecules replicate (see Figure 13.10). The key experimental method was the separation of DNA that contained ^{14}N ("light" DNA) from DNA that contained ^{15}N ("heavy" DNA), using an ultracentrifuge to create a density gradient of cesium chloride. DNA molecules suspended in the cesium chloride formed bands within the gradient, at their own density (**FIGURE A**).

FIGURE B shows the results of the experiment. Each sample contained the same number of bacteria, so the total amount of DNA in each panel was the same. The photos show the DNA bands in the tubes after centrifugation, and the plots show quantitative analysis of the bands, where height indicates amount of DNA.

QUESTION 1

Use the heights of the peaks to estimate the percent of total DNA that was heavy, intermediate, and light at each generational stage. Create a table summarizing these calculations and discuss whether they support the authors' conclusions.

QUESTION 2

What would the data look like if the bacteria had been allowed to divide for three more generations?

QUESTION 3

If Meselson and Stahl had done their experiment starting with light DNA and then added ^{15}N for succeeding generations, what would the bands look like? Draw them alongside the actual data, above.

QUESTION 4

What would the data look like if conservative replication were the correct model? What would the data look like if dispersive replication were correct? Draw these alongside the actual data above.

FIGURE A

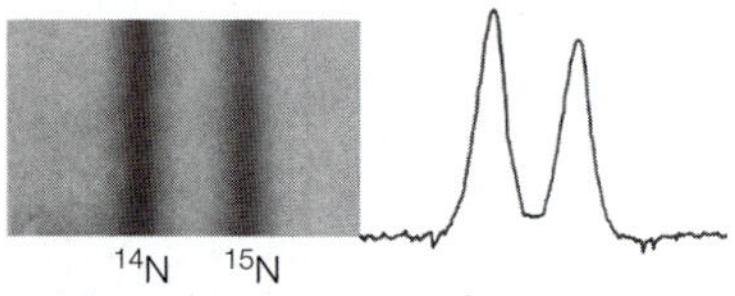

^{14}N ^{15}N

FIGURE B

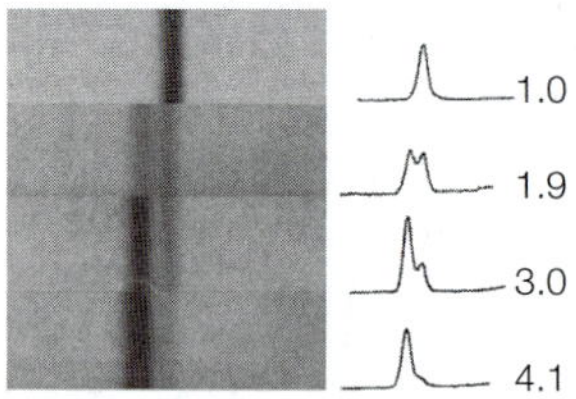

Generations
1.0
1.9
3.0
4.1

Go to **BioPortal** for all WORKING WITH**DATA** exercises

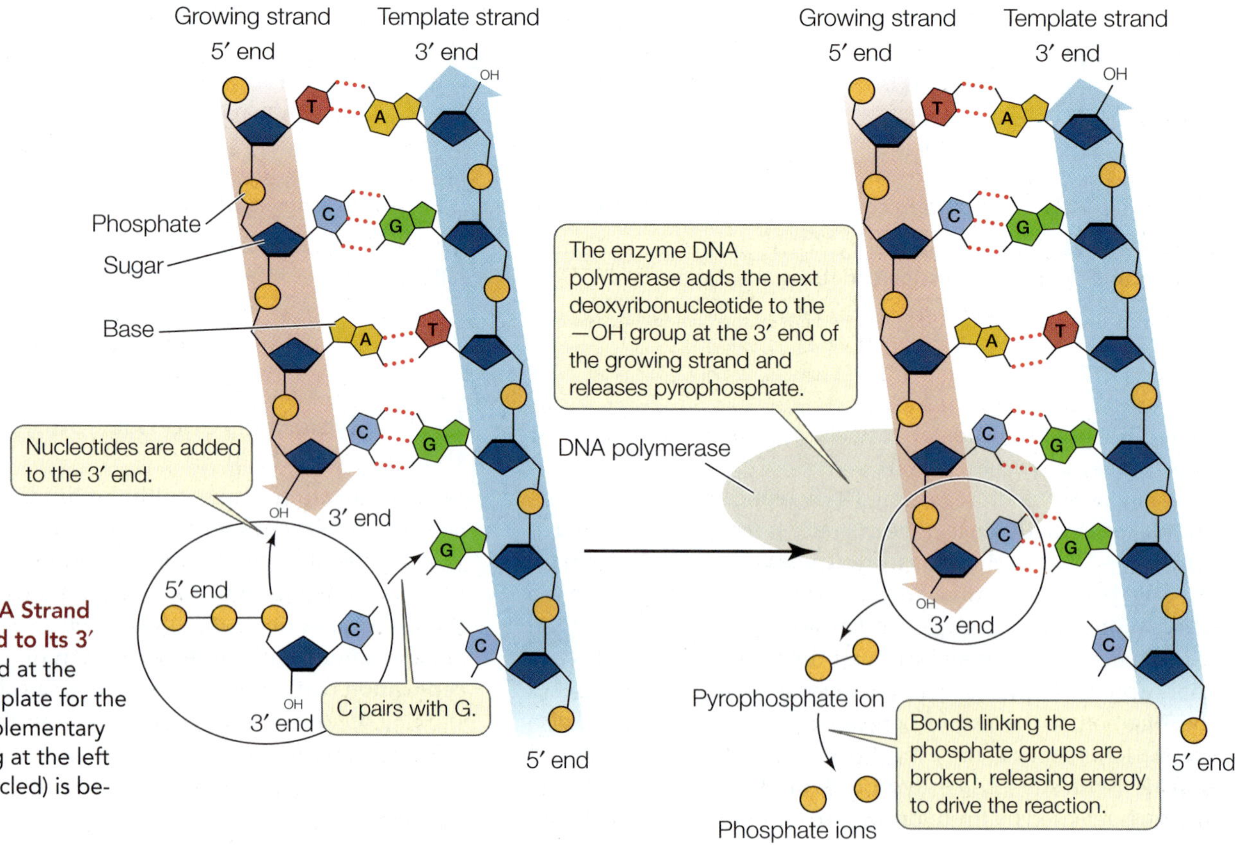

Growing strand
5′ end
Template strand
3′ end
OH
Phosphate
Sugar
Base
T
A
C
G
Nucleotides are added to the 3′ end.
OH
3′ end
5′ end
3′ end
C pairs with G.
5′ end
The enzyme DNA polymerase adds the next deoxyribonucleotide to the —OH group at the 3′ end of the growing strand and releases pyrophosphate.
DNA polymerase
Pyrophosphate ion
Bonds linking the phosphate groups are broken, releasing energy to drive the reaction.
Phosphate ions

13.11 Each New DNA Strand Grows from Its 5′ End to Its 3′ End The DNA strand at the right (blue) is the template for the synthesis of the complementary strand that is growing at the left (pink). Here dCTP (circled) is being added.

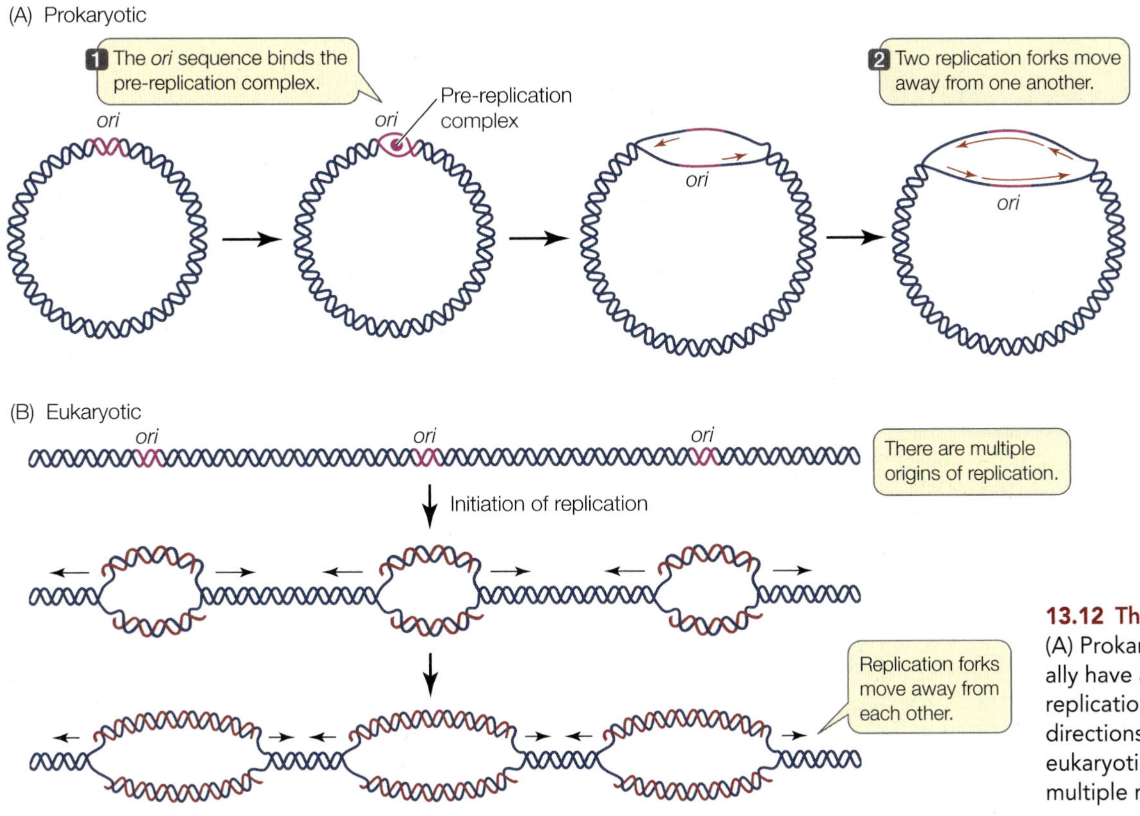
(A) Prokaryotic
1 The ori sequence binds the pre-replication complex.
ori
ori
Pre-replication complex
ori
2 Two replication forks move away from one another.
ori
(B) Eukaryotic
ori
ori
ori
There are multiple origins of replication.
Initiation of replication
Replication forks move away from each other.

13.12 The Origin of DNA Replication (A) Prokaryotic chromosomes usually have a single origin where DNA replication starts and proceeds in both directions. (B) The usually much larger eukaryotic chromosome typically has multiple replication origins.

The 245 bp *ori* sequence is at a particular location on the chromosome. Once the pre-replication complex binds to it, the DNA is unwound and replication proceeds in both directions around the circle, forming two **replication forks** (Figure 13.12A). The replication rate in *E. coli* is approximately 1,000 bp per second, so it takes about 40 minutes to fully replicate the chromosome (with two replication forks). Rapidly dividing *E. coli* cells divide every 20 minutes. In these cells, new rounds of replication begin at the *ori* of each new chromosome before the first chromosome has fully replicated. In this way the cells can divide more frequently than the time needed to finish replicating the original chromosome.

Eukaryotic chromosomes are typically much longer than those of prokaryotes—up to a billion bp—and are linear, not circular. If replication occurred from a single *ori* with two forks growing away from each other, it would take weeks to fully replicate a chromosome. So eukaryotic chromosomes have multiple origins of replication, scattered at intervals of 10,000 to 40,000 bp (Figure 13.12B).

DNA REPLICATION BEGINS WITH A PRIMER A DNA polymerase elongates a polynucleotide strand by covalently linking new nucleotides to a preexisting strand. However, it cannot start this process without a short "starter" strand, called a **primer**. In most organisms this primer is a short single strand of RNA (Figure 13.13), but in some organisms it is DNA. The primer is complementary to the DNA template and is synthesized one nucleotide at a time by an enzyme called a **primase**. The DNA polymerase then adds nucleotides to the 3′ end of the primer and continues until the replication of that section of DNA has been completed. Then the RNA primer is degraded, DNA is added in its place, and the resulting DNA fragments are connected by the action of other enzymes. When DNA replication is complete, each new strand consists only of DNA.

DNA POLYMERASES ARE LARGE DNA polymerases are much larger than their substrates (the dNTPs) and the template DNA, which is very thin. Molecular models of the enzyme–substrate–template complex from bacteria show that the enzyme is shaped like an open right hand with a palm, a thumb, and fingers (Figure 13.14). Within the "palm" is the active site of the enzyme, which brings together each dNTP substrate and the template. The "finger" regions have precise shapes that can recognize the different shapes of the four nucleotide bases. They bind to the bases by hydrogen bonding and rotate inward. Most cells contain more than one kind of DNA polymerase, but only one of them is responsible for chromosomal DNA replication. The others are involved in primer removal and DNA repair. Fifteen DNA polymerases have been identified in humans, whereas the bacterium *E. coli* has five DNA polymerases.

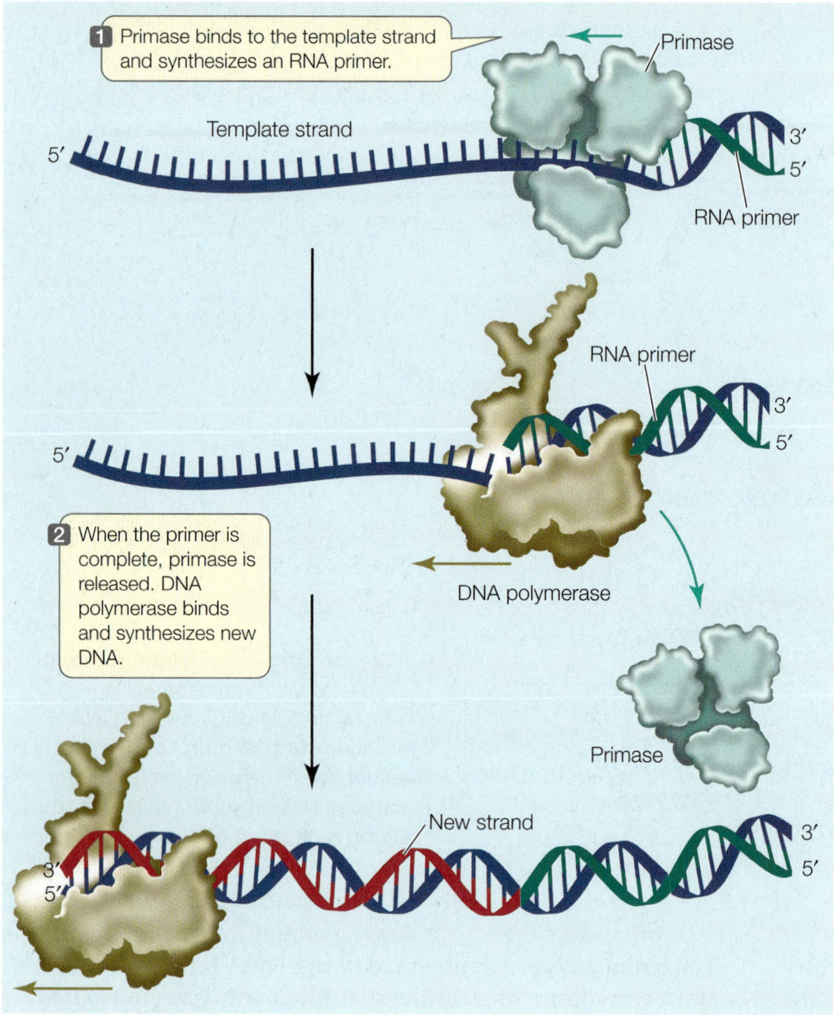

1 Primase binds to the template strand and synthesizes an RNA primer.
Primase
Template strand
5′
3′
5′
RNA primer
RNA primer
5′
3′
5′
2 When the primer is complete, primase is released. DNA polymerase binds and synthesizes new DNA.
DNA polymerase
Primase
New strand
3′
5′
3′
5′

13.13 DNA Forms with a Primer DNA polymerases require a primer—a "starter" strand of DNA or RNA to which they can add new nucleotides.

Many other proteins assist with DNA polymerization

Various other proteins play roles in other replication tasks; some of these are shown in **Figure 13.15**. The first event at the origin of replication is the localized unwinding and separation (denaturation) of the DNA strands. As we discussed in Section 13.2, the two strands are held together by hydrogen bonds and Van der Waals forces. An enzyme called **DNA helicase** uses energy from ATP hydrolysis to unwind and separate the strands, and **single-strand binding proteins** bind to the unwound strands to keep them from reassociating into a double helix. This process makes each of the two template strands available for complementary base pairing.

The two DNA strands grow differently at the replication fork

The DNA at the replication fork—the site where DNA unwinds to expose the bases so that they can act as templates—opens up like a zipper in one direction. Study **Figure 13.16** and try to imagine what is happening over a short period of time. Remember that the two DNA strands are antiparallel; that is, the 3′ end of one strand is paired with the 5′ end of the other.

- One newly replicating strand (the **leading strand**) is oriented so that it can grow continuously at its 3′ end as the fork opens up.
- The other new strand (the **lagging strand**) is oriented so that as the fork opens up, its exposed 3′ end gets farther and farther away from the fork, and an unreplicated gap is formed. This gap would get bigger and bigger if there were not a special mechanism to overcome this problem.

Synthesis of the lagging strand requires the synthesis of relatively small, discontinuous stretches of DNA (100–200 nucleotides in eukaryotes; 1,000–2,000 nucleotides in prokaryotes). These discontinuous stretches are synthesized just as the

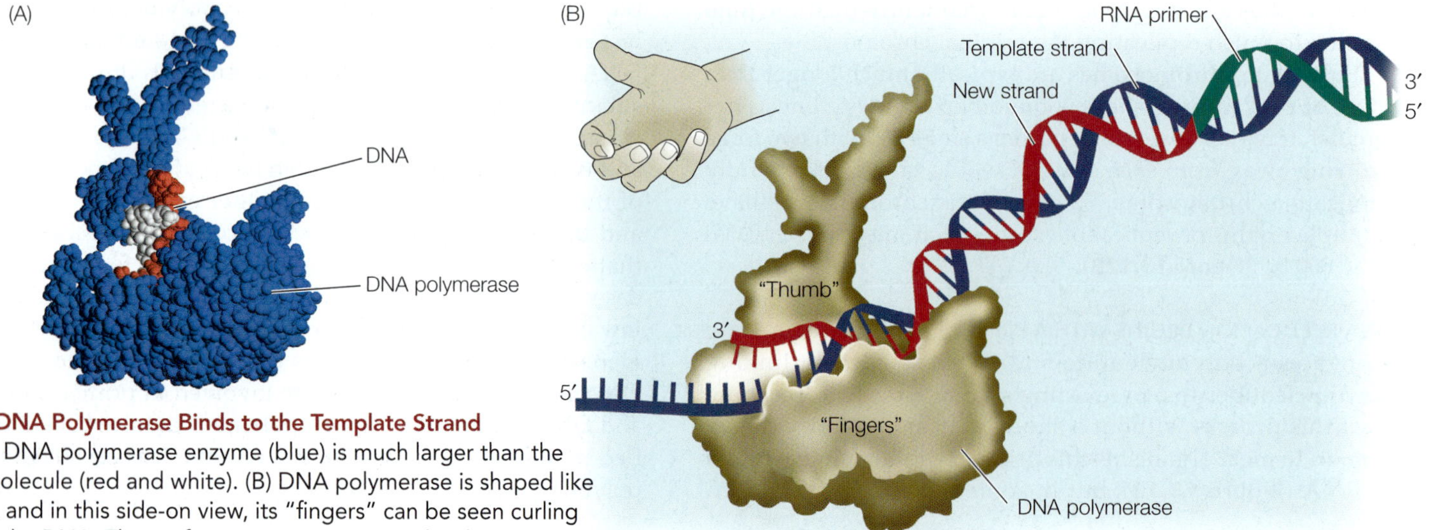

(A)
DNA
DNA polymerase
(B)
RNA primer
Template strand
New strand
3′
5′
"Thumb"
3′
5′
"Fingers"
DNA polymerase

13.14 DNA Polymerase Binds to the Template Strand (A) The DNA polymerase enzyme (blue) is much larger than the DNA molecule (red and white). (B) DNA polymerase is shaped like a hand, and in this side-on view, its "fingers" can be seen curling around the DNA. These "fingers" can recognize the distinctive shapes of the four bases.

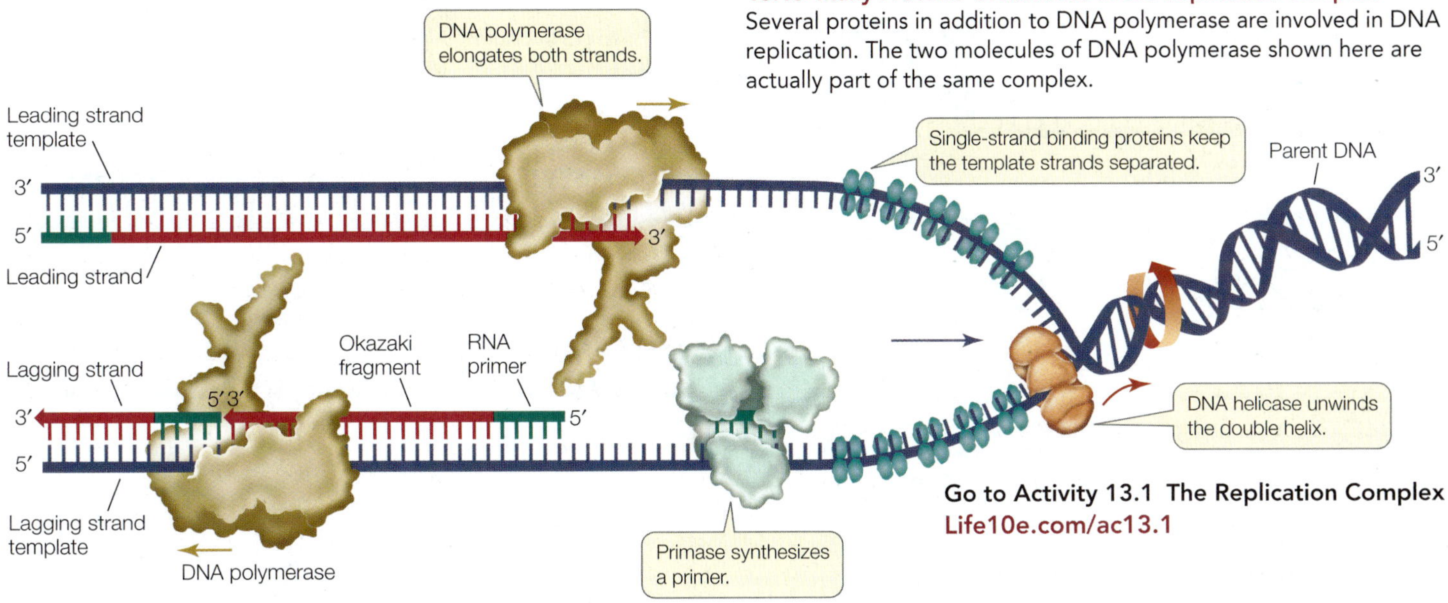
DNA polymerase elongates both strands.
Leading strand template
3′
5′
Leading strand
3′
Single-strand binding proteins keep the template strands separated.
Parent DNA
3′
5′
Lagging strand
Okazaki fragment
RNA primer
3′
5′3′
5′
5′
Lagging strand template
DNA polymerase
Primase synthesizes a primer.
DNA helicase unwinds the double helix.

13.15 Many Proteins Collaborate in the Replication Complex Several proteins in addition to DNA polymerase are involved in DNA replication. The two molecules of DNA polymerase shown here are actually part of the same complex.

Go to Activity 13.1 The Replication Complex
Life10e.com/ac13.1

leading strand is, by the addition of new nucleotides one at a time to the 3′ end of the new strand, but the synthesis of this new strand moves in the direction opposite to that in which the replication fork is moving. These stretches of new DNA are called **Okazaki fragments** (after their discoverer, the Japanese biochemist Reiji Okazaki). While the leading strand grows continuously "forward," the lagging strand grows in shorter, "backward" stretches with gaps between them.

A single primer is needed for synthesis of the leading strand, but each Okazaki fragment requires its own primer to be synthesized by the primase. In bacteria, DNA polymerase III then synthesizes an Okazaki fragment by adding nucleotides to one primer until it reaches the primer of the previous fragment. At this point, DNA polymerase I removes the old primer and replaces it with DNA. Left behind is a tiny nick—the final phosphodiester linkage between the adjacent Okazaki fragments is missing. The enzyme **DNA ligase** catalyzes the formation of that bond, linking the fragments and making the lagging strand whole (**Figure 13.17**).

Working together, DNA helicase, the two DNA polymerases, primase, DNA ligase, and the other proteins of the pre-replication complex do the job of DNA synthesis with a speed and accuracy that are almost unimaginable. In *E. coli*, the replication complex makes new DNA at a rate in excess of 1,000 base pairs per second, committing errors in fewer than one base in a million.

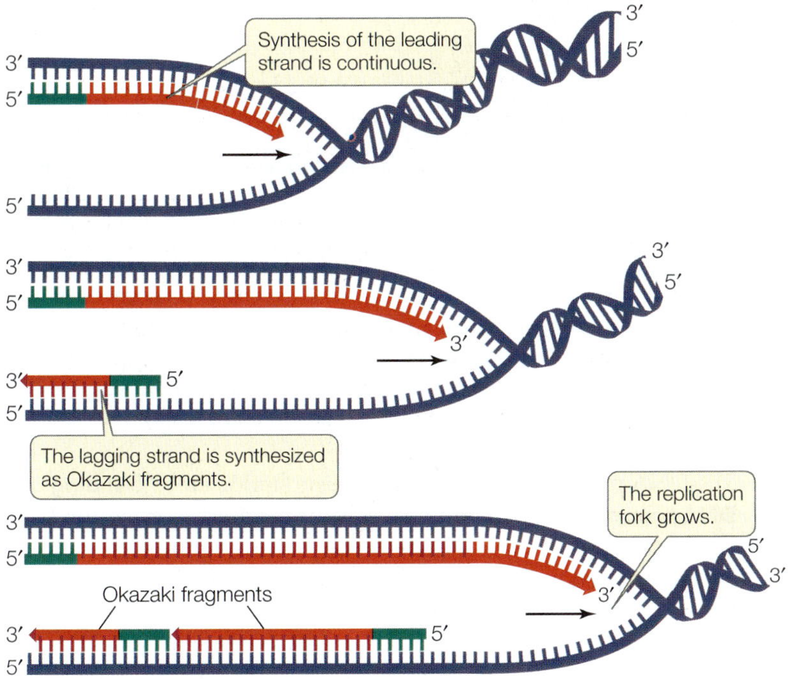
Synthesis of the leading strand is continuous.
3′
5′
5′
3′
5′
3′
5′
3′
3′
5′
5′
The lagging strand is synthesized as Okazaki fragments.
The replication fork grows.
3′
5′
5′
3′
3′
Okazaki fragments
3′
5′
5′

13.16 The Two New Strands Form in Different Ways As the parent DNA unwinds, both new strands are synthesized in the 5′-to-3′ direction, although their template strands are antiparallel. The leading strand grows continuously forward, but the lagging strand grows in short discontinuous stretches called Okazaki fragments. Eukaryotic Okazaki fragments are hundreds of nucleotides long, with gaps between them.

Go to Animated Tutorial 13.4
Leading and Lagging Strand Synthesis
Life10e.com/at13.4

A SLIDING CLAMP INCREASES THE RATE OF DNA REPLICATION

How do DNA polymerases work so fast? We saw in Section 8.3 that an enzyme catalyzes a chemical reaction:

Substrate binds to enzyme → one product is formed → enzyme is released → cycle repeats

DNA replication would not proceed as rapidly as it does if it went through such a cycle for each nucleotide. Instead, DNA polymerases are **processive**—that is, they catalyze the formation of many phosphodiester linkages each time they bind to a DNA molecule:

Substrates bind to enzyme → many products are formed → enzyme is released → cycle repeats

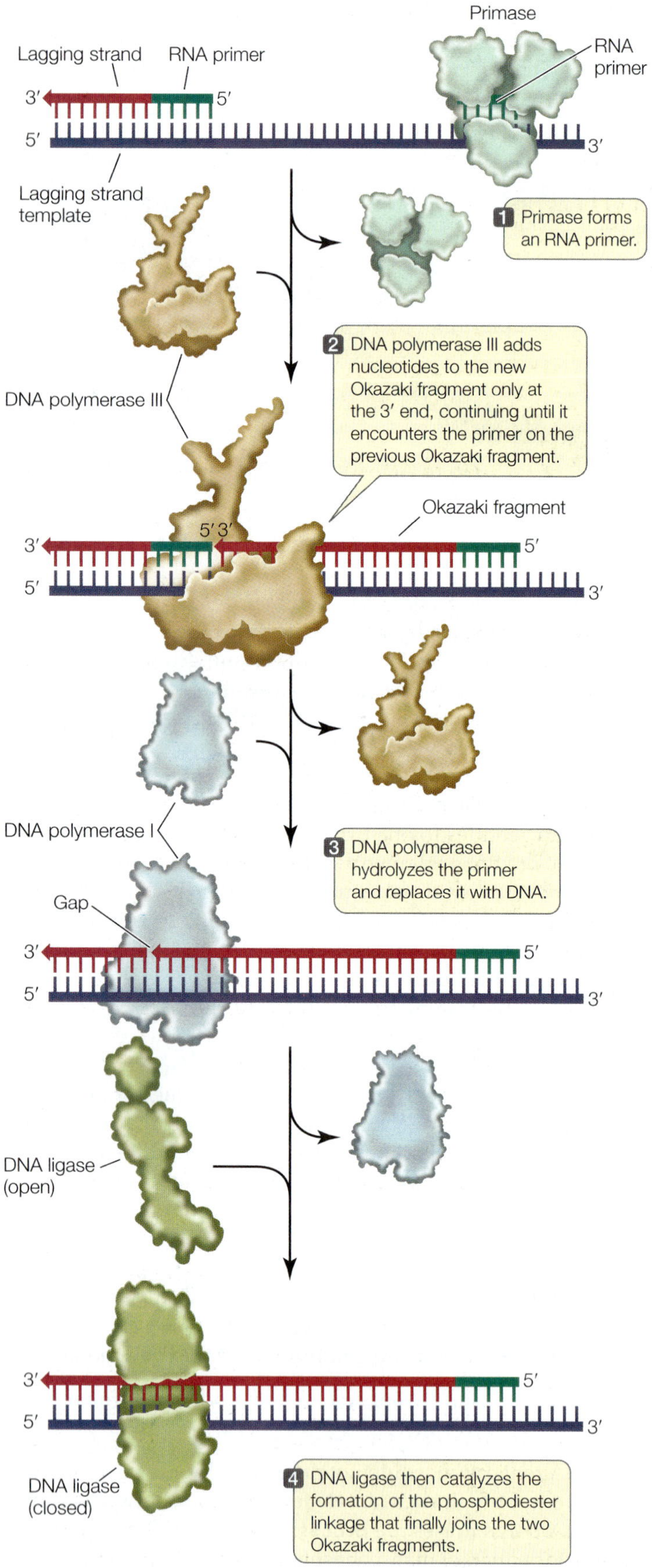

Primase
Lagging strand
RNA primer
RNA primer
3′
5′
5′
3′
Lagging strand template
1 Primase forms an RNA primer.
2 DNA polymerase III adds nucleotides to the new Okazaki fragment only at the 3′ end, continuing until it encounters the primer on the previous Okazaki fragment.
DNA polymerase III
Okazaki fragment
5′ 3′
3′
5′
5′
3′
DNA polymerase I
3 DNA polymerase I hydrolyzes the primer and replaces it with DNA.
Gap
3′
5′
5′
3′
DNA ligase (open)
3′
5′
5′
3′
DNA ligase (closed)
4 DNA ligase then catalyzes the formation of the phosphodiester linkage that finally joins the two Okazaki fragments.

13.17 The Lagging Strand Story In bacteria, DNA polymerase I and DNA ligase cooperate with DNA polymerase III to complete the complex task of synthesizing the lagging strand.

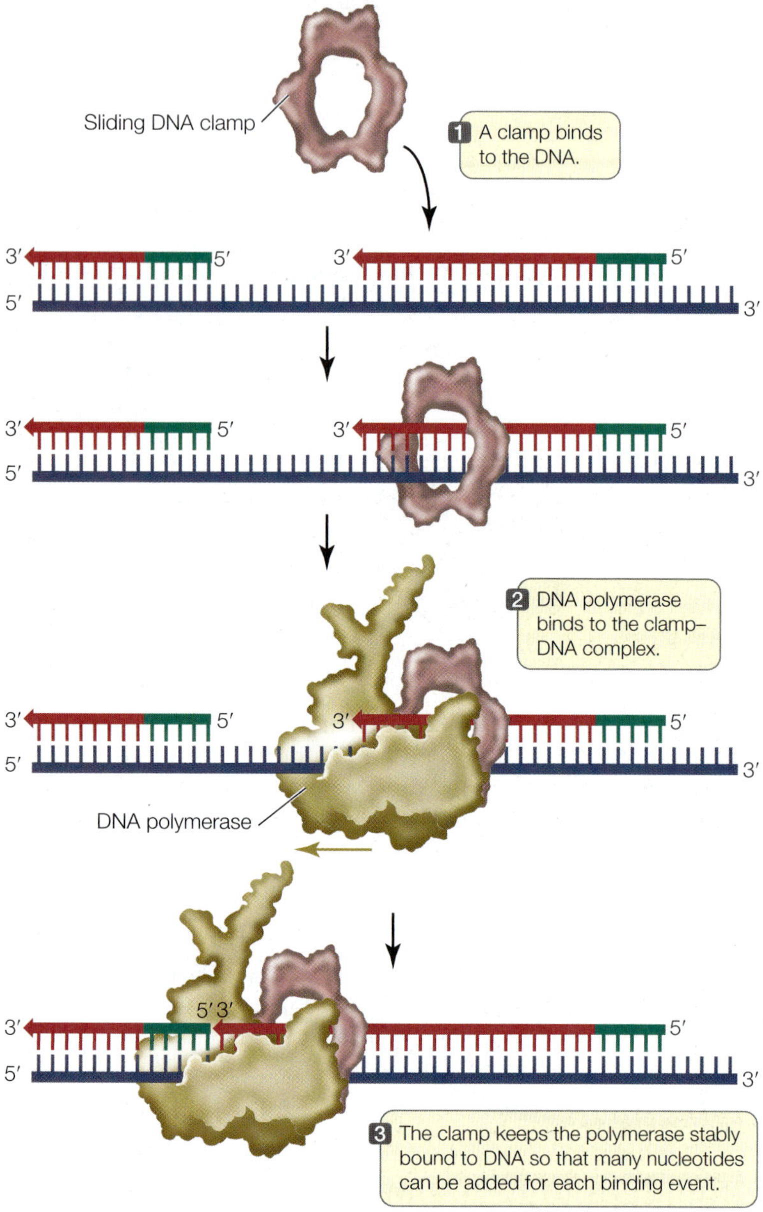

Sliding DNA clamp
1 A clamp binds to the DNA.
3′
5′
5′
3′
5′
3′
2 DNA polymerase binds to the clamp–DNA complex.
DNA polymerase
5′ 3′
3 The clamp keeps the polymerase stably bound to DNA so that many nucleotides can be added for each binding event.

13.18 A Sliding DNA Clamp Increases the Efficiency of DNA Polymerization The clamp increases the efficiency of polymerization by keeping the enzyme bound to the substrate, so the enzyme does not have to repeatedly bind to template and substrate.

The DNA polymerase–DNA complex is stabilized by a **sliding DNA clamp**, which has multiple identical subunits assembled into a doughnut shape (Figure 13.18). The doughnut's "hole" is just large enough to encircle the DNA double helix, along with a thin layer of water molecules for lubrication. The clamp binds to the DNA polymerase–DNA complex, keeping the enzyme and the DNA associated tightly with each other. If the clamp is absent, DNA polymerase dissociates from DNA after forming 20 to 100 phosphodiester linkages. With the clamp, it can polymerize up to 50,000 nucleotides before it detaches.

DNA IS THREADED THROUGH A REPLICATION COMPLEX Until recently, DNA replication was always depicted to look like a locomotive (the replication complex) moving along a railroad track (the DNA). While this does occur in some organisms,

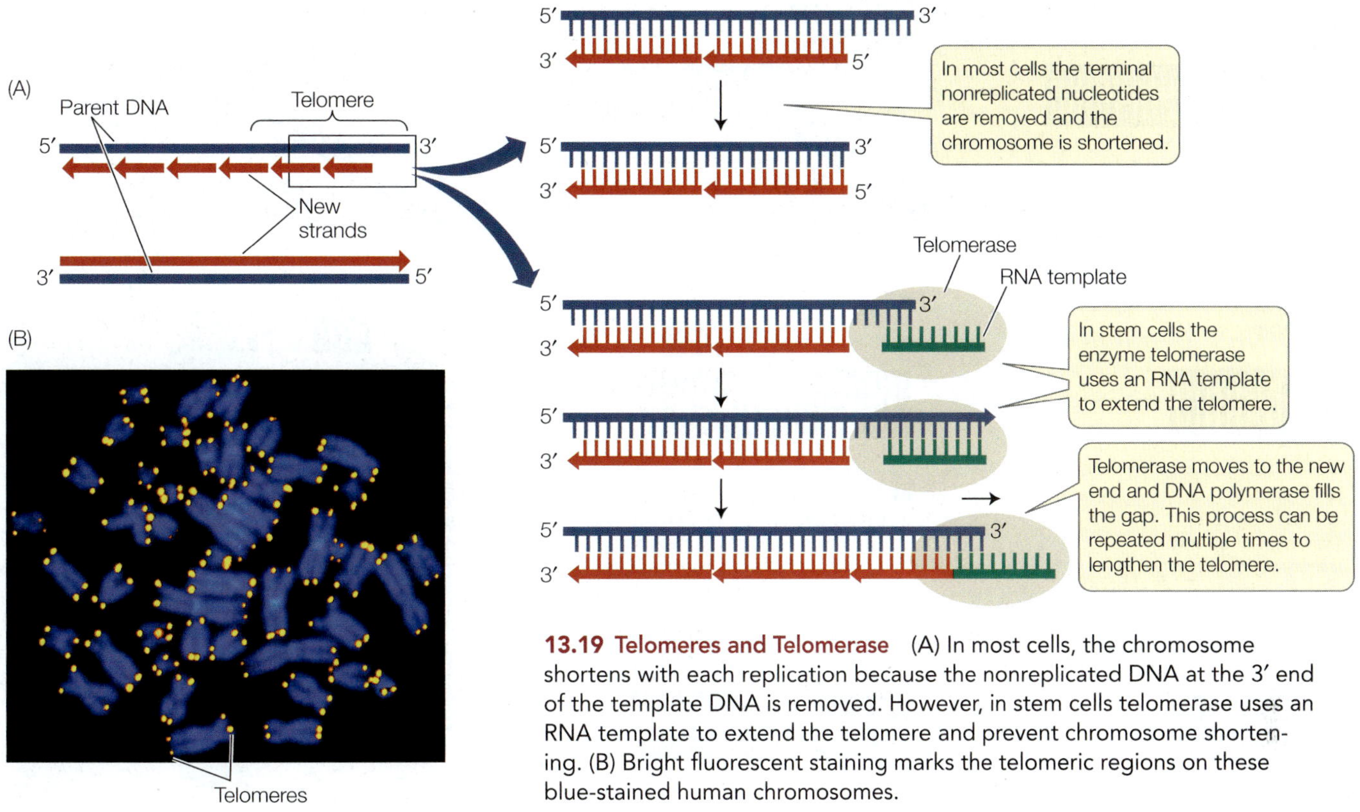

(A)
Parent DNA
Telomere
5′
3′
New strands
3′
5′
In most cells the terminal nonreplicated nucleotides are removed and the chromosome is shortened.
Telomerase
RNA template
In stem cells the enzyme telomerase uses an RNA template to extend the telomere.
Telomerase moves to the new end and DNA polymerase fills the gap. This process can be repeated multiple times to lengthen the telomere.
(B)
Telomeres

13.19 Telomeres and Telomerase (A) In most cells, the chromosome shortens with each replication because the nonreplicated DNA at the 3′ end of the template DNA is removed. However, in stem cells telomerase uses an RNA template to extend the telomere and prevent chromosome shortening. (B) Bright fluorescent staining marks the telomeric regions on these blue-stained human chromosomes.

most commonly in eukaryotes the replication complexes seem to be stationary, attached at specific positions on the nuclear matrix: a dynamic network of protein fibers within the nucleus. It is the DNA that moves, essentially sliding into the replication complex as one double-stranded molecule and emerging as two double-stranded molecules.

Telomeres are not fully replicated and are prone to repair

As we will discuss in Section 13.4, DNA may be damaged by radiation or chemicals. When this happens DNA repair mechanisms are activated, and breaks in DNA are rejoined via a combination of DNA synthesis and DNA ligase activity. So the ends of chromosomes are a potential problem: the DNA repair system might recognize the ends as breaks, and join two chromosomes together. This would create havoc with genomic integrity.

In many eukaryotes, there are repetitive sequences at the ends of chromosomes called **telomeres**. In humans, the telomere sequence is TTAGGG, and it is repeated about 2,500 times at each chromosome end. These repeats bind special proteins that prevent the DNA repair system from recognizing the chromosome ends as breaks. In addition, the repeats may form loops that have a similar protective role.

But there is another problem with chromosome ends. As we have discussed, replication of the lagging strand occurs by the addition of Okazaki fragments to RNA primers. When the terminal RNA primer is removed, no DNA can be synthesized to replace it because there is no 3′ end to extend. In most cells, the short piece of single stranded DNA at each end of the chromosome is removed. Thus the chromosome becomes slightly shorter with each cell division (**Figure 13.19**).

Each human chromosome can lose 50 to 200 base pairs of telomeric DNA after each round of DNA replication and cell division. After many cell divisions, the genes near the ends of the chromosomes can be lost, and the cell dies. This phenomenon explains, in part, why many cell lineages do not last the entire lifetime of the organism: their telomeres are lost. Continuously dividing cells, such as bone marrow stem cells and gamete-producing cells, have a special mechanism for maintaining their telomeric DNA. An enzyme called **telomerase** catalyzes the addition of any lost telomeric sequences in these cells (see Figure 13.19). Telomerase contains an RNA sequence that acts as a template for the telomeric DNA repeat sequence.

Telomerase is expressed in more than 90 percent of human cancers, and may be an important factor in the ability of cancer cells to divide continuously. Since most normal cells do not have this ability, telomerase is an attractive target for drugs designed to attack tumors specifically.

There is also interest in telomerase and aging. When cultured human cells are transformed with a telomerase gene that is expressed at high levels, their telomeres do not shorten. Instead of living 20 to 30 cell generations and then dying, the cells become immortal. It remains to be seen how this finding relates to the aging of a whole organism.

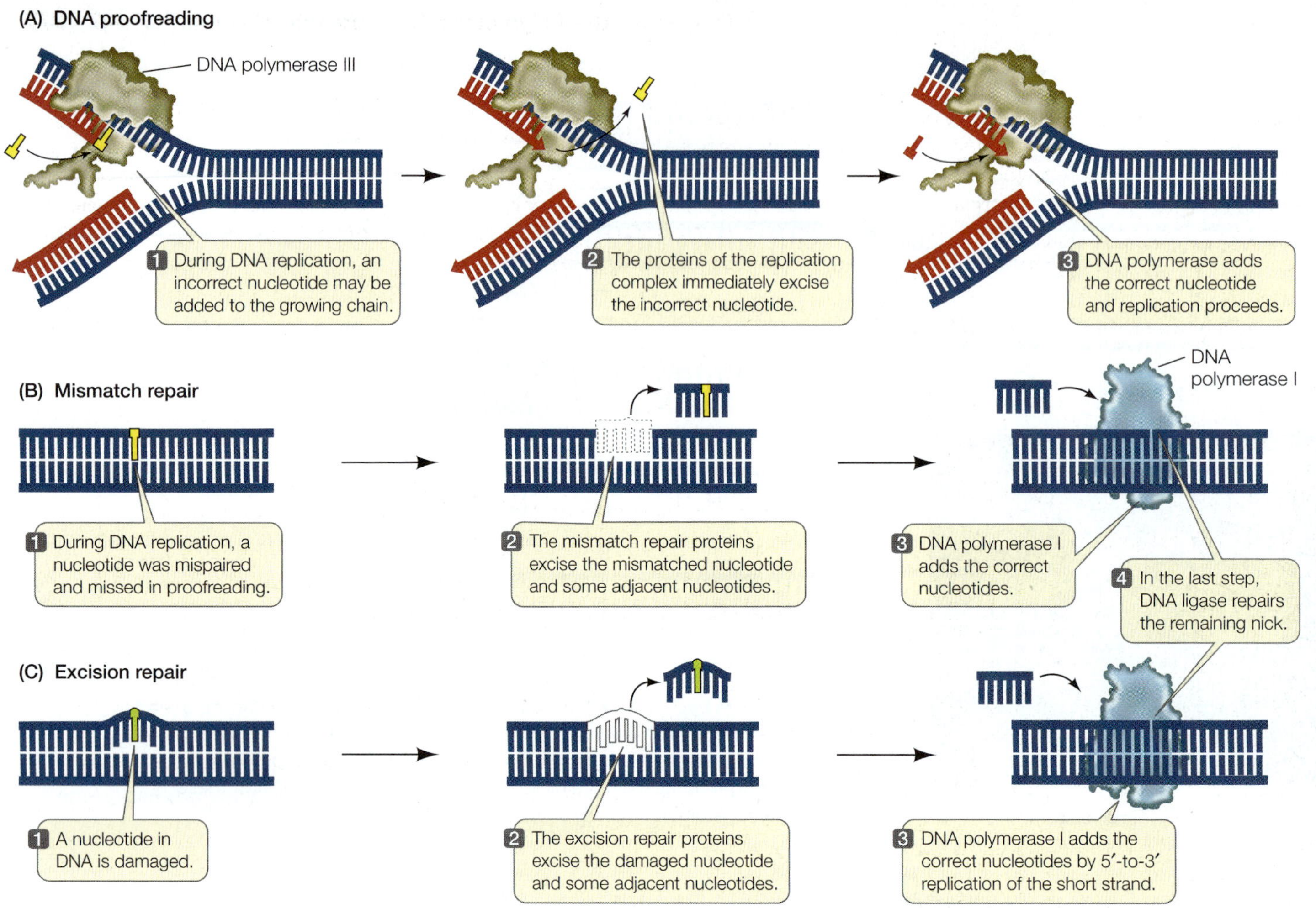
(A) DNA proofreading
DNA polymerase III
1 During DNA replication, an incorrect nucleotide may be added to the growing chain.
2 The proteins of the replication complex immediately excise the incorrect nucleotide.
3 DNA polymerase adds the correct nucleotide and replication proceeds.
(B) Mismatch repair
DNA polymerase I
1 During DNA replication, a nucleotide was mispaired and missed in proofreading.
2 The mismatch repair proteins excise the mismatched nucleotide and some adjacent nucleotides.
3 DNA polymerase I adds the correct nucleotides.
4 In the last step, DNA ligase repairs the remaining nick.
(C) Excision repair
1 A nucleotide in DNA is damaged.
2 The excision repair proteins excise the damaged nucleotide and some adjacent nucleotides.
3 DNA polymerase I adds the correct nucleotides by 5′-to-3′ replication of the short strand.

13.20 DNA Repair Mechanisms The proteins of the replication complex function in DNA repair mechanisms, reducing the rate of errors in the replicated DNA. Another mechanism (excision repair) repairs damage to existing DNA molecules.

RECAP 13.3

Meselson and Stahl showed that DNA replication is semiconservative: each parent DNA strand serves as a template for a new strand. A complex of proteins, most notably DNA polymerase, is involved in replication. New DNA is polymerized in one direction only, and since the two strands are antiparallel, one strand is made continuously and the other is synthesized in short Okazaki fragments that are eventually joined.

- How did the Meselson–Stahl experiment differentiate between the three models for DNA replication? See p. 268 and Figures 13.9, 13.10
- Name five enzymes needed for DNA replication. What are their roles? See pp. 271–274 and Figures 13.13–13.17
- Why is the leading strand of DNA replicated continuously while the lagging strand must be replicated in fragments? See pp. 272–273 and Figure 13.16

The complex process of DNA replication is amazingly accurate, but it is not perfect. What happens when things go wrong?

13.4 How Are Errors in DNA Repaired?

DNA must be accurately replicated and faithfully maintained. This is essential for the proper functioning of every cell, whether a prokaryote or a cell in a complex, multicellular organism. Yet the replication of DNA is not perfectly accurate, and DNA is subject to damage by chemicals and other environmental agents. In the face of these threats, how has life gone on for so long?

DNA repair mechanisms help preserve life. DNA polymerases initially make significant numbers of mistakes in assembling polynucleotide strands. Without DNA repair, the observed error rate of one for every 10^5 bases replicated would result in about 60,000 mutations every time a human cell divided. Fortunately, our cells can repair damaged nucleotides and correct DNA replication errors. Cells have at least three DNA repair mechanisms at their disposal:

- A **proofreading** mechanism corrects errors in replication as DNA polymerase makes them.
- A **mismatch repair** mechanism scans DNA immediately after it has been replicated and corrects any base-pairing mismatches.
- An **excision repair** mechanism removes abnormal bases that have formed because of chemical damage and replaces them with functional bases.

Most DNA polymerases perform a proofreading function each time they introduce a new nucleotide into a growing DNA strand (**Figure 13.20A**). When a DNA polymerase recognizes a mispairing of bases, it removes the improperly introduced nucleotide and tries again. (Other proteins in the replication complex also play roles in proofreading.) The error rate for this process is only about 1 in 10,000 repaired base pairs, and it lowers the overall error rate for replication to about one error in every 10^{10} bases replicated.

After the DNA has been replicated, a second set of proteins surveys the newly replicated molecule and looks for mismatched base pairs that were missed in proofreading (**Figure 13.20B**). For example, this mismatch repair mechanism might detect an A-C base pair instead of an A-T pair. The repair system can make the correct choice out of two options: remove the C and replace it with T, or remove the A and replace it with G. When mismatch repair fails, DNA sequences are altered. One form of colon cancer arises in part from a failure of mismatch repair.

DNA molecules can also be damaged during the life of a cell (for example, when it is in G1). High-energy radiation, chemicals from the environment, and random spontaneous chemical reactions can all damage DNA. For example, when adjacent thymines on the same DNA strand absorb ultraviolet light (at about 260 nm), they form a covalent bond between the bases, making a thymine dimer. These dimers interfere with base pairing during replication, leading to mutations. This is the primary cause of skin cancer in humans. Excision repair mechanisms deal with these kinds of damage (**Figure 13.20C**). Individuals who suffer from a condition known as xeroderma pigmentosum lack an excision repair mechanism that normally corrects the damage caused by ultraviolet radiation. They can develop skin cancers after even a brief exposure to sunlight.

RECAP 13.4

DNA replication is not perfect. In addition, DNA may be altered or damaged by environmental factors. Repair mechanisms detect and repair mismatched or damaged DNA.

- Explain the roles of DNA proofreading, mismatch repair, and excision repair. **See Figure 13.20**

Understanding how DNA is replicated and repaired has allowed scientists to develop techniques for studying genes. We'll look at just one of those techniques next.

13.5 How Does the Polymerase Chain Reaction Amplify DNA?

The principles underlying DNA replication in cells have been used to develop an important laboratory technique that has been vital in analyzing genes and genomes. This technique allows researchers to make multiple copies of short DNA sequences.

Go to Animated Tutorial 13.5
Polymerase Chain Reaction
Life10e.com/at13.5

The polymerase chain reaction makes multiple copies of DNA sequences

In order to study DNA and perform genetic manipulations, it is often necessary to make multiple copies of a DNA sequence. This is necessary because the amount of DNA isolated from a biological sample is often too small to work with. The **polymerase chain reaction** (**PCR**) technique essentially automates this replication process by copying a short region of DNA many times in a test tube. This process is referred to as DNA amplification.

The PCR reaction mixture contains:

- a sample of double-stranded DNA from a biological sample, to act as the template;
- two short, artificially synthesized primers that are complementary to the ends of the sequence to be amplified;
- the four dNTPs (dATP, dTTP, dCTP, and dGTP);
- a DNA polymerase that can tolerate high temperatures without becoming degraded; and
- salts and a buffer to maintain a near-neutral pH.

The PCR amplification is a cyclic process in which a sequence of steps is repeated over and over again (**Figure 13.21**):

- The first step involves heating the reaction mixture to near boiling point, to separate (denature) the two strands of the DNA template.
- The reaction is then cooled to allow the primers to bind (or anneal) to the template strands.
- Next, the reaction is warmed to an optimum temperature for the DNA polymerase to catalyze the production of the complementary new strands.

A single cycle takes a few minutes to produce two copies of the target DNA sequence, leaving the new DNA in the double-stranded state. Repeating the cycle many times leads to an exponential increase in the number of copies of the DNA sequence.

The PCR technique requires that the base sequences at the 3′ end of each strand of the target DNA sequence be known, so that complementary primers, usually 15 to 30 bases long, can be made in the laboratory. Because of the uniqueness of DNA sequences, a pair of primers this length will usually bind to only a single region of DNA in an organism's genome. This specificity, despite the incredible diversity of DNA sequences, is a key to the power of PCR.

One initial problem with PCR was its temperature requirements. To denature DNA, it must be heated to more than 90°C—a temperature that destroys most DNA polymerases. When the PCR technique was first being developed, new enzyme had to be added after denaturation in each cycle, which made the technique impractical for widespread use.

This problem was solved by nature: in the hot springs at Yellowstone National Park, as well as in other high-temperature locations, there lives a bacterium called, appropriately, *Thermus aquaticus* ("hot water"). The means by which this organism survives temperatures of up to 95°C was investigated by Thomas Brock and his colleagues at the University of Wisconsin, Madison. They discovered that *T. aquaticus* has an entire metabolic

RESEARCHTOOLS

13.21 The Polymerase Chain Reaction The steps in this cyclic process are repeated many times to produce multiple identical copies of a DNA fragment. This makes enough DNA for chemical analysis and genetic manipulations.

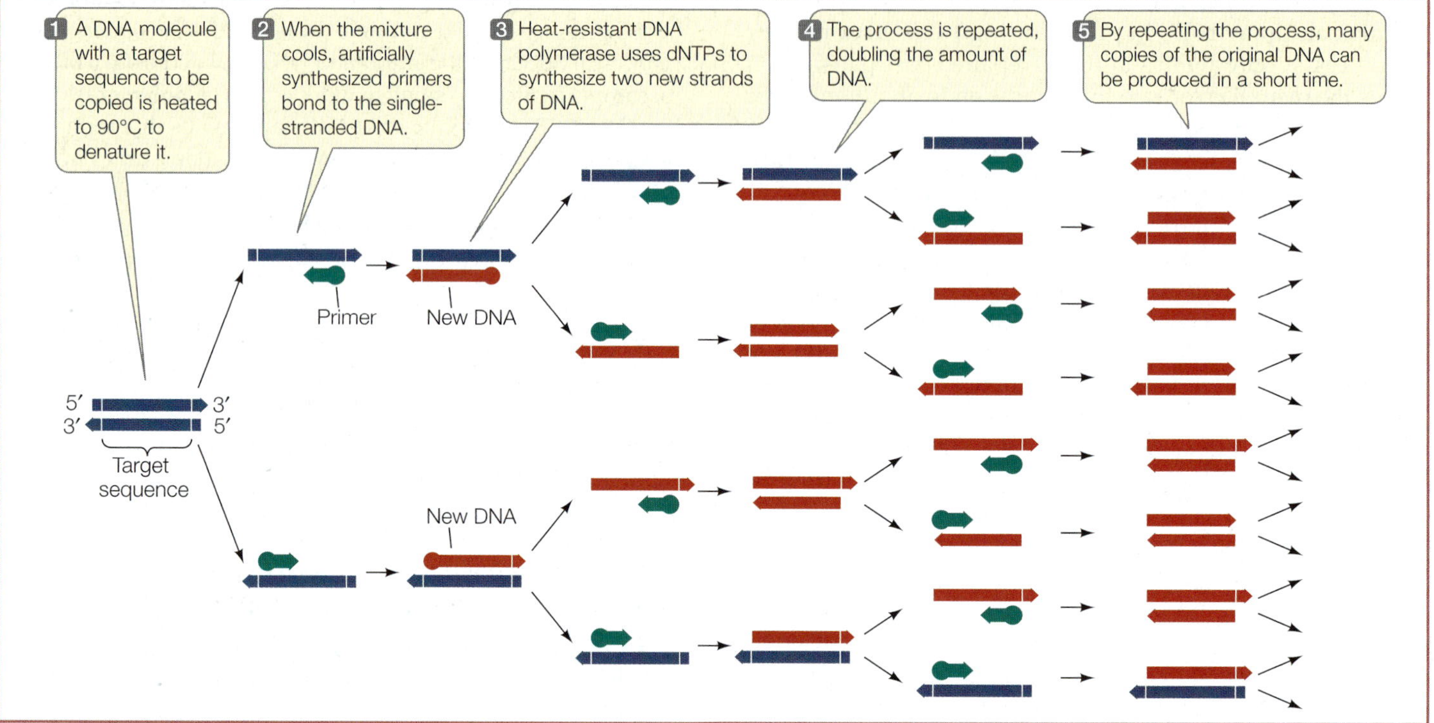
1 A DNA molecule with a target sequence to be copied is heated to 90°C to denature it.
2 When the mixture cools, artificially synthesized primers bond to the single-stranded DNA.
3 Heat-resistant DNA polymerase uses dNTPs to synthesize two new strands of DNA.
4 The process is repeated, doubling the amount of DNA.
5 By repeating the process, many copies of the original DNA can be produced in a short time.
Primer
New DNA
5′
3′
3′
5′
Target sequence
New DNA

machinery that is heat-resistant, including a DNA polymerase that does not denature at these high temperatures.

Scientists pondering the problem of copying DNA by PCR read Brock's basic research articles and got a clever idea: why not use *T. aquaticus* DNA polymerase in the PCR technique? It could withstand the 90°C denaturation temperature and would not have to be added during each cycle. The idea worked, and it earned biochemist Kary Mullis a Nobel prize. PCR has had an enormous impact on genetic research. Some of its most striking applications will be described in Chapters 15–18. These applications range from amplifying DNA in order to identify an individual person or organism, to detection of diseases.

RECAP 13.5

Knowledge of the mechanisms of DNA replication led to the development of a technique for making multiple copies of DNA sequences.

- What is the role of primers in PCR? **See p. 277 and Figure 13.21**

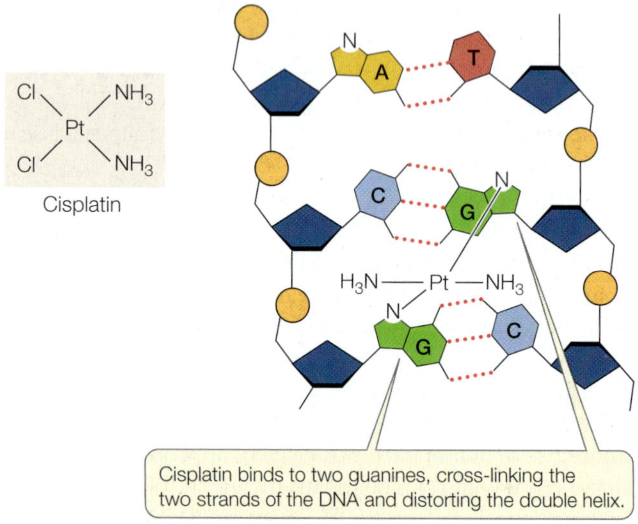
Cl
Cl
Pt
NH_3
NH_3
Cisplatin
N
A
T
C
G
N
H_3N — Pt — NH_3
N
G
C
Cisplatin binds to two guanines, cross-linking the two strands of the DNA and distorting the double helix.

13.22 Cisplatin: A Small but Lethal Molecule

How does cisplatin work?

ANSWER

Cisplatin contains a platinum atom bonded to two amino groups and two chlorine atoms (**Figure 13.22**). In Rosenberg's experiments, this compound was formed when the platinum electrode reacted with salts in the surrounding solution. The bonds between the platinum atom and the chlorine atoms are weak, and the latter can be displaced by electron-rich substances (you may know from chemistry that these are called nucleophiles). In DNA, one of the nitrogen atoms of guanine (see Figure 4.1) displaces one of the chlorines, forming a strong covalent bond. If there is a nearby guanine on the opposite DNA strand, it displaces the other chlorine and the DNA becomes cross-linked. This type of DNA lesion is not repaired by any of the cell's usual DNA repair mechanisms.

CHAPTERSUMMARY 13

13.1 What Is the Evidence that the Gene Is DNA?

- Griffith's experiments in the 1920s demonstrated that some substance in cells can cause heritable changes in other cells. **Review Figure 13.1**
- The location and quantity of DNA in the cell suggested that DNA might be the genetic material. Avery and his colleagues isolated the transforming principle from bacteria and identified it as DNA. **Review Figure 13.2**
- The Hershey–Chase experiments established conclusively that DNA (and not protein) is the genetic material, by tracing the DNA of radioactively labeled viruses, with which they infected bacterial cells. **Review Figure 13.4, ANIMATED TUTORIAL 13.1**
- Genetic transformation of eukaryotic cells is often called **transfection**. Transformation and transfection can be studied with the aid of a **genetic marker** gene that confers a known and observable phenotype. **Review Figure 13.5**

13.2 What Is the Structure of DNA?

- Chargaff's rule states that the amount of **adenine** in DNA is equal to the amount of **thymine**, and that the amount of **guanine** is equal to the amount of **cytosine**; thus the total abundance of purines (A + G) equals the total abundance of pyrimidines (T + C).
- X-ray crystallography showed that the DNA molecule is a double helix. Watson and Crick proposed that the two strands in DNA are **antiparallel**. **Review Figure 13.7**
- **Complementary base pairing** between A and T and between G and C accounts for Chargaff's rule. The bases are held together by hydrogen bonding.
- Reactive groups are exposed in the paired bases, allowing for recognition by other molecules such as proteins. **Review Figure 13.8**

13.3 How Is DNA Replicated?
See ANIMATED TUTORIAL 13.2

- Meselson and Stahl showed that DNA undergoes **semiconservative replication**. Each parent strand acts as a **template** for the synthesis of a new strand; thus the two replicated DNA molecules each contain one parent strand and one newly synthesized strand. **Review Figure 13.10, ANIMATED TUTORIAL 13.3**
- In DNA replication, the enzyme **DNA polymerase** catalyzes the addition of nucleotides to the 3′ end of each strand. Which nucleotides are added is determined by complementary base pairing with the template strand. **Review Figure 13.11**
- The pre-replication complex is a huge protein complex that attaches to the chromosome at the **origin of replication** (*ori*).
- Replication proceeds from the origin of replication on both strands in the 5′-to-3′ direction, forming a **replication fork**. **Review Figure 13.12**
- **Primase** catalyzes the synthesis of a short RNA **primer** to which nucleotides are added by DNA polymerase. **Review Figure 13.13**
- Many proteins assist in DNA replication. **DNA helicase** separates the strands, and **single-strand binding proteins** keep the strands from reassociating. **Review Figure 13.15, ACTIVITY 13.1**
- The **leading strand** is synthesized continuously and the **lagging strand** in pieces called **Okazaki fragments**. The fragments are joined together by **DNA ligase**. **Review Figures 13.16, 13.17, ANIMATED TUTORIAL 13.4**
- The speed with which DNA polymerization proceeds is attributed to the **processive** nature of DNA polymerases, which can catalyze many polymerizations at a time. A **sliding DNA clamp** helps ensure the stability of this process. **Review Figure 13.18**
- At the ends of eukaryotic chromosomes are regions of repetitive DNA sequence called **telomeres**. Unless the enzyme **telomerase** is present, a short segment at the end of each telomere is lost each time the DNA is replicated. After multiple cell cycles, the telomeres shorten enough to cause chromosome instability and cell death. **Review Figure 13.19**

13.4 How Are Errors in DNA repaired?

- DNA polymerases make about one error in 10^5 bases replicated. DNA is also subject to natural alterations and chemical damage. DNA can be repaired by at least three different mechanisms, including **proofreading**, **mismatch repair**, and **excision repair**. **Review Figure 13.20**

13.5 How Does the Polymerase Chain Reaction Amplify DNA?

- The **polymerase chain reaction** technique uses DNA polymerase to make multiple copies of DNA in the laboratory. **Review Figure 13.21, ANIMATED TUTORIAL 13.5**

Go to the Interactive Summary to review key figures, Animated Tutorials, and Activities
Life10e.com/is13

CHAPTERREVIEW

REMEMBERING

1. In the Hershey–Chase experiment,
   a. DNA from parent bacteriophages appeared in progeny bacteriophages.
   b. most of the phage DNA never entered the bacteria.
   c. more than three-fourths of the phage protein appeared in progeny phages.
   d. DNA was labeled with radioactive sulfur.
   e. DNA formed the coat of the bacteriophages.

2. Which statement about complementary base pairing is not true?
   a. Complementary base pairing plays a role in DNA replication.
   b. In DNA, T pairs with A.
   c. Purines pair with purines, and pyrimidines pair with pyrimidines.
   d. In DNA, C pairs with G.
   e. The base pairs are of equal length.

3. In semiconservative replication of DNA,
   a. the original double helix remains intact and a new double helix forms.
   b. the strands of the double helix separate and act as templates for new strands.
   c. polymerization is catalyzed by RNA polymerase.
   d. polymerization is catalyzed by a double-helical enzyme.
   e. DNA is synthesized from amino acids.

4. The primer used for DNA replication
   a. is a short strand of RNA added to the 3′ end.
   b. is needed only once on a leading strand.
   c. remains on the DNA after replication.
   d. ensures that there will be a free 5′ end to which nucleotides can be added.
   e. is added to only one of the two template strands.

5. The role of DNA ligase in DNA replication is to
   a. add more nucleotides to the growing strand one at a time.
   b. open up the two DNA strands to expose template strands.
   c. ligate base to sugar to phosphate in a nucleotide.
   d. bond Okazaki fragments to one another.
   e. remove incorrectly paired bases.

6. What is the correct order for the following events in excision repair of DNA? (1) DNA polymerase I adds correct nucleotides by 5′-to-3′ replication; (2) damaged nucleotides are recognized; (3) DNA ligase seals the new strand to existing DNA; (4) part of a single strand is excised.
   a. 1, 2, 3, 4
   b. 2, 1, 3, 4
   c. 2, 4, 1, 3
   d. 3, 4, 2, 1
   e. 4, 2, 3, 1

UNDERSTANDING & APPLYING

7. One strand of DNA has the sequence 5′-ATTCCG-3′ The complementary strand for this is
   a. 5′-TAAGGC-3′
   b. 5′-ATTCCG-3′
   c. 5′-ACCTTA-3′
   d. 5′-CGGAAT-3′
   e. 5′-GCCTTA-3′

8. Using the following information, calculate the number of origins of DNA replication on a human chromosome: DNA polymerase adds nucleotides at 3,000 base pairs per minute in one direction; replication is bidirectional; S phase lasts 300 minutes; there are 120 million base pairs per chromosome. In a typical chromosome 3 micrometers (μm) long, how many origins are there per micrometer?

9. The drug dideoxycytidine, used to treat certain viral infections, is a nucleotide made with 2′,3′-dideoxyribose. This sugar lacks —OH groups at both the 2′ and the 3′ positions. Explain why this drug stops the growth of a DNA chain when added to DNA.

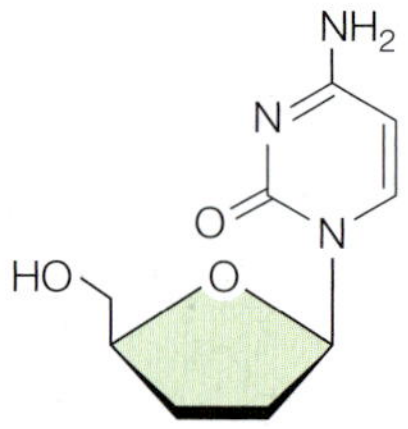
NH2
N
O
N
HO
O
Dideoxycytidine

ANALYZING & EVALUATING

10. Suppose that Meselson and Stahl had continued their experiment on DNA replication for another ten bacterial generations. Would there still have been any ^{14}N–^{15}N hybrid DNA present? Would it still have appeared in the centrifuge tube? Explain.

11. Outline a series of experiments using radioactive isotopes (such as ^{32}P and ^{35}S) to show that during bacterial conjugation it is DNA and not protein that moves from the donor cell to the recipient cell and is responsible for bacterial transformation.

Go to BioPortal at **yourBioPortal.com** for Animated Tutorials, Activities, LearningCurve Quizzes, Flashcards, and many other study and review resources.

14 From DNA to Protein: Gene Expression

CHAPTER**OUTLINE**

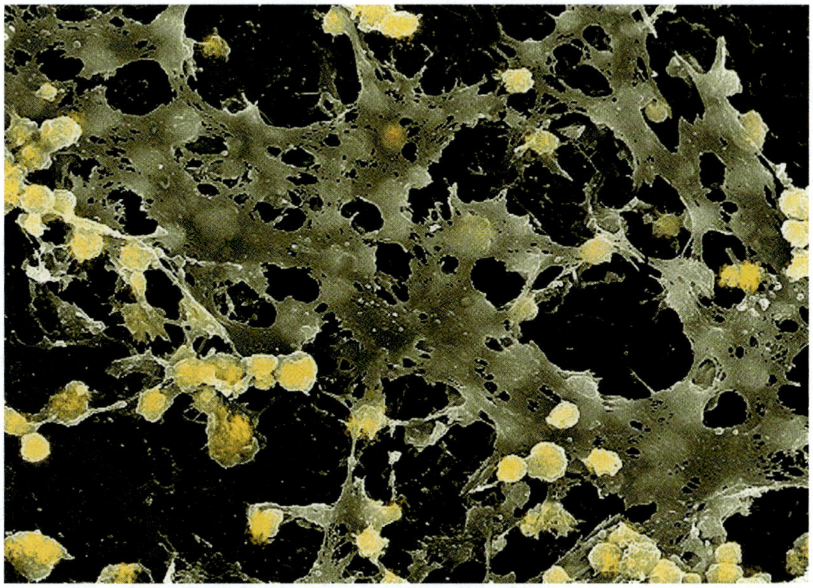

MRSA Methicillin-resistant *Staphylococcus aureus*, a major cause of serious illness and death in the United States and Europe, is treated with antibiotics that target its gene expression.

AT AGE 87, Janet enjoyed her life at the Sunshine Senior Citizen's Rest Home. Her only medical problem was periodic loss of bladder control, but this was alleviated when her nurse placed a plastic tube called a catheter into her bladder with an unobtrusive external bag. Things were going well until one day, unbeknown to Janet, some *Staphylococcus aureus* bacteria from the environment got into the bag and catheter. Finding a hospitable environment on the inner surface of the plastic, the bacteria attached to it via extracellular polysaccharides. Gradually, they divided and formed a colony, recruiting other free-living *S. aureus* cells that happened by. Within a few weeks, a slimy bacteria-laden coating called a biofilm (similar to dental plaque) had formed.

In time, some of the bacteria from the biofilm entered Janet's body and began to reproduce. Her advanced age and weak immune system permitted a significant infection to develop in her bladder and lungs. Fever, chills, shortness of breath, and the beginnings of kidney failure raged in her body. Racing for a treatment, her doctor first got a sample of the bacteria that Janet coughed up and sent it to a pathology lab for testing with antibiotics. The mainstay of treatment of "Staph" has been a penicillin-related antibiotic, methicillin, that binds to a bacterial protein needed to make new cell walls after cell division. Unfortunately, the bacteria had a mutation that made the target protein resistant to the antibiotic.

The lab made a diagnosis of MRSA: methicillin-resistant *S. aureus*. These "superbugs" are now common in hospitals and nursing homes and cause about 20,000 deaths a year in the United States.

Finally, Janet's doctor tried the antibiotic tetracycline. This molecule prevents gene expression in *S. aureus* and many other bacteria, but does not affect gene expression in eukaryotic cells. It has this specificity because it binds to a bacteria-specific protein in the ribosome, preventing the attachment of transfer RNA that carries amino acids to the ribosome for protein synthesis. Eukaryotic ribosomes do not have a binding site for tetracycline. Fortunately, the antibiotic killed the MRSA that infected Janet's bladder and lungs, and she is happy and healthy. But there are strains of MRSA that have developed resistance to tetracycline and other antibiotics as well. Scientists are now working to find new ways to combat this very serious problem.

Can new treatments focused on gene expression control MRSA?

See answer on p. 301.

14.1 What Is the Evidence that Genes Code for Proteins?

Chapter 4 introduced DNA and its role in gene expression. Chapter 13 described how DNA is the carrier of genetic information and how DNA is replicated. Here we will focus on the evidence for proteins as major products of gene expression, and we will describe how a gene is expressed as protein.

The molecular basis of phenotypes was actually discovered before it was known that DNA was the genetic material. Scientists had studied the chemical differences between individuals carrying wild-type and mutant alleles in organisms as diverse as humans and bread molds. They found that the major phenotypic differences resulted from differences in specific proteins.

Observations in humans led to the proposal that genes determine enzymes

The identification of a gene product as a protein began with a mutation. In the early twentieth century, the English physician Archibald Garrod saw several children with a rare disease. One symptom of the disease was that the urine turned dark brown when exposed to air. This was especially noticeable on the infants' diapers. The disease was given the descriptive name alkaptonuria ("black urine").

Garrod noticed that the disease was most common in children whose parents were first cousins. Mendelian genetics had just been "rediscovered," and Garrod realized that because first cousins share on average 1/8 of their alleles, the children of first cousins might inherit a rare mutant allele from both parents. He proposed that alkaptonuria was a phenotype caused by a recessive, mutant allele.

Garrod took the analysis further by identifying the biochemical abnormality in the affected children. He isolated from them an unusual substance, homogentisic acid, which accumulated in blood, joints (where it crystallized and caused severe pain), and urine (where it turned black). The chemical structure of homogentisic acid is similar to that of the amino acid tyrosine:

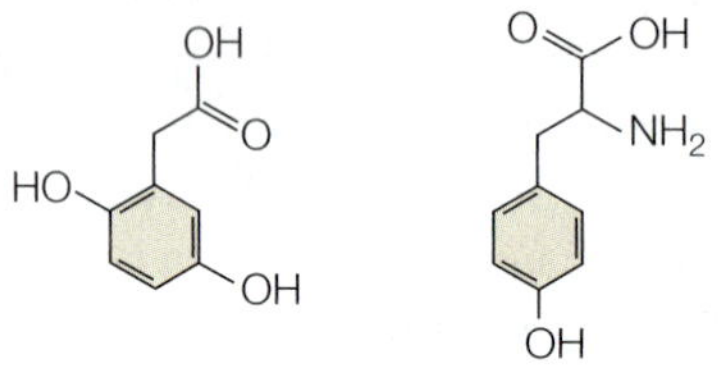
OH
O
HO
OH
O
OH
NH2
OH

Homogentisic acid Tyrosine

Enzymes as biological catalysts had just been discovered. Garrod proposed that homogentisic acid was a breakdown product of tyrosine. Normally, homogentisic acid would be converted to a harmless product. According to Garrod, there was a normal human allele that determined the synthesis of an enzyme that catalyzed this conversion:

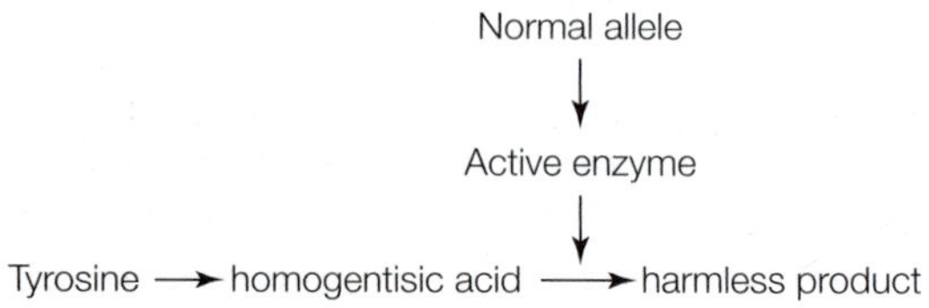
Normal allele
Active enzyme
Tyrosine
homogentisic acid
harmless product

When the allele was mutated, the enzyme was inactive and homogentisic acid accumulated instead:

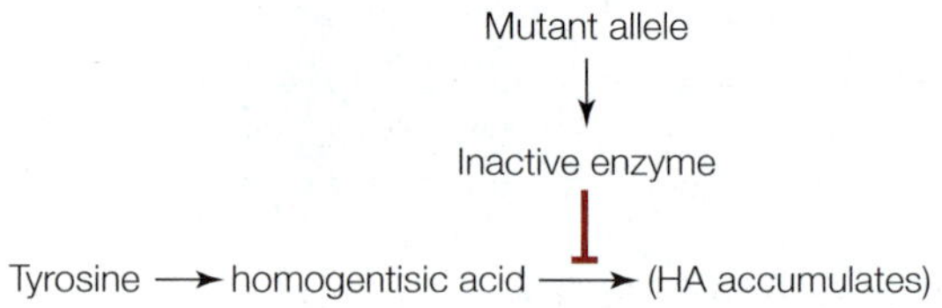
Mutant allele
Inactive enzyme
Tyrosine
homogentisic acid
(HA accumulates)

Therefore Garrod correlated *one gene to one enzyme* and coined the term "inborn error of metabolism" to describe this genetically determined biochemical disease. But his hypothesis needed direct confirmation by the identification of the specific enzyme and specific gene mutation involved. This did not occur until the enzyme, homogentisic acid oxidase, was described as active in healthy people and inactive in alkaptonuria patients in 1958, and the specific DNA mutation was described in 1996.

To relate genes and enzymes more generally, biologists turned to simpler organisms that could be manipulated in the laboratory.

Experiments on bread mold established that genes determine enzymes

As they work to explain the principles that govern life, biologists often turn to organisms that are easy to manipulate experimentally. Such **model organisms** have certain characteristics that make them attractive experimental subjects. For example:

- They are easy to grow in the laboratory or greenhouse.
- They have short generation times.
- They are easy to manipulate genetically, by crossing or by other methods.
- They often produce large numbers of progeny.

Biologists have used model organisms to develop principles of genetics that can then be applied more generally to other organisms. You have seen some of these organisms in previous chapters:

- Pea plants (*Pisum sativum*) were used by Mendel in his genetics experiments.
- Fruit flies (*Drosophila*) were used by Morgan in his genetics experiments.
- *Escherichia coli* was used by Meselson and Stahl to study DNA replication.

To this list we now add the bread mold *Neurospora*. *Neurospora* is an ascomycete fungus (see Chapter 30). This mold is haploid for most of its life, so there are no dominant or recessive alleles: all alleles are expressed phenotypically and are not masked by a heterozygous condition. *Neurospora* is simple to grow in the laboratory. Biologists at Stanford University led by George Beadle and Edward Tatum undertook studies to biochemically define the phenotypes in *Neurospora*.

Like Garrod, Beadle and Tatum hypothesized that the expression of a specific gene results in the activity of a specific enzyme. Now, they set out to test this hypothesis directly. They grew *Neurospora* on a nutritional medium containing sucrose, minerals, and biotin, which is the only vitamin that wild-type *Neurospora* cannot synthesize itself. Using this minimal medium, the enzymes of wild-type *Neurospora* could catalyze all the metabolic reactions needed for growth.

The scientists then treated the wild-type *Neurospora* with X rays, which function as a mutagen. A **mutagen** is something that damages DNA, causing **mutations**: heritable alterations in the DNA sequence. After the X-ray treatment, some *Neurospora* strains could no longer grow on the minimal medium. These mutant strains grew only if they were supplied with specific additional nutrients, such as particular vitamins. Beadle and Tatum hypothesized that these genetic strains had mutations in the genes that code for production of enzymes needed to synthesize the additional nutrients. For each mutant strain, the scientists were able to find a single compound that, when added to the minimal medium, supported the growth of that strain. These results suggested that mutations have simple effects, and that each mutation causes a defect in only one enzyme in a metabolic pathway. These conclusions confirmed Garrod's **one-gene, one-enzyme hypothesis**.

Mutations provide a powerful way to determine cause and effect in biology. Nowhere has this been more evident than in the elucidation of biochemical pathways. Such pathways consist of sequential events (chemical reactions) in which each event is dependent on the occurrence of the preceding event. The general reasoning is as follows:

- *Observation*. A particular gene (*a*) is present and a particular reaction catalyzed by a particular enzyme (A) occurs; the two are correlated.
- *Hypothesis*. Gene *a* determines the synthesis of the enzyme A.
- *Test of hypothesis*. Mutate gene *a*. Prediction: no functional enzyme is made, and the reaction does not occur.

Two colleagues of Beadle and Tatum, Adrian Srb and Norman Horowitz, used this experimental approach to isolate *Neurospora* mutants that could not survive without the amino acid arginine in their growth medium. By adding particular compounds to the medium, Srb and Horowitz were able to identify a series of steps in the biochemical pathway leading to the synthesis of arginine (**Figure 14.1**).

INVESTIGATING**LIFE**

14.1 One Gene, One Enzyme Srb and Horowitz had several mutant strains of *Neurospora* that could not make arginine (*arg*). Several compounds are needed for arginine synthesis. By testing these compounds in the growth media for the mutant strains, the researchers deduced that each mutant strain was deficient in one enzyme along a biochemical pathway.[a]

HYPOTHESIS Each gene determines an enzyme in a biochemical pathway.

Method Place spores (single cells that divide to produce mold colonies) of each arg mutant strain on a minimal nutritional medium with and without supplements.

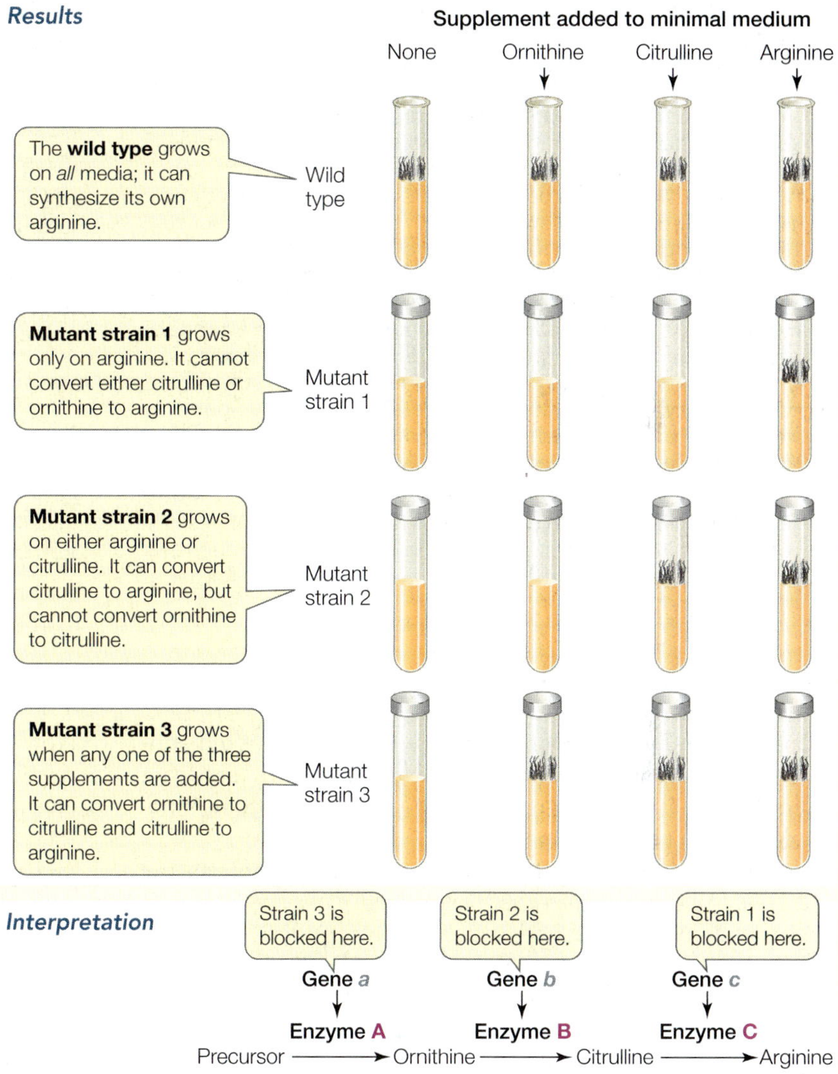

Results
Supplement added to minimal medium
None
Ornithine
Citrulline
Arginine
The wild type grows on all media; it can synthesize its own arginine.
Wild type
Mutant strain 1 grows only on arginine. It cannot convert either citrulline or ornithine to arginine.
Mutant strain 1
Mutant strain 2 grows on either arginine or citrulline. It can convert citrulline to arginine, but cannot convert ornithine to citrulline.
Mutant strain 2
Mutant strain 3 grows when any one of the three supplements are added. It can convert ornithine to citrulline and citrulline to arginine.
Mutant strain 3
Interpretation
Strain 3 is blocked here.
Strain 2 is blocked here.
Strain 1 is blocked here.
Gene a
Gene b
Gene c
Enzyme A
Enzyme B
Enzyme C
Precursor
Ornithine
Citrulline
Arginine

If an organism cannot convert one particular compound to another, it presumably lacks an enzyme required for conversion, and the mutation is in the gene that codes for that enzyme.

CONCLUSION Each gene specifies a particular enzyme.

Go to **BioPortal** for discussion and relevant links for all INVESTIGATING**LIFE** figures.

[a]Srb, A. M. and N. H. Horowitz. 1944. *Journal of Biological Chemistry* 154: 129–139.

One gene determines one polypeptide

The one-gene, one-enzyme relationship has undergone several modifications in light of our current knowledge of molecular biology. Many proteins, including many enzymes, are composed of more than one polypeptide chain, or subunit (that is, they have a quaternary structure; see Section 3.2). Look at the illustration of hemoglobin in Figure 3.11.

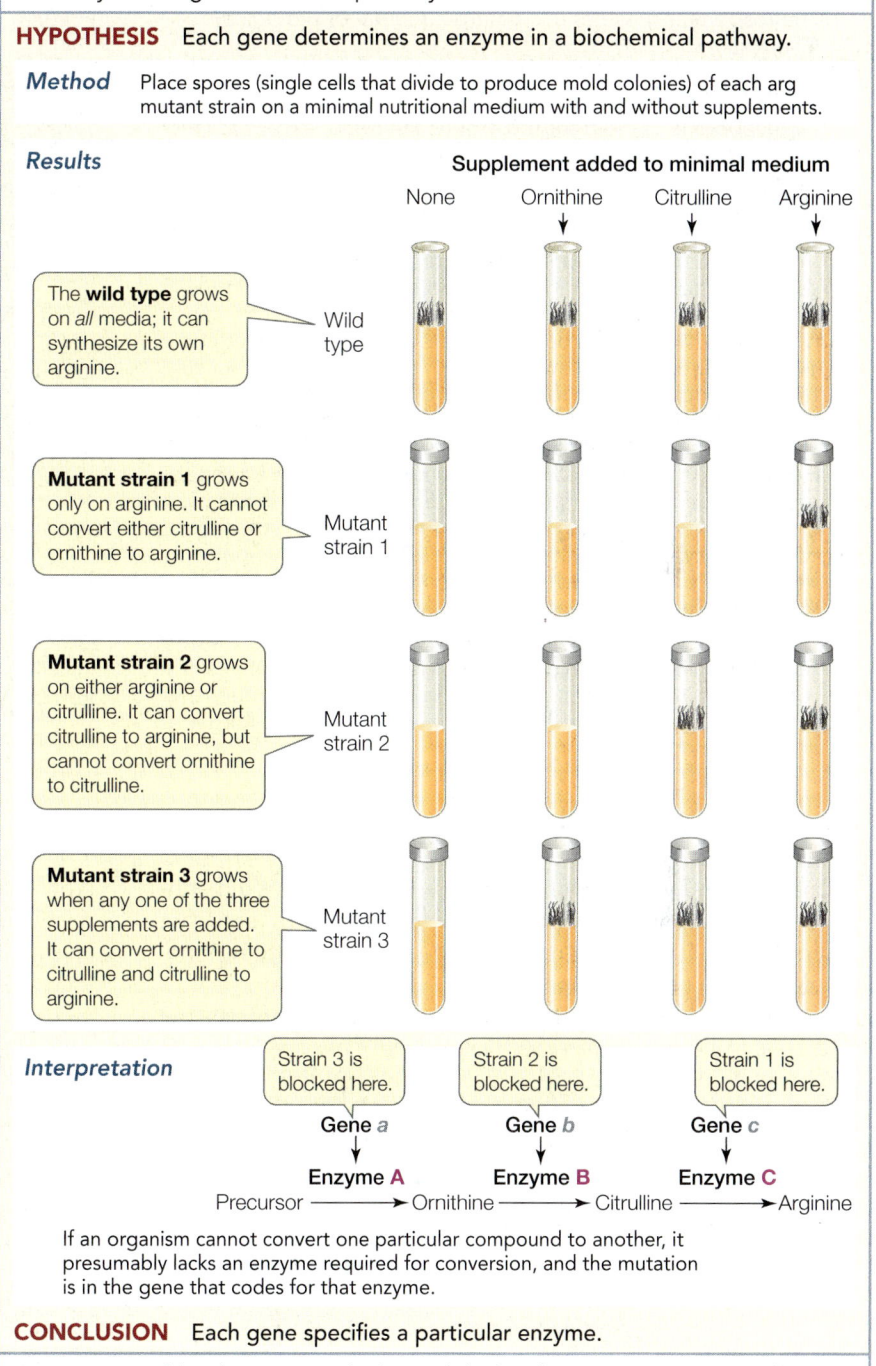

WORKING WITH **DATA:**

One Gene, One Enzyme

Original Paper

Srb, A. M. and N. H. Horowitz. 1944. The ornithine cycle in *Neurospora* and its genetic control. *Journal of Biological Chemistry* 154: 129–139.

Analyze the Data

Neurospora is haploid for most of its life cycle, except for the formation of a diploid cell when it undergoes mating; the cell then undergoes meiosis to form haploid spores. Beadle and Tatum used X rays to cause genetic mutations in *Neurospora*. They isolated mutant strains that were unable to grow on minimal medium, but were able to grow if the medium was supplemented with particular compounds. Their colleagues Adrian Srb and Norman Horowitz analyzed 15 mutant strains (the *arg* mutants) that could not synthesize arginine, but could grow on medium supplemented with arginine. The scientists tested various compounds and found two, ornithine and citrulline, that could be used instead of arginine to support the growth of some of the mutant strains (see Figure 14.1). Srb and Horowitz concluded that ornithine and citrulline are intermediates in the biochemical pathway leading to the synthesis of arginine. The various *arg* strains had mutations in genes that encode enzymes that catalyzed different steps in this pathway. The results of this and similar experiments supported the "one-gene, one-enzyme" hypothesis.

Fifteen mutant strains that required arginine for growth were tested for growth in the presence of the other substances. The results for three of the strains are shown in the first three rows of the table, with growth expressed as dry weight of fungal material after growth for 5 days.

QUESTION 1

Based on the biochemical pathway for arginine synthesis shown in Figure 14.1, which enzyme (A, B, or C) was mutated in each strain?

QUESTION 2

Why was there some growth in strains 34105 and 33442 even when there were no additions to the growth medium?

QUESTION 3

Nineteen other amino acids were tested as substitutes for arginine in the three strains. In all cases, there was no growth. Explain these results.

QUESTION 4

Sexual reproduction in *Neurospora* was used to create double mutants, which carried the mutations from both parental strains. A double mutant derived from strains 33442 and 36703 had the growth characteristics shown in the last row of the table. Explain these data in terms of the genes, mutations, and biochemical pathway.

Strain	Arginine added	Citrulline added	Ornithine added	No addition
34105	33.2	30.0	25.5	1.1
33442	43.8	42.7	2.5	2.3
36703	20.4	0.0	0.0	0.0
Double mutant	22.0	0.0	0.0	0.0

Go to **BioPortal** for all WORKING WITH**DATA** exercises

This protein has four polypeptides—two α and two β subunits, and the different subunits are encoded by separate genes. Thus it is more correct to speak of a **one-gene, one-polypeptide relationship**.

So far we have seen that in terms of protein synthesis, the *function of a gene is to prescribe the production of a single, specific polypeptide*. But not all genes code for polypeptides. As we will see below and in Chapter 16, there are many DNA sequences that are transcribed to RNA molecules that are not translated into polypeptides, but instead have other functions.

RECAP 14.1

Studies of mutations in humans and bread molds led to our understanding of the one-gene, one-polypeptide relationship. In most cases, the function of a gene is to code for a specific polypeptide.

- What is a model organism, and why is *Neurospora* a good model for studying biochemical genetics? **See p. 282**
- How were the experiments on *Neurospora* set up to determine the order of steps in a biochemical pathway? **See pp. 282–283 and Figure 14.1**
- Explain the distinction between the phrases "one-gene, one-enzyme" and "one-gene, one-polypeptide." **See pp. 283–284**

Now that we have established the one-gene, one-polypeptide relationship, how does it work? That is, how is the information encoded in DNA used to produce a particular polypeptide?

14.2 How Does Information Flow from Genes to Proteins?

As we discussed in Chapter 13 and Section 14.1, two of the greatest biological discoveries of the twentieth century were that DNA is the hereditary material, and that DNA codes for proteins. We now know that the human genome contains about 21,000 protein-coding genes along with thousands of other genes that are transcribed into noncoding RNA molecules, which themselves have various functions in the cell. In the remainder of this chapter we will focus on the processes that occur when a protein-coding gene is expressed. You will see that certain noncoding RNAs have important roles in this process. Gene expression was briefly outlined in Section 4.1. To review, this process occurs in two major steps:

- During **transcription**, the information in a DNA sequence (a gene) is copied into a complementary RNA sequence.
- During **translation**, this RNA sequence is used to create the amino acid sequence of a polypeptide.

DNA
Transcription
RNA
Translation
Polypeptide

Francis Crick and James Watson deciphered the structure of DNA in 1953 (see Section 13.2), and it was they who first proposed this model for gene expression. They took the concept further by suggesting that gene expression can only go in one direction: DNA can be used to create a protein, but a protein can never be used to create DNA. At the time, Crick called this "the **central dogma** of molecular biology."

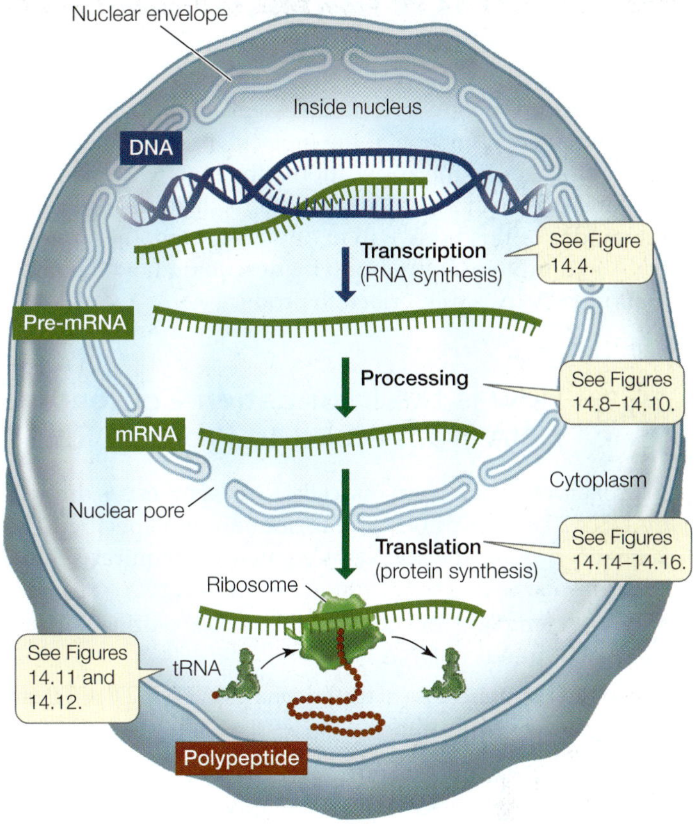

Nuclear envelope
Inside nucleus
DNA
Transcription
(RNA synthesis)
See Figure 14.4.
Pre-mRNA
Processing
See Figures 14.8–14.10.
mRNA
Cytoplasm
Nuclear pore
Translation
(protein synthesis)
See Figures 14.14–14.16.
Ribosome
See Figures 14.11 and 14.12.
tRNA
Polypeptide

14.2 From Gene to Protein This diagram summarizes the processes of gene expression in eukaryotes.

Go to Activity 14.1 Eukaryotic Gene Expression
Life10e.com/ac14.1

Three types of RNA have roles in the information flow from DNA to protein

There are numerous types of RNA. Three of them have vital roles in gene expression.

MESSENGER RNA AND TRANSCRIPTION When a particular gene is expressed, one of the two DNA strands in the gene is transcribed to produce a complementary RNA strand, which is then processed to produce **messenger RNA** (**mRNA**). In eukaryotic cells, the mRNA travels from the nucleus to the cytoplasm, where it is translated into a polypeptide (**Figure 14.2**). The nucleotide sequence of the mRNA determines the ordered sequence of amino acids in the polypeptide chain, which is built by a ribosome.

RIBOSOMAL RNA AND TRANSLATION The ribosome is essentially a protein synthesis factory composed of multiple proteins and several **ribosomal RNAs** (**rRNAs**). One of the rRNAs catalyzes peptide bond formation between amino acids, to form a polypeptide.

TRANSFER RNA MEDIATES BETWEEN mRNA AND PROTEIN Another RNA, called **transfer RNA** (**tRNA**), can both bind a specific amino acid and recognize specific sequences of nucleotides in mRNA. It is the tRNA that recognizes which amino acid should be added next to a growing polypeptide chain (see Figure 14.2).

Go to Media Clip 14.1
Protein Synthesis: An Epic on a Cellular Level
Life10e.com/mc14.1

In some cases, RNA determines the sequence of DNA

Certain viruses present exceptions to the general process of gene expression outlined above. As we saw in Section 13.1, a virus is a non-cellular infectious particle that reproduces inside cells. Many viruses, such as the tobacco mosaic virus, influenza viruses, and poliovirus, have RNA rather than DNA as their genetic material. With its nucleotide sequence, RNA could potentially act as an information carrier and be expressed as a protein. But if RNA is usually single-stranded, how do these viruses replicate? They generally solve this problem by transcribing from RNA to RNA, making an RNA strand that is complementary to their genomes. This "opposite" strand is then used to make multiple copies of the viral genome by transcription:

RNA
RNA
Polypeptide

Human immunodeficiency viruses (HIV) and certain rare tumor viruses also have RNA as their genomes, but do not replicate by transcribing from RNA to RNA. Instead, after infecting a host cell, such a virus makes a DNA copy of its genome, which becomes incorporated into the host's genome. The virus relies on the host cell's transcription machinery to make more RNA. This RNA can be either translated to produce viral proteins, or incorporated as the viral genome into new viral particles. Synthesis of DNA from RNA is called **reverse transcription**, and not surprisingly, such viruses are called **retroviruses**.

DNA
RNA
Polypeptide

RECAP 14.2

When protein-coding genes are expressed, the DNA code is used to produce RNA, and the RNA sequence determines the order of amino acids in a polypeptide. Transcription is the process by which the information in DNA is copied into RNA. Translation is the process by which this information is converted into a polypeptide chain.

- What is the central dogma of molecular biology? **See p. 285**
- What are the roles of mRNA, rRNA, and tRNA in gene expression? **See p. 285 and Figure 14.2**

Understanding gene expression is essential for understanding how organisms function at the molecular level. This understanding is key to the application of biology to human welfare, in areas such as agriculture and medicine. Much of the remainder of this book will in one way or another involve DNA and proteins. Let's begin by describing how the information in DNA is transcribed to produce RNA.

14.3 How Is the Information Content in DNA Transcribed to Produce RNA?

In prokaryotic and eukaryotic cells, RNA synthesis is directed by DNA. Transcription—the formation of a specific RNA sequence from a specific DNA sequence—requires several components:

- A DNA template for complementary base pairing; one of the two strands of DNA
- The four ribonucleoside triphosphates ATP, GTP, CTP, and UTP, to act as substrates
- An RNA polymerase enzyme
- Salts and a pH buffer to create an appropriate chemical environment for RNA polymerase if done in a test tube

mRNA is not the only molecule produced by transcription. The same process is responsible for the synthesis of tRNA and rRNA, whose important roles in protein synthesis will be described below. Like polypeptides, these RNAs are encoded by specific genes. Eukaryotes also make many kinds of small RNAs, including small nuclear RNA (snRNA), microRNA (miRNA), and small interfering RNA (siRNA), which are also transcribed. **Table 14.1** summarizes some of the RNAs found in eukaryotic cells. We will discuss the roles of miRNA and siRNA in Chapter 16.

RNA polymerases share common features

RNA polymerases from both prokaryotes and eukaryotes catalyze the synthesis of RNA from the DNA template. There is only one kind of RNA polymerase in bacteria, whereas there are several kinds in eukaryotes; however, they all share a common structure (**Figure 14.3**). Like DNA polymerases, RNA polymerases catalyze the addition of nucleotides in a 5′-to-3′ direction and are processive; that is, a single enzyme–template binding event results in the polymerization of hundreds of RNA bases. But unlike DNA polymerases (see Figure 13.13), RNA polymerases *do not require a primer*.

TABLE 14.1 Some RNAs in Eukaryotic Cells

RNA Type	Location of Activity	Role
Ribosomal RNA (rRNA)	Cytoplasm (ribosome)	Binding of mRNA and tRNA and protein synthesis
Messenger RNA (mRNA)	Cytoplasm	Carrier of gene sequence
Transfer RNA (tRNA)	Cytoplasm	Adaptor between mRNA and protein sequences
MicroRNA (miRNA)	Nucleus and cytoplasm	Regulates transcription and translation
Small interfering RNA (siRNA)	Nucleus and cytoplasm	Regulates other RNAs
Small nuclear RNA (snRNA)	Nucleus	Mediates mRNA processing

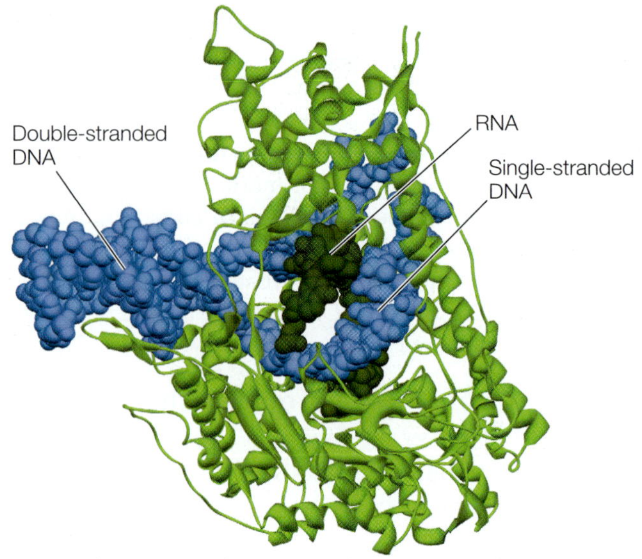
Double-stranded DNA
RNA
Single-stranded DNA

14.3 RNA Polymerase This enzyme from bacteriophage T7 is similar to most other RNA polymerases. Note the size relationship between enzyme and DNA. See Figure 14.4 for details.

Transcription occurs in three steps

Transcription can be divided into three distinct processes: initiation, elongation, and termination. You can follow these processes in **Figure 14.4**.

INITIATION Transcription begins when RNA polymerase binds to a special sequence of DNA called a **promoter** (see Figure 14.4A). Eukaryotic genes generally have one promoter each, whereas in prokaryotes and viruses, several genes often share one promoter. Promoters are important control sequences that "tell" the RNA polymerase two things:

- Where to start transcription
- Which strand of DNA to transcribe

A promoter reads in a particular direction, so it orients the RNA polymerase and thus "aims" it at the correct strand to use as a template. Part of each promoter is the **initiation site**, where transcription begins. Groups of nucleotides lying "upstream" from the initiation site (5′ on the non-template strand and 3′ on the template strand) help the RNA polymerase bind. Other proteins, which can bind to specific DNA sequences and to RNA polymerase, help direct the polymerase onto the promoter. These proteins, called **sigma factors** and **transcription factors**, help determine which specific genes are expressed at a particular time in the cell.

Although every gene has a promoter, not all promoters are identical. Some are more effective at transcription initiation than others. Furthermore, there are differences between transcription initiation in prokaryotes and in eukaryotes. We will discuss promoters and their roles in the regulation of gene expression in Chapter 16.

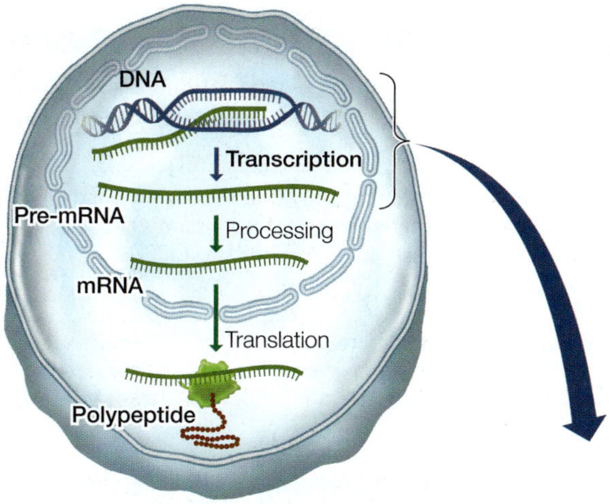
DNA
Transcription
Pre-mRNA
Processing
mRNA
Translation
Polypeptide

14.4 DNA Is Transcribed to Form RNA DNA is locally unwound by RNA polymerase to serve as a template for RNA synthesis. The RNA transcript is formed and then peels away, allowing the DNA that has already been transcribed to rewind into a double helix. Three distinct processes—initiation, elongation, and termination—constitute DNA transcription. RNA polymerase is much larger in reality than indicated here, covering about 50 base pairs.

Go to Animated Tutorial 14.1
Transcription
Life10e.com/at14.1

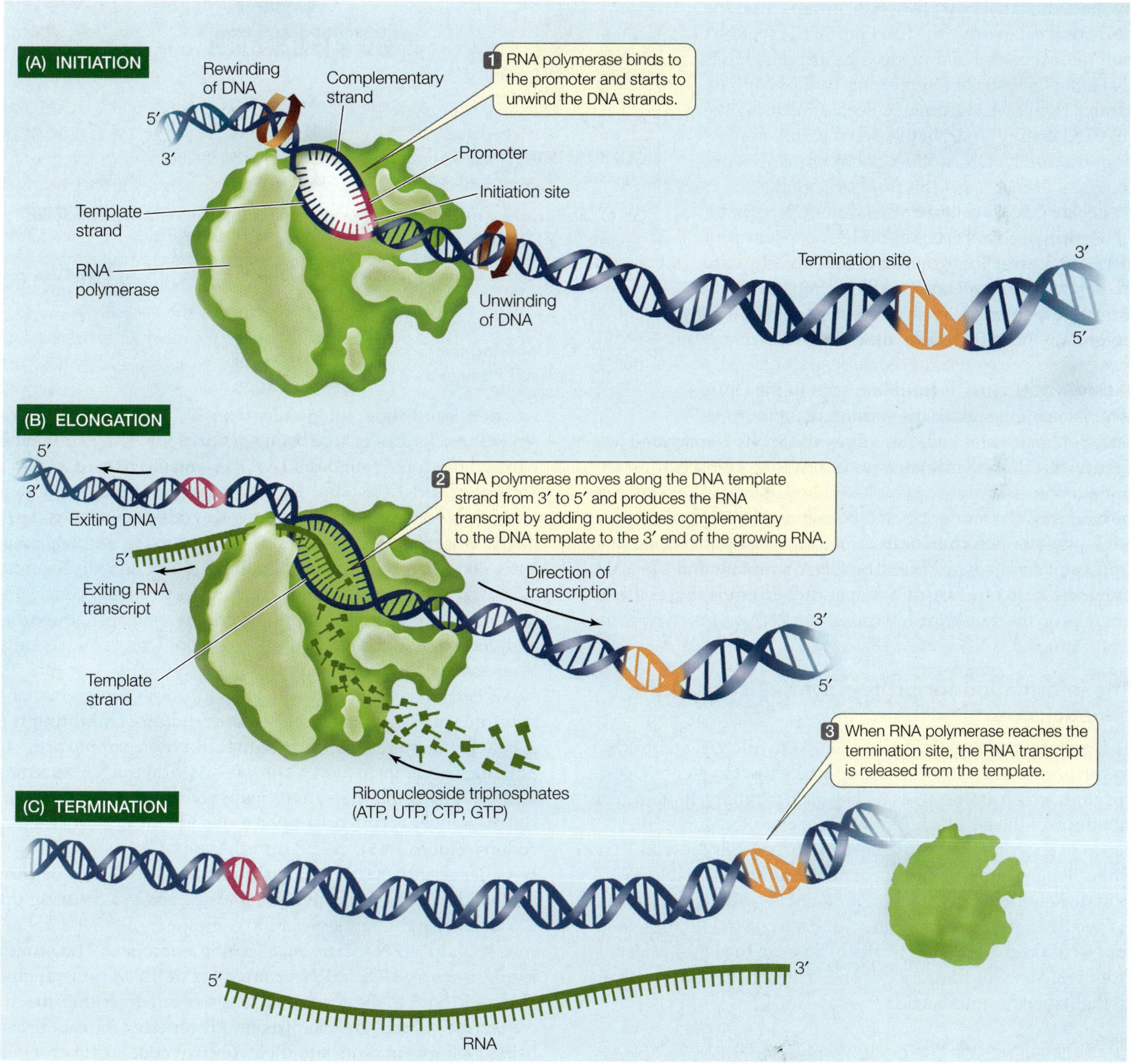
(A) INITIATION
Rewinding of DNA
Complementary strand
1 RNA polymerase binds to the promoter and starts to unwind the DNA strands.
5′
3′
Promoter
Initiation site
Template strand
RNA polymerase
Unwinding of DNA
Termination site
3′
5′
(B) ELONGATION
5′
3′
Exiting DNA
2 RNA polymerase moves along the DNA template strand from 3′ to 5′ and produces the RNA transcript by adding nucleotides complementary to the DNA template to the 3′ end of the growing RNA.
5′
Exiting RNA transcript
Direction of transcription
3′
5′
Template strand
Ribonucleoside triphosphates (ATP, UTP, CTP, GTP)
3 When RNA polymerase reaches the termination site, the RNA transcript is released from the template.
(C) TERMINATION
5′
3′
RNA

ELONGATION After RNA polymerase has bound to the promoter, it begins the process of **elongation** (see Figure 14.4B). RNA polymerase unwinds the DNA about 10 base pairs at a time and reads the template strand in the 3′-to-5′ direction. The first nucleotide in the new RNA forms its 5′ end, and subsequent nucleotides complementary to the DNA template are added to its 3′ end. Thus the RNA transcript is antiparallel to the DNA template strand.

Recall from Section 13.3 that DNA polymerase uses dNTPs (deoxyribonucleoside triphosphates) as substrates, and forms covalent bonds between each incoming dNTP and the 3′ end of the growing polynucleotide chain (see Figure 13.11). Energy released by the removal of two phosphate groups from the dNTP is used to drive the reaction. Similarly, RNA polymerase uses (ribo)nucleoside triphosphates (NTPs) as substrates, removing two phosphate groups from each substrate molecule and using the released energy to drive the polymerization reaction. But unlike DNA, RNA polymerase does not require a primer to get this process started.

Because RNA polymerases do not proofread, transcription errors occur at a rate of one for every 10^4 to 10^5 bases. Because many copies of RNA are made, however, and because they often have only a relatively short life span, these errors are not as potentially harmful as mutations in DNA.

INVESTIGATING**LIFE**

14.5 Deciphering the Genetic Code Nirenberg and Matthaei used a test tube protein synthesis system to determine the amino acids specified by synthetic mRNAs of known compositions.[a]

HYPOTHESIS An artificial mRNA containing only one repeating base will direct the synthesis of a protein containing only one repeating amino acid.

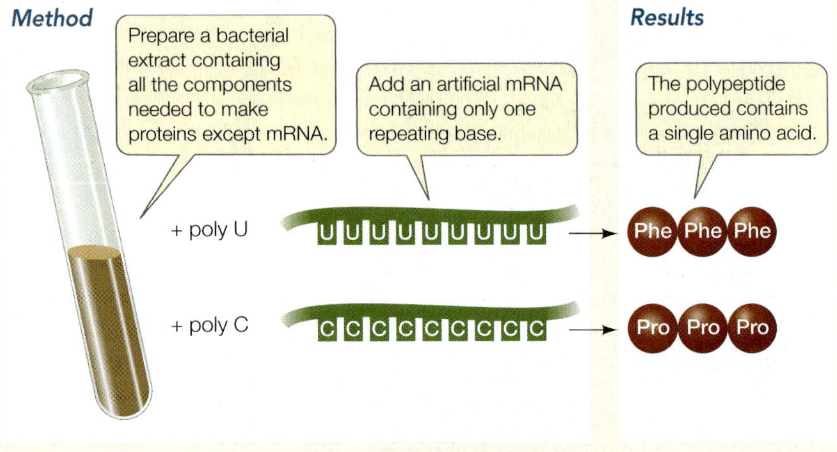
Method
Results
Prepare a bacterial extract containing all the components needed to make proteins except mRNA.
Add an artificial mRNA containing only one repeating base.
The polypeptide produced contains a single amino acid.
+ poly U
U U U U U U U U U
Phe Phe Phe
+ poly C
C C C C C C C C C
Pro Pro Pro

CONCLUSION Poly U contains codons for phenylalanine.
Poly C contains codons for proline.

Go to **BioPortal** for discussion and relevant links for all INVESTIGATING**LIFE** figures.

[a]Nirenberg, M. and H. Matthaei. 1961. *Proceedings of the National Academy of Sciences USA* 47: 1588–1602.

Go to Animated Tutorial 14.2
Deciphering the Genetic Code
Life10e.com/at14.2

TERMINATION Just as initiation sites in the DNA template strand specify the starting point for transcription, particular base sequences specify its **termination** (see Figure 14.4C). The mechanisms controlling transcription termination in eukaryotes are not well understood. There are two mechanisms for transcription termination in bacteria. For some genes, the newly formed transcript forms a loop that causes the transcript to fall away from the DNA template and the RNA polymerase. In other cases, a helper protein binds to specific sequences on the transcript and causes the RNA to detach from the DNA template.

The information for protein synthesis lies in the genetic code

The **genetic code** relates genes (DNA) to mRNA, and mRNA to the amino acids that make up proteins. The genetic code specifies which amino acids will be used to build a protein. You can think of the genetic information in an mRNA molecule as a series of sequential, nonoverlapping three-letter "words." The three "letters" are three adjacent nucleotide bases in the mRNA polynucleotide. Each three-letter "word" is called a **codon**, and it specifies a particular amino acid. Each codon is complementary to the corresponding triplet of bases in the DNA molecule from which it was transcribed. The genetic code relates codons to their specific amino acids.

CHARACTERISTICS OF THE CODE Molecular biologists "broke" the genetic code in the early 1960s. The problem they addressed was perplexing: how could more than 20 "code words" be written with an "alphabet" consisting of only four "letters"? In other words, how could four bases (A, U, G, and C) code for 20 different amino acids?

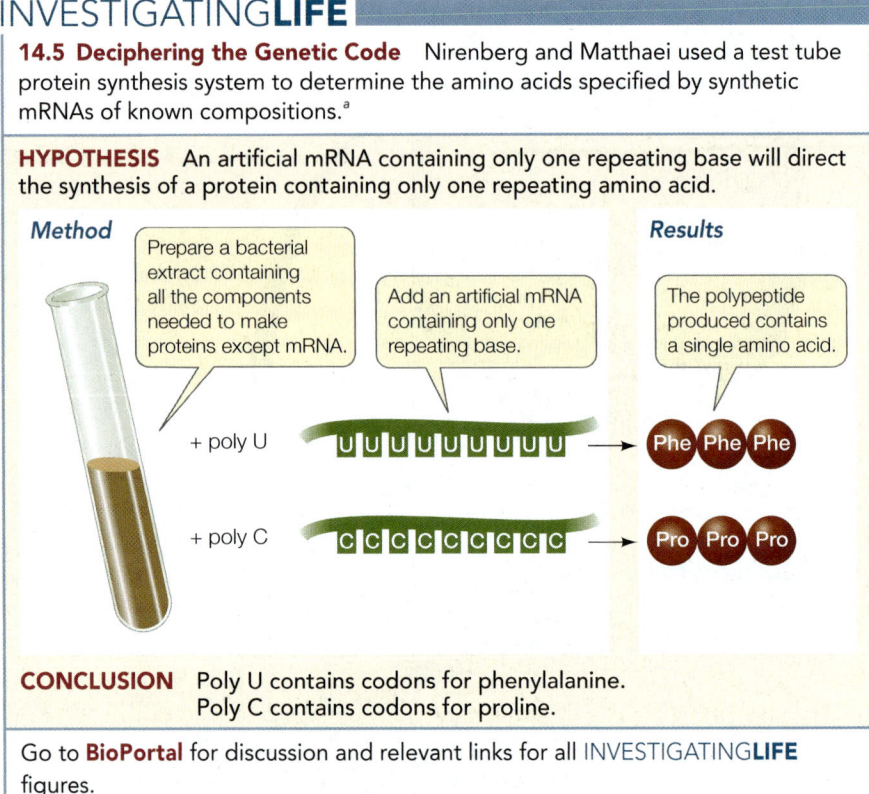

A triplet code, based on three-letter codons, was considered likely. Since there are only four letters (A, G, C, and U), a one-letter code clearly could not unambiguously encode 20 amino acids; it could encode only four of them. A two-letter code could have only $4 \times 4 = 16$ unambiguous codons—still not enough. But a triplet code could have $4 \times 4 \times 4 = 64$ codons, more than enough to encode the 20 amino acids.

Marshall W. Nirenberg and J. H. Matthaei, at the U.S. National Institutes of Health, made the first decoding breakthrough in 1961 when they realized that they could use a simple artificial polynucleotide instead of a complex natural mRNA as a messenger. They could then identify the polypeptide that the artificial messenger encoded. This led to the identification of the first codons (**Figure 14.5**).

Other scientists later found that artificial mRNAs only three nucleotides long—each amounting to one codon—could bind to a ribosome, and that the resulting complex could then bind to a corresponding tRNA carrying a specific amino acid. Thus, for example, a simple UUU mRNA caused the tRNA carrying phenylalanine to bind to the ribosome. To discover which amino acid a codon represented, the scientists simply repeated the experiment using a sample of artificial mRNA for that codon, and observed which amino acid became bound to it.

First letter	Second letter: U	Second letter: C	Second letter: A	Second letter: G	Third letter
U	UUU Phenylalanine	UCU Serine	UAU Tyrosine	UGU Cysteine	U
U	UUC Phenylalanine	UCC Serine	UAC Tyrosine	UGC Cysteine	C
U	UUA Leucine	UCA Serine	UAA Stop codon	UGA Stop codon	A
U	UUG Leucine	UCG Serine	UAG Stop codon	UGG Tryptophan	G
C	CUU Leucine	CCU Proline	CAU Histidine	CGU Arginine	U
C	CUC Leucine	CCC Proline	CAC Histidine	CGC Arginine	C
C	CUA Leucine	CCA Proline	CAA Glutamine	CGA Arginine	A
C	CUG Leucine	CCG Proline	CAG Glutamine	CGG Arginine	G
A	AUU Isoleucine	ACU Threonine	AAU Asparagine	AGU Serine	U
A	AUC Isoleucine	ACC Threonine	AAC Asparagine	AGC Serine	C
A	AUA Isoleucine	ACA Threonine	AAA Lysine	AGA Arginine	A
A	AUG Methionine; start codon	ACG Threonine	AAG Lysine	AGG Arginine	G
G	GUU Valine	GCU Alanine	GAU Aspartic acid	GGU Glycine	U
G	GUC Valine	GCC Alanine	GAC Aspartic acid	GGC Glycine	C
G	GUA Valine	GCA Alanine	GAA Glutamic acid	GGA Glycine	A
G	GUG Valine	GCG Alanine	GAG Glutamic acid	GGG Glycine	G

14.6 The Genetic Code Genetic information is encoded in mRNA in three-letter units—codons—made up of nucleoside monophosphates with the bases uracil (U), cytosine (C), adenine (A), and guanine (G) and is read in a 5′-to-3′ direction on mRNA. To decode a codon, find its first letter in the left column, then read across the top to its second letter, then read down the right column to its third letter. The amino acid the codon specifies is given in the corresponding row. For example, AUG codes for methionine, and GUA codes for valine.

Go to Activity 14.2 The Genetic Code
Life10e.com/ac14.2

The complete genetic code is shown in **Figure 14.6**. Notice that there are many more codons than there are different amino acids in proteins. All possible combinations of the four available "letters" (the bases) give 64 (4^3) different three-letter codons, yet these codons determine only 20 amino acids. AUG, which codes for methionine, is also the **start codon**, the initiation signal for translation. Three of the codons (UAA, UAG, UGA) are **stop codons**, or termination signals for translation. When the translation machinery reaches one of these codons, translation stops, and the polypeptide is released from the translation complex.

THE GENETIC CODE IS REDUNDANT BUT NOT AMBIGUOUS The 60 codons that are not start or stop codons are far more than enough to code for the other 19 amino acids. Indeed, there is more than one codon for almost all amino acids. Thus we say that the genetic code is redundant (or degenerate). For example, leucine is represented by six different codons (see Figure 14.6). Only methionine and tryptophan are represented by just one codon each.

A *redundant* code should not be confused with an *ambiguous* code. If the code were ambiguous, a single codon could specify either of two (or more) different amino acids, and there would be doubt about which amino acid should be incorporated into a growing polypeptide chain. Redundancy in the code simply means that there is more than one clear way to say "Put leucine here." The genetic code is not ambiguous: a given amino acid may be encoded by more than one codon, but a codon can code for only one amino acid.

THE GENETIC CODE IS (NEARLY) UNIVERSAL The same basic genetic code is used by all the species on our planet. Thus the code must be an ancient one that has been maintained intact throughout the evolution of living organisms. Exceptions are known: for example, within mitochondria and chloroplasts, the code differs slightly from that in prokaryotes and in the nuclei of eukaryotic cells; and in one group of protists, UAA and UAG code for glutamine rather than functioning as stop codons. The significance of these differences is not yet clear. What is clear is that the exceptions are few.

The common genetic code means that there is also a common language for evolution. Natural selection acts on phenotypic variations that result from genetic variation. The genetic code probably originated early in the evolution of life. As we saw in Chapter 4, simulation experiments indicate the plausibility of individual nucleotides and nucleotide polymers arising spontaneously on the primeval Earth. The common code also has profound implications for genetic engineering, as we will see in Chapter 18, since it means that the code for a human gene is the same as that for a bacterial gene. It is therefore impressive, but not surprising, that a human gene can be expressed in *E. coli* via laboratory manipulations, since these cells speak the same "molecular language."

The codons in Figure 14.6 are mRNA codons. The base sequence of the DNA strand that is transcribed to produce the mRNA is complementary and antiparallel to these codons. Thus, for example,

- 3′-AAA-5′ in the template DNA strand corresponds to phenylalanine (which is encoded by the mRNA codon 5′-UUU-3′)
- 3′-ACC-5′ in the template DNA corresponds to tryptophan (which is encoded by the mRNA codon 5′-UGG-3′)

The non-template strand of DNA has the same sequence as the mRNA (but with T's instead of U's), and is often referred to as the "coding strand." By convention, DNA sequences are usually shown beginning with the 5′ end of the coding sequence.

RECAP 14.3

Transcription, which is catalyzed by an RNA polymerase, proceeds in three steps: initiation, elongation, and termination. The genetic code relates the information in mRNA (as a linear sequence of codons) to protein (a linear sequence of amino acids).

- What are the steps of gene transcription that produce mRNA? **See pp. 286–288 and Figure 14.4**
- How do RNA polymerases work? **See pp. 287–288**
- How was the genetic code elucidated? **See p. 288 and Figure 14.5**

The general features of transcription that we have described were first elucidated in model prokaryotes, such as *E. coli*. Biologists then used the same methods to analyze this process in eukaryotes, and although the basics are the same, there are some notable (and important) differences. We will now turn to a more detailed description of eukaryotic gene expression.

14.4 How Is Eukaryotic DNA Transcribed and the RNA Processed?

Since the genetic code is the same, you might expect the process of gene expression to be the same in eukaryotes as it is in prokaryotes. And basically it is. However, there are significant differences in gene structure between prokaryotes and eukaryotes, that is, there are differences in the organization of the nucleotide sequences in the genes. In addition, in eukaryotes but not prokaryotes, a nucleus separates transcription and translation (**Table 14.2**).

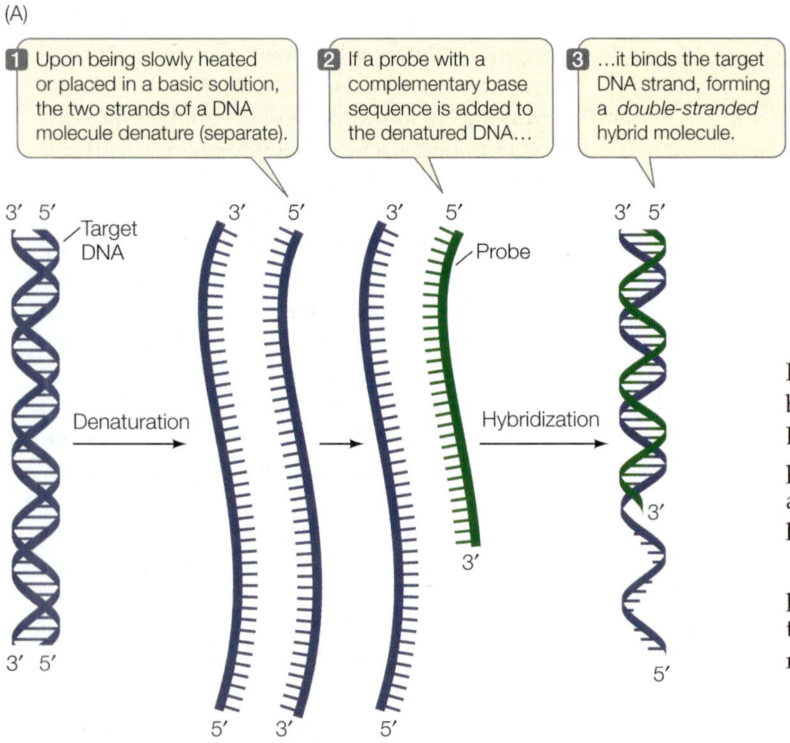

(A)
1 Upon being slowly heated or placed in a basic solution, the two strands of a DNA molecule denature (separate).
2 If a probe with a complementary base sequence is added to the denatured DNA…
3 …it binds the target DNA strand, forming a *double-stranded* hybrid molecule.
3′ 5′
Target DNA
Denaturation
3′ 5′
5′ 3′
3′ 5′
Probe
5′
3′
Hybridization
3′ 5′
3′
5′

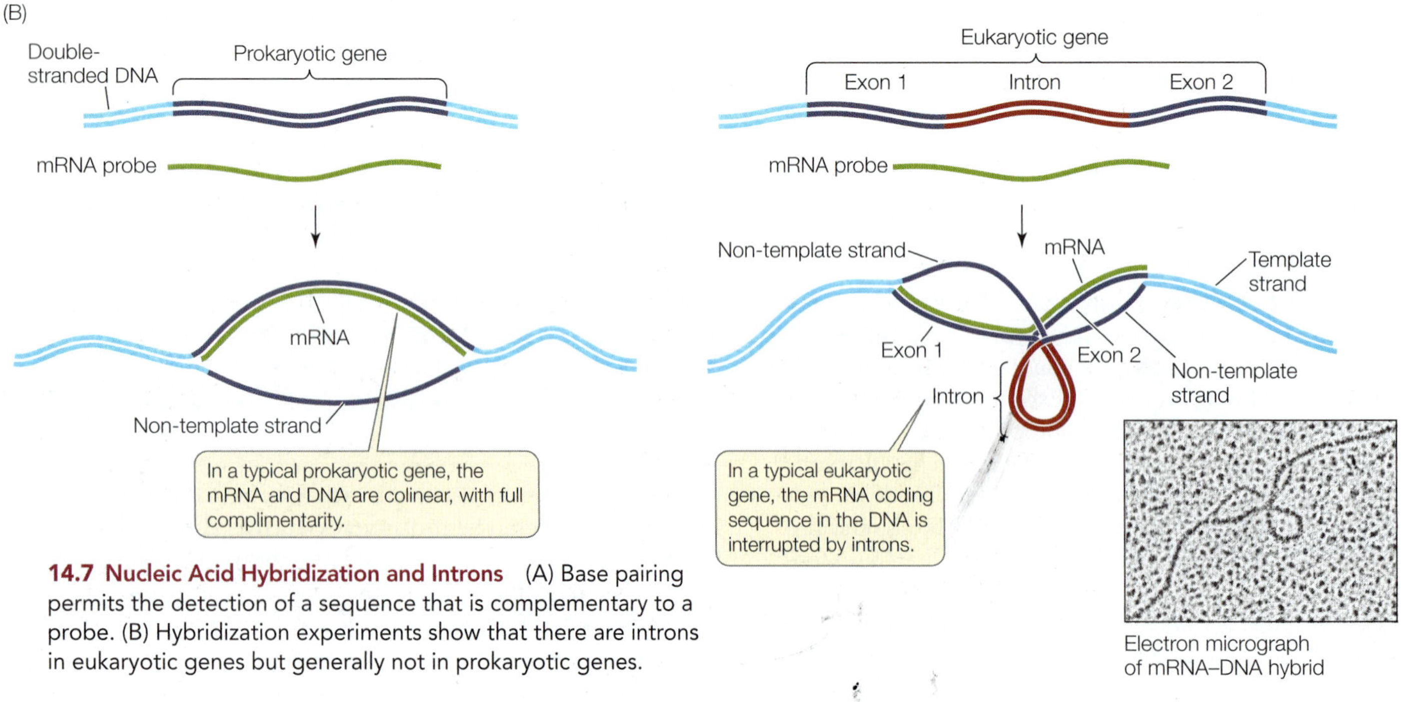

(B)
Double-stranded DNA
Prokaryotic gene
mRNA probe
mRNA
Non-template strand
In a typical prokaryotic gene, the mRNA and DNA are colinear, with full complimentarity.
Eukaryotic gene
Exon 1
Intron
Exon 2
mRNA probe
Non-template strand
mRNA
Template strand
Exon 1
Exon 2
Non-template strand
Intron
In a typical eukaryotic gene, the mRNA coding sequence in the DNA is interrupted by introns.
Electron micrograph of mRNA–DNA hybrid

14.7 Nucleic Acid Hybridization and Introns (A) Base pairing permits the detection of a sequence that is complementary to a probe. (B) Hybridization experiments show that there are introns in eukaryotic genes but generally not in prokaryotic genes.

Thus far we have examined features of gene transcription that are common to prokaryotes and eukaryotes. We will look in more detail at prokaryotic gene structure in Chapter 16, where we describe how transcription is controlled in prokaryotes. Let's look now at the distinctive structures of eukaryotic genes and their transcription.

Many eukaryotic genes are interrupted by noncoding sequences

The sequence of an mRNA that reaches the ribosome is complementary to the sequence of a gene that is part of DNA. One way to show this is by the technique of **nucleic acid hybridization**, shown in **Figure 14.7A**. This technique involves two steps:

1. A sample of chromosomal DNA containing the gene is denatured to break the hydrogen bonds between the base pairs and separate the two strands.
2. The single-stranded mRNA (called a **probe**) is incubated with the denatured DNA. If the probe has a base sequence complementary to the target DNA, a probe–target double helix forms by hydrogen bonding between the bases. Because the two strands are from different sources, the resulting double-stranded region is called a hybrid.

Hybridization experiments can be performed with various combinations of DNA and RNA (RNA as target and DNA as probe; DNA as both target and probe, etc.). In many hybridization experiments, the probe is labeled in some way so that its binding to a specific target sequence can be detected. The double-stranded hybrids can also be viewed by electron microscopy.

Now let's examine what happens when mRNA probes from prokaryotes and eukaryotes are incubated with their respective chromosomal DNAs and then viewed under an electron microscope.

TABLE **14.2**

Differences between Prokaryotic and Eukaryotic Gene Expression

Characteristic	Prokaryotes	Eukaryotes
Transcription and translation occurrence	At the same time in the cytoplasm	Transcription in the nucleus, then translation in the cytoplasm
Gene structure	DNA sequence is read in the same order as the amino acid sequence	Noncoding introns within coding sequence
Modification of mRNA after initial transcription but before translation	Usually none	Introns spliced out; 5′ cap and 3′ poly A tail added

- In prokaryotes (**Figure 14.7B, left**), there is usually a 1:1 linear complementarity between the base sequence of the mRNA at the ribosome and that of the chromosomal DNA.
- In eukaryotes (**Figure 14.7B, right**), one or more DNA loops are often observed within the mRNA–DNA hybrid, indicating that there are stretches of DNA sequence that do not have a complementary sequence in the mRNA that is translated at the ribosome.

One question that immediately arises from the eukaryotic result is, does this "extra" DNA actually get transcribed and then removed at the RNA level before mRNA arrives at the ribosome, or does transcription just "skip" these "extra" non-translated sequences? The answer comes from an experiment in which the initial mRNA transcript in the cell nucleus—the **precursor mRNA**, or **pre-mRNA**—is hybridized with chromosomal DNA. In this case, there is full, linear, loop-free hybridization. So the intervening regions (**introns**) of DNA actually get transcribed and then spliced out of the pre-mRNA in the nucleus, leaving expressed sequences (**exons**) in the mRNA that reaches the ribosome (**Figure 14.8**).

Introns *interrupt, but do not scramble*, the DNA sequence of a gene. The base sequences of the exons in the template strand, if joined and taken in order, form a continuous sequence that is complementary to that of the mature mRNA. In some cases, the separated exons often encode different functional regions, or **domains**, of the protein. For example, the globin polypeptides that make up hemoglobin each have two domains: one for binding to a nonprotein pigment called heme, and another for binding to the other globin subunits. These two domains are encoded by different exons in the globin genes. Most (but not all) eukaryotic genes contain introns, and in rare cases, introns are also found in prokaryotes. The largest human gene encodes a muscle protein called titin; it has 363 exons, which together code for 38,138 amino acids.

Eukaryotic gene transcripts are processed before translation

The primary transcript of a eukaryotic gene is modified in several ways before it leaves the nucleus: both ends of the pre-mRNA are modified, and the introns are removed.

MODIFICATION AT BOTH ENDS Two steps in the processing of pre-mRNA take place in the nucleus, one at each end of the molecule (**Figure 14.9**):

- A **5′ cap** is added to the 5′ end of the pre-mRNA as it is transcribed. The 5′ cap is a chemically modified molecule of guanosine triphosphate (GTP). It facilitates the binding of mRNA to the ribosome for translation, and it protects

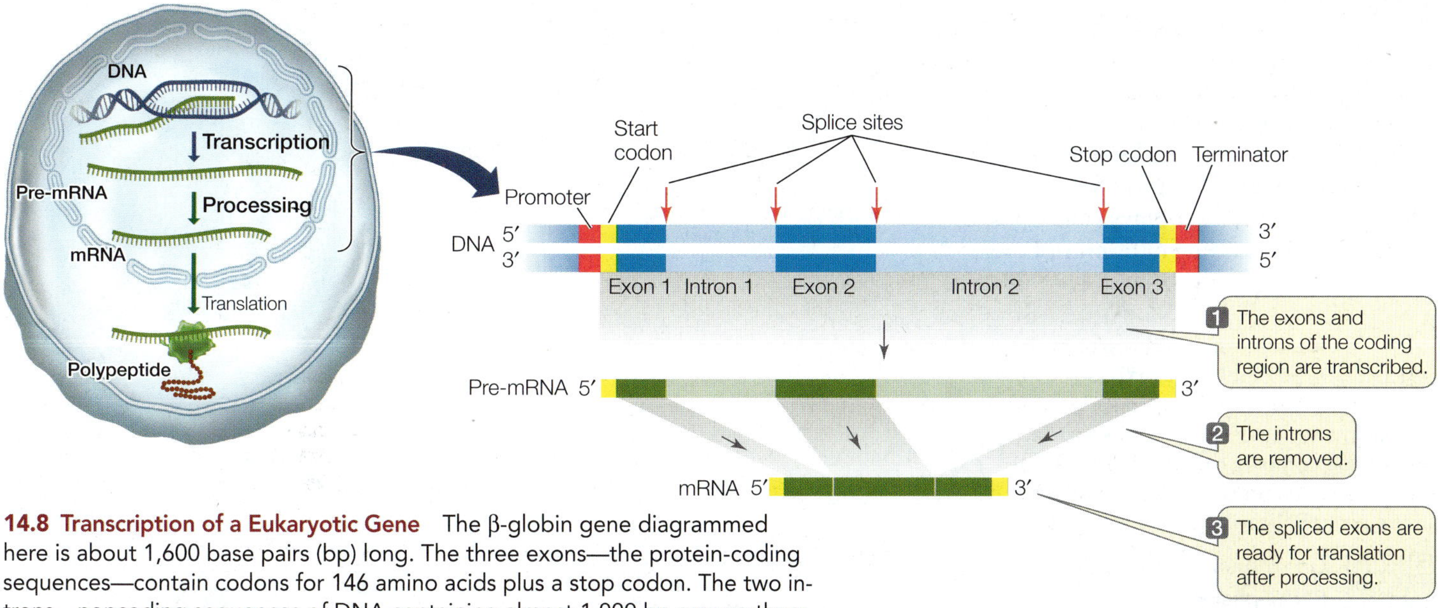
DNA
Transcription
Pre-mRNA
Processing
mRNA
Translation
Polypeptide
Start codon
Splice sites
Stop codon
Terminator
Promoter
DNA
5′
3′
3′
5′
Exon 1
Intron 1
Exon 2
Intron 2
Exon 3
1 The exons and introns of the coding region are transcribed.
Pre-mRNA 5′
3′
2 The introns are removed.
mRNA 5′
3′
3 The spliced exons are ready for translation after processing.

14.8 Transcription of a Eukaryotic Gene The β-globin gene diagrammed here is about 1,600 base pairs (bp) long. The three exons—the protein-coding sequences—contain codons for 146 amino acids plus a stop codon. The two introns—noncoding sequences of DNA containing almost 1,000 bp among them—are initially transcribed but are spliced out of the pre-mRNA transcript.

14.9 Processing the Ends of Eukaryotic Pre-mRNA Modifications at each end of the pre-mRNA transcript—the 5′ cap and the poly A tail—are important for mRNA function.

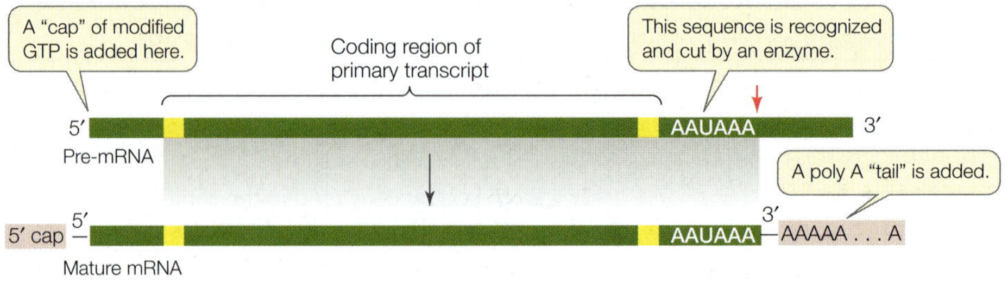
A "cap" of modified GTP is added here.
Coding region of primary transcript
This sequence is recognized and cut by an enzyme.
5′
AAUAAA
3′
Pre-mRNA
A poly A "tail" is added.
5′ cap
5′
3′
AAUAAA
AAAAA . . . A
Mature mRNA

the mRNA from being digested by ribonucleases that break down RNAs.

- A **poly A tail** is added to the 3′ end of the pre-mRNA at the end of transcription. Transcription ends downstream of the termination codon: along the coding strand of the DNA in the 3′ direction. In eukaryotes, there is usually a "polyadenylation" sequence (AAUAAA) near the 3′ end of the pre-mRNA, after the last codon. This sequence acts as a signal for an enzyme to cut the pre-mRNA. Immediately after this cleavage, another enzyme adds 100 to 300 adenine nucleotides (the poly A tail) to the 3′ end of the pre-mRNA. This tail may assist in the export of mature mRNA from the nucleus and is important for mRNA stability.

SPLICING TO REMOVE INTRONS The next step in the processing of eukaryotic pre-mRNA within the nucleus is removal of the introns. If these RNA sequences were not removed, a very different amino acid sequence, and possibly a nonfunctional protein, would result. A process called **RNA splicing** removes the introns and splices the exons together.

As soon as the pre-mRNA is transcribed, several **small nuclear ribonucleoprotein particles (snRNPs)** bind to the ends of each intron. Each snRNP has an RNA component—the snRNA (see Table 14.1). There are several types of these RNA–protein particles in the nucleus.

At the boundaries between introns and exons are **consensus sequences**—short stretches of DNA that appear, with little variation, in many different genes. (The polyadenylation sequence mentioned above is another example of a consensus sequence. Most promoters also contain consensus sequences for the binding of transcription factors and RNA polymerases.) The RNA in one of the snRNPs has a stretch of bases that is complementary to the consensus sequence at the 5′ exon–intron boundary, and it binds to the pre-mRNA by complementary base pairing. Another snRNP binds to the pre-mRNA near the 3′ intron–exon boundary (**Figure 14.10**).

Next, using energy from adenosine triphosphate (ATP), proteins are added to form a large RNA–protein complex called a **spliceosome**. This complex cuts the pre-mRNA at exon–intron boundaries, releases the introns, and joins the ends of the exons together to produce mature mRNA.

Molecular studies of human genetic diseases have provided insights into intron consensus sequences and splicing machinery. For example, people with the genetic disease β-thalassemia have a defect in the production of one of the hemoglobin subunits. These people suffer from severe anemia because they have an inadequate supply of red blood cells. In some cases, the genetic mutation that causes the disease occurs at an intron consensus sequence in the β-globin gene. Consequently, β-globin pre-mRNA cannot be spliced

14.10 The Spliceosome: An RNA Splicing Machine The binding of snRNPs to consensus sequences bordering the introns on the pre-mRNA results in a series of proteins binding and forming a large complex called a spliceosome. This structure determines the exact position of each cut in the pre-mRNA with great precision.

Go to Animated Tutorial 14.3
RNA Splicing
Life10e.com/at14.3

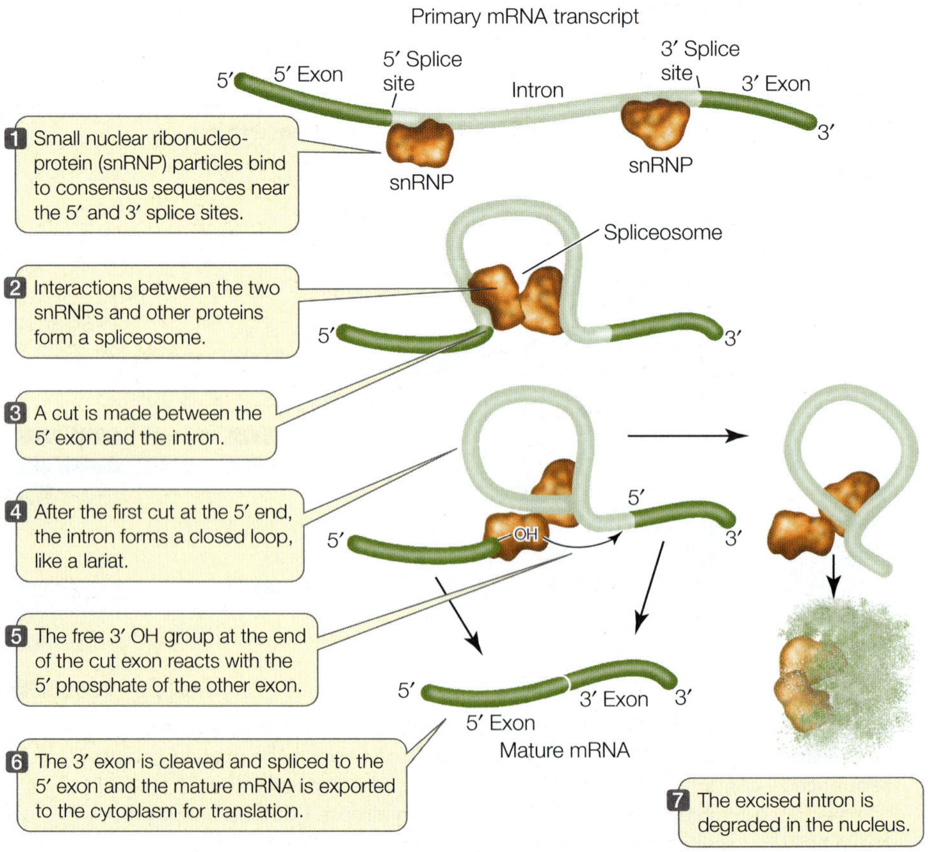
Primary mRNA transcript
5′
5′ Exon
5′ Splice site
Intron
3′ Splice site
3′ Exon
3′
snRNP
snRNP
1 Small nuclear ribonucleoprotein (snRNP) particles bind to consensus sequences near the 5′ and 3′ splice sites.
Spliceosome
2 Interactions between the two snRNPs and other proteins form a spliceosome.
5′
3′
3 A cut is made between the 5′ exon and the intron.
4 After the first cut at the 5′ end, the intron forms a closed loop, like a lariat.
5′
OH
5′
3′
5 The free 3′ OH group at the end of the cut exon reacts with the 5′ phosphate of the other exon.
5′
3′ Exon
3′
5′ Exon
Mature mRNA
6 The 3′ exon is cleaved and spliced to the 5′ exon and the mature mRNA is exported to the cytoplasm for translation.
7 The excised intron is degraded in the nucleus.

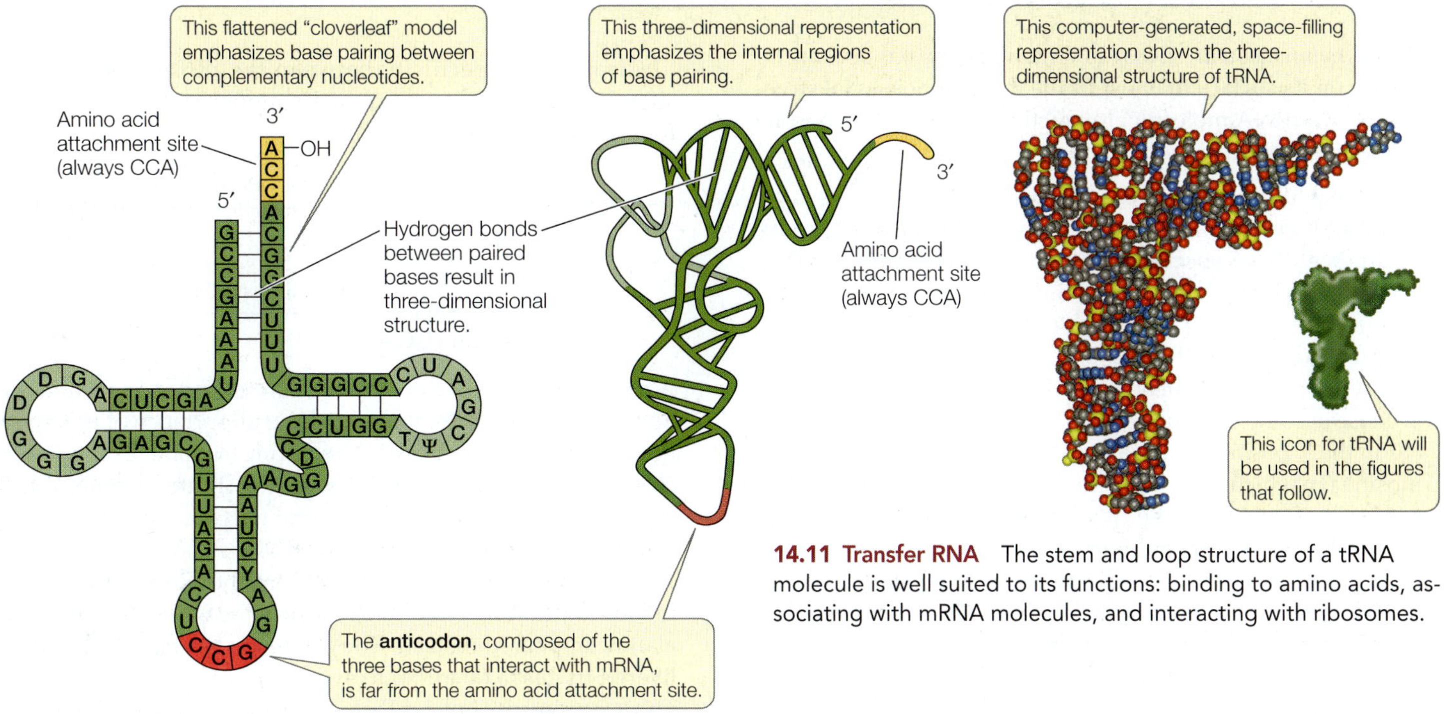
This flattened "cloverleaf" model emphasizes base pairing between complementary nucleotides.
This three-dimensional representation emphasizes the internal regions of base pairing.
This computer-generated, space-filling representation shows the three-dimensional structure of tRNA.
Amino acid attachment site (always CCA)
3′
5′
OH
Hydrogen bonds between paired bases result in three-dimensional structure.
Amino acid attachment site (always CCA)
This icon for tRNA will be used in the figures that follow.
The anticodon, composed of the three bases that interact with mRNA, is far from the amino acid attachment site.

14.11 Transfer RNA The stem and loop structure of a tRNA molecule is well suited to its functions: binding to amino acids, associating with mRNA molecules, and interacting with ribosomes.

correctly, and β-globin mRNA that encodes a nonfunctional polypeptide is made. This finding offers another example of how biologists can use mutations to elucidate biological processes.

After processing is completed in the nucleus, the mature mRNA moves out into the cytoplasm through the nuclear pores. In the nucleus, a protein complex called the cap-binding complex binds to the 5′ cap of the processed mRNA. This complex is recognized by a receptor at the nuclear pore. Together, these proteins lead the mRNA through the pore. Unprocessed or incompletely processed pre-mRNAs remain in the nucleus.

RECAP 14.4

Most eukaryotic genes contain noncoding sequences called introns, which are removed from the pre-mRNA transcript.

- Describe the method of nucleic acid hybridization and how it shows introns in eukaryotic genes. **See pp. 290–291 and Figure 14.7**
- How is the pre-mRNA transcript modified at the 5′ and 3′ ends? **See pp. 291–292 and Figure 14.9**
- How does RNA splicing happen? What are the consequences if it does not happen correctly? **See p. 292 and Figure 14.10**

Transcription and posttranscriptional events produce an mRNA that is ready to be translated into a sequence of amino acids in a polypeptide. We will turn now to the events of translation.

14.5 How Is RNA Translated into Proteins?

The translation of mRNA into proteins requires a molecule that links the information contained in mRNA codons with specific amino acids in proteins. This function is performed by transfer RNA (tRNA). Two key events must take place to ensure that the protein made is the one specified by the mRNA:

- The tRNAs must read mRNA codons correctly.
- The tRNAs must deliver the amino acids that correspond to each mRNA codon.

Once the tRNAs "decode" the mRNA and deliver the appropriate amino acids, components of the ribosome catalyze the formation of peptide bonds between amino acids. We will now turn to these two steps.

Go to Animated Tutorial 14.4
Translation
Life10e.com/at14.4

Transfer RNAs carry specific amino acids and bind to specific codons

There is at least one specific tRNA molecule for each of the 20 amino acids. Each tRNA has three functions that are fulfilled by its structure and base sequence (**Figure 14.11**):

- *tRNAs bind to particular amino acids.* Each tRNA binds to a specific enzyme that attaches it to only 1 of the 20 amino acids. This covalent attachment is at the 3′ end of the tRNA. We will describe the details of this vital process in the next section. When it is carrying an amino acid, the tRNA is said to be "charged."
- *tRNAs bind to mRNA.* At about the midpoint on the tRNA polynucleotide chain there is a triplet of bases called the **anticodon**, which is complementary to the mRNA codon for the particular amino acid that the tRNA carries. Like the two strands of DNA, the codon and anticodon bind together via noncovalent hydrogen bonds. For example, the mRNA codon for arginine is 5′-CGG-3′, and the complementary tRNA anticodon is 3′-GCC-5′.

- *tRNAs interact with ribosomes.* The ribosome has several sites on its surface that just fit the three-dimensional structure of a tRNA molecule. Interaction between the ribosome and the tRNA is noncovalent.

Recall that 61 different codons encode the 20 amino acids in proteins (see Figure 14.6). Does this mean that the cell must produce 61 different tRNA species, each with a different anticodon? No. The cell gets by with about two-thirds of that number of tRNA species because the specificity for the base at the 3′ end of the codon (and the 5′ end of the anticodon) is not always strictly observed. This phenomenon is called wobble, and it is possible because in some cases unusual or modified nucleotide bases occur in the 5′ position of the anticodon. One such unusual base is inosine (I), which can pair with A, C, and U. For example, the presence of inosine in the tRNA with the anticodon 3′-CGI-5′ allows it to recognize and bind to three of the alanine codons: GCA, GCC, and GCU. Wobble occurs in some matches but not in others; of most importance, it does not allow the genetic code to be ambiguous. That is, *each mRNA codon binds to just one tRNA species, carrying a specific amino acid.*

Each tRNA is specifically attached to an amino acid

The charging of each tRNA with its correct amino acid is achieved by a family of enzymes known as aminoacyl-tRNA synthetases. Each enzyme is specific for one amino acid and for its corresponding tRNA. The reaction uses ATP, forming a high-energy bond between the amino acid and the tRNA (**Figure 14.12**). The energy in this bond is later used in the formation of peptide bonds between amino acids in a growing polypeptide chain.

Clearly, the specificity between the tRNA and its corresponding amino acid is extremely important. These reactions, for example, are highly specific:

$$\text{Cysteine} + \underset{\text{(anticodon ACA)}}{\text{tRNAcys}} \xrightarrow{\text{Cys-tRNA synthetase}} \text{cys-tRNA}$$

$$\text{Alanine} + \underset{\text{(anticodon CGA)}}{\text{tRNAala}} \xrightarrow{\text{Ala-tRNA synthetase}} \text{ala-tRNA}$$

A clever experiment by Seymour Benzer and his colleagues at Purdue University demonstrated the importance of this specificity. They took the cys-tRNA molecule (see above) and chemically modified the cysteine, converting it into alanine. Which component—the amino acid or the tRNA—would be recognized when this hybrid charged tRNA was put into a protein synthesizing system? The answer was the tRNA. Everywhere in the synthesized protein where cysteine was supposed to be, alanine appeared instead. The cysteine-specific tRNA had delivered its cargo (alanine) to every mRNA codon for cysteine. This experiment showed that the protein synthesis machinery recognizes the anticodon of the charged tRNA, not the amino acid attached to it.

The ribosome is the workbench for translation

The **ribosome** is the molecular workbench where the task of translation is accomplished. Its structure enables it to hold mRNA and charged tRNAs in the correct positions, thus allowing a polypeptide chain to be assembled efficiently. A given ribosome does not specifically produce just one kind

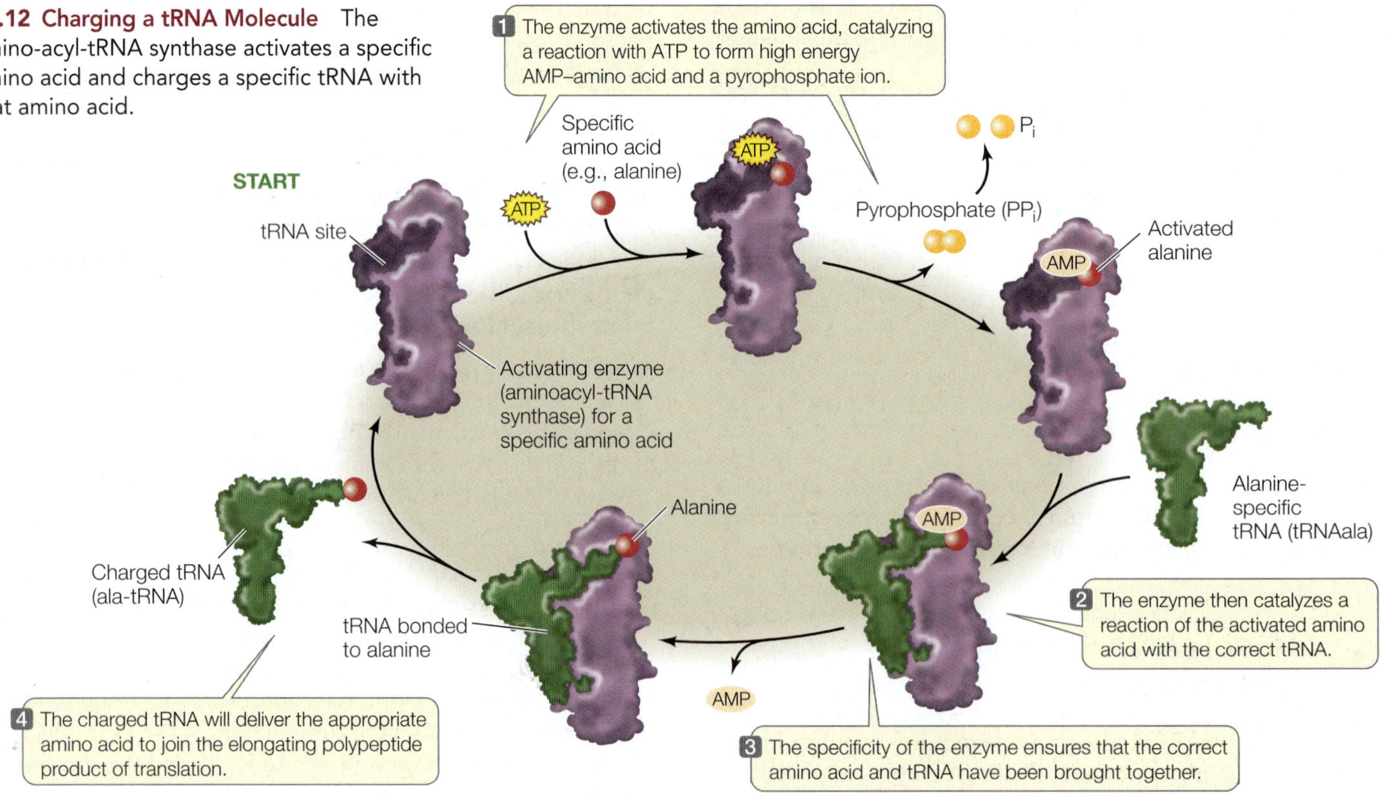
1 The enzyme activates the amino acid, catalyzing a reaction with ATP to form high energy AMP–amino acid and a pyrophosphate ion.
START
tRNA site
ATP
Specific amino acid (e.g., alanine)
ATP
P_i
Pyrophosphate (PP_i)
AMP
Activated alanine
Activating enzyme (aminoacyl-tRNA synthase) for a specific amino acid
Alanine-specific tRNA (tRNAala)
Alanine
AMP
Charged tRNA (ala-tRNA)
tRNA bonded to alanine
2 The enzyme then catalyzes a reaction of the activated amino acid with the correct tRNA.
AMP
4 The charged tRNA will deliver the appropriate amino acid to join the elongating polypeptide product of translation.
3 The specificity of the enzyme ensures that the correct amino acid and tRNA have been brought together.

14.12 Charging a tRNA Molecule The amino-acyl-tRNA synthase activates a specific amino acid and charges a specific tRNA with that amino acid.

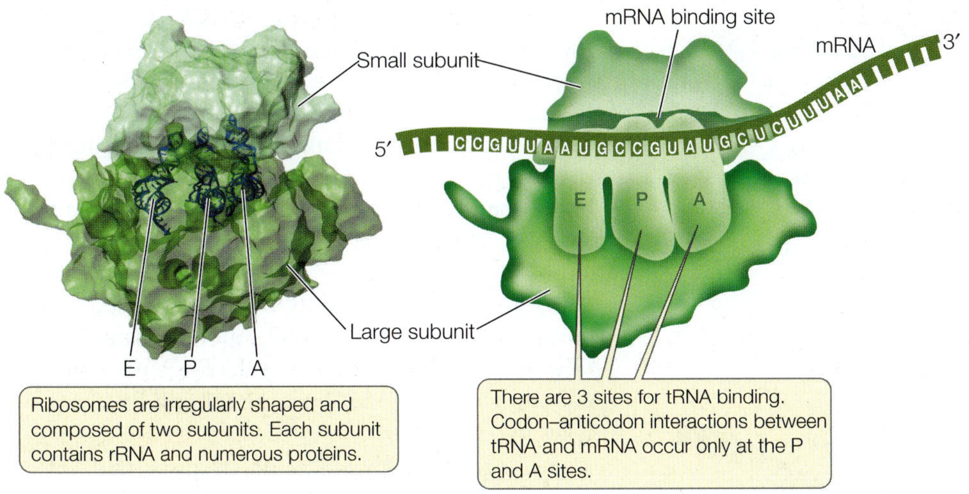
mRNA binding site
Small subunit
mRNA
3′
5′
E
P
A
Large subunit
Ribosomes are irregularly shaped and composed of two subunits. Each subunit contains rRNA and numerous proteins.
There are 3 sites for tRNA binding. Codon–anticodon interactions between tRNA and mRNA occur only at the P and A sites.

14.13 Ribosome Structure Each ribosome consists of a large and a small subunit. The subunits remain separate when they are not in use for protein synthesis.

of protein. A ribosome can use any mRNA and all species of charged tRNAs, and thus can be used to make many different polypeptide products. Ribosomes can be used over and over again, and there are thousands of them in a typical cell.

Although ribosomes are small in contrast to other cellular structures, their mass of several million daltons makes them large in comparison with charged tRNAs. Each ribosome consists of two subunits, a large one and a small one (**Figure 14.13**). The two subunits and several dozen other molecules interact noncovalently. In fact, when hydrophobic interactions between the proteins and RNAs are disrupted, the ribosome falls apart. The two subunits separate and all the RNAs and proteins separate from one another. If the disrupting agent is removed, the complex structure self-assembles perfectly! When not active in the translation of mRNA, the ribosomes exist as two separate subunits.

In eukaryotes, the large subunit consists of three different molecules of ribosomal RNA (rRNA) and 49 different protein molecules, arranged in a precise configuration. The small subunit consists of 1 rRNA molecule and 33 different protein molecules. The ribosomes of prokaryotes are somewhat smaller than those of eukaryotes, and their ribosomal proteins and RNAs are different. These differences explain why antibiotics that target the prokaryotic ribosome (such as the tetracycline used to cure Janet's bladder infection in the opening story) can be used to kill bacteria without harming a patient's cells. Mitochondria and chloroplasts also contain ribosomes, some of which are similar to those of prokaryotes (see Chapter 5).

On the large subunit of the ribosome there are three sites to which a tRNA can bind, and these are designated A, P, and E (see Figure 14.13). The mRNA and ribosome move in relation to one another, and as they do so, a charged tRNA traverses these three sites in order:

- The *A (aminoacyl tRNA) site* is where the charged tRNA anticodon binds to the mRNA codon, thus lining up the correct amino acid to be added to the growing polypeptide chain.
- The *P (peptidyl tRNA) site* is where the tRNA adds its amino acid to the polypeptide chain.
- The *E (exit) site* is where the tRNA, having given up its amino acid, resides before being released from the ribosome and going back to the cytosol to pick up another amino acid and begin the process again.

The ribosome has a fidelity function that ensures that the mRNA–tRNA interactions are accurate; that is, that a charged tRNA with the correct anticodon (e.g., 3′-UAC-5′) binds to the appropriate codon in mRNA (e.g., 5′-AUG-3′). When proper binding occurs, hydrogen bonds form between the paired bases. The rRNA of the small ribosomal subunit plays a role in validating the three-base-pair match. If hydrogen bonds have not formed between all three base pairs, the tRNA must be the wrong one for that mRNA codon, and the incorrect tRNA is ejected from the ribosome.

Translation takes place in three steps

Translation is the process by which the information in mRNA (derived from DNA) is used to specify and link a specific sequence of amino acids, producing a polypeptide. Like transcription, translation occurs in three steps: initiation, elongation, and termination.

INITIATION The translation of mRNA begins with the formation of an **initiation complex**, which consists of a charged tRNA and a small ribosomal subunit, both bound to the mRNA (**Figure 14.14**).

In prokaryotes, the rRNA of the small ribosomal subunit first binds to a complementary ribosome binding site (AGGAGG; known as the Shine–Dalgarno sequence) on the mRNA. This sequence is less than 10 bases upstream of the actual start codon but lines up the start codon so that it is adjacent to the P site of the large subunit:

mRNA 5′ . . . A G G A G G . . . (start codon) . . . 3′
rRNA 3′ . . . U C C U C C . . . (P site) . . . 5′

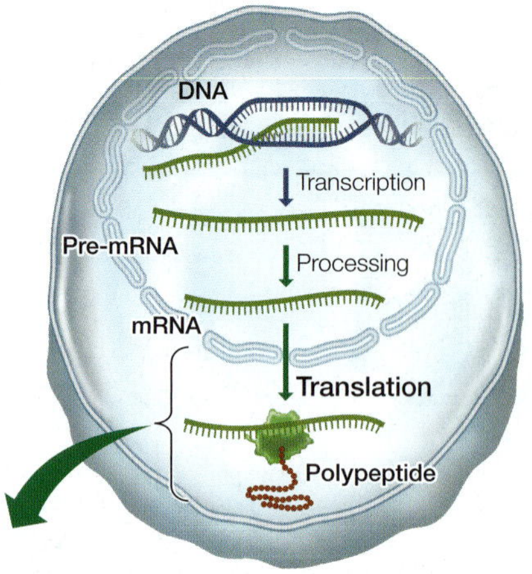
DNA
Transcription
Pre-mRNA
Processing
mRNA
Translation
Polypeptide

14.14 The Initiation of Translation
Translation begins with the formation of an initiation complex. In prokaryotes, the small ribosomal subunit binds to the Shine–Dalgarno sequence to begin the process, whereas in eukaryotes, it binds to the 5′ cap.

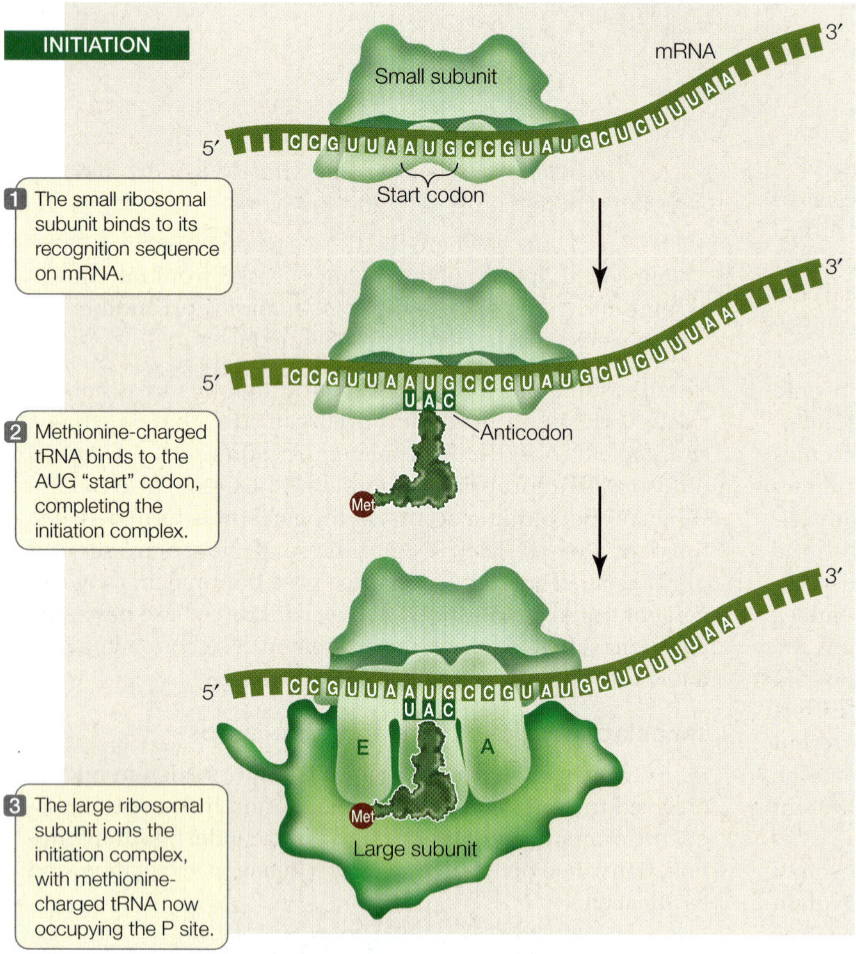
INITIATION
Small subunit
mRNA
5′
3′
CCGUUAAUGCCGUAUGCUCUUUAA
Start codon
1 The small ribosomal subunit binds to its recognition sequence on mRNA.
UAC
Anticodon
Met
2 Methionine-charged tRNA binds to the AUG "start" codon, completing the initiation complex.
E
A
Large subunit
3 The large ribosomal subunit joins the initiation complex, with methionine-charged tRNA now occupying the P site.

Eukaryotes load the mRNA onto the ribosome somewhat differently: the small ribosomal subunit binds to the 5′ cap on the mRNA and then moves along the mRNA until it reaches the start codon.

Recall that the mRNA start codon in the genetic code is AUG (see Figure 14.6). The anticodon (UAG) of a methionine-charged tRNA binds to this start codon by complementary base pairing to complete the initiation complex. Thus the first amino acid in a polypeptide chain is always methionine. However, not all mature proteins have methionine as their N-terminal amino acid. In many cases, the initial methionine is removed by an enzyme after translation.

After the methionine-charged tRNA has bound to the mRNA, the large subunit of the ribosome joins the complex. The methionine-charged tRNA now lies in the P site of the ribosome, and the A site is aligned with the second mRNA codon. These ingredients—mRNA, two ribosomal subunits, and methionine-charged tRNA—are assembled by a group of proteins called initiation factors.

ELONGATION A charged tRNA whose anticodon is complementary to the second codon of the mRNA now enters the open A site of the large ribosomal subunit **Figure 14.15**. The large subunit then catalyzes two reactions:

- It breaks the bond between the tRNA and its amino acid in the P site.
- It catalyzes the formation of a peptide bond between that amino acid and the one attached to the tRNA in the A site.

Because the large ribosomal subunit performs these two actions, it is said to have **peptidyl transferase** activity. In this way, methionine (the amino acid in the P site) becomes the N terminus of the new protein. The second amino acid is now bound to methionine but remains attached to its tRNA at the A site.

How does the large ribosomal subunit catalyze peptide bond formation? Harry Noller and his colleagues at the University of California at Santa Cruz did a series of experiments and found that:

- If they removed almost all of the proteins from the large subunit, it still catalyzed peptide bond formation.
- If the rRNA was extensively modified, peptidyl transferase activity was destroyed.

Thus *rRNA is the catalyst*. The purification and crystallization of ribosomes has allowed scientists to examine their structure in detail, and the catalytic role of rRNA in peptidyl transferase activity has been confirmed. This supports the hypothesis that RNA, and catalytic RNA in particular, evolved before DNA (see Section 4.3).

After the first tRNA releases its methionine, it moves to the E site and is then dissociated from the ribosome, returning to the cytosol to become charged with another methionine. The second tRNA, now bearing a dipeptide (a two-amino acid chain), is shifted to the P site as the ribosome moves one codon along the mRNA in the 5′-to-3′ direction.

The elongation process continues, and the polypeptide chain grows, as these steps are repeated. Follow the process in Figure 14.15. All these steps are assisted by ribosomal proteins called elongation factors.

TERMINATION The elongation cycle ends, and translation is terminated, when a stop codon—UAA, UAG, or UGA—enters

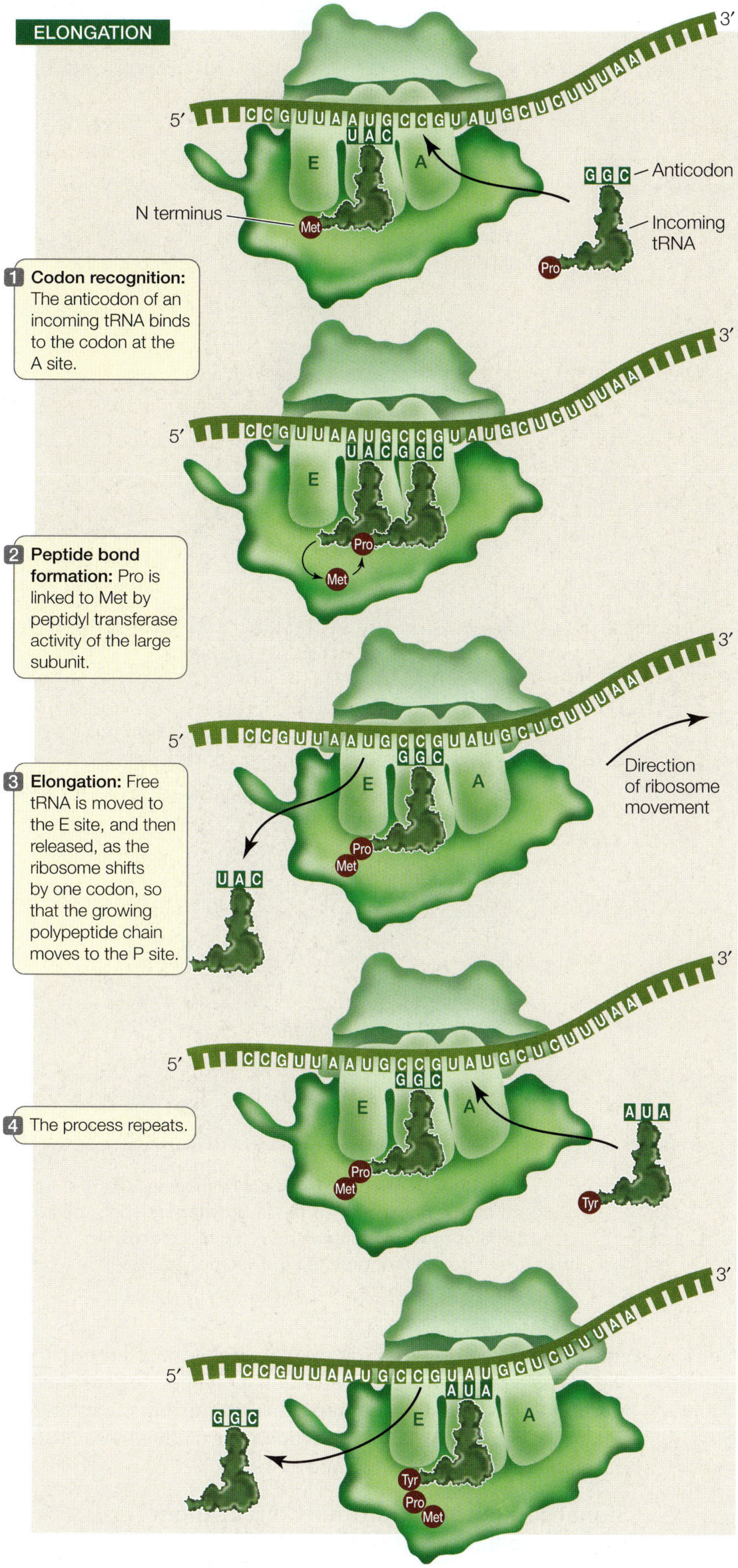
ELONGATION
5′
3′
E
A
Anticodon
Incoming tRNA
N terminus
Met
Pro
1 Codon recognition: The anticodon of an incoming tRNA binds to the codon at the A site.
2 Peptide bond formation: Pro is linked to Met by peptidyl transferase activity of the large subunit.
3 Elongation: Free tRNA is moved to the E site, and then released, as the ribosome shifts by one codon, so that the growing polypeptide chain moves to the P site.
Direction of ribosome movement
4 The process repeats.
Tyr

14.15 The Elongation of Translation The polypeptide chain elongates as the mRNA is translated.

TABLE 14.3 Signals that Start and Stop Transcription and Translation

	Transcription	Translation
Initiation	Promoter DNA	AUG start codon in the mRNA
Termination	Terminator DNA	UAA, UAG, or UGA in the mRNA

the A site (**Figure 14.16**). These codons do not correspond with any amino acids, nor do they bind any tRNAs. Rather, they bind a protein release factor, which allows hydrolysis of the bond between the polypeptide chain and the tRNA in the P site. The newly completed polypeptide thereupon separates from the ribosome. Its C terminus is the last amino acid to join the chain. Its N terminus, at least initially, is methionine, as a consequence of the AUG start codon. In its amino acid sequence, it contains information specifying its conformation, as well as its ultimate cellular destination.

Table 14.3 summarizes the nucleic acid signals for initiation and termination of transcription and translation.

Polysome formation increases the rate of protein synthesis

Several ribosomes can work simultaneously at translating a single mRNA molecule, producing multiple polypeptides at the same time. As soon as the first ribosome has moved far enough from the site of translation initiation, a second initiation complex can form, then a third, and so on. An assemblage consisting of a strand of mRNA with its beadlike ribosomes and their growing polypeptide chains is called a **polyribosome**, or **polysome** (**Figure 14.17**). Cells that are actively synthesizing proteins contain large numbers of polysomes and few free ribosomes or ribosomal subunits.

RECAP 14.5

A key step in protein synthesis is the attachment of an amino acid to its proper tRNA. This attachment is carried out by an activating enzyme. Translation of the genetic information from mRNA into protein occurs at the ribosome. Multiple ribosomes may act on a single mRNA to make multiple copies of the protein that it encodes.

- How is an amino acid attached to a specific tRNA, and why is specificity in this process so important? See p. 294 and Figure 14.12
- Describe the events of initiation, elongation, and termination of translation. See pp. 295–297 and Figures 14.14–14.16

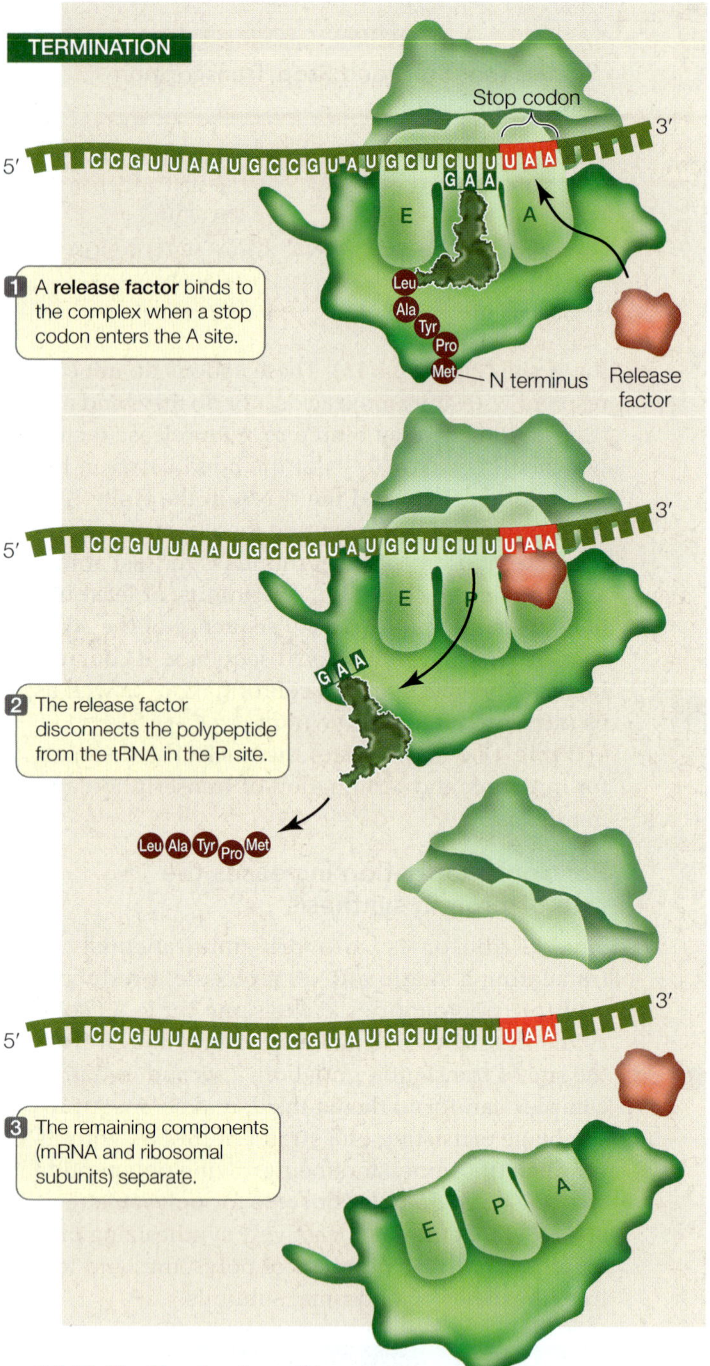
TERMINATION
Stop codon
5′
3′
CCGUUAAUGCCGUAUGCUCUUUAA
GAA
E
A
Leu
Ala
Tyr
Pro
Met
N terminus
Release factor
1 A release factor binds to the complex when a stop codon enters the A site.
E
P
GAA
2 The release factor disconnects the polypeptide from the tRNA in the P site.
Leu Ala Tyr Pro Met
3 The remaining components (mRNA and ribosomal subunits) separate.
E
P
A

14.16 The Termination of Translation Translation terminates when the A site of the ribosome encounters a stop codon on the mRNA.

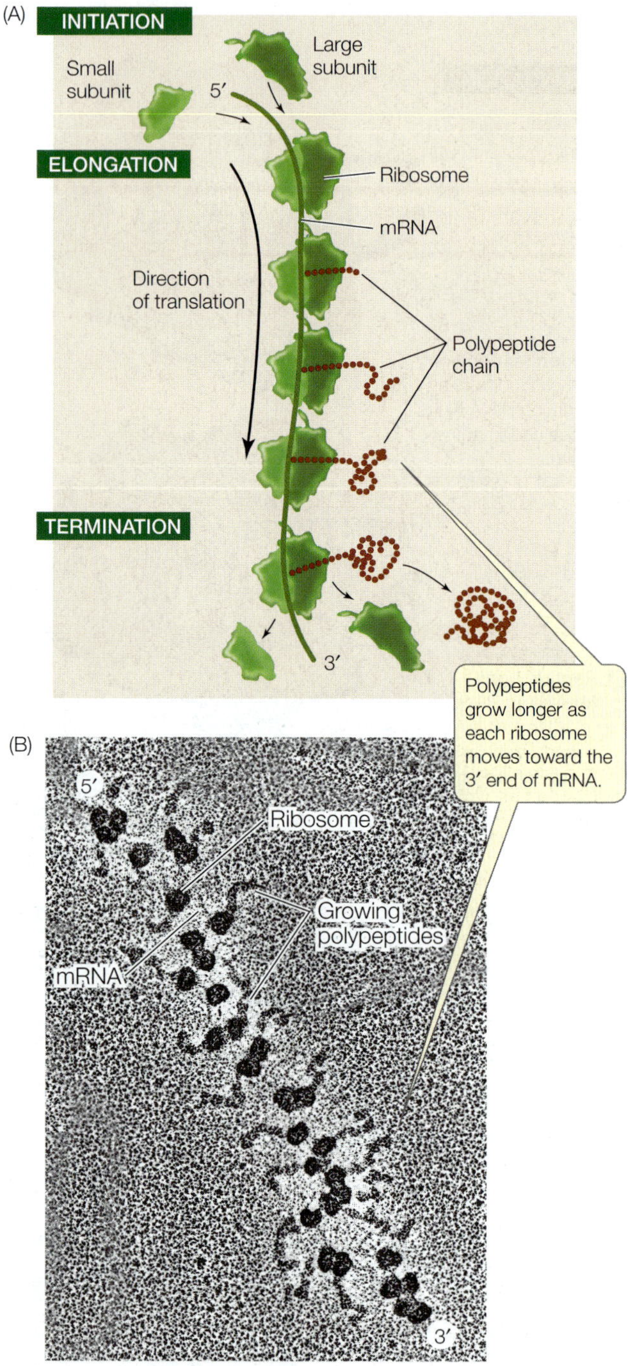
(A)
INITIATION
Small subunit
Large subunit
5′
ELONGATION
Ribosome
mRNA
Direction of translation
Polypeptide chain
TERMINATION
3′
Polypeptides grow longer as each ribosome moves toward the 3′ end of mRNA.
(B)
5′
Ribosome
Growing polypeptides
mRNA
3′

14.17 A Polysome (A) A polysome consists of multiple ribosomes and their growing polypeptide chains moving along an mRNA molecule. (B) An electron micrograph of a polysome.

The polypeptide chain that is released from the ribosome is not necessarily a functional protein. Let's look at some of the posttranslational changes that can affect the fate and function of a polypeptide.

14.6 What Happens to Polypeptides after Translation?

The site of a polypeptide's function may be far away from its point of synthesis in the cytoplasm. This is especially true for eukaryotes. The polypeptide may be moved into an organelle, or even out of the cell. In addition, polypeptides are often modified by the addition of new chemical groups that have functional significance. In this section we examine these posttranslational aspects of protein synthesis.

Signal sequences in proteins direct them to their cellular destinations

As a polypeptide chain emerges from the ribosome it may simply fold into its three-dimensional shape and perform its cellular role. However, a newly formed polypeptide may

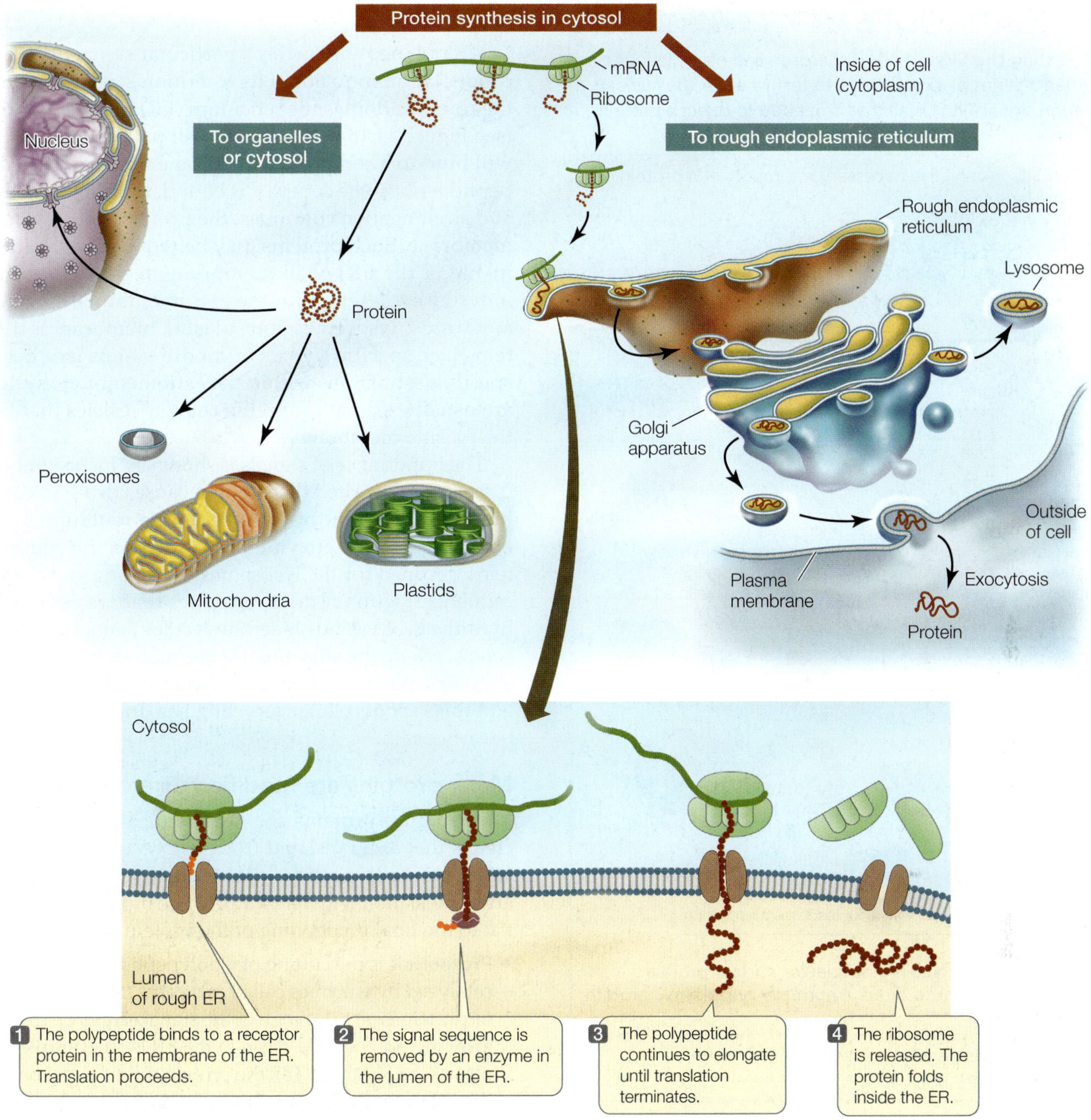

Protein synthesis in cytosol
mRNA
Ribosome
Inside of cell (cytoplasm)
Nucleus
To organelles or cytosol
To rough endoplasmic reticulum
Rough endoplasmic reticulum
Lysosome
Protein
Golgi apparatus
Peroxisomes
Mitochondria
Plastids
Outside of cell
Plasma membrane
Exocytosis
Protein
Cytosol
Lumen of rough ER
1 The polypeptide binds to a receptor protein in the membrane of the ER. Translation proceeds.
2 The signal sequence is removed by an enzyme in the lumen of the ER.
3 The polypeptide continues to elongate until translation terminates.
4 The ribosome is released. The protein folds inside the ER.

14.18 Destinations for Newly Translated Polypeptides in a Eukaryotic Cell Signal sequences on newly synthesized polypeptides bind to specific receptor proteins on the outer membranes of the organelles to which they are "addressed." Once the protein has bound to it, the receptor forms a channel in the membrane, and the protein enters the organelle.

contain a **signal sequence** (or **signal peptide**)—a short stretch of amino acids that indicates where in the cell the polypeptide belongs. Proteins destined for different locations have different signals.

Protein synthesis always begins on free ribosomes, and the "default" location for a protein is the cytosol. In the absence of a signal sequence, the protein will remain in the same cellular compartment where it was synthesized. Some proteins contain signal sequences that "target" them to the nucleus, mitochondria, plastids, or peroxisomes (**Figure 14.18**). A signal sequence binds to a specific receptor protein at the surface of the organelle. Once it has bound, a channel forms in the organelle membrane, allowing the targeted protein to move into the organelle. For example, here is a nuclear localization signal (NLS):

-Pro-Pro-Lys-Lys-Lys-Arg-Lys-Val-

How do we know this? The function of this peptide was established using experiments like the one illustrated in **Figure 14.19**. Proteins were made in the laboratory with or without the peptide, and then tested by injecting them into cells. Only proteins with the NLS were found in the nucleus.

INVESTIGATINGLIFE

14.19 Testing the Signal A. Richardson and his colleagues performed a series of experiments to test whether the nuclear localization signal (NLS) is all that is needed to direct a protein to the nucleus.[a]

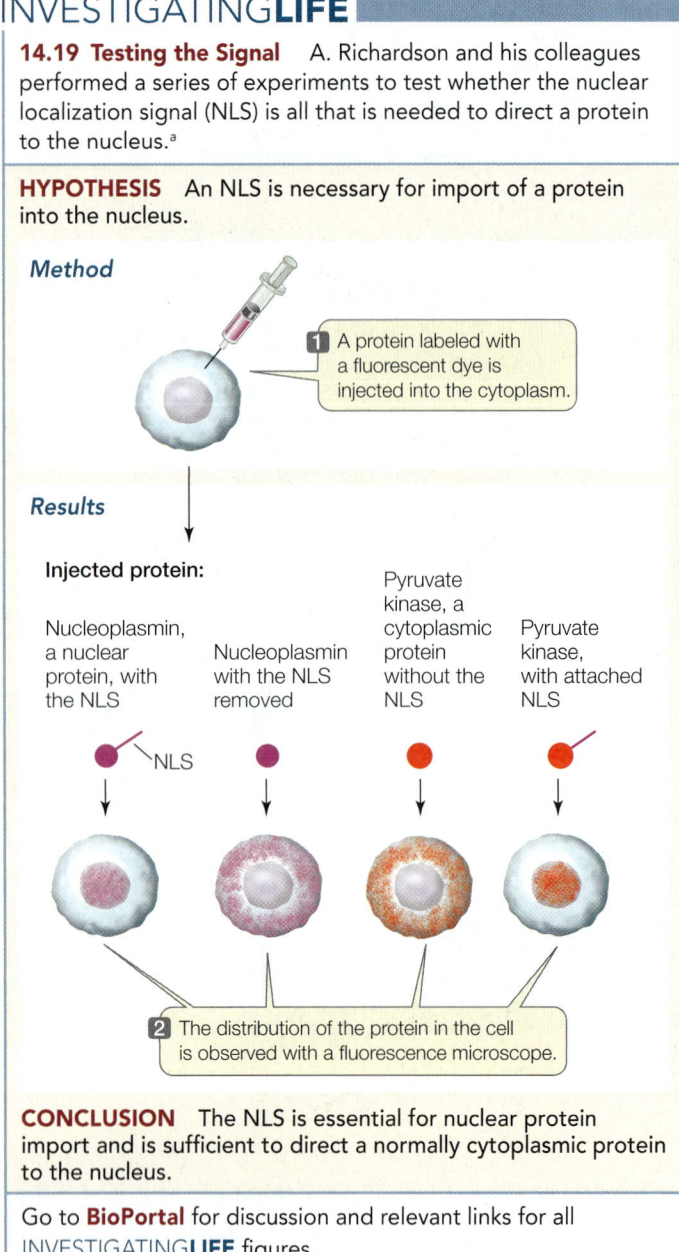

HYPOTHESIS An NLS is necessary for import of a protein into the nucleus.

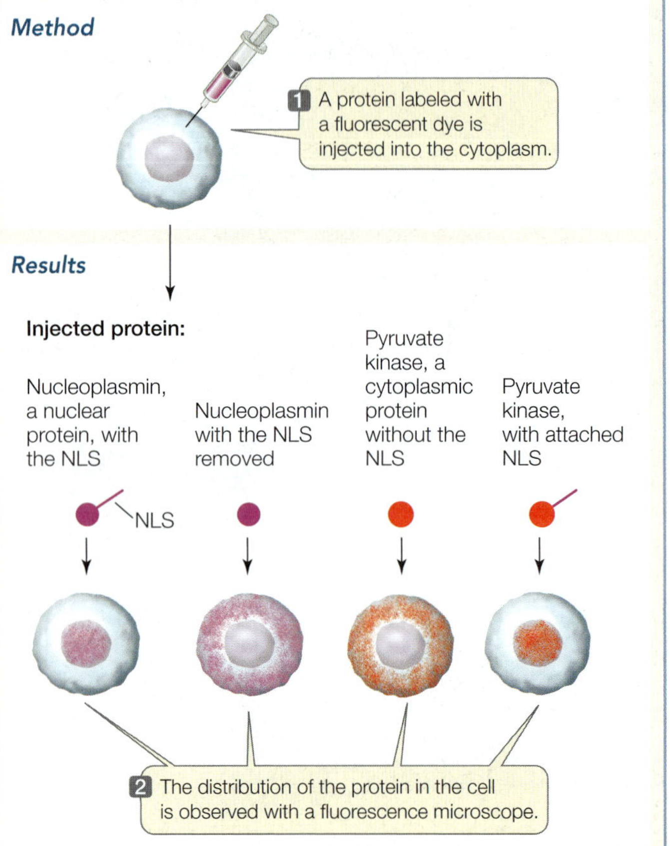
Method
1 A protein labeled with a fluorescent dye is injected into the cytoplasm.
Results
Injected protein:
Nucleoplasmin, a nuclear protein, with the NLS
Nucleoplasmin with the NLS removed
Pyruvate kinase, a cytoplasmic protein without the NLS
Pyruvate kinase, with attached NLS
NLS
2 The distribution of the protein in the cell is observed with a fluorescence microscope.

CONCLUSION The NLS is essential for nuclear protein import and is sufficient to direct a normally cytoplasmic protein to the nucleus.

Go to **BioPortal** for discussion and relevant links for all INVESTIGATINGLIFE figures.

[a]Dingwall, C. et al. 1988. *The Journal of Cell Biology* 107: 841–849.

If a polypeptide carries a particular signal of about 20 hydrophobic amino acids at its N terminus, it will be directed to the rough endoplasmic reticulum (ER) for further processing (see Figure 14.18). Translation will pause, and the ribosome will bind to a receptor at the ER membrane. Once the polypeptide–ribosome complex is bound, translation will resume, and as elongation continues, the protein will traverse the ER membrane. Such proteins may be retained in the lumen (the inside of the ER) or in membrane of the ER, or they may move elsewhere within the endomembrane system (Golgi apparatus, lysosomes, and plasma membrane). If the proteins lack specific signals or modifications (see below) that specify destinations within the endomembrane system, they are usually secreted from the cell via vesicles that fuse with the plasma membrane.

The importance of signals is shown by Inclusion-cell (I-cell) disease, an inherited disease that causes death in early childhood. People with this disease have a mutation in the gene encoding a Golgi enzyme that adds specific sugars to proteins destined for the lysosomes. These sugars act like signal sequences; without them, enzymes that are essential for the hydrolysis of various macromolecules cannot reach the lysosomes, where the enzymes are normally active. Without these enzymes, the macromolecules accumulate in the lysosomes, and this lack of cellular recycling has drastic effects, resulting in early death.

Many proteins are modified after translation

Most mature proteins are not identical to the polypeptide chains that are translated from mRNA on the ribosomes. Instead, most polypeptides are modified in any of several ways after translation (**Figure 14.20**). These modifications are essential to the final functioning of the protein.

- **Proteolysis** is the cutting of a polypeptide chain, a reaction catalyzed by enzymes called proteases (also called peptidases or proteinases). Cleavage of the signal sequence from the growing polypeptide chain in the ER is an example of proteolysis (see Figure 14.18); the protein might move back out of the ER through the membrane channel if the signal sequence were not cut off. Some proteins are actually made from polyproteins (long polypeptides) that are cut into final products

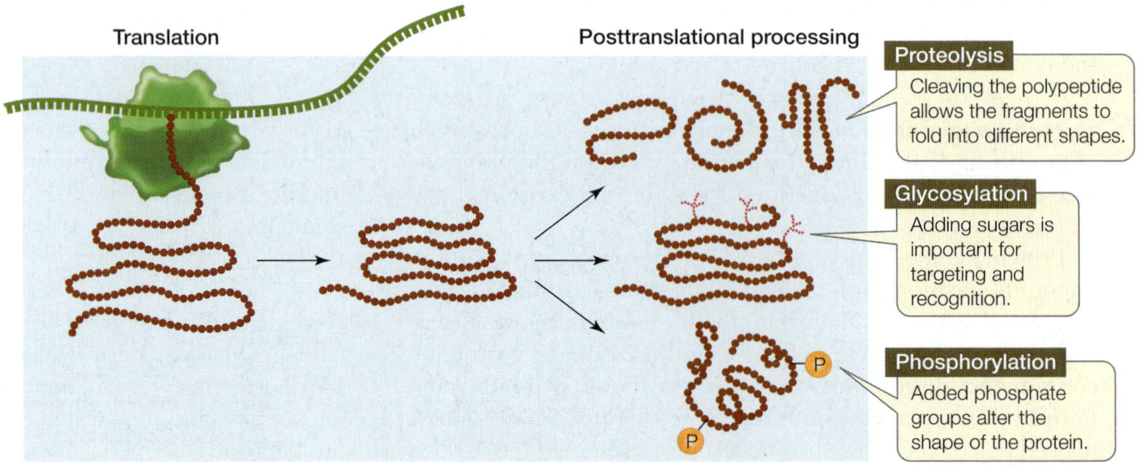
Translation
Posttranslational processing
Proteolysis
Cleaving the polypeptide allows the fragments to fold into different shapes.
Glycosylation
Adding sugars is important for targeting and recognition.
Phosphorylation
Added phosphate groups alter the shape of the protein.
P
P

14.20 Posttranslational Modifications of Proteins Most polypeptides must be modified after translation in order to become functional proteins.

by proteases. These proteases are essential to some viruses, including human immunodeficiency virus (HIV), because the large viral polyprotein cannot fold properly unless it is cut. Certain drugs used to treat acquired immune deficiency syndrome (AIDS) work by inhibiting the HIV protease, thereby preventing the formation of proteins needed for viral reproduction.

- **Glycosylation** is the addition of sugars to proteins to form glycoproteins. In both the ER and the Golgi apparatus, resident enzymes catalyze the addition of various sugars or short sugar chains to certain amino acid R groups on proteins. One such type of "sugar coating" is essential for directing proteins to lysosomes, as mentioned above. Other types are important in the conformation of proteins and their recognition functions at the cell surface (see Section 6.2). Other attached sugars help stabilize extracellular proteins, or proteins stored in vacuoles in plant seeds.
- **Phosphorylation** is the addition of phosphate groups to proteins, and is catalyzed by protein kinases. The charged phosphate groups change the conformation of a protein, often exposing the active site of an enzyme or the binding site for another protein. We have seen the important role of phosphorylation in cell signaling (see Chapter 7) and the cell cycle (see Chapter 11).

RECAP 14.6

Signal sequences in polypeptides direct them to their appropriate destinations inside or outside the cell. Many polypeptides are modified after translation.

- How do signal sequences determine where a protein will go after it is made? **See pp. 299–300 and Figure 14.18**
- What are some ways in which posttranslational modifications alter protein structure and function? **See pp. 300–301 and Figure 14.20**

All of the processes we have just described result in a functional protein, but only if the amino acid sequence of that protein is correct. If the sequence is not correct, cellular dysfunction may result. Changes in the DNA—mutations—are a major source of errors in amino acid sequences. This is the subject of the next chapter.

Can new treatments focused on gene expression control MRSA?

ANSWER

Help may be on the way in the form of a different type of antibiotic that is being developed by Paul Dunman at the University of Rochester, New York. Dr. Dunman and his colleagues have described an important step in gene expression in bacteria, namely the breakdown of mRNA. Bacteria often exist in rapidly changing environments, and so they must adapt quickly by expressing new sets of genes appropriate for the new conditions. They are able to do this in part because they break down mRNA as soon as it is used, so that its monomers can be used in new mRNA molecules. The scientists screened a library of commercially available small molecules and identified one that targets a protein in this RNA breakdown machinery. Bacteria treated with this molecule cannot recycle old mRNA molecules to provide the monomers needed for new ones. Without the ability to adapt, the bacteria soon die. MRSA are particularly sensitive to this antibiotic.

CHAPTER SUMMARY 14

14.1 What Is the Evidence that Genes Code for Proteins?

- Experiments on metabolic enzymes in the bread mold *Neurospora* led to the **one-gene, one-enzyme hypothesis**. We now know that there is a **one-gene, one-polypeptide relationship**. Review Figure 14.1

14.2 How Does Information Flow from Genes to Proteins?

- The **central dogma** of molecular biology states that DNA encodes RNA, and RNA encodes proteins. Proteins do not encode proteins, RNA, or DNA.
- The process by which the information in DNA is copied to RNA is called **transcription**. The process by which a protein is built from the information in RNA is called **translation**. Review Figure 14.2, ACTIVITY 14.1
- A product of transcription is **messenger RNA (mRNA)**. **Transfer RNAs (tRNAs)** translate the genetic information in the mRNA into a corresponding sequence of amino acids to produce a polypeptide.
- Certain RNA viruses are exceptions to the central dogma. For example, **retroviruses** synthesize DNA from RNA in a process called **reverse transcription**.

14.3 How Is the Information Content in DNA Transcribed to Produce RNA?

- In a given gene, only one of the two strands of DNA (the template strand) acts as a template for transcription. **RNA polymerase** is the catalyst for transcription.
- RNA transcription from DNA proceeds in three steps: **initiation**, **elongation**, and **termination**. Review Figure 14.4, ANIMATED TUTORIAL 14.1
- Initiation requires a **promoter**, to which RNA polymerase binds. Part of each promoter is the **initiation site**, where transcription begins.
- Elongation of the RNA molecule proceeds from the 5′ to 3′ end.

continued

- Particular base sequences specify termination, at which point transcription ends and the RNA transcript separates from the DNA template.
- The **genetic code** is a "language" of triplets of mRNA nucleotide bases (**codons**) corresponding to 20 specific amino acids; there are **start** and **stop codons** as well. The code is redundant (an amino acid may be represented by more than one codon) but not ambiguous (no single codon represents more than one amino acid). **Review Figures 14.5, 14.6, ANIMATED TUTORIAL 14.2, ACTIVITY 14.2**

14.4 How Is Eukaryotic DNA Transcribed and the RNA Processed?

- Unlike prokaryotes, where transcription and translation occur in the cytoplasm and are coupled, in eukaryotes transcription occurs in the nucleus and translation occurs later in the cytoplasm.
- Eukaryotic genes contain **introns**, which are noncoding sequences within the transcribed regions of genes. **Review Figures 14.7B, 14.8**
- The initial transcript of a eukaryotic protein-coding gene is modified with a **5′ cap** and a 3′ **poly A tail**. **Review Figure 14.9**
- Pre-mRNA introns are removed in the nucleus via **RNA splicing** by the **small nuclear ribonucleoprotein particles**. Then the mRNA passes through a nuclear pore into the cytoplasm, where it is translated through **ribosomes**. **Review Figure 14.10, ANIMATED TUTORIAL 14.3**

14.5 How Is RNA Translated into Proteins?

See ANIMATED TUTORIAL 14.4

- During translation, amino acids are linked together in the order specified by the codons in the mRNA. This task is achieved by tRNAs, which bind to (are charged with) specific amino acids.
- Each tRNA species has an amino acid attachment site as well as an **anticodon** complementary to a specific mRNA codon. A specific activating enzyme charges each tRNA with its specific amino acid. **Review Figures 14.11, 14.12**
- The **ribosome** is the molecular workbench where translation takes place. It has one large and one small subunit, both made of **ribosomal RNA** and proteins.
- Three sites on the large subunit of the ribosome interact with tRNA anticodons. The A site is where the charged tRNA anticodon binds to the mRNA codon; the P site is where the tRNA adds its amino acid to the growing polypeptide chain; and the E site is where the tRNA is released. **Review Figure 14.13**
- Translation occurs in three steps: **initiation**, **elongation**, and **termination**. The **initiation complex** consists of tRNA bearing the first amino acid, the small ribosomal subunit, and mRNA. A specific complementary sequence on the small subunit rRNA binds to the transcription initiation site on the mRNA. **Review Figure 14.14**
- The growing polypeptide chain is elongated by the formation of peptide bonds between amino acids, catalyzed by the rRNA. **Review Figure 14.15**
- When a stop codon reaches the A site, it terminates translation by binding a release factor. **Review Figure 14.16**
- In a **polysome**, more than one ribosome moves along a strand of mRNA at one time. **Review Figure 14.17**

14.6 What Happens to Polypeptides after Translation?

- **Signal sequences** of amino acids direct polypeptides to their cellular destinations. **Review Figures 14.18, 14.19**
- Destinations in the cytoplasm include organelles, which proteins enter after being recognized and bound by surface receptors.
- Proteins "addressed" to the ER bind to a receptor protein in the ER membrane. **Review Figure 14.18**
- Posttranslational modifications of polypeptides include **proteolysis**, in which a polypeptide is cut into smaller fragments; **glycosylation**, in which sugars are added; and **phosphorylation**, in which phosphate groups are added. **Review Figure 14.20**

Go to the Interactive Summary to review key figures, Animated Tutorials, and Activities Life10e.com/is14

CHAPTER**REVIEW**

REMEMBERING

1. The adapters that allow translation of the four-letter nucleic acid language into the 20-letter protein language are called
   a. aminoacyl-tRNA synthase.
   b. transfer RNAs.
   c. ribosomal RNAs.
   d. messenger RNAs.
   e. ribosomes.

2. Which of the following does *not* occur after eukaryotic mRNA is transcribed?
   a. Binding of a sigma factor to the promoter
   b. Capping of the 5′ end
   c. Addition of a poly A tail to the 3′ end
   d. Splicing out of the introns
   e. Transport to the cytosol

3. Transcription
   a. produces only mRNA.
   b. requires ribosomes.
   c. requires tRNAs.
   d. produces RNA growing from the 5′ end to the 3′ end.
   e. takes place only in eukaryotes.

4. Which statement about translation is *not* true?
   a. Translation is RNA-directed polypeptide synthesis.
   b. An mRNA molecule can be translated by only one ribosome at a time.
   c. The same genetic code operates in almost all organisms and organelles.
   d. Energy is used in the formation of the bond between a tRNA and an amino acid.
   e. There are both start and stop codons.

5. Which statement about RNA is *not* true?
   a. Transfer RNA functions in translation.
   b. Ribosomal RNA functions in translation.
   c. RNAs are produced by transcription.
   d. Messenger RNAs are produced on ribosomes.
   e. DNA codes for mRNA, tRNA, and rRNA.

6. The genetic code
   a. is different for prokaryotes and eukaryotes.
   b. has changed during the course of recent evolution.
   c. has 64 codons that code for amino acids.
   d. has more than one codon for many amino acids.
   e. is ambiguous.

UNDERSTANDING & APPLYING

7. Normally, *Neurospora* can synthesize all 20 amino acids. A certain strain of this mold cannot grow in minimal nutritional medium, but grows only when the amino acid leucine is added to the medium. This strain
   a. is dependent on leucine for energy.
   b. has a mutation affecting a biochemical pathway leading to the synthesis of carbohydrates.
   c. has a mutation affecting the biochemical pathways leading to the synthesis of all 20 amino acids.
   d. has a mutation affecting the biochemical pathway leading to the synthesis of leucine.
   e. has a mutation affecting the biochemical pathways leading to the syntheses of 19 of the 20 amino acids.

8. An mRNA has the sequence 5′-AUGAAAUCCUAG-3′. What is the template DNA strand for this sequence?
   a. 5′-TACTTTAGGATC-3′
   b. 5′-ATGAAATCCTAG-3′
   c. 5′-GATCCTAAAGTA-3′
   d. 5′-TACAAATCCTAG-3′
   e. 5′-CTAGGATTTCAT-3′

9. In rats, a gene 1,440 base pairs long codes for an enzyme made up of 192 amino acids. Discuss this apparent discrepancy.

10. Errors in transcription occur about 100,000 times as often as errors in DNA replication. Why can this high rate be tolerated in RNA synthesis but not in DNA synthesis?

ANALYZING & EVALUATING

11. Har Gobind Khorana at the University of Wisconsin synthesized artificial mRNAs such as poly CA (CACA ...) and poly CAA (CAACAACAA ...). He found that poly CA codes for a polypeptide consisting of alternating threonine (Thr) and histidine (His) residues. There are two possible codons in poly CA: CAC and ACA. One of these must encode histidine and the other threonine—but which is which? The answer comes from results with poly CAA, which produces three different polypeptides: poly Thr, poly Gln (glutamine), and poly Asn (asparagine). (An artificial mRNA can be read, inefficiently, beginning at any point in the chain; there is no specific initiation signal. Thus poly CAA can be read as a polymer of CAA, of ACA, or of AAC.) Compare the results of the poly CA and poly CAA experiments, and determine which codon corresponds with threonine and which with histidine.

12. Beadle and Tatum's experiments showed that a biochemical pathway could be deduced from mutant strains. In bacteria, the biosynthesis of the amino acid tryptophan (T) from the precursor chorismate (C) involves four intermediate chemical compounds, which we will call D, E, F, and G. Here are the phenotypes of various mutant strains. Each strain has a mutation in a gene for a different enzyme; + means growth with the indicated compound added to the medium, and 0 means no growth. Based on these data, order the compounds (C, D, E, F, G, and T) and enzymes (1, 2, 3, 4, and 5) in a biochemical pathway.

	Addition to medium					
Mutant strain	C	D	E	F	G	T
1	0	0	0	0	+	+
2	0	+	+	0	+	+
3	0	+	0	0	+	+
4	0	+	+	+	+	+
5	0	0	0	0	0	+

Go to BioPortal at **yourBioPortal.com** for Animated Tutorials, Activities, LearningCurve Quizzes, Flashcards, and many other study and review resources.

15 Gene Mutation and Molecular Medicine

CHAPTER OUTLINE

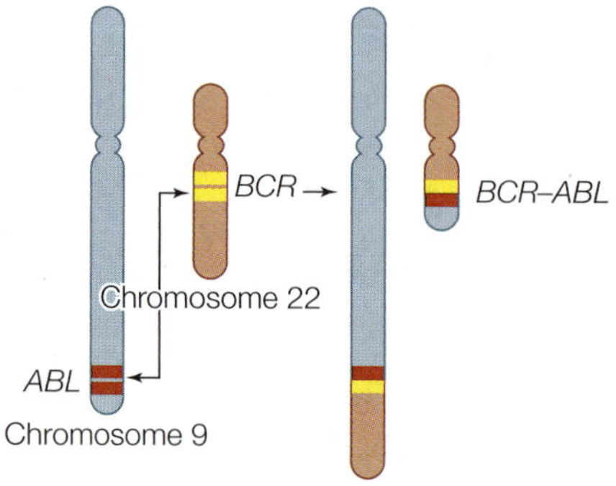

BCR
BCR–ABL
Chromosome 22
ABL
Chromosome 9

Shuffling the Genetic Deck in Cancer A mutation in a bone marrow cell involves a swap of the ends of two chromosomes. The result is a new gene, encoding a new protein that stimulates cell division. This mutation can cause leukemia.

KAREEM ABDUL-JABBAR had a very successful career as a basketball player. He led his university team (UCLA) to three national championships, and later he led his professional teams to six championships. But in late 2008, with persistent hot flashes and sweats, he went to see his doctor, who sent his blood for testing. A blood smear, where cells are examined in the microscope, showed a high proportion of white blood cells (normally scarce compared with red blood cells). A few days later, there was a diagnosis: chronic myelogenous leukemia (CML), a type of cancer in which immature white blood cells divide continuously. If untreated, the white blood cells crowd out the red blood cells (causing anemia), the platelets (causing clotting disorders), and other types of white blood cells that are part of the immune system (causing infections).

If Abdul-Jabbar had been diagnosed ten years earlier, his CML would have been treated with toxic drugs to stop cancer cell division. He might have expected to live about five years after diagnosis. But a revolutionary treatment has dramatically improved survival in CML. Like many other patients with this cancer, Abdul-Jabbar now has no detectable tumor cells and can expect to live a normal life span. The new treatment came from knowledge of a mutation involved in CML, the abnormal protein produced by that mutation, and a drug targeted to that protein.

The mutation that causes CML occurs in a developing white blood cell in the bone marrow, in which parts of chromosomes 9 and 22 are swapped (translocated). As a result, part of a gene on chromosome 9 (*ABL*) becomes fused with part of a gene on chromosome 22 (*BCR*). The swap usually occurs within the coding regions of the two genes, and the altered chromosome 22 can end up with a new gene made up of the first part of *BCR* and the last part of *ABL*.

The *BCR–ABL* gene fusion happens to be an oncogene; it makes a protein product that stimulates cell division and causes cancer. Immature white blood cells containing the gene fusion divide continuously instead of differentiating into mature white blood cells. The BCR–ABL protein has been crystallized and its structure determined. Its unique structure allowed scientists to design and screen through millions of chemicals to find one that would specifically bind to and inhibit the protein. After a promising chemical was identified, it was modified to improve its binding ability, resulting in the drug imatinib (sold as Gleevec). This drug has been a highly successful targeted therapy for CML and has saved the lives of many patients, including Abdul-Jabbar.

This approach—describing a mutation and its protein phenotype, and then designing a drug for that protein—is at the heart of a new branch of healing called molecular medicine.

? Are there other targeted therapies directed to specific types of cancer?

See answer on p. 325.

15.1 What Are Mutations?

In Chapter 12 we described mutations as inherited changes in genes, and we saw that different alleles may produce different phenotypes (short pea plants versus tall, for example). In the following two chapters we described the chemical nature of genes as DNA sequences, and how they are expressed as phenotypes (in particular, proteins). As we mentioned in Section 14.1, a **mutation** is a change in the nucleotide sequence of DNA that can be passed on from one cell, or organism, to another.

As an example of just one cause of mutations, recall from Chapter 13 that DNA polymerases make errors. Repair systems such as proofreading are in place to correct them. But some errors escape being corrected and are passed on to the daughter cells.

Mutations in multicellular organisms can be divided into two types:

- **Somatic mutations** are those that occur in somatic (body) cells. These mutations are passed on to the daughter cells during mitosis, and to the offspring of those cells in turn, but are not passed on to sexually produced offspring. For example, a mutation in a single human skin cell could result in a patch of skin cells that all have the same mutation, but it would not be passed on to the person's children. The mutation that results in chronic myelogenous leukemia (CML; see the opening story) is a somatic mutation in white blood cells.
- **Germ line mutations** are those that occur in the cells of the germ line—the specialized cells that give rise to gametes. A gamete with the mutation passes it on to a new organism at fertilization. The new organism will have the mutation in every cell of its body and will pass the mutation on to all of its progeny.

In either case, the mutations may or may not have phenotypic effects.

Mutations have different phenotypic effects

Phenotypically, we can understand mutations in terms of their effects on proteins and their function (**Figure 15.1**).

- A **silent mutation** does not usually affect protein function (see Figure 15.1B). It can be a mutation in a region of DNA that does not encode a protein, or it can be in the coding region of a gene but not affect the amino acid sequence. Because of the redundancy of the genetic code, a base change in a coding region will not always cause a change in the amino acid sequence when the altered mRNA is translated (see Figure 15.2). Silent mutations are common, and they usually result in genetic diversity that is not expressed as phenotypic differences. We say "usually" here because some silent mutations within coding regions actually do affect protein function. A silent mutation can affect mRNA stability, or it can affect the rate of translation because of differences in the abundance of specific tRNAs. In either case, the abundance of the protein can be affected, and this in turn can affect phenotype.

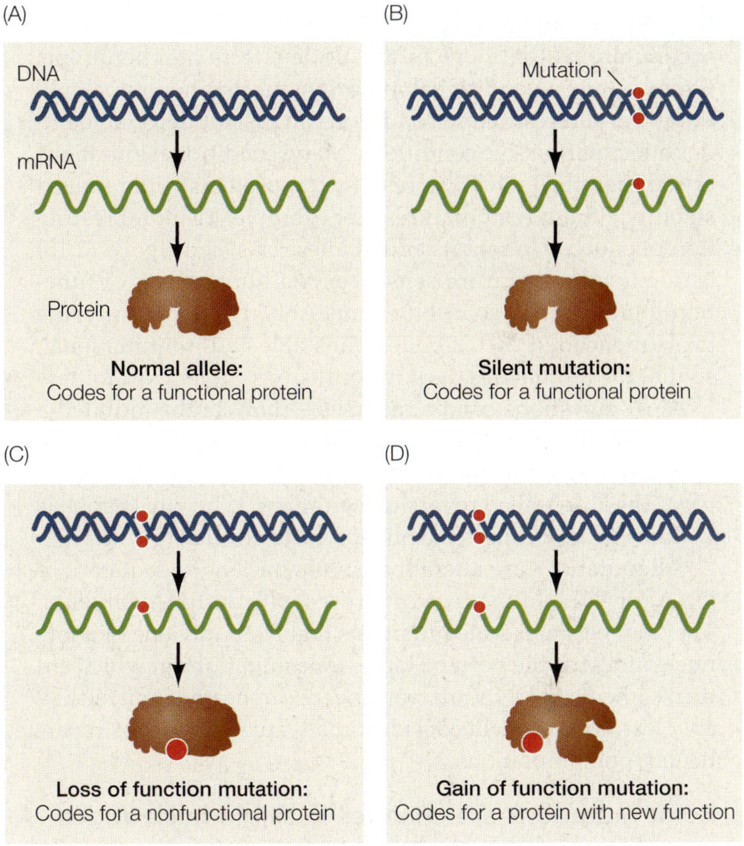
(A)
(B)
DNA
mRNA
Protein
Normal allele:
Codes for a functional protein
Mutation
Silent mutation:
Codes for a functional protein
(C)
(D)
Loss of function mutation:
Codes for a nonfunctional protein
Gain of function mutation:
Codes for a protein with new function

15.1 Mutation and Phenotype Mutations may or may not affect the protein phenotype.

Go to Animated Tutorial 15.1
Gene Mutations
Life10e.com/at15.1

- A **loss of function mutation** affects protein function (see Figure 15.1C). Such a mutation may cause a gene to not be expressed at all, or the gene may be expressed but produce a protein that no longer plays its cellular role, such as its catalytic function if it is an enzyme. Loss of function mutations almost always show recessive inheritance in diploid organisms, because the presence of one wild-type allele will usually result in sufficient functional protein for the cell. For example, recall from Section 12.1 that the familiar wrinkled-seed allele in pea plants, originally studied by Mendel, is due to a recessive loss of function mutation in the *SBE1* (*s*tarch *b*ranching *e*nzyme). Normally the protein made by this gene catalyzes the branching of starch as seeds develop. In the mutant, the SBE1 protein is not functional, and that leads to osmotic changes, causing the wrinkled appearance.
- A **gain of function mutation** leads to a protein with an altered function (see Figure 15.1D). This kind of mutation usually shows dominant inheritance, because the presence of the wild-type allele does not prevent the mutant allele from functioning. This type of mutation is common in cancer. For example, as we saw in the opening story, the *BCR–ABL* gene fusion is a gain of function mutation that causes uncontrolled cell division.

In addition to the broad categories shown in Figure 15.1, there are mutations that have more subtle effects on phenotype. For example, a **conditional mutation** affects phenotype only under certain restrictive conditions and is not detectable under other, permissive conditions. Many conditional mutations are temperature-sensitive, resulting in proteins with reduced stability at high temperatures. For example, the point restriction phenotype in rabbits and Siamese cats (see Figure 12.15) is due to a temperature-sensitive (conditional, loss of function) mutation in a coat color gene. At body temperature, the protein encoded by the gene is unstable and nonfunctional, so that the animal has dark fur only in its cooler extremities.

Most mutations can be reversed—they can be mutated a second time so that the DNA reverts to its original sequence or to a coding sequence that results in the non-mutant phenotype. These are called **reversion mutations**. When this happens within a gene, it causes the phenotype to go back to wild type.

All mutations are alterations in the nucleotide sequence of DNA (or RNA in the case of viruses with an RNA genome). They can be small-scale mutations that alter only one or a few nucleotides, or they can be large-scale mutations in which entire segments of DNA are rearranged, duplicated, or irretrievably lost. Next we will consider small-scale mutations, in particular, point mutations.

Point mutations are changes in single nucleotides

A **point mutation** is the addition or subtraction of a single nucleotide, or the substitution of one nucleotide base for another. There are two kinds of base substitution:

- A **transition** is the substitution of one purine for the other purine, or one pyrimidine for the other:

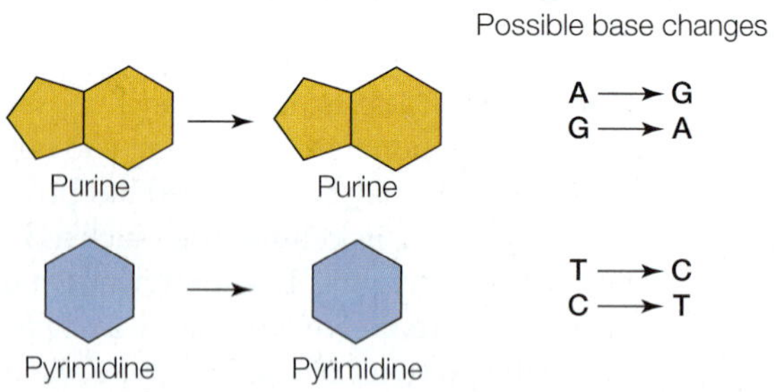
Possible base changes
Purine
Purine
A → G
G → A
Pyrimidine
Pyrimidine
T → C
C → T

- A **transversion** is the substitution of a purine for a pyrimidine, or vice versa:

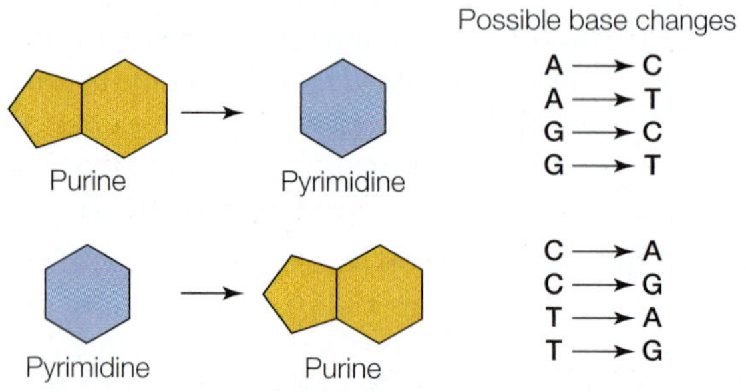
Possible base changes
Purine
Pyrimidine
A → C
A → T
G → C
G → T
Pyrimidine
Purine
C → A
C → G
T → A
T → G

A point mutation in the coding region of a gene will result in an alteration in the mRNA sequence, but a change in the mRNA may or may not result in a change in the protein. As we have already discussed, a silent mutation has no effect on the amino acid sequence of an encoded polypeptide. By contrast, missense, nonsense, and frameshift mutations result in changes in the protein, some of them drastic (**Figure 15.2**).

MISSENSE MUTATIONS Some base substitutions change the genetic code such that one amino acid substitutes for another in a protein. These are called **missense mutations** (see Figure 15.2C). A specific example is the mutation that causes sickle-cell disease, a serious heritable blood disorder. The disease occurs in people who carry two copies of the sickle allele of the gene for β-globin—a subunit of hemoglobin, the protein in human blood that carries oxygen. The sickle allele differs from the normal allele by one base pair, resulting in a polypeptide that differs by one amino acid from the normal protein. Individuals who are homozygous for this recessive allele have defective, sickle-shaped red blood cells (**Figure 15.3**).

A missense mutation may result in a defective protein, but often it has no effect on the protein's function. For example, a hydrophilic amino acid may be substituted for another hydrophilic amino acid, so that the shape of the protein is unchanged. Or a missense mutation might reduce the functional efficiency of a protein rather than completely inactivating it. Therefore individuals homozygous for a missense mutation in a protein essential for life may survive if enough of the protein's function is retained.

In some cases, a gain of function missense mutation occurs. An example is a mutation in the human *TP53* gene, which codes for a tumor suppressor—a protein that inhibits the cell cycle (see Section 11.7). Certain mutations of the *TP53* gene cause this protein to no longer inhibit cell division, but to promote it and prevent programmed cell death. So like the BCR–ABL fusion protein described in the opening story, this kind of mutation results in a TP53 protein that has gained an oncogenic (cancer-causing) function.

NONSENSE MUTATIONS A **nonsense mutation** involves a base substitution that causes a stop codon (for translation) to form somewhere in the mRNA (see Figure 15.2D). A nonsense mutation results in a shortened protein, since translation does not proceed beyond the point where the mutation occurred. For example, a common mutation causing thalassemia (another blood disorder affecting hemoglobin) in Mediterranean populations is a nonsense mutation that drastically shortens the β-globin subunit. Shortened proteins are usually not functional; however, if the nonsense mutation occurs near the 3′ end of the gene, it may have no effect on function.

FRAME-SHIFT MUTATIONS Not all point mutations are base substitutions. One or two nucleotides may be inserted into, or deleted from, a sequence of DNA. Such mutations in coding sequences are known as **frame-shift mutations** because they alter the reading frame in which the three-base codons are read during translation (see Figure 15.2E). Think again of codons as three-letter words, each corresponding to a particular amino acid. Translation proceeds codon by codon; if a nucleotide is added to the mRNA or subtracted from it, then the three-letter

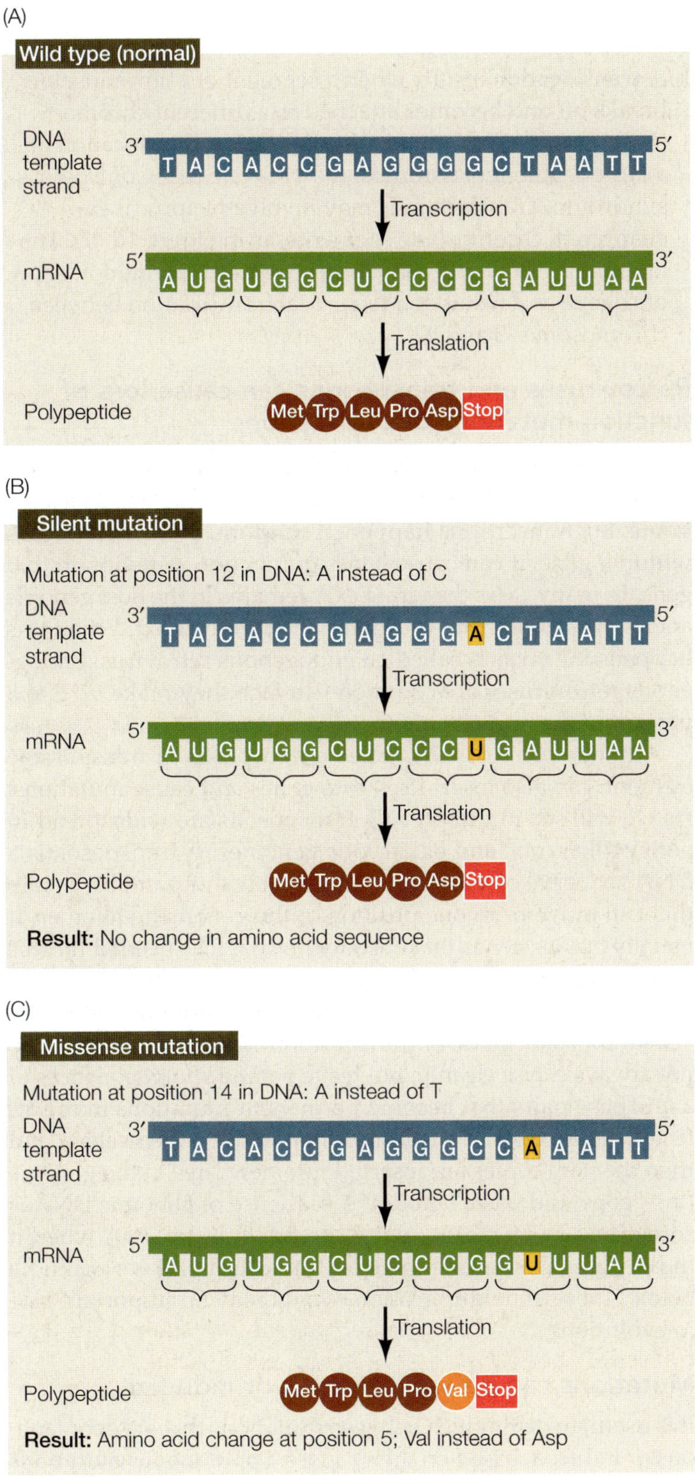
(A)
Wild type (normal)
DNA template strand
3′ T A C A C C G A G G G G C T A A T T 5′
Transcription
mRNA
5′ A U G U G G C U C C C C G A U U A A 3′
Translation
Polypeptide
Met Trp Leu Pro Asp Stop
(B)
Silent mutation
Mutation at position 12 in DNA: A instead of C
DNA template strand
3′ T A C A C C G A G G G A C T A A T T 5′
Transcription
mRNA
5′ A U G U G G C U C C C U G A U U A A 3′
Translation
Polypeptide
Met Trp Leu Pro Asp Stop
Result: No change in amino acid sequence
(C)
Missense mutation
Mutation at position 14 in DNA: A instead of T
DNA template strand
3′ T A C A C C G A G G G G C A A A T T 5′
Transcription
mRNA
5′ A U G U G G C U C C C C G U U U A A 3′
Translation
Polypeptide
Met Trp Leu Pro Val Stop
Result: Amino acid change at position 5; Val instead of Asp

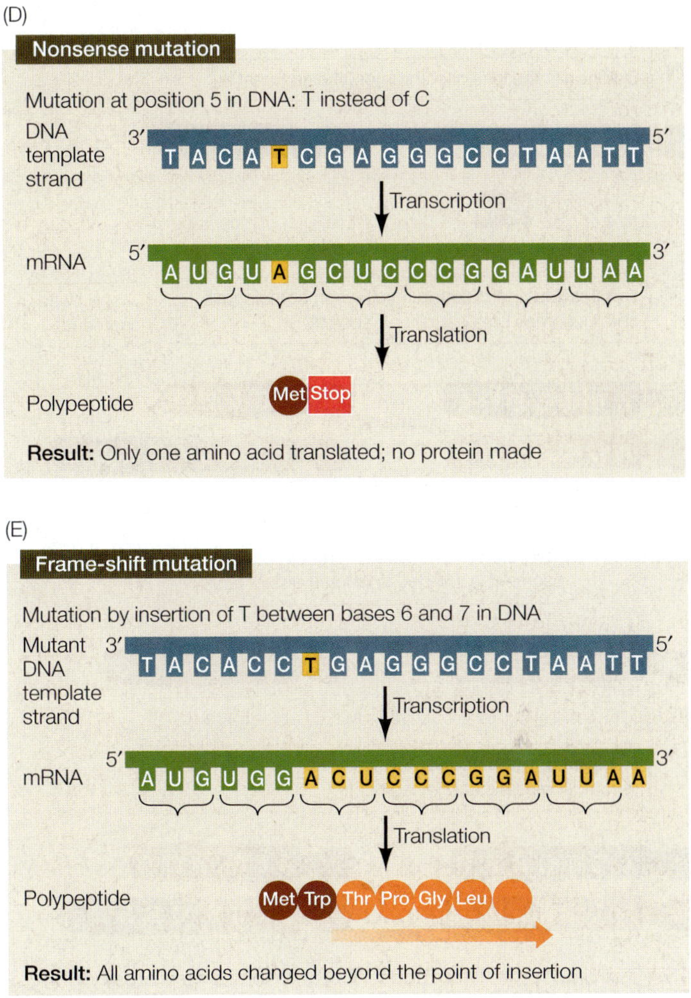
(D)
Nonsense mutation
Mutation at position 5 in DNA: T instead of C
DNA template strand
3′ T A C A T C G A G G G G C T A A T T 5′
Transcription
mRNA
5′ A U G U A G C U C C C C G A U U A A 3′
Translation
Polypeptide
Met Stop
Result: Only one amino acid translated; no protein made
(E)
Frame-shift mutation
Mutation by insertion of T between bases 6 and 7 in DNA
Mutant DNA template strand
3′ T A C A C C T G A G G G G C T A A T T 5′
Transcription
mRNA
5′ A U G U G G A C U C C C C G A U U A A 3′
Translation
Polypeptide
Met Trp Thr Pro Gly Leu
Result: All amino acids changed beyond the point of insertion

15.2 Point Mutations When they occur in the coding regions of proteins, single base changes can cause silent, missense, nonsense, or frame-shift mutations.

"words" are altered as translation proceeds beyond that point, and the result is a completely different amino acid sequence. Frame-shift mutations almost always lead to the production of nonfunctional proteins.

Chromosomal mutations are extensive changes in the genetic material

Changes in single nucleotides are not the most dramatic changes that can occur in the genetic material. Whole DNA molecules can break and rejoin, grossly disrupting the sequence of genetic information. There are four types of such

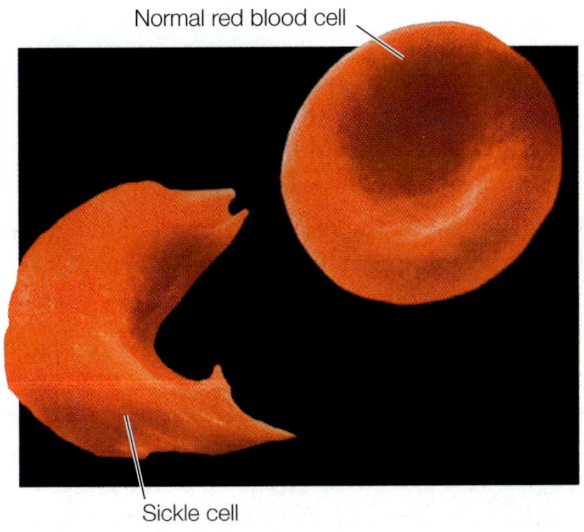
Normal red blood cell
Sickle cell

15.3 Sickle and Normal Red Blood Cells The misshapen red blood cell on the left is caused by a missense mutation and an incorrect amino acid in one of the two polypeptides of hemoglobin.

Go to Media Clip 15.1
Sickle Cells: Deformed by a Mutation
Life10e.com/mc15.1

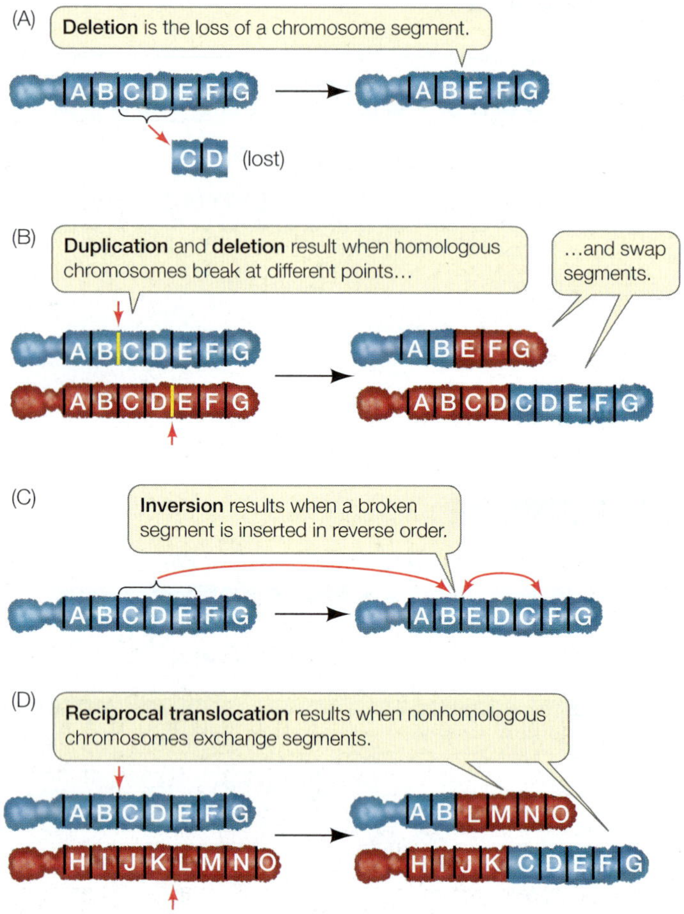
(A)
Deletion is the loss of a chromosome segment.
A B C D E F G
A B E F G
C D (lost)
(B)
Duplication and deletion result when homologous chromosomes break at different points...
...and swap segments.
A B C D E F G
A B C D E F G
A B E F G
A B C D C D E F G
(C)
Inversion results when a broken segment is inserted in reverse order.
A B C D E F G
A B E D C F G
(D)
Reciprocal translocation results when nonhomologous chromosomes exchange segments.
A B C D E F G
H I J K L M N O
A B L M N O
H I J K C D E F G

15.4 Chromosomal Mutations Chromosomes may break during replication, and parts of chromosomes may then rejoin incorrectly. This can result in deletions (A and B), duplications (B), inversions (C), or reciprocal translocations (D). Note that the letters on these illustrations represent large segments of the chromosomes. Because chromosomes contain regions of noncoding DNA, each segment may include anywhere from zero to hundreds or thousands of genes.

chromosomal mutations: deletions, duplications, inversions, and translocations. These mutations can be caused by severe damage to chromosomes resulting from mutagens or by drastic errors in chromosome replication.

- A **deletion** occurs by the removal of part of the genetic material and can happen if a chromosome breaks at two points and then rejoins, leaving out the DNA between the breaks (**Figure 15.4A**).
- A **duplication** can be produced at the same time as a deletion and can occur if homologous chromosomes break at different positions and then reconnect to the wrong partners (**Figure 15.4B**). One of the two chromosomes ends up with a deleted segment, and the other has two copies (a duplication) of the same segment.
- An **inversion** can also result from the breaking and rejoining of a chromosome, and can occur if a segment of DNA becomes "flipped," so that it runs in the opposite direction from its original orientation (**Figure 15.4C**).
- A **translocation** results when a segment of a chromosome breaks off and becomes attached to a different chromosome. As we mentioned in Section 11.5, a translocation of a large segment of chromosome 21 is one cause of Down syndrome. Translocations may involve reciprocal exchanges of chromosome segments, as in **Figure 15.4D**. The opening story of this chapter described the formation of an oncogene as a result of a reciprocal translocation between chromosomes 9 and 22.

Retroviruses and transposons can cause loss of function mutations or duplications

In Section 14.2 we mentioned that certain viruses called retroviruses can insert their genetic material into the host cell's genome. Such insertions happen at random, and if one occurs within a gene, it can cause a loss of function mutation in that gene. In many cases the viral DNA remains in the host genome and is passed on from one generation to the next. When this happens the virus is called an endogenous retrovirus. Endogenous retroviruses are common—in fact, they make up 5 to 8 percent of the human genome!

Another form of DNA, called a transposon or transposable element, can also insert itself into genes and cause mutations. As we will see in Chapter 17, transposons are widespread in both prokaryotic and eukaryotic genomes. A transposon is a DNA sequence of a few hundred to a few thousand base pairs that can move from one position in the genome to another. It usually carries genes that encode the enzymes needed for this movement. Some transposons cut themselves out from their positions in the genome and then insert into other sites (the "cut and paste" mode of transposition). These transposons do not always excise cleanly, but leave behind short sequences of a few base pairs that become permanent mutations in the affected genes. Other transposons first replicate themselves, and then the new copies are inserted into new sites in the genome (the "copy and paste" mode). A sequence of genomic DNA is sometimes carried along with the transposon DNA when it moves, and this results in gene duplication. As we will note below, these gene duplication events play an important role in evolution.

Mutations can be spontaneous or induced

It is useful to distinguish between mutations that are spontaneous or induced, based on their causes. **Spontaneous mutations** are permanent changes in the genetic material that occur without any outside influence. The movement of transposons is an example of spontaneous mutation. Spontaneous mutations can also occur because cellular processes are imperfect, and may occur by several mechanisms:

- *The four nucleotide bases of DNA can have different structures, leading to mistakes during replication.* Each base can exist in two different forms (called tautomers), one of which is common and one rare. When a base temporarily forms its rare tautomer, it can pair with the wrong base. For example, C normally pairs with G, but if C is in its rare tautomer at the time of DNA replication, it pairs with (and DNA

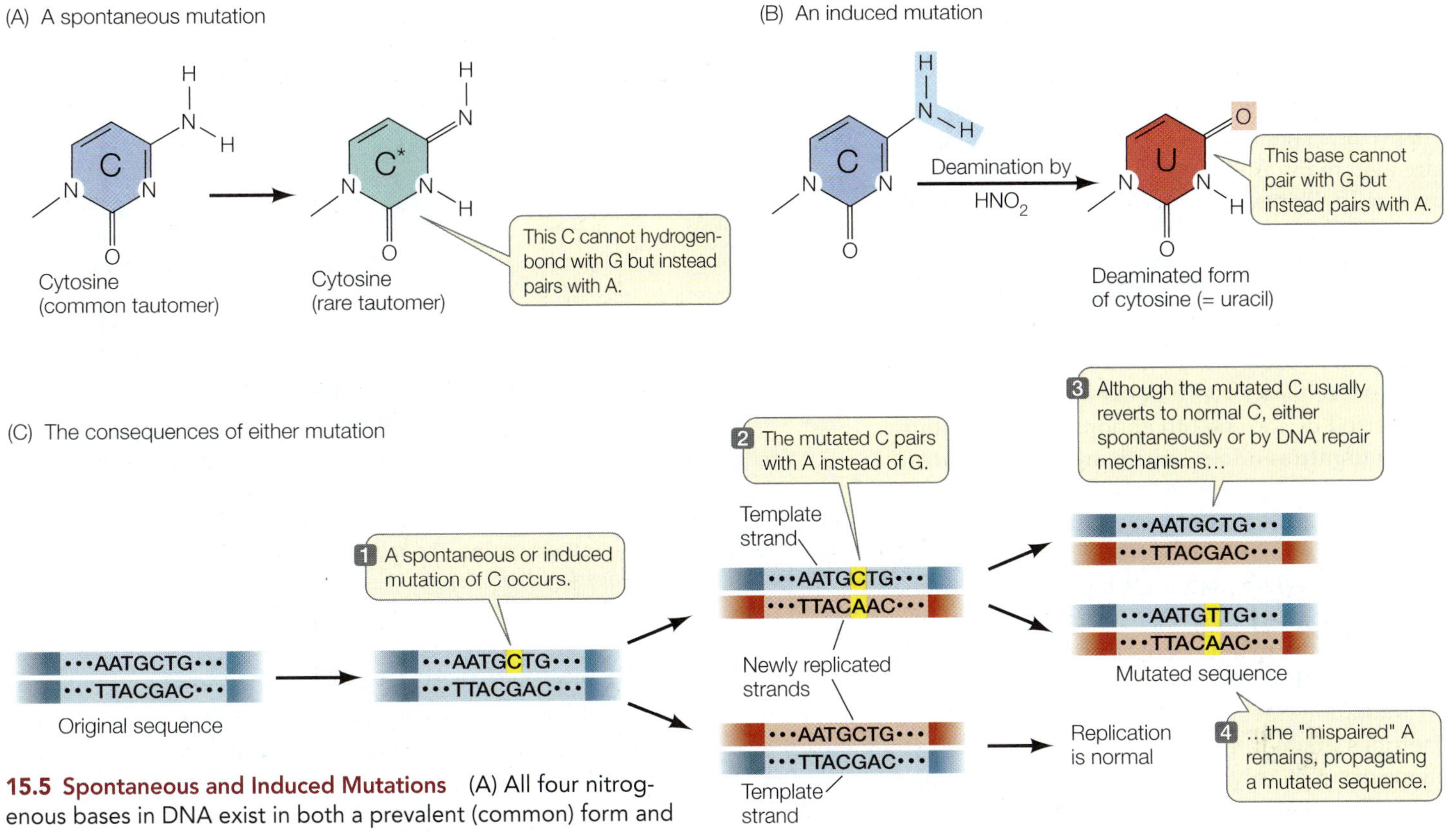

(A) A spontaneous mutation
Cytosine (common tautomer)
Cytosine (rare tautomer)
This C cannot hydrogen-bond with G but instead pairs with A.
(B) An induced mutation
Deamination by HNO2
Deaminated form of cytosine (= uracil)
This base cannot pair with G but instead pairs with A.
(C) The consequences of either mutation
•••AATGCTG•••
•••TTACGAC•••
Original sequence
1 A spontaneous or induced mutation of C occurs.
2 The mutated C pairs with A instead of G.
Template strand
•••AATGCTG•••
•••TTACAAC•••
Newly replicated strands
•••AATGCTG•••
•••TTACGAC•••
Template strand
3 Although the mutated C usually reverts to normal C, either spontaneously or by DNA repair mechanisms...
•••AATGTTG•••
•••TTACAAC•••
Mutated sequence
Replication is normal
4 ...the "mispaired" A remains, propagating a mutated sequence.

15.5 Spontaneous and Induced Mutations (A) All four nitrogenous bases in DNA exist in both a prevalent (common) form and a rare form. When a base spontaneously forms its rare tautomer, it can pair with a different base. (B) Mutagens such as nitrous acid (HNO_2) can induce changes in the bases. (C) The results of both spontaneous and induced mutations are permanent changes in the DNA sequence following replication.

polymerase will insert) an A. The result is a point mutation: G → A (**Figure 15.5A and C**).

- *Bases in DNA may change because of a chemical reaction*—for example, loss of an amino group in cytosine (a reaction called deamination). If this occurs in a DNA molecule, the error will usually be repaired. However, since the repair mechanism is not perfect, the altered nucleotide will sometimes remain during replication. In these cases, DNA polymerase will add an A (which base-pairs with U) instead of G (which normally pairs with C).
- *DNA polymerase can make errors in replication* (see Section 13.4)—for example, by inserting a T opposite a G. Most of these errors are repaired by the proofreading function of the replication complex, but some errors escape detection and become permanent.
- *Meiosis is not perfect*. Nondisjunction—the failure of homologous chromosomes to separate during meiosis—can occur, leading to one too many chromosomes or one too few (see Figure 11.20). Random chromosome breakage and rejoining can produce deletions, duplications, inversions, or translocations.

Induced mutations occur when some agent from outside the cell—a **mutagen**—causes a permanent change in DNA. As we mentioned above, retroviruses can function as mutagens. In addition, certain chemicals and radiation can cause mutations:

- *Some chemicals can alter nucleotide bases*. For example, nitrous acid (HNO_2) and similar molecules can react with cytosine and convert it to uracil by deamination. More specifically, they convert an amino group on the cytosine (—NH_2) into a keto group (—C=O) (**Figure 15.5B**). This alteration has the same result as spontaneous deamination: instead of a G, DNA polymerase inserts an A (see Figure 15.5C).
- *Some chemicals add groups to the bases*. For instance, benzopyrene, a component of cigarette smoke, adds a large chemical group to guanine, making it unavailable for base pairing. When DNA polymerase reaches such a modified guanine, it inserts any one of the four bases at random. Three-fourths of the time the inserted base is not cytosine, and a mutation results.
- *Radiation damages the genetic material*. Radiation can damage DNA in two ways. First, ionizing radiation (including X rays, gamma rays, and radiation from unstable isotopes) produces highly reactive chemicals called free radicals. Free radicals can change bases in DNA to forms that are not recognized by DNA polymerase. Ionizing radiation can also break the sugar–phosphate backbone of DNA, causing chromosomal abnormalities. Second, ultraviolet radiation (from the sun or a tanning lamp) is absorbed by thymine, causing it to form covalent bonds with adjacent bases. This, too, plays havoc with DNA replication by distorting the double helix.

Mutagens can be natural or artificial

Many people associate mutagens with materials made by humans, but just as there are many human-made chemicals that cause mutations, there are also many mutagenic substances that occur naturally. Plants (and to a lesser extent animals) make thousands of small molecules with a variety of functions, such as defense against pathogens (see Chapter 39). Some of these are mutagenic and potentially carcinogenic. Examples of human-made mutagens are nitrites, which are used to preserve meats. Once in mammals, nitrites get converted by the smooth endoplasmic reticulum (SER) to nitrosamines, which are strongly mutagenic because they cause deamination of cytosine (see above). An example of a naturally occurring mutagen is aflatoxin, which is made by the mold *Aspergillus*. When mammals ingest the mold, the aflatoxin is converted by the ER into a product that, like benzopyrene from cigarette smoke, binds to guanine; this also causes mutations.

Radiation can also be human-made or natural. Some of the isotopes made in nuclear reactors and nuclear bomb explosions are certainly harmful. For example, extensive studies have shown increased mutations in the survivors of the atom bombs dropped on Japan in 1945. As previously mentioned, natural ultraviolet radiation in sunlight also causes mutations, in this case by affecting thymine and, to a lesser extent, other bases in DNA.

Biochemists have estimated how much DNA damage occurs in the human genome under normal circumstances: among the genome's 3.2 billion base pairs, there are about 16,000 DNA-damaging events per cell per day, of which 80 percent are repaired.

Some base pairs are more vulnerable than others to mutation

In certain regions of DNA, many of the cytosine residues have methyl groups added at their 5′ positions, forming 5′–methylcytosine. This methylation plays an important role in gene regulation (see Section 16.4). DNA sequencing has revealed that mutation "hot spots" are often located where cytosines have been methylated.

As we discussed above, unmethylated cytosine can lose its amino group, either spontaneously or because of a chemical mutagen, to form uracil (see Figure 15.5). This type of error is usually detected by the cell and repaired, because uracil is recognized as inappropriate for DNA (it occurs only in RNA). However, when 5′–methylcytosine loses its amino group, the product is thymine, a natural base for DNA (**Figure 15.6**). The DNA repair mechanism ignores this thymine. During replication, however, the mismatch repair mechanism recognizes that G-T is a mismatched pair, although it cannot tell which base is incorrect. Half of the time it matches a new C to the G, but the other half of the time it matches a new A to the T, resulting in a mutation.

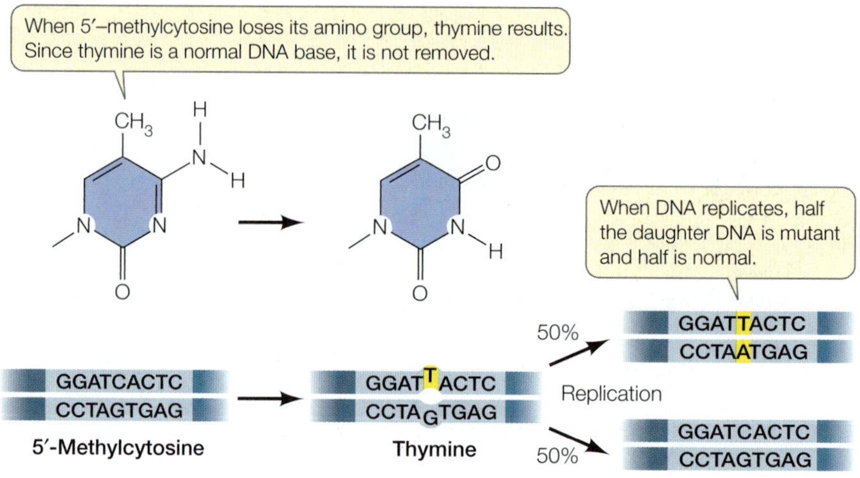
When 5′–methylcytosine loses its amino group, thymine results. Since thymine is a normal DNA base, it is not removed.
When DNA replicates, half the daughter DNA is mutant and half is normal.
GGATCACTC
CCTAGTGAG
5′-Methylcytosine
GGATTACTC
CCTAGTGAG
Thymine
50%
Replication
50%
GGATTACTC
CCTAATGAG
GGATCACTC
CCTAGTGAG

15.6 5′-Methylcytosine in DNA Is a "Hot Spot" for Mutations If cytosine has been methylated to 5′-methylcytosine, the mutation is unlikely to be repaired and a C-G base pair is replaced with a T-A pair.

Mutations have both benefits and costs

As we will see in Part Seven of this book, mutations are the raw material of evolution: they provide the genetic diversity that makes natural selection possible. This diversity can be beneficial in two ways. First, a mutation in a somatic cell may benefit the organism immediately. Second, a mutation in a germ line cell may have no immediate selective advantage to the organism, but it may cause a phenotypic change in the organism's offspring. If the environment changes in a later generation, this mutation may be advantageous, and thus selected for, under the new conditions.

We have seen that gene duplication may arise through either chromosomal rearrangements or through the movements of transposons. Gene duplication is not always harmful and is an important source of genetic variation. In a duplicated pair of genes, one gene may continue to play its original role in the cell, while the other may acquire a gain of function mutation that produces a new phenotype. As for any other mutation, this may be of immediate benefit for the organism, or it may provide a later generation with a selective advantage.

By contrast, mutations in genes whose products are needed for normal cellular processes are often deleterious, especially if they occur in germ line cells. In such cases, some offspring can inherit harmful recessive alleles in the homozygous condition. In their extreme form, such mutations produce phenotypes that are lethal, killing the organism during early development.

Mutations in somatic cells can also have dramatic harmful consequences for an organism. In Chapter 11 we described how mutations in oncogenes can result in uncontrolled cell division, whereas loss of function mutations in tumor suppressor genes can prevent the inhibition of cell division in cells that normally do not divide. Both kinds of mutations can lead to cancer, and they can be either spontaneous or induced. While spontaneous mutagenesis is not in our control, we can certainly try to avoid mutagenic substances and radiation. Not surprisingly, many things that cause cancer (carcinogens) are also mutagens. A good example is benzopyrene (discussed above), which is found in coal tar, car exhaust fumes, and charbroiled foods, as well as in cigarette smoke.

A major public policy goal is to reduce the effects of both human-made and natural mutagens on human health. Here are two examples:

- The Montreal Protocol is the only international environmental agreement signed and adhered to by all members of the United Nations. It bans chlorofluorocarbons and other substances that cause depletion of the ozone layer in the upper atmosphere of Earth. Such depletion can result in increased ultraviolet radiation reaching Earth's surface. This would cause more somatic mutations that lead to skin cancer.
- Bans on cigarette smoking have rapidly spread throughout the world. Cigarette smoking causes cancer because of increased exposure of somatic cells in the lungs and throat to benzopyrene and other carcinogens.

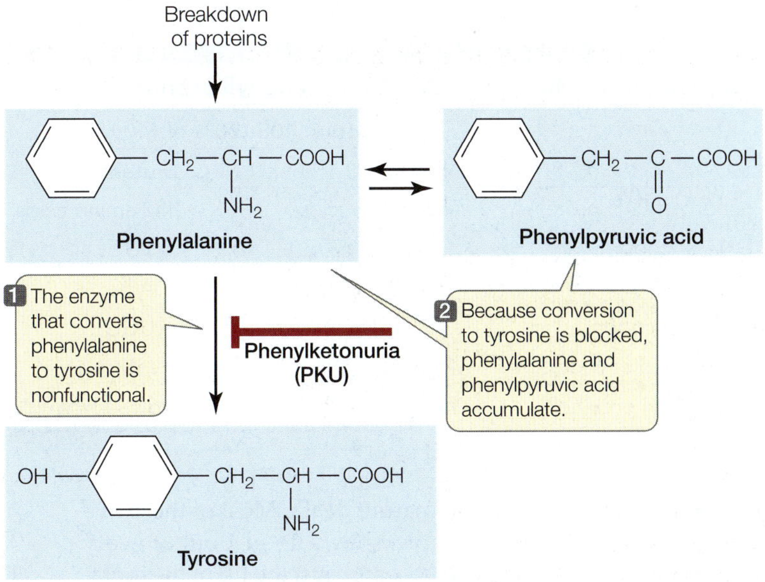
Breakdown of proteins
CH_2—CH—COOH
NH_2
Phenylalanine
CH_2—C—COOH
O
Phenylpyruvic acid
1 The enzyme that converts phenylalanine to tyrosine is nonfunctional.
Phenylketonuria (PKU)
2 Because conversion to tyrosine is blocked, phenylalanine and phenylpyruvic acid accumulate.
OH
CH_2—CH—COOH
NH_2
Tyrosine

15.7 One Gene, One Enzyme Phenylketonuria is caused by an abnormality in a specific enzyme that metabolizes the amino acid phenylalanine. Knowing the molecular causes of such single-gene, single-enzyme metabolic diseases can aid researchers in developing screening tests as well as treatments.

RECAP 15.1

Mutations are alterations in the nucleotide sequence of DNA. They may be changes in single nucleotides or extensive rearrangements of chromosomes. If they occur in somatic cells, they will be passed on to daughter cells; if they occur in germ line cells, they will be passed on to offspring.

- What are the various kinds of point mutations? **See p. 306 and Figure 15.2**
- What distinguishes the various kinds of chromosomal mutations: deletions, duplications, inversions, and translocations? **See p. 308 and Figure 15.4**
- Explain the difference between spontaneous and induced mutagenesis. Give an example of each. **See pp. 308–309 and Figure 15.5**
- Why do many mutations involve G-C base pairs? **See p. 310 and Figure 15.6**

We have seen that there are many different ways in which DNA can be altered, in terms of both the types of changes and the mechanisms by which they occur. We will turn now to the ways in which mutations can cause disease.

15.2 What Kinds of Mutations Lead to Genetic Diseases?

The biochemistry that relates genotype (DNA) and phenotype (proteins) has been most completely described for model organisms, such as the prokaryote *E. coli* and the eukaryotes yeast and *Drosophila*. While the details vary, there is great similarity in the fundamental processes among these forms of life. These similarities have permitted the application of knowledge and methods discovered using these model organisms to the study of human biochemical genetics. Our focus in this chapter is mutations that affect human phenotypes, leading to diseases.

Genetic mutations may make proteins dysfunctional

Genetic mutations are often expressed phenotypically as proteins that differ from normal (wild-type) proteins. Abnormalities in enzymes, receptor proteins, transport proteins, structural proteins, and most of the other functional classes of proteins have all been implicated in genetic diseases.

LOSS OF ENZYME FUNCTION In 1934, the urine of two mentally retarded young siblings was found to contain phenylpyruvic acid, an unusual by-product of the metabolism of the amino acid phenylalanine. It was not until two decades later that the complex clinical phenotype of the disease that afflicted these children, called phenylketonuria (PKU), was traced back to its molecular cause. The disease results from an abnormality in a single enzyme, phenylalanine hydroxylase (PAH), which catalyzes the conversion of dietary phenylalanine to tyrosine (**Figure 15.7**). This enzyme is not active in the livers of PKU patients, leading to excesses of phenylalanine and phenylpyruvic acid in the blood. Since then, the nucleotide sequence of the *PAH* gene has been compared between healthy people and those with the PKU disease, and more than 400 different disease-causing mutations have been found. A common one is a missense mutation that results in tryptophan instead of arginine at position 408 in the polypeptide chain (**Table 15.1**). As is often the case with loss of function mutations, the mutant alleles are recessive, because one functional allele is all that is needed to produce enough functional PAH to prevent the disease.

Hundreds of human genetic diseases that result from enzyme abnormalities have been discovered, some of which lead

TABLE 15.1
Two Common Mutations That Cause Phenylketonuria

	Codon 408 (20% of PKU Cases)		Codon 280 (2% of PKU Cases)	
	Normal	Mutant	Normal	Mutant
Length of PAH protein	452 amino acids	452 amino acids	452 amino acids	452 amino acids
DNA at codon	...CGG...	...TGG...	...GAA...	...AAA...
	...GCC...	...ACC...	...CTT...	...TTT...
mRNA at codon	...CGG...	...UGG...	...GAA...	...AAA...
Amino acid at codon	Arginine	Tryptophan	Glutamic acid	Lysine
Active PAH enzyme?	Yes	No	Yes	No

to mental retardation and premature death. Most of these diseases are rare; PKU, for example, shows up in 1 out of every 12,000 newborns. But these diseases are just the tip of the mutation iceberg. Some mutations result in amino acid changes that have no obvious clinical effects (see Figure 15.1). As we have just seen, there can be numerous alleles of a gene. Some produce proteins that function normally, whereas others produce variants that cause disease.

ABNORMAL HEMOGLOBIN As we mentioned in Section 15.1, sickle-cell disease is caused by a recessive, missense mutation. This blood disorder most often afflicts people whose ancestors came from the tropics or from the Mediterranean region. About 1 in 655 African-Americans is homozygous for the sickle allele and has the disease.

Recall that human hemoglobin contains four globin subunits—two α chains and two β chains—as well as the pigment heme (see Figure 3.11). In sickle-cell disease, one of the 146 amino acids in the β-globin polypeptide chain is abnormal: at position 6, glutamic acid is replaced by valine. This replacement changes the charge of the protein (glutamic acid is negatively charged and valine is neutral), causing it to form long, needlelike aggregates in the red blood cells. The phenotypic result is sickle-shaped red blood cells (see Figure 15.3) and an impaired ability of the blood to carry oxygen. The sickled cells tend to block narrow blood capillaries, resulting in tissue damage and eventually death by organ failure.

Because hemoglobin is easy to isolate and study, its variations in the human population have been extensively documented (**Figure 15.8**). Hundreds of single amino acid alterations in β-globin have been reported. For example, at the same position that is mutated in sickle-cell disease (resulting in hemoglobin S), glutamic acid may be replaced by lysine, causing hemoglobin C disease. In this case, the resulting anemia is usually not severe. Many alterations of hemoglobin do not affect the protein's function. In fact, about 5 percent of all humans carry at least one missense point mutation in a β-globin allele.

Some of the more common examples of inherited diseases caused by specific protein defects are listed in **Table 15.2**. These mutations can be dominant, codominant, or recessive, and some are sex-linked.

Disease-causing mutations may involve any number of base pairs

Disease-causing mutations may involve a single base pair, a long stretch of DNA, multiple segments of DNA, or even entire chromosomes (as we saw for Down syndrome in Section 11.5).

POINT MUTATIONS There are many examples of point mutations in human genetic diseases. In some cases, all of the people with the disease have the same genetic mutation. This is the case with sickle-cell anemia. In other cases, many different loss of function point mutations in one gene can lead to the same disease, as we saw above for PKU. This makes sense if you think about the three-dimensional structure of an enzyme protein and the many amino acid changes that could affect its activity.

LARGE DELETIONS Larger mutations may involve many base pairs of DNA. For example, deletions in the X chromosome that

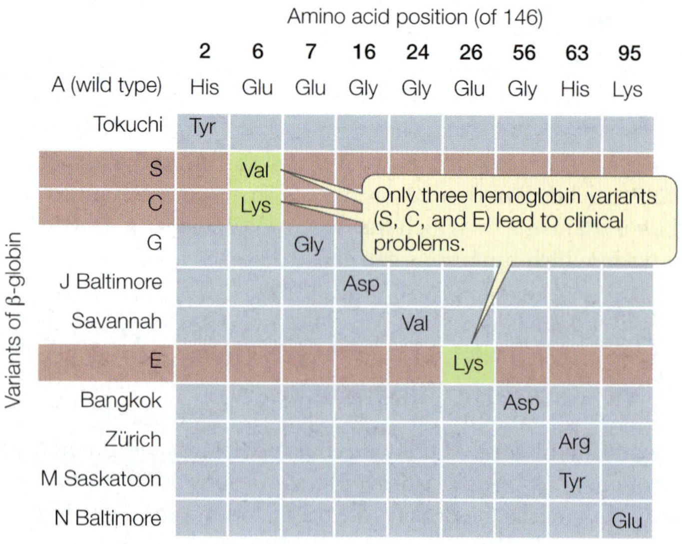

Amino acid position (of 146)
2 6 7 16 24 26 56 63 95
A (wild type) His Glu Glu Gly Gly Glu Gly His Lys
Tokuchi Tyr
S Val
C Lys
G Gly
J Baltimore Asp
Savannah Val
E Lys
Bangkok Asp
Zürich Arg
M Saskatoon Tyr
N Baltimore Glu
Variants of β-globin
Only three hemoglobin variants (S, C, and E) lead to clinical problems.

15.8 Hemoglobin Polymorphism Each of these mutant alleles codes for a protein with a single amino acid change in the 146-amino acid chain of β-globin. Only three of the hundreds of known variants of β-globin, shown on the left, are known to lead to clinical abnormalities. "S" is the sickle-cell anemia allele.

TABLE 15.2 Some Human Genetic Diseases

Disease Name	Inheritance Pattern; Births Frequency	Gene Mutated; Protein Product	Clinical Phenotype
Familial hypercholesterolemia	Autosomal codominant; 1 in 500 heterozygous	*LDLR*; low-density lipoprotein receptor	High blood cholesterol, heart disease
Cystic fibrosis	Autosomal recessive; 1 in 4,000	*CFTR*; chloride ion channel in membrane	Immune, digestive, and respiratory illness
Duchenne muscular dystrophy	Sex-linked recessive; 1 in 3,500 males	*DMD*; the muscle membrane protein dystrophin	Muscle weakness
Hemophilia A	Sex-linked recessive; 1 in 5,000 males	*HEMA*; factor VIII blood clotting protein	Inability to clot blood after injury, hemorrhage

include the gene for the protein dystrophin result in Duchenne muscular dystrophy. Dystrophin is important in organizing the structure of muscles, and people who have only the abnormal form have severe muscle weakness. Sometimes only part of the dystrophin gene is missing, leading to an incomplete but partly functional protein and a mild form of the disease. In other cases, deletions span the entire sequence of the gene, so that the protein is missing entirely, resulting in a severe form of the disease. In yet other cases, deletions involve millions of base pairs and cover not only the dystrophin gene but adjacent genes as well; the result may be several diseases in the same person.

CHROMOSOMAL ABNORMALITIES Chromosomal abnormalities also cause human diseases. Such abnormalities result from the gain or loss of complete chromosomes (aneuploidy) (see Figure 11.20), or from the gain or loss of chromosomal segments (see Figure 15.4). About 1 newborn in 200 has a chromosomal abnormality. This may be inherited from a parent who also has the abnormality, or it may result from an error in meiosis during the formation of gametes in one of the parents. One example is the fragile-X syndrome, which is a restriction in the tip of the X chromosome that can result in mental retardation (**Figure 15.9**). About 1 male in 1,500 and 1 female in 2,000 are affected. Although the basic pattern of inheritance is that of an X-linked recessive trait, there are departures from this pattern. Not all people with the fragile-X chromosomal abnormality are mentally retarded, as we will see.

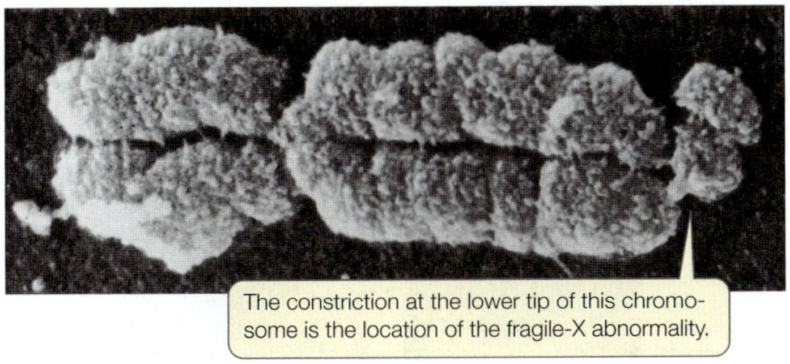
The constriction at the lower tip of this chromosome is the location of the fragile-X abnormality.

15.9 A Fragile-X Chromosome at Metaphase The chromosomal abnormality associated with fragile-X syndrome shows up under the microscope as a constriction in the chromosome. This occurs during preparation of the chromosome for microscopy.

Expanding triplet repeats demonstrate the fragility of some human genes

About one-fifth of all males who have the fragile-X chromosomal abnormality are phenotypically normal, as are most of their daughters. But many of those daughters' sons are mentally retarded. In a family in which the fragile-X syndrome appears, later generations tend to show earlier onset and more severe symptoms of the disease. It is almost as if the abnormal allele itself is changing—and getting worse. And that's exactly what is happening.

The gene responsible for fragile-X syndrome (*FMR1*) contains a repeated triplet, CGG, at a certain point in the promoter region (**Figure 15.10**). In normal people, this triplet is repeated 6 to 54 times (the average is 29). In mentally retarded people with fragile-X syndrome, the CGG sequence is repeated 200 to 2,000 times.

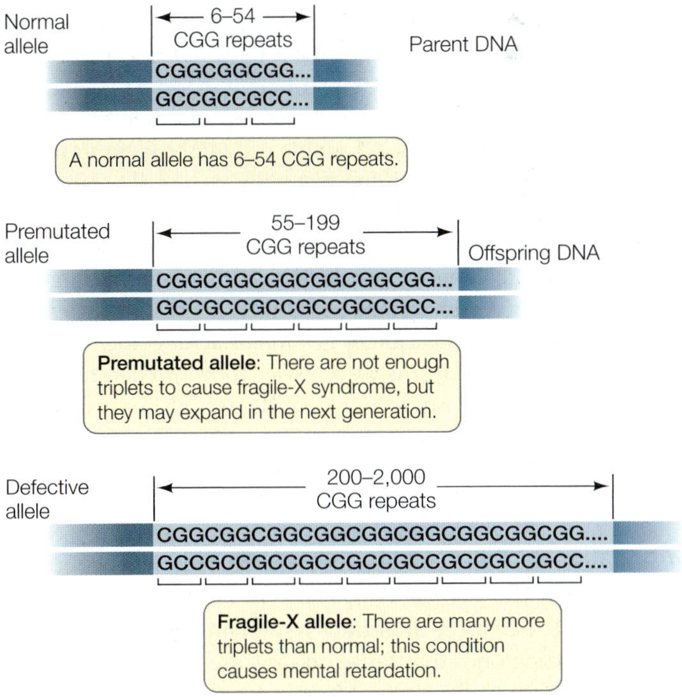
Normal allele
6–54 CGG repeats
Parent DNA
CGGCGGCGG...
GCCGCCGCC...
A normal allele has 6–54 CGG repeats.
Premutated allele
55–199 CGG repeats
Offspring DNA
CGGCGGCGGCGGCGGCGG...
GCCGCCGCCGCCGCCGCC...
Premutated allele: There are not enough triplets to cause fragile-X syndrome, but they may expand in the next generation.
Defective allele
200–2,000 CGG repeats
CGGCGGCGGCGGCGGCGGCGGCGGCGG....
GCCGCCGCCGCCGCCGCCGCCGCCGCC....
Fragile-X allele: There are many more triplets than normal; this condition causes mental retardation.

15.10 The CGG Repeats in the *FMR1* Gene Expand with Each Generation The genetic defect in fragile-X syndrome is caused by 200 or more repeats of the CGG triplet.

Males carrying a moderate number of repeats (55–199) show no symptoms and are called premutated. These repeats become more numerous as the daughters of these men pass the chromosome on to their children. With 200 or more repeats, increased methylation of the cytosines in the CGG triplets is likely, which inhibits transcription of the *FMR1* gene. The normal role of the protein product of this gene is to bind to mRNAs involved in neuron function and to regulate their translation at the ribosome. When the FMR1 protein is not made in adequate amounts, these mRNAs are not properly translated, and nerve cells die. Their loss often results in mental retardation.

This phenomenon of **expanding triplet repeats** has been found in more than a dozen other diseases, such as myotonic dystrophy (involving repeated CTG triplets) and Huntington's disease (in which CAG is repeated). Such repeats, which may be found within a protein-coding region or outside it, appear to be present in many other genes without causing harm. How the repeats expand is not known; one hypothesis is that DNA polymerase may slip after copying a repeat and then fall back to copy it again.

Cancer often involves somatic mutations

As we saw in the opening story of this chapter and discussed in Chapter 11, a mutation in a somatic cell can result in cancer. Many chromosomal and point mutations have been described in cancer cells. Such mutations affect oncogenes, whose products stimulate cell division, or tumor suppressor genes, whose products inhibit cell division.

More than two gene mutations are usually needed for full-blown cancer. Because colon cancer progresses to full malignancy slowly, it has been possible to identify the gene mutations that lead to each stage. Figure 15.11 outlines the "molecular biography" of this form of cancer. At least three tumor suppressor genes and one oncogene must be mutated in sequence in a cell in the colon lining for cancer to develop. Although the occurrence of all of these events in a single cell might seem unlikely, remember that the colon lining has millions of cells, that these cells arise from stem cells that are constantly dividing, and that these changes take place over many years of exposure to natural and synthetic substances in foods, which may act as mutagens.

The description of somatic mutations (as well as the rarer germ line mutations) in cancer is a major achievement of molecular medicine. Cancer is responsible for more deaths in the Western world that any other disease except heart disease. Mutational analysis allows for screening and diagnosis, and as we saw in the opening story, knowledge of the mutated gene product may point the way to targeted therapy.

Most diseases are caused by multiple genes and environment

The example of cancer illustrates how many common phenotypes, including ones that cause disease, are **multifactorial**; that is, they are caused by the interactions of many genes and proteins with one or more factors in the environment. When studying genetics, we tend to call individuals either normal (wild type) or abnormal (mutant); however, in reality every individual contains thousands or millions of genetic variations that arose through

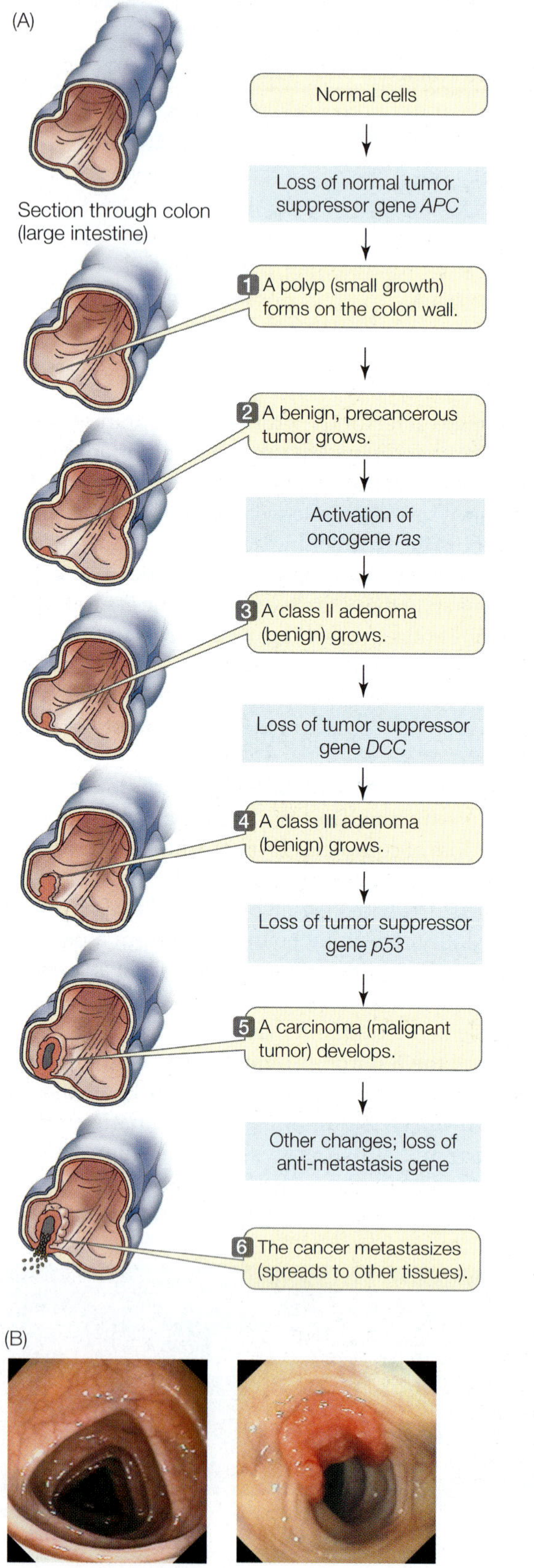
(A)
Section through colon (large intestine)
Normal cells
Loss of normal tumor suppressor gene *APC*
1 A polyp (small growth) forms on the colon wall.
2 A benign, precancerous tumor grows.
Activation of oncogene *ras*
3 A class II adenoma (benign) grows.
Loss of tumor suppressor gene *DCC*
4 A class III adenoma (benign) grows.
Loss of tumor suppressor gene *p53*
5 A carcinoma (malignant tumor) develops.
Other changes; loss of anti-metastasis gene
6 The cancer metastasizes (spreads to other tissues).
(B)

15.11 Multiple Somatic Mutations Transform a Normal Colon Epithelial Cell into a Cancer Cell (A) At least five genes must be mutated in a single cell to produce colon cancer. (B) These images from a screening test reveal a normal colon (left) and colon cancer (right).

mutations. Our susceptibility to disease is often determined by complex interactions between these genotypes and factors in the environment, such as the foods we eat or the pathogens we encounter. For example, a complex set of genotypes determines who among us can eat a high-fat diet and not experience a heart attack, or who will succumb to disease when exposed to infectious bacteria. Estimates suggest that up to 60 percent of all people are affected by diseases that are genetically influenced. Identifying these genetic influences is another major task of molecular medicine and human genome sequencing.

RECAP 15.2

Many genetic mutations are expressed as nonfunctional enzymes, structural proteins, or membrane proteins. Human genetic diseases may be inherited in dominant, codominant, or recessive patterns, and they may be sex-linked.

- Describe an example of an abnormal protein in humans that results from a genetic mutation and causes a disease. **See pp. 311–313**
- Describe an example of an abnormal protein in humans that results from a genetic mutation and does *not* cause a disease. **See p. 312**
- How do expanding repeats cause genetic diseases? **See pp. 313–314 and Figure 15.10**
- How do somatic mutations cause cancer? **See p. 314 and Figure 15.11**

In the previous section we described the ways in which mutations can lead to human disease. We will turn now to the ways that biologists detect mutations in DNA.

15.3 How Are Mutations Detected and Analyzed?

A challenge for biologists studying mutations is to precisely describe the DNA changes that lead to specific protein changes—an area of research called molecular genetics. Of course, the most direct and comprehensive way to analyze DNA is to determine its sequence of bases. DNA sequencing technologies are continually improving, and the entire genomes of many organisms have now been sequenced completely. Furthermore, the genomes of closely related organisms have been compared in order to identify mutations. We will discuss sequencing technology in Chapter 17. In this section we will look at some of the techniques that are used in combination with DNA sequencing to study DNA, and to identify mutations that cause disease.

Restriction enzymes cleave DNA at specific sequences

All organisms, including bacteria, must have ways of dealing with their enemies. As we saw in Section 13.1, bacteria can be attacked by viruses called bacteriophages. These viruses inject their genetic material into the host cell and turn it into a virus-producing factory, eventually killing the cell. Some bacteria defend themselves against such invasions by producing **restriction enzymes** (also known as restriction endonucleases), which

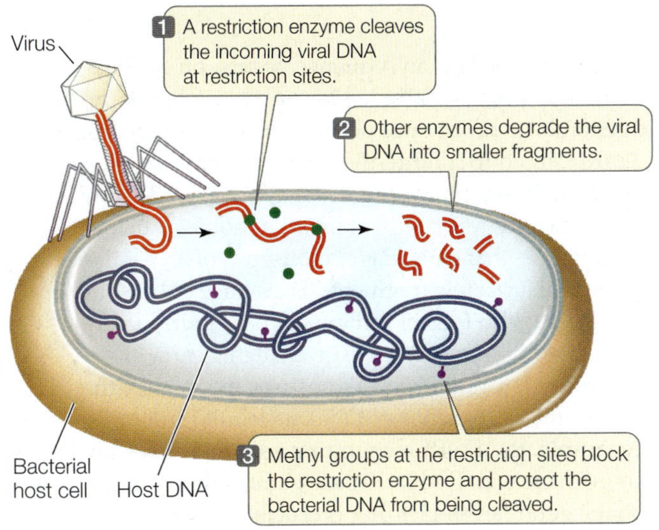
Virus
1 A restriction enzyme cleaves the incoming viral DNA at restriction sites.
2 Other enzymes degrade the viral DNA into smaller fragments.
Bacterial host cell
Host DNA
3 Methyl groups at the restriction sites block the restriction enzyme and protect the bacterial DNA from being cleaved.

15.12 Bacteria Fight Invading Viruses by Making Restriction Enzymes

cut double-stranded DNA molecules—such as those injected by bacteriophages—into smaller, noninfectious fragments (**Figure 15.12**). These enzymes break the bonds of the DNA backbone between the 3′ hydroxyl group of one nucleotide and the 5′ phosphate group of the next nucleotide. This cutting process is called **restriction digestion**.

There are many such restriction enzymes, each of which cleaves DNA at a specific sequence of bases called a **recognition sequence** or a **restriction site**. Most recognition sequences are four to six base pairs long. Because each sequence of bases has a unique structure (see Section 13.2), it can be specifically recognized by a particular restriction enzyme. Cells protect themselves from being digested by their own enzymes by modifying their DNA, often with methyl groups, to prevent binding by the restriction enzymes.

Restriction enzymes can be isolated from the cells that make them and used as biochemical reagents in the laboratory to give information about the nucleotide sequences of DNA molecules from other organisms. If DNA from any organism is incubated in a test tube with a restriction enzyme (along with buffers and salts that help the enzyme function), that DNA will be cut wherever the restriction site occurs. A specific sequence of bases defines each restriction site. For example, the enzyme *Eco*RI (named after its source strain of the bacterium *E. coli*) cuts DNA only where it encounters the following paired sequence in the DNA double helix:

5′...GAATTC...3′
3′...CTTAAG...5′

Note that this sequence is palindromic, like the word "mom." This means that both strands have the same sequence when they are read from their 5′ (or their 3′) ends. The *Eco*RI enzyme has two identical active sites on its two subunits, which cleave the two strands simultaneously between the G and the A of each strand:

5′...G↓AATTC...3′	→	5′...G	AATTC...3′
3′...CTTAA↑G...5′		3′...CTTAA	G...5′

The *Eco*RI recognition sequence occurs, on average, about once in every 4,000 base pairs in a typical prokaryotic genome, or about once per four prokaryotic genes. So *Eco*RI can chop a large piece of DNA into smaller pieces containing, on average, just a few genes. Using *Eco*RI in the laboratory to cut small genomes, such as those of viruses that have tens of thousands of base pairs, may result in just a few fragments. For a huge eukaryotic chromosome with tens of millions of base pairs, a very large number of fragments will be created.

Of course, "on average" does not mean that the enzyme cuts all stretches of DNA at regular intervals. For example, the *Eco*RI recognition sequence does not occur even once in the 40,000 base pairs of the T7 bacteriophage genome—a fact that is crucial to the survival of this virus, since its host is *E. coli*. Fortunately for *E. coli*, the *Eco*RI recognition sequence does appear in the DNA of other bacteriophages.

Gel electrophoresis separates DNA fragments

Restriction enzyme digestion is used to manipulate DNA in the laboratory and to identify and analyze mutations. After a laboratory sample of DNA has been cut with a restriction enzyme, the DNA is in fragments, which must be separated to identify (map) where the cuts were made. Because the recognition sequence does not occur at regular intervals, the fragments are not all the same size, and this property provides a way to separate them from one another. Separating the fragments is necessary to determine the number and molecular sizes (in base pairs) of the fragments produced, or to identify and purify an individual fragment for further analysis or for use in an experiment.

A convenient way to separate or purify DNA fragments is by **gel electrophoresis**. Samples containing the fragments are placed in wells at one end of a semisolid gel (usually made of agarose or polyacrylamide), and an electric field is applied to the gel (**Figure 15.13**). Because of its phosphate groups, DNA is negatively charged at neutral pH; therefore, because opposite charges attract, the DNA fragments move through the gel toward the positive end of the field. Because the spaces between the polymers of the gel are small, small DNA molecules can move through the gel faster than larger ones. Thus DNA fragments of different sizes separate from one another, forming bands that can be detected with a dye. This provides three types of information:

- *The number of fragments*. The number of fragments produced by digestion of a DNA sample with a given restriction enzyme depends on how many times that enzyme's recognition sequence occurs in the sample. Thus gel electrophoresis can provide some information about the presence of specific DNA sequences (the restriction sites) in the DNA sample.

RESEARCH**TOOLS**

15.13 Separating Fragments of DNA by Gel Electrophoresis
A mixture of DNA fragments is placed in a gel, and an electric field is applied across the gel. The negatively charged DNA moves toward the positive end of the field, with smaller molecules moving faster than larger ones. After minutes to hours for separation, the electric power is shut off and the separated fragments can be analyzed.

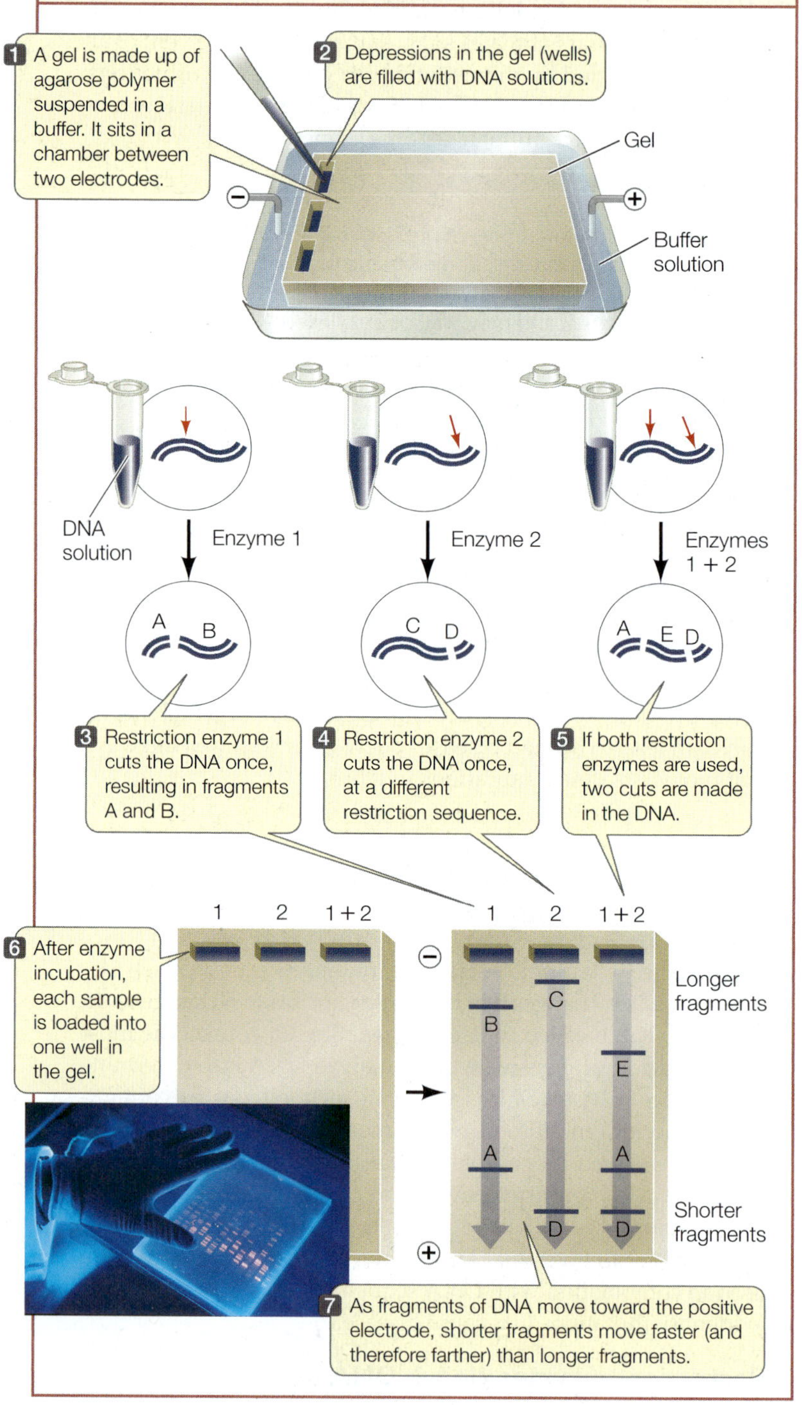

1 A gel is made up of agarose polymer suspended in a buffer. It sits in a chamber between two electrodes.
2 Depressions in the gel (wells) are filled with DNA solutions.
Gel
Buffer solution
DNA solution
Enzyme 1
Enzyme 2
Enzymes 1 + 2
A B
C D
A E D
3 Restriction enzyme 1 cuts the DNA once, resulting in fragments A and B.
4 Restriction enzyme 2 cuts the DNA once, at a different restriction sequence.
5 If both restriction enzymes are used, two cuts are made in the DNA.
1 2 1 + 2
6 After enzyme incubation, each sample is loaded into one well in the gel.
Longer fragments
Shorter fragments
B C E A A D D
7 As fragments of DNA move toward the positive electrode, shorter fragments move faster (and therefore farther) than longer fragments.

Go to Animated Tutorial 15.2
Gel Electrophoresis
Life10e.com/at15.2

- *The sizes of the fragments*. DNA fragments of known size (size markers) are often placed in one well of the gel to provide a standard for comparison. The size markers are used to determine the sizes of the DNA fragments in samples in the other wells. By comparing the fragment sizes obtained with two or more restriction enzymes, the locations of their recognition sites relative to one another can be worked out (mapped).

- *The relative abundance of a fragment*. In many experiments, the investigator is interested in how much DNA is present. The relative intensity of a band produced by a specific fragment can indicate the amount of that fragment.

DNA fingerprinting combines PCR with restriction analysis and electrophoresis

The methods we have just described are used in **DNA fingerprinting**, which identifies individuals based on differences in their DNA sequences. DNA fingerprinting works best with sequences that are highly polymorphic—that is, sequences that have multiple alleles (because of many point mutations during the evolution of the organism) and are therefore likely to be different in different individuals. Two types of polymorphisms are especially informative:

- **Single nucleotide polymorphisms** (**SNPs**; pronounced "snips") are inherited variations involving a single nucleotide base—they are point mutations. These polymorphisms have been mapped for many organisms. If one parent is homozygous for the base A at a certain point on the genome, and the other parent is homozygous for a G at that point, the offspring will be heterozygous: one chromosome will have A at that point and the other will have G. If a SNP occurs in a restriction enzyme recognition site, such that one variant is recognized by the enzyme and the other isn't, then individuals can be distinguished from one another very easily using the polymerase chain reaction (PCR) (see Section 13.5). A fragment containing the polymorphic sequence is amplified by PCR from samples of total DNA isolated from each individual. The fragments are then cut with the restriction enzyme and analyzed by gel electrophoresis.

  Go to Activity 15.1 Allele-Specific Cleavage
  Life10e.com/ac15.1

- **Short tandem repeats** (**STRs**) are short, repetitive DNA sequences that occur side by side on the chromosomes, usually in the noncoding regions. These repeat patterns, which contain one to five base pairs, are also inherited. For example, at a particular locus on chromosome 15 there may be an STR of "AGG." An individual may inherit an allele with six copies of the repeat (AGGAGGAGGAGGAGGAGG) from her mother and an allele with two copies (AGGAGG) from her father. Again, PCR is used to amplify DNA fragments containing these repeated sequences, and then the amplified fragments, which have different sizes because of the different lengths of the repeats, are distinguished by gel electrophoresis (**Figure 15.14A**).

The method of DNA fingerprinting used most commonly today involves STR analysis. The Federal Bureau of Investigation in the United States uses 13 STR loci in its Combined DNA Index System (CODIS) database (**Table 15.3**). An analysis of these loci in your DNA would reveal your particular DNA fingerprint. Looking at Table 15.3, you might inherit:

From your mother: allele 72 from chromosome 4; allele 23 from chromosome 7; allele 14 from chromosome 11; and allele 12 from chromosome 18

From your father: allele 56 from chromosome 4; allele 22 from chromosome 7; allele 16 from chromosome 11; and allele 12 from chromosome 18

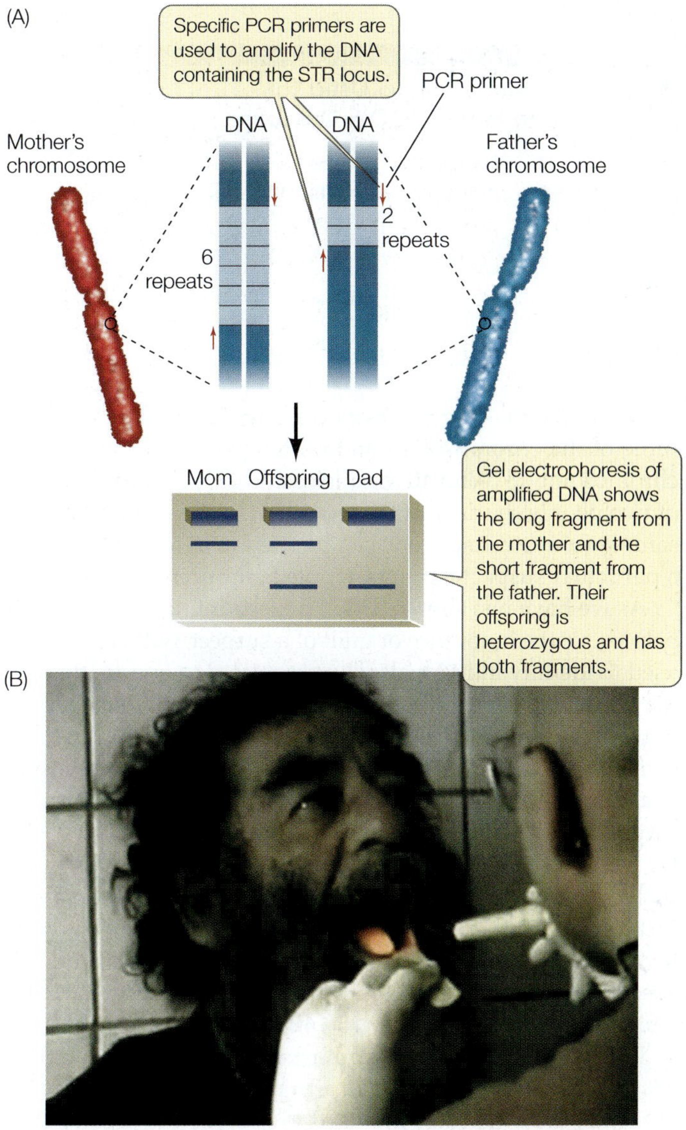
(A)
Specific PCR primers are used to amplify the DNA containing the STR locus.
PCR primer
DNA
DNA
Mother's chromosome
Father's chromosome
6 repeats
2 repeats
Mom
Offspring
Dad
Gel electrophoresis of amplified DNA shows the long fragment from the mother and the short fragment from the father. Their offspring is heterozygous and has both fragments.
(B)

15.14 DNA Fingerprinting with Short Tandem Repeats (A) A particular STR locus can be analyzed to determine the number of repeat sequences that were inherited by an individual from each parent. The two alleles can be identified in an electrophoresis gel on the basis of their sizes. When several STR loci are analyzed, the pattern can constitute a definitive identification of an individual. (B) When the dictator Saddam Hussein was captured in Iraq, a sample of his cheek epithelial cells was taken for DNA fingerprinting. A comparison with DNA fingerprints of relatives provided military scientists with evidence that the man in question was indeed Saddam Hussein.

TABLE 15.3 Four of the Genetic Loci Used for Identification in the CODIS Database

Human Chromosome	Locus Name	Repeated Sequence	Number of Alleles
4	FGA	CTTT	80
7	D7S820	GATA	30
11	TH01	TCAT	20
18	S18S51	AGAA	51

Note that in this case you are heterozygous for the alleles on three of the chromosomes and homozygous for the allele on chromosome 18. With all the alleles and 13 loci, the probability of two people sharing the same alleles is very small. So a DNA sample from a crime scene can be used to determine whether a particular suspect left that sample at the scene.

As we have just shown, DNA fingerprinting can be used to help prove the innocence or guilt of a suspect, but it can also be used to identify individuals who are related to one another. On May 2, 2011, Osama bin Laden was killed by U.S. soldiers at his home in Pakistan. He was identified at the scene by comparison with photographs, a wife who pointed him out, and instant analysis using a digital camera with facial recognition software. In addition, DNA fingerprinting was used. Bin Laden's son Khalid was also killed in the raid, and a sister had previously died in the U.S. Analyses of their DNA along with that of Osama indicated that the three shared many polymorphisms and were highly likely to be closely related. The same methods were also used to identify Saddam Hussein, who was captured in 2003 and later executed in Iraq (**Figure 15.14B**).

DNA analysis with genetic markers such as SNPs and STRs has applications throughout all areas of biological research. For example, these markers are used to analyze the organization of genomes, to identify species or individuals within species, to compare species or organisms to see how closely related they are, and to analyze particular genes and the phenotypes associated with them. In the remainder of this chapter we will focus on the use of these markers and other technologies that are used to study and treat genetic diseases.

Reverse genetics can be used to identify mutations that lead to disease

We have seen for diseases such as PKU and sickle-cell anemia that the clinical phenotypes of inherited diseases could be traced to individual proteins, and that the genes could then be identified. With the advent of new ways to identify DNA variations, a new pattern of human genetic analysis has emerged. In these cases, the clinical phenotype is first related to a DNA variation, and then the protein involved is identified. This pattern of discovery is called **reverse genetics**, because it proceeds in the opposite direction to genetic analyses done before the mid-1980s. For example, in sickle-cell anemia, the protein abnormality in hemoglobin was described first (a single amino acid change), and then the gene for β-globin was isolated and the DNA mutation was pinpointed.

Clinical phenotype → protein phenotype → gene

By contrast, for cystic fibrosis (see Table 15.2), a mutant version of the gene *CFTR* was isolated first, and then the protein was characterized:

Clinical phenotype → gene → protein phenotype

Whichever approach is used, final identification of the protein(s) involved in a disease is important in designing specific therapies.

Genetic markers can be used to find disease-causing genes

To identify a mutant gene by reverse genetics, close linkage to a marker sequence is used, in a process called linkage analysis. To understand this type of analysis, imagine an astronaut looking down from space, trying to find her son on a park bench on Chicago's North Shore. The astronaut first picks out reference points—landmarks that will lead her to the park. She recognizes the shape of North America, then moves to Lake Michigan, then the Willis Tower, and so on. Once she has zeroed in on the North Shore Park, she can use advanced optical instruments to find her son.

The reference points for gene isolation are **genetic markers**. Genetic markers such as STRs and SNPs can be used as landmarks to find a gene of interest, if the latter also has multiple alleles (for example, normal and disease-causing alleles). The key to this method is the well-established observation that if two genes are located near each other on the same chromosome, they are usually passed on together from parent to offspring (see Section 12.4). The same holds true for any pair of DNA genetic markers. In the case of linkage analysis, the idea is to find markers that are progressively closer to the gene of interest.

As noted above, SNPs and STRs are widespread in eukaryotic genomes. There is roughly one SNP for every 1,330 base pairs in the human genome, and many regions of the genome also contain repetitive DNA sequences such as those found in STRs. SNPs and STRs can be analyzed using the PCR techniques mentioned above. SNPs can also be detected using sophisticated chemical methods such as mass spectrometry.

To narrow down the location of a gene, a scientist must find a genetic marker that is *always inherited with the gene*. To do this, family medical histories are taken and pedigrees are constructed. If a genetic marker and a genetic disease are inherited together in many families, then they must be near each other on the same chromosome (**Figure 15.15**). This narrows down the location of the gene to a few hundred thousand base pairs.

Once a linked DNA region is identified, many methods are available to identify the actual gene responsible for a genetic disease. The complete sequence of the region can be searched for candidate genes, using information available from databases of genome sequences. With luck a scientist can make an educated guess, based on biochemical or physiological information about the disease, along with information about the functions of candidate genes, as to which gene is responsible

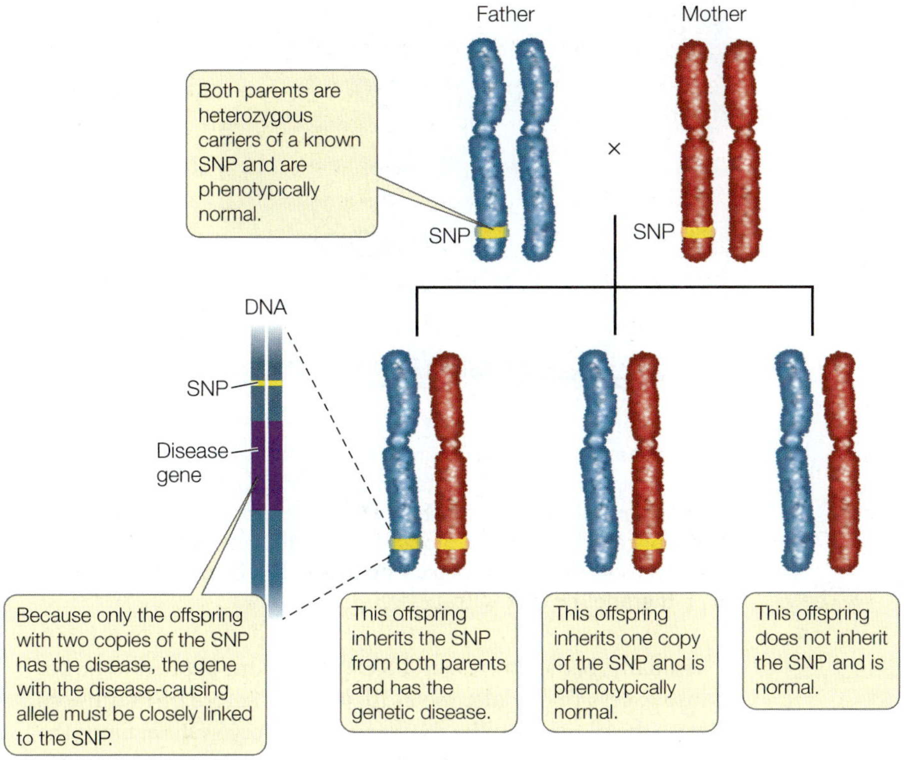
Father
Mother
×
Both parents are heterozygous carriers of a known SNP and are phenotypically normal.
SNP
SNP
DNA
SNP
Disease gene
Because only the offspring with two copies of the SNP has the disease, the gene with the disease-causing allele must be closely linked to the SNP.
This offspring inherits the SNP from both parents and has the genetic disease.
This offspring inherits one copy of the SNP and is phenotypically normal.
This offspring does not inherit the SNP and is normal.

15.15 DNA Linkage Analysis Linkage of an SNP to a disease-causing allele in many families narrows down the location of the defective gene, making its isolation and identification possible. In the example shown here, the disease-causing allele is recessive.

for the disease. The identification of DNA polymorphisms within candidate genes that correlate with the presence or absence of disease can also help narrow down the search. A variety of techniques, such as analyzing mRNA levels of candidate genes in diseased and healthy individuals, are used to confirm that the correct gene has been identified.

The DNA barcode project aims to identify all organisms on Earth

One of the most exciting aspects of DNA technology for biologists is its potential to identify species, varieties, and even individual organisms from their DNA. In order to repeat experiments and report scientific results, it is essential that biologists know exactly what species or varieties they are studying. However, different organisms can sometimes look very much alike in nature. About 1.8 million species have been named and described, but about ten times that number probably have yet to be identified. A proposal to use DNA technology to identify known species and detect the unknown ones has been endorsed by a large group of scientific organizations known as the Consortium for the Barcode of Life (CBOL).

Evolutionary biologist Paul Hebert at the University of Guelph in Ontario, Canada, was walking down the aisle of a supermarket in 1998 when he noticed the barcodes on all the packaged foods. This gave him an idea to identify each species with a "DNA barcode" that is based on a short sequence from a single gene. The gene he chose is the cytochrome oxidase gene, a component of the respiratory chain that is present in most organisms. Because this gene mutates readily, there are many allelic differences among species. A fragment of 650 to 750 base pairs in this gene is being sequenced for all organisms, and so far sufficient variation has been detected to make it diagnostic for each species (**Figure 15.16**).

Once the DNA of the targeted gene fragment has been sequenced for all known species, a simple device can be used in the field to analyze DNA from an organism and identify its species. The barcode project has the potential to advance biological research on evolution, to track species diversity in ecologically significant areas, to help identify new species, and even to detect undesirable microbes or bioterrorism agents.

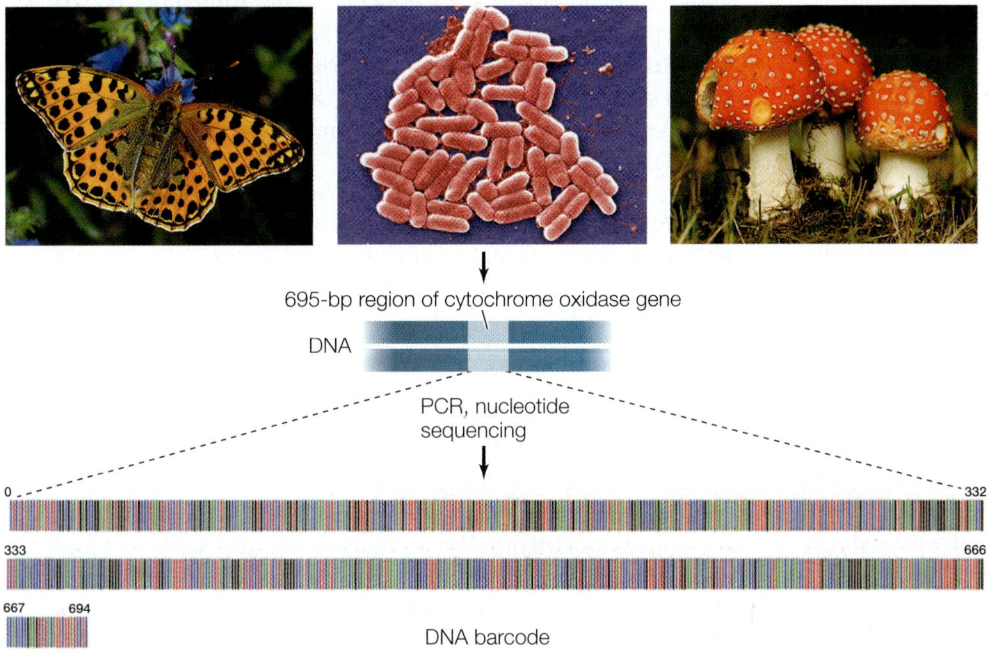
695-bp region of cytochrome oxidase gene
DNA
PCR, nucleotide sequencing
0
332
333
666
667
694
DNA barcode

Go to Media Clip 15.2
Barcode of Life
Life10e.com/mc15.2

15.16 A DNA Barcode A 650- to 750-base-pair region of the cytochrome oxidase gene can be amplified by PCR from any organism and then sequenced. This knowledge is used to make a bar code in which each of the four DNA bases is represented as a different color. Such a species barcode permits accurate and rapid identification of a particular species for experimental, ecological, or evolutionary studies.

RECAP 15.3

Large DNA molecules can be cut into smaller pieces by restriction digestion and then sorted by gel electrophoresis. PCR is used to amplify sequences of interest from complex samples. These techniques are used in DNA fingerprinting to analyze DNA polymorphisms for the purpose of identifying individuals. Genes involved in disease can be identified by first detecting the abnormal DNA sequence and then the protein that the wild-type allele encodes. Scientists hope to be able to identify all species using DNA analysis.

- How does a restriction enzyme recognize a restriction site on DNA? **See p. 315**
- How does gel electrophoresis separate DNA fragments? **See p. 316 and Figure 15.13**
- What are STRs, and how are they used to identify individuals? **See pp. 317–318 and Figure 15.14**
- How can a gene mutation that causes a disease be mapped and detected before its protein product is known? **See p. 318 and Figure 15.15**

The determination of the precise molecular phenotypes and genotypes of various human genetic diseases has made it possible to diagnose these diseases even before symptoms appear. Let's take a detailed look at some of these genetic screening techniques.

15.4 How Is Genetic Screening Used to Detect Diseases?

Genetic screening is the use of a test to identify people who have, are predisposed to, or are carriers of a genetic disease. It can be done at many times of life and used for many purposes.

- *Prenatal screening* can be used to identify an embryo or fetus with a disease so that medical intervention can be applied or decisions can be made about whether or not to continue the pregnancy.
- *Newborn babies* can be screened so that proper medical intervention can be initiated quickly for those babies who need it.
- *Asymptomatic people* who have relatives with a genetic disease can be screened to determine whether they are carriers of the disease-associated allele or are likely to develop the disease themselves.

Genetic screening can be done at the level of either the phenotype or the genotype.

Screening for disease phenotypes involves analysis of proteins and other chemicals

Genetic screening can involve examining a protein or other chemical that is relevant to a phenotype associated with a particular disease. Perhaps the best example is the test for phenylketonuria (PKU), which has made it possible to identify the disease in newborns, so that treatment can be started immediately. It is very likely that you were screened as a newborn for PKU.

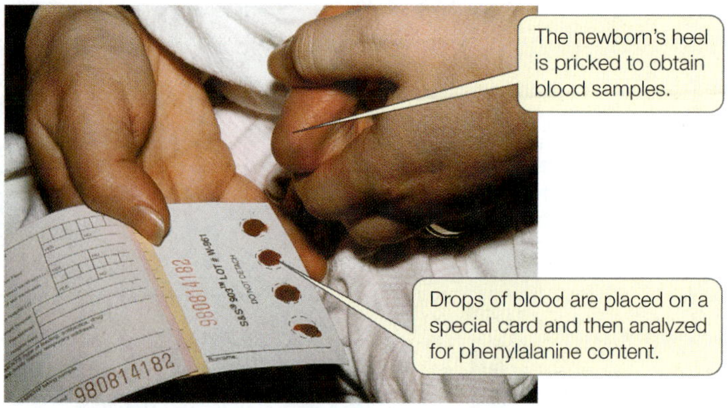

The newborn's heel is pricked to obtain blood samples.
Drops of blood are placed on a special card and then analyzed for phenylalanine content.

15.17 Genetic Screening of Newborns for Phenylketonuria A blood test is used to screen newborns for phenylketonuria. Small samples of blood are taken from a newborn's heel. The samples are placed in a machine that measures the phenylalanine concentration in the blood. Early detection means that the symptoms of the condition can be prevented by putting the baby on a therapeutic diet.

Initially, babies born with PKU have a normal phenotype because excess phenylalanine in their blood before birth diffuses across the placenta to the mother's circulatory system. Since the mother is almost always heterozygous, and therefore has adequate phenylalanine hydroxylase activity, her body metabolizes the excess phenylalanine from the fetus. After birth, however, the baby begins to consume protein-rich food (milk) and to break down some of his or her own proteins. Phenylalanine begins to accumulate in the blood. After a few days, the phenylalanine level in the baby's blood may be ten times higher than normal. Within days, the developing brain is damaged, and untreated children with PKU become severely mentally retarded. But if detected early, PKU can be treated with a special diet low in phenylalanine to avoid the brain damage that would otherwise result. Thus early detection is imperative.

Newborn screening for PKU and other diseases began in 1963 with the development of a simple, rapid test for the presence of excess phenylalanine in blood serum (Figure 15.17). This method uses dried blood spots from newborn babies and can be automated so that a screening laboratory can process many samples in a day. Newborn babies' blood is now screened for up to 35 genetic diseases. Some are common, such as congenital hypothyroidism, which occurs about once in 4,000 births and causes reduced growth and mental retardation because of low levels of thyroid hormone. With early intervention, many of these infants can be successfully treated. So it is not surprising that newborn screening is legally mandatory in many countries, including the United States and Canada.

DNA testing is the most accurate way to detect abnormal genes

The level of phenylalanine in the blood is an indirect measure of phenylalanine hydroxylase activity in the liver. But how can we screen for genetic diseases that are not detectable by blood tests? What if blood is difficult to obtain, as it is in a fetus? How are genetic abnormalities in heterozygotes, who express the normal protein at some level, identified?

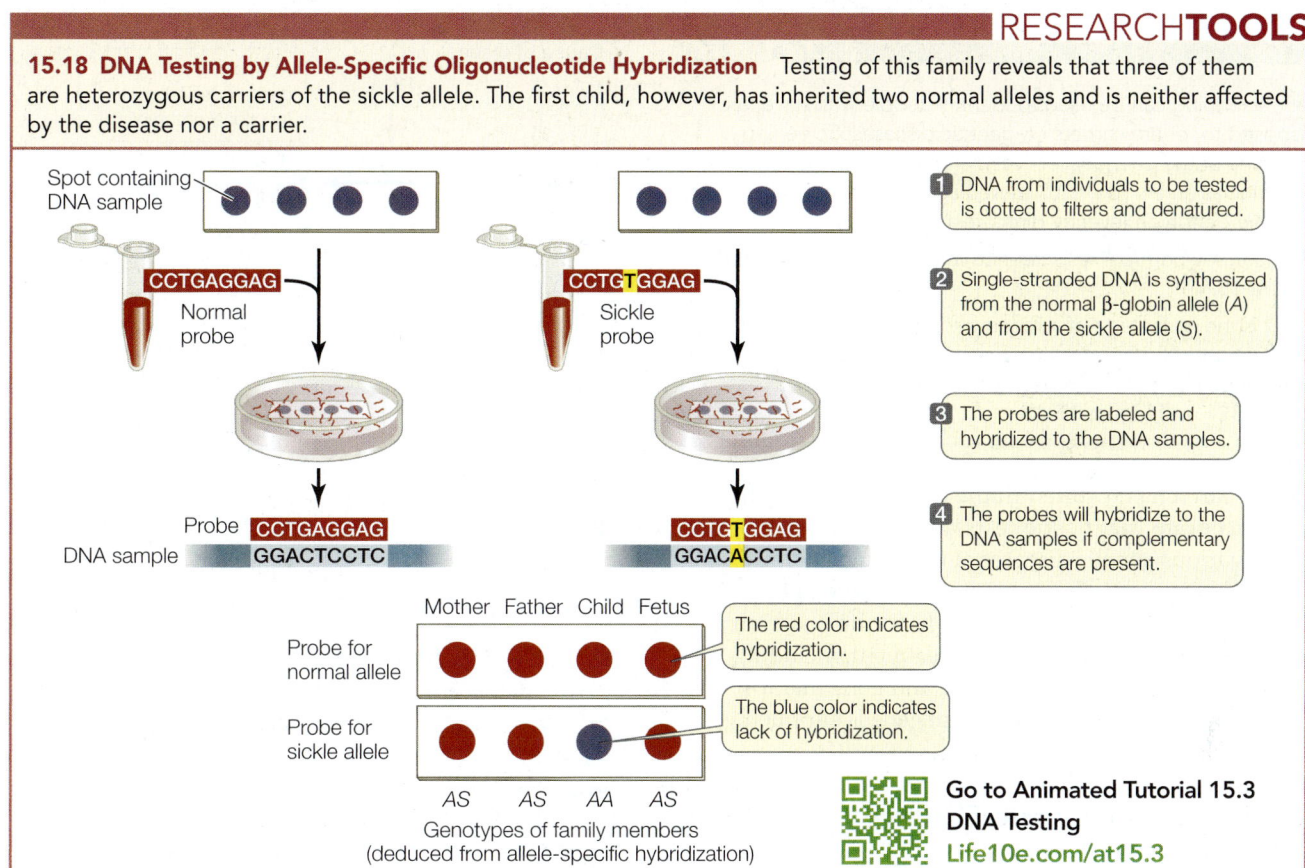

RESEARCH**TOOLS**

15.18 DNA Testing by Allele-Specific Oligonucleotide Hybridization Testing of this family reveals that three of them are heterozygous carriers of the sickle allele. The first child, however, has inherited two normal alleles and is neither affected by the disease nor a carrier.

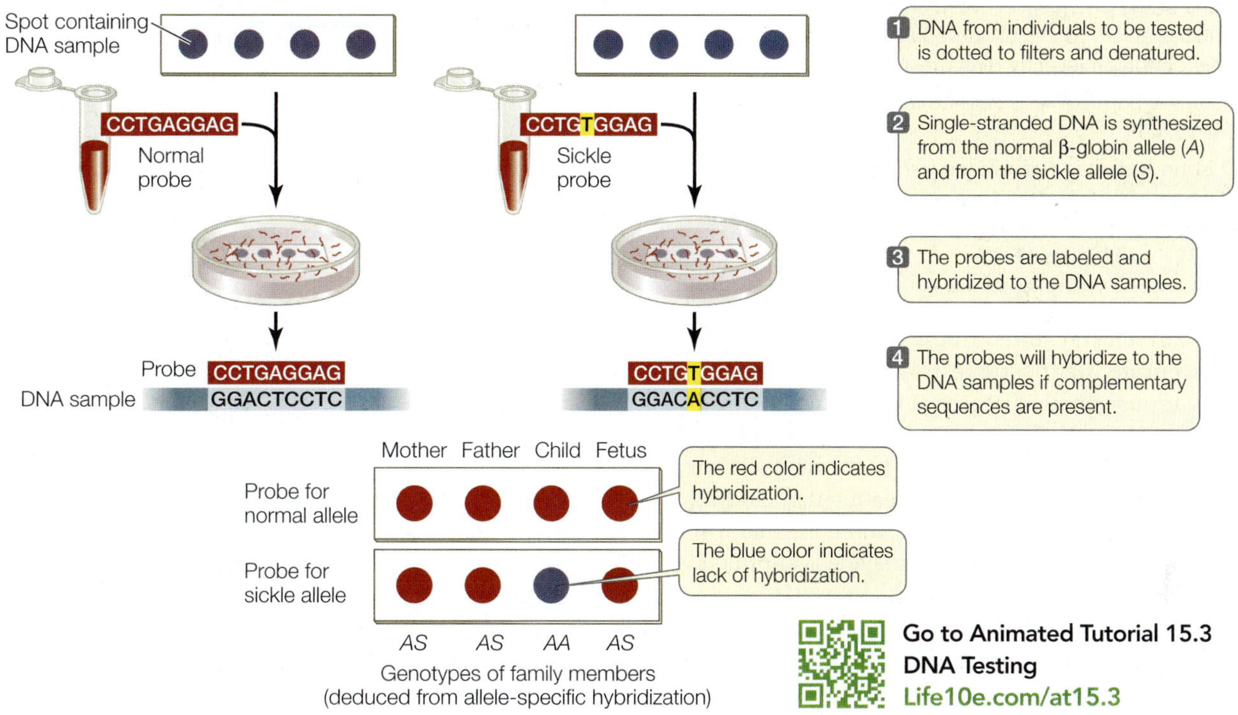

Spot containing DNA sample
CCTGAGGAG
Normal probe
CCTGTGGAG
Sickle probe
Probe
CCTGAGGAG
DNA sample
GGACTCCTC
CCTGTGGAG
GGACACCTC
1 DNA from individuals to be tested is dotted to filters and denatured.
2 Single-stranded DNA is synthesized from the normal β-globin allele (A) and from the sickle allele (S).
3 The probes are labeled and hybridized to the DNA samples.
4 The probes will hybridize to the DNA samples if complementary sequences are present.
Mother Father Child Fetus
Probe for normal allele
Probe for sickle allele
The red color indicates hybridization.
The blue color indicates lack of hybridization.
AS AS AA AS
Genotypes of family members (deduced from allele-specific hybridization)

Go to Animated Tutorial 15.3
DNA Testing
Life10e.com/at15.3

DNA testing is the direct analysis of DNA for a mutation, and it offers the most direct and accurate way of detecting an abnormal allele. Now that the mutations responsible for many human diseases have been identified, any cell in the body can be examined at any time of life for mutations. The amplification power of PCR means that only one or a few cells are needed for testing. These methods work best for diseases caused by only one or a few different mutations.

Consider, for example, two parents who are both heterozygous for the cystic fibrosis allele but who want to have a normal child. If treated with the appropriate hormones, the mother can be induced to "superovulate," releasing several eggs. An egg can be injected with a single sperm from her husband and the resulting zygote allowed to divide to the eight-cell stage. If one of these embryonic cells is removed, it can be tested for the presence of the cystic fibrosis allele. If the test is negative, the remaining seven-cell embryo can be implanted in the mother's womb where with luck, it will develop normally.

Such preimplantation screening is performed only rarely. More typical are analyses of fetal cells after normal fertilization and implantation in the womb. Fetal cells can be analyzed at about the tenth week of pregnancy by chorionic villus sampling, or during the thirteenth to seventeenth weeks by amniocentesis. In either case, only a few fetal cells are necessary to perform DNA testing. Recently, very sensitive methods were developed so that DNA testing can be done with the few fetal cells that are released into the mother's blood. A 10-milliliter blood sample from a pregnant woman has enough fetal cells for the analysis of many disorders, including Down syndrome and cystic fibrosis. This relatively noninvasive procedure could replace amniocentesis and chorionic villus sampling—which both carry a slight risk of causing a miscarriage—in the near future.

DNA testing can also be performed with newborns. The blood samples used for screening for PKU and other disorders contain enough of the baby's blood cells to permit DNA analysis using PCR-based techniques. DNA analysis is now being used to screen for sickle-cell disease and cystic fibrosis; similar tests for other diseases will surely follow. Of the numerous methods of DNA testing available, we will describe DNA hybridization, using sickle-cell anemia as an example.

Allele-specific oligonucleotide hybridization can detect mutations

Nucleic acid hybridization (see Figure 14.7) can be used to detect the presence of a specific DNA sequence, such as a sequence containing a particular mutation. Samples of DNA are collected from people who may or may not carry the mutation, and PCR is used to amplify the region of DNA where the mutation may occur. Short synthetic DNA strands called oligonucleotide probes are hybridized with the denatured PCR products. The probe is labeled in some way (e.g., with radioactivity or a fluorescent dye) so that hybridization can be readily detected (**Figure 15.18**).

Detection of a mutation by DNA screening can be used for diagnosis of a genetic disease, so that appropriate treatment can begin. In addition, DNA screening provides a person with important information about his or her genome.

RECAP 15.4

Genetic screening can be used to identify people who have, are predisposed to, or are carriers of, genetic diseases. Screening can be done at the phenotype level by identifying an abnormal protein such as an enzyme with altered activity. It can also be done at the genotype level by direct testing of DNA.

- How are newborn babies screened for PKU? See p. 320 and Figure 15.17
- What is the advantage of screening for genetic mutations by allele-specific oligonucleotide hybridization relative to screening phenotype differences in enzyme activity? See p. 321 and Figure 15.18

Ongoing research has resulted in the development of increasingly accurate diagnostic tests and a better understanding of various genetic diseases at the molecular level. This knowledge is now being applied to the development of new treatments for genetic diseases. In the next section we will survey various approaches to treatment, including modifications of the mutant phenotype and gene therapy, in which the normal version of a mutant gene is supplied.

15.5 How Are Genetic Diseases Treated?

Most treatments for genetic diseases simply try to alleviate the patient's symptoms. But to effectively treat these diseases—whether they affect all cells, as in inherited disorders such as PKU, or only somatic cells, as in cancer—physicians must be able to diagnose the disease accurately, understand how the disease works at the molecular level, and intervene early, before the disease ravages or kills the individual. There are two main approaches to treating genetic diseases:

- modifying the disease phenotype
- replacing the defective gene (modifying the genotype)

Genetic diseases can be treated by modifying the phenotype

Altering the phenotype of a genetic disease so that it no longer harms an individual is commonly done in one of three ways: by restricting the substrate of a deficient enzyme, by inhibiting a harmful metabolic reaction, or by supplying a missing protein product (Figure 15.19).

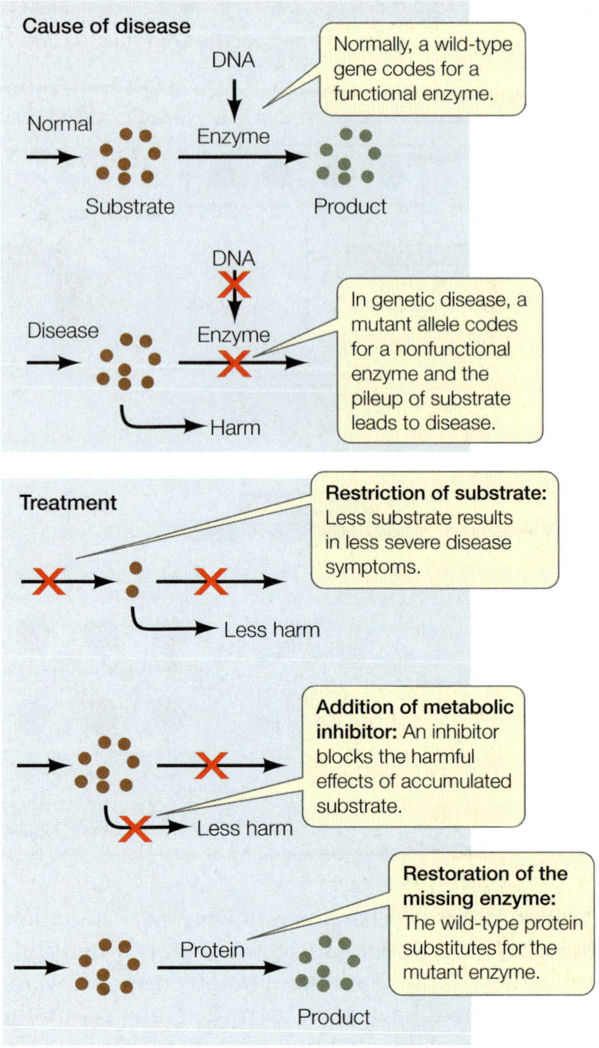
Cause of disease
DNA
Normally, a wild-type gene codes for a functional enzyme.
Normal
Enzyme
Substrate
Product
DNA
In genetic disease, a mutant allele codes for a nonfunctional enzyme and the pileup of substrate leads to disease.
Disease
Enzyme
Harm
Treatment
Restriction of substrate: Less substrate results in less severe disease symptoms.
Less harm
Addition of metabolic inhibitor: An inhibitor blocks the harmful effects of accumulated substrate.
Less harm
Restoration of the missing enzyme: The wild-type protein substitutes for the mutant enzyme.
Protein
Product

15.19 Strategies for Treating Genetic Diseases

RESTRICTING THE SUBSTRATE Restricting the substrate of a deficient enzyme is the approach taken when a newborn is diagnosed with PKU. In this case, the deficient enzyme is phenylalanine hydroxylase, and the substrate is phenylalanine (see Figure 15.7). The infant's inability to break down phenylalanine in food leads to a buildup of the substrate, which causes the clinical symptoms. So the infant is immediately put on a special diet that contains only enough phenylalanine for immediate use. Lofenelac, a milk-based product that is low in phenylalanine, is fed to these infants just like formula. Later, certain fruits, vegetables, cereals, and noodles low in phenylalanine can be added to the diet. Meat, fish, eggs, dairy products, and bread, which contain high amounts of phenylalanine, must be avoided, especially during childhood, when brain development is most rapid. The artificial sweetener aspartame must also be avoided because it is made of two amino acids, one of which is phenylalanine.

People with PKU are generally advised to stay on a low-phenylalanine diet for life. Although maintaining these dietary restrictions may be difficult, it is effective. Numerous follow-up studies since newborn screening was initiated have shown that people with PKU who stay on the diet are no different from the rest of the population in terms of mental ability. This is an impressive achievement in public health, given the severity of mental retardation in untreated patients.

METABOLIC INHIBITORS In Section 11.7 we described how drugs that are inhibitors of various cell cycle processes are used to treat cancer. Drugs are also used to treat the symptoms of many genetic diseases. As biologists have gained insight into the molecular characteristics of these diseases and the specific proteins involved, a more specific approach to treatment is taking shape. This approach—called molecular medicine—was used to develop the inhibitor used to treat Kareem Abdul-Jabar's chronic myelogenous leukemia (see the opening story).

SUPPLYING THE MISSING PROTEIN An obvious way to treat a disease caused by the lack of a functional protein is to supply that protein. This approach is used to treat hemophilia A, a disease in which blood factor VIII is missing and blood clotting is impaired (see Table 15.2). At first the missing protein was obtained from blood and was sometimes contaminated with viruses (e.g., HIV) or other pathogens that could harm the recipient. Now, however, human clotting proteins are produced by recombinant DNA technology (see Chapter 18), making it possible to provide the protein in a much purer form.

Unfortunately, the phenotypes of many diseases caused by genetic mutations are very complex. In these cases, simple interventions like those we have just described do not work. Indeed, a recent survey of 351 diseases caused by single-gene mutations showed that current therapies increased patients' life spans by an average of only 15 percent.

Gene therapy offers the hope of specific treatments

If a cell lacks an allele that encodes a functional product, an optimal treatment would be to provide a functional allele. The objective of **gene therapy** is to add a new gene that will be expressed in appropriate cells in a patient. What cells should be targeted? There are two approaches:

- Germ line gene therapy, in which the new gene is inserted into a gamete (usually an egg) or the fertilized egg. In this case, all cells of the adult will carry the new gene. Ethical considerations preclude its use in humans.
- Somatic cell gene therapy, in which the new gene is inserted into somatic cells involved in the disease. This method is being tried for numerous diseases, ranging from inherited genetic disorders to cancer.

There are two approaches to somatic cell gene therapy:

- In *ex vivo* gene therapy, target cells are removed from the patient, given the new gene, and then reinserted into the patient. This approach is being used, for example, for diseases caused by defects in genes that are expressed in white blood cells.
- In *in vivo* gene therapy, the gene is actually inserted directly into a patient, targeted to the appropriate cells. An example is a treatment for lung cancer in which a solution with a therapeutic gene is actually squirted onto a tumor.

Armed with knowledge of how genes are expressed (see Chapter 14) and regulated (see Chapter 16), physicians can design a therapeutic gene that contains not only a normal protein-coding sequence but also other sequences—such as an appropriate promotor—required for the gene's expression in targeted cells.

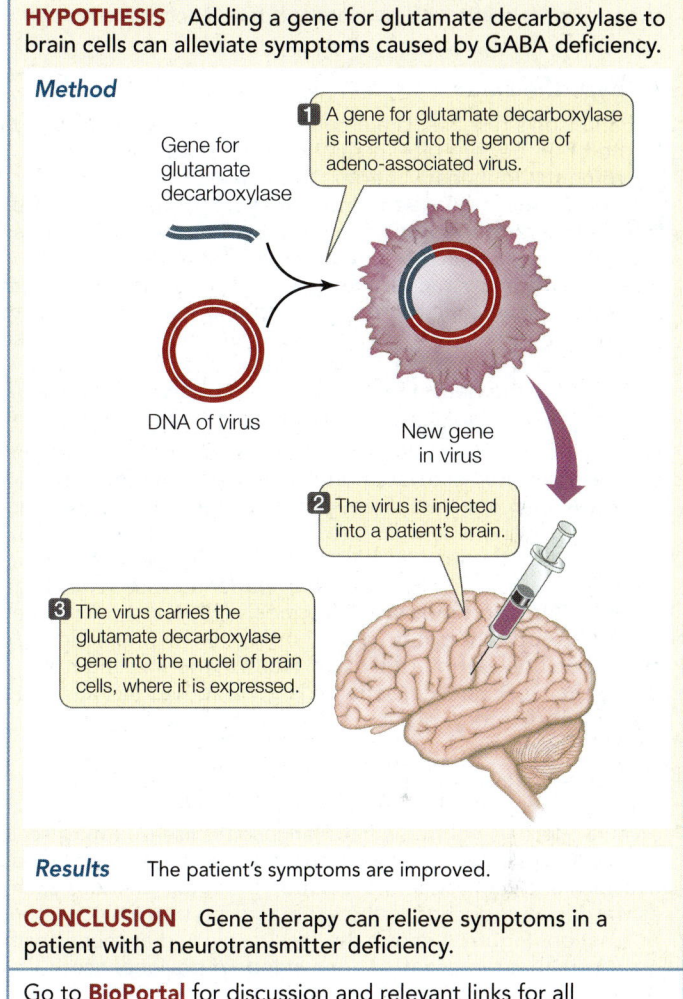

INVESTIGATING**LIFE**

15.20 Gene Therapy Andrew Feigin and his colleagues showed that a virus can be used to insert a therapeutic gene into the brains of patients with Parkinson's disease.[a] The gene they used encodes glutamate decarboxylase, the enzyme that catalyzes the synthesis of the neurotransmitter GABA.

HYPOTHESIS Adding a gene for glutamate decarboxylase to brain cells can alleviate symptoms caused by GABA deficiency.

Method

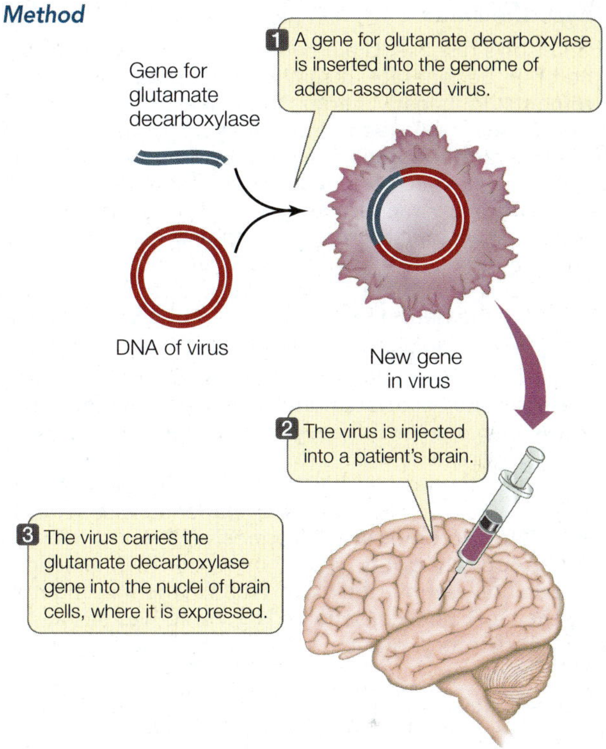
1 A gene for glutamate decarboxylase is inserted into the genome of adeno-associated virus.
Gene for glutamate decarboxylase
DNA of virus
New gene in virus
2 The virus is injected into a patient's brain.
3 The virus carries the glutamate decarboxylase gene into the nuclei of brain cells, where it is expressed.

Results The patient's symptoms are improved.

CONCLUSION Gene therapy can relieve symptoms in a patient with a neurotransmitter deficiency.

Go to **BioPortal** for discussion and relevant links for all INVESTIGATING**LIFE** figures.

[a]LeWitt, P. A. et al. 2011. *Lancet Neurology* 10: 309–319.

A major challenge has been getting the therapeutic gene into cells. Uptake of DNA into eukaryotic cells is a rare event, and once the DNA is inside a cell, its entry into the nucleus and expression are rarer still. One solution to these problems is to insert the gene into a carrier virus that can infect human cells but has been altered genetically to prevent replication. An example is the DNA virus called **adeno-associated virus**, which has been widely used in human gene therapy clinical trials. This virus has a small genome that allows splicing in of a human gene; infects most human cells, including nondividing cells such as neurons; is harmless to humans; does not provoke rejection by the immune system; and enters the cell nucleus, where its DNA with the new gene can be expressed.

A recent success in human gene therapy using adeno-associated virus is shown in **Figure 15.20**. Parkinson's disease is a neurological condition that affects about 1 in 200 persons

WORKING WITH DATA:

Gene Therapy for Parkinson's Disease

Original Paper

LeWitt, P. A. and 20 additional authors. 2011. AAV2-GAD gene therapy for advanced Parkinson's disease: a double-blind, sham-surgery controlled randomized trial. *Lancet Neurology* 10: 309–319.

Analyze the Data

In Parkinson's disease, brain cells (neurons) in specific regions of the brain degenerate, resulting in a loss of control of movements. In its early stages, the disease can be treated with drugs that replace the molecules no longer supplied by the degenerated neurons. However, as the disease progresses, these drugs become less effective and have unacceptable side effects. An alternate approach to treatment in such patients is gene therapy to supply an enzyme whose activity produces a molecule that can reduce symptoms. In this case, the enzyme is glutamate decarboxylase and the molecule produced is the neurotransmitter γ-aminobutyric acid (GABA).

Feigin and his colleagues inserted the glutamate decarboxylase gene into a virus that normally infects human cells, in this case, adeno-associated virus (see Figure 15.20). This virus was then injected directly into the region of the brain involved in Parkinson's disease. Sixteen patients received a dose of gene therapy: an injection of a saline solution containing the virus. Twenty-one other patients, with similar ages (average 60) and times since diagnosis (10 years), were in a control group that got a "sham" operation—an injection of saline solution without the virus. Neither the investigators doing the work nor the patients knew who was in the gene therapy and control groups, so this was a "double-blind" study.

At several times after injection, the patients in both groups were examined for changes in motor function. They were each given a score using the unified Parkinson's disease rating scale (UPDRS), which combines both the patient's monitoring of daily activities and a physician's evaluation of motor functions such as walking. A reduction in the score indicates an improvement of function. **Figure A** shows the results, plotted as means with error bars that represent +/– one standard deviation (see Appendix B).

QUESTION 1

Compare the control (sham) group with the gene therapy group. Were there any differences in their UPDRS scores? If so, when were the differences initially apparent? Were the differences statistically significant? How can you tell?

FIGURE A

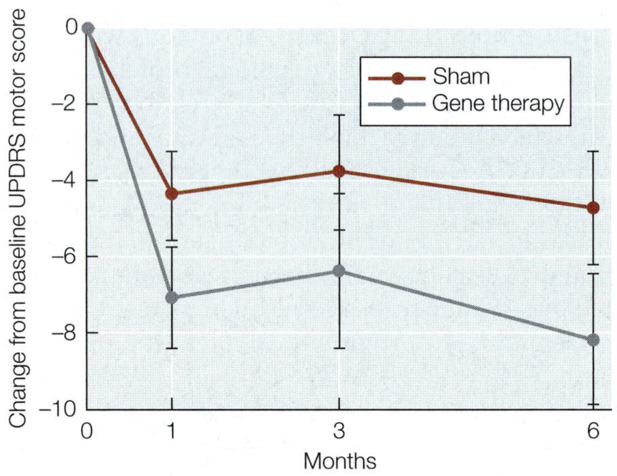
Change from baseline UPDRS motor score
0
–2
–4
–6
–8
–10
0
1
3
6
Months
Sham
Gene therapy

FIGURE B

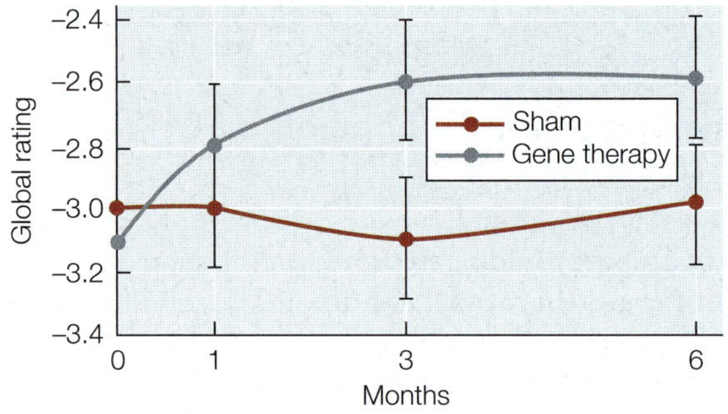
Global rating
–2.4
–2.6
–2.8
–3.0
–3.2
–3.4
0
1
3
6
Months
Sham
Gene therapy

QUESTION 2

Why do you think the score for the control group changed after the sham treatment?

QUESTION 3

A second way to assess symptoms is the global rating of Parkinsonism, in which the patients are asked to evaluate their own symptoms. In this case, an increase in the score indicates improvement. The results are shown in **Figure B**. How did the two groups of patients compare in their own evaluation of their symptoms? How does the fact that this was a double-blind study influence the strength of your conclusions?

Go to **BioPortal** for all WORKING WITH**DATA** exercises

in their 60s, and 1 in 25 persons in their 80s. Its symptoms include muscle stiffness and shaking and—in about half of the patients—dementia. In patients with Parkinson's disease, there is inadequate production of the neurotransmitter γ-amino butyric acid (GABA; see Section 45.3) in a particular part of the brain, because of degeneration of the neurons that normally produce GABA. This results in poor coordination of movement. Synthesis of GABA is catalyzed by the enzyme glutamate decarboxylase. A team led by Dr. Andrew Feigin at the Feinstein Institute in New York has successfully treated patients with Parkinson's disease by using a glutamate decarboxylase gene packaged into adeno-associated virus and injected into the affected part of the brain. After six months, the patients who received the virus-encapsulated gene had increased levels of GABA and significant improvement in their disease symptoms. Other diseases being treated with gene therapy using this virus include cystic fibrosis, hemophilia, and muscular dystrophy.

RECAP 15.5

Treatment of a human genetic disease may involve an attempt to modify the abnormal phenotype by restricting the substrate of a deficient enzyme, inhibiting a harmful metabolic reaction, or supplying a missing protein. By contrast, gene therapy aims to address a genetic defect by inserting a normal allele into a patient's cells.

- How do metabolic inhibitors used in chemotherapy function in treating cancer? **See pp. 322–323 and Figure 15.19**
- How does *in vivo* gene therapy work? Can you give an example? **See pp. 323–324 and Figure 15.20**

In this chapter we dealt with mutations in general, focusing on DNA changes that affect phenotypes through specific protein products. But there is much more to molecular genetics than the sequences of genes and proteins. Determining which genes will be expressed when and where is a major function of the genome. In Chapter 16 we will turn to gene regulation.

?

Are there other targeted therapies directed to specific types of cancer?

ANSWER

As the somatic mutations in various cancers are described, specific therapies are being developed. For example, some breast cancer cells overexpress the receptor for the steroid hormone estrogen. This makes the cancer cells abnormally sensitive to estrogen, which stimulates them to divide. Tamoxifen, an analog of estrogen that binds but does not activate the estrogen receptor, is being used to treat breast cancers of this type. Other cancer cells express the receptor for epidermal growth factor, another signal that stimulates cell division. In these cases, treatment with erlotinib, a drug targeted to that receptor, may be effective in slowing cancer growth.

CHAPTER SUMMARY 15

15.1 What Are Mutations?

- **Mutations** are heritable changes in DNA. **Somatic mutations** are passed on to daughter cells, but only **germ line mutations** are passed on to sexually produced offspring. **Review ANIMATED TUTORIAL 15.1**
- **Point mutations** result from alterations in single base pairs of DNA. **Silent mutations** can occur in noncoding DNA or in coding regions of genes and do not affect the amino acid sequences of proteins. **Missense**, **nonsense**, and **frame-shift** mutations all cause changes in protein sequences. **Review Figure 15.2**
- Chromosomal mutations (**deletions**, **duplications**, **inversions**, and **translocations**) involve large regions of chromosomes. **Review Figure 15.4**
- **Spontaneous mutations** occur because of instabilities in DNA or chromosomes. **Induced mutations** occur when a mutagen damages DNA. **Review Figure 15.5**
- Mutations can occur in hot spots where cytosine has been methylated to 5′-methylcytosine. **Review Figure 15.6**
- Mutations, although often detrimental to an individual organism, are the raw material of evolution.

15.2 What Kinds of Mutations Lead to Genetic Diseases?

- Abnormalities in proteins have been implicated in genetic diseases.
- While a single amino acid difference can be the cause of disease, amino acid variations have been detected in many functional proteins. **Review Figures 15.7, 15.8**
- **Point mutations**, **deletions**, and chromosome abnormalities are associated with genetic diseases. **Review Figure 15.9**
- The effects of fragile-X syndrome worsen with each generation. This pattern is the result of an **expanding triplet repeat**. **Review Figure 15.10**
- A series of genetic mutations can lead to colon cancer. **Review Figure 15.11**
- **Multifactorial** diseases are caused by the interactions of many genes and proteins with the environment. They are much more common than diseases caused by mutations in a single gene.

15.3 How Are Mutations Detected and Analyzed?

- **Restriction enzymes**, which are made by microorganisms as a defense against viruses, bind to and cut DNA at specific **recognition sequences** (also called **restriction sites**). These enzymes can be used to produce small fragments of DNA for study, a technique known as restriction digestion. **Review Figure 15.12**
- DNA fragments can be separated by size using **gel electrophoresis**. **Review Figure 15.13, ANIMATED TUTORIAL 15.2**
- **DNA fingerprinting** is used to distinguish among specific individuals or to reveal which individuals are most closely related to one another. It involves the detection of DNA polymorphisms, including **single nucleotide polymorphisms** (**SNPs**) and **short tandem repeats** (**STRs**). **Review Figure 15.14, ACTIVITY 15.1**
- It is possible to isolate both the mutant genes and the abnormal proteins responsible for human diseases. **Review Figure 15.15**
- The goal of the DNA barcoding project is to sequence a single region of DNA in all species for identification purposes.

15.4 How Is Genetic Screening Used to Detect Diseases?

- **Genetic screening** is used to detect human genetic diseases, alleles predisposing people to those diseases, or carriers of those disease alleles.
- Genetic screening can be done by looking for abnormal protein expression. **Review Figure 15.17**
- **DNA testing** is the direct identification of mutant alleles. Any cell can be tested at any time in the life cycle. **Review Figure 15.18, ANIMATED TUTORIAL 15.3**

continued

15.5 How Are Genetic Diseases Treated?

- There are three ways to modify the phenotype of a genetic disease: restrict the substrate of a deficient enzyme, inhibit a harmful metabolic reaction, or supply a missing protein. **Review Figure 15.19**
- Cancer sometimes can be treated with treated with metabolic inhibitors.
- In **gene therapy**, a mutant gene is replaced with a normal gene. Both *ex vivo* and *in vivo* therapies are being developed. **Review Figure 15.20**

Go to the Interactive Summary to review key figures, Animated Tutorials, and Activities Life10e.com/is15

CHAPTER**REVIEW**

REMEMBERING

1. Phenylketonuria is an example of a genetic disease in which
   a. a single enzyme is not functional.
   b. inheritance is sex-linked.
   c. two parents without the disease cannot have a child with the disease.
   d. mental retardation always occurs, regardless of treatment.
   e. a transport protein does not work properly.

2. Mutations of the gene for β-globin
   a. are usually lethal.
   b. occur only at amino acid position 6.
   c. number in the hundreds.
   d. always result in sickling of red blood cells.
   e. can always be detected by gel electrophoresis.

3. Multifactorial (complex) diseases
   a. are less common than single-gene diseases.
   b. involve the interaction of many genes with the environment.
   c. affect less than 1 percent of humans.
   d. involve the interactions of several mRNAs.
   e. are exemplified by sickle-cell disease.

4. Mutational "hot spots" in human DNA
   a. always occur in genes that are transcribed.
   b. are common at cytosines that have been modified to 5′–methylcytosine.
   c. involve long stretches of nucleotides.
   d. occur only where there are long repeats.
   e. are very rare in genes that code for proteins.

5. Genetic diagnosis by DNA testing
   a. detects only mutant and not normal alleles.
   b. can be done only on eggs or sperm.
   c. involves hybridization to rRNA.
   d. often uses restriction enzymes and a polymorphic restriction site.
   e. cannot be done with PCR.

6. Current treatments for human genetic diseases include all of the following *except*
   a. restricting a dietary substrate.
   b. replacing the mutant gene in all cells.
   c. alleviating the patient's symptoms.
   d. inhibiting a harmful metabolic reaction.
   e. supplying a protein that is missing.

UNDERSTANDING & APPLYING

7. Compare the following:
   a. Loss of function mutation versus gain of function mutation
   b. Missense mutation versus nonsense mutation
   c. Spontaneous versus induced mutation

8. In the past, it was common for people with phenylketonuria (PKU) who were placed on a low-phenylalanine diet after birth to be allowed to return to a normal diet during their teenage years. Although the levels of phenylalanine in their blood were high, their brains were thought to be beyond the stage when they could be harmed. If a woman with PKU becomes pregnant, however, a problem arises. Typically, the fetus is heterozygous but is unable, at early stages of development, to metabolize the high levels of phenylalanine that arrive from the mother's blood. (a) Why is the fetus likely to be heterozygous? (b) What do you think would happen to the fetus during this "maternal PKU" situation? (c) What would be your advice to a woman with PKU who wants to have a child?

ANALYZING & EVALUATING

9. Cystic fibrosis is an autosomal recessive disease in which thick mucus is produced in the lungs and airways. The gene responsible for this disease encodes a protein composed of 1,480 amino acids. In most patients with cystic fibrosis, the protein has 1,479 amino acids: a phenylalanine is missing at position 508. A baby is born with cystic fibrosis. He has an older brother who is not affected. How would you test the DNA of the older brother to determine whether he is a carrier for cystic fibrosis? How would you design a gene therapy protocol to "cure" the cells in the younger brother's lungs and airways?

10. Several efforts are under way to identify human genetic polymorphisms that correlate with multifactorial diseases such as diabetes, heart disease, and cancer. What would be the uses of such information? What concerns do you think are being raised about this kind of genetic testing?

11. Tay-Sachs disease is caused by a recessively inherited mutation in the gene coding for the enzyme hexosaminidase A (HEXA), which normally breaks down a lipid called GM2 ganglioside. Accumulation of this lipid in the brain leads to progressive deterioration of the nervous system and death, usually by age 4. HEXA activity in blood serum is 0 to 6 percent in homozygous recessives and 7 to 35 percent in heterozygous carriers, compared with non-carriers (100 percent). The most common mutation in the *HEXA* gene is an insertion of four base pairs, which presumably leads to a premature stop codon. How would you do genetic screening for carriers of this disease by enzyme testing and by DNA testing? What are the advantages of DNA testing? How would you investigate the premature stop codon hypothesis?

12. Alkaptonuria is an inborn error of metabolism, caused by defects in an enzyme in the pathway that breaks down tyrosine (see Section 14.1). Humans who are homozygous for one of these mutations make nonfunctional enzyme and accumulate the enzyme's substrate, homogentisic acid, which causes their disease symptoms. In 1996, researchers in Spain cloned and sequenced the gene for the enzyme, and characterized several mutant alleles. Here is the wild-type coding strand sequence for part of the gene, with the amino acid sequence of the protein below:

... TTG ATA CCC ATT GCC ...

... Leu Ile Pro Ile Ala ...

Here is the sequence for one of the mutant alleles:

... TTG ATA TCC ATT GCC ...

a. What is the amino acid sequence produced by the mutant allele? What type of mutation is this: silent, nonsense, missense, or frame-shift?
b. Why is this mutation likely to affect the function of the enzyme? (Hint: see Section 3.2, especially Table 3.2.)
c. The mutation creates a new recognition site for the restriction enzyme *Eco*RV (5′-GATATC-3′). How would you take advantage of this new site to screen individuals for the mutant allele?

Go to BioPortal at **yourBioPortal.com** for Animated Tutorials, Activities, LearningCurve Quizzes, Flashcards, and many other study and review resources.

16 Regulation of Gene Expression

CHAPTER**OUTLINE**

Stress Extreme physical or emotional stress during pregnancy can have negative effects on a child many years later. A little relaxation, on the other hand, never hurt anyone.

ANY TEENAGER OR PARENT OF A TEENAGER will tell you that those years can be rough. But some young people have a rougher time than others. Teens whose mothers were extremely stressed during pregnancy often have more behavioral problems than those whose mothers had calmer pregnancies. A recent study led by psychologist Thomas Elbert and evolutionary biologist Axel Meyer at the University of Konstanz in Germany implicates the regulation of gene expression in these behavioral differences. The study examined a gene for the glucocorticoid receptor, which is involved in regulating hormonal responses to stress. The researchers found that teenagers whose mothers had suffered physical abuse during pregnancy had higher rates of methylation in the promoter of this gene than teenagers whose mothers had not suffered such abuse.

A major control point for gene expression is the promoter, a sequence of DNA adjacent to the coding region of a gene where proteins bind and control the rate of transcription. The ability of these transcription factors to bind to the promoter is affected by the level of DNA methylation in the promoter. As we mentioned in Section 15.1, in certain regions of DNA many of the cytosine residues have methyl groups added at their 5′ positions, forming 5′-methylcytosine. If a gene promoter has a high degree of methylation, some transcription factors can't bind to it. Instead, specific repressor proteins bind to the methylated DNA and prevent expression of the gene. DNA methylation plays an important role in gene regulation and is a normal part of development. But the level of methylation can change over time and can vary among individuals, as was the case for the teenagers in Elbert and Meyer's study. Their finding is interesting because it shows a correlation between maternal stress in humans and DNA methylation in their offspring. However, a lot more work is needed to show how the mother's stress brought about this DNA change, and how this change affects behavior.

Other studies show a link between early life experiences, promoter methylation in the glucocorticoid receptor gene, and behavior. People who are abused as children have altered hormonal stress responses and an increased risk of suicide. One study showed that suicide victims who were abused as children had more methylation of the glucocorticoid receptor gene promotor than in people who were not abused, and this led to lower expression of the gene.

Such studies have spawned the new field of behavioral epigenetics. Epigenetics is the study of heritable changes in gene expression that do not involve changes in the DNA sequence. A deeper understanding of the regulation of gene expression may lead to better treatments of important social problems.

> Question: Can epigenetic changes be manipulated?
>
> See answer on p. 349.

16.1 How Is Gene Expression Regulated in Prokaryotes?

Prokaryotes conserve energy and resources by making certain proteins only when they are needed. The protein content of a bacterium can change rapidly when conditions warrant. There are several ways in which a prokaryotic cell could shut off the supply of an unneeded protein. The cell could:

- downregulate the transcription of mRNA for that protein;
- hydrolyze the mRNA after it is made, thereby preventing translation;
- prevent translation of the mRNA at the ribosome;
- hydrolyze the protein after it is made; or
- inhibit the function of the protein.

Whichever mechanism is used, it must be both responsive to environmental signals and efficient. The earlier the cell intervenes in the process of protein synthesis, the less energy it wastes. Selective blocking of transcription is far more efficient than transcribing the gene, translating the message, and then degrading or inhibiting the protein. While all five mechanisms for regulating protein levels are found in nature, prokaryotes generally use the most efficient one: transcriptional regulation.

As we described in Chapter 14, gene expression begins at the promoter, where RNA polymerase binds to initiate transcription. In a genome with many genes, not all promoters are active at a given time—there is selective gene transcription. The "decision" regarding which genes to activate involves two types of regulatory proteins that bind to DNA: repressor proteins and activator proteins. In both cases, these proteins bind to the promoter to regulate the gene (Figure 16.1):

- In **negative regulation**, binding of a repressor protein prevents transcription.
- In **positive regulation**, an activator protein binds DNA to stimulate transcription.

You will see examples of these mechanisms, or combinations of them, as we examine regulation in prokaryotes, eukaryotes, and viruses.

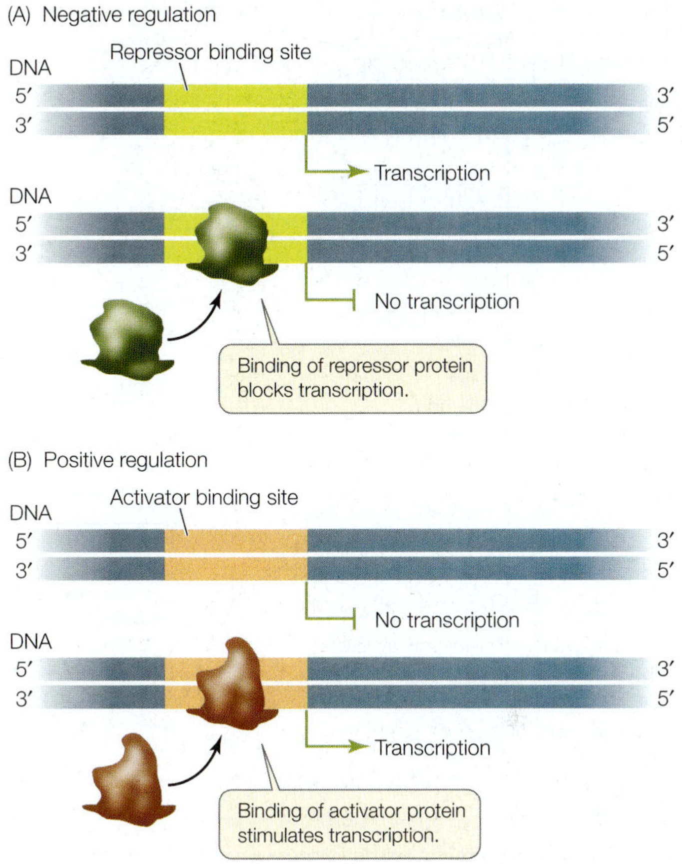

(A) Negative regulation
Repressor binding site
DNA
5′
3′
3′
5′
Transcription
DNA
5′
3′
3′
5′
No transcription
Binding of repressor protein blocks transcription.
(B) Positive regulation
Activator binding site
DNA
5′
3′
3′
5′
No transcription
DNA
5′
3′
3′
5′
Transcription
Binding of activator protein stimulates transcription.

16.1 Positive and Negative Regulation Proteins regulate gene expression by binding to DNA and preventing or allowing RNA polymerase to bind DNA at the promotor region to control transcription of the gene.

Regulating gene transcription conserves energy

As a normal inhabitant of the human intestine, *E. coli* must be able to adjust to sudden changes in its chemical environment. Its host may present it with one foodstuff one hour (e.g., glucose in fruit juice) and another the next (e.g., lactose in milk). Such changes in nutrients present the bacterium with a metabolic challenge. Glucose is its preferred energy source, and is the easiest sugar to metabolize. Lactose is a β-galactoside—a disaccharide containing galactose β-linked to glucose (see Section 3.3). Three proteins are involved in the initial uptake and metabolism of lactose by *E. coli*:

- β-Galactoside permease is a carrier protein in the bacterial plasma membrane that moves the sugar into the cell.
- β-Galactosidase is an enzyme that hydrolyses lactose to glucose and galactose.
- β-Galactoside transacetylase transfers acetyl groups from acetyl CoA to certain β-galactosides. Its role in the metabolism of lactose is not clear.

When *E. coli* is grown on a medium that contains glucose but no lactose or other β-galactosides, the levels of these three proteins are extremely low—the cell does not waste energy and materials making the unneeded enzymes. But if the environment changes such that lactose is the predominant sugar available and very little glucose is present, the bacterium promptly begins making all three enzymes after a short lag period. There are only a few molecules of β-galactosidase present in an *E. coli* cell when glucose is present in the medium. But when glucose is absent, the addition of lactose can induce the synthesis of about 1,500 times more molecules of β-galactosidase per cell (Figure 16.2A)!

What is behind this dramatic increase? An important clue comes from measuring the amount of mRNA for β-galactosidase. The mRNA level dramatically increases during the lag period after lactose is added to the medium, and this mRNA is presumably translated into protein (Figure 16.2B). Moreover, the high mRNA level depends on the presence of

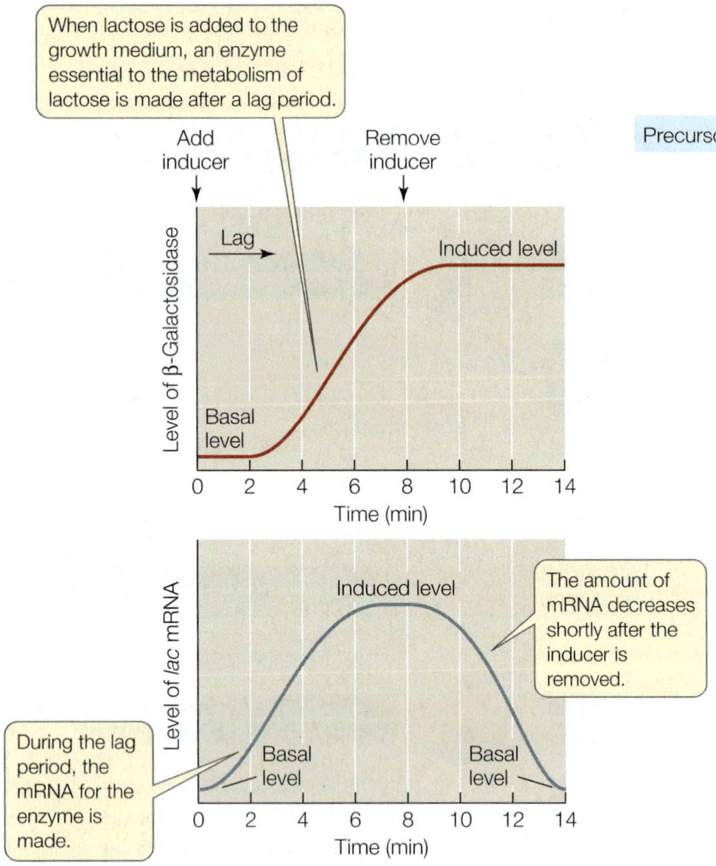
When lactose is added to the growth medium, an enzyme essential to the metabolism of lactose is made after a lag period.
Add inducer
Remove inducer
Level of β-Galactosidase
Lag
Induced level
Basal level
0 2 4 6 8 10 12 14
Time (min)
Level of lac mRNA
Induced level
The amount of mRNA decreases shortly after the inducer is removed.
During the lag period, the mRNA for the enzyme is made.
Basal level
Basal level
0 2 4 6 8 10 12 14
Time (min)

16.2 An Inducer Stimulates the Expression of a Gene for an Enzyme (A) When lactose is added to the growth medium for the bacterium *E. coli*, the synthesis of β-galactosidase begins only after an initial lag. (B) There is a lag because the mRNA for β-galactosidase has to be made before the protein can be made. The amount of mRNA decreases rapidly after the lactose is removed, indicating that transcription is no longer occurring. These changes in mRNA levels indicate that the mechanism of induction by lactose is transcriptional regulation.

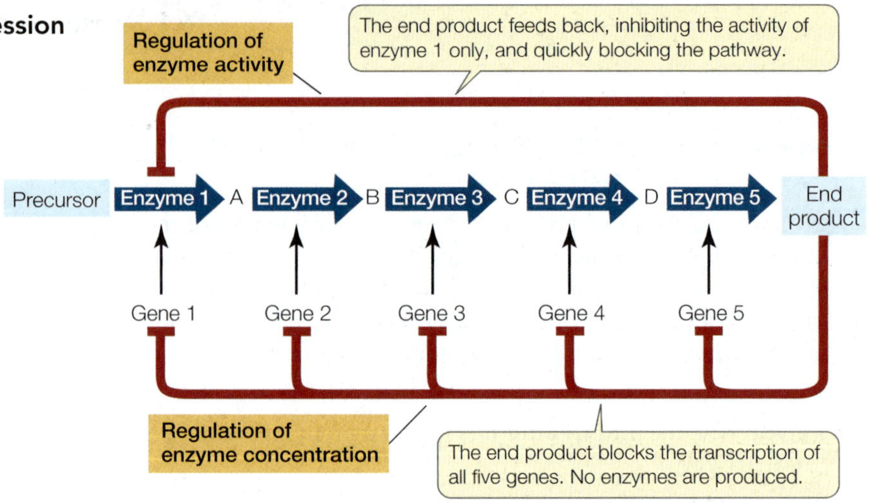
Regulation of enzyme activity
The end product feeds back, inhibiting the activity of enzyme 1 only, and quickly blocking the pathway.
Precursor
Enzyme 1
A
Enzyme 2
B
Enzyme 3
C
Enzyme 4
D
Enzyme 5
End product
Gene 1
Gene 2
Gene 3
Gene 4
Gene 5
Regulation of enzyme concentration
The end product blocks the transcription of all five genes. No enzymes are produced.

16.3 Two Ways to Regulate a Metabolic Pathway Feedback from the end product of a metabolic pathway can block enzyme activity (allosteric regulation), or it can stop the transcription of genes that code for the enzymes in the pathway (transcriptional regulation).

lactose, because if the lactose is removed, the mRNA level goes down. The response of the bacterial cell to lactose is clearly at the level of transcription.

Compounds that, like lactose, stimulate the synthesis of a protein are called **inducers**. The proteins that are produced are called **inducible proteins**, whereas proteins that are made all the time at a constant rate are called **constitutive proteins**. (Think of the constitution of a country, a document that does not change under normal circumstances.)

We have now seen two basic ways of regulating the rate of a metabolic pathway. In Section 8.5 we described the allosteric regulation of enzyme activity, which allows the rapid fine-tuning of metabolism. Regulation of protein synthesis—that is, regulation of the concentration of enzymes—is slower but results in greater savings of energy and resources. Protein synthesis is a highly endergonic process, since assembling mRNA, charging tRNA, and moving the ribosomes along mRNA all require the hydrolysis of nucleoside triphosphates such as ATP. **Figure 16.3** compares these two modes of regulation.

Operons are units of transcriptional regulation in prokaryotes

The genes that encode the three enzymes for processing lactose in *E. coli* are **structural genes**; they specify the primary structures (the amino acid sequences) of protein molecules. Structural genes are genes that can be transcribed into mRNA. The three genes lie adjacent to one another on the *E. coli* chromosome. This arrangement is no coincidence: the genes share a single promoter, and their DNA is transcribed into a single, continuous molecule of mRNA. Because this particular mRNA governs the synthesis of all three lactose-metabolizing enzymes, either all or none of these enzymes are made, depending on whether their common message—their mRNA—is present in the cell.

A cluster of genes with a single promoter is called an **operon**, and the operon that encodes the three lactose-metabolizing enzymes in *E. coli* is called the *lac* operon. The *lac* operon promoter can be very efficient (the maximum rate of mRNA synthesis can be high), but mRNA synthesis can be shut down when the enzymes are not needed. In addition to the promoter, an operon has other **regulatory sequences** that are not transcribed. A typical operon consists of a promoter, an operator, and two or more structural genes (**Figure 16.4**). The **operator** is a short stretch of DNA that lies between the promoter and the structural genes. It can bind very tightly with regulatory proteins that either activate or repress transcription.

There are numerous mechanisms to control the transcription of operons; here we will focus on three examples:

- An inducible operon regulated by a repressor protein
- A repressible operon regulated by a repressor protein
- An operon regulated by an activator protein

Operator–repressor interactions control transcription in the *lac* and *trp* operons

The *lac* operon contains a promoter, to which RNA polymerase binds to initiate transcription, and an operator, to which a **repressor** protein can bind. The gene that encodes this repressor is located near the *lac* operon on the *E. coli* chromosome.

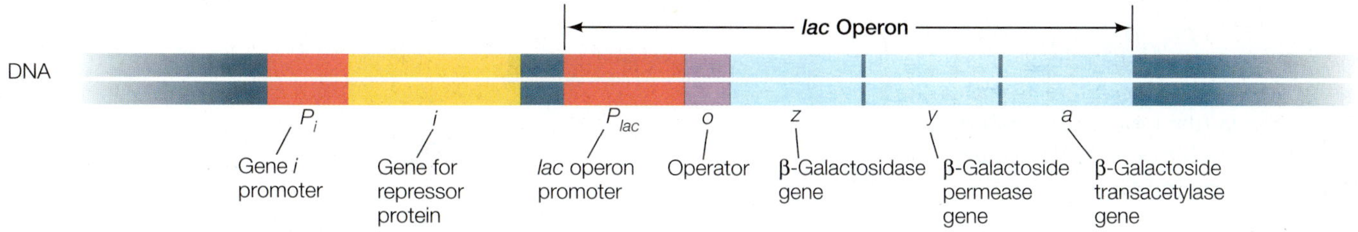
lac Operon
DNA
P_i
i
P_{lac}
o
z
y
a
Gene i promoter
Gene for repressor protein
lac operon promoter
Operator
β-Galactosidase gene
β-Galactoside permease gene
β-Galactoside transacetylase gene

16.4 The *lac* Operon of *E. coli* The *lac* operon of *E. coli* is a segment of DNA that includes a promoter, an operator, and the three structural genes that code for lactose-metabolizing enzymes.

When the repressor is bound, transcription of the operon is blocked. This example of negative regulation was elegantly worked out by Nobel Prize winners François Jacob and Jacques Monod.

The repressor protein has two binding sites: one for the operator and the other for the inducer. The environmental signal that induces the *lac* operon (for example, in the human digestive tract) is lactose, but the actual inducer is allolactose, a molecule that forms from lactose once it enters the cell. In the absence of the inducer, the repressor protein fits into the major groove of the operator DNA and recognizes and binds to a specific nucleotide base sequence. This prevents the binding of RNA polymerase to the promoter, and the operon is not transcribed (**Figure 16.5A**). When the inducer is present, it binds to the repressor and changes the shape of the repressor. This change in three-dimensional structure (conformation) prevents the repressor from binding to the operator. As a result, RNA polymerase can bind to the promoter and start transcribing the structural genes of the *lac* operon (**Figure 16.5B**).

You can see from this example that a key to transcriptional control of gene expression is the presence of regulatory sequences that do not code for proteins, but are binding sites for regulatory proteins and other proteins involved in transcription.

In contrast to the inducible system of the *lac* operon, other operons in *E. coli* are repressible; that is, they are repressed only under specific conditions. In such a system, the repressor is not normally bound to the operator. But if another molecule called a **co-repressor** binds to the repressor, the repressor changes shape and binds to the operator, thereby inhibiting transcription. An example is the *trp* operon, whose structural genes catalyze the synthesis of the amino acid tryptophan:

$$\text{Precursor molecules} \rightarrow \rightarrow \rightarrow \rightarrow \rightarrow \text{tryptophan} \quad (\text{5 Enzyme-catalyzed reactions})$$

When tryptophan is present in the cell in adequate concentrations, it is advantageous to stop making the enzymes for tryptophan synthesis. To do this, the cell uses a repressor that binds to an operator in the *trp* operon. But the repressor of the *trp* operon is not normally bound to the operator; it only binds when its shape is changed by binding to tryptophan, the co-repressor.

To summarize the differences between these two types of operons:

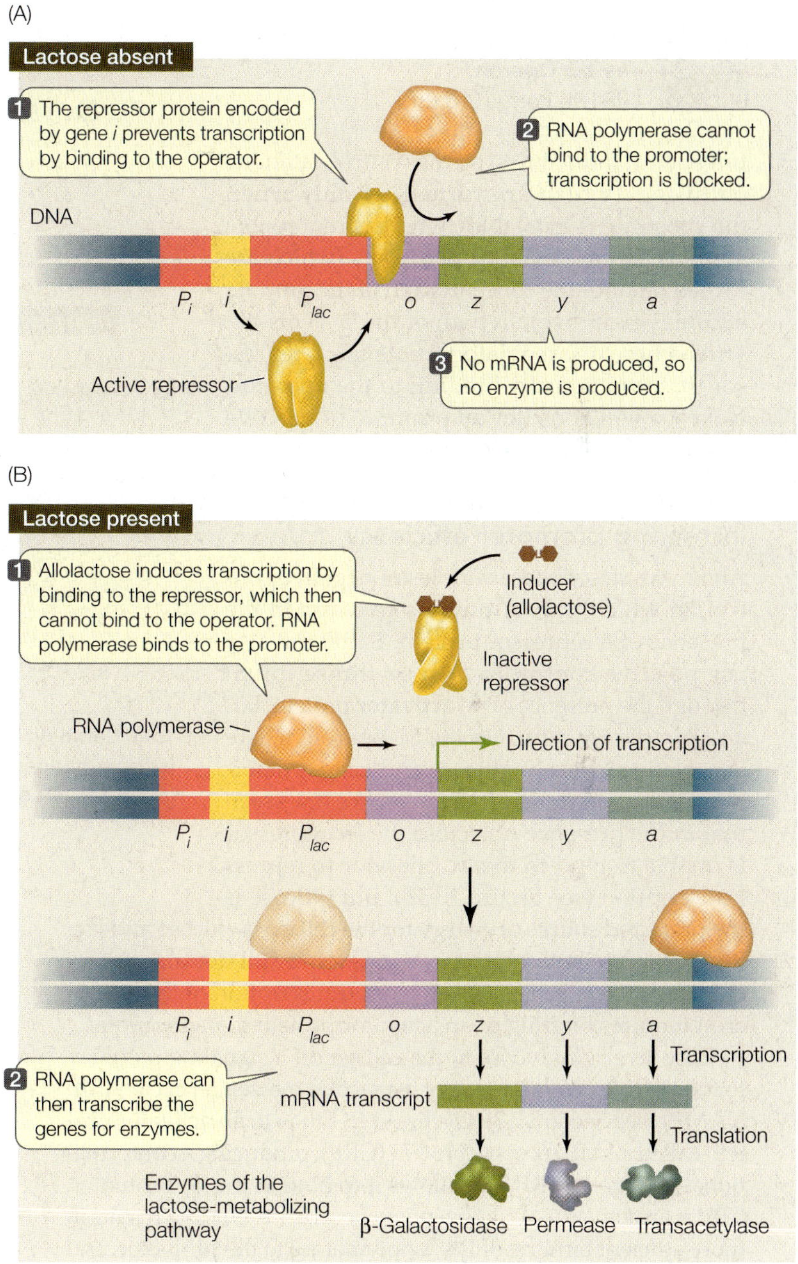
(A)
Lactose absent
1 The repressor protein encoded by gene i prevents transcription by binding to the operator.
2 RNA polymerase cannot bind to the promoter; transcription is blocked.
DNA
P_i i P_{lac} o z y a
Active repressor
3 No mRNA is produced, so no enzyme is produced.
(B)
Lactose present
1 Allolactose induces transcription by binding to the repressor, which then cannot bind to the operator. RNA polymerase binds to the promoter.
Inducer (allolactose)
Inactive repressor
RNA polymerase
Direction of transcription
P_i i P_{lac} o z y a
Transcription
2 RNA polymerase can then transcribe the genes for enzymes.
mRNA transcript
Translation
Enzymes of the lactose-metabolizing pathway
β-Galactosidase
Permease
Transacetylase

16.5 The *lac* Operon: An Inducible System (A) When lactose is absent, the synthesis of enzymes for its metabolism is inhibited. (B) Lactose (the inducer) leads to synthesis of the enzymes in the lactose-metabolizing pathway by binding to the repressor protein and preventing its binding to the operator.

Go to Animated Tutorial 16.1
The *lac* Operon
Life10e.com/at16.1

- In *inducible* systems, the substrate of a metabolic pathway (the inducer) interacts with a regulatory protein (the repressor), rendering the repressor incapable of binding to the operator and thus allowing transcription.
- In *repressible* systems, the product of a metabolic pathway (the co-repressor) binds to a regulatory protein, which is then able to bind to the operator and block transcription.

Go to Animated Tutorial 16.2
The *trp* Operon
Life10e.com/at16.2

In general, inducible systems control catabolic pathways (which are turned on only when the substrate is available), whereas repressible systems control anabolic pathways (which are turned on until the concentration of the product becomes excessive). In both of the systems described here, the regulatory protein is a repressor that functions by binding to the operator. Next we will consider an example of positive control involving an activator.

Protein synthesis can be controlled by increasing promoter efficiency

Above we described examples of negative control, in which transcription is *decreased* in the presence of a repressor protein. *E. coli* can also use positive control to *increase* transcription through the presence of an **activator** protein. For an example we return to the *lac* operon, where the relative levels of glucose and lactose determine the amount of transcription. We have seen that in the presence of lactose the *lac* repressor is unable to bind to the *lac* operator to repress transcription (see Figure 16.5B). But glucose is the preferred source of energy for the cell, so if glucose and lactose levels are both high, the *lac* operon is still not transcribed efficiently. This is because efficient transcription of the *lac* operon requires binding of an activator protein to its promoter.

Low levels of glucose in the cell set off a signaling pathway that leads to increased levels of the second messenger cyclic AMP (cAMP) (see Section 7.3). Cyclic AMP binds to an activator protein called *c*AMP *r*eceptor *p*rotein (CRP), producing a conformational change in CRP that allows it to bind to the *lac* promoter. CRP is an activator of transcription, because its binding results in more efficient binding of RNA polymerase to the promoter, and thus increased transcription of the structural genes (**Figure 16.6**). In the presence of abundant glucose, cAMP levels are low, CRP does not bind to the promoter, and the efficiency of transcription of the *lac* operon is reduced. This is an example of **catabolite repression**, a system of gene regulation in which the presence of the preferred energy source represses other catabolic pathways. The mechanisms controlling positive and negative regulation of the *lac* operon are summarized in **Table 16.1**.

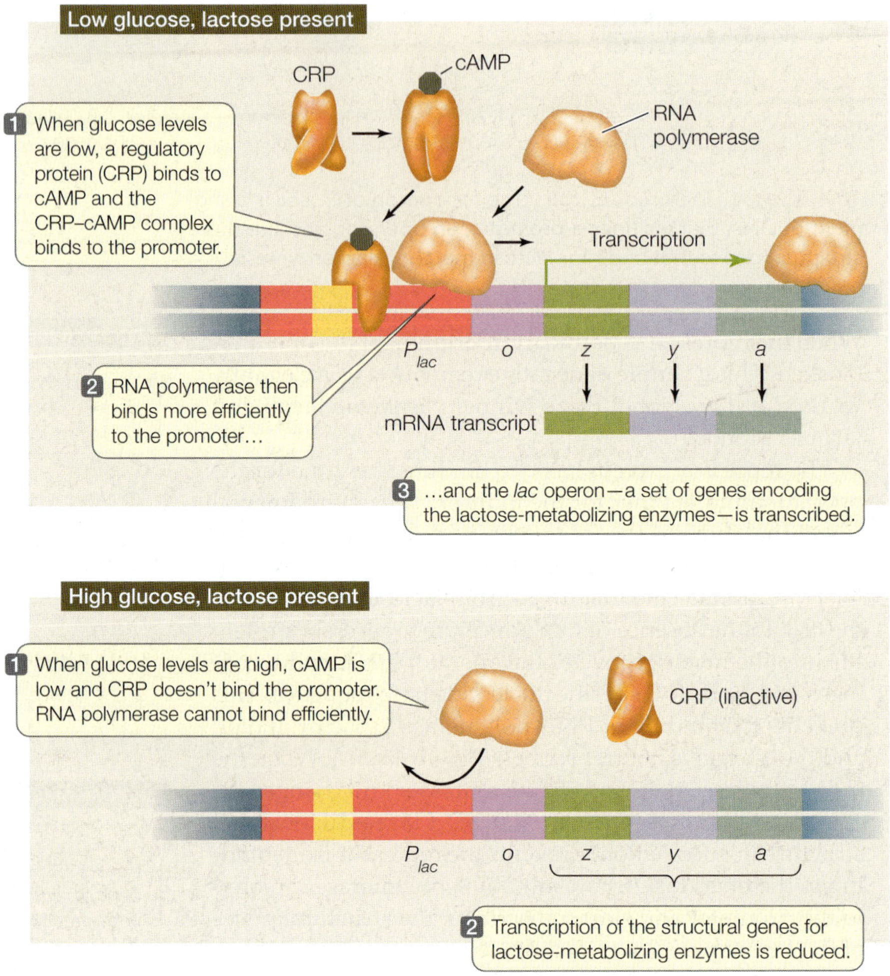
Low glucose, lactose present
CRP
cAMP
RNA polymerase
1 When glucose levels are low, a regulatory protein (CRP) binds to cAMP and the CRP–cAMP complex binds to the promoter.
Transcription
P_{lac}
o
z
y
a
2 RNA polymerase then binds more efficiently to the promoter...
mRNA transcript
3 ...and the *lac* operon—a set of genes encoding the lactose-metabolizing enzymes—is transcribed.
High glucose, lactose present
1 When glucose levels are high, cAMP is low and CRP doesn't bind the promoter. RNA polymerase cannot bind efficiently.
CRP (inactive)
P_{lac}
o
z
y
a
2 Transcription of the structural genes for lactose-metabolizing enzymes is reduced.

16.6 Catabolite Repression Regulates the *lac* Operon The promoter for the *lac* operon does not function efficiently in the absence of cAMP, as occurs when glucose levels are high. High glucose levels thus repress the enzymes that metabolize lactose.

RNA polymerases can be directed to particular classes of promoters

Thus far we have described a promoter as a specific DNA sequence located upstream of a transcription initiation site. It binds and orients RNA polymerase so that the polymerase transcribes the correct DNA strand. As we mentioned in Chapter 14, not all promoters are identical. However, they share **consensus sequences** that allow them to be recognized by the RNA polymerase and other proteins. (Recall from Section 14.4 that a consensus sequence is a short stretch of DNA that appears, with little variation, in many different genes.)

Prokaryotic promoters generally have two sites for consensus sequences, which begin 10 and 35 base pairs upstream of the transcription start site (the –10 element and the –35 element). Different classes of promoters have different consensus sequences at these two sites. The largest class consists of promoters for "housekeeping genes," which are all the genes that are normally expressed in actively growing cells. In these genes, the consensus –10 element is 5′-TATAAT-3′, and the

TABLE 16.1

Positive and Negative Regulation in the *lac* Operon

Glucose	cAMP Levels	RNA Polymerase Binding to Promoter	Lactose	*lac* Repressor	Transcription of *lac* Genes?	Lactose Used by Cells?
Present	Low	Absent	Absent	Active and bound to operator	No	No
Present	Low	Present, not efficient	Present	Inactive and not bound to operator	Low level	No
Absent	High	Present, very efficient	Present	Inactive and not bound to operator	High level	Yes
Absent	High	Absent	Absent	Active and bound to operator	No	No

consensus –35 element is 5′-TTGACAT-3′ (N stands for any nucleotide):

```
5′-NNNTTGACATNNNNNNNNNNNNNNNNNTATAATNNNN*NNNNNNNNNNNN-3′
3′-NNNAACTGTANNNNNNNNNNNNNNNNNATATTANNNN*NNNNNNNNNNNN-5′
     –35 Element                 –10 Element   Transcription start site
```

Other classes of genes have different consensus sequences at their –10 and –35 sites. The different classes of consensus sequences are recognized by regulatory proteins called sigma factors.

Sigma factors are proteins in prokaryotic cells that bind to RNA polymerase and direct it to specific classes of promoters. The RNA polymerase must be bound to a sigma factor before it can recognize a promoter and begin transcription. Genes that encode proteins with related functions may be at different locations in the genome, but because they share consensus sequences in their promoters, they are recognized by a particular sigma factor. The sigma-70 factor is active most of the time and binds to the consensus sequences of housekeeping genes; other sigma factors are activated only under specific conditions. For example, when *E. coli* cells experience stress conditions such as DNA damage or osmotic stress, the sigma-38 factor is activated, and it directs RNA polymerase to the promoters of various genes that are expressed under stress conditions. *E. coli* has seven sigma factors; this number varies in other prokaryotes.

As we will see in the next section, this form of global gene regulation by proteins binding to RNA polymerase is also common in eukaryotes.

RECAP 16.1

Gene expression in prokaryotes is most commonly regulated through control of transcription. An operon consists of a set of closely linked structural genes and the DNA sequences (promoter and operator) that control their transcription. Operons can be regulated by both negative and positive controls. Sigma factors control the expression of specific classes of prokaryotic genes that share consensus sequences in their promoters.

- What is the difference between positive and negative regulation of gene expression? **See Figure 16.1 and Table 16.1**
- Describe the molecular conditions at the *lac* operon promoter in the presence versus absence of lactose. **See p. 331 and Figure 16.5**
- What are the key differences between an inducible system and a repressible system? **See p. 332**
- How do sigma factors and consensus sequences act to affect the expressions of classes of genes? **See pp. 332–333**

Studies of bacteria have provided a basic understanding of mechanisms that regulate gene expression and of the roles of regulatory proteins in both positive and negative regulation. We will now turn to the transcriptional control of gene expression in eukaryotes.

16.2 How Is Eukaryotic Gene Transcription Regulated?

For the normal development of a multicellular organism from fertilized egg to adult, and for each cell to acquire and maintain its proper specialized function, certain proteins must be made at just the right times and in just the right cells; these proteins must not be made at other times in other cells. Here are two examples from humans:

- In human pancreatic exocrine cells, the digestive enzyme procarboxypeptidase A makes up 7.6 percent of all the protein in the cell; in other cell types it is usually undetectable.
- In human breast duct cells, alpha-lactalbumin, a protein in breast milk, is made only late in pregnancy and during lactation. Alpha-lactalbumin is not made in any other cell types.

Clearly the expression of eukaryotic genes must be precisely regulated.

As in prokaryotes, eukaryotic gene expression can be regulated at several different points in the process of transcribing and translating the gene into a protein (**Figure 16.7**). In this section we will describe the mechanisms that result in the selective transcription of specific genes. The mechanisms for regulating gene expression in eukaryotes have similar themes as in those of prokaryotes. Both types of cells use DNA–protein interactions and negative and positive control. However, there are many differences, some of them dictated by the presence of a nucleus, which physically separates transcription and translation (**Table 16.2**).

General transcription factors act at eukaryotic promoters

As in prokaryotes, a promoter in a eukaryotic gene is a sequence of DNA near the 5′ end of the coding region, where RNA polymerase binds and initiates transcription. Although eukaryotic promoters are much more diverse than those of prokaryotes, many contain an element similar to the –10 element in prokaryotic promoters. This element is usually located close to the transcription start site and is called the **TATA box** because it

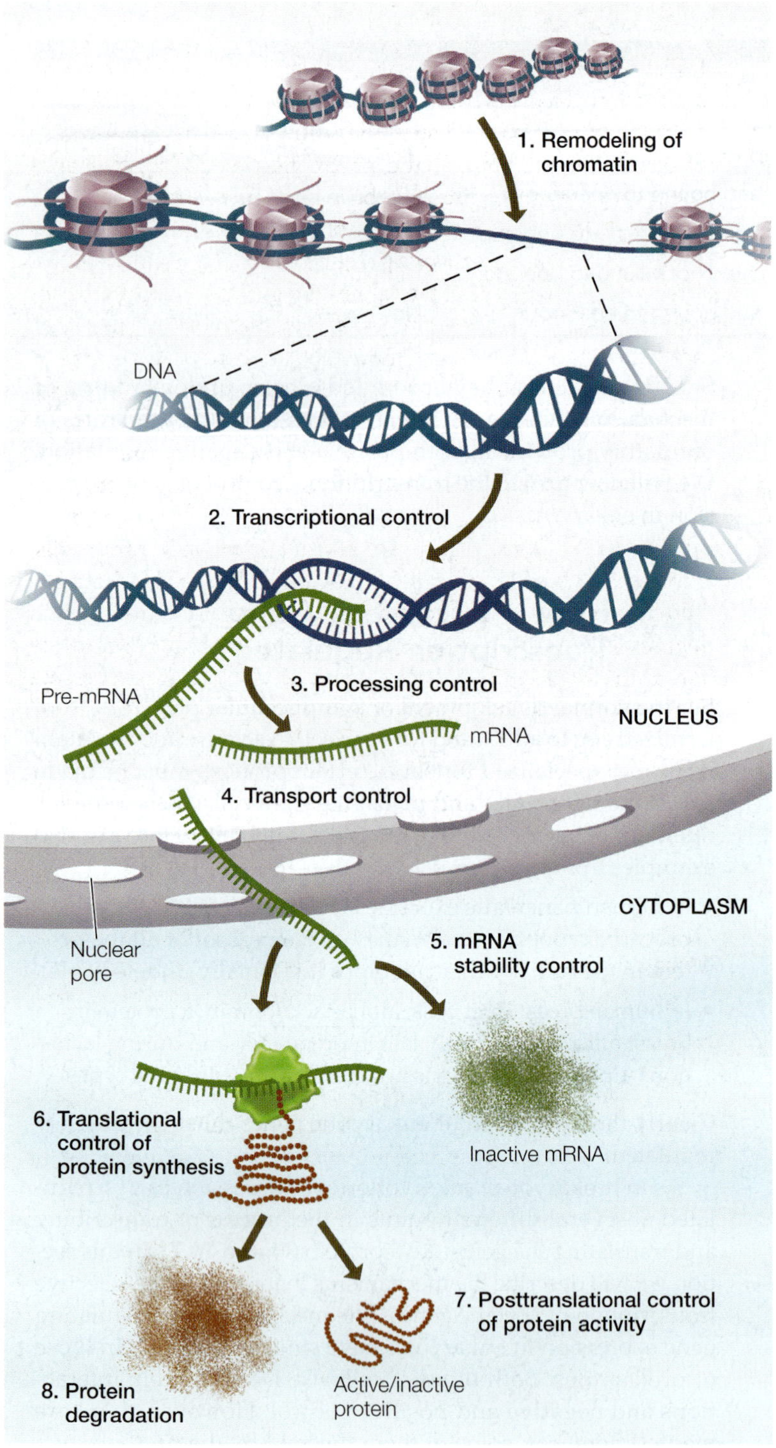
1. Remodeling of chromatin
DNA
2. Transcriptional control
Pre-mRNA
3. Processing control
mRNA
NUCLEUS
4. Transport control
CYTOPLASM
Nuclear pore
5. mRNA stability control
6. Translational control of protein synthesis
Inactive mRNA
7. Posttranslational control of protein activity
8. Protein degradation
Active/inactive protein

16.7 Potential Points for the Regulation of Gene Expression
Gene expression can be regulated before transcription (1), during transcription (2, 3), after transcription but before translation (4, 5), at translation (6), or after translation (7).

Go to Activity 16.1 Eukaryotic Gene Expression Control Points
Life10e.com/ac16.1

is rich in AT base pairs. The TATA box is the site where DNA begins to denature so that the template strand can be exposed. In addition to the TATA box, eukaryotic promoters typically include multiple regulatory sequences that are recognized and

TABLE 16.2 Transcription in Prokaryotes and Eukaryotes

	Prokaryotes	Eukaryotes
Locations of functionally related genes	Often clustered in operons	Often distant from one another with separate promoters
RNA polymerases	One	Three: I transcribes rRNA II transcribes mRNA III transcribes tRNA and small RNAs
Promoters and other regulatory sequences	Few	Many
Initiation of transcription	Binding of RNA polymerase to promoter	Binding of many proteins, including RNA polymerase

bound by **transcription factors**: regulatory proteins that help control transcription.

Like the prokaryotic RNA polymerase, eukaryotic RNA polymerase II cannot simply bind to the promoter and initiate transcription. Rather, it does so only after various **general transcription factors** have assembled on the chromosome (Figure 16.8). First, the protein TFIID ("TF" stands for transcription factor) binds to the TATA box. Binding of TFIID changes both its own shape and that of the DNA, presenting a new surface that attracts the binding of other general transcription factors to form a transcription initiation complex. RNA polymerase II binds only after several other proteins have bound to this complex.

Each general transcription factor has a role in gene expression:

- TFIIB binds both DNA polymerase and TFIID, and helps identify the transcription initiation site.
- TFIIF prevents nonspecific binding of the complex to DNA and helps recruit RNA polymerase to the complex; it is similar in function to a bacterial sigma factor.
- TFIIE binds to the promoter and stabilizes the denaturation of the DNA.
- TFIIH opens up the DNA for transcription.

Some regulatory DNA sequences, such as the TATA box, are common to the promoters of many eukaryotic genes and are recognized by general transcription factors that are found in all the cells of an organism. Other regulatory sequences are present in only a few genes and are recognized by specific transcription factors. These factors may be found only in certain types of cells or at certain stages of the cell cycle, or they may be activated by signaling pathways in response to cellular or environmental signals (see Chapter 7). Specific transcription factors play an important role in cell differentiation—the structural and functional specialization of cells during development.

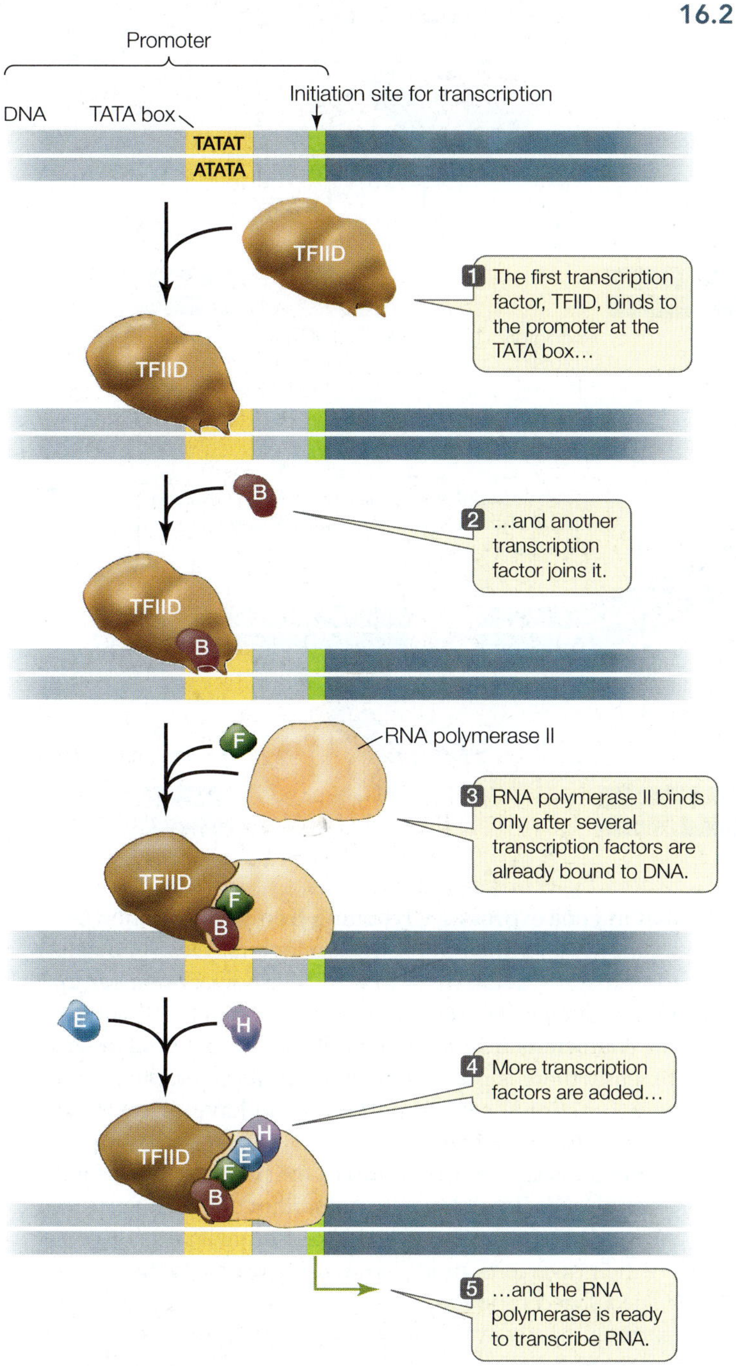
Promoter
Initiation site for transcription
DNA
TATA box
TATAT
ATATA
TFIID
1 The first transcription factor, TFIID, binds to the promoter at the TATA box...
B
2 ...and another transcription factor joins it.
F
RNA polymerase II
3 RNA polymerase II binds only after several transcription factors are already bound to DNA.
E
H
4 More transcription factors are added...
5 ...and the RNA polymerase is ready to transcribe RNA.

16.8 The Initiation of Transcription in Eukaryotes Apart from TFIID, which binds to the TATA box, each general transcription factor in this transcription complex has binding sites only for the other proteins in the complex, and does not bind directly to DNA. B, E, F, and H are general transcription factors.

Go to Animated Tutorial 16.3
Initiation of Transcription
Life10e.com/at16.3

Specific proteins can recognize and bind to DNA sequences and regulate transcription

Some regulatory DNA sequences are positive elements termed **enhancers**. They bind transcription factors that either activate transcription or increase the rate of transcription. Other regulatory elements are **silencers**: they bind factors that repress transcription. Most of the regulatory elements needed for correct expression of a gene are found within a few hundred base pairs of the transcription start site. For example, the mouse albumin gene promoter contains all the information needed for liver cell–specific expression within 170 base pairs upstream of the transcription start site. Other regulatory elements may be located thousands of base pairs away (as far as 20,000), and they may affect the expression of several nearby genes. When transcription factors bind to these elements, they interact with the RNA polymerase complex, causing the DNA to bend (**Figure 16.9**).

Often many transcription factors are involved, and *the combination of factors present determines the rate of transcription*. For example, the immature red blood cells in bone marrow make large amounts of β-globin. At least 13 different transcription factors are involved in regulating transcription of the β-globin gene in these cells. Not all of these factors are present or active in other cells, such as the immature white blood cells produced by the same bone marrow. As a result, the β-globin gene is not transcribed in those cells. So although the same genes are present in all cells, the fate of the cell is determined by which of its genes are expressed. How do transcription factors recognize specific DNA sequences?

Specific protein–DNA interactions underlie binding

As we have seen, transcription factors with specific DNA-binding domains are involved in the activation and inactivation of specific genes. There are several common structural themes in the protein domains that bind to DNA. These themes, or **structural motifs**, consist of different combinations of structural elements (protein conformations) and may include special components such as zinc. One of the common structural motifs is the helix-turn-helix, in which two α helices are connected via a nonhelical turn. The interior-facing "recognition" helix interacts with the bases inside the DNA. The exterior-facing helix sits on the sugar–phosphate backbone, ensuring that the interior helix is presented to the bases in the correct configuration:

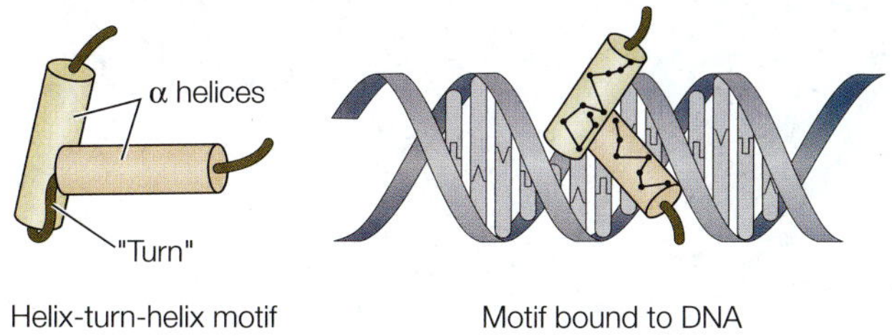
α helices
"Turn"
Helix-turn-helix motif
Motif bound to DNA

How does a protein recognize a sequence in DNA? As pointed out in Section 13.2, the complementary bases in DNA not only form hydrogen bonds with each other, but also can form additional hydrogen bonds with proteins, particularly at points exposed in the major and minor grooves. In this way, an intact DNA double helix can be recognized by a protein motif whose structure:

- fits into the major or minor groove;
- has amino acids that can project into the interior of the double helix; and
- has amino acids that can form hydrogen bonds with the interior bases.

16.9 Transcription Factors and Transcription Initiation The actions of many proteins determine whether and where RNA polymerase II will transcribe DNA.

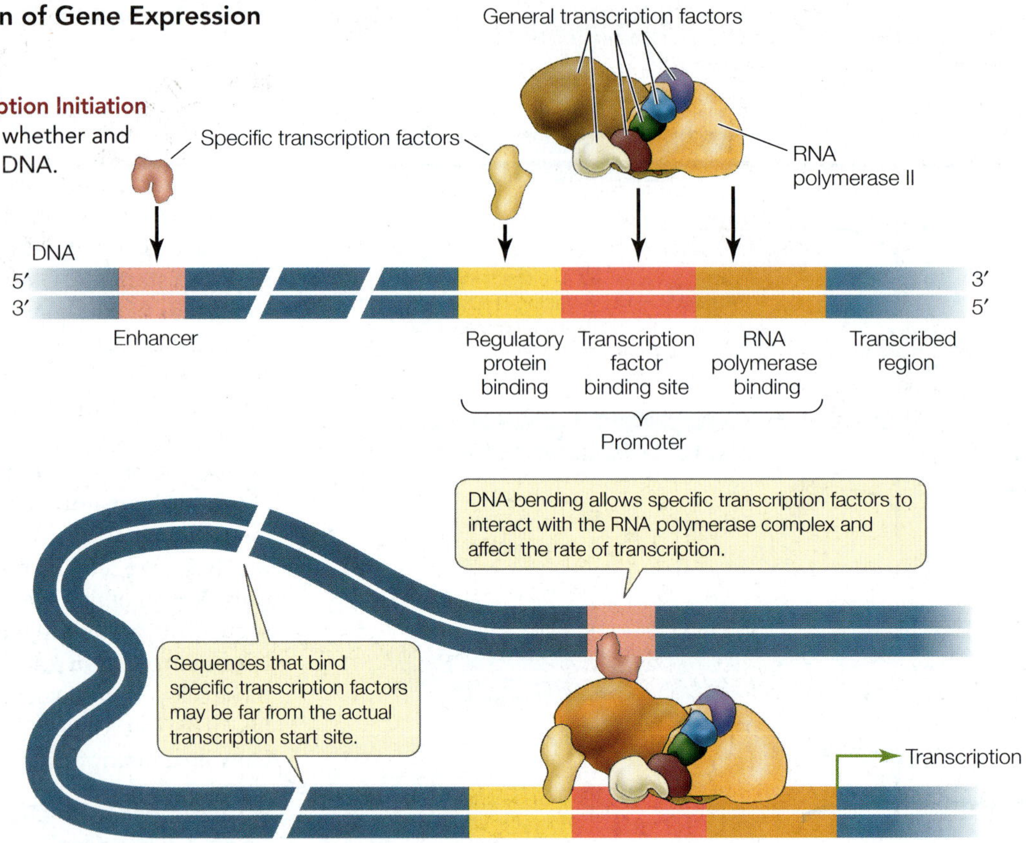
General transcription factors
Specific transcription factors
RNA polymerase II
DNA
5′
3′
3′
5′
Enhancer
Regulatory protein binding
Transcription factor binding site
RNA polymerase binding
Transcribed region
Promoter
DNA bending allows specific transcription factors to interact with the RNA polymerase complex and affect the rate of transcription.
Sequences that bind specific transcription factors may be far from the actual transcription start site.
Transcription

The helix-turn-helix motif fits these three criteria. Many repressor proteins, including the bacterial *lac* repressor, have this helix-turn-helix motif in their structure:

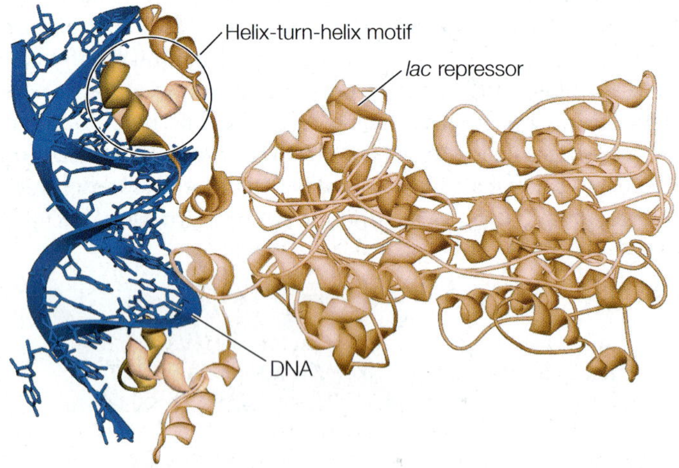
Helix-turn-helix motif
lac repressor
DNA

These proteins often form dimers, as shown here (one molecule is shown in white, the other in gray).

Repressors can inhibit transcription in several different ways. They can prevent the binding of transcriptional activators to DNA, or they can interact with other DNA-binding proteins to decrease the rate of transcription.

The expression of transcription factors underlies cell differentiation

During the development of a complex organism from fertilized egg to adult, cells become more and more differentiated (specialized). Differentiation is mediated in many cases by changes in gene expression, resulting from the activation (and inactivation) of various transcription factors. We will discuss this topic in more detail in Chapter 19. For now, remember that all differentiated cells contain the entire genome, and that their specific characteristics arise from differential gene expression.

Currently there is great interest in cellular therapy: providing new, functional cells to patients who have diseases that involve the degeneration of certain cell types. An example is Alzheimer's disease, which involves the degeneration of neurons in the brain. Because of the possibility of immune system rejection (see Chapter 42), it would be optimal if patients could receive their own cells, modified in some way to be functional. Since specialized functions are mediated by transcription factors, turning readily available cells into a particular desired cell type might be achieved by altering transcription factor expression. Marius Wernig and his colleagues at Stanford University have made important progress toward this goal (**Figure 16.10**). They took skin fibroblasts from mice and manipulated the expression of transcription factors in the cells to change them into neurons. By repeating their experiments on human fibroblasts, they have brought cellular therapy closer to reality.

The expression of sets of genes can be coordinately regulated by transcription factors

How do eukaryotic cells coordinate the regulation of several genes whose transcription must be turned on at the same time? Prokaryotes solve this problem by arranging multiple genes in an operon that is controlled by a single promoter, and by using sigma factors to recognize particular classes of promoters. Most eukaryotic genes have their own separate promoters, and genes that are coordinately regulated may be far apart. In these

INVESTIGATING**LIFE**

16.10 Expression of Specific Transcription Factors Turns Fibroblasts into Neurons Fibroblasts are cells that secrete abundant extracellular matrix and contribute to the structural integrity of organs. Neurons are highly specialized cells in the nervous system. Marius Wernig and his colleagues performed a series of experiments to find out whether expressing neuronal transcription factors in fibroblasts would be sufficient to cause the fibroblasts to become neurons.[a]

HYPOTHESIS Expression of neuron-specific transcription factors in fibroblasts will turn the latter into neurons.

Method

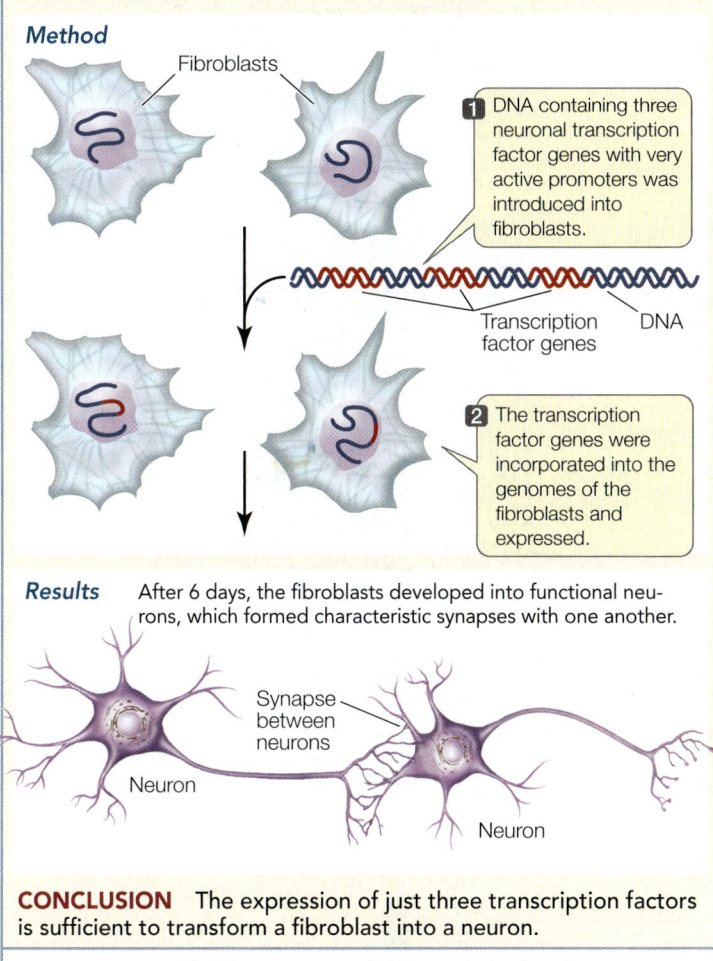

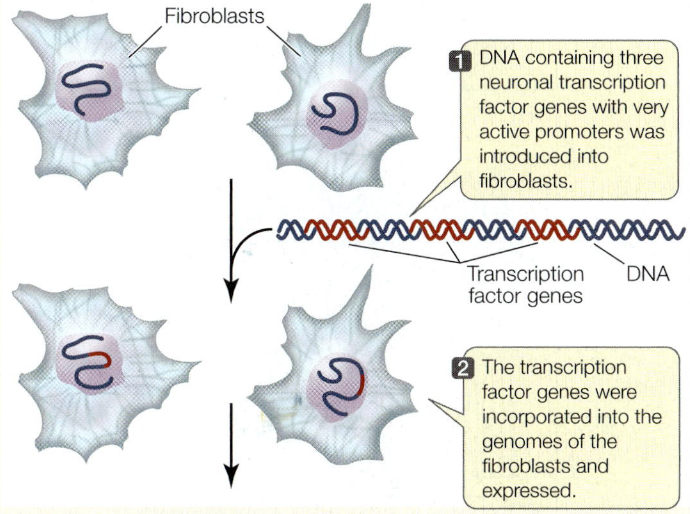
Fibroblasts
1 DNA containing three neuronal transcription factor genes with very active promoters was introduced into fibroblasts.
Transcription factor genes
DNA
2 The transcription factor genes were incorporated into the genomes of the fibroblasts and expressed.

Results After 6 days, the fibroblasts developed into functional neurons, which formed characteristic synapses with one another.

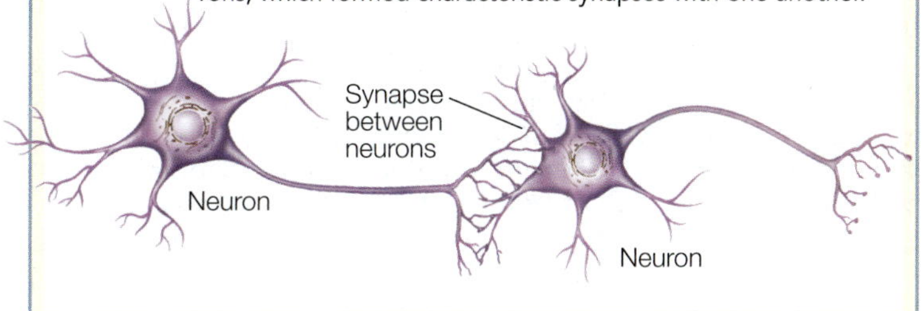
Synapse between neurons
Neuron
Neuron

CONCLUSION The expression of just three transcription factors is sufficient to transform a fibroblast into a neuron.

Go to **BioPortal** for discussion and relevant links for all INVESTIGATING**LIFE** figures.

[a]Vierbuchen, T. et al. 2010. *Nature* 463: 1035–1041.

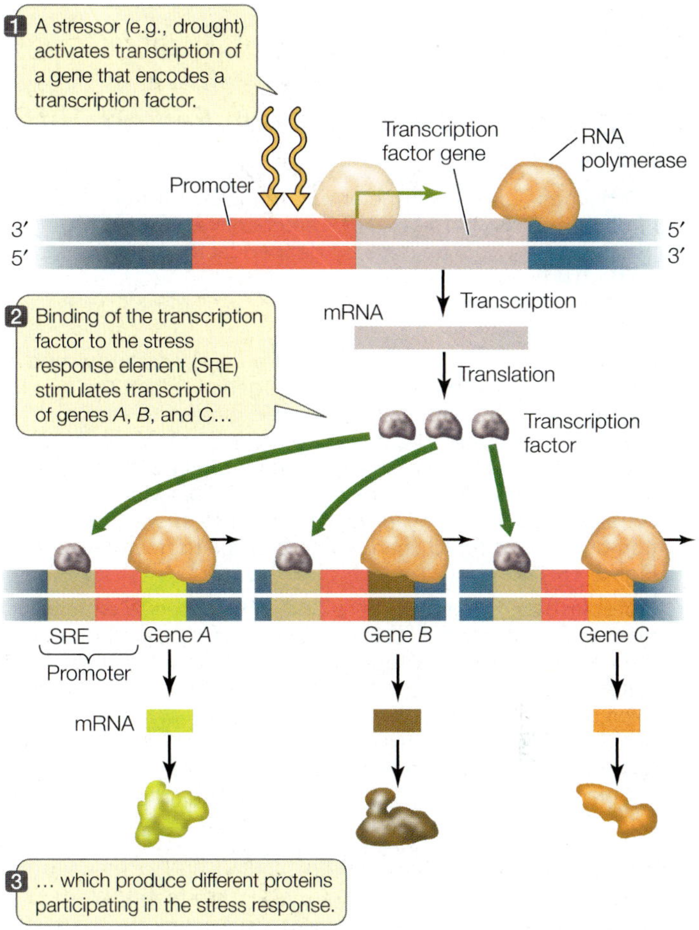
1 A stressor (e.g., drought) activates transcription of a gene that encodes a transcription factor.
Transcription factor gene
RNA polymerase
Promoter
3′
5′
5′
3′
2 Binding of the transcription factor to the stress response element (SRE) stimulates transcription of genes A, B, and C...
mRNA
Transcription
Translation
Transcription factor
SRE
Promoter
Gene A
Gene B
Gene C
mRNA
3 ... which produce different proteins participating in the stress response.

16.11 Coordinating Gene Expression A single environmental signal, such as drought stress, causes the synthesis of a transcription factor that acts on many genes.

cases, the expression of genes can be coordinated if they share regulatory sequences that bind the same transcription factors.

This type of coordination is used by organisms to respond to stress—for example, by plants in response to drought. Under conditions of drought stress, a plant must simultaneously synthesize several proteins whose genes are scattered throughout the genome. The synthesis of these proteins comprises the stress response. To coordinate expression, each of these genes has a specific regulatory sequence near its promoter called the stress response element (SRE). A transcription factor binds to this element and stimulates mRNA synthesis (**Figure 16.11**). The stress response proteins not only help the plant conserve water, but also protect the plant against excess salt in the soil and freezing. This finding has considerable importance for agriculture because crops are often grown under less than optimal conditions or are affected by weather.

RECAP 16.2

A number of general transcription factors must bind to a eukaryotic promoter before RNA polymerase will bind to it and begin transcription. Other, specific transcription factors bind to regulatory DNA sequences and interact with the RNA polymerase complex to control differential gene expression. This provides several ways to increase or decrease transcription.

- Describe some of the different ways in which transcription factors regulate gene transcription. **See pp. 334–336 and Figure 16.9**
- How can more than one gene be regulated at the same time? **See pp. 336–337 and Figure 16.11**

We have seen how prokaryotes and eukaryotes regulate the transcription of their genes and operons. In the next section we will see how prokaryotic and eukaryotic viruses are able to hijack transcription mechanisms in order to complete their life cycles.

WORKING WITH**DATA:**

Expression of Transcription Factors Turns Fibroblasts into Neurons

Original Paper

Vierbuchen, T., A. Ostermeier, Z. P. Pang, Y. Kokubu, T. C. Südhof, and M. Wernig. 2010. Direct conversion of fibroblasts to functional neurons by defined factors. *Nature* 463: 1035–1041.

Analyze the Data

Different cell types in an organism are usually distinguished by differences in the expression of their genes. When the regulatory regions of differentially expressed genes are analyzed, they are found to bind different transcription factors. Would expression of a specific set of transcription factors be sufficient to change one cell type into another? Marius Wernig and his colleagues at Stanford University set out to answer this question by attempting to change fibroblasts into neurons (see Figure 16.10).

Fibroblasts occur widely in the body. They secrete extracellular matrix materials and infiltrate various organs, contributing to the organs' structural integrity. These cells divide actively when cultured in the laboratory. Neurons have highly specialized roles in the nervous system (see Chapter 45) and do not divide in culture. Wernig and his colleagues found 19 transcription factors that were strongly expressed in mouse neurons and not in fibroblasts. When five of these transcription factors—Ascl1, Brn2, Mytl1, Zic1, and Olig2—were introduced into fibroblasts and expressed from very strong promoters, the fibroblasts became neurons. Subsequent experiments showed that a combination just three of these—Ascl1, Brn2, and Mytl1—was sufficient to cause efficient transformation. In a further study, the researchers were able to repeat these experiments with human cells, using four transcription factors to convert fibroblasts into neurons. Since fibroblasts are easily isolated from human skin, the possibility of creating neurons from them that can be used for cellular therapy in diseases where neurons degenerate (e.g., Alzheimer's disease) is exciting.

Three main criteria were used to determine that the transformed cells were neurons: morphology, electrical excitability, and lack of cell division. The transformed cells clearly showed the distinctive cellular architecture of neurons (see Figure 45.1). Wernig and his colleagues conducted additional experiments to test for the other two properties of neurons in their transformed cells.

QUESTION 1:

As you will see in Chapter 45, neurons respond to electrical stimulation by generating an action potential. The electrical activity of a stimulated transformed fibroblast cell is shown in **FIGURE A**—8, 12, and 20 days after addition of the transcription factors. What is the magnitude of the action potential of the transformed cell in millivolts (mv)? Look up Figure 45.10. How does this compare?

QUESTION 2:

The rate of cell division in the population of transformed cells was measured by the incorporation of the labeled nucleotide BrdU into their DNA. The percentage of labeled—and hence dividing—cells is shown in **FIGURE B**. Did cell division stop in the transformed cells? Explain your answer.

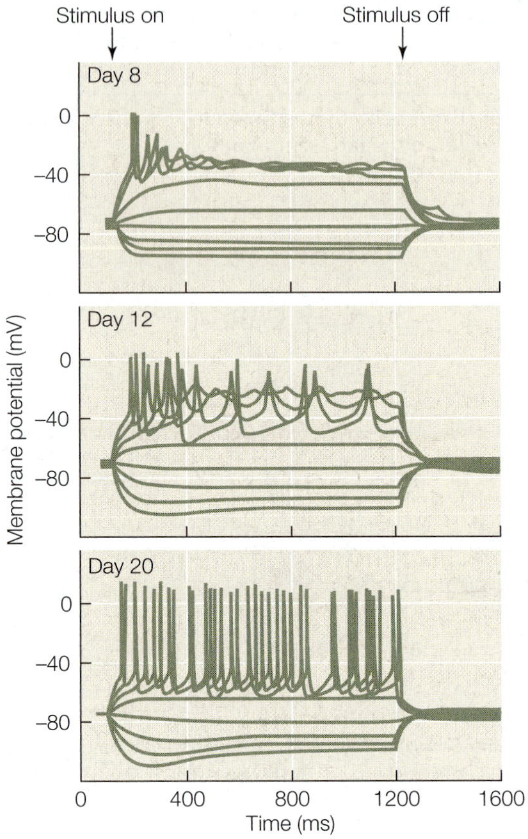
Stimulus on
Stimulus off
Day 8
Day 12
Day 20
0
–40
–80
Membrane potential (mV)
0
400
800
1200
1600
Time (ms)

FIGURE A Response of Transformed Cells to Electrical Stimulation The different traces show the cell's response to different amounts of stimulation, some of which were too small to trigger the production of action potentials (upward spikes).

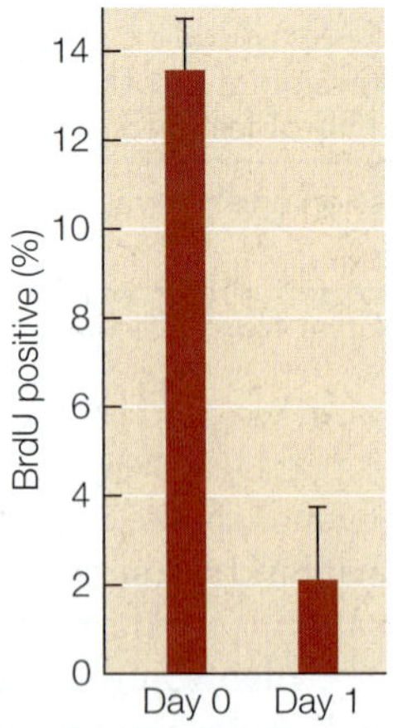
BrdU positive (%)
0
2
4
6
8
10
12
14
Day 0
Day 1

FIGURE B Cell Division Labeled BrdU was added to the transformed cells at the time transcription factors were added (Day 0) or one day later (Day 1). The number of labeled cells was assessed 13 days after the transcription factors were added.

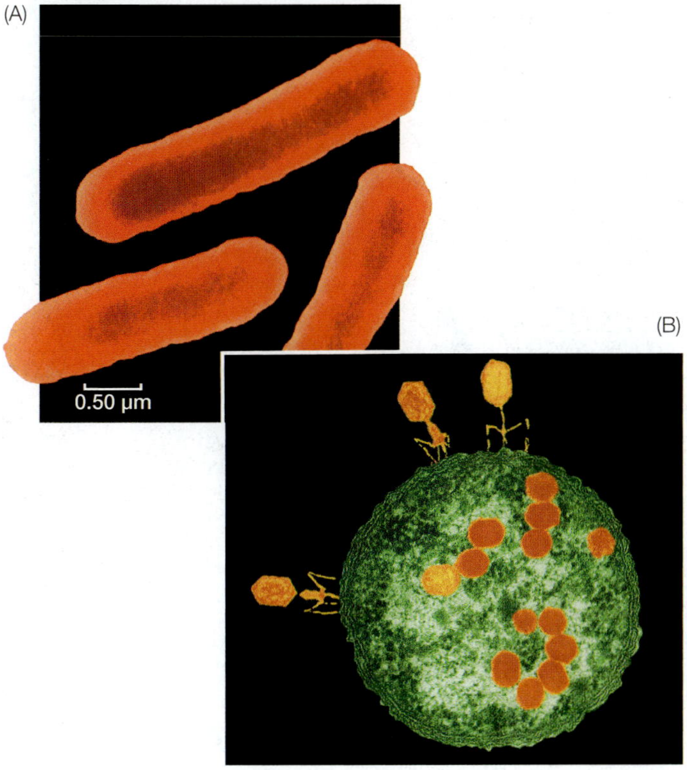
(A)
0.50 μm
(B)

16.12 Bacteriophage and Host (A) *E. coli* cells (viewed here from the side) are the host for bacteriophage T2. (B) Bacteriophages have attached to this *E. coli* cell, and the reproductive cycle is underway, producing new phage particles. The cell is viewed in transverse section.

16.3 How Do Viruses Regulate Their Gene Expression?

"A virus is a piece of bad news wrapped in protein." This quote from immunologist Sir Peter Medawar is certainly true for the cells that viruses infect. As we described in Chapter 13, a bacterial virus (**bacteriophage**) injects its genetic material into a host cell and turns that cell into a virus factory (see Figure 13.3). Other viruses enter cells intact and then shed their coats and take over the cell's replication machinery. Viral life cycles can be very efficient. An example is the poliovirus: a single poliovirus infecting a mammalian cell can produce more than 100,000 new virus particles!

Viruses are small infectious agents that infect cellular organisms and that cannot reproduce outside their host cells. Most virus particles, called **virions**, consist of only two or three components: the genetic material made up of DNA or RNA, a protein coat that protects the genetic material, and in some cases, an envelope of lipids that surrounds the protein coat. As we will see in this section, viral genomes include sequences that encode regulatory proteins. These proteins "hijack" the host cells' transcriptional machinery, allowing the viruses to complete their life cycles.

Many bacteriophages undergo a lytic cycle

The Hershey–Chase experiment (see Figure 13.4) involved a typical **lytic** viral reproductive cycle, so named because soon after infection, the host cell bursts (lyses), releasing progeny viruses. In this cycle, the viral genetic material takes over the host's synthetic machinery for its own reproduction immediately after infection. In the case of some bacteriophages, the process is extremely rapid—within 15 minutes, new phage particles appear in the bacterial cell (**Figure 16.12**). Ten minutes later, the "game is over," and these particles are released from the lysed cell. What happens during this rapid life cycle?

At the molecular level, the reproductive cycle of a typical lytic virus has two stages: early and late, as illustrated in **Figure 16.13**. The reproductive cycle involves both positive

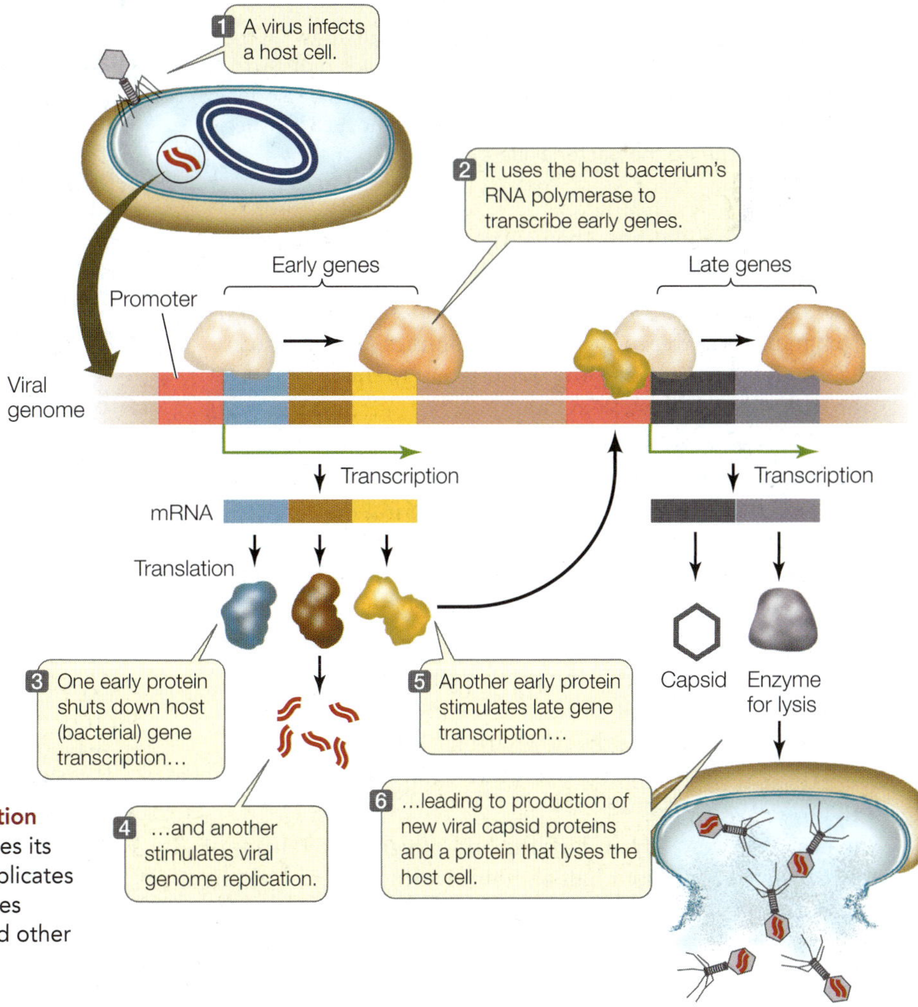
1 A virus infects a host cell.
2 It uses the host bacterium's RNA polymerase to transcribe early genes.
Early genes
Late genes
Promoter
Viral genome
Transcription
Transcription
mRNA
Translation
3 One early protein shuts down host (bacterial) gene transcription...
4 ...and another stimulates viral genome replication.
5 Another early protein stimulates late gene transcription...
Capsid
Enzyme for lysis
6 ...leading to production of new viral capsid proteins and a protein that lyses the host cell.

16.13 The Lytic Cycle: A Strategy for Viral Reproduction In a host cell infected with a virus, the viral genome uses its early genes to shut down host transcription while it replicates itself. Once the viral genome is replicated, its late genes produce capsid proteins that package the genome and other proteins that lyse the host cell.

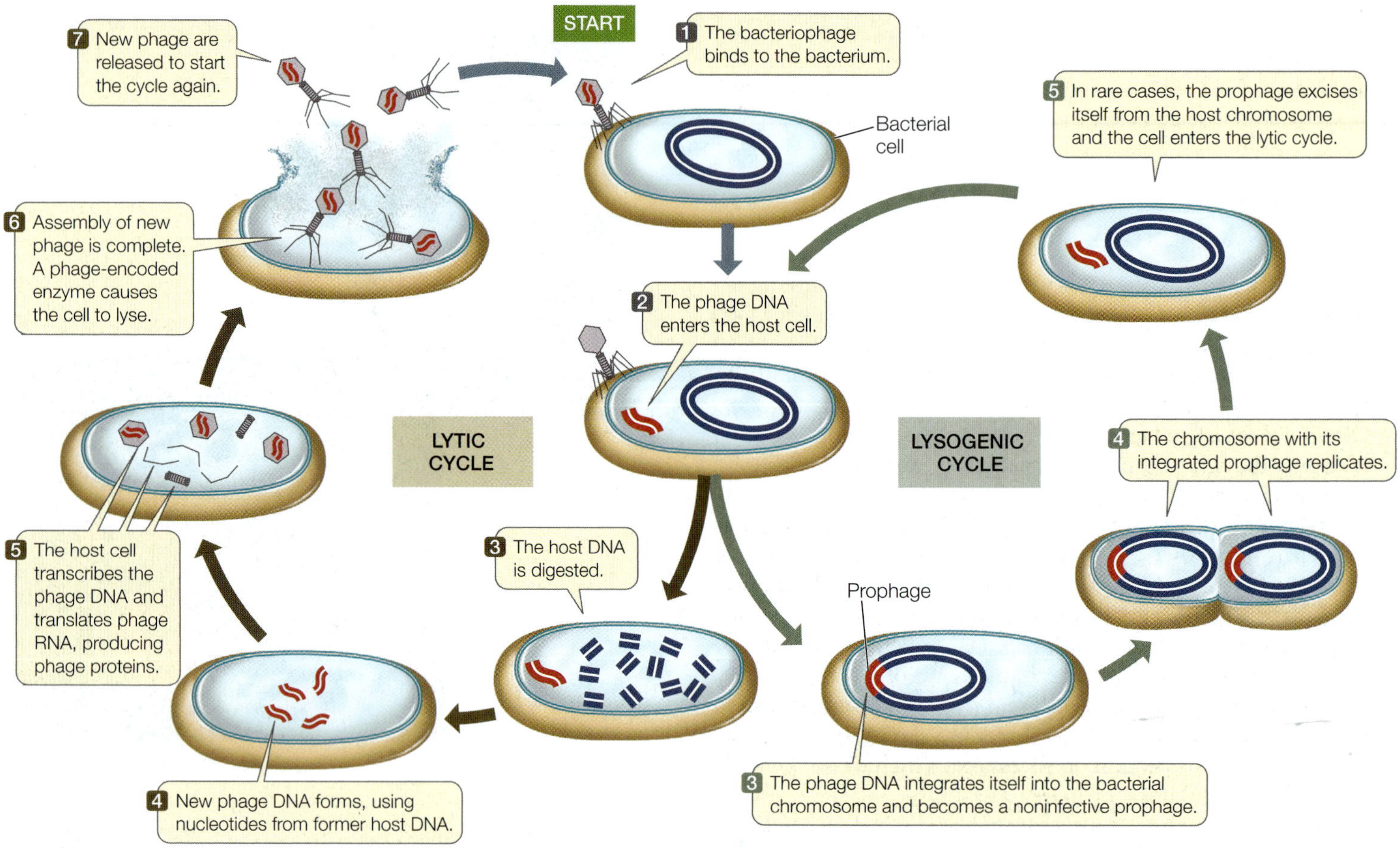

START
1 The bacteriophage binds to the bacterium.
Bacterial cell
2 The phage DNA enters the host cell.
LYTIC CYCLE
3 The host DNA is digested.
4 New phage DNA forms, using nucleotides from former host DNA.
5 The host cell transcribes the phage DNA and translates phage RNA, producing phage proteins.
6 Assembly of new phage is complete. A phage-encoded enzyme causes the cell to lyse.
7 New phage are released to start the cycle again.
LYSOGENIC CYCLE
Prophage
3 The phage DNA integrates itself into the bacterial chromosome and becomes a noninfective prophage.
4 The chromosome with its integrated prophage replicates.
5 In rare cases, the prophage excises itself from the host chromosome and the cell enters the lytic cycle.

16.14 The Lytic and Lysogenic Cycles of Bacteriophages In the lytic cycle, infection of a bacterium by viral DNA leads directly to multiplication of the virus and lysis of the host cell. In the lysogenic cycle, an inactive prophage is integrated into the host DNA where it is replicated during the bacterial life cycle.

and negative regulation, which stimulate and inhibit, respectively, gene expression:

- The viral genome contains a promoter that binds host RNA polymerase. In the early stage (1–2 minutes after phage DNA entry), viral genes that lie adjacent to this promoter are transcribed (positive regulation).
- These early genes often encode proteins that shut down host transcription (negative regulation) and stimulate viral genome replication and transcription of viral late genes (positive regulation). Three minutes after DNA entry, viral nuclease enzymes digest the host's chromosome, providing nucleotides for the synthesis of viral genomes.
- In the late stage, viral late genes are transcribed (positive regulation); they encode the proteins that make up the **capsid** (the outer shell of the virus) and enzymes that lyse the host cell to release the new virions. This begins 9 minutes after DNA entry and 6 minutes before the first new phage particles appear.

The entire process—from binding and infection to release of new phage—takes about half an hour. During this period, the sequence of transcriptional events is carefully controlled to produce complete, infective virions.

Some bacteriophages can undergo a lysogenic cycle

Like all nucleic acid genomes, those of viruses can mutate and evolve by natural selection. Some viruses have evolved an advantageous process called **lysogeny** that postpones the lytic cycle. In lysogeny, the viral DNA becomes integrated into the host DNA and becomes a **prophage** (Figure 16.14). As the host cell divides, the viral DNA gets replicated along with that of the host. The prophage can remain inactive within the bacterial genome for thousands of generations, producing many copies of the original viral DNA.

However, if the host cell is not growing well, the virus "cuts its losses." It switches to a lytic cycle, in which the prophage excises itself from the host chromosome and reproduces. In other words, the virus is able to enhance its chances of multiplication and survival by inserting its DNA into the host chromosome, where it sits as a silent passenger until conditions are right for lysis.

Uncovering the regulation of gene expression that underlies the lysis/lysogeny switch was a major achievement of molecular biologists. Here we present just an outline of the process to give you an idea of the positive and negative regulatory mechanisms involved (Figure 16.15). The model virus bacteriophage λ (lambda) has been used extensively to study the lysogenic mechanism.

So how does a prophage "know" when to switch to the lytic cycle? The virus genome includes genes that encode the regulatory proteins cI and Cro, which compete for specific promoters

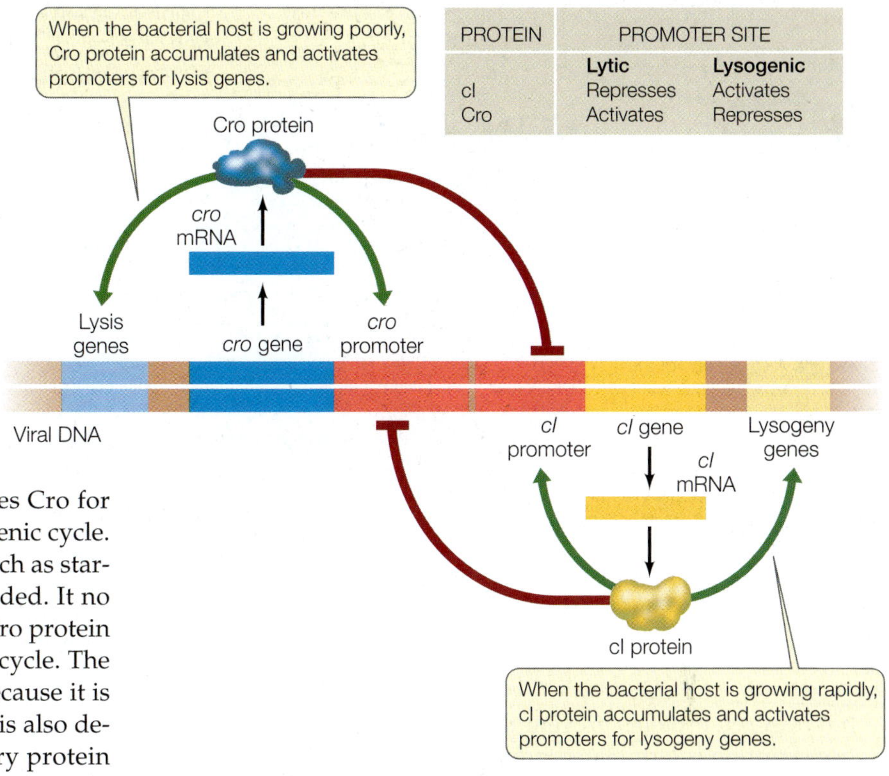
When the bacterial host is growing poorly, Cro protein accumulates and activates promoters for lysis genes.
PROTEIN
PROMOTER SITE
Lytic
Lysogenic
cI
Represses
Activates
Cro
Activates
Represses
Cro protein
cro mRNA
Lysis genes
cro gene
cro promoter
Viral DNA
cI promoter
cI gene
cI mRNA
Lysogeny genes
cI protein
When the bacterial host is growing rapidly, cI protein accumulates and activates promoters for lysogeny genes.

16.15 Control of Bacteriophage λ Lysis and Lysogeny Two regulatory proteins, Cro and cI, compete to control expression of one another and genes for viral lysis and lysogeny.

on the viral DNA—including their own promoters (see Figure 16.15). Cro and cI have opposite effects on each promoter: cI represses the expression of genes involved in the lytic cycle and promotes expression of genes involved in lysogeny, whereas Cro has the opposite effect.

The outcome of a bacteriophage infection depends on the relative abundance of these two regulatory proteins. Under favorable bacterial growth conditions, cI accumulates and outcompetes Cro for DNA binding, and the bacteriophage enters a lysogenic cycle. But if the host cell undergoes stressful conditions such as starvation, exposure to toxins, or radiation, cI is degraded. It no longer represses transcription of the *Cro* gene, the Cro protein is produced, and the bacteriophage enters the lytic cycle. The cI protein is degraded under stressful conditions because it is structurally similar to an *E. coli* protein (LexA) that is also degraded under such conditions. LexA is a regulatory protein that represses DNA repair mechanisms under normal conditions, but it is degraded by other proteins when the cell becomes stressed.

The reproductive cycle of bacteriophage λ is a paradigm for our understanding of viral life cycles in general. This system has served as a model to help us understand how the complicated reproductive cycles of other viruses, including HIV, are controlled.

Eukaryotic viruses can have complex life cycles

Eukaryotes are susceptible to infection by various kinds of viruses whose genomes may consist of RNA or DNA. A subgroup of RNA viruses are called retroviruses (see also Section 26.4):

- *DNA viruses.* Many viral particles contain double-stranded DNA. However, some contain single-stranded DNA, and a complementary strand is made after the viral genome enters the host cell. Like some bacteriophages, DNA viruses that infect eukaryotes are capable of undergoing both lytic and lysogenic life cycles. Examples include the herpes viruses and papillomaviruses (which cause warts).
- *RNA viruses.* Some viral genomes are made up of RNA that is usually, but not always, single-stranded. The RNA is translated by the host's machinery to produce viral proteins, some of which are involved in replication of the RNA genome. The influenza virus has an RNA genome.
- *Retroviruses.* As we described in Section 14.2, a **retrovirus** is an RNA virus that carries a gene for **reverse transcriptase**, a protein that synthesizes DNA from an RNA template. The retrovirus uses this protein to make a DNA copy of its genome, which then becomes integrated into the host genome. The integrated DNA acts as a template for both mRNA and new viral genomes. Human immunodeficiency virus (HIV) is a retrovirus that infects cells of the immune system and causes acquired immune deficiency syndrome (AIDS).

HIV gene regulation occurs at the level of transcription elongation

As we have discussed so far, many instances of gene regulation occur at the level of transcription *initiation*, involving both activator and repressor proteins that bind to the promoters of genes. However, studies of HIV and other viruses have revealed that transcription can also be controlled at the *elongation* stage.

HIV is an **enveloped virus**; it is enclosed within a phospholipid membrane derived from its host cells (a specific type of immune system cell) (**Figure 16.16**). During infection, proteins in this membrane interact with proteins on the host cell surface, and the viral envelope fuses with the host plasma membrane. After the virus enters the cell, its capsid is broken down. The viral reverse transcriptase then uses the virus's RNA template to produce a complementary DNA (cDNA) strand, while at the same time degrading the viral RNA. The enzyme then makes a complementary copy of the cDNA, and the resulting double-stranded DNA is inserted into the host's chromosome by the enzyme integrase. The integrated DNA is referred to as the **provirus**. Both the reverse transcriptase and the integrase are carried inside the HIV virion, along with other proteins needed at the very early stages of infection.

The provirus resides permanently in the host chromosome, and can remain in a latent (inactive) state for many years. During this time transcription of the viral DNA is initiated,

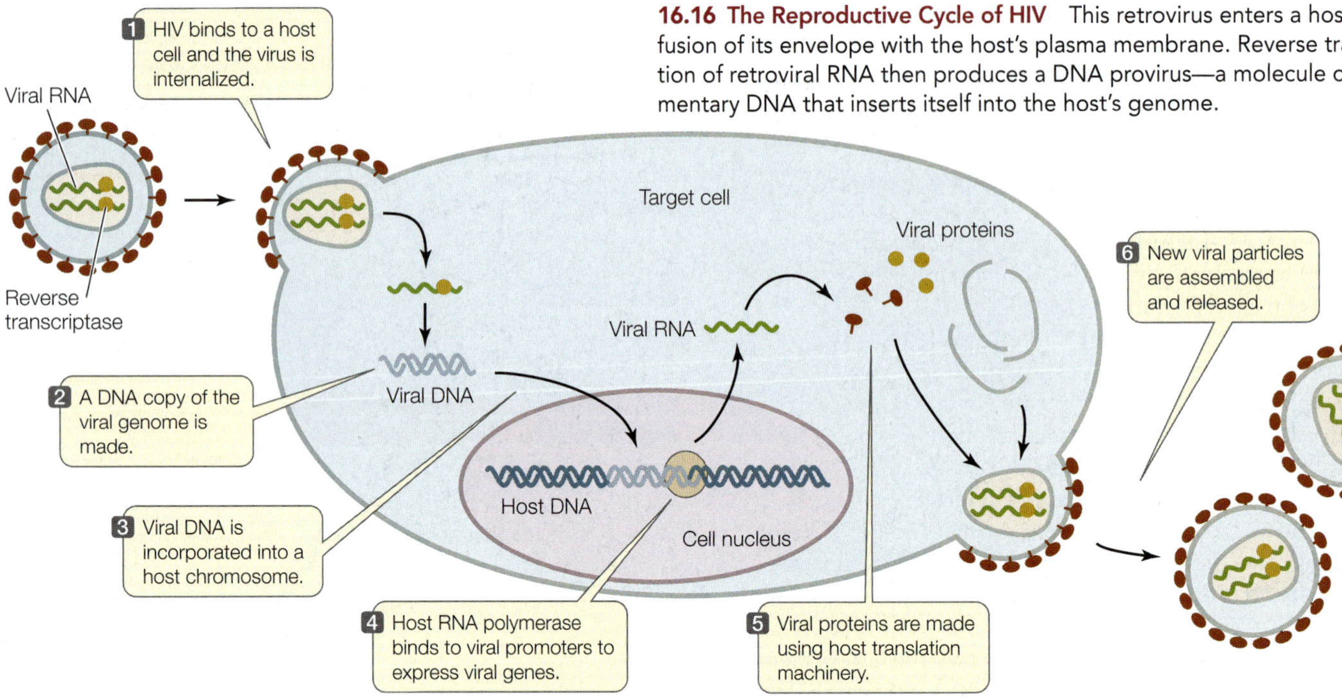

1 HIV binds to a host cell and the virus is internalized.
Viral RNA
Reverse transcriptase
Target cell
Viral proteins
Viral RNA
Viral DNA
2 A DNA copy of the viral genome is made.
Host DNA
Cell nucleus
3 Viral DNA is incorporated into a host chromosome.
4 Host RNA polymerase binds to viral promoters to express viral genes.
5 Viral proteins are made using host translation machinery.
6 New viral particles are assembled and released.

16.16 The Reproductive Cycle of HIV This retrovirus enters a host cell via fusion of its envelope with the host's plasma membrane. Reverse transcription of retroviral RNA then produces a DNA provirus—a molecule of complementary DNA that inserts itself into the host's genome.

but host cell proteins prevent the RNA from elongating, and transcription is terminated prematurely (**Figure 16.17A**). Under some circumstances, such as when the host immune cell is activated, the level of transcription initiation increases and some viral RNA is made. One of these viral genes encodes a protein called tat (*t*rans*a*ctivator of *t*ranscription), which binds to a stem-and-loop structure at the 5′ end of the viral RNA. As a result of tat binding, the production of full-length viral RNA is dramatically increased (**Figure 16.17B**), and the rest of the viral reproductive cycle is able to proceed. It was only after the discovery of this mechanism in HIV and similar viruses that researchers found that many eukaryotic genes are regulated at the level of transcription elongation.

Almost every step in the reproductive cycle of HIV is, in principle, a potential target for drugs to treat AIDS. The classes of anti-HIV drugs currently in use include:

- Reverse transcriptase inhibitors that block viral DNA synthesis from RNA (step 2 in Figure 16.16)
- Integrase inhibitors that block the incorporation of viral DNA into the host chromosome (step 3)
- Protease inhibitors that block the posttranslational processing of viral proteins (step 5)

Combinations of drugs from these classes have been spectacularly successful in treating HIV infection. But a problem with HIV treatment is the rapid emergence of drug-resistant strains,

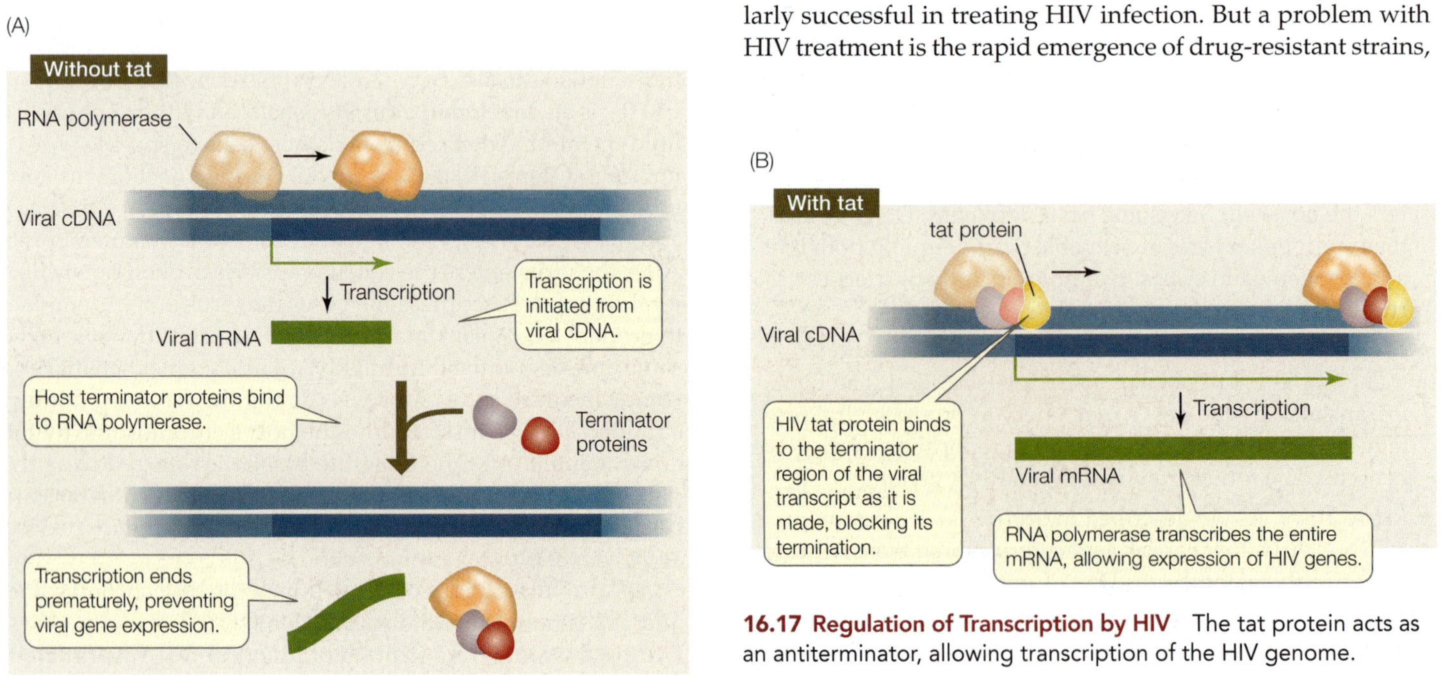

(A)
Without tat
RNA polymerase
Viral cDNA
Transcription
Viral mRNA
Transcription is initiated from viral cDNA.
Host terminator proteins bind to RNA polymerase.
Terminator proteins
Transcription ends prematurely, preventing viral gene expression.
(B)
With tat
tat protein
Viral cDNA
Transcription
Viral mRNA
HIV tat protein binds to the terminator region of the viral transcript as it is made, blocking its termination.
RNA polymerase transcribes the entire mRNA, allowing expression of HIV genes.

16.17 Regulation of Transcription by HIV The tat protein acts as an antiterminator, allowing transcription of the HIV genome.

and drugs targeted to other steps in the HIV reproductive cycle are being developed. These include drugs that interact with the binding of viral particles to host cells, and drugs that interfere with tat activity.

RECAP 16.3

A virus consists of nucleic acids, a few proteins, and in some cases, a lipid envelope. Viruses require host cells to reproduce. Viral life cycles can include lytic and lysogenic stages. Like its prokaryotic host, bacteriophage λ uses both positive and negative regulators of transcription initiation. Studies of HIV revealed a new mechanism for gene regulation: the regulation of transcription elongation.

- What are the lytic and lysogenic cycles of bacteriophages? **See pp. 339–341 and Figure 16.14**
- Describe positive and negative regulation of gene expression in the bacteriophage λ and HIV life cycles. **See pp. 340–342 and Figures 16.15, 16.17**

So far we have discussed mechanisms that cells and viruses use to control gene transcription. These mechanisms usually involve the interaction of regulatory proteins with specific DNA sequences. However, there are other mechanisms for controlling gene expression that do not depend on specific DNA sequences. We will discuss these mechanisms in the next section.

16.4 How Do Epigenetic Changes Regulate Gene Expression?

In the mid-twentieth century, the great developmental biologist Conrad Hal Waddington coined the term "epigenetics" and defined it as "that branch of biology which studies the causal interactions between genes and their products which bring the phenotype into being." Today **epigenetics** is defined more specifically, referring to the study of changes in gene expression that occur without changes in the DNA sequence. These changes are reversible but sometimes are stable and heritable. You saw an example of this phenomenon in the opening story of this chapter. Epigenetic changes include two processes: DNA **methylation** and chromosomal protein alterations.

DNA methylation occurs at promoters and silences transcription

Depending on the organism, from 1 to 5 percent of cytosine residues in the DNA are chemically modified by the addition of a methyl group ($—CH_3$) to the 5′ carbon, to form 5′-methylcytosine (Figure 16.18). This covalent addition is catalyzed by the enzyme **DNA methyltransferase** and, in mammals, usually occurs in C residues that are adjacent to G residues. DNA regions rich in these doublets are called **CpG islands**, and are especially abundant in promoters.

This covalent change in DNA is heritable: when DNA is replicated, a **maintenance methylase** catalyzes the formation of 5′-methylcytosine in the new DNA strand. However, the pattern of cytosine methylation can also be altered, because methylation is reversible: a third enzyme, appropriately called **demethylase**, catalyzes the removal of the methyl group from cytosine (see Figure 16.18).

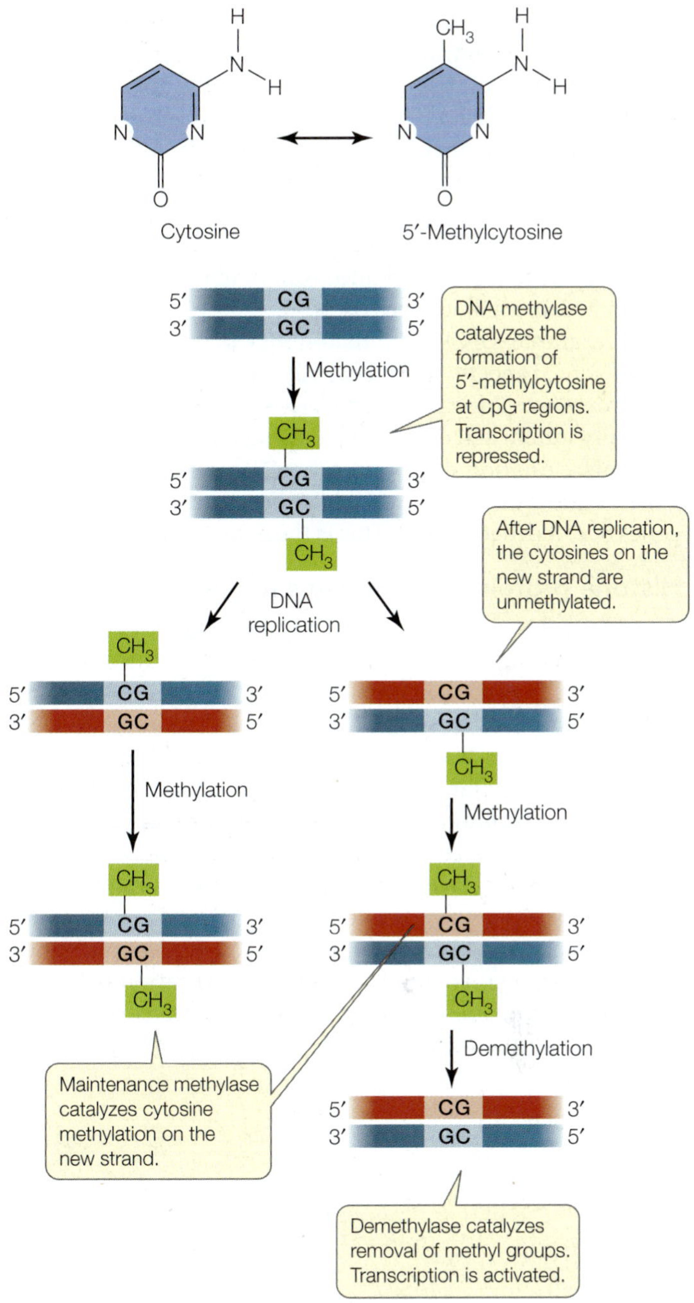

Cytosine
5′-Methylcytosine
CH3
DNA methylase catalyzes the formation of 5′-methylcytosine at CpG regions. Transcription is repressed.
Methylation
DNA replication
After DNA replication, the cytosines on the new strand are unmethylated.
Methylation
Maintenance methylase catalyzes cytosine methylation on the new strand.
Demethylation
Demethylase catalyzes removal of methyl groups. Transcription is activated.

16.18 DNA Methylation: An Epigenetic Change The reversible formation of 5′-methylcytosine in DNA can alter the rate of transcription.

What is the effect of DNA methylation? During replication and transcription, 5′-methylcytosine behaves just like plain cytosine: it base-pairs with guanine. But extra methyl groups in a promoter attract proteins that bind methylated DNA. These proteins are generally involved in the repression of gene transcription; thus heavily methylated genes tend to be inactive. This form of genetic regulation is epigenetic because it affects gene expression patterns without altering the DNA sequence.

DNA methylation is important in development from egg to embryo. For example, when a mammalian sperm enters an egg, many genes in first the male and then the female genome become demethylated. Thus many genes that are usually inactive are expressed during early development. As the embryo develops and its cells become more specialized, genes whose products are not needed in particular cell types become methylated. These methylated genes are "silenced"; their transcription is repressed. However, unusual or abnormal events can sometimes turn silent genes back on.

For example, DNA methylation may play roles in the genesis of some cancers. In cancer cells, oncogenes get activated and promote cell division, and tumor suppressor genes (which normally inhibit cell division) are turned off (see Chapter 11). This misregulation can occur when the promoters of oncogenes become demethylated while those of tumor suppressor genes become methylated. This is the case in colorectal cancer.

Histone protein modifications affect transcription

Another mechanism for epigenetic gene regulation is the alteration of chromatin structure, or chromatin remodeling. As we saw in Chapter 11, DNA is packaged with histone proteins into nucleosomes (see Figure 11.9), which can make DNA physically inaccessible to RNA polymerase and the rest of the transcription apparatus. Each histone protein has a "tail" of approximately 20 amino acids at its N terminus that sticks out of the compact structure and contains certain positively charged amino acids (notably lysine). Ordinarily there is strong ionic attraction between the positively charged histone proteins and DNA, which is negatively charged because of its phosphate groups. However, enzymes called histone acetyltransferases can add acetyl groups to these positively charged amino acids, thus changing their charges:

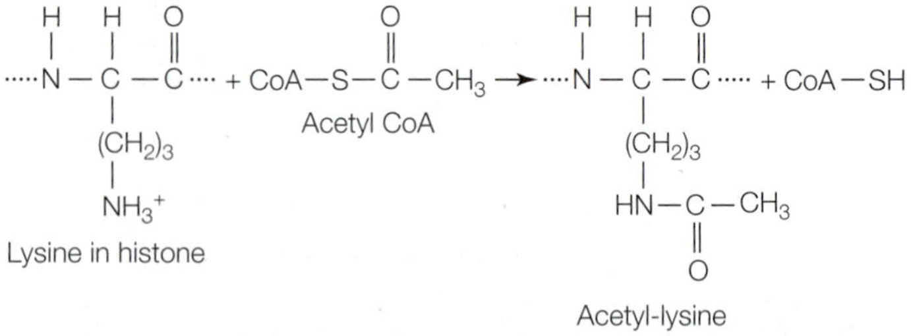

Acetyl CoA
Lysine in histone
Acetyl-lysine

Reducing the positive charges of the histone tails reduces the affinity of the histones for DNA, opening up the compact nucleosome (**Figure 16.19**). Additional chromatin remodeling proteins can bind to the loosened nucleosome–DNA complex, opening up the DNA for gene expression. Histone acetyltransferases can thus activate transcription.

Another kind of chromatin remodeling protein, histone deacetylase, can remove the acetyl groups from histones and thereby repress transcription. Histone deacetylases are targets for drug development to treat some forms of cancer. As noted above, certain genes block cell division in normal specialized tissues. In some cancers these genes are less active than in normal cells, and the histones near them show excessive levels of deacetylation. Theoretically, a drug acting as a histone deacetylase inhibitor could tip the balance toward acetylation, and this might activate genes that normally inhibit cell division.

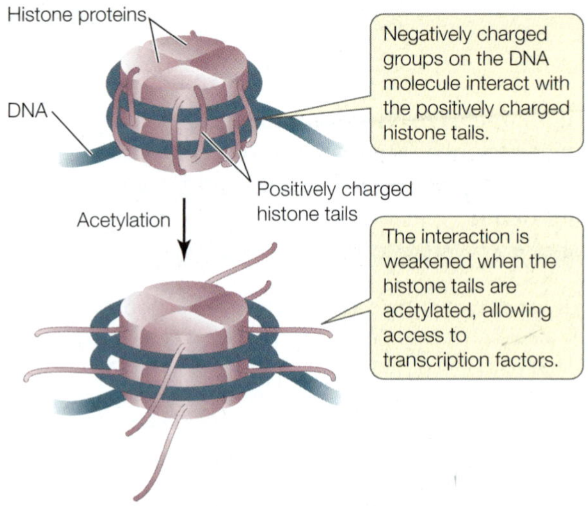

Histone proteins
DNA
Negatively charged groups on the DNA molecule interact with the positively charged histone tails.
Positively charged histone tails
Acetylation
The interaction is weakened when the histone tails are acetylated, allowing access to transcription factors.

16.19 Epigenetic Remodeling of Chromatin for Transcription Initiation of transcription requires that nucleosomes change their structure, becoming less compact. This chromatin remodeling makes DNA accessible to the transcription initiation complex (see Figure 16.8).

Other types of histone modification can affect gene activation and repression. For example, histone methylation (not to be confused with DNA methylation) is associated with gene inactivation, and histone phosphorylation also affects gene expression, the specific effects depending on which amino acids are modified. All of these effects are reversible, and so the activity of a eukaryotic gene may be determined by very complex patterns of histone modification.

Epigenetic changes can be induced by the environment

Despite the fact that they are reversible, many epigenetic changes such as DNA methylation and histone modification can permanently alter gene expression patterns in a cell. If the cell is a germ line cell that forms gametes, the epigenetic changes can be passed on to the next generation. But what determines these epigenetic changes? A clue comes from a recent study of monozygotic (identical) twins.

Monozygotic twins come from a single fertilized egg that divides to produce two separate cells; each of these goes on to develop a separate individual. Monozygotic twins thus have identical genomes. But are they identical in their epigenomes? A comparison of DNA in hundreds of such twin pairs shows that in tissues of three-year-olds, the DNA methylation patterns are virtually the same. But by age 50, by which time the twins have usually been living apart in different environments for decades, the patterns are quite different. This indicates that the *environment plays an important role in epigenetic modifications*, and thus in the regulation of genes that these modifications affect.

Go to Media Clip 16.1
The Surprising Epigenetics of Identical Twins
Life10e.com/mc16.1

DNA methylation can result in genomic imprinting

In mammals specific patterns of methylation develop for each sex during gamete formation. This happens in two stages: first, the existing methyl groups are removed from the 5′-methylcytosines by a demethylase, and then a DNA methylase adds methyl groups to a new set of cytosines. When the gametes form, they carry this new pattern of methylation.

The DNA methylation pattern in male gametes (sperm) differs from that in female gametes (eggs) at about 200 genes in the mammalian genome. That is, a given gene in this group may be methylated in eggs but unmethylated in sperm (Figure 16.20). In this case the offspring would inherit a maternal gene that is transcriptionally inactive (methylated) and a paternal gene that is transcriptionally active (demethylated). This is called **genomic imprinting**.

An example of imprinting is found in a region on human chromosome 15 called 15q11. This region is imprinted differently during the formation of male and female gametes, and offspring normally inherit both the paternally and maternally derived patterns. In rare cases there is a chromosome deletion in one of the gametes, and the newborn baby inherits just the male or the female imprinting pattern in this particular chromosome region:

- If the male pattern is the only one present (female region deleted), the baby develops Angelman syndrome, characterized by epilepsy, tremors, and constant smiling.
- If the female pattern is the only one present (male region deleted), the baby develops a quite different phenotype called Prader-Willi syndrome, marked by muscle weakness and obesity.

Note that the gene sequences are the same in both cases: it is the epigenetic patterns that are different.

Imprinting of specific genes occurs primarily in mammals and flowering plants. Most imprinted genes are involved with embryonic development. An embryo must have both the paternally and maternally imprinted gene patterns to develop properly. In fact, attempts to make an embryo that has chromosomes from only one sex (for example, by chemically treating an egg cell to double its chromosomes) usually fail. So imprinting has an important lesson for genetics: *males and females may be the same genetically (except for the X and Y chromosomes), but they differ epigenetically.*

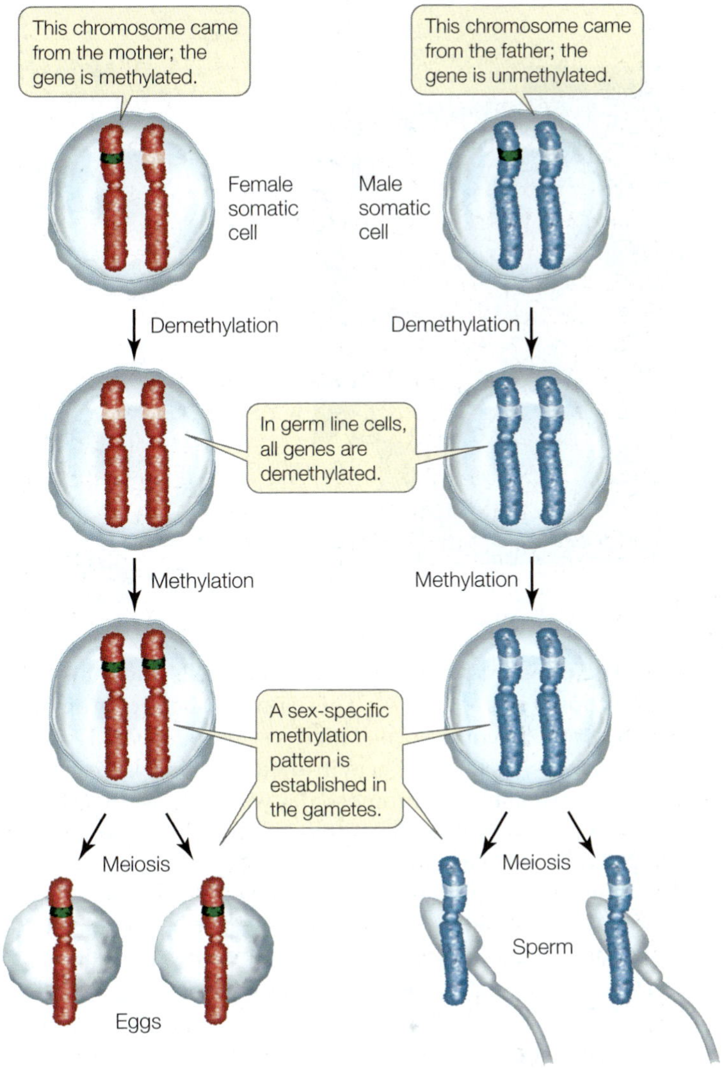
This chromosome came from the mother; the gene is methylated.
This chromosome came from the father; the gene is unmethylated.
Female somatic cell
Male somatic cell
Demethylation
Demethylation
In germ line cells, all genes are demethylated.
Methylation
Methylation
A sex-specific methylation pattern is established in the gametes.
Meiosis
Meiosis
Eggs
Sperm

16.20 Genomic Imprinting For some genes, epigenetic DNA methylation differs in male and female gametes. As a result, an individual might inherit an allele from the female parent that is transcriptionally silenced; but the same allele from the male parent would be expressed.

Global chromosome changes involve DNA methylation

Like single genes, large regions of chromosomes or even entire chromosomes can have distinct patterns of DNA methylation. Under a microscope, two kinds of chromatin can be distinguished in the stained interphase nucleus: **euchromatin** and **heterochromatin**. The euchromatin appears diffuse and stains lightly; it contains the DNA that is transcribed into mRNA. Heterochromatin is condensed and stains darkly; any genes it contains are generally not transcribed.

Perhaps the most dramatic example of heterochromatin is the inactive X chromosome of mammals. A normal female mammal has two X chromosomes; a normal male has an X and a Y (see Section 12.4). The X and Y chromosomes probably arose from a pair of autosomes (non–sex chromosomes) about 300 million years ago. Over time, mutations in the Y chromosome resulted in maleness-determining genes, and the Y chromosome gradually lost most of the genes it once shared with its X homolog. As a result, females and males differ greatly in the "dosage" of X-linked genes. Each female cell has two copies of each gene on the X chromosome and therefore has the potential to produce twice as much of each protein product. Nevertheless, for 75 percent of the genes on the X chromosome, transcription is generally the same in males and in females. How does this happen?

During early embryonic development, one of the X chromosomes in each cell of a female is largely inactivated with regard to transcription. The same X chromosome remains inactive in all that cell's descendants. In a given embryonic cell, the "choice" of which X in the pair to inactivate is random. Recall that one X in a female comes from her father and one from her mother. Thus in one embryonic cell the paternal X might be the one remaining transcriptionally active, but in a neighboring cell the maternal X might be active.

The inactivated X chromosome is identifiable within the nucleus because it is very compact, even during interphase. Typically, a nuclear structure called a Barr body (after its discoverer,

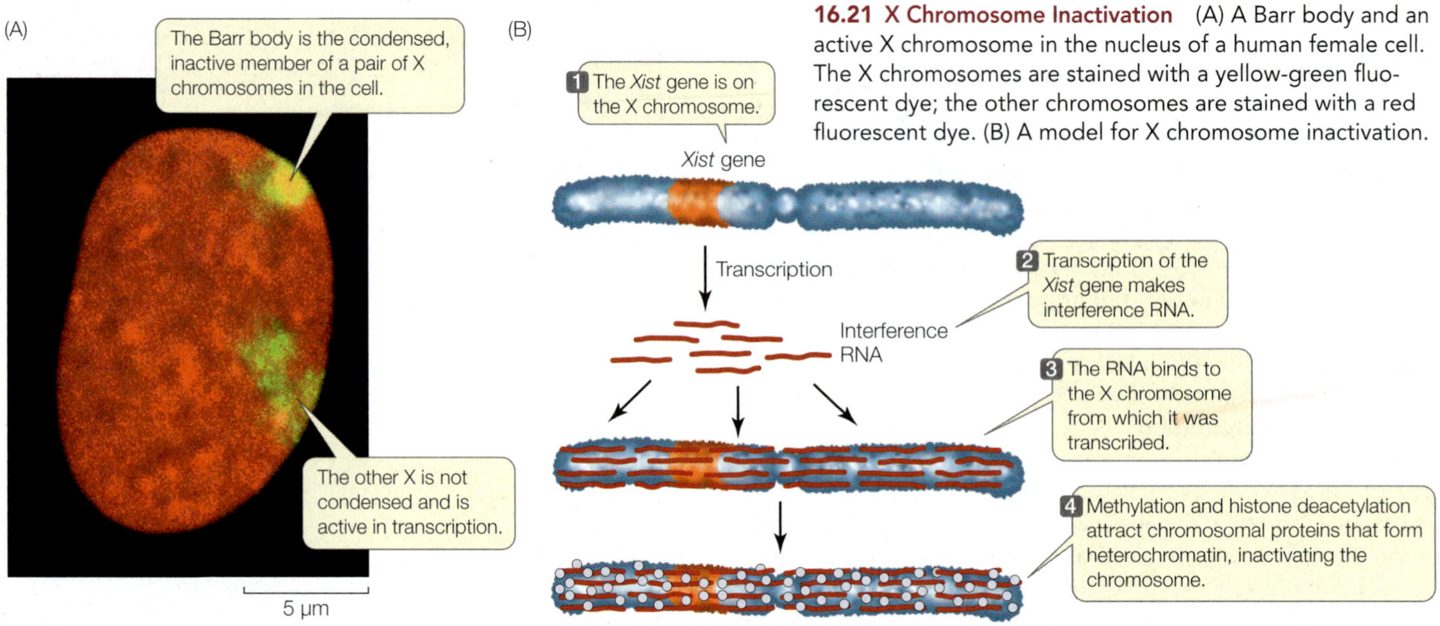

(A)
The Barr body is the condensed, inactive member of a pair of X chromosomes in the cell.
The other X is not condensed and is active in transcription.
5 μm
(B)
1 The *Xist* gene is on the X chromosome.
Xist gene
Transcription
Interference RNA
2 Transcription of the *Xist* gene makes interference RNA.
3 The RNA binds to the X chromosome from which it was transcribed.
4 Methylation and histone deacetylation attract chromosomal proteins that form heterochromatin, inactivating the chromosome.

16.21 X Chromosome Inactivation (A) A Barr body and an active X chromosome in the nucleus of a human female cell. The X chromosomes are stained with a yellow-green fluorescent dye; the other chromosomes are stained with a red fluorescent dye. (B) A model for X chromosome inactivation.

Murray Barr) can be seen in human female cells under the light microscope (**Figure 16.21A**). This clump of heterochromatin, which is not present in normal males, is the inactivated X chromosome, and it consists of heavily methylated DNA. A female with the normal two X chromosomes will have one Barr body, whereas a rare female with three Xs will have two, and an XXXX female will have three. Males that are XXY will have one. These observations suggest that the interphase cells of each person, male or female, have a single active X chromosome, and thus a constant dosage of expressed X chromosome genes.

Condensation of the inactive X chromosome makes its DNA sequences physically unavailable to the transcriptional machinery. Most of the genes of the inactive X are heavily methylated. However, one gene, *Xist* (for *X i*nactivation-specific *t*ranscript), is only lightly methylated and is transcriptionally active. On the active X chromosome, *Xist* is heavily methylated and not transcribed. The RNA transcribed from *Xist* binds to the X chromosome from which it is transcribed, and this binding leads to a spreading of inactivation along the chromosome. The *Xist* RNA transcript is an example of **interference RNA** (**Figure 16.21B**).

RECAP 16.4

Epigenetics describes stable changes in gene expression that do not involve changes in DNA sequences. These changes involve modifications of DNA (cytosine methylation) or of histone proteins bound to DNA. Epigenetic changes can be affected by the environment, and can also result in genome imprinting, in which expression of some genes depends on their parental origin.

- How are DNA methylation patterns established, and how do they affect gene expression? **See p. 343 and Figure 16.18**
- How do histone modifications affect transcription? **See p. 344 and Figure 16.19**
- Why and how does X chromosome inactivation occur? **See pp. 345–346**

Gene expression involves transcription and then translation. So far we have described how gene expression is regulated at the transcriptional level. But as Figure 16.7 shows, there are many points at which regulation can occur after the initial gene transcript is made.

16.5 How Is Eukaryotic Gene Expression Regulated after Transcription?

Eukaryotic gene expression can be regulated both in the nucleus prior to mRNA export, and after the mRNA leaves the nucleus. Posttranscriptional control mechanisms include alternative splicing of pre-mRNA, gene silencing by miRNAs and siRNAs, repression of translation, and regulation of protein breakdown in the proteasome.

Different mRNAs can be made from the same gene by alternative splicing

Most primary mRNA transcripts contain several introns (see Figure 14.7). Before the RNA is exported from the nucleus, a splicing mechanism recognizes the boundaries between exons and introns and converts pre-mRNA, which has the introns, into mature mRNA, which does not:

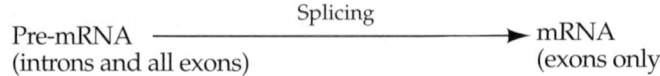

$$\text{Pre-mRNA (introns and all exons)} \xrightarrow{\text{Splicing}} \text{mRNA (exons only)}$$

For many genes, alternative splicing can occur, where some exons are spliced out along with the introns (**Figure 16.22**). This mechanism generates a family of different proteins, with different functions, from a single gene. Before the human genome was sequenced, most scientists estimated that there were 80,000 to 150,000 protein-coding genes. You can imagine their surprise when the actual sequence revealed only about 21,000! In fact, there are many more human mRNAs than there are human genes, and most of this variation comes from alternative

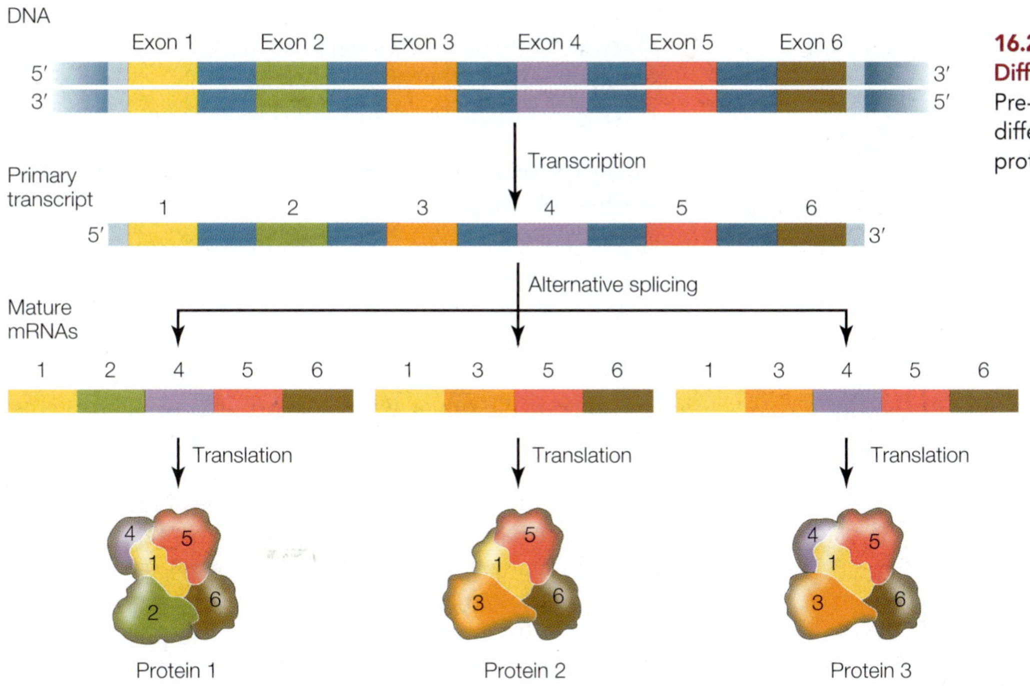

DNA
Exon 1
Exon 2
Exon 3
Exon 4
Exon 5
Exon 6
5′
3′
3′
5′
Transcription
Primary transcript
1
2
3
4
5
6
5′
3′
Alternative splicing
Mature mRNAs
1 2 4 5 6
1 3 5 6
1 3 4 5 6
Translation
Translation
Translation
Protein 1
Protein 2
Protein 3

16.22 Alternative Splicing Results in Different Mature mRNAs and Proteins Pre-mRNA can be spliced differently in different tissues, resulting in different proteins.

splicing. Indeed, recent surveys show that about half of all human genes are alternatively spliced. Alternative splicing may be a key to the differences in levels of complexity among organisms. For example, although humans and chimpanzees have similar-sized genomes, there is more alternative splicing in the human brain than in the brain of a chimpanzee.

Alternative RNA splicing is controlled both by regulatory elements in the RNA sequence that bind specific proteins (similar to regulatory sequences in DNA) and by secondary RNA structures that form by hybridization between nucleotides in the single-stranded RNA molecule.

Small RNAs are important regulators of gene expression

As we will discuss in Chapter 17, less than 5 percent of the genome in most plants and animals codes for proteins. Some of the genome encodes ribosomal RNA and transfer RNAs, but until recently biologists thought that the rest of the genome was not transcribed; some even called it "junk." Recent investigations, however, have shown that some of these noncoding regions are transcribed. The RNAs produced from these regions are often very small and therefore difficult to detect. In both prokaryotes and eukaryotes, these tiny RNA molecules are called **microRNA** (**miRNA**).

The first miRNA sequences were found in the worm *Caenorhabditis elegans*. This model organism, which has been studied extensively by developmental biologists, goes through several larval stages. Victor Ambros at the University of Massachusetts found mutations in two genes that had different effects on progress through these stages:

- *lin-14* mutations (named for abnormal cell *lin*eage) caused the larvae to skip the first stage and go straight to the second stage. Thus the gene's normal role is to facilitate events in the first larval stage.
- *lin-4* mutations caused certain cells in later larval stages to repeat a pattern of development normally shown in the first larval stage. It was as if the cells were stuck in that stage. So the normal role of this gene is to negatively regulate *lin-14*, turning its expression off so the cells can progress to the next stage.

Not surprisingly, further investigation showed that *lin-14* encodes a transcription factor that affects the transcription of genes involved in larval cell progression. It was originally expected that *lin-4*, the negative regulator, would encode a protein that downregulates genes activated by the lin-14 protein. But this turned out to be incorrect. Instead, *lin-4* encodes a 22-base miRNA that inhibits *lin-14* expression posttranscriptionally, by binding to its mRNA.

More than 5,000 miRNAs have now been described in eukaryotes. The human genome has about 1,000 miRNA-encoding regions. Each miRNA is about 22 bases long and usually has dozens of mRNA targets because the base pairing between the miRNA and the target mRNA doesn't have to be perfect. MicroRNAs are transcribed as longer precursors that fold into double-stranded RNA molecules and are then processed through a series of steps into single-stranded miRNAs. A protein complex guides the miRNA to its target mRNA, where translation is inhibited (**Figure 16.23A**). The remarkable conservation of the miRNA gene-silencing mechanism indicates that it is evolutionarily ancient and biologically important.

In addition to miRNAs, the eukaryotic RNA silencing mechanism also recognizes a similar class of molecules called **small interfering RNAs** (**siRNAs**). These often arise from viral infections, when two complementary strands of a viral genome are transcribed. Large double-stranded RNAs are formed, and as with miRNAs, these are converted into shorter single-stranded sequences; these bind to the target RNA and cause its

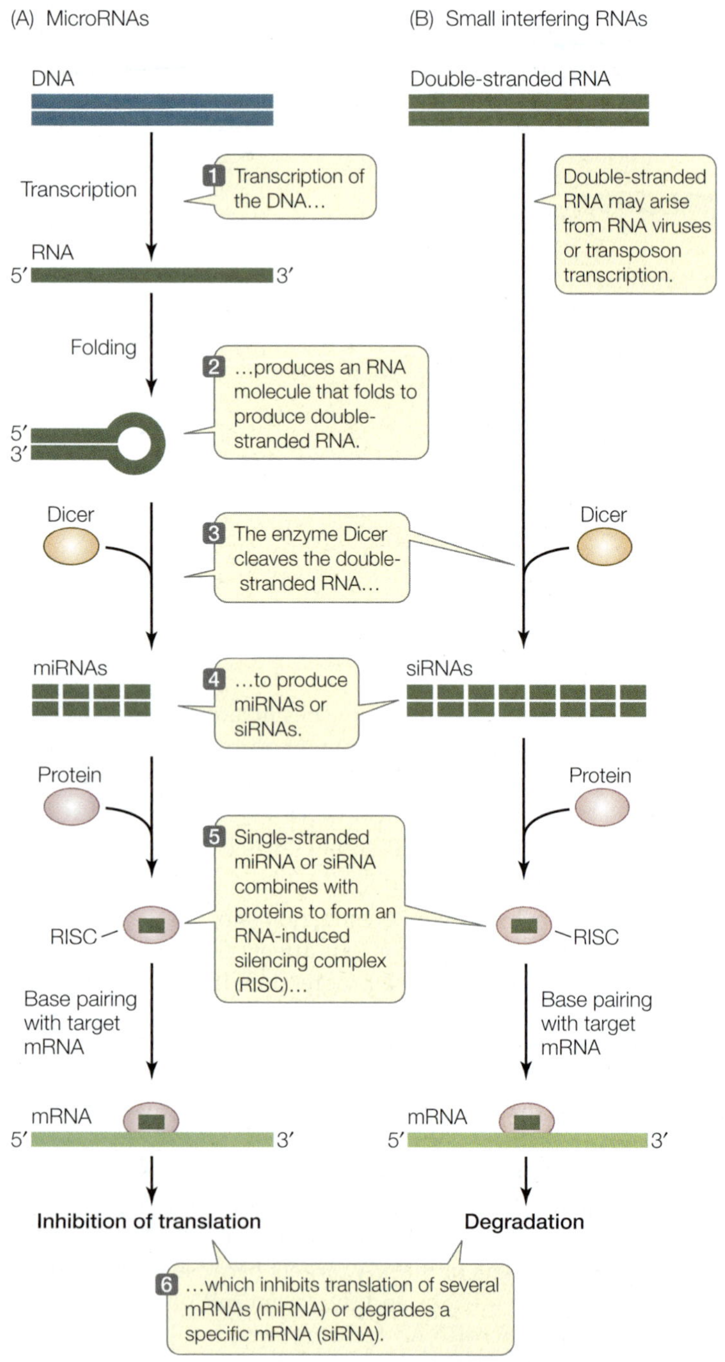

(A) MicroRNAs
(B) Small interfering RNAs
DNA
Double-stranded RNA
Transcription
1 Transcription of the DNA...
Double-stranded RNA may arise from RNA viruses or transposon transcription.
RNA
5′
3′
Folding
2 ...produces an RNA molecule that folds to produce double-stranded RNA.
5′
3′
Dicer
3 The enzyme Dicer cleaves the double-stranded RNA...
Dicer
miRNAs
4 ...to produce miRNAs or siRNAs.
siRNAs
Protein
Protein
5 Single-stranded miRNA or siRNA combines with proteins to form an RNA-induced silencing complex (RISC)...
RISC
RISC
Base pairing with target mRNA
Base pairing with target mRNA
mRNA
5′
3′
mRNA
5′
3′
Inhibition of translation
Degradation
6 ...which inhibits translation of several mRNAs (miRNA) or degrades a specific mRNA (siRNA).

16.23 mRNA Inhibition by RNAs MicroRNAs and small interfering RNAs can inhibit translation by binding to target mRNAs.

degradation (**Figure 16.23B**). Small interfering RNAs are also derived from transposon sequences, which are widespread in eukaryotic genomes (see Section 15.1). Therefore it is likely that gene silencing involving siRNAs evolved as a defense mechanism to prevent the translation of viral and transposon sequences. MicroRNAs and siRNAs are similar molecules that are processed by the same cellular enzymes. A major difference between them is that:

- miRNAs are synthesized from DNA sequences separate from their target, whereas
- siRNAs are targeted to their sequence of origin.

Translation of mRNA can be regulated by proteins and riboswitches

Is the amount of a protein in a cell determined by the amount of its mRNA? Scientists have examined the relationship between mRNA abundance and protein abundance in yeast cells.

For about one-third of the many genes surveyed, there was a clear correlation between mRNA and protein: more of one led to more of the other. But for two-thirds of the proteins, there was no apparent relationship between the two: sometimes there was lots of mRNA and little or no protein, or lots of protein and little mRNA. The concentrations of these proteins must therefore have been determined by factors acting after the mRNA was made. Cells have two major ways to control the amount of a protein after transcription: (1) they can regulate translation of the protein's mRNA and (2) they can regulate how long a newly synthesized protein persists in the cell (protein longevity).

REGULATION OF TRANSLATION There are a variety of ways in which the translation of mRNA can be regulated. One way, as we saw in the previous section, is to inhibit translation with siRNAs and miRNAs. A second way involves modification of the guanosine triphosphate cap on the 5′ end of the mRNA (see Section 14.4). An mRNA that is capped with an unmodified GTP molecule is not translated. For example, stored mRNAs in the egg cells of the tobacco hornworm moth are capped with unmodified GTP molecules and are not translated. After the egg is fertilized, however, the caps are modified, allowing the mRNA to be translated to produce the proteins needed for early embryonic development.

In another system, repressor proteins directly block translation. For example, in mammalian cells the protein ferritin binds free iron ions (Fe^{2+}). When iron is present in excess, ferritin synthesis rises dramatically, but the amount of ferritin mRNA remains constant, indicating that the increase in ferritin synthesis is due to an increased rate of mRNA translation. Indeed, when the iron level in the cell is low, a translational repressor protein binds to the 5′ noncoding region of ferritin mRNA and prevents its translation by blocking its attachment to a ribosome. When the iron level rises, some of the excess Fe^{2+} ions bind to the repressor and alter its three-dimensional structure, causing the repressor to detach from the mRNA and allowing translation to proceed (**Figure 16.24**).

The binding site for the translational repressor on mRNA is a stem-and-loop region with sufficient three-dimensional structure for recognition by a protein or metabolite. This regulatory mechanism occurs widely, and the RNA region that is bound is called a **riboswitch**.

REGULATION OF PROTEIN LONGEVITY The protein content of a cell at any given time is a function of both protein synthesis and protein degradation. Certain proteins can be targeted for destruction in a chain of events that begins when an enzyme attaches a 76-amino acid protein called **ubiquitin** (so named because it is ubiquitous, or widespread) to a lysine residue of the protein to be destroyed. Other ubiquitins then attach to the primary one, forming a polyubiquitin chain. The protein–polyubiquitin complex then binds to a huge protein complex called a **proteasome** (from *protease* and *soma*, "body") (Figure 16.25). Upon entering the proteasome, the polyubiquitin is removed and ATP energy is used to unfold the target protein. Three different proteases then digest the protein into small peptides and amino acids.

You may recall from Section 11.2 that cyclins are proteins that regulate the activities of key enzymes at specific points in the cell cycle. Cyclins must be broken down at just the right time, and this is done by attaching ubiquitin to them and degrading them in the proteasomes. Viruses can hijack this system. For example, some strains of the human papillomavirus (HPV) add ubiquitin to the p53 and retinoblastoma proteins, targeting them for proteasomal degradation. These proteins normally inhibit the cell cycle, so the result of this HPV activity is unregulated cell division (cancer).

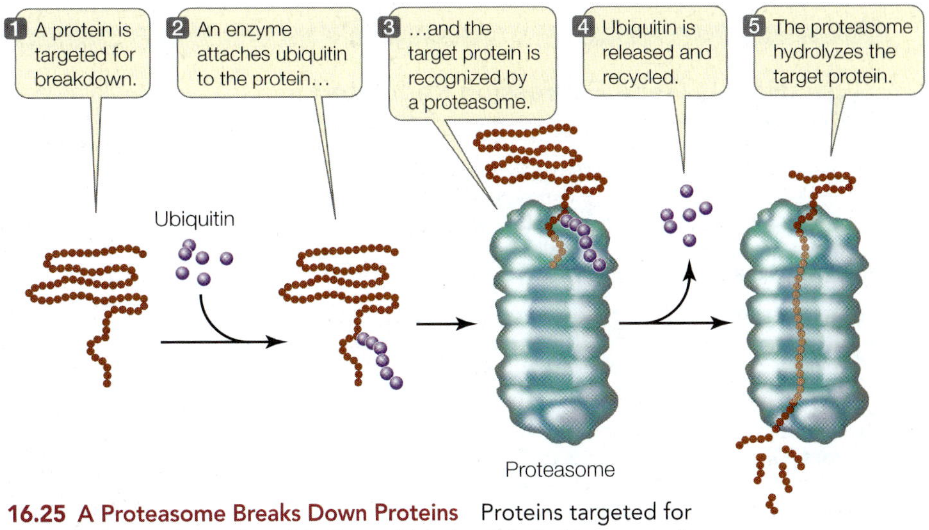

1 A protein is targeted for breakdown.
2 An enzyme attaches ubiquitin to the protein...
3 ...and the target protein is recognized by a proteasome.
4 Ubiquitin is released and recycled.
5 The proteasome hydrolyzes the target protein.
Ubiquitin
Proteasome

16.25 A Proteasome Breaks Down Proteins Proteins targeted for degradation are bound to ubiquitin, which then binds the targeted protein to a proteasome. The proteasome is a complex structure where proteins are digested by several powerful proteases.

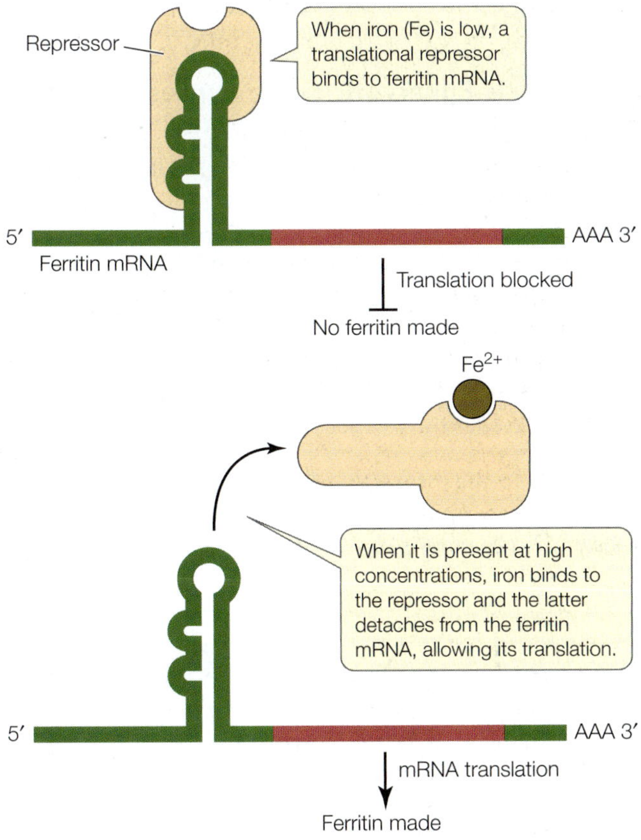

Repressor
When iron (Fe) is low, a translational repressor binds to ferritin mRNA.
5′
AAA 3′
Ferritin mRNA
Translation blocked
No ferritin made
Fe2+
When it is present at high concentrations, iron binds to the repressor and the latter detaches from the ferritin mRNA, allowing its translation.
5′
AAA 3′
mRNA translation
Ferritin made

16.24 Translational Repressor Can Repress Translation
Binding of a protein to a target mRNA can inhibit its translation.

RECAP 16.5

One of the most important means of posttranscriptional regulation is alternative RNA splicing, which allows more than one protein to be made from a single gene. The stability of mRNA in the cytoplasm can also be regulated. MicroRNAs, siRNAs, mRNA modifications, and translational repressors can prevent mRNA translation. Proteins in the cell can be targeted for breakdown by ubiquitin and then hydrolyzed in proteasomes.

- How can a single pre-mRNA sequence encode several different proteins? **See pp. 346–347 and Figure 16.22**
- How do miRNAs and siRNAs regulate gene expression? **See pp. 347–348 and Figure 16.23**
- What is a riboswitch? **See p. 348 and Figure 16.24**
- Explain the role of the proteasome. **See p. 349 and Figure 16.25**

Can epigenetic changes be manipulated?

ANSWER

Epigenetic changes often involve the addition of methyl (–CH_3) groups. Nutrients in the diet such as folic acid and SAM-e (*S*-adenosyl methinione) contain methyl groups and participate in reactions that modify DNA. Experiments with mice have shown that feeding young animals a diet enriched with these nutrients causes changes in epigenetic patterns and gene expression that remain throughout life. It remains to be seen what the effects of such a diet on human infants would be. But the results of these experiments raise the possibility of altering gene expression by diet.

CHAPTER SUMMARY 16

16.1 How Is Gene Expression Regulated in Prokaryotes?

- Some proteins are synthesized only when they are needed. Proteins that are made only in the presence of a particular compound—an **inducer**—are **inducible proteins**. Proteins that are made at a constant rate regardless of conditions are **constitutive proteins**. **Review Figure 16.2**
- An **operon** consists of a promoter, an **operator**, and two or more **structural genes**. Promoters and operators do not code for proteins, but serve as binding sites for regulatory proteins. **Review Figure 16.4**
- Regulatory genes code for regulatory proteins, such as **repressors**. When a repressor binds to an operator, transcription of the structural gene is inhibited. **Review Figure 16.5, ANIMATED TUTORIALS 16.1, 16.2**
- The *lac* operon is an example of an inducible system, in which the presence of an inducer (lactose) keeps the repressor from binding the operator, allowing the transcription of structural genes for lactose metabolism.
- Transcription can be enhanced by the binding of an **activator** protein to the promoter. **Review Figure 16.6**
- **Catabolite repression** is the inhibition of a catabolic pathway for one energy source by a different, preferred energy source.

16.2 How Is Eukaryotic Gene Transcription Regulated?

- Eukaryotic gene expression is regulated both during and after transcription. **Review Figure 16.7, ACTIVITY 16.1**
- **Transcription factors** and other proteins bind to DNA and affect the rate of initiation of transcription at the promoter. **Review Figures 16.8, 16.9, ANIMATED TUTORIAL 16.3**
- The interactions of these proteins with DNA are highly specific and depend on protein domains and DNA sequences.
- Genes at distant locations from one another can be coordinately regulated by transcription factors and promoter elements. **Review Figure 16.11**

16.3 How Do Viruses Regulate Their Gene Expression?

- **Viruses** are not cells, and rely on host cells to reproduce.
- The basic unit of a virus is a **virion**, which consists of a nucleic acid genome (DNA or RNA) and a protein coat, called a **capsid**.
- **Bacteriophages** are viruses that infect bacteria.
- Viruses undergo a **lytic** cycle, which causes the host cell to burst, releasing new virions.
- Some viruses have promoters that bind host RNA polymerase, which they use to transcribe their own genes and proteins. **Review Figure 16.13**
- Some viruses can also undergo **lysogeny**, in which a molecule of their DNA, called a **prophage**, is inserted into the host chromosome, where it replicates for generations. **Review Figure 16.14**
- The cellular environment determines whether a phage undergoes a lytic or a lysogenic cycle. Regulatory proteins that compete for promoters on phage DNA control the switch between the two life cycles. **Review Figure 16.15**
- A **retrovirus** uses **reverse transcriptase** to generate a cDNA **provirus** from its RNA genome. The provirus is incorporated into the host's DNA and can be activated to produce new virions. **Review Figure 16.16**

16.4 How Do Epigenetic Changes Regulate Gene Expression?

- **Epigenetics** refers to changes in gene expression that do not involve changes in DNA sequences.
- **Methylation** of cytosine residues generally inhibits transcription. **Review Figure 16.18**
- Modifications of histone proteins in nucleosomes make transcription either easier or more difficult. **Review Figure 16.19**
- Epigenetic changes can occur because of the environment.
- DNA methylation can explain **genomic imprinting**, where the expression of a gene depends on its parental origin. **Review Figure 16.20**

16.5 How Is Eukaryotic Gene Expression Regulated after Transcription?

- Alternative splicing of pre-mRNA can produce different proteins. **Review Figure 16.22**
- Small RNAs do not code for proteins but regulate the translation and longevity of mRNA. **Review Figure 16.23**
- The translation of mRNA to proteins can be regulated by translational repressors.
- The **proteasome** can break down proteins, thus affecting protein longevity. **Review Figure 16.25**

See ACTIVITY 16.2 for a concept review of this chapter.

Go to the Interactive Summary to review key figures, Animated Tutorials, and Activities Life10e.com/is16

CHAPTERREVIEW

REMEMBERING

1. Which of the following statements about the *lac* operon is *not* true?
   a. When the inducer binds to the repressor, the repressor can no longer bind to the operator.
   b. When the inducer binds to the operator, transcription is stimulated.
   c. When the repressor binds to the operator, transcription is inhibited.
   d. When the inducer binds to the repressor, the shape of the repressor is changed.
   e. The repressor has binding sites for both DNA and the inducer.

2. In the lysogenic cycle of bacteriophage λ,
   a. a repressor, cI, blocks the lytic cycle.
   b. the bacteriophage carries DNA between bacterial cells.
   c. both early and late viral genes are transcribed.
   d. the viral genome is made into RNA, which stays in the host cell.
   e. many new viruses are made immediately, regardless of host health.

3. An operon is
   a. a molecule that can turn genes on and off.
   b. an inducer bound to a repressor.
   c. a series of regulatory sequences controlling transcription of protein-coding genes.
   d. any long sequence of DNA.
   e. a promoter, an operator, and a group of linked structural genes.

4. Which of the following is true of both positive and negative gene regulation?
   a. They directly reduce the rate of transcription of certain genes.
   b. They involve transcription factors (or RNA) binding to DNA.
   c. They involve transcription of all genes in the genome.
   d. They are not both active in the same organism or virus.
   e. They act away from the promoter.

5. Which statement about selective gene transcription in eukaryotes is *not* true?
   a. Transcription factors can bind at a site on DNA distant from the promoter.
   b. Transcription requires transcription factors.
   c. Genes are usually transcribed as groups called operons.
   d. Both positive and negative regulation occur.
   e. Many proteins bind at the promoter.

6. Control of gene expression in eukaryotes includes all of the following except
   a. alternative RNA splicing.
   b. binding of proteins to DNA.
   c. transcription factors.
   d. stabilization of mRNA by miRNA.
   e. DNA methylation.

UNDERSTANDING & APPLYING

7. Compare the life cycles of a lysogenic bacteriophage and HIV (Figures 16.14 and 16.16) with respect to:
   a. how the virus enters the cell.
   b. how the new viruses are released from the cell.
   c. how the viral genome is replicated.
   d. how new viruses are produced.

8. Compare the roles of proteins and DNA sequences in the initiation of transcription in a prokaryote gene and a eukaryotic gene.

9. A protein-coding gene in a eukaryote has three introns. How many different proteins could be made by alternative splicing of the pre-mRNA from this gene?

ANALYZING & EVALUATING

10. The repressor protein that acts on the *lac* operon of *E. coli* is encoded by a regulatory gene. The repressor is made in small quantities and at a constant rate. Would you surmise that the promoter for this repressor protein is efficient or inefficient? Is synthesis of the repressor constitutive, or is it inducible and under environmental control?

11. In colorectal cancer, tumor suppressor genes are not active. This important factor results in uncontrolled cell division. Two possible explanations for the inactive genes are (a) mutations in the coding regions, resulting in inactive proteins, or (b) epigenetic silencing at the promoters of the genes, resulting in reduced transcription. How would you investigate these two possibilities?

Promoter
Coding region
3′
5′
5′
3′
Epigenetic silencing
CH_3
Point mutations

Go to BioPortal at **yourBioPortal.com** for Animated Tutorials, Activities, LearningCurve Quizzes, Flashcards, and many other study and review resources.

17 Genomes

CHAPTER**OUTLINE**

Variation in Dogs The Chihuahua (bottom) and the Brazilian mastiff (top) are the same species, *Canis lupus familiaris*, yet show great variation in size. Genome sequencing has revealed insights into how size is controlled by genes.

CANIS LUPUS FAMILIARIS, the dog, was domesticated by humans from the gray wolf thousands of years ago. While there are many kinds of wolves, they all look about the same. Not so with "man's best friend." The American Kennel Club recognizes about 155 different breeds, which not only look different, but also vary greatly in size. For example, an adult Chihuahua weighs just 1.5 kg, whereas a Scottish deerhound weighs 70 kg. No other mammal shows such large phenotypic variation. Also, there are hundreds of genetic diseases in dogs, many of which have counterparts in humans. To find out about the genes behind the phenotypic variation, and to elucidate the relationships between genes and diseases, the Dog Genome Project was started in the late 1990s.

Two dogs—a boxer and a poodle—were the first to have their entire genomes sequenced. The dog genome contains 2.8 billion base pairs of DNA in 39 pairs of chromosomes. There are 19,000 protein-coding genes, most of them with close counterparts in other mammals, including humans. The whole genome sequence made it easy to create a map of genetic markers—specific nucleotides or short sequences of DNA at particular locations on the genome that differ among individual dogs or breeds.

Genetic markers are being used to map the locations of (and thus identify) genes that control particular traits. For example, Dr. Elaine Ostrander and her colleagues at the National Institutes of Health studied Portuguese water dogs to identify genes that control size. Taking samples of cells for DNA isolation was relatively easy: a cotton swab was swept over the inside of the cheek. As Dr. Ostrander said, the dogs "didn't care, especially if they thought they were going to get a treat or if there was a tennis ball in our other hand." It turned out that the gene for *i*nsulin-like *g*rowth *f*actor 1 (IGF-1) is important in determining size: large breeds have an allele that codes for an active IGF-1, and small breeds have a different allele that codes for a less active IGF-1.

Inevitably, some scientists have set up companies to test dogs for genetic variations, using DNA supplied by anxious owners and breeders. Some traditional breeders disapprove, but others say it will improve the breeds and give more joy (and prestige) to owners. In other words, the issues surrounding the Dog Genome Project are not unlike those arising from the Human Genome Project.

?

What does dog genome sequencing reveal about other animals?

See answer on p. 370.

17.1 How Are Genomes Sequenced?

Genome sequencing involves determining the nucleotide base sequence of the entire genome of an organism. For a prokaryotic organism with a single chromosome, the genome sequence is one continuous series of base pairs (bp). In the case of a diploid, sexually reproducing species with multiple autosomes and a pair of sex chromosomes (see Section 12.4), the "sequenced genome" usually means the sequence of all the bases in one set of autosomes and in each of the two sex chromosomes. With advances in the technology for DNA sequencing, there has been an explosion of genetic information that scientists can use in a variety of ways:

- The genomes of different species can be compared to find out how they differ at the DNA level, and this can be used to trace evolutionary relationships.
- The sequences of individuals within a species can be compared to identify mutations that affect particular phenotypes.
- The sequence information can be used to identify genes for particular traits, such as genes associated with diseases.

The notion of sequencing the entire genome of a complex organism was not contemplated until 1986. The Nobel laureate Renato Dulbecco and others proposed at that time that the world scientific community be mobilized to undertake the sequencing of the entire human genome. One motive was to detect DNA damage in people who had survived the atomic bomb attacks and been exposed to radiation in Japan during World War II. But in order to detect changes in the human genome, scientists first needed to know its normal sequence.

The result was the publicly funded **Human Genome Project**, an enormous undertaking that was successfully completed in 2003. This effort was aided and complemented by privately funded groups. The project benefited from the development of many new methods that were first developed to sequence smaller genomes—those of prokaryotes and simple eukaryotes, the model organisms you have encountered in earlier chapters of this book. Many of these methods are still applied widely, and powerful new methods for sequencing genomes have emerged. These methods are complemented by new ways to examine phenotypic diversity in a cell's proteins and in the metabolic products of the cell's enzymes.

New methods have been developed to rapidly sequence DNA

Many prokaryotes have a single chromosome, whereas eukaryotes have many. Because of their differing sizes, chromosomes can be separated from one another, identified, and experimentally manipulated. It might seem that the most straightforward way to sequence a chromosome would be to start at one end and simply sequence the DNA molecule one nucleotide at a time. The task is somewhat simplified because only one of the two strands needs to be sequenced, the other being complementary. However, this large-polymer approach is not practical, since at most only several hundred bp can be sequenced at a time using current methods.

As you will see, the key to determining genome sequences is to perform many sequencing reactions simultaneously, after first breaking the DNA up into millions of small, overlapping fragments.

In the 1970s Frederick Sanger and his colleagues invented a way to sequence DNA by using chemically modified nucleotides that were originally developed to stop cell division in cancer. This method, or a variation of it, was used to obtain the first human genome sequence as well as those of several model organisms. However, it was relatively slow, expensive, and labor-intensive. The first decade of the new millennium saw the development of faster and less expensive methods, often referred to under the general term **high-throughput sequencing**. These methods use miniaturization techniques first developed for the electronics industry, as well as the principles of DNA replication, often in combination with the polymerase chain reaction (PCR).

High-throughput sequencing methods are rapidly evolving. Just one of the many approaches is outlined here and illustrated in **Figure 17.1**. First the DNA is prepared for sequencing by attaching it to a solid surface and amplifying the DNA by PCR (see Figure 17.1A):

1. A large molecule of DNA is cut into small fragments of about 100 bp each. This can be done physically, using mechanical forces to shear (break up) the DNA, or by using enzymes that hydrolyze the phosphodiester bonds between nucleotides at intervals in the DNA backbone.
2. The DNA is denatured by heat, breaking the hydrogen bonds that hold the two strands together. Each single strand acts as a template for the synthesis of new, complementary DNA.
3. Short, synthetic oligonucleotides are attached to each end of each fragment, and these are attached to a solid support. The support can be a microbead or a flat surface.
4. The DNA is amplified by PCR (see Section 13.5) using primers complementary to the synthetic oligonucleotides attached to the ends of each DNA. The multiple (approximately 1,000) copies of the DNA at a single location allow for easy detection of added nucleotides during the sequencing steps.

Once the DNA has been attached to a solid substrate and amplified, it is ready for sequencing (see Figure 17.1B):

1. At the beginning of each sequencing cycle, the fragments are heated to denature them. A solution containing a universal primer (complementary to one of the same synthetic oligonucleotides used for the PCR amplification step), DNA polymerase, and the four deoxyribonucleoside triphosphates (dNTPs: dATP, dGTP, dCTP, and dTTP) is then added to the DNA. Recall that dNTPs are the substrates that the DNA polymerase uses in DNA synthesis (see Section 13.3). Each of the four kinds of dNTP is tagged with a different colored fluorescent dye.
2. The DNA synthesis reaction is set up so that only one nucleotide is added to the new DNA strand in each sequencing cycle. After each addition, the unincorporated dNTPs are removed.

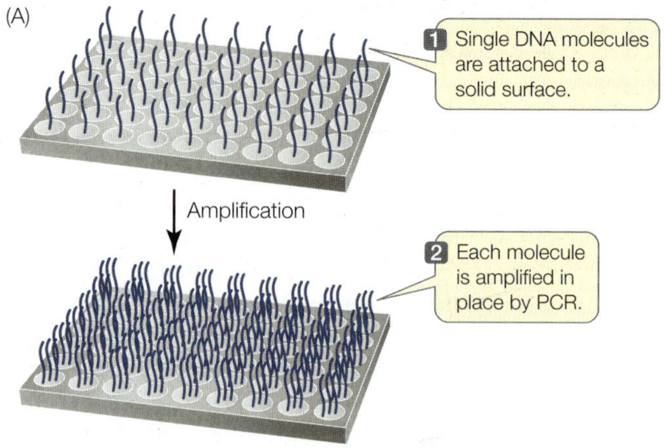
(A)
1 Single DNA molecules are attached to a solid surface.
Amplification
2 Each molecule is amplified in place by PCR.

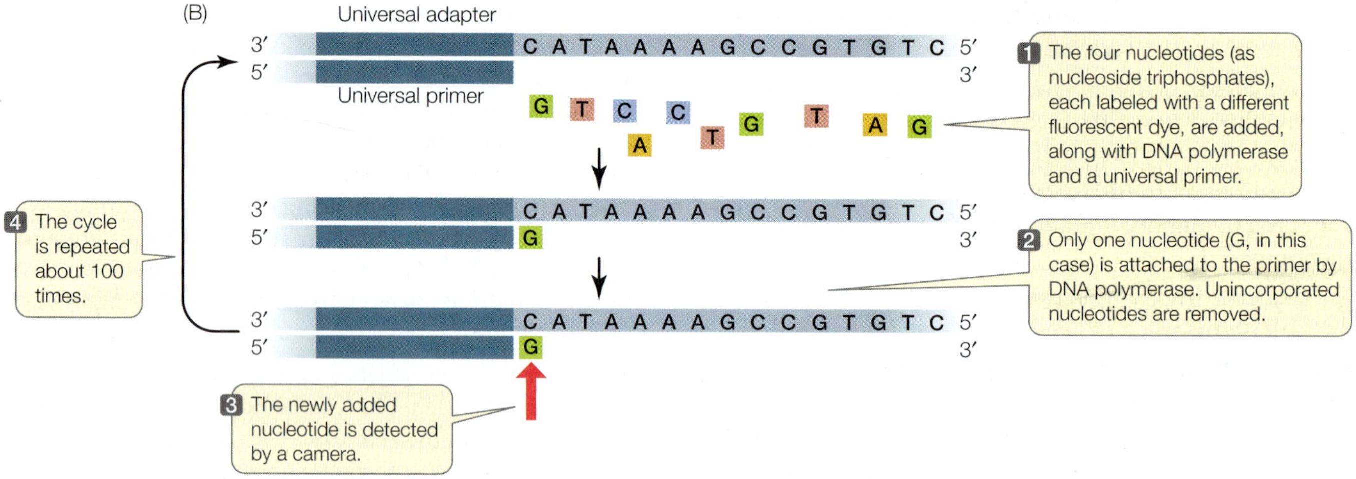
(B)
Universal adapter
3′ 5′
C A T A A A A G C C G T G T C
5′ 3′
Universal primer
G T C C A T G T A G
1 The four nucleotides (as nucleoside triphosphates), each labeled with a different fluorescent dye, are added, along with DNA polymerase and a universal primer.
2 Only one nucleotide (G, in this case) is attached to the primer by DNA polymerase. Unincorporated nucleotides are removed.
3 The newly added nucleotide is detected by a camera.
4 The cycle is repeated about 100 times.

17.1 DNA Sequencing High-throughput sequencing involves (A) the chemical amplification of DNA fragments and (B) the synthesis of complementary strands using fluorescently labeled nucleotides.

Go to Animated Tutorial 17.1
Sequencing the Genome
Life10e.com/at17.1

Go to Animated Tutorial 17.2
High-Throughput Sequencing
Life10e.com/at17.2

3. The fluorescence of the new nucleotide at each location is detected with a camera. The color of the fluorescence indicates which of the four nucleotides was added.
4. The fluorescent tag is removed from the nucleotide that is already attached, and then the DNA synthesis cycle is repeated. Images are captured after each nucleotide is added. The series of colors at each location indicate the sequence of nucleotides in the growing DNA strand at that location.

The power of this method derives from the fact that:

- It is fully automated and miniaturized.
- Millions of different fragments are sequenced at the same time.
- It is an inexpensive way to sequence large genomes. For example, at the time of this writing, a complete human genome could be sequenced in a few days for several thousand dollars. This is in contrast to the Human Genome Project, which took 13 years and $2.7 billion to sequence one genome!

The technology used to sequence millions of short DNA fragments is only half the story, however. Once these sequences have been determined, the problem becomes how to put them together. In other words, how are they arranged in the chromosomes from which they came? Imagine if you cut out every word in this book (there are more than half a million of them), put them on a table, and tried to arrange them in their original order! The enormous task of determining DNA sequences is possible because the original DNA fragments are overlapping.

Let's illustrate the process using a single 10-bp DNA molecule. (This is a double-stranded molecule, but for convenience we show only the sequence of the noncoding strand.) The molecule is cut three ways (for example, using three different restriction enzymes). Cutting with the first enzyme generates the fragments:

TG, ATG, and CCTAC

Cutting the same molecule with the second enzyme generates the fragments:

AT, GCC, and TACTG

Cutting with the third enzyme results in:

CTG, CTA, and ATGC

Can you put the fragments in the correct order? (The answer is ATGCCTACTG.) For genome sequencing, the sequence fragments are called "reads" (**Figure 17.2**). Of course, the problem of ordering 2.5 million fragments from human chromosome 1 (246 million bp) is more challenging than our 10-bp example above! The field of **bioinformatics** was developed to analyze DNA sequences using sophisticated mathematics and computer programs to handle the large amounts of data generated in genome sequencing.

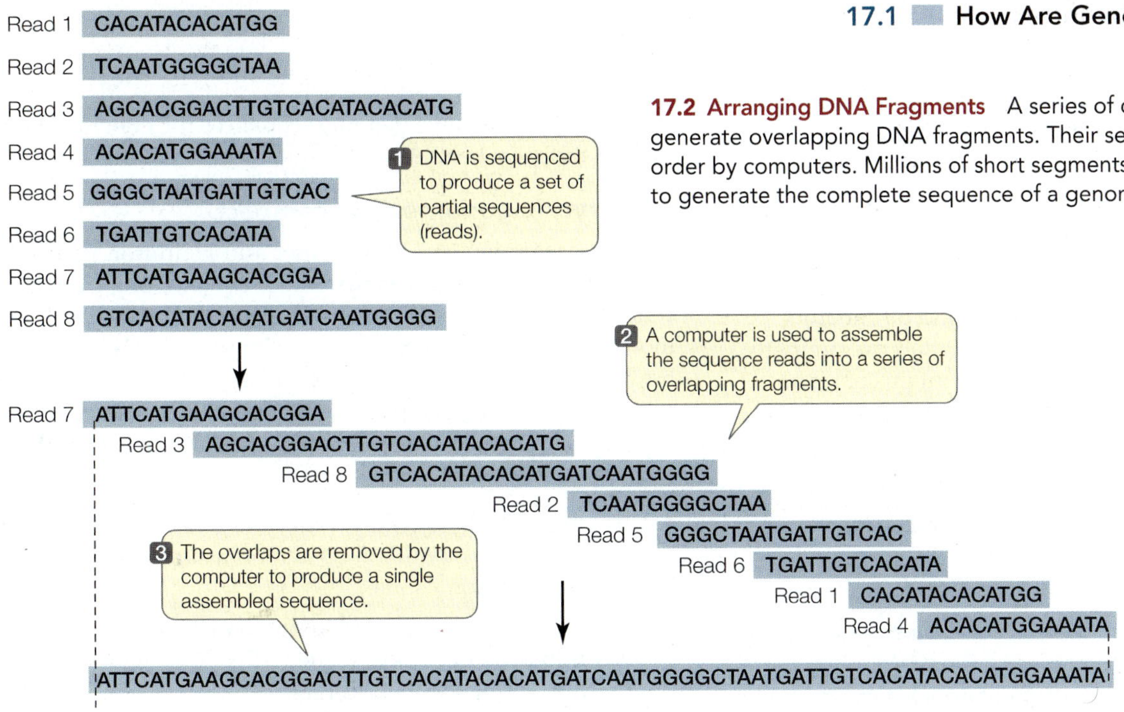
Read 1 CACATACACATGG
Read 2 TCAATGGGGCTAA
Read 3 AGCACGGACTTGTCACATACACATG
Read 4 ACACATGGAAATA
Read 5 GGGCTAATGATTGTCAC
Read 6 TGATTGTCACATA
Read 7 ATTCATGAAGCACGGA
Read 8 GTCACATACACATGATCAATGGGG
1 DNA is sequenced to produce a set of partial sequences (reads).
2 A computer is used to assemble the sequence reads into a series of overlapping fragments.
Read 7 ATTCATGAAGCACGGA
Read 3 AGCACGGACTTGTCACATACACATG
Read 8 GTCACATACACATGATCAATGGGG
Read 2 TCAATGGGGCTAA
Read 5 GGGCTAATGATTGTCAC
Read 6 TGATTGTCACATA
Read 1 CACATACACATGG
Read 4 ACACATGGAAATA
3 The overlaps are removed by the computer to produce a single assembled sequence.
ATTCATGAAGCACGGACTTGTCACATACACATGATCAATGGGGCTAATGATTGTCACATACACATGGAAATA

17.2 Arranging DNA Fragments A series of different cuts is used to generate overlapping DNA fragments. Their sequences are arranged in order by computers. Millions of short segments are arranged in this way to generate the complete sequence of a genome.

Genome sequences yield several kinds of information

New genome sequences are being published at an accelerating pace, creating a torrent of biological information (Figure 17.3). This information is used in two related fields of research, both focused on studying genomes. In **functional genomics**, biologists use sequence information to identify the functions of various parts of genomes. These parts include:

- *Open reading frames*, which are the coding regions of genes. For protein-coding genes, these regions can be recognized by the start and stop codons for translation, and by consensus sequences that indicate the locations of

17.3 The Genomic Book of Life Genome sequences contain many features, some of which are summarized in this overview. Sifting through all the information contained in a genome sequence can help us understand how an organism functions and what its evolutionary history might be.

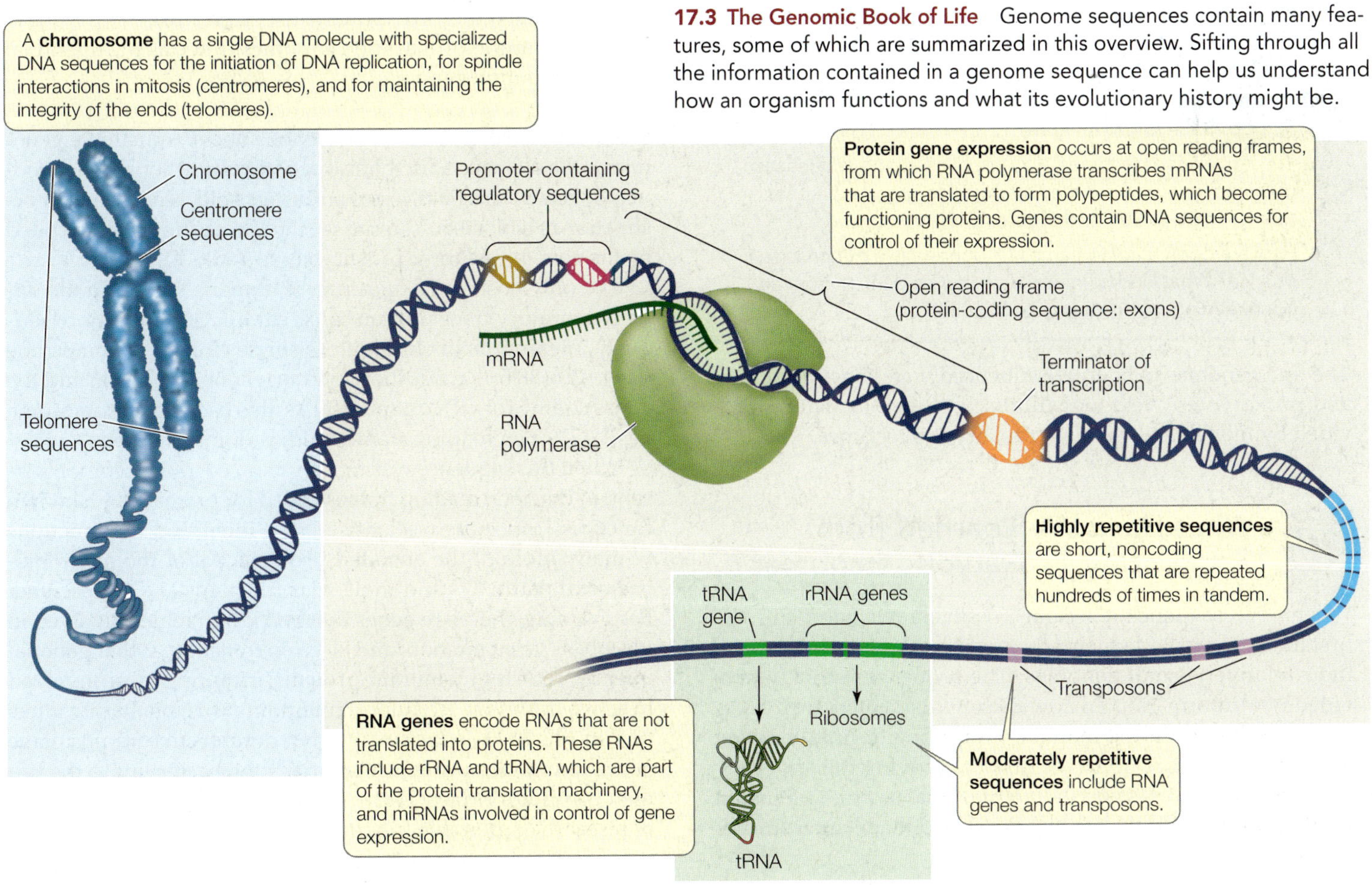
A chromosome has a single DNA molecule with specialized DNA sequences for the initiation of DNA replication, for spindle interactions in mitosis (centromeres), and for maintaining the integrity of the ends (telomeres).
Chromosome
Centromere sequences
Telomere sequences
Promoter containing regulatory sequences
mRNA
RNA polymerase
Protein gene expression occurs at open reading frames, from which RNA polymerase transcribes mRNAs that are translated to form polypeptides, which become functioning proteins. Genes contain DNA sequences for control of their expression.
Open reading frame (protein-coding sequence: exons)
Terminator of transcription
Highly repetitive sequences are short, noncoding sequences that are repeated hundreds of times in tandem.
tRNA gene
rRNA genes
Ribosomes
tRNA
Transposons
RNA genes encode RNAs that are not translated into proteins. These RNAs include rRNA and tRNA, which are part of the protein translation machinery, and miRNAs involved in control of gene expression.
Moderately repetitive sequences include RNA genes and transposons.

introns. A major goal of functional genomics is to understand the function of every open reading frame in each genome.

- *Amino acid sequences* of proteins, which can be deduced by applying the genetic code to the DNA sequences of open reading frames.
- *Regulatory sequences*, such as promoters and terminators for transcription. These are identified by their proximity to open reading frames and because they contain consensus sequences for the binding of specific transcription factors.
- *RNA genes*, including rRNA, tRNA, small nuclear RNA, and microRNA genes.
- *Other noncoding sequences* that can be classified into various categories, including centromeric and telomeric regions, transposons, and other repetitive sequences.

Sequence information is also used in **comparative genomics**: the comparison of a newly sequenced genome (or parts thereof) with sequences from other organisms. This can provide further information about the functions of sequences and can be used to trace evolutionary relationships among different organisms.

RECAP 17.1

The sequencing of genomes involves cutting large chromosomes into fragments, sequencing the fragments, and then assembling the fragment sequences into continuous sequences for entire chromosomes. Current sequencing methods use automation and powerful computers. They use labeled nucleotides that are detected at the ends of growing polynucleotide chains.

- Describe one high-throughput method for DNA sequencing. **See pp. 353–354 and Figure 17.1**
- For sequencing genomes, why are overlapping sequences obtained, and how are they arranged to give the final sequence? **See p. 354 and Figure 17.2**
- How are open reading frames recognized in a genomic sequence? What kind of information can be derived from an open reading frame? **See pp. 355–356**

The first genomes to be fully sequenced were those of viruses and prokaryotes. Next we will discuss the information provided by the relatively simple prokaryotic genomes.

17.2 What Have We Learned from Sequencing Prokaryotic Genomes?

When DNA sequencing became possible in the late 1970s, the first life forms to be sequenced were the simplest viruses with their relatively small genomes. The sequences quickly provided new information on how these viruses infect their hosts and reproduce. But the manual sequencing techniques used on viruses were not up to the task of studying the larger genomes of prokaryotes and eukaryotes. The newer, automated sequencing techniques we just described made such studies possible. We now have genome sequences for many prokaryotes, to the great benefit of microbiology and medicine.

Prokaryotic genomes are compact

In 1995 a team led by Craig Venter and Hamilton Smith determined the first complete genomic sequence of a free-living cellular organism, the bacterium *Haemophilus influenzae*. Many more prokaryotic sequences have followed, revealing not only how prokaryotes apportion their genes to perform different cellular functions, but also how their specialized functions are carried out. There are several notable features of bacterial and archaeal genomes:

- They are relatively small. Prokaryotic genomes range from about 160,000 to 12 million bp and are usually organized into a single chromosome.
- They are compact. Typically, more than 85 percent of the DNA is in protein-coding sequences or RNA genes, with only short sequences between genes.
- The genes usually do not have introns. An exception is the rRNA and tRNA genes of archaea, which frequently contain introns.
- In addition to the main chromosome, prokaryotes often have smaller, circular molecules of DNA called plasmids, which may be transferred between cells (see Chapter 12).

Beyond these similarities, there is great diversity among these single-celled organisms, reflecting the huge variety of environments in which they are found (see Chapter 26).

Let's examine prokaryotic genomes in terms of functional and comparative genomics.

FUNCTIONAL GENOMICS As described above, functional genomics is the biological discipline that assigns functions to the products of genes. This field, less than 20 years old, is now a major occupation of biologists. You can see the various functions encoded by the genomes of three prokaryotes in **Table 17.1**.

The only host for *H. influenzae* is humans. It lives in the upper respiratory tract and can cause ear infections or, more seriously, meningitis in children. Its single circular chromosome has 1,830,138 bp. In addition to its origin of replication and the genes coding for rRNAs and tRNAs, this bacterial chromosome has 1,727 open reading frames with promoters nearby.

When this sequence was first announced, only 1,007 (58 percent) of the open reading frames coded for proteins with known functions. Since then, scientists have identified the functions of many more of the encoded proteins. All of the major biochemical pathways and molecular functions are represented. For example, there are genes that encode enzymes involved in glycolysis, fermentation, and electron transport. Other gene sequences code for membrane proteins, including those involved in active transport. An important finding was that highly infective strains of *H. influenzae*, but not noninfective strains, have genes for surface proteins that attach the bacterium to the human respiratory tract. These surface proteins are now a focus of research on possible treatments for *H. influenzae* infections.

TABLE 17.1
Gene Functions in Three Bacteria

	Number of Genes in:		
Category	*E. coli*	*H. influenzae*	*M. genitalium*
Total protein-coding genes	4,288	1,727	482
Biosynthesis of amino acids	131	68	1
Biosynthesis of cofactors	103	54	5
Biosynthesis of nucleotides	58	53	19
Cell envelope proteins	237	84	17
Energy metabolism	243	112	31
Intermediary metabolism	188	30	6
Lipid metabolism	48	25	6
DNA replication, recombination, and repair	115	87	32
Protein folding	9	6	7
Regulatory proteins	178	64	7
Transcription	55	27	12
Translation	182	141	101
Uptake of molecules from the environment	427	123	34

COMPARATIVE GENOMICS Soon after the sequence of *H. influenzae* was announced, smaller (*Mycoplasma genitalium*: 580,073 bp) and larger (*E. coli*: 4,639,221 bp) prokaryotic sequences were completed. Thus began the era of comparative genomics. Scientists can identify genes that are present in one bacterium and missing in another, allowing them to relate these genes to bacterial function.

M. genitalium, for example, lacks enzymes needed to synthesize amino acids, whereas *E. coli* and *H. influenzae* both possess such enzymes. This finding reveals that *M. genitalium* must obtain all its amino acids from its environment (usually the human urogenital tract). Furthermore, *E. coli* has 55 genes that encode transcriptional activators, whereas *M. genitalium* has only 7. This relative lack of control over gene expression suggests that the biochemical flexibility of *M. genitalium* must be limited compared with that of *E. coli*.

The sequencing of prokaryotic and viral genomes has many potential benefits

Prokaryotic genome sequencing is providing insights into microorganisms that are important for agriculture and medicine. Scientists who analyze the sequences have discovered previously unknown genes and proteins that can be targeted for isolation and functional study. They have also discovered surprising relationships between some organisms, suggesting that genes may be transferred between different species.

- *Rhizobium* species are bacteria that form symbiotic associations with plants, living inside the roots of legumes such as beans, peas, and clover. The bacteria fix atmospheric nitrogen from the air and convert it into forms usable by the plants, reducing the need for nitrogen-containing fertilizers. Genome sequences from several *Rhizobium* species have been used to identify the genes involved in successful symbiosis, and this knowledge is being used both to improve the efficiency of this process and to broaden the range of plants that can form these beneficial associations.
- *E. coli* strain O157:H7 causes illness (sometimes severe) in at least 70,000 people a year in the United States. Its genome has 5,416 genes, of which 1,387 are different from those in the familiar (and harmless) laboratory strains of this bacterium. Many of these unique genes are also present in other pathogenic bacteria, such as *Salmonella* and *Shigella*. This finding suggests that there is extensive genetic exchange among these species, and that "superbugs" that have acquired multiple genes for antibiotic resistance may be on the horizon.
- *Severe acute respiratory syndrome* (*SARS*) was first detected in southern China in 2002 and rapidly spread in 2003. There is no effective treatment, and 10 percent of infected people die. Isolation of the causative agent, a virus, and the rapid sequencing of its genome revealed several novel proteins that are possible targets for antiviral drugs or vaccines. Research is underway on both fronts, since further outbreaks are anticipated.

Genome sequencing also provides insights into organisms involved in global ecological cycles (see Chapter 58). In addition to carbon dioxide, another important gas that contributes to global warming is methane (CH_4; see Figure 2.7). Some bacteria, such as *Methanococcus*, produce methane in the stomachs of cows. Others, such as *Methylococcus*, remove methane from the air and use it as an energy source. The genomes of both of these bacteria have been sequenced. Understanding the genes involved in methane production and consumption may help us slow the progress of global warming.

Metagenomics allows us to describe new organisms and ecosystems

If you take a microbiology laboratory course you will learn how to identify various prokaryotes on the basis of their growth on particular artificial media. For example, staphylococci are a group of bacteria that infect skin and nasal passages. When grown on a medium called blood agar, they form round, raised colonies. Microorganisms can also be identified by their nutritional requirements or the conditions under which they will grow (for example, aerobic versus anaerobic). Such culture methods have been the mainstay of microbial identification for more than a century and are still useful and important. However, scientists can now use PCR and DNA sequencing to identify microbes without culturing them in the laboratory.

In 1985 Norman Pace, then at Indiana University, came up with the idea of isolating DNA directly from environmental samples. He used PCR to amplify specific sequences from the samples to determine whether particular microbes were present. The PCR products were sequenced to explore their

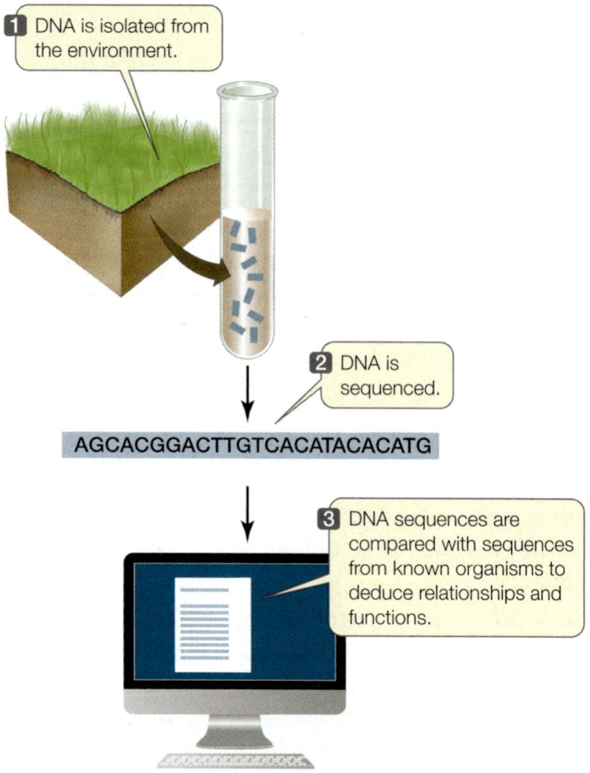
1 DNA is isolated from the environment.
2 DNA is sequenced.
AGCACGGACTTGTCACATACACATG
3 DNA sequences are compared with sequences from known organisms to deduce relationships and functions.

17.4 Metagenomics Microbial DNA extracted from the environment can be sequenced and analyzed. This has led to the description of many new genes and species.

diversity. The term **metagenomics** was coined to describe this approach of analyzing genes without isolating the intact organism. It is now possible to sequence DNA samples from almost any environment. The sequences can be used to detect the presence of both known microbes and heretofore unidentified organisms (Figure 17.4). For example:

- Sequencing of DNA from 200 liters of seawater indicated that it contained 5,000 different viruses and 2,000 different bacteria, many of which had not been described previously.
- One kilogram of marine sediment contained 1 million different viruses, most of them new.
- Water runoff from a mine contained many new species of prokaryotes thriving in this apparently inhospitable environment. Some of these organisms exhibited metabolic pathways that were previously unknown to biologists. These organisms and their capabilities may be useful in cleaning up pollutants from the water.

These and other discoveries are truly extraordinary and potentially very important. It is estimated that 90 percent of the microbial world has been invisible to biologists and is only now being revealed by metagenomics. Entirely new ecosystems of bacteria and viruses are being discovered in which, for example, one species produces a molecule that another metabolizes. It is hard to overemphasize the importance of such an increase in our knowledge of the hidden world of microbes. This knowledge will help us understand natural ecological processes, and has the potential to help us find better ways to manage environmental catastrophes such as oil spills, or to remove toxic heavy metals from soil.

Some sequences of DNA can move about the genome

As we mentioned in Section 15.1, **transposable elements** (or **transposons**) are segments of DNA that can move from place to place in the genome. Genome sequencing has allowed scientists to study these elements more broadly, and they are now known to be widespread in both prokaryotes and eukaryotes. Prokaryotic transposable elements are often short sequences of 1,000 to 2,000 bp, and they can be found in both chromosomes and in plasmids. A transposable element might be at one location in the genome of one *E. coli* strain, and at a different location in another strain. The insertion of this movable DNA sequence from elsewhere in the genome into the middle of a protein-coding gene disrupts that gene (Figure 17.5A). Any mRNA expressed from the disrupted gene will contain the extra sequence, and the protein it encodes will be altered and almost certainly nonfunctional. So transposable elements can produce significant phenotypic effects. Sometimes, two transposable elements located near one another (within a few thousand bp) will transpose together and carry the intervening DNA sequence with them. These are referred to as **composite transposons** (Figure 17.5B). Genes for

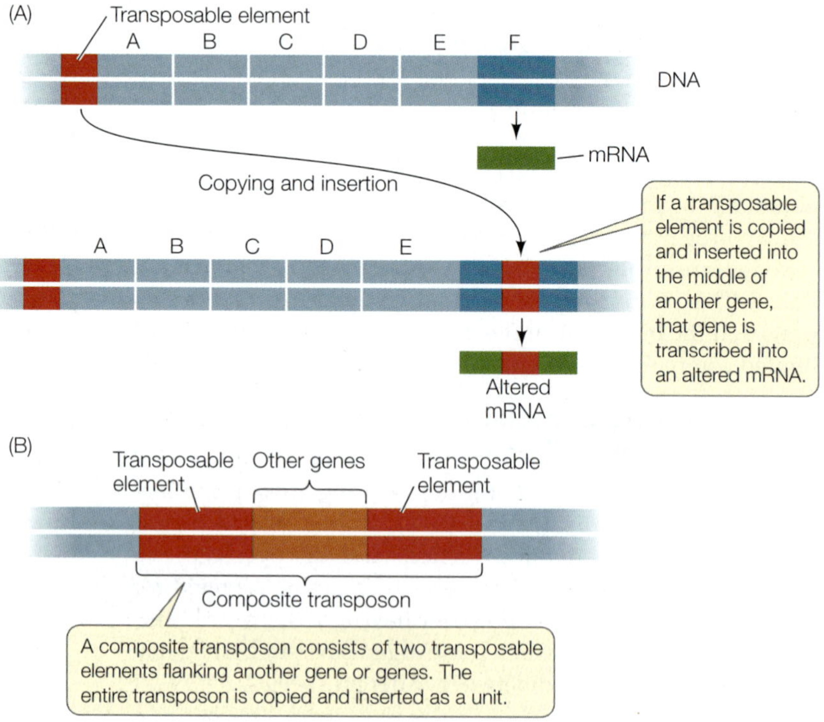
(A)
Transposable element
A B C D E F
DNA
mRNA
Copying and insertion
A B C D E
If a transposable element is copied and inserted into the middle of another gene, that gene is transcribed into an altered mRNA.
Altered mRNA
(B)
Transposable element
Other genes
Transposable element
Composite transposon
A composite transposon consists of two transposable elements flanking another gene or genes. The entire transposon is copied and inserted as a unit.

17.5 DNA Sequences That Move Transposable elements are DNA sequences that move from one location to another. (A) In one method of transposition ("copy and paste"), the DNA sequence is replicated and the copy inserts elsewhere in the genome. (B) Composite transposons contain additional genes flanked by two transposable elements.

INVESTIGATING**LIFE**

17.6 Using Transposon Mutagenesis to Determine the Minimal Genome *Mycoplasma genitalium* has one of the smallest known genomes of any prokaryote. But are all of its genes essential to life? By inactivating the genes one by one, scientists determined which of them are essential for the cell's survival. This research may lead to the construction of artificial cells with customized genomes, designed to perform functions such as degrading oil and making plastics.[a]

HYPOTHESIS Only some of the genes in a bacterial genome are essential for cell survival.

Method

Results

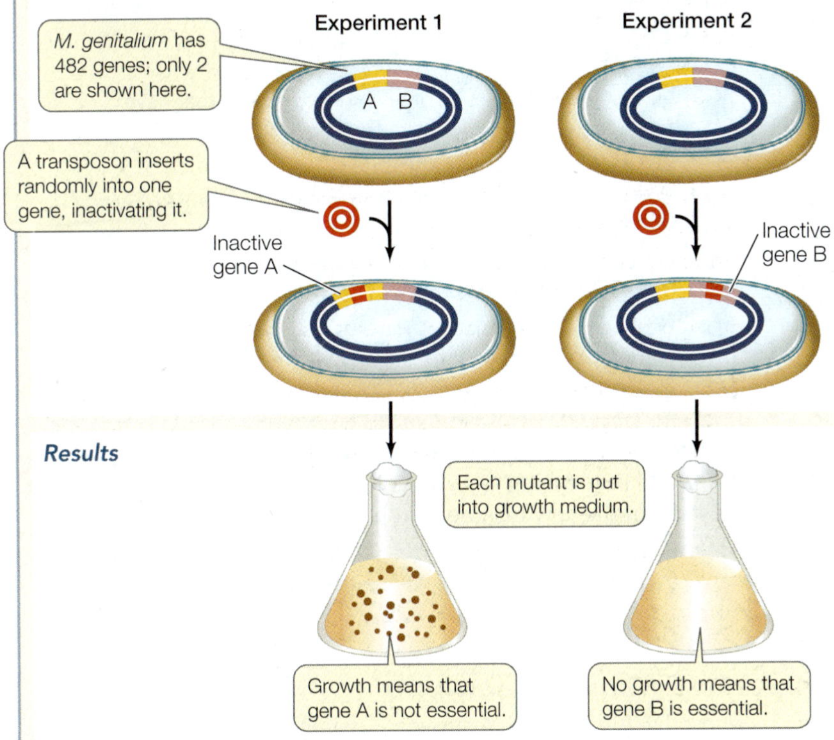

Experiment 1
Experiment 2
M. genitalium has 482 genes; only 2 are shown here.
A B
A transposon inserts randomly into one gene, inactivating it.
Inactive gene A
Inactive gene B
Each mutant is put into growth medium.
Growth means that gene A is not essential.
No growth means that gene B is essential.

CONCLUSION If each gene is inactivated in turn, a "minimal essential genome" can be determined.

Go to **BioPortal** for discussion and relevant links for all INVESTIGATING**LIFE** figures.

[a]Hutchison, C. et al. 1999. *Science* 286: 2165–2169, and Glass, J. I. et al. 2006. *Proceedings of the National Academy of Sciences USA* 103: 425–430.

antibiotic resistance can be multiplied and transferred between bacteria in this way: composite transposons carrying genes for antibiotic resistance can insert into a plasmid that then moves between bacteria by conjugation (see Section 12.6).

As we mentioned in Chapter 15, the mechanisms that allow transposable elements to move vary. For example, a transposable element may be replicated, and then the copy can be inserted into another site in the genome (the "copy and paste" mode). Or the element might splice out of one location and move to another location ("cut and paste"). The elements usually carry genes for enzymes such as transposases, which catalyze the reactions needed for transposition. Often the elements are flanked by inverted repeat DNA sequences that are recognized by these enzymes.

Will defining the genes required for cellular life lead to artificial life?

When the genomes of prokaryotes and eukaryotes are compared, a striking conclusion arises: certain genes are present in all organisms (universal genes). Not surprisingly, these include genes whose products are involved in DNA replication, transcription, and RNA translation to form proteins. There are also some (nearly) universal gene segments that are present in many genes in many organisms; for example, the sequence that codes for an ATP binding site in a protein. These findings suggest that there is some ancient, minimal set of DNA sequences that is common to all cells. One way to identify these sequences is to look for them in computer analyses of sequenced genomes.

Another way to define the minimal genome is to take an organism with a simple genome and deliberately mutate one gene at a time to see what happens. *M. genitalium* has one of the smallest known genomes—only 482 protein-coding genes. Even so, some of its genes are dispensable under some circumstances. For example, *M. genitalium* has genes for metabolizing both glucose and fructose, but it can survive in the laboratory on a medium containing only one of these sugars. Under such conditions it doesn't need the genes for metabolizing the other sugar.

What about other genes? A team led by Craig Venter has addressed this question with experiments involving the use of transposons as mutagens. When transposons in the bacterium are activated, they insert themselves into genes at random, mutating and inactivating them (**Figure 17.6**). The mutated bacteria are tested for growth and survival, and DNA from interesting mutants is sequenced to find out which genes contain transposons. The astonishing result of these studies is that *M. genitalium* can survive in the laboratory with a minimal genome of only 382 protein-coding genes!

One goal of this research is to design new life forms with specific purposes, such as bacteria that will clean up oil spills. The next step toward this goal is to create an artificial genome and insert it into bacterial cells. Venter's team recently synthesized the entire genome of *Mycoplasma mycoides*, using a computer-directed blueprint that included some extra sequences that could be used to track the survival of the genome in multiplying cells. They then transplanted this genome into empty cells of a different species, *Mycoplasma capricolum*, whose own DNA had been hydrolyzed. The new DNA directed the cell to perform all the biochemical functions of life, including reproduction (**Figure 17.7**). Since the new cell's genome had some extra sequences, it was an entirely new organism, called *Mycoplasma mycoides JCV1-syn.1.0*.

At the time of DNA transfer, the recipient cell still had its original small molecules and proteins, but after approximately 30 cell divisions, all of the proteins in the new cell colony had been made using the synthetic genome's sequences. The cells had used substances in their environment to make their own small and large molecules. They were new individuals of a new organism.

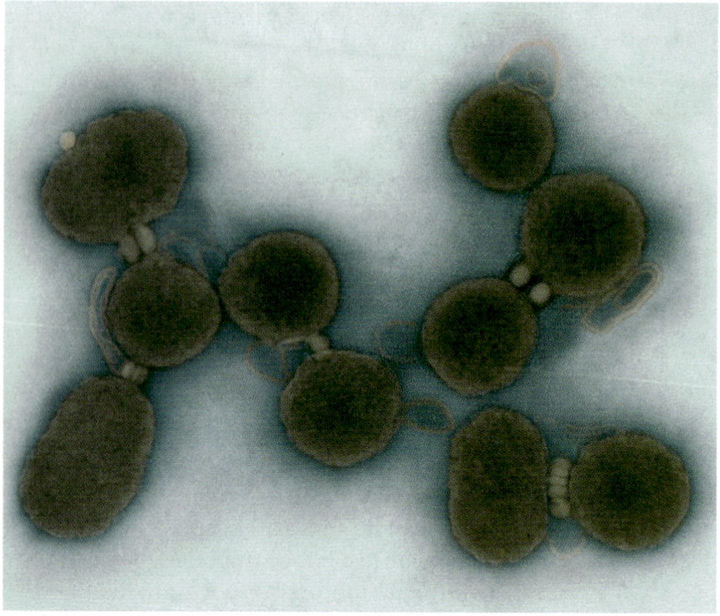

17.7 Synthetic Cells Cells of *Mycoplasma mycoides JCVI-syn 1.0*, the first synthetic organism, are shown in this false-colored micrograph.

RECAP 17.2

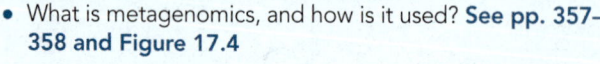

DNA sequencing is used to study the genomes of prokaryotes that are important to humans and to ecosystems. Functional genomics uses gene sequences to determine the functions of the gene products. Comparative genomics compares gene sequences from different organisms to help identify their functions and evolutionary relationships. Transposable elements, including composite transposons, move from one place to another in the genome. Studies of the minimal genome may lead to the creation of artificial species.

- Give some examples of prokaryotic genomes that have been sequenced. What have the sequences shown? **See pp. 356–357**
- What is metagenomics, and how is it used? **See pp. 357–358 and Figure 17.4**
- How do transposons move about the genome? **See pp. 358–359 and Figure 17.5**
- How are selective inactivation studies being used to determine the minimal genome, and what are possible practical applications of this? **See p. 359 and Figures 17.6 and 17.7**

Advances in DNA sequencing and sequence analysis have led to the rapid sequencing of eukaryotic genomes. We will now look at some of the new insights that have come from these studies.

WORKING WITH**DATA:**

Using Transposon Mutagenesis to Determine the Minimal Genome

Original Papers

Hutchison, C., S. N. Peterson, S. R. Gill, R. T. Cline, O. White, C. M. Fraser, H. O. Smith, and J. C. Venter. 1999. Global transposon mutagenesis and a minimal *Mycoplasma* genome. *Science* 286: 2165–2169.

Glass, J. I., N. Assad-Garcia, N. Alperovich, S. Yooseph, M. R. Lewis, M. Maruf, C. A. Hutchison III, H. O. Smith, and J. C. Venter. 2006. Essential genes of a minimal bacterium. *Proceedings of the National Academy of Sciences USA* 103: 425–430.

Analyze the Data

Mycoplasma genitalium is a tiny parasitic bacterium. Its genome was the second bacterial genome to be sequenced, and the genome's size (580,073 bp) and number of protein-coding genes (482) made it the smallest known bacterial genome at the time. For this reason, it was a useful organism for exploring the question of how many genes are necessary for life. An approach to answering this question is transposon mutagenesis. A transposon is a movable sequence of DNA that can insert within the coding region of a gene. If the gene is essential to life, a cell harboring the transposon will not survive. A team led by Craig Venter used this approach to identify 382 *M. genitalium* genes that are essential for its survival in the laboratory (see Figure 17.6). They went on to chemically synthesize all of these genes, put them together to make a chromosome, and put the chromosome into a *Mycoplasma* cell whose DNA had been destroyed. They thereby created a new cell with just the synthetic minimal genome (see Figure 17.7).

The growth of *M. genitalium* strains with insertions in genes (intragenic regions) was compared with the growth of strains with insertions in noncoding (intergenic) regions of the genome. The results are shown in the table.

QUESTION 1

Explain these data in terms of genes essential for growth and survival. Are all of the genes in *M. genitalium* essential for growth? If not, how many are essential? Why did some of the insertions in intergenic regions prevent growth?

QUESTION 2

If a transposon inserts into the following regions of a gene, there might be no effect on the phenotype. Explain in each case:

a. near the 3′ end of a coding region

b. within a gene coding for rRNA (see Section 17.3)

How does this affect your answer to Question 1?

Type of insertion	Number of different genes/regions	Number that grew
Intragenic	482	100
Intergenic	199	184

Go to **BioPortal** for all WORKING WITH**DATA** exercises

17.3 What Have We Learned from Sequencing Eukaryotic Genomes?

As genomes have been sequenced and described, a number of major differences have emerged between eukaryotic and prokaryotic genomes. For example, in Table 17.2 compare the bacterial genomes with those of yeasts, plants, and animals—all eukaryotes. Key differences include the following:

- *Eukaryotic genomes are larger than those of prokaryotes*, and they have more protein-coding genes. This difference is not surprising given that multicellular organisms have many cell types with specialized functions. As we saw above, the simple prokaryote *Mycoplasma* has several hundred protein-coding genes in a genome of 0.58 million bp. A rice plant, in contrast, has almost 43,000 genes!
- *Eukaryotic genomes have more regulatory sequences*—and many more regulatory proteins—than prokaryotic genomes. The greater complexity of eukaryotes requires much more regulation, which is evident in the many points of control associated with the expression of eukaryotic genes (see Figure 16.13).
- *Much of eukaryotic DNA is noncoding*. Distributed throughout many eukaryotic genomes are various kinds of DNA sequences that are not transcribed into mRNA, most notably introns and gene control sequences. As we discussed in Chapter 16, some noncoding sequences are transcribed into microRNAs. In addition, eukaryotic genomes contain various kinds of repeated sequences. These features are rare in prokaryotes.
- *Eukaryotes have multiple chromosomes*, whereas prokaryotes often have a single, circular chromosome. As we have described in previous chapters, eukaryotic chromosomes have multiple origins of replication, a centromere region that holds the replicated chromosomes together before mitosis, and a telomeric sequence at each end of the chromosome that maintains chromosome integrity.

Model organisms reveal many characteristics of eukaryotic genomes

Most of our information about eukaryotic genomes has come from model organisms that have been studied extensively: the yeast *Saccharomyces cerevisiae*, the nematode (roundworm) *Caenorhabditis elegans*, the fruit fly *Drosophila melanogaster*, and the thale cress plant (*Arabidopsis thaliana*). Model organisms have been chosen because they are relatively easy to grow and study in a laboratory, their genetics are well studied, and they exhibit characteristics that represent a larger group of organisms.

YEAST: THE BASIC EUKARYOTIC MODEL Yeasts are single-celled eukaryotes. Like most eukaryotes, they have membrane-enclosed organelles, such as the nucleus and endoplasmic reticulum, and a life cycle that alternates between haploid and diploid generations (see Figure 11.15).

TABLE 17.2 Representative Sequenced Genomes

Organism	Haploid Genome Size (Mb)	Number of Protein-coding Genes	Percent of Genome That Codes for Proteins
Bacteria			
Mycoplasma genitalium	0.58	482	88
Haemophilus influenzae	1.8	1,727	89
Escherichia coli	4.6	4,288	88
Yeasts			
Saccharomyces cerevisiae	12.2	6,275	70
Schizosaccharomyces pombe	13.8	4,824	60
Plants			
Arabidopsis thaliana	125	25,498	25
Oryza sativa (rice)	420	42,653	12
Animals			
Caenorhabditis elegans (nematode)	100	20,470	25
Drosophila melanogaster (fruit fly)	140	15,016	13
Tetraodon nigroviridis (pufferfish)	340	27,918	10
Gallus gallus (chicken)	1,060	~20,000	3
Homo sapiens (human)	3,200	~21,000	1.2

Mb = millions of base pairs

TABLE 17.3 Comparison of the Genomes of *E. coli* and *S. cerevisiae*

	E. coli	Yeast
Genome length (base pairs)	4,640,000	12,157,000
Number of protein-coding genes	4,288	6,275
Proteins with roles in:		
Metabolism	650	650
Energy production/storage	240	175
Membrane transport	280	250
DNA replication/repair/recombination	115	175
Transcription	230	400
Translation	182	350
Protein targeting/secretion	35	430
Cell structure	180	250

Whereas the prokaryote *E. coli* has a single circular chromosome with about 4.6 million bp and 4,288 protein-coding genes, budding yeast (*Saccharomyces cerevisiae*) has 16 linear chromosomes and a haploid content of about 12.2 million bp, with 6,275 protein-coding genes. Gene inactivation studies similar to those carried out for *M. genitalium* (see Figure 17.6) indicate that fewer than 20 percent of the yeast's genes are essential to survival.

The most striking difference between the yeast genome and that of *E. coli* is the number of genes for targeting proteins to organelles (**Table 17.3**). Both of these single-celled organisms appear to use about the same number of genes to perform the basic functions of cell survival. It is the compartmentalization of the eukaryotic yeast cell into organelles that requires it to have many more genes. This finding is direct, quantitative confirmation of something we have known for a century: the eukaryotic cell is structurally more complex than the prokaryotic cell.

THE NEMATODE: UNDERSTANDING EUKARYOTIC DEVELOPMENT In 1965 Sydney Brenner was fresh from being part of the team that first isolated mRNA. He was looking for a simple organism in which to study multicellularity, and settled on *Caenorhabditis elegans*, a 1-millimeter-long nematode (roundworm) that normally lives in the soil. It can also live in the laboratory, where it has become a favorite model organism of developmental biologists (see Section 19.4). The nematode has a transparent body that develops over 3 days from a fertilized egg to an adult worm made up of nearly 1,000 cells. In spite of its small number of cells, the nematode has a nervous system, digests food, reproduces sexually, and ages. So it is not surprising that an intense effort was made to sequence the genome of this model organism.

The *C. elegans* genome (100 million bp) is eight times larger than that of the yeast *Saccharomyces cerevisiae* and has 3.3 times as many protein-coding genes (20,470). Gene inactivation studies have shown that the worm can survive in laboratory cultures with only 10 percent of these genes. So the minimal genome of a worm is about twice the size of that of the yeast, which in turn is about three times the size of the minimal genome for *Mycoplasma*. What do these extra genes do?

All cells must have genes for survival, growth, and division. In addition, the cells of multicellular organisms must have genes for holding cells together to form tissues, for cell differentiation, and for intercellular communication. Looking at **Table 17.4**, you will recognize functions that we discussed in earlier chapters, including gene regulation (see Chapter 16) and cell communication (see Chapter 7).

***DROSOPHILA MELANOGASTER*: RELATING GENETICS TO GENOMICS** The fruit fly *Drosophila melanogaster* is a famous model organism. Studies of fruit flies resulted in the formulation of many basic principles of genetics (see Section 12.4). More than 2,500 mutations of *D. melanogaster* had been described by the 1990s when genome sequencing began, and this fact alone was a good reason for sequencing the fruit fly's DNA. The fruit fly is a much larger organism than *C. elegans*, both in size (it has ten times more cells) and complexity, and it undergoes complicated developmental transformations from egg to larva to pupa to adult.

Not surprisingly, the fly's genome (about 140 million bp) is larger than that of *C. elegans*. But as we mentioned earlier, genome size does not necessarily correlate with the number of genes encoded. In this case, the larger fruit fly genome contains fewer genes (15,016) than the smaller nematode genome. **Figure 17.8** summarizes the functions of the *Drosophila* genes that have been characterized so far; this distribution is typical of complex eukaryotes.

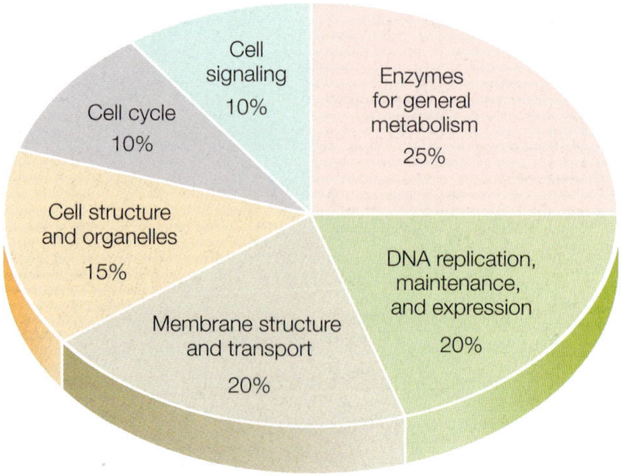
Cell signaling 10%
Enzymes for general metabolism 25%
Cell cycle 10%
Cell structure and organelles 15%
DNA replication, maintenance, and expression 20%
Membrane structure and transport 20%

17.8 Functions of the Eukaryotic Genome The distribution of gene functions in *Drosophila melanogaster* shows a pattern that is typical of many complex organisms.

TABLE 17.4
C. elegans Genes Essential to Multicellularity

Function	Protein/ Domain	Number of Genes
Transcription control	Zinc finger; homeobox	540
RNA processing	RNA binding domains	100
Nerve impulse transmission	Gated ion channels	80
Tissue formation	Collagens	170
Cell interactions	Extracellular domains; glycotransferases	330
Cell–cell signaling	G protein–linked receptors; protein kinases; protein phosphatases	1,290

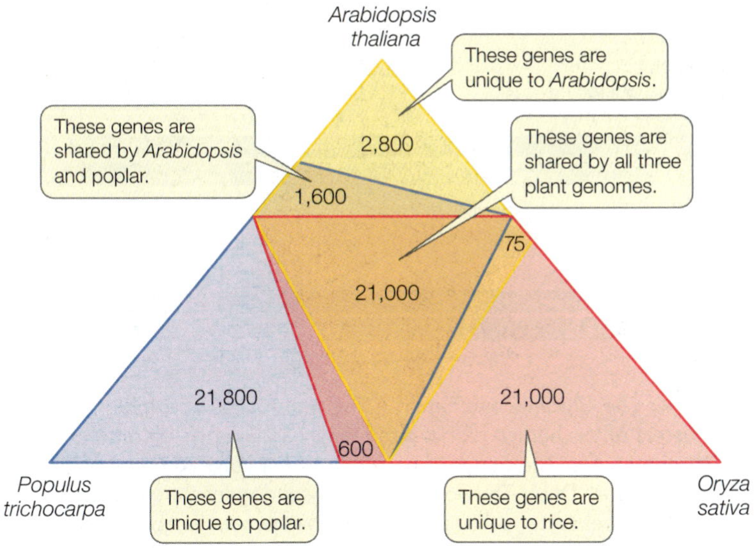

Arabidopsis thaliana
These genes are unique to Arabidopsis.
These genes are shared by Arabidopsis and poplar.
2,800
These genes are shared by all three plant genomes.
1,600
75
21,000
21,800
21,000
600
Populus trichocarpa
These genes are unique to poplar.
These genes are unique to rice.
Oryza sativa

17.9 Plant Genomes Three plant genomes share a common set of approximately 21,000 genes that appear to comprise the "minimal" plant genome.

***ARABIDOPSIS*: STUDYING THE GENOMES OF PLANTS** About 250,000 species of flowering plants dominate the land and fresh water. But in the context of the history of life, the flowering plants are fairly young, having evolved only about 200 million years ago. The genomes of some plants are huge—for example, the genome of corn is about 3 billion bp, and that of wheat is 16 billion bp. So although we are naturally most interested in the genomes of plants we use as food and fiber, it is not surprising that scientists first chose to sequence a simpler flowering plant.

Arabidopsis thaliana, thale cress, is a member of the mustard family and has long been a favorite model organism of plant biologists. It is small (hundreds could grow and reproduce in the space occupied by this page), it is easy to manipulate, and it has a relatively small genome of 125 million bp.

The *Arabidopsis* genome has an estimated 25,498 protein-coding genes, but remarkably, many of these genes are duplicates and probably originated by chromosomal rearrangements. When these duplicate genes are subtracted from the total, about 15,000 unique genes are left—similar to the number of genes found in fruit flies. Indeed, many of the genes found in these animals have homologs (related genes) in *Arabidopsis* and other plants, suggesting that plants and animals have a common ancestor.

But *Arabidopsis* has some genes that distinguish it as a plant (**Table 17.5**). These include genes involved in photosynthesis, in the transport of water into the root and throughout the plant, in the assembly of the cell wall, in the uptake and metabolism of inorganic substances from the environment, and in the synthesis of specific molecules used for defense against microbes and herbivores (organisms that eat plants). Plants cannot escape their enemies or other adverse conditions as animals can, and so they must cope with the situation where they are. The ability to make tens of thousands of unusual molecules helps them fight their enemies and adapt to the environment (see Chapter 39).

TABLE 17.5
Arabidopsis Genes Unique to Plants

Function	Number of Genes
Cell wall and growth	42
Water channels	300
Photosynthesis	139
Defense and metabolism	94

The plant-specific genes in *Arabidopsis* are also found in the genomes of other plants, including rice (*Oryza sativa*), the first major crop plant whose sequence has been determined. Rice is the world's most important crop; it is a staple in the diet of 3 billion people. Despite its larger genome, rice has a set of genes remarkably similar to those of *Arabidopsis*. More recently the genome of the poplar tree *Populus trichocarpa* was sequenced. This rapidly growing tree is widely used for manufacturing paper and is a potential source of fixed carbon for making fuel. A comparison of the three genomes shows many genes in common, comprising the basic minimal plant genome (**Figure 17.9**).

Eukaryotes have gene families

About half of all eukaryotic protein-coding genes exist as only one copy in the haploid genome (two copies in somatic cells). The rest are present in multiple copies, which arose from gene duplications. Over evolutionary time, different copies of genes have undergone separate mutations, giving rise to groups of closely related genes called **gene families**. Some gene families, such as those encoding the globin proteins that make up hemoglobin, contain only a few members. Other families, such as the genes encoding the immunoglobulins that make up antibodies, have hundreds of members. In the human genome, there are about 21,000 protein-coding genes, but 16,000 of these are members of gene families. So only about one-third of the human genes are unique.

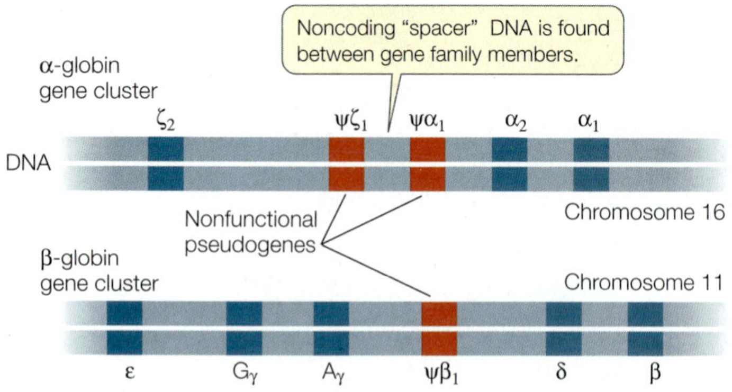
Noncoding "spacer" DNA is found between gene family members.
α-globin gene cluster
ζ2
ψζ1
ψα1
α2
α1
DNA
Nonfunctional pseudogenes
Chromosome 16
β-globin gene cluster
Chromosome 11
ε
Gγ
Aγ
ψβ1
δ
β

17.10 The Globin Gene Family The α-globin and β-globin clusters of the human globin gene family are located on different chromosomes. The genes of each cluster are separated by noncoding "spacer" DNA. The nonfunctional pseudogenes are indicated by the Greek letter psi (ψ). The γ gene has two variants, A_γ and G_γ.

The DNA sequences in a gene family are usually different from one another. As long as at least one member encodes a functional protein, the other members may mutate in ways that change the functions of the proteins they encode. During evolution, the availability of multiple copies of a gene allows for selection of mutations that provide advantages under certain circumstances. If a mutated gene is useful, it may be selected for in succeeding generations. If the mutated gene is a total loss, the functional copy is still there to carry out its role.

The family of genes that encode globins is a good example of a gene family in vertebrates. The globins are components of hemoglobin and myoglobin (an oxygen-binding protein present in muscle). The globin genes all arose long ago from a single common ancestral gene. In humans there are three functional members of the α-globin cluster and five in the β-globin cluster (**Figure 17.10**). In adults, each hemoglobin molecule is a tetramer containing two identical α-globin subunits, two identical β-globin subunits, and four heme pigments (see Figure 3.11).

During human development, different members of the globin gene cluster are expressed at different times and in different tissues. This differential gene expression has great physiological significance. For example, hemoglobin that contains γ-globin, a subunit found in the hemoglobin of the human fetus, binds O_2 more tightly than adult hemoglobin does. This specialized form of hemoglobin ensures that in the placenta, O_2 will be transferred from the mother's blood to the developing fetus's blood. Just before birth the liver stops synthesizing fetal hemoglobin and the bone marrow cells take over, making the adult form (2 α and 2 β). Thus hemoglobins with different binding affinities for O_2 are provided at different stages of human development.

In addition to genes that encode proteins, many gene families include nonfunctional **pseudogenes**, which are designated with the Greek letter psi (ψ) (see Figure 17.10). These pseudogenes result from mutations that cause a loss of function rather than an enhanced or new function. The DNA sequence of a pseudogene may not differ greatly from that of other family members. It may simply lack a promoter, for example, and thus fail to be transcribed. Or it may lack a recognition site needed for the removal of an intron, so that the transcript it makes is not correctly processed into a useful mature mRNA. In some gene families pseudogenes outnumber functional genes. In such cases, there appears to be no evolutionary advantage for the deletion of the pseudogenes, even though they have no apparent function.

Eukaryotic genomes contain many repetitive sequences

Eukaryotic genomes contain numerous repetitive DNA sequences that do not code for polypeptides. These include highly repetitive sequences, moderately repetitive sequences, and transposons.

Highly repetitive sequences are short (less than 100 bp) sequences that are repeated thousands of times in tandem (side-by-side) arrangements in the genome. They are not transcribed. Their proportion in eukaryotic genomes varies, from 10 percent in humans to about half the genome in some species of fruit flies. Often they are associated with heterochromatin, the densely packed, transcriptionally inactive part of the genome. Other highly repetitive sequences are scattered around the genome. For example, short tandem repeats (STRs) of 1 to 5 bp can be repeated up to 100 times at a particular chromosomal location. The copy number of an STR at a particular location varies among individuals and is inherited. In Chapter 15 we described how STRs can be used in the identification of individuals (DNA fingerprinting).

Moderately repetitive sequences are repeated 10 to 1,000 times in the eukaryotic genome. These sequences include the genes that are transcribed to produce tRNAs and rRNAs, which are used in protein synthesis. The cell makes tRNAs and rRNAs constantly, but even at the maximum rate of transcription, single copies of the tRNA and rRNA genes would be inadequate to supply the large amounts of these molecules needed by most cells. Thus the genome has multiple copies of these genes.

In mammals, four different rRNA molecules make up the ribosome: the 18S, 5.8S, 28S, and 5S rRNAs. (The S stands for Svedberg unit, which is a measure of size.) The 18S, 5.8S, and 28S rRNAs are transcribed together as a single precursor RNA molecule (**Figure 17.11**). As a result of several posttranscriptional steps, the precursor is cut into the final three rRNA products, and the noncoding "spacer" RNA is discarded. The sequence encoding these RNAs is moderately repetitive in humans: a total of 280 copies of the sequence are located in clusters on five different chromosomes.

Apart from the RNA genes, most moderately repetitive sequences are transposons, which, like the prokaryotic transposons we discussed earlier, can move about in the genome. Transposons make up more than 40 percent of the human genome and about 50 percent of the corn genome, although the percentage is smaller (3–10 percent) in many other eukaryotes.

Table 17.6 summarizes the four main types of transposons in eukaryotes. Three types—long terminal repeats (LTRs), long interspersed elements (LINEs), and short interspersed elements (SINEs)—are retrotransposons. Retrotransposons move about

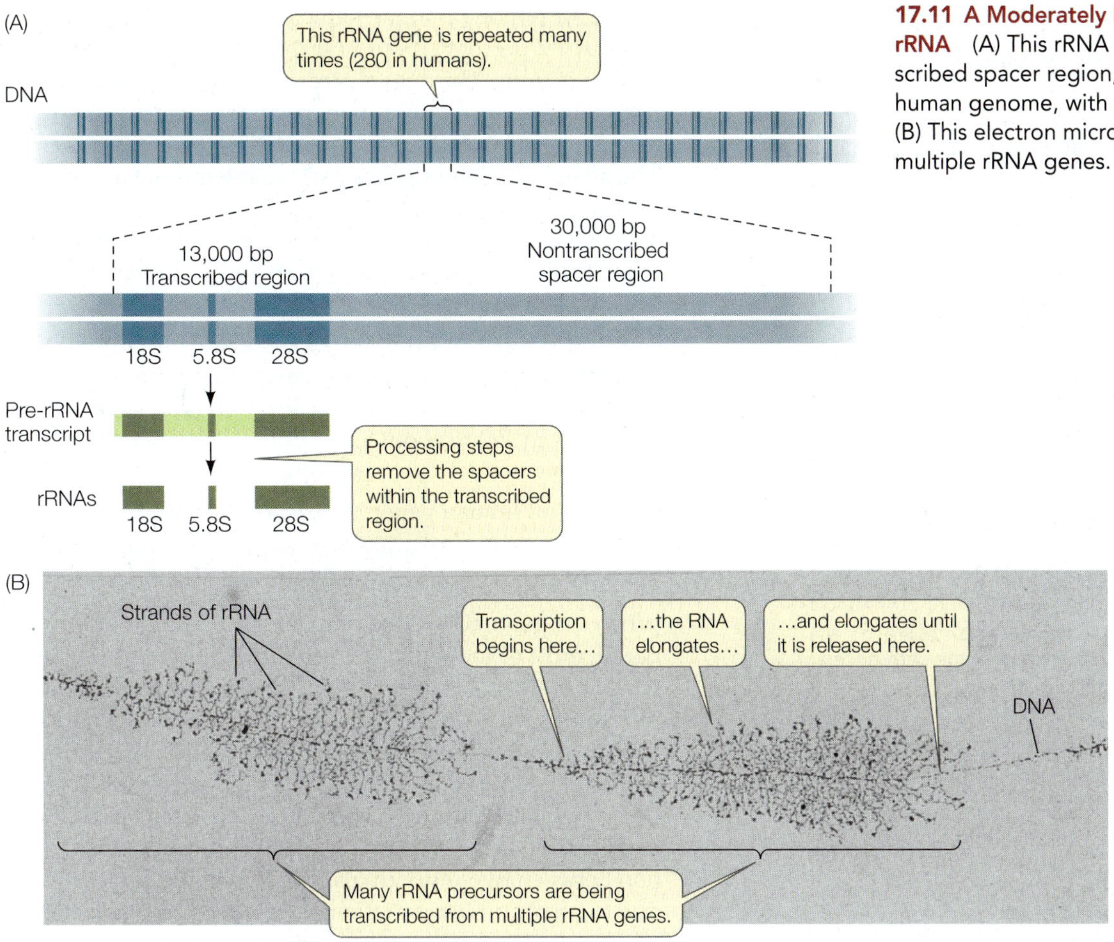
(A)
This rRNA gene is repeated many times (280 in humans).
DNA
13,000 bp
Transcribed region
30,000 bp
Nontranscribed spacer region
18S 5.8S 28S
Pre-rRNA transcript
rRNAs
18S 5.8S 28S
Processing steps remove the spacers within the transcribed region.
(B)
Strands of rRNA
Transcription begins here...
...the RNA elongates...
...and elongates until it is released here.
DNA
Many rRNA precursors are being transcribed from multiple rRNA genes.

17.11 A Moderately Repetitive Sequence Codes for rRNA (A) This rRNA gene, along with its nontranscribed spacer region, is repeated 280 times in the human genome, with clusters on five chromosomes. (B) This electron micrograph shows transcription of multiple rRNA genes.

the genome in a distinctive way: they are transcribed into RNA, which then acts as a template for new DNA. The new DNA becomes inserted at a new location in the genome. This "copy and paste" mechanism results in two copies of the transposon: one at the original location and the other at a new location. A single type of SINE retrotransposon, the 300-bp Alu element, accounts for 11 percent of the human genome; it is present in a million copies!

Transposons of the fourth type, the DNA transposons, do not use RNA intermediates. Like some prokaryotic transposable elements, they are excised from the original location and become inserted at a new location without being replicated (a "cut and paste" mechanism).

What role do these moving sequences play in the cell? The best answer so far seems to be that transposons are simply cellular parasites that can be replicated. The insertion of a transposon at a new location can have important consequences. For example, the insertion of a transposon into the coding region of a gene results in a mutation (see Figure 17.5). This phenomenon accounts for rare forms of several genetic diseases in humans, including hemophilia and muscular dystrophy. If the insertion of a transposon takes place in the germ line, a gamete with a new mutation results. If the insertion takes place in a somatic cell, cancer may result.

Sometimes an adjacent gene can be replicated along with a transposon, resulting in a gene duplication. A transposon can carry a gene, or a part of it, to a new location in the genome, shuffling the genetic

TABLE 17.6 Major Transposable Element Groups in the Eukaryotic Genome

Class	Element	Length (bp)	Number in Human Genome	Percent of Human Genome
Retrotransposon	LTRs (long terminal repeats)	100–5,000	450,000	8
	LINEs (long, interspersed elements)	6,000–8,000	850,000	17
	SINEs (short interspersed elements)	<300	1,500,000	15
DNA transposon		2,000–3,000	300,000	3

material and creating new genes. Clearly, transposition stirs the genetic pot in the eukaryotic genome and thus contributes to genetic variation.

Transposons also may have played a role in endosymbiosis, the process by which chloroplasts and mitochondria are thought to have descended from once free-living prokaryotes (see Section 5.5). In modern eukaryotes the chloroplasts and mitochondria contain some DNA, but the nucleus contains most of the genes that encode the organelles' proteins. If the organelles were once independent, they must originally have contained all of those genes. How did the genes move to the nucleus? They may have done so by DNA transpositions between organelles and the nucleus, which still occur today. The DNA that remains in the organelles may be the remnants of more complete prokaryotic genomes.

RECAP 17.3

The genomes of eukaryotes contain more genes than those of prokaryotes. Some of these genes encode functions associated with the compartmentalization of eukaryotic cells; others are needed to support multicellularity. The genome sequences of model organisms have been used to identify common features of the eukaryotic genome, including the presence of abundant regulatory sequences, repetitive sequences, and noncoding DNA. Some eukaryotic genes are in families, which may include members that are mutated and nonfunctional.

- What are the major differences between prokaryotic and eukaryotic genomes? **See p. 361**
- Describe one class of proteins found in *C. elegans* that has few counterparts in yeasts. **See p. 362 and Table 17.4**
- What is the evolutionary role of eukaryotic gene families? **See pp. 363–364 and Figure 17.10**
- Why are there multiple copies of sequences coding for rRNA in the mammalian genome? **See p. 364**
- What effects can transposons have on a genome? **See pp. 365–366**

The analysis of eukaryotic genomes has resulted in an enormous amount of useful information, as we have seen. In the next section we will look more closely at the human genome.

17.4 What Are the Characteristics of the Human Genome?

Since the first human genome sequence was completed early in the first decade of this millennium, the haploid genomes of numerous other individuals have been sequenced and published. With the rapid development of sequencing technologies, the time is approaching when a human genome can be sequenced for less than $1,000.

The human genome sequence held some surprises

The following are just some of the interesting facts that we have learned about the human genome:

- Of the 3.2 billion bp in the haploid human genome, an estimated 1.2 percent (about 21,000 genes) make up protein-coding regions. This was a surprise. Before sequencing began, the diversity of human proteins suggested there might be 80,000 to 150,000 genes in the human genome. The actual number of genes—not many more than in a fruit fly—means that posttranscriptional mechanisms (such as alternative splicing) must account for the observed number of proteins in humans. That is, the average human gene must code for several different proteins.
- The average gene has 27,000 bp. Gene sizes vary greatly, from about 1,000 bp to 2.4 million bp. Variation in gene size was expected given that human proteins (and RNAs) vary in size, from 100 to about 5,000 amino acids per polypeptide chain.
- Virtually all human genes have many introns.
- About half of the genome is made up of transposons (see Table 17.6) and other highly repetitive sequences.
- When the genomes of two unrelated individuals are compared, most of the sequence—about 99.5 percent—is identical. Despite this apparent homogeneity, there are many differences, and as more genomes are sequenced, more variants are found. Current estimates suggest that each haploid genome contains about 3.3 million single nucleotide polymorphisms (SNPs; pronounced "snips") (see Section 15.3), so these account for about one-fifth of the variation between two individuals. The remaining four-fifths are due to copy number variation: differences in sequence copy number that have arisen through chromosomal deletions, duplications, or translocations (see Figure 15.4) or through duplications caused by transposons.
- Genes are not evenly distributed over the genome. Chromosome 19 is packed densely with genes, whereas chromosome 8 has long stretches without coding regions. The Y chromosome has the fewest genes (about 230), and chromosome 1 has the most (about 3,000).

Go to Media Clip 17.1
A Big Surprise from Genomics
Life10e.com/mc17.1

Comparative genomics reveals the evolution of the human genome

Comparisons between sequenced genomes from prokaryotes and eukaryotes have revealed some of the evolutionary relationships between genes. Some genes are present in both prokaryotes and eukaryotes, others are only in eukaryotes, and still others are only in animals or only in vertebrates (**Figure 17.12**).

The genomes of various other primates, including all of the great apes, have now been sequenced. The search is on for a set of human genes that differ from those found in other primates and that make us unique. Chimpanzees are our closest living relatives, sharing nearly 99 percent of our DNA sequence. Gorillas and orangutans are next closest, with genomes that are about 98 percent and 97 percent similar to ours, respectively.

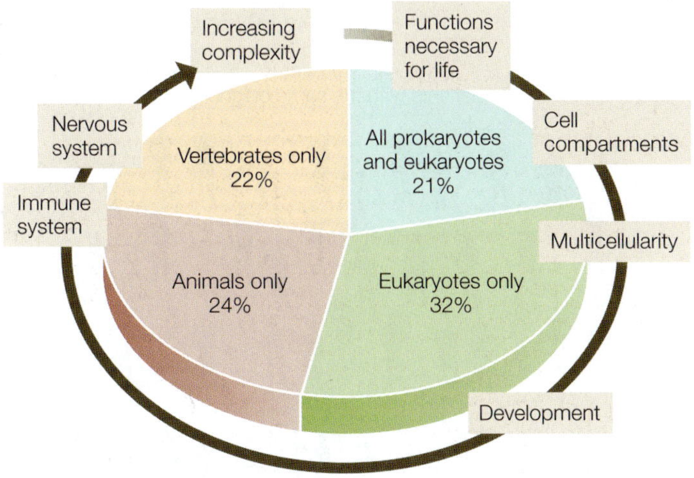
Increasing complexity
Functions necessary for life
Nervous system
Immune system
Cell compartments
Multicellularity
Development
Vertebrates only 22%
All prokaryotes and eukaryotes 21%
Animals only 24%
Eukaryotes only 32%

17.12 Evolution of the Genome A comparison of the human and other genomes has revealed how genes with new functions have been added over the course of evolution. Each percentage number refers to genes in the human genome. Thus 21 percent of human genes have homologs in prokaryotes and other eukaryotes, 32 percent of human genes occur only in other eukaryotes, and so on.

The gorilla sequence was completed in 2012 and was the last of the great ape genomes to be sequenced. The international team of researchers who published the sequence have identified about 500 protein-coding genes that have undergone accelerated evolution in humans, chimpanzees, and gorillas, including genes involved in hearing and brain development. Further analyses of these sequences may reveal genes that distinguish us from other apes, and that "make humans human."

Other clues about "human" genes have come from sequencing the genomes of ancient human relatives. An international team of scientists led by Svante Pääbo at the Max Planck Institute has extracted and sequenced DNA from the bones of Neanderthals, who lived in Europe up to 50,000 years ago. The entire Neanderthal genome has been sequenced. It is more than 99 percent identical to our human DNA, justifying the classification of Neanderthals as part of the same genus, *Homo*.

Comparisons of humans and Neanderthals with regard to specific genes and mutations are ongoing and have already revealed several interesting facts:

- The gene *MC1R* is involved in skin and hair pigmentation. A point mutation found in Neanderthals but not humans caused lower activity of the MC1R protein when it was introduced into cell cultures. Such lower activity of MC1R is known to result in fair skin and red hair in humans. So it appears that at least some Neanderthals may have had pale skin and red hair (**Figure 17.13**).
- The gene *FOXP2* is involved in vocalization in many organisms, including birds and mammals. Mutations in this gene result in severe speech impairment in humans. The Neanderthal *FOXP2* gene is identical to that of humans, whereas that of chimpanzees is slightly different. This has led to speculation that Neanderthals were capable of speech.
- While the human and Neanderthal genome sequences are very similar, there are differences in many point mutations and larger chromosomal arrangements. There are distinctive "human" DNA sequences, and also distinctive "Neanderthal" sequences. There is some mixture of the two, indicating that humans and Neanderthals interbred, with transfer of DNA between the two.

17.13 A Neanderthal Child Genome sequencing and analyses have led to this reconstruction of a Neanderthal child who lived about 60,000 years ago.

Human genomics has potential benefits in medicine

Most complex phenotypes are determined not by single genes but by multiple genes interacting with the environment. The single-allele explanations of phenylketonuria and sickle-cell anemia (see Chapter 15) do not apply to such common disorders as diabetes, heart disease, and Alzheimer's disease. To understand the genetic bases of these diseases, biologists are now using rapid genotyping technologies to create "haplotype maps," which are used to identify SNPs that are linked to genes involved in disease.

HAPLOTYPE MAPPING The SNPs that differ among individuals are not inherited as independent alleles. Rather, a set of SNPs that are present on a segment of chromosome are usually inherited as a unit. This linked piece of a chromosome is called a **haplotype**. You can think of the chromosome as a sentence, the haplotype as a word, and the SNP as a letter in the word. Analyzing SNPs is faster and less expensive than sequencing whole genomes, so haplotype mapping provides a shortcut for identifying the locations of genes and mutations involved in particular diseases (see Section 15.3). By comparing the haplotypes of individuals with and without a particular genetic disease, the genetic loci associated with the disease can be identified (**Figure 17.14**).

New technologies are continually being developed to analyze thousands or millions of SNPs in the genomes of individuals. Such technologies include rapid sequencing methods and

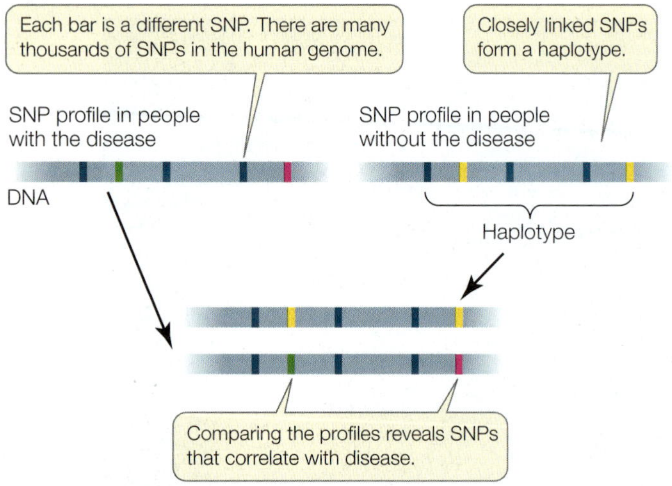
Each bar is a different SNP. There are many thousands of SNPs in the human genome.
Closely linked SNPs form a haplotype.
SNP profile in people with the disease
SNP profile in people without the disease
DNA
Haplotype
Comparing the profiles reveals SNPs that correlate with disease.

17.14 SNP Genotyping and Disease Scanning the genomes of people with and without particular diseases reveals correlations between SNPs and complex diseases.

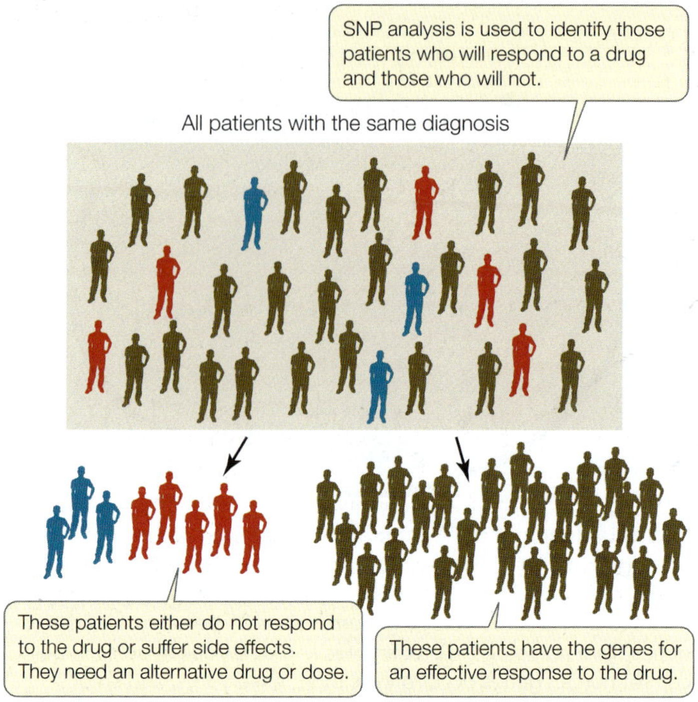
SNP analysis is used to identify those patients who will respond to a drug and those who will not.
All patients with the same diagnosis
These patients either do not respond to the drug or suffer side effects. They need an alternative drug or dose.
These patients have the genes for an effective response to the drug.

17.15 Pharmacogenomics Correlations between genotypes and responses to drugs will help physicians develop personalized medical care. The different colors indicate individuals with different SNPs.

DNA microarrays (see Chapter 18) that depend on DNA hybridization to identify specific SNPs. For example, a microarray of 500,000 SNPs has been used to analyze thousands of people to find out which SNPs are associated with specific diseases. The amount of data is prodigious: 500,000 SNPs, thousands of people, thousands of medical records. With so much natural variation, statistical measures of association between a haplotype and a disease need to be very rigorous.

GENOTYPING TECHNOLOGY AND PERSONAL GENOMICS Association tests like the one described above have revealed particular haplotypes or alleles that are associated with modestly increased risks for such diseases as breast cancer, diabetes, arthritis, obesity, and coronary heart disease (**Table 17.7**). Private companies can now scan a human genome for these variants—and the price for this service keeps getting lower. However, at this point it is unclear what a person without symptoms should do with the information, since multiple genes, environmental influences, and epigenetic effects all contribute to the development of these diseases.

Of course, the most comprehensive way to analyze a person's genome is by actually sequencing it. Once the cost of genome sequencing is within an affordable range, SNP testing may be superseded.

TABLE 17.7
SNP Human Genome Scans and Diseases

Disease	Location of SNP (Chromosome Number)	Percent of Increased Risk	
		Heterozygotes	Homozygotes
Breast cancer	8	20	63
Coronary heart disease	9	20	56
Heart attack	9	25	64
Obesity	16	32	67
Diabetes	10	65	277
Prostate cancer	8	26	58

PHARMACOGENOMICS Genetic variation can affect how an individual responds to a particular drug. For example, a drug may be chemically modified in the liver to make it more or less active. Consider an enzyme that catalyzes the following reaction:

$$\text{Active drug} \rightarrow \text{less active drug}$$

A mutation in the gene that encodes this enzyme may make the enzyme less active. For a given dose of the drug, a person with the mutation would have more active drug in the bloodstream than a person without the mutation. So the effective dose of the drug would be lower in the person with the mutation.

Now consider a different case, in which the liver enzyme is needed to make the drug active:

$$\text{Inactive drug} \rightarrow \text{active drug}$$

A person carrying a mutation in the gene encoding this liver enzyme would not be affected by the drug, since the activating enzyme is not present.

The study of how an individual's genome affects his or her response to drugs or other outside agents is called **pharmacogenomics**. Just as it is possible to identify haplotypes or SNPs that are associated with particular disease susceptibilities, it is also possible to identify SNPs that are associated with specific drug responses. This type of analysis makes it possible to predict whether a drug will be effective. The objective is to personalize drug treatment so that a physician can know in advance whether an individual will benefit from a particular drug (**Figure 17.15**). This approach might also be used to reduce the incidence of adverse drug reactions by identifying individuals who will metabolize a drug slowly, which can lead to a dangerously high level of the drug in the body.

RECAP 17.4

The haploid human genome has 3.2 billion bp, but about 1.2 percent of the genome codes for proteins. Most human genes are subject to alternative splicing; this may account for the fact that there are more proteins than genes. Haplotype mapping to find correlations of specific SNPs with disease and drug susceptibility holds promise for personalized medicine.

- What are some of the major characteristics of the human genome? See p. 366
- How does haplotype mapping work in personalized medicine? See pp. 367–368 and Figures 17.14, 17.15

Genome sequencing has advanced our understanding of biology enormously. High-throughput technologies are now being applied to other components of the cell: proteins and metabolites. We will now turn to the results of these studies.

17.5 What Do the New Disciplines of Proteomics and Metabolomics Reveal?

"The human genome is the book of life." Statements like this were common at the time the human genome sequence was first revealed. They reflect "genetic determinism," that a person's phenotype is determined by his or her genotype. But is an organism just a product of gene expression? We know that it is not. The proteins and small molecules present in any cell at a given point in time reflect not just gene expression but also modifications by the intracellular and extracellular environment. Two new fields have emerged to complement genomics and take a more complete snapshot of a cell and organism: proteomics and metabolomics.

The proteome is more complex than the genome

As mentioned above, many genes encode more than a single protein (**Figure 17.16A**). Alternative splicing leads to different combinations of exons in the mature mRNAs transcribed from a single gene (see Figure 16.22). Posttranslational modifications also increase the number of protein variants that can be derived from one gene (see Figure 14.20). However, in a multicellular organism many proteins are produced only by certain cells under specific conditions. Even single-celled organisms express only a subset of their genes at any particular time. The **proteome** is the sum total of the proteins produced by a cell, tissue, or organism at a given time, under defined conditions.

Two methods are commonly used to analyze the proteome:

- Because of their unique amino acid compositions (primary structures), most proteins have unique combinations of electric charge and size. On the basis of these two properties, they can be separated by two-dimensional gel electrophoresis. Thus isolated, individual proteins can be analyzed, sequenced, and studied (**Figure 17.16B**).
- Mass spectrometry uses electromagnets to identify proteins by the masses of their atoms. Each protein has a unique pattern of these atomic masses.

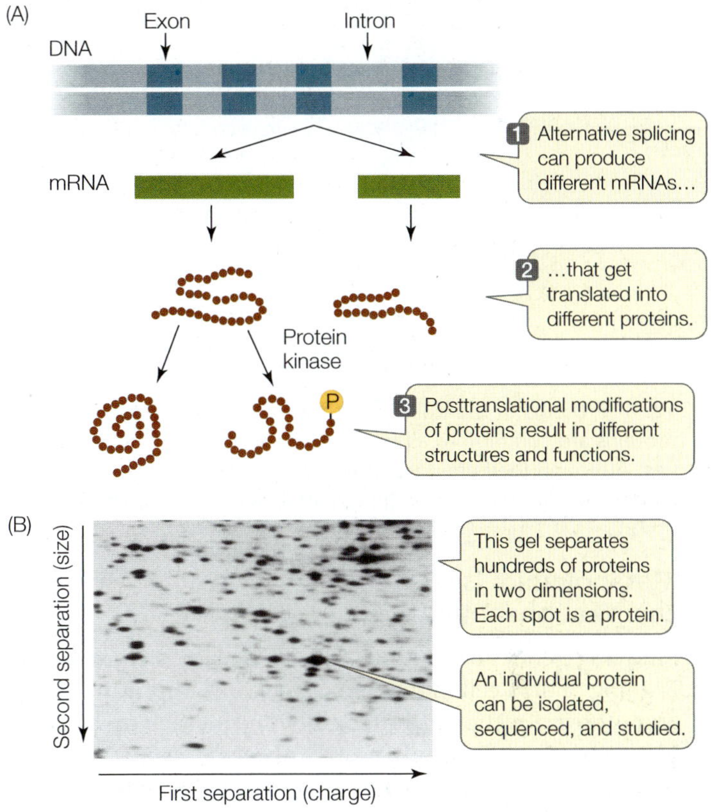

(A)
DNA
Exon
Intron
mRNA
1 Alternative splicing can produce different mRNAs…
2 …that get translated into different proteins.
Protein kinase
P
3 Posttranslational modifications of proteins result in different structures and functions.
(B)
Second separation (size)
First separation (charge)
This gel separates hundreds of proteins in two dimensions. Each spot is a protein.
An individual protein can be isolated, sequenced, and studied.

17.16 Proteomics (A) A single gene can code for multiple proteins. (B) A cell's proteins can be separated on the basis of charge and size by two-dimensional gel electrophoresis. The two separations can distinguish most proteins from one another.

The ultimate aim of proteomics is just as ambitious as that of genomics. Whereas genomics seeks to describe the genome and its expression, proteomics seeks to identify and characterize all of the expressed proteins.

Comparisons of the proteomes of humans and other eukaryotic organisms have revealed a common set of proteins that can be categorized into groups with similar amino acid sequences and similar functions. When considered on a whole organism basis, 46 percent of the yeast proteome, 43 percent of the nematode proteome, and 61 percent of the fruit fly proteome are shared by the human proteome. Functional analyses indicate that this set of 1,300 proteins provides the basic metabolic functions of a eukaryotic cell, such as glycolysis, the citric acid cycle, membrane transport, protein synthesis, DNA replication, and so on (**Figure 17.17**).

Of course, these are not the only human proteins. There are many more, which presumably distinguish us as *human* eukaryotic organisms. As we have mentioned before, proteins have different functional regions called domains (for example, a domain for binding a substrate, or a domain for spanning a membrane). While a particular organism may have many unique proteins, those proteins are often just unique combinations of domains that exist in other organisms. *This reshuffling of the genetic deck is a key to evolution.*

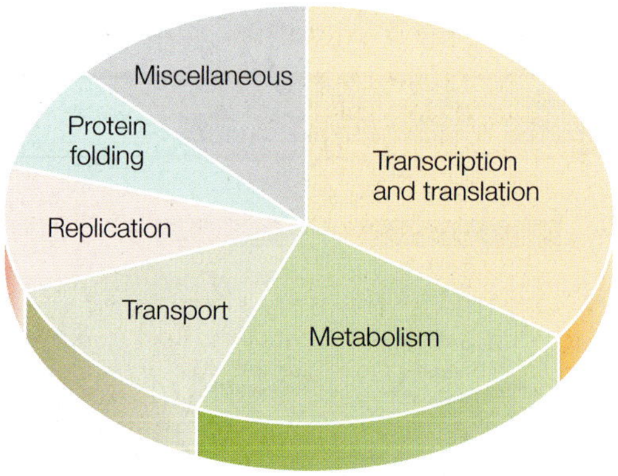
Miscellaneous
Protein folding
Transcription and translation
Replication
Transport
Metabolism

17.17 Proteins of the Eukaryotic Proteome About 1,300 proteins are common to all eukaryotes and fall into these categories. Although their amino acid sequences may differ to a limited extent, they perform the same essential functions in all eukaryotes.

Metabolomics is the study of chemical phenotype

Studying genes and proteins gives a limited picture of what is going on in a cell. But as we have seen, both gene function and protein function are affected by the internal and external environments of the cell. Many proteins are enzymes, and their activities affect the concentrations of their substrates and products. So as the proteome changes, so will the abundances of small molecules called metabolites. The **metabolome** is the complete set of small molecules present in a cell, tissue, or organism under defined conditions. These include:

- *Primary metabolites* involved in normal processes, such as intermediates in pathways such as glycolysis. This category also includes hormones and other signaling molecules.
- *Secondary metabolites*, which are often unique to particular organisms or groups of organisms. They are often involved in special responses to the environment. Examples are antibiotics made by microbes, and the many chemicals made by plants that are used in defense against pathogens and herbivores.

Not surprisingly, measuring metabolites involves sophisticated analytical instruments. If you have studied organic or analytical chemistry, you may be familiar with gas chromatography and high-performance liquid chromatography, which separate molecules, and mass spectrometry and nuclear magnetic resonance spectroscopy, which are used to identify them. These measurements result in "chemical snapshots" of cells or organisms, which can be related to physiological states.

There has been some progress in defining the human metabolome. A database created by David Wishart and colleagues at the University of Alberta contains more than 6,500 metabolite entries. The challenge now is to relate levels of these substances to physiology. For example, you probably know that high levels of glucose in the blood are associated with diabetes. But what about the early stages of heart disease? There may be a pattern of metabolites that is diagnostic of this disease. This could aid in early diagnosis and treatment.

Plant biologists are far ahead of medical researchers in the field of metabolomics. Over the years, tens of thousands of secondary metabolites have been identified in plants, many of them made in response to environmental challenges. Some of these will be discussed in Chapter 39. The metabolome of the model organism *Arabidopsis thaliana* is being described, and will give insight into how a plant copes with stresses such as drought or pathogen attack. This knowledge could be helpful in optimizing plant growth for agriculture.

RECAP 17.5

The proteome is the total of all proteins produced by a cell, tissue, or organism under specific conditions. The metabolome is the total content of small molecules such as intermediates in primary metabolism, hormones, and secondary metabolites. The proteome and the metabolome can be analyzed using chemical methods that separate and identify molecules.

- How is the proteome analyzed? See p. 369 and Figure 17.17
- Explain the differences between genome, protoeome, and metabolome.

?

What does dog genome sequencing reveal about other animals?

ANSWER

Some breeds of dogs have leg muscles that are highly developed and permit them to run very fast. Genome analyses have shown that the gene for myostatin, a protein that inhibits muscle growth, is mutated and leads to an inactive protein in these dogs. The lack of myostatin leads to bulkier muscles (**Figure 17.18**). Comparative genomics reveals that the myostatin gene is also mutated in a cattle breed that has overdeveloped muscles. In humans there is potential for manipulating myostatin to treat muscle-wasting diseases such as muscular dystrophy. As might be expected, athletes anxious to have bulkier muscles are also very interested in this gene and its protein product.

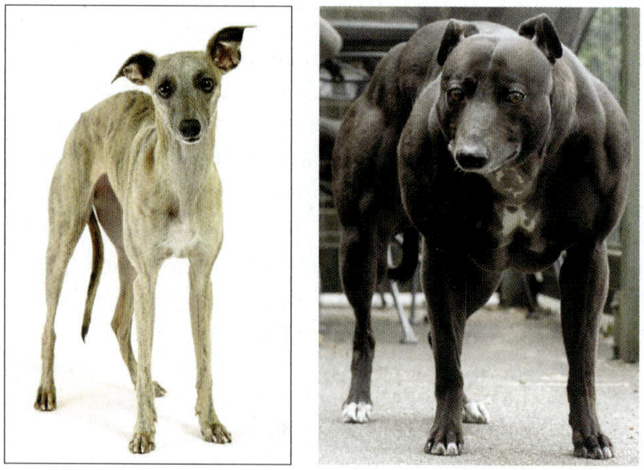

17.18 Muscular Gene These dogs are both whippets, but the muscle-bound dog (right) has a mutation in the myostatin gene.

CHAPTER**SUMMARY** 17

17.1 How Are Genomes Sequenced?

- New methods of DNA sequencing involve miniaturization and computerized analysis. **Review Figure 17.1, ANIMATED TUTORIALS 17.1, 17.2**
- Genomes are sequenced in overlapping fragments, and then the fragments are lined up to give the final sequence. **Review Figure 17.2**
- The analysis of genome sequences gives information about protein-coding and noncoding regions. **Review Figure 17.3**

17.2 What Have We Learned from Sequencing Prokaryotic Genomes?

- DNA sequencing is used to study the genomes of prokaryotes that are important to humans and ecosystems.
- **Metagenomics** is the identification of DNA sequences without first isolating, growing, and identifying the organisms present in an environmental sample. Many of these sequences are from prokaryotes that were heretofore unknown to biologists. **Review Figure 17.4**
- **Transposable elements** and **composite transposons** can move about the genome. **Review Figure 17.5**
- Transposon mutagenesis can be used to inactivate genes one by one. The mutated organism can be tested for survival and an artificial genome created based on a minimal set of essential genes. **Review Figures 17.6, 17.7**

17.3 What Have We Learned from Sequencing Eukaryotic Genomes?

- Genome sequences from model organisms have demonstrated some common features of the eukaryotic genome. In addition, there are specialized genes for cellular compartmentalization, development, and features unique to plants. **Review Tables 17.2–17.5, Figures 17.8, 17.9**
- Some eukaryotic genes exist as members of **gene families**. Proteins may be made from these closely related genes at different times and in different tissues. Some members of gene families may be nonfunctional **pseudogenes**. **Review Figure 17.10**
- Repeated sequences are present in the eukaryotic genome. **Review Table 17.6**
- **Moderately repetitive sequences** include those coding for rRNA and transposons. **Review Figure 17.11**

17.4 What Are the Characteristics of the Human Genome?

- The haploid human genome has 3.2 billion bp.
- Only about 1.2 percent of the genome codes for proteins; the rest consists of repeated sequences and noncoding DNA.
- Virtually all human genes have introns, and alternative splicing leads to the production of more than one protein per gene.
- SNP genotyping (haplotype mapping) correlates variations in the genome with diseases or drug sensitivity. It may lead to personalized medicine. **Review Figure 17.14**
- **Pharmacogenomics** is the analysis of genetics as applied to drug metabolism. **Review Figure 17.15**

17.5 What Do the New Disciplines of Proteomics and Metabolomics Reveal?

- The **proteome** is the total protein content of an organism.
- There are more proteins than there are protein-coding genes in the genome.
- The proteome can be analyzed using chemical methods that separate and identify proteins. These include two-dimensional electrophoresis and mass spectrometry. **See Figure 17.16**
- The **metabolome** is the total content of small molecules, such as intermediates in primary metabolism, hormones, and secondary metabolites.

See ACTIVITY 17.1 for a concept review of this chapter

Go to the Interactive Summary to review key figures, Animated Tutorials, and Activities Life10e.com/is17

CHAPTER**REVIEW**

REMEMBERING

1. Genome sequencing is done
   a. one chromosome at a time.
   b. on entire chromosomal DNA molecules.
   c. on overlapping fragments.
   d. manually, with large electrophoresis gels.
   e. using radioactive isotopes to label DNA.

2. The minimal genome of *Mycoplasma genitalium*
   a. has 100 genes.
   b. has been used to create new species.
   c. is made of RNA.
   d. is larger than the genome of *E. coli*.
   e. was derived from point mutations in each gene.

3. Which is *not* true of metagenomics?
   a. It has been done with bacteria.
   b. It has revealed many new species.
   c. It has revealed many new metabolic capacities.
   d. It involves extracting DNA from the environment.
   e. It cannot be done on seawater.

4. Transposons
   a. always use RNA for replication.
   b. are approximately 50 bp long.
   c. are made up of either DNA or RNA.
   d. do not contain genes coding for proteins.
   e. make up more than 40 percent of the human genome.

5. Vertebrate gene families
   a. contain only active genes.
   b. include the globins.
   c. are not produced by gene duplications.
   d. increase the number of unique genes in the genome.
   e. are not transcribed.

6. The human genome
   a. contains very few repeated sequences.
   b. has 3.2 billion bp.
   c. is 10 percent protein-coding sequences.
   d. has genes evenly distributed along chromosomes.
   e. has few genes with introns.

UNDERSTANDING & APPLYING

7. Eukaryotic protein-coding genes differ from their prokaryotic counterparts in that eukaryotic genes
   a. are double-stranded.
   b. are present in only a single copy.
   c. contain introns.
   d. have promoters.
   e. are transcribed into mRNA.

8. A comparison of the genomes of yeasts and bacteria shows that only yeasts have many genes for
   a. energy metabolism.
   b. cell wall synthesis.
   c. intracellular protein targeting.
   d. DNA-binding proteins.
   e. RNA polymerase.

9. The genomes of the fruit fly and the nematode are similar to those of yeasts, except that the former organisms have many genes for
   a. intercellular signaling.
   b. synthesis of polysaccharides.
   c. cell cycle regulation.
   d. intracellular protein targeting.
   e. transposable elements.

10. Why is identifying an organism's proteome and metabolome more complex than defining its genome?

ANALYZING & EVALUATING

11. The genomes of rice, wheat, and corn are similar to one another and to that of *Arabidopsis* in many ways. Discuss how these plants might nevertheless have very different proteins.

12. It is the year 2025. You are taking care of a patient who is concerned about having an early stage of kidney cancer. His mother died from this disease.
    a. Assume that the SNPs linked to genes involved in the development of this type of cancer have been identified. How would you determine if this man has a genetic predisposition for developing kidney cancer? Explain how you would do the analysis.
    b. How might you develop a metabolomic profile for kidney cancer and then use it to determine whether your patient has kidney cancer?
    c. If the patient was diagnosed with cancer by the methods in (a) and (b), how would you use pharmacogenomics to choose the right medications to treat the tumor in this patient?

Go to BioPortal at **yourBioPortal.com** for Animated Tutorials, Activities, LearningCurve Quizzes, Flashcards, and many other study and review resources.

18 Recombinant DNA and Biotechnology

CHAPTER OUTLINE

Oil on the Beach A massive oil spill from a drilling rig in the Gulf of Mexico in 2010 released oil that reached the seashore. On this beach in Louisiana, huge piles of contaminated sand await cleaning. Naturally occurring bacteria broke down a lot of the oil, but biotechnology may be able to improve the process.

IN 2010 THE DEEPWATER HORIZON oil rig in the Gulf of Mexico exploded, and crude oil began gushing out of the well below it. The oil flowed unabated for three months until the wellhead was finally capped; a total of 210 million gallons were released. The oil slick was visible from space—it damaged marine habitats and killed wildlife, washed up on beaches, and shut down fishing and tourism in the area. Efforts to remove the giant slick included collecting the oil, burning it, and dispersing it chemically. In addition, scientists identified bacteria living in the Gulf waters that were able to digest and break down components of the crude oil. Despite the clean-up efforts and the actions of these bacteria, much of the oil remains in the Gulf today.

Bioremediation—the use of microorganisms to remove pollutants in the environment—is an attractive option for cleaning up these spills. One approach is to encourage the growth of natural microorganisms that digest components of the crude oil. After the oil tanker *Exxon Valdez* ran aground near the Alaskan shore in 1989, nitrogen fertilizers were applied to nearby beaches to encourage the growth of oil-eating bacteria. Similar approaches have been tried in Kuwait, where destruction of wells during the 1991 Gulf War led to a massive release of oil. But the success of such methods has been limited because the naturally occurring bacteria digest only certain components of the crude oil, and because of technical difficulties in bringing the organisms into contact with the oil. Within the scientific community, there is great interest in producing genetically modified organisms that can clean up oil more rapidly and effectively.

The possibility of genetically changing bacteria for bioremediation began in 1971. Ananda Chakrabarty, at the General Electric Research Center in New York, used genetic crosses to develop a strain of the bacterium *Pseudomonas* with multiple genes for the breakdown of various hydrocarbons in oil. He and his company applied for a patent to legally protect their discovery and profit from it. In a landmark case, the U.S. Supreme Court ruled in 1980 that "a live, human-made microorganism is patentable" under the U.S. Constitution.

The biotechnology industry has flourished since then. Many genetically modified organisms (GMOs) have been developed and approved for use in agriculture, medicine, and other industries. At the same time, concerns about the safety of GMOs have been raised. In particular, there are well-founded fears concerning the environmental consequences of releasing genetically modified "superbugs" into the environment. Such approaches to environmental cleanup remain at the experimental stage.

?

Are there other uses for microorganisms in environmental cleanup?

See answer on p. 389.

18.1 What Is Recombinant DNA?

Recombinant DNA is a DNA molecule that has been made in the laboratory using at least two different sources of DNA. In Chapter 15 we discussed the use of restriction enzymes (restriction endonucleases) to detect mutations. These bacterial enzymes are also used in the laboratory to cut up (cleave) DNA. In the late 1960s scientists discovered other enzymes that act on DNA. One such enzyme is **DNA ligase**, which catalyzes the joining of DNA fragments; one of its functions is to join Okazaki fragments during DNA replication (see Section 13.3).

With these enzymes in hand, scientists could cut up DNA molecules and then splice the fragments together in new combinations ("recombine" the DNA fragments). In 1973 Stanley Cohen at Stanford University, Herbert Boyer at the University of California, and their colleagues did just that. They isolated two different plasmids (circular DNA molecules that replicate independently in bacterial cells; see Figure 12.24) from *Escherichia coli*. These plasmids contained different antibiotic resistance genes. The scientists cut the two plasmids with restriction enzymes, mixed the fragments together, and then used DNA ligase to rejoin the fragments. The products of this ligation reaction were inserted into new *E. coli* cells, and the cells were grown on solid medium containing both antibiotics. (See Section 18.2 for a discussion of how DNA is inserted into cells.) A few of the bacteria had been transformed with a recombinant plasmid containing both of the antibiotic resistance genes and could grow on the medium (**Figure 18.1**). With this experiment, **recombinant DNA technology** was born.

Let's look more closely at what happens when DNA is cut with a restriction enzyme and then rejoined with DNA ligase. Some restriction enzymes recognize palindromic DNA sequences—sequences that read the same way in both directions. For example, you can read the DNA recognition sequence for the restriction enzyme *Eco*R1 from 5′ to 3′ as GAATTC on both strands:

5′.......GAATTC......3′

3′.......CTTAAG......5′

Some restriction enzymes cut the DNA straight through the middle of the palindrome, generating "blunt-ended" fragments. Others, such as *Eco*RI, make staggered cuts—they cut the phosphodiester bond of one strand of the double helix several bases away from where they cut the other (**Figure 18.2**). After *Eco*RI makes its two cuts in the complementary strands, the ends of the strands are held together only by the hydrogen bonds between four base pairs. At warm temperatures (above room temperature) these hydrogen bonds are too weak to hold the two strands together, so the DNA separates into fragments when it is warmed. Each fragment carries a single-stranded "overhang" at the location of each cut. These overhangs are called **sticky ends** because they have specific base sequences that can bind by base pairing with complementary sticky ends.

Any two sticky ends that are complementary can form hydrogen bonds with one another. The sticky ends of the original molecule may rejoin, or two different fragments may join (see Figure 18.2). Indeed, a fragment from one source, such as a human, can be joined to a fragment from another source, such as a bacterium. Initially the fragments are held together only by weak hydrogen bonds, but the enzyme ligase catalyzes the formation of covalent bonds between adjacent nucleotides at the ends of the fragments, joining them to form a single, larger molecule.

Besides *Eco*RI, there are hundreds of other restriction enzymes, each with its own recognition sequence. With these tools—restriction enzymes and DNA ligase—scientists can cut and rejoin different DNA molecules from any and all sources,

INVESTIGATING**LIFE**

18.1 Recombinant DNA With the discovery of restriction enzymes and DNA ligase, it became possible to combine DNA fragments from different sources in the laboratory. But would such "recombinant DNA" be functional when inserted into a living cell? The results of this experiment completely changed the scope of genetic research, increasing our knowledge of gene structure and function, and ushered in the new field of biotechnology.[a]

HYPOTHESIS Biologically functional recombinant plasmids can be made in the laboratory.

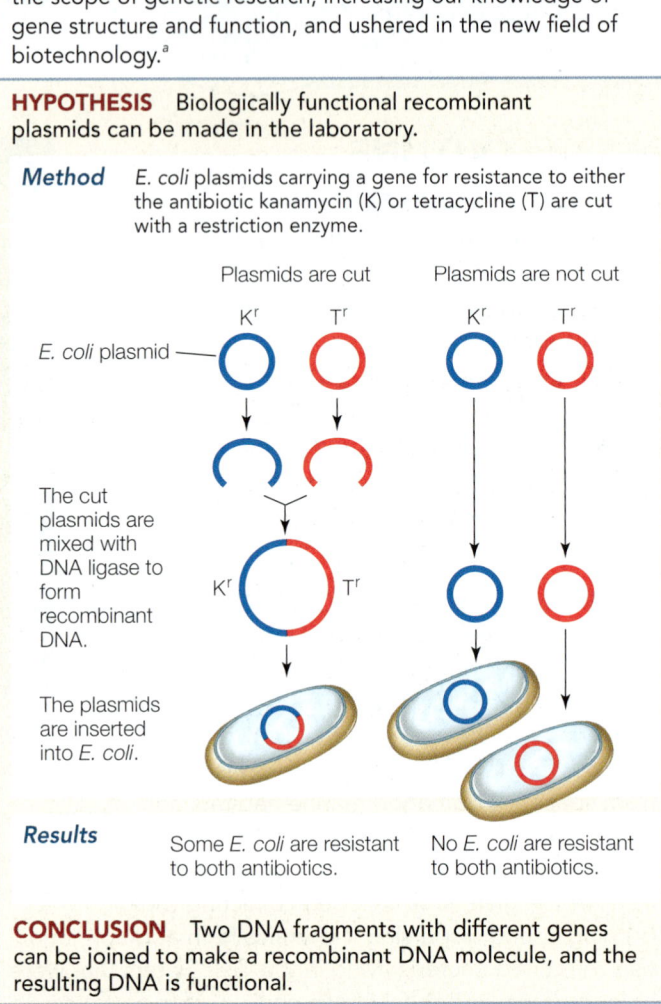

Method *E. coli* plasmids carrying a gene for resistance to either the antibiotic kanamycin (K) or tetracycline (T) are cut with a restriction enzyme.

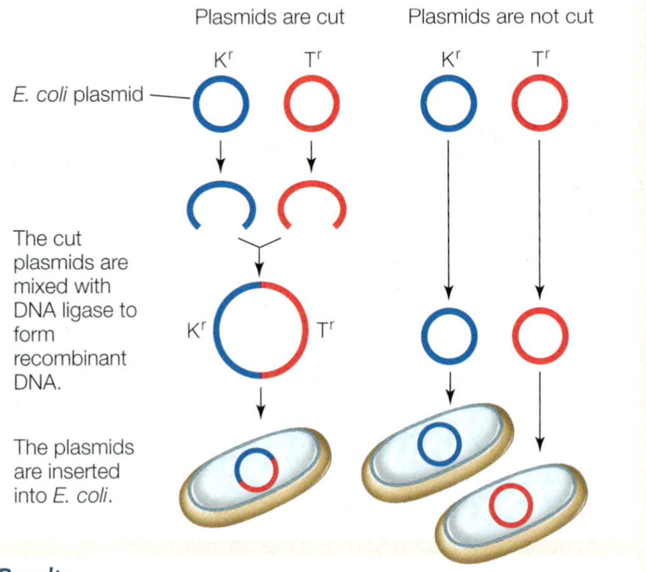

Plasmids are cut
Plasmids are not cut
K^r
T^r
K^r
T^r
E. coli plasmid
The cut plasmids are mixed with DNA ligase to form recombinant DNA.
K^r
T^r
The plasmids are inserted into E. coli.

Results Some *E. coli* are resistant to both antibiotics. No *E. coli* are resistant to both antibiotics.

CONCLUSION Two DNA fragments with different genes can be joined to make a recombinant DNA molecule, and the resulting DNA is functional.

Go to **BioPortal** for discussion and relevant links for all INVESTIGATING**LIFE** figures.

[a]Cohen, S. N. et al. 1973. *Proceedings of the National Academy of Sciences USA* 70: 3240–3244.

Go to Media Clip 18.1
Striking Views of Recombinant DNA Being Made
Life10e.com/mc18.1

WORKING WITH**DATA:**

Recombinant DNA

Original Paper

Cohen, S. N., A. C. Y. Chang, H. W. Boyer, and R. B. Helling. 1973. Construction of biologically functional bacterial plasmids *in vitro*. *Proceedings of the National Academy of Sciences USA* 70: 3240–3244.

Analyze the Data

In the 1960s scientists discovered restriction enzymes and DNA ligase. In 1973 Stanley Cohen, Hebert Boyer, and colleagues used these two enzymes as chemical reagents to show that biologically functional recombinant plasmids can be constructed in the laboratory (see Figure 18.1). Specifically, they used the restriction enzyme *Eco*RI to cut two *E. coli* plasmids, one containing a resistance gene for kanamycin and the other containing a resistance gene for tetracycline. The fragments were mixed together, and DNA ligase was used to join them in random combinations. The ligated molecules were then inserted back into *E. coli*. Some of the transformed cells were resistant to both antibiotics, indicating that they had been transformed with a recombinant plasmid that contained both the tetracycline and kanamycin resistance genes. For this landmark experiment on recombinant DNA, Cohen was awarded the Nobel Prize in 1986.

QUESTION 1

Two plasmids were used in this study. One, pSC101, had a gene for tetracycline resistance, and the other, pSC102, had a gene for kanamycin resistance. In one experiment, some pSC101 was cut with the restriction enzyme *Eco*RI but not sealed up with DNA ligase. Cut or intact pSC101 were used to transform *E. coli* cells sensitive to antibiotics. The transformed cells were plated on media containing tetracycline, kanamycin, or chloramphenicol (another antibiotic). The results are shown in **TABLE A**. What can you conclude from this experiment?

QUESTION 2

In another experiment, pSC101 and pSC102 were mixed and treated in various ways. The mixtures were used to transform antibiotic-sensitive *E. coli*. The three treatments were: intact ("None"), cut with *Eco*RI, and cut with *Eco*RI then sealed with DNA ligase. The results are shown in **TABLE B**. Did treatment with DNA ligase improve the efficiency of genetic transformation by the cut plasmids? What is the quantitative evidence for your statement?

QUESTION 3

How did the antibiotic-resistant bacteria arise in the "None" DNA treatment?

QUESTION 4

Did the *Eco*RI + DNA ligase treatment result in an increase in doubly-resistant bacteria over controls? What data provide evidence for your statement?

QUESTION 5

For the *Eco*RI + DNA ligase treatment, compare the number of transformants that were resistant to either tetracycline or kanamycin alone to the number that were doubly resistant. What accounts for the large difference? (Hint: look at Figure 18.2.)

TABLE A

	Transformants per µg DNA	
Plasmid	Tetracycline	Kanamycin
Intact pSC101	3×10^5	None
*Eco*RI-treated pSC101	2.8×10^4	None

TABLE B

	Transformants per µg DNA		
Treatment of DNA	Tetracycline	Kanamycin	Tetracycline and Kanamycin
None	2×10^5	1×10^5	2×10^2
*Eco*RI	1×10^4	1.1×10^3	7×10^1
*Eco*RI + DNA ligase	1.2×10^4	1.3×10^3	5.7×10^2

Go to **BioPortal** for all WORKING WITH**DATA** exercises

including artificially synthesized DNA sequences. Recently, new tools have been developed for making recombinant DNA. Methods based on the polymerase chain reaction (PCR) allow the joining of any two DNA molecules without the need for conveniently placed restriction enzyme sites. There are even methods for the construction of multiple artificial genes using programmable microchips. Despite these advances, restriction enzymes and DNA ligase are still used routinely for recombinant DNA construction in biology labs.

RECAP 18.1

DNA fragments from different sources can be linked together to make recombinant DNA.

- How did Cohen and Boyer make the first recombinant DNA? **See Figure 18.1**
- How does a staggered cut in DNA create a "sticky end"? **See p. 374 and Figure 18.2**

Recombinant DNA has no biological significance until it is inserted inside a living cell, which can replicate and transcribe the transplanted genetic information. How can recombinant DNA made in the laboratory be inserted and expressed in living cells?

18.2 How Are New Genes Inserted into Cells?

One goal of recombinant DNA technology is to **clone**—that is, to produce many identical copies of—a particular gene. We have seen the term "clone" used in the context of whole cells or organisms (see Chapters 11 and 12) that are genetically identical to one another. A gene can be cloned by inserting it into a bacterial cell such as *E. coli*. The bacterium is allowed to reproduce and multiply into millions of identical cells, all carrying copies of the gene. Cloning might be done for analysis,

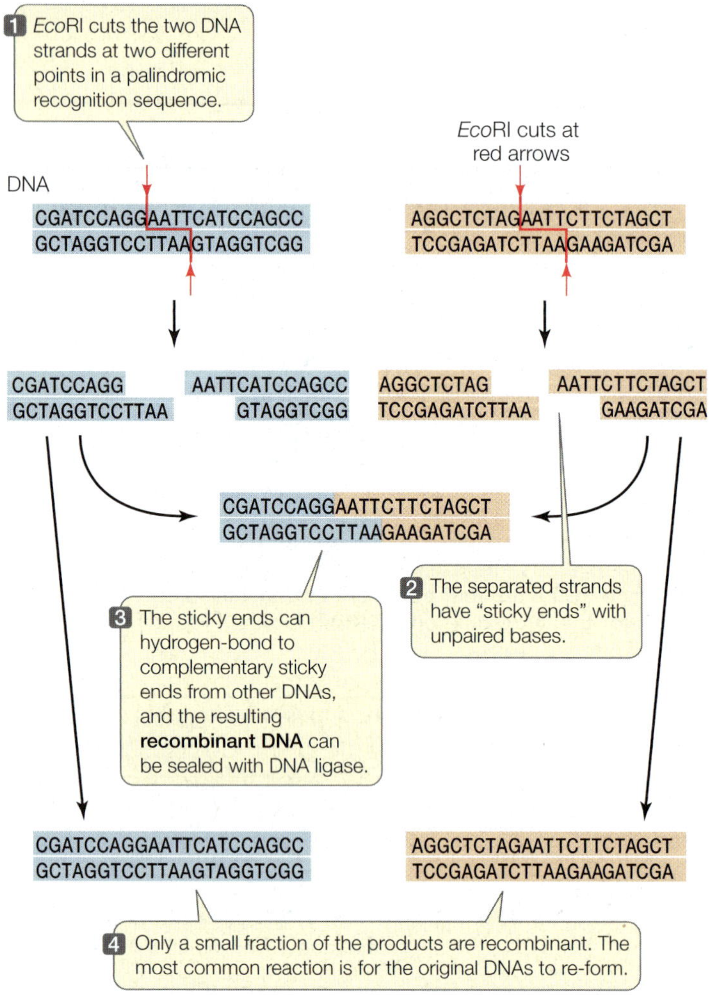
1 EcoRI cuts the two DNA strands at two different points in a palindromic recognition sequence.
EcoRI cuts at red arrows
DNA
CGATCCAGGAATTCATCCAGCC
GCTAGGTCCTTAAGTAGGTCGG
AGGCTCTAGAATTCTTCTAGCT
TCCGAGATCTTAAGAAGATCGA
CGATCCAGG
GCTAGGTCCTTAA
AATTCATCCAGCC
GTAGGTCGG
AGGCTCTAG
TCCGAGATCTTAA
AATTCTTCTAGCT
GAAGATCGA
CGATCCAGGAATTCTTCTAGCT
GCTAGGTCCTTAAGAAGATCGA
2 The separated strands have "sticky ends" with unpaired bases.
3 The sticky ends can hydrogen-bond to complementary sticky ends from other DNAs, and the resulting recombinant DNA can be sealed with DNA ligase.
CGATCCAGGAATTCATCCAGCC
GCTAGGTCCTTAAGTAGGTCGG
AGGCTCTAGAATTCTTCTAGCT
TCCGAGATCTTAAGAAGATCGA
4 Only a small fraction of the products are recombinant. The most common reaction is for the original DNAs to re-form.

18.2 Cutting, Splicing, and Joining DNA Some restriction enzymes (*Eco*RI is shown here) make staggered cuts in DNA. *Eco*RI can be used to cut two different DNA molecules (blue and orange). The exposed bases can hydrogen-bond with complementary exposed bases on other DNA fragments, forming recombinant DNA. DNA ligase stabilizes the recombinant molecule by forming covalent bonds in the DNA backbone.

to produce a protein product in quantity, or as a step toward creating an organism with a new phenotype.

Recombinant DNA is cloned by inserting it into host cells in a process known as **transformation**—or **transfection** if the host cells are derived from an animal. A host cell or organism that contains recombinant DNA is referred to as a **transgenic** cell or organism, and the non-native DNA is called a transgene. Later in this chapter we will encounter many examples of transgenic organisms, including yeast, mice, wheat plants, and even cows.

Various methods are used to create transgenic cells. Generally these methods are inefficient in that only a few of the cells that are exposed to the recombinant DNA actually become transformed with it. In order to select for the transgenic cells, **selectable marker** genes, such as genes that confer resistance to antibiotics, are often included as part of the recombinant DNA molecule. When an antibiotic resistance gene is used as the selectable marker, the cells from the transformation experiment are grown in the presence of the antibiotic; the antibiotic kills all nontransgenic cells, leaving only the transgenic cells. Antibiotic resistance genes were the markers used in Cohen and Boyer's experiment (see Figure 18.1).

Genes can be inserted into prokaryotic or eukaryotic cells

In theory, any cell or organism can act as a host for the introduction of recombinant DNA. Most research has been done using model organisms:

- *Bacteria* are easily grown and manipulated in the laboratory. Much of their molecular biology is known, especially for well-studied bacteria such as *E. coli*. Furthermore, bacteria contain plasmids, which are easily manipulated to carry recombinant DNA into the cell. But since the processes of transcription, translation, and posttranslational modification proceed differently in prokaryotes than they do in eukaryotes, bacteria might not be suitable as hosts to express eukaryotic genes.
- *Yeasts* such as *Saccharomyces cerevisiae* are commonly used as eukaryotic hosts for recombinant DNA studies. The advantages of using yeasts include rapid cell division (a life cycle completed in 2–4 hours), ease of growth in the laboratory, and a relatively small genome size (see Table 17.2). In addition, yeast cells have most of the characteristics of other eukaryotic cells, with a notable exception being those associated with multicellularity.
- *Plant cells* are good hosts because of their ability to make stem cells (unspecialized totipotent cells) from mature plant tissues. The unspecialized cells can be transformed with recombinant DNA and then studied in culture, or grown into new plants. There are also methods for making whole transgenic plants without going through the cell culture step. These methods result in plants that carry the recombinant DNA in all their cells, including the germ line cells.
- *Cultured animal cells* can be used to study expression of human or animal genes, for example for medical purposes. Whole transgenic animals can also be created.

A variety of methods are used to insert recombinant DNA into host cells

Methods for inserting DNA into host cells vary. The cells may be chemically treated to make their outer membranes more permeable, and then mixed with the DNA so that it can diffuse into the cells. Another approach is called electroporation: a short electric shock is used to create temporary pores in the membranes through which the DNA can enter. Viruses can be altered so that they carry recombinant DNA into cells. A common method for transforming plants involves a specific bacterium that inserts DNA into plant cells. Transgenic animals can be produced by injecting recombinant DNA into the nuclei of fertilized eggs. There are even "gene guns" that "shoot" the host cells with tiny metal particles coated with the DNA.

The challenge of inserting new DNA into a cell lies not only in getting it into the host cell, but also in getting it to replicate as the host cell divides. DNA polymerase does not bind to and copy just any sequence. If the new DNA is to be replicated, it must become part of a segment of DNA that contains an origin of replication (see Figure 13.12). Such a DNA molecule is called a **replicon**, or replication unit.

There are two general ways in which the newly introduced DNA can become part of a replicon within the host cell:

- It may be inserted into a host chromosome. Although the site of insertion is usually random, this is nevertheless a common method of integrating new genes into host cells.
- It can enter the host cell as part of a carrier DNA sequence, called a **vector**, and can then either integrate into a host chromosome or be replicated from its own origin of replication.

Several types of vectors are used to get DNA into cells, some of which we will now describe in more detail.

PLASMIDS AS VECTORS As we described in Chapter 12, plasmids are small, circular DNA molecules that replicate autonomously in many prokaryotic cells. Several characteristics make plasmids useful as transformation vectors:

- Plasmids are relatively small (an *E. coli* plasmid usually has 2,000–6,000 base pairs) and are therefore easy to manipulate in the laboratory.
- A typical plasmid has one or more restriction enzyme recognition sequences that each occur only once in the plasmid sequence. These sites make it easy to insert additional DNA into the plasmid before it is used to transform host cells.
- Many plasmids contain genes that confer resistance to antibiotics, which can serve as selectable markers.
- Plasmids have a bacterial origin of replication (*ori*) and can replicate independently of the host chromosome. It is not uncommon for a bacterial cell to contain hundreds of copies of a recombinant plasmid. For this reason, the power of bacterial transformation to amplify a gene is extraordinary. A one-liter culture of bacteria harboring the human β-globin gene in a typical plasmid has as many copies of that gene as there are cells in a typical adult human (10^{14}).

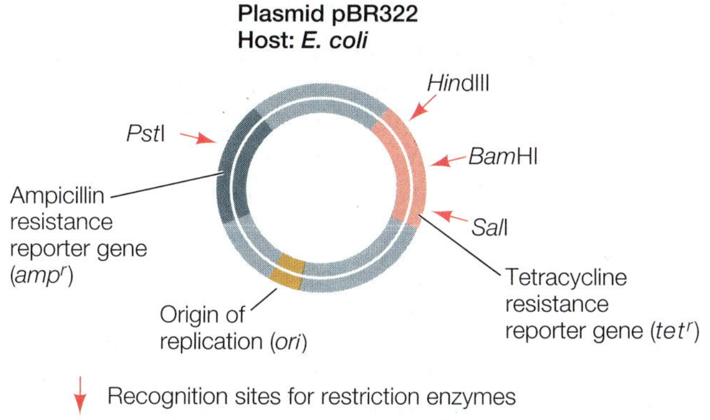
Plasmid pBR322
Host: E. coli
HindIII
PstI
BamHI
Ampicillin resistance reporter gene (amp^r)
SalI
Tetracycline resistance reporter gene (tet^r)
Origin of replication (ori)
Recognition sites for restriction enzymes

The plasmids used as vectors in the laboratory have been extensively altered to include convenient features: multiple cloning sites, often with 20 or more unique restriction enzyme sites for cloning purposes; origins of replication that will function in a variety of host cells; and various kinds of reporter genes (see below), such as selectable marker genes.

PLASMID VECTORS FOR PLANTS An important vector for carrying new DNA into many types of plants is a plasmid found in the bacterium *Agrobacterium tumefaciens*. This bacterium lives in the soil, infects plants, and causes a disease called crown gall, which is characterized by the presence of growths (or tumors) on the plant. *Agrobacterium* contains a plasmid called Ti (for *t*umor-*i*nducing). When the bacterium infects a plant cell, a region of the Ti plasmid called the T DNA is inserted into the cell, where it becomes incorporated into one of the plant's chromosomes. The Ti plasmid carries the genes needed for this transfer and incorporation of the T DNA. The T DNA carries genes that are expressed by the host cell, causing the growth of tumors and the production of specific sugars that the bacterium uses as sources of energy. Scientists have exploited this remarkable natural "genetic engineer" to insert foreign DNA into the genomes of plants.

When used as a vector for plant transformation, the tumor-inducing and sugar-producing genes on the T DNA are removed and replaced with foreign DNA. The recombinant Ti plasmids are first used to transform *Agrobacterium* cells from which the original Ti plasmids have been removed. Then the *Agrobacterium* cells are used to infect plant cells.

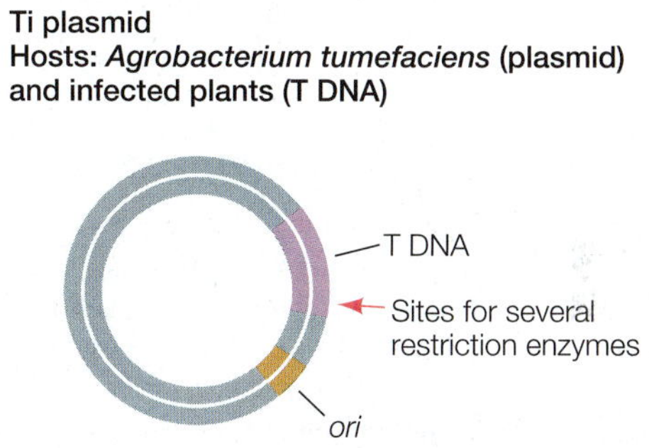
Ti plasmid
Hosts: Agrobacterium tumefaciens (plasmid) and infected plants (T DNA)
T DNA
Sites for several restriction enzymes
ori

VIRUSES AS VECTORS Constraints on plasmid replication limit the size of the new DNA that can be inserted into a plasmid to about 10,000 base pairs. Although many prokaryotic genes may be smaller than this, most eukaryotic genes—with their introns and extensive flanking sequences—are bigger. A vector that accommodates larger DNA inserts is needed for these genes.

Both prokaryotic and eukaryotic viruses are used as vectors for eukaryotic DNA. Bacteriophage λ, which infects *E. coli*, has a DNA genome of about 45,000 base pairs. About 20,000 base pairs are not necessary for the bacteriophage to complete its life cycle (see Figure 16.14). These 20,000 base pairs can be deleted and replaced with DNA from another organism, which then gets replicated along with the virus DNA. Because viruses infect cells naturally, they offer a great advantage over plasmids, which often require artificial means to coax them to enter host cells.

Reporter genes help select or identify host cells containing recombinant DNA

Even when a population of host cells is exposed to an appropriate vector, only a small proportion of the cells actually take up the

RESEARCH**TOOLS**

18.3 Selection for Recombinant DNA Selectable marker (reporter) genes are used by biologists to select for bacteria that have taken up a plasmid. In a typical experiment, most of the bacteria will not take up any DNA. Of those that do, only a small fraction will take up recombinant DNA.

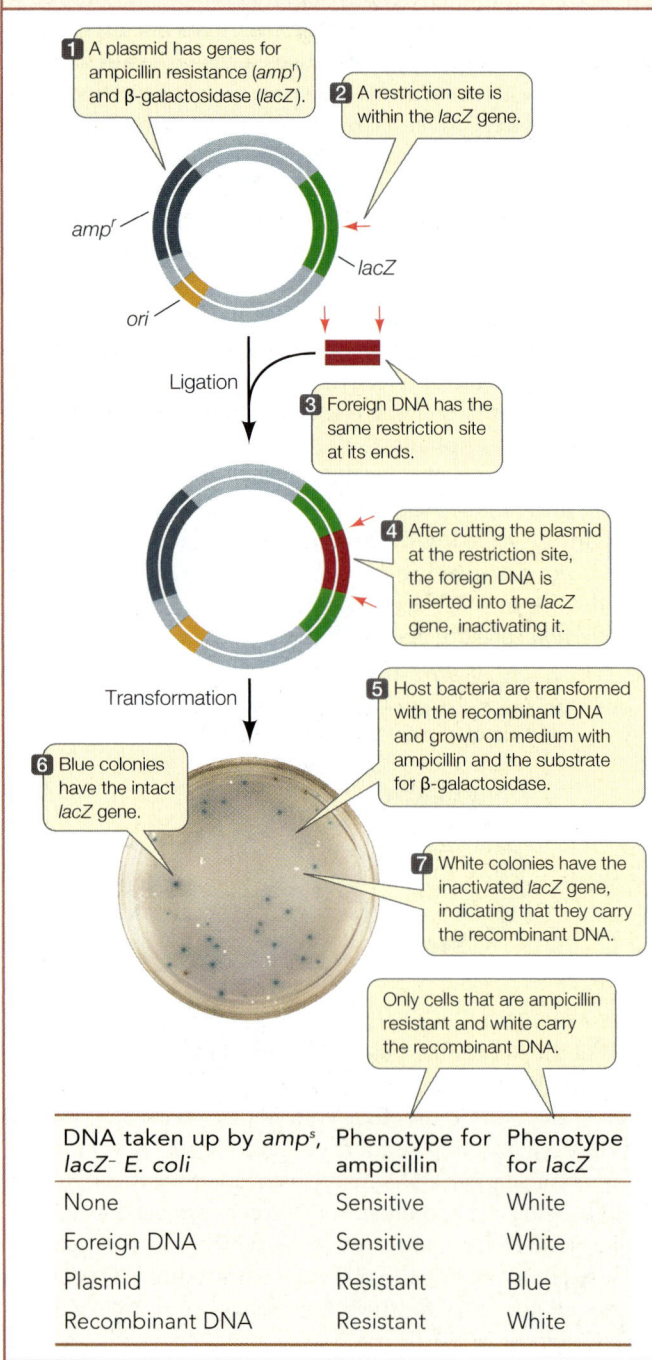

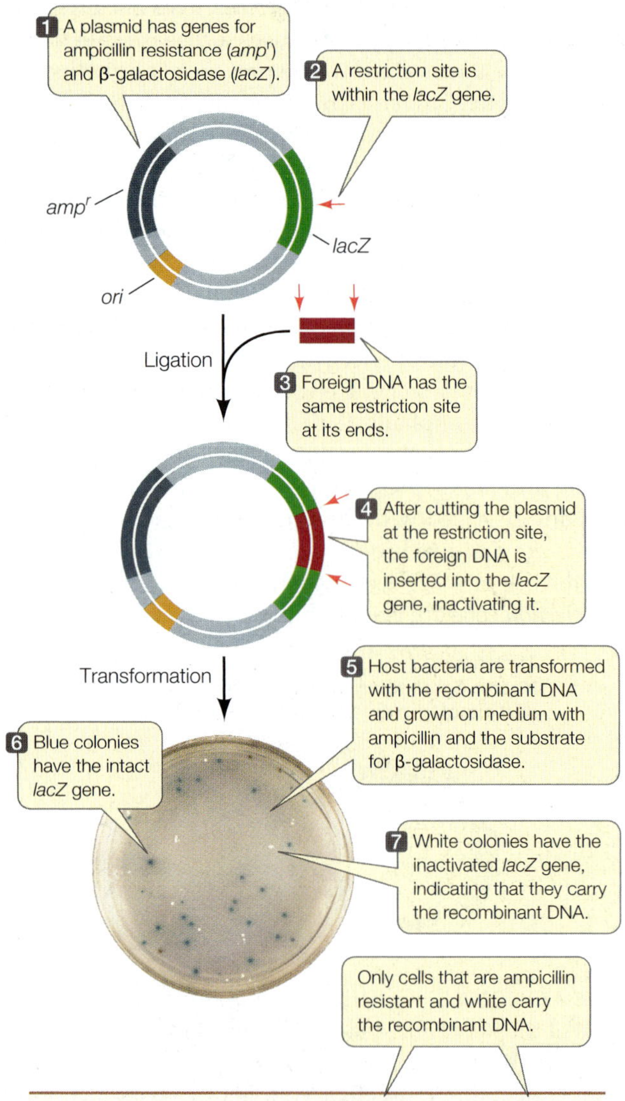
1 A plasmid has genes for ampicillin resistance (amp^r) and β-galactosidase (*lacZ*).
2 A restriction site is within the *lacZ* gene.
amp^r
lacZ
ori
Ligation
3 Foreign DNA has the same restriction site at its ends.
4 After cutting the plasmid at the restriction site, the foreign DNA is inserted into the *lacZ* gene, inactivating it.
Transformation
5 Host bacteria are transformed with the recombinant DNA and grown on medium with ampicillin and the substrate for β-galactosidase.
6 Blue colonies have the intact *lacZ* gene.
7 White colonies have the inactivated *lacZ* gene, indicating that they carry the recombinant DNA.
Only cells that are ampicillin resistant and white carry the recombinant DNA.

DNA taken up by amp^s, $lacZ^-$ *E. coli*	Phenotype for ampicillin	Phenotype for *lacZ*
None	Sensitive	White
Foreign DNA	Sensitive	White
Plasmid	Resistant	Blue
Recombinant DNA	Resistant	White

vector. Furthermore, the process of making recombinant DNA is far from perfect. During a ligation reaction, DNA molecules can combine in various ways, many of which do not produce the desired recombinant molecule (see Figures 18.1 and 18.2). Methods have been developed to improve the chance that a desired combination will occur. A simple approach is to cut the vector with two different restriction enzymes that have sites near each other. This produces a molecule with incompatible sticky ends, which is much less likely to simply recircularize during the ligation reaction. The desired insert molecule is cut with the same two enzymes, so that theoretically a functional circular plasmid can only be produced if the vector and insert ligate with one another. Even so, it is often the case that only a small proportion of the ligation products have the desired recombinant sequence.

How can we identify or select the host cells that contain the desired sequence? One way is to use selectable markers, such as those used in early experiments with recombinant DNA (see Figure 18.1). Selectable markers are one type of **reporter gene**, which is any gene whose expression is easily observed. There are several types of reporter genes:

- An antibiotic resistance gene in a plasmid or other vector allows the selection of transformed host cells. If the host cells are normally sensitive to the antibiotic, they will grow on medium containing the antibiotic only if they have been transformed by the vector. This approach is used in the selection of transgenic prokaryotic and eukaryotic cells, including those of plants and animals.
- The β-galactosidase (*lacZ*) gene in the *E. coli lac* operon (see Figure 16.4) codes for an enzyme that can convert the artificial substrate X-Gal into a bright blue product. Many plasmids contain the *lacZ* gene with a multiple cloning site (i.e., multiple unique restriction enzyme sites) within its sequence. These plasmids also carry genes for antibiotic resistance. To clone a specific DNA fragment, the foreign DNA and the plasmid are each cut with one of the restriction enzymes that can cut within the cloning site. The vector and the foreign DNA are then combined in a ligation reaction, and the ligation products are used to transform bacteria. Bacterial colonies containing the plasmid are selected on a solid medium containing the antibiotic. X-Gal is also included in the medium. If a bacterial colony contains a recombinant plasmid that carries the foreign DNA inserted into the *lacZ* gene, it will not make β-galactosidase and the colony will be white. Clones that contain the original plasmid with no insert express the *lacZ* gene and make blue colonies (**Figure 18.3**).
- Green fluorescent protein (GFP), which normally occurs in the jellyfish *Aequorea victoria*, emits visible green light when exposed to ultraviolet light. The gene for this protein has been isolated and incorporated into vectors. GFP is now widely used as a reporter gene (**Figure 18.4**). GFP has also been modified to emit other colors when exposed to ultraviolet light, and these new variants are widely used by molecular biologists.

RECAP 18.2

Recombinant DNA can be cloned by using a vector to insert it into a suitable host cell. The vector often has a selectable marker or other reporter gene that gives the host cell a phenotype by which transgenic cells can be identified.

- List the characteristics of a plasmid that make it suitable for introducing new DNA into a host cell. **See p. 377**
- How are cells harboring a vector that carries recombinant DNA identified? **See p. 378 and Figure 18.3**

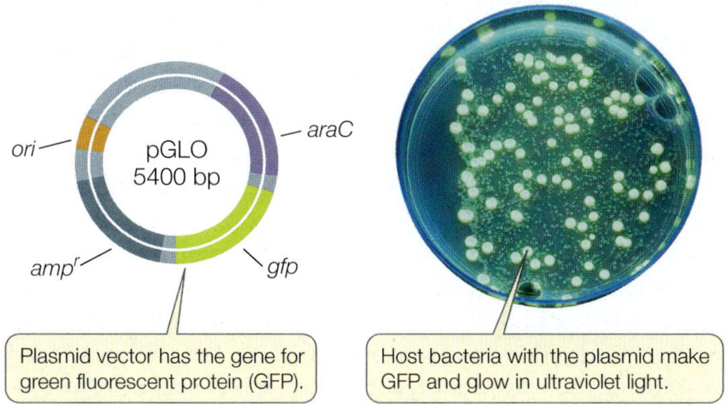
ori
pGLO
5400 bp
araC
ampr
gfp
Plasmid vector has the gene for green fluorescent protein (GFP).
Host bacteria with the plasmid make GFP and glow in ultraviolet light.

18.4 Green Fluorescent Protein as a Reporter The presence of a plasmid with the gene for green fluorescent protein (GFP) is readily apparent in transgenic cells because the GFP that the cells produce glows under ultraviolet light. This allows the identification of cells carrying a plasmid without the use of selection on antibiotics. That is, no cells are killed during the selection process.

We have described how DNA can be cut with restriction enzymes, inserted into a vector, and introduced into host cells. We have also seen how host cells carrying recombinant DNA can be identified. Now let's consider where the genes or DNA fragments used in these procedures come from.

18.3 What Sources of DNA Are Used in Cloning?

A major goal of molecular cloning experiments is to learn the functions of the DNA sequences and the proteins they encode. The DNA fragments used in cloning procedures are obtained from several sources. They include random fragments of chromosomes that are maintained as gene libraries, complementary DNA obtained by reverse transcription from mRNA, products of the polymerase chain reaction (PCR), and artificially synthesized or mutated DNA.

Libraries provide collections of DNA fragments

A **genomic library** is a collection of DNA fragments that together comprise the genome of an organism. Restriction enzyme digestion or other means, such as mechanical shearing, can be used to break chromosomes into smaller pieces. These smaller DNA fragments still constitute a genome (Figure 18.5A), but the information is now in many smaller "volumes." Each fragment is inserted into a vector, which is then taken up by a host cell. Proliferation of a single transformed cell produces a colony of cells, each of which harbors many copies of the same fragment of DNA.

When plasmids are used as vectors, about 700,000 separate fragments are required to make a library of the human genome. By using bacteriophage λ, which can carry about four times as much DNA as a plasmid, the number of "volumes" in the library can be reduced to about 160,000. Although this still seems like a large number, a single petri plate can hold thousands of phage colonies, or plaques, which are easily screened for the presence of a particular DNA sequence by hybridization to an appropriate DNA probe.

cDNA is made from mRNA transcripts

A much smaller DNA library—one that includes only the genes transcribed in a particular tissue—can be made from **complementary DNA**, or **cDNA** (Figure 18.5B). This involves isolating mRNA from cells, then making cDNA copies of that mRNA by complementary base pairing. An enzyme, **reverse transcriptase**, catalyzes this reaction.

A collection of cDNAs from a particular tissue at a particular time in the life cycle of an organism is called a **cDNA**

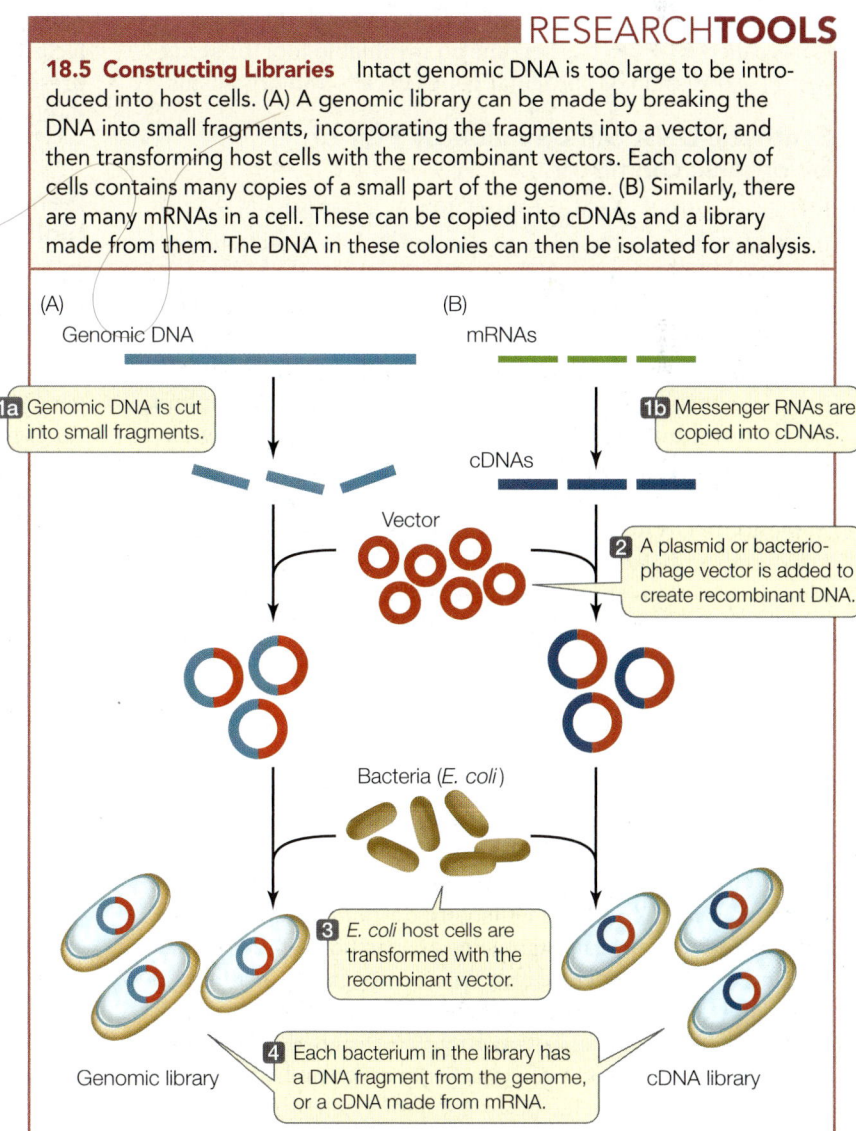

RESEARCH**TOOLS**

18.5 Constructing Libraries Intact genomic DNA is too large to be introduced into host cells. (A) A genomic library can be made by breaking the DNA into small fragments, incorporating the fragments into a vector, and then transforming host cells with the recombinant vectors. Each colony of cells contains many copies of a small part of the genome. (B) Similarly, there are many mRNAs in a cell. These can be copied into cDNAs and a library made from them. The DNA in these colonies can then be isolated for analysis.

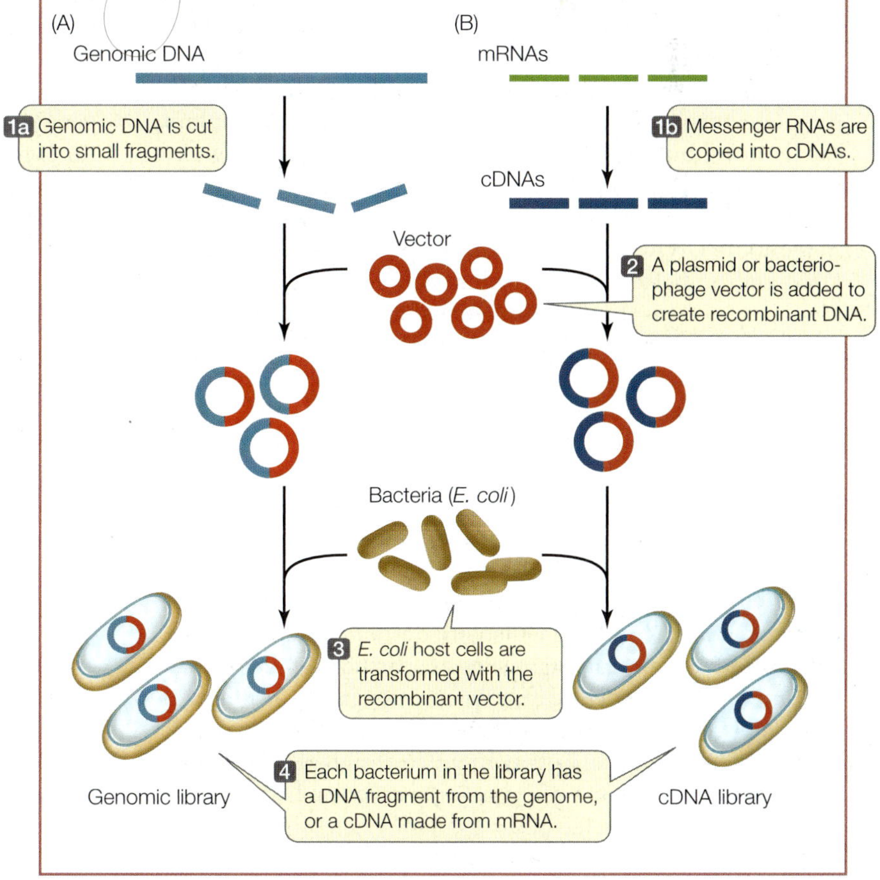
(A)
Genomic DNA
(B)
mRNAs
1a Genomic DNA is cut into small fragments.
1b Messenger RNAs are copied into cDNAs.
cDNAs
Vector
2 A plasmid or bacteriophage vector is added to create recombinant DNA.
Bacteria (E. coli)
3 E. coli host cells are transformed with the recombinant vector.
Genomic library
4 Each bacterium in the library has a DNA fragment from the genome, or a cDNA made from mRNA.
cDNA library

library. Because mRNAs do not last long in the cytoplasm, the types and amounts of mRNAs present in a cell are a good representation of the transcription rates of all the genes in that cell. Thus a cDNA library is a "snapshot" that captures the transcription pattern of a set of cells at a given point in time. cDNA libraries have been invaluable for comparing gene expression in different tissues at different stages of development. For example, researchers have found that up to one-third of all the genes of an animal are expressed only during development. cDNA is also a good starting point for cloning eukaryotic genes, because the clones contain only the coding sequences of the genes (the introns having been spliced out; see Figure 14.8). Also, if a eukaryotic gene is highly expressed in a particular tissue, a cDNA library made from that tissue will be enriched for the gene, making it easier to identify and clone the gene.

Reverse transcriptase along with PCR (see below) can also be used to create and amplify a specific cDNA sequence without the need to construct a library. In this case, RNA is isolated from an organism or tissue, and reverse transcriptase is used to make cDNA from the RNA. Then PCR is used to amplify a specific sequence directly from the sample of cDNA. This procedure, called **RT-PCR**, has become an invaluable tool for studying the expression of particular genes in cells and organisms.

Synthetic DNA can be made by PCR or by organic chemistry

PCR is a method of amplifying DNA in a test tube, based on the same processes used to replicate DNA in a cell (see Figure 13.21). PCR can begin with as little as 10^{-12} g of DNA (a picogram). Any fragment of DNA can be amplified by PCR as long as appropriate primers are available. You will recall that DNA replication (by PCR or in a cell) requires not just a template onto which DNA polymerase adds complementary nucleotides, but also a short oligonucleotide primer where replication begins (see Figure 13.13). If the appropriate primers (two are needed—one for each strand of DNA) are added to template DNA in a PCR reaction, millions of copies of the DNA region between the primers can be produced in just a few hours. This amplified DNA can then be inserted into a vector to create recombinant DNA and be cloned in host cells.

Artificial DNA can be synthesized using organic chemistry to link nucleotides together in a specified sequence. This process is now fully automated, and a service laboratory can make large numbers of short- to medium-length sequences overnight. PCR primers, for example, are made in this way. The synthesis of artificial DNA does not require a template, so DNA with any sequence can be made. This flexibility is useful for creating DNA fragments with desirable characteristics, such as convenient restriction sites or specific mutations. Longer synthetic sequences can be pieced together to construct completely artificial genes that have been designed for specific purposes. For example, a gene might be designed to be highly expressed in a particular cell type, or to encode a highly active enzyme.

RECAP 18.3

DNA for cloning can be obtained using genomic libraries, cDNA, or artificially synthesized DNA fragments.

- How are genomic DNA and cDNA libraries made and used? **See pp. 379–380 and Figure 18.5**
- Explain how RT-PCR is used to amplify a specific gene sequence. **See p. 380**

We've explored the various sources of DNA that can be used to make recombinant DNA molecules, and how organisms are transformed with recombinant DNA. We will now turn to some of the ways that recombinant DNA and transformation methods can be used to study the functions of genes and proteins.

18.4 What Other Tools Are Used to Study DNA Function?

So far in this chapter we have seen how recombinant DNA is made and how organisms are transformed with recombinant DNA. In this section we will examine several additional techniques for studying DNA. These approaches include expression of genes in different biological systems, mutagenesis, methods to block gene expression, and DNA microarrays to analyze large numbers of nucleotide sequences.

Genes can be expressed in different biological systems

One way to study a gene and its protein product is to express it in cells that do not normally express that gene. Often a scientist will want to express a gene derived from one kind of organism in another, very different organism—for example, a human gene in a bacterium, or a bacterial gene in a plant. For successful expression, the gene of interest must be combined with a promoter and other regulatory sequences from the host organism: a bacterial promoter will not function in a plant cell, for example. The coding region of the gene of interest is inserted between a promoter and a transcription termination sequence derived from the host organism, or from an organism that uses similar mechanisms for gene regulation.

Another way to study a gene is to overexpress it in cells where it is normally expressed at lower levels, so that much more of its protein product is made. To achieve this, a copy of the coding region is inserted downstream of a different, stronger promoter, and cells are transformed with the recombinant DNA.

There are many thousands of examples of how such experiments have shed light on the functions of genes and their protein products. One example involves a genetic system that probably evolved to prevent inbreeding in plants. Most plants produce flowers with both male and female parts, but many plant species are self-incompatible; they cannot self-pollinate (see Sections 22.3 and 38.1). Their flowers produce a protein that recognizes "self" pollen, and prevents the pollen from fertilizing egg cells in the same flower (see Figure 38.5). Genetic crosses suggested that a particular multi-allelic gene, called the *S* gene, was responsible for self-incompatibility. Definitive proof was obtained when plants were transformed with recombinant DNA containing an *S* allele different from their own *S* alleles. The transgenic plants rejected not only their own pollen but also pollen from flowers that naturally carried the foreign *S* allele.

DNA mutations can be created in the laboratory

Mutations that occur in nature have been important in demonstrating cause-and-effect relationships for a specific gene. However, mutations in nature are rare events. Recombinant DNA technology allows us to ask "what if" questions by creating mutations artificially. Because synthetic DNA can be made with any desired sequence, it can be manipulated to create specific mutations, the consequences of which can be observed when the mutant DNA is expressed in host cells.

Genes can be inactivated by homologous recombination

As we have just discussed, a gene can be studied by expressing it in cells where it is not normally expressed or by expressing it at higher than normal levels. Another way to study a gene is to inactivate it, so that it is not transcribed and translated into a functional protein. An example of this approach is the use of transposon mutagenesis to define the minimal genome (see Section 17.2). In animals, such a manipulation is called a **knockout** experiment. Methods have been developed to knock out genes in model organisms such as mice, fruit flies, and nematodes. We will focus here on the technique used in mice.

In mice, **homologous recombination** can be used to knock out a specific gene (Figure 18.6). As we saw in Chapter 11, homologous recombination occurs when a pair of homologous chromosomes line up during meiosis (see Figures 11.17 and 11.18). Homologous recombination also occurs in cells that are not undergoing meiosis, as part of their DNA repair processes. A key feature of homologous recombination is that it involves an exchange of DNA between molecules with identical, or nearly identical, sequences.

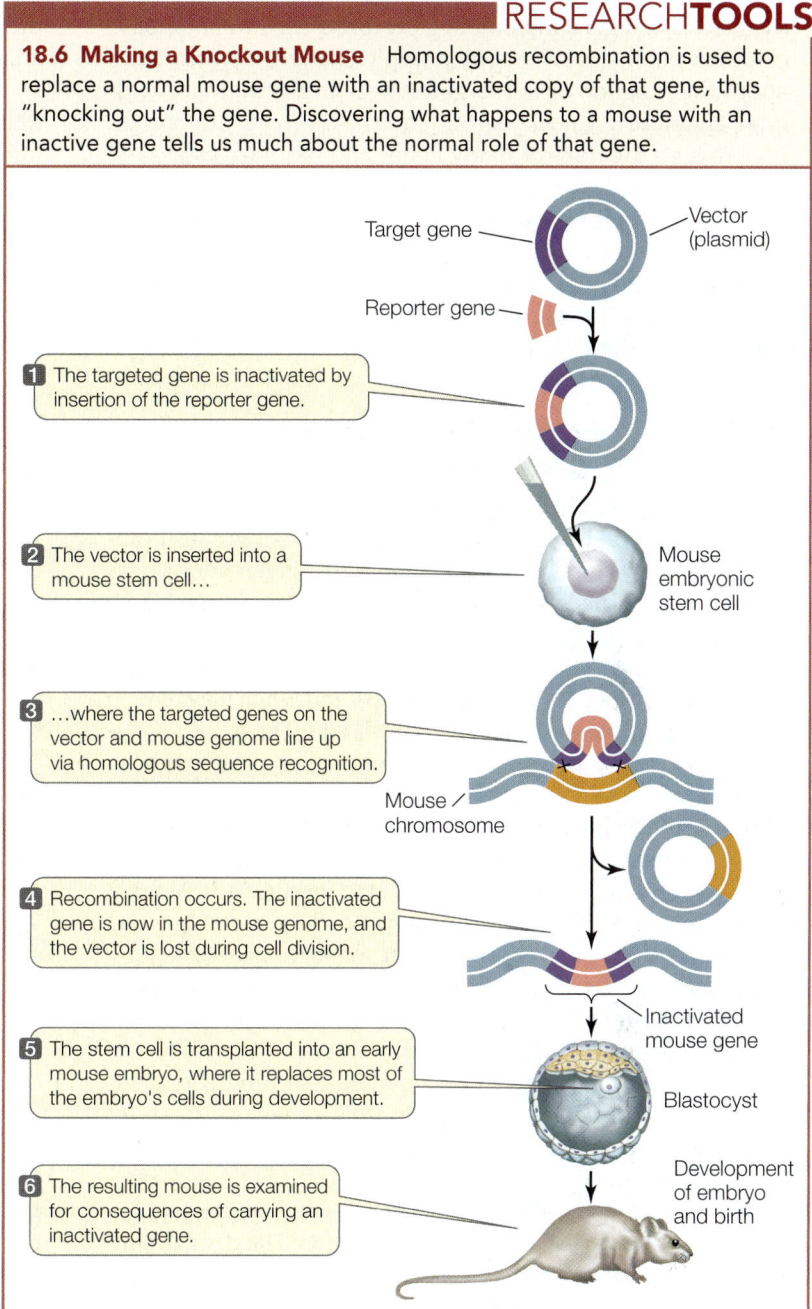

RESEARCH**TOOLS**

18.6 Making a Knockout Mouse Homologous recombination is used to replace a normal mouse gene with an inactivated copy of that gene, thus "knocking out" the gene. Discovering what happens to a mouse with an inactive gene tells us much about the normal role of that gene.

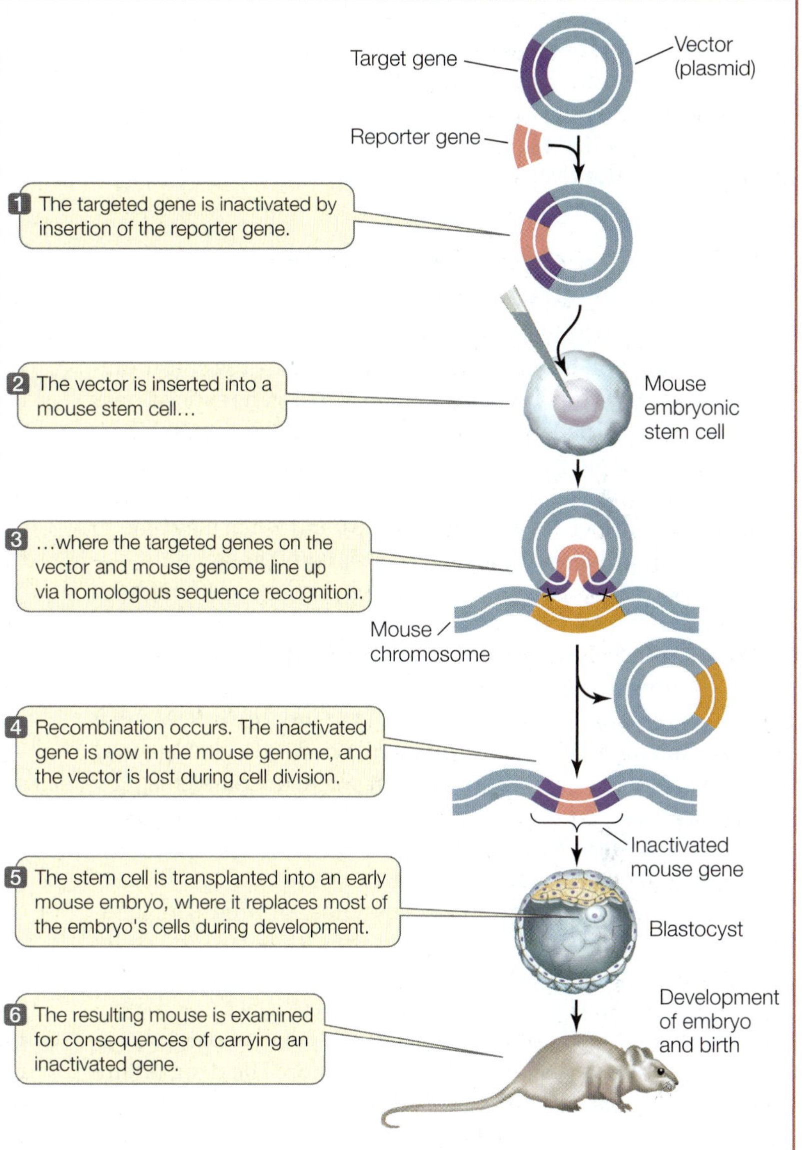
Target gene
Vector (plasmid)
Reporter gene
1 The targeted gene is inactivated by insertion of the reporter gene.
2 The vector is inserted into a mouse stem cell…
Mouse embryonic stem cell
3 …where the targeted genes on the vector and mouse genome line up via homologous sequence recognition.
Mouse chromosome
4 Recombination occurs. The inactivated gene is now in the mouse genome, and the vector is lost during cell division.
Inactivated mouse gene
5 The stem cell is transplanted into an early mouse embryo, where it replaces most of the embryo's cells during development.
Blastocyst
Development of embryo and birth
6 The resulting mouse is examined for consequences of carrying an inactivated gene.

To create a knockout mouse, the normal allele of a gene of interest is inserted into a plasmid. Restriction enzymes are then used to insert a fragment containing a selectable marker gene into the middle of the mouse gene. This mutated version of the gene will eventually replace the wild-type gene in a living mouse. The extra DNA interferes with transcription and translation of the targeted gene, making it nonfunctional.

Once the recombinant plasmid has been made, it is used to transfect mouse embryonic stem cells. (A **stem cell** is an unspecialized cell that divides and differentiates into specialized cells.) Much of the targeted gene is still present in the plasmid (although in two separated regions), and these sequences tend to line up with their homologous sequences in the normal allele on the mouse chromosome. Sometimes recombination occurs, and the plasmid's nonfunctional allele is "swapped" with the functional allele in the host cell. The nonfunctional allele is inserted permanently into the host cell's genome and the normal allele is lost, because the plasmid cannot replicate in mouse cells. The active marker gene in the insert is used to select those stem cells carrying the nonfunctional allele.

A transfected stem cell is now transplanted into a mouse embryo at an early stage of development. If the mouse that develops from this embryo has the mutant allele in its germ line cells, its progeny will have the allele in every cell in their bodies. Such mice are inbred to create knockout mice carrying the inactivated gene in homozygous form. The mutant mouse can then be observed for phenotypic changes, to find clues about the function of the targeted gene in the normal (wild-type) animal. The knockout technique has been very important in assessing the roles of many genes and has been especially valuable in studying human genetic diseases. Many of these diseases, such as phenylketonuria, have knockout mouse models—mouse strains produced by homologous recombination that are affected by an analogous disease. These models can be used to study a disease and to test potential treatments. Mario Capecchi, Martin Evans, and Oliver Smithies shared the Nobel Prize in 2007 for developing the knockout mouse technique.

Complementary RNA can prevent the expression of specific genes

Another way to study the expression of a specific gene is to block the translation of its mRNA. This is an example of scientists imitating nature. As we described in Section 16.5, gene expression is sometimes regulated by the production of double-stranded RNA molecules that are processed to produce short, single-stranded microRNA (miRNA). These miRNAs are complementary to specific mRNA sequences (see Figure 16.23), and when they bind to their target mRNAs, they inhibit translation. The hybrid molecules tend to break down rapidly in the cytoplasm. Although the target gene continues to be transcribed, translation does not take place. Scientists have used the idea of hybrid RNAs to develop methods for blocking the expression of particular genes (**Figure 18.7**). An organism can be transformed to produce mRNAs that are complementary to and thus bind to specific endogenous mRNAs. Or cells can be injected with synthetic complementary sequences (see Figure 18.7B). Such a complementary molecule is called **antisense RNA** because it binds by base pairing to the "sense" bases on the mRNA.

There is much interest in the possibility of producing antisense drugs to treat cancer. For example, the gene *bcl2* codes for a protein that blocks apoptosis (programmed cell death; see Section 11.6). In some forms of cancer, *bcl2* is activated inappropriately through mutation. These cells fail to undergo apoptosis, continue to divide, and form a tumor. The drug oblimersen is an antisense RNA that binds to *bcl2* mRNA, prevents production of the protein, and leads to apoptosis of tumor cells and shrinkage of the tumor.

A technique related to antisense RNA takes advantage of **RNA interference** (**RNAi**), a natural mechanism for inhibiting mRNA translation that involves small interfering RNA (siRNA; see Figure 16.23). siRNAs bind to complementary regions on their target mRNAs, which are then degraded. Since the discovery of RNAi in the late 1990s, scientists have synthesized artificial double-stranded siRNAs to inhibit the expression of known genes (see Figure 18.7C). Because these double-stranded siRNAs are more stable than antisense RNAs, the use of siRNAs is the preferred approach for blocking translation. An RNAi-based therapy is being developed to treat macular degeneration, an eye disease that results in near-blindness when blood vessels proliferate in the eye. The signaling molecule that stimulates vessel proliferation is a growth factor. An siRNA that targets this growth factor's mRNA shows promise in stopping and even reversing the progress of the disease.

Although medical applications for RNAi are still at the experimental stage, antisense RNA and RNAi have been widely used to test cause-and-effect relationships in biological research.

DNA microarrays reveal RNA expression patterns

The science of genomics faces two major quantitative challenges. First, there are very large numbers of genes in eukaryotic genomes. Second, there are myriad distinct patterns of gene expression in different tissues at different times. For example, the cells of a skin cancer at its early stage may have a unique mRNA "fingerprint" that differs from those of normal skin cells and cells from a more advanced skin cancer. In such a case, the pattern of gene expression could provide invaluable information to a clinician trying to characterize a patient's tumor (see below).

To find such patterns, scientists could isolate mRNA from cells and measure the amount of each gene's mRNA one gene at a time by hybridization or RT-PCR. But that would involve many steps and take a very long time. It is far simpler to do these hybridizations all in one step. This is possible with **DNA microarray** technology, which provides large arrays of sequences for hybridization experiments.

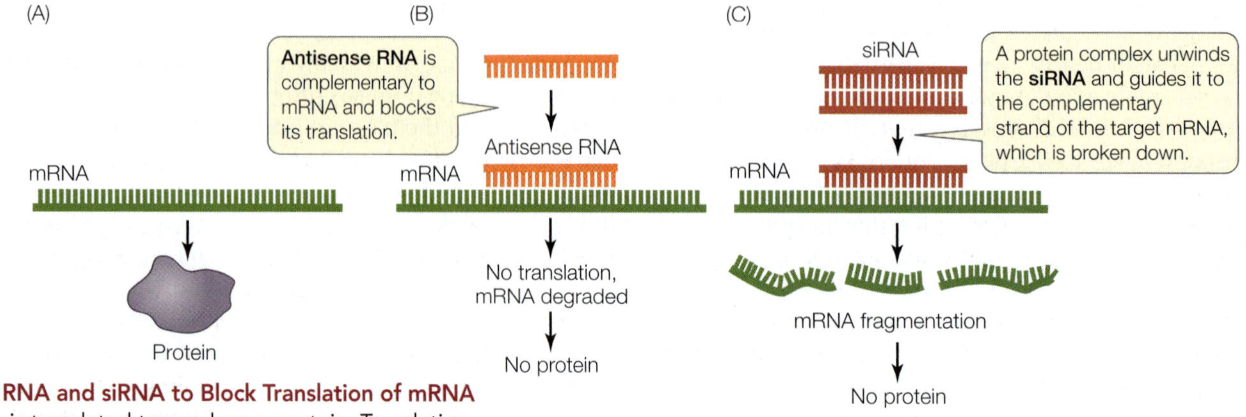
(A)
mRNA
Protein
(B)
Antisense RNA is complementary to mRNA and blocks its translation.
Antisense RNA
mRNA
No translation, mRNA degraded
No protein
(C)
siRNA
A protein complex unwinds the siRNA and guides it to the complementary strand of the target mRNA, which is broken down.
mRNA
mRNA fragmentation
No protein

18.7 Using Antisense RNA and siRNA to Block Translation of mRNA
(A) Normally an mRNA is translated to produce a protein. Translation of a target mRNA can be prevented with (B) an antisense RNA or (C) a small interfering RNA (siRNA) that is complementary to the target mRNA.

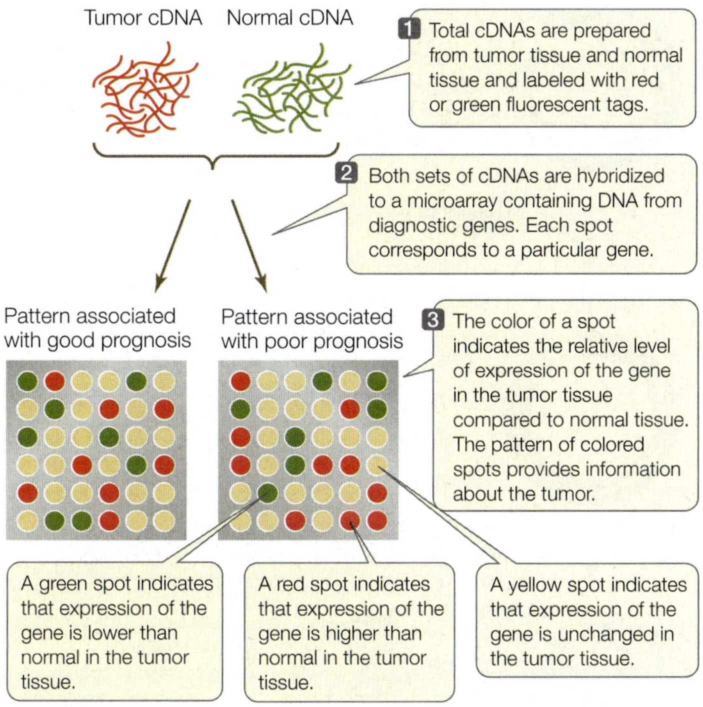
Tumor cDNA
Normal cDNA
1 Total cDNAs are prepared from tumor tissue and normal tissue and labeled with red or green fluorescent tags.
2 Both sets of cDNAs are hybridized to a microarray containing DNA from diagnostic genes. Each spot corresponds to a particular gene.
Pattern associated with good prognosis
Pattern associated with poor prognosis
3 The color of a spot indicates the relative level of expression of the gene in the tumor tissue compared to normal tissue. The pattern of colored spots provides information about the tumor.
A green spot indicates that expression of the gene is lower than normal in the tumor tissue.
A red spot indicates that expression of the gene is higher than normal in the tumor tissue.
A yellow spot indicates that expression of the gene is unchanged in the tumor tissue.

18.8 DNA Microarray for Medical Diagnosis The pattern of expression of 70 genes in tumor tissues indicates whether breast cancer is likely to recur. Actual arrays have more spots than shown here.

Go to Animated Tutorial 18.1
DNA Microarray Technology
Life10e.com/at18.1

DNA microarrays ("gene chips") contain a series of DNA sequences attached to a glass slide in a precise order. The slide is divided into a grid of microscopic spots, or "wells." Each spot contains thousands of copies of a particular oligonucleotide of 20 or more bases. A computer controls the addition of these oligonucleotide sequences in a predetermined pattern. Each oligonucleotide can hybridize with only one DNA or RNA sequence and thus is a unique identifier of a gene. Many thousands of different oligonucleotides can be placed on a single microarray.

Microarrays can be used to examine patterns of gene expression in different tissues and under different conditions, and they can be used to identify individual organisms with particular mutations. You can see the concept of microarray analysis by following the example illustrated in **Figure 18.8**. Most women with breast cancer are treated with surgery to remove the tumor, and then treated with radiation soon afterward to kill cancer cells that the surgery may have missed. But a few cancer cells may still survive in some patients, and these eventually form tumors in the breast or elsewhere in the body. The challenge for physicians is to develop criteria to identify patients likely to have surviving cancer cells so that they can be treated aggressively with tumor-killing chemotherapy. Scientists at the Netherlands Cancer Institute used medical records to identify patients whose cancer did or did not recur. Then they extracted mRNA from the patients' tumors, which had been stored after surgery, and made cDNA from the samples. The cDNAs were hybridized to microarrays containing sequences derived from 1,000 human genes. The researchers found 70 genes whose expression differed dramatically between tumors from patients whose cancers recurred and tumors from patients whose cancers did not recur. From this information the Dutch group identified gene expression signatures that are useful in clinical decision-making: patients with a good prognosis can avoid unnecessary chemotherapy, whereas those with a poor prognosis can receive more aggressive treatment.

RECAP 18.4

Researchers can study the function of a gene by expressing it in cells where it is not normally expressed, by overexpressing it, or by knocking out the gene in a living organism. Antisense RNAs and siRNAs prevent gene expression by selectively blocking mRNA translation. DNA microarrays allow the simultaneous analysis of many different mRNA transcripts.

- How is a gene "knocked out" in a living organism? **See p. 381 and Figure 18.6**
- How do antisense RNA and siRNA molecules affect gene expression? **See p. 382 and Figure 18.7**

Now that you've seen how DNA can be fragmented, recombined, manipulated, and put back into living organisms, let's see some examples of how these techniques are used to make useful products.

18.5 What Is Biotechnology?

Biotechnology is the use of cells or whole living organisms to produce materials useful to people, such as foods, medicines, and chemicals. People have been using various forms of biotechnology for a long time. For example, the use of yeasts to brew beer and wine dates back at least 8,000 years, and the use of bacterial cultures to make cheese and yogurt is a technique many centuries old. For a long time people exploited these biochemical transformations without being aware of the organisms and genes involved.

About 100 years ago it became clear that specific bacteria, yeasts, and other microbes could be used as biological converters to make certain products. Alexander Fleming's discovery that the mold *Penicillium* makes the antibiotic penicillin led to the large-scale commercial culture of microbes to produce antibiotics as well as other useful chemicals. Today microbes are grown in vast quantities to make much of the industrial-grade alcohol, glycerol, butyric acid, and citric acid that are used as is or as starting materials in the manufacture of other products.

In contrast, the commercial harvesting of proteins, including hormones and enzymes, was until recently limited by the (often) minuscule amounts that could be extracted from organisms that produce them naturally. Yields were low, and purification was difficult and costly. Gene cloning has changed all this. Now that almost any gene can be inserted into bacteria or yeasts and the cells can be induced to make and export the gene product in large amounts, these microbes have become versatile factories for important products. Today there is interest in producing nutritional supplements and pharmaceuticals in whole transgenic animals and harvesting them in large quantities from, for example, the milk of cows or the eggs of chickens. Key to this boom in biotechnology has been the development of specialized vectors that not only carry genes into cells, but also make those cells express the genes at high levels.

Expression vectors can turn cells into protein factories

If a eukaryotic gene is inserted into a typical plasmid and used to transform *E. coli*, little if any of the gene product will be made unless other key prokaryotic DNA sequences are included with the gene. A bacterial promoter, a signal for transcription termination, and a special sequence that is necessary for binding of bacterial ribosomes to the mRNA must all be included in the transformation vector if the gene is to be expressed in a bacterial cell.

To solve this kind of problem, scientists make **expression vectors** that have all the characteristics of cloning vectors, as well as the extra sequences needed for the foreign gene (also called a transgene) to be expressed in the host cell. For bacterial hosts, these additional sequences include the elements named above (Figure 18.9). For eukaryotes, they include the poly A–addition sequence and a promoter that contains all the elements needed for expression in a eukaryotic cell. An expression vector can be designed to deliver transgenes to any class of prokaryotic or eukaryotic host and may include additional features:

- An *inducible promoter*, which responds to a specific signal, can be included. For example, a promoter that responds to hormonal stimulation can be used so that the transgene will be expressed at high levels only when the hormone is added.
- A *tissue-specific promoter*, which is expressed only in a certain tissue at a certain time, can be used if localized expression is desired. For example, many seed proteins are expressed only in the plant embryo. Coupling a transgene to a seed-specific promoter will allow the gene to be expressed only in seeds.
- *Signal sequences* can be added so that the gene product is directed to an appropriate destination. For example, when a protein is made by yeast or bacterial cells in a liquid medium, it is economical to include a signal directing the protein to be secreted into the extracellular medium for easier recovery.

RECAP 18.5

Expression vectors maximize the expression of transgenes inserted into host cells.

- How do expression vectors work? **See p. 384 and Figure 18.9**

This chapter has introduced many of the methods that are used in biotechnology. Let's turn now to the ways biotechnology is being applied to meet some specific human needs.

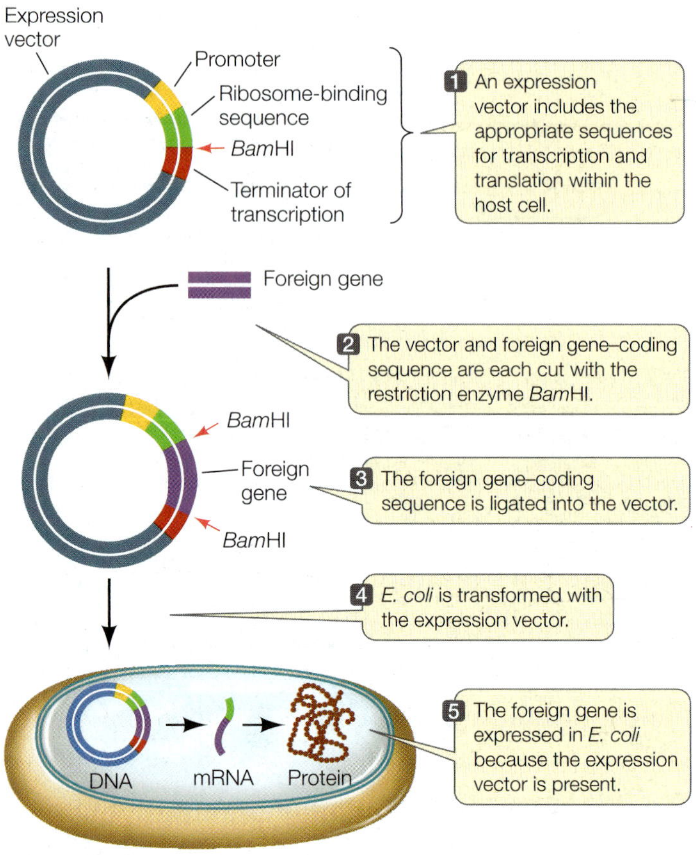
Expression vector
Promoter
Ribosome-binding sequence
BamHI
Terminator of transcription
1 An expression vector includes the appropriate sequences for transcription and translation within the host cell.
Foreign gene
2 The vector and foreign gene–coding sequence are each cut with the restriction enzyme BamHI.
BamHI
Foreign gene
BamHI
3 The foreign gene–coding sequence is ligated into the vector.
4 E. coli is transformed with the expression vector.
DNA
mRNA
Protein
5 The foreign gene is expressed in E. coli because the expression vector is present.

18.9 Expression of a Transgene in a Host Cell Produces Large Amounts of Its Protein Product To be expressed in *E. coli*, a gene derived from a eukaryote requires bacterial sequences for transcription initiation (promoter), transcription termination, and ribosome binding. Expression vectors contain these additional sequences, enabling the eukaryotic protein to be synthesized in the prokaryotic cell.

Go to Activity 18.1 Expression Vectors
Life10e.com/ac18.1

18.6 How Is Biotechnology Changing Medicine and Agriculture?

Huge potential for improvements in health and agriculture derive from recent developments in biotechnology. We now have the ability to make virtually any protein by recombinant DNA technology and to insert transgenes into many kinds of host cells. With these revolutionary developments, concerns have been raised about ethics and safety. We will now turn to the promises and problems of biotechnology that uses DNA manipulation.

Medically useful proteins can be made using biotechnology

Many medically useful products are being made using biotechnology (Table 18.1), and more are in various stages of development. A good illustration of a medical application of biotechnology is the manufacture of tissue plasminogen activator (TPA).

When a wound begins bleeding, a blood clot soon forms to stop the flow. Later, as the wound heals, the clot dissolves.

TABLE 18.1 Some Medically Useful Products of Biotechnology

Product	Use
Colony-stimulating factor	Stimulates production of white blood cells in patients with cancer and AIDS
Erythropoietin	Prevents anemia in patients undergoing kidney dialysis and cancer therapy
Factor VIII	Replaces clotting factor missing in patients with hemophilia A
Growth hormone	Replaces missing hormone in people of short stature
Insulin	Stimulates glucose uptake from blood in people with insulin-dependent (Type I) diabetes
Platelet-derived growth factor	Stimulates wound healing
Tissue plasminogen activator	Dissolves blood clots after heart attacks and strokes
Vaccine proteins: Hepatitis B, herpes, influenza, Lyme disease, meningitis, pertussis, etc.	Prevent and treat infectious diseases

How does the blood perform these conflicting functions at the right times? Mammalian blood contains an enzyme called plasminogen. When activated, it is converted into plasmin and catalyzes the dissolution of the clotting proteins. The conversion of plasminogen into plasmin is catalyzed by the enzyme TPA, which is produced by cells lining the blood vessels:

$$\underset{\text{(Inactive)}}{\text{Plasminogen}} \xrightarrow{\text{TPA}} \underset{\text{(Active)}}{\text{plasmin}}$$

Heart attacks and strokes can be caused by blood clots that form in major blood vessels leading to the heart or the brain, respectively. In the 1970s a bacterial enzyme called streptokinase was found to stimulate the dissolution of the clots in some patients. Treatment with this enzyme saved lives, but being a foreign protein, it triggered the body's immune system to react against it. Worse, the drug sometimes prevented clotting throughout the entire circulatory system, leading to a dangerous situation in which blood could not clot where needed; uncontrolled bleeding was a serious concern.

When TPA was discovered, it had many advantages over streptokinase: it bound specifically to clots, and it did not provoke an immune reaction because it is an enzyme that is normally found in the body. But the amounts of TPA that could be harvested from human tissues were tiny, certainly not enough to inject at the site of a clot in the emergency room.

Recombinant DNA technology solved this problem. TPA mRNA was isolated and used to make cDNA, which was then inserted into an expression vector and used to transform *E. coli* (Figure 18.10). The transgenic bacteria made the protein in quantity, and it soon became available commercially. This drug has had considerable success in dissolving blood clots in people experiencing strokes and heart attacks.

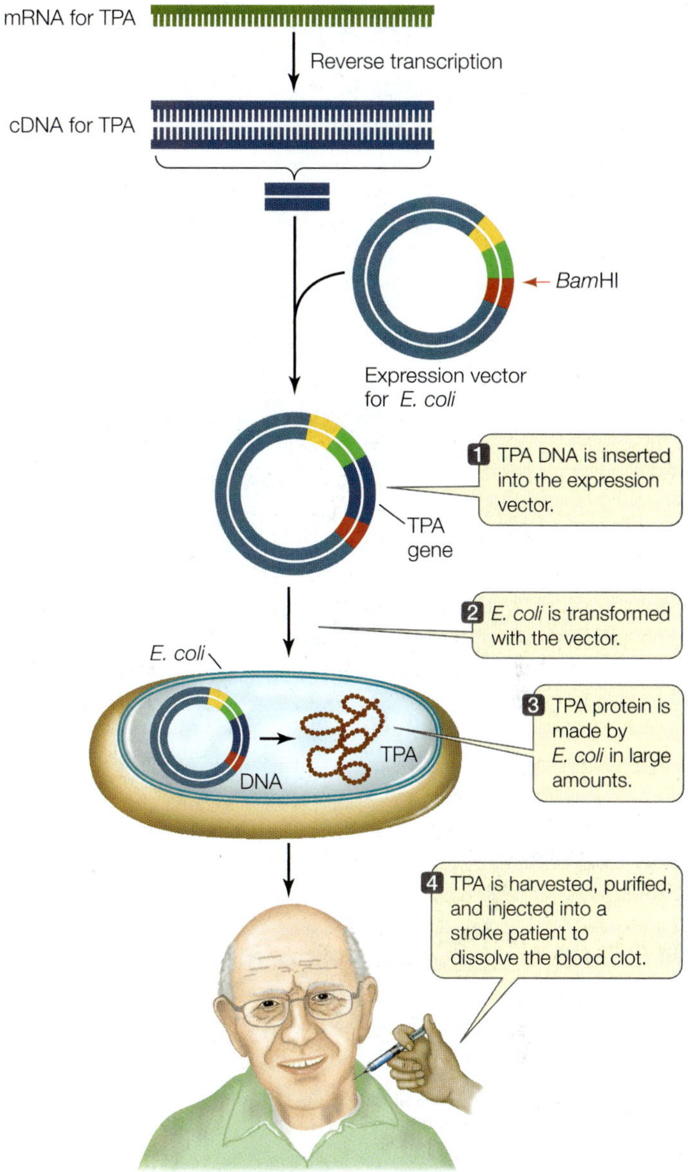

mRNA for TPA
Reverse transcription
cDNA for TPA
BamHI
Expression vector for E. coli
TPA gene
1 TPA DNA is inserted into the expression vector.
2 E. coli is transformed with the vector.
E. coli
DNA
TPA
3 TPA protein is made by E. coli in large amounts.
4 TPA is harvested, purified, and injected into a stroke patient to dissolve the blood clot.

18.10 Tissue Plasminogen Activator TPA is a naturally occurring human protein that dissolves blood clots. It is used to treat patients suffering from blood clotting in heart attacks or strokes, and is manufactured using recombinant DNA technology.

Another way of making medically useful products in large amounts is **pharming**: the production of pharmaceuticals in farm animals or plants. For example, a gene encoding a useful protein might be placed next to the promoter of the gene that encodes lactoglobulin, an abundant milk protein. Transgenic animals carrying this recombinant DNA will secrete large amounts of the foreign protein into their milk. These natural "bioreactors" can produce abundant supplies of the protein, which can be separated easily from the other components of the milk (Figure 18.11).

Pharming is being used to produce human growth hormone (hGH), a protein made in the pituitary gland in the brain that has

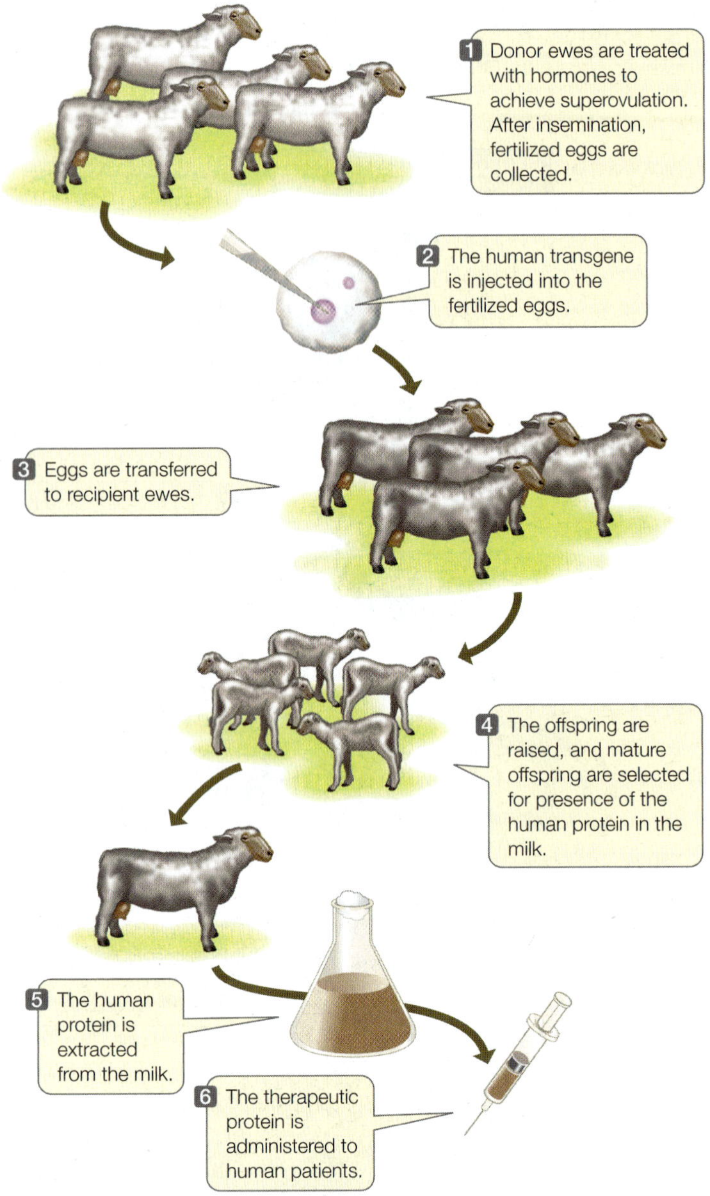
1 Donor ewes are treated with hormones to achieve superovulation. After insemination, fertilized eggs are collected.
2 The human transgene is injected into the fertilized eggs.
3 Eggs are transferred to recipient ewes.
4 The offspring are raised, and mature offspring are selected for presence of the human protein in the milk.
5 The human protein is extracted from the milk.
6 The therapeutic protein is administered to human patients.

18.11 Pharming An expression vector carrying a desired gene can be put into an animal egg, which is implanted into a surrogate mother. The transgenic offspring produce the new protein in their milk. The milk is easily harvested and the protein isolated, purified, and made clinically available to patients.

many effects, especially in growing children (see Chapter 41). Children with growth hormone deficiency have short stature as well as other abnormalities. In the past they were treated with hGH isolated from the pituitary glands of organ donors, but the supply was too limited to meet demand. Recombinant DNA technology was used to coax bacteria to make the protein, but the cost of treatment was high ($30,000 a year per person). In 2004 a team led by Daniel Salamone at the University of Buenos Aires made a transgenic cow that secretes hGH in her milk. The yield is prodigious: only 15 such cows are needed to meet the needs worldwide of children suffering from this type of dwarfism.

DNA manipulation is changing agriculture

The cultivation of plants and the husbanding of animals provide the world's oldest examples of biotechnology, dating back more than 10,000 years. Over the centuries, people have adapted crops and farm animals to their needs, producing organisms with desirable characteristics such as large seeds, high fat content in milk, or resistance to disease.

Until recently, the most common way to improve crop plants and farm animals was to identify individuals with desirable phenotypes that existed as a result of natural variation. Through many deliberate crosses—a process called selective breeding—the genes responsible for the desirable trait could be introduced into a widely used variety or breed of that organism.

Despite some spectacular successes, among them the breeding of high-yielding varieties of wheat, rice, and hybrid corn, such deliberate crossing can be a hit-or-miss affair. Many desirable traits are controlled by multiple genes, and it is hard to predict the results of a cross or to maintain a prized combination as a pure-breeding variety year after year. In sexual reproduction, combinations of desirable genes are quickly separated by meiosis. Furthermore, traditional breeding takes a long time: many plants and animals take years to reach maturity and then can reproduce only once or twice a year—a far cry from the rapid reproduction of bacteria.

Modern recombinant DNA technology has several advantages over traditional methods of breeding (**Figure 18.12**):

- *The ability to identify specific genes*. The development of genetic markers allows breeders to select for specific desirable genes, making the breeding process more precise and rapid.
- *The ability to introduce any gene from any organism into a plant or animal species*. This ability, combined with mutagenesis techniques, vastly expands the range of possible new traits.
- *The ability to generate new organisms quickly*. Manipulating cells in the laboratory and regenerating a whole plant or animal by cloning is much faster than traditional breeding.

Consequently, recombinant DNA technology has found many applications in agriculture (**Table 18.2**). We will describe a few examples to demonstrate the approaches that plant scientists have used to improve crop plants.

PLANTS THAT MAKE THEIR OWN INSECTICIDES Plants are subject to infections by viruses, bacteria, and fungi, but probably the

TABLE 18.2
Agricultural Applications of Biotechnology under Development

Goal	Technology/Genes
Improving the environmental adaptations of plants	Genes for drought tolerance, salt tolerance
Improving nutritional traits	High-lysine seeds; β-carotene in rice
Improving crops after harvest	Delay of fruit ripening; sweeter vegetables
Using plants as bioreactors	Plastics, oils, and drugs produced in plants

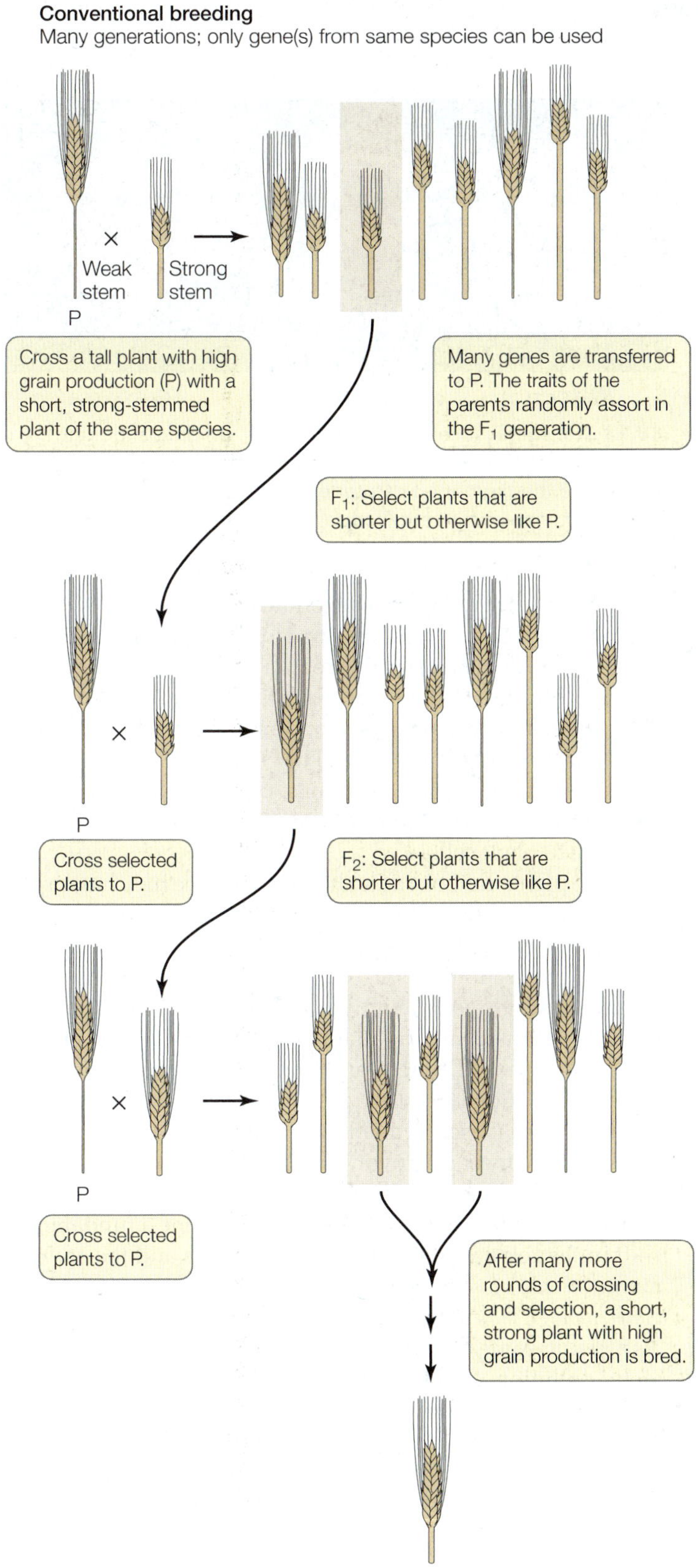
Conventional breeding
Many generations; only gene(s) from same species can be used
Weak stem
Strong stem
P
Cross a tall plant with high grain production (P) with a short, strong-stemmed plant of the same species.
Many genes are transferred to P. The traits of the parents randomly assort in the F_1 generation.
F_1: Select plants that are shorter but otherwise like P.
P
Cross selected plants to P.
F_2: Select plants that are shorter but otherwise like P.
P
Cross selected plants to P.
After many more rounds of crossing and selection, a short, strong plant with high grain production is bred.

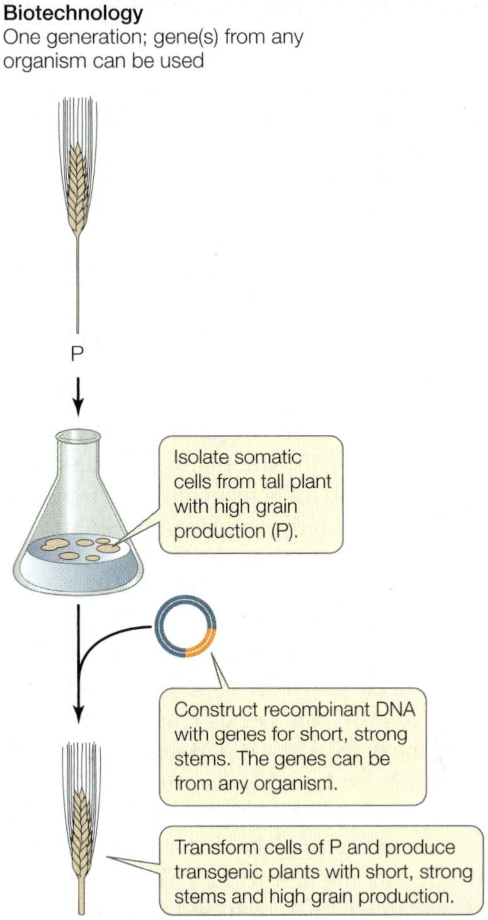
Biotechnology
One generation; gene(s) from any organism can be used
P
Isolate somatic cells from tall plant with high grain production (P).
Construct recombinant DNA with genes for short, strong stems. The genes can be from any organism.
Transform cells of P and produce transgenic plants with short, strong stems and high grain production.

18.12 Genetic Modification of Plants versus Conventional Plant Breeding Plant biotechnology offers many potential advantages over conventional breeding. In the example here, the objective is to transfer gene(s) for short, strong stems into a wheat plant that has high grain production but a tall, weak stem.

most important crop pests are herbivorous insects. From the locusts of biblical (and modern) times to the cotton boll weevil, insects have continually eaten the crops people grow.

The development of insecticides has improved the situation, but insecticides have their own problems. Many, including the organophosphates, are relatively nonspecific and kill beneficial insects in the broader ecosystem as well as crop pests. Some pesticides even have toxic effects on other groups of organisms, including people. What's more, many insecticides persist in the environment for a long time.

Some bacteria protect themselves by producing proteins that can kill insects. For example, the bacterium *Bacillus thuringiensis* produces a protein that is toxic to the insect larvae that prey on it.

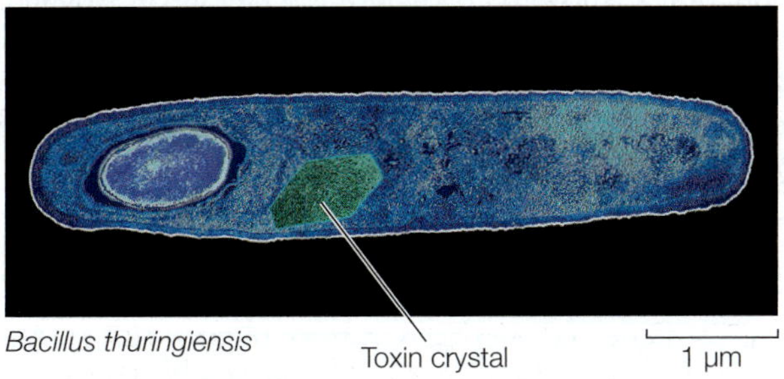
Bacillus thuringiensis
Toxin crystal
1 μm

The toxicity of this protein is 80,000 times greater than that of a typical commercial insecticide. When a hapless larva eats the bacteria, the toxin becomes activated and binds specifically to the insect's gut, producing holes and killing the insect. Dried preparations of *B. thuringiensis* have been sold for decades as safe insecticides that break down rapidly in the environment. But the biodegradation of these preparations is their limitation, because it means that the dried bacteria must be applied repeatedly during the growing season.

A longer-acting approach is to have the crop plants themselves make the toxin, and this is exactly what plant scientists have done. The toxin gene from *B. thuringiensis* has been isolated, cloned, and extensively modified by the addition of a plant promoter and other regulatory sequences. Transgenic corn, cotton, soybeans, tomatoes, and other crops are now being grown successfully with this added gene. Farmers growing these transgenic crops use far less of other pesticides.

CROPS THAT ARE RESISTANT TO HERBICIDES Herbivorous insects are not the only threat to agriculture. Weeds may grow in fields and compete with crop plants for water and soil nutrients. Glyphosate is a widely used and effective herbicide, or weed killer, that works only on plants. It inhibits an enzyme system in the chloroplast that is involved in the synthesis of amino acids. Glyphosate is a broad-spectrum herbicide that kills most weeds, but unfortunately it also kills crop plants. One solution to this problem is to use it to rid a field of weeds before the crop plants start to grow. But as any gardener knows, when the crop begins to grow, the weeds reappear. If the crop were not affected by the herbicide, the herbicide could be applied to the field at any time without harming the crop.

Scientists have used expression vectors to make plants that synthesize a different form of the target enzyme for glyphosate that is unaffected by the herbicide. The gene for this enzyme has been inserted into corn, cotton, and soybean plants, making them resistant to glyphosate. This technology has expanded rapidly, and a large proportion of cotton and soybean plants now carry this gene.

GRAINS WITH IMPROVED NUTRITIONAL CHARACTERISTICS To remain healthy, humans must consume adequate amounts of β-carotene, which the body converts into vitamin A. About 400 million people worldwide suffer from vitamin A deficiency, which makes them susceptible to infections and blindness. One reason is that rice grains, which do not contain β-carotene, make up a large part of their diets. Rice grains lack the two-enzyme biochemical pathway that synthesizes β-carotene.

Plant biologists Ingo Potrykus and Peter Beyer isolated one of the genes for the β-carotene pathway from the bacterium *Erwinia uredovora* and the other from daffodil plants. They added a promoter and other signals for expression in the developing rice grain and then transformed rice plants with the two genes. The resulting rice plants produce grains that look yellow because of their high β-carotene content. A newer variety with a corn gene replacing the one from daffodils makes even more β-carotene and is golden in color (**Figure 18.13**). A daily intake of about 150 grams of this cooked rice can supply all the β-carotene a person needs. This new transgenic strain has been crossed with strains adapted for various local environments, in the hope of improving the diets of millions of people.

CROPS THAT ADAPT TO THE ENVIRONMENT Agriculture depends on ecological management—tailoring the environment to the needs of crop plants and animals. A farm field is an unnatural, human-designed system that must be carefully managed to maintain optimal conditions for crop growth. For example, excessive irrigation can cause increases in soil salinity. The Fertile Crescent, the region between the Tigris and Euphrates rivers in the Middle East where agriculture probably originated 10,000 years ago, is no longer fertile. It is now a desert, largely because the soil has a high salt concentration. Few plants can grow on salty soils, partly because of osmotic effects that result in wilting, and partly because excess salt ions are toxic to plant cells.

Wild type
Golden rice 1
Golden rice 2

18.13 Transgenic Rice Rich in β-Carotene Middle and right: The grains from these transgenic rice strains are colored because they make the pigment β-carotene, which is converted to vitamin A in the human body. Left: Normal rice grains do not contain β-carotene.

Some plants can tolerate salty soils because they have a protein that transports Na^+ ions out of the cytoplasm and into the vacuole, where the ions can accumulate without harming plant growth (see Section 5.3 for a description of the plant vacuole). Scientists developed a highly active version of this gene and used it to transform crop plants that are less tolerant to salt, including rapeseed, wheat, and tomatoes. When this gene was added to tomato plants, they grew in water that was four times as salty as the typical lethal level (**Figure 18.14**). This finding raises the prospect of growing crops on what were previously unproductive soils.

This example illustrates what could become a fundamental shift in the relationship between crop plants and the environment. *Instead of manipulating the environment to suit the plant, biotechnology may allow us to adapt the plant to the environment.* As a result, some of the negative effects of agriculture, such as water pollution, could be lessened.

There is public concern about biotechnology

Concerns have been raised about the safety and wisdom of genetically modifying crops and other organisms. These concerns are centered on three claims:

- Genetic manipulation is an unnatural interference with nature.
- Genetically altered foods are unsafe to eat.
- Genetically altered crop plants are dangerous to the environment.

Advocates of biotechnology tend to agree with the first claim. However, they point out that all crops are unnatural in the sense that they come from artificially bred plants growing in

(A)

(B)

18.14 Salt-Tolerant Tomato Plants Transgenic plants containing a gene for salt tolerance thrive in salty water (A), whereas plants without the transgene die (B). This technology may allow crops to be grown on salty soils.

a manipulated environment (a farmer's field). Recombinant DNA technology just adds another level of sophistication to these technologies.

To counter the concern about whether genetically engineered crops are safe for human consumption, biotechnology advocates point out that only single genes are added and that these genes are specific for plant function. For example, the *B. thuringiensis* toxin produced by transgenic plants has no effect on people. However, as plant biotechnology moves from adding genes that improve plant growth to adding genes that affect human nutrition, such concerns will become more pressing.

Various negative environmental impacts have been envisaged. There is concern about the possible "escape" of transgenes from crops to other species. If the gene for herbicide resistance, for example, were inadvertently transferred from a crop plant to a closely related weed, that weed could thrive in herbicide-treated areas. Another negative impact is the possibility that increased use of an herbicide will select for weeds with naturally occurring mutations that make them resistant to that herbicide. This is indeed occurring. Widespread use of glyphosate on fields of glyphosate-resistant crops has resulted in the selection of rare mutations in weeds that make them resistant to glyphosate. More than ten resistant weed species have appeared in the U.S.

As we mentioned in the opening story, scientists have developed microorganisms that are able to break down components of crude oil. These have not been released into the environment because of the unknown effects that such organisms might have on natural ecosystems. However, these organisms might provide a way to rapidly clean up catastrophic oil spills.

Because of the potential benefits of biotechnology, scientists believe that it is wise to proceed, albeit with caution.

RECAP 18.6

Biotechnology has been used to produce medicines and to develop transgenic plants with improved agricultural and nutritional characteristics.

- What are the advantages of using biotechnology for plant breeding compared with traditional methods? **See Figure 18.12**
- What are some of the concerns that people might have about biotechnology? **See pp. 388–389**

?

Are there other uses for microorganisms in environmental cleanup?

ANSWER

Microorganisms currently play a critical role in cleaning up human-made waste products in the environment. For example, composting involves the use of bacteria to break down large molecules, including carbon-rich polymers and proteins in waste products such as wood chips, paper, straw, and kitchen scraps. Other bacteria are used in wastewater treatment to break down human waste and household chemicals in wastewater treatment plants. As we mentioned in the opening story, much research effort has gone into the development of genetically modified bacteria that could be used for environmental cleanup. An interesting example is the radiation-resistant bacterium *Deinococcus radiodurans*, which has been engineered to precipitate heavy metals and to break down components of crude oil. This organism may be useful for bioremediation at radioactively contaminated sites.

CHAPTER SUMMARY 18

18.1 What Is Recombinant DNA?

- **Recombinant DNA** is formed by the combination of two DNA sequences from different sources. Review Figure 18.1
- Many restriction enzymes make staggered cuts in the two strands of DNA, creating fragments that have **sticky ends** with unpaired bases.
- DNA fragments with sticky ends can be used to create recombinant DNA. DNA molecules from different sources can be cut with the same restriction enzyme and spliced together using **DNA ligase**. Review Figure 18.2

18.2 How Are New Genes Inserted into Cells?

- One goal of recombinant DNA technology is to **clone** a particular gene, either for analysis or to produce its protein product in quantity.
- Bacteria, yeasts, and cultured plant and animal cells are commonly used as hosts for recombinant DNA. The insertion of foreign DNA into host cells is called **transformation** or (for animal cells) **transfection**. Transformed or transfected cells are called **transgenic** cells.
- Various methods are used to get recombinant DNA into cells. These include chemical or electrical treatment of the cells, the use of viral vectors, and injection. *Agrobacterium tumefaciens* is often used to insert DNA into plant cells.
- To identify host cells that have taken up a foreign gene, the inserted sequence can be tagged with one or more **reporter genes**, which are genetic markers with easily identifiable phenotypes. **Selectable markers** allow for the selective growth of transgenic cells. Review Figures 18.3, 18.4
- Replication of the foreign gene in the host cell requires that it become part of a segment of DNA that contains a **replicon** (origin and terminus of replication).
- **Vectors** are DNA sequences that can carry new DNA into host cells. Plasmids and viruses are commonly used as vectors.

18.3 What Sources of DNA Are Used in Cloning?

- DNA fragments from a genome can be inserted into host cells to create a **genomic library**. Review Figure 18.5A
- The mRNAs produced in a certain tissue at a certain time can be extracted and used to create **complementary DNA (cDNA)** by **reverse transcription**. Review Figure 18.5B
- PCR products can be used for cloning. Synthetic DNA containing any desired sequence can be made in the laboratory.

18.4 What Other Tools Are Used to Study DNA Function?

- Homologous recombination can be used to **knock out** a gene in a living organism. Review Figure 18.6
- Gene silencing techniques can be used to inactivate the mRNA transcript of a gene, which may provide clues to the gene's function. Artificially created **antisense RNA** or siRNA can be added to a cell to prevent translation of a specific mRNA. Review Figure 18.7
- **DNA microarray** technology permits the screening of thousands of cDNA sequences at the same time. Review Figure 18.8, ANIMATED TUTORIAL 18.1

18.5 What Is Biotechnology?

- **Biotechnology** is the use of living cells to produce materials useful to people. Recombinant DNA technology has resulted in a boom in biotechnology.
- **Expression vectors** allow a transgene to be expressed in a host cell. Review Figure 18.9, ACTIVITY 18.1

18.6 How Is Biotechnology Changing Medicine and Agriculture?

- Recombinant DNA techniques have been used to make medically useful proteins. Review Figure 18.10
- **Pharming** is the use of transgenic plants or animals to produce pharmaceuticals. Review Figure 18.11
- Because recombinant DNA technology has several advantages over traditional agricultural biotechnology, it is being extensively applied to agriculture. Review Figure 18.12
- Transgenic crop plants can be adapted to their environments, rather than vice versa.
- There is public concern about the application of recombinant DNA technology to food production.

Go to the Interactive Summary to review key figures, Animated Tutorials, and Activities
Life10e.com/is18

CHAPTER**REVIEW**

REMEMBERING

1. Which of the following is used as a reporter gene in recombinant DNA work with bacteria as host cells?
   a. rRNA
   b. Green fluorescent protein
   c. Antibiotic sensitivity
   d. Ability to make ornithine
   e. Vitamin synthesis

2. Which feature is *not* desirable in a vector for gene cloning?
   a. An origin of DNA replication
   b. Genetic markers for the presence of the vector
   c. Many recognition sequences for the restriction enzyme to be used
   d. One recognition sequence each for one or more different restriction enzymes
   e. Genes other than the target gene for transfection

3. RNA interference (RNAi) inhibits
   a. DNA replication.
   b. neither transcription nor translation of specific genes.
   c. recognition of the promoter by RNA polymerase.
   d. transcription of all genes.
   e. translation of specific mRNAs.

4. An expression vector requires all of the following except
   a. genes for ribosomal RNA.
   b. a reporter gene.
   c. a promoter of transcription.
   d. an origin of DNA replication.
   e. restriction enzyme recognition sequences.

5. Pharming is a term that describes
   a. the use of animals in transgenic research.
   b. plants making genetically altered foods.
   c. synthesis of recombinant drugs by bacteria.
   d. large-scale production of cloned animals.
   e. synthesis of a drug by a transgenic plant or animal.

6. Which of the following could *not* be used to test whether expression of a particular gene is necessary for a particular biological function?
   a. RNAi
   b. Knockout technology
   c. Antisense
   d. Mutant tRNA
   e. Transposon mutagenesis

UNDERSTANDING & APPLYING

7. Assume you are using a plasmid vector that contains genes encoding ampicillin resistance and the green fluorescent protein, with a restriction site in the latter gene only. Outline the sequence of steps for inserting a piece of foreign DNA into this plasmid, introducing the recombinant plasmid into bacteria, and verifying that the plasmid and the foreign gene are both present in the bacteria:
   1. Transform host cells.
   2. Select colonies for antibiotic resistance.
   3. Select colonies that do not glow under ultraviolet light.
   4. Digest vector and foreign DNA with a restriction enzyme.
   5. Ligate the digested plasmid together with the foreign DNA.

   a. 4, 5, 1, 3, 2
   b. 4, 5, 1, 2, 3
   c. 1, 3, 4, 2, 5
   d. 3, 2, 1, 4, 5
   e. 1, 3, 2, 5, 4

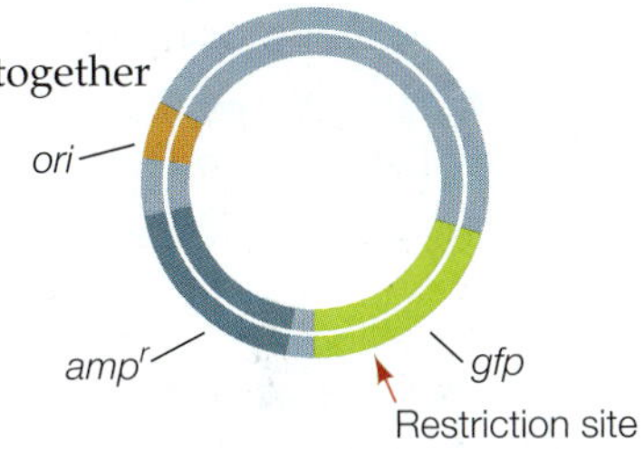
ori
amp^r
gfp
Restriction site

8. In a genomic library of frog DNA in *E. coli* bacteria,
   a. all bacterial cells have the same sequences of frog DNA.
   b. all bacterial cells have different sequences of frog DNA.
   c. each bacterial cell has a random fragment of frog DNA.
   d. each bacterial cell has many fragments of frog DNA.
   e. the frog DNA is always transcribed into mRNA in the bacterial cells.

9. Compare PCR (see Section 13.5) and cloning as methods to amplify a gene. What are the requirements, benefits, and drawbacks of each method?

10. Compare traditional genetic methods with recombinant DNA methods for producing genetically altered plants. For each case, describe (a) sources of new genes; (b) numbers of genes transferred; and (c) how long the process takes.

ANALYZING & EVALUATING

11. As specifically as you can, outline the steps you would take to (a) insert and express the gene for a new, nutritious seed protein in wheat, and (b) insert and express a gene for a human enzyme in sheep's milk.

12. What are the major public concerns about biotechnology as applied to food? In your locality or country, look up regulations on either producing or labeling foods made using biotechnology. What were the scientific and philosophical bases for these regulations?

Go to BioPortal at **yourBioPortal.com** for Animated Tutorials, Activities, LearningCurve Quizzes, Flashcards, and many other study and review resources.

19 Differential Gene Expression in Development

CHAPTER OUTLINE

Bartolo Colon Stem cells helped repair damage to his tendons, and he was able to pitch—and win—again.

IT WAS A MASTERFUL PERFORMANCE. Baseball star Bartolo Colon pitched a five-hit shutout for the New York Yankees as they defeated the Oakland Athletics. His journey to that day on the mound had been eventful. Growing up poor in the Dominican Republic, he was spotted by baseball talent agents as a teenager, and at age 20 he signed a contract with a U.S. major league team. By age 23 he had won the Cy Young Award as professional baseball's top pitcher. But then Colon partially tore the group of muscles and tendons in the elbow and shoulder of his pitching arm. For several years his performance deteriorated significantly, and he even missed an entire season.

Colon came back as good as ever, thanks to stem cell therapy, which he received in the Dominican Republic. A surgeon extracted fat and bone marrow from Colon's body and isolated mesenchymal stem cells from these tissues. Stem cells are actively dividing, unspecialized cells that have the potential to produce different cell types depending on the signals they receive from the body. Mesenchymal stem cells are able to differentiate into various kinds of connective tissue, including bone, cartilage, blood vessels, tendons, and muscle. Colon's stem cells were injected into his elbow and shoulder, and months later he was apparently healed.

Scientists see much promise in the use of stem cell treatments to heal a wide variety of medical conditions, including cancer, Parkinson's disease, brain and spinal cord injuries, heart and other muscular damage, diabetes, blindness—even baldness. The basic idea is to inject stem cells into damaged tissues, where they will differentiate and form new, healthy tissues. Many procedures are being tried experimentally in the U.S., but so far the only kind of stem cell therapy approved by the U.S. government is hematopoietic stem cell transplantation (see Section 19.5). Other countries, including China, Mexico, Ukraine, and the Dominican Republic, already have clinics that offer various stem cell treatments.

The processes by which an unspecialized stem cell proliferates and forms specialized cells and tissues with distinctive appearances and functions are similar to the developmental processes that occur in the embryo. Much of our knowledge of developmental biology has come from studies on model organisms such as fruit flies, nematodes, frogs, sea urchins, mice, and the small flowering plant *Arabidopsis thaliana*. Eukaryotes share many similar genes, and the cellular and molecular principles underlying their development also turn out to be similar. Thus discoveries from one organism can aid us in understanding other organisms, including ourselves.

?

What are other uses of stem cells derived from fat?

See answer on p. 410.

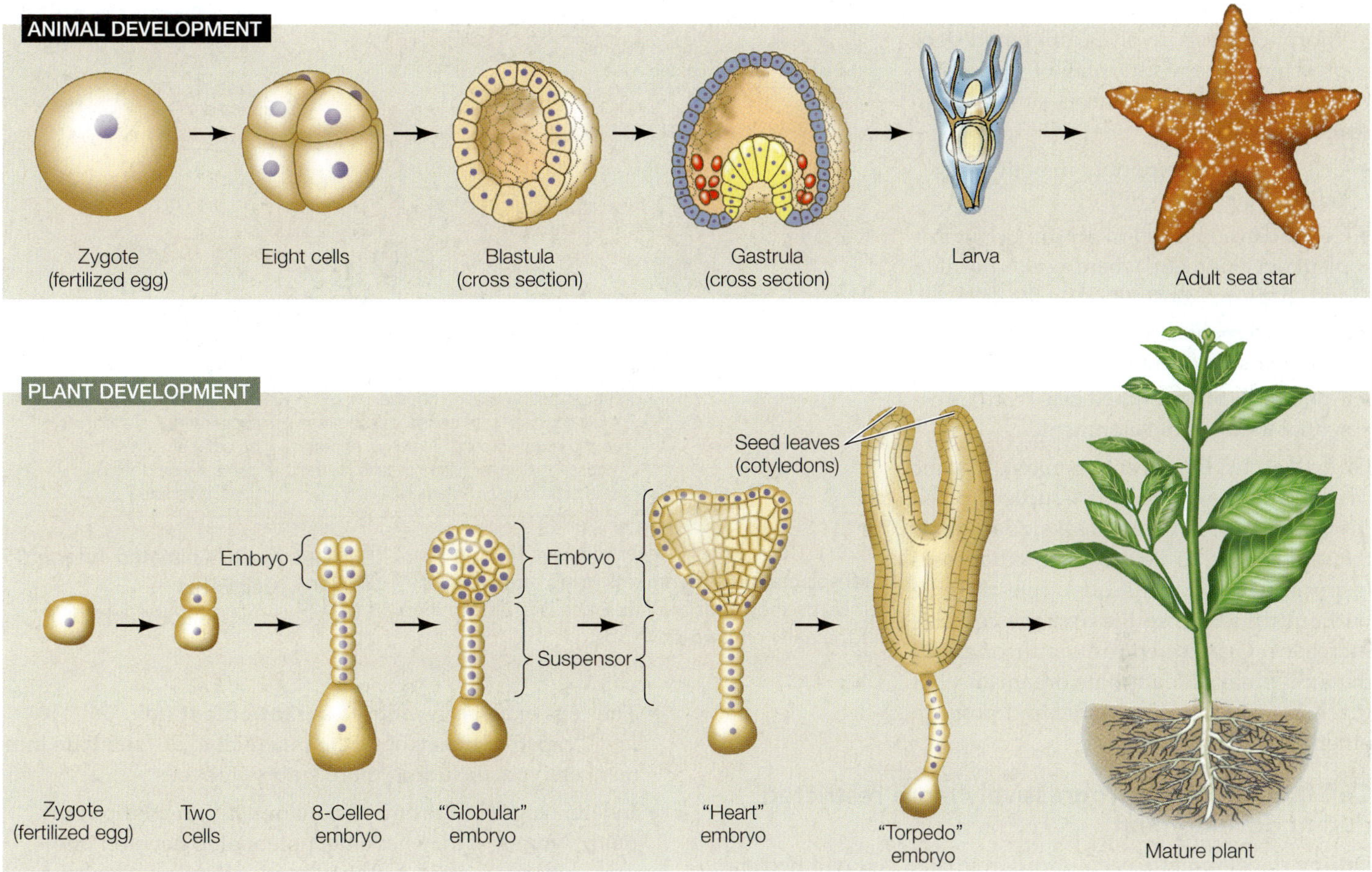

ANIMAL DEVELOPMENT
Zygote
(fertilized egg)
Eight cells
Blastula
(cross section)
Gastrula
(cross section)
Larva
Adult sea star
PLANT DEVELOPMENT
Seed leaves
(cotyledons)
Embryo
Embryo
Suspensor
Zygote
(fertilized egg)
Two
cells
8-Celled
embryo
"Globular"
embryo
"Heart"
embryo
"Torpedo"
embryo
Mature plant

19.1 From Fertilized Egg to Adult The stages of development from zygote to maturity are shown for an animal and for a plant. The blastula is a hollow sphere of cells; the gastrula has three cell layers.

Go to Activity 19.1 Stages of Development
Life10e.com/ac19.1

19.1 What Are the Processes of Development?

Development is the process by which a multicellular organism, beginning with a single cell, goes through a series of changes, taking on the successive forms that characterize its life cycle (Figure 19.1). After the egg is fertilized it is called a zygote, and in the earliest stages of development a plant or animal is called an **embryo**. Sometimes the embryo is contained within a protective structure such as a seed coat, an eggshell, or a uterus. An embryo does not photosynthesize or feed itself. Instead, it obtains its food from its mother either directly (via the placenta) or indirectly (by way of nutrients stored in a seed or egg). A series of embryonic stages precedes the birth of the new, independent organism. Many organisms continue to develop throughout their life cycles, with development ceasing only at death.

Development involves distinct but overlapping processes

The developmental changes an organism undergoes as it progresses from a fertilized egg to a mature adult involve four processes:

- **Determination** sets the developmental *fate* of a cell—what type of cell it will become—even before any characteristics of that cell type are observable. For example, the mesenchymal stem cells described in the opening story look unspecialized, but their fate to become connective tissue cells has already been determined.
- **Differentiation** is the process by which different types of cells arise, leading to cells with specific structures and functions. For example, mesenchymal stem cells differentiate to become muscle, fat, tendon, or other connective tissue cells.
- **Morphogenesis** (Greek for "origin of form") is the organization and spatial distribution of differentiated cells into the multicellular body and its organs.
- **Growth** is the increase in size of the body and its organs by cell division and cell enlargement.

Determination and differentiation occur largely because of differential gene expression. The cells that arise from repeated mitoses in the early embryo may look the same superficially, but they soon begin to differ in terms of which of the thousands of genes in the genome are expressed.

Morphogenesis involves differential gene expression and the interplay of signals between cells. Morphogenesis can occur in several ways:

- Cell division is important in both plants and animals.
- Cell expansion is especially important in plant development, where a cell's position and shape are constrained by the cell wall.
- Cell movements are very important in animal morphogenesis.
- Apoptosis (programmed cell death) is essential in organ development.

Growth occurs by cell enlargement. In some cases, cell enlargement is coupled to cell division, so the average cell size remains the same as the tissue grows; in other cases (especially in plant tissues), cells enlarge without dividing, so the average cell size increases. Growth continues throughout the individual's life in some organisms, but reaches a more or less stable end point in others.

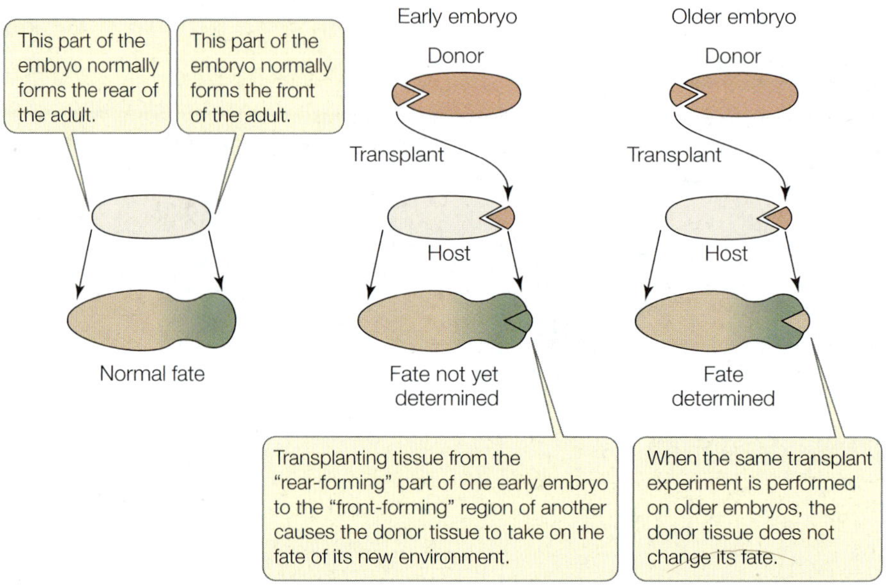

This part of the embryo normally forms the rear of the adult.
This part of the embryo normally forms the front of the adult.
Normal fate
Early embryo
Donor
Transplant
Host
Fate not yet determined
Older embryo
Donor
Transplant
Host
Fate determined
Transplanting tissue from the "rear-forming" part of one early embryo to the "front-forming" region of another causes the donor tissue to take on the fate of its new environment.
When the same transplant experiment is performed on older embryos, the donor tissue does not change its fate.

19.2 A Cell's Fate Is Determined in the Embryo Transplantation experiments using amphibian embryos show that the fate of cells is determined as the early embryo develops.

Go to Animated Tutorial 19.1 Cell Fates
Life10e.com/at19.1

Cell fates become progressively more restricted during development

During development, each undifferentiated cell will become part of a particular type of tissue—this is referred to as **cell fate**. Cell fate determination occurs as the embryo develops. The timing of this determination varies with the organism, but it is typically quite early. One way to find out the timing is to transplant cells from one embryo to a different region of a recipient embryo (Figure 19.2). Do the transplanted cells adopt the differentiation pattern of their new surroundings, or do they continue on their own path, with their fate already sealed?

In amphibian embryos determination happens early. If the donor tissue is from an early-stage embryo (blastula), it adopts the fate of the new surroundings. In this case, cell fate has not been determined and is influenced by the extracellular environment. But if the donor tissue is from an older embryo (gastrula), it continues on its original developmental path. In this case, cell fate has already been determined and is no longer influenced by the extracellular environment.

Cell fate determination is influenced by changes in gene expression as well as the extracellular environment. Determination is not something that is visible under the microscope—cells do not change their appearance when they become determined. Determination is followed by differentiation—the actual changes in biochemistry, structure, and function that result in different cell types. *Determination is a commitment; the final realization of that commitment is differentiation.*

During animal development, cell fate becomes progressively more restricted. This can be thought of in terms of **cell potency**, which is a cell's potential to differentiate into other cell types:

- The cells of an early embryo are **totipotent** (*toti*, "all"; *potent*, "capable"); they have the potential to differentiate into any cell type, including more embryonic cells.
- In later stages of the embryo, many cells are **pluripotent** (*pluri*, "many"); they have the potential to develop into most other cell types, but they cannot form new embryos.
- Through later developmental stages, including adulthood, certain stem cells are **multipotent**; they can differentiate into several different, related cell types. Mesenchymal stem cells (see the opening story) are one kind of multipotent stem cell.
- Many cells in the mature organism are **unipotent**; they can produce only one cell type—their own.

As you will see in Section 19.5, many plant cell types can be manipulated in the laboratory to dedifferentiate, form embryos, and develop into new plants. Much more recently, researchers have found ways to manipulate some mammalian cells to make them dedifferentiate and then redifferentiate into new tissues. So even though cell fate becomes progressively more restricted during normal development, this process can be altered in the laboratory.

RECAP 19.1

Development takes place via the processes of determination, differentiation, morphogenesis, and growth. Cells in the very early embryo have not yet had their fates determined; as development proceeds, their potential fates become more and more restricted.

- What are the four processes of development? See p. 393
- What did the experiments of the type illustrated in Figure 19.2 tell us about how cell fates become determined? See p. 394
- Describe the differences between totipotent, pluripotent, and multipotent cells. See p. 394

We have discussed the basic processes that occur during development, and seen that cell fate determination occurs before cells differentiate and become specialized. We will now turn to the mechanisms of cell fate determination.

19.2 How Is Cell Fate Determined?

The fertilized egg undergoes many cell divisions to produce the many differentiated cells in the body (such as liver, muscle, and nerve cells). How can one cell produce so many different cell types? There are two ways that determination occurs:

- **Cytoplasmic segregation** (unequal cytokinesis). A factor within an egg, zygote, or precursor cell may be unequally distributed in the cytoplasm. After cell division, the factor ends up in some daughter cells or regions of cells, but not others.
- **Induction** (cell-to-cell communication). A factor is actively produced and secreted by certain cells to induce other cells to become determined.

Cytoplasmic segregation can determine polarity and cell fate

Some differences in gene expression patterns are the result of cytoplasmic differences among cells. One such cytoplasmic difference is the emergence of distinct "top" and "bottom" ends of an organism or structure; such a difference is called **polarity**. Many examples of polarity are observed as development proceeds. Our heads are distinct from our rear ends, and the distal (far) ends of our arms and legs (wrists, ankles, fingers, toes) differ from the proximal (near) ends (shoulders and hips). Polarity may develop early; even within the fertilized egg, the yolk and other factors are often distributed asymmetrically. During early development in animals, polarity is specified by an animal pole at the top of the zygote and a vegetal pole at the bottom. This polarity can lead to determination of cell fates at a very early stage of development. For example, sea urchin embryos can be bisected at the eight-cell stage in two ways:

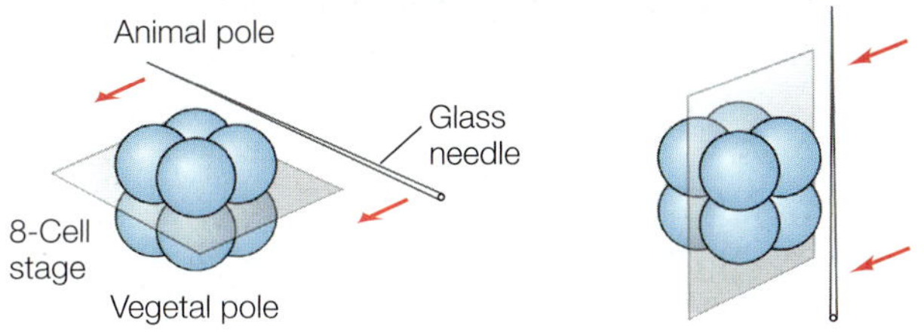
Animal pole
Glass needle
8-Cell stage
Vegetal pole

If the two halves of these embryos (each with four cells) are allowed to develop, the results are dramatically different for the two different cuts:

- If the embryo is cut into a top half and a bottom half (left, above), the bottom half develops into a small sea urchin and the top half does not develop at all.
- If the embryo is cut into two side halves (right, above), both halves develop into small sea urchins.

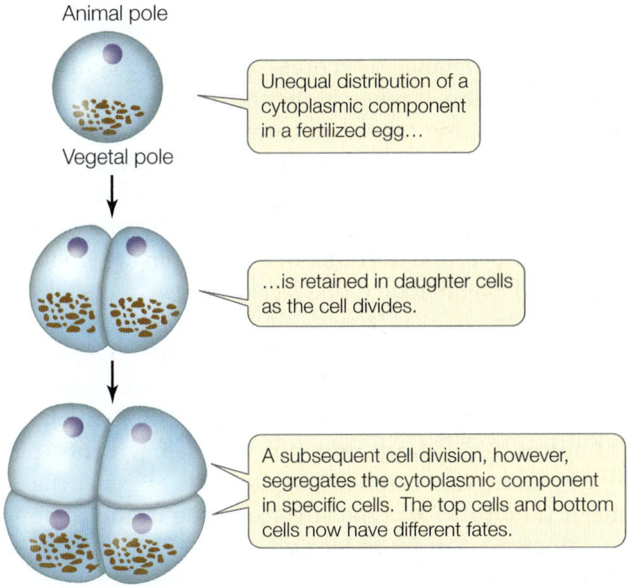
Animal pole
Vegetal pole
Unequal distribution of a cytoplasmic component in a fertilized egg...
...is retained in daughter cells as the cell divides.
A subsequent cell division, however, segregates the cytoplasmic component in specific cells. The top cells and bottom cells now have different fates.

19.3 The Principle of Cytoplasmic Segregation The unequal distribution of some component in the cytoplasm of a cell may determine the fates of its descendants.

Go to Animated Tutorial 19.2
Early Asymmetry in the Embryo
Life10e.com/at19.2

These results indicate that the top and bottom halves of an eight-celled sea urchin embryo have already developed distinct fates. Such observations led to the model of cytoplasmic segregation shown in Figure 19.3. The model states that certain materials called **cytoplasmic determinants** are distributed unequally in the egg cytoplasm. Cytoplasmic determinants include specific proteins, small regulatory RNAs, and mRNAs, and they play roles in directing the embryonic development of many organisms. What accounts for the unequal distribution of these determinants?

The cytoskeleton contributes to the asymmetric distribution of cytoplasmic determinants in the egg. Recall from Section 5.3 that an important function of the microtubules and microfilaments in the cytoskeleton is to help move materials around in the cell. Two properties allow these structures to accomplish this:

- Microtubules and microfilaments have polarity—they grow by adding subunits to the plus end.
- The cytoskeletan can bind motor proteins that are used to transport the cytoplasmic determinants.

For example, in the sea urchin egg there is a protein that binds to both the growing (+) end of a microfilament and to an mRNA encoding a cytoplasmic determinant. As the microfilament grows toward one end of the cell, it carries the mRNA along with it. The asymmetrical distribution of the mRNA leads to a similar distribution of the protein it encodes.

Inducers passing from one cell to another can determine cell fates

The term "induction" has different meanings in different contexts. In biology it can be used broadly to refer to the initiation of, or cause of, a change or process. But in the context of cellular differentiation, it refers to the signaling events by which cells in a developing organism communicate and influence one another's developmental fate. Induction involves chemical signals and signal transduction mechanisms. We will describe two examples of this form of induction: one in the developing vertebrate eye, and the other in a developing reproductive structure of the nematode *Caenorhabditis elegans*.

LENS DETERMINATION IN THE VERTEBRATE EYE The development of the lens in the vertebrate eye is a classic example of induction. In a frog embryo, the developing forebrain bulges out at both sides to form the optic vesicles, which expand until they come into contact with the cells at the surface of the head (**Figure 19.4**). The surface tissue in the region of contact thickens, forming a lens placode—tissue that will ultimately form the lens. The lens placode bends inward, folds over on itself, and ultimately detaches from the surface tissue to produce a structure that will develop into the lens. If the growing optic vesicle is cut away before it contacts the surface cells, no lens forms. Placing an impermeable barrier between the optic vesicle and the surface cells also prevents the lens from forming. These observations suggest that the surface tissue begins to develop into a lens when it receives a signal from the optic vesicle. Such a signal is termed an **inducer** (signaling factor).

Inducers trigger sequences of gene expression in the responding cells. How cells switch on different sets of genes that govern development and direct the formation of body plans is of great interest to developmental and evolutionary biologists. They use model organisms to investigate the major principles governing these processes.

VULVAL DETERMINATION IN THE NEMATODE The genome of the nematode *Caenorhabditis elegans* was one of the first eukaryotic genomes to be sequenced (see Section 17.3). It develops from fertilized egg to larva in only about 8 hours, and the worm reaches the adult stage in just 3.5 days. The process is easily observed using a low-magnification dissecting microscope because the body covering is transparent (**Figure 19.5A**).

The adult nematode is hermaphroditic, containing both male and female reproductive organs. It lays eggs through a pore called the vulva on the ventral (lower) surface. During development, a single cell, called the anchor cell, induces the vulva to form from six cells on the worm's ventral surface. In this case there are two molecular signals: the primary (1°) inducer and the secondary (2°) inducer. Each of the six ventral cells has three possible fates: it may become a primary vulval precursor cell, a secondary vulval precursor cell, or simply become part of the worm's skin—an epidermal cell. You can follow the sequence of events in **Figure 19.5B**. The concentration gradient of the primary inducer, LIN-3, is key. (LIN stands for abnormal cell *lin*eage.) The anchor cell produces the LIN-3 protein, which diffuses out of the cell and forms a concentration gradient with respect to adjacent cells. Three cells receive more LIN-3 than the others and become vulval precursor cells; cells farther from the anchor cell receive less LIN-3 and become epidermal cells. The cell closest to the anchor cell receives the most LIN-3—enough LIN-3 to turn on expression of the secondary inducer. The secondary inducer then acts on the two adjacent cells. This second induction event results in the two classes of vulval precursor cells: primary and secondary. Induction leads to the activation or inactivation of specific sets of genes through signal transduction cascades. This differential gene expression leads to cell differentiation.

RECAP 19.2

Cell fate determination involves cytoplasmic segregation and induction. Cytoplasmic segregation is the unequal distribution of gene products in the egg, zygote, or early embryo. Induction occurs when one cell or tissue sends a chemical signal to another.

- How does cytoplasmic segregation result in polarity in a fertilized egg, and how does polarity affect cell fate determination? **See p. 395 and Figure 19.3**
- Describe how induction influences tissue formation in the vertebrate eye. **See pp. 395–396 and Figure 19.4**
- Describe the process of induction during nematode vulva development. **See p. 396 and Figure 19.5**

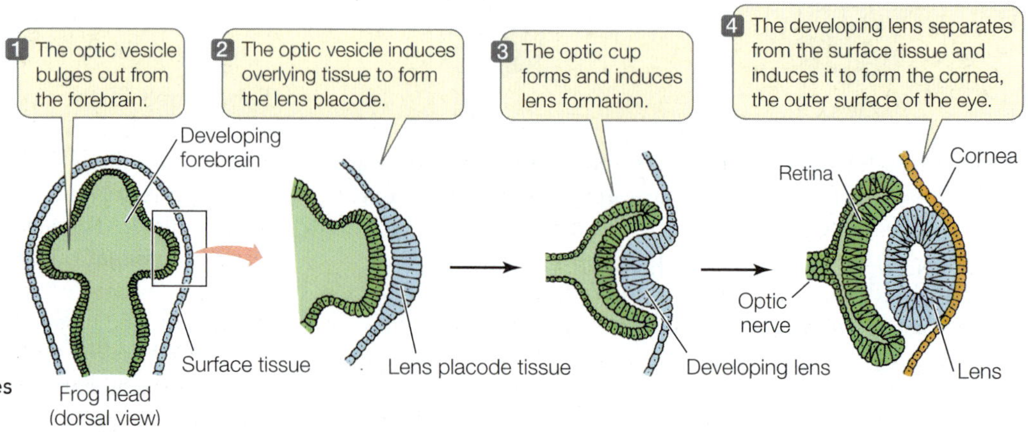
1 The optic vesicle bulges out from the forebrain.
2 The optic vesicle induces overlying tissue to form the lens placode.
3 The optic cup forms and induces lens formation.
4 The developing lens separates from the surface tissue and induces it to form the cornea, the outer surface of the eye.
Developing forebrain
Surface tissue
Frog head (dorsal view)
Lens placode tissue
Developing lens
Retina
Cornea
Optic nerve
Lens

19.4 Embryonic Inducers in Vertebrate Eye Development The eye of a frog develops as different cells induce changes in neighboring cells.

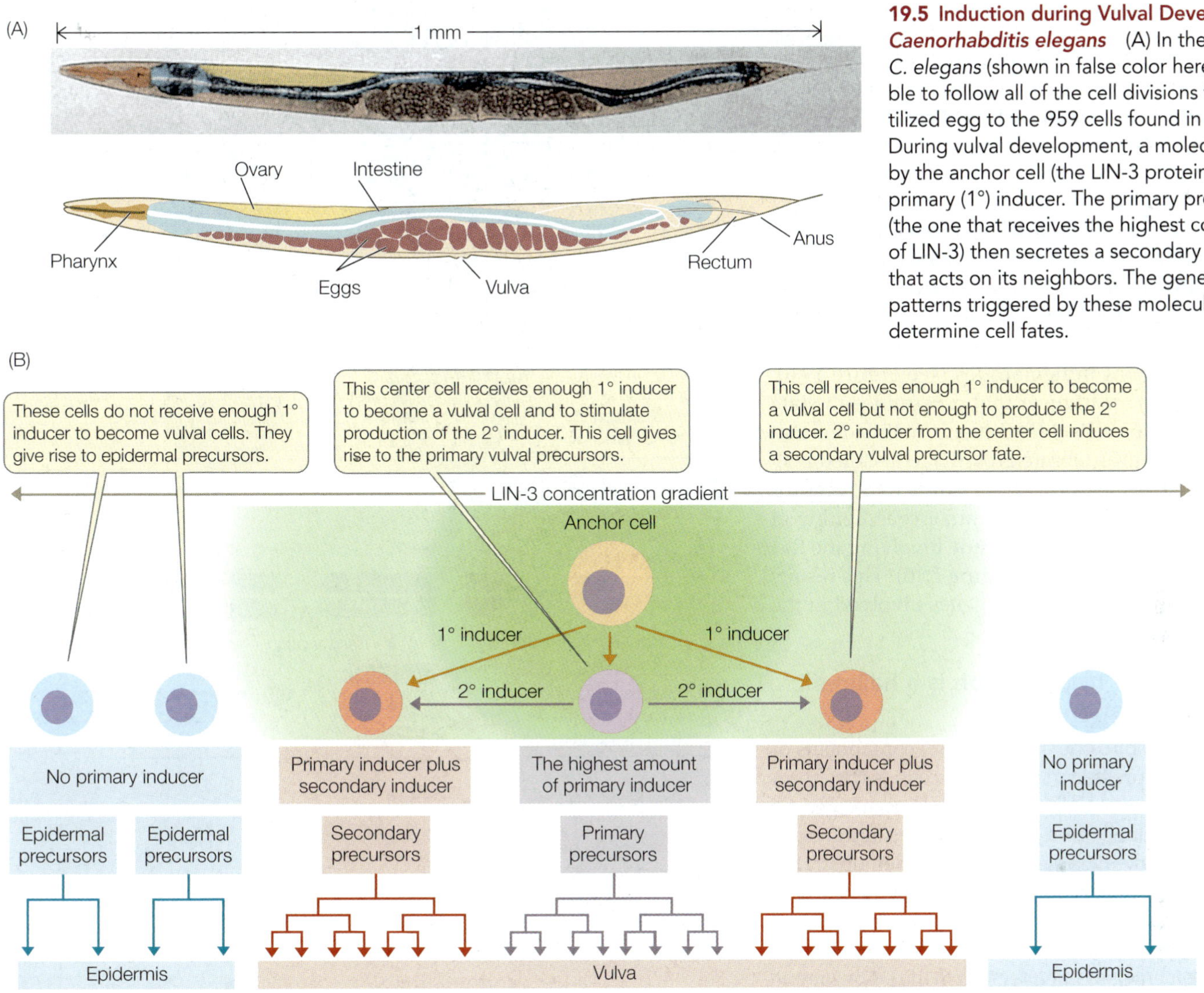

(A)
1 mm
Ovary
Intestine
Pharynx
Eggs
Vulva
Rectum
Anus
(B)
These cells do not receive enough 1° inducer to become vulval cells. They give rise to epidermal precursors.
This center cell receives enough 1° inducer to become a vulval cell and to stimulate production of the 2° inducer. This cell gives rise to the primary vulval precursors.
This cell receives enough 1° inducer to become a vulval cell but not enough to produce the 2° inducer. 2° inducer from the center cell induces a secondary vulval precursor fate.
LIN-3 concentration gradient
Anchor cell
1° inducer
1° inducer
2° inducer
2° inducer
No primary inducer
Primary inducer plus secondary inducer
The highest amount of primary inducer
Primary inducer plus secondary inducer
No primary inducer
Epidermal precursors
Epidermal precursors
Secondary precursors
Primary precursors
Secondary precursors
Epidermal precursors
Epidermis
Vulva
Epidermis

19.5 Induction during Vulval Development in *Caenorhabditis elegans* (A) In the nematode *C. elegans* (shown in false color here), it is possible to follow all of the cell divisions from the fertilized egg to the 959 cells found in the adult. (B) During vulval development, a molecule secreted by the anchor cell (the LIN-3 protein) acts as the primary (1°) inducer. The primary precursor cell (the one that receives the highest concentration of LIN-3) then secretes a secondary (2°) inducer that acts on its neighbors. The gene expression patterns triggered by these molecular switches determine cell fates.

We have seen that cytoplasmic segregation and induction both influence cell fate determination. We have seen two examples of how induction leads to organ formation in developing multicellular organisms. Next we will take a closer look at how gene expression affects cell fate determination and differentiation.

19.3 What Is the Role of Gene Expression in Development?

Although every cell contains all the genes needed to produce every protein encoded by an organism's genome, each cell expresses only selected genes. For example, certain cells in hair follicles produce keratin, the protein that makes up hair, whereas other cell types in the body do not. What determines whether a cell will produce keratin? Chapter 16 described a number of ways in which cells regulate gene expression and protein production—by controlling transcription, translation, and posttranslational protein modifications. The mechanisms that control gene expression during cell fate determination and cell differentiation generally work at the level of transcription.

Cell fate determination involves signal transduction pathways that lead to differential gene expression

As we have seen, cell fate determination can occur by the process of induction. When an inducer molecule binds to its specific receptor on the surface of a cell, a signal transduction pathway leads to the activation of one or more transcription factors. Recall that transcription factors are DNA binding proteins that regulate the expression of specific genes (see Section 16.2). **Figure 19.6** illustrates the induction of one cell (the cell on the left), which is exposed to a high concentration of inducer. This results in the activation of a transcription factor in the cytoplasm, causing it to enter the nucleus and switch on the expression of a specific gene. The cell on the right is exposed to a lower concentration of the inducer, and as a result, gene expression is not activated.

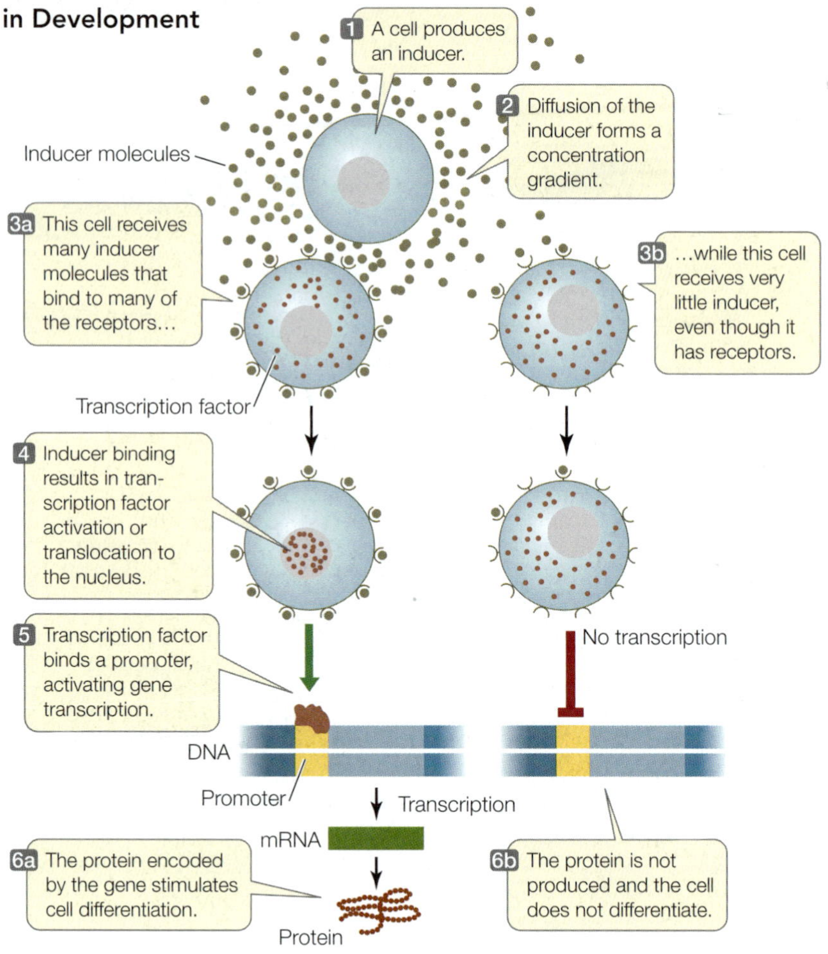
1 A cell produces an inducer.
2 Diffusion of the inducer forms a concentration gradient.
Inducer molecules
3a This cell receives many inducer molecules that bind to many of the receptors...
3b ...while this cell receives very little inducer, even though it has receptors.
Transcription factor
4 Inducer binding results in transcription factor activation or translocation to the nucleus.
5 Transcription factor binds a promoter, activating gene transcription.
No transcription
DNA
Promoter
Transcription
mRNA
6a The protein encoded by the gene stimulates cell differentiation.
Protein
6b The protein is not produced and the cell does not differentiate.

19.6 Induction The concentration of an inducer directly affects the degree to which a transcription factor is activated. The inducer acts by binding to a receptor on the target cell. This binding is followed by signal transduction involving transcription factor activation or translocation from the cytoplasm to the nucleus. In the nucleus it acts to stimulate the expression of genes involved in cell differentiation.

It turns out that development is often controlled by these kinds of molecular switches, which allow a cell to proceed down one of two alternative paths. One challenge for developmental biologists is to find these switches and determine how they work. In the case of vulval development in nematodes (see Figure 19.5), LIN-3 is the primary inducer that determines the fate of vulval precursor cells. LIN-3 is a growth factor that is homologous to a vertebrate growth factor called EGF (*e*pidermal *g*rowth *f*actor). LIN-3 binds to a receptor on the surfaces of the vulval precursor cells, setting in motion a signal transduction cascade involving the Ras protein and MAP kinases (see Figure 7.10). This results in increased transcription of the genes involved in the differentiation of vulval cells.

Differential gene transcription is a hallmark of cell differentiation

The gene for β-globin, one of the protein components of hemoglobin, is expressed in red blood cells as they form in the bone marrow of mammals. That this same gene is also present—but unexpressed—in neurons in the brain (which do not make hemoglobin) can be demonstrated by nucleic acid hybridization. Recall that in nucleic acid hybridization, a probe made of single-stranded DNA or RNA of known sequence is added to denatured DNA to reveal complementary coding regions in the DNA sample (see Figure 14.7). A probe for the β-globin gene can be applied to DNA from brain cells and immature red blood cells (recall that mature mammalian red blood cells lose their nuclei during development). In both cases, the probe finds its complement, showing that the β-globin gene is present in both types of cells. However, if the β-globin probe is applied to mRNA, rather than DNA, from the two cell types, it finds β-globin mRNA only in the red blood cells, not in the brain cells. This result shows that the gene is expressed in only one of the two cell types.

What leads to this differential gene expression? One well-studied example of cell differentiation is the conversion of undifferentiated muscle precursor cells into cells that are destined to form muscle (**Figure 19.7**). In the vertebrate embryo these cells come from a tissue layer called the mesoderm (see Section 44.3). A key event in the commitment of these cells to become

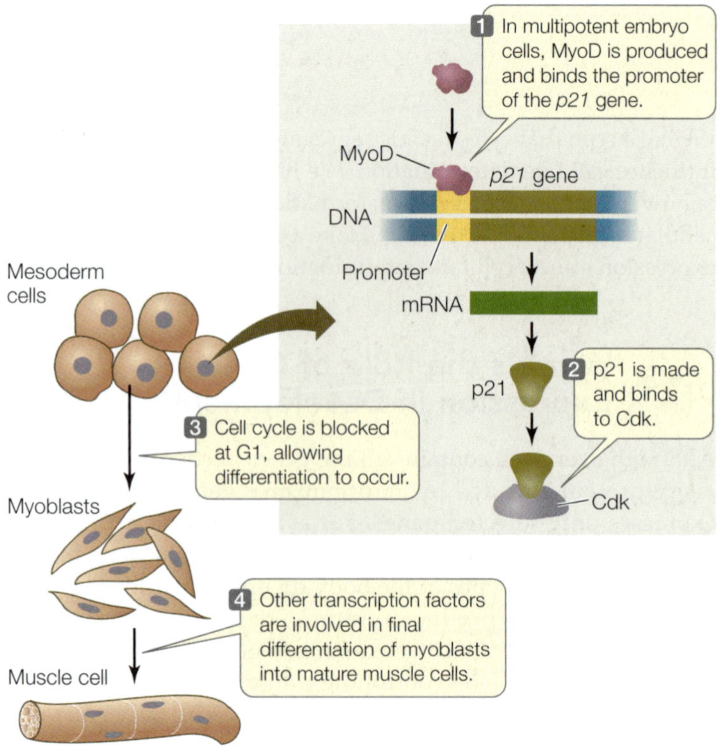
1 In multipotent embryo cells, MyoD is produced and binds the promoter of the *p21* gene.
MyoD
p21 gene
DNA
Promoter
mRNA
p21
2 p21 is made and binds to Cdk.
Cdk
Mesoderm cells
3 Cell cycle is blocked at G1, allowing differentiation to occur.
Myoblasts
4 Other transcription factors are involved in final differentiation of myoblasts into mature muscle cells.
Muscle cell

19.7 Transcription and Differentiation in the Formation of Muscle Cells Production of the transcription factor MyoD is important in muscle cell differentiation.

muscle is that they stop dividing. Indeed, in many parts of the embryo, *cell division and cell differentiation are mutually exclusive.* Cell signaling activates the gene for a transcription factor called **MyoD** (for *myo*blast-*d*etermining gene); this in turn activates the gene for p21, which is an inhibitor of the cyclin-dependent kinases (Cdk's) that normally stimulate the cell cycle at G1 (see Figure 11.6). Expression of the *p21* gene causes the cell cycle to stop, and other transcription factors then enter the picture so that differentiation can proceed. Interestingly, *myoD* is also activated in the stem cells that are present in adult muscle, indicating a role of this transcription factor in the repair of muscle tissue as it gets damaged and worn out.

Genes such as *myoD* that direct the most fundamental decisions in development (often by regulating other genes on other chromosomes) usually encode transcription factors. In some cases a single transcription factor can cause a cell to differentiate in a certain way. In others, complex interactions between genes and proteins determine a sequence of transcriptional events that leads to differentiation.

RECAP 19.3

Cell fate determination involves the activation of signal transduction pathways that lead to differential gene expression. Differentiation involves selective gene expression, controlled at the level of transcription by transcription factors.

- How do inducer molecules cause changes in gene expression? **See pp. 397–398 and Figure 19.6**
- What techniques could you use to identify genes expressed during cell differentiation? **See p. 398**
- Describe the roles of transcription factors in controlling differentiation. **See pp. 398–399 and Figure 19.7**

We have seen how cell fate is determined, and we have looked at the roles of gene expression in cell fate determination and differentiation. We will now take a closer look at how gene expression affects differentiation and morphogenesis.

19.4 How Does Gene Expression Determine Pattern Formation?

Pattern formation is the process that results in the spatial organization of a tissue or organism. It is inextricably linked to morphogenesis, the creation of body form. You might expect morphogenesis to involve a lot of cell division, followed by differentiation—and it does. But what you might not expect is the amount of programmed cell death—apoptosis—that occurs during morphogenesis.

Multiple proteins interact to determine developmental programmed cell death

We noted in Section 11.6 that apoptosis is a programmed series of events that leads to cell death. Apoptosis is an integral part of the normal development and life of an organism. For example, in an early human embryo, the hands and feet look like tiny paddles: the tissues that will become fingers and toes are joined by connective tissue. Between days 41 and 56 of development, the cells between the digits die, freeing the individual fingers and toes:

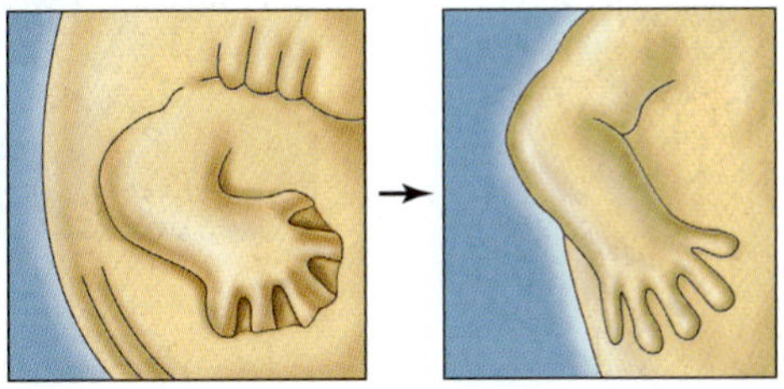

Day 41
Day 56

Many cells and structures form and then disappear during development, in processes involving apoptosis.

Model organisms have been very useful in studying the genes and proteins involved in apoptosis. Mutants with altered cell death phenotypes have been used to identify the genes and proteins involved. For example, the nematode worm *C. elegans* produces precisely 1,090 somatic cells as it develops from a fertilized egg into an adult, but 131 of those cells die, leaving 959 cells in the adult worm. The sequential activation of two proteins called CED-4 and CED-3 (for *ce*ll *d*eath) appears to control this programmed cell death (**Figure 19.8A**). A third protein called CED-9, which is bound to the outside of the mitochondrion, inhibits apoptosis in cells that are not programmed to die. In these cells, CED-9 binds CED-4 and prevents it from activating CED-3. If the cell receives a signal for apoptosis, CED-9 releases CED-4, which then activates CED-3, a protease (an enzyme that breaks down proteins).

A similar system controls apoptosis during human development. The apoptosis pathway in humans involves a class of proteases called caspases (see Figure 11.23). Caspases are similar in amino acid sequence to CED-3 in *C. elegans*. Furthermore, the human proteins Bcl-2 and Apaf1 are similar in

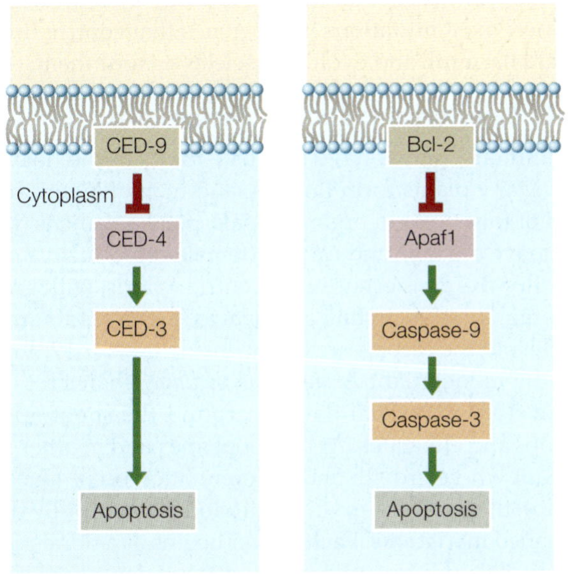

(A) C. elegans
(B) Human neuron
CED-9
Cytoplasm
CED-4
CED-3
Apoptosis
Bcl-2
Apaf1
Caspase-9
Caspase-3
Apoptosis

19.8 Pathways for Apoptosis In the worm *C. elegans* (A) and humans (B), similar pathways for apoptosis are controlled by genes with similar sequences and functions.

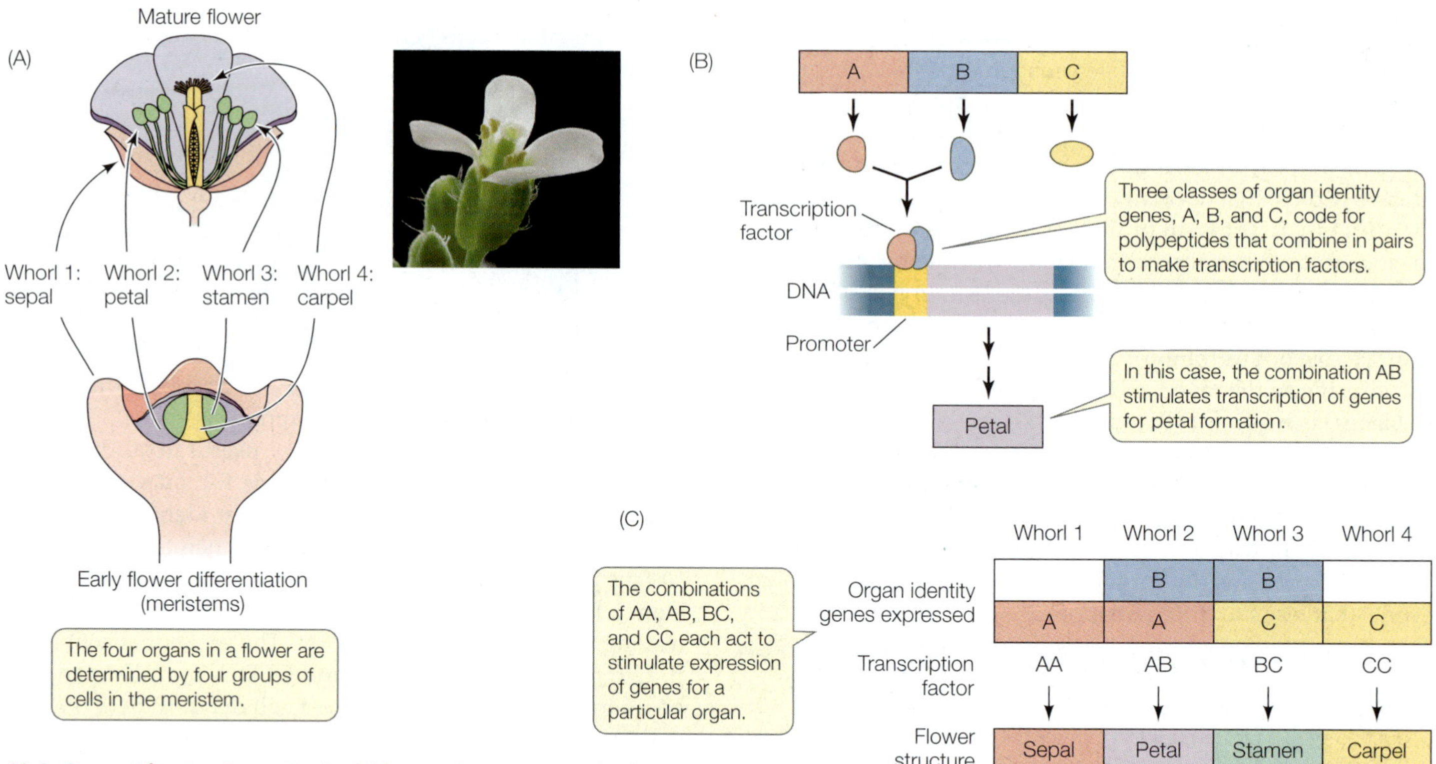
(A)
Mature flower
Whorl 1: sepal
Whorl 2: petal
Whorl 3: stamen
Whorl 4: carpel
Early flower differentiation (meristems)
The four organs in a flower are determined by four groups of cells in the meristem.
(B)
A
B
C
Transcription factor
DNA
Promoter
Petal
Three classes of organ identity genes, A, B, and C, code for polypeptides that combine in pairs to make transcription factors.
In this case, the combination AB stimulates transcription of genes for petal formation.
(C)
The combinations of AA, AB, BC, and CC each act to stimulate expression of genes for a particular organ.
Whorl 1
Whorl 2
Whorl 3
Whorl 4
Organ identity genes expressed
B
B
A
A
C
C
Transcription factor
AA
AB
BC
CC
Flower structure
Sepal
Petal
Stamen
Carpel

19.9 Organ Identity Genes in *Arabidopsis* Flowers (A) The four organs of a flower—carpels (yellow), stamens (green), petals (purple), and sepals (pink)—grow in whorls that develop from the floral meristem. (B) Floral organs are determined by three classes of organ identity genes whose polypeptide products combine in pairs to form transcription factors. (C) Combinations of polypeptide subunits in transcription factors activate gene expression for specific organs.

structure and function to the *C. elegans* proteins CED-9 and CED-4, respectively (Figure 19.8B). So humans and nematodes, two species separated by more than 600 million years of evolutionary history, have similar genes controlling programmed cell death. The conservation of this pathway indicates its importance: most mutations in the genes that control this pathway are harmful, and evolution selects against them.

Plants have organ identity genes

Like animals, plants have organs—for example, leaves and roots. Many plants form flowers, and many flowers are composed of four types of organs: sepals, petals, stamens (male reproductive organs), and carpels (female reproductive organs). These floral organs occur in concentric whorls, with groups of each organ type encircling a central axis. The sepals are on the outside and the carpels are on the inside (Figure 19.9A).

In the model plant *Arabidopsis thaliana* (thale cress), flowers develop in a radial pattern around the shoot apex as it develops and elongates. At the shoot apex and in other parts of the plant where growth and differentiation occur (such as the root tip), there are groups of undifferentiated, rapidly dividing cells called **meristems**. Each flower begins as a floral meristem of about 700 undifferentiated cells arranged in a dome, and the four whorls develop from this meristem. How is the identity of a particular whorl determined? Three classes of genes called **organ identity genes** encode proteins that act in combination to produce specific whorl features (Figure 19.9B and C):

- Genes in class A are expressed in whorls 1 and 2 (which form sepals and petals, respectively).
- Genes in class B are expressed in whorls 2 and 3 (which form petals and stamens).
- Genes in class C are expressed in whorls 3 and 4 (which form stamens and carpels).

These genes encode transcription factors that are active as dimers, that is, proteins with two polypeptide subunits. The composition of the dimer determines which genes the transcription factor activates. For example, a dimer made up of two class A monomers activates transcription of the genes that make sepals; a dimer made up of A and B monomers results in petals, and so forth. A common feature of the A, B, and C proteins, as well as many other plant transcription factors, is a DNA-binding domain called the **MADS box**. The name "MADS" comes from the initials of four genes encoding proteins with this domain.

Two lines of experimental evidence support this model for floral organ determination:

- *Loss-of-function mutations*: for example, a mutation in a class A gene results in no sepals or petals.
- *Gain-of-function mutations*: for example, a promoter for a class C gene can be artificially coupled to a class A gene. In

this case, the class A gene is expressed in all four whorls, resulting in only sepals and petals. In any organism, the replacement of one organ for another is called homeosis, and this type of mutation is a **homeotic mutation**.

Transcription of the floral organ identity genes is controlled by other gene products, including the LEAFY protein. Plants with loss-of-function mutations in the *LEAFY* gene make stems instead of flowers, with increased numbers of modified leaves called bracts. The wild-type LEAFY protein is a transcription factor that stimulates expression of the class A, B, and C genes so that they produce flowers. This finding has practical applications. It usually takes 6 to 20 years for a citrus tree to produce flowers and fruits. Scientists have made transgenic orange trees expressing the *LEAFY* gene coupled to a strongly expressed promoter. These trees flower and fruit years earlier than normal trees.

Morphogen gradients provide positional information

During development, the key cellular question "What will I be?" is often answered in part by "Where am I?" Think of the cells in the developing nematode, which can develop into different parts of the vulva depending on their positions relative to the anchor cell (see Figure 19.5). This spatial "sense" is called **positional information**. Positional information often comes in the form of an inducer called a **morphogen**, which diffuses from one cell or group of cells to surrounding cells, setting up a concentration gradient (as we saw for LIN-3 in *C. elegans* vulval induction). There are two requirements for a signal to be considered a morphogen:

- It must directly affect target cells, rather than triggering a secondary signal that affects target cells.
- Different concentrations of the signal must cause different effects.

Developmental biologist Lewis Wolpert uses the "French flag model" to explain morphogens (**Figure 19.10**). This model can be applied to the differentiation of the vulva in *C. elegans* and to the development of a vertebrate limb.

The vertebrate limb develops from a paddle-shaped limb bud (**Figure 19.11**). The cells that develop into different digits must receive positional information; if they do not, the limb will be totally disorganized—imagine a hand with only thumbs or only little fingers. A group of cells at the posterior base of the limb bud, just where it joins the body wall, is called the zone of polarizing activity (ZPA). The cells of the ZPA secrete a protein morphogen called Sonic *h*edge*h*og (Shh). Shh forms a gradient that determines the posterior–anterior (little finger to thumb) axis of the developing limb. In humans and other primates, the cells exposed to the highest dose of Shh form the little finger; those that receive the lowest dose develop into the thumb.

A cascade of transcription factors establishes body segmentation in the fruit fly

Perhaps the best-studied example of how morphogens determine cell fate is body segmentation in the fruit fly *Drosophila melanogaster*. The body segments of this model organism are clearly different from one another. The adult fly has an anterior head (composed of several fused segments), three different thoracic segments, and eight abdominal segments at the posterior end. Each segment develops into different body parts: for example, antennae and eyes develop from head segments, wings from the thorax, and so on.

The life cycle of *Drosophila* from fertilized egg to adult takes about 2 weeks at room temperature. The egg hatches into a larva, which then forms a pupa, which finally is transformed

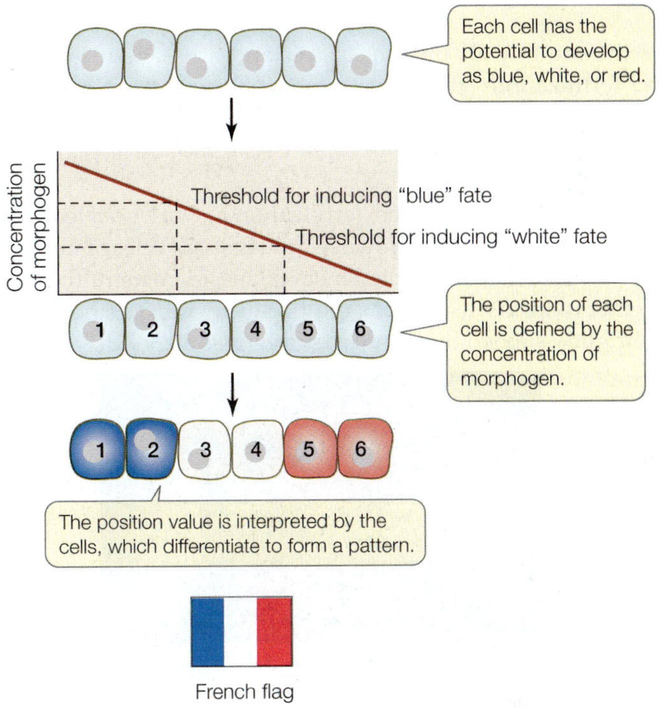

Each cell has the potential to develop as blue, white, or red.
Concentration of morphogen
Threshold for inducing "blue" fate
Threshold for inducing "white" fate
1 2 3 4 5 6
The position of each cell is defined by the concentration of morphogen.
1 2 3 4 5 6
The position value is interpreted by the cells, which differentiate to form a pattern.
French flag

19.10 The French Flag Model In the "French flag" model, a concentration gradient of a diffusible morphogen signals each cell to specify its position.

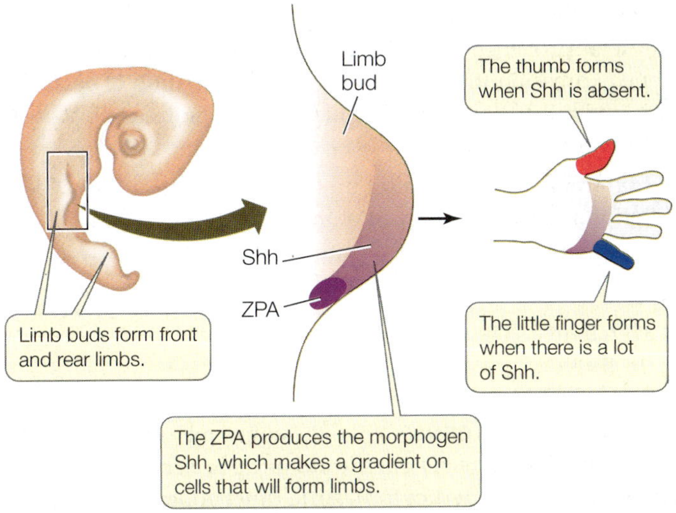

Limb bud
The thumb forms when Shh is absent.
Shh
ZPA
Limb buds form front and rear limbs.
The little finger forms when there is a lot of Shh.
The ZPA produces the morphogen Shh, which makes a gradient on cells that will form limbs.

19.11 Specification of the Vertebrate Limb and the French Flag Model The zone of polarizing activity (ZPA) in the limb bud of the embryo secretes the morphogen Sonic *h*edge*h*og (Shh). Cells in the bud form different digits depending on the concentration of Shh.

into the adult fly. By the time a larva appears—about 24 hours after fertilization—there are recognizable segments. The thoracic and abdominal segments all look similar, *but the fates of their cells to become different adult segments have already been determined.*

As with other organisms, fertilization in *Drosophila* leads to a rapid series of mitoses. However, the first 12 cycles of nuclear division are not accompanied by cytokinesis. So a multinucleate embryo forms instead of a multicellular embryo (the nuclei are brightly stained in the micrographs below):

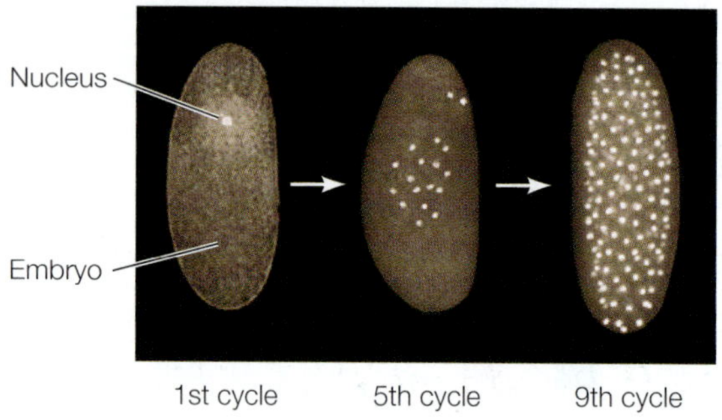
Nucleus
Embryo
1st cycle
5th cycle
9th cycle

With no cell membranes to cross, morphogens can diffuse easily within the embryo. We focus here on the determination events that occur in the first 24 hours after fertilization.

The events leading to cell fate determination in *Drosophila* were elucidated using experimental genetics:

- First, developmental mutations were identified. For example, a mutant strain might produce larvae with two heads or no segments.
- Then the mutant was compared with wild-type flies, and the gene responsible for the developmental mistake, and the gene's protein product (if appropriate), were isolated.
- Finally, experiments with the gene (making transgenic flies) and protein (injecting the protein into an egg or into an embryo) were done to confirm their roles in the proposed developmental pathway.

These approaches revealed a sequential pattern (cascade) of gene expression that results in the determination of each segment within 24 hours after fertilization. Several classes of genes are involved:

- **Maternal effect genes** set up the major axes (anterior–posterior and dorsal–ventral) of the egg.
- **Segmentation genes** determine the boundaries and polarity of each segment.
- **Hox genes** determine which organ will be made at a given location.

MATERNAL EFFECT GENES Like the eggs and early embryos of sea urchins, *Drosophila* eggs and larvae are characterized by unevenly distributed cytoplasmic determinants (see Figure 19.3). These molecular determinants are the products of specific maternal effect genes. The genes are transcribed in the cells of the mother's ovary, and the mRNAs are passed to the egg by cytoplasmic bridges. Two maternal effect genes called *bicoid* and *nanos* help determine the anterior–posterior axis of the egg. (The dorsal–ventral axis is determined by other maternal effect genes that will not be described here.)

The mRNAs for *bicoid* and *nanos* diffuse from the mother's cells into what will be the anterior end of the egg. The *bicoid* mRNA is translated into Bicoid protein, a transcription factor that diffuses away from the anterior end, establishing a gradient in the egg cytoplasm (**Figure 19.12A**). Meanwhile, the egg's cytoskeleton transports the *nanos* mRNA from the anterior end of the egg, where it was deposited, to the posterior end, where it is translated (**Figure 19.12B**).

The actions of Bicoid and Nanos establish a gradient of yet another protein, called Hunchback, which determines the anterior and posterior ends of the embryo. Initially, the *hunchback* mRNA is evenly distributed in the embryo, but Nanos inhibits its translation, thus preventing Hunchback protein accumulation at the posterior end of the embryo (**Figure 19.12C**). Meanwhile, at the anterior end of the embryo, Bicoid stimulates increased transcription of the *hunchback* gene, thus increasing the amount of *hunchback* mRNA (and thus Hunchback protein) and further strengthening the Hunchback gradient.

How did biologists elucidate these pathways? Let's look more closely at the experimental approaches used in this case.

- Females that are homozygous for a particular *bicoid* mutation produce larvae with no head and no thorax; thus the Bicoid protein must be needed for the anterior structures to develop.
- If the eggs of these *bicoid* mutants are injected at the anterior end with cytoplasm from the anterior region of a wild-type egg, the injected eggs develop into normal larvae. This experiment also shows that the Bicoid protein is involved in the development of anterior structures.
- If cytoplasm from the anterior region of a wild-type egg is injected into the posterior region of another egg, anterior structures develop there. The degree of induction depends on how much cytoplasm is injected.
- Eggs from homozygous *nanos* mutant females develop into larvae with missing abdominal segments.
- If cytoplasm from the posterior region of a wild-type egg is injected into the posterior region of a *nanos* mutant egg, it will develop normally.

These and other experiments led scientists to understand the cascade of events that determines cell fates.

The events involving *bicoid*, *nanos*, and *hunchback* begin before fertilization and continue after it, during the multinucleate stage, which lasts a few hours. At this stage the embryo looks like a bunch of indistinguishable nuclei under the light microscope. But the cell fates have already begun to be determined. After the anterior and posterior ends have been established, the next step in pattern formation is the determination of segment number and locations.

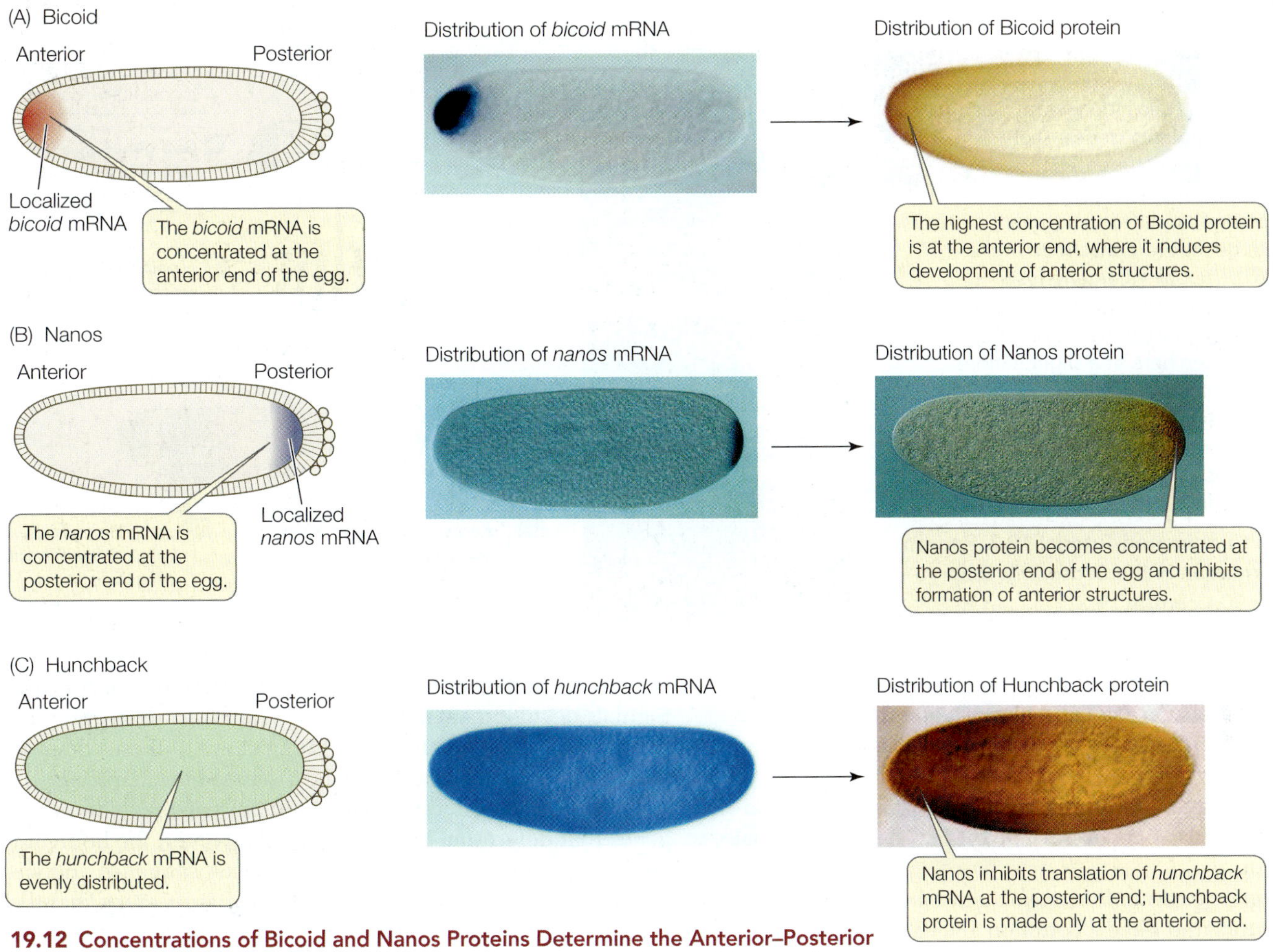

(A) Bicoid
Anterior
Posterior
Localized bicoid mRNA
The bicoid mRNA is concentrated at the anterior end of the egg.
Distribution of bicoid mRNA
Distribution of Bicoid protein
The highest concentration of Bicoid protein is at the anterior end, where it induces development of anterior structures.
(B) Nanos
Anterior
Posterior
The nanos mRNA is concentrated at the posterior end of the egg.
Localized nanos mRNA
Distribution of nanos mRNA
Distribution of Nanos protein
Nanos protein becomes concentrated at the posterior end of the egg and inhibits formation of anterior structures.
(C) Hunchback
Anterior
Posterior
The hunchback mRNA is evenly distributed.
Distribution of hunchback mRNA
Distribution of Hunchback protein
Nanos inhibits translation of hunchback mRNA at the posterior end; Hunchback protein is made only at the anterior end.

19.12 Concentrations of Bicoid and Nanos Proteins Determine the Anterior–Posterior Axis The anterior–posterior axis of *Drosophila* arises from gradients of the morphogens encoded by (A) *bicoid* and (B) *nanos*. Together, Bicoid and Nanos establish a concentration gradient of Hunchback (C).

SEGMENTATION GENES The number and polarity of the *Drosophila* larval segments are determined by the segmentation genes. These genes are expressed when there are about 6,000 nuclei in the embryo (about 3 hours after fertilization). Three classes of segmentation genes act one after the other to regulate finer and finer details of the segmentation pattern:

- **Gap genes** organize broad areas along the anterior–posterior axis. Mutations in gap genes result in gaps in the body plan—the omission of several consecutive larval segments.
- **Pair rule genes** divide the embryo into units of two segments each. Mutations in pair rule genes result in embryos missing every other segment.
- **Segment polarity genes** determine the boundaries and anterior–posterior organization of the individual segments. Mutations in segment polarity genes can result in segments in which posterior structures are replaced by reversed (mirror-image) anterior structures.

The expression of these genes is sequential (**Figure 19.13**). The products of the gap genes activate pair rule genes, and the pair rule gene products activate segment polarity genes. By the end of this cascade, nuclei throughout the embryo "know" which segment they will be part of in the adult fly.

The next set of genes in the cascade determines the form and function of each segment.

HOX GENES Hox (for "Homeobox") genes encode a family of transcription factors that are expressed in different combinations along the length of the embryo, and help determine cell fate within each segment. Hox gene expression tells the cells of a segment in the head to make eyes, those of a segment in the thorax to make wings, and so on. The *Drosophila* Hox genes occur in two clusters on chromosome 3, in the same order as the segments whose function they determine (**Figure 19.14**). By the time the fruit fly larva hatches, its segments are completely determined. Hox genes are shared by all animals and are homeotic genes—that is, a mutation in a Hox gene can result in one organ being replaced by another.

In *Drosophila*, the maternal effect genes, segmentation genes, and Hox genes interact to "build" a larva step by step, beginning with the unfertilized egg. How do we know that the Hox genes determine segment identity? A clue comes from

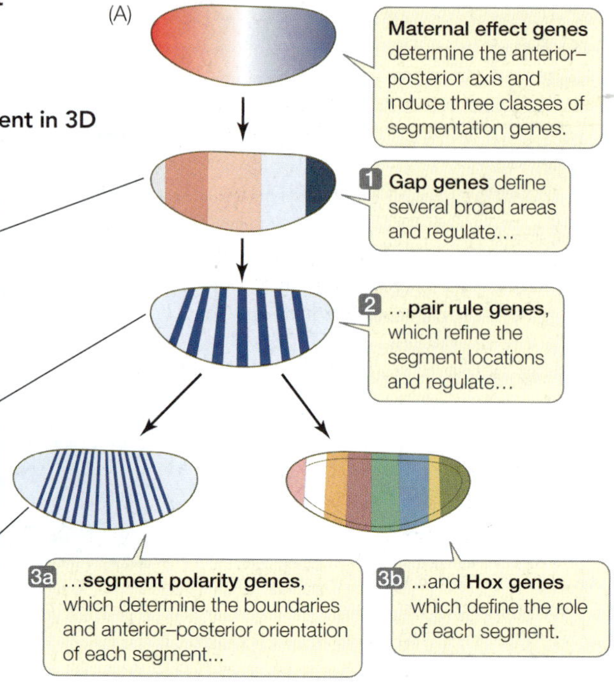
(A)
Maternal effect genes determine the anterior–posterior axis and induce three classes of segmentation genes.
1 Gap genes define several broad areas and regulate…
2 …pair rule genes, which refine the segment locations and regulate…
3a …segment polarity genes, which determine the boundaries and anterior–posterior orientation of each segment…
3b …and Hox genes which define the role of each segment.

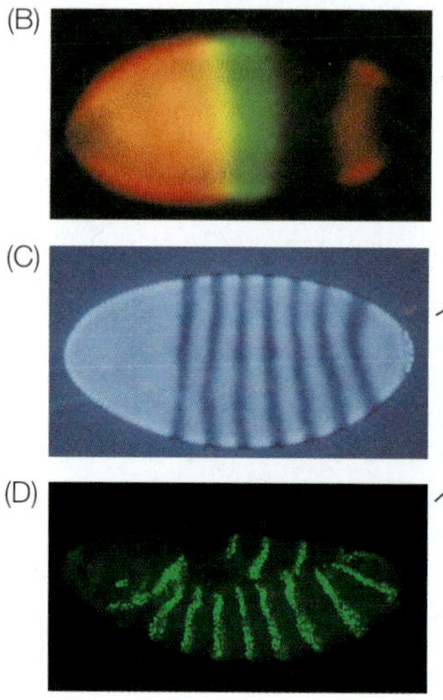
(B)
(C)
(D)

19.13 A Gene Cascade Controls Pattern Formation in the *Drosophila* Embryo
(A) Maternal effect genes induce gap, pair rule, and segment polarity genes—collectively referred to as segmentation genes. (B) Expression of two gap genes, *hunchback* (orange) and *Krüppel* (green), overlaps; both genes are transcribed in the yellow area. (C) The pair rule gene *fushi tarazu* is transcribed in the dark blue areas. (D) The segment polarity gene *engrailed* (bright green) is seen here at a slightly more advanced stage than is depicted in (A). By the end of this cascade, a group of nuclei at the anterior of the embryo, for example, is determined to become the first head segment in the adult fly.

Go to Media Clip 19.1 Spectacular Fly Development in 3D
Life10e.com/mc19.1

Go to Animated Tutorial 19.3 Pattern Formation in the *Drosophila* Embryo
Life10e.com/at19.3

homeotic mutations. A mutation in the Hox gene *antennapedia* causes legs to grow on the head in place of antennae (**Figure 19.15**). When another Hox gene, *ultrabithorax* is mutated, an extra pair of wings grows in a thoracic segment where wings do not normally occur (see Figure 20.3). So the normal (wild-type) functions of Hox genes must be to "tell" a segment what organ to form.

The *antennapedia* and *ultrabithorax* genes both encode transcription factors and have a common 180 base pair sequence called the **homeobox**. It encodes a 60 amino acid sequence called the **homeodomain**. The homeodomain recognizes and binds to a specific DNA sequence in the promoters of its target genes. This protein domain is found in transcription factors that regulate development in many other animals with an anterior–posterior axis. The evolutionary significance of these common pathways for development will be discussed in Chapter 20.

As we have seen, both plants and animals have homeotic genes that determine organ identity. However, the homeotic genes in plants and animals differ in DNA sequence and encoded protein structure. This is not surprising, given that the

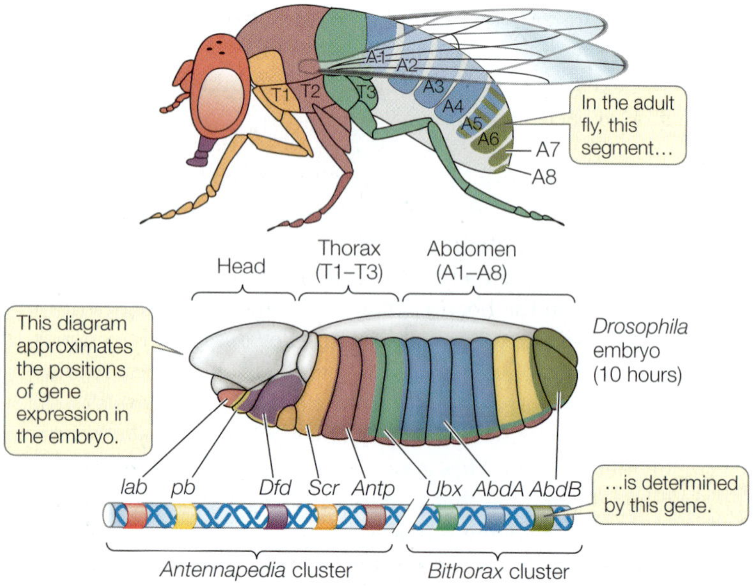
A1
A2
A3
A4
A5
A6
A7
A8
T1
T2
T3
In the adult fly, this segment…
Head
Thorax (T1–T3)
Abdomen (A1–A8)
This diagram approximates the positions of gene expression in the embryo.
Drosophila embryo (10 hours)
lab
pb
Dfd
Scr
Antp
Ubx
AbdA
AbdB
…is determined by this gene.
Antennapedia cluster
Bithorax cluster

19.14 Hox Genes in *Drosophila* Determine Segment Identity
Two clusters of Hox genes on chromosome 3 (center) determine segment function in the adult fly (top). These genes are expressed in the embryo (bottom) long before the structures of the segments actually appear.

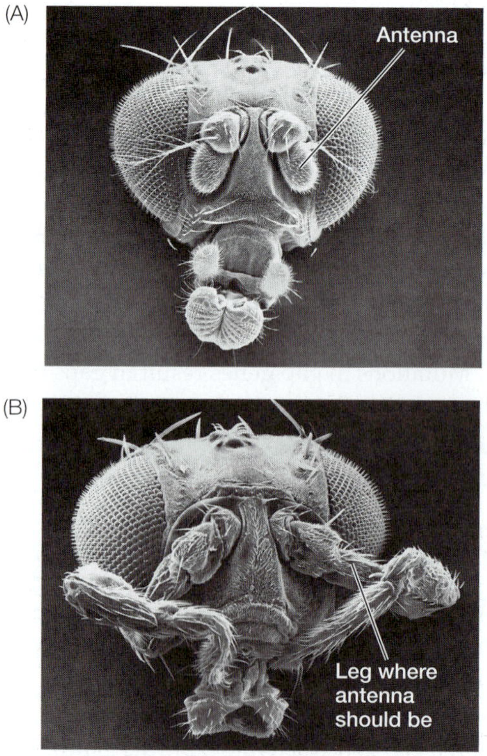
(A)
Antenna
(B)
Leg where antenna should be

19.15 A Homeotic Mutation in *Drosophila* Mutations of the Hox genes cause body parts to form on inappropriate segments. (A) A wild-type fruit fly. (B) An *antennapedia* mutant fruit fly.

last common ancestor of plants and animals was unicellular, and therefore multicellularity evolved independently in plants and animals.

RECAP 19.4

Cascades of transcription factors govern pattern formation and the subsequent development of animal and plant organs. Often these transcription factors create or respond to morphogen gradients. In plants, cell fate is often determined by MADS box genes; in animal embryos, cell fate is determined in part by Hox genes.

- How is apoptosis crucial in shaping the developing embryo? **See p. 399**
- How do organ identity genes act in *Arabidopsis*? **See pp. 400–401 and Figure 19.9**
- List the key attributes of a morphogen. How does the Bicoid protein fit this definition? **See pp. 401–402 and Figure 19.12**
- How is segment identity established in the *Drosophila* embryo? **Review pp. 401–403 and Figure 19.13**

Is a mesophyll cell in a plant leaf or a liver cell in a human irrevocably committed to that specialization? Under the right experimental circumstances, differentiation is reversible in some cells. The next section will describe how some cells can be manipulated to express different sets of genes used in differentiation.

19.5 Is Cell Differentiation Reversible?

A zygote has the ability to give rise to every type of cell in the organism; in other words, it is totipotent. Its genome contains instructions for all of the structures and functions that will arise throughout the life cycle of the organism. Later in development, the cellular descendants of the zygote lose their totipotency and become determined. These determined cells then differentiate into specialized cells. The human liver cell and the leaf mesophyll cell generally retain their differentiated forms and functions throughout their lives. But this does not necessarily mean that they have irrevocably lost their totipotency. Most of the differentiated cells in an animal or plant have nuclei containing the entire genome of the organism, and they therefore have the genetic capacity for totipotency. We will explore here several examples of how this capacity has been demonstrated experimentally.

Plant cells can be totipotent

A carrot root cell normally faces a dark future. It cannot photosynthesize and generally does not give rise to new carrot plants. However, in 1958 Frederick Steward at Cornell University showed that if he isolated cells from a carrot root and maintained them in a suitable nutrient medium, he could induce them to dedifferentiate—to lose their differentiated characteristics. The cells could divide and give rise to masses of undifferentiated cells called calli (singular "callus"), which could be maintained in culture indefinitely. Furthermore, if they were provided with the right chemical cues, the cells could develop into embryos and eventually into complete new plants (Figure 19.16). Since the new plants were genetically identical to the cells from which they came, they were clones of the original carrot plant.

The ability to clone an entire carrot plant from a differentiated root cell indicated that the cell contained a functional, complete carrot genome, and that under the right conditions,

INVESTIGATING**LIFE**

19.16 Cloning a Plant When cells were removed from a plant and put into a medium with nutrients and hormones, they lost many of their specialized features—in other words, they dedifferentiated. Did these cells retain the ability to differentiate again? Frederick Steward found that a cultured carrot cell did indeed retain the ability to develop into an embryo and a new plant.[a]

HYPOTHESIS Differentiated plant cells can be totipotent and can be induced to generate all types of the plant's cells.

Method

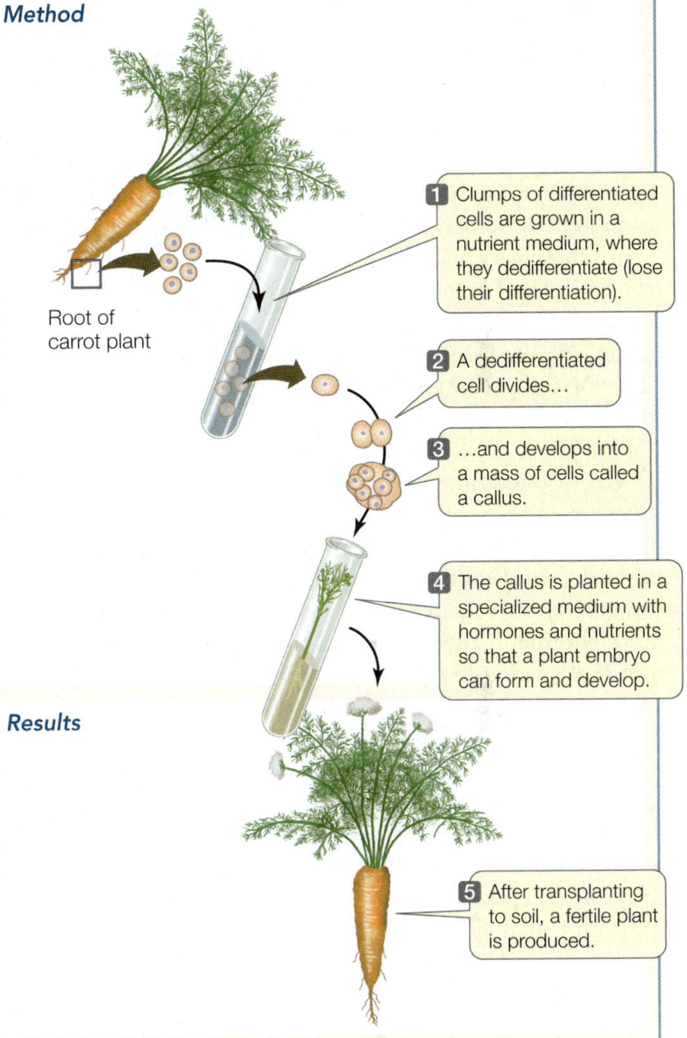

Root of carrot plant
1 Clumps of differentiated cells are grown in a nutrient medium, where they dedifferentiate (lose their differentiation).
2 A dedifferentiated cell divides…
3 …and develops into a mass of cells called a callus.
4 The callus is planted in a specialized medium with hormones and nutrients so that a plant embryo can form and develop.
Results
5 After transplanting to soil, a fertile plant is produced.

CONCLUSION Differentiated plant cells can be totipotent.

Go to **BioPortal** for discussion and relevant links for all INVESTIGATING**LIFE** figures.

[a]Steward, F. C., et al. 1958. *American Journal of Botany* 45: 705–709.

the cell and its descendants could express the appropriate genes in the right sequence to form a new plant. Many types of cells from other plant species show similar behavior in the laboratory. This ability to generate a whole plant from a single cell has been invaluable in agriculture and forestry. For example, trees from planted forests are used in making paper, lumber, and other products. To replace the trees reliably, forestry companies regenerate new trees from the leaves of selected trees with desirable traits. The characteristics of these clones are more uniform and predictable than those of trees grown from seeds.

Nuclear transfer allows the cloning of animals

Animal somatic cells cannot be manipulated as easily as plant cells can. However, experiments such as the one shown on page 395 have demonstrated the totipotency of early embryonic cells from animals. In humans this totipotency permits both genetic screening (see Section 15.4) and certain assisted reproductive technologies (see Section 43.4). A human embryo can be isolated in the laboratory and one or a few cells removed and examined to determine whether a certain genetic condition is present. Because of their totipotency, the remaining cells can develop into a complete embryo, which can be implanted into the mother's uterus, where it develops into a normal fetus and infant.

An isolated animal embryo cell generally won't develop into a complete organism, but the *nucleus* of such a cell has the genetic potential to do so. Nuclear transfer experiments have shown that the genetic information from a single animal cell can be used to create cloned animals. Robert Briggs and Thomas King performed the first such experiments in the 1950s using frog embryos. First they removed the nuclei from unfertilized eggs, forming enucleated eggs. Then, with very fine glass needles, they punctured cells from early embryos and drew up parts of their contents, including the nuclei. Each nucleus was injected into an enucleated egg. They stimulated the eggs to divide, and many went on to form embryos, and eventually frogs, which were clones from the original implanted nuclei. These experiments led to two important conclusions:

- No information is lost from the nuclei of cells as they pass through the early stages of embryonic development. This fundamental principle of developmental biology is known as genomic equivalence.
- The cytoplasmic environment around a cell nucleus can modify its fate.

More recent studies have demonstrated that a cell from a fully developed animal can be induced to dedifferentiate and give rise to an entire new individual. In 1996 Ian Wilmut and his colleagues at the Roslin Institute in Edinburgh cloned the first mammal by somatic cell nuclear transfer. This method involves the fusion of a somatic (nonreproductive) cell from an adult animal, containing the donor nucleus, with an enucleated egg. The fully differentiated donor cells were isolated from a Finn Dorset ewe's udder and starved of nutrients for a week, halting the cells in the G1 phase of the cell cycle. One of these cells was fused with an enucleated egg from a Scottish Blackface ewe, and this fused cell began to divide. After several cell divisions, the resulting early embryo was transplanted into the womb of a surrogate mother. Eventually, a lamb named Dolly was born (**Figure 19.17**). Dolly showed all the characteristics of a Finn Dorset sheep: she carried the same genetic material as the nuclear donor, and thus was a clone of that donor.

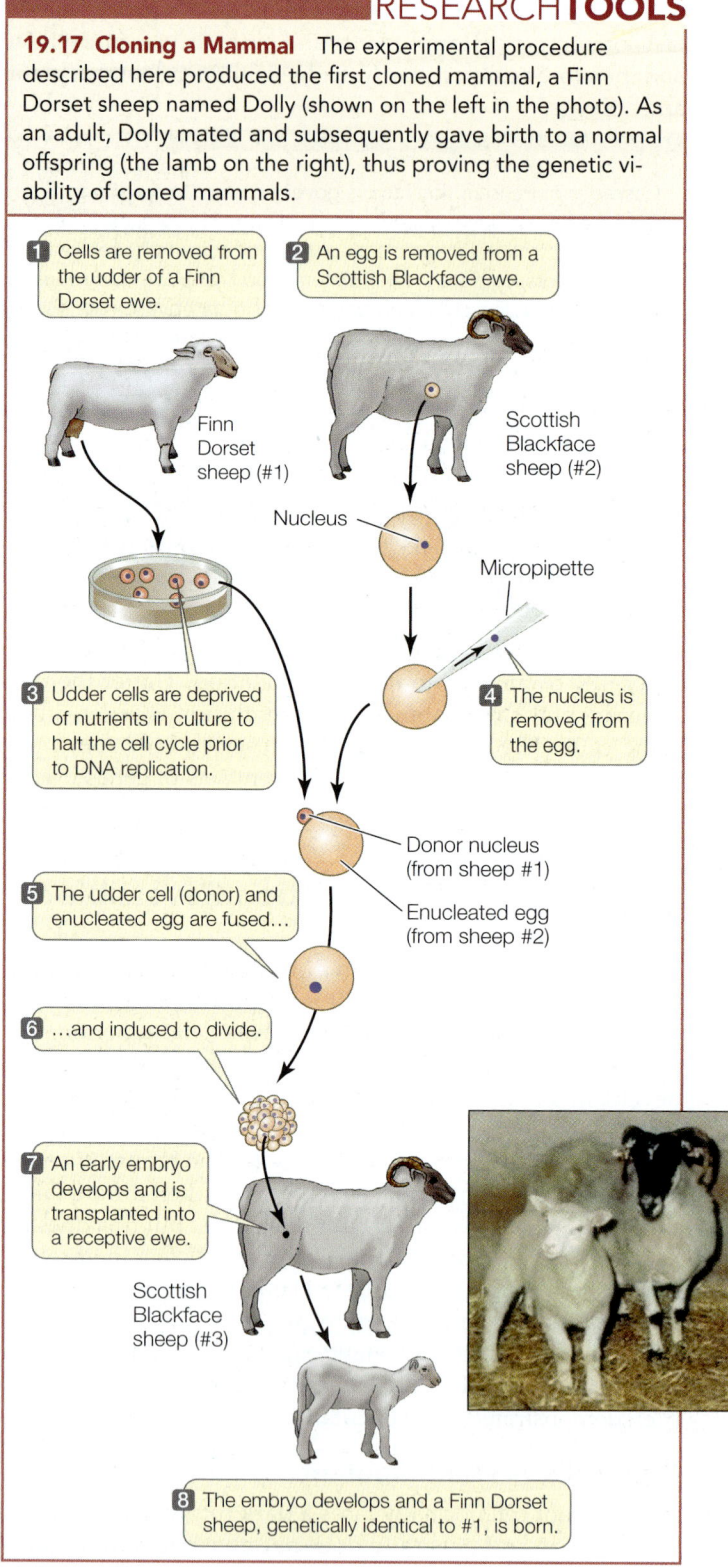

RESEARCH**TOOLS**

19.17 Cloning a Mammal The experimental procedure described here produced the first cloned mammal, a Finn Dorset sheep named Dolly (shown on the left in the photo). As an adult, Dolly mated and subsequently gave birth to a normal offspring (the lamb on the right), thus proving the genetic viability of cloned mammals.

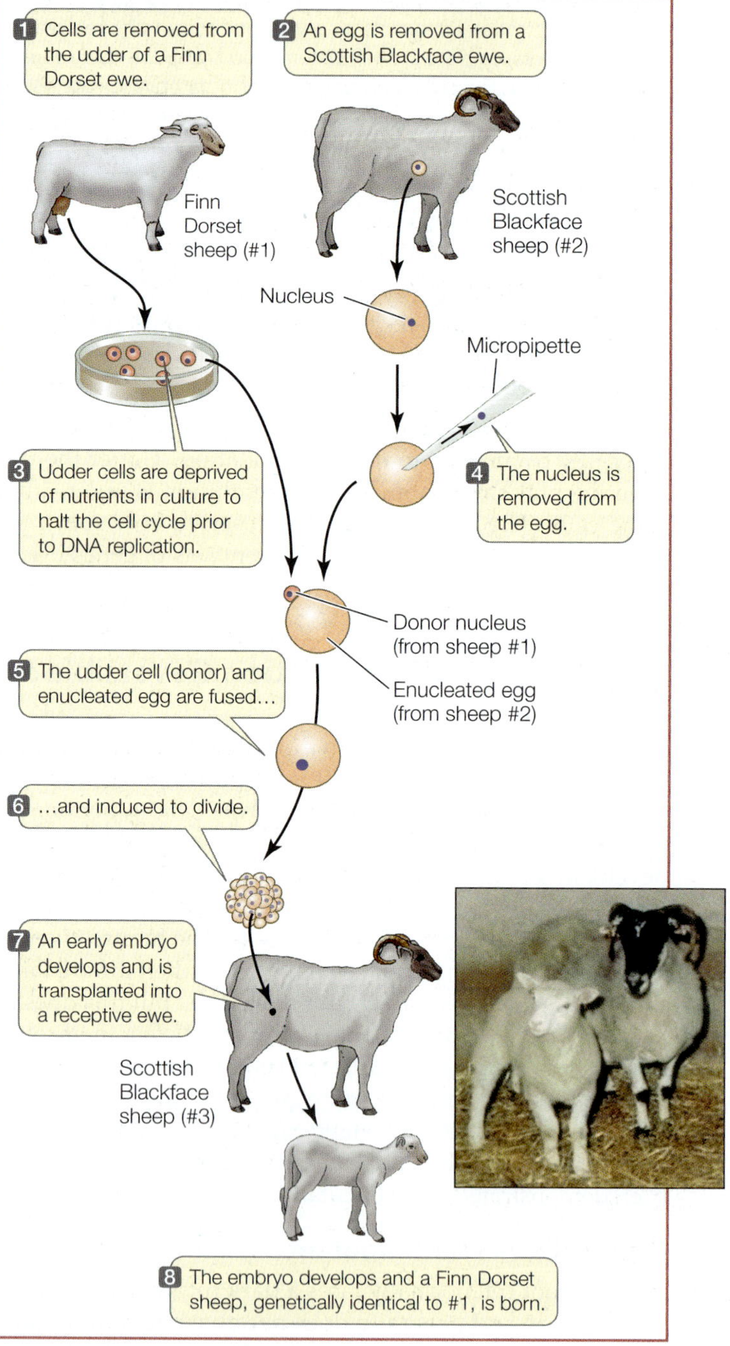
1 Cells are removed from the udder of a Finn Dorset ewe.
2 An egg is removed from a Scottish Blackface ewe.
Finn Dorset sheep (#1)
Scottish Blackface sheep (#2)
Nucleus
Micropipette
3 Udder cells are deprived of nutrients in culture to halt the cell cycle prior to DNA replication.
4 The nucleus is removed from the egg.
Donor nucleus (from sheep #1)
Enucleated egg (from sheep #2)
5 The udder cell (donor) and enucleated egg are fused...
6 ...and induced to divide.
7 An early embryo develops and is transplanted into a receptive ewe.
Scottish Blackface sheep (#3)
8 The embryo develops and a Finn Dorset sheep, genetically identical to #1, is born.

WORKING WITH**DATA:**

Cloning a Mammal

Original Paper

Wilmut, I., A. E. Schnieke, J. McWhir, A. J. Kind, and K. H. S. Campbell. 1997. Viable offspring derived from fetal and adult mammalian cells. *Nature* 385: 810–813.

Analyze the Data

In 1997 Ian Wilmut and colleagues announced the first successful cloning of a mammal by somatic cell nuclear transfer (SCNT; see Figure 19.17). The team fused mammary epithelium cells from a donor sheep (a Finn Dorset) with enucleated eggs from sheep of a different breed (Scottish Blackface). The eggs were induced to divide, and the resulting embryos were implanted into the uteruses of recipient ewes (surrogate mother sheep; also Scottish Blackface). This resulted in the birth of one live lamb, Dolly, who was genetically identical to the ewe that donated the mammary cells. This work demonstrated that, under appropriate circumstances, animal cells are totipotent.

Aside from several ethical issues surrounding the idea of cloning mammals, the use of the SCNT technique itself raises scientific and medical concerns. One concern is the risk of premature aging. Initially Dolly appeared to be a healthy sheep, but at the age of four she developed severe arthritis, a condition usually associated with older animals. In 1999, research published in the journal Nature suggested that Dolly may have been susceptible to premature aging because of the shortened telomeres in her cells. (See Section 13.3 for a discussion of telomeres and their roles in DNA replication and aging.) Dolly was euthanized at age 6 years, which is approximately half the normal life span of a sheep.

QUESTION 1

In addition to mammary epithelium (ME) cells, Wilmut's team also attempted cloning by nuclear transfer from fetal fibroblasts (FB) and embryo-derived cells (EC). The results are shown in the table. What can you conclude about the efficiency of this cloning process?

QUESTION 2

Compare the efficiencies of cloning using nuclear donors from different sources. What can you conclude about the ability of different nuclei to be reprogrammed?

	Number of attempts that progressed to each stage		
Stage	ME	FB	EC
Egg fusions	277	172	385
Embryos transferred to recipient ewes	29	34	72
Pregnancies	1	4	14
Live lambs	1	2	4

QUESTION 3

Polymorphic DNA markers were used to analyze Dolly's genetic make-up. The data for four short tandem repeat (STR) markers (FCB 11, FCB 304, MAF 33, and MAF 209) are shown in the figure. (See Section 15.3 and Figure 15.14A for an explanation of STR analysis.) In the electrophoresis gels, different genotypes produce DNA bands of different sizes. A sample of Dolly's DNA was compared with samples from her nuclear donor (mammary cells from a Finn Dorset ewe) and from the recipient (her surrogate mother, a Scottish Blackface ewe). Are the DNA bands from Dolly the same sizes as those from her nuclear donor or from her surrogate mother? What does this indicate about Dolly's genetic makeup?

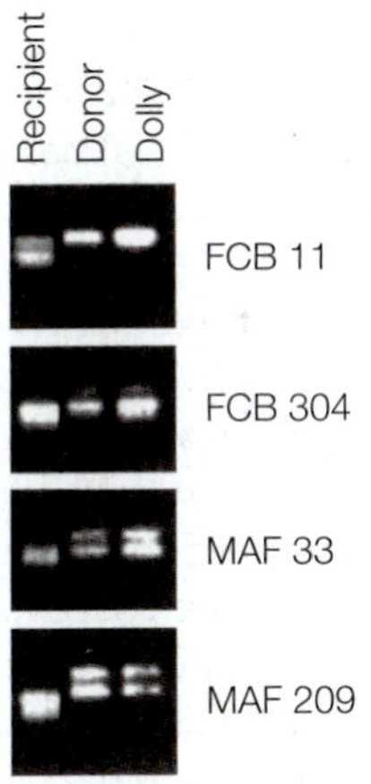
Recipient
Donor
Dolly
FCB 11
FCB 304
MAF 33
MAF 209

Go to **BioPortal** for all WORKING WITH**DATA** exercises

The production of Dolly demonstrated that a fully differentiated cell from a mature organism can revert to a totipotent state, and that this cell can be used to create a new animal. Many other animal species, including cats, dogs, horses, pigs, rabbits, and mice, have since been cloned by nuclear transfer. The cloning of animals has practical uses and has given us important information about developmental biology. There are several reasons to clone animals:

- *Expansion of the numbers of valuable animals*: One goal of Wilmut's experiments was to develop a method for cloning transgenic animals with useful phenotypes. For example, as we mentioned in Section 18.6, a cow was genetically engineered to make human growth hormone in her milk. This animal was then cloned to produce additional cows that do the same thing. Only 15 such cows would supply the world's need for this medication, which is used to treat short stature caused by growth hormone deficiency.
- *Preservation of endangered species*: The banteng, a relative of the cow, was the first endangered animal to be cloned. An enucleated egg from a cow, and a cow surrogate mother were used. Cloning may be the only way to save some endangered species with low rates of natural reproduction, such as the giant panda.
- *Perpetuation of pets*: Many people get great personal benefit from pets, and the death of a pet can be devastating. Companies have been set up to clone cats and dogs from cells provided by their owners. Of course, the behavioral characteristics of the beloved pet, which are certainly derived in part from the environment, may not be the same in the cloned pet as in its genetic parent.

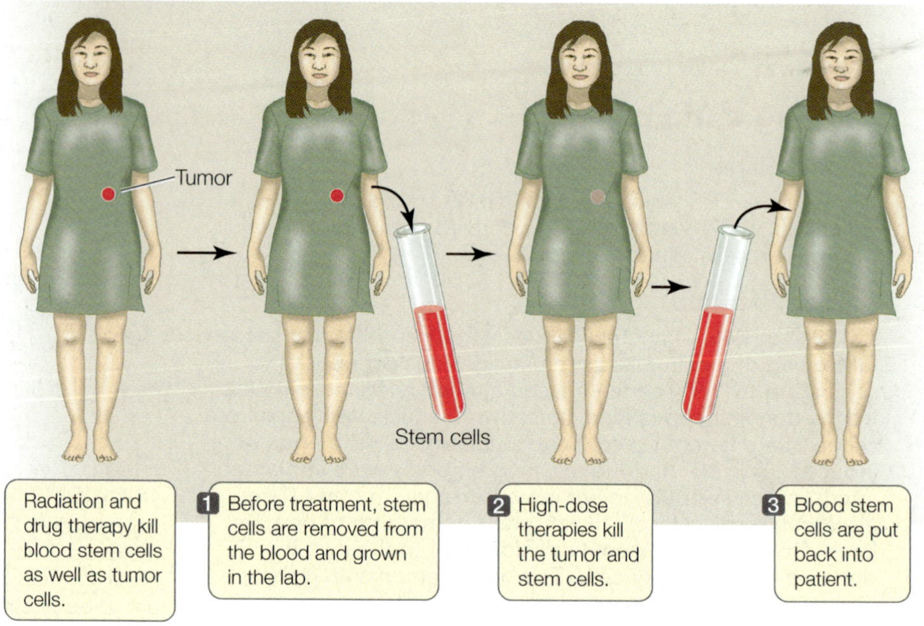
Tumor
Stem cells
Radiation and drug therapy kill blood stem cells as well as tumor cells.
1 Before treatment, stem cells are removed from the blood and grown in the lab.
2 High-dose therapies kill the tumor and stem cells.
3 Blood stem cells are put back into patient.

19.18 Stem Cell Transplantation Multipotent stem cells can be used in hematopoietic stem cell transplantation to replace stem cells destroyed by cancer therapy.

Multipotent stem cells differentiate in response to environmental signals

Stem cells have been mentioned several times already in this chapter (for example, in the opening story). **Stem cells** are rapidly dividing, undifferentiated cells that can differentiate into diverse cell types.

In plants, stem cells occur in the meristem. In general, plants have far fewer (15–20) broad cell types than animals (as many as 200). In mammals, stem cells are found in adult tissues that need frequent cell replacement, such as the skin, the inner lining of the intestine, and the bone marrow, where blood cells and other types of cells are formed. Canadian cell biologists Ernest McCulloch and James Till discovered mammalian stem cells in the early 1960s when they injected bone marrow cells into adult mice. They noticed that the recipient mice developed small clumps of tissue in their spleens. When they looked more carefully at the clumps, they found that each was composed of undifferentiated stem cells. Before this, stem cells were believed to be present only in animal embryos.

The stem cells found in adult animals are not totipotent, because their ability to differentiate is limited to a relatively few cell types. In other words, they are multipotent (see Section 19.1). For example, there are two types of multipotent stem cells in the bone marrow. Hematopoietic stem cells produce red and white blood cells, whereas mesenchymal stem cells (like those used to treat Bartolo Colon's injury in the opening story) produce bone and connective tissues, including muscle.

The proliferation and differentiation of multipotent stem cells are "on demand." For example, hematopoietic stem cells proliferate in the bone marrow in response to growth factors, and the extra stem cells are released into the blood. This is the basis of an important therapy called hematopoietic stem cell transplantation (**Figure 19.18**). Some cancer treatments kill all dividing cells in the body, so hematopoietic stem cells can be depleted in patients given these treatments. To circumvent this problem, stem cells are harvested from the bone marrow or blood of the patient (prior to cancer treatment) or of a donor, and then the cells are injected back into the patient after cancer treatment.

Signals from adjacent cells can stimulate stem cell differentiation. Many controlled experiments have shown that damaged animal tissues (for example, hearts and tendons) that are injected with stem cells can heal more effectively than tissues that don't receive this treatment. The mechanisms by which this occurs are still not clear. There is some evidence that the injected cells can actually insert themselves into the damaged tissue and differentiate into new cells of that tissue. Alternatively, injected cells may secrete growth factors and other molecules that induce the cells in the surrounding tissue to regenerate into healthy tissue. Whatever the mechanisms by which multipotent stem cells contribute to the healing of damaged tissues, their use in treating diseases is very promising.

Pluripotent stem cells can be obtained in two ways

As stated earlier, totipotent stem cells that can differentiate into any cell type are found only in very early embryos. In both mice and humans, a slightly later embryonic stage is a hollow sphere of cells called a **blastocyst** (see Figure 44.4). A group of cells within the blastocyst is pluripotent: they can differentiate into most cell types but cannot give rise to a complete organism. In mice, these **embryonic stem cells** (**ESCs**) can be removed from the blastocyst and grown in laboratory culture almost indefinitely, if provided with the right conditions. When cultured mouse ESCs are injected back into another mouse blastocyst, the stem cells mix with the resident cells and differentiate to form all the cell types in the mouse. This indicates that the ESCs do not lose any of their developmental potential while growing in the laboratory.

ESCs growing in the laboratory can also be induced to differentiate in a particular way if the right signal is provided (**Figure 19.19A**). For example, treatment of mouse ESCs with a derivative of vitamin A causes them to form neurons, whereas other growth factors induce them to form blood cells. Such experiments demonstrate both the cells' developmental potential and the roles of environmental signals. This finding raises the possibility of using ESC cultures as sources of differentiated cells to repair specific tissues, such as a damaged pancreas in diabetes, or a brain that malfunctions in Parkinson's disease.

ESCs can be harvested from human embryos conceived by in vitro ("under glass"—in the laboratory) fertilization, with the consent of the donors. Since more than one embryo is usually conceived in this procedure, embryos not used for reproduction might be available for embryonic stem cell isolation. These cells could then be grown in the laboratory and used as sources of tissues for transplantation into patients with tissue damage. There are two problems with this approach:

- Some people object to the destruction of human embryos for this purpose.
- The stem cells, and tissues derived from them, would provoke an immune response in a recipient (see Chapter 42).

Shinya Yamanaka and coworkers at Kyoto University in Japan developed another way to produce pluripotent stem cells that gets around these two problems (**Figure 19.19B**). Instead of extracting ESCs from blastocysts, they make pluripotent stem cells from skin cells. They developed this method systematically:

1. First they used microarrays (see Figure 18.8) to compare the genes expressed in ESCs with non-stem cells. They found several genes that were uniquely expressed at high levels in ESCs. These genes were believed to be essential to the undifferentiated state and function of stem cells.
2. Next they isolated the genes, coupled them to highly expressing promoters, and inserted them into skin cells (see Section 18.5). They found that the skin cells now expressed the newly added genes at high levels.
3. Finally, they showed that the altered skin cells were pluripotent and could be induced to differentiate into many tissues. They called these cells **induced pluripotent stem cells (iPS cells)**.

Because the iPS cells can be made from skin cells of the individual who is to be treated, an immune response may be avoided. Such cells have already been used for cell therapy in animals for diseases similar to human Parkinson's disease (a brain disorder), diabetes, and sickle cell anemia. If it can be shown conclusively that iPS cells have the same properties as ESCs, human uses are sure to follow.

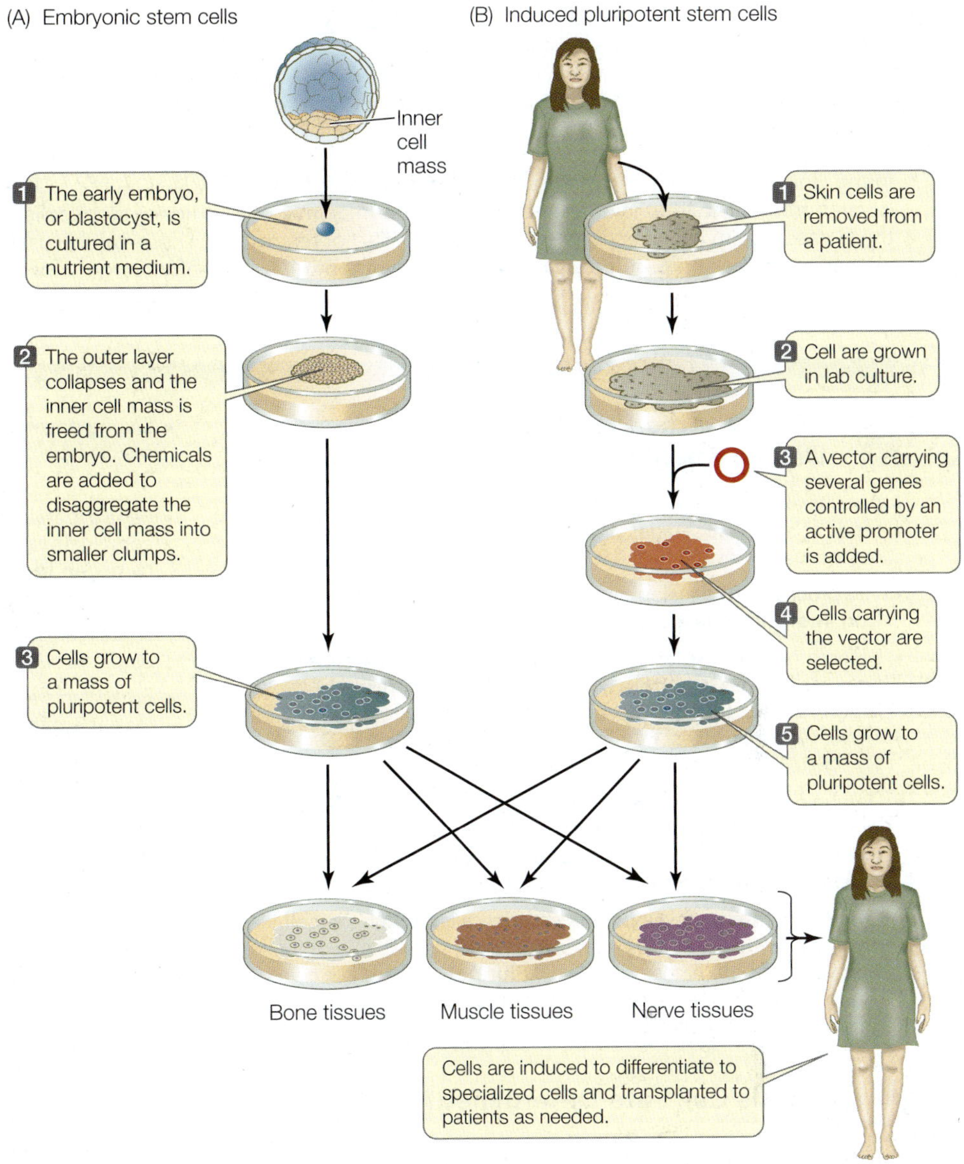
(A) Embryonic stem cells
(B) Induced pluripotent stem cells
Inner cell mass
1 The early embryo, or blastocyst, is cultured in a nutrient medium.
2 The outer layer collapses and the inner cell mass is freed from the embryo. Chemicals are added to disaggregate the inner cell mass into smaller clumps.
3 Cells grow to a mass of pluripotent cells.
1 Skin cells are removed from a patient.
2 Cell are grown in lab culture.
3 A vector carrying several genes controlled by an active promoter is added.
4 Cells carrying the vector are selected.
5 Cells grow to a mass of pluripotent cells.
Bone tissues
Muscle tissues
Nerve tissues
Cells are induced to differentiate to specialized cells and transplanted to patients as needed.

19.19 Two Ways to Obtain Pluripotent Stem Cells Pluripotent stem cells can be obtained either from human embryos (A) or by adding highly expressed genes to skin cells to transform them into stem cells (B).

Go to Animated Tutorial 19.4
Embryonic Stem Cells
Life10e.com/at19.4

RECAP 19.5

Differentiated cells retain their ability to differentiate into other cell types, given appropriate chemical signals. This has made cloning and stem cell technologies possible.

- How are stem cells found in adult body tissues different from embryonic stem cells? **See p. 408**
- What are the two ways to produce pluripotent stem cells? **See pp. 408–409 and Figure 19.19**

What are other uses of stem cells derived from fat?

ANSWER

In the United States, veterinarians use multipotent stem cells derived from fat to treat injuries and osteoarthritis in animals. In these cases, stem cell injections improve the healing of tendon injuries in horses and ease the symptoms of osteoarthritis in dogs. A procedure has been developed to isolate large quantities of stem cells from human patients in the operating room. These can be used immediately to repair tissues, for example, after surgery for breast cancer.

CHAPTER SUMMARY 19

19.1 What Are the Processes of Development?

- A multicellular organism begins its development as an embryo. A series of embryonic stages precedes the birth of an independent organism. **Review Figure 19.1, ACTIVITY 19.1**
- The processes of development are **determination**, **differentiation**, **morphogenesis**, and **growth**.
- Differential gene expression is responsible for the differences among cell types. **Cell fate** is determined by environmental factors, such as the cell's position in the embryo, as well as by intracellular influences. **Review Figure 19.2, ANIMATED TUTORIAL 19.1**
- Determination is followed by differentiation, the actual changes in biochemistry, structure, and function that result in cells of different types. Determination is a commitment; differentiation is the realization of that commitment.
- Over the course of development, embryo cells decrease in **cell potency**. **Totipotent** cells (such as a zygote) are capable of forming every cell type in the adult body. **Pluripotent** cells can give rise to most cell types, **multipotent** cells to several cell types, and **unipotent** cells to only one cell type.

19.2 How Is Cell Fate Determined?

- **Cytoplasmic segregation**—the unequal distribution of **cytoplasmic determinants** in the egg, zygote, or early embryo—can establish **polarity** and lead to cell fate determination. **Review Figure 19.3, ANIMATED TUTORIAL 19.2**
- **Induction** is a process by which embryonic animal tissues direct the development of neighboring cells and tissues by secreting chemical signals, called **inducers**. **Review Figures 19.4, 19.5**

19.3 What Is the Role of Gene Expression in Development?

- Inducers act through signaling pathways to determine cell fate. **Review Figure 19.6**
- Differential gene expression results in cell differentiation. Transcription factors are especially important in regulating gene expression during differentiation.
- Complex interactions of many genes and their products are responsible for differentiation during development. **Review Figure 19.7**

19.4 How Does Gene Expression Determine Pattern Formation?

- **Pattern formation** is the process that results in the spatial organization of a tissue or organism.
- During development, selective elimination of cells by apoptosis results from the expression of specific genes. **Review Figure 19.8**
- Sepals, petals, stamens, and carpels form in plants as a result of combinatorial interactions between transcription factors encoded by **organ identity genes**. **Review Figure 19.9**
- The transcription factors encoded by floral organ identity genes contain an amino acid sequence called the **MADS box** that can bind to DNA.
- Both plants and animals use **positional information** as a basis for pattern formation. Positional information usually comes in the form of a signal called a **morphogen**. Different concentrations of the morphogen cause different effects. **See Figures 19.10, 19.11**
- In the fruit fly *D. melanogaster*, a cascade of transcriptional activation sets up the axes of the embryo, the development of the segments, and finally the determination of cell fate in each segment. The cascade involves the sequential expression of maternal effect genes, gap genes, pair rule genes, segment polarity genes, and Hox genes. **Review Figures 19.13, 19.14, ANIMATED TUTORIAL 19.3**
- Hox genes help determine cell fate in the embryos of all animals. The **homeobox** is a DNA sequence found in Hox genes and other genes that code for transcription factors. The sequence of amino acids encoded by the homeobox is called the **homeodomain**.

19.5 Is Cell Differentiation Reversible?

- The ability to create clones from differentiated cells demonstrates the principle of genomic equivalence. **Review Figures 19.16, 19.17**
- **Stem cells** produce daughter cells that differentiate when provided with appropriate intercellular signals. Some multipotent stem cells in the adult body can differentiate into a limited number of cell types to replace dead cells and maintain tissues. **Review Figure 19.18**
- **Embryonic stem cells** are pluripotent and can be cultured in the laboratory. Under suitable environmental conditions, these cells can differentiate into almost any tissue type. **Induced pluripotent stem cells** have similar characteristics. This has led to technologies designed to replace cells or tissues damaged by injury or disease. **Review Figure 19.19, ANIMATED TUTORIAL 19.4**

Go to the Interactive Summary to review key figures, Animated Tutorials, and Activities
Life10e.com/is19

CHAPTER**REVIEW**

REMEMBERING

1. Which statement about determination is true?
   a. Differentiation precedes determination.
   b. All cells are determined after two cell divisions in most organisms.
   c. A determined cell will keep its determination no matter where it is placed in an embryo.
   d. A cell changes its appearance when it becomes determined.
   e. A differentiated cell has the same pattern of transcription as a determined cell.

2. Cloning experiments on sheep, frogs, and mice have shown that
   a. nuclei of adult cells are not pluripotent.
   b. nuclei of embryonic cells can be totipotent.
   c. nuclei of differentiated cells have different genes than zygote nuclei have.
   d. differentiation is fully reversible in all cells of a frog.
   e. differentiation involves permanent changes in the genome.

3. The term "induction" describes a process in which a cell or cells
   a. influence the development of another cell or group of cells.
   b. trigger cell movements in an embryo.
   c. stimulate the transcription of their own genes.
   d. organize the egg cytoplasm before fertilization.
   e. inhibit the movement of the embryo.

4. Which statement about induction is *not* true?
   a. One group of cells induces adjacent cells to develop in a certain way.
   b. It triggers a sequence of gene expression in target cells.
   c. Single cells cannot form an inducer.
   d. A tissue may be induced as well as make an inducer.
   e. The chemical identification of specific inducers has not been achieved.

5. Homeotic mutations
   a. are often severe and result in structures at inappropriate places.
   b. cause subtle changes in the forms of larvae or adults.
   c. occur only in prokaryotes.
   d. do not affect the animal's DNA.
   e. are confined to the zone of polarizing activity.

6. Which statement about the homeobox is *not* true?
   a. It is transcribed and translated.
   b. It is found only in animals.
   c. Proteins containing the homeodomain bind to DNA.
   d. It is a sequence of DNA shared by more than one gene.
   e. It occurs in Hox genes.

UNDERSTANDING & APPLYING

7. In fruit flies, the following genes are used to determine segment polarity: (k) gap genes; (l) Hox genes; (m) maternal effect genes; (n) pair rule genes. In what order are these genes expressed during development?
   a. k, l, m, n
   b. l, k, n, m
   c. m, k, n, l
   d. n, k, m, l
   e. n, m, k, l

8. Molecular biologists can attach genes to active promoters and insert them into cells (see Section 18.5). What would happen if the following were inserted and overexpressed? Explain your answers.
   a. *ced-9* in embryonic neuron precursors of *C. elegans*
   b. *myoD* in undifferentiated myoblasts
   c. the gene for Sonic hedgehog in a chick limb bud
   d. *nanos* at the anterior end of the *Drosophila* embryo

9. A powerful method to test for the function of a gene in development is to generate a "knockout" organism, in which the gene in question is inactivated (see Section 18.4). What do you think would happen in each of the following cases?
   a. a knockout of *ced-9* in *C. elegans*
   b. a knockout of *nanos* in *Drosophila*

10. If you wanted a rose flower with only petals, what kind of homeotic mutation would you seek in the rose genome?

ANALYZING & EVALUATING

11. During development, an animal cell's potential for differentiation becomes ever more limited. In the normal course of events, most cells in the adult animal have the potential to be only one or a few cell types. On the basis of what you have learned in this chapter, discuss possible mechanisms for the progressive limitation of the cell's potential.

12. Cloning involves considerable reprogramming of gene expression in a differentiated cell so that it acts like an egg cell. How would you investigate this reprogramming?

20 Genes, Development, and Evolution

CHAPTER**OUTLINE**

Beak Diversity Can Evolve through Changes in Development Genes expressed during embryonic development affect beak length and depth in birds. Changes in the expression patterns of these genes contributed to the evolution of the distinctive beaks of the Australian pelican (*Pelecanus conspicillatus*) and the silver gull (*Chroicocephalus novaehollandiae*).

AN ICONIC ILLUSTRATION of Charles Darwin's theory of evolution by natural selection is the finches of the Galápagos. Each island in the archipelago is home to a combination of species of these small, dull-colored birds. Their beaks are of different shapes and sizes, ranging from the thick, short, strong beaks of seed-eating species to the thin, long beaks of insect-eating species. Darwin wrote that "Seeing this gradation and diversity of structure in one small intimately related group of birds, one might really fancy that from an original paucity of birds in this archipelago, one species had been taken and modified for different ends."

Darwin had no idea of the genetic basis for such modification. Now we do, but recent discoveries have changed much of our thinking about certain aspects of evolutionary genetics. Based on Mendelian genetics and the central dogma that genes code for proteins, most explanations of evolutionary change have focused on the effects of gene mutations on the structural proteins that make up individual organisms. We now know that mechanisms of evolution can depend as much on changes in the regulatory sequences that control the timing, amount, and location of gene expression as on mutations that change the structural genes themselves.

The beak of the finch develops from tissues at the anterior of the embryo that will form the facial bones. Cell divisions in this embryonic tissue are controlled by signaling proteins, one of which is called bone morphogenetic protein 4 (BMP4); another is the protein calmodulin. If BMP4 is present early and in large amounts, the beak becomes broad and deep. If calmodulin is present early and in large amounts, the beak grows longer and thinner. Thus beak structure is affected by changes in the timing of protein production as well as the amount of protein made.

Beak differences among Galápagos finches are impressive, but consider the diverse beaks seen in other birds—hummingbirds, flamingos, toucans, and pelicans, to name only a few. Could such dramatic differences in beak size and shape also be a result of changes in expression patterns of BMP4 and calmodulin? The realization that major evolutionary change can be the result of subtle changes in spatial, temporal, and quantitative distribution of signaling molecules and changes in the noncoding regions of the DNA that control gene expression has revolutionized the study of evolution, producing the new field of evolutionary developmental biology.

?

How are gene expression patterns involved in the shaping of the diverse beaks of birds?

See answer on p. 424.

20.1 How Can Small Genetic Changes Result in Large Changes in Phenotype?

Genetic mutations are the source of variation for evolution. But most genetic mutations change only a single nucleotide in a very large genome. How can such small changes lead to the huge diversity we see in living organisms? This question has become acute in recent years with the realization that the actual number of protein-coding genes in many species is many fewer than expected (only about 21,000 in humans, for example). Furthermore, the genomes of related species that appear very different can be very similar, as is the case for the genomes of humans and chimpanzees, which are more than 95 percent identical. Part of the answer to this question has been found in the many and subtle ways that gene expression is controlled in time, space, and amount during development. Study of the relationship between development and evolution has given rise to the new field of **evolutionary developmental biology**, or **evo-devo**. The basic principles of evo-devo are:

- Organisms share similar molecular mechanisms for development that include a "toolkit" of regulatory molecules that control the expression of genes.
- Toolkit regulatory molecules are able to act independently in different tissues and regions of the body, enabling modular evolutionary change.
- Developmental differences can arise from changes in the timing of regulatory molecule action, the location of its action, or the quantity of its action.
- Differences among species can arise from alterations in the expression of developmental genes.
- Developmental changes can arise from environmental influences on developmental processes.

The development of a multicellular organism from a fertilized egg—a single cell—involves an intricate pattern of sequential gene expression. When developmental biologists began to describe the events responsible for the differentiation and controlled proliferation of cells and tissues at the molecular level, they found common regulatory genes and pathways in organisms that don't appear similar at all, such as those that direct eye development in fruit flies and mice.

Developmental genes in distantly related organisms are similar

About a dozen major kinds of eyes are found among the different animals, including the camera-like eyes of humans and the compound eyes of insects (see Section 46.4). Although the eyes of insects and vertebrates evolved independently, a remarkable discovery in the 1990s showed that common developmental pathways were involved in the origin of eyes in both groups. Swiss developmental biologists Rebecca Quiring and Walter Gehring were using lines of mutant fruit flies to identify the transcription factors involved in fly development. One of these mutations, appropriately named *eyeless*, resulted in flies with no eyes. Quiring and Gehring isolated the protein product of the *eyeless* gene, which they determined was a transcription factor that controls the genes responsible for eye development. By making recombinant DNA constructs that allowed the *eyeless* gene to be expressed in different embryonic tissues of transgenic flies, they were able to produce flies with extra eyes on various body parts such as the legs, the antennae, and under the wings.

A big surprise came when database searches revealed that the *eyeless* gene sequence was similar to that of the *Pax6* gene in mice; *Pax6*, when mutated, leads to the development of abnormally small eyes. Could the extremely different eyes of flies and mice be variations on a common developmental theme? To test for functional similarity between the insect and mammalian genes, Gehring and colleagues repeated their experiments on flies, but using the mouse *Pax6* gene instead of the fly *eyeless* gene. Once again, eyes developed at various sites on the transgenic flies. Thus a gene whose expression normally leads to the development of a mammalian eye now led to the development of the very different insect eye.

We have to look very far back in evolutionary time for a common ancestor of fruit flies and mice. Yet the *eyeless* and *Pax6* genes contain sequences that are highly conserved, not only in these two model species but in others as well (Figure 20.1). Biologists call such genes **homologous**, meaning that they evolved from a gene present in a common ancestor. In recent years a large number of homologous genes (the regulatory "toolkit") have been shown to control development in distantly related species.

An even more dramatic example of homology in genes that control development in animals is the Hox gene cluster discussed in Chapter 19. This set of genes codes for transcription factors that provide positional information and control the pattern of development in the different body segments of the embryo (see Figure 19.14). The genes in the Hox cluster share a homologous sequence, the homeobox. The similarity among the repeated gene sequences of the homeobox suggests that the Hox genes arose through duplication of an ancestral gene, which then diverged to take on new functions (see Section 24.3). Duplication originally provided protection from loss-of-function mutations in these critical genes. One copy of the gene could provide the critical function even if the other copy acquired mutations that altered its function. New functions in duplicated genes could then be favored through natural selection. Thus duplication and divergence allowed for "evolutionary experiments" using sets of duplicated genes such as the Hox cluster.

The duplication-and-divergence hypothesis derives support from the deep evolutionary history (that is, very early divergences on the tree of life) of the Hox genes. Two clusters of Hox genes are found in cnidarians (jellyfish and their relatives), one expressed in the anterior of the larvae and one expressed in the posterior of the larvae. The number of Hox genes increases among animals that have bilateral symmetry (see Section 31.2) as complexity of the patterning of the body axis increases. The increase in Hox genes reaches a maximum in vertebrates, where there are not only more Hox genes per cluster but there are four clusters, each on a different chromosome. Comparing

Mouse *Pax6* gene:
DNA GTATCCAACGGTTGTGTGAGTAAAATTCTGGGCAGGTATTACGAGACTGGCTCCATCAGA
Amino acids V S N G C V S K I L G R Y Y E T G S I R
Fly *eyeless* gene:
77% GTATCAAATGGATGTGTGAGCAAAATTCTCGGGAGGTATTATGAAACAGGAAGCATACGA
100% V S N G C V S K I L G R Y Y E T G S I R
Shark eye control gene:
85% GTGTCCAACGGTTCTGTCAGTAAAATCCTGGGCAGATACTATGAAACAGGATCCATCAGA
100% V S N G C V S K I L G R Y Y E T G S I R
Squid eye control gene:
78% GTCTCCAACGGCTGCGTTAGCAAGATTCTCGGACGGTACTATGAGACGGGCTCCATAAGA
100% V S N G C V S K I L G R Y Y E T G S I R

20.1 DNA Sequence Similarity in Eye Development Genes Genes controlling eye development contain regions that are highly conserved, even among species with very different eyes. These sequences, from a conserved region of the *Pax6* gene and its homologs in other species, are similar at the DNA level (top sequence in each pair) and identical at the amino acid level (bottom sequence). The percentages beside the sequences represent the percent match with the corresponding DNA and protein sequences in the mouse.

these gene clusters in such distantly related species as *Drosophila* and mice, the genes are arranged in similar clusters and are expressed in similar patterns along the body axes of their embryos (**Figure 20.2**).

The homology of the Hox genes across animal species is remarkable. Over the millions of years that have elapsed since cnidarians, insects, and vertebrates last shared a common ancestor, the genes for these transcription factors have been conserved. This conservation extends beyond similarities of gene sequence and organization to similarities in protein structure and function during development. These similarities led biologists to form one of the fundamental principles of evo-devo: Shared developmental mechanisms controlled by specific DNA sequences comprise a **genetic toolkit** that has been modified and reshuffled to produce the remarkable diversity of plants, animals, and other organisms we know today. Small changes in the application of the genetic toolkit—when, where, and how much the transcription factor genes are expressed—influence the development of the organism and produce the variation upon which natural selection can work.

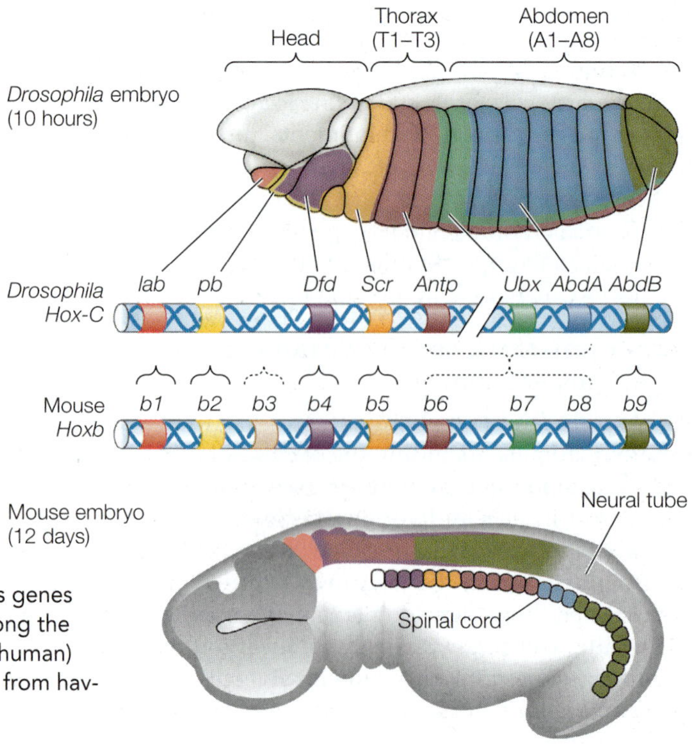
Head
Thorax (T1–T3)
Abdomen (A1–A8)
Drosophila embryo (10 hours)
lab
pb
Dfd
Scr
Antp
Ubx
AbdA
AbdB
Drosophila Hox-C
b1
b2
b3
b4
b5
b6
b7
b8
b9
Mouse *Hoxb*
Mouse embryo (12 days)
Neural tube
Spinal cord

20.2 Regulatory Genes Show Similar Expression Patterns Homologous genes encoding similar transcription factors are expressed in similar patterns along the anterior–posterior axes of both insects and vertebrates. The mouse (and human) Hox genes are present in multiple copies; this prevents a single mutation from having drastic effects.

RECAP 20.1

Changes in development underlie many changes in morphology that result in evolution of body form and function. Study of the evolutionary aspects of development has revealed that a genetic toolkit consisting of highly conserved regulatory genes governs pattern formation in multicellular organisms.

- Describe what evolutionary developmental biologists mean by a genetic toolkit. **See pp. 413 and 414**
- How does the story of the eye-determining genes *Pax6* and *eyeless* support the claim that genes controlling development are highly conserved? **See p. 413 and Figure 20.1**
- Using Hox genes as an example, explain how duplication and divergence have enabled the evolution of complexity in body form. **See pp. 413–414 and Figure 20.2**

We saw in Chapter 19 that many developmental mutations in fruit flies result in striking abnormalities that affect only a single structure, segment, or region (e.g., a head segment that forms a leg; see Figure 19.15). The rest of the embryo is often unaffected. How is this possible?

20.2 How Can Mutations with Large Effects Change Only One Part of the Body?

Development involves interactions between gene products that produce a sequence of transcriptional events resulting in differential gene expression. The examples of homeotic mutations revealed that these processes can be localized in different tissues or regions of the embryo. Thus the embryo can be thought of as consisting of **modules** that encompass both the genes and various signaling pathways that determine specific physical structures such as eyes, body segments, wings, or legs. Because developmental genes can be controlled separately in the different modules, structures that arise from different modules can change independently of one another in both developmental and evolutionary time.

Go to Animated Tutorial 20.1
Modularity
Life10e.com/at20.1

Genetic switches govern how the genetic toolkit is used

Developmental modules based on a common set of genetic instructions can evolve separately within a species because **genetic switches** control how the toolkit is used. These switches include gene promoters and the transcription factors that bind to promoters, as well as the enhancers and repressors that can modulate the interactions of transcription factors and promoters (see Section 16.2). The signal cascades resulting from the interaction of transcription factors, enhancers, and repressors with gene promoters determine where, when, and to what extent genes are turned on and off. Multiple switches control each gene, creating different expression patterns in different locations. In this way, elements of the genetic toolkit can be involved in multiple developmental processes and still allow individual modules to develop and evolve independently.

Genetic switches integrate positional information in the developing embryo and play key roles in determining the developmental pathways of different modules. In *Drosophila*, for example, each Hox gene codes for a transcription factor that is expressed in a particular segment or appendage of the developing fly. The pattern and function of each segment depends on the unique Hox gene or combination of Hox genes that are expressed in that segment. As an example, let's look at wing development in insects.

Drosophila species are members of the insect group Diptera, which means "two wings"—that is, they have a single pair of wings, whereas most insects have two pairs of wings (i.e., four wings). The single pair of wings of dipterans develops on the second thoracic segment, where the Hox gene *antennapedia* (*Antp*) is expressed. *Antp* is also expressed in the third thoracic segment, but in that segment a pair of balancing organs called halteres develops in dipterans. A critical difference between thoracic segments 2 and 3 is that another Hox gene, *ultrabithorax* (*Ubx*), is expressed along with *Antp* in segment 3. *Ubx* represses the expression of *Antp* in dipterans. If *Ubx* is inactivated by mutation, a second pair of wings forms in thoracic segment 3, as is typical of many other insect groups (**Figure 20.3**). Thus some major morphological differences among groups of animals can result from relatively small changes in gene expression.

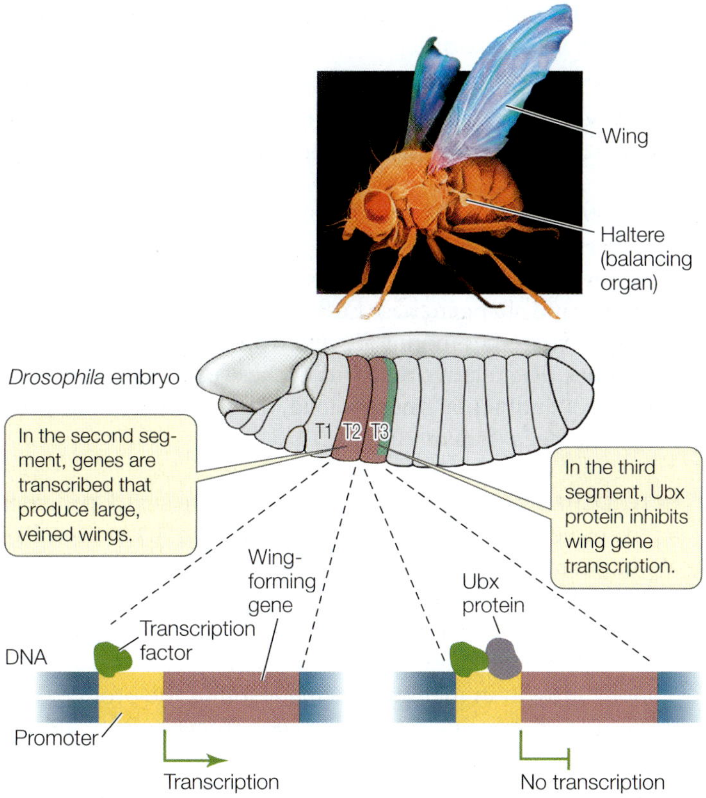

Wing
Haltere (balancing organ)
Drosophila embryo
T1
T2
T3
In the second segment, genes are transcribed that produce large, veined wings.
In the third segment, Ubx protein inhibits wing gene transcription.
Wing-forming gene
Ubx protein
Transcription factor
DNA
Promoter
Transcription
No transcription

20.3 Segments Differentiate under Control of Genetic Switches The binding of a single protein, Ultrabithorax (Ubx), determines whether a thoracic segment produces full wings or halteres.

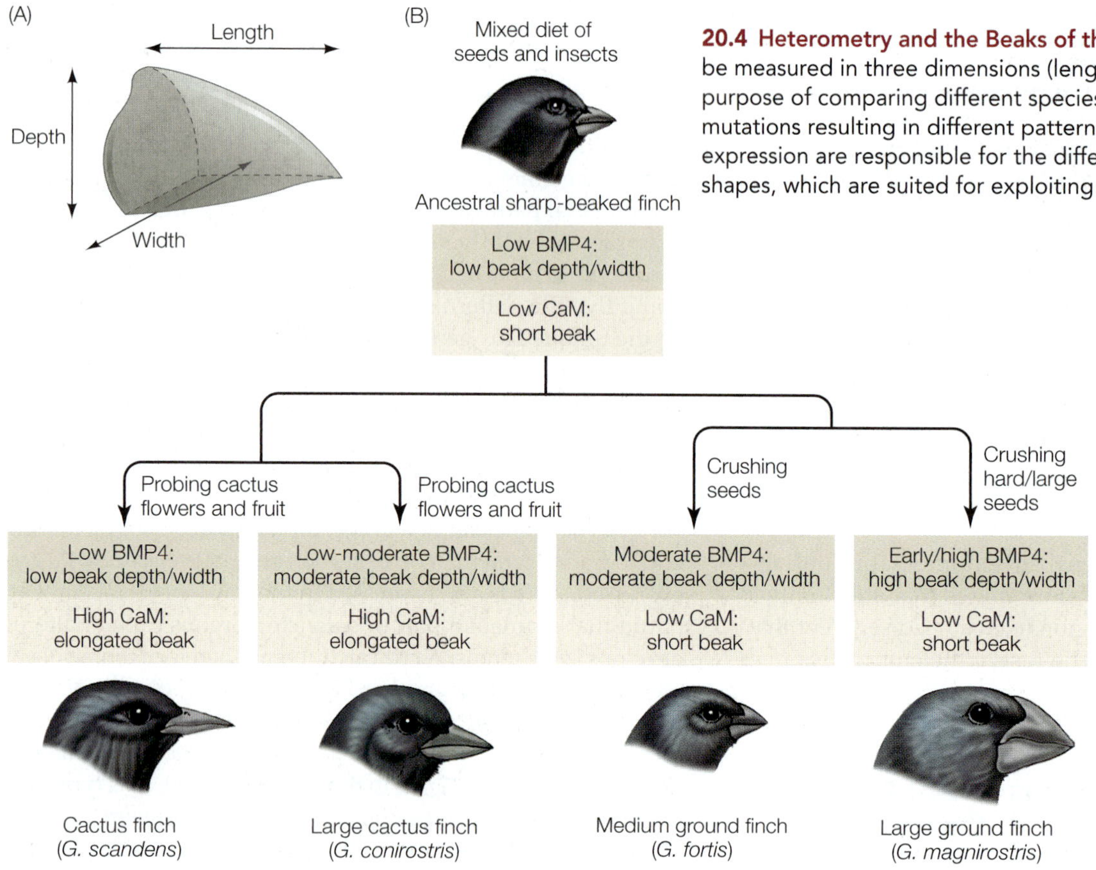
(A)
Length
Depth
Width
(B)
Mixed diet of seeds and insects
Ancestral sharp-beaked finch
Low BMP4: low beak depth/width
Low CaM: short beak
Probing cactus flowers and fruit
Probing cactus flowers and fruit
Crushing seeds
Crushing hard/large seeds
Low BMP4: low beak depth/width
High CaM: elongated beak
Low-moderate BMP4: moderate beak depth/width
High CaM: elongated beak
Moderate BMP4: moderate beak depth/width
Low CaM: short beak
Early/high BMP4: high beak depth/width
Low CaM: short beak
Cactus finch (*G. scandens*)
Large cactus finch (*G. conirostris*)
Medium ground finch (*G. fortis*)
Large ground finch (*G. magnirostris*)

20.4 Heterometry and the Beaks of the Finches (A) Bird beaks can be measured in three dimensions (length, width, and depth) for the purpose of comparing different species. (B) Among Galápagos finches, mutations resulting in different patterns of BMP4 and calmodulin (CaM) expression are responsible for the different species' beak sizes and shapes, which are suited for exploiting different types of food.

Modularity allows for differences in the patterns of gene expression

Because of modularity, transcription factors and other genes of the toolkit can regulate the expression of structural genes in different amounts, at different times, and in different locations.

HETEROMETRY We saw an example of **heterometry** ("different measure") at the opening of this chapter when we described beak development in Galápagos finches. Studies revealed that beak size and shape in these birds is influenced by the level of expression (i.e., the amount of protein produced) of two regulatory genes. The relative amounts of these proteins determines whether an individual's beak is long, thin, and narrow or short, thick, and deep (Figure 20.4).

HETEROCHRONY The evolution of the giraffe's neck provides an example of **heterochrony** ("different time"). Giraffes, like all mammals except manatees and sloths, have seven cervical (neck) vertebrae. So giraffes did not get longer necks by adding more vertebrae. However, the cervical vertebrae of giraffes are much longer than those of other mammals (Figure 20.5).

Bone growth in mammals is the result of the proliferation of cartilage-producing cells called chondrocytes. Bone growth

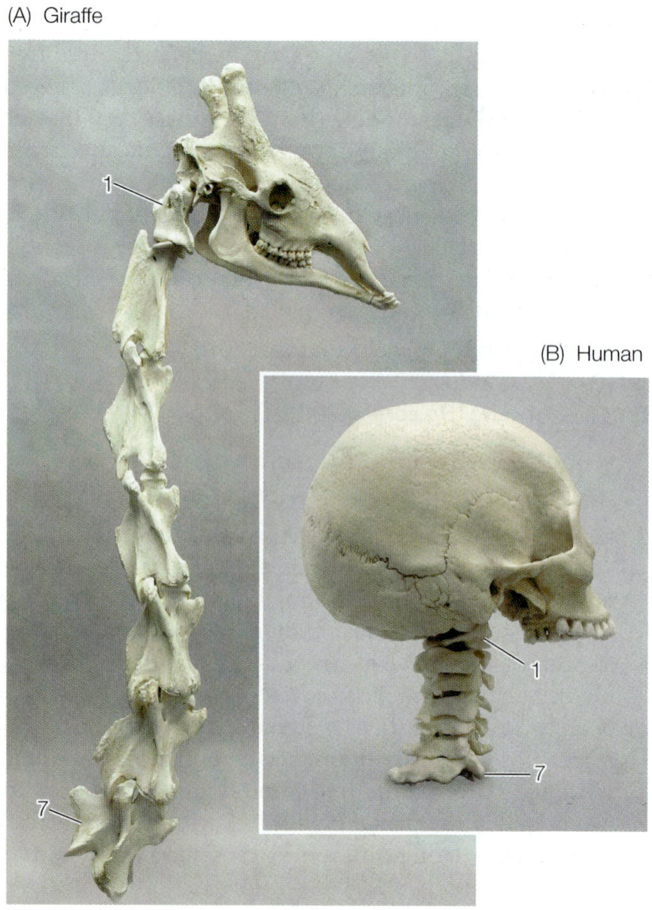
(A) Giraffe
1
7
(B) Human
1
7

20.5 Heterochrony in the Development of a Longer Neck There are seven vertebrae in the neck of the giraffe (left) and human (right; not to scale). But the vertebrae of the giraffe are much longer (25 cm compared to 1.5 cm) because during development, growth continues for a longer period of time. This timing difference is called heterochrony.

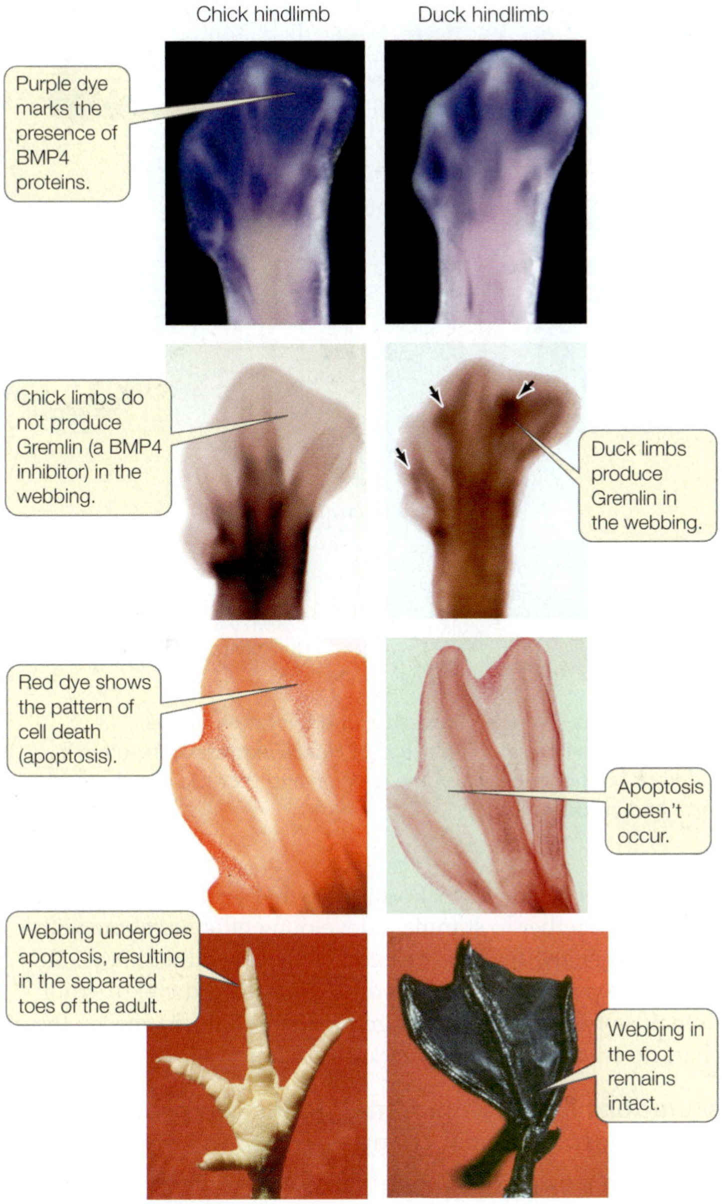
Chick hindlimb
Duck hindlimb
Purple dye marks the presence of BMP4 proteins.
Chick limbs do not produce Gremlin (a BMP4 inhibitor) in the webbing.
Duck limbs produce Gremlin in the webbing.
Red dye shows the pattern of cell death (apoptosis).
Apoptosis doesn't occur.
Webbing undergoes apoptosis, resulting in the separated toes of the adult.
Webbing in the foot remains intact.

20.6 Changes in Gremlin Expression Correlate with Changes in Hindlimb Structure The left column of photos shows foot development in a chicken; the right column shows foot development in a duck. Gremlin protein in the webbing of the duck foot inhibits BMP4 signaling, thus preventing the embryonic webbing from undergoing apoptosis.

is stopped by a genetic signal that results in apoptosis, or cell death, of chondrocytes and calcification of the bone matrix (see Section 48.3). In giraffes this signaling process is delayed in the cervical vertebrae, so that these vertebrae grow longer. Thus the evolution of longer necks resulted from *changes in the timing* of expression of the genes that control bone formation.

HETEROTOPY Spatial differences in the expression of a developmental gene are known as **heterotopy** ("different place"), exemplified by the different development of feet in ducks and chickens. The feet of all bird embryos have webs of skin that connect their toes. This webbing is retained in adult ducks (and other aquatic birds) but not in adult chickens (and other nonaquatic birds). The loss of webbing is controlled by the BMP4 signaling protein, which as we have noted is also involved in beak development (another example of use of a common genetic toolkit to produce different kinds of change).

BMP4 protein instructs the cells that produce webbing to undergo apoptosis and thus eliminates the webbing between the toes. The hindlimbs of both duck and chicken embryos express the *BMP4* gene in the webbing between the toes; however, they differ in expression of the *Gremlin* gene, which encodes a protein that inhibits *BMP4* expression (**Figure 20.6**). In ducks, but not in chickens, *Gremlin* is expressed in the webbing cells and Gremlin protein inhibits *BMP4* expression. With no BMP4 protein to stimulate apoptosis, a webbed foot develops. If chick hindlimbs are experimentally exposed to Gremlin during development, the adult chicken will have ducklike webbed feet (**Figure 20.7**).

INVESTIGATING**LIFE**

20.7 Changing the Form of an Appendage Ducks have webbed feet and chickens do not—a major difference in the adaptations of these species. Webbing is initially present in the chick embryo, but undergoes apoptosis that is stimulated by the protein BMP4. In ducks another protein, Gremlin, binds to BMP4 and inhibits it, preventing apoptosis and resulting in webbed feet. Juan Hurle and his colleagues at the Universidad de Cantabria in Spain asked what would happen if Gremlin were put onto a developing chick foot.[a]

HYPOTHESIS Adding Gremlin protein (a BMP4 inhibitor) to a developing chicken foot will transform it into a ducklike foot.

Method Open a small window in chick egg shell and carefully add Gremlin-secreting beads to the webbing of embryonic chicken hindlimbs. Add beads that do not contain Gremlin to other hindlimbs (controls). Close the eggs and observe limb development.

Results In the hindlimbs in which Gremlin was secreted, the webbing does not undergo apoptosis, and the hindlimb resembles that of a duck. The control hindlimbs develop the normal chicken form.

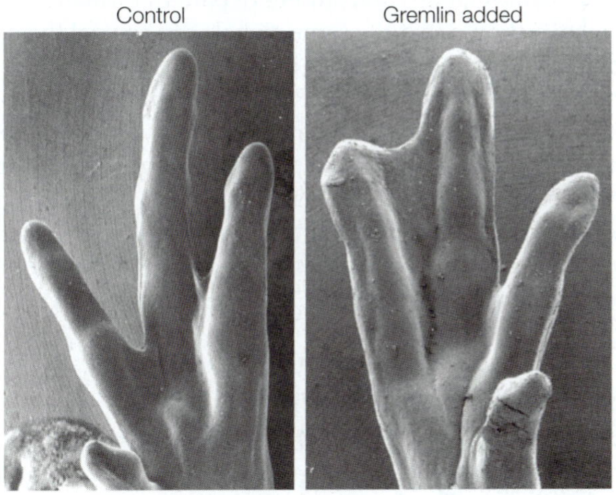
Control
Gremlin added

CONCLUSION Changes in *Gremlin* gene expression account for major morphological differences in the webbing of bird feet.

Go to **BioPortal** for discussion and relevant links for all INVESTIGATING**LIFE** figures.

[a]Merino, R. et al. 1999. *Development* 126: 5515–5522.

RECAP 20.2

Genetic toolkit genes act independently among modules of a developing embryo. There are various ways in which the expression of these developmental genes can differ among species, which results in major morphological differences. They can differ in amount of expression (heterometry), in the timing of expression (heterochrony), or in the location of expression (heterotopy).

- What are genetic switches, and how can they alter the way a particular gene is expressed? **See p. 415 and Figure 20.3**
- Explain how heterometry and natural selection produced the beak variations we see in today's Galápagos finches. **See p. 416 and Figure 20.4**
- How is heterochrony involved in the evolution of the long neck of giraffes? **See pp. 416–417 and Figure 20.5**
- Explain why the differences in webbing of duck feet and chicken feet are an example of heterotopy. **See p. 417 and Figure 20.6**

Genetic manipulations and studies of pattern formation in embryos have shown that the same signals can control the development of different structures in an individual organism. As we have noted, the protein BMP4 promotes apoptosis between developing digits in bird feet and is also involved in the formation of bird beaks. Perhaps not surprisingly, BMP4—bone morphogenetic protein 4—is also involved in the formation of bone. Might the processes that generate multiple structures *within* an organism also explain how different structures develop in different species?

20.3 How Can Developmental Changes Result in Differences among Species?

We have seen in the case of the Galápagos finches how subtle changes in the expression of developmental genes can alter the morphology of the beak leading to adaptations for exploiting different types of food (see Figure 20.4). Small differences in the expression of developmental genes can produce even more dramatic differences between species.

Differences in Hox gene expression patterns result in major differences in body plans

A diagnostic feature of vertebrates is their vertebral columns, but the vertebral columns exhibit considerable variation across the groups of vertebrates. For example, most mammals have 7 cervical vertebrae, but birds can have many more; chickens have 14 and swans have 25. Mammals have 13 thoracic vertebrae with ribs, birds have fewer, and snakes have many more. These significant differences result from the spatial patterns of Hox gene expression that govern the transitions from one region to another (**Figure 20.8**). The anterior limit of expression of *Hoxc6*, for example, always falls at the boundary between the cervical and thoracic vertebrae in mammals and birds. In addition, the anteriormost segment that expresses *Hoxc6* is the segment where the forelimbs will develop. Posterior to that segment, *Hoxc6* and *Hoxc8* act together to stimulate development of vertebrae with ribs—that is, the thoracic vertebrae. Over evolutionary time, genetic changes that expanded or contracted the expression domains of the different Hox genes resulted in changes in the characteristic numbers of different vertebrae.

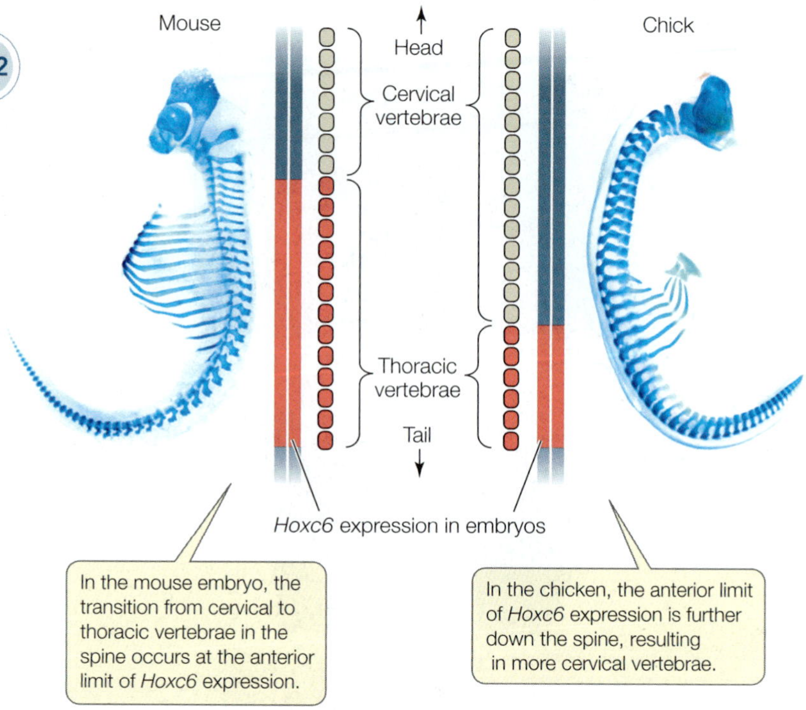

Mouse
Chick
Head
Cervical vertebrae
Thoracic vertebrae
Tail
Hoxc6 expression in embryos
In the mouse embryo, the transition from cervical to thoracic vertebrae in the spine occurs at the anterior limit of *Hoxc6* expression.
In the chicken, the anterior limit of *Hoxc6* expression is further down the spine, resulting in more cervical vertebrae.

20.8 Changes in Gene Expression and Evolution of the Spine Differences in the pattern of *Hoxc6* expression result in a different boundary between the cervical and thoracic vertebrae in mice and chicks.

Mutations in developmental genes can produce major morphological changes

Sometimes a major developmental change is due to an alteration in the regulatory molecule itself rather than a change in where, when, or how much it is expressed. This is called **heterotypy** ("different type"). An excellent example of heterotypy is a gene that controls the number of legs in arthropods. Arthropods all have head, thoracic, and abdominal regions with variable numbers of segments. Insects, such as *Drosophila*, have three pairs of legs on their three thoracic segments, whereas centipedes have many legs on both thoracic and abdominal segments. All arthropods express a gene called *Distalless* (*Dll*) that controls segmental leg development. In insects, *Dll* expression is repressed in abdominal segments by the Hox gene *Ubx*. *Ubx* is expressed in the abdominal segments of all arthropods, but it has different effects in different species. In centipedes, Ubx protein activates expression of the *Dll* gene to promote the formation of legs. During the evolution of insects, a change in the *Ubx* gene sequence resulted in a modified Ubx protein that *represses Dll* expression in abdominal segments. A phylogenetic tree of arthropods shows that this change in *Ubx* occurred in the ancestor of insects, at the same time that abdominal legs were lost (**Figure 20.9**).

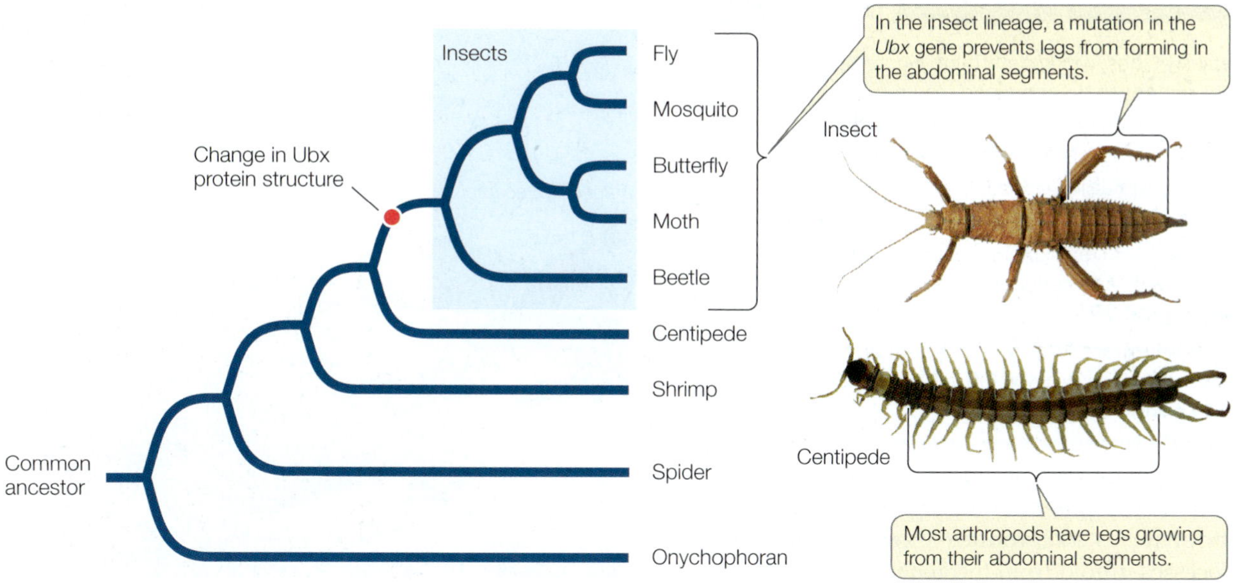

Insects
Fly
Mosquito
Butterfly
Moth
Beetle
Centipede
Shrimp
Spider
Onychophoran
Common ancestor
Change in Ubx protein structure
In the insect lineage, a mutation in the *Ubx* gene prevents legs from forming in the abdominal segments.
Insect
Centipede
Most arthropods have legs growing from their abdominal segments.

20.9 A Mutation in a Hox Gene Changed the Number of Legs in Insects In the insect lineage (blue box) of the arthropods, a change to the *Ubx* gene resulted in a protein that inhibits the *Dll* gene, which is required for legs to form. Because insects express this modified *Ubx* gene in their abdominal segments, no legs grow from these segments. Other arthropods, such as centipedes, produce an unmodified Ubx protein and do grow legs from their abdominal segments.

An example of heterotypy that has a huge impact on our food supply is a gene that is partly responsible for the easily accessible edible kernels of domestic corn. The wild relative of domesticated corn—a plant called teosinte—has kernels encased in tough shells (glumes) that develop under control of a gene called *Tga1*. The "liberation" of the edible kernels from these casings that revolutionized this grain's importance to human agriculture is in part based on a mutation in *Tga1* that results in a protein different from the ancestral form by a single amino acid. Genetic experiments have shown that when the domesticated corn *Tga1* gene is inserted into teosinte, teosinte kernels break free of their glumes; conversely, when the teosinte *Tga1* gene is inserted into domestic corn, many of the corn kernels become encased in hard glumes (**Figure 20.10**).

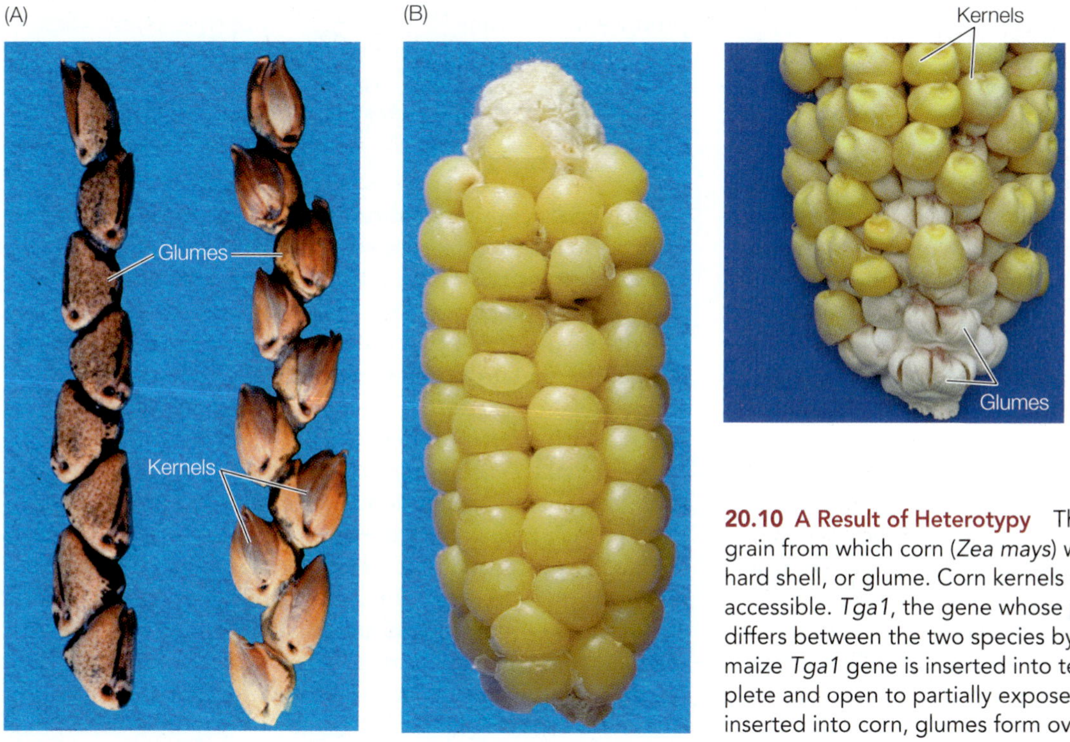

(A)
Glumes
Kernels
(B)
Kernels
Glumes

20.10 A Result of Heterotypy The edible kernels of teosinte, the grain from which corn (*Zea mays*) was domesticated, are encased in a hard shell, or glume. Corn kernels have no glumes and thus are easily accessible. *Tga1*, the gene whose product controls glume formation, differs between the two species by one amino acid. (A) If the domestic maize *Tga1* gene is inserted into teosinte, the glumes are incomplete and open to partially expose the kernels. (B) If teosinte *Tga1* is inserted into corn, glumes form over the kernels.

RECAP 20.3

Changes in genetic switches, which determine where, when, and to what extent genes will be expressed, can result in major morphological changes. Natural selection can result in fixation of these differences and subsequent speciation.

- How can Hox gene expression explain the differences in the vertebral columns of different vertebrate groups? **See p. 418**
- How does the activity of *Ubx* explain the fact that insects have three pairs of legs and centipedes have many? **See p. 418 and Figure 20.9**
- Explain how heterotypy accounts for a major morphological difference between corn and its related wild relative, teosinte. **See p. 419 and Figure 20.10**

This chapter has focused on how modular genetic signaling cascades control the development of an organism and how changes in genetic switches can produce differences among species. These processes unfold from the genetic information contained in the fertilized egg, but information from the environment can also influence the genetic signaling cascades and thereby alter the form of the organism.

Trachemys scripta
Percent male
100
80
60
40
20
0
25
26
27
28
29
30
Temperature (°C)

20.11 Hot Females, Cool Males The sex of a red-eared slider turtle depends on the temperature at which the egg is incubated. Higher temperatures produce only females and lower temperatures produce only males, apparently due to temperature sensitivity in the synthesis of sex hormones.

20.4 How Can the Environment Modulate Development?

In some cases environmental signals produce developmental changes that affect the morphology (phenotype) of an organism. The ability of an organism to modify its development in response to environmental conditions is called **developmental plasticity** or **phenotypic plasticity**; it means that a single genotype has the capacity to produce two or more different phenotypes.

Go to Media Clip 20.1
Predator-Induced Development
Life10e.com/mc20.1

Temperature can determine sex

Section 12.4 described how genetic mechanisms can determine sex. In some reptiles, however, sex is determined not by genetic differences between individuals, but by the temperature at which the eggs are incubated. Research in the laboratory of David Crews at the University of Texas has shown that if eggs of the red-eared slider turtle (*Trachemys scripta*) are incubated at temperatures below 28.6°C, all the hatchlings are males, whereas eggs incubated above 29.4°C all produce females. In between these two temperatures—a range of less than 1°C—a clutch of eggs will produce both males and females (**Figure 20.11**). In other species with temperature-dependent sex determination, the incubation temperatures that produce males and females may be different. Thus the effects of temperature on sex determination can vary among species. But how can temperature control developmental plasticity?

In all vertebrate embryos, the development of male and female organs is controlled by the actions of sex steroid hormones. This is the case whether the organism's sex determination is controlled by its genotype or by temperature. Sex steroid biosynthesis in both males and females begins with cholesterol and goes through many chemical reactions to produce the male sex steroids (androgens) and the female sex steroids (estrogens). In this biosynthetic sequence, the step that produces the first androgen—testosterone—precedes the step that produces estrogens; therefore both males and females produce testosterone.

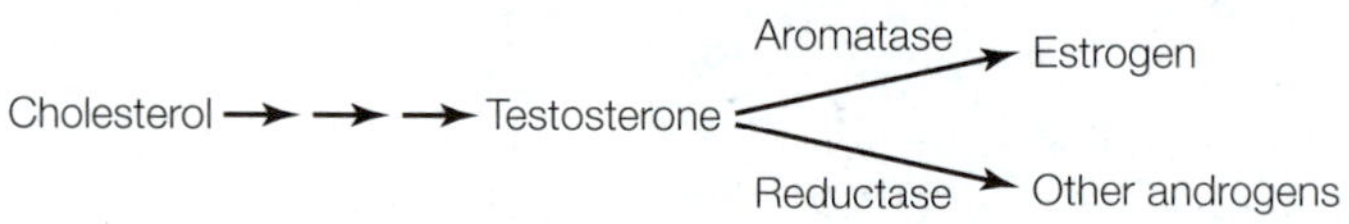
Cholesterol
Testosterone
Aromatase
Estrogen
Reductase
Other androgens

In animals with temperature-controlled sex determination, incubation temperature controls the expression of the enzyme aromatase, which converts testosterone to estrogen. If aromatase is abundantly expressed, estrogens are dominant and female organs develop. If aromatase is not expressed, testosterone is dominant and male organs develop. Applying estrogen to eggs results in the development of females, even at the male-inducing temperature.

What is the evolutionary advantage of this sex determination mechanism? Incubation temperature can affect other aspects of the phenotype of the developing organism besides its sex, such as its growth rate and eventual adult body size. For example, red-eared slider females are larger than the males, and the largest females produce many more eggs than smaller females. But even small males can produce enough sperm to fertilize all the eggs of the larger females. For these kinds of reasons, incubation temperature may have a differential effect

INVESTIGATING**LIFE**

20.12 Temperature-Dependent Sex Determination Can Be Associated with Sex-Specific Fitness Differences In some reptiles, sex is determined by the incubation temperature of the developing embryo. This led to the hypothesis that male-inducing temperatures during development result in males with higher reproductive fitness. Daniel Warner and Rick Shine at the University of Sydney, Australia tested this hypothesis by using a drug to block estrogen synthesis, so that males developed instead of females at high and low temperatures. These males had much lower reproductive fitness than males that developed at the normal male-inducing temperature (which is intermediate). In contrast, females showed highest fitness when they developed from eggs incubated at higher temperatures.[a]

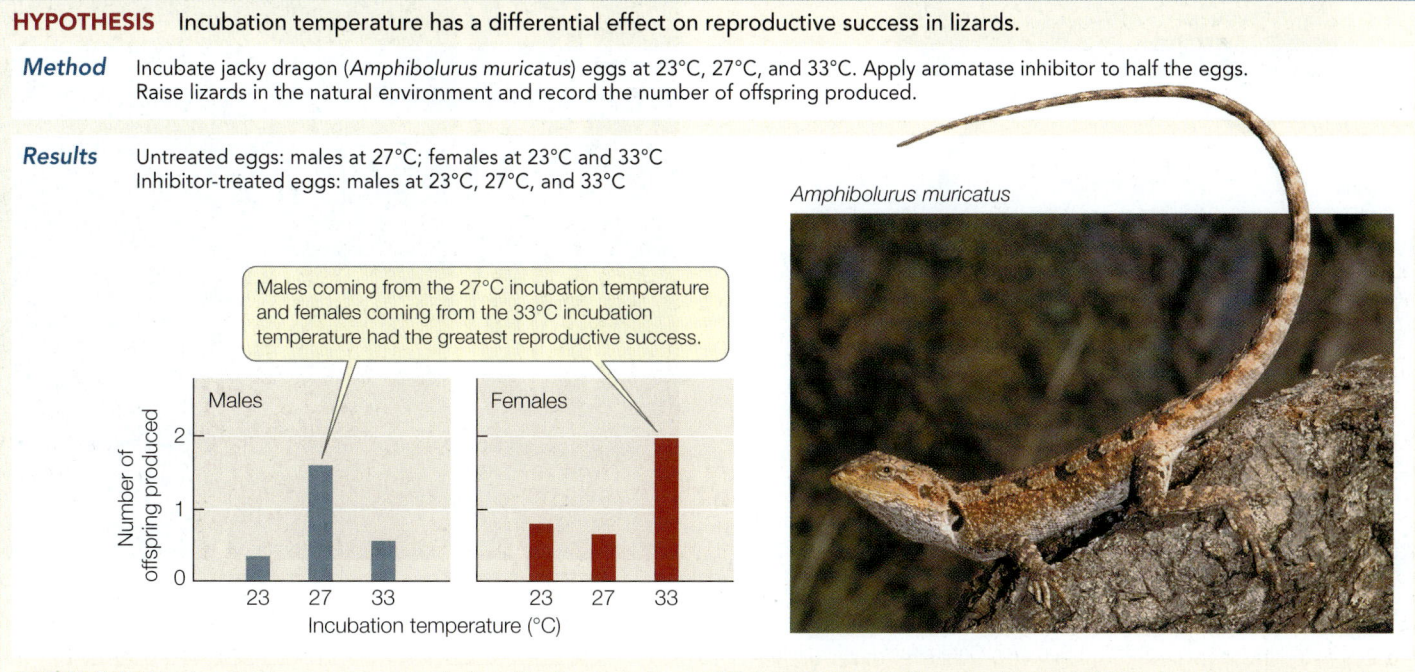

HYPOTHESIS Incubation temperature has a differential effect on reproductive success in lizards.

Method Incubate jacky dragon (*Amphibolurus muricatus*) eggs at 23°C, 27°C, and 33°C. Apply aromatase inhibitor to half the eggs. Raise lizards in the natural environment and record the number of offspring produced.

Results Untreated eggs: males at 27°C; females at 23°C and 33°C
Inhibitor-treated eggs: males at 23°C, 27°C, and 33°C

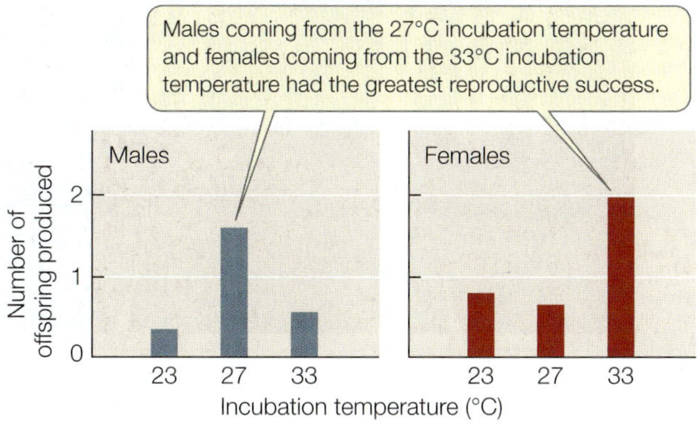
Males coming from the 27°C incubation temperature and females coming from the 33°C incubation temperature had the greatest reproductive success.
Males
Females
Number of offspring produced
0
1
2
23
27
33
23
27
33
Incubation temperature (°C)

Amphibolurus muricatus

CONCLUSION Incubation temperature during development has a differential effect on fitness of male and female jacky dragons, and thus could be a selective pressure leading to the pattern of temperature determination of sex in this species.

Go to **BioPortal** for discussion and relevant links for all INVESTIGATING**LIFE** figures.

[a]Warner, D. A. and R. Shine. 2008. *Nature* 451: 566–568.

on the reproductive success of males and females in a population. Figure 20.12 describes an experiment that clearly suggests incubation temperature differentially affects the reproductive successes of males and females in one lizard species.

Dietary information can be a predictor of future conditions

In some species with short life spans, an individual may encounter only one of several distinct but predictable seasonal environments. Developmental plasticity allows such individuals to develop the phenotype that enhances their survival in the environmental they will encounter as adults. An excellent example is the moth *Nemoria arizonaria*, which produces two generations each year. Caterpillars from eggs that have overwintered and hatch in the spring feed on the oak tree flowers, or catkins, that are present at that time of year. These caterpillars complete their development and transform into adult moths in summer. The summer moths lay their eggs on oak leaves, and the caterpillars that hatch eat the mature leaves. When these caterpillars transform into adult moths, they lay eggs that overwinter and hatch the following spring when the catkins are once again in bloom. Although newly hatched caterpillars all look similar, the body form of spring caterpillars resembles the catkins (Figure 20.13A), whereas the body form of summer caterpillars resembles young oak branches (Figure 20.13B). Their different larval diets—that is, the different biochemistry of the molecules in oak catkins versus those in mature oak leaves—trigger developmental changes that result in two different phenotypes, each of which is cryptic in the respective environments. This phenotypic plasticity results in lower predation and increased evolutionary fitness.

A variety of environmental signals influence development

In addition to temperature and diet, other environmental signals may initiate developmental change. One ubiquitous source of environmental information is sunlight, which provides predictive information about seasonal changes. Outside the equatorial region, lengthening days herald spring and summer whereas shortening days indicate oncoming winter. Many insects use day length to enter or exit a period of developmental or reproductive arrest called **diapause**, which enables them to better survive harsh conditions. Deer, moose, and elk use day length to time the development and the dropping of antlers,

Summer adult
Nemoria arizonaria caterpillars
Catkins
(A)
(B)
Spring: Caterpillars feed on catkins
Summer: Caterpillars feed on leaves

20.13 Spring and Summer Forms of a Caterpillar (A) Spring caterpillars of the moth *Nemoria arizonaria* resemble the oak catkins on which they feed. They develop into adults (shown above), which lay eggs on oak leaves. (B) The summer caterpillars feed on oak leaves and are camouflaged by their resemblance to oak twigs.

In the dark, stems elongate and leaf growth is reduced.
In the light, stems are shorter and leaves are bigger.

20.14 Light Seekers The bean plants on the left were grown under low light levels. The plant's cells have elongated in response to the low light, and the plants have become spindly. The control plants on the right were grown under normal light conditions.

and many organisms use day length to optimize the timing of reproduction or migration. Many plants initiate flowering in response to the length of the night (an absence of light), and in some plant species developmental changes are induced by certain wavelengths of light.

You may wonder why processes such as antler growth, reproductive timing, and seasonal migration are considered in a chapter on development. Development encompasses more than the events that occur before an organism reaches maturity. Development includes changes in body form and function that can occur throughout the life of the organism. For example, giant redwoods that are thousands of years old still have undifferentiated tissues called meristems that produce new differentiated tissues—stems, leaves, reproductive structures, and so on—throughout the life of the tree. These developmental processes are not a simple read-out of a genetic program; they are adjusted to optimize plant form in the environment in which the organism grows. Light, which plants need for photosynthesis, is an important environmental signal in plant development. Dim light stimulates the cells of the stems to elongate, so plants growing in the shade become tall and spindly (**Figure 20.14**). This developmental plasticity is adaptive because a tall, spindly plant is more likely to reach a patch of brighter light than a plant that remains compact and bushy. A plant in bright light does not need to grow tall and so can put more energy into producing leaves.

RECAP 20.4

Developmental plasticity enables some developing organisms to adjust their forms to fit the environments in which they live. Organisms respond to environmental signals that are accurate predictors of future conditions. Development continues throughout life, and can result in adaptive changes in the forms and functions of adult organisms.

- Describe several examples of how phenotype can be a response to environmental signals. **See pp. 420–422 and Figures 20.11, 20.13, and 20.14**
- How would you determine whether or not an environmental effect on development is adaptive? **See pp. 420–421 and Figure 20.12**

Appropriate responses to new environmental conditions are likely to evolve over time, but what are the limits of such evolution? To what extent do developmental genes dictate the structures and forms that are possible?

20.5 How Do Developmental Genes Constrain Evolution?

Four decades ago, the French geneticist François Jacob made the analogy that evolution works like a tinker, assembling new structures by combining and modifying the available materials, and not like an engineer, who is free to develop dramatically different designs (a jet engine to replace a propeller-driven engine, for example). We have seen that morphological evolution is not usually governed by the acquisition of radically new genes, but proceeds primarily by "tinkering" with the expression patterns of existing genes. Thus developmental genes and their expression constrain evolution in two major ways:

- Nearly all evolutionary innovations are modifications of previously existing structures.
- The basic set of regulatory genes that control development is broadly conserved, changing only slowly over the course of evolution.

Evolution usually proceeds by changing what's already there

The features of organisms usually evolve from preexisting features in their ancestors. New "wing genes" did not suddenly appear in insects, birds, and bats; instead, wings arose as modifications of existing structures. Wings evolved independently in insects and vertebrates—once in insects, and in three independent instances among the vertebrates (**Figure 20.15**). Vertebrate wings are modified forelimbs and share many of the same basic skeletal elements. They have a humerus that connects the wing to the rest of the skeleton; two longer bones (radius and ulna) that project away from the humerus; and metacarpals and phalanges (digits). Although these elements are present in all the various vertebrate wings, the independent origins of vertebrate wings is clear from the different role these skeletal elements play in each wing type. During development these bones grow to different lengths and weights in different organisms, and evolution proceeded in different paths in producing bird, bat, and pterosaur wings.

Developmental regulatory genes also influence how organisms lose structures. The ancestors of present-day snakes lost their forelimbs as a result of changes in the segmental expression of Hox genes. The snake lineage subsequently lost its hindlimbs by the loss of expression of the *Sonic hedgehog* gene in the limb bud tissue. Some snakes, such as the boas and pythons, have rudimentary pelvic and upper leg bones.

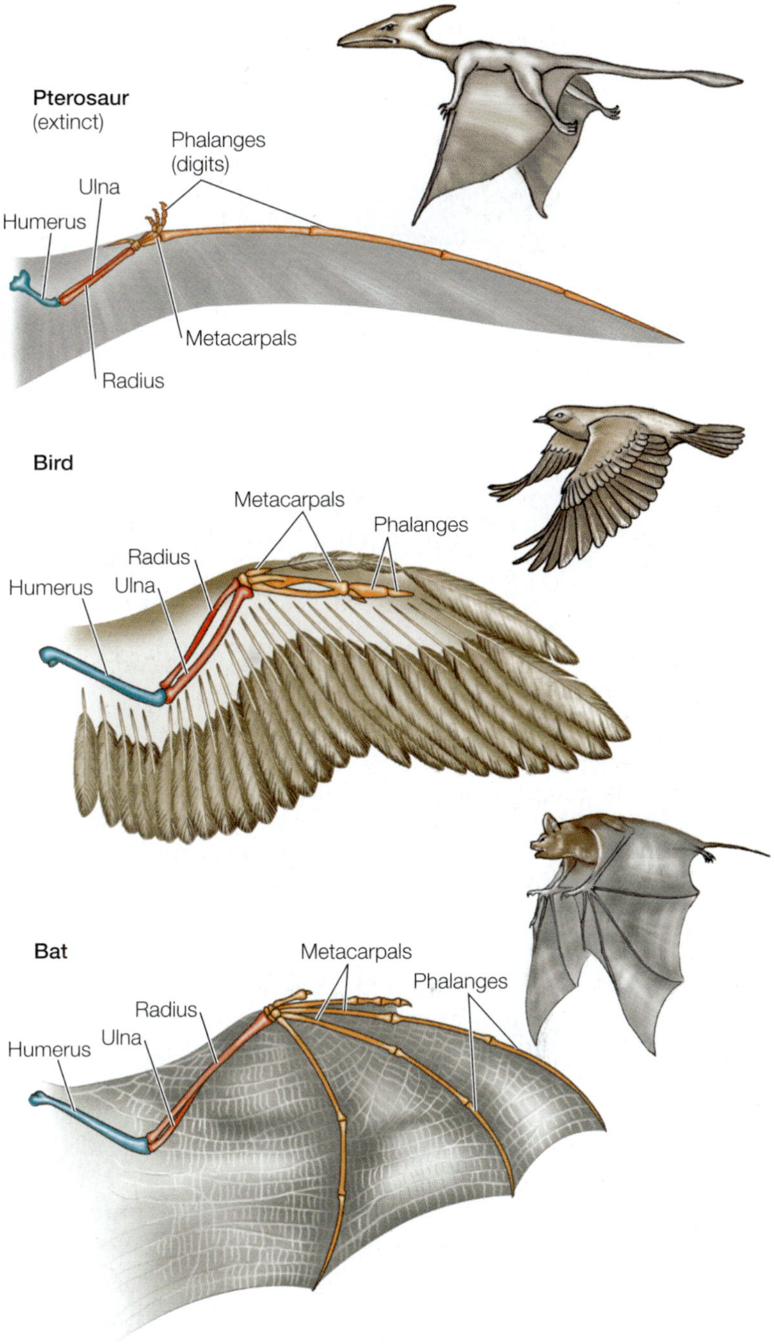
Pterosaur
(extinct)
Phalanges
(digits)
Ulna
Humerus
Metacarpals
Radius
Bird
Metacarpals
Phalanges
Radius
Humerus
Ulna
Bat
Metacarpals
Phalanges
Radius
Humerus
Ulna

20.15 Wings Evolved Three Times in Vertebrates The wings of pterosaurs (the earliest flying vertebrates, which lived from 265 to 220 million years ago), birds, and bats are all modified forelimbs and are constructed from the same skeletal components. These components, however, have different forms in the different groups, supporting the independent evolution of wings in each group.

Conserved developmental genes can lead to parallel evolution

The existence of highly conserved developmental genes makes it likely that similar traits will evolve repeatedly, especially among closely related species. This process is known as **parallel evolution**, and a good example is provided by a small fish, the three-spined stickleback (*Gasterosteus aculeatus*).

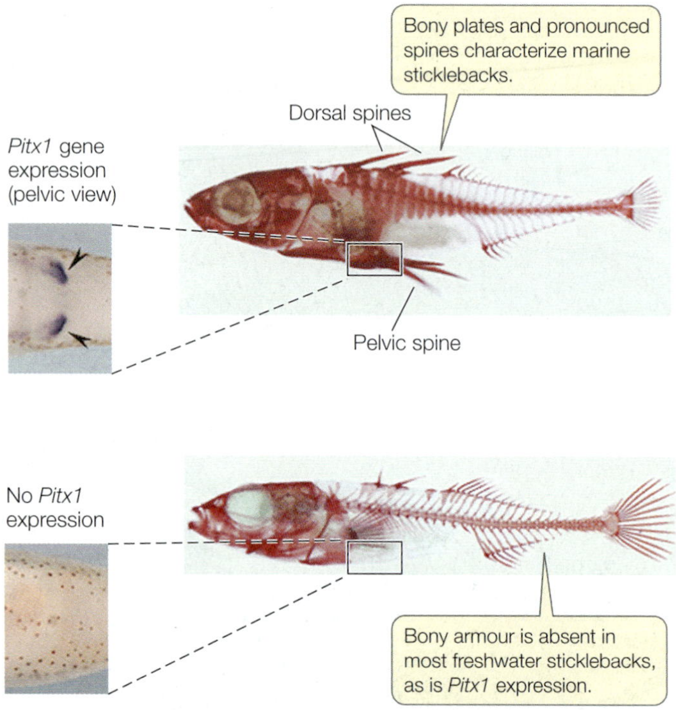
Bony plates and pronounced spines characterize marine sticklebacks.
Dorsal spines
Pitx1 gene expression (pelvic view)
Pelvic spine
No *Pitx1* expression
Bony armour is absent in most freshwater sticklebacks, as is *Pitx1* expression.

20.16 Parallel Phenotypic Evolution in Sticklebacks
A developmental gene, *Pitx1*, encodes a transcription factor that stimulates the production of plates and spines. This gene is active in marine sticklebacks, but mutated and inactive in various freshwater populations of the fish. The fact that this mutation is found in geographically distant and isolated freshwater populations is evidence for parallel evolution.

Sticklebacks are widely distributed throughout the Atlantic and Pacific oceans; they are also found in many freshwater lakes and rivers. Marine sticklebacks spend most of their lives at sea but return to fresh water to breed. Members of freshwater populations that are isolated in lakes never encounter salt water. Genetic evidence shows that freshwater populations have arisen from marine populations many times, and independently. Marine sticklebacks have structures that protect them from predatory marine fish; these are bony plates and well-developed pelvic bones with pelvic spines that lacerate the mouths of predators. Freshwater sticklebacks do not face such predatory dangers; their body armor is greatly reduced, and their dorsal and pelvic spines are shorter or even lacking (**Figure 20.16**).

The differences between marine and freshwater sticklebacks are not induced by environmental conditions. Marine species reared in fresh water still grow armor and spines. The differences are due to the expression of a developmental regulatory gene, *Pitx1*. This gene codes for a transcription factor normally expressed in regions of the developing embryo that in marine sticklebacks form the head, trunk, tail, and pelvis. However, in separate and long-isolated populations of freshwater sticklebacks from Japan, British Columbia, California, and Iceland, the *Pitx1* gene is no longer expressed in the pelvis, and spines do not develop. This same change in regulatory gene expression resulted in similar phenotypic changes *independently* in several different populations, and is thus a good example of parallel evolution. Is there a common selective mechanism at work in these cases, and what could that mechanism be? One reasonable hypothesis is that decreased predation pressure in the freshwater environment allowed increased reproductive success for sticklebacks that invest less energy in the development of unnecessary protective structures.

RECAP 20.5

Nearly all evolutionary innovations are modifications of previously existing structures. The conservation of many developmental regulatory genes makes it likely that similar traits will evolve repeatedly.

- How have diverse body forms evolved by means of modifications to existing structures? **See p. 423 and Figure 20.15**
- How do the differences between marine and freshwater sticklebacks exemplify parallel evolution via changes in gene regulation? **See p. 424 and Figure 20.16**

Many novel traits have arisen during the course of evolution, but most of them failed to persist beyond even a single generation. Part Six of this book will examine the processes of evolution—the powerful forces that influence the survival and reproductive success of various life forms. We will examine how different adaptations become prevalent in different environments, resulting in the extraordinary diversity of life on Earth today, which we will describe in further detail in Part Seven.

?

How are gene expression patterns involved in the shaping of the diverse beaks of birds?

ANSWER

As described in Figure 20.4, studies of Galápagos finches indicate that high levels of BMP4 protein expression are associated with the short, robust beaks of the ground finches and that high calmodulin expression is associated with the narrow, sharp beaks of the cactus finches. The timing and level of expression of the genes for these two proteins, under the control of transcription factors and their promotors, enhancers, and repressors—the genetic toolkit—result in morphological modifications of the birds' beaks.

Consider also the difference between the short, narrow, sharp beaks of chickens and the long, thick, broad beaks of ducks. In other studies, analysis of BMP4 in the developing beak tissues of chickens and ducks reveals that its expression is higher in duck embryos. Using molecular genetic means of increasing expression of BMP4 in the developing beak tissue of chicks results in bigger, broader, more robust beaks in the adult chicken. Blocking BMP4 expression in the developing beak tissue of chickens results in adult birds with very small beaks. Thus across these two distantly related groups of birds, the same developmental regulatory genes are involved in shaping the morphology of the beak.

CHAPTER**SUMMARY** 20

20.1 How Can Small Genetic Changes Result in Large Changes in Phenotype?

- **Evolutionary developmental biology**, or **evo-devo**, is the study of the evolutionary aspects of development. This field focuses on the molecular mechanisms that underlie the development of phenotypic diversity.
- Changes in development underlie evolutionary changes in morphology that produce differences in body forms.
- Similarities in the basic mechanisms of development between widely divergent organisms reflect common ancestry. **Review Figure 20.1**
- Evolutionary diversity is produced using a modest number of regulatory genes.
- Genes encoding the transcription factors and other regulatory proteins that govern pattern formation in the developing bodies of multicellular organisms can be thought of as a **genetic toolkit**. These regulatory genes have been highly conserved throughout evolution. **Review Figure 20.2**

20.2 How Can Mutations with Large Effects Change Only One Part of the Body?

- The bodies of developing and mature organisms are organized into self-contained units, or **modules**, that can be modified independently. **See ANIMATED TUTORIAL 20.1**
- The genetic toolkit involves **genetic switches**—promoters, enhancers, and repressors—that can alter the expression of developmental genes in different modules independently of one another.
- Developmental genes can be expressed in a modular fashion in different amounts (**heterometry**), at different times (**heterochrony**), or in different locations (**heterotopy**). **Review Figures 20.3–20.6**

20.3 How Can Developmental Changes Result in Differences among Species?

- Changes in genetic switches that determine where, when, and to what extent a set of genes will be expressed underlie both the transformation of an individual from egg to adult and the evolution of differences among species.
- Morphological differences among species can result from mutations in the genes that regulate the deveolpment of modules such as body segments or wings. **Review Figures 20.8, 20.9**

20.4 How Can the Environment Modulate Development?

- The ability of an organism to modify its development in response to environmental conditions is called **developmental**, or **phenotypic**, **plasticity**.
- In many species of reptiles, sex development is determined by incubation temperature, which acts through genes that control the production, modification, and action of sex hormones. **Review Figure 20.11**
- The adaptive significance of developmental plasticity is not always obvious, but experiments can test for effects on reproductive success. **Review Figure 20.12**
- Some environmental cues, such as those that anticipate seasons, are highly regular and can reliably drive seasonal adaptations in body form and function. **Review Figure 20.13**
- Environmental cues that trigger developmental change are diverse and can act at any stage of the life of an organism.

20.5 How Do Developmental Genes Constrain Evolution?

- Virtually all evolutionary innovations are modifications of preexisting structures. **Review Figure 20.15**
- Because many genes that govern development have been highly conserved, similar traits are likely to evolve repeatedly, especially among closely related species. This process is called **parallel phenotypic evolution**. **Review Figure 20.16**

See ACTIVITY 20.1 for a concept review of this chapter.

Go to the Interactive Summary to review key figures, Animated Tutorials, and Activities
Life10e.com/is20

CHAPTER**REVIEW**

REMEMBERING

1. Which of the following is *not* one of the principles of evolutionary developmental biology (evo-devo)?
   a. Animal groups share similar molecular mechanisms for morphogenesis.
   b. Changes in the timing of gene expression are important in the evolution of new structures.
   c. Evolution of development is not responsive to the environment.
   d. Changes in the locations of gene expression in the embryo can lead to new structures.
   e. Evolution often occurs by modification of existing developmental genes and pathways.

2. The developmental control pathway that determines the segmental body plan in *Drosophila*
   a. has similar gene sequences and chromosomal arrangements in the mouse.
   b. is unique to insects.
   c. determines only those organs that arise in head segments.
   d. induces the development of wings in each thoracic segment.
   e. arose through new genes that had not existed before in any form.

3. The process whereby changes in the timing of developmental gene expression can change the form of an organism is called
   a. heterochrony.
   b. developmental plasticity.
   c. adaptation.
   d. modularity.
   e. heterometry.

4. Modularity is important for development because it
   a. guarantees that all units of a developing embryo will change in a coordinated way.
   b. coordinates the establishment of the anterior–posterior axis of the developing embryo.
   c. allows changes in developmental genes to change one part of the body without affecting other parts.
   d. guarantees that the timing of gene expression is the same in all parts of a developing embryo.
   e. allows organisms to be built up one module at a time.

5. Organisms often respond to environmental signals that accurately predict future conditions by
   a. stopping development until the signal changes.
   b. altering their development to adapt to the future environment.
   c. altering their development such that the resulting adult can produce offspring adapted to the future environment.
   d. producing new mutants.
   e. developing normally because the predicted conditions may not last long.

UNDERSTANDING & APPLYING

6. What do you predict would happen if you blocked the expression of Gremlin in the feet of duck embryos?

7. If you applied an inhibitor of aromatase to red-eared slider turtle eggs being incubated at warm and at cool temperatures, what would be the outcome?

8. In the mouse, *Hoxc6* is first expressed in the posterior part of the twelfth body segment (somite) and marks the boundary between the cervical and thoracic regions of the vertebral column. *Hoxc8* is expressed a little more posteriorly and is coexpressed with *Hoxc6* for most of the length of the thoracic region (where ribs form). In contrast, in the developing python, *Hoxc6* and *Hoxc8* are expressed together along most of the length of the embryo. What hypotheses can you formulate about the roles of these Hox genes in the development of features of the body axis in mice and snakes?

ANALYZING & EVALUATING

9. In a series of experiments on chick embryos, researchers applied different concentrations of BMP4 to the embryo's beak growth region; they then measured the size of the beak cartilage at a later stage of development. Based on their data in the table below, what would you conclude about the role of BMP4 in beak growth?

Amount of BMP4	Cartilage diameter (mm)
None (control)	0.5
0.1 units	0.7
0.3 units	1.0
1.0 units	1.8

10. *Plasmodium vivax* is a protist that causes a form of malaria. When *P. vivax* enters the blood, it attaches to a glycoprotein on the red blood cells. The same glycoprotein is found on a variety of other cells as well. Some human populations in Africa are immune to *P. vivax* because they lack this particular glycoprotein, but only on their red blood cells. The transcription of this glycoprotein is under the influence of a number of enhancers, and the one that is expressed in red blood cell precursors is mutated in the population that is immune to this form of malaria. What does this case illustrate in evolutionary developmental terms?

Go to BioPortal at **yourBioPortal.com** for Animated Tutorials, Activities, LearningCurve Quizzes, Flashcards, and many other study and review resources.

PART SIX
The Patterns and Processes of Evolution

21 Mechanisms of Evolution

CHAPTER OUTLINE

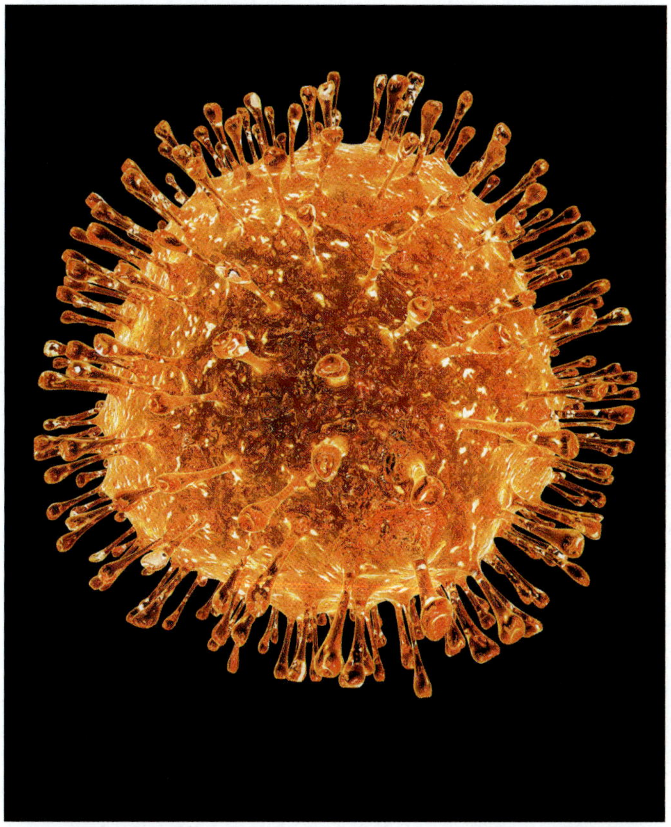

A Deadly Pathogen An artist's image of the H1N1 influenza virus that was the target of a recent flu vaccine. The spikes represent viral surface proteins, whose rapid evolution allows flu viruses to escape the host's immune system.

ON NOVEMBER 11, 1918, an armistice agreement signed in France signaled the end of World War I. But the death toll from four years of war was soon surpassed by the casualties of a massive influenza epidemic that began in the spring of 1918 among soldiers in a U.S. Army barracks. Over the next 18 months, this particular strain of flu virus spread across the globe, killing more than 50 million people worldwide—more than twice the number of combat-related deaths.

The 1918–1919 pandemic was noteworthy because the death rate among young adults—who are usually less likely to die from influenza than are the elderly or the very young—was 20 times higher than in flu epidemics before or since. Why was that particular virus so deadly, especially to typically hardy individuals? The 1918 flu strain triggered an especially intense reaction in the human immune system. This overreaction meant that people with strong immune systems were likely to be more severely affected.

In most cases, however, our immune system helps us fight viruses; this response is the basis of vaccination. Since 1945, programs to administer flu vaccines have helped keep the number and severity of influenza outbreaks in check. Last year's vaccine, however, will probably not be effective against this year's virus. New strains of flu virus are evolving continuously, ensuring genetic variation in the virus population. If these viruses did not evolve, we would become resistant to them, and annual vaccination would become unnecessary. But because the viruses do evolve, biologists must develop a new and different flu vaccine each year.

Vertebrate immune systems recognize proteins on the viral surface, so changes in these proteins mean that the virus can escape detection. Those virus strains with the greatest number of changes to their surface proteins are most likely to avoid detection and infect their hosts, and thus have an advantage over other strains. Biologists can observe evolution in action by following changes in influenza virus proteins from year to year.

We learn a great deal about the processes of evolution by examining rapidly evolving organisms such as viruses, and these studies contribute to the development of evolutionary theory. Evolutionary theory, in turn, is put to many practical uses, such as the development of better strategies for combating deadly diseases.

?

How do biologists use evolutionary theory to develop better flu vaccines?

See answer on p. 446.

21.1 What Is the Relationship between Fact and Theory in Evolution?

Biological populations change in their genetic makeup over time. This change in the genetic composition of populations over time is called **evolution**. We can, and do, observe evolutionary change on a regular basis, both in laboratory experiments and in natural populations. We measure the rates at which new mutations arise, observe the spread of new genetic variants through a population, and see the effects of genetic change on the form and function of organisms. In the fossil record, we observe the long-term morphological changes that are the result of underlying genetic changes. These underlying changes in the genetic makeup of populations drive the origin and extinction of species and fuel the diversification of life.

In addition to observing and recording physical changes over evolutionary time, biologists have accumulated a large body of evidence that has shown us *how* these changes occur and *what* evolutionary changes have occurred in the past. The resulting understanding of the mechanisms of evolutionary change is known as **evolutionary theory**.

We constantly apply evolutionary theory to the prevention and treatment of diseases, to the development of better agricultural crops and practices, and to the development of industrial processes that produce new molecules with useful properties. At a more basic level, evolutionary theory allows biologists to understand how life diversifies and how species interact. It also helps us make predictions about the biological world.

In everyday speech, people tend to use the word "theory" to mean an untested hypothesis, or even a guess. But "evolutionary theory" does not refer to any single hypothesis, and it certainly is not guesswork. A few scientists grasped the concept of evolutionary change even before Charles Darwin so clearly described his observations, presented his conclusions, and articulated an explanation for evolution in *On the Origin of Species*. The rediscovery of Gregor Mendel's experiments and the subsequent establishment of the principles of genetic inheritance early in the 1900s set the stage for vast amounts of research. By the end of the twentieth century, findings from many fields of biology firmly upheld Darwin's basic premises about the common ancestry of life and the role of natural selection as an important mechanism of evolution. Today a vast array of geological, morphological, and molecular data all support the factual basis of evolution.

When we refer to evolutionary theory, we are referring to our understanding of the mechanisms that result in genetic changes in populations over time. In many cases we can observe evolution directly, as with the influenza viruses described at the opening of this chapter. Over much longer periods, we can observe evolutionary changes in the fossil record. It is evolutionary theory, however, that allows us to apply our understanding of evolution to problems in medicine, agriculture, industry, and throughout biology.

Go to Media Clip 21.1
Watching Evolution in Real Time
Life10e.com/mc21.1

Darwin and Wallace introduced the idea of evolution by natural selection

In the early 1800s, it was not yet evident to many people that life evolves. But several biologists had suggested that the species living on Earth had changed over time—that is, that evolution had taken place. Jean-Baptiste Lamarck, for one, presented strong evidence for the fact of evolution in 1809, but his ideas about *how* evolution occurred were not convincing. At that time, no one had yet envisioned a viable mechanism for evolution.

In the 1820s, a young Charles Darwin became passionately interested in the subjects of geology (with its new sense of Earth's great age) and natural history (the scientific study of how different organisms function and carry out their lives in nature). Despite these interests, he planned, at his father's behest, to become a doctor. But surgery conducted without anesthesia nauseated Darwin, and he gave up medicine to study at Cambridge University for a career as a clergyman in the Church of England. Always more interested in science than in theology, he gravitated toward scientists on the faculty, especially the botanist John Henslow. In 1831, Henslow recommended Darwin for a position on HMS *Beagle*, a Royal Navy vessel that was preparing for a survey voyage around the world (**Figure 21.1**).

Whenever possible during the five-year voyage, Darwin went ashore to study rocks and to observe and collect plants and animals. He noticed striking differences between the species he saw in South America and those of Europe. He observed that the species of the temperate regions of South America (Argentina and Chile) were more similar to those of tropical South America (Brazil) than they were to temperate European species. When he explored the islands of the Galápagos archipelago, west of Ecuador, he noted that most of the animals were endemic (found nowhere else) to the islands, although they were similar to animals found on the mainland of South America. Darwin also observed that the fauna of the Galápagos differed from island to island. He postulated that in earlier times some animals from mainland South America had arrived on the archipelago and had subsequently undergone different and distinctive changes on each of the islands. He wondered what might account for these changes.

When he returned to England in 1836, Darwin continued to ponder his observations. His ruminations were strongly influenced by the geologist Charles Lyell, who a few years earlier had popularized the idea that Earth had been shaped by slow-acting forces that were still at work. Darwin reasoned that similar thinking could be applied to the living world. Over the next decade, he developed the framework of an explanatory theory for evolutionary change based on three major propositions:

- Species are not immutable; they change over time.
- Divergent species share a common ancestor and have diverged from one another gradually through time (a concept Darwin termed **descent with modification**).
- Changes in species over time can be explained by **natural selection**: the differential survival and reproduction of individuals based on variation in their traits.

Charles Robert Darwin

21.1 Darwin and the Voyage of the *Beagle* The mission of HMS *Beagle* was to chart the oceans and collect oceanographic and biological information from around the world. The world map indicates the ship's path; the inset map shows the Galápagos Islands, whose organisms were an important source of Darwin's ideas on natural selection. The portrait is of Charles Darwin at age 27, shortly after the *Beagle* returned to England.

Go to Activity 21.1 Darwin's Voyage
Life10e.com/ac21.1

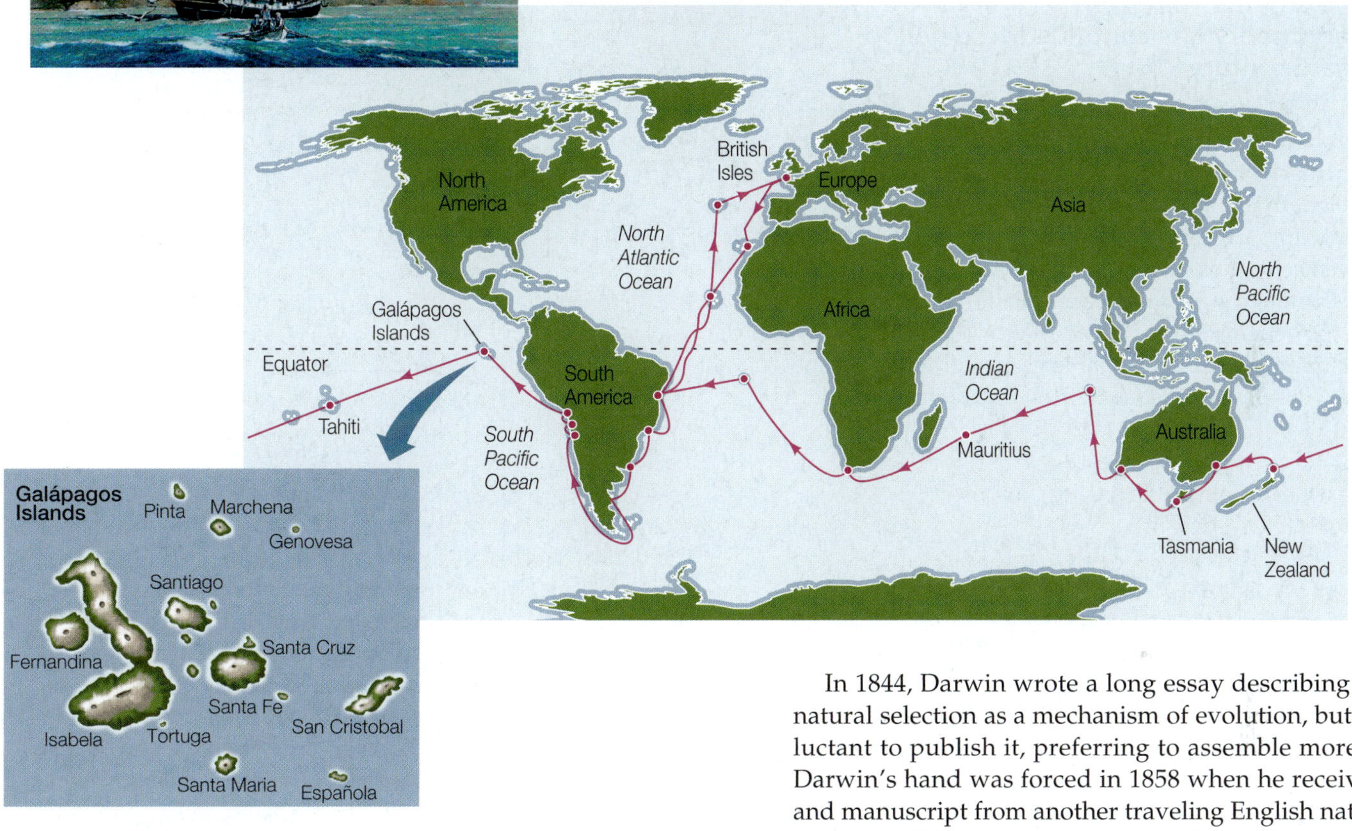
North America
British Isles
Europe
Asia
North Atlantic Ocean
Africa
North Pacific Ocean
Galápagos Islands
Equator
South America
Indian Ocean
Tahiti
South Pacific Ocean
Mauritius
Australia
Tasmania
New Zealand
Galápagos Islands
Pinta
Marchena
Genovesa
Santiago
Santa Cruz
Fernandina
Santa Fe
San Cristobal
Isabela
Tortuga
Santa Maria
Española

The first of these propositions was not unique to Darwin; several earlier authors had argued for the fact of evolution. A more revolutionary idea was his second proposition, that *divergent species are related to one another through common descent*. But Darwin is probably best known for his third proposition, that of natural selection.

Darwin realized that many more individuals of most species are born than survive to reproduce. He also knew that, although offspring usually resemble their parents, offspring are not identical to one another or to either parent. Finally, he was well aware of the fact that human breeders of plants and animals often selected their breeding stock based on the occurrence of particular traits. Over time, this selection resulted in dramatic changes in the appearance of the descendants of those plants or animals. In natural populations, wouldn't the individuals with the best chances of survival and reproduction be similarly "selected," and thus pass their traits on to the next generation? Darwin's simple but powerful idea was that nature did the selecting in natural populations on the basis of traits that resulted in greater survival and, eventually, greater likelihood of reproduction.

In 1844, Darwin wrote a long essay describing the role of natural selection as a mechanism of evolution, but he was reluctant to publish it, preferring to assemble more evidence. Darwin's hand was forced in 1858 when he received a letter and manuscript from another traveling English naturalist, Alfred Russel Wallace, who was studying the biota of the Malay Archipelago. Wallace asked Darwin to evaluate his manuscript, which included an explanation of natural selection almost identical to Darwin's. Darwin was at first dismayed, believing Wallace to have preempted his idea. Parts of Darwin's 1844 essay, together with Wallace's manuscript, were presented to the Linnean Society of London on July 1, 1858, thereby crediting both men for the idea of natural selection. Darwin then worked quickly to finish *On the Origin of Species*, which was published the following year.

Although Darwin and Wallace independently articulated the concept of natural selection, Darwin developed his ideas first. Furthermore, *On the Origin of Species* proved to be a stunning work of scholarship that provided exhaustive evidence from many fields supporting both the premise of evolution itself and the role of natural selection as a mechanism of evolution. Thus both concepts are more closely associated with Darwin than with Wallace.

Go to Animated Tutorial 21.1
Natural Selection
Life10e.com/at21.1

The publication of *On the Origin of Species* in 1859 stirred considerable interest (and controversy) among scientists and the public alike. Scientists spent much of the rest of the nineteenth century amassing biological and paleontological data to test evolutionary ideas and to document the history of life on Earth. By 1900, the fact of biological evolution (defined at that time as change in the physical characteristics of populations over time) was established beyond any reasonable doubt. But the *genetic* basis of evolutionary change was not yet understood.

Evolutionary theory has continued to develop over the past century

Shortly after 1900, several individuals rediscovered Mendel's work (which had been published in 1866 but rarely read or cited) and the basic mechanisms of genetic inheritance began to be unraveled. In the first decades of the twentieth century, Thomas Hunt Morgan's studies on fruit flies led to his discovery of the role of chromosomes in inheritance. In the 1920s and early 1930s, the major principles of population genetics were established, the genetic basis of new variation (i.e., mutations) began to be understood, and mechanisms of evolution such as genetic drift were described (see Section 21.2). This work set the stage for a "modern synthesis" of genetics and evolution that took place over the period 1936–1947. Some of the major contributors to this synthesis and a few of their books are listed in **Figure 21.2**.

Although chromosomes were soon understood to be the basis of genetic transmission in eukaryotes, their molecular structure remained a mystery until soon after the modern synthesis. Then, in 1953, James Watson and Francis Crick published their paper on the structure of DNA, opening the door to

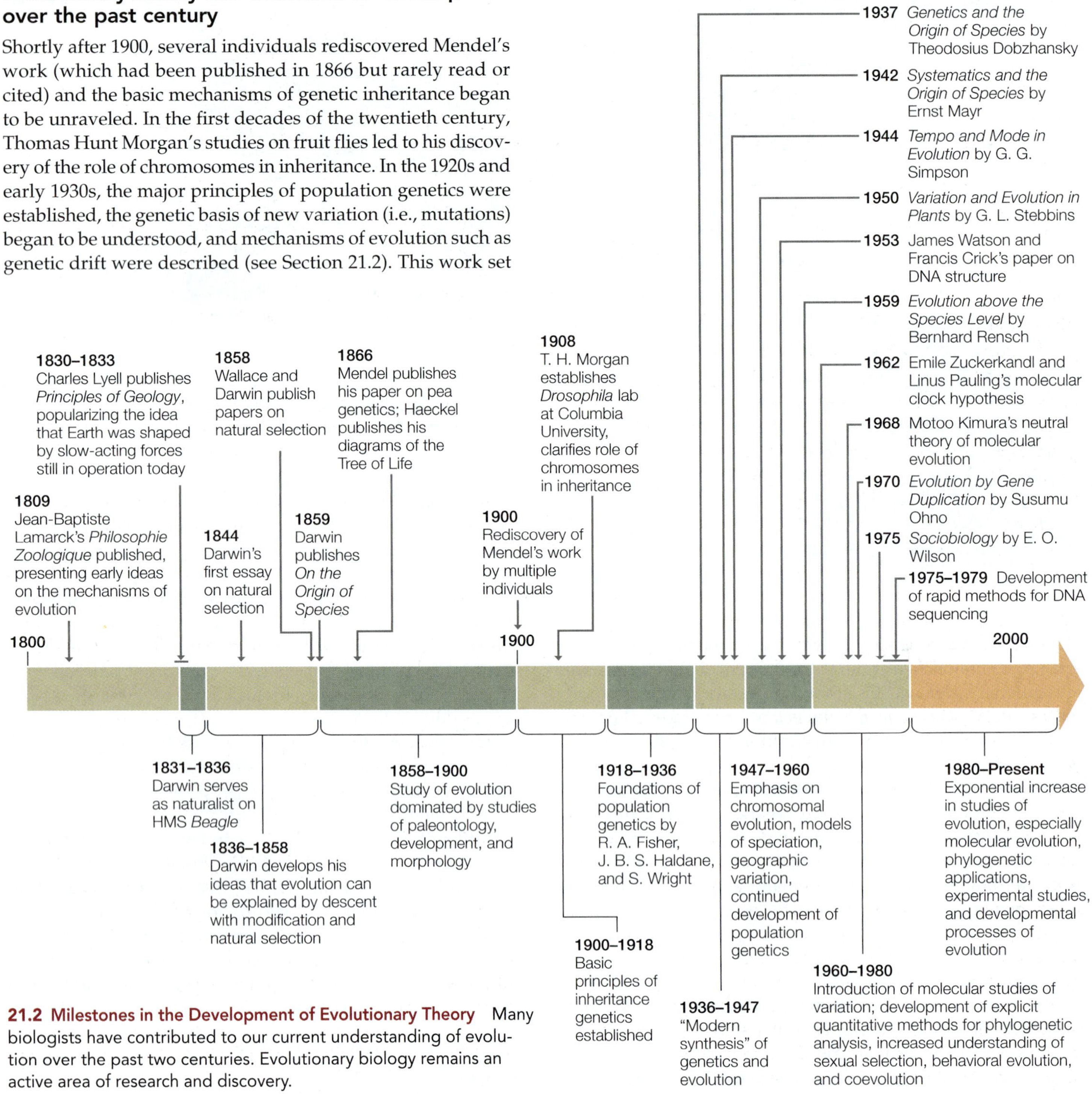

1937 Genetics and the Origin of Species by Theodosius Dobzhansky
1942 Systematics and the Origin of Species by Ernst Mayr
1944 Tempo and Mode in Evolution by G. G. Simpson
1950 Variation and Evolution in Plants by G. L. Stebbins
1953 James Watson and Francis Crick's paper on DNA structure
1959 Evolution above the Species Level by Bernhard Rensch
1962 Emile Zuckerkandl and Linus Pauling's molecular clock hypothesis
1968 Motoo Kimura's neutral theory of molecular evolution
1970 Evolution by Gene Duplication by Susumu Ohno
1975 Sociobiology by E. O. Wilson
1975–1979 Development of rapid methods for DNA sequencing
1830–1833 Charles Lyell publishes Principles of Geology, popularizing the idea that Earth was shaped by slow-acting forces still in operation today
1858 Wallace and Darwin publish papers on natural selection
1866 Mendel publishes his paper on pea genetics; Haeckel publishes his diagrams of the Tree of Life
1908 T. H. Morgan establishes Drosophila lab at Columbia University, clarifies role of chromosomes in inheritance
1809 Jean-Baptiste Lamarck's Philosophie Zoologique published, presenting early ideas on the mechanisms of evolution
1844 Darwin's first essay on natural selection
1859 Darwin publishes On the Origin of Species
1900 Rediscovery of Mendel's work by multiple individuals
1800
1900
2000
1831–1836 Darwin serves as naturalist on HMS Beagle
1836–1858 Darwin develops his ideas that evolution can be explained by descent with modification and natural selection
1858–1900 Study of evolution dominated by studies of paleontology, development, and morphology
1900–1918 Basic principles of inheritance genetics established
1918–1936 Foundations of population genetics by R. A. Fisher, J. B. S. Haldane, and S. Wright
1936–1947 "Modern synthesis" of genetics and evolution
1947–1960 Emphasis on chromosomal evolution, models of speciation, geographic variation, continued development of population genetics
1960–1980 Introduction of molecular studies of variation; development of explicit quantitative methods for phylogenetic analysis, increased understanding of sexual selection, behavioral evolution, and coevolution
1980–Present Exponential increase in studies of evolution, especially molecular evolution, phylogenetic applications, experimental studies, and developmental processes of evolution

21.2 Milestones in the Development of Evolutionary Theory Many biologists have contributed to our current understanding of evolution over the past two centuries. Evolutionary biology remains an active area of research and discovery.

our current detailed understanding of molecular evolutionary mechanisms. By the 1960s, biologists could study and document changes in allele frequencies in populations over time (see Section 21.3). Most of this early work necessarily focused on variants of proteins that differed within and between populations and species because, even though the molecular structure of DNA was known, it was not yet practical to sequence long stretches of DNA. Nonetheless, many important advances occurred in evolutionary theory during this time. These advances were not focused solely on a genetic understanding of evolution. E. O. Wilson's 1975 book *Sociobiology*, for example, invigorated studies of the evolution of behavior (a subject that had fascinated Darwin).

In the late 1970s, several techniques were developed that allowed the rapid sequencing of long stretches of DNA, which in turn allowed researchers to ascertain the amino acid sequences of proteins. This capability opened a new door for evolutionary biologists, who could now explore the structures of genes and proteins and document evolutionary changes within and between species in ways never before possible. In the past three decades, well over a quarter of a million scientific papers on evolutionary observations, experiments, and theory have been published.

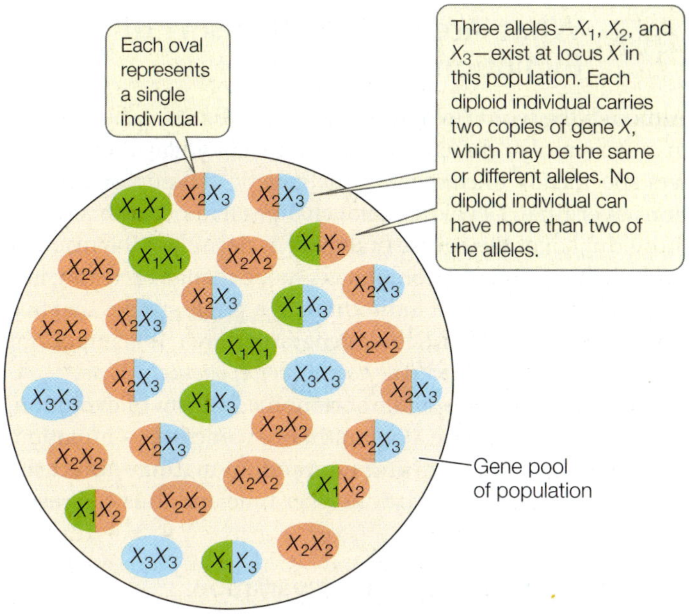
Each oval represents a single individual.
Three alleles—X_1, X_2, and X_3—exist at locus X in this population. Each diploid individual carries two copies of gene X, which may be the same or different alleles. No diploid individual can have more than two of the alleles.
Gene pool of population

21.3 A Gene Pool A gene pool is the sum of all the alleles found in a population or at a particular locus in that population. This figure shows the gene pool for one locus, X, in a population of diploid organisms. The allele frequencies in this case are 0.20 for X_1, 0.50 for X_2, and 0.30 for X_3 (see Figure 21.10).

Genetic variation contributes to phenotypic variation

For a population to evolve, its members must possess heritable genetic variation, which is the raw material on which mechanisms of evolution act. In everyday life, we do not directly observe the genetic composition of organisms. What we see are **phenotypes**, the physical expressions of organisms' genes (including interactions among those genes). The features of a phenotype are its **characters**—eye color, for example. The specific form of a character, such as brown eyes, is a **trait**. A **heritable trait** is a trait that is at least partly determined by the organism's genes. The genetic constitution that governs a character is called its **genotype**. *A population evolves when individuals with different genotypes survive or reproduce at different rates.*

Different forms of a gene, known as **alleles**, may exist at a **locus** (a particular site on a chromosome). At any given locus, a single diploid individual carries no more than two of all the alleles found in the population. The sum of all copies of all alleles at all loci found in a population constitutes that population's **gene pool**. We can also refer to the "gene pool" for a particular locus (**Figure 21.3**). The gene pool contains the genetic variation that produces the phenotypic traits on which natural selection acts. Evolution can be defined as changes in the proportions of alleles in the gene pool over time. Thus, to understand evolution and the roles of various evolutionary mechanisms, we need to know how much genetic variation populations have, what the sources of that genetic variation are, and how genetic variation changes in populations over space and time.

The study of the genetic basis of evolution is made more difficult by the fact that genotypes alone do not determine all phenotypes. When one allele is dominant to another, for example, a particular phenotype can be produced by more than one genotype (e.g., *AA* and *Aa* individuals may be phenotypically identical). In addition, as we described in Section 20.4, a given genotype can produce different phenotypes depending on the environmental conditions encountered during development. For example, the cells of all the leaves on a tree or shrub are usually genetically identical, yet leaves of the same plant often differ in shape and size depending, for example, on the amount of ambient light they receive.

RECAP 21.1

Evolutionary change is directly observable in biological populations. Natural selection is one of the major mechanisms that results in evolution. It acts on genetic variation, which is required for evolutionary change to occur.

- How would you respond to someone who said that evolution was "just a theory"? **See pp. 428–430**
- Articulate the principle of natural selection and explain how natural selection leads to evolutionary change. **See pp. 428–429**
- What do you think were the most important of the discoveries and technological breakthroughs that have provided biologists with a more thorough understanding of evolution since Darwin's time? **See pp. 430–431 and Figure 21.2**

Although the importance of natural selection to evolution has been confirmed in many thousands of scientific studies, it is not the only process that drives evolution. In the next section we'll consider a more complete view of evolutionary processes and how they operate.

21.2 What Are the Mechanisms of Evolutionary Change?

Although the word "evolution" is often used in a general sense to mean simply "change," evolution in a biological context refers specifically to changes in the genetic makeup of populations over time. Developmental changes that occur in a single individual over the course of the life cycle are not the result of evolutionary change. Evolution is genetic change occurring in a **population**—a group of individuals of a single species that live and interbreed in a particular geographic area. It is important to remember that *individuals do not evolve; populations do.*

Natural selection is an important mechanism of evolution, but it does not act alone. Four additional processes—mutation, gene flow, genetic drift, and nonrandom mating—affect the genetic makeup of populations over time and thus can result in evolution.

Mutation generates genetic variation

The source of genetic variation is mutation. As described in Section 15.1, a **mutation** is any change in the nucleotide sequence of an organism's DNA. The process of DNA replication is not perfect, and some changes appear almost every time a genome is replicated. Mutations occur randomly with respect to their costs or benefits to the organism; it is natural selection acting on this random variation that results in adaptation. Most mutations are either harmful to their bearers (deleterious mutations) or have no effect (neutral mutations). But a few mutations are beneficial, and even previously deleterious or neutral alleles may become advantageous if environmental conditions change. In addition, mutation can restore genetic variation that other evolutionary mechanisms have removed. Thus mutation both creates and helps maintain genetic variation in populations.

Mutation rates can be very high, particularly in viruses, some of which have mutation rates as high as 10^{-3} changes per nucleotide per generation. The rapid evolution of influenza viruses challenges effective vaccine production, as we saw at the opening of this chapter. In some genes of eukaryotes, the mutation rate is much lower (on the order of 10^{-8} to 10^{-9} changes per base pair per generation). Even low overall mutation rates, however, create considerable genetic variation because each of a large number of genes may change and because many populations contain large numbers of individuals. For example, if the probability of a point mutation (an addition, deletion, or substitution of a single base) were 10^{-9} per base pair per generation, then each human gamete—the DNA of which contains 3×10^9 base pairs—would average three new point mutations ($3 \times 10^9 \times 10^{-9} = 3$), and each diploid zygote would carry an average of six new mutations. The current human population of about 7 billion people would thus be expected to carry about 42 billion new mutations (i.e., changes in the nucleotide sequences of their DNA that were not present one generation earlier). So even though the mutation rate in humans is low, human populations still contain enormous genetic variation on which other evolutionary mechanisms can act.

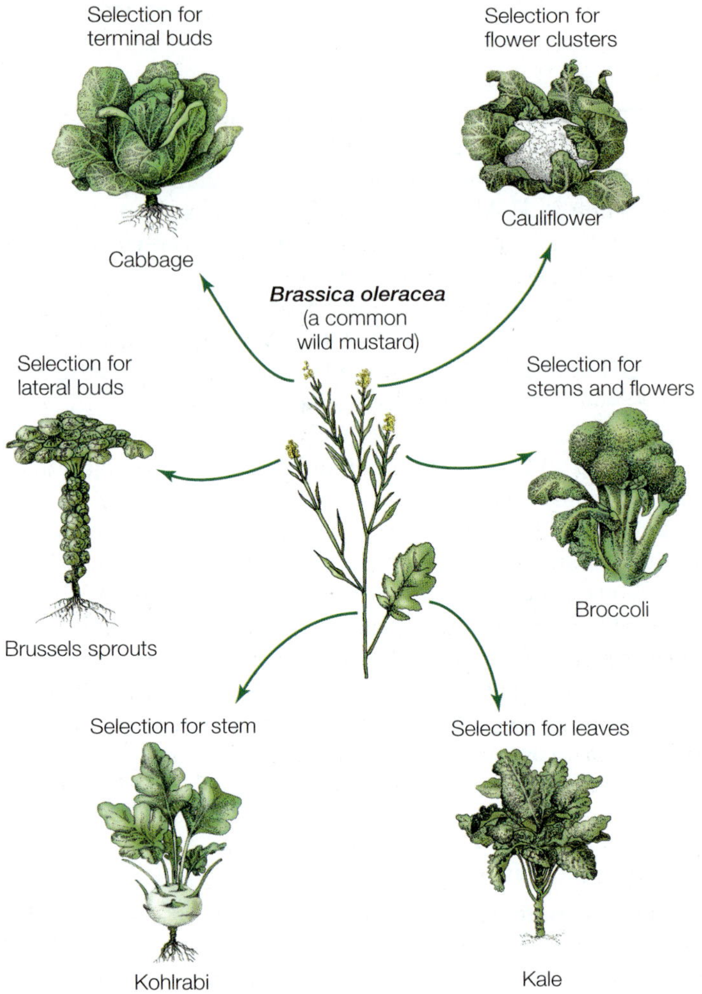
Selection for terminal buds
Cabbage
Selection for flower clusters
Cauliflower
Brassica oleracea (a common wild mustard)
Selection for lateral buds
Brussels sprouts
Selection for stems and flowers
Broccoli
Selection for stem
Kohlrabi
Selection for leaves
Kale

21.4 Many Vegetables from One Species All the crop plants shown here derive from a single wild mustard species. European agriculturalists produced these crop species by selecting and breeding plants with unusually large buds, stems, leaves, or flowers. The results substantiate the vast amount of variation present in the gene pool of the ancestral species.

Recall from Section 21.1 that the gene pool is the sum of the genetic variation in a population. Mutation adds new alleles to the gene pool. Biologists use two simple measures to characterize the variation in a given gene pool. The proportion of each allele in the gene pool is its **allele frequency**. Similarly, the proportion of each genotype among the individuals in the population is its **genotype frequency**. The calculations of allele and genotype frequencies in a population allow biologists to measure evolutionary change, as will be described in Section 21.3.

Selection acting on genetic variation leads to new phenotypes

As a result of mutation, the gene pools of nearly all populations contain variation for many characters. Selection on this variation can produce rapid evolutionary change, and evolution can proceed in many different directions. For example, **artificial selection**—the purposeful selection of specific phenotypes by humans—on different characters in a single European species of wild mustard has resulted in many different crop plants (**Figure 21.4**). Agriculturalists were able to achieve these results because the gene pool of the original mustard population

21.5 Artificial Selection Charles Darwin, who raised pigeons as a hobby, noted similar forces at work in artificial and natural selection. The "fancy" pigeons shown here represent 3 of the more than 300 varieties derived from the wild rock pigeon (*Columba livia*; left) by artificial selection on characters such as color and feather distribution.

contained variation in the alleles for the characters of interest (such as stem thickness or number of leaves).

As we noted in Section 21.1, Darwin compared this artificial selection by animal and plant breeders with natural selection. Many of Darwin's observations on the nature of variation and selection came from domesticated plants and animals. Darwin bred pigeons and thus knew firsthand the astonishing diversity in color, size, form, and behavior that breeders could achieve in these birds (**Figure 21.5**). He recognized close parallels between selection by breeders and selection in nature. Natural selection resulted in traits that helped organisms survive and reproduce more effectively; artificial selection resulted in traits that were preferred by the human breeders, for whatever reason.

Laboratory experiments also demonstrate the potential for selection to result in evolutionary change, often resulting in descendant populations containing phenotypes that were not present in the ancestral populations. In one such experiment, investigators, starting with a population of the fruit fly *Drosophila melanogaster* with intermediate numbers of abdominal bristles, bred two new populations. One population was selected for high and the other for low bristle numbers. After 35 generations, all flies in both the high- and low-bristle lineages had bristle numbers that fell well outside the range seen in the original population (**Figure 21.6**). In this experiment, artificial selection resulted in new combinations of the many different genes that were present in the original population, so that the phenotypic variation in subsequent generations fell outside that of the original population.

Natural selection works in much the same way. Slight trait differences among individuals increase the chance that a given individual will survive and reproduce, which increases the frequency of the favored trait in the next generation. A favored trait that evolves through natural selection is known as an **adaptation**; this word is used to describe both the trait itself and the process that produces the trait. Biologists regard an organism as being adapted to a particular environment when they can demonstrate that a slightly different organism reproduces and survives less well in that environment. To understand adaptation, biologists compare the performances of individuals that differ in their traits.

One consequence of natural seection is the purging of deleterious mutations from populations. Individuals with deleterious mutations are less likely to survive and reproduce, so they are less likely to pass their alleles on to the next generation. Biologists often distinguish between two broad categories of selection: **positive selection** (selection for beneficial changes) and **purifying selection** (selection against deleterious changes).

Gene flow may change allele frequencies

Few populations are completely isolated from other populations of the same species. Migration of individuals and movements of gametes (in pollen, for example) between populations—a phenomenon called **gene flow**—can change allele frequencies in a population. If the arriving individuals survive

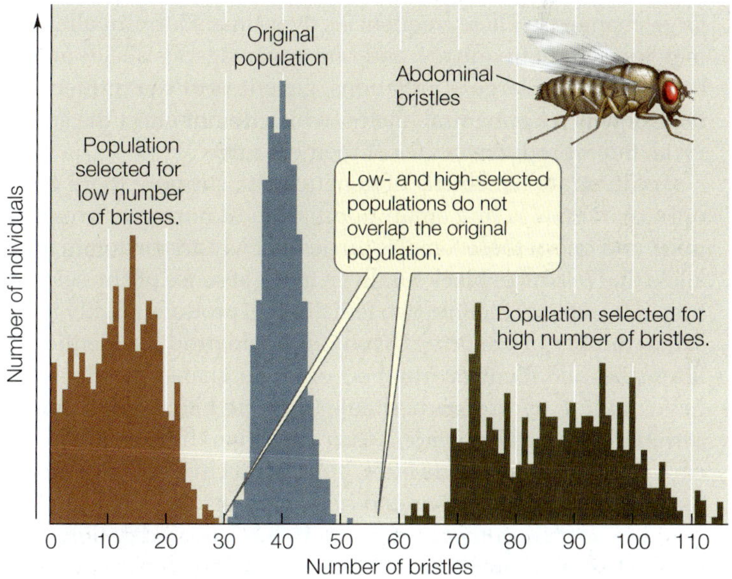

Original population
Abdominal bristles
Population selected for low number of bristles.
Low- and high-selected populations do not overlap the original population.
Population selected for high number of bristles.
Number of individuals
0
10
20
30
40
50
60
70
80
90
100
110
Number of bristles

21.6 Artificial Selection Reveals Genetic Variation When investigators subjected *Drosophila melanogaster* to artificial selection for abdominal bristle number, that character evolved rapidly. The graph shows the number of flies with different numbers of bristles in the original population and after 35 generations of artificial selection for low and for high bristle numbers.

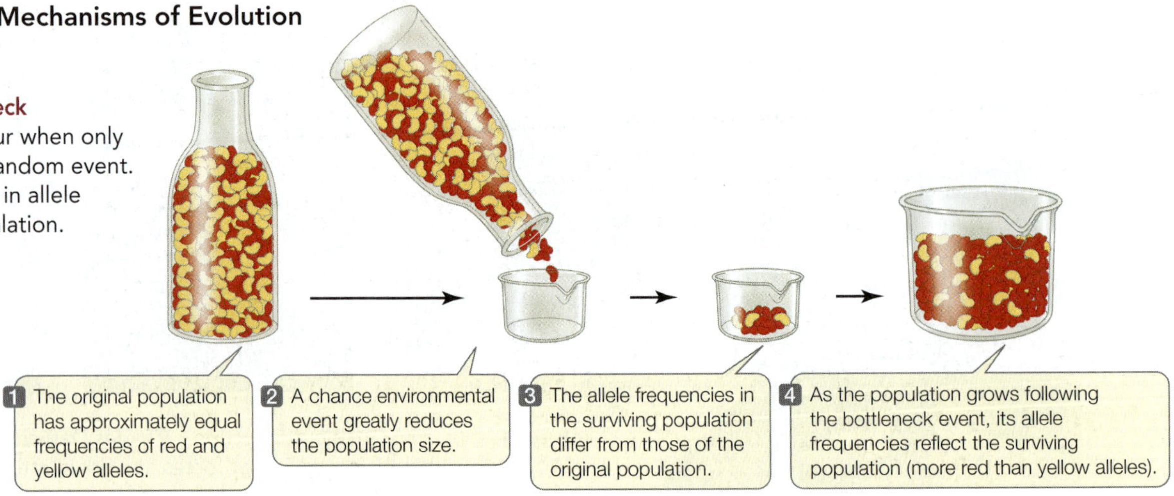
1 The original population has approximately equal frequencies of red and yellow alleles.
2 A chance environmental event greatly reduces the population size.
3 The allele frequencies in the surviving population differ from those of the original population.
4 As the population grows following the bottleneck event, its allele frequencies reflect the surviving population (more red than yellow alleles).

21.7 A Population Bottleneck Population bottlenecks occur when only a few individuals survive a random event. The result may be a change in allele frequencies within the population.

and reproduce in their new location, they may add new alleles to the population's gene pool, or they may change the frequencies of alleles present in the original population.

About 35,000 years ago, modern humans expanded their range into the range of Neanderthals in Europe and western Asia. Although modern humans largely replaced the Neanderthals over the next 7,000 years, some interbreeding between these populations occurred, as evidenced by the presence of a small percentage of Neanderthal genes in modern non-African human populations. This incorporation of Neanderthal genes is an example of gene flow. Traits such as red hair (which was common in Neanderthal populations) may have entered modern human populations in this manner.

Genetic drift may cause large changes in small populations

In small populations, **genetic drift**—random changes in allele frequencies from one generation to the next—may produce large changes in allele frequencies over time. Harmful alleles may increase in frequency, and rare advantageous alleles may be lost. Even in large populations, genetic drift can influence the frequencies of neutral alleles (which do not affect the survival and reproductive rates of their bearers).

To illustrate the effects of genetic drift, suppose there are only two females in a small population of normally brown mice, and one of these females carries a newly arisen dominant allele that produces black fur. Even in the absence of any selection, it is unlikely that the two females will produce exactly the same number of offspring. Even if they do produce identical litter sizes and identical numbers of litters, chance events that have nothing to do with heritable traits are likely to result in differential mortality among their offspring. If, for example, each female produces one litter, but a flood envelops the black female's nest and kills her and her offspring, the novel allele could be lost from the population in just one generation. In contrast, if the brown female's litter is lost, then the frequency of the newly arisen allele (and phenotype) for black fur will rise dramatically in just one generation

Genetic drift is particularly potent when a population is reduced dramatically in size. Even populations that are normally large may occasionally pass through environmental conditions that only a small number of individuals survive, a situation known as a **population bottleneck**. The effect of genetic drift in such a situation is illustrated in **Figure 21.7**, in which red and yellow beans represent two alleles of a gene. Most of the beans in the small sample of the "population" that "survives" the bottleneck event are, just by chance, red, so the new population has a much higher frequency of red beans than the previous generation had. In a real population, the red and yellow allele frequencies would be described as having "drifted."

A population forced through a bottleneck is likely to lose much of its genetic variation. For example, when Europeans first arrived in North America, millions of greater prairie-chickens (*Tympanuchus cupido*) inhabited the midwestern prairies. As a result of hunting and habitat destruction by the new settlers, the Illinois population of this species plummeted from about 100 million birds in 1900 to fewer than 50 individuals in the 1990s. A comparison of DNA from birds collected during the mid-twentieth century with DNA from the population surviving in the 1990s showed that Illinois prairie-chickens had lost most of their genetic diversity. Loss of genetic variation in small populations is one of the problems facing biologists who attempt to protect endangered species.

Genetic drift can have similar effects when a few pioneering individuals colonize a new region. Because of its small size, the colonizing population is unlikely to possess all the alleles found in the gene pool of its source population. The resulting reduction in genetic variation, called a **founder effect**, is equivalent to that in a large population reduced by a bottleneck. When a few humans migrated across the Bering Strait to colonize the Americas, for example, they brought with them a small sample of the genetic diversity that was present in Asian populations.

Go to Animated Tutorial 21.2
Genetic Drift
Life10e.com/at21.2

Nonrandom mating can change genotype or allele frequencies

Mating patterns often alter genotype frequencies because the individuals in a population do not choose mates at random. For example, self-fertilization is common in many groups of organisms, especially plants. Any time individuals mate preferentially with other individuals of the same genotype (including themselves), homozygous genotypes will increase, and heterozygous genotypes will decrease, in frequency over time. The

Euplectes progne

21.8 What Is the Advantage? The extensive tail of the male African long-tailed widowbird inhibits its ability to fly. Darwin attributed the evolution of this seemingly nonadaptive trait to sexual selection.

INVESTIGATING**LIFE**

21.9 Sexual Selection in Action Behavioral ecologist Malte Andersson tested Darwin's hypothesis that excessively long tails evolved in male widowbirds because female preference for longer-tailed males increased their mating and reproductive success.[a]

HYPOTHESIS Female widowbirds prefer to mate with the male that displays the longest tail; longer-tailed males thus are favored by sexual selection because they will father more offspring.

Method

1. Capture males and artificially lengthen or shorten tails by cutting or gluing on feathers. In a control group, cut and replace tails to their normal length (to control for the effects of tail-cutting).
2. Release the males to establish their territories and mate.
3. Count the nests with eggs or young on each male's territory.

Results Male widowbirds with artificially shortened tails established and defended display sites sucessfully but fathered fewer offspring than did control or unmanipulated males. Males with artificially lengthened tails fathered the most offspring.

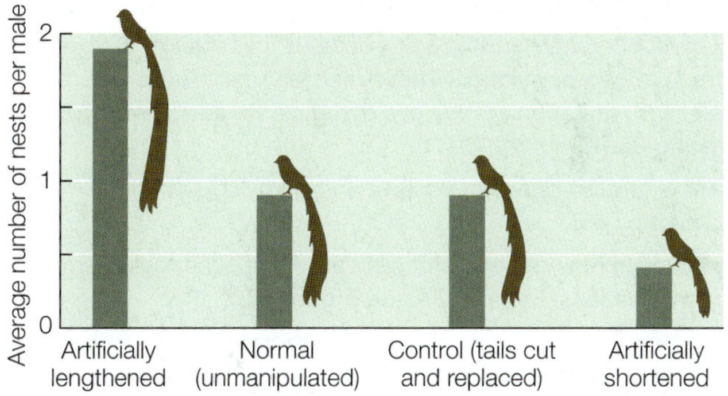

Average number of nests per male
2
1
0
Artificially lengthened
Normal (unmanipulated)
Control (tails cut and replaced)
Artificially shortened

CONCLUSION Sexual selection in *Euplectes progne* has favored the evolution of long tails in the male.

Go to **BioPortal** for discussion and relevant links for all INVESTIGATING**LIFE** figures.

[a]Andersson, M. 1982. *Nature* 299: 818–820

opposite effect (more heterozygotes, fewer homozygotes) is expected when individuals mate primarily or exclusively with individuals of different genotypes.

Nonrandom mating systems that do not affect the relative reproductive success of individuals produce changes in genotype frequencies but not in allele frequencies, and thus do not, by themselves, result in evolutionary change in a population. However, nonrandom mating systems that result in differential reproductive success among individuals do produce allele frequency changes from one generation to the next. **Sexual selection** occurs when individuals of one sex mate preferentially with particular individuals of the opposite sex rather than at random.

Sexual selection was first suggested by Charles Darwin, who developed the idea to explain the evolution of conspicuous traits that would appear to inhibit survival, such as bright colors, long tails, and elaborate courtship displays in males of many species. He hypothesized that these features either improved the ability of their bearers to compete for access to mates (intrasexual selection) or made their bearers more attractive to members of the opposite sex (intersexual selection). The concept of sexual selection was either ignored or questioned for many decades, but recent investigations have demonstrated its importance.

Darwin argued that while natural selection typically favors traits that enhance the survival of their bearers or their descendants, sexual selection is primarily about successful reproduction. An animal that survives but fails to reproduce makes no contribution to the next generation. Thus sexual selection may favor traits that enhance an individual's chances of reproduction even if they reduce its chances of survival. For example, females may be more likely to see or hear males with a conspicuous trait (and thus be more likely to mate with those males), even though the conspicuous trait may increase the chances that the male will be seen or heard by a predator. The sexual signal may also indicate a successful genotype in the male. In many species of frogs, for example, females prefer males with low-frequency calls. Male calls vary with body size, and a low-frequency call is indicative of a large-bodied frog. Frogs exhibit indeterminate growth—that is, they continue to grow indefinitely—so a large frog is a long-lived frog. In this case, the male's call represents what is known as an honest signal of the male's ability to survive in the local environment.

One trait that Darwin attributed to sexual selection is the remarkable tail of the male African long-tailed widowbird, which is longer than the bird's head and body combined (**Figure 21.8**) and impairs its ability to fly. Male widowbirds normally select and defend a territory where they perform courtship displays to attract females. Malte Andersson investigated whether sexual selection drove the evolution of widowbird tails (**Figure 21.9**). He captured male widowbirds and clipped the tails of some individuals while artificially lengthening the tails of others (by gluing on additional feathers). He then cut and re-glued the tail feathers of still other males, which served as controls. When released, both short- and long-tailed males successfully defended their display territories, indicating that a long tail does not confer an advantage in male–male competition.

However, males with artificially elongated tails attracted about four times more females than did males with shortened tails.

Why do female widowbirds prefer males with long tails? Biologists have developed several hypotheses. One possibility is that the ability to grow and maintain a costly feature such as a long tail may indicate that the male bearing it is vigorous and healthy. If so, then females that are attracted to long tails are indirectly attracted to vigorous, healthy males, which probably carry beneficial genes that should lead to high survivorship among their offspring. Another possibility is that females judge a male's body size by its overall length and they prefer larger males. It is also possible that a long tail stimulates the female visual system for reasons that are unrelated to any aspect of male health or vigor; the reasons why females prefer sexually selected traits are not always clear. What *is* clear is that female preferences often lead to the evolution of dramatic male adornments and mating displays.

RECAP 21.2

Evolutionary mechanisms are processes that change the genetic structure of a population. These mechanisms include natural selection, mutation, gene flow, genetic drift, and nonrandom mating.

- Distinguish between positive and purifying selection. **See p. 433**
- Explain how genetic drift can cause large changes in small populations. **See p. 434 and Figure 21.7**
- Describe some other examples (beyond those described in this book) of traits that may have evolved through sexual selection.

RESEARCHTOOLS

21.10 Calculating Allele and Genotype Frequencies Allele and genotype frequencies for a gene locus with two alleles in the population can be calculated using the equations in panel 1. When the equations are applied to two populations (panel 2), we find that the *frequencies of alleles A and a in the two populations* are the same, but the alleles are distributed differently between heterozygous and homozygous genotypes.

1 In any population, where N is the total number of individuals in the population:

$$\text{Frequency of allele } A = p = \frac{2N_{AA} + N_{Aa}}{2N} \qquad \text{Frequency of allele } a = q = \frac{2N_{aa} + N_{Aa}}{2N}$$

Frequency of genotype $AA = N_{AA}/N$
Frequency of genotype $Aa = N_{Aa}/N$
Frequency of genotype $aa = N_{aa}/N$

2 Compute the allele and genotype frequencies for two separate populations of $N = 200$:

Population 1 (mostly homozygotes)	Population 2 (mostly heterozygotes)
$N_{AA} = 90$, $N_{Aa} = 40$, and $N_{aa} = 70$	$N_{AA} = 45$, $N_{Aa} = 130$, and $N_{aa} = 25$
$p = \frac{180 + 40}{400} = 0.55$	$p = \frac{90 + 130}{400} = 0.55$
$q = \frac{140 + 40}{400} = 0.45$	$q = \frac{50 + 130}{400} = 0.45$
Freq. AA = 90/200 = 0.45	Freq. AA = 45/200 = 0.225
Freq. Aa = 40/200 = 0.20	Freq. Aa = 130/200 = 0.65
Freq. aa = 70/200 = 0.35	Freq. aa = 25/200 = 0.125

The evolutionary mechanisms discussed so far act by changing the frequencies of alleles and genotypes in populations. How are these changes measured by biologists? In other words, how do we know that evolution is occurring?

21.3 How Do Biologists Measure Evolutionary Change?

Much of evolution occurs through gradual changes in the relative frequencies of different alleles in a population from one generation to the next. As we'll see in Chapter 24, major genetic changes can also be sudden, as happens, for example, when two formerly separated populations merge and hybridize. But in most cases, we can measure evolution by looking at gradual changes in allele and genotype frequencies (or of their respective phenotypes) in populations.

Evolutionary change can be measured by allele and genotype frequencies

To measure allele frequencies in a population precisely, we would need to count every allele at every locus in every individual in the population. Fortunately, we do not need to make such complete measurements because we can reliably estimate allele frequencies for a given locus by counting alleles in a sample of individuals from the population. The sum of all allele frequencies at a locus is equal to 1, so measures of allele frequency range from 0 to 1.

An allele's frequency is calculated using the following formula:

$$p = \frac{\text{number of copies of the allele in the population}}{\text{total number of copies of all alleles in the population}}$$

If only two alleles (we'll call them A and a) are found among the members of a diploid population, those alleles can combine to form three different genotypes: AA, Aa, and aa. A population with more than one allele at a locus is said to be polymorphic ("many forms") at that locus. Applying the formula above, as shown in **Figure 21.10**, we can calculate the relative frequencies of alleles A and a in a population of N individuals:

- Let N_{AA} be the number of individuals that are homozygous for the A allele (AA).
- Let N_{Aa} be the number that are heterozygous (Aa).
- Let N_{aa} be the number that are homozygous for the a allele (aa).

Note that $N_{AA} + N_{Aa} + N_{aa} = N$, the total number of individuals in the population, and that the total number of copies of both alleles present in the population is $2N$, because each individual is diploid. Each AA individual has two copies of the A allele, and each Aa individual has one copy of the A allele. Therefore, the total number of A alleles in the population is $2N_{AA} + N_{Aa}$. Similarly, the total number of a alleles in the population is $2N_{aa} + N_{Aa}$. If p represents the frequency of A and q represents the frequency of a, then

$$q = \frac{2N_{aa} + N_{Aa}}{2N}$$

and

$$p = \frac{2N_{AA} + N_{Aa}}{2N}$$

The calculations in Figure 21.10 demonstrate two important points. First, notice that for each population, $p + q = 1$, which means that $q = 1 - p$. So when there are only two alleles at a given locus in a population, we can calculate the frequency of one allele and obtain the second allele's frequency by subtraction. If there is only one allele at a given locus in a population, its frequency is 1: The population is then monomorphic ("one form") at that locus, and the allele is said to be **fixed**.

The second thing to notice is that population 1 (consisting mostly of homozygotes) and population 2 (consisting mostly of heterozygotes) have the same allele frequencies for A and a; thus they have the same gene pool for this locus. Because the alleles in the gene pool are distributed differently among individuals, however, the genotype frequencies of the two populations differ.

The frequencies of the different alleles at each locus and the frequencies of the different genotypes in a population describe that population's **genetic structure**. Allele frequencies measure the amount of genetic variation in a population; genotype frequencies show how a population's genetic variation is distributed among its members. Other measures, such as the proportion of loci that are polymorphic, are also used to measure variation in populations. With these measurements, it becomes possible to consider how the genetic structure of a population changes or remains the same over generations—that is, to measure evolutionary change.

Evolution will occur unless certain restrictive conditions exist

In 1908 the British mathematician Godfrey Hardy and the German physician Wilhelm Weinberg independently deduced the conditions that must prevail if the genetic structure of a population is to remain the same over time. If the conditions they identified do not exist, then evolution will occur. The resulting principle is known as **Hardy–Weinberg equilibrium**. Hardy–Weinberg equilibrium constitutes a model in which allele frequencies do not change across generations, and genotype frequencies can be predicted from allele frequencies. The expectations of Hardy–Weinberg equilibrium apply only to sexually reproducing organisms. Several conditions must be met for a population to be at Hardy–Weinberg equilibrium (which, you should notice, correspond inversely to the five principal mechanisms of evolution discussed in Section 21.2):

- *There is no mutation*. The alleles present in the population do not change and no new alleles are added to the gene pool.
- *There is no selection among genotypes*. Individuals with different genotypes have equal probabilities of survival and equal rates of reproduction.
- *There is no gene flow*. There is no movement of individuals or gametes into or out of the population or reproductive contact with other populations.
- *Population size is infinite*. The larger a population, the smaller will be the effect of genetic drift.
- *Mating is random*. Individuals do not preferentially choose mates with certain genotypes.

If these idealized conditions hold, two major consequences follow. First, the frequencies of alleles at a locus remain constant from generation to generation—that is, no evolutionary change occurs in the population. Second, following one generation of random mating, the genotypes occur at the following frequencies:

Genotype	AA	Aa	aa
Frequency	p^2	$2pq$	q^2

To understand why these consequences are important, start by considering a population that is *not* in Hardy–Weinberg equilibrium, such as generation I in **Figure 21.11**. This could occur, for example, if the initial population is founded by migrants from several other populations, thus violating the Hardy–Weinberg assumption of no gene flow. In this example, generation I has more homozygous individuals and fewer heterozygous individuals than would be expected under Hardy–Weinberg equilibrium (a condition known as heterozygote deficiency).

Even with a starting population that is not in Hardy–Weinberg equilibrium, we can predict that after a single generation of random mating, if the other Hardy–Weinberg assumptions are not violated, the allele frequencies will remain unchanged, but the genotype frequencies will return to Hardy–Weinberg expectations. Let's explore why this is true.

In generation I of Figure 21.11, the frequency of the A allele (p) is 0.55. Because we assume that individuals select mates at random, without regard to their genotype, gametes carrying A or a combine at random—that is, as predicted by the allele frequencies p and q. In this example, the probability that a particular sperm or egg will bear an A allele is 0.55. In other words, 55 out of 100 randomly sampled sperm or eggs will bear an A allele. Because $q = 1 - p$, the probability that a sperm or egg will bear an a allele is $1 - 0.55 = 0.45$.

To obtain the probability of two A-bearing gametes coming together at fertilization, we multiply the two independent probabilities of their occurrence:

$$p \times p = p^2 = (0.55)^2 = 0.3025$$

Therefore 0.3025, or 30.25 percent, of the offspring in generation II will have homozygous genotype AA. Similarly, the probability of two a-bearing gametes coming together is

$$q \times q = q^2 = (0.45)^2 = 0.2025$$

which means that 20.25 percent of generation II will have the aa genotype.

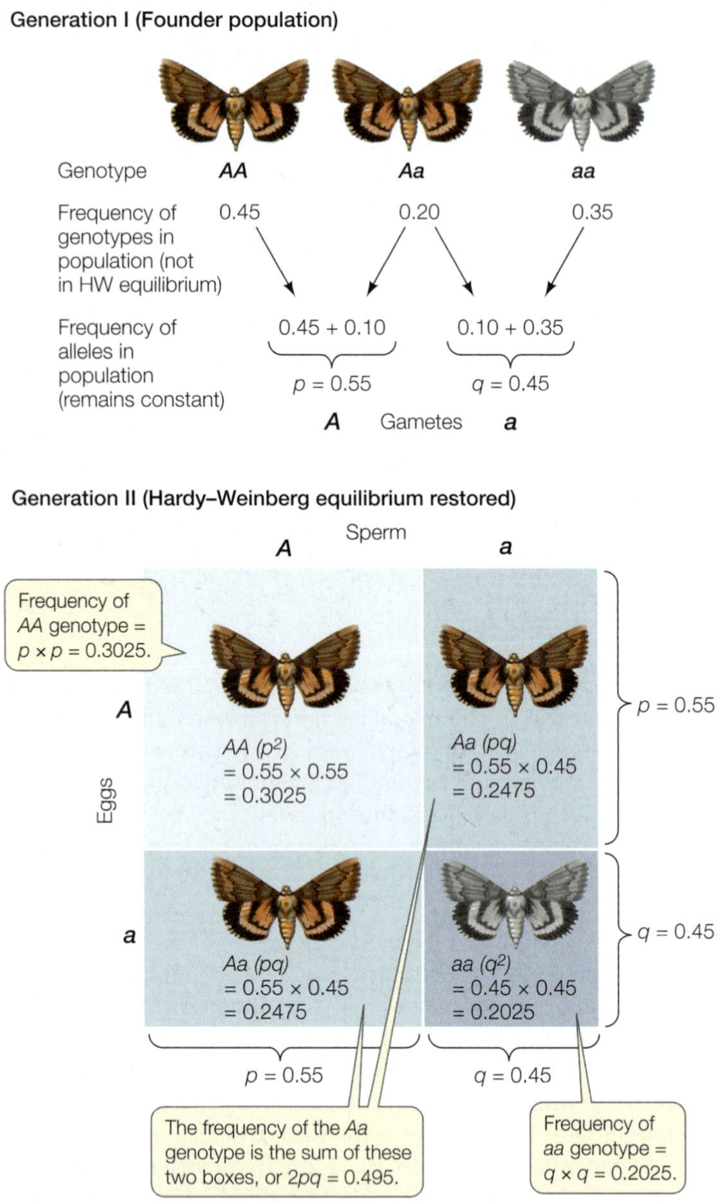
Generation I (Founder population)
Genotype
AA
Aa
aa
Frequency of genotypes in population (not in HW equilibrium)
0.45
0.20
0.35
Frequency of alleles in population (remains constant)
0.45 + 0.10
0.10 + 0.35
p = 0.55
q = 0.45
A
Gametes
a
Generation II (Hardy–Weinberg equilibrium restored)
Sperm
A
a
Eggs
A
a
Frequency of AA genotype = p × p = 0.3025.
AA (p2) = 0.55 × 0.55 = 0.3025
Aa (pq) = 0.55 × 0.45 = 0.2475
Aa (pq) = 0.55 × 0.45 = 0.2475
aa (q2) = 0.45 × 0.45 = 0.2025
p = 0.55
q = 0.45
p = 0.55
q = 0.45
The frequency of the Aa genotype is the sum of these two boxes, or 2pq = 0.495.
Frequency of aa genotype = q × q = 0.2025.

21.11 One Generation of Random Mating Restores Hardy–Weinberg Equilibrium Generation I of this population is made up of migrants from several source populations and so is not initially in Hardy–Weinberg equilibrium. After one generation of random mating, the allele frequencies are unchanged, and the genotype frequencies return to Hardy–Weinberg expectations. The lengths of the sides of each rectangle are proportional to the allele frequencies in the population; the areas of the rectangles are proportional to the genotype frequencies.

There are two ways of producing a heterozygote. An *A* sperm may combine with an *a* egg, the probability of which is $p \times q$; or an *a* sperm may combine with an *A* egg, the probability of which is $q \times p$. Consequently, the overall probability of obtaining a heterozygote is $2pq$, or 0.495 in this example. The frequencies of the *AA*, *Aa*, and *aa* genotypes in generation II of Figure 21.11 now meet Hardy–Weinberg expectations, and the frequencies of the two alleles (p and q) have not changed from generation I.

Under the assumptions of Hardy–Weinberg equilibrium, allele frequencies p and q remain constant from generation to generation. If Hardy–Weinberg assumptions are violated and the genotype frequencies in the parental generation are altered (say, by the loss of a large number of *AA* individuals from the population), then the allele frequencies in the next generation will be altered. However, based on the new allele frequencies, another generation of random mating will be sufficient to restore the genotype frequencies to Hardy–Weinberg equilibrium.

Go to Animated Tutorial 21.3
Hardy–Weinberg Equilibrium
Life10e.com/at21.3

Deviations from Hardy–Weinberg equilibrium show that evolution is occurring

You probably have realized that populations in nature never meet the stringent conditions necessary to be at Hardy–Weinberg equilibrium—which explains why all biological populations evolve. Why, then, is this model considered so important for the study of evolution? There are two reasons. First, the expectations of Hardy–Weinberg equilibrium are useful for predicting the approximate genotype frequencies of a population from its allele frequencies. Second—and crucially—the model allows biologists to evaluate which mechanisms of evolution are acting on a particular population. Specific patterns of deviation from Hardy–Weinberg equilibrium can help us identify the various mechanisms of evolutionary change.

Natural selection acts directly on phenotypes

Although evolution is defined as changes in the genetic makeup of a population from one generation to the next, natural selection acts directly on the phenotype—that is, on the physical features expressed by an organism with a given genotype—and therefore acts only indirectly on the genotype. The reproductive contribution of a phenotype to subsequent generations relative to the contributions of other phenotypes is called its **fitness**.

Changes in reproductive rate do not necessarily change the genetic structure of a population. For example, if all individuals in a population experience the same increase in reproductive rate (during an environmentally favorable year, for instance), the genetic structure of the population will not change. Changes in numbers of offspring are responsible for increases and decreases in the *size* of a population, but only changes in the *relative* success of different phenotypes in a population will lead to changes in allele frequencies from one generation to the next. The fitness of individuals of a particular phenotype is a function of the probability of those individuals surviving multiplied by the average number of offspring they produce over their lifetimes. In other words, the *fitness of a phenotype is determined by the relative rates of survival and reproduction of individuals with that phenotype.*

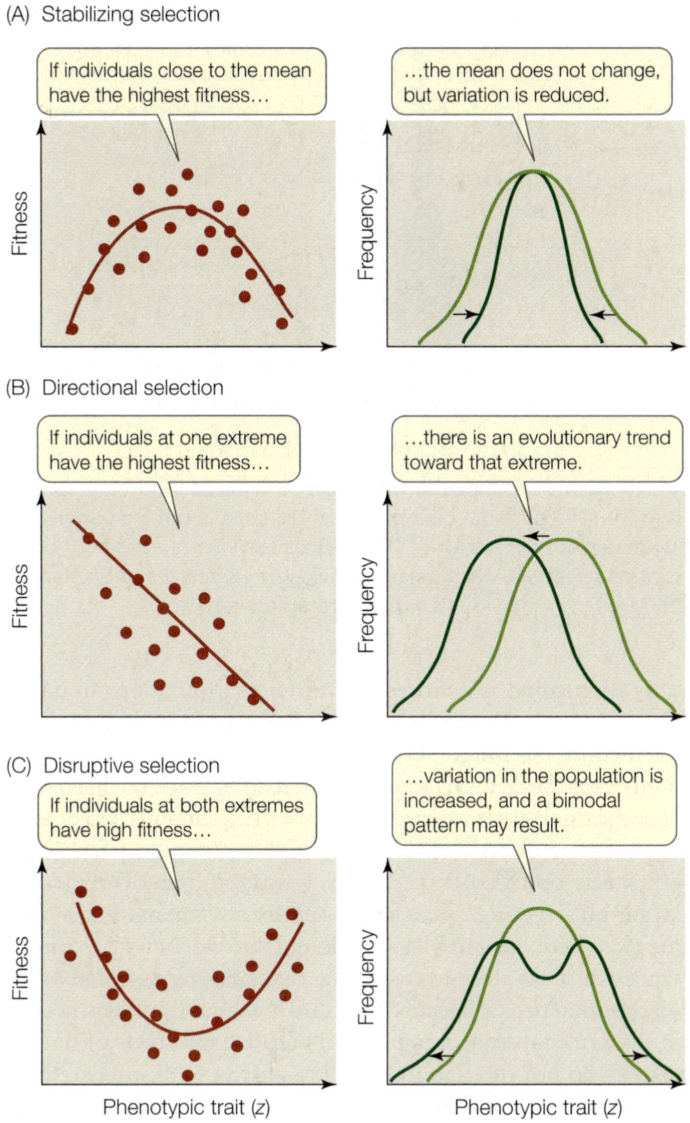

(A) Stabilizing selection
If individuals close to the mean have the highest fitness...
...the mean does not change, but variation is reduced.
Fitness
Frequency
(B) Directional selection
If individuals at one extreme have the highest fitness...
...there is an evolutionary trend toward that extreme.
Fitness
Frequency
(C) Disruptive selection
If individuals at both extremes have high fitness...
...variation in the population is increased, and a bimodal pattern may result.
Fitness
Frequency
Phenotypic trait (z)
Phenotypic trait (z)

21.12 Natural Selection Can Operate in Several Ways The graphs in the left-hand column show the fitness of individuals with different phenotypes for the same character. The right-hand graphs show the distribution of the phenotypes in the population before (light green) and after (dark green) the influence of selection.

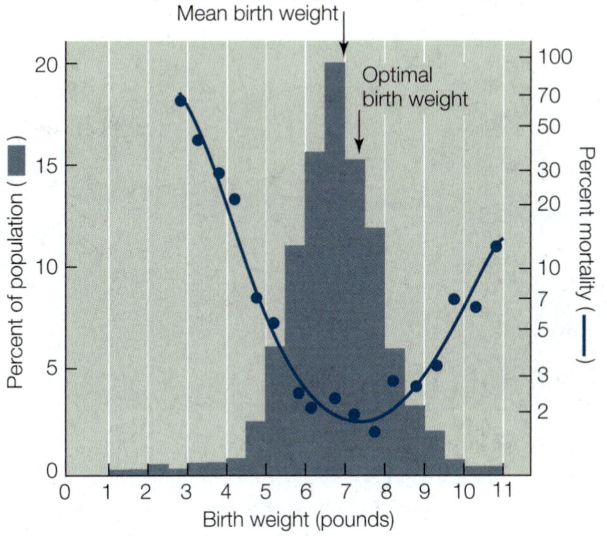

Mean birth weight
Optimal birth weight
Percent of population
Percent mortality
Birth weight (pounds)
0
5
10
15
20
100
70
50
30
20
10
7
5
3
2
0 1 2 3 4 5 6 7 8 9 10 11

21.13 Human Birth Weight Is Influenced by Stabilizing Selection Babies that weigh more or less than average are more likely to die soon after birth than are babies with weights closer to the population mean.

Natural selection can change or stabilize populations

Until now, our discussion has focused on changes in alleles at a single genetic locus. Phenotypic traits controlled by a single locus are often distinguished by discrete qualities (black versus white, or smooth versus wrinkled), and are called qualitative traits. Many traits, however, are influenced by alleles at more than one locus. Such traits are likely to show continuous, quantitative variation rather than discrete qualitative variation, and so are known as quantitative traits. Body size, for example, is influenced by genes at many loci as well as by the environment (nutrition, for example). Therefore the distribution of body sizes of individuals in a population is likely to resemble a continuous bell-shaped curve.

Natural selection can act on characters with quantitative variation in any one of several different ways, producing quite different results (**Figure 21.12**):

- **Stabilizing selection** preserves the average characteristics of a population by favoring average individuals.
- **Directional selection** changes the characteristics of a population by favoring individuals that vary in one direction from the mean of the population.
- **Disruptive selection** changes the characteristics of a population by favoring individuals that vary in both directions from the mean of the population.

STABILIZING SELECTION If the smallest and largest individuals in a population contribute fewer offspring to the next generation than do individuals closer to the average body size, then stabilizing selection is operating on body size (see Figure 21.12A). Stabilizing selection reduces variation in populations, but it does not change the mean. Natural selection frequently acts in this way, countering increases in variation brought about by sexual recombination, mutation, or gene flow. Rates of phenotypic change in many species are slow because natural selection is often stabilizing. Stabilizing selection operates, for example, on human birth weight. Babies who are lighter or heavier at birth than the population mean die at higher rates than babies whose weights are close to the mean (**Figure 21.13**).

DIRECTIONAL SELECTION Directional selection is operating when individuals at one extreme of a character distribution contribute more offspring to the next generation than other individuals do, shifting the average value of that character in the population toward that extreme. We noted in Section 21.2 that, in the case of a single locus, selection for a particular genetic variant is referred to as positive selection. Positive selection for many genetic variants at many loci, or for new combinations of those variants, results in the overall directional selection of a quantitative trait. By favoring one phenotype over another, directional selection results in an increase of the frequencies of alleles that produce the favored phenotype (as with the surface proteins of influenza discussed at the opening of this chapter).

21.14 A Result of Directional Selection In the American Southwest, long horns were advantageous for defending calves from attacks by predators, so cows with longer horns were more likely to raise calves successfully. As a result, horn length in feral herds of cattle increased between the early 1500s and the 1860s, leading to the Texas Longhorn breed. This evolutionary trend has been maintained by modern ranchers practicing artificial selection.

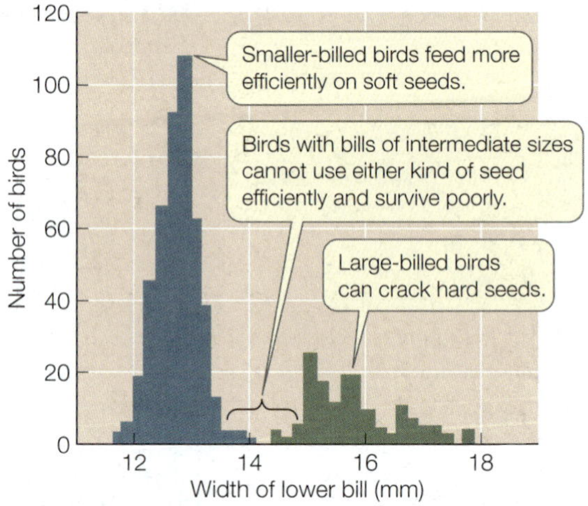

Smaller-billed birds feed more efficiently on soft seeds.
Birds with bills of intermediate sizes cannot use either kind of seed efficiently and survive poorly.
Large-billed birds can crack hard seeds.
Number of birds
0
20
40
60
80
100
120
12
14
16
18
Width of lower bill (mm)

21.15 Disruptive Selection Results in a Bimodal Character Distribution The bimodal distribution of bill sizes in the black-bellied seedcracker of West Africa (*Pyrenestes ostrinus*) is a result of disruptive selection, which favors individuals with larger and smaller bill sizes over individuals with intermediate-sized bills.

If directional selection operates over many generations, an evolutionary trend is seen in the population (see Figure 21.12B). Evolutionary trends can be reversed if the environment changes so that different phenotypes are favored. Or they can be halted when an optimal phenotype is reached or when trade-offs oppose further change (see Section 21.5). The character then undergoes stabilizing selection.

The long horns of Texas Longhorn cattle (**Figure 21.14**) are an example of a trait that has evolved through directional selection. Texas Longhorns are descendants of cattle brought to the New World by Christopher Columbus, who picked up a few cattle in the Canary Islands and brought them to the island of Hispaniola in 1493. The cattle multiplied, and their descendants were taken to the mainland of Mexico. Spaniards exploring what would become Texas and the southwestern United States brought these cattle with them, some of which escaped and formed feral herds. Populations of feral cattle increased greatly over the next few hundred years, but they faced heavy predation from bears, mountain lions, and wolves, especially on the young calves. Cows with longer horns were more successful in protecting their calves against attacks. Over a few hundred years, the average horn length in the feral herds increased considerably. In addition, these cattle evolved resistance to the endemic diseases of the Southwest, as well as higher fecundity and longevity. Texas Longhorns often live and produce calves well into their twenties—about twice as long as many breeds of cattle that have been artificially selected by humans for traits such as high fat content or high milk production (which are examples of artificial directional selection).

DISRUPTIVE SELECTION Disruptive selection is operating when individuals at opposite extremes of a character distribution contribute more offspring to the next generation than do individuals close to the mean. Directional selection increases variation in the population (see Figure 21.12C).

The strikingly bimodal (two-peaked) distribution of bill sizes in the black-bellied seedcracker, a West African finch, illustrates how disruptive selection can influence populations in nature (**Figure 21.15**). The seeds of two types of sedges (marsh plants) are the most abundant food source for these finches during part of the year. Birds with large bills can readily crack the hard seeds of the sedge *Scleria verrucosa*. Birds with small bills can crack *S. verrucosa* seeds only with difficulty; however, they feed more efficiently on the soft seeds of *S. goossensii* than do birds with larger bills. Young finches whose bills deviate markedly from the two predominant bill sizes do not survive as well as finches whose bills are close to one of the two sizes represented by the distribution peaks. Because there are few abundant food sources in the finches' environment, and because the seeds of the two sedges do not overlap in hardness, birds with intermediate-sized bills are less efficient in using either one of the species' principal food sources. Disruptive selection therefore maintains a bimodal bill-size distribution.

RECAP 21.3

Hardy–Weinberg equilibrium describes the theoretical conditions for a non-evolving population. Deviations from Hardy–Weinberg expectations provide information about how evolution is occurring in a given population. Natural selection can both change and stabilize phenotypes within populations.

- Why is the concept of Hardy–Weinberg equilibrium important even though the assumptions on which it is based are never completely met in nature? **See pp. 437–438**
- Explain why natural selection that acts on a phenotype results in changes in genotype frequencies. **See p. 438**
- Describe the differences between stabilizing, directional, and disruptive selection, giving examples of each. **See pp. 439–440 and Figure 21.12**

Genetic drift, stabilizing selection, and directional selection all tend to reduce genetic variation within populations. Nevertheless, as we have seen, most populations harbor considerable genetic variation. What processes produce and maintain genetic variation within populations?

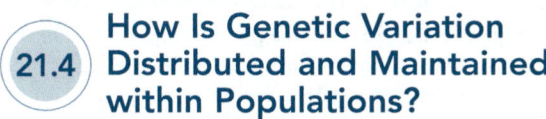

21.4 How Is Genetic Variation Distributed and Maintained within Populations?

Genetic variation is the raw material on which mechanisms of evolution act. In this section we will discuss several factors—neutral mutations, sexual recombination, frequency-dependent selection, and heterozygote advantage—that affect how genetic variation is established, how it is distributed among individuals, and how it is maintained within populations.

Neutral mutations accumulate in populations

An allele that does not affect the fitness of an organism—that is, an allele that is no better or worse than alternative alleles at the same locus—is called a **neutral allele**. Neutral alleles are added to a population over time through mutation, providing the population with considerable genetic variation. The frequencies of neutral alleles are not affected directly by natural selection. Even in large populations, neutral alleles may be lost, or may increase in frequency, purely by genetic drift.

Much of the phenotypic variation we are able to observe is not neutral. However, modern techniques enable us to measure neutral variation at the molecular level and provide the means to distinguish it from adaptive variation. Section 24.2 will describe how variation in neutral molecular traits can be used to study divergence among genes, populations, and species.

Sexual recombination amplifies the number of possible genotypes

In asexually reproducing organisms, each new individual is genetically identical to its parent unless there has been a mutation. When organisms reproduce sexually, however, offspring differ from their parents not only because they result from the combination of genetic material from two different gametes, but also through crossing over and independent assortment of chromosomes during meiosis, as described in Section 11.5. Sexual recombination generates an endless variety of genotypic combinations that increase the evolutionary potential of populations—a long-term advantage of sex. Although many species reproduce asexually most of the time, few are strictly asexual over long periods of evolutionary time. Almost all have some means of achieving genetic recombination.

The evolution of the mechanisms of meiosis and sexual recombination were crucial events in the history of life. Exactly how these attributes arose is puzzling, however, because sex has at least three striking disadvantages in the short term:

- Recombination breaks up adaptive combinations of genes.
- Sex reduces the rate at which females pass genes on to their offspring.
- Dividing offspring into separate sexes greatly reduces the overall reproductive rate.

To see why this last disadvantage exists, consider an asexual female that produces the same number of offspring as a sexual female. Let's assume that both females produce two offspring, but that 50 percent of the sexual female's offspring will be males (and thus only contribute sperm). In this next (F_1) generation, both asexual females will produce two more offspring each, but there is only one sexual F_1 female to produce offspring. Thus the effective reproductive rate of the asexual lineage is twice that of the sexual lineage.

The evolutionary problem is to identify the advantages of sex that can overcome such short-term disadvantages. Several hypotheses have been proposed to explain the existence of sex, none of which are mutually exclusive. One is that sexual recombination facilitates repair of damaged DNA, because breaks and other errors in DNA on one chromosome can be repaired by copying the intact sequence from the homologous chromosome.

Another advantage is that sexual reproduction permits the elimination of deleterious mutations. As Section 13.4 described, DNA replication is not perfect. Errors are introduced in every generation, and most of these errors result in lower fitness. Asexual organisms have no mechanism to eliminate deleterious mutations. Hermann J. Muller noted that the accumulation of deleterious mutations in a nonrecombining genome is like a genetic ratchet. The mutations accumulate—"ratchet up"—at each replication. A mutation occurs and is passed on when the genome replicates, then two new mutations occur in the next replication, so three mutations are passed on, and so on. Over time, the least-mutated class of individuals is lost from the population as new mutations occur. Deleterious mutations cannot be eliminated except by the death of the lineage or a rare back mutation. This accumulation of deleterious mutations in lineages that lack genetic recombination is known as **Muller's ratchet**. In sexual species, on the other hand, genetic recombination produces some individuals with more of these deleterious mutations and some with fewer. The individuals with fewer deleterious mutations are more likely to survive. Thus sexual reproduction allows natural selection to eliminate particular deleterious mutations from the population over time.

Another advantage of sex is the great variety of genetic combinations it creates in each generation. Sexual recombination does not directly influence the frequencies of alleles; rather, *it generates new combinations of alleles on which natural selection can act*. It expands variation in a character influenced by alleles at many loci by creating new genotypes. For example, genetic variation can be a defense against pathogens and parasites. Most pathogens and parasites have much shorter life cycles than their hosts and can rapidly evolve counteradaptations to host defenses. Sexual recombination can give the host's defenses a chance to keep up.

Frequency-dependent selection maintains genetic variation within populations

Natural selection often preserves variation as a polymorphism (the presence of two or more variants of a character in the same population). When the fitness of a given phenotype depends on its frequency in a population, a polymorphism may be maintained by a process known as **frequency-dependent selection**. *Perissodus microlepis*, a small fish that lives in Lake Tanganyika in East Africa, provides an example of frequency-dependent selection.

P. microlepis feeds on the scales of other fish, approaching its prey from behind and dashing in to bite off several scales

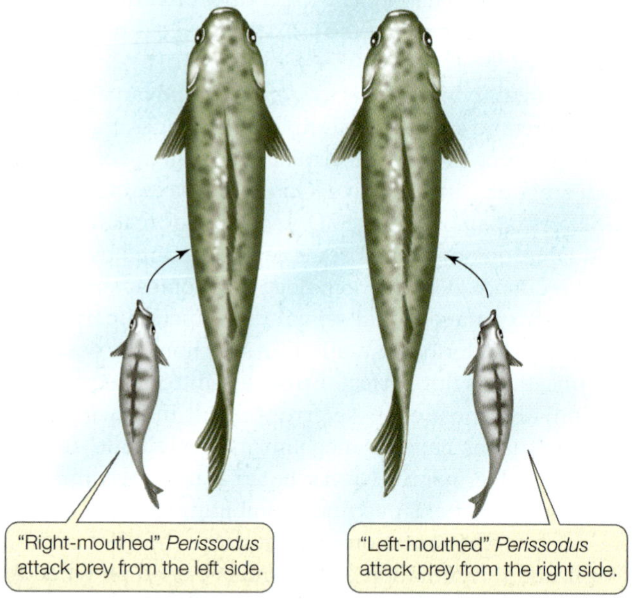

"Right-mouthed" *Perissodus* attack prey from the left side.
"Left-mouthed" *Perissodus* attack prey from the right side.

21.16 A Stable Polymorphism Frequency-dependent selection maintains equal proportions of left- and right-mouthed individuals of the scale-eating fish *Perissodus microlepis*.

from the prey's flank. Because of an asymmetrical jaw joint, the mouth of this scale-eating species opens either to the right or to the left; the direction is genetically determined (**Figure 21.16**). "Right-mouthed" individuals always attack from the victim's left, and "left-mouthed" individuals always attack from the victim's right. The distorted mouth enlarges the area of teeth in contact with the prey's flank, but only if the scale-eater attacks from the appropriate side.

Prey fish are alert to approaching scale-eaters, so attacks are more likely to be successful if the prey must watch both flanks. Vigilance by prey thus favors equal numbers of right- and left-mouthed scale-eaters in a population, because if attacks from one side were more common than the other, prey fish would pay more attention to potential attacks from that side. Over an 11-year study of *P. microlepis* in Lake Tanganyika, the genetic polymorphism was found to be stable, and the two phenotypes of the scale-eaters remained at about equal frequencies.

Heterozygote advantage maintains polymorphic loci

In many cases, different alleles of a particular gene are advantageous under different environmental conditions. Most organisms experience a wide variety of environmental conditions over time. A night is dramatically different from the preceding day. A cold, cloudy day differs from a clear, hot one. Day length and temperature change seasonally. For many genes, a single allele is unlikely to perform well under all these conditions. In such cases, heterozygous individuals (with two different alleles) are likely to outperform individuals that are homozygous (with only one of those two alleles).

INVESTIGATING**LIFE**

21.17 A Heterozygote Mating Advantage Among butterflies of the genus *Colias*, males that are heterozygous for two alleles of the PGI enzyme can fly farther under a broader range of temperatures than males that are homozygous for either allele. Does this ability give heterozygous males a mating advantage?[a]

HYPOTHESIS Heterozygous male *Colias* will have proportionally greater mating success than homozygous males.

Method

1. For each of two *Colias* species, capture mated female butterflies in the field. In the laboratory, allow them to lay eggs.
2. Determine the genotypes of the females and their offspring, and thus the genotypes of the fathers.
3. Compare the frequency of heterozygotes among successfully mating males with the frequency of heterozygotes among all viable males (i.e., males captured flying with females).

Results For both species, the proportion of heterozygotes among the males that mated successfully was higher than the proportion of heterozygotes among all viable males.

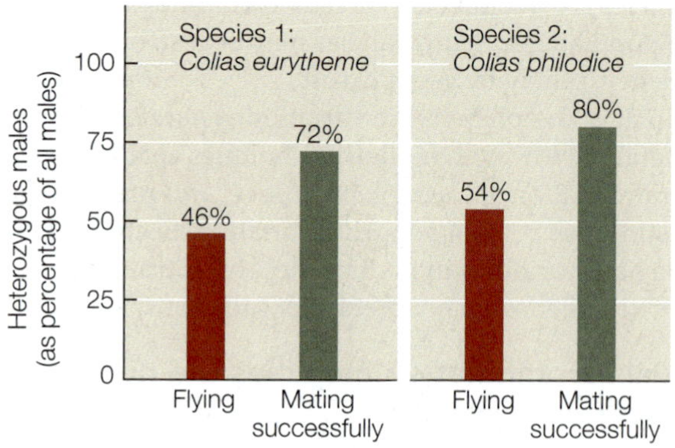

Species 1: Colias eurytheme
Species 2: Colias philodice
Heterozygous males (as percentage of all males)
100
75
50
25
0
46%
72%
54%
80%
Flying
Mating successfully
Flying
Mating successfully

CONCLUSION Heterozygous *Colias* males have a mating advantage over homozygous males.

Go to **BioPortal** for discussion and relevant links for all INVESTIGATING**LIFE** figures.

[a]Watt, W. B. et al. 1985. *Genetics* 109:157–175.

Colias butterflies of the Rocky Mountains live in environments where dawn temperatures are often too cold, and afternoon temperatures too hot, for the butterflies to fly. Populations of these butterflies are polymorphic for a gene that encodes the enzyme phosphoglucose isomerase (PGI), which influences how well a butterfly flies at different temperatures. Butterflies with certain PGI genotypes can fly better during the cold hours of early morning; others perform better during midday heat. The optimal body temperature for flight is 35°C–39°C, but some butterflies can fly with body temperatures as low as 29°C or as high as 40°C. During spells of unusually hot weather, heat-tolerant genotypes are favored; during spells of unusually cool weather, cold-tolerant genotypes are favored.

Heterozygous *Colias* butterflies can fly over a greater range of temperatures than homozygous individuals, which should give them an advantage in foraging and finding mates. A test of this prediction found that heterozygous males did indeed have a mating advantage, and further, that this advantage maintains the polymorphism in the population (**Figure 21.17**).

WORKING WITH**DATA:**

Do Heterozygous Males Have a Mating Advantage?

Original Paper

Watt, W. B., P. A. Carter and S. M. Blower. 1985. Adaptation at specific loci. IV. Differential mating success among glycolytic allozyme genotypes of *Colias* butterflies. *Genetics* 109: 157–175.

Analyze the Data

Ward Watt and his colleagues tested the hypothesis that males with two different alleles for the PGI enzyme (heterozygotes) were more likely to mate successfully with females than were homozygous males. They reasoned that the heterozygous males could fly farther under a broader range of temperatures than could homozygous males, and that this ability would give heterozygous males greater access to receptive females. To test this hypothesis, they needed to know the frequency of heterozygotes among successfully mating males, and they needed to compare that frequency with the frequency of heterozygotes among males in the general population (i.e. all the potential mates available to females). To estimate the frequency of heterozygotes among mating males, Watt et al. collected mated female butterflies in the field and allowed them to lay eggs in the laboratory. They hatched the eggs and determined the genotypes of the offspring, as well as the genotypes of the females. Using this information, they could determine the genotypes of the males that fathered the larvae. They then compared the estimated frequency of heterozygotes among the successful fathers with the frequency of heterozygotes among all viable males in the population. Samples of their data are given in the table.

QUESTION 1

If we assume that the proportions of each genotype among mating males should be the same as the proportions seen among all viable males, what is the number of *mating males* expected to be heterozygous in each sample?

QUESTION 2

Use a chi-square test (see Appendix B) to evaluate the significance of the difference in the observed and expected numbers of heterozygous and homozygous individuals among the mating males. The critical value ($P = 0.05$) of the chi-square distribution with one degree of freedom is 3.841. Are the observed numbers of genotypes among mating males significantly different ($P < 0.05$) from the expected numbers in these samples?

QUESTION 3

The investigators determined the genotypes of enough larvae from each batch of eggs to judge the genotype of the father with 99% certainty. How many larvae did they need to measure to achieve that level of certainty?

Hint: If the female is homozygous—say, genotype *ii*—the number needed is small. However, if a female is heterozygous—say, genotype *ij*—and only *ii* and *ij* progeny are found among her offspring, more larvae need to be genotyped. In this particular case, the father can be only *ii* or *ij*. If he were *ij*, the probability that any one offspring is *not jj* = 0.75, so the chance of getting only *ii* and *ij* among *n* offspring is 0.75^n. What value of *n* is required to reduce the probability of error in determining the father's genotype to 0.01?

	All viable males[a]		Mating males	
Species	Heterozygous/ total	Percent heterozygous	Heterozygous/ total	Percent heterozygous
C. philodice	32/74	43.2	31/50	62.0
C. eurytheme	44/92	47.8	45/59	76.3

[a]"Viable males" are males captured flying with females (hence with the potential to mate)

Go to **BioPortal** for all WORKING WITH**DATA** exercises

Of course, the heterozygous genotype can never become fixed in the population, because the offspring of two heterozygotes will always include both classes of homozygotes in addition to heterozygotes.

Genetic variation within species is maintained in geographically distinct populations

Much of the genetic variation within species is preserved as differences among members living in different places (populations). Populations often vary genetically because they are subjected to different selective pressures in different environments. Environmental conditions may vary significantly even over short distances. For example, in the Northern Hemisphere, temperature and soil moisture differ dramatically between north- and south-facing mountain slopes. In the Rocky Mountains of Colorado, the proportion of ponderosa pines (*Pinus ponderosa*) that are heterozygous for a particular peroxidase enzyme is particularly high on south-facing slopes, where temperatures fluctuate dramatically, often on a daily basis. This heterozygous genotype performs well over a broad range of temperatures. On north-facing slopes and at higher elevations, where temperatures are cooler and fluctuate less strikingly, a peroxidase homozygote, which has a lower optimal temperature, is much more frequent.

Plant species may also vary geographically in the chemicals they synthesize to defend themselves against herbivores. Some individuals of the white clover (*Trifolium repens*) produce the poisonous chemical cyanide. Poisonous individuals are less appealing to herbivores—particularly mice and slugs—than are nonpoisonous individuals. However, clover plants that produce cyanide are more likely to be killed by frost, because freezing damages cell membranes and releases cyanide into the plant's own tissues.

In European populations of *Trifolium repens*, the frequency of cyanide-producing individuals increases gradually from north to south and from east to west (**Figure 21.18**). Such a pattern of gradual change in phenotype across a geographic gradient is known as **clinal variation**. In the white clover cline,

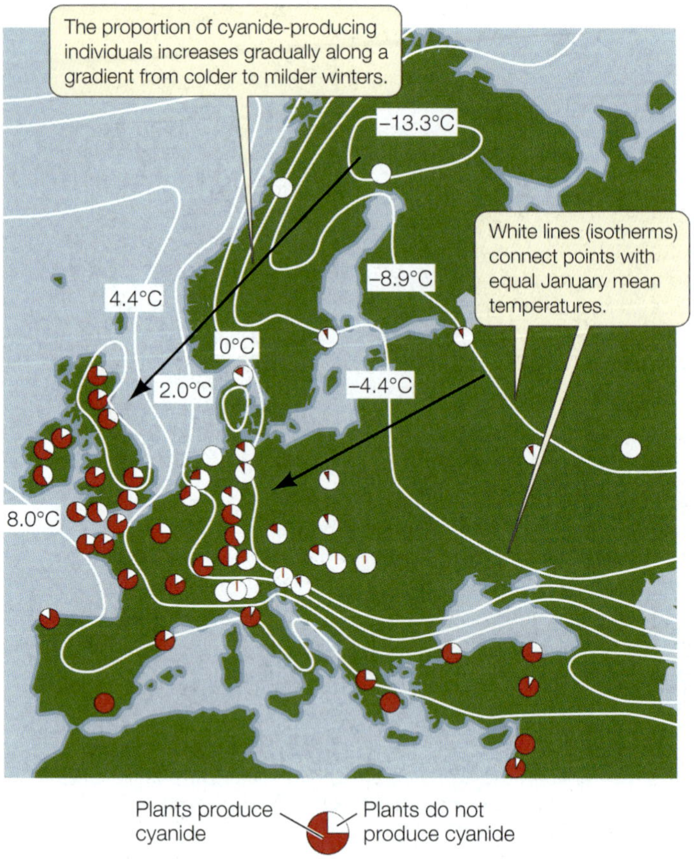
The proportion of cyanide-producing individuals increases gradually along a gradient from colder to milder winters.
–13.3°C
White lines (isotherms) connect points with equal January mean temperatures.
–8.9°C
4.4°C
0°C
2.0°C
–4.4°C
8.0°C
Plants produce cyanide
Plants do not produce cyanide

21.18 Geographic Variation in a Defensive Chemical The proportion of cyanide-producing individuals in European populations of white clover (*Trifolium repens*) depends on winter temperatures.

poisonous plants make up a large proportion of populations only in areas where winters are mild. Cyanide-producing individuals are rare where winters are cold, even though herbivores graze clovers heavily in those areas.

RECAP 21.4

Neutral mutations, sexual recombination, frequency-dependent selection, and heterozygote advantage all act to maintain considerable genetic variation in most populations. Variation within species is also maintained among geographically distinct populations.

- Why is sexual reproduction is so prevalent in nature, despite its having at least three short-term evolutionary disadvantages? **See p. 441**
- How does frequency-dependent selection act to maintain genetic variation in a population? **See pp. 441–442**

The mechanisms of evolution have produced a remarkable variety of organisms. There are organisms that have adapted to nearly every environment on Earth. This natural variation, along with the success of breeders attempting to produce desired traits in domesticated plants and animals, suggests that evolution can produce a wide variety of adaptive traits. But are there limits to the adaptations evolution can produce?

21.5 What Are the Constraints on Evolution?

We would be mistaken to assume that evolutionary mechanisms can produce any trait we might imagine. Evolution is constrained in many ways. Lack of appropriate genetic variation, for example, prevents the development of many potentially favorable traits. If the allele for a given trait does not exist in a population, that trait cannot evolve, even if it would be highly favored by natural selection. Most possible combinations of genes and genotypes have never existed in any population and so have never been tested under natural selection.

In addition, constraints are imposed on organisms by the dictates of physics and chemistry. The size of cells, for example, is constrained by the stringencies of surface area-to-volume ratios (see Figure 5.2). The ways in which proteins can fold are limited by the bonding capacities of their constituent molecules (see Section 2.2). And the energy transfers that fuel life must operate within the laws of thermodynamics (see Section 8.1). Keep in mind that evolution works within the boundaries of these universal constraints as well as the constraints described in this section.

Developmental processes constrain evolution

As Section 20.5 explained, developmental constraints on evolution are paramount because *all evolutionary innovations are modifications of previously existing structures*. Human engineers seeking to power an airplane can start "from scratch" to design a completely new type of engine (powered by jet propulsion) to replace the previous type (powered by propellers). Evolutionary changes, however, cannot happen in this way. Current phenotypes of organisms are constrained by historical conditions and past selective pressures.

A striking example of such developmental constraints is provided by the evolution of fishes that spend most of their time on the sea bottom, where a ventrally flattened body is advantageous. One such lineage, the bottom-dwelling skates and rays, shares a common ancestor with sharks, whose bodies are already somewhat ventrally flattened and whose skeletal frame is made of flexible cartilage. Skates and rays evolved a body type that further flattened their bellies, allowing them to swim along the ocean floor (**Figure 21.19A**).

By contrast, plaice, sole, and flounder are bottom-dwelling descendants of deep-bellied, laterally flattened ancestors with bony skeletons. The only way these fishes can lie flat is to flop over on their sides. Their ability to swim is thus curtailed, but their bodies can lie still and are well camouflaged. During development, one eye of these flatfishes moves so that both eyes are positioned on the same side of the body (**Figure 21.19B**). Such shifts in eye position have evolved several times, and shifts have happened in both directions (that is, both left- and right-eyed flatfishes have evolved independently). Small shifts in the position of one eye probably helped ancestral flatfishes see better, resulting in the body forms found today. This path to producing a flattened body may not be optimal, but the fishes' developmental capabilities constrain the pathways that evolution can take.

(A) *Taeniura lymma*

(B) *Bothus lunatus*

21.19 Two Solutions to a Single Problem (A) This stingray, whose ancestors were dorsoventrally flattened, lies on its belly. Stingrays' bodies are symmetrical around the dorsal backbone. (B) The flounder, whose ancestors were laterally flattened, lies on its side. (The backbone of this individual is at the right.) Flounders' eyes migrate during development so that both are on the same side of the body.

Trade-offs constrain evolution

Adaptations frequently impose both costs and benefits. For an adaptation to be favored, the fitness benefits it confers must exceed the fitness costs it imposes—in other words, the **trade-off** must be worthwhile. For example, there are metabolic costs associated with developing and maintaining certain conspicuous features (such as antlers or horns) that males use to compete with other males for access to females. The fact that these features are common in many species suggests that the benefits derived from possessing them must outweigh the costs.

As a result of trade-offs, many traits that are adaptive in one context may be maladaptive in another. Consider the rough-skinned newt and one of its predators, the common garter snake (**Figure 21.20A**). The newt sequesters a potent neurotoxin called tetrodotoxin (TTX) in its skin. TTX paralyzes nerves and muscles by blocking sodium channels (see Section 6.3). Most vertebrates—including many garter snakes—will die if they eat a rough-skinned newt. But some garter snakes can eat rough-skinned newts and survive: TTX-resistant sodium channels have evolved in the nerves and muscles of such individuals. However, for several hours after eating a newt, TTX-resistant snakes can move only slowly, and they never move as fast as nonresistant snakes. Resistant snakes are thus more vulnerable to their own predators than are TTX-sensitive snakes. This vulnerability leads to selection *against* TTX-resistant sodium channels in garter snake populations that occur outside the range of rough-skinned newts, even though there is selection *for* TTX resistance in many areas where newts are present (**Figure 21.20B**).

(A)

(B)

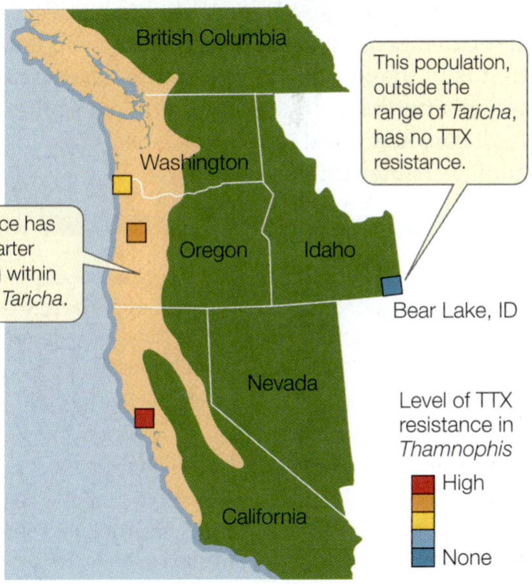
British Columbia
Washington
Oregon
Idaho
Nevada
California
This population, outside the range of *Taricha*, has no TTX resistance.
TTX resistance has evolved in garter snakes living within the range of *Taricha*.
Bear Lake, ID
Level of TTX resistance in *Thamnophis*
High
None

21.20 Resistance to a Toxin Comes at a Cost (A) Garter snakes (*Thamnophis sirtalis*) prey on rough-skinned newts (*Taricha granulosa*). Rough-skinned newts defend themselves by sequestering a neurotoxin, TTX, in their skin. In turn, TTX-resistant sodium channels have evolved in some snake populations, allowing the snakes to eat toxic newts but resulting in slower movement by the snakes. (B) High TTX resistance in garter snakes is found only in regions where snake and newt populations overlap (tan area).

Go to Animated Tutorial 21.4
Assessing the Costs of Adaptation
Life10e.com/at21.4

Short-term and long-term evolutionary outcomes sometimes differ

The short-term changes in allele frequencies within populations that we have emphasized in this chapter, often termed **microevolutionary** changes, are an important focus of study for evolutionary biologists. These changes can be observed directly, they can be manipulated experimentally, and they demonstrate the actual processes by which evolution occurs. By themselves, however, they may not be sufficient to predict long-term, or **macroevolutionary**, changes.

Long-term patterns of evolutionary change can be strongly influenced by events that occur so infrequently (such as a meteorite impact) or so slowly (such as continental drift) that they are unlikely to be observed during short-term studies. The evolutionary mechanisms at work may change over time with changing environmental conditions. Even among the descendants of a single ancestral species, different lineages may evolve in different directions. Additional types of evidence—evidence demonstrating the effects of rare and unusual events on trends in the fossil record—must be gathered if we wish to understand the course of evolution over billions of years. We will consider these long-term aspects of evolution in the remaining chapters of this section.

RECAP 21.5

Developmental processes constrain evolution because all evolutionary innovations are modifications of previously existing structures. An adaptation can evolve only if the fitness benefits it confers exceed the fitness costs it imposes.

- Describe an example of an evolutionary trade-off in which the advantages of an adaptation outweigh its costs in the long run. See p. 445 and Figure 21.19
- How could the presence of a great deal of genetic variation within a population increase the chances that some members of the population would survive an unprecedented environmental change? Why is there no guarantee that this would be the case?

?

How do biologists use evolutionary theory to develop better flu vaccines?

ANSWER

Many different strains of influenza virus circulate among human populations and other vertebrate hosts each year, but only a few of those strains survive and produce descendants. One of the ways in which influenza strains differ is in the configuration of proteins on their surface. These surface proteins are the targets of recognition by the host immune system. When changes occur in the surface proteins of an influenza virus, the host immune system may no longer recognize the invading virus, so that virus is more likely to replicate successfully. Those virus strains with the greatest number of changes to their surface proteins are most likely to escape detection by the host immune system, and are therefore most likely to spread among the host population and result in future flu epidemics. In other words, there is positive selection for change in the surface proteins of influenza viruses.

By comparing the survival and proliferation rates of influenza virus strains that have different gene sequences coding for surface proteins, biologists can study the adaptation of the viruses over time. If biologists can predict which of the currently circulating flu virus strains are most likely to escape host detection, then they can identify the strains that are most likely to be involved in upcoming influenza epidemics and can target those strains for vaccine production.

How can biologists make such predictions? By determining the ratio of synonymous to nonsynonymous substitutions in genes that encode viral surface proteins, biologists can detect which codon changes (i.e., mutations) are under positive selection (using methods we will discuss in Section 24.2). They can then assess which of the currently circulating flu strains show the greatest number of changes in these positively selected codons. It is these flu strains that are most likely to survive, proliferate, and lead to the flu epidemics of the future, so they are the logical targets for new vaccines. This practical application of evolutionary theory leads to more effective flu vaccines—and thus fewer illnesses and influenza-related deaths each year.

CHAPTER SUMMARY 21

21.1 What Is the Relationship between Fact and Theory in Evolution?

- **Evolution** is genetic change in populations over time. Evolution can be observed directly in living populations as well as in the fossil record of life.
- **Evolutionary theory** refers to our understanding of the mechanisms of evolutionary change.
- Charles Darwin in best known for his ideas on the common ancestry of divergent species and on **natural selection** (the differential survival and reproduction of individuals based on variation in their traits) as a mechanism of evolution. See ANIMATED TUTORIAL 21.1, ACTIVITY 21.1
- Since Darwin's time, many biologists have contributed to the development of evolutionary theory, and rapid progress in our understanding continues today. Review Figure 21.2
- For a population to evolve, its members must possess heritable genetic variation.

21.2 What Are the Mechanisms of Evolutionary Change?

- **Mutation** is the source of the genetic variation on which mechanisms of evolution act.
- The term **adaptation** refers both to a trait that evolves through natural selection and to the process that produces such traits.

continued

- Within populations, natural selection acts to increase the frequency of beneficial alleles (**positive selection**) and to decrease the frequency of deleterious alleles (**purifying selection**).
- Movement of individuals or gametes between populations results in **gene flow**.
- In small populations, **genetic drift**—the random loss of individuals and the alleles they possess from one generation to the next—may produce large changes in allele frequencies over time and greatly reduce genetic variation. See ANIMATED TUTORIAL 21.2
- **Population bottlenecks** occur when only a few individuals survive a random event, resulting in a drastic shift in allele frequencies within the population and the loss of genetic variation. Similarly, a population established by a small number of individuals colonizing a new region may lose genetic variation via a **founder effect**. Review Figure 21.7
- Nonrandom mating may result in changes in genotype and allele frequencies in a population.
- **Sexual selection** results from differential reproductive success based on individuals' phenotypes. Review Figure 21.9

21.3 How Do Biologists Measure Evolutionary Change?

- Allele frequencies measure the amount of genetic variation in a population. Genotype frequencies show how a population's genetic variation is distributed among its members. Together, allele and genotype frequencies describe a population's **genetic structure**. Review Figure 21.10
- **Hardy–Weinberg equilibrium** predicts genotype frequencies from allele frequencies in the absence of evolution. Deviation from these frequencies indicates that evolutionary mechanisms are at work. Review Figure 21.11, ANIMATED TUTORIAL 21.3
- Natural selection can act on characters with quantitative variation in three different ways. Review Figure 21.12
- **Stabilizing selection** acts to reduce variation without changing the mean value of a trait.
- **Directional selection** acts to shift the mean value of a trait toward one extreme.
- **Disruptive selection** favors both extremes of trait values, resulting in a bimodal character distribution.

21.4 How Is Genetic Variation Distributed and Maintained within Populations?

- Neutral mutations, sexual recombination, frequency-dependent selection, and heterozygote advantage can all maintain genetic variation within populations.
- **Neutral alleles** do not affect the fitness of an organism, are not affected by natural selection, and may accumulate or be lost by genetic drift.
- Despite its short-term disadvantages, sexual reproduction generates countless genotypic combinations that increase the evolutionary potential and survivorship of populations.
- A polymorphism may be maintained by **frequency-dependent selection** when the fitness of a genotype depends on its frequency in a population.
- Genetic variation within species may be maintained by the existence of genetically distinct populations over geographic space. A gradual change in phenotype across a geographic gradient is known as **clinal variation**. Review Figure 21.18

21.5 What Are the Constraints on Evolution?

- Developmental processes constrain evolution because all evolutionary innovations are modifications of previously existing structures.
- Most adaptations impose costs as well as benefits. An adaptation can evolve only if the benefits it confers exceed the costs it imposes. Review Figure 21.20, ANIMATED TUTORIAL 21.4

Go to the Interactive Summary to review key figures, Animated Tutorials, and Activities
Life10e.com/is21

CHAPTER**REVIEW**

REMEMBERING

1. Long-horned cattle have greater difficulty moving through heavily forested areas compared with cattle that have short or no horns, but long-horned cattle are better able to defend their young against predators. This contrast is an example of
   a. an adaptation.
   b. genetic drift.
   c. natural selection.
   d. a trade-off.
   e. none of the above

2. Which statement about allele frequencies is *not* true?
   a. The sum of all allele frequencies at a locus is always 1.
   b. If there are two alleles at a locus and we know the frequency of one of them, we can obtain the frequency of the other by subtraction.
   c. If an allele is missing from a population, its frequency in that population is 0.
   d. If two populations have the same allele frequencies at a locus, they must have the same proportion of homozygotes at that locus.
   e. If there is only one allele at a locus, its frequency is 1.

3. Which of the following is *not* an assumption of Hardy–Weinberg equilibrium?
   a. There is no migration between populations.
   b. Natural selection is not acting on the alleles in the population.
   c. Mating is random.
   d. Multiple alleles must be present at every locus.
   e. All of the above

4. Laboratory selection experiments with fruit flies have demonstrated that
   a. bristle number is not genetically controlled.
   b. bristle number is not genetically controlled, but changes in bristle number are caused by the environment in which the fly is raised.
   c. bristle number is genetically controlled, but there is little variation on which natural selection can act.
   d. bristle number is genetically controlled, but selection cannot result in flies having more bristles than any individual in the original population had.
   e. bristle number is genetically controlled, and selection can result in flies having more, or fewer, bristles than any individual in the original population had.

5. Disruptive selection maintains a bimodal distribution of bill size in the black-bellied seedcracker because
   a. bills of intermediate sizes are difficult to form.
   b. the species' two major food sources differ markedly in size and hardness.
   c. males use their large bills in displays.
   d. migrants introduce different bill sizes into the population each year.
   e. older birds need larger bills than younger birds.

UNDERSTANDING & APPLYING

6. In what ways does artificial selection by humans differ from natural selection? Can you give some examples of a trait that might be favored by artificial selection in agriculture, but selected against by natural selection in a wild population?

7. As far as we know, natural selection cannot adapt organisms to future events. Yet many organisms appear to respond to natural events before they happen. For example, many mammals go into hibernation while it is still quite warm. Similarly, many birds leave the temperate zone for their southern wintering grounds long before winter has arrived. How do you think such "anticipatory" behaviors evolve?

8. As more humans live longer, many people face degenerative conditions such as Alzheimer's disease that (in most cases) are linked to advancing age. Assuming that some individuals may be genetically predisposed to successfully combat these conditions, is it likely that natural selection alone would act to favor such a predisposition in human populations? Why or why not?

ANALYZING & EVALUATING

9. The following sample lists the genotype at locus *A* for 10 individuals in a diploid population. Based on this sample, answer the questions that follow.
   a. Sample: *AA*, *AA*, *Aa*, *Aa*, *Aa*, *Aa*, *aa*, *aa*, *aa*, *aa*
   b. What is the observed frequency of allele *a*? The observed frequency of allele *A*?
   c. What are the observed frequencies of genotypes *aa*, *Aa*, and *AA*?
   d. After one generation of random mating, what would be the Hardy–Weinberg expectations for the frequencies of genotypes *aa*, *Aa*, and *AA*?
   e. What are some of the reasons you might expect the observed genotype frequencies to differ from the Hardy–Weinberg expectations?

10. Imagine you are studying a color polymorphism in a species of mice; in this species, some individuals have black coats and some have white coats. You want to know if the mice are mating randomly or if there is mate selection based on coat color. You decide to examine genotype frequencies at *a locus that is unrelated to coat color*. You collect the following data from mice sampled in a single location that potentially represents a single breeding population.
    a. In a sample of 25 white-coated mice: 1 mouse has genotype *aa*, 4 have genotype *Aa*, and 20 have genotype *AA*.
    b. In a sample of 25 black-coated mice: 24 mice have genotype *aa* and 1 has genotype *Aa*.

    Do your data support random mating between black mice and white mice? Can you think of other explanations for your data that do not involve nonrandom mate selection? How might you test and distinguish these various hypotheses?

Go to BioPortal at **yourBioPortal.com** for Animated Tutorials, Activities, LearningCurve Quizzes, Flashcards, and many other study and review resources.

22 Reconstructing and Using Phylogenies

CHAPTER OUTLINE

Multiple Fluorescences The reef-building coral *Acropora millepora* shows both cyan and red fluorescences. This photograph was taken under a microscope that affects the colors we see; the colors are perceived differently by marine animals in their natural environment.

GREEN FLUORESCENT PROTEIN (GFP) was discovered in 1962 when Osamu Shimomura, an organic chemist and marine biologist, led a team that was able to extract the protein from the tissues of the bioluminescent jellyfish *Aequorea victoria* and purify it. Some 30 years after its initial discovery, Martin Chalfie had the idea (and the technology) to link the gene for GFP to other protein-coding genes so that the expression of specific genes of interest could be visualized in glowing green within cells and tissues of living organisms (see Figure 18.4). This work was extended by Roger Tsien, who changed some of the amino acids within GFP to create fluorescent proteins of several distinct colors. Different-colored fluorescent proteins meant that the expression of a number of different proteins could be visualized and studied in the same organism at the same time. These three scientists were awarded the 2008 Nobel Prize in Chemistry for the isolation of GFP and its development for visualizing gene expression.

Although Tsien was able to produce different-colored proteins, he could not produce a *red* protein. This was frustrating; a red fluorescent protein would have been particularly useful to biologists because red light penetrates tissues more easily than do other colors. Tsien's work inspired Mikhail Matz to look for new fluorescent proteins in corals (which are relatives of the jellyfishes). Among the different coral species he studied, Matz found proteins that fluoresced in various shades of green, cyan (blue-green)—and red.

How had fluorescent red pigments evolved among the corals, given that the necessary molecular changes had eluded Tsien? To answer this question, Matz sequenced the genes of the fluorescent proteins and used these sequences to reconstruct the evolutionary history of the amino acid changes that produced different colors in different species of corals.

Matz's work showed that the ancestral fluorescent protein in corals was green, and that red fluorescent proteins evolved in a series of gradual steps. His analysis of evolutionary relationships allowed him to retrace these steps. Such an evolutionary history, often depicted as a branching diagram of relationships among lineages, is called a phylogeny.

The evolution of many aspects of an organism's biology can be studied using phylogenetic methods. This information is used in all fields of biology to understand the structure, function, and behavior of organisms.

> ? How are phylogenetic methods used to resurrect protein sequences from extinct organisms?
>
> See answer on p. 464.

22.1 What Is Phylogeny?

Phylogeny is the history of evolutionary relationships among organisms or their genes. A **phylogenetic tree** is a diagram that portrays a reconstruction of that history. Phylogenetic trees are commonly used to depict the evolutionary histories of species, populations, and genes. Each branching point (or **node**) in a phylogenetic tree represents a point at which lineages diverged in the past. In the case of species, these splits represent past speciation events, when one lineage divided into two. Thus a phylogenetic tree can trace evolutionary relationships from the ancient common ancestor of a group of species through the various speciation events (when lineages split) up to the present populations of the organisms (Figure 22.1). Over the past several decades, phylogenetic trees have become important tools for studying and describing evolutionary patterns and for applying evolutionary theory throughout biology. You will need to understand phylogenetic trees to comprehend many articles and books about biology, including this one.

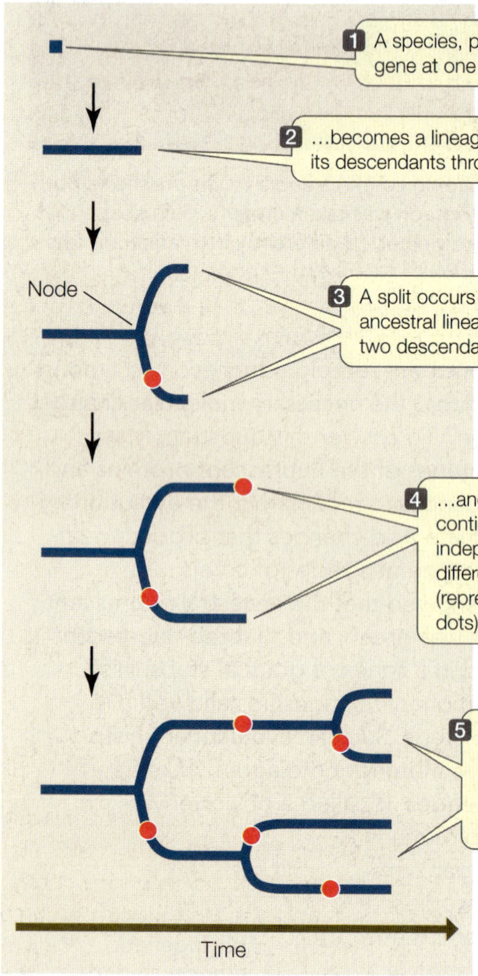
1 A species, population, or gene at one point in time...
2 ...becomes a lineage as we follow its descendants through time.
Node
3 A split occurs when the ancestral lineage divides into two descendant lineages...
4 ...and each lineage continues to evolve independently as different traits (represented by red dots) arise.
5 The lineages continue to split, and a phylogenetic tree emerges.
Time

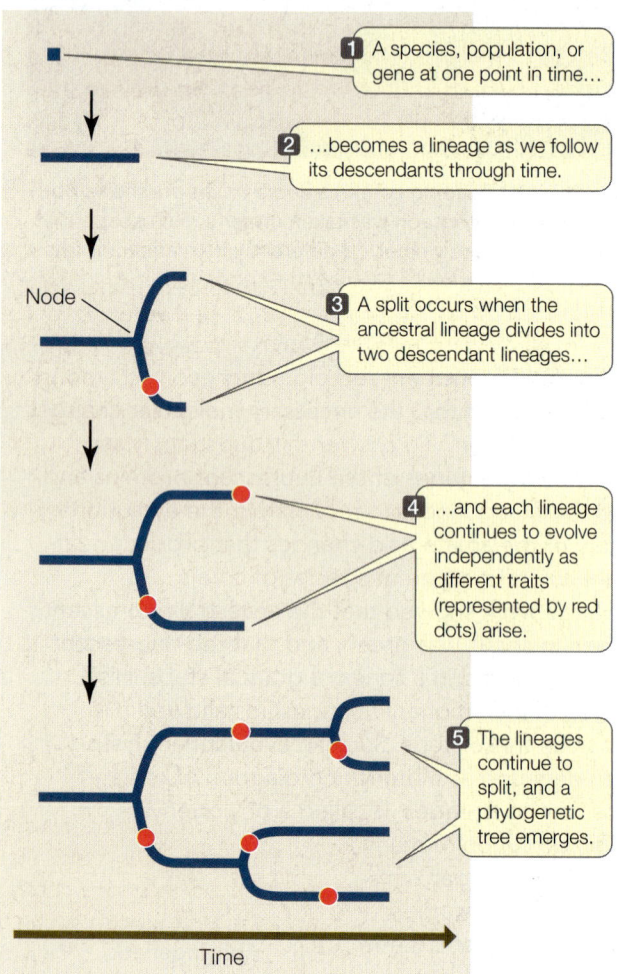

22.1 The Components of a Phylogenetic Tree Evolutionary relationships among organisms can be represented in a treelike diagram.

A phylogenetic tree may be used to portray the evolutionary history of all life forms; of a major evolutionary group (such as the insects); of a small group of closely related species; or in some cases, even of individuals, populations, or genes within a species. The common ancestor of all the organisms in the tree forms the **root** of the tree.

The phylogenetic trees in this book depict time flowing from left (earliest) to right (most recent) (Figure 22.2A). It is also common practice to draw trees with the earliest times at the bottom. The timing of a splitting event in a lineage is shown by the position of a node on the time axis, (sometimes called the divergence axis). These splits represent events in which one lineage diverged into two, such as a speciation event (for a tree of species), a gene duplication event (for a tree of genes), or a transmission event (for a tree of viral lineages transmitted through a host population). The divergence axis may have an explicit scale or simply show the relative timing of splitting events.

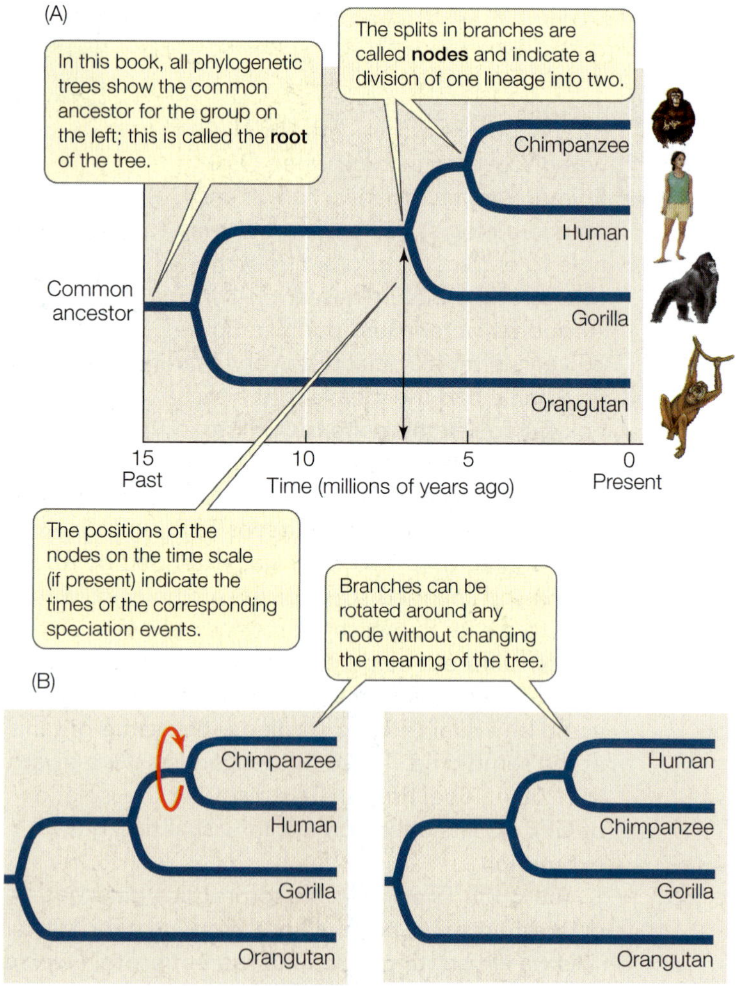
(A)
In this book, all phylogenetic trees show the common ancestor for the group on the left; this is called the root of the tree.
The splits in branches are called nodes and indicate a division of one lineage into two.
Chimpanzee
Human
Gorilla
Orangutan
Common ancestor
15 Past
10
5
0 Present
Time (millions of years ago)
The positions of the nodes on the time scale (if present) indicate the times of the corresponding speciation events.
Branches can be rotated around any node without changing the meaning of the tree.
(B)
Chimpanzee
Human
Gorilla
Orangutan
Human
Chimpanzee
Gorilla
Orangutan

22.2 How to Read a Phylogenetic Tree (A) Phylogenetic trees can be produced with time scales, as shown here, or with no indication of time. If no time scale is shown, then the trees are only meant to depict the relative order of divergence events. (B) Lineages can be rotated around a given node, so the vertical order of taxa is largely arbitrary.

In this book, the order of nodes along the horizontal (time) axis has meaning, but the vertical distance between the branches does not. Vertical distances are adjusted for legibility and clarity of presentation; they do not correlate with the degree of similarity or difference between groups. Note too that lineages can be rotated around nodes in the tree, so the vertical order of lineages is also largely arbitrary (Figure 22.2B). The important information in the tree is the branching order along the time axis, as this indicates when the various lineages last shared a common ancestor.

Any species or group of species that we designate or name is called a **taxon** (plural taxa). Some examples of familiar taxa include humans, primates, mammals, and vertebrates (note that in this series, each taxon in the list is also a member of the next, more inclusive taxon). Any taxon that consists of an ancestor and all of its evolutionary descendants is called a **clade**. Clades can be identified by picking any point on a phylogenetic tree and then tracing all the descendant lineages to the tips of the terminal branches (Figure 22.3). Two species that are each other's closest relatives are called **sister species**; and any two clades that are each other's closest relatives are called **sister clades**.

Before the 1980s, phylogenetic trees tended to be seen only in the literature on evolutionary biology, especially in the area of **systematics**: the study and classification of biodiversity. But

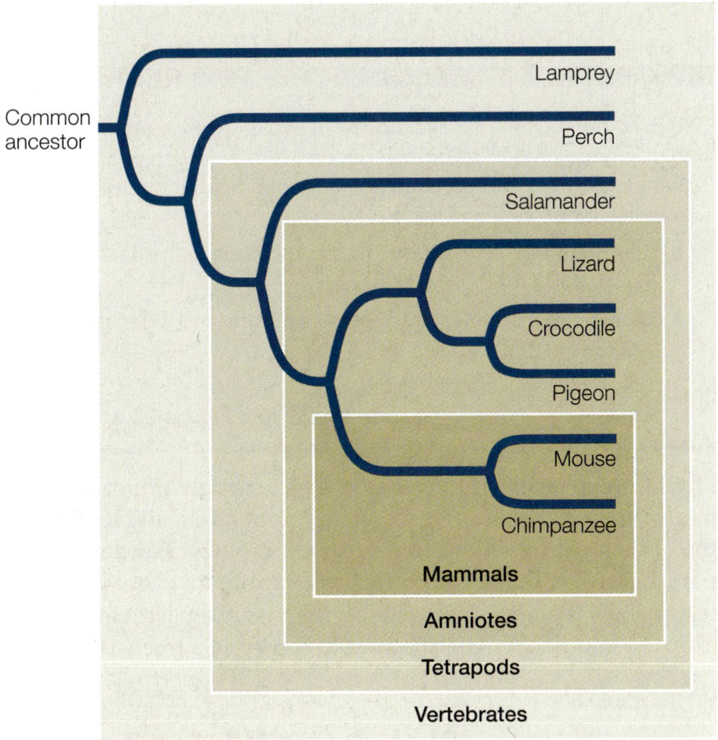
Common ancestor
Lamprey
Perch
Salamander
Lizard
Crocodile
Pigeon
Mouse
Chimpanzee
Mammals
Amniotes
Tetrapods
Vertebrates

22.3 Clades Represent an Ancestor and All of Its Evolutionary Descendants All clades are subsets of larger clades, with all of life as the most inclusive taxon. In this example, the groups called mammals, amniotes, tetrapods, and vertebrates represent successively larger clades. Only a few species within each clade are represented on the tree.

almost every journal in the life sciences published during the last few years contains phylogenetic trees. Trees are widely used in molecular biology, biomedicine, physiology, behavior, ecology, and virtually all other fields of biology. Why have phylogenetic studies become so important?

All of life is connected through evolutionary history

In biology, we study life at all levels of organization—from genes, cells, organisms, populations, and species to the major divisions of life. In most cases, however, no individual gene or organism (or other unit of study) is exactly like any other gene or organism that we investigate.

Consider the individuals in your biology class. We recognize each person as an individual, but we know that no two are exactly alike. If we knew everyone's family tree in detail, the genetic similarity of any pair of students would be more predictable. We would find that more closely related students have more traits in common (from the color of their hair to their susceptibility or resistance to diseases). Similarly, biologists use phylogenies to make comparisons and predictions about shared traits across genes, populations, and species.

One of the great unifying concepts in biology is that all life is connected through its evolutionary history. The complete evolutionary history of life is known as the **tree of life**. Biologists estimate that there are tens of millions of species on Earth. Only about 1.8 million have been formally described and named. New species are being discovered and named all the time, and phylogenetic trees are continually being reviewed and revised. Thus our knowledge of the tree of life is far from complete, even for known species. Yet knowledge of evolutionary relationships is essential for making comparisons in biology, so biologists construct phylogenetic trees for groups of interest as the need arises. The evolutionary relationships among species, as shown in the tree of life, also form the basis for biological classification. This evolutionary framework allows biologists to make many predictions about the behavior, ecology, physiology, genetics, and morphology of species that have not yet been studied in detail.

Comparisons among species require an evolutionary perspective

When biologists make comparisons among species, they observe traits that differ within the group of interest and try to ascertain when those traits evolved. In many cases, investigators are interested in how the evolution of a trait depends on environmental conditions or selection pressures. For instance, scientists have used phylogenetic analyses to discover changes in the genome of HIV that confer resistance to particular drug treatments. The association of a particular genetic change in HIV with a particular drug treatment provides a hypothesis about the evolution of resistance that can be tested experimentally.

Any features shared by two or more species that have been inherited from a common ancestor are said to be **homologous**. Homologous features may be any heritable traits, including DNA sequences, protein structures, anatomical structures, and even some behavior patterns. Traits that are shared across a group of interest are likely to have been inherited from a common ancestor. For example, all living vertebrates have a vertebral column, and all known fossil vertebrates had a vertebral column; thus the vertebral column is judged to be homologous in all vertebrates.

In tracing the evolution of a trait, biologists distinguish between ancestral and derived traits. A trait that was present in the ancestor of a group is known as an **ancestral trait** for that group. A trait found in a descendant that differs from its ancestral form is a **derived trait**. Derived traits that are shared among a group of organisms and are viewed as evidence of the common ancestry of that group are called **synapomorphies** (*syn*, "shared"; *apo*, "derived"; *morph*, "form," referring to the "form" of a trait). Thus the vertebral column is considered a synapomorphy—a shared, derived trait—of the vertebrates.

A particular trait may be ancestral or derived, depending on our phylogenetic point of reference. For example, all birds have feathers, which are highly modified scales. We infer from this fact that feathers were present in the common ancestor of modern birds, and therefore we consider the presence of feathers to be an *ancestral* trait for any group of modern birds (such as the songbirds). Feathers are not present in any other living animals, although there is fossil evidence for the presence of feathers in many extinct species of theropod dinosaurs. If we were reconstructing the phylogeny of all vertebrates, the presence of feathers would be a *derived* trait that informs us about the close evolutionary relationships between birds and their extinct theropod relatives.

Not all similar traits are evidence of relatedness, however. Similar traits can develop in distantly related groups of organisms for either of the following reasons:

- Independently evolved traits subjected to similar selection pressures may become superficially similar, a phenomenon called **convergent evolution**. For example, although the wing bones of bats and birds are homologous, having been inherited from a common ancestor, the wings of bats and the wings of birds—both adaptations for flight—are not homologous because they evolved independently from the forelimbs of different nonflying ancestors (Figure 22.4).
- A character may revert from a derived state back to an ancestral state in an event called an **evolutionary reversal**. For example, most frogs lack teeth in the lower jaw, but the ancestor of frogs did have such teeth. Teeth have been regained in the lower jaw of one South American species and thus represent an evolutionary reversal in that species.

Similar traits in distantly related taxa generated by convergent evolution or by evolutionary reversals are called homoplastic traits, or **homoplasies**.

Go to Media Clip 22.1
Morphing Arachnids
Life10e.com/mc22.1

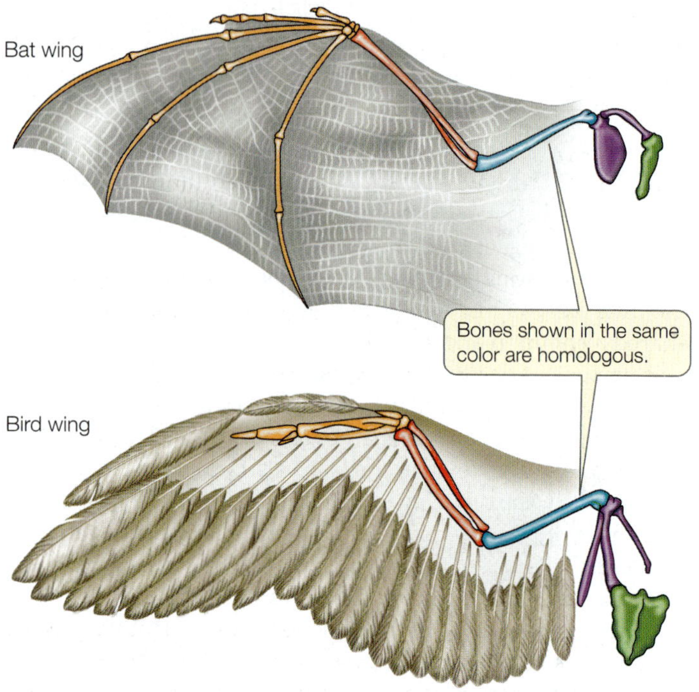
Bat wing
Bones shown in the same color are homologous.
Bird wing

22.4 The Bones Are Homologous, the Wings Are Not
The supporting bone structures of both bat wings and bird wings are derived from a common four-limbed ancestor and are thus homologous. However, the wings themselves—an adaptation for flight—evolved independently in the two groups.

RECAP 22.1

A phylogenetic tree is a description of evolutionary relationships among organisms or their genes. All living organisms share a common ancestor and are related through the phylogenetic tree of life.

- Describe the different elements of a phylogenetic tree. See p. 450 and Figure 22.2
- Explain the difference between an ancestral and a derived trait. See p. 450
- How might similar traits arise independently in species that are only distantly related? See p. 452 and Figure 22.4

Phylogenetic analyses have become increasingly important to many types of biological research in recent years, and they are the basis for the comparative nature of biology. For the most part, however, evolutionary history cannot be observed directly. How, then, do biologists reconstruct the past? One way is by using phylogenetic analyses to construct a tree.

22.2 How Are Phylogenetic Trees Constructed?

To illustrate how a phylogenetic tree is constructed, consider the eight vertebrate animals listed in Table 22.1: lamprey, perch, salamander, lizard, crocodile, pigeon, mouse, and chimpanzee.

TABLE **22.1**

Eight Vertebrates and the Presence or Absence of Some Shared Derived Traits

	Derived Trait							
Taxon	Jaws	Lungs	Claws or nails	Gizzard	Feathers	Fur	Mammary glands	Keratinous scales
Lamprey (outgroup)	–	–	–	–	–	–	–	–
Perch	+	–	–	–	–	–	–	–
Salamander	+	+	–	–	–	–	–	–
Lizard	+	+	+	–	–	–	–	+
Crocodile	+	+	+	+	–	–	–	+
Pigeon	+	+	+	+	+	–	–	+
Mouse	+	+	+	–	–	+	+	–
Chimpanzee	+	+	+	–	–	+	+	–

We will assume initially that any given derived trait evolved only once during the evolution of these animals (that is, there has been no convergent evolution) and that no derived traits were lost from any of the descendant groups (there has been no evolutionary reversal). For simplicity, we have selected traits that are either present (+) or absent (–).

In a phylogenetic analysis, the group of organisms of primary interest is called the **ingroup**. As a point of reference, an ingroup is compared with an **outgroup**, a species or group that is closely related to the ingroup but is known to be phylogenetically outside it; the root of the tree is located between the ingroup and the outgroup. Any trait that is present in both the ingroup and the outgroup must have evolved before the origin of the ingroup and thus must be ancestral for the ingroup. In contrast, traits that are present only in some members of the ingroup must be derived traits within that ingroup. As we will see in Chapter 33, a group of jawless fishes called the lampreys is thought to have separated from the lineage leading to the other vertebrates before the jaw arose. Therefore we have included the lamprey as the outgroup for our analysis. Because derived traits are traits acquired by other members of the vertebrate lineage *after* they diverged from the outgroup, any trait that is present in both the lamprey and the other vertebrates is judged to be ancestral.

We begin by noting that the chimpanzee and mouse share two derived traits—mammary glands and fur—that are absent in both the outgroup and the other species of the ingroup. We then infer that mammary glands and fur are derived traits that evolved in a common ancestor of chimpanzees and mice after that lineage separated from the lineages leading to the other vertebrates. In other words, we provisionally assume that mammary glands and fur evolved only once among the animals in our ingroup. These traits are synapomorphies that unite chimpanzees and mice (as well as all other mammals, although we have not included other mammalian species in this example). By the same reasoning, we can infer that the other shared derived traits are synapomorphies for the various groups in which they are expressed. For instance, keratinous scales are a synapomorphy of the lizard, crocodile, and pigeon.

Table 22.1 also tells us that, among the animals in our ingroup, the pigeon has a unique trait: the presence of feathers. As we discussed in Section 22.1, feathers are a synapomorphy of birds and some of their extinct dinosaur relatives. But since we only have one bird (and no extinct species) in this example, the presence of feathers provides no clues concerning relationships among the eight species of vertebrates we have sampled. Gizzards, however, are found in both birds and crocodiles, so this trait is evidence of a close relationship between birds and crocodilians.

By combining information about the various synapomorphies, we can construct a phylogenetic tree. We infer, for example, that mice and chimpanzees, the only two animals that share fur and mammary glands in our example, share a more recent common ancestor with each other than they do with pigeons and crocodiles. Otherwise we would need to assume that the ancestors of pigeons and crocodiles also had fur and mammary glands but subsequently lost them—unnecessary additional assumptions.

Figure 22.5 shows a phylogenetic tree for the vertebrates listed in Table 22.1, based on the shared derived traits we examined and the assumption that each derived trait evolved only once. This particular tree was easy to construct because the animals and characters we chose met the assumptions that derived traits appeared only once and were never lost after they appeared. Had we included a snake in the group, our second assumption would have been violated, because we know that the lizard ancestors of snakes had limbs that were subsequently lost. We would need to examine additional characters to determine that the lineage leading to snakes separated from the one leading to lizards long after the lineage leading to lizards separated from the others. In fact, the analysis of several characters shows that snakes evolved from burrowing lizards that became adapted to a subterranean existence.

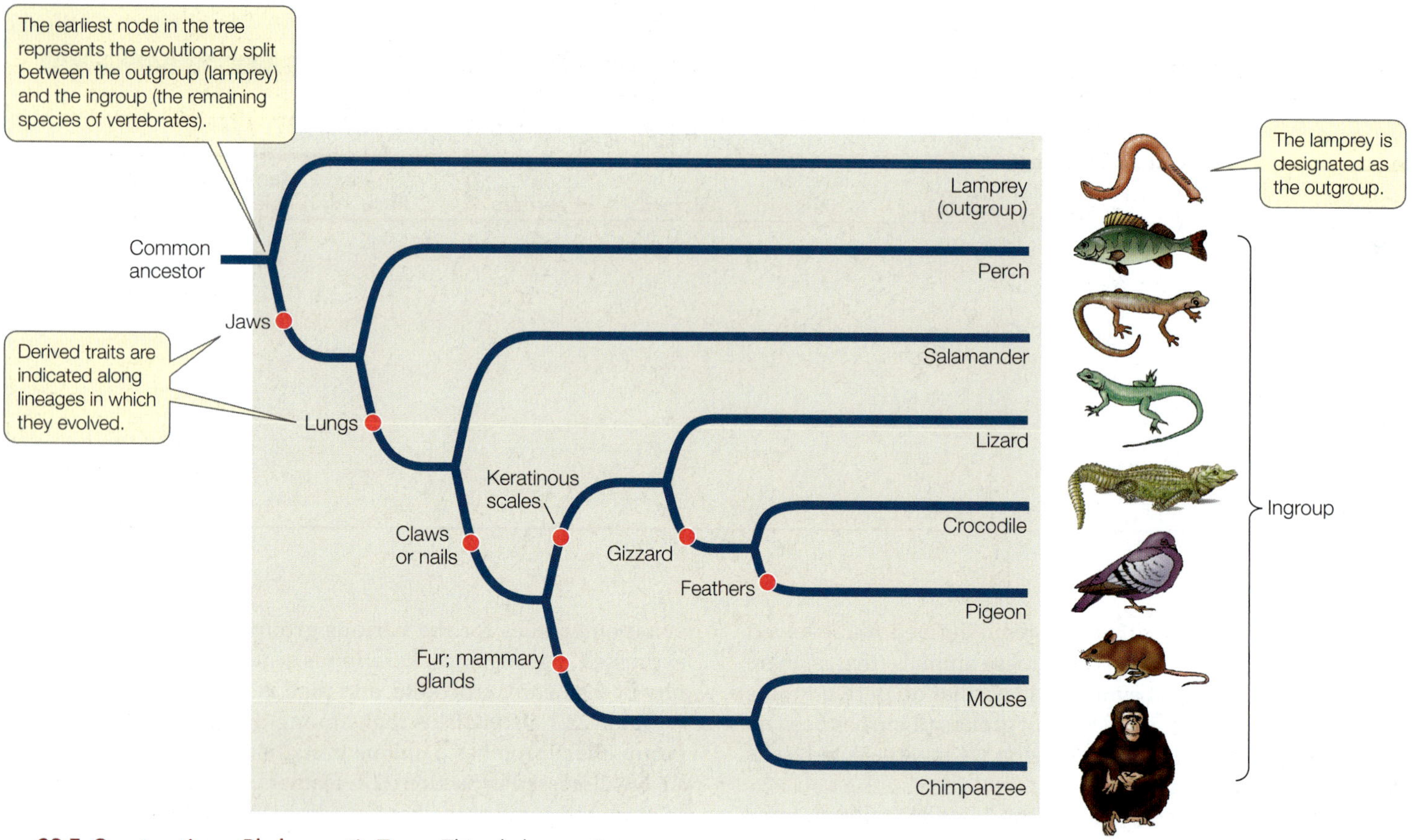

The earliest node in the tree represents the evolutionary split between the outgroup (lamprey) and the ingroup (the remaining species of vertebrates).
The lamprey is designated as the outgroup.
Common ancestor
Jaws
Derived traits are indicated along lineages in which they evolved.
Lungs
Claws or nails
Keratinous scales
Gizzard
Feathers
Fur; mammary glands
Lamprey (outgroup)
Perch
Salamander
Lizard
Crocodile
Pigeon
Mouse
Chimpanzee
Ingroup

22.5 Constructing a Phylogenetic Tree This phylogenetic tree was constructed from the information given in Table 22.1 using the parsimony principle. Each clade in the tree is supported by at least one shared derived trait, or synapomorphy.

Go to Activity 22.1 Constructing a Phylogenetic Tree
Life10e.com/ac22.1

Parsimony provides the simplest explanation for phylogenetic data

The phylogenetic tree shown in Figure 22.5 is based on only a very small sample of traits. Typically, biologists construct phylogenetic trees using hundreds or thousands of traits. With larger data sets, we would expect to observe some traits that have changed more than once, and thus we would expect to see some convergence and evolutionary reversal. How do we determine which traits are synapomorphies and which are homoplasies? One way is to invoke the principle of parsimony.

In its most general form, the principle of **parsimony** states that the preferred explanation of our observations is the simplest explanation. Applying the parsimony principle to the construction of phylogenetic trees entails minimizing the number of evolutionary changes that need to be assumed over all characters in all taxa in the tree. In other words, the best hypothesis under the parsimony principle is the one that requires the fewest homoplasies. This application of parsimony is a specific case of a general principle of logic called Occam's razor: the best explanation is the one that fits the data best while making the fewest assumptions.

We apply the parsimony principle in constructing phylogenetic trees not because all evolutionary change occurs parsimoniously, but because it is logical to adopt the simplest explanation that can account for the observed data. More complicated explanations are accepted only when the evidence requires them. Phylogenetic trees represent our best estimates about evolutionary relationships. They are continually modified as additional evidence becomes available.

Phylogenies are reconstructed from many sources of data

Naturalists have constructed various forms of phylogenetic trees for more than 150 years. In fact, the only figure in the first edition of *On the Origin of Species* was a phylogenetic tree. Tree construction has been revolutionized, however, by the advent of computer software for trait analysis and tree construction, allowing us to consider far more data than could ever before be processed. By combining these methods with the massive comparative data sets being generated through studies of genomes, biologists are learning details about the tree of life at a remarkable pace (see Appendix A: The Tree of Life).

Any trait that is genetically determined, and therefore heritable, can be used in a phylogenetic analysis. Evolutionary relationships can be revealed through studies of morphology, development, the fossil record, behavioral traits, and molecular traits such as DNA and protein sequences. Let's take a closer look at the types of data used in modern phylogenetic analyses.

Go to Animated Tutorial 22.1 Using Phylogenetic Analysis to Reconstruct Evolutionary History
Life10e.com/at22.1

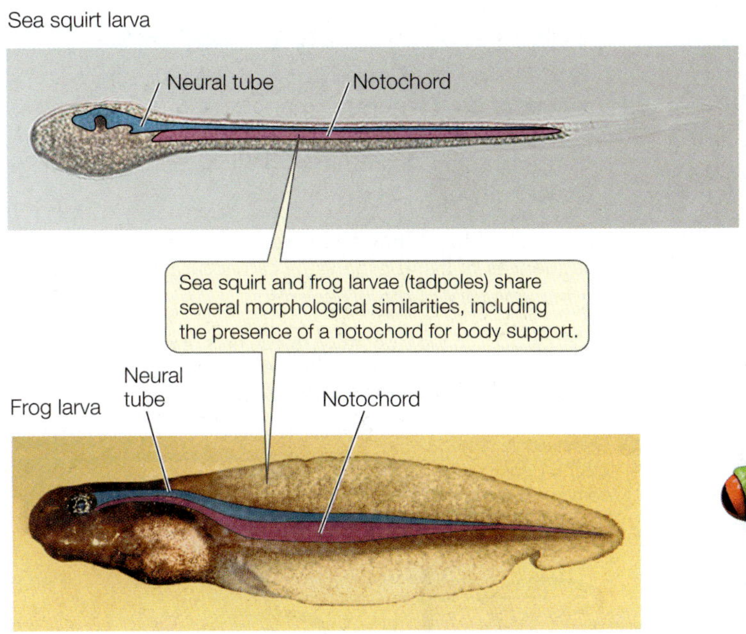
Sea squirt larva
Neural tube
Notochord
Sea squirt and frog larvae (tadpoles) share several morphological similarities, including the presence of a notochord for body support.
Frog larva
Neural tube
Notochord

Adult
Adult
Despite the similarity of their larvae, the morphology of adult frogs and sea squirts provides little evidence of the common ancestry of these two groups.

22.6 Development Reveals the Evolutionary Relationship between Sea Squirts and Vertebrates All chordates—a taxonomic group that includes sea squirts and frogs—have notochords at some stage of their development. The larvae share similarities that are not apparent in the adults. Such similarities in development can provide useful evidence of evolutionary relationships. The notochord is lost in adult sea squirts. In adult frogs, as in all vertebrates, the vertebral column replaces the notochord as the support structure.

MORPHOLOGY An important source of phylogenetic information is **morphology**: the presence, size, shape, and other attributes of body parts. Since living organisms have been observed, depicted, collected, and studied for millennia, we have a wealth of recorded morphological data as well as extensive museum and herbarium collections of organisms whose traits can be measured. New technological tools, such as the electron microscope and computed tomography (CT) scans, enable systematists to examine and analyze the structures of organisms at much finer scales than was formerly possible.

Most species are described and known primarily by their morphology, and morphology provides the most comprehensive data set available for many taxa. The features of morphology that are important for phylogenetic analysis are often specific to a particular group of organisms. For example, the presence, development, shape, and size of various features of the skeletal system are important for the study of vertebrate phylogeny, whereas floral structures are important for studying the relationships among flowering plants.

Morphological approaches to phylogenetic analysis have some limitations, however. Some taxa exhibit little morphological diversity despite great species diversity. For example, the phylogeny of the leopard frogs of North and Central America would be difficult to infer from morphological differences alone because the many species look very similar, despite important differences in their behavior and physiology. At the other extreme, few morphological traits can be compared across distantly related species (earthworms and mammals, for instance). Furthermore, some morphological variation has an environmental (rather than a genetic) basis and so must be excluded from phylogenetic analyses. An accurate phylogenetic analysis often requires information beyond that supplied by morphology.

DEVELOPMENT Similarities in developmental patterns may reveal evolutionary relationships. Some organisms exhibit similarities in early developmental stages only. The larvae of marine creatures called sea squirts, for example, have a flexible gelatinous rod in the back—the notochord—that disappears as the larvae develop into adults. All vertebrate animals also have a notochord at some time during their development (**Figure 22.6**). This shared structure is one of the reasons for inferring that sea squirts are more closely related to vertebrates than would be suspected if only adult sea squirts were examined.

PALEONTOLOGY The fossil record is another important source of information on evolutionary history. Fossils show us where and when organisms lived in the past and give us an idea of what they looked like. Fossils provide important evidence that helps us distinguish ancestral from derived traits. The fossil record can also reveal when lineages diverged and began their independent evolutionary histories. Furthermore, in groups with few species that have survived to the present, information on extinct species is often critical to an understanding of the large divergences among the surviving species. The fossil record does have limitations, however. Few or no fossils have been found for some groups, and the fossil record for many groups is fragmentary.

BEHAVIOR Some behavioral traits are culturally transmitted and some are inherited. If a particular behavior is culturally transmitted, it may not accurately reflect evolutionary relationships (but may nonetheless reflect cultural connections). Bird songs, for instance, are often learned and may be inappropriate traits for phylogenetic analysis. Frog calls, however, are genetically determined and appear to be acceptable sources of information for reconstructing phylogenies.

MOLECULAR DATA All heritable variation is encoded in DNA, so the complete genome of an organism contains an enormous set of traits (the individual nucleotide bases of DNA) that can be used in phylogenetic analyses. In recent years, DNA sequences have become among the most widely used sources of data for constructing phylogenetic trees. Comparisons of nucleotide sequences are not limited to the DNA in the cell nucleus. Eukaryotes have genes in their mitochondria as well as in their nuclei; plant cells also have genes in their chloroplasts. The chloroplast genome (cpDNA), which is used extensively in phylogenetic studies of plants, has changed slowly over evolutionary time, so it is often used to study relatively ancient phylogenetic relationships. Most animal mitochondrial DNA (mtDNA) has changed more rapidly, so mitochondrial genes have been used extensively to study evolutionary relationships among closely related animal species (the mitochondrial genes of plants evolve more slowly). Many nuclear gene sequences are also commonly analyzed, and now that many entire genomes have been sequenced, they too are used to construct phylogenetic trees. Information on gene products (such as the amino acid sequences of proteins) is also widely used for phylogenetic analyses, as we will see in Chapter 24.

INVESTIGATING**LIFE**

22.7 Testing the Accuracy of Phylogenetic Analysis To test whether analysis of gene sequences can accurately reconstruct evolutionary phylogeny, we must have an unambiguously known phylogeny to compare against the reconstruction. Will the observed phylogeny match the reconstruction?[a]

HYPOTHESIS A phylogeny reconstructed by analyzing the DNA sequences of living organisms can accurately match the known evolutionary history of the organisms.

Method In the laboratory, researchers produced an unambiguous phylogeny of nine viral lineages, enhancing the mutation rate to increase variation among the lineages.

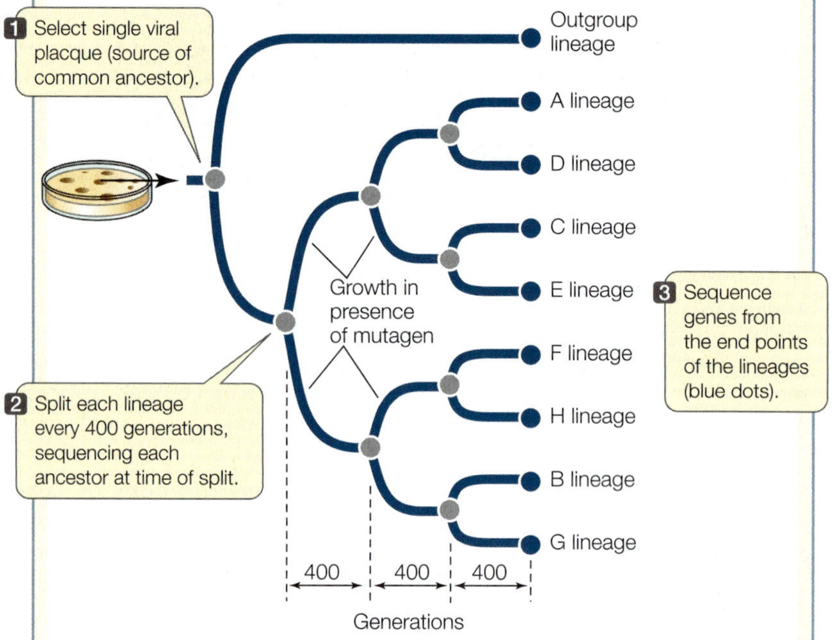

1 Select single viral plaque (source of common ancestor).
2 Split each lineage every 400 generations, sequencing each ancestor at time of split.
3 Sequence genes from the end points of the lineages (blue dots).
Outgroup lineage
A lineage
D lineage
C lineage
E lineage
F lineage
H lineage
B lineage
G lineage
Growth in presence of mutagen
400
400
400
Generations

Viral sequences from the end points of each lineage (blue dots) were subjected to phylogenetic analysis by investigators who were unaware of the history of the lineages or the gene sequences of the ancestral viruses. These investigators reconstructed the phylogeny and ancestral DNA sequences based solely on their analyses of the descendants' genomes.

Results The true phylogeny and ancestral DNA sequences were accurately reconstructed solely from the DNA sequences of the viruses at the tips of the tree branches.

CONCLUSION Phylogenetic analysis of DNA sequences can accurately reconstruct evolutionary history.

Go to **BioPortal** for discussion and relevant links for all INVESTIGATING**LIFE** figures.

[a]Hillis, D. M. et al. 1992. *Science* 255: 589–592.

Mathematical models expand the power of phylogenetic reconstruction

As biologists began to use DNA sequences to construct phylogenetic trees in the 1970s and 1980s, they developed explicit mathematical models describing how DNA sequences change over time. These models account for multiple changes at a given position in a DNA sequence. They also take into account different rates of change at different positions in a gene, at different positions in a codon, and among different nucleotides (see Section 24.1). For example, transitions (changes between two purines or between two pyrimidines) are usually more likely than are transversions (changes between a purine and pyrimidine).

Mathematical models can be used to compute how a tree might evolve given the observed data. **Maximum likelihood** methods identify the tree that is most likely to have produced the observed data, given the assumptions of the model. Maximum likelihood methods can be used for any kind of characters, but they are most often used with molecular data, for which explicit mathematical models of evolutionary change are easier to develop. The principal advantages of maximum likelihood analyses are that they incorporate more information about evolutionary change than do parsimony methods, and they are easier to treat in a statistical framework. The principal disadvantages are that they are computationally intensive and require explicit mathematical models of evolutionary change (which are difficult to develop for some kinds of characters).

WORKING WITH**DATA**:

Does Phylogenetic Analysis Correctly Reconstruct Evolutionary History?

Original Papers

Hillis, D. M., J. J. Bull, M. E. White, M. R. Badgett, and I. J. Molineux. 1992. Experimental phylogenetics: Generation of a known phylogeny. *Science* 255: 589–592.

Bull, J. J., C. W. Cunningham, I. J. Molineux, M. R. Badgett, and D. M. Hillis. 1993. Experimental molecular evolution of bacteriophage T7. *Evolution* 47: 993–1007.

Analyze the Data

Refer to the description of Hillis and colleagues' experiment with T7 virus below and in Figure 22.7. The full DNA sequences for the viral lineages produced in this experiment are thousands of nucleotides long. However, 23 of the nucleotide positions are shown in the table below, and you can use these data to repeat the researchers' analysis. Each nucleotide position represents a separate character.

QUESTION 1

Construct a phylogenetic tree from the nucleotide positions using the parsimony principle (see Section 22.2 and the examples in Table 22.1 and Figure 22.5). Use the outgroup to root your tree. Assume that all changes among nucleotides are equally likely.

QUESTION 2

Using your tree from Question 1, can you reconstruct the DNA sequences of the ancestral lineages?

QUESTION 3

Why did the investigators use a blind study design, in which the true identities of the viral lineages were not revealed until the analyses were complete? What potential for bias were they avoiding?

QUESTION 4

Transitions are mutations that change one purine to the other (G ↔ A) or one pyrimidine to the other (C ↔ T), whereas transversions exchange a purine for a pyrimidine or vice versa (e.g., A → C or T; C → A or G). Which kind of mutation predominates in this phylogeny? Why might this be the case?

	Character at position																						
Lineage	1	2	3	4	5	6	7	8	9	10	11	12	13	14	15	16	17	18	19	20	21	22	23
A	T	C	G	G	G	C	C	C	C	C	C	C	A	A	C	C	G	A	T	A	C	A	A
B	C	C	G	G	G	T	C	C	C	T	C	C	G	A	T	T	A	G	C	G	T	G	G
C	C	C	G	G	G	C	C	C	T	C	C	T	A	A	C	C	G	G	T	A	C	A	A
D	T	C	A	G	G	C	C	C	C	C	C	C	A	A	C	C	G	A	T	A	C	A	A
E	C	T	G	G	G	C	C	C	C	C	C	T	A	A	C	C	G	G	T	A	C	A	A
F	C	T	G	A	A	C	C	C	C	C	C	C	G	A	C	T	G	G	C	G	C	G	G
G	C	C	G	G	G	T	T	C	C	T	C	C	G	A	T	T	A	G	C	G	C	G	G
H	C	C	G	G	A	C	C	C	C	C	C	C	G	C	C	T	G	G	C	G	C	G	G
Outgroup	C	C	G	G	G	C	C	T	C	C	T	C	G	A	C	C	G	G	C	A	C	G	G

Go to **BioPortal** for all WORKING WITH**DATA** exercises

The accuracy of phylogenetic methods can be tested

If phylogenetic trees represent reconstructions of past events, and if many of these events occurred before any humans were around to witness them, how can we test the accuracy of phylogenetic methods? Biologists have conducted experiments both in living organisms and with computer simulations that have demonstrated the effectiveness and accuracy of phylogenetic reconstruction methods.

In one such experiment, David Hillis, James Bull, and their colleagues at the University of Texas used a single viral culture of bacteriophage T7 as a starting point and allowed lineages to evolve from this ancestral virus in the laboratory (**Figure 22.7**). The initial culture was split into two separate lineages, one of which became the ingroup for analysis; the other lineage became the outgroup used for rooting the tree. Mutagens were added to the viral cultures to increase the mutation rate so that the amount of change and the degree of homoplasy would be typical of the organisms analyzed in average phylogenetic analyses. The lineages in the ingroup were split in two after every 400 generations and samples of the virus were saved for analysis at each of these branching points. The lineages were allowed to evolve until there were eight lineages in the ingroup. The investigators then sequenced samples from the end points of the eight lineages as well as from the ancestors at the branching points. They then gave the sequences from the end points to other investigators to analyze, without revealing the known history of the lineages or the sequences of the ancestral viruses.

After the phylogenetic analysis was completed, the investigators asked two questions: Did phylogenetic methods reconstruct the known history correctly, and were the sequences of the ancestral viruses reconstructed accurately? The answer in both cases was yes: the branching order of the lineages was reconstructed exactly as it had occurred, more than 98 percent of the nucleotide positions of the ancestral viruses were reconstructed correctly, and 100 percent of the amino acid changes in the viral proteins were reconstructed correctly.

The experiment shown in Figure 22.7 demonstrated that phylogenetic analysis was accurate under the conditions tested, but it did not examine all possible conditions. Other experimental studies have taken other factors into account, such as the sensitivity of phylogenetic analysis to convergence under similar environments or to highly variable rates of evolutionary change.

Computer simulations based on mathematical models of evolutionary change have also been used extensively to study the effectiveness of phylogenetic analysis. These studies too have confirmed the accuracy of phylogenetic methods and have been used to refine those methods and extend them to new applications.

RECAP 22.2

Phylogenetic trees can be constructed by using the parsimony principle to find the simplest explanation for phylogenetic data. Maximum likelihood methods incorporate more explicit mathematical models of evolutionary change to reconstruct evolutionary history.

- Describe the process of reconstructing a phylogenetic tree. **See p. 453 and Figure 22.5**
- What are two methods biologists have used to test whether phylogenetic trees provide accurate reconstructions of evolutionary history? **See pp. 457–458 and Figure 22.7**

Biologists in many fields now routinely reconstruct the phylogenetic relationships of organisms. Let's examine some of the many uses of these phylogenetic trees.

22.3 How Do Biologists Use Phylogenetic Trees?

Information about the evolutionary relationships among organisms is useful to scientists investigating a wide variety of biological questions. In this section we will illustrate how phylogenetic trees can be used to ask questions about the past and to compare aspects of the biology of organisms in the present.

Phylogenetic trees can be used to reconstruct past events

Reconstruction of past events is important for understanding many biological processes. In the case of zoonotic diseases (diseases caused by infectious organisms transmitted to humans from another animal host), for example, it is important to understand when, where, and how the disease first entered a human population. Human immunodeficiency virus (HIV) is the cause of such a zoonotic disease: acquired immunodeficiency syndrome, or AIDS. Phylogenetic analyses have become important for studying the transmission of viruses such as HIV. They are also important for understanding the present global diversity of such viruses and for determining their origins in human populations.

A broad phylogenetic analysis of immunodeficiency viruses shows that humans acquired these viruses from two different hosts: HIV-1 from chimpanzees, and HIV-2 from sooty mangabeys (**Figure 22.8**). HIV-1 is the common form of the virus in human populations in central Africa, where chimpanzees are hunted for food, and HIV-2 is the common form in human populations in western Africa, where sooty mangabeys are hunted for food. Thus it seems likely that these viruses entered human populations through hunters who cut themselves while skinning chimpanzees or sooty mangabeys. The relatively recent global pandemic of AIDS occurred when these infections in local African populations rapidly spread through human populations around the world.

In recent years, phylogenetic analysis has become important in forensic investigations that involve viral transmission events. For example, phylogenetic analysis was critical for a criminal investigation of a physician who was accused of purposefully injecting blood from one of his HIV-positive patients into his former girlfriend in an attempt to kill her. The phylogenetic analysis revealed that the HIV strains present in the girlfriend were a subset of those present in the physician's patient (**Figure 22.9**). Other evidence was needed, of course, to connect the physician to this purposeful transmission event, but the phylogenetic analysis was important to support the contention that the virus had been transmitted from the patient to the victim.

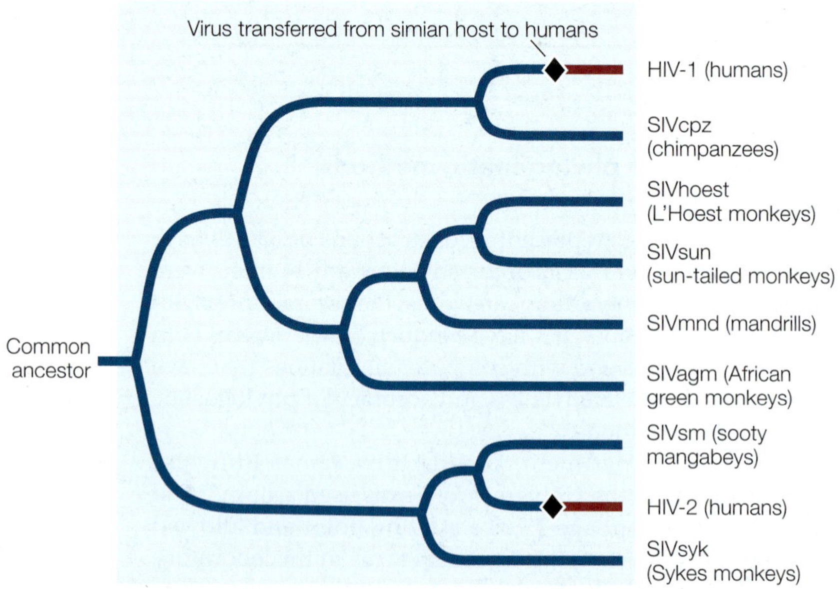
Virus transferred from simian host to humans
Common ancestor
HIV-1 (humans)
SIVcpz (chimpanzees)
SIVhoest (L'Hoest monkeys)
SIVsun (sun-tailed monkeys)
SIVmnd (mandrills)
SIVagm (African green monkeys)
SIVsm (sooty mangabeys)
HIV-2 (humans)
SIVsyk (Sykes monkeys)

22.8 Phylogenetic Tree of Immunodeficiency Viruses The evolutionary relationships of immunodeficiency viruses show that these viruses have been transmitted to humans from two different simian hosts: HIV-1 from chimpanzees and HIV-2 from sooty mangabeys. (SIV stands for simian immunodeficiency virus.)

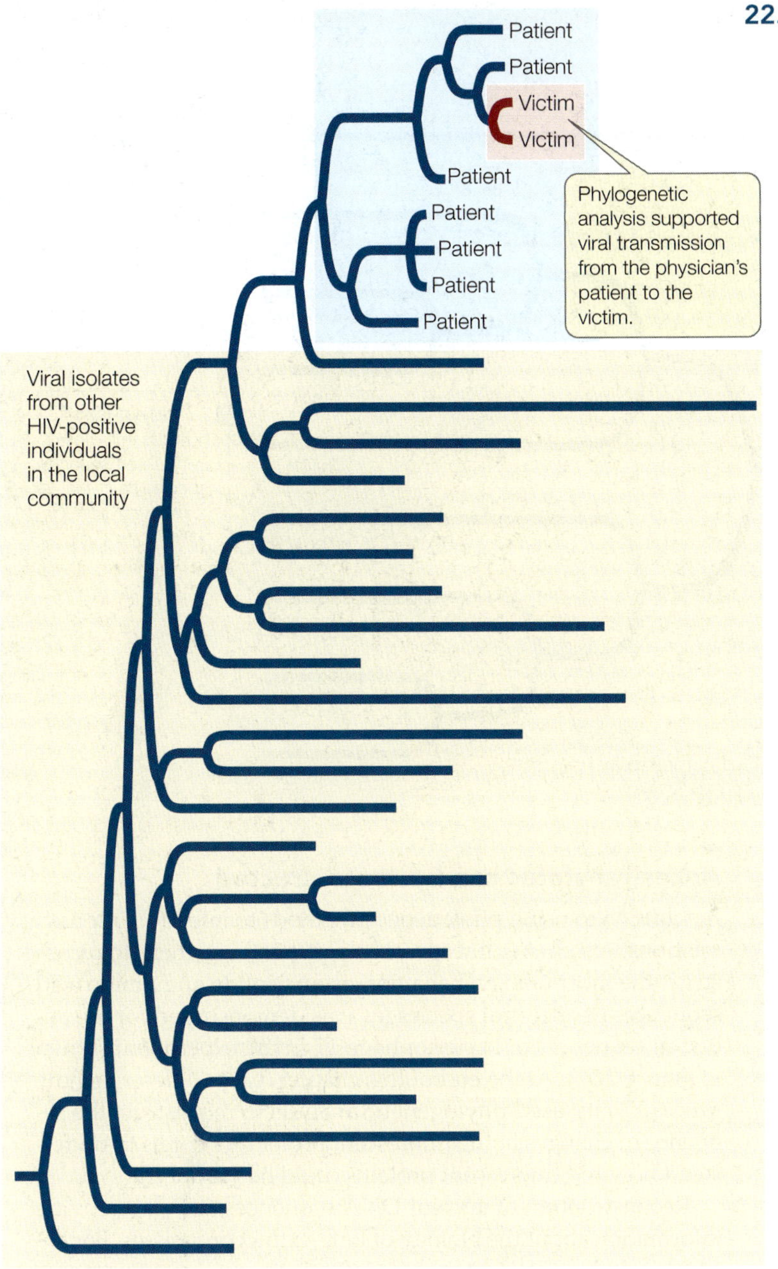
Patient
Patient
Victim
Victim
Patient
Patient
Patient
Patient
Patient
Phylogenetic analysis supported viral transmission from the physician's patient to the victim.
Viral isolates from other HIV-positive individuals in the local community

22.9 A Forensic Application of Phylogenetic Analysis This phylogenetic analysis demonstrated that strains of HIV virus present in a victim (shown in red) were a phylogenetic subset of viruses isolated from a physician's patient (shown in blue). This analysis was part of the evidence used to show that the physician drew blood from his HIV-positive patient and injected it into the victim in an attempt to kill her. The physician was found guilty of attempted murder by the jury.

Phylogenies allow us to compare and contrast living organisms

Male swordtails—a group of fishes in the genus *Xiphophorus*—have a long, colorful tail extension, and their reproductive success is closely associated with this appendage. Males with a long sword are more likely to mate successfully than are males with a short sword (an example of sexual selection; see Sections 21.2 and 23.5). Several explanations have been advanced for the evolution of this structure, including the hypothesis that the sword simply exploits a preexisting bias in the female sensory system—i.e., that female swordtails had a preference for males with long tails even before the tails evolved (perhaps because females assess the size of males by their total body length—including the tail—and prefer larger males).

To test this sensory exploitation hypothesis, phylogenetic reconstruction was used to identify the relatives of swordtails that had split most recently from their lineage before the evolution of swords. These closest relatives turned out to be the platyfishes, another group of *Xiphophorus*. Even though male platyfish do not normally have swords, when researchers attached artificial sword-like structures to the tails of some male platyfish, female platyfish preferred those males, thus providing support for the sensory exploitation hypothesis (**Figure 22.10**).

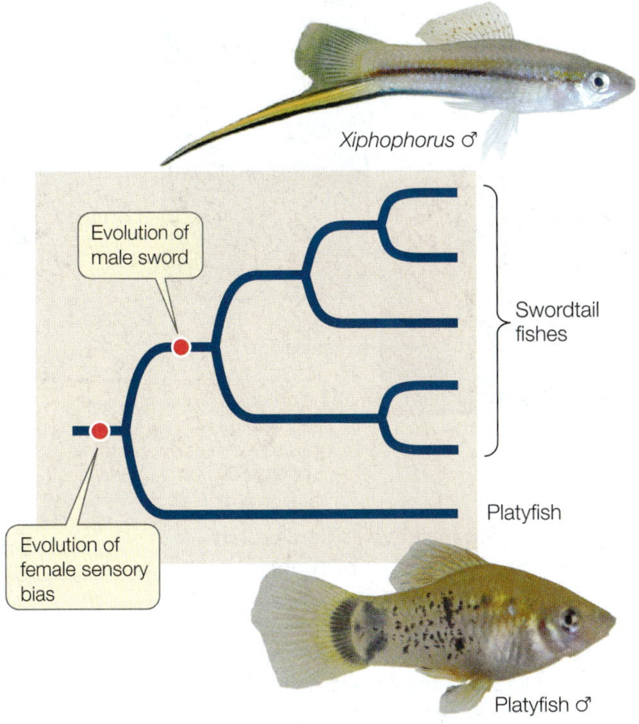
Xiphophorus ♂
Evolution of male sword
Swordtail fishes
Platyfish
Evolution of female sensory bias
Platyfish ♂

22.10 The Origin of a Sexually Selected Trait The tail extension of male swordtails (genus *Xiphophorus*) apparently evolved through sexual selection, as females mated preferentially with males that had long "swords." Phylogenetic analysis reveals that the platyfishes split from the swordtails before the evolution of the sword. The independent finding that female platyfish prefer male platyfish with an artificial sword further supports the idea that this appendage evolved as a result of a preexisting preference in females.

Phylogenies can reveal convergent evolution

Like most animals, many flowering plants (angiosperms) reproduce by mating with another individual of the same species. But in many angiosperm species, the same individual produces both male and female gametes (contained within pollen and ovules, respectively). Self-incompatible plant species have mechanisms to prevent fertilization of the ovule by the

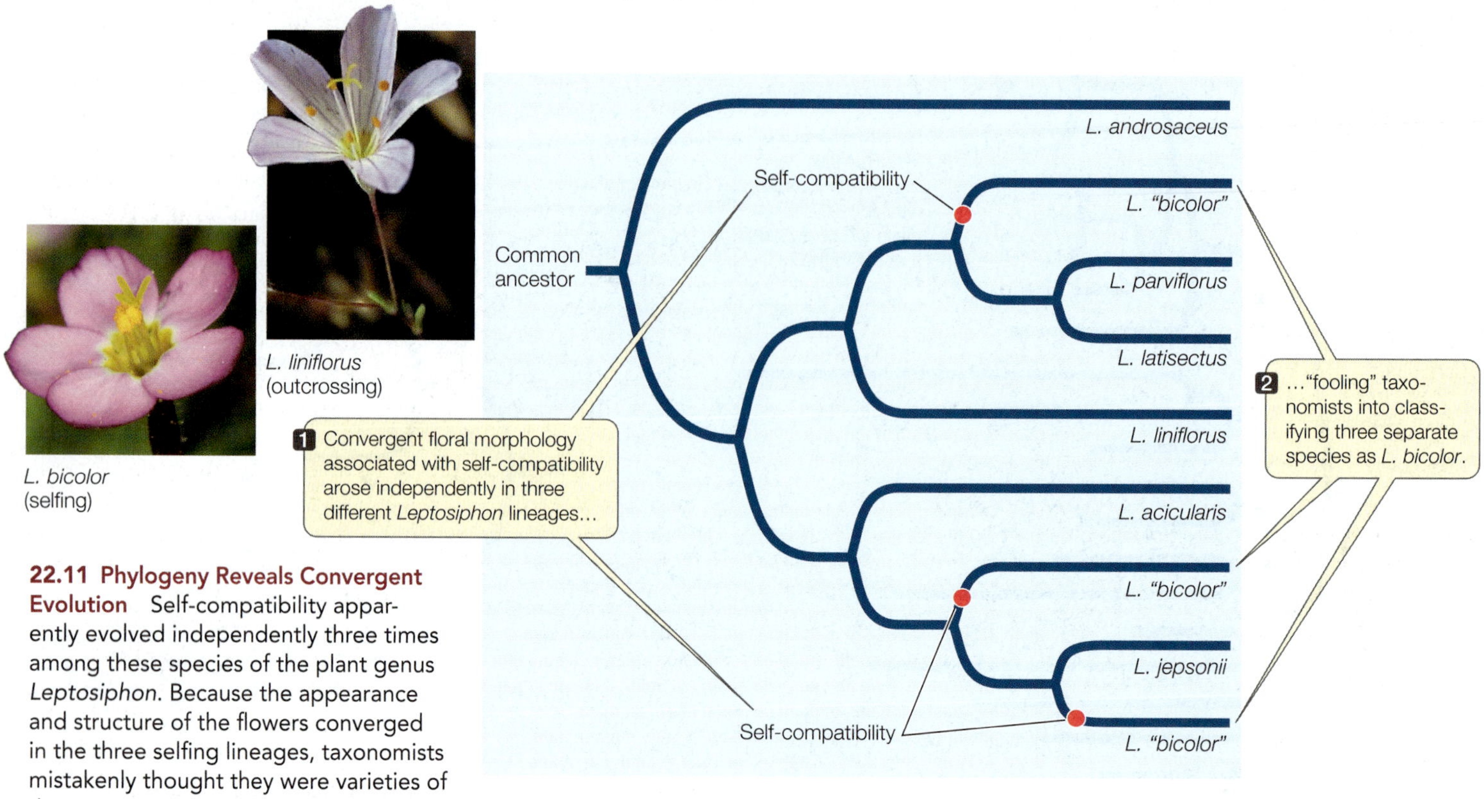
L. liniflorus
(outcrossing)
L. bicolor
(selfing)
Common ancestor
Self-compatibility
L. androsaceus
L. "bicolor"
L. parviflorus
L. latisectus
L. liniflorus
L. acicularis
L. "bicolor"
L. jepsonii
L. "bicolor"
Self-compatibility
1 Convergent floral morphology associated with self-compatibility arose independently in three different Leptosiphon lineages…
2 …"fooling" taxonomists into classifying three separate species as L. bicolor.

22.11 Phylogeny Reveals Convergent Evolution Self-compatibility apparently evolved independently three times among these species of the plant genus *Leptosiphon*. Because the appearance and structure of the flowers converged in the three selfing lineages, taxonomists mistakenly thought they were varieties of the same species.

individual's own pollen, and so must reproduce by outcrossing with another individual. Individuals of some species, however, regularly fertilize their ovules using their own pollen; they are referred to as self-fertilizing, or selfing, species, and their gametes as self-compatible.

The evolution of angiosperm fertilization mechanisms was examined in *Leptosiphon*, a genus in the phlox family that exhibits a diversity of mating systems and pollination mechanisms. The self-incompatible (outcrossing) species of *Leptosiphon* have long petals and are pollinated by long-tongued flies. In contrast, the self-pollinating species have short petals and do not require insect pollinators to reproduce successfully. Using nuclear ribosomal DNA sequences, investigators reconstructed the phylogeny of a subgroup of this genus (**Figure 22.11**). They then determined whether the gametes of each species were self-compatible by artificially pollinating flowers with the plant's own pollen or with pollen from other individuals and observing whether viable seeds formed.

The reconstructed phylogeny suggests that self-incompatibility is the ancestral state and that self-compatibility evolved three times within this group of *Leptosiphon*. The change to self-compatibility eliminated the plants' dependence on pollinators and was accompanied by the evolution of reduced petal size. Indeed, the striking morphological similarity of the flowers in the self-compatible taxa led to their being classified as members of a single species (*L. bicolor*). Phylogenetic analysis, however, showed them to be members of three distinct lineages. From this information, we can infer that self-compatibility and its associated floral structure are the result of convergent evolution in the three independent lineages that had been called *L. bicolor*.

Ancestral states can be reconstructed

In addition to using phylogenetic methods to infer evolutionary relationships among lineages, biologists can use them to reconstruct the morphology, behavior, or nucleotide and amino acid sequences of ancestral species (as was demonstrated for the ancestral sequences of bacteriophage T7 in the experiment shown in Figure 22.7). At the end of this chapter, we will describe how Mikhail Matz used phylogenetic analysis to reconstruct the sequence of changes in the fluorescent proteins of corals to understand how red fluorescent proteins could be produced.

Reconstruction of ancient DNA sequences can also provide information about the biology of long-extinct organisms. For example, phylogenetic analysis was used to reconstruct an opsin protein found in the ancestral archosaur (the most recent common ancestor of birds, dinosaurs, and crocodiles). Opsins are pigment proteins involved in vision; different opsins (with different amino acid sequences) are excited by different wavelengths of light. A team of investigators used a phylogenetic analysis of opsins from living vertebrates to estimate the amino acid sequence of the visual pigment that existed in the ancestral archosaur. A protein with that sequence was then constructed in the laboratory. The investigators tested the reconstructed opsin and found a significant shift toward the red end of the spectrum in the light sensitivity of this protein compared with that of most modern opsins. Modern species that exhibit similar opsin sensitivity are adapted for nocturnal vision, so the investigators inferred that the ancestral archosaur might have been active at night. Thus, reminiscent of the movie *Jurassic Park*, phylogenetic analyses are being used to reconstruct extinct species, one protein at a time.

Go to Animated Tutorial 22.2
Phylogeny and Molecular Evolution
Life10e.com/at22.2

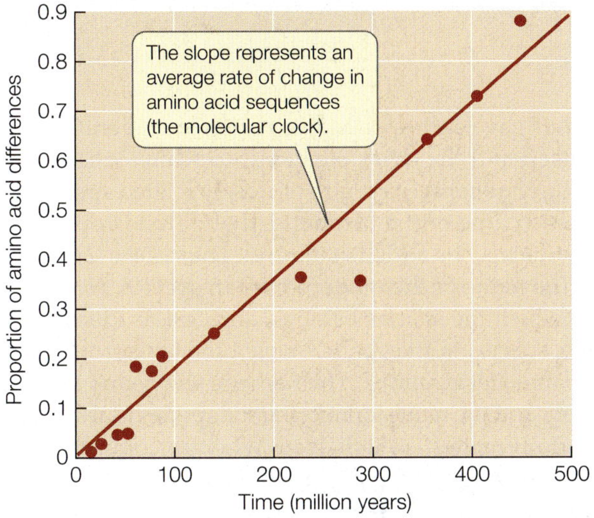
The slope represents an average rate of change in amino acid sequences (the molecular clock).
Proportion of amino acid differences
0.9
0.8
0.7
0.6
0.5
0.4
0.3
0.2
0.1
0
0
100
200
300
400
500
Time (million years)

22.12 A Molecular Clock for the Protein Hemoglobin Amino acid replacements in hemoglobin have occurred at a relatively constant rate over nearly 500 million years of evolution. The graph shows the relationship between the time of divergence and the proportion of amino acids that have changed for 13 pairs of vertebrate hemoglobin proteins. The average rate of change represents the molecular clock for hemoglobin in vertebrates.

Molecular clocks help date evolutionary events

For many applications, biologists want to know not only the order in which evolutionary lineages diverged, but also the timing of those splits. In 1965, Emile Zuckerkandl and Linus Pauling hypothesized that rates of molecular change were constant enough that they could be used to predict evolutionary divergence times—an idea that has become known as the molecular clock hypothesis.

Of course, different genes evolve at different rates, and there are also differences in evolutionary rates among species related to generation times, environments, efficiencies of DNA repair systems, and other biological factors. Nonetheless, among closely related species, a given gene usually evolves at a reasonably constant rate, so the protein encoded by that gene does as well (Figure 22.12). A **molecular clock** is the average rate at which a given gene or protein accumulates changes, and this rate of change can be used to gauge the time of a particular split in the phylogeny. Molecular clocks must be calibrated using independent data, including the fossil record, known times of divergence, or biogeographic dates (such as the dates for separations of continents). Using such calibrations, times of divergence have been estimated for many groups of species that have diverged over millions of years.

Molecular clocks are used not only to date ancient events but also to study the timing of comparatively recent events. For example, most samples of HIV-1 have been collected from humans only since the early 1980s, although a few isolates from medical biopsies are available from as early as the 1950s. But biologists were able to use the observed changes in HIV-1 over the past several decades to extrapolate back to the common ancestor of a group of HIV-1 samples and estimate when HIV-1 first entered human populations from chimpanzees (Figure 22.13). Their molecular clock was calibrated using samples from the 1980s and

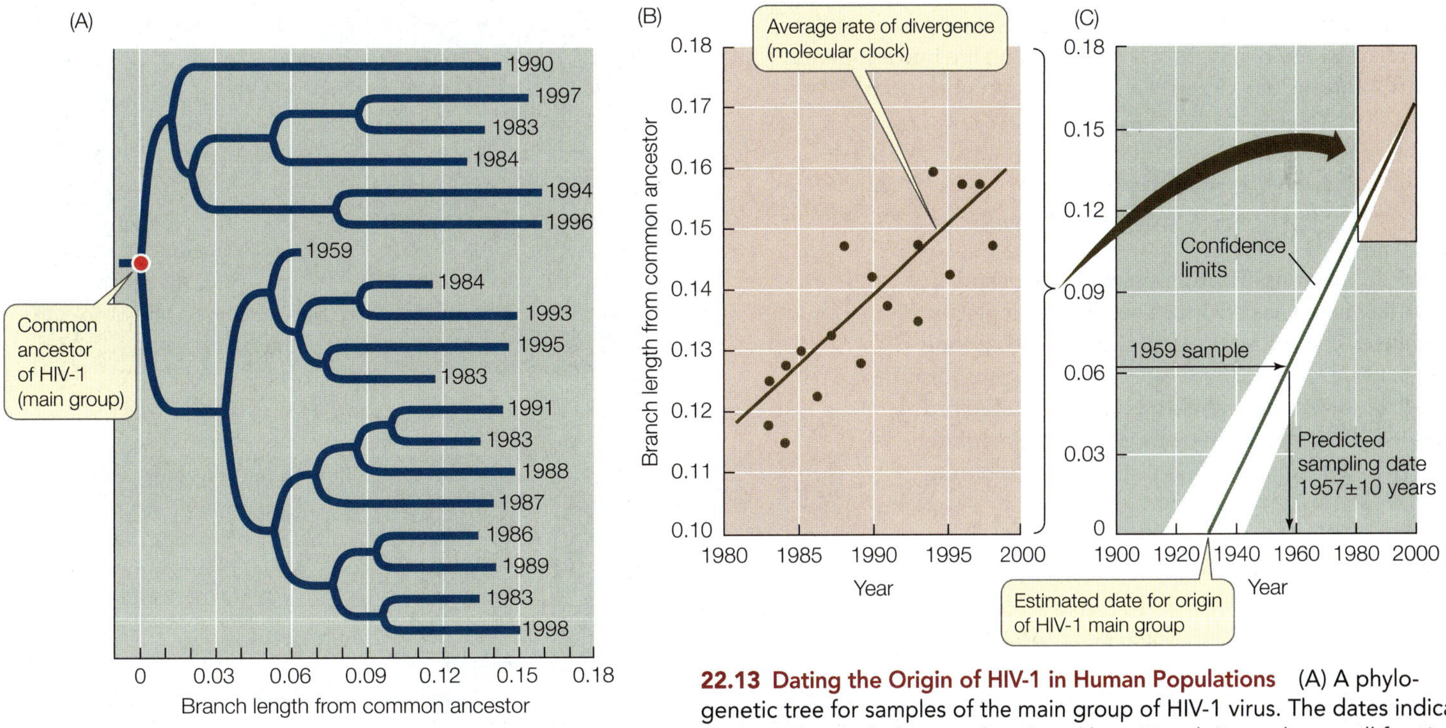
(A)
1990
1997
1983
1984
1994
1996
1959
1984
1993
1995
1983
1991
1983
1988
1987
1986
1989
1983
1998
Common ancestor of HIV-1 (main group)
0
0.03
0.06
0.09
0.12
0.15
0.18
Branch length from common ancestor
(B)
Average rate of divergence (molecular clock)
Branch length from common ancestor
0.18
0.17
0.16
0.15
0.14
0.13
0.12
0.11
0.10
1980
1985
1990
1995
2000
Year
(C)
0.18
0.15
0.12
0.09
0.06
0.03
0
Confidence limits
1959 sample
Predicted sampling date 1957±10 years
1900
1920
1940
1960
1980
2000
Year
Estimated date for origin of HIV-1 main group

22.13 Dating the Origin of HIV-1 in Human Populations (A) A phylogenetic tree for samples of the main group of HIV-1 virus. The dates indicate the years in which the samples were taken. (For clarity, only a small fraction of the samples that were examined in the original study are shown.) (B) A plot of sample year versus genetic divergence from the common ancestor provided an average rate of divergence, or a molecular clock. (C) The molecular clock was used to date a sample taken in 1959 (as a test of the clock) and to estimate the date of origin of the HIV-1 main group (about 1930).

1990s, then tested using samples from the 1950s. As shown in Figure 22.13C, a sample from 1959 was dated by the molecular clock at 1957 ± 10 years. Extrapolation back to the common ancestor of the samples suggested a date of origin for this group of viruses of about 1930. Although AIDS was unknown to Western medicine until the 1980s, this analysis shows that HIV-1 was present (probably at a very low frequency) in human populations in Africa for at least a half-century before its emergence as a global pandemic. Biologists have used similar analyses to conclude that immunodeficiency viruses have been transmitted repeatedly into human populations from multiple primates for more than a century (see also Figure 22.8).

RECAP 22.3

Phylogenetic trees are used to reconstruct the past history of lineages, to determine when and where traits arose, and to make biological comparisons among genes, populations, and species. They can also be used to reconstruct ancestral traits and to estimate the timing of evolutionary events.

- Explain how phylogenetic trees can help determine the number of times a particular trait evolved. **See pp. 459–460 and Figure 22.11**
- How did the reconstruction of ancestral traits help biologists explain the evolution of visual pigment proteins? **See p. 460**
- How do molecular clocks add a time dimension to phylogenetic trees? **See p. 461 and Figure 22.12**

All of life is connected through evolutionary history, and the relationships among organisms provide a natural basis for making biological comparisons. For these reasons, biologists use phylogenetic relationships as the basis for organizing life into a coherent classification system, described in the next section.

22.4 How Does Phylogeny Relate to Classification?

The biological classification system in widespread use today is derived from a system developed by the Swedish biologist Carolus Linnaeus in the mid-1700s. Linnaeus developed a naming system called **binomial nomenclature** that has allowed scientists throughout the world to refer unambiguously to the same organisms by the same names (**Figure 22.14**).

(A) Campanula rotundifolia
(B) Endymion non-scriptus
(C) Mertensia virginica

22.14 Many Different Plants Are Called Bluebells
All three of these distantly related plant species are called "bluebells." Binomial nomenclature allows us to avoid the ambiguity of such common names and communicate exactly what is being described. (A) *Campanula rotundifolia*, found on the North American Great Plains, belongs to a larger group of bellflowers. (B) *Endymion non-scriptus*, English bluebell, is related to hyacinths. (C) *Mertensia virginica*, Virginia bluebell, belongs in a very different group of plants known as borages.

Linnaeus gave each species two names, one identifying the species itself and the other the genus to which it belongs. A **genus** (plural **genera**) is a group of closely related species. Optionally, the name of the taxonomist who first proposed the species name may be added at the end. Thus *Homo sapiens* Linnaeus is the name of the modern human species. *Homo* is the genus to which the species belongs, and *sapiens* identifies the particular species in the genus *Homo*; Linnaeus proposed the species name *Homo sapiens*. The name of the genus is always capitalized, and the name identifying the species is always lowercased. Both names are italicized, whereas common names (humans in this case) of organisms are not. Rather than repeating the name of a genus that is used several times in the same discussion, biologists often spell it out only once and abbreviate it to the initial letter thereafter (*D. melanogaster* rather than *Drosophila melanogaster*, for example).

As noted earlier, any group of organisms that is treated as a unit in a biological classification system, such as the genus *Drosophila*, or all insects, is called a taxon. In the Linnaean system, species and genera are further grouped into a hierarchical system of ranked taxonomic categories. The taxon above the genus in the Linnaean system is the **family**. The names of animal families end in the suffix "–idae." Thus Formicidae is the family that contains all ant species, and the family Hominidae contains humans and our recent fossil relatives as well as our closest living relatives, the chimpanzees and gorillas. Family names are based on the name of a member genus but are not italicized; Formicidae is based on the genus *Formica*, and Hominidae is based on *Homo*. The same rules are used in classifying plants, except that the suffix "-aceae" is used for plant family names instead of "-idae." Thus Rosaceae is the family that includes the genus roses (*Rosa*) and other genera closely related to *Rosa*.

In the Linnaean system, families are grouped into **orders**, orders into **classes**, classes into **phyla** (singular phylum), and phyla into **kingdoms**. However, Linnaean classification is subjective—there are no explicit criteria for deciding if a particular taxon should be treated as an order or a family, for example—and today Linnaean terms are used largely for convenience. Although families are always grouped within orders, orders within classes, and so forth, there is nothing that makes a "family" in one group equivalent (in number of genera or in evolutionary age, for instance) to a "family" in another group.

Linnaeus recognized the overarching hierarchy of life, but he developed his system before evolutionary thought had become widespread. Biologists today recognize the tree of life as the basis for biological classification, and they often name taxa without placing them in any Linnaean rank. But regardless of whether they rank organisms in the Linnaean hierarchy or use unranked taxon names, modern biologists use evolutionary relationships as the basis for distinguishing and naming taxa.

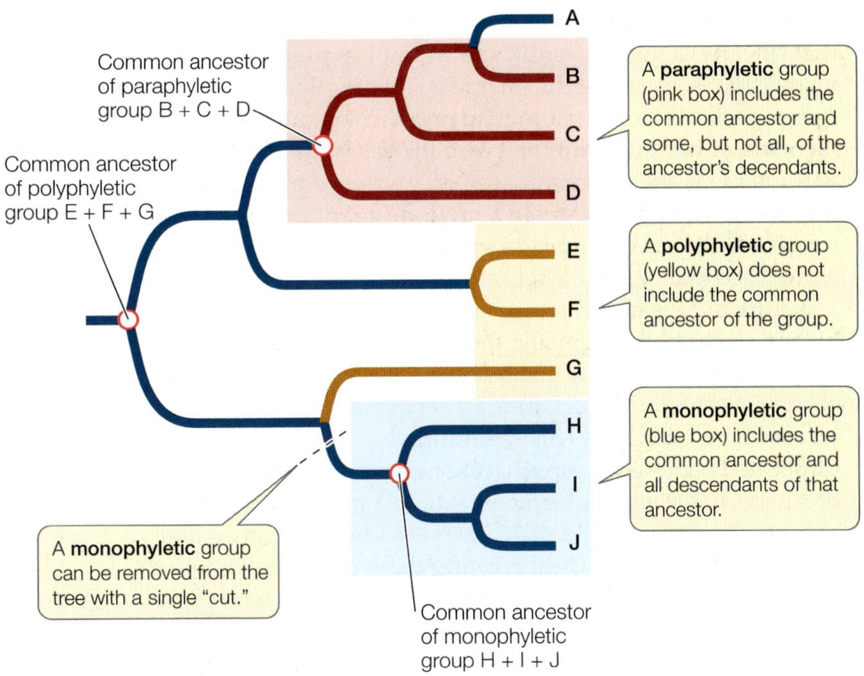
A
B
C
D
E
F
G
H
I
J
Common ancestor of paraphyletic group B + C + D
Common ancestor of polyphyletic group E + F + G
A **paraphyletic** group (pink box) includes the common ancestor and some, but not all, of the ancestor's decendants.
A **polyphyletic** group (yellow box) does not include the common ancestor of the group.
A **monophyletic** group (blue box) includes the common ancestor and all descendants of that ancestor.
A **monophyletic** group can be removed from the tree with a single "cut."
Common ancestor of monophyletic group H + I + J

22.15 Monophyletic, Polyphyletic, and Paraphyletic Groups Monophyletic groups are the basis of taxa in modern biological classifications. Polyphyletic and paraphyletic groups are not appropriate for use in classifications because they do not accurately reflect evolutionary history.

Go to Activity 22.2 Types of Taxa Life10e.com/ac22.2

Evolutionary history is the basis for modern biological classification

Biological classification systems are used to express relationships among organisms. The kind of relationship we wish to express influences which features we use to classify organisms. If, for instance, we were interested in a system that would help us decide what plants and animals were desirable as food, we might devise a classification system based on tastiness, ease of capture, and the number of edible parts each organism possessed. Early Hindu classifications of organisms were designed according to these criteria. Such systems served the needs of the people who developed them, but they are not adequate for formal scientific classification.

Biologists today use systems of classification to express the evolutionary relationships of organisms. Taxa are expected to be **monophyletic**, meaning that the taxon contains an ancestor and all descendants of that ancestor, and no other organisms (**Figure 22.15**). In other words, every taxon should be a complete branch on the tree of life (a clade; see Figure 22.3). Although biologists seek to describe and name only monophyletic taxa, the detailed phylogenetic information needed to do so is not always available. A group that does not include its common ancestor is called a **polyphyletic** group. A group that does not include all the descendants of a common ancestor is called a **paraphyletic** group.

A true monophyletic group (i.e., a clade) can be removed from a phylogenetic tree by a single "cut" in the tree, as shown in Figure 22.15. Note that there are many monophyletic groups on any phylogenetic tree, and that these groups are successively smaller subsets of larger monophyletic groups. This hierarchy of taxa, with all of life as the most inclusive taxon and many smaller taxa within larger taxa down to individual species, is the modern basis for biological classification.

Virtually all taxonomists now agree that polyphyletic and paraphyletic groups are inappropriate as taxonomic units because they do not correctly reflect evolutionary history. The classifications used today still contain such groups, however, because some organisms have not been evaluated phylogenetically. As mistakes in prior classifications are detected, taxonomic names are revised and polyphyletic and paraphyletic groups are eliminated from the classifications.

Several codes of biological nomenclature govern the use of scientific names

Several sets of explicit rules govern the use of scientific names for organisms. Biologists around the world follow these rules voluntarily to facilitate communication and dialogue. Although there may be dozens of common names for an organism in many different languages, the rules of biological nomenclature are designed so that there is only one correct scientific name for any single recognized taxon and so that (ideally) a given scientific name applies to only a single taxon (that is, each scientific name is unique). Sometimes the same species is named more than once (when more than one taxonomist has taken up the task); the rules specify that the valid name is the first name that was proposed. If the same name is inadvertently given to two different species, then a replacement name must be given to the species that was named second.

Because of the historical separation of the fields of zoology, botany (including, originally, the study of fungi), and microbiology, different sets of taxonomic rules were developed for each of these groups. Yet another set of rules for classifying viruses emerged later. This separation has resulted in many duplicated names in groups that are governed by different sets of rules: *Drosophila*, for instance, is both a genus of fruit flies and a genus of fungi, and there are species in both groups that have identical names. Until recently these duplicated names caused little confusion, since traditionally biologists who studied fruit flies were unlikely to read the literature on fungi (and vice versa). Today, however, given the use of large, universal biological databases (such as GenBank, which includes DNA sequences from across the tree of life), it is increasingly important that each taxon have a unique name. Taxonomists are now working to develop common sets of rules that can be applied across all living organisms. For example, a universal system, known as the PhyloCode, that emphasizes the hierarchical and unified nature of the tree of life has been proposed for classifying all species of life.

RECAP 22.4

Biologists organize and classify life by identifying and naming monophyletic groups. Several sets of rules govern the use of scientific names so that each species and higher taxon can be identified and named unambiguously.

- Explain the difference between monophyletic, paraphyletic, and polyphyletic groups. **See p. 463 and Figure 22.15**
- Why do biologists prefer monophyletic groups in formal classifications? **See p. 463**

Now that we have seen how evolution occurs and how phylogenies can be used to study evolutionary relationships, we are ready to consider the process of speciation. Speciation is what leads to the branching events on the tree of life, and it is the process that results in the millions of species that constitute biodiversity.

?

How are phylogenetic methods used to resurrect protein sequences from extinct organisms?

ANSWER

Most genes and proteins of organisms that lived millions of years ago have decomposed in the fossil remains of these species. Nonetheless, the sequences of many ancient genes and proteins can be reconstructed by the methods described in this chapter. As we saw in Section 22.3, just as we can reconstruct the morphological features of a clade's ancestors, we can reconstruct their DNA and protein sequences—if we have enough information about the genomes of their descendants. Biologists have reconstructed gene sequences from species that have been extinct for millions of years. Using this information, a laboratory can reconstruct real proteins that correspond to those sequences. This is how Mikhail Matz and his colleagues were able to resurrect fluorescent proteins from the extinct ancestors of modern corals, then visualize the colors produced by these proteins in the laboratory (**Figure 22.16**).

Biologists have even used phylogenetic analysis to reconstruct some protein sequences that were present in the common ancestor of life mentioned in Section 22.1. These hypothetical protein sequences were then made into actual proteins in the laboratory. When biologists measured the temperature optima for these resurrected proteins, they found that the proteins functioned best in the range of 55°C–65°C. This result is consistent with hypotheses that life evolved in a high-temperature environment.

To reconstruct protein sequences from species that have been extinct for millions or even billions of years, biologists use detailed mathematical models that take into account much of what we have learned about molecular evolution, as described in Section 22.2. These models incorporate information on rates of replacement among different amino acids in proteins, information on different substitution rates among nucleotides, and changes in the rate of molecular evolution among the major lineages of life.

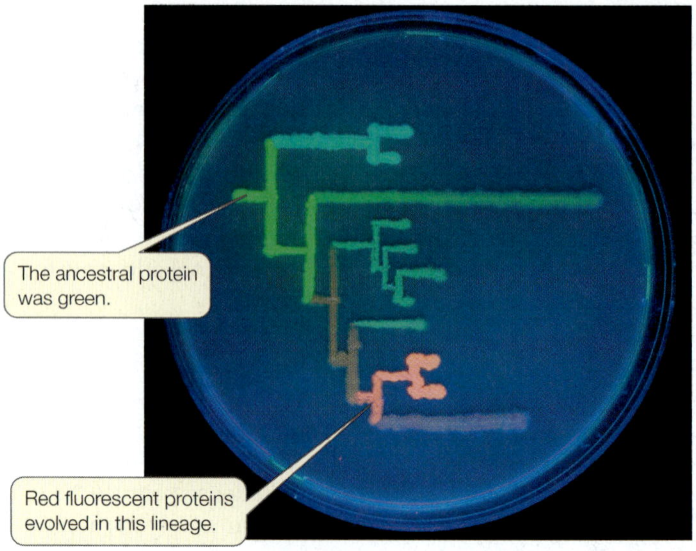
The ancestral protein was green.
Red fluorescent proteins evolved in this lineage.

22.16 Evolution of Fluorescent Proteins of Corals Mikhail Matz and his colleagues used phylogenetic analysis to reconstruct the sequences of fluorescent proteins that were present in the extinct ancestors of modern corals. They then expressed these proteins in bacteria and plated the bacteria in the form of a phylogenetic tree to show how the colors evolved over time.

CHAPTER SUMMARY 22

22.1 What Is Phylogeny?

- **Phylogeny** is the history of evolutionary relationships among organisms or their genes. Groups of evolutionarily related species are represented as branches in a **phylogenetic tree**. Review Figures 22.1, 22.2
- Named species and groups of species are called **taxa**. A taxon that consists of an ancestor and all of its evolutionary descendants is called a **clade**. Review Figure 22.3
- **Homologies** are similar traits that have been inherited from a common ancestor. Review Figure 22.4
- A derived trait that is shared by two or more taxa and is inherited from their common ancestor is called a **synapomorphy**.
- Distantly related species may show similar traits that do not result from common ancestry. **Convergent evolution** and **evolutionary reversals** can give rise to such traits, which are called **homoplasies**.

22.2 How Are Phylogenetic Trees Constructed?

- Phylogenetic trees can be constructed from synapomorphies using the logic of **parsimony**. Review Figure 22.5, ACTIVITY 22.1, ANIMATED TUTORIAL 22.1
- Sources of phylogenetic information include **morphology**, patterns of development, the fossil record, behavioral traits, and molecular traits such as DNA and protein sequences.
- Phylogenetic trees can also be constructed with **maximum likelihood** methods, which find the tree most likely to have generated the observed data.
- Phylogenetic methods have been tested in both experimental and simulation studies, and have been shown to be accurate under a wide variety of conditions.

22.3 How Do Biologists Use Phylogenetic Trees?

- Phylogenetic trees are used to make comparisons among living organisms. Review Figure 22.10
- Phylogenetic trees are used to reconstruct the past and to understand the origin of traits. Review Figure 22.11
- Biologists can use phylogenetic trees to reconstruct ancestral states. See ANIMATED TUTORIAL 22.2
- Phylogenetic trees may include estimates of divergence times of lineages determined by **molecular clock** analysis. Review Figure 22.13

22.4 How Does Phylogeny Relate to Classification?

- Biologists use phylogenetic relationships to organize life into a coherent classification system.
- Taxa in modern classifications are expected to be **monophyletic** groups. **Paraphyletic** and **polyphyletic** groups are not considered appropriate taxonomic units. Review Figure 22.15, ACTIVITY 22.2
- Several sets of rules govern the use of scientific names, with the goal of providing unique and universal names for taxa.

Go to the Interactive Summary to review key figures, Animated Tutorials, and Activities Life10e.com/is22

CHAPTER REVIEW

REMEMBERING

1. Phylogenetic trees may be constructed for
   a. genes.
   b. species.
   c. major evolutionary groups.
   d. viruses.
   e. all of the above.

2. A shared derived trait, used as the basis for identifying a monophyletic group, is called
   a. a synapomorphy.
   b. a homoplasy.
   c. a parallel trait.
   d. a convergent trait.
   e. a phylogeny.

3. Convergent evolution and evolutionary reversal are two sources of
   a. homology.
   b. parsimony.
   c. synapomorphy.
   d. monophyly.
   e. homoplasy.

4. Taxonomists strive to describe and name only taxa that are
   a. monophyletic.
   b. paraphyletic.
   c. polyphyletic.
   d. homoplastic.
   e. monomorphic.

5. Which of the following groups have separate sets of rules for nomenclature?
   a. Animals
   b. Plants and fungi
   c. Bacteria
   d. Viruses
   e. All of the above

6. If two scientific names are proposed for the same species, how do taxonomists decide which name should be used?
   a. The name that provides the most accurate description of the organism is used.
   b. The name that was proposed most recently is used.
   c. The name that was used in the most recent taxonomic revision is used.
   d. The first name to be proposed is used.
   e. Taxonomists use whichever name they prefer.

UNDERSTANDING & APPLYING

7. What is the problem with the classification shown below? How could the limits of Genus A and/or Genus B be modified so that both genera are monophyletic?

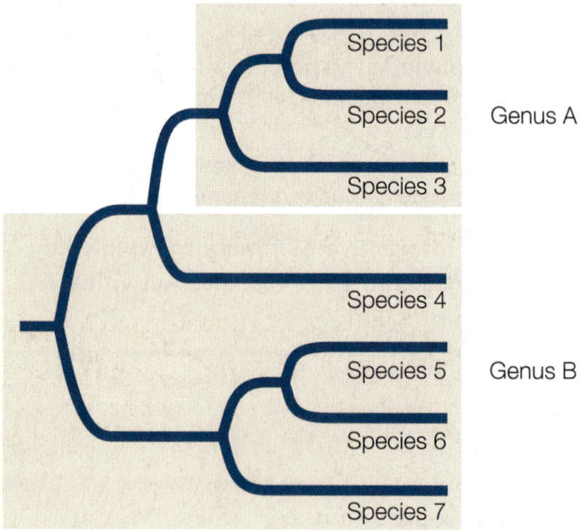
Species 1
Species 2
Species 3
Genus A
Species 4
Species 5
Species 6
Species 7
Genus B

8. How are fossils helpful in identifying ancestral and derived traits of organisms? Describe an example.

9. In the figure shown below, what is the estimated average rate of change (expressed as the proportion of amino acid change per million years)?

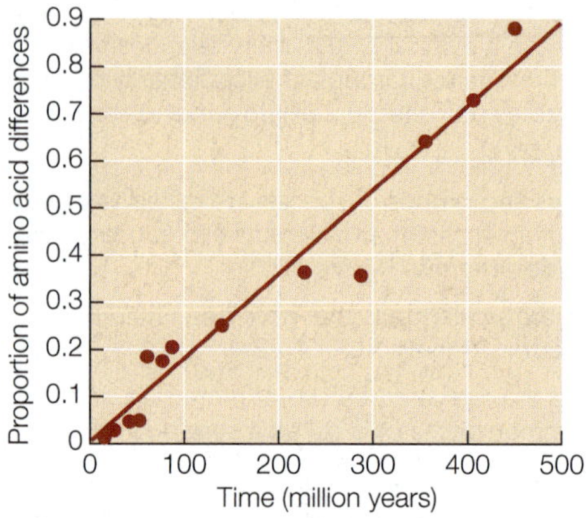
Proportion of amino acid differences
0.9
0.8
0.7
0.6
0.5
0.4
0.3
0.2
0.1
0
0
100
200
300
400
500
Time (million years)

ANALYZING & EVALUATING

10. West Nile virus kills birds of many species and can cause fatal encephalitis (inflammation of the brain) in humans and horses. The virus was first isolated in Africa (where it is thought to be endemic) in the 1930s, and by the 1990s it had been found throughout much of Eurasia. West Nile virus was not found in North America until 1999 (when it was first detected in New York), but since that time it has spread rapidly across most of the United States. Use the phylogenetic tree of West Nile virus isolates shown below to construct a hypothesis about the origin of the virus lineage that was introduced into the United States. The isolates are identified by their place and date of isolation.

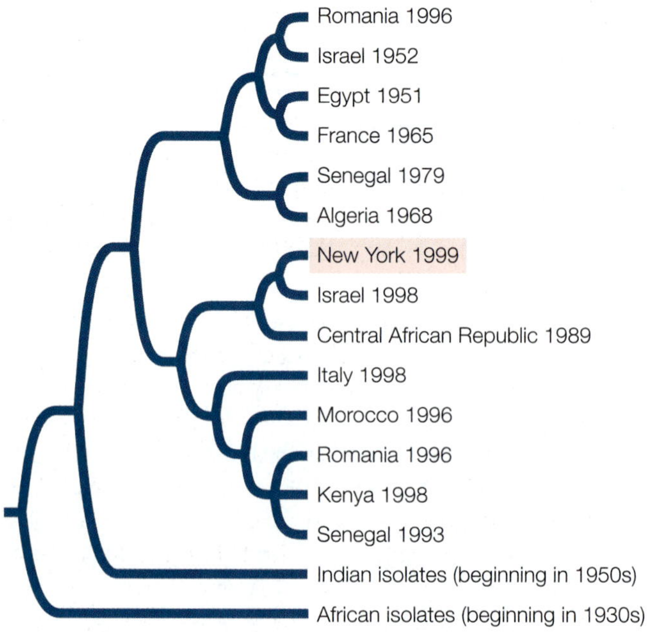
Romania 1996
Israel 1952
Egypt 1951
France 1965
Senegal 1979
Algeria 1968
New York 1999
Israel 1998
Central African Republic 1989
Italy 1998
Morocco 1996
Romania 1996
Kenya 1998
Senegal 1993
Indian isolates (beginning in 1950s)
African isolates (beginning in 1930s)

Go to BioPortal at **yourBioPortal.com** for Animated Tutorials, Activities, LearningCurve Quizzes, Flashcards, and many other study and review resources.

23 Speciation

CHAPTER OUTLINE

Many Species from One This composite photograph shows a few of the nearly 1,000 species of haplochromine cichlids that are endemic to Lake Malawi, all of which arose from a single founder species.

NOT QUITE 2 MILLION YEARS AGO, a tectonic split in the Great Rift Valley of East Africa led to the formation of Lake Malawi, which lies between the modern countries of Malawi, Tanzania, and Mozambique. A few fish species entered the newly formed lake, including individuals of one species known as a haplochromine cichlid. Today the descendants of these "founding fish" include nearly a thousand distinct species of haplochromine cichlids. All of them are endemic to this single large and deep freshwater lake—they are not found anywhere else in the world. This vast array of cichlid species makes Lake Malawi the world's most diverse lake in terms of its fish fauna. How did so many different species arise from a single ancestral species in less than 2 million years?

By studying the history and timing of speciation events in Lake Malawi, biologists have pieced together some of the processes that led to so many cichlid species. The earliest haplochromine cichlids to enter the new lake encountered diverse habitats, as some of its shores were rocky while others were sandy. Different populations of the original cichlid species quickly adapted; fish in rocky habitats adapted to breeding and living in rocky conditions, and those in sandy habitats evolved specializations for life over sand. These changes resulted in an early speciation event.

Within each of these two major habitat types, there were numerous opportunities for diet specialization. Some populations of cichlids became rock scrapers, others became bottom feeders, fish predators, scale-eaters, pelagic (open water) zooplankton eaters, or plant-feeding specialists. Each of these feeding specializations required different mouth morphology. The offspring of fish that bred with fish of similar morphology were more likely to survive than were fish with two very different parental morphologies. These differences in fitness led to the formation of many more new species, each adapted to a different feeding mode.

The Lake Malawi cichlids continued to diverge and form new species. Male cichlids competed for the attention of females through their bright body colors. Diversification of the body colors of males, and of the preferences of females for different body colors, led to more and more new species, each isolated from the other by their sexual preferences. Today biologists are studying the genomes of Lake Malawi cichlids to understand the details of the genetic changes that have given rise to so many species over so little time.

?

Can biologists study the processes of speciation in the laboratory?

See answer on p. 482.

23.1 What Are Species?

Biological diversity does not vary in a smooth, incremental way. People have long recognized groups of similar organisms that mate with one another, and they have noticed that there are usually distinct morphological differences between these groups. A group of organisms that can mate with one another and produce fertile offspring is commonly called a **species** (note that this is both the plural and singular form of the word). Species are the result of the process of **speciation**: the divergence of biological lineages and the emergence of reproductive isolation between those lineages.

Although "species" is a useful and common term, its usage varies among biologists who are interested in different aspects of speciation. Different biologists think about species differently because they ask different questions. How can we recognize and identify a species? How do new species arise? How do species remain distinct, especially from their recent close relatives? Why do rates of speciation differ among groups of organisms? In answering these questions, biologists focus on different attributes of species, leading to several ways of thinking about what species are and how they form. Most of the various **species concepts** proposed by biologists are simply different ways of approaching the question "What are species?"

We can recognize many species by their appearance

Someone who is knowledgeable about a group of organisms, such as birds or flowering plants, can usually distinguish the different species found in a particular area simply by looking at them. Standard field guides to birds, mammals, insects, and wildflowers are possible only because many species change little in appearance over large geographic distances (Figure 23.1A).

More than 250 years ago, Carolus Linnaeus developed the system of binomial nomenclature by which species are named today (see Section 22.4). Linnaeus described and named thousands of species, but because he knew nothing about the genetics or the mating behavior of the organisms he was naming, he classified them on the basis of their appearance alone. In other words, Linnaeus used a **morphological species concept**, a construct that assumes that a species comprises individuals that "look alike" and that individuals that do not look alike belong to different species. Although Linnaeus could not have known it, the members of most of the groups he classified as species look alike because they share many alleles of the genes that code for their morphological features.

Using morphology to define species has limitations. Members of the same species do not always look alike. For example, males, females, and young individuals do not always resemble one another closely (Figure 23.1B). Furthermore, morphology is of little use in the case of cryptic species—instances in which two or more species are morphologically indistinguishable but do not interbreed (Figure 23.2). Biologists therefore cannot rely on appearance alone in determining whether individual organisms are members of the same or different species. Today, biologists use several additional types of information—especially behavioral and genetic data—to differentiate species.

Reproductive isolation is key

The most important factor in the divergence of sexually reproducing lineages from one another is the evolution of **reproductive isolation**, a state in which two groups of organisms can no longer exchange genes. If individuals of group A mate and reproduce only with one another, then group A constitutes a distinct species within which genes recombine. In other words, group A is an independent evolutionary lineage—a separate branch on the tree of life.

It was his recognition of the importance of reproductive isolation that brought evolutionary biologist Ernst Mayr to propose the **biological species concept**: *"Species are groups of actually or potentially interbreeding natural populations which are reproductively isolated from other such groups."* The phrase "actually or potentially" is an important element of this definition.

(A)
(B)
Aix sponsa
Male, Florida
Aix sponsa
Male, California
Aix sponsa
Female

23.1 Not All Members of the Same Species Look Alike (A) It is easy to identify these two male wood ducks as members of the same species, even though they are found on opposite coasts 2,000 miles apart. Despite their geographic separation, the two individuals are morphologically very similar. (B) Wood ducks are sexually dimorphic, which means the female's appearance is quite different from that of the male.

(A)
Hyla versicolor

(B)
Hyla chrysoscelis

23.2 Cryptic Species Look Alike but Do Not Interbreed These two species of gray treefrogs cannot be distinguished by their external morphology, but they do not interbreed even when they occupy the same geographic range. *Hyla versicolor* is a tetraploid species (four sets of chromosomes), whereas *H. chrysoscelis* is diploid (two sets of chromosomes). And, although they look alike, the males have distinctive mating calls; female frogs recognize and mate with males of their own species based on these calls.

"Actually" says that the individuals live in the same area and interbreed with one another. "Potentially" says that even though the individuals do not live in the same area, and therefore do not interbreed, other information suggests that they *would* do so if they were able to get together. This widely used species concept does not apply to organisms that reproduce asexually, and it is limited to a single point in evolutionary time.

The lineage approach takes a long-term view

Evolutionary biologists often think of species as branches on the tree of life. This idea can be termed a **lineage species concept**. In this framework for thinking about species, one species splits into two descendant species, which thereafter evolve as distinct lineages. A lineage species concept allows biologists to consider species over evolutionary time.

A **lineage** is an ancestor–descendant series of populations followed over time. Each species has a history that starts with a speciation event by which one lineage is split into two and ends either with extinction or with another speciation event, at which time the species produces two descendant species. The process of lineage splitting is usually gradual, taking thousands of generations to complete. At the other extreme, an ancestral lineage may be split in two within a few generations (as happens with polyploidy, which we'll discuss in Section 23.3). The gradual nature of most splitting events means that at a single point in time, the final outcome of the process may not be clear. In these cases, it may be difficult to predict whether the incipient species will continue to diverge and become fully isolated from one another, or whether the two lineages will merge again in the future.

The different species concepts are not mutually exclusive

Many named variants of these three major classes of species concepts exist. These various concepts are not incompatible; they simply emphasize different aspects of species or speciation. The morphological species concept emphasizes the practical aspects of recognizing species, although it sometimes results in underestimation or overestimation of actual number of species. Mayr's biological species concept emphasizes that reproductive isolation is what allows sexual species to evolve independently of one another. The lineage species concept embraces the idea that sexual species are maintained by reproductive isolation, but extends the concept of a species as a lineage over evolutionary time. The lineage species concept is also able to accommodate species that reproduce asexually.

Virtually all species exhibit some degree of genetic recombination among individuals, even if recombination events are relatively rare. Significant reproductive isolation between species is therefore necessary for lineages to remain distinct over evolutionary time. Furthermore, reproductive isolation is responsible for the morphological distinctness of most species because mutations that result in morphological changes cannot spread between reproductively isolated species. Therefore, no matter which species concept we emphasize, the evolution of reproductive isolation is important for understanding the origin of species.

RECAP 23.1

Species are distinct lineages on the tree of life. Speciation is usually a gradual process as one lineage divides into two. Over time, lineages of sexual species remain distinct from one another because they have become reproductively isolated.

- Explain how the various species concepts emphasize different attributes of species. See pp. 468–469
- Why is the biological species concept not applicable to asexually reproducing organisms? See pp. 468–469
- Explain the role of reproductive isolation in each of the species concepts discussed in this section? See pp. 468–469

Although Charles Darwin titled his groundbreaking book *On the Origin of Species*, it included very little about speciation as we understand it today. Darwin devoted most of his attention to demonstrating that individual species are altered over time by natural selection. The remaining sections of this chapter discuss the many aspects of speciation that biologists have learned about since Darwin's time.

23.2 What Is the Genetic Basis of Speciation?

Not all evolutionary changes result in new species. A single lineage may change over time without giving rise to a new species. Speciation requires the interruption of gene flow within a species whose members formerly exchanged genes. But if a genetic change prevents reproduction between individuals of a species, how can such a change spread through a species in the first place?

Incompatibilities between genes can produce reproductive isolation

If a new allele that causes reproductive incompatibility arises in a population, it cannot spread through the population because no other individuals will be reproductively compatible with the individual that carries the new allele. So how can one reproductively cohesive lineage ever split into two reproductively isolated species? Several early geneticists, including Theodosius Dobzhansky and Hermann Joseph Muller, developed a genetic model to explain this apparent conundrum (**Figure 23.3**).

The Dobzhansky–Muller model is quite simple. First, assume that a single ancestral population is subdivided into two separate populations by some barrier to gene flow (by the formation of a new mountain range, for instance), and that these two populations then evolve as independent lineages. In one of the two populations, a new allele (*A*) arises and becomes fixed. In the other population, another new allele (*B*) becomes fixed *at a different gene locus*. Neither new allele at either locus results in any loss of reproductive compatibility. However, the two new forms of these two different genes have never occurred together in the same individual or population. Recall that the products of many genes must work together in an organism. It is possible that the new protein forms encoded by the two new alleles will not be compatible with each other. If individuals from the two populations come back together after these genetic changes, they may still be able to interbreed. However, the hybrid offspring may have a new combination of genes that is functionally inferior to that of either parent, or even lethal. This will not happen with all new combinations of genes, but over time, isolated populations will accumulate many allele differences at many gene loci. Some combinations of these differentiated genes will not function well together in hybrids. Thus genetic incompatibility between the two isolated populations will develop over time.

Many empirical examples support the Dobzhansky–Muller model. This model works not only for pairs of individual genes, but also for some kinds of chromosomal rearrangements. Bats of the genus *Rhogeessa*, for example, exhibit considerable variation in centric fusions of their chromosomes. The chromosomes of the various species contain the same basic chromosomal arms, but in some species two acrocentric (one-armed) chromosomes have fused at the centromere to form larger, metacentric (two-armed) chromosomes. A polymorphism in centric fusion causes few, if any, problems in meiosis because the respective chromosomes can still align and assort normally. Therefore, a given centric fusion can become fixed in a lineage. However, if a *different* centric fusion becomes fixed in a second lineage, then hybrids between individuals of each lineage will not be able to produce normal gametes in meiosis (**Figure 23.4**). Most of the closely related species of *Rhogeessa* display different combinations of these centric fusions and are thereby reproductively isolated from one another.

Reproductive isolation develops with increasing genetic divergence

As pairs of species diverge genetically, they become increasingly reproductively isolated (**Figure 23.5**). Both the rate at which reproductive isolation develops and the mechanisms that produce it vary from group to group, as we'll see in the next two sections of this chapter. Reproductive incompatibility has been shown to develop gradually in many groups of plants, animals, and fungi, reflecting the slow pace at which incompatible genes accumulate in each lineage. In some cases, complete reproductive isolation may take millions of years. In other cases (as with the chromosomal fusions of *Rhogeessa* described above), reproductive isolation can develop over just a few generations.

Partial reproductive isolation has evolved in strains of *Phlox drummondii* artificially isolated by humans. In 1835, Thomas Drummond, after whom this species of garden plant is named, collected seeds in Texas and distributed them to nurseries in

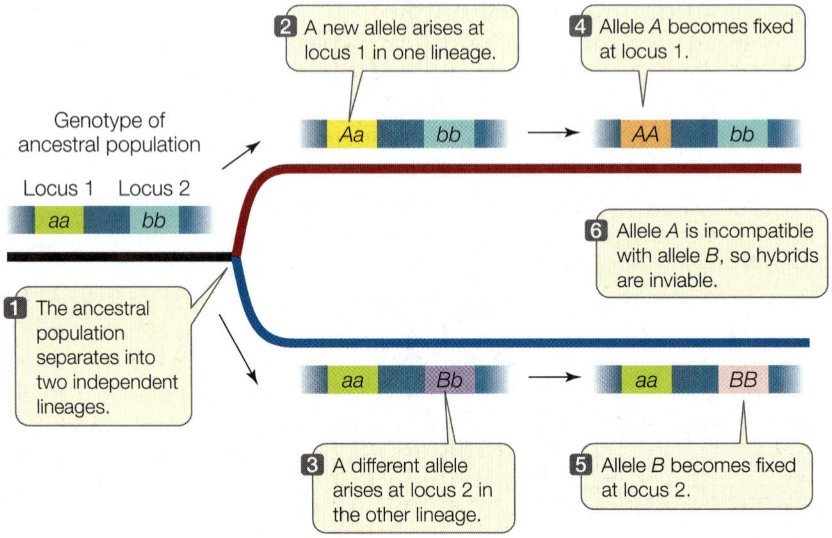
Genotype of ancestral population
Locus 1
Locus 2
aa
bb
1 The ancestral population separates into two independent lineages.
2 A new allele arises at locus 1 in one lineage.
Aa
bb
4 Allele A becomes fixed at locus 1.
AA
bb
6 Allele A is incompatible with allele B, so hybrids are inviable.
aa
Bb
3 A different allele arises at locus 2 in the other lineage.
aa
BB
5 Allele B becomes fixed at locus 2.

23.3 The Dobzhansky–Muller Model In this simple two-locus version of the model, two lineages from the same ancestral population become physically separated from each other and evolve independently. A new allele becomes fixed in each descendant population, but at a different locus. Neither of the new alleles is incompatible with the ancestral alleles, but the two new alleles in the two different genes are incompatible with each other. Thus the two descendant lineages are reproductively incompatible.

Go to Animated Tutorial 23.1
Speciation Simulation
Life10e.com/at23.1

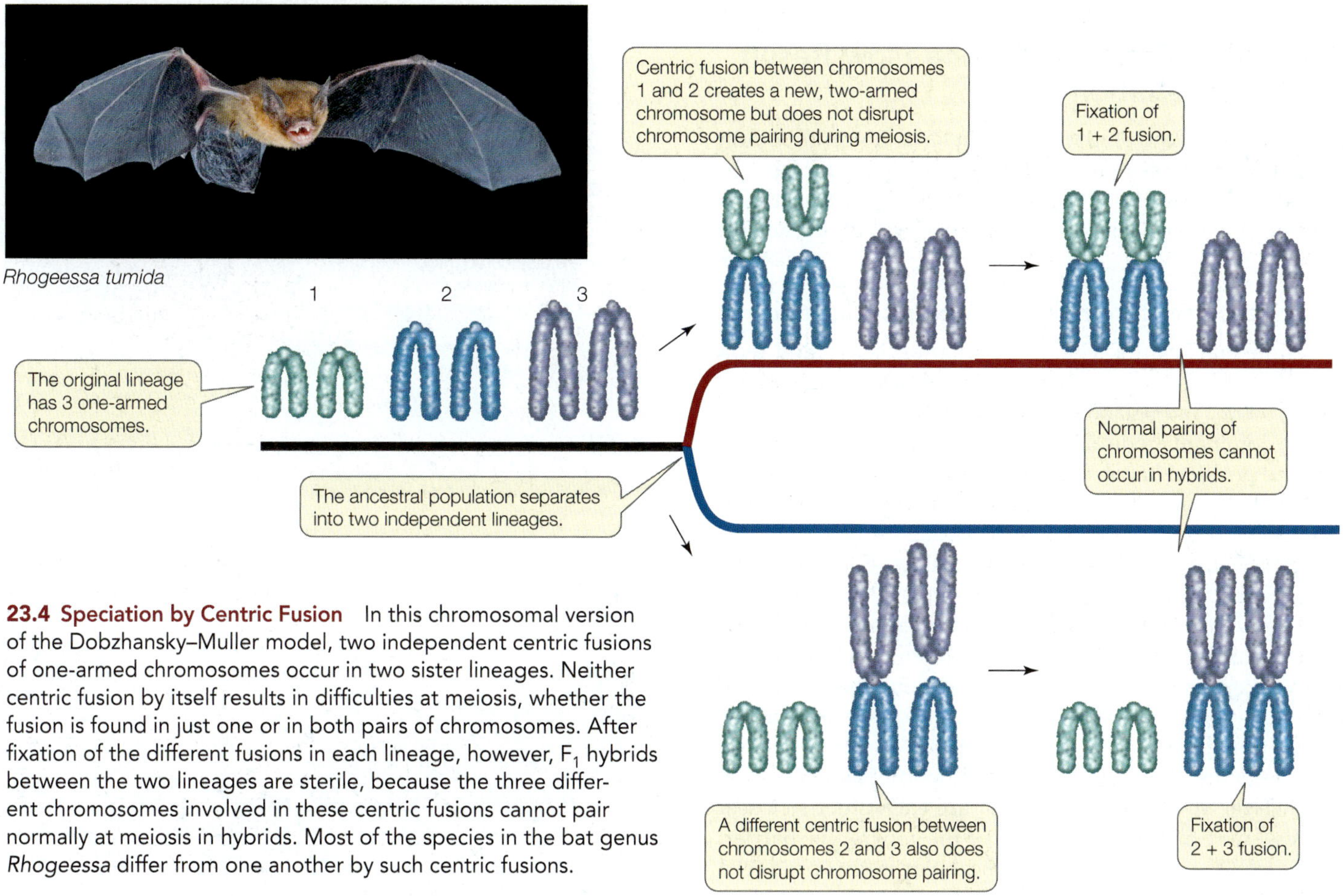

Rhogeessa tumida
1
2
3
The original lineage has 3 one-armed chromosomes.
The ancestral population separates into two independent lineages.
Centric fusion between chromosomes 1 and 2 creates a new, two-armed chromosome but does not disrupt chromosome pairing during meiosis.
Fixation of 1 + 2 fusion.
Normal pairing of chromosomes cannot occur in hybrids.
A different centric fusion between chromosomes 2 and 3 also does not disrupt chromosome pairing.
Fixation of 2 + 3 fusion.

23.4 Speciation by Centric Fusion In this chromosomal version of the Dobzhansky–Muller model, two independent centric fusions of one-armed chromosomes occur in two sister lineages. Neither centric fusion by itself results in difficulties at meiosis, whether the fusion is found in just one or in both pairs of chromosomes. After fixation of the different fusions in each lineage, however, F_1 hybrids between the two lineages are sterile, because the three different chromosomes involved in these centric fusions cannot pair normally at meiosis in hybrids. Most of the species in the bat genus *Rhogeessa* differ from one another by such centric fusions.

Europe. The European nurseries established more than 200 true-breeding strains of *P. drummondii* that differed in flower size, flower color, and plant growth form. The breeders did not select directly for reproductive incompatibility between strains, but in subsequent experiments in which strains were crossed, biologists found that reproductive compatibility between strains (as measured by seed production) had been reduced by 14 to 50 percent, depending on the cross—even though the strains had been isolated from one another for less than two centuries.

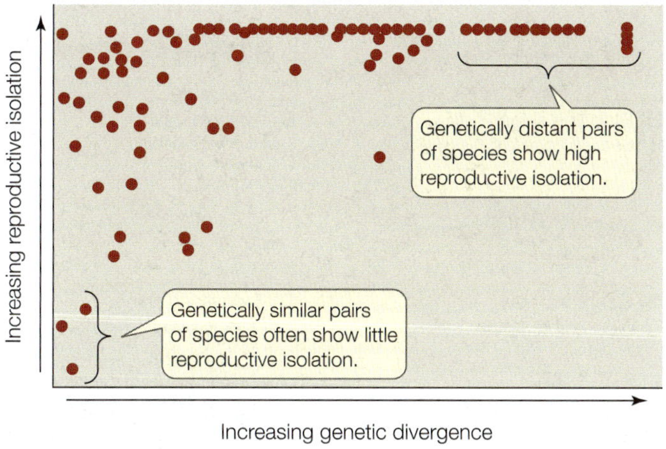

Increasing reproductive isolation
Genetically distant pairs of species show high reproductive isolation.
Genetically similar pairs of species often show little reproductive isolation.
Increasing genetic divergence

23.5 Reproductive Isolation Increases with Genetic Divergence Among pairs of *Drosophila* species, the more the species differ genetically, the greater their reproductive isolation from each other. Each dot represents a comparison of one species pair. Such positive relationships between genetic distance and reproductive isolation have been observed in many groups of plants, animals, and fungi.

RECAP 23.2

When two parts of a population become isolated from each other by some barrier to gene flow, they begin to diverge genetically. The Dobzhansky-Muller model describes how new alleles or chromosomal arrangements that arise in the two descendent lineages can lead to genetically incompatibility, and hence reproductive isolation, of the two lineages.

- How can centric fusions occur in one *Rhogeessa* lineage without causing any reproductive difficulties, when different centric fusions in two different lineages of *Rhogeessa* lead to disruption of meiosis in hybrid individuals? **See p. 470 and Figure 23.4**
- What empirical evidence can you cite in support of the idea that genetic divergence of populations leads to loss of reproductive compatibility? **See pp. 470–471 and Figure 23.5**

We have now seen how the splitting of an ancestral population leads to genetic divergence and reproductive incompatibility in the two descendant lineages. How do populations become separated in the first place?

23.3 What Barriers to Gene Flow Result in Speciation?

Many biologists who study speciation have concentrated on geographic processes that can result in the division of an ancestral species. Splitting of the geographic range of a species is one obvious way of achieving such a division, but it is not the only way.

Physical barriers give rise to allopatric speciation

Speciation that results when a population is divided by a physical barrier is known as **allopatric speciation** (Greek *allos*, "other"; *patria*, "homeland"). Allopatric speciation is thought to be the dominant mode of speciation in most groups of organisms. The physical barrier that divides the range of a species may be a body of water or a mountain range for terrestrial organisms or dry land for aquatic organisms—in other words, any type of habitat that is inhospitable to the species. Such barriers can form when continents drift, sea levels rise and fall, glaciers advance and retreat, or climates change. The populations separated by such barriers are often, but not always, initially large. The lineages that descend from these founding populations evolve differences for a variety of reasons, including mutation, genetic drift, and adaptation to different environments in the two areas. As a result, many pairs of closely related **sister species**—species that are each other's closest relatives—may exist on either side of the geographic barrier. An example of a physical geographical barrier that produced many pairs of sister species was the Pleistocene glaciation that isolated freshwater streams in the eastern highlands of the Appalachian Mountains from streams in the Ozark and Ouachita Mountains (**Figure 23.6**). This splitting event resulted in many parallel speciation events among isolated lineages of stream-dwelling organisms.

Allopatric speciation may also result when some members of a population cross an existing barrier and establish a new, isolated population. Many of the more than 800 species of *Drosophila* found in the Hawaiian Islands are restricted to a single island. We know that these species are the descendants of new populations founded by individuals dispersing among the islands when we find that the closest relative of a species on one island is a species on a neighboring island rather than a species on the same island. Biologists who have studied the chromosomes of these fruit flies estimate that speciation in this group of *Drosophila* has resulted from at least 45 such founder events (**Figure 23.7**).

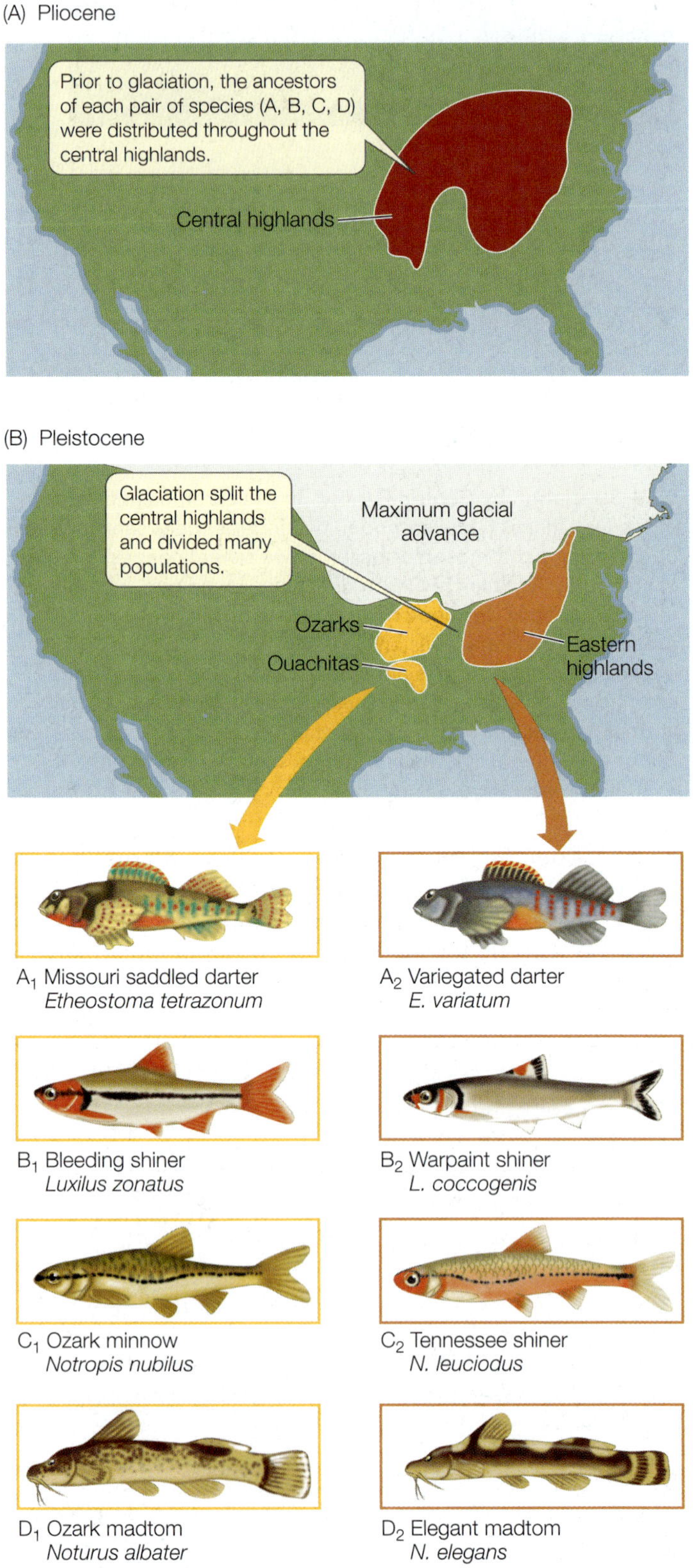
(A) Pliocene
Prior to glaciation, the ancestors of each pair of species (A, B, C, D) were distributed throughout the central highlands.
Central highlands
(B) Pleistocene
Glaciation split the central highlands and divided many populations.
Maximum glacial advance
Ozarks
Ouachitas
Eastern highlands
A_1 Missouri saddled darter *Etheostoma tetrazonum*
A_2 Variegated darter *E. variatum*
B_1 Bleeding shiner *Luxilus zonatus*
B_2 Warpaint shiner *L. coccogenis*
C_1 Ozark minnow *Notropis nubilus*
C_2 Tennessee shiner *N. leuciodus*
D_1 Ozark madtom *Noturus albater*
D_2 Elegant madtom *N. elegans*

23.6 Allopatric Speciation Allopatric speciation may result when an ancestral population is divided into two separate populations by a physical barrier and those populations then diverge. (A) Many species of freshwater stream fishes were distributed throughout the central highlands of North America in the Pliocene epoch (about 3–5 million years ago). (B) During the Pleistocene, glaciers advanced and isolated fish populations in the Ozark and Ouachita Mountains to the west from fish populations in the highlands of the Appalachian Mountains to the east. Numerous species diverged as a result of this separation, including the ancestors of the four pairs of sister species shown here.

Go to Animated Tutorial 23.2
Speciation Mechanisms
Life10e.com/at23.2

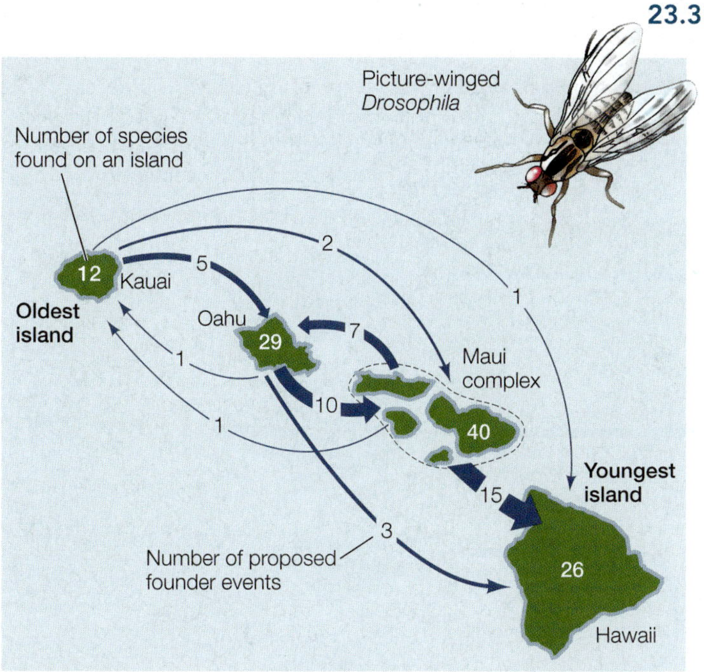

Picture-winged
Drosophila
Number of species
found on an island
12
Kauai
Oldest
island
Oahu
29
Maui
complex
40
Youngest
island
26
Hawaii
5
2
1
7
1
10
1
15
3
Number of proposed
founder events

23.7 Founder Events Lead to Allopatric Speciation The large number of species of picture-winged *Drosophila* in the Hawaiian Islands is the result of founder events: the founding of new populations by individuals dispersing among the islands. The islands, which were formed in sequence as Earth's crust moved over a volcanic "hot spot," vary in age.

Go to Animated Tutorial 23.3
Founder Events and Allopatric Speciation
Life10e.com/at23.3

The 13 species of finches found in the islands of the Galápagos archipelago, some 1,000 km off the coast of Ecuador, are one of the most famous examples of allopatric speciation. Darwin's finches (as they are usually called, because Darwin was the first scientist to study them) arose in the Galápagos from a single South American finch species that colonized the islands. Today the Galápagos species differ strikingly not only from their closest mainland relative, but also from one another (**Figure 23.8**). The islands are sufficiently far apart that the birds move among them only infrequently. In addition, environmental conditions differ widely from island to island. Some islands are relatively flat and arid; others have forested mountain slopes. Over millions of years, finch lineages on the different islands have differentiated to the point that when occasional immigrants do arrive from other islands, they either do not breed with the residents or, if they do, the resulting offspring do not survive as well as the offspring of established residents. The genetic distinctness of each finch species from the others and the genetic cohesiveness of the individual species are thus maintained.

Sympatric speciation occurs without physical barriers

Although geographic isolation is usually required for speciation, speciation can also occur in the absence of a physical barrier. Speciation without physical isolation is called **sympatric speciation** (Greek *sym*, "together with"). But how can such speciation happen? Given that speciation is usually a gradual process, how can reproductive isolation develop when individuals have frequent opportunities to mate with one another?

DISRUPTIVE SELECTION Sympatric speciation may occur with some forms of disruptive selection (see Section 21.3) if individuals with different genotypes have a preference for distinct microhabitats where mating takes place. For example, sympatric speciation via disruptive selection appears to be taking place in the apple maggot fly (*Rhagoletis pomonella*) of eastern North America. Until the mid-1800s, *Rhagoletis* flies courted, mated, and deposited their eggs only on hawthorn fruits. About 150 years ago, European immigrants introduced apple trees into the region, and some flies began to lay their eggs on apples. Apple trees are closely related to hawthorns, but the smell of the fruits differs, and the apple fruits appear earlier in the season than those of hawthorns. Some early-emerging female *Rhagoletis* laid their eggs on apples, and over time, a genetic preference for the smell of apples evolved among early-emerging insects. When the offspring of these flies sought out apple trees for mating and egg deposition, they mated with other flies reared on apples, which shared the same genetic preferences.

Today the two groups of *Rhagoletis pomonella* in eastern North America appear to be on the way to becoming distinct species. One group mates and lays eggs primarily on hawthorn fruits, the other on apples. The incipient species are partially reproductively isolated because they mate primarily with individuals raised on the same fruit and which emerge at the same of year. In addition, the larvae of the apple-feeding flies now grow more rapidly on apples than they originally did. Sympatric speciation that arises from such host-plant specificity may be widespread among insects, many of which feed on only one plant species.

POLYPLOIDY The most common means of sympatric speciation is **polyploidy**, which results from the duplication of sets of chromosomes within individuals. Polyploidy can arise either from chromosome duplication in a single species (**autopolyploidy**) or from the combining of the chromosomes of two different species (**allopolyploidy**).

An autopolyploid individual originates when (for example) two accidentally unreduced diploid gametes (each with two sets of chromosomes) combine to form a tetraploid individual (with four sets of chromosomes). Tetraploid and diploid individuals of the same species are reproductively isolated because their hybrid offspring are triploid and thus are usually sterile; those offspring rarely produce viable gametes because their chromosomes do not segregate evenly during meiosis (**Figure 23.9**). So a tetraploid individual usually cannot produce viable offspring by mating with a diploid individual—but it *can* do so if it self-fertilizes or mates with another tetraploid. Thus polyploidy can result in complete reproductive isolation in two generations—an important exception to the general rule that speciation is a gradual process.

Allopolyploids may be produced when individuals of two different (but closely related) species interbreed. Such hybridization often disrupts normal meiosis, which can result in chromosomal doubling. Allopolyploids are often fertile because

23.8 Allopatric Speciation among Darwin's Finches The descendants of the ancestral finch that colonized the Galápagos archipelago several million years ago evolved into at least 13 different species whose beaks are variously adapted to feed on buds, seeds, and insects.

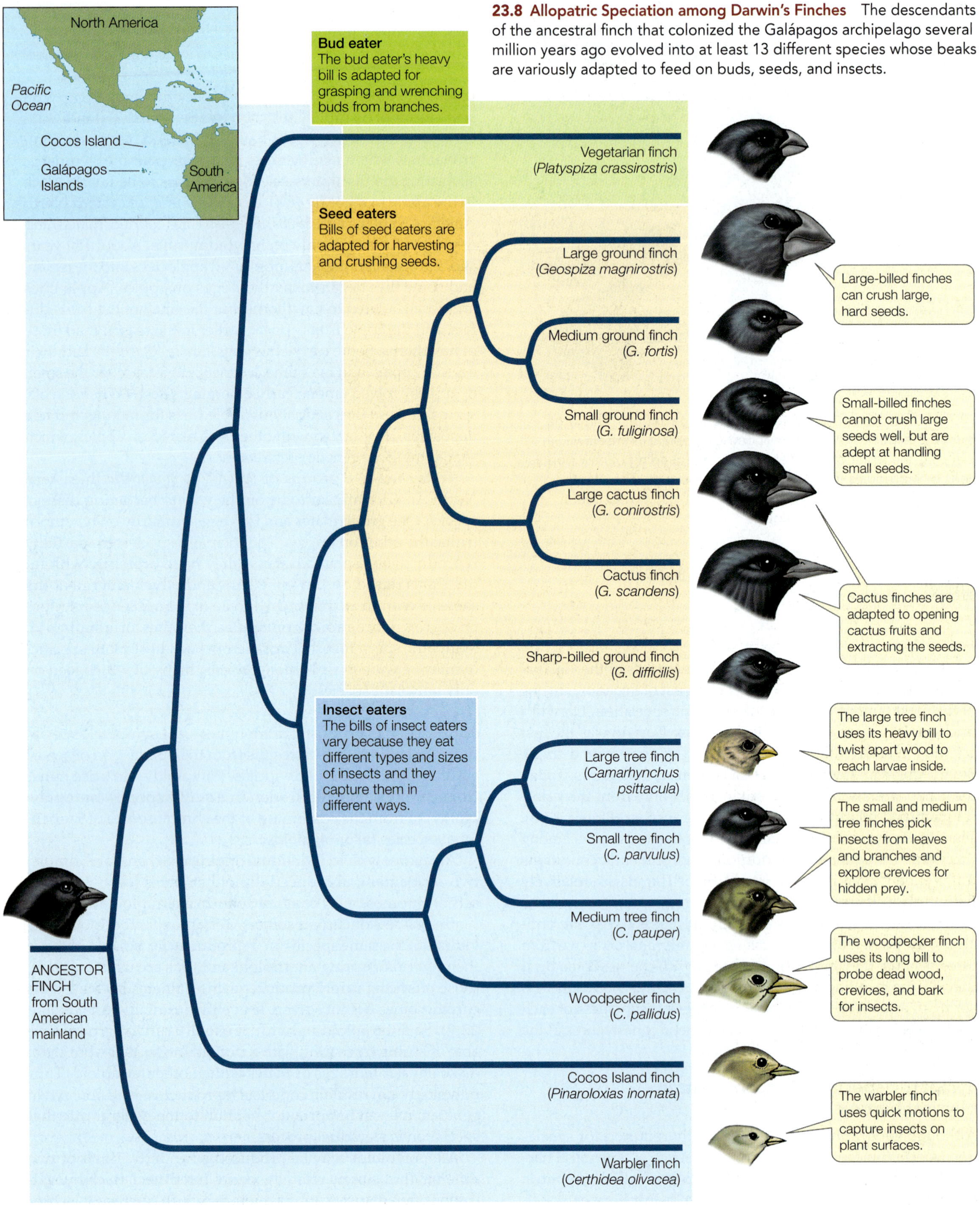
North America
Pacific Ocean
Cocos Island
Galápagos Islands
South America
Bud eater
The bud eater's heavy bill is adapted for grasping and wrenching buds from branches.
Vegetarian finch (Platyspiza crassirostris)
Seed eaters
Bills of seed eaters are adapted for harvesting and crushing seeds.
Large ground finch (Geospiza magnirostris)
Medium ground finch (G. fortis)
Small ground finch (G. fuliginosa)
Large cactus finch (G. conirostris)
Cactus finch (G. scandens)
Sharp-billed ground finch (G. difficilis)
Large-billed finches can crush large, hard seeds.
Small-billed finches cannot crush large seeds well, but are adept at handling small seeds.
Cactus finches are adapted to opening cactus fruits and extracting the seeds.
Insect eaters
The bills of insect eaters vary because they eat different types and sizes of insects and they capture them in different ways.
Large tree finch (Camarhynchus psittacula)
Small tree finch (C. parvulus)
Medium tree finch (C. pauper)
Woodpecker finch (C. pallidus)
Cocos Island finch (Pinaroloxias inornata)
Warbler finch (Certhidea olivacea)
The large tree finch uses its heavy bill to twist apart wood to reach larvae inside.
The small and medium tree finches pick insects from leaves and branches and explore crevices for hidden prey.
The woodpecker finch uses its long bill to probe dead wood, crevices, and bark for insects.
The warbler finch uses quick motions to capture insects on plant surfaces.
ANCESTOR FINCH from South American mainland

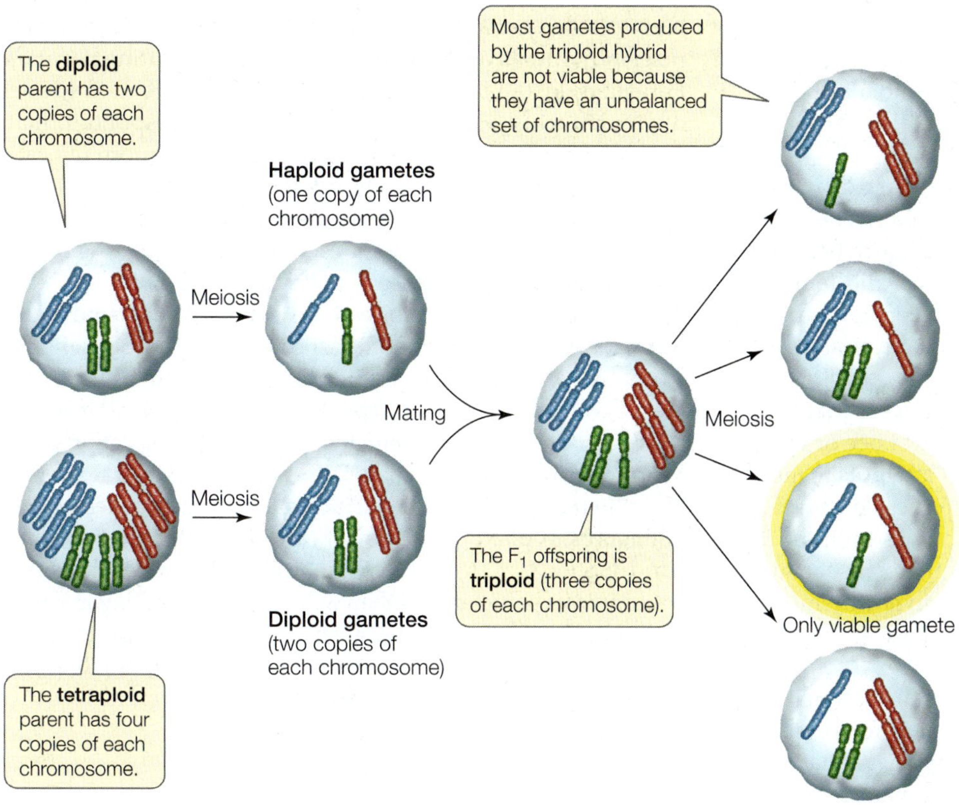

The diploid parent has two copies of each chromosome.
Haploid gametes (one copy of each chromosome)
Meiosis
Meiosis
Mating
Most gametes produced by the triploid hybrid are not viable because they have an unbalanced set of chromosomes.
Meiosis
The F1 offspring is triploid (three copies of each chromosome).
Only viable gamete
Diploid gametes (two copies of each chromosome)
The tetraploid parent has four copies of each chromosome.

23.9 Tetraploids Are Reproductively Isolated from Their Diploid Ancestors Even if the triploid offspring of a diploid and a tetraploid parent survives and reaches sexual maturity, most of the gametes it produces have aneuploid (unbalanced) numbers of chromosomes. Such triploid individuals are effectively sterile. (For simplicity, the diagram shows only three homologous chromosomes. Most species have many more chromosomes, so viable gametes are extremely rare.)

each of the chromosomes has a nearly identical partner with which to pair during meiosis.

Speciation by polyploidy has been particularly important in the evolution of plants, although it has contributed to speciation in animals as well (such as the treefrogs in Figure 23.2). New species arise by polyploidy more easily among plants than among animals because many species of plants can reproduce by self-fertilization. In addition, if polyploidy arises in several offspring of a single parent, the siblings can fertilize one another. Botanists estimate that about 70 percent of flowering plant species and 95 percent of fern species are the result of recent polyploidization. Some of these species arose from hybridization between two species followed by chromosomal duplication and self-fertilization. Other species diverged from polyploid ancestors, so that the new species share their ancestors' duplicated sets of chromosomes.

RECAP 23.3

Allopatric speciation results from the separation of populations by geographic barriers; it is the dominant mode of speciation among most groups of organisms. Sympatric speciation may result from disruptive selection that results in ecological isolation, but polyploidy is the most common cause of sympatric speciation among plants.

- Explain why an effective barrier to gene flow for one species may not effectively isolate another species. See p. 472
- What are some obstacles to sympatric speciation? See p. 473
- What is the difference between allopolyploidy and autopolyploidy? See pp. 473–474

Most populations separated by a physical barrier become reproductively isolated only slowly and gradually. If two incipient species once again come into contact with each other, what keeps them from merging back into a single species?

23.4 What Happens When Newly Formed Species Come into Contact?

As we saw in Section 23.2, once a barrier to gene flow is established, reproductive isolation will begin to develop through genetic divergence. Over many generations, differences accumulate in the isolated lineages, reducing the probability that individuals from each lineage will mate successfully with individuals in the other when they come back into contact. In this way, reproductive isolation can evolve as a by-product of the genetic changes in the two diverging lineages.

Reproductive isolation may be incomplete when incipient species come back into contact, however, in which case some hybridization will occur. If hybrid individuals are less fit than non-hybrids, selection will favor parents that do not produce hybrid offspring. Under these conditions, selection will result in the strengthening, or **reinforcement**, of mechanisms that prevent hybridization.

Mechanisms that prevent hybridization from occurring are called **prezygotic isolating mechanisms**. Mechanisms that reduce the fitness of hybrid offspring are called **postzygotic isolating mechanisms**. Postzygotic isolating mechanisms result in selection against hybridization, which in turn leads to the reinforcement of prezygotic isolating mechanisms.

Prezygotic isolating mechanisms prevent hybridization

Prezygotic isolating mechanisms, which come into play before fertilization, can prevent hybridization in several ways.

MECHANICAL ISOLATION Differences in the sizes and shapes of reproductive organs may prevent the union of gametes from different species. With animals, there may be a match between the shapes of the reproductive organs of males and females of the same species, so that reproduction between individuals with mismatched reproductive structures is not physically possible. In plants, mechanical isolation may involve a pollinator. For example, some orchid species produce flowers that look and smell like the females of particular species of bee or wasp (**Figure 23.10**). When a male insect visits and attempts to mate with a flower (thinking it is a female of his species), his mating behavior results in the transfer of pollen to and from his body by appropriately configured anthers and stigmas on the orchid. Other insects, which may visit the flower but do not attempt to mate with it, do not trigger this transfer of pollen.

TEMPORAL ISOLATION Many organisms have distinct mating seasons. If two closely related species breed at different times of the year (or different times of day), they may never have an opportunity to hybridize. For example, in sympatric populations of three closely related leopard frog species, each species breeds at a different time of year (**Figure 23.11**). Although there is some overlap in the breeding seasons, the opportunities for hybridization are minimized.

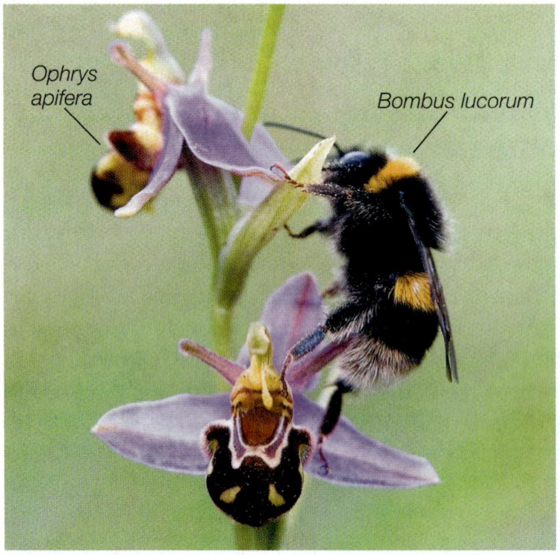

Ophrys apifera
Bombus lucorum

23.10 Mechanical Isolation through Mimicry Many orchid species maintain reproductive isolation by means of flowers that look and smell like females of of one—and only one—bee or wasp species. A male insect of the correct species must land on the flower and attempt to mate with it; only males of this particular species are physically configured to collect and transfer the orchid's pollen. The constraints of this method of pollen transfer reproductively isolate the plant from related orchid species that attract different insect pollinators. The species shown here are the two players in one such interspecific relationship; see Figure 56.11 for another example.

BEHAVIORAL ISOLATION Individuals may reject, or fail to recognize, individuals of other species as potential mating partners. For example, the mating calls of male frogs of related species diverge quickly (**Figure 23.12**). Female frogs respond to mating calls from males of their own species but ignore the calls of other species, even closely related ones. The evolution of female preferences for certain male coloration patterns among the cichlids of Lake Malawi, described at the opening of this chapter, is another example of behavioral isolation.

Sometimes the mate choice of one species is mediated by the behavior of individuals of other species. For example, whether or not two plant species hybridize may depend on the food preferences of their pollinators. The floral traits of plants, including their color and shape, can enhance reproductive isolation either by influencing which pollinators are attracted to their flowers or by influencing where pollen is deposited on the bodies of their pollinators. A plant whose flowers are pendant (hanging downward; **Figure 23.13A**) will be pollinated by an animal with different physical characteristics than will a plant whose flowers grow upright (**Figure 23.13B**). Because each pollinator prefers (and is adapted to) a different type of flower, pollinators will rarely transfer pollen from one plant species to the other.

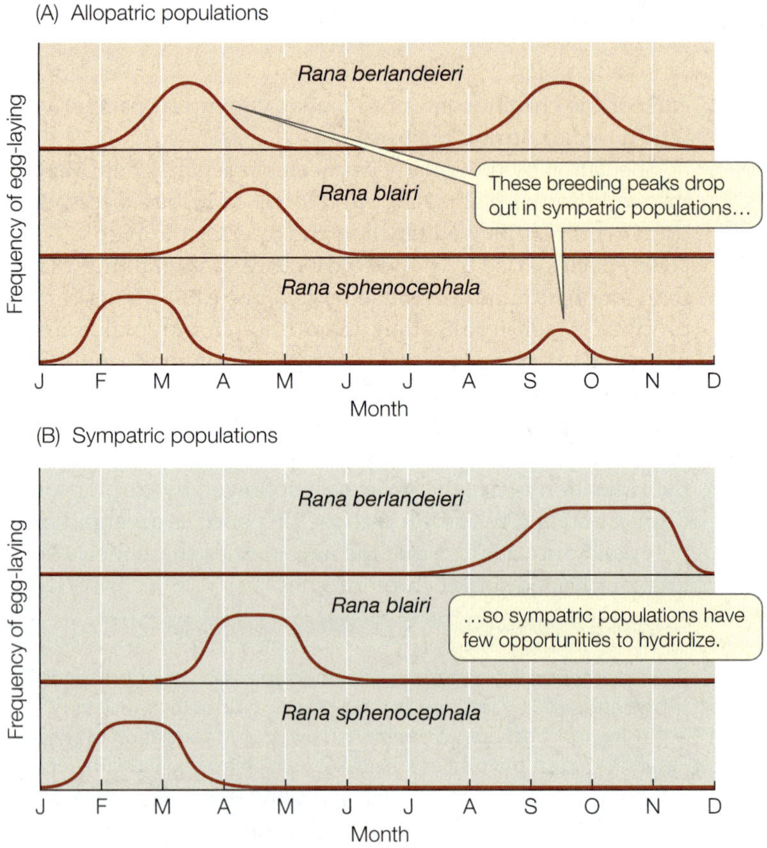

(A) Allopatric populations
Frequency of egg-laying
Rana berlandeieri
Rana blairi
Rana sphenocephala
These breeding peaks drop out in sympatric populations...
J F M A M J J A S O N D
Month
(B) Sympatric populations
Frequency of egg-laying
Rana berlandeieri
Rana blairi
Rana sphenocephala
...so sympatric populations have few opportunities to hydridize.
J F M A M J J A S O N D
Month

23.11 Temporal Isolation of Breeding Seasons (A) The peak breeding seasons of three species of leopard frogs (*Rana*) overlap when the species are physically separated (allopatry). (B) Where two or more species of *Rana* live together (sympatry), overlap between their peak breeding seasons is greatly reduced or eliminated. Selection against hybridization in areas of sympatry helps reinforce this prezygotic isolating mechanism.

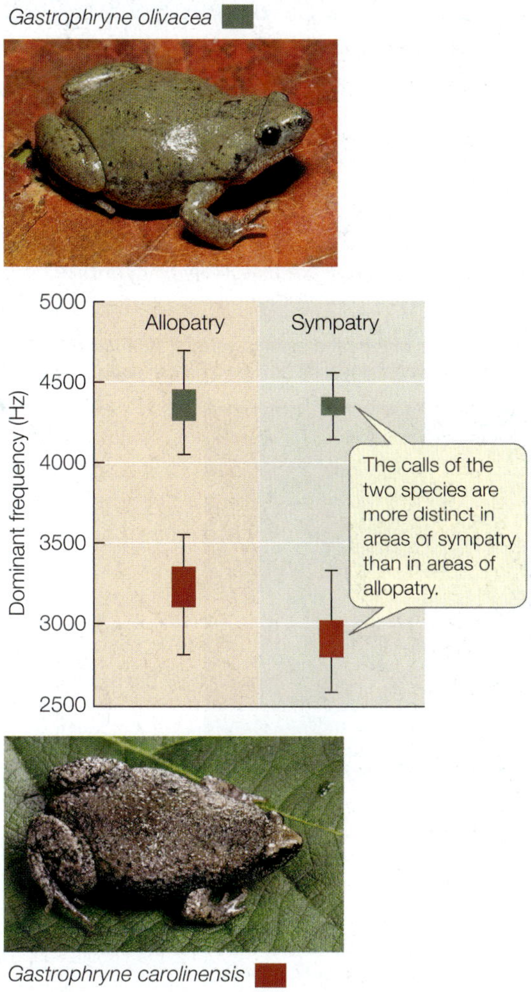
Gastrophryne olivacea
5000
4500
4000
3500
3000
2500
Dominant frequency (Hz)
Allopatry
Sympatry
The calls of the two species are more distinct in areas of sympatry than in areas of allopatry.
Gastrophryne carolinensis

23.12 Behavioral Isolation in Mating Calls The males of most frog species produce species-specific calls. The calls of the two closely related frog species shown here differ in their dominant frequency (a high-frequency sound wave results in a high-pitched sound; a low frequency results in a low-pitched sound). Female frogs are attracted to the calls of males of their own species.

Go to Media Clip 23.1
Narrowmouth Toads Calling for Mates
Life10e.com/mc23.1

(A)
(B)
(C) Aquilegia formosa
(D) A. pubescens

23.13 Reproductive Isolating Mechanisms May Be Mediated by Species Interactions (A) This hummingbird's morphology and behavior are adapted for feeding on nectar from pendant flowers. (B) The nectar-extracting proboscis of this hawkmoth is adapted to flowers that grow upright. (C) *Aquilegia formosa* flowers are normally pendant and are pollinated by hummingbirds. (D) Flowers of *A. pubescens* are normally upright and are pollinated by hawkmoths. In addition, their long floral spurs appear to restrict access by some other potential pollinators.

Such isolation by pollinator behavior is seen in the mountains of California in two sympatric species of columbines (*Aquilegia*) that have diverged in flower color, structure, and orientation. *Aquilegia formosa* (**Figure 23.13C**) has pendant flowers with short spurs (spikelike, nectar-containing structures) and is pollinated by hummingbirds. *A. pubescens* (**Figure 23.13D**) has upright, lighter-colored flowers with long spurs and is pollinated by hawkmoths. The difference in pollinators means that these two species are effectively reproductively isolated even though they populate the same geographic range.

HABITAT ISOLATION When two closely related species evolve preferences for living or mating in different habitats, they may never come into contact during their respective mating periods. The *Rhagoletis* flies discussed in Section 23.3 experienced such habitat isolation, as did the cichlid fishes that first adapted to rocky or sandy habitats upon entering Lake Malawi, as described at the opening of this chapter.

GAMETIC ISOLATION The sperm of one species may not attach to the eggs of another species because the eggs do not release the appropriate attractive chemicals, or the sperm may be unable to penetrate the egg because the two gametes are chemically incompatible. Thus, even though the gametes of two species may come into contact, the gametes never fuse into a zygote.

Gametic isolation is extremely important for many aquatic species that spawn (release their gametes directly into the environment). It has been extensively studied in sea urchins. A protein known as bindin is found in sea urchin sperm and functions in attaching ("binding") the sperm to eggs. All sea urchin species studied produce this egg-recognition protein, but the bindin gene sequence diverges so rapidly that it has become species-specific—that is, sperm can attach only to eggs of the same species, so no interspecific hybridization occurs.

Postzygotic isolating mechanisms result in selection against hybridization

Genetic differences that accumulate between two diverging lineages may reduce the survival and reproductive rates of hybrid offspring in any of several ways:

- *Low hybrid zygote viability*. Hybrid zygotes may fail to mature normally, either dying during development or developing phenotypic abnormalities that prevent them from becoming reproductively capable adults.
- *Low hybrid adult viability*. Hybrid offspring may have lower survivorship than non-hybrid offspring.
- *Hybrid infertility*. Hybrids may mature into infertile adults. For example, the offspring of matings between horses and donkeys—mules—are sterile. Although otherwise healthy, mules produce no descendants.

Natural selection does not directly favor the evolution of postzygotic isolating mechanisms. But if hybrids have low fitness, then individuals that breed only within their own species will leave more surviving offspring than will individuals that interbreed with another species. Therefore, individuals that avoid interbreeding with members of other species will have a selective advantage, and any trait that contributes to such avoidance will be favored.

Donald Levin of the University of Texas has studied reinforcement of prezygotic isolating mechanisms in flowers of the genus *Phlox*. Levin noticed that most individuals of *P. drummondii* in most of the range of the species in Texas have pink flowers. However, where *P. drummondii* is sympatric with its close relative, the pink-flowered *P. cuspidata*, most *P. drummondii* have red flowers. No other *Phlox* species has red flowers. Levin performed an experiment whose results showed that reinforcement may explain why red flowers are favored where the two species are sympatric (**Figure 23.14**).

Likely cases of reinforcement are often detected by comparing sympatric and allopatric populations of potentially hybridizing species, as in the case of *Phlox*. If reinforcement is occurring, then sympatric populations of closely related species are expected to evolve more effective prezygotic reproductive barriers than do allopatric populations of the same species. As Figure 23.11 shows, the breeding seasons of sympatric populations of different leopard frog species overlap much less than do those of allopatric populations. Similarly, the frequencies of the frog mating calls illustrated in Figure 23.12 are more divergent in sympatric populations than in allopatric populations. In both cases, there appears to have been natural selection against hybridization in areas of sympatry.

INVESTIGATING**LIFE**

23.14 Flower Color Reinforces a Reproductive Barrier in *Phlox* Most *Phlox drummondii* flowers are pink, but in regions where they are sympatric with *P. cuspidata*—which is always pink—most *P. drummondii* individuals are red. Most pollinators preferentially visit flowers of one color or the other. In this experiment, Donald Levin explored whether flower color reinforces a prezygotic reproductive barrier, lessening the chances of interspecific hybridization.[a]

HYPOTHESIS Red-flowered *P. drummondii* are less likely to hybridize with *P. cuspidata* than are pink-flowered *P. drummondii*.

Method 1. Introduce equal numbers of red- and pink-flowered *P. drummondii* individuals into an area with many pink-flowered *P. cuspidata*.

P. cuspidata
P. drummondii

2. After the flowering season ends, measure hybridization by assessing the genetic composition of the seeds produced by *P. drummondii* plants of both colors.

Results Of the seeds produced by pink-flowered *P. drummondii*, 38% were hybrids with *P. cuspidata*. Only 13% of the seeds produced by red-flowered individuals were genetic hybrids.

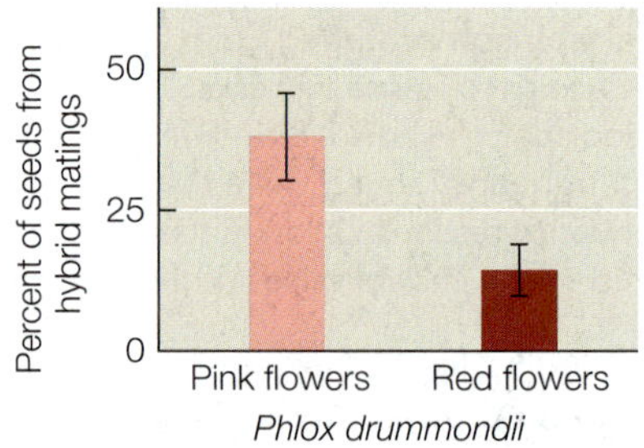
Percent of seeds from hybrid matings
50
25
0
Pink flowers
Red flowers
Phlox drummondii

CONCLUSION *P. drummondii* and *P. cuspidata* are less likely to hybridize if the flowers of the two species differ in color.

Go to **BioPortal** for discussion and relevant links for all INVESTIGATING**LIFE** figures.

[a]Levin, D. A. 1985. *Evolution* 39: 1275–1281.

Hybrid zones may form if reproductive isolation is incomplete

Unless reproductive isolation is complete, closely related species may hybridize in areas where their ranges overlap, resulting in the formation of a hybrid zone. When a hybrid zone first forms, most hybrids are offspring of crosses

WORKING WITH**DATA:**

Does Flower Color Act as a Prezygotic Isolating Mechanism?

Original Paper

Levin, D. A. 1985. Reproductive character displacement in *Phlox*. *Evolution* 39: 1275–1281.

Analyze the Data

Donald Levin proposed that *Phlox drummondii* has red flowers only in locations where it is sympatric with pink-flowered *Phlox cuspidata* because having red flowers decreases interspecific hybridization. To test this hypothesis, Levin introduced equal numbers of red- and pink-flowered *P. drummondii* individuals into an area with many pink-flowered *P. cuspidata*. At the end of the flowering season, he assessed the genetic composition of the seeds produced by *P. drummondii*. The results are shown in the table (below, right).

QUESTION 1

Check the 95% confidence intervals for the proportion of hybrid seeds in red- and pink-flowered *P. drummondii* (shown graphically in Figure 23.14). There are many websites available for calculating confidence intervals; a good one is the Vassar College statistical computation site, VassarStats.net. You can go to this site and select "Proportions" from the left-hand menu, then select "The Confidence Interval of a Proportion." What are the numerical values of the 95% confidence intervals?

QUESTION 2

You can see that the proportions of hybrids among the seeds of red- versus pink-flowered samples are significantly different because the 95% confidence intervals do not overlap. To quantify the significance of this difference, use the website suggested in Question 1, but select "Significance of the Difference between Two Independent Proportions" from the "Proportions" menu. What null hypothesis are you testing in this case? (See Appendix B if you need help.) What is the *P*-value of getting results at least as different as these two samples if your null hypothesis is true?

QUESTION 3

How would you extend or improve the experimental design of this study? What kinds of additional test sites or conditions would you want to examine? How might replicate or control sites make the study more convincing?

	Number of seeds (progeny)		
Morph (flower color)	*P. drummondii*	Hybrid	Total
Red	181 (87%)	27 (13%)	208
Pink	86 (62%)	53 (38%)	139

Go to **BioPortal** for all WORKING WITH**DATA** exercises

between purebred individuals of the two hybridizing species. However, subsequent generations include a variety of individuals with varying proportions of their genes derived from the original two species, so hybrid zones often contain recombinant individuals resulting from many generations of hybridization.

Detailed genetic studies can tell us much about why narrow hybrid zones may persist for long periods between the ranges of two species. In Europe, the hybrid zone between two toad species of the genus *Bombina* has been studied intensively. The fire-bellied toad (*B. bombina*) lives in eastern Europe; the closely related yellow-bellied toad (*B. variegata*) lives in western and southern Europe. The ranges of the two species overlap in a long but very narrow zone stretching 4,800 kilometers from eastern Germany to the Black Sea (**Figure 23.15**). Hybrids between the two species suffer from a range of defects, many of which are lethal. Those hybrids that survive often have skeletal abnormalities, such as misshapen mouths, ribs that are fused to vertebrae, and a reduced number of vertebrae. By following the fates of thousands of toads from the hybrid zone, investigators found that a hybrid toad, on average, is only half as fit as a purebred individual of either species. The hybrid zone remains narrow because there is strong selection against hybrids and because adult toads do not move over long distances. The zone has persisted for hundreds of years, however, because individuals of both species continue to move short distances into it, continually replenishing the hybrid population.

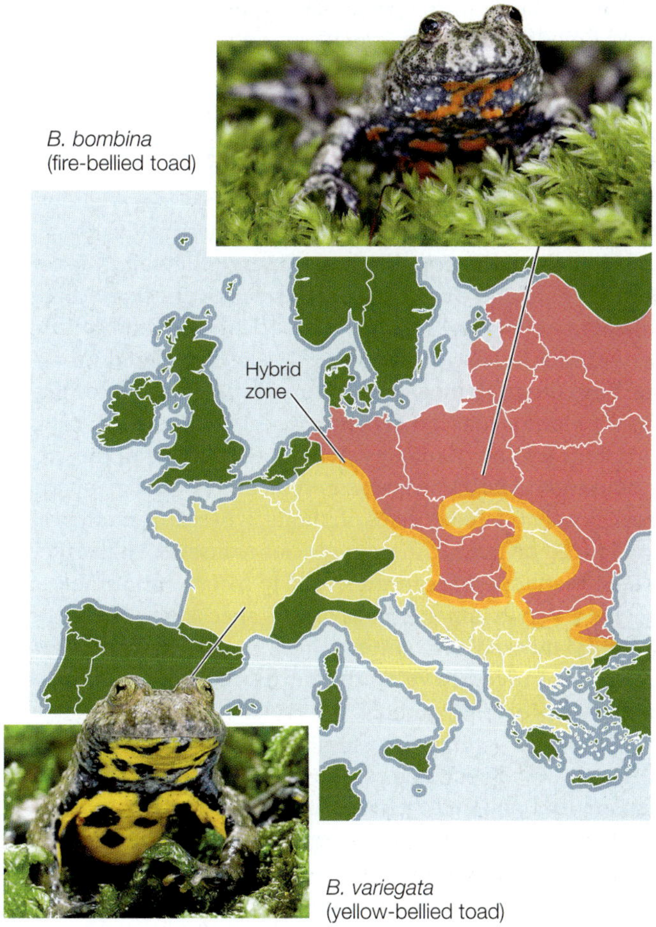
B. bombina
(fire-bellied toad)
Hybrid
zone
B. variegata
(yellow-bellied toad)

23.15 A Hybrid Zone The narrow zone (shown in orange) in which fire-bellied toads meet and hybridize with yellow-bellied toads has been stable for hundreds of years.

RECAP 23.4

Reproductive isolation may result from prezygotic or postzygotic isolating mechanisms. Lower fitness of hybrids can lead to the reinforcement of prezygotic isolating mechanisms.

- Distinguish among five types of prezygotic isolating mechanisms. **See pp. 476–477**
- Why are postzygotic isolating mechanisms said to *reinforce* prezygotic isolating mechanisms? **See pp. 478–479**
- Why don't most narrow hybrid zones, such as the one between *Bombina bombina* and *B. variegata*, get wider over time? **See p. 479 and Figure 23.15**

Some groups of organisms have many species, others only a few. Hundreds of species of *Drosophila* evolved in the small area of the Hawaiian Islands over about 20 million years. In contrast, there are only a few species of horseshoe crabs in the world, and only one species of ginkgo tree, even though these latter groups have persisted for hundreds of millions of years. Why do different groups of organisms have such different rates of speciation?

23.5 Why Do Rates of Speciation Vary?

Many factors influence the likelihood that a lineage will split to form two or more species. Therefore, rates of speciation (the proportion of existing species that split to form new species over a given period) vary greatly among groups of organisms. What are some of the factors that influence the probability of a given lineage splitting into two?

Several ecological and behavioral factors influence speciation rates

DIET SPECIALIZATION Populations of species that have specialized diets may be more likely to diverge than are those with more generalized diets. To investigate the effects of diet specialization on rates of speciation, Charles Mitter and colleagues compared species richness in some closely related groups of true bugs (hemipterans). The common ancestor of these groups was a predator that fed on other insects, but a dietary shift to herbivory (eating plants) evolved at least twice in the groups under study. Herbivorous bugs typically specialize on one or a few closely related species of plants, whereas predatory bugs tend to feed on many different species of insects. High diversity of host plant species can thus lead to a correspondingly high species diversity among herbivorous specialists. The Mitter et al. study showed that among these insects, the herbivorous groups do indeed have many more species than the related predatory groups (**Figure 23.16**).

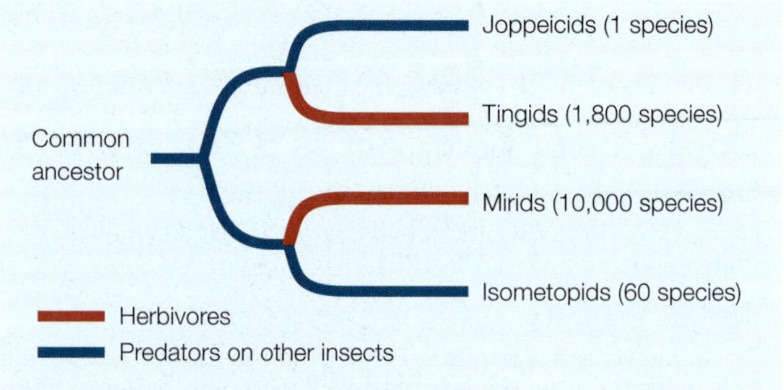
Joppeicids (1 species)
Tingids (1,800 species)
Common ancestor
Mirids (10,000 species)
Isometopids (60 species)
Herbivores
Predators on other insects

23.16 Dietary Shifts Can Promote Speciation Herbivorous groups of hemipteran insects have speciated several times faster than closely related predatory groups.

POLLINATION Speciation rates are faster in animal-pollinated than in wind-pollinated plants. Animal-pollinated groups have, on average, 2.4 times as many species as related groups pollinated by wind. Among animal-pollinated plants, speciation rates are correlated with pollinator specialization. In columbines (*Aquilegia*), the rate of evolution of new species has been about three times faster in lineages that have long nectar spurs than in lineages that lack spurs. Why do nectar spurs increase the speciation rate? Apparently it is because spurs restrict the number of pollinator species that visit the flowers, thus increasing opportunities for reproductive isolation (see Figure 23.13).

SEXUAL SELECTION It appears that the mechanisms of sexual selection (see Section 21.2) result in high rates of speciation, as we saw in the case of the cichlids of Lake Malawi. Some of the most striking examples of sexual selection are found in birds with polygynous mating systems. Bird-watchers travel thousands of miles to Papua New Guinea to witness the mating displays of male birds of paradise, which have long, brightly colored tail feathers (**Figure 23.17A**) and look distinctly different

(A) *Paradisaea minor*
(B) *Manucodia comrii*

23.17 Sexual Selection Can Lead to Higher Speciation Rates (A) Birds of paradise and (B) manucodes are closely related bird groups of the South Pacific. Speciation rates are much higher among the sexually dimorphic, polygynous birds of paradise (33 species) than among the sexually monomorphic, monogamous manucodes (5 species).

from females of their species—a phenomenon called sexual dimorphism. Males assemble at display grounds called leks, and females come there to choose a mate. After mating, the females leave the display grounds, build their nests, lay their eggs, and feed their offspring with no help from the males. The males remain at the lek to court more females.

The closest relatives of the birds of paradise are the manucodes (**Figure 23.17B**). Male and female manucodes differ only slightly in size and plumage (they are sexually monomorphic). They form monogamous pair bonds, and both sexes contribute to raising the young. There are only 5 species of manucodes, compared with 33 species of birds of paradise. By itself, this one comparison would not be convincing evidence that sexually dimorphic clades have higher rates of speciation than do monomorphic clades. However, when biologists examined all the examples of birds in which one clade is sexually dimorphic and the most closely related clade is sexually monomorphic, the sexually dimorphic clades were significantly more likely to contain more species. But why would sexual dimorphism be associated with a higher rate of speciation?

Animals with complex sexually selected behaviors are likely to form new species at a high rate because they make sophisticated discriminations among potential mating partners. They distinguish members of their own species from members of other species, and they make subtle discriminations among members of their own species on the basis of size, shape, appearance, and behavior. Such discriminations can greatly influence which individuals are most successful in mating and producing offspring, so they may lead to rapid evolution of behavioral isolating mechanisms among populations.

DISPERSAL ABILITY Speciation rates are usually higher in groups with poor dispersal abilities than in groups with good dispersal abilities because even narrow barriers can be effective in dividing a species whose members are highly sedentary. Until recently, the Hawaiian Islands had about 1,000 species of land snails, many of which were restricted to a single valley. Because snails move only short distances, the high ridges that separate the valleys were effective barriers to their dispersal. Unfortunately, introductions of other species and changes in habitat have resulted in the recent extinction of most of these unique Hawaiian land snails.

Rapid speciation can lead to adaptive radiation

The rapid proliferation of a large number of descendant species from a single ancestor species is called an **evolutionary radiation**. Evolutionary radiations often occur when a species colonizes a new area, such as an island archipelago that contains no other closely related species, because of the large number of open ecological niches. If such a rapid proliferation of species results in an array of species that live in a variety of environments and differ in the characteristics they use to exploit those environments, it is referred to as an **adaptive radiation**.

Several remarkable adaptive radiations have occurred in the Hawaiian Islands. In addition to its 1,000 species of land snails, the native Hawaiian biota includes 1,000 species of flowering plants, 10,000 species of insects, and more than 100 bird species. However, there were no amphibians, no terrestrial reptiles, and only one native terrestrial mammal (a bat) on the islands until humans introduced additional species. The 10,000 known native species of insects on Hawaii are believed to have evolved from about 400 immigrant species; only 7 immigrant species are believed to account for all the native Hawaiian land birds. Similarly, as we saw earlier in this chapter, an adaptive radiation in the Galápagos archipelago resulted in the 13 species of Darwin's finches, which differ strikingly in the size and shape of their bills and, accordingly, in the food resources they use (see Figure 23.8).

The 28 species of Hawaiian sunflowers called silverswords are an impressive example of an adaptive radiation in plants (**Figure 23.18**). DNA sequences show that these species share a relatively recent common ancestor with a species of tarweed from the Pacific coast of North America. Whereas all mainland

Madia sativa (tarweed)
Argyroxiphium sandwicense
Wilkesia gymnoxiphium
Dubautia menziesii

23.18 Rapid Evolution among Hawaiian Silverswords The Hawaiian silverswords, three closely related genera of the sunflower family, are believed to have descended from a single common ancestor (a plant similar to the tarweed *Madia sativa*) that colonized Hawaii from the Pacific coast of North America. The four plants shown here are more closely related than they appear to be based on their morphology.

tarweeds are small, upright herbs (non-woody plants such as *Madia sativa*; see Figure 23.18), the silverswords include shrubs, trees, and vines as well as both upright and ground-hugging herbs. Silversword species occupy nearly all the habitats of the Hawaiian Islands, from sea level to above the timberline in the mountains. Despite their extraordinary morphological diversification, all silverswords are genetically very similar.

The Hawaiian silverswords are more diverse in size and shape than the mainland tarweeds because their tarweed ancestors first arrived on islands that harbored very few plant species. In particular, there were few trees and shrubs because such large-seeded plants rarely disperse to oceanic islands. Trees and shrubs have evolved from non-woody ancestors on many oceanic islands. On the mainland, however, tarweeds live in ecological communities that contain many tree and shrub species in lineages with long evolutionary histories. In those environments, opportunities to exploit the "tree" way of life have already been preempted.

RECAP 23.5

Dietary specialization, pollinator specialization, sexual selection, and poor dispersal abilities are correlated with high rates of speciation. Open ecological niches present opportunities for adaptive radiations.

- How can pollinator specialization in plants and sexual selection in animals increase rates of speciation? **See pp. 480–481**
- Why do adaptive radiations often occur when a founder species invades an isolated geographic area? **See p. 481**

23.19 Evolution in the Laboratory For their experiments on the evolution of prezygotic isolating mechanisms in *Drosophila melanogaster*, Rice and Salt built an elaborate system of varying habitats contained within vials inside a large fly enclosure. Some groups of flies developed preferences for widely divergent habitats and became reproductively isolated within 35 generations.

The processes described in this chapter, operating over billions of years, have produced a world in which life is organized into millions of species, each adapted to live in a particular environment and to use environmental resources in a particular way. In the next chapter we consider how species evolve at the level of their genes and genomes.

?

Can biologists study the process of speciation in the laboratory?

ANSWER

Although speciation usually takes thousands or millions of years, and although it is typically studied in natural settings such as Lake Malawi, some aspects of speciation can be studied and observed in controlled laboratory experiments. Most such experiments use organisms with short generation times, in which evolution is expected to be relatively rapid.

William Rice and George Salt conducted an experiment in which fruit flies (*Drosophila melanogaster*) were allowed to choose food sources in different habitats. The habitats—where mating also took place—were vials in different parts of an experimental cage (**Figure 23.19**). The vials differed in three environmental factors: (1) light; (2) the direction (up or down) in which the fruit flies had to move to reach food; and (3) the concentrations of two aromatic chemicals, ethanol and acetaldehyde. In just 35 generations, the two groups of flies that chose the most divergent habitats had become reproductively isolated from each other, having evolved distinct preferences for the different habitats.

The experiment by Rice and Salt (see *American Naturalist* 131: 911–917, 1988) demonstrated an example of habitat isolation as a prezygotic isolating mechanism. Even though the different habitats were in the same cage, and individual fruit flies were capable of flying from one habitat to the other, habitat preferences were inherited by offspring from their parents, and populations from the two divergent habitats did not interbreed. Similar habitat isolation is thought to have resulted in the early split between cichlids that preferred the rocky versus the sandy shores of Lake Malawi. In controlled experiments like this one, biologists can observe many aspects of the process of speciation directly.

CHAPTER SUMMARY 23

23.1 What Are Species?

- **Speciation** is the divergence of biological lineages and the emergence of reproductive isolation between those lineages.
- The **morphological species concept** distinguishes species on the basis of physical similarities and differences.
- The **biological species concept** distinguishes species on the basis of **reproductive isolation.**
- The **lineage species concept** recognizes evolutionarily independent lineages as species, allowing biologists to consider species over evolutionary time.

23.2 What Is the Genetic Basis of Speciation?

- Speciation usually results from the interruption of gene flow within a population.
- The Dobzhansky–Muller model describes how reproductive isolation between two physically isolated populations can develop through the accumulation of incompatible genes or chromosomal arrangements. **Review Figures 23.3, 23.4, ANIMATED TUTORIAL 23.1**
- Reproductive isolation increases with increasing genetic divergence between populations. **Review Figure 23.5**

23.3 What Barriers to Gene Flow Result in Speciation?

See ANIMATED TUTORIAL 23.2

- **Allopatric speciation**, which results when populations are separated by a physical barrier, is the dominant mode of speciation in most groups of organisms. This type of speciation may follow founder events, in which some members of a population cross a barrier and found a new, isolated population. **Review Figures 23.6–23.8, ANIMATED TUTORIALS 23.2 and 23.3**
- **Sympatric speciation** results when the genomes of two groups diverge in the absence of physical isolation. Such divergence can result from disruptive selection if individuals with different genotypes prefer distinct microhabitats.
- Sympatric speciation can occur within two generations via **polyploidy**. Polyploidy may arise from chromosome duplications within a species (**autopolyploidy**) or from hybridization that results in combining the chromosomes of two species (**allopolyploidy**). **Review Figure 23.9**

23.4 What Happens When Newly Formed Species Come into Contact?

- **Prezygotic isolating mechanisms** prevent hybridization; **postzygotic isolating mechanisms** reduce the fitness of hybrids.
- Postzygotic isolating mechanisms lead to **reinforcement** of prezygotic isolating mechanisms by natural selection. **Review Figures 23.11, 23.12, 23.14**
- Hybrid zones may form and persist if reproductive isolation between species is incomplete. **Review Figure 23.15**

23.5 Why Do Rates of Speciation Vary?

- Dietary specialization, pollinator specialization, sexual selection, and dispersal ability all influence speciation rates. **Review Figure 23.16**
- **Evolutionary radiation** refers to the rapid proliferation of descendant species from a single ancestor species. This often occurs following colonization, when new species may rapidly move into unoccupied ecological niches in a process known as **adaptive radiation**.

See ACTIVITY 23.1 for a concept review of this chapter

Go to the Interactive Summary to review key figures, Animated Tutorials, and Activities Life10e.com/is23

CHAPTER REVIEW

REMEMBERING

1. Which of the following is *not* a condition expected to favor allopatric speciation?
   a. Continents drift apart and separate previously connected lineages.
   b. A mountain range separates formerly connected populations.
   c. Different environments on two sides of a barrier cause populations to diverge.
   d. The range of a species is separated by loss of intermediate habitat.
   e. Tetraploid individuals arise in one part of the range of a species.

2. Which of the following is *not* a potential prezygotic reproductive barrier?
   a. Temporal segregation of breeding seasons
   b. Differences in chemicals that attract mates
   c. Hybrid infertility
   d. Spatial segregation of mating sites
   e. Sperm that cannot penetrate an egg

3. A common means of sympatric speciation is
   a. polyploidy.
   b. hybrid infertility.
   c. temporal segregation of breeding seasons.
   d. spatial segregation of mating sites.
   e. imposition of a geographic barrier.

4. Which of the following is often associated with higher rates of speciation?
   a. Sexual dimorphism in birds
   b. Diet specialization in insects
   c. Poor dispersal ability
   d. Animal pollination in plants
   e. All of the above

UNDERSTANDING & APPLYING

5. The Dobzhansky–Muller model of speciation suggests that divergence among alleles at *different* gene loci leads to genetic incompatibility between species. Why is genetic incompatibility between two alleles at the *same* locus considered less likely?

6. Why do some combinations of chromosomal centric fusions cause problems in meiosis? Can you diagram what would happen at meiosis in a hybrid of the divergent lineages shown in Figure 23.4?

7. Assume that the reproductive isolation seen in the *Phlox* strains discussed in Section 23.2 results from lethal combinations of incompatible alleles at several loci among the various strains. Given this assumption, why might the reproductive isolation seen among these strains be partial rather than complete?

8. If allopatric speciation is the most prevalent mode of speciation, what do you predict about the geographic distributions of many closely related species?

ANALYZING & EVALUATING

9. In each of the species of columbine shown in Figure 23.13, the orientation of the flowers and the length of flower spurs are associated with a particular type of pollinator (hummingbirds or hawkmoths). Columbine flowers vary in other ways as well; for example, they differ in color, and probably in odor. What experiments could you design to determine the traits that various pollinators use to distinguish among the flowers of different columbine species?

10. The different finch species in the phylogeny shown in Figure 23.8 have all evolved on islands of the Galápagos archipelago within the past 3 million years. Molecular clock analysis (see Chapter 24) has been used to determine the dates of the various speciation events in that phylogeny. Geological techniques for dating rock samples (see Chapter 25) have been used to determine the ages of the various Galápagos islands. The table below shows the number of species of Darwin's finches and the number of islands that have existed in the archipelago at several times during the past 4 million years.

Time (mya)	Number of islands	Number of finch species
0.25	18	14
0.50	18	9
0.75	9	7
1.00	6	5
2.00	4	3
3.00	4	1
4.00	3	0

a. Plot the number of species of Darwin's finches and the number of islands in the Galápagos archipelago (dependent variables) against time (independent variable).

b. Are the data consistent with the hypothesis that isolation of populations on newly formed islands is related to speciation in this group of birds? Why or why not?

c. If no more islands form in the Galápagos archipelago, do you think that speciation by geographic isolation will continue to occur among Darwin's finches? Why or why not? What additional data could you collect to test your hypothesis (without waiting to see if speciation occurs)?

Go to BioPortal at **yourBioPortal.com** for Animated Tutorials, Activities, LearningCurve Quizzes, Flashcards, and many other study and review resources.

24 Evolution of Genes and Genomes

CHAPTER**OUTLINE**

Adapting to a Murky Environment The elephant-nose fish (*Gnathonemus petersi*), a river-dwelling species from West Africa, is one of many fishes in which weakly discharging electric organs have evolved via duplications and modifications in sodium channel proteins.

SOME FISHES GENERATE HIGH-VOLTAGE electric discharges (up to 650 volts) that they use to stun their prey; perhaps the best known examples are the electric eels of Central and South America. Several other fish species produce weaker electric discharges. Most of these weakly electric fishes live in murky water, where visual cues are limited. They use electric signals to locate (but not to stun) their prey. Electric signals also allow them to communicate with other individuals of their own species.

Electric organs have evolved independently in several fish lineages. How did these organs evolve? Consider first the physical basis of the electric discharges. Voltage-gated sodium channels are large protein complexes that underlie the generation and propagation of rapid electric signals in nerve, muscle, and heart tissues. Electric signals are transmitted along nerves to muscles as the sodium channels embedded in cell membranes are stimulated to open. These channels control the concentration of positively charged sodium ions (Na^+) on the inside relative to the outside of cells, resulting in an electric charge that is transmitted across the surface of the muscle, leading to muscular contraction.

Most vertebrates have multiple copies of the genes encoding the several proteins that make up the sodium channel. These copies arose in the distant past through a series of gene duplications. Such duplications allowed the differentiation and specialization of protein function, making it possible for different sodium channels to exist in different types of tissue. In the case of electric fishes, one of the sodium channel genes expressed in muscle diverged, and a new functional protein evolved. Changes in a relatively small number of nucleotide positions in the gene resulted in modified sodium channels, allowing the development of a new organ with a unique function—the generation of externally transmitted electric energy.

The repeated evolution of electric organs from muscle tissue has been facilitated by the relative simplicity of the molecular changes required. Gene duplication has expedited the process, since redundant genes allow for such specialization in protein function. Finally, interspecific differences in sodium channel function have arisen from additional changes in the nucleotide sequences of the genes. These small differences allow different species to use different communication signals, which improves intraspecific communication while reducing interspecific interference.

?

How do evolutionary studies of sodium channel genes help us understand some human genetic disorders?

See answer on p. 502.

24.1 How Are Genomes Used to Study Evolution?

An organism's **genome** is the full set of genes it contains, as well as any noncoding regions of the DNA (or, in the case of some viruses, RNA). Most of the genes of eukaryotic organisms are found on chromosomes in the nucleus, but genes are also present in chloroplasts and mitochondria. In organisms that reproduce sexually, both males and females contribute nuclear genes, but mitochondrial and chloroplast genes are usually transmitted only via the cytoplasm of one of the two gametes (usually from the female parent).

Genomes must be replicated to be transmitted from parents to offspring. DNA replication does not occur without error, however. Mistakes in DNA replication—mutations—provide much of the raw material for evolutionary change. Mutations are essential for the long-term survival of life because they are the initial source of the genetic variation that permits species to evolve in response to changes in their environment.

A particular allele of a gene will not be passed on to successive generations unless an individual carrying that allele survives and reproduces. The allele must function in combination with many other genes in the genome or it will quickly be selected against. Moreover, the degree and timing of a gene's expression are affected by its location in the genome. For these reasons, the genes of an individual organism can be viewed as interacting members of a group, among which there are divisions of labor but also strong interdependencies.

A genome, then, is not simply a random collection of genes in a random order along chromosomes. Rather, it is a complex set of integrated genes, regulatory sequences, and structural elements, interspersed with vast stretches of noncoding DNA that may have little direct function. Both the positions of genes and their sequences are subject to evolutionary change, as are the extent and location of noncoding DNA. All of these changes can affect the phenotype of an organism.

Biologists have now sequenced the complete genomes of a large number of organisms, including humans. The information in these sequences is helping us understand how and why organisms differ, how they function, and how they have evolved.

Evolution of genomes results in biological diversity

The field of **molecular evolution** investigates the mechanisms and consequences of the evolution of macromolecules—particularly nucleic acids (DNA and RNA) and proteins. Molecular evolutionists study relationships between the structures of genes and proteins and the functions of organisms. They also examine molecular variation to reconstruct evolutionary history and to study the mechanisms and consequences of evolution. Students of this field ask questions such as: What does molecular variation tell us about a gene's function? Why do the genomes of different organisms vary in size? What evolutionary forces shape patterns of variation among genomes? And a crucial question from an evolutionary perspective: How do genomes acquire new functions? Investigations into the evolution of particular nucleic acids and proteins are instrumental in reconstructing the evolutionary histories of genes. Ultimately, molecular evolutionary biologists hope to explain the molecular basis of biological diversity.

The evolution of nucleic acids and proteins depends on genetic variation introduced by mutations. One of several ways in which genes evolve is by means of nucleotide substitutions (the incorporation of point mutations in populations). In genes that encode proteins, nucleotide substitutions sometimes result in amino acid replacements that can change the charge, the structure, and other chemical and physical properties of the encoded protein. Phenotypic changes in a protein molecule often affect the way that protein functions in the organism.

Evolutionary changes in genes and proteins can be identified by comparing nucleotide or amino acid sequences from different organisms. The longer two sequences have been evolving separately, the more differences they accumulate (bearing in mind that different genes in the same species evolve at different rates). Determining how long ago changes in nucleotide or amino acid sequences occurred is a useful step toward inferring their causes. Knowledge of the pattern and rate of evolutionary change in a given macromolecule is useful in reconstructing the evolutionary history of groups of organisms.

To compare genes or proteins from Tdifferent organisms, biologists need a way to identify homologous parts of macromolecules. (Recall from Section 22.1 that homologous features are those shared by two or more species that have been inherited from a common ancestor.) Homologous parts of a protein can be identified by their homologous amino acid sequences. And, since nucleotide sequences encode amino acid sequences, the concept of homology extends down to the level of individual nucleotide positions. Therefore one of the first steps in studying the evolution of genes or proteins is to align homologous positions in the amino acid or nucleotide sequence of interest.

Genes and proteins are compared through sequence alignment

Once the nucleotide or amino acid sequences of molecules from different organisms have been determined, they can be compared. Homologous positions can be identified only if we first pinpoint the locations of deletions and insertions that have occurred in the molecules of interest in the time since the organisms diverged from a common ancestor. A simple hypothetical example illustrates this **sequence alignment** technique. In **Figure 24.1** we compare two amino acid sequences from homologous proteins in different organisms. The two sequences at first appear to differ in both the number and identity of their amino acids. If we insert a gap after the first amino acid in sequence 2 (after leucine), however, the similarities in the two sequences become obvious. This gap represents the occurrence of one of two evolutionary events: an insertion of an amino acid in the longer protein or a deletion of an amino acid in the shorter protein. Having adjusted for this insertion or deletion event, we can see that the two sequences differ by only one amino acid at position 6 (serine or phenylalanine).

Go to Activity 24.1 Amino Acid Sequence Alignment
Life10e.com/ac24.1

RESEARCH**TOOLS**

24.1 Amino Acid Sequence Alignment Amino acid sequence alignment is a way of arranging protein sequences to identify regions of homology between the sequences. Gaps are inserted between the amino acid residues to align similar residues in columns. Differences and similarities between each pair of aligned sequences are then summarized in a similarity matrix. Homologous DNA sequences can be aligned in a similar manner.

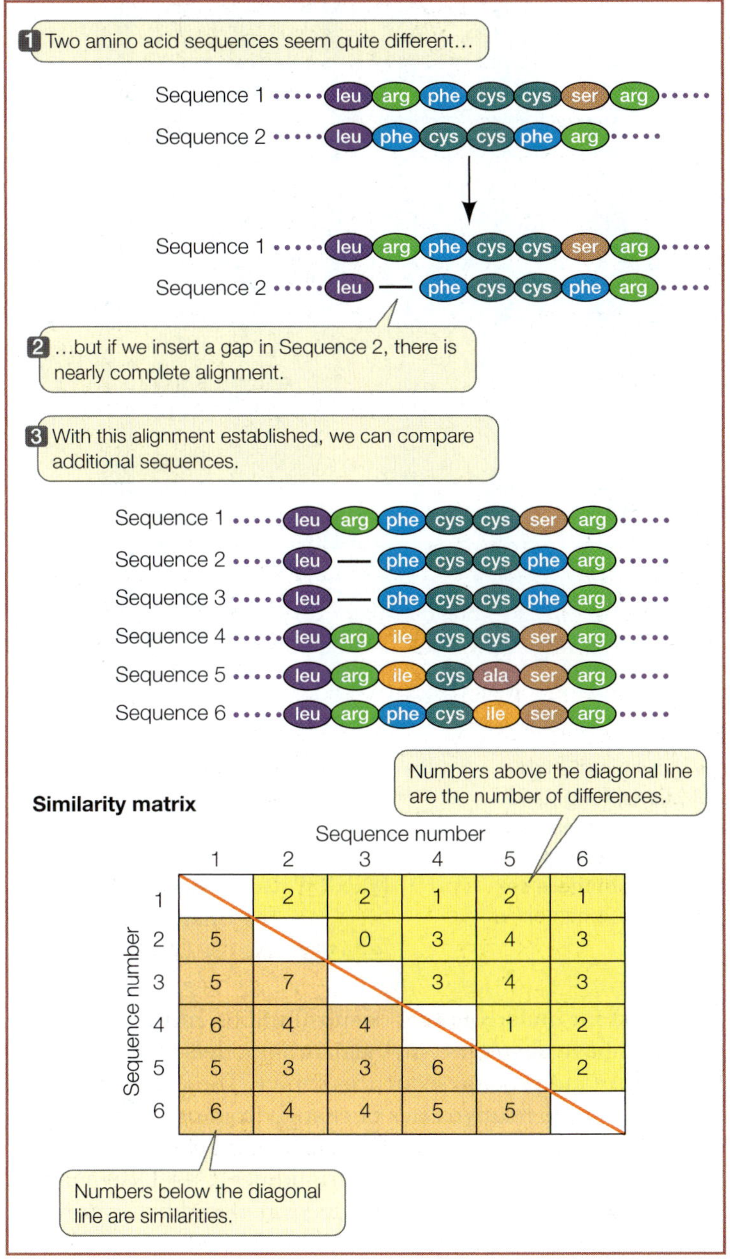

1 Two amino acid sequences seem quite different…
Sequence 1 leu arg phe cys cys ser arg
Sequence 2 leu phe cys cys phe arg
Sequence 1 leu arg phe cys cys ser arg
Sequence 2 leu — phe cys cys phe arg
2 …but if we insert a gap in Sequence 2, there is nearly complete alignment.
3 With this alignment established, we can compare additional sequences.
Sequence 1 leu arg phe cys cys ser arg
Sequence 2 leu — phe cys cys phe arg
Sequence 3 leu — phe cys cys phe arg
Sequence 4 leu arg ile cys cys ser arg
Sequence 5 leu arg ile cys ala ser arg
Sequence 6 leu arg phe cys ile ser arg
Similarity matrix
Numbers above the diagonal line are the number of differences.
Sequence number
1 2 3 4 5 6
Sequence number
1: 2 2 1 2 1
2: 5 0 3 4 3
3: 5 7 3 4 3
4: 6 4 4 1 2
5: 5 3 3 6 2
6: 6 4 4 5 5
Numbers below the diagonal line are similarities.

Adding a single gap—that is, identifying a deletion or an insertion—aligns the two sequences in Figure 24.1. Additional sequences can now be added to the alignment in a similar manner. Longer amino acid sequences, and those that have diverged more extensively, require more elaborate adjustments. Explicit models (incorporated into computer algorithms) have been developed to account for the relative probabilities of deletions, insertions, and particular amino acid replacements.

Having aligned the sequences, we can compare them by counting the number of nucleotides or amino acids that differ between them. Summing the numbers of the same and different amino acids in each pair of sequences allows us to construct a **similarity matrix**, which gives us a measure of the minimum number of changes that have occurred since the divergence of each pair of organisms (see Figure 24.1).

Go to Activity 24.2 Similarity Matrix Construction
Life10e.com/ac24.2

Models of sequence evolution are used to calculate evolutionary divergence

The sequence comparison procedure illustrated in Figure 24.1 gives a simple count of the number of similarities and differences between the proteins of two species. In the context of two aligned DNA sequences, we can count the number of differences at homologous nucleotide positions, and this count indicates the minimum number of nucleotide changes that must have occurred since the two sequences diverged from a common ancestral sequence.

Although it is useful in determining a *minimum* number of changes between two DNA sequences, the count provided by sequence alignment almost certainly underestimates the *actual* number of changes that have occurred since the sequences diverged. Any given change counted in a similarity matrix of DNA sequences may result from multiple substitution events that occurred at a given nucleotide position over time. As illustrated in **Figure 24.2**, any of the following events may have

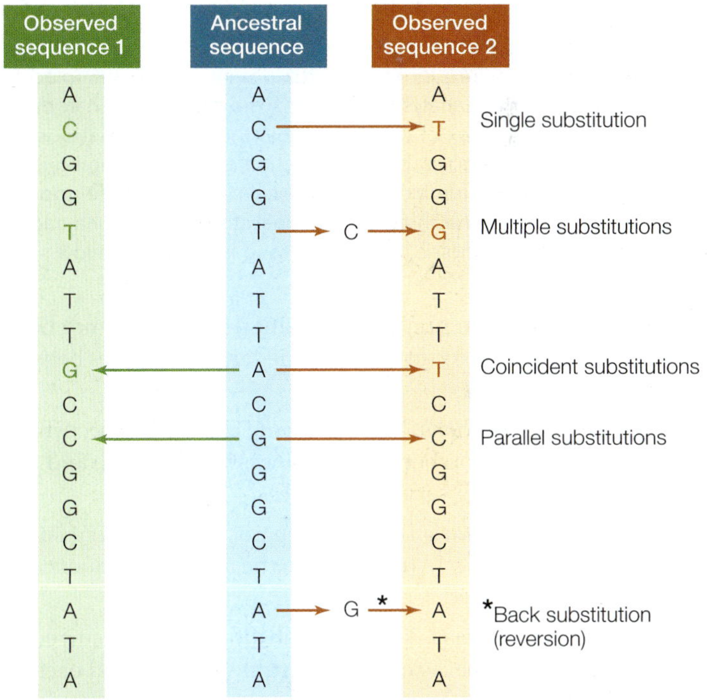

Observed sequence 1
Ancestral sequence
Observed sequence 2
A C G G T A T T G C C G G C T A T A
A C G G T A T T A C G G G C T A T A
A T G G G A T T T C C G G C T A T A
C
G
Single substitution
Multiple substitutions
Coincident substitutions
Parallel substitutions
*Back substitution (reversion)

24.2 Multiple Substitutions Are Not Reflected in Pairwise Sequence Comparisons Two observed sequences descended from a common ancestral sequence (center) have undergone a series of substitutions. Although the two observed sequences differ by only three nucleotides (colored letters), these three differences result from a total of nine substitutions (arrows).

Tuna

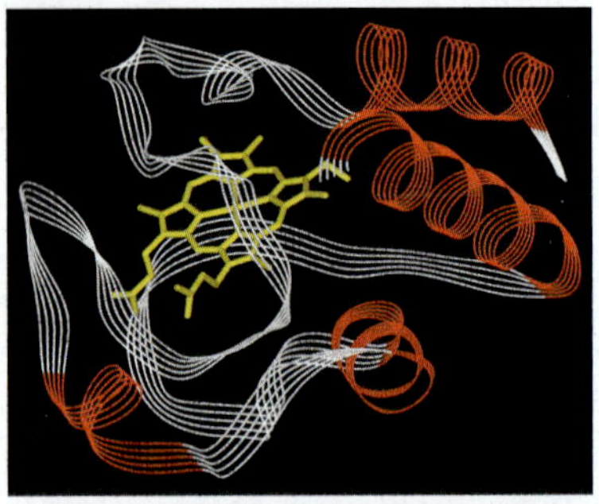

Rice

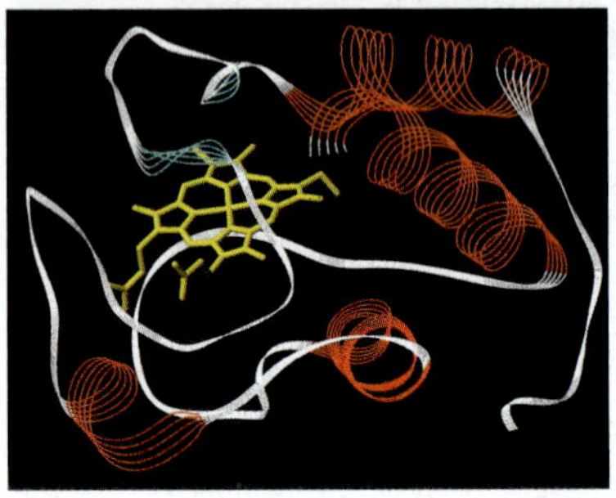

Acidic side chains

- D Aspartic acid
- E Glutamic acid

Basic side chains

- H Histidine
- K Lysine
- R Arginine

Hydrophobic side chains

- F Phenylalanine
- I Isoleucine
- L Leucine
- M Methionine
- V Valine
- Y Tyrosine
- W Tryptophan
- A Alanine

Other

- C Cysteine
- P Proline
- Q Glutamine
- N Asparagine
- S Serine
- T Threonine
- G Glycine

The number 1 indicates an invariant position in the cytochrome *c* molecule (i.e., all the organisms have the same amino acid in this position). Such a position is probably under strong purifying selection.

Amino acids at positions marked by red arrowheads have side chains that interact with the heme group.

Position in sequence	1				5					10					15					20					25					30				
(red arrowheads)														▼			▼	▼																
Number of amino acids at the position	1	3	5	5	5	1	3	3	4	1	4	3	2	1	3	3	1	1	2	4	3	4	2	3	4	2	1	4	1	1	2	1	5	1
Human, chimpanzee	G	D	V	E	K	G	K	K	I	F	I	M	K	C	S	Q	C	H	T	V	E	K	G	G	K	H	K	T	G	P	N	L	H	G
Rhesus monkey	G	D	V	E	K	G	K	K	I	F	I	M	K	C	S	Q	C	H	T	V	E	K	G	G	K	H	K	T	G	P	N	L	H	G
Horse	G	D	V	E	K	G	K	K	I	F	V	Q	K	C	A	Q	C	H	T	V	E	K	G	G	K	H	K	T	G	P	N	L	H	G
Donkey	G	D	V	E	K	G	K	K	I	F	V	Q	K	C	A	Q	C	H	T	V	E	K	G	G	K	H	K	T	G	P	N	L	H	G
Cow, pig, sheep	G	D	V	E	K	G	K	K	I	F	V	Q	K	C	A	Q	C	H	T	V	E	K	G	G	K	H	K	T	G	P	N	L	H	G
Dog	G	D	V	E	K	G	K	K	I	F	V	Q	K	C	A	Q	C	H	T	V	E	K	G	G	K	H	K	T	G	P	N	L	H	G
Rabbit	G	D	V	E	K	G	K	K	I	F	V	Q	K	C	A	Q	C	H	T	V	E	K	G	G	K	H	K	T	G	P	N	L	H	G
Gray whale	G	D	V	E	K	G	K	K	I	F	V	Q	K	C	A	Q	C	H	T	V	E	K	G	G	K	H	K	T	G	P	N	L	H	G
Gray kangaroo	G	D	V	E	K	G	K	K	I	F	V	Q	K	C	A	Q	C	H	T	V	E	K	G	G	K	H	K	T	G	P	N	L	N	G
Chicken, turkey	G	D	I	E	K	G	K	K	I	F	V	Q	K	C	S	Q	C	H	T	V	E	K	G	G	K	H	K	T	G	P	N	L	H	G
Pigeon	G	D	I	E	K	G	K	K	I	F	V	Q	K	C	S	Q	C	H	T	V	E	K	G	G	K	H	K	T	G	P	N	L	H	G
Pekin duck	G	D	V	E	K	G	K	K	I	F	V	Q	K	C	S	Q	C	H	T	V	E	K	G	G	K	H	K	T	G	P	N	L	H	G
Snapping turtle	G	D	V	E	K	G	K	K	I	F	V	Q	K	C	A	Q	C	H	T	V	E	K	G	G	K	H	K	T	G	P	N	L	N	G
Rattlesnake	G	D	V	E	K	G	K	K	I	F	T	M	K	C	S	Q	C	H	T	V	E	K	G	G	K	H	K	T	G	P	N	L	H	G
Bullfrog	G	D	V	E	K	G	K	K	I	F	V	Q	K	C	A	Q	C	H	T	C	E	K	G	G	K	H	K	V	G	P	N	L	Y	G
Tuna	G	D	V	A	K	G	K	K	T	F	V	Q	K	C	A	Q	C	H	T	V	E	N	G	G	K	H	K	V	G	P	N	L	W	G
Dogfish	G	D	V	E	K	G	K	K	V	F	V	Q	K	C	A	Q	C	H	T	V	E	N	G	G	K	H	K	T	G	P	N	L	S	G
Samia cynthia (moth)	G	N	A	E	N	G	K	K	I	F	V	Q	R	C	A	Q	C	H	T	V	E	A	G	G	K	H	K	V	G	P	N	L	H	G
Tobacco hornworm moth	G	N	A	D	N	G	K	K	I	F	V	Q	R	C	A	Q	C	H	T	V	E	A	G	G	K	H	K	V	G	P	N	L	H	G
Screwworm fly	G	D	V	E	K	G	K	K	I	F	V	Q	R	C	A	Q	C	H	T	V	E	A	G	G	K	H	K	V	G	P	N	L	H	G
Drosophila (fruit fly)	G	D	V	E	K	G	K	K	L	F	V	Q	R	C	A	Q	C	H	T	V	E	A	G	G	K	H	K	V	G	P	N	L	H	G
Baker's yeast	G	S	A	K	K	G	A	T	L	F	K	T	R	C	E	L	C	H	T	V	E	K	G	G	P	H	K	V	G	P	N	L	H	G
Candida krusei (yeast)	G	S	A	K	K	G	A	T	L	F	K	T	R	C	A	E	C	H	T	I	E	A	G	G	P	H	K	V	G	P	N	L	H	G
Neurospora crassa (mold)	G	D	S	K	K	G	A	N	L	F	K	T	R	C	A	E	C	H	–	–	E	–	N	L	T	Q	K	I	G	P	A	L	H	G
Wheat	G	N	P	D	A	G	A	K	I	F	K	T	K	C	A	Q	C	H	T	V	D	A	G	A	–	H	K	Q	G	P	N	L	H	G
Sunflower	G	D	P	T	T	G	A	K	I	F	K	T	K	C	A	Q	C	H	T	V	E	K	G	A	–	H	K	Q	G	P	N	L	N	G
Mung bean	G	D	S	K	S	G	E	K	I	F	K	T	K	C	A	Q	C	H	T	V	D	K	G	A	–	H	K	Q	G	P	N	L	N	G
Rice	G	N	P	K	A	G	E	K	I	F	K	T	K	C	A	Q	C	H	T	V	D	K	G	A	–	H	K	Q	G	P	N	L	N	G
Sesame	G	D	V	K	S	G	E	K	I	F	K	T	K	C	A	Q	C	H	T	V	D	K	G	A	–	H	K	Q	G	P	N	L	N	G

Gaps indicate insertion and/or deletion events.

24.3 Amino Acid Sequences of Cytochrome *c* The amino acid sequences shown in the table were obtained from analyses of the enzyme cytochrome *c* from 33 species of plants, fungi, and animals. Note the lack of variation across the sequences at positions 70–80, suggesting that this region is under strong stabilizing selection and that changing its amino acid sequence would impair the protein's function. The molecular models at the upper left are created from these sequences and show the three-dimensional structures of tuna and rice cytochrome *c*. Alpha helixes are in red, and the molecule's heme group is shown in yellow.

Go to Media Clip 24.1
The Ubiquitous Protein
Life10e.com/mc24.1

occurred at a given nucleotide position that would not be revealed by a simple count of similarities and differences between two DNA sequences:

- *Multiple substitutions.* More than one change has occurred at a given position between the ancestral sequence and at least one of the observed sequences.
- *Coincident substitutions.* At a given position, different substitutions have occurred between the ancestral sequence and each observed sequence.
- *Parallel substitutions.* The same substitution has occurred independently between the ancestral sequence and each observed sequence.
- *Back substitutions* (also called reversions). In a variation on multiple substitutions, after a change at a given position, a subsequent substitution has changed the position back to the ancestral state.

To correct for undercounting of substitutions, molecular evolutionists have developed mathematical models that describe how DNA (and protein) sequences evolve. These models take into account the relative rates of change from one nucleotide to another; for example, transitions (changes between the two purines, A ↔ G, or between the pyrimidines, C ↔ T) are more frequent than transversions (a purine is replaced by a pyrimidine, or vice versa). These models also include parameters such as the different rates of substitution across different parts of a gene and the proportions of each nucleotide present in a given sequence. Once such parameters have been estimated, the model is used to correct for multiple substitutions, coincident substitutions, parallel substitutions, and back substitutions. The revised estimate accounts for the *total* number of substitutions likely to have occurred between two sequences, which is almost always greater than the observed number of differences.

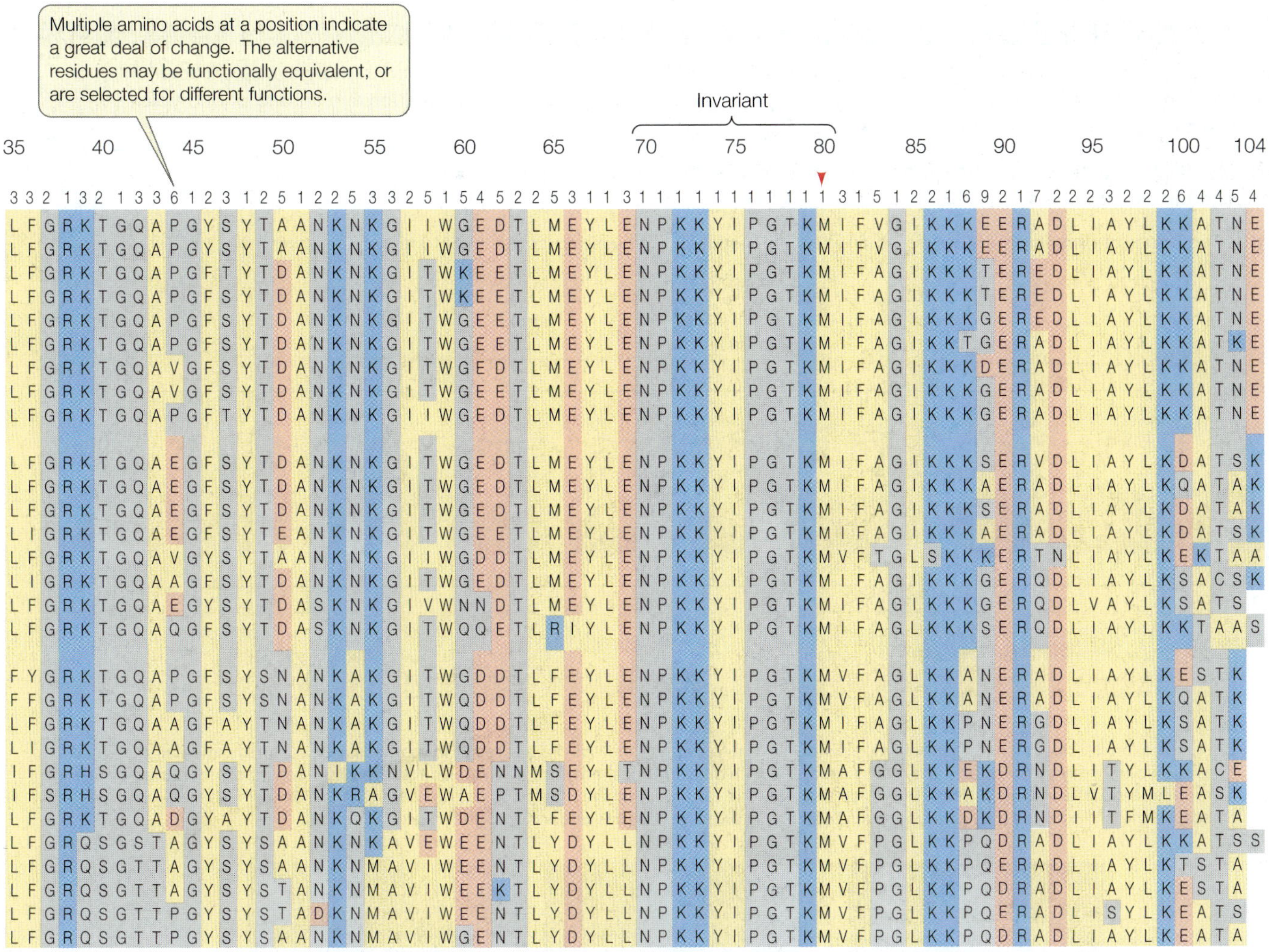
Multiple amino acids at a position indicate a great deal of change. The alternative residues may be functionally equivalent, or are selected for different functions.
Invariant
35 40 45 50 55 60 65 70 75 80 85 90 95 100 104

As sequence information becomes available for more and more genes in an ever-expanding database, sequence alignments can be extended across multiple homologous sequences, and the minimum number of insertions, deletions, and substitutions can be summed across homologous genes of an entire group of organisms. Similar databases have been constructed for homologous proteins. **Figure 24.3** shows aligned data for cytochrome *c* protein sequences in 33 species of animals, plants, and fungi. Such information is used extensively in determining evolutionary relationships among species.

Experimental studies examine molecular evolution directly

Although molecular evolutionists are often interested in naturally evolved genes and proteins, molecular evolution can also be observed directly in the laboratory. Increasingly, evolutionary biologists are studying evolution experimentally. Because substitution rates are related to generation time rather than to absolute time, most of these experiments use unicellular organisms or viruses with short generations. Viruses, bacteria, and unicellular eukaryotes (such as the yeasts) can be cultured in large populations in the laboratory, and many of these organisms can evolve rapidly. In the case of some RNA viruses, the natural substitution rate may be as high as 1 substitution per 1,000 nucleotides per generation. Therefore in a virus of a few thousand nucleotides, one or more substitutions are expected (on average) every generation, and these changes can easily be determined by sequencing the entire viral genome (because of its small size). Generation time may be only tens of minutes (rather than years or decades, as in many animals), so biologists can directly observe substantial molecular evolution in a controlled population over the course of days, weeks, or months.

An example of an experimental evolutionary study is shown in **Figure 24.4**. Paul Rainey and Michael Travisano wanted to examine a potential cause of adaptive radiations, which are a major source of biological diversity (see Section 23.5). While Rainey and Travisano clearly couldn't experimentally manipulate animals over many millions of years, they could test the idea that heterogeneous environments with unoccupied niches lead to adaptive radiation by experimentally manipulating a bacterial lineage.

Rainey and Travisano inoculated several flasks containing culture medium with the same strain of the bacterium *Pseudomonas fluorescens*. They then shook some of the cultures to maintain a constantly uniform environment. They left others alone (static cultures), allowing them to develop spatially

INVESTIGATINGLIFE

24.4 Evolution in a Heterogeneous Environment Paul Rainey and Michael Travisano cultured the rapidly evolving bacterium *Pseudomonas fluorescens* in homogeneous and heterogeneous environments to examine the relationship among phenotypic diversity, molecular divergance, and environmental variability.[a]

HYPOTHESIS Heterogeneous environments are more conducive to the evolution of phenotypic diversity than are homogeneous environments.

Method One colony of *Pseudomonas fluorescens* (all of a single genotype) was used to inoculate many replicate cultures.

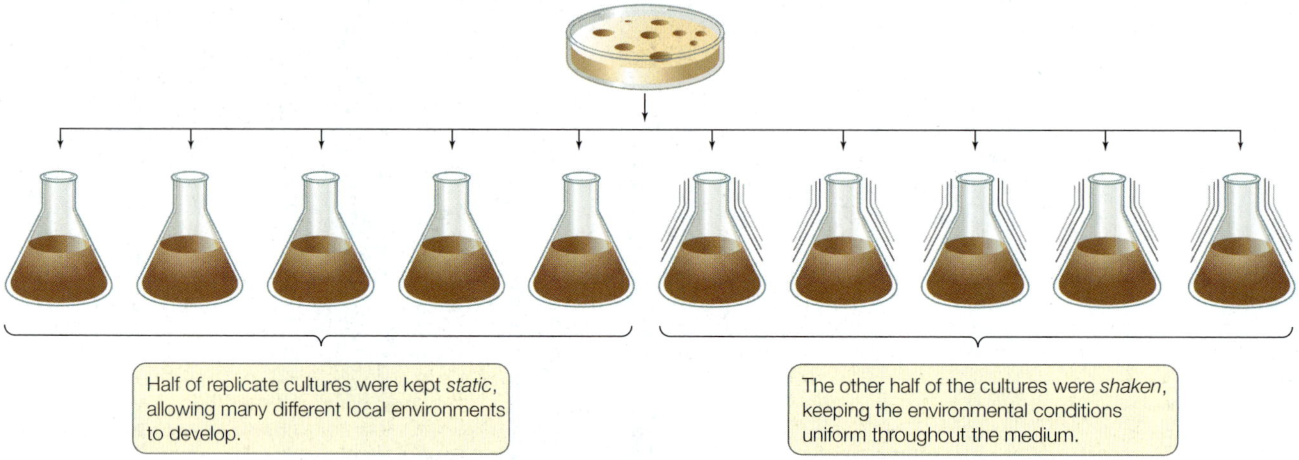
Half of replicate cultures were kept *static*, allowing many different local environments to develop.
The other half of the cultures were *shaken*, keeping the environmental conditions uniform throughout the medium.

Results In the shaken flasks, the ancestral morph persisted; the uniform environment did not result in morphological diversification. In the static flasks, two new morphs regularly arose, each adapted to a different local environment. Molecular analysis revealed that the mutations that produce these phenotypes arose in both shaken and static cultures, but the mutations did not persist in the uniform (shaken) environment because the phenotypes they produced were selectively disadvantageous under homogeneous conditions.

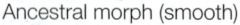

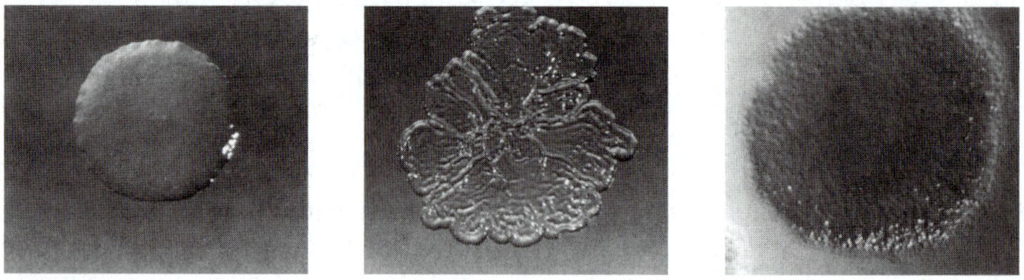

Ancestral morph (smooth) "Wrinkly spreader" morph "Fuzzy spreader" morph

CONCLUSION Phenotypic change and diversification are enhanced in a heterogeneous environment.

Go to **BioPortal** for discussion and relevant links for all INVESTIGATING**LIFE** figures.

[a]Rainey, P. B. and M. Travisano. 1998. *Nature* 394: 69–72.

distinct environments. In the static cultures, the environment on the surface of the medium differed from that on the walls of the flasks and from parts of the culture not touching any surfaces.

When the cultures were started, the ancestral phenotype of the bacterium produced a smooth colony phenotype, which the investigators called a "smooth morph." In just a few days, however, the static cultures consistently and independently developed two other morphs: a "wrinkly spreader" and a "fuzzy spreader." The researchers determined that the two new morphs had a genetic basis and were adaptively superior in certain environments found within the static cultures. For example, the "wrinkly spreader" cells adhered firmly to one another as well as to surfaces and thus were able to form a mat across the surface of the medium, where they could compete successfully for oxygen.

DNA sequencing of the genomes of these morphs showed that the same phenotypes had evolved repeatedly in different static cultures and that many different substitutions could produce the same phenotypes. The homogeneous shaken cultures, in contrast, showed no changes in phenotype. The same mutations occurred in the shaken cultures, but they did not persist because the novel phenotypes they produced were selectively disadvantageous (i.e., less fit) under the "shaken" environmental conditions.

Experimental molecular evolutionary studies are used for a wide variety of purposes and have greatly expanded the ability of evolutionary biologists to test evolutionary concepts and principles. Biologists now routinely study evolution in the laboratory and, as we will see later in this chapter, they can use in vitro evolutionary techniques to produce novel molecules that perform new functions with industrial and pharmaceutical uses.

RECAP 24.1

The genomes of all organisms evolve over time. Evolutionary changes can be detected by comparing the nucleic acid and protein sequences of different species. Experimental studies of molecular evolution allow biologists to study many processes of evolution directly under controlled conditions.

- How do biologists align nucleotide and amino acid sequences they wish to compare, and how do they determine the minimum number of changes that have occurred between pairs of aligned sequences? **See pp. 486–487 and Figure 24.1**
- Explain why a simple count of nucleotide differences between two sequences underestimates the actual number of nucleotide substitutions since the sequences diverged. **See Figure 24.2**

We have seen that molecular evolutionists can directly observe the evolution of genomes over time, compare the genomes of different organisms, and reconstruct the changes that have occurred during their evolution. Let's turn now to the question of how genomes change and examine some of the consequences of those changes.

24.2 What Do Genomes Reveal about Evolutionary Processes?

A mutation, as we saw in Section 15.1, is any change in the genetic material. A nucleotide substitution is the product of one type of mutation, incorporated into a population. Many nucleotide substitutions have no effect on phenotype, even if the change occurs in a gene that encodes a protein, because most amino acids are specified by more than one codon (see Figure 14.6). A substitution that does not change the encoded amino acid is known as a **synonymous substitution** or **silent substitution** (**Figure 24.5A**). Synonymous substitutions do not affect the functioning of a protein (although they may have other effects, as described in Section 15.1) and are therefore less likely than other types of substitutions to be subject to natural selection.

A nucleotide substitution that *does* change the amino acid sequence encoded by a gene is known as a **nonsynonymous substitution** or **missense substitution** (**Figure 24.5B**). In general, nonsynonymous substitutions are likely to be deleterious to the organism. But not every amino acid replacement alters a protein's shape and charge (and hence its functional properties), so some nonsynonymous substitutions may also be selectively neutral (or nearly so). Conversely, an amino acid replacement that confers an advantage to the organism would result in positive selection for the corresponding nonsynonymous substitution.

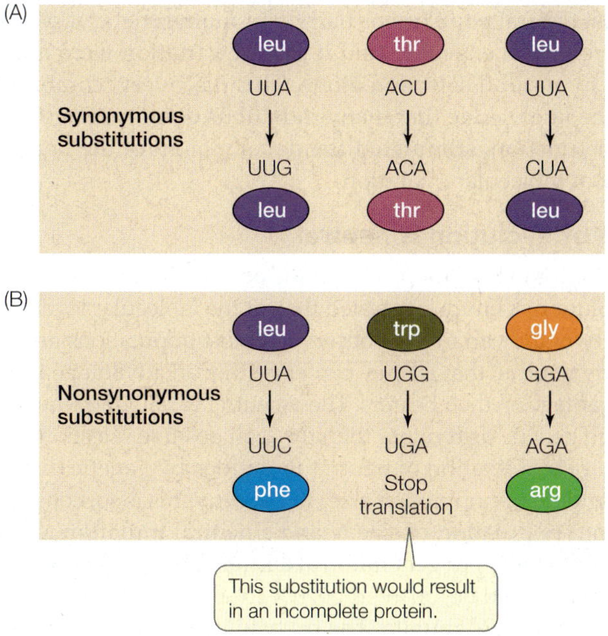

(A)
Synonymous substitutions
leu
UUA
UUG
leu
thr
ACU
ACA
thr
leu
UUA
CUA
leu
(B)
Nonsynonymous substitutions
leu
UUA
UUC
phe
trp
UGG
UGA
Stop translation
gly
GGA
AGA
arg
This substitution would result in an incomplete protein.

24.5 When One Nucleotide Does or Doesn't Make a Difference (A) Synonymous (silent) substitutions do not change the amino acid specified and do not affect protein function; such substitutions are unlikely to be subject to natural selection. (B) Nonsynonymous (missense) substitutions do change the amino acid sequence and are likely to have an effect (often deleterious) on protein function; such substitutions are targets for natural selection.

Investigators have measured the average rate of nonsynonymous nucleotide substitutions in several mammalian protein-coding genes at about 0.9 substitutions per position per billion years. Synonymous substitutions in these genes have occurred about five times more frequently than nonsynonymous substitutions. In other words, *substitution rates are highest at nucleotide positions that do not change the amino acid being expressed* (**Figure 24.6**). Substitution rates are even higher in **pseudogenes**, which are duplicate, nonfunctional copies of genes.

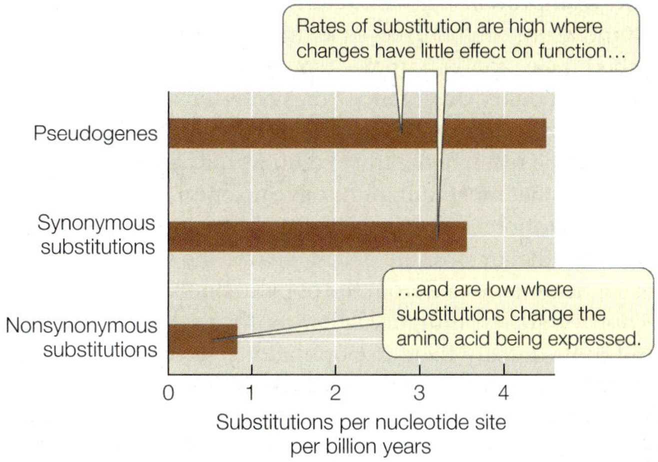

Rates of substitution are high where changes have little effect on function...
Pseudogenes
Synonymous substitutions
Nonsynonymous substitutions
...and are low where substitutions change the amino acid being expressed.
0
1
2
3
4
Substitutions per nucleotide site per billion years

24.6 Rates of Substitution Differ Rates of nonsynonymous substitution typically are much lower than rates of synonymous substitution and the substitution rate in pseudogenes. This pattern reflects differing levels of functional constraints.

Most natural populations harbor far more genetic variation than we would expect to find if genetic variation were influenced by natural selection alone. This discovery, combined with the knowledge that many mutations do not change molecular function, stimulated the development of the neutral theory of molecular evolution.

Much of evolution is neutral

In 1968, Motoo Kimura proposed the **neutral theory** of molecular evolution. Kimura suggested that, at the molecular level, the majority of the variants we observe in most populations are selectively neutral; that is, they confer neither an advantage nor a disadvantage on their bearers. These neutral variants accumulate through genetic drift rather than through positive selection.

The rate of fixation of neutral mutations by genetic drift is independent of population size. To see why this is so, consider a diploid population of size N and a neutral mutation rate μ (mu) per gamete per generation at a locus. The number of new mutations would be, on average, $\mu \times 2N$, because $2N$ gene copies are available to mutate. The probability that a given mutation will be fixed by drift alone is its frequency, which equals $1/(2N)$ for a newly arisen mutation. We can multiply these two terms to get the rate of fixation of neutral mutations (m) in a given population of N individuals:

$$m = 2N\mu \frac{1}{2N}$$

Therefore the rate of fixation m of neutral mutations depends only on the neutral mutation rate μ and is independent of population size. A given mutation is more likely to appear in a large population than in a small one, but any mutation that does appear is more likely to become fixed in a small population. These two influences of population size cancel each other out, so the rate of fixation of neutral mutations is equal to the mutation rate (i.e., $m = \mu$).

As as long as the underlying mutation rate is constant, macromolecules evolving in separate populations should diverge from one another in neutral changes at a constant rate. Investigators have confirmed that the rate of evolution of particular genes and proteins is often relatively constant over time and can therefore can be used as a "molecular clock." As we described in Section 22.3, molecular clocks can be used to calculate evolutionary divergence times between species.

Although much of the genetic variation we observe in populations is the result of neutral evolution, the neutral theory does not imply that most mutations have no effect on the organism. Many mutations are never observed in populations because they are lethal or strongly detrimental to the organism and are thus quickly removed from the population through natural selection. Similarly, mutations that confer a selective advantage tend to be quickly fixed in populations, so they do not result in variation at the population level either. Nonetheless, in any population, some amino acid positions will remain constant under purifying selection, others will vary through neutral genetic drift, and still others will differ between species as a result of positive selection for change. How can these evolutionary processes be distinguished?

Positive and purifying selection can be detected in the genome

As we have just seen, substitutions in a protein-coding gene can be either synonymous or nonsynonymous, depending on whether they change the resulting amino acid sequence of the protein. The relative rates of synonymous and nonsynonymous substitutions are expected to differ in regions of genes that are evolving neutrally, under positive selection for change, or staying unchanged under purifying selection.

- If a given amino acid in a protein can be one of many alternatives (without changing the protein's function), then an amino acid replacement is *neutral* with respect to the fitness of an organism. In this case, the rates of synonymous and nonsynonymous substitutions in the corresponding DNA sequences are expected to be very similar, so the ratio of the two rates should be close to 1.
- If a given amino acid position is under *positive* selection for change, the observed rate of nonsynonymous substitutions is expected to exceed the rate of synonymous substitutions in the corresponding DNA sequences.
- If a given amino acid position is under *purifying* selection, then the observed rate of synonymous substitutions is expected to be much higher than the rate of nonsynonymous substitutions in the corresponding DNA sequences.

By comparing the gene sequences that encode homologous proteins from many species, scientists can determine the history and timing of synonymous and nonsynonymous substitutions. This information can be mapped on a phylogenetic tree, as we saw in Chapter 22. Regions of genes that are evolving under neutral, purifying, or positive selection can be identified by comparing the nature and rates of substitutions across the phylogenetic tree.

A study of the evolution of lysozyme illustrates how and why particular amino acid positions might be under different modes of selection (**Figure 24.7**). The enzyme lysozyme (see Figure 3.9) is found in almost all animals. It is produced in the tears, saliva, and milk of mammals and in the albumen (whites) of bird eggs. Lysozyme digests the cell walls of bacteria, rupturing and killing them. As a result, it plays an important role as a first line of defense against invading bacteria. Most animals defend themselves against bacteria by digesting them, which is probably why most animals have lysozyme. Some animals also use lysozyme in the digestion of food.

Among mammals, a mode of digestion called foregut fermentation has evolved twice. In mammals with this mode of digestion, the foregut—the posterior esophagus or the stomach—has been converted into a chamber in which bacteria break down ingested plant matter by fermentation. Foregut fermenters can extract nutrients from the otherwise indigestible cellulose that makes up a large proportion of plant tissue. Foregut fermentation evolved independently in ruminants (a group of hoofed mammals that includes cattle) and in certain leaf-eating monkeys, such as langurs. Caro-Beth Stewart knew that these evolutionary events were independent because both langurs and ruminants have close relatives that are not foregut fermenters.

INVESTIGATING**LIFE**

24.7 Convergent Molecular Evolution Langurs (a group of monkeys) and cattle are only distantly related but both have evolved foregut fermentation. They uniquely express the enzyme lysozyme in their stomachs (foreguts) to aid in breaking down bacteria that are involved in fermentation. Stewart and colleagues compared the gene sequences of lysozyme in mammals with and without foregut fermentation to see if there is convergence in the independently evolved amino acid sequences of lysozyme in langurs and cattle.[a]

HYPOTHESIS Similar selective conditions in distantly related mammals have resulted in convergence of adaptations for foregut fermentation in the amino acid sequences of lysozyme.

Method 1. Isolate and sequence lysozyme from two distantly related mammal species with foregut fermentation (langur and cattle) as well as other mammals that are more closely related to either langurs or to cattle but lack foregut fermentation.

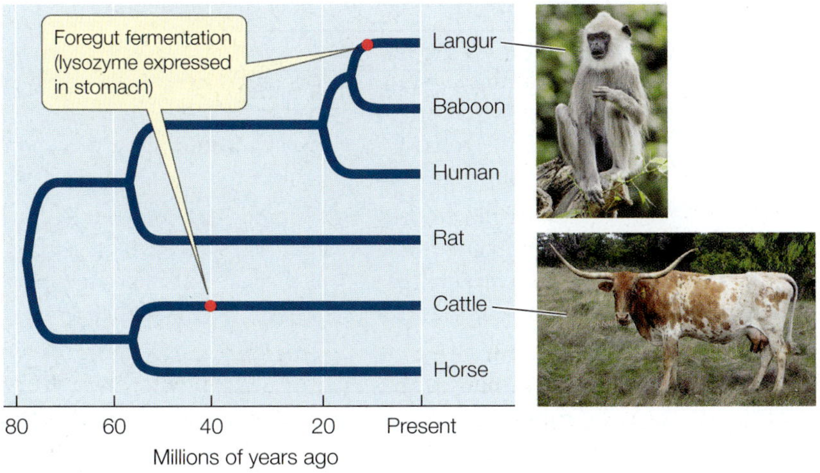

Foregut fermentation (lysozyme expressed in stomach)
Langur
Baboon
Human
Rat
Cattle
Horse
80
60
40
20
Present
Millions of years ago

2. Tabulate the pairwise differences in the amino acid sequences. Plot the amino acid changes on the phylogenetic tree and count the number of convergent similarities between each pair of species. The results can then be plotted as a matrix.

Results The matrix shows the number of *all* pairwise amino acid differences above the diagonal and the number of *convergent* similarities below the diagonal.

	Langur	Baboon	Human	Rat	Cattle	Horse
Langur		14	18	38	32	65
Baboon	0		14	33	39	65
Human	0	1		37	41	64
Rat	0	1	0		55	64
Cattle	5	0	0	0		71
Horse	0	0	0	0	1	

The lysozymes of langurs and cattle are convergent for 5 amino acids.

CONCLUSION The lysozyme sequences of the two species with foregut fermentation account for the majority of the convergent amino acid replacements observed among these species, demonstrating molecular convergence associated with the independent evolution of foregut fermentation.

Go to **BioPortal** for discussion and relevant links for all INVESTIGATING**LIFE** figures.

[a]Stewart, C.-B. et al. 1987. *Nature* 330: 401–404.

In both foregut-fermenting lineages, lysozyme has been modified to play a new, nondefensive role. This lysozyme ruptures some of the bacteria that live in the foregut, releasing nutrients metabolized by the bacteria, which the mammal then absorbs. How many changes in the lysozyme molecule were needed to allow it to perform this function amid the digestive enzymes and acidic conditions of the mammalian foregut? To answer this question, Stewart and her colleagues compared the lysozyme-coding sequences in foregut fermenters with those in several of their nonfermenting relatives. They determined which amino acids differed and which were shared among the species, as well as the rates of synonymous and nonsynonymous substitutions in lysozyme genes across the evolutionary history of the sampled species.

For many of the amino acid positions of lysozyme, the rate of synonymous substitutions in the corresponding gene sequence was much higher than the rate of nonsynonymous substitutions. This observation indicates that many of the amino acids that make up lysozyme are evolving under purifying selection. In other words, there is selection against change in the protein at these positions, and the observed amino acids must therefore be critical for lysozyme function. At other positions, several different amino acids function equally well, and the corresponding gene sequences had similar rates of synonymous and nonsynonymous substitutions. The most striking finding was that amino acid replacements in lysozyme happened at a much higher rate in the lineage leading to langurs than in any other primate lineage. The high rate of nonsynonymous substitutions in the langur lysozyme gene shows that lysozyme went through a period of rapid change in adapting to the foregut of langurs. Moreover, the lysozymes of langurs and cattle share five unique amino acid replacements, all of which lie on the surface of the lysozyme molecule, well away from the enzyme's active site. Two of these shared replacements involve changes from arginine to lysine, which makes the proteins more resistant to attack by the stomach enzyme pepsin. By understanding the functional significance of amino acid replacements, molecular evolutionists can explain the observed changes in amino acid sequences in terms of changes in the functioning of the protein.

A large body of fossil, morphological, and molecular evidence shows that langurs and ruminants do not share a recent common ancestor. However, langur and ruminant lysozymes share several amino acids that neither mammal shares with the lysozymes of its own closer relatives. The lysozymes of these two mammals

WORKING WITH DATA:

Detecting Convergence in Lysozyme Sequences

Original Paper

Stewart, C.-B., J. W. Schilling, and A. C. Wilson. 1987. Adaptive evolution in the stomach lysozymes of foregut fermenters. *Nature* 330: 401–404.

Analyze the Data

Caro-Beth Stewart and her colleagues collected lysozyme sequences from six species of mammals. A small sample of their data is shown in the table. The phylogeny of these six species is well supported from analysis of many genes and much morphological data. Using the phylogenetic tree (see Figure 24.7), plot the amino acid changes across the phylogeny of the six mammals. Assume that the ancestral state is the amino acid present at the base of the tree.

QUESTION 1

Which amino acid positions show unique convergence between the langur and cattle lineages (i.e., the derived state is found *only* in cattle and langurs)?

QUESTION 2

Which additional position is convergent between cattle and the ancestor of langurs and baboons?

QUESTION 3

Did you detect any other convergent amino acid changes between any other pair of lineages? What does this suggest about the convergent changes you observed between cattle and langurs?

	Amino acid position										
Species	2	14	17	21	50	63	75	87	117	118	130
Langur	I	K	L	K	E	Y	D	N	Q	N	V
Baboon	I	R	L	R	Q	Y	N	D	Q	N	V
Human	V	R	M	R	R	Y	N	D	Q	N	V
Rat	T	R	M	Y	Q	Y	N	D	K	N	V
Cattle	V	K	L	K	E	W	D	N	R	D	L
Horse	V	A	M	G	G	W	N	E	K	D	L
Ancestral state	V	R	M	R	Q	W	N	D	K	N	V

Go to **BioPortal** for all WORKING WITH**DATA** exercises

have undergone convergent evolution at some amino acid positions despite their very different ancestry. The amino acids they share give these lysozymes the ability to lyse the bacteria that ferment plant material in the foregut.

The hoatzin, an unusual leaf-eating South American bird and the only known avian foregut fermenter, offers another remarkable example of the convergent evolution of lysozyme. Many birds have an enlarged esophageal chamber called a crop. The crop of the hoatzin contains lysozyme and bacteria and acts as a fermenting chamber. Many of the amino acid replacements that occurred in the adaptation of hoatzin crop lysozyme are identical to those that evolved in ruminants and langurs. Thus even though the hoatzin and foregut-fermenting mammals have not shared a common ancestor in hundreds of millions of years, they have all evolved similar adaptations in their lysozymes that enable them to recover nutrients from their fermenting bacteria.

Genome size also evolves

We know that genome size varies tremendously among organisms. Across broad taxonomic categories, there is some correlation between genome size and organismal complexity. The genome of the tiny bacterium *Mycoplasma genitalium* has only 470 genes. *Rickettsia prowazekii*, the bacterium that causes typhus, has 634 genes. *Homo sapiens*, by contrast, has about 21,000 protein-coding genes. **Figure 24.8** shows the numbers of genes in a sample of organisms whose genomes have been fully sequenced, arranged by their evolutionary relationships.

As this figure reveals, a larger genome does not always indicate greater complexity (compare rice with the other plants, for example). It is not surprising that more complex genetic instructions are needed for building and maintaining a large multicellular organism than a small single-celled bacterium. What *is* surprising is that some multicellular organisms, such as lungfishes, some salamanders, and lilies, have about 40 times as much DNA as humans do. Structurally, a lungfish or a lily is not 40 times more complex than a human. So why does genome size vary so much?

Differences in genome size are not so great if we take into account only the portion of the DNA that actually encodes RNAs or proteins. Although the organisms with the largest total amounts of nuclear DNA (some ferns and flowering plants) have 80,000 times as much DNA as do the bacteria with the smallest genomes, no species has more than about 100 times as many protein-coding genes as a bacterium. Therefore much of the variation in genome size lies not in the number of functional genes but in the amount of noncoding DNA (**Figure 24.9**).

Why do the cells of most eukaryotic organisms have so much noncoding DNA? Does this noncoding DNA have a function, or is it "junk"? Although some of this DNA does not appear to have a direct function, it can alter the expression of the surrounding genes. The degree or timing of a gene's expression can be changed dramatically depending on the gene's position relative to noncoding sequences. Other regions of noncoding DNA consist of pseudogenes that are carried in the genome simply because the cost of doing so is very small. These

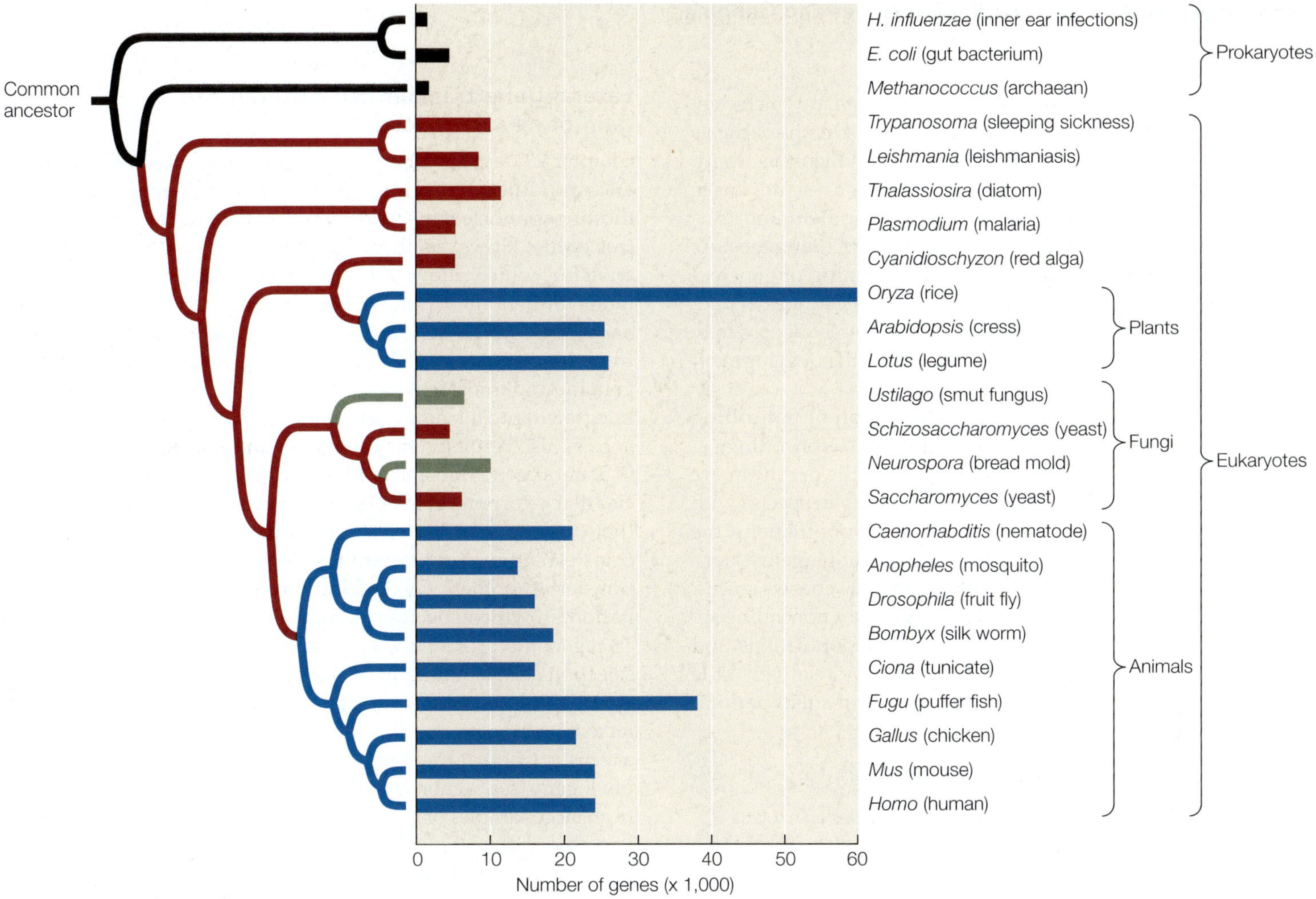
Common ancestor
H. influenzae (inner ear infections)
E. coli (gut bacterium)
Methanococcus (archaean)
Prokaryotes
Trypanosoma (sleeping sickness)
Leishmania (leishmaniasis)
Thalassiosira (diatom)
Plasmodium (malaria)
Cyanidioschyzon (red alga)
Oryza (rice)
Arabidopsis (cress)
Lotus (legume)
Plants
Ustilago (smut fungus)
Schizosaccharomyces (yeast)
Neurospora (bread mold)
Saccharomyces (yeast)
Fungi
Eukaryotes
Caenorhabditis (nematode)
Anopheles (mosquito)
Drosophila (fruit fly)
Bombyx (silk worm)
Ciona (tunicate)
Fugu (puffer fish)
Gallus (chicken)
Mus (mouse)
Homo (human)
Animals
0
10
20
30
40
50
60
Number of genes (x 1,000)

24.8 Genome Size Varies Widely This tree shows the numbers of genes from a sample of organisms whose genomes have been fully sequenced, arranged by their evolutionary relationships. Bacteria and archaea typically have fewer genes than most eukaryotes. Among eukaryotes, multicellular organisms with tissue organization (plants and animals; blue branches) have more genes than single-celled organisms (red branches) or multicellular organisms that lack pronounced tissue organization (green branches).

pseudogenes may become the raw material for the evolution of new genes with novel functions. Some noncoding DNA functions solely in maintaining chromosomal structure. Still other sequences consist of "selfish" transposable elements that proliferate because they reproduce faster than the host genome.

DNA does not just accumulate in genomes over time; noncritical nucleotide sequences are also lost from genomes. Some species differ so much in genome size because they lose noncritical sequences at very different rates. Investigators can use retrotransposons to estimate the rates at which species lose DNA. Retrotransposons are transposable elements (see Figure 17.5) that copy themselves through an RNA intermediate. The most common type of retrotransposon carries duplicated sequences at each end, called long terminal repeats, or LTRs. Occasionally, LTRs recombine in the host genome in such a way that the DNA between them is excised. When this happens, one recombined LTR is left behind. The number of such "orphaned" LTRs in a genome is a measure of how many retrotransposons have been lost. By comparing the number of LTRs in the genomes of Hawaiian crickets (*Laupala*) and fruit flies (*Drosophila*), investigators found that *Laupala* loses DNA more than 40 times more slowly than does *Drosophila*. Therefore it is not surprising that the genome of *Laupala* is much larger than that of *Drosophila*.

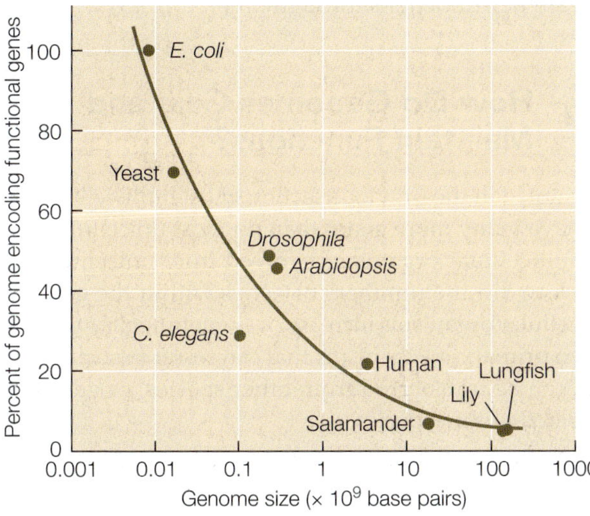
Percent of genome encoding functional genes
100
80
60
40
20
0
E. coli
Yeast
Drosophila
Arabidopsis
C. elegans
Human
Lungfish
Lily
Salamander
0.001
0.01
0.1
1
10
100
1000
Genome size (× 10^9 base pairs)

24.9 A Large Proportion of DNA Is Noncoding Most of the DNA of bacteria and yeasts encodes RNAs or proteins, but a large percentage of the DNA of multicellular species is noncoding.

Why do species differ so greatly in the rate at which they gain or lose apparently functionless DNA? One hypothesis is that genome size is related to the rate at which the organism develops, which may be under selection pressure. Large genomes can slow down the rate of development and thus alter the relative timing of expression of particular genes. As discussed in Section 20.2, changes in the timing of gene expression—heterochrony—can produce major changes in phenotype. Thus although some noncoding DNA sequences may have no direct function, they may still affect the development of the organism.

Another hypothesis is that the proportion of noncoding DNA is related primarily to population size. Noncoding sequences that are only slightly deleterious to the organism are likely to be purged by selection most efficiently in species with large population sizes. In species with small populations, the effects of genetic drift can overwhelm selection against noncoding sequences that have small deleterious consequences. Therefore selection against the accumulation of noncoding sequences is most effective in species with large populations, and such species (such as bacteria and yeasts) have relatively little noncoding DNA compared with species with small populations (see Figure 24.9).

RECAP 24.2

The neutral theory of molecular evolution provides an explanation for the relatively constant rate of molecular change seen in many species. By examining the relative rates of synonymous and nonsynonymous substitutions in genes over time, biologists can distinguish the evolutionary mechanisms acting on individual genes.

- Describe how the ratio of synonymous to nonsynonymous substitutions can be used to determine whether a particular gene region is evolving neutrally, under positive selection, or under purifying selection. See pp. 492–493
- Contrast two hypotheses proposed to explain the differences in genome size among different organisms. See pp. 494–495

We have examined some of the ways in which organisms can lose DNA without losing gene functions. But how do organisms gain new functions through time?

24.3 How Do Genomes Gain and Maintain Functions?

As we noted in the previous section, most multicellular organisms have many more genes than do most unicellular species. But multicellular organisms evolved from unicellular ancestors. How did the numbers of genes within the genomes of multicellular organisms increase over evolutionary time? There are two primary mechanisms that can result in such increases: genes can be transferred from other species, or genes can be duplicated within species.

Lateral gene transfer can result in the gain of new functions

Chapter 23 described how, through the process of speciation, ancestral lineages divide into descendant lineages, and it is those speciation events that are captured by the branches in the tree of life. However, there are also processes of **lateral gene transfer**, which allow individual genes, organelles, or fragments of genomes to move horizontally from one lineage to another. Some species may pick up fragments of DNA directly from the environment. Other genes may be picked up in a viral genome and transferred to a new host when the virus becomes integrated into the new host's genome. Hybridization between species also results in the transfer of large numbers of genes.

Lateral gene transfer can be highly advantageous to a species that incorporates novel genes from a distant relative. Genes that confer antibiotic resistance, for example, are commonly transferred among different species of bacteria. Lateral gene transfer is another way, in addition to mutation and recombination, in which species can increase their genetic variation. That genetic variation then provides the raw material on which selection acts, resulting in evolution.

A phylogenetic tree constructed from a single laterally transferred genome fragment is likely to reflect only the evolutionary history of that fragment, rather than the overall organismal phylogeny (see Section 26.3). Most biologists prefer to build trees from large samples of genes or their products, so that the underlying species tree (as well as any lateral gene transfer events) can be reconstructed. Depictions of lateral gene transfer events on the underlying species tree are known as **reticulations**.

The degree to which lateral gene transfer events occur in various parts of the tree of life is a matter of considerable current investigation and debate. Lateral gene transfer appears to be relatively uncommon among most eukaryote lineages, although the two major endosymbioses that gave rise to mitochondria and chloroplasts can be viewed as lateral transfers of entire bacterial genomes to the eukaryote lineage. Some groups of eukaryotes, most notably some plants, are subject to relatively high levels of hybridization among closely related species. Hybridization leads to the exchange of many genes among recently separated lineages of plants. The greatest degree of lateral transfer, however, appears to occur among bacteria. Many bacterial genes have been transferred repeatedly among lineages of bacteria, to the point that relationships among bacterial species are often hard to decipher. Nonetheless, the broad relationships of the major groups of bacteria can still be determined (as we will discuss in Part Seven of this book). Lateral transfer of genes also makes it difficult to identify the boundaries of bacterial species, which is one reason why fewer bacterial species have been named than are known to exist.

Most new functions arise following gene duplication

Gene duplication is yet another way in which genomes can acquire new functions. When a gene is duplicated, one copy of that gene is potentially freed from having to perform its original function. The initially identical copies of a duplicated gene can have any one of four subsequent fates:

- Both copies of the gene may retain their original function (which can result in a change in the amount of gene product that is produced by the organism).
- Both copies of the gene may retain the ability to produce the original gene product, but the expression of the genes may diverge in different tissues or at different times in development.
- One copy of the gene may be incapacitated by the accumulation of deleterious substitutions and become a nonfunctional pseudogene, or may be eliminated from the genome altogether.
- One copy of the gene may retain its original function while the second copy accumulates enough substitutions that it can perform a different function.

How often do gene duplications arise, and which of these four outcomes is most likely? Investigators have found that rates of gene duplication are fast enough for a yeast or *Drosophila* population to acquire several hundred duplicate genes over the course of a million years. They have also found that most of the duplicated genes in these organisms are very young. Many extra genes are lost from a genome within 10 million years (which is rapid on an evolutionary time scale).

Some genes may be duplicated many times, resulting in large numbers of related pseudogenes scattered throughout the genome. In the human genome, the functional copy of the ribosomal protein gene *RPL21* is located on chromosome pair 13, but pseudogenes derived from it are found on most of the other chromosome pairs (**Figure 24.10**). Although not all genes are represented by pseudogenes, there are nearly as many known pseudogenes in the human genome as there are functional protein-coding genes.

Although many extra genes disappear rapidly, some duplication events lead to the evolution of genes with new functions. Several successive rounds of duplication and mutation may result in a **gene family**: a group of homologous genes with related functions, often arrayed in tandem along a chromosome. An example of this process is provided by the globin gene family (see Figure 17.10). The globins were among the first proteins to be sequenced and compared. Comparisons of their amino acid sequences strongly suggest that the different globins arose via gene duplications. These comparisons also allow us to estimate how long the globins have been evolving separately because differences among these proteins have accumulated with time.

Hemoglobin, a tetramer (four-subunit molecule) consisting of two α-globin and two β-globin polypeptide chains, carries oxygen in blood. Myoglobin, a monomer, is the primary O_2 storage protein in muscle. Myoglobin's affinity for O_2 is much higher than that of hemoglobin, but hemoglobin has evolved to be more diversified in its roles. Hemoglobin binds O_2 in the lungs or gills, where the O_2 concentration is relatively high, transports it to deep body tissues, where the O_2 concentration is low, and releases it in those tissues. With its more complex tetrameric structure, hemoglobin is able to carry four molecules of O_2, as well as hydrogen ions and carbon dioxide, in the blood.

To estimate the time of the globin gene duplication that gave rise to the α- and β-globin gene clusters, we can create a **gene tree**—a phylogenetic tree that describes the evolutionary history of particular genes or gene families, in this case the gene sequences that encode the various globins (**Figure 24.11**). The rate of molecular evolution of globin genes has been estimated from other studies, using the divergence times of groups of vertebrates that are well documented in the fossil record. These studies indicate an average rate of divergence for globin genes of about 1 nucleotide substitution every 2 million years. By applying this rate to the globin gene tree, we can estimate the divergence time of the two globin gene clusters at about 450 million years ago.

Many gene duplications affect only one or a few genes at a time, but entire genomes are duplicated in polyploid organisms (which include many plants). When all the genes are duplicated, there are massive opportunities for new functions to evolve. That is exactly what appears to have happened in the evolution of vertebrates. The genomes of the jawed vertebrates appear to have four diploid sets of many major genes, which has led biologists to conclude that two genome-wide duplication events occurred in the ancestor of these species.

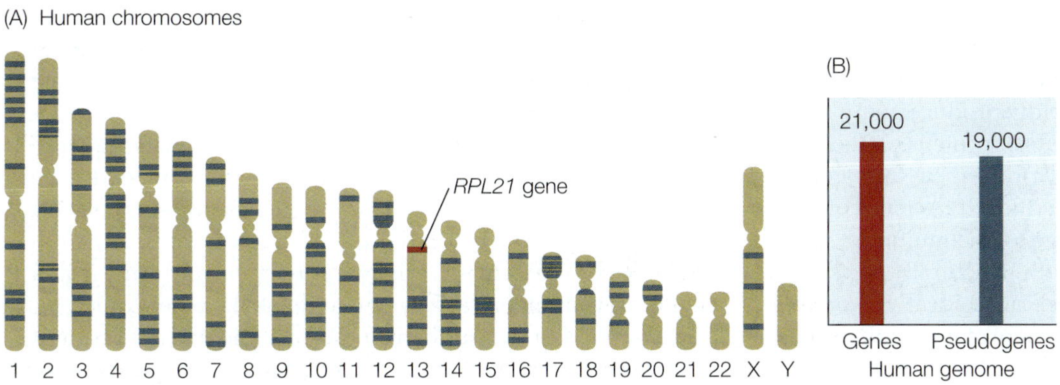
(A) Human chromosomes
RPL21 gene
1 2 3 4 5 6 7 8 9 10 11 12 13 14 15 16 17 18 19 20 21 22 X Y
(B)
21,000
19,000
Genes
Pseudogenes
Human genome

24.10 Some Functional Genes are Duplicated Many Times as Nonfunctional Pseudogenes (A) The functional gene that encodes ribosomal protein *RPL21* is located on human chromosome 13 (indicated in red). In addition, there are many nonfunctional pseudogenes of *RPL21* in the human genome, produced through repeated duplication events (indicated in blue). (B) Although *RPL21* represents a relatively extreme example of pseudogene duplication, there are almost as many known pseudogenes in the human genome as there are functional genes.

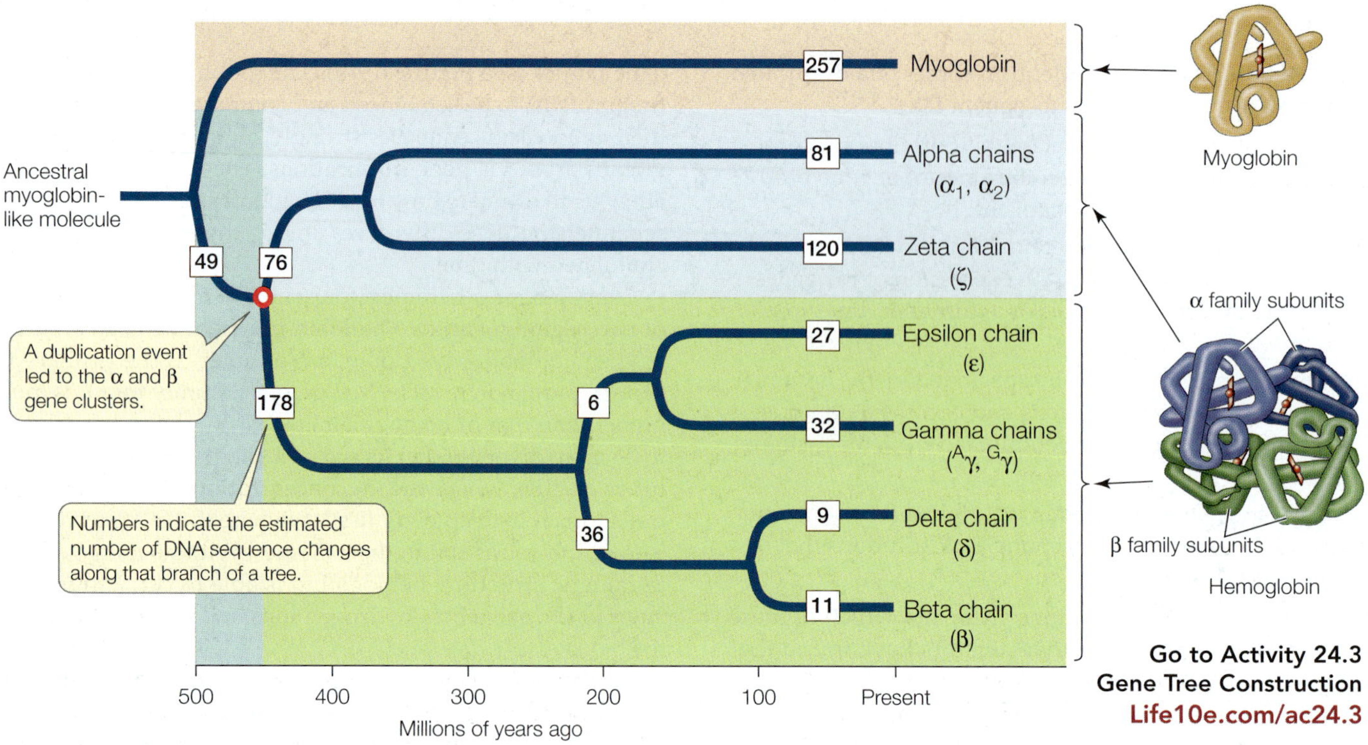

Ancestral myoglobin-like molecule
257
Myoglobin
81
Alpha chains (α_1, α_2)
120
Zeta chain (ζ)
49
76
A duplication event led to the α and β gene clusters.
178
6
27
Epsilon chain (ε)
32
Gamma chains ($^A\gamma$, $^G\gamma$)
36
9
Delta chain (δ)
11
Beta chain (β)
Numbers indicate the estimated number of DNA sequence changes along that branch of a tree.
500
400
300
200
100
Present
Millions of years ago
Myoglobin
α family subunits
β family subunits
Hemoglobin

Go to Activity 24.3 Gene Tree Construction Life10e.com/ac24.3

24.11 A Globin Family Gene Tree A molecular clock analysis suggests that the α-globin (blue) and β-globin (green) gene clusters diverged about 450 million years ago, soon after the origin of the vertebrates.

These duplications have allowed considerable specialization of individual vertebrate genes, many of which are now highly tissue-specific in their expression. A good example is the duplication of sodium channel genes, which allowed the evolution of the electric organs of electric fishes described at the opening of this chapter.

Some gene families evolve through concerted evolution

Although the members of the globin gene family have diversified in form and function, the members of many other gene families do not evolve independently of one another. For instance, almost all organisms have many copies (up to thousands) of the ribosomal RNA genes. Ribosomal RNA (rRNA) is the principal structural element of ribosomes and, as such, has a primary role in protein synthesis. Every living species needs to synthesize proteins, often in large amounts (especially during early development). Having many copies of the rRNA genes ensures that organisms can rapidly produce many ribosomes and thereby maintain a high rate of protein synthesis.

Like all portions of the genome, ribosomal RNA genes evolve, and differences accumulate in the rRNA genes of different species. But within any one species, the multiple copies of the rRNA genes are very similar, both structurally and functionally. This similarity makes sense because, ideally, every ribosome in a species should synthesize proteins in the same way. In other words, within a given species, the multiple copies of these rRNA genes are evolving in concert with one another, a phenomenon called **concerted evolution**.

How does concerted evolution occur? Two different mechanisms appear to be responsible. The first of these is **unequal crossing over**. When DNA is replicated during meiosis in a diploid species, the homologous chromosome pairs align and recombine by crossing over (see Section 11.4). In the case of highly repeated genes, however, it is easy for genes to become displaced in alignment, since so many copies of the same genes are present on the chromosomes (**Figure 24.12A**). The end result is that one chromosome may gain extra copies of the repeat and the other chromosome may have fewer copies of the repeat. If a new substitution arises in one copy of the repeat, it can spread to new copies (or be eliminated) through unequal crossing over. Thus, over time, a novel substitution will either become fixed or it will be lost entirely. In either case, all copies of the repeat will remain very similar to one another.

The second mechanism that produces concerted evolution is **biased gene conversion**. This mechanism can be much faster than unequal crossing over and has been shown to be the primary mechanism for concerted evolution of rRNA genes. DNA strands are frequently broken and repaired (see Section 13.4). At many times during the cell cycle, the genes for ribosomal RNA are clustered close together. If damage occurs to one of the genes, a copy of the rRNA gene on another chromosome may be used to repair the damaged copy, and the sequence that is used as a template can thereby replace the original sequence (**Figure 24.12B**). In many cases, this repair system appears to be biased in favor of using particular sequences as templates for repair, and thus the favored sequence rapidly spreads across all copies of the gene. In this way, changes may appear in a single copy and then rapidly spread to all the other copies.

Regardless of the mechanism responsible, the net result of concerted evolution is that the copies of a highly repeated gene do not evolve independently of one another. Mutations still occur, but once they arise in one copy, they either spread rapidly

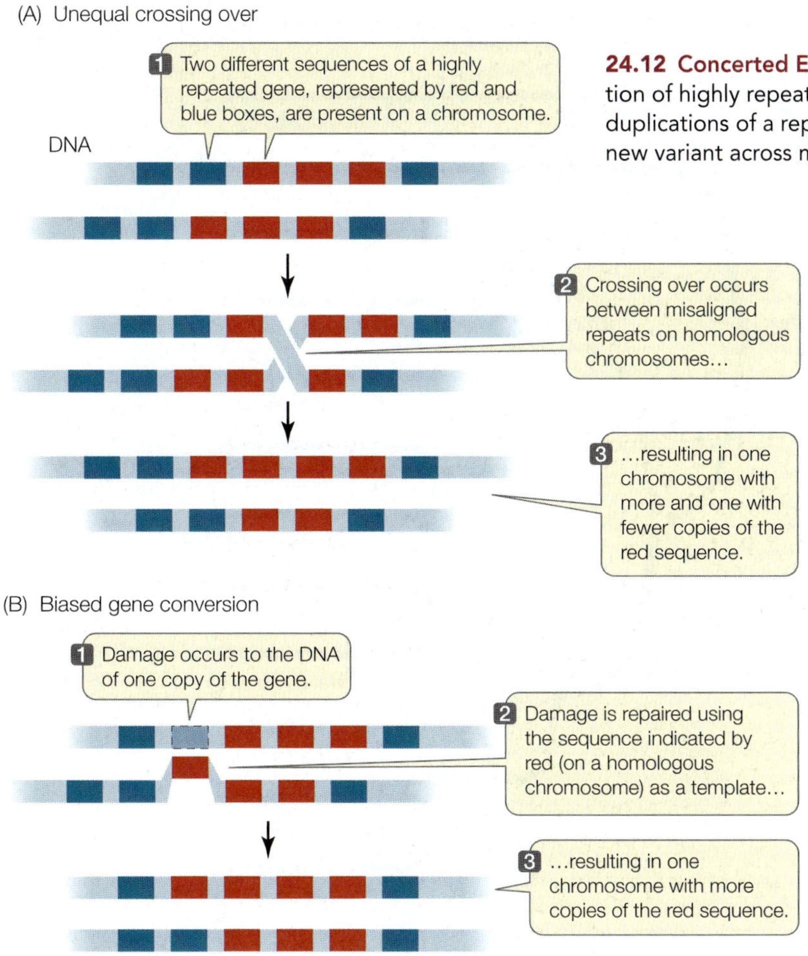

(A) Unequal crossing over
1 Two different sequences of a highly repeated gene, represented by red and blue boxes, are present on a chromosome.
DNA
2 Crossing over occurs between misaligned repeats on homologous chromosomes…
3 …resulting in one chromosome with more and one with fewer copies of the red sequence.
(B) Biased gene conversion
1 Damage occurs to the DNA of one copy of the gene.
2 Damage is repaired using the sequence indicated by red (on a homologous chromosome) as a template…
3 …resulting in one chromosome with more copies of the red sequence.

24.12 Concerted Evolution Two mechanisms can produce concerted evolution of highly repeated genes. (A) Unequal crossing over results in deletions and duplications of a repeated gene. (B) Biased gene conversion can rapidly spread a new variant across multiple copies of a repeated gene.

across all the copies or are lost from the genome completely. This process allows the products of each copy to remain similar over time in both sequence and function.

Go to Animated Tutorial 24.1
Concerted Evolution
Life10e.com/at24.1

RECAP 24.3

Lateral gene transfer can result in the transfer of genetic functions between even distantly related species. Gene duplication can lead to the evolution of new functions. Some highly repeated genes undergo concerted evolution, which maintains uniform functionality.

- Explain the potential advantages of lateral gene transfer. **See p. 496**
- What are four possible outcomes of gene duplication? **See p. 497**
- Describe the pattern of concerted evolution among highly repeated genes and the mechanisms that lead to concerted evolution. **See p. 498 and Figure 24.12**

We have seen how the principles and methods of molecular evolution have opened new vistas in evolutionary biology. Next we will consider some of the practical applications of this field.

24.4 What Are Some Applications of Molecular Evolution?

Studies of molecular evolution have practical applications throughout biology, from understanding basic aspects of biological function to studies of human health.

Molecular sequence data are used to determine the evolutionary history of genes

A gene tree can show the evolutionary relationships of a single gene in different species, or it can trace the evolution of members of a gene family (as in Figure 24.11). The methods for constructing a gene tree are the same as those we described in Section 22.2 for building phylogenetic trees of species. The process involves identifying differences between genes and using those differences to reconstruct the evolutionary history of the genes. Gene trees are often used to construct phylogenetic trees of species, but the two types of trees are not necessarily equivalent. Processes such as gene duplication can give rise to differences between the phylogenetic trees of genes and species. From a gene tree, biologists can reconstruct the history and timing of gene duplication events and learn how gene diversification has resulted in the evolution of new protein functions.

All the genes of a particular gene family have similar sequences because they have a common ancestry. As we discussed in Section 22.1, features that are similar as a result of common ancestry are said to be homologous. When discussing gene trees, however, we usually need to distinguish between two forms of homology. Homologous genes that are found in different species and whose divergence we can trace to the speciation events that gave rise to those species are called **orthologs**. Homologous genes in the same or different species that are related through gene duplication events are called **paralogs**. When we examine a gene tree, the questions we wish to address determine whether we should compare orthologous or paralogous genes. If we wish to reconstruct the evolutionary history of the species that contain the genes, then our comparison should be restricted to orthologs (because they will reflect the history of speciation events). If we are interested in the changes in function that have resulted from gene duplication events, however, then the appropriate comparison is among paralogs (because they will reflect the history of gene duplication events). If our focus is on the diversification of a gene family through both processes, then we will want to include both paralogs and orthologs in our analysis.

Figure 24.13 depicts a gene tree for the members of a gene family called *engrailed* (its members encode transcription

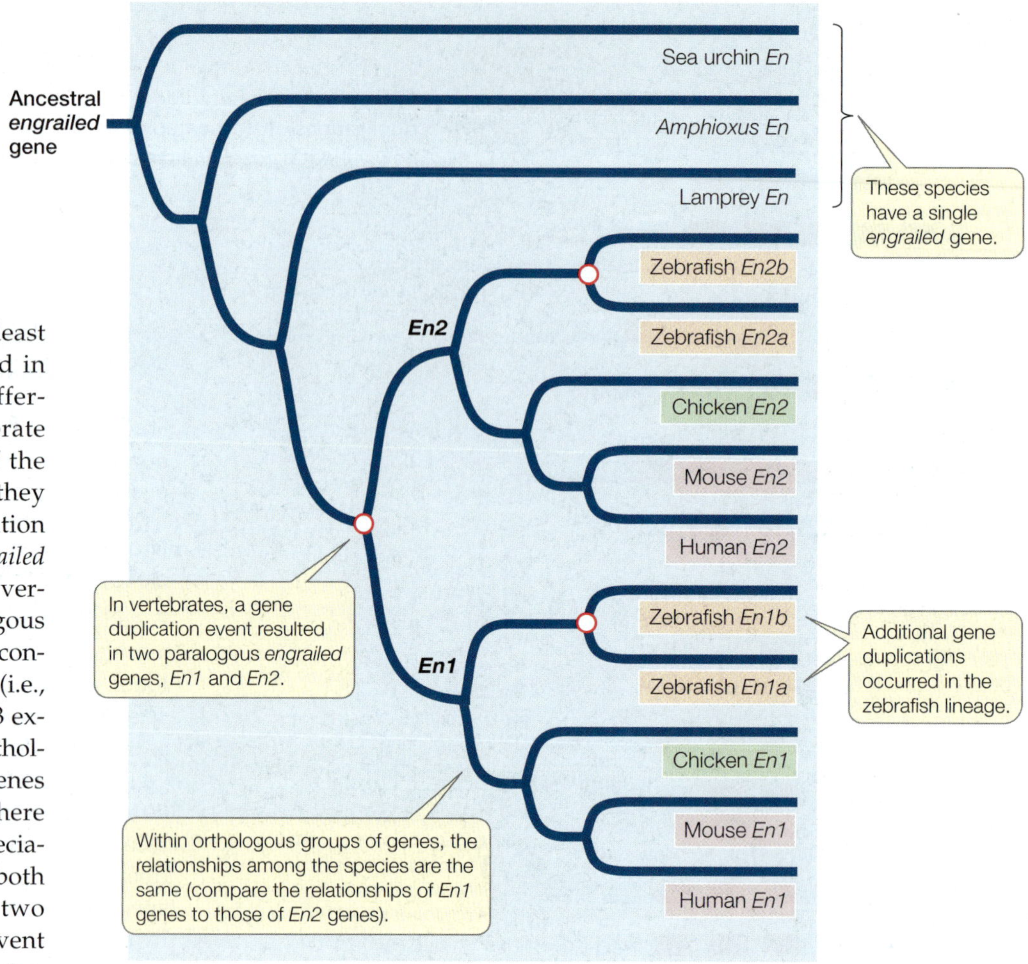
Ancestral
engrailed
gene
Sea urchin *En*
Amphioxus En
Lamprey *En*
These species have a single *engrailed* gene.
Zebrafish *En2b*
Zebrafish *En2a*
En2
Chicken *En2*
Mouse *En2*
Human *En2*
In vertebrates, a gene duplication event resulted in two paralogous *engrailed* genes, *En1* and *En2*.
Zebrafish *En1b*
Zebrafish *En1a*
En1
Additional gene duplications occurred in the zebrafish lineage.
Chicken *En1*
Mouse *En1*
Human *En1*
Within orthologous groups of genes, the relationships among the species are the same (compare the relationships of *En1* genes to those of *En2* genes).

24.13 Phylogeny of the *engrailed* Genes
The *engrailed* genes are homologous because they share a common ancestor. Speciation events have generated orthologous *engrailed* genes, and gene duplication events (open circles) have generated paralogous *engrailed* genes among bony vertebrates.

factors that regulate development). At least three gene duplications have occurred in this family, resulting in up to four different *engrailed* genes (*En*) in some vertebrate species (such as the zebrafish). All of the *engrailed* genes are homologs because they have a common ancestor. Gene duplication events have generated paralogous *engrailed* genes (*En1* and *En2*) in some lineages of vertebrates. We could compare the orthologous sequences of the *En1* group of genes to reconstruct the history of the bony vertebrates (i.e., all the vertebrate species in Figure 24.13 except the lamprey), or we could use the orthologous sequences of the *En2* group of genes and expect the same answer (because there is only one history of the underlying speciation events). All bony vertebrates have both groups of *engrailed* genes because the two groups arose from a gene duplication event in the common ancestor of bony vertebrates. If we wanted to focus on the diversification that occurred as a result of this duplication, then the appropriate comparison would be between the paralogous genes of the *En1* versus *En2* groups.

Gene evolution is used to study protein function

Earlier in this chapter we discussed the ways in which biologists can detect regions of genes that are under positive selection for change. What are the practical uses of this information? Consider the evolution of the family of genes encoding voltage-gated sodium channels, which we introduced at the opening of this chapter. Sodium channels have many functions, including the control of nerve impulses in the nervous system. Sodium channels can be blocked by various toxins, such as tetrodotoxin (TTX), a neurotoxin present in the tissues of some puffer fishes and several other animals. A human who eats those tissues of a puffer fish that contain TTX can become paralyzed and die because the toxin blocks sodium channels and prevents nerves and muscles from functioning.

But puffer fishes have sodium channels too, so why doesn't the TTX cause paralysis in the puffer fish itself? The sodium channels of puffer fishes (and other animals that sequester TTX, such as the rough-skinned newt shown in Figure 21.20) have evolved to become resistant to the toxin. Nucleotide substitutions in the puffer fish genome have resulted in changes in the proteins that make up the sodium channels, and those changes prevent TTX from binding to the sodium channel pore.

Several different substitutions that result in TTX resistance have evolved in the various duplicated sodium channel genes of the many species of puffer fish. Many other changes that have nothing to do with the evolution of TTX resistance have occurred in these genes as well. Biologists who study the function of sodium channels can learn a great deal about how the channels work (and about neurological diseases that are caused by mutations in the sodium channel genes) by understanding which changes have been selected for TTX resistance. They do this by comparing the rates of synonymous and nonsynonymous substitutions across the genes in various lineages that have evolved TTX resistance. In a similar manner, molecular evolutionary principles are used to understand function and diversification of function in many other proteins.

In vitro evolution is used to produce new molecules

As biologists studied the relationships among selection, evolution, and function in macromolecules, they realized that molecular evolution could be used in a controlled laboratory environment to produce new molecules with novel and useful functions. Thus were born the applications of **in vitro evolution**.

Living organisms produce thousands of compounds that humans have found useful. The search for such naturally occurring compounds, which can be used for pharmaceutical, agricultural, or industrial purposes, has been termed "bioprospecting." These compounds are the result of millions of years of molecular evolution across millions of species of living organisms. Yet biologists can also imagine molecules that *could* have evolved but, lacking the right combination of selection pressures and opportunities, have not. For instance, we might want to have a molecule that binds a particular environmental contaminant so that the contaminant can be isolated and extracted from the environment. But if the contaminant is synthetic (i.e., not produced naturally), it is unlikely that any living organism will have evolved a molecule with the function we desire. This problem was the inspiration for the field of in vitro evolution.

The principles of in vitro evolution are based on the principles of molecular evolution that we have learned from the natural world. Consider the evolution of a new RNA molecule that was produced in the laboratory using the principles of mutation and selection. This molecule's intended function was to join two other RNA molecules (acting as a ribozyme with a function similar to that of the naturally occurring DNA ligase described in Section 13.3, but for RNA molecules). The process started with a large pool of random RNA sequences (10^{15} different sequences, each about 300 nucleotides long), which were then selected for any ligase activity (**Figure 24.14**). None were very effective ligases, but some were slightly better than others. The best of the ribozymes were selected and reverse-transcribed into cDNA (using the enzyme reverse transcriptase). The cDNA molecules were then amplified using the polymerase chain reaction (PCR; see Figure 13.21).

PCR amplification is not perfect, and it introduced many new mutations into the pool of sequences. These sequences were then transcribed back into RNA molecules using RNA polymerase, and the process was repeated. The ligase activity of the RNAs evolved quickly; after 10 rounds of in vitro evolution, it had increased by about 7 million times. Similar techniques have since been used to create a wide variety of molecules with novel enzymatic and binding functions.

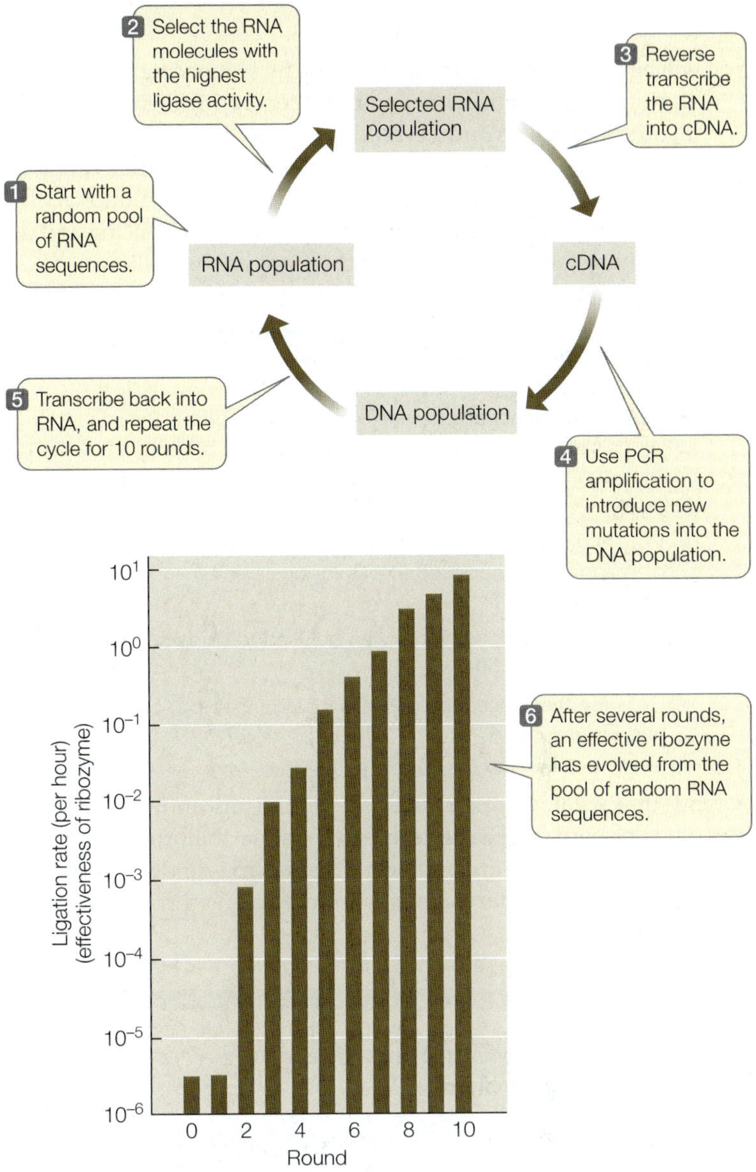

1 Start with a random pool of RNA sequences.
RNA population
2 Select the RNA molecules with the highest ligase activity.
Selected RNA population
3 Reverse transcribe the RNA into cDNA.
cDNA
4 Use PCR amplification to introduce new mutations into the DNA population.
DNA population
5 Transcribe back into RNA, and repeat the cycle for 10 rounds.
6 After several rounds, an effective ribozyme has evolved from the pool of random RNA sequences.
Ligation rate (per hour) (effectiveness of ribozyme)
10^{1}
10^{0}
10^{-1}
10^{-2}
10^{-3}
10^{-4}
10^{-5}
10^{-6}
0
2
4
6
8
10
Round

24.14 In Vitro Evolution Starting with a large pool of random RNA sequences, investigators produced a new ribozyme through rounds of mutation and selection for the ability to ligate RNA sequences.

Molecular evolution is used to study and combat diseases

Many of the most problematic human diseases are caused by living, evolving organisms that present a moving target for modern medicine. Recall the example of influenza described at the opening of Chapter 21 and that of HIV in Chapter 22. The control of these and many other human diseases depends on techniques that can track the evolution of pathogenic organisms over time.

The transportation advances of the past century have allowed humans to move around the world with unprecedented speed and frequency. Unfortunately, this mobility has allowed pathogens to be transmitted among human populations at increasing rates, which has led to the global emergence of many "new" diseases. Most of these emerging diseases are caused by viruses. Virtually all new viral diseases have been identified by evolutionary comparison of their genomes with those of known viruses. In recent years, for example, rodent-borne hantaviruses have been identified as the source of widespread respiratory illnesses, and the virus (and its host) that causes Sudden Acute Respiratory Syndrome (SARS) was identified using evolutionary comparisons of genes. Studies of the origins, the timing of emergence, and the global diversity of many human pathogens depend on the principles of molecular evolution, as do the efforts to develop and use effective vaccines against these pathogens. For example, the techniques used to develop modern polio vaccines, as well as the methods used to track their effectiveness in human populations, rely on molecular evolutionary approaches.

In the future, molecular evolution will become even more critical to the identification of human (and other) diseases. Once biologists have collected data on the genomes of enough organisms, it will be possible to identify an infection by sequencing a portion of the infecting organism's genome and comparing this sequence with other sequences on an evolutionary tree. At present it is difficult to identify many common viral infections (those that cause "colds," for instance). As genomic databases and evolutionary trees increase, however, automated methods of sequencing and rapid phylogenetic comparison of the sequences will allow us to identify and treat a much wider array of human illnesses.

RECAP 24.4

Molecular evolution has provided biologists with new tools to understand the functions of macromolecules and how those functions can change over time. These tools can be used to develop synthetic molecules and to identify and combat human diseases.

- Why might a biologist limit a particular investigation to orthologous (as opposed to paralogous) genes? **See pp. 499–500**
- Explain how gene evolution can be used to study protein function. **See p. 500**
- Describe the process of in vitro evolution. **See pp. 500–501 and Figure 24.14**

Now that we have discussed how organisms and biological molecules evolve, we are ready to consider the broader evolutionary history of life on Earth. Chapter 25 will describe the long-term evolutionary changes that have given rise to all of life's diversity.

?

How do evolutionary studies of sodium channel genes help us understand some human genetic disorders?

ANSWER

Our understanding of how genes and proteins function is largely based on comparative evolutionary analyses of homologous genes across many species. These analyses reveal the parts of the genes that are conserved across species and evolutionary time, as well as the parts of the genes that covary with certain functions (such as the generation of electric signals, as in the case of the sodium channel genes discussed in the chapter opening story). These evolutionary patterns allow biologists to correlate sequence variation with function. These correlations, in turn, allow biologists to predict the causes (and in many cases, suggest treatments) for genetic disorders found in humans. For example, certain mutations in sodium channel genes can cause the inability to feel pain. Other mutations may result in extreme sensitivity to pain, heart disease, or sudden infant death. Understanding the relationship between sequence variation and function is the first step in understanding these disorders, as well as a necessary step in finding treatments for people who suffer from these disorders.

CHAPTER SUMMARY 24

24.1 How Are Genomes Used to Study Evolution?

- A **genome** is an organism's full set of genes, regulatory sequences, and structural elements as well as noncoding DNA.
- The field of **molecular evolution** concerns relationships between the structures of genes and proteins and the functions of organisms.
- **Sequence alignments** of proteins or nucleic acids from different organisms allow us to compare the sequences and identify homologous positions. **Review Figure 24.1, ACTIVITY 24.1**
- The minimum number of changes between sequences can be calculated from a **similarity matrix.** Models of sequence evolution can be used to account for changes that cannot be observed directly. **Review Figure 24.2, ACTIVITY 24.2**

24.2 What Do Genomes Reveal about Evolutionary Processes?

- **Nonsynonymous substitutions** of nucleotides result in amino acid replacements in proteins, but **synonymous substitutions** do not. **Review Figure 24.5**
- The **neutral theory** of molecular evolution states that much of the molecular change in nucleotide sequences does not change genome function. The rate of fixation of neutral mutations is independent of population size and is equal to the mutation rate.
- Positive selection for change in a protein-coding gene may be detected by a higher rate of nonsynonymous than synonymous substitutions. The reverse is true of purifying selection.
- Common selective constraints can lead to convergent evolution of amino acid sequences in distantly related species. **Review Figure 24.7**
- The total size of genomes varies much more widely across multicellular species than does the number of functional genes. **Review Figures 24.8, 24.9**

24.3 How Do Genomes Gain and Maintain Functions?

- **Lateral gene transfer** can result in the rapid acquisition of new functions from distantly related species.
- **Gene duplications** can result in increased production of the gene's product, in nonfunctional **pseudogenes**, or in new gene functions. Several rounds of gene duplication can give rise to multiple genes with related functions, collectively known as a **gene family**. **Review Figures 24.10, 24.11.**
- **Gene trees** describe the evolutionary history of particular genes or gene families. **See ACTIVITY 24.3**
- Some highly repeated genes undergo **concerted evolution**, in which the multiple copies within the genome maintain their similarity, even as the genes diverge among species. **Review Figure 24.12, ANIMATED TUTORIAL 24.1**

continued

24.4 What Are Some Applications of Molecular Evolution?

- **Orthologs** are genes that are related through speciation events, whereas **paralogs** are genes that are related through gene duplication events. Review Figure 24.13
- Protein function can be studied by examining gene evolution. Detection of positive selection can be used to identify molecular changes that have resulted in functional changes.
- **In vitro evolution** is used to produce synthetic molecules with particular desired functions. Review Figure 24.14
- Many diseases are identified, studied, and combated through molecular evolutionary investigations.

Go to the Interactive Summary to review key figures, Animated Tutorials, and Activities
Life10e.com/is24

CHAPTERREVIEW

REMEMBERING

1. A higher rate of synonymous than nonsynonymous substitutions in a protein-coding gene is expected under
   a. purifying selection.
   b. positive selection.
   c. neutral evolution.
   d. concerted evolution.
   e. none of the above

2. Before nucleotide and amino acid sequences can be compared in an evolutionary framework, they must be aligned to account for
   a. deletions and insertions.
   b. selection and neutrality.
   c. parallelisms and convergences.
   d. gene families.
   e. all of the above

3. The rate of fixation of neutral mutations, $m = 2N\mu \frac{1}{2N}$, is
   a. independent of population size.
   b. higher in small populations than in large populations.
   c. higher in large populations than in small populations.
   d. slower than the rate of fixation of deleterious mutations.
   e. none of the above

4. When a gene is duplicated, which of the following may occur?
   a. Production of the gene's product may increase.
   b. The two copies may become expressed in different tissues.
   c. One copy of the gene may accumulate deleterious substitutions and become functionless.
   d. The two copies may diverge and acquire different functions.
   e. All of the above

5. Genome size varies widely among different multicellular organisms (see graph below). What is the greatest contributing cause of this variation?
   a. The number of protein-coding genes
   b. The amount of noncoding DNA
   c. The number of duplicated genes
   d. The degree of concerted evolution
   e. The amount of positive selection for change in protein-coding genes

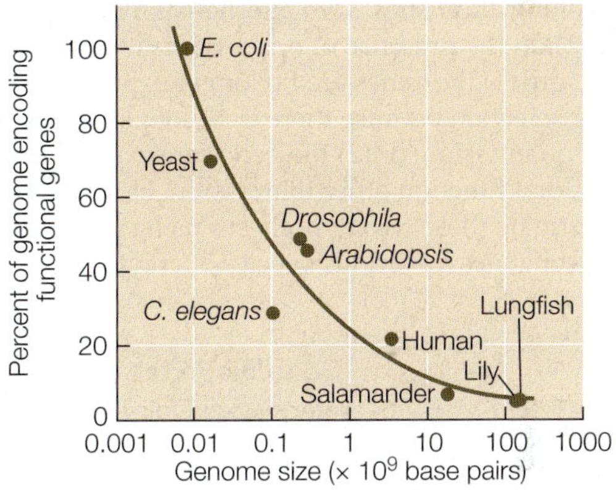

Percent of genome encoding functional genes
100
80
60
40
20
0
E. coli
Yeast
Drosophila
Arabidopsis
C. elegans
Human
Lungfish
Lily
Salamander
0.001
0.01
0.1
1
10
100
1000
Genome size (× 10^9 base pairs)

6. Paralogous genes are genes that trace back to a common
   a. speciation event.
   b. substitution event.
   c. insertion event.
   d. deletion event.
   e. duplication event.

UNDERSTANDING & APPLYING

7. Rates of evolutionary change differ among different molecules, and different species differ widely in generation times and population sizes. How does this variation limit how and in what ways we can use the concept of a molecular clock to help us answer questions about the evolution of both molecules and organisms?

8. Based on what you have learned about evolutionary processes, what modifications could you introduce into in vitro evolution experiments (such as the one shown in Figure 24.14) to increase the diversity or rate of evolution of new molecules?

ANALYZING & EVALUATING

9. Over evolutionary history, many groups of organisms that inhabit caves have lost the organs of sight. For instance, although surface-dwelling crayfishes have functional eyes, several crayfish species that are restricted to underground habitats lack eyes. Opsins are a group of light-sensitive proteins known to have an important function in vision (see Chapter 46), and opsin genes are expressed in eye tissues. Opsin genes are present in the genomes of eyeless, cave-dwelling crayfishes. Two alternative hypotheses that might explain the presence of opsin genes in an eyeless organism are (1) the genes are no longer experiencing purifying selection because there is no longer selection for function in vision; or (2) the genes are experiencing selection for a function other than vision. How would you investigate these two hypotheses using the sequences of the opsin genes in various species of crayfishes?

10. Analysis of synonymous and nonsynonymous substitutions in protein-coding genes can be used to distinguish neutral evolution, positive selection, and purifying selection. An investigator compared many gene sequences encoding surface proteins from influenza viruses sampled over time and collected the data shown in the table. Use the table to answer the following questions.

Amino acid position	Synonymous substitutions	Nonsynonymous substitutions
12	0	7
15	1	9
61	0	12
80	7	0
137	12	1
156	24	2
165	3	4
226	38	3

a. Which positions encode amino acids that have probably changed as a result of positive selection? Why?
b. Which positions encode amino acids that have probably changed as a result of purifying selection? Why?

(Hint: To calculate rates of each substitution type, you will need to consider the number of synonymous and nonsynonymous substitutions present *relative to the number of possible substitutions of each type*. There are approximately 3 times as many possible nonsynonymous substitutions as there are synonymous substitutions.)

Go to BioPortal at **yourBioPortal.com** for Animated Tutorials, Activities, LearningCurve Quizzes, Flashcards, and many other study and review resources.

25 The History of Life on Earth

CHAPTER**OUTLINE**

Dragonflies as Big as Hawks *Meganeuropsis permiana* is shown here in an artist's reconstruction from fossils. Except for its size, this giant from the Permian period was similar to modern dragonflies (depicted in the inset at the same scale).

ALMOST ANYONE who has spent time around freshwater ponds is familiar with dragonflies. Their bright colors and transparent wings stimulate our visual senses on bright summer afternoons as they fly about their business of devouring mosquitoes, mating, and laying their eggs. The largest dragonflies alive today have wingspans that can be covered by a human hand. Three hundred million years ago, however, dragonflies such as *Meganeuropsis permiana* had wingspans of more than 70 centimeters—well over 2 feet, matching or exceeding the wingspans of many modern birds of prey. These dragonflies were the largest flying predators of their time.

No flying insects alive today are anywhere near this size. But during the Carboniferous and Permian geological periods, 350–250 million years ago, many groups of flying insects contained gigantic species. *Meganeuropsis* probably ate huge mayflies and other giant flying insects that shared its home in the Permian swamps. These enormous insects were themselves eaten by giant amphibians. None of these insects or amphibians would be able to survive on Earth today. The oxygen concentrations in Earth's atmosphere were about 50 percent higher then than they are now, and those high oxygen concentrations are thought to have been necessary to support giant insects and their huge amphibian predators.

Paleontologists have uncovered fossils of *Meganeuropsis permiana* in the rocks of Kansas. How do we know the age of these fossils, and how can we know how much oxygen that long-vanished atmosphere contained? The layering of rocks allows us to tell their ages relative to one another, but it does not by itself indicate a given layer's absolute age.

One of the remarkable scientific achievements of the twentieth century was the development of sophisticated techniques that use the decay rates of various radioisotopes, the ratios of certain molecules in rocks and fossils, and changes in Earth's magnetic field to infer conditions and events in the remote past and to date them accurately. It is those methods that allow us to age the fossils of *Meganeuropsis* and to calculate the concentration of oxygen in Earth's atmosphere at the time.

Earth is about 4.5 billion years old, and life has existed on it for about 3.8 billion of those years. That means human civilizations have occupied Earth for less than 0.0003 percent of the history of life. Discovering what happened before humans were around is an ongoing and exciting area of science.

?

Can modern experiments test hypotheses about the evolutionary impact of ancient environmental changes?

See answer on p. 522.

TABLE 25.1
Earth's Geological History

Eon	Era	Period	Onset	Major Physical Changes on Earth
Phanerozoic (~0.5 billion years long)	Cenozoic	Quaternary (Q)	2.6 mya	Cold/dry climate; repeated glaciations
		Tertiary (T)	65.5 mya	Continents near current positions; climate cools
	Mesozoic	Cretaceous (K)	145.5 mya	Laurasian continents attached to one another; Gondwana begins to drift apart; meteorite strikes near current Yucatán Peninsula at end of period
		Jurassic (J)	201.6 mya	Two large continents form: Laurasia (north) and Gondwana (south); climate warm
		Triassic (Tr)	251.0 mya	Pangaea begins to drift apart; hot/humid climate
	Paleozoic	Permian (P)	299 mya	Extensive lowland swamps; O_2 levels 50% higher than present; by end of period continents aggregate to form Pangaea, and O_2 levels drop rapidly
		Carboniferous (C)	359 mya	Climate cools; marked latitudinal climate gradients
		Devonian (D)	416 mya	Continents collide at end of period; giant meteorite probably strikes Earth
		Silurian (S)	444 mya	Sea levels rise; two large land masses emerge; hot/humid climate
		Ordovician (O)	488 mya	Massive glaciation; sea level drops 50 meters
		Cambrian (Є)	542 mya	Atmospheric O_2 levels approach current levels
Proterozoic	Collectively called the Precambrian (~4 billion years long)		2.5 bya	Atmospheric O_2 levels increase from negligible to about 18%; "snowball Earth" from about 750 to 580 mya
Archean			3.8 bya	Earth accumulates more atmosphere (still almost no O_2); meteorite impacts greatly reduced
Hadean			4.5–4.6 bya	Formation of Earth; cooling of Earth's surface; atmosphere contains almost no free O_2; oceans form; Earth under almost continuous bombardment from meteorites

Note: mya, million years ago; bya, billion years ago.

25.1 How Do Scientists Date Ancient Events?

Some evolutionary changes happen rapidly enough to be studied directly and manipulated experimentally. Plant and animal breeding by agriculturalists and the evolution of surface proteins in influenza viruses are examples of rapid, short-term evolution that we have discussed in previous chapters. Other evolutionary changes, such as the appearance of new species and evolutionary lineages, usually take place over a **geological time scale** (Table 25.1).

To understand long-term patterns of evolutionary change, we must not only think in time scales spanning many millions of years, but also consider events and conditions very different from those we observe today. Earth of the distant past was so unlike our present Earth that it would seem like a foreign planet inhabited by strange organisms. The continents were not where they are now, and climates were sometimes dramatically different from those of today. We know this because much of Earth's history is recorded in its rocks.

We cannot tell the ages of rocks just by looking at them, but we can determine the ages of rocks *relative to one another*. The first person to formally recognize this method of relative dating was the seventeenth century Danish physician Nicolaus Steno. Steno realized that in undisturbed sedimentary rocks (rocks formed by the accumulation of sediments), the oldest layers of rock, or **strata** (singular *stratum*), lie at the bottom, and successively higher strata are progressively younger.

Geologists subsequently combined Steno's insight with their observations of fossils contained in sedimentary rocks to establish the following principles of **stratigraphy**:

- Fossils of similar organisms are found in widely separated places on Earth.
- Certain fossils are always found in younger strata, and certain other fossils are always found in older strata.
- Organisms found in younger strata are more similar to modern organisms than are those found in older strata.

These patterns revealed much about the relative ages of sedimentary rocks and the fossils they contained, as well as patterns in the evolution of life. But geologists still could not tell the absolute age of these rocks. A method for *absolute* dating of rocks—that is, determining their actual age rather than just their age relative to one another)—did not become available

Major Events in the History of Life
Humans evolve; many large mammals become extinct
Diversification of birds, mammals, flowering plants, and insects
Dinosaurs continue to diversify; mass extinction at end of period (~76% of species lost)
Diverse dinosaurs; radiation of ray-finned fishes; first fossils of flowering plants
Early dinosaurs; first mammals; marine invertebrates diversify; mass extinction at end of period (~65% of species lost)
Reptiles diversify; giant amphibians and flying insects present; mass extinction at end of period (~96% of species lost)
Extensive fern/horsetail/giant club moss forests; first reptiles; insects diversify
Jawed fishes diversify; first insects and amphibians; mass extinction at end of period (~75% of marine species lost)
Jawless fishes diversify; first ray-finned fishes; plants and animals colonize land
Mass extinction at end of period (~75% of species lost)
Rapid diversification of multicellular animals; diverse photosynthetic protists
Origin of photosynthesis, multicellular organisms, and eukaryotes
Origin of life; prokaryotes flourish
Life not yet present

until after radioactivity was discovered at the beginning of the twentieth century.

Radioisotopes provide a way to date fossils and rocks

Radioactive isotopes of elements—**radioisotopes**—decay in a predictable pattern over long periods (see Section 2.1). Over a specific time interval, known as a **half-life**, half of the atoms in a radioisotope decay to become a different, stable (nonradioactive) isotope (**Figure 25.1A**). The use of this knowledge to date fossils and rocks is known as **radiometric dating**.

To use a radioisotope to date a past event, we must know or estimate the concentration of that isotope at the time of that event, and we must know the radioisotope's half-life. For example, the production of carbon-14 (^{14}C, a radioisotope of carbon) in the upper atmosphere—by the reaction of neutrons with nitrogen-14 (^{14}N, a stable isotope of nitrogen)—just balances the natural radioactive decay of ^{14}C into ^{14}N. Therefore the ratio of ^{14}C to the more common stable isotope of carbon, carbon-12 (^{12}C), is relatively constant in living organisms and in their environment. As soon as an organism dies, however, it ceases to exchange carbon compounds with its environment.

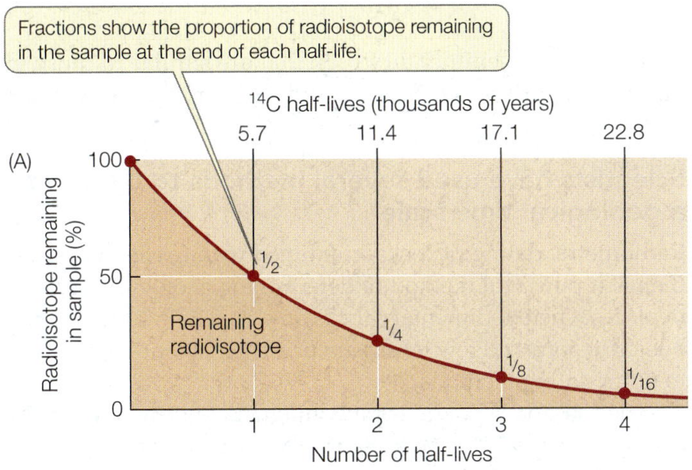

Fractions show the proportion of radioisotope remaining in the sample at the end of each half-life.
(A)
^{14}C half-lives (thousands of years)
5.7
11.4
17.1
22.8
Radioisotope remaining in sample (%)
100
50
0
1/2
1/4
1/8
1/16
Remaining radioisotope
1
2
3
4
Number of half-lives

(B)

Radioisotope	Decay product	Half-life (years)	Useful dating range (years)
Carbon-14 (^{14}C)	Nitrogen-14 (^{14}N)	5,700	100 – 60,000
Uranium-234 (^{234}U)	Thorium-230 (^{230}Th)	80,000	10,000 – 500,000
Uranium-235 (^{235}U)	Lead-207 (^{207}Pb)	704 million	200,000 – 4.5 billion
Potassium-40 (^{40}K)	Argon-40 (^{40}Ar)	1.3 billion	10 million – 4.5 billion

25.1 Radioactive Isotopes Allow Us to Date Ancient Rocks The decay of radioactive isotopes into stable isotopes happens at a steady rate. A half-life is the time it takes for half of the remaining atoms to decay in this way. (A) The graph demonstrates the principle of half-life using carbon-14 (^{14}C) as an example. The half-life of ^{14}C is 5,700 years. (B) Different radioisotopes have different characteristic half-lives that allow us to estimate the ages of rocks.

Its decaying ^{14}C is no longer replenished, and the ratio of ^{14}C to ^{12}C in its remains decreases over time. Paleontologists can use the ratio of ^{14}C to ^{12}C in fossil material to date fossils that are less than 60,000 years old (and thus the sedimentary rocks that contain those fossils). If fossils are older than that, so little ^{14}C remains that the limits of detection using this particular isotope are reached.

Radiometric dating methods have been expanded and refined

Sedimentary rocks are formed from materials that existed for varying lengths of time before being weathered, fragmented, and transported, sometimes over long distances, to the site of their deposition. Therefore the radioisotopes in sedimentary rock do not contain reliable information about the date of its formation. Radiometric dating of rocks older than 60,000 years requires estimating radioisotope concentrations in **igneous rock**, which is formed when molten material cools. To date sedimentary strata, geologists search for places where volcanic ash or lava flows have intruded into the sedimentary rock.

A preliminary estimate of the age of an igneous rock determines which radioisotopes can be used to date it (**Figure 25.1B**). The decay of potassium-40 (which has a half-life of 1.3 billion years) to argon-40, for example, has been used to date many of the ancient events in the evolution of life. Fossils in the adjacent sedimentary rock that are similar to those in other rocks of known ages provide additional clues to the rock's age.

Scientists have used several methods to construct a geological time scale

Radiometric dating of rocks, combined with fossil analysis, is the most powerful method of determining geological age. But in places where sedimentary rocks do not contain suitable igneous intrusions and few fossils are present, paleontologists turn to other dating methods.

One method, known as paleomagnetic dating, relates the ages of rocks to patterns in Earth's magnetism, which change over time. Earth's magnetic poles move and occasionally reverse themselves. Both sedimentary and igneous rocks preserve a record of Earth's magnetic field at the time they were formed, and that record can be used to determine the ages of those rocks. Other methods of dating events in life's history use information about continental drift, information about sea level changes, and molecular clocks (a method that was described in Section 22.3).

Geologists used all of these methods to develop a geological time scale (see Table 25.1). They divided the broad history of life into four **eons**. The Hadean eon refers to the time on Earth before life evolved. The early history of life occurred in the Archean eon, which ended about the time that photosynthetic organisms first appeared on Earth. Prokaryotic life diversified rapidly in the Proterozoic eon, and the first eukaryotes in the fossil record date from this time. These three eons are sometimes referred to collectively as Precambrian time, or simply the **Precambrian**. The Precambrian lasted for approximately four billion years and thus accounts for the vast majority of geological time. It was in the Phanerozoic eon, however—a mere 542-million-year time span—that multicellular eukaryotes rapidly diversified. To emphasize the events of the Phanerozoic, Table 25.1 shows the subdivision of this eon into eras and periods. The boundaries between these divisions of time are based largely on the striking differences geologists observe in the assemblages of fossil organisms contained in successive strata. This geological record of life reveals a remarkable story of a world in which the continents and biological communities are constantly changing.

RECAP 25.1

Fossils in sedimentary rock strata enabled geologists to determine the relative ages of organisms, but absolute dating was not possible until the discovery of radioactivity. Geologists divide the history of life into eons, eras, and periods, based on assemblages of fossil organisms found in successive layers of rocks.

- What observations suggested that fossils could be used to determine the relative ages of rocks? See p. 506
- How is the rate of decay of radioisotopes used to estimate the absolute ages of rocks? See p. 507 and Figure 25.1

As geologists began to develop accurate ways to measure the age of Earth, they began to understand that Earth is far older than anyone had previously understood. During its 4.5-billion-year history, Earth has undergone massive physical changes. These changes have influenced the evolution of life, and life, in its turn, has influenced Earth's physical environment.

25.2 How Have Earth's Continents and Climates Changed over Time?

As we saw in the previous section, the Phanerozoic eon has been notable for the rapid diversification of multicellular eukaryotes. But the diversity of multicellular organisms has not simply increased steadily through time. New species have arisen, and species have gone extinct, throughout the history of life. But there have been times during which extinction rates have increased dramatically over the background levels (**Figure 25.2**). These mass extinction events are the cause of some of the striking differences in fossil assemblages that geologists use to divide the Phanerozoic eon into eras and periods. After each mass extinction, the diversity of life rebounded, although recovery took millions of years. In this section we will discuss some of the physical changes on Earth that have resulted in such dramatic changes in life's diversity.

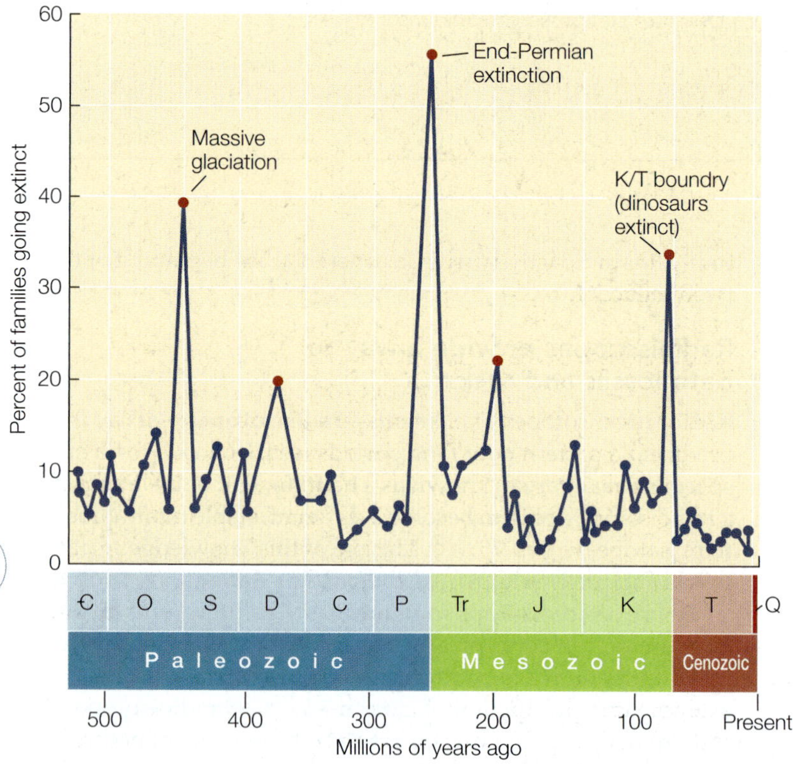

60
50
40
30
20
10
0
Percent of families going extinct
End-Permian extinction
Massive glaciation
K/T boundry (dinosaurs extinct)
Є
O
S
D
C
P
Tr
J
K
T
Q
Paleozoic
Mesozoic
Cenozoic
500
400
300
200
100
Present
Millions of years ago

25.2 Periodic Mass Extinctions Mark Many Geologic Boundaries Five sharp rises (marked by red dots) above the background extinction rate have occurred throughout the Phanerozoic. The most sweeping of these events, the end-Permian extinction, was associated with dramatic drops in sea level (see Figure 25.4), global temperature, and atmospheric oxygen level (see Figure 25.8).

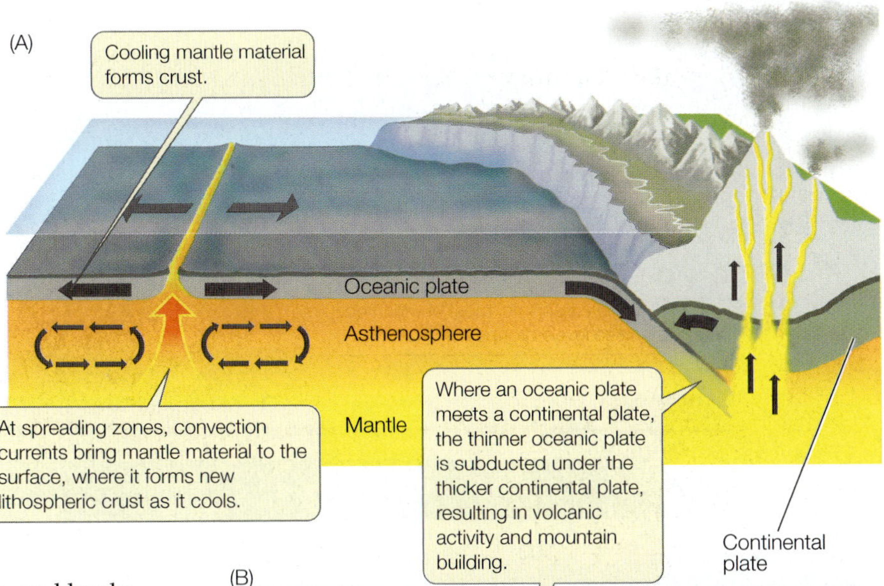
(A)
Cooling mantle material forms crust.
Oceanic plate
Asthenosphere
Mantle
At spreading zones, convection currents bring mantle material to the surface, where it forms new lithospheric crust as it cools.
Where an oceanic plate meets a continental plate, the thinner oceanic plate is subducted under the thicker continental plate, resulting in volcanic activity and mountain building.
Continental plate

(B)

25.3 Plate Tectonics and Continental Drift (A) The heat of Earth's core generates convection currents in the viscous mantle material of the asthenosphere. These currents push the continental plates, along with the land masses they carry, together or apart. Where plates collide, one may slide under the other, creating mountain ranges and often volcanoes. (B) The Cascade Range of the Pacific Northwest of North America is an example of a mountain chain produced by subduction of an oceanic plate under a continental plate.

Go to Media Clip 25.1
Lava Flows and Magma Explosions
Life10e.com/mc25.1

The continents have not always been where they are today

The globes and maps that adorn our walls, shelves, and books give an impression of a static Earth. It would be easy for us to assume that the continents have always been where they are. But we would be wrong. The idea that Earth's land masses have changed their positions over the millennia, and that they continue to do so, was put forth in 1912 by the German meteorologist and geophysicist Alfred Wegener. His idea, known as **continental drift**, was initially met with skepticism and resistance. By the 1960s, however, physical evidence and increased understanding of **plate tectonics**—the geophysics of the movement of major land masses—had convinced virtually all geologists of the reality of Wegener's vision. Plate tectonics provided the geological mechanism that explained Wegener's hypothesis of continental drift.

Earth's crust consists of several solid plates, which collectively make up the solid **lithosphere** ("stone sphere"). Thick continental and thinner oceanic lithospheric plates overlie a viscous, malleable layer of Earth's mantle, known as the **asthenosphere** ("weak sphere"). Heat produced by radioactive decay deep in Earth's core sets up large-scale convection currents in the mantle. New crust is formed as mantle material rises between diverging plates, pushing them apart.

Where oceanic plates and continental plates converge, the thinner oceanic plate is forced underneath the thicker continental plate, a process known as **subduction**. Subduction results in volcanism and mountain building on the continental boundary (**Figure 25.3A**). For example, in the Pacific Northwest of North America, a series of volcanoes formed the Cascade mountain range as the Juan de Fuca oceanic plate has been subducted beneath a portion of the continental North American Plate (**Figure 25.3B**). When two oceanic plates collide, one is also subducted below the other, producing a deep oceanic trench and associated volcanic activity. The deepest part of the world's oceans—the Mariana Trench in the western Pacific—formed where two oceanic plates collided. Volcanic activity associated with the subduction at the Mariana Trench produced the nearby Mariana Islands.

When two thick continental plates collide, neither plate is subducted. Instead, the plates push up against one another, forming high mountain chains. The highest mountain chain in the world, the Himalayas, was formed this way when the Indian Plate collided with the Eurasian Plate. When continental plates diverge, new crust forms in the intervening spaces, resulting in deep clefts called rift valleys in which large freshwater lakes typically form. The Great Rift Valley lakes of eastern Africa, including Lake Malawi (discussed at the opening of Chapter 23), were formed in this way. Two plates can also slide past one another, forming a transform fault boundary (such as the San Andreas Fault that produces violent seismic activity in parts of California).

Many physical conditions on Earth have oscillated in response to plate tectonic processes. We now know that the movement of the plates has sometimes brought continents together and at other times has pushed them apart, as seen in the maps across the top of Figure 25.14. The positions and sizes of the continents influence oceanic circulation patterns, global climates, and sea levels. Sea level is influenced directly by plate tectonic processes (which can influence the depth of ocean basins) and indirectly by oceanic circulation patterns, which affect patterns of glaciation. As climates cool, glaciers form and tie up water over land masses; as climates warm, glaciers melt and release water.

Go to Animated Tutorial 25.1
Movement of the Continents
Life10e.com/at25.1

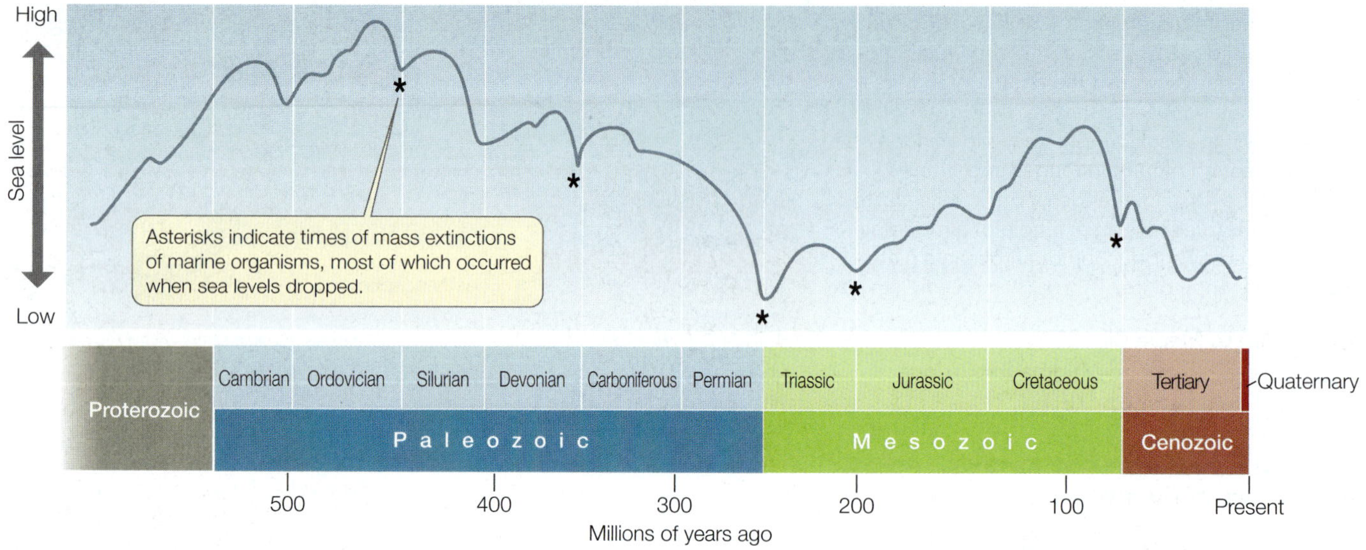
High
Sea level
Low
Asterisks indicate times of mass extinctions of marine organisms, most of which occurred when sea levels dropped.
Proterozoic
Cambrian
Ordovician
Silurian
Devonian
Carboniferous
Permian
Paleozoic
Triassic
Jurassic
Cretaceous
Mesozoic
Tertiary
Quaternary
Cenozoic
500
400
300
200
100
Present
Millions of years ago

25.4 Sea Levels Have Changed Repeatedly Rapid drops in sea level are associated with periods of globally cooler temperatures and increased glaciation. Most mass extinctions of marine organisms have coincided with low sea levels.

Earth's climate has shifted between hot and cold conditions

Through much of its history, Earth's climate was considerably warmer than it is today, and temperatures decreased more gradually toward the poles. At other times, Earth was colder than it is today. Rapid drops in sea level near the ends of the Ordovician, Devonian, Permian, Triassic, and Cretaceous periods, and most recently in the Quaternary period, resulted mainly from increased global glaciation (**Figure 25.4**). Many of these drops in sea level were accompanied by mass extinctions—particularly of marine organisms, which could not survive the disappearance of the shallow seas that covered vast areas of the continental shelves.

Earth's cold periods were separated by long periods of milder climates. Because we are living in one of the colder periods, it is difficult for us to imagine the mild climates that were found at high latitudes during much of the history of life. The Quaternary period has been marked by a series of glacial advances, interspersed with warmer interglacial intervals during which the glaciers retreated.

"Weather" refers to the daily events at a given location, such as individual storms and the high and low temperatures on a given day. "Climate" refers to long-term average expectations over the various seasons at a given location. Weather often changes rapidly; climates typically change slowly. However, major climate shifts have taken place over periods as short as 5,000 to 10,000 years, primarily as a result of changes in Earth's orbit around the sun. A few climate shifts have been even more rapid: during one Quaternary interglacial period, the ice-locked Antarctic Ocean became nearly ice-free in less than 100 years. Some climate changes have been so rapid that the extinctions caused by them appear to be nearly instantaneous in the fossil record. Such rapid changes are usually caused by sudden shifts in ocean currents.

We are currently living in a time of rapid climate change thought to be caused by a buildup of atmospheric CO_2, primarily from the burning of fossil fuels by human populations. We are reversing the energy transformations that occurred with the massive burial and decomposition of organic material that occurred (especially) in the Carboniferous, Permian, and Triassic, which gave rise to the fossil fuels we are using today. But we are burning these fuels over a few hundred years, rather than the many millions of years over which those deposits accumulated. The current rate of increase in atmospheric CO_2 is unprecedented in Earth's history. A doubling of the atmospheric CO_2 concentration—which may happen during the current century—is expected to increase the average temperature of Earth, change rainfall patterns, melt glaciers and ice caps, and raise sea levels.

Volcanoes have occasionally changed the history of life

Most volcanic eruptions produce only local or short-lived effects, but a few large volcanic eruptions have had major consequences for life. When Krakatau (a volcanic island in the Sunda Strait off Indonesia) erupted in 1883, it ejected more than 25 cubic kilometers of ash and rock as well as large quantities of sulfur dioxide gas (SO_2). The SO_2 was ejected into the stratosphere and carried by high-altitude winds around the planet. Its presence led to high concentrations of sulfurous acid (H_2SO_3) in high-altitude clouds, creating a "parasol effect" that reduced the amount of sunlight reaching Earth's surface. Global temperatures dropped by 1.2°C in the year following the eruption, and global weather patterns showed strong effects for another 5 years. More recently, the eruption of Mount Pinatubo in the Philippines in 1991 (**Figure 25.5**) temporarily reduced global temperatures by about 0.5°C.

Although these individual volcanoes had only relatively short-term effects on global temperatures, they suggest that the simultaneous eruption of many volcanoes could have a much stronger effect on Earth's climate. What would cause many volcanoes to erupt at the same time? The collision of

25.5 Volcanic Eruptions Can Cool Global Temperatures When Mount Pinatubo erupted in 1991, it increased the concentrations of sulfurous acid in high-altitude clouds, which temporarily lowered global temperatures by about 0.5°C.

Iridium-rich layer at the Cretaceous-Tertiary (K/T) boundary

25.6 Evidence of a Meteorite Impact The white layers of rock are Cretaceous in age; the layers at the upper left were deposited in the Tertiary. Between the two is a thin, dark layer of clay that contains large amounts of iridium, a metal common in some meteorites but rare on Earth. Its high concentration in this sediment layer, deposited about 65.5 million years ago, suggests the impact of a large meteorite at that time.

continents during the Permian period, about 275 million years ago (mya), formed a single, gigantic land mass and caused a multitude of massive volcanic eruptions. Emissions from these eruptions blocked considerable sunlight, contributing to the advance of glaciers and a consequent drop in sea level (see Figure 25.4). Thus volcanic eruptions were probably at least part of the explanation for the greatest mass extinction in Earth's history.

Extraterrestrial events have triggered changes on Earth

At least 30 meteorites of sizes between tennis balls and soccer balls strike Earth each year. Collisions with larger meteorites or comets are rare, but such collisions have probably been responsible for several mass extinctions. Several types of evidence tell us about these collisions. Their craters, and the dramatically disfigured rocks that result from their impact, are found in many places. Geologists have discovered compounds in these rocks that contain helium and argon with isotope ratios characteristic of meteorites, which are very different from the ratios found elsewhere on Earth.

A meteorite caused or contributed to a mass extinction at the end of the Cretaceous period (about 65.5 mya). The first clue that a meteorite was responsible came from the abnormally high concentrations of the element iridium found in a thin layer separating rocks deposited during the Cretaceous from rocks deposited during the Tertiary (**Figure 25.6**). Iridium is abundant in some meteorites, but it is exceedingly rare on Earth's surface. When scientists discovered a circular crater 180 km in diameter buried beneath the northern coast of Mexico's Yucatán Peninsula, they constructed the following scenario. When the meteorite that formed that crater collided with Earth, it released energy equivalent to that of 100 million megatons of high explosives, creating great tsunamis. A massive plume of debris rose into the atmosphere, spread around Earth, and descended. The descending debris heated the atmosphere to several hundred degrees and ignited massive fires. It also blocked the sun, preventing plants from photosynthesizing. The settling debris formed the iridium-rich layer. About a billion tons of soot with a composition matching that of smoke from forest fires were also deposited. These events had devastating effects on biodiversity. Many fossil species (including non-avian dinosaurs) that are found in Cretaceous rocks are not found in the Tertiary rocks of the next stratum.

Oxygen concentrations in Earth's atmosphere have changed over time

As the continents have moved over Earth's surface, the world has experienced other physical changes, including large increases and decreases in atmospheric oxygen concentrations. The atmosphere of early Earth probably contained little or no free oxygen gas (O_2). The increase in atmospheric O_2 came in two big steps more than a billion years apart.

The first step occurred about 2.5 billion years ago (bya), when certain bacteria gained the ability to use water as the source of hydrogen ions for photosynthesis. By chemically splitting H_2O, these bacteria generated O_2 as a waste product. They also made electrons available for reducing CO_2 to form the carbohydrate end-products of photosynthesis (see Section 10.3). The O_2 they produced dissolved in water and reacted with dissolved iron. The reaction product then precipitated

25.7 Banded Iron Formations Indicate Early Photosynthesis The alternating red and dark layers in this 2.25-billion-year-old sedimentary rock formation from Lake Superior resulted from a reaction between dissolved iron and the atmospheric oxygen produced by Earth's first photosynthetic organisms. The chemical reaction produced nearly pure iron oxide, or hematite, which forms the gray, metallic layers in this sample. The red bands are jasper tinged with much smaller amounts of iron oxide.

as iron oxide, which accumulated in alternating layers of red and dark rock known as banded iron formations (**Figure 25.7**). These formations provide evidence for the earliest photosynthetic organisms. As photosynthetic organisms continued to release O_2, oxygen gas began to accumulate in the atmosphere.

The second step occurred about a billion years later, when some of these photosynthetic bacteria became endosymbionts within eukaryotic cells, leading to the eventual evolution of chloroplasts in plants and other photosynthetic eukaryotes. This change resulted in continued accumulation of O_2 in Earth's atmosphere (**Figure 25.8**).

One group of photosynthetic bacteria, the cyanobacteria, formed rocklike structures called stromatolites, which are abundantly preserved in the fossil record. To this day, cyanobacteria still form stromatolites in a few very salty bodies of water (**Figure 25.9**). Cyanobacteria liberated enough O_2 to open the way for the evolution of oxidation reactions as the energy source for the synthesis of ATP.

Thus the evolution of life irrevocably changed the physical nature of Earth. Those physical changes, in turn, influenced the evolution of life. When it first appeared in the atmosphere, O_2 was toxic to most of the anaerobic prokaryotes that inhabited Earth at the time. Over millennia, however, prokaryotes that evolved the ability to tolerate and use O_2 not only survived but gained the advantage. Aerobic metabolism proceeds more rapidly, and harvests energy more efficiently, than anaerobic metabolism. Organisms with aerobic metabolism replaced anaerobes in most of Earth's environments.

An atmosphere rich in O_2 also made possible larger and more complex organisms. Small single-celled aquatic organisms can obtain enough oxygen by simple diffusion even when dissolved oxygen concentrations in the water are very low. Larger single-celled organisms, however, have lower surface area-to-volume ratios; to obtain enough oxygen by simple diffusion, they must live in an environment with a relatively high oxygen concentration. Bacteria can thrive at 1 percent of the current oxygen concentration; eukaryotic cells require levels that are at least 2–3 percent of the current concentration. For concentrations of dissolved oxygen in the oceans to have reached these levels, much higher atmospheric concentrations were needed.

Probably because it took many millions of years for Earth to develop an oxygenated atmosphere, only single-celled prokaryotes lived on Earth for more than 2 billion years. About 1.5 bya, atmospheric O_2 concentrations became high enough for larger eukaryotic cells to flourish. Further increases in atmospheric O_2 concentrations in the late Precambrian enabled several groups of multicellular organisms to evolve (see Figure 25.8).

Oxygen concentrations increased again during the Carboniferous and Permian periods because of the evolution of large vascular plants. These plants lived in the expansive lowland swamps that existed at the time (see Table 25.1). Massive amounts of organic material were buried in these swamps as the plants died, leading to the formation of Earth's vast coal deposits. Because the buried organic material was not subject to oxidation as it decomposed, and because the living plants were producing large quantities of O_2, atmospheric O_2 increased to

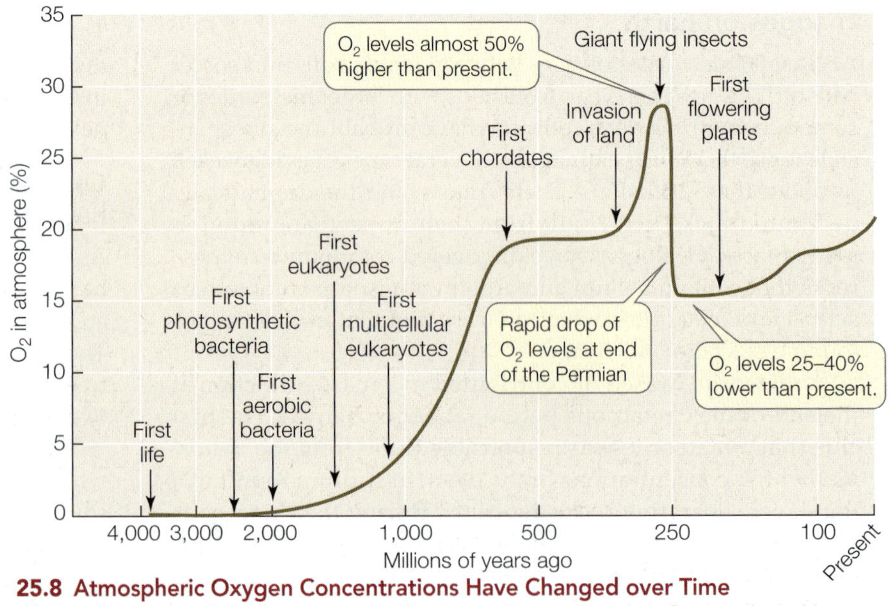

O_2 in atmosphere (%)
35
30
25
20
15
10
5
0
O_2 levels almost 50% higher than present.
Giant flying insects
First flowering plants
Invasion of land
First chordates
First eukaryotes
First photosynthetic bacteria
First multicellular eukaryotes
Rapid drop of O_2 levels at end of the Permian
O_2 levels 25–40% lower than present.
First aerobic bacteria
First life
4,000 3,000 2,000
1,000
500
250
100
Present
Millions of years ago

25.8 Atmospheric Oxygen Concentrations Have Changed over Time Changes in atmospheric oxygen concentrations have strongly influenced, and have been influenced by, the evolution of life. (Note that the horizontal axis of the graph is on a logarithmic scale.)

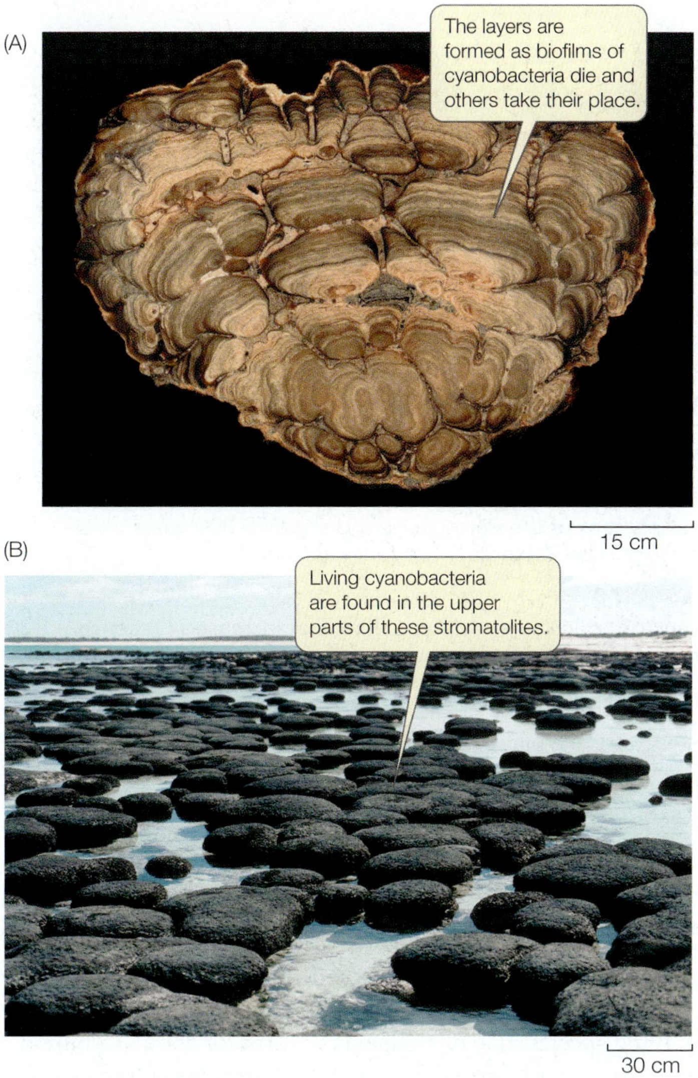

(A)
The layers are formed as biofilms of cyanobacteria die and others take their place.
15 cm
(B)
Living cyanobacteria are found in the upper parts of these stromatolites.
30 cm

25.9 Stromatolites (A) A vertical section through a fossil stromatolite. (B) These rocklike structures are living stromatolites that thrive in the very salty waters of Shark Bay in Western Australia.

INVESTIGATING**LIFE**

25.10 Atmospheric Oxygen Concentrations and Body Size in Insects C. Jaco Klok and his colleagues asked whether insects raised in hyperoxic conditions would evolve to be larger than their counterparts raised under today's atmospheric conditions. They raised strains of fruit flies (*Drosophila melanogaster*) under both conditions to test the effects of increased O_2 concentrations on the evolution of body size.[a]

HYPOTHESIS In hyperoxic conditions, increased partial pressure of oxygen results in evolution of increased body size in flying insects.

Method

1. Separate a population of fruit flies into multiple lineages.
2. Raise half the lineages in current atmospheric (control) conditions; raise the other lineages in hyperoxic (experimental) conditions. Continue all lineages for seven generations.
3. Raise the F_8 individuals of all lineages under identical (current) atmospheric conditions.
4. Weigh 50 flies from each of the replicate lines and test for statistical differences in body weight.

Results The average body mass of F_8 individuals of both sexes raised under hyperoxic conditions was significantly ($P < 0.001$) greater than that of individuals in the control lineages.

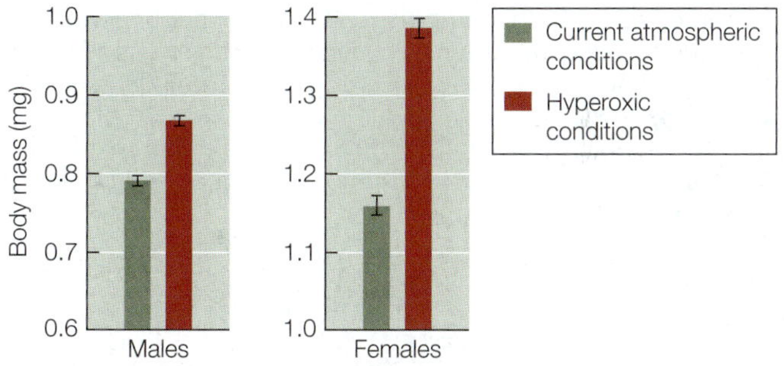

Body mass (mg)
1.0
0.9
0.8
0.7
0.6
Males
1.4
1.3
1.2
1.1
1.0
Females
Current atmospheric conditions
Hyperoxic conditions

CONCLUSION Increased O_2 concentrations led to evolution of larger body size in fruit flies, consistent with the trends seen among other flying insects in the fossil record.

Go to **BioPortal** for discussion and relevant links for all INVESTIGATING**LIFE** figures.

[a]Klok, C. J. et al. 2009. *Journal of Evolutionary Biology* 22: 2496–2504.

concentrations that have not been reached again in Earth's history (see Figure 25.8). As mentioned at the opening of this chapter, these high concentrations of atmospheric O_2 allowed the evolution of giant flying insects and large amphibians that could not survive in today's atmosphere.

The drying of the lowland swamps at the end of the Permian reduced burial of organic matter as well as the production of O_2, so atmospheric O_2 concentrations dropped rapidly. Over the past 200 million years, with the diversification of flowering plants, O_2 concentrations have again increased, but not to the levels that characterized the Carboniferous and Permian periods.

Biologists have conducted experiments that demonstrate the changing selection pressures that can accompany changes in atmospheric O_2 concentrations. When fruit flies (*Drosophila*) were raised in hyperoxic conditions (i.e., with artificially increased atmospheric concentrations of O_2), they evolved larger body sizes in just a few generations (**Figure 25.10**). The present atmospheric O_2 concentrations appear to constrain body size in these flying insects; increases in O_2 appear to relax those constraints. This experiment demonstrates that the stabilizing selection on body size at present O_2 concentrations can quickly switch to directional selection when environmental conditions change (see Section 21.3).

RECAP 25.2

Physical conditions on Earth have changed dramatically over time. Changes in Earth's climate and sea levels have had major effects on biological evolution. Continental drift, volcanic eruptions, and large meteorite strikes have contributed to major climate changes during Earth's history, and many of these climate shifts have resulted in mass extinction events. Changes in atmospheric concentrations of O_2 have also influenced the evolution of life.

- How have plate tectonic processes, volcanic eruptions, and meteorite strikes resulted in mass extinction events? **See pp. 510–511 and Figure 25.4**
- Describe how increases in atmospheric concentrations of O_2 affected the evolution of multicellular organisms. **See pp. 511–512 and Figure 25.8**

WORKING WITH**DATA**:

The Effects of Oxygen Concentration on Insect Body Size

Original Papers

Harrison, J. F. and G. G. Haddad. 2011. Effects of oxygen on growth and size: Synthesis of molecular, organismal and evolutionary studies with *Drosophila melanogaster*. *Annual Review of Physiology* 73: 13.1–13.9.

Harrison, J. F., A. Kaiser and J. M. VandenBrooks. 2010. Atmospheric oxygen level and the evolution of insect body size. *Proceedings of the Royal Society of London B* 277: 1937–1946.

Klok, C. J., A. J. Hubb and J. F. Harrison. 2009. Single and multigenerational responses of body mass to atmospheric oxygen concentration in *Drosophila melanogaster*: Evidence for roles of plasticity and evolution. *Journal of Evolutionary Biology* 22: 2496–2504.

Analyze the Data

In the data shown in Figure 25.10, the body mass of individuals in the experimental population of *Drosophila* increased, on average, about 2 percent per generation under hyperoxic conditions (although the rate of increase was not constant over the experiment). Here you will extrapolate from Harrison et al.'s study to determine whether the observed rate of increase in body mass per generation these researchers observed in *Drosophila* is sufficient to account for the giant dragonflies of the Permian period.

QUESTION 1

Assume that the average rate of increase in dragonfly size during the Permian was much slower than the rate observed in the experiment in Figure 25.10. We'll assume that the actual rate of increase for dragonflies was just 0.01 percent per generation, rather than the 2 percent observed over a few generations for *Drosophila*. We'll assume further that dragonflies complete only one generation per year (as opposed to 40 or more generations for *Drosophila*). Starting with an average body mass of 1 gram, calculate the projected increase in dragonfly body mass over 50,000 years.*

QUESTION 2

What percent of the Permian period does 50,000 years represent? Use Table 25.1 for your calculation.

QUESTION 3

Given your calculations, do you think that increased O_2 concentrations during the Permian were sufficient to account for the evolution of giant dragonflies? Why or why not?

*This calculation is similar to computing compound interest for a savings account. Use the formula $W = S(1 + R)^N$ where W = the final mass, S = the starting mass, R = the rate of increase per generation (0.0001 in this case), and N = the number of generations.)

Go to **BioPortal** for all WORKING WITH**DATA** exercises

The many dramatic physical events in Earth's history have influenced the nature and timing of evolutionary changes among Earth's living organisms. We now will look more closely at some of the major events that characterize the history of life on Earth.

25.3 What Are the Major Events in Life's History?

How do we know about the effects of the physical changes described in the previous section on the evolution of life? To reconstruct life's history, scientists rely heavily on the fossil record. As we have seen, geologists divided Earth's history into eons, eras, and periods based on distinct fossil assemblages (see Table 25.1). Biologists refer to the assemblage of all organisms of all kinds living at a particular time or place as a **biota**. All of the plants living at a particular time or place are its **flora**; all of the animals are its **fauna**.

About 300,000 species of fossil organisms have been described and named, and the number steadily grows. The number of named species, however, is only a tiny fraction of the species that have ever lived. We do not know how many species lived in the past, but we have ways of making reasonable estimates. Of the present-day biota, about 1.8 million species have been named. The actual number of living species is estimated to be in the tens of millions, and possibly much higher, because many species have not yet been discovered and described by biologists. So the number of described fossil species is only about 3 percent of the estimated minimum number of living species. Life has existed on Earth for about 3.8 billion years. Many species last only a few million years before undergoing speciation or going extinct; therefore Earth's biota must have turned over many times during geological history. So the total number of species that have lived over evolutionary time must vastly exceed the number living today. Why have only about 300,000 of these tens of millions of species been described from fossils to date?

Several processes contribute to the paucity of fossils

Only a tiny fraction of organisms ever become fossils, and only a tiny fraction of fossils are ever discovered by paleontologists. Most organisms live and die in oxygen-rich environments, in which they quickly decompose. Organisms are not likely to become fossils unless they are transported by wind or water to sites that lack oxygen, where decomposition proceeds slowly or not at all. Furthermore, geological processes transform many rocks, destroying the fossils they contain, and many fossil-bearing rocks are deeply buried and inaccessible. Paleontologists have studied only a tiny fraction of the sites that contain fossils, although they find and describe many new ones every year.

The fossil record is most complete for marine animals that had hard skeletons (which resist decomposition). Among the nine major animal groups with hard-shelled members, approximately 200,000 species have been described from

25.11 Insect Fossils Chunks of amber—fossilized tree resin—often contain insects such as this fly. Insects were preserved when they became trapped in the sticky resin.

fossils—roughly twice the number of living marine species in these same groups. Paleontologists lean heavily on these groups in their interpretations of the evolution of life. Insects and spiders are also relatively well represented in the fossil record because they are numerically abundant and have hard exoskeletons (**Figure 25.11**). The fossil record, though incomplete, is good enough to document clearly the factual history of the evolution of life.

By combining evidence of physical changes during Earth's history with evidence from the fossil record, scientists have composed portraits of what Earth and its inhabitants may have looked like at different times. We know in general where the continents were and how life changed over time, but many of the details are poorly known, especially for events in the more remote past.

Precambrian life was small and aquatic

Life first appeared on Earth about 3.8 bya (**Figure 25.12**). The fossil record of organisms that lived prior to the Phanerozoic is fragmentary, but it is good enough to establish that the total number of species and individuals increased dramatically in the late Precambrian.

For most of its history, life was confined to the oceans, and all organisms were small. For more than 3 billion years, all organisms lived in shallow seas. These seas slowly began to teem with microscopic prokaryotes. After the first eukaryotes appeared about 1.5 billion years ago, during the Proterozoic, unicellular eukaryotes and small multicellular animals fed on the microorganisms. Small floating organisms, known collectively as **plankton**, were strained from the water and eaten by slightly larger filter-feeding animals. Other animals ingested sediments on the seafloor and digested the remains of organisms within them. But it still took nearly a billion years before eukaryotes began to diversify rapidly into the many different morphological forms that we know today.

What limited the diversity of multicellular eukaryotes (in terms of their size and shape) for much of their early existence? It is likely that a combination of factors was responsible. We have already noted that O_2 levels increased throughout the Proterozoic, and it is likely that high atmospheric and dissolved O_2 concentrations were needed to support large multicellular organisms. In addition, geologic evidence points to a series of intensely cold periods during the late Proterozoic, which would have resulted in seas that were largely covered by

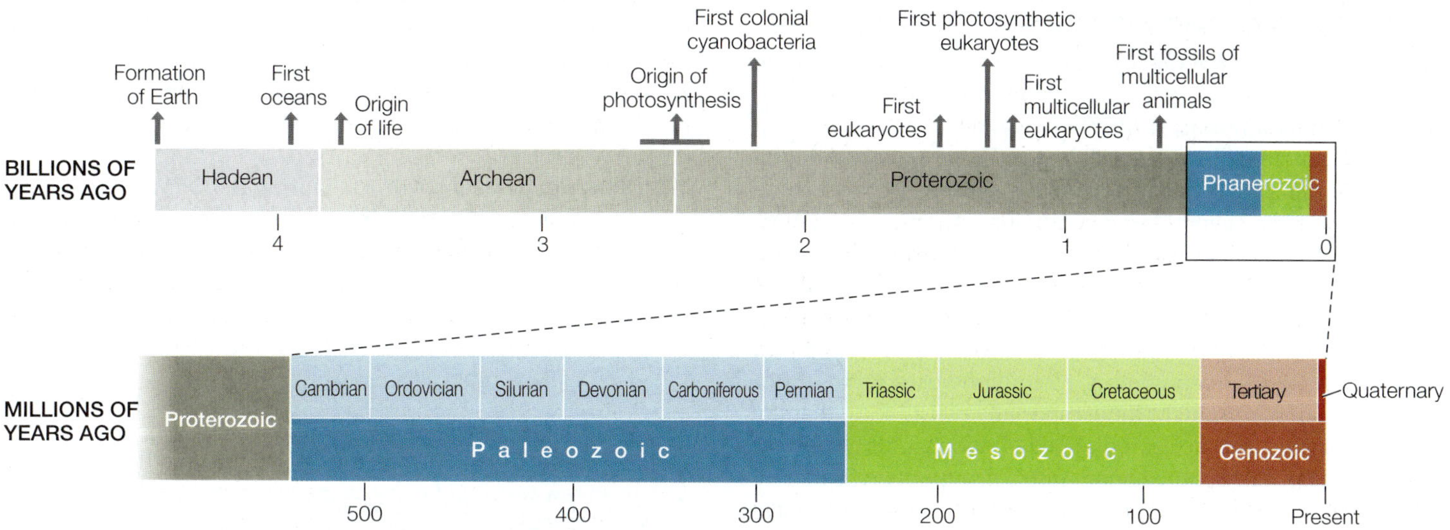
Formation of Earth
First oceans
Origin of life
Origin of photosynthesis
First colonial cyanobacteria
First eukaryotes
First photosynthetic eukaryotes
First multicellular eukaryotes
First fossils of multicellular animals
BILLIONS OF YEARS AGO
Hadean
Archean
Proterozoic
Phanerozoic
4
3
2
1
0
MILLIONS OF YEARS AGO
Proterozoic
Cambrian
Ordovician
Silurian
Devonian
Carboniferous
Permian
Triassic
Jurassic
Cretaceous
Tertiary
Quaternary
Paleozoic
Mesozoic
Cenozoic
500
400
300
200
100
Present

25.12 A Sense of Life's Time The top timeline shows the 4.5-billion-year history of Earth. Most of this history is accounted for by the Precambrian, a 3.4-billion-year time span that saw the origin of life and the evolution of cells, photosynthesis, and multicellularity. The final 600 million years are expanded in the bottom timeline and detailed in Figure 25.14.

25.13 Diversification of Multicellular Organisms: The "Cambrian Explosion" Shortly after the end of Proterozoic glaciations (about 580 mya), several major radiations of multicellular organisms appear in the fossil record. (A) These microscopic fossils from the Doushantuo rock formation of China are the remains of tiny one-, two-, four-, and eight-celled stages of multicellular organisms. (B) Unusual soft-bodied marine invertebrates, unlike any animals alive at present, characterize the fossilized fauna preserved at Ediacara in southern Australia. (C) By the early Phanerozoic, fossilized faunas such as those preserved in Canada's Burgess Shale include extinct representatives of some of the major animal groups alive today.

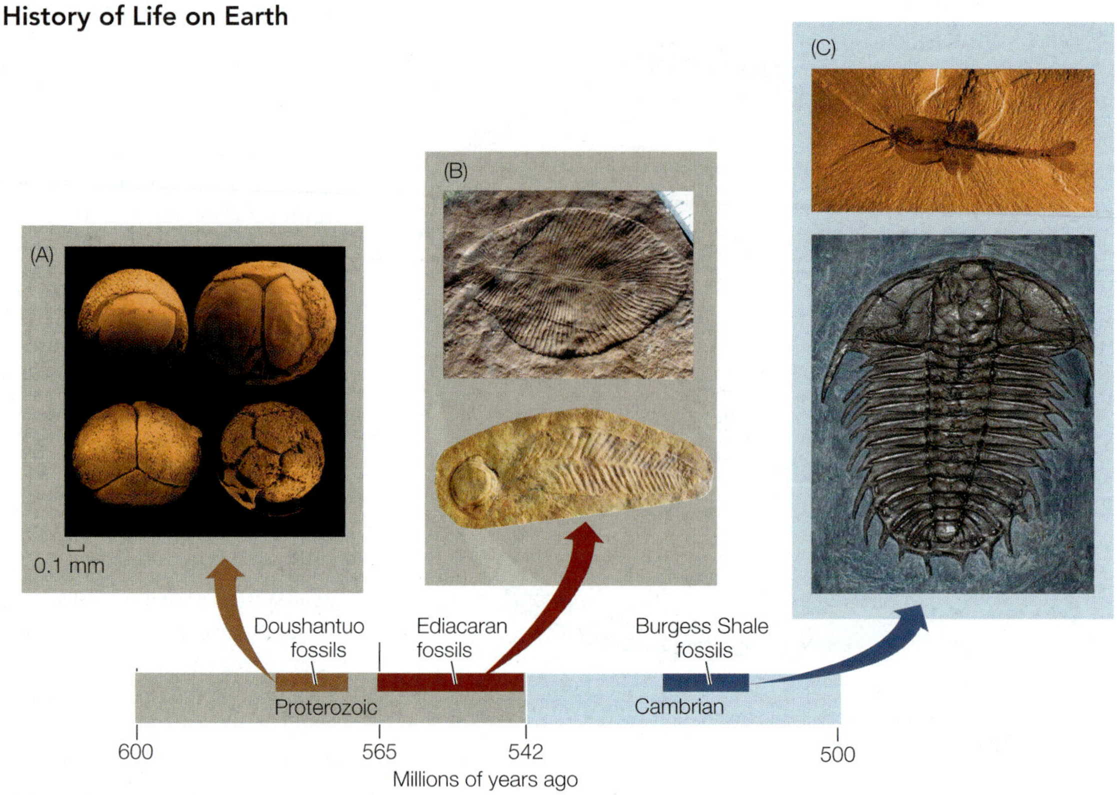
(A)
0.1 mm
(B)
(C)
Doushantuo fossils
Ediacaran fossils
Burgess Shale fossils
Proterozoic
Cambrian
600
565
542
500
Millions of years ago

ice and continents that were covered by glaciers. The "snowball Earth" hypothesis suggests that cold conditions confined life to warm places such as hot springs, deep thermal vents, and perhaps a few equatorial oceans that avoided ice cover. The last of these Proterozoic glaciations ended about 580 million years ago, just before several major radiations of multicellular eukaryotes appear in the fossil record (**Figure 25.13**). Many of the multicellular organisms known from the late Proterozoic and early Phanerozoic were very different from any animals living today and may be members of groups that left no living descendants.

Life expanded rapidly during the Cambrian period

The Cambrian period (542–488 mya) marks the beginning of the Paleozoic, the first era of the Phanerozoic. The O_2 concentration in the Cambrian atmosphere was approaching the current level, and the glaciations of the late Proterozoic had ended nearly 40 million years earlier. Earth's land masses had come together to form several large continents. A rapid diversification of life took place that is called the **Cambrian explosion**. This name is somewhat misleading, as the series of radiations it refers to actually began before the start of the Cambrian and continued for about 60 million years into the early Cambrian (see Figure 25.13). Nonetheless, 60 million years represents a relatively short amount of time, especially considering that the first eukaryotes had appeared about a billion (= 1,000 million) years earlier. Many of the major animal groups represented by species alive today first appeared during these evolutionary radiations. **Figure 25.14** provides an overview of the numerous continental and biotic innovations that have characterized the Phanerozoic.

For the most part, fossils tell us only about the hard parts of organisms, but in three known Cambrian fossil beds—the Burgess Shale in British Columbia, Sirius Passet in northern Greenland, and the Chengjiang site in southern China—the soft parts of many animals were preserved. Crustacean arthropods (crabs, shrimps, and their relatives) are the most diverse group in the Chinese fauna; some of them were large carnivores. Multicellular life was largely or completely aquatic during the Cambrian. If there was life on land at this time, it was probably restricted to microorganisms.

Many groups of organisms that arose during the Cambrian later diversified

Geologists divide the remainder of the Paleozoic era into the Ordovician, Silurian, Devonian, Carboniferous, and Permian periods. Each period is characterized by the diversification of specific groups of organisms. Mass extinctions marked the ends of the Ordovician, Devonian, and Permian.

THE ORDOVICIAN (488–444 MYA) During the Ordovician period, the continents, which were located primarily in the Southern Hemisphere, still lacked multicellular life. Evolutionary radiation of marine organisms was spectacular during the early Ordovician, especially among animals, such as brachiopods and mollusks, that lived on the seafloor and filtered small prey

25.14 A Brief History of Multicellular Life on Earth The geologically rapid "explosion" of life shortly before and during the Cambrian saw the rise of many major animal groups that have representatives surviving today. The following three pages depict life's history through the Phanerozoic. The movements of the major continents during the past half-billion years are shown in the maps of Earth, and associated biotas for each time period are depicted. The artists' reconstructions are based on fossils such as those shown in the photographs.

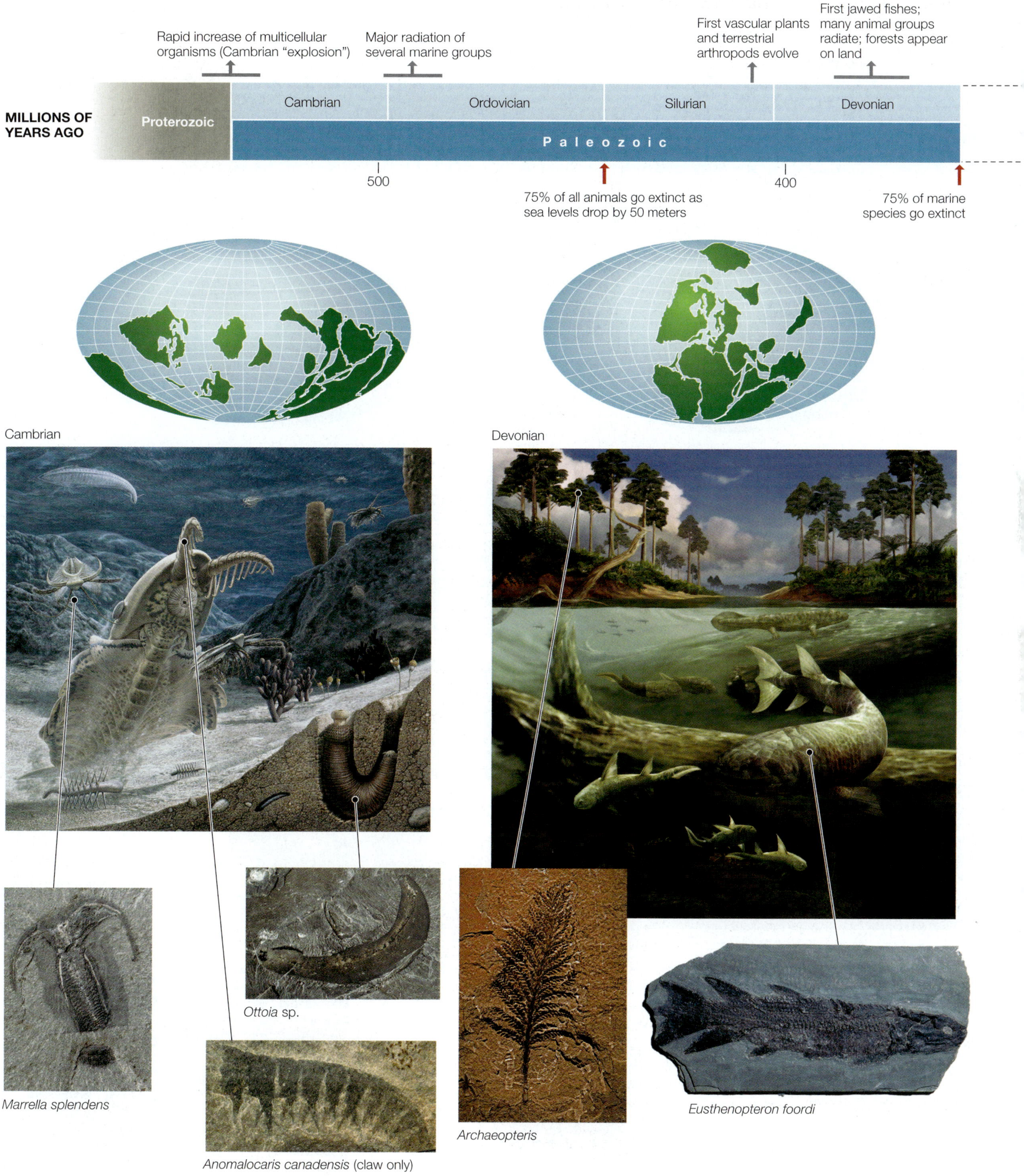

Rapid increase of multicellular organisms (Cambrian "explosion")
Major radiation of several marine groups
First vascular plants and terrestrial arthropods evolve
First jawed fishes; many animal groups radiate; forests appear on land
MILLIONS OF YEARS AGO
Proterozoic
Cambrian
Ordovician
Silurian
Devonian
Paleozoic
500
400
75% of all animals go extinct as sea levels drop by 50 meters
75% of marine species go extinct
Cambrian
Devonian
Marrella splendens
Ottoia sp.
Anomalocaris canadensis (claw only)
Archaeopteris
Eusthenopteron foordi

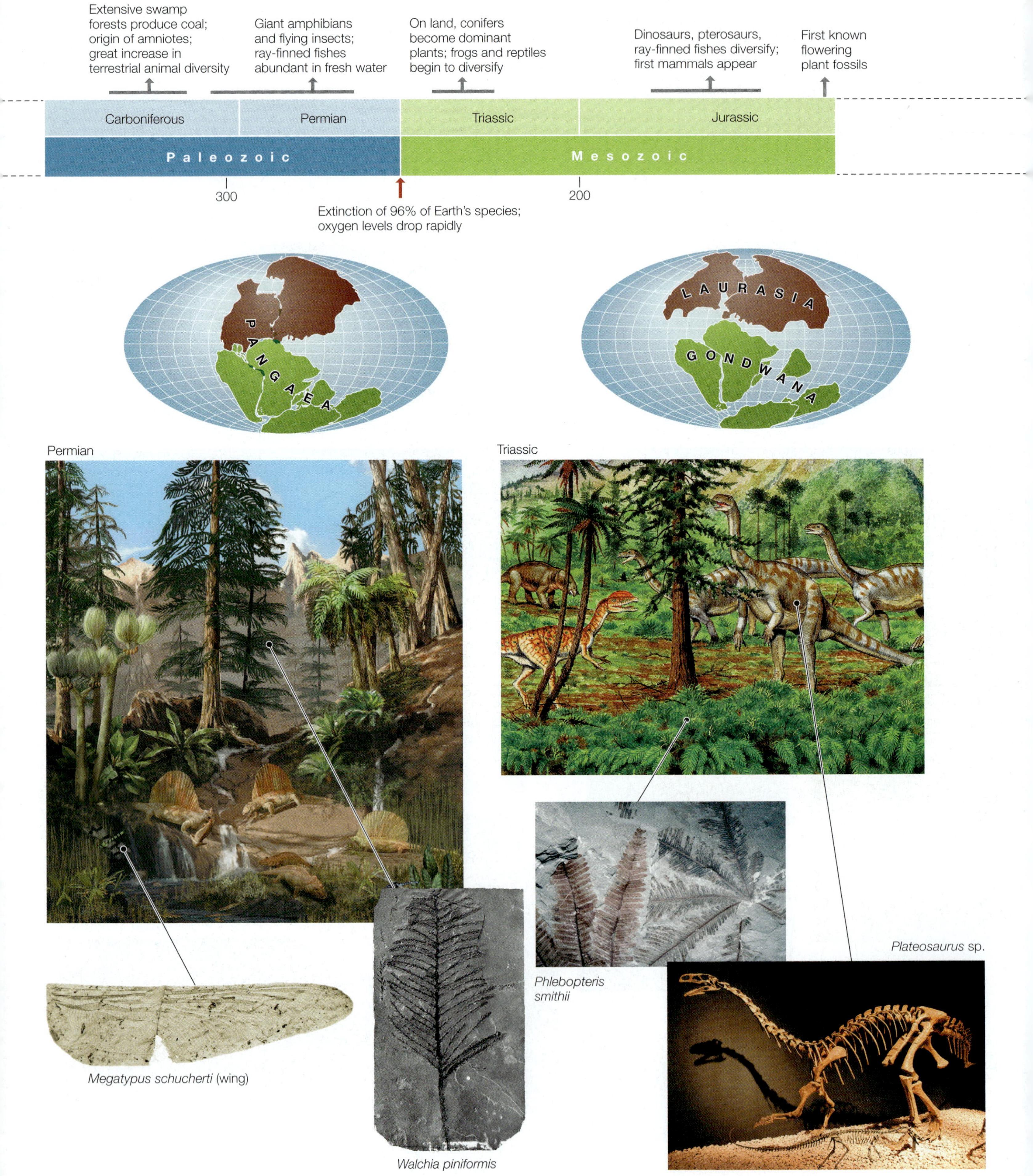

Extensive swamp forests produce coal; origin of amniotes; great increase in terrestrial animal diversity
Giant amphibians and flying insects; ray-finned fishes abundant in fresh water
On land, conifers become dominant plants; frogs and reptiles begin to diversify
Dinosaurs, pterosaurs, ray-finned fishes diversify; first mammals appear
First known flowering plant fossils
Carboniferous
Permian
Triassic
Jurassic
Paleozoic
Mesozoic
300
200
Extinction of 96% of Earth's species; oxygen levels drop rapidly
PANGAEA
LAURASIA
GONDWANA
Permian
Triassic
Megatypus schucherti (wing)
Walchia piniformis
Phlebopteris smithii
Plateosaurus sp.

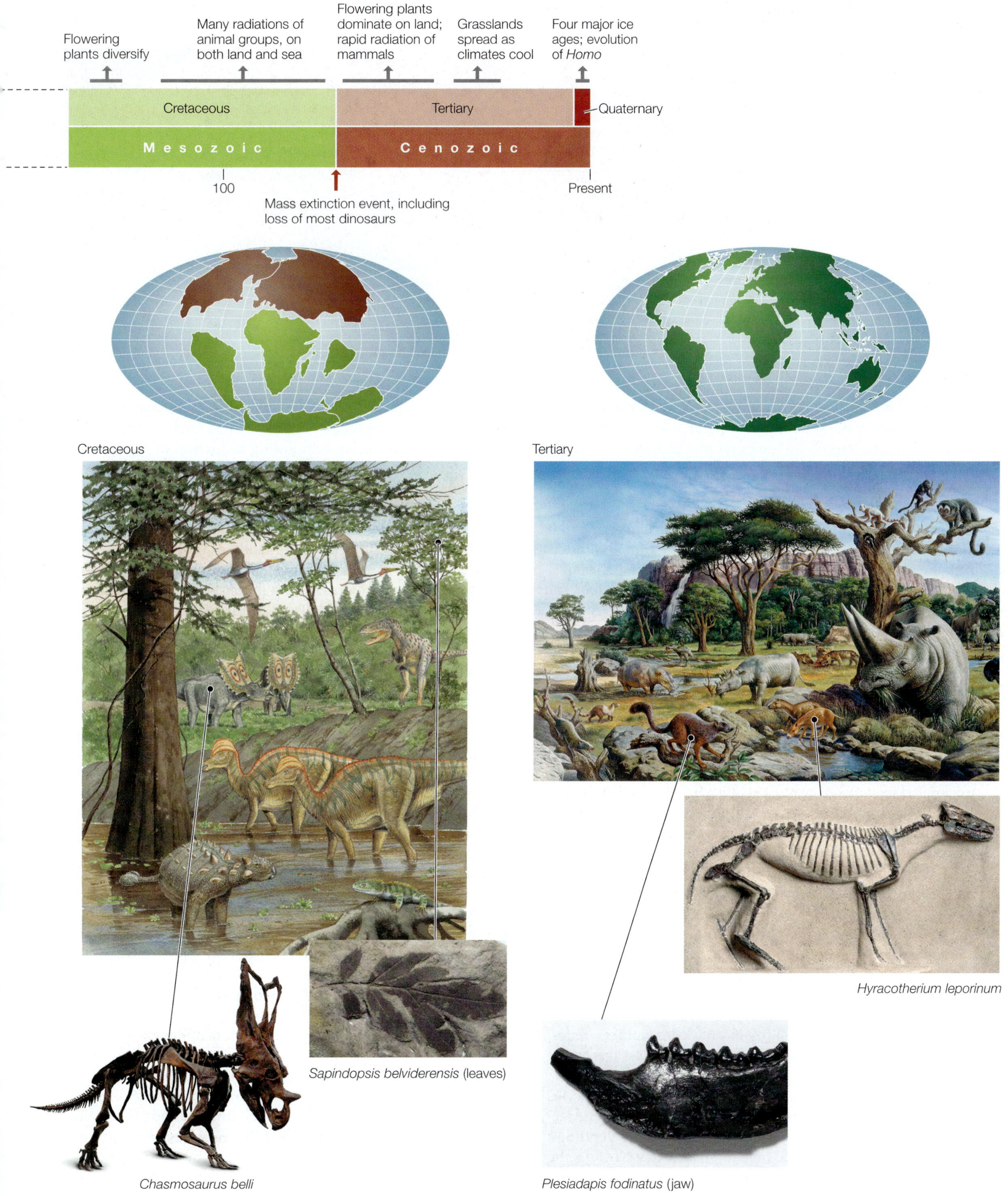

Flowering plants diversify
Many radiations of animal groups, on both land and sea
Flowering plants dominate on land; rapid radiation of mammals
Grasslands spread as climates cool
Four major ice ages; evolution of *Homo*
Cretaceous
Tertiary
Quaternary
Mesozoic
Cenozoic
100
Present
Mass extinction event, including loss of most dinosaurs
Cretaceous
Tertiary
Hyracotherium leporinum
Sapindopsis belviderensis (leaves)
Chasmosaurus belli
Plesiadapis fodinatus (jaw)

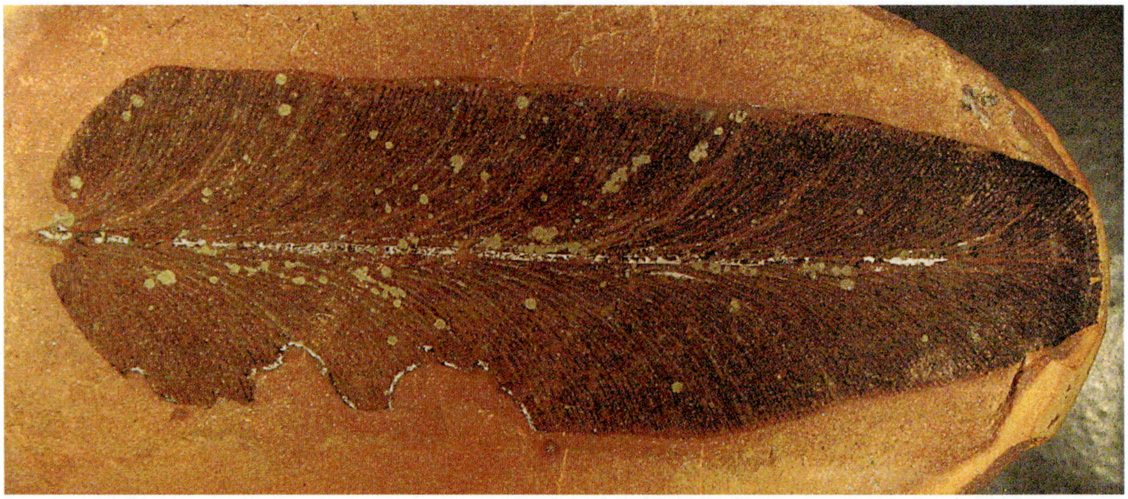

25.15 Evidence of Insect Diversification The margins of this fossil fern leaf from the Carboniferous have been chewed by insects.

from the water. At the end of the Ordovician, as massive glaciers formed over the southern continents, sea levels dropped about 50 meters, and ocean temperatures dropped. About 75 percent of all animal species became extinct, probably because of these environmental changes.

THE SILURIAN (444–416 MYA) During the Silurian period, the continents began to cluster together. Marine life rebounded from the mass extinction at the end of the Ordovician. Animals able to swim in open water and feed above the ocean floor appeared for the first time. Jawless fishes diversified, and the first fishes with supporting rays in their fins appeared. The tropical sea was uninterrupted by land barriers, and most marine organisms were widely distributed. On land, the first vascular plants evolved late in the Silurian (about 420 mya). The first terrestrial arthropods—scorpions and millipedes—evolved at about the same time.

THE DEVONIAN (416–359 MYA) Rates of evolutionary change accelerated in many groups of organisms during the Devonian period. The major land masses continued to move slowly toward each other. In the oceans there were great evolutionary radiations of corals and of shelled, squidlike cephalopod mollusks. Fishes diversified as jawed forms replaced jawless ones and as bony armor gave way to the less rigid scales of modern fishes.

Terrestrial communities changed dramatically during the Devonian. Club mosses, horsetails, and ferns became common; some attained the size of large trees. Their roots accelerated the weathering of rocks, resulting in the development of the first forest soils. The first plants to produce seeds appeared in the Devonian. The earliest fossil centipedes, spiders, mites, and insects date to this period, as do the earliest terrestrial vertebrates.

A mass extinction of about 75 percent of all marine species marked the end of the Devonian. Paleontologists are uncertain about its cause, but two large meteorites that collided with Earth at about that time (one in present-day Nevada, the other in Western Australia) may have been responsible, or at least a contributing factor. The continued coalescence of the continents, with the corresponding reduction in the area of continental shelves, may have also contributed to this mass extinction.

THE CARBONIFEROUS (359–299 MYA) Large glaciers formed over high-latitude portions of the southern land masses during the Carboniferous period, but extensive swamp forests grew on the tropical continents. These forests were dominated by giant tree ferns and horsetails with small leaves. Their fossilized remains formed the coal we now mine for energy. In the seas, crinoids (sea lilies and feather stars) reached their greatest diversity, forming "meadows" on the seafloor.

The diversity of terrestrial animals increased greatly during the Carboniferous. Snails, scorpions, centipedes, and insects were abundant and diverse. Insects evolved wings, becoming the first animals to fly. Flight gave herbivorous insects easy access to tall plants; plant fossils from this period show evidence of chewing by insects (**Figure 25.15**). The terrestrial vertebrates split into two lineages. The amphibians became larger and better adapted to terrestrial existence, while the sister lineage led to the amniotes: vertebrates with well-protected eggs that can be laid in dry places.

THE PERMIAN (299–251 MYA) During the Permian period, the continents coalesced into a single supercontinent called **Pangaea**. Permian rocks contain representatives of many of the major groups of insects we know today. By the end of the period the amniotes had split into two lineages: the reptiles, and a second lineage that would lead to the mammals. Ray-finned fishes became common in the fresh waters of Pangaea.

Toward the end of the Permian, conditions for life deteriorated. Massive volcanic eruptions resulted in outpourings of lava that covered large areas of Earth. The ash and gases produced by the volcanoes blocked sunlight and cooled the climate. The death and decay of the massive Permian forests rapidly used up atmospheric oxygen, and the loss of photosynthetic organisms meant that relatively little new atmospheric oxygen was produced. In addition, much of Pangaea was located close to the South Pole by the end of the Permian. All of these factors combined to produce the most extensive continental glaciers since the "snowball Earth" times of the late Proterozoic. Atmospheric oxygen concentrations gradually dropped from about 30 percent to 15 percent. At such low concentrations, most animals would have been unable to survive at elevations above 500 meters; thus about half of the land area would have been uninhabitable at the end of the Permian. The combination of these changes resulted in the most drastic mass extinction in Earth's history. Scientists estimate that about 96 percent of all multicellular species became extinct at the end of the Permian.

Geographic differentiation increased during the Mesozoic era

The few organisms that survived the Permian mass extinction found themselves in a relatively empty world at the start of the Mesozoic era (251 mya). As Pangaea slowly began to break apart, the biotas of the newly separated continents began to diverge. The oceans rose and once again flooded the continental shelves, forming huge, shallow inland seas. Atmospheric oxygen concentrations gradually rose. Life once again proliferated and diversified, but different groups of organisms came to the fore. The three groups of phytoplankton (photosynthetic floating organisms) that dominate today's oceans—dinoflagellates, coccolithophores, and diatoms—became ecologically important at this time, and their remains are the primary origin of the world's oil deposits. Seed-bearing plants replaced the trees that had ruled the Permian forests.

The Mesozoic era is divided into three periods: the Triassic, Jurassic, and Cretaceous. The Triassic and Cretaceous were terminated by mass extinctions, probably caused by meteorite impacts.

THE TRIASSIC (251–201.6 MYA) Pangaea began to break apart during the Triassic period. Many invertebrate groups diversified, and many burrowing animals evolved from groups living on the surfaces of seafloor sediments. On land, conifers and seed ferns were the dominant trees. The first frogs and turtles appeared. A great radiation of reptiles began, which eventually gave rise to crocodilians, dinosaurs, and birds. The end of the Triassic was marked by a mass extinction that eliminated about 65 percent of the species on Earth.

THE JURASSIC (201.6–145.5 MYA) During the Jurassic period, Pangaea became completely divided into two large continents: **Laurasia**, which drifted northward, and **Gondwana** in the south. Ray-finned fishes rapidly diversified in the oceans. The first lizards appeared, and flying reptiles (pterosaurs) evolved. Most of the large terrestrial predators and herbivores of the period were dinosaurs. Several groups of mammals made their first appearance, and the earliest known fossils of flowering plants are from late in this period.

THE CRETACEOUS (145.5–65.5 MYA) By the early Cretaceous period, Laurasia and Gondwana had begun to break apart into the continents we know today. A continuous sea encircled the tropics. Sea levels were high, and Earth was warm and humid. Life proliferated both on land and in the oceans. Marine invertebrates increased in diversity. On land, the reptile radiation continued as dinosaurs diversified further and the first snakes appeared. Early in the Cretaceous, flowering plants began the radiation that led to their current dominance of the land. By the end of the period, many groups of mammals had appeared.

As described in Section 25.2, another meteorite-caused mass extinction took place at the end of the Cretaceous. In the seas, many planktonic organisms and bottom-dwelling invertebrates became extinct. On land, almost all animals larger than about 25 kg in body weight became extinct. Many species of insects died out, perhaps because the growth of their food plants was greatly reduced following the impact. Some species in northern North America and Eurasia survived in areas that were not subjected to the devastating fires that engulfed most low-latitude regions.

Modern biotas evolved during the Cenozoic era

By the early Cenozoic era (65.5 mya), the positions of the continents resembled those of today, but Australia was still attached to Antarctica, and the Atlantic Ocean was much narrower. The Cenozoic was characterized by an extensive radiation of mammals, but other groups were also undergoing important changes.

Flowering plants diversified extensively and came to dominate world forests except in the coolest regions, where the forests were composed primarily of gymnosperms. Mutations of two genes in one group of plants (the legumes) allowed these plants to use atmospheric nitrogen directly by forming symbioses with a few species of nitrogen-fixing bacteria. The evolution of this symbiosis, which can be thought of as the first "green revolution," dramatically increased the amount of nitrogen available for terrestrial plant growth; this symbiosis remains fundamental to the ecological base of life as we know it today.

The Cenozoic era is divided into the Tertiary and the Quaternary periods, which are commonly subdivided into **epochs** (**Table 25.2**).

TABLE **25.2** Subdivisions of the Cenozoic Era

Period	Epoch	Onset (mya)
Quaternary	Holocene (Recent)	0.01 (~10,000 years ago)
	Pleistocene	2.6
Tertiary	Pliocene	5.3
	Miocene	23
	Oligocene	34
	Eocene	55.8
	Paleocene	65

THE TERTIARY (65.5–2.6 MYA) During the Tertiary period, Australia began its northward drift. By 20 mya it had nearly reached its current position. The early Tertiary was a hot and humid time, and the ranges of many plants shifted into higher latitudes. The tropics were probably too hot to support rainforest vegetation and instead were clothed in low-lying vegetation. In the middle of the Tertiary, however, Earth's climate became considerably cooler and drier. Many lineages of flowering plants evolved herbaceous (nonwoody) forms, and grasslands spread over much of Earth.

By the start of the Cenozoic era, invertebrate faunas had already come to resemble those of today. It is among the terrestrial vertebrates that evolutionary changes during the Tertiary were most rapid. Frogs, snakes, lizards, birds, and mammals all underwent extensive radiations during this period. Three waves of mammals dispersed from Asia to North America across one of the several land bridges that have intermittently connected the two continents during the past 55 million years. Rodents, marsupials, primates, and hoofed mammals appeared in North America for the first time.

THE QUATERNARY (2.6 MYA TO PRESENT) We are living in the Quaternary period. It is commonly subdivided into two epochs, the Pleistocene and the Holocene (also known as the Recent).

The Pleistocene was a time of drastic cooling and climatic fluctuations. During 4 major and about 20 minor "ice ages," massive glaciers spread across the continents, and the ranges of animal and plant populations shifted toward the equator. The last of these glaciers retreated from temperate latitudes less than 15,000 years ago. Organisms are still adjusting to this change. Many high-latitude ecological communities have occupied their current locations for no more than a few thousand years.

It was during the Pleistocene epoch that divergence within one group of mammals, the primates, resulted in the evolution of the hominoid lineage. Subsequent hominoid radiation eventually led to the species *Homo sapiens*—modern humans. Many large bird and mammal species became extinct in Australia and in the Americas when *H. sapiens* arrived on those continents about 45,000 and 15,000 years ago, respectively. Many paleontologists believe these extinctions were the result of hunting and other influences of *Homo sapiens.*

The tree of life is used to reconstruct evolutionary events

The fossil record reveals broad patterns in life's evolution. To reconstruct major events in the history of life, biologists also rely on the phylogenetic information in the tree of life. We can use phylogeny, in combination with the fossil record, to reconstruct the timing of such major events as the acquisition of mitochondria in the ancestral eukaryotic cell, the several independent origins of multicellularity, and the movement of life onto dry land. We can also follow major changes in the genomes of organisms, and we can even reconstruct many gene sequences of species that are long extinct, as described in Section 22.3.

Changes in Earth's physical environment have clearly influenced the diversity of organisms we see on the planet today. To study the evolution of that diversity, biologists examine the evolutionary relationships among species. Deciphering phylogenetic relationships is an important step in understanding how life has diversified on Earth. The next part of this book will explore the major groups of life and the different solutions these groups have evolved to meet major challenges such as reproduction, energy acquisition, dispersal, and escape from predation.

RECAP 25.3

Life evolved in the oceans about 3.8 billion years ago. It diversified as atmospheric oxygen approached its current level. Numerous climate changes and rearrangements of the continents, as well as meteorite impacts, contributed to five major mass extinctions.

- Why have so few of the multitudes of organisms that have existed over millennia become fossilized? **See pp. 514–515**
- What do we mean when we refer to the "Cambrian explosion"? **See p. 516**
- What are the major changes that have occurred in terrestrial biotic communities over the course of the Phanerozoic? **See pp. 516, 520–522 and Figure 25.14**

Can modern experiments test hypotheses about the evolutionary impact of ancient environmental changes?

ANSWER

Several experiments have been conducted to test the link between O_2 concentrations and evolution of body size in flying insects (one of these is discussed in Figure 25.10). Results of these experiments are consistent with the evolution of larger body size in flying insects in hyperoxic (high-oxygen) environments.

Experiments have also been conducted under hypoxic (low-oxygen) conditions, such as existed at the end of the Permian. Results of these experiments suggest that the evolution of body size is constrained under hypoxic conditions, even under strong artificial selection for larger body size. These latter results are consistent with the extinction of many of the large flying insects at the end of the Permian, the result of rapidly decreasing O_2 concentrations. Giant flying insects simply could not have survived the lower O_2 concentrations that existed at that time. The mass extinction at the end of the Permian is the only known mass extinction that involved considerable loss of insect diversity.

CHAPTERSUMMARY 25

25.1 How Do Scientists Date Ancient Events?

- The relative ages of organisms can be determined by the **strata** of **sedimentary rocks** in which their fossils are found.
- Paleontologists use a variety of **radioisotopes** with different **half-lives** to date events at different times in the remote past. Review Figure 25.1
- Geologists divide the history of life into **eons**, **eras**, and **periods**. These divisions are based largely on major differences in the fossil assemblages found in successive layers of rocks. Review Table 25.1

25.2 How Have Earth's Continents and Climates Changed over Time?

- **Plate tectonic** processes result in **continental drift** as well as volcanism and mountain building. Changes in the positions and sizes of the continents affect oceanic circulation patterns, climate, and sea levels. Review Figure 25.3, ANIMATED TUTORIAL 25.1
- Major physical events on Earth, such as the collision of continents that formed the single gigantic land mass **Pangaea**, have affected Earth's surface, climate, and atmosphere. In addition, extraterrestrial events such as meteorite strikes have created sudden and dramatic climate shifts. Some dramatic changes in physical conditions on Earth have caused mass extinctions. Review Figure 25.4
- Oxygen-generating cyanobacteria released enough O_2 to open the door to oxidation reactions in metabolic pathways. Aerobic prokaryotes were able to harvest more energy than anaerobic organisms and began to predominate. Increases in atmospheric O_2 levels also supported the evolution of large eukaryotic cells. Review Figure 25.8

25.3 What Are the Major Events in Life's History?

- Paleontologists use the fossil record and evidence of geological changes to determine what Earth and its **biota** may have looked like at different times. Review Figure 25.12
- During most of its history, life was confined to the oceans. Multicellular life diversified extensively during the **Cambrian explosion**. Review Figure 25.13
- The remaining periods of the Paleozoic era were each characterized by the diversification of specific groups of organisms. The Paleozoic ended with the most drastic mass extinction in Earth's history, at the end of the Permian. Review Figure 25.14
- During the Mesozoic era, distinct terrestrial biotas evolved on each continent. Dinosaurs diversified to become the dominant large predators and herbivores. The era ended with a mass extinction event caused by the collision of a giant meteorite with Earth.
- The Cenozoic era is divided into the Tertiary and the Quaternary periods, which in turn are subdivided into **epochs**. This era saw the emergence of the modern biota as mammals radiated extensively and the angiosperms (flowering plants) became dominant. Review Table 25.2

See ACTIVITY 25.1 for a concept review of this chapter

Go to the Interactive Summary to review key figures, Animated Tutorials, and Activities
Life10e.com/is25

CHAPTERREVIEW

REMEMBERING

1. In undisturbed strata of sedimentary rock, the oldest rocks
   a. lie at the top.
   b. lie at the bottom.
   c. are in the middle.
   d. are distributed among the strata of younger rocks.
   e. None of the above

2. The concentration of oxygen in Earth's atmosphere
   a. has increased steadily over time.
   b. has decreased steadily over time.
   c. has been both higher and lower in the past than at present.
   d. was lower during most of the Permian than at present.
   e. was at its highest levels in the Cambrian.

3. Many of the coal beds we now mine for energy are largely the remains of
   a. plants that grew in swamps during the Carboniferous period.
   b. algae that grew in marshes during the Devonian period.
   c. giant insects and amphibians of the Permian period.
   d. plants that grew in the oceans during the Carboniferous period.
   e. None of the above

4. The mass extinction at the end of the Ordovician period was probably caused by
   a. the collision of Earth with a large meteorite.
   b. massive volcanic eruptions.
   c. massive glaciation on the southern continents and associated climate changes.
   d. the coming together of the continents to form Pangaea.
   e. changes in Earth's orbit.

5. The cause of the mass extinction at the end of the Mesozoic era probably was
   a. continental drift.
   b. the collision of Earth with a large meteorite.
   c. changes in Earth's orbit.
   d. massive glaciation.
   e. changes in the salt concentration of the oceans.

6. Which of the following times was marked by the largest mass extinction of life in the history of Earth?
   a. The end of the Cretaceous
   b. The end of the Devonian
   c. The end of the Permian
   d. The end of the Triassic
   e. The end of the Silurian

UNDERSTANDING & APPLYING

7. Scientists date ancient events using a variety of methods, but nobody was present to witness or record those events. Accepting those dates requires us to understand the accuracy and appropriateness of indirect measurement techniques. What other basic scientific concepts are also based on the results of indirect measurement techniques?

8. Why is it useful to be able to date past events absolutely as well as relatively?

9. What conditions may have favored the evolution of multicellular groups of organisms near the end of the Precambrian?

ANALYZING & EVALUATING

10. The experiment in Figure 25.10 showed that the body size of insects may evolve quickly following changes in atmospheric oxygen concentrations. What other experiments could you devise to test the effects of changing atmospheric oxygen concentrations?

Go to BioPortal at **yourBioPortal.com** for Animated Tutorials, Activities, LearningCurve Quizzes, Flashcards, and many other study and review resources.

26 Bacteria, Archaea, and Viruses

CHAPTER**OUTLINE**

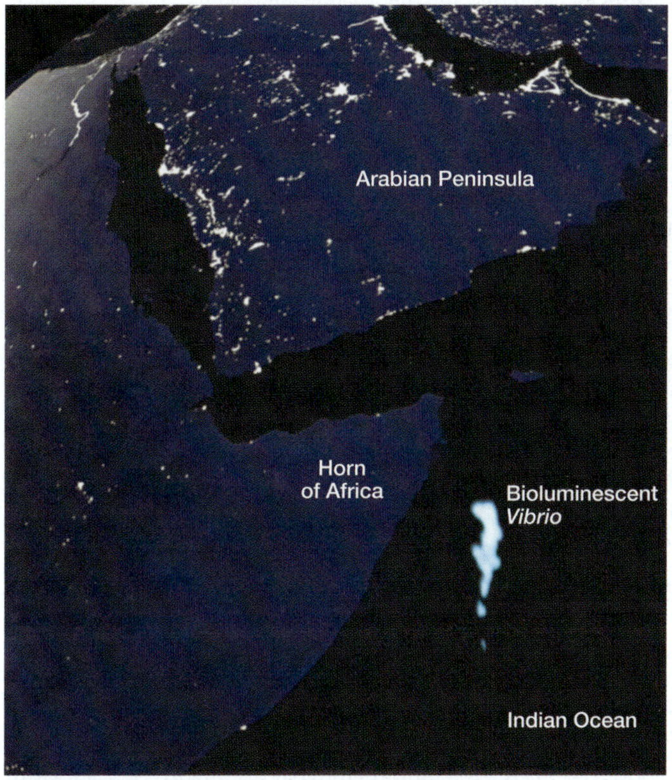
Arabian Peninsula
Horn of Africa
Bioluminescent *Vibrio*
Indian Ocean

Bacteria Seen from Space A satellite image reveals vast expanses of bioluminescence in the Indian Ocean. Spreading over thousands of square kilometers, this glowing "milky sea" is produced by dense populations of *Vibrio* bacteria.

ONE NIGHT IN JANUARY of 1995, the British merchant vessel *Lima* was off the coast of Somalia, near the Horn of Africa. This area is infamous for bands of pirates, so the crew was keeping a watchful eye on the seas when they spotted an eerie, whitish glow on the horizon. It was directly in their path, and there was no way to avoid it. Was this strange sight the result of some strange trick of piracy?

Within 15 minutes of first sighting the glow, the *Lima* was surrounded by shining waters for as far as her crew could see. The ship's log recorded that "it appeared as though the ship was sailing over a field of snow or gliding over the clouds." Fortunately for the crew, the glow had nothing to do with pirates.

For centuries, mariners in this part of the world had reported occasional "milky seas" in which the sea surface produced a strange glow at night, extending from horizon to horizon. Scientists up to that point had never been able to confirm the reality or the cause of such phenomena. It was well established, however, that many organisms emit light by bioluminescence, a complex, enzyme-catalyzed biochemical reaction that emits light but not heat.

What kind of organism could cause the vast expanse of bioluminescence observed by the *Lima*? Some marine organisms emit flashes of light when they are disturbed, but they could not have produced the sustained and uniform glow seen in milky seas. The only organisms known to produce the level of sustained bioluminescence consistent with milky seas are certain prokaryotes, such as bacteria of the genus *Vibrio*. Using information supplied by the *Lima*, biologists scanned satellite images of the Indian Ocean for the specific light wavelengths emitted by *Vibrio*. The satellite images clearly identified thousands of square kilometers of *Vibrio*-produced milky seas.

Vibrio's bioluminescence requires a critical concentration of a specific chemical signal produced by the bacteria, so at low densities, free-living *Vibrio* populations do not glow. But as a colony establishes itself on phytoplankton, the bacteria's population density increases and concentrations of the luminescence signal build up. Eventually bacterial density—and concentrations of the signal—become high enough for the huge colony to produce light at a rate of about 10^3 photons per second per *Vibrio* cell. Such chemical-induced action among bacterial cells is referred to as quorum sensing.

> What adaptive advantage does bioluminescence provide to *Vibrio* bacteria?
>
> See answer on p. 546.

26.1 Where Do Prokaryotes Fit into the Tree of Life?

You may think that you have little in common with a bacterium. But all multicellular eukaryotes—including you—share many attributes with bacteria and archaea, together called **prokaryotes**. For example, all organisms, whether eukaryotes or prokaryotes,

- have plasma membranes and ribosomes (see Chapters 5 and 6).
- have a common set of metabolic pathways (see Chapters 8 and 9).
- replicate DNA semiconservatively (see Chapter 13).
- use DNA as the genetic material to encode proteins, and use similar genetic codes to produce those proteins by transcription and translation (see Chapter 14).

These shared features support the conclusion that all living organisms share a common ancestor. If life had multiple origins, there would be little reason to expect all organisms to use overwhelmingly similar genetic codes or to share structures as unique as ribosomes. Furthermore, similarities in the DNA sequences of universal genes (such as those that encode the structural components of ribosomes) confirm the monophyly of life.

Despite these commonalities, major differences have also evolved across the diversity of life. Based on the differences in cell structure and biochemical functioning that they have observed, many biologists now recognize three domains (primary divisions) of life, two prokaryotic and one eukaryotic (**Figure 26.1**).

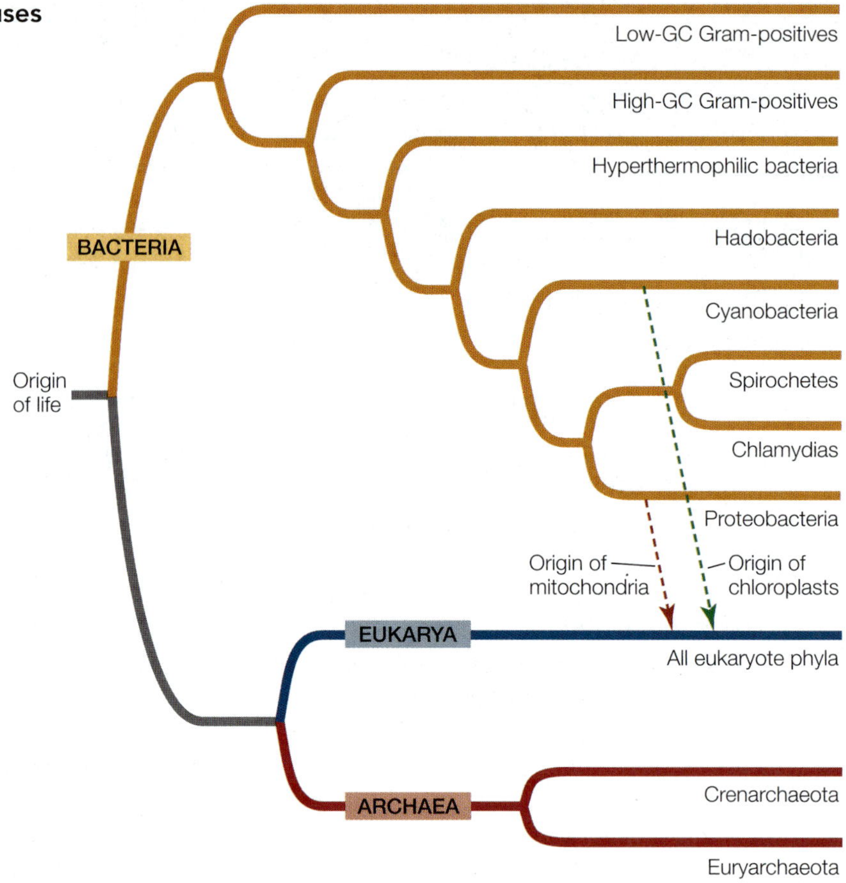

Low-GC Gram-positives
High-GC Gram-positives
Hyperthermophilic bacteria
Hadobacteria
Cyanobacteria
Spirochetes
Chlamydias
Proteobacteria
BACTERIA
Origin of life
Origin of mitochondria
Origin of chloroplasts
EUKARYA
All eukaryote phyla
ARCHAEA
Crenarchaeota
Euryarchaeota

26.1 The Three Domains of the Living World This summary classification of the domains Bacteria and Archaea shows their relationships to each other and to Eukarya. The relationships among the many clades of bacteria, not all of which are listed here, are incompletely resolved at this time.

Go to Animated Tutorial 26.1
The Evolution of the Three Domains
Life10e.com/at26.1

All prokaryotic organisms are unicellular, although they may form large, coordinated colonies or communities consisting of many individuals. The domain Eukarya, by contrast, encompasses both unicellular and multicellular life forms. As we saw in Chapter 5, prokaryotic cells differ from eukaryotic cells in some important ways:

- *Prokaryotic cells do not divide by mitosis.* Instead, after replicating their DNA, prokaryotic cells divide by their own method, binary fission (see Section 11.1).
- *The organization of the genetic material differs.* The DNA of the prokaryotic cell is not organized within a membrane-enclosed nucleus. DNA molecules in prokaryotes are often circular. Many (but not all) prokaryotes have only one main chromosome and are effectively haploid, although many have additional smaller DNA molecules, called plasmids (see Section 12.6).
- *Prokaryotes have none of the membrane-enclosed cytoplasmic organelles that are found in most eukaryotes.* However, the cytoplasm of a prokaryotic cell may contain a variety of infoldings of the plasma membrane and photosynthetic membrane systems that are not found in eukaryotes.

Although the study and classification of eukaryotic organisms goes back centuries, much of our knowledge of the evolutionarily ancient prokaryotic domains is extremely recent. Not until the final quarter of the twentieth century did advances in molecular genetics and biochemistry enable the research that revealed the deep-seated distinctions between the domains Bacteria and Archaea.

The two prokaryotic domains differ in significant ways

A glance at **Table 26.1** will show you that there are major differences between the two prokaryotic domains (most of which cannot be seen even under an electron microscope). In some ways archaea are more like eukaryotes; in other ways they are more like bacteria. (Note that we use lowercase when referring to members of these domains and initial capitals when referring to the domains themselves.) The basic unit of an archaeon (the term for a single archaeal organism) or bacterium (a single bacterial organism) is the prokaryotic cell. Each single-celled prokaryote contains a full complement of genetic and protein-synthesizing systems, including DNA, RNA, and all

TABLE **26.1**
The Three Domains of Life

	Domain		
Characteristic	Bacteria	Archaea	Eukarya
Membrane-enclosed nucleus	Absent	Absent	Present
Membrane-enclosed organelles	Few	Absent	Many
Peptidoglycan in cell wall	Present	Absent	Absent
Membrane lipids	Ester-linked	Ether-linked	Ester-linked
	Unbranched	Branched	Unbranched
Ribosomes[a]	70S	70S	80S
Initiator tRNA	Formylmethionine	Methionine	Methionine
Operons	Yes	Yes	Rare
Plasmids	Yes	Yes	Rare
Number of RNA polymerases[b]	One	One	Three
Ribosomes sensitive to chloramphenicol and streptomycin	Yes	No	No
Ribosomes sensitive to diphtheria toxin	No	Yes	Yes

[a]70S ribosomes are smaller than 80S ribosomes.
[b]The structure of archaeal RNA polymerase is similar to that of eukaryotic polymerases.

the enzymes needed to transcribe and translate genetic information into proteins. The prokaryotic cell also contains at least one system for generating the ATP it needs.

Genetic studies clearly indicate that all three domains of life had a single common ancestor. Across a large portion of their genome, eukaryotes share a more recent common ancestor with archaea than they do with bacteria (see Figure 26.1). However, the mitochondria of eukaryotes (as well as the chloroplasts of photosynthetic eukaryotes, such as plants) originated through endosymbiosis with bacteria (see Section 5.5). Some biologists prefer to view the origin of eukaryotes as a fusion of two equal partners (one ancestor that was related to modern archaea and another that was more closely related to modern bacteria). Others view the divergence of the early eukaryotes from the archaea as an event separate from and earlier than the later endosymbioses. In either case, some eukaryote genes are most closely related to those of archaea, whereas others are most closely related to those of bacteria. The tree of life therefore contains some merging of lineages as well as the predominant divergence of lineages.

The last common ancestor of the three domains probably lived about 3 billion years ago. It probably had DNA as its genetic material, as well as machinery for transcription and translation that produced RNAs and proteins, respectively. This ancestor almost certainly had a circular chromosome. Archaea, Bacteria, and Eukarya are all the products of billions of years of mutation, natural selection, and genetic drift, and they are all well adapted to present-day environments. The earliest prokaryote fossils, which date back at least 3.5 billion years, indicate that there was considerable diversity among the prokaryotes even during those earliest days of life.

The small size of prokaryotes has hindered our study of their evolutionary relationships

Until about 300 years ago, nobody had even *seen* an individual prokaryote; these organisms remained invisible to humans until the invention of the first simple microscope. Prokaryotes are so small, however, that even the best light microscopes don't reveal much about them. It took advanced microscopic equipment and modern molecular techniques to open up the microbial world. (Microscopic organisms—both prokaryotes and eukaryotes—are often collectively referred to as "microbes.")

Before DNA sequencing became practical, taxonomists based prokaryote classification on observable phenotypic characters such as shape, color, motility, nutritional requirements, and sensitivity to antibiotics. One of the characters most widely used to classify prokaryotes is the structure of their cell walls.

The cell walls of almost all bacteria contain **peptidoglycan**, a cross-linked polymer of amino sugars that produces a firm, protective, meshlike structure around the cell. Peptidoglycan is a substance unique to bacteria; its absence from the cell walls of archaea is a key difference between the two prokaryotic domains. Peptidoglycan is also an excellent target for combating pathogenic (disease-causing) bacteria because it has no counterpart in eukaryotic cells. Antibiotics such as penicillin and ampicillin, as well as other agents that specifically interfere with the synthesis of peptidoglycan-containing cell walls, tend to have little, if any, effect on the cells of humans and other eukaryotes.

The **Gram stain** is a technique that can be used to separate most types of bacteria into two distinct groups. A smear of bacterial cells on a microscope slide is soaked in a violet dye and treated with iodine; it is then washed with alcohol and counterstained with a red dye called safranin. **Gram-positive bacteria** retain the violet dye and appear blue to purple (Figure

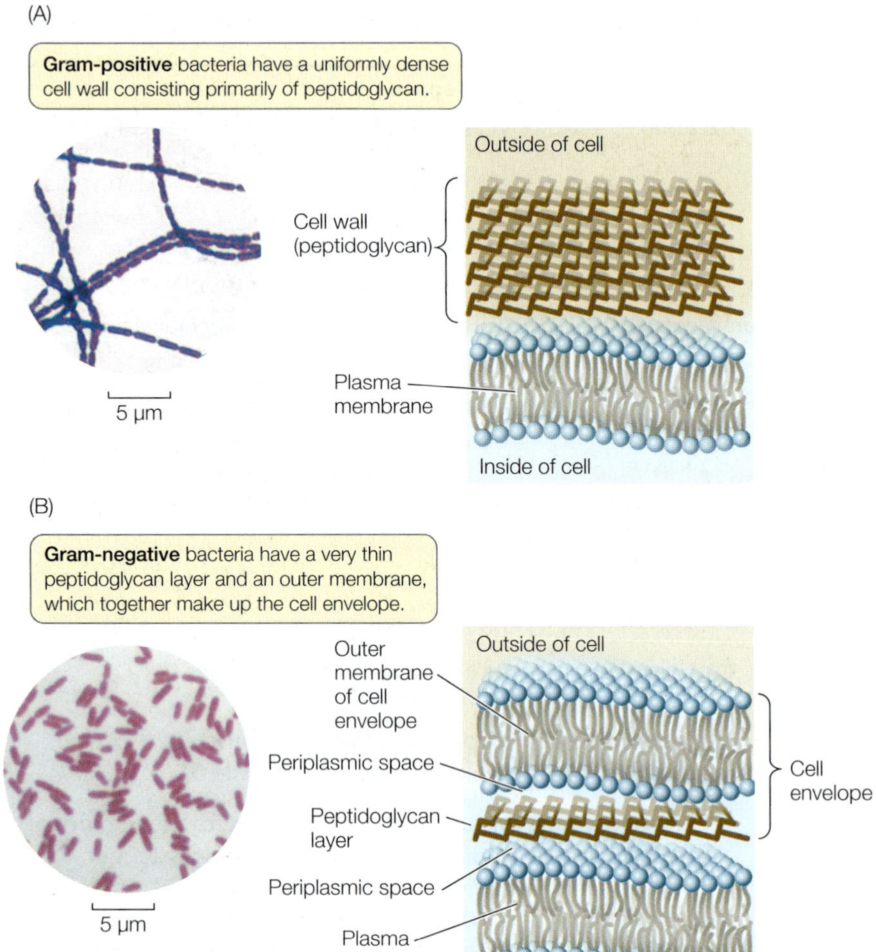
(A)
Gram-positive bacteria have a uniformly dense cell wall consisting primarily of peptidoglycan.
5 µm
Outside of cell
Cell wall (peptidoglycan)
Plasma membrane
Inside of cell
(B)
Gram-negative bacteria have a very thin peptidoglycan layer and an outer membrane, which together make up the cell envelope.
5 µm
Outside of cell
Outer membrane of cell envelope
Periplasmic space
Peptidoglycan layer
Periplasmic space
Plasma membrane
Cell envelope
Inside of cell

Go to Activity 26.1 Gram Stain and Bacteria
Life10e.com/ac26.1

26.2 The Gram Stain and the Bacterial Cell Wall When treated with Gram-staining reagents, the cell walls of bacteria react in one of two ways. (A) Gram-positive bacteria have a thick peptidoglycan cell wall that retains the violet dye and appears deep blue or purple. (B) Gram-negative bacteria have a thin peptidoglycan layer that does not retain the violet dye, but picks up the counterstain and appears pink to red.

26.2A). The alcohol washes the violet stain out of **Gram-negative bacteria**, which then pick up the safranin counterstain and appear pink to red (Figure 26.2B). For most bacteria, the effect of the Gram stain is determined by the chemical structure of the cell wall:

- A *Gram-negative cell wall* usually has a thin peptidoglycan layer, which is surrounded by a second, outer membrane quite distinct in chemical makeup from the plasma membrane. Together the cell wall and the outer membrane are called the cell envelope. Between the plasma membrane and the outer membrane is a **periplasmic space**. This space contains proteins that are important in digesting some materials, transporting others, and detecting chemical gradients in the environment.
- *A Gram-positive cell wall* usually has about five times as much peptidoglycan as a Gram-negative cell wall. Its thick peptidoglycan layer is a meshwork that may serve some of the same purposes as the periplasmic space of the Gram-negative cell envelope.

Shape is another phenotypic characteristic that is useful for the basic identification of bacteria. The three most common shapes are spheres, rods, and spiral forms (Figure 26.3). Many bacterial names are based on these shapes. A spherical bacterium is called a **coccus** (plural *cocci*). Cocci may live singly or may associate in two- or three-dimensional arrays such as chains, plates, blocks, or clusters of cells. A rod-shaped bacterium is called a **bacillus** (plural *bacilli*). A spiral bacterium (shaped like a corkscrew) is called a **spirillum** (plural *spirilla*). Bacilli and spirilla may be single, form chains, or gather in regular clusters. Among the other bacterial shapes are long filaments and branched filaments.

Less is known about the shapes of archaea because many of these organisms have never been seen. Many archaea are known only from samples of DNA from the environment. However, the species whose morphologies are known include cocci, bacilli, and even triangular and square species; the last grow on surfaces, arranged like sheets of postage stamps.

The nucleotide sequences of prokaryotes reveal their evolutionary relationships

Analyses of the nucleotide sequences of ribosomal RNA (rRNA) genes provided the first comprehensive evidence of evolutionary relationships among prokaryotes. Comparisons of rRNA genes from a great many organisms have revealed probable phylogenetic relationships throughout the tree of life. Databases such as GenBank contain rRNA gene sequences from hundreds of thousands of species—more than any other type of gene sequence.

For several reasons, rRNA is particularly useful for phylogenetic studies of living organisms:

- rRNA was present in the common ancestor of all life and is therefore evolutionarily ancient.
- No free-living (i.e., not parasitic) organism lacks rRNA, so rRNA genes can be compared across the tree of life.
- rRNA plays a critical role in translation in all organisms, so lateral transfer of rRNA genes among distantly related species is unlikely.
- rRNA has evolved slowly enough that gene sequences from even distantly related species can be aligned and analyzed.

Although studies of rRNA genes reveal much about the evolutionary relationships of prokaryotes, they don't always reveal the entire evolutionary history of these organisms. In some groups of prokaryotes, analyses of multiple gene sequences

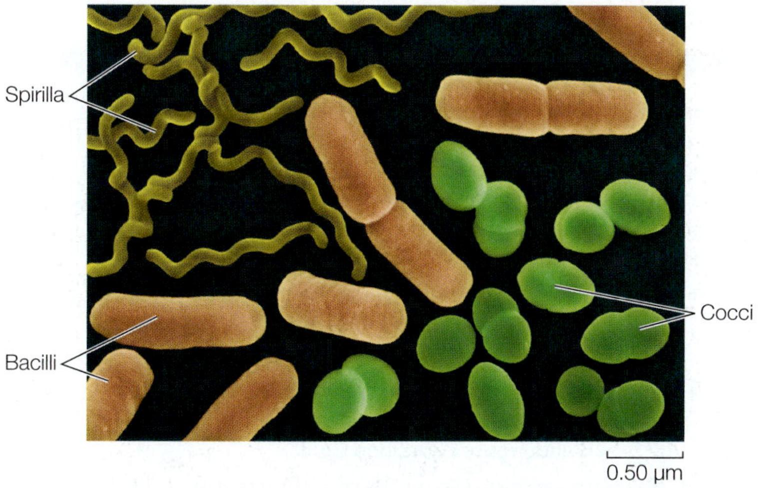
Spirilla
Cocci
Bacilli
0.50 μm

26.3 Bacterial Cell Shapes This composite, colorized micrograph shows the three most common bacterial shapes. Spherical cells are called cocci; those pictured are a species of *Enterococcus* from the mammalian gut. Rod-shaped cells are called bacilli; these *Escherichia coli* also reside in the gut. The helix-shaped spirilla are *Leptospira interrogans*, a human pathogen.

have suggested several different phylogenetic patterns. How could such differences among gene sequences arise in the same organisms? Studies of whole prokaryotic genomes have revealed that even distantly related prokaryotes sometimes exchange genetic material.

Lateral gene transfer can lead to discordant gene trees

As noted earlier, prokaryotes reproduce by binary fission. If we could follow these divisions back through evolutionary time, we would be tracing the complete tree of life for bacteria and archaea. This underlying tree of evolutionary relationships (represented in highly abbreviated form in Appendix A) is called the organismal (or species) tree. Because binary fission is an asexual process that replicates whole genomes, we would expect phylogenetic trees constructed from most gene sequences (see Chapter 22) to reflect these same relationships.

There are other processes, however—including transformation, conjugation, and transduction—that allow the exchange of genetic information between some prokaryotes without reproduction. Thus prokaryotes can exchange and recombine their DNA with that of other individuals (this is sex in the genetic sense of the word), but this genetic exchange is not directly linked to reproduction, as it is in most eukaryotes.

From early in evolution to the present day, some genes have been moving "sideways" from one prokaryote species to another, a phenomenon known as **lateral gene transfer**. Lateral gene transfers are well documented among closely related species, and some have been documented even across the domains of life. Consider, for example, the genome of *Thermotoga maritima*, a bacterium that can survive extremely high temperatures. By comparing the 1,869 gene sequences of *T. maritima* with sequences encoding the same proteins in other species, investigators found that some of this bacterium's genes have their closest relationships not with the genes of other bacterial species, but with the genes of archaea that live in similar extreme environments.

When genes involved in lateral transfer events are sequenced and analyzed, the resulting *gene trees* will not match the organismal tree in every respect (**Figure 26.4**). The gene trees will vary because the history of lateral transfer events may be different for different genes. Biologists can reconstruct the underlying organismal phylogeny by comparing multiple genes to produce a consensus tree, or by concentrating only on genes that are unlikely to be involved in lateral gene transfer

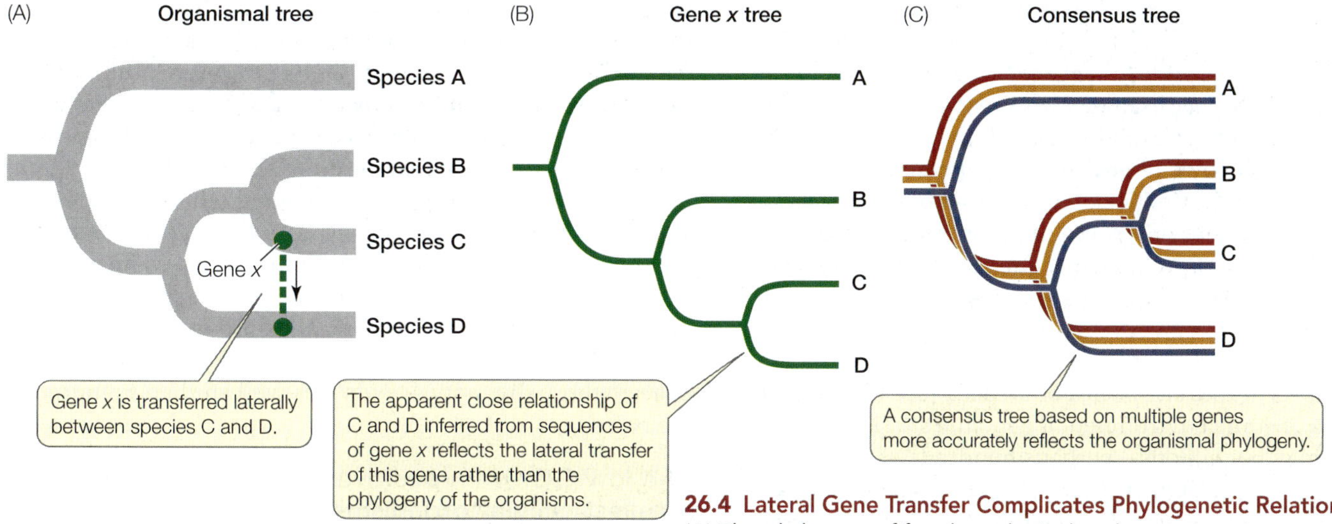
(A)
Organismal tree
Species A
Species B
Species C
Gene x
Species D
Gene x is transferred laterally between species C and D.
(B)
Gene x tree
A
B
C
D
The apparent close relationship of C and D inferred from sequences of gene x reflects the lateral transfer of this gene rather than the phylogeny of the organisms.
(C)
Consensus tree
A
B
C
D
A consensus tree based on multiple genes more accurately reflects the organismal phylogeny.

26.4 Lateral Gene Transfer Complicates Phylogenetic Relationships (A) The phylogeny of four hypothetical prokaryote species, two of which have been involved in a lateral transfer of gene *x*. (B) A tree based only on gene *x* shows the phylogeny of the laterally transferred gene, rather than the organismal phylogeny. (C) A consensus tree based on multiple genes is more likely to reflect the true organismal phylogeny, especially if those genes come from a stable core of genes involved in fundamental processes.

events. For example, genes that are involved in fundamental cellular processes (such as the rRNA genes discussed above) are unlikely to be replaced by the same genes from other species because functional, locally adapted copies of these genes are already present.

What kinds of genes are most likely to be involved in lateral gene transfer? Genes that result in a new adaptation that confers higher fitness on a recipient species are most likely to be retained. For example, genes that produce antibiotic resistance are often transferred among bacterial species on plasmids, especially under strong selection pressure such as that imposed by modern antibiotic medications. Improper or overly frequent use of antibiotics can select for resistant strains of pathogenic bacteria that are much harder to treat. This phenomenon explains why informed physicians have become more careful in prescribing antibiotics.

It is debatable whether lateral gene transfer has seriously complicated our attempts to resolve the tree of prokaryotic life. Recent work suggests that it has not; although lateral gene transfer complicates studies in some individual species (and makes the boundaries of "species" of bacteria difficult to determine), it need not present problems at higher taxonomic levels. It is now possible to make nucleotide sequence comparisons of entire genomes, and these studies are revealing a stable core of fundamental genes whose phylogenies are uncomplicated by lateral gene transfer. Gene trees based on this stable core more accurately reveal the organismal phylogeny. The problem that remains is that only a very small proportion of the prokaryotic world has been described and studied.

The great majority of prokaryote species have never been studied

Most prokaryotes have defied all attempts to grow them in pure culture, causing biologists to wonder how many species, and possibly even large groups of species, we might be missing. A window onto this problem was opened with the introduction of a new way of examining nucleic acid sequences. When biologists are unable to work with the whole genome of a single prokaryote species, they can instead examine individual genes collected from a random sample of the environment. This technique is known as **environmental genomics**.

Biologists now routinely isolate gene sequences, or even whole genomes, from environmental samples such as soil and seawater. Comparing such sequences with previously known ones has revealed that an extraordinary number of the sequences represent new, previously unrecognized species. Biologists have described only about 10,000 species of bacteria and only a few hundred species of archaea (see Figure 1.7). The results of some environmental genomic studies suggest that there may be millions—perhaps hundreds of millions—of prokaryote species. Other biologists put the estimate much lower, arguing that the high dispersal ability of many bacterial species greatly reduces endemism (i.e., the number of species restricted to a small geographic area). Only the magnitude of these estimates differs, however; all sides agree that we have just begun to uncover Earth's bacterial and archaeal diversity.

RECAP 26.1

Bacteria and Archaea are distinct prokaryotic domains of the tree of life. The small size of prokaryotes, combined with their potential for lateral gene transfer, has hindered our ability to understand their evolutionary relationships. Environmental genomic studies have suggested a much higher diversity of prokaryotes than previous studies had revealed.

- What findings led to the establishment of Bacteria and Archaea as separate domains? **See pp. 526–527 and Table 26.1**
- How did biologists classify bacteria before it became possible to determine their nucleotide sequences? **See pp. 527–528 and Figures 26.2 and 26.3**
- Why are nucleotide sequences of rRNA genes particularly useful for phylogenetic studies of prokaryotes? **See p. 528**
- How does lateral gene transfer complicate phylogenetic studies of prokaryotes? **See pp. 528–529 and Figure 26.4**

Despite the challenges of reconstructing prokaryote phylogeny, taxonomists are starting to establish evolutionary classification systems for these organisms. With the understanding that new information will necessitate periodic revision of these classifications, we next apply one current system to our discussion of prokaryote diversity.

26.2 Why Are Prokaryotes So Diverse and Abundant?

The prokaryotes were alone on Earth for a very long time, adapting to new and changing environments. They are still with us, in massive numbers and incredible diversity, and they are found everywhere. If success is measured by numbers of individuals, the prokaryotes are the most successful organisms on Earth. Individual bacteria and archaea in the oceans have been estimated to number more than 3×10^{28}—perhaps 100 million times more than the number of stars in the visible universe. Closer to home, the individual bacteria living in your intestinal tract outnumber all the humans who have ever lived.

Given our still-fragmentary knowledge of prokaryote diversity, it is not surprising that there are several different hypotheses about the relationships of the major groups of prokaryotes. In this book we use a widely accepted classification system that has considerable support from nucleotide sequence data. In this section we will discuss the eight bacterial groups that have the broadest phylogenetic support and have received the most study (see Figure 26.1). We will then describe the archaea, whose diversity is even less well studied than that of the bacteria.

The low-GC Gram-positives include some of the smallest cellular organisms

The **low-GC Gram-positives**, also known as Firmicutes, derive the first part of their name from the relatively low ratio of G-C to A-T nucleotide base pairs in their DNA. The second part of their name is less accurate: some of the low-GC Gram-positives

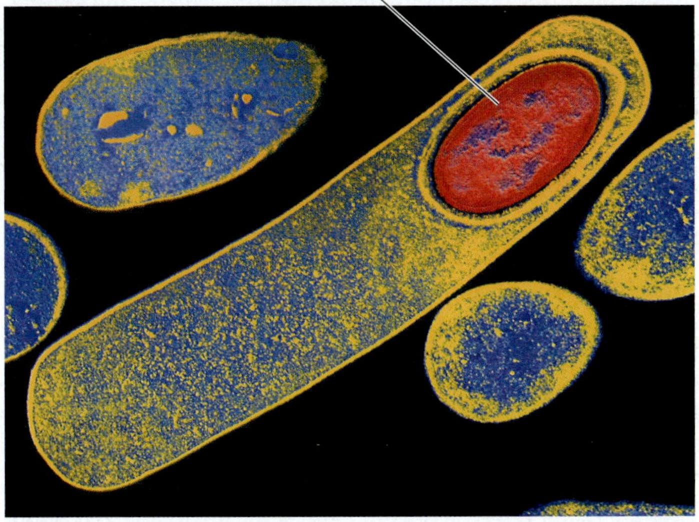
Endospore
Clostridium difficile
0.3 μm

26.5 A Structure for Waiting Out Bad Times Under harsh conditions, some low-GC Gram-positive bacteria can replicate their DNA and encase it in an endospore. The parent cell then breaks down, and the endospore survives in a dormant state until conditions improve.

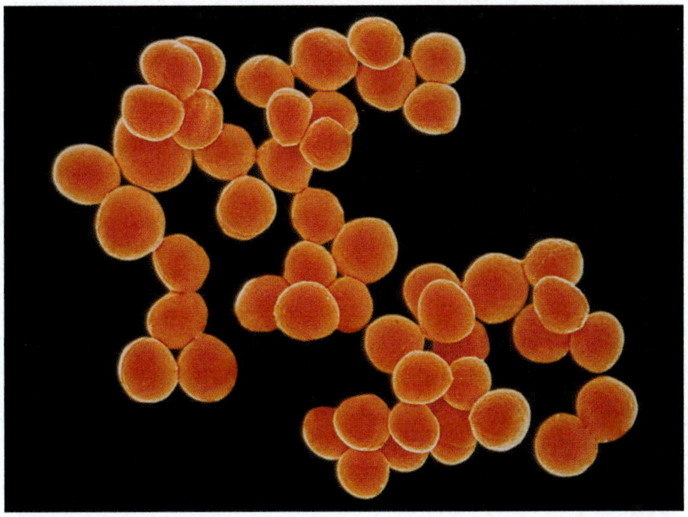
Staphylococcus aureus
1 μm

26.6 Staphylococci "Grape clusters" are the usual arrangement of these low-GC Gram-positive coccal bacteria, often the cause of skin or wound infections.

are in fact Gram-negative, and some have no cell wall at all. Despite these differences, phylogenetic analyses of DNA sequences support the monophyly of this bacterial group.

One group of low-GC Gram-positives can produce resting structures called **endospores** (Figure 26.5). When a key nutrient such as nitrogen or carbon becomes scarce, the bacterium replicates its DNA and encapsulates one copy, along with some of its cytoplasm, in a tough cell wall heavily thickened with peptidoglycan and surrounded by a spore coat. The parent cell then breaks down, releasing the endospore. Endospore production is not a reproductive process; the endospore merely replaces the parent cell. The endospore, however, can survive harsh environmental conditions that would kill the parent cell, such as high or low temperatures or drought, because it is dormant—its normal metabolic activity is suspended. Later, if it encounters favorable conditions, the endospore becomes metabolically active and divides, forming new cells that are descendants of the parent cell. Members of this endospore-forming group of low-GC Gram-positives include the many species of *Clostridium* and *Bacillus*. Some of their endospores can be reactivated after more than a thousand years of dormancy. There are even credible claims of reactivation of *Bacillus* endospores millions of years old.

Endospores of *Bacillus anthracis* are the cause of anthrax. Anthrax is primarily a disease of cattle and sheep, but it can be fatal in humans. When the endospores sense macrophages in mammalian blood, they reactivate and release toxins into the bloodstream. *Bacillus anthracis* has been used as a bioterrorism agent because it is relatively easy to transport large quantities of its endospores and release them among human populations, where they may be inhaled or ingested.

Low-GC Gram-positives of the genus *Staphylococcus*—the **staphylococci** (Figure 26.6)—are abundant on the human body surface; they are responsible for boils and many other skin problems. *Staphylococcus aureus* is the best-known human pathogen in this genus; it is present in 20 to 40 percent of normal adults (and in 50 to 70 percent of hospitalized adults). In addition to skin diseases, *S. aureus* can cause respiratory, intestinal, and wound infections.

Another interesting group of low-GC Gram-positives, the **mycoplasmas**, lack cell walls, although some have a stiffening material outside the plasma membrane. The mycoplasmas are among the smallest cellular organisms known (Figure 26.7). The smallest mycoplasmas have a diameter of about 0.2 μm. They are small in another crucial sense as well: they have less than half as much DNA as most other prokaryotes. It has been speculated

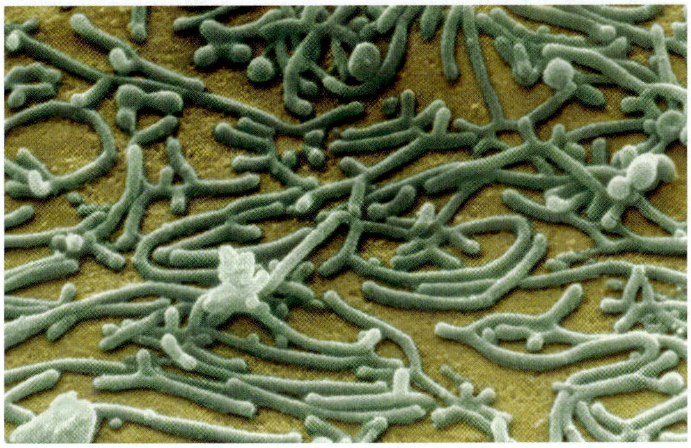
Mycoplasma sp.
0.7 μm

26.7 Tiny Cells With about one-fifth as much DNA as *E. coli*, mycoplasmas are among the smallest known bacteria.

that the DNA in a mycoplasma, which codes for fewer than 500 proteins, may be close to the minimum amount required to encode the essential properties of a living cell (see Figure 17.6).

Some high-GC Gram-positives are valuable sources of antibiotics

High-GC Gram-positives, also known as actinobacteria, have a higher ratio of G–C to A–T nucleotide base pairs than do the low-GC Gram-positives. These bacteria develop an elaborately branched system of filaments (**Figure 26.8**) that resembles the filamentous growth habit of fungi, albeit at a smaller scale. Some high-GC Gram-positives reproduce by forming chains of spores at the tips of the filaments. In species that do not form spores, the branched, filamentous growth ceases and the structure breaks up into typical cocci or bacilli, which then reproduce by binary fission.

The high-GC Gram-positives include several medically important bacteria. *Mycobacterium tuberculosis* causes tuberculosis, which kills 3 million people each year. Genetic data suggest that this bacterium arose 3 million years ago in East Africa, making it the oldest known human bacterial pathogen. The genus *Streptomyces* produces streptomycin as well as hundreds of other antibiotics. We derive most of our antibiotics from members of the high-GC Gram-positives.

Hyperthermophilic bacteria live at very high temperatures

Several lineages of bacteria and archaea are **extremophiles**: they thrive under extreme conditions that would kill most other organisms. The hyperthermophilic bacteria, for example, are thermophiles (Greek, "heat-lovers"). Genera such as *Aquifex* live near volcanic vents and in hot springs, sometimes at temperatures near the boiling point of water. Some species of *Aquifex* need only hydrogen, oxygen, carbon dioxide, and mineral salts to live and grow. Species of the genus *Thermotoga* live deep underground in oil reservoirs as well as in other high-temperature environments.

Biologists have hypothesized that high temperatures characterized the ancestral conditions for life, given that most environments on early Earth were much hotter than those of today. Reconstructions of ancestral bacterial genes have supported this hypothesis by showing that the ancestral sequences functioned best at elevated temperatures. The monophyly of the hyperthermophilic bacteria, however, is not well established.

Hadobacteria live in extreme environments

The **hadobacteria**, including such genera as *Deinococcus* and *Thermus*, are another group of thermophilic extremophiles. The group's name is derived from Hades, the ancient Greek name for the underworld. *Deinococcus* are resistant to radiation and can consume nuclear waste and other toxic materials. They can also survive extremes of cold as well as hot temperatures. Another member of this group, *Thermus aquaticus*, was the source of the thermally stable DNA polymerase that was critical for the development of the polymerase chain reaction. *Thermus aquaticus* was originally isolated from a hot spring, but it can be found wherever hot water occurs (including many residential hot water heaters).

Cyanobacteria were the first photosynthesizers

Cyanobacteria, sometimes called *blue-green bacteria* because of their pigmentation, are photosynthetic. They use chlorophyll *a* for photosynthesis and release oxygen gas (O_2); many species also fix nitrogen. The production of oxygen by these bacteria transformed the atmosphere of early Earth (see Section 25.2).

Cyanobacteria carry out the same type of photosynthesis that is characteristic of eukaryotic photosynthesizers. They contain elaborate and highly organized internal membrane systems called **photosynthetic lamellae**. As mentioned in Section 26.1, the chloroplasts of photosynthetic eukaryotes are derived from an endosymbiotic cyanobacterium.

Cyanobacteria may live free as single cells or associate in multicellular colonies. Depending on the species and on growth conditions, these colonies may range from flat sheets one cell thick to filaments to spherical balls of cells. Some filamentous colonies of cyanobacteria differentiate into three specialized cell types: vegetative cells, spores, and heterocysts (**Figure 26.9**). **Vegetative cells** photosynthesize, **spores** are resting stages that can survive harsh environmental conditions and eventually develop into new filaments, and **heterocysts** are cells specialized for nitrogen fixation. All of the known cyanobacteria with heterocysts fix nitrogen. Heterocysts also have a role in reproduction: when filaments break apart to reproduce, the heterocyst may serve as a breaking point.

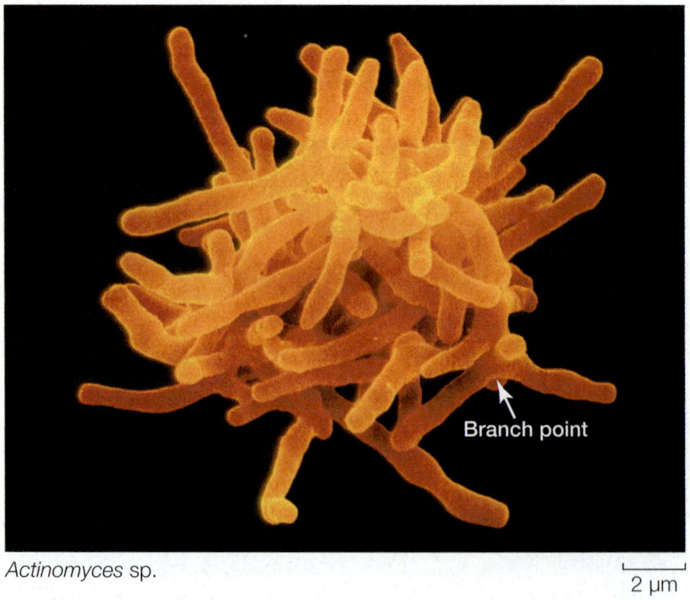
Branch point
Actinomyces sp.
2 μm

26.8 Actinobacteria Are High-GC Gram-Positives The tangled, branching filaments seen in this scanning electron micrograph are typical of this medically important bacterial group.

Go to Media Clip 26.1
Cyanobacteria
Life10e.com/mc26.1

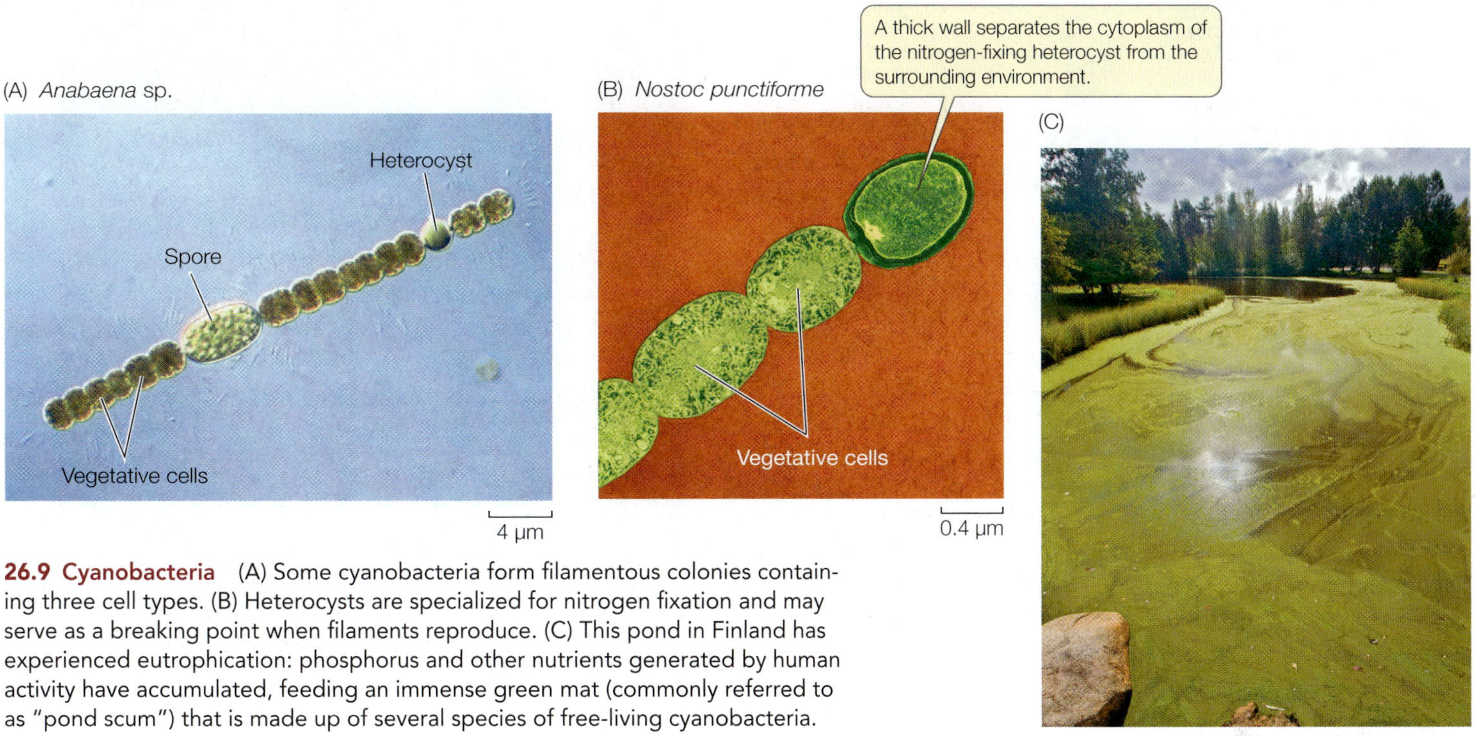
(A) Anabaena sp.
Heterocyst
Spore
Vegetative cells
4 μm
(B) Nostoc punctiforme
A thick wall separates the cytoplasm of the nitrogen-fixing heterocyst from the surrounding environment.
Vegetative cells
0.4 μm
(C)

26.9 Cyanobacteria (A) Some cyanobacteria form filamentous colonies containing three cell types. (B) Heterocysts are specialized for nitrogen fixation and may serve as a breaking point when filaments reproduce. (C) This pond in Finland has experienced eutrophication: phosphorus and other nutrients generated by human activity have accumulated, feeding an immense green mat (commonly referred to as "pond scum") that is made up of several species of free-living cyanobacteria.

Spirochetes move by means of axial filaments

Spirochetes are Gram-negative, motile bacteria characterized by unique structures called axial filaments (**Figure 26.10A**), which are modified flagella running through the periplasmic space. The cell body is a long cylinder coiled into a helix (**Figure 26.10B**). The axial filaments begin at either end of the cell and overlap in the middle. Motor proteins connect the axial filaments to the cell wall, enabling the corkscrew-like movement of the bacterium. Many spirochetes are parasites of humans; a few are pathogens, including those that cause syphilis and Lyme disease. Others live free in mud or water.

Chlamydias are extremely small parasites

Chlamydias are among the smallest bacteria (0.2–1.5 μm in diameter). These tiny, Gram-negative cocci can live only as parasites in the cells of other organisms. It was once believed that their obligate parasitism resulted from an inability to produce ATP—that chlamydias were "energy parasites." However, genome sequencing indicates that chlamydias have the genetic capacity to produce at least some ATP. They can augment this capacity by using an enzyme called a translocase, which allows them to take up ATP from the cytoplasm of their host in exchange for ADP from their own cells.

Chlamydias are unique among prokaryotes because of their complex life cycle, which involves two different forms of cells, elementary bodies and reticulate bodies (**Figure 26.11**). Various strains of chlamydias cause eye infections, sexually transmitted diseases, and some forms of pneumonia in humans.

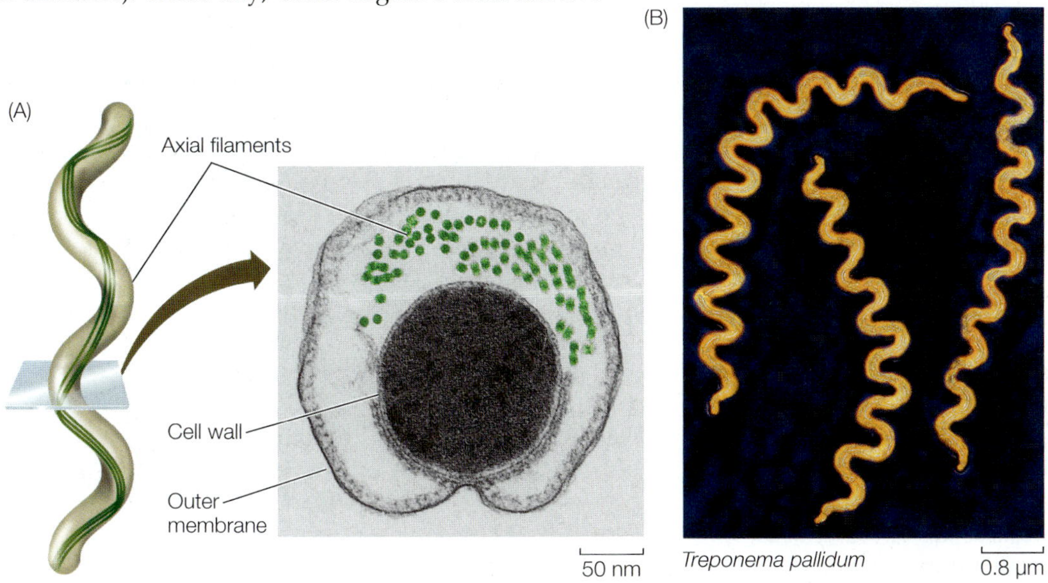
(A)
Axial filaments
Cell wall
Outer membrane
50 nm
(B)
Treponema pallidum
0.8 μm

26.10 Spirochetes Get Their Shape from Axial Filaments
(A) A spirochete from the gut of a termite, seen in cross section, shows the axial filaments these helical prokaryotes use to produce a corkscrew-like movement. (B) This spirochete species causes syphilis in humans.

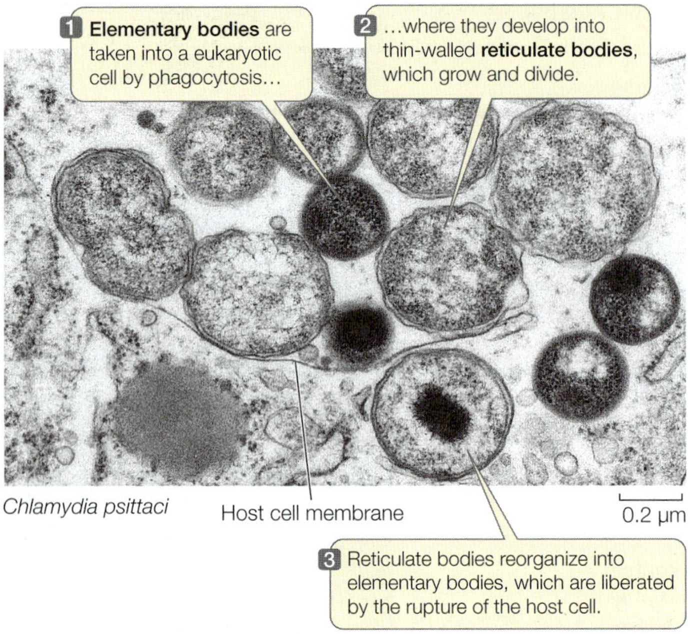
1 Elementary bodies are taken into a eukaryotic cell by phagocytosis...
2 ...where they develop into thin-walled reticulate bodies, which grow and divide.
Chlamydia psittaci
Host cell membrane
0.2 µm
3 Reticulate bodies reorganize into elementary bodies, which are liberated by the rupture of the host cell.

26.11 Chlamydias Change Form Elementary bodies and reticulate bodies are the two cell forms of the chlamydia life cycle.

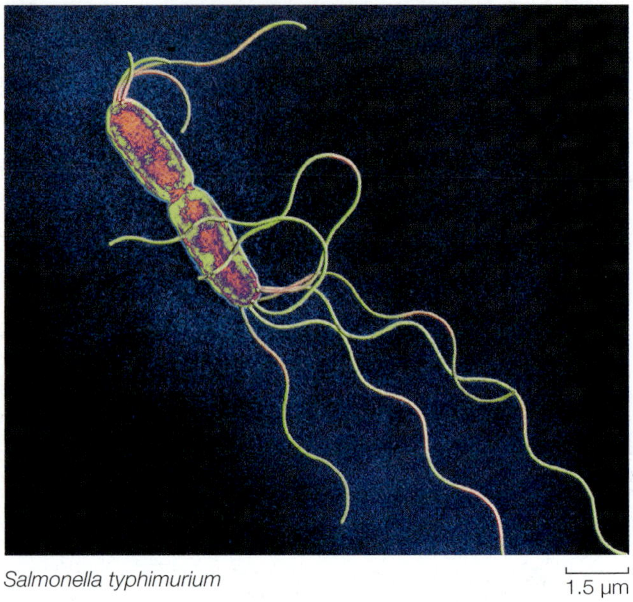
Salmonella typhimurium
1.5 µm

26.12 Proteobacteria Include Human Pathogens These conjugating cells of *Salmonella typhimurium* are exchanging genetic material. This pathogen causes a wide range of gastrointestinal illnesses in humans.

The proteobacteria are a large and diverse group

By far the largest bacterial group, in terms of numbers of described species, is the **proteobacteria**. The proteobacteria include many species of Gram-negative photoautotrophs that use light-driven reactions to metabolize sulfur, as well as dramatically diverse bacteria that bear no phenotypic resemblance to the photoautotrophic species. Genetic and morphological evidence indicates that the mitochondria of eukaryotes were derived from a proteobacterium by endosymbiosis.

Among the proteobacteria are some nitrogen-fixing genera such as *Rhizobium*, and other bacteria that contribute to the global nitrogen and sulfur cycles. *Escherichia coli*, one of the most studied organisms on Earth, is a proteobacterium. So too are many of the most famous human pathogens, such as *Yersinia pestis* (the cause of bubonic plague), *Vibrio cholerae* (cholera), and *Salmonella typhimurium* (gastrointestinal disease; **Figure 26.12**).

Go to Media Clip 26.2
A Swarm of *Salmonella*
Life10e.com/mc26.2

The bioluminescent *Vibrio* we discussed at the opening of this chapter are also proteobacteria. There are many potential applications of the genes that encode bioluminescent proteins in bacteria; these genes are already being inserted into the genomes of other species, with the resulting bioluminescence used as a marker of gene expression. Futuristic proposals for making use of bioluminescence in bioengineered organisms include crop plants that glow when they become water-stressed and need to be irrigated; and glowing trees that could illuminate highways, replacing electric lights.

Although most plant diseases are caused by fungi and viruses, about 200 known plant diseases are of bacterial origin. *Crown gall*, with its characteristic tumors, is one of the most striking (**Figure 26.13**). The causal agent of crown gall is *Agrobacterium tumefaciens*, a proteobacterium that harbors a plasmid often used in recombinant DNA studies as a vehicle for inserting genes into new plant hosts.

Gene sequencing enabled biologists to differentiate the domain Archaea

The original identification of Archaea as a domain separate from Bacteria and Eukarya was based on phylogenetic relationships determined from rRNA gene sequences. This separation was supported when biologists sequenced the first complete archaeal genome, which consisted of 1,738 genes—more than half of which were unlike any genes ever found in the other two domains.

26.13 Crown Gall Crown gall, a type of tumor shown here growing on the trunk of a white oak, is caused by the proteobacterium *Agrobacterium tumefaciens*.

INVESTIGATING**LIFE**

26.14 What Is the Highest Temperature Compatible with Life?

Can any organism thrive at temperatures above 120°C? This is the temperature used for sterilization, known to destroy all previously described organisms. Kazem Kashefi and Derek Lovley isolated an unidentified prokaryote from water samples taken near a hydrothermal vent and found it survived and even multiplied at 121°C. The organism was dubbed "Strain 121," and its gene sequencing results indicate that it is an archaeal species.[a]

HYPOTHESIS Some prokaryotes can survive at temperatures above the 120°C threshold of sterilization.

Method

1. Seal samples of unidentified, iron-reducing, thermal vent prokaryotes in tubes with a medium containing Fe^{3+} as an electron acceptor. Control tubes contain Fe^{3+} but no organisms.
2. Hold both tubes in a sterilizer at 121°C for 10 hours. If the iron-reducing organisms are metabolically active, they will reduce the Fe^{3+} to Fe^{2+} (as magnetite, which can be detected with a magnet).

Results

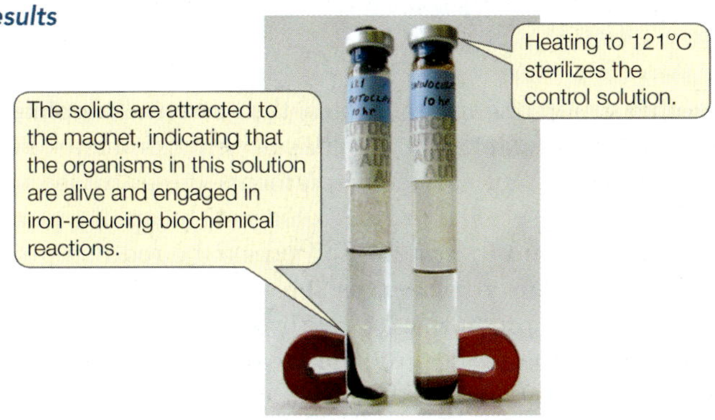

Heating to 121°C sterilizes the control solution.
The solids are attracted to the magnet, indicating that the organisms in this solution are alive and engaged in iron-reducing biochemical reactions.

CONCLUSION Archaea of "Strain 121" can survive at temperatures above the previously defined sterilization limit.

Go to **BioPortal** for discussion and relevant links for all INVESTIGATING**LIFE** figures.

[a]Kashefi, K. and D. R. Lovley. 2003. *Science* 301: 934.

WORKING WITH**DATA:**

A Relationship between Temperature and Growth in an Archaean

Original Paper

Kashefi, K. and D. R. Lovley. 2003. Extending the upper temperature limit for life. *Science* 301: 934.

Analyze the Data

After Strain 121 was isolated, its growth was examined at various temperatures. The table below shows generation time (time between cell divisions) at nine temperatures.

Temperature (°C)	Generation time (hr)
85	10
90	4
95	3
100	2.5
105	2
110	4
115	6
120	20
130	No growth, but cells not killed

QUESTION 1

Make a graph from these data showing time as a function of temperature.

QUESTION 2

Which temperature appears to be closest to optimum for the growth of Strain 121?

QUESTION 3

Note that no growth occurred at 130°C, but that the cells were not killed. How would you demonstrate that these cells were still alive?

Go to **BioPortal** for all WORKING WITH**DATA** exercises

Archaea are known for living in extreme habitats such as those with high salinity (salt content), low oxygen concentrations, high temperatures, or high or low pH (**Figure 26.14**). Many archaea are not extremophiles, however—they are common in soil, for example. Perhaps the largest numbers of archaea live in the ocean depths.

One current classification scheme divides Archaea into two principal groups, **Crenarchaeota** and **Euryarchaeota**. Less is known about two more recently discovered groups, **Korarchaeota** and **Nanoarchaeota**. In fact, we know relatively little about the phylogeny of archaea, in part because the study of these prokaryotes is still in its early stages.

Two characteristics shared by all archaea are the absence of peptidoglycan in their cell walls and the presence of lipids of distinctive composition in their cell membranes (see Table 26.1). The unusual lipids in the membranes of archaea are found in all archaea and in no bacteria or eukaryotes. Most lipids in bacterial and eukaryotic membranes contain unbranched long-chain fatty acids connected to glycerol molecules by **ester linkages**:

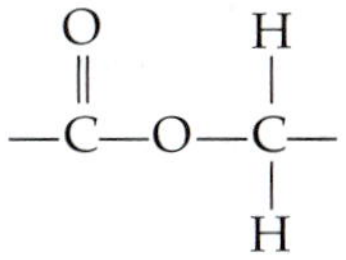

In contrast, some lipids in archaeal membranes contain long-chain hydrocarbons connected to glycerol molecules by **ether linkages**:

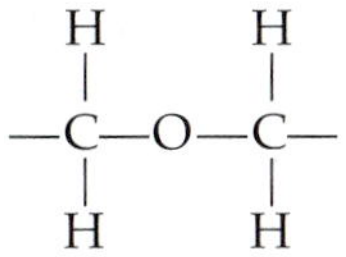

These ether linkages are a synapomorphy of archaea. In addition, the long hydrocarbon chains in the lipids of archaea are branched. One class of archaeal lipids, with hydrocarbon chains 40 carbon atoms in length, contains glycerol at *both* ends of the hydrocarbons (**Figure 26.15**). These lipids form a *lipid monolayer* structure that is unique to archaea. They still fit into a biological membrane because they are twice as long as the typical lipids in the bilayers of other membranes. Lipid monolayers

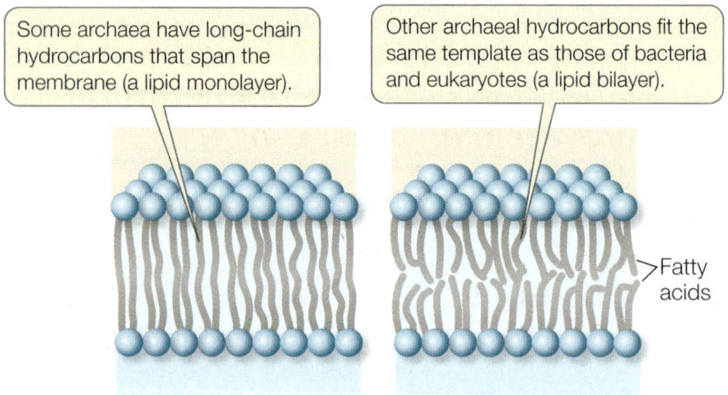
Some archaea have long-chain hydrocarbons that span the membrane (a lipid monolayer).
Other archaeal hydrocarbons fit the same template as those of bacteria and eukaryotes (a lipid bilayer).
Fatty acids

26.15 Membrane Architecture in Archaea The long-chain hydrocarbons of many archaeal lipids have glycerol molecules at both ends, so that the membranes they form consist of a lipid monolayer. In contrast, the membranes of other archaea, bacteria, and eukaryotes consist of a lipid bilayer.

and bilayers are both found among the archaea. The effects, if any, of these structural features on membrane performance are unknown. In spite of this striking difference in their lipids, the membranes of all three domains have similar overall structures, dimensions, and functions.

Most crenarchaeotes live in hot or acidic places

Most known crenarchaeotes are either thermophilic, acidophilic (acid loving), or both. Members of the genus *Sulfolobus* live in hot sulfur springs at temperatures of 70°C to 75°C. They become metabolically inactive at 55°C (131°F). Hot sulfur springs are also extremely acidic. *Sulfolobus* grows best in the range from pH 2 to pH 3, but some members of this genus readily tolerate pH values as low as 0.9. Most acidophilic thermophiles maintain an internal pH of 5.5 to 7 (close to neutral) in spite of their acidic environment. These and other crenarchaeotes thrive where very few other organisms can even survive (**Figure 26.16**).

26.16 Some Crenarchaeotes Like It Hot Thermophilic crenarchaeotes can thrive in the intense heat of volcanic hot sulfur springs such as these in Nevada's Black Rock Desert.

Euryarchaeotes are found in surprising places

Some species of Euryarchaeota are **methanogens**: they produce methane (CH_4) by reducing carbon dioxide as the key step in their energy metabolism. All of the methanogens are obligate anaerobes (see Section 26.3). Comparison of their rRNA gene sequences has revealed a close evolutionary relationship among these methanogenic species, which were previously assigned to several different groups of bacteria.

Methanogenic euryarchaeotes release approximately 2 billion tons of methane gas into Earth's atmosphere each year, accounting for 80 to 90 percent of the methane that enters the atmosphere, including that produced in some mammalian digestive systems. Approximately a third of this methane comes from methanogens living in the guts of ruminants such as cattle, sheep, and deer, and another large fraction comes from methanogens living in the guts of termites and cockroaches. Methane is increasing in Earth's atmosphere by about 1 percent per year and contributes to the greenhouse effect (see Section 58.3). Part of that increase is due to increases in cattle and rice farming and the methanogens associated with both.

Another group of euryarchaeotes, the **extreme halophiles** (salt lovers), live exclusively in very salty environments. Because they contain pink carotenoid pigments, these archaea are sometimes easy to see (**Figure 26.17**). Extreme halophiles grow in the Dead Sea and in brines of all types; the reddish pink spots that can occur on pickled fish are colonies of halophilic archaea. Few other organisms can live in the saltiest homes that the extreme halophiles occupy; most would "dry" to death, losing too much water to the hypertonic environment. Extreme halophiles have been found in lakes with pH values as high as 11.5—the most alkaline environment inhabited by living organisms, and almost as alkaline as household ammonia.

Some of the extreme halophiles have a unique system for trapping light energy and using it to form ATP—without using any form of chlorophyll—when oxygen is in short supply.

26.17 Extreme Halophiles Highly saline environments such as these commercial seawater evaporating ponds in San Francisco Bay are home to extreme halophiles. The archaea are easily visible here because of the rich red coloration of their carotenoid pigments.

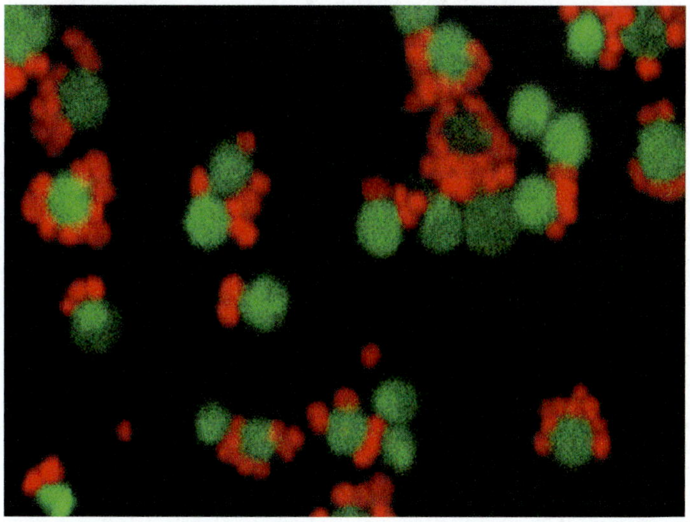
1 μm

26.18 A Nanoarchaeote Growing in Mixed Culture with a Crenarchaeote *Nanoarchaeum equitans* (red), discovered living near deep-sea hydrothermal vents, is the only representative of the nano-archaeote group so far identified. This tiny organism lives attached to cells of the crenarchaeote *Ignicoccus* (green). For this confocal laser micrograph, the two species were visually differentiated by fluorescent dye "tags" that are specific to their distinct gene sequences.

They use the pigment retinal (also found in the vertebrate eye) combined with a protein to form a light-absorbing molecule called microbial rhodopsin.

Another member of the Euryarchaeota, *Thermoplasma*, has no cell wall. It is thermophilic and acidophilic, its metabolism is aerobic, and it lives in coal deposits. Its genome of 1,100,000 base pairs is among the smallest (along with that of the mycoplasmas) found in any free-living organism, although some parasitic organisms have even smaller genomes.

Korarchaeotes and nanoarchaeotes are less well known

The korarchaeotes are known only from DNA isolated directly from marine hydrothermal vents and freshwater hot springs. No korarchaeote has been successfully grown in pure culture.

Another distinctive archaeal lineage has been discovered at a deep-sea hydrothermal vent off the coast of Iceland. It is the first representative of the group christened Nanoarchaeota because of their minute size (Greek *nanos*, "dwarf"). This organism is a parasite that lives on cells of *Ignicoccus*, a crenarchaeote. Because of their association, the two species can be grown together in culture (**Figure 26.18**).

RECAP 26.2

Bacteria and Archaea are highly diverse groups that survive in almost every imaginable habitat on Earth. Many can survive and even thrive in habitats where no eukaryotes can live, including extremely hot, acidic, or saline conditions.

- How does the diversity of environments occupied by prokaryotes compare with the diversity of environments occupied by multicellular organisms with which you are familiar?
- Explain why Gram staining is of limited use in understanding the evolutionary relationships of bacteria. **See pp. 530–531**
- What makes the membranes of archaea unique? **See pp. 535–536 and Figure 26.15**

Prokaryotes are found almost everywhere on Earth and live in a wide variety of ecosystems. In the next section we will examine the contributions of prokaryotes to the functioning of those ecosystems.

26.3 How Do Prokaryotes Affect Their Environments?

Many people think of prokaryotes primarily as disease-causing organisms, but only a small percentage of bacteria, and no archaea, are known to be pathogens. Prokaryotes play roles in ecosystems that reach far beyond human sickness and health.

Prokaryotes have diverse metabolic pathways

Bacteria and archaea outdo the eukaryotes in terms of metabolic diversity. Although they are much more diverse in size and shape, eukaryotes draw on fewer metabolic mechanisms for their energy needs. In fact, much of eukaryotes' energy metabolism is carried out in organelles—mitochondria and chloroplasts—that are endosymbiotic descendants of bacteria. The long evolutionary history of bacteria and archaea, during which they have had time to adapt to a wide variety of habitats, has led to the extraordinary diversity of their metabolic "lifestyles"—their use or nonuse of oxygen, their varied sources of energy and carbon atoms, and the materials they release as waste products. This diversity of metabolic pathways, some of them unique to prokaryotes, makes prokaryotes a key component of the cycling of materials through ecosystems.

ANAEROBIC VERSUS AEROBIC METABOLISM The presence of oxygen is poisonous to some prokaryotes, so these **obligate anaerobes** can live only by anaerobic metabolism. At the other extreme from the obligate anaerobes, some prokaryotes are **obligate aerobes**. They require oxygen for cellular respiration and are unable to survive for extended periods in the *absence* of oxygen. Other prokaryotes can shift their metabolism between anaerobic and aerobic modes and thus are called **facultative anaerobes**. Many facultative anaerobes alternate between anaerobic metabolic processes (such as fermentation) and cellular respiration, as conditions dictate. **Aerotolerant anaerobes** are not damaged by oxygen when it is present, but they cannot conduct cellular respiration. By definition, an anaerobe does not use oxygen as an electron acceptor for respiration.

NUTRITIONAL CATEGORIES All organisms face the same two nutritional challenges: they must synthesize energy-rich compounds such as ATP to power their life-sustaining metabolic reactions, and they must obtain carbon atoms to build their

TABLE 26.2 How Organisms Obtain Their Energy and Carbon

Nutritional Category	Energy Source	Carbon Source
Photoautotrophs (some bacteria, some eukaryotes)	Light	Carbon dioxide
Photoheterotrophs (some bacteria)	Light	Organic compounds
Chemoautotrophs (some bacteria, many archaea)	Inorganic substances	Carbon dioxide
Chemoheterotrophs (found in all three domains)	Usually organic compounds; sometimes inorganic substances	Organic compounds

own organic molecules. Biologists recognize four broad nutritional categories of organisms: photoautotrophs, photoheterotrophs, chemoautotrophs, and chemoheterotrophs. Prokaryotes are represented in all four groups (**Table 26.2**).

Photoautotrophs perform photosynthesis. They use light as their energy source and carbon dioxide (CO_2) as their carbon source. The cyanobacteria, like green plants and other photosynthetic eukaryotes, use chlorophyll *a* as their key photosynthetic pigment and produce oxygen gas (O_2) as a by-product of noncyclic electron transport.

There are other photoautotrophs among the bacteria, but these organisms use bacteriochlorophyll rather than chlorophyll *a* as their key photosynthetic pigment, and they do not produce O_2. Instead, some of these photosynthesizers produce particles of pure sulfur because hydrogen sulfide (H_2S), rather than H_2O, is their electron donor for photophosphorylation. Many proteobacteria fit into this category. Bacteriochlorophyll molecules absorb light of longer wavelengths than do the chlorophyll molecules used by other photosynthesizing organisms. As a result, bacteria using this pigment can grow in water under fairly dense layers of algae, using light of wavelengths that are not absorbed by the algae (**Figure 26.19**).

Photoheterotrophs use light as their energy source but must obtain their carbon atoms from organic compounds made by other organisms. Their "food" consists of organic compounds such as carbohydrates, fatty acids, and alcohols. For example, some photoheterotrophs take up compounds released from plant roots (as in rice paddies) or from decomposing photosynthetic bacteria in hot springs and metabolize them to form building blocks for other compounds. Sunlight provides them with the energy necessary for ATP formation through photophosphorylation.

Chemoautotrophs obtain their energy by oxidizing inorganic substances, and they use some of that energy to fix carbon. Some chemoautotrophs use reactions identical to those of the typical photosynthetic cycle, but others use alternative pathways for carbon fixation. Some bacteria oxidize ammonia or nitrite ions to form nitrate ions. Others oxidize hydrogen gas, hydrogen sulfide, sulfur, and other materials. Many archaea are chemoautotrophs.

Finally, **chemoheterotrophs** obtain carbon atoms from one or more complex organic compounds that have been synthesized by other organisms, and usually obtain energy from breaking down these organic compounds as well. Most known bacteria and archaea are chemoheterotrophs—as are all animals and fungi and many protists.

Although most chemoheterotrophs rely on the breakdown of organic compounds for energy, some chemoheterotrophic prokaryotes obtain their energy by breaking down inorganic substances. Organisms that obtain energy from oxidizing inorganic substances (both chemoautotrophs as well as some chemoheterotrophs) are also known as **lithotrophs** (Greek, "rock consumers").

Prokaryotes play important roles in element cycling

The metabolic diversity of the prokaryotes makes them key players in the cycles that keep chemical elements moving through ecosystems. Many prokaryotes are **decomposers**: organisms that metabolize organic compounds in dead organic material and return the products to the environment as inorganic substances. Prokaryotes, along with fungi, return tremendous quantities of carbon to the atmosphere as carbon dioxide, thus carrying out a key step in the carbon cycle.

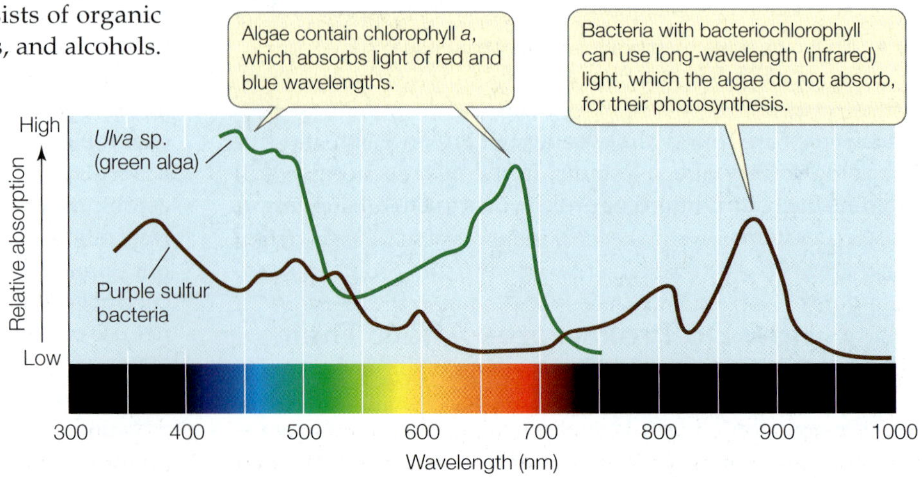

Algae contain chlorophyll *a*, which absorbs light of red and blue wavelengths.
Bacteria with bacteriochlorophyll can use long-wavelength (infrared) light, which the algae do not absorb, for their photosynthesis.
High
Low
Relative absorption
Ulva sp. (green alga)
Purple sulfur bacteria
300
400
500
600
700
800
900
1000
Wavelength (nm)

26.19 Bacteriochlorophyll Absorbs Long-Wavelength Light The green alga *Ulva* contains chlorophyll, which absorbs no light of wavelengths longer than 750 nm. Purple sulfur bacteria, which contain bacteriochlorophyll, can conduct photosynthesis using longer infrared wavelengths. As a result, these bacteria can grow under layers of algae.

The key metabolic reactions of many prokaryotes involve nitrogen or sulfur. For example, some bacteria carry out respiratory electron transport without using oxygen as an electron acceptor. These organisms use oxidized inorganic ions such as nitrate, nitrite, or sulfate as electron acceptors. Examples include the **denitrifiers**, which release nitrogen to the atmosphere as nitrogen gas (N_2). These normally aerobic bacteria, mostly species of the genera *Bacillus* and *Pseudomonas*, use nitrate (NO_3^-) as an electron acceptor in place of oxygen under anaerobic conditions:

$$2\ NO_3^- + 10\ e^- + 12\ H^+ \rightarrow N_2 + 6\ H_2O$$

Denitrifiers play a key role in the cycling of nitrogen through ecosystems. Without denitrifiers, which convert nitrate ions into nitrogen gas, all forms of nitrogen would leach from the soil and end up in lakes and oceans; the resulting deficit of nitrogen on land would make terrestrial life much more difficult.

Nitrogen fixers convert atmospheric nitrogen gas into ammonia (NH_3), a chemical form that is usable by the nitrogen fixers themselves as well as by other organisms:

$$N_2 + 6\ H \rightarrow 2\ NH_3$$

All organisms require nitrogen in order to build proteins, nucleic acids, and other important compounds. Nitrogen fixation is thus vital to life as we know it. This all-important biochemical process is carried out by a wide variety of archaea and bacteria (including cyanobacteria) but by no other organisms, so we depend on these prokaryotes for our very existence.

Ammonia is oxidized to nitrate in soil and in seawater by chemoautotrophic bacteria and archaea called **nitrifiers**. Bacteria of two genera, *Nitrosomonas* and *Nitrosococcus*, convert ammonia (NH_3) to nitrite ions (NO_2^-), and *Nitrobacter* oxidize nitrite to nitrate (NO_3^-), the form of nitrogen most easily used by many plants. What do the nitrifiers get out of these reactions? Their metabolism is powered by the energy released by the oxidation of ammonia or nitrite. For example, by passing the electrons from nitrite through an electron transport system, *Nitrobacter* can make ATP, and using some of this ATP, can also make NADH. With this ATP and NADH, the bacterium can convert CO_2 and H_2O into glucose.

We have already seen the importance of the cyanobacteria in the cycling of oxygen: in ancient times, the oxygen generated by their photosynthesis converted Earth's atmosphere from an anaerobic to an aerobic environment. Other prokaryotes—both bacteria and archaea—contribute to the cycling of sulfur. Deep-sea hydrothermal vent ecosystems depend on chemoautotrophic prokaryotes that are incorporated into large communities of crabs, mollusks, and giant worms, all living at a depth of 2,500 meters—below any hint of sunlight. These bacteria obtain energy by oxidizing hydrogen sulfide and other substances released in the near-boiling water flowing from volcanic vents in the ocean floor.

Many prokaryotes form complex communities

Prokaryotes do not usually live in isolation. Rather, they live in communities of many different microbial species, which often include microscopic eukaryotes as well as myriad prokaryotic species. While some microbial communities are harmful to humans, others provide important services. For example, microbial communities help us digest our food, break down municipal waste, and recycle organic matter and chemical elements in the environment.

Some microbial communities form layers in sediments; others form clumps a meter or more in diameter. Many microbial communities form dense **biofilms**. Upon contacting a solid surface, the cells bind to that surface and secrete a sticky, gel-like, polysaccharide matrix that traps other cells (Figure 26.20). Once a biofilm forms, the cells become more difficult to kill.

Biofilms are found in many places, and in some of those places they cause problems for humans. The material on our teeth that we call dental plaque is a biofilm. Pathogenic bacteria are difficult for the immune system—and modern medicine—to combat once they form a biofilm, which may be impermeable to antibiotics. Worse, some drugs stimulate the bacteria in a biofilm to lay down more matrix, making the film even more impermeable. Biofilms may form on just about any available surface, including contact lenses and artificial joint replacements. They foul metal pipes and cause corrosion, a major problem in steam-driven electricity generation plants. Fossil stromatolites—large, rocky structures made up of alternating layers of fossilized biofilm and calcium carbonate—are among the oldest remnants of life on Earth (see Figure 25.9).

Some biologists are studying the chemical signals that prokaryotes use to communicate with one another and trigger density-linked activities such as biofilm formation. We saw one example of this type of communication—called **quorum sensing**—in the chapter-opening discussion of biominescent *Vibrio*. In the case of health-threatening bacteria, researchers hope to be able to block the quorum-sensing signals that lead to the production of the matrix polysaccharides, thus preventing pathological biofilms from forming.

Prokaryotes live on and in other organisms

Prokaryotes engage in many kinds of mutually beneficial relationships with eukaryotes. As we have seen, the mitochondria and chloroplasts of eukaryotes are descended from what were once free-living bacteria. Much later in evolutionary history, some plants became associated with bacteria to form cooperative nitrogen-fixing nodules on their roots.

Many animals harbor a variety of bacteria and archaea in their digestive tracts. Cattle depend on prokaryotes to perform important steps in digestion. Like most animals, cattle cannot produce cellulase, the enzyme needed to start the digestion of the cellulose that makes up the bulk of their plant food. However, bacteria living in a special section of the gut, called the rumen, produce enough cellulase to process the daily diet for the cattle. Human health also depends on many hundreds of species of symbiotic bacteria.

Microbiomes are critical to human health

Although only a few bacterial species are pathogens, popular notions of bacteria as "germs" and fear of the consequences

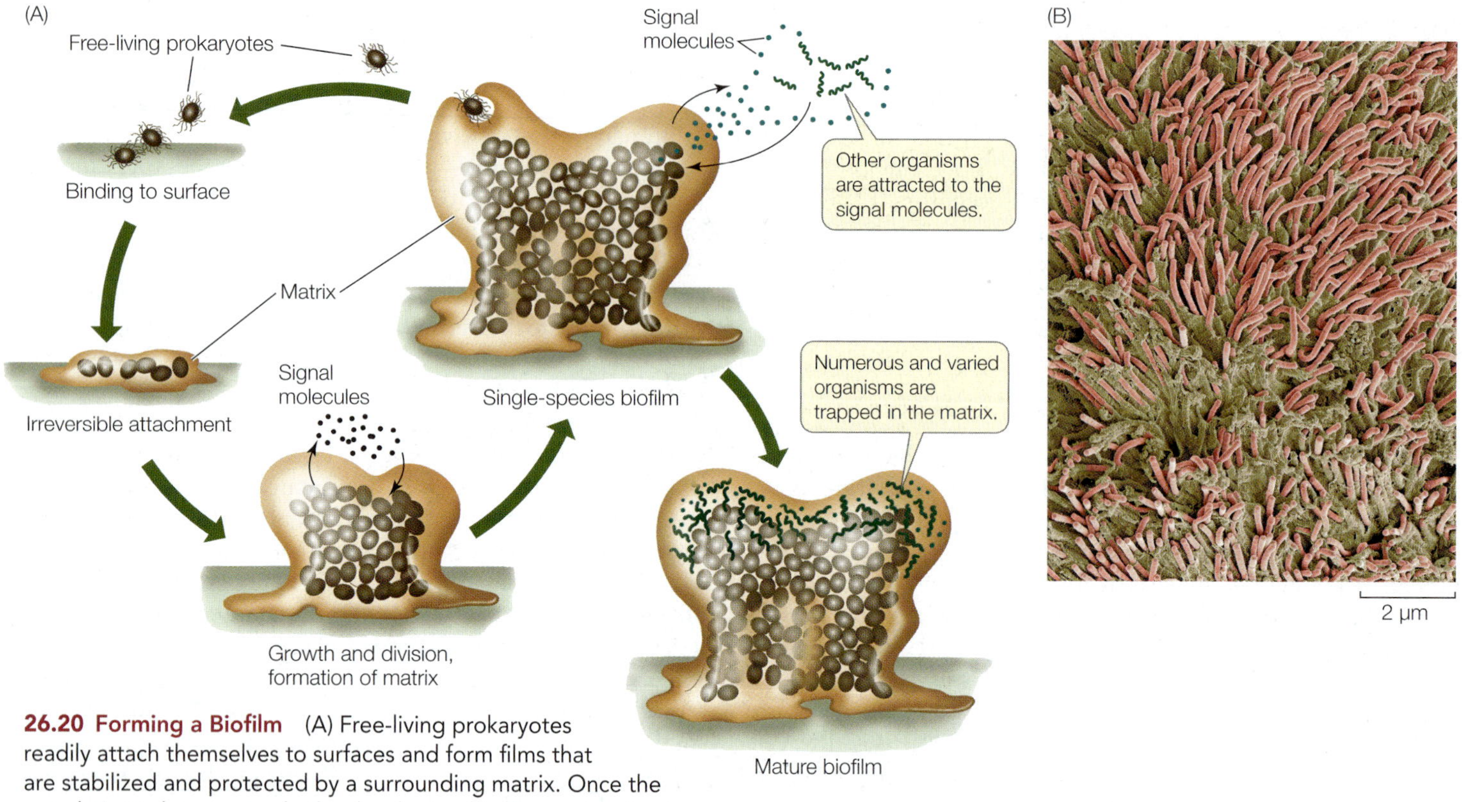
(A)
Free-living prokaryotes
Binding to surface
Irreversible attachment
Matrix
Signal molecules
Growth and division, formation of matrix
Single-species biofilm
Signal molecules
Other organisms are attracted to the signal molecules.
Numerous and varied organisms are trapped in the matrix.
Mature biofilm
(B)
2 μm

26.20 Forming a Biofilm (A) Free-living prokaryotes readily attach themselves to surfaces and form films that are stabilized and protected by a surrounding matrix. Once the population is large enough, the developing biofilm can send out chemical signals that attract other microorganisms. (B) Scanning electron micrography reveals a biofilm of dental plaque. The bacteria (red) are embedded in a matrix consisting of proteins from both bacterial secretions and saliva.

of infection cause many people to assume that most bacteria are harmful. Increasingly, however, biologists are discovering that human health depends in many ways on the health of our **microbiomes**: communities of bacteria that live in and on our bodies (**Figure 26.21**).

Every surface of your body is covered with diverse communities of bacteria. A 2009 study identified more than 1,000 species of bacteria that live on human skin. Inside your body, your digestive system teems with bacteria. If you count up all the cells in a human body, only about 10 percent of them are human cells. The rest are microbes—mostly bacteria, along with some archaea and microscopic eukaryotes.

Biologists are discovering that many complex health problems are linked to the disruption of our microbiomes. These diverse microbial communities affect the expression of our genes and play a critical role in the development and maintenance of a healthy immune system. When our microbiomes contain an appropriate community of beneficial bacteria, our bodies function normally. But these communities are strongly affected by our life experiences, by the food we eat, by the medicines we take, and by our exposure to various environmental toxins. When our microbial communities are disrupted, they must be restored before the body can function normally. The recent rapid increase in occurrences of autoimmune diseases—diseases in which our immune systems begin to attack our bodies (see Chapter 42)—has been linked to the changing diversity and composition of our microbiomes.

The early acquisition of an appropriate microbiome is critical for lifelong health. Normally, a human infant acquires much of its microbiome at birth, from the microbiome in its mother's vagina. Other components of the microbiome are also acquired from the mother, especially through breast feeding. Recent studies have shown that babies born by cesarean section, as well as babies that are bottle-fed on artificial milk formula, typically acquire microbes from a wider variety of sources. Many of the bacteria acquired in this way are not well suited for human health. Biologists have discovered that the incidence of many autoimmune diseases is much higher in people who were born by cesarean section and in those who were fed on formula as infants, compared with individuals who were born vaginally and breast-fed. The difference appears to be related to the composition of the individual's original microbiome.

Our microbiomes may be related to many other health concerns. For example, physicians have long noted a connection between autism and gastrointestinal disorders. In 2012, microbiologists discovered that children with autism have high levels of bacteria of the genus *Sutterella* adhering to their intestinal walls. These bacteria are absent or rare in children without autism. It is not yet known if *Sutterella* bacteria cause the gastrointestinal problems or if they are merely its symptoms, but it appears clear that the intestinal microbiomes of children with autism are distinctive.

Humans use some of the metabolic products—especially vitamins B_{12} and K—produced by the microorganisms living in the large intestine. Communities of bacteria line our intestines with a dense biofilm that is in intimate contact with the mucosal lining of the gut. This biofilm facilitates nutrient transfer from the intestine into the body, functioning like a specialized "tissue" that is essential to our health. This biofilm has a complex ecology that scientists have just begun to explore in

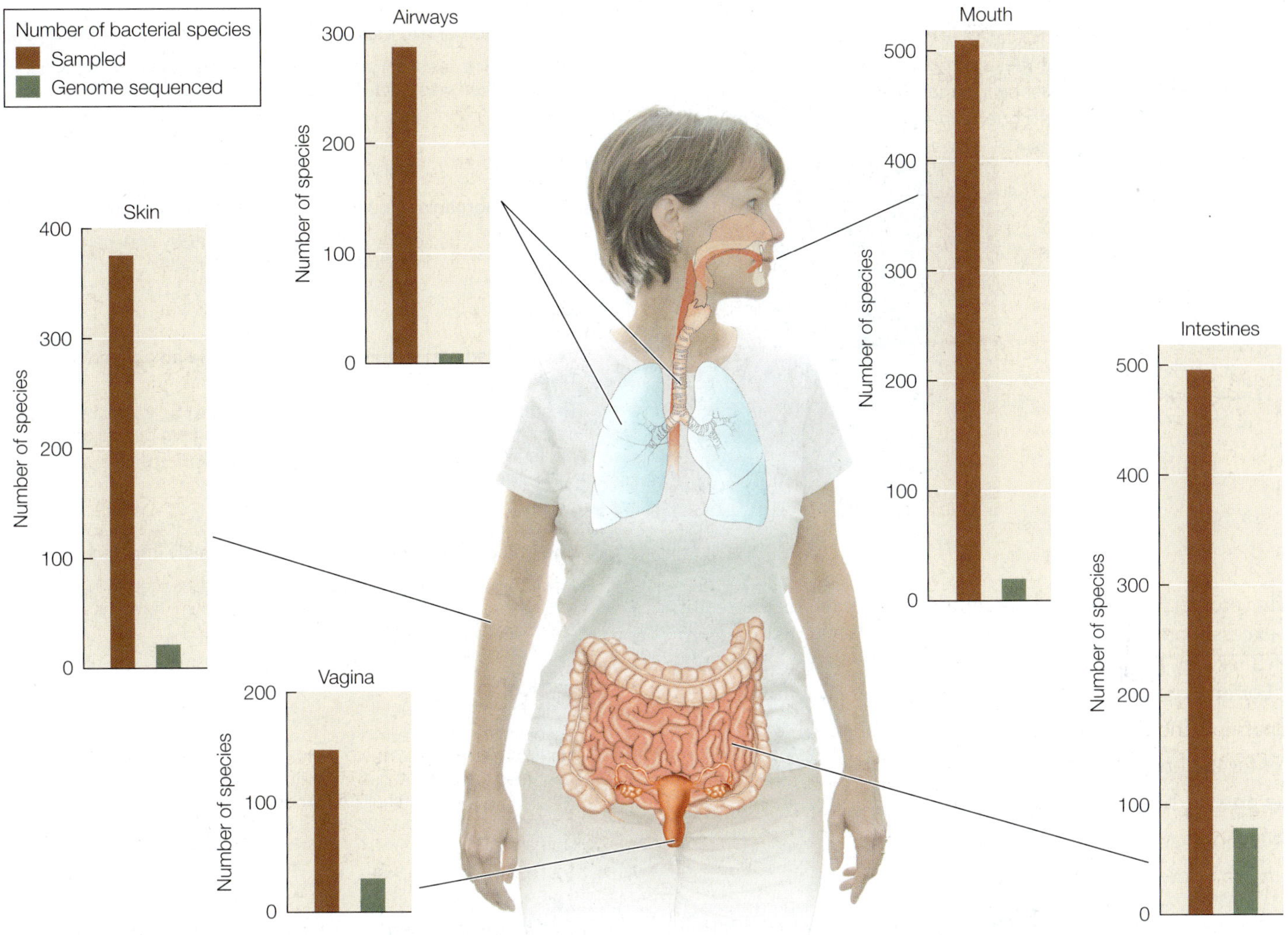

Number of bacterial species
Sampled
Genome sequenced
Airways
Skin
Mouth
Intestines
Vagina
Number of species
0
100
200
300
400
500

26.21 The Body's Microbiome Is Critical to the Maintenance of Health Surveys of the human microbiome have shown that this community includes thousands of diverse bacterial species that are adapted to grow in or on various parts of the body. Although we now know that the composition of this microbiome is closely associated with many aspects of human health, most of the component species are poorly characterized and remain largely unstudied by biologists. What has become clear is that, although the "subcommunities" in different parts of the body share similarities, each is a site-specific assemblage of many distinctive species.

detail—including the possibility that the species composition of an individual's gut microbiome may contribute to obesity (or the resistance to it).

A small minority of bacteria are pathogens

The tiny percentage of prokaryotes that are pathogens all fall into the domain Bacteria. Many archaea live in association with humans and other eukaryotes but are not known to cause any diseases. Why are there no pathogenic archaea? One clue is that different groups of viruses and plasmids infect bacteria than infect archaea, and pathogenesis in bacteria is largely coupled with traits carried by these plasmids and viruses. Differences in the external structures of bacteria and archaea make cross-domain infection by viruses and plasmids highly unlikely. Pathogenic traits may evolve relatively rarely, and they do not appear to have evolved in the viruses or plasmids that infect archaea.

How can we know if a particular microbe is responsible for a disease? The late nineteenth century was a productive era in the history of medicine—a time when bacteriologists, chemists, and physicians proved that many diseases are caused by microbial agents. During this time, the German physician Robert Koch laid down a set of four rules for establishing that a particular microorganism causes a particular disease:

1. The microorganism is always found in individuals with the disease.
2. The microorganism can be taken from the host and grown in pure culture.
3. A sample of the culture produces the same disease when injected into a new, healthy host.
4. The newly infected host yields a new, pure culture of microorganisms identical to those obtained in the second step.

These rules, called **Koch's postulates**, were important tools in a time when it was not widely understood that microorganisms cause disease. Although modern medical science has more powerful diagnostic tools, Koch's postulates remain useful. For example, physicians were taken aback in the 1990s

26.22 Satisfying Koch's Postulates Robin Warren and Barry Marshall of the University of Western Australia won the 2005 Nobel Prize in Medicine for showing that ulcers are caused not by the action of stomach acid but by infection with the bacterium *Helicobacter pylori*.

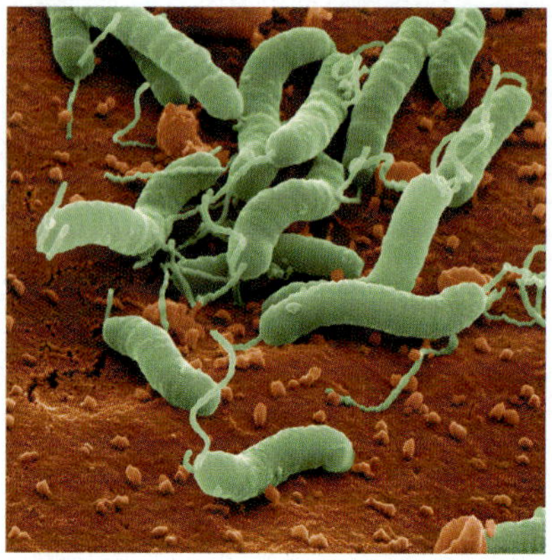

Helicobacter pylori 1.5 μm

Marshall and Warren set out to satisfy Koch's postulates:

Test 1

The microorganism must be present in every case of the disease.

Results: Biopsies from the stomachs of many patients revealed that the bacterium was always present if the stomach was inflamed or ulcerated.

Test 2

The microorganism must be cultured from a sick host.

Results: The bacterium was isolated from biopsy material and eventually grown in culture media in the laboratory.

Test 3

The isolated and cultured bacteria must be able to induce the disease.

Results: Marshall was examined and found to be free of bacteria and inflammation in his stomach. After drinking a pure culture of the bacterium, he developed stomach inflammation (gastritis).

Test 4

The bacteria must be recoverable from newly infected individuals.

Results: Biopsy of Marshall's stomach 2 weeks after he ingested the bacteria revealed the presence of the bacterium, now christened *Helicobacter pylori*, in the inflamed tissue.

Conclusion

Antibiotic treatment eliminated the bacteria and the inflammation in Marshall's stomach. The experiment was repeated on healthy volunteers, and many patients with gastric ulcers were cured with antibiotics. Thus Marshall and Warren demonstrated that the stomach inflammation leading to ulcers is caused by *H. pylori* infections in the stomach.

when stomach ulcers—long accepted and treated as the result of excess stomach acid—were shown by Koch's postulates to be caused by the bacterium *Helicobacter pylori* (**Figure 26.22**).

For an organism to be a successful pathogen, it must:

- arrive at the body surface of a potential host;
- enter the host's body;
- evade the host's defenses;
- reproduce inside the host; and
- infect a new host.

Failure to complete any of these steps ends the reproductive career of a pathogenic organism. Yet in spite of the many defenses available to potential hosts (see Chapters 39 and 42), pathogenic bacteria are often surprisingly difficult to combat, even with today's arsenal of antibiotics. One source of this difficulty, as we have seen, is their ability to form biofilms.

For the host, the consequences of a bacterial infection depend on several factors. One is the **invasiveness** of the pathogen: its ability to multiply in the host's body. Another is its **toxigenicity**, or ability to produce toxins (chemical substances that are harmful to the host's tissues). *Corynebacterium diphtheriae*, the agent that causes diphtheria, has low invasiveness and multiplies only in the throat, but its toxigenicity is so great that the entire body is affected. In contrast, *Bacillus anthracis*, which causes anthrax, has low toxigenicity, but is so invasive that the entire bloodstream ultimately teems with the bacteria.

There are two general types of bacterial toxins, exotoxins and endotoxins. **Endotoxins** are lipopolysaccharides (complexes consisting of a polysaccharide and a lipid component) that form part of the outer membrane of certain Gram-negative bacteria. They are released when these bacteria lyse (burst). Endotoxins are rarely fatal to the host; they normally cause fever, vomiting, and diarrhea. Among the endotoxin producers are some strains of the proteobacteria *Salmonella* and *Escherichia*.

Exotoxins are soluble proteins released by living, multiplying bacteria. They are highly toxic—often fatal—to the host. Human diseases induced by bacterial exotoxins include tetanus (*Clostridium tetani*), cholera (*Vibrio cholerae*), and bubonic plague (*Yersinia pestis*). Anthrax is caused by three exotoxins produced by *Bacillus anthracis*. Botulism is caused by exotoxins produced by the obligate anaerobe *Clostridium botulinum* that are among the most poisonous ever discovered. The lethal dose for humans of one exotoxin of *C. botulinum* is about one-millionth of a gram. Nonetheless, much smaller doses of this exotoxin are marketed under various trade names (the best known being Botox®) and are used to treat muscle spasms as well as for cosmetic purposes (temporary wrinkle reduction in the skin).

RECAP 26.3

Many prokaryotes are beneficial and even necessary to other forms of life. Most animals, including humans, depend on a complex community of prokaryotes—a microbiome—to maintain health, especially of the immune and digestive systems. Pathogenic bacteria are the direct causes of diseases.

- How are the four nutritional categories of prokaryotes distinguished? **See p. 539 and Table 26.2**
- Why is nitrogen metabolism in the prokaryotes vital to other organisms? **See p. 538**
- How do biofilms form, and why are they of special interest to researchers? **See p. 539 and Figure 26.20**

Before moving on to discuss the diversity of eukaryotic life, it is appropriate to consider another category of life that includes some pathogens: the viruses.

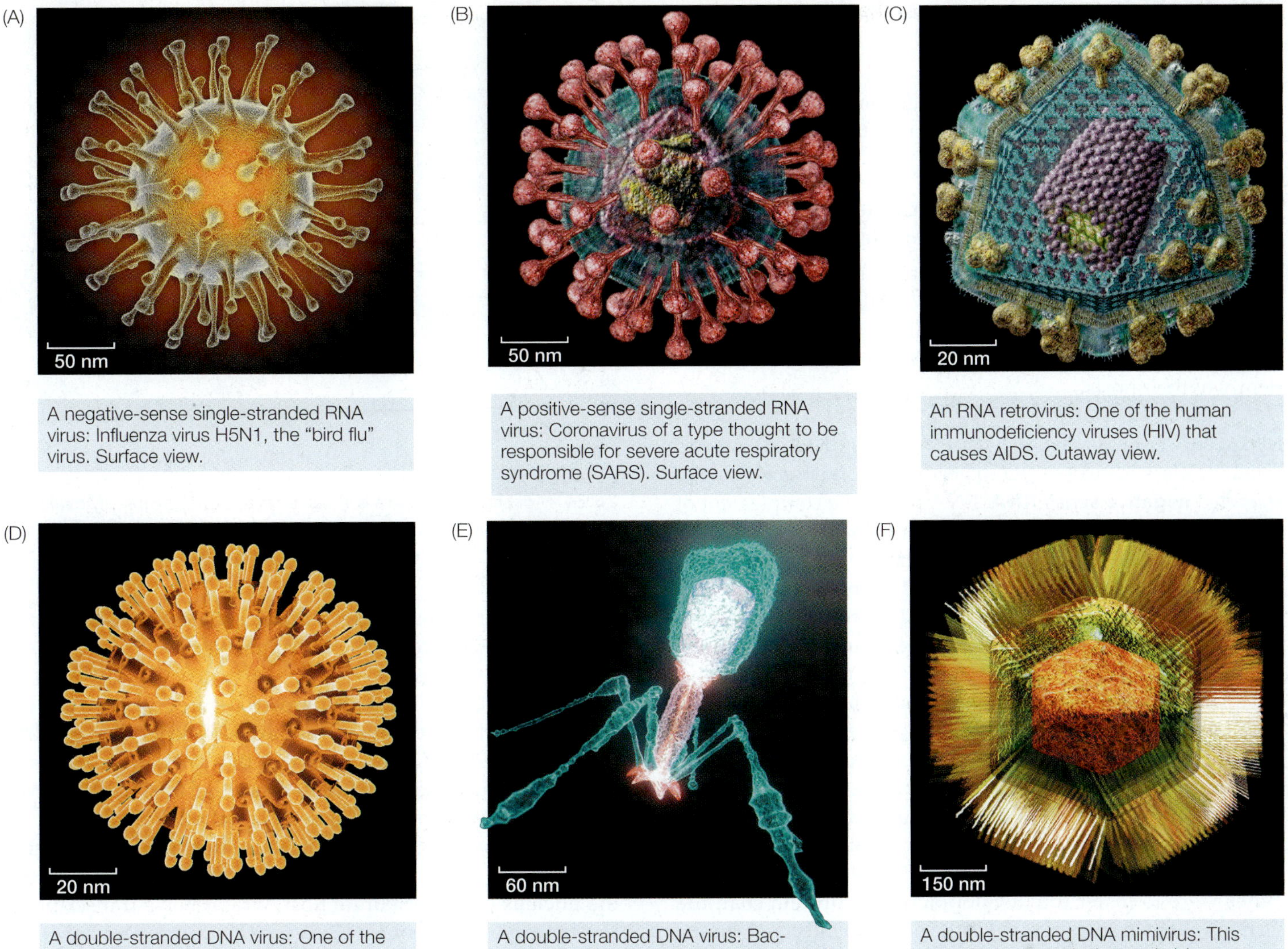
(A)
50 nm
(B)
50 nm
(C)
20 nm
(D)
20 nm
(E)
60 nm
(F)
150 nm

(A) A negative-sense single-stranded RNA virus: Influenza virus H5N1, the "bird flu" virus. Surface view.

(B) A positive-sense single-stranded RNA virus: Coronavirus of a type thought to be responsible for severe acute respiratory syndrome (SARS). Surface view.

(C) An RNA retrovirus: One of the human immunodeficiency viruses (HIV) that causes AIDS. Cutaway view.

(D) A double-stranded DNA virus: One of the many herpes viruses (Herpesviridae). Different herpes viruses are responsible for many human infections, including chicken pox, shingles, cold sores. and genital herpes (HSV1/2). Surface view.

(E) A double-stranded DNA virus: Bacteriophage T4. Viruses that infect bacteria are referred to as bacteriophage (or simply phage). T4 attaches leglike fibers to the outside of its host cell and injects its DNA into the cytoplasm through its "tail" (pink structure in this rendition).

(F) A double-stranded DNA mimivirus: This *Acanthamoeba polyphaga* mimivirus (APMV) has the largest diameter of all known viruses and a genome larger than some prokaryote genomes. It is named for its host, an amoeba. Cutaway view.

26.23 Viruses Are Diverse Relatively small genomes and rapid evolutionary rates make it difficult to reconstruct phylogenetic relationships among some classes of viruses. Instead, viruses are classified largely by general characteristics of their genomes. The images here are computer artists' reconstructions based on cryo-electron micrographs.

26.4 How Do Viruses Relate to Life's Diversity and Ecology?

Although they are not cellular, viruses are numerically among the most abundant forms of life on Earth, and their effects on other organisms are enormous. Where did viruses come from, and how do they fit into the tree of life? Biologists are still working to answer these questions.

Some biologists do not think of viruses as living organisms, primarily because they are not cellular and must depend on cellular organisms for basic life functions such as replication and metabolism (see Section 16.1). But viruses are derived from the cells of living organisms. They use the same forms of genetic information storage and transmission as do cellular organisms. Viruses infect all cellular forms of life—bacteria, archaea, and eukaryotes. They replicate, mutate, evolve, and interact with other organisms, often causing serious diseases in their hosts. Finally, viruses clearly evolve independently of other organisms, so it is almost impossible not to treat them as a part of life.

Several factors make viral phylogeny difficult to resolve. First, the tiny size of many viral genomes restricts the phylogenetic analyses that can be conducted to relate them to the genomes of cellular organisms. Second, their rapid mutation rate, which results in rapid evolution of viral genomes, tends to cloud their evolutionary relationships over long periods. Third, there are no known fossil viruses—viruses are too small and delicate to fossilize—so the paleontological record offers no clues to virus origins. Finally, viruses are highly diverse (**Figure 26.23**). Several lines of evidence support the hypothesis that viruses have evolved repeatedly within each of the major groups of life. The difficulty in resolving their deep evolutionary relationships makes a phylogeny-based classification of viruses difficult. Instead, viruses are placed in one of several groups on the basis of the structure of their genomes, although most of these defined groups are not thought to represent monophyletic taxa.

Many RNA viruses probably represent escaped genomic components of cellular life

Although viruses are obligate parasites of cellular species, many viruses may once have been cellular components involved in basic cellular functions—that is, they may be "escaped" components of cellular life that now evolve independently of their hosts.

NEGATIVE-SENSE SINGLE-STRANDED RNA VIRUSES A case in point is a class of viruses whose genome is composed of single-stranded **negative-sense RNA**: RNA that is the complement of the mRNA needed for protein translation. Many of these negative-sense single-stranded RNA viruses have only a few genes, including one for an RNA-dependent RNA polymerase that allows them to make complementary mRNA from their negative-sense RNA genome. Modern cellular organisms cannot generate mRNA in this manner (at least in the absence of viral infections), but scientists speculate that single-stranded RNA genomes may have been common in the distant past, before DNA became the primary molecule for genetic information storage.

A self-replicating RNA polymerase gene that began to replicate independently of a cellular genome could conceivably acquire a few additional protein-coding genes through recombination with its host's DNA. If one or more of these genes were to foster the development of a protein coat, the virus might then survive outside the host and infect new hosts. It is believed that this scenario has been repeated many times independently across the tree of life, given that many of the negative-sense single-stranded RNA viruses that infect organisms from bacteria to humans are not closely related to one another. In other words, negative-sense single-stranded RNA viruses do not represent a distinct taxonomic group, but rather exemplify a particular process of cellular escape that probably happened many different times.

Familiar examples of negative-sense single-stranded RNA viruses include the viruses that cause measles, mumps, rabies, and influenza (see Figure 26.23A).

POSITIVE-SENSE SINGLE-STRANDED RNA VIRUSES The genome of another type of single-stranded RNA virus is composed of positive-sense RNA. Positive-sense genomes are already set for translation; no replication of the genome to form a complement strand is needed before protein translation can take place. Positive-sense single-stranded RNA viruses (see Figure 26.23B) are the most abundant and diverse class of viruses. Most of the viruses that cause diseases in crop plants are members of this group. These viruses kill patches of cells in the leaves or stems of plants, leaving live cells amid a patchwork of discolored dead tissue (giving them the name of mosaic or mottle viruses; Figure 26.24). Other viruses in this group infect bacteria, fungi, and animals. Human diseases caused by positive-sense single-stranded RNA viruses include polio, hepatitis C, and the common cold. As is true of the other functionally defined groups of viruses, these viruses appear to have evolved multiple times across the tree of life from different groups of cellular ancestors.

Yellow areas are dead leaf cells, killed by the mosaic virus.

26.24 Mosaic Viruses Are a Problem for Agriculture Mosaic, or "mottle," viruses are the most diverse class of viruses. This leaf is from an apple tree infected with a mosaic virus.

RNA RETROVIRUSES The RNA retroviruses are best known as the group that includes the human immunodeficiency viruses (HIV; see Figure 26.23C). Like the previous two categories of viruses, RNA retroviruses have genomes composed of single-stranded RNA and probably evolved as escaped cellular components.

Retroviruses are so named because they regenerate themselves by reverse transcription. When the retrovirus enters the nucleus of its vertebrate host, viral reverse transcriptase produces complementary DNA (cDNA) from the viral RNA genome, then replicates that single-stranded cDNA to produce double-stranded DNA. Another virally encoded enzyme, called integrase, catalyzes the integration of the new piece of double-stranded viral DNA into the host's genome. The viral genome is then replicated along with the host cell's DNA; the integrated retroviral DNA is known as a **provirus**.

Retroviruses are only known to infect vertebrates, although genomic elements that resemble portions of these viruses are a component of the genomes of a wide variety of organisms, including bacteria, plants, and many animals. Several retroviruses are associated with various forms of cancer, as cells infected with these viruses are likely to undergo uncontrolled replication.

DOUBLE-STRANDED RNA VIRUSES Double-stranded RNA viruses may have evolved repeatedly from single-stranded RNA ancestors—or perhaps vice versa. These viruses, which are not closely related to one another, infect organisms from throughout the tree of life. Many plant diseases are caused by double-stranded RNA viruses. Other viruses of this type cause many cases of infant diarrhea in humans.

Some DNA viruses may have evolved from reduced cellular organisms

Another class of viruses is composed of viruses that have a double-stranded DNA genome (see Figure 26.23D–F). This group is also almost certainly polyphyletic (with many independent origins). Many of the common phage that infect bacteria are double-stranded DNA viruses, as are the viruses that cause smallpox and herpes in humans.

Some biologists think that at least some of the DNA viruses may represent highly reduced parasitic organisms that have lost their cellular structure as well as their ability to survive as free-living species. For example, the mimiviruses, which are some of the largest DNA viruses (see Figure 26.23F), have a genome in excess of a million base pairs of DNA that encodes more than 900 proteins. This genome is similar in size to the genomes of many parasitic bacteria and about twice as large as the genomes of the smallest bacteria (**Figure 26.25**). Phylogenetic analyses of these DNA viruses suggest that they have evolved repeatedly from cellular organisms. Furthermore, recombination among different viruses may have allowed the exchange of various genetic modules, further complicating the history and origins of these viruses.

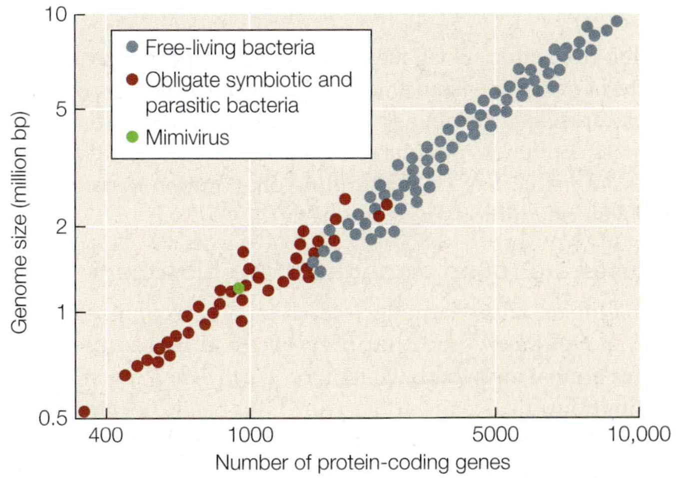
Free-living bacteria
Obligate symbiotic and parasitic bacteria
Mimivirus
Genome size (million bp)
10
5
2
1
0.5
400
1000
5000
10,000
Number of protein-coding genes

26.25 Mimiviruses Have Genomes Similar in Size to Those of Many Parasitic Bacteria The genome of *Acanthamoeba polyphaga* mimivirus contains 1,181,404 base pairs and encompasses 911 protein-encoding genes. This observation is consistent with the hypothesis that this virus evolved from a parasitic bacterium.

Vertebrate genomes contain endogenous retroviruses

In introducing the RNA retroviruses, we noted that they insert their genomes into the genomes of their vertebrate hosts. As these incorporated retroviral genomes evolve over time, many become nonfunctional copies that are no longer expressed as functional viruses. These sequences may provide a record of ancient viral infections that plagued our ancestors. Humans, for example, carry about 100,000 fragments of endogenous retroviruses in our genome. These fragments make up about 8 percent of our DNA—a considerably larger fraction of our genome than the fraction that comprises all our protein-coding genes (about 1.2 percent of our genome).

In 2006 a French biologist named Thierry Heidmann examined the variants of one such endogenous retrovirus in a number of different people. He reasoned that each of these sequences might contain small numbers of changes from the original sequence. If so, then a consensus of the variants might resemble the original, functional sequence of a retrovirus. Heidmann constructed the consensus sequence in the lab and inserted it into human cells in culture. The cultured cells produced functional retroviruses, which could then infect other cells. Heidmann named this reconstructed virus "Phoenix" after the mythical bird that rose from the dead. Since then, biologists have resurrected other retroviruses from our genomes, and they have then identified and studied the genes that our cells use to disable these retroviral sequences and keep them from producing functional viruses. In this way, we are beginning to understand how our bodies fight retroviral infections.

Although many endogenous retroviral fragments in our genomes represent nonfunctional "ghost sequences," in some cases we appear to have derived new functions from captured retroviral sequences. For example, when a developing fetus forms a placenta, the cells in the outer layer of the placenta fuse together. This fusion is accomplished by the expression of a protein on the cell's surface, which allows the cells to attach and merge with one another. The protein that is responsible for this fusion is encoded in an endogenous retroviral sequence. Thus vertebrates appear to have co-opted some critical functions from retroviral genomes that were inserted during ancient infections.

Viruses can be used to fight bacterial infections

Although some viruses cause devastating diseases, other viruses have been used to fight disease. Most bacterial diseases are treated today with antibiotics. But antibiotics were first discovered in the 1930s, and they were not widely used to treat bacterial diseases until the 1940s; antibiotics were not available during World War I, when bacterial infections plagued the battlefields. Battlefield wounds were often infected by bacteria, and in the absence of antibiotics, these infections often led to the loss of limbs and lives. While trying to find a way to combat this problem, a physician named Felix d'Herelle discovered the first evidence of viruses that attack bacteria. He named these viruses **bacteriophages**, or "eaters of bacteria." d'Herelle extracted bacteriophages from the stool of infected patients. He then used these extracts to treat patients with deadly bacterial infections, including dysentery, cholera, and bubonic plague. This practice became known as phage therapy. After the war, phage therapy was widely used among the general public to treat bacterial infections of the skin and intestines.

Go to Media Clip 26.3
Bacteriophages Attack *E. coli*
Life10e.com/mc26.3

Phage therapy was mostly replaced by the use of antibiotics in the 1930s and 1940s as physicians grew concerned about treating patients with live viruses. Phage therapy continued to be used in the Soviet Union but largely disappeared from Western medical practice. Today, however, many antibiotics are losing their effectiveness as bacterial pathogens evolve resistance to these drugs. Phage therapy is once again an active area of research, and it is likely that bacteriophages will become increasingly important as weapons against bacterial diseases. One advantage that bacteriophages may have over antibiotics is that, like bacteria, bacteriophages can evolve. As bacteria evolve resistance to a strain of bacteriophages, biologists can select for new strains of bacteriophages that retain their effectiveness against the pathogens. In this way, biologists are using their understanding of evolution to combat the problem of antibiotic-resistant bacteria.

Viruses are found throughout the biosphere

As biologists have learned to search for and recognize viruses, they have discovered that they occur in incredible abundance almost anywhere we look on Earth. They are abundant throughout the oceans of the world, for example, with an estimated 10^{30} individual viruses in marine environments. This means there are about 15 times more viruses in the oceans than there are cellular organisms, including all bacteria, archaea, and eukaryotes combined. Many of these viruses are bacteriophages, and they have an enormous effect on the ecology of the oceans. Every day, about half of the bacteria in the oceans are killed by viruses. Huge marine blooms of bacteria, such as the *Vibrio* bloom that produced the milky seas described at the opening of this chapter, do not last long because viral blooms soon follow the initial bacterial bloom. As the bacteriophages increase, they begin to kill bacteria faster than the bacteria can reproduce. Another species of *Vibrio* causes cholera in humans, and cholera epidemics also fade as bacteriophages increase and control these *Vibrio* population booms.

Many nutrient cycles, such as the carbon cycle (see Section 58.3), are strongly influenced by viral populations. Photosynthetic bacteria and algae produce about half of the oxygen in Earth's atmosphere. During photosynthesis, these organisms also take up carbon dioxide, which they use to produce organic carbon compounds. As these photosynthetic organisms are killed by viruses, their remains settle to the ocean bottom. These remains are a primary source of the vast oil reserves that occur under oceanic sediments. The constant growth and death of these photosynthetic organisms controlled largely by viruses. Because of their importance in controlling these populations, viruses have an enormous impact on Earth's climate and ecosystems.

RECAP 26.4

Viruses are highly diverse and appear to have evolved independently from many different cellular organisms within each of the major groups of life. Some viruses appear to have evolved from escaped components of cellular organisms, whereas other viruses may have evolved from parasitic cellular ancestors.

- Why is it difficult to place viruses precisely within the tree of life? **See p. 543**
- What are the two main hypotheses of viral origins? **See pp. 544–545**
- How can viruses be used to treat some human diseases? **See pp. 545–546**
- What are some of the ways that viruses can affect Earth's ecosystems? **See p. 546**

It may be best to view viruses as "spin-offs" from the various branches on the tree of life—sometimes evolving independently of cellular genomes, sometimes recombining with them. One way to think of viruses is as the "bark" on the tree of life: an important component all across the tree, but not quite like the main branches. The third of those main branches—the eukaryotes—will be our focus in the rest of Part 7.

?

What adaptive advantage does bioluminescence provide to *Vibrio* bacteria?

ANSWER

Although these marine *Vibrio* are able to live independently, they truly thrive inside the guts of fish and other marine animals. Inside a fish, *Vibrio* cells attach themselves to food particles, including phytoplankton, and are often expelled into the ocean as waste. How can they get back into their preferred environment? The bioluminescent glow produced by a dense colony of free-living *Vibrio* growing on phytoplankton attracts fish, which consume the phytoplankton and thus ingest the bacteria—which gets the bacteria into a new host fish.

Margaret McFall-Ngai and her colleagues have studied one species of bioluminescent *Vibrio*, *V. fischeri*, in which a symbiotic relationship with the Hawaiian bobtail squid (*Euprymna scolopes*) has evolved. Chemicals produced by a developing squid embryo specifically "recruit" *V. fischeri* from the surrounding seawater. The bacteria then preferentially migrate to specific tissues that develop into a bioluminescent "light organ" in the belly of the adult squid (**Figure 26.26**). The tiny (about 3 cm long) adult squid feed while floating near the sea surface at night. The soft glow produced by the bioluminescent bacteria mimics the moonlight above, so the squid are less visible to potential predators coming at them from below.

Vibrio fischeri live symbiotically inside the squid's light organ.

26.26 Bioluminescent Bacterial Symbionts *Vibrio* bacteria within the light organ emit bioluminescence downward as the Hawaiian bobtail squid floats near the ocean surface to feed. At night, this allows the squid to blend in with moonlight or starlight rather than becoming a target for a predator from below.

CHAPTER**SUMMARY** 26

26.1 Where Do Prokaryotes Fit into The Tree of Life?

- Two of life's three domains, Bacteria and Archaea, are prokaryotic. They are distinguished from Eukarya in several ways, including their lack of a nucleus and of membrane-enclosed organelles. **Review Table 26.1**
- Eukaryotes are related to both Archaea and Bacteria and appear to have originated through endosymbiosis between members of these two lineages. The last common ancestor of all three domains probably lived about 3 billion years ago. **Review Figure 26.1, ANIMATED TUTORIAL 26.1**
- Bacteria can be classified into two groups by the **Gram stain. Gram-negative bacteria** have a periplasmic space between the plasma membrane and a distinct outer membrane. **Gram-positive bacteria** have a thick cell wall containing about five times as much peptidoglycan as a Gram-negative wall. **Review Figure 26.2, ACTIVITY 26.1**
- The three most common bacterial shapes are **cocci** (spheres), **bacilli** (rods), and **spirilla** (helices). **Review Figure 26.3**
- Phylogenetic classification of prokaryotes is now based principally on the nucleotide sequences of rRNA and other genes involved in fundamental cellular processes.
- Although **lateral gene transfer** has occurred throughout prokaryotic evolutionary history, elucidation of many aspects of prokaryote phylogeny is still possible. **Review Figure 26.4**

26.2 Why Are Prokaryotes So Diverse and Abundant?

- Prokaryotes are the most numerous organisms on Earth.
- The **low-GC Gram-positives** include the **mycoplasmas**, which are among the smallest cellular organisms ever discovered.
- Some **high-GC Gram-positives** produce important antibiotics.
- The photosynthetic **cyanobacteria** release oxygen into the atmosphere. Cyanobacteria may live free as single cells or associate in multicellular colonies. **Review Figure 26.9**
- **Spirochetes** have unique structures called axial filaments that allow them to move in a corkscrew-like manner. **Review Figure 26.10**
- The **proteobacteria** embrace the largest number of known species of bacteria. Smaller groups include the **hyperthermophilic bacteria**, **hadobacteria,** and **chlamydias**.
- Many archaea are **extremophiles**. **Review Figure 26.14**
- **Ether linkages** in the branched long hydrocarbon chains of the lipids that make up the cell membranes are a synapomorphy of Archaea. **Review Figure 26.15**

26.3 How Do Prokaryotes Affect Their Environments?

- Prokaryotes form complex communities, of which **biofilms** are one example. **Review Figure 26.19**
- Prokaryote metabolism is very diverse. Some prokaryotes are anaerobic, others are aerobic, and still others can shift between these modes.
- Prokaryotes fall into four broad nutritional categories: **photoautotrophs**, **photoheterotrophs, chemoautotrophs**, and **chemoheterotrophs**. **Review Table 26.2**
- Prokaryotes play key roles in the cycling of elements such as nitrogen, oxygen, sulfur, and carbon.
- Diverse communities of bacteria and archaea live on and in most animals. The composition of these **microbiomes** is often closely associated with the animal's health. **Review Figure 26.21**
- **Koch's postulates** establish the criteria by which an organism may be classified as a pathogen. Relatively few bacteria—and no archaea—are known to be pathogens. **Review Figure 26.22**

26.4 How Do Viruses Relate to Life's Diversity and Ecology?

- Viruses have evolved many times from many different groups of cellular organisms. They are placed in groups according to the structure of their genomes, but these groups are not thought to represent monophyletic taxa. **Review Figure 26.23**
- Some viruses are probably derived from escaped components of cellular organisms; others are thought to have evolved as highly reduced parasites. **Review Figure 26.25**
- A large fraction of vertebrate (including human) genomes consists of incorporated remains of retroviral genomes.
- Bacteriophages have been used to treat bacterial infections in humans.
- Viruses are found in virtually all of Earth's environments and have a huge impact on the planet's ecosystems.

Go to the Interactive Summary to review key figures, Animated Tutorials, and Activities Life10e.com/is26

CHAPTER**REVIEW**

REMEMBERING

1. The division of the living world into three domains
   a. is based on the number of cells in organisms of each group.
   b. is based mostly on major morphological differences between archaea and bacteria.
   c. emphasizes the greater importance of eukaryotes.
   d. was proposed by the early microscopists.
   e. is based on phylogenetic relationships determined from nucleotide sequences of rRNA and other genes.

2. Which statement about nitrogen metabolism is *not* true?
   a. Certain prokaryotes reduce atmospheric N_2 to ammonia.
   b. Some nitrifiers are soil bacteria.
   c. Denitrifiers are obligate anaerobes.
   d. Nitrifiers obtain energy by oxidizing ammonia and nitrite.
   e. Without nitrifiers, terrestrial organisms would lack a nitrogen supply.

3. All photosynthetic bacteria
   a. use chlorophyll *a* as their photosynthetic pigment.
   b. use bacteriochlorophyll as their photosynthetic pigment.
   c. release oxygen gas (O_2).
   d. produce particles of sulfur.
   e. are photoautotrophs.

4. Gram-negative bacteria
   a. appear blue to purple following Gram staining.
   b. appear pink to red following Gram staining.
   c. are all either bacilli or cocci.
   d. contain no peptidoglycan in their cell walls.
   e. are all photosynthetic.

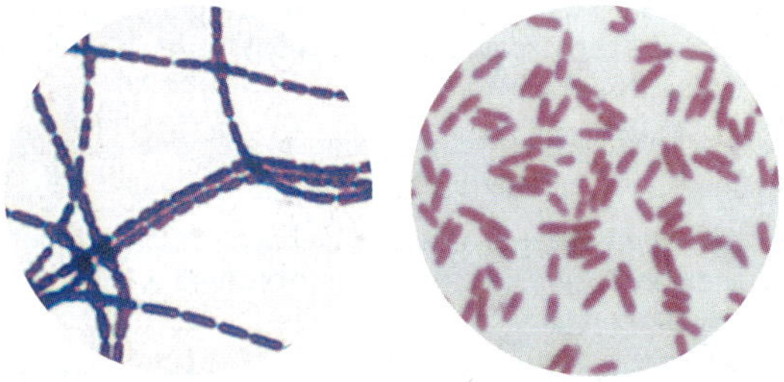

5. Archaea
   a. have cytoskeletons.
   b. have distinctive lipids in their plasma membranes.
   c. survive only at moderate temperatures and near neutral pH.
   d. all produce methane.
   e. have substantial amounts of peptidoglycan in their cell walls.

6. Genetic evidence suggests that viruses
   a. are most closely related to Bacteria.
   b. are most closely related to Archaea.
   c. are most closely related to Eukarya.
   d. have evolved multiple times from many different cellular species.
   e. evolved from the fusion of a bacterial and an archaeal species.

UNDERSTANDING & APPLYING

7. Why do systematic biologists find rRNA sequence data more useful than data on metabolism or cell structure for classifying prokaryotes?

8. The figure below shows an organismal tree in which gene *x* has undergone a lateral transfer event. Draw the phylogenetic tree you would expect based on gene *x*, as well as the phylogenetic tree you would expect based on a consensus of non-transferred genes.

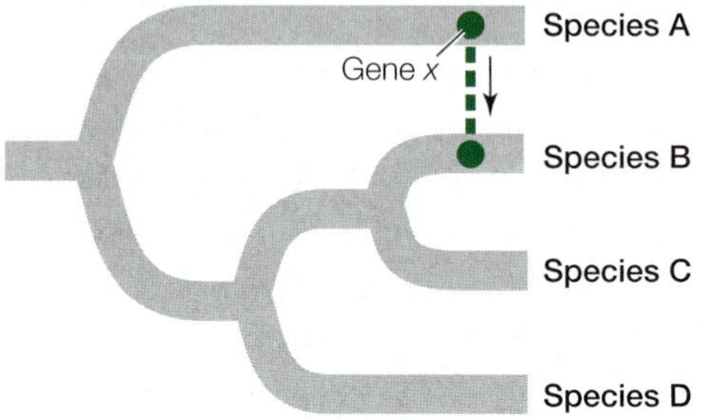
Gene *x*
Species A
Species B
Species C
Species D

9. Do you consider viruses to be living organisms? Why or why not?

ANALYZING & EVALUATING

10. Kashefi and Lovley were able to grow an unnamed archaeal species at temperatures above 120°C only because they used Fe_3^+ as an electron acceptor—no other electron acceptor they tried allowed the archaean to grow (see Figure 26.14). How might you explore the same or other high-temperature environments for other hyperthermophilic organisms not detected by Kashefi and Lovley using Fe_3^+?

27 The Origin and Diversification of Eukaryotes

CHAPTER**OUTLINE**

A Toxic Sea A bloom of dinoflagellates of the genus *Noctiluca* was responsible for this red tide in Puget Sound in the U. S. state of Washington.

IN THE SUMMER of 2005, a devastating red tide crippled the shellfish industry along the Atlantic coast of North America from Canada to Massachusetts. This red tide was produced by a bloom of dinoflagellates of the genus *Alexandrium*. These protists produce a powerful toxin that accumulates in clams, mussels, and oysters. A person who eats a mollusk contaminated with the toxin can experience a syndrome known as paralytic shellfish poisoning. Many people were sickened by eating mollusks that were harvested before the problem was diagnosed, and losses to the shellfish industry in 2005 were estimated at $50 million.

Several species of dinoflagellates produce toxic red tides in many parts of the world. Along the Gulf of Mexico, red tides caused by dinoflagellates of the genus *Karenia* produce a neurotoxin that affects the central nervous systems of fish, which become paralyzed and cannot respire effectively. Huge numbers of dead fish wash up on Gulf Coast beaches during a *Karenia* red tide. In addition, wave action can produce aerosols of the *Karenia* toxin, and these aerosols often cause asthma-like symptoms in humans on shore.

After the losses that resulted from the 2005 red tide, biologists at the Woods Hole Oceanographic Institution (WHOI) on Cape Cod began to monitor and model dinoflagellate populations off the New England coast. If biologists could accurately forecast future blooms, people in the area could be made aware of the problem in advance and adjust the shellfish harvest (and their eating habits) accordingly.

Biologists from WHOI monitored counts of dinoflagellates in the water and in seafloor sediments. They also monitored river runoff, water currents, water temperature and salinity, winds, and tides. Another environmental factor they considered was the "nor'easter" storms common along the New England coast. By correlating their measurements of these environmental factors with dinoflagellate counts, biologists produced a model that predicted growth of dinoflagellate populations.

In spring 2008, the WHOI team determined that all the factors were in place to produce another red tide like the one of 2005—if a nor'easter occurred to blow the dinoflagellates toward the coast. A nor'easter did occur at just the wrong time, and another red tide materialized in summer 2008, just as predicted. But this time, people were warned. Shellfish harvesters adjusted their harvest, and many fewer people were harmed by eating toxic mollusks.

?

Can dinoflagellates be beneficial, as well as harmful, to marine ecosystems?

See answer on p. 566.

27.1 How Did the Eukaryotic Cell Arise?

We easily recognize trees, mushrooms, and insects as plants, fungi, and animals, respectively. But there is a dazzling assortment of other eukaryotic organisms—mostly microscopic—that do not fit into these three groups. Eukaryotes that are not plants, animals, or fungi have traditionally been called **protists**. But phylogenetic analyses reveal that many of the groups we commonly refer to as protists are not, in fact, closely related. Thus the term "protist" does not describe a formal taxonomic group, but is a convenience term for "all the eukaryotes that are not plants, animals, or fungi."

The unique characteristics of the eukaryotic cell lead scientists to conclude that the eukaryotes are monophyletic and that a single eukaryotic ancestor diversified into the many different protist lineages as well as giving rise to the plants, fungi, and animals. As we saw in Section 26.1, eukaryotes are generally thought to be more closely related to Archaea than to Bacteria. The mitochondria and chloroplasts of eukaryotes, however, are clearly derived from bacterial lineages (see Figure 26.1).

Biologists traditionally hypothesized that the split of Eukarya from Archaea was followed by endosymbioses with bacterial lineages that led to the origin of mitochondria and chloroplasts. Some biologists prefer to view the origin of eukaryotes as the *fusion* of lineages from the two prokaryote groups. This difference is largely a semantic one that hinges on the point at which we deem the eukaryote lineage to have become definitively "eukaryotic." In either case, we can make some reasonable inferences about the events that led to the evolution of a new cell type, bearing in mind that the environment underwent an enormous change—from low to high availability of atmospheric oxygen—during the course of these events.

The modern eukaryotic cell arose in several steps

At least five events were significant in the origin of the eukaryotic cell (**Figure 27.1**):

- The origin of a flexible cell surface
- The origin of a cytoskeleton
- The origin of a nuclear envelope, which enclosed a genome organized into chromosomes

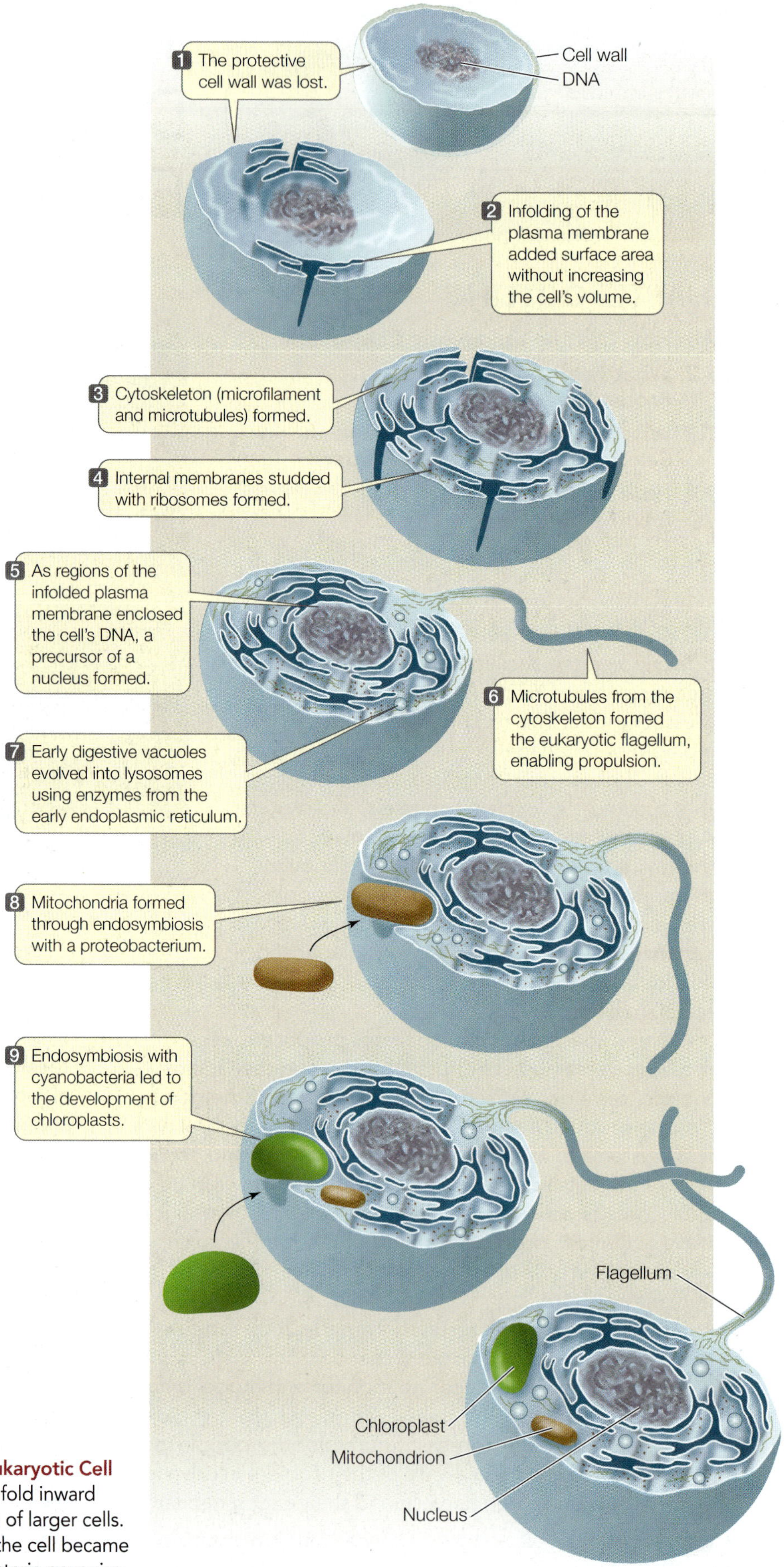

1 The protective cell wall was lost.
Cell wall
DNA
2 Infolding of the plasma membrane added surface area without increasing the cell's volume.
3 Cytoskeleton (microfilament and microtubules) formed.
4 Internal membranes studded with ribosomes formed.
5 As regions of the infolded plasma membrane enclosed the cell's DNA, a precursor of a nucleus formed.
6 Microtubules from the cytoskeleton formed the eukaryotic flagellum, enabling propulsion.
7 Early digestive vacuoles evolved into lysosomes using enzymes from the early endoplasmic reticulum.
8 Mitochondria formed through endosymbiosis with a proteobacterium.
9 Endosymbiosis with cyanobacteria led to the development of chloroplasts.
Flagellum
Chloroplast
Mitochondrion
Nucleus

27.1 A Hypothetical Sequence for the Evolution of the Eukaryotic Cell The loss of a firm cell wall allowed the plasma membrane to fold inward and create more surface area, which facilitated the evolution of larger cells. As cells grew larger, cytoskeletal complexity increased, and the cell became increasing compartmentalized. Endosymbioses involving bacteria gave rise to mitochondria and (in photosynthetic eukaryotes) to chloroplasts.

- The appearance of digestive vacuoles
- The acquisition of mitochondria and chloroplasts via endosymbiosis

FLEXIBLE CELL SURFACE We presume that ancient prokaryotic organisms, like most present-day prokaryotic cells, had firm cell walls. The first step toward the eukaryotic condition was the loss of the cell wall.

Consider the possibilities open to a flexible cell without a firm wall, starting with cell size. As a cell grows larger, its surface area-to-volume ratio decreases (see Figure 5.2). Unless the surface area can be increased, the cell's volume will reach an upper limit. If the cell's surface is flexible, however, it can fold inward, creating more surface area for gas and nutrient exchange. With a surface flexible enough to allow infolding, the cell can exchange materials with its environment rapidly enough to sustain a larger volume and more rapid metabolism (Figure 27.1, steps 1–2). Furthermore, a flexible surface can pinch off bits of the environment and bring them into the cell by endocytosis. These infoldings of the cell surface (which also exist in some modern prokaryotes) were important for the evolution of large eukaryotic cells.

CHANGES IN CELL STRUCTURE AND FUNCTION Other early steps that were important for the evolution of the eukaryotic cell are likely to have included three advances: the formation of ribosome-studded internal membranes, some of which surrounded the DNA; development of a complex cytoskeleton; and the evolution of digestive vacuoles (Figure 27.1, steps 3–7).

Until a few years ago, biologists thought that cytoskeletons were restricted to eukaryotes. Improved imaging technology and molecular analyses have now revealed homologs of many cytoskeletal proteins in prokaryotes, showing that simple cytoskeletons evolved before the origin of eukaryotes. The cytoskeleton of a eukaryote, however, is much more developed and complex than that of a prokaryote. This greater development of microfilaments and microtubules supports the eukaryotic cell and allows it to manage changes in shape, to distribute daughter chromosomes, and to move materials from one part of the large cell to other parts. In addition, the presence of microtubules in the cytoskeleton could have given rise in some cells to the characteristic eukaryotic flagellum.

The DNA of a prokaryotic cell is attached to a site on its plasma membrane. If that region of the plasma membrane were to fold into the cell, the first step would be taken toward the evolution of a nucleus, a primary feature of the eukaryotic cell.

The nuclear envelope appeared early in the eukaryote lineage. The next step was probably phagocytosis—the ability to engulf and digest other cells in digestive vacuoles. These digestive vacuoles eventually incorporated digestive enzymes to form lysosomes. Other infoldings of the plasma membrane developed into the endoplasmic reticulum and Golgi apparatus.

ENDOSYMBIOSIS At the same time the processes outlined above were taking place, cyanobacteria were generating oxygen gas as a product of photosynthesis. Increasing concentrations of O_2 in the oceans, and eventually in the atmosphere, had disastrous consequences for most organisms of the time, which were unable to tolerate the newly oxidizing environment. But some prokaryotes evolved strategies to utilize the increasing oxygen and—fortunately for us—so did some of the new phagocytic eukaryotes.

At about this time, endosymbioses began to play a role in eukaryote evolution (Figure 27.1, steps 8–9). The theory of endosymbiosis proposes that certain organelles are the descendants of prokaryotes engulfed, but not digested, by early eukaryotic cells. One crucial event in the history of eukaryotes was the incorporation of a proteobacterium that evolved into the mitochondrion. Initially, the new organelle's primary function was probably to detoxify O_2 by reducing it to water. Later, this reduction became coupled with the formation of ATP in cellular respiration. Upon completion of this step, the essential modern eukaryotic cell was complete.

Photosynthetic eukaryotes are the result of yet another endosymbiotic step: the incorporation of a prokaryote related to today's cyanobacteria, which became the chloroplast.

Chloroplasts have been transferred among eukaryotes several times

Eukaryotes in several different groups possess chloroplasts, and groups with chloroplasts appear in several distantly related eukaryote clades. Some of these groups differ in the photosynthetic pigments their chloroplasts contain. And not all chloroplasts are limited to a pair of surrounding membranes like those found in plants—in some microbial eukaryotes, they are surrounded by three or more membranes. We now view these observations as evidence of a remarkable series of endosymbioses. This conclusion is supported by extensive evidence from electron microscopy and nucleic acid sequence comparisons.

All chloroplasts trace their ancestry back to the engulfment of one cyanobacterium by a larger eukaryotic cell. This event, the step that first gave rise to the photosynthetic eukaryotes, is known as **primary endosymbiosis** (Figure 27.2A). The cyanobacterium, a Gram-negative bacterium, had both an inner and an outer membrane (see Figure 26.2B). Thus the original chloroplasts had two surrounding membranes: the inner and outer membranes of the cyanobacterium. Remnants of the peptidoglycan-containing cell wall of the bacterium are present in the form of a bit of peptidoglycan between the chloroplast membranes of glaucophytes, the first eukaryote group to branch off following primary endosymbiosis (as we will see in Chapter 28). Primary endosymbiosis also gave rise to the chloroplasts of the red algae, green algae, and land plants. The red algal chloroplast retains certain pigments of the original cyanobacterial endosymbiont that are absent in green algal chloroplasts.

Almost all remaining photosynthetic eukaryotes are the result of additional rounds of endosymbiosis. For example, the photosynthetic euglenids derived their chloroplasts from **secondary endosymbiosis** (Figure 27.2B). Their ancestor took up a unicellular green alga, retaining its chloroplast and eventually losing the rest of the constituents of the alga. This history explains why the photosynthetic euglenids have the same photosynthetic pigments as the green algae and land plants. It also

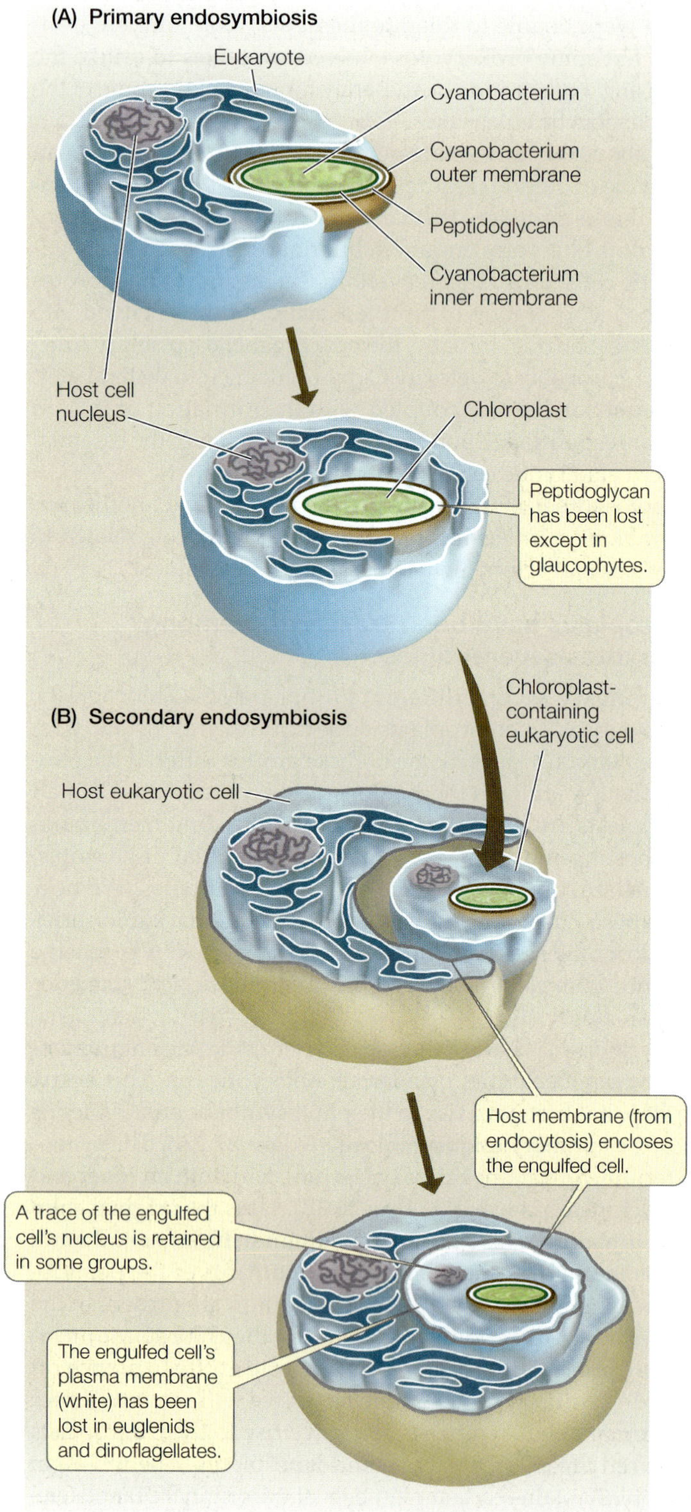
(A) Primary endosymbiosis
Eukaryote
Cyanobacterium
Cyanobacterium outer membrane
Peptidoglycan
Cyanobacterium inner membrane
Host cell nucleus
Chloroplast
Peptidoglycan has been lost except in glaucophytes.
(B) Secondary endosymbiosis
Chloroplast-containing eukaryotic cell
Host eukaryotic cell
Host membrane (from endocytosis) encloses the engulfed cell.
A trace of the engulfed cell's nucleus is retained in some groups.
The engulfed cell's plasma membrane (white) has been lost in euglenids and dinoflagellates.

27.2 Endosymbiotic Events in the Evolution of Chloroplasts
(A) A single instance of primary endosymbiosis ultimately gave rise to all of today's chloroplasts. (B) Secondary endosymbiosis—the uptake and retention of a chloroplast-containing cell by another eukaryotic cell—took place several times, independently.

Go to Animated Tutorial 27.1
Family Tree of Chloroplasts
Life10e.com/at27.1

accounts for the third membrane of the euglenid chloroplast, which is derived from the euglenid's plasma membrane (as a result of endocytosis). An additional round—**tertiary endosymbiosis**—occurred when a dinoflagellate apparently lost its chloroplast and took up another protist that had acquired its chloroplast through secondary endosymbiosis.

RECAP 27.1

The modern eukaryotic cell probably arose from an ancestral prokaryote in several steps, including the origin of a flexible cell surface, the enclosure of the genetic material in a nucleus, and endosymbiosis.

- What were the steps that led to the evolution of the eukaryotic cell? **See pp. 550–551 and Figure 27.1**
- Why was the development of a flexible cell surface a key event for eukaryote evolution? **See p. 551**
- Explain how increased availability of atmospheric oxygen (O_2) could have influenced the evolution of the eukaryotic cell. **See p. 551**

The features that eukaryotes gained from archaea and bacteria have allowed them to exploit many different environments. This led to the evolution of great diversity among eukaryotes, beginning with a radiation that started in the Precambrian.

27.2 What Features Account for Protist Diversity?

Most eukaryotes can be placed in one of eight major clades that began to diversify about 1.5 billion years ago: alveolates, stramenopiles, rhizaria, excavates, plants, amoebozoans, fungi, and animals (Figure 27.3). Plants, fungi, and animals each have close protist relatives (such as the choanoflagellate relatives of animals), which we will discuss along with those major multicellular eukaryote groups in Chapters 28–33.

Each of the five major groups of eukaryotes covered in this chapter consists of organisms with enormously diverse body forms and lifestyles. Some protists are motile, whereas others do not move; some are photosynthetic, others heterotrophic; most are unicellular, but some are multicellular. Most are microscopic, but a few are huge (giant kelps, for example, can grow to half the length of a football field). We refer to the unicellular species of protists as **microbial eukaryotes**, but keep in mind that there are large, multicellular protists as well.

Multicellularity has arisen dozens of times across the evolutionary history of eukaryotes. Four of the origins of multicellularity resulted in large organisms that are familiar to most people: plants, animals, fungi, and brown algae (the last are a group of stramenopiles). In addition, there are dozens of smaller and less familiar groups among the eukaryotes that include multicellular species. Recent experimental studies have shown that artificial selection for multicellularity can produce repeated, convergent evolution of multicellular forms over just a few months in some normally unicellular eukaryotic species. In addition, many unicellular species retain individual identities but nonetheless associate in large multicellular colonies.

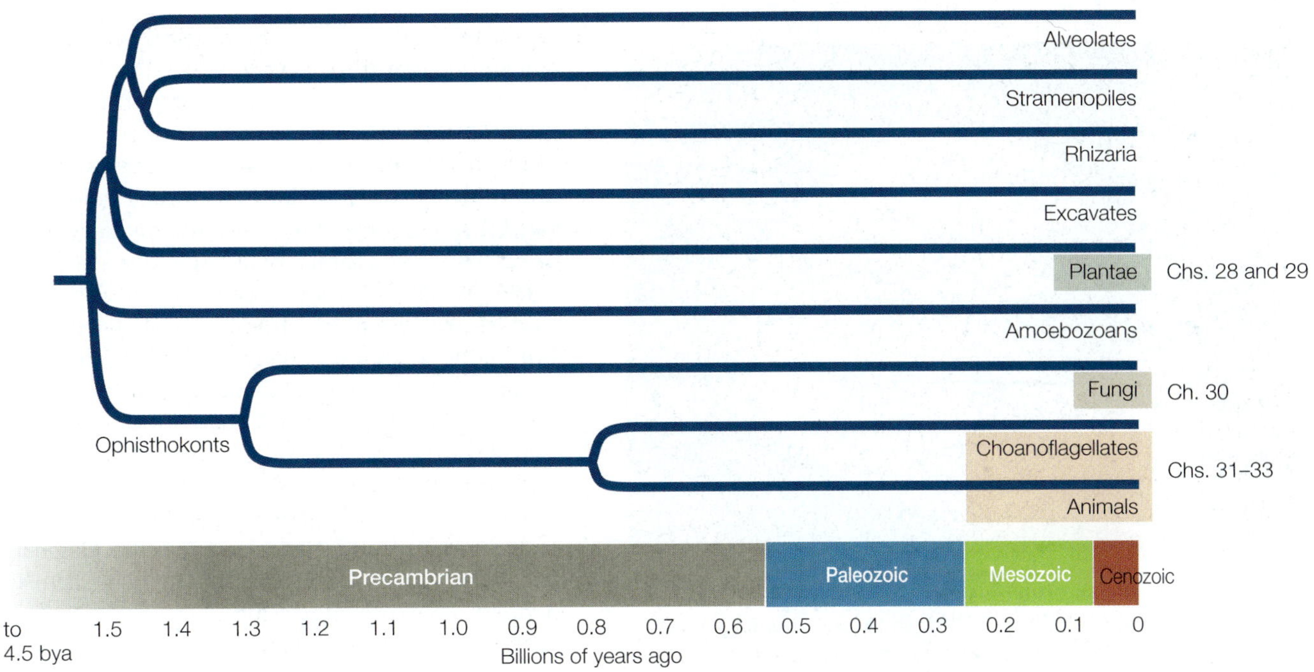

Alveolates
Stramenopiles
Rhizaria
Excavates
Plantae
Chs. 28 and 29
Amoebozoans
Fungi
Ch. 30
Ophisthokonts
Choanoflagellates
Chs. 31–33
Animals
Precambrian
Paleozoic
Mesozoic
Cenozoic
to 4.5 bya
1.5
1.4
1.3
1.2
1.1
1.0
0.9
0.8
0.7
0.6
0.5
0.4
0.3
0.2
0.1
0
Billions of years ago

27.3 Precambrian Divergence of Major Eukaryote Groups
A phylogenetic tree shows one current hypothesis and estimated time line for the origin of the major groups of eukaryotes. The rapid divergence of major lineages between 1.5 and 1.4 billion years ago makes reconstruction of their precise relationships difficult. The major multicellular groups (tinted boxes) will be covered in subsequent chapters.

There is a near-continuum between fully integrated, multicellular organisms on the one hand and loosely integrated multicellular colonies of cells on the other. Biologists do not always agree on where to draw the line between the two.

Biologists used to classify protists largely on the basis of their life histories and reproductive features. In recent years, however, electron microscopy and gene sequencing have revealed many new patterns of evolutionary relatedness among these groups. Analyses of slowly evolving gene sequences are making it possible to explore evolutionary relationships among eukaryotes in ever greater detail and with greater confidence. Nonetheless, some substantial areas of uncertainty remain, and lateral gene transfer may complicate our efforts to reconstruct the evolutionary history of protists (as is also true for prokaryotes; see Section 26.1). Today we recognize great diversity among the many distantly related protist clades.

Alveolates have sacs under their plasma membranes

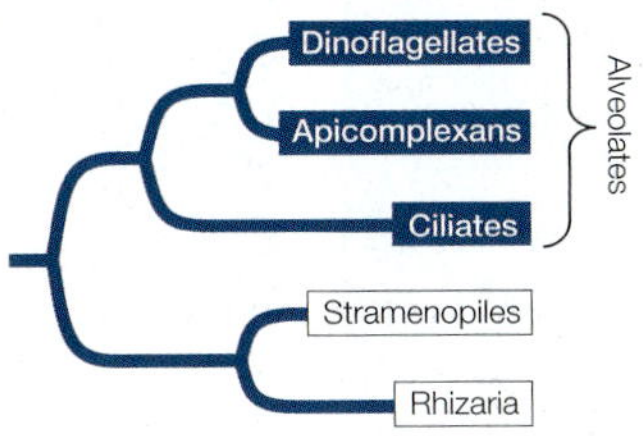

Dinoflagellates
Apicomplexans
Ciliates
Alveolates
Stramenopiles
Rhizaria

Alveolates are so named because they possess sacs, called alveoli, just beneath their plasma membranes, which may play a role in supporting the cell surface. All alveolates are unicellular, and most are photosynthetic, but they are diverse in body form. The alveolate groups we will consider in detail here are the dinoflagellates, apicomplexans, and ciliates.

DINOFLAGELLATES Most **dinoflagellates** are marine and photosynthetic; they are important primary producers of organic matter in the oceans. Although fewer *species* of dinoflagellates live in fresh water, individuals can be abundant in freshwater environments. The dinoflagellates are of great ecological, evolutionary, and morphological interest. A distinctive mixture of photosynthetic and accessory pigments gives their chloroplasts a golden brown color. Some dinoflagellate species cause red tides, as discussed at the opening of this chapter. Other species are photosynthetic endosymbionts that live within the cells of other organisms, including invertebrate animals (such as corals; see Figure 27.21) and other marine protists (see Figure 27.12A). Still others are nonphotosynthetic and live as parasites within other marine organisms.

Dinoflagellates have a distinctive appearance. They generally have two flagella, one in an equatorial groove around the cell, the other starting near the same point as the first and passing down a longitudinal groove before extending into the surrounding medium (**Figure 27.4**). Some dinoflagellates can take on different forms, including amoeboid ones, depending on environmental conditions. It has been claimed that the dinoflagellate *Pfiesteria piscicida* can occur in at least two dozen distinct forms, although this claim is highly controversial. In any case, this remarkable dinoflagellate, when present in large enough numbers, is harmful to fish and can both stun and feed on them.

APICOMPLEXANS The exclusively parasitic **apicomplexans** derive their name from the apical complex, a mass of organelles contained in the apical end (the tip) of the cell. These organelles help the apicomplexan invade its host's tissues. For example, the apical complex enables *Plasmodium*, the causative agent of malaria, to enter its target cells in the human body after transmission by a mosquito.

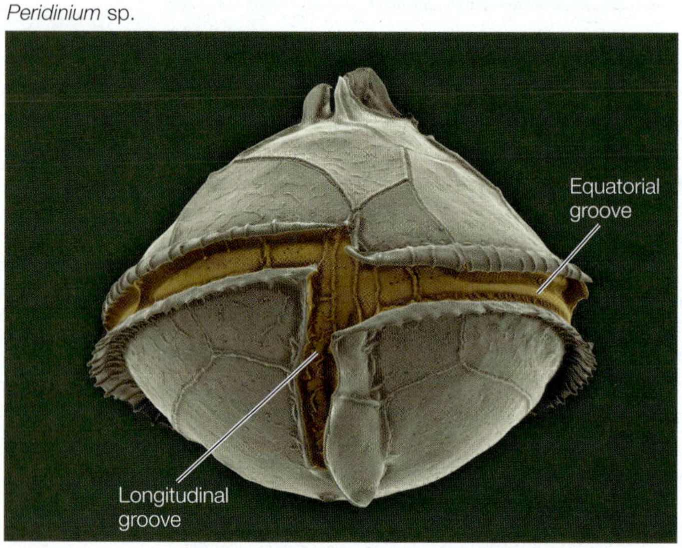
Peridinium sp.
Equatorial groove
Longitudinal groove

27.4 A Dinoflagellate The presence of two flagella is characteristic of many dinoflagellates, although these appendages are contained within deep grooves and thus are seldom visible. One flagellum lies within the equatorial groove and provides forward thrust and spin to the organism. The second flagellum originates in the longitudinal groove and acts like the rudder of a boat.

Go to Media Clip 27.1
A Dinoflagellate Shows Off Its Flagellum
Life10e.com/mc27.1

Like many obligate parasites, apicomplexans have elaborate life cycles featuring asexual and sexual reproduction through a series of very dissimilar life stages (see Figure 27.20). In many species, these life stages are associated with two different types of host organisms, as is the case with *Plasmodium*. Another apicomplexan, *Toxoplasma*, alternates between cats and rats to complete its life cycle. A rat infected with *Toxoplasma* loses its fear of cats, which makes it more likely to be eaten by, and thus transfer the parasite to, a cat.

CILIATES The **ciliates** are named for their numerous hairlike cilia, which are shorter than, but otherwise identical to, eukaryotic flagella. The ciliates are much more complex in body form than are most other unicellular eukaryotes (**Figure 27.5**). Their definitive characteristic is the possession of two types of nuclei (whose roles we will describe in Section 27.3 when we discuss protist reproduction). Almost all ciliates are heterotrophic, although a few contain photosynthetic endosymbionts.

Paramecium, a frequently studied ciliate genus, exemplifies the complex structure and behavior of ciliates (**Figure 27.6**). The slipper-shaped cell is covered by an elaborate pellicle, a structure composed principally of an outer membrane and an inner layer of closely packed, membrane-enclosed sacs (the alveoli) that surround the bases of the cilia. Defensive organelles called trichocysts are also present in the pellicle. In response to a threat, a microscopic explosion expels the trichocysts in a few milliseconds, and they emerge as sharp darts, driven forward at the tip of a long, expanding filament.

The cilia provide *Paramecium* with a form of locomotion that is generally more precise than locomotion by flagella. A *Paramecium* can coordinate the beating of its cilia to propel itself either forward or backward in a spiraling manner. It can also back off swiftly when it encounters a barrier or a negative stimulus. The coordination of ciliary beating is probably the result of a differential distribution of ion channels in the plasma membrane near the two ends of the cell.

Organisms living in fresh water are hypertonic to their environment. Many freshwater protists, including *Paramecium*, address this problem by means of specialized **contractile vacuoles** that excrete the excess water the organisms constantly take in by osmosis. The excess water collects in the contractile vacuoles, which then contract and expel the water from the cell.

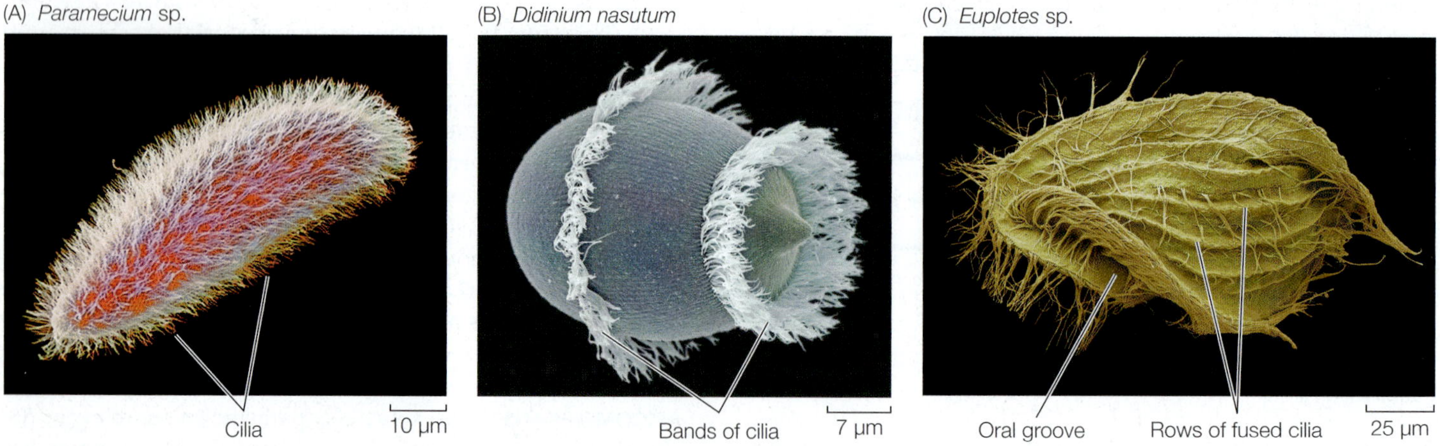
(A) Paramecium sp.
(B) Didinium nasutum
(C) Euplotes sp.
Cilia
10 μm
Bands of cilia
7 μm
Oral groove
Rows of fused cilia
25 μm

27.5 Diversity among the Ciliates (A) A free-swimming organism, this *Paramecium* belongs to a ciliate group whose members have many cilia of uniform length. (B) The barrel-shaped *Didinium nasutum* feeds on other ciliates, including *Paramecium*. Its cilia occur in two separate bands. (C) Some of the cilia in *Euplotes* fuse into flat sheets that direct food particles into an oral groove.

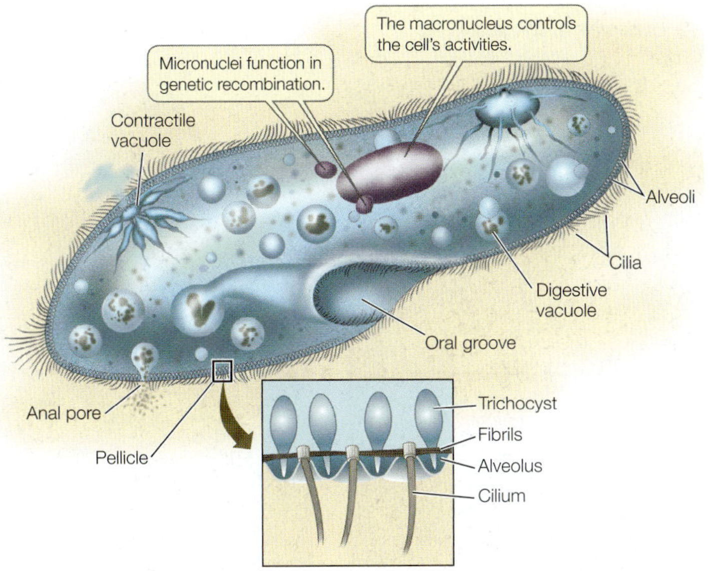
The macronucleus controls the cell's activities.
Micronuclei function in genetic recombination.
Contractile vacuole
Alveoli
Cilia
Digestive vacuole
Oral groove
Anal pore
Pellicle
Trichocyst
Fibrils
Alveolus
Cilium

27.6 Anatomy of *Paramecium* *Paramecium*, with its many specialized organelles, exemplifies the complex body form of ciliates.

Go to Activity 27.1 Anatomy of *Paramecium*
Life10e.com/ac27.1

Paramecium and many other protists engulf solid food by endocytosis, forming a **digestive vacuole** within which the food is digested (**Figure 27.7**). Smaller vesicles containing digested food pinch away from the digestive vacuole and enter the cytoplasm. These tiny vesicles provide a large surface area across which the products of digestion can be absorbed by the rest of the cell.

Go to Animated Tutorial 27.2
Digestive Vacuoles
Life10e.com/at27.2

Stramenopiles typically have two flagella of unequal length

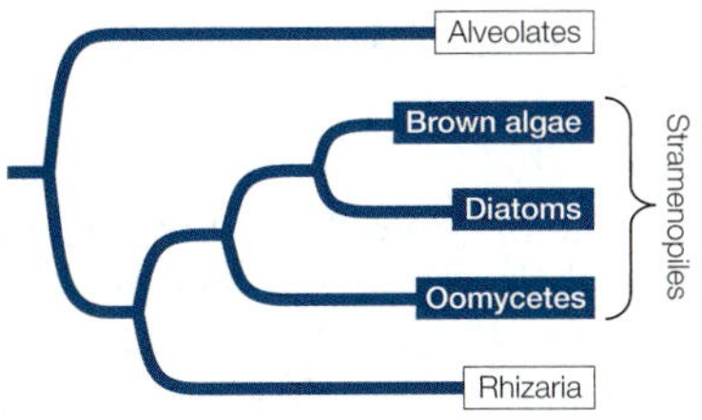
Alveolates
Brown algae
Diatoms
Oomycetes
Rhizaria
Stramenopiles

A morphological synapomorphy of most **stramenopiles** is the possession of two flagella of unequal length, with rows of tubular hairs on the longer of the two. Some stramenopiles lack flagella, but they are descended from ancestors that possessed flagella. The stramenopiles include the diatoms and the brown algae, which are photosynthetic, and the oomycetes, which are not.

DIATOMS All of the **diatoms** are unicellular, although some species associate in filaments. Many have sufficient carotenoids in their chloroplasts to give them a yellow or brownish color. All of them synthesize carbohydrates and oils as photosynthetic storage products. Diatoms lack flagella except in male gametes.

Architectural magnificence on a microscopic scale is the hallmark of the diatoms. Almost all diatoms deposit silica (hydrated silicon dioxide) in their cell walls. The cell wall of a diatom is constructed in two pieces, with the top overlapping the bottom like the top of a petri dish. The silica-impregnated walls have intricate patterns unique to each species (**Figure 27.8**). Despite their remarkable morphological diversity, all diatoms are symmetrical—either bilaterally (with "right" and "left" halves) or radially (with the type of symmetry possessed by a circle).

Diatoms reproduce both sexually and asexually. Asexual reproduction by binary fission is somewhat constrained by the stiff cell wall. Both the top and bottom of the "petri dish" become tops of new "dishes" without changing appreciably in size; as a result, the new cell made from the former bottom is smaller than the parent cell. If this process continued indefinitely, one cell line would simply vanish, but sexual reproduction largely solves this

INVESTIGATING**LIFE**

27.7 The Role of Vacuoles in Ciliate Digestion An acidic environment is known to aid digestion in many multicellular organisms. Do ciliates also use acid to obtain nutrients?[a]

HYPOTHESIS The digestive vacuoles of *Paramecium* produce an acidic environment that allows the organism to digest food particles.

Method

1. Feed *Paramecium* yeast cells stained with Congo red, a dye that is red at neutral or basic pH but turns green at acidic pH.
2. Under a light microscope, observe the formation and degradation of digestive vacuoles within the *Paramecium*. Note time and sequence of color (i.e., acid level) changes.

Results

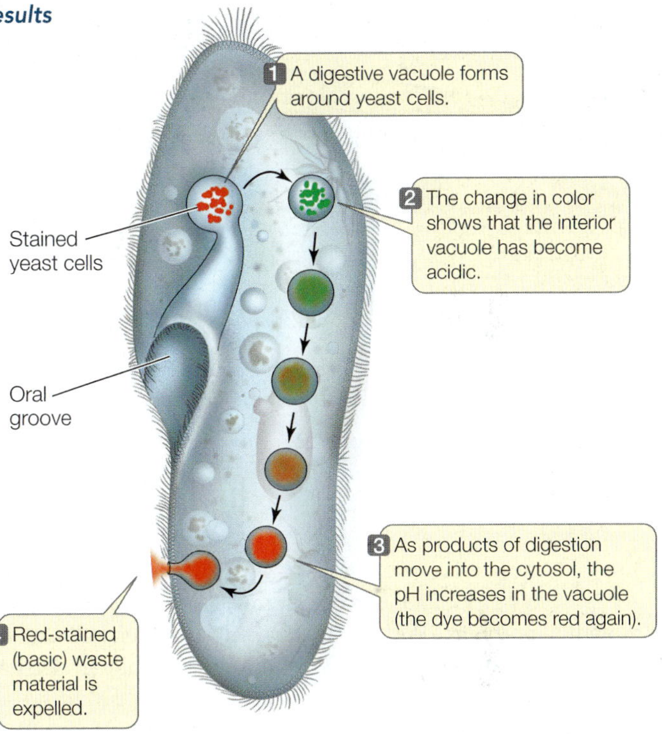
1 A digestive vacuole forms around yeast cells.
2 The change in color shows that the interior vacuole has become acidic.
Stained yeast cells
Oral groove
3 As products of digestion move into the cytosol, the pH increases in the vacuole (the dye becomes red again).
4 Red-stained (basic) waste material is expelled.

CONCLUSION Some ciliates acidify digestive vacuoles to assist in the breakdown of food.

Go to **BioPortal** for discussion and relevant links for all INVESTIGATING**LIFE** figures.

[a]Mast, S. O. 1947. *Biological Bulletin* 92: 31–72.

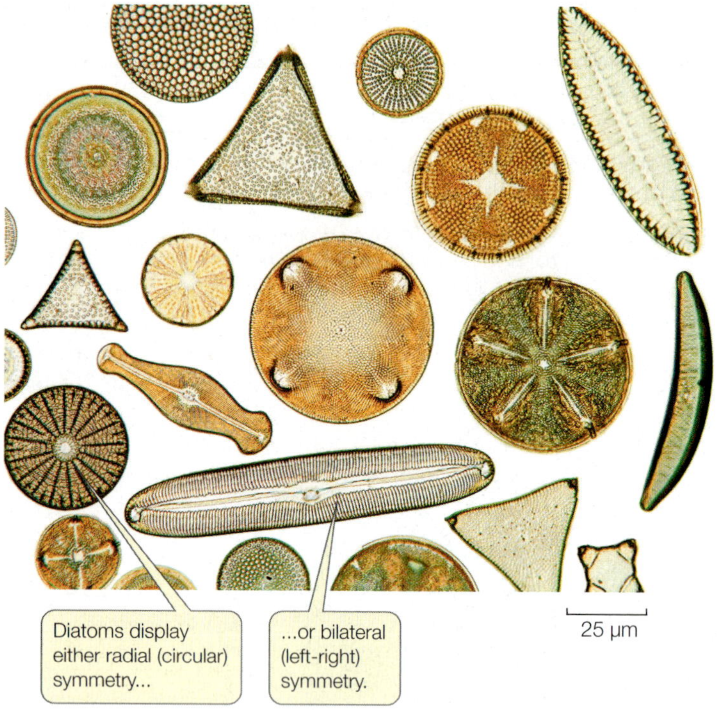
Diatoms display either radial (circular) symmetry...
...or bilateral (left-right) symmetry.
25 µm

27.8 Diatom Diversity This bright-field micrograph illustrates the variety of species-specific forms found among the diatoms.

Go to Media Clip 27.2
Diatoms In Action
Life10e.com/mc27.2

(A) *Himanthalia elongata*
(B) *Postelsia palmiformis*
Holdfasts

27.9 Brown Algae (A) This seaweed illustrates the filamentous growth form of the brown algae. (B) Sea palms exemplify the leaf-like growth form of brown algae. Sea palms and many other brown algal species are "glued" to the rocks by tough, branched structures called holdfasts that can withstand the pounding of the surf.

potential problem. Gametes are formed, shed their cell walls, and fuse. The resulting zygote then grows substantially in size before a new cell wall is laid down.

Diatoms are found in all the oceans and are frequently present in great numbers. They are major photosynthetic producers in coastal waters and are among the dominant organisms in the dense "blooms" of phytoplankton that occasionally appear in the open ocean (see Section 27.4). Diatoms are also common in fresh water and even occur on the wet surfaces of terrestrial mosses.

BROWN ALGAE The **brown algae** obtain their namesake color from the carotenoid **fucoxanthin**, which is abundant in their chloroplasts. The combination of this yellow-orange pigment with the green of chlorophylls *a* and *c* yields a brownish tinge. All brown algae are multicellular, and some are extremely large. Giant kelps such as those of the genus *Macrocystis* may be up to 60 meters long (see Figure 54.9).

Go to Media Clip 27.3
A Kelp Forest
Life10e.com/mc27.3

The brown algae are almost exclusively marine. They are composed either of branched filaments (**Figure 27.9A**) or of leaflike growths (**Figure 27.9B**). Some float in the open ocean; the most famous example is the genus *Sargassum*, which forms dense mats in the Sargasso Sea in the mid-Atlantic. Most brown algae, however, attach themselves to rocks near the shore. A few thrive only where they are regularly exposed to heavy surf. All of the attached forms develop a specialized structure, called a holdfast, that literally glues them to the rocks. The "glue" of the holdfast is alginic acid, a gummy polymer found in the walls of many brown algal cells. In addition to its function in holdfasts, alginic acid cements algal cells and filaments together. It is harvested and used by humans as an emulsifier in ice cream, cosmetics, and other products.

OOMYCETES The **oomycetes** are the water molds and their terrestrial relatives. Water molds are filamentous and stationary. They are **absorptive heterotrophs**—that is, they secrete enzymes that digest large food molecules into smaller molecules that they can absorb. They are all aquatic and **saprobic**—meaning they feed on dead organic matter. If you have seen a whitish, cottony mold growing on dead fish or dead insects in water, it was probably a water mold of the common genus *Saprolegnia* (**Figure 27.10**).

Some other oomycetes, such as the downy mildews, are terrestrial. Although most of the terrestrial oomycetes are harmless or helpful decomposers of dead matter, a few are plant parasites that attack crops such as avocados, grapes, and potatoes.

Oomycetes were once classified as fungi. However, we now know that their similarity to fungi is only superficial, and that the oomycetes are more distantly related to the fungi than are many other eukaryote groups, including humans (see Figure 27.3). For example, the cell walls of oomycetes are typically made of cellulose, whereas those of fungi are made of chitin.

Saprolegnia sp.

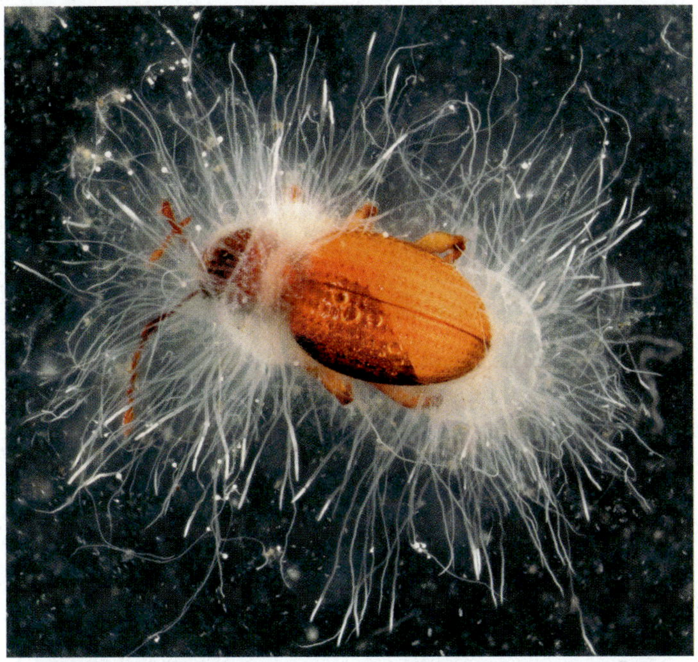
3 mm

27.10 An Oomycete The filaments of a water mold radiate from the carcass of a beetle.

1 mm

27.11 Building Blocks of Limestone Some foraminiferans secrete calcium carbonate to form shells. The shells of different species have distinctive shapes. Over millions of years, the shells of foraminiferans have accumulated to form limestone deposits.

Rhizaria typically have long, thin pseudopods

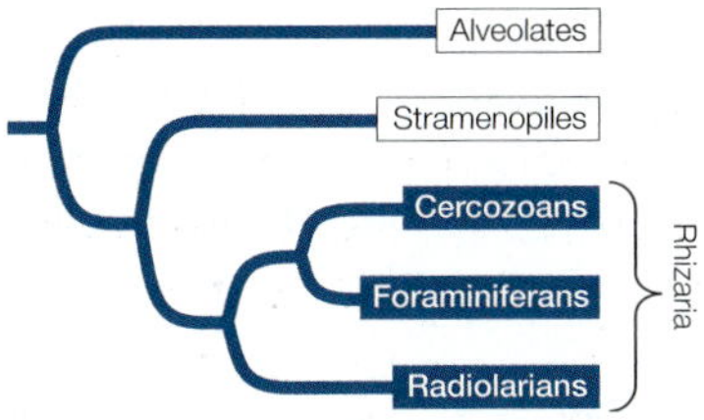
Alveolates
Stramenopiles
Cercozoans
Foraminiferans
Radiolarians
Rhizaria

The three primary groups of **Rhizaria**—cercozoans, foraminiferans, and radiolarians—are unicellular and mostly aquatic. The rhizaria have contributed their shells to ocean sediments, some of which have become terrestrial features over the course of geological history.

CERCOZOANS The **cercozoans** are a diverse group with many forms and habitats. Some are aquatic; others live in soil. One group of cercozoans possesses chloroplasts derived from a green alga by secondary endosymbiosis, and those chloroplasts contain a trace of the alga's nucleus.

FORAMINIFERANS Some **foraminiferans** secrete external shells of calcium carbonate (**Figure 27.11**). These shells have accumulated over time to produce much of the world's limestone. Some foraminiferans live as plankton; others live on the seafloor. Living foraminiferans have been found 10,896 meters down in the western Pacific's Challenger Deep—the deepest point in the world's oceans. At that depth, however, they cannot secrete normal shells because the surrounding water is too poor in calcium carbonate.

In living planktonic foraminiferans, long, threadlike, branched pseudopods extend through numerous microscopic apertures in the shell and interconnect to create a sticky, reticulated net, which the foraminiferans use to catch smaller plankton. In some foraminiferan species, the pseudopods provide locomotion.

RADIOLARIANS **Radiolarians** are recognizable by their thin, stiff pseudopods, which are reinforced by microtubules (**Figure 27.12A**). These pseudopods greatly increase the surface area of the cell, and they help the cell stay afloat in its marine environment.

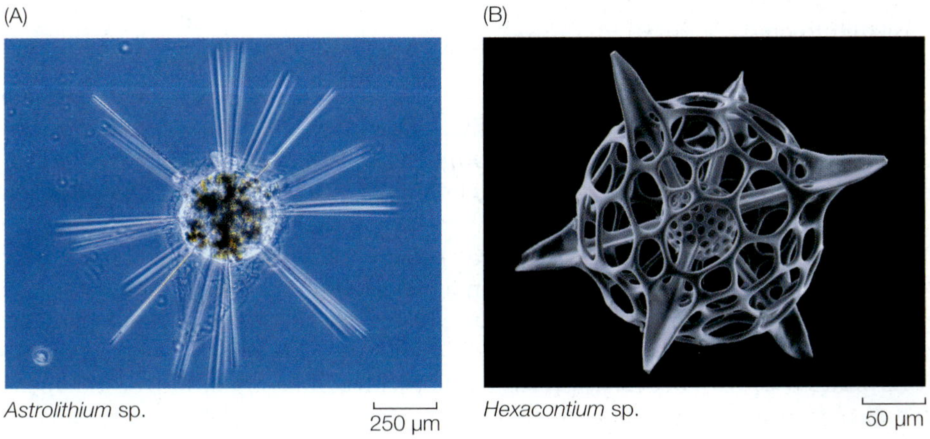
(A)
Astrolithium sp.
250 μm
(B)
Hexacontium sp.
50 μm

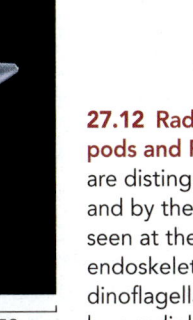

27.12 Radiolarians Exhibit Distinctive Pseudopods and Radial Symmetry (A) The radiolarians are distinguished by their thin, stiff pseudopods and by their radial symmetry. The pigmentation seen at the center of this radiolarian's glassy endoskeleton is imparted by endosymbiotic dinoflagellates. (B) The endoskeleton secreted by a radiolarian.

Radiolarians also are immediately recognizable by their distinctive radial symmetry. Almost all radiolarian species secrete glassy endoskeletons (internal skeletons). The skeletons of the different species are as varied as snowflakes, and many have elaborate geometric designs (**Figure 27.12B**). A few radiolarians are among the largest of the unicellular eukaryotes, measuring several millimeters across.

Excavates began to diversify about 1.5 billion years ago

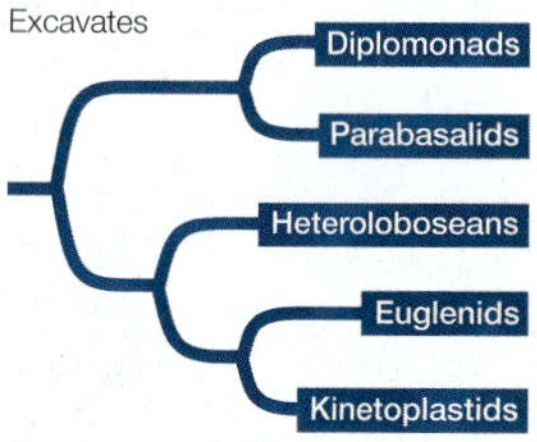
Excavates
Diplomonads
Parabasalids
Heteroloboseans
Euglenids
Kinetoplastids

The **excavates** include a number of diverse groups that began to split from one another soon after the origin of eukaryotes. Several groups of excavates lack mitochondria, an absence that once led to the view that these groups might represent early-diverging eukaryotes that diversified before the evolution of mitochondria. However, the discovery of genes in the nucleus that are normally associated with mitochondria suggests that the absence of mitochondria is a derived condition in these organisms. In other words, ancestors of these excavate groups probably possessed mitochondria that were lost or reduced over the course of evolution. The existence of these organisms today shows that eukaryotic life is possible without mitochondria.

DIPLOMONADS AND PARABASALIDS The **diplomonads** and the **parabasalids** are unicellular and lack mitochondria (although they have reduced organelles that are derived from mitochondria). The parasitic *Giardia lamblia*, a diplomonad, causes the intestinal disease giardiasis. *Giardia* infections may result from contact with contaminated water; in the United States, such infections are most common among hikers and campers using spring or stream water in recreational areas. This tiny organism has a cytoskeleton and multiple flagella, and contains two nuclei bounded by nuclear envelopes (**Figure 27.13A**).

In addition to flagella and a cytoskeleton, the parabasalids have undulating membranes that also contribute to the cell's locomotion. *Trichomonas vaginalis* (**Figure 27.13B**) is a parabasalid responsible for a sexually transmitted disease in humans. Infection of the male urethra, where it may occur without symptoms, is less common than infection of the vagina.

HETEROLOBOSEANS The amoeboid body form appears in several protist groups that are only distantly related to one another. The body forms of **heteroloboseans**, for example, resemble those of loboseans, an amoebozoan group that is not at all closely related to heteroloboseans (see the next section). Amoebas of the free-living heterolobosean genus *Naegleria*, some of which can enter the human body and cause a fatal disease of the nervous system, usually have a two-stage life cycle, in which one stage has amoeboid cells and the other flagellated cells.

EUGLENIDS AND KINETOPLASTIDS The **euglenids** and **kinetoplastids** together constitute a clade of unicellular excavates with flagella. Their mitochondria contain distinctive disc-shaped cristae, and their flagella contain a crystalline rod not found in other organisms. They reproduce primarily asexually.

(A) *Giardia* sp.

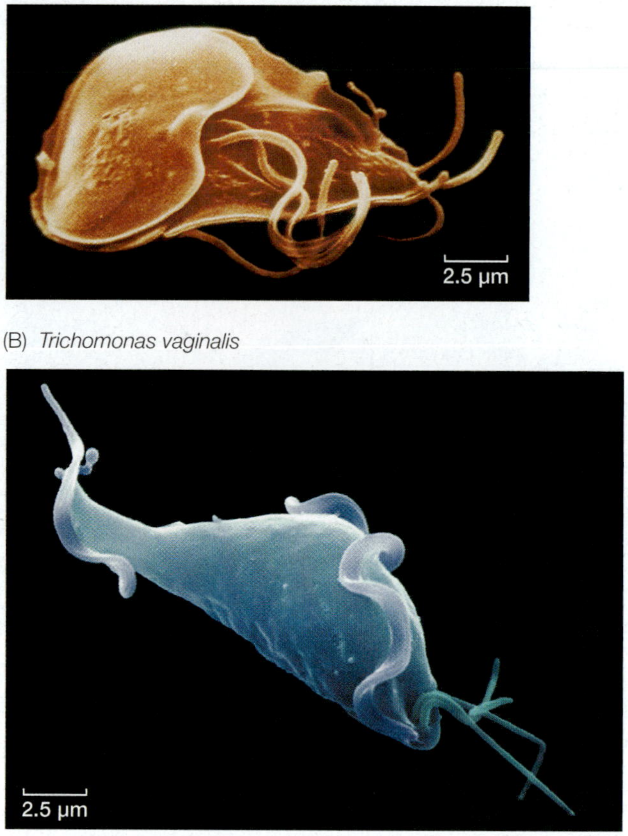
2.5 μm
(B) Trichomonas vaginalis
2.5 μm

27.13 Some Excavate Groups Lack Mitochondria (A) *Giardia*, a diplomonad, has flagella and two nuclei. (B) *Trichomonas*, a parabasalid, has flagella and undulating membranes. Neither of these organisms possesses mitochondria.

The flagella of euglenids arise from a pocket at the anterior end of the cell. Spiraling strips of proteins under the plasma membrane control the cell's shape. Some euglenids are photosynthetic. **Figure 27.14** depicts a typical cell of the genus *Euglena*. This common freshwater organism has a complex cell structure. It propels itself through the water with the longer of its two flagella, which may also serve as an anchor to hold the organism in place. The second flagellum is often rudimentary.

Go to Media Clip 27.4
Euglenids
Life10e.com/mc27.4

The euglenids have diverse nutritional requirements. Many species are always heterotrophic. Other species, including species of *Euglena*, are fully autotrophic in sunlight, using chloroplasts to synthesize organic compounds through photosynthesis. When kept in the dark, these euglenids lose their photosynthetic pigment and begin to feed exclusively on dissolved organic material in the water around them. A "bleached" *Euglena* resynthesizes its photosynthetic pigment when it is returned to the light and becomes autotrophic again. But *Euglena* cells treated with certain antibiotics or mutagens lose their photosynthetic pigment completely; neither they nor

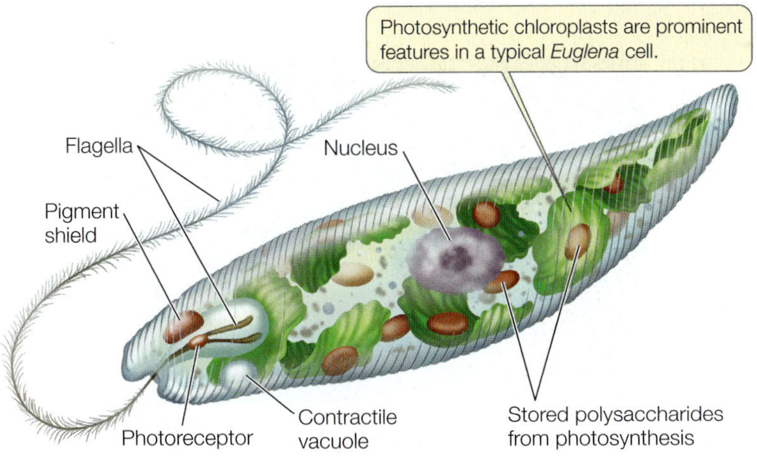
Photosynthetic chloroplasts are prominent features in a typical *Euglena* cell.
Flagella
Nucleus
Pigment shield
Photoreceptor
Contractile vacuole
Stored polysaccharides from photosynthesis

27.14 A Photosynthetic Euglenid In the *Euglena* species illustrated in this drawing, the second flagellum is rudimentary. Note that the primary flagellum originates at the anterior of the organism and trails toward its posterior.

their descendants are ever autotrophs again. However, those descendants function well as heterotrophs.

The kinetoplastids are unicellular parasites with two flagella and a single, large mitochondrion. The mitochondrion contains a kinetoplast, a unique structure housing multiple circular DNA molecules and associated proteins. Some of these DNA molecules encode "guide proteins" that edit mRNA within the mitochondrion.

The kinetoplastids include several medically important species of pathogenic trypanosomes (**Table 27.1**). Some of these organisms are able to change their cell surface recognition molecules frequently. This ability allows them to evade our best attempts to kill them and thus eradicate the diseases they cause.

Amoebozoans use lobe-shaped pseudopods for locomotion

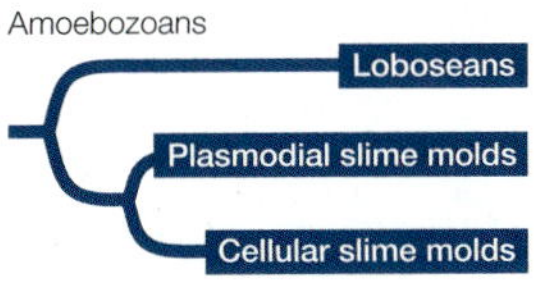
Amoebozoans
Loboseans
Plasmodial slime molds
Cellular slime molds

Amoebozoans appear to have diverged from other eukaryotes about 1.5 billion years ago (see Figure 27.3). It is not yet clear whether they are more closely related to opisthokonts (which include fungi and animals) or to other major groups of eukaryotes. The lobe-shaped pseudopods of amoebozoans are a hallmark of the amoeboid body form (**Figure 27.15**). Amoebozoan pseudopods differ in form and function from the slender pseudopods of rhizaria. We consider three amoebozoan groups here: the loboseans and two groups known as slime molds.

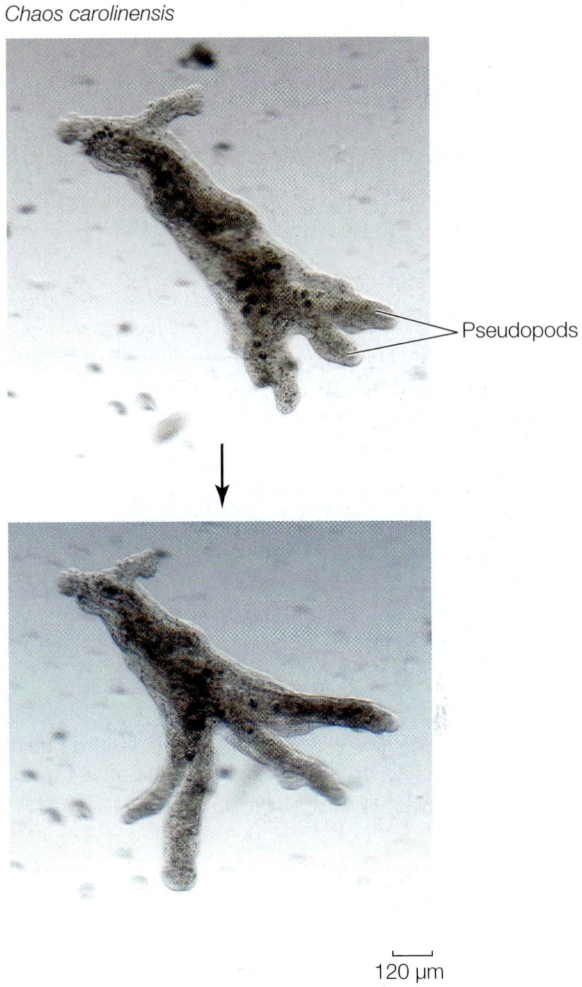
Chaos carolinensis
Pseudopods
120 µm

27.15 An Amoeba in Motion The flowing pseudopods of this "chaos amoeba" (a lobosean) are constantly changing shape as it moves and feeds.

Go to Media Clip 27.5
Amoeboid Movement
Life10e.com/mc27.5

TABLE 27.1
Three Pathogenic Trypanosomes

	Trypanosoma brucei	*Trypanosoma cruzi*	*Leishmania major*
Human disease	Sleeping sickness	Chagas disease	Leishmaniasis
Insect vector	Tsetse fly	Assassin bugs (many species)	Sand fly
Vaccine or effective cure	None	None	None
Strategy for survival	Changes surface recognition molecules frequently	Causes changes in surface recognition molecules on host cell	Reduces effectiveness of macrophage hosts
Site in human body	Bloodstream; in final stages, attacks nerve tissue	Enters cells, especially muscle cells	Enters cells, primarily macrophages
Approximate number of deaths per year	50,000	45,000	60,000

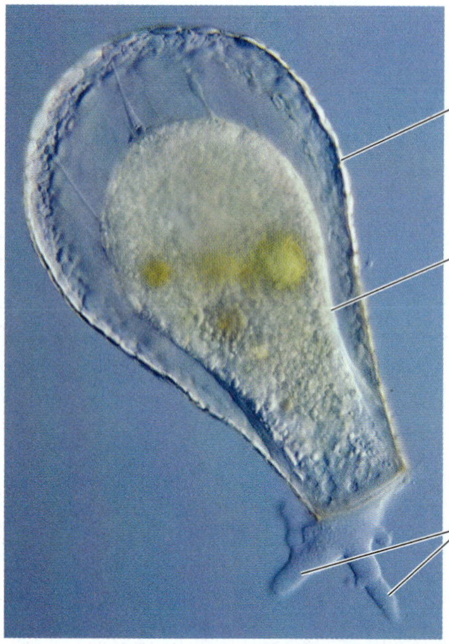
Nebela collaris
Shell (test) made of sand grains
Plasma membrane of amoeba
Pseudopods of amoeba
18 µm

27.16 Life in a Glass House This testate amoeba has built a lightbulb-shaped shell, or test, by gluing sand grains together. Its pseudopods extend through the single aperture in the test.

(A)
30 mm
(B)
1.5 mm

27.17 A Plasmodial Slime Mold (A) The plasmodial form of the slime mold *Hemitrichia serpula* covers rocks, decaying logs, and other objects as it engulfs bacteria and other food items; it is also responsible for the organism's common name of "pretzel mold." (B) Fruiting structures of *Hemitrichia*.

Go to Media Clip 27.6
Plasmodial Slime Mold Growth
Life10e.com/mc27.6

LOBOSEANS **Loboseans** are small amoebozoans that feed on other small organisms and particles of organic matter by phagocytosis, engulfing them with pseudopods. Many loboseans are adapted for life on the bottoms of lakes, ponds, and other bodies of water. Their creeping locomotion and their manner of engulfing food particles fit them for life close to a relatively rich supply of sedentary organisms or organic particles. Most loboseans exist as predators, parasites, or scavengers. Members of one group of loboseans, the testate amoebas, live inside shells. Some of these amoebas produce casings by gluing sand grains together (**Figure 27.16**); other testate amoebas have shells secreted by the organism itself.

PLASMODIAL SLIME MOLDS If the nucleus of an amoeba began rapid mitotic division, accompanied by a tremendous increase in cytoplasm and organelles, but no cytokinesis, the resulting organism would resemble the multinucleate mass of a **plasmodial slime mold**. During its vegetative (feeding, nonreproductive) stage, a plasmodial slime mold is a wall-less mass of cytoplasm with numerous diploid nuclei. This mass streams very slowly over its substrate in a remarkable network of strands called a plasmodium (**Figure 27.17A**). The plasmodium is an example of a **coenocyte**: many nuclei enclosed in a single plasma membrane. The outer cytoplasm of the plasmodium (closest to the environment) is normally less fluid than the interior cytoplasm and thus provides some structural rigidity.

Plasmodial slime molds provide a dramatic example of movement by **cytoplasmic streaming**. The outer cytoplasm of the plasmodium becomes more fluid in places, and cytoplasm rushes into those areas, stretching the plasmodium. This streaming reverses its direction every few minutes as cytoplasm rushes into a new area and drains away from an older one, moving the plasmodium over its substrate. Sometimes an entire wave of plasmodium moves across a surface, leaving strands behind. Microfilaments and a contractile protein called myxomyosin interact to produce the streaming movement. As it moves, the plasmodium engulfs food particles by endocytosis—predominantly bacteria, yeasts, spores of fungi, and other small organisms as well as decaying animal and plant remains.

A plasmodial slime mold can grow almost indefinitely in its plasmodial stage as long as the food supply is adequate and other conditions, such as moisture and pH, are favorable. If conditions become unfavorable, however, one of two things can happen. In one case, the plasmodium can form an irregular mass of hardened, cell-like components called a **sclerotium**. This resting structure rapidly becomes a plasmodium again when favorable conditions are restored.

Alternatively, the plasmodium can transform itself into spore-bearing **fruiting structures** (Figure 27.17B). These stalked or branched structures rise from heaped masses of plasmodium. They derive their rigidity from walls that form and thicken between their nuclei. The diploid nuclei of the plasmodium divide by meiosis as the fruiting structure develops. One or more knobs, called sporangia, develop on the end of the stalk. Within a sporangium, haploid nuclei become surrounded by walls to form spores. Eventually, as the fruiting structure dries, it sheds its spores.

The spores germinate into wall-less, haploid cells called **swarm cells**, which can either divide mitotically to produce more haploid swarm cells or function as gametes. Swarm cells can live as separate individual cells that move by means of flagella or pseudopods, or they can become walled and resistant resting cysts when conditions are unfavorable; when conditions improve again, the cysts release swarm cells. Two swarm cells can also fuse to form a diploid zygote, which divides by mitosis (but without a wall forming between the nuclei) and thus forms a new coenocytic plasmodium.

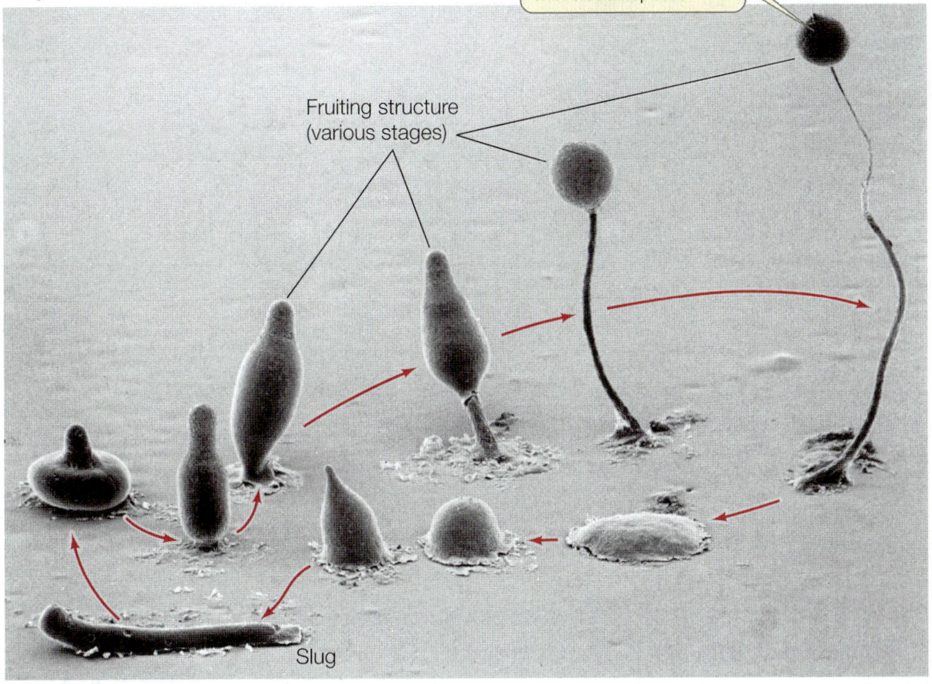
Dictyostelium discoideum
The sporangium of the mature fruiting structure will release spores.
Fruiting structure (various stages)
Slug
0.25 mm

27.18 A Cellular Slime Mold This composite micrograph shows the life cycle of the slime mold *Dictyostelium*.

Go to Media Clip 27.7
Cellular Slime Mold Aggregation
Life10e.com/mc27.7

CELLULAR SLIME MOLDS Whereas the plasmodium is the basic vegetative unit of the plasmodial slime molds, a single amoeboid cell is the vegetative unit of the **cellular slime molds** (Figure 27.18). Cells called **myxamoebas**, which have single haploid nuclei, swarm together as they engulf bacteria and other food particles by endocytosis and reproduce by mitosis and fission. This simple life cycle stage, consisting of swarms of independent, isolated cells, can persist indefinitely as long as food and moisture are available.

When conditions become unfavorable, the cellular slime molds form fruiting structures, as do their plasmodial counterparts. The individual myxamoebas aggregate into a mass called a **slug** or **pseudoplasmodium**. Unlike the true plasmodium of the plasmodial slime molds, this structure is not simply a giant sheet of cytoplasm with many nuclei; the individual myxamoebas within the slug retain their plasma membranes and therefore their identity.

A slug may migrate over a substrate for several hours before becoming motionless and reorganizing to construct a delicate, stalked fruiting structure. Cells at the top of the fruiting structure develop into thick-walled spores, which are eventually released. Later, under favorable conditions, the spores germinate, releasing myxamoebas.

The cycle from myxamoebas through slug and spores to new myxamoebas is asexual. Cellular slime molds also have a sexual cycle, in which two myxamoebas fuse. The product of this fusion develops into a spherical structure that ultimately germinates, releasing new haploid myxamoebas.

RECAP 27.2

The major lineages of eukaryotes began to diverge about 1.5 billion years ago. Major groups of eukaryotes are highly diverse in their habitat, nutrition, locomotion, and body form. Many protists are photosynthetic autotrophs, but heterotrophic lineages have evolved repeatedly. Although most protists are unicellular, multicellularity has arisen independently many times.

- Contrast the major distinctive features of alveolates, excavates, stramenopiles, rhizaria, and amoebozoans.
- Give examples of alveolates, stremenoplies, and excavates that are important for medical or agricultural reasons. See pp. 553–558 and Table 27.1.

The ancient origins of the major eukaryote lineages and the adaptation of these lineages to a wide variety of lifestyles and environments resulted in enormous protist diversity. It is not surprising, then, that reproductive modes among protists are also highly diverse.

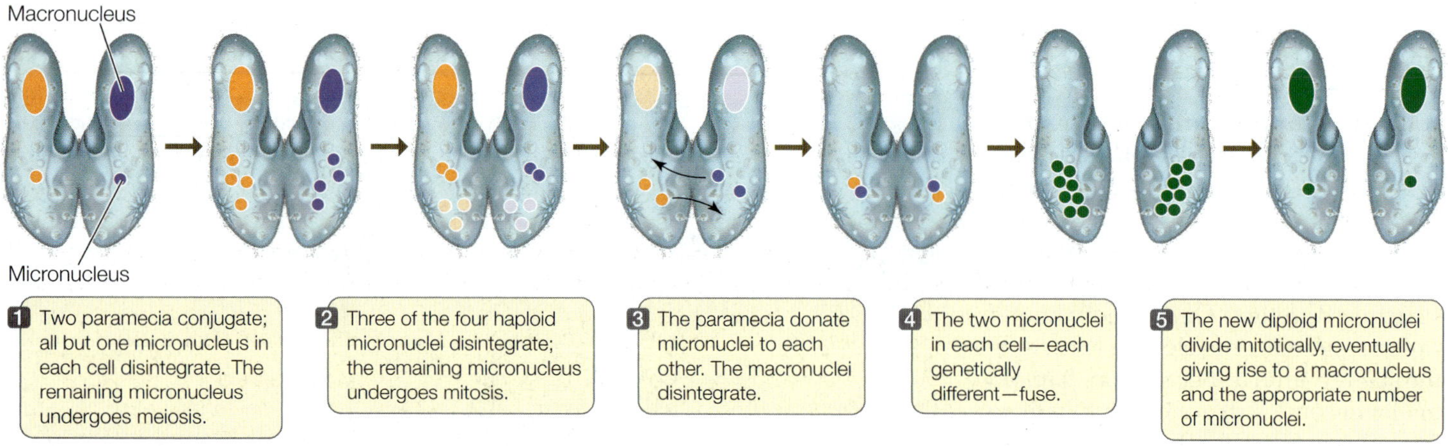
Macronucleus
Micronucleus
1 Two paramecia conjugate; all but one micronucleus in each cell disintegrate. The remaining micronucleus undergoes meiosis.
2 Three of the four haploid micronuclei disintegrate; the remaining micronucleus undergoes mitosis.
3 The paramecia donate micronuclei to each other. The macronuclei disintegrate.
4 The two micronuclei in each cell—each genetically different—fuse.
5 The new diploid micronuclei divide mitotically, eventually giving rise to a macronucleus and the appropriate number of micronuclei.

27.19 Conjugation in Paramecia The exchange of micronuclei by two conjugating *Paramecium* individuals results in genetic recombination. After conjugation, the cells separate and continue their lives as two individuals.

27.3 What Is the Relationship between Sex and Reproduction in Protists?

Although most protists engage in both asexual and sexual reproduction, sexual reproduction has yet to be observed in some groups. In some protists, as in all prokaryotes, the acts of sex and reproduction are not directly linked. Several asexual reproductive processes have been observed among the protists:

- The equal splitting of one cell into two by mitosis followed by cytokinesis
- The splitting of one cell into multiple (i.e., more than two) cells
- The outgrowth of a new cell from the surface of an old one (known as **budding**)
- The formation of specialized cells (spores) that are capable of developing into new individuals (know as **sporulation**)

Asexual reproduction results in offspring that are genetically nearly identical to their parents (they differ only by new mutations that may arise during DNA replication). Such asexually reproduced groups of nearly identical organisms are known as **clonal lineages**, or **clones**.

Sexual reproduction among the protists takes various forms. In some protists, as in animals, the gametes are the only haploid cells. In others, the zygote is the only diploid cell. In still others, both diploid and haploid cells undergo mitosis, giving rise to alternating multicellular diploid and haploid life stages.

Some protists reproduce without sex and have sex without reproduction

As noted in Section 27.2, members of the genus *Paramecium* are ciliates, which commonly have two types of nuclei in a single cell (one macronucleus and from one to several micronuclei; see Figure 27.6). The micronuclei are typical eukaryotic nuclei and are essential for genetic recombination. The macronucleus contains many copies of the genetic information, packaged in units containing only a few genes each. The macronuclear DNA is transcribed and translated to regulate the life of the cell.

When paramecia reproduce asexually, all of the nuclei are copied before the cell divides. Paramecia (and many other protists) also have an elaborate sexual behavior called **conjugation**, in which two individuals line up tightly against each other and fuse in the oral groove region of the body. Nuclear material is extensively reorganized and exchanged over the next several hours (**Figure 27.19**). Each cell ends up with two haploid micronuclei, one of its own and one from the other cell, which fuse to form a new diploid micronucleus. A new macronucleus develops from that micronucleus through a series of dramatic chromosomal rearrangements. The exchange of nuclei is fully reciprocal: each of the two paramecia gives and receives an equal amount of DNA. The two organisms then separate and go their own ways, each equipped with new combinations of alleles.

Conjugation in *Paramecium* is a sexual process (i.e., a process of genetic recombination), but it is not a reproductive process. Two cells begin conjugation and two cells are there at the end, so no new cells are created. As a rule, each asexual clone of paramecia must conjugate periodically. Experiments have shown that if some species are not permitted to conjugate, the clones can live through only about 350 cell divisions before dying out.

Some protist life cycles feature alternation of generations

Alternation of generations is a feature of the life cycles of many multicellular protists, all land plants, and some fungi. In these life cycles, a multicellular, diploid, spore-producing stage gives rise to a multicellular, haploid, gamete-producing stage (see Figure 28.6). When two haploid gametes fuse, a diploid organism is produced. The haploid organism, the diploid organism, or both may also reproduce asexually. Note that alternation of generations is distinct from the familiar reproductive system of animals, in which the only haploid stages are unicellular gametes produced by multicellular, diploid adults.

The two alternating (spore-producing and gamete-producing) generations differ genetically (one has diploid cells, the other haploid cells), but they may or may not differ morphologically. In **heteromorphic** alternation of generations, the two generations differ morphologically; in **isomorphic** alternation of generations, they do not. Examples of both heteromorphic and isomorphic alternation of generations are found among the brown algae.

The gamete-producing generation does not produce gametes by meiosis because the gamete-producing organism is already haploid. Instead, specialized cells of the diploid spore-producing organism, called **sporocytes**, divide meiotically to produce four haploid *spores*. The spores may eventually germinate and divide mitotically to produce the multicellular haploid generation, which then produces gametes by mitosis and cytokinesis.

Gametes, unlike spores, can produce new organisms only by fusing with other gametes. The fusion of two gametes produces a diploid zygote, which then undergoes mitotic divisions to produce a diploid organism. The diploid organism's sporocytes then undergo meiosis and produce haploid spores, starting the cycle anew.

RECAP 27.3

Protists reproduce both asexually and sexually, although sex occurs independently of reproduction in some species. Some multicellular protists exhibit alternation of generations, alternating between multicellular haploid and diploid life stages.

- Why is conjugation between paramecia considered a sexual process but not a reproductive process? **See p. 562 and Figure 27.19**
- Although most diploid animals have haploid stages (for example, eggs and sperm), their life cycles are not considered an example of alternation of generations. Why not? **See pp. 562–563**

Given the diversity of protists and of the environments in which they live, it is not surprising that they influence their environments in numerous ways.

27.4 How Do Protists Affect Their Environments?

As we have seen, many microbial eukaryotes are food for aquatic animals, while others poison those animals or act as pathogens. The remains of some form the sands of many modern beaches, and others are a major source of the oil that sometimes fouls those beaches.

Phytoplankton are primary producers

A single protist clade, the diatoms, performs about one-fifth of all photosynthetic carbon fixation on Earth—about the same amount as all of Earth's rainforests. These spectacular unicellular organisms (see Figure 27.8) are the predominant component of the oceanic phytoplankton, but the phytoplankton include many other protists that also contribute heavily to global photosynthesis. Like green plants on land, these "floating photosynthesizers" are the gateway for energy from the sun into the rest of the living world; in other words, they are **primary producers**. These autotrophs are eaten by heterotrophs, including animals and many other protists. Those consumers, in turn, are eaten by other consumers. Most aquatic heterotrophs (with the exception of some species in the deep sea) depend on photosynthesis performed by phytoplankton for their energy supply.

Some microbial eukaryotes are deadly

Some microbial eukaryotes are pathogens that cause serious diseases in humans and other vertebrates. The best-known pathogenic protists are members of the genus *Plasmodium*, a highly specialized group of apicomplexans that spend part of their complex life cycle as parasites in human red blood cells. *Plasmodium* parasites cause malaria, one of the world's three most serious infectious diseases; it infects more than 350 million people, and kills more than 1 million people, each year. On average, about two people die from malaria every minute of every day—most of them in sub-Saharan Africa, although malaria occurs in more than 100 countries.

Mosquitoes of the genus *Anopheles* transmit *Plasmodium* to humans when an infected female mosquito penetrates the person's skin. The parasites enter the human bloodstream and travel to cells in the liver and the lymphatic system, where they change form, multiply, and reenter the bloodstream to invade red blood cells, where they continue to multiply. Eventually the blood cells lyse (burst), releasing new swarms of *Plasmodium*. These episodes of cell lysis coincide with the primary symptoms of malaria, which include fever, shivering, vomiting, joint pain, and convulsions.

If another *Anopheles* bites the victim, the mosquito ingests *Plasmodium* cells along with blood. Some of these cells develop into gametes that unite in the mosquito, forming zygotes. The zygotes lodge in the mosquito's gut, divide several times, and move into its salivary glands, from which they can be passed on to another human host (**Figure 27.20**). Thus *Plasmodium* is an extracellular parasite in the mosquito vector and an intracellular parasite in the human host. Such an organism—that is, a parasite that requires more than one host—is said to have a **complex life cycle**.

Plasmodium is a singularly difficult pathogen for humans to combat. Its life cycle is best broken by removing stagnant water, in which mosquitoes breed. Using insecticides to reduce the *Anopheles* population can also be effective, but the benefits must be weighed against the ecological, economic, and health risks posed by the insecticides themselves.

Even some of the phytoplankton that are such important primary producers can be deadly, as described at the opening of this chapter. Some diatoms and dinoflagellates reproduce in enormous numbers when environmental conditions are favorable for their growth. In the resulting "red tides," the concentration of dinoflagellates may reach 60 million per liter of ocean water and produce potent nerve toxins that harm or kill many vertebrates, especially fish.

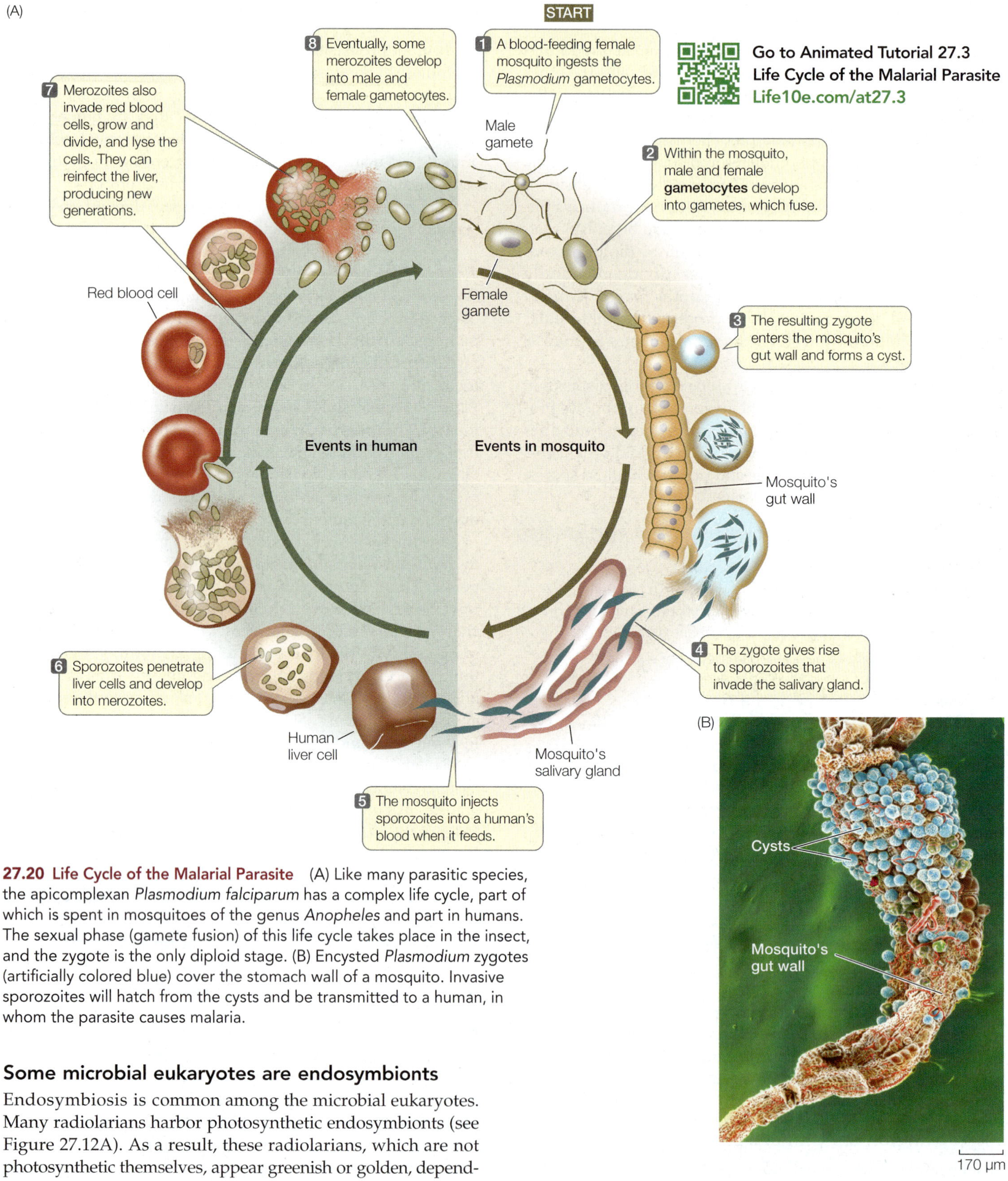
(A)
START
1 A blood-feeding female mosquito ingests the *Plasmodium* gametocytes.
Male gamete
2 Within the mosquito, male and female **gametocytes** develop into gametes, which fuse.
Female gamete
3 The resulting zygote enters the mosquito's gut wall and forms a cyst.
Mosquito's gut wall
4 The zygote gives rise to sporozoites that invade the salivary gland.
Mosquito's salivary gland
5 The mosquito injects sporozoites into a human's blood when it feeds.
Human liver cell
6 Sporozoites penetrate liver cells and develop into merozoites.
Red blood cell
7 Merozoites also invade red blood cells, grow and divide, and lyse the cells. They can reinfect the liver, producing new generations.
8 Eventually, some merozoites develop into male and female gametocytes.
Events in human
Events in mosquito
(B)
Cysts
Mosquito's gut wall
170 μm

Go to Animated Tutorial 27.3
Life Cycle of the Malarial Parasite
Life10e.com/at27.3

27.20 Life Cycle of the Malarial Parasite (A) Like many parasitic species, the apicomplexan *Plasmodium falciparum* has a complex life cycle, part of which is spent in mosquitoes of the genus *Anopheles* and part in humans. The sexual phase (gamete fusion) of this life cycle takes place in the insect, and the zygote is the only diploid stage. (B) Encysted *Plasmodium* zygotes (artificially colored blue) cover the stomach wall of a mosquito. Invasive sporozoites will hatch from the cysts and be transmitted to a human, in whom the parasite causes malaria.

Some microbial eukaryotes are endosymbionts

Endosymbiosis is common among the microbial eukaryotes. Many radiolarians harbor photosynthetic endosymbionts (see Figure 27.12A). As a result, these radiolarians, which are not photosynthetic themselves, appear greenish or golden, depending on the type of endosymbiont they contain. This arrangement is often mutually beneficial: the radiolarian can make use of the carbon compounds produced by its photosynthetic endosymbiont, and the endosymbiont may in turn make use of metabolites made by the host or receive physical protection. In some cases, the endosymbiont is exploited for its photosynthetic products while receiving little or no benefit itself.

Dinoflagellates are common endosymbionts and can be found in both animals and other protists. Most, but not all, dinoflagellate endosymbionts are photosynthetic. Some dinoflagellates live endosymbiotically in the cells of corals, contributing the products of their photosynthesis to the partnership. Their importance to the corals is demonstrated when the dinoflagellates die or are expelled by the corals as a result of changing environmental conditions such as rising water temperatures or increased water turbidity. This phenomenon, known as **coral bleaching**, reduces the corals' food supply. Unless the corals can acquire new endosymbionts, they are unlikely to survive (**Figure 27.21**).

We rely on the remains of ancient marine protists

Diatoms are lovely to look at, but their importance to us goes far beyond aesthetics, and even beyond their role as primary producers. Diatoms store oil as an energy reserve and to keep themselves afloat at the correct depth in the ocean. Over millions of years, untold numbers of diatoms have died and sunk to the ocean floor, where their bodies have undergone chemical changes. In this way, diatoms have become a major source of petroleum and natural gas, two of our most important fossil fuels and political concerns.

Because the silica-containing cell walls of dead diatoms resist decomposition, some sedimentary rocks are composed almost entirely of diatom skeletons that sank to the seafloor over time. Diatomaceous earth obtained from such rocks has many industrial uses, such as insulation, filtration, toothpaste, and metal polishing. It has also been used as an "Earth-friendly" insecticide that clogs the tracheae (breathing structures) of insects.

Other ancient marine protists have also contributed to the rocks of today. Some foraminiferans, as we have seen, secrete shells of calcium carbonate. After they reproduce (by mitosis and cytokinesis), the daughter cells abandon the parent shell and make new shells of their own. The discarded shells of ancient foraminiferans form a layer hundreds to thousands of meters deep over millions of square kilometers of ocean bottom. The extensive limestone deposits seen in various parts of the world are the result of tectonic processes that have raised these layers above sea level. Foraminiferan shells also make up much of the sand of some beaches. A single gram of such sand may contain as many as 50,000 foraminiferan shells and shell fragments.

The shells of individual foraminiferans are easily preserved as fossils in marine sediments. Each geological period is characterized by a distinctive assemblage of foraminiferan species. Because the shells of foraminiferan species have such distinctive shapes (see Figure 27.11) and, because they are so abundant, the remains of foraminiferans are especially valuable in classifying and dating sedimentary rocks. In addition, analyses of the chemical makeup of foraminiferan shells can be used to estimate the global temperatures prevalent at the time when the shells were formed.

INVESTIGATING**LIFE**

27.21 Can Corals Reacquire Dinoflagellate Endosymbionts Lost to Bleaching? Some corals lose their chief nutritional source when their photosynthetic endosymbionts die, often as a result of changing environmental conditions. This experiment by Cynthia Lewis and Mary Alice Coffroth investigated the ability of corals to acquire new endosymbionts after bleaching.[a]

HYPOTHESIS Bleached corals can acquire new photosynthetic endosymbionts from their environment.

Method

1. Count numbers of *Symbiodinium*, a photosynthetic dinoflagellate, living symbiotically in samples of a coral (*Briareum* sp.).
2. Stimulate bleaching by maintaining all *Briareum* colonies in darkness for 12 weeks.
3. After 12 weeks of darkness, count numbers of *Symbiodinium* in the coral samples; then return all colonies to light.
4. In some of the bleached colonies (the experimental group), introduce *Symbiodinium* strain B211—dinoflagellates that contain a unique molecular marker. Do not expose the others (the control group) to strain B211. Maintain both groups in the light for 6 weeks.

Results

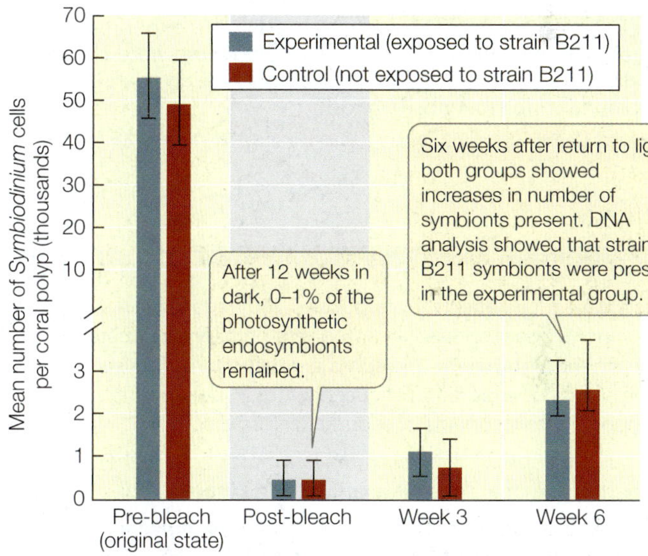
Mean number of Symbiodinium cells per coral polyp (thousands)
70
60
50
40
30
20
10
3
2
1
0
Experimental (exposed to strain B211)
Control (not exposed to strain B211)
After 12 weeks in dark, 0–1% of the photosynthetic endosymbionts remained.
Six weeks after return to light, both groups showed increases in number of symbionts present. DNA analysis showed that strain B211 symbionts were present in the experimental group.
Pre-bleach (original state)
Post-bleach
Week 3
Week 6

Pre-bleach

Post-bleach

CONCLUSION Corals can acquire new endosymbionts from their environment following bleaching.

Go to **BioPortal** for discussion and relevant links for all INVESTIGATING**LIFE** figures.

[a]Lewis, C. L. and M. A. Coffroth. 2004. *Science* 304: 1490–1492.

WORKING WITH DATA:

Uptake of Endosymbionts After Coral Bleaching

Original Paper

Lewis, C. L. and M. A. Coffroth. 2004. The acquisition of exogenous algal symbionts by an octocoral after bleaching. *Science* 304: 1490–1492.

Analyze the Data

The data shown in the table at right come from DNA analyses of *Symbiodinium* strains found in the experimental and control colonies of corals (*Briareum*) before and after bleaching. *Symbiodinium* strain B211 (which was not present before bleaching) was introduced to the experimental colonies after bleaching. Use these data to answer the questions below.

	Symbiodinium strain present (% of colonies)			
	Non-B211	B211	None*	Colony died
Experimental colonies (strain B211 added)				
Pre-bleach	100	0	0	0
Post-bleach	58	0	42	0
Week 3	0	92	0	8
Week 6	8	58	8	25
Control colonies (no strain B211)				
Pre-bleach	100	0	0	0
Post-bleach	67	0	33	0
Week 3	67	0	33	0
Week 6	67	0	17	17

*Colonies remained alive but no *Symbiodinium* were detected.

QUESTION 1

Are new strains of *Symbiodinium* taken up only by coral colonies that have lost all their original endosymbionts?

QUESTION 2

Does the acquisition of a new *Symbiodinium* strain always result in survival of a recovering *Briareum* colony?

QUESTION 3

In week 3, only strain B211 was detected in the experimental colonies, but in week 6, non-B211 *Symbiodinium* were detected in 8 percent of the experimental colonies. Can you suggest an explanation for this observation?

Go to **BioPortal** for all WORKING WITH DATA exercises

RECAP 27.4

Protists have many effects, both positive and negative, on their environment. Some species are important primary producers, many are endosymbionts, and some are pathogens. Protists are among the most important producers of fossil fuels, and they are important components of sedimentary rocks.

- What is the role of female *Anopheles* mosquitoes in the transmission of malaria? **See p. 563 and Figure 27.20**
- Explain the roles of dinoflagellates in the two very different phenomena of coral bleaching and red tides. **See pp. 549 and 564–565**
- What are some of the ways in which diatoms are important to human society? **See p. 565**

The next six chapters will explore the major evolutionary radiations of multicellular eukaryotes, along with the protist ancestors from which they arose. Chapters 28 and 29 will describe the origin and diversification of plants, Chapter 30 will present the fungi, and Chapters 31–33 will provide a brief overview of the animals.

Can dinoflagellates be beneficial, as well as harmful, to marine ecosystems?

ANSWER

Not all dinoflagellate blooms produce problems for other species. Dinoflagellates are important components of many ecosystems, as we have seen throughout this chapter. Photosynthetic dinoflagellates also produce much of the atmospheric oxygen that most animals need to survive.

Corals and many other species depend on symbiotic dinoflagellates for food (see Figure 27.21). In addition, as photosynthetic organisms, free-living planktonic dinoflagellates are among the most important primary producers in aquatic food webs. They are a major component of the phytoplankton and provide an important food source for many species (see Section 27.4).

Some dinoflagellates produce a beautiful bioluminescence. Unlike the bioluminescent bacteria described at the start of Chapter 26, however, dinoflagellates cannot generate a steady bioluminescence, but produce flashes of light when disturbed, as people who swim in the ocean at night in certain regions often observe.

CHAPTERSUMMARY 27

27.1 How Did the Eukaryotic Cell Arise?

- The term **protist** does not describe a formal taxonomic group. It is shorthand for "all eukaryotes that are not plants, animals, or fungi."
- Early events in the evolution of the eukaryotic cell probably included the loss of the firm cell wall and infolding of the plasma membrane. Such infolding probably led to segregation of the genetic material in a membrane-enclosed nucleus. **Review Figure 27.1**
- Mitochondria evolved by endosymbiosis with a proteobacterium.
- **Primary endosymbiosis** of a eukaryote and a cyanobacterium gave rise to the first chloroplasts. **Secondary endosymbiosis** and **tertiary endosymbiosis** between chloroplast-containing eukaryotes and other eukaryotes gave rise to the distinctive chloroplasts of euglenids, dinoflagellates, and other groups. **Review Figure 27.2, ANIMATED TUTORIAL 27.1**

27.2 What Features Account for Protist Diversity?

- Most eukaryotes can be placed in one of eight major clades that diverged about 1.5 billion years ago: alveolates, stramenopiles, rhizaria, excavates, plants, amoebozoans, fungi, and animals. **Review Figure 27.3**
- Most, but not all, protists are unicellular.
- **Alveolates** are unicellular organisms with sacs (alveoli) beneath their plasma membranes. Alveolate clades include the marine **dinoflagellates**, the parasitic **apicomplexans**, and the diverse, highly motile **ciliates**. **See ACTIVITY 27.1, ANIMATED TUTORIAL 27.2**
- **Stramenopiles** typically have two flagella of unequal length, the longer one bearing rows of tubular hairs. Among the stramenopiles are the unicellular **diatoms**, the multicellular **brown algae**, and the nonphotosynthetic **oomycetes**, many of which are **saprobic**.
- **Rhizaria** are unicellular and aquatic. They include the **cercozoans**; the **foraminiferans**, which secrete shells of calcium carbonate; and the **radiolarians,** which have thin, stiff pseudopods and glassy endoskeletons.
- The **excavates** include parasitic as well as free-living species. The **diplomonads** and **parabasalids** lack typical mitochondria. **Heteroloboseans** have an amoeboid body form and a two-stage life cycle. **Euglenids** have anterior flagella; some are photosynthetic. The **kinetoplastids**, which include several human pathogens, have a single, large mitochondrion.
- The **amoebozoans** move by means of lobe-shaped pseudopods. A **lobosean** consists of a single amoeboid cell. **Plasmodial slime molds** are amoebozoans whose vegetative stage is a **coenocyte** that moves by cytoplasmic streaming. In **cellular slime molds**, the individual cells maintain their identity at all times but aggregate to form fruiting structures.

27.3 What is the Relationship between Sex and Reproduction in Protists?

- Asexual reproduction gives rise to **clonal lineages** of organisms.
- **Conjugation** in *Paramecium* is a sexual process but not a reproductive one. **Review Figure 27.19**
- **Alternation of generations**, which includes a multicellular diploid stage and a multicellular haploid stage, is a feature of many multicellular protist life cycles (as well as those of some fungi and all land plants). The alternating generations may be **heteromorphic** or **isomorphic**.

27.4 How Do Protists Affect Their Environments?

- The diatoms are responsible for about one-fifth of the photosynthetic carbon fixation on Earth. They and other members of the phytoplankton are important **primary producers** in the marine environment. Ancient diatoms are a major source of today's petroleum and natural gas deposits.
- Some protists are pathogens of humans and other vertebrates. **Review Figure 27.20, ANIMATED TUTORIAL 27.3**
- Endosymbiotic relationships are common among microbial protists and typically benefit both the endosymbionts and their protist or animal partners. **Review Figure 27.21**

Go to the Interactive Summary to review key figures, Animated Tutorials, and Activities Life10e.com/is27

CHAPTERREVIEW

REMEMBERING

1. Which statement about eukaryotic phytoplankton is *not* true?
   a. Some are important primary producers.
   b. Some contributed to the formation of petroleum.
   c. Some form toxic "red tides."
   d. Some are food for marine animals.
   e. They constitute a clade.

2. The chloroplasts of photosynthetic protists
   a. are structurally identical.
   b. gave rise to mitochondria.
   c. are all descended from a once free-living cyanobacterium.
   d. all have exactly two surrounding membranes.
   e. are all descended from a once free-living red alga.

3. Reproduction in protists
   a. is sexual in some species
   b. is asexual in some species.
   c. can occur through both asexual and sexual processes in some species.
   d. can occur independently of sex in some species.
   e. All of the above

UNDERSTANDING & APPLYING

4. For each pair of groups below, describe how you could recognize members of the two groups and differentiate them from each another. Then describe features that the two groups in each pair share.
   a. Foraminiferans and radiolarians
   b. Ciliates and dinoflagellates
   c. Diatoms and brown algae
   d. Plasmodial slime molds and cellular slime molds

5. Given that sex and reproduction are independent of each other in the ciliates, what does that suggest about the role of sex in maintenance of populations?

ANALYZING & EVALUATING

Background for Questions 6–7:

In most temperate regions of the oceans, there is a spring bloom of phytoplankton. Although the red tide blooms described at the opening of this chapter are harmful, phytoplankton blooms can also be beneficial for marine communities. In fact, many species of marine life depend on these blooms for their survival. The dates of spring phytoplankton blooms near the coast of Nova Scotia, Canada, were determined by examining remote satellite images. The table below presents these dates as deviations from the mean date of the spring bloom in this region; it also gives the survival index for larval haddock (an important commercial fish) for the year after each bloom. The survival index is the ratio of the mass of juvenile fish to the mass of mature fish; higher values indicate better survival of larval fish.

Year	Deviation in bloom date* (days)	Survival index
1	+5	1.9
2	+11	2.2
3	–15	6.8
4	+5	1.9
5	–4	4.9
6	–20	10.3
7	+6	2.1
8	+14	1.9

*Negative values indicate blooms occurring earlier than the mean date; positive values indicate later blooms.

6. Plot the survival index of larval haddock against the deviation in the date of the spring phytoplankton bloom. Calculate a correlation coefficient for their relationship (see Appendix B).

7. Formulate one or more hypotheses to explain your results. Keep in mind that larval haddock include phytoplankton in their diet, and that phytoplankton blooms also provide some cover in which larval fish can hide from potential predators.

Background for Questions 8–10:

Ribosomal RNA (rRNA) genes are present in the nuclear genome of eukaryotes. There are also rRNA genes in the genomes of mitochondria and chloroplasts. Therefore photosynthetic eukaryotes have three different sets of rRNA genes, which encode the structural RNA of three separate sets of ribosomes. Translation of each genome takes place on its own set of ribosomes. The gene tree shows the evolutionary relationships among rRNA gene sequences isolated from the nuclear genomes of humans, yeast, and corn; from an archaeon (*Halobacterium*), a proteobacterium (*E. coli*), and a cyanobacterium (*Chlorobium*); and from the mitochondrial and chloroplast genomes of corn. Use the gene tree to answer the following questions.

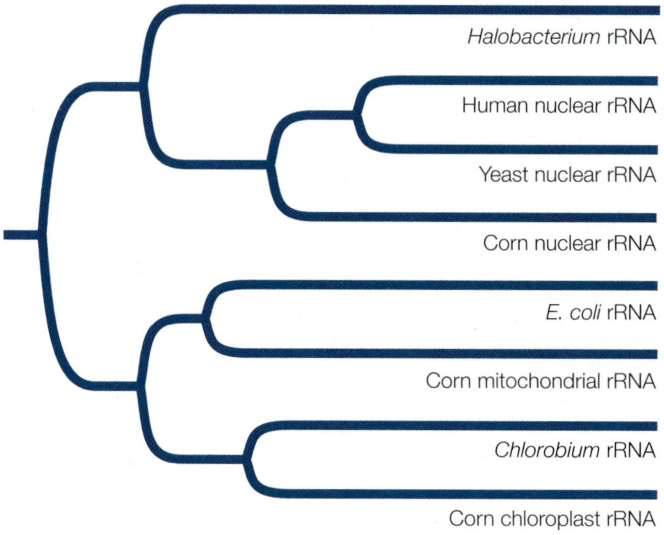
Halobacterium rRNA
Human nuclear rRNA
Yeast nuclear rRNA
Corn nuclear rRNA
E. coli rRNA
Corn mitochondrial rRNA
Chlorobium rRNA
Corn chloroplast rRNA

8. Why aren't the three rRNA genes of corn one another's closest relatives?

9. How would you explain the closer relationship of the mitochondrial rRNA gene of corn to the rRNA gene of *E. coli* than to the nuclear rRNA genes of other eukaryotes? Can you explain the relationship of the rRNA gene from the chloroplast of corn to the rRNA gene of the cyanobacterium?

10. If you were to sequence the rRNA genes from human and yeast mitochondrial genomes, where would you expect these two sequences to fit on the gene tree?

Go to BioPortal at **yourBioPortal.com** for Animated Tutorials, Activities, LearningCurve Quizzes, Flashcards, and many other study and review resources.

28 Plants without Seeds: From Water to Land

CHAPTER OUTLINE

Fireball on the Gulf A "blowout"—an uncontrolled release of petroleum—ignited a fireball above the drilling rig *Deepwater Horizon*. Fueled by gushing oil and natural gas, the fire could not be extinguished, and crews were forced to sink the rig without containing the blowout. Oil continued to flow from the deep-sea well for 3 months.

IN THE GULF OF MEXICO, about 60 kilometers south of the Louisiana coast, the oil rig *Deepwater Horizon* was drilling an exploratory oil well in the seafloor beneath about 1,500 meters of water when, on April 20, 2010, an explosive blowout occurred and could not be contained. Over the next 3 months, almost 5 million barrels of petroleum flowed from the well into the Gulf, making this event the worst marine oil spill in history. The spill caused massive mortality among marine life, as well as considerable damage along the coast as the oil floated to the surface and washed ashore.

Why was oil to be found so deep beneath the Gulf, and what led geologists to expect to find oil there? Most people know that petroleum is a fossil fuel, meaning that it is derived from the ancient remains of once-living organisms. Fewer people know that most petroleum is derived largely from the remains of phytoplankton, including many species of green algae (as well as other microbial groups, as discussed in Chapter 27). These algae produce complex hydrocarbons through photosynthesis. They accumulate hydrocarbons both as an energy reserve and as a way to increase their buoyancy in water. When these algae die, they drop to the bottom of the ocean, and over many millions of years, their buried remains decompose into petroleum deposits.

Today there is great interest in using solar power to help meet human energy needs. But unicellular eukaryotes first incorporated tiny solar energy converters into their cells about 1.5 billion years ago, when they formed partnerships with photosynthetic cyanobacteria. These endosymbionts—which over time would become the chloroplasts of modern plants—allowed many eukaryotes to use solar energy to drive the reactions that convert carbon dioxide into organic carbon compounds. Over many millions of years, the carbon compounds produced in the cells of marine algae accumulated in ocean sediments. Today humans are tapping that trapped solar energy in the form of petroleum and other fossil fuels.

> ?
>
> Given that petroleum is produced naturally from green algae, can humans use green algae to produce oil commercially?
>
> See answer on p. 585.

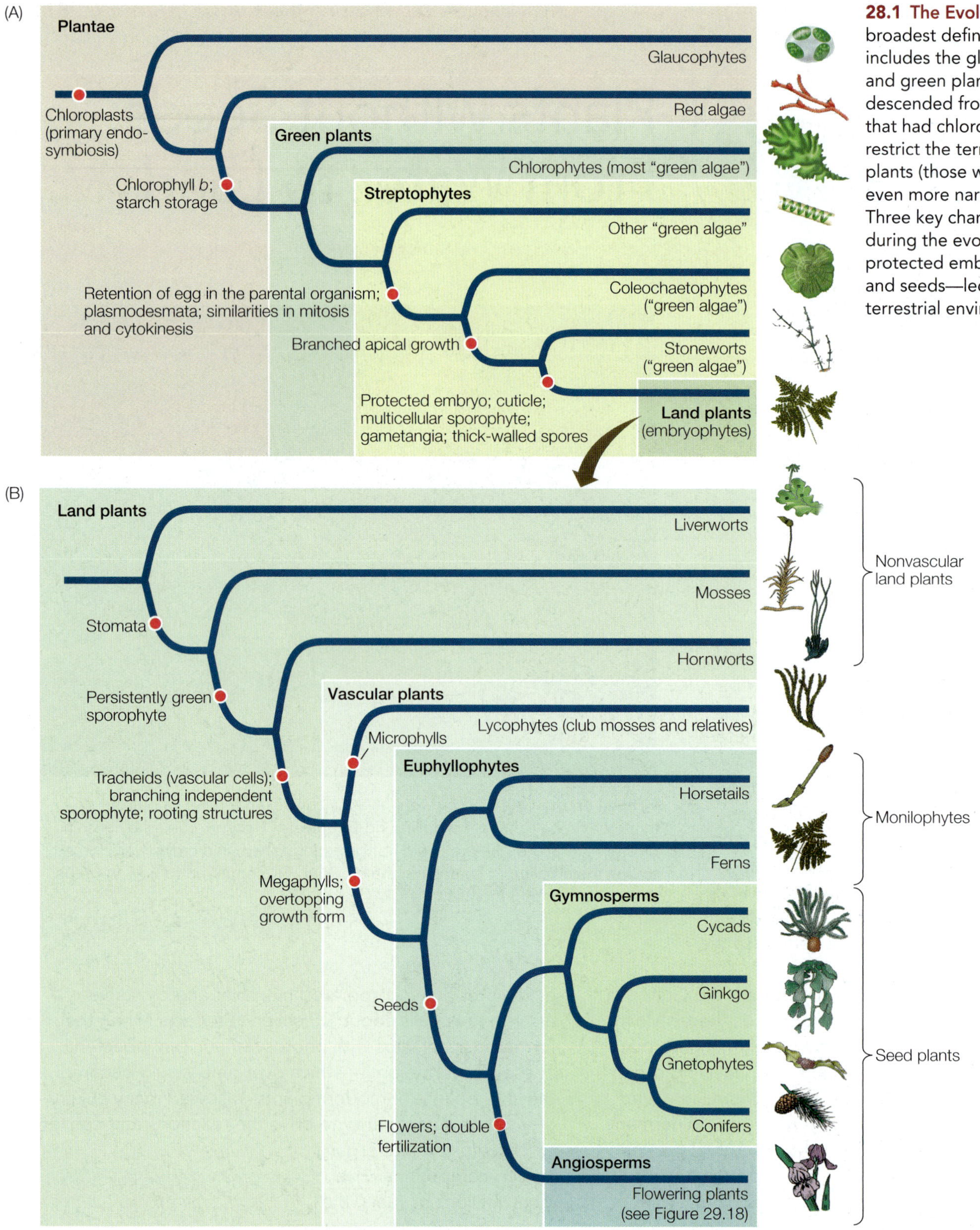

(A)
Plantae
Glaucophytes
Chloroplasts (primary endo-symbiosis)
Red algae
Green plants
Chlorophyll b; starch storage
Chlorophytes (most "green algae")
Streptophytes
Other "green algae"
Retention of egg in the parental organism; plasmodesmata; similarities in mitosis and cytokinesis
Coleochaetophytes ("green algae")
Branched apical growth
Stoneworts ("green algae")
Protected embryo; cuticle; multicellular sporophyte; gametangia; thick-walled spores
Land plants (embryophytes)
(B)
Land plants
Liverworts
Mosses
Stomata
Hornworts
Persistently green sporophyte
Vascular plants
Lycophytes (club mosses and relatives)
Microphylls
Euphyllophytes
Tracheids (vascular cells); branching independent sporophyte; rooting structures
Horsetails
Ferns
Megaphylls; overtopping growth form
Gymnosperms
Cycads
Ginkgo
Seeds
Gnetophytes
Conifers
Flowers; double fertilization
Angiosperms
Flowering plants (see Figure 29.18)
Nonvascular land plants
Monilophytes
Seed plants

28.1 The Evolution of Plants In its broadest definition, the term "plant" includes the glaucophytes, red algae, and green plants—all the groups descended from a common ancestor that had chloroplasts. Some biologists restrict the term "plant" to the green plants (those with chlorophyll *b*) or, even more narrowly, to the land plants. Three key characteristics that emerged during the evolution of land plants—protected embryos, vascular tissues, and seeds—led to their success in the terrestrial environment.

28.1 How Did Photosynthesis Arise in Plants?

More than a billion years ago, when a cyanobacterium was first engulfed by an early eukaryote, the history of life was altered radically. The chloroplasts that resulted from primary endosymbiosis of this cyanobacterium were obviously important for the evolution of plants and other photosynthetic eukaryotes, but they were also critical to the evolution of all life on land. Until photosynthetic plants were able to move onto land, there was very little there to support multicellular animals or fungi, and almost all life was restricted to the oceans and fresh waters.

Primary endosymbiosis is a shared derived trait—a synapomorphy—of the group known as **Plantae** (**Figure 28.1**). Although *Plantae* is Latin for "plants," in everyday language—and throughout this book—the unmodified common name "plants"

WORKING WITH**DATA:**

The Phylogeny of Land Plants

Original Paper

Qiu, Y.-L. et al. 2006. *Proceedings of the National Academy of Sciences USA* 103: 15511–15516.

Analyze the Data

In addition to the morphological characters of land plants shown on the phylogeny in Figure 28.1, DNA sequences are widely used to study and reconstruct the evolutionary history of plants. These sequences are many tens of thousands of nucleotides long and have been collected from a large number of species. The full data set used by Yin-Long Qiu and his colleagues (available at treebase.org) includes DNA sequences from 67 genes. The table below provides sample sequences from a chloroplast gene that has been used to reconstruct the relationships of representative plant species; the table shows 27 nucleotide positions for 10 species.

QUESTION 1

Construct a phylogenetic tree of these 10 species using the parsimony method (see Section 22.2 and the examples in Table 22.1 and Figure 22.5 for instructions). Use the outgroup to root your tree. Assume that all changes among nucleotides are equally likely.

QUESTION 2

How many changes (from one nucleotide to another) occur along each branch on your tree?

QUESTION 3

Which nucleotide positions (i.e., which character states) exhibit homoplasy (convergence or reversal of the character state)?

QUESTION 4

Which group on your tree represents the streptophytes? The land plants? The vascular plants? The euphyllophytes?

	Nucleotide position (character state)																										
Species	1	2	3	4	5	6	7	8	9	10	11	12	13	14	15	16	17	18	19	20	21	22	23	24	25	26	27
Outgroup (Chlorophyte alga)	T	A	T	T	A	T	G	A	T	T	C	C	A	A	A	T	A	T	T	A	T	A	A	T	C	T	A
Stonewort	T	A	T	T	T	A	A	A	T	T	A	C	T	A	A	T	A	A	T	A	T	A	A	T	C	T	A
Liverwort	A	C	T	T	T	T	A	A	T	G	A	T	T	C	A	G	A	A	T	A	T	A	A	T	C	T	A
Moss	A	C	T	T	T	T	A	A	T	A	T	T	T	T	A	A	T	A	T	A	A	A	A	T	C	T	T
Hornwort	A	C	T	T	T	T	A	A	T	G	T	T	T	T	A	A	T	A	C	A	G	A	A	A	C	T	T
Lycophyte	A	C	T	C	C	C	G	G	T	G	T	T	C	T	G	A	T	A	C	A	A	G	G	A	C	C	T
Fern	C	C	T	C	C	G	A	G	C	G	T	T	C	T	T	A	G	A	T	A	A	G	G	A	C	C	T
Pine tree	A	C	C	C	C	G	C	G	C	G	T	T	C	T	G	A	T	G	C	G	A	G	G	A	T	C	T
Rice	A	C	C	C	C	G	C	G	C	G	T	T	C	T	G	A	T	G	C	G	A	G	G	A	T	A	T
Tobacco	A	C	C	A	C	G	C	G	C	G	T	T	C	T	G	A	T	G	C	G	A	G	G	A	T	A	T

Go to **BioPortal** for all WORKING WITH**DATA** exercises

is usually used to refer only to the land plants. However, the first several clades that branch off the tree of life after primary endosymbiosis are all aquatic. Most aquatic photosynthetic eukaryotes (other than those secondarily derived from land plants) are known by the common name **algae**. This name, however, is just a convenient way to refer to these groups, which are not all closely related. Many of the photosynthetic groups discussed in Chapter 27 (which acquired chloroplasts through secondary endosymbiosis) are also commonly called algae.

Several distinct clades of algae were among the first photosynthetic eukaryotes

The ancestor of Plantae was unicellular and may have been similar in general form to the modern **glaucophytes** (**Figure 28.2**). These microscopic freshwater algae are thought to be the sister group of the rest of Plantae (see Figure 28.1A). The chloroplast of glaucophytes is unique in containing a small amount of peptidoglycan between its inner and outer membranes—the same arrangement found in cyanobacteria. Peptidoglycan has been lost from the remaining photosynthetic eukaryotes.

In contrast to the glaucophytes, almost all **red algae** are multicellular (**Figure 28.3**). Their characteristic color is a result of the accessory photosynthetic pigment **phycoerythrin**, which is found in relatively large amounts in the chloroplasts of many red algae. In addition to phycoerythrin, red algal chloroplasts contain chlorophyll *a* and the accessory pigments phycocyanin and carotenoids.

The red algae include species that grow in the shallowest tide pools as well as the photosynthesizers found deepest in the ocean (as deep as 260 meters if nutrient conditions are right and the water is clear enough to permit light to penetrate). A few red algae inhabit fresh water. Most grow attached to a substrate by a holdfast.

Glaucocystis

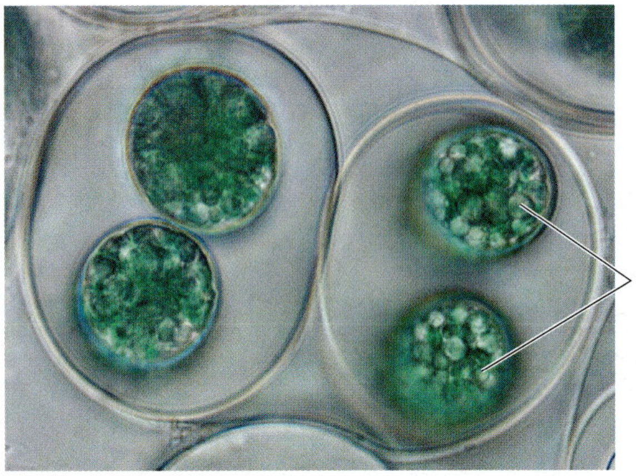
Chloroplasts
20 μm

28.2 Glaucophytes May Resemble Some of the Earliest Plantae The large chloroplasts of unicellular glaucophytes differ from chloroplasts of other Plantae in retaining a layer of peptidoglycan. This feature is thought to have been retained from the endosymbiotic cyanobacteria that gave rise to the chloroplasts of Plantae. The photograph shows a colony of two individuals, each with two chloroplasts.

Despite their name, red algae don't always appear red in color. The ratio of two pigments—phycoerythrin (red) and chlorophyll *a* (green)—depends largely on the intensity of light that reaches the alga. In deep water, where light is dim, algae accumulate large amounts of phycoerythrin and have red coloration. But many species growing near the surface contain a higher concentration of chlorophyll *a* and are bright green.

The remaining algal groups in Plantae are the various "green algae." Like land plants, the green algae contain both chlorophylls *a* and *b* and store their reserve of photosynthetic products as starch in chloroplasts. All the groups that share these features are commonly called **green plants** because both of their photosynthetic pigments are green.

The largest clade of "green algae" is the **chlorophytes**. There are more than 17,000 species of chlorophytes, most of which are aquatic (some are marine, though more are freshwater forms), although there are a few terrestrial forms that live in moist environments. Chlorophytes range in size from microscopic unicellular forms to multicellular forms many centimeters long and display an incredible variety of shapes and body forms. Surprisingly large and well-formed colonies of cells are found in some unicellular freshwater groups, such as the genus *Volvox* (**Figure 28.4A**). Certain cells in these colonies are specialized for reproduction. The cells in these colonies are not differentiated into specialized tissues and organs, as in land plants and animals, but they show vividly how the preliminary step of this great evolutionary innovation might have been taken.

Volvox is a colonial unicellular chlorophyte, but there are also many true multicellular species of chlorophytes. Some of these are filamentous. Others, like species in the genus *Ulva* (**Figure 28.4B**), grow into thin, membranous sheets up to 30 centimeters in width.

Two groups of green algae are the closest relatives of land plants

All green algae other than the chlorophytes form a group together with the land plants known as **streptophytes** (see Figure 28.1A). Several microscopic structural features, backed by clear evidence from molecular studies, indicate that the closest relatives of the land plants are two groups of aquatic green algae, the **coleochaetophytes** (**Figure 28.5A**) and the **stoneworts** (**Figure 28.5B**). Both of these multicellular algal groups retain their eggs in the parental organism, as land plants do. As in land plants, the cytoplasm of adjacent cells in these algal groups is connected through structures called plasmodesmata; they also share similarities in the details of mitosis and cytokinesis. Of these two groups, stoneworts are thought to be the sister group of land plants (see Figure 28.1A). The growth form of stoneworts is branching and apical (new growth occurs at the tips of branches), as in most land plants. Phylogenetic

(A) *Ceramium* sp.

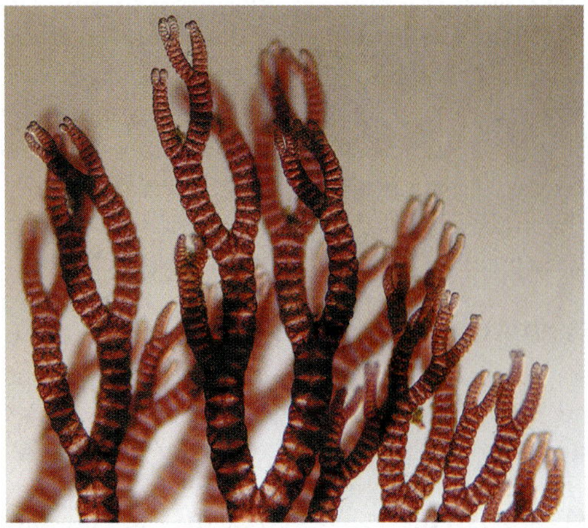
1.5 mm

(B) *Calliarthron sp.*

7.5 mm

28.3 Red Algae Contain a Red Accessory Photosynthetic Pigment (A) Differential contrast light microscopy reveals the rich red color of the pigment phycoerythrin. (B) Coralline red alga is named for its coral-like appearance.

Parent colony
Somatic cells
Daughter colonies produced by reproductive cells
(A) *Volvox* sp.
120 μm
(B) *Ulva rigida*
3 cm

28.4 Chlorophytes Display a Wide Diversity of Forms (A) *Volvox* colonies are precisely spaced arrangements of individual cells. Specialized reproductive cells produce daughter colonies, which will eventually release new individuals. (B) Sea lettuce grows in marine waters and intertidal areas.

analysis of gene sequences has confirmed the close relationships of coleochaetophytes and stoneworts to the land plants.

There are ten major groups of land plants

One of the key synapomorphies of the **land plants** is development from an embryo that is protected by tissues of the parent plant. For this reason, land plants are sometimes called **embryophytes** (*phyton*, "plant"). The green plants, the streptophytes, and the land plants have each been called "the plant kingdom" by different authorities; others take an even broader view and include red algae and glaucophytes as "plants." To avoid confusion in this chapter, we will use modifying terms ("land plants" or "green plants," for example) to refer to the various clades of Plantae shown in Figure 28.1.

The land plants that exist today fall naturally into ten major clades (listed by their common names in the center column of Table 28.1). Members of seven of those clades possess well-developed vascular systems that transport materials throughout the plant body. We call these seven groups, collectively, the **vascular plants**, or **tracheophytes**, because they all possess fluid-conducting cells called **tracheids**. The remaining three clades (liverworts, mosses, and hornworts) lack tracheids and are referred to collectively as **nonvascular land plants**. Note, however, that (unlike the vascular plants, which *are* a clade) *the three groups of nonvascular land plants do not form a clade.*

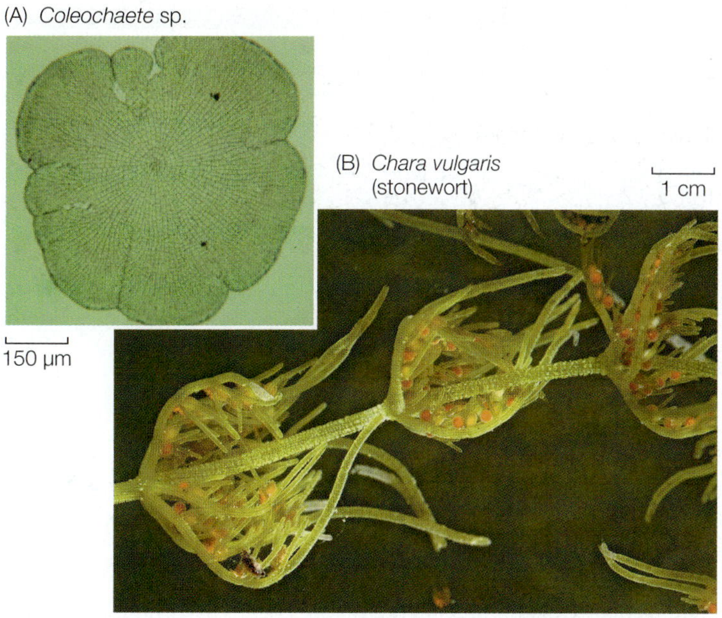
(A) *Coleochaete* sp.
150 μm
(B) *Chara vulgaris* (stonewort)
1 cm

28.5 The Closest Relatives of Land Plants (A) This species is a representative of the coleochaetophytes, the sister clade of stoneworts plus land plants. (B) The land plants probably evolved from a common ancestor shared with the stoneworts, an abundant group of multicellular green algae often found in freshwater pools and lakes (although a few species are found in marine environments). A species in the common genus *Chara* is shown here.

Go to Media Clip 28.1
Reproductive Structures of *Chara*
Life10e.com/mc28.1

RECAP 28.1

Primary endosymbiosis is a synapomorphy of the Plantae. The glaucophytes, the sister clade of the other Plantae, are unicellular algae that are similar to some of the earliest photosynthetic eukaryotes. The green plants contain chlorophyll *b* in addition to the chlorophyll *a* found in all Plantae.

- Explain the different possible uses of the term "plant." **See pp. 570–571 and Figure 28.1**
- Why doesn't the name "algae" designate a formal taxonomic group? **See pp. 571–572**
- What are some of the key differences between glaucophytes, red algae, and the various clades of green algae? **See pp. 571–572**
- What evidence supports the phylogenetic relationship between land plants and the various groups of green algae? **See pp. 572–573**

The green algal ancestors of the land plants lived at the margins of ponds or marshes, ringing them with a mat of dense green. It was from such a marginal habitat, which was sometimes wet and sometimes dry, that early plants made the transition onto land.

TABLE 28.1
Classification of Land Plants

Group	Common Name	Characteristics
Nonvascular land plants		
Hepatophyta	Liverworts	No stomata; gametophyte flat or leafy
Bryophyta	Mosses	Filamentous stage; gametophyte leafy; sporophyte grows apically (at the tip)
Anthocerophyta	Hornworts	Embedded archegonia; sporophyte grows basally (i.e., from the ground)
Vascular plants		
Lycopodiophyta	Lycophytes: Club mosses and allies	Microphylls in spirals; sporangia in leaf axils
Monilophyta	Horsetails, ferns	Simple leaves in whorls or frondlike compound leaves
SEED PLANTS		
Gymnosperms		
Cycadophyta	Cycads	Compound leaves; swimming sperm; seeds on modified leaves
Ginkgophyta	Ginkgo	Deciduous; fan-shaped leaves; swimming sperm
Gnetophyta	Gnetophytes	Vessels in vascular tissue; opposite, simple leaves
Coniferophyta	Conifers	Seeds in cones; needlelike or scalelike leaves
Angiosperms	Flowering plants	Endosperm; carpels; gametophytes much reduced; seeds contained within fruits

28.2 When and How Did Plants Colonize Land?

How did the land plants arise? To address this question, we can compare land plants with their closest relatives among the green algae. The features that differ between the two groups include the adaptations that allowed the first land plants to survive in the terrestrial environment.

Adaptations to life on land distinguish land plants from green algae

Land plants first appeared in the terrestrial environment between 450 and 500 million years ago. How did they survive in an environment that differed so dramatically from the aquatic environment of their ancestors? While the water essential for life is everywhere in the aquatic environment, water is difficult to obtain and retain in the terrestrial environment.

No longer bathed in fluid, organisms on land faced potentially lethal desiccation (drying). Large terrestrial organisms had to develop ways to transport water to body parts distant from the source of the water. And whereas water provides aquatic organisms with support against gravity, a plant living on land must either have some other support system or sprawl unsupported on the ground. A land plant must also use different mechanisms for dispersing its gametes and progeny than its aquatic relatives, which can simply release them into the water.

Survival on land was facilitated by the evolution among plants of numerous adaptations, including:

- The *cuticle*, a membrane covered in waxes to retard water loss
- *Stomata*, small openings in leaves and stems that open and close to regulate gas exchange and water loss
- *Gametangia*, multicellular organs that enclose plant gametes and prevent them from drying out
- *Embryos*, young plants contained within a protective structure
- Certain *pigments* that afford protection against the mutagenic ultraviolet radiation that bathes the terrestrial environment
- Thick *spore walls* containing a polymer that protects the spores from desiccation and resists decay
- A *mutually beneficial association with fungi* (mycorrhizae) that promotes nutrient uptake from the soil

The cuticle may be the most important—and the earliest—of these features. Composed of several unique waxy lipids that coat the leaves and stems of land plants, the cuticle has several functions, the most obvious and important of which is to keep water from evaporating from the plant body.

As ancient plants colonized the land, they not only adapted to the terrestrial environment, they also modified it by contributing to the formation of soil. Acids secreted by plants helped break down rock, and the organic compounds produced by the breakdown of dead plants contributed nutrients to the soil. Such effects are repeated today wherever plants colonize and grow in new areas.

Life cycles of land plants feature alternation of generations

A universal feature of the life cycles of land plants is alternation of generations. Recall from Section 27.3 the two hallmarks of alternation of generations:

- The life cycle includes both a multicellular diploid stage and a multicellular haploid stage.

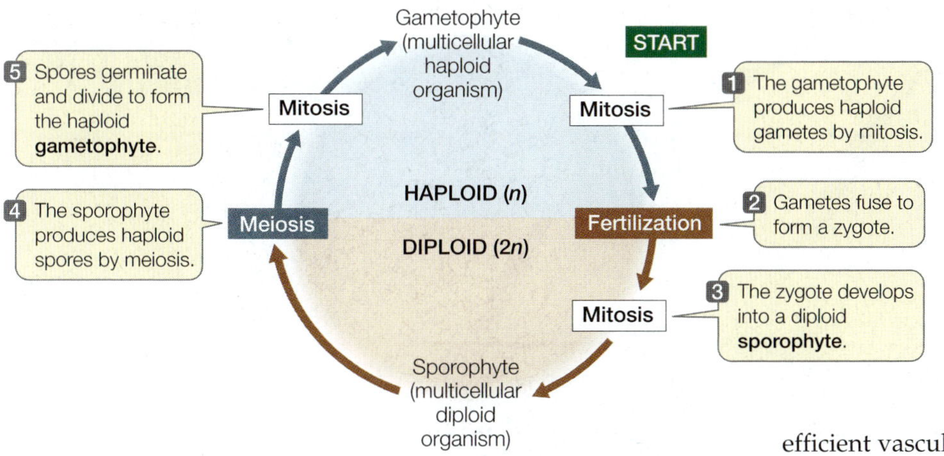
Gametophyte (multicellular haploid organism)
START
5 Spores germinate and divide to form the haploid gametophyte.
Mitosis
Mitosis
1 The gametophyte produces haploid gametes by mitosis.
HAPLOID (n)
4 The sporophyte produces haploid spores by meiosis.
Meiosis
Fertilization
2 Gametes fuse to form a zygote.
DIPLOID (2n)
Mitosis
3 The zygote develops into a diploid sporophyte.
Sporophyte (multicellular diploid organism)

28.6 Alternation of Generations in Land Plants A multicellular diploid sporophyte generation that produces spores by meiosis alternates with a multicellular haploid gametophyte generation that produces gametes by mitosis.

- Gametes are produced by mitosis, not by meiosis. Meiosis produces **spores** that develop into multicellular haploid organisms.

If we begin looking at the land plant life cycle at the single-cell stage—the diploid zygote—then the first phase of the cycle is the formation, by mitosis and cytokinesis, of a multicellular **embryo**, which eventually grows into a mature diploid plant. This multicellular diploid plant is called the **sporophyte** ("spore plant").

Cells contained within specialized reproductive organs of the sporophyte, called **sporangia** (singular *sporangium*), undergo meiosis to produce haploid, unicellular spores. By mitosis and cytokinesis, a spore develops into a haploid plant. This multicellular haploid plant, called the **gametophyte** ("gamete plant"), produces haploid gametes by mitosis. The fusion of two gametes (fertilization) forms a single diploid cell—the zygote—and the cycle is repeated (Figure 28.6).

The sporophyte generation extends from the zygote through the adult multicellular diploid plant and sporangium formation; the gametophyte generation extends from the spore through the adult multicellular haploid plant to the gametes. The transitions between the generations are accomplished by fertilization and by meiosis. In all land plants, the sporophyte and the gametophyte differ genetically: the sporophyte has diploid cells, and the gametophyte has haploid cells.

There is a trend toward reduction of the gametophyte generation in plant evolution. In the nonvascular land plants, the gametophyte is larger, longer-lived, and more self-sufficient than the sporophyte. In those groups that appeared later in plant evolution, however, the sporophyte is the larger, more conspicuous, longer-lived, and more self-sufficient generation.

Nonvascular land plants live where water is readily available

The living species of nonvascular land plants are the liverworts, mosses, and hornworts. These three groups are thought to be similar in many ways to the earliest land plants. Most of these plants grow in dense mats, usually in moist habitats. Even the largest of these species are only about half a meter tall, and most are only a few centimeters tall or long. Why have they not evolved to be taller? The probable answer is that they lack an efficient vascular system for transporting water and minerals from the soil to distant parts of the plant body.

The nonvascular land plants lack the true leaves, stems, and roots that characterize the vascular plants, although they have structures analogous to each. Their growth form allows water to move through the mats of plants by capillary action. They have leaflike structures that readily catch and hold any water that splashes onto them. They are small enough that minerals can be distributed throughout their bodies by diffusion. As in all land plants, layers of maternal tissue protect their embryos from desiccation. Many nonvascular land plants also have a cuticle, although it is often very thin (even absent in some species) and thus is not highly effective in retarding water loss.

Most nonvascular land plants live on the soil or on vascular plants, but some grow on bare rock, on dead and fallen tree trunks, and even on buildings. Their ability to grow on such marginal surfaces results from a mutualistic association with fungi. The earliest association of land plants with fungi dates back at least 460 million years. This mutualism probably facilitated the absorption of water and minerals, especially phosphorus, from the first soils.

Nonvascular land plants are widely distributed over six continents and even exist (albeit very locally) on the coast of the seventh, Antarctica. Most are terrestrial. Although a few species live in fresh water, these aquatic species are descended from terrestrial ones. None live in the oceans.

The sporophytes of nonvascular land plants are dependent on the gametophytes

In the nonvascular land plants, the conspicuous green structure visible to the naked eye is the gametophyte. The gametophyte is photosynthetic and is therefore nutritionally independent; the sporophyte may or may not be photosynthetic, but it is always nutritionally dependent on the gametophyte and remains permanently attached to it.

Figure 28.7 illustrates the life cycle of a moss, which is typical of the life cycles of nonvascular land plants. A sporophyte produces unicellular haploid spores as products of meiosis within a sporangium. When a spore germinates, it gives rise to a multicellular haploid gametophyte whose cells contain chloroplasts and are thus photosynthetic. Eventually gametes form within specialized sex organs, called the **gametangia**. The **archegonium** is a multicellular, flask-shaped female sex organ with a long neck

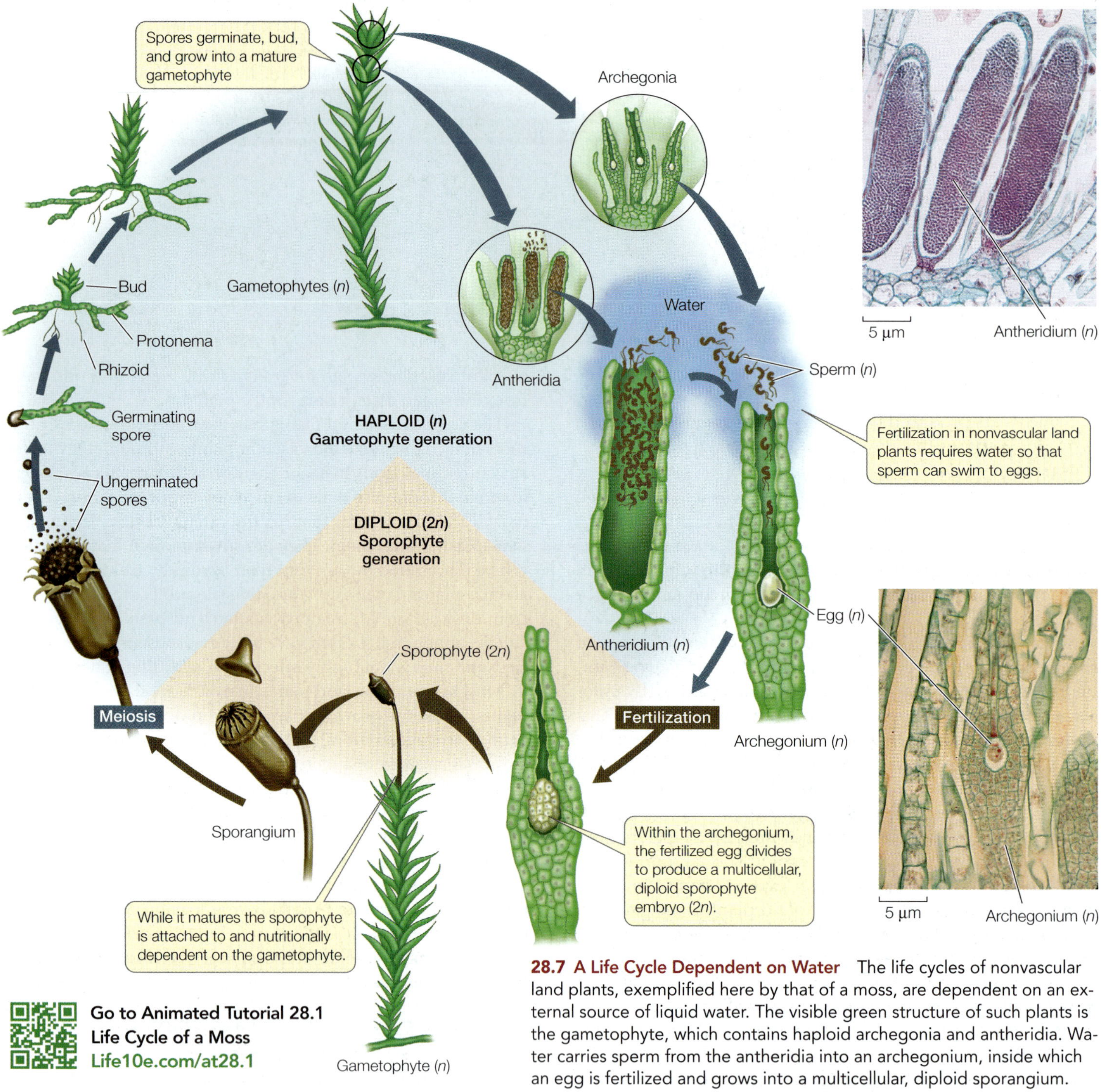

Spores germinate, bud, and grow into a mature gametophyte
Archegonia
Bud
Gametophytes (n)
Protonema
Rhizoid
Water
Sperm (n)
5 µm
Antheridium (n)
Antheridia
Germinating spore
HAPLOID (n) Gametophyte generation
Fertilization in nonvascular land plants requires water so that sperm can swim to eggs.
Ungerminated spores
DIPLOID (2n) Sporophyte generation
Egg (n)
Sporophyte (2n)
Antheridium (n)
Meiosis
Fertilization
Archegonium (n)
Sporangium
Within the archegonium, the fertilized egg divides to produce a multicellular, diploid sporophyte embryo (2n).
5 µm
Archegonium (n)
While it matures the sporophyte is attached to and nutritionally dependent on the gametophyte.
Gametophyte (n)

28.7 A Life Cycle Dependent on Water The life cycles of nonvascular land plants, exemplified here by that of a moss, are dependent on an external source of liquid water. The visible green structure of such plants is the gametophyte, which contains haploid archegonia and antheridia. Water carries sperm from the antheridia into an archegonium, inside which an egg is fertilized and grows into a multicellular, diploid sporangium.

Go to Animated Tutorial 28.1
Life Cycle of a Moss
Life10e.com/at28.1

Go to Media Clip 28.2
Bryophyte Reproduction
Life10e.com/mc28.2

and a swollen base; it produces a single egg. The **antheridium** is a male sex organ in which sperm, each bearing two flagella, are produced in large numbers. Archegonia and antheridia are produced on the same individual in many species, so each individual has both male and female reproductive structures. Adjacent individuals often fertilize one another's gametes, however, which helps maintain genetic diversity in the population.

Once released from the antheridium, the sperm must swim or be splashed by raindrops to a nearby archegonium on the same or a neighboring plant—a constraint that reflects the aquatic origins of the nonvascular land plants' ancestors. The sperm are aided on their journey by chemical attractants released by the egg or the archegonium. Before sperm can enter the archegonium, however, certain cells in the neck of the archegonium must break down, leaving a water-filled canal through which the sperm can swim to complete their journey. Notice that *all of these events require liquid water*.

Once sperm arrive at an egg, the nucleus of a sperm fuses with the egg nucleus to form a diploid zygote. Mitotic divisions of the zygote produce a multicellular, diploid sporophyte embryo. After the sporophyte grows out of the

These cups contain gemmae—small, lens-shaped outgrowths of the plant body, each capable of developing into a new plant.
The banana-like structures bear archegonia.

(A) *Bazzania trilobata* 2 cm (B) *Marchantia* sp. 0.3 cm (C) *Marchantia polymorpha* 2.5 cm

28.8 Liverwort Diversity (A) The gametophyte of a leafy liverwort. (B) The gametophytes of the thalloid liverwort *Marchantia* lie flat to the ground. (C) *Marchantia* gametophytes bearing archegonia.

archegonium it produces a single sporangium, within which meiotic divisions produce spores and thus the next gametophyte generation.

Liverworts are the sister clade of the remaining land plants

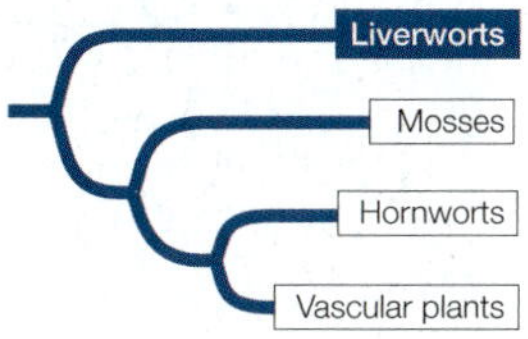
Liverworts
Mosses
Hornworts
Vascular plants

There are about 9,000 species of **liverworts**. Most liverworts have leafy gametophytes (**Figure 28.8A**). Some have thalloid gametophytes: green, leaflike layers that lie flat on the ground (**Figure 28.8B,C**). The simplest liverwort gametophytes are flat plates of cells, a centimeter or so long, that produce antheridia or archegonia on their upper surfaces and rhizoids (rootlike filaments) on their lower surfaces.

Liverwort sporophytes are shorter than those of mosses and hornworts, rarely exceeding a few millimeters. The liverwort sporophyte has a stalk that raises the sporangium above the gametophyte. In most species, the stalk elongates by expansion of cells throughout its length. This elongation raises the sporangium above ground level, allowing the spores to be dispersed more widely. The sporangia of liverworts are simple: a globular sporangium wall surrounds a mass of spores. In some species of liverworts, spores are not released by the sporophyte until the surrounding sporangium wall rots. In other liverworts, however, the spores are thrown from the sporangium by structures that shorten and compress as they dry out. When the stress becomes sufficient, the compressed structure snaps back to its resting position, throwing spores in all directions.

Among the most familiar thalloid liverworts are species of the genus *Marchantia*. *Marchantia* is easily recognized by the characteristic structures on which its male and female gametophytes bear their antheridia (Figure 28.8B) and archegonia (Figure 28.8C). Like most liverworts, *Marchantia* also reproduces asexually by simple fragmentation of the gametophyte. In addition, *Marchantia* and some other liverworts and mosses reproduce asexually by means of gemmae (singular gemma), which are lens-shaped clumps of cells. In a few liverworts, the gemmae are held in structures called gemmae cups, which promote dispersal of the gemmae by raindrops.

Go to Media Clip 28.3
Liverwort Life Cycle
Life10e.com/mc28.3

Water and sugar transport mechanisms emerged in the mosses

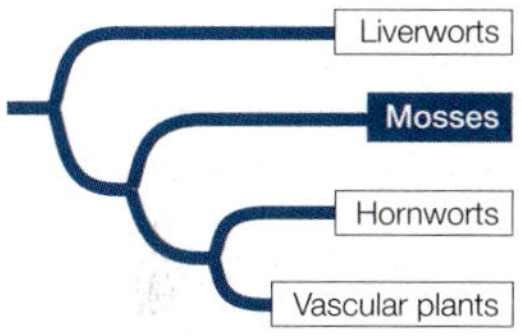
Liverworts
Mosses
Hornworts
Vascular plants

The most familiar of the nonvascular land plants are the **mosses**. These hardy little plants, of which there are about 15,000 species, are found in almost every terrestrial environment. They are often found on damp, cool ground, where they form thick mats (**Figure 28.9**). The mosses are the sister clade of the vascular plants plus the hornworts (see Figure 28.1).

The mosses, along with the hornworts and vascular plants, share an advance over the liverworts in their adaptation to life on land: they have openings called **stomata**, which allow CO_2 to enter the plant body and allow water and O_2 to leave it. Stomata are a synapomorphy of mosses and all other land plants except liverworts.

In mosses, the gametophyte begins its development following spore germination as a branched, filamentous structure called a protonema (see Figure 28.7). Although the protonema looks a bit like a filamentous green alga, this structure is unique to the mosses. Some of the filaments contain chloroplasts and are photosynthetic; others, called rhizoids, are nonphotosynthetic and anchor the protonema to the substratum. After a period of linear growth, cells close to the tips of the photosynthetic filaments divide rapidly in three dimensions to form buds. The buds eventually develop a distinct tip, or apex, and produce the familiar leafy moss shoot with leaflike structures arranged spirally. These leafy shoots produce antheridia or archegonia (see Figure 28.7).

(A)
Sporophytes
Gametophytes
(B) Polytrichum sp.

28.9 Mosses Often Cover the Ground in Dense Mats (A) Dense layers of moss carpet a field of solidified volcanic lava in Iceland. (B) A close-up view of moss growing on a forest floor in Michigan.

(A)
Sporophyte
Gametophyte
Sphagnum sp.
(B)

28.10 Sphagnum Moss (A) *Sphagnum* bogs are extremely dense growths of the moss shown here in a close-up view. (B) A farmer mines a bog for peat, a fossil fuel formed from decomposing *Sphagnum* mosses.

Some moss gametophytes are too large to transport enough water through their bodies solely by diffusion. Gametophytes and sporophytes of many mosses contain a type of cell called a hydroid, which dies and leaves a tiny channel through which water can travel. The hydroid is functionally similar to the tracheid, the characteristic water-conducting cell of the vascular plants, but it lacks lignin and the cell wall structure found in tracheids. The possession of hydroids and of a limited system for transport of sugar in some mosses shows that the term "nonvascular plant" is somewhat misleading when applied to these plants. Despite their simple system of internal transport, however, the mosses are not considered vascular plants because they lack tracheids or other components of xylem and phloem.

Mosses of the genus *Sphagnum* (Figure 28.10A) often grow in cool, swampy places, where the plants begin to decompose in water after they die. Rapidly growing upper layers of moss compress the deeper-lying, decomposing layers. Partially decomposed plant matter is called **peat**. In some parts of the world, people derive the majority of their fuel from peat bogs (Figure 28.B). *Sphagnum*-dominated peatlands cover a total area approximately half the size of the United States—more than 1 percent of Earth's surface. Millions of years ago, continued compression of peat composed primarily of other seedless plants gave rise to coal.

Hornworts have distinctive chloroplasts and stalkless sporophytes

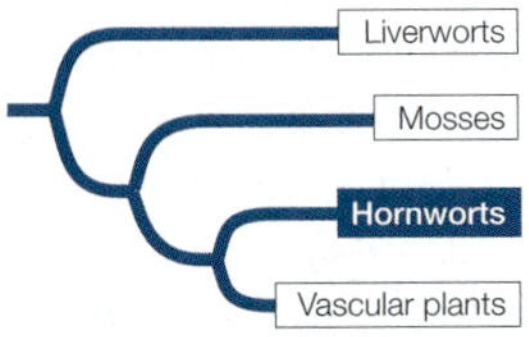
Liverworts
Mosses
Hornworts
Vascular plants

The approximately 100 species of **hornworts** are so named because their sporophytes look like little horns (Figure 28.11). Hornworts appear at first glance to be liverworts with very simple gametophytes. Their gametophytes are flat plates of cells a few cells thick.

Hornworts have several characteristics that distinguish them from liverworts and mosses. First, the cells of hornworts each contain a single large, platelike chloroplast, whereas the cells of the other two groups contain numerous small, lens-shaped

The sporophytes of hornworts can reach 20 cm in height.
Gametophytes are flat plates a few cells thick.
Anthoceros sp.

28.11 Hornworts Get Their Name from Their Hornlike Sporophytes Unlike liverworts or mosses, the sporophytes of hornworts are persistently green. They share this trait with the vascular plants.

chloroplasts. Second, of the sporophytes in all three groups, those of the hornworts come closest to being capable of growth without a set limit. Liverwort and moss sporophytes have a stalk that stops growing as the sporangium matures, so elongation of the sporophyte is strictly limited. The hornwort sporophyte, however, has no stalk, and it is persistently green (a trait shared with vascular plants). A basal region of the sporangium remains capable of indefinite cell division, continuously producing new spore-bearing tissue above. The sporophytes of some hornworts growing in mild and continuously moist conditions can become as tall as 20 centimeters. Eventually, however, the sporophyte's growth is limited by the lack of a transport system.

Hornworts have a symbiotic relationship that promotes their growth by providing them with access to nitrogen, which is often a limiting resource. Hornworts have internal cavities filled with mucilage; these cavities are often populated by cyanobacteria that convert atmospheric nitrogen gas into a form usable by their host plant.

RECAP 28.2

The transition of plants to land required numerous adaptations, including the cuticle, stomata, gametangia, protected embryos, and mutually beneficial associations with fungi. Nonvascular land plants rely on liquid water for reproduction.

- Explain what is meant by alternation of generations. **See pp. 574–576 and Figure 28.6**
- Describe several adaptations of plants to the terrestrial environment, and describe the distribution of those adaptations among the liverworts, mosses, and hornworts.

New features appeared in plants as they continued to adapt to the terrestrial environment. One of the most important of these was vascular tissues, the characteristic that defines the vascular plants.

28.3 What Features Allowed Land Plants to Diversify in Form?

The first plants possessing vascular tissues did not arise until tens of millions of years after the earliest nonvascular plants had colonized the land. But once vascular tissues arose, their ability to transport water and food throughout the plant body allowed the vascular plants to spread to new terrestrial environments and to diversify rapidly.

Vascular tissues transport water and dissolved materials

The key synapomorphy of the vascular plants is a well-developed vascular system containing two types of tissues that are specialized for the transport of materials from one part of the plant to another. One type of vascular tissue, the **xylem**, conducts water and minerals from the soil to aerial parts of the plant. Because some of its cell walls contain a stiffening substance called lignin, xylem also provides support against gravity in the terrestrial environment. The other type of vascular tissue, the **phloem**, conducts the products of photosynthesis from sites where they are produced or released to sites where they are used or stored. (Xylem and phloem will be discussed in detail in Chapters 34 and 35.)

Although the vascular plants are an extraordinarily large and diverse group, one particular event was critical to their evolution. Sometime during the Paleozoic era, probably in the mid-Silurian (430 mya), a new cell type—the tracheid—evolved in sporophytes of the earliest vascular plants. The tracheid is the principal water-conducting element of the xylem in all vascular plants except the angiosperms (flowering plants) and gnetophytes—and tracheids persist even in these groups, along with a more specialized and efficient system derived from them.

The evolution of tracheids set the stage for the complete and permanent invasion of land by plants. First, these cells provided a pathway for the transport of water and mineral nutrients from a source of supply to regions of need in the plant body. Second, the cell walls of tracheids, stiffened by lignin, provided rigid structural support. This support is a crucial factor in a terrestrial environment because it allows plants to grow upward and thus compete for sunlight. A taller plant can intercept more direct sunlight (and thus conduct photosynthesis more readily) than a shorter plant, which may be shaded by the taller one. Increased height also improves the dispersal of spores.

The vascular plants featured another evolutionary novelty: a branching, independent sporophyte. A branching sporophyte body can produce more spores than an unbranched body, and it can develop in complex ways. The sporophyte of a vascular plant is nutritionally independent of the gametophyte at maturity. Among the vascular plants, the sporophyte is the large and obvious plant one normally notices in nature, in contrast to the relatively small, dependent sporophytes typical of most nonvascular land plants.

28.12 Artist's Reconstruction of an Ancient Forest Forests of the Carboniferous period were characterized by abundant vascular plants such as club mosses, ferns, and horsetails, some of which reached heights of 40 meters. Huge flying insects (see the opening of Chapter 25) thrived in these forests, which are the source of modern coal deposits.

Sporangia
Dichotomous branching
Rhizoids
Rhizome

28.13 An Ancient Relative of the Vascular Plants The extinct rhyniophyte *Aglaophyton major* lacked roots and leaves. It had a central column of xylem running through its stems but no true tracheids. A horizontal underground stem called a rhizome anchored the plant. The dichotomously branching aerial stems were less than 50 centimeters tall. Some stems were topped by sporangia.

Vascular plants allowed herbivores to colonize the land

The initial absence of herbivores (plant-eating animals) on land helped make the first vascular plants successful. By the late Silurian period (about 425 mya), vascular plants were being preserved as fossils that we can study today. The proliferation of these plants made the terrestrial environment more hospitable to animals. Arthropods, vertebrates, and other animals moved onto land only after vascular plants became established there.

Trees of various kinds appeared in the Devonian period and dominated the landscape of the Carboniferous period (359–299 mya). Forests of lycophytes (club mosses) up to 40 meters tall, along with horsetails and tree ferns, flourished in the tropical swamps of what would become North America and Europe (Figure 28.12). Plant material from those forests sank into the swamps and was gradually covered by layers of sediment. Over millions of years, as the buried plant material was subjected to intense pressure and elevated temperatures, it was transformed into coal. Today that coal provides over half of our electricity. The world's coal deposits, although huge, are not infinite, and humans are burning coal deposits at a far faster rate than they were produced.

In the subsequent Permian period, when the continents came together to form Pangaea, the continental interior became warmer and drier. The 200-million-year reign of the lycophyte–fern forests came to an end as they were replaced by forests of early gymnosperms.

The closest relatives of vascular plants lacked roots

The closest relatives of living vascular plants belonged to several extinct groups called **rhyniophytes** (Figure 28.13). The rhyniophytes were one of a very few types of land plants in the Silurian period. The landscape at that time probably consisted mostly of bare ground, with mats of nonvascular plants and stands of rhyniophytes in low-lying moist areas. Early versions of the structural features of the vascular plant groups appeared in the rhyniophytes of that time. These shared features strengthen the case for the origin of all vascular plants from a common nonvascular land plant ancestor.

Rhyniophytes did not have roots. They were apparently anchored in the soil by horizontal portions of stem called **rhizomes**, which bore water-absorbing unicellular filaments called **rhizoids**. These plants also bore aerial branches, and sporangia—homologous to the sporangia of mosses—were found at the tips of those branches. Their branching pattern was **dichotomous**; that is, the apex (tip) of the shoot divided to produce two equivalent new branches, with each pair of branches diverging at approximately the same angle from the original stem.

The lycophytes are sister to the other vascular plants

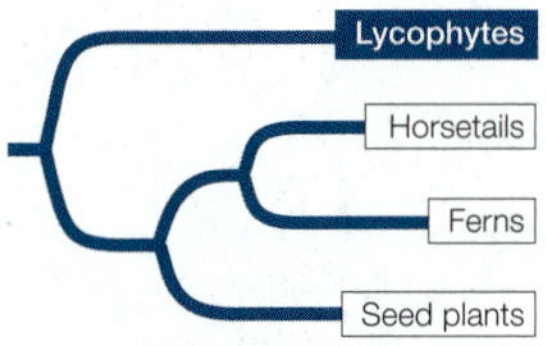
Lycophytes
Horsetails
Ferns
Seed plants

The club mosses and their relatives, the spike mosses and quillworts, are collectively called **lycophytes**. The lycophytes are the sister clade to the remaining vascular plants (see Figure 28.1B). There are relatively few (just over 1,200) surviving species of lycophytes.

The lycophytes have true roots that branch dichotomously. The arrangement of vascular tissue in their stems is simpler than that in other vascular plants. They bear simple leaflike structures called **microphylls**, which are arranged spirally on the stem. Growth in lycophytes comes entirely from apical cell division. Branching in the stems, which is also dichotomous, occurs by division of an apical cluster of dividing cells.

The sporangia of many club mosses are aggregated in conelike structures called **strobili** (singular *strobilus*; Figure 28.14A), which are clusters of spore-bearing microphylls attached to the end of the stem. Other club mosses lack strobili and bear their sporangia on (or adjacent to) the upper surfaces of specialized microphylls.

Horsetails and ferns constitute a clade

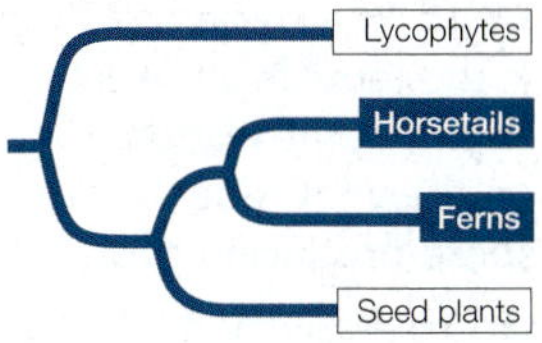
Lycophytes
Horsetails
Ferns
Seed plants

The horsetails and ferns were once thought to be only distantly related. From analysis of gene sequences we now know that they form a clade, the **monilophytes**. In the monilophytes—as in the seed plants, to which they are the sister clade (see Figure 28.1)—there is differentiation between a main stem and side branches (including the leaves derived from these branches). This pattern contrasts with the dichotomous branching characteristic of the lycophytes and rhyniophytes, in which each split gives rise to two branches of similar size.

Today there are only about 15 species of **horsetails**, all in the genus *Equisetum*. The horsetails have reduced true leaves that form in distinct whorls (circles) around the stem (Figure 28.14B). Horsetails are sometimes called "scouring rushes" because rough silica deposits found in their cell walls once made them useful for cleaning. They have true roots that branch irregularly. Horsetails have a large sporophyte and a small gametophyte, both independent of each other.

Strobilus
Microphylls

(A) *Lycopodium annotinum*

Leaves
(in whorls)
Sporangia-bearing
structure

(B) *Equisetum pratense*

(C) *Marsilea* sp. *Salvinia* sp.

(D) *Dicksonia antarctica*

28.14 Lycophytes and Monilophytes
(A) Club mosses have microphylls arranged spirally on their stems. Strobili are visible at the tips of these stems. (B) Horsetails have a distinctive growth pattern in which the stem grows in segments above each whorl of leaves. These are fertile shoots with sporangia-bearing structures at the apex. (C) The leaves of two species of water ferns. (D) Tree ferns dominate this forest on the island of Tasmania, Australia.

28.15 Life Cycle of a Fern The most conspicuous stage in the fern life cycle is the mature diploid sporophyte, shown at the bottom of this diagram. The inset shows sori on the underside of a fern leaf. Each sorus contains many spore-producing sporangia.

Go to Activity 28.1 The Fern Life Cycle
Life10e.com/ac28.1

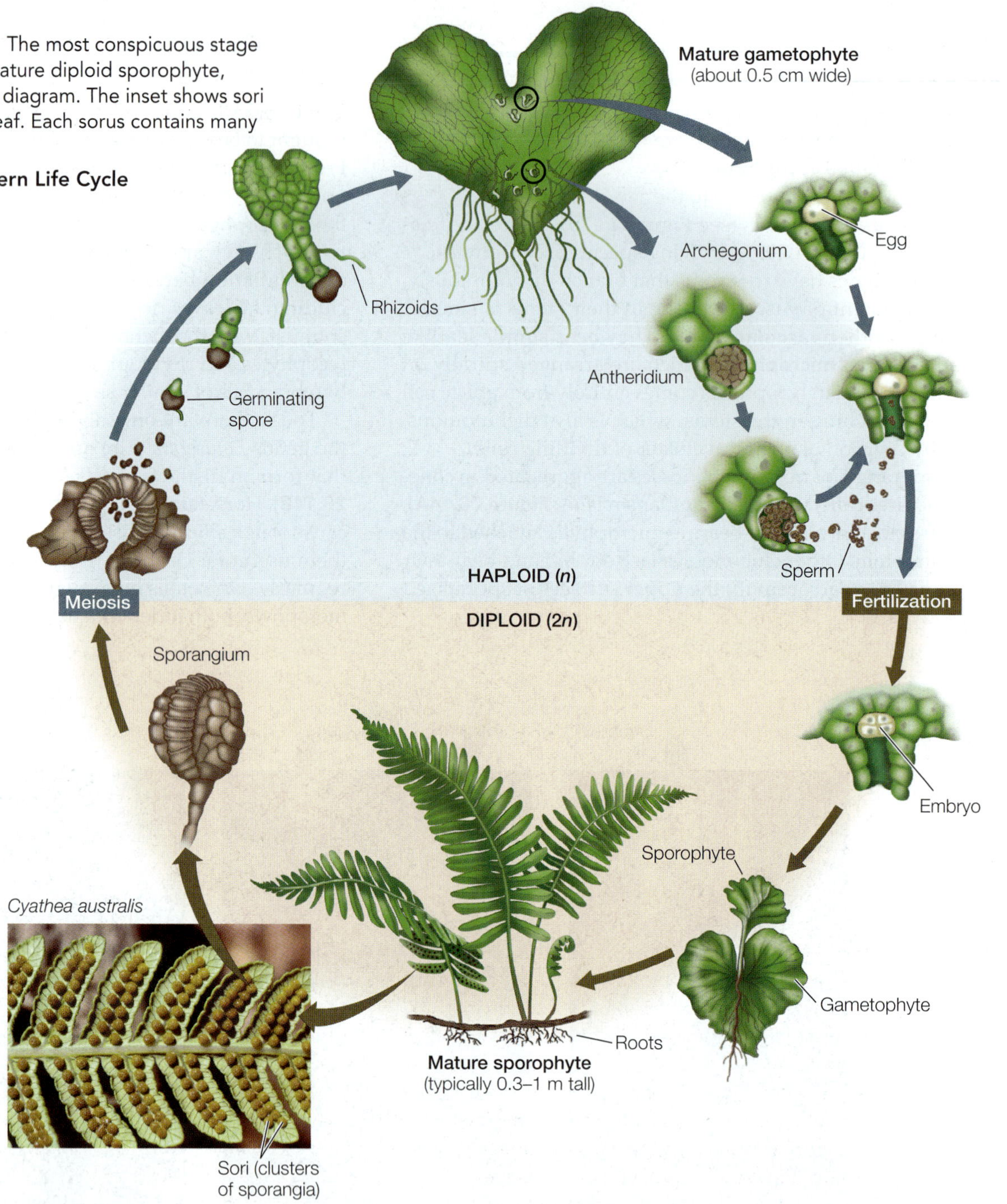
Mature gametophyte
(about 0.5 cm wide)
Archegonium
Egg
Rhizoids
Antheridium
Germinating spore
Sperm
HAPLOID (n)
DIPLOID (2n)
Meiosis
Fertilization
Sporangium
Embryo
Sporophyte
Gametophyte
Cyathea australis
Roots
Mature sporophyte
(typically 0.3–1 m tall)
Sori (clusters of sporangia)

The first ferns appeared during the Devonian period; today this group comprises more than 12,000 species. Analyses of gene sequences indicate that a few species traditionally allied with ferns may in fact be more closely related to horsetails than to ferns. Nonetheless, the majority of ferns form a monophyletic group.

Although most ferns are terrestrial, a few species live in shallow fresh water (**Figure 28.14C**). Terrestrial ferns are characterized by large leaves with branching vascular strands (**Figure 28.14D**). Some fern leaves become climbing organs and may grow to be as long as 30 meters.

In the alternating generations of a fern, the gametophyte is small, delicate, and short-lived, but the sporophyte can be very large and can sometimes survive for hundreds of years (**Figure 28.15**). Ferns require liquid water for the transport of the male gametes to the female gametes, so most ferns inhabit shaded, moist woodlands and swamps. The sporangia of ferns are typically borne on a stalk in clusters called **sori** (singular *sorus*). The sori are found on the undersurfaces of the leaves, sometimes covering the entire undersurface and sometimes located at the edges.

The vascular plants branched out

Several features that were new to the vascular plants evolved in lycophytes and monilophytes. Roots probably had their evolutionary origins as branches, either of a rhizome or of the

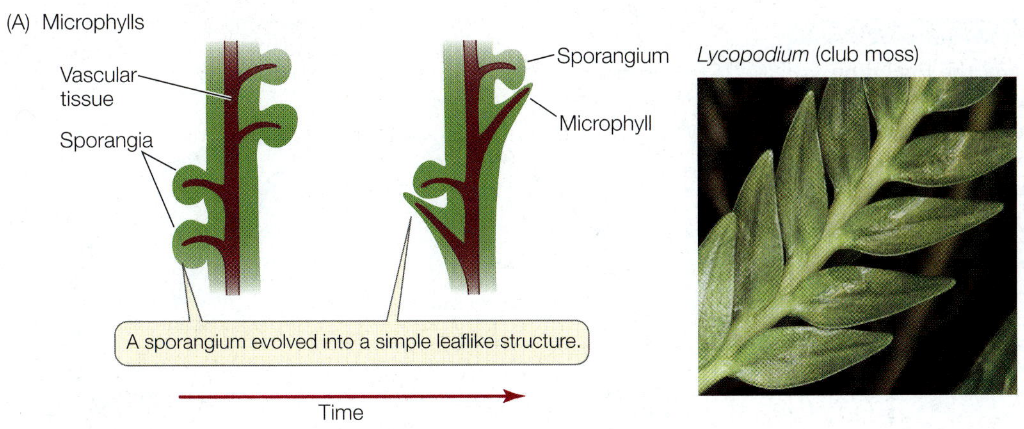
(A) Microphylls
Vascular tissue
Sporangia
Sporangium
Microphyll
A sporangium evolved into a simple leaflike structure.
Time
Lycopodium (club moss)

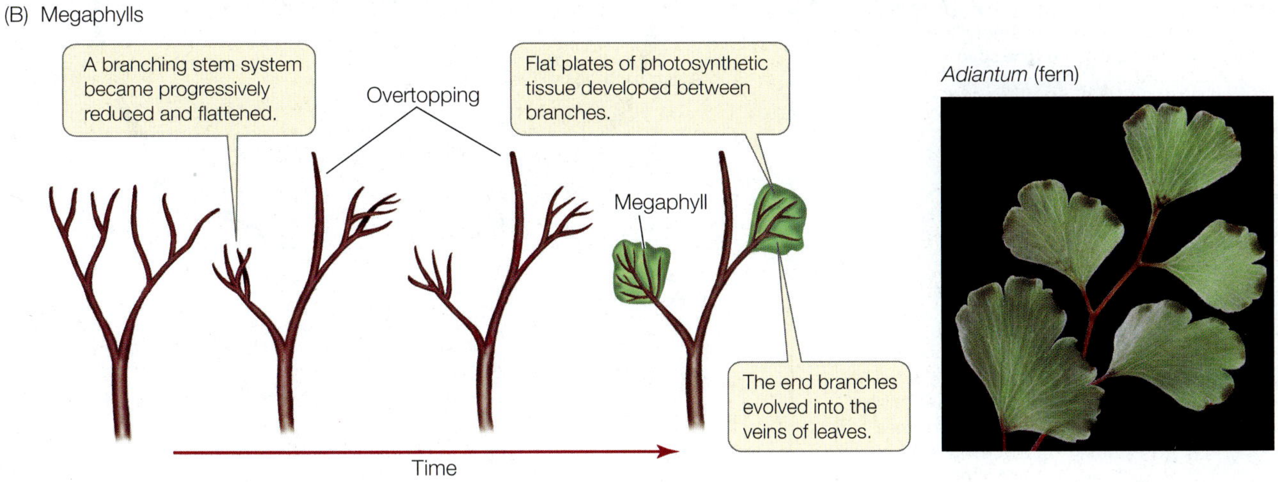
(B) Megaphylls
A branching stem system became progressively reduced and flattened.
Overtopping
Flat plates of photosynthetic tissue developed between branches.
Megaphyll
The end branches evolved into the veins of leaves.
Time
Adiantum (fern)

28.16 Evolution of Leaves (A) Microphylls are thought to have evolved from sterile sporangia. (B) The megaphylls of monilophytes and seed plants may have arisen as photosynthetic tissue developed between branch pairs that were "left behind" as dominant branches overtopped them.

aboveground portion of a stem. These branches presumably penetrated the soil and branched further. The underground portions could anchor the plant firmly, and even in this primitive condition, they could absorb water and minerals.

The microphylls of lycophytes were probably the first leaflike structures to evolve among the vascular plants. Microphylls are usually small and only rarely have more than a single vascular strand, at least in existing species. Some biologists believe that microphylls had their evolutionary origins as sterile sporangia (**Figure 28.16A**). A typical feature of this type of leaf is a vascular strand that departs from the vascular system of the stem in such a way that the structure of the stem's vascular system is scarcely disturbed. This pattern was evident even in the lycophyte trees of the Carboniferous period, many of which had microphylls many centimeters long.

The monilophytes and seed plants constitute a clade called the **euphyllophytes** (*eu*, "true"; *phyllon*, "leaf"). An important synapomorphy of the euphyllophytes is **overtopping**, a growth pattern in which one branch differentiates from and grows beyond the others (**Figure 28.16B**). Overtopping would have given these plants an advantage in the competition for light, enabling them to shade their dichotomously branching competitors. The overtopping growth of the euphyllophytes also allowed a new type of leaflike structure to evolve. This larger, more complex leaf is called a **megaphyll**. The megaphyll is thought to have arisen from the flattening of a portion of a branching stem system that exhibited overtopping growth. This change was followed by the development of photosynthetic tissue between the members of overtopped groups of branches, which had the advantage of increasing the photosynthetic surface area of those branches.

The first megaphylls, which were very small, appeared in the Devonian period. We might expect that evolution should have led swiftly to the appearance of more and larger megaphylls because of their greater photosynthetic capacity. However, it took some 50 million years, until the Carboniferous period, for large megaphylls to become common. Why should this have been so, especially given that other advances in plant structure were taking place during that time?

According to one theory, the high concentration of CO_2 in the atmosphere during the Devonian period reduced selection for the stomata that allow a leaf to take up CO_2 for use in photosynthesis. With more CO_2 available, fewer stomata were needed. In the Devonian, larger leaves with a limited number of stomata would have absorbed heat from sunlight, but they would have been unable to lose heat fast enough by evaporation of water through their stomata. The resulting overheating would have been lethal. Recent research has supported this hypothesis, indicating that larger megaphylls evolved only as CO_2 concentrations dropped over millions of years (**Figure 28.17**).

INVESTIGATINGLIFE

28.17 Atmospheric CO_2 Concentrations and the Evolution of Megaphylls High concentrations of atmospheric CO_2 during the first part of the Devonian may have limited the evolution of leaf size. C. P. Osborne and colleagues compared the leaf sizes of fossil plants against estimates of CO_2 concentrations in the atmosphere at the time the plants were alive.[a]

HYPOTHESIS High atmospheric CO_2 concentrations during the early Devonian, and the resulting lack of selection for more stomata, kept leaf sizes small.

Method

1. Analyze 300 plant fossils from the Devonian and Carboniferous periods and measure the sizes of their leaves.
2. Compare the pattern of change in leaf size with that of the estimated change in atmospheric CO_2 concentrations over the same time span.

Results

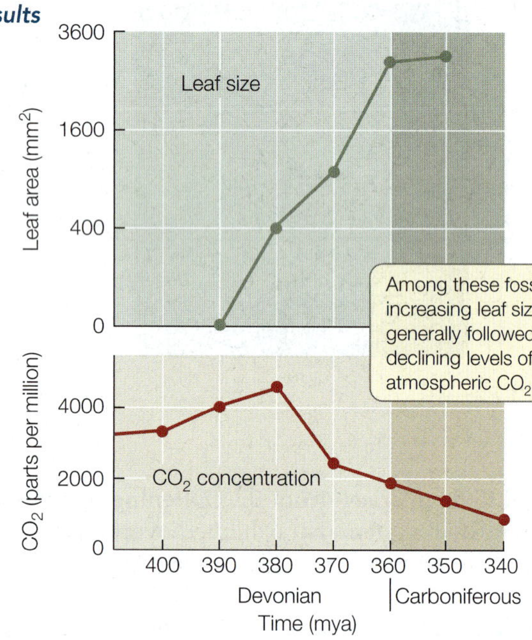
Leaf size
Leaf area (mm²)
3600
1600
400
0
Among these fossils, increasing leaf sizes generally followed declining levels of atmospheric CO₂.
CO₂ (parts per million)
4000
2000
0
CO₂ concentration
400 390 380 370 360 350 340
Devonian
Carboniferous
Time (mya)

CONCLUSION Leaf size increased as levels of CO_2 in the atmosphere decreased.

Go to **BioPortal** for discussion and relevant links for all INVESTIGATING**LIFE** figures.

[a]Osborne, C. P. et al. 2004. *Proceedings of the National Academy of Sciences USA* 101: 10360–10362.

Heterospory appeared among the vascular plants

In the lineages of present-day, seedless vascular plants that are most similar to their ancestors, the gametophyte and the sporophyte are independent, and both are usually photosynthetic. The spores produced by the sporophyte are of a single type and develop into a single type of gametophyte that bears both female and male reproductive organs (see Figure 28.15). Such plants, which bear a single type of spore, are said to be **homosporous** (Figure 28.18A).

A system with two distinct types of spores evolved somewhat later. Plants of this type are said to be **heterosporous**

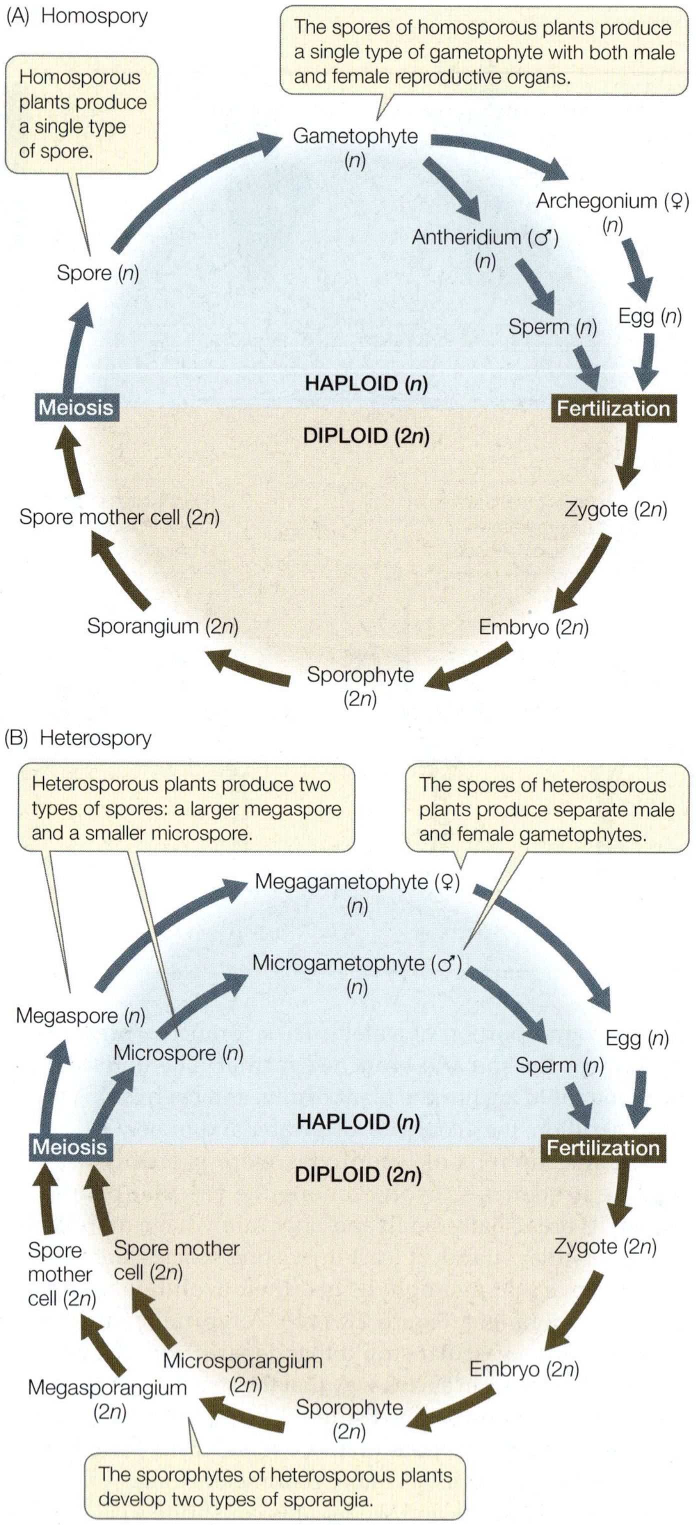
(A) Homospory
Homosporous plants produce a single type of spore.
The spores of homosporous plants produce a single type of gametophyte with both male and female reproductive organs.
Gametophyte (n)
Archegonium (♀) (n)
Antheridium (♂) (n)
Spore (n)
Sperm (n)
Egg (n)
HAPLOID (n)
Meiosis
Fertilization
DIPLOID (2n)
Spore mother cell (2n)
Zygote (2n)
Sporangium (2n)
Embryo (2n)
Sporophyte (2n)
(B) Heterospory
Heterosporous plants produce two types of spores: a larger megaspore and a smaller microspore.
The spores of heterosporous plants produce separate male and female gametophytes.
Megagametophyte (♀) (n)
Microgametophyte (♂) (n)
Megaspore (n)
Microspore (n)
Egg (n)
Sperm (n)
HAPLOID (n)
Meiosis
Fertilization
DIPLOID (2n)
Spore mother cell (2n)
Spore mother cell (2n)
Zygote (2n)
Microsporangium (2n)
Megasporangium (2n)
Embryo (2n)
Sporophyte (2n)
The sporophytes of heterosporous plants develop two types of sporangia.

28.18 Homospory and Heterospory (A) Homosporous plants bear a single type of spore. Each gametophyte has two types of sex organs, antheridia (male) and archegonia (female). (B) Heterosporous plants bear two types of spores that develop into distinctly male and female gametophytes.

Go to Activities 28.2 Heterospory and 28.3 Homospory
Life10e.com/ac28.2
Life10e.com/ac28.3

(Figure 28.18B). In heterospory, one type of spore—the **megaspore**—develops into a specifically female gametophyte (a **megagametophyte**) that produces only eggs. The other type, the **microspore**, is smaller and develops into a male gametophyte (a **microgametophyte**) that produces only sperm. The sporophyte produces megaspores in small numbers in **megasporangia** and microspores in large numbers in **microsporangia**. Heterospory affects not only the spores and the gametophytes but also the sporophyte plant itself, which must develop two types of sporangia.

The earliest vascular plants were all homosporous, but heterospory evidently evolved several times independently among later groups of vascular plants. The fact that heterospory evolved repeatedly suggests that it affords selective advantages. Subsequent evolution in the land plants featured ever greater specialization of the heterosporous condition.

RECAP 28.3

Vascular plants are characterized by a vascular system specialized for the transport of materials from one part of the plant to another. A new type of cell, the tracheid, marked the origin of this group. Later evolutionary events included the appearance of roots, leaves, and heterospory.

- How do xylem and phloem serve the vascular plants? See p. 579
- Describe the evolution and distribution of different kinds of leaves and roots among the vascular plants. See pp. 581–583 and Figure 28.16

All of the vascular plant groups we have discussed thus far disperse by means of spores. The embryos of these seedless vascular plants develop directly into sporophytes, which either survive or die, depending on environmental conditions. The spores of some seedless plants may remain dormant and viable for long periods, but the embryos of seedless plants are relatively unprotected. Greater protection of the embryo evolved in the seed plants, which we will consider in the next chapter.

28.19 Biodiesel from Algae In this vertical-growth algal cultivation system for biofuel production, algae are grown in sheets of clear plastic, exposed to sunlight on all sides.

? Given that petroleum is produced naturally from green algae, can humans use green algae to produce oil commercially?

ANSWER

Scientists are developing new methods for growing green algae for the production of biofuels (fuels produced directly from living organisms, such as biodiesel). Some species of green algae can produce up to 60 percent of their dry weight in oil. So biofuels can certainly be produced from green algae, although the process is not yet commercially viable. Like conventional fossil fuels, biofuels release carbon dioxide into the atmosphere when burned. In the production of biofuels, however, algae remove carbon dioxide from the atmosphere, so the use of these fuels is more sustainable, and results in less accumulation of CO_2 in the atmosphere over time, than the use of fossil fuels.

The primary commercial limitations to growing algae for biofuels include the need to establish efficient growing facilities, water needs, costs of fertilizers, costs and difficulties associated with harvest and refining, and labor expenses. Many new methods for growing and harvesting algae are being developed (Figure 28.19), and algae production in brackish water as well as in wastewater is being explored. Once some of the technical difficulties have been overcome, it is possible that commercial production of biofuels from algae will provide a significant source of energy for humans.

CHAPTER SUMMARY 28

28.1 How Did Photosynthesis Arise in Plants?

- Primary endosymbiosis gave rise to chloroplasts and the subsequent diversification of the **Plantae**. The descendants of the first photosynthetic eukaryote include **glaucophytes**, **red algae**, several groups of green algae, and **land plants**, all of which contain chlorophyll *a*. **Review Figure 28.1**
- **Green plants**, which include the green algae and the land plants, are characterized by the presence of chlorophyll *b* (in addition to chlorophyll *a*). **Review Figure 28.1**
- Land plants, also known as **embryophytes**, arose from an aquatic green algal ancestor related to today's **stoneworts**. Land plants develop from embryos that are protected by parental tissue. **Review Figure 28.1**

28.2 When and How Did Plants Colonize Land?

- The acquisition of a **cuticle**, **stomata**, **gametangia**, a protected **embryo**, protective pigments, thick spore walls with a protective polymer, and mutualistic associations with fungi were all adaptations of land plants to terrestrial life.
- All land plant life cycles feature alternation of generations, in which a multicellular diploid **sporophyte** alternates with a multicellular haploid **gametophyte**. **Review Figure 28.6**
- The **nonvascular land plants** comprise the **liverworts**, **mosses**, and **hornworts**. These groups lack specialized vascular tissues for the conduction of water or nutrients through the plant body.
- The life cycles of nonvascular land plants depend on liquid water. The sporophyte is usually smaller than the gametophyte and depends on it for water and nutrition. **Review Figure 28.7, ANIMATED TUTORIAL 28.1**
- Liverworts lack stomata, but they are present in mosses, hornworts, and vascular plants. Hornworts have a persistently green sporophyte, a characteristic shared with vascular plants.

28.3 What Features Allowed Land Plants to Diversify in Form?

- The **vascular plants** have a **vascular system** consisting of **xylem** and **phloem** that conducts water, minerals, and products of photosynthesis through the plant body. The vascular system includes cells called **tracheids**.
- The **rhyniophytes**, the earliest known vascular plants, are known to us only in fossil form. They lacked true roots and leaves but possessed **rhizomes** and **rhizoids**. **Review Figure 28.13**
- The **lycophytes** (club mosses and relatives) have only small, simple leaflike structures (**microphylls**). **Monilophytes** (which include **horsetails** and **ferns**) have true leaves, and so together with seed plants are known as **euphyllophytes**.
- Unlike nonvascular land plants, the diploid sporophyte is the more conspicuous life stage of lycophytes and monilophytes. **Review Figure 28.15, ACTIVITY 28.1**
- Microphylls probably evolved from sterile sporangia. **Megaphylls** (true leaves) may have resulted from the flattening and reduction of a portion of a stem system with **overtopping** growth. **Review Figure 28.16**
- The earliest-diverging groups of vascular plants are **homosporous**, but **heterospory**—the production of distinct **megaspores** and **microspores**—has evolved several times. Megaspores develop into female **megagametophytes**; microspores develop into male **microgametophytes**. **Review Figure 28.18, ACTIVITIES 28.2 and 28.3**

Go to the Interactive Summary to review key figures, Animated Tutorials, and Activities Life10e.com/is28

CHAPTER REVIEW

REMEMBERING

1. Which statement about alternation of generations in land plants is *not* true?
   a. The gametophyte and sporophyte differ in appearance.
   b. Meiosis occurs in sporangia.
   c. Gametes are always produced by meiosis.
   d. The zygote is the first cell of the sporophyte generation.
   e. The gametophyte and sporophyte differ in chromosome number.

2. Which of the following provide evidence for the close relationship between land plants and stoneworts?
   a. Phylogenetic analysis of DNA sequences
   b. Similar mechanics of mitosis and cytokinesis
   c. Similarities in chloroplast structure
   d. Similarities of their biochemistry
   e. All of the above

3. Liverworts, mosses, and hornworts
   a. lack a sporophyte generation.
   b. grow in dense masses, allowing capillary movement of water.
   c. possess xylem and phloem.
   d. possess true leaves.
   e. possess true roots.

4. Which statement about ferns is *not* true?
   a. The sporophyte is larger than the gametophyte.
   b. Most are heterosporous.
   c. The young sporophyte can grow independently of the gametophyte.
   d. The leaf is a megaphyll.
   e. The gametophytes produce archegonia and antheridia.

5. The ferns
   a. lack a gametophyte generation.
   b. have a large and prominent gametophyte but a much smaller sporophyte.
   c. are more closely related to club mosses than to horsetails.
   d. are monilophytes.
   e. produce seeds.

UNDERSTANDING & APPLYING

6. Contrast microphylls with megaphylls in terms of structure, evolutionary origin, and occurrence among plants.

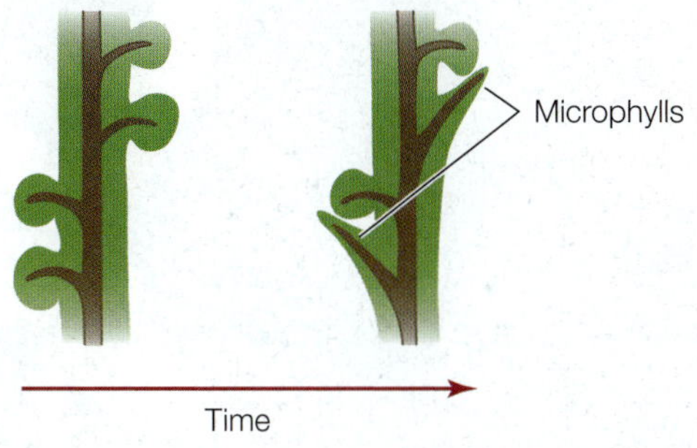
Microphylls
Time

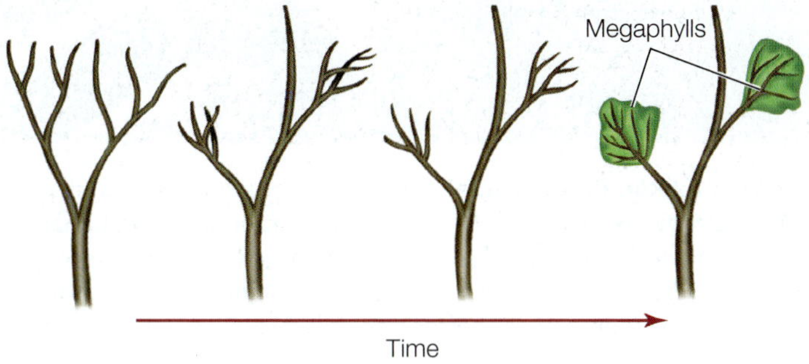
Megaphylls
Time

7. Heterospory is thought to have evolved multiple times within plants, suggesting an advantage to the heterosporous condition. Why might heterospory provide an evolutionary advantage over homospory?

8. Contrast the life cycles of mosses (see Figure 28.7) and ferns (see Figure 28.15). How are these life cycles similar, and how do they differ?

ANALYZING & EVALUATING

9. Are the morphological characters plotted on the phylogeny in Figure 28.1 consistent with your analysis of DNA sequences from the Working with Data exercise on p. 571? What characteristics of plants that we have discussed in this chapter appear to have evolved more than once in the history of land plants?

10. The findings of Osborne and co-workers on the evolution of megaphylls (see Figure 28.17) support the idea that large megaphylls became common only after the atmospheric CO_2 level had dropped, so that more stomata were favored, allowing evaporation from the stomata to cool larger leaves. How might you extend that work to confirm the involvement of temperature as a factor limiting leaf size?

Go to BioPortal at **yourBioPortal.com** for Animated Tutorials, Activities, LearningCurve Quizzes, Flashcards, and many other study and review resources.

29 The Evolution of Seed Plants

CHAPTER OUTLINE

Darwin's Ever-Fascinating Orchids On examining a specimen of the orchid *Angraecum sesquipedale*, Charles Darwin noted its exceptionally long nectar tube and predicted the existence of a pollinator with a correspondingly long proboscis. This pollinator, the sphinx moth *Xanthopan morgani*, was not discovered until after Darwin's death.

IN THE EARLY 1860s, much of middle- and upper-class England was caught up in an "orchid frenzy." Amateur gardeners and professional botanists alike were enchanted with these beautiful flowers and devoted considerable effort and money to collecting and raising them. Following the publication of *On The Origin of Species* in 1859, Charles Darwin wrote his next book on this group of plants. *On the Various Contrivances by which British and Foreign Orchids are Fertilised by Insects*, which he referred to proudly as "my little book on the fertilisation of orchids," appeared in 1862.

There are more than 25,000 species of orchids, which makes them one of the most diverse plant groups. Darwin wanted to know why orchids had experienced such rapid diversification and was particularly impressed with the role that insect pollinators might have played in this process. Seeking examples to demonstrate the power of natural selection, Darwin found such examples in abundance among the orchids.

Orchids show an impressive variety of specialized pollination mechanisms, many of which demonstrate that they have coevolved with their pollinators. For example, Darwin observed a South American orchid of the genus *Catasetum* shooting a packet of pollen at an insect that landed on its flower. When he was shown *Angraecum sesquipedale*, an orchid from Madagascar with a nectar tube more than a foot long, Darwin hypothesized that there must be a moth with a proboscis of unprecedented length that fed from and pollinated that flower. Many people scoffed at his vision, but the moth he described was eventually discovered—21 years after his death.

In 1836 the explorer Robert Schomburgk shook the botanical world with a report that he had seen flowers described as belonging to three different genera of orchids—*Catasetum*, *Monachanthus*, and *Myanthus*—growing together on a single plant. The English botanist John Lindley remarked that this observation would "shake to the foundation all our ideas of the stability of genera and species." Orchid enthusiasts were befuddled by their efforts to grow specimens of *Myanthus*, only to have them flower with the more common blooms of *Catasetum*. Darwin knew that he needed to find the explanation for these odd observations, for otherwise he would have to conclude that individual plants were able to change their specific identity, something that did not fit with his explanations of the evolution of diversity.

> ? What was Darwin's explanation for the three distinct flowers growing on a single orchid plant?
>
> See answer on p. 605.

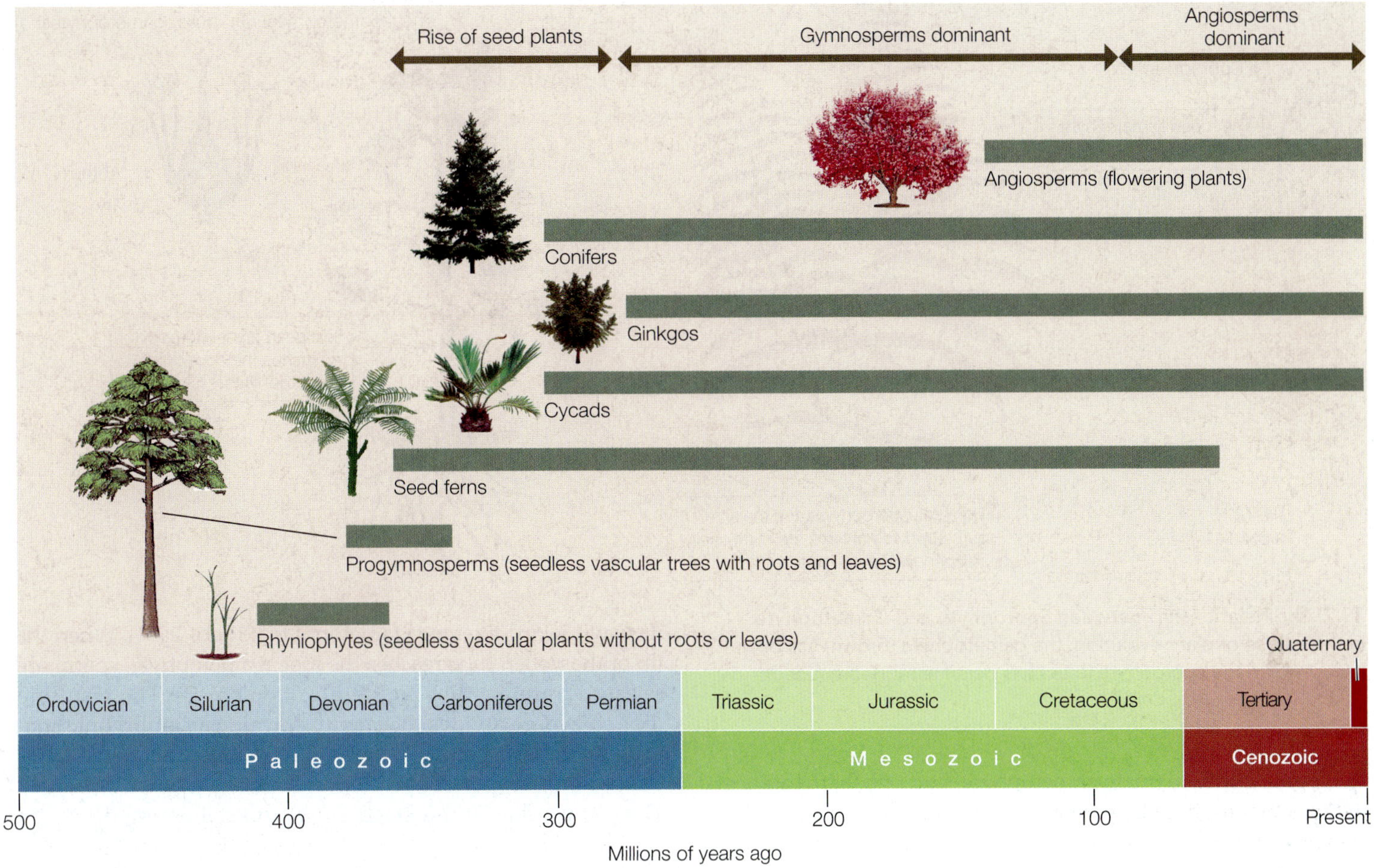

Rise of seed plants
Gymnosperms dominant
Angiosperms dominant
Angiosperms (flowering plants)
Conifers
Ginkgos
Cycads
Seed ferns
Progymnosperms (seedless vascular trees with roots and leaves)
Rhyniophytes (seedless vascular plants without roots or leaves)
Quaternary
Ordovician
Silurian
Devonian
Carboniferous
Permian
Triassic
Jurassic
Cretaceous
Tertiary
Paleozoic
Mesozoic
Cenozoic
500
400
300
200
100
Present
Millions of years ago

29.1 The Fossil Record of Seed Plant Evolution Woody growth evolved in the seedless progymnosperms. The now-extinct seed ferns had woody growth, fernlike foliage, and seeds attached to their leaves. New lineages of seed plants arose during the Carboniferous, but the earliest known fossils of flowering plants are from near the Jurassic–Cretaceous boundary.

29.1 How Did Seed Plants Become Today's Dominant Vegetation?

By the late Devonian period, more than 360 million years ago, Earth was home to a great variety of land plants, many of which we discussed in Chapter 28. The land plants shared the hot, humid terrestrial environment with insects, spiders, centipedes, and fishlike amphibians (early tetrapods). These plants and animals evolved together, each acting as agents of natural selection on the other. In the Devonian, a new innovation appeared when some plants developed extensively thickened woody stems. Among the first plants with this adaptation were seedless vascular plants called **progymnosperms**, all species of which are now extinct. The progymnosperms included many large trees.

Another innovation, the **seed**, arose in the seed plants. Seeds provide a secure and lasting dormant stage for the embryo. A plant embryo may safely wait within its seed (in some cases for many years, or even centuries) until conditions are right for germination.

The earliest fossil evidence of seed plants is found in late Devonian rocks. Like the progymnosperms, these now-extinct **seed ferns** were woody. They possessed fernlike foliage but had seeds attached to their leaves. By the end of the Permian, other groups of seed plants had become dominant (Figure 29.1). Today's living seed plants fall into two major groups, the **gymnosperms** (such as pines and cycads) and the hugely diverse group known as the **angiosperms** (flowering plants).

Features of the seed plant life cycle protect gametes and embryos

In Section 28.2 we described a major trend in land plant evolution: the sporophyte became less dependent on the gametophyte, which became smaller in relation to the sporophyte. This trend continued with the seed plants, whose gametophyte generation is reduced even further than that of the ferns (Figure 29.2). The haploid seed plant gametophyte develops partly or entirely while attached to and nutritionally dependent on the diploid sporophyte.

Among the seed plants, only a few groups of gymnosperms (including modern cycads and ginkgos) have swimming sperm. Even in these groups, sperm is transferred via pollen grains, so fertilization does not require liquid water outside the plant body. This adaptation, along with the advent of the seed, gave seed plants the opportunity to colonize drier areas and spread over the terrestrial environment.

Seed plants are heterosporous (see Figure 28.18B)—that is, they produce two types of spores, one that becomes a

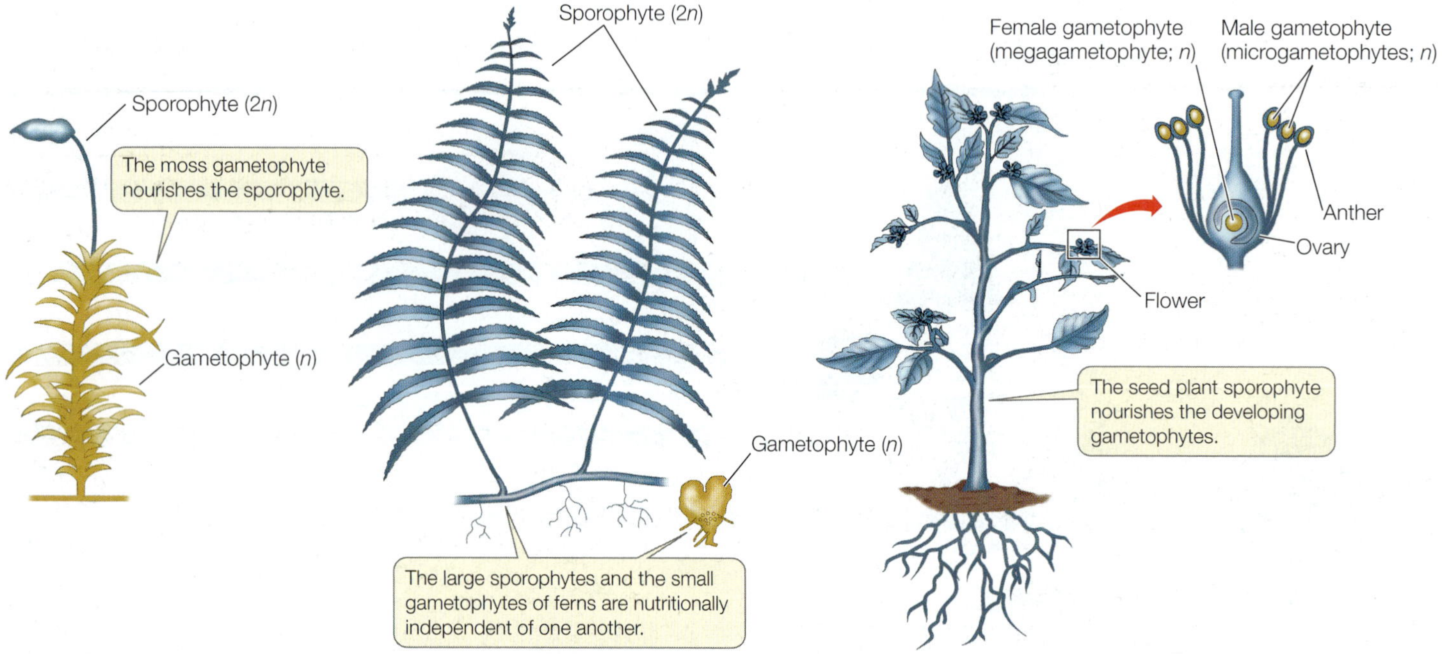
Sporophyte (2n)
The moss gametophyte nourishes the sporophyte.
Gametophyte (n)
Sporophyte (2n)
Gametophyte (n)
The large sporophytes and the small gametophytes of ferns are nutritionally independent of one another.
Female gametophyte (megagametophyte; n)
Male gametophyte (microgametophytes; n)
Anther
Ovary
Flower
The seed plant sporophyte nourishes the developing gametophytes.

29.2 The Relationship between Sporophyte and Gametophyte In the course of plant evolution, the gametophyte (brown) has been reduced and the sporophyte (blue) has become more prominent.

microgametophyte (male gametophyte) and one that becomes a megagametophyte (female gametophyte). These plants form separate microsporangia and megasporangia on structures that are grouped on short stems.

Within the microsporangium, the meiotic products are microspores. Within its spore wall, a microspore divides mitotically one or a few times to form a multicellular male gametophyte called a **pollen grain**. Pollen grains are released from the microsporangium to be distributed by wind or by an animal pollinator (Figure 29.3). As in seedless land plants, the spore wall surrounding the pollen grain contains sporopollenin, the most chemically resistant biological compound known, which protects the pollen grain against dehydration and chemical damage—another advantage in terms of survival in the terrestrial environment.

In contrast to the microspores, the megaspores of seed plants are not shed. Instead, they develop into female gametophytes within the megasporangia. These megagametophytes are dependent on the sporophyte for food and water.

In most seed plant species, only one of the meiotic products in a megasporangium survives. The surviving haploid nucleus divides mitotically, and the resulting cells divide again to produce a multicellular female gametophyte. The megasporangium is surrounded by an **integument**: a layer of sporophytic tissue that protects the megasporangium and its contents. Together, the megasporangium and integument constitute the **ovule**, which will develop into a seed after fertilization.

The arrival of a pollen grain at an appropriate landing point, close to a female gametophyte on a sporophyte of the same species, is called **pollination**. A pollen grain that reaches this point produces a slender **pollen tube** that elongates and usually digests its way toward the megagametophyte (Figure 29.4). When the tip of the pollen tube reaches the megagametophyte, sperm are released from the tube and fertilization occurs. The resulting diploid zygote divides repeatedly, forming an embryonic sporophyte. After a period of embryonic development, growth is temporarily suspended (the embryo enters a dormant stage). The end product at this stage is the multicellular seed.

29.3 Blown on the Wind Pollen grains are the male gametophytes of seed plants. The male flowers of this hazel tree release pollen-containing spores that are dispersed by the wind and may land near female gametophytes of the same or other hazel plants.

Go to Media Clip 29.1
Pollen Transfer by Wind
Life10e.com/mc29.1

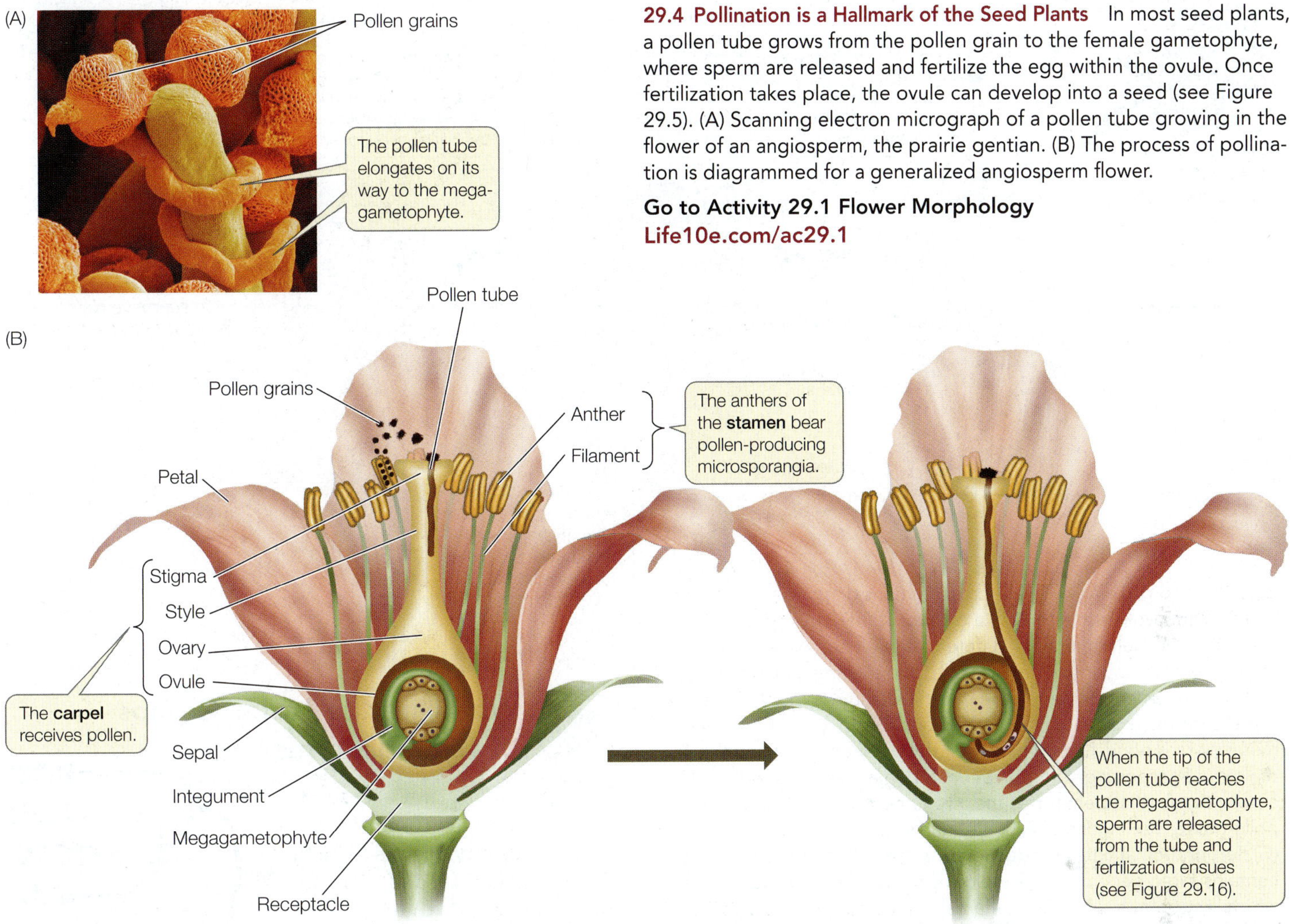
(A)
Pollen grains
The pollen tube elongates on its way to the mega-gametophyte.
(B)
Pollen tube
Pollen grains
Petal
Anther
Filament
The anthers of the **stamen** bear pollen-producing microsporangia.
Stigma
Style
Ovary
Ovule
The **carpel** receives pollen.
Sepal
Integument
Megagametophyte
Receptacle
When the tip of the pollen tube reaches the megagametophyte, sperm are released from the tube and fertilization ensues (see Figure 29.16).

29.4 Pollination is a Hallmark of the Seed Plants In most seed plants, a pollen tube grows from the pollen grain to the female gametophyte, where sperm are released and fertilize the egg within the ovule. Once fertilization takes place, the ovule can develop into a seed (see Figure 29.5). (A) Scanning electron micrograph of a pollen tube growing in the flower of an angiosperm, the prairie gentian. (B) The process of pollination is diagrammed for a generalized angiosperm flower.

Go to Activity 29.1 Flower Morphology
Life10e.com/ac29.1

The seed is a complex, well-protected package

A seed contains tissues from three generations (**Figure 29.5**). A seed coat develops from the integument—the tissues of the diploid sporophyte parent that surround the megasporangium. Within the megasporangium is haploid tissue from the female gametophyte, which contains a supply of nutrients for the developing embryo. (This tissue is fairly extensive in most gymnosperm seeds. In angiosperm seeds it is greatly reduced, and nutrition for the embryo is supplied instead by a tissue called endosperm.) In the center of the seed is the third generation, the embryo of the new diploid sporophyte.

The seed is a well-protected resting stage. The seeds of some species may remain dormant but stay viable (capable of growth and development) for many years, germinating only when conditions are favorable for the growth of the sporophyte. During the dormant stage, the seed coat protects the embryo from excessive drying and may also protect it against potential predators that would otherwise consume the embryo and its nutrient reserves. Many seeds have structural adaptations that promote their dispersal by wind or, more often, by animals. When the young sporophyte resumes growth, it draws on the food reserves in the seed. The possession of seeds is a major reason for the enormous evolutionary success of the seed plants, which are the dominant life forms of most modern terrestrial floras.

A change in stem anatomy enabled seed plants to grow to great heights

Fossils of the closest relatives of seed plants (progymnosperms) and the earliest seed plants (seed ferns) are found in late Devonian rocks (see Figure 29.1). These plants had thickened woody stems, developed through the proliferation of xylem. This type of growth, which increases the diameter of stems and roots in many modern seed plants, is called **secondary growth**. Its product is secondary xylem, or wood.

The younger portion of the wood produced by secondary growth is well adapted for water transport, but older wood becomes clogged with resins or other materials. Although no longer functional in transport, the older wood continues to provide support for the plant. This support allows woody plants to grow taller than other plants around them and thus capture more light for photosynthesis.

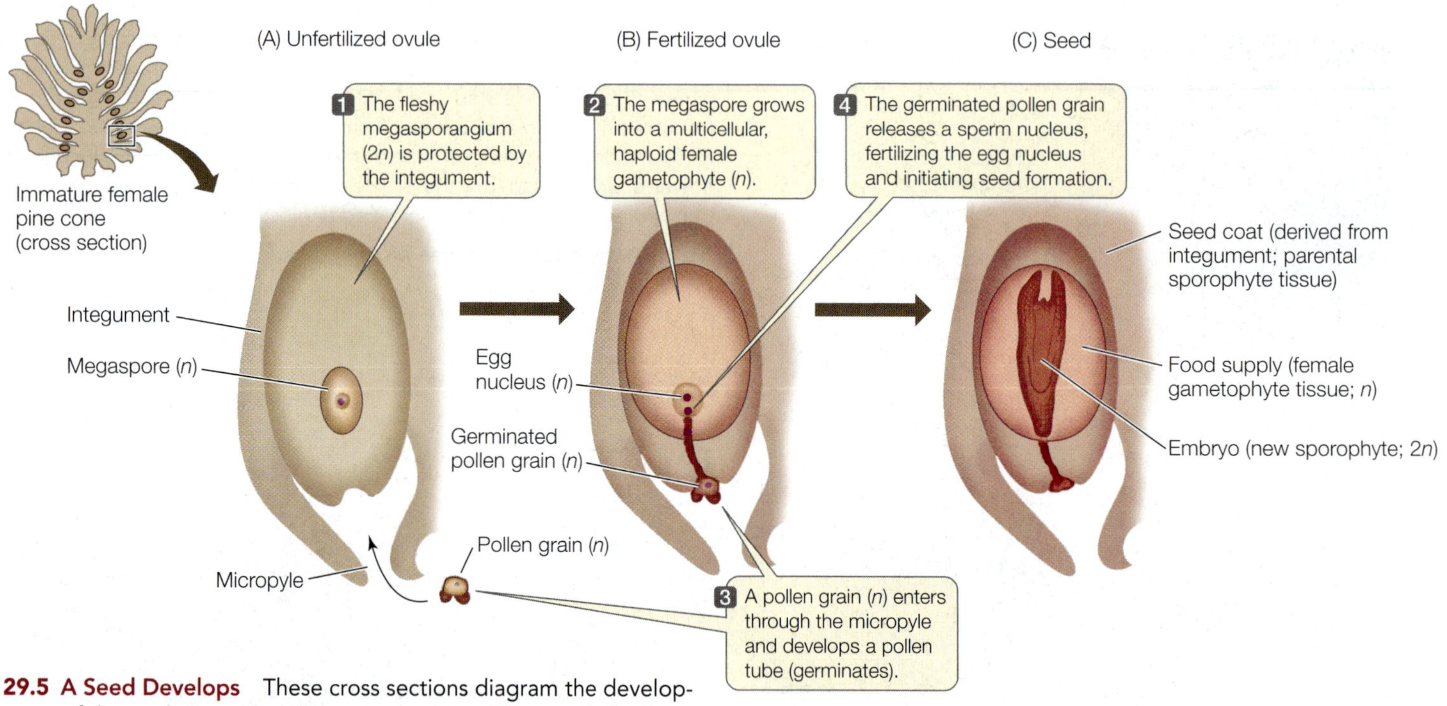
(A) Unfertilized ovule
(B) Fertilized ovule
(C) Seed
1 The fleshy megasporangium (2n) is protected by the integument.
2 The megaspore grows into a multicellular, haploid female gametophyte (n).
4 The germinated pollen grain releases a sperm nucleus, fertilizing the egg nucleus and initiating seed formation.
Immature female pine cone (cross section)
Integument
Megaspore (n)
Egg nucleus (n)
Germinated pollen grain (n)
Pollen grain (n)
Micropyle
Seed coat (derived from integument; parental sporophyte tissue)
Food supply (female gametophyte tissue; n)
Embryo (new sporophyte; 2n)
3 A pollen grain (n) enters through the micropyle and develops a pollen tube (germinates).

29.5 A Seed Develops These cross sections diagram the development of the ovule into a seed in a gymnosperm (*Pinus* sp.). Angiosperm seed development has differences (e.g., angiosperm integuments have two layers rather than one, and the angiosperm embryo is nourished by specialized tissue called endosperm) but follows the same principle (compare Figures 29.8 and 29.16). (A) The haploid megaspore is nourished by tissues of the parental sporophyte (the diploid megasporangium). (B) The mature megaspore is fertilized by a pollen grain that penetrates the integument, germinates (grows a pollen tube; see Figure 29.4A), and releases a sperm nucleus. (C) Fertilization initiates production of a seed. A mature seed contains three generations: a diploid embryo (the new sporophyte), which is surrounded by haploid female gametophyte tissue that supplies nutrition, which is in turn surrounded by the seed coat (diploid parental sporophyte tissue).

Not all seed plants are woody. In the course of seed plant evolution, many groups lost the woody growth habit; however, other advantageous attributes helped them become established in an astonishing variety of places.

RECAP 29.1

Pollen grains, seeds, and wood are major evolutionary innovations of the seed plants. Protection of gametes and embryos is a hallmark of seed plants.

- Distinguish between the roles of the megagametophyte and the pollen grain. See p. 590 and Figure 29.4
- Explain the importance of pollen in freeing seed plants from dependence on liquid water. See p. 590 and Figure 29.4
- What are some of the advantages afforded by seeds? By wood? See pp. 591–592 and Figure 29.5

The seed ferns have long been extinct, but the surviving seed plants have been remarkable successes. After the seed ferns, the gymnosperms were the next group of plants to dominate terrestrial environments.

29.2 What Are the Major Groups of Gymnosperms?

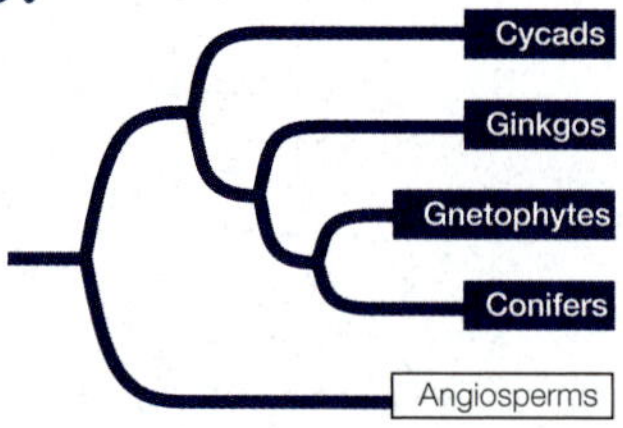
Cycads
Ginkgos
Gnetophytes
Conifers
Angiosperms

The gymnosperms are seed plants that do not form flowers or fruits. Gymnosperms (which means "naked-seeded") are so named because their ovules and seeds, unlike those of angiosperms, are not protected by ovary or fruit tissue. Although there are probably fewer than 1,200 living species of gymnosperms, these plants are second only to the angiosperms in their dominance of the terrestrial environment.

There are four major groups of living gymnosperms

The living gymnosperms can be divided into four major groups (see Figure 28.1B):

- **Cycads** are palmlike plants of the tropics and subtropics (**Figure 29.6A**). Of the present-day gymnosperms, the cycads are probably the earliest-diverging clade. There are about 300 species, some of which grow as tall as 20 meters. The tissues of many species are highly toxic to humans if ingested.
- **Ginkgos**, common during the Mesozoic era, are represented today by a single species, *Ginkgo biloba*, the maidenhair tree (**Figure 29.6B**). There are both male (microsporangiate) and female (megasporangiate) maidenhair trees. The difference is determined by X and Y sex chromosomes, as in humans; few other plants have distinct sex chromosomes.

(A) *Encephalartos sp.*

(B) *Ginkgo biloba*

(C) *Welwitschia mirabilis*

(D) *Pinus longaeva*

29.6 Diversity among the Gymnosperms
(A) Many cycads have growth forms that resemble both ferns and palms, although cycads are not closely related to either group. (B) The characteristic broad leaves of the maidenhair tree. (C) The straplike leaves of *Welwitschia*, a gnetophyte, grow throughout the life of the plant, breaking and splitting as they grow. (D) Conifers dominate many types of landscapes in the Northern Hemisphere. Bristlecone pines such as these are the longest-lived individual trees known.

- **Gnetophytes** number about 90 species in three very different genera, which share certain characteristics analogous to ones found in the angiosperms. One of the gnetophytes is *Welwitschia* (**Figure 29.6C**), a long-lived desert plant with straplike leaves that sprawl on the sand and can grow as long as 3 meters.
- **Conifers** are by far the most abundant of the gymnosperms. There are about 700 species of these cone-bearing plants, including the pines and redwoods (**Figure 29.6D**).

With the exception of the gnetophytes, the living gymnosperm groups have only tracheids as water-conducting and support cells within the xylem; they lack the vessel elements and fibers (cells specialized for water conduction and support, respectively) that are found in angiosperms. While the gymnosperm water-transport and support system may seem somewhat less efficient than that of the angiosperms, it serves some of the largest trees known. The coastal redwoods of California are the tallest gymnosperms; the largest are well over 100 meters tall.

During the Permian, as environments became warmer and dryer, the conifers and cycads flourished. Gymnosperm forests changed over time as the gymnosperm groups evolved. Gymnosperms dominated most of the Mesozoic era, during which the continents drifted apart and large dinosaurs lived. Gymnosperms were the principal trees in all forests until about 90 million years ago, and even today conifers are the dominant trees in many forests. The oldest living single organism on Earth today is a gymnosperm in California—a bristlecone pine that germinated some 4,800 years ago, at about the time the ancient Egyptians were starting to develop writing.

Conifers have cones and no swimming sperm

The great Douglas fir and cedar forests found in northwestern North America, the massive boreal forests of pine, fir, and spruce of the northern regions of the Northern Hemisphere, and the alpine forests on the upper slopes of mountain ranges everywhere rank among the great forests of the world. All these forests are dominated by trees belong to one group of gymnosperms: the conifers, or cone-bearers.

Male and female **cones** contain the reproductive structures of conifers. The female (seed-bearing) cone is known as a

(A) Pinus contorta
Woody scale
Seed
Central axis
Cross section of a megastrobilus
Female cones, or megastrobili
(B) Pinus contorta
Herbaceous scale
Pollen-containing microspores
Central axis
Cross section of a microstrobilus
Male cones, or microstrobili

29.7 Female and Male Cones (A) The woody scales of female cones (megastrobili) are modified branches. (B) The herbaceous scales of male cones (microstrobili) are modified leaves.

megastrobilus (plural *megastrobili*); an example is the familiar woody pine cone. The seeds in a megastrobilus are protected by a tight cluster of woody scales, which are modifications of branches extending from a central axis (**Figure 29.7A**). The typically much smaller male (pollen-bearing) cone is known as a **microstrobilus**. The microstrobilus is typically herbaceous rather than woody, as its scales are composed of modified leaves, beneath which are the pollen-bearing microsporangia (**Figure 29.7B**).

The life cycle of a pine illustrates reproduction in gymnosperms (**Figure 29.8**). The production of male gametophytes in the form of pollen grains frees the plant completely from dependence on liquid water for fertilization. Wind assists conifer pollen grains in their travel from the microstrobilus to near the female gametophyte inside a megastrobilus. A pollen tube provides the sperm with the means for the last stage of travel by growing through maternal sporophytic tissue. When the pollen tube reaches the female gametophyte, it releases two sperm, one of which degenerates after the other unites with an egg. Union of sperm and egg results in a zygote; mitotic divisions and further development of the zygote result in an embryo.

The megasporangium, in which the female gametophyte will form, is enclosed in a layer of sporophytic tissue—the integument—that will eventually develop into the seed coat that protects the embryo (see Figure 29.5). The integument, the megasporangium inside it, and the tissue attaching it to the maternal sporophyte constitute the ovule. The pollen grain enters the ovule through a small opening in the integument at the tip of the ovule, the **micropyle**.

Most conifer ovules, which will develop into seeds after fertilization, are borne exposed on the upper surfaces of the scales of the megastrobilus. The only protection of the ovules comes from the scales, which are tightly pressed against one another within the cone. Some pines, such as the lodgepole pine, have such tightly closed cones that only fire suffices to split them open and release the seeds. These species are said to be fire-adapted, and fire is essential to their reproduction. A fire devastated lodgepole pine forests in Yellowstone National Park in 1988, but also released large numbers of seeds from cones. As a result, large numbers of lodgepole pine seedlings are now emerging in the burn area (**Figure 29.9**).

About half of all conifer species have soft, fleshy tissues that envelop their seeds. Some of these are fleshy, fruitlike cones, as in junipers. Others are fruitlike extensions of the seeds, called arils, as in yews. These tissues, although often mistaken for "berries," are not true fruits. As we will see in the next section, true fruits are the plant's ripened ovaries, which are absent in gymnosperms. Nonetheless, the fleshy tissues that surround many conifers serve a similar purpose to that of the fruits of flowering plants, acting as an enticement for seed-dispersing animals. Animals eat these fleshy tissues and disperse the seeds in their feces, often depositing the seeds considerable distances away from the parent plant.

Go to Animated Tutorial 29.1 Life Cycle of a Conifer
Life10e.com/at29.1

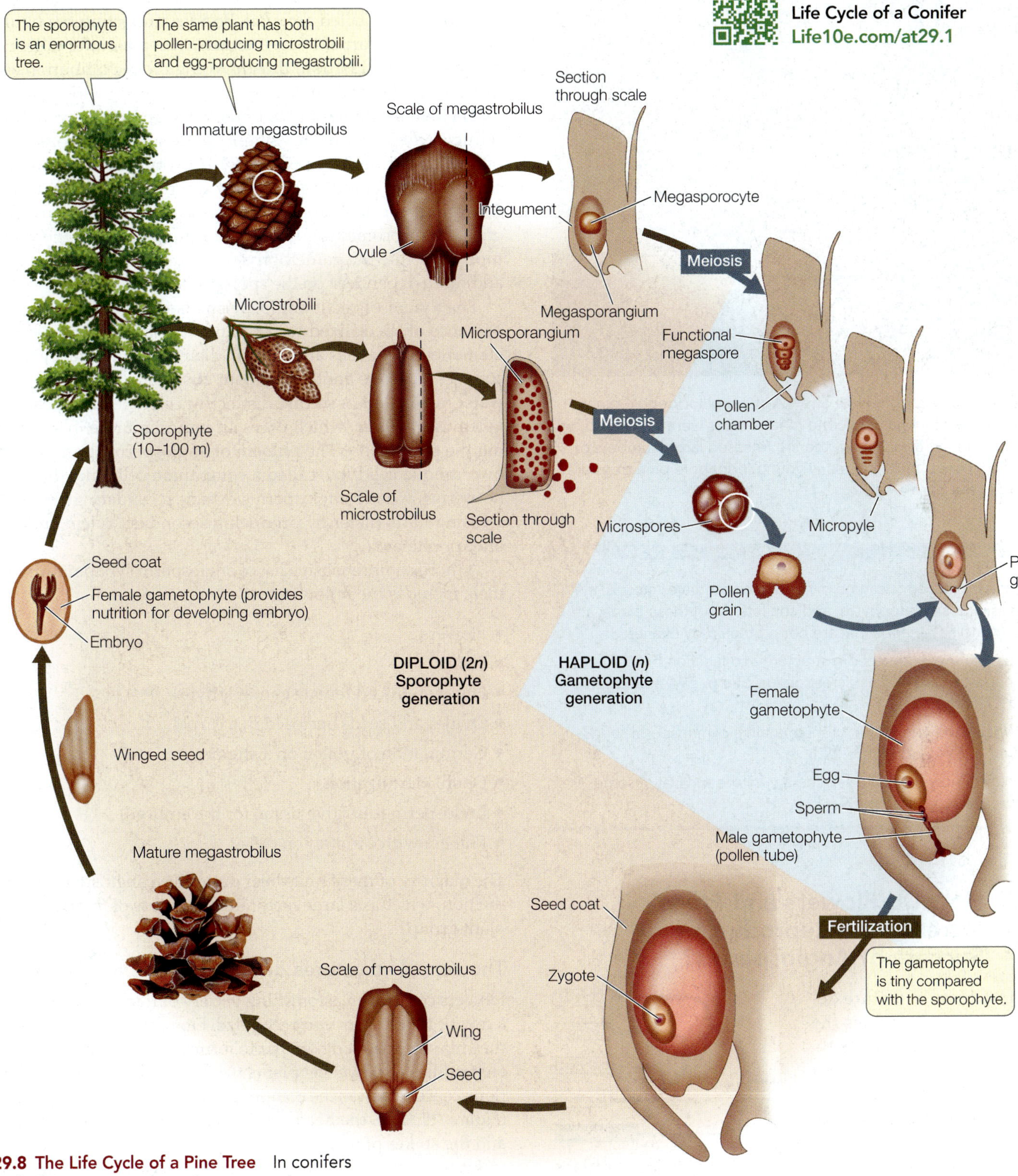
The sporophyte is an enormous tree.
The same plant has both pollen-producing microstrobili and egg-producing megastrobili.
Immature megastrobilus
Scale of megastrobilus
Section through scale
Integument
Megasporocyte
Ovule
Meiosis
Megasporangium
Microstrobili
Microsporangium
Functional megaspore
Pollen chamber
Meiosis
Sporophyte (10–100 m)
Scale of microstrobilus
Section through scale
Microspores
Micropyle
Pollen grain
Pollen grain
Seed coat
Female gametophyte (provides nutrition for developing embryo)
Embryo
DIPLOID (2n) Sporophyte generation
HAPLOID (n) Gametophyte generation
Female gametophyte
Winged seed
Egg
Sperm
Male gametophyte (pollen tube)
Mature megastrobilus
Seed coat
Fertilization
The gametophyte is tiny compared with the sporophyte.
Scale of megastrobilus
Zygote
Wing
Seed

29.8 The Life Cycle of a Pine Tree In conifers and other gymnosperms, the gametophytes are small and nutritionally dependent on the sporophyte generation.

Go to Activity 29.2 Life Cycle of a Conifer
Life10e.com/ac29.2

29.9 From Devastation, New Life A stand of lodgepole pines in Yellowstone National Park. The mature trees were destroyed by a forest fire in 1988. However, the fire released large numbers of seeds from cones, and now many young lodgepole pine trees are growing in the burn area.

RECAP 29.2

Living gymnosperms can be divided into four major groups: cycads, ginkgos, gnetophytes, and conifers. All of these plants are woody and have seeds that are not protected by ovaries.

- Explain the differences in structure and function between a megastrobilus and a microstrobilus. See p. 594 and Figures 29.7 and 29.8
- What is the role of the integument in a gymnosperm seed? See p. 594 and Figure 29.8
- Explain how fire can be necessary for the survival of some plant species. See p. 594

29.3 How Do Flowers and Fruits Increase the Reproductive Success of Angiosperms?

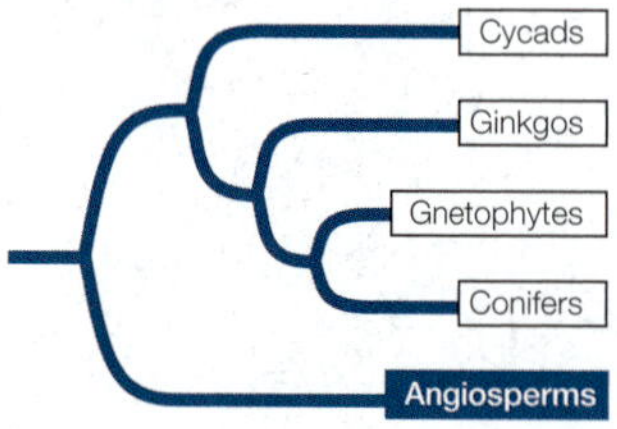
Cycads
Ginkgos
Gnetophytes
Conifers
Angiosperms

The most obvious feature defining the angiosperms is the **flower**, which is their sexual structure. Production of **fruits** is also a synapomorphy (shared derived trait) of angiosperms. After fertilization, the ovary of a flower (together with the seeds it contains) develops into a fruit that protects the seeds and can promote seed dispersal. As we will see, both flowers and fruits give angiosperms major reproductive advantages.

Angiosperms have many shared derived traits

The name *angiosperm* ("enclosed seed") is drawn from another distinctive synapomorphy of these plants that is related to the formation of fruits: the ovules and seeds are enclosed in a modified leaf called a **carpel**. Besides protecting the ovules and seeds, the carpel often interacts with incoming pollen to prevent self-pollination, thus favoring cross-pollination and increasing genetic diversity.

The female gametophyte of the angiosperms is even more reduced than that of the gymnosperms, usually consisting of only seven cells (see Figure 29.16). Thus the angiosperms represent the current extreme of the trend we have traced throughout the evolution of the vascular plants: the sporophyte generation becomes larger and more independent of the gametophyte, while the gametophyte generation becomes smaller and more dependent on the sporophyte.

The xylem of most angiosperms is distinguished by the presence of specialized water-transporting cells called **vessel elements**. These cells are larger in diameter than tracheids and connect with one another without obstruction, allowing easy water movement. A second distinctive cell type in angiosperm xylem is the **fiber**, which plays an important role in supporting the plant body. The phloem of angiosperms possesses its own unique cell type, called a companion cell. Like the gymnosperms, woody angiosperms exhibit secondary growth, increasing in diameter by producing secondary xylem and secondary phloem.

A more comprehensive list of angiosperm synapomorphies, then, includes the following:

- Flowers
- Fruits
- Highly reduced female gametophytes
- Ovules and seeds enclosed in a carpel
- Germination of pollen on a stigma
- Double fertilization
- Endosperm (nutritive tissue for the embryo)
- Phloem with companion cells

The majority of these traits bear directly on angiosperm reproduction, which is a large factor in the success of this dominant plant group.

The sexual structures of angiosperms are flowers

Flowers come in an astonishing variety of forms—just think of a few of the flowers you recognize. Flowers may be single, or they may be grouped together to form an **inflorescence**. Different families of flowering plants have characteristic types of inflorescences, such as the compound umbels of the carrot family (Figure 29.10A), the heads of the aster family (Figure 29.10B), and the spikes of many grasses (Figure 29.10C).

If you examine any familiar flower, you will notice that the outer parts look somewhat like leaves. In fact, all the parts of a flower *are* modified leaves. The diagram in Figure 29.4B represents a generalized flower, for which there is no exact counterpart in nature. The structures bearing microsporangia are called **stamens**. Each stamen is composed of a **filament** bearing an **anther** that contains the pollen-producing microsporangia. The structures bearing megasporangia are called carpels. A structure

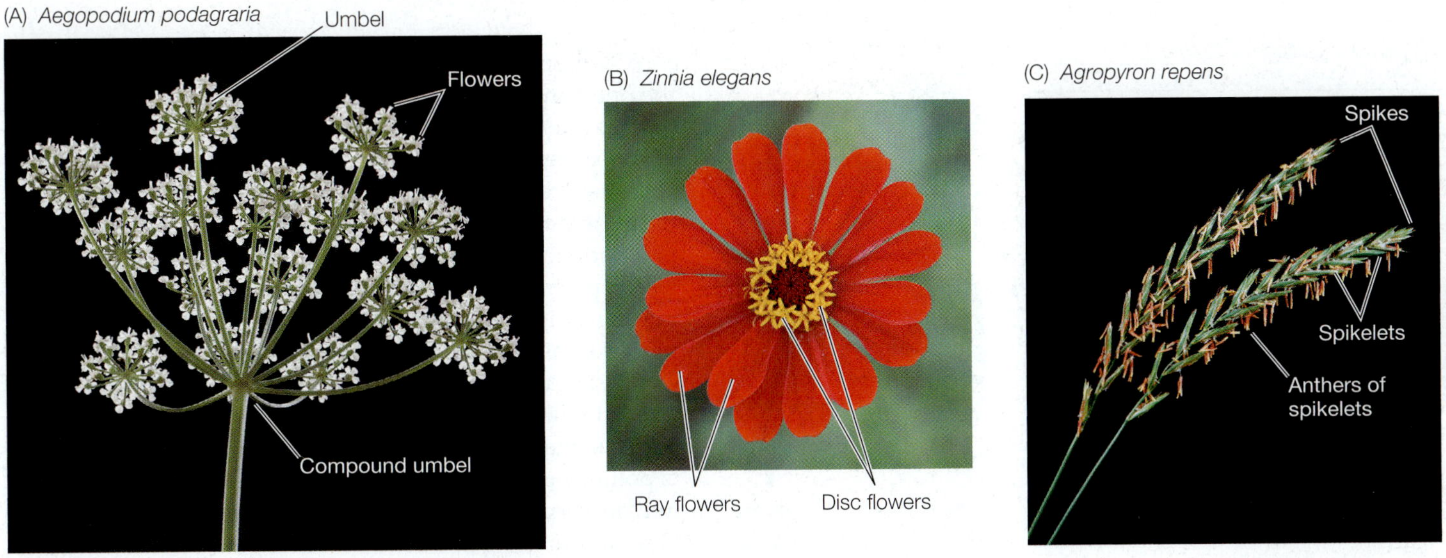
(A) *Aegopodium podagraria*
Umbel
Flowers
Compound umbel
(B) *Zinnia elegans*
Ray flowers
Disc flowers
(C) *Agropyron repens*
Spikes
Spikelets
Anthers of spikelets

29.10 Inflorescences (A) The inflorescence of bishop's goutweed, a member of the carrot family, is a compound umbel. Each umbel bears flowers on stalks that arise from a common center. (B) Zinnias are members of the aster family; their inflorescence is a head. Within the head, each of the long, petal-like structures is a ray flower; the central portion of the head consists of dozens to hundreds of disc flowers. (C) Some grasses, such as quack grass, have inflorescences called spikes, which are composed of many smaller groups of flowers, or spikelets.

composed of one carpel or two or more fused carpels is called a **pistil**. The swollen base of the pistil, containing one or more ovules (each containing a megasporangium surrounded by two protective integuments), is called the **ovary**. The apical stalk of the pistil is the **style**, and the terminal surface that receives pollen grains is the **stigma**.

In addition, many flowers contain specialized sterile (non-spore-bearing) leaves. The inner ones are called **petals** (collectively, the corolla) and the outer ones **sepals** (collectively, the calyx). The corolla and calyx (collectively, the perianth) can be quite showy and often play roles in attracting animal pollinators to the flower. The calyx more commonly protects the immature flower in bud. From base to apex, these floral organs—sepals, petals, stamens, and carpels—are usually positioned in circular arrangements or whorls and attached to a central stalk.

The generalized flower in Figure 29.4B has both functional megasporangia and functional microsporangia; such flowers are referred to as **perfect** (or hermaphroditic). Many angiosperms produce two types of flowers, one with only megasporangia and the other with only microsporangia. Consequently, either the stamens or the carpels are nonfunctional or absent in a given flower, and the flower is referred to as **imperfect** (see Figure 38.1).

Species such as corn or birch, in which both megasporangiate (female) and microsporangiate (male) flowers occur on the same plant, are said to be **monoecious** ("one-housed"—but, it must be added, one house with separate rooms). Complete separation of imperfect flowers occurs in some other angiosperm species, such as willows and date palms; in these species, an individual plant produces either flowers with stamens or flowers with carpels, but never both. Such species are said to be **dioecious** ("two-housed").

Flower structure has evolved over time

The flowers of the earliest-diverging clades of angiosperms have a large and variable number of tepals (undifferentiated sepals and petals), carpels, and stamens (**Figure 29.11A**). Evolutionary change within the angiosperms has included some striking modifications of this early condition: reductions in the number of each type of floral organ to a fixed number, differentiation of petals from sepals, and changes in symmetry from radial (as in a lily or magnolia) to bilateral (as in the orchid shown at the opening of this chapter), often accompanied by an extensive fusion of parts (**Figure 29.11B**).

(A) *Nymphaea* sp.
(B) *Viola tricolor*

29.11 Flower Form and Evolution (A) A water lily shows the major features of early flowers: it is radially symmetrical, and the individual tepals, stamens, and carpels are separate, numerous, and attached at their bases. (B) Violets such as this "Johnny jump-up" have a bilaterally symmetrical structure that evolved much later than radial flower symmetry.

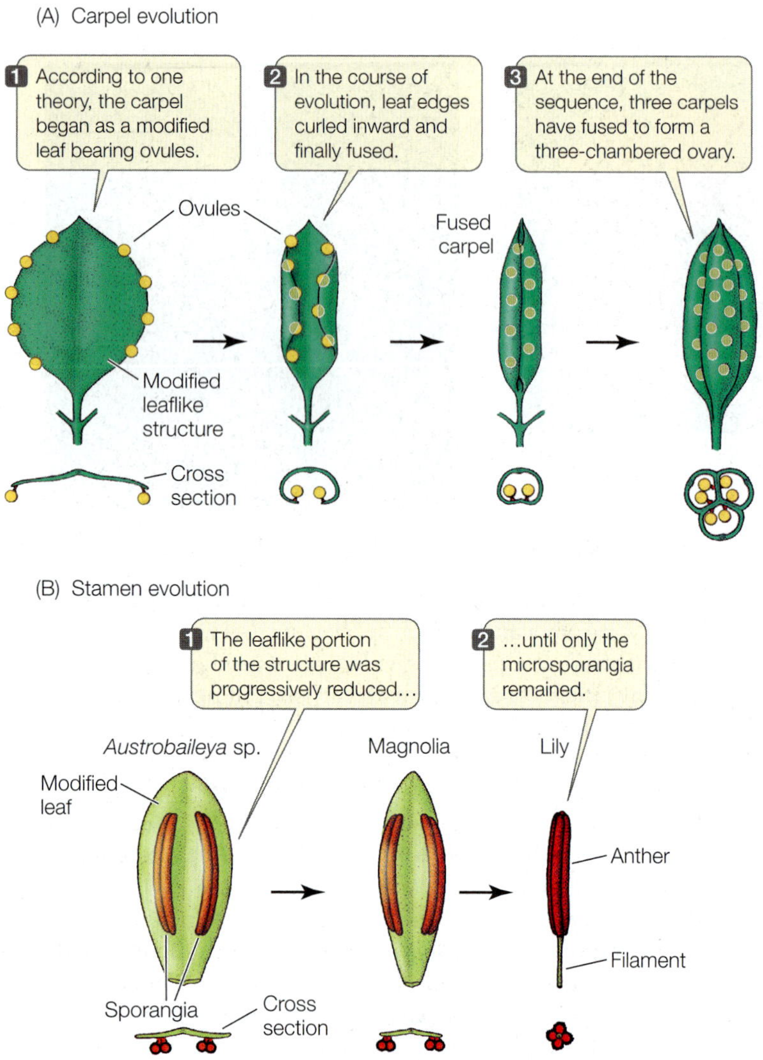

(A) Carpel evolution
1 According to one theory, the carpel began as a modified leaf bearing ovules.
2 In the course of evolution, leaf edges curled inward and finally fused.
3 At the end of the sequence, three carpels have fused to form a three-chambered ovary.
Ovules
Fused carpel
Modified leaflike structure
Cross section
(B) Stamen evolution
1 The leaflike portion of the structure was progressively reduced…
2 …until only the microsporangia remained.
Austrobaileya sp.
Magnolia
Lily
Modified leaf
Anther
Filament
Sporangia
Cross section

29.12 Carpels and Stamens Evolved from Leaflike Structures (A) Possible stages in the evolution of a carpel from a more leaflike structure. (B) The stamens of three modern plants show three possible stages in the evolution of that organ. (It is *not* implied that these species evolved from one another; their structures simply illustrate the possible stages.)

According to one hypothesis, the first carpels to evolve were leaves with marginal ovules, folded but incompletely closed. Early in angiosperm evolution, carpels fused with one another, forming a single, multichambered ovary (**Figure 29.12A**). In some flowers, the other floral organs are attached at the top of the ovary rather than at the bottom as in Figure 29.4B. The stamens of the most ancient flowers may have been leaflike (**Figure 29.12B**), with little resemblance to the stamens of the generalized flower seen in Figure 29.4B.

Why do so many flowers have pistils with long styles and anthers with long filaments? Natural selection has favored length in both of these floral organs, probably because length increases the likelihood of successful pollination. Long filaments may bring the anthers into contact with insect bodies, or they may place the anthers in a better position to catch the wind. Similar arguments apply to long styles.

A perfect flower represents a compromise of sorts. On the one hand, by attracting a pollinating bird or insect, the plant is attending to both its female and male functions with a single flower type, whereas plants with imperfect flowers must create that attraction twice—once for each type of flower. On the other hand, the perfect flower can favor self-pollination, which is usually disadvantageous. Another potential problem is that the female and male functions might interfere with each other—for example, the stigma might be placed so as to make it difficult for pollinators to reach the anthers, thus reducing the export of pollen to other flowers.

Might there be a way around these problems? One solution is seen in the bush monkeyflower (*Mimulus aurantiacus*), which is pollinated by hummingbirds. Its flower has a stigma that initially serves as a screen, hiding the anthers (**Figure 29.13**). Once a hummingbird touches the stigma, however, one of the stigma's two lobes is retracted, so that subsequent hummingbird visitors pick up pollen from the previously screened anthers. Thus the first bird to visit the flower transfers pollen from another plant to the stigma. Later visitors pick up pollen from the now-accessible anthers, fulfilling the flower's male function. **Figure 29.14** describes the experiment that revealed the function of this mechanism.

Angiosperms have coevolved with animals

Whereas most gymnosperms are pollinated by wind, most angiosperms are pollinated by animals. The many different

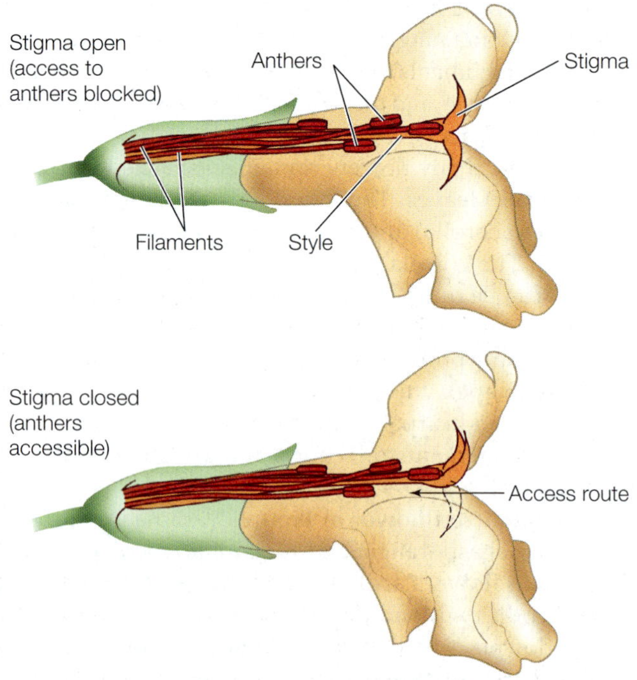

Stigma open (access to anthers blocked)
Anthers
Stigma
Filaments
Style
Stigma closed (anthers accessible)
Access route

29.13 An Unusual Way to Prevent Selfing Both long stamens and long styles facilitate cross-pollination, but if these male and female structures are too close to each other, the likelihood of (disadvantageous) self-pollination increases. In *Mimulus aurantiacus*, the stigma is initially open, blocking access to the anthers. A hummingbird's touch as it deposits pollen on the stigma causes one lobe of the stigma to retract, creating a path to the anthers and allowing pollen dispersal by subsequent hummingbird visitors.

INVESTIGATINGLIFE

29.14 The Effect of Stigma Retraction in Monkeyflowers

Elizabeth Fetscher's experiments showed that the unusual stigma retraction response to pollination in monkeyflowers (illustrated in Figure 29.13) enhances the dispersal of pollen to other flowers.[a]

HYPOTHESIS The stigma-retraction response in *M. aurantiacus* increases the likelihood that an individual flower's pollen will be exported to another flower once pollen from another flower has been deposited on its stigma.

Method

1. Set up three groups of monkeyflower arrays. Each array consists of one pollen-donor flower and multiple pollen-recipient flowers (with the anthers removed to prevent pollen donation).
2. In control arrays, the stigma of the pollen donor is allowed to function normally.
3. In one group of experimental arrays, the stigma of the pollen donor is permanently propped open (blocking access to the anthers).
4. In a second group of experimental arrays, the stigma of the pollen donor is artificially sealed closed (allowing access to anthers).
5. Allow hummingbirds to visit the arrays, then count the pollen grains transferred from each donor flower to the recipient flowers in the same array.

Results

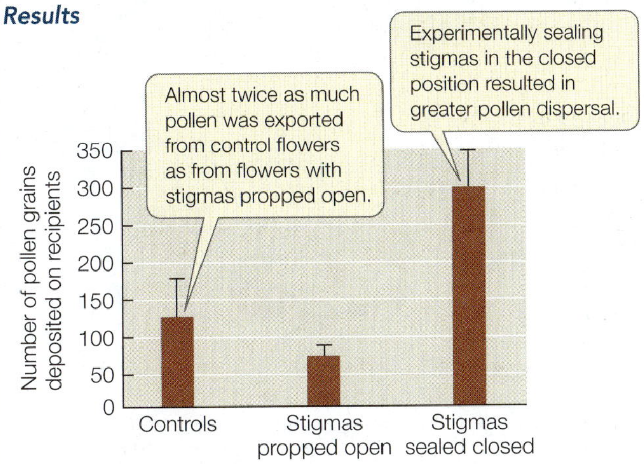
Almost twice as much pollen was exported from control flowers as from flowers with stigmas propped open.
Experimentally sealing stigmas in the closed position resulted in greater pollen dispersal.
Number of pollen grains deposited on recipients
350
300
250
200
150
100
50
0
Controls
Stigmas propped open
Stigmas sealed closed

CONCLUSION The stigma-retraction response enhances the male function of the flower (dispersal of pollen) once the female function (receipt of pollen) has been performed.

Go to **BioPortal** for discussion and relevant links for all INVESTIGATING**LIFE** figures.

[a]Fetscher, A. E. 2001. *Proceedings of the Royal Society B* 268: 525–529.

Rubeckia fulgida

29.15 See Like a Bee To normal human vision (above), the petals of a black-eyed Susan appear solid yellow. Ultraviolet photography reveals patterns that attract bees to the central region, where pollen and nectar are located.

mutualistic pollination relationships between plants and animals are vital to both parties. We mentioned coevolution between insects and orchids at the opening of this chapter, and we'll revisit plant–pollinator coevolution in more detail when we discuss mutualisms in Chapter 56, but we'll consider a few important aspects of the plant–pollinator relationship here.

Many flowers entice animals to visit them by providing food rewards. Pollen grains themselves sometimes serve as food for animals. In addition, some flowers produce a sugary fluid called nectar, and some of these flowers have specialized structures to store and distribute it, as we saw at the opening of this chapter. In the process of visiting flowers to obtain nectar or pollen, animals often carry pollen from one flower to another or from one plant to another. Thus, in their quest for food, the animals contribute to the genetic diversity of the plant population. Insects, especially bees, are among the most important pollinators; birds and some species of bats are also major pollinators.

Go to Media Clip 29.2
Pollen Transfer by a Bat
Life10e.com/mc29.2

For more than 150 million years, angiosperms and their animal pollinators have coevolved in the terrestrial environment. Animals have affected the evolution of the plants, and plants have affected the evolution of the animals. Flower structure has become incredibly diverse under these selection pressures. Some of the products of coevolution are highly specific; for example, the flowers of some yucca species are pollinated by only one species of yucca moth, and that moth may exclusively pollinate just one species of yucca (see Figure 56.12). Such specific relationships provide plants with a reliable mechanism for transferring pollen only to members of their own species.

Most plant–pollinator interactions are much less specific; that is, many different animal species pollinate the same plant species, and the same animal species pollinates many different plant species. However, even these less specific interactions have developed some specialization (see Table 56.1). For example, many bird-pollinated flowers are red and odorless, whereas many insect-pollinated flowers have characteristic odors. Some bee-pollinated flowers have markings, called nectar guides, that are conspicuous only to animals, such as bees, that can see colors in the ultraviolet region of the spectrum (**Figure 29.15**).

29.16 The Life Cycle of an Angiosperm Double fertilization results in triploid endosperm in most species of angiosperms. One sperm nucleus fertilizes the egg to form the zygote, while the other combines with the two polar nuclei to form the endosperm.

Go to Animated Tutorial 29.2
Life Cycle of an Angiosperm
Life10e.com/at29.2

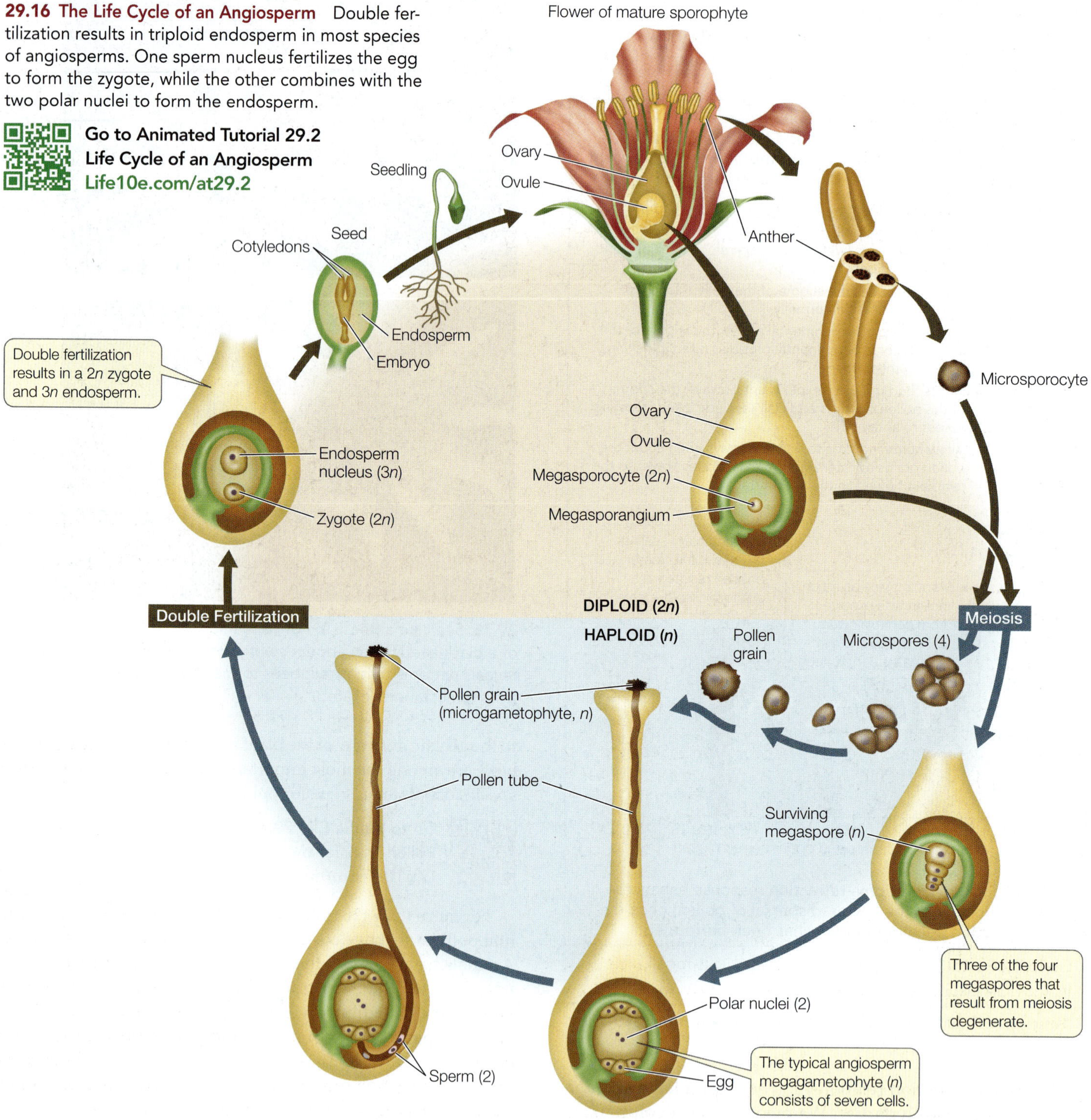

Flower of mature sporophyte
Ovary
Ovule
Anther
Seedling
Seed
Cotyledons
Endosperm
Embryo
Microsporocyte
Double fertilization results in a 2n zygote and 3n endosperm.
Endosperm nucleus (3n)
Zygote (2n)
Ovary
Ovule
Megasporocyte (2n)
Megasporangium
DIPLOID (2n)
HAPLOID (n)
Double Fertilization
Meiosis
Pollen grain
Microspores (4)
Pollen grain (microgametophyte, n)
Pollen tube
Surviving megaspore (n)
Three of the four megaspores that result from meiosis degenerate.
Polar nuclei (2)
Egg
The typical angiosperm megagametophyte (n) consists of seven cells.
Sperm (2)

The angiosperm life cycle produces diploid zygotes nourished by triploid endosperms

Like all seed plants, angiosperms are heterosporous. As we have seen, their ovules are contained within carpels rather than being exposed on the surfaces of scales, as in most gymnosperms. The male gametophytes, as in the gymnosperms, are pollen grains.

Pollination in the angiosperms consists of the arrival of a microgametophyte—a pollen grain—on the receptive surface of a flower (the stigma). As in the gymnosperms, pollination is the first in a series of events that results in the formation of a seed. The next event is the growth of a pollen tube extending to the megagametophyte. The third event is a fertilization process that, in detail, is unique to the angiosperms (**Figure 29.16**).

In angiosperms, *two* male gametes, contained in a single microgametophyte, participate in fertilization. The nucleus of one sperm combines with that of the egg to produce a diploid zygote—the first cell of the sporophyte generation. In

(A)
(B)
(C)
(D)
(E)
(F)

29.17 Fruits Come in Many Forms (A) The single seeds inside the simple fruits of peaches are dispersed by animals. (B) Each seed of the horse chestnut is covered by a hard, woody fruit that allows it to survive drought. Although such fruits are commonly called "nuts," this is a culinary rather than a biological term. (C) The highly reduced simple fruits of dandelions are dispersed by wind. (D) A multiple fruit, the jackfruit (*Artocarpus heterophyllus*) of tropical Asia, is the largest tree-borne fruit in the world. (E) An aggregate fruit (blackberry). (F) An accessory fruit (pear).

most angiosperms, the other sperm nucleus combines with two other haploid nuclei of the female gametophyte to form a cell with a *triploid* ($3n$) nucleus. That cell in turn gives rise to a triploid tissue called the **endosperm**, which nourishes the embryonic sporophyte during its early development. This process, in which two fertilization events take place, is known as **double fertilization**. In some angiosperms, additional haploid nuclei are incorporated to form even higher ploidy levels in the endosperm, or the second sperm fuses with only one haploid nucleus, resulting in diploid endosperm.

The angiosperm zygote develops into an embryo, which consists of an embryonic axis that will become a stem and a root and one or two **cotyledons**, or "seed leaves." The cotyledons have different fates in different plants. In many, they serve as absorptive organs that take up and digest the endosperm. In others, they enlarge and become photosynthetic when the seed germinates. Often they play both roles.

The ovule develops into a seed containing the products of the double fertilization that characterizes angiosperms: the diploid zygote and a triploid endosperm (see Figure 29.16).

Fruits aid angiosperm seed dispersal

As mentioned at the start of this section, the production of seed-protecting fruits is an important synapomorphy. Fruits may attach to or be eaten by an animal. Fruits are not necessarily fleshy; they can be hard and woody, or small with modified structures that allow them to be dispersed by wind or water (**Figure 29.17**).

A fruit may consist of only the mature ovary and its seeds, or it may include other parts of the flower or structures associated with it. A **simple fruit** is one that develops from a single carpel or several fused carpels, such as a plum or peach. A raspberry is an example of an **aggregate fruit**—one that develops from several separate carpels of a single flower. Pineapples and figs are examples of **multiple fruits**, formed from a cluster of flowers (an inflorescence). Fruits derived from parts in addition to the carpel and seeds are called **accessory fruits**; examples are apples, pears, and strawberries.

Go to Media Clip 29.3
Flower and Fruit Formation
Life10e.com/mc29.3

Recent analyses have revealed the phylogenetic relationships of angiosperms

Figure 29.18 shows the relationships among the major angiosperm clades. The two largest clades—the **monocots** and the **eudicots**—include the great majority of angiosperm species. The monocots are so called because they have a single

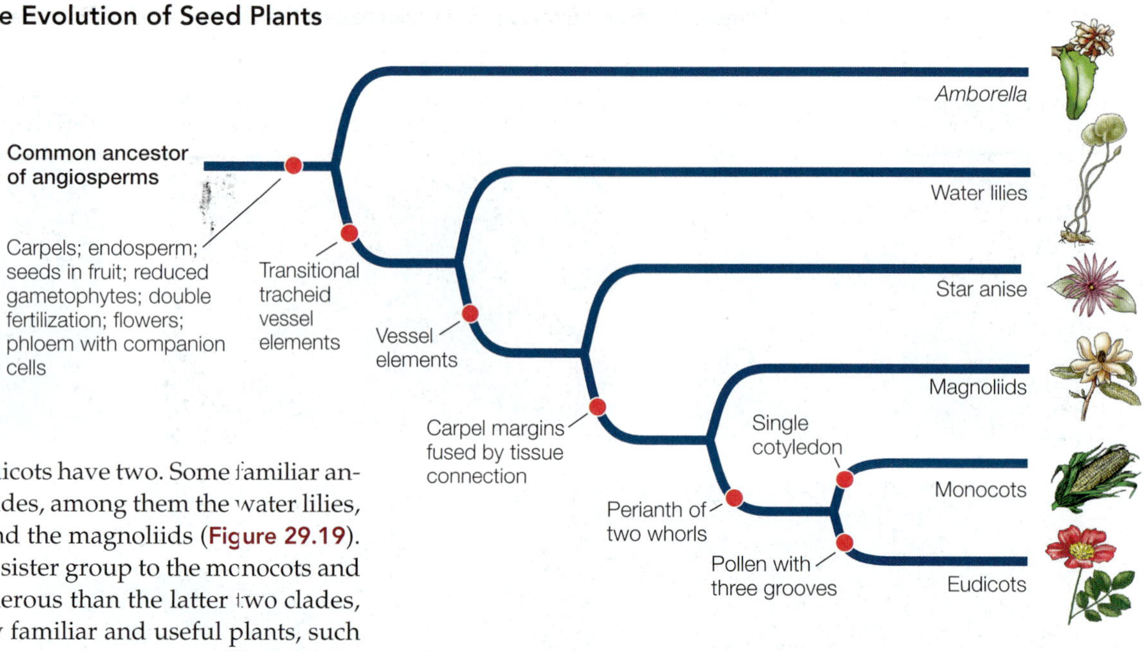
Common ancestor of angiosperms
Carpels; endosperm; seeds in fruit; reduced gametophytes; double fertilization; flowers; phloem with companion cells
Transitional tracheid vessel elements
Vessel elements
Carpel margins fused by tissue connection
Perianth of two whorls
Single cotyledon
Pollen with three grooves
Amborella
Water lilies
Star anise
Magnoliids
Monocots
Eudicots

29.18 Evolutionary Relationships among the Angiosperms Recent analyses of many angiosperm genes have clarified the relationships among the major groups.

embryonic cotyledon; the eudicots have two. Some familiar angiosperms belong to other clades, among them the water lilies, star anise and its relatives, and the magnoliids (Figure 29.19). The **magnoliids** are the likely sister group to the monocots and eudicots. Although less numerous than the latter two clades, the magnoliids include many familiar and useful plants, such as avocados, cinnamon, black pepper, and magnolias.

The root of the evolutionary tree of flowering plants was once a matter of great controversy. A fundamental challenge was identifying the group that is sister to the remaining angiosperms, and members of the magnoliid clade were leading candidates for the position. At the close of the twentieth century, however, an impressive convergence of molecular and morphological evidence led to the conclusion that the sister group to the remaining flowering plants is the single species of the genus *Amborella* (see Figure 29.19A). This woody shrub, with cream-colored flowers, lives only on New Caledonia, an island in the South Pacific. Its 5 to 8 carpels have a spiral arrangement,

(A) *Amborella trichopoda*
Sterile stamens
(B) *Victoria amazonica*
(C) *Illicium floridanum*
(D) *Magnolia* sp.
(E) *Aristolochia littoralis*

29.19 Monocots and Eudicots Are Not the Only Surviving Angiosperms (A) *Amborella*, a shrub, is sister to the remaining extant angiosperms. Notice the sterile stamens on this female flower, which may serve to lure insects that are searching for pollen. (B) The water lily clade was the next to diverge after *Amborella*. (C) Star anise and its relatives belong to another early-diverging angiosperm clade. (D, E) The largest clade other than the monocots and eudicots is the magnoliid complex, which includes magnolias and the group known as "Dutchman's pipe."

(A)

(B) *Saccharum* sp.

(C) *Posidonia oceanica*

(D) *Phoenix dactylifera*

29.20 Monocots (A) Monocots include many popular garden flowers such as these hyacinths (*Muscari armeniacum*; blue), tulips (*Tulipa* sp.; red) and daffodils (*Narcissus* sp.; yellow). (B) Monocot grasses such as this sugarcane feed the world; wheat, rice, and maize (corn) are also grasses. (C) Seagrasses such as this Neptune's grass form "meadows" in the shallow, sunlit waters of the world's oceans. (D) Palms are among the few monocot trees. Date palms like these are a major food source in some areas of the world.

and it has 30 to 100 stamens. The xylem of *Amborella* lacks vessel elements, which evolved after this deepest split in the angiosperm evolutionary tree.

Representatives of the two largest angiosperm clades are everywhere. The monocots (**Figure 29.20**) include grasses, cattails, lilies, orchids, and palms. The eudicots (**Figure 29.21**) include the vast majority of familiar seed plants, including most herbs (i.e., nonwoody plants), vines, trees, and shrubs. Among the eudicots are such diverse plants as oaks, willows, beans, snapdragons, roses, and sunflowers.

(A) *Delonix regia*

(B) *Passiflora caerulea*

Filaments of corona

(C) *Echinocereus reichenbachii*

(D) *Rafflesia arnoldii*

29.21 Eudicots (A) The royal poinciana, or "flame tree," is native to Madagascar. Humans have introduced this ornamental flowering tree throughout the world's tropical regions. (B) Passionflower vines are found throughout the tropics and subtropics. Their flowers are distinguished by elaborate coronas (the purple filaments on this flower). (C) Cacti comprise a large group of eudicots, with about 1,500 species in the Americas. Many, such as this black lace cactus, bear large flowers for a brief period of each year. (D) *Rafflesia arnoldii*, found in the rainforests of Indonesia, bears the largest flower in the world. The flower lives as a parasite on tropical vines and has lost its leaf, stem, and even root structures. It smells like decaying meat, which attracts its fly pollinators.

RECAP 29.3

The synapomorphies of angiosperms include flowers, fruit, carpels, double fertilization, and endosperm. Most angiosperms also possess distinctive cells in the xylem and phloem. The largest angiosperm clades are the monocots and the eudicots.

- Explain the difference between pollination and fertilization. **See pp. 598–601**
- What are the respective roles of the two sperm in double fertilization in angiosperms? **See pp. 600–601 and Figure 29.16**
- What are the different functions of flowers, fruits, and seeds?

The remarkable diversity of the seed plants has been shaped by both biotic and abiotic components of the environments to which they have adapted. In turn, land plants—and seed plants in particular—shape their environments.

TABLE 29.1 Some Medicinal Plants and Their Products

Product	Plant Source	Medical Application
Atropine	Belladonna	Dilate pupils for eye examination
Bromelain	Pineapple stem	Control tissue inflammation
Digitalin	Foxglove	Strengthen heart muscle contraction
Ephedrine	*Ephedra*	Ease nasal congestion
Menthol	Japanese mint	Relieve coughing
Morphine	Opium poppy	Relieve pain
Quinine	*Cinchona* bark	Treat malaria
Taxol	Pacific yew	Treat ovarian and breast cancers
Tubocurarine	Curare plant	Muscle relaxant (used in surgery)
Vincristine	Periwinkle	Treat leukemia and lymphoma

29.4 How Do Plants Benefit Human Society?

Plants make profound contributions to ecosystem services—processes by which the environment maintains resources that benefit human society. Plants produce oxygen and remove carbon dioxide from the atmosphere, and they play important roles in forming soils and renewing soil fertility. Plant roots help hold soil in place, providing a defense against erosion by wind and water (**Figure 29.22**). They also moderate local climate in various ways, such as by increasing humidity, providing shade, and blocking wind.

Seed plants have been sources of medicine since ancient times

One of the oldest human professions is that of medicine man (shaman) or "wise woman"—a person who cures others with medicines, most of which are derived from plants. It is said that in 2700 BCE, a legendary Chinese emperor understood the use of some 365 medicinal plants. Although we also use medicines derived from fungi, lichens, and actinobacteria, seed plants are the source of many of our medications; a few examples are shown in **Table 29.1**. Even in synthetic pharmaceuticals, the chemical structures of the active ingredients are often based on the biochemistry of substances isolated from plants.

How are plant-based medicines discovered? These days, many are found by systematic testing of tremendous numbers of plants from all over the world, a process that began in the 1960s. One example of a medicine discovered in this way is taxol, an important anticancer drug. Among the myriad plant samples that had been tested by 1962, extracts of the bark of the Pacific yew (*Taxus brevifolia*) showed anti-tumor activity in tests against rodent tumors. The active ingredient, taxol, was isolated in 1971 and tested against human cancers in 1977. After another 16 years, the U.S. Food and Drug Administration approved it for human use, and taxol is now widely used in treating breast and ovarian cancers as well as several other types of cancers.

Widespread screening of plant samples was eventually deemphasized in favor of a purely chemical approach. Using automation and miniaturization, pharmaceutical laboratories generate vast numbers of compounds, which are then screened just as plant materials had been screened in the search for taxol. Today,

29.22 Plants Prevent Erosion When forest vegetation was cleared on these hillsides in Malaysia, landslides and extensive soil erosion quickly followed. Adjacent forested areas did not have landslides.

however, the screening of plant materials is getting renewed interest. Both screening approaches are based on trial and error.

The other leading source of plant-based medicines is work by ethnobotanists, who study how people use and view plants in their local environments. This work proceeds all over the globe today. An older example of this approach is the discovery of quinine as a treatment for malaria. In the sixteenth century, Spanish priests in Peru became aware that the native population used the bark of local *Cinchona* trees to treat fevers. The priests successfully used the bark to treat malaria. Word of the medicine spread to Europe, where it is said to have been in use in Rome by 1631. The active ingredient of *Cinchona* bark—quinine—was identified in 1820, and quinine remained the standard malarial remedy well into the twentieth century.

29.23 Rice Feeds Much of the World's Human Population
These rice fields, or "paddies," are on the island of Luzon in the Philippines. Rice has been cultivated in this manner for thousands of years.

Seed plants are our primary food source

Plants are primary producers; that is, they trap energy and carbon by means of photosynthesis, making those resources available not only for their own needs, but also for the herbivores and omnivores that consume them, for the carnivores and omnivores that eat the herbivores, and for the prokaryotes and fungi that complete food webs. The earliest steps in human civilization involved cultivating angiosperms to provide a reliable food supply.

Today, twelve species of seed plants stand between the human race and starvation: rice, coconut, wheat, corn (maize), potato, sweet potato, cassava (also called tapioca or manioc), sugarcane, sugar beet, soybean, common bean (*Phaseolus vulgaris*), and banana. Hundreds of other seed plants are cultivated for food, but none rank with these twelve in importance. Indeed, more than half of the world's human population derives the bulk of its food energy from the seeds of a single plant, *Oryza sativa*, better known as rice. Rice is particularly important in eastern Asia, where it has been cultivated for more than 8,000 years (**Figure 29.23**). People also use rice straw in many ways, such as thatching for roofs, food and bedding for livestock, and clothing. Even rice hulls have many uses, ranging from fertilizer to fuel.

What was Darwin's explanation for the three distinct flowers growing on a single orchid plant?

ANSWER

After obtaining specimens of the plant in question and dissecting the flowers, Darwin was able to demonstrate that the orchid was a single species (*Catasetum macrocarpum*) that bore three distinct types of flowers: megasporangiate (female), microsporangiate (male), and perfect (hermaphroditic). The three types of flowers were remarkable in their morphological differences, which were great enough to have misled botanists into describing the different flower types as species in different genera. Most individual plants were either male (specimens identified as *Catasetum*) or female (specimens identified as *Monachanthus*), but some individuals that bore predominately male or female flowers also produced perfect flowers (specimens identified as *Myanthus*).

The case of *C. macrocarpum* demonstrates that some plants blur the lines between the categories we have discussed in this chapter. Most flowers on a *C. macrocarpum* individual are either male or female, but some plants can bear some perfect flowers as well.

Why do the male and female flowers of *C. macrocarpum* look so different? Part of the explanation is their different roles in pollination. Recall Darwin's observation of a *Catasetum* flower shooting a packet of pollen at an insect that landed on its flower. The pollinia (pollen packets) and associated structures in male flowers of *Catasetum* are coiled like springs and are released suddenly when disturbed by an insect. This release forcefully propels the pollinia precisely into position on the back of the insect. The insect pollinator of *C. macrocarpum* is a specific bee species, the males of which are attracted to the odor of the flowers. The flowers produce no nectar reward, but the male bee does gather the chemical that produces the scent. The bee then moves on to another flower. When the bee visits a female flower on a different *C. macrocarpum* individual (again attracted by the same scent), no such "loaded spring" awaits. Instead, the morphology of the female flower enhances the removal of the pollinia from the insect's body. In this way, floral morphology makes cross-fertilization more likely and reduces the chances of self-pollination.

Orchids were important in forming Darwin's ideas about the mechanisms of evolution, for they showed that even aspects of the coloration and form of flowers evolved in response to natural selection. This conclusion ran counter to the thinking of the day. For example, Thomas Huxley, one of Darwin's earliest and strongest supporters, doubted that the beauty of color in plants and animals could be explained on the basis of their importance to function. Darwin showed that the beauty of flowers is indeed connected to their reproductive success and is a key element in explaining the great diversity of plants.

CHAPTERSUMMARY 29

29.1 How Did Seed Plants Become Today's Dominant Vegetation?

- Fossils of woody **seed ferns** are the earliest evidence of seed plants. The surviving groups of seed plants are the **gymnosperms** and **angiosperms**. **Review Figure 29.1**
- All seed plants are heterosporous, and their gametophytes are much smaller than (and dependent on) their sporophytes. **Review Figure 29.2**
- Seed plants do not require liquid water for fertilization. **Pollen grains**, the microgametophytes of seed plants, are carried to a megagametophyte by wind or by animals.
- An **ovule** consists of the seed plant megagametophyte and the **integument** of sporophytic tissue that protects it.
- Following **pollination**, a **pollen tube** emerges from the pollen grain, elongates, and usually delivers gametes to the megagametophyte. **Review Figure 29.4, ACTIVITY 29.1**
- The ovule develops into a **seed** that contains an embryo (the new sporophyte generation). Seeds are well protected and are often capable of long periods of dormancy, germinating only when conditions are favorable. **Review Figure 29.5**

29.2 What Are the Major Groups of Gymnosperms?

- The gymnosperms produce ovules and seeds that are not protected by ovary or fruit tissues.
- The major gymnosperm groups are the **cycads**, **ginkgos**, **gnetophytes**, and **conifers**. **Review Figure 29.6**
- The megaspores of conifers are produced in woody **cones** called **megastrobili**; the microspores are produced in herbaceous cones called **microstrobili**. **Review Figures 29.7 and 29.8, ACTIVITY 29.2, ANIMATED TUTORIAL 29.1**

29.3 How Do Flowers and Fruits Increase the Reproductive Success of Angiosperms?

- **Flowers** and **fruits** are unique to the angiosperms, distinguishing them from the gymnosperms.
- The xylem of most angiosperms is more complex than that of the gymnosperms. It contains two specialized cell types: **vessel elements**, which function in water transport, and **fibers**, which play an important role in structural support.
- The ovules and seeds of angiosperms are enclosed in and protected by **carpels**.
- The floral organs, from the base to the apex of the flower, are the **sepals**, **petals**, **stamens**, and **pistil**. Stamens bear microsporangia in **anthers**. The pistil (consisting of one or more carpels) includes an **ovary** containing ovules. The **stigma** is the receptive surface of the pistil.
- A flower with both megasporangia and microsporangia is referred to as **perfect**; a flower with only one or the other is **imperfect**.
- A **monoecious** species has megasporangiate and microsporangiate flowers on the same plant. A **dioecious** species is one in which megasporangiate and microsporangiate flowers occur on different plants.
- The carpels and stamens of flowers probably evolved from leaflike structures. **Review Figure 29.12**
- Some plants with perfect flowers have adaptations to prevent self-fertilization. **Review Figure 29.13**
- Many angiosperms have coevolved with their animal pollinators.
- Angiosperms exhibit **double fertilization**, usually resulting in the production of a diploid zygote and triploid **endosperm**. **Review Figure 29.16, ANIMATED TUTORIAL 29.2**
- The oldest evolutionary split among the angiosperms is between the clade represented by the single species in the genus *Amborella* and all the remaining flowering plants. **Review Figure 29.18**
- The most species-rich angiosperm clades are the **monocots** and the **eudicots**. The **magnoliids** are likely the sister group to the monocots and eudicots.

29.4 How Do Plants Benefit Human Society?

- Plants provide ecosystem services that affect soil, water, air quality, and climate.
- Plants are primary producers and as such are the foundation of terrestrial food webs.
- Plants provide humans with many important medicinal products.

Go to the Interactive Summary to review key figures, Animated Tutorials, and Activities Life10e.com/is29

CHAPTERREVIEW

REMEMBERING

1. Which of the following statements about seed plants is *true*?
   a. Seeds are produced only by flowering plants (angiosperms).
   b. The sporophyte generation is more reduced in seed plants than in the ferns.
   c. The gametophytes of seed plants are independent of the sporophytes.
   d. All seed plant species are heterosporous.
   e. The zygote of seed plants divides repeatedly to form the gametophyte.

2. Most angiosperms
   a. have seeds enclosed in a carpel.
   b. have haploid endosperm that nourishes the developing embryo.
   c. lack secondary growth.
   d. bear two kinds of cones.
   e. lack flowers.

3. Which statement about flowers is *not* true?
   a. Pollen is produced in the anthers.
   b. Pollen is received on the stigma.
   c. An inflorescence is a cluster of flowers.
   d. A species having female and male flowers on the same plant is dioecious.
   e. A flower with both megasporangia and microsporangia is said to be perfect.

4. Which statement about the angiosperm pollen grain is *not* true?
   a. It is the male gamete.
   b. It is haploid.
   c. It produces a long tube.
   d. It is multicellular.
   e. It is produced in microsporangia.

5. The eudicots
   a. include many herbs, vines, shrubs, and trees.
   b. along with the monocots are the only extant angiosperm clades.
   c. are not a clade.
   d. include the magnolias.
   e. include orchids and palm trees.

UNDERSTANDING & APPLYING

6. Not all flowers possess all of the following floral organs: sepals, petals, stamens, and carpels. Which floral organ or organs do you think might be found in the flowers that have the smallest number of floral organ types? Discuss the possibilities, both for a single flower and for a species.

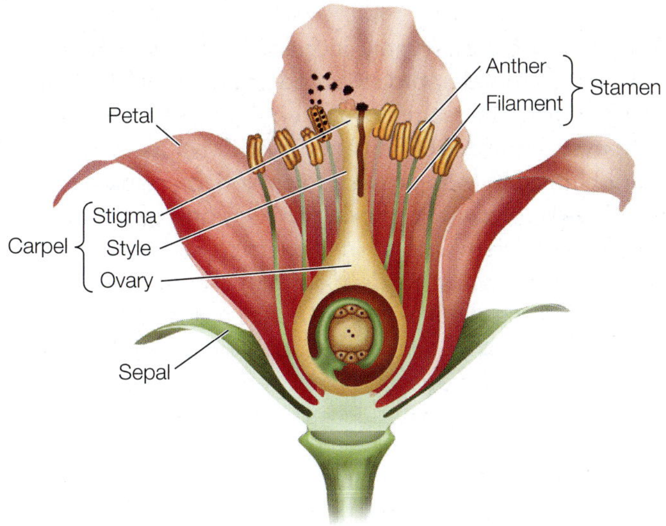
Anther
Filament
Stamen
Petal
Carpel
Stigma
Style
Ovary
Sepal

7. The origin of the angiosperms has long been "an abominable mystery," as Charles Darwin once put it. The earliest known angiosperm fossils are from the late Jurassic or early Cretaceous, but fossils of their sister group, the gymnosperms, are known from as early as the late Carboniferous (about 150 million years earlier; see Figure 29.1). Given that these two sister groups are thought to have arisen from a single split in the seed plant lineage, what might explain the lack of earlier angiosperm fossils?

ANALYZING & EVALUATING

In 1879, W. J. Beal began an experiment that he could not hope to finish in his lifetime. He prepared 20 lots of seeds for long-term storage. Each lot consisted of 50 seeds from each of 23 species. He mixed each lot of seeds with sand and placed the mixture in an uncapped bottle, then buried all the bottles on a sandy knoll. At 10-year intervals over the next century, biologists have excavated a bottle and checked the viability of its contents. The table below shows the number of germinating seeds (of the original 50) from three of the species in years 50–100 of this ongoing experiment. Use these data to answer Questions 8–10.

	Years after burial					
Species	50	60	70	80	90	100
Oenothera biennis (Evening primrose)	19	12	7	5	0	0
Rumex crispus (Curly dock)	26	2	7	1	0	0
Verbascum blattaria (Moth mullein)	31	34	37	35	10	21

8. Calculate the percent of viable seeds for these three species in years 50–100 and graph seed survivorship as a function of time buried.

9. No seeds of the first two species were viable after 90 years of the experiment. Assume 100 percent seed viability at the start of the experiment (year 0), and predict from your graph the approximate year when you think the last of the *Verbascum blattaria* seeds will germinate.

10. What factors do you think might influence the differences among the species in long-term seed viability?

Go to BioPortal at **yourBioPortal.com** for Animated Tutorials, Activities, LearningCurve Quizzes, Flashcards, and many other study and review resources.

30 The Evolution and Diversity of Fungi

CHAPTER**OUTLINE**

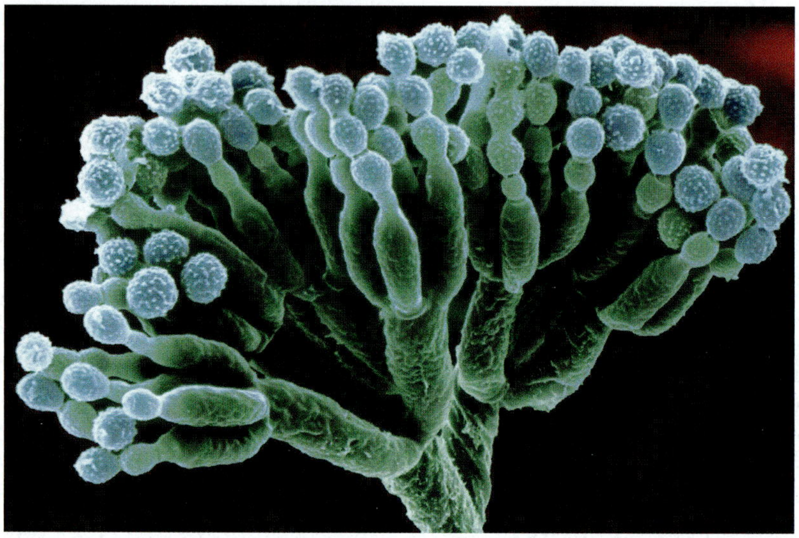

Source of a "Miracle Drug" All species of the fungus *Penicillium* are recognizable by their dense, spore-bearing structures. The derivation of the antibiotic agent penicillin from these fungi was one of the most important achievements in medical history.

ALEXANDER FLEMING WAS ALREADY A famous scientist in 1928, but his laboratory was often a mess. That year he was studying the properties of *Staphylococcus* bacteria, the agents of dangerous staph infections. In August he took a long vacation with his family. When he returned in early September, he found that some of his petri dishes of *Staphylococcus* had become infested with a fungus that killed many of the bacteria.

Many scientists would have sighed at the loss, thrown out the petri dishes, and started new cultures of bacteria. But when Fleming looked at the dishes, he saw something exciting. Around each colony of fungi was a ring within which all the bacteria were dead.

Fleming hypothesized that the bacteria-free rings around the fungal colonies were produced by a substance excreted from the fungi, which he initially called "mould juice." He identified the fungi as members of the genus *Penicillium* and eventually named the antibacterial substance they produced "penicillin." Fleming published his discovery in 1929, but initially it received very little attention.

Over the next decade, Fleming produced small quantities of penicillin for testing as an antibacterial agent. Some of the tests showed promise, but many were inconclusive, and eventually Fleming gave up on the research. But his tests had shown enough promise to attract the attention of several chemists, who worked out the practical problems of producing a stable form of the substance. Clinical trials of this stable form of penicillin were extremely successful, and by 1945 it was being produced and distributed as an antibiotic on a large scale. That same year, Fleming and two of the chemists, Howard Florey and Ernst Chain, received the Nobel Prize in Medicine for their work on penicillin.

The development of penicillin was one of the most important achievements in modern medicine. Until the introduction of modern antibiotics, the most widespread agents of human death included bacterial infections such as gangrene, tuberculosis, and syphilis. Penicillin proved to be highly effective in curing such infections, and its success led to the creation of the modern pharmaceutical industry. Soon many additional antibiotic compounds were isolated from other fungi or synthesized in the laboratory, leading to a "golden age" of human health.

?

Have antibiotics derived from fungi eliminated the danger of bacterial diseases in human populations?

See answer on p. 627.

30.1 What Is a Fungus?

Fungi are organisms that digest their food outside their bodies. They secrete digestive enzymes to break down large food molecules in the environment, then absorb the breakdown products through the plasma membranes of their cells in a process known as **absorptive heterotrophy**. This mode of nutrition allows them to be successful in a wide variety of environments. Many fungi are **saprobes**, which means that they absorb nutrients from dead organic matter. Others are parasites, absorbing nutrients from living hosts. Still others are mutualists, living in intimate associations with other organisms that benefit both partners.

Modern fungi are believed to have evolved from a unicellular protist ancestor that had a flagellum. The probable common ancestor of the animals was also a flagellated protist much like the living choanoflagellates (see Figure 31.2). Current evidence, including the sequences of many genes, suggests that the fungi, choanoflagellates, and animals share a common ancestor not shared by other eukaryotes. These three lineages form a group known as the **opisthokonts** (Figure 30.1). A synapomorphy of the opisthokonts is a flagellum that, if present, is posterior, as in animal sperm. The flagella of all other eukaryotes are anterior.

Synapomorphies that distinguish the fungi as a group among the opisthokonts include absorptive heterotrophy and the presence of **chitin**, a nitrogen-containing structural polysaccharide, in their cell walls. The fungi represent one of the four independent evolutionary origins of large multicellular organisms (plants, brown algae, and animals are the other three).

Unicellular yeasts absorb nutrients directly

Most fungi are multicellular, but single-celled species are found in most fungal groups. Unicellular, free-living fungi are referred to as **yeasts** (Figure 30.2). Some fungi have both a yeast life stage and a multicellular life stage. Thus the term "yeast" does not refer to a single taxonomic group, but rather to a lifestyle that has evolved multiple times. Yeasts live in liquid or moist environments and absorb nutrients directly across their cell surfaces.

The ease with which many yeasts can be cultured, combined with their rapid growth rates, has made them ideal model organisms for study in the laboratory. They present many of the same advantages to laboratory investigators as do many bacteria, but because they are eukaryotes, their genome structures and cells are much more like those of humans and other eukaryotes than are those of bacteria.

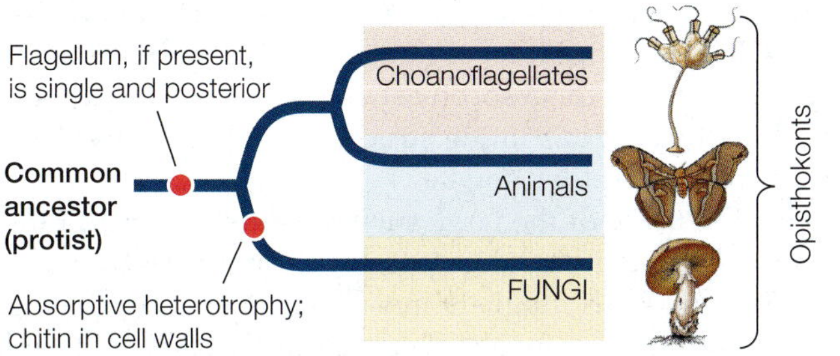
Flagellum, if present, is single and posterior
Common ancestor (protist)
Absorptive heterotrophy; chitin in cell walls
Choanoflagellates
Animals
FUNGI
Opisthokonts

30.1 Fungi in Evolutionary Context Absorptive heterotrophy and the presence of chitin in their cell walls distinguish the fungi from other opisthokonts.

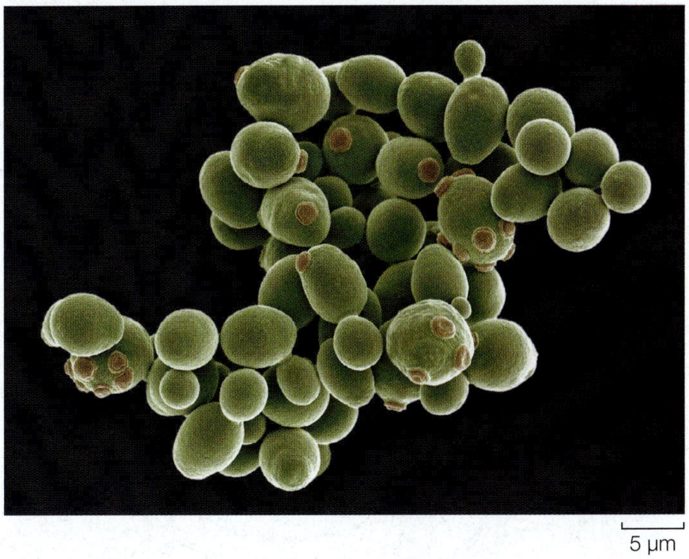
Saccharomyces cerevisiae
5 μm

30.2 Yeasts Unicellular, free-living fungi are known as yeasts. Many yeasts reproduce by budding—mitosis followed by asymmetrical cell division—as illustrated here.

Multicellular fungi use hyphae to absorb nutrients

The body of a multicellular fungus is called a **mycelium** (plural *mycelia*). A mycelium is composed of a mass of individual tubular filaments called **hyphae** (singular *hypha*; Figure 30.3A), in which absorption of nutrients takes place. The cell walls of the hyphae are greatly strengthened by microscopic fibrils of chitin. In some species of fungi, the hyphae are subdivided into cell-like compartments by incomplete cross-walls called **septa** (singular *septum*); these hyphae are referred to as **septate**. Septa do not completely close off compartments in the hyphae. Pores at the centers of the septa allow organelles—sometimes even nuclei—to move in a controlled way between compartments (Figure 30.3B). In other species of fungi, the hyphae lack septa but may contain hundreds of nuclei; these hyphae are referred to as **coenocytic**. The coenocytic condition results from repeated nuclear divisions without cytokinesis.

Certain modified hyphae, called **rhizoids**, anchor some fungi to their substrate (i.e., the dead organism or other matter on which they are feeding). These rhizoids are not homologous to the rhizoids of plants, and they are not specialized to absorb nutrients and water.

Fungi can grow very rapidly. In some species, the total hyphal growth of a fungal mycelium (not the growth of an individual hypha) may exceed 1 kilometer a day! The hyphae may be widely dispersed to forage for nutrients over a large area, or they may clump together in a cottony mass to exploit a rich nutrient source. The familiar mushrooms you may notice in the environment are spore-producing fruiting structures (Figure 30.3C). In the fungal species that produce these structures, the

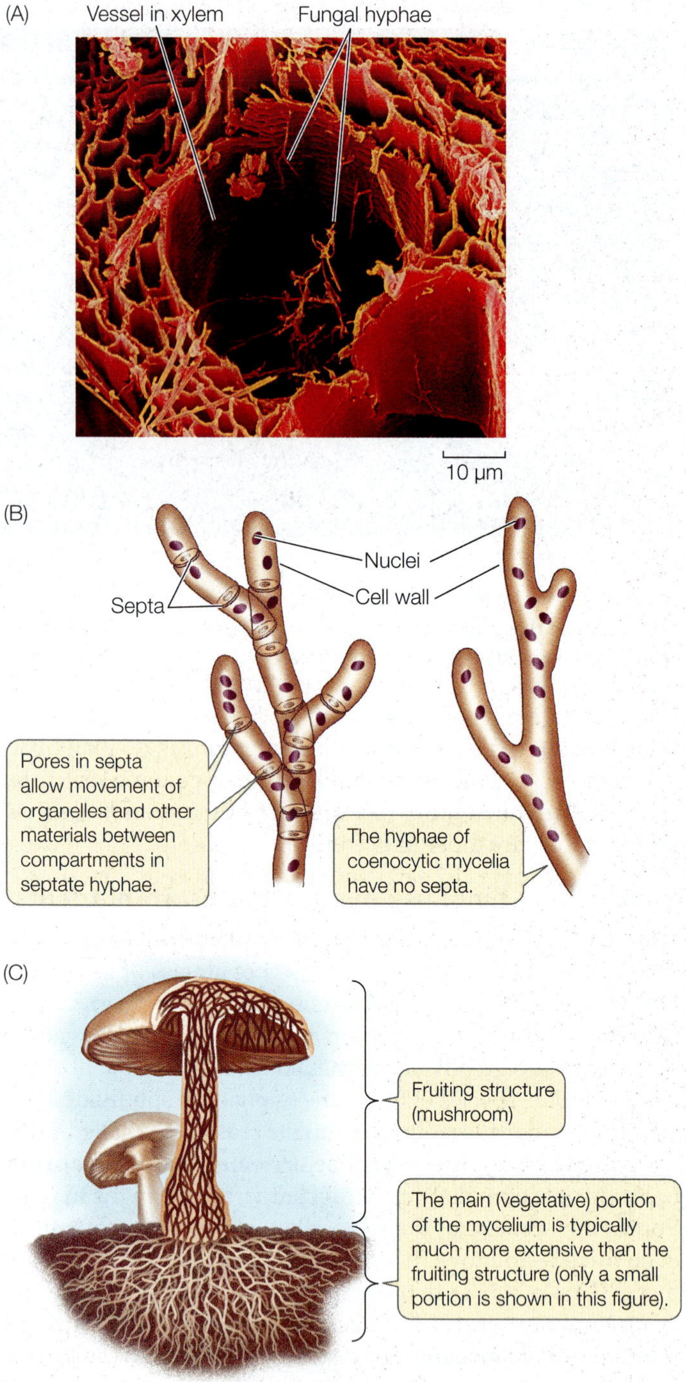
(A)
Vessel in xylem
Fungal hyphae
10 μm
(B)
Septa
Nuclei
Cell wall
Pores in septa allow movement of organelles and other materials between compartments in septate hyphae.
The hyphae of coenocytic mycelia have no septa.
(C)
Fruiting structure (mushroom)
The main (vegetative) portion of the mycelium is typically much more extensive than the fruiting structure (only a small portion is shown in this figure).

30.3 Mycelia Are Made Up of Hyphae (A) The minute individual hyphae of fungal mycelia can penetrate small spaces. In this artificially colored micrograph, hyphae (yellow structures) of a dry-rot fungus are penetrating the xylem tissues of a log. (B) The hyphae of septate fungal species are divided into organelle-containing compartments by porous septa. The hyphae of coenocytic fungal species have no septa. (C) The fruiting structure of a club fungus is short-lived, but the filamentous, nutrient-absorbing mycelium can be long-lived and cover large areas.

mycelial mass is often far larger than the visible mushroom. The mycelium of one individual fungus discovered in Oregon covers almost 900 hectares underground and weighs considerably more than a blue whale (the largest animal). Aboveground, this individual is evident only as isolated clumps of mushrooms.

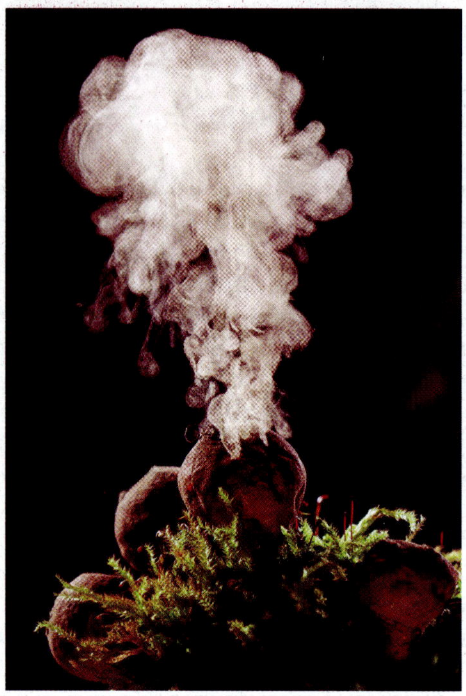
Lycoperdon perlatum

30.4 Spores Galore Puffballs (a type of club fungus) disperse trillions of spores in great bursts. Few of the spores travel very far, however; some 99 percent of them fall within 100 meters of the parent puffball.

What happens when a fungus faces a dwindling food supply? A common strategy is to reproduce rapidly and abundantly. When conditions are good, fungi produce great quantities of spores, but the rate of spore production is commonly even higher when nutrient supplies go down. The spores may then remain dormant until conditions improve, or they may be dispersed to areas where nutrient supplies are higher.

Not only are fungal spores abundant in number, but they are extremely tiny and easily spread by wind or water (**Figure 30.4**). These attributes virtually ensure that they will be scattered over great distances and that at least some of them will find conditions suitable for growth. The air we breathe contains as many as 10,000 fungal spores per cubic meter. No wonder we find fungi just about everywhere.

Fungi are in intimate contact with their environment

The filamentous hyphae of a fungus give it a unique relationship with its physical environment. The fungal mycelium has an enormous surface area-to-volume ratio compared with that of most large multicellular organisms. This large ratio is a marvelous adaptation for absorptive heterotrophy. Throughout the mycelium (except in fruiting structures), all of the hyphae are very close to their food source.

The downside of the large surface area-to-volume ratio of the mycelium is its tendency to lose water rapidly in a dry environment. Thus fungi are most common in moist environments. You have probably observed the tendency of molds, toadstools, and other fungi to appear in damp places.

Another characteristic of some fungi is a tolerance for highly hypertonic environments (those with a solute concentration higher than their own; see Section 6.3). Many fungi are more resilient than bacteria in hypertonic surroundings. Jelly in the refrigerator, for example, will not become a growth medium for bacteria because it is too hypertonic to those organisms, but it may eventually harbor mold colonies. Mold in the refrigerator illustrates yet another trait of many fungi: tolerance of temperature extremes. Many fungi grow in temperatures as low as –6°C, and some can tolerate temperatures higher than 50°C.

RECAP 30.1

Fungi, like animals, are opisthokonts. Fungi are distinguished from other opisthokonts by absorptive heterotrophy and by the presence of chitin in their cell walls. Unicellular fungi called yeasts absorb nutrients directly across their cell surfaces. The body form of multicellular fungi—a mycelium made up of rapidly growing hyphae—allows them to practice absorptive heterotrophy efficiently in a variety of moist environments.

- Describe the relationship between fungal structure and absorptive heterotrophy. **See pp. 609–610 and Figure 30.3**
- What are the advantages and disadvantages to multicellular fungi of the large surface area-to-volume ratio of the mycelium? **See p. 610**

Fungi are important components of healthy ecosystems. They interact with other organisms in many ways, some of which are harmful and some beneficial to those other organisms.

30.2 How Do Fungi Interact with Other Organisms?

Without the fungi, our planet would be very different. Picture Earth with only a few stunted plants and watery environments choked with the remains of dead organisms. Fungi do much of Earth's garbage disposal. Fungi not only help clean up the landscape and form soil, but also play key roles in the recycling of mineral nutrients. Furthermore, the colonization of the terrestrial environment was made possible in large part by associations fungi formed with land plants and other organisms.

Saprobic fungi are critical to the planetary carbon cycle

Saprobic fungi, along with bacteria, are the major decomposers on Earth. These organisms contribute to the decay of nonliving organic matter and thus to the recycling of the elements required by living things. In forests, for example, fungi digest and absorb nutrients from fallen trees, thus decomposing their wood. Fungi are the principal decomposers of cellulose and lignin, the main components of plant cell walls (which most bacteria cannot break down). Other fungi produce enzymes that decompose keratin and thus break down animal structures such as hair and nails.

Were it not for the fungal decomposers, Earth's carbon cycle would fail: great quantities of carbon atoms would remain trapped forever on forest floors and elsewhere. Instead, those carbon atoms are returned to the atmosphere in the form of CO_2 by fungal respiration, where they are again available for photosynthesis by plants.

There was a time in Earth's history when populations of saprobic fungi declined dramatically. Vast tropical swamps existed during the Carboniferous period, as we saw in Chapter 25. When plants in these swamps died, they began to form peat. Peat formation led to acidification of the swamps; that acidity, in turn, drastically reduced the fungal population. The result? With the decomposers largely absent, large quantities of peat remained on the swamp floor and over time were converted into coal.

In contrast to their decline during the Carboniferous, fungi did very well at the end of the Permian, about 250 million years ago, when the aggregation of the continents produced volcanic eruptions that triggered a global mass extinction. The fossil record shows that even as 96 percent of all multicellular species became extinct, fungi flourished—demonstrating both their hardiness and their role in recycling the elements in dead plants and animals.

Simple sugars and the breakdown products of complex polysaccharides are the favored source of carbon for saprobic fungi. Most fungi obtain nitrogen from proteins or the products of protein breakdown. Many fungi can use nitrate (NO_3^-) or ammonium (NH_4^+) ions as their sole source of nitrogen. No known fungus can get its nitrogen directly from inorganic nitrogen gas, however, as can some bacteria and plant–bacteria associations (that is, fungi cannot fix nitrogen; see Section 26.3).

Go to Media Clip 30.1
Fungal Decomposers
Life10e.com/mc30.1

Some fungi engage in parasitic or predatory interactions

Whereas saprobic fungi obtain their energy, carbon, and nitrogen directly from dead organic matter, other species of fungi obtain their nutrition from parasitic—and even predatory—interactions.

PARASITIC FUNGI **Mycologists** (biologists who study fungi) distinguish between two classes of parasitic fungi based on their degree of dependence on their host. **Facultative parasites** can grow on living organisms but can also grow independently (including on artificial media). **Obligate parasites** can grow only on a specific, living host. The fact that their growth depends on a living host shows that obligate parasites have specialized nutritional requirements.

Plants and insects are the most common hosts of parasitic fungi. The filamentous structure of fungal hyphae is especially well suited to a life of absorbing nutrients from living plants. The slender hyphae of a parasitic fungus can invade a plant through stomata, through wounds, or in some cases, by direct penetration of epidermal cell walls (**Figure 30.5A**). Once inside the plant, the hyphae branch out to expand the mycelium. Some hyphae produce **haustoria**, branching projections that push through cell walls into living plant cells, absorbing the nutrients within those cells. The haustoria do not break

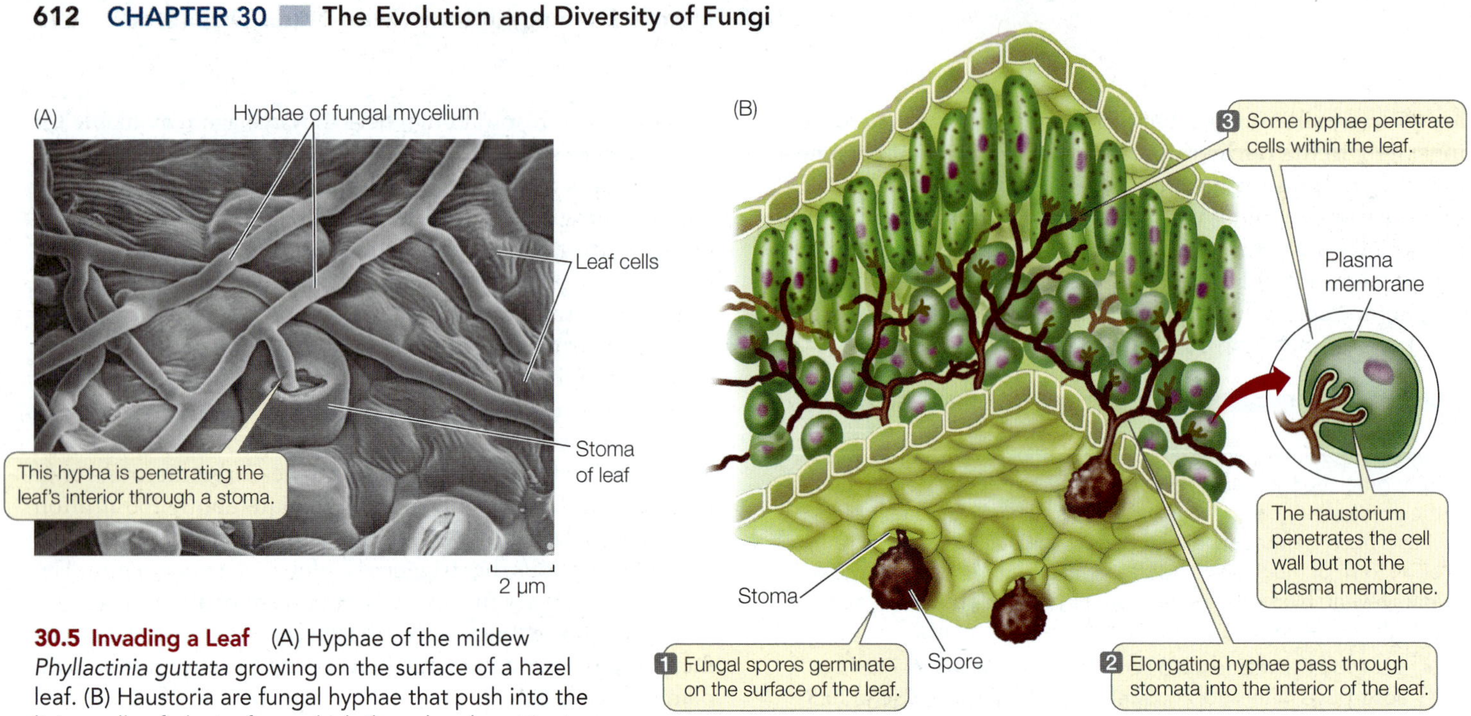
(A)
Hyphae of fungal mycelium
Leaf cells
Stoma of leaf
This hypha is penetrating the leaf's interior through a stoma.
2 μm
(B)
3 Some hyphae penetrate cells within the leaf.
Plasma membrane
The haustorium penetrates the cell wall but not the plasma membrane.
Stoma
Spore
1 Fungal spores germinate on the surface of the leaf.
2 Elongating hyphae pass through stomata into the interior of the leaf.

30.5 Invading a Leaf (A) Hyphae of the mildew *Phyllactinia guttata* growing on the surface of a hazel leaf. (B) Haustoria are fungal hyphae that push into the living cells of plants, from which they absorb nutrients.

through the plasma membranes inside the cell walls; they simply invaginate into the membranes, so that the plasma membrane fits them like a glove (**Figure 30.5B**). Fruiting structures may form, either within the plant body or on its surface.

Some parasitic fungi live in a close physical (symbiotic) relationship with a plant host that is usually not lethal to the plant. Others are *pathogenic*, sickening or even killing the host from which they derive nutrition.

Go to Media Clip 30.2
Mind-Control Killer Fungi
Life10e.com/mc30.2

PATHOGENIC FUNGI Although most human diseases are caused by bacteria or viruses, fungal pathogens are a major cause of death among people with compromised immune systems. Most people with AIDS die of fungal diseases, including the pneumonia caused by *Pneumocystis jirovecii*. Even *Candida albicans* and certain other yeasts that are normally part of a healthy microbiome can cause severe diseases, such as esophagitis (which impairs swallowing), in individuals with AIDS and in individuals taking immunosuppressive drugs. Various fungi cause other, less threatening human diseases, such as ringworm and athlete's foot. Our limited understanding of the basic biology of these fungi hampers our ability to treat the diseases they cause. As a result, fungal diseases are a growing international health problem.

The worldwide decline of amphibian species has been linked to the spread of a chytrid fungus, *Batrachochytrium dendrobatidis*. Genetic analyses indicate that the populations of this fungus that are attacking amphibian populations around the world are genetically almost identical, which suggests a recent introduction of the fungus across the globe. This chytrid appears to be native to southern Africa, and its spread around the world may have been initiated in the 1930s with exports of the African clawed frog (*Xenopus laevis*), which was once widely used in human pregnancy tests.

Fungi are by far the most important plant pathogens, much more so than bacteria and viruses. Pathogenic fungi cause crop losses amounting to billions of dollars each year. Major fungal diseases of crop plants include black stem rust of wheat and other diseases of wheat, corn, and oats. The agent of black stem rust is *Puccinia graminis*, which has a complicated life cycle that involves two plant hosts (wheat and barberry). In an epidemic in 1935, *P. graminis* was responsible for the loss of about one-fourth of the wheat crop in Canada and the United States.

PREDATORY FUNGI Some fungi have adaptations that enable them to function as active predators, trapping nearby microscopic protists or animals. The most common predatory strategy seen in fungi is to secrete sticky substances from the hyphae so that passing organisms stick to them. The hyphae then quickly invade the trapped prey, growing and branching within it, spreading through its body, absorbing nutrients, and eventually killing it.

A more dramatic adaptation for predation is the constricting ring formed by some species of soil fungi (**Figure 30.6**). When nematodes (tiny roundworms) are present in the soil, these fungi form three-celled rings with a diameter that just fits a nematode. A nematode crawling through one of these rings stimulates the fungus, causing the cells of the ring to swell and trap the worm. Fungal hyphae quickly invade and digest the unlucky victim.

Mutualistic fungi engage in relationships that benefit both partners

Certain relationships between fungi and other organisms have nutritional consequences for both partners. Two relationships of this type are **symbiotic** (the partners live in close, permanent contact with each other) as well as **mutualistic** (the relationship benefits both partners; see Section 56.3).

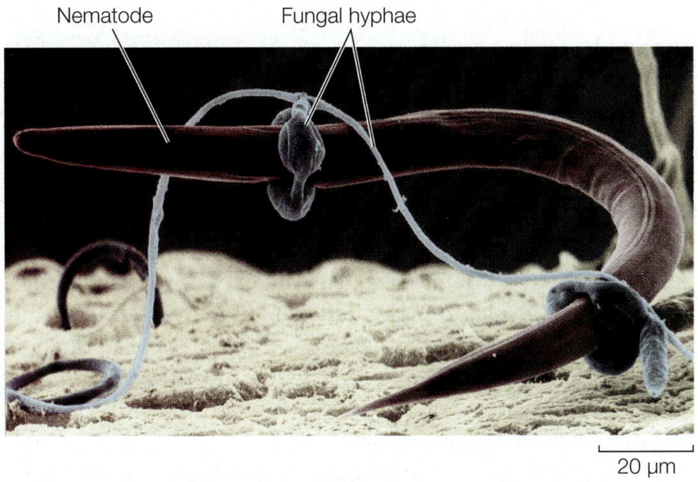
Nematode
Fungal hyphae
20 μm

30.6 Fungus as Predator A nematode is trapped by hyphal rings of the soil-dwelling fungus *Arthrobotrys dactyloides*.

LICHENS A **lichen** is not a single organism, but rather a meshwork of two radically different species: a fungus and a photosynthetic microorganism. Together, the organisms that constitute a lichen can survive some of the harshest environments on Earth. The biota of Antarctica, for example, features more than a hundred times as many species of lichens as of plants. Relatively little experimental work has focused on lichens, perhaps because they grow so slowly—typically less than 1 centimeter in a year.

There are nearly 30,000 described "species" of lichens, each of which is assigned the name of its fungal component. These fungal components may constitute as many as 20 percent of all fungal species. Most of them are sac fungi (Ascomycota). Some of them are able to grow independently without a photosynthetic partner, but most have never been observed in nature other than in a lichen association. The photosynthetic component of a lichen is most often a unicellular green alga, but it can be a cyanobacterium, or may even include both.

Lichens are found in all sorts of exposed habitats: on tree bark, on open soil, and on bare rock. Reindeer moss (not a moss at all, but the lichen *Cladonia subtenuis*) covers vast areas in Arctic, sub-Arctic, and boreal regions, where it is an important part of the diets of reindeer and other large mammals.

The body forms of lichens fall into three principal categories. **Crustose** (crustlike) lichens adhere tightly to their substrate (**Figure 30.7A**). **Foliose** (leafy) lichens are loosely attached and grow parallel to their substrate (**Figure 30.7B**) **Fruticose** lichens are highly branched and can grow upward like shrubs or hang in long strands from tree branches or rocks (**Figure 30.7C**).

A cross section of a typical foliose lichen reveals a tight upper region of fungal hyphae; a layer of photosynthetic cyanobacteria or algae; a loose hyphal layer; and finally hyphal rhizoids that attach the structure to its substrate (**Figure 30.8**). The meshwork of fungal hyphae takes up mineral nutrients needed by the photosynthetic cells and also holds water tenaciously, providing a suitably moist environment. The fungus obtains fixed carbon from the photosynthetic products of the algal or cyanobacterial cells.

Within the lichen, fungal hyphae are tightly pressed against the photosynthetic cells and sometimes invade them without breaching the plasma membrane (similar to the haustoria in parasitic fungi; see Figure 30.5). The photosynthetic cells not only survive these intrusions but continue to grow. Algal cells in a lichen "leak" photosynthetic products at a greater rate than do similar cells growing on their own, and photosynthetic cells taken from lichens grow more rapidly on their own than when associated with a fungus. On the basis of these observations, we could consider lichen fungi to be parasitic on their photosynthetic partners. In many places where lichens grow, however, the photosynthetic cells could not grow at all on their own.

Lichens can reproduce simply by fragmentation of the vegetative body (the **thallus**) or by means of specialized structures called **soredia** (singular *soredium*). Soredia consist of one or a few photosynthetic cells bound by fungal hyphae. They become detached from the lichen, are dispersed by air currents, and upon arriving at a favorable location, develop into a new

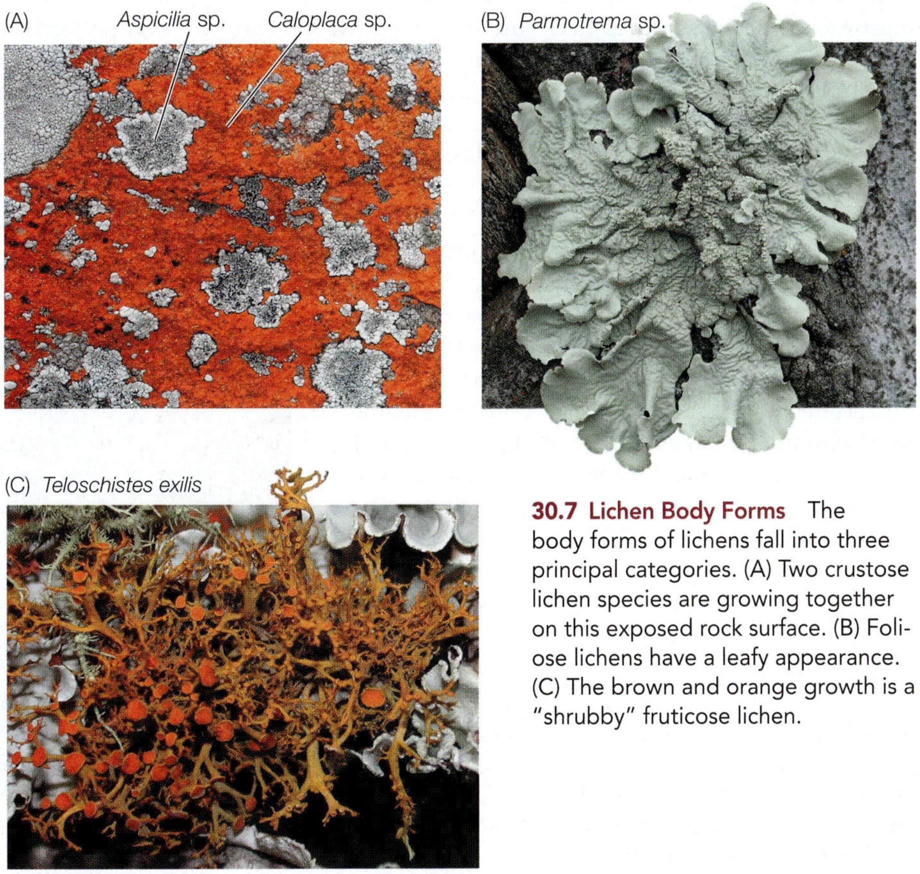
(A) *Aspicilia* sp. *Caloplaca* sp.
(B) *Parmotrema* sp.
(C) *Teloschistes exilis*

30.7 Lichen Body Forms The body forms of lichens fall into three principal categories. (A) Two crustose lichen species are growing together on this exposed rock surface. (B) Foliose lichens have a leafy appearance. (C) The brown and orange growth is a "shrubby" fruticose lichen.

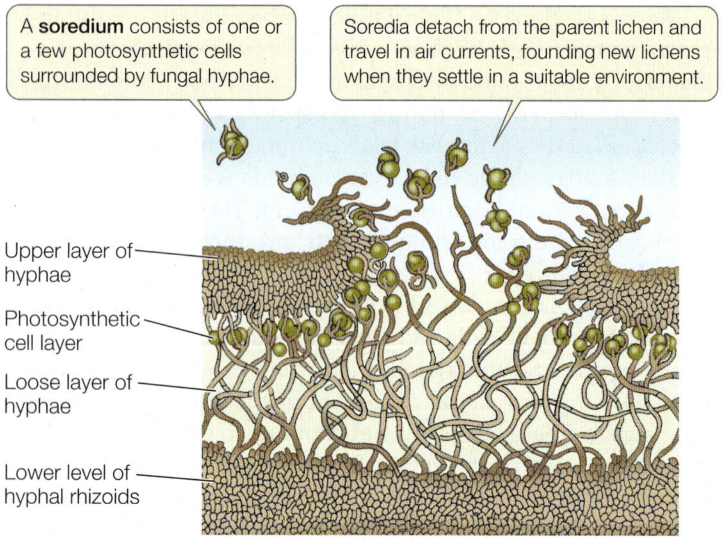
A **soredium** consists of one or a few photosynthetic cells surrounded by fungal hyphae.
Soredia detach from the parent lichen and travel in air currents, founding new lichens when they settle in a suitable environment.
Upper layer of hyphae
Photosynthetic cell layer
Loose layer of hyphae
Lower level of hyphal rhizoids

30.8 Lichen Anatomy Cross section showing the layers of a foliose lichen and the release of soredia.

lichen thallus. Alternatively, the fungal partner may go through its sexual reproductive cycle, producing haploid spores. When these spores are discharged, however, they disperse alone, unaccompanied by the photosynthetic partner.

Lichens are often the first colonists on new areas of bare rock. They get most of the mineral nutrients they need from the air and rainwater, augmented by minerals absorbed from dust. A lichen begins to grow shortly after a rain, as it begins to dry. As it grows, the lichen acidifies its environment slightly, and this acidity contributes to the slow breakdown of rocks, an early step in soil formation. With further drying, the lichen's photosynthesis ceases. The water content of the lichen may drop to less than 10 percent of its dry weight, at which point it becomes highly insensitive to extremes of temperature.

MYCORRHIZAE Many vascular plants depend on a symbiotic association with fungi. This ancient association between plants and fungi was critical to the successful exploitation of the terrestrial environment by plants. Unassisted, the root hairs of many plants often do not take up enough water or minerals to sustain their growth. However, the roots of such plants usually become infected with fungi, forming an association called a **mycorrhiza**. (We'll describe the infection process in detail in Section 36.4.) There are two types of mycorrhizae, distinguished by whether or not the fungal hyphae penetrate the plant cell walls.

In **ectomycorrhizae**, the fungus wraps around the root, and its mass is often as great as that of the root itself (Figure 30.9A). The fungal hyphae wrap around individual cells in the root but do not penetrate the cell walls. An extensive web of hyphae penetrates the soil in the area around the root, so that up to 25 percent of the soil volume near the root may be fungal hyphae. The hyphae attached to the root increase its surface area for the absorption of water and minerals, and the mass of hyphae in the soil acts like a sponge to hold water in the neighborhood of the root. Infected roots are short, swollen, and club-shaped, and they lack root hairs.

The fungal hyphae of **arbuscular mycorrhizae** enter the root and penetrate the cell walls of the root cells, forming arbuscular (treelike) structures inside the cell wall but outside the plasma membrane. These structures, like the haustoria of parasitic fungi and the contact regions of fungal hyphae and photosynthetic cells in lichens, become the primary site of exchange between plant and fungus (Figure 30.9B). As in the ectomycorrhizae, the fungus forms a vast web of hyphae leading from the root surface into the surrounding soil.

The mycorrhizal association is important to both partners. The fungus obtains needed organic compounds, such as sugars and amino acids, from the plant. In return, the fungus, because of its high surface area-to-volume ratio and its ability to penetrate the fine structure of the soil, greatly increases the plant's ability to absorb water and minerals (especially phosphorus). The fungus may also provide the plant with certain growth hormones and may protect it against attack by disease-causing microorganisms.

Plants that have active arbuscular mycorrhizae typically are a deeper green, exhibit higher growth rates, and may resist drought and temperature extremes better than plants of the same species that have little mycorrhizal development. Attempts to introduce some plant species to new areas have failed

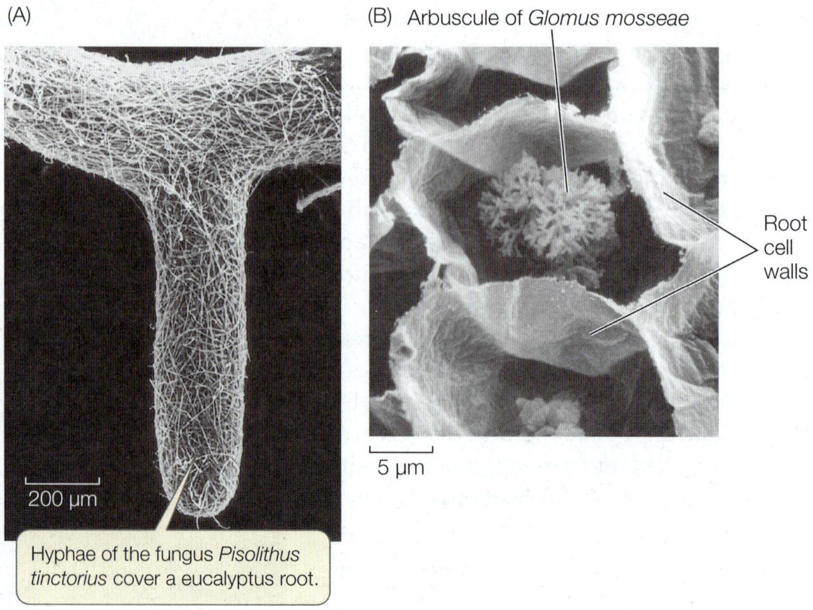
(A)
200 μm
Hyphae of the fungus *Pisolithus tinctorius* cover a eucalyptus root.
(B) Arbuscule of *Glomus mosseae*
Root cell walls
5 μm

30.9 Mycorrhizal Associations (A) Ectomycorrhizal fungi wrap themselves around a plant root, increasing the area available for absorption of water and minerals. (B) Hyphae of arbuscular mycorrhizal fungi infect the root internally and penetrate the root cell walls, branching within the cells and forming a treelike structure, the arbuscule. (The cell cytoplasm has been removed to better visualize the arbuscule.)

until a bit of soil from the native area (presumably containing the fungus necessary to establish mycorrhizae) was provided. Trees without ectomycorrhizae do not grow well in the absence of abundant nutrients and water, so the health of our forests depends on the presence of ectomycorrhizal fungi. Many agricultural crops require inoculation of seeds with appropriate mycorrhizal fungi prior to planting. Without these fungi, the plants are unlikely to grow well, or in some cases at all. Certain plants that live in nitrogen-poor habitats, such as cranberry bushes and orchids, invariably have mycorrhizae. Orchid seeds will not germinate in nature unless they are already infected by the fungus that will form their mycorrhizae. Plants that lack chlorophyll always have mycorrhizae, which they often share with the roots of green, photosynthetic plants. In effect, these plants without chlorophyll are feeding on nearby green plants, using the fungus as a bridge.

Endophytic fungi protect some plants from pathogens, herbivores, and stress

In a tropical rainforest, 10,000 or more fungal spores may land on a single leaf each day. Some are plant pathogens, some do not affect the plant at all, and some invade the plant in a beneficial way. Fungi that live within aboveground parts of plants without causing obvious deleterious symptoms are called **endophytic fungi**. Recent research has shown that endophytic fungi are abundant in plants in all terrestrial environments.

Among the grasses, individual plants with endophytic fungi are more resistant to pathogens and to insect and mammalian herbivores than are plants lacking endophytes. The fungi produce alkaloids (nitrogen-containing compounds) that are toxic to animals. The alkaloids do not harm the host plant; in fact, some plants produce alkaloids (such as nicotine) themselves. The fungal alkaloids also increase the ability of grasses to resist stresses of various types, including drought (water shortage) and salty soils. Such resistance is useful in agriculture.

The role, if any, of endophytic fungi in most broad-leaved plants is unclear. They may convey protection against pathogens, or they may simply occupy space within leaves without conferring any benefit, but also without doing harm. The benefit, in fact, might be all for the fungus.

RECAP 30.2

Fungi interact with other organisms in many ways, both harmful and beneficial. Saprobic fungi play critical roles in the recycling of elements required by living organisms. Lichens are mutualistic associations of a fungus with algae or cyanobacteria. Mycorrhizae are associations of fungi and the roots of plants; they are essential for the survival of most plant species.

- What is the role of fungi in Earth's carbon cycle? See p. 611
- Describe the nature and benefits of the lichen association. See pp. 613–614 and Figure 30.8
- Why do plants grow better in the presence of mycorrhizal fungi? See pp. 614–615 and Figure 30.9

Before molecular techniques clarified their phylogenetic relationships, one criterion used for assigning fungi to taxonomic groups was the nature of their life cycles—including the types of fruiting structures they produced. The next section will take a closer look at life cycles in the six major groups of fungi.

30.3 How Do Major Groups of Fungi Differ in Structure and Life History?

Major fungal groups were originally distinguished by their structures and processes for sexual reproduction and, to a lesser extent, by other morphological differences. Although fungal life cycles are even more diverse than was once realized, specific types of life cycles generally characterize the six major groups of fungi: microsporidia, chytrids, zygospore fungi (Zygomycota), arbuscular mycorrhizal fungi (Glomeromycota), sac fungi (Ascomycota), and club fungi (Basidiomycota). Figure 30.10 diagrams the evolutionary relationships of these groups as they are understood today.

The chytrids and the zygospore fungi may not represent monophyletic groups, as they each consist of several distantly related lineages that retain some ancestral features. The clades

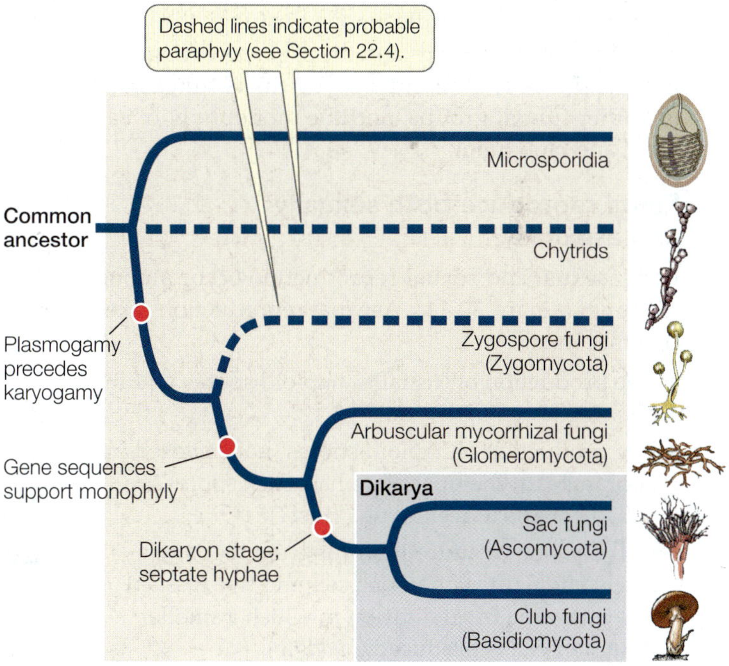
Dashed lines indicate probable paraphyly (see Section 22.4).
Common ancestor
Microsporidia
Chytrids
Plasmogamy precedes karyogamy
Zygospore fungi (Zygomycota)
Gene sequences support monophyly
Arbuscular mycorrhizal fungi (Glomeromycota)
Dikarya
Dikaryon stage; septate hyphae
Sac fungi (Ascomycota)
Club fungi (Basidiomycota)

30.10 A Phylogeny of the Fungi Microsporidia are reduced, parasitic fungi whose relationships among the fungi are uncertain. They may be the sister group of most other fungi, or they may be more closely related to particular groups of chytrids or zygospore fungi. The dashed lines indicate that chytrids and zygospore fungi are thought to be paraphyletic; the relationships of the lineages within these two informal groups (see Table 30.1) are not yet well resolved. The sac fungi and club fungi together form the clade Dikarya.

Go to Activity 30.1 Fungal Phylogeny
Life10e.com/ac30.1

TABLE 30.1
Classification of the Fungi

Group	Common Name	Features
Microsporidia	Microsporidia	Intracellular parasites of animals; greatly reduced, among smallest eukaryotes known; polar tube used to infect hosts
Chytrids (paraphyletic)[a] Chytridiomycota Neocallimastigomycota Blastocladiomycota	Chytrids	Mostly aquatic and microscopic; zoospores and gametes have flagella
Zygomycota (paraphyletic)[a] Entomophthoromycotina Kickxellomycotina Mucoromycotina Zoopagomycotina	Zygospore fungi	Reproductive structure is a unicellular zygospore with many diploid nuclei; hyphae coenocytic; no fleshy fruiting body
Glomeromycota	Arbuscular mycorrhizal fungi	Form arbuscular mycorrhizae in plant roots; only asexual reproduction is known
Ascomycota	Sac fungi	Sexual reproductive saclike structure known as an ascus, which contains haploid ascospores; hyphae septate; dikaryon
Basidiomycota	Club fungi	Sexual reproductive structure is a basidium, a swollen cell at the tip of a specialized hypha that supports haploid basidiospores; hyphae septate; dikaryon

[a]The formally named groups within the chytrids and Zygomycota are each thought to be monophyletic, but their relationships to one another (and to microsporidia) are not yet well resolved.

that are thought to be monophyletic within these two informal groupings are listed in Table 30.1. Recent evidence from DNA analyses has established the placement of the microsporidia among the fungi, the likely paraphyly of the chytrids and the zygospore fungi, the independence of arbuscular mycorrhizal fungi from the other fungal groups, and the monophyly of sac fungi and club fungi.

Fungi reproduce both sexually and asexually

Both asexual and sexual reproduction occur among the fungi (Figure 30.11). Asexual reproduction takes several forms:

- The production of (usually) haploid spores within sporangia
- The production of haploid spores (not enclosed in sporangia) at the tips of hyphae; such spores are called **conidia** (Greek *konis*, "dust")
- Cell division by unicellular fungi—either a relatively equal division of one cell into two (*fission*) or an asymmetrical division in which a smaller daughter cell is produced (*budding*)
- Simple breakage of the mycelium

Sexual reproduction is rare (or even unknown) in some groups of fungi but is common in others. Sexual reproduction may not occur in some species, or it may occur so rarely that mycologists have never observed it. Species in which no sexual stage has been observed were once placed in a separate taxonomic group because knowledge of the sexual life cycle was considered necessary for classifying fungi. Now, however, these species can be related to other species of fungi through analysis of their DNA sequences.

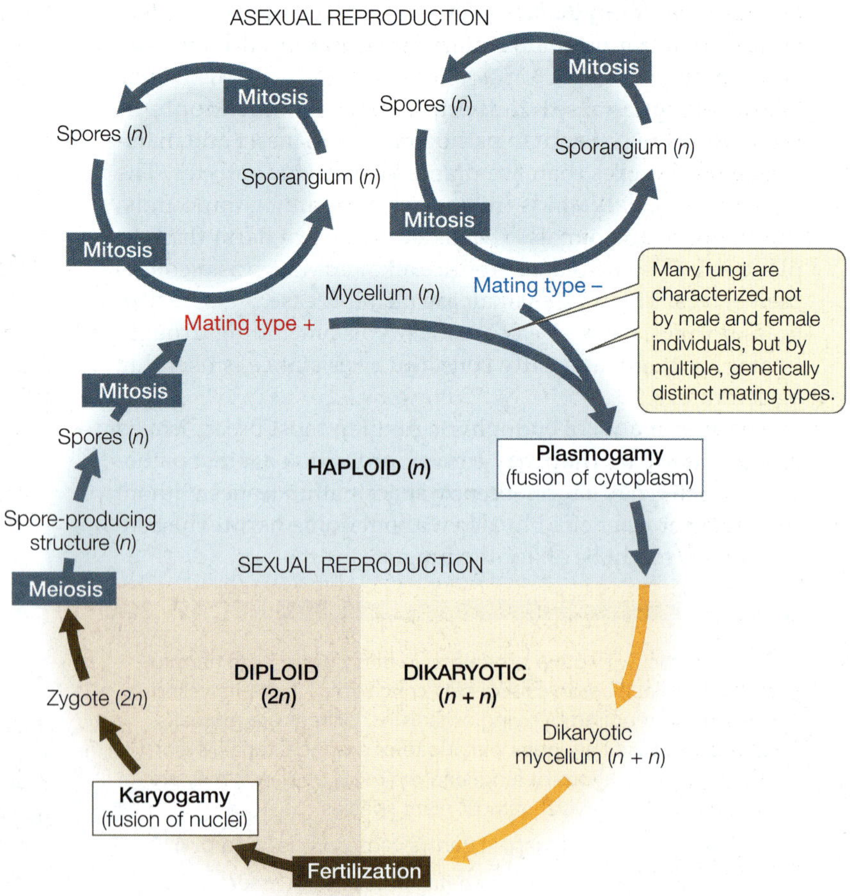
ASEXUAL REPRODUCTION
Mitosis
Spores (n)
Sporangium (n)
Mitosis
Mitosis
Spores (n)
Sporangium (n)
Mitosis
Mycelium (n)
Mating type +
Mating type –
Many fungi are characterized not by male and female individuals, but by multiple, genetically distinct mating types.
Mitosis
Spores (n)
HAPLOID (n)
Plasmogamy
(fusion of cytoplasm)
Spore-producing structure (n)
SEXUAL REPRODUCTION
Meiosis
DIPLOID (2n)
DIKARYOTIC (n + n)
Zygote (2n)
Dikaryotic mycelium (n + n)
Karyogamy
(fusion of nuclei)
Fertilization

30.11 A Generalized Fungal Life Cycle Environmental conditions may determine which mode of reproduction—sexual or asexual—takes place at a given time.

Sexual reproduction in most fungi features an interesting twist: There is no morphological distinction between female and male structures, or between female and male individuals, in most groups of fungi. Rather, there is a genetically determined distinction between two *or more* **mating types**. Individuals of the same mating type cannot mate with each other, but they can mate with individuals of another mating type within the same species, thus avoiding self-fertilization. Individuals of different mating types differ genetically but are often visually and physiologically indistinguishable.

Microsporidia are highly reduced, parasitic fungi

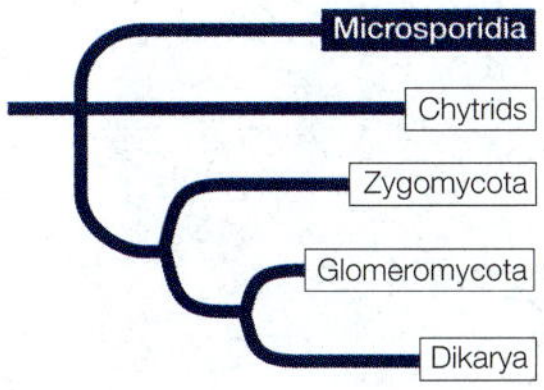
Microsporidia
Chytrids
Zygomycota
Glomeromycota
Dikarya

Microsporidia are unicellular parasitic fungi. They are among the smallest eukaryotes known, with infective spores that are only 1 to 40 micrometers (μm) in diameter. About 1,500 species have been described, but many more species are thought to exist. Their relationships among the eukaryotes have puzzled biologists for many decades.

Microsporidia lack true mitochondria, although they have reduced structures, known as **mitosomes**, that are derived from mitochondria. Unlike mitochondria, however, mitosomes contain no DNA; the mitochondrial genome has been completely transferred to the nucleus. Because microsporidia lack mitochondria, biologists initially suspected that they represented an early lineage of eukaryotes that diverged before the endosymbiotic event from which mitochondria evolved. The presence of mitosomes, however, indicates that this hypothesis is incorrect. DNA sequence analysis, along with the fact that their cell walls contain chitin, has confirmed that the microsporidia are in fact highly reduced, parasitic fungi, although their exact placement among the fungal lineages is still being investigated.

Microsporidia are obligate intracellular parasites of animals, especially of insects, crustaceans, and fishes. Some species are known to infect mammals, including humans. Most infections by microsporidia cause chronic disease in the host, with effects that include weight loss, reduced fertility, and a shortened life span. The host cell is penetrated by a polar tube that grows from the microsporidian spore. The function of the polar tube is to inject the contents of the spore, the sporoplasm, into the host (**Figure 30.12**). The sporoplasm replicates within the host cell and produces new infective spores. The life cycle of some species is complex and involves multiple hosts, whereas other species infect a single host. In some insects, parasitic microsporidia are transmitted vertically (i.e., from parent to offspring). Reproduction is thought to be strictly asexual in some microsporidians, but it includes poorly understood asexual and sexual cycles in other species.

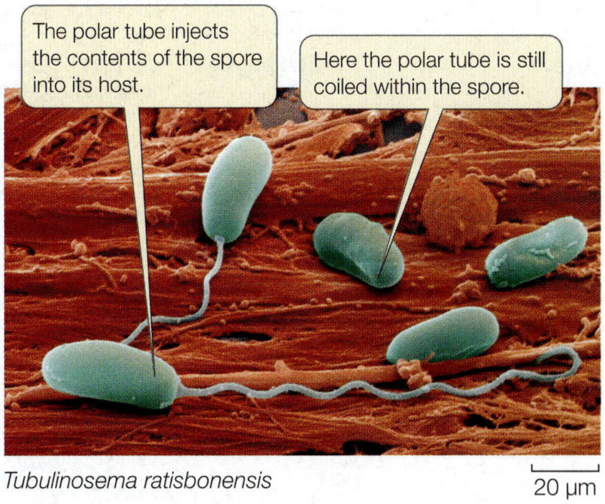
The polar tube injects the contents of the spore into its host.
Here the polar tube is still coiled within the spore.
Tubulinosema ratisbonensis
20 μm

30.12 Invasion of the Microsporidia The spores of microsporidia grow polar tubes that transfer the contents of the spores into the host's cells. The species shown here infects many animals, including humans.

Most chytrids are aquatic

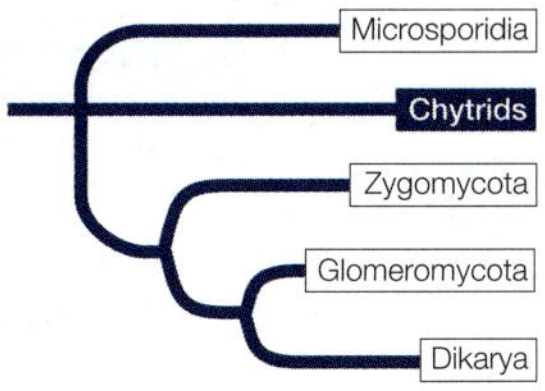
Microsporidia
Chytrids
Zygomycota
Glomeromycota
Dikarya

The **chytrids** (**Figure 30.13**) include several distinct lineages of aquatic microorganisms that were once classified with the protists. However, morphological evidence (cell walls that consist primarily of chitin) and molecular evidence support the classification of the chytrids as early-diverging fungi. In this book we use the term "chytrid" to refer to all three of the formally named clades listed as chytrids in Table 30.1, but some mycologists use this term to refer to only one of those clades, the Chytridiomycota. There are fewer than 1,000 described species among the three groups of chytrids.

Chytrids reproduce both sexually and asexually. Like the animals, chytrids that reproduce sexually possess flagellated gametes. The retention of this trait reflects the aquatic environment in which fungi first evolved. Chytrids are the only fungi that include species with flagella at any life cycle stage. Both the spores (called zoospores) and the gametes are flagellated (**Figure 30.14A**).

The alternation between multicellular haploid (n) and multicellular diploid ($2n$) generations that exists in plants and certain protist groups is seen in many fungi as well. For fungal groups other than the chytrids, this cycle differs from the usual system known as "alternation of generations" (see Figure 28.6) in that

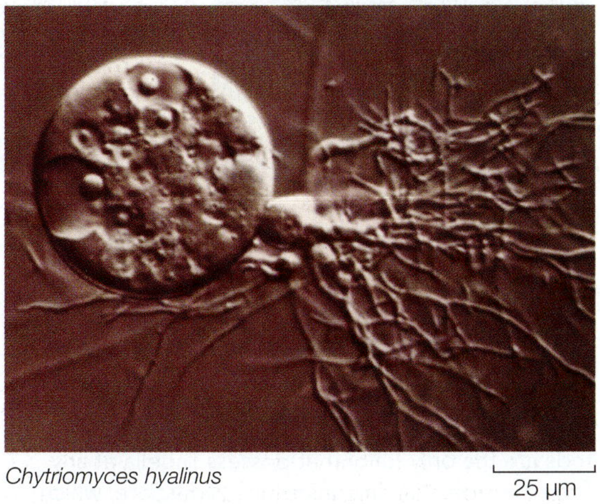
Chytriomyces hyalinus
25 μm

30.13 A Chytrid Branched rhizoids emerge from the sporangium of a mature chytrid.

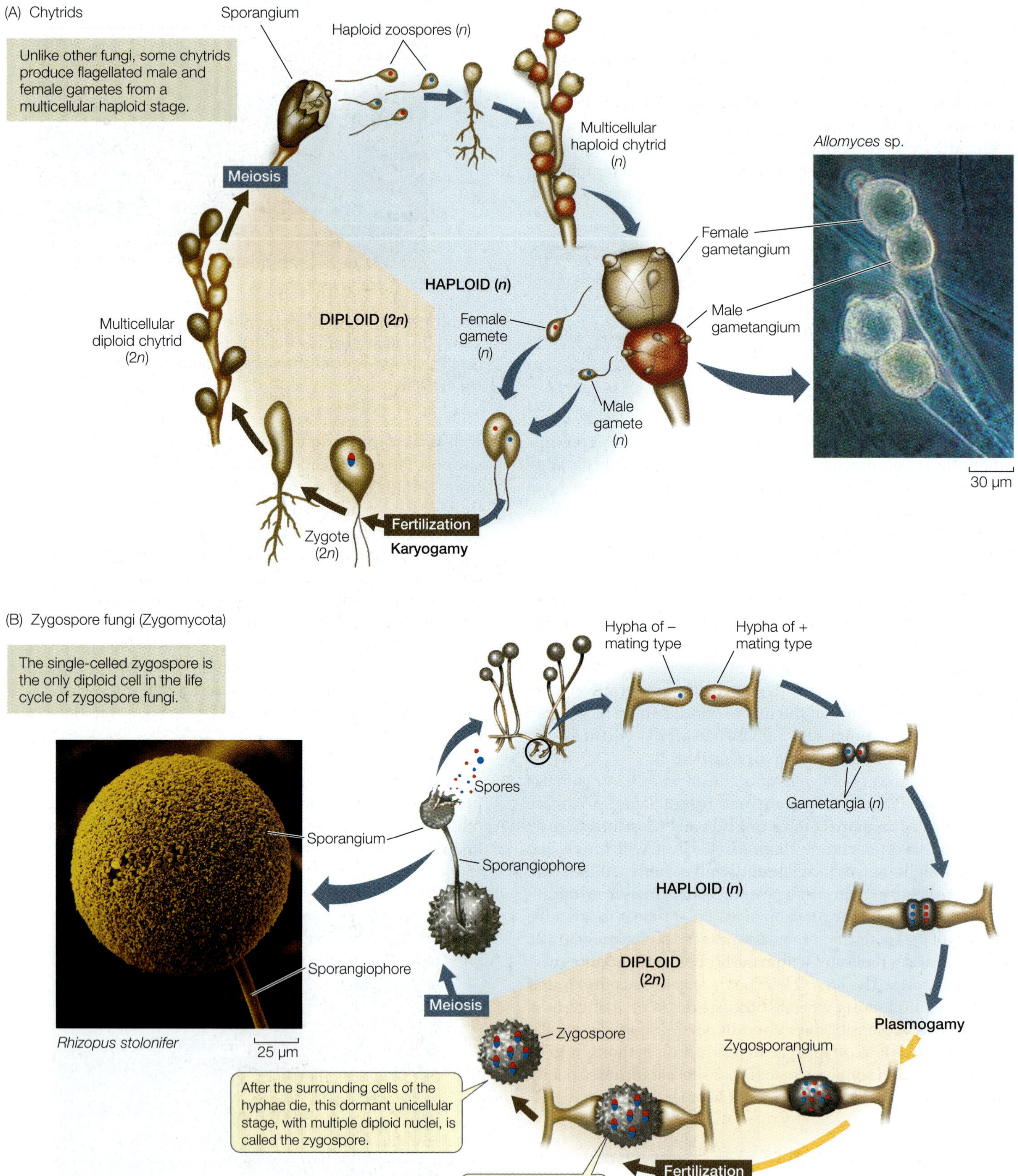
(A) Chytrids
Unlike other fungi, some chytrids produce flagellated male and female gametes from a multicellular haploid stage.
Sporangium
Haploid zoospores (n)
Multicellular haploid chytrid (n)
Allomyces sp.
Meiosis
Female gametangium
HAPLOID (n)
DIPLOID (2n)
Female gamete (n)
Male gametangium
Multicellular diploid chytrid (2n)
Male gamete (n)
30 µm
Fertilization
Zygote (2n)
Karyogamy
(B) Zygospore fungi (Zygomycota)
The single-celled zygospore is the only diploid cell in the life cycle of zygospore fungi.
Hypha of – mating type
Hypha of + mating type
Spores
Gametangia (n)
Sporangium
Sporangiophore
HAPLOID (n)
Sporangiophore
DIPLOID (2n)
Meiosis
Plasmogamy
Rhizopus stolonifer
25 µm
Zygospore
Zygosporangium
After the surrounding cells of the hyphae die, this dormant unicellular stage, with multiple diploid nuclei, is called the zygospore.
Multiple diploid nuclei form within the zygosporangium.
Fertilization
Karyogamy

30.14 Sexual Life Cycles of Chytrids and Zygospore Fungi (A) Chytrids are the only fungi that possess flagella at any stage of the life cycle. (B) The unicellular zygospore, which contains multiple diploid nuclei, is a resting structure unique to the zygospore fungi.

the multicellular haploid stages do not produce specialized male and female gametes that are independent of the hyphae. Instead, in most fungi, cells of haploid hyphae from different mating types (of which there can be many more than two) fuse to form the diploid stage. But in sexually reproducing chytrids, the multicellular haploid stage produces independent male and female gametes, both of which are flagellated (see Figure 30.14A). The male gamete is distinguished from the female gamete by its smaller size; otherwise the two gametes are very similar in form.

The chytrids are diverse in form; some are unicellular, others have rhizoids, and still others have coenocytic hyphae. They may be parasitic (on organisms such as algae, mosquito larvae, nematodes, and amphibians) or saprobic. Some have complex mutualistic relationships with foregut-fermenting animals such as cattle and deer. Many chytrids live in freshwater habitats or in moist soil, but some are marine.

Sporangia
Sporangiophores
Pilobolus sp.
150 μm

30.15 Zygospore Fungi Produce Sporangiophores These transparent structures are sporangiophores produced by a zygospore fungus growing on decomposing animal dung. The sporangiophores grow toward the light and end in tiny sporangia, which the stalked sporangiophores can eject as far as 2 meters. Animals ingest sporangia and then disseminate the spores in their feces.

Go to Animated Tutorial 30.1
Life Cycle of a Zygospore Fungus
Life10e.com/at30.1

Some fungal life cycles feature separate fusion of cytoplasms and nuclei

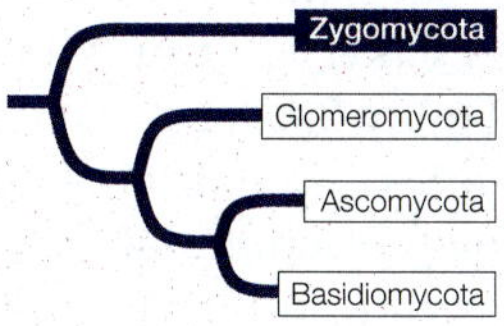
Zygomycota
Glomeromycota
Ascomycota
Basidiomycota

Most members of the remaining four groups of fungi are terrestrial. Although the terrestrial fungi grow in moist places, they do not have motile gametes, and they do not release gametes into the environment, so liquid water is not required for fertilization. Instead, the cytoplasms of two individuals of different mating types fuse (a process called **plasmogamy**) before their nuclei fuse (a process called **karyogamy**; see Figure 30.11). Sexual species of terrestrial fungi include some zygospore fungi, sac fungi, and club fungi.

Zygospore fungi reproduce sexually when adjacent hyphae of two different mating types release chemical signals that cause them to grow toward each other. These hyphae produce gametangia, which are specialized cells for reproduction that are retained as part of the hyphae. In the gametangia, nuclei replicate without cell division, resulting in multiple haploid nuclei in both gametangia. The two gametangia then fuse to form a zygosporangium that contains many haploid nuclei of each mating type (**Figure 30.14B**). Haploid nuclei of different mating types then pair up to form multiple diploid nuclei within the zygosporangium. A thick, multilayered cell wall forms around the zygosporangium to form a well-protected resting stage that can remain dormant for months. In harsh environmental conditions, this resting stage may be the only cell that survives as the surrounding cells of the hyphae die. At this stage the single surviving cell is known as a **zygospore**, which is the basis of the name of the zygospore fungi. When environmental conditions improve, the nuclei within the zygospore undergo meiosis and one or more stalked **sporangiophores** sprout, each bearing a sporangium. Each sporangium contains the products of meiosis: haploid nuclei that are incorporated into spores. These spores disperse and germinate to form a new generation of haploid hyphae.

The zygospore fungi include four major lineages of terrestrial fungi that live on soil as saprobes, as parasites of insects and spiders, or as mutualists of other fungi and invertebrate animals. They produce no cells with flagella, and only one diploid cell—the zygospore—appears in the entire life cycle. Their hyphae are coenocytic. These species do not form a fleshy fruiting structure; rather, the hyphae spread in a radial pattern from the spore, with occasional stalked sporangiophores reaching up into the air (**Figure 30.15**).

More than 1,000 species of zygospore fungi have been described. One species you may have seen is *Rhizopus stolonifer*, the black bread mold. *Rhizopus* produces many stalked sporangiophores, each bearing a single sporangium containing hundreds of minute spores (see Figure 30.14B).

Arbuscular mycorrhizal fungi form symbioses with plants

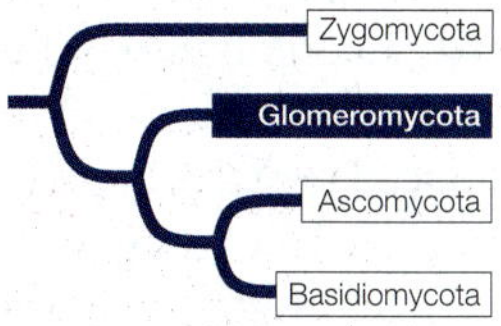
Zygomycota
Glomeromycota
Ascomycota
Basidiomycota

Arbuscular mycorrhizal fungi (Glomeromycota) are terrestrial fungi that associate with plant roots in a symbiotic, mutualistic relationship (see Figure 30.9B). Fewer than 200 species have been described, but 80 to 90 percent of all plants have associations with them. Molecular systematic studies have suggested that arbuscular mycorrhizal fungi are the sister group of the Dikarya (sac fungi and club fungi).

The hyphae of arbuscular mycorrhizal fungi are coenocytic. These fungi use glucose from their plant partners as their

30.16 Sexual Life Cycles among the Dikarya (A) In sac fungi, the products of meiosis are borne in a microscopic sac called an ascus. The fleshy fruiting structure, the ascoma, consists of both dikaryotic and haploid hyphae. (B) The basidium is the characteristic sexual reproductive structure of the club fungi. The fruiting structures, called basidiomata, consist solely of dikaryotic hyphae, and the dikaryotic phase can last a long time.

Go to Activity 30.2 Life Cycle of a Dikaryotic Fungus
Life10e.com/ac30.2

primary energy source, converting it into other, fungus-specific sugars that cannot return to the plant. Arbuscular mycorrhizal fungi reproduce asexually; there is not yet any direct evidence that they reproduce sexually.

The dikaryotic condition is a synapomorphy of sac fungi and club fungi

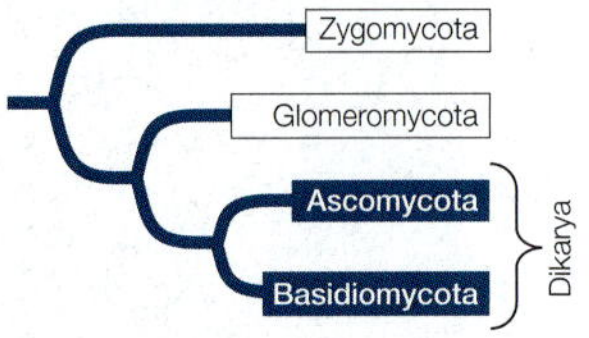
Zygomycota
Glomeromycota
Ascomycota
Basidiomycota
Dikarya

In the two remaining groups of fungi—the sac fungi and the club fungi—some stages have a nuclear configuration other than the familiar haploid or diploid states (see Figure 30.11). In these fungi, karyogamy (fusion of nuclei) occurs long after plasmogamy (fusion of cytoplasm), so that *two genetically different haploid nuclei coexist and divide within each cell of the mycelium* (**Figure 30.16**). This stage of the life cycle is called a **dikaryon** ("two nuclei"), and its ploidy is indicated as $n + n$. The dikaryon is a synapomorphy of the sac fungi and club fungi, which are placed together in a clade called **Dikarya**.

Eventually, specialized fruiting structures form, within which pairs of genetically dissimilar nuclei—one from each parent—fuse, giving rise to zygotes long after the original "mating." The diploid zygote nucleus then undergoes meiosis, producing four haploid nuclei. The mitotic descendants of those nuclei become spores, which germinate to give rise to the next haploid generation.

A life cycle with a dikaryon stage has several unusual features. First, there are no gamete *cells*, only gamete *nuclei*. Second, the only true diploid structure is the zygote, although for a long period the genes of both parents are present in the dikaryon and can be expressed. In effect, the dikaryon is neither diploid ($2n$) nor haploid (n); rather, it is dikaryotic ($n + n$). Therefore a harmful recessive mutation in one nucleus may be compensated for by a normal allele on the same chromosome in the other nucleus, and dikaryotic hyphae often have characteristics that are different from their n or $2n$ products. The dikaryotic condition is perhaps the most distinctive of the genetic peculiarities of the fungi.

The sexual reproductive structure of sac fungi is the ascus

The **sac fungi** (Ascomycota) are found in terrestrial, marine, and freshwater habitats. There are approximately 64,000 known species, nearly half of which are the fungal partners in lichens. The hyphae of sac fungi are segmented by more or less regularly spaced septa. A large pore in each septum permits movement of nuclei and organelles from one segment to the next.

Sac fungi are distinguished by the production of sacs called **asci** (singular *ascus*), which at maturity contain sexually produced haploid **ascospores** (see Figure 30.16A). The ascus is the characteristic sexual reproductive structure of the sac fungi. In the past, the sac fungi were classified on the basis of whether or not the asci are contained within a specialized fruiting structure known as an ascoma (plural *ascomata*) and on differences in the morphology of that fruiting structure. DNA sequence analyses have resulted in a revision of these traditional groupings, however.

SAC FUNGUS YEASTS Some species of sac fungi are unicellular yeasts. The 1,000 or so species in this group are among the most important domesticated fungi. Perhaps the best known is baker's, or brewer's, yeast (*Saccharomyces cerevisiae*; see Figure 30.2 and Section 30.4), which metabolizes glucose obtained from its environment into ethanol and carbon dioxide by fermentation. Other sac fungus yeasts live on fruits such as figs and grapes and play an important role in the making of wine. Many others are associated with insects; in the guts of some insects, they provide enzymes that break down materials that are otherwise difficult for the insects to digest, especially cellulose.

Sac fungus yeasts reproduce asexually by budding. Sexual reproduction takes place when two adjacent haploid cells of dissimilar mating types fuse. In some species, the resulting zygote buds to form a diploid cell population. In others, the zygote nucleus undergoes meiosis immediately; when this happens, the entire cell becomes an ascus. Depending on whether the products of meiosis then undergo mitosis, a yeast ascus contains either eight or four ascospores, which germinate to become haploid cells. The sac fungus yeasts have lost the dikaryon stage.

FILAMENTOUS SAC FUNGI Most sac fungi are filamentous species, such as the cup fungi (**Figure 30.17**), in which the ascomata are cup-shaped and can be as large as several centimeters across (although most are much smaller). The inner surfaces of the ascomata, which are covered with a mixture of specialized hyphae and asci, produce huge numbers of spores. The edible ascomata of some species, including morels and truffles, are regarded by humans as gourmet delicacies (and can sell at prices higher than gold). The underground ascomata of truffles have a strong odor that attracts mammals such as pigs, which then eat the fungi and disperse the spores.

The sexual reproductive cycle of filamentous sac fungi includes the formation of a dikaryon, although this stage is relatively brief compared with that in club fungi. Many filamentous sac fungi form multinucleate mating structures (see Figure 30.16A). Mating structures of two different mating types fuse and produce a dikaryotic mycelium, containing nuclei from both mating types. The dikaryotic mycelium often forms a cup-shaped ascoma, which bears the asci. Only after the formation of asci do the nuclei from the two mating types finally fuse.

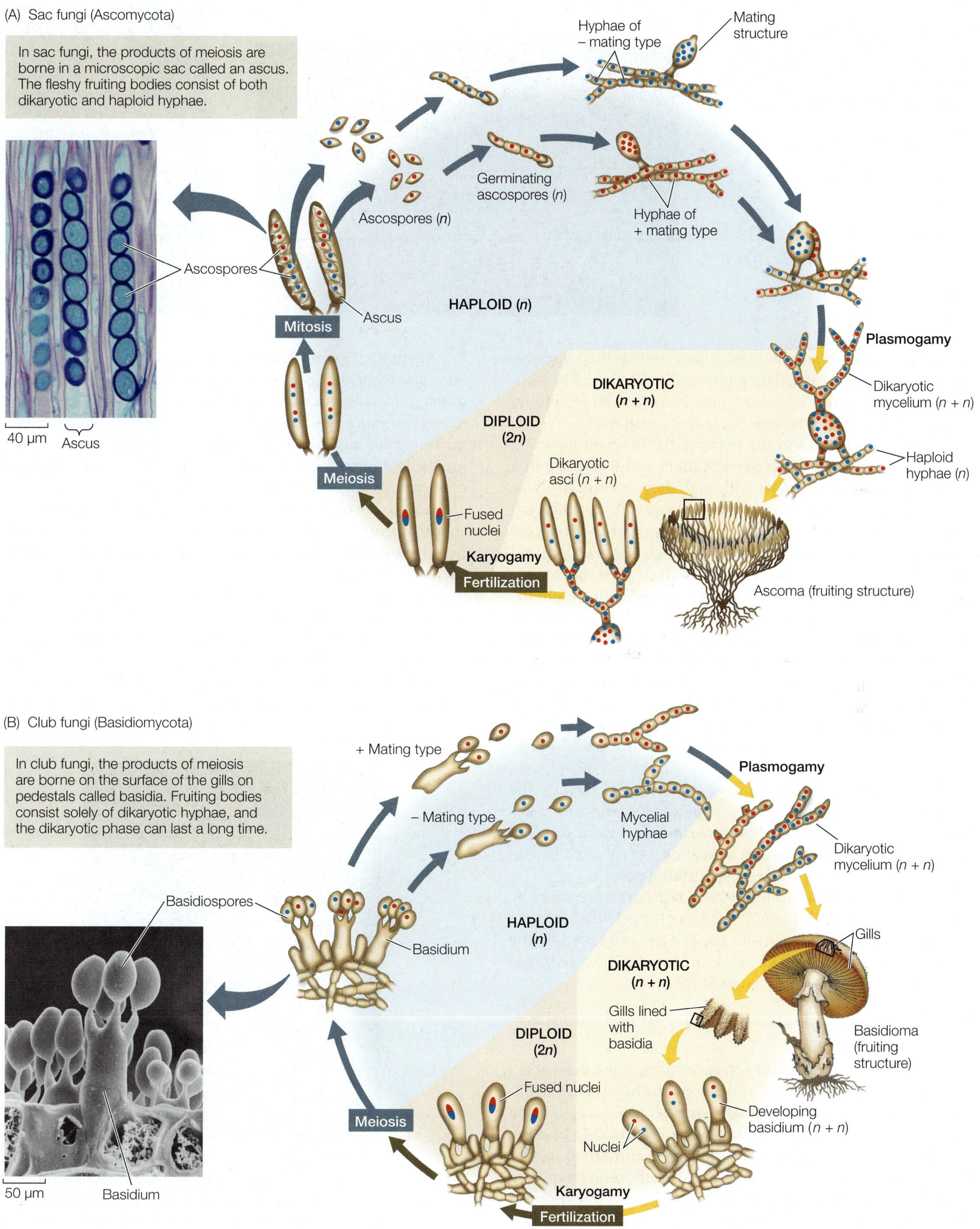
(A) Sac fungi (Ascomycota)
In sac fungi, the products of meiosis are borne in a microscopic sac called an ascus. The fleshy fruiting bodies consist of both dikaryotic and haploid hyphae.
Ascospores
40 µm
Ascus
Hyphae of – mating type
Mating structure
Germinating ascospores (n)
Ascospores (n)
Hyphae of + mating type
HAPLOID (n)
Ascus
Mitosis
Plasmogamy
Dikaryotic mycelium (n + n)
DIKARYOTIC (n + n)
DIPLOID (2n)
Haploid hyphae (n)
Dikaryotic asci (n + n)
Meiosis
Fused nuclei
Karyogamy
Fertilization
Ascoma (fruiting structure)
(B) Club fungi (Basidiomycota)
In club fungi, the products of meiosis are borne on the surface of the gills on pedestals called basidia. Fruiting bodies consist solely of dikaryotic hyphae, and the dikaryotic phase can last a long time.
+ Mating type
Plasmogamy
– Mating type
Mycelial hyphae
Dikaryotic mycelium (n + n)
Basidiospores
Basidium
HAPLOID (n)
Gills
DIKARYOTIC (n + n)
Gills lined with basidia
DIPLOID (2n)
Basidioma (fruiting structure)
Fused nuclei
Developing basidium (n + n)
Meiosis
Nuclei
Karyogamy
Fertilization
50 µm
Basidium

(A) *Aleuria aurantia*

(B) *Morchella esculenta*

30.17 Sac Fungi (A) These brilliant red cups are the ascomata of a cup fungus. (B) Morels, which have a spongelike ascoma and a subtle flavor, are considered a culinary delicacy by humans.

Both nuclear fusion and the subsequent meiosis that produces haploid ascospores take place within individual asci. The ascospores are ultimately released (sometimes shot off forcefully) by the ascus to begin the new haploid generation.

The sac fungi also include many of the filamentous fungi known as molds. **Molds** consist of filamentous hyphae that do not form large ascomata, although they can still produce asci and ascospores. Many molds are parasites of flowering plants. Chestnut blight and Dutch elm disease are both caused by molds. The chestnut blight fungus, which was introduced to the United States in the 1890s, had destroyed the American chestnut as a commercial tree species by 1940. Before the blight, this species accounted for more than half the trees in eastern U.S. forests. Another familiar story is that of the American elm. Sometime before 1930, the Dutch elm disease fungus (first discovered in the Netherlands but native to Asia) was introduced into the United States on infected elm logs from Europe. Spreading rapidly—sometimes by way of connected root systems—the fungus destroyed great numbers of American elm trees.

Other plant pathogens among the sac fungi include the powdery mildews that infect cereal crops, lilacs, and roses, among many other plants. Mildews can be a serious problem to farmers and gardeners, and a great deal of research has focused on ways to control these agricultural pests.

The filamentous sac fungi can also reproduce asexually by means of conidia that form at the tips of specialized hyphae (**Figure 30.18**). Small chains of conidia are produced by the millions and can survive for weeks in nature. The conidia are what give molds their characteristic colors.

The sexual reproductive structure of club fungi is the basidium

Club fungi (Basidiomycota) produce some of the most spectacular fruiting structures found among the fungi. These fruiting structures, called **basidiomata** (singular *basidioma*), include mushrooms of all kinds, puffballs (see Figure 30.4), and the bracket fungi often encountered on trees and fallen logs in a damp forest. About 30,000 species of club fungi have been described. They include about 4,000 species of mushrooms, including both poisonous and edible species (**Figure 30.19A**). Bracket fungi (**Figure 30.19B**) play an important role in the carbon cycle by breaking down wood; they also do great economic damage to both cut lumber and timber stands. Some of the most economically damaging plant pathogens are club fungi, including the rust fungi and smut fungi that parasitize cereal grains. In contrast, other club fungi contribute to the survival of plants as fungal partners in ectomycorrhizae.

The hyphae of club fungi characteristically have septa with small, distinctive pores. As they grow, haploid hyphae of different mating types meet and fuse, forming dikaryotic hyphae, each cell of which contains two nuclei, one from each parent hypha. The dikaryotic mycelium grows and eventually, when triggered by rain or another environmental cue, produces a basidioma. The dikaryon stage may persist for years—some club fungi live for decades or even centuries. This pattern contrasts with the life cycle of the sac fungi, in which the dikaryon is found only in the stages leading up to formation of the asci.

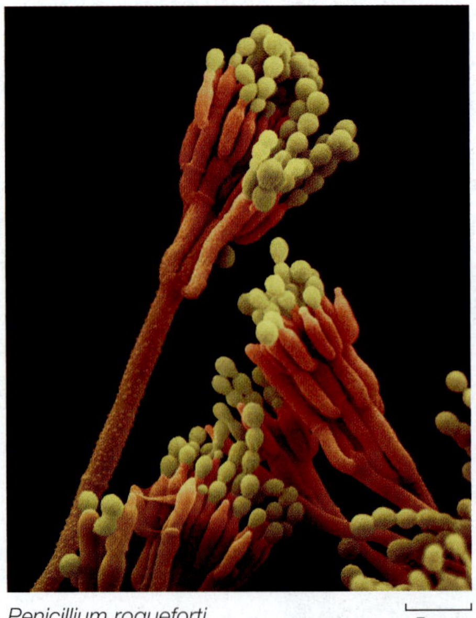
5 μm

Penicillium roqueforti

30.18 Conidia Chains of conidia (yellow) are developing at the tips of specialized hyphae arising from a *Penicillium roqueforti* mold. This species is used to produce "blue" cheese.

(A) *Armillaria* sp.

(B) *Laetiporus sulphureus*

30.19 Club Fungus Basidiomata
(A) These edible mushrooms are the fruiting structures of a honey fungus. Mycelia of this genus form connected underground masses that cover many hectares and are among the largest and longest-lived organisms in the world. (B) A bracket fungus growing parasitically on a tree. Although this particular species is edible, many similar-appearing bracket fungi are poisonous.

The **basidium** (plural *basidia*), a swollen cell at the tip of a specialized hypha, is the characteristic sexual reproductive structure of the club fungi (see Figure 30.16B). In mushroom-forming club fungi, the basidia typically form on specialized structures of the basidiomata known as gills. The basidium is the site of nuclear fusion and meiosis and thus plays the same role in the club fungi as the ascus does in the sac fungi and the zygosporangium does in the zygospore fungi.

After nuclei fuse in the basidium, the resulting diploid nucleus undergoes meiosis, and the four resulting haploid nuclei are incorporated into haploid **basidiospores**, which form on tiny stalks on the outside of the basidium. A single basidioma of the common bracket fungus *Ganoderma applanatum* can produce as many as 4.5 *trillion* basidiospores in one growing season. Basidiospores typically are forcibly discharged from their basidia and then germinate, and give rise to hyphae with haploid nuclei.

RECAP 30.3

Sexual reproduction is common in some groups of fungi but has never been observed in others. Many fungal species have two or more genetically distinct mating types. The sac fungi and club fungi share a dikaryotic condition, in which two genetically different haploid nuclei coexist in the same cell.

- Explain how microsporidia infect the cells of their animal hosts. **See p. 617 and Figure 30.12**
- What is the role of the zygospore in the life cycle of zygospore fungi? **See p. 619 and Figure 30.14B**
- Explain the dikaryon stage in terms of plasmogamy and karyogamy. **See p. 620 and Figure 30.16**
- What distinguishes the fruiting bodies of sac fungi from those of club fungi? **See pp. 620–623 and Figure 30.16**

Fungi are of special interest to biologists because of the roles they play in interactions with other organisms. But they are also useful as tools for studying many kinds of biological problems and for finding solutions to those problems.

30.4 What Are Some Applications of Fungal Biology?

We've briefly noted the important part that fungi play in the production of human foods and beverages. We have also described the diverse roles that fungi play in natural ecosystems, from decomposers to pathogens to mutualistic partners. These diverse ecological roles have led to the use of fungi in studies of environmental change and in remediation of environmental pollution. Many fungi are also important model organisms for laboratory investigations of basic biological process. Others, as we saw at the opening of this chapter, have given us treatments for human diseases.

Fungi are important in producing food and drink

Grains from grasses provide most of the world's food supply for humans. But in most cases, we do not eat these grains directly as they are produced by the plants. Instead, we use them as a source of starch. To make the starch more pleasing and digestible for human consumption, we usually convert it to more complex and tasty forms of food and drink, often with the help of fungi.

Baker's (or brewer's) yeast (*Saccharomyces cerevisiae*) converts the starch from grain into ethanol. This process also forms carbon dioxide bubbles in bread dough, causing it to rise, which gives baked bread its light texture. The ethanol and carbon dioxide are baked away in bread making (which produces the pleasant aroma of baking bread). In contrast, the ethanol and carbon dioxide are retained when yeast is used to ferment grain into beer. The carbon dioxide gives beer its fizz, and the alcohol contributes to the taste and appeal of beer to those who enjoy it. Sugars, especially from fruit such as grapes, are also converted into alcohol and carbon dioxide by yeasts in the production of wine (although the carbon dioxide is not retained in finished wine, as it is in beer). Many different strains of *S. cerevisiae* are used in wine production, which contributes to the distinctive nature of wine from different regions and wineries. Many other species of local, native yeasts are also used in producing distinctive local wines and beers. For example, fission yeast (*Schizosaccharomyces pombe*) was first isolated from African millet beer; it takes its specific name (*pombe*) from the Swahili word for beer.

Go to Media Clip 30.3
Time Lapse of Beer Formation
Life10e.com/mc30.3

Brown molds of the genus *Aspergillus* are important in some human diets. *Aspergillus tamarii* acts on soybeans in the production of soy sauce, and *A. oryzae* is used in brewing the Japanese alcoholic beverage sake from rice. *Aspergillus niger* is the source of most commercial citric acid production; citric acid gives food and soft drinks a tart taste and is also used as a food preservative. But some species of *Aspergillus* that grow on grains and on nuts such as peanuts and pecans produce extremely carcinogenic (cancer-inducing) compounds called aflatoxins. Aflatoxins can occur in high concentrations in foods such as peanut butter. In the United States and most other industrialized countries, moldy grain infected with *Aspergillus* is typically thrown out. In Africa, where food is scarcer, the grain gets eaten, moldy or not, and causes severe health problems, including high levels of certain cancers.

Penicillium is a genus of green molds, of which some species produce the antibiotic penicillin, as described in the beginning of this chapter. But several species of *Penicillium* are important for food production as well. For example, *P. camembertii* and *P. roqueforti* are the organisms responsible for the characteristic strong flavors of Camembert and Roquefort cheeses, respectively.

Many fungi serve directly as a human food source. Mushroom enthusiasts seek out the delicious fruiting structures of a wide variety of edible sac and club fungi. In the United States, relatively few species of mushrooms are grown commercially, and wild mushrooms are collected mostly for personal consumption. But in many parts of the world, a wide variety of wild mushrooms are collected for sale and consumption. Fungi used for food are not limited to fruiting bodies such as mushrooms, however. Various species of lichens are eaten in Arctic regions as well as in parts of North America and Asia. In southwestern China, for example, several species of lichens are used as a primary ingredient in cooking (**Figure 30.20**).

30.20 Some Lichens Are Edible In southwestern China, several species of lichens that grow on tree bark serve as a primary ingredient in the local cuisine.

Fungi record and help remediate environmental pollution

Each year, biologists deposit samples of many groups of organisms in the collections of natural history museums. These museum collections serve many purposes, one of which is to document changes in the biota of our planet over time.

Collections of fungi made over many decades or centuries provide a record of the environmental pollutants that were present when the fungi were growing. Biologists can analyze these historical samples to see how different sources of pollutants were affecting our environment before anyone thought to take direct measurements. These long-term records are also useful for analyzing the effectiveness of cleanup efforts and regulatory programs for controlling environmental pollutants.

We have already seen that fungi are critical to the planetary carbon cycle because of their role in breaking down dead organic matter. Fungi are also used in remediation efforts to help clean up sites that have been polluted by oil spills or contaminated with toxic petroleum-derived hydrocarbons. Many herbicides, pesticides, and other synthetic hydrocarbons are broken down primarily through the action of fungi.

Lichen diversity and abundance are indicators of air quality

Lichens can live in many harsh environments where few other species can survive, as we saw in Section 30.2. In spite of their hardiness, however, lichens are highly sensitive to air pollution because they are unable to excrete any toxic substances they absorb. This sensitivity means that lichens are good biological indicators of air pollution levels. It also explains why they are not commonly found in heavily industrialized regions or in large cities.

Sensitive biological indicators of pollution such as lichen growth allow biologists to monitor air quality without the use of specialized equipment. Monitoring the diversity and abundance of lichens growing on trees is a practical and inexpensive system for gauging air quality around cities (**Figure 30.21**). Maps of lichen diversity provide environmental biologists with a tool for tracking the spatial distribution of air pollutants and their effects. Lichens can also provide a long-term measure of the effects of air pollution across many seasons and years.

Fungi are used as model organisms in laboratory studies

Much of what we know about many basic aspects of cell and molecular biology comes from the study of model organisms. Among the eukaryotes, some fungi have numerous advantages over model plant and animals systems for laboratory investigations.

Of particular importance as model organisms are several species of sac fungi: *Aspergillus nidulans* (a brown mold), *Neurospora crassa* (a red bread mold), *Saccharomyces cerevisiae* (baker's, or brewer's, yeast), and *Schizosaccharomyces pombe* (fission yeast). These species can be cultured in large numbers in small spaces, and they have short generation times, so that genetic

WORKING WITH**DATA:**

Using Fungi to Study Environmental Contamination

Original Paper

Flegal, A. R., C. Gallon, S. Hibdon, Z. E. Kuspa, and L. F. Laporte. 2010. Declining—but persistent—atmospheric contamination in central California from the resuspension of historic leaded gasoline emissions as recorded in the lace lichen (*Ramalina menziesii* Taylor) from 1892 to 2006. *Environmental Science and Technology* 44: 5613–5618.

Analyze the Data

A. Russell Flegal and his colleagues analyzed over 100 years' worth of museum samples of lace lichens collected near San Francisco, California for evidence of lead contamination. They measured concentrations of lead (Pb) as well as the ratios of two lead isotopes, ^{206}Pb and ^{207}Pb. The isotope ratio was used to determine the source of lead contamination. Possible sources included a lead smelter that operated in the area from 1885 to 1971 (which produced emissions with a $^{206}Pb/^{207}Pb$ ratio of about 1.15–1.17); leaded gasoline in use from the 1930s to the early 1980s, peaking in 1970 (which produced automobile emissions with a $^{206}Pb/^{207}Pb$ ratio of 1.18–1.23); and resuspension of historic lead contamination as atmospheric aerosols in recent decades (with an intermediate $^{206}Pb/^{207}Pb$ ratio of about 1.16–1.19).

Before analyzing the data, use the information in the preceding paragraph to formulate hypotheses about the following two questions: What trends in atmospheric lead concentrations would you expect to see? What $^{206}Pb/^{207}Pb$ ratios would you expect to find at different times from the late 1800s to the early 2000s? After formulating your hypotheses, plot lead concentration in the lichen samples against year of sample collection. Make a second plot, this one of $^{206}Pb/^{207}Pb$ ratio against year of sample collection.

Sample	Year collected	Lead concentration (µg of Pb/g lichen)	$^{206}Pb/^{207}Pb$ ratio
1	1892	11.9	1.165
2	1894	4.0	1.155
3	1906	13.7	1.154
4	1907	22.9	1.157
5	1945	49.9	1.187
6	1957	34.2	1.185
7	1978	50.9	1.221
8	1982	10.0	1.215
9	1983	4.6	1.224
10	1987	1.0	1.198
11	1988	1.3	1.199
12	1995	1.9	1.202
13	2000	0.4	1.184
14	2006	1.8	1.184

QUESTION 1

Do your analyses support the hypotheses you formulated?

QUESTION 2

Are your hypotheses consistent with your analyses of trends in both lead concentrations and $^{206}Pb/^{207}Pb$ ratios through time? If not, how would you modify your hypotheses, and what additional tests can you design to test your ideas?

Go to **BioPortal** for all WORKING WITH**DATA** exercises

(A) (B)

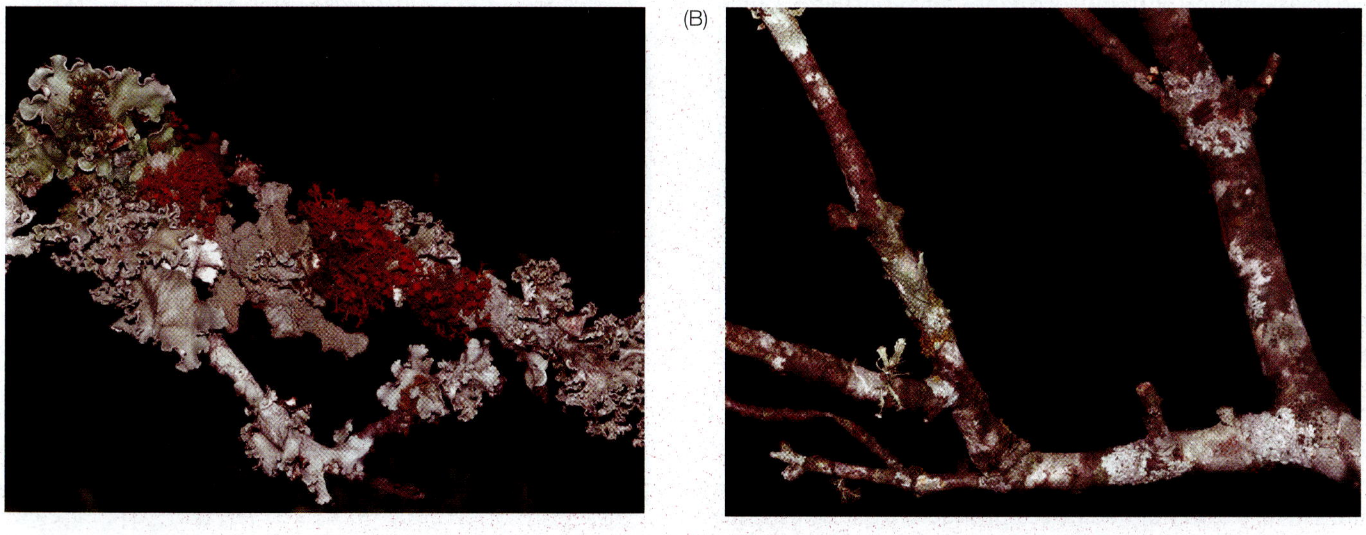

30.21 More Lichens, Better Air Lichen abundance and diversity are excellent indicators of air quality. (A) In suitable environments with few pollutants in the air, many lichen species show luxuriant growth on trees. These lichens are growing on an oak tree 150 kilometers west of Austin, Texas, in an area relatively free of human habitation and industry. (B) As air quality declines, so do the number and diversity of lichens. These lichens are growing on an oak tree inside the city of Austin, where air quality is reduced (primarily by automobile exhaust).

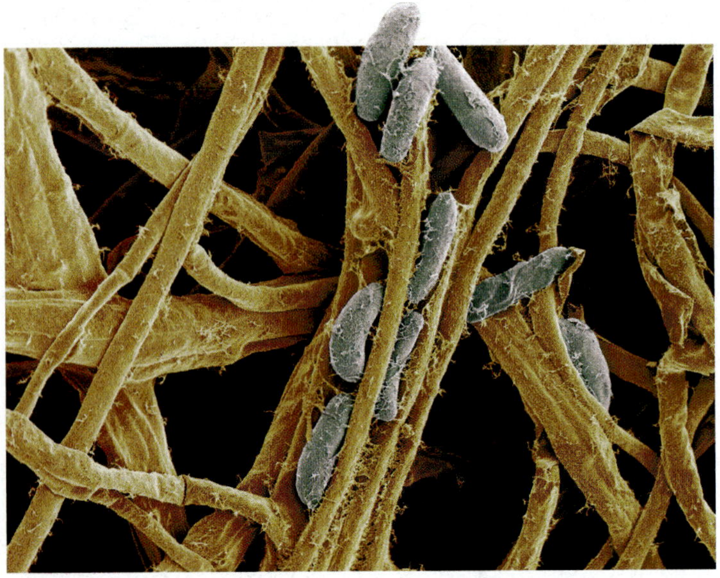

30.22 A Pathogenic Fungus Attacks a Parasitic Plant The fungus *Fusarium oxysporum* is a potent pathogen of witchweed (*Striga*), a parasitic plant that attacks crops. The fungal spores are shown in blue; fungal filaments are in tan. Both colors were added to enhance this scanning electron micrograph.

investigations can be conducted in days rather than years. Furthermore, their genomes are relatively small and encode relatively few genes compared with those of most plants and animals, so it is easier to elucidate the functions of the fungal genes responsible for basic biological functions.

Reforestation may depend on mycorrhizal fungi

When a forest is cut down, it is not just the trees that are lost. A forest is an ecosystem that depends on the interactions of many species. As we have discussed, many plants depend on close relationships with mycorrhizal fungal partners. When trees are removed from a site, the populations of mycorrhizal fungi there decline rapidly. If we wish to restore the forest on the site, we cannot simply replant it with trees and other plants and expect them to survive. The mycorrhizal fungal community must be reestablished as well. For large forest restoration projects, a planned succession of plant growth and soil improvement is often necessary before forest trees can be replanted. As the community of soil fungi gradually recovers, trees that have been inoculated with appropriate mycorrhizal fungi in tree nurseries can be planted to reintroduce greater diversity to the soil fungal community.

Fungi provide important weapons against diseases and pests

We started this chapter with the story of the discovery of penicillin. The discovery of antibiotics produced by fungi revolutionized medical treatment of bacterial diseases in humans and their domestic animals. Live strains of fungi are also used to combat various pest species of plants and animals.

In Africa, the parasitic plant witchweed (*Striga*) causes crop losses of about $7 billion every year. A group of Canadian biologists discovered that a strain of the mold *Fusarium oxysporum* could be applied to crops to control witchweed without harming the crop plants (Figure 30.22). Other strains of *Fusarium* that preferentially attack coca plants, the source of cocaine, have been proposed to combat illicit drug production. Still other fungi are used to attack various animal pests, such as termites and aphids, and even malaria-carrying mosquitoes, as we'll see below.

RECAP 30.4

Fungi are important to humans in many ways. Some species are consumed directly as food, while others are important in food production. Fungi serve as important indicators of ecosystem health and are critical in reforestation and in pollution remediation efforts. Several species are important model organisms for studies of eukaryote cell and molecular biology. Fungi are widely used to combat diseases and pests.

- What are some of the ways in which fungi or fungal products contribute to the human food supply? **See pp. 623–624**
- What are some advantages of using surveys of lichen diversity and museum collections of lichens to measure long-term changes in air quality, compared with direct measurements of atmospheric pollutants? **See pp. 624–625 and Figure 30.21**
- Why are some fungi particularly appropriate as model organisms for the study of eukaryote cell and molecular biology? **See pp. 625–626**

Whether living on their own or in symbiotic associations, fungi have spread successfully over much of Earth since their origin from a protist ancestor. An earlier ancestor of fungi also gave rise to the choanoflagellates and the animals, which we will describe in the following three chapters.

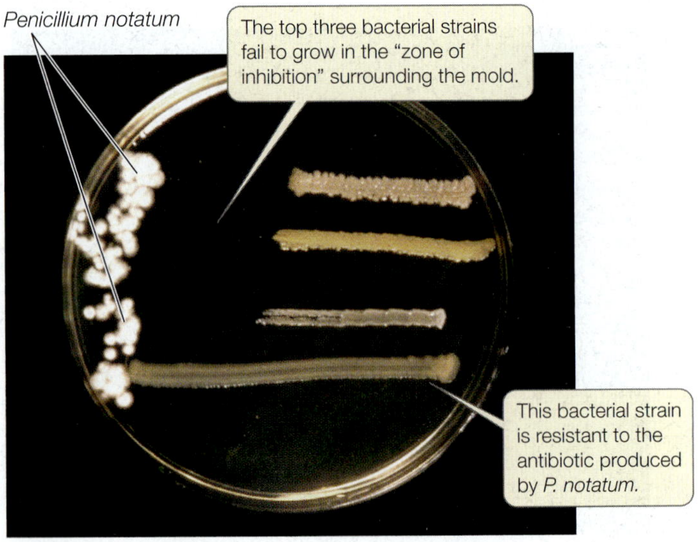
Penicillium notatum
The top three bacterial strains fail to grow in the "zone of inhibition" surrounding the mold.
This bacterial strain is resistant to the antibiotic produced by *P. notatum*.

30.23 Penicillin Resistance In a petri dish similar to those in Alexander Fleming's lab, four strains of a pathogenic bacterium have been cultured along with *Penicillium* mold. One strain is resistant to the mold's antibiotic substance, as is evidenced by its growth up to the mold.

Have antibiotics derived from fungi eliminated the danger of bacterial diseases in human populations?

ANSWER

Beginning in the 1940s, antibiotics derived from fungi ushered in a "golden age" of freedom from bacterial infections. Today, however, many of these antibiotics are losing their effectiveness as pathogenic bacteria evolve resistance to them (**Figure 30.23**)

Most medical antibiotics are chemically modified forms of substances that are found naturally in fungi and other organisms. Fungi naturally produce antibiotic compounds to defend themselves against bacterial growth and to reduce competition from bacteria for nutritional resources. These naturally occurring compounds are usually chemically modified to increase their stability, to improve their effectiveness, and to facilitate synthetic production. From the late 1950s to the late 1990s, no new major classes of antibiotics were discovered. In recent years, however, three new classes of antibiotics have been synthesized based on information learned from naturally occurring, fungally derived antibiotics, leading to improved treatment of some formerly resistant strains of bacteria.

Fungi have also been used to combat non-bacterial diseases. One of the more unusual applications of fungi is in the war against malaria. Biologists have discovered that two species of fungi, *Beauveria bassiana* and *Metarhizium anisopliae,* can kill malaria-causing mosquitoes when applied to mosquito netting. Mosquitoes have not yet shown evidence of developing resistance to these biological pathogens.

CHAPTER**SUMMARY** 30

30.1 What Is a Fungus?

- Fungi are distinguished from other **opisthokonts** by **absorptive heterotrophy** and by the presence of **chitin** in their cell walls. **Review Figure 30.1**
- Some fungi are **saprobes**, others are parasites, and some are mutualists.
- **Yeasts** are unicellular, free-living fungi.
- The body of a multicellular fungus is a **mycelium**—a meshwork of filaments called **hyphae.** Hyphae may be **septate** (having **septa**) or **coenocytic** (multinucleate). **Review Figure 30.3**

30.2 How Do Fungi Interact with Other Organisms?

- Saprobic fungi, which act as decomposers, make crucial contributions to the recycling of elements, especially carbon.
- Many fungi are parasites, harvesting nutrients from host cells by means of **haustoria**. **Review Figure 30.5**
- Certain fungi have relationships with other organisms that are **symbiotic** and **mutualistic**.
- Some fungi associate with unicellular green algae, cyanobacteria, or both to form **lichens**, which live on exposed surfaces of rocks, trees, and soil. **Review Figure 30.8**
- **Mycorrhizae** are mutualistic associations of fungi with plant roots. They improve a plant's ability to take up nutrients and water.
- **Endophytic fungi** live within plants and may provide their hosts with protection from herbivores and pathogens.

30.3 How Do Major Groups of Fungi Differ in Structure and Life History?

- The **microsporidia**, **chytrids**, and **zygospore fungi** diversified early in fungal evolution. The arbuscular mycorrhizal fungi, sac fungi, and club fungi form a monophyletic group, and the latter two groups form the clade Dikarya. **Review Figure 30.10, Table 30.1, ACTIVITY 30.1**
- Many species of fungi reproduce both sexually and asexually. In many fungi, sexual reproduction occurs between individuals of different **mating types**. **Review Figure 30.11**
- The **microsporidia** are highly reduced unicellular fungi. They are obligate intracellular parasites of animals.
- The three distinct lineages of **chytrids** all include species with flagellated gametes. **Review Figure 30.14A**
- In the sexual reproduction of terrestrial fungi, **plasmogamy** (fusion of cytoplasm) precedes **karyogamy** (fusion of nuclei).
- **Zygospore fungi** have a resting stage known as a **zygospore**, which contains many diploid nuclei. Their spores are dispersed from simple stalked **sporangiophores**. **Review Figure 30.14B, ANIMATED TUTORIAL 30.1**
- **Arbuscular mycorrhizal fungi** form symbiotic associations with plant roots. They are only known to reproduce asexually. Their hyphae are coenocytic.
- In sac fungi and club fungi, a mycelium containing two genetically different haploid nuclei, called a **dikaryon**, is formed. The **dikaryotic** ($n + n$) condition is unique to the fungi. **Review Figure 30.16, ACTIVITY 30.2**
- **Sac fungi** have septate hyphae with large pores; their sexual reproductive structures are **asci**. Some sac fungi are unicellular yeasts. Many filamentous sac fungi produce fleshy fruiting structures called **ascomata**. The dikaryon stage in the sac fungus life cycle is relatively brief. **Review Figure 30.16A**
- **Club fungi** have septate hyphae with distinctive small pores. Their fruiting structures are called **basidiomata**, and their sexual reproductive structures are **basidia**. The dikaryon stage may last for years. **Review Figure 30.16B**

30.4 What Are Some Applications of Fungal Biology?

- Some fungi are consumed as food by humans; other fungi are critical in baking, fermentation, and flavoring food.
- Fungi play important roles in cleaning up environmental pollutants such as synthetic petroleum-derived hydrocarbons.

continued

- The diversity and abundance of lichen growth on trees is a sensitive indicator of air quality.
- Reforestation projects require restoration of the mycorrhizal fungal community.
- Several species of fungi are important model organisms.
- Fungi provide important weapons against diseases and pests.

Go to the Interactive Summary to review key figures, Animated Tutorials, and Activities
Life10e.com/is30

CHAPTER REVIEW

REMEMBERING

1. The absorptive heterotrophy of fungi is aided by
   a. dikaryon formation.
   b. spore formation.
   c. the fact that they are all parasites.
   d. their large surface area-to-volume ratio.
   e. their possession of chloroplasts.

2. Which of the following is a reason that lichens can be useful indicators of environmental change?
   a. Lichens excrete the toxic substances that they absorb, and these excretions can be used to measure local contamination.
   b. Because lichens retain toxins, historical collections of lichens can be used to measure past atmospheric conditions.
   c. Atmospheric pollutants (such as ^{206}Pb) stimulate lichen growth, so heavy lichen growth is a sign of air pollution.
   d. Atmospheric pollution stimulates mycorrhizal activity, which inhibits lichen growth.
   e. None of the above

3. Which statement about the dikaryon stage is *not* true?
   a. The cytoplasm of two cells fuses before their nuclei fuse.
   b. The two haploid nuclei are genetically different.
   c. The two nuclei are of the same mating type.
   d. The dikaryon stage ends when the two nuclei fuse.
   e. Not all fungi have a dikaryon stage.

4. Microsporidia
   a. lack true mitochondria.
   b. are parasites of animals.
   c. contain mitosomes.
   d. are among the smallest eukaryotes known.
   e. All of the above

5. Which statement about lichens is *not* true?
   a. They can reproduce by fragmentation of the vegetative body.
   b. They are often the first colonists in a new area.
   c. They render their environment more basic (alkaline).
   d. They contribute to soil formation.
   e. They may contain less than 10 percent water by weight.

UNDERSTANDING & APPLYING

6. You are shown an object that looks superficially like a pale green mushroom. Describe at least three criteria (including anatomical and chemical traits) that would enable you to tell whether the object is a piece of a plant or a piece of a fungus.

7. Many fungi are dikaryotic during part of their life cycle. Why are dikaryons described as $n + n$ instead of $2n$?

8. If all the fungi on Earth were suddenly to die, how would the surviving organisms be affected?

ANALYZING & EVALUATING

9. Consider the following data for lichens at five survey sites:

Site number	Number of lichen species	Tree branches covered in lichens (%)
1	5	38
2	1	2
3	3	15
4	8	75
5	13	100

   Predict the relative order of the sites with respect to their distance from the center of a large city. Other factors (besides distance to city center, such as prevailing wind direction) might affect your prediction. Can you think of two other major factors that might influence these results?

10. When biologists isolate DNA from whole-plant samples to study plant genomes, the investigators sometimes find that some of the genes they isolate appear to be more closely related to fungal genes than they are to the genes of close relatives of the plants they are studying. What is a likely explanation of this observation? How could you test your hypothesis?

Go to BioPortal at yourBioPortal.com for Animated Tutorials, Activities, LearningCurve Quizzes, Flashcards, and many other study and review resources.

31 Animal Origins and the Evolution of Body Plans

CHAPTER OUTLINE

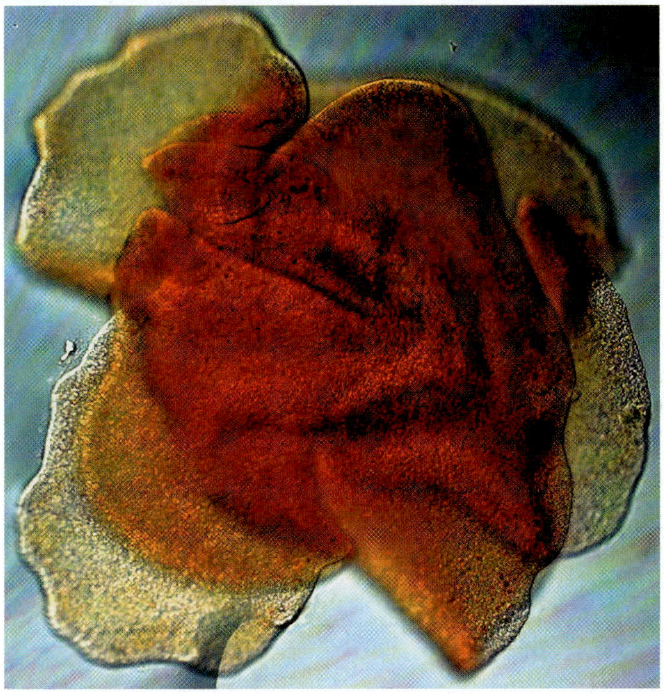

Did Placozoans Diverge at the Root of the Animal Tree? Answering that question requires an understanding of animal phylogeny. Although placozoans are morphologically simple and have the smallest genome of any animal studied to date, they may have descended from ancestors with more complex body plans.

IN 1883 THE ZOOLOGIST FRANZ SCHULZE noticed something unusual in his Austrian laboratory: transparent, flattened organisms were crawling on the sides of his saltwater aquarium. These organisms, which Schulze had collected accidentally along with the sponges that were his primary interest, appeared to be animals, but they were unlike any animals previously described—especially since they continually changed shape as they moved.

Further examination revealed that the new organisms were indeed animals. Structurally, however, they were among the simplest animals that Schulze—or anyone else—had ever observed, being made up of only four types of cells. He named the new species *Trichoplax adhaerens*, which means "sticky hairy plate," and argued that the new species had no close relationships with other major animal groups. For decades, however, most biologists dismissed Schulze's findings, insisting that the transparent organisms must be larval forms of other, well-known, animals.

In the 1960s more detailed studies confirmed the distinctive nature of *Trichoplax*. Even then, this odd animal continued to be known almost exclusively from aquariums. Only in the past decade have biologists located natural populations of *T. adhaerens* growing on hard surfaces in tropical and subtropical coastal regions. A few additional closely related species have been discovered (although most have not yet been formally named). Collectively, these species are known as placozoans (Greek, "flat animals").

The more biologists have studied *Trichoplax*, the odder this animal appears. It has the smallest genome of any animal studied to date. The mature stages lack body symmetry and have no mouth, gut, or nervous system. Is *Trichoplax* a relict representative of a group of animals that appeared early in animal evolution?

Which groups of organisms are at the root of the animal tree is currently a subject of considerable investigation and debate. Several possible hypotheses of relationships are being explored with genomic analyses. The structural simplicity of *Trichoplax* is now considered by most biologists to be an evolutionary reversal from a more complex body form. Most genomic studies point to other groups of animals as forming the earliest split with the remaining species.

This chapter explores the earliest branches on the animal tree and shows how a few fundamental "body plans" have been modified to yield the remarkable variety of animal forms described in this and the following two chapters.

> ? Which animal groups are involved in the earliest split in the animal tree?
>
> See answer on p. 648.

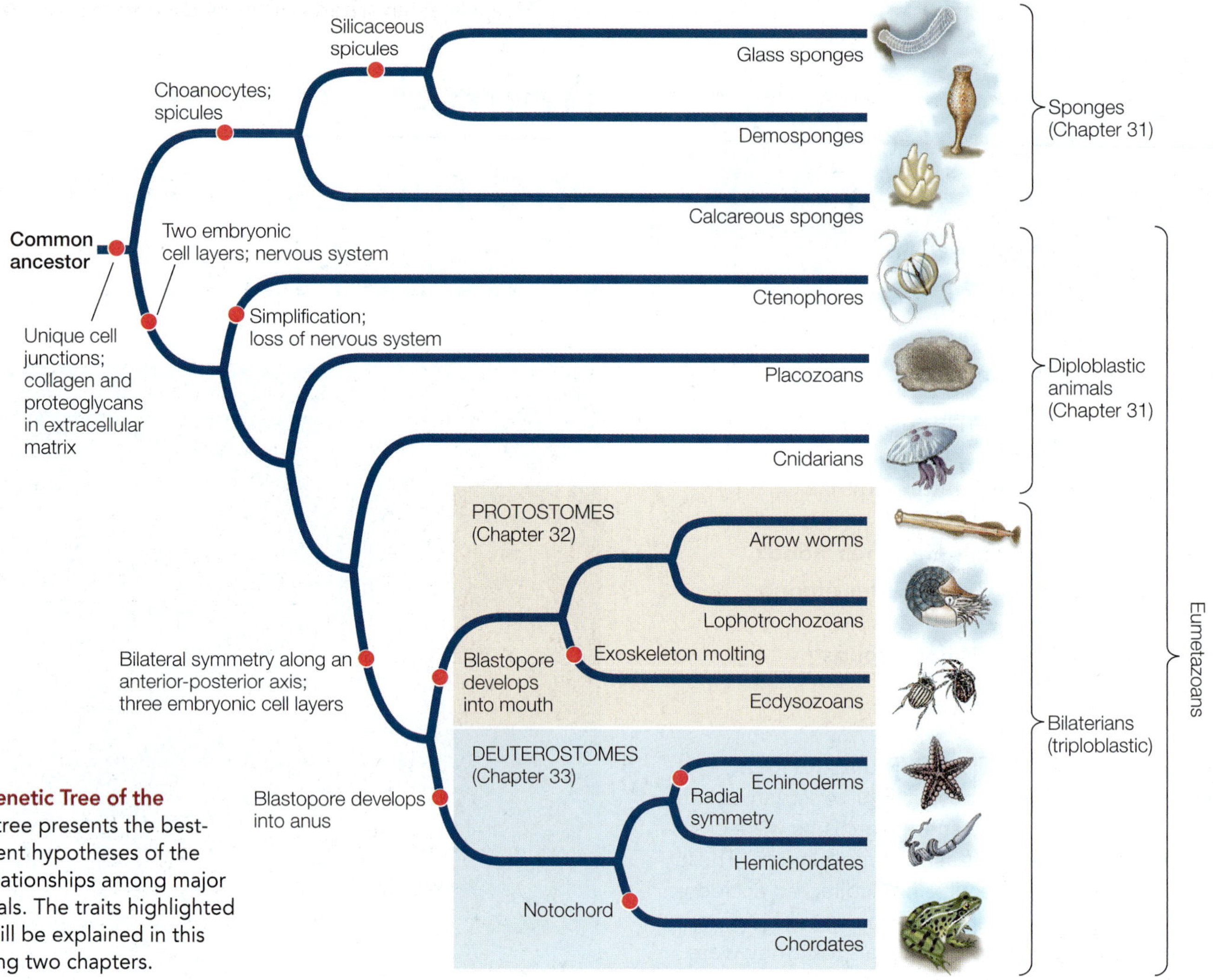

Silicaceous spicules
Choanocytes; spicules
Glass sponges
Demosponges
Calcareous sponges
Sponges (Chapter 31)
Common ancestor
Two embryonic cell layers; nervous system
Unique cell junctions; collagen and proteoglycans in extracellular matrix
Ctenophores
Simplification; loss of nervous system
Placozoans
Cnidarians
Diploblastic animals (Chapter 31)
PROTOSTOMES (Chapter 32)
Arrow worms
Lophotrochozoans
Exoskeleton molting
Blastopore develops into mouth
Ecdysozoans
Bilateral symmetry along an anterior-posterior axis; three embryonic cell layers
DEUTEROSTOMES (Chapter 33)
Echinoderms
Radial symmetry
Blastopore develops into anus
Hemichordates
Notochord
Chordates
Bilaterians (triploblastic)
Eumetazoans

31.1 A Phylogenetic Tree of the Animals This tree presents the best-supported current hypotheses of the evolutionary relationships among major groups of animals. The traits highlighted by red circles will be explained in this and the following two chapters.

31.1 What Characteristics Distinguish the Animals?

How do we recognize an organism as an animal? That may seem obvious for many familiar animals, but less so for groups such as sponges, which were once thought to be plants. Some of the general characteristics we associate with animals include:

- *Multicellularity.* In contrast to the Bacteria, Archaea, and most protists (see Chapters 26 and 27), all animals are multicellular. Animal life cycles feature complex patterns of development from a single-celled zygote into a multicellular adult.
- *Heterotrophic metabolism.* In contrast to most plants, all animals are heterotrophs. Animals are able to synthesize very few organic molecules from inorganic chemicals, so they must obtain the necessary organic molecules from their environment.
- *Internal digestion.* Although the fungi are also heterotrophs (see Chapter 30), animals and fungi digest their food differently. While fungi digest food outside their bodies, most animals use internal processes to break down materials from their environment into the organic molecules they need. Most animals ingest food into an internal **gut** that is continuous with the outside environment and in which digestion takes place.
- *Movement and nervous systems.* In contrast to the majority of plants and fungi, most animals can move their bodies. This movement is often coordinated through a well-developed nervous system, which also typically functions as a sensory system. Animals must move to find food or bring food to them. Muscle tissue and nervous systems are unique to animals, and many animal body plans are specialized for movement and detection of prey.

Although these general features help us recognize animals, none is diagnostic for all animals. Some animals do not move, at least during certain life stages, and some plants and fungi do have limited movement. Some animals lack a centralized nervous system, and some lack a gut. Many multicellular organisms are not animals. On what basis do we group all animals together in a single clade?

Animal monophyly is supported by gene sequences and morphology

The most convincing evidence that all the organisms considered to be animals share a common ancestor comes from phylogenetic analyses of their gene sequences. Relatively few complete animal genomes are available, but more are being sequenced each year. Analyses of these genomes, as well as of many individual gene sequences, have shown that the animals are indeed monophyletic. The best-supported phylogenetic

WORKING WITH**DATA**:

Reconstructing Animal Phylogeny

Original Paper

Dunn, C. W. and 17 others 2008. Broad phylogenomic sampling improves resolution of the Animal Tree of Life. *Nature* 452: 745–749.

Analyze the Data

Several breakthroughs in our understanding of animal phylogeny have occurred in recent years as the sequences of genes and proteins have been compared across species. Casey Dunn and his colleagues compared sequences from many different proteins across a wide variety of animal groups. The table below is a sample of their data that can be used to reconstruct the relationships among these representative species. In the original paper, Dunn and colleagues reported on 11,234 amino acid positions among 77 species of animals. Twenty-seven of these amino acid positions for ten of those species are shown in the table.

QUESTION 1

As you did in Chapter 22 (see Section 22.2), construct a phylogenetic tree of these ten species using the parsimony method. Use the outgroup (a choanoflagellate) to root your tree. Assume that all changes from one amino acid to another are equally likely.

QUESTION 2

How many character state changes (i.e., changes from one amino acid to another) occur along each branch on your tree?

QUESTION 3

Which amino acid positions (i.e., which character numbers) exhibit homoplasy (convergence or reversal of the character state)?

QUESTION 4

Which group on your tree represents the bilaterian animals? The protostomes? The deuterostomes?

	Character state (amino acid at position)																										
Species	1	2	3	4	5	6	7	8	9	10	11	12	13	14	15	16	17	18	19	20	21	22	23	24	25	26	27
Outgroup (choanoflagellate)	Y	G	L	G	Q	D	P	N	F	P	K	S	F	S	V	A	L	T	V	I	R	Q	N	L	V	I	L
Clam	Y	S	T	G	L	H	E	N	Y	A	R	A	M	R	I	A	L	T	I	V	K	L	S	I	V	I	L
Earthworm	Y	A	T	G	L	H	E	N	Y	P	H	A	M	R	I	A	L	T	I	V	K	L	S	I	V	M	L
Tardigrade	Y	A	T	G	L	H	E	H	Y	K	R	A	M	R	V	A	T	S	I	V	R	L	N	L	V	L	L
Fruit fly	F	A	T	G	L	H	E	N	Y	K	R	A	M	R	I	A	L	S	I	V	S	L	D	L	V	L	L
Sea urchin	Y	A	T	G	L	L	E	N	Y	P	N	A	M	R	I	A	L	T	V	I	R	Q	N	L	T	V	K
Human	W	A	A	G	L	R	E	H	Y	P	K	A	I	R	I	S	V	T	V	I	R	Q	N	L	T	V	K
Chicken	W	A	A	G	L	R	E	H	Y	P	R	A	I	R	I	A	V	T	V	I	R	Q	N	L	T	V	K
Lancelet	Y	A	T	G	L	R	E	H	Y	P	K	A	M	R	I	A	V	T	V	I	R	L	N	L	T	V	K
Sponge	Y	G	L	S	L	R	P	N	F	P	K	S	M	S	V	A	L	T	V	I	R	Q	N	L	V	I	L

Go to **BioPortal** for all WORKING WITH**DATA** exercises

tree for the major animal groups is shown in **Figure 31.1**. **Table 31.1** summarizes the living members of those groups.

Although animals were considered to belong to a single clade long before gene sequencing became possible, surprisingly few morphological features are shared across all species of animals. Two morphological synapomorphies have been identified that distinguish the animals:

- A common set of extracellular matrix molecules, including collagen and proteoglycans (see Figure 5.22)
- Unique types of junctions between cells (tight junctions, desmosomes, and gap junctions; see Figure 6.7)

Although some animals in a few groups lack one or the other of these traits, it is believed that these traits were possessed by the ancestor of all animals and subsequently lost in those groups. Similarities among animals in the organization and function of Hox and other developmental genes (see Chapter 20) provide additional evidence of developmental mechanisms shared by a common animal ancestor.

The common ancestor of animals was probably a colonial flagellated protist similar to existing colonial choanoflagellates. Choanoflagellate colonies have clear similarities to the multicellular sponges (**Figure 31.2**). The best-supported hypothesis of animal origins postulates a choanoflagellate-like lineage in which certain cells in the colony became specialized—some for movement, others for nutrition, others for reproduction, and so on. Once this functional specialization had begun, cells could have continued to differentiate. Coordination among groups of cells could have improved by means of specific regulatory and signaling molecules that guided differentiation and migration of cells in developing embryos. Such coordinated groups of cells eventually could have evolved into the larger and more complex organisms that we call animals.

Nearly 80 percent of the 1.8 million named species of living organisms are animals, and millions of additional animal species await discovery. Evidence for the evolutionary relationships among animal groups can be found in fossils, in patterns of embryonic development, in the morphology

TABLE 31.1

Summary of Living Members of the Major Animal Groups

Group	Approximate Number of Living Species Described	Major Subgroups, Other Names, and Notes
Sponges	8,500	Demosponges, glass sponges, calcareous sponges
Ctenophores	250	Comb jellies
Placozoans	2	Additional species have been discovered but not yet formally named
Cnidarians	12,500	Anthozoans: corals, sea anemones Hydrozoans: hydras and hydroids Scyphozoans: jellyfish Myxozoans: parasitic mucous animals; sometimes placed in group distinct from cnidarians
Orthonectids	45	Microscopic wormlike parasites of marine invertebrates; relationships uncertain
Rhombozoans	125	Tiny (0.5–7 mm) parasites of cephalopods; relationships uncertain
PROTOSTOMES		
Arrow worms	180	Glass worms
Lophotrochozoans		
Bryozoans	5,500	Moss animals
Entoprocts	170	Sessile aquatic animals, 0.1–7 mm long, superficially similar to bryozoans
Flatworms	30,000	Free-living flatworms; flukes and tapeworms (all parasitic); monogeneans (ectoparasites of fishes)
Gastrotrichs	800	"Hairy backs"
Rotifers and relatives	3,000	Rotifers, spiny-headed worms, and jaw worms
Ribbon worms	1,200	Proboscis worms
Brachiopods	450	Lampshells
Phoronids	10	Sessile marine filter feeders
Annelids	19,000	Polychaetes (generally marine; may not be monophyletic) Clitellates: earthworms, freshwater worms, leeches
Mollusks	117,000	Monoplacophorans Chitons Bivalves: clams, oysters, mussels Gastropods: snails, slugs, limpets Cephalopods: squids, octopuses, nautiloids
Ecdysozoans		
Kinorhynchs	180	Mud dragons
Loriciferans	30	Brush heads
Priapulids	20	Penis worms
Nematodes	25,000	Roundworms
Horsehair worms	350	Gordian worms
Tardigrades	1,200	Water bears
Onychophorans	180	Velvet worms
Arthropods		
Chelicerates	114,000	Horseshoe crabs, pycnogonids, and arachnids (scorpions, harvestmen, spiders, mites, ticks)
Myriapods	12,000	Millipedes, centipedes
Crustaceans	67,000	Crabs, shrimps, lobsters and crayfishes, barnacles, copepods
Hexapods	1,020,000	Insects and their wingless relatives
DEUTEROSTOMES		
Xenoturbellids	2	Secondarily simple marine worms; relationships uncertain
Acoels	400	Very small (mostly <2 mm) flattened marine worms; relationships uncertain
Echinoderms	7,500	Crinoids (sea lilies and feather stars); brittle stars; sea stars; sea daisies; sea urchins; sea cucumbers
Hemichordates	120	Acorn worms and pterobranchs
Tunicates	2,800	Sea squirts (ascidians), salps, and larvaceans
Lancelets	35	Cephalochordates
Vertebrates	65,000	Hagfishes, lampreys, cartilaginous fishes, ray-finned fishes, coelacanths, lungfishes, amphibians, reptiles (including birds), and mammals

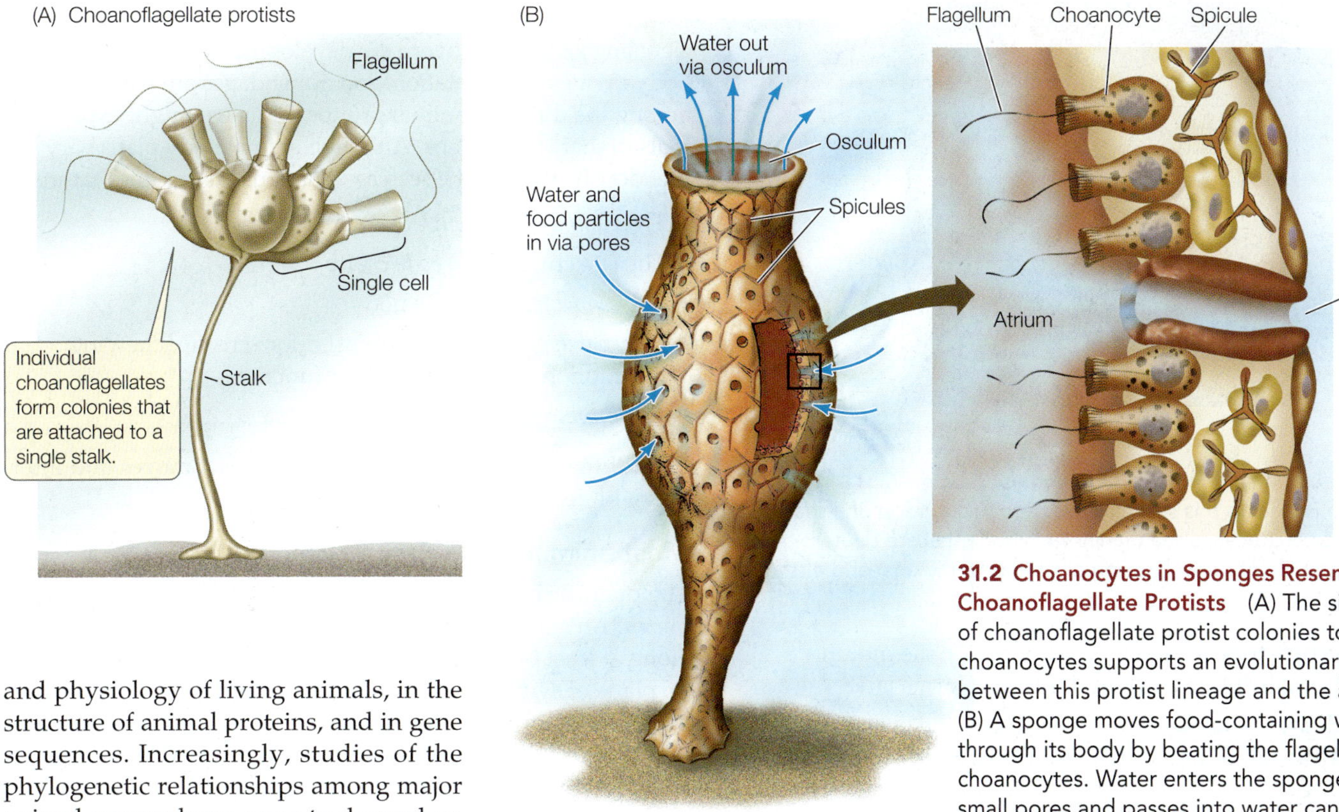
(A) Choanoflagellate protists
Flagellum
Single cell
Individual choanoflagellates form colonies that are attached to a single stalk.
Stalk
(B)
Water out via osculum
Osculum
Water and food particles in via pores
Spicules
Flagellum
Choanocyte
Spicule
Pore
Atrium

31.2 Choanocytes in Sponges Resemble Choanoflagellate Protists (A) The similarity of choanoflagellate protist colonies to sponge choanocytes supports an evolutionary link between this protist lineage and the animals. (B) A sponge moves food-containing water through its body by beating the flagella of its choanocytes. Water enters the sponge through small pores and passes into water canals or an open atrium, where the choanocytes capture food particles from the water. The spicules are supportive, skeletal structures.

and physiology of living animals, in the structure of animal proteins, and in gene sequences. Increasingly, studies of the phylogenetic relationships among major animal groups have come to depend on genomic sequence comparisons.

A few basic developmental patterns differentiate major animal groups

Differences in patterns of embryonic development have until recently provided many of the important clues to animal phylogeny. Analyses of gene sequences, however, are now showing that some developmental patterns are more evolutionarily variable than previously thought. Here we describe the basic developmental patterns that vary among the major animal clades.

The early cell divisions of an embryo are known as **cleavage**. As described in Section 44.1, several different patterns of cleavage exist among animals. Although these patterns can be useful for characterizing major animal groups, genomic analyses have shown that many changes have occurred in cleavage patterns throughout animal evolution.

Cleavage patterns are influenced by the configuration of the yolk, the acellular nutritive material that nourishes the growing embryo. The eggs of many animal groups contain a small amount of yolk that is evenly distributed throughout the egg cytoplasm. In some of these groups, the zygote and its descendant cells divide completely and evenly in a pattern known as **radial cleavage**. Radial cleavage is thought to be the ancestral condition for the animals other than sponges, as it is widely distributed among the major lineages. **Spiral cleavage**—a complicated permutation of radial cleavage—is found among many lophotrochozoans (a group that includes earthworms and clams). Lophotrochozoans with spiral cleavage are thus sometimes known as spiralians. The early branches of the ecdysozoans (molting animals, such as insects and nematodes) have radial cleavage, but most ecdysozoans have an idiosyncratic cleavage pattern that is neither radial nor spiral in organization (see Figure 44.3C). In reptiles, the presence of a large body of yolk within the fertilized egg creates an incomplete cleavage pattern in which the dividing cells form an embryo on top of the yolk mass.

Distinct layers of cells form during the early development of most animals. These cell layers differentiate into specific organs and organ systems as development continues. The embryos of **diploblastic** animals have two cell layers: an outer **ectoderm** and an inner **endoderm**. Embryos of **triploblastic** animals have, in addition to ectoderm and endoderm, a third distinct cell layer, **mesoderm**, which lies between the ectoderm and the endoderm. The existence of three cell layers in embryos is a synapomorphy of the triploblastic animals (which form a clade), whereas the diploblastic animals (ctenophores, placozoans, and cnidarians, which are not a clade) exhibit the ancestral condition (see Figure 31.1). Some biologists consider sponges to be diploblastic, but since they do not have clearly differentiated tissue types or embryonic cell layers, the term is not usually applied to them.

During early development in many animals, in a process known as gastrulation, a hollow ball one cell thick indents to form a cup-shaped structure. The opening of the cavity formed by this indentation is called the blastopore (**Figure 31.3**). The process of gastrulation is detailed in Section 44.2; the point to remember here is that the *overall pattern* of gastrulation immediately after formation of the blastopore divides the triploblastic animals into two major groups:

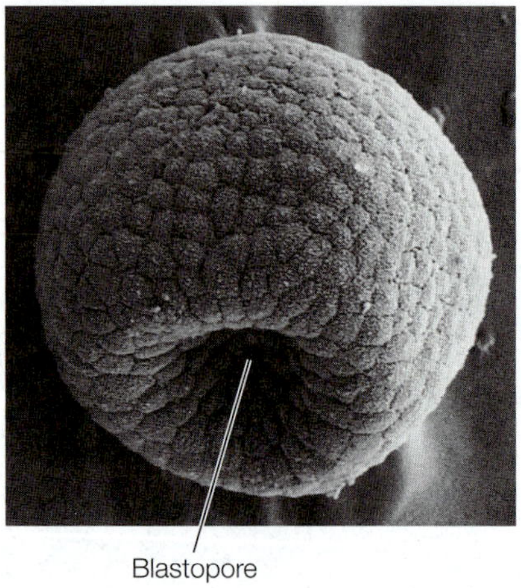
Blastopore

31.3 Gastrulation Illuminates Evolutionary Relationships The blastopore is clear in this scanning electron micrograph of a sea urchin gastrula. Because sea urchins (echinoderms) are deuterostomes, this blastopore will eventually become the anal end of the animal's gut.

- In the **protostomes** (Greek, "mouth first"), the mouth arises from the blastopore, and the anus forms later.
- In the **deuterostomes** ("mouth second"), the blastopore becomes the anus, and the mouth forms later.

Although the developmental patterns of animals are more varied than suggested by this simple dichotomy, sequencing data indicate that the protostomes and deuterostomes represent distinct animal clades. Together these two groups are known as the **bilaterians** (named for their usual bilateral symmetry), and they account for the vast majority of animal species.

RECAP 31.1

The animals are thought to be monophyletic because they share several derived traits, especially among their gene sequences. Major developmental differences also provide evidence of their evolutionary relationships.

- What general features of animals distinguish this group from other living organisms? See p. 630
- Describe the difference between diploblastic and triploblastic embryos. See p. 633
- Describe the difference between protostomes and deuterostomes. See p. 634

We will begin our exploration of animal diversity by discussing the general features of animal body plans. Later in this chapter we will describe several groups of animals that diverged before the origin of the bilaterians. We will devote Chapter 32 to the protostomes and Chapter 33 to the deuterostomes.

31.2 What Are the Features of Animal Body Plans?

The general structure of an animal, the arrangement of its organ systems, and the integrated functioning of its parts are referred to as its **body plan**. As Chapter 20 described, the regulatory and signaling genes that govern the development of body symmetry, body cavities, segmentation, and appendages are widely shared among the different animal groups. Thus we might expect animals to share body plans. Although animal body plans vary tremendously, they can be seen as variations on five key features:

- The *symmetry* of the body
- The structure of the *body cavity*
- The *segmentation* of the body
- *External appendages* that are used for sensing, chewing, locomotion, mating, and other functions
- The development of the *nervous system*

Each of these features affects how an animal moves and interacts with its environment.

Most animals are symmetrical

The overall shape of an animal can be described by its **symmetry**. An animal is said to be symmetrical if it can be divided along at least one plane into similar halves. Animals that have no plane of symmetry are said to be asymmetrical. Placozoans and many sponges are asymmetrical, but most other animals have some kind of symmetry, which is governed by the expression of regulatory genes during development.

The simplest form of symmetry is **spherical symmetry**, in which body parts radiate out from a central point. An infinite number of planes passing through the central point can divide a spherically symmetrical organism into similar halves. Spherical symmetry is widespread among unicellular protists, but most animals possess other forms of symmetry.

In organisms with **radial symmetry**, body parts are arranged around one main axis at the body's center (**Figure 31.4A**). Ctenophores (comb jellies) are radially symmetrical, as are many cnidarians (such as sea anemones and jellyfishes) and echinoderms. A perfectly radially symmetrical animal can be divided into similar halves by any plane that contains the main axis. However, most radially symmetrical animals—including the adults of echinoderms such as sea stars—are slightly modified, so that only some planes can divide them into identical halves. Some radially symmetrical animals are sessile (they remain fixed in one place) or drift with water currents. Others move about slowly but can move equally well in any direction.

Bilateral symmetry is characteristic of animals that have a distinct front end, which typically precedes the rest of the body as the animal moves. A bilaterally symmetrical animal can be divided into mirror-image (left and right) halves by a single plane that passes through the midline of its body. This plane runs from the front, or **anterior**, end of the body, to the rear, or **posterior**, end (**Figure 31.4B**). A plane at right angles to the midline divides the body into two dissimilar sides. The back of a bilaterally symmetrical animal is its **dorsal** surface; the underside is its **ventral** surface.

Bilateral symmetry is strongly correlated with **cephalization** (Greek *kephalos*, "head"), which is the concentration of sensory organs and nervous tissues at the anterior end of the animal.

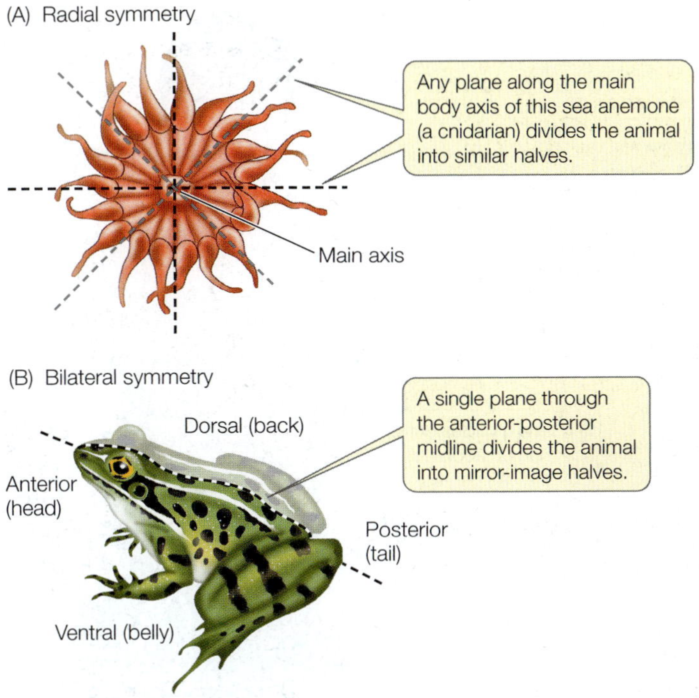
(A) Radial symmetry
Any plane along the main body axis of this sea anemone (a cnidarian) divides the animal into similar halves.
Main axis
(B) Bilateral symmetry
Dorsal (back)
A single plane through the anterior-posterior midline divides the animal into mirror-image halves.
Anterior (head)
Posterior (tail)
Ventral (belly)

31.4 Body Symmetry Most animals are either (A) radially or (B) bilaterally symmetrical.

Cephalization has been favored by natural selection because the anterior end of a bilaterally symmetrical animal typically encounters new environments first.

The structure of the body cavity influences movement

The body plans of triploblastic animals can be divided into three types based on the presence and structure of an internal, fluid-filled **body cavity**.

- **Acoelomate** animals such as flatworms lack an enclosed, fluid-filled body cavity. Instead, the space between the gut (derived from endoderm) and the muscular body wall (derived from mesoderm) is filled with masses of cells called **mesenchyme** (Figure 31.5A). These animals typically move by beating cilia.
- **Pseudocoelomate** animals have a body cavity called a pseudocoel, a fluid-filled space lying between the mesoderm and endoderm. Many of the internal organs are suspended in the pseudocoel, which is enclosed by muscles (mesoderm) only on its outside; there is no inner layer of mesoderm surrounding the internal organs (Figure 31.5B).
- **Coelomate** animals have a body cavity, the coelom, that develops within the mesoderm. The coelom is lined with a thin layer of tissue called the **peritoneum**, which also surrounds the internal organs. The coelom is thus completely enclosed by mesoderm (Figure 31.5C).

The structure of an animal's body cavity strongly influences the ways in which it can move. The body cavities of many animals function as **hydrostatic skeletons**. Fluids are relatively

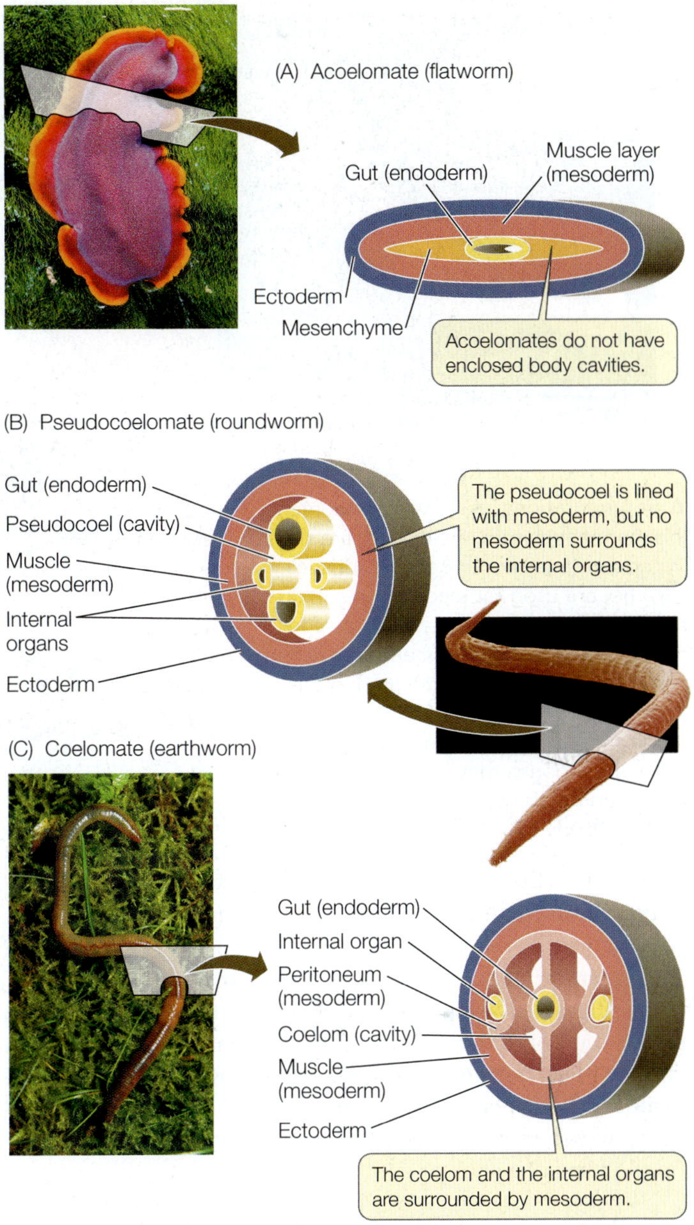
(A) Acoelomate (flatworm)
Gut (endoderm)
Muscle layer (mesoderm)
Ectoderm
Mesenchyme
Acoelomates do not have enclosed body cavities.
(B) Pseudocoelomate (roundworm)
Gut (endoderm)
Pseudocoel (cavity)
Muscle (mesoderm)
Internal organs
Ectoderm
The pseudocoel is lined with mesoderm, but no mesoderm surrounds the internal organs.
(C) Coelomate (earthworm)
Gut (endoderm)
Internal organ
Peritoneum (mesoderm)
Coelom (cavity)
Muscle (mesoderm)
Ectoderm
The coelom and the internal organs are surrounded by mesoderm.

31.5 Animal Body Cavities (A) Acoelomates do not have enclosed body cavities. (B) Pseudocoelomates have a body cavity enclosed by mesoderm only on its outside. (C) Coelomates have a body cavity that is enclosed by mesoderm on both its inside and its outside.

Go to Activity 31.1 Animal Body Cavities
Life10e.com/ac31.1

incompressible, so when the muscles surrounding a fluid-filled body cavity contract, fluids shift to another part of the cavity. If the body tissues around the cavity are flexible, fluids squeezed out of one region can cause some other region to expand. The moving fluids can thus move specific body parts. (You can see how a hydrostatic skeleton works by watching a snail emerge from its shell.) A coelomate animal has better control over the movement of the fluids in its body cavity than a pseudocoelomate animal does. An animal that has longitudinal muscles (running along the length of the body) as well as

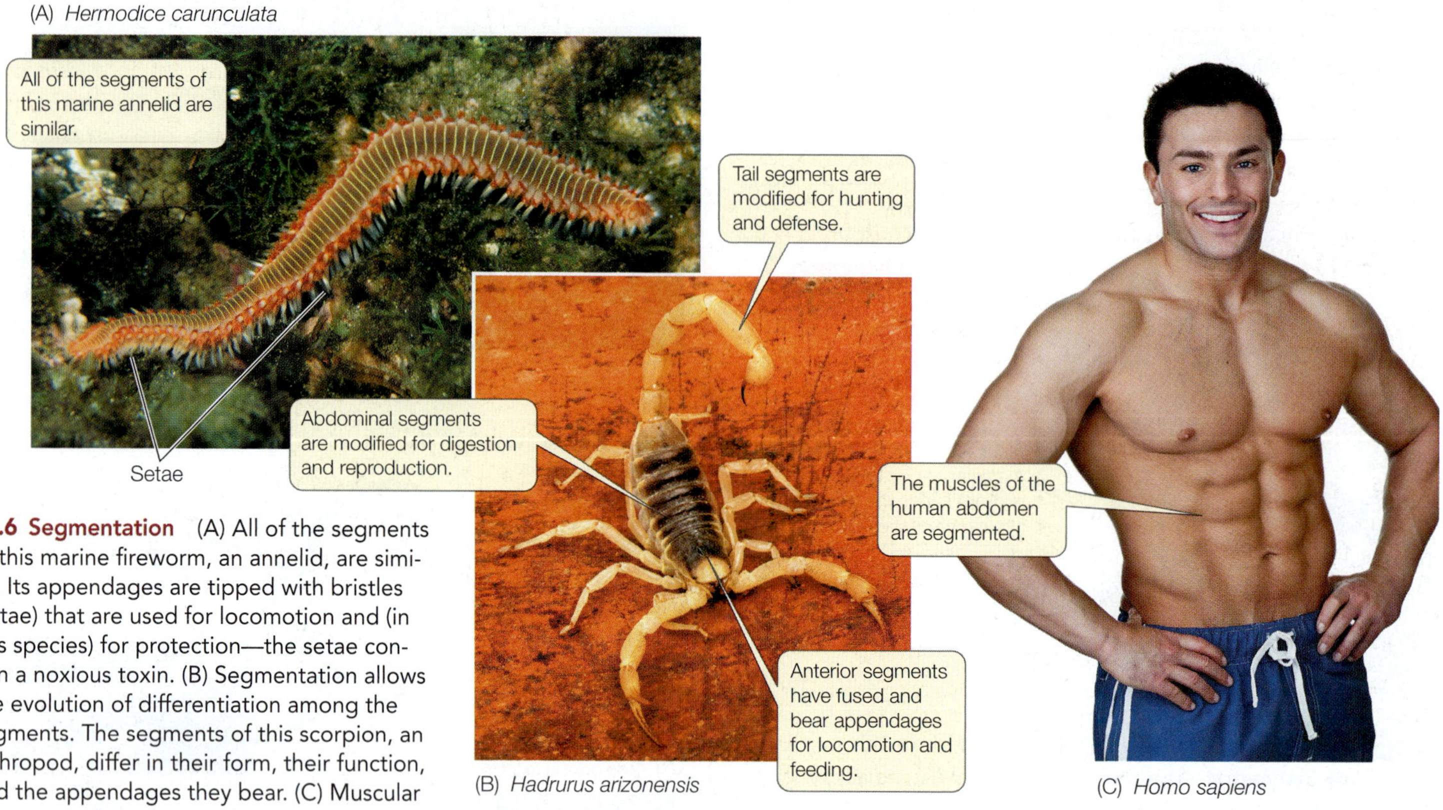
(A) Hermodice carunculata
All of the segments of this marine annelid are similar.
Setae
Abdominal segments are modified for digestion and reproduction.
Tail segments are modified for hunting and defense.
Anterior segments have fused and bear appendages for locomotion and feeding.
(B) Hadrurus arizonensis
The muscles of the human abdomen are segmented.
(C) Homo sapiens

31.6 Segmentation (A) All of the segments of this marine fireworm, an annelid, are similar. Its appendages are tipped with bristles (setae) that are used for locomotion and (in this species) for protection—the setae contain a noxious toxin. (B) Segmentation allows the evolution of differentiation among the segments. The segments of this scorpion, an arthropod, differ in their form, their function, and the appendages they bear. (C) Muscular segmentation is clearly visible in the abdomen of this body builder.

circular muscles (encircling the body cavity) has even greater control over its movement.

In terrestrial environments, the hydrostatic function of fluid-filled body cavities applies mostly to relatively small, soft-bodied organisms. Most larger animals (as well as many smaller ones) have hard skeletons that provide protection and facilitate movement. Muscles are attached to those firm structures, which may be inside the animal or on its outer surface (in the form of a shell or cuticle).

Segmentation improves control of movement

Segmentation—the division of the body into segments—is seen in many animal groups. Segmentation facilitates specialization of different body regions. It also allows an animal to alter the shape of its body in complex ways and to control its movements precisely. If an animal's body is segmented, muscles in each individual segment can change the shape of that segment independently of the others. In only a few segmented animals is the body cavity separated into discrete compartments, but even partly separated compartments allow better control of movement. As we will see in Chapters 32 and 33, segmentation occurs in several groups of protostomes and deuterostomes.

In many animals, such as annelids (earthworms and their relatives), similar body segments are repeated many times (**Figure 31.6A**). In other animals, including most arthropods, segments differ strikingly from one another (**Figure 31.6B**). As we'll describe in Chapter 32, the dramatic evolutionary radiation of the arthropods (including the insects, spiders, centipedes, and crustaceans) was based on modifications of a segmented body plan that features muscles attached to the inner surface of an external skeleton, including a variety of external appendages that move these animals. In some animals, distinct body segments are not apparent externally (as with the segmented vertebrae of vertebrates, including humans). Nonetheless, muscular segmentation is clearly visible in humans with well-defined, muscular bodies (**Figure 31.6C**).

Appendages have many uses

Getting around under their own power is important to many animals. It allows them to obtain food, to avoid predators, and to find mates. Even some species that are sessile as adults, such as sea anemones, have larval stages that use cilia to swim, thus increasing the animal's chances of finding a suitable habitat.

Appendages that project from the body greatly enhance an animal's ability to move around. Many echinoderms, including sea urchins and sea stars, have myriad tube feet that allow them to move slowly across the substrate (see Figure 33.3B). Animals whose appendages have become modified into specialized limbs are capable of better controlled, more rapid movement. The presence of jointed limbs has been a prominent factor in the evolutionary success of the arthropods and the vertebrates. In four independent instances—among the arthropod insects and among the vertebrate pterosaurs, birds, and bats—body plans emerged in which limbs were modified into wings, allowing these animals to use powered flight.

Appendages also include many structures that are not used for locomotion. Many animals have antennae, which are specialized appendages used for sensing the environment. Other appendages (such as the claws and mouthparts of many arthropods) are adaptations for capturing prey or chewing food. In some species, appendages are used for reproductive purposes, such as sperm transfer or egg incubation.

Nervous systems coordinate movement and allow sensory processing

The bilaterian animals have a well-coordinated central nervous system. More diffuse nervous systems, called **nerve nets**, are present in some other animals, such as cnidarians and ctenophores. Nervous systems appear to be completely absent in a few groups, such as sponges and placozoans.

The central nervous system of bilaterians coordinates the actions of muscles, which allows coordinated movement of appendages and body parts. This coordination of muscles permits highly effective and efficient movement on land, in water, or through the air. The central nervous system is also essential for the processing of sensory information gathered from a wide variety of sensory systems. Many animals have sensory systems for detecting light, for forming images of their environment (sight), for mechanical touch, for detecting movement, for detecting sounds (hearing), for detecting electric fields, and for chemical detection (e.g., taste and smell). These sensory systems allow animals to find food, and the ability of animals to move allows them to capture or collect food from their environment. These same abilities also allow most animals to move to avoid potential predators or to search for suitable mates. Most animals can also assess the suitability of different environments and move appropriately in response to that information.

Go to Media Clip 31.1
Nervous Systems Lead to Efficient Predators
Life10e.com/mc31.1

RECAP 31.2

The body plans of animals are variations on patterns of symmetry, body cavity structure, segmentation, appendages, and nervous systems.

- Describe the main types of symmetry found in animals and explain how an animal's symmetry can influence the way it moves. **See p. 634 and Figure 31.4**
- Explain several ways in which body cavities, segmentation, and centralized nervous systems improve control over movement. **See pp. 635–637**

Many of the modifications to the body plans of animals affect their ways of finding, capturing, and processing food. Evolutionary changes in symmetry, body cavities, appendages, segmentation, and sensory systems have played key roles in enabling animals to obtain food from their environment as well as helping them avoid becoming food for other animals.

31.3 How Do Animals Get Their Food?

As noted in Section 31.1, animals are heterotrophs, or "other-feeders." Although some animals rely on photosynthetic endosymbionts to nourish them (see Figure 27.21), most animals must expend energy to obtain an outside source of nutrition, otherwise known as food.

The need to locate food has favored the evolution of sensory structures that provide animals with detailed information about their environment as well as nervous systems that can receive, process, and coordinate that information. Furthermore, in order to acquire food, animals must either move through the environment to where food is located, or move the environment and the food it contains to them. Animals that move from one place to another are **motile**; animals that stay in one place are **sessile**.

The principal feeding strategies that animals use fall into five broad categories:

- **Filter feeders** (or **suspension feeders**) strain small organisms from their environment.
- **Herbivores** eat plants or parts of plants.
- **Predators** capture and eat other animals.
- **Parasites** live in or on other, generally much larger, organisms, from which they obtain energy and nutrients.
- **Detritivores** feed on dead organic material.

Each of these strategies can be found in many different animal groups, and none of them is limited to a single group. Individuals of some species employ more than one of these feeding strategies, and some animals employ different feeding strategies at different stages of their life cycle. The constant and ongoing need to obtain food, the variety of nutrient sources available in a given environment, and the necessity of competing with other animals to obtain food means that a variety of feeding strategies can be found among all the major animal groups.

Filter feeders capture small prey

Air and water often contain small organisms and organic molecules that are potential food for animals. Moving air and water may carry those items to an animal that positions itself in a good location. Other animals can move through the environment, filtering out prey items as they move. In either case, filter feeders use some kind of straining device to filter the food from the environment.

Many sessile aquatic animals rely on water currents to bring prey to them (Figure 31.7A). Some sessile filter feeders (such as sponges; see Figure 31.2) expend energy to move water past their food-capturing devices. Motile filter feeders move their bodies to the nutrient source. Flamingos, for example, use their serrated beaks to filter small organisms out of the muddy mixture they pick up as they wade through shallow water (Figure 31.7B). Blue whales—the largest animals that have ever lived—are filter feeders that strain tiny crustaceans from the water column as they swim.

Go to Media Clip 31.2
Filter Feeders
Life10e.com/mc31.2

Herbivores eat plants

An individual plant has many different structures—leaves, wood, roots, sap, flowers, fruits, nectar, and seeds—that animals can consume. Not surprisingly, then, many different kinds of herbivores—animals that feed only on plants—may

(A) *Spirobranchus* sp.

(B) *Phoenicopterus ruber*

31.7 Filter-Feeding Strategies (A) Sessile marine filter feeders such as this Christmas tree worm, a polychaete, allow the ocean currents to bring their food—plankton—to them. (B) The greater flamingo of South America is a motile filter feeder. It uses its appendages (legs) to stir up mud as it wades through ocean lagoons and salty lakes. The bird then uses its beak (inset) to strain small organisms out of the muddy mixture.

feed on a single kind of plant, consuming different parts of the plant or feeding on the same part in different ways. Whereas an individual animal that is captured by a predator is likely to die, herbivores often feed on plants without killing them.

Animals do not need to expend energy subduing and killing plants in order to feed on them. However, plant matter can be difficult to digest and can pose special challenges to terrestrial herbivores because the dominant land plants tend to have several different kinds of tissues, many of which are tough or fibrous. Herbivorous animals typically have long, complex guts to accomplish the tasks involved in digesting plants (see Section 51.2). Animals also must expend energy to detoxify plants' defensive chemicals.

Predators and omnivores capture and subdue prey

Predators possess features that enable them to capture and subdue other animals (referred to as their **prey**). Many vertebrate predators have sensitive sensory organs that enable them to locate prey, as well as sharp teeth or claws that allow them to capture and subdue prey (Figure 31.8). Predators may stalk and pursue their prey or wait (often camouflaged) for their prey to come to them.

Omnivores ("all-devouring") are animals, such as raccoons and humans, that eat both plants and other animals. The diets of some omnivores differ at different life stages; many songbirds, for example, eat fruit or seeds as adults but feed insects to their young.

Parasites live in or on other organisms

Parasites obtain nutrients from another organism—a **host**—by living on or within the host. Some animal parasites consume parts of the host itself (such as ticks that suck body fluids); others highjack nutrients the host would otherwise consume (such as tapeworms that may live in the intestines of mammals).

(A) *Haliaeetus leucocephalus*

(B) *Tropidolaemus wagleri*

31.8 Active and Sit-and-Wait Predators (A) The appendages (legs and wings) of the bald eagle, along with its strong beak, are adaptations to the life of an active predatory hunter. (B) Many snakes, such as Wagler's pit viper, rely on camouflage to conceal it from potential prey. These snakes typically sit motionless, waiting in one spot for unsuspecting prey to walk within striking range.

Most animal parasites are much smaller than their hosts, and many parasites can consume parts of their host without killing it. To set up residence within a host, a parasite must first overcome the host's defenses. Parasites often have complex life cycles that rely on multiple hosts, as we will see in Section 31.4.

Parasites that live inside their hosts are called endoparasites, and these are often morphologically very simple. Endoparasites often function without a digestive system, absorbing their food directly from the host's gut or body tissues. Many flatworms are endoparasites of humans and other mammals, as we will describe in Chapter 32.

Parasites that live outside their hosts are called ectoparasites; they are generally more complex morphologically than endoparasites. Ectoparasites have digestive tracts and mouthparts that enable them to pierce the host's tissues or suck on the host's body fluids. Fleas and ticks are ectoparasitic arthropods that feed on many vertebrates, including humans.

Detritivores live on the remains of other organisms

Detritivores feed on the dead bodies or waste products of other organisms, organic matter known as **detritus**. Detritivores (sometimes called decomposers) perform an important ecosystem function by breaking down dead organic matter and returning the nutrients it contains to the environment in a form that can be used by other organisms. Detritivores are common in any soil with high organic content, as well as on the ocean floor. Well-known detritivores include earthworms and other annelids, millipedes, and many insects and crustaceans.

Charles Darwin became fascinated with earthworms and wrote a book called *The Formation of Vegetable Mould through the Action of Worms*. He was particularly impressed by the importance of earthworms in soil formation. Darwin conducted many interesting experiments to establish how quickly earthworms break down organic matter and build up rich soils.

Go to Media Clip 31.3
Detritivores
Life10e.com/mc31.3

RECAP 31.3

Animals are heterotrophs that must expend energy to acquire food from their environment. Most animals either move through the environment to where food is located or move the environment and the food it contains to them.

- How can you distinguish among filter feeders, predators, and parasites—all of which may feed on other animals? **See pp. 637–639**
- What adaptations are necessary for animals that eat plants? What adaptations are needed for a predatory lifestyle? **See pp. 638–639**

As an animal grows from a single-celled zygote into a larger, more complex adult, its body structure, its diet, and the environment in which it lives may all change. In the next section we will describe some animal life cycles and discuss why they are so varied.

31.4 How Do Life Cycles Differ among Animals?

The life cycle of an animal encompasses its embryonic development, birth, growth to maturity, reproduction, and death. During its life an individual animal ingests food, grows, interacts with other individuals of the same and other species, and reproduces.

Many animal life cycles feature specialized life stages

In some groups of animals, newborns look much like miniature versions of the adults (a pattern called **direct development**). Newborns of most animal species, however, differ dramatically from adults. Many animal species have a life stage called a **larva** (plural *larvae*), which is an immature form that the animal takes early in its life before assuming an adult form. Some of the most striking life cycle changes are found among insects such as beetles, flies, moths, butterflies, and bees, which undergo radical change (called **metamorphosis**) between their larval and adult stages (**Figure 31.9**). In these animals, one stage may be specialized for feeding and the other for reproduction. Adults

(A)
(B)
(C)

31.9 A Life Cycle with Complete Metamorphosis (A) The larval stage (caterpillar) of the monarch butterfly (*Danaus plexippus*) is specialized for feeding. (B) The pupa is the stage during which the transformation to the adult form occurs. (C) The adult butterfly is specialized for dispersal and reproduction.

of most moth species, for example, do not eat. In some animal species, individuals eat during all life cycle stages, but what they eat changes with the stage. For example, butterfly larvae (known as caterpillars) eat leaves and flowers, whereas most adult butterflies eat only nectar. Having different life cycle stages that are specialized for different activities may increase the efficiency with which an animal performs those activities.

Most animal life cycles have at least one dispersal stage

At some time during their lives, most animals move, or are moved, so few animals die exactly where they were born. Movement of organisms away from a parent organism or from an existing population is called **dispersal**.

Animals that are sessile as adults typically disperse as eggs or larvae. Most sessile marine animals discharge their eggs and sperm into the water, where fertilization takes place. A larva soon hatches and floats freely in the plankton, where it filters small food items from the water. Many animals that live on the seafloor, including polychaete worms and mollusks, have a radially symmetrical larval form known as a **trochophore** (**Figure 31.10A**). Other animals, such as crustaceans, have a bilaterally symmetrical larval form called a **nauplius** (**Figure 31.10B**). Both types of larvae feed for some time in the plankton and may travel long distances before settling on the ocean floor and transforming into adults.

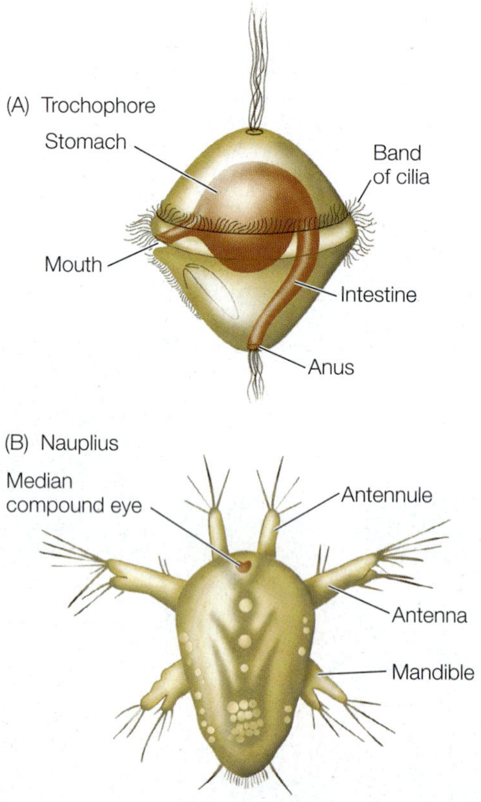
(A) Trochophore
Stomach
Band of cilia
Mouth
Intestine
Anus
(B) Nauplius
Median compound eye
Antennule
Antenna
Mandible

31.10 Planktonic Larval Forms of Marine Animals (A) The trochophore ("wheel-bearer") is a distinctive larval form found in several marine animal clades with spiral cleavage, most notably the polychaete worms and the mollusks. (B) This nauplius larva will mature into a crustacean with a segmented body and jointed appendages.

Other animal species that are motile as adults disperse when they are mature. A caterpillar, for example, may spend its entire larval stage feeding on a single plant, but after it metamorphoses into a flying adult—a butterfly—it may fly to and lay eggs on other plants located far from the one where it spent its caterpillar days. In some species, individuals disperse during several different life cycle stages.

Parasite life cycles facilitate dispersal and overcome host defenses

Animals that live as endoparasites are bathed in the nutritious tissues of their host or in the digested food that fills their host's digestive tract. Thus they may not need to exert much energy to obtain food, but to survive they must overcome the host's defenses. Furthermore, either they or their offspring must disperse to new hosts while their host is still living, because they die when their host dies.

The fertilized eggs of some parasites are voided with the host's feces and later ingested directly by other host individuals. Most parasite species, however, have complex life cycles involving one or more intermediate hosts and several larval stages (**Figure 31.11**). Some intermediate hosts transport individual parasites directly between other host species. Others house and support the parasite until another host ingests them. Complex life cycles may thus facilitate the transfer of individual parasites among hosts.

Some animals form colonies of genetically identical, physiologically integrated individuals

Most people tend to view the distinction between individuals and populations as clear-cut. In several groups of animals, however, asexual reproduction without fission leads to the formation of colonies composed of many physiologically integrated individuals. At first appearance, these colonies may look much like a single integrated organism. The individuals in the colony are clonal copies of one another, so they are genetically homogeneous.

Coloniality has arisen several times among animal groups, with widely varying levels of integration and specialization among the individuals. In some species, colonies are composed of loosely connected but integrated individuals that all function alike (**Figure 31.12**). In other colonial species, the individuals may become specialized for different functions, just as different cell types in multicellular organisms do. The Portuguese man-of-war (a cnidarian; see Figure 31.19) is an example of such a colonial animal, as it is composed of many individuals of four different specialized body forms, all integrated and functioning together. The individuals in the colony are themselves multicellular, however, unlike the cells of a single multicellular organism.

Mature tapeworm
Final hosts (fish-eating mammals)
START
1 The zygote, which has developed in a host mammal's gut, is passed with its feces.
2 The embryo develops in water.
First larval stage (free-swimming)
3 The larva hatches.
Second larval stage
4 The free-swimming first larval stage is ingested by a copepod (first intermediate host).
5 The tapeworm develops into the second larval stage and is passed on when a perch eats the copepod.
6 The larva moves to the muscles of the perch (second intermediate host).
7 The perch is eaten by a larger fish (third intermediate host).
Third larval stage
8 The fish is eaten by a mammalian host; the tapeworm matures and reproduces in the mammal's gut.

31.11 Reaching a New Host by a Complex Route The broad fish tapeworm *Diphyllobothrium latum* must pass through the bodies of a copepod (a type of crustacean) and at least one fish before it can reinfect its primary host, a mammal. Such complex life cycles assist the parasite's colonization of new host individuals, but they also provide opportunities for humans to break the cycle with hygienic measures (such as thoroughly cooking food to kill the parasites).

No life cycle can maximize all benefits

A common saying, "a jack-of-all-trades is master of none," suggests why there are constraints on the evolution of life cycles. The characteristics an animal has in any one life cycle stage may improve its performance in one activity but reduce its performance in another—a situation known as a trade-off. An animal that is good at filtering small food items from the water, for example, is unlikely to be good at capturing large prey. Similarly, energy devoted to building protective structures such as shells cannot be used for growth.

Some major trade-offs can be seen in animal reproduction. Some animals produce large numbers of small eggs, each with a small energy store (**Figure 31.13A**). Other animals produce a small number of large eggs, each with a large energy store (**Figure 31.13B**). With a fixed amount of energy available for reproduction, a female animal can produce many small eggs or a few large eggs, but she cannot produce many large eggs. Thus there is a trade-off between

Plumatella repens
The individual animals...
...secrete a mineralized shell that connects the colony.
1 mm

31.12 Colonial Animals This bryozoan colony consists of many asexually reproducing, genetically homogeneous, physiologically interacting individuals. The colony looks much like a single individual with many parts, but in fact is a group of individuals acting together.

(A) *Rana sylvatica*

(B) *Aptenodytes patagonicus*

31.13 Many Small or Few Large Allocation of energy to eggs requires trade-offs. (A) This wood frog has divided her reproductive energy among a large number of small eggs. (B) This King penguin has invested all of her reproductive energy in one large egg.

the number of offspring produced and the energy resources each offspring receives from its mother.

The larger the energy store in an egg, the longer an offspring can develop before it must either find its own food or be fed by its parents. Birds of all species lay relatively small numbers of relatively large eggs, but incubation periods vary. In some species, eggs hatch when the young are still helpless (**Figure 31.14A**). Such **altricial** young must be fed and cared for until they can feed themselves; parents can provide for only a small number of altricial offspring. In contrast, some bird species incubate their eggs longer, and the hatchlings are developed to the point that they are able to forage for themselves almost immediately (**Figure 31.14B**). The young of such species are called **precocial**.

RECAP 31.4

Many animals have life cycle stages that differ from one another morphologically. In some animals, the larval form is a dispersal stage; in other species, the adults are more likely to disperse than are larvae. In several groups of organisms, asexual reproduction without fission leads to coloniality.

- How do trade-offs constrain the evolution of life cycles? **See pp. 641–642**
- Explain the difference between multicellularity and coloniality. **See p. 641 and Figure 31.12**

Variations in body symmetry, body cavity structure, life cycles, patterns of development, and survival strategies differentiate

(A) *Alcedo atthis*

(B) *Anser anser*

31.14 Helpless or Independent (A) The altricial young of the common kingfisher are essentially helpless when they hatch. Their parents feed and care for them for several weeks. (B) Grey goose hatchlings are precocial, ready to swim and feed independently almost immediately after hatching.

millions of animal species. In the remainder of this chapter and in Chapters 32 and 33, we will become acquainted with the major animal groups and learn how the general characteristics we have just described apply to each of them.

31.5 What Are the Major Groups of Animals?

The bilaterians make up a large monophyletic group embracing all animals other than sponges, ctenophores, placozoans, cnidarians, and a few poorly known groups of parasitic animals (see Figure 31.1). Some of the traits that support the monophyly of bilaterians are the presence of three distinct cell layers in embryos (triploblasty) and the presence of at least seven Hox genes (see Chapters 19 and 20). Although bilateral symmetry is often viewed as a synapomorphy of bilaterians (and the trait gives the group its name), some groups of cnidarians are also bilaterally symmetrical. Recent studies have shown that the genetic basis of bilateral symmetry in bilaterians and in those cnidarian groups that have this trait is similar, so this feature may have been present in the ancestor of both groups.

Bilaterian animals can be divided into the two major clades mentioned earlier, the protostomes and the deuterostomes. These two groups have been diversifying separately for more than 500 million years—ever since the "Cambrian explosion" that we described in Chapter 25. We will describe the protostomes in Chapter 32 and the deuterostomes in Chapter 33.

Go to Activity 31.2 Sponge and Diploblast Classificaton
Life10e.com/ac31.2

The remainder of this chapter describes those animal groups that are not bilaterians. The simplest animals, the sponges, have no distinct tissue types. All other animals are usually known as **eumetazoans**. Two groups treated here as eumetazoans, the ctenophores and placozoans, have weakly differentiated tissue layers (placozoans also lack a nervous system), so some biologists exclude these two groups from Eumetazoa as well. Most eumetazoans have some form of body symmetry, a gut, and tissues organized into distinct organs (although there have been secondary losses of some of these features in some eumetazoans).

Sponges are loosely organized animals

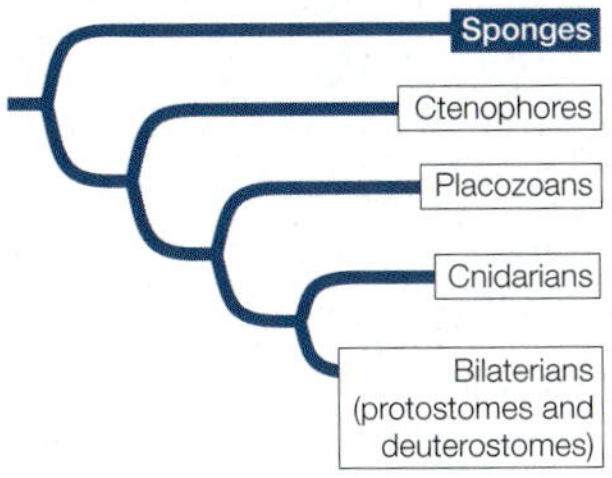
Sponges
Ctenophores
Placozoans
Cnidarians
Bilaterians (protostomes and deuterostomes)

Sponges are the simplest animals. Although they have some specialized cells, they have no distinct embryonic cell layers and no true organs. Early naturalists classified sponges as plants because they were sessile and lacked body symmetry.

Sponges have hard skeletal elements called **spicules**, which may be small and simple or large and complex (see Figure 31.2B). Three major groups of sponges, which separated soon after the split between sponges and the rest of the animals, are distinguished by their spicules. Members of two groups (glass sponges and demosponges) have skeletons composed of silicaceous spicules made of hydrated silicon dioxide (**Figure 31.15A,B**). These spicules are remarkable in having greater flexibility and toughness than synthetic glass rods of similar length. Members of the third group, the calcareous sponges,

(A) *Xestospongia testudinaria*
(B) *Euplectella aspergillum*
(C) *Sycon* sp.
Spicules

31.15 Sponge Diversity (A) The majority of sponge species are demosponges, such as these Pacific barrel sponges. The system of pores and water canals (see Figure 31.2) that is typical of the sponge body plan is apparent in this photograph. (B) The supporting structures of both demosponges and glass sponges are silicaceous spicules, seen here in the skeleton of a glass sponge. (C) The spicules of calcareous sponges are made of calcium carbonate.

take their name from their calcium carbonate spicules (Figure 31.15C). There is some question about the monophyly of sponges. Analyses of some gene sequences have suggested that calcareous sponges are actually more closely related to the eumetazoans than to the other groups of sponges. However, genomic analyses that combine information from many genes support the monophyly of sponges.

The body plan of sponges of all three groups—even large ones, which may reach 1 meter or more in length—is an aggregation of cells built around a water canal system. Sponges bring water into their bodies by beating the flagella of their specialized feeding cells, called **choanocytes** (see Figure 31.2B). Water, along with any food particles it contains, enters the sponge by way of small pores and passes into the water canals or a central atrium, where the choanocytes capture food particles. (You may recall from Section 31.1 that the choanocytes are similar in structure to protists known as choanoflagellates; that similarity provides evidence for the close relationship of animals to choanoflagellates.)

A skeleton of simple or branching spicules, often combined with a complex network of elastic fibers, supports the body of most sponges. Sponges also produce an extracellular matrix, composed of collagen, adhesive glycoproteins, and other molecules, that holds the cells together. Most species are filter feeders; a few species are predators that trap prey on hook-shaped spicules that protrude from the body surface.

Most of the 8,500 species of sponges are marine animals; only about 50 species live in fresh water. Sponges come in a wide variety of sizes and shapes that are adapted to different movement patterns of water. Sponges living in intertidal or shallow subtidal environments with strong wave action are firmly attached to the substrate. Most sponges that live in slowly flowing water are flattened and are oriented at right angles to the direction of current flow. They intercept water and the food items it contains as it flows past them.

Sponges reproduce both sexually and asexually. In most species, a single individual produces both eggs and sperm, but individuals do not self-fertilize. Water currents carry sperm from one individual to another. Sponges also reproduce asexually by budding and fragmentation.

Ctenophores are radially symmetrical and diploblastic

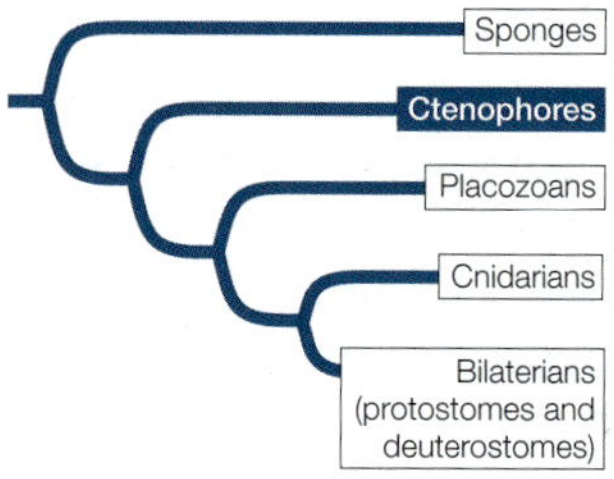
Sponges
Ctenophores
Placozoans
Cnidarians
Bilaterians (protostomes and deuterostomes)

Ctenophores, also known as comb jellies, were until recently thought to be most closely related to the cnidarians (jellyfishes, corals, and their relatives). But ctenophores lack most of the Hox genes found in all other eumetazoans, and recent studies of their genomes have indicated that ctenophores were probably among the earliest lineages to split from the remaining animals. Most recent phylogenetic analyses place the ctenophores as the sister group of all other animals except sponges.

Ctenophores have a radially symmetrical, diploblastic body plan. The two cell layers are separated by an inert, gelatinous extracellular matrix called **mesoglea**. Ctenophores, unlike sponges, have a **complete gut**: food enters through a mouth, and wastes are eliminated through two anal pores.

Ctenophores move by beating cilia rather than by muscular contractions. Most of the 250 known species have eight comblike rows of cilia-bearing plates, called **ctenes** (Figure 31.16). The feeding tentacles of ctenophores are covered with cells that discharge adhesive material when they contact prey. After capturing its prey, a ctenophore retracts its tentacles to bring

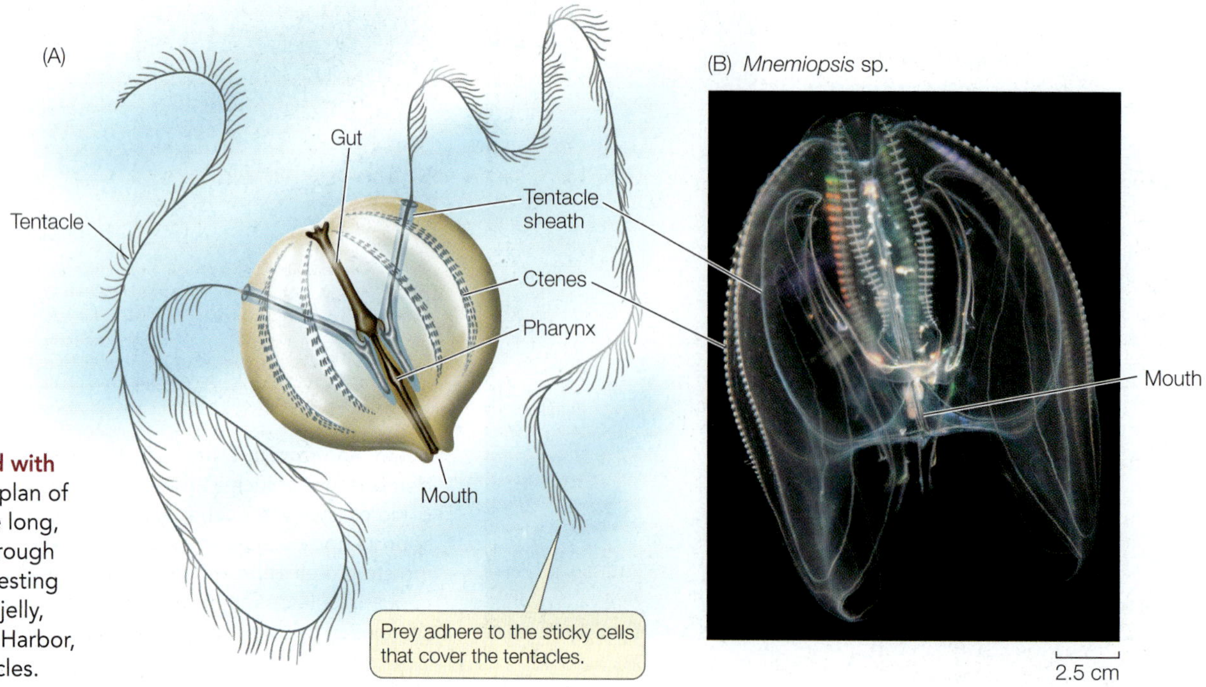
(A)
Gut
Tentacle
Tentacle sheath
Ctenes
Pharynx
Mouth
Prey adhere to the sticky cells that cover the tentacles.
(B) *Mnemiopsis* sp.
Mouth
2.5 cm

31.16 Comb Jellies Feed with Tentacles (A) The body plan of a typical ctenophore. The long, sticky tentacles sweep through the water, efficiently harvesting small prey. (B) This comb jelly, photographed in Sydney Harbor, Australia, has short tentacles.

the food to its mouth. In some species, the entire surface of the body is coated with sticky mucus that captures prey. Most ctenophores eat small planktonic organisms, although some eat other ctenophores. They are common in open seas and can become abundant in coastal bays, where large populations of ctenophores may inhibit the growth of other organisms.

Ctenophore life cycles are uncomplicated. Gametes are released into the gut and discharged through the mouth or the anal pores. Fertilization takes place in open seawater. In nearly all species, the fertilized egg develops directly into a miniature ctenophore, which gradually grows into an adult.

Go to Media Clip 31.4
Ctenophores
Life10e.com/mc31.4

Placozoans are abundant but rarely observed

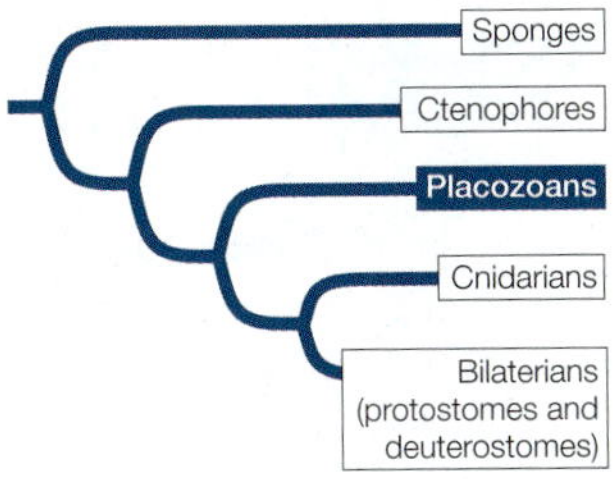
Sponges
Ctenophores
Placozoans
Cnidarians
Bilaterians (protostomes and deuterostomes)

As discussed at the start of this chapter, **placozoans** are structurally very simple animals with only a few distinct cell types (Figure 31.17A). Individuals in the mature, asymmetrical life stage are usually observed adhering to surfaces (such as the glass of aquariums, where they were first discovered, or to rocks and other hard substrates in nature). Their structural simplicity—they have no mouth, gut, or nervous system—initially led biologists to suspect they might be the sister group of all other animals. Most phylogenetic analyses have not supported this hypothesis, however, and some aspects of the placozoans' structural simplicity may be secondarily derived. They are generally considered to have a diploblastic body plan, with upper and lower surface layers that sandwich a layer of contractile fiber cells.

Recent studies have found that placozoans have a pelagic (open-ocean) life stage that is capable of swimming (Figure 31.17B), but the life history of placozoans is incompletely known. Most studies have focused on the larger adherent stages that are most easily observed in aquariums. The transparent nature and small size of placozoans make them very difficult to observe in nature. Nonetheless, it is known that placozoans can reproduce both asexually and sexually, although the details of their sexual reproduction are poorly understood. As we noted at the opening of this chapter, placozoans have been studied mainly in aquariums, where they appear after being inadvertently collected with other marine organisms, although we now know that pelagic-stage placozoans are abundant in warm seas around the world.

Cnidarians are specialized predators

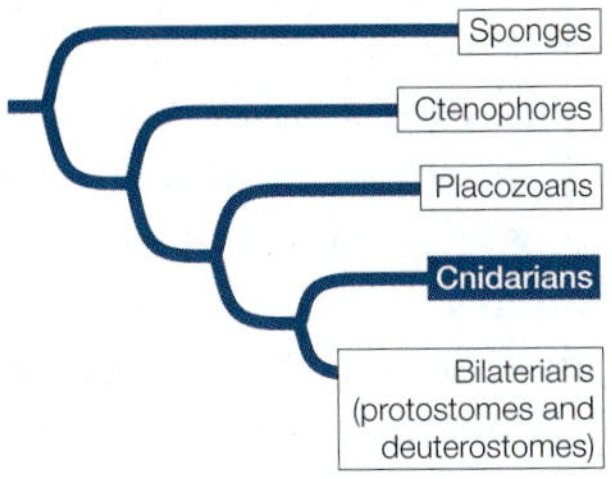
Sponges
Ctenophores
Placozoans
Cnidarians
Bilaterians (protostomes and deuterostomes)

The **cnidarians** (jellyfishes, sea anemones, corals, and hydrozoans) make up the largest and most diverse group of non-bilaterian animals. The mouth of a cnidarian is connected to a blind sac called the **gastrovascular cavity** (a cnidarian thus does not have a complete gut). The gastrovascular cavity functions in digestion, circulation, and gas exchange, and it also acts as a hydrostatic skeleton. The single opening serves as both mouth and anus.

The life cycle of many cnidarians has two distinct stages, one sessile and the other motile (Figure 31.18), although one or the other of these stages is absent in some groups. In the sessile **polyp** stage, a cylindrical stalk attaches the animal to the substrate. The motile **medusa** (plural *medusae*) is a free-swimming stage shaped like a bell or an umbrella. It typically floats with its mouth and feeding tentacles facing downward.

Mature polyps produce medusae by asexual budding. Medusae then reproduce sexually, producing eggs or sperm by meiosis and releasing the gametes into the water. A fertilized egg develops into a free-swimming, ciliated larva called a **planula**, which eventually settles to the bottom and develops into a polyp.

Cnidarians are specialized predators adapted for capturing and subduing relatively large and complex prey. Their tentacles are covered with specialized cells that contain stinging organelles called **nematocysts**, which inject toxins into their prey (Figure 31.19). Some cnidarians, including many corals and anemones, gain additional nutrition from photosynthetic endosymbionts that live in their tissues.

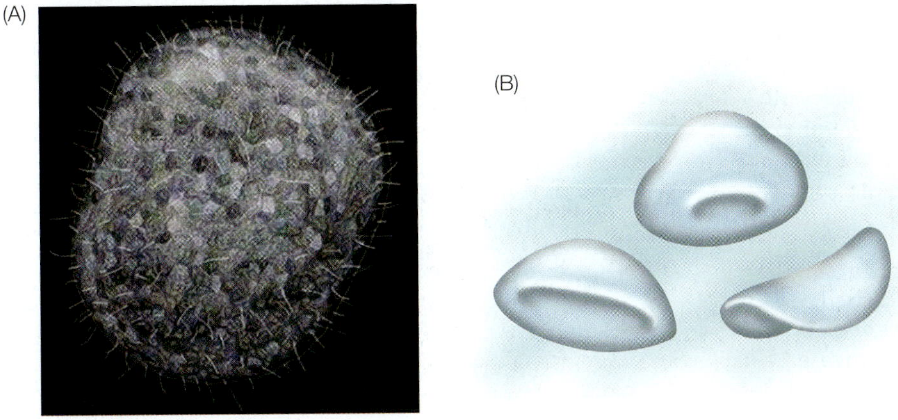
(A)
(B)

31.17 Placozoan Simplicity (A) As seen in this artist's rendition, adult placozoans are tiny (1–2 mm across), flattened, asymmetrical animals. (B) Recent studies have found a symmetrical, weakly swimming pelagic stage of placozoan to be abundant in many warm tropical and subtropical seas.

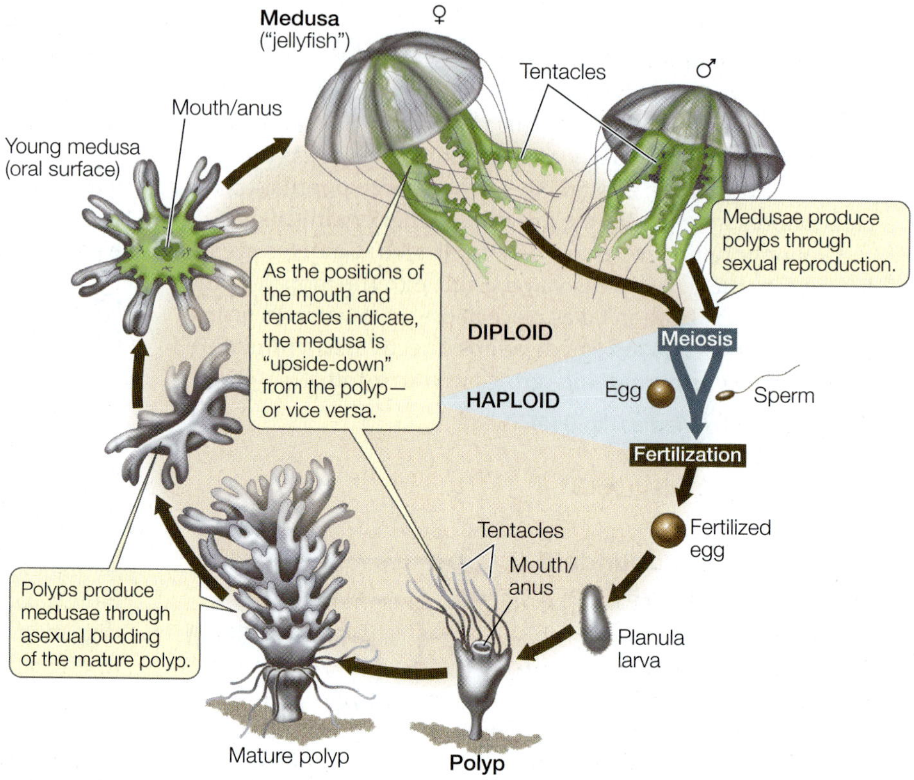
Medusa
("jellyfish")
♀
♂
Tentacles
Mouth/anus
Young medusa
(oral surface)
Medusae produce polyps through sexual reproduction.
As the positions of the mouth and tentacles indicate, the medusa is "upside-down" from the polyp—or vice versa.
DIPLOID
HAPLOID
Meiosis
Egg
Sperm
Fertilization
Fertilized egg
Tentacles
Mouth/anus
Planula larva
Polyps produce medusae through asexual budding of the mature polyp.
Mature polyp
Polyp

31.18 The Life Cycle of Most Cnidarians Has Two Stages The life cycle of a scyphozoan (jellyfish) exemplifies the typical cnidarian body forms: the sessile, asexual polyp and the motile, sexual medusa. Some species of cnidarians have life cycles that lack polyps or medusae.

Go to Animated Tutorial 31.1
Life Cycle of a Cnidarian
Life10e.com/at31.1

Cnidarians have cells containing muscle fibers whose contractions enable the animals to move, as well as simple nerve nets that integrate the body's activities. Their bodies also contain specialized structural molecules (collagen, actin, and myosin). Yet cnidarians, like ctenophores, are largely made up of inert mesoglea. They have low metabolic rates and can survive in environments where they encounter prey only infrequently.

Of the roughly 12,500 living cnidarian species, all but a few live in the oceans (**Figure 31.20**). The smallest cnidarians can barely be seen without a microscope. One small group, known as myxozoans, consists of tiny parasites, usually with a two-host life cycle that includes a fish and an annelid worm or a bryozoan. The largest known jellyfish is 2.5 meters in diameter, and some colonial hydrozoans (which include the Portuguese man-of-war; see Figure 31.19) can reach lengths in excess of 30 meters. Here we describe the three clades of cnidarians that contain the most species: anthozoans, scyphozoans, and hydrozoans.

ANTHOZOANS Members of the **anthozoan** clade include sea anemones, sea pens, and corals. Sea anemones (see Figure 31.20A), all of which are solitary, are widespread in both warm and cold ocean waters. Sea pens (see Figure 31.20B), by contrast, are colonial. Each colony consists of two or more different kinds of polyps. The primary polyp has a lower portion anchored in the bottom sediment and a branched upper portion that projects above the substrate. Along the upper portion, the primary polyp produces smaller secondary polyps by budding. Some of these secondary polyps differentiate into feeding polyps; in some species, other secondary polyps differentiate to circulate water through the colony.

The common names of coral groups—brain corals, staghorn corals, and organ pipe corals, among others—often describe their appearance (**Figure 31.21A**). Corals are sessile and colonial. The polyps of most species form a skeleton by secreting a matrix of organic molecules on which they then deposit calcium carbonate. As the colony grows, old polyps die, but their calcium carbonate skeletons remain. Living corals form a layer on top of a growing bank of skeletal remains, eventually forming chains of islands and reefs (**Figure 31.21B**). The Great Barrier Reef along the northeastern

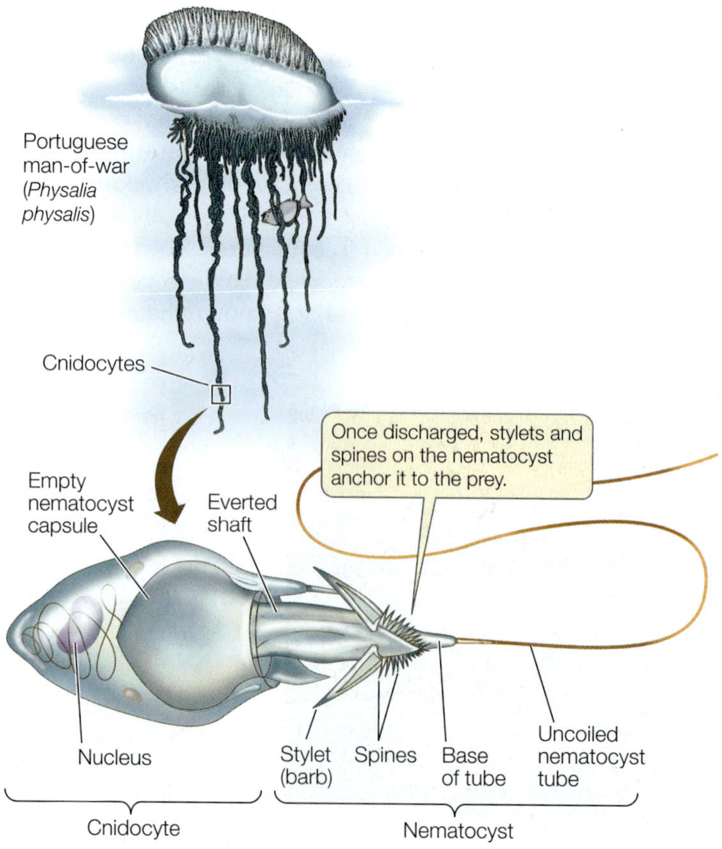
Portuguese man-of-war (*Physalia physalis*)
Cnidocytes
Empty nematocyst capsule
Everted shaft
Once discharged, stylets and spines on the nematocyst anchor it to the prey.
Nucleus
Stylet (barb)
Spines
Base of tube
Uncoiled nematocyst tube
Cnidocyte
Nematocyst

31.19 Nematocysts Are Potent Weapons The tentacles of the Portuguese man-of-war, a hydrozoan, are rife with specialized cells called cnidocytes. These cells contain stinging organelles called nematocysts, which inject toxins into their prey. The Portuguese man-of-war is a colonial organism, composed of many physiologically integrated individuals with specialized functions.

(A) Sagartia modesta
(B) Pteroeides sp.
(C) Gonionemus vertens
(D) Polyorchis penicillatus

31.20 Diversity among Cnidarians (A) Sea anemones are sessile, living attached to marine substrates. Water currents carry prey into the nematocyst-studded tentacles. (B) The sea pen is a colonial cnidarian that lives in soft bottom sediments and projects polyps above the substrate. (C) This jellyfish illustrates the complexity of a scyphozoan medusa. (D) The internal structure of the medusa of a North Atlantic colonial hydrozoan is visible here.

coast of Australia is a system of coral formations more than 2,000 kilometers long, which is about the distance from New York City to St. Louis. A single coral reef in the Red Sea has been calculated to contain more material than all the buildings in the major cities of North America combined.

Corals flourish at shallow depths in clear, nutrient-poor tropical waters. They grow well in such environments because unicellular photosynthetic dinoflagellates live endosymbiotically within their cells. These dinoflagellates provide the corals with products of photosynthesis; the corals, in turn, provide the dinoflagellates with nutrients and a place to live. This endosymbiotic relationship explains why reef-forming corals are restricted to clear surface waters, where light levels are high enough to support photosynthesis.

Coral reefs throughout the world are threatened by rising CO_2 levels (which result in increased ocean temperatures) and acidification of ocean waters. Polluted runoff from development on adjacent shorelines is an additional threat to corals. Warmer temperatures lead to the loss of coral endosymbionts (known as coral bleaching; see Figure 27.21), and acidification can cause coral skeletons to dissolve. An overabundance of nitrogen in runoff is advantageous to algae, which overgrow and eventually smother the corals.

SCYPHOZOANS The several hundred species of **scyphozoans** are all marine. The mesoglea of their medusae is thick and firm, giving rise to their common name of jellyfish (or sea jellies). The medusa rather than the polyp dominates the life cycle of scyphozoans. An individual medusa is male or female, releasing eggs or sperm into the open sea. A fertilized egg develops into a small planula larva that quickly settles on a substrate and develops into a small polyp. This polyp feeds and grows and may produce additional polyps by budding. After a period of growth, the polyp begins to bud off small medusae, which feed, grow, and transform into adult medusae (see Figures 31.18 and 31.20C).

HYDROZOANS The polyp typically dominates the life cycle of **hydrozoans**, but some species have only medusae (see Figure 31.19) and others have only polyps. Most hydrozoans are colonial. A single planula larva eventually gives rise to a colony of

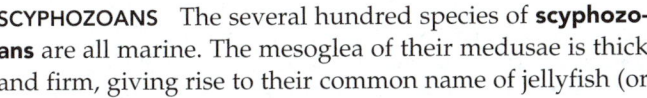

(A) *Diploria labyrinthiformis*

(B)

31.21 Corals (A) The descriptive common name of this Caribbean coral is "brain coral." (B) Many different coral species form this reef in the Red Sea between Egypt and the Arabian Peninsula.

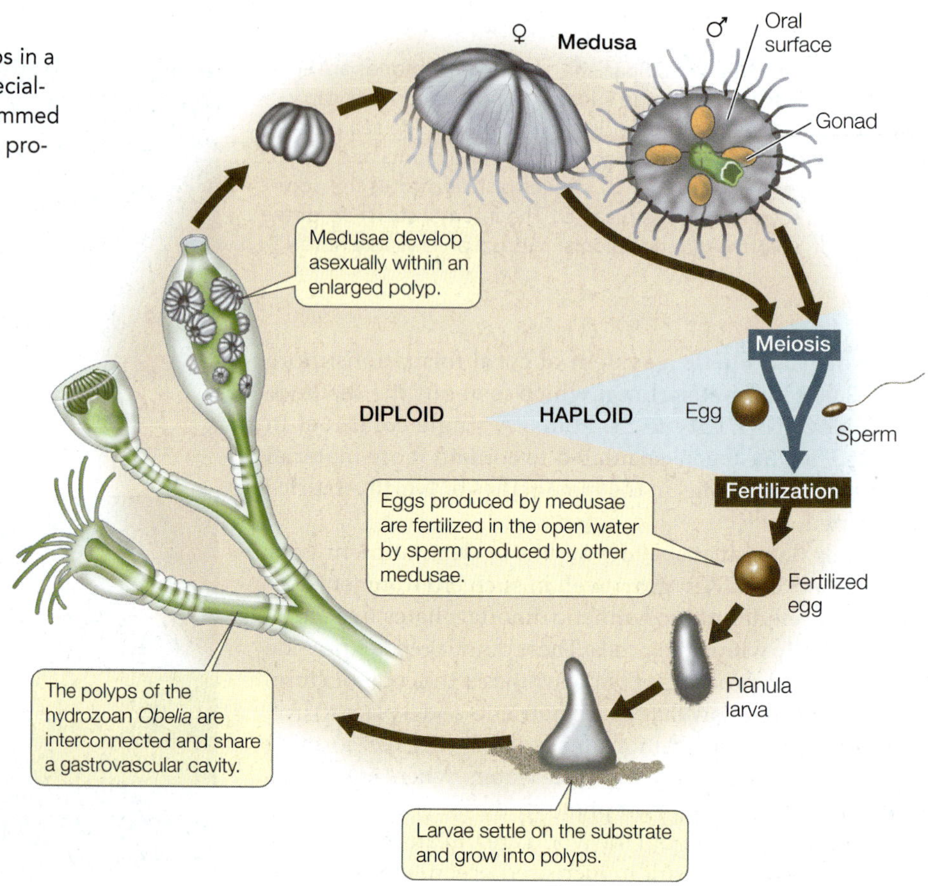

♀
Medusa
♂
Oral surface
Gonad
Medusae develop asexually within an enlarged polyp.
Meiosis
DIPLOID
HAPLOID
Egg
Sperm
Fertilization
Eggs produced by medusae are fertilized in the open water by sperm produced by other medusae.
Fertilized egg
Planula larva
The polyps of the hydrozoan *Obelia* are interconnected and share a gastrovascular cavity.
Larvae settle on the substrate and grow into polyps.

31.22 Many Hydrozoans Are Colonial The polyps in a hydrozoan colony may differentiate to perform specialized tasks. In the species whose life cycle is diagrammed here, the medusa is the sexual reproductive stage, producing eggs and sperm in organs called gonads.

many polyps, all interconnected and sharing a continuous gastrovascular cavity (**Figure 31.22**). Within such a colony, some polyps have tentacles with many nematocysts; they capture prey for the colony. Other individuals lack tentacles and are unable to feed, but are specialized for the asexual production of medusae. Still others are fingerlike and defend the colony with their nematocysts.

Some small groups of parasitic animals may be the closest relatives of bilaterians

Two small groups of tiny marine parasites are listed in Table 31.1 but are not depicted in the phylogeny in Figure 31.1: the orthonectids and the rhombozoans. Recent genomic analyses suggest that these groups may be among the closest surviving relatives of the bilaterians, although their exact phylogenetic placement is uncertain. Both groups are highly reduced parasites that lack many of the structures that traditionally have been used to study animal relationships. As their genomes become more completely known, the relationships of these two groups to other animals should become clearer. Two other small groups (also listed in Table 31.1 but not shown in Figure 31.1) have been proposed as also falling just outside the bilaterians: the xenoturbellids and acoels. Recent genomic analyses, however, have suggested that these animals are actually highly specialized deuterostomes.

RECAP 31.5

Bilaterian animals are classified into two major clades, protostomes and deuterostomes. The non-bilaterian animals comprise the sponges, ctenophores, placozoans, cnidarians, and some small groups of parasitic animals.

- Why are sponges considered to be animals even though they lack the complex body structures found among most other animal groups? **See pp. 643–644**
- Describe some major features of the following groups: sponges, ctenophores, placozoans, and cnidarians.

?

Which animal groups are involved in the earliest split in the animal tree?

ANSWER

Most data support the placement of sponges as the sister group of all other animals. Morphologically, sponges are the most similar to the choanoflagellates, and most genomic analyses continue to place sponges at the base of the animal tree. The ctenophores share some superficial similarities with cnidarians and were traditionally considered their sister group, but more recent studies increasingly support the placement of the ctenophores closer to the base of the animal tree. A few studies even suggest that the ctenophores are the sister group of all other animals, including sponges, although most studies place the ctenophores as the sister group of all animals except sponges, and this is the phylogeny we show in Figure 31.1.

Although placozoans, with only four cell types and no true organs, are less structurally complex than ctenophores, this structural simplicity is now thought to represent an evolutionary reversal in the placozoan lineage. The alternative possibility, which seems less likely, is that the organ systems of ctenophores evolved independently from those of the cnidarians and bilaterians.

CHAPTERSUMMARY 31

31.1 What Characteristics Distinguish the Animals?

- Animals share a set of derived traits not found in other groups of organisms. These traits include similarities in the sequences of many of their genes, the structure of their cell junctions, and the components of their extracellular matrix.
- Patterns of embryonic development provide clues to the evolutionary relationships among animals. **Diploblastic** animals, which include the ctenophores, placozoans, and cnidarians, develop two embryonic cell layers. **Triploblastic** animals develop three cell layers. **Review Figure 31.1**
- Differences in their patterns of early development characterize two major triploblastic clades, the **protostomes** and the **deuterostomes**.

31.2 What Are the Features of Animal Body Plans?

- Animal **body plans** can be described in terms of symmetry, **body cavity** structure, **segmentation**, types of appendages, and nervous system development.
- A few animals have no symmetry, but most animals have either **radial symmetry** or **bilateral symmetry**. **Review Figure 31.4**
- Many bilaterally symmetrical animals exhibit **cephalization**: the concentration of sensory organs and nervous tissues in an anterior head.
- On the basis of their body cavity structure, animals can be described as **acoelomates**, **pseudocoelomates**, or **coelomates**. **Review Figure 31.5, ACTIVITY 31.1**
- Segmentation, which takes many forms, improves control of movement, as do appendages. **Review Figure 31.6**
- The development of a nervous system is important for the coordination of muscular movement and the processing of sensory information.

31.3 How Do Animals Get Their Food?

- **Motile** animals can move to find food; **sessile** animals stay in one place, but may expend energy to move the environment and the food it contains to them
- **Filter feeders** strain small organisms and organic molecules from their environment.
- **Herbivores** consume plants, usually without killing them.
- **Predators** have morphological features such as sharp teeth, beaks, and claws that enable them to capture and subdue animal **prey**.
- **Parasites** live in or on other organisms and obtain nutrition from those **host** individuals.
- **Detritivores** consume dead organic matter and return the nutrients it contains to the ecosystem.

31.4 How Do Life Cycles Differ among Animals?

- The stages of an animal's life cycle may be specialized for different activities. An immature stage whose morphology is dramatically different from that of the adult stage is called a **larva**.
- Most animal life cycles have at least one **dispersal** stage. Many sessile marine animals can be grouped by the presence of one of two distinct larval dispersal stages: **trochophore** or **nauplius**. **Review Figure 31.10**
- A characteristic of an animal or a life cycle stage may improve the animal's performance in one activity but reduce its performance in another, a situation known as a trade-off.
- Parasites have complex life cycles that may involve one or more hosts and several larval stages. **Review Figure 31.11**
- In some groups of animals, asexual reproduction without fission leads to the formation of colonies composed of many genetically homogeneous, physiologically integrated individuals.

31.5 What Are the Major Groups of Animals?

- **Eumetazoans** include all animals except sponges. Animals other than sponges, ctenophores, placozoans, and cnidarians—that is, the triploblastic protostomes and deuterostomes—belong to a large monophyletic group called **bilaterians**. **Review Figure 31.1, ACTIVITY 31.2**
- **Sponges** are simple animals that lack differentiated cell layers and true organs. They have skeletons made up of siliceous or calcareous **spicules**. They create water currents and capture food with flagellated feeding cells called **choanocytes**. **Review Figure 31.2**
- **Ctenophores** are radially symmetrical and have two cell layers separated by an inert extracellular matrix called **mesoglea**. **Review Figure 31.16**
- **Placozoans** are asymmetrical as adults. They have only a few cell types and lack true organs, although their simplicity may be secondarily derived.
- The life cycle of most **cnidarians** has two distinct stages: a sessile **polyp** stage and a motile **medusa** stage that reproduces sexually. A fertilized egg develops into a free-swimming **planula** larva, which settles to the bottom and develops into a polyp. **Review Figures 31.18, 31.22, ANIMATED TUTORIAL 31.1**

Go to the Interactive Summary to review key figures, Animated Tutorials, and Activities
Life10e.com/is31

CHAPTER REVIEW

REMEMBERING

1. A bilaterally symmetrical animal can be divided into mirror-image halves by
   a. any plane through the midline of its body.
   b. any plane from its anterior to its posterior end.
   c. any plane from its dorsal to its ventral surface.
   d. a single plane through the midline of its body from its anterior to its posterior end.
   e. a single plane that divides it into dorsal and ventral halves.

2. In the common ancestor of eumetazoans, the pattern of early cleavage was
   a. spiral.
   b. radial.
   c. biradial.
   d. deterministic.
   e. haphazard.

3. A fluid-filled body cavity can function as a hydrostatic skeleton because
   a. fluids are moderately compressible.
   b. fluids are highly compressible.
   c. fluids are relatively incompressible.
   d. fluids have the same density as body tissues.
   e. fluids can be moved by ciliary action.

4. Many parasites evolved complex life cycles because
   a. they are too simple to disperse readily.
   b. they are poor at recognizing new hosts.
   c. they were driven to it by host defenses.
   d. complex life cycles increase the probability of a parasite's transfer to a new host.
   e. their nonparasitic ancestors had complex life cycles and they simply retained them.

5. The endosymbiotic dinoflagellates present in many corals
   a. provide the corals with products of photosynthesis.
   b. allow corals to flourish in clear, nutrient-poor tropical waters.
   c. can be lost when environmental conditions change.
   d. All of the above
   e. None of the above

UNDERSTANDING & APPLYING

6. Differentiate among the members of each of the following sets of related terms:

   radial symmetry/bilateral symmetry
   protostome/deuterostome
   diploblastic/triploblastic
   coelomate/pseudocoelomate/acoelomate

7. In this chapter we listed some of the traits shared by all animals that have convinced most biologists that all animals are descendants of a single common ancestral lineage. In your opinion, which of these traits provides the most compelling evidence that animals are monophyletic?

8. Why is bilateral symmetry strongly associated with cephalization, the concentration of sensory organs in an anterior head?

9. How does a slow metabolic rate enable an animal to live in an unproductive environment?

ANALYZING & EVALUATING

10. The discoveries that the pelagic stages of placozoans are abundant in warm seas and that the mature stages settle on smooth surfaces suggest how these organisms might be collected and surveyed. What sampling procedures might you use to discover whether placozoans occur at a particular location along a coast?

Go to BioPortal at **yourBioPortal.com** for Animated Tutorials, Activities, LearningCurve Quizzes, Flashcards, and many other study and review resources.

32 Protostome Animals

CHAPTER OUTLINE

How Many Species? In an important study, entomologist Terry Erwin fogged tree canopies in a Panamanian rainforest with insecticide. Using the number of species represented among the fallen insects as a base, he extrapolated to estimate the total number of insect species on Earth.

OF THE 1.8 MILLION ANIMAL SPECIES that have been discovered and named by biologists, a large majority are protostomes. One group of protostomes, the insects, accounts for more than 1 million of these species, or more than half of all known species of living organisms. Although these numbers may seem incredibly large, they represent a relatively small fraction of the total protostome diversity that is thought to exist on Earth.

As recently as the 1980s, many biologists thought that about half of existing insect species had been described, but today they think that the number of described insect species may be a much smaller fraction of the total number of living species. Why did they change their minds?

A simple but important field study suggested that the number of existing insect species had been significantly underestimated. Knowing that the insects of tropical rainforests—the most species-rich habitat on Earth—were poorly known, entomologist Terry Erwin made a comprehensive sample of one group of insects, the beetles, in the canopies of a single species of tropical forest tree, *Luehea seemannii*, in Panama. Erwin fogged the canopies of 19 large *L. seemannii* trees with an insecticide and collected the insects that fell from the trees in collection nets. He collected about 1,200 species of beetles—many of them previously undescribed—from this one species of tree.

Erwin then used a set of assumptions to estimate the total number of insect species in tropical rainforests. His assumptions included estimates of the number of species of trees in these forests; the proportion of beetles that specialize on a specific species of host tree; the relative proportion of beetles to other insect groups; and the proportion of beetles that live in trees versus those that live in leaf litter on the ground. From this and similar studies, Erwin estimated that there may be 30 million or more species of insects on Earth. Although recent tests of Erwin's assumptions suggest that 30 million was an overestimate, it is clear that the vast majority of insect species remain to be discovered.

Erwin's pioneering study highlighted the fact that we live on a poorly known planet, most of whose species have yet to be named and described. Much of that undiscovered diversity occurs among several groups of protostomes.

?

Which groups of protostomes are thought to contain the most undiscovered species?

See answer on p. 675.

32.1 What Is a Protostome?

You may recall that the embryos of diploblastic animals (the ctenophores, placozoans, and cnidarians, which we discussed in Chapter 31) have two cell layers: an outer ectoderm and an inner endoderm (see Section 31.1). Sometime after the origin of the diploblastic animals, a third embryonic cell layer evolved: the mesoderm, which lies between the ectoderm and the endoderm. Mesoderm is found in the two major triploblastic animal clades, the protostomes and the deuterostomes. If we were to judge solely on the basis of numbers, both of species and of individuals, the protostomes would emerge as by far the more successful of the two groups.

As noted in Section 31.1, the name "protostome" means "mouth first." In protostomes, the embryonic blastopore becomes the mouth as the animal develops. In contrast, in deuterostomes ("mouth second"), the blastopore becomes the anal opening of the gut. The protostomes are extremely varied, but they are all bilaterally symmetrical animals whose bodies exhibit two major derived traits:

- An anterior brain that surrounds the entrance to the digestive tract
- A ventral nervous system consisting of paired or fused longitudinal nerve cords

Other aspects of protostome body organization differ widely from group to group (**Table 32.1**). Before gene sequences were available for phylogenetic analysis, biologists considered the structure of the body cavity to be a critical feature in animal classification. But the results of genetic analyses have shown that body cavity forms have undergone considerable convergence in the course of protostome evolution. Although the common ancestor of the protostomes had a coelom, subsequent modifications of the coelom distinguish many protostome lineages. In some lineages (such as the flatworms and entoprocts), the coelom has been lost (that is, these groups reverted to an acoelomate state). Some lineages are characterized by a pseudocoel, a body cavity lined with mesoderm in which the internal organs are suspended (see Figure 31.5B). In two of the most prominent protostome clades, the coelom has been highly modified:

- Arthropods lost the ancestral condition of the coelom over the course of evolution. Their internal body cavity has become a **hemocoel**, or "blood chamber," in which fluid from an open circulatory system bathes the internal organs before returning to blood vessels.
- Most mollusks have an open circulatory system with some of the attributes of the hemocoel, but they retain vestiges of an enclosed coelom around their major organs.

The protostomes can be divided into two major clades—the lophotrochozoans and the ecdysozoans—largely on the basis of DNA sequence analysis (**Figure 32.1**).

Go to Activity 32.1 Features of the Protostomes
Life10e.com/ac32.1

TABLE 32.1 Anatomical Characteristics of Some Major Protostome Groups

Group	Body Cavity	Digestive Tract	Circulatory System
Arrow worms	Coelom	Complete	None
LOPHOTROCHOZOANS			
Bryozoans	Coelom	Complete	None
Entoprocts	None	Complete	None
Flatworms	None	Blind gut	None
Rotifers	Pseudocoel	Complete	None
Gastrotrichs	Pseudocoel	Complete	None
Ribbon worms	Coelom	Complete	Closed
Brachiopods	Coelom	Complete in most	Open
Phoronids	Coelom	Complete	Closed
Annelids	Coelom	Complete	Closed or open
Mollusks	Reduced coelom	Complete	Open except in cephalopods
ECDYSOZOANS			
Nematodes	Pseudocoel	Complete	None
Horsehair worms	Pseudocoel	Greatly reduced	None
Arthropods	Hemocoel	Complete	Open

Cilia-bearing lophophores and trochophores evolved among the lophotrochozoans

Lophotrochozoans derive their name from two different ciliated features: a feeding structure known as a lophophore and a free-living larval form known as a trochophore. Neither the lophophore nor the trochophore is universal to all lophotrochozoans, however.

Several distantly related groups of lophotrochozoans (including bryozoans, entoprocts, brachiopods, and phoronids) have a **lophophore**, a circular or U-shaped ring of ciliated, hollow tentacles around the mouth (**Figure 32.2**). This complex organ is used for both food collection and gas exchange. Biologists once grouped taxa that have lophophores together as "lophophorates," but it is now clear that they are not one another's closest relatives. The lophophore appears to have evolved independently at least twice, or else it is an ancestral feature of lophotrochozoans and has been lost in many groups. Nearly all animals with a lophophore are sessile as adults. They use the tentacles and cilia of the lophophore to capture small floating organisms from the water. Other sessile lophotrochozoans have less well developed tentacles that they use for the same purpose.

Go to Media Clip 32.1
Feeding with a Lophophore
Life10e.com/mc32.1

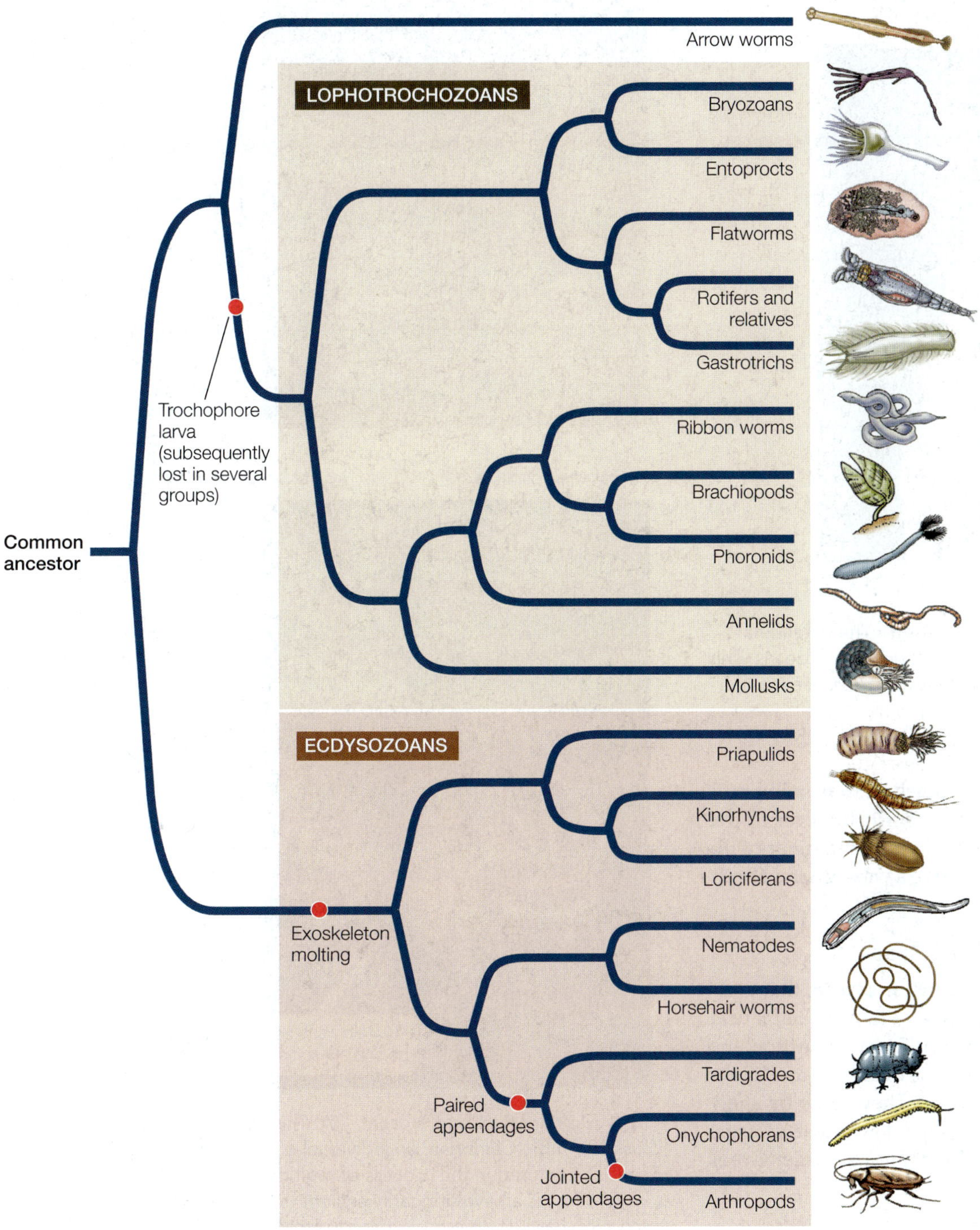
Arrow worms
LOPHOTROCHOZOANS
Bryozoans
Entoprocts
Flatworms
Rotifers and relatives
Gastrotrichs
Ribbon worms
Brachiopods
Phoronids
Annelids
Mollusks
Trochophore larva (subsequently lost in several groups)
Common ancestor
ECDYSOZOANS
Priapulids
Kinorhynchs
Loriciferans
Nematodes
Horsehair worms
Tardigrades
Onychophorans
Arthropods
Exoskeleton molting
Paired appendages
Jointed appendages

32.1 Phylogenetic Tree of Protostomes Two major lineages, the lophotrochozoans and the ecdysozoans, dominate the protostome tree. Some small groups are not included in this tree. The phylogenetic relationships shown here are supported mainly by genomic sequence data. Although genomic studies are contributing greatly to our knowledge of animal phylogeny, most species of protostomes have yet to be studied in detail.

Go to Activity 32.2 Protosome Classification
Life10e.com/ac32.2

Some lophotrochozoans, especially in their larval form, use cilia for locomotion. The larval form known as a **trochophore** moves by beating a band of cilia (see Figure 31.10A). This movement of cilia also brings plankton closer to the larva, where it can capture and ingest them (its cilia are therefore similar in function to the cilia of the lophophore). Trochophore larvae are found among many of the major groups of lophotrochozoans, including the mollusks, annelids, ribbon worms, entoprocts, and bryozoans. This larval form was probably present in the common ancestor of lophotrochozoans but has been subsequently lost in several lineages.

As we discussed in Chapter 31, some lophotrochozoans (including flatworms, ribbon worms, annelids, and mollusks) exhibit a form of cleavage in early development known as spiral cleavage. Some biologists group these taxa together as "spiralians," although phylogenetic analyses of gene sequences do not support a spiralian monophyly. Nonetheless, spiral cleavage may have been present in the lophotrochozoan ancestor and subsequently lost in several descendant lineages.

Many lineages of lophotrochozoans have a wormlike body form, which means that they are bilaterally symmetrical, legless, soft-bodied, and at least several times longer than they

Bryozoans can oscillate, rotate, and retract their lophophore tentacles.
Plumatella repens
100 μm

32.2 Bryozoans Use the Lophophore to Feed The extended lophophore dominates the anatomy of the colonial bryozoans. This species inhabits fresh water, although most bryozoans are marine.

are wide. A wormlike body form enables animals to burrow efficiently through marine sediment or soil. However, as we will see in Section 32.2, the mollusks—the most familiar of the lophotrochozoans to many people—have a very different body organization.

Ecdysozoans must shed their cuticles

Ecdysozoans have an external covering, or **cuticle**, that is secreted by the underlying epidermis (the outermost cell layer). The cuticle provides these animals with both protection and support. Once formed, however, the cuticle cannot grow. How, then, can ecdysozoans increase in size? They do so by shedding, or **molting**, the cuticle and replacing it with a new, larger one. This molting process gives the clade its name (Greek *ecdysis*, "to get out of").

Go to Media Clip 32.2
Molting a Cuticle
Life10e.com/mc32.2

A fossil Cambrian arthropod preserved in the process of molting shows that molting evolved more than 500 million years ago (Figure 32.3A). An increasingly rich array of molecular and genetic evidence, including a set of Hox genes shared by all ecdysozoans, suggests they have a single common ancestor. Thus molting of a cuticle is a trait that may have evolved only once during animal evolution.

Before an ecdysozoan molts, a new cuticle is already forming underneath the old one. Once the old cuticle is shed, the new one expands and hardens. Until it has hardened, though, the animal is vulnerable to its enemies, both because its outer surface is easy to penetrate and because an individual with a soft cuticle can move only slowly or not at all (Figure 32.3B).

In many ecdysozoans that have wormlike bodies, the cuticle is relatively thin and flexible; it offers the animal some protection but provides only modest body support. A thin cuticle allows the exchange of gases, minerals, and water across the body surface, but it restricts the animal to moist habitats. Many species of ecdysozoans with thin cuticles live in marine sediments from which they obtain food, either by ingesting sediments and extracting

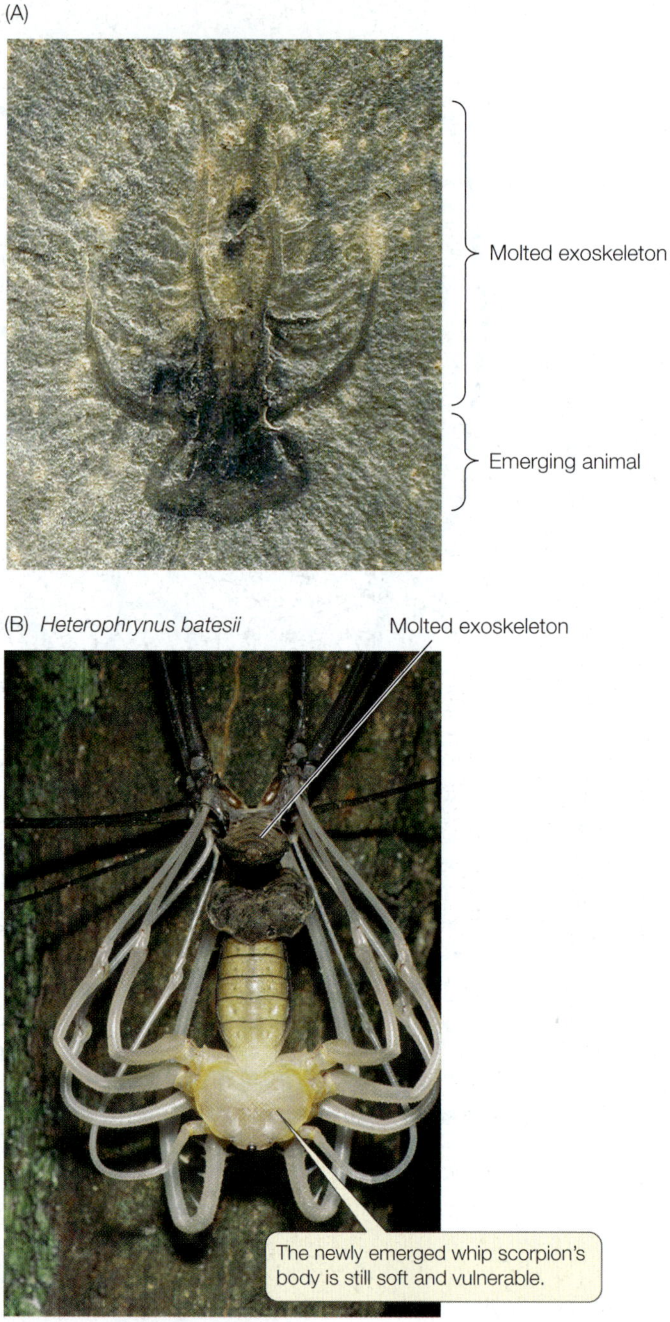
(A)
Molted exoskeleton
Emerging animal
(B) Heterophrynus batesii
Molted exoskeleton
The newly emerged whip scorpion's body is still soft and vulnerable.

32.3 Molting: Past and Present (A) This 500-million-year-old fossil from the Cambrian, an individual of a long-extinct arthropod species captured in the process of molting, shows that the molting process is an evolutionarily ancient trait. (B) This tailless whip scorpion has just emerged from its discarded exoskeleton. It will be highly vulnerable until its new cuticle has hardened.

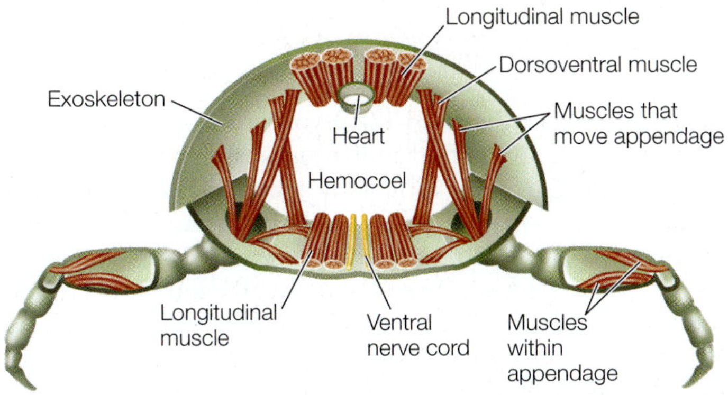
Longitudinal muscle
Dorsoventral muscle
Exoskeleton
Muscles that move appendage
Heart
Hemocoel
Longitudinal muscle
Ventral nerve cord
Muscles within appendage

32.4 Arthropod Skeletons Are Rigid and Jointed This cross section through a thoracic segment of a generalized arthropod illustrates the arthropod body plan, which is characterized by a rigid exoskeleton with jointed appendages.

organic material from them or by capturing larger prey using a toothed **pharynx** (a muscular organ at the anterior end of the digestive tract). Some freshwater species absorb nutrients directly through their thin cuticles, as do parasitic species that live within their hosts (endoparasites). Many wormlike ecdysozoans are predators, eating protists and small animals.

The cuticles of other ecdysozoans, mainly arthropods, function as external skeletons, or **exoskeletons**. These exoskeletons are thickened by layers of protein and a strong, waterproof polysaccharide called **chitin**. An animal with a rigid, chitin-reinforced exoskeleton can neither move in a wormlike manner nor use cilia for locomotion. A hard exoskeleton also impedes the passage of oxygen and nutrients into the animal, presenting new challenges in other areas besides growth. Thus new mechanisms of locomotion and gas exchange evolved in those ecdysozoans with hard exoskeletons.

To move rapidly, an animal with a rigid exoskeleton must have body extensions that can be manipulated by muscles. Such appendages evolved in the late Precambrian, leading to the **arthropod** ("jointed foot") clade. Arthropod appendages exist in an amazing variety of forms. They serve many functions, including walking and swimming, gas exchange, food capture and manipulation, copulation, and sensory perception. Arthropods grasp food with their mouths and associated appendages and digest it internally. Their muscles are attached to the inside of the exoskeleton. Each segment has muscles that operate that segment and the appendages attached to it (**Figure 32.4**).

The arthropod exoskeleton has had a profound influence on the evolution of these animals. Encasement within a rigid body covering provides support for walking on dry land, and the waterproofing provided by chitin keeps the animal from dehydrating in dry air. Thus aquatic arthropods were, in short, excellent candidates to invade terrestrial environments. As we will see, they did so several times.

Arrow worms retain some ancestral developmental features

Nearly all triploblastic animal groups can be readily classified as either protostomes or deuterostomes, but the evolutionary relationships of one small group, the **arrow worms**, were debated for many years. Although the early development of arrow worms seems similar to that of deuterostomes, it is now known that arrow worms merely retain developmental features that are ancestral to triploblastic animals in general. Recent studies of gene sequences clearly identify arrow worms as protostomes. There is still some question as to whether they are the closest relatives of the lophotrochozoans (as shown in Figure 32.1) or possibly the sister group of all other protostomes.

The arrow worm body is divided into three compartments: head, trunk, and tail (**Figure 32.5**). The body is transparent or translucent. Most arrow worms swim in the open sea. A few species live on the seafloor. Their abundance as fossils indicates that they were common more than 500 million years ago. The 180 or so living species of arrow worms are small enough—ranging from 3 millimeters to 12 centimeters in length—that their gas exchange and waste excretion requirements are met by diffusion through the body surface. They lack a circulatory system; wastes and nutrients are moved around the body in the coelomic fluid, which is propelled by cilia that line the coelom.

Arrow worms are hermaphroditic; that is, each individual produces both male and female gametes. Eggs are fertilized internally following elaborate courtship between two individuals. Miniature adults hatch directly from the eggs; these animals have no distinct larval stage.

Arrow worms are stabilized in the water by means of one or two pairs of lateral fins and a tail fin. They are major predators of planktonic organisms in the open ocean, ranging in size from

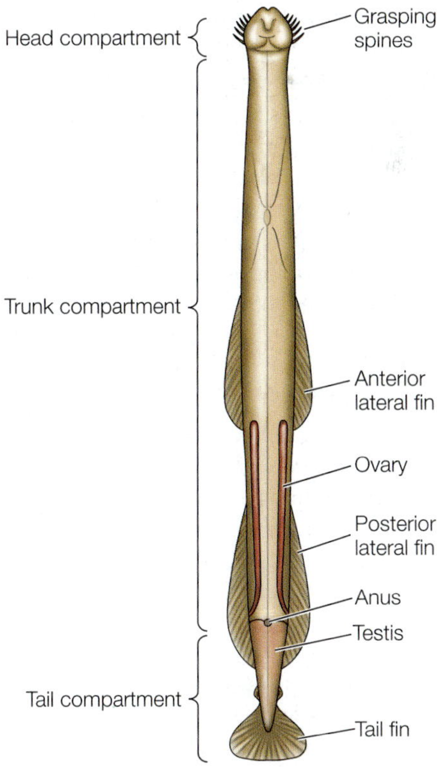
Head compartment
Grasping spines
Trunk compartment
Anterior lateral fin
Ovary
Posterior lateral fin
Anus
Testis
Tail compartment
Tail fin

32.5 An Arrow Worm Arrow worms have a three-part body organization. Their fins and grasping spines are adaptations for a predatory lifestyle. Individuals are hermaphroditic, producing both eggs in an ovary and sperm in a testis.

small protists to young fish as large as the arrow worms themselves. An arrow worm typically lies motionless in the water until water movement signals the approach of prey. The arrow worm then darts forward and uses the stiff spines adjacent to its mouth to grasp its prey.

RECAP 32.1

The shared derived traits of protostomes include a blastopore that develops into a mouth, an anterior brain, and a ventral nervous system. Several lophotrochozoan groups are characterized by a filter-feeding structure known as a lophophore or by cilia-bearing larvae known as trochophores. Ecdysozoans, which have a body covering known as a cuticle, must molt periodically in order to grow.

- How does an animal's body covering influence the way it exchanges gases, feeds, and moves? **See pp. 654–655**
- What features made arthropods well adapted for colonizing terrestrial environments? **See p. 655**

In the next section we continue our survey of the protostomes with a more detailed look at the major groups of lophotrochozoans and the diverse body forms that are found among them.

32.2 What Features Distinguish the Major Groups of Lophotrochozoans?

Lophotrochozoans come in a variety of sizes and shapes, ranging from relatively simple animals with a blind gut (that is, a gut with only one opening) and no internal transport system to animals with a complete gut (having separate entrance and exit openings) and a complex internal transport system. They include some species-rich groups, such as flatworms, annelids, and mollusks. A number of these groups have wormlike bodies, but the lophotrochozoans encompass a wide variety of morphologies, including a few groups with external shells. Some lophotrochozoan groups have only recently been discovered by biologists.

Most bryozoans and entoprocts live in colonies

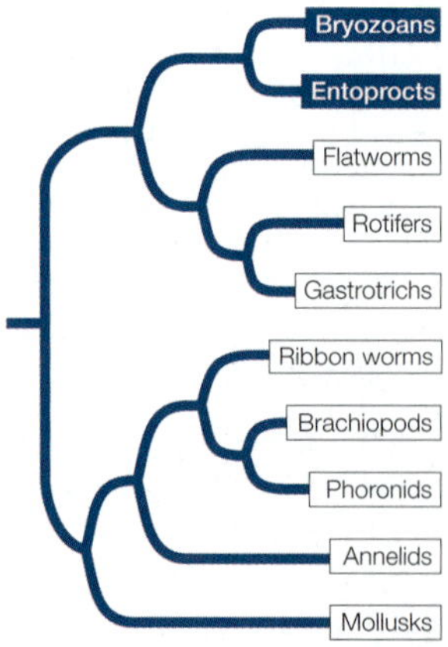
Bryozoans
Entoprocts
Flatworms
Rotifers
Gastrotrichs
Ribbon worms
Brachiopods
Phoronids
Annelids
Mollusks

Most of the 5,500 species of **bryozoans** ("moss animals") and 170 species of **entoprocts** (meaning "anus inside") are colonial animals that live in a "house" made of material secreted by the external body wall. The colonial species are sessile, but the few solitary species can slowly move around in their environment. Almost all bryozoans and entoprocts are marine, although a few species occur in fresh or brackish water.

A bryozoan colony consists of many small (1–2 mm) individuals connected by strands of tissue along which nutrients can be moved (see Figure 31.12). Bryozoan colonies can grow to contain more than 2 million individuals, all stemming from the asexual reproduction of the colony's founder. Rocks in coastal regions in many parts of the world are covered with luxuriant growths of bryozoans. Some bryozoans create miniature reefs in shallow waters. In some species, the individual colony members are differentially specialized for feeding, reproduction, defense, or support. Individual bryozoans in a colony are able to oscillate their lophophore to increase contact with prey. They can also retract it into their "house" (see Figure 32.2).

Bryozoans can reproduce sexually by releasing sperm into the water, which carries the sperm to other individuals. Eggs are fertilized internally; developing embryos are brooded before they exit as larvae to seek suitable sites for attachment to the substrate. Entoprocts can also reproduce both sexually and asexually. Some species of entoprocts release unfertilized eggs into the water for fertilization, whereas other species brood their developing young as bryozoans do.

Bryozoans and entoprocts differ in the placement of the anus. In bryozoans, the anus is located outside the ring of tentacles that make up the lophophore, whereas the anus of entoprocts is located in the center of this ring. The lophophores of the two groups also function differently: food particles are carried from the tips to the bases of the tentacles in bryozoans, but from the bases to the tips of the tentacles in entoprocts. Entoprocts lack a coelom, whereas bryozoans have a three-part coelom.

Flatworms, rotifers, and gastrotrichs are structurally diverse relatives

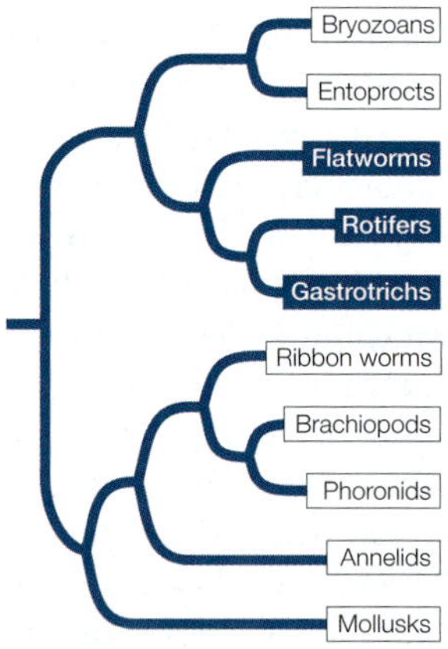
Bryozoans
Entoprocts
Flatworms
Rotifers
Gastrotrichs
Ribbon worms
Brachiopods
Phoronids
Annelids
Mollusks

Flatworms, rotifers, gastrotrichs, and their close relatives are a structurally diverse group of organisms whose relationships to one another has been hypothesized only recently. If recent genomic studies prove correct, this monophyletic lophotrochozoan group includes both acoelomate subgroups (e.g., the flatworms) and pseudocoelomate subgroups (e.g., the rotifers and gastrotrichs), and yet the closest relatives of this group—the bryozoans and entoprocts—are coelomate and acoelomate, respectively. Thus this group provides an example of the evolutionary convergence in body cavity form that we described in Section 32.1.

FLATWORMS **Flatworms** lack specialized organs for transporting oxygen to their internal tissues. In the absence of a gas transport system, each cell must be near a body surface, a requirement met by the dorsoventrally flattened body form that gives these animals their common name. The digestive tract of a flatworm consists of a mouth opening into a blind gut. The gut is often highly branched, forming intricate patterns that increase the surface area available for the absorption of nutrients. Some small free-living flatworms are cephalized, with a

(A) *Eurylepta californica*, a free-living flatworm

(B) Diagram of a typical parasitic flatworm

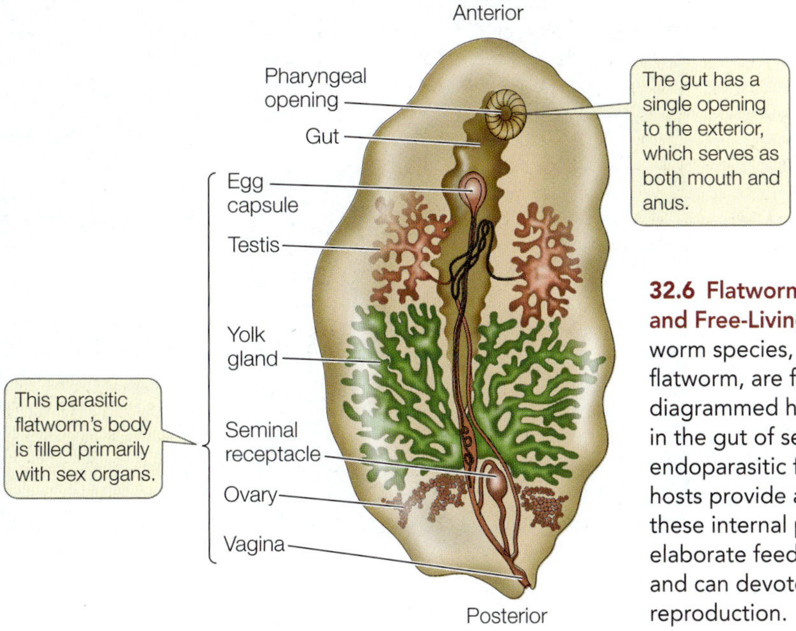
Anterior
Pharyngeal opening
Gut
Egg capsule
Testis
Yolk gland
Seminal receptacle
Ovary
Vagina
Posterior
The gut has a single opening to the exterior, which serves as both mouth and anus.
This parasitic flatworm's body is filled primarily with sex organs.

32.6 Flatworms Include Both Parasites and Free-Living Forms (A) Some flatworm species, such as this Pacific marine flatworm, are free-living. (B) The fluke diagrammed here lives endoparasitically in the gut of sea urchins and is typical of endoparasitic flatworms. Because their hosts provide all the nutrition they need, these internal parasites do not require elaborate feeding or digestive organs and can devote most of their bodies to reproduction.

head bearing chemoreceptor organs, two simple eyes, and a tiny brain composed of anterior thickenings of the longitudinal nerve cords. Free-living flatworms glide over surfaces, powered by broad bands of cilia (**Figure 32.6A**).

Although many flatworms are free-living, most flatworm species are parasites. Of the parasitic species, most are endoparasites. There are also flatworms that feed externally on animal tissues (living or dead), and some graze on plants. A likely evolutionary transition was from feeding on dead organisms to feeding on the body surfaces of dying hosts to invading and consuming parts of healthy hosts.

Most of the 30,000 species of living flatworms are tapeworms and flukes; members of these two groups are endoparasites, particularly of vertebrates (**Figure 32.6B**). Because they absorb digested food from the digestive tracts of their hosts, many endoparasitic flatworms lack digestive tracts of their own. Some cause serious human diseases, such as schistosomiasis, which is common in parts of Asia, Africa, and South America. The species that causes this devastating disease has a complex life cycle involving both freshwater snails and mammals as hosts. Members of another flatworm group, the monogeneans, are ectoparasites of fishes and other aquatic vertebrates. The turbellarians include most of the free-living species.

ROTIFERS Most species of **rotifers** are tiny (50–500 μm long)—smaller than some ciliate protists—but they have specialized internal organs (**Figure 32.7A and B**). A complete gut passes from an anterior mouth to a posterior anus; the body cavity is a pseudocoel that functions as a hydrostatic skeleton. Rotifers typically propel themselves through the water by means of rapidly beating cilia rather than by muscular contraction.

The most distinctive organ of rotifers is a conspicuous ciliated organ called the corona, which surmounts the head of many species. Coordinated beating of the cilia sweeps particles of organic matter from the water into the animal's mouth and down to a complicated structure called the mastax, in which food is ground into small pieces. By contracting muscles around the pseudocoel, a few rotifer species that prey on protists and small animals can protrude the mastax through the mouth and seize small objects with it.

Go to Media Clip 32.3
Rotifer Feeding
Life10e.com/mc32.3

Most of the known species of rotifers live in fresh water. Some species rest on the surfaces of mosses or lichens in a desiccated, inactive state until it rains. When rain falls, they absorb water and become mobile, feeding in the films of water that temporarily cover the plants. Most rotifers live no longer than a few weeks.

Both males and females are found in some species of rotifers, but only females are known among the bdelloid rotifers (the b in "bdelloid" is silent). Biologists have concluded that the bdelloid rotifers may have existed for tens of millions of years without regular sexual reproduction. Lack of genetic recombination generally leads to the buildup of deleterious mutations, so long-term asexual reproduction typically leads to extinction (see Section 21.4). Recent studies, however, have indicated that bdelloid rotifers may avoid this problem by picking up fragments of genes from their environment during the desiccation–rehydration cycle, which allows genetic recombination among individuals in the absence of direct sexual exchange.

A few highly reduced lineages appear to have descended from the free-living rotifers. The spiny-headed worms are parasites with complex life cycles, often parasitizing several animal hosts (**Figure 32.7C**). The jaw worms are tiny marine organisms that glide between sand grains in shallow marine environments. Although spiny-headed worms and jaw worms are

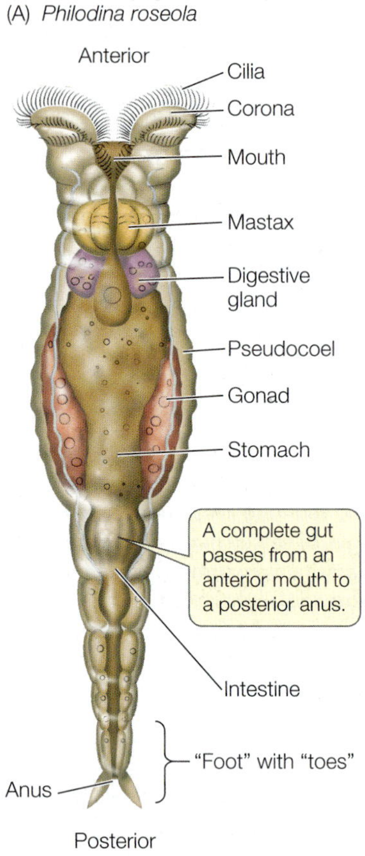
(A) *Philodina roseola*
Anterior
Cilia
Corona
Mouth
Mastax
Digestive gland
Pseudocoel
Gonad
Stomach
A complete gut passes from an anterior mouth to a posterior anus.
Intestine
"Foot" with "toes"
Anus
Posterior

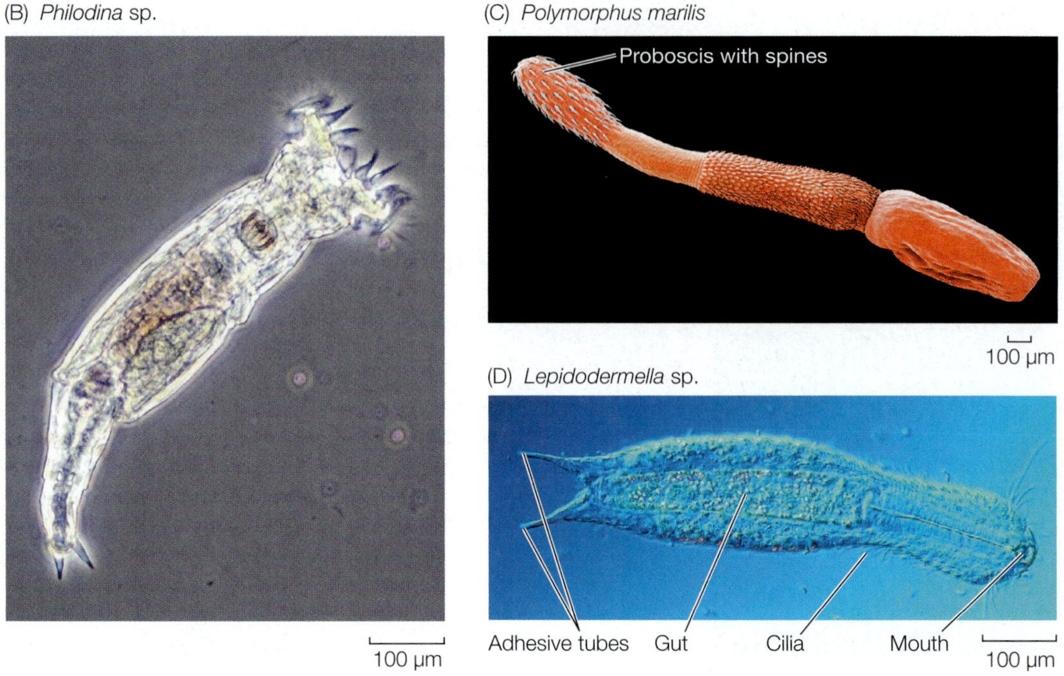
(B) *Philodina* sp.
100 µm
(C) *Polymorphus marilis*
Proboscis with spines
100 µm
(D) *Lepidodermella* sp.
Adhesive tubes
Gut
Cilia
Mouth
100 µm

32.7 Rotifers and Gastrotrichs (A) The rotifer diagrammed here reflects the general structure of many rotifers. (B) A micrograph reveals the internal complexity of the microscopic rotifers. (C) Spiny-headed worms are parasitic rotifer relatives. The spines of the proboscis anchor the animal to the organs of its host. (D) Gastrotrichs superficially resemble rotifers but have flattened ventral surfaces covered with cilia, as flatworms do.

structurally quite distinct, molecular analyses have revealed that both groups are essentially highly modified rotifers.

GASTROTRICHS The 800 species of **gastrotrichs** (also called "hairy backs") are abundant, tiny (0.05–3 mm) animals that live in marine sediments, in fresh waters, and in the water films that surround grains of soil. Their transparent bodies have a flat ventral surface that is covered with cilia (Figure 32.7D). Most species are simultaneous hermaphrodites, with both male and female reproductive organs, although the male organs have been greatly reduced or lost in some species that reproduce asexually.

Ribbon worms have a long, protrusible feeding organ

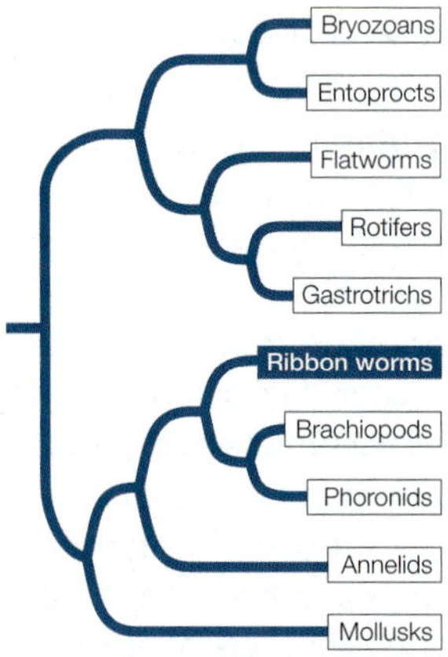
Bryozoans
Entoprocts
Flatworms
Rotifers
Gastrotrichs
Ribbon worms
Brachiopods
Phoronids
Annelids
Mollusks

Ribbon worms (nemerteans) have simple nervous and excretory systems similar to those of flatworms. Unlike flatworms, however, they have a complete digestive tract with a mouth at one end and an anus at the other. Small ribbon worms move slowly by beating their cilia. Larger ones employ waves of muscle contraction to move over the surface of sediments or to burrow into them.

Within the body of nearly all of the 1,200 species of ribbon worms is a fluid-filled cavity called the rhynchocoel, within which lies a hollow, muscular **proboscis**. The proboscis, which is the worm's feeding organ, may extend much of the length of the body. Contraction of the muscles surrounding the rhynchocoel causes the proboscis to evert explosively through an anterior pore (Figure 32.8A). The proboscis may be armed with sharp stylets that pierce prey and discharge paralysis-causing toxins into the wound.

Go to Media Clip 32.4
Explosive Extrusion of Ribbon Worm Proboscis
Life10e.com/mc32.4

Most ribbon worm species are marine, although there are species that live in fresh water or on land. Most species are less than 20 centimeters long, but individuals of some species reach 20 meters or more. Some genera feature species that are conspicuous and brightly colored (Figure 32.8B). Recent molecular analyses suggest that ribbon worms may be most closely related to the brachiopods and phoronids.

Brachiopods and phoronids use lophophores to extract food from the water

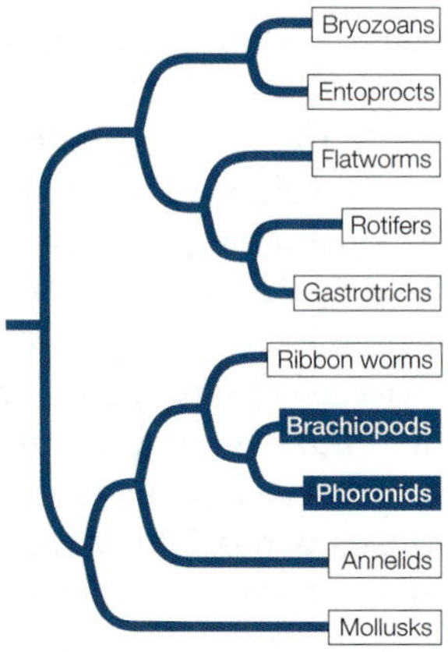
Bryozoans
Entoprocts
Flatworms
Rotifers
Gastrotrichs
Ribbon worms
Brachiopods
Phoronids
Annelids
Mollusks

Recall that the bryozoans and entoprocts use a lophophore to feed. Brachiopods and phoronids also feed using a lophophore, but this structure may have evolved separately in these groups. Although neither the brachiopods nor the phoronids are represented by many living species, the brachiopods (which have hard external shells and thus leave an excellent

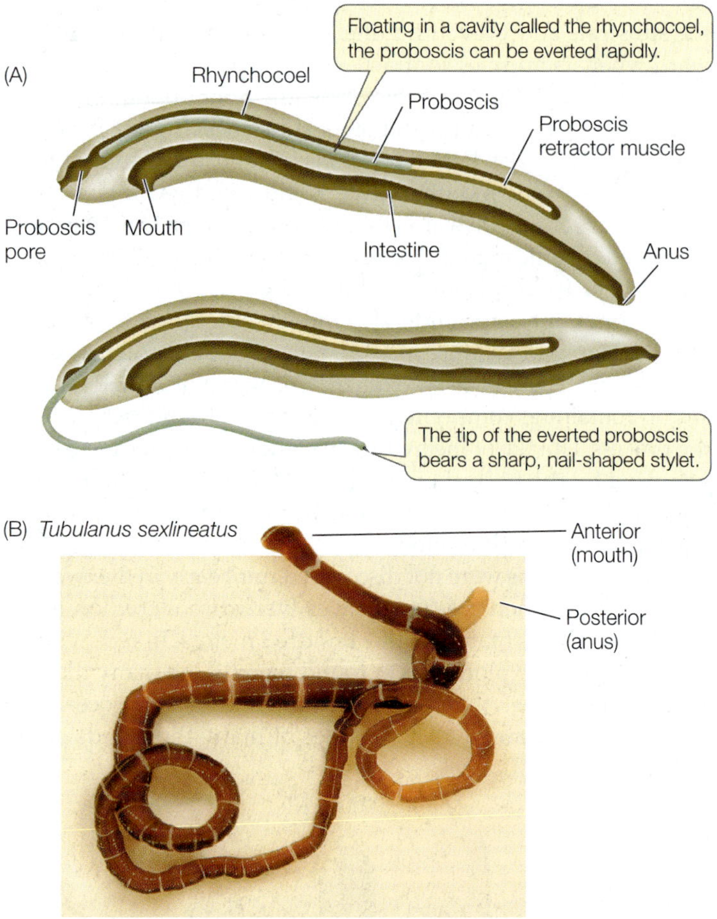
(A)
Rhynchocoel
Floating in a cavity called the rhynchocoel, the proboscis can be everted rapidly.
Proboscis
Proboscis retractor muscle
Proboscis pore
Mouth
Intestine
Anus
The tip of the everted proboscis bears a sharp, nail-shaped stylet.
(B) Tubulanus sexlineatus
Anterior (mouth)
Posterior (anus)

32.8 Ribbon Worms (A) The proboscis is the ribbon worm's feeding organ. (B) This large marine nemertean is found in harbors and bays along the Pacific Coast of North America. Its proboscis is not everted in this photograph.

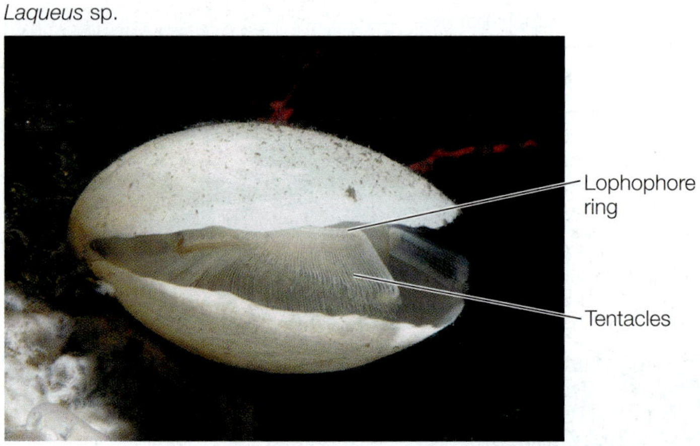
Laqueus sp.
Lophophore ring
Tentacles

32.9 A Brachiopod's Lophophore The lophophore of this North Pacific brachiopod can be seen between the valves of its shell.

fossil record) are known to have been much more abundant in the past.

BRACHIOPODS **Brachiopods** (lampshells) are solitary marine animals. They have a rigid shell that is divided into two parts connected by a ligament (**Figure 32.9**). The two halves can be pulled shut to protect the soft body. Brachiopods superficially resemble bivalve mollusks, but shells have evolved independently in the two groups. The two halves of the brachiopod shell are dorsal and ventral, rather than lateral as in bivalves. The lophophore is located within the shell. The beating of cilia on the lophophore draws water into the slightly opened shell. Food is trapped in the lophophore and directed to a ridge, along which it is transferred to the mouth.

Most brachiopods are 4 to 6 centimeters long. They live attached to a solid substrate or embedded in soft sediments. Most species are attached by means of a short, flexible stalk that holds the animal above the substrate. Gases are exchanged across body surfaces, especially the tentacles of the lophophore. Most brachiopods release their gametes into the water, where they are fertilized. The larvae remain among the plankton for only a few days before they settle and develop into adults.

Brachiopods reached their peak abundance and diversity in Paleozoic and Mesozoic times. More than 26,000 fossil species have been described. Only about 450 species survive, but they remain common in some marine environments.

PHORONIDS The ten known species of **phoronids** are small (5–25 cm long), sessile worms that live in muddy or sandy sediments or attached to rocky substrates. Phoronids are found in marine waters from the intertidal zone to about 400 meters deep. They secrete tubes made of chitin, within which they live, and have a U-shaped gut with the anus located outside the lophophore (**Figure 32.10**). Their cilia drive water into the top of the lophophore, and the water exits through the narrow spaces between the tentacles. Suspended food particles are caught and transported to the mouth by ciliary action. Some species release eggs into the water, where they are fertilized, but other species produce large eggs that are fertilized internally and retained in the parent's body, where they are brooded until they hatch.

Annelids have segmented bodies

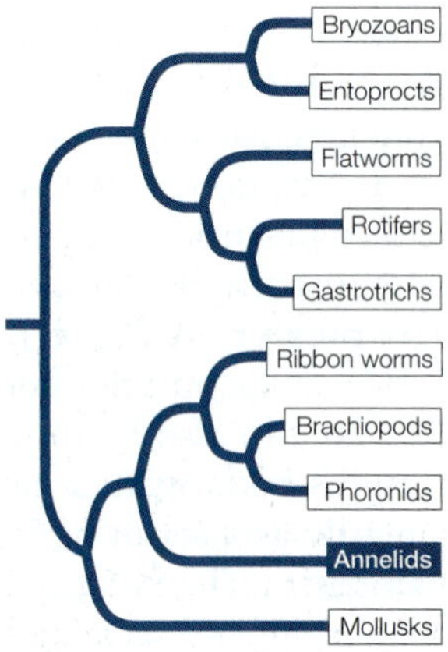
Bryozoans
Entoprocts
Flatworms
Rotifers
Gastrotrichs
Ribbon worms
Brachiopods
Phoronids
Annelids
Mollusks

The wormlike bodies of **annelids** are clearly segmented. As described in Section 31.2, segmentation allows an animal to move different parts of its body independently, giving it much better control of its movement. The earliest segmented worms, preserved as fossils from the middle Cambrian, were burrowing marine annelids.

In most large annelids, the coelom in each segment is isolated from those in other segments (**Figure 32.11**). A separate nerve center called a ganglion (plural ganglia) controls each segment; nerve cords that connect the ganglia coordinate their

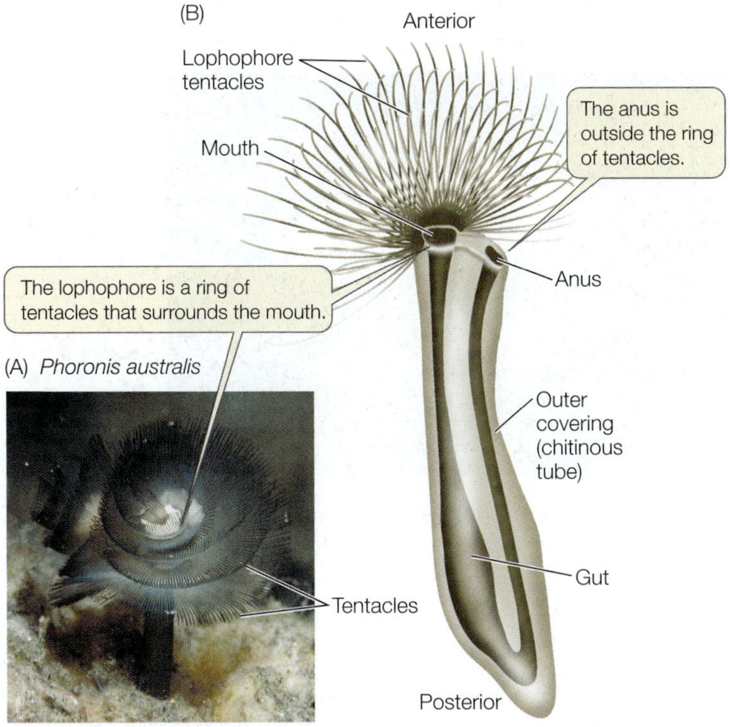
(B)
Anterior
Lophophore tentacles
Mouth
The anus is outside the ring of tentacles.
Anus
The lophophore is a ring of tentacles that surrounds the mouth.
(A) Phoronis australis
Outer covering (chitinous tube)
Gut
Tentacles
Posterior

32.10 Phoronids (A) The tentacles of this phoronid's lophophore form a spiral. (B) The phoronid gut is U-shaped, as seen in this generalized diagram.

the substrate and preventing the animal from slipping backward when its muscles contract.

Members of one polychaete clade, the **pogonophorans**, secrete tubes made of chitin and other substances, in which they live (**Figure 32.12B**). Pogonophorans have lost their digestive tract (they have no mouth or gut). So how do they obtain nutrition? Part of the answer is that pogonophorans can take up dissolved organic matter directly from the sediments in which they live or from the surrounding water. Much of their nutrition, however, is provided by endosymbiotic bacteria that the pogonophorans house in a specialized organ known as the trophosome. These bacteria oxidize hydrogen sulfide and other sulfur-containing compounds, fixing carbon from methane in the process. Uptake of the hydrogen sulfide, methane, and oxygen used by the bacteria is facilitated by hemoglobin in the pogonophorans' tentacles. It is this hemoglobin that gives the tentacles their red coloration.

Pogonophorans were not discovered until early in the twentieth century, when the first species were discovered on the seafloor at depths of up to a few hundred meters. In recent decades, deep-sea explorers have found them living many thousands of meters below the ocean surface. In these deep oceanic sediments, they may reach densities of many thousands per

functioning. Most annelids lack a rigid external protective covering; instead they have a thin, permeable body wall that serves as a general surface for gas exchange. Most annelids are thus restricted to moist environments because they lose body water rapidly in dry air. The approximately 19,000 described species live in marine, freshwater, and moist terrestrial environments.

POLYCHAETES More than half of all annelid species are commonly known as **polychaetes** ("many hairs"), although this is a descriptive term rather than the name of a single clade. Recent molecular studies indicate that polychaetes are paraphyletic with respect to the remaining annelids. Most polychaetes are marine, and many live in burrows in soft sediments. Most of them have one or more pairs of eyes and one or more pairs of tentacles, with which they capture prey or filter food from the surrounding water, at the anterior end of the body (**Figure 32.12A**; see also Figure 31.6A). In some species, the body wall of most segments extends laterally as a series of thin outgrowths called parapodia. The parapodia function in gas exchange, and some species use them to move. Stiff bristles called setae protrude from each parapodium, forming temporary contact with

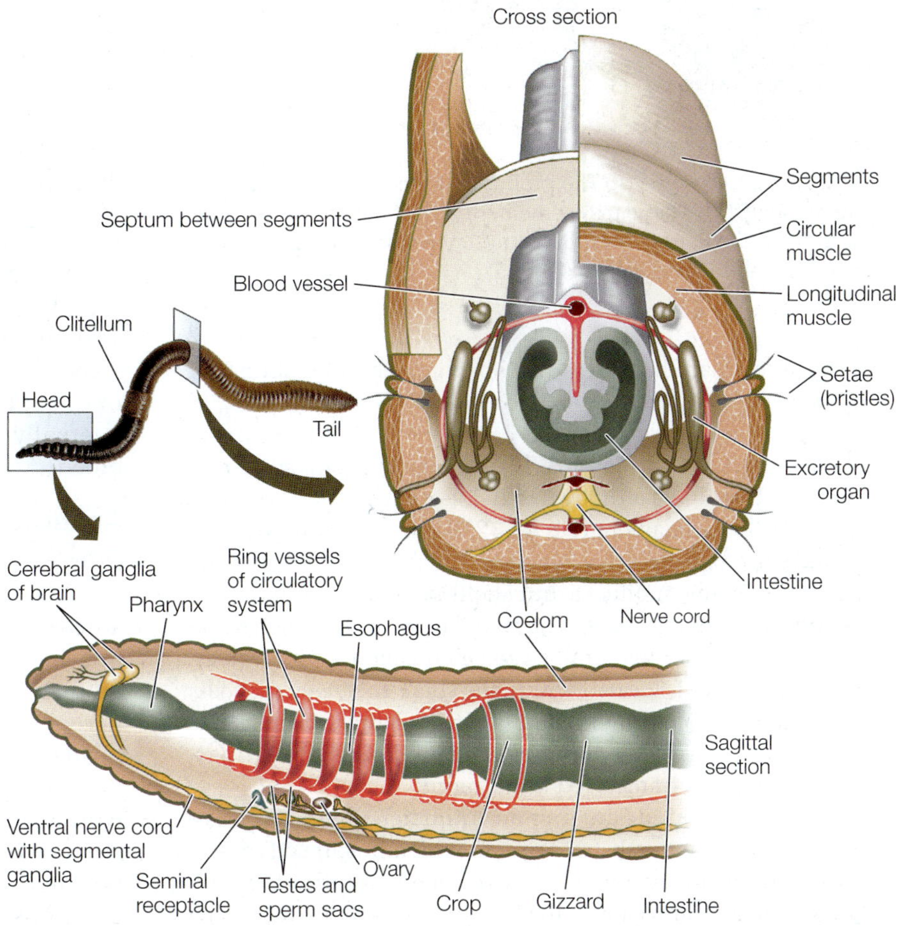
Cross section
Segments
Septum between segments
Circular muscle
Blood vessel
Longitudinal muscle
Clitellum
Setae (bristles)
Head
Tail
Excretory organ
Intestine
Cerebral ganglia of brain
Ring vessels of circulatory system
Pharynx
Esophagus
Coelom
Nerve cord
Sagittal section
Ventral nerve cord with segmental ganglia
Seminal receptacle
Testes and sperm sacs
Ovary
Crop
Gizzard
Intestine

32.11 Annelids Have Many Body Segments The segmented structure of annelids such as this earthworm is apparent both externally and internally. Many organs are repeated serially.

(A) Spirographis spallanzanii
(B) Riftia sp.
(C) Lumbricus terrestris
(D) Hirudo medicinalis

32.12 Diversity among the Annelids (A) "Fan worms" or "feather duster worms" are sessile marine polychaetes that grow in masses, filtering food from the water with their tentacles. This individual has been removed from its chitinous tube. (B) Pogonophorans live around hydrothermal vents deep in the ocean. Their tentacles can be seen protruding from their chitinous tubes. (C) Earthworms, like all oligochaetes, are hermaphroditic; when they copulate, each individual donates and receives sperm. (D) The medicinal leech has been a tool of physicians and healers for centuries. Even today, leeches have uses in clinical practice.

square meter. About 160 species have been described. The largest and most remarkable pogonophorans are 2 meters or more in length and live near deep-sea hydrothermal vents—volcanic openings in the seafloor through which hot, sulfide-rich water pours. The methane and hydrogen sulfide from these vents provide the raw materials for carbon fixation by the pogonophorans' endosymbiotic bacteria.

CLITELLATES The approximately 3,000 described species of **clitellates**, which form a well-supported clade within the annelids, are found in freshwater, marine, and terrestrial environments. The clitellates appear to be phylogenetically nested among various groups of polychaetes, although the exact relationships are not yet clear. There are two major groups of clitellates, the oligochaetes and the leeches.

Oligochaetes ("few hairs") have no parapodia, eyes, or anterior tentacles, and they have only four pairs of setae bundles per segment. Earthworms—the most familiar oligochaetes—burrow in and ingest soil, from which they extract food particles. All oligochaetes are hermaphroditic. Sperm are exchanged simultaneously between two copulating individuals (**Figure 32.12C**). Eggs and sperm are deposited in a cocoon outside the adult's body. Fertilization occurs within the cocoon after it is shed, and when development is complete, miniature worms emerge and immediately begin independent life.

Leeches, like oligochaetes, lack parapodia and tentacles. The coelom of leeches is not divided into compartments; the coelomic space is largely filled with undifferentiated tissue. Groups of segments at each end of the body are modified to form suckers, which serve as temporary anchors that aid the leech in its movement. With its posterior sucker attached to a substrate, the leech extends its body by contracting its circular muscles. The anterior sucker is then attached, the posterior one detached, and the leech shortens itself by contracting its longitudinal muscles. Leeches live in freshwater or terrestrial habitats.

Most leeches are ectoparasites that feed by making an incision in a host, from which blood flows. A leech can ingest so much blood in a single feeding that its body may enlarge severalfold. The leech secretes an anticoagulant into the wound that keeps the host's blood flowing. For centuries, medical practitioners employed leeches to draw blood to treat diseases they believed were caused by an excess of blood or by "bad blood." Although most leeching practices (such as inserting a leech in a person's throat to alleviate swollen tonsils) have been abandoned, *Hirudo medicinalis* (the medicinal leech; **Figure 32.12D**) is used today to reduce fluid pressure and prevent

blood clotting in damaged tissues, to eliminate pools of coagulated blood, and to prevent scarring. The anticoagulants of certain other leech species also contain anesthetics and blood vessel dilators and are being studied for possible medical uses.

Go to Media Clip 32.5
Leeches Feeding on Blood
Life10e.com/mc32.5

Mollusks have undergone a dramatic evolutionary radiation

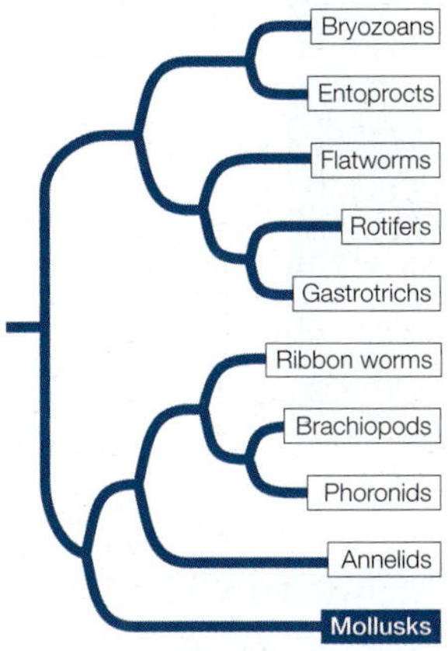

Bryozoans
Entoprocts
Flatworms
Rotifers
Gastrotrichs
Ribbon worms
Brachiopods
Phoronids
Annelids
Mollusks

Mollusks are the most diverse group of lophotrochozoans, both in numbers of species and in the environments they occupy. Although the major groups of mollusks differ dramatically in morphology, they all share the same three major body components: a foot, a visceral mass, and a mantle (Figure 32.13A).

- The molluscan **foot** is a large, muscular structure that originally was both an organ of locomotion and a support for the internal organs. In squids and octopuses, the foot has been modified to form arms and tentacles borne on a head with complex sensory organs. In other groups, such as clams, the foot is a burrowing organ. In some groups the foot is greatly reduced.
- The heart and the digestive, excretory, and reproductive organs are concentrated in a centralized, internal **visceral mass**.
- The **mantle** is a fold of tissue that covers the organs of the visceral mass. The mantle secretes the hard, calcareous shell that is typical of many mollusks.

In most mollusks, the mantle extends beyond the visceral mass to form a mantle cavity. Within this cavity lie gills that are used for gas exchange. When cilia on the gills beat, they create a current of water. The tissue of the gills, which is highly vascularized (contains many blood vessels), takes up oxygen from the water and releases carbon dioxide. Many mollusk species use their gills as filter-feeding devices, whereas others feed using a rasping structure known as a **radula** to scrape algae from rocks. In some mollusks, such as the marine cone snails, the radula has been modified into a drill or poison dart.

In all mollusks except cephalopods, the blood vessels do not form a closed system. Blood and other fluids empty into a large, fluid-filled hemocoel, through which fluids move and deliver oxygen to the internal organs. Eventually the fluids reenter the blood vessels and are moved by a heart.

Monoplacophorans were the most abundant mollusks during the Cambrian period, 500 million years ago, but only a few species survive today. In monoplacophorans, in contrast to all other living mollusks, the gas exchange organs, muscles, and excretory pores are repeated over the length of the body.

The four major clades of living mollusks are the chitons, gastropods, bivalves, and cephalopods. Each of these groups is readily identifiable and distinct, even though they share variations on a common body plan.

CHITONS Eight overlapping calcareous plates, surrounded by a structure known as the girdle, protect the internal organs and muscular foot of **chitons** (Figure 32.13B). The chiton body is bilaterally symmetrical, and the internal organs, particularly the digestive and nervous systems, are relatively simple. Most chitons are marine omnivores that scrape algae, bryozoans, and other organisms from rocks with a sharp radula. An adult chiton spends most of its life clinging tightly to rock surfaces with its large, muscular, mucus-covered foot. It moves slowly by means of rippling waves of muscular contraction in the foot. Fertilization in most chitons takes place in the water, but in a few species fertilization is internal and embryos are brooded within the body. There are approximately 1,000 living species of chitons.

GASTROPODS **Gastropods** (Figure 32.13C) are the most species-rich and widely distributed mollusks, with about 85,000 living species. Snails, whelks, limpets, slugs, nudibranchs (sea slugs), and abalones are all gastropods. Most species move by gliding on the muscular foot, but in a few species—the sea butterflies and heteropods—the foot is a swimming organ with which the animal moves through open ocean waters.

Marine nudibranchs and terrestrial slugs are gastropods that have lost their protective shell over the course of evolution (Figure 32.14). Without a shell, these groups rely on other forms of protection from predation. The coloration of many nudibranchs is aposematic, meaning that it serves to warn potential predators of toxicity. Other nudibranch species and most terrestrial slugs exhibit camouflaged coloration.

Shelled gastropods have one-piece shells. The only mollusks that live in terrestrial environments—land snails and slugs—are gastropods. In these terrestrial species, the mantle tissue is modified into a highly vascularized lung.

BIVALVES Clams, oysters, scallops, and mussels are all familiar **bivalves**. The approximately 30,000 species are found in both marine and freshwater environments. Bivalves have a hinged, two-part shell that extends over the sides of the body as well as the top (Figure 32.13D). Many clams use the foot to burrow into mud and sand. Bivalves feed by taking in water through an opening called an incurrent siphon and filtering food from the water with their large gills, which are also the main sites of gas exchange. Water and gametes exit through the excurrent siphon. Fertilization takes place in open water in most species.

CEPHALOPODS The **cephalopods**—squids, cuttlefish, octopuses, and nautiluses—first appeared near the beginning of the Cambrian period. By the Ordovician period a variety of types were present. Today there are about 800 living species. In these mollusks the excurrent siphon is modified to

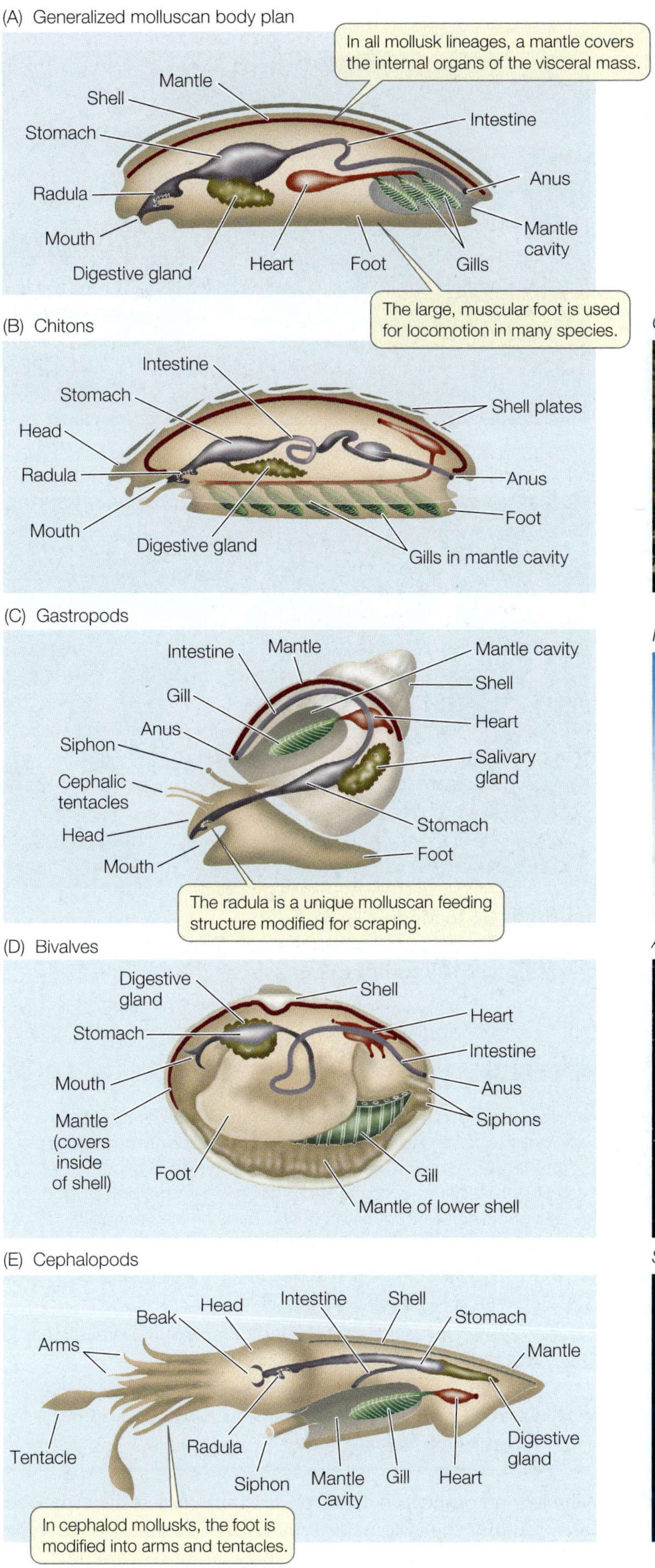
(A) Generalized molluscan body plan
In all mollusk lineages, a mantle covers the internal organs of the visceral mass.
Mantle
Shell
Stomach
Intestine
Anus
Radula
Mouth
Mantle cavity
Digestive gland
Heart
Foot
Gills
The large, muscular foot is used for locomotion in many species.
(B) Chitons
Intestine
Stomach
Shell plates
Head
Radula
Anus
Foot
Mouth
Digestive gland
Gills in mantle cavity
(C) Gastropods
Intestine
Mantle
Mantle cavity
Shell
Gill
Heart
Anus
Siphon
Salivary gland
Cephalic tentacles
Head
Stomach
Mouth
Foot
The radula is a unique molluscan feeding structure modified for scraping.
(D) Bivalves
Digestive gland
Shell
Heart
Stomach
Intestine
Mouth
Anus
Siphons
Mantle (covers inside of shell)
Foot
Gill
Mantle of lower shell
(E) Cephalopods
Head
Intestine
Shell
Beak
Stomach
Arms
Mantle
Tentacle
Radula
Digestive gland
Siphon
Mantle cavity
Gill
Heart
In cephalod mollusks, the foot is modified into arms and tentacles.

32.13 Organization and Diversity of Molluscan Bodies (A) The major molluscan groups display different variations on a body plan that includes three major components: a foot, a visceral mass of internal organs, and a mantle. In many species, the mantle secretes a calcareous shell. (B) Chitons have eight overlapping calcareous plates surrounded by a girdle. (C) Most gastropods have a single dorsal shell, into which they can retreat for protection. (D) Bivalves get their name from their two hinged shells, which can be tightly closed. (E) Cephalopods are active predators; they use their arms and tentacles to capture prey. This cuttlefish has an internal shell but no external shell.

Chaetopleura angulata
Helix pomatia
Argopecten irradians
Sepia sp.

(A) *Flabellina iodinea*

(B) *Ariolimax columbianus*

(C) *Octopus macropus*

32.14 Mollusks in Some Groups Have Lost Their Shells (A) Nudibranchs ("naked gills"), also called sea slugs, are shell-less gastropods. This species is brightly colored, alerting potential predators of its toxicity. (B) Banana slugs are terrestrial, shell-less gastropods that feed on decomposing vegetation on the damp forest floor. (C) Octopuses have neither an external nor an internal shell, which allows these cephalopods to squeeze through tight spaces.

allow the animal to control the water content of the mantle cavity (**Figure 32.13E**). The modification of the mantle into a device for forcibly ejecting water from the cavity through the siphon enables these animals to move rapidly through the water by "jet propulsion." With their greatly enhanced mobility, cephalopods became the major predators in the open waters of the Devonian oceans. They remain important marine predators today.

As is typical of active, rapidly moving predators, cephalopods have a head with complex sensory organs—most notably eyes that are comparable to those of vertebrates in their ability to resolve images (see Figure 46.14). The head is closely associated with a large, branched foot that bears the arms and/or tentacles and a siphon. Arms are distinguished by the presence of suckers along most of their length. Tentacles, in contrast, have suckers only near the tips or lack suckers altogether. Octopuses typically have eight arms and no tentacles, whereas squids and cuttlefishes have eight arms plus two tentacles. Cephalopods use their arms and tentacles to capture and subdue prey; octopuses also use their arms to move over the substrate. The large, muscular mantle provides a solid external supporting structure. The gills hang in the mantle cavity.

Many early cephalopods had an external chambered shell divided by partitions. The only surviving cephalopods with such shells are the nautiluses (genus *Nautilus*). The chambers inside nautilus shells are connected by a strand of tissue that runs through ducts in the partitions. Blood in this tissue carries water *from* the chambers and gases *into* the chambers, thus providing buoyancy. Most cephalopods retain an internal shell that functions for internal support and, in some species, is also chambered and buoyant. Octopuses have completely lost their shells, which allows them to compress their bodies through very small openings.

Go to Media Clip 32.6
Octopuses Can Pass through Small Openings
Life10e.com/mc32.6

RECAP 32.2

Lophotrochozoans include animals with diverse body types. Wormlike forms include some flatworms, ribbon worms, phoronids, and annelids. There has been convergent evolution of lophophores (in bryozoans and entoprocts versus brachiopods and phoronids) and of external two-part shell coverings (in brachiopods versus bivalve mollusks).

- How do flatworms survive without an internal transport system? **See pp. 656–657 and Figure 32.6**
- Why are most annelids restricted to moist environments? **See p. 660**
- Briefly describe how the basic body organization of mollusks has been modified to yield a wide variety of body forms. **See pp. 662–664 and Figure 32.13**

The second of the two major protostome clades, the ecdysozoans, contains the vast majority of Earth's animal species. What evolutionary innovations led to this massive diversity?

32.3 What Features Distinguish the Major Groups of Ecdysozoans?

Many ecdysozoans are wormlike in form, although others—the arthropods, onychophorans, and tardigrades—have limbs. In this section we will look at the two clades of wormlike ecdysozoans: the priapulids, kinorhynchs, and loriciferans in one clade and the nematodes and horsehair worms in the other. Section 32.4 will be devoted to the most diverse ecdysozoans—the arthropods and their relatives—and the many forms their appendages take.

Several marine ecdysozoan groups have relatively few species

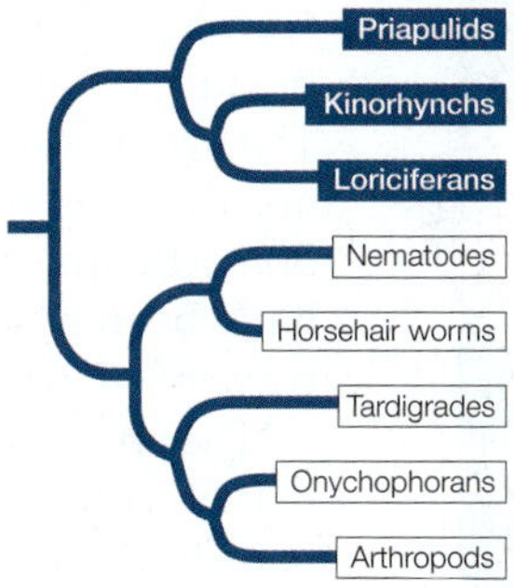

Priapulids
Kinorhynchs
Loriciferans
Nematodes
Horsehair worms
Tardigrades
Onychophorans
Arthropods

Members of several species-poor groups of wormlike marine ecdysozoans—the priapulids, kinorhynchs, and loriciferans—have relatively thin cuticles that are molted periodically as the animals grow to full size. In 2004, embryos of a fossil species related to these ecdysozoans were discovered in sediments laid down in China about 500 million years ago. This remarkable discovery shows that the ancestors of these animals developed directly from an egg to the adult form, as most of their modern descendants do.

The 20 species of **priapulids** are cylindrical, unsegmented, wormlike animals with a three-part body plan consisting of a proboscis, trunk, and caudal appendage ("tail"). It should be clear from their appearance why they were named after the Greek fertility god Priapus (Figure 32.15A). Priapulids range in length from 0.5 millimeters to 20 centimeters. They live in burrows in fine marine sediments and prey on soft-bodied invertebrates such as polychaetes, which they capture with a toothed, muscular pharynx that they evert through the mouth and then withdraw into the body together with the grasped prey. Fertilization is external, and most species have a larval form that also lives in the mud.

About 180 species of **kinorhynchs** have been described. They live in marine sands and muds and are virtually microscopic; no kinorhynchs are longer than 1 millimeter. Their bodies are divided into 13 segments, each covered with a separate cuticular plate (Figure 32.15B). These plates are periodically molted during growth. Kinorhynchs feed by ingesting sediments through a retractable proboscis (the group name means "movable snout"). They then digest the organic material found in the sediment, which may include living algae as well as dead matter. Kinorhynchs have no distinct larval stage; fertilized eggs develop directly into juveniles, which emerge from their egg cases with 11 of the 13 body segments already formed.

Loriciferans are also minute animals less than 1 millimeter long. They were not discovered until 1983. About 100 living species are known to exist, although only about 30 of these have been formally described to date. The body is divided into a head, neck, thorax, and abdomen and is covered by six plates, from which the loriciferans get their name (*Latin lorica,*

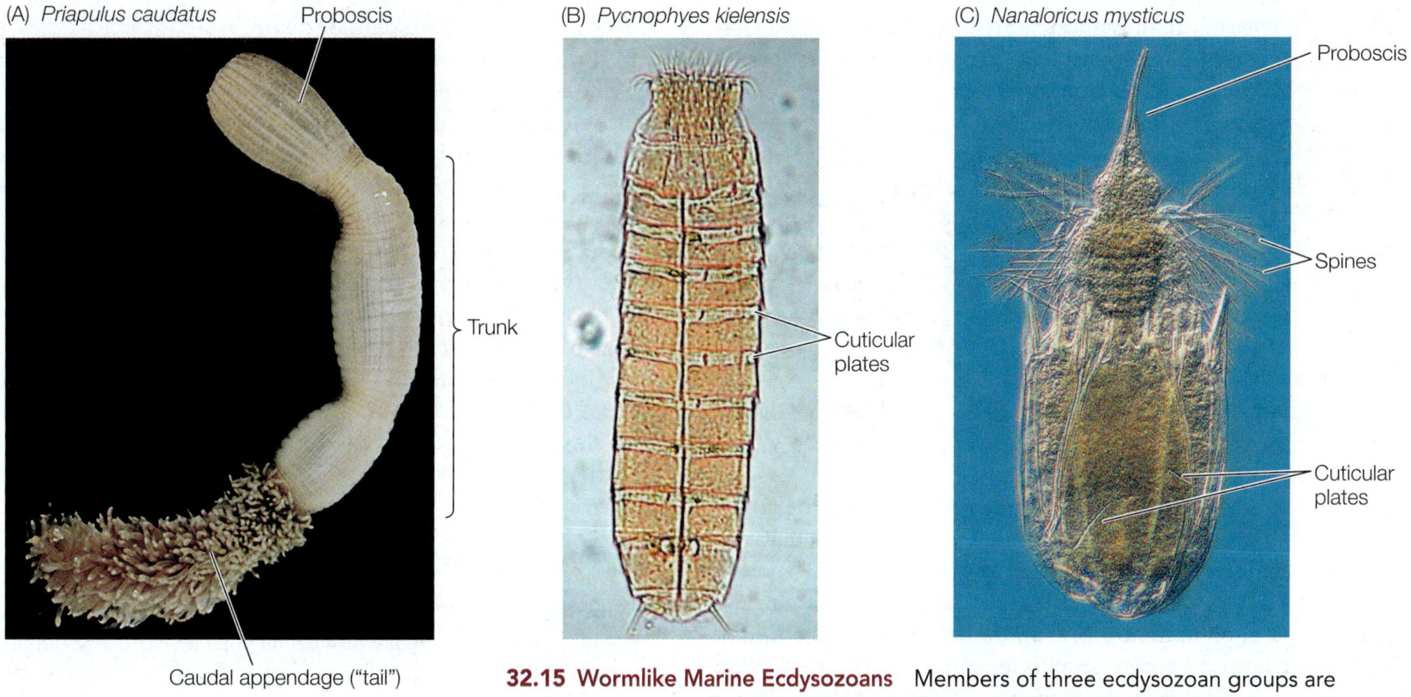

(A) *Priapulus caudatus*
Proboscis
Trunk
Caudal appendage ("tail")
(B) *Pycnophyes kielensis*
Cuticular plates
(C) *Nanaloricus mysticus*
Proboscis
Spines
Cuticular plates

32.15 Wormlike Marine Ecdysozoans Members of three ecdysozoan groups are marine bottom-dwellers. (A) Most priapulid species live in burrows on the ocean floor, extending the proboscis to feed. (B) Kinorhynchs are virtually microscopic. The cuticular plates that cover their bodies are molted periodically. (C) Six cuticular plates form a "corset" around the minute loriciferan body.

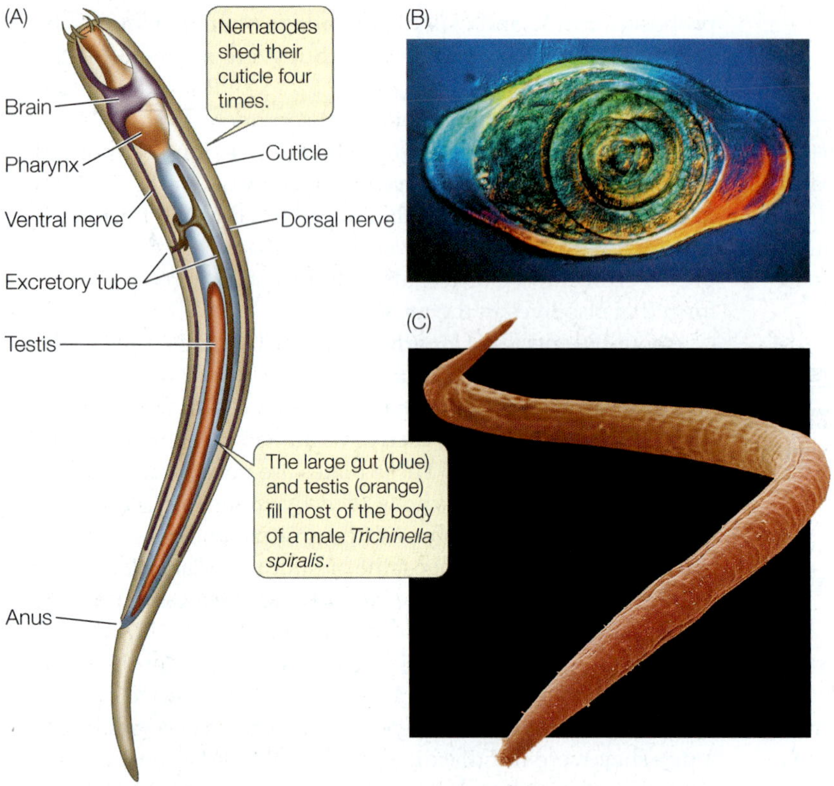
(A)
Brain
Pharynx
Ventral nerve
Excretory tube
Testis
Anus
Nematodes shed their cuticle four times.
Cuticle
Dorsal nerve
The large gut (blue) and testis (orange) fill most of the body of a male *Trichinella spiralis*.
(B)
(C)

32.16 Nematodes (A) The body plan of *Trichinella spiralis*, which causes trichinosis, is typical of parasitic nematodes. (B) This polarized light micrograph shows a cyst of *T. spiralis* in the muscle tissue of a host. (C) This free-living nematode lives in freshwater environments.

"corset"). The plates around the base of the neck bear anterior-directed spines of unknown function (**Figure 32.15C**). Loriciferans live in coarse marine sediments. Little is known about what they eat, but some species apparently eat bacteria.

Nematodes and their relatives are abundant and diverse

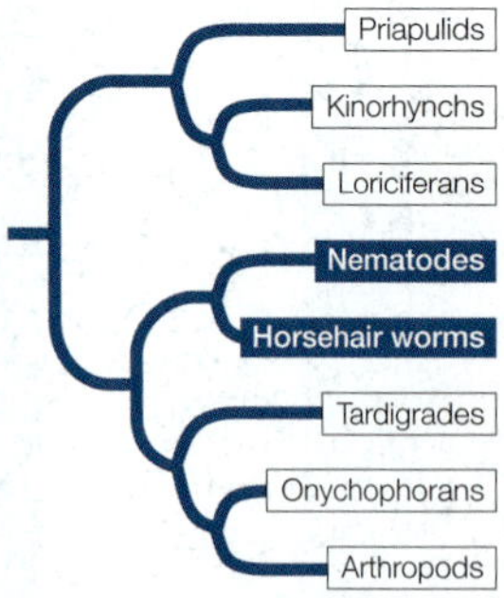
Priapulids
Kinorhynchs
Loriciferans
Nematodes
Horsehair worms
Tardigrades
Onychophorans
Arthropods

Nematodes (roundworms) have a thick, multilayered cuticle that gives their unsegmented body its shape (**Figure 32.16**). As a nematode grows, it sheds its cuticle four times. Nematodes exchange oxygen and nutrients with their environment through both the cuticle and the gut wall, which is only one cell layer thick. Materials are moved through the gut by rhythmic contraction of a highly muscular pharynx. Nematodes move by contracting their longitudinal muscles.

Nematodes are probably the most abundant and universally distributed of all animal groups. Many nematodes are microscopic; the largest known nematode, which reaches a length of 9 meters, is a parasite in the placentas of sperm whales. About 25,000 species have been described, but the actual number of living species may be more than 1 million. Countless nematodes live as scavengers in the upper layers of the soil, on the bottoms of lakes and streams, and in marine sediments. The topsoil of rich farmland may contain from 3 to 9 billion nematodes per acre. A single rotting apple may contain as many as 90,000 individuals.

One soil-inhabiting nematode, *Caenorhabditis elegans*, serves as a model organism in the laboratories of geneticists and developmental biologists. It is ideal for such research because it is easy to cultivate, matures in 3 days, and has a fixed number of body cells. Its genome has been completely sequenced.

Many nematodes are predators, feeding on protists and small animals (including other roundworms). Most significant to humans, however, are the many species that parasitize plants and animals. The nematodes that parasitize humans (causing serious diseases such as trichinosis and elephantiasis), domestic animals, and economically important plants have been studied intensively in an effort to find ways of controlling them.

The structure of parasitic nematodes is similar to that of free-living species, but the life cycles of many parasitic species have special stages that facilitate the transfer of individuals among hosts. *Trichinella spiralis*, the species that causes the human disease trichinosis, has a relatively simple life cycle. A person may become infected by eating the flesh of an animal (usually a pig) that has *Trichinella* larvae encysted in its muscles (see Figure 32.16B). The larvae are activated in the person's digestive tract, emerge from their cysts, and attach to the intestinal wall, where they feed. Later they bore through the intestinal wall and are carried in the bloodstream to muscles, where they form new cysts. If present in great numbers, these cysts can cause severe pain or death.

About 350 species of the unsegmented **horsehair worms** have been described. As their name implies, these animals are extremely thin in diameter; horsehair worms range from a few millimeters up to 1 meter in length. Most adult worms live in fresh water, among leaf litter and algal mats near the edges of streams and ponds. A few species live in damp soil.

Horsehair worm larvae are endoparasites of freshwater crayfishes and of terrestrial and aquatic insects (**Figure 32.17**). An adult horsehair worm has no mouth, and its gut is greatly reduced and probably nonfunctional. Some species may feed only as larvae, absorbing nutrients from their hosts across the body wall. But other species continue to shed their cuticles and grow after they have left their hosts, suggesting that adult worms may also absorb nutrients from their environment.

RECAP 32.3

Priapulids, kinorhynchs, and loriciferans are relatively small, poorly known groups of wormlike marine ecdysozoans. Nematodes and horsehair worms have unsegmented wormlike bodies. Nematodes are probably the most abundant and universal animal group.

- Describe at least three different ways in which nematodes have a significant impact on humans. **See p. 666**

An adult horsehair worm leaves the body of the wood cricket it parasitized during its larval development.

32.17 Horsehair Worm Larvae Are Parasitic The larva of this horsehair worm (*Paragordius tricuspidatus*) can manipulate its host's behavior. The hatching worm causes the cricket to jump into water, where the worm emerges from the insect's body to continue its life cycle as a free-living adult. The cricket, having delivered its parasitic burden, drowns.

We will next turn to the animals that not only dominate the ecdysozoan clade but also constitute the most diverse group of animals on Earth.

32.4 Why Are Arthropods So Diverse?

Arthropods and their relatives are ecdysozoans with paired appendages. Arthropods are the most diverse group of animals in numbers of species (more than 1.2 million have been described, and many more remain to be discovered). Furthermore, the number of individual arthropods alive at any one time is estimated to be about 10^{18}, or 1 billion billion. Among the animals, only the nematodes are thought to exist in greater numbers.

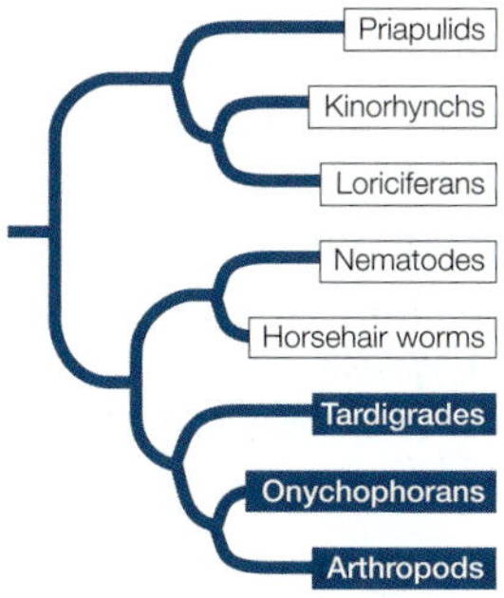
Priapulids
Kinorhynchs
Loriciferans
Nematodes
Horsehair worms
Tardigrades
Onychophorans
Arthropods

Several key features have contributed to the success of the arthropods. As we have seen, their muscles are attached to the inside of their rigid exoskeletons. Their bodies are segmented, and each segment has muscles that operate that segment and the jointed appendages attached to it (see Figure 32.4). Jointed appendages permit complex movements, and different appendages are specialized for different functions. Encasement of the body within a rigid exoskeleton provides the animal with support for walking in the water or on dry land and provides some protection against predators. The waterproofing provided by chitin keeps the animal from dehydrating in dry air.

The four major arthropod groups living today are all species-rich: chelicerates (including the arachnids—spiders, scorpions, mites, and their relatives), myriapods (millipedes and centipedes), crustaceans (including shrimps, crabs, and barnacles), and hexapods (insects and their wingless relatives). The latter three groups are together known as mandibulates.

The jointed appendages of arthropods gave the clade its name, from the Greek words *arthron*, "joint," and *podos*, "foot" or "limb." Arthropods evolved from ancestors with simple, unjointed appendages. The exact forms of those ancestors are unknown, but some arthropod relatives with segmented bodies and unjointed appendages survive today. Before we describe the modern arthropods, we will discuss those arthropod relatives, as well as an early clade that went extinct but left an important fossil record.

Arthropod relatives have fleshy, unjointed appendages

The two living groups most closely related to the arthropods provide us with clues about the likely appearance of ancestral arthropod appendages. **Tardigrades** (water bears) have fleshy, unjointed legs and use their fluid-filled body cavities as hydrostatic skeletons (Figure 32.18A). Tardigrades are tiny (0.5–1.5 mm long) and lack both a circulatory system and gas exchange organs. The 1,200 known extant species live in marine sands and on temporary water films on plants. When these films dry out, the animals also lose water and shrink to small, barrel-shaped objects that can survive for at least a decade in

(A) Echiniscus sp.
50 μm
(B) Macroperipatus torquatus
1 cm

32.18 Arthropod Relatives with Unjointed Appendages (A) Tardigrades ("water bears") can be abundant on the wet surfaces of mosses and plants and in temporary pools of water. (B) Onychophorans ("velvet worms") have unjointed legs and use the body cavity as a hydrostatic skeleton. They are sometimes referred to as "living fossils," meaning they are an ancient group that has changed very little over millennia.

a dormant state. Tardigrades have been found at densities as high as 2 million per square meter of moss.

Until fairly recently, biologists debated whether the **onychophorans** (velvet worms) were more closely related to annelids or to arthropods, but molecular evidence clearly links them to the latter. Indeed, with their soft, fleshy, unjointed, claw-bearing legs and elongate bodies, onychophorans may be similar in appearance to the ancestors of arthropods (**Figure 32.18B**). The 180 species of onychophorans live in leaf litter in humid tropical environments. They have soft, segmented bodies that are covered by a thin, flexible cuticle that contains chitin. Like the tardigrades, they use their fluid-filled body cavities as hydrostatic skeletons. Fertilization is internal, and the large, yolky eggs are brooded within the body of the female.

Jointed appendages appeared in the trilobites

The **trilobites** flourished in Cambrian and Ordovician seas, but they disappeared in the great Permian extinction at the close of the Paleozoic era (251 mya). Because they had heavy exoskeletons that readily fossilized, they left behind an abundant record of their existence (**Figure 32.19**). About 10,000 species have been described.

The trilobites are the earliest known arthropods to have had jointed appendages. The body segmentation and appendages of trilobites followed a relatively simple, repetitive plan, but some of their appendages were modified for different functions. This specialization of appendages is a theme in the continuing evolution of the arthropods.

Chelicerates have pointed, nonchewing mouthparts

In the **chelicerates**, the head bears two pairs of pointed appendages modified to form mouthparts, called chelicerae, that are used to grasp (rather than chew) prey. Chelicerates typically have a two-part body plan, with anterior segments fused to form a cephalothorax, and rear segments fused to form an abdomen. In some groups, such as mites and ticks, there is no clear distinction between these two body parts. Most chelicerates have four pairs of walking legs. The 114,000 described species are grouped into three major clades: pycnogonids, horseshoe crabs, and arachnids.

The pycnogonids, or sea spiders, make up a poorly known group of about 1,000 marine species (**Figure 32.20A**). Most are small, with leg spans less than 1 cm, but some deep-sea species have leg spans up to 60 cm. A few pycnogonids eat algae, but most are carnivorous, eating a variety of small invertebrates.

There are only four living species of horseshoe crabs, but many close relatives are known from fossils. Horseshoe crabs, which have changed very little morphologically over their long history, have a large horseshoe-shaped covering over most of the body. They are common in shallow waters along the eastern coast of North America and the southern and eastern coasts of Asia, where they scavenge and prey on bottom-dwelling

(A) *Endeis* sp.

(B) *Limulus polyphemus*

32.20 Two Small Chelicerate Groups (A) Although they are not spiders, it is easy to see how sea spiders got their common name. (B) A spawning aggregation of horseshoe crabs. Horseshoe crabs, like the onychophorans (see Figure 32.18B), are an example of "living fossils."

Cheirurus ingricus

32.19 A Trilobite Fossil The relatively simple, repetitive segments of the now-extinct trilobites are illustrated by a fossil trilobite from the shallow seas of the Ordovician period, some 450 million years ago.

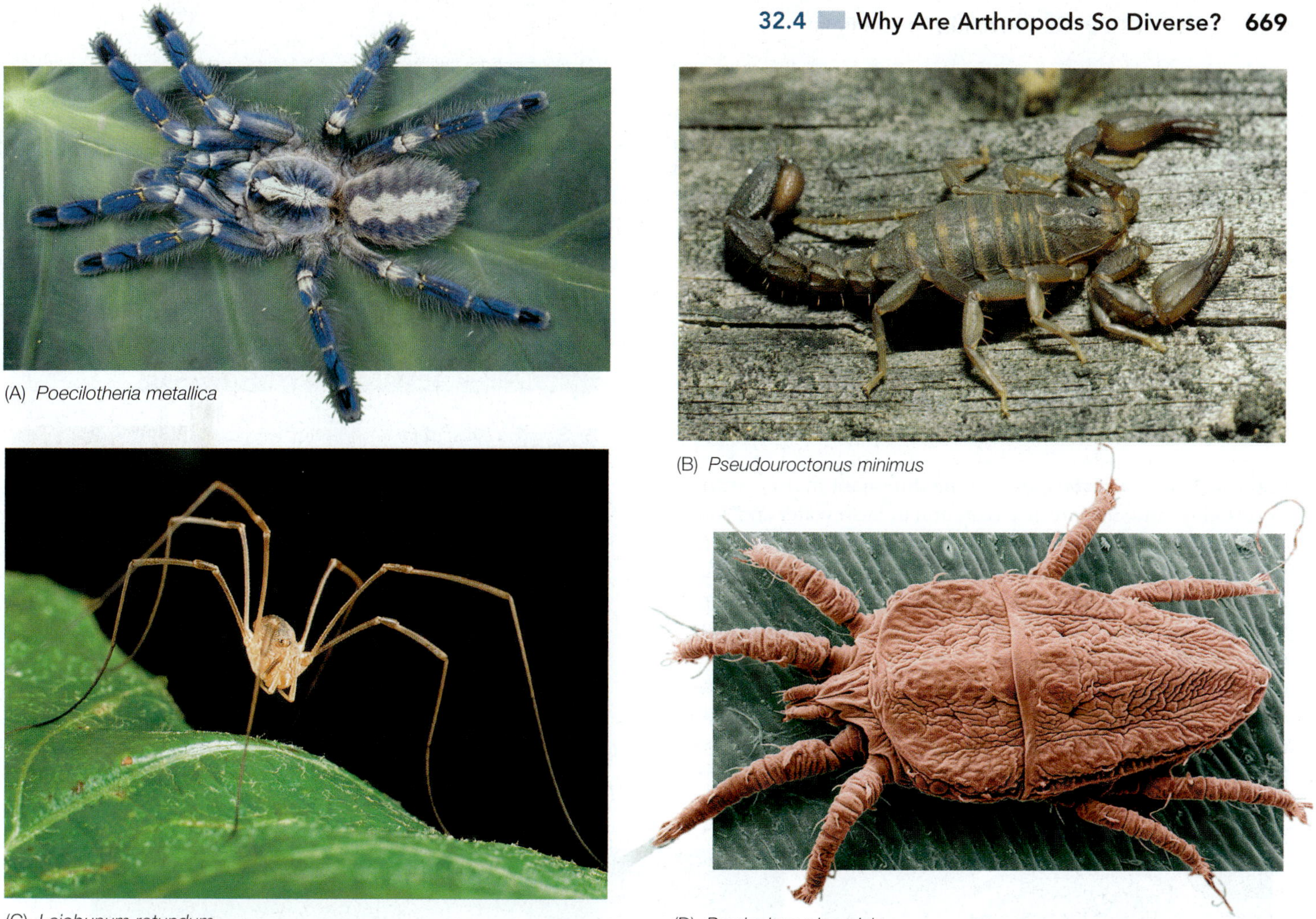

(A) *Poecilotheria metallica*

(B) *Pseudouroctonus minimus*

(C) *Leiobunum rotundum*

(D) *Brevipalpus phoenicis*

32.21 Arachnid Diversity (A) Tarantulas encompass several hundred species of hairy, ground-dwelling spiders, some of which can grow to the size of a dinner plate. Their venomous bite, although painful, is usually not dangerous to humans. (B) Scorpions are nocturnal predators. (C) Harvestmen, also called daddy longlegs, are scavengers. (D) Mites include many free-living species as well as blood-sucking ectoparasites.

animals. Periodically they crawl into the intertidal zone in large numbers to mate and lay eggs (**Figure 32.20B**).

Arachnids are abundant in terrestrial environments. Most arachnids have a simple life cycle in which miniature adults hatch from internally fertilized eggs and begin independent lives almost immediately. Some arachnids retain their eggs during development and give birth to live young.

The most species-rich and abundant arachnids are the spiders, scorpions, harvestmen, mites, and ticks (**Figure 32.21**). More than 60,000 described species of mites and ticks live in soil, leaf litter, mosses, and lichens, under bark, and as parasites of plants and animals. Mites are vectors for wheat and rye mosaic viruses; they cause mange in domestic animals and skin irritation in humans.

Spiders, of which about 50,000 species have been described, are important terrestrial predators with hollow chelicerae, which they use to inject venom into their prey. Some have excellent vision that enables them to chase and seize their prey. Others spin elaborate webs made of protein threads in which they snare prey. The threads are produced by modified abdominal appendages connected to internal glands that secrete the proteins, which solidify on contact with air. The webs of different groups of spiders are strikingly varied, and this variation enables the spiders to position their snares in many different environments for many different types of prey.

Mandibles and antennae characterize the remaining arthropod groups

The remaining three arthropod groups—the myriapods, crustaceans, and hexapods—have mouthparts that are mandibles, rather than chelicerae, so they are collectively called **mandibulates**. Mandibles can be used for chewing as well as for biting and holding food. Another distinctive characteristic of the mandibulates is the presence of sensory antennae on the head.

MYRIAPODS The **myriapods** comprise the centipedes, millipedes, and their close relatives. Centipedes and millipedes have a well-formed head that bears the mandibles and antennae characteristic of mandibulates. Their distinguishing feature is a long, flexible, segmented trunk that bears many pairs of legs. Centipedes, which have one pair of legs per segment (**Figure 32.22A**), prey on insects and other small animals. In

(A) *Scolopendra hardwicki*

(B) *Sigmoria trimaculata*

32.22 Myriapods (A) Centipedes have modified appendages that function as poisonous fangs for capturing active prey. They have one pair of legs per segment. (B) Millipedes are scavengers and plant eaters; they have smaller jaws and legs than centipedes do, and they have two pairs of legs per segment.

millipedes, two adjacent segments are fused so that each fused segment has two pairs of legs (**Figure 32.22B**). Millipedes scavenge and eat plants. More than 3,000 species of centipedes and 9,000 species of millipedes have been described; many more species probably remain unknown. Although most myriapods are less than a few centimeters long, some tropical species are ten times that size.

CRUSTACEANS **Crustaceans** are the dominant marine arthropods today, and they are also common in fresh water and some terrestrial environments. The most familiar crustaceans are the shrimps, lobsters, crayfishes, and crabs (all decapods; **Figure 32.23A**) and the sow bugs (isopods; **Figure 32.23B**). Additional species-rich groups include the amphipods, ostracods, branchiopods (**Figure 32.23C**), and copepods (**Figure 32.23D**), all of which are found in freshwater and marine environments.

Barnacles are unusual crustaceans that are sessile as adults (**Figure 32.23E**). Adult barnacles look more like mollusks than like other crustaceans, but as the zoologist Louis Agassiz

(A) *Randallia ornata*

(C) *Triops longicaudatus*

(B) *Armadillidium vulgare*

(D) Cyclopoid copepod

(E) *Lepas pectinata*

32.23 Crustacean Diversity (A) This decapod crustacean, a purple sand crab, is also called a "globe crab" based on its body shape. (B) This pillbug, a terrestrial isopod, can roll into a tight ball when threatened. (C) This tadpole shrimp, a branchiopod, is common in seasonal pools of the southwestern United States. (D) This minute copepod from a freshwater pond is brooding eggs. (E) Gooseneck barnacles attach to a substrate by their muscular stalks and feed by protruding and retracting their feeding appendages.

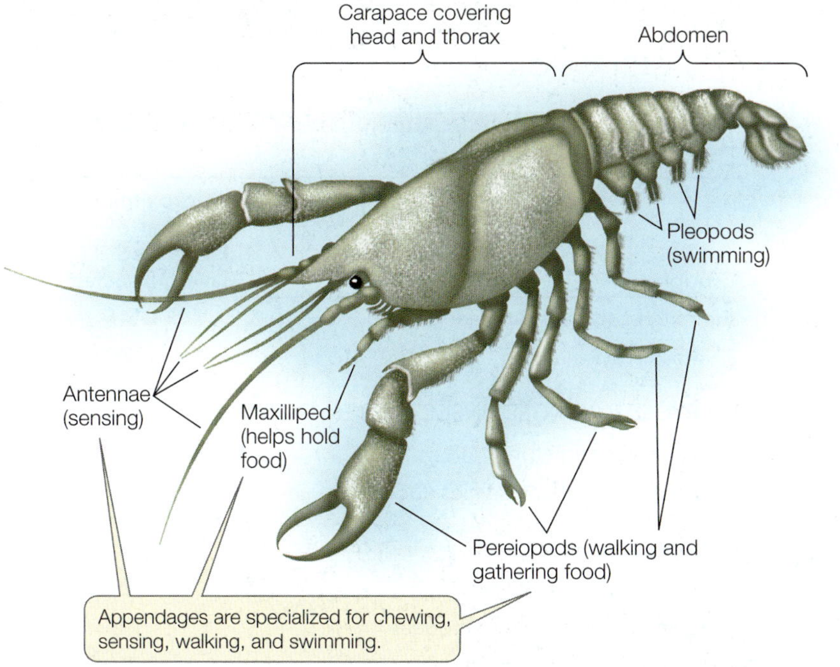
Carapace covering head and thorax
Abdomen
Pleopods (swimming)
Antennae (sensing)
Maxilliped (helps hold food)
Pereiopods (walking and gathering food)
Appendages are specialized for chewing, sensing, walking, and swimming.

32.24 Crustacean Body Plan The bodies of crustaceans are divided into three regions: the head, thorax, and abdomen. Each body region bears specialized appendages. A shell-like carapace covers the head and thorax.

remarked more than a century ago, a barnacle is "nothing more than a little shrimp-like animal, standing on its head in a limestone house and kicking food into its mouth."

Go to Media Clip 32.7
Barnacles Feeding
Life10e.com/mc32.7

Most of the 67,000 described species of crustaceans have a body that is divided into three regions: head, thorax, and abdomen (**Figure 32.24**). The segments of the head are fused together, and the head bears five pairs of appendages. Each of the multiple thoracic and abdominal segments usually bears one pair of appendages. The appendages on different parts of the body are specialized for different functions, such as gas exchange, chewing, capturing food, sensing, walking, and swimming. In many species, a fold of the exoskeleton, the carapace, extends dorsally and laterally back from the head to cover and protect some of the other segments.

The fertilized eggs of most crustacean species are attached to the outside of the female's body, where they remain during their early development (see Figure 32.23D). At hatching, the young of some species are released as larvae; those of other species are released as juveniles that are similar in form to the adults. Still other species release eggs into the water or attach them to an object in the environment.

More than half of all described species are insects

During the Devonian period, more than 400 million years ago, some mandibulates colonized terrestrial environments. Of the several groups (including some crustacean isopods and decapods) that successfully colonized the land, none is more prominent today than the six-legged **hexapods**, which include the **insects** and their wingless relatives.

The wingless relatives of the insects—the springtails, two-pronged bristletails, and proturans—are probably the most similar of living forms to insect ancestors (**Figure 32.25**). These hexapods have a simple life cycle: they hatch from eggs as miniature adults. They differ from insects in having internal mouthparts. Springtails can be extremely abundant (up to 200,000 per m^2) in soil, leaf litter, and on vegetation and are the most abundant hexapods in the world in terms of number of individuals (as opposed to number of species).

As we saw at the opening of this chapter, more than 1 million of the 1.8 million described living species on Earth are insects. Like crustaceans, insects have a body with three regions: head, thorax, and abdomen. They have a single pair of antennae on the head and three pairs of legs attached to the thorax. In most groups of insects, the thorax also bears two pairs of wings. Unlike other arthropods, insects have no appendages growing from their abdominal segments (**Figure 32.26**). Insects are distinguished from springtails and other hexapods by their external mouthparts and by antennae that contain a motion-sensitive receptor called Johnston's organ. In addition, insects have a derived mechanism for gas exchange in air: a system of air sacs and tubular channels called tracheae (singular trachea) that extend from external openings called spiracles inward to tissues throughout the body (see Figure 49.4).

Table 32.2 lists the major insect groups. Two groups—the jumping bristletails and silverfish—are wingless and have simple life cycles, like the springtails and other non-insect hexapods. The remaining groups are all pterygote insects. **Pterygotes** have two pairs of wings, except in some groups in which one or both pairs of wings have been secondarily lost. These secondarily wingless groups include the parasitic lice and fleas, some beetles, and the worker individuals of many ant species.

Hatchling pterygotes do not look like adults; they undergo substantial changes at each molt. The immature stages of insects between molts are called **instars**. A substantial change that

Tomocerus minor

0.5 mm

32.25 Wingless Hexapods The wingless hexapods, such as this springtail, have a simple life cycle. They hatch looking like miniature adults, then grow by successive molts of the cuticle.

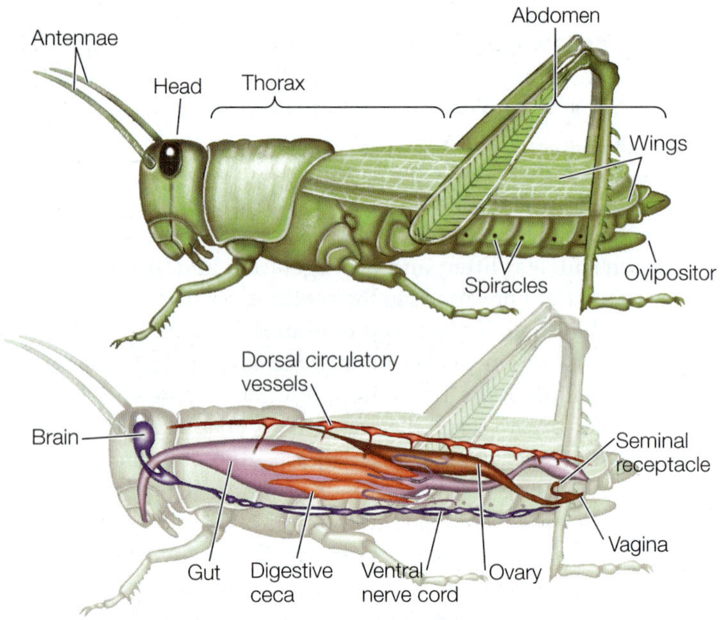
Antennae
Head
Thorax
Abdomen
Wings
Spiracles
Ovipositor
Dorsal circulatory vessels
Brain
Seminal receptacle
Vagina
Gut
Digestive ceca
Ventral nerve cord
Ovary

32.26 Insect Body Plan Like those of crustaceans, the bodies of insects are divided into three regions: head, thorax, and abdomen. In insects, however, the thorax bears three pairs of legs and, in most groups, two pairs of wings. Unlike other arthropods, insects have no appendages growing from their abdominal segments.

occurs between one developmental stage and another is called **metamorphosis**. If the changes between stages are gradual, an insect is said to have **incomplete metamorphosis**. If the change between at least some stages is dramatic, an insect is said to have **complete metamorphosis**. In many insects with complete metamorphosis, different stages are specialized for different functions and use different food sources. In many species the larvae are specialized for feeding and growing, whereas the adults are specialized for reproduction and dispersal.

Go to Media Clip 32.8
Complete Metamorphosis
Life10e.com/mc32.8

The adults of most flying insects have two pairs of stiff, membranous wings attached to the thorax. True flies, however, have one pair of wings and a pair of stabilizers called halteres. In winged beetles, one pair of wings—the forewings—forms heavy, hardened wing covers.

Two groups of pterygotes, the mayflies and dragonflies (Figure 32.27A), cannot fold their wings against their bodies. This is the ancestral condition for pterygote insects, and the mayflies and dragonflies are not closely related to one another. Members of both groups have predatory or herbivorous aquatic larvae that transform into flying adults after they crawl out of the water. Dragonflies (and their relatives the damselflies) are active predators as adults. In contrast, adult mayflies lack functional digestive tracts. Mayflies live only about a day, just long enough to mate and lay eggs.

All other pterygote insects—the **neopterans**—can tuck their wings out of the way upon landing and crawl into crevices and other tight places. Some neopteran groups undergo incomplete metamorphosis, so hatchlings of these insects are sufficiently

TABLE **32.2**
The Major Insect Groups[a]

Group	Approximate Number of Described Living Species
Jumping bristletails (Archaeognatha)	515
Silverfish (Thysanura)	560
PTERYGOTE (WINGED) INSECTS (PTERYGOTA)	
Mayflies (Ephemeroptera)	3,250
Dragonflies and damselflies (Odonata)	5,900
Neopterans (Neoptera)[b]	
Ice-crawlers (Grylloblattodea)	35
Gladiators (Mantophasmatodea)	15
Stoneflies (Plecoptera)	3,750
Webspinners (Embioptera)	465
Angel insects (Zoraptera)	40
Earwigs (Dermaptera)	2,000
Grasshoppers and crickets (Orthoptera)	24,000
Stick insects (Phasmida)	3,000
Cockroaches (Blattodea)	4,650
Termites (Isoptera)	2,700
Mantids (Mantodea)	2,400
Booklice and barklice (Psocoptera)	5,750
Thrips (Thysanoptera)	6,000
Lice (Phthiraptera)	5,100
True bugs, cicadas, aphids, leafhoppers (Hemiptera)	104,000
Holometabolous neopterans (Holometabola)[c]	
Ants, bees, wasps (Hymenoptera)	117,000
Beetles (Coleoptera)	388,000
Strepsipterans (Strepsiptera)	610
Lacewings, ant lions, and mantidflies (Neuroptera)	5,900
Dobsonflies, alderflies, and fishflies (Megaloptera)	350
Snakeflies (Raphidoptera)	250
Scorpionflies (Mecoptera)	760
Fleas (Siphonaptera)	2,100
True flies (Diptera)	155,000
Caddisflies (Trichoptera)	14,300
Butterflies and moths (Lepidoptera)	158,000

[a]The hexapod relatives of insects include the springtails (Collembola; 3,000 spp.), two-pronged bristletails (Diplura; 600 spp.), and proturans (Protura; 10 spp.). All are wingless and have internal mouthparts.
[b]Neopteran insects can tuck their wings close to their bodies.
[c]Holometabolous insects are neopterans that undergo complete metamorphosis.

WORKING WITH**DATA**:

How Many Species of Insects Exist on Earth?

Original Papers

Erwin, T. L. 1988. The tropical forest canopy: The heart of biotic diversity, In E. O. Wilson, ed., *Biodiversity*, 123–129. National Academy Press, Washington, D.C.,

Erwin, T. L. 1997. Biodiversity at its utmost: Tropical forest beetles. In M. L. Reaka-Kudla, D. E. Wilson, and E. O. Wilson, eds., *Biodiversity II*, 27–40. Joseph Henry Press, Washington, D.C.

Analyze the Data

The data in the table were used by entomologist Terry Erwin to estimate the undescribed diversity of insects. Review the design of Erwin's experiment in the opening story of this chapter. Then use Erwin's data to answer the questions.

QUESTION 1

From the data in the table, estimate the number of insect species in an average hectare of Panamanian forest. Assume that the data for beetles on *L. seemannii* are representative of the other tree species, and that all the species of beetles that are *not* host-specific were collected in the original sample. Remember to sum your estimates of the number of (a) host-specific beetle species in the forest canopy; (b) non-host-specific beetle species in the forest canopy; (c) beetle species on the forest floor; and (d) species of all insects other than beetles.

QUESTION 2

There are about 50,000 species of tropical forest trees. Assume that the data for beetles on *L. seemannii* are representative of other species of tropical trees and calculate the number of host-specific beetles found on these trees. Add an estimated 1 million species of non-host-specific beetles that are expected across different species of trees (including those in temperate regions). Estimate the number of ground-dwelling beetle species based on the percentage used in Question 1. Now use this information to estimate the number of insect species on Earth, based on the percentage of beetles among all insect species.

Approximate number of beetle species collected from *Luehea seemannii* trees	1,200
Estimated number of host-specific beetles in this sample	163
Number of tree species per hectare of Panamanian forest	70
Percent of beetle species living in tree canopy (as opposed to ground-dwelling species)	75%
Percent of beetles among all insect species	40%

Go to **BioPortal** for all WORKING WITH**DATA** exercises

similar in form to adults to be recognizable. Examples include the grasshoppers (**Figure 32.27B**), roaches, mantids, stick insects, termites, stoneflies, earwigs, thrips, true bugs (**Figure 32.27C**), aphids, cicadas, and leafhoppers. These groups acquire adult organ systems, such as wings and compound eyes, gradually through several juvenile instars.

More than 80 percent of all insects belong to a subgroup of the neopterans called the **holometabolous** insects (see Table 32.2), which undergo complete metamorphosis (**Figure 32.27D**). The many species of beetles account for almost half of this group (**Figure 32.27E**). Also included are lacewings and their relatives; caddisflies; butterflies and moths (**Figure 32.27F**); sawflies; true flies (**Figure 32.27G**); and bees, wasps, and ants, some species of which display unique and highly specialized social behaviors (**Figure 32.27H**).

Molecular data suggest that insects began to diversify about 450 million years ago, about the time of the appearance of the first land plants. These early hexapods evolved in a terrestrial environment that lacked any other similar organisms, which in part accounts for their remarkable success. But the success of the insects is also due to their wings. Pterygote insects were the first animals in evolutionary history to achieve the ability to fly. Homologous genes control the development of insect wings and crustacean appendages, suggesting that the insect wing evolved from a dorsal branch of a crustacean-like limb (**Figure 32.28**). The dorsal limb branch of crustaceans is used for gas exchange. Thus the insect wing probably evolved from a gill-like structure that had a gas exchange function.

Flight opened up many new lifestyles and feeding opportunities that only the insects could exploit, such as pollination of (and coevolution with) flowering plants. Flight is almost certainly one of the reasons for the remarkable numbers of both insect species and individual insects, and for their unparalleled evolutionary success.

RECAP 32.4

All arthropods have segmented bodies. Muscles in each segment operate that segment and the appendages attached to it. Jointed, specialized appendages permit complex patterns of movement, including, in insects, the ability to fly. With flight, insects took advantage of new feeding and lifestyle opportunities, which contributed to the unparalleled evolutionary success of this group.

- What features have contributed to making arthropods among the most abundant animals on Earth, both in number of species and number of individuals? **See p. 667**
- Describe the difference between incomplete and complete metamorphosis. **See pp. 671–672**

An Overview of Protostome Evolution

The protostomes encompass a staggering number of different body forms and lifestyles. The following aspects of protostome evolution have contributed to this enormous diversity:

- The evolution of *segmentation* permitted some groups of protostomes to move different parts of the body independently of one another. Species in some groups gradually evolved the ability to move rapidly over and through the substrate, through water, and through air.

(A) *Libellula quadrimaculata*
(B) *Phymateus morbillosus*
(C) *Coquerelia ventralis*
(D) *Limnephilus* sp. (larva)
(E) *Eupholus magnificus*
(F) *Argema mittrei*
(G) *Lucilia caesar*
(H) *Polistes nympha*

32.27 Diverse Pterygotes (A) Unlike most flying insects, a dragonfly cannot fold its wings over its back. (B) Orthopteran insects such as grasshoppers have incomplete metamorphosis: they undergo several molts, but the juvenile instars resemble small adults (incomplete metamorphosis). (C) Hemipterans such as this Madagascan shield bug are known as "true" bugs. (D–H) Holometabolous insects undergo complete metamorphosis. (D) A larval caddisfly (right) emerges from its dark pupal case. (E) The beetles (Coleoptera) comprise the largest insect group; beetles such as this New Guinea weevil account for more than half of all holometabolous species. (F) Butterflies and moths are the lepidopterans, whose phases of complete metamorphosis are familiar to many (see Figure 31.9). (G) Blowflies are among the "true" flies, the Dipterans. Adult blowflies feed on pollen or nectar, but lay their eggs on carrion, upon which the larvae feed. (H) These paper wasps are hymenopterans, a group in which most members display social behaviors.

- *Complex life cycles* with dramatic changes in form between one stage and another allow individuals of different stages to specialize on different resources.
- *Parasitism* has evolved repeatedly, and many protostome groups parasitize plants and animals.
- The evolution of *diverse feeding structures* allowed protostomes to specialize on many different food sources. Specialization on food sources undoubtedly contributed to reproductive isolation and further diversification.
- Predation was a major selection pressure favoring the development of *hard external body coverings* (exoskeletons and

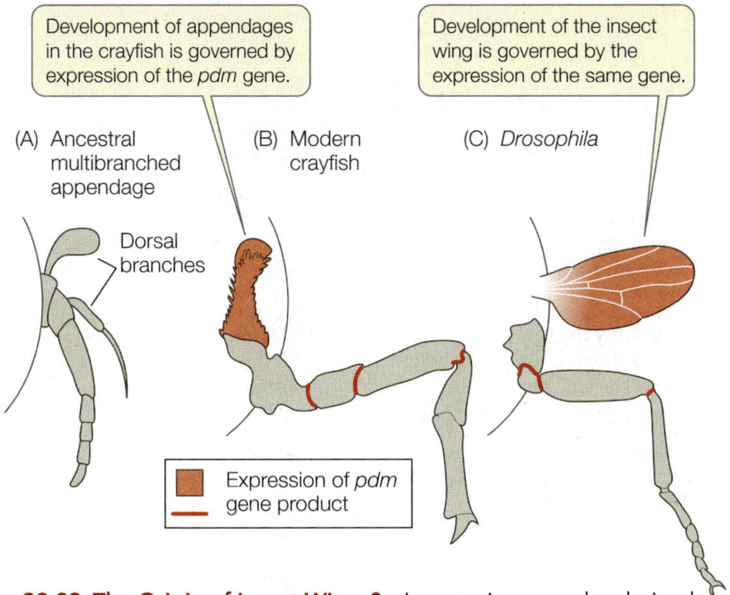
Development of appendages in the crayfish is governed by expression of the *pdm* gene.
Development of the insect wing is governed by the expression of the same gene.
(A) Ancestral multibranched appendage
(B) Modern crayfish
(C) *Drosophila*
Dorsal branches
Expression of *pdm* gene product

32.28 The Origin of Insect Wings? Insect wings may be derived from an ancestral appendage similar to that of modern crustaceans. (A) A diagram of the ancestral, multibranched arthropod limb. The uppermost dorsal branch may have been used for gas exchange. (B, C) The *pdm* gene, a Hox gene, is expressed throughout the dorsal limb branch and walking leg of the thoracic limb of a crayfish (B) and in the wings and legs of *Drosophila* (C).

shells). Such coverings evolved independently in many lophotrochozoan and ecdysozoan groups. In addition to providing protection, these coverings became key elements in the development of new systems of locomotion.

- *Better locomotion* permitted prey to escape from predators, but also allowed predators to pursue their prey more effectively. Thus the evolution of animals has been, and continues to be, a complex "arms race" among predators and prey.

Go to Animated Tutorial 32.1
An Overview of the Protostomes
Life10e.com/at32.1

Many major evolutionary trends among the protostomes are shared by the deuterostomes, which include the chordates, the group to which humans belong. We turn to the deuterostomes in the next chapter.

?

Which groups of protostomes are thought to contain the most undiscovered species?

ANSWER

It is perhaps easier to list the groups of protostomes for which a nearly complete inventory of living species has been done than to list all the groups for which many new species remain to be described. Among the insects, the best-studied group, in terms of species, is the butterflies, which are widely collected and studied. There are still many species of other lepidopterans (such as moths), however, remaining to be discovered. Most other major insect groups contain many undescribed species.

New species discovery and description rates remain high for almost all other major groups of protostomes. Second to the insects, and perhaps even rivaling the insects in undiscovered diversity, are the nematodes. Although known nematode diversity is only about one-fortieth of known insect diversity (in terms of number of described species), the taxonomy of nematodes has been much more poorly studied than that of insects. Some biologists think there are likely to be species-specific parasitic nematodes specializing on most other species of multicellular organisms. If so, then there may be as many species of nematodes as there are of plants, fungi, and other animals combined.

Most of the other diverse groups of protostomes also contain many as yet undetected species, judging from the rate of new species descriptions. In particular, flatworms (especially the parasitic flukes and tapeworms), marine annelids, mollusks, crustaceans, myriapods, and chelicerates all contain large numbers of undescribed species.

CHAPTER SUMMARY 32

32.1 What Is a Protostome?

- Protostomes ("mouth first") are bilaterally symmetrical animals with an anterior brain that surrounds the entrance to the digestive tract and a ventral nervous system. The embryonic blastopore of protostomes develops into a mouth.
- There are two major clades of protostomes, the lophotrochozoans and the ecdysozoans. **Review Figure 32.1, Table 32.1, ACTIVITIES 32.1, 32.2, ANIMATED TUTORIAL 32.1**
- **Lophotrochozoans** include a wide variety of body forms. Within this group, **lophophores** (complex organs for both food collection and gas exchange), free-living **trochophore** larvae, and spiral cleavage evolved. Some of these features were subsequently lost in some lineages (or evolved convergently).
- **Ecdysozoans** have a body covering known as the **cuticle**, which they must **molt** in order to grow. Some ecdysozoans have a relatively thin cuticle. Others, especially the **arthropods**, have a rigid cuticle reinforced with **chitin** that functions as an **exoskeleton**. **Review Figure 32.4**
- **Arrow worms** may be most closely related to lophotrochozoans, or they may be the sister group of all other protostomes. **Review Figure 32.5**

32.2 What Features Distinguish the Major Groups of Lophotrochozoans?

- Lophotrochozoans range from animals with a blind gut and no internal transport system to animals with complete digestive tracts and complex internal transport systems. **Review Figure 32.6**
- Most species of **bryozoans** and **entoprocts** live in colonies produced through asexual reproduction. Individuals of both groups feed using a lophophore.
- **Flatworms**, **rotifers**, **gastrotrichs**, and their close relatives form a structurally diverse clade of ciliated lophotrochozoans. **Review Figure 32.7**
- **Ribbon worms** feed using a long, protrusible proboscis. **Review Figure 32.8**

continued

- The shelled **brachiopods** and wormlike **phoronids** use a lophophore to feed; this lophophore may have evolved independently of the lophophore in bryozoans and entoprocts. **Review Figures 32.9, 32.10**
- **Annelids** are a diverse group of segmented worms that live in moist terrestrial and aquatic environments. **Review Figure 32.11**
- **Mollusks** underwent a dramatic evolutionary radiation based on a body plan consisting of three major components: a **foot**, a **mantle**, and a **visceral mass**. The four major living molluscan clades—**chitons**, **bivalves**, **gastropods**, and **cephalopods**—demonstrate the diversity that evolved from this three-part body plan. **Review Figure 32.13**

32.3 What Features Distinguish the Major Groups of Ecdysozoans?

- Members of several species-poor groups of wormlike marine ecdysozoans—**priapulids**, **kinorhynchs**, and **loriciferans**—have thin cuticles.
- **Nematodes** have a thick, multilayered cuticle. Nematodes are among the most abundant and universally distributed of all animal groups. **Review Figure 32.16**
- **Horsehair worms** are extremely thin; many are endoparasites as larvae.

32.4 Why Are Arthropods So Diverse?

- One major ecdysozoan clade, the arthropods, has evolved jointed, paired appendages that have a wide diversity of functions. Collectively, arthropods are the dominant animals on Earth in number of described species, and among the most abundant in number of individuals.
- Encasement within a rigid exoskeleton provides arthropods with support for walking as well as some protection from predators. The waterproofing provided by chitin keeps arthropods from dehydrating in dry air.
- Jointed appendages permit complex movement patterns. Each arthropod segment has muscles attached to the inside of the exoskeleton that operate that segment and the appendages attached to it.
- The **onychophorans** and the **tardigrades** are arthropod relatives that have simple, unjointed appendages. **Trilobites,** the first arthropods known to have had jointed appendages, disappeared in the Permian mass extinction.
- **Chelicerates** have a two-part body and pointed mouthparts that grasp prey; most chelicerates have four pairs of walking legs.
- Mandibles and antennae are synapomorphies of the **mandibulates**, which include the myriapods, crustaceans, and hexapods.
- The bodies of **myriapods** have two regions: a head with mandibles and antennae, and a segmented trunk that bears many pairs of legs.
- **Crustaceans** have segmented bodies that are divided into three regions—head, thorax, and abdomen—with different, specialized appendages in each region. **Review Figure 32.24**
- **Hexapods**—insects and their relatives—are the dominant terrestrial arthropods. They have the same three body regions as crustaceans, but no appendages form in their abdominal segments. **Review Figure 32.26, Table 32.2**
- Wings and the ability to fly first evolved among the **pterygote** insects, allowing them to exploit new lifestyles. **Review Figure 32.28**

Go to the Interactive Summary to review key figures, Animated Tutorials, and Activities
Life10e.com/is32

CHAPTER**REVIEW**

REMEMBERING

1. Which of the following is *not* part of the molluscan body plan?
   a. Mantle
   b. Foot
   c. Radula
   d. Visceral mass
   e. Jointed appendages

2. The outer covering of ecdysozoans
   a. is always hard and rigid.
   b. is always thin and flexible.
   c. is hard and rigid in larvae but thin in adults.
   d. ranges from very thin to hard and rigid, depending on the species.
   e. grows throughout life to accommodate a growing body.

3. Which groups are arthropod relatives with unjointed legs?
   a. Trilobites and onychophorans
   b. Onychophorans and tardigrades
   c. Trilobites and tardigrades
   d. Onychophorans and chelicerates
   e. Tardigrades and chelicerates

4. The body plan of insects comprises which of the following three regions?
   a. Head, abdomen, and trachea
   b. Head, abdomen, and cephalothorax
   c. Cephalothorax, abdomen, and trachea
   d. Head, thorax, and abdomen
   e. Abdomen, trachea, and mantle

5. Insects whose hatchlings are sufficiently similar in form to adults to be recognizable are said to have
   a. instars.
   b. neopterous development.
   c. accelerated development.
   d. incomplete metamorphosis.
   e. complete metamorphosis.

6. Factors that may have contributed to the remarkable evolutionary success of insects include
   a. the lack of any other similar organisms in the terrestrial environments colonized by insects.
   b. the ability to fly.
   c. complete metamorphosis.
   d. a new mechanism for delivering oxygen to their internal tissues.
   e. All of the above

UNDERSTANDING & APPLYING

7. Segmentation either has arisen several times during animal evolution, or else arose early in animal evolution and was subsequently lost multiple times. What advantages does segmentation provide? Given these advantages, why might some animals have lost their segmentation?

8. Major structural novelties have arisen only infrequently during the course of evolution. Which of the features of protostomes do you think are major evolutionary novelties? Which of these features may have led to major evolutionary radiations?

9. There are more described and named species of insects than of all other species on Earth combined. However, only a very few insect species live in marine environments, and those species are restricted to the intertidal zone or the ocean surface. What factors may have contributed to the insects' lack of success in the oceans?

ANALYZING & EVALUATING

10. In the Working with Data exercise on page 673, you were asked to make many assumptions to estimate the number of species of insects on Earth. Do you think these assumptions are reasonable? Why or why not? Would you argue for a different set of assumptions? How do you think these changes in assumptions would affect your calculations? Can you think of ways to test these assumptions?

Go to BioPortal at **yourBioPortal.com** for Animated Tutorials, Activities, LearningCurve Quizzes, Flashcards, and many other study and review resources.

33 Deuterostome Animals

CHAPTER**OUTLINE**

32.1 What Is a Deuterostome?

33.2 What Features Distinguish the Echinoderms, Hemichordates, and Their Relatives?

33.3 What New Features Evolved in the Chordates?

33.4 How Did Vertebrates Colonize the Land?

33.5 What Traits Characterize the Primates?

Good Parent or Cannibal? Young frogs emerge from the digestive tract of a female *Rheobatrachus silus*, one of the now-extinct gastric brooding frogs of Australia. In this unique form of gestation, eggs hatch and tadpoles develop within the protected environment of the mother's stomach.

WHY WOULD A FROG swallow its own offspring? That may not sound like a good parenting skill, but female gastric brooding frogs of Australia shut down their digestive system to brood their tadpoles in their stomach. This strategy provides the tadpoles with a safe haven from predators, making it much more likely that they will survive to metamorphosis. That was the case, at least, until gastric brooding frogs went extinct in the early 1980s, after humans introduced pathogenic fungi into their native range.

Not all adult frogs are involved in raising their young. Female bullfrogs, for example, lay thousands of eggs each year and provide no parental care for their offspring. The eggs are fertilized by a male bullfrog and left to develop on their own. The eggs hatch into tadpoles, which transform into tiny frogs—if they aren't eaten first by aquatic predators. Out of the tens of thousands of tadpoles an adult bullfrog produces in its lifetime, an average of only two offspring will survive to reproduce.

Among other species of frogs, complex behaviors associated with parental care change these long odds. Rather than producing huge numbers of offspring, each with a minimal chance of surviving, some frogs invest more energy in each individual offspring and care for the young as they grow. This strategy increases the chances that any one offspring will survive and reproduce—but it also means that far fewer offspring can be produced.

Frogs' strategies for parental care include many behaviors in addition to gastric brooding. The females of many species guard their eggs until they hatch. Other females carry their tadpoles around with them on their backs, or even in special brood pouches. Males often provide parental care as well. Males of many frog species guard egg masses or carry young, sometimes in unusual ways. In Darwin's frog of South America, for example, the tadpoles develop within the male's vocal sacs.

Parental care can extend beyond protection to feeding the young. Some female poison frogs of the tropical Americas carry each of their tadpoles to one of the many tiny pools of water that collect in bromeliad plants growing on trees. The female then returns to each bromeliad "pond" and lays unfertilized eggs as food for the single tadpole developing there.

Frogs and other vertebrates constitute one of the major groups of deuterostome animals. Deuterostomes are of particular interest to biologists not only because of their often complex behaviors, their importance in many ecosystems, and their widespread use as models in developmental biology and genetics, but because we are deuterostomes.

> ?
>
> How has the evolution of complex behaviors affected the diversification of some major groups of deuterostomes?
>
> See answer on p. 705.

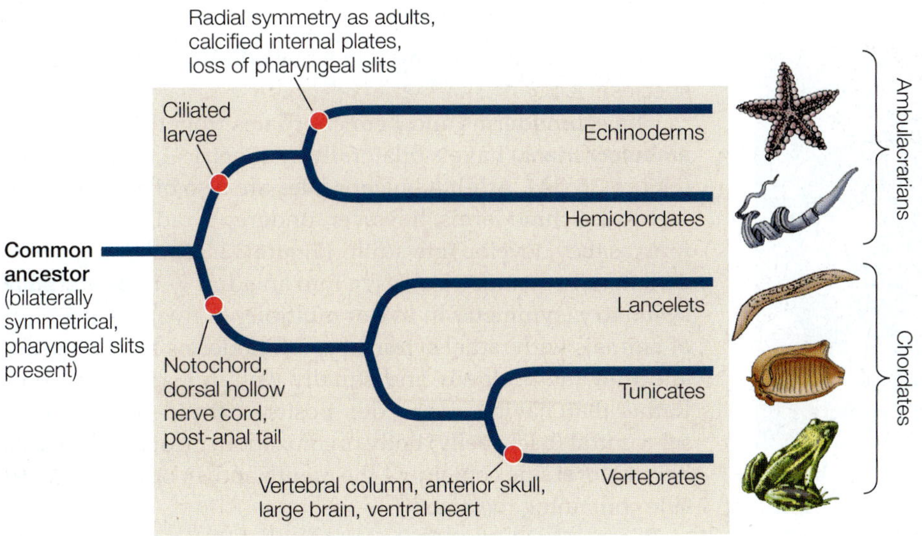
Radial symmetry as adults, calcified internal plates, loss of pharyngeal slits
Ciliated larvae
Echinoderms
Hemichordates
Common ancestor (bilaterally symmetrical, pharyngeal slits present)
Lancelets
Notochord, dorsal hollow nerve cord, post-anal tail
Tunicates
Vertebrates
Vertebral column, anterior skull, large brain, ventral heart
Ambulacrarians
Chordates

33.1 Phylogeny of the Deuterostomes The three principal groups of deuterostomes are the echinoderms, the hemichordates, and the chordates, which include the lancelets, tunicates, and vertebrates. The echinoderms and the vertebrates contain most of the described species.

Go to Activity 33.1 Deuterostome Phylogeny
Life10e.com/ac33.1

33.1 What Is a Deuterostome?

It may surprise you to learn that both you and a sea urchin are deuterostomes. Adult sea stars, sea urchins, and sea cucumbers—the most familiar echinoderms—look so different from adult vertebrates (fishes, frogs, lizards, birds, and mammals) that it may be difficult to believe all these animals are closely related. The evidence that all deuterostomes share a common ancestor that is not shared with the protostomes includes early developmental patterns and phylogenetic analysis of gene sequences, factors that are not apparent in the forms of the adult animals.

Deuterostomes share early developmental patterns

Historically, the deuterostomes were distinguished by three early developmental patterns:

- Radial cleavage
- Development of the blastopore into the anus and formation of the mouth at the opposite end of the embryo from the blastopore (the pattern that gives the deuterostomes their name)
- Development of a coelom from mesodermal pockets that bud off from the cavity of the gastrula rather than by splitting of the mesoderm, as occurs among protostomes

These distinctions, however, are not the strongest evidence for the monophyly of the deuterostomes. Radial cleavage is not exclusive to deuterostomes, and as noted in Section 31.1, it is now thought to be the ancestral condition for all bilaterians. In fact, some of the groups now known to be protostomes were once thought to be deuterostomes because their developmental patterns are similar to those of echinoderms and chordates. The development of the blastopore into an anus does characterize the deuterostomes, but it may be the ancestral condition for bilaterians rather than a derived feature of deuterostomes. Today the strongest support for the shared evolutionary relationships of the deuterostomes comes from phylogenetic analyses of DNA sequences of many different genes.

There are three major deuterostome clades

All deuterostomes are triploblastic, coelomate animals (see Figure 31.5C). Skeletal elements, where present, are internal rather than external. Some species have segmented bodies, but the segments are less obvious than those of annelids and arthropods. Although there are far fewer species of deuterostomes than of protostomes (see Table 31.1), we have a special interest in deuterostomes because we are members of that clade. The deuterostomes are also of interest because they include many large animals that strongly influence the characteristics of ecosystems. Many deuterostome species have been intensively studied in all fields of biology. Complex behaviors, such as the parenting behaviors described at the opening of this chapter, are especially well developed among some deuterostomes.

The major groups of living deuterostomes comprise three distinct clades (**Figure 33.1**):

- **Echinoderms**: sea stars (starfish), sea urchins, and their relatives
- **Hemichordates**: acorn worms and pterobranchs
- **Chordates**: sea squirts, lancelets, and vertebrates

In addition, some recent genomic analyses suggest that two poorly known groups, the xenoturbellids and the acoels, may also be deuterostomes.

Go to Animated Tutorial 33.1
An Overview of the Deuterostomes
Life10e.com/at33.1

Fossils shed light on deuterostome ancestors

Scientists are learning much about the ancestors of modern deuterostomes from fossils recently discovered in 520-million-year-old rocks in China. Some of these early deuterostomes had skeletons similar to those of echinoderms but, unlike modern adult echinoderms, they had bilateral symmetry and a pharynx with slits through which water flowed. Another early deuterostome group, the yunnanozoans, was discovered in China's Yunnan Province. The well-preserved fossils reveal animals that had large mouths, six pairs of external gills, and a segmented posterior body section bearing a light cuticle (**Figure 33.2**). The features of these fossil animals, together with

Yunnanozoon lividum
Mouth
Esophagus
External gills
Segments

33.2 Ancestral Deuterostomes Had External Gills The extinct yunnanozoans may be ancestral deuterostomes. This fossil, which dates from the Cambrian, shows the six pairs of external gills and segmented posterior body that characterized these animals.

findings from phylogenetic analyses of living species, show that the earliest deuterostomes were bilaterally symmetrical, segmented animals with pharyngeal slits. The adult forms of the living echinoderms with their unique symmetry (in which the body parts are arranged along five radial axes) evolved much later. Other deuterostomes retained the ancestral bilateral symmetry.

RECAP 33.1

The three major clades of deuterostomes are the echinoderms, the hemichordates, and the chordates. The common ancestry of these groups is supported by early developmental similarities and by phylogenetic analyses of DNA sequences.

- Why is radial cleavage no longer considered to be evidence for the monophyly of deuterostomes? See p. 679
- What three traits did the earliest deuterostomes have in common? See pp. 679–680

We will begin our survey of the deuterostomes with the echinoderms and hemichordates, the most distant of our relatives within that clade.

33.2 What Features Distinguish the Echinoderms, Hemichordates, and Their Relatives?

About 13,000 species of echinoderms in 23 major groups have been described from their fossil remains. They are probably only a small fraction of the echinoderm species that have ever lived. Only 6 of the 23 major groups known from fossils are represented by species that survive today; many clades were lost during the periodic mass extinctions that have occurred throughout Earth's history. Nearly all of the 7,500 extant species of echinoderms live in marine environments. There are far fewer species of living hemichordates (about 120 known species).

The echinoderms and hemichordates (together known as **ambulacrarians**) have a bilaterally symmetrical, ciliated larva (**Figure 33.3A**). Adult hemichordates are also bilaterally symmetrical. Echinoderms, however, undergo a radical change in form as they develop into adults (**Figure 33.3B**), changing from a bilaterally symmetrical larva into an adult with **pentaradial symmetry** (symmetry in five or multiples of five). As is typical of animals with radial symmetry, echinoderms have no head, and they move slowly and equally well in many directions. Rather than having an anterior–posterior (head–tail) and dorsal–ventral (back–belly) body organization, most echinoderms have an **oral** side containing the mouth and an opposite **aboral** side containing the anus.

Recent genomic analyses suggest that two groups of small, highly reduced , soft-bodied marine organisms, the xenoturbellids and the acoels, are the sister group of the ambulacrarians. The two known species of **xenoturbellids** are wormlike organisms up to 4 cm long that feed on or parasitize mollusks in the northern Atlantic Ocean. They have a very simple body plan, with almost no well-defined organ systems. The mostly tiny (<2 mm) **acoels** are also highly reduced, wormlike organisms that live as plankton, between grains of sediment, or on other organisms such as corals (**Figure 33.4**). They are among the simplest of bilaterian animals, with no gut, circulatory system, respiratory system, gonads, or excretory system. They feed by forming a vacuole around tiny food items. They are hermaphrodites, and their gametes fill the body between the epidermis and digestive vacuole. There are about 400 known species of acoels.

Echinoderms have unique structural features

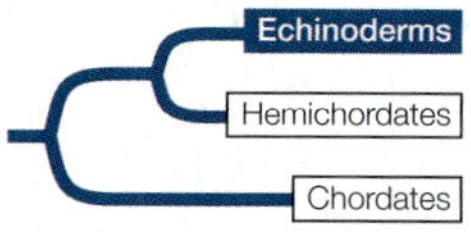
Echinoderms
Hemichordates
Chordates

In addition to having pentaradial symmetry, adult echinoderms have two unique structural features. One is a system of calcified internal plates covered by thin layers of skin and some muscles. The calcified plates of most echinoderms are thick, and they fuse inside the entire body, forming an **internal skeleton**. The other unique feature of this group is a **water vascular system**, a network of water-filled canals leading to extensions called **tube feet** (see Figure 33.3B). This system functions in gas exchange, locomotion, and feeding. Seawater enters the system through a perforated structure called a madreporite. A calcified canal leads from the madreporite to the ring canal, which surrounds the esophagus (the tube leading from the mouth to the stomach). Radial canals branch off from the ring canal, extending through the arms (in species that have arms) and connecting with the tube feet. These structural innovations have been modified in many ways, resulting in a striking array of very different animals.

Members of one major extant echinoderm clade, the crinoids (sea lilies and feather stars), were more abundant and species-rich 300 to 500 million years ago than they are today. There are some 80 described living sea lily species, most of which are

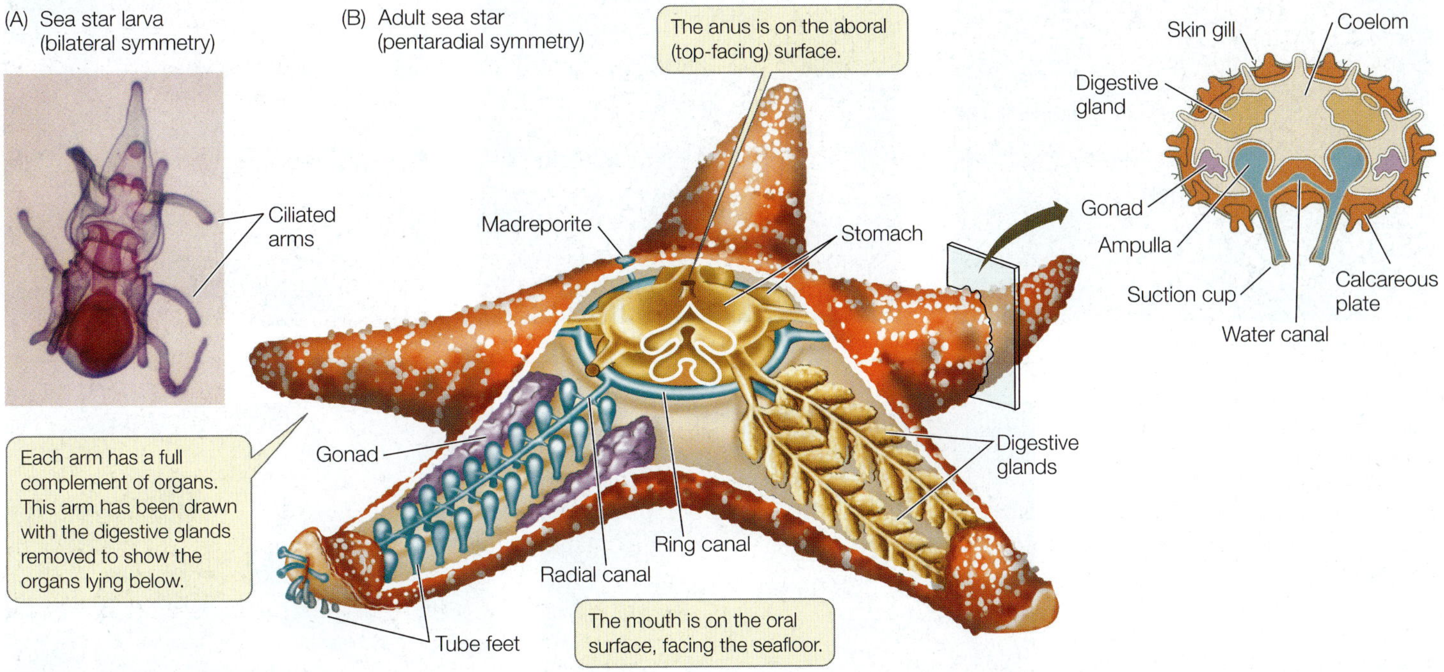
(A) Sea star larva (bilateral symmetry)
Ciliated arms
(B) Adult sea star (pentaradial symmetry)
The anus is on the aboral (top-facing) surface.
Madreporite
Stomach
Skin gill
Coelom
Digestive gland
Gonad
Ampulla
Suction cup
Water canal
Calcareous plate
Each arm has a full complement of organs. This arm has been drawn with the digestive glands removed to show the organs lying below.
Gonad
Digestive glands
Ring canal
Radial canal
Tube feet
The mouth is on the oral surface, facing the seafloor.

33.3 Echinoderms Are Bilaterally Symmetrical as Larvae but Radially Symmetrical as Adults (A) The ciliated larva of a sea star has bilateral symmetry. Hemichordates have a similar larval form. (B) An adult sea star displays the pentaradial symmetry of adult echinoderms. The canals and tube feet of the water vascular system, as well as the calcified internal skeleton, are shown in this diagram. The body's orientation is oral–aboral rather than anterior–posterior.

Wamionoa sp.

33.4 Highly Reduced Acoels Are Probably Relatives of the Ambulacrarians Acoels (yellow) are seen here living on bubble coral (white). Acoels ("without coelom") feed by enveloping food particles in a vacuole within which nutrients are digested. These hermaphroditic animals can reproduce rapidly and may become problematic in saltwater aquariums.

sessile organisms attached to the substrate by a stalk. Feather stars (**Figure 33.5A**) grasp the substrate with specialized flexible appendages that allow for limited movement. About 600 living species of feather stars have been described.

Unlike the mostly sessile crinoids, most surviving echinoderms are motile. The two main groups of motile echinoderms are the echinozoans (sea urchins and sea cucumbers) and the asterozoans (sea stars and brittle stars). Sea urchins are hemispherical in shape and lack arms (**Figure 33.5B**). They are covered with spines that are attached to the underlying skeleton with ball-and-socket joints. These joints enable the spines to be moved so they can converge toward a point that has been touched. The spines, which vary among species in size and shape, can be used for locomotion; a few produce toxic substances. They provide effective protection for the urchin, as many a scuba diver has found out the hard way. Sand dollars are flattened, disc-shaped relatives of sea urchins.

Sea cucumbers also lack arms, and their bodies are oriented in an atypical manner for an echinoderm (**Figure 33.5C**). The mouth is anterior and the anus is posterior (front and rear), in contrast to the oral–aboral (top and bottom) orientation of other echinoderms. Sea cucumbers can use most of their tube feet to move, but they use them primarily for attaching to the substrate.

Sea stars, popularly called starfish, are the most familiar echinoderms (**Figure 33.5D**). Their gonads and digestive organs are located in the arms, as seen in Figure 33.3B. Their tube feet serve as organs of locomotion, gas exchange, and attachment. Each tube foot of a sea star consists of an internal ampulla connected by a muscular tube to an external suction cup that can stick to the substrate. The tube foot is moved by expansion and contraction of the circular and longitudinal muscles of the tube. Brittle stars are similar in structure to sea stars, but their flexible arms are composed of jointed, hard plates (**Figure 33.5E**).

(A) *Oxycomanthus bennetti*

(B) *Sphaerechinus granularis*

(C) *Synaptula* sp.

(D) *Marthasterias glacialis*

(E) *Ophiopholis aculeata*

33.5 Echinoderm Diversity (A) The flexible arms of this golden feather star (a crinoid) are clearly visible. (B) Sea urchins are important grazers on algae in the intertidal zones of the world's oceans. (C) Sea cucumbers are unique among echinoderms in having an anterior–posterior, rather than an oral–aboral, orientation of the mouth and anus. (D) Sea stars are important predators on bivalve mollusks such as mussels and clams. Suction tips on its tube feet allow a sea star to grasp both shells of the bivalve and pull them open. (E) The arms of the brittle star are composed of hard but jointed plates.

Echinoderms use their tube feet in a great variety of ways to capture prey. Sea lilies, for example, feed by orienting their arms in passing water currents. Food particles then strike and stick to the tube feet, which are covered with mucus-secreting glands. The tube feet transfer these particles to grooves in the arms, where ciliary action carries the food to the mouth.

Most sea urchins capture phytoplankton with their tube feet or scrape algae from rocks with a complex rasping structure. Sea cucumbers capture food with their anterior tube feet, which are modified into large, feathery, sticky tentacles that can be protruded from the mouth. Periodically, a sea cucumber withdraws the tentacles, wipes off the material that has adhered to them, and digests it.

Many sea stars use their tube feet to capture large prey such as polychaetes, gastropod and bivalve mollusks, small crustaceans such as crabs, and fishes. With hundreds of tube feet acting simultaneously, a sea star can grasp a bivalve in its arms, anchor the arms with its tube feet, and by steady contraction of the muscles in its arms, gradually exhaust the muscles the bivalve uses to keep its shell closed (see Figure 33.5D). To feed on the bivalve, the sea star can push its stomach out through its mouth and then through the narrow space between the two halves of the bivalve's shell. The sea star's stomach then secretes enzymes that digest the prey.

Go to Media Clip 33.1
Sea Star Feeding on a Bivalve
Life10e.com/mc33.1

Most of the 2,000 species of brittle stars ingest particles from the upper layers of sediments and assimilate the organic material from them, although some species filter suspended food particles from the water, and others capture small animals.

Hemichordates are wormlike marine deuterostomes

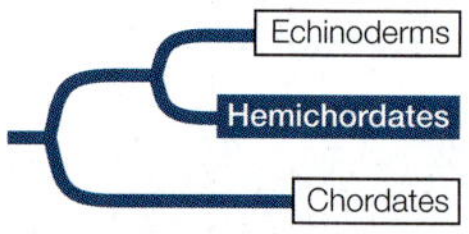
Echinoderms
Hemichordates
Chordates

Hemichordates—acorn worms and pterobranchs—have a bilaterally symmetrical, wormlike body organized in three major parts: a proboscis, a collar (which bears the mouth), and a trunk (which contains the other body parts). The 90 known species of acorn worms range up to 2 meters long (**Figure 33.6A**). They live in burrows in muddy and sandy marine sediments. The digestive tract of an acorn worm consists of a mouth behind which are a muscular pharynx and an intestine. The pharynx opens to the outside through several pharyngeal slits through which water can exit. Highly vascularized tissue surrounding the pharyngeal slits serves as a gas exchange apparatus. Acorn worms respire by pumping water into the mouth and out through the pharyngeal slits. They capture prey with the large proboscis, which is coated with sticky mucus to which small organisms in the sediment stick. The mucus and its attached prey are conveyed by cilia to the mouth. In the esophagus, the food-laden mucus is compacted into a ropelike mass that is moved through the digestive tract by ciliary action.

The 30 living species of pterobranchs are sedentary marine animals up to 12 millimeters long that live in a tube secreted

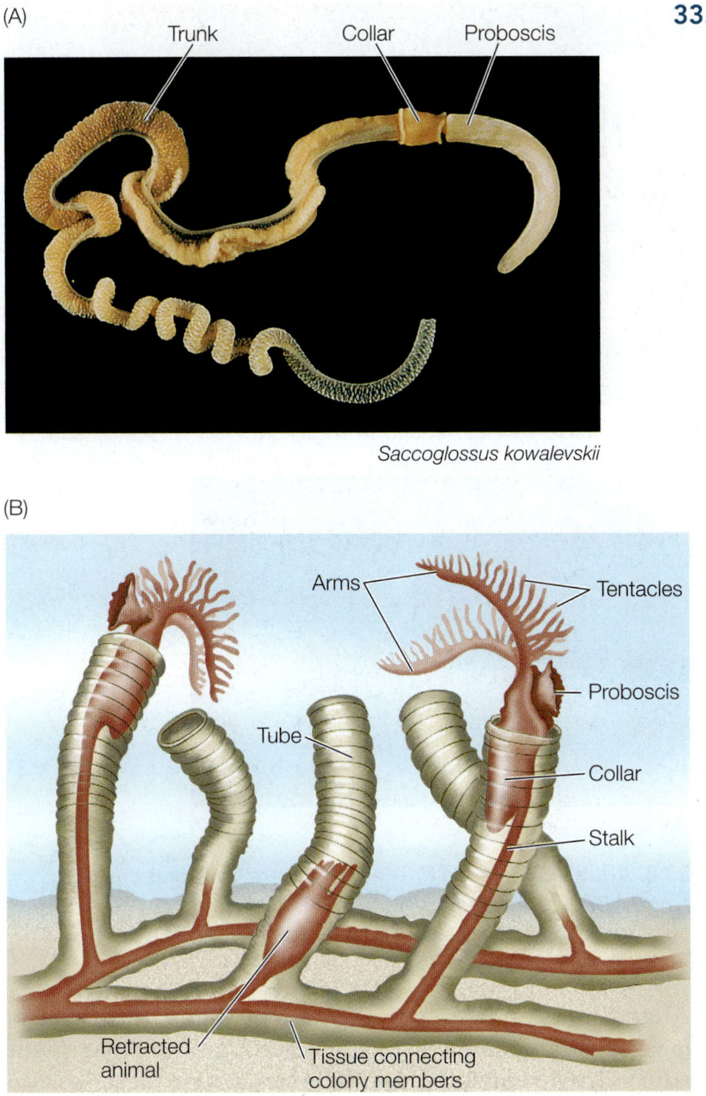
(A)
Trunk
Collar
Proboscis
Saccoglossus kowalevskii
(B)
Arms
Tentacles
Proboscis
Tube
Collar
Stalk
Retracted animal
Tissue connecting colony members

33.6 Hemichordates (A) The proboscis of an acorn worm is modified for burrowing. (B) Some pterobranch species form colonies.

by the proboscis. Some species are solitary; others form colonies of individuals joined together (**Figure 33.6B**). Behind the proboscis is a collar with anywhere from one to nine pairs of arms. The arms bear long tentacles that capture prey and function in gas exchange.

RECAP 33.2

Echinoderms are characterized by pentaradial symmetry, an internal skeleton of calcified plates, and a unique water vascular system. Hemichordates have a bilaterally symmetrical body divided into three parts: proboscis, collar, and trunk.

- How does the body form of echinoderm larvae differ from that of adults? **See p. 681 and Figure 33.3**
- Describe some of the ways that echinoderms use their tube feet to obtain food. **See p. 682**
- Explain how hemichordates obtain food. **See p. 682**

Having described the deuterostome groups that are most distantly related to us, we will next turn our attention to the unique features that evolved in the chordates, a clade dominated by the vertebrates.

33.3 What New Features Evolved in the Chordates?

As we have seen, it is not obvious from examining adult animals that echinoderms and chordates share a common ancestor. The evolutionary relationships among some chordate groups are not immediately apparent either. The features that reveal all of these evolutionary relationships are seen primarily in the larvae—in other words, it is during the early developmental stages that these evolutionary relationships are evident.

There are three principal chordate clades: the **lancelets** (also called cephalochordates), the **tunicates** (also called urochordates), and the **vertebrates** (see Figure 33.1). Adult chordates vary greatly in form, but all chordates display the following derived structures at some stage in their development (**Figure 33.7**):

- A *dorsal hollow nerve cord*
- A *tail* that extends beyond the anus
- A dorsal supporting rod, the *notochord*

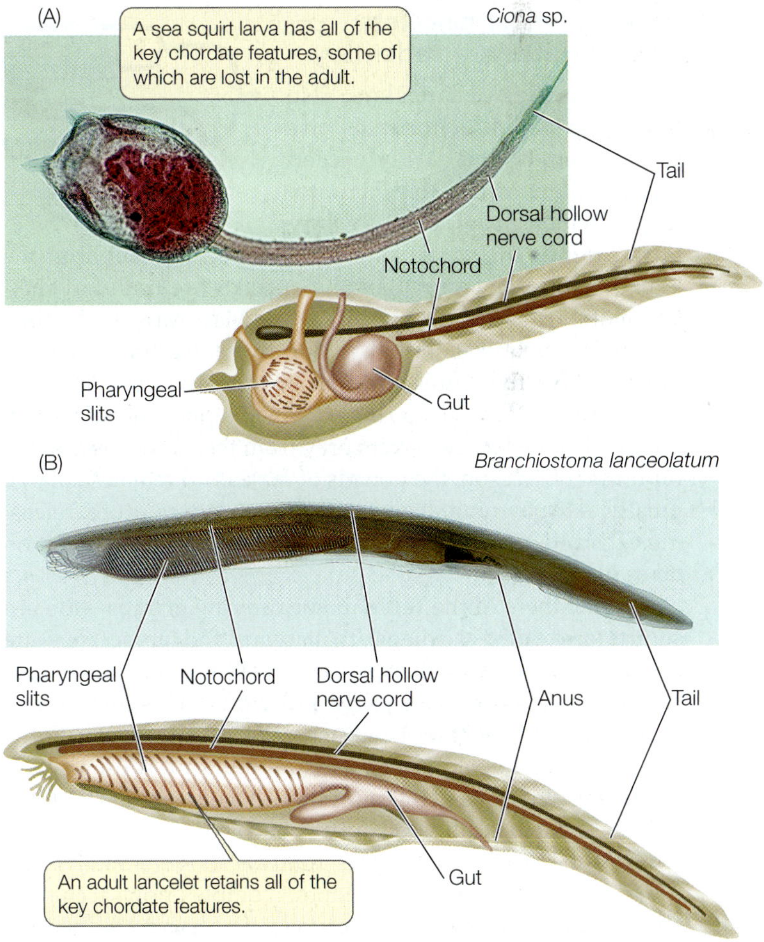
(A)
A sea squirt larva has all of the key chordate features, some of which are lost in the adult.
Ciona sp.
Tail
Dorsal hollow nerve cord
Notochord
Pharyngeal slits
Gut
(B)
Branchiostoma lanceolatum
Pharyngeal slits
Notochord
Dorsal hollow nerve cord
Anus
Tail
Gut
An adult lancelet retains all of the key chordate features.

33.7 Key Features May Be Most Apparent in Early Development (A) The sea squirt larva (but not the adult) has all three key features of chordates: a dorsal hollow nerve cord, a post-anal tail, and a notochord. (B) All three chordate synapomorphies are retained in the adult lancelet.

The **notochord** is the most distinctive derived chordate trait. It is composed of a core of large cells with turgid fluid-filled vacuoles, which make it rigid but flexible. In the tunicates the notochord is lost during metamorphosis to the adult stage. In most vertebrate species it is replaced during development by skeletal structures that provide support for the body.

The pharyngeal slits found in the common ancestor of deuterostomes are present at some developmental stage in all chordates but are often lost or greatly modified in adults. In chordates, the pharyngeal slits are separated and supported by structural elements called pharyngeal arches. In tunicates and lancelets, the pharynx functions as a straining device to filter small food particles. In fishes and larval amphibians, some of the pharyngeal arches develop into gill arches, which support the respiratory gills and are often used as feeding structures as well. Developmentally, some pharyngeal arches also develop into elements of the vertebrate jaw, as well as parts of the tongue, larynx, trachea, and middle ear of tetrapods (four-legged vertebrates). Some of the pharyngeal slits are modified in tetrapods to form the eustachian tube and middle ear chamber.

Adults of most lancelets and tunicates are sedentary

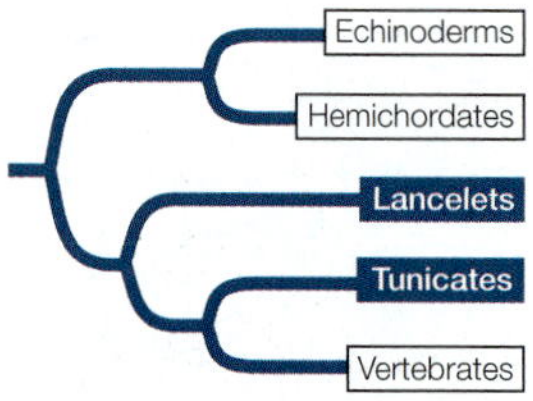
Echinoderms
Hemichordates
Lancelets
Tunicates
Vertebrates

The 35 species of lancelets, also known as cephalochordates, are small animals that rarely exceed 5 centimeters in length. The notochord, which provides body support, extends the entire length of the body throughout the animal's life (see Figure 33.7B). Lancelets are found in shallow marine and brackish waters worldwide. Most of the time they lie covered in sand with their head protruding above the sediment, but they can swim. The pharynx has been enlarged and modified to form a structure called a pharyngeal basket, with which the lancelet filters prey from the water. During the reproductive season, the gonads of males and females enlarge greatly. At spawning, the walls of the gonads rupture, releasing eggs and sperm into the water column, where fertilization takes place.

All members of the three major tunicate groups—the sea squirts (also called ascidians), thaliaceans, and larvaceans—are marine animals. More than 90 percent of the 2,800 known species of tunicates are sea squirts. Individual sea squirts range in length from less than 1 millimeter to 60 centimeters. Some sea squirts form colonies by asexual budding from a single founder. Colonies may measure several meters across. The baglike body of an adult sea squirt is enclosed in a tough tunic, which is the basis for the name "tunicate" (**Figure 33.8A**). The tunic is composed of proteins and a complex polysaccharide secreted by epidermal cells. The sea squirt pharynx is enlarged into a pharyngeal basket that filters prey from the water passing through it.

In addition to its pharyngeal slits, a sea squirt larva has a dorsal hollow nerve cord and a notochord that is restricted mostly to the tail region (see Figure 33.7A). Bands of muscle that surround the notochord provide support for the body.

(A) *Clavelina dellavallei*
1 cm
(B) *Pegea* sp.
1 cm

33.8 Adult Tunicates (A) The transparent tunic and the pharyngeal basket are clearly visible in this sea squirt. (B) A chainlike colony of thaliaceans (salps) floats in tropical waters.

After a short time swimming in the plankton, the larvae of most species settle on the seafloor and transform into sessile adults. The swimming, tadpolelike larvae suggest a close evolutionary relationship between tunicates and vertebrates (see Figure 22.6).

Thaliaceans (salps and their relatives) are tunicates that can live singly or in chainlike colonies up to several meters long (**Figure 33.8B**). They float in tropical and subtropical oceans at depths down to 1,500 meters. Larvaceans are solitary planktonic animals that retain the notochord and dorsal hollow nerve cord throughout their lives. Most larvaceans are less than 5 millimeters long, but some species that live near the bottom of deep ocean waters build delicate casings of mucus that may be more than a meter wide. They snare sinking organic particles (their primary food source) with elaborate filters built into their mucus "houses." When the old "house" gets clogged with excess debris, the animal builds a new one.

A dorsal supporting structure replaces the notochord in vertebrates

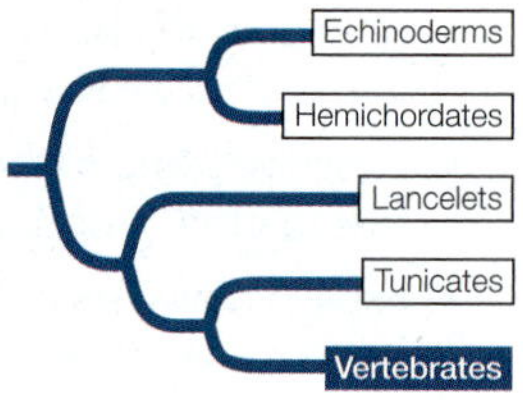
Echinoderms
Hemichordates
Lancelets
Tunicates
Vertebrates

In one chordate group, the vertebrates, a new dorsal supporting structure evolved. This group takes its name from the jointed, dorsal **vertebral column** that replaces the notochord during early development as the primary supporting structure. The individual elements in the vertebral column are called vertebrae. Four other key features characterize the vertebrates as well (**Figure 33.9**):

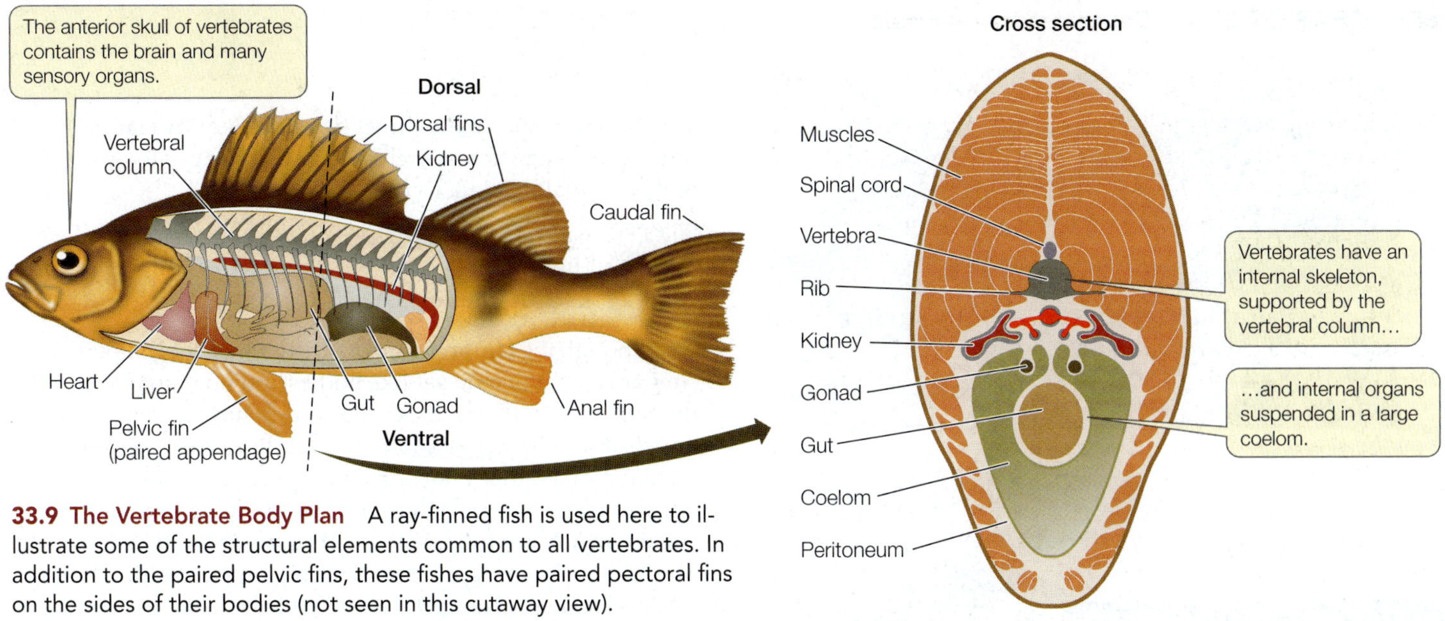
The anterior skull of vertebrates contains the brain and many sensory organs.
Dorsal
Dorsal fins
Vertebral column
Kidney
Caudal fin
Heart
Liver
Gut
Gonad
Anal fin
Pelvic fin (paired appendage)
Ventral
Cross section
Muscles
Spinal cord
Vertebra
Rib
Kidney
Gonad
Gut
Coelom
Peritoneum
Vertebrates have an internal skeleton, supported by the vertebral column…
…and internal organs suspended in a large coelom.

33.9 The Vertebrate Body Plan A ray-finned fish is used here to illustrate some of the structural elements common to all vertebrates. In addition to the paired pelvic fins, these fishes have paired pectoral fins on the sides of their bodies (not seen in this cutaway view).

- An anterior *skull* enclosing a large brain
- A rigid internal *skeleton* supported by the vertebral column
- Internal organs *suspended in a coelom*
- A well-developed *circulatory system*, driven by contractions of a ventral *heart*

These structural features can support large, active animals. The internal skeleton provides support for an extensive muscular system, which receives oxygen from the circulatory system and is controlled by the central nervous system. The evolution of these features allowed many vertebrates to become large, active predators, which in turn allowed the vertebrates to diversify widely (**Figure 33.10**).

All of the nonvertebrate deuterostomes (acoels and xenoturbellids, echinoderms, hemichordates, lancelets, and tunicates) live in marine environments. The lineage that led to the vertebrates is also thought to have evolved in the oceans, although probably in an estuarine environment (where fresh water meets salt water). The first vertebrates appeared in the Cambrian; since then they have radiated into marine, freshwater, terrestrial, and aerial environments worldwide. There are about 65,000 species of living vertebrates.

The phylogenetic relationships of jawless fishes are uncertain

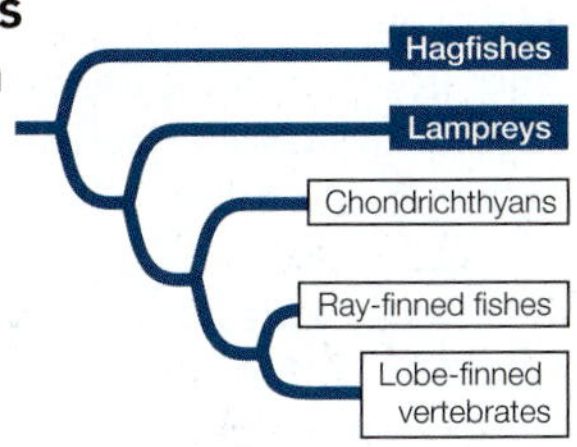
Hagfishes
Lampreys
Chondrichthyans
Ray-finned fishes
Lobe-finned vertebrates

The **hagfishes** are thought by many to be the sister group to the remaining vertebrates (see Figure 33.10). Hagfishes (**Figure 33.11A**) have a weak circulatory system

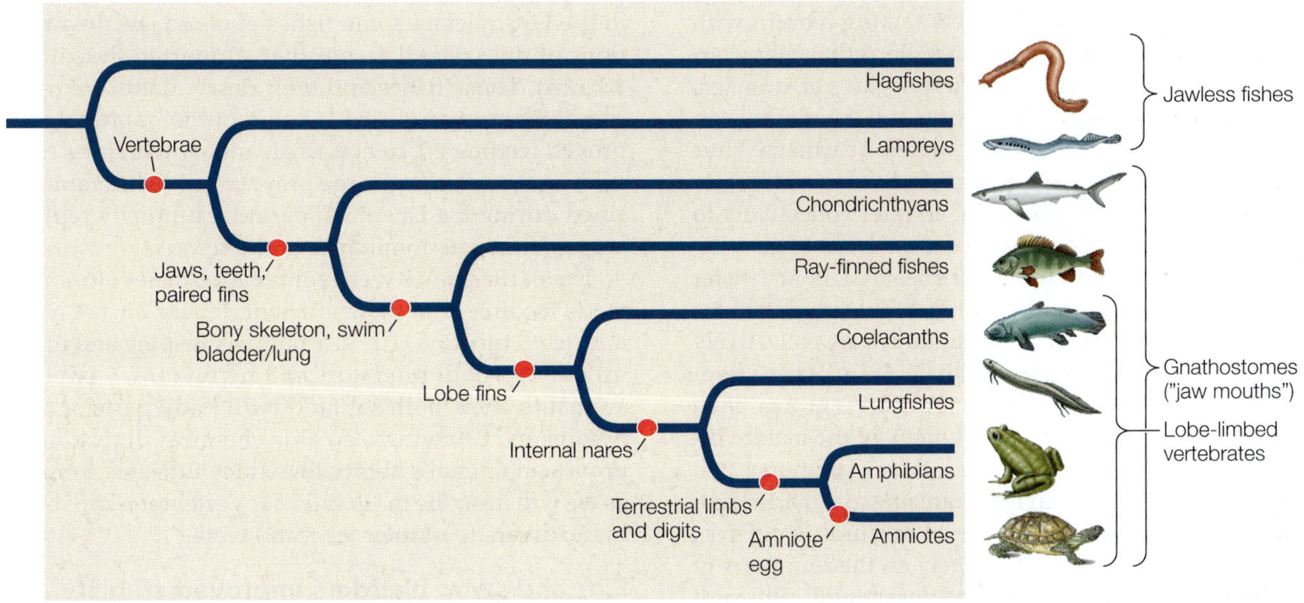
Hagfishes
Lampreys
Chondrichthyans
Ray-finned fishes
Coelacanths
Lungfishes
Amphibians
Amniotes
Vertebrae
Jaws, teeth, paired fins
Bony skeleton, swim bladder/lung
Lobe fins
Internal nares
Terrestrial limbs and digits
Amniote egg
Jawless fishes
Gnathostomes ("jaw mouths")
Lobe-limbed vertebrates

33.10 Phylogeny of the Living Vertebrates This phylogenetic tree shows the evolution of some of the key innovations among the major groups of vertebrates.

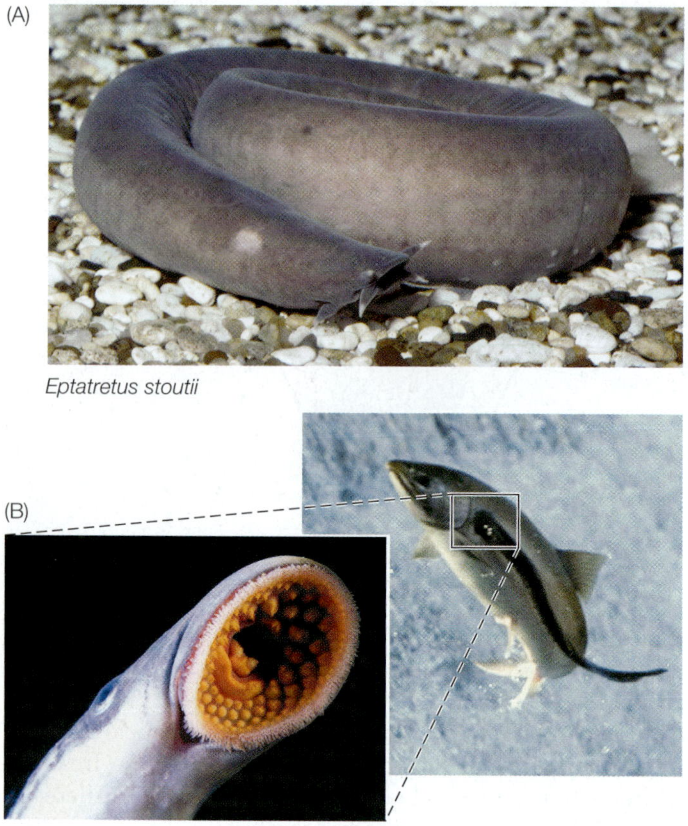
(A)
Eptatretus stoutii
(B)
Petromyzon marinus

33.11 Modern Jawless Fishes (A) Hagfishes burrow in the ocean mud, from which they extract small prey. They also scavenge on dead or dying fish. Hagfishes have degenerate eyes, which has led to their being called (inaccurately) "blind eels." (B) Sea lampreys are ectoparasites that attach to the bodies of living fish and use their large, jawless mouths to suck blood and flesh. They can survive in both fresh and salt water, as this individual attached to a salmon returning to its spawning ground will do.

with three small accessory hearts (rather than a single, large heart); a partial cranium, or skull (containing a brain with no cerebrum or cerebellum, two main regions that characterize the brains of other vertebrates); and no jaws or stomach. They also lack separate, jointed vertebrae and have a skeleton composed of a firm but pliable material called cartilage. Thus some biologists do not consider hagfishes to be vertebrates and instead use the term "craniates" to refer collectively to the hagfishes and the vertebrates. Some analyses of gene sequences suggest, however, that hagfishes may be the sister group of the **lampreys** (**Figure 33.11B**); in this phylogenetic arrangement, the hagfishes and the lampreys are collectively called the cyclostomes ("circle mouths"). If in fact the hagfishes and lampreys do form a monophyletic group, then hagfishes must have secondarily lost many of the major vertebrate morphological features during their evolution.

The 80 known species of hagfishes are unusual marine animals that produce copious quantities of mucus as a defense. They are virtually blind and rely largely on the four pairs of sensory tentacles around the mouth to detect food. Although they have no jaws, hagfishes have a tonguelike structure equipped with toothlike rasps that they use to tear apart dead organisms and to capture their principal prey, polychaete worms. Hagfishes have direct development (no larvae), and individuals may actually change sex from year to year (from male to female and vice versa).

Go to Media Clip 33.2
Hagfish Slime
Life10e.com/mc33.2

Although the lampreys and hagfishes may look superficially similar (with elongate eel-like bodies and no paired fins), they differ greatly in their biology. Lampreys have a complete skull and distinct and separate (although rudimentary) vertebrae, all cartilaginous rather than bony. Lampreys undergo a complete metamorphosis from filter-feeding larvae, known as ammocoetes, which are morphologically similar to adult lancelets. The adults of many species of lampreys are parasitic, although several lineages of lampreys evolved to become nonfeeding as adults. These nonfeeding adults survive for only a few weeks after metamorphosis—just long enough to breed. In the species that are parasitic as adults, the round mouth is a rasping and sucking organ that is used to attach to prey and rasp at the flesh (see Figure 33.11B).

The nearly 50 species of lampreys either live permanently in fresh water or are anadromous—meaning they live in coastal salt water and move into fresh water to breed. Some species of lampreys are critically endangered because of recent habitat changes and losses.

Jaws and teeth improved feeding efficiency

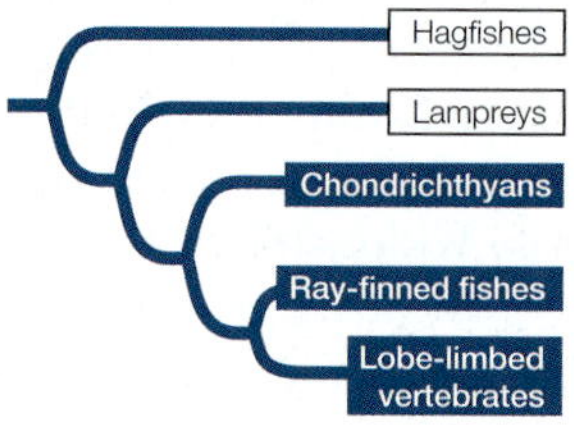
Hagfishes
Lampreys
Chondrichthyans
Ray-finned fishes
Lobe-limbed vertebrates

Many kinds of jawless fishes were found in the seas, estuaries, and fresh waters of the Ordovician, Silurian, and Devonian periods, but hagfishes and lampreys are the only jawless fishes that survived beyond the Devonian. Late in the Ordovician, some fishes evolved jaws via modifications of the skeletal arches that supported the gills (**Figure 33.12A**). Those fishes and their descendants are referred to as **gnathostomes** (Greek, "jaw mouths"). Jaws greatly improved feeding efficiency, as an animal with jaws can grasp, subdue, and swallow large prey. Jawed fishes rapidly diversified during the Devonian period, eventually replacing the jawless fishes in dominance of the seas.

The earliest jaws were simple, but the evolution of teeth made feeding even more efficient (**Figure 33.12B**). In predators, teeth function crucially both in grasping and in breaking up prey. In both predators and herbivores, teeth enable an animal to chew both soft and hard body parts of their food organisms. Chewing also aids chemical digestion and improves an animal's ability to extract nutrients from its food, as we will describe in Chapter 51. Vertebrates are remarkable in the diversity of their jaws and teeth.

Fins and swim bladders improved stability and control over locomotion

Most jawed fishes have a pair of pectoral fins just behind the gill slits and a pair of pelvic fins anterior to the anus (see

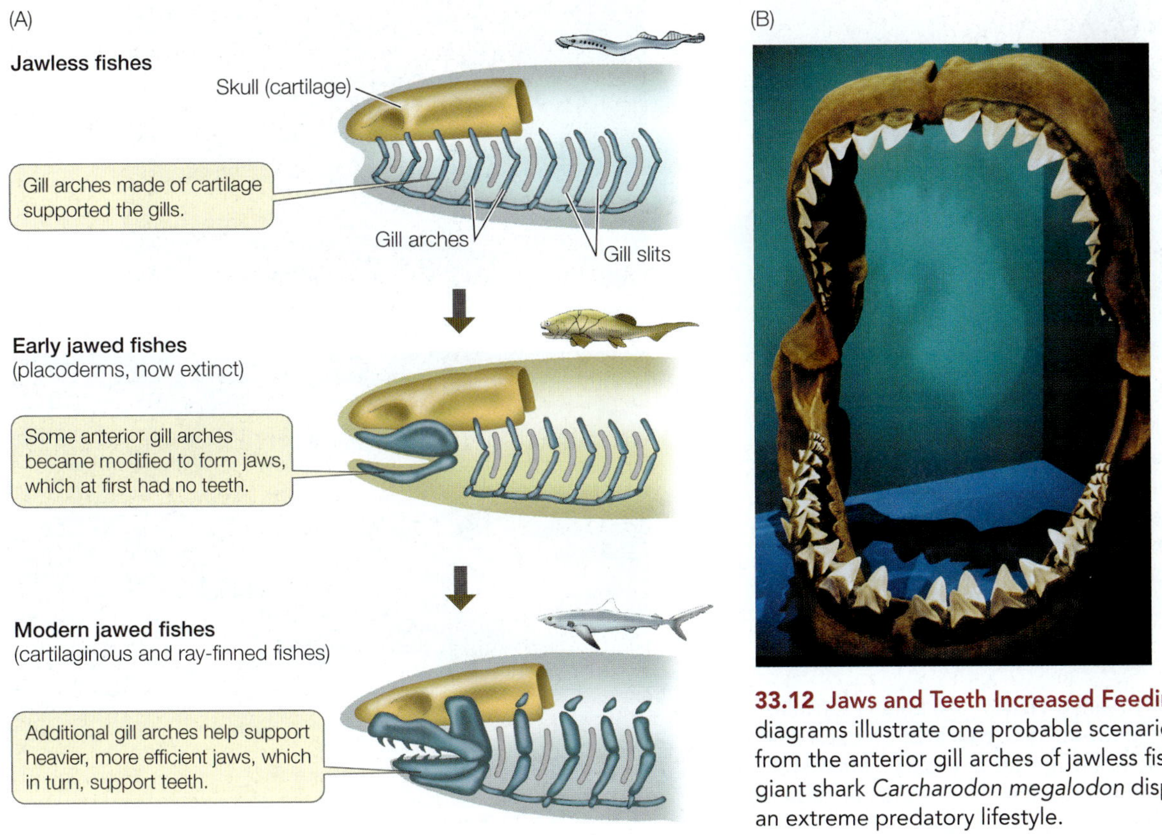
(A)
Jawless fishes
Skull (cartilage)
Gill arches made of cartilage supported the gills.
Gill arches
Gill slits
Early jawed fishes
(placoderms, now extinct)
Some anterior gill arches became modified to form jaws, which at first had no teeth.
Modern jawed fishes
(cartilaginous and ray-finned fishes)
Additional gill arches help support heavier, more efficient jaws, which in turn, support teeth.
(B)

33.12 Jaws and Teeth Increased Feeding Efficiency (A) These diagrams illustrate one probable scenario for the evolution of jaws from the anterior gill arches of jawless fishes. (B) Jaws of the extinct giant shark *Carcharodon megalodon* display the teeth that indicate an extreme predatory lifestyle.

Figures 33.9 and 33.13A). These paired fins stabilize the fish's position in water (and in some cases, help propel it). Median dorsal and anal fins also stabilize the fish, or may be used for propulsion in some species. In many fishes, the caudal (tail) fin helps propel the animal and enables it to turn rapidly.

Several groups of jawed fishes became abundant during the Devonian. Among them were the **chondrichthyans**—sharks, skates, and rays (about 1,000 living species) and chimaeras (40 living species). Like hagfishes and lampreys, these fishes have a skeleton composed entirely of cartilage. Their skin is flexible and leathery, sometimes bearing scales that give it the consistency of sandpaper. Sharks move forward by means of lateral (side-to-side) undulations of the body and caudal fin (**Figure 33.13A**). Skates and rays propel themselves by means of vertical undulating movements of their greatly enlarged pectoral fins (**Figure 33.13B**).

Most sharks are predators, but some feed by straining plankton from the water. Most skates and rays live on the ocean floor, where they feed on mollusks and other animals buried in the sediments. Nearly all chondrichthyans live in the oceans, but a few are estuarine or migrate into lakes and rivers. One group of stingrays is found in river systems of South America. The less familiar chimaeras (**Figure 33.13C**) live in deep-sea or cold waters.

One lineage of aquatic gnathostomes gave rise to the bony vertebrates, which soon split into two main lineages—the **ray-finned fishes** and the **lobe-limbed vertebrates**. Bony vertebrates have internal skeletons of calcified, rigid bone rather than flexible cartilage. In early bony vertebrates, gas-filled sacs supplemented the gas exchange function of the gills by giving the animals access to atmospheric oxygen. These features enabled these fishes to live where oxygen was periodically in short supply, as it often is in freshwater environments. In the ray-finned fishes, these lunglike sacs evolved into **swim bladders**, which are organs of buoyancy. By adjusting the amount of gas in its swim bladder, a ray-finned fish can control the depth at which it remains suspended in the water while expending very little energy to maintain its position.

The outer body surface of most species of ray-finned fishes is covered with thin, flat, lightweight scales that provide protection or enhance movement through the water. The gills open into a single chamber covered by a hard flap, called an operculum. Movement of the operculum increases the flow of water over the gills, where gas exchange takes place.

Ray-finned fishes began to diversify during the Mesozoic era and continued to radiate extensively throughout the Tertiary period. Today there are about 32,000 known living species, encompassing a remarkable variety of sizes, shapes, and lifestyles (**Figure 33.14**). The smallest are less than 1 centimeter long; the largest weigh as much as 900 kilograms. Ray-finned fishes exploit nearly all types of aquatic food sources. In the oceans they filter plankton from the water, rasp algae from rocks, eat corals and other soft-bodied colonial animals, dig animals from soft sediments, and prey on virtually all kinds of other fishes. In fresh water they eat plankton, devour insects, eat fruits that fall into the water, and prey on other aquatic vertebrates and, occasionally,

(A) *Charcharodon charcharis*

Dorsal fin
Caudal fin
Pelvic fin
Pectoral fin

(B) *Myliobatis australis*

Pectoral fins

Dorsal fin
Pectoral fin
Pelvic fin

(C) *Hydrolagus colliei*

33.13 Chondrichthyans (A) Most sharks are active marine predators, as epitomized by the great white shark seen here. (B) Skates and rays, represented here by an eagle ray, feed on the ocean bottom. Their modified pectoral fins are used for propulsion; their other fins are greatly reduced. (C) A chimaera, or ratfish. Many of these deep-sea fishes possess modified dorsal fins that contain toxins.

33.14 Diversity among the Ray-Finned Fishes (A) Eels such as this moray have the large teeth and powerful jaws typical of predatory fishes. (B) There are more than 500 described species of wrasses. Many species, such as this flame fairy wrasse, inhabit coral reefs. (C) Another large ray-fin clade, the serranids, includes the sea basses and groupers. Panther groupers such as this one are endangered by the loss of Pacific coral reef habitat. (D) A unique structure that resembles a fishing lure has evolved among the anglerfishes. Deep-sea anglerfishes such as this one live below the level of light penetration; their lures are bioluminescent.

(A) *Gymnothorax meleagris*

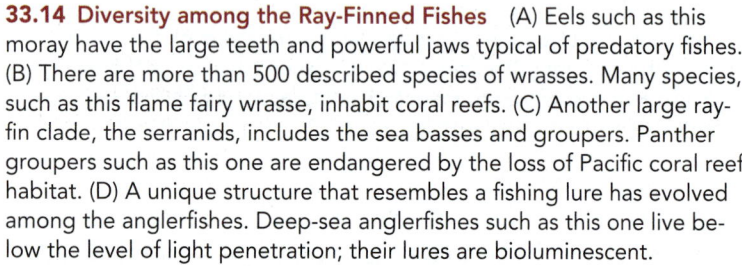

(B) *Cirrhilabrus jordani*

(C) *Cromileptes altivelis*

(D) *Gigantactis vanhoeffeni* luring prey

terrestrial vertebrates. Many ray-finned fishes are solitary, but in open water others form large aggregations called schools. Many species perform complicated behaviors to maintain schools, build nests, court mates, and care for their young.

Although ray-finned fishes can readily control their position in open water using their fins and swim bladder, their eggs tend to sink. Some species produce small eggs that are buoyant enough to complete their development in open water, but many marine fishes move to food-rich shallow waters to lay their eggs. That is why coastal waters and estuaries are so important in the life cycles of many marine fishes. Some ray-finned fishes, such as salmon, are anadromous, moving from the ocean to the fresh waters in which they breed.

RECAP 33.3

Chordates are characterized by a dorsal hollow nerve chord, a post-anal tail, and a dorsal supporting rod called a notochord at some point during the life cycle. Specialized structures for support (a vertebral column), locomotion (such as fins), and feeding (jaws and teeth) evolved among the vertebrates.

- What synapomorphies characterize the chordates and the vertebrates, respectively? **See pp. 683–685 and Figures 33.7 and 33.9**
- How do the hagfishes differ from the lampreys in morphology and life history? Why do some biologists contend that hagfishes are not vertebrates? **See pp. 685–686**

In the lobe-limbed vertebrates, the gas-filled sacs that gave rise to swim bladders in ray-finned fishes became specialized for another purpose: breathing air. That adaptation set the stage for the vertebrates to move onto the land.

33.4 How Did Vertebrates Colonize the Land?

The evolution of lunglike sacs in fishes set the stage for the vertebrate invasion of the land. Some early ray-finned fishes probably used those sacs to supplement their gills when oxygen levels in the water were low, as many groups of ray-finned fishes do today. But with their unjointed fins, those fishes could only flop around when out of water. Changes in the structure of the fins first allowed lobe-limbed vertebrates to support themselves better in shallow water and, later, to move better on land.

Jointed limbs enhanced support and locomotion on land

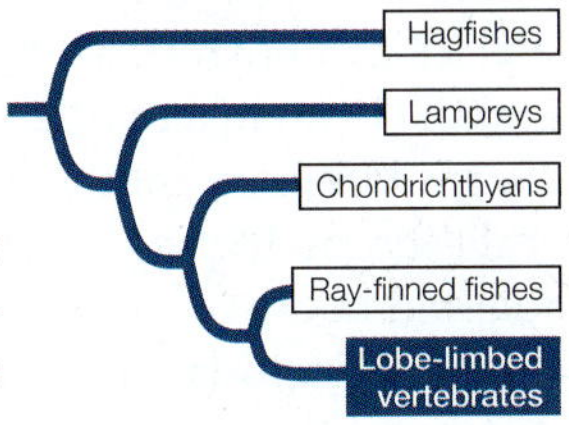
Hagfishes
Lampreys
Chondrichthyans
Ray-finned fishes
Lobe-limbed vertebrates

In the lobe-limbed vertebrates, the paired pelvic and pectoral fins developed into more muscular fins that were joined to the body by a single enlarged bone. The modern representatives of these lobe-limbed vertebrates include the coelacanths, lungfishes, and tetrapods.

The coelacanths flourished from the Devonian until about 65 million years ago, when they were thought to have become extinct. But in 1938 a commercial fisherman caught a living coelacanth off South Africa. Since that time, hundreds of individuals of this extraordinary fish, *Latimeria chalumnae*, have been collected. A second species, *L. menadoensis*, was discovered in 1998 off the Indonesian island of Sulawesi. *Latimeria*, a predator of other fishes, reaches a length of about 1.8 meters and weighs up to 82 kilograms (**Figure 33.15A**). Its skeleton is composed mostly of cartilage, not bone. The

(A) *Latimeria chalumnae*
(B) *Protopterus annectens*
Tiktaalik's pectoral fins show some of the skeletal structures of tetrapod limbs.
(C) *Tiktaalik roseae*

33.15 The Closest Relatives of Tetrapods (A) The coelacanth, discovered in deep waters of the Indian Ocean off the South African coast, represents one of two surviving species of a group that was once thought to be extinct. (B) All surviving lungfish species, such as this African lungfish, live in the Southern Hemisphere. (C) *Tiktaalik*, a fossil lobe-limbed vertebrate from the Devonian, is believed to represent a transitional species intermediate between the finned fishes and the limbed tetrapods.

cartilaginous skeleton is a derived feature in this clade because it had bony ancestors.

Go to Media Clip 33.3
Coelacanths in the Deep Seas
Life10e.com/mc33.3

Lungfishes were important predators in shallow-water habitats in the Devonian, but most lineages died out. The six surviving species live in stagnant swamps and muddy waters in South America, Africa, and Australia (**Figure 33.15B**). Lungfishes have lungs derived from the lunglike sacs of their ancestors as well as gills. When ponds dry up, individuals of most species can burrow deep into the mud and survive for many months in an inactive state while breathing air.

It is believed that some early aquatic lobe-limbed vertebrates began to use terrestrial food sources, became more fully adapted to life on land, and eventually evolved to become ancestral **tetrapods** ("four legs"). How was this transition from an animal that swam in water to one that walked on land accomplished? Early in 2006, scientists reported the discovery of a Devonian fossil lobe-limbed vertebrate, since then named *Tiktaalik*, which possessed intermediate appendages between the fins of fishes and the limbs of terrestrial tetrapods (**Figure 33.15C**). It appears that limbs able to prop up a large fish and make the front-to-rear movements necessary for walking evolved while these animals still lived in water. These limbs appear to have functioned in holding the animals upright in shallow water, perhaps even allowing them to hold their head above the water's surface. These same structures were then co-opted for movement on land, at first probably for foraging on brief trips out of water.

Among the lobe-limbed vertebrates, limbs capable of movement on land evolved from the short, muscular fins of aquatic ancestors (**Figure 33.16**). The resulting four terrestrial limbs give the tetrapods their name. The basic skeletal elements of those limbs can be traced through major changes in limb form and function among the terrestrial vertebrates.

An early split in the tetrapod tree led to two main groups of terrestrial vertebrates: **amphibians**, most of which remained tied to moist environments; and the **amniotes**, many of which adapted to much drier conditions.

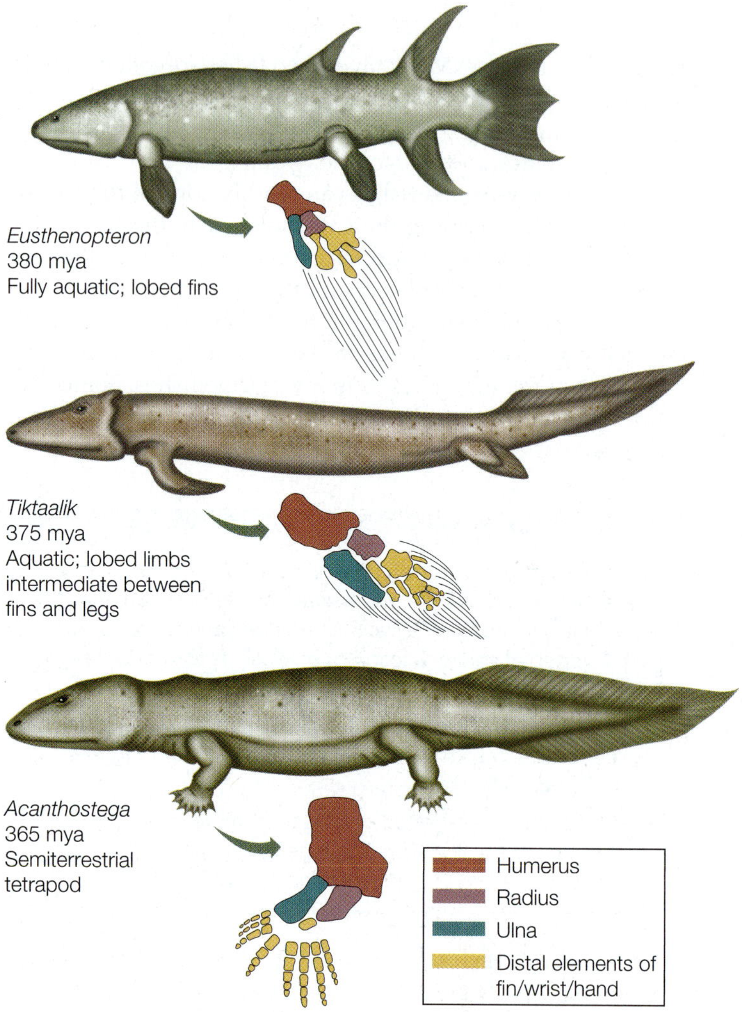

Eusthenopteron
380 mya
Fully aquatic; lobed fins
Tiktaalik
375 mya
Aquatic; lobed limbs intermediate between fins and legs
Acanthostega
365 mya
Semiterrestrial tetrapod
Humerus
Radius
Ulna
Distal elements of fin/wrist/hand

33.16 Tetrapod Limbs Are Modified Fins The major skeletal elements of the tetrapod limb were already present in aquatic lobe-limbed fishes some 380 million years ago. The relative sizes and positions of these elements changed as lobe-limbed vertebrates moved to a terrestrial environment, where limbs were needed to support and move the animal's body on land.

Amphibians usually require moist environments

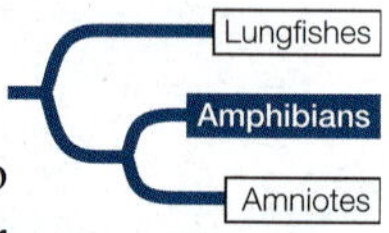

Lungfishes
Amphibians
Amniotes

Most modern amphibians are confined to moist environments because they lose water rapidly through the skin when exposed to dry air. In addition, their eggs are enclosed within delicate membranous envelopes that cannot prevent water loss in dry conditions. In some amphibian species, adults live mostly on land but return to fresh water to lay and fertilize their eggs (**Figure 33.17**). The fertilized eggs give rise to larvae that live in water until they undergo metamorphosis to become terrestrial adults. However, many amphibians (especially those in tropical and subtropical areas) have evolved a wide variety of additional reproductive modes and types of parental care, as described in the opening of this chapter. Internal fertilization, for example, evolved many times among amphibian species. Many species develop directly into adultlike forms from fertilized eggs laid on land or carried by the parents. Other species of amphibians are entirely aquatic, never leaving the water at any stage of their lives, and many of these species retain a larval-like morphology.

The more than 7,000 known species of amphibians living today belong to three major groups: the wormlike, limbless, tropical, burrowing, or aquatic caecilians (**Figure 33.18A**), the tailless frogs and toads (collectively called anurans; **Figure 33.18B**), and the tailed salamanders (**Figure 33.18C and D**).

Anurans are most diverse in wet tropical and warm temperate regions, although a few are found at very high latitudes. There are far more anurans than any other amphibians, with well over 6,000 described species and more being discovered every year. Some anurans have tough skins and other adaptations that enable them to live for long periods in deserts, whereas others live in moist terrestrial and arboreal environments. Some species are completely aquatic as adults. All anurans have a short vertebral column and a pelvic region that is

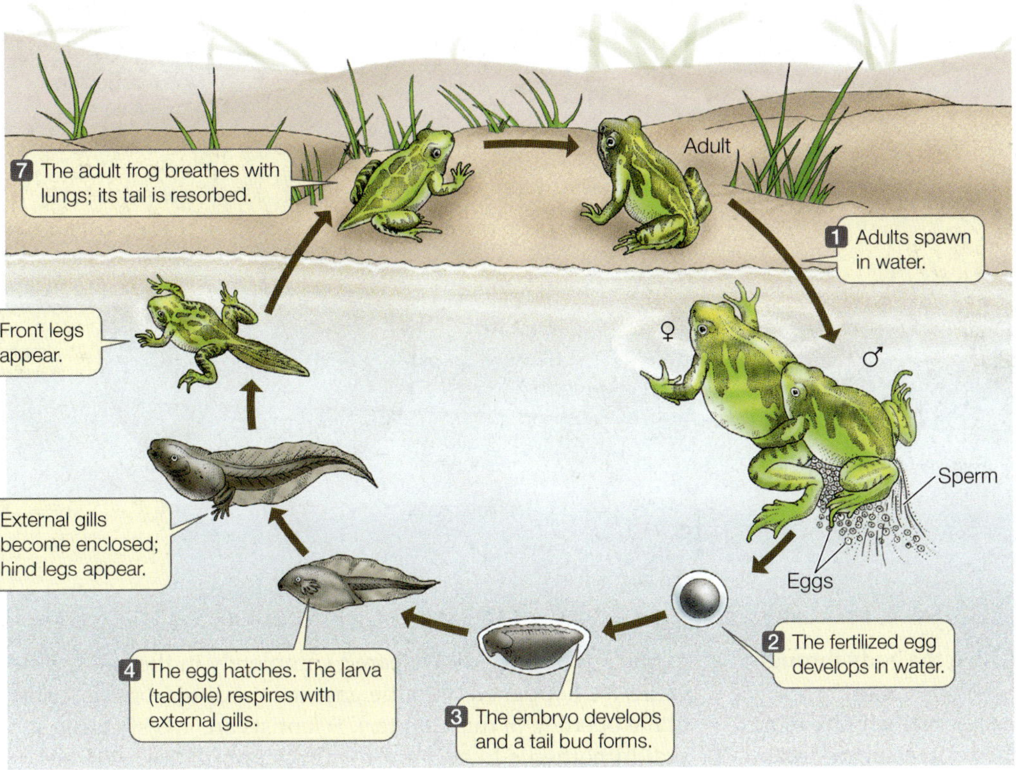
7 The adult frog breathes with lungs; its tail is resorbed.
Adult
1 Adults spawn in water.
6 Front legs appear.
♀
♂
Sperm
5 External gills become enclosed; hind legs appear.
Eggs
2 The fertilized egg develops in water.
4 The egg hatches. The larva (tadpole) respires with external gills.
3 The embryo develops and a tail bud forms.

33.17 In and Out of the Water Many amphibian species have life cycles like the one diagrammed here, in which the early stages take place in water and the aquatic tadpole transforms into a terrestrial adult through metamorphosis. Some species of amphibians, however, have direct development (with no aquatic larval stage), and others are aquatic throughout life.

Go to Animated Tutorial 33.2 Life Cycle of a Frog
Life10e.com/at33.2

(A) *Siphonops annulatus*

(B) *Bufo periglenes*

(C) *Ambystoma mavortium*

(D) *Eurycea waterlooensis*

33.18 Diversity among the Amphibians (A) Burrowing caecilians superficially look more like worms than like amphibians. (B) Male golden toads in the cloud forest of Monteverde, Costa Rica. This species has recently become extinct, one of many amphibian species to do so in the past few decades. (C) An adult barred tiger salamander. (D) This Austin blind salamander's life cycle is completely aquatic; it has no adult terrestrial stage. The eyes of this cave dweller have become greatly reduced.

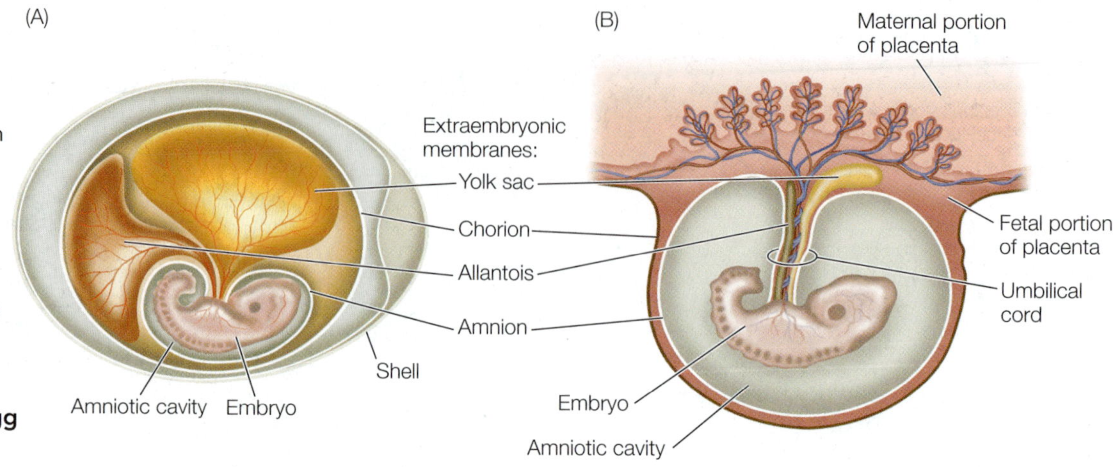
(A)
(B)
Maternal portion of placenta
Extraembryonic membranes:
Yolk sac
Chorion
Allantois
Amnion
Shell
Amniotic cavity
Embryo
Fetal portion of placenta
Umbilical cord
Embryo
Amniotic cavity

33.19 The Amniote Egg (A) The evolution of the amniote egg, with its water-retaining shell, four extraembryonic membranes, and embryo-nourishing yolk, was a major step in adaptation to the terrestrial environment. A chick egg is shown here. (B) In mammals, the developing embryo is retained inside the mother's body, with which it exchanges nutrients and wastes via the placenta. Note the correspondence between the various membranes in (A) and (B).

Go to Activity 33.2 The Amniote Egg
Life10e.com/ac33.2

modified for leaping, hopping, or propelling the body through water by kicking the hind legs.

The more than 600 described species of salamanders are most diverse in temperate regions of the Northern Hemisphere and in cool, moist environments in the mountains of Central America, although a few species penetrate into tropical regions. Many salamanders live in rotting logs or moist soil. One major group has lost the lungs, and these species exchange gases entirely through the skin and mouth lining—body parts that all amphibians use, in addition to their lungs, for gas exchange. A completely aquatic lifestyle has evolved several times among the salamanders (see Figure 33.18D). These aquatic species have arisen through a developmental process known as **neoteny**, or the retention of juvenile traits (in this case, gills) by delayed somatic development. Most species of salamanders have internal fertilization, which is usually achieved through the transfer of a small, jellylike, sperm-embedded capsule called a spermatophore.

Many amphibians have complex social behaviors. Most male anurans utter loud, species-specific calls to attract females of their own species (and sometimes to defend breeding territories), and they compete for access to females that arrive at the breeding sites. Many amphibians lay large numbers of eggs, which they abandon once they are deposited and fertilized. As described in the opening of this chapter, however, other amphibians lay only a few eggs, which are fertilized and then cared for. A few species of frogs, salamanders, and caecilians are viviparous, meaning that they give birth to well-developed young that have received nutrition from the female during gestation.

Go to Media Clip 33.4
Answering a Mating Call
Life10e.com/mc33.4

Amphibians are the focus of much attention today because populations of many species are declining rapidly, especially in mountainous regions of western North America, Central and South America, and northeastern Australia. Worldwide, about one-third of amphibian species are now threatened with extinction or have disappeared completely in the last few decades (as happened with the gastric brooding frogs described at the opening of this chapter). Scientists are investigating several hypotheses to account for these population declines, as described in Chapters 1 and 30.

Amniotes colonized dry environments

Several key innovations for conserving water contributed to the ability of the amniotes to exploit a wide range of terrestrial habitats. The **amniote egg** (which gives the group its name) is relatively impermeable to water and allows the embryo to develop in a contained aqueous environment (**Figure 33.19A**). Its leathery or brittle, calcium-impregnated shell retards evaporation of the fluids inside but permits passage of oxygen and carbon dioxide. The amniote egg also stores large quantities of food in the form of **yolk**, allowing the embryo to attain a relatively advanced state of development before it hatches. Within the shell are **extraembryonic membranes** that protect the embryo from desiccation and assist it with gas exchange and excretion of nitrogenous waste products of metabolism.

In several different groups of amniotes, modifications of the amniote egg allowed the embryo to develop inside (and exchange nutrients and wastes with) its mother's body (**Figure 33.19B**). In most mammals the egg lost its shell entirely while the functions of the extraembryonic membranes were retained and expanded; we will examine the roles of these membranes in detail in Section 44.5.

Other innovations evolved in the organs of terrestrial adults. A tough, impermeable skin, covered with scales or modifications of scales such as hair and feathers, greatly reduced water loss. Adaptations of the vertebrate excretory organs, called kidneys, allowed amniotes to excrete concentrated urine, ridding the body of nitrogenous wastes without losing a large amount of water in the process (see Chapter 52).

During the Carboniferous, the amniotes split into two major groups: the **reptiles** and the lineage that eventually led to the **mammals** (**Figure 33.20**).

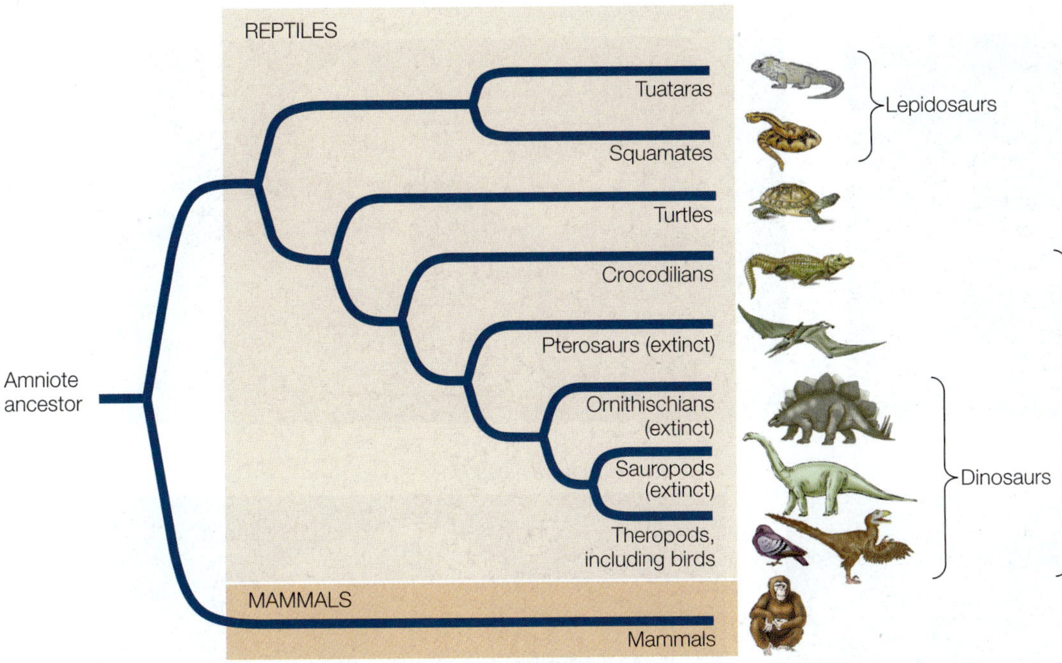
REPTILES
Tuataras
Squamates
Lepidosaurs
Turtles
Crocodilians
Pterosaurs (extinct)
Ornithischians (extinct)
Sauropods (extinct)
Theropods, including birds
Dinosaurs
Archosaurs
Amniote ancestor
MAMMALS
Mammals

33.20 Phylogeny of Amniotes This tree of amniote relationships shows the primary split between mammals and reptiles. The reptile portion of the tree shows that one lineage led to the lepidosaurs (snakes, lizards, and tuataras), another to the turtles, and a third to the archosaurs (the crocodiles, several extinct groups, and the birds).

Reptiles adapted to life in many habitats

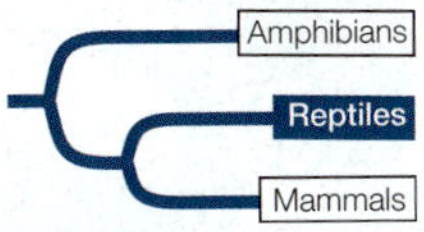
Amphibians
Reptiles
Mammals

The lineage leading to modern reptiles began to diverge from other amniotes more than 300 million years ago. More than 19,000 species of reptiles exist today, more than half of which are birds. Birds are the only living representatives of the otherwise extinct dinosaurs, the dominant terrestrial predators of the Mesozoic.

The **lepidosaurs** constitute the second most species-rich clade of living reptiles. This group is composed of the **squamates** (lizards, snakes, and amphisbaenians—the last a group of mostly legless, wormlike, burrowing reptiles with greatly reduced eyes) and the **tuataras**, which superficially resemble lizards but differ from them in tooth attachment and several internal anatomical features. Many species related to the tuataras lived during the Mesozoic era, but today only two species survive (Figure 33.21A), and these are restricted to a few islands off New Zealand.

The skin of a lepidosaur is covered with horny scales that greatly reduce loss of water from the body surface. These scales, however, make the skin unavailable as an organ of gas exchange. Gases are exchanged almost entirely via the lungs, which are proportionally much larger in surface area than those of amphibians. A lepidosaur forces air into and out of its lungs by bellows-like movements of its ribs. The three-chambered lepidosaur heart partially separates oxygenated blood from the lungs from deoxygenated blood returning from the body. With this type of heart, lepidosaurs can generate high blood pressure and can sustain a relatively high metabolism.

Most lizards are insectivores, although some are herbivores and a few prey on other vertebrates. Most lizards walk on four limbs (Figure 33.21B), although limblessness has evolved repeatedly among the lizards, especially in burrowing and grassland species. The largest lizard is the predaceous Komodo dragon of the East Indies, which grows as long as 3 meters and can weigh more than 150 kilograms.

Go to Media Clip 33.5
Komodo Dragons Bring Down Prey
Life10e.com/mc33.5

The major group of limbless squamates is the snakes (Figure 33.21C). All snakes are carnivores, and many can swallow objects much larger than themselves. Several snake groups evolved venom glands and the ability to inject venom rapidly into their prey.

The **turtles** comprise a reptilian group that has changed relatively little since the early Mesozoic. In these reptiles, dorsal and ventral bony plates form a shell into which the head and limbs can be withdrawn in many species (Figure 33.21D). The dorsal shell is a modification of the ribs. It is a mystery how the pectoral girdles evolved to be inside the ribs of turtles, making them unlike any other vertebrates. Most turtles live in aquatic environments, but several groups, such as tortoises and box turtles, are terrestrial. Sea turtles spend their entire lives at sea except when they come ashore to lay eggs. Human exploitation of sea turtles and their eggs has resulted in worldwide declines of these species, all of which are now endangered. A few species of turtles are strict herbivores or carnivores, but most species are omnivores that eat a variety of aquatic and terrestrial plants and animals.

Crocodilians and birds share their ancestry with the dinosaurs

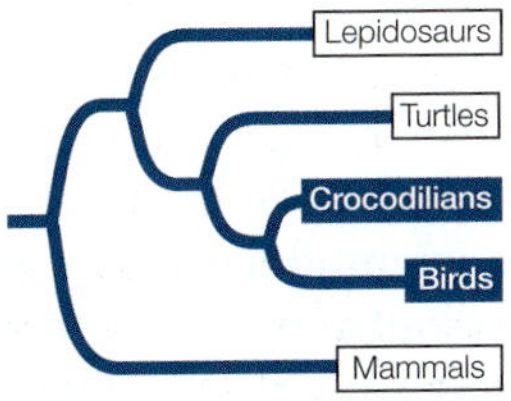
Lepidosaurs
Turtles
Crocodilians
Birds
Mammals

Another reptilian clade, the **archosaurs**, includes the crocodilians, pterosaurs, dinosaurs, and birds. Only the crocodilians and birds are represented by living species today. Modern **crocodilians**—crocodiles, caimans, gharials, and alligators—are

(A) *Sphenodon punctatus*

(B) *Eublepharis macularius*

(C) *Diadophis punctatus*

(D) *Chelonoidis nigra abingdonii*

33.21 Reptilian Diversity (A) This tuatara represents one of only two surviving species in its lineage. (B) The leopard gecko, a desert-dwelling lizard native to Afghanistan, Pakistan, and northwestern India. (C) The ringneck snake of North America is nonvenomous. It coils its tail to reveal a bright orange underbelly, which distracts potential predators from the vital head region. (D) Galápagos tortoises are the largest turtles and among the largest reptiles. They have been documented to live for more than 100 years in the wild.

confined to tropical and warm temperate environments (**Figure 33.22A**). All crocodilians are carnivorous; they eat vertebrates of all kinds, including large mammals. Crocodilians spend much of their time in water, but they lay their eggs in nests they build on land or on floating piles of vegetation. The eggs are warmed by heat generated by decaying organic matter that the female places in the nest. The female provides other forms of parental care as well: typically she guards the eggs until they hatch, and in some species she continues to guard and communicate with her offspring after they hatch.

Dinosaurs rose to prominence about 215 million years ago and dominated terrestrial environments for about 150 million years. However, only one group of dinosaurs, the **birds**, survived the mass extinction at the end of the Cretaceous. During the Mesozoic, most terrestrial animals more than a meter long were dinosaurs. Many were agile and could run rapidly; they had special muscles that enabled the lungs to be filled and emptied while the limbs moved. We can infer the existence of such muscles in dinosaurs from the structure of the vertebral column in fossils. Some of the largest dinosaurs weighed as much as 70,000 kilograms.

Biologists have long accepted the phylogenetic position of birds among the reptiles, although birds clearly have many unique, derived morphological features. In addition to the strong morphological evidence for this placement, fossil and molecular data emerging over the last few decades have provided definitive supporting evidence. Birds are a specialized group of **theropods**, a clade of predatory dinosaurs that shared such traits as a bipedal stance, hollow bones, a furcula ("wishbone"), elongated metatarsals with three-fingered feet, elongated forelimbs with three fingers, and a pelvis that points backward. Modern birds are endothermic, meaning that they regulate their body temperatures by producing and retaining metabolic heat, rather than by absorbing heat from their external environment (see Section 40.5). Although we cannot directly assess this physiological trait in extinct species, many fossil theropods share morphological traits that suggest they may have been endothermic as well.

The living bird species fall into two major groups that diverged about 80 to 90 million years ago from a flying ancestor. The few modern descendants of one lineage include a group of secondarily flightless and weakly flying birds, some of which are very large. This group, called the palaeognaths, includes the South and Central American tinamous and several large flightless birds of the southern continents—the rheas, emu, kiwis, cassowaries, and the world's largest bird, the ostrich (**Figure 33.22B**). The

(A) *Crocodylus porosus*
(B) *Struthio camelus*

33.22 Archosaurs The two surviving groups of archosaurs are very different. (A) The crocodilians live in tropical and warm temperate climates. This crocodile species lives in saltwater and estuarine environments along Australia's coast. (B) Birds are the only other living archosaur group, represented here by the winged but flightless ostrich.

second lineage, the neognaths, has left a much larger number of descendants, most of which have retained the ability to fly.

Feathers allowed birds to fly

Fossil theropods discovered recently in early Cretaceous deposits in Liaoning Province, in northeastern China, show that the scales of some small predatory dinosaurs were highly modified to form **feathers**. Initially these feathers were simply a body covering that probably provided insulation and enhanced coloration. But the feathers of some later dinosaurs, such as *Microraptor gui*, were structurally similar to those of modern birds (**Figure 33.23A**).

Another theropod that was even more closely related to modern birds, *Archaeopteryx*, lived about 150 million years ago. *Archaeopteryx* had teeth (unlike modern birds), but it was covered with feathers that are virtually identical to those of birds (**Figure 33.23B**). It also had well-developed wings, a long tail, and a furcula to which some of the flight muscles were probably attached. *Archaeopteryx* had clawed fingers on its forelimbs, but it also had typical perching bird claws on its hindlimbs. It probably lived in trees and shrubs and used the fingers to assist it in clambering over branches. It probably glided or flew weakly. The descendants of *Archaeopteryx* and similar Mesozoic theropods were the modern birds, most of which are accomplished fliers.

Go to Media Clip 33.6
Falcons in Flight
Life10e.com/mc33.6

The evolution and specialization of feathers was a major force for diversification. Feathers are lightweight, but they are strong and structurally complex (**Figure 33.24**). The stiff central shaft of the flight feathers on a bird's wings arise from the skin of the forelimbs to create the flying surfaces. Other strong feathers sprout like a fan from the shortened tail and serve as stabilizers during flight. The feathers that cover the body, along with an underlying layer of down feathers, provide birds with insulation that helps them survive in virtually all of Earth's climates.

The bones of theropod dinosaurs, including birds, are hollow with internal struts that increase their strength. Hollow bones would have made early theropods lighter and more mobile; later they facilitated the evolution of flight. The sternum (breastbone) of flying birds forms a large, vertical keel to which the flight muscles are attached.

Flight is metabolically expensive. A flying bird consumes energy at a rate about 15 to 20 times faster than a running lizard of the same weight. Because birds have such high metabolic rates, they generate large amounts of heat. They control the rate of heat loss using their feathers, which may be held close to the body or elevated to alter the amount of insulation they provide. The lungs of birds allow air to flow through unidirectionally rather than pumping air in and out (see Section 49.2). This

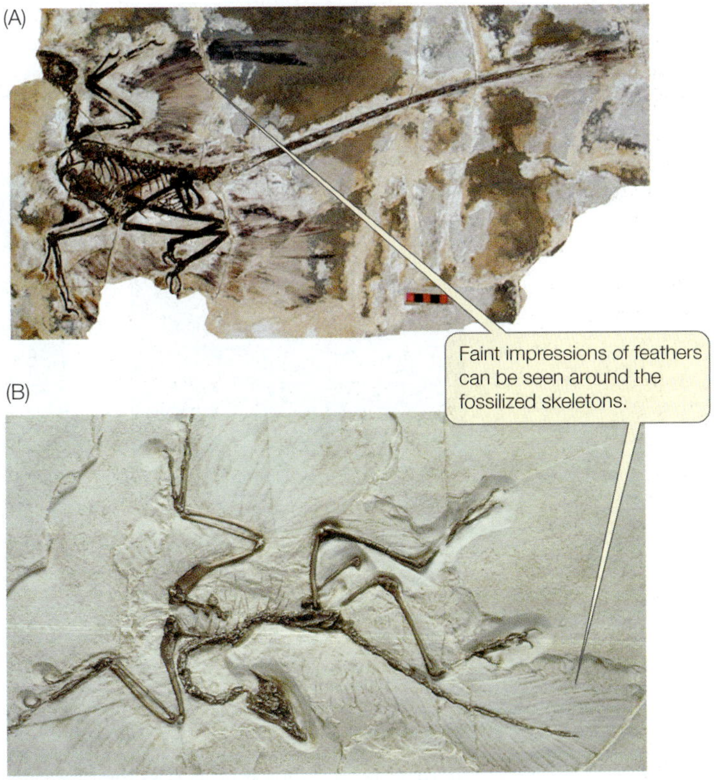
(A)
(B)
Faint impressions of feathers can be seen around the fossilized skeletons.

33.23 Mesozoic Bird Relatives Fossils support the evolution of birds from other theropods. (A) *Microraptor gui* was a feathered theropod from the early Cretaceous (about 140 mya). (B) *Archaeopteryx* was even more closely related to modern birds.

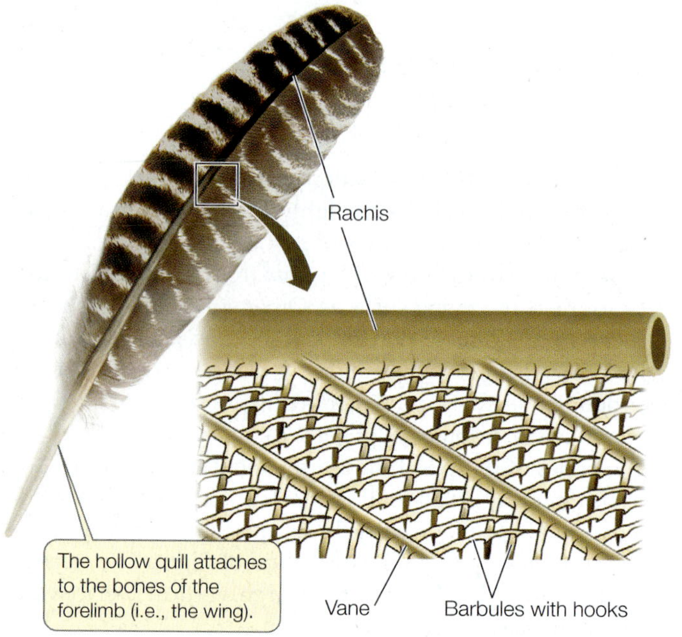
Rachis
The hollow quill attaches to the bones of the forelimb (i.e., the wing).
Vane
Barbules with hooks

33.24 Feather Anatomy The flight feathers of birds are attached to the wing's skin by the hollow portion, or quill, of a stiff central shaft. The rachis is the solid portion of the shaft from which radiate fine branches (vanes) with interlocking hooks and barbules. Overall, this structure represents a major evolutionary innovation: a strong, lightweight surface that enables flight.

flow-through structure of the lungs increases the efficiency of gas exchange and thereby supports a high metabolic rate.

There are about 10,000 species of living birds, which range in size from the 150-kilogram ostrich to a tiny hummingbird weighing only 2 grams (**Figure 33.25**). The teeth so prominent among other dinosaurs were secondarily lost in the ancestral birds, but birds nonetheless eat almost all types of animal and plant material. Insects and fruits are the most important dietary items for terrestrial species. Birds also eat seeds, nectar and pollen, leaves and buds, carrion, and other vertebrates. By eating the fruits and seeds of plants, birds serve as major agents of seed dispersal.

Mammals radiated after the extinction of non-avian dinosaurs

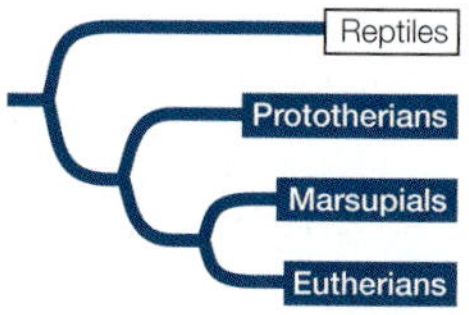
Reptiles
Prototherians
Marsupials
Eutherians

Small and medium-sized mammals coexisted with the large dinosaurs throughout most of the Mesozoic era, and most of the major groups of mammals that are alive today arose in the Cretaceous. After the non-avian dinosaurs disappeared during the mass extinction at the end of the Cretaceous, mammals increased dramatically in numbers, diversity, and size. Today mammals range in size from tiny shrews and bats weighing only about 2 grams to the blue whale, the largest animal on Earth, which measures up to 33

(A) *Bombycilla cedrorum*

(B) *Trichoglossus haematodus*

(C) *Tyto alba*

(D) *Eudocimus ruber*

33.25 Diversity among the Birds (A) Perching, or passeriform, birds such as this cedar waxwing constitute the most species-rich of all bird groups. (B) Some 375 species of parrots, macaws, parakeets, and lorikeets such as the one shown here are another large bird group. (C) The barn owl is a nighttime predator that can find prey using its sensitive auditory "sonar" system. (D) The scarlet ibis of South America and the Caribbean is one of many species of wading birds.

meters long and can weigh as much as 160,000 kilograms. Mammals have far fewer, but more highly differentiated, teeth than do fishes, amphibians, or reptiles. Differences among mammals in the number, type, and arrangement of teeth reflect their varied diets (see Figure 51.6).

Four key features distinguish the mammals:

- *Sweat glands*, which secrete sweat that evaporates and thereby cools an animal
- *Mammary glands*, which in females secrete a nutritive fluid (milk) on which newborn individuals feed
- *Hair*, which provides a protective and insulating covering
- A *four-chambered heart* that completely separates the oxygenated blood coming from the lungs from the deoxygenated blood returning from the body (this last characteristic is convergent with the archosaurs, including modern birds and crocodiles)

Mammalian eggs are fertilized within the female's body, and in nearly all mammalian groups the resulting embryos undergo a period of development inside the female's body in an organ called the uterus. In the uterus, the embryo is contained in an amniotic sac that is homologous to one of the three membranes found in the amniote egg (see Figure 33.19). The embryo is connected to the wall of the uterus by an organ called a placenta. The placenta allows for nutrient and gas exchange, as well as waste elimination from the developing embryo, via the female's circulatory system. Most mammals develop a covering of hair (fur), which is luxuriant in some species but has been greatly reduced in others, including cetaceans (whales and dolphins) and humans. Thick layers of insulating fat (blubber) replace hair as a heat-retention mechanism in the cetaceans; humans learned to use clothing for this purpose when they dispersed from warm tropical areas.

The approximately 5,700 species of living mammals are divided into two primary groups: the **prototherians** and the **therians** (Table 33.1). Members of the therian clade are further divided into the **marsupials** and the **eutherians**.

(A) *Tachyglossus aculeatus*

(B) *Ornithorhynchus anatinus*

33.26 Prototherians (A) The short-beaked echidna is one of four surviving species of echidnas. (B) The duck-billed platypus lives in freshwater streams in eastern Australia.

PROTOTHERIANS Only five species of prototherians are known, and they are found only in Australia and New Guinea. These mammals, the duck-billed platypus and four species of echidnas, differ from other mammals in laying shelled eggs and having sprawling legs (Figure 33.26). Prototherians supply milk for their young, but they have no nipples on their mammary glands; the milk simply oozes out and is lapped off the fur by the offspring.

MARSUPIALS Females of most marsupial species have a ventral pouch in which they carry and feed their offspring (see Figure 33.27A). Gestation (pregnancy) in marsupials is brief; the young are born tiny but with well-developed forelimbs, with which they climb to the pouch. They attach to a nipple but cannot suck. The mother ejects milk into the tiny offspring until it grows large enough to suckle. Once her offspring have left the uterus, a female marsupial may become sexually receptive again. She can then carry fertilized eggs that are capable of initiating development to replace the offspring in her pouch should something happen to them.

At one time marsupials were found on all continents, but the approximately 350 living species are now restricted to Australasia (**Figure 33.27A and B**) and the Americas (especially South America; **Figure 33.27C**). Of the seven major groups of marsupials shown in Table 33.1, only the New World opossums, the shrew opossums, and the diminutive monito del monte are found in the Americas. Only one species, the Virginia opossum, is found in North America north of Mexico. Marsupials radiated to become herbivores, insectivores, and carnivores, but no marsupials live in the oceans. None can fly, although some arboreal (tree-dwelling) marsupials are gliders. The largest living marsupials are the kangaroos of Australia, which can weigh up to 90 kilograms. Much larger marsupials existed in Australia until humans exterminated them soon after reaching that continent about 40,000 years ago.

EUTHERIANS The majority of mammals are eutherians ("true" therians). Eutherians are sometimes called placental mammals, but this name is inappropriate because some marsupials also have placentas. Eutherians are more developed at birth than are marsupials; no external pouch houses them after they are born.

The more than 5,300 species of living eutherians are divided into 20 major groups (see Table 33.1). The relationships of these groups to one another have been difficult to determine because most of the major groups diverged within a short time during an explosive adaptive radiation. Modern genomic analyses have elucidated these relationships, however (**Figure**

TABLE 33.1 Major Groups of Living Mammals

Group	Number of Described Species	Examples
PROTOTHERIANS		
Monotremes (Monotremata)	5	Echidnas, duck-billed platypus
THERIANS		
Marsupials		
Diprotodonts (Diprotodontia)	146	Kangaroos, wallabies, possums, koala, wombats
New World opossums (Didelphimorphia)	93	Opossums
Carnivorous marsupials (Dasyuromorphia)	72	Quolls, dunnarts, numbat, Tasmanian devil
Omnivorous marsupials (Peramelemorphia)	22	Bandicoots and bilbies
Shrew opossums (Paucituberculata)	7	Andean rat opossums
Marsupial moles (Notoryctemorphia)	2	Southern and northern marsupial moles
Microbiothere (Microbiotherea)	1	Monito del monte
Eutherians		
Rodents (Rodentia)	2,337	Rats, mice, squirrels, woodchucks, ground squirrels, beaver, capybara
Bats (Chiroptera)	1,171	Fruit bats, echo-locating bats
Even-toed hoofed mammals and cetaceans (Cetartiodactyla)	469	Deer, sheep, goats, cattle, antelopes, giraffes, camels, swine, hippopotamus, whales, dolphins
Shrews, moles, and relatives (Soricomorpha)	428	Shrews, moles, solenodons
Primates (Primates)	396	Lemurs, monkeys, apes, humans
Carnivores (Carnivora)	284	Wolves, dogs, bears, cats, weasels, pinnipeds (seals, sea lions, walruses)
Rabbits and relatives (Lagomorpha)	92	Rabbits, hares, pikas
African insectivores (Afrosoricida)	50	Tenrecs, golden moles
Hedgehogs (Erinaceomorpha)	24	European hedgehog
Armadillos (Cingulata)	21	Giant armadillo, nine-banded armadillo
Tree shrews (Scandentia)	20	Pygmy tree shrew, pen-tailed tree shrew
Odd-toed hoofed mammals (Perissodactyla)	16	Horses, zebras, tapirs, rhinoceroses
Elephant shrews (Macroscelidea)	15	Elephant shrews, jumping shrews, sengis
Anteaters, sloths (Pilosa)	10	Anteaters, tamanduas, two- and three-toed sloths
Pangolins (Pholidota)	8	Asian and African pangolins
Hyraxes and relatives (Hyracoidea)	5	Hyraxes, dassies
Sirenians (Sirenia)	4	Manatees, dugongs
Elephants (Proboscidea)	3	African and Indian elephants
Colugos (Dermoptera)	2	Flying lemurs
Aardvark (Tubulidentata)	1	Aardvark

33.28). These studies have revealed that the major early splits in eutherian lineages are closely associated with the breakup of the continents during the Mesozoic (see Figure 25.14), after which the major groups of mammals radiated independently in Laurasia, Africa, and South America. The reconnection of South America and North America via the Panamanian land bridge about 3 million years ago resulted in a huge faunal exchange between those continents, which is particularly evident among the mammals. South American groups such as armadillos moved north into North America, and Laurasian groups such as carnivores and odd- and even-toed hoofed animals moved south into South America.

(B) *Macrotis lagotis*
(A) *Macropus giganteus*
(C) *Didelphis virginiana*

33.27 Diversity among the Marsupials (A) Australia's eastern gray kangaroo is among the largest living marsupials. This female carries her young offspring in the characteristic marsupial pouch. (B) The greater bilby of Australia's arid regions is a nocturnal omnivore The bilby does not require drinking water, obtaining the moisture it needs from its various food items. (C) The North American opossum is the only marsupial found north of Mexico.

Eutherians are extremely varied in their form and ecology (**Figure 33.29**). The extinction of the non-avian dinosaurs at the end of the Cretaceous may have made it possible for them to diversify and radiate into a large range of ecological niches. Many eutherian species grew large, and some assumed the roles of dominant terrestrial predators previously occupied by the large dinosaurs. Among these predators, social hunting behavior evolved in several species, including members of the carnivore and primate clades.

The two most diverse groups of eutherians are the rodents and the bats, which together comprise about two-thirds of the species. Rodents are traditionally defined by the unique morphology of their teeth, which are adapted for gnawing through substances such as wood. The bats probably owe much of their

33.28 Major Groups of Eutherians Diversified as the Continents Drifted Apart This phylogenetic tree shows the relationships among most of the major terrestrial eutherian groups, the location of the earliest fossils found for each group (also indicated in the color distinctions of the various branches), and the current distributions of the groups. The major splits in eutherian evolution correspond in large degree to the tectonic history of the major continents (see Figures 25.14 and 54.12).

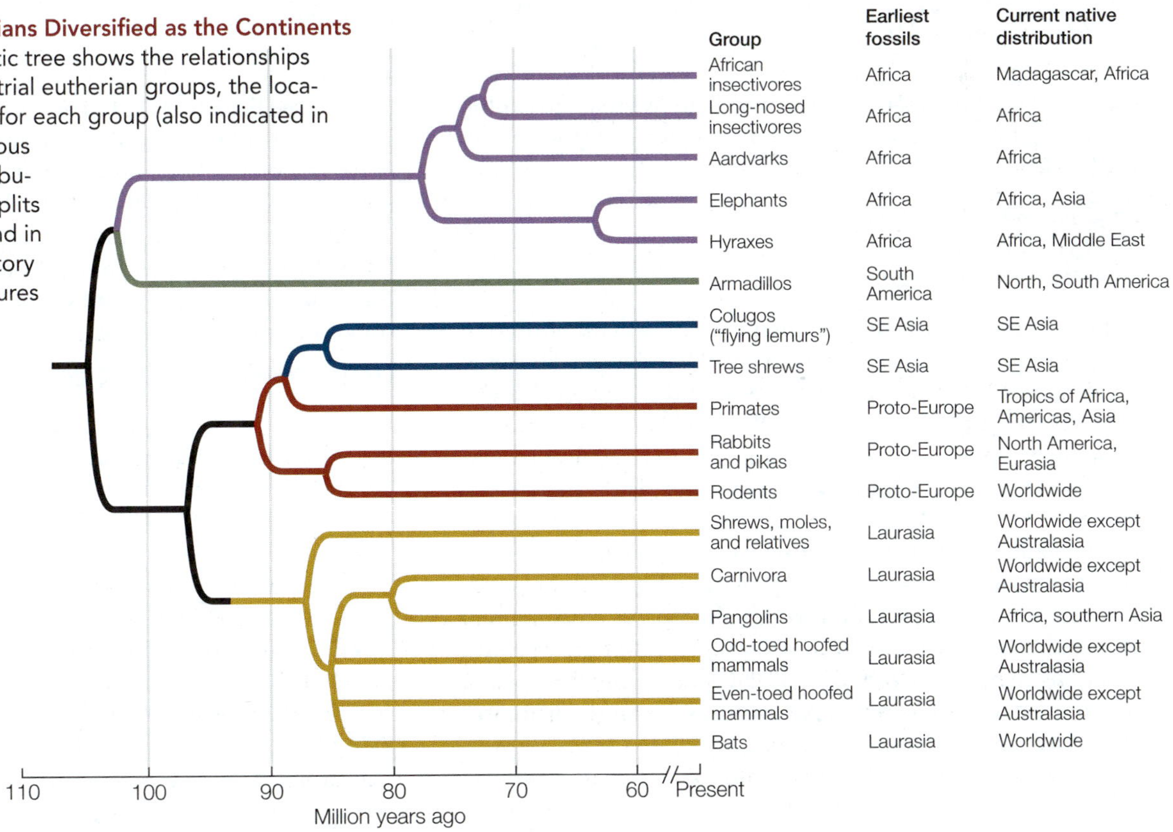
Group
Earliest fossils
Current native distribution
African insectivores | Africa | Madagascar, Africa
Long-nosed insectivores | Africa | Africa
Aardvarks | Africa | Africa
Elephants | Africa | Africa, Asia
Hyraxes | Africa | Africa, Middle East
Armadillos | South America | North, South America
Colugos ("flying lemurs") | SE Asia | SE Asia
Tree shrews | SE Asia | SE Asia
Primates | Proto-Europe | Tropics of Africa, Americas, Asia
Rabbits and pikas | Proto-Europe | North America, Eurasia
Rodents | Proto-Europe | Worldwide
Shrews, moles, and relatives | Laurasia | Worldwide except Australasia
Carnivora | Laurasia | Worldwide except Australasia
Pangolins | Laurasia | Africa, southern Asia
Odd-toed hoofed mammals | Laurasia | Worldwide except Australasia
Even-toed hoofed mammals | Laurasia | Worldwide except Australasia
Bats | Laurasia | Worldwide
110
100
90
80
70
60
Present
Million years ago

(A) *Castor canadensis*

(B) *Macroderma gigas*

(C) *Rangifer tarandus*

(D) *Stenella longirostris*

33.29 Diversity among the Eutherians (A) The North American beaver exhibits the gnawing teeth that characterize rodents. Almost half of all eutherians are rodents. (B) Flight evolved in the ancestor of bats. This ghost bat is endemic to Australia; bats and rodents are the only eutherian groups native to that continent. (C) Large hoofed mammals such as reindeer are important herbivores in terrestrial environments. Although this bull is grazing by himself, reindeer are usually found in huge herds. (D) Spinner dolphins are cetaceans, a cetartiodactyl group that returned to the marine environment.

success to the evolution of flight, which allows them to exploit a variety of food sources and colonize remote locations with relative ease.

Go to Media Clip 33.7
Bats Feeding in Flight
Life10e.com/mc33.7

Grazing and browsing by members of several eutherian groups helped transform the terrestrial landscape. Herds of grazing herbivores fed on open grasslands, whereas browsers fed on shrubs and trees. The effects of these herbivores on plant life favored the evolution of the spines, tough leaves, and difficult-to-eat growth forms found in many plants. In turn, adaptations in the teeth and digestive systems of many herbivore lineages allowed these species to consume many plants despite such defenses—a striking example of coevolution. A large animal can survive on food of lower quality than a small animal can, and large size evolved in several groups of grazing and browsing mammals (see Figure 33.29C). The evolution of large herbivores, in turn, favored the evolution of large carnivores able to attack and overpower them.

Several lineages of terrestrial eutherians subsequently returned to the aquatic environments their ancestors had left behind (see Figure 33.29D). The completely aquatic cetaceans—whales and dolphins—evolved from even-toed hoofed ancestors (whales are closely related to the hippopotamuses). The seals, sea lions, and walruses also returned to the marine environment, and their limbs became modified into flippers. Weasel-like otters retain their limbs but have also returned to aquatic environments, colonizing both fresh and salt water. The manatees and dugongs colonized estuaries and shallow seas.

RECAP 33.4

The initial vertebrate colonization of dry land was facilitated by the evolution of lunglike sacs and jointed limbs. The amniotes evolved impermeable body coverings, efficient kidneys, and the amniote egg, which resists desiccation.

- What are the similarities and differences between a shelled amniote egg laid by a reptile and the retained amniotic sac of a mammal? **See p. 692 and Figure 33.19**
- How has the diversification of mammals been influenced by mass extinction events and continental drift? **See pp. 697–698 and Figure 33.28**

The biology of one eutherian group—the primates—has been the subject of extensive research. The behavior, ecology, physiology, and molecular biology of the primates are of special interest to us because this lineage includes humans.

33.5 What Traits Characterize the Primates?

One lineage of small, arboreal, insectivorous eutherians underwent extensive evolutionary radiation to become the **primates** (Figure 33.30). A nearly complete fossil of an early primate, *Carpolestes*, found in Wyoming, was dated at 56 million years ago; it had grasping feet with an opposable big toe that had a nail rather than a claw. Grasping limbs with opposable digits, an adaptation to arboreal life, are one of the major features that distinguish primates from other mammals.

Two major lineages of primates split late in the Cretaceous

About 90 million years ago, late in the Cretaceous period, the primates split into two clades: the prosimians and the anthropoids. **Prosimians**—lemurs, lorises, and galagos—once lived on all continents, but today they are restricted to Africa, Madagascar, and tropical Asia. All mainland prosimian species are arboreal and nocturnal. On the island of Madagascar, however, the site of a remarkable radiation of lemurs, there are also diurnal and terrestrial species (Figure 33.31). Tarsiers were once considered prosimians as well, although today we know that they are more closely related to monkeys and apes than to lemurs, lorises, and galagos.

The second primate lineage, the **anthropoids**—tarsiers, New World monkeys, Old World monkeys, and apes—began to diversify shortly after the mass extinction event at the end of the Cretaceous, in Africa or Asia. New World monkeys diverged from Old World monkeys and apes slightly later, but early enough that they may have originated in Africa and reached South America when those two continents were still close to each other. New World monkeys now live only in South and Central America, and all of them are arboreal (Figure 33.32A). Many of them have a long, prehensile tail with which they can grasp branches. Many Old World monkeys are arboreal as well, but several species are terrestrial (Figure 33.32B). No Old World monkey has a prehensile tail.

About 35 million years ago, a lineage that led to the modern apes separated from the Old World monkeys. Between 22 and 5.5 million years ago, dozens of species of apes lived in Europe, Asia, and Africa. The

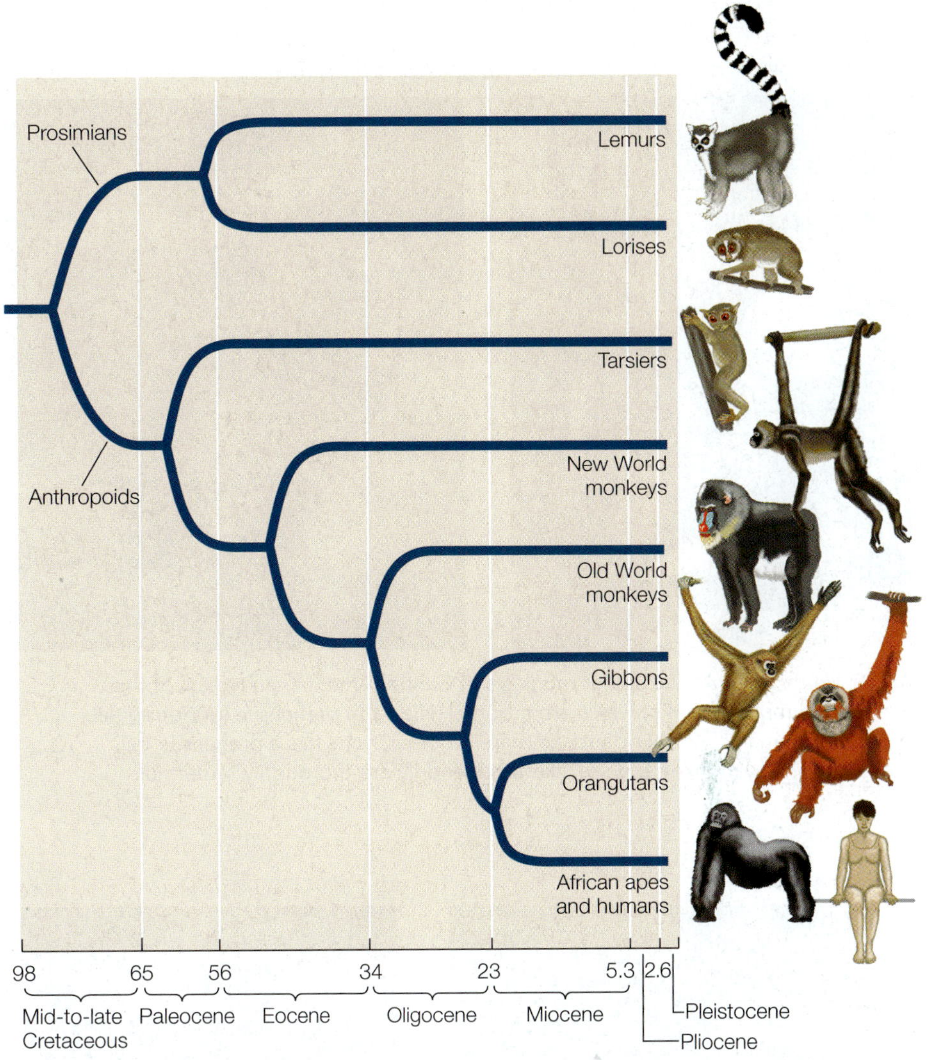
Prosimians
Anthropoids
Lemurs
Lorises
Tarsiers
New World monkeys
Old World monkeys
Gibbons
Orangutans
African apes and humans
98
65
56
34
23
5.3
2.6
Mid-to-late Cretaceous
Paleocene
Eocene
Oligocene
Miocene
Pleistocene
Pliocene

33.30 Phylogeny of the Primates The phylogeny of primates is among the best studied of any major group of mammals. This tree is based on evidence from many genes, morphology, and fossils.

Propithecus diadema

33.31 A Prosimian The diademed sifaka is one of the many lemur species found in Madagascar, where it is part of a unique assemblage of endemic plants and animals. Sifakas live in groups of up to a dozen animals and defend their territories.

(A) *Ateles geoffroyi* (B) *Mandrillus sphinx*

33.32 Monkeys (A) The spider monkeys of Central America are typical of the New World monkeys, all of which are arboreal. Note the prehensile (gripping) tail. (B) Although many Old World monkeys are arboreal, none has a prehensile tail. Many Old World monkey species, like this mandrill, are thoroughly terrestrial.

Asian apes—gibbons and orangutans (**Figure 33.33A and B**)—descended from two of these ape lineages. Orangutans are the closest living sister group of the modern African apes: gorillas (**Figure 33.33C**), chimpanzees (**Figure 33.33D**), and humans.

Bipedal locomotion evolved in human ancestors

About 6 million years ago in Africa, a lineage split occurred that would lead to the chimpanzees on the one hand and to the **hominin** clade, which includes modern humans and their extinct close relatives, on the other.

The earliest protohominins, known as ardipithecines, had distinct morphological adaptations for **bipedal locomotion** (walking on two legs). Bipedal locomotion frees the forelimbs to manipulate objects and to carry them while walking. It also elevates the eyes, enabling the animal to see over tall vegetation to spot predators and prey. Bipedal locomotion is also energetically more economical than

(A) *Hylobates lar* (B) *Pongo pygmaeus* (C) *Gorilla gorilla*

(D) *Pan troglodytes*

33.33 Apes (A) The several genera of gibbons are all smaller in size than the other apes. Gibbons are found throughout southeastern Asia. (B) Orangutans are also native to Asia, living in the forests of Sumatra and Borneo. (C) Gorillas—the largest apes—are restricted to humid African forests. This male is a lowland gorilla. (D) Chimpanzees, our closest relatives, are found in forested regions of Africa.

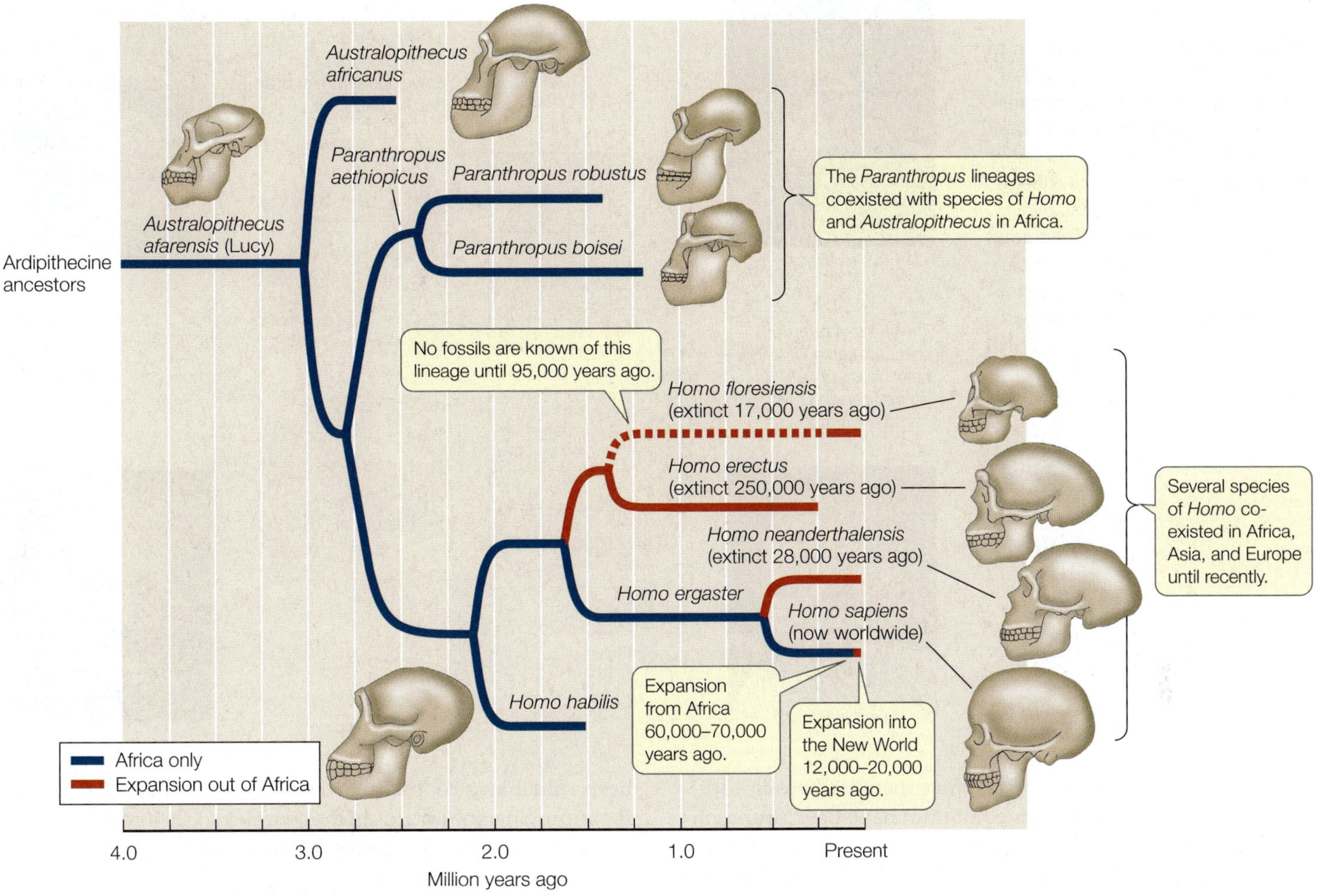

Australopithecus africanus
Paranthropus aethiopicus
Paranthropus robustus
Paranthropus boisei
Australopithecus afarensis (Lucy)
Ardipithecine ancestors
The Paranthropus lineages coexisted with species of Homo and Australopithecus in Africa.
No fossils are known of this lineage until 95,000 years ago.
Homo floresiensis (extinct 17,000 years ago)
Homo erectus (extinct 250,000 years ago)
Homo neanderthalensis (extinct 28,000 years ago)
Homo ergaster
Homo sapiens (now worldwide)
Homo habilis
Several species of Homo co-existed in Africa, Asia, and Europe until recently.
Expansion from Africa 60,000–70,000 years ago.
Expansion into the New World 12,000–20,000 years ago.
Africa only
Expansion out of Africa
4.0
3.0
2.0
1.0
Present
Million years ago

33.34 A Phylogenetic Tree of Hominins At times in the past, more than one hominin species lived on Earth at the same time. Originating in Africa, hominins spread to Europe and Asia multiple times. All but one of those species are now extinct, but that one species, modern *Homo sapiens*, has colonized nearly every corner of the planet.

quadrupedal locomotion (walking on four legs). All three advantages were probably important for the ardipithecines and their descendants, the australopithecines (**Figure 33.34**).

The first australopithecine skull was found in South Africa in 1924. Since then australopithecine fossils have been found at many sites in Africa. The most complete fossil skeleton yet found was discovered in Ethiopia in 1974. The skeleton, approximately 3.5 million years old, was that of a young female who has since become known to the world as "Lucy." Lucy was assigned to the species *Australopithecus afarensis*. Fossil remains of more than 100 *A. afarensis* individuals have since been discovered, and there have been recent discoveries of fossils of other australopithecine species that lived in Africa 4 to 5 million years ago.

Experts disagree over how many species are represented by australopithecine fossils, but it is clear that multiple species of hominins lived together over much of eastern Africa several million years ago. A lineage of larger species (weighing about 40 kilograms) is represented by *Paranthropus robustus* and *P. boisei*, both of which died out between 1 and 1.5 million years ago. A lineage of smaller australopithecines gave rise to the genus *Homo*.

Early members of the genus *Homo* lived contemporaneously with *Paranthropus* in Africa for about a million years. Some 2-million-year-old fossils of an extinct species called *H. habilis* were discovered in the Olduvai Gorge, Tanzania. Other fossils of *H. habilis* have been found in Kenya and Ethiopia. Associated with these fossils were tools that these early hominins used to obtain food.

Another extinct hominin species, *Homo erectus*, arose in Africa about 1.6 million years ago. Soon thereafter it had spread as far as eastern Asia, becoming the first hominin to leave Africa. Members of *H. erectus* were nearly as large as modern people, but their brains were smaller and they had comparatively thick skulls. The cranium, which had thick, bony walls, may have been an adaptation to protect the brain, ears, and eyes from impacts caused by a fall or a blow from a blunt object. What would have been the source of such blows? Fighting with other *H. erectus* individuals is a possible answer.

Homo erectus used fire for cooking and for hunting large animals, and made characteristic stone tools that have been found in many parts of Africa and Asia. Populations of *H. erectus* survived until at least 250,000 years ago, although more recent fossils may also be attributable to this species. In 2004 some 18,000-year-old fossil remains of a small *Homo* were found on

The chimp skull undergoes much more dramatic change in development than does the human skull.
Pan
The skulls of juvenile chimpanzees and humans are quite similar to each other.
Homo

33.35 Neoteny in the Evolution of Humans The skulls and heads of juvenile chimpanzees and humans start out relatively similar in shape. The grid over the skulls shows how the various bony elements change in their relative proportions over the course of maturation. The adult human skull retains a shape closer to its juvenile shape, resulting in a brain that is much larger relative to other parts of the skull (notably the jaw).

the island of Flores in Indonesia. Since then, numerous additional fossils of this diminutive hominin have been found on Flores, dating from 95,000 to 17,000 years ago. Many anthropologists think that this small species, named *H. floresiensis*, was most closely related to *H. erectus*.

Human brains became larger as jaws became smaller

In another hominin lineage that diverged from *H. erectus* and *H. floresiensis*, the brain increased rapidly in size, and the jaw muscles, which were large and powerful in earlier hominins, dramatically decreased in size. These two changes were simultaneous, which suggests that they might have been developmentally linked. These changes are another example of evolution by neoteny (which, you may recall from our discussion of amphibians, is the retention of juvenile traits through delayed somatic development). Human and chimpanzee skulls are similar in shape at birth, but chimpanzee skulls undergo a dramatic change in shape as the animals mature (**Figure 33.35**). In particular, the jaw grows considerably in relation to the brain case. As human skulls grow, relative proportions much closer to those of the juvenile skull are retained, which results in a large brain case and small jaw compared with those of chimpanzees. A mutation in a regulatory gene that is expressed only in the head may have removed a barrier that had previously prevented this remodeling of the human cranium.

The striking enlargement of the brain relative to body size in the hominin lineage was probably favored by an increasingly complex social life. Any features that allowed group members to communicate more effectively with one another would have been valuable in cooperative hunting and gathering as well as for improving one's status in the complex social interactions that must have characterized early human societies, just as they do ours today.

Several *Homo* species coexisted during the mid-Pleistocene, from about 1.5 million to about 250,000 years ago. All were skilled hunters of large mammals, but plants were important components of their diets as well. During this period another distinctly human trait emerged: rituals and a concept of life after death. Deceased individuals were buried with tools and clothing, supplies for their presumed existence in the next world.

One species, *Homo neanderthalensis*, was widespread in Europe and Asia between about 500,000 and 28,000 years ago. Neanderthals were short, stocky, and powerfully built. Their massive skull housed a brain somewhat larger than our own. They manufactured a variety of tools and hunted large mammals, which they probably ambushed and subdued in close combat. Early modern humans (*H. sapiens*) expanded out of Africa between 70,000 and 60,000 years ago. Then, about 35,000 years ago, *H. sapiens* moved into the range of *H. neanderthalensis* in Europe and western Asia, so the two species must have interacted with each other. Neanderthals abruptly disappeared about 28,000 years ago. Many anthropologists believe that Neanderthals were exterminated by those early modern humans. Scientists have been able to isolate large parts of the genome of *H. neanderthalensis* from recent fossils and compare it with our own. These studies suggest that there was some limited interbreeding between the two species while they occupied the same range. As a result, in humans with Eurasian ancestry, 1 to 4 percent of the genes in their genomes may be derived from Neanderthal ancestors.

Early modern humans made and used a variety of sophisticated tools. They created the remarkable paintings of large mammals, many of them showing scenes of hunting, found in European caves. The animals they depicted were characteristic of the cold steppes and grasslands that occupied much of Europe during periods of glacial expansion. Early modern humans also spread across Asia, reaching North America perhaps as early as 20,000 years ago, although the date of their arrival in the Americas is still uncertain. Within a few thousand years, they had spread southward through North America to the southern tip of South America.

Humans developed complex language and culture

As our ancestors evolved larger brains, their behavioral capabilities increased, especially the capacity for language. Most animal communication consists of a limited number of signals, which refer mostly to immediate circumstances and are associated with charged emotional states induced by those circumstances. Human language is far richer in its symbolic character than other animal vocalizations. Our words can refer to past and future times and to distant places. We are capable of learning thousands of words, many of them referring to abstract concepts. We can rearrange words to form sentences with complex meanings.

The expanded mental abilities of humans enabled the development of a complex culture, in which knowledge and traditions are passed along from one generation to the next by teaching and observation. Cultures can change rapidly because genetic changes are not necessary for a cultural trait to spread through a population. Cultural norms, however, are not transferred automatically and must be deliberately taught to each generation.

Go to Media Clip 33.8
Humans Develop Complex Social Behaviors
Life10e.com/mc33.8

Cultural transmission greatly facilitated the domestication of plants and animals and the resultant conversion of most human societies from ones in which food was obtained by hunting and gathering to ones in which pastoralism (herding large animals) and agriculture provided most of the food. The development of agriculture led to an increasingly sedentary life, the growth of cities, greatly expanded food supplies, rapid increases in the human population, and the appearance of occupational specializations, such as artisans, shamans, and teachers.

RECAP 33.5

Grasping limbs with opposable digits distinguish primates from other mammals. Bipedal locomotion and large brains evolved in the primate ancestors that led to humans.

- Describe the differences between Old World and New World monkeys. **See p. 701 and Figures 33.32 and 33.33**
- Explain how neoteny resulted in the development of humans with relatively large brains and small jaws. **See p. 704 and Figure 33.35**

?

How has the evolution of complex behaviors affected the diversification of some major groups of deuterostomes?

ANSWER

Speciation in many groups of vertebrates is closely associated with the diversification of behaviors that are used for mate recognition or mate selection. These behavioral displays can be visual, olfactory, or auditory. Many vertebrate species have elaborate and stereotypical courtship displays, usually performed by males. In some cases these displays involve male-to-male combat, and the winning male mates with any attending females. In other cases a female may watch the displays of several different males and select a mate based on her perception of the quality of the display.

Some vertebrates, including most mammals, rely on olfactory cues to determine mate receptivity and to select mates. Humans have greatly reduced olfactory senses compared with most other mammals, although research has shown that we use more olfactory cues than we realize in selecting potential mates. This is the basis for the $27.5-billion-per-year perfume industry, which attempts to create smells that other humans will find sexually attractive.

In other vertebrate groups, such as frogs, mating calls play a primary role in mate selection. The calls are species-specific, but there is often some within-species variation in the calls as well. In many frog species, females can use the pitch of the calls to select larger, older males as mates. Males that are older have demonstrated that they have successful combinations of genes that have allowed them to survive and thrive in the local environment.

Whether mate selection is based on auditory, visual, or olfactory cues, divergence in the signaling system can lead to rapid speciation. Closely related species of frogs may look very similar but have distinctly different calls. Since the females use the calls to select appropriate mates of their own species, evolution of call differences among different populations can lead to rapid reproductive isolation and speciation.

CHAPTER SUMMARY 33

33.1 What Is a Deuterostome?

- Deuterostomes vary greatly in adult form, but based on the distinctive patterns of early development they share and on phylogenetic analyses of their gene sequences, they are judged to be monophyletic.
- There are far fewer species of deuterostomes than of protostomes, but many deuterostomes are large and ecologically important.
- The deuterostomes comprise three major clades: the **echinoderms**, **hemichordates**, and **chordates**. **Review Figure 33.1, ACTIVITY 33.1, ANIMATED TUTORIAL 33.1**

33.2 What Features Distinguish the Echinoderms, Hemichordates, and Their Relatives?

- Echinoderms and hemichordates, together called **ambulacrarians**, have bilaterally symmetrical, ciliated larvae. Adult echinoderms have **pentaradial symmetry** and an **oral–aboral** body orientation. **Review Figure 33.3**
- The **xenoturbellids** and **acoels** are reduced, soft-bodied wormlike marine animals with few distinct organ systems. Their relationships are uncertain, but recent analyses suggest that they may be the sister group of the ambulacrarians.
- Echinoderms have an **internal skeleton** of calcified plates and a unique **water vascular system** connected to extensions called **tube feet**. **Review Figure 33.3**
- Hemichordates are bilaterally symmetrical and have a three-part body divided into a proboscis, collar, and trunk. **Review Figure 33.6**

33.3 What New Features Evolved in the Chordates?

- Chordates fall into three principal clades: **lancelets**, **tunicates**, and **vertebrates**.
- At some stage in their development, all chordates have a dorsal hollow nerve cord, a post-anal tail, and a **notochord**. Lancelets have all three key chordate features as adults. Tunicates have these features as larvae but lose them as adults. **Review Figure 33.7**
- The vertebrate body is characterized by an internal skeleton, which is supported by a **vertebral column** that replaces the notochord. It is also characterized by internal organs suspended in a coelom, a ventral heart, and an anterior skull enclosing a large brain. **Review Figure 33.9**
- From estuarine ancestors, vertebrates diversified into many lineages of marine and freshwater fishes. One of these lineages, the lobe-limbed vertebrates, later radiated into terrestrial environments. **Review Figure 33.10**
- In the **gnathostomes**, jaws evolved from gill arches. Jaws enabled these vertebrates to grasp large prey and, together with teeth, allowed them to cut food into small pieces. **Review Figure 33.12**
- **Chondrichthyans** have skeletons of cartilage; almost all species are marine. The skeletons of **ray-finned fishes** are made of bone; these fishes have colonized all aquatic environments.

33.4 How Did Vertebrates Colonize the Land?

- Lungs and jointed appendages enabled one lineage of **lobe-limbed vertebrates** to colonize the land. This lineage gave rise to the **tetrapods**. **Review Figure 33.16**
- The earliest split in the tetrapod tree is between the **amphibians** and the **amniotes** (**reptiles** and **mammals**).
- Most modern amphibians are confined to moist environments because their bodies and their eggs lose water rapidly. **Review Figure 33.17, ANIMATED TUTORIAL 33.2**
- An impermeable skin, efficient kidneys, and an **amniote egg** that could resist desiccation evolved in the amniotes. **Review Figure 33.19, ACTIVITY 33.2**
- The major living reptile groups are the **lepidosaurs** (**tuataras**, along with the **squamates**, which include lizards, snakes, and amphisbaenians), the **turtles**, and the **archosaurs** (**crocodilians** and **birds**). **Review Figure 33.20**
- Birds evolved from a group of active, predatory dinosaurs known as theropods. Feathers arose among the theropods, originally for insulation and to enhance coloration, but eventually developed into adaptations for flight in birds. **Review Figures 33.23, 33.24**
- Mammals are unique among animals in supplying their young with a nutritive fluid (milk) secreted by mammary glands. There are two primary mammalian clades: the **prototherians** (of which there are only five species) and the species-rich **therians**. The therian clade is further subdivided into the **marsupials** and the **eutherians**. **Review Table 33.1**
- Mammalian phylogeny is strongly associated with the breakup of the major continents during the Mesozoic. Major lineages of eutherians diversified in Laurasia, Africa, and South America. **Review Figure 33.28**

33.5 What Traits Characterize the Primates?

- Grasping limbs with opposable digits distinguish **primates** from other mammals. The **prosimian** clade includes the lemurs, lorises, and galagos; the **anthropoid** clade includes tarsiers, monkeys, and apes. **Review Figure 33.30**
- The ancestors of **hominins** were terrestrial apes that developed efficient **bipedal locomotion**. **Review Figure 33.34**
- In the lineage leading to *Homo*, brains became larger as jaws became smaller; the two events appear to be developmentally linked and are an example of evolution via **neoteny**. **Review Figure 33.35**

See ACTIVITY 33.3 for a concept review of this chapter.

Go to the Interactive Summary to review key figures, Animated Tutorials, and Activities Life10e.com/is33

CHAPTER**REVIEW**

REMEMBERING

1. Which of the following are *not* deuterostomes?
   a. Acorn worms
   b. Sea stars
   c. Tunicates
   d. Brachiopods
   e. Lancelets

2. The structure used by adult tunicates to capture food is a
   a. pharyngeal basket.
   b. proboscis.
   c. lophophore.
   d. mucus net.
   e. radula.

3. In most ray-finned fishes, lunglike sacs evolved into
   a. pharyngeal gill slits.
   b. true lungs.
   c. coelomic cavities.
   d. swim bladders.
   e. none of the above

4. The horny scales that cover the skin of reptiles prevent them from
   a. using their skin as an organ of gas exchange.
   b. sustaining high levels of metabolic activity.
   c. laying their eggs in water.
   d. flying.
   e. crawling into small spaces.

5. Which statement about bird feathers is *not* true?
   a. They are highly modified reptilian scales.
   b. They provide insulation for the body.
   c. They exist in two layers.
   d. They help birds fly.
   e. They are important sites of gas exchange.

6. The relatively large brain case and small jaw of humans relative to chimpanzees is an example of
   a. genetic drift.
   b. neoteny.
   c. concerted evolution.
   d. cultural transfer.
   e. none of the above

UNDERSTANDING & APPLYING

7. The body plan of most vertebrates is based on four appendages. What are the varied forms that these appendages take, and how are they used? In which lineages have two or more of these appendages been lost?

8. Amphibians have survived and prospered for millions of years, but today many species are disappearing, and populations of others are declining seriously. What features of their life histories might make amphibians especially vulnerable to the kinds of environmental changes now happening on Earth?

9. In the not-too-distant past, the idea that birds were reptiles met with skepticism. Explain how fossils, morphology, and molecular evidence now support the position of birds among the reptiles.

ANALYZING & EVALUATING

10. On the phylogenetic tree of amniotes (below), map the evolutionary origin of the following traits: endothermy, hair, and feathers. Which of these traits appears to have evolved convergently in more than one lineage? What is a likely functional relationship among these traits? What are some reasons why many paleontologists think many extinct theropods were endothermic?

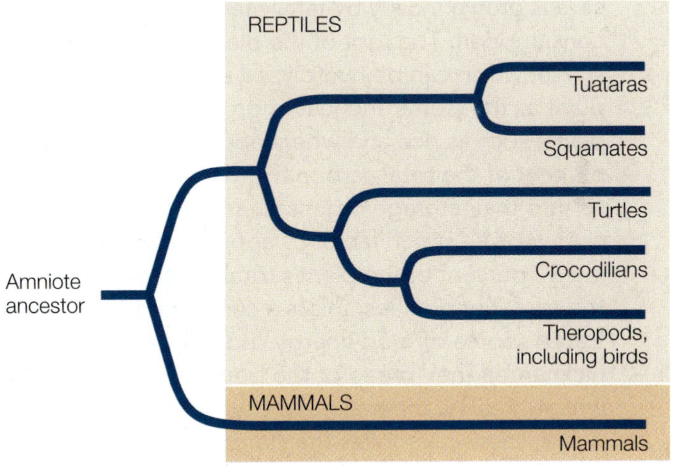
REPTILES
Tuataras
Squamates
Turtles
Crocodilians
Theropods, including birds
Amniote ancestor
MAMMALS
Mammals

PART EIGHT
Flowering Plants: Form and Function

34 The Plant Body

CHAPTER OUTLINE

Cassava The roots of this plant supply food energy to hundreds of millions of people.

TO NORTH AMERICANS AND EUROPEANS, it is tapioca. To Central and South Americans, it is yuca. To Africans, it is manioc or bananku. The roots of the cassava plant, *Manihot esculenta*, are important in the diets of more than 800 million people. Cassava is grown mostly by family farmers for their own consumption. The root of the plant is a store of starch, which can be hydrolyzed and used by the plant as the stems and leaves grow. Other crop plants such as rice and wheat apportion about 35 percent of the total carbon fixed in photosynthesis into their storage organs (the grains), whereas cassava is a "starch factory," apportioning an astounding 80 percent of the plant's total photosynthate into the root. For humans, this is a convenient and concentrated source of food energy. Indeed, cassava has been nicknamed the "bread of the tropics" because, just as wheat bread is the major starchy food in the Western world, cassava supplies the starch in tropical areas.

Most cassava plants are clones. The plants have wide adaptability and grow well in dry soils, in both hot and cool climates. Typically, the farmer breaks off some pieces of stem and plants them. Some of the stem cells dedifferentiate and form roots, while others become growing shoots. A whole new plant develops, and the roots are ready to eat 6 months to 2 years later. It's a fairly easy process, much simpler than the work needed to grow rice, for example. As we pointed out in Chapter 19, totipotency is a remarkable property of plant cells, one that distinguishes them from animal cells.

There are countless recipes for preparing cassava as food. Cassava does not, however, offer complete nutrition. Although rich in carbohydrates, cassava is a relatively poor source of protein, a requirement for the human diet. A diet based on cassava presents other difficulties as well. Cassava roots and leaves contain cyanogenic compounds, which are converted to cyanide by digestive enzymes and gut flora. Cyanide is highly toxic and potentially lethal because it blocks electron transport in the mitochondria. Therefore it is essential to soak, cook, or ferment cassava, to break down the cyanogenic compounds and eliminate the cyanide, before it can be eaten safely. In all probability, the plant uses cyanide production as a protection against predators that eat it.

We open with the topic of cassava because it offers a preview of a wide range of studies embraced by the discipline of plant physiology. Plant physiology is a broad subject, covering photosynthesis, transport, plant nutrition, regulation of growth and development, reproduction, and the interactions between plants and their environments.

> ? How might plant physiologists improve the cassava plant for human use?
>
> See answer on p. 724.

34.1 What Is the Basic Body Plan of Plants?

Plants live by harvesting energy from sunlight and by collecting water and mineral nutrients from the atmosphere and the soil. Because these resources are sometimes limited, plants must collect them from large areas, both above and below ground. The plant is further challenged by its inability to move; a plant cannot, for example, relocate from a dry, shady location to one that is wet and sunny.

The plant body plan allows plants to respond to these challenges:

- Stems, leaves, and roots enable a plant anchored to one spot to capture scarce resources effectively, both above and below the ground.
- Plants can grow throughout their lifetimes, enabling them to respond to environmental cues. A plant can redirect its growth to exploit opportunities in its immediate environment; for example, it can extend its roots toward a water supply.

In Chapters 28 and 29 we saw how modern plants arose from aquatic ancestors, giving rise to simple land plants and then to vascular plants. Despite their obvious differences in size and form, all vascular plants have essentially the same simple structural organization. This chapter describes the basic architecture of the largest group of vascular plants, the angiosperms (flowering plants), and shows how so much diversity can literally grow out of such a simple, basic form.

As we saw in Figure 29.1, angiosperms first appeared about 145 million years ago. They radiated explosively over a period of about 60 million years and eventually became the dominant form of plant life on Earth. Today there are more than 250,000 angiosperm species. Flowers are the main distinguishing feature of angiosperms; they consist of modified leaves and stems and carry the organs for sexual reproduction. We will examine the structures and functions of flowers in detail in Section 38.1. In this chapter we'll focus on the three kinds of vegetative (nonsexual) organs that angiosperms possess: roots, stems, and leaves. Each of these vegetative organs can be understood in terms of its structure. By structure we mean both the overall form of the organ (its morphology) and the arrangement of its cells and tissues (its anatomy).

Plant organs are organized into two systems (**Figure 34.1**):

- The **root system** anchors the plant in place, absorbs water and dissolved minerals, and stores the products of photosynthesis from the shoot system. The extreme branching of plant roots and their high surface area-to-volume ratios allow them to absorb water and mineral nutrients from the soil efficiently.
- The **shoot system** of a plant consists of the stems, leaves, and flowers. Broadly speaking, the **leaves** are the chief organs of photosynthesis. The **stems** hold and display the leaves to the sun and provide connections for the transport of materials between roots and leaves.

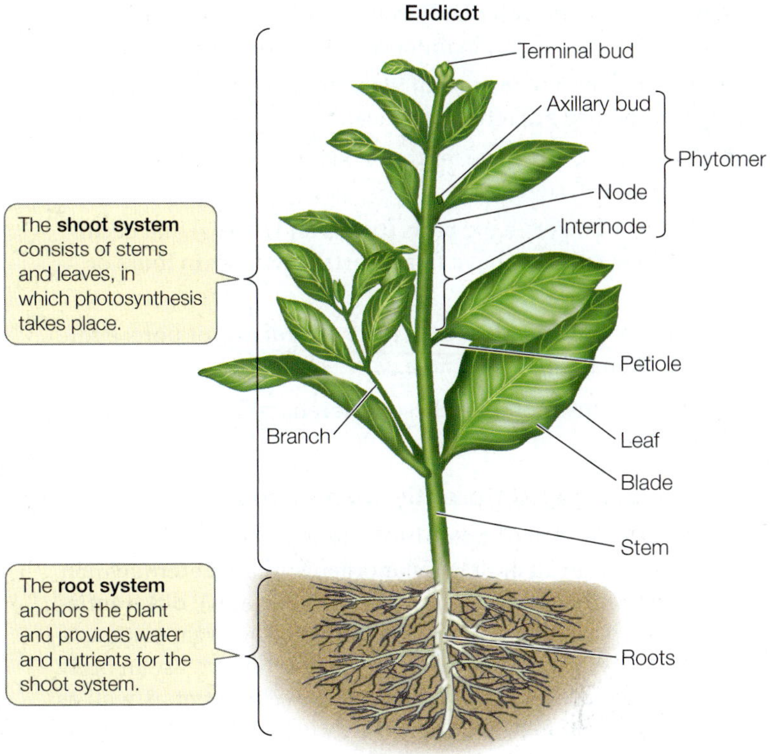
Eudicot
Terminal bud
Axillary bud
Phytomer
Node
Internode
The **shoot system** consists of stems and leaves, in which photosynthesis takes place.
Petiole
Branch
Leaf
Blade
Stem
The **root system** anchors the plant and provides water and nutrients for the shoot system.
Roots

34.1 Vegetative Organs and Systems The basic plant body plan, with root and shoot systems, and the principal vegetative organs.

Shoots and roots are composed of repeating modules called **phytomers**. Each phytomer in the shoot consists of a **node** carrying one or more leaves; an **internode**, which is the interval of stem between two nodes; and one or more **axillary buds**, each of which forms in the angle (axil) where a leaf meets the stem. A **bud** is an undeveloped shoot that can develop further to produce another leaf, a phytomer, a flower, or a flowering stem. The axillary buds (also called lateral buds) are distinguished from the bud at the end of a stem or branch, which is called a **terminal bud**. If it becomes active, an axillary bud can develop into a new branch, or an extension of the shoot system. The arrangement of leaves along the stem (called the phyllotaxy) is characteristic of the plant species.

Plant roots also have a modular construction. In the roots, each phytomer consists of a root segment between two branches.

Most angiosperms are either monocots or eudicots

As we saw in Section 29.3, most angiosperms belong to one of two major clades. *Monocots* are generally narrow-leaved flowering plants such as grasses, lilies, orchids, and palms. *Eudicots* are broad-leaved flowering plants such as soybeans, roses, sunflowers, and maples. These two clades, which account for 97 percent of flowering plant species, differ in several basic characteristics:

- Monocots have one **cotyledon** (leaf in the embryo), whereas eudicots have two.

- In eudicots, the vascular bundles in the stem are arranged in concentric circles; in monocots they are scattered.
- In monocots, the major leaf veins are usually parallel; in eudicots they are reticulate, meaning they form a network.
- Eudicots usually have taproot systems; monocots have fibrous root systems.
- Monocot flowers have parts (petals and sepals) that occur in threes; eudicots have floral parts that occur in fours or fives.
- Monocot pollen grains each have one furrow or pore; eudicot pollen grains have three.

We will discuss some of these differences in more detail in Section 34.3.

Plants develop differently than animals

As we discussed in Chapter 19, the four processes that govern development in all multicellular organisms are **determination** (the commitment of cells to their ultimate fates); **differentiation** (cell specialization); **morphogenesis** (the organization of cells into tissues and organs); and **growth** (increase in body size). These processes govern plant development as well as animal development, but they are influenced by four unique properties of plants: apical meristems, totipotency, vacuoles, and cell walls.

APICAL MERISTEMS Animals have stem cells to replace tissues lost through damage or apoptosis. Plants have **meristems**: regions of undifferentiated cells where cell division occurs. **Apical meristems** are found at the tips of shoots and roots and allow plants to continue growing throughout their lives. We will discuss apical meristems in more detail below.

TOTIPOTENCY During normal animal development, only the early embryonic cells are totipotent: they can differentiate into any type of cell in the body (see Section 19.1). In contrast, some differentiated plant cells can dedifferentiate and become totipotent (see Figure 19.16). This means that a plant can readily repair damage wrought by the environment or herbivores.

VACUOLES Mature plant cells usually contain a single **central vacuole**, which may account for up to 90 percent of a cell's volume (see Figure 5.13). The vacuole is a watery sac containing a high concentration of solutes, including enzymes, amino acids, and sugars produced by photosynthesis. Many of these solutes are pumped into the vacuole by transporter proteins located in the **tonoplast**, the vacuolar membrane. This active accumulation of solutes provides the osmotic force for water uptake into the vacuole, as we will see in Section 35.1. As the vacuole expands, it exerts turgor pressure on the cell wall (see Chapter 6). Turgor pressure keeps plants upright and is essential for plant growth.

CELL WALLS Each plant cell is surrounded by a cell wall, which is interrupted by membrane-lined cytoplasmic channels called plasmodesmata (see Section 5.4). This rigid extracellular matrix makes it impossible for cells to move from place to place, as they do in animal development. Instead, plant morphogenesis is controlled by the planes of cell division, which determine the direction in which a piece of tissue will grow (**Figure 34.2**). In addition, unequal cell division can occur when the cytoplasm contains differentiation signals that are localized in one part of a cell (see Section 19.2). Plant cytokinesis (the division of the cytoplasm) occurs along a cell plate laid down by membranous vesicles produced by the Golgi apparatus (see Figure 11.13B). Unlike animal cells, in which the location of cytokinesis depends on the location of the middle of the mitotic spindle, the location of the plant cell plate is determined earlier—as early as mitotic prophase.

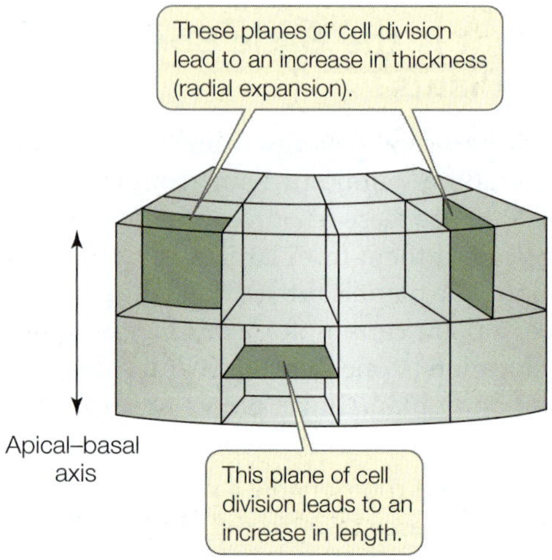
These planes of cell division lead to an increase in thickness (radial expansion).
Apical–basal axis
This plane of cell division leads to an increase in length.

34.2 Cytokinesis and morphogenesis The plane of cell division can determine the growth pattern of a plant's organs, as in this section of a shoot.

One of the major ways that plants grow is by cell expansion. Some cells can increase in volume by 100,000 to 1,000,000 times! As a growing plant cell takes up water, it exerts turgor pressure on the cell wall, which resists cell expansion. For the cell to expand, the wall must expand too. Proteins called expansins reside in the cell wall and help loosen it by disrupting the noncovalent bonds between cellulose microfibrils and other polysaccharides in the cell wall. This is followed by the assembly of new polysaccharides and microfibrils, allowing the cell wall to grow:

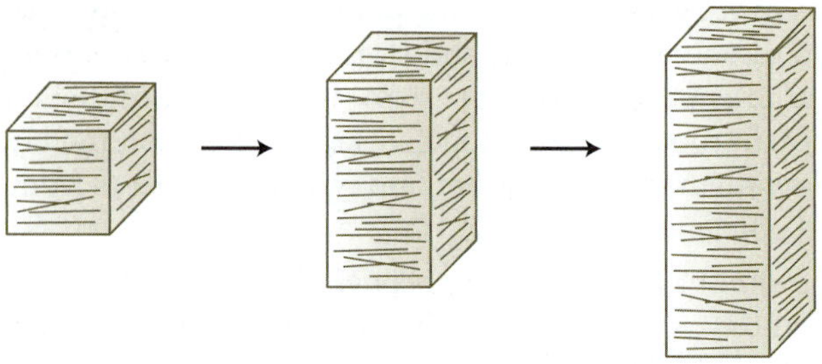

The wall of a growing plant cell is called the **primary cell wall**. When cell expansion stops, some types of plant cells deposit one or more additional cellulosic layers to form a thick, rigid **secondary cell wall** that is internal to the primary cell wall:

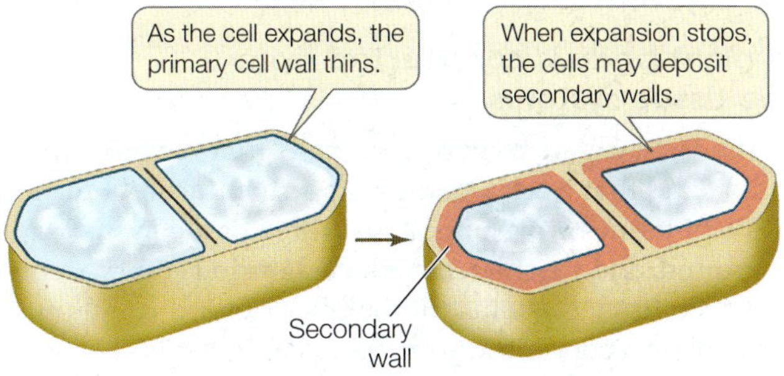
As the cell expands, the primary cell wall thins.
When expansion stops, the cells may deposit secondary walls.
Secondary wall

Secondary cell walls cannot expand. Instead they provide the mechanical support that allows some plants to produce large stems. The secondary wall contains layers of ordered cellulose microfibrils embedded in a remarkable substance called **lignin**. Lignin is a major component of wood; it is a complex, carbon-containing polymer that forms a hydrophobic matrix. This matrix is strong, waterproof, and resistant to digestion by animals. After cellulose, lignin is the most abundant biological polymer on Earth, accounting for 20 to 35 percent of the dry weight of wood.

Apical–basal polarity and radial symmetry are characteristics of the plant body

Two basic patterns are established early in plant embryogenesis (embryo formation) (**Figure 34.3**):

- The *apical–basal axis*: the arrangement of cells and tissues along the main axis from root to shoot
- The *radial axis*: the concentric arrangement of the tissue systems. (which we will describe in Section 34.2)

Both axes are best understood in developmental terms. We will focus here on embryogenesis in *Arabidopsis thaliana*. As we have seen in previous chapters, *Arabidopsis* is a model eudicot

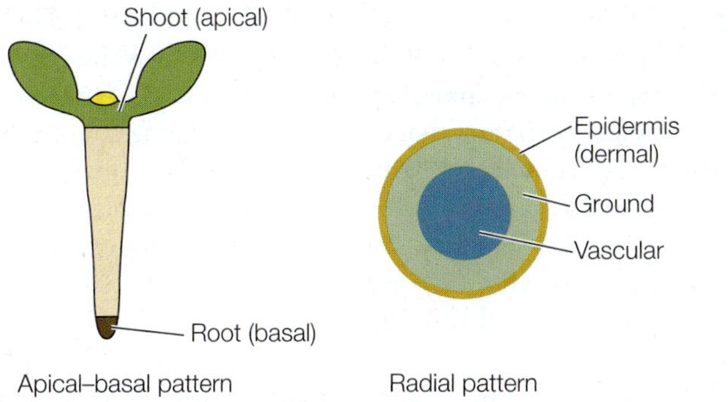
Shoot (apical)
Root (basal)
Apical–basal pattern
Epidermis (dermal)
Ground
Vascular
Radial pattern

34.3 Two Patterns for Plant Morphogenesis (A) The apical–basal pattern is the arrangement of tissues along a main axis from root to shoot. (B) The radial pattern determines the concentric arrangement of tissues as organs grow in thickness.

that has been studied extensively by plant physiologists and geneticists.

The first step in the formation of an *Arabidopsis* embryo is a mitotic division of the zygote that gives rise to two daughter cells (**Figure 34.4, Step 1**). An asymmetrical plane of cell division results in an uneven distribution of cytoplasm between these two cells, and this determines their different fates. Signals in the smaller, upper daughter cell cause it to produce the embryo proper, and the other, larger daughter cell produces a supporting structure called the **suspensor** (**Figure 34.4, Step 2**). This division not only establishes the apical–basal axis of the new plant but also determines its polarity (which end is the tip, or apex, and which is the base). A long, thin suspensor and a more spherical or globular embryo are distinguishable after just four mitotic divisions (see Figure 19.1). The suspensor soon ceases to elongate.

In eudicots such as *Arabidopsis*, the initially globular embryo develops into the characteristic heart stage as the cotyledons start to grow (**Figure 34.4, Step 3**). Further elongation of the cotyledons and of the main axis of the embryo gives rise to the

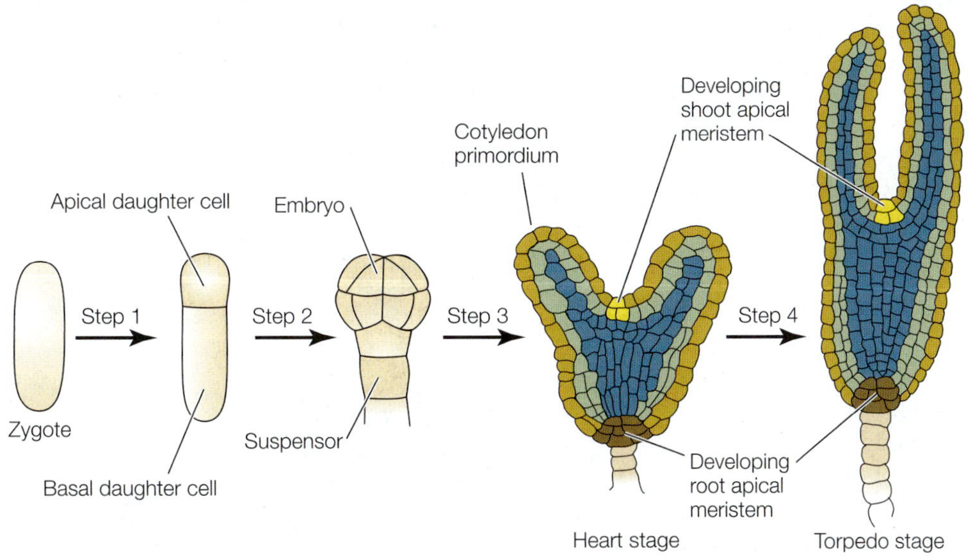
Apical daughter cell
Zygote
Step 1
Basal daughter cell
Embryo
Step 2
Suspensor
Cotyledon primordium
Step 3
Developing shoot apical meristem
Step 4
Developing root apical meristem
Heart stage
Torpedo stage

34.4 Plant Embryogenesis The basic body plant of the model eudicot *Arabidopsis thaliana* is established in several steps. By the heart stage, the three tissue systems are established: the dermal (gold), ground (light green), and vascular (blue) tissue systems.

torpedo stage, during which some of the internal tissues begin to differentiate (**Figure 34.4, Step 4**). Between the cotyledons is the **shoot apical meristem**; at the other end of the axis is the **root apical meristem**. Each of these meristems contains undifferentiated cells that will continue to divide to give rise to the organs that will develop during the life of the plant.

As shown in Step 2 of Figure 34.4, the plant embryo is first a sphere and later a cylinder. The root and stem retain this cylindrical shape throughout the plant's life. You can see this most easily in the trunk (mature stem) of a tree. By the end of embryogenesis, the radial symmetry of the plant has been established (see Figure 34.3). The embryonic plant contains three tissue systems, arranged concentrically, which will give rise to the tissues of the adult plant body. We will discuss these tissue systems in the next section.

RECAP 34.1

The vegetative plant body consists of a root system and a shoot system. The plant body is modular, made up of repeated units called phytomers. Plant cells are characterized by apical meristems, totipotency, vacuoles, and cell walls. Plants have apical–basal and radial patterns. Most angiosperms are either monocots or eudicots, which differ in several basic ways.

- How do plants explore their environment for resources even though they cannot move? **See p. 709**
- How does plant development differ from animal development? **See pp. 710–711**
- How do apical–basal and radial patterns develop? **See pp. 711–712 and Figures 34.3 and 34.4**

The dramatic differences between plant and animal body plans are not surprising given that the multicellular forms of plants and animals evolved independently from entirely distinct protist ancestors (see Figure 1.7). In the next two sections we will look more closely at the unique characteristics of the plant body, by following its development from a zygote into an adult.

34.2 What Are the Major Tissues of Plants?

By the end of embryogenesis, the radial axis of the plant has been established. Unlike complex animals that can have dozens of different tissues (for example, in humans there are three kinds of muscle tissue alone), plants have just three major tissue systems, each of which has specialized cells. These three tissue systems are arranged concentrically in the embryo and give rise to the tissues of the adult plant body. In turn, these tissues form the major vegetative plant organs: roots, stems, and leaves.

The plant body is constructed from three tissue systems

A tissue is an organized group of cells that have features in common and that work together as a structural and functional unit. Plant tissues are grouped into three **tissue systems**: dermal, ground, and vascular. Established during embryogenesis, these three tissue systems ultimately extend throughout the plant body in a concentric arrangement (**Figure 34.5**). Each tissue system has distinct functions and is composed of different mixtures of cell types.

DERMAL TISSUE SYSTEM The **dermal tissue system** forms the **epidermis** (outer covering) of a plant and usually consists of a single cell layer. The stems and roots of woody plants develop a dermal tissue called **periderm**.

During plant development, the epidermis grows to cover the expanding plant body. The cells of the epidermis are initially small and round and usually have a small central vacuole or none at all. Once cell division ceases in the epidermis of an organ, the epidermal cells expand. Some epidermal cells differentiate to form one of three specialized structures:

- Stomatal guard cells, which form stomata (pores) for gas exchange in leaves

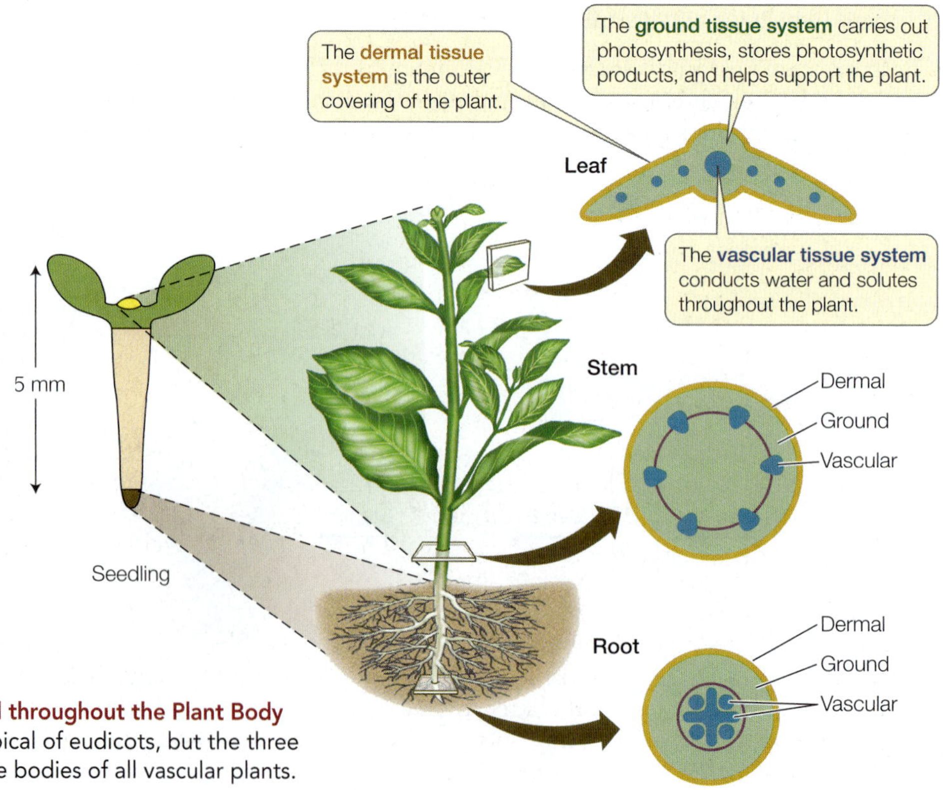
The dermal tissue system is the outer covering of the plant.
The ground tissue system carries out photosynthesis, stores photosynthetic products, and helps support the plant.
Leaf
The vascular tissue system conducts water and solutes throughout the plant.
5 mm
Seedling
Stem
Dermal
Ground
Vascular
Root
Dermal
Ground
Vascular

34.5 Three Tissue Systems Extend throughout the Plant Body The arrangement shown here is typical of eudicots, but the three tissue systems are continuous in the bodies of all vascular plants.

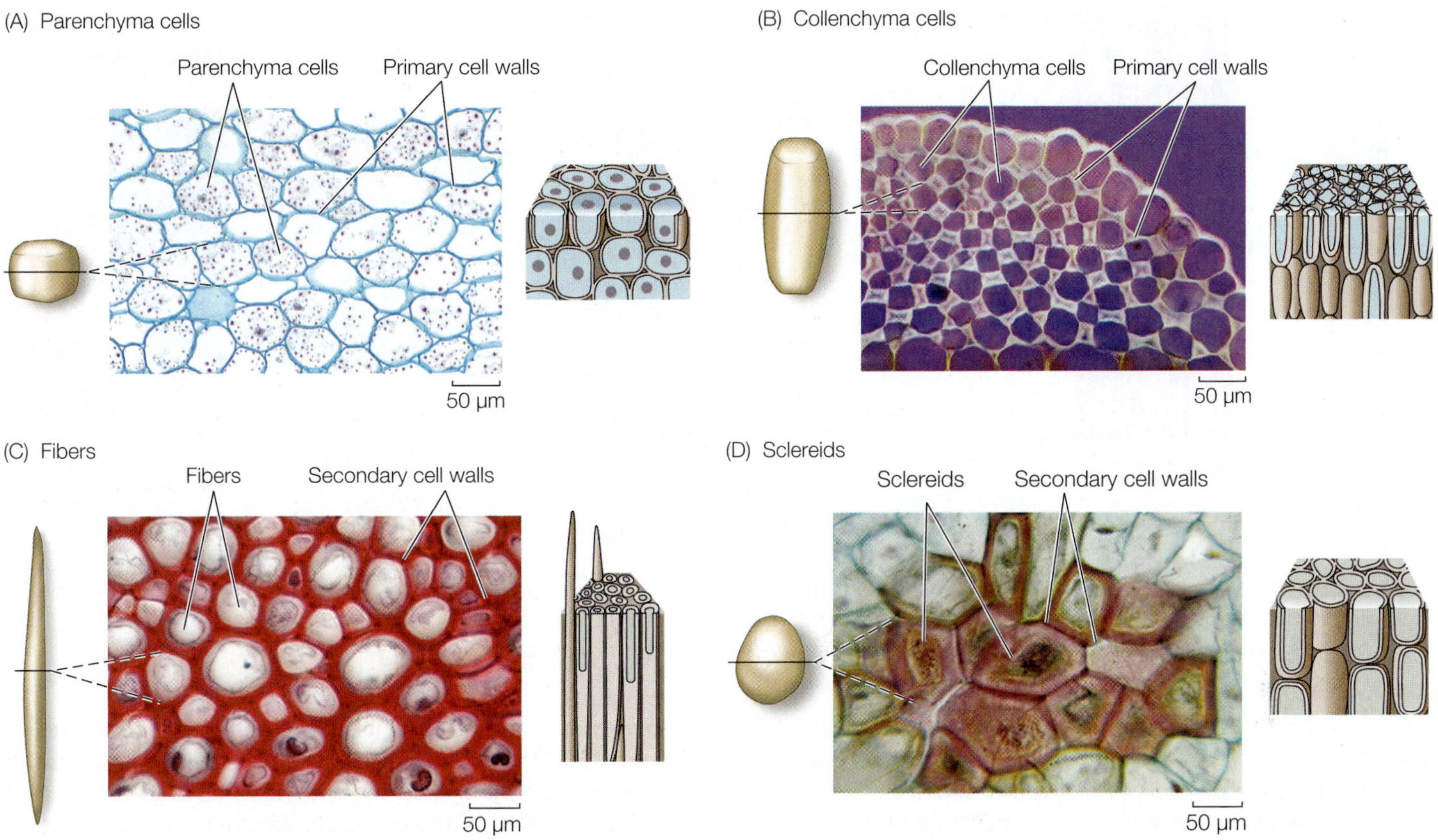
(A) Parenchyma cells
Parenchyma cells
Primary cell walls
50 µm
(B) Collenchyma cells
Collenchyma cells
Primary cell walls
50 µm
(C) Fibers
Fibers
Secondary cell walls
50 µm
(D) Sclereids
Sclereids
Secondary cell walls
50 µm

34.6 Ground Tissue Cell Types (A) Parenchyma cells in the petiole of *Coleus*. Note the thin, uniform cell walls. (B) Collenchyma cells make up the outer cell layers of this spinach leaf vein. Their walls are thick at the corners of the cells and thin elsewhere. (C) Sclerenchyma: fibers in a sunflower stem (*Helianthus*). The thick secondary walls are stained red. (D) Sclerenchyma: sclereids. The extremely thick secondary walls of sclereids are laid down in layers. They provide support and a hard texture to structures such as nuts and seeds.

- Trichomes, or leaf hairs, which provide protection against insects and damaging solar radiation
- Root hairs, which greatly increase root surface area, thus providing more surface for the uptake of water and mineral nutrients

The aboveground epidermal cells secrete a protective extracellular layer called a **cuticle**. The cuticle is made up of cutin (a polymer composed of long chains of fatty acids), a complex mixture of waxes, and cell wall polysaccharides. The cuticle limits water loss, reflects potentially damaging solar radiation, and serves as a barrier against pathogens.

GROUND TISSUE SYSTEM Virtually all the tissue lying between the dermal tissue and the vascular tissue in both shoots and roots is part of the **ground tissue system**. Therefore the ground tissues make up most of the plant body. The ground tissues function primarily in storage, support, and photosynthesis. To fulfill these diverse functions, ground tissues contain three cell types that are classified according to their cell wall structure: parenchyma, collenchyma, and sclerenchyma.

The most common cell type in plants is the **parenchyma** cell (**Figure 34.6A**). Parenchyma cells have large vacuoles and thin walls consisting only of a primary wall and the shared middle lamella. The **middle lamella** is a layer of pectin that cements adjacent plant cells together (see Figure 5.21). Parenchyma cells play important roles in photosynthesis (mainly in the leaves); proteins, starch, fats, or oils may be stored in parenchyma cells of the seeds and/or roots. Many retain the capacity to divide and give rise to new cells, as when a wound results in cell proliferation.

Collenchyma cells resemble parenchyma cells that have been modified to provide flexible support. They are generally elongated, and their primary walls are characteristically thick at the corners of the cells (**Figure 34.6B**). In these cells the primary wall thickens in part because of the deposition of pectins, but no secondary wall forms. Collenchyma cells provide support to leaf petioles, nonwoody stems, and growing organs. Tissue made of collenchyma cells is flexible, permitting stems and petioles to sway in the wind without snapping. The familiar "strings" in celery consist primarily of collenchyma cells.

Sclerenchyma cells have thickened secondary walls that enable the cells to perform their major function: support. Many sclerenchyma cells undergo programmed cell death (apoptosis; see

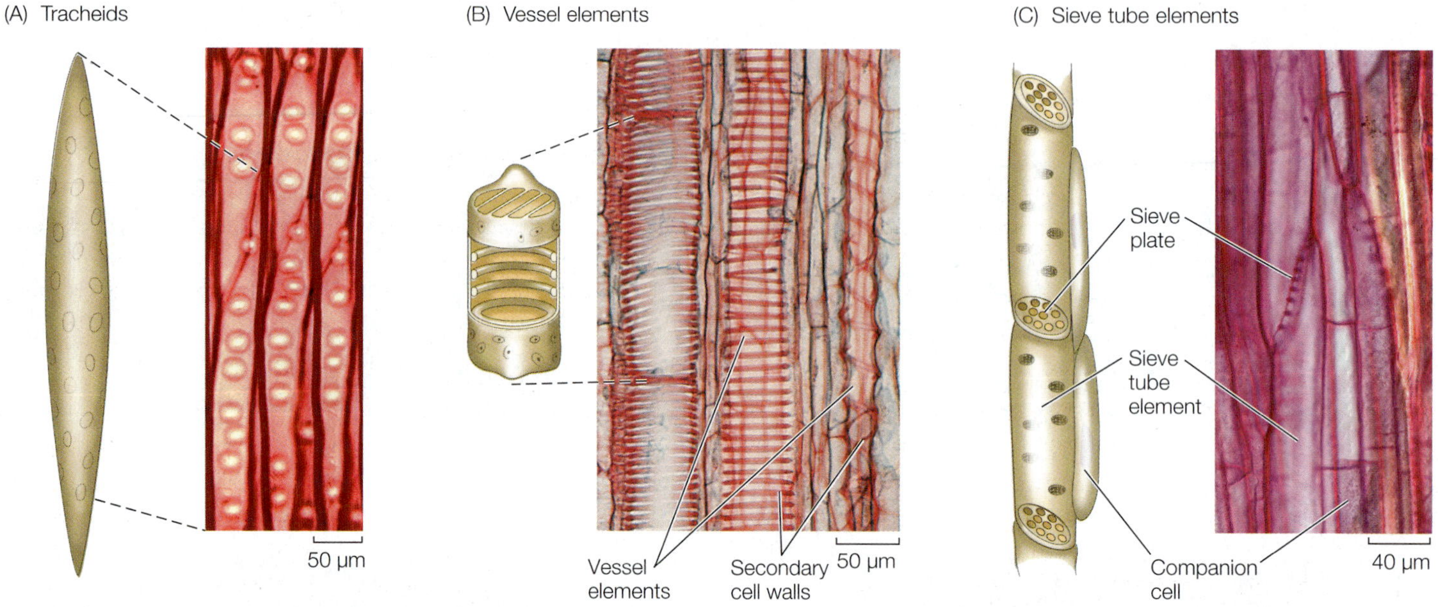
(A) Tracheids
50 µm
(B) Vessel elements
Vessel elements
Secondary cell walls
50 µm
(C) Sieve tube elements
Sieve plate
Sieve tube element
Companion cell
40 µm

34.7 Vascular Tissue Cell Types (A, B) Tracheary elements: (A) Tracheids in pinewood. The thick secondary walls are stained dark red. (B) Vessel elements in the stem of a squash. The secondary walls are stained red; note the different patterns of thickening, including rings and spirals. (C) Sieve tube elements and companion cells in the stem of a cucumber.

Section 11.6) after developing the lignified secondary walls, and thus perform their supporting function when dead. There are two types of sclerenchyma cells: elongated **fibers** and variously shaped **sclereids**. Fibers are often organized into bundles and provide relatively rigid support to wood, bark, and other parts of the plant (Figure 34.6C). Sclereids may pack together densely, as in a nut's shell or in some seed coats (Figure 34.6D). Isolated clumps of sclereids, called stone cells, occur in pears and some other fruits and give them their characteristic gritty texture.

VASCULAR TISSUE SYSTEM The **vascular tissue system** is the plant's plumbing, or transport system—the distinguishing feature of vascular plants. Its two constituent tissues, the xylem and phloem, distribute materials throughout the plant. The **xylem** distributes water and mineral ions taken up by the roots to all the cells of the stems and leaves. **Phloem** performs a variety of functions, including transport, support, and storage. All the living cells of the plant body require a source of energy and chemical building blocks. The phloem meets these needs by transporting carbohydrates away from the sites of production, which are called **sources** (primarily leaves). The carbohydrates are transported to sites of utilization or storage, called **sinks**. Sinks include growing tissues, storage organs, and developing flowers.

Let's take a closer look at the structures of the diverse cell types that make up these vascular tissues. In Chapter 35 we will see how they transport water and materials throughout the plant body.

Cells of the xylem transport water and dissolved minerals

Xylem tissue contains conducting cells called **tracheary elements** that have secondary cell walls and undergo apoptosis before assuming their function of transporting water and dissolved minerals. There are two types of tracheary elements: tracheids and vessel elements. The spindle-shaped **tracheids** (Figure 34.7A) are evolutionarily more ancient than vessel elements and are the major cell type in the wood of gymnosperms (see Section 29.2). When tracheids die, their internal components disintegrate and pits remain between the cells. Pits are cavities in the secondary cell walls spanned by porous structures that allow water and minerals to move between tracheids and thus through the xylem tissue.

Flowering plants have evolved a water-conducting system made up of vessels, which are formed from individual cells called **vessel elements**. These cells are laid down end-to-end. Like tracheids, vessel elements have pits in their cell walls, but their pits are generally larger in diameter than those of tracheids. Before they undergo apoptosis, the end walls of vessel elements partially break down, forming a continuous hollow tube that functions as an open pipeline for water conduction (Figure 34.7B). In the course of angiosperm evolution, vessel elements have become shorter and wider, and their end walls have become less oblique and less obstructed. These adaptations have presumably increased the efficiency of water transport through the vessels. The xylem of many angiosperms includes both tracheids and vessels.

Cells of the phloem transport the products of photosynthesis

The transport cells of the phloem, unlike those of the mature xylem, are living cells. In flowering plants the characteristic cells of the phloem are called **sieve tube elements** (Figure 34.7C). Like vessel elements, these cells meet end-to-end. They form long sieve tubes, which transport carbohydrates and many other materials from their sources (usually leaves) to tissues that consume or store them (for example, roots).

Unlike in vessel elements, which break down their end walls, the end walls of sieve tube elements contain plasmodesmata, which enlarge to form pores. The end walls look and function like sieves and are thus called sieve plates. Although the sieve tube elements remain alive, some of their components, including the nucleus, ribosomes, and vacuole, break down during development. The sieve tube elements are, however, closely connected via plasmodesmata to **companion cells**—specialized parenchyma cells that retain all their organelles and function as "life support systems" for the sieve tube elements.

RECAP 34.2

The three concentric tissue systems of the plant embryo—dermal, ground, and vascular—give rise to the tissues and organs of the adult plant. These tissue systems have unique combinations of specialized cells that carry out the various functions necessary for plant life.

- What distinguishes the three tissue systems in terms of their location and functions? See pp. 712–714 and Figure 34.5
- What structural differences make tissues consisting of collenchyma cells more flexible than those consisting primarily of sclerenchyma cells? See pp. 713–714
- Outline the differences between tracheids and vessel elements. See p. 714 and Figures 34.6 and 34.7

After the plant embryo has formed, it is encased in a seed coat and is ready to germinate. We will discuss aspects of seed germination in the chapters that follow. For now, let's consider the processes by which the embryo grows into a mature plant.

34.3 How Do Meristems Build a Continuously Growing Plant?

As noted at the beginning of this chapter, plants and animals develop and function differently. While animals use their mobility to forage for food, plants are sessile (rooted in one place) and grow toward scarce resources, both above and below the ground. Therefore plants grow in two directions: the shoots grow toward sunlight, and the roots grow toward water and dissolved minerals in the soil.

In most animals, growth is **determinate**: it ceases when the adult state is reached. Determinate growth is also characteristic of some plant organs, such as leaves, flowers, and fruits. The growth of shoots and roots, however, is a lifelong process. Such open-ended growth is **indeterminate**.

Plants increase in size through primary and secondary growth

Plant growth can occur in either of two ways:

- **Primary growth** is characterized by cell division followed by cell enlargement. It results in the proliferation and lengthening of shoots and roots. All seed plants have a primary plant body, which consists of all the nonwoody parts of the plant. Many herbaceous plants consist entirely of a primary plant body.
- **Secondary growth** increases plant thickness. Woody plants, such as trees and shrubs, have a secondary plant body consisting of wood and bark. As the tissues of the secondary plant body are laid down, the stems and roots thicken.

A hierarchy of meristems generates the plant body

As we have already mentioned, meristems are localized regions of undifferentiated cells that are the sources of all new growth in the adult plant. Even before seed germination, the plant embryo has two meristems: a shoot apical meristem near the end of the embryonic shoot, and a root apical meristem at the end of the embryonic root (see Figure 34.4).

Meristematic cells are small and closely packed, with very small vacuoles and thin primary cell walls. They are undifferentiated and retain the ability to produce new cells indefinitely. The cells that perpetuate the meristems, called **initials**, are comparable to animal stem cells (discussed in Section 19.1). When the initials divide, some of the daughter cells develop into new initials, and some differentiate into more specialized cells.

Several types of meristem contribute to the growth and development of the adult plant:

- Apical meristems in the root and shoot (**Figure 34.8**) orchestrate primary growth, ultimately giving rise to every cell in the primary plant body.
- When the initials of apical meristems divide, some of their daughter cells differentiate and become the **primary meristems**. Three kinds of primary meristem (see below) give rise to the three major tissue systems (dermal, ground, and vascular) described in Section 34.2.
- **Lateral meristems** (also called secondary meristems) orchestrate secondary growth (see Figure 34.8). Two lateral meristems, vascular cambium and cork cambium, contribute to the secondary plant body.

Indeterminate primary growth originates in apical meristems

Because apical meristems can perpetuate themselves indefinitely, a shoot or root can continue to lengthen and grow indefinitely; in other words, growth of the shoot or root is indeterminate. All plant organs arise ultimately from cell divisions in apical meristems, followed by cell expansion and differentiation. Several types of apical meristems play roles in organ formation:

- Shoot apical meristems supply the cells for new leaves and stems. In addition to the main stem of the plant, each branch has its own shoot apical meristem. Shoot apical meristems are also called **vegetative meristems**, because they give rise to vegetative tissues (leaves, stems, and roots).
- When the plant is ready to flower, one or more of its shoot apical meristems are transformed into **inflorescence meristems**, and these in turn develop **floral meristems**. See Section 38.2 for more on floral development.
- Root apical meristems supply the cells that extend roots, enabling the plant to penetrate and explore the soil for water and minerals. Each type of root (i.e., the taproot, a lateral root, or an adventitious root; see below) has its own root apical meristem.

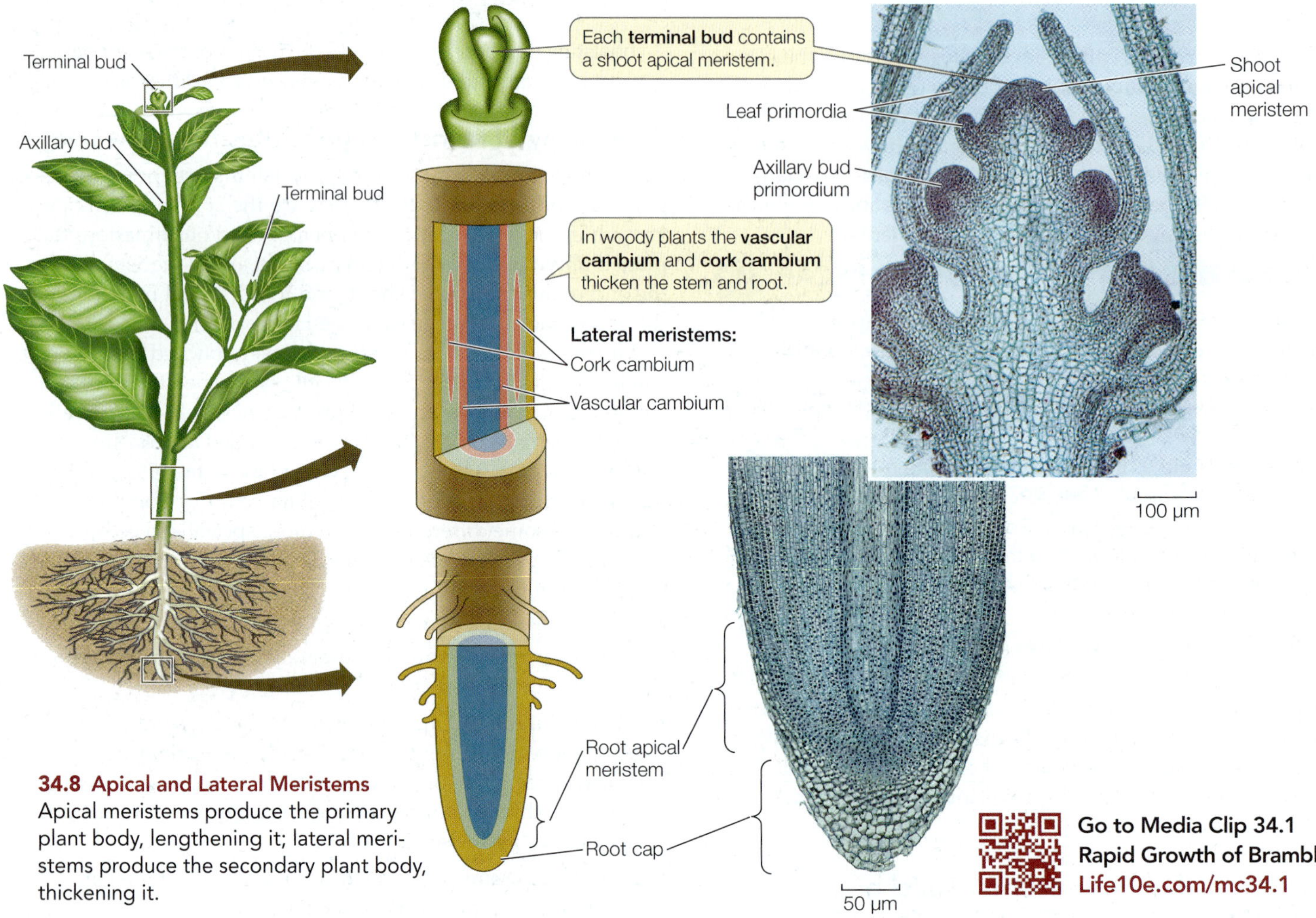
Terminal bud
Axillary bud
Terminal bud
Each terminal bud contains a shoot apical meristem.
Leaf primordia
Axillary bud primordium
Shoot apical meristem
In woody plants the vascular cambium and cork cambium thicken the stem and root.
Lateral meristems:
Cork cambium
Vascular cambium
100 µm
Root apical meristem
Root cap
50 µm

34.8 Apical and Lateral Meristems
Apical meristems produce the primary plant body, lengthening it; lateral meristems produce the secondary plant body, thickening it.

Go to Media Clip 34.1
Rapid Growth of Brambles
Life10e.com/mc34.1

Apical meristems in both the shoot and the root give rise to a set of **primary meristems**, which produce the tissues of the primary plant body. From the outside to the inside of the shoot or root, the primary meristems are the **protoderm**, the **ground meristem**, and the **procambium** (see Figure 34.8). These meristems, in turn, give rise to the three tissue systems:

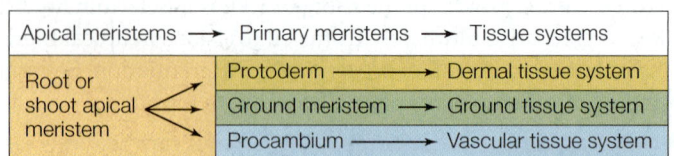

Apical meristems →	Primary meristems →	Tissue systems
Root or shoot apical meristem	Protoderm →	Dermal tissue system
	Ground meristem →	Ground tissue system
	Procambium →	Vascular tissue system

Because meristems can continue to produce new organs throughout the lifetime of the plant, the plant body is much more variable in form than the animal body, which produces each organ only once.

Let's look more closely at how the root apical meristem produces the root system.

The root apical meristem gives rise to the root cap and the root primary meristems

The root apical meristem produces all the cells that contribute to the growth and development of the root (Figure 34.9A). Some of the daughter cells from the apical (tip) end of the root apical meristem contribute to a **root cap**, which protects the delicate growing region of the root as it pushes through the soil. The root cap secretes a mucopolysaccharide (slime) that acts as a lubricant. Even so, the cells of the root cap are often damaged or scraped away and must therefore be replaced constantly. The root cap is also the structure that detects the pull of gravity and thus controls the downward growth of roots.

In the middle of the root apical meristem is a quiescent center, in which cell divisions are rare. The quiescent center can become more active when needed—following injury, for example. The daughter cells produced above the quiescent center (that is, away from the root cap) become the three primary meristems.

The apical and primary meristems constitute the **zone of cell division**, the source of all the cells of the root's primary tissues. Just above this zone is the **zone of cell elongation**, where the newly formed cells are elongating and thus pushing the root farther into the soil. Above that zone is the **zone of maturation**, where the cells are differentiating, taking on specialized forms and functions. These three zones grade imperceptibly into one another; there is no abrupt line of demarcation.

The products of the root's primary meristems become root tissues

The products of the three primary meristems (the protoderm, ground meristem, and procambium) are the tissue systems of the mature root. The differing arrangement of the three tissue

34.9 Tissues and Regions of the Root Tip (A) Extensive cell division creates the complex structure of the root. (B) Root hairs, seen with a scanning electron microscope.

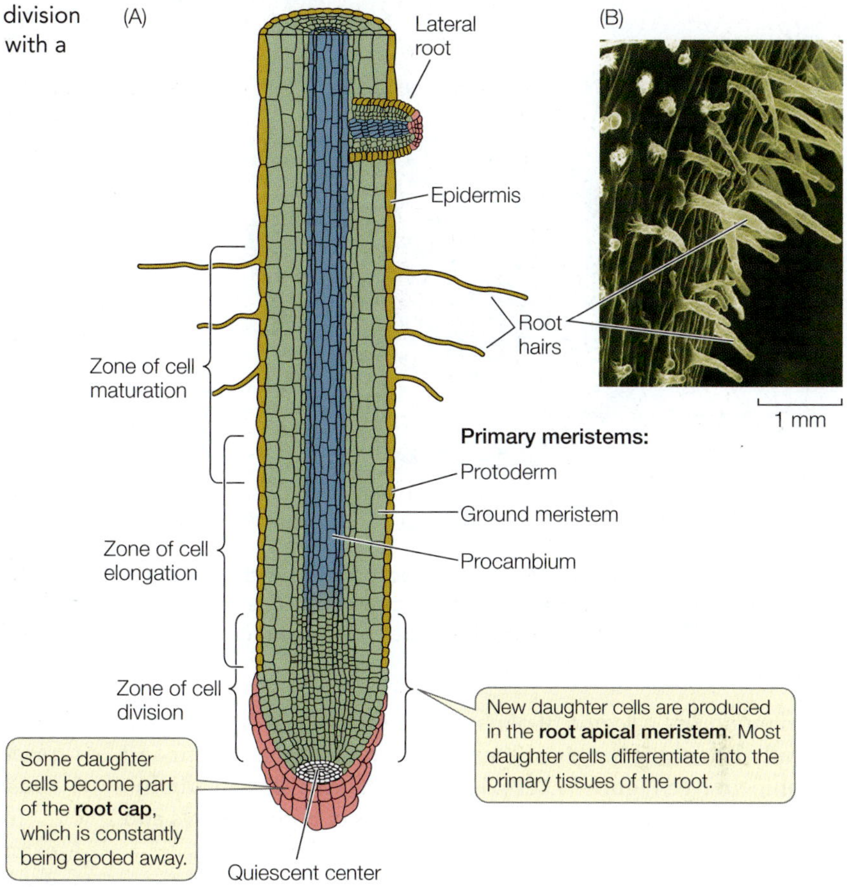

(A)
Lateral root
Epidermis
Root hairs
Zone of cell maturation
Primary meristems:
Protoderm
Ground meristem
Procambium
Zone of cell elongation
Zone of cell division
New daughter cells are produced in the **root apical meristem**. Most daughter cells differentiate into the primary tissues of the root.
Some daughter cells become part of the **root cap**, which is constantly being eroded away.
Quiescent center
(B)
1 mm

systems in the roots of eudicots and monocots is one of the ways in which the two clades are distinguished (Figure 34.10).

The protoderm gives rise to the epidermis, the outer layer of cells that is adapted for protection of the root and absorption of mineral ions and water. Many of the epidermal cells produce long, delicate **root hairs**, which vastly increase the surface area of the root (Figure 34.9B). Root hairs grow out among the soil particles, probing nooks and crannies and taking up water and minerals.

Internal to the epidermis, the ground meristem gives rise to a region of ground tissue that is many cells thick, called the **cortex**. The cells of the cortex are relatively unspecialized and often serve as storage depots. The innermost layer of the cortex is the **endodermis**. Unlike other cortical cells, the endodermal cells contain a waterproof substance called suberin in their primary cell walls. The suberin forms a cylindrical ring around the inside of the endodermis, which allows the endodermal cells to control the movement of water and dissolved mineral ions into and out of the vascular tissue system.

Within the endodermis is the vascular cylinder, or **stele**, produced by the procambium. The stele consists of three tissues: pericycle, xylem, and phloem. The **pericycle** consists of one or more layers of relatively undifferentiated cells. It has three important functions:

- It is the tissue within which lateral roots arise (Figure 34.11).
- It can contribute to secondary growth by giving rise to lateral meristems that thicken the root.
- Its cells contain membrane transport proteins that export nutrient ions into the cells of the xylem.

At the very center of the root of a eudicot lies the xylem. Seen in cross section, it typically has the shape of a star with a variable number of points (see Figure 34.10A). Between the points are bundles of phloem.

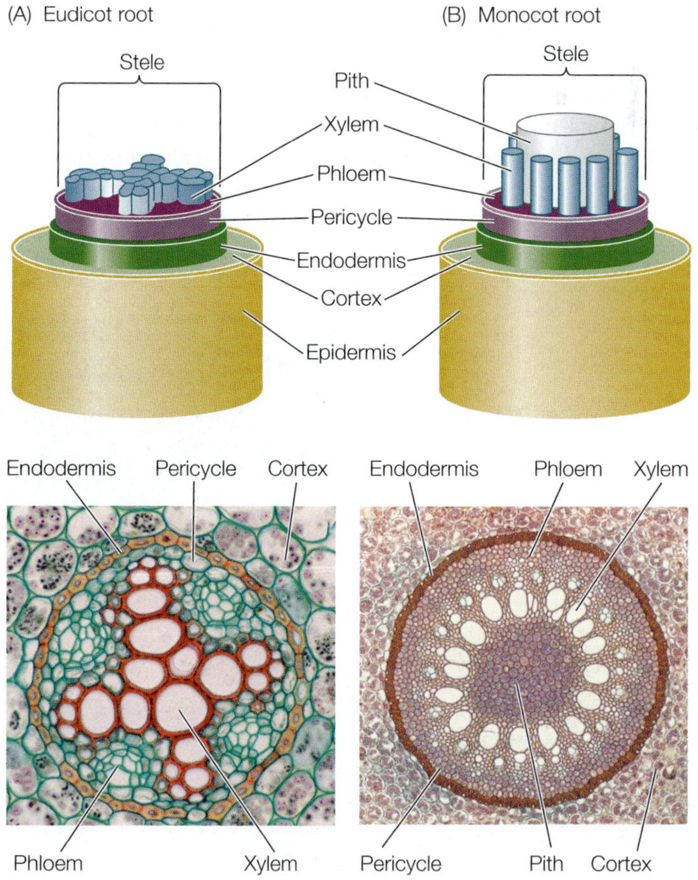

(A) Eudicot root
(B) Monocot root
Stele
Stele
Pith
Xylem
Phloem
Pericycle
Endodermis
Cortex
Epidermis
Endodermis
Pericycle
Cortex
Endodermis
Phloem
Xylem
Phloem
Xylem
Pericycle
Pith
Cortex

34.10 Products of the Root's Primary Meristems The protoderm gives rise to the outermost layer (epidermis). The ground meristem produces the cortex, the innermost layer of which is the endodermis. The primary vascular tissues of the root are found in the stele, which is the product of the procambium. The arrangement of tissues in the stele differs in the roots of (A) eudicots and (B) monocots. The photomicrographs show cross sections of the stele of a representative eudicot (the buttercup, *Ranunculus*) and a representative monocot (corn, *Zea mays*), showing the arrangement of the primary root tissues.

Go to Activity 34.1 Eudicot Root Life10e.com/ac34.1

Go to Activity 34.2 Monocot Root Life10e.com/ac34.2

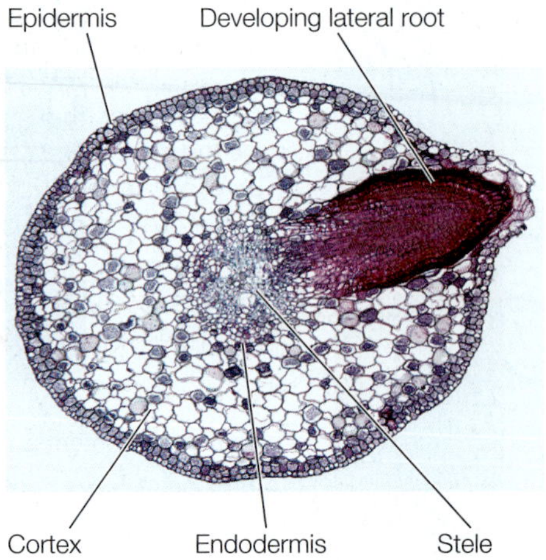
Epidermis
Developing lateral root
Cortex
Endodermis
Stele

34.11 Lateral Root Anatomy Cross section through the tip of a lateral root in a willow tree. Cells in the pericycle divide and the products differentiate, forming the tissues of a lateral root.

In monocots, a region of parenchyma cells called the **pith** typically lies in the center of the root, surrounded by xylem and phloem (see Figure 34.10B). Pith, which often stores carbohydrate reserves, is also found in the stems of both eudicots and monocots.

The root system anchors the plant and takes up water and dissolved minerals

Water and minerals enter most plants through the root system, which is located in the soil. Because light does not penetrate the soil, roots typically lack the capacity for photosynthesis. Although hidden from view, the root system is often larger than the visible shoot system. For example, the root system of a 4-month-old winter rye plant (*Secale cereale*) was found to be 130 times longer in total than the shoot system, with almost 13 million branches that had a cumulative length of more than 500 kilometers!

Angiosperm root systems develop from the embryonic root, called the **radicle**. From this common starting point, the root systems of monocots and eudicots develop differently. Following seed germination, the radicle of most eudicots develops as a primary root called the **taproot**, which extends downward by tip growth and outward by initiating **lateral roots**. The taproot and the lateral roots form a **taproot system**, which can take a variety of forms. For example, the taproot itself often functions as a nutrient storage organ, as in carrots (*Daucus carota*), sugar beets (*Beta vulgaris*), and sweet potato (*Ipomoea batatas*) (**Figure 34.12A**).

In contrast, the primary root of monocots (and some eudicots) is short-lived. Because they originate from the stem at ground level or just below, the roots of a typical monocot are called **adventitious** ("arriving from outside") **roots**, and they form a **fibrous root system** composed of numerous thin roots that are all roughly equal in diameter (**Figure 34.12B**). Many fibrous root systems have large surface areas for the absorption of water and minerals. A fibrous root system clings to soil very well. The fibrous root systems of grasses, for example, may protect steep hillsides where runoff from rain would otherwise cause erosion.

In some plants—corn, banyan, and pandanus trees, for example—adventitious roots grow down from above the ground and function as props to help support the shoot system (**Figure 34.12C**). These **prop roots** have evolved for different reasons in different species. For example, some monocots may develop prop roots because they are unable to support aboveground growth through the thickening of their stems. Pandanus trees often grow near coastal beaches, where their prop roots help provide support in very sandy soils. Banyans begin life as epiphytes (plants that grow on other plants) and then develop woody prop roots, which enable them to grow into huge trees.

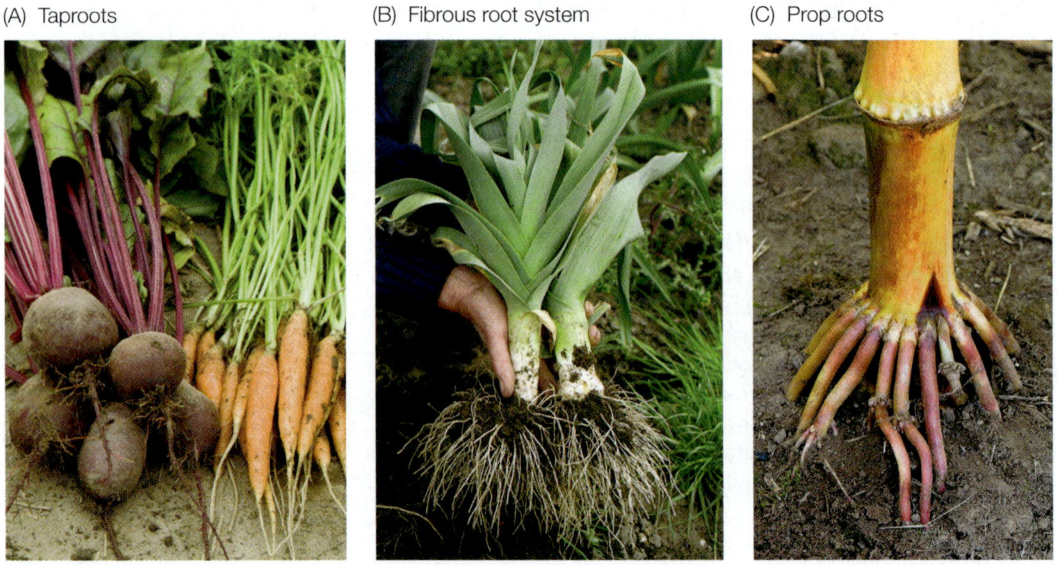
(A) Taproots
(B) Fibrous root system
(C) Prop roots

34.12 Root Systems of Eudicots and Monocots (A) The taproot systems of eudicots, such as carrots, sugar beets, and sweet potato, contrast with (B) the fibrous root system of a leek and (C) the adventitious prop roots of corn.

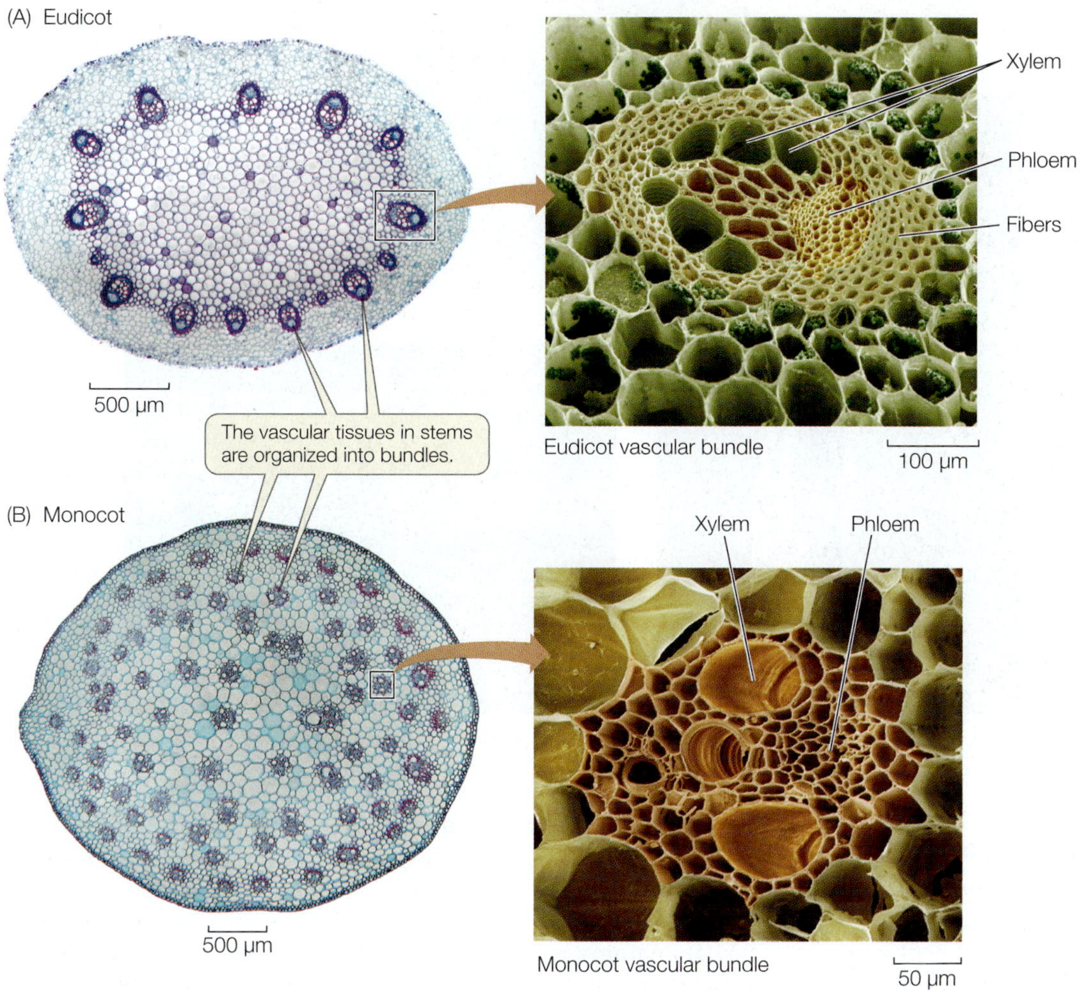
(A) Eudicot
500 µm
Xylem
Phloem
Fibers
Eudicot vascular bundle
100 µm
The vascular tissues in stems are organized into bundles.
(B) Monocot
Xylem
Phloem
500 µm
Monocot vascular bundle
50 µm

34.13 Vascular Bundles in Stems
(A) In herbaceous eudicot stems, the vascular bundles are arranged in a cylinder, with pith in the center and the cortex outside the cylinder. (B) A scattered arrangement of vascular bundles is typical of monocot stems.

Go to Activity 34.3 Eudicot Stem
Life10e.com/ac34.3

Go to Activity 34.4 Monocot Stem
Life10e.com/ac34.4

The products of the stem's primary meristems become stem tissues

Recall that shoots and roots are composed of repeating modules called phytomers. A phytomer in the shoot system consists of a node with its attached leaf or leaves, the internode below the node, and axillary buds, each of which forms in the angle between a leaf and the stem (see Figure 34.1). Shoots grow by adding new phytomers. Initially these form only on the primary stem of the plant, but if an axillary bud develops into a branch, it does so by producing new phytomers. The new phytomers originate from cells in the shoot apical meristems, which are present in the terminal buds of each branch and the main stem.

The shoot apical meristem, like the root apical meristem, forms three primary meristems: protoderm, ground meristem, and procambium. These primary meristems, in turn, give rise to the three shoot tissue systems.

As the shoot grows and extends, bulges called **leaf primordia** develop on the sides of the shoot apical meristem at regular intervals. These primordia are made up of primary meristematic tissues, which go on to develop into the mature tissues of the leaf. In addition, bud primordia form at the bases of the leaf primordia. These have the potential to become new apical meristems and initiate new shoots. The sites where leaf primordia form become nodes on the developing stem. The regions between the nodes (the internodes) lengthen initially via cell division in the primary meristematic tissues, and later by cell elongation in the mature stem tissues. The growing stem has no protective structure analogous to the root cap, but the leaf primordia can act as a protective covering for the shoot apical meristem.

The vascular systems of stems differ from those of roots. In a root, the vascular tissue lies deep in the interior, with the xylem at or near the center (see Figure 34.10). The vascular tissue of a young stem, however, is divided into discrete **vascular bundles** (**Figure 34.13**). Each vascular bundle contains both xylem and phloem. In eudicots the vascular bundles generally form a cylinder, whereas in monocots they are scattered throughout the stem.

In addition to the vascular tissues, the stem contains other important storage and supportive tissues. In eudicots the pith lies inside the ring of vascular bundles and also extends between them, forming regions called pith rays. To the outside lies the cortex, which may contain supportive collenchyma

(A)
Branches
Tuber (modified stem)
(B)
"Barrel" (enlarged stem)
Spines (modified leaves)
(C)
Shoots
Runner (horizontal stem)

34.14 Modified Stems (A) A potato is a modified stem called a tuber; the sprouts that grow from its "eyes" are shoots, not roots. (B) The stem of this barrel cactus is enlarged to store water. Its highly modified leaves serve as thorny spines. Most of this plant's photosynthesis occurs in the stem. (C) The runners of ground ivy are horizontal stems that produce roots and shoots at intervals. Rooted portions of the plant can live independently if the runner is cut.

cells with thickened walls. The pith and cortex constitute the ground tissue system of the stem. The outermost cell layer of the young stem is the epidermis.

The stem supports leaves and flowers

The central function of stems is to elevate and support the photosynthetic organs (leaves) as well as the reproductive organs (flowers). Various modifications of stems are seen in nature. The tuber of a potato, for example—the part of the plant eaten by humans—is not a root but rather an underground stem. The "eyes" of a potato are depressions containing axillary buds—in other words, a sprouting potato is just a branching stem (**Figure 34.14A**). Many desert plants have enlarged, water-retaining stems (**Figure 34.14B**). The runners of strawberry plants are horizontal stems that develop adventitious roots some distance from the main stem (**Figure 34.14C**); if the links between the rooted portions are broken, independent plants can develop on each side of the break—a form of vegetative (asexual) reproduction (see Section 38.3).

Leaves are determinate organs produced by shoot apical meristems

For most of its life, a plant produces leaves from apical meristems. As we mentioned earlier, apical meristems that produce leaves are called vegetative meristems. Leaves originate from the edges of the apical meristem as initial cells that differentiate into leaf primordia. A highly simplified way to think of the development of a leaf from a leaf primordium is to imagine leaves as flattened stems. However, there are two important differences:

- Unlike the growth of the stem, which is indeterminate, the growth of a leaf is determinate.
- Whereas the tissues of the stem are arranged in a radial pattern, the leaf, as a flat organ, has a distinct top side and bottom side.

Leaf anatomy is beautifully adapted to carry out photosynthesis, and to support that process by exchanging the gases O_2 and CO_2 with the environment, while at the same time limiting evaporative water loss. Its extensive vascular system supplies the leaf with water and mineral nutrients and exports the products of photosynthesis to the rest of the plant. **Figure 34.15A** shows a section of a typical eudicot leaf in three dimensions.

The photosynthetic parenchyma tissue in leaves is called the **mesophyll** (which means "middle of the leaf"). Most eudicot leaves have two zones of mesophyll: an upper layer of elongated cells called the palisade mesophyll, and a lower layer of irregularly shaped cells called the spongy mesophyll. Within the mesophyll is a great deal of air space through which CO_2 can diffuse to photosynthesizing cells.

Vascular tissue branches extensively throughout the leaf, forming a network of veins (**Figure 34.15B**). Typically, monocot leaves have a parallel pattern of major veins; for an example, look at the grass blades on a nearby lawn. Dicot leaves, in contrast, have major veins in a netlike pattern, as in a maple or oak leaf. Veins extend to within a few cell diameters of all the cells of the leaf, ensuring that the mesophyll cells are well supplied with water and minerals. The products of photosynthesis are loaded into the veins for export to the rest of the plant.

The epidermis covers the entire surface of the leaf and is made up of nonphotosynthetic cells. The epidermal cells secrete a waxy cuticle that is impermeable to water. Although this impermeability prevents excessive water loss, it also poses

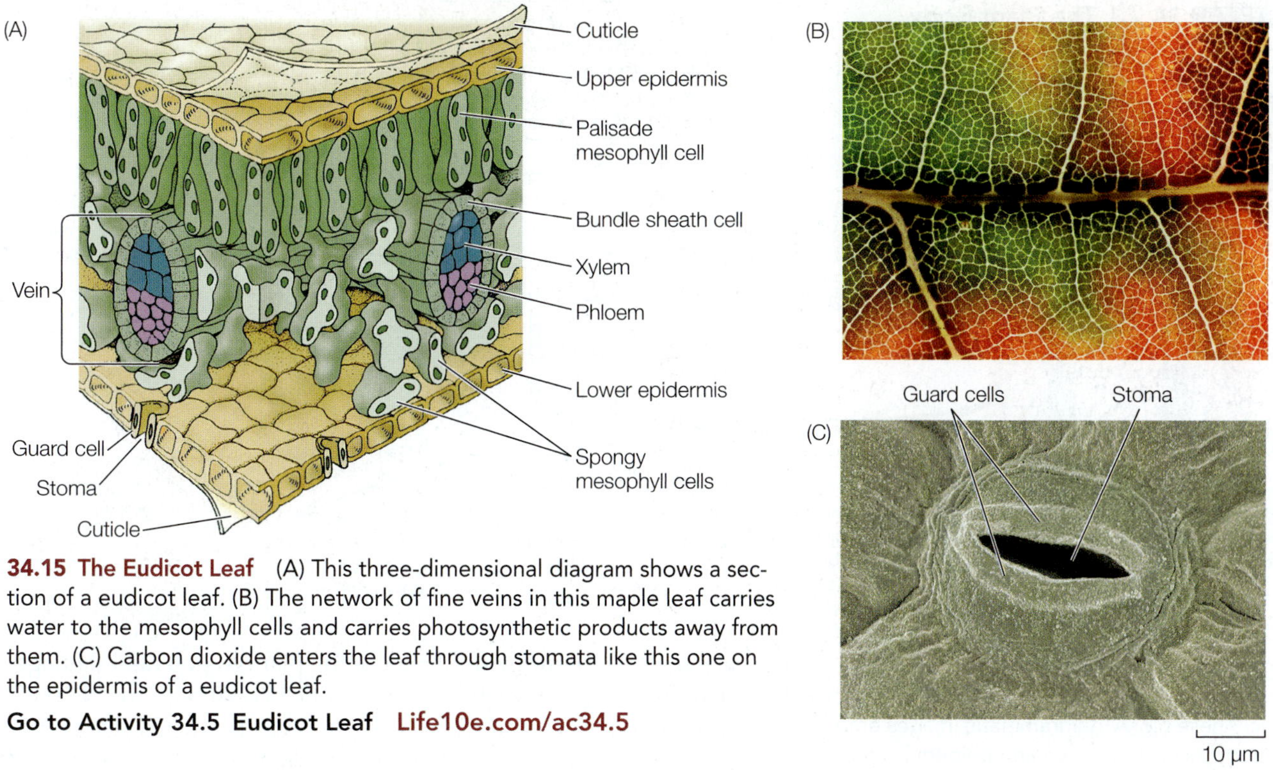

(A)
Cuticle
Upper epidermis
Palisade mesophyll cell
Bundle sheath cell
Xylem
Phloem
Vein
Lower epidermis
Guard cell
Stoma
Spongy mesophyll cells
Cuticle
(B)
(C)
Guard cells
Stoma
10 μm

34.15 The Eudicot Leaf (A) This three-dimensional diagram shows a section of a eudicot leaf. (B) The network of fine veins in this maple leaf carries water to the mesophyll cells and carries photosynthetic products away from them. (C) Carbon dioxide enters the leaf through stomata like this one on the epidermis of a eudicot leaf.

Go to Activity 34.5 Eudicot Leaf Life10e.com/ac34.5

a problem: while the epidermis keeps water in the leaf, it also keeps out CO_2—the other raw material of photosynthesis.

The problem of balancing water retention and carbon dioxide availability is solved by an elegant regulatory system that will be discussed in more detail in Section 35.3. Stomatal guard cells are modified epidermal cells that can change their shape, thereby opening or closing pores called **stomata** (singular *stoma*). The stomata serve as passageways between the environment and the leaf's interior (**Figure 34.15C**). When the stomata are open, carbon dioxide can enter and oxygen can leave, but water can also be lost.

Many eudicot stems and roots undergo secondary growth

As we have seen, the roots and stems of some eudicots develop a secondary plant body, the tissues of which we commonly refer to as wood and bark. These tissues are derived by secondary growth from the two lateral meristems, the vascular cambium and the cork cambium.

The **vascular cambium** is a cylindrical layer of tissue consisting predominantly of elongated cells that divide frequently. It supplies the cells of the secondary xylem and secondary phloem, which eventually become wood and bark. The **cork cambium** produces mainly waxy-walled protective cells. It supplies some of the cells that become bark.

Each year, deciduous trees lose their leaves and have bare branches and twigs over the winter (**Figure 34.16**). The apical meristems of the

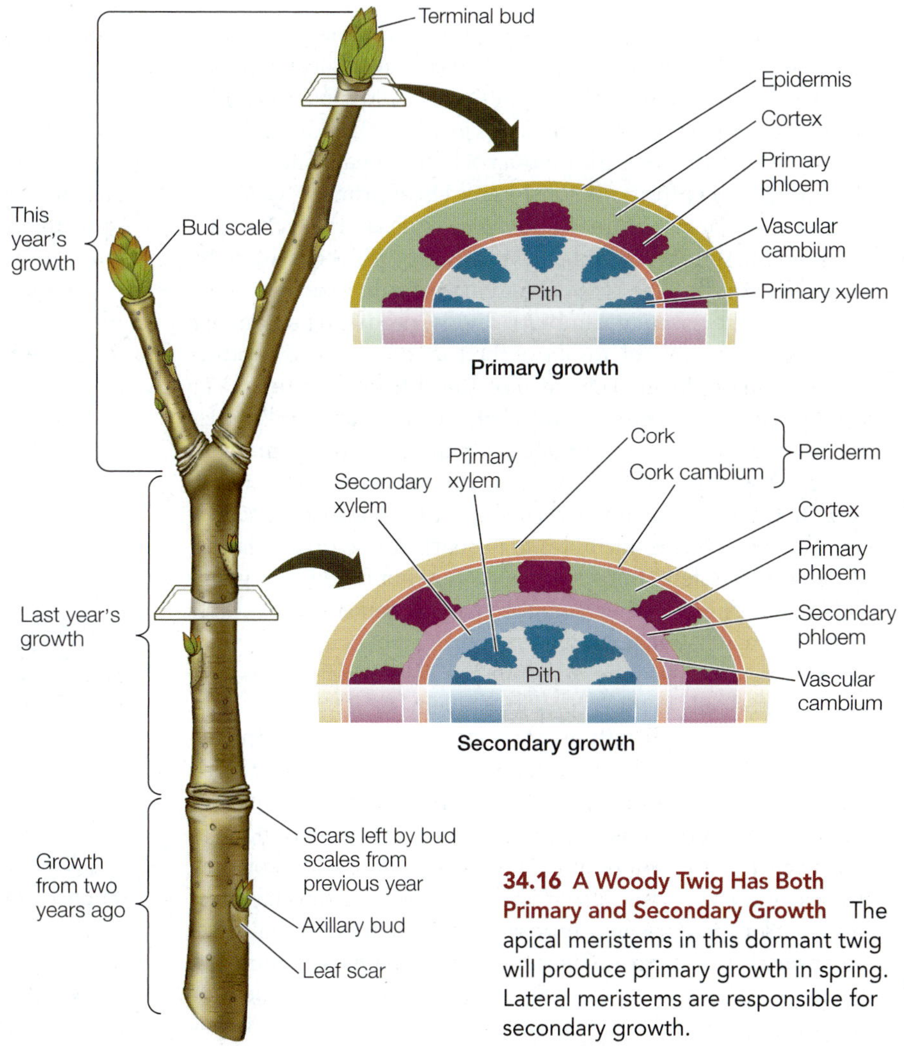

Terminal bud
Epidermis
Cortex
Primary phloem
Vascular cambium
Primary xylem
Pith
Primary growth
This year's growth
Bud scale
Cork
Cork cambium
Periderm
Primary xylem
Secondary xylem
Cortex
Primary phloem
Secondary phloem
Vascular cambium
Pith
Secondary growth
Last year's growth
Scars left by bud scales from previous year
Growth from two years ago
Axillary bud
Leaf scar

34.16 A Woody Twig Has Both Primary and Secondary Growth The apical meristems in this dormant twig will produce primary growth in spring. Lateral meristems are responsible for secondary growth.

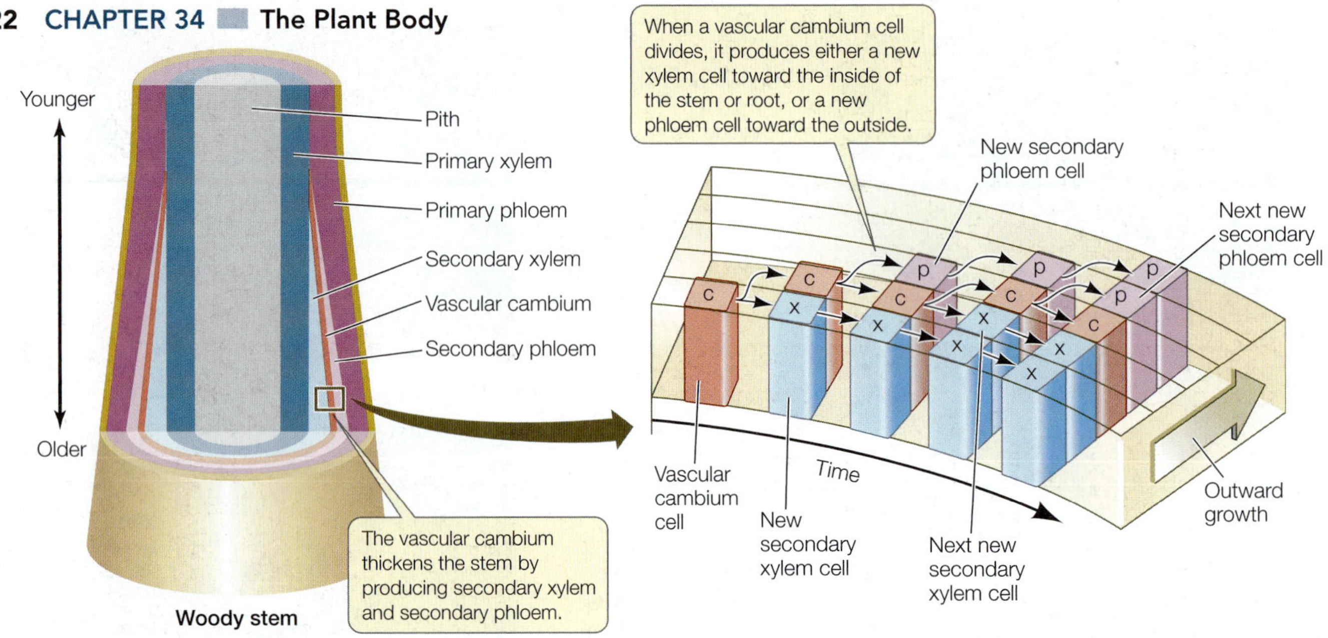
Younger
Older
Pith
Primary xylem
Primary phloem
Secondary xylem
Vascular cambium
Secondary phloem
Woody stem
The vascular cambium thickens the stem by producing secondary xylem and secondary phloem.
When a vascular cambium cell divides, it produces either a new xylem cell toward the inside of the stem or root, or a new phloem cell toward the outside.
New secondary phloem cell
Next new secondary phloem cell
Vascular cambium cell
New secondary xylem cell
Time
Next new secondary xylem cell
Outward growth
c
x
p

34.17 Vascular Cambium Thickens Stems and Roots Stems and roots grow thicker because a thin layer of cells, the vascular cambium, remains meristematic. These highly diagrammatic images emphasize the pattern of deposition of secondary xylem and phloem by the vascular cambium.

Go to Animated Tutorial 34.1
Secondary Growth: The Vascular Cambium
Life10e.com/at34.1

twigs are enclosed in buds protected by bud scales. When the buds begin to grow in spring, the scales fall away, leaving scars that show where the bud was. These scars allow us to identify the parts of the twig from each year's growth. The dormant twig shown in Figure 34.16 is the product of both primary and secondary growth. Only the buds consist entirely of primary tissues.

The vascular cambium is initially a single layer of cells lying between the primary xylem and the primary phloem within the vascular bundles. The root or stem increases in diameter when the cells of the vascular cambium divide, producing secondary xylem cells toward the inside of the root or stem and producing secondary phloem cells toward the outside (**Figure 34.17**). In the stem, cells in the pith rays between the vascular bundles also divide, forming a continuous cylinder of vascular cambium running the length of the stem. This cylinder, in turn, gives rise to complete cylinders of secondary xylem (the **wood**) and secondary phloem, which contributes to the bark. It also produces vascular rays for lateral transport, a structure not found in primary xylem and phloem. Therefore the vascular cambium produces vessel elements, tracheids, and supportive fibers in the secondary xylem; and sieve tube elements, companion cells, fibers, and parenchyma cells in the secondary phloem.

As secondary growth of stems or roots continues, the expanding vascular tissue stretches and breaks the epidermis and the outer layers of the cortex, which ultimately flake away. Before these dermal tissues are broken away, cells lying near the surface of the secondary phloem begin to divide, forming a cork cambium. This meristematic tissue produces layers of cork, a protective tissue composed of cells with thick walls waterproofed with suberin. The cork soon becomes the outermost tissue of the stem or root (see Figure 34.16). Without the activity of the cork cambium, the sloughing off of the outer primary tissues would expose the plant to potential damage, such as excessive water loss or invasion by microorganisms. Sometimes the cork cambium produces cells toward the inside as well as the outside; these cells constitute a tissue known as the phelloderm.

The cork cambium, cork, and phelloderm constitute the secondary dermal tissue called periderm. As the vascular cambium continues to produce secondary vascular tissue, these corky layers are lost, but the continuous formation of new cork cambia in the underlying secondary phloem gives rise to new corky layers. The periderm and the secondary phloem—that is, all the tissues external to the vascular cambium—constitute the **bark**.

When periderm forms on stems or roots, the underlying tissues still need to release carbon dioxide and take up oxygen for cellular respiration. Spongy regions in the periderm called lenticels allow such gas exchange (**Figure 34.18**).

Cross sections of most trunks (mature stems) of trees in temperate-zone forests show annual rings of wood (**Figure 34.19**), which result from seasonal environmental conditions.

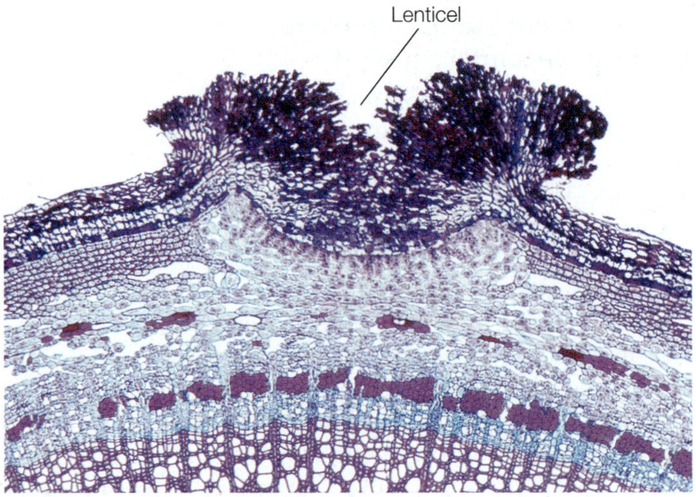
Lenticel

34.18 Lenticels Allow Gas Exchange through the Periderm The region of periderm that appears to be broken open is a lenticel in a year-old elderberry (*Sambucus*) twig; note the spongy tissue that constitutes the lenticel.

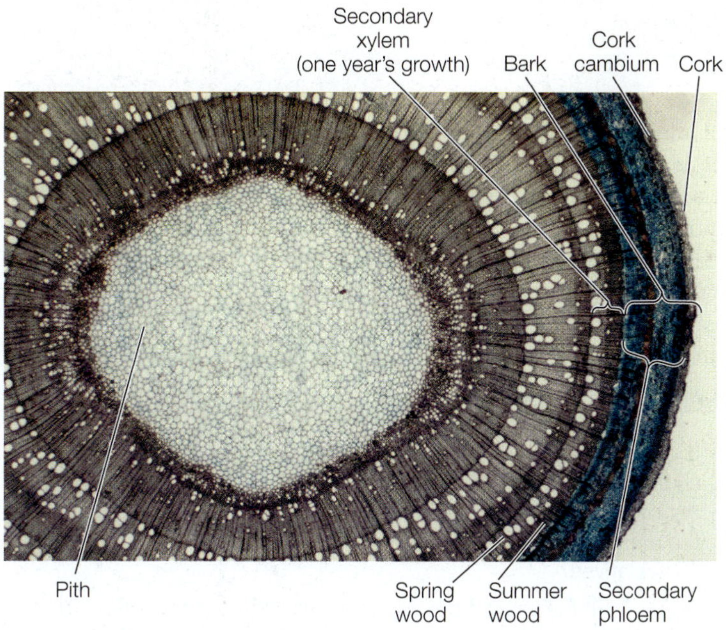
Secondary xylem (one year's growth)
Bark
Cork cambium
Cork
Pith
Spring wood
Summer wood
Secondary phloem

34.19 Annual Rings Rings of secondary xylem are the most noticeable feature of this cross section from a tree trunk.

In spring, when water is relatively plentiful, the tracheids or vessel elements produced by the vascular cambium tend to be large in diameter and thin-walled. Such wood is well adapted for transporting water and minerals. As water becomes less available during the summer, narrower cells with thicker walls are produced, making this summer wood darker and perhaps more dense than the wood formed in spring. Thus each growing season is usually recorded in a tree trunk by a clearly visible annual ring. Trees in the moist tropics do not undergo seasonal growth, so they do not lay down such obvious regular rings. Variations in temperature or water supply can lead to the formation of more than one ring in a single year, but most commonly a single annual ring is formed.

Only eudicots and other non-monocot angiosperms, along with many gymnosperms, have a vascular cambium and a cork cambium and thus undergo secondary growth. The few monocots that form thickened stems—palms, for example—do so without secondary growth. Palms have a very wide apical meristem that produces a wide stem, and dead leaf bases add to the diameter of the stem. All monocots grow in essentially this way, as do other angiosperms that lack secondary growth.

RECAP 34.3

Meristems are localized regions of cell division that are the sources of all new organs in the adult plant. Apical meristems are responsible for primary growth, which is associated with the lengthening and branching of shoots and roots. Lateral meristems increase plant thickness and form wood and bark in many eudicots.

- Explain how an apical meristem can be maintained for years while continuing to form leaves. **See p. 715 and Figure 34.8**
- What cells are derived from the root apical meristem, and what is the general process of root growth? **See pp. 716–717 and Figures 34.9 and 34.10**
- How does the vascular cambium give rise to thicker stems and roots? **See pp. 721–722 and Figures 34.16 and 34.17**

The building of the plant body by meristems allows a plant to respond to its environment by redirecting its growth. Thus individual plants of the same species can vary greatly in form. What underlies this variation, and how have we humans used it to our advantage?

34.4 How Has Domestication Altered Plant Form?

We have seen in this chapter that a very simple, modular plant body plan underlies the remarkable diversity of the flowering plants that cover our planet. Differences in plant form among species are not surprising, given the high levels of genetic diversity among plant species. However, members of the same species can also show remarkable diversity in form. From a genetic perspective, this suggests that minor differences in gene content or gene regulation can underlie dramatic differences in plant form.

It is hard to believe that modern corn was domesticated from the wild grass teosinte, which still grows in the hills of Mexico (**Figure 34.20**). One of the most conspicuous differences is that teosinte, like other wild grasses, is highly branched, whereas domesticated corn has a single shoot. This morphological difference is due in large part to the activity of a single gene called *teosinte branched 1* (*tb1*). The protein product of *tb1* regulates the growth of axillary buds (see Figure 34.1). The allele of *tb1* in domesticated corn represses branching, whereas the allele in teosinte permits branching.

Teosinte
Corn

34.20 Modern Corn Was Domesticated from the Wild Grass Teosinte Each teosinte plant has multiple branches. Beginning more than 8,000 years ago in Mexico, farmers favored plants with minimal branching. Reducing the number of branches results in fewer ears per plant but allows each ear to grow larger and produce more seeds.

Even harder to believe is that a single species, *Brassica oleracea* (wild mustard), is the ancestor of many familiar and morphologically diverse crops, including kale, broccoli, Brussels sprouts, and cabbage (see Figure 21.4). But each of these familiar vegetable crops has the same basic body plan. Starting with morphologically diverse populations of the wild ancestor, humans selected and planted the seeds of variants with traits they found desirable. For example, Brussels sprout plants were selected for their enlarged axillary buds, cabbage plants were selected for their enlarged terminal buds and short internodes, and broccoli and cauliflower plants were selected for their large clusters of flower buds. Many generations of such artificial selection produced the crops that fill the produce section of the supermarket or the stands of the farmers' market.

Just as they were for ancient farmers, the genomes of plants are priceless resources today. The genetic variation in crop plants and their wild relatives can be used to improve our crop plants or adapt them to changing conditions. The improvement of crop plants is a work in progress that is being carried out in plant breeding programs worldwide. In fact, these programs are more important than ever. Increased human activity is dramatically changing our planet and leading to the extinction of more and more plant species. For this reason, various organizations around the world have developed seed banks, where seeds of diverse species, and variants within species, are stored.

RECAP 34.4

Crop domestication involves artificial selection of certain desirable traits found in wild plant populations. By understanding the basic body plan of plants, one can more easily understand the morphological relationship between a crop plant and its wild relatives.

- Why are the seeds from wild relatives of crop plants valuable? See p. 725

How might plant physiologists improve the cassava plant for human use?

ANSWER

Many people depend on cassava roots for food. Unfortunately, however, the varieties that are traditionally grown have inadequate levels of protein, iron, and β-carotene. (The human body converts β-carotene into vitamin A.) Furthermore, cassava plants have high levels of cyanogens that are converted to toxic cyanide when eaten. BioCassava Plus is an international consortium of scientists who are using biotechnology to develop cassava varieties with improved nutritional contents. Another team has developed cassava plants that contain reduced levels of cyanogens. The anatomy of the plant is also being changed. Cassava has been crossed with a treelike relative, *Manihot glaziovii*, and the resulting plants produce roots that are not only fleshy and edible but that grow deep into the soil, where they can tap into water supplies far below the surface in dry climates. This crop is being investigated for use in sub-Saharan Africa.

CHAPTER SUMMARY 34

34.1 What Is the Basic Body Plan of Plants?

- The vegetative organs of flowering plants are roots, which form a **root system**, and stems and leaves, which form a **shoot system**. Review Figure 34.1
- Plant development differs from animal development in that plants have apical meristems, cell walls, vacuoles, and in some cases, totipotent cells.
- Plants have apical–basal and radial axes of symmetry. Review Figures 34.3, 34.4

34.2 What Are the Major Tissues of Plants?

- Three **tissue systems**, arranged concentrically, extend throughout the plant body: the vascular tissue, dermal tissue, and ground tissue systems. Review Figure 34.5
- The **dermal tissue system** protects the plant body surface. Dermal cells form the **epidermis** and, in woody plants, the **periderm**.
- The **ground tissue system** contains cells of three types. Some **parenchyma** cells carry out photosynthesis; others store starch. **Collenchyma** cells provide flexible support. **Sclerenchyma** cells include **fibers** and **sclereids** that provide strength and mechanical support. Review Figures 34.6, 34.7
- The **vascular tissue system** includes **xylem**, which conducts water and minerals absorbed by the roots, and **phloem**, which conducts the products of photosynthesis throughout the plant body.
- **Tracheary elements** include **tracheids** and **vessel elements**, which are the conducting cells of the xylem. **Sieve tube elements** are the conducting cells of the phloem.

34.3 How Do Meristems Build a Continuously Growing Plant?

- All seed plants possess a primary plant body consisting of nonwoody tissues. Woody plants also possess a secondary plant body consisting of wood and bark. **Apical meristems** generate the primary plant body, and **lateral meristems** generate the secondary plant body. Review Figure 34.8
- Apical meristems are responsible for **primary growth** (lengthening of roots and shoots). Apical meristems at the tips of stems and roots give rise to three **primary meristems** (**protoderm**, **ground meristem**, and **procambium**), which in turn produce the three tissue systems of the primary plant body.
- The root apical meristem gives rise to the **root cap** and to three primary meristems. Root tips have overlapping **zones of cell division**, **cell elongation**, and **cell maturation**. Review Figure 34.9

continued

- The vascular tissue of roots is contained within the **stele**. It is arranged differently in eudicot and monocot roots. **Review Figures 34.10, 34.11, ACTIVITIES 34.1, 34.2**
- In nonwoody stems, the vascular tissue is divided into **vascular bundles**, each containing both xylem and phloem. **Review Figure 34.13, ACTIVITIES 34.3, 34.4**
- Eudicot leaves have two zones of photosynthetic **mesophyll** that are supplied by veins with water and minerals. Veins also carry the products of photosynthesis to other parts of the plant body. A waxy **cuticle** limits water loss from the leaf. Guard cells control openings called **stomata** in the leaf that allow CO_2 to enter, but also allow some water to escape. **Review Figure 34.15, ACTIVITY 34.5**
- Two lateral meristems, the **vascular cambium** and **cork cambium**, are responsible for secondary growth. The vascular cambium produces secondary xylem (wood) and secondary phloem. The cork cambium produces a protective tissue called cork. **Review Figures 34.16, 34.17, ANIMATED TUTORAL 34.1**

34.4 How Has Domestication Altered Plant Form?

- The plant body plan is simple, yet it can be changed dramatically by minor differences in genes, as evidenced by the natural diversity of wild plants.
- Crop domestication involves artificial selection of certain desirable traits found in wild populations. **Review Figure 34.20**

Go to the Interactive Summary to review key figures, Animated Tutorials, and Activities Life10e.com/is34

CHAPTER**REVIEW**

REMEMBERING

1. Roots
   a. always form a fibrous root system that holds the soil.
   b. possess a root cap at their tip.
   c. form branches from axillary buds.
   d. are commonly photosynthetic.
   e. do not show secondary growth.

2. Which statement about parenchyma cells is *not* true?
   a. They are alive when they perform their functions.
   b. They typically lack a secondary wall.
   c. They often function as storage depots.
   d. They are the most common cell type in the plant body.
   e. They are found only in stems and roots.

3. Which statement about meristems is *not* true?
   a. They are formed during embryogenesis.
   b. They have secondary cell walls.
   c. Their cells have small central vacuoles.
   d. They are clusters of undifferentiated cells.
   e. They retain the ability to produce new cells indefinitely.

4. The pericycle
   a. is the innermost layer of the cortex.
   b. is the tissue within which lateral roots arise.
   c. consists of highly differentiated cells.
   d. forms a star-shaped structure at the very center of the root.
   e. is waterproofed by suberin.

5. Which statement about leaf anatomy is *not* true?
   a. Opening of stomata is controlled by guard cells.
   b. The cuticle is secreted by the epidermis.
   c. The veins contain xylem and phloem.
   d. The cells of the mesophyll are packed together, minimizing air space.
   e. The spines of cacti are actually modified leaves.

UNDERSTANDING & APPLYING

6. Which of these statements is true of secondary growth but *not* of primary growth?
   a. It occurs in eudicots and monocots.
   b. It involves the proliferation of roots and shoots through branching.
   c. It derives from the vascular cambium and the cork cambium.
   d. It occurs in palms.
   e. It derives from the shoot apical meristem.

7. Which of the following is a difference between monocots and eudicots?
   a. Only eudicots have phytomers.
   b. Only monocots have shoot and root apical meristems.
   c. Monocot stems do not undergo secondary growth.
   d. The vascular bundles of monocot stems are commonly arranged as a cylinder.
   e. Eudicot embryos commonly have one cotyledon.

8. Compare sclerenchyma cells and collenchyma cells in terms of structure and function.

9. Compare primary and secondary growth. Do all angiosperms undergo secondary growth? Explain.

ANALYZING & EVALUATING

10. When a young oak was 5 meters tall, a thoughtless person carved his initials in its trunk at a height of 1.5 meters above the ground. Today that tree is 10 meters tall. How high above the ground are those initials? Explain your answer in terms of plant growth.

11. Take a walk through a farmers' market or the produce section of a supermarket. Use your knowledge of plant growth and form to figure out what desirable trait was selected to produce some of your favorite salad vegetables.

Go to BioPortal at **yourBioPortal.com** for Animated Tutorials, Activities, LearningCurve Quizzes, Flashcards, and many other study and review resources.

35 Transport in Plants

CHAPTER OUTLINE

Thirsty Rice Cultivation of rice, the most important food crop in Asia, requires large quantities of water.

EVERYONE KNOWS THAT PLANTS NEED WATER to grow. However, it may come as a surprise that the cultivation of crop plants consumes far more water than all other human activities combined. The worldwide demand for fresh water is increasing at a greater rate than the supply. This situation makes it imperative that we understand how plants use water so that we can select or breed plants that use it more efficiently. Much of the mass that plants acquire as they grow is due to their net fixation of atmospheric CO_2 into carbohydrates through photosynthesis. But plants need to take up a lot of water to grow. The ratio of net photosynthetic carbon fixation to water uptake is known as a plant's water-use efficiency.

Droughts and dwindling water supplies are challenging farmers all over the world. One of the least water-efficient of all crop plants is, unfortunately, one of our most important: rice. Rice plants use up to three times more water per unit of growth than crops such as wheat and corn. The precariousness of heavily water-dependent rice farming was dramatically demonstrated in eastern India between 1997 and 2003, when drought reduced rice production by more than 5 million tons—some farmers lost up to 50 percent of their crops.

A strain of rice requiring less water yet producing the same amount of grain would both make the world supply of rice less vulnerable to drought and help conserve water for other uses. A team of molecular biologists, plant physiologists, and crop scientists led by Andrew Pereira at Virginia Polytechnic began a quest for such a strain of rice by studying an entirely different plant—the model organism *Arabidopsis thaliana* (thale cress). They searched for genetic variants of *Arabidopsis* that had superior water-use efficiency. One variant they studied was particularly hard to pull out of the ground because of its extensive root system (indicating higher capacity for water uptake) and had thick leaves with abundant photosynthetic tissue (indicating prolific photosynthesis). Molecular and physiological characterization of this *Arabidopsis* strain showed that its improved water usage was linked to a mutation in a single gene that codes for a transcription factor. When this gene (called *HARDY*) was isolated and put into rice plants using recombinant DNA technology, the transformed rice plants not only had higher water-use efficiency than wild-type rice plants but were more tolerant of dry soil as well.

Many laboratories around the world are using *Arabidopsis* to isolate genes involved in water usage and other important physiological processes. The knowledge gained from these studies may lead to various improvements of crop plants.

What other methods are used to reduce water loss in agriculture?

See answer on p. 738.

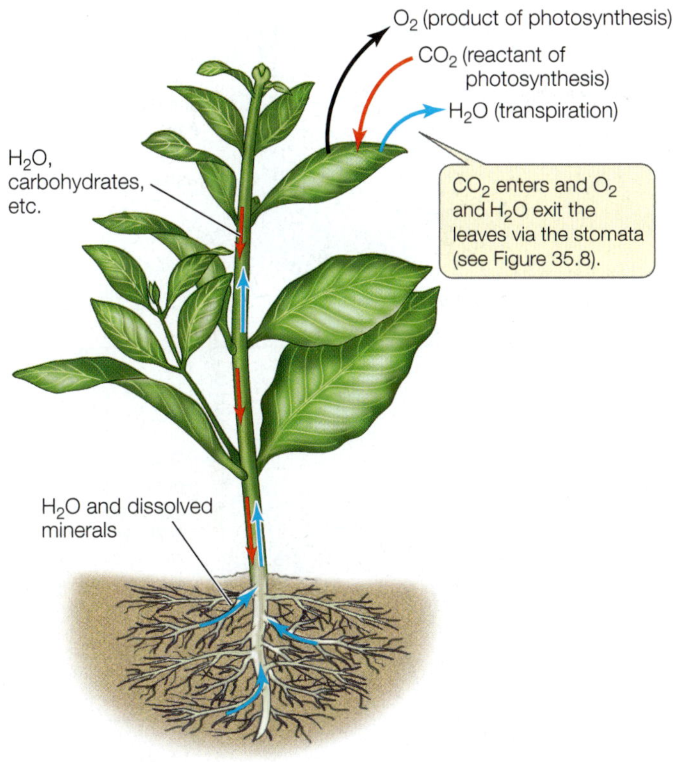
O_2 (product of photosynthesis)
CO_2 (reactant of photosynthesis)
H_2O (transpiration)
H_2O, carbohydrates, etc.
CO_2 enters and O_2 and H_2O exit the leaves via the stomata (see Figure 35.8).
H_2O and dissolved minerals

35.1 The Pathways of Water and Solutes in a Plant
Water travels from the soil to the atmosphere, with only a small fraction used within the plant.

35.1 How Do Plants Take Up Water and Solutes?

Terrestrial plants must obtain both water and mineral nutrients from the soil, usually through their roots. The roots, in turn, obtain carbohydrates and other important materials from the leaves (**Figure 35.1**). Water is required for photosynthesis in the leaves (see Section 10.1), for transporting solutes between plant organs, for cooling the plant, and for developing the internal pressure that supports the plant body.

The minerals that a plant needs are transported along with the water. Several steps in water and mineral transport will be considered in this chapter. In this section we will focus on the first part of the journey—the uptake of water and minerals into the roots and their transport into the xylem.

Water potential differences govern the direction of water movement

To enter a root cell, a solution must pass through the cell's plasma membrane. In Section 6.3 we described osmosis, the movement of water through a selectively permeable membrane from a region of lower solute concentration (higher water potential) to a region of higher solute concentration (lower water potential). Plant biologists define **water potential** (Ψ, **psi**) as the tendency of a solution (water plus solutes) to take up water from pure water across a membrane. By definition, the water potential of pure water is zero. Any solution that has a water potential less than zero has a tendency to take up water from pure water. The lower (more negative) the water potential, the greater the driving force for water movement across the membrane.

Water potential has two major components:

- **Solute potential** (Ψ_s): Solutes affect the osmotic behavior of a solution. The solute potential of pure water is zero, so any solution will create a negative solute potential. The greater the concentration of solutes, the lower the water potential, and the lower (more negative) the solute potential.
- **Pressure potential** (Ψ_p): As plant cells take up water, they tend to swell. However, the presence of the cell wall provides resistance to swelling (see Figure 6.9). The result is an increase in pressure inside the cell (**turgor pressure**), which decreases the tendency of the cell to take up more water. Therefore the pressure potential within a plant cell is usually positive.

A solution's water potential is the sum of its (negative) solute potential (Ψ_s) and its (usually positive) pressure potential (Ψ_p):

$$\Psi = \Psi_s + \Psi_p$$

We measure solute potential, pressure potential, and water potential in megapascals (MPa), a unit of pressure. Atmospheric pressure, "one atmosphere," is about 0.1 MPa, or 14.7 pounds per square inch; a typical pressure in an automobile tire is about 0.2 MPa.

Whenever water moves by osmosis, the following important rule applies: *water always moves across a selectively permeable membrane toward the region of lower (more negative) water potential.* In the left-hand tube in **Figure 35.2A**, the solution has a negative solute potential and a pressure potential of zero (relative to the atmosphere), so water moves across the membrane into the solution. In the right-hand tube, a piston is used to increase the pressure potential in the solution. When the negative solute potential is balanced by the positive pressure potential there is no net movement of water across the membrane.

In a plant cell immersed in pure water (**Figure 35.2B**), turgor pressure is comparable to the pressure potential exerted by the piston in Figure 35.2A. Water enters the cell by osmosis until the pressure potential exactly balances the solute potential and the water potential is zero. At this point the cell is **turgid**—that is, it has a significantly positive pressure potential. The cells in a plant are not surrounded by pure water, and their water potential is dependent on the water potential in the soil. But because turgid cells have a positive pressure potential, there is no net movement of water into them. The physical structures of many plants are maintained by the positive pressure potential of the water in their cells. If the pressure potential drops (for example, if the plant does not have enough water), the plant wilts (**Figure 35.3**).

Within living plant tissues, the movement of water from cell to cell follows a gradient of water potential. Over long distances, in unobstructed tubes such as xylem vessels and phloem sieve tubes, the flow of water and dissolved solutes is driven by a *gradient of pressure potential*, not a gradient of water potential. The movement of a solution from a region of higher pressure

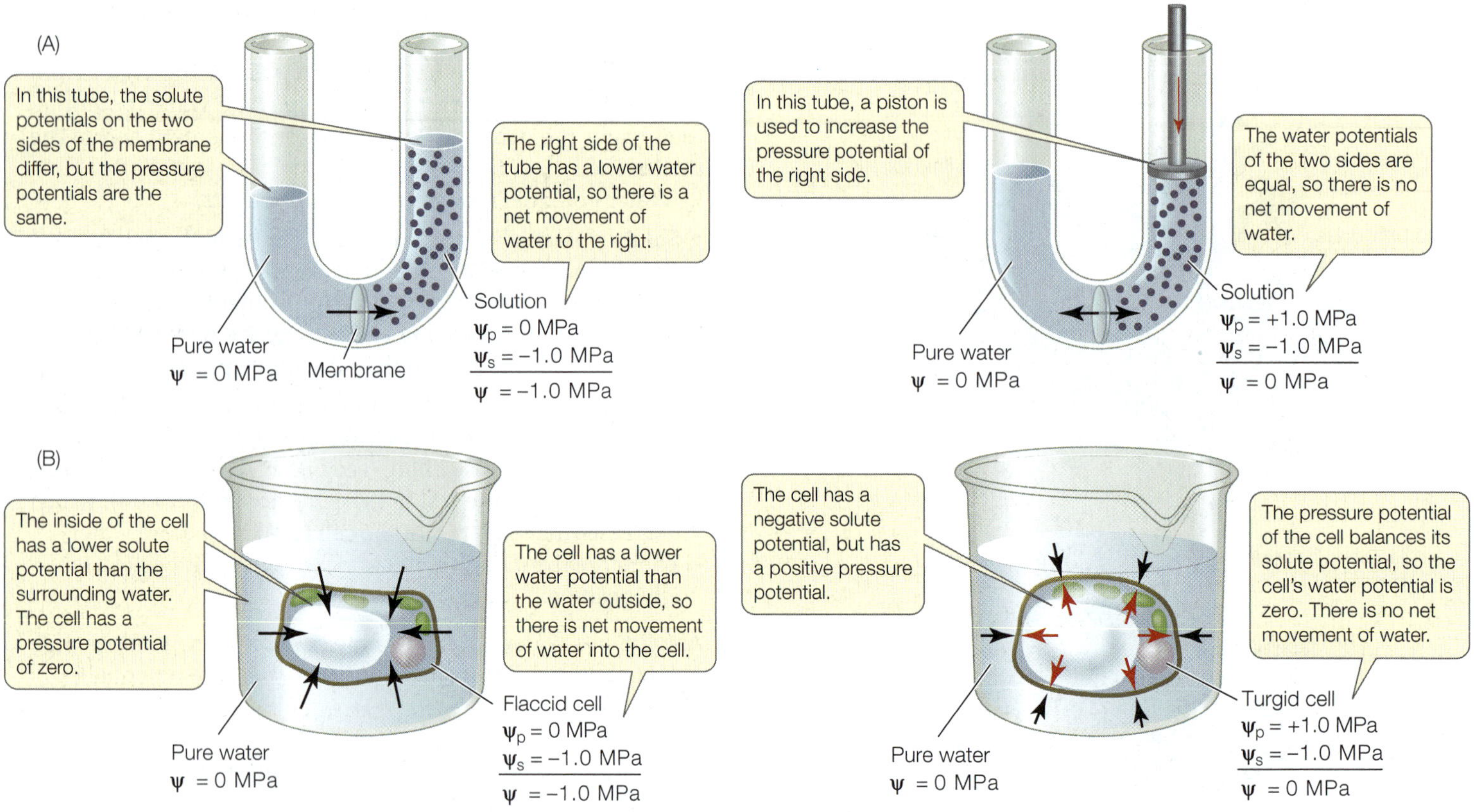

(A)
In this tube, the solute potentials on the two sides of the membrane differ, but the pressure potentials are the same.
The right side of the tube has a lower water potential, so there is a net movement of water to the right.
Pure water
ψ = 0 MPa
Membrane
Solution
ψ_p = 0 MPa
ψ_s = −1.0 MPa
ψ = −1.0 MPa
In this tube, a piston is used to increase the pressure potential of the right side.
The water potentials of the two sides are equal, so there is no net movement of water.
Pure water
ψ = 0 MPa
Solution
ψ_p = +1.0 MPa
ψ_s = −1.0 MPa
ψ = 0 MPa
(B)
The inside of the cell has a lower solute potential than the surrounding water. The cell has a pressure potential of zero.
The cell has a lower water potential than the water outside, so there is net movement of water into the cell.
Pure water
ψ = 0 MPa
Flaccid cell
ψ_p = 0 MPa
ψ_s = −1.0 MPa
ψ = −1.0 MPa
The cell has a negative solute potential, but has a positive pressure potential.
The pressure potential of the cell balances its solute potential, so the cell's water potential is zero. There is no net movement of water.
Pure water
ψ = 0 MPa
Turgid cell
ψ_p = +1.0 MPa
ψ_s = −1.0 MPa
ψ = 0 MPa

35.2 Water Potential, Solute Potential, and Pressure Potential A theoretical illustration of water potential. (B) The effect of differences in water potential on a plant cell.

Go to Animated Tutorial 35.1 Water Uptake in Plants
Life10e.com/at35.1

potential to a region of lower pressure potential is called **bulk flow**. As we will see later in this chapter, bulk flow in the xylem is between regions of differing *negative* pressure potentials (tension). By contrast, bulk flow in the phloem is between regions of differing *positive* pressure potentials (turgor pressure).

Water and ions move across the root cell plasma membrane

The movement of water and mineral ions across a root cell plasma membrane can be impeded for two reasons:

- The membrane is hydrophobic, whereas water and mineral ions are polar.
- Some mineral ions must be moved against their concentration gradients.

However, as we saw in Chapter 6, membrane proteins assist with the movement of materials across membranes:

- *Aquaporins.* Aquaporins (see Figure 6.11) are located in both the plasma membrane and the tonoplast (vacuolar membrane) of a plant cell. Aquaporins allow water to diffuse rapidly across these membranes. The abundance of aquaporins in a particular cell depends on that cell's need to obtain and retain water, and can vary with environmental conditions. The permeability of some aquaporins also can be regulated. Alterations in aquaporin abundance and permeability change the *rate* of osmosis across the membrane. Note that water movement through aquaporins is always

The cells of this plant have low turgor pressure and the plant is wilted.
The water potential of cells of this plant is zero because the negative solute potential is balanced by an equally positive pressure potential. The plant is upright because its cells are turgid.

35.3 A Wilted Plant A plant wilts when the pressure potential in its cells (the turgor pressure) is low.

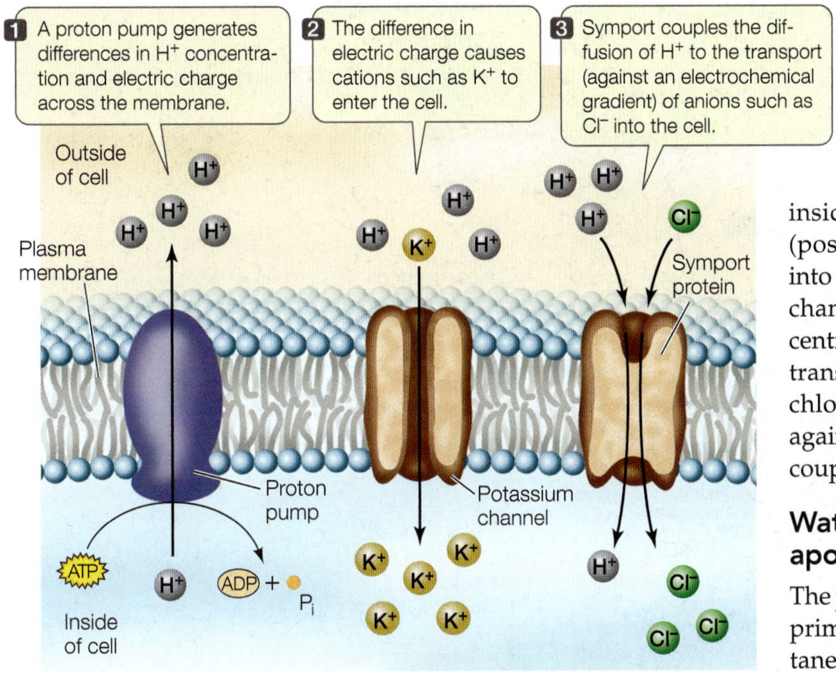
1 A proton pump generates differences in H+ concentration and electric charge across the membrane.
2 The difference in electric charge causes cations such as K+ to enter the cell.
3 Symport couples the diffusion of H+ to the transport (against an electrochemical gradient) of anions such as Cl− into the cell.
Outside of cell
Plasma membrane
Symport protein
Proton pump
Potassium channel
ATP
ADP + Pi
Inside of cell

35.4 The Proton Pump in Transport of K^+ and Cl^- The active transport of hydrogen ions (H^+) out of the cell by the proton pump (1) drives the movement of both cations (2) and anions (3) into the cell.

passive: from a region of higher water potential to one of lower water potential.

- *Ion channels and pumps.* When the concentration of a charged ion in the soil is greater than that in the plant, transport proteins can move the ions into the plant by facilitated diffusion, which is a passive process (see Section 6.3). The concentrations of most ions in the soil solution, however, are lower than those inside the plant. In these cases the plant must actively take up ions *against* their concentration gradients—a process that requires energy (Section 6.4).

Electric charge differences also play a role in the uptake of mineral ions. For example, a negatively charged ion that moves into a negatively charged compartment is moving against an *electrical gradient*, and this requires energy. Concentration and electrical gradients combine to form an electrochemical gradient (see Section 45.2). Uptake against an electrochemical gradient involves active transport, which requires energy and specific transport proteins.

Unlike animals, plants do not have a sodium–potassium pump (see Section 6.4) to drive active transport. Rather, plants have a **proton pump**, which uses energy obtained from ATP to move protons out of the cell against a proton concentration gradient (**Figure 35.4, step 1**). Because protons (H^+) are positively charged, their accumulation outside the cell has two results:

- An electrical gradient is created, with the region outside the cell more positively charged than the inside.
- A proton concentration gradient develops, with more protons outside the cell than inside.

Both the electrical gradient and the concentration gradient assist with the movement of other ions into the cell. Because the inside of the cell is more negative than the outside, cations (positively charged ions) such as potassium (K^+) can move into the cell by facilitated diffusion through specific membrane channels (**Figure 35.4, step 2**). In addition, the proton concentration gradient can be harnessed to drive secondary active transport, in which anions (negatively charged ions) such as chloride (Cl^-) are moved into the cell. These ions can move against the electrochemical gradient because symport proteins couple their movement with that of H^+ (**Figure 35.4, step 3**).

Water and ions pass to the xylem by way of the apoplast and symplast

The journey from the soil through the roots to the xylem occurs primarily by one of two pathways, either separately or simultaneously: the fast lane (called the apoplast) and the slow(er) lane (called the symplast) (**Figure 35.5**):

- The **apoplast** (Greek *apo*, "away from"; *plast*, "living material") consists of the cell walls, which lie outside the plasma membranes, and the intercellular spaces (spaces between cells) that are common in many plant tissues. The apoplast is a continuous meshwork through which water and dissolved substances can flow without ever having to cross a membrane. Movement of materials through the apoplast is thus unregulated and rapid.
- The **symplast** (Greek *sym*, "together with") is the continuous cytoplasm of the living cells, which are connected by plasmodesmata (see Figure 7.19). The selectively permeable plasma membranes of the root cells control access to the symplast, so movement of water and dissolved substances into the symplast is tightly regulated.

Water and minerals that pass from the soil solution through the apoplast can travel as far as the endodermis, the innermost

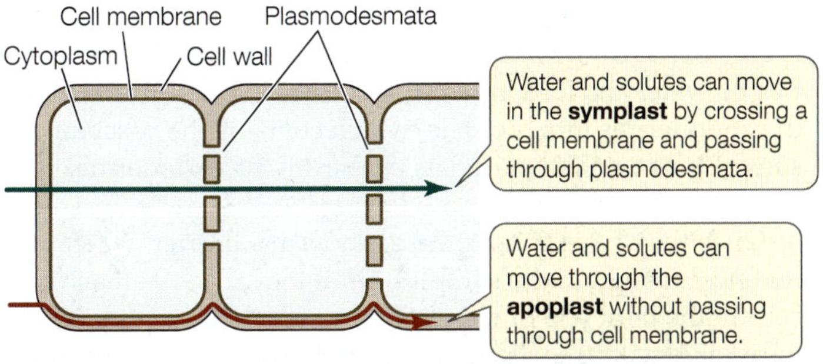
Cell membrane
Plasmodesmata
Cytoplasm
Cell wall
Water and solutes can move in the **symplast** by crossing a cell membrane and passing through plasmodesmata.
Water and solutes can move through the **apoplast** without passing through cell membrane.

35.5 Apoplast and Symplast Plant cell walls and intercellular spaces constitute the apoplast. The symplast comprises the living cells, which are connected by plasmodesmata. To enter the symplast, water and solutes must pass through a plasma membrane. No such selective barrier limits movement through the apoplast.

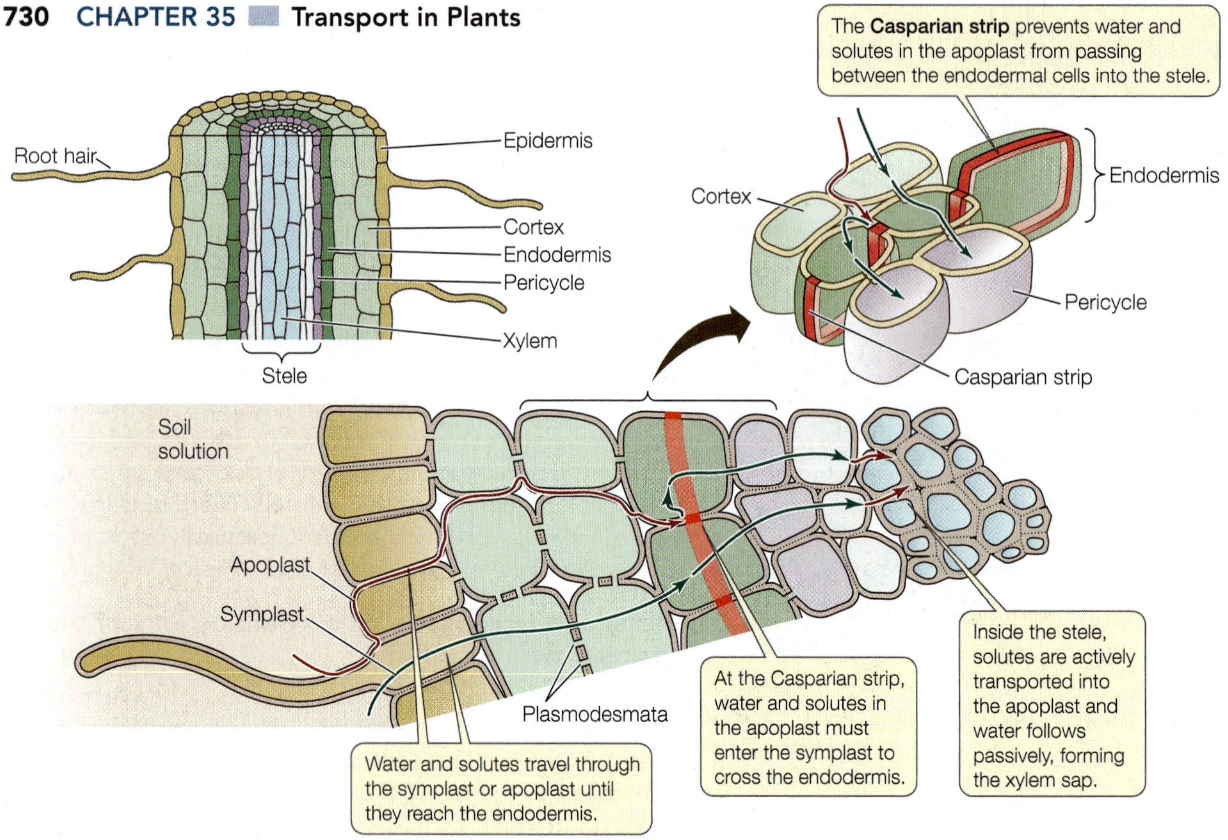

The **Casparian strip** prevents water and solutes in the apoplast from passing between the endodermal cells into the stele.
Root hair
Epidermis
Cortex
Endodermis
Pericycle
Xylem
Stele
Cortex
Endodermis
Pericycle
Casparian strip
Soil solution
Apoplast
Symplast
Plasmodesmata
Water and solutes travel through the symplast or apoplast until they reach the endodermis.
At the Casparian strip, water and solutes in the apoplast must enter the symplast to cross the endodermis.
Inside the stele, solutes are actively transported into the apoplast and water follows passively, forming the xylem sap.

35.6 Pathways to the Root Xylem Water and solutes can move into the root through the symplast or the apoplast until they reach the endodermis (shown in dark green); there the water and solutes must enter the symplast to bypass the Casparian strip (red), a region of the endodermal cell wall that is impregnated with the water-repelling substance suberin. Inside the stele, the water and solutes enter the xylem (blue).

Go to Activity 35.1 Apoplast and Symplast of the Root
Life10e.com/ac35.1

layer of the root cortex (**Figure 35.6**; see Section 34.3). The endodermis is distinguished from the rest of the ground tissue by the presence of the **Casparian strip**. This waxy, suberin-impregnated region of the endodermal cell wall forms a hydrophobic belt around each endodermal cell where it is in contact with other endodermal cells. The Casparian strip acts as a seal that prevents water and ions from moving through apoplastic spaces between the endodermal cells (see Figure 35.6). Therefore all water and ions must enter the symplast in order to cross the endodermis into the stele, which contains the vascular tissues of the root. The materials pass from the endodermal cells to cells in the stele via plasmodesmata.

Once they have passed the endodermal barrier, water and minerals remain in the symplast until they reach parenchyma cells in the pericycle or xylem. These cells then actively export mineral ions into the apoplast of the stele. As the concentrations of mineral ions in the apoplast increase, its water potential becomes more negative. Consequently, water moves out of the cells and into the apoplast by osmosis. In other words, ions are transported actively, and water follows passively. The end result is that water and minerals end up in the xylem, where they constitute the **xylem sap**.

RECAP 35.1

Differences in water potential govern the osmotic flow of water from the soil into the plant stele; this is a passive process. Uptake of minerals from the soil that occurs along an electrochemical gradient is an active process requiring energy and membrane transport proteins. Water and minerals can move into the root through either the apoplast or the symplast, but must enter and leave via the symplast to reach the xylem.

- What distinguishes water potential, solute potential, and pressure potential? **See p. 727 and Figure 35.2**
- Why is the cell wall important in determining the direction of water movement and plant form? **See pp. 727–728**
- What are aquaporins, and why are they needed? **See p. 728**
- What are the differences between the apoplast and the symplast? **See p. 729 and Figures 35.5 and 35.6**

So far we've described the movement of water and minerals into plant roots and their entry into the root xylem. How does the xylem sap move once it is in the xylem?

35.2 How Are Water and Minerals Transported in the Xylem?

Water has arrived in the xylem—and must travel "uphill" from there. Before considering the ascent of water and minerals to the leaves, reacquaint yourself with the cells that make up the xylem: the tracheids and vessel elements (see Figure 34.7A and B). Recall that these xylem cells are dead and lack all cell contents. When fused end to end, they form long tubular "straws" of lignified cell walls called **xylem vessels**. These vessels provide both structural support and the rigidity needed to maintain a gradient of pressure.

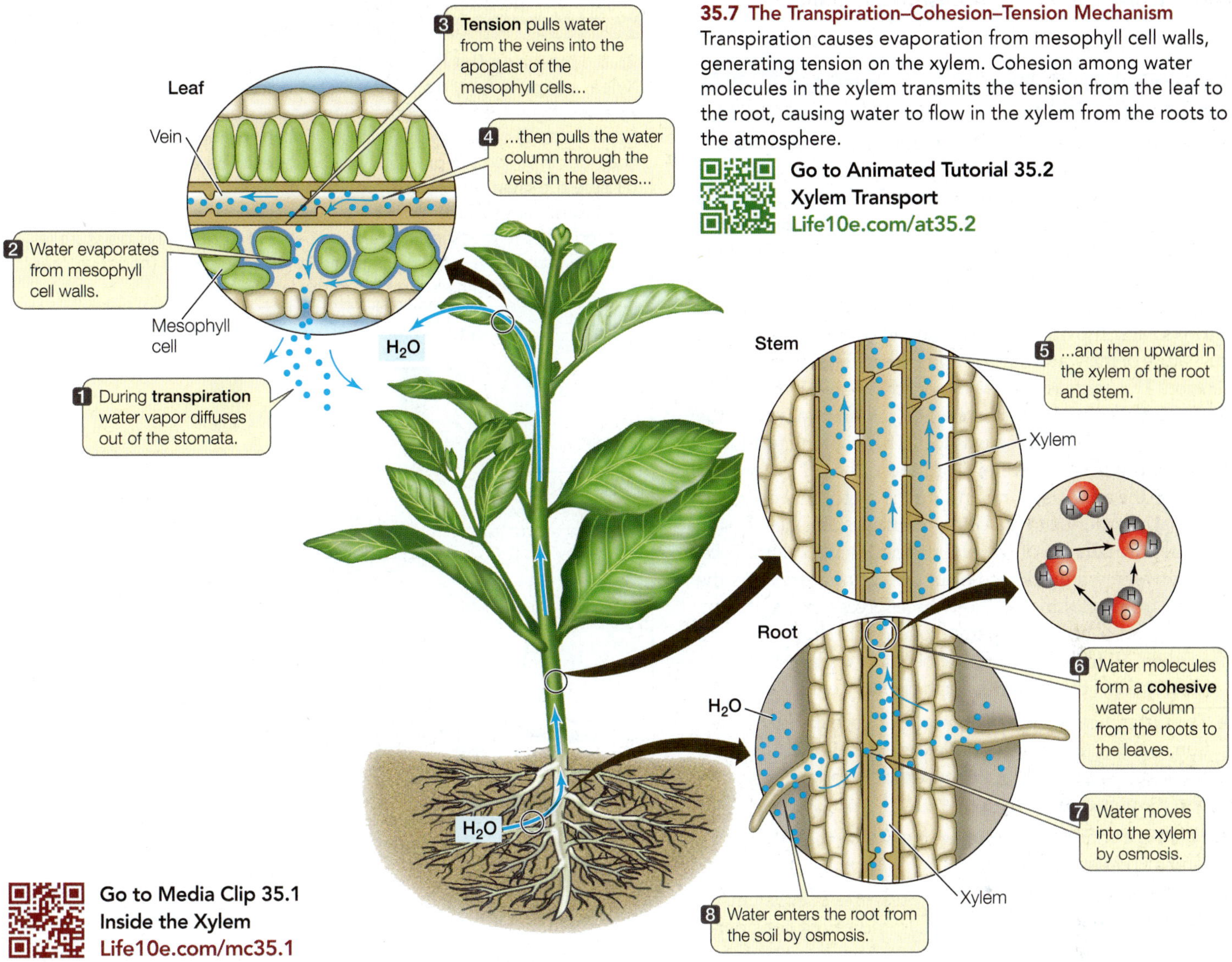

Leaf
Vein
Mesophyll cell
3 Tension pulls water from the veins into the apoplast of the mesophyll cells...
4 ...then pulls the water column through the veins in the leaves...
2 Water evaporates from mesophyll cell walls.
1 During transpiration water vapor diffuses out of the stomata.
H_2O
Stem
5 ...and then upward in the xylem of the root and stem.
Xylem
Root
H_2O
6 Water molecules form a cohesive water column from the roots to the leaves.
7 Water moves into the xylem by osmosis.
8 Water enters the root from the soil by osmosis.
Xylem
H_2O

35.7 The Transpiration–Cohesion–Tension Mechanism Transpiration causes evaporation from mesophyll cell walls, generating tension on the xylem. Cohesion among water molecules in the xylem transmits the tension from the leaf to the root, causing water to flow in the xylem from the roots to the atmosphere.

Go to Animated Tutorial 35.2
Xylem Transport
Life10e.com/at35.2

Go to Media Clip 35.1
Inside the Xylem
Life10e.com/mc35.1

Consider the magnitude of what the xylem accomplishes when it transports large amounts of water over great distances within a plant. A single maple tree 15 meters tall was estimated to have some 177,000 leaves, with a total leaf surface area of 675 square meters—about one and a half times the area of a basketball court. During a summer day, that tree loses 220 liters of water *per hour* to the atmosphere by evaporation from the leaves. So to prevent wilting, the xylem needs to transport 220 liters of water up to 15 meters from the roots to the leaves every hour. (By comparison, a 50-gallon drum holds 189 liters.)

Until the twentieth century, scientists proposed two possible mechanisms for moving water through the xylem: upward pressure by living cells, and capillary action. Both of these possibilities were largely ruled out by experiments:

- A simple experiment in 1893 ruled out the hypothesis that root cells might initiate a *pumping mechanism* to propel water upward. A tree was cut at its base and the sawed-off part was placed in a vat containing a solution of poison that killed living cells. The poison rose up the trunk, killing any living cells it encountered along the way. The experiment demonstrated that a living pump in the root is not necessary to push the xylem sap up a tree. Because the roots were absent, it was clear that they are not involved in xylem movement. Furthermore, when the poison sap reached the leaves, they died and all upward movement of the solution stopped, showing that living leaves are necessary for water to move in the xylem.
- Because of its surface tension (see Section 2.4) and adhesive forces between water and its container, water will move up a narrow column by a mechanism called *capillary action*. Capillary action was ruled out as a primary mechanism for upward xylem sap transport when calculations showed that xylem vessels (at 100 micrometers in diameter) are too wide to get water to the top of a 15-meter tree in this fashion. In fact, the maximum height for a water column raised by capillary action alone in a 100-micrometer tube would be only 0.15 meters.

The transpiration–cohesion–tension mechanism accounts for xylem transport

The current model of xylem transport involves three processes (**Figure 35.7**):

- *Transpiration* of water molecules from the leaves by evaporation
- *Tension* in the xylem sap resulting from transpiration from the leaves

- *Cohesion* of water molecules in the xylem sap, from the leaves to the roots

The relative amount of water vapor in the atmosphere is lower than that in the leaf. Because of this difference, water vapor diffuses from the intercellular spaces of the leaf to the outside air, in a process called **transpiration**. Within the leaf blade, water evaporates from the moist walls of the mesophyll cells and enters the intercellular spaces. As water evaporates from the aqueous film coating each cell, the film shrinks back into tiny spaces in the cell walls, increasing the curvature of the water surface and thus increasing its surface tension. This increased *tension* (negative pressure potential) in the surface film draws more water into the walls from the cells, replacing that which was lost. The resulting tension in the mesophyll draws water from the xylem of the nearest vein into the apoplast surrounding the mesophyll cells. The removal of water from the veins, in turn, establishes tension on the entire column of water contained in the xylem. *Cohesion* between water molecules in the column prevents the column from breaking. So these three forces—transpiration, tension, and cohesion—operate together to draw water up the xylem, all the way from the roots to the leaves.

Each part of this theory is supported by evidence:

- The difference in water potential between the soil solution and the air is huge, on the order of –100 MPa. This difference should generate more than enough tension to pull a water column up a tree.
- There is a continuous column of water in the xylem, which is caused by cohesion.
- Actual measurements of xylem pressures in cut stems show negative pressure potentials, indicating considerable tension in the xylem (**Figure 35.8**).

The transpiration–cohesion–tension mechanism accounts for the movement of water through the xylem. Dissolved mineral ions are carried along with the water to all of the plant's living tissues, where the ions are used for various cellular processes (see Chapter 36 for more on plant nutrition). In addition to promoting the transport of minerals, transpiration has an added benefit of cooling a plant's leaves. The evaporation of water from mesophyll cells consumes heat, thereby decreasing the leaf temperature. A farmer can hold a leaf between thumb and forefinger to estimate its temperature; if the leaf doesn't feel cool, that means transpiration is not occurring and it must be time to water.

RESEARCH**TOOLS**

35.8 Measuring the Pressure of Xylem Sap with a Pressure Chamber Xylem sap pulls away from a cut stem because the pressure in the intact xylem is lower than that of the atmosphere. The negative pressure potential originally present in the plant can be measured in a pressure chamber in which the pressure can be raised. The cut surface remains outside the chamber. As gas pressure increases, the xylem sap is pushed back to the cut surface. When the sap first becomes visible again at the cut surface, the pressure in the chamber is recorded. This pressure is equal in magnitude but opposite in sign to the tension (negative pressure potential) originally present in the xylem.

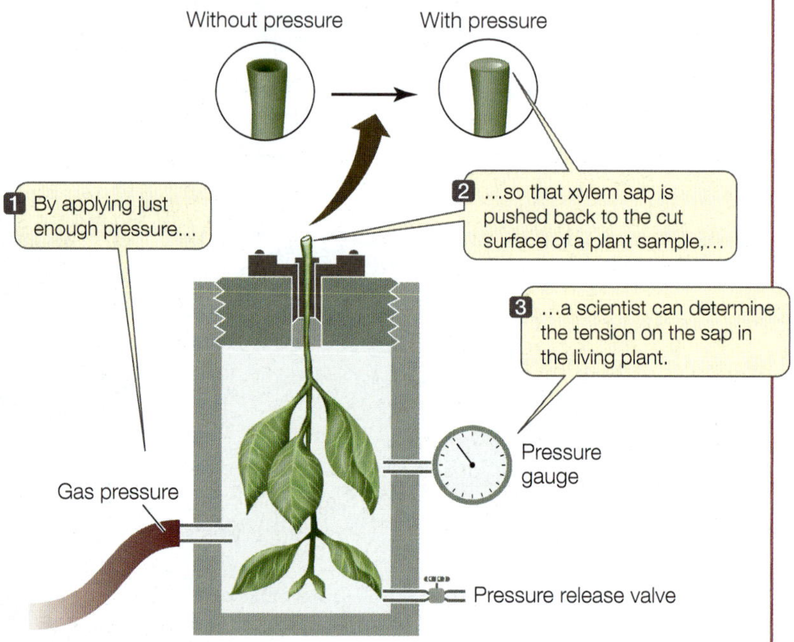
Without pressure
With pressure
1 By applying just enough pressure…
2 …so that xylem sap is pushed back to the cut surface of a plant sample,…
3 …a scientist can determine the tension on the sap in the living plant.
Pressure gauge
Gas pressure
Pressure release valve

Although transpiration provides the driving force for the transport of water and minerals in the xylem, it also results in the loss of tremendous quantities of water from the plant. How plants control this loss will be the subject of the next section.

RECAP 35.2

The transpiration–cohesion–tension mechanism explains the ascent of xylem sap. Transpiration draws water out of leaves, resulting in tension that pulls water from the xylem. Because of cohesion between water molecules, water is pulled passively through the xylem vessels in continuous columns, always toward a region with lower pressure potential.

- What are the roles of transpiration, cohesion, and tension in xylem transport? **See pp. 731–732 and Figure 35.7**
- What experiment rules out the role of pressure from the roots in the upward flow of water in xylem? **See p. 731**

35.3 How Do Stomata Control the Loss of Water and the Uptake of CO_2?

The epidermis of leaves and stems secretes a waxy cuticle, which is impermeable to water and thus helps minimize the loss of water from transpiration. However, the cuticle is also impermeable to carbon dioxide. The cuticle poses a dilemma: how can the plant balance its need to retain water with its need to obtain CO_2 for photosynthesis?

An elegant compromise has evolved in plants in the form of pores called **stomata** (singular *stoma*) in the epidermis of their leaves. A pair of specialized epidermal cells, called **guard cells**, controls the opening and closing of each stoma (**Figure 35.9A**). When the stomata are open, CO_2 can enter the leaf by diffusion—but water vapor diffuses out of the leaf at the same time. Closed stomata prevent water loss but also exclude CO_2 from the leaf.

Most plants open their stomata only when the light intensity is sufficient to maintain a moderate rate of photosynthesis. At night, when darkness precludes photosynthesis, their stomata are closed; no CO_2 is needed, and water is conserved. Even during the day, the stomata close if water is being lost at too rapid a rate.

Stomata are ancient structures; they have been found in plant fossils that are more than 400 million years old. For this reason they are thought to predate the evolution of leaves. Stomata are found in all vascular plants and in many nonvascular plants, including mosses (but not liverworts; see Chapter 28).

The stoma and guard cells seen in Figure 35.9A are typical of eudicots. Monocots typically have specialized epidermal cells associated with their guard cells. However, the principle of operation, which we will now describe in more detail, is the same for both monocot and eudicot stomata.

The guard cells control the size of the stomatal opening

The opening and closing of stomata are regulated by several environmental factors, including light, CO_2 levels, temperature, and water availability. For example, light causes the stomata of most plants to open, admitting CO_2 for photosynthesis. A low level of CO_2 in the intercellular spaces of the leaf will trigger the opening of the stomata as well, allowing the uptake of more CO_2. However, to conserve water, plants will close their stomata on hot days and when water availability is limited.

Stomata can respond to these environmental stimuli in a matter of minutes. How does this important biological process happen so rapidly? Stomata open and close rapidly in response to changes in turgor pressure in the guard cells. These changes in turgor pressure arise in response to changes in K^+ concentrations in the guard cells. These concentration changes are triggered by environmental cues.

In sunlight, a pigment in the guard cell membrane absorbs blue light, which activates a proton pump that actively transports H^+ out of the guard cells and into the apoplast of the surrounding epidermis. The resulting electrochemical gradient drives K^+ into the guard cells, making the water potential of the guard cells more negative (**Figure 35.9B**). Negatively charged chloride ions and organic ions move into and out of the guard cells along with the potassium ions, functioning to maintain the electrical balance and contributing to the change in the solute potential of the guard cells. Water enters by osmosis (guard cell membranes are particularly rich in aquaporin protein channels), which increases the pressure potential of the guard cells. The cellulose microfibrils in the guard cell walls are arranged so that the shape of the cells changes in response to the increase in pressure potential, and a gap—the stoma—appears between them.

In the absence of sunlight, the stomata of most plants close. Lacking blue light, the proton pump becomes less active, potassium and chloride ions diffuse passively out of the guard cells, and water follows by osmosis (see Figure 35.9B, step 3). These changes lower the pressure potential in the guard cells so that they sag together, closing the stoma.

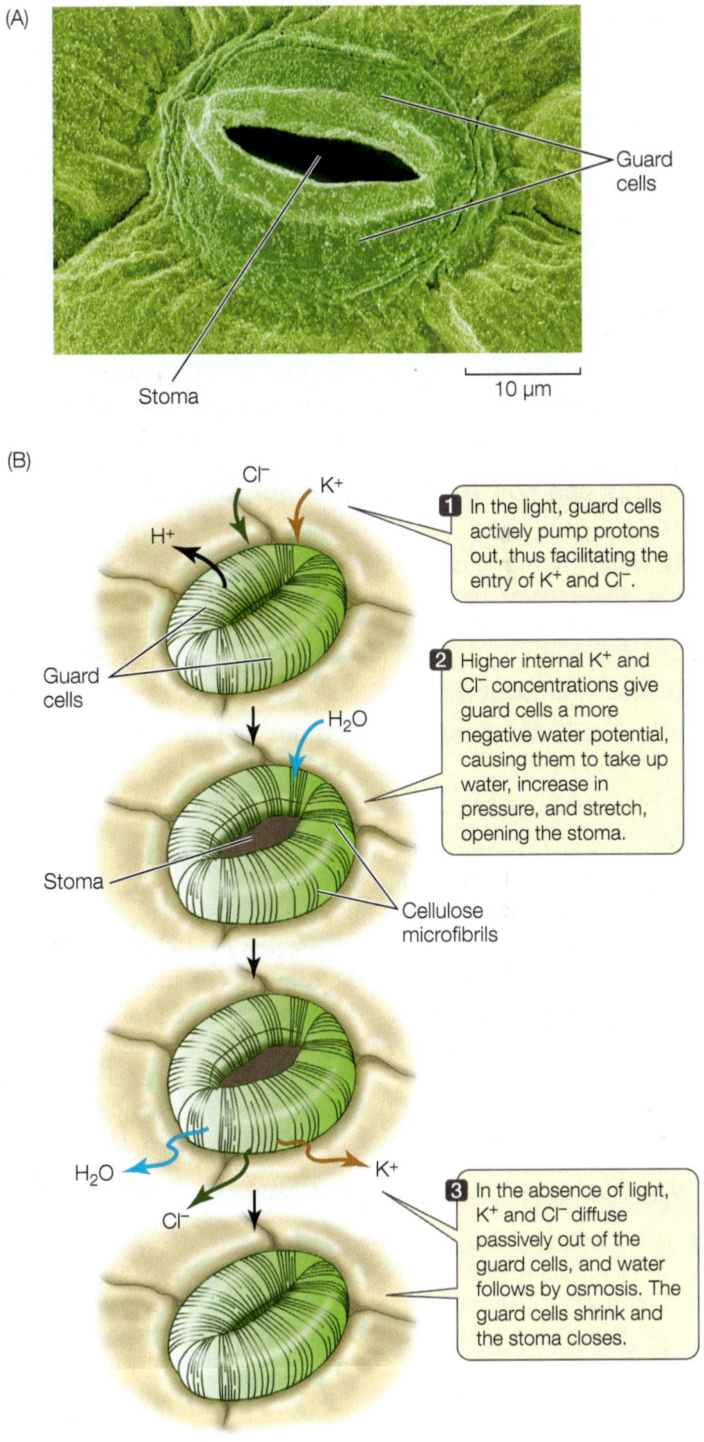

(A)
Guard cells
Stoma
10 μm
(B)
Cl^-
K^+
H^+
1 In the light, guard cells actively pump protons out, thus facilitating the entry of K^+ and Cl^-.
Guard cells
H_2O
2 Higher internal K^+ and Cl^- concentrations give guard cells a more negative water potential, causing them to take up water, increase in pressure, and stretch, opening the stoma.
Stoma
Cellulose microfibrils
H_2O
K^+
Cl^-
3 In the absence of light, K^+ and Cl^- diffuse passively out of the guard cells, and water follows by osmosis. The guard cells shrink and the stoma closes.

35.9 Stomata (A) Scanning electron micrograph of an open stoma formed by two sausage-shaped guard cells. (B) Potassium ion concentrations affect the water potential of the guard cells, controlling the opening and closing of stomata. Negatively charged ions (e.g., Cl^-) that accompany K^+ maintain electrical balance and contribute to the changes in water potential that open and close the stomata.

Stomata respond not only to light but to water availability. Water stress is a common problem for plants, especially on hot, windy days. On hot, dry days plants will close their stomata even when the sun is shining. The water potential of the mesophyll cells is the cue for this protective response. If the mesophyll is too dehydrated—that is, if its water potential is too negative—its cells release a plant hormone called abscisic acid. Abscisic acid causes the guard cells to close the stomata and prevent further drying of the leaf. Although this protective response reduces the rate of photosynthesis, it protects the plant from wilting.

Not all plants respond to environmental cues in the same way. CAM plants such as cacti (see Section 10.4) are often found in the desert. To avoid excess water loss, these plants close their stomata during the day and open them at night. They have a metabolic adaptation that allows them to accumulate CO_2 at night when their stomata are open, then release the CO_2 inside the plant for photosynthesis during the day.

Plants can control their total numbers of stomata

Individual stomata are tiny, yet plants can lose large amounts of water. A single corn plant can lose 2 quarts of water per day. Plant water loss can be great because there can be a huge numbers of stomata on every leaf: up to 250,000 per square inch of leaf surface! Plants limit water loss by controlling stomata in two very different ways:

- By regulating stomatal opening and closing (described above)
- By controlling the total number of stomata

Whereas the process of opening and closing stomata can occur in minutes, as we saw above, the process of controlling the total number of stomata occurs over days or weeks. Trees can reduce their total number of stomata by shedding some of their leaves. Other plants reduce the numbers of stomata on new leaves that develop during a drought.

CO_2 levels can also trigger changes in stomatal density. If the model plant *Arabidopsis* is exposed to high CO_2 levels, its new leaves have fewer stomata than they would under normal conditions. Why do you think this might be advantageous?

RECAP 35.3

CO_2, which is needed for photosynthesis, enters leaves via tiny pores called stomata. Stomata also permit the loss of water by transpiration. Guard cells open or close stomata in response to a variety of environmental cues. Plants control their total numbers of stomata by shedding leaves or by altering the density of stomata on new leaves.

- What is the role of K^+ ions in the functioning of guard cells? **See p. 733 and Figure 35.9**
- Describe how environmental cues (such as CO_2 level and water availability) can affect stomatal function and density during the life of a plant. **See p. 734**

Stomata are normally open during daylight hours, allowing photosynthesis—the production of carbohydrates from CO_2 and water. In the next section we'll see how the products of photosynthesis are delivered to other parts of the plant, supporting plant growth.

35.4 How Are Substances Translocated in the Phloem?

Photosynthesis occurs primarily in the leaf (see Figure 10.1). The carbohydrate products of photosynthesis (mainly sucrose) diffuse to the nearest small vein (composed of xylem and phloem), where they are actively transported into sieve tube elements of the phloem. The movement of carbohydrates and other solutes through the phloem is called **translocation**. The products of photosynthesis are called **photosynthates**, and the content of the phloem is called the **phloem sap**.

The photosynthates and other substances are translocated from sources to sinks.

- A **source** is an organ (such as a mature leaf or a storage root) that *produces*, by photosynthesis or by digestion of stored reserves, more sugars than it requires.
- A **sink** is an organ (such as a root, flower, developing fruit, or immature leaf) that *consumes* sugars for its own growth and storage needs of the plant.

A given organ can be a sink at some times and a source at others. An example is the roots of maple trees, which store sugars sent down from the leaves during one growing season, then send sugars back upward to support the emergence of new leaves the following spring.

Sucrose and other solutes are carried in the phloem

Evidence that the phloem carries sucrose and other solutes was first obtained in the 1600s when the Italian scientist Marcello Malpighi removed a ring of bark from the trunk of a tree—that is, he "girdled" the tree. The bark contained the phloem, while the xylem in the underlying wood remained intact. Over time, the bark in the region above the girdle swelled:

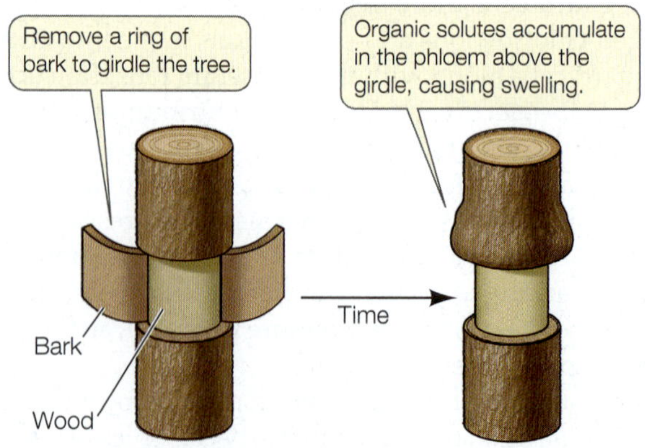
Remove a ring of bark to girdle the tree.
Organic solutes accumulate in the phloem above the girdle, causing swelling.
Bark
Wood
Time

Malpighi correctly concluded that a solution coming from the leaves above the girdle was trapped in the bark. Later the bark below the girdle died, presumably because it no longer received nutrients from the leaves. Eventually the roots, and then

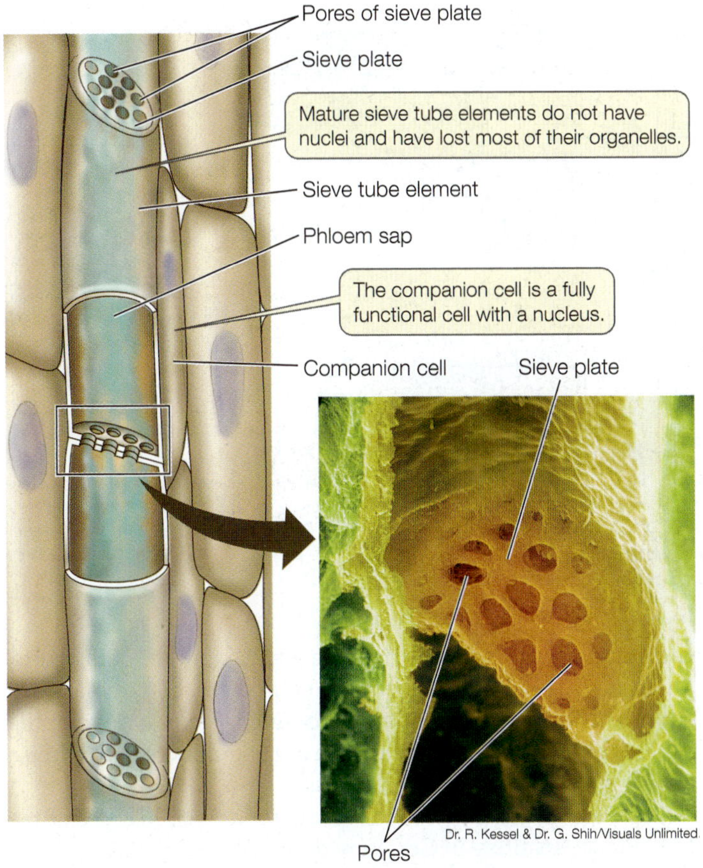
Pores of sieve plate
Sieve plate
Mature sieve tube elements do not have nuclei and have lost most of their organelles.
Sieve tube element
Phloem sap
The companion cell is a fully functional cell with a nucleus.
Companion cell
Sieve plate
Dr. R. Kessel & Dr. G. Shih/Visuals Unlimited
Pores

35.10 Sieve Tubes Individual sieve tube elements join together to form long tubes that transport carbohydrates and other nutrient molecules throughout the plant body in the phloem. Sieve plates form at the ends of each sieve tube element, and phloem sap passes through the pores in the sieve plate.

the entire tree, died—suggesting that sugar transport might occur in the phloem.

The cells that make up the phloem's conducting tubes are sieve tube elements (see Figure 34.7C). Like the vessel elements in the xylem, sieve tube elements meet end to end. However, unlike vessel elements, whose end walls are broken down as they mature, sieve tube elements retain their end walls. Communication between sieve tube elements is achieved by plasmodesmata in their end walls. During sieve tube development, the diameter of these plasmodesmata increases 10- to 100-fold, resulting in pores that allow the flow of phloem sap between neighboring cells. Because the end walls of sieve tube elements look and function like sieves, they are called **sieve plates** (**Figure 35.10**).

What happens next is truly remarkable and makes sieve tube elements among the most unusual cell types in nature. As the holes in the sieve plates expand, most of the cell contents are lost, including the nucleus, Golgi apparatus, and most of the ribosomes and cytoskeleton. Despite this, sieve tube elements live for an entire growing season in deciduous trees, and for decades in some other plants. How can sieve tube elements live for so long with no nucleus? The answer is that each sieve tube element has one or more **companion cells** (see Figure 35.10). Companion cells are produced as daughter cells along with the sieve tube elements when parent cells divide. Numerous plasmodesmata link a companion cell with its neighboring sieve tube element. Companion cells retain all their organelles and provide all the components needed to maintain the sieve tube elements—they may be thought of as the "life support systems" of the sieve tube elements.

Plant biologists in the twentieth century used aphids to precisely analyze the contents of the phloem. Aphids are insects that feed on plants by drilling into sieve tube elements with a specialized organ, the stylet. The pressure potential in the sieve tube is higher than that outside the plant, so the phloem contents are forced through the stylet into the aphid's digestive tract. So great is the pressure that some of the liquid is forced through the insect's body and out its anus. If an aphid is frozen in the act of feeding, its body can be chopped off the plant stem, leaving the stylet intact. Phloem sap continues to flow from the stylet for hours, and can be collected for analysis.

Sieve tube element
The aphid's stylet has successfully penetrated the sieve tube.
Sap droplet
Longistigma caryae (aphid)
Stylet

These and other experiments led to several important observations:

- Sucrose makes up 90 percent of the phloem sap solutes. The phloem sap also contains hormones, small molecules such as amino acids, mineral nutrients, and viruses.
- The flow rate can be very high, as much as 100 centimeters per hour.
- Different sieve tube elements conduct their contents in different directions—for example, up or down the stem. Therefore the overall movement in the phloem is bidirectional.
- The movement of phloem sap requires living cells, in contrast to movement in the xylem.

The pressure flow model appears to account for translocation in the phloem

As noted above, phloem sap flows under positive pressure through the sieve tubes. It moves by bulk flow (i.e., from a region of higher pressure potential to a region of lower pressure potential) from one sieve tube element to the next. We need to

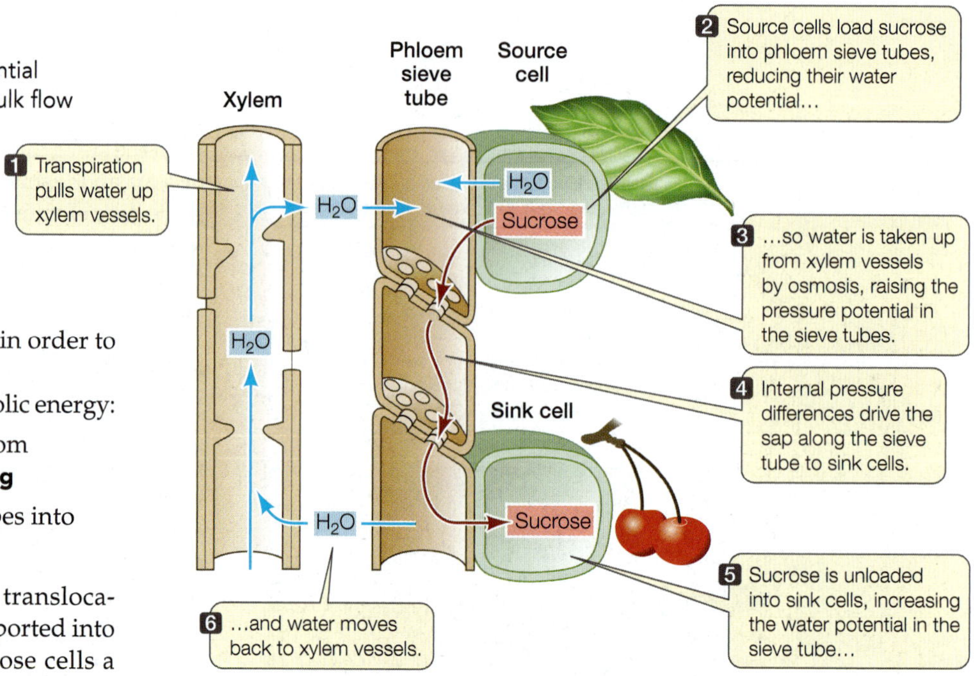
Xylem
Phloem sieve tube
Source cell
Sink cell
H₂O
Sucrose
1 Transpiration pulls water up xylem vessels.
2 Source cells load sucrose into phloem sieve tubes, reducing their water potential...
3 ...so water is taken up from xylem vessels by osmosis, raising the pressure potential in the sieve tubes.
4 Internal pressure differences drive the sap along the sieve tube to sink cells.
5 Sucrose is unloaded into sink cells, increasing the water potential in the sieve tube...
6 ...and water moves back to xylem vessels.

35.11 The Pressure Flow Model Water potential differences produce a pressure gradient and bulk flow of phloem sap from sources to sinks.

Go to Animated Tutorial 35.3 The Pressure Flow Model
Life10e.com/at35.3

understand how this pressure is generated in order to understand translocation in the phloem.

Two steps in translocation require metabolic energy:

- Transport of sucrose and other solutes from sources into the sieve tubes; called **loading**
- Removal of the solutes from the sieve tubes into sinks; called **unloading**

According to the **pressure flow model** of translocation in the phloem, sucrose is actively transported into sieve tube elements at a source, giving those cells a greater sucrose concentration than the surrounding cells. Therefore the cells develop a lower solute potential—a more negative Ψ_s—at which point water enters the sieve tube elements from xylem vessels by osmosis. The entry of this water causes a greater pressure potential (turgor pressure) at the source end of the sieve tube, so that the entire fluid content of the sieve tube is pushed toward the sink end of the tube—in other words, the sap moves by bulk flow in response to a pressure gradient (Figure 35.11). In the sink, the sucrose is unloaded both passively and by active transport, and water moves back to the xylem vessels. In this way the gradients of solute potential and pressure potential, which are needed for movement of the phloem sap, are maintained.

Sugars and other solutes move from the mesophyll cells in the leaf to the phloem by two general routes: apoplastic and symplastic. The exact details of these routes vary widely among plant species. In many plants, sugars and other solutes follow the **apoplastic pathway**: solutes move out of the mesophyll cells and then diffuse through the apoplast to the sieve tubes. Specific sugars and amino acids are then actively transported into the sieve tube elements. Active transport across membranes (see Section 6.4) allows plants to regulate which specific substances enter the phloem. In contrast, in the **symplastic pathway**, solutes remain within the symplast and pass through plasmodesmata all the way from the mesophyll cells to the sieve tube cells. Because no membranes are crossed in the symplastic pathway, the solutes are loaded into the phloem by mechanisms other than active membrane transport.

The pressure flow model of translocation in the phloem is contrasted with the transpiration–cohesion–tension model of xylem transport in Table 35.1.

Upon arriving in sink regions, solutes are actively transported *out* of the sieve tube elements and into the surrounding tissues. This unloading serves several purposes: it helps maintain the gradient of solute potential and hence of pressure potential in the sieve tubes; it supplies carbohydrates and amino acids to developing organs, where they are used for growth; and it helps build up high concentrations of proteins and carbohydrates in storage organs such as storage roots, fruits, and seeds. The demand for fixed carbon can be very large in some tissues, such as a rapidly growing stem or a developing potato tuber. These tissues receive a lot of phloem sap, while a slowly growing organ receives less.

An experiment by Lothar Willmitzer and colleagues at the Max Planck Institute in Germany demonstrated the importance of sink strength in potato tubers (Figure 35.12). Sink strength is the relative ability of an organ to attract photosynthates. A potato tuber is formed from an underground stem that changes its function from transport to synthesis and storage of starch. Because the source of glucose to make this starch is sucrose from the phloem, there is a high demand for phloem unloading into the developing potato tuber. The enzyme invertase catalyzes the hydrolysis of sucrose to glucose and fructose, and the fructose is quickly converted to glucose. Willmitzer and colleagues transformed potato plants by introducing a gene that caused invertase to accumulate in the apoplast of the tuber tissue, where the enzyme accumulated to high levels. In these plants, the sucrose that was transported out of the phloem into the apoplast was rapidly hydrolyzed. The resulting glucose was taken up by the tuber cells and converted to starch. The action of the invertase in the apoplast lowered the sucrose concentration in the potato tissue and led to a high rate of phloem unloading (see Figure 35.11). As a result, the potato tubers

TABLE 35.1 Mechanisms of Sap Flow in Plant Vascular Tissues

	Xylem	Phloem
Driving force for bulk flow	Transpiration from leaves	Active transport of sucrose at source and sink
Site of bulk flow	Nonliving vessel elements and tracheids	Living sieve tube elements
Pressure potential in sap	Negative (pull from top; tension)	Positive (push from source; pressure)

INVESTIGATING**LIFE**

35.12 Manipulating Sucrose Transport from the Phloem

Does a change in the concentration of sucrose in a sink tissue affect the unloading of sucrose from the phloem? Lothar Willmitzer and colleagues genetically modified potato plants so that sucrose in the developing tubers was immediately hydrolyzed. This lowered the effective sucrose concentration in the "sink" for phloem unloading. The researchers then tested whether this affected tuber development, a reflection of sucrose unloading from phloem.[a]

HYPOTHESIS Reducing the sucrose concentration in a sink organ will increase the transport and unloading of sucrose from the phloem.

Method Plants were transformed with a gene for invertase, an enzyme that hydrolyzes sucrose.

The potato plant has underground tubers, modified stems that store starch.

Results

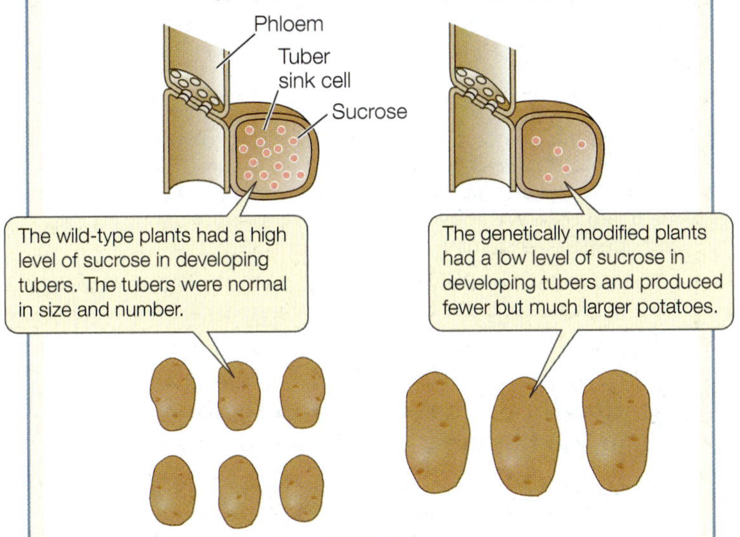

Wild-type plants
Transgenic plants
Phloem
Tuber sink cell
Sucrose
The wild-type plants had a high level of sucrose in developing tubers. The tubers were normal in size and number.
The genetically modified plants had a low level of sucrose in developing tubers and produced fewer but much larger potatoes.

CONCLUSION Increasing sink strength increases sucrose transport into developing tissues.

Go to **BioPortal** for discussion and relevant links for all INVESTIGATING**LIFE** figures.

[a]Sonnewald, U. et al. 1997. *Nature Biotechnology* 15: 794–797.

that developed on the transformed plants were larger than those of wild-type plants. This experiment demonstrated that transport into the apoplast is an efficient way for phloem cells to unload sucrose, and that the sucrose concentration gradient between the phloem and the apoplast influences the rate of unloading.

WORKING WITH**DATA**:

Manipulating Sucrose Transport from the Phloem

Original Paper

Sonnewald, U., N.-R. Hajirezaei, J. Kossmann, A. Heyer, R. Tretheway, and L. Willmitzer. 1997. Increased potato tuber size from apoplastic expression of yeast invertase. *Nature Biotechnology* 15: 794–797.

Analyze the Data

Plant tissues can be exporters (sources) or importers (sinks) of carbohydrates, mainly in the form of sucrose. Many of the foods grown for human consumption are sink tissues; we eat the crops' carbohydrates, mostly as starch. As a potato tuber develops from an underground stem, or stolon, it requires a lot of carbohydrate, which it obtains as sucrose from the phloem. Lothar Willmitzer and colleagues at the Max Planck Institute in Germany hypothesized that the rate of sucrose unloading from phloem to tuber is affected by the sucrose concentration difference (gradient) between the phloem and the developing tuber. To test their hypothesis, they used genetic engineering to express the gene for yeast invertase in developing tuber cells (see Figure 35.12).

Normal potato plants have invertase but do not express it at high levels in tubers. The yeast invertase gene used to transform the plants was coupled to a plant promoter that caused high expression of the enzyme in the developing tubers. Invertase hydrolyzes sucrose, so the researchers predicted that the concentration of sucrose would be lower in the tubers of transgenic plants (and the gradient steeper) than in wild-type plants. Experiments performed on the transgenic plants confirmed that increasing the sucrose concentration difference between the phloem and the tubers led to increased sucrose unloading from the phloem.

QUESTION 1

To test whether the introduced yeast invertase was having the predicted effects on sucrose levels, the researchers measured invertase activity and sugars in tubers of wild-type and transgenic plants. The results are in the first three rows of the table (mean ± standard deviation). Which group of plants had higher invertase activity? Carry out a statistical test to evaluate your conclusion (see Appendix B). How do the sugar content measurements relate to invertase activity?

QUESTION 2

The researchers estimated sucrose unloading from phloem by comparing the sizes of the tubers that developed on normal and transgenic plants, assuming that larger tubers must have unloaded more sucrose from the phloem. The tuber yields are shown in the last two rows of the table (mean ± standard deviation). What can you conclude about the distribution of sucrose to developing tubers?

	Wild type	Transgenic
Invertase activity (units/mg protein)	9 ± 4	598 ± 73
Sucrose (mg/g tuber weight)	15.8 ± 0.7	1.2 ± 0.2
Glucose (mg/g tuber weight)	2.1 ± 0.4	38.0 ± 4.3
Tubers per plant	6.4 ± 0.4	2.6 ± 0.1
Tuber weight (g/tuber)	18	57.3

Go to **BioPortal** for all WORKING WITH**DATA** exercises

The control of phloem sap travel from source to sink has great importance for humans. The parts of plants that we use for food, such as seeds (e.g., the rice in the opening story), fruits, and storage organs, are mostly sinks. Increasing the flow of sucrose into these organs as they develop can increase food production in crop plants. The experiment on potatoes, described above, is an important step toward understanding and manipulating phloem transport.

RECAP 35.4

Carbohydrates produced by photosynthesis are translocated from source to sink through the phloem by a pressure flow mechanism.

- Explain the difference between a source and a sink. See p. 734
- How does loading of sucrose at the source result in bulk flow toward the sink? See pp. 735–736 and Figure 35.11

?

What other methods are used to reduce water loss in agriculture?

ANSWER

Optimal plant growth depends on the presence of an appropriate amount of soil water near the roots during a plant's life cycle. In addition to using plants that are more water-efficient, plant growers can manipulate the environment to reduce water losses that are due to evaporation and soil runoff. Simply putting a pan out in the field and measuring the rate of evaporation can be a guide as to when to increase irrigation. Cheap sensors in the soil can measure water content and alert the grower to deficits or excess. Farmers can use laser technology to grade (shape) fields to reduce water runoff during irrigation.

CHAPTER SUMMARY 35

35.1 How Do Plants Take Up Water and Solutes?

- Water moves through biological membranes by osmosis, always moving toward regions with a more negative water potential. The **water potential** (Ψ) of a cell or solution is the sum of the **solute potential** (Ψ_s) and the **pressure potential** (Ψ_p). **Review Figure 35.2, ANIMATED TUTORIAL 35.1**
- **Turgid** plant cells have significant positive pressure potential because the rigid cell wall limits expansion of the cell. This positive pressure (**turgor pressure**) maintains the physical structure of many plant cells; if the pressure potential drops, the plant wilts.
- The movement of a solution due to a difference in pressure potential between two parts of a plant is called **bulk flow**.
- Aquaporins are channel proteins that facilitate movement of water molecules through biological membranes.
- Mineral uptake requires transport proteins. Some minerals enter the plant passively by facilitated diffusion; others enter by active transport. A **proton pump** provides energy for the active transport of many mineral ions across membranes in plants. **Review Figure 35.4**
- Water and minerals pass from the soil into the root by way of the **apoplast** and **symplast**, but must pass through the symplast to cross the endodermis and enter the xylem. The **Casparian strip** in the endodermis blocks movement of water and minerals through the apoplast. **Review Figures 35.5, 35.6, ACTIVITY 35.1**

35.2 How Are Water and Minerals Transported in the Xylem?

- Experiments proved that neither a root pump nor capillary action can alone account for the ascent of xylem sap in trees.
- Water transport in the xylem results from the combined effects of **transpiration**, cohesion, and tension—the transpiration–cohesion–tension mechanism. Evaporation from the leaf produces tension in the mesophyll cells, which pulls a column of water—held together by cohesion—up through the xylem from the root. **Review Figure 35.7, ANIMATED TUTORIAL 35.2**
- Transport in the xylem is passive. It does not require the expenditure of energy by the plant.

35.3 How Do Stomata Control the Loss of Water and the Uptake of CO_2?

- The waxy cuticle of plant epidermis is impermeable to both water and carbon dioxide. **Stomata** allow for carbon dioxide uptake (when open) while minimizing transpirational water loss (when closed).
- A pair of **guard cells** controls the size of the stomatal opening. A light-activated proton pump moves protons out of the guard cells to the walls of surrounding epidermal cells, setting up an electrochemical gradient that drives the transport of potassium ions into the guard cells. Water follows osmotically, swelling the guard cells and opening the stomata. **Review Figure 35.9**
- When threatened by dehydration, mesophyll cells release abscisic acid, which causes guard cells to close the stomata, even in the light.

35.4 How Are Substances Translocated in the Phloem?

- Products of photosynthesis, as well as some minerals, are translocated through sieve tubes in the phloem by way of living sieve tube elements. **Review Figure 35.10**
- **Translocation** in the phloem can proceed in both directions in the stem. Translocation requires a supply of ATP.
- Translocation in the phloem is explained by the **pressure flow model**: the difference in solute concentration between **sources** and **sinks** creates a difference in (positive) pressure potential along the sieve tubes, resulting in bulk flow. **Review Figure 35.11, Table 35.1, ANIMATED TUTORIAL 35.3**

Go to the Interactive Summary to review key figures, Animated Tutorials, and Activities
Life10e.com/is35

CHAPTER**REVIEW**

REMEMBERING

1. Osmosis
   a. requires ATP.
   b. results in the bursting of plant cells placed in pure water.
   c. can cause a cell to become turgid.
   d. is independent of solute concentrations.
   e. continues until the pressure potential equals the water potential.

2. Water potential
   a. is the difference between the solute potential and the pressure potential.
   b. is analogous to the air pressure in an automobile tire.
   c. is the movement of water through a membrane.
   d. determines the direction of water movement between cells.
   e. is defined as 1.0 MPa for pure water under no applied pressure.

3. Which statement about proton pumping across the plasma membrane of plants is *not* true?
   a. It requires ATP.
   b. The region inside the membrane becomes positively charged with respect to the region outside.
   c. It enhances the movement of K^+ ions into the cell.
   d. It moves protons out of the cell against a proton concentration gradient.
   e. It can drive the secondary active transport of negatively charged ions.

4. Which statement is *not* true?
   a. The symplast consists of the interconnected cytoplasm of living cells.
   b. Water can enter the stele without entering the symplast.
   c. The Casparian strips prevent water from moving through the apoplast between endodermal cells.
   d. The endodermis is a cell layer in the cortex.
   e. Water can move freely in the apoplast without entering cells.

5. Which of the following is *not* part of the transpiration–cohesion–tension mechanism?
   a. Water evaporates from the walls of mesophyll cells.
   b. Removal of water from the xylem exerts a pull on the water column.
   c. Water is remarkably cohesive.
   d. The wider the tube, the greater the tension its water column can withstand.
   e. At each step, water moves to a region with a more strongly negative water potential.

6. Which statement about phloem transport is *not* true?
   a. It takes place in sieve tubes.
   b. It depends on mechanisms for loading solutes into the phloem at sources.
   c. It stops if the phloem is killed by heat.
   d. A high pressure potential is maintained in the sieve tubes.
   e. At sinks, solutes are actively transported into sieve tube elements.

UNDERSTANDING & APPLYING

7. Shoot epidermal cells protect against excess water loss. How do they perform this function? What differences might you expect to find in the structure of the epidermis in stems, roots, and leaves?

8. Compare sources to sinks. Give examples of each. How might the distribution of sources and sinks change in the course of a year?

9. What is the minimum number of plasma membranes a water molecule would have to cross in order to get from the soil solution to the atmosphere by way of the stele? To get from the soil solution to a mesophyll cell in a leaf?

ANALYZING & EVALUATING

10. In the story that opened this chapter we saw that a mutation in the *HARDY* gene resulted in *Arabidopsis* plants with a more extensive root system and thicker leaves than wild-type plants. When the *HARDY* gene was isolated, it was found to encode a transcription factor that stimulates expression of genes for increased water-use efficiency. What phenotype due to a mutation in the *HARDY* gene would cause *Arabidopsis* plants to use water more efficiently? How would you investigate the effect of the *HARDY* mutation on stomata? What results would you expect?

11. Here are measurements of water potential (Ψ) in a 100-meter-tall tree and its surroundings:

Region	Ψ (MPa)
Soil water	–0.3
Xylem of root	–0.6
Xylem of trunk	–1.2
Inside of leaf	–2.0
Outside air	–58.5

    Gravity exerts a force of –0.01 MPa per meter of height above ground.
    a. Is the water potential in the leaf sufficiently low to draw water to the top of the tree?
    b. Would transpiration continue if soil water potential decreased to –1.0?
    c. What would you expect to happen to the xylem water potential if all of the stomata closed?

Go to BioPortal at **yourBioPortal.com** for Animated Tutorials, Activities, LearningCurve Quizzes, Flashcards, and many other study and review resources.

36 Plant Nutrition

CHAPTER**OUTLINE**

Nitrogen Sipper Corn plants, such as those being fertilized here, extract a lot of nitrogen compounds from the soil. Excess nitrogen the corn plants do not take up gets left behind in the soil, where it can be a pollutant.

CROPS SUCH AS RICE, WHEAT, AND CORN supply more than half of the human diet. But like other organisms, the plants themselves require good nutrition in order to grow and produce our food. One of the nutrients most often in short supply in soils is nitrogen. Fertilizer or manure can be used to overcome this deficiency, producing a spectacular effect on plant growth and grain production. The production yield of many crops increased significantly in the past century, in part because of the application of more fertilizer.

But nitrogen fertilizer is expensive in two ways. First, the century-old process of manufacturing ammonium from hydrogen and nitrogen gases is very energy-intensive; when the price of oil goes up, the price of fertilizer also rises. In 2012 it cost farmers in the U.S. about $8 billion a year to spread nitrogen fertilizer on 90 million acres of cornfields. This was three times as much as it cost in 2000. Imagine what this means if you are a poor farmer without the money to pay the higher price for fertilizer needed for your crop to grow.

Nitrogen fertilizer is also environmentally expensive. When it rains excessively, nitrogen fertilizer can be lost from farm fields and end up in lakes, rivers, or groundwater. When nitrogen-laden rivers enter the sea, excessive growth of marine algae is likely to result. Eventually the algae die, and the organisms that decompose the algae use up so much oxygen that there is not enough left to support animal life. Nitrogen fertilizer runoff has thus resulted in vast "dead zones" in waters near the mouths of major rivers, including the Mississippi River Delta in the Gulf of Mexico. An additional environmental cost is the conversion of some nitrogen fertilizer to nitrous oxide gas (N_2O), which contributes to global warming.

Scientists are working on several strategies by which nitrogen fertilizer might be used efficiently. One strategy is to improve farming practices, to apply optimal rates of nitrogen to crops while reducing losses to the environment. The other strategy is to alter the genetics of crop plants to improve their uptake and assimilation of nitrogen.

Several companies in the U.S. are working to modify corn genetics to increase the nitrogen use efficiency. Many processes are involved in a plant's use of nitrogen from the soil, such as uptake into the roots, transport to other organs in the vascular system, and incorporation of nitrogen into molecules such as amino acids and nucleotides. Each process involves many genes, and so inheritance is complex. An understanding of these processes, however, will yield improvements in nitrogen use efficiency.

?

What progress has been made in improving the nitrogen use efficiency of corn?

See answer on p. 753.

36.1 What Nutrients Do Plants Require?

Every living thing—and a plant is no exception—must obtain raw materials from its environment. These **nutrients** include the major ingredients of macromolecules: carbon, hydrogen, oxygen, and nitrogen. Plants are autotrophs, and obtain both carbon and oxygen from atmospheric carbon dioxide through the reactions of photosynthesis (see Chapter 10). Hydrogen comes mainly from water, so it is plentiful when there is an adequate water supply. Nitrogen, as you will see later in this chapter, enters most plants from the soil. The activities of microorganisms are important in converting organic nitrogen and nitrogen gas into inorganic forms that are usable by plants.

In addition to nitrogen, organisms require other **mineral nutrients**: inorganic elements that are used for various cellular processes. For example, proteins contain sulfur (S), nucleic acids contain phosphorus (P), chlorophyll contains magnesium (Mg), cytochromes contain iron (Fe), and cellular signaling can involve calcium (Ca). Most plants obtain these nutrients from the soil. Within the soil, minerals dissolve in water as ions, forming a solution—called the **soil solution**—that contacts the roots of plants.

All plants require specific macronutrients and micronutrients

A plant nutrient is called an **essential element** if the plant fails to complete its life cycle or grows abnormally when the element is absent or insufficient. In addition to needing the nonmineral elements carbon, oxygen, and hydrogen, plants require many mineral nutrients (**Table 36.1**). Essential elements fall roughly into two categories—macronutrients and micronutrients—based on the amounts required by plants.

- A plant needs **macronutrients** in concentrations of at least 1 gram per kilogram of the plant's dry matter.
- A plant needs **micronutrients** in concentrations of less than 100 milligrams per kilogram of the plant's dry matter.

How do we know if a plant is getting enough of a particular nutrient?

TABLE **36.1**
Mineral Elements Required by Plants

Element (Abbreviation; Absorbed Form)	Typical Amount in Plant (g/kg)	Major Functions	Deficiency Symptoms
MACRONUTRIENTS			
Nitrogen (N; NO_3^- and NH_4^+)	15	In proteins, nucleic acids	Oldest leaves turn yellow and die prematurely; plant is stunted
Phosphorus (P; $H_2PO_4^-$ and HPO_4^{2-})	2	In nucleic acids, ATP, phospholipids	Plant is dark green with purple veins and is stunted
Potassium (K; K^+)	10	Enzyme activation; water balance; ion balance; stomatal opening	Older leaves have dead edges
Sulfur (S; SO_2^{4-})	1	In proteins and coenzymes	Young leaves are yellow to white with yellow veins
Calcium (Ca; Ca^{2+})	5	Affects the cytoskeleton, membranes, and many enzymes; second messenger	Growing points die back; young leaves are yellow and crinkly
Magnesium (Mg; Mg^{2+})	2	In chlorophyll; required by many enzymes; stabilizes ribosomes	Older leaves have yellow stripes between veins
MICRONUTRIENTS			
Iron (Fe; Fe^{2+} and Fe^{3+})	0.1	In active site of many redox enzymes and electron carriers; chlorophyll synthesis	Young leaves are white or yellow
Chlorine (Cl; Cl^-)	0.1	Photosynthesis; ion balance	Leaf tips wilt; leaves turn yellow and die
Manganese (Mn; Mn^{2+})	0.05	Activation of many enzymes	Younger leaves are pale with green veins
Boron [B; $B(OH)_3$]	0.02	Required for proper cell wall formation and expansion	Poor growth of leaves and roots
Zinc (Zn; Zn^{2+})	0.02	Enzyme activation; auxin synthesis	Young leaves are abnormally small; older leaves have many dead spots
Copper (Cu; Cu^{2+})	0.006	In active site of many redox enzymes and electron carriers	New leaves are dark green, may have dead spots
Nickel (Ni; Ni^{2+})	0.00005	Activation of the enzyme urease	Leaf tips die; deficiency is rare
Molybdenum (Mo; MoO_4^{2-})	0.0001	Nitrate reduction	Leaves turn yellow between veins; older leaves die

Healthy
Iron-deficient
Nitrogen-deficient

36.1 Mineral Nutrient Deficiency Symptoms The plants on the left were grown with a full complement of essential nutrients. The center plants were deprived of iron, whereas the plants on the right were deprived of nitrogen.

Go to Animated Tutorial 36.1
Nitrogen and Iron Deficiences
Life10e.com/at36.1

Deficiency symptoms reveal inadequate nutrition

When a plant is deficient in an essential element, it displays characteristic **deficiency symptoms**. Table 36.1 lists some of these symptoms, and plants deficient in iron and nitrogen are also shown in Figure 36.1. Such symptoms help growers diagnose mineral nutrient deficiencies in plants. With proper diagnosis, the missing nutrient(s) can be provided in the form of a **fertilizer** (an added source of mineral nutrients).

We know that the elements listed in Table 36.1 are essential to the life of all plants. How did biologists discover which elements are essential?

Hydroponic experiments identified essential elements

The essential elements for plants were identified by growing plants **hydroponically**—that is, with their roots suspended in nutrient solutions instead of soil. Growing plants in this manner allows for greater control of nutrient availability than is possible in a complex medium such as soil.

Plant
Air
Root
Culture solution

INVESTIGATING**LIFE**

36.2 Is Nickel an Essential Element for Plant Growth?
Using highly purified salts in growth media, Patrick Brown and his colleagues tested whether barley can complete its life cycle in the absence of nickel.[a] Other investigators showed that no other element could substitute for nickel.

HYPOTHESIS Nickel is an essential element for a plant to complete its life cycle.

Method

1. Grow barley plants for 3 generations in nutrient solutions containing 0, 0.6, and 1.0 μM NiSO4.
2. Harvest seeds from 5–6 third-generation plants in each of the groups.
3. Determine the nickel concentration in seeds from each plant.
4. Germinate other seeds from the same plants on nickel-free medium and plot the success of germination against nickel concentration.

Results There was a positive correlation between seed germination and seed nickel concentration. There was significantly less germination at the lowest nickel concentrations.

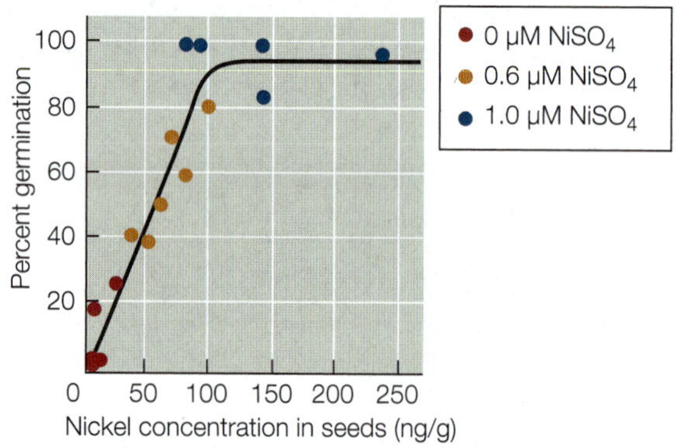
100
80
60
40
20
Percent germination
0
50
100
150
200
250
Nickel concentration in seeds (ng/g)
0 μM $NiSO_4$
0.6 μM $NiSO_4$
1.0 μM $NiSO_4$

CONCLUSION Barley seeds require nickel in order to survive and germinate, and therefore nickel is an essential element for plant nutrition.

Go to **BioPortal** for discussion and relevant links for all INVESTIGATING**LIFE** figures.

[a]Brown, P. H., R. M. Welch, and E. E. Cary. 1987. *Plant Physiology* 85: 801–803.

In the first experiments of this type, performed a century and a half ago, plants seemed to grow normally in solutions containing only calcium nitrate [$Ca(NO_3)_2$], magnesium sulfate ($MgSO_4$), and potassium phosphate (KH_2PO_4). A solution missing any of these compounds could not support normal growth. Tests with other compounds that included various combinations of these elements soon established the existence of six essential elements: calcium, nitrogen, magnesium, sulfur, potassium, and phosphorus. These are now known as the essential mineral macronutrients.

Iron was the first micronutrient to be clearly established as essential, in the 1840s. The last micronutrient to be listed as essential was nickel, in 1983. That experiment is described in Figure 36.2. Identifying essential micronutrients proved to be more difficult than identifying macronutrients because of the small amounts involved. Sufficient amounts of micronutrients can be present in the environment used to grow plants or in the plants themselves. A seed may contain enough of a micronutrient to supply the embryo and the entire plant throughout its lifetime. There might even be enough left over to pass on to third-generation

WORKING WITH**DATA**:

Is Nickel an Essential Element for Plant Growth?

Original Paper

Brown, P. H., R. M. Welch, and E. E. Cary. 1987. Nickel: A micronutrient essential for higher plants. *Plant Physiology* 85: 801–803.

Analyze the Data

For an element to be classified as an essential nutrient in plants, two criteria must be satisfied. Specifically, one must demonstrate that a plant cannot complete its life cycle in the absence of the element, and that no other element can substitute for the test element. To determine if nickel is an essential element for plants, Patrick Brown and colleagues tested whether barley could complete its life cycle in the absence of nickel. The researchers grew barley plants for three generations in nutrient solutions containing either 0, 0.6, or 1.0 μM $NiSO_4$, then tested the seeds for nickel content and germination. The results indicated a positive correlation between seed germination and nickel concentration (see Figure 36.2). Importantly, no germination was observed at the lowest nickel concentration. Together with results from another study showing that no other element could substitute for nickel, the investigators concluded that nickel is an essential micronutrient.

As shown in Figure 36.2, Ni depletion reduced the germination of barley grains. This could be caused by a reduced ability of the parental plant to produce grains, or, alternatively, a reduced germination ability of the grain itself. To investigate these two possibilities, Brown and colleagues measured grain production by twenty plants for each nickel treatment. The results are shown in **TABLE A**.

QUESTION 1

Did Ni depletion reduce the ability of the parental plants to produce grain? What data support your answer?

QUESTION 2

The data are expressed as mean ± standard deviation. What statistical test would you perform to determine whether the means are significantly different? Are they? (Refer to Appendix B if needed.)

QUESTION 3

To determine whether the reduced germination ability could be corrected by adding Ni, dry seeds from the three groups in the table were soaked in water with or without added 1.0 μM $NiSO_4$. The results are shown in **TABLE B**. Did adding Ni solution to dry seeds affect germination?

QUESTION 4

Plant biologists define the critical value as that concentration of a mineral nutrient in plant tissue that results in a 15 percent reduction in the optimum yield of the plant. Using maximum germination percentage in the graph in Figure 36.2 as the optimum yield, what is the critical value for nickel?

TABLE A

Ni in nutrient solution (μM)	Total grain weight (g)	Grain number (per plant)
0	7.3 ± 1.3	175 ± 26
0.6	7.5 ± 0.9	179 ± 35
1.0	8.4 ± 1.5	195 ± 41

TABLE B

Ni supplied to maternal plant (μM)	Germination after soaking seed with or without nickel (%)	
	0 μM $NiSO_4$	1 μM $NiSO_4$
0	9.1 ± 10	11.5 ± 10.3
0.6	55.2 ± 17	51.2 ± 20.3
1.0	98.2 ± 6.7	97.1 ± 9.1

Go to **BioPortal** for all WORKING WITH**DATA** exercises

plants. Because of such difficulties, nutrition experiments must be performed in tightly controlled laboratories with special air filters that exclude microscopic salt particles in the air, and must use only the purest available chemicals.

RECAP 36.1

Plants are autotrophs that obtain carbon and oxygen by photosynthesis, and mineral nutrients and water from the soil. Nutrients required by plants are classified as either macronutrients or micronutrients depending on the amount needed. Micronutrients are often needed in such minute amounts that only sophisticated chemical experiments can determine their essentiality.

- What are some specific mineral deficiency symptoms seen in plants? **See pp. 741–742 and Table 36.1**
- Why do plants need phosphorus? Why do they need nitrogen? **See p. 741**
- Outline an experimental method for determining whether an element is essential to a plant. **See Figure 36.2**

As we have seen, all plants require a specific set of nutrients for growth. Unlike many other organisms, however, plants can't move around to find nutrients. Let's look at how a plant finds and takes up nutrients from its environment.

36.2 How Do Plants Acquire Nutrients?

Many organisms can move from place to place to find the nutrients they need. But a plant cannot change its location (it is sessile), and so must obtain nutrients from its immediate environment. With the exception of the carbon and oxygen in CO_2, a plant's supply of nutrients is strictly local, and a plant may use up the water and mineral nutrients in its local environment as it grows. How does a plant cope with the problem of scarce nutrient supplies?

Plants rely on growth to find nutrients

As discussed in Chapter 34, plants differ fundamentally from animals in that they grow throughout their lifetimes. In fact,

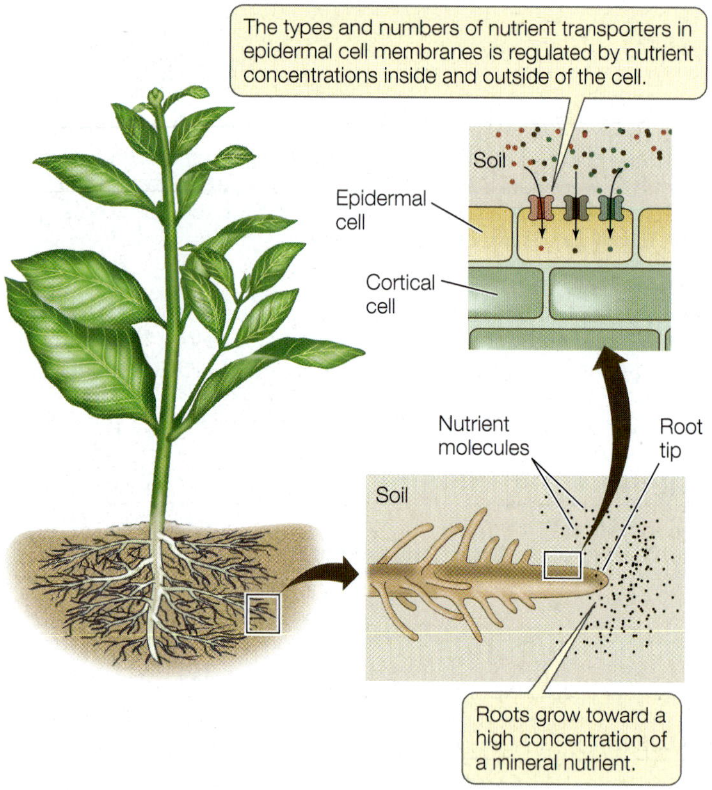
The types and numbers of nutrient transporters in epidermal cell membranes is regulated by nutrient concentrations inside and outside of the cell.
Soil
Epidermal cell
Cortical cell
Nutrient molecules
Root tip
Soil
Roots grow toward a high concentration of a mineral nutrient.

36.3 Plants Regulate Their Nutrition Plant roots branch and grow toward nutrients. Nutrients are taken up by transport proteins in the epidermal cell membrane. The number of transporters for a given nutrient can be regulated in response to nutrient availability.

growth is a plant's version of movement. For example, roots obtain most of the mineral nutrients plants need. By growing through the soil, roots mine it for new sources of mineral nutrients and water. The growth of stems and leaves helps a plant secure light and carbon dioxide, which in turn allows the roots to continue their growth through the soil. Deficiencies in water or specific mineral nutrients can stimulate plants to grow more roots, to improve the plants' chances of finding the nutrients they need.

As it grows, a plant—or even a single root—must deal with a variable environment. Animal droppings create high local concentrations of nitrogen. A particle of calcium carbonate may make a tiny area of the soil alkaline, while dead organic matter may make a nearby area acidic. Such microenvironments encourage or discourage the proliferation of a root system and help direct its growth. A major effort is underway to identify the signals in the soil and signaling pathways in the root that result in growth toward a source of nutrients (**Figure 36.3**).

Nutrient uptake and assimilation are regulated

Nutrients must cross the plasma membranes of cells in order to be used or assimilated into larger molecules. Polar molecules, including mineral ions, cross the membrane via specialized transport systems. In Chapter 35 we discussed the movement of water and ions into plant roots by way of the apoplast (through cell walls and intercellular spaces) or symplast (directly through cells). The Casparian strip prevents water and ions from entering the xylem tissues of the roots (see Section 35.1); therefore these nutrients must enter the symplast before they can be transported to other tissues. In most cases, ions are actively transported across the plasma membrane of epidermal cells into the symplast because their concentrations in the soil solution are generally lower than their concentrations inside cells (see Figure 35.4).

Plants have specialized transport systems for the uptake of specific ions (see Figure 36.3). For example, *Arabidopsis thaliana* has more than 50 genes that encode nitrate (NO_3^-) transporters, 6 genes encoding ammonium (NH_4^+) transporters, and at least 4 genes for phosphate (PO_4^{3-}) transporters. Nutrient uptake is highly regulated because the levels of ions inside cells must be maintained at constant levels. The genes for ion transporters are regulated at the transcriptional level by the amounts of each nutrient inside cells: low nutrient levels stimulate transcription, whereas high levels repress transcription. In addition, the transporter proteins themselves are regulated (for example, by phosphorylation) to control their ion transport activity.

The assimilation of nutrients into more complex molecules is also regulated according to the plant's needs. The enzymes involved in assimilating nitrate and ammonium into amino acids are regulated at the transcriptional and posttranscriptional levels to increase assimilation when available nitrogen is abundant. The uptake and assimilation of nitrogen are also stimulated by photosynthesis, and this ensures that the nitrogen status in the plant is coordinated with its carbon status.

RECAP 36.2

Both the uptake and assimilation of nutrients are regulated according to a plant's needs.

- How does the ability to grow throughout their lifetime allow plants to seek out nutrients? **See p. 744**
- Describe how plants control the uptake and assimilation of mineral nutrients. **See Figure 36.3**

Plants acquire many essential elements from the soil. As we will see in the next section, soils have complex structures that affect the availability of nutrients for plants.

36.3 How Does Soil Structure Affect Plants?

Most terrestrial plants grow in soil. Soils provide:

- Mechanical support
- Mineral nutrients and water from the soil solution
- O_2 for root respiration

Soils also harbor many bacteria and other organisms; some of these are beneficial to plant life, but others are harmful. Some soils contain toxic levels of metal ions such as cadmium, chromium, and lead (see Section 39.4).

Soils are modified by natural phenomena such as rain, temperature extremes, and the activities of plants and animals. They are also modified by the activities of humans, particularly in agriculture. In this section we will examine the composition, structure, and formation of soils, as well as their role in plant nutrition.

Soils are complex in structure

Soils have living and nonliving components (**Figure 36.4**). The living components include plant roots as well as populations of bacteria, fungi, protists, and animals such as earthworms and insects. The nonliving portion of the soil includes rock fragments ranging in size from large stones to sand to silt, and finally to tiny particles of clay that are 2 micrometers (μm) or less in diameter. Soil also contains water and dissolved mineral nutrients, air spaces, and dead organic matter. The air spaces in soil contain O_2.

Although soils vary greatly, almost all of them have a soil profile consisting of several recognizable horizontal layers, called **horizons**, lying on top of one another. Soil scientists recognize three major horizons—termed A, B, and C—in the profile of a typical soil (**Figure 36.5**).

- The **A horizon** is the **topsoil** that supports the plant's nutrient needs. It contains most of the soil's living and dead organic matter.
- The **B horizon** is the **subsoil**, which accumulates materials from the topsoil above it and from the parent rock below.
- The **C horizon** is the **parent rock**, also called bedrock, that is in the process of breaking down to form soil.

Soil fertility is a soil's ability to support plant growth. A topsoil's fertility is determined by several factors. Topsoils vary greatly in their proportions of sand, silt, and clay, and this influences their ability to support plant growth. For example, mineral nutrients tend to be **leached** from the upper soil horizons—dissolved in rain or irrigation water and carried to deeper horizons, where they are unavailable to plant roots. Dissolved minerals are readily leached from sandy soil because sand particles are relatively large and cannot hold water. Clay, by contrast, binds more water than sand does, and the charged surfaces of clay particles bind mineral ions that plant roots ultimately take up. But clay particles are tiny and pack tightly together, leaving little space for air. A **loam** is a soil that is an optimal mixture of sand, silt, and clay, and thus has sufficient levels of air, water, and available nutrients for plants. Loams also usually contain organic matter. Most of the best topsoils for agriculture are loams.

In addition to mineral particles, soils contain dead organic matter, largely from plants. Soil organisms break down dead leaves and other plant organs on the ground into a substance called **humus**. This material is used as a food source by microbes that break down complex organic molecules and release simpler molecules into the soil solution. Humus also provides air spaces that increase O_2 availability to plant roots.

Soils form through the weathering of rock

Rocks are broken down into soil particles—**weathered**—in two ways. First there is *mechanical weathering*, which is the physical breakdown of materials by wetting, drying, and freezing. Second there is *chemical weathering*, the alteration of the chemistry of the materials in the rocks. Several

A horizon
Topsoil
B horizon
Subsoil
C horizon
Weathering parent rock

36.5 A Soil Profile The A, B, and C horizons can sometimes be seen at construction sites such as this one in Massachussets. The upper layer (the A horizon) is home to most of the living organisms in the soil.

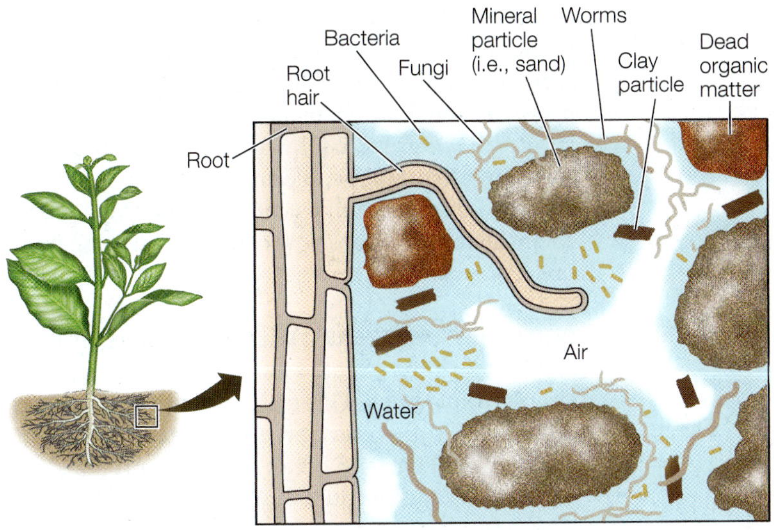

Mineral particle (i.e., sand)
Worms
Bacteria
Dead organic matter
Root hair
Fungi
Clay particle
Root
Air
Water

36.4 The Complexity of Soil Soils favorable for plant growth contain both clay and larger mineral particles, as well as water, air, and organic matter. Other organisms are also present.

types of chemical weathering occur, all of which influence the availability of mineral nutrients:

- Oxidation by atmospheric oxygen
- Hydrolysis (reaction with water)
- Reaction with acids (particularly carbonic acid)

The parent rock and the weathering it undergoes determine the basic structure and chemical composition of a soil. However, a key soil characteristic for plants is the availability of nutrients, which must be dissolved in the soil solution for uptake by the plant. Chemical weathering often results in clay particles that are covered with negatively charged chemical groups, which bind positively charged mineral nutrients. How might roots obtain these mineral nutrients?

Soils are the source of plant nutrition

Humus and clay particles often carry negative charges. These particles form ionic attractions (see Section 2.2) with the positively charged ions (cations) of many minerals that are important for plant nutrition, such as potassium (K^+), magnesium (Mg^{2+}), and calcium (Ca^{2+}). To become available to plants or other organisms, these cations must be detached from the clay particles.

Recall that the root surface is covered with root hair cells (see Figure 34.8). Transporters in the plasma membranes of these cells actively pump protons (H^+) out of the cell. In addition, cellular respiration in the roots releases CO_2, which dissolves in the soil water and reacts with it to form carbonic acid. This acid ionizes to form bicarbonate and free protons:

$$CO_2 + H_2O \rightleftharpoons H_2CO_3 \rightleftharpoons H^+ + HCO_3^-$$

Proton-pumping by the root and ionization of carbonic acid both act to increase the proton concentration (lower the pH) in the soil surrounding the root. The protons bind more strongly to clay particles than do mineral cations; in essence, they trade places with the cations in a process called **cation exchange** (Figure 36.6). Cation exchange releases important cations into the soil solution, where they are available to be taken up by the roots. Soil fertility is determined in part by the soil's ability to provide nutrients in this manner.

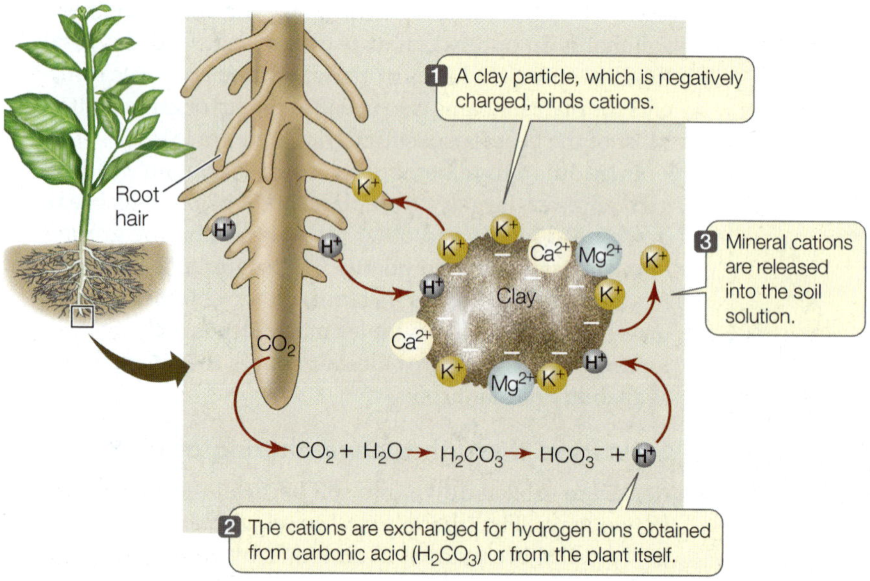
1 A clay particle, which is negatively charged, binds cations.
Root hair
H^+
K^+
Ca^{2+}
Mg^{2+}
Clay
CO_2
3 Mineral cations are released into the soil solution.
$CO_2 + H_2O \rightarrow H_2CO_3 \rightarrow HCO_3^- + H^+$
2 The cations are exchanged for hydrogen ions obtained from carbonic acid (H_2CO_3) or from the plant itself.

36.6 Cation Exchange Plants obtain mineral nutrients from the soil primarily in the form of positive ions; potassium (K^+) is the example shown here.

Some soil particles, such as ones containing oxides of iron or aluminium, are positively charged under acid conditions and can exchange anions in a process similar to cation exchange. However, soil pH is rarely low enough for anion exchange to occur. As a result, important anions such as nitrate (NO_3^-) and sulfate (SO_4^{2-})—direct sources of nitrogen and sulfur, respectively—may leach rapidly from the A horizon.

As we have just seen, soil fertility is affected by soil pH. The proton concentration affects the binding of cations and anions to soil particles, and can affect the solubility of other nutrients, such as iron. In addition, soil pH affects the absorption of nutrients by plant roots. The pH level of a soil depends on its mineral and organic contents and can be altered by various factors, including rainfall, weathering, plant growth, and fertilizer applications. The optimal soil pH for most plants is in the range 6 to 7.5, but some plants, such as blueberries and cranberries, prefer pH levels of 4.5 to 5.

Fertilizers can be used to add nutrients to soil

Leaching and the harvesting of crops can deplete a soil of its nutrients, so that new crops grow poorly on that soil. Soil fertility can be restored or increased in various ways, including shifting agriculture to another location or applying organic or chemical fertilizers.

SHIFTING AGRICULTURE In the past, when the soil could no longer support a level of plant growth sufficient for agricultural purposes, people simply moved to another location. The nutrients in the soil of a field allowed to lie fallow will be replenished gradually through the addition of organic matter from the growth and death of plants naturally present, and by the weathering of the parent rock. Both processes take a long time, which is not a problem as long as a lot of land is available. Today, however, the food needs of a large human population are too great to allow land to be left vacant for a long time, and people are disinclined to move away from settled homesteads. As a consequence, chemical fertilizers are now commonly used to improve soil fertility.

ORGANIC FERTILIZERS Microorganisms in the soil break down organic molecules into smaller, simpler molecules. These simpler molecules can dissolve in soil water and enter plant roots. For example, soil bacteria break down the proteins in dead leaves and produce ammonium ions (NH_4^+), which in turn are converted into nitrate (NO_3^-). Both ammonium and nitrate can be taken up and used by plants:

$$\text{Proteins in leaves} \xrightarrow{\text{Bacteria}} NH_4^+ \xrightarrow{\text{Bacteria}} NO_3^-$$

Farmers can increase the nutrient content of soil by adding organic materials such as compost (partially decomposed plant materials) or manure (waste from farm animals). Manure is a particularly good source of nitrogen. In either case, the addition of these **organic fertilizers** adds nutrients to the soil much more rapidly than weathering or the gradual addition of organic matter from natural vegetation. Organic fertilizers allow for a slow release of ions as the materials decompose.

INORGANIC FERTILIZERS Organic fertilizers may act too slowly to restore fertility if a soil is to be used every year. **Inorganic fertilizers** supply mineral nutrients in forms that can be taken up immediately by plants or that are rapidly converted to usable forms in the soil. Inorganic fertilizers are easily transported and handled, and allow farmers to control the amount of a particular nutrient that is supplied to each crop. Particular fertilizers are used in varying amounts, depending on the needs of the crop and the type of soil. For example, much higher amounts of nitrogen are applied to cornfields than to soybean fields. Inorganic fertilizers come in many forms. Common ones include ammonia (NH_3), urea (NH_2-CO-NH_2), and salts formed from positive and negative ions such as ammonium (NH_4^+), potassium (K^+), nitrate (NO_3^-), phosphate (PO_4^{3-}), and sulfate (SO_4^{2-}).

The use of inorganic fertilizers became widespread in many countries during the twentieth century. This contributed (along with improved genetic strains and other technological advances) to the "green revolution"—a massive increase in world food production during the second half of the century. However, as we saw in the opening story, excessive fertilizer use has led to environmental problems. Agricultural scientists are developing methods for determining the exact amount of a particular nutrient that a crop requires, and for applying fertilizer at an optimal time to maximize yield while reducing losses to the environment.

RECAP 36.3

Land plants live anchored in the soil and obtain water and mineral nutrients from it. Soils are complex in structure and vary in fertility. Farmers can add fertilizers to improve the nutrient contents of soils.

- Explain how mechanical and chemical weathering form soil from rock. See pp. 745–746
- How is soil fertility enhanced by the process of cation exchange? See p. 746 and Figure 36.6
- What are the differences between organic and inorganic fertilizers in terms of plant nutrition? See pp. 746–747

Thus far we have focused on the uptake of nutrients in the soil by plant roots. An understanding of how plants acquire nutrients from the soil would be incomplete, however, without taking into account the involvement of soil microbes, including fungi and bacteria. In the next section we will focus on the intimate interactions of plants with these organisms, which are essential to the success of most terrestrial plants.

36.4 How Do Fungi and Bacteria Increase Nutrient Uptake by Plant Roots?

One gram of soil can contain 6,000 to 50,000 bacterial *species* and up to 200 *meters* of fungal hyphae (the long branching cells of fungi), although both are largely invisible to the naked eye. In Chapter 39 we will describe the strategies plants use to prevent infection by harmful soil microbes. But plants actively encourage a few species of fungi and bacteria to infect their roots and even invade root cells. In this section we will describe the mutually beneficial relationships in which products are exchanged between the plants and these special soil microbes.

Plants send signals for colonization

In Chapter 30 we described mycorrhizae, the association of fungi with plant roots—an interaction that occurs in more than 90 percent of terrestrial plants. Our example in Chapter 30 was ectomycorrhizal fungi, which wrap themselves around a plant root (see Figure 30.9). In this chapter we will examine arbuscular mycorrhizae, in which fungal hyphae penetrate root cells. We will also describe here the close association between the roots of some plants and rhizobia, a group of nitrogen-fixing bacteria (bacteria that convert atmospheric N_2 into a more biologically useful form). We will see that mycorrhizal and rhizobial associations are both initiated by signals sent by plant roots that attract the soil organisms, and that the development of these associations involves similar genes and cellular pathways.

FORMATION OF MYCORRHIZAL ASSOCIATIONS The events in the formation of arbuscular mycorrhizae are shown in Figure 36.7A. Plant roots produce compounds called **strigolactones** that stimulate rapid growth of fungal hyphae toward the root. In response, the fungi produce signals that stimulate expression of plant symbiosis-related genes. The products of some of these genes give rise to the prepenetration apparatus (PPA), which guides the growth of the fungal hyphae into the root cortex. The sites of nutrient exchange between fungus and plant are the arbuscules, which form within root cortical cells. Despite the intimacy of this association, the plant and fungal cytoplasms never mix—they are separated by two membranes, the fungal plasma membrane and the periarbuscular membrane (PAM), which is continuous with the plant plasma membrane.

FORMATION OF NITROGEN-FIXING NODULES A group of plants called legumes (members of the plant family Fabaceae) can form symbioses with soil bacteria in several genera collectively known as rhizobia. The legume roots release flavonoids and other chemical signals that attract the rhizobia to the vicinity of the roots (Figure 36.7B). The flavonoids also trigger the transcription of bacterial *nod* genes, the products of which synthesize Nod (nodulation) factors. These factors, when secreted by the bacteria, cause cells in the root cortex to divide, leading to the formation of a primary nodule meristem. This meristem gives rise to the plant tissue that constitutes the **root nodule**.

36.7 Roots Send Signals for Colonization Plant roots send chemical signals to arbuscular mycorrhizal fungi (A) and nitrogen-fixing bacteria (B) to stimulate colonization.

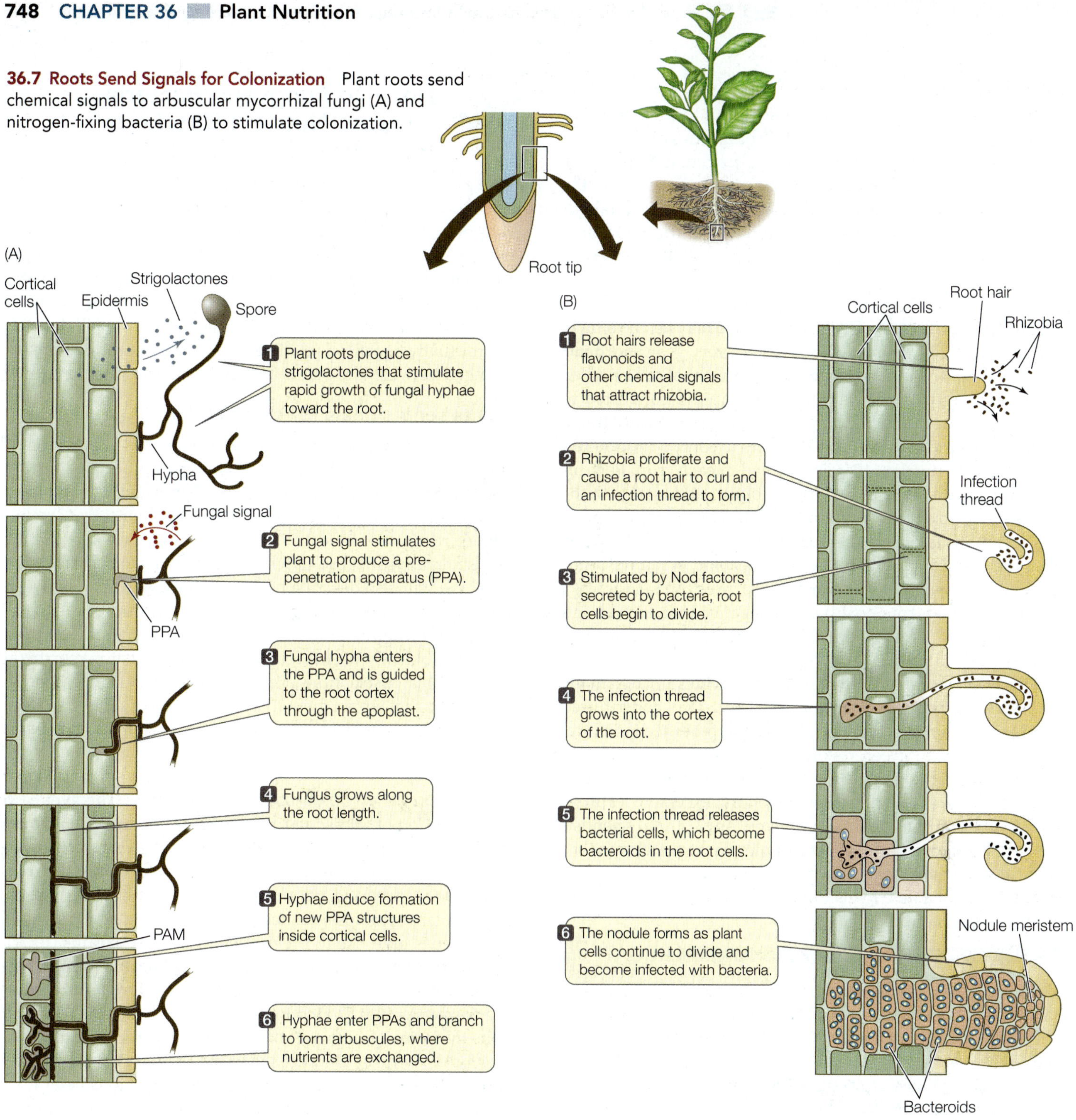
Root tip
(A)
Cortical cells
Epidermis
Strigolactones
Spore
1 Plant roots produce strigolactones that stimulate rapid growth of fungal hyphae toward the root.
Hypha
Fungal signal
2 Fungal signal stimulates plant to produce a pre-penetration apparatus (PPA).
PPA
3 Fungal hypha enters the PPA and is guided to the root cortex through the apoplast.
4 Fungus grows along the root length.
PAM
5 Hyphae induce formation of new PPA structures inside cortical cells.
6 Hyphae enter PPAs and branch to form arbuscules, where nutrients are exchanged.
(B)
Cortical cells
Root hair
Rhizobia
1 Root hairs release flavonoids and other chemical signals that attract rhizobia.
2 Rhizobia proliferate and cause a root hair to curl and an infection thread to form.
Infection thread
3 Stimulated by Nod factors secreted by bacteria, root cells begin to divide.
4 The infection thread grows into the cortex of the root.
5 The infection thread releases bacterial cells, which become bacteroids in the root cells.
6 The nodule forms as plant cells continue to divide and become infected with bacteria.
Nodule meristem
Bacteroids

Bacteria enter the root via an infection thread, analogous to the PPA in mycorrhizal associations, and eventually reach cells inside the root nodule. There the bacteria are released into the cytoplasm of the nodule cells, enclosed in membrane vesicles similar to the PAM. Inside the vesicles, the bacteria differentiate into **bacteroids**—the form of the bacteria that can fix nitrogen.

A COMMON MECHANISM There is increasing evidence that nodule formation depends on some of the same genes and mechanisms that allow mycorrhizae to develop. For example, both processes involve invagination of the plasma membrane to allow entry of the fungal hyphae or rhizobia. The similarities in the structures formed during the development of mycorrhizae and nodules are especially striking, considering that the symbioses involve members of two different kingdoms (fungi and bacteria) (Figure 36.8).

Mycorrhizae expand the root system

In many cases the roots of vascular plants cannot nutritionally support plant growth alone—they simply cannot reach all the

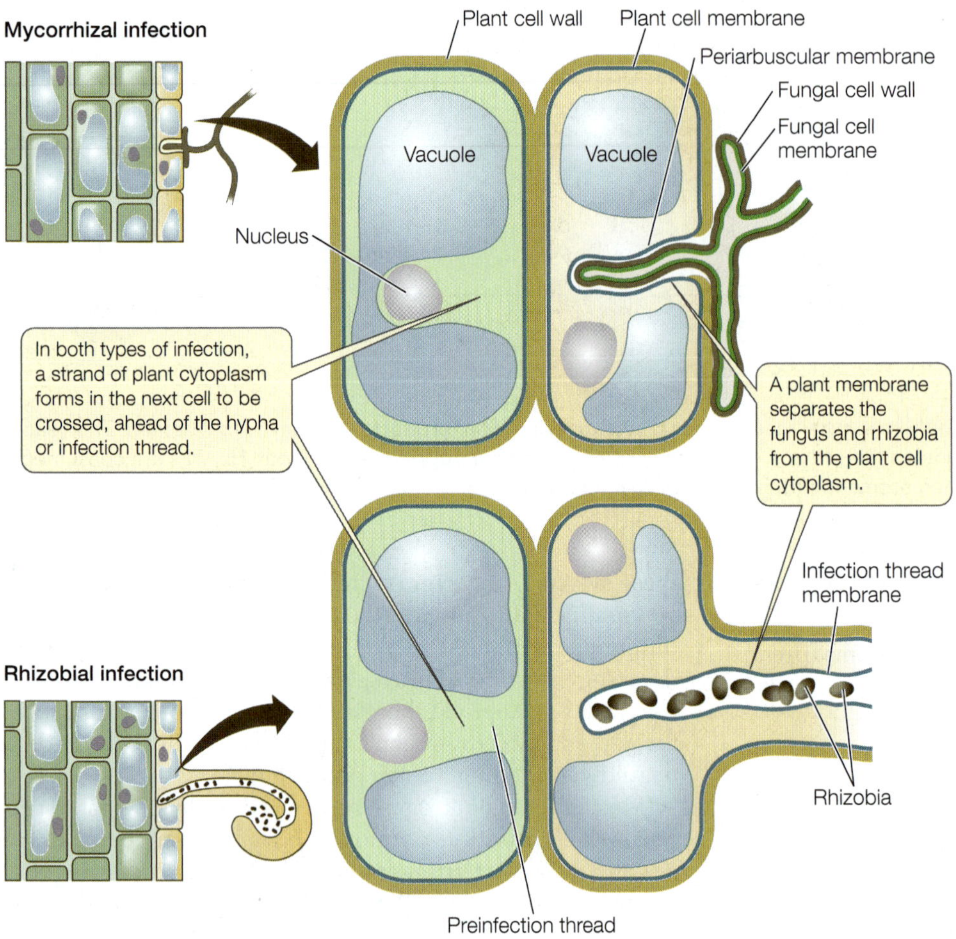

Mycorrhizal infection
Plant cell wall
Plant cell membrane
Periarbuscular membrane
Fungal cell wall
Fungal cell membrane
Vacuole
Vacuole
Nucleus
In both types of infection, a strand of plant cytoplasm forms in the next cell to be crossed, ahead of the hypha or infection thread.
A plant membrane separates the fungus and rhizobia from the plant cell cytoplasm.
Infection thread membrane
Rhizobial infection
Rhizobia
Preinfection thread

36.8 Intracellular Structures in Plant–Fungus and Plant–Rhizobium Symbioses Several steps in the development of mycorrhizae and nodules involve similar structures.

nutrients available in the soil. Mycorrhizae expand the root surface area 10-fold to 1,000-fold, increasing the amount of soil that can be mined for nutrients. In addition, because fungal hyphae are much finer than root hairs, they can get into pores in the soil that are inaccessible to roots. In this way, mycorrhizae probe a vast quantity of soil for nutrients and deliver them into root cortical cells.

The primary nutrient that the plant obtains from a mycorrhizal interaction is phosphorus. In exchange, the fungus obtains an energy source: the products of photosynthesis. In fact, up to 20 percent of the photosynthate of terrestrial plants is directed to and consumed by arbuscular mycorrhizal fungi. Such associations are excellent examples of mutualism, an interaction between two species in which both species benefit (further discussed in Chapter 56). Mutualism is a type of symbiosis, in which two different species live in close contact for a significant portion of their life cycles.

Soil bacteria are essential in getting nitrogen from air to plant cells

The essential mineral nutrient most commonly in short supply, in both natural and agricultural situations, is nitrogen. This is surprising because elemental nitrogen (N_2) makes up almost four-fifths of Earth's atmosphere. However, plants cannot use N_2 directly as a nutrient. The triple bond linking the two nitrogen atoms is extremely stable, and a great deal of energy is required to break it; thus N_2 is a highly unreactive substance.

Some prokaryotes have an enzyme that enables them to convert N_2 into a more reactive and biologically useful form by a process called **nitrogen fixation**:

$$N_2 + 6\,H \rightarrow 2\,NH_3$$

Nitrogen fixers, including those present in root nodules, fix approximately 170 million metric tons of nitrogen per year. Humans use industrial methods to fix about 80 million metric tons per year. In addition, about 20 million metric tons per year are fixed in the atmosphere by nonbiological means such as lightning, volcanic eruptions, and forest fires. Rain brings these atmospherically formed products to the ground.

Two types of organisms can fix nitrogen:

- Free-living organisms living in soil and water (e.g., *Azotobacter* bacteria and *Nostoc* cyanobacteria [sometimes called blue-green algae])
- Symbiotic organisms living in other organisms (e.g., rhizobia in roots of legumes, and *Anabaena* cyanobacteria in aquatic ferns)

Nitrogenase catalyzes nitrogen fixation

Nitrogen fixation is the reduction (see Section 9.1) of nitrogen gas. It proceeds by the stepwise addition of three pairs of

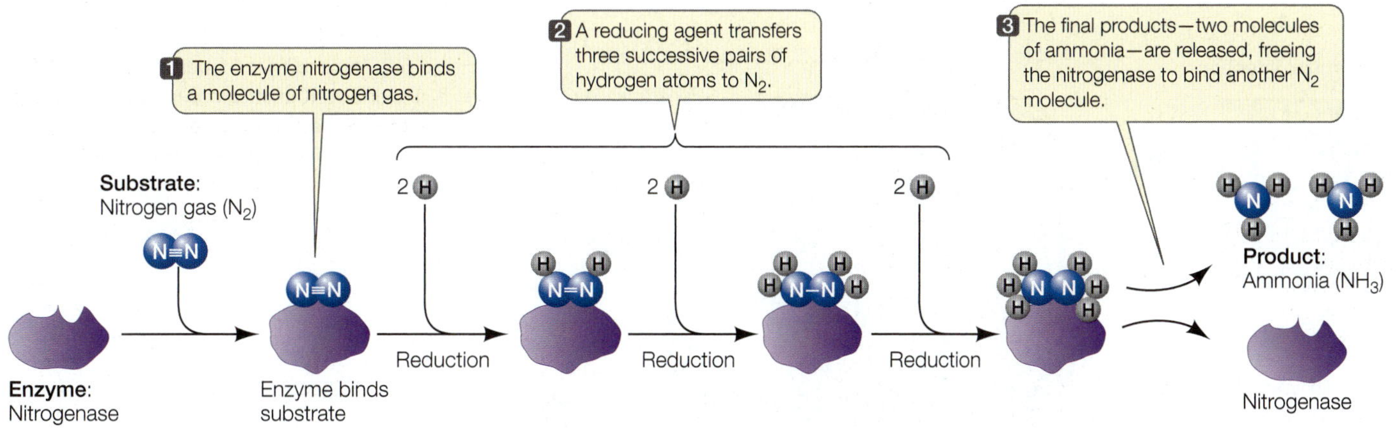

1 The enzyme nitrogenase binds a molecule of nitrogen gas.
2 A reducing agent transfers three successive pairs of hydrogen atoms to N2.
3 The final products—two molecules of ammonia—are released, freeing the nitrogenase to bind another N2 molecule.
Substrate: Nitrogen gas (N2)
Enzyme: Nitrogenase
Enzyme binds substrate
Reduction
Reduction
Reduction
Product: Ammonia (NH3)
Nitrogenase

36.9 Nitrogenase Fixes Nitrogen Throughout the chemical reactions of nitrogen fixation, the reactants are bound to the enzyme nitrogenase. A reducing agent transfers hydrogen atoms to nitrogen, and eventually the final product—ammonia—is released. This reaction requires a large input of energy: about 16 ATPs are consumed per reaction.

hydrogen atoms to N_2 (**Figure 36.9**). In addition to N_2, these reactions require three things:

- A strong reducing agent to transfer hydrogen atoms (protons and electrons) to N_2 and to the intermediate products of the reaction
- A great deal of energy, which is supplied by ATP
- The enzyme **nitrogenase**, which catalyzes the reaction

Depending on the species of nitrogen fixer, either respiration or photosynthesis provides the necessary reducing agent and ATP.

Nitrogenase is strongly inhibited by oxygen, and many nitrogen fixers are anaerobes that live in environments with little or no O_2. But rhizobia are aerobic and fix nitrogen in aerobic plant roots. How can nitrogenase function under these circumstances?

Plants typically house nitrogen-fixing bacteria in root nodules. Within a nodule, O_2 is maintained at a low level that is sufficient to support respiration, but not so high as to inactivate nitrogenase. This is possible because the cytoplasm of nodule cells contains a plant-produced protein called **leghemoglobin**, which is an O_2 carrier. Leghemoglobin is a close relative of hemoglobin, the red, oxygen-carrying pigment of animals, and is thus an evolutionarily ancient molecule. Some plant nodules contain enough of it to be bright pink inside. Leghemoglobin, with its iron-containing heme groups, transports enough oxygen to the nitrogen-fixing bacteria to support their respiration, while keeping free oxygen concentrations low enough to protect nitrogenase.

Biological nitrogen fixation does not always meet agricultural needs

Crop rotation systems have been used for hundreds or thousands of years by many human civilizations. In these systems, each field is used to grow different crops in different years, with legumes (such as alfalfa, clover, peas, and beans) included in the rotation. The rotation may also include periods of grazing by farm animals. Because of their association with nitrogen-fixing bacteria, legumes can replace all or some of the nitrogen removed by grain crops such as wheat and corn. Even with these systems, however, bacterial nitrogen fixation is not always sufficient to support the needs of agriculture. Some traditional farmers used to plant dead fish along with corn; the decaying fish released nitrogen that the developing corn could use. Today farmers use inorganic nitrogen fertilizers produced through industrial nitrogen fixation to meet the food needs of a rapidly expanding human population.

Most industrial nitrogen fixation is done by the Haber process, which produces ammonia (NH_3) from methane (CH_4) gas and molecular nitrogen (N_2). The Haber process requires a great deal of energy. (Biological nitrogen fixation also consumes lots of energy in the form of ATP—about 16 ATP per N fixed.) At present in the United States, the manufacture of nitrogen-containing fertilizer takes more energy—primarily natural gas and hydroelectric—than does any other aspect of crop production. The rising cost of energy sources raise serious questions about the sustainability of this approach to fertilizer production.

Plants and bacteria participate in the global nitrogen cycle

Nitrogen moves through the biosphere in a **global nitrogen cycle** (**Figure 36.10**), which includes four key steps:

1. *Fixation* of atmospheric N_2 to NH_3 and NH_4^+ by bacteria and by abiotic processes
2. *Nitrification* of these molecules to nitrate by bacteria
3. *Nitrate reduction* by plants
4. *Denitrification* of nitrate by bacteria back to N_2, which is then released to the atmosphere to begin another cycle

Nitrogen fixation results in the formation of ammonia (NH_3), most of which is rapidly ionized to form ammonium (NH_4^+). The balance between ammonia and ammonium depends on pH; neutral and acidic pH levels favor the ionic form. Ammonia can be toxic to plants, but ammonium ions are taken up safely. Soil bacteria called **nitrifiers** oxidize ammonia to nitrate ions (NO_3^-)—another form that plants can take up—by the process of **nitrification**. Soil pH affects which form of nitrogen is taken up by plants: nitrate ions are taken up preferentially under more basic conditions, and ammonium

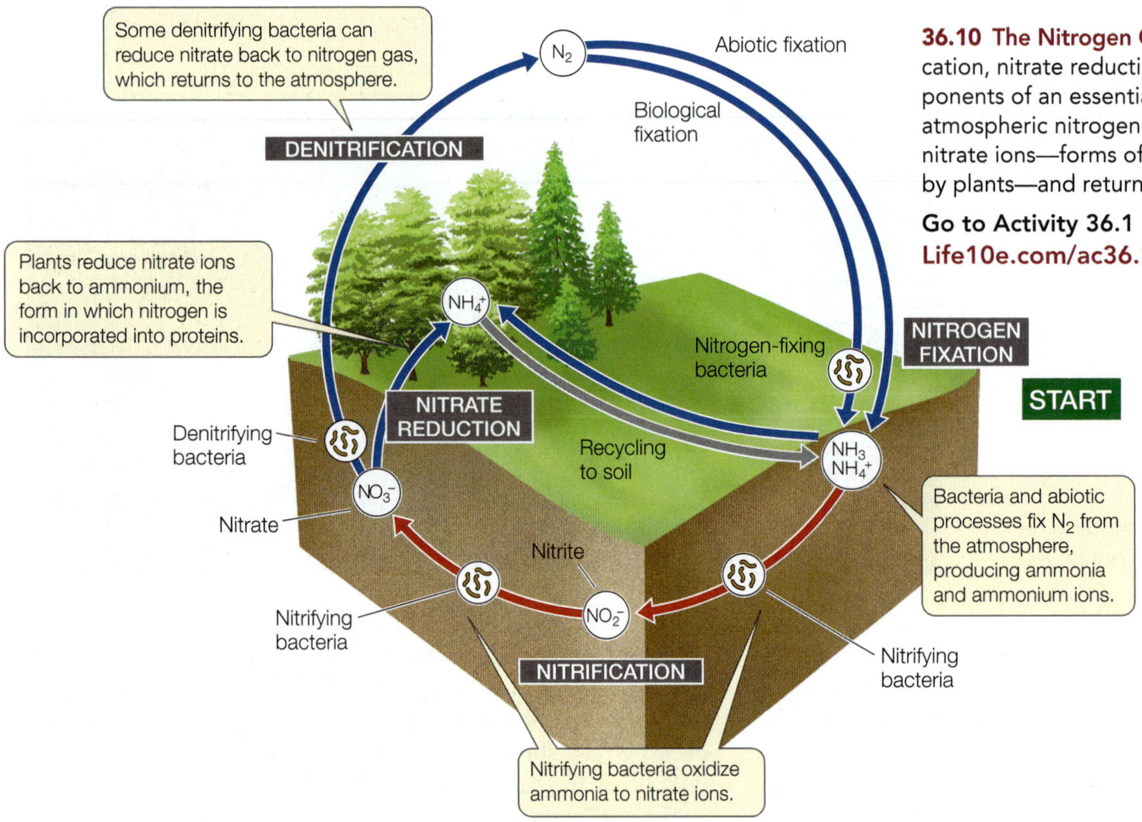
Some denitrifying bacteria can reduce nitrate back to nitrogen gas, which returns to the atmosphere.
DENITRIFICATION
N_2
Abiotic fixation
Biological fixation
Plants reduce nitrate ions back to ammonium, the form in which nitrogen is incorporated into proteins.
NH_4^+
NITROGEN FIXATION
Nitrogen-fixing bacteria
START
NITRATE REDUCTION
Recycling to soil
Denitrifying bacteria
NH_3 NH_4^+
NO_3^-
Nitrate
Nitrite
Bacteria and abiotic processes fix N_2 from the atmosphere, producing ammonia and ammonium ions.
Nitrifying bacteria
NO_2^-
NITRIFICATION
Nitrifying bacteria
Nitrifying bacteria oxidize ammonia to nitrate ions.

36.10 The Nitrogen Cycle Nitrogen fixation, nitrification, nitrate reduction, and denitrification are components of an essential chemical cycle that converts atmospheric nitrogen gas into ammonium ions and nitrate ions—forms of nitrogen that can be taken up by plants—and returns N_2 to the atmosphere.

Go to Activity 36.1 The Nitrogen Cycle Life10e.com/ac36.1

ions under more acidic ones. To use nitrate, a plant must first reduce it to ammonium in a process called **nitrate reduction**, which occurs in two enzyme-catalyzed steps. First, nitrate is converted to nitrite (NO_2^-) in the cytoplasm, and then nitrite is converted to ammonium in the plastids. The plant uses the ammonium to manufacture amino acids, from which the plant's proteins and all its other nitrogen-containing compounds are formed. Animals cannot reduce nitrogen and so depend on plants to supply them with reduced nitrogenous compounds.

The nitrogen cycle is essential for life on Earth: nitrogen-containing compounds constitute 5 to 30 percent of a plant's total dry weight. The nitrogen content of animals is even higher, and all of it arrives there by way of the plant kingdom.

RECAP 36.4

Two mutualistic interactions with soil microbes are critical to the success of terrestrial plants. Fungi and plants form mycorrhizae, which greatly increase the soil volume that roots can scavenge for nutrients. Bacteria in soils and root nodules fix atmospheric nitrogen into forms that plants and ultimately animals can use. Denitrification returns nitrogen from dead organisms and animal waste back to the atmosphere, completing the global nitrogen cycle.

- How is the formation of a root nodule on a legume similar to the formation of an arbuscular mycorrhiza? **See p. 747–749 and Figures 36.7 and 36.8**
- What is exchanged between plants and fungi in mycorrhizae? **See p. 749**
- What, besides nitrogenase, is required to reduce nitrogen gas to a form plants can use? **See p. 750 and Figure 36.9**

Let's turn now to some special mechanisms for obtaining nutrients that have evolved in plant species with unusual lifestyles.

36.5 How Do Carnivorous and Parasitic Plants Obtain a Balanced Diet?

Most plants obtain their mineral nutrients from the soil solution. Carnivorous and parasitic plants are examples of plants that obtain nutrients from other sources.

Carnivorous plants supplement their mineral nutrition

Some plants augment their nitrogen supply by capturing and digesting flies and other insects. There are about 500 of these **carnivorous plant** species, the best known of which are Venus flytraps (genus *Dionaea*; **Figure 36.11A**), sundews (genus *Drosera*; **Figure 36.11B**), and pitcher plants (genus *Sarracenia*).

Carnivorous plants are typically found in boggy habitats that are acidic and nutrient deficient. To obtain extra nitrogen, these plants capture animals, digest their proteins, and absorb the amino acids. Pitcher plants have pitcher-shaped leaves that collect small amounts of rainwater. Insects and even small rodents are lured into the pitchers by bright colors or attractive scents and are prevented from leaving by stiff, downward-pointing hairs. The animals eventually die and are digested by a combination of plant enzymes and bacteria in the water. Sundews have leaves covered with hairs that secrete a sticky, sugary liquid. Insects become stuck to these hairs, and more hairs curve over to further entrap them. Enzymes secreted by the plant digest the insects. Venus flytraps have specialized leaves with two halves that fold together. When an insect touches special hairs called trigger hairs on the

(A) Dionaea muscipula
(B) Drosera rotundifolia

36.11 Carnivorous Plants Some plants have adapted to nitrogen-poor environments by becoming carnivorous. (A) The Venus flytrap obtains nitrogen from the bodies of insects trapped inside the plant when its hinges snap shut. (B) Sundews trap insects on sticky hairs. Secreted enzymes will digest the carcass externally.

Go to Media Clip 36.1
A Venus Flytrap "Snaps to It"
Life10e.com/mc36.1

leaf, its two halves snap together. The spiny margins interlock and imprison the insect before it can escape. The leaf then secretes enzymes that digest its prey.

The closing of the Venus flytrap's leaf is one of the fastest movements in the plant world, requiring only 0.1 seconds. To find out how this happens, Dr. Lakshminarayanan Mahadevan and colleagues at Harvard University painted fluorescent dots on the surface of the flytrap's leaf surface and used high-speed cameras to record the trap snapping shut when its trigger hairs were touched. They then used computer analysis of the recorded dot movements to generate a mathematical model to help explain the movement. The researchers found that the first step is the osmosis-driven elongation of cells on the outer surface of the leaf. The expansion of only one side of the leaf causes it to snap from a convex into a concave shape, much like a contact lens flipping inside out.

Carnivorous plants photosynthesize and extract soil nutrients just like other plants, but eating insects helps them grow faster in their natural habitats. They use the additional nitrogen from the insects to make more proteins, chlorophyll, and other nitrogen-containing compounds.

Parasitic plants take advantage of other plants

Approximately 1 percent of flowering plant species derive some or all of their water, mineral nutrients, and sometimes even photosynthate from other plants. These **parasitic plants** have evolved absorptive organs called **haustoria**, which invade the host and tap into the vascular tissues in the root or stem.

Parasitic plants are divided into two broad classes based on their nutritional interactions with their hosts. **Hemiparasites** can still photosynthesize but derive water and mineral nutrients from the living bodies of other plants. Perhaps the most familiar hemiparasites are the several genera of mistletoes. Mistletoes are green and carry on some photosynthesis, but they parasitize other plants for water and mineral nutrients and may derive photosynthetic products from them as well. Dwarf mistletoe (*Arceuthobium americanum*) is a serious parasite in forests of the western United States, destroying more than 3 billion board feet of lumber per year.

Holoparasites are completely parasitic and do not perform photosynthesis. They are taxonomically and morphologically diverse. Some, such as members of the dodder family, are plantlike in appearance, with small leaf remnants and flowers (**Figure 36.12**). Some holoparasites do not have leaves or stems because they spend most of their life cycle underground and only break the surface to flower.

Several parasitic plant species lack many of the genes normally present in the chloroplast genome (which in turn is only a remnant of the genome of the original endosymbiont from which the chloroplast evolved; see Sections 5.5 and 27.1). These genes, which are needed for photosynthesis, have been lost because there is no evolutionary pressure to retain them. Thus while the parasitic lifestyle can be viewed as a free ride, for

The host goldenrod has scars from prior attachment sites.
Dodder flower buds
Tendrils of dodder
Host stem

36.12 A Parasitic Plant Tendrils of dodder (genus *Cuscuta*) wrap around a goldenrod (genus *Solidago*). The parasitic dodder obtains water, sugars, and other nutrients through tiny, rootlike protuberances that penetrate the surface of the host plant.

some plants it is also a one-way ticket, with no possibility of return to self-sufficiency.

The plant–parasite relationship is similar to plant–fungus and plant–bacteria associations

Plant–bacteria and plant–fungus associations both involve reciprocal signaling between the two species (see Figure 36.7). Parasitic plants also need to detect signals from nearby plants so they can grow toward them and obtain their nutrients, but obviously this is to the disadvantage of the potential host plant. In one interesting case, a parasitic plant has evolved the ability to recognize the chemical signals produced by plants to attract beneficial fungi.

The holoparasite *Striga* (witchweed) is a serious pest of cereal crops in Africa. In Section 36.4 we saw that arbuscular fungi are attracted to plant roots by compounds called strigolactones. One of these same molecules induces the seed germination of some parasitic plants, including *Striga*. Scientists strongly suspect that this is no coincidence. The mycorrhizal interaction is ancient (more than 400 million years old) and predates the evolution of parasitic plants. For this reason scientists hypothesize that a mechanism evolved in the ancestors of modern *Striga* to recognize a compound that was already produced by plants to attract soil microbes.

In *Striga* we thus find an example of "opportunistic evolution"—that is, the repurposing of preexisting processes rather than the development of new processes. This is not the first time we have encountered this phenomenon in this chapter. Recall that the formation of nodules by rhizobia uses some of the same mechanisms used by arbuscular fungi to establish residence inside plant cells (see Figure 36.7). This implies an evolutionary connection between the two symbioses.

RECAP 36.5

Carnivorous plants supplement their nutrition by extracting materials from animals. Rapid reflexes have evolved in some of these plants for trapping their prey. Parasitic plants, by contrast, get some or all of their sustenance from other plants. Holoparasites cannot function as autotrophs, having lost chloroplast genes coding for photosynthetic machinery. At least one parasitic plant responds to the same signaling molecule that the host plant uses to attract beneficial fungi.

- Why do carnivorous plants eat animals? See p. 752
- How do the needs of holoparasitic plants differ from those of carnivorous plants? See p. 752
- What characteristics are shared among plant–parasite, plant–fungus, and plant–bacteria associations? See p. 753

?

What progress has been made in improving the nitrogen use efficiency of corn?

ANSWER

There has been significant progress in developing corn varieties with improved nitrogen use efficiency, using both conventional plant breeding (genetics) and biotechnology. The recent publication of the corn genome sequence will provide useful information about genes involved in nitrogen use. For example, the seed company Pioneer Hi-Bred is developing a strain of corn that produces a more efficient version of the enzyme glutamine synthetase (GS). GS adds ammonia to glutamate to form glutamine, and therefore plays an important role in nitrogen assimilation in plants. The new corn strain uses up to 20 percent less fertilizer to produce the same yields as other corn varieties.

CHAPTER SUMMARY 36

36.1 What Nutrients Do Plants Require?

- Plants are photosynthetic autotrophs that can produce all their organic molecules from carbon dioxide, water, and minerals, including a nitrogen source.
- **Mineral nutrients** are obtained from the **soil solution**.
- Plants require 14 **essential elements**, 6 of which are **macronutrients** and 8 of which are **micronutrients**. **Deficiency symptoms** suggest what essential element a plant lacks. **Review Table 36.1, Figure 36.1, ANIMATED TUTORIAL 36.1**
- The essential elements were discovered by growing plants **hydroponically**, meaning in solutions that lacked individual elements. **Review Figure 36.2**

36.2 How Do Plants Acquire Nutrients?

- Root growth allows plants, which are sessile, to search for mineral resources.
- Plants can regulate the uptake of nutrients by increasing the number or activity of active transport proteins in root epidermal cells. **Review Figure 36.3**

36.3 How Does Soil Structure Affect Plants?

- **Soils** contain water, air, and inorganic and organic substances. Soils have living (biotic) and nonliving (abiotic) components. **Review Figure 36.4**
- A soil typically consists of two or three horizontal zones called **horizons**. **Topsoil** forms the uppermost or **A horizon**. Topsoil tends to lose mineral nutrients through **leaching**. **Loams** are excellent agricultural topsoils, with a good balance of sand, silt, clay, and organic matter. **Review Figure 36.5**
- Soils form by mechanical and chemical **weathering** of rock. Chemical weathering imparts mineral nutrients to clay particles. Plant litter and other organic matter decompose to form **humus**. Plants obtain some mineral nutrients through **cation exchange** between the soil solution and the surface of clay particles. **Review Figure 36.6**
- Farmers use **fertilizers** to make up for deficiencies in soil mineral nutrient content.

continued

36.4 How Do Fungi and Bacteria Increase Nutrient Uptake by Plant Roots?

- Mycorrhizae are symbiotic root–fungus associations that greatly increase a plant's absorption of water and minerals, especially phosphorus. They occur in more than 90 percent of terrestrial plant species.
- The arbuscules are the sites of nutrient exchange between the fungus and plant. **Review Figure 36.7**
- In the earliest stages of mycorrhiza formation, the hyphae of arbuscular fungi grow toward **strigolactones**, compounds that are produced by the plant roots.
- Some **nitrogen fixers** live free in soil or water; others live symbiotically as **bacteroids** within plant roots. The formation of a root nodule requires interaction between the root system of a legume and rhizobia. **Review Figure 36.7**
- Several steps in the formation of root nodules and arbuscules are similar and probably involve some of the same plant genes. **Review Figure 36.8**
- In **nitrogen fixation**, nitrogen gas (N_2) is reduced to ammonia (NH_3) or ammonium ions (NH_4^+) in a reaction catalyzed by **nitrogenase**. **Review Figure 36.9**
- Plants and bacteria interact in the **global nitrogen cycle**, which involves a series of reductions and oxidations of nitrogen-containing molecules. **Review Figure 36.10, ACTIVITY 36.1**

36.5 How Do Carnivorous and Parasitic Plants Obtain a Balanced Diet?

- **Carnivorous plants** are autotrophs that supplement a low nitrogen supply by feeding on insects or other small animals.
- **Parasitic plants** draw on other plants to meet their needs, which may include minerals, water, or the products of photosynthesis.
- **Hemiparasites**, such as mistletoes, can still photosynthesize. **Holoparasites** cannot function as autotrophs because they have lost chloroplast genes that code for components of the photosynthetic apparatus (which they no longer need).
- A strigolactone—a compound in the same category of compounds plants use to attract mycorrhizal fungi—also induces the germination of some parasitic plants, including *Striga*. Scientists hypothesize that a mechanism evolved in the ancestors of modern *Striga* to recognize a compound that was already produced by plants to attract arbuscular fungi.

Go to the Interactive Summary to review key figures, Animated Tutorials, and Activities Life10e.com/is36

CHAPTER**REVIEW**

REMEMBERING

1. Macronutrients
   a. are so called because they are more essential than micronutrients.
   b. include manganese, boron, and zinc, among others.
   c. function as catalysts.
   d. are required in concentrations of at least 1 gram per kilogram of plant dry matter.
   e. are obtained by the process of photosynthesis.

2. Which of the following is *not* an essential mineral element for plants?
   a. Potassium
   b. Magnesium
   c. Calcium
   d. Lead
   e. Phosphorus

3. Fertilizers
   a. always have a defined chemical composition.
   b. are not required if crops are removed frequently enough.
   c. restore needed mineral nutrients to the soil.
   d. are needed to provide carbon, hydrogen, and oxygen to plants.
   e. are needed to destroy soil pests.

4. In a typical soil,
   a. the topsoil tends to lose mineral nutrients by leaching.
   b. there are four or more horizons.
   c. the C horizon consists primarily of loam.
   d. the dead and decaying organic matter gathers in the B horizon.
   e. more clay means more air space and thus more oxygen for roots.

5. Nitrogen fixation is
   a. performed only by plants.
   b. the oxidation of nitrogen gas.
   c. catalyzed by the enzyme nitrogenase.
   d. a single-step chemical reaction.
   e. possible because N_2 is a highly reactive substance.

6. Which of the following is true of the formation of *both* arbuscules and root nodules?
   a. Invasion of a plant root by a fungus
   b. Invasion of a plant root by a bacterium
   c. Strigolactones produced by the root are recognized by the microbe
   d. Root cells are invaded but there is no direct contact between plant and microbe cell contents
   e. Root cells are invaded and there is direct contact between plant and microbe cell contents

UNDERSTANDING & APPLYING

7. Methods for determining whether a particular element is essential have been known for more than a century. Since these methods are so well established, why was the essentiality of some elements discovered only recently?

8. Soils are dynamic systems. What changes might result when land is subjected to heavy irrigation for agriculture after being relatively dry for many years? What changes in the soil might result when a virgin deciduous forest is cut down and replaced by crops that are harvested each year?

9. The biosphere of Earth as we know it depends on the existence of a few species of nitrogen-fixing prokaryotes. What do you think might happen if one of these species were to become extinct? If all of them were to disappear? (See also Figure 58.13.)

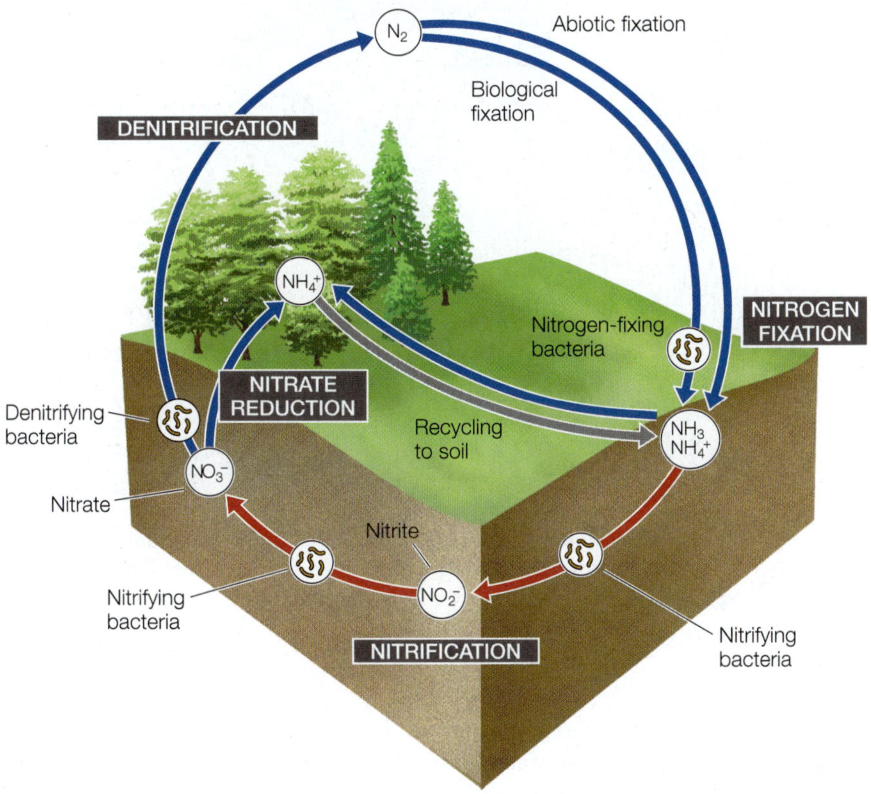

N_2
Abiotic fixation
Biological fixation
DENITRIFICATION
NH_4^+
Nitrogen-fixing bacteria
NITROGEN FIXATION
NITRATE REDUCTION
Denitrifying bacteria
Recycling to soil
NH_3
NH_4^+
NO_3^-
Nitrate
Nitrite
Nitrifying bacteria
NO_2^-
NITRIFICATION
Nitrifying bacteria

ANALYZING & EVALUATING

10. Some mutant *Arabidopsis* plants that are very bushy (their shoots are more highly branched than wild-type plants) cannot make strigolactones because of a mutation in a gene necessary for strigolactone biosynthesis. If an investigator applies strigolactones to the plants, they grow normally. What does this experiment suggest about the role of strigolactones in plant growth? How does this add to the story of strigolactones as signals for arbuscules and parasitic plants?

11. Holoparasitic plants have lost many of the morphological and genetic traits necessary for an autotrophic lifestyle. From an evolutionary point of view, how do you think this happened? (Hint: think about selection pressures.)

Go to BioPortal at **yourBioPortal.com** for Animated Tutorials, Activities, LearningCurve Quizzes, Flashcards, and many other study and review resources.

37 Regulation of Plant Growth

CHAPTER**OUTLINE**

Norman Borlaug Seen here in a field of semi-dwarf wheat, plant geneticist Norman Borlaug carried out a program of genetic crosses that led to high-yielding varieties and saved millions from starvation.

AGRICULTURAL SCIENTISTS are constantly searching for ways to help farmers produce more food for a growing population. One way is to breed crop plants whose physiology allows them to produce more grain per plant (resulting in higher yields). The drawback of this approach is that the sheer weight of the load of seeds may cause the stem to bend over. The problem is made worse when fertilizer makes the plants grow taller. Harvesting seeds on the ground is very difficult; think of how hard it would be to pick up seeds one by one, when some have already sprouted.

During World War II, the island nation of Japan was blockaded and could not import food or other supplies. Food was rationed and many people were hungry, but there were no major famines in Japan during that period. How were the Japanese able to produce enough grain to feed their population? One answer to this question lay in the fields: the Japanese had bred genetic strains of rice and wheat with short, strong stems that could bear high yields of grain without bending over. An agricultural advisor to the occupying American army sent samples of the grains to the U.S.

A decade later, the American plant geneticist Norman Borlaug, who was working in Mexico at the time, began making genetic crosses between the Japanese wheat and other varieties that had genes conferring rapid growth, adaptability to varying climates, and resistance to fungal diseases. The results were "semi-dwarf" wheat varieties that gave record yields. The varieties were grown first in Mexico, and later in India and Pakistan during the 1960s. At about the same time and using a similar strategy, scientists in the Philippines developed semi-dwarf rice with equally spectacular results. People who had lived on the edge of starvation now produced enough food. Countries that had relied on food from other countries were now able to grow more than enough grain, and export the surplus. The development of these semi-dwarf grains began what was called the "Green Revolution." Borlaug was awarded the Nobel Peace Prize for his research on wheat, which is estimated to have saved a billion lives.

> **?** What changes in growth patterns made the new strains of wheat and rice successful?
>
> See answer on p. 775.

37.1 How Does Plant Development Proceed?

As described in Chapter 34, plants are sessile organisms that must seek out resources above and below the ground. A number of unique features enable plants to obtain the resources they need to grow and reproduce:

- *Meristems*. Plants have permanent collections of stem cells (undifferentiated, constantly dividing cells) that allow them to continue growing throughout their lifetimes (see Section 34.3).
- *Post-embryonic organ formation*. Unlike animals, plants can initiate development of new organs such as leaves and flowers throughout their lifetimes.
- *Differential growth*. Plants can allocate their resources so that they grow more of the organs that will benefit them most—for example, more leaves to harvest more sunlight or more roots to obtain more water and nutrients.

Plants must continuously monitor their ever-changing environments and redirect their growth appropriately. For example, the amount of available light changes from day to night and from season to season. In addition, other plants are often vying for what light there is, and plants modulate their growth to compete with their neighbors for this precious resource. As you will see in this chapter, several mechanisms have evolved in plants that enable them to sense changes in their environments, and trigger appropriate growth responses.

The development of a plant—the series of progressive changes that take place throughout its life—is regulated in many ways. Key factors involved in regulating plant growth and development are:

- *Environmental cues*, such as day length, water availability, and various chemicals in the environment
- *Receptors* that allow a plant to sense environmental cues, such as photoreceptors that absorb light, and chemoreceptors that signal the presence of pathogens (see Chapter 39)
- *Hormones*—chemical signals that mediate the effects of the environmental cues, including those sensed by receptors
- *Regulatory proteins and enzymes* that catalyze the biochemical reactions of development

We will explore these regulatory mechanisms in more detail later in this chapter. But first let's look at the initial steps of plant development—from seed to seedling—and the types of internal and external cues that guide them.

In early development, the seed germinates and forms a growing seedling

Chapter 38 will describe the events of plant reproduction and development that lead to the formation of seeds. Here we begin with the seed, the structure that contains the early embryo. Unlike most animal embryos, plant seeds may be held in "suspended animation," with the development of the embryo halted, for long periods. If development stops even when external conditions (such as water supply) are adequate for development, the seed is said to be **dormant**.

DORMANCY Seed dormancy may last for weeks, months, or years. Plants use several mechanisms to maintain dormancy:

- *Exclusion of water or oxygen* from the embryo by an impermeable seed coat
- *Mechanical restraint* of the embryo by a tough seed coat
- *Chemical inhibition* of germination by growth regulators
- *Photodormancy*: some seeds need a period of light or dark before they can germinate
- *Thermodormancy*: some seeds need high or low temperatures to germinate

Dormancy can be broken by factors that overcome these mechanisms. For example, the seed coat may be damaged by passage through an animal's digestive system, or heavy rains may wash away chemical inhibitors. There are many unusual methods to overcome dormancy. One interesting example is the breaking of dormancy by components of smoke. *Emmenanthe penduliflora* is a common plant in dry chaparral of the southwestern U.S., an area that is prone to wildfires.

Emmenanthe penduliflora seeds germinate after exposure to smoke.

These plants germinate rapidly after a fire. John Keeley of Occidental College in Los Angeles found that dormancy in seeds of this plant is broken not by heat but by smoke—in particular, by the nitrogen oxides found in smoke. Other molecules in smoke have been identified that regulate seed germination.

Plant biologists distinguish between seed *dormancy*, which prevents germination under conditions that are suitable for plant growth, and seed *quiescence*, which occurs when a seed fails to germinate because conditions are unfavorable for growth. Some seeds may remain quiescent, yet viable, for centuries: botanists have germinated a 1,300-year-old lotus seed recovered from a dry lake bed in China.

Seed dormancy and quiescence are common, so they must provide selective advantages for plants. Dormancy ensures that the seed will germinate at a time suitable for the plant to complete its life cycle. For example, some seeds require exposure to a long cold period (winter) before they germinate in the spring; this ensures that the plant has the entire growing season to mature and set new seeds. Dormancy and quiescence also help seeds survive droughts or long-distance dispersal, allowing plants to colonize new territory.

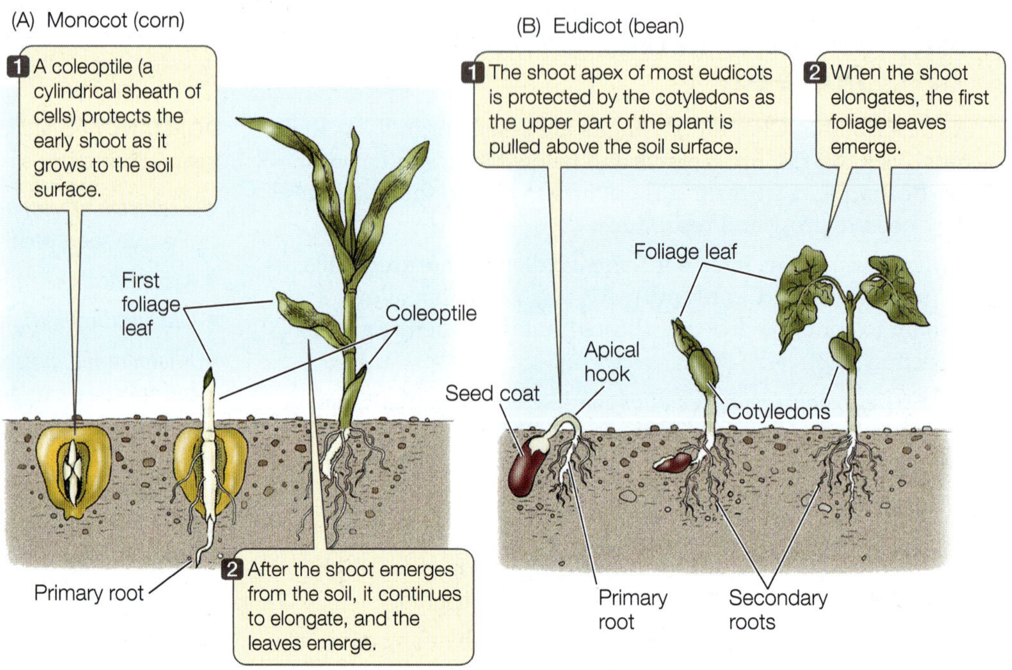

(A) Monocot (corn)
1 A coleoptile (a cylindrical sheath of cells) protects the early shoot as it grows to the soil surface.
First foliage leaf
Coleoptile
Primary root
2 After the shoot emerges from the soil, it continues to elongate, and the leaves emerge.
(B) Eudicot (bean)
1 The shoot apex of most eudicots is protected by the cotyledons as the upper part of the plant is pulled above the soil surface.
2 When the shoot elongates, the first foliage leaves emerge.
Foliage leaf
Apical hook
Seed coat
Cotyledons
Primary root
Secondary roots

37.1 Patterns of Early Shoot Development (A) In grasses and some other monocots, growing shoots are protected by a coleoptile until they reach the soil surface. (B) In most eudicots, the growing point of the shoot is protected within the cotyledons.

Go to Activity 37.1 Monocot Shoot Development
Life10e.com/ac37.1

Go to Activity 37.2 Eudicot Shoot Development
Life10e.com/ac37.2

GERMINATION Seeds begin to **germinate**, or sprout, when dormancy is broken and environmental conditions are satisfactory. The first step in germination is the uptake of water, called **imbibition** (from *imbibe*, "to drink in"). Before germination, a seed contains very little water: only 5 to 15 percent of its weight is water, compared with 80 to 95 percent for most other plant parts. Seeds also contain polar macromolecules, such as cellulose and starch, which attract and bind polar water molecules. Consequently a seed has a very negative water potential (see Chapter 35) and will take up water if the seed coat is permeable to water. The force exerted by imbibing seeds, which expand several-fold in volume, demonstrates the magnitude of their water potential; for example, imbibing cocklebur seeds can exert a pressure of up to 1,000 atmospheres.

As a seed takes up water, it undergoes metabolic changes: enzymes are activated upon hydration, RNA and then proteins are synthesized, the rate of cellular respiration increases, and other metabolic pathways are activated. In many seeds, cell division is not initiated during the early stages of germination. Instead, growth results solely from the expansion of small, preformed cells.

As germination proceeds, starch, proteins, and lipids that are stored in the seed are hydrolyzed to provide metabolic energy and chemical building blocks—carbohydrates, amino acids, and lipid monomers—for the growing embryo. These reserves are stored in the cotyledons (the first leaf or leaves of the embryo) or in the endosperm (the non-embryonic storage tissue of the seed). Germination is completed when the **radicle** (embryonic root) emerges from the seed coat. The plant is then called a **seedling**.

If the seed germinates underground, the new seedling must elongate rapidly (in the right direction!) and cope with a period of life in darkness or dim light. Photoreceptors that sense light and specialized cells that sense gravity direct this stage of development and prepare the seedling for growth in the light.

The pattern of early shoot development varies among the flowering plants. **Figure 37.1** shows the shoot development patterns of monocots and eudicots. In monocots, the growing shoot is protected by a sheath of cells called the **coleoptile** as it pushes its way through the soil. In dicots, the shoot is protected by the cotyledons.

Several hormones and photoreceptors help regulate plant growth

The above description of the early stages of plant development illustrates some of the environmental cues that influence plant growth. A plant's responses to these cues involve signal transduction pathways. Various mechanisms are used by the plant to sense changes in the environment, and these mechanisms activate signal transduction pathways that result in the synthesis and activation of specific plant hormones. In turn, these hormones act as signals that trigger pathways resulting in changes in plant growth. In some cases these changes involve alterations in the expression of specific genes.

Hormones are chemical signals that act at very low concentrations at sites often distant from where they are produced. Most plant hormones are very different from their counterparts in animals (**Table 37.1**). Each plant hormone plays multiple

TABLE **37.1**
Comparison of Plant and Animal Hormones

Characteristic	Plant Hormones	Animal Hormones
Size, chemistry	Small organic molecules	Peptides, proteins, small molecules
Site of synthesis	Many locations in the plant	Specialized glands or cells
Site of action	Local or distant	Distant, transported
Effects	Often diverse	Often specific
Regulation	By biochemical feedback	By feedback and by central nervous system

TABLE 37.2
Plant Growth Hormones

Hormone	Structure	Typical Activities
Abscisic acid*	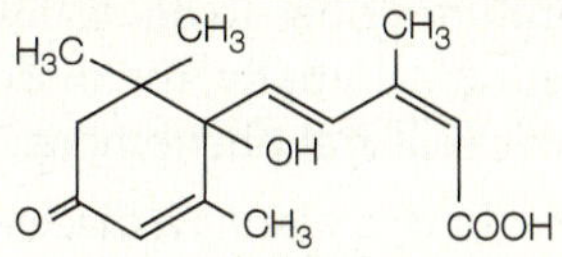	Maintains seed dormancy; closes stomata
Auxins (mainly indole-3-acetic acid)	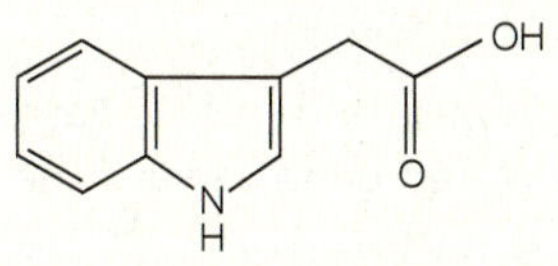	Promote stem elongation, adventitious root initiation, and fruit growth; inhibit axillary bud outgrowth, leaf abscission, and root elongation
Brassinosteroids	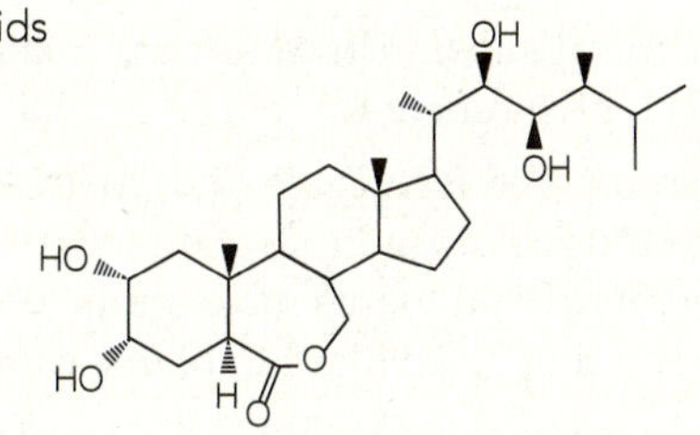	Promote stem and pollen tube elongation; promote vascular tissue differentiation
Cytokinins	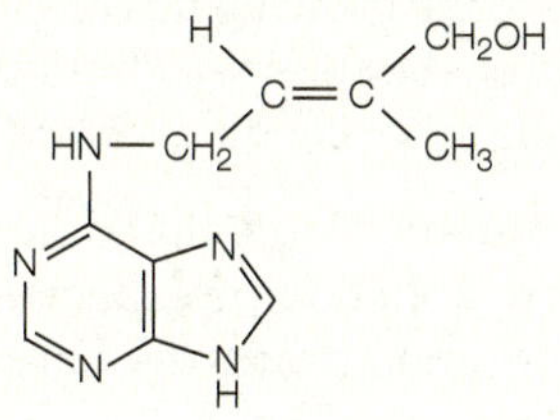	Inhibit leaf senescence; promote cell division and axillary bud outgrowth; affect root growth
Ethylene	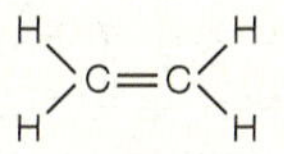	Promotes fruit ripening and leaf abscission; inhibits stem elongation and gravitropism
Gibberellins (e.g., gibberellic acid)	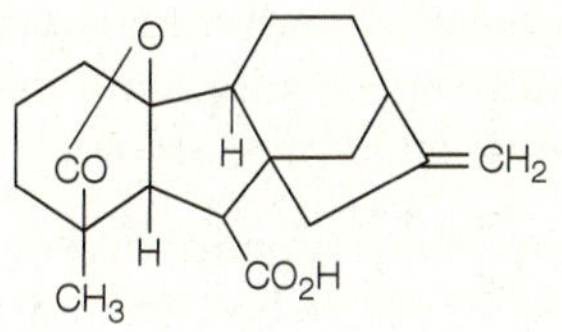	Promote seed germination, stem growth, and ovule and fruit development; break winter dormancy; mobilize nutrient reserves in grass seeds

*See Chapter 38.

regulatory roles, and interactions among them can be complex. Several hormones regulate the growth and development of plants from seedling to adult (**Table 37.2**). Other hormones (for example, jasmonic acid and salicylic acid) are involved in the plant's defenses against herbivores and microorganisms, as we will discuss in Chapter 39.

Perhaps the most important environmental cue for a plant is light: the source of energy for photosynthesis. Plants have an abundance of **photoreceptors** that detect changes in the quality and direction of light as well as the timing of light availability (daylength). Photoreceptors are often proteins associated with pigments. Light acts directly on photoreceptors, which in turn regulate developmental processes that need to be responsive to light, such as the many changes that occur as a seedling emerges from the soil.

Genetic screens have increased our understanding of plant signal transduction

In Chapter 19 we described how genetic studies can be used to identify the steps along a developmental pathway. The reasoning behind these experiments is that if a mutation in a specific gene disrupts a developmental process, then the product of that gene must be involved in that process. Similarly, genetic studies can be used to analyze pathways of receptor activation and signal transduction in plants: if proper signaling does not occur in a mutant plant, then the mutant gene must be involved in the signal transduction process. Mapping the mutant gene and identifying its function is a starting point for understanding the signaling pathway. *Arabidopsis thaliana* has been a major model organism for plant biologists investigating signal transduction.

One technique for identifying the genes involved in a plant signal transduction pathway is illustrated in **Figure 37.2**. This technique, called a **genetic screen**, involves creating a large, random collection of mutant plants and identifying those individuals that are likely to have a defect in the pathway of interest. Plant genes can be randomly mutated in a variety of ways, including treatment with a chemical mutagen or the insertion of transposons (see Section 17.2) randomly in the genome. After treatment, the plants are grown and then examined for a specific phenotype, usually a characteristic that is influenced by the pathway of interest. Once mutant plants have been selected, their genotypes are compared with those of wild-type plants. *Arabidopsis* mutants with altered developmental patterns have provided a wealth of new information about the mechanisms of hormone and receptor (particularly photoreceptor) action.

RECAP 37.1

Plant development is under the control of external cues in the environment and is mediated by hormones. These signals activate pathways that may result in changes in gene expression. Genetic screens have been useful in describing signal transduction pathways in the model plant *Arabidopsis thaliana.* Seed dormancy often precedes seed germination.

- Under what circumstances is seed dormancy advantageous? **See p. 757**
- Describe how monocots and eudicots differ in early development. **See p. 758 and Figure 37.1**
- What is a genetic screen, and how can it be used to analyze the regulation of plant development? **See pp. 759–760 and Figure 37.2**

RESEARCH**TOOLS**

37.2 A Genetic Screen Genetics of the model plant *Arabidopsis thaliana* can be used to identify the steps of a signal transduction pathway. If a mutant strain does not respond to a hormone (in this case, ethylene), the corresponding wild-type gene must be essential for the pathway (in this case, ethylene response). This method has been instrumental to scientists in understanding plant growth regulation.

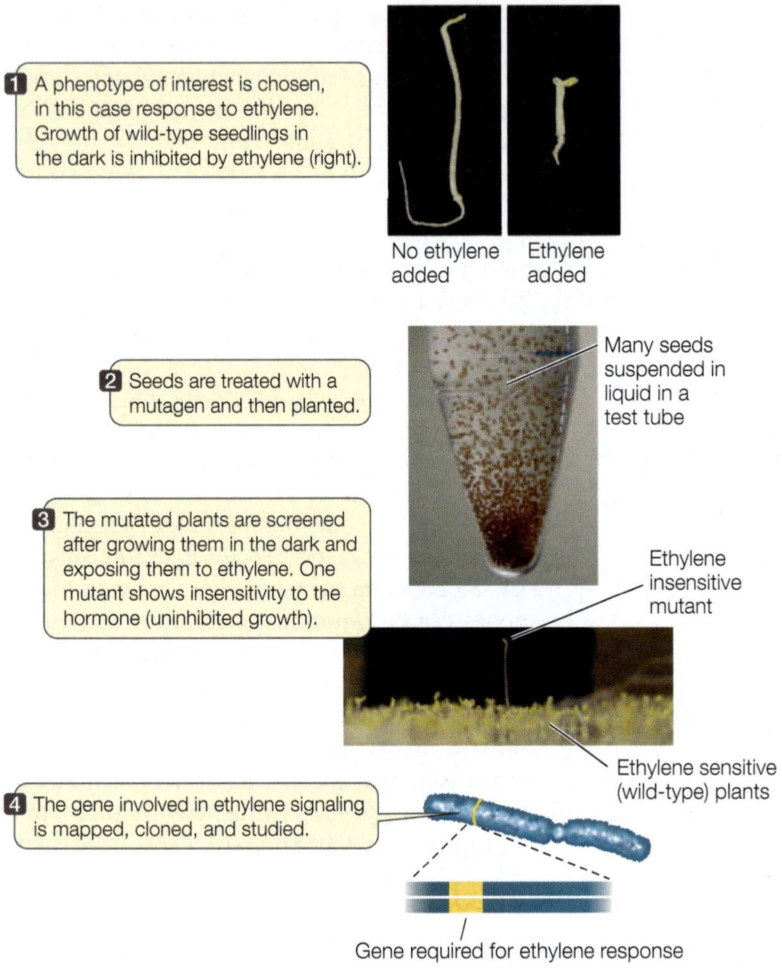
1 A phenotype of interest is chosen, in this case response to ethylene. Growth of wild-type seedlings in the dark is inhibited by ethylene (right).
No ethylene added
Ethylene added
2 Seeds are treated with a mutagen and then planted.
Many seeds suspended in liquid in a test tube
3 The mutated plants are screened after growing them in the dark and exposing them to ethylene. One mutant shows insensitivity to the hormone (uninhibited growth).
Ethylene insensitive mutant
Ethylene sensitive (wild-type) plants
4 The gene involved in ethylene signaling is mapped, cloned, and studied.
Gene required for ethylene response

You have now seen the early stages of plant development and growth, and how the environment influences these processes. We will now turn to the subject of plant hormones, which are central to the internal regulation of development. We will describe how hormones were discovered and what physiological effects they have on plants. We will emphasize how genetic screens and other methods have led to a deeper molecular understanding of the action of plant hormones.

37.2 What Do Gibberellins and Auxin Do?

The discovery of two key plant hormones exemplifies the experimental approaches that plant biologists have used to investigate the mechanisms of plant development. **Gibberellins** (there are several active forms) and **auxin** (there are several forms, but one predominates) were the first plant hormones to be discovered, early in the twentieth century. Initially, the discoveries came from observations of natural phenomena:

- *Gibberellins*: In rice plants, a disease caused by the fungus *Gibberella fujikori* resulted in plants that grew overly tall and spindly.
- *Auxin*: Biologists and indoor gardeners noted that seedlings would bend toward the light when placed near a light source.

A chemical substance was then isolated that could cause each phenomenon:

- Gibberellic acid (see Table 37.2) made by the *G. fujikori* fungus caused rice plants to overgrow. Later it was found that plants make gibberellic acid as well, and that applying it to plants caused growth.
- Indole-3-acetic acid (see Table 37.2) applied asymmetrically to the growing tips of seedlings caused cell elongation on the side away from the light, which resulted in the shoot bending toward the light.

Finally, mutant plants that do not make each hormone exhibit a phenotype expected in the absence of the hormone, and adding the hormone reverses that phenotype (**Figure 37.3**):

- Tomato plants that do not make gibberellic acid are very short; supplying them with the hormone results in normal growth.
- *Arabidopsis thaliana* individuals that do not make auxin are also short; supplying them with that hormone reverses that phenotype.

Note that the phenotype involved—short stature, or dwarfism—is similar in both mutant plants. This observation exemplifies a concept that is important to keep in mind when studying plant hormones: their actions are not unique and specific, as is the case with animal hormones (see Table 37.1).

The approaches outlined above—observation, hormone isolation, hormone treatment, and analysis of mutants—are just some of the methods used to identify plant hormones and understand their roles in plant development. Plant biologists have also studied hormones using chemical inhibitors and using plant transformation experiments that alter hormone levels or the plants' responses to hormones.

Gibberellins have many effects on plant growth and development

The functions of gibberellins can be inferred from the effects of experimentally decreasing gibberellins or blocking their action at various points in plant development. Such experiments reveal that gibberellins have multiple roles in regulating plant growth.

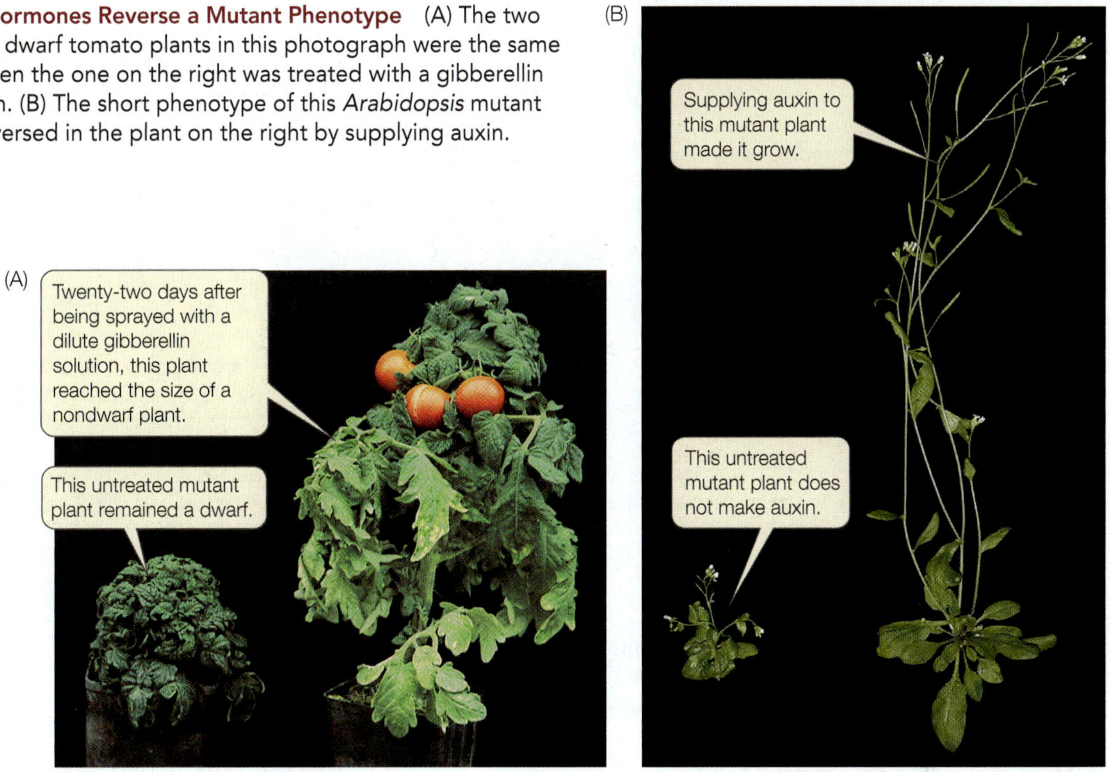

(A)
Twenty-two days after being sprayed with a dilute gibberellin solution, this plant reached the size of a nondwarf plant.
This untreated mutant plant remained a dwarf.
(B)
Supplying auxin to this mutant plant made it grow.
This untreated mutant plant does not make auxin.

37.3 Hormones Reverse a Mutant Phenotype (A) The two mutant dwarf tomato plants in this photograph were the same size when the one on the right was treated with a gibberellin solution. (B) The short phenotype of this *Arabidopsis* mutant was reversed in the plant on the right by supplying auxin.

STEM ELONGATION The effects of gibberellins on wild-type plants are not as dramatic as those seen on dwarf plants. However, gibberellins are indeed active in wild-type plants, because inhibitors of gibberellin synthesis cause a reduction in stem elongation. Such inhibitors can be put to practical uses. For example, plants such as chrysanthemums that are grown in greenhouses tend to get tall, but leggy plants do not appeal to consumers. Flower growers spray such plants with gibberellin synthesis inhibitors to control their height. Some wheat crops are similarly sprayed to keep them short, so they do not fall over when they produce grain; the result is chemically produced semi-dwarfs similar to the genetically produced varieties described in the opening story of this chapter. In some plants, such as cabbage, the normal growth habit is to be a squat, leafy head near the ground. When environmental signals are right, however, the plant "bolts," quickly producing a tall stem with flowers—a response that can be mediated by gibberellins.

FRUIT GROWTH Gibberellins and other hormones regulate the growth of fruits. Grapevines that produce seedless grapes develop smaller fruits than varieties that produce seed-bearing grapes. Biologists wanting to explain this phenomenon removed seeds from immature seeded grapes and found that this prevented normal fruit growth, suggesting that the seeds are sources of a growth regulator. Biochemical studies showed that developing seeds produce gibberellins, which diffuse into the immature fruit tissue. Spraying young seedless grapes with a gibberellin solution causes them to grow as large as seeded ones, and this is now a standard commercial practice (**Figure 37.4**).

MOBILIZATION OF SEED RESERVES Early in seed germination, hydrolytic enzymes are produced to break down stored reserves of starch, proteins, and lipids. Just after imbibition in germinating seeds of barley and other cereals, the embryo secretes gibberellins. The hormones diffuse through the endosperm to a surrounding tissue called the **aleurone layer**, which lies underneath the seed coat. The gibberellins trigger a cascade of events in the aleurone layer, causing it to synthesize and secrete enzymes that hydrolyze proteins and starch stored in the endosperm (**Figure 37.5**). These observations have practical importance: in the beer brewing industry, gibberellins are used to enhance the "malting" (germination) of barley and the breakdown of its endosperm, producing sugar that is fermented to alcohol.

37.4 Gibberellins and Fruit Growth Spraying developing seedless grapes with gibberellins (right) increases their size compared with untreated fruit (left).

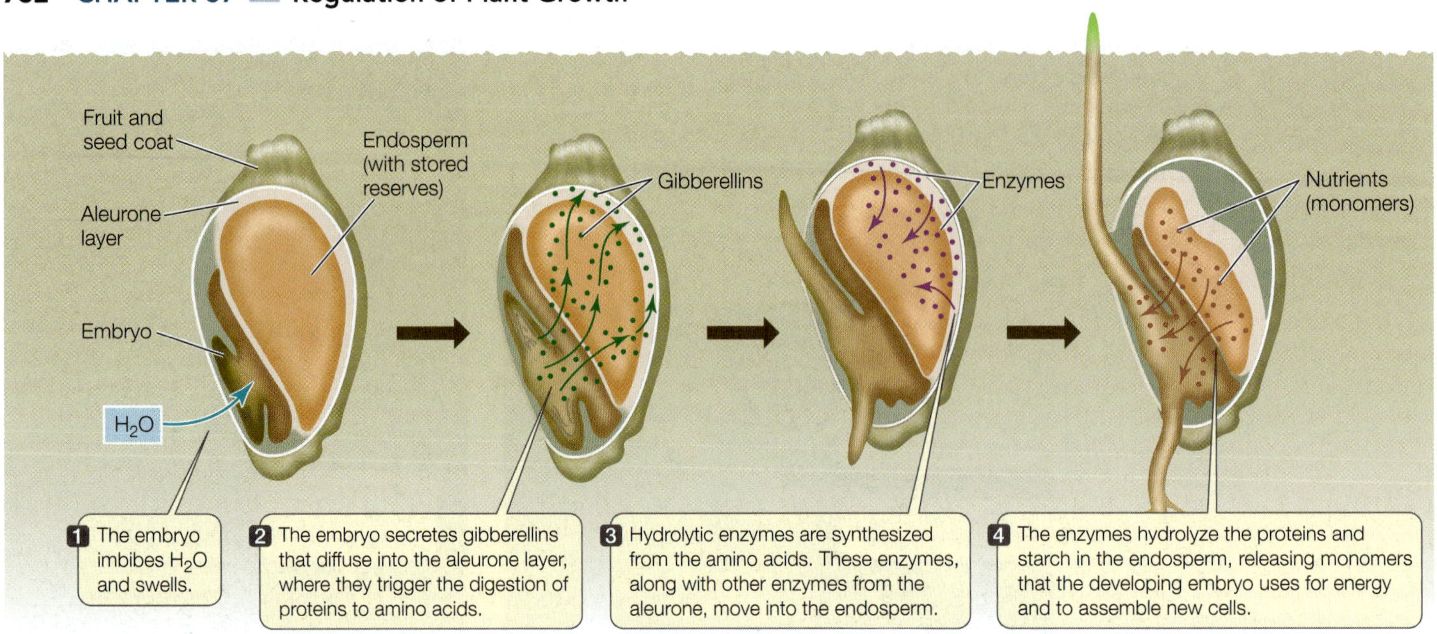
Fruit and seed coat
Aleurone layer
Endosperm (with stored reserves)
Embryo
H_2O
Gibberellins
Enzymes
Nutrients (monomers)
1 The embryo imbibes H_2O and swells.
2 The embryo secretes gibberellins that diffuse into the aleurone layer, where they trigger the digestion of proteins to amino acids.
3 Hydrolytic enzymes are synthesized from the amino acids. These enzymes, along with other enzymes from the aleurone, move into the endosperm.
4 The enzymes hydrolyze the proteins and starch in the endosperm, releasing monomers that the developing embryo uses for energy and to assemble new cells.

37.5 Embryos Mobilize Their Reserves During seed germination in cereal grasses, gibberellins trigger a cascade of events that result in the conversion of starch and protein reserves into monomers that can be used by the developing embryo.

Go to Activity 37.3 Events of Seed Germination
Life10e.com/ac37.3

Another hormone, abscisic acid (ABA) (see Table 37.2) plays an antagonistic role with gibberellins in seed dormancy and germination. As we will describe in Section 38.1, ABA levels are high in dormant seeds and fall off during germination. Thus ABA plays a role in maintaining seed dormancy, whereas gibberellins function to break dormancy and promote germination.

Auxin plays a role in differential plant growth

Auxin was first discovered for its role in the bending of plants toward light. Subsequent research has shown that auxin is involved in many other aspects of plant growth and development.

IDENTIFYING AUXIN AND ITS TRANSPORT Auxin (from the Latin "to increase") was discovered in the context of **phototropism**: a response to light in which plant stems bend toward a light source. This was a familiar observation by biologists and home gardeners when the ever-curious Charles Darwin and his son Francis investigated this phenomenon and concluded that a signal was made in the shoot apex and diffused down the shoot in a polar (unidirectional) fashion, stimulating cell elongation.

Go to Animated Tutorial 37.1 Tropisms
Life10e.com/at37.1

The Darwins worked with canarygrass (*Phalaris canariensis*) seedlings grown in the dark. While underground, a young grass seedling is protected by a coloeptile (see Figure 37.1). The coleoptiles of grasses are phototropic—they grow toward the light.

To find the light-receptive region of the coleoptile, the Darwins "blindfolded" the coleoptiles of dark-grown canarygrass seedlings in various places and then illuminated them from one side (**Figure 37.6**). The coleoptile grew toward the light whenever its tip was exposed. If the top millimeter or more of the coleoptile was covered, however, the coleoptile showed no phototropic response. The Darwins concluded that the tip contains the photoreceptor that responds to light. The actual bending toward the light, however, takes place in a growing region a few millimeters below the tip. Therefore, the Darwins reasoned, *some type of signal must travel from the tip of the coleoptile to the growing region.*

Others showed that placing the coleoptile tip (the source of the growth signal) on a decapitated coleoptile led to bending, even when a block of gelatin separated the tip and coleoptile.

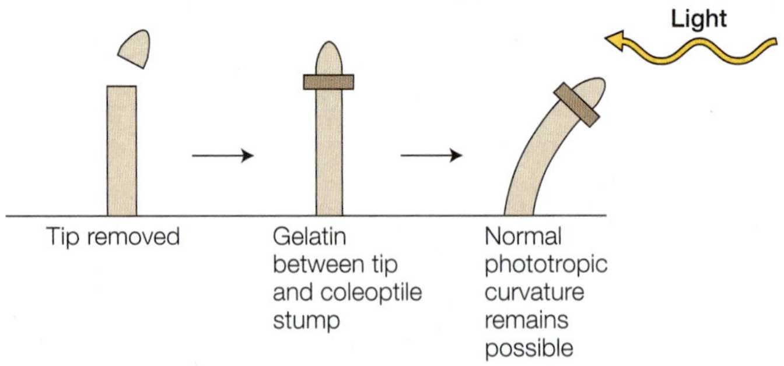
Light
Tip removed
Gelatin between tip and coleoptile stump
Normal phototropic curvature remains possible

In these experiments, the growth signal moved from the coleoptile tip into the gelatin and then down into the decapitated coleoptile; later the growth signal was isolated from such gelatin blocks and identified as indole-3-acetic acid (see Table 37.2), the major form of auxin.

Go to Animated Tutorial 37.2 Went's Experiment
Life10e.com/at37.2

INVESTIGATING**LIFE**

37.6 The Darwins' Phototropism Experiment Charles Darwin and his son Francis wanted to know how plants bend toward the light. They grew canarygrass seedlings (coleoptiles) in the dark. To discover what part of the coleoptile responds to light, they covered up ("blindfolded") different regions of each coleoptile and then exposed the seedlings to light from one side. The Darwins discovered that the tip of the seedling senses the light and that growth occurs below the tip. Their observations led them to hypothesize the existence of a growth-promoting signal produced by the coleoptile tip.[a]

HYPOTHESIS Only part of the coleoptile senses the light that triggers phototropism.

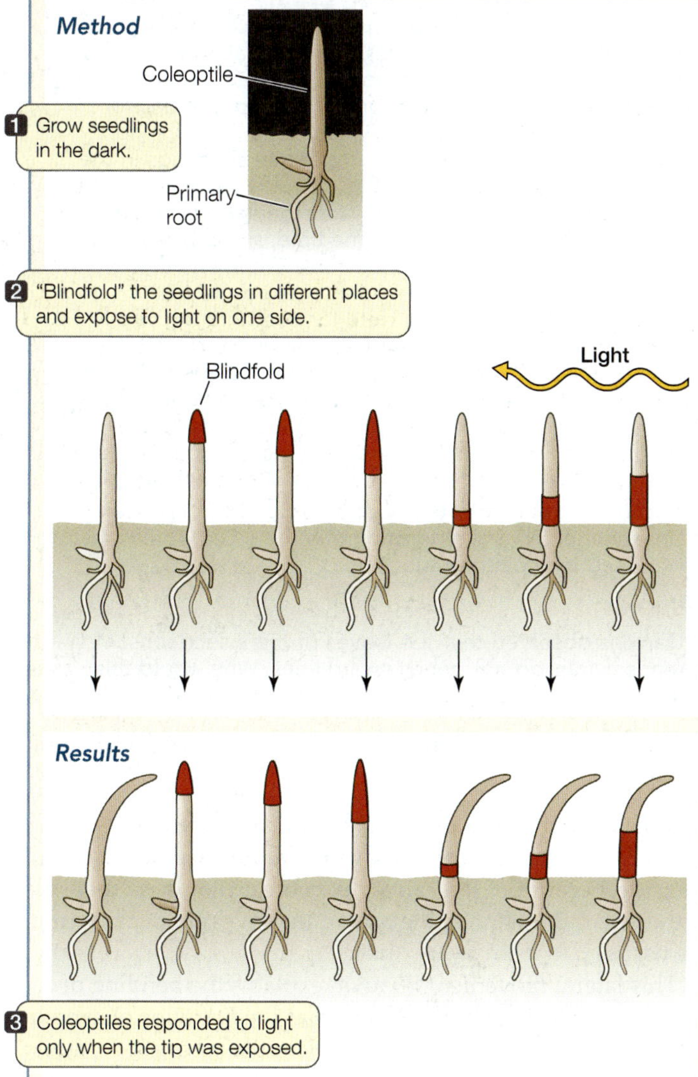

Method
Coleoptile
1 Grow seedlings in the dark.
Primary root
2 "Blindfold" the seedlings in different places and expose to light on one side.
Blindfold
Light
Results
3 Coleoptiles responded to light only when the tip was exposed.

CONCLUSION The part of the coleoptile that senses light is in the tip, and it sends a signal from the tip to the growing region.

Go to **BioPortal** for discussion and relevant links for all INVESTIGATING**LIFE** figures.

[a]Darwin, C. R. 1880. *The Power of Movement in Plants*. London, John Murray.

MECHANISM OF AUXIN TRANSPORT The movement of auxin down a coleoptile is an example of polar (apical-to-basal) transport. Polar transport of auxin (**Figure 37.7**) depends on four biochemical processes that may be familiar from earlier chapters:

- *Diffusion across a plasma membrane*. Polar molecules (in the chemical sense) diffuse across plasma membranes less readily than nonpolar molecules (see Chapter 6).
- *Membrane protein asymmetry*. Active transport carriers (see Chapter 6) for auxin are located only in the portion of the plasma membrane at the basal (bottom) end of the cell.
- *Proton pumping/chemiosmosis*. A proton pump (see Chapter 36) moves H^+ from the cytoplasm to the cell wall, thereby increasing the intracellular pH and decreasing the pH in the cell wall. Proton pumping also sets up an electrochemical gradient, which provides potential energy to drive the transport of auxin by the carriers mentioned above (see Chapter 9).
- *Ionization of a weak acid*. The main form of auxin, indole-3-acetic acid, is a weak acid:

$$A^- + H^+ \rightleftharpoons HA$$

When the pH is low, this reaction is driven to the right, and HA (non-ionized auxin) is the predominant form. When the pH is higher, there is more A^- (ionized auxin).

While polar auxin transport distributes the hormone along the *longitudinal* axis of the plant, *lateral* (side-to-side) redistribution

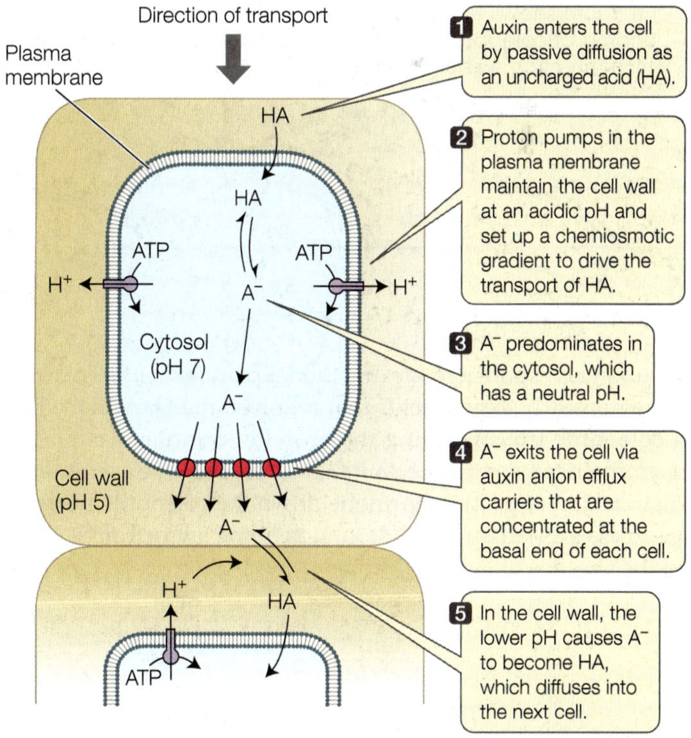

Direction of transport
Plasma membrane
HA
HA
ATP
ATP
H+
H+
A–
Cytosol (pH 7)
A–
Cell wall (pH 5)
A–
H+
HA
ATP
1 Auxin enters the cell by passive diffusion as an uncharged acid (HA).
2 Proton pumps in the plasma membrane maintain the cell wall at an acidic pH and set up a chemiosmotic gradient to drive the transport of HA.
3 A– predominates in the cytosol, which has a neutral pH.
4 A– exits the cell via auxin anion efflux carriers that are concentrated at the basal end of each cell.
5 In the cell wall, the lower pH causes A– to become HA, which diffuses into the next cell.

37.7 Polar Transport of Auxin Proton pumps set up a chemiosmotic gradient directing ionized auxin (A^-) toward the basally placed active transport carriers for auxin, which leads to a net movement of auxin in a basal direction.

WORKING WITH**DATA**:

The Darwins' Phototropism Experiment

Original Paper

Darwin, C. R. 1880. *The Power of Movement in Plants.* Chapter IX: Sensitiveness of plants to light: Its transmitted effects. London, John Murray.

Analyze the Data

Fascinated by experiments showing that light induces plants to bend toward it and that this is due to increased growth on the "dark side" of the stem, Charles Darwin and his son Francis performed experiments to learn more about how plants respond to light. Their experiments were done on canarygrass (*Phalaris canariensis*) seedlings grown in the dark to simulate events that occur when the seed germinates in the soil. These seedlings have a sheath—the coleoptile—that covers the developing leaf. When placed in a box with an opening at a window (unidirectional light), the seedlings bent toward the light source. This bending seemed to be led by the very tip of the coleoptile, and the Darwins hypothesized that the light-sensing mechanism might reside in the coleoptile, while growth occurs in tissue below the tip. Their hypothesis was validated when they repeated their experiments with the coleoptile tip cut off: bending did not occur. Because it was possible that the cut plants did not bend because they had been irreversibly damaged in some way, the Darwins repeated their initial experiment with undamaged seedlings whose coleoptile tips had been covered with a thin glass tube cap that was blacked out with India ink. In this case the plants did not bend. The Darwins' conclusion that the tip of the coloptile senses the light and transmits a message to tissue below the tip to grow was initially controversial but ultimately led to the identification of auxin, the first plant growth hormone.

QUESTION 1

The figure shows a drawing from the Darwins' book of the bending of coleoptiles after 8 hours of light exposure. From which direction was the light shone?

Fig. 181.

QUESTION 2

The Darwins reported:

> "Seven cotyledons [Note: The Darwins used the term "cotyledon" for what is now called a coleoptile.] had their tips cut off for lengths varying between 0.1 and 0.16 of an inch, and these, when left exposed all day to a lateral light, remained upright. In another set of 7 cotyledons, the tips were cut off for a length of only about 0.05 of an inch (1.27 mm) and these became bowed towards a lateral light, but not nearly so much as the many other seedlings in the same pots."

What do these data indicate about the possible role of the tip and about the possibility that injury in cutting blocks the bending response?

QUESTION 3

The Darwins describe their further experiments:

> "The summits of nine cotyledons, differing somewhat in height, were enclosed for rather less than half their lengths in uncoloured or transparent tubes; and these were then exposed before a south-west window on a bright day for 8 h. All of them became strongly curved towards the light, in the same degree as the many other free seedlings in the same pots; so that the glass-tubes certainly did not prevent the cotyledons from bending towards the light. Nineteen other cotyledons were, at the same time, similarly enclosed in tubes thickly painted with Indian ink. On five of them, the paint, to our surprise, contracted after exposure to the sunlight, and very narrow cracks were formed, through which a little light entered; and these five cases were rejected. Of the remaining 14 cotyledons, the lower halves of which had been fully exposed to the light for the whole time, 7 continued quite straight and upright; 1 was considerably bowed to the light, and 6 were slightly bowed, but with the exposed bases of most of them almost or quite straight."

What do these data indicate about the role of the tip? Can you explain why there was slight bending in the 6 coleoptiles that were covered with painted tubes?

QUESTION 4

The Darwins observed that the leaves of the insect-consuming Venus flytrap do not bend toward light. Why would this response not be important to an insectivorous plant?

Go to BioPortal for all WORKING WITHDATA exercises

of auxin is responsible for directional plant growth. This was shown in early experiments that followed the Darwins' when a coleoptile tip containing the growth hormone was placed asymmetrically on the decapitated coleoptile. The asymmetric distribution of growth hormone down the coleoptile resulted in excess growth on that side, and bending away from it, even in the absence of light:

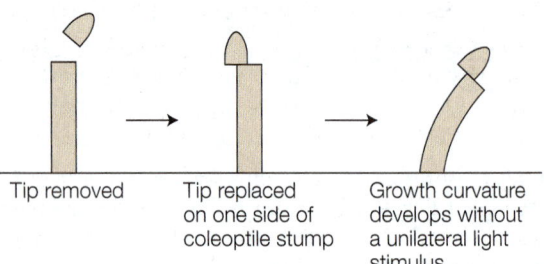

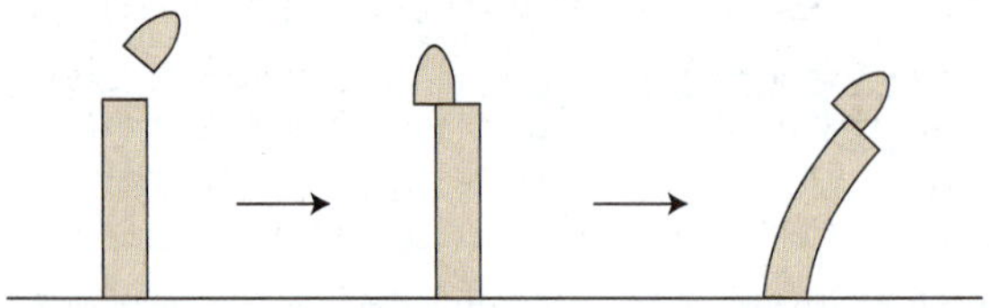
Tip removed
Tip replaced on one side of coleoptile stump
Growth curvature develops without a unilateral light stimulus

The redistribution of auxin to one side is carried out by auxin carrier proteins that move from the base of the cell to one side; because of this, auxin exits the cell only on that side of the cell, rather than at the base, and moves sideways within the tissue.

This lateral movement of auxin explains the bending of canarygrass seedlings toward light that the Darwins observed. When light strikes a canarygrass coleoptile on one side, auxin at the tip moves laterally toward the shaded side. The asymmetry thus established is maintained as polar transport moves auxin down the coleoptile, so that in the growing region below, the auxin concentration is highest on the shaded side. Cell elongation is thus speeded up on that side, causing the coleoptile to bend toward the light (**Figure 37.8A**).

Light is not the only signal that can cause the redistribution of auxin. Auxin moves to the lower side of a shoot that has been

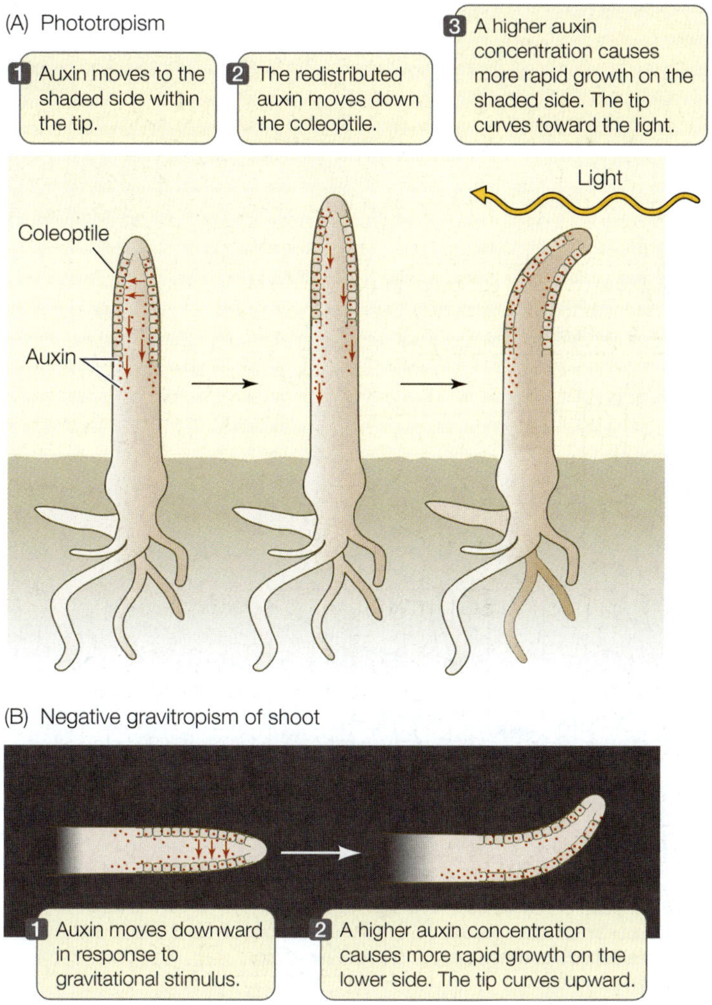
(A) Phototropism
1 Auxin moves to the shaded side within the tip.
2 The redistributed auxin moves down the coleoptile.
3 A higher auxin concentration causes more rapid growth on the shaded side. The tip curves toward the light.
Light
Coleoptile
Auxin
(B) Negative gravitropism of shoot
1 Auxin moves downward in response to gravitational stimulus.
2 A higher auxin concentration causes more rapid growth on the lower side. The tip curves upward.

37.8 Plants Respond to Light and Gravity Phototropism (A) and gravitropism (B) occur in shoot apices in response to a redistribution of auxin.

tipped sideways, causing more rapid growth in the lower side and hence an upward bending of the shoot. Such growth in a direction determined by gravity is called **gravitropism** (**Figure 37.8B**). The upward gravitropic response of shoots is defined as negative gravitropism; that of roots, which bend downward, is positive gravitropism. Gravitropism in roots also involves differential growth caused by lateral movement of auxin, but the details of the mechanism differ between the root and the shoot.

Auxin affects plant growth in several ways

Like the gibberellins, auxin has many roles in plant development. It affects the vegetative and reproductive growth of plants in several ways.

ROOT INITIATION Cuttings from the shoots of some plants can produce roots and develop into entire new plants. For this to occur, certain undifferentiated cells in the interior of the shoot, originally destined to function only in food storage, must set off on a new mission: they must change their cell fate and become organized into the apical meristem of a new root. These changes are similar to those that take place in the pericycle of a root when a lateral root forms (see Figure 34.10). Shoot cuttings of many species can be made to develop roots by dipping the cut surfaces into an auxin solution. These observations suggest that in an intact plant, the plant's own auxin plays a role in the initiation of lateral roots. Commercial preparations that enhance the rooting of plant cuttings typically contain synthetic auxins.

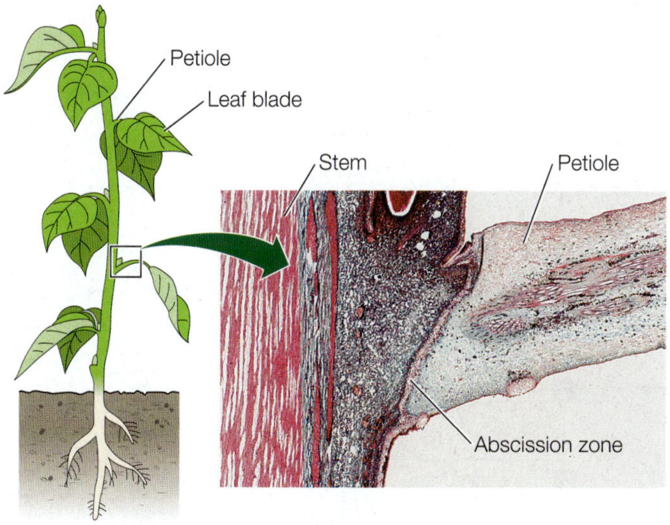
Petiole
Leaf blade
Stem
Petiole
Abscission zone

37.9 Changes Occur When a Leaf Is about to Fall The breakdown of cells in the abscission zone of the petiole causes the leaf to fall.

LEAF ABSCISSION In contrast to its stimulatory effect on root initiation, auxin inhibits the detachment of old leaves from stems. This detachment process, called **abscission**, is the cause of autumn leaf fall. Many leaves consist of a blade and a petiole that attaches the blade to the stem. Abscission results from the breakdown of a specific part of the petiole, the abscission zone (**Figure 37.9**). If the blade of a leaf is cut off, the petiole falls from the plant more rapidly than if the leaf had remained intact. If the cut surface is treated with an auxin solution, however, the petiole remains attached to the plant, often longer than an intact leaf would have. The timing of leaf abscission in nature appears to be determined in part by a decrease in the movement through the petiole of auxin produced in the blade.

APICAL DOMINANCE Auxin helps maintain **apical dominance**, a phenomenon in which apical buds inhibit the growth of axillary buds (see Figure 34.1), resulting in the growth of a single main stem with minimal branching. Apical dominance can be demonstrated by an experiment with a young seedling. If the plant remains intact, the stem elongates and the axillary buds remain inactive. Removal of the apical bud—the major site of auxin production—results in growth of the axillary buds. If the cut surface of the stem is treated with auxin, however, the axillary buds do not grow. The apical buds of branches also exert apical dominance: the axillary buds on the branch are inactive unless the apex of the branch is removed. That is why gardeners prune shrubs to encourage branching.

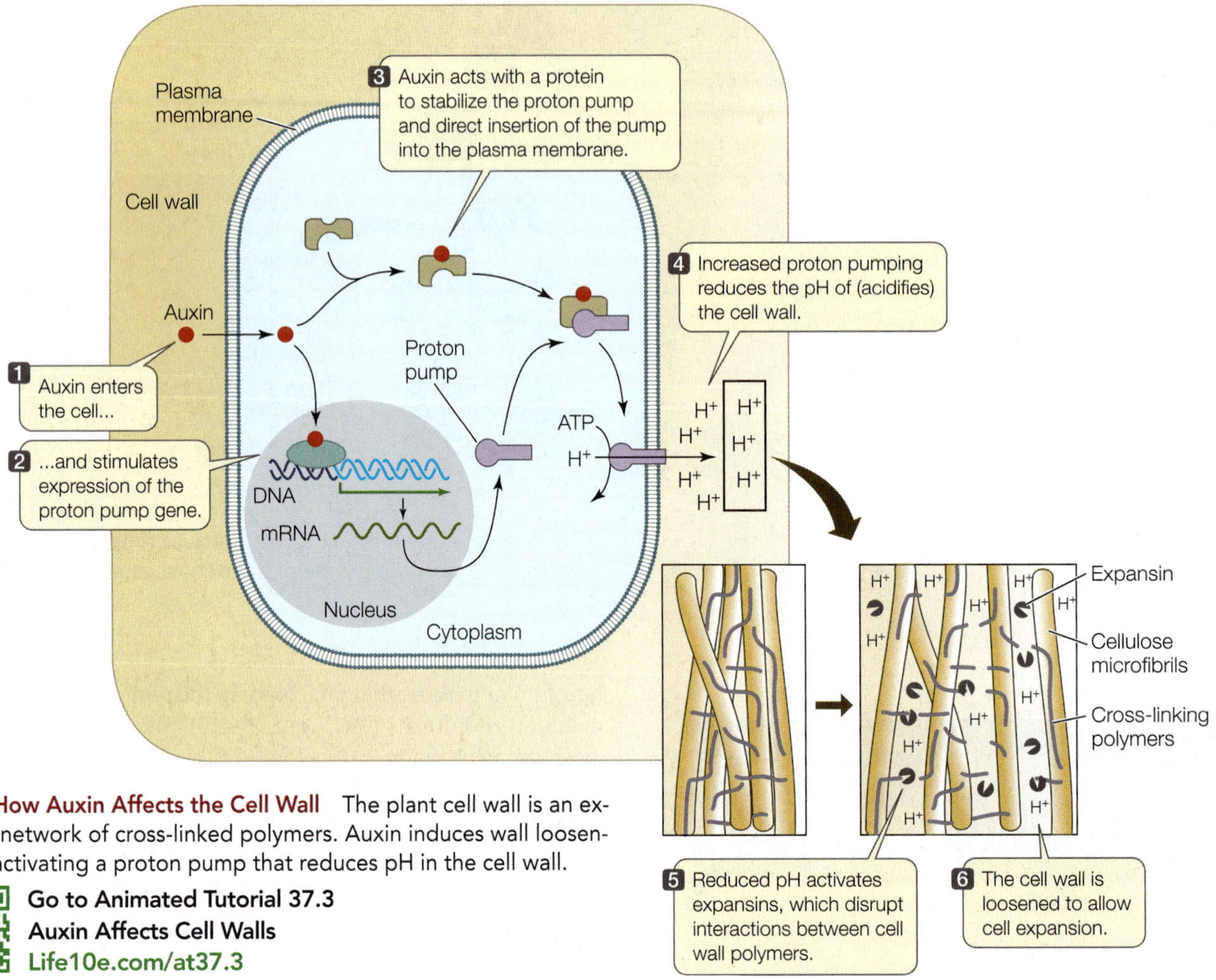

Plasma membrane
Cell wall
Auxin
1 Auxin enters the cell...
2 ...and stimulates expression of the proton pump gene.
3 Auxin acts with a protein to stabilize the proton pump and direct insertion of the pump into the plasma membrane.
Proton pump
ATP
H+
DNA
mRNA
Nucleus
Cytoplasm
4 Increased proton pumping reduces the pH of (acidifies) the cell wall.
Expansin
Cellulose microfibrils
Cross-linking polymers
5 Reduced pH activates expansins, which disrupt interactions between cell wall polymers.
6 The cell wall is loosened to allow cell expansion.

37.10 How Auxin Affects the Cell Wall The plant cell wall is an extensive network of cross-linked polymers. Auxin induces wall loosening by activating a proton pump that reduces pH in the cell wall.

Go to Animated Tutorial 37.3
Auxin Affects Cell Walls
Life10e.com/at37.3

In the experiments on leaves and stems just discussed, removal of a particular part of the plant elicits a response—abscission or loss of apical dominance—and that response is prevented by treatment with auxin. These results are consistent with other data showing that the excised part of the leaf or stem is an auxin source and that auxin in the intact plant delays the abscission of leaves and helps maintain apical dominance.

FRUIT DEVELOPMENT Fruit development normally depends on prior fertilization of the ovule (egg), but in many species treatment of an unfertilized ovary with auxin or gibberellins causes **parthenocarpy**—fruit formation without fertilization. Parthenocarpic fruits form spontaneously in some cultivated varieties of plants, including seedless grapes, bananas, and some cucumbers.

CELL EXPANSION Cell division followed by cell expansion is what causes plant growth. Because the plant cell wall normally prevents expansion of the cell contents inside the plasma membrane (see Section 34.1), the cell wall plays a key role in controlling the rate and direction of plant cell growth. Auxin acts on cell walls to regulate this process.

The expansion of a plant cell is driven primarily by the uptake of water, which enters the cytoplasm of the cell and accumulates in its central vacuole (see Section 35.1). Growth of the vacuole accounts for most of the increase in volume of a growing cell, and the vacuole often makes up more than 90 percent of the volume of a mature cell. As the vacuole expands, it presses the cytoplasm against the cell wall, and the wall resists this force (the basis of turgor pressure). The cell wall is an extensively cross-linked network of polysaccharides and proteins, dominated by cellulose microfibrils. If the cell is to expand, some adjustments must be made in the wall structure to allow the wall to "give" under turgor pressure. Think of a balloon (the cell surrounded by a membrane) inside a box (the cell wall). How does the cell wall "box" loosen to allow expansion?

The **acid growth hypothesis** explains auxin-induced cell expansion (Figure 37.10). The hypothesis holds that protons (H^+) are pumped from the cytoplasm into the cell wall, lowering the pH of the wall and activating enzymes called expansins that catalyze changes in the cell wall structure such that the polysaccharides adhere to each other less strongly. This loosens the cell wall, making it easier to stretch as the cell expands. Auxin has two roles in this process: to increase the synthesis of the proton

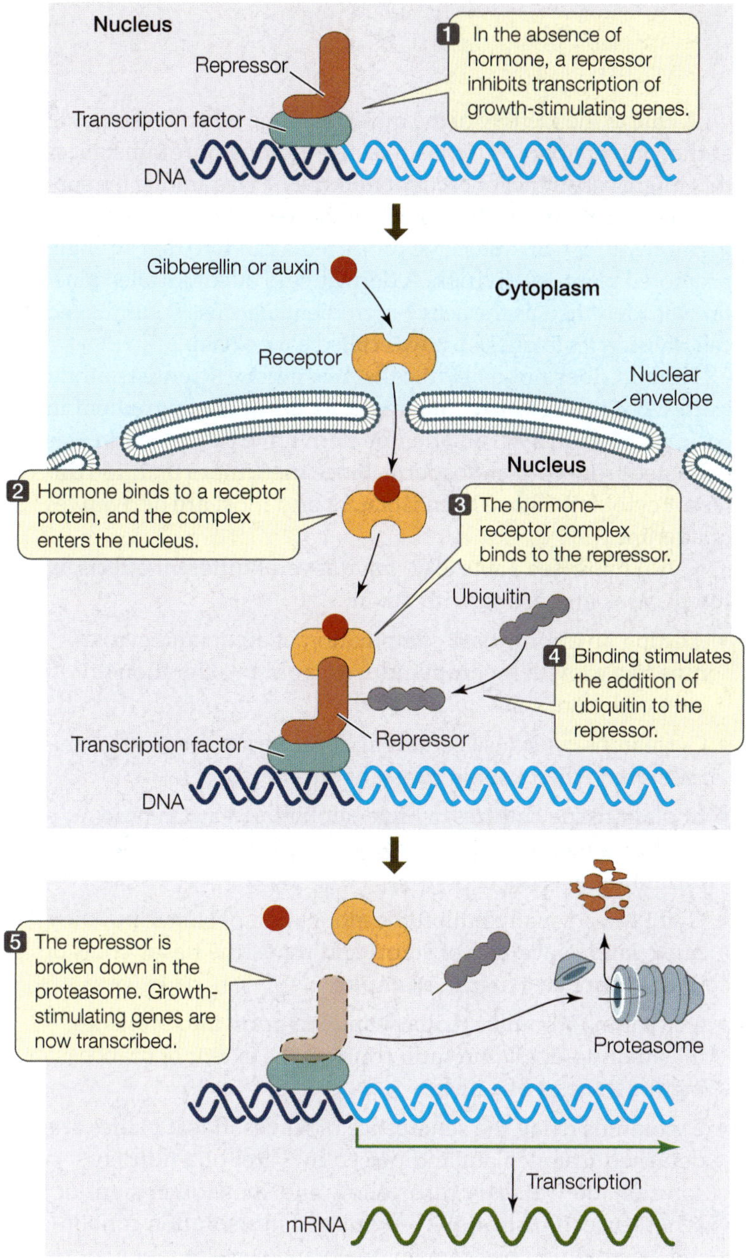

Nucleus
Repressor
Transcription factor
DNA
1 In the absence of hormone, a repressor inhibits transcription of growth-stimulating genes.
Gibberellin or auxin
Cytoplasm
Receptor
Nuclear envelope
Nucleus
2 Hormone binds to a receptor protein, and the complex enters the nucleus.
3 The hormone–receptor complex binds to the repressor.
Ubiquitin
4 Binding stimulates the addition of ubiquitin to the repressor.
Transcription factor
Repressor
DNA
5 The repressor is broken down in the proteasome. Growth-stimulating genes are now transcribed.
Proteasome
Transcription
mRNA

37.11 Gibberellins and Auxin Have Similar Signal Transduction Pathways Although the specific proteins involved are different, both hormones act to stimulate gene transcription by inactivating a repressor protein.

Go to Media Clip 37.1
Gibberellin Binding to Its Receptor
Life10e.com/mc37.1

pumps, and to guide their insertion into the plasma membrane. Auxin may also increase the activity of proton pump proteins already in the plasma membrane. Several lines of evidence support the acid growth hypothesis. For example, adding acid to the cell wall to lower the pH stimulates cell expansion even in the absence of auxin. Conversely, when a buffer is used to prevent the wall from becoming more acidic, auxin-induced cell expansion is blocked. The acid growth hypothesis holds more or less well depending on species; in some plants, auxin stimulates secretion of new cell wall components quickly enough to account for even rapid changes in growth rate.

At the molecular level, auxin and gibberellins act similarly

The molecular mechanisms underlying both auxin and gibberellin action have been worked out with the help of genetic screens (see Figure 37.2). Biologists started by identifying mutant plants whose growth and development are insensitive to the hormones—that is, plants that are *not* affected by added hormone. Such mutants fall into two general categories:

- *Excessively tall plants.* These plants resemble wild-type plants given an excess of hormone and grow no taller when given extra hormone. They grow tall even when treated with inhibitors of hormone synthesis. Their hormone response is always "on," even in the absence of the hormone. In such cases, it is presumed that the normal allele for the mutant gene codes for an inhibitor of the hormone signal transduction pathway. In wild-type plants, that pathway is "off," but in the mutant plants, the pathway is "on" and the plant grows tall.
- *Dwarf plants.* These plants resemble dwarf plants that are deficient in hormone synthesis (see Figure 37.3), but they do not respond to added hormone. In these mutants the hormone response is always "off," regardless of the presence of the hormone.

Remarkably, some mutations of both types turned out to affect the same protein, which turns out to be a repressor of a transcription factor that stimulates the expression of growth-promoting genes. The repressor protein has two important domains, which explains how mutations in the same protein can have seemingly opposite effects:

- *One region of the repressor protein binds to the transcription complex to inhibit transcription of growth-promoting genes.* This is the region mutated in the excessively tall plants: the growth-promoting genes are always "on" because the repressor does not bind to the transcription complex.
- *Another region of the repressor protein causes it to be removed from the transcription complex.* This is the region mutated in the dwarf plants: the growth-promoting genes are always "off" because the repressor is always bound to the complex.

These observations allowed biologists to figure out how auxin and gibberellins work in wild-type plants. Of course, the repressor proteins involved in responding to the two hormones are different, but the actions of both hormones are similar: they *act by removing the repressor from the transcription complex* (**Figure 37.11**). The hormones do this by binding to a receptor protein, which in turn binds to the repressor. Binding of the hormone–receptor complex stimulates polyubiquitination of the repressor, targeting it for breakdown in the proteasome (see Figure 16.25). The receptors contain or associate with a region called an F-box that facilitates protein–protein interactions necessary for causing polyubiquitination of a target protein. Whereas animal genomes have few F-box-containing proteins, plant genomes have hundreds, an indication that this type of gene regulation is common in plants.

RECAP 37.2

Gibberellins are plant hormones that affect stem growth, fruit size, seed germination, and many other aspects of plant development; the effects vary from species to species. Auxin regulates cell expansion and thus mediates phototropism and gravitropism; it also plays roles in apical dominance, leaf abscission, fruit development, and root initiation. The acid growth hypothesis explains auxin-induced cell wall loosening. Similar molecular mechanisms explain the effects of auxin and gibberellins on gene expression.

- How were gibberellins and auxin shown to be plant hormones? See p. 760
- How do gibberellins contribute to the germination of barley seeds? See Figure 37.5
- What is the evidence for polar transport of auxin, and how does it occur? See pp. 762–763 and Figures 37.6 and 37.7
- Explain why, even though auxin moves *away* from the lighted side of a coleoptile tip, the coleoptile bends *toward* the light. See p. 764 and Figure 37.8
- How does auxin cause cell wall loosening? See p. 766 and Figure 37.10
- What is the general signal transduction pathway for auxin and gibberellin? See p. 767 and Figure 37.11

How can a single hormone, such as auxin or a gibberellin, have so many effects? As we have seen, a single signal transduction pathway may affect more than one gene. We will learn about other important plant hormones in the next section, and they too have multiple effects.

37.3 What Are the Effects of Cytokinins, Ethylene, and Brassinosteroids?

Like animal cells, plant cells differentiate after they form from undifferentiated stem cells (called meristem cells in plants). But unlike animal cells, which generally do not divide after differentiation, plant cells retain the ability to divide. For example, in leaf abscission (see Figure 37.9) differentiated parenchyma cells in the petiole resume division, forming a specialized, weak layer of cells. Also, cells of the phloem and cortex can resume division and form secondary meristems. What stimulates these cells to divide? An answer came from studies of cells isolated from plants and cultured in the laboratory.

Cytokinins are active from seed to senescence

Like bacteria and yeasts, plant cells such as parenchyma cells can be grown in a liquid or solidified growth medium containing sugars and salts. The cells will divide continuously until they run out of nutrients. In the early days of plant cell culturing, scientists experimented with many supplements to determine the optimal chemical environment for growth. The best supplement was coconut milk, the fluid that surrounds the developing embryo in coconut fruit. Investigators suspected that a molecule in the fluid must stimulate plant cell division.

A clue to the identity of the molecule came when Folke Skoog at the University of Wisconsin tested various pure substances that might substitute for coconut milk. DNA was among the substances tested, and it did not work; however, heating DNA at high pressure in an autoclave produced a mixture that strongly promoted plant cell division. A derivative of adenine called kinetin was identified as the active ingredient. Because it stimulated cell division (cytokinesis), it was called a **cytokinin**.

Kinetin does not exist in cells, but it gave scientists a hint as to what type of molecule might be the active ingredient in coconut milk. In 1963 an adenine derivative called **zeatin** was extracted from corn endosperm. Since then, more than 150 different cytokinins have been isolated, and most are derivatives of adenine.

Cytokinins (see Table 37.2) have several different effects, in many cases interacting with auxin:

- Adding an appropriate combination of auxin and cytokinins to a growth medium induces rapid proliferation of cultured plant cells.
- Cytokinins can cause certain light-requiring seeds to germinate even when kept in constant darkness.
- In plant tissue cultures, a high cytokinin-to-auxin ratio promotes the formation of shoots; a low ratio promotes the formation of roots.
- Cytokinins usually inhibit the elongation of stems, but they cause lateral swelling of stems and roots (the fleshy roots of radishes are an extreme example).
- Cytokinins stimulate axillary buds to grow into branches; the auxin-to-cytokinin ratio controls the extent of branching (bushiness) of a plant.
- Cytokinins delay the senescence of leaves. If leaf blades are detached from a plant and placed in water or a nutrient solution, they quickly turn yellow and show other signs of senescence. If instead they are placed in a solution containing a cytokinin, they remain green and senesce much more slowly. Roots contain abundant cytokinins, and cytokinin transport to the leaves delays senescence.

This leaf was treated with cytokinin, which delays senescence.
This leaf was not treated.

Cytokinin signaling appears to act through a two-component pathway (a type of signal transduction pathway common in bacteria):

- A *receptor* that can act as a protein kinase, phosphorylating itself as well as a target protein

- A target protein, generally a transcription factor, that can act as an *effector*

Genetic screens in *Arabidopsis* for defects in the response to cytokinin have identified the receptor (AHK; *Arabidopsis* *h*istidine *k*inase) and target effector (ARR; *Arabidopsis* *r*esponse *r*egulator), the latter acting as a transcription factor when phosphorylated. The signal transduction pathway also includes a third protein (AHP; *Arabidopsis* *h*istidine *p*hosphotransfer protein), which transfers phosphates from the receptor to the effector (**Figure 37.12**). The *Arabidopsis* genome has more than 20 genes that are expressed in response to this signaling pathway.

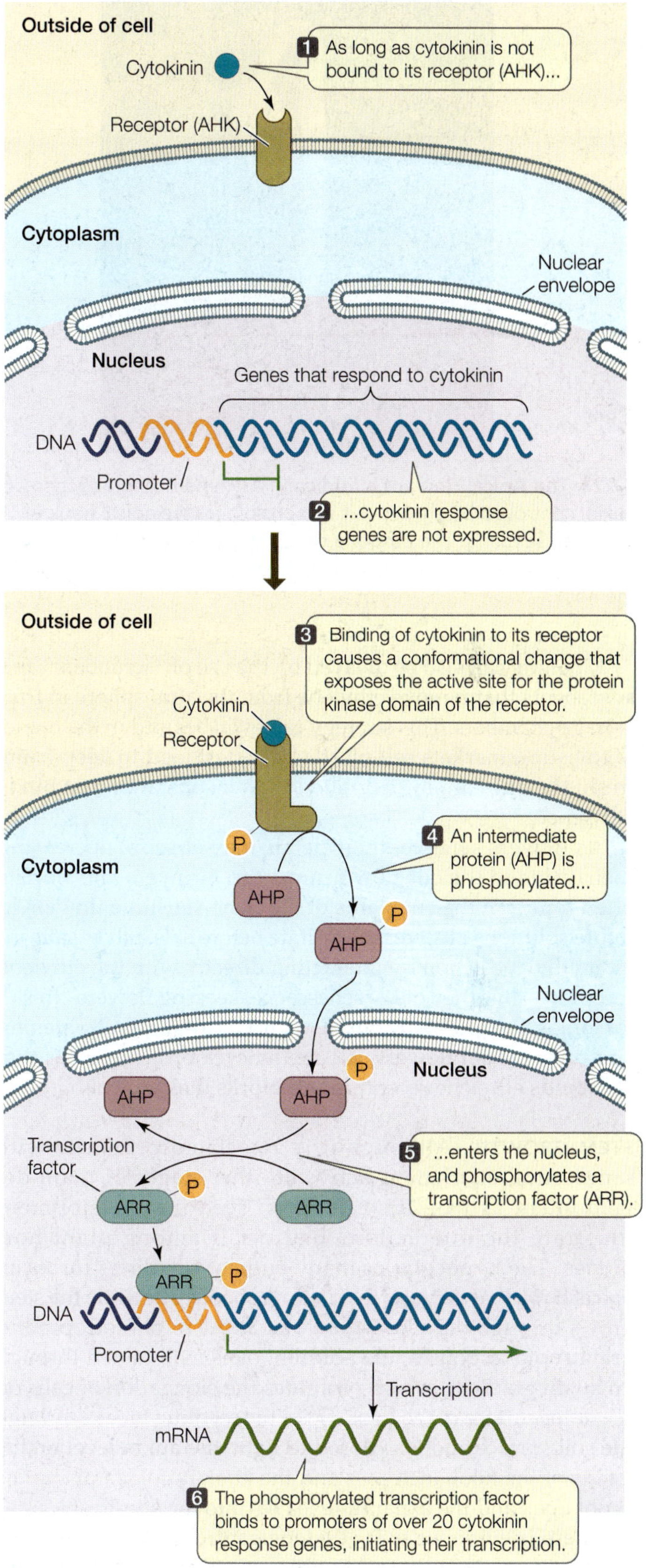
Outside of cell
Cytokinin
1 As long as cytokinin is not bound to its receptor (AHK)...
Receptor (AHK)
Cytoplasm
Nuclear envelope
Nucleus
Genes that respond to cytokinin
DNA
Promoter
2 ...cytokinin response genes are not expressed.
Outside of cell
3 Binding of cytokinin to its receptor causes a conformational change that exposes the active site for the protein kinase domain of the receptor.
Cytokinin
Receptor
P
Cytoplasm
4 An intermediate protein (AHP) is phosphorylated...
AHP
AHP
P
Nuclear envelope
Nucleus
AHP
AHP
P
Transcription factor
5 ...enters the nucleus, and phosphorylates a transcription factor (ARR).
ARR
P
ARR
ARR
P
DNA
Promoter
Transcription
mRNA
6 The phosphorylated transcription factor binds to promoters of over 20 cytokinin response genes, initiating their transcription.

37.12 The Cytokinin Response Pathway Plant cells respond to cytokinins using a signal transduction pathway involving a receptor and an effector protein.

Ethylene is a gaseous hormone that hastens leaf senescence and fruit ripening

Whereas the cytokinins delay senescence, another plant hormone promotes it: the gas **ethylene** (see Table 37.2), which is sometimes called the senescence hormone. Ethylene can be produced by all parts of the plant, and like all plant hormones, it has several effects.

Back when streets were lit by gas rather than by electricity, leaves on trees near street lamps dropped earlier than those on trees farther from the lamps. We now know why: ethylene, a combustion product of the illuminating gas, caused the early abscission. Whereas auxin delays leaf abscission, ethylene strongly promotes it; thus the balance of auxin and ethylene controls abscission.

FRUIT RIPENING By promoting senescence, ethylene also speeds the ripening of fruit. As a fruit ripens, it loses chlorophyll and its cell walls break down; ethylene promotes both of these processes. Ethylene also causes an increase in its own production. Thus once ripening begins, more and more ethylene forms, and because it is a gas, it diffuses readily throughout the fruit and even to neighboring fruits on the same or other plants. The old saying "one rotten apple spoils the barrel" is true. That rotten apple is a rich source of ethylene, which speeds the ripening and subsequent rotting of the other fruit in a barrel or other confined space.

Farmers used to poke holes in developing figs to make them ripen faster. We now know that wounding causes an increase in ethylene production by the fruit, and that the raised ethylene level promotes ripening in many fruits, including apples, bananas, melons, apricots, and tomatoes. Today commercial shippers and storers of such fruit hasten ripening by adding ethylene to storage chambers. This use of ethylene is the single most important use of a natural plant hormone in agriculture and commerce.

These stored tomatoes were treated with ethylene to promote ripening.
These tomatoes were not treated with ethylene.

Apical hook

37.13 The Apical Hook of a Eudicot Asymmetrical production of auxin, controlled by ethylene, is responsible for the apical hook of this bean seedling. The ethylene concentration was highest on the right side, so more rapid growth on the left caused and maintained the hook.

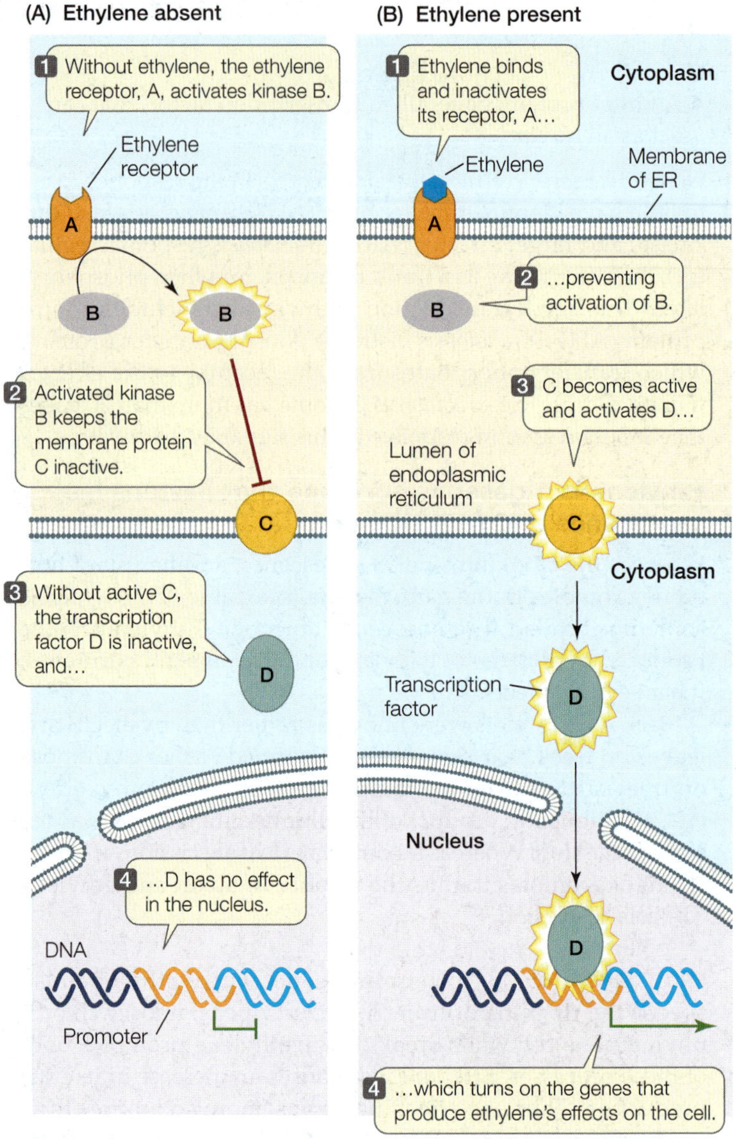
(A) Ethylene absent
(B) Ethylene present
1 Without ethylene, the ethylene receptor, A, activates kinase B.
Ethylene receptor
A
B
B
2 Activated kinase B keeps the membrane protein C inactive.
C
3 Without active C, the transcription factor D is inactive, and…
D
4 …D has no effect in the nucleus.
DNA
Promoter
1 Ethylene binds and inactivates its receptor, A…
Cytoplasm
Ethylene
Membrane of ER
A
2 …preventing activation of B.
B
3 C becomes active and activates D…
Lumen of endoplasmic reticulum
C
Cytoplasm
Transcription factor
D
Nucleus
D
4 …which turns on the genes that produce ethylene's effects on the cell.

37.14 The Signal Transduction Pathway for Ethylene This diagram shows the roles of four proteins (A, B, C, and D) in the signal transduction pathway through which ethylene exerts its many effects.

Ripening can also be delayed by the use of "scrubbers" and adsorbents that remove ethylene from the atmosphere in fruit storage chambers. This strategy can even be used in the home. Many supermarkets sell plastic bags designed to keep fruits fresh; the bags are impregnated with a substance that binds ethylene.

As flowers senesce, their petals may abscise, decreasing their value in the cut-flower industry. Growers and florists often immerse the cut stems of ethylene-sensitive flowers in dilute solutions of silver thiosulfate before sale. Silver salts inhibit ethylene action by interacting directly with the ethylene receptor—thus they delay senescence, keeping flowers "fresh" for longer. An alternative product, used to delay fruit ripening and preserve cut flowers, is 1-methylcyclopropene, a gas that also binds ethylene receptors and blocks their function.

STEM GROWTH Although it is associated primarily with senescence, ethylene is active at other stages of plant development as well. Its effects on seedling development illustrate the interactions that occur among plant hormones. The hypocotyl of many eudicot seedlings forms an **apical hook** that protects the delicate shoot apex while the stem grows through the soil (**Figure 37.13**). As in phototropic and gravitropic responses, the apical hook is maintained through an auxin gradient, which promotes the elongation of cells on the outer surface of the hook. Once the seedling breaks through the soil surface and is exposed to light, the auxin level on the inside of the hook increases and the hook unfolds, raising the shoot apex and the expanding leaves into the sun.

There is evidence that ethylene controls the formation of the auxin gradient during seedling development. Treatment of dark-grown seedlings of some species with ethylene results in what is called the "triple response": an exaggeration of the apical hook and a thickening and shortening of the hypocotyl and root (this response was exploited in a genetic screen for ethylene response mutants; see Figure 37.2). It has been shown that ethylene affects both auxin synthesis and transport during apical hook development.

THE ETHYLENE SIGNAL TRANSDUCTION PATHWAY The mechanism of ethylene action has been worked out by analyzing *Arabidopsis* mutants that have ethylene-related defects. Some of these mutants do not respond to applied ethylene, and others act as if they have been exposed to ethylene even though they have not. Researchers studied the mutant genes and compared their protein products with other known proteins; thus they worked out some of the details of the signal transduction pathway through which ethylene acts (**Figure 37.14**).

The pathway includes two membrane proteins in the endoplasmic reticulum. The first is an ethylene receptor (labeled A in the figure), and the second is a channel-like protein (C). In the absence of ethylene, receptor A activates a protein kinase (B) that keeps C inactive by phosphorylation. When receptor A binds ethylene, it stops activating B. Without B to inactivate it, C activates a transcription factor (D), which then moves into the nucleus, where it turns on the genes that produce ethylene's effects in the cell. In other words, ethylene turns off the "off" signal.

Brassinosteroids are plant steroid hormones

In animals, steroid hormones such as cortisol and estrogen are formed from cholesterol (see Figure 41.2). Initially, biologists isolated a plant steroid hormone from the pollen of rape, a member of the Brassicaceae (mustard family). When applied to various plant tissues, this **brassinosteroid** (see Table 37.2) stimulated cell elongation, pollen tube elongation, and vascular tissue differentiation, but it inhibited root elongation. Since then, dozens of chemically related, growth-affecting brassinosteroids have been found in most plants.

Mutant plants that either do not make brassinosteroids or have defects in brassinosteroid reception and signal transduction are usually dwarf, infertile, and slow to develop. These effects can be reversed by adding small amounts of brassinosteroids, indicating that brassinosteroids are true hormones. These hormones have diverse effects, which vary among plants. Brassinosteroids can:

- promote xylem differentiation
- promote growth of pollen tubes during reproduction
- promote seed germination
- promote apical dominance and leaf senescence
- enhance cell elongation and cell division in shoots

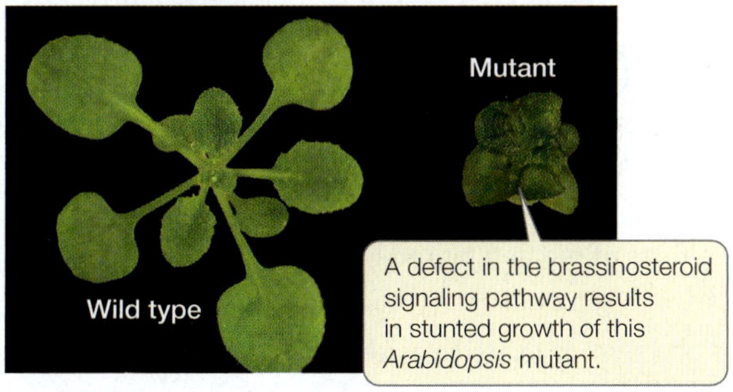
Mutant
Wild type
A defect in the brassinosteroid signaling pathway results in stunted growth of this *Arabidopsis* mutant.

The signaling pathway for these plant steroids differs sharply from those for steroid hormones in animals. In animals, steroids diffuse through the plasma membrane and bind to receptors in the cytoplasm. In contrast, the receptor for brassinosteroids is an integral protein in the plasma membrane.

RECAP 37.3

Cytokinins, ethylene, and brassinosteroids work in concert with auxin and gibberellins to mediate plant development. Their signaling pathways vary from a simple two-component receptor–effector system (cytokinin) to inhibition of an inhibitor of an effector (ethylene).

- How do cytokinins interact with auxin to regulate a plant's development? **See p. 768**
- What is the role of ethylene in fruit ripening? How is this knowledge used commercially? **See p. 769**
- How do the signaling pathways for cytokinins and ethylene differ? **See pp. 769–770 and Figures 37.12 and 37.14**

A plant's response to light—the energy source for photosynthesis—is crucial to its survival. We saw how the Darwins' pioneering investigations of phototropism led to the discovery of auxin. Let's now look more closely at how plants sense and respond to light.

37.4 How Do Photoreceptors Participate in Plant Growth Regulation?

As we pointed out in Section 37.1, plants respond to many different environmental cues, and light is possibly the most important of these cues. Much has been learned about the receptors that plants use to sense light, and we will focus on those receptors here.

Plants respond to two aspects of light: (1) its *quality*—that is, the wavelengths of light that can be absorbed by molecules in the plant; and (2) its *quantity*—that is, the intensity and duration of light exposure.

Chapter 10 described photosynthesis: how chlorophyll and other pigments absorb light at certain wavelengths (quality), and how light intensity affects photosynthetic rate (quantity). Here we will consider how light affects plant development. Earlier in this chapter we described phototropism and how auxin mediates a plant stem's bending toward light. In addition to phototropism, light influences seed germination, shoot elongation, the initiation of flowering, and many other important aspects of plant development. Several photoreceptors take part in these processes. Three or more types of **blue-light receptors** mediate the effects of higher-intensity blue light, and phytochrome mediates the effects of red light.

Phototropins, cryptochromes, and zeaxanthin are blue-light receptors

Charles and Francis Darwin showed that the apical tip of a growing coleoptile receives light as a signal and then redistributes auxin to stimulate cell elongation below the tip on the shaded side. You may recall from Chapter 10 that an action spectrum involves exposing plants to different wavelengths of light to determine what wavelengths are most effective in driving a given process (e.g., photosynthesis). For photosynthesis, such studies showed that the most effective wavelengths are those absorbed by chlorophylls (see Figure 10.5). When an action spectrum was obtained for phototropism of coleoptiles, blue light (peak 436 nm) was found to be the most effective at inducing the coleoptile to curve (**Figure 37.15**). What is the blue-light-absorbing receptor/pigment? Biologists have used a genetic approach to answer this question, once again employing the model plant *Arabidopsis*.

Researchers recovered blue-light-insensitive *Arabidopsis* mutants from a genetic screen and identified the gene for a blue-light receptor protein located in the plasma membrane called **phototropin**. Phototropin protein has a flavin mononucleotide associated with it that absorbs blue light, leading to a change in the shape of the protein. This change exposes an active site for a protein kinase, which in turn initiates a signal transduction cascade that ultimately results in the stimulation of cell elongation by auxin.

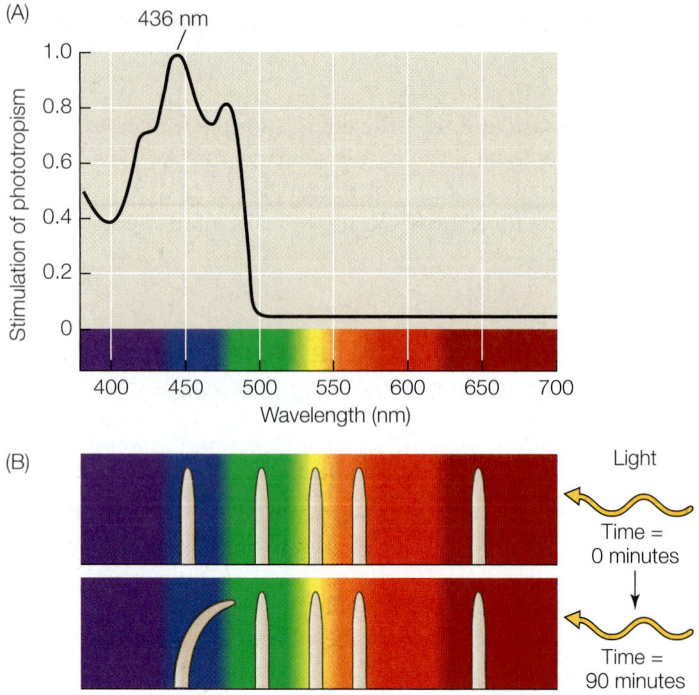
(A)
436 nm
1.0
0.8
0.6
0.4
0.2
0
Stimulation of phototropism
400
450
500
550
600
650
700
Wavelength (nm)
(B)
Light
Time =
0 minutes
Time =
90 minutes

37.15 Action Spectrum for Phototropism (A) The action spectrum for bending of a coleoptile toward light is similar to the absorption spectrum for the receptor, phototropin. (B) After 90 minutes, only the coleoptiles exposed to blue light bend.

Phototropin is also involved in chloroplast movements in relation to light, and participates with another type of blue-light receptor, the plastid pigment **zeaxanthin**, in the light-induced opening of stomata (see Figure 35.8).

Yet another class of blue-light receptors is the **cryptochromes**, which absorb blue and ultraviolet light. These yellow pigments are located primarily in the plant cell nucleus and affect seedling development and flowering. The exact mechanism of cryptochrome action is not yet known. Strong blue light inhibits cell elongation through the action of cryptochromes, although the most rapid responses are mediated by phototropins.

Phytochromes mediate the effects of red and far-red light

Photomorphogenesis refers to a number of physiological and developmental events in plants that are controlled by light. For example:

- A bean seedling germinating below ground has an elongated stem, a pale yellow, folded leaf, and a hook that protects the first leaves (see Figures 37.1 and 37.13)—it is **etiolated**. As the seedling reaches the surface of the soil, it undergoes several light-induced changes: the apical hook straightens, the rudimentary leaves unfold, and chlorophyll is made, so that photosynthesis can begin. Even very dim light will induce these changes.
- Lettuce seeds spread on the soil will germinate only in response to light. Even just a flash of dim light will suffice.
- Adult cocklebur plants flower when they are exposed to long nights. If there is a brief light flash in the middle of the night, they do not flower.

Action spectra of the above processes show that they are induced by red light (650–680 nm). This indicates that plants must have a photoreceptor pigment that absorbs red light and initiates photomorphogenesis.

What is especially remarkable about these red light responses is that *they are reversible by far-red light* (710–740 nm). For example, if lettuce seeds are exposed to brief, alternating periods of red and far-red light in close succession, they respond only to the final exposure. If it is red, they germinate; if it is far-red, they remain dormant (**Figure 37.16**). This reversibility of the effects of red and far-red light regulates many other aspects of plant development, including flowering and seedling growth.

The basis for the effects of red and far-red light resides in the bluish photoreceptor pigment protein in the cytosol of plants called **phytochrome**. Phytochrome exists in two

INVESTIGATING**LIFE**

37.16 Sensitivity of Seeds to Red and Far-Red Light Lettuce seeds will germinate if exposed to a brief period of light. An action spectrum indicated that red light was most effective in promoting germination, but far-red light would reverse the stimulation if presented right after the red light flash. Harry Borthwick and his colleagues asked what would be the effect of repeated alternating flashes of red and far-red light. In each case, the final exposure determined the germination response.[a] This observation led to the conclusion that a single, photoreversible molecule was involved. That molecule turned out to be phytochrome.

HYPOTHESIS The effects of red and far-red light on lettuce seed germination are mutually reversible.

Method Expose lettuce seeds to alternate periods of red light R for 1 minute and far-red light FR for 4 minutes.

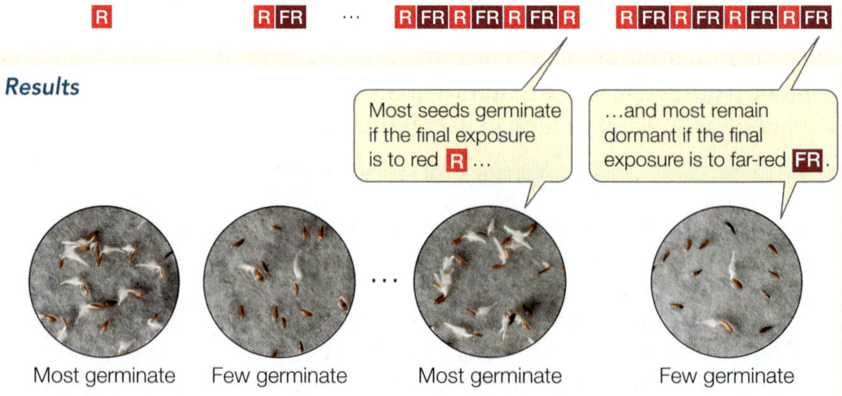
R
R FR
. . .
R FR R FR R FR R
R FR R FR R FR R FR
Results
Most seeds germinate if the final exposure is to red R . . .
. . .and most remain dormant if the final exposure is to far-red FR.
Most germinate
Few germinate
Most germinate
Few germinate

CONCLUSION Red light and far-red light reverse each other's effects.

Go to **BioPortal** for discussion and relevant links for all INVESTIGATING**LIFE** figures.

[a]Borthwick, H. A. et al. 1952. *Proceedings of the National Academy of Sciences USA* 38: 662–666.

interconvertible "isoforms," or states. The molecule undergoes a conformational change upon absorbing light at particular wavelengths. The default or "ground" state, which absorbs principally red light, is called $\mathbf{P_r}$. When P_r absorbs a photon of red light it is converted into $\mathbf{P_{fr}}$. P_{fr} is the active form of phytochrome—the form that triggers important biological processes in various plants.

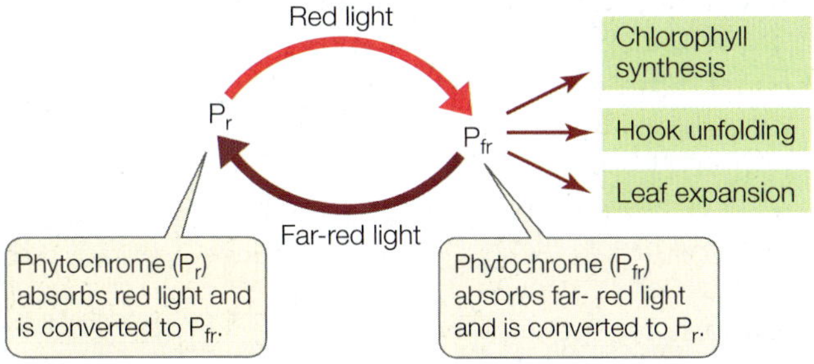
Red light
Far-red light
P_r
P_{fr}
Chlorophyll synthesis
Hook unfolding
Leaf expansion
Phytochrome (P_r) absorbs red light and is converted to P_{fr}.
Phytochrome (P_{fr}) absorbs far- red light and is converted to P_r.

The part of phytochrome that absorbs red and far red light is a covalently attached pigment called a chromophore (**Figure 37.17A**). The chromophore of P_r preferentially absorbs red light; when it does so, it changes conformation and the phytochrome is converted to the P_{fr} form. When the chromophore of P_{fr} absorbs far red light, the phytochrome is converted back to the P_r form. If you know organic chemistry, this reaction is a familiar *cis-trans* isomerization.

The absorption spectra for P_r and P_{fr} correlate with their action spectra (**Figure 37.17B**). As we have seen, these processes include seed germination, shoot development after etiolation, and flowering. In *Arabidopsis* there is a gene family that encodes five slightly different phytochromes, each functioning in different photomorphogenic responses.

For a plant in nature, the ratio of red to far-red light determines whether a phytochrome-mediated response will occur. For example, during daylight the ratio is about 1.2:1; because there is more red than far-red light, the P_{fr} form predominates. But for a plant growing in the shade of other plants, the ratio is as low as 0.13:1, and phytochrome is mostly in the P_r form. The low ratio of red to far-red light in the shade results from absorption of red light by chlorophyll in the leaves overhead, so less of the red light gets through to the plants below. Shade-intolerant species respond by stimulating cell elongation in the stem and thus growing taller to escape the shade. Shade cast by other plants also prevents germination of seeds that require red light to germinate (see Figure 37.16). The reflective properties of the soil can also affect the red to far-red ratio—and thus plant behavior. For example, cotton seedlings grow more slowly on soils (such as clay) that reflect more red than far-red light.

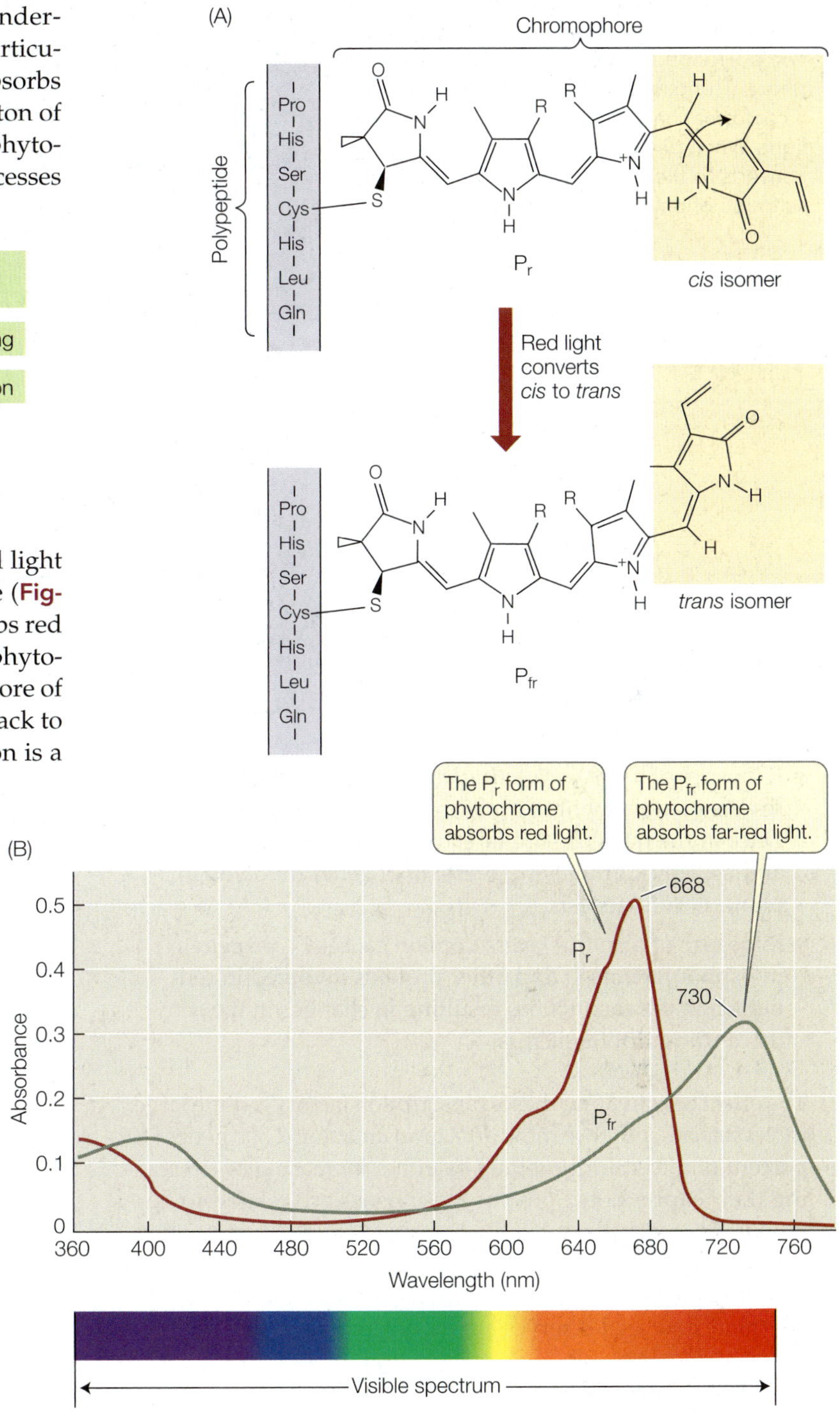
(A)
Chromophore
Polypeptide
Pro
His
Ser
Cys
His
Leu
Gln
P_r
cis isomer
Red light converts *cis* to *trans*
P_{fr}
trans isomer
(B)
The P_r form of phytochrome absorbs red light.
The P_{fr} form of phytochrome absorbs far-red light.
668
730
P_r
P_{fr}
Absorbance
Wavelength (nm)
Visible spectrum

37.17 Phytochrome Exists in Two Forms Absorption spectra of phytochrome reveal two interconvertible forms. (A) The *cis* and *trans* isomers of phytochrome's chromophore. (B) The P_r form absorbs red light; the P_{fr} form absorbs far-red light.

Phytochrome stimulates gene transcription

How does phytochrome, or more specifically P_{fr}, work? Phytochrome has two subunits (**Figure 37.18**), each composed of a protein chain and a chromophore. Gene transcription is stimulated when P_r is converted to the P_{fr} isoform. When P_r absorbs red light and the chromophore changes shape (see Figure 37.17A), this leads to a change in the conformation of the protein itself, from the P_r form to the P_{fr} form. Conversion to the P_{fr} form exposes two important regions of the phytochrome protein (see Figure 37.18), both of which affect transcriptional activity:

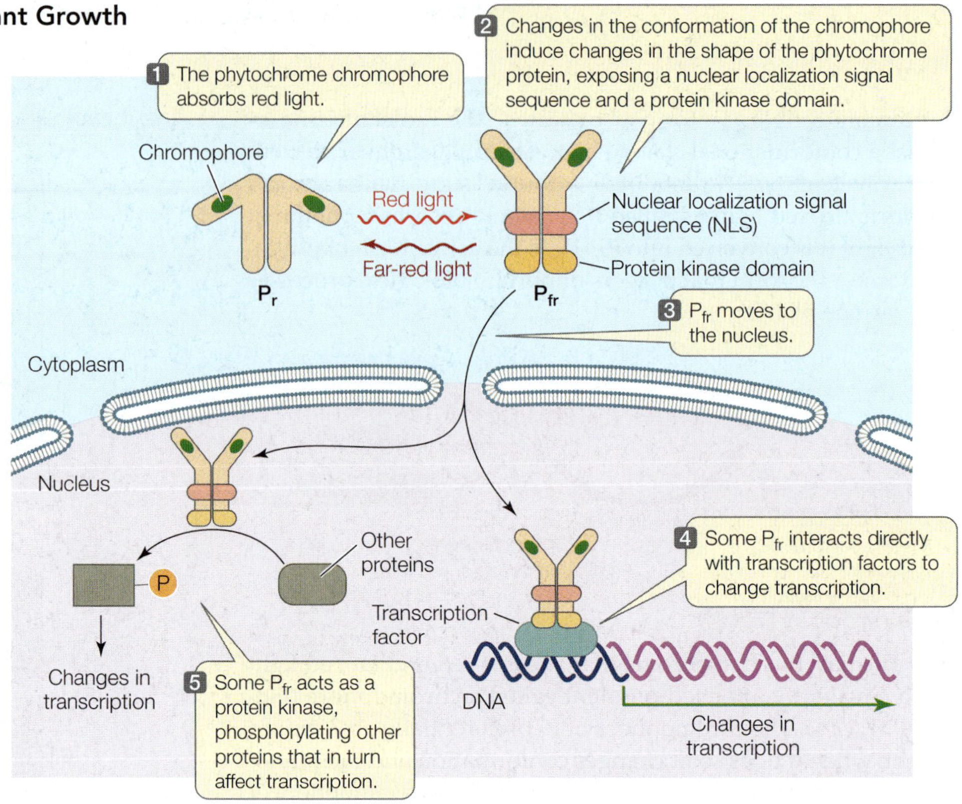
1 The phytochrome chromophore absorbs red light.
2 Changes in the conformation of the chromophore induce changes in the shape of the phytochrome protein, exposing a nuclear localization signal sequence and a protein kinase domain.
Chromophore
Red light
Far-red light
Nuclear localization signal sequence (NLS)
Protein kinase domain
P_r
P_{fr}
3 P_{fr} moves to the nucleus.
Cytoplasm
Nucleus
Other proteins
P
Transcription factor
4 Some P_{fr} interacts directly with transcription factors to change transcription.
Changes in transcription
5 Some P_{fr} acts as a protein kinase, phosphorylating other proteins that in turn affect transcription.
DNA
Changes in transcription

37.18 Phytochrome Stimulates Gene Transcription Phytochrome is composed of two polypeptide chains, each with a chromophore. This pair of polypeptides undergoes a conformational change upon absorbing light. When phytochrome absorbs red light, it converts to the P_{fr} form, which activates transcription of phytochrome-responsive genes.

- Exposure of a *nuclear localization sequence* (see Figure 14.19) results in movement of P_{fr} from the cytosol to the nucleus. Once in the nucleus, P_{fr} binds to transcription factors and thereby stimulates expression of genes involved in photomorphogenesis.
- Exposure of a *protein kinase* domain causes P_{fr} protein to phosphorylate itself and other proteins involved in red-light signal transduction, resulting in changes in the activity of transcription factors.

The effect of activating these transcription factors is quite large: in *Arabidopsis*, phytochrome affects an amazing 2,500 genes (10 percent of the entire genome!) by either increasing or decreasing their expression. Some of these genes are related to hormones. For example, when P_{fr} is formed in seed germination, genes for gibberellin synthesis are activated and genes for gibberellin breakdown are repressed. As a result, gibberellins accumulate and seed reserves are mobilized.

Circadian rhythms are entrained by light reception

The timing and duration of biological activities in living organisms are governed in all eukaryotes and some prokaryotes by what is commonly called a "biological clock"—an oscillator within cells that alternates back and forth between two states at roughly 12-hour intervals. The major outward manifestations of this clock are known as **circadian rhythms** (Latin *circa*, "about"; *dies*, "day"). Think of your own life: in all probability you sleep at night and are awake during the day. The circadian rhythms of animals are discussed in Section 53.3. In plants, circadian rhythms influence, for example, the opening (during the day) and closing (at night) of stomata in *Arabidopsis*, and the raising toward the sun (during the day) and lowering (at night) of leaves in bean plants. From these two examples, it is obvious that circadian rhythms are ecologically useful adaptations, in that they relate the plant's physiology to its environment.

Two qualities characterize circadian rhythms, as well as other regular biological cycles: the **period** is the length of one cycle, and the **amplitude** is the magnitude of the change over the course of a cycle. The circadian rhythms of plants have several noteworthy characteristics:

- The period of a circadian rhythm is remarkably insensitive to temperature, although lowering the temperature may drastically reduce the amplitude.
- Circadian rhythms are highly persistent; they may continue for days, even in the absence of environmental cues, such as light–dark periods.
- Circadian rhythms can be entrained, within limits, by light–dark cycles that do not exactly correspond to 24 hours. That is, the period of a rhythm can be made to coincide (within limits) with that of the light–dark cycle to which the organism is exposed.

Consider what happens when a person abruptly moves across many time zones: what was the night becomes the day, and gradually the person's sleep–wakefulness circadian rhythm entrains to the new environmental cues. Similar entrainment occurs in plants adapting to day length as the seasons progress during the year. The action spectrum for plant entrainment indicates that phytochrome (and to a lesser extent, blue-light receptors) is very likely involved. At sundown phytochrome is mostly in the active P_{fr} form. But as the night progresses, P_{fr} gradually gets converted back to the inactive P_r form. By dawn phytochrome is mostly in the P_r state, but as daylight begins, it rapidly converts to P_{fr}. The switch to the P_{fr} state resets the plant's biological clock. However long the night, the clock is still reset at dawn every day. Thus while the total period measured by the clock is consistent, the clock adjusts to changes in day length over the course of the year.

RECAP 37.4

Light controls several physiological and developmental events in plants, a process called photomorphogenesis. Pigment photoreceptors such as phototropin, cryptochromes, and phytochrome mediate the effects of light on plant growth and development. Phytochrome exists in two interconvertible states; conversion from one state to the other is controlled by the ratio of red to far-red light. Circadian rhythms are influenced by light reception.

- Give the evidence for blue-light receptors in plants. **See pp. 771–772 and Figure 37.15**
- Why does red light affect seed germination differently than far-red light does? **See pp. 772–773 and Figure 37.16**
- What are circadian rhythms, and how are they related to photoreception? **See p. 774**

Photoreceptors also play a regulatory role in flowering. In addition to light, another environmental cue—temperature—regulates flowering. We will examine these topics and others in the next chapter, which focuses on reproduction in flowering plants.

?

What changes in growth patterns made the new strains of wheat and rice successful?

ANSWER

In normal wheat plants, gibberellins stimulate stem elongation. But in the semi-dwarf plants, a mutation affects the signal transduction mechanism for gibberellins so that the stem cells do not respond to it and growth is reduced. In rice, the mutation is in the gene for an enzyme in the biochemical pathway for the synthesis of gibberellins. Without the hormone, the stem does not elongate. The lives of countless people have been saved by intentional disruptions of hormone signaling.

CHAPTER**SUMMARY** 37

37.1 How Does Plant Development Proceed?

- As sessile organisms, plants maximize their ability to grow by using meristems, forming new organs, and growing throughout life.
- The environment, photoreceptors, hormones, and the plant's genome all regulate plant development.
- Seed **dormancy**, which has adaptive advantages, is maintained by a variety of mechanisms. In nature, dormancy is broken by, for example, abrasion, fire, leaching, and low temperatures. When dormancy ends and the seed **imbibes** water, it **germinates** and develops into a **seedling**. **Review Figure 37.1, ACTIVITIES 37.1, 37.2**
- Plant hormones differ in structure and physiology from animal hormones. **Review Table 37.1**
- Plants have several hormones, each of which regulates multiple aspects of development. Interactions among these hormones are often complex. **Review Table 37.2**
- **Genetic screens** using the model organism *Arabidopsis thaliana* have contributed greatly to our understanding of signaling in plants. **Review Figure 37.2**

37.2 What Do Gibberellins and Auxin Do?

- Both **gibberellins** and **auxin** can induce growth in plants otherwise genetically destined to be dwarfs. **Review Figure 37.3**
- Gibberellins have many effects that vary among different plants, including cell elongation, fruit ripening, and mobilization of seed storage polymers. **Review Figures 37.3–37.5, ACTIVITY 37.3**
- Auxin was discovered in the context of stem and coleoptile growth, in particular **phototropism**. In the shoot, it is made in the growing tip and transported down to stimulate cell elongation. **Review Figures 37.6, 37.7, ANIMATED TUTORIALS 37.1, 37.2**
- According to the **acid growth hypothesis**, auxin stimulates cell elongation through the release of protons into the cell wall (acidification of the cell wall). **Review Figure 37.10, ANIMATED TUTORIAL 37.3**
- Both auxin and gibberellins act through the breakdown of transcriptional repressors.

37.3 What Are the Effects of Cytokinins, Ethylene, and Brassinosteroids?

- **Cytokinins** are adenine derivatives that promote plant cell division, promote seed germination in some species, inhibit stem elongation, promote lateral swelling of stems and roots, stimulate the growth of axillary buds, promote the expansion of leaf tissue, and delay leaf senescence.
- A balance between auxin and **ethylene** controls leaf abscission. Ethylene promotes senescence and fruit ripening. It indirectly causes the formation of a protective **apical hook** in eudicot seedlings. In stems, it inhibits elongation, promotes lateral swelling, and causes a loss of gravitropic sensitivity.
- Ethylene acts on cells by a protein kinase pathway located in the endoplasmic reticulum. **Review Figure 37.14**
- Dozens of different **brassinosteroids** affect cell elongation, pollen tube elongation, vascular tissue differentiation, and root elongation. These steroids act at a plasma membrane receptor.

continued

37.4 How Do Photoreceptors Participate in Plant Growth Regulation?

- **Phototropins** are blue-light photoreceptors for phototropism and chloroplast movements. **Zeaxanthin** acts in conjunction with the phototropins to mediate the light-induced opening of stomata. **Cryptochromes** are blue-light photoreceptors that control seedling development, stem elongation, and floral initiation.
- **Phytochrome** exists in the cytosol in two interconvertible forms, P_r and P_{fr}. The relative amounts of these two forms are a function of the ratio of red to far-red light. Phytochrome affects seedling growth, flowering, and etiolation. **Review Figure 37.16**
- The phytochrome signal transduction pathway affects transcription in two different ways; the P_{fr} form interacts directly with some transcription factors, and influences transcription indirectly through interactions with protein kinases. **Review Figure 37.18**
- **Circadian rhythms** are activities that occur on a near-24-hour cycle. Light can entrain these activities through photoreceptors such as phytochrome.

Go to the Interactive Summary to review key figures, Animated Tutorials, and Activities Life10e.com/is37

CHAPTER**REVIEW**

REMEMBERING

1. Which of the following is *not* an advantage of seed dormancy?
   a. It makes the seed more likely to be digested by birds that disperse it.
   b. It counters the effects of year-to-year variations in the environment.
   c. It increases the likelihood that a seed will germinate in the right place.
   d. It favors dispersal of the seed.
   e. It may result in germination at a favorable time of year.

2. The gibberellins
   a. are responsible for phototropism and gravitropism.
   b. are gases at room temperature.
   c. are produced only by fungi.
   d. cause flowering in plants.
   e. inhibit the synthesis of digestive enzymes by barley seeds.

3. In coleoptile tissue, auxin
   a. is transported from base to tip.
   b. is transported from tip to base.
   c. can be transported toward either the tip or the base, depending on the orientation of the coleoptile with respect to gravity.
   d. is transported by simple diffusion, with no preferred direction.
   e. is not transported, because auxin is used where it is made.

4. Signal transduction for both auxin and gibberellins involves
   a. binding of the hormone to a nuclear receptor.
   b. degradation of a repressor of gene transcription.
   c. production of a small molecule second messenger.
   d. light absorption followed by chemical changes.
   e. breakdown of the hormone.

5. Ethylene
   a. causes the triple response in seedlings growing underground.
   b. is liquid at room temperature.
   c. delays the ripening of fruits.
   d. generally promotes stem elongation.
   e. inhibits the swelling of stems, in opposition to cytokinin's effects.

6. Phytochrome
   a. is the only photoreceptor pigment in plants.
   b. exists in two forms that are interconvertible by light.
   c. is a pigment that is colored red or far-red.
   d. is a green-light receptor.
   e. is the photoreceptor for phototropism in coleoptiles.

UNDERSTANDING & APPLYING

7. Describe the circumstances under which it would be advantageous for a species to have the dormancy of its seeds broken by fire.

8. Cocklebur fruits contain two seeds each that are kept dormant by two different mechanisms. Why might having two mechanisms of dormancy be advantageous to cockleburs?

9. Supermarkets sell plastic bags that are impregnated with activated charcoal, which binds gases. The bags are designed to keep fruit fresh. How do they work?

ANALYZING & EVALUATING

10. Corn stunt spiroplasma (a bacterium) causes a great reduction in the growth rate of infected corn plants. Diseased plants take on a dwarfed form. Since their appearance is reminiscent of genetically dwarfed corn, you suspect that the bacterium may inhibit the synthesis of gibberellins by corn plants. Describe two experiments you might conduct to test this hypothesis, only one of which should require chemical measurement.

A corn plant infected with corn stunt spiroplasma

11. The semi-dwarf wheat and rice plants that led to the Green Revolution described in the chapter opening have mutations in the signal transduction pathway for gibberellins. You wish to use genetic engineering to make corn plants that are semi-dwarf.
    a. How would you do a genetic screen to identify the genes in corn involved in gibberellin signaling?
    b. Assuming that the signal transduction pathway is similar to that in *Arabidopsis*, what gene would you select for inactivation?
    c. Besides short stature, what other effects would you expect for the signal transduction mutant strain? How would you use other hormones to overcome them?

Go to BioPortal at **yourBioPortal.com** for Animated Tutorials, Activities, LearningCurve Quizzes, Flashcards, and many other study and review resources.

38 Reproduction in Flowering Plants

CHAPTER**OUTLINE**

Dandelion This plant can reproduce by forming seeds without sex.

YOU MAY NOT CONSIDER A BOTANIST specializing in plant reproduction to be "cool," but Arthur Hemmings, a fictional plant biologist at the famous Kew Botanical Gardens in London, might change your mind. In addition to being a plant biologist, Hemmings is an undercover agent for Britain's Office of Food Security. Hemmings is the protagonist in the recent novel *Day of the Dandelion* by Paul Pringle. The theft of a packet of seeds from the lab of an Oxford University geneticist, coupled with the disappearance of the scientist and his lab assistant, prompts the government to assign Hemmings to the case. The stolen seeds are unusual and highly valued. The seeds will be a hybrid between corn (*Zea mays*) and its close relative gamagrass (*Tripsacum dactyloides*), having the high grain production and disease-resistance characteristics of modern corn and an ability to reproduce asexually in a method called apomixis (thanks to a gene from gamagrass). The race to find the country (China? Russia?) or company (a multinational corporation that wants to dominate the food supply?) that has stolen the seeds takes the fictional scientist–spy around the world. It is an exciting tale, with a basis in reality because scientists at the U.S. Department of Agriculture have indeed bred a corn–*Tripsacum* hybrid that shows apomixis, although its productivity is low.

As we described in Chapter 12, modern farmers typically plant hybrid corn, created by seed companies from crosses between homozygous inbred strains. The heterozygous offspring of these crosses show hybrid vigor, with valuable characteristics such as high grain production. If a farmer allows these plants to reproduce among themselves, however, the result will be a collection of plants with some homozygosity and some heterozygosity (think of the results from this cross: AaBbCc × AaBbCc), and hybrid vigor will be lost. So crosses between inbred lines must be done every time hybrid seeds are needed.

Some species of plants, such as dandelions, blackberries, and gamagrass, have genes for apomixis, which prevents meiosis in the cells that form the female gametes during sexual reproduction. Instead of forming haploid egg cells, these plants can form diploid egg cells that don't need to be fertilized by male gametes. The egg cells go on to form clone diploid offspring: plants that are genetically identical to the mother plant. For these plant species, apomixis provides an evolutionary advantage in that it allows for reproduction without pollination, and for the propagation of well-adapted genotypes. However, a disadvantage is that the genetic variation that results from sexual reproduction is lost.

For the plant breeder, apomixis could be a boon. Apomixis would allow for the rapid propagation of hybrid plants without the need to make crosses and recreate the hybrids every time they are grown. The potential profits are staggering—thus the excitement of *Day of the Dandelion*.

?

By what genetic mechanism is apomixis brought about?

See answer on p. 794.

38.1 How Do Angiosperms Reproduce Sexually?

Most angiosperms (flowering plants) have evolved to reproduce sexually because this strategy has the selective advantage of producing the genetic diversity that is the raw material for evolution. Sexual reproduction in angiosperms involves mitosis, meiosis, and the alternation of haploid and diploid generations (see Figure 11.15). There are several important differences between sexual reproduction in angiosperms and in vertebrate animals (the latter also discussed in Chapter 43):

- Meiosis in plants produces spores, after which mitosis produces gametes; in animals, meiosis usually produces gametes directly.
- In most plants, there are multicellular diploid (sporophyte) and haploid (gametophyte) life stages (alternation of generations); in animals, there is no multicellular haploid stage.
- In plants, the cells that will form gametes are determined in the adult organism, usually in response to environmental conditions; in animals, the germline cells are determined before birth.

The flower is an angiosperm's structure for sexual reproduction

The plant life cycle typically involves the alternation of haploid and diploid generations (see Chapter 11):

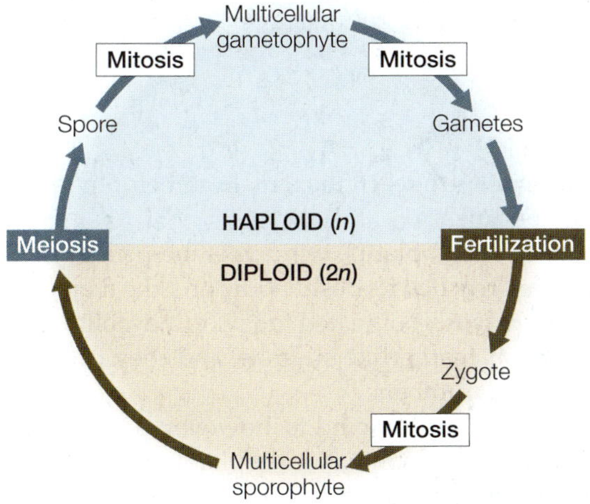
Multicellular gametophyte
Mitosis
Mitosis
Spore
Gametes
HAPLOID (n)
Meiosis
Fertilization
DIPLOID (2n)
Zygote
Mitosis
Multicellular sporophyte

In angiosperms, the plant that we see in nature is a sporophyte, and the male and/or female gametophytes are contained in the flowers (see Section 29.3 for a description of flower parts and floral evolution). A complete flower consists of four concentric groups of organs arising from modified leaves: the carpels, stamens, petals, and sepals.

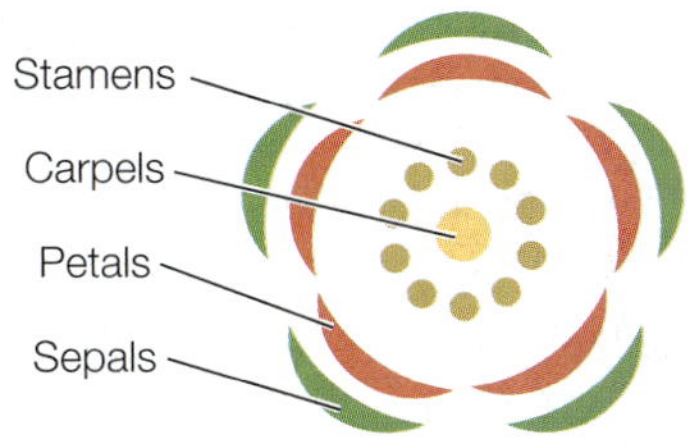
Stamens
Carpels
Petals
Sepals

The parts of the flower are usually borne on a stem tip, and derive from a meristem.

- The carpels are the female sex organs that contain the developing female gametophytes.
- The stamens are the male sex organs that contain the developing male gametophytes.

The differentiation of the meristem into the various organs of the flower is controlled by specific transcription factors (see Figure 19.9).

Most angiosperm species are **hermaphroditic**, and have flowers with both stamens and carpels; such flowers are termed *perfect* (**Figure 38.1A**). *Imperfect* flowers, by contrast, are those with only male or only female sex organs. Male flowers have stamens but not carpels, and female flowers have carpels but not stamens. Some plants, such as corn, bear both male and female flowers on an individual plant; such species are called **monoecious** ("one house") (**Figure 38.1B**). In **dioecious** species, individual plants bear either male-only or female-only flowers; an example is American holly (**Figure 38.1C**).

Flowering plants have microscopic gametophytes

Figure 38.2 offers a detailed look at the gametophytes central to angiosperm reproduction. The haploid gametophytes—the gamete-producing structures—develop from haploid spores in the flower:

- Each female gametophyte (megagametophyte) is called an **embryo sac**, and it develops inside an ovule. One or more ovules are contained within the ovary, which is the lower part of the carpel.
- Male gametophytes (microgametophytes), which are called **pollen grains**, develop inside the anther, which is part of the stamen.

FEMALE GAMETOPHYTE Of the four haploid megaspores resulting from meiosis, three undergo apoptosis (programmed cell death). The remaining megaspore undergoes three mitotic divisions without cytokinesis, producing eight haploid nuclei, all initially contained within a single cell—three nuclei at one end, three at the other, and two in the middle. Subsequent cell wall formation leads to an elliptical, seven-celled megagametophyte with a total of eight nuclei:

- At one end of the megagametophyte are three small cells: the *egg cell* and two cells called *synergids*. The egg cell is the female gamete, and the synergids participate in fertilization by attracting the pollen tube. The pollen tube enters one of the synergids before the sperm cells are released for fertilization.
- At the opposite end of the megagametophyte are three antipodal cells, which eventually degenerate.
- In the large central cell are two **polar nuclei**.

The megagametophyte, or embryo sac, is the entire seven-cell, eight-nucleus structure.

(A) Perfect: lily (*Lilium* sp.)
Carpel
Stamens
(C) Imperfect dioecious: American holly (*Ilex opaca*)
Male flower with stamens
Female flower with carpels

(B) Imperfect monoecious: corn
Male flower with stamens
Female flower with carpels

38.1 Perfect and Imperfect Flowers (A) A lily is an example of a perfect flower, meaning one that has both male and female sex organs. (B) Imperfect flowers are either male or female. Corn is a monoecious species: both types of imperfect flowers are borne on the same plant. (C) American holly is a dioecious species; some American holly plants bear male imperfect flowers whereas others bear female imperfect flowers.

MALE GAMETOPHYTE The four haploid products of meiosis (the microspores) each develop a cell wall and undergo a single mitotic division, producing four two-celled pollen grains that are released into the environment. The two cells in a pollen grain have different roles:

- After pollination (see below) the generative cell divides by mitosis to form two sperm cells that participate in fertilization.
- The tube cell forms the elongating pollen tube that delivers the sperm to the embryo sac.

These events occur after the pollen grain is transferred to a stigma (part of the female reproductive organ)—a process called **pollination**.

Pollination in the absence of water is an evolutionary adaptation

As Chapter 28 described, the union of gametes in aquatic plants is accomplished in the water. Fertilization of mosses and ferns also requires at least a film of water for the movement of gametes. While there are mechanisms to ensure that fertilization occurs if and when the two gametes meet, fertilization is clearly a low-probability event. The evolution of pollen made it possible for male gametes to reach the female gametophyte without an aqueous conduit.

In the first seed plants, wind was the primary vehicle by which pollen reached its destination, and the majority of gymnosperms are wind-pollinated today. Wind-pollinated flowers have sticky or featherlike stigmas, and they produce pollen grains in great numbers.

Pollen transport by wind is, however, a relatively chancy means of achieving pollination, explaining why about 90 percent of all angiosperms still rely on animals—including insects, birds, and bats—for pollen transport. Pollen transport by animals greatly increases the probability that pollen will get to the female gametophyte. As described in Sections 29.3 and 56.3, the structures of flowers have coevolved with their animal pollinators to enhance the plants' chances of successful pollination. Suitably pigmented, shaped, and scented flowers attract the pollinating animal, resulting in a pollen transfer from flower to flower within the same plant species (**Figure 38.3**).

A pollen tube delivers sperm cells to the embryo sac

When a functional pollen grain lands on the stigma of a compatible stigma, it germinates. A key event is water uptake

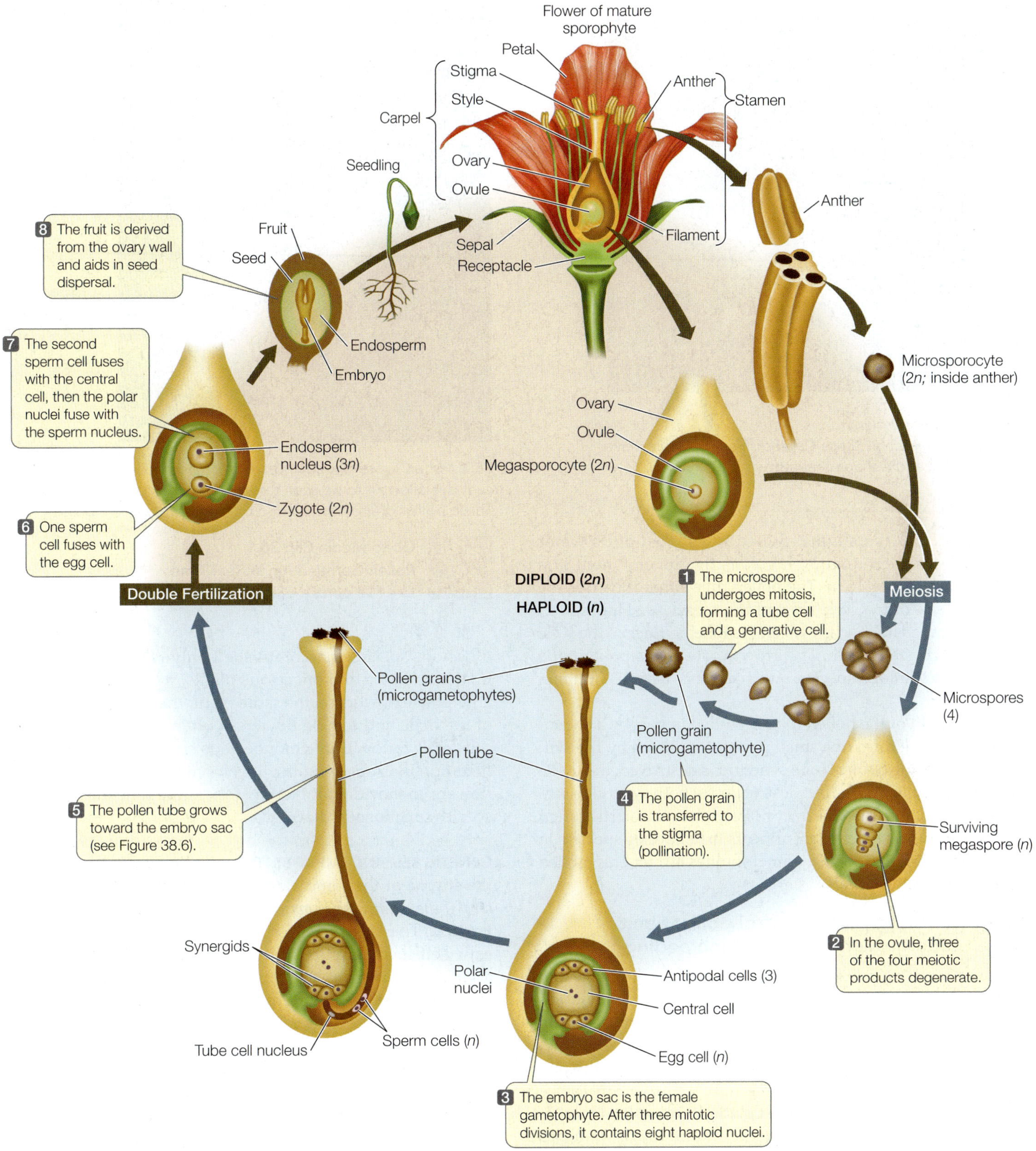

Flower of mature sporophyte
Petal
Stigma
Style
Carpel
Ovary
Ovule
Anther
Stamen
Filament
Sepal
Receptacle
Seedling
Anther
Microsporocyte (2*n*; inside anther)
8 The fruit is derived from the ovary wall and aids in seed dispersal.
Fruit
Seed
Endosperm
Embryo
7 The second sperm cell fuses with the central cell, then the polar nuclei fuse with the sperm nucleus.
Endosperm nucleus (3*n*)
Zygote (2*n*)
6 One sperm cell fuses with the egg cell.
Double Fertilization
Ovary
Ovule
Megasporocyte (2*n*)
DIPLOID (2*n*)
HAPLOID (*n*)
1 The microspore undergoes mitosis, forming a tube cell and a generative cell.
Meiosis
Microspores (4)
Pollen grains (microgametophytes)
Pollen grain (microgametophyte)
Pollen tube
4 The pollen grain is transferred to the stigma (pollination).
5 The pollen tube grows toward the embryo sac (see Figure 38.6).
Surviving megaspore (*n*)
2 In the ovule, three of the four meiotic products degenerate.
Synergids
Polar nuclei
Antipodal cells (3)
Central cell
Egg cell (*n*)
Tube cell nucleus
Sperm cells (*n*)
3 The embryo sac is the female gametophyte. After three mitotic divisions, it contains eight haploid nuclei.

38.2 Sexual Reproduction in Angiosperms The embryo sac is the female gametophyte; the pollen grain is the male gametophyte. The male and female cells meet and fuse within the embryo sac. Angiosperms have double fertilization, in which a zygote and an endosperm form from separate fusion events. The zygote forms by the fusion of one sperm cell with the egg cell. The endosperm forms after the other sperm cell fuses with the central cell, which contains two nuclei. The three nuclei fuse, forming a triploid cell.

Go to Activity 38.1
Sexual Reproduction in Angiosperms
Life10e.com/ac38.1

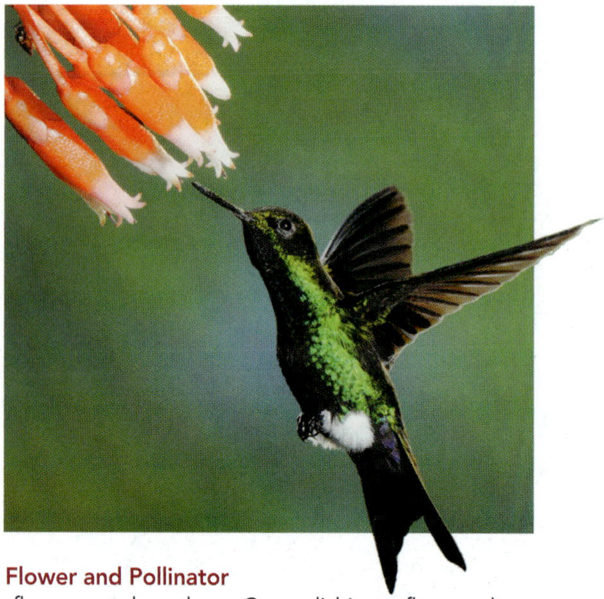

38.3 Flower and Pollinator
Some flowers, such as these *Cavendishia* sp. flowers, have red pigments and a shape that attracts certain birds.

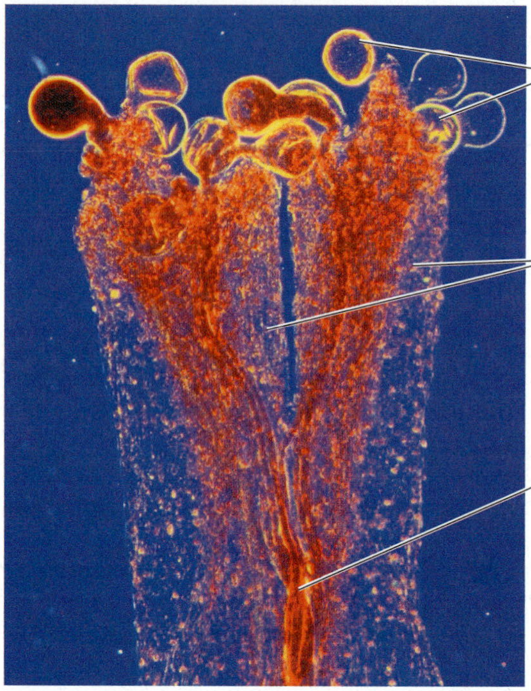
Pollen grains on stigmas
Styles
Pollen tube

38.4 Pollen Tubes Begin to Grow Staining pollen with a fluorescent dye allows it to be seen through a fluorescence microscope. These pollen grains have landed on the stigmas of a crocus.

Go to Media Clip 38.1
Pollen Germination in Real Time
Life10e.com/mc38.1

by pollen from the stigma: pollen loses most of its water as it matures. Germination involves the development of a **pollen tube** (**Figure 38.4**). The pollen tube either traverses the spongy tissue of the style (part of the carpel; see Figure 38.2) or, if the style is hollow, grows on the inner surface of the style until it reaches an ovule. The growth rate of the pollen tube varies greatly among species, but it can be as fast as 3 millimeters an hour.

The growth of the pollen tube is guided in part by a chemical signal in the form of a small protein produced by the synergids within the ovule. If one synergid is destroyed, the ovule still attracts pollen tubes, but destruction of both synergids renders the ovule unable to attract pollen tubes, and fertilization does not occur. The attractant appears to be species-specific: in some cases, isolated female gametophytes attract only pollen tubes of the same species.

Many flowering plants control pollination or pollen tube growth to prevent inbreeding

You may recall from discussions of Mendel's work (see Section 12.1) that some plants can reproduce sexually by both cross-pollination and self-pollination. Self-pollination increases the chances of successful pollination but leads to homozygosity, which reduces genetic diversity. Because diversity is the raw material of evolution by natural selection, homozygosity can be selectively disadvantageous. Most plants have evolved mechanisms that prevent self-fertilization. A multitude of strategies exist (for example, see Figure 29.13). Two primary means to prevent self-fertilization are (1) physical separation of male and female gametophytes and (2) genetic self-incompatibility.

SEPARATION OF MALE AND FEMALE GAMETOPHYTES Self-fertilization is prevented in dioecious species, which bear only male or female flowers on a particular plant. Pollination in dioecious species is accomplished only when one plant pollinates another. In monoecious plants, which bear both male and female flowers on the same plant, the physical separation of the male and female flowers is often sufficient to prevent self-fertilization. Some monoecious species prevent self-fertilization by staggering the development of male and female flowers so they do not bloom at the same time, making these species functionally dioecious.

GENETIC SELF-INCOMPATIBILITY A pollen grain that lands on the stigma of the same plant will fertilize the female gamete *only if the plant is self-compatible,* meaning capable of self-pollination. To prevent self-fertilization, many plants are genetically self-incompatible. **Self-incompatibility** depends on the ability of a plant to determine whether pollen is genetically similar or genetically different from itself. Rejection of "same-as-self" pollen prevents self-fertilization. How does it occur?

Self-incompatibility in plants is controlled by a cluster of tightly linked genes called the *S* locus (for self-incompatibility). The *S* locus encodes proteins in the pollen and style that interact during the recognition process. A self-incompatible species typically has many alleles of the *S* locus. The pollen phenotype may be determined by its own haploid genotype or by the diploid genotype of its parent plant. In either case, if the pollen expresses an allele that matches either of the alleles expressed in the recipient pistil, the pollen is rejected. Depending on the type of self-incompatibility system, the rejected pollen either fails to germinate or is prevented from growing through the style (**Figure 38.5**); either way, self-fertilization is prevented.

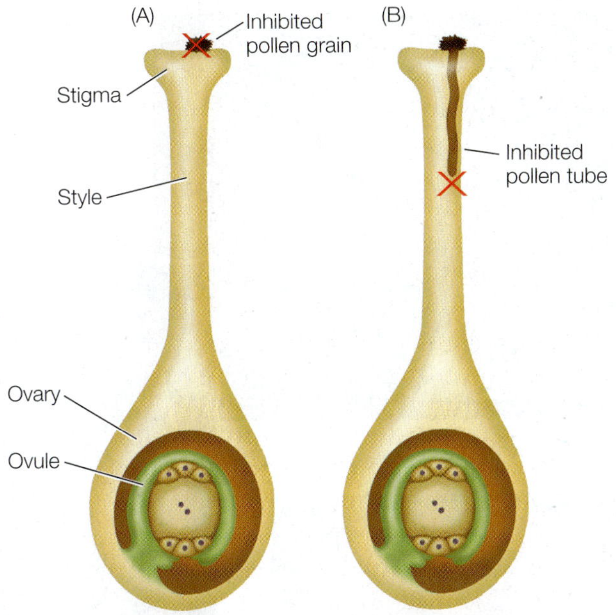

(A)
Inhibited pollen grain
(B)
Stigma
Inhibited pollen tube
Style
Ovary
Ovule

38.5 Self-Incompatibility In a self-incompatible plant, pollen is rejected if it expresses an *S* allele that matches one of the *S* alleles of the stigma and style. Self pollen may (A) fail to germinate or (B) its pollen tube may die before reaching an ovule. In either case, the egg cannot be fertilized by a sperm from the same plant.

Angiosperms perform double fertilization

In most angiosperm species, the mature pollen grain consists of two cells: a large tube cell enclosing a much smaller generative cell. When a compatible pollen grain lands on the stigma of a plant of the same species, it germinates, and the pollen tube grows through the style tissue to the embryo sac. During this process, the generative cell undergoes one mitotic division and cytokinesis to produce two haploid **sperm cells** (Figure 38.6, steps 1 and 2).

Two fertilization events now occur. One of the two synergids degenerates when the pollen tube arrives and the two sperm cells are released into its remains. (Figure 38.6, step 3). Each sperm cell then fuses with a different cell of the embryo sac (Figure 38.6, steps 4 and 5). One sperm cell fuses with the egg cell, and the two nuclei fuse, producing the diploid zygote. The other sperm cell fuses with the central cell, and its nucleus fuses with the two polar nuclei, forming a **triploid (3n) cell**. Immediately after fertilization the triploid nucleus undergoes rapid mitotic divisions to form a specialized nutritive tissue, the **endosperm**. In most species the endosperm nucleus initially divides without cytokinesis, forming a large, multinucleate cell, and cell walls form later between the nuclei. After the endosperm begins developing, the zygote undergoes mitotic division to form the new sporophyte embryo. The developing embryo uses the endosperm tissue as a source of nutrients, energy, and carbon-based anabolic building blocks. In some cases the endosperm persists until germination and is used as a source of nutrients by the developing seedling. This source of nutrients is important because the seedling often begins its development underground and cannot perform photosynthesis right away.

The remaining cells of the male and female gametophytes—the antipodal cells, the remaining synergid, and the pollen tube nucleus—degenerate as the embryo begins to develop.

Double fertilization is so named because it involves two cell fusion events:

- One sperm cell fuses with the egg cell.
- The other sperm cell fuses with the central cell.

The fusion of a sperm cell with the central cell to form the triploid endosperm nucleus is one of the defining characteristics of angiosperms.

Go to Animated Tutorial 38.1 Double Fertilization
Life10e.com/at38.1

38.6 Double Fertilization Two sperm are involved in two cell fusion events, hence the term "double fertilization." One sperm is involved in the formation of the diploid zygote, and the other results in the formation of the triploid endosperm cell that divides to form endosperm. Double fertilization is a characteristic feature of angiosperm reproduction.

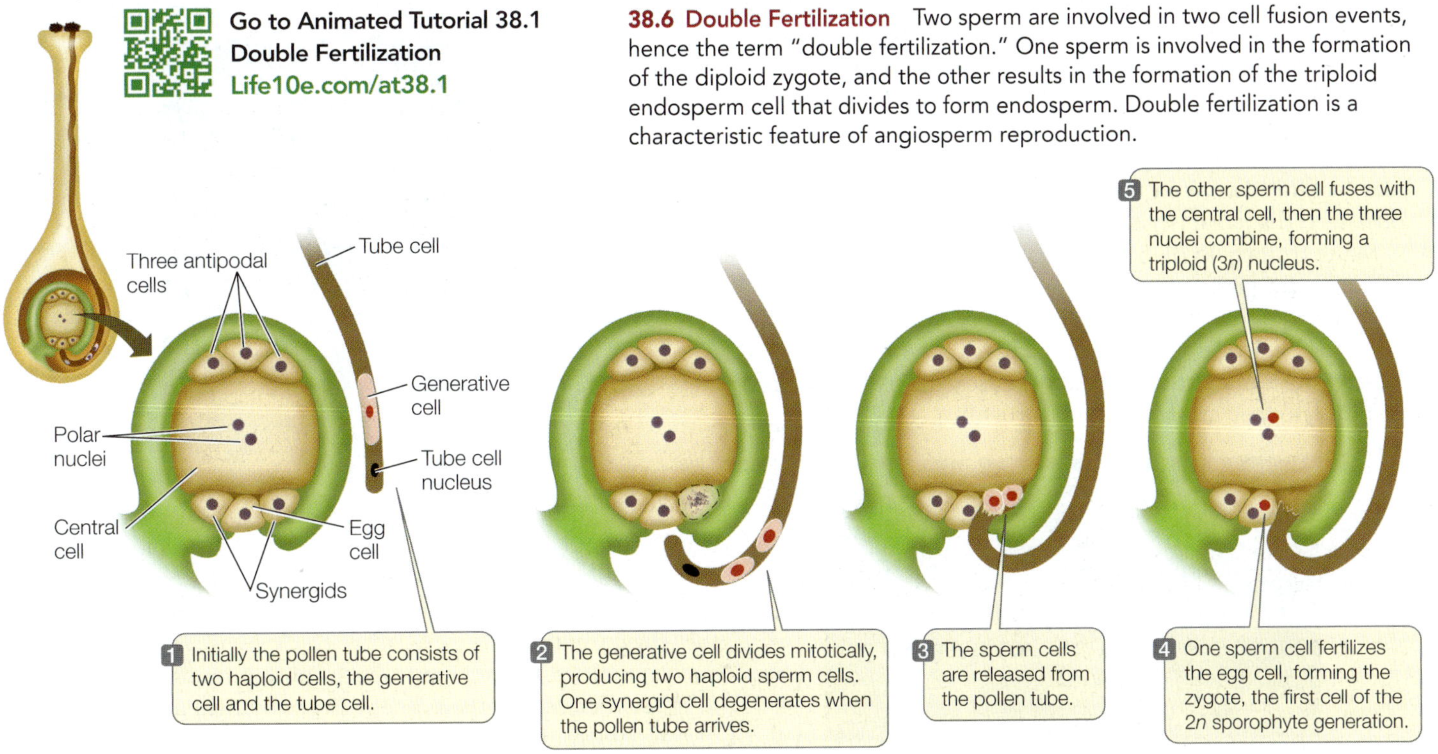

Three antipodal cells
Tube cell
Generative cell
Polar nuclei
Tube cell nucleus
Central cell
Egg cell
Synergids
1 Initially the pollen tube consists of two haploid cells, the generative cell and the tube cell.
2 The generative cell divides mitotically, producing two haploid sperm cells. One synergid cell degenerates when the pollen tube arrives.
3 The sperm cells are released from the pollen tube.
4 One sperm cell fertilizes the egg cell, forming the zygote, the first cell of the 2n sporophyte generation.
5 The other sperm cell fuses with the central cell, then the three nuclei combine, forming a triploid (3n) nucleus.

Embryos develop within seeds contained in fruits

Fertilization initiates the highly coordinated growth and development of the embryo, endosperm, integuments, and carpel. The integuments—protective tissue layers surrounding the ovule—develop into the seed coat, and the ovary wall becomes the outer layers of the fruit that encloses the seed (see Figure 38.2).

In Chapter 37 we described the events in plant embryonic development and its hormonal control. As seeds develop, they prepare for dispersal and dormancy by losing up to 95 percent of their water content. You can see this desiccation by comparing corn grains (e.g., popcorn) with ripe corn from the cob or a can. A dry seed is still alive; it has protective proteins that keep its cells in a viscous state.

In angiosperms, the ovary—together with the seeds it contains—develops into a fruit after fertilization has occurred. Fruits have two main functions:

- They protect the seed from damage by animals and infection by microbial pathogens.
- They aid in seed dispersal.

A **fruit** may consist of only the mature ovary and seeds, or it may include other parts of the flower. Some species produce fleshy, edible fruits, such as peaches and tomatoes, whereas the fruits of other species are dry or inedible (**Figure 38.7**).

(A)
Seed with embryo
Fruit (ovary wall)
(B)
Seed with embryo
Fruit (ovary wall)
(C)
Seed with embryo
Fruit (ovary wall)

38.7 Angiosperm Fruits There are a variety of fruits, but all have seed containing the embryo, surrounded by a fruit that comes from the wall of the ovary. (A) Garden pea. (B) Tomato. (C) Corn.

The diverse forms of fruits reflect the varied strategies plants use to disperse their progeny. Because plants cannot move, their progeny need some mechanism for separating themselves from their parents. Wide dispersal of progeny may not always be advantageous, however. If a plant has successfully grown and reproduced, its location is likely to be favorable for the next generation too. Some offspring do indeed stay near their parents. This is the case in many tree species, whose seeds simply fall to the ground. However, this strategy has several potential disadvantages. If the species is a perennial, the offspring that germinate near their parents will be competing with their parents for resources, which may be too limited to support a dense population. Furthermore, even though local conditions were good enough for the parent to produce at least some seeds, there is no guarantee that conditions will still be good the next year, or that they won't be better elsewhere. Thus in many cases seed dispersal is vital to a species' survival.

Many fruits help disperse seeds over substantial distances, increasing the probability that at least a few of the seeds will find suitable conditions for germination and growth to sexual maturity. Some wind-dispersed seeds have fruits with "wings," like those of the familiar maple. In other cases, the fruits and seeds are tiny, and they include feathery structures, such as this thistle:

Other fruits, such as these burs, attach themselves to animals (or to your clothes and shoes):

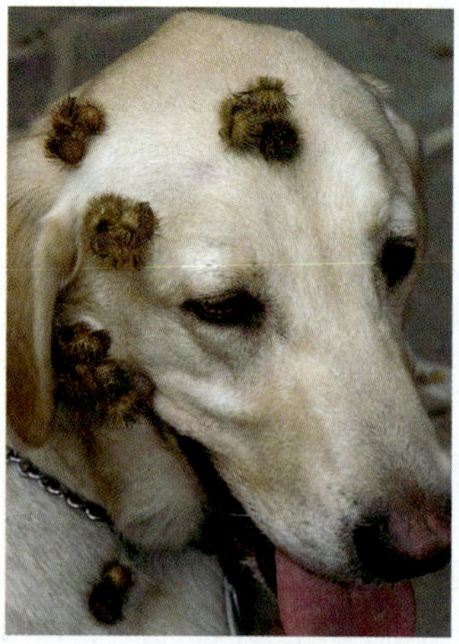

Water disperses some fruits; coconuts have been known to float thousands of miles between islands. Seeds swallowed whole by an animal along with fruits such as berries travel through the animal's digestive tract and are deposited some distance from the parent plant.

Biologists are beginning to understand the relationships between seed development (from ovules) and fruit development (from carpels). Some seedless fruits, such as varieties of watermelons and grapes, develop when fertilization occurs but the embryo then aborts. In other cases, such as bananas and pineapples, the fruit develops without fertilization. In most cases, however, fruit development does not occur in the absence of fertilization. Several years ago a farmer in Spain who grows sugar apples (*Annona squamosa*) noticed a seedless fruit and brought it to the attention of scientists. A single gene was subsequently identified whose mutated form results in this seedless phenotype. The wild-type version of this gene encodes a transcription factor important to the development of the outer tissues of the ovule. The gene is present in the genomes of all angiosperms examined. In the future, therefore, it may be possible to produce other seedless fruits by engineering mutations in this gene.

Seed development is under hormonal control

Chapter 37 described the role of the gibberellin hormones in the mobilization of stored macromolecules in the seed endosperm during germination. The development of seeds is under the control of a different hormone, **abscisic acid** (**ABA**). (Unfortunately, its name is misleading, because it does not directly control leaf abscission.) Most plant tissues make ABA, and like other plant hormones, it has multiple effects (see Table 37.2). For example, ABA plays a role in stomatal closing under water stress conditions (see Section 35.3). During early seed development the ABA level is low, and it rises as the seed matures. This increase stimulates the endosperm to synthesize seed storage proteins. It also stimulates the synthesis of proteins that prevent cell death as the seeds dry.

ABA also keeps the developing seed from germinating on the plant before it dries. Premature germination, termed **vivipary**, is undesirable in seed crops (such as wheat) because the grain is damaged if it starts to sprout. Viviparous seedlings are also unlikely to survive if they remain attached to the parent plant and are unable to establish themselves in the soil. Mutants of corn that are insensitive to ABA have viviparous seeds, indicating the importance of ABA in preventing precocious germination.

The general effect of ABA in preventing germination extends to seed dormancy. Seeds stay dormant if their ABA levels are high and germinate when the levels go down. This usually occurs as dormancy is broken.

RECAP 38.1

Flowers contain the organs for sexual reproduction in angiosperms. Plants that use pollen for reproduction have several selective advantages, among them the ability to accomplish fertilization without water. After fertilization, the flower develops into seed(s) and fruit. The selective advantages of seeds and fruits include long-term viability and multiple modes of dispersal.

- What are the relationships between an ovule and an ovary, and between a fruit and a seed? **See p. 779 and Figure 38.2**
- How do plants prevent self-pollination? **See p. 782 and Figure 38.5**
- Describe the roles of the two sperm cells in double fertilization. **See p. 783 and Figure 38.6**
- How is plant development controlled by the hormone abscisic acid? **See p. 785**

We have now traced the sexual life cycle of angiosperms from the flower to the gametophytes, pollination, fertilization, and the dispersal of seeds. We discussed seed germination and seedling development in Chapter 37, and indeterminate, vegetative plant growth in Chapter 34. The next section will cover the rest of the angiosperm life cycle—the transition from the vegetative to the flowering state—and how this transition is regulated.

38.2 What Determines the Transition from the Vegetative to the Flowering State?

Flowering is one of the major events in a plant's life. Flowering requires a reallocation of energy and materials away from making more plant parts (vegetative growth) to making flowers and gametes (reproductive growth). Once a plant is old enough, it can respond to internal or external signals to initiate reproduction. Flowering can happen right at maturity as part of a predetermined developmental program (as in a dandelion plant in the summer) or in response to environmental cues such as light or temperature (as with most ornamental flowers).

Plants fall into three categories depending on when they mature and initiate flowering, and what happens after they flower:

- **Annuals** complete their lives in one year. This class includes many crops important to the human diet, such as corn, wheat, rice, and soybean. When the environment is suitable, these plants grow rapidly, with little or no secondary growth. After flowering, they use most of their materials and energy to develop seeds and fruits, and the rest of the plant withers away.
- **Biennials** take 2 years to complete their lives. They are much less common than annuals and include carrots, cabbage, and onions. Typically, biennials produce only vegetative growth during the first year and store carbohydrates in underground roots (carrot) and stems (onion). In the second year they use most of the stored carbohydrates to produce flowers and seeds rather than vegetative growth, and the plant dies after seeds form.
- **Perennials** live 3 or more—sometimes many more—years. Maple trees can live up to 400 years. Perennials include many trees and shrubs, as well as wildflowers. Typically these plants flower every year but stay alive and keep growing for another season; the reproductive cycle repeats each year. However, some perennials (e.g., century plant) grow vegetatively for many years, flower once, and die.

No matter what type of life cycle they have, angiosperms all make the transition to flowering. This transition entails significant developmental changes, to which we now turn.

Shoot apical meristems can become inflorescence meristems

The first visible sign of a transition to the flowering state may be a change in one or more apical meristems in the shoot

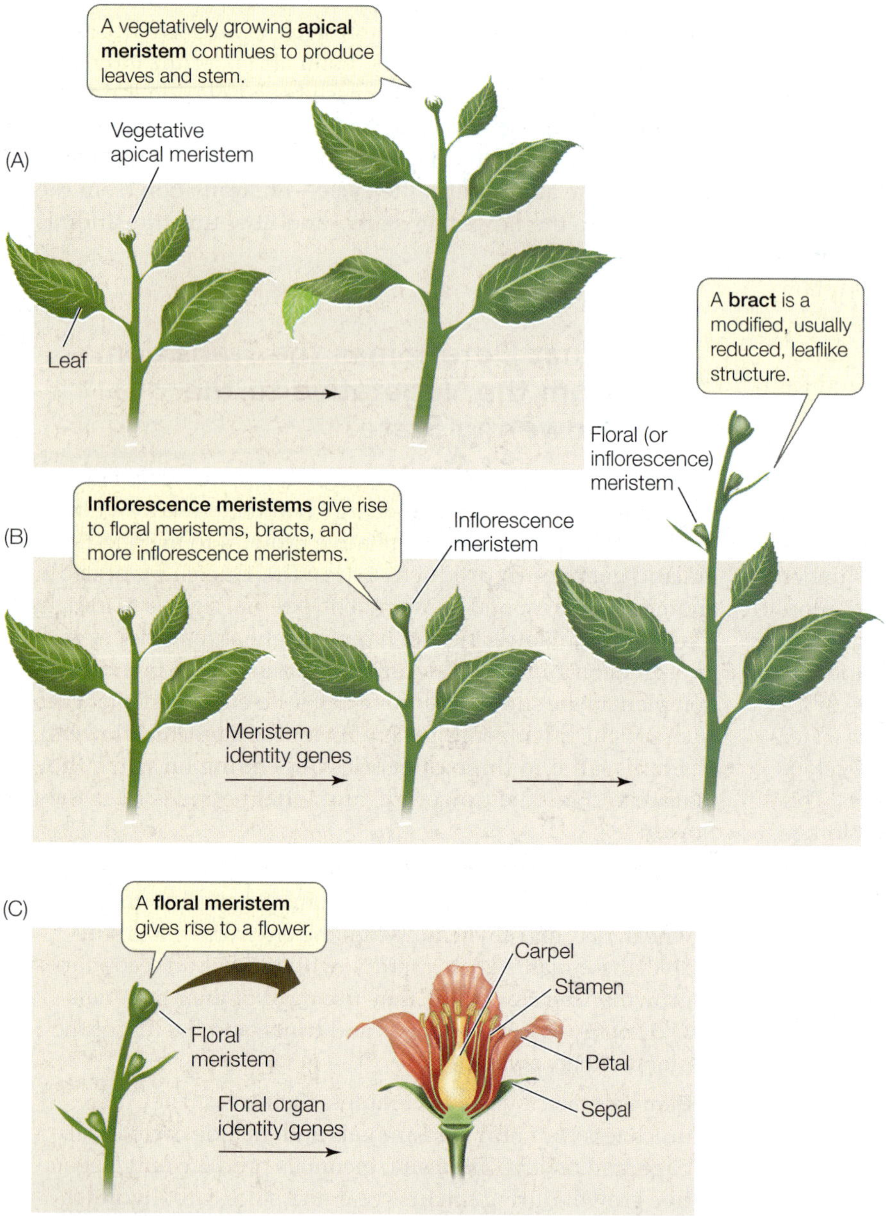
A vegetatively growing **apical meristem** continues to produce leaves and stem.
Vegetative apical meristem
(A)
Leaf
A **bract** is a modified, usually reduced, leaflike structure.
Floral (or inflorescence) meristem
Inflorescence meristems give rise to floral meristems, bracts, and more inflorescence meristems.
Inflorescence meristem
(B)
Meristem identity genes
(C)
A **floral meristem** gives rise to a flower.
Carpel
Stamen
Floral meristem
Petal
Floral organ identity genes
Sepal

38.8 Flowering and the Apical Meristem A vegetative apical meristem (A) grows without producing flowers. Once the transition to the flowering state is made (B), inflorescence meristems give rise to bracts and to floral meristems (C), which become the flowers.

system. As described in Chapter 34, meristems have a pool of undetermined cells. During vegetative growth, a shoot apical meristem continually produces leaves, axillary buds, and stem tissues (**Figure 38.8A**) in a kind of unrestricted growth called *indeterminate growth* (see Section 34.3).

Flowers may appear singly or in an orderly cluster that constitutes an **inflorescence**. If a vegetative apical meristem becomes an **inflorescence meristem**, it ceases production of leaves and axillary buds and produces other structures: smaller leafy structures called bracts, as well as new meristems in the angles between the bracts and the stem (**Figure 38.8B**). These new meristems may also be inflorescence meristems, or they may be **floral meristems**, each of which gives rise to a flower.

Each floral meristem typically produces four consecutive whorls or spirals of organs—the sepals, petals, stamens, and carpels discussed earlier in the chapter—separated by very short internodes, keeping the flower compact (**Figure 38.8C**). In contrast to vegetative apical meristems and some inflorescence meristems, floral meristems are responsible for *determinate growth*—growth of limited duration.

A cascade of gene expression leads to flowering

The genes that determine the transition from shoot apical meristems to inflorescence meristems and from inflorescence meristems to floral meristems have been studied in model organisms such as *Arabidopsis*.

MERISTEM IDENTITY GENES Expression of two **meristem identity genes** initiates a cascade of further gene expression that leads to flower formation. These genes encode the transcription factors LEAFY and APETALA1, which together are necessary and sufficient for determining the transition to flowering. Evidence for the roles of these factors comes from both genetic and plant transformation experiments. For example, a mutant allele of the *APETALA1* gene leads to continued vegetative growth, even if conditions are suitable for flowering. However, if the wild-type *APETALA1* gene is coupled with a constitutive (always on) promoter and used to transform *Arabidopsis* plants, the plants will flower prematurely, regardless of environmental conditions. This is powerful evidence that APETALA1 plays a role in switching meristem cells from a vegetative to a reproductive fate (see Figure 38.8B).

FLORAL ORGAN IDENTITY GENES The products of the meristem identity genes trigger the expression of **floral organ identity genes**. As described in Section 19.4, these genes work in concert to specify the successive whorls of the flower (see Figure 19.9). The floral organ identity genes encode transcription factors that determine whether cells in the floral meristem will be sepals, petals, stamens, or carpels. For example, *AGAMOUS* is a class C gene that causes florally determined cells to form stamens and carpels.

Depending on the species, plants initiate these gene expression changes, and the events that follow, in response to either internal or external cues. The most well studied external cues are photoperiod (day length) and temperature. We will begin with photoperiod.

38.9 Mammoth Plant Wild-type tobacco (left) is much smaller than the Maryland Mammoth mutant of the same age (right), which does not respond to an environmental cue to stop growing and flower.

Photoperiodic cues can initiate flowering

The study of how light affects the transition to flowering began with two observations in the early twentieth century:

- Normally, tobacco grows to about 1.5 meters tall before flowering in the summer, but a variety called Maryland Mammoth grows to 5 meters (**Figure 38.9**). Farmers in Virginia were frustrated because they could not easily get seeds from this luxuriant plant for successive crops. Instead of flowering, it continued to grow until the late fall frost killed it.
- Because of improvements in agricultural techniques, soybean yields became so great that it was hard for farmers to harvest all the plants at once. Hoping to stagger the harvests, farmers tried planting the seeds in groups several weeks apart, but all resulting plants nevertheless formed flowers and seeds at the same time.

The explanation for both of these observations was the same: the signal that set the plants' shoot apical meristems on the path to flowering was the length of daylight, or **photoperiod**. When soybeans experience days of a certain length, they flower, regardless of how "old" they are. Maryland Mammoth tobacco *can* flower, but it doesn't do so in Virginia because it dies when the weather there gets cold. When the plant is grown in a greenhouse to prevent freezing, however, it flowers in December, when the days are short. Maryland Mammoth is now grown commercially in Florida.

Scientists used greenhouse experiments to measure the day length required for different plant species to flower. Maryland Mammoth tobacco did not flower if exposed to more than 14 hours of light per day; flowering was only initiated once day length became shorter than 14 hours, as it does in December. Other plants (such as soybeans and henbane) flowered only when the days were long (**Figure 38.10**). Control of an organism's responses by the length of day or night is called **photoperiodism**.

Plants vary in their responses to photoperiodic cues

Plants that flower in response to photoperiodic stimuli fall into two main classes:

- **Short-day plants (SDPs)** flower only when the day is shorter than a critical maximum. They include poinsettias and chrysanthemums, as well as Maryland Mammoth tobacco. Thus, for example, we see chrysanthemums in nurseries in the fall and poinsettias in winter.
- **Long-day plants (LDPs)** flower only when the day is longer than a critical minimum. Spinach and clover are examples of LDPs. For example, spinach tends to flower in the summer, so it is normally planted in early spring.

While there are variations on these two patterns, photoperiodic control of flowering serves an important role: it synchronizes the flowering of plants of the same species in a local population, and this promotes cross-pollination and successful reproduction.

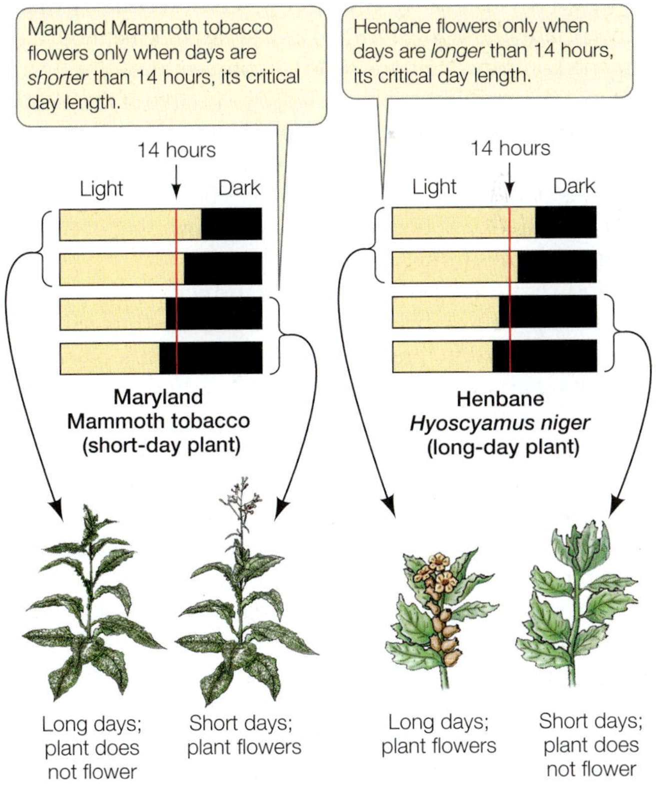

Maryland Mammoth tobacco flowers only when days are *shorter* than 14 hours, its critical day length.
Henbane flowers only when days are *longer* than 14 hours, its critical day length.
14 hours
Light
Dark
14 hours
Light
Dark
Maryland Mammoth tobacco (short-day plant)
Henbane *Hyoscyamus niger* (long-day plant)
Long days; plant does not flower
Short days; plant flowers
Long days; plant flowers
Short days; plant does not flower

38.10 Day Length and Flowering Flowering of Maryland Mammoth tobacco is initiated when the days become shorter than a critical length. Maryland Mammoth tobacco is thus called a short-day plant. Henbane, a long-day plant, shows an inverse pattern of flowering.

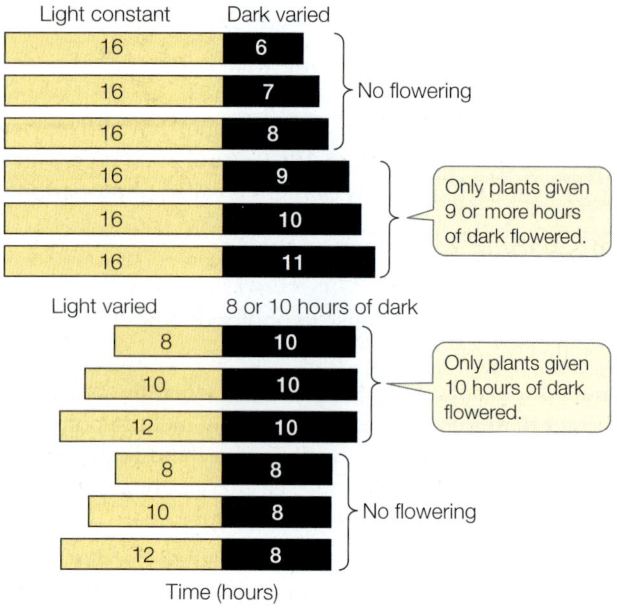
Light constant
Dark varied
16
6
16
7
16
8
No flowering
16
9
16
10
16
11
Only plants given 9 or more hours of dark flowered.
Light varied
8 or 10 hours of dark
8
10
10
10
12
10
Only plants given 10 hours of dark flowered.
8
8
10
8
12
8
No flowering
Time (hours)

38.11 Night Length and Flowering Greenhouse experiments using cocklebur, a short-day plant, showed that night length, not day length, is the environmental cue that initiates flowering.

Night length is a key photoperiodic cue that determines flowering

The terms "short-day plant" and "long-day plant" imply that *day length* is the environmental cue that triggers flowering. Actually, the important cue is *night length*, as a series of greenhouse experiments confirmed (**Figure 38.11**). In a greenhouse, the overall length of a day or night can be varied, irrespective of the 24-hour natural cycle. For example, if cocklebur, an SDP, is exposed to several long periods of light (16 hours each), it will still flower as long as the dark period between them is 9 hours or longer. This 9-hour inductive dark period also induces flowering even if the light period varies from 8 hours to 12 hours.

Biologists noticed that when the inductive dark period was interrupted by a brief period of light, the flowering signal generated by the long night disappeared. It took several days of long nights for the plant to recover and initiate flowering. Interrupting the day with a dark period had no effect on flowering. A clue as to what occurred in the plant when the flash of light was given came when biologists determined the action spectrum for the wavelengths of light that were effective. As with lettuce seed germination (see Figure 37.16), red light was most effective at breaking the "night" stimulus, and its effect was reversible by far-red light (**Figure 38.12**). Later, the photoreceptor involved was identified as a phytochrome.

One way that plants sense night length is through phytochrome activity. As explained in Section 37.4, during the day there is more light in the red than in the far-red range, and the inactive P_r isoform of phtyochrome is converted to the active P_{fr} isoform. At night there is a gradual, spontaneous conversion of this P_{fr} back to P_r, so at the end of the night the phytochrome is predominantly in its inactive form. But a short burst of red light in the middle of the night will disrupt this rhythm by converting the P_r to P_{fr} again.

INVESTIGATING**LIFE**

38.12 Interrupting the Night Knowing that plants measure night duration, the question became whether the dark hours to which a plant is exposed must be continuous. Using SDPs and LDPs as test subjects, Karl Hamner and James Bonner demonstrated that this was the case by interrupting the night with short bursts of light. Sterling Hendricks and William Siegelman repeated the experiments using light of different wavelengths to gain information about the photoreceptor involved.[a]

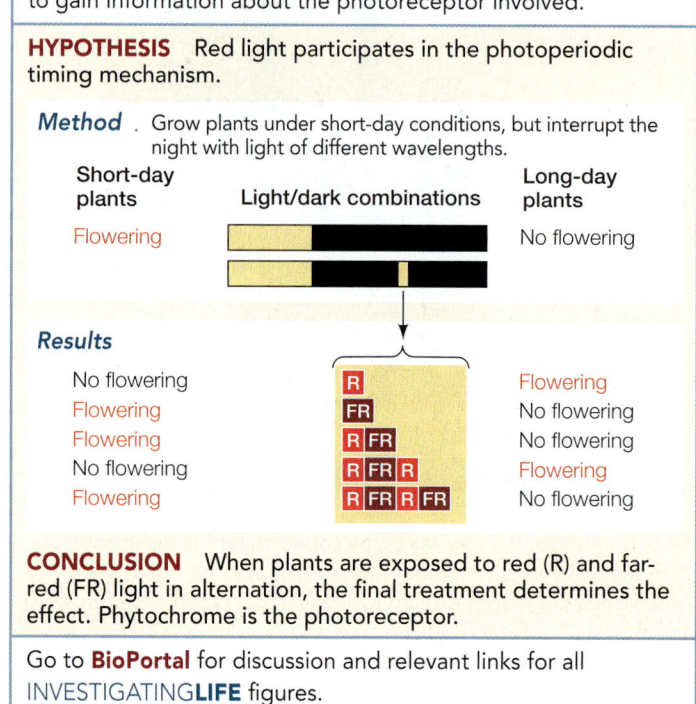

HYPOTHESIS Red light participates in the photoperiodic timing mechanism.

Method Grow plants under short-day conditions, but interrupt the night with light of different wavelengths.

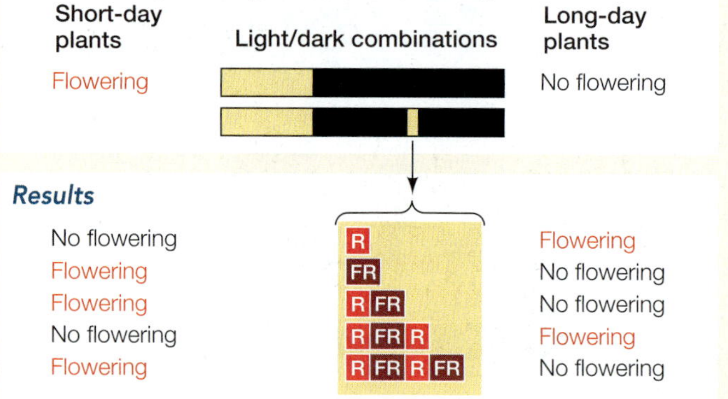
Short-day plants
Light/dark combinations
Long-day plants
Flowering
No flowering
Results
No flowering
Flowering
Flowering
No flowering
Flowering
R
FR
R FR
R FR R
R FR R FR
Flowering
No flowering
No flowering
Flowering
No flowering

CONCLUSION When plants are exposed to red (R) and far-red (FR) light in alternation, the final treatment determines the effect. Phytochrome is the photoreceptor.

Go to **BioPortal** for discussion and relevant links for all INVESTIGATING**LIFE** figures.

[a]Hendricks, S. B. and H. W. Siegelman. 1967. *Comparative Biochemistry* 27: 211–235.

Go to Animated Tutorial 38.2
The Effect of Interrupted Days and Nights
Life10e.com/at38.2

As described in Section 37.4, phytochromes and a blue-light receptor function together to "entrain" a circadian rhythm in plants. These photoreceptors cycle through active and inactive phases over repeated 24-hour periods, and in doing so they activate signaling pathways that result in regular cycles in the expression of specific genes. One gene whose expression follows a circadian rhythm is *CONSTANS* (*CO*), which encodes a transcriptional regulator that controls the expression of flowering genes.

Experiments with *Arabidopsis*, an LDP, have shown that photoperiodic flowering times are determined by interactions between photoreceptors and the CO protein. *CO* gene expression goes through regular 24-hour cycles with peak expression late in the day. On long days, this occurs in daylight, late in the afternoon, but on short days the peak expression occurs after dark. On long days, the active forms of phytochrome and a blue-light receptor activate pathways that stabilize the CO protein, which promotes flowering. This process does not occur on short days.

The flowering stimulus originates in a leaf

Early experiments indicated that reception of the photoperiodic stimulus occurs within the leaf. For example, in the LDP spinach, flowering occurred if the leaves were exposed to long-day periods of light, while the shoot apical meristem was masked to simulate short days. Flowering could *not* occur when the

INVESTIGATINGLIFE

38.13 The Flowering Signal Moves from Leaf to Bud The receptors for photoperiod are in the leaf, but the transition to flowering occurs in the shoot apical meristem. To investigate whether there is a diffusible substance that travels from leaf to bud, Hamner and Bonner exposed only the leaf to the photoperiodic stimulus[a]

HYPOTHESIS The leaves measure the photoperiod.

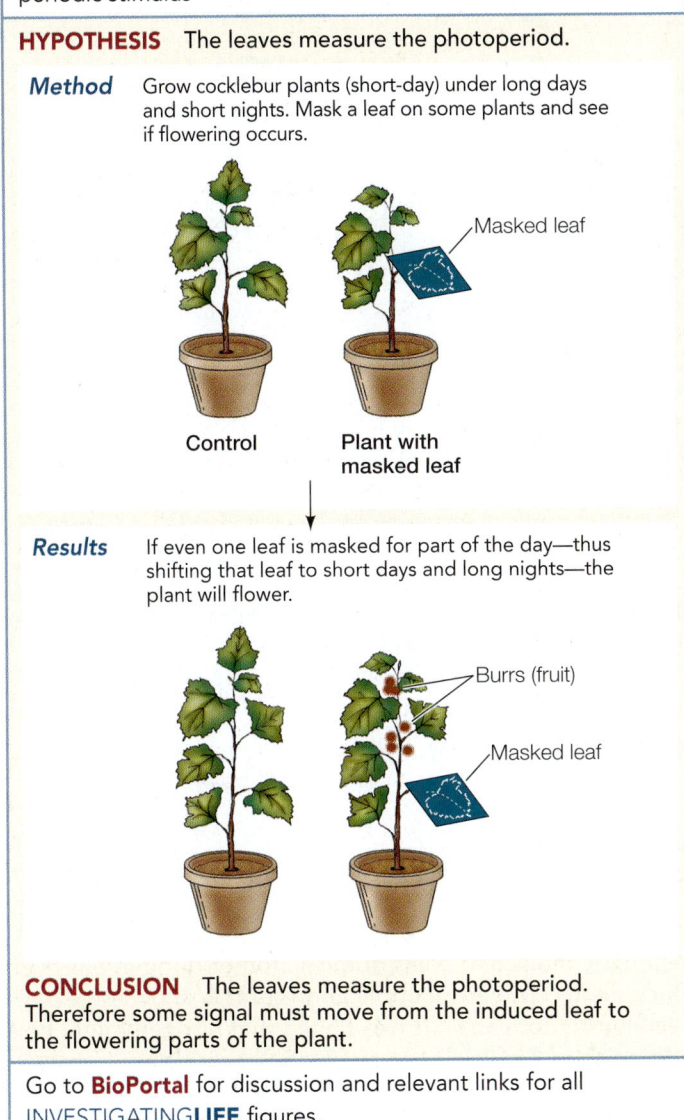

Method Grow cocklebur plants (short-day) under long days and short nights. Mask a leaf on some plants and see if flowering occurs.

Masked leaf
Control
Plant with masked leaf

Results If even one leaf is masked for part of the day—thus shifting that leaf to short days and long nights—the plant will flower.

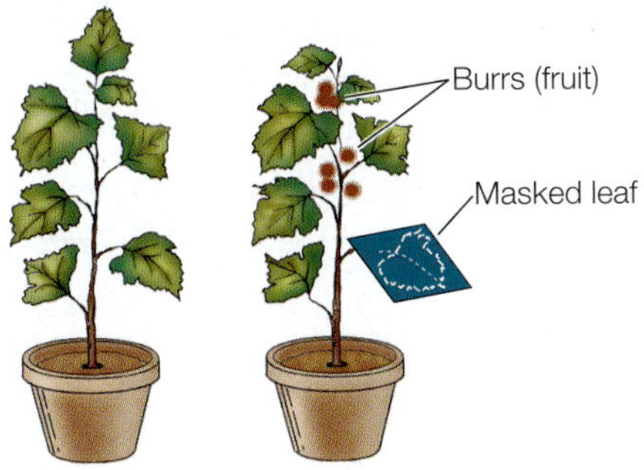
Burrs (fruit)
Masked leaf

CONCLUSION The leaves measure the photoperiod. Therefore some signal must move from the induced leaf to the flowering parts of the plant.

Go to **BioPortal** for discussion and relevant links for all INVESTIGATING**LIFE** figures.

[a]Hamner, K. C. and J. Bonner. 1938. *Botanical Gazette* 100: 388–431.

leaves were masked to simulate short days, while the bud was exposed to long-day periods of light.

These "masking" experiments were extended to SDP plants as well (**Figure 38.13**). Because the receptor of the stimulus (in the leaf) is physically separated from the tissue on which the stimulus acts (the bud meristem), the inference can be drawn that a systemic signal travels from the leaf through the plant's tissues to the bud meristem. Other evidence that a diffusible chemical travels from the leaf to the bud meristem signal includes the following:

- If a photoperiodically induced leaf is immediately removed from a plant after the inductive dark period, the plant does not flower. If, however, the induced leaf remains attached to the plant for several hours, the plant will flower. This

WORKING WITH**DATA:**

The Flowering Signal Moves from Leaf to Bud

Original Paper

Hamner, K. C. and J. Bonner. 1938. Photoperiodism in relation to hormones as factors in floral initiation and development. *Botanical Gazette* 100: 388–431.

Analyze the Data

In 1938 Karl Hamner at the University of Chicago was working on the role of plant nutrition in flowering. The plant he studied, cocklebur, is a short-day plant that requires 16 hours of darkness to flower. When the plants were kept in 6 hours darkness (16 hours light) in a greenhouse, they did not flower. One day Hamner came to the lab to find all the plants flowering. It turned out that there had been a power outage, and the plants had received a single inductive short day (long night). Realizing that this provided a simple system to study flowering, Hamner invited a major scientist in the field, James Bonner from Caltech, to join him for the summer. The two biologists carried out a series of experiments using the single inductive period that showed that flowering is induced by night length as opposed to day length (see Figure 38.12) and that the flowering signal is received by the leaf from which it travels to a bud, inducing flowering. A Russian plant physiologist, Mikhail Chailakhyan, named this signal florigen. More recently, the molecular nature of this signal was described.

QUESTION 1

Intact cocklebur plants (6) or plants with their leaves removed (6) were placed in the inductive (short day) photoperiod. After 14 days, the researchers obtained the results in **TABLE A**. Based on these data, which part of the plant senses the photoperiod?

QUESTION 2

Cocklebur plants were treated so that a single leaf was exposed to the inductive (short) photoperiod while the rest of the plant received the long photoperiod. After 18 days, the results were as shown in **TABLE B**. What do these results indicate about the location of the receptor for flowering?

QUESTION 3

What do the data tell you about the signal generated by the plant in response to photoperiod and that induces flowering in the apical meristem?

TABLE A

Treatment	Number of plants that flowered
No inductive period, intact plant	0
Inductive period, intact plant	6
Inductive period, leaves removed	0

TABLE B

Treatment	Number of plants	Result
Untreated	6	Vegetative
Treated, one leaf	32	Flower

Go to **BioPortal** for all WORKING WITH**DATA** exercises

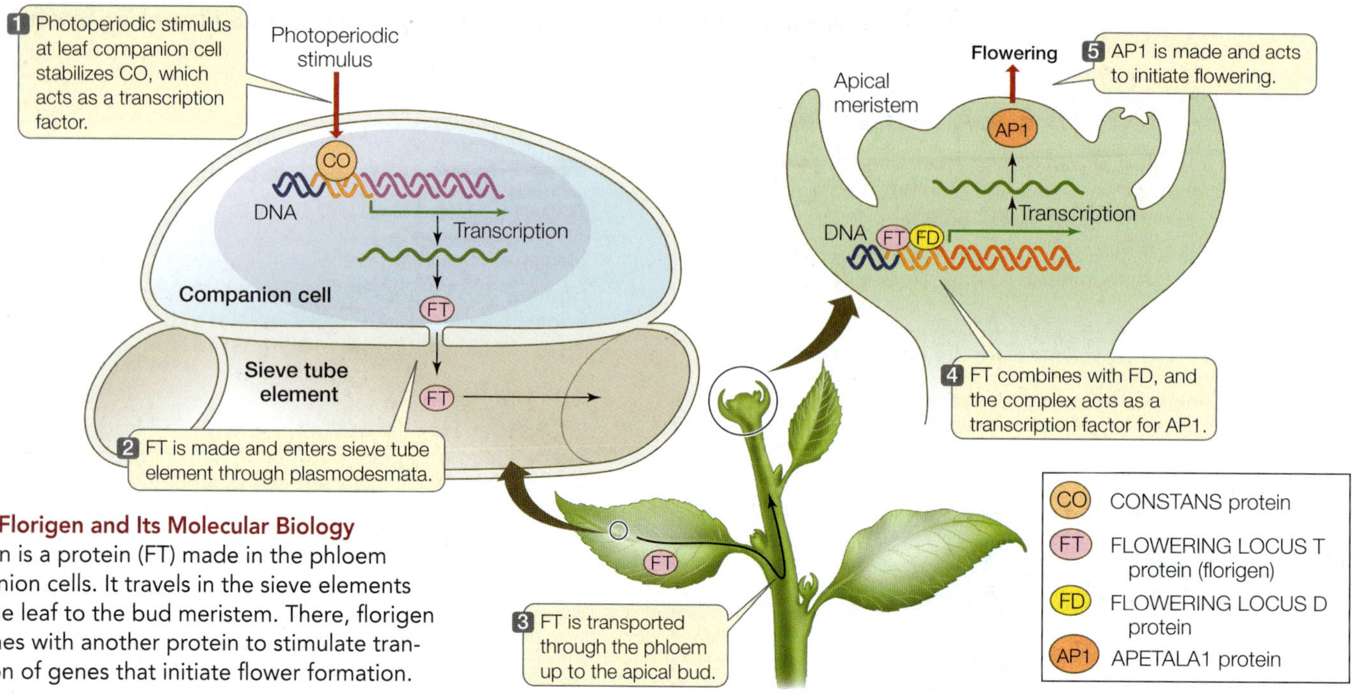
1 Photoperiodic stimulus at leaf companion cell stabilizes CO, which acts as a transcription factor.
Photoperiodic stimulus
CO
DNA
Transcription
Companion cell
FT
Sieve tube element
FT
2 FT is made and enters sieve tube element through plasmodesmata.
FT
3 FT is transported through the phloem up to the apical bud.
Flowering
5 AP1 is made and acts to initiate flowering.
Apical meristem
AP1
Transcription
DNA
FT
FD
4 FT combines with FD, and the complex acts as a transcription factor for AP1.
CO CONSTANS protein
FT FLOWERING LOCUS T protein (florigen)
FD FLOWERING LOCUS D protein
AP1 APETALA1 protein

38.14 Florigen and Its Molecular Biology
Florigen is a protein (FT) made in the phloem companion cells. It travels in the sieve elements from the leaf to the bud meristem. There, florigen combines with another protein to stimulate transcription of genes that initiate flower formation.

result suggests that something is synthesized in the leaf in response to the inductive dark period, and then moves out of the leaf to induce flowering.

- If two cocklebur plants are grafted together, and if only one of the plants is exposed to inductive long nights, both plants flower.
- In several species, if an induced leaf from one species is grafted onto another, noninduced plant of a different species, the recipient plant flowers.

The transmissible signal was given a name, **florigen** ("flower inducing"), in 1937, but its chemical nature has been described only in the past decade.

Florigen is a small protein

The characterization of florigen was made possible by genetic and molecular studies of the model plant *Arabidopsis*. Three genes are involved in the signaling response for flowering (**Figure 38.14**).

- ***FT*** (***FLOWERING LOCUS T***) *codes for florigen.* FT is a small (20 kDa) protein that can travel through plasmodesmata. FT is synthesized in the phloem companion cells of the leaf and then diffuses into the adjacent sieve elements. It then is carried through the phloem to the apical meristem. If the *FT* gene is coupled to an active promoter and expressed at high levels in the shoot meristem, flowering is induced even in the absence of an appropriate photoperiodic stimulus.
- ***CO*** (***CONSTANS***) *codes for the transcription factor that activates the synthesis of FT.* As described above, *CO* expression follows a circadian rhythm, and stabilization of the CO protein by photoreceptors allows it to function. Like *FT*, *CO* is expressed in leaf companion cells. If *CO* is experimentally overexpressed in the leaf, flowering is induced. However, if *CO* is overexpressed in the apical meristem, flowering is not induced, indicating that *CO* functions in the leaf.
- ***FD*** (***FLOWERING LOCUS D***) *encodes a protein that binds to FT protein when it arrives in the apical meristem.* The FD protein is a transcription factor that, when bound to FT, activates promoters for meristem identity genes, such as *APETALA1* (see Figure 38.8). The expression of FD primes meristem cells to change from a vegetative fate to a reproductive fate once FT arrives.

Before FT was isolated, grafting experiments indicated that many different plant species could be induced to flower by the same chemical signal. Results of molecular experiments confirmed that the *FT* gene is involved in photoperiod signaling in many species:

- Transgenic plants (e.g., tobacco and tomato) that express the *Arabidopsis FT* gene at high levels flower regardless of day length.
- Transgenic *Arabidopsis* plants that express *FT* homologs from other plants (e.g., rice and tomato) flower regardless of day length.

While the molecular basis of the action of florigen has been elucidated, commercial applications of this knowledge have been harder to realize. It was hoped that florigen might be a very small molecule, like an auxin or gibberellin, that could be sprayed on economically important plants to induce flowering at will. The fact that florigen is a protein that cannot readily enter cells from the outside environment makes the development of commercial florigen treatments unlikely.

We have considered the photoperiodic regulation of flowering, from photoreceptors in the leaf to florigen that travels from the induced leaf to the sites of flower formation. In some plants, however, flowering is induced by other stimuli. These additional stimuli can function with photoperiodism or independently of it.

Flowering can be induced by temperature or gibberellin

Whereas some plants use the environmental cue of day length to induce flowering, other plants use different mechanisms. These include temperature (another environmental cue) and gibberellin (an internal, hormonal cue).

Winter-annual *Arabidopsis* without vernalization
Winter-annual *Arabidopsis* with vernalization

38.15 Vernalization A genetic strain of *Arabidopsis* (winter-annual *Arabidopsis*) requires vernalization for flowering. Without it, the plant is large and vegetative (left), but with the cold period it is smaller and flowers (right).

TEMPERATURE In some plant species, notably certain cereal grains, the environmental signal for flowering is cold temperature, a phenomenon called **vernalization** (Latin *vernus*, "spring"). In both wheat and rye, we distinguish two categories of flowering behavior. Spring wheat, for example, is a typical annual plant: it is sown in the spring and flowers in the same year. Winter wheat is sown in the fall, grows to a seedling, overwinters (often covered by snow), and flowers the following summer. If winter wheat is not exposed to cold in its first year, it will not flower normally the next year.

How vernalization leads to flowering has been elucidated using model organisms such as *Arabidopsis*. In strains of *Arabidopsis* that require vernalization to flower (**Figure 38.15**), a gene called *FLC* (*FLOWERING LOCUS C*) encodes a transcription factor that blocks the FT–FD florigen pathway (see Figure 38.14) by inhibiting expression of FT and FD. Cold temperature inhibits the synthesis of FLC protein, allowing the FT and FD proteins to be expressed, and flowering to proceed. Similar proteins control some steps in vernalization in cereals.

Epigenetics (see Section 16.4) plays an important role in the inhibition of *FLC* gene expression by cold temperature. Before vernalization, the chromatin at the promoter of the *FLC* gene is in a relaxed configuration, with histone protein acetylation lowering the ionic attraction of these proteins for DNA, which allows transcription (**Figure 38.16A**). During vernalization, a gene is expressed whose protein product is involved in the deacetylation of histones on the *FLC* gene. Deactylation causes the chromatin to be more compact, which blocks *FLC* gene expression (**Figure 38.16B**).

GIBBERELLIN *Arabidopsis* plants do not flower if they are genetically deficient in the hormone gibberellin, or if they are treated with an inhibitor of gibberellin synthesis. These observations implicate gibberellins in flowering. Direct application of gibberellins to *Arabidopsis* buds results in activation of the

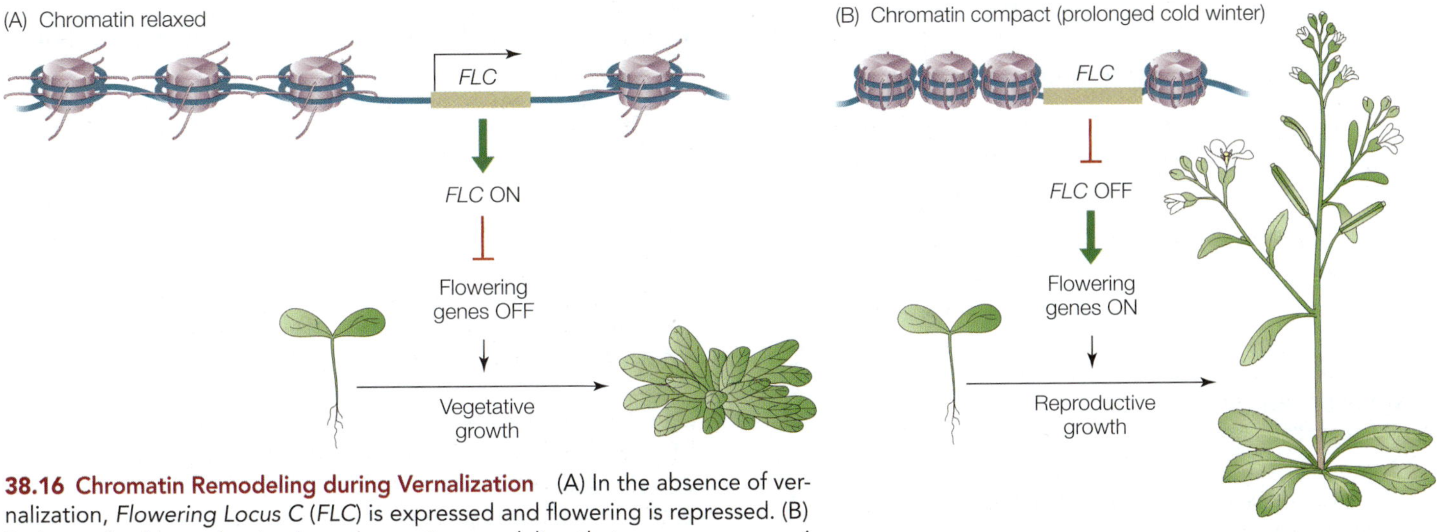
(A) Chromatin relaxed
FLC
FLC ON
Flowering genes OFF
Vegetative growth
(B) Chromatin compact (prolonged cold winter)
FLC
FLC OFF
Flowering genes ON
Reproductive growth

38.16 Chromatin Remodeling during Vernalization (A) In the absence of vernalization, *Flowering Locus C* (*FLC*) is expressed and flowering is repressed. (B) Prolonged cold weather leads to chromatin remodeling that represses expression of *FLC*. The absence of FLC protein allows flowering genes to be expressed.

meristem identity gene *LEAFY*, which in turn promotes the transition to flowering.

Some plants do not require an environmental cue to flower

Several plant species and strains do not require a photoperiod or vernalization to flower, but instead flower on cue from an "internal clock." For example, flowering in some strains of tobacco will be initiated in the terminal bud when the stem has grown four phytomers in length (recall that stems are composed of repeating units called phytomers; see Figure 34.1). If such a bud with a single adjacent phytomer is removed and planted, the cutting will flower because the bud has already received the cue for flowering. But the rest of the shoot below the bud that has been removed will not flower because it is only three phytomers long. After it grows an additional phytomer, it will flower. These results suggest that there is something about the *position* of the bud (atop four phytomers of stem) that determines its transition to flowering.

The bud might "know" its position by the concentration of some substance that forms a positional gradient along the length of the plant. Such a gradient could be formed if the root makes a diffusible inhibitor of flowering whose concentration diminishes with plant height. When the plant reaches a certain height, the concentration of the inhibitor would become sufficiently low at the tip of the shoot to allow flowering. What this inhibitor might be is unclear, but there is evidence that it acts by decreasing the amount of FLC, allowing the FT–FD pathway to proceed (just as cold acts on FLC in vernalization). A positional gradient that acts on FLC would be consistent with other mechanisms affecting flowering, which all converge on *LEAFY* and *APETALA1*:

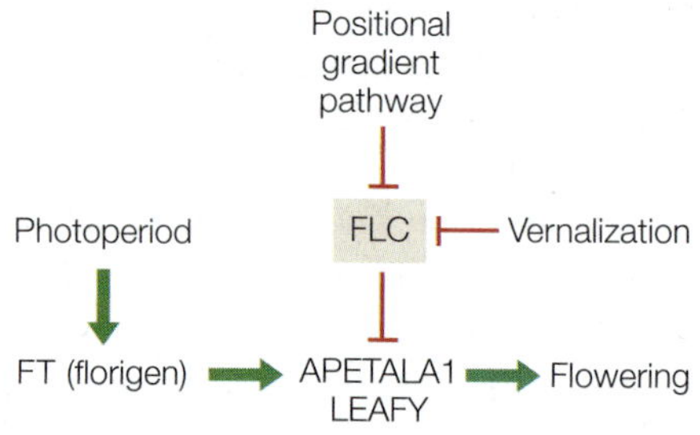
Positional gradient pathway
Photoperiod
FLC
Vernalization
FT (florigen)
APETALA1
LEAFY
Flowering

RECAP 38.2

Flowering of some angiosperms is controlled by night length, a phenomenon called photoperiodism. Low temperatures can induce flowering in some species (vernalization). Some species flower when their stems have grown by a certain amount, independent of environmental cues. All pathways to flowering converge on the meristem identity genes.

- What are the differences between apical meristems, inflorescence meristems, and floral meristems? What genes control the transitions between them? **See p. 786 and Figure 38.8**
- Explain why "short-day plant" is a misleading term. **See p. 788 and Figure 38.11**
- What is the evidence for florigen? What is its molecular mechanism of action? **See pp. 788–790 and Figures 38.13 and 38.14**

We have seen how environmental factors interact with genes to control flowering in angiosperms. The function of flowers is sexual reproduction, which maintains beneficial genetic variation in a population. Many angiosperms, however, also benefit from being able to reproduce asexually.

38.3 How Do Angiosperms Reproduce Asexually?

Although sexual reproduction takes up most of the space in this chapter, asexual reproduction accounts for many of the individual plants present on Earth. This fact suggests that in some circumstances asexual reproduction must be advantageous.

We have noted that genetic recombination is one of the advantages of sexual reproduction. Self-fertilization is a form of sexual reproduction, but offers fewer opportunities for genetic recombination than does cross-fertilization. A diploid, self-fertilizing plant that is heterozygous for a certain locus can produce both kinds of homozygotes for that locus plus the heterozygote among its progeny, but it cannot produce any progeny carrying alleles that it does not itself possess. Nevertheless many self-fertilizing plant species produce viable and vigorous offspring.

Asexual reproduction eliminates genetic recombination altogether. A plant that reproduces asexually produces progeny genetically identical to the parent (clones). What, then, is the advantage of asexual reproduction? If a plant is well adapted to its environment, asexual reproduction allows it to pass on to all its progeny a superior combination of alleles, which might otherwise be separated by sexual recombination.

Many forms of asexual reproduction exist

Stems, leaves, and roots are considered vegetative organs and are distinguished from flowers, the reproductive parts of the plant. Asexual reproduction is often accomplished through the modification of a vegetative organ, which is why the term **vegetative reproduction** is sometimes used to describe asexual reproduction in plants. Another type of asexual reproduction, apomixis (see the opening story), involves flowers but no fertilization.

Often the stem is the organ that is modified for vegetative reproduction. Strawberries, for example, produce horizontal stems, called stolons or runners, which grow along the soil surface, form roots at intervals, and establish potentially independent plants. Asexual reproduction by shoot tips is accomplished when the tips of upright branches sag to the ground and develop roots, as in blackberry and forsythia.

Some plants, such as potatoes, form enlarged fleshy tips of underground stems, called tubers, that can produce new plants (from the "eyes"). Rhizomes are horizontal underground stems that can give rise to new shoots. Bamboo is a striking example of a plant that reproduces vegetatively by means of rhizomes. A single bamboo plant can give rise to a stand—even a forest—of plants constituting a single, physically connected entity (**Figure 38.17A**).

(A)
These bamboo shoots all arise from the same underground stem.
(B)
Each clove of garlic is a bulb that can give rise to a new plant.
(C)
The plantlets forming on the margin of this *Kalanchoe* leaf will fall to the ground and become independent plants.

38.17 Vegetative Organs Modified for Reproduction (A) The rhizomes of bamboo are underground stems that produce plants at intervals. (B) Bulbs are short stems with large leaves that store nutrients and can give rise to new plants. (C) In *Kalanchoe*, new plantlets can form on leaves.

Whereas stolons and rhizomes are horizontal stems, bulbs and corms are short, vertical, underground stems. Lilies and garlic form bulbs (**Figure 38.17B**), short stems with many fleshy, highly modified leaves that store nutrients. These storage leaves make up most of the bulb. They can give rise to new plants by dividing or by producing new bulbs from axillary buds. Crocuses, gladioli, and many other plants produce corms, underground stems that function very much as bulbs do. Corms are disclike and consist primarily of stem tissue; they lack the fleshy modified leaves that are characteristic of bulbs.

Stems are not the only vegetative organs modified for asexual reproduction. Leaves may also be the source of new plantlets, as in some succulent plants of the genus *Kalanchoe* (**Figure 38.17C**). Many kinds of angiosperms, ranging from grasses to trees such as aspens and poplars, form interconnected, genetically homogeneous populations by means of suckers—shoots produced by roots. What appears to be a whole stand of aspen trees, for example, may be a clone derived from a single tree by suckers. This is why the leaves of a whole stand of aspens typically turn yellow at the same time.

Plants that reproduce vegetatively often grow in physically unstable environments such as eroding hillsides. Plants with stolons or rhizomes, such as beach grasses, rushes, and sand verbena, are common pioneers on coastal sand dunes. Rapid vegetative reproduction enables these plants, once introduced, not only to multiply but also to survive burial by the shifting sand; in addition, the dunes are stabilized by the extensive network of rhizomes or stolons that develops. Vegetative reproduction is also common in some deserts, where the environment is often not suitable for seed germination and the establishment of seedlings.

Vegetative reproduction has a disadvantage

Vegetative reproduction is highly efficient in an environment that is stable over the long term. A change in the environment, however, can leave an asexually reproducing species at a disadvantage.

A striking example is provided by the demise of the English elm, *Ulmus procera*, which was apparently introduced into England as a clone by the ancient Romans. This tree reproduces asexually by suckers and is incapable of sexual reproduction. In 1967 Dutch elm disease first struck the English elms. After two millennia of clonal growth, the population lacked genetic diversity, and no individuals carried genes that would protect them against the disease. Today the English elm is all but gone from England.

Vegetative reproduction is important in agriculture

One of the oldest methods of vegetative reproduction used in agriculture consists of simply making cuttings of stems, inserting them in soil, and waiting for them to form roots and become autonomous plants. The cuttings are usually encouraged to root by treatment with auxin, a plant hormone.

Woody plants can be propagated asexually by **grafting**: attaching a bud or a piece of stem from one plant to a root or root-bearing stem of another plant. The part of the resulting plant that comes from the root-bearing "host" is called the **stock**; the part grafted on is the **scion** (**Figure 38.18**). The vascular cambium of the scion associates with that of the stock, forming a continuous cambium that produces xylem and phloem. The cambium allows the transport of water and minerals to the scion and of photosynthate to the stock. Much of the fruit grown for market in the United States is produced on grafted trees, as are wine grapes.

Another method widely used for asexual plant propagation is **meristem culture**, in which pieces of shoot apical meristem are cultured on growth media to generate plantlets, which can then be planted in the field. This strategy is vital when uniformity is desired, as in forestry, or when virus-free plants are the goal, as with strawberries and potatoes.

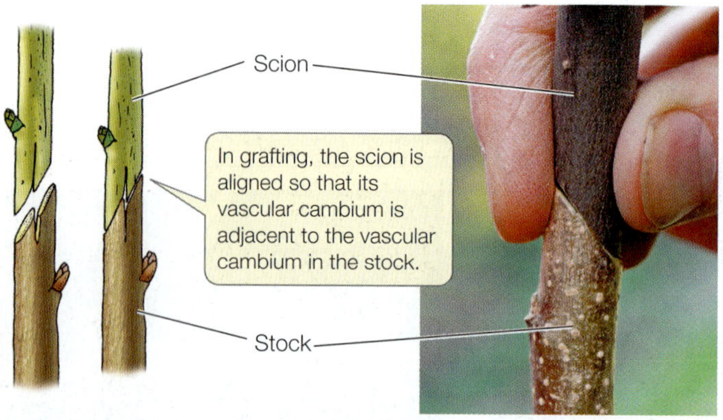
Scion
In grafting, the scion is aligned so that its vascular cambium is adjacent to the vascular cambium in the stock.
Stock

38.18 Grafting Grafting—attaching a piece of a plant to the root or root-bearing stem of another plant—is a common horticultural technique. The "host" root or stem is the stock; the upper grafted piece is the scion. In the photo, a scion of one apple variety is being grafted onto a stock of another variety.

RECAP 38.3

Angiosperms may reproduce asexually by means of modified stems, roots, or leaves. Asexual reproduction is advantageous when a plant has a superior genotype well adapted to its environment, but it decreases the genetic diversity of plant populations.

- What are the advantages and disadvantages of asexual reproduction for a plant? See pp. 792–793
- Explain how vegetative reproduction of plants is advantageous to humans. See pp. 793–794

We have seen how angiosperms reproduce sexually and asexually. A disadvantage of asexual reproduction is that its genetic inflexibility may leave a population unable to cope with new challenges. In the next chapter we will focus on the mechanisms that have evolved in plants to cope with biological and physical challenges in their environment.

By what genetic mechanism is apomixis brought about?

ANSWER

Apomixis involves several processes, including abnormal meiosis, followed by the development of an embryo within a seed. *Arabidopsis* plants with a mutation in the *SWII* gene show apomixis but no viable seeds. In normal plants, the expression of *SWII* is essential for chromosome pairing during meiosis I (see Figure 11.16). But in the mutant strain, meiosis I resembles mitosis, and the chromosomes replicate again before what would be meiosis II. The search is now on for genes that promote embryo and seed development in plants that exhibit apomixis. Scientists are trying to isolate and transfer such genes into corn and other cereal crops with the hope that plant breeders can use apomixis to propagate plants with desirable traits (such as high yields and disease- and insect-resistance) without compromising their hybrid vigor.

CHAPTER SUMMARY 38

38.1 How Do Angiosperms Reproduce Sexually?

- Sexual reproduction promotes genetic diversity in a population. The flower is an angiosperm's structure for sexual reproduction.
- Flowering plants have microscopic gametophytes. The megagametophyte is the **embryo sac**, which typically contains eight nuclei in a total of seven cells. The microgametophyte is the **pollen grain**, which usually contains two cells. Review Figure 38.2, ACTIVITY 38.1
- Following **pollination**, the pollen grain delivers **sperm cells** to the embryo sac by means of a **pollen tube**. Review ANIMATED TUTORIAL 38.1
- Plants have both physical and genetic methods of preventing inbreeding. Physical separation of the gametophytes and genetic **self-incompatibility** prevent self-pollination. Review Figure 38.5
- Most angiosperms exhibit **double fertilization**: one sperm cell fertilizes the egg cell, forming a zygote, and the other sperm cell fertilizes the central cell, where its nucleus unites with the two **polar nuclei** to form a triploid **endosperm**. Review Figure 38.6
- Ovules develop into seeds, and the ovary wall and the enclosed seeds develop into a **fruit**.
- The hormone **abscisic acid** promotes seed development and dormancy.

38.2 What Determines the Transition from the Vegetative to the Flowering State?

- In **annuals** and **biennials**, flowering and seed formation usually leads to death of the rest of the plant. **Perennials** live a long time and typically reproduce repeatedly.
- For a vegetatively growing plant to flower, an apical meristem in the shoot system must become an **inflorescence meristem**, which in turn must give rise to one or more **floral meristems**. These events are under the influence of **meristem identity genes** and **floral organ identity genes**. Review Figure 38.8

continued

- Some plants flower in response to **photoperiod**. **Short-day plants** flower when the nights are longer than a critical night length specific to each species; **long-day plants** flower when the nights are shorter than a critical night length. **Review Figure 38.11**
- The mechanism of photoperiodic control involves phytochromes and a biological clock. **Review Figure 38.12, ANIMATED TUTORIAL 38.2**
- A flowering signal, called **florigen**, is formed in a photoperiodically induced leaf and is translocated to the sites where flowers will form. **Review Figures 38.13, 38.14**
- In some angiosperm species, exposure to low temperatures—**vernalization**—is required for flowering; in others, internal signals (one of which is gibberellin in some plants) induce flowering. **Review Figures 38.15, 38.16**

38.3 How Do Angiosperms Reproduce Asexually?

- Asexual reproduction allows rapid multiplication of organisms that are well suited to their environment.
- **Vegetative reproduction** involves the modification of a vegetative organ—usually the stem—for reproduction. **Review Figure 38.17**
- Horticulturists often **graft** different plants together to take advantage of favorable properties of both **stock** and **scion**. **Review Figure 38.18**

Go to the Interactive Summary to review key figures, Animated Tutorials, and Activities Life10e.com/is38

CHAPTER**REVIEW**

REMEMBERING

1. The typical angiosperm female gametophyte
   a. is called a microspore.
   b. has eight nuclei.
   c. has eight cells.
   d. is called a pollen grain.
   e. is carried to the male gametophyte by wind or animals.

2. Pollination in angiosperms
   a. always requires wind.
   b. never occurs within a single flower.
   c. always requires help by animal pollinators.
   d. is also called fertilization.
   e. makes most angiosperms independent of external water for reproduction.

3. Which statement about double fertilization is *not* true?
   a. It is found in most angiosperms.
   b. It includes fusion between the microsporocyte and the megasporocyte.
   c. One of its products is a triploid nucleus.
   d. One sperm cell fuses with the egg cell.
   e. One sperm cell fuses with the central cell.

4. Which statement about photoperiodism is *not* true?
   a. It is related to the biological clock.
   b. Phytochrome plays a role in the timing process.
   c. It is based on measurement of the length of the night.
   d. Some plants do not flower in response to photoperiod.
   e. It is limited to plants.

5. Florigen is
   a. produced in the leaves and transported to the apical bud.
   b. produced in the roots and transported to the shoots.
   c. produced in the apical meristem of the stem and transported to the base.
   d. the same as gibberellin.
   e. activated by prolonged (more than a month) high temperature.

6. Which statement about vernalization is *not* true?
   a. It decreases the abundance of an inhibitor of flowering.
   b. It involves exposure to cold temperatures.
   c. It only occurs in crop plants such as cereals.
   d. It inhibits synthesis of the FLC protein.
   e. If winter wheat is not exposed to cold, it will not flower.

UNDERSTANDING & APPLYING

7. Thompson Seedless grapes are produced by vines that are triploid. Think about the consequences of this chromosomal condition for meiosis in the flowers. Why are these grapes seedless? Describe the role played by the flower in fruit formation when no seeds are being formed. How do you suppose Thompson Seedless grapes are propagated?

8. Poinsettias are popular ornamental plants that typically bloom just before Christmas. Their flowering is photoperiodically controlled. Are they long-day or short-day plants? Explain.

9. You plan to induce the flowering of a crop of long-day plants in the field by using artificial light. Is it necessary to keep the lights on continuously from sundown until the point at which the critical day length is reached? Why or why not?

ANALYZING & EVALUATING

10. Describe the proteins and mutations that could be involved in the following observations:
    a. A mutant plant flowers without its normal inductive dark period. When a leaf from the mutant plant is grafted onto an unexposed wild-type plant, the recipient plant flowers.
    b. A mutant plant does not flower when exposed to the normal inductive dark period. When a leaf from a mutant plant that has been exposed to the inductive dark period is grafted onto an unexposed wild-type plant, the recipient plant flowers.
    c. A plant flowers only after exposure to cold.
    d. If a gene is experimentally overexpressed in the leaf, flowering is induced. Overexpression of the gene in the shoot apical meristem does not, however, induce flowering.
11. The isolation of a mutation in the *Arabidopsis SWII* gene that results in abnormal meiosis has offered insights into apomixis. How would you try to identify other genes or mutations that function along with the gene product of *SWII*, with the goal of producing fully fertile apomictic plants?

Go to BioPortal at **yourBioPortal.com** for Animated Tutorials, Activities, LearningCurve Quizzes, Flashcards, and many other study and review resources.

39 Plant Responses to Environmental Challenges

CHAPTER OUTLINE

A New Way to Fight Malaria Sweet wormwood (*Artemisia annua*) grows in forests throughout the world. It synthesizes a defensive chemical, called artemisinin, that is now being used to treat people with malaria.

ARTEMISIA ANNUA, or sweet wormwood, is a fernlike shrub about 2 meters tall that is native to Asia but grows all over the world. Unlike animals, which can sometimes escape their enemies, plants must confront their enemies in place. Over time, plants have evolved elaborate mechanisms to fight off attackers. One such mechanism is the production of defensive chemicals. *Artemisia* produces a chemical called artemisinin that is toxic to certain cells, including those of the parasite that causes malaria.

Malaria has infected humans for more than 50,000 years. Today about 400 million cases arise worldwide and close to 1 million people die from malaria each year. The disease is caused by a mosquito-borne parasite that infects liver and blood cells in the human body.

Many cultures have long histories of extracting medicines from plants growing in their environments. About 1,700 years ago the Chinese herbalist Ge Hong noted that drinking a tea made from *Artemisia* was useful in treating malaria. Meanwhile, the indigenous people of Peru were using tinctures from the bark of a different plant—cinchona—to treat malaria. In the 1600s Jesuit priests noted how effective the cinchona treatment was, and brought it back to Europe. During the nineteenth century the active ingredient of cinchona was identified as quinine, and quinine became the most widely used medicine for both the prevention and treatment of malaria. But during the twentieth century, the supply of quinine could not keep up with demand, especially for soldiers fighting in the world wars. This led to the development of various synthetic antimalarial drugs.

As more modern drugs were developed to treat malaria, traditional herbs such as *Artemisia* fell into disuse. But gradually the parasite developed resistance to the drugs, and doctors turned back to *Artemisia*. Indeed, during the Vietnam War, drinking *Artemisia* tea helped Vietnamese soldiers cope with the quinine-resistant malaria that struck American soldiers. Chinese scientists isolated the active ingredient of the antimalarial tea and called it artemisinin.

The exact mechanism of action of artemisinin is still unclear. It appears that the drug reacts with iron in red blood cells, forming free radicals that damage lipids and DNA in the infecting parasite. Because of political differences between China and the rest of the world, news of the discovery of artemisinin, and access to the drug, were restricted until the early 1980s. Now it is widely available.

Until recently, hundreds of such plant chemicals were contemplated only in the context of plant biochemistry (how plants make them). Today we view them as adaptations arising from a plant's interactions with its environment.

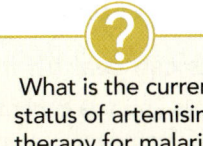

> ? What is the current status of artemisinin therapy for malaria?
> See answer on p. 812.

39.1 How Do Plants Deal with Pathogens?

Botanists know of thousands of diseases that can affect plants. Each is caused by a different strain of pathogen. Plant pathogens—which include bacteria, fungi, protists, nematodes, and viruses—are part of nature, and for that reason alone they merit our study in biology. For example, many diseases affect tomato plants, some of which may be familiar to you from growing tomatoes in your backyard or shopping for them (Figure 39.1). Like some human infectious diseases such as pneumonia, which may be bacterial or viral, plant diseases are named for the symptoms rather than the agent that causes them. For example, the term "blight," characterized by browning and death of plant tissues, can be caused by bacteria, fungi, or oomycetes. Just as medical schools have departments of pathology, many universities in agricultural regions have departments of plant pathology.

Successful infection by a pathogen can have significant effects on a plant, reducing photosynthesis and causing massive cell and tissue death. Like the responses of the human immune system (see Chapter 42), the responses by which plants fight off disease are varied and fascinating. Plants and pathogens have evolved together in a continuing "arms race": pathogens have evolved ways to attack plants, and plants have evolved ways to defend themselves against those attacks.

What determines the outcome of a battle between a plant and a pathogen? The key to success for the plant is to respond to information from the pathogen quickly and massively. Plants use both mechanical and chemical defenses in this effort. These defenses can either be:

- **Constitutive**, always present in the plant, or
- **Induced**, produced in reaction to damage or stress.

Fungus:
anthracnose
Bacterium:
spotted wilt
Fungus:
powdery mildew
Virus:
tomato mosaic
virus
Fungus:
Fusarium wilt
Bacterium:
bacterial
speck
Nematode:
root knot nematode
Bacterium:
crown gall tumor

39.1 Diseases of Tomato Plants A wide variety of disease agents cause a variety of symptoms.

Physical barriers form constitutive defenses

A plant's first line of defense is its outer surfaces, which can prevent the entry of pathogens. As Chapter 34 described, the organs of a growing plant that are exposed to the outside environment are largely covered with cutin, suberin, and waxes. These substances not only prevent water loss by evaporation but can also prevent fungal spores and bacteria from entering the underlying tissues. Some fungi get around this defense by secreting enzymes that hydrolyze these substances, breaking them down to gain entry.

Much more important to the plant are the induced resistance mechanisms, summarized in Figure 39.2, that are initiated when a pathogen lands on a plant.

Plants can seal off infected parts to limit damage

Whereas animals generally repair tissues that have been damaged by pathogens, plants do not. Instead, plants seal off and sacrifice damaged tissues so that the rest of the plant does not become infected. Plants have the option of discarding damaged tissues because most plants, unlike most animals, can replace damaged parts by growing new ones.

Before we look at the details of the defensive process, note that a key response by plant cells to invasion by pathogens is the rapid deposition of additional polymers on the inside of the cell wall. These macromolecules not only reinforce the mechanical barrier formed by the cell wall, but also block the plasmodesmata, limiting the ability of viral pathogens to move from cell to cell. The polysaccharides also serve as a base on which lignin may be laid down. Lignin enhances the mechanical barrier, and the toxicity of lignin precursor chemicals makes the cell inhospitable to some pathogens. These lignin building blocks are only one example of the toxic substances that plants use as chemical defenses.

Induced responses to pathogens are controlled by receptors. Plant pathogens cause the host plant to activate various chemical defense responses. A wide range of molecules called **elicitors** have been identified that trigger these defenses. These molecules vary in character, from peptides made by bacteria to cell wall fragments from fungi. Elicitors can also be derived from fragments of plant cell wall components broken down by pathogens.

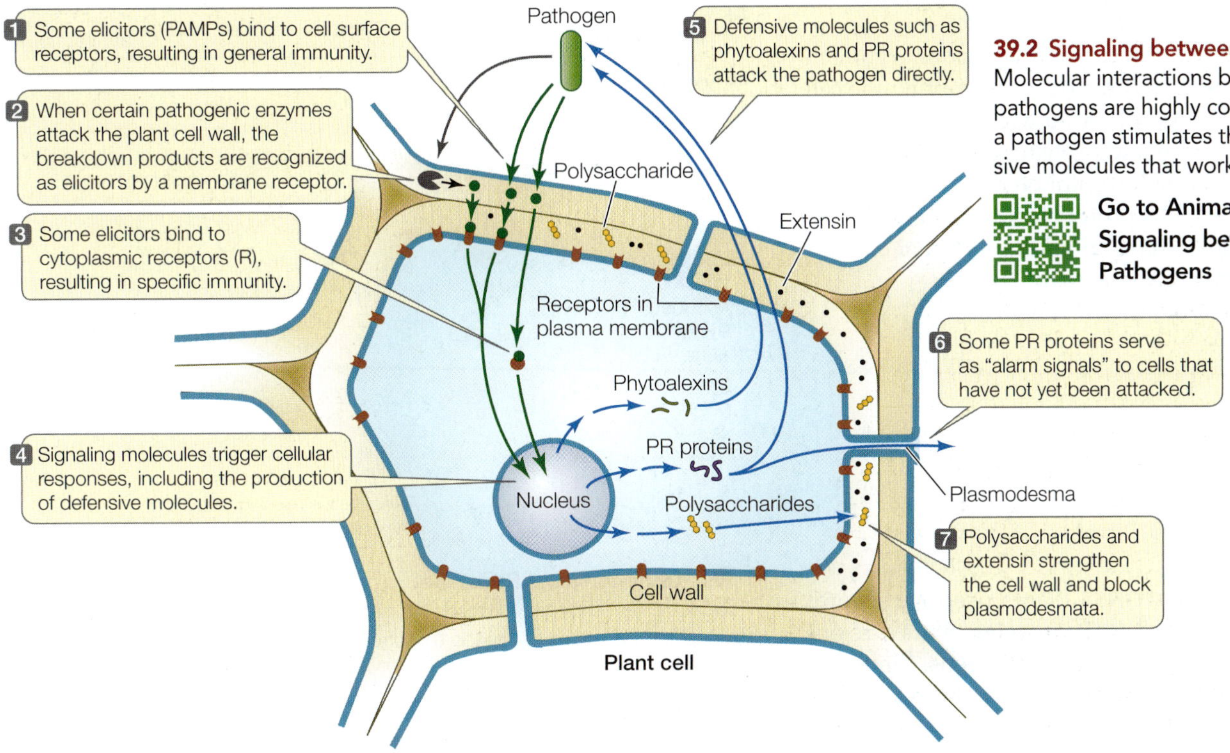

1 Some elicitors (PAMPs) bind to cell surface receptors, resulting in general immunity.
2 When certain pathogenic enzymes attack the plant cell wall, the breakdown products are recognized as elicitors by a membrane receptor.
3 Some elicitors bind to cytoplasmic receptors (R), resulting in specific immunity.
4 Signaling molecules trigger cellular responses, including the production of defensive molecules.
Pathogen
5 Defensive molecules such as phytoalexins and PR proteins attack the pathogen directly.
Polysaccharide
Extensin
Receptors in plasma membrane
Phytoalexins
PR proteins
Nucleus
Polysaccharides
Cell wall
Plant cell
6 Some PR proteins serve as “alarm signals” to cells that have not yet been attacked.
Plasmodesma
7 Polysaccharides and extensin strengthen the cell wall and block plasmodesmata.

39.2 Signaling between Plants and Pathogens Molecular interactions between plants and pathogens are highly coevolved. The presence of a pathogen stimulates the plant to produce defensive molecules that work in many different ways.

Go to Animated Tutorial 39.1 Signaling between Plants and Pathogens Life10e.com/at39.1

The responses of plants to elictors can be described in terms of the “plant immune system.” Two forms of immunity are recognized (see Figure 39.2):

- General immunity is triggered by general elicitors called pathogen associated molecular patterns (PAMPs). PAMPs are usually molecules that are produced by entire classes of pathogens, such as flagellin (found in bacterial flagella) or chitin (found in fungal cell walls). Thus general immunity is an overall response rather than a response that is triggered by a specific pathogen in a particular plant. PAMPs are recognized by transmembrane receptors called pattern recognition receptors, which activate signaling pathways that lead to general immunity (also called PAMP-triggered immunity, or PTI).
- Specific immunity is triggered by specific elicitors called effectors. Effectors include a wide variety of specific pathogen-produced molecules that enter the plant cell. Once inside the cell, effectors bind to cytoplasmic receptors called R proteins that trigger the specific immunity response (also called effector-triggered immunity, or ETI).

General and specific immunity both involve multiple responses

Many of the signaling pathways associated with general and specific immunity are the same, although the latter is specific for particular pathogens in particular plants, and is much stronger than general immunity. Both forms of immunity involve signaling pathways that are triggered by binding between the elicitors (PAMPs or effectors) with their receptors. These pathways lead to various responses:

- *Formation of reactive oxygen and NO*: Receptor binding triggers the rapid production of nitric oxide (NO) and reactive oxygen species such as superoxide and hydrogen peroxide. These reactive molecules are toxic to some pathogens, and they are components of signal transduction pathways leading to local and systemic (plantwide) defenses.
- *Callose deposition*: The β-1,3-glucan polymer callose is deposited on the inside of the cell wall to strengthen the wall and seal off the cell.
- *Hormone signaling*: Some pathways result in the production of plant hormones, including salicylic acid and jasmonic acid. We will describe the roles of these hormones in immunity later in the chapter.
- *Changes in gene expression*: Signal transduction cascades lead to changes in gene expression. The upregulated genes include pathogenesis-related (PR) genes and genes encoding the production of antimicrobial substances called phytoalexins.

PHYTOALEXINS **Phytoalexins** are antibiotics that are produced by infected plants and are toxic to many fungi and bacteria. Most are small molecules, and each is made by only a few plant species. They are produced by infected cells and their immediate neighbors within hours of the onset of infection. Because their antimicrobial activity is nonspecific, phytoalexins can destroy many species of fungi and bacteria in addition to the one that originally triggered their production. Some phytoalexins can also kill the plant cells that produced them, thus sealing off the infection site (the hypersensitive response; see below).

Phytoalexins are an example of an induced plant defense: they are not normally present in plants but are synthesized rapidly when a bacterial or fungal infection occurs. Physical injuries and viral infections can also induce the production of phytoalexins.

Camalexin, a phytoalexin made by the model organism *Arabidopsis thaliana*, appears to function by disrupting the cell membranes of invading fungal or bacterial pathogens. Its production is induced by a conserved protein kinase cascade (see Section 7.3) that is triggered by receptor binding in either general or specific immunity. This protein kinase cascade results in the upregulation of genes encoding enzymes that convert the amino acid tryptophan to camalexin:

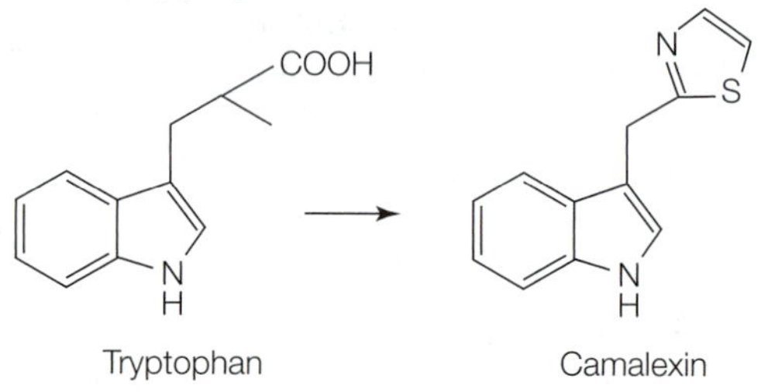
COOH
N
H
Tryptophan
N
S
N
H
Camalexin

PATHOGENESIS-RELATED PROTEINS Plants produce several types of **pathogenesis-related (PR) proteins**. Some are enzymes that break down the cell walls of pathogens. Chitinase, for example, is a PR protein that breaks down chitin, which is found in many fungal cell walls. In some cases the breakdown products of the pathogen cell walls serve as elicitors that trigger further defensive responses.

Another class of PR proteins are the plant defensins, which are similar to defensins produced by animals (see Section 42.2). These small peptides bind to fungal membranes and are toxic to a wide range of fungal targets, but they are not toxic to plant or animal cells.

Other PR proteins may serve as alarm signals to plant cells that have not yet been attacked. In general, PR proteins appear not to be rapid-response weapons; rather, they act more slowly, perhaps after other, faster responses have blunted the pathogen's attack.

Specific immunity involves gene-for-gene resistance

As mentioned above, pathogen effectors are molecules that are secreted inside plant cells by fungal and bacterial pathogens. These molecules are often proteins whose function is to inhibit some aspect of plant immunity. For example, many effectors are proteases that break down specific plant proteins involved in immune responses. Thus effectors enable pathogens to overcome general immunity and invade the plant, causing disease. The genes that encode effectors have evolved as part of the "arms race" between plants and their pathogens.

In response to the evolution of effectors, plants have evolved intracellular receptors that recognize specific effectors. These receptors are called R (for *r*esistance) proteins. When an R protein binds its ligand (a specific effector), it activates the signal transduction pathways of the specific immunity response. As we mentioned above, specific immunity is stronger than general immunity, and it enables the plant to prevent growth of the pathogen and remain healthy.

R proteins are encoded by **resistance (*R*) genes**. During the middle of the twentieth century, the plant pathologist Harold

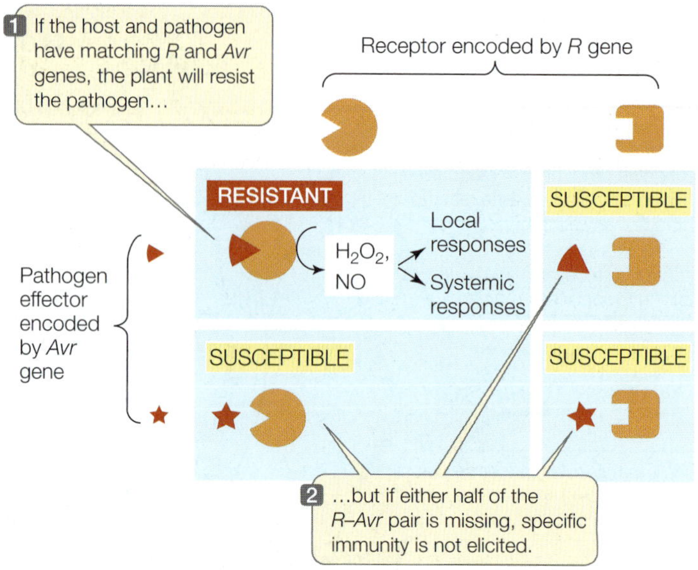
1 If the host and pathogen have matching *R* and *Avr* genes, the plant will resist the pathogen...
Receptor encoded by *R* gene
RESISTANT
H_2O_2, NO
Local responses
Systemic responses
SUSCEPTIBLE
Pathogen effector encoded by *Avr* gene
SUSCEPTIBLE
SUSCEPTIBLE
2 ...but if either half of the *R*–*Avr* pair is missing, specific immunity is not elicited.

39.3 Gene-for-Gene Resistance If a gene in a pathogen that codes for an elicitor "matches" a gene in a plant that codes for a receptor, the receptor binds the elicitor, and a defensive response results.

Henry Flor, at North Dakota State University, realized that there is a special relationship between *R* genes and certain pathogen genes named ***Avirulence* (*Avr*) genes**. Flor studied strains of the rust fungus *Melampsora lini* and the flax plant (*Linum usitatissimum*). He found that specific strains of rust fungus (with particular *Avr* genes) were able to cause disease only on some varieties of flax—those that *didn't* carry specific *R* genes. If a flax variety carried the right *R* gene, it was resistant to those strains of rust fungus. Thus a particular *Avr* gene rendered the fungus *avirulent* on flax plants with the corresponding *R* gene. Flor named this the **gene-for-gene concept**. We now know that *Avr* genes encode effector molecules that bind to receptors encoded by the *R* genes, resulting in the specific immunity response (Figure 39.3). Hundreds of *R* genes and their corresponding *Avr* genes have been identified. A major goal of plant breeders for the past 50 years has been to breed new *R* genes into crops to make them more resistant to pathogens.

Specific immunity usually leads to the hypersensitive response

Many signaling pathways and plant responses are shared between general and specific immunity, although in the latter these responses are accelerated and amplified. Furthermore, specific immunity usually leads to a form of programmed cell death called the **hypersensitive response**. When this occurs, plant cells at and near the site of pathogen infection die, forming a necrotic lesion (Figure 39.4). This cell death deprives the pathogen of nutrients and prevents the spread of the infection.

As described above, specific immunity is triggered by binding between specific pathogen effectors (*Avr* gene products) and R receptor proteins. The pathways that lead to cell death have not been worked out in detail, but the generation of nitric oxide and reactive oxygen species is an important component.

39.4 Sealing Off the Pathogen and the Damage These necrotic lesions on the leaves of a broad bean plant are a response to "chocolate spot" fungus, *Botrytis fabae*.

These reactive molecules may contribute to cell death directly—for example, by disrupting cellular membranes. They also function as signaling molecules that activate pathways leading to cell death.

Systemic acquired resistance is a form of long-term immunity

Thus far we have described the events that occur in plant cells at or near the site of invasion by a pathogen. In both general and specific immunity, the infected cells also send hormonal signals to the rest of the plant, stimulating a systemic response. **Systemic acquired resistance** is a general increase in the resistance of the entire plant to a wide range of pathogens. It is not limited to the pathogen that originally triggered it, or to the site of the original infection, and its effect may last as long as an entire growing season.

This defensive response is initiated by the plant hormone salicylic acid.

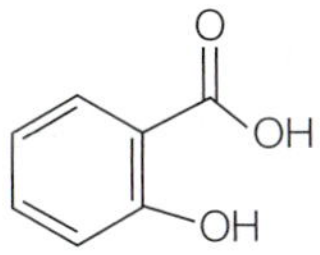

Salicylic acid

Salicylic acid production is triggered by receptor binding in both general and specific immunity; salicylic acid then functions as a signal that mediates various defense responses. For example, salicylic acid triggers the production of reactive oxygen species and the induction of genes that encode PR proteins. There are many classes of PR proteins, which function in various ways to protect plants against insect attack and against invasion by fungi, bacteria, viruses, and nematodes.

These responses occur at the site of infection and, to a lesser extent, throughout the plant. A derivative of salicylic acid can form a gas that can travel through the air, carrying the defense signal not only to other parts of the same plant but to other nearby plants.

Another type of systemic acquired resistance is a more specific defense against viruses with RNA genomes. The plant uses its own enzymes to convert some of the single-stranded RNA of the invading virus into double-stranded RNA (dsRNA) and to chop that dsRNA into small interfering RNAs (siRNAs) (see Section 16.5). Simultaneously, some of the viral RNA is transcribed, forming mRNAs that advance the infection. However, the siRNAs interact with another cellular component to degrade those mRNAs, blocking viral replication. The siRNAs spread quickly throughout the entire plant through plasmodesmata, providing systemic resistance.

RECAP 39.1

Plants protect themselves against pathogens with constitutive and induced defenses. General and specific immunity involve several common signaling pathways, but specific immunity is a stronger response that usually leads to hypersensitive cell death. Systemic acquired resistance provides a longer-lasting, more general immunity throughout the plant.

- Name two types of defensive compounds that plant cells produce when they are infected by bacteria or fungi. **See pp. 799–800 and Figure 39.2**
- How do *R* and *Avr* genes determine which pathogens a plant can resist? **See p. 800 and Figure 39.3**
- How do infected plant cells signal infection to other parts of the plant, or to other plants? **See p. 801**

Not all biological threats to plants come from pathogens. Another threat comes from the many animals that eat plants.

39.2 How Do Plants Deal with Herbivores?

Herbivores—animals that eat plants—depend on plants for energy and nutrients. Their foraging activities cause physical damage to plants, and they often spread disease among plants as well. While the majority of herbivores are insects (**Figure 39.5**), every major class of vertebrates includes at least a few herbivores (see also Section 56.2, which discusses herbivory in the ecological context of species interactions). Plants cannot evade their consumers by running away, but they have many other ways of protecting themselves against herbivory.

Mechanical defenses against herbivores are widespread

Plants have both constitutive and induced mechanical defenses against herbivores. Constitutive anatomical barriers include trichomes (specialized hairs; see p. 807) and thorns, spines that are specialized for defense. An example of an induced mechanical defense is the production of latex. Some plants, such as *Euphorbia* species, produce a thick, white aqueous suspension of cellular debris, oils, and resins called latex when they are injured by an herbivore. Insects trapped by this sticky substance starve to death.

TABLE 39.1
Secondary Metabolites Used in Defense

Class	Type	Role	Example
Nitrogen-containing 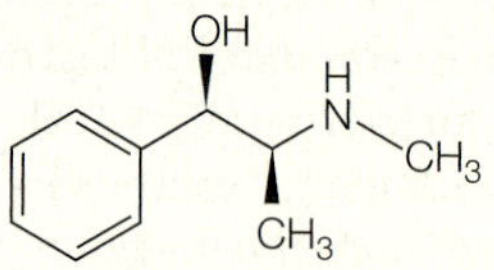 Ephedrine (an alkaloid)	Alkaloids	Neurotoxin	Nicotine in tobacco
	Glycosides	Inhibit electron transport	Dhurrin in sorghum
	Nonprotein amino acids	Disrupt protein structure	Canavanine in jack bean
Nitrogen–sulfur-containing 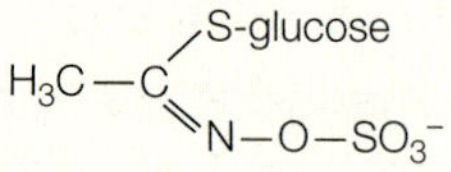 Methylglucosinolide	Glucosinolates	Inhibit respiration	Methylglucosinolate in cabbage
Phenolics 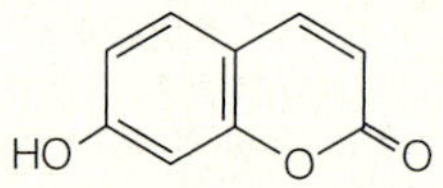 Umbelliferone	Coumarins	Block cell division	Umbelliferone in carrots
	Flavonoids	Phytoalexins	Capsidol in peppers
	Tannins	Inhibit enzymes	Gallotannin in oak trees
Terpenes 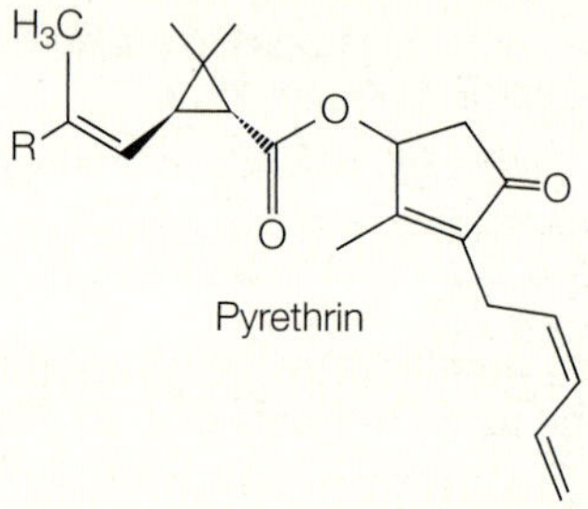 Pyrethrin	Monoterpenes	Neurotoxins	Pyrethrin in chrysanthemums
	Diterpenes	Disrupt reproduction and muscle function	Gossypol in cotton
	Triterpenes	Inhibit ion transport	Digitalis in foxglove
	Sterols	Block animal hormones	Spinasterol in spinach
	Polyterpenes	Deter feeding	Latex in *Euphorbia*

Plants produce constitutive chemical defenses against herbivores

Plants attract, resist, and inhibit other organisms with a wide range of chemicals known as secondary metabolites. You learned about one of these chemicals, artemisinin, in the opening of this chapter.

Primary metabolites are substances such as proteins, nucleic acids, carbohydrates, lipids, and their building blocks, which are produced and used by all living organisms, including plants. Primary metabolites are used in basic cellular processes such as photosynthesis, respiration, and nutrient uptake. **Secondary metabolites** are substances that are not used for basic cellular processes. Each is found in only certain organisms or groups of organisms.

The more than 10,000 known plant secondary metabolites range in molecular mass from about 70 to more than 390,000 daltons, but most have a low molecular mass (Table 39.1). Some are produced by only a single plant species, whereas

(A) Locusta migratoria
(B) Manduca sexta

39.5 Insect Herbivores The great majority of herbivores are insects. (A) Some herbivores, such as this locust, are generalists that will attack nearly any plant. (B) Others are specialists, like this tobacco hornworm, which feeds only on tobacco and related plants.

INVESTIGATING**LIFE**

39.6 Nicotine Is a Defense against Herbivores The secondary metabolite nicotine, made by tobacco plants, is an insecticide, yet most commercial varieties of tobacco are susceptible to insect attack. Ian Baldwin demonstrated that a tobacco strain with a reduced nicotine concentration was more susceptible to insect damage.[a]

HYPOTHESIS Nicotine helps protect tobacco plants against insects.

Method

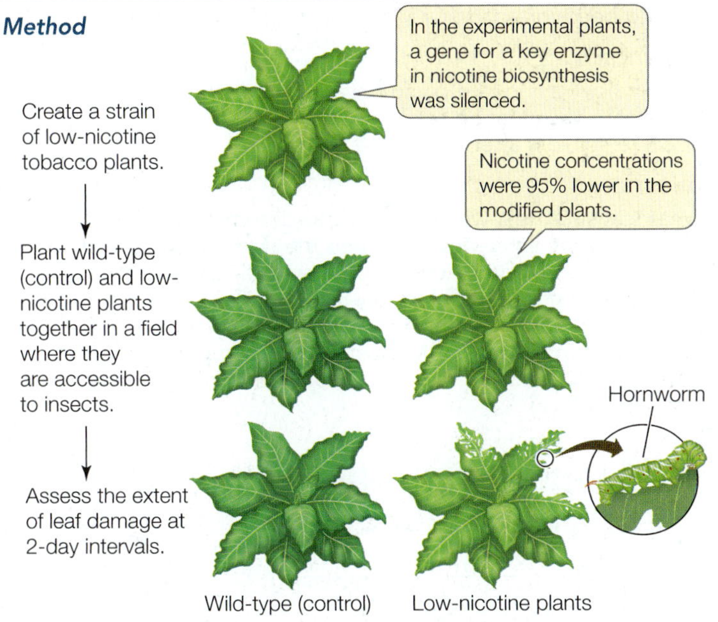
Create a strain of low-nicotine tobacco plants.
In the experimental plants, a gene for a key enzyme in nicotine biosynthesis was silenced.
Nicotine concentrations were 95% lower in the modified plants.
Plant wild-type (control) and low-nicotine plants together in a field where they are accessible to insects.
Hornworm
Assess the extent of leaf damage at 2-day intervals.
Wild-type (control)
Low-nicotine plants

Results The low-nicotine plants suffered more than twice as much leaf damage as did the wild-type controls.

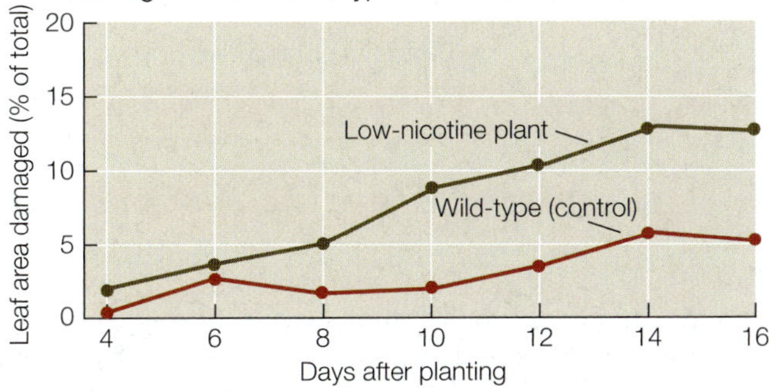
Leaf area damaged (% of total)
20
15
10
5
0
4
6
8
10
12
14
16
Days after planting
Low-nicotine plant
Wild-type (control)

CONCLUSION Nicotine provides tobacco plants with at least some protection against insects.

Go to **BioPortal** for discussion and relevant links for all INVESTIGATING**LIFE** figures.

[a]Steppuhn, A. et al. 2004. *PLoS Biology* 2(8): e217.

others are characteristic of entire genera or even families. The effects of defensive secondary metabolites on animals are diverse. Some act on the nervous systems of herbivorous insects, mollusks, or mammals. Others mimic the natural hormones of insects, causing some larvae to fail to develop into adults. Still others damage the digestive tracts of herbivores. Some secondary metabolites are toxic to fungal pathogens. For example, as we saw at the opening of this chapter, humans make use of secondary metabolites as medicines.

The secondary metabolite nicotine was one of the first insecticides to be used by farmers and gardeners. This molecule kills insects by inhibiting the functioning of the nervous system. Yet commercial varieties of tobacco and related plants that produce nicotine are still attacked, with moderate damage, by pests such as the tobacco hornworm (see Figure 39.5B). Given that observation, does nicotine really deter herbivores? Biologists answered this question conclusively with a study that used tobacco plants in which an enzyme involved in nicotine biosynthesis had been silenced, lowering the nicotine concentration in the plants by more than 95 percent. These low-nicotine plants suffered much more damage from insect herbivory than did normal plants (**Figure 39.6**).

Some secondary metabolites play multiple roles

Canavanine is a secondary metabolite whose role is defensive and is based on its chemical structure. Canavanine is an amino acid that is not found in proteins, but is very similar to the amino acid arginine, which is found in almost all proteins:

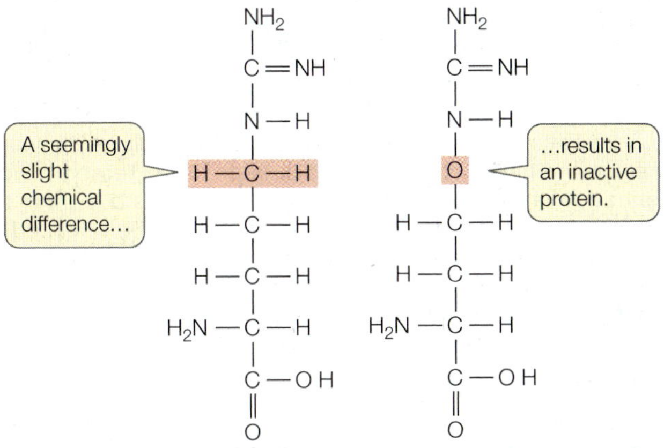
A seemingly slight chemical difference…
…results in an inactive protein.

Arginine **Canavanine**

When an insect larva consumes canavanine-containing plant tissue, the canavanine is incorporated into the insect's proteins in some of the places where the mRNA codes for arginine, because the enzyme that charges the tRNA specific for arginine fails to discriminate accurately between the two amino acids (see Section 14.5). The structure of canavanine, however, is different enough from that of arginine that some of the resulting proteins end up with a modified tertiary structure, and hence reduced biological activity. These defects in protein structure and function lead to developmental abnormalities that kill the insect.

In plants that produce them, canavanine and other secondary metabolites are constitutive defenses—that is, they are present regardless of whether the plant is under attack. Other chemical defenses come into play only when an herbivore strikes.

Plants respond to herbivory with induced defenses

In Section 39.1 we described the defenses that are induced in plants in response to pathogen attack. Plants also respond to wounding and herbivory with induced defenses involving signal transduction pathways. Less is known about the elicitors and receptors involved in these responses, but several classes of chemical elicitors have been identified. These elicitors are either derived from the herbivores themselves or are products of the digestion of plant tissues. For example, the enzymes

WORKING WITH**DATA:**

Nicotine Is a Defense against Herbivores

Original Paper

Steppuhn, A., K. Gase, B. Krock, R. Halitschke, and I. T. Baldwin. 2004. Nicotine's defensive function in nature. *PLoS Biology* 2(8): e217.

Analyze the Data

In 2004 Anke Steppuhn, Ian Baldwin, and colleagues at the Max Planck Institute for Chemical Ecology in Germany tested the hypothesis that nicotine helps protect tobacco plants against insects. They generated a line of low-nicotine transgenic plants by modifying the gene for putrescine *N*-methyl transferase, a key regulatory enzyme in the nicotine biosynthesis pathway. Both the low-nicotine and wild-type tobacco plants were transplanted into a field plantation where they were accessible to naturally occurring herbivores. The extent of leaf damage by insects was then measured at 2-day intervals for a period of 16 days. Results showed that the low-nicotine plants lost more than twice as much of their total leaf area as did the wild-type controls, indicating that nicotine provides tobacco plants with protection against insects (see Figure 39.6).

In a separate experiment, Baldwin and his colleagues showed that treatment with jasmonic acid (jasmonate) increased the concentration of nicotine in wild-type tobacco plants but not in the low-nicotine transgenic plants. The researchers then planted a group of wild-type and low-nicotine tobacco plants and treated them with jasmonate, a plant hormone, 7 days after planting. The plants were assessed for herbivore damage every 2 days after being planted. The results are shown in the figure.

QUESTION 1

Compare these data to those in Figure 39.6 for the untreated plants. What was the effect of jasmonate treatment on the resistance of wild-type plants to herbivore damage? What was the effect on low-nicotine plants?

QUESTION 2

What do these data reveal about the role of nicotine in preventing herbivore damage?

QUESTION 3

Explain how jasmonate could have had the effect it did on the transgenic low-nicotine plants, even though their nicotine levels were still low after jasmonate treatment.

QUESTION 4

What statistical test would you use to determine the possible significance of differences between the jasmonate-treated and untreated plants? At 10 days, the mean damage ± standard deviation for untreated low-nicotine plants was 6.0 ± 1.5 percent ($n = 36$); for treated low-nicotine plants it was 2.2 ± 0.6 percent ($n = 28$). Run a statistical test comparing the two results, calculate the *P* value, and comment on significance (see Appendix B).

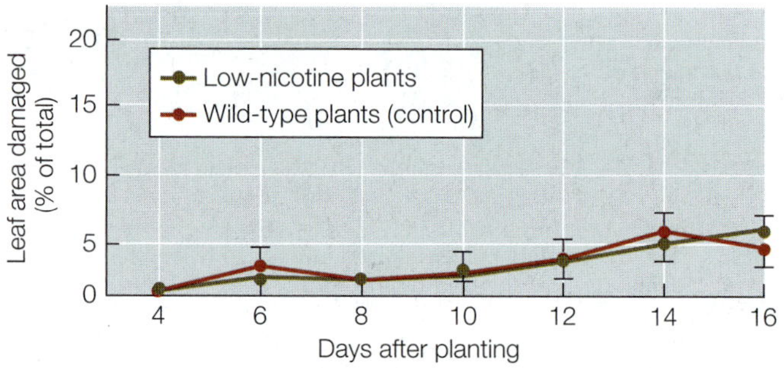
Low-nicotine plants
Wild-type plants (control)
Leaf area damaged (% of total)
20
15
10
5
0
4
6
8
10
12
14
16
Days after planting

Go to **BioPortal** for all WORKING WITH**DATA** exercises

that insects use to digest plant carbohydrates and lipids can elicit defensive responses. Some elicitors are produced when plant material passes through an herbivore's digestive tract; these elicitors are composed of a fatty acid derived from the plant and an amino acid derived from the insect gut. One such elicitor, produced by insects feeding on corn plants, has been named volicitin for its ability to induce production of volatile signals that can travel to other plant parts—and to neighboring corn plants—and stimulate their defense responses. In addition, many herbivorous insects lay their eggs on plants, and some components of the fluids secreted during egg laying have been identified as elicitors.

The signal transduction pathways that are activated by herbivory or wounding involve several key components, some of which are shared by the pathways that are induced by pathogens:

- *Membrane signaling:* The plasma membrane is the part of the plant cell that is in contact with the environment. Within the first minute after an herbivore strikes, changes in the electric potential of the plasma membrane occur in the damaged area. As we will describe in our chapter on the animal nervous system (see Section 45.2), such changes can be rapidly transmitted as a signal along the plasma membrane. In the case of plants responding to herbivory, the continuity of the symplast (see Figure 35.5) ensures that the signal travels over much of the plant within 10 minutes.
- *Reactive oxygen species:* Both wounding and herbivory trigger the production of reactive oxygen species (such as superoxide and hydrogen peroxide), which act as signaling molecules in pathways that lead to changes in gene expression.

Hormone signaling: Herbivory induces the production of several hormones that stimulate various plant responses. The most important of these is **jasmonic acid (jasmonate)**, which triggers systemic defenses against herbivores (**Figure 39.7**).

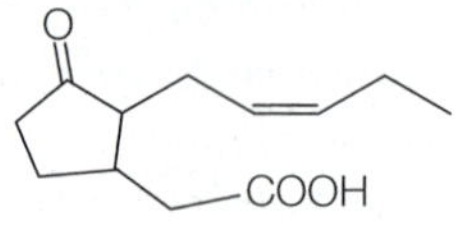
O
COOH

Jasmonic acid

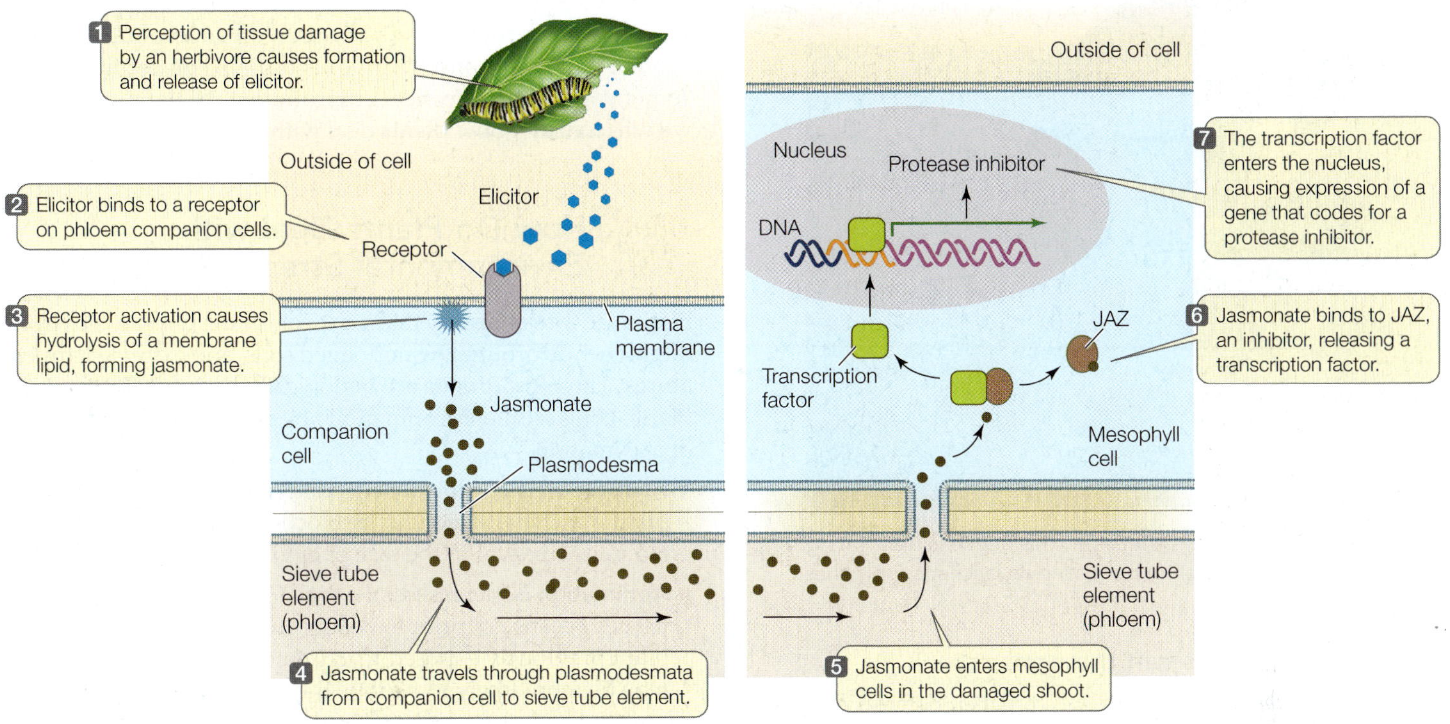
1 Perception of tissue damage by an herbivore causes formation and release of elicitor.
Outside of cell
Elicitor
2 Elicitor binds to a receptor on phloem companion cells.
Receptor
3 Receptor activation causes hydrolysis of a membrane lipid, forming jasmonate.
Plasma membrane
Jasmonate
Companion cell
Plasmodesma
Sieve tube element (phloem)
4 Jasmonate travels through plasmodesmata from companion cell to sieve tube element.
Outside of cell
Nucleus
Protease inhibitor
DNA
7 The transcription factor enters the nucleus, causing expression of a gene that codes for a protease inhibitor.
JAZ
6 Jasmonate binds to JAZ, an inhibitor, releasing a transcription factor.
Transcription factor
Mesophyll cell
Sieve tube element (phloem)
5 Jasmonate enters mesophyll cells in the damaged shoot.

39.7 A Signal Transduction Pathway for Induced Defenses
The chain of events initiated by herbivory that leads to the production of a defensive chemical can consist of many steps. These steps may include the synthesis of one or two hormones, binding of receptors, gene activation, and finally, synthesis of defensive compounds.

Jasmonates trigger a range of responses to wounding and herbivory

When the plant senses an herbivore-produced elicitor, it makes jasmonate and a variety of jasmonate derivatives. These molecules trigger many plant defenses both at the site of herbivore attack and throughout the plant (see Figure 39.7). These defenses include the production of specific secondary metabolites and defensive proteins. Jasmonates induce changes in gene expression by binding a transcriptional inhibitor called a JAZ protein. After binding by jasmonate, the JAZ protein is targeted for degradation, and the previously inhibited genes can be expressed.

Protease inhibitors are an important group of defensive proteins that are synthesized in response to insect attack. Once inside an insect's gut, these inhibitors interfere with the digestion of proteins and thus stunt the insect's growth.

Jasmonates can also "call for help" by triggering the formation of volatile compounds that attract insects that prey on the herbivores attacking the plant.

Why don't plants poison themselves?

Why don't the defensive chemicals that are so toxic to herbivores and pathogens kill the plants that produce them? In some cases the defensive chemicals are directed at organs or systems that are not found in plants, such as the nervous, digestive, or endocrine systems of animals. In addition, plants that produce toxic defensive chemicals use one or more other measures to protect themselves.

COMPARTMENTALIZATION Isolation of the toxic substance is a common means of avoiding exposure. Plants store their toxins in vacuoles if the toxins are water-soluble. If they are hydrophobic, the toxins may be dissolved in latex (see p. 806) and stored in a specialized compartment, or they may be dissolved in waxes on the epidermal surface. Such compartmentalized storage keeps the toxins away from the mitochondria, chloroplasts, and other parts of the plant's metabolic machinery.

STORAGE OF PRECURSORS Some plants store the precursors of toxic substances in one type of tissue, such as the epidermis, and store the enzymes that convert those precursors into the active toxin in another type, such as the mesophyll. When an herbivore chews part of the plant, cells are ruptured, the enzymes come into contact with the precursors, and the toxin is produced. The only part of the plant that is damaged by the toxin is that which was already damaged by the herbivore. Plants such as sorghum and some legumes, which respond to herbivory by producing cyanide (a potent inhibitor of cellular respiration), are among those that use this type of protective measure.

MODIFIED PROTEINS The plant has modified proteins that do not react with the toxin. As noted above, canavanine resembles arginine and therefore plays havoc with protein synthesis in insect larvae. In plants that make canavanine, the plant enzyme that charges the arginine tRNA discriminates correctly between arginine and canavanine, so canavanine is not incorporated into the plant's proteins.

39.8 Disarming a Plant's Defenses This beetle is inactivating a milkweed's defense system by cutting its laticifer supply lines.

Plants don't always win the arms race

Milkweeds such as *Asclepias syriaca* store their defensive chemicals in latex in specialized tubes called **laticifers**, which run alongside the veins in the leaves. When damaged, a milkweed releases copious amounts of toxic latex from its laticifers. Field studies have shown that most insects that feed on neighboring plants of other species do not attack laticiferous plants, but there are exceptions. One population of beetles that feeds on *A. syriaca* exhibits a remarkable prefeeding behavior: these beetles cut a few veins in the leaves before settling down to dine. Cutting the veins causes massive latex leakage from the adjacent laticifers and interrupts the latex supply to a downstream portion of the leaf. The beetles then move to the relatively latex-free portion and eat their fill (**Figure 39.8**).

We have described just one of many ways that herbivores circumvent plant defenses. A successful plant defense exerts strong selection pressure on herbivores to get around it somehow; a successful herbivore, in turn, exerts strong selection pressure on plants to develop new defensive strategies. One can imagine, for example, that over time *A. syriaca* might evolve to have thicker walls around the base of its laticifers, such that the beetles can no longer cut them, or to produce a different toxin that does not depend on laticifers.

RECAP 39.2

Many plants use secondary metabolites as constitutive defenses against herbivory. Other defenses are induced by herbivory through signal transduction pathways. The plant hormone jasmonate and its derivatives stimulate local and systemic responses to herbivores.

- Describe one example of a secondary metabolite and how it affects herbivores. **See pp. 802–803**
- What role do jasmonates play in plant defense? **See p. 805 and Figure 39.7**
- What are three ways in which a plant avoids being poisoned by its own defensive chemicals? **See p. 805**

A plant's survival depends not only on successful defenses against pathogens and herbivores but also on coping with a sometimes hostile physical environment. In the next section we will consider how plants deal with environmental stresses.

39.3 How Do Plants Deal with Environmental Stresses?

Plants are threatened by many aspects of the physical environment, such as drought, waterlogged soils, and extreme temperatures. These conditions are biologically harmful (**Table 39.2**). Plants cope with environmental stresses through adaptation or acclimation.

- **Adaptation** is genetically encoded resistance to stress. A plant may have structures or biochemical properties that aid in its survival in the face of environmental challenges.
- **Acclimation** is increased tolerance for environmental extremes because of prior exposure to them. An individual plant previously exposed to extreme cold, for example, may be more likely to survive the subsequent winter.

Go to Media Clip 39.1
Leaves for Every Environment
Life10e.com/mc39.1

Some plants have special adaptations to live in very dry conditions

Many plants, especially those living in deserts, must cope with extremely limited water supplies. A variety of anatomical and life-cycle adaptations allow plants to survive under these conditions. Many of these adaptations are ways to avoid, reduce, or cope with the inevitable water loss through transpiration that occurs during active photosynthesis. Other adaptations help plants tolerate the excessive light and heat that are often found in deserts.

DROUGHT AVOIDERS Some desert plants have no special structural adaptations for water conservation. Instead, these desert annuals, called drought avoiders, simply evade periods of drought. Drought avoiders carry out their entire life cycle—from seed to seed—during a brief period in which rainfall has

TABLE 39.2 Environmental Stresses on Plants

Condition	Effect on Plants
Drought	Reduced water potential, dehydration
Flooding	Reduced O_2 and respiration
High temperature	Changes in membrane fluidity and in proteins
Low temperature	Changes in membrane fluidity, damage by ice crystals
Salinity	Reduced water potential, dehydration
Metal element toxicity	Disruption of metabolism

39.9 Desert Annuals Avoid Drought The seeds of many desert annuals lie dormant for long periods, awaiting conditions appropriate for germination. When they do receive enough moisture to germinate, they grow and reproduce rapidly before the short wet season ends. During the long dry spells, only dormant seeds remain alive.

made the surrounding desert soil sufficiently moist for growth and reproduction (Figure 39.9). A different drought avoidance strategy is seen in some African and South American deciduous perennial plants, which shed their leaves in response to drought as a way to conserve water. These plants remain dormant until conditions are again favorable for growth, much as deciduous trees in temperate climates shed their leaves in fall and are dormant until the following spring.

LEAF STRUCTURES Most desert plants are not drought avoiders, but rather grow in their dry environment year-round. Plants adapted to dry environments are called **xerophytes** (Greek *xeros*, "dry"). Three structural adaptations are found in the leaves of many xerophytes:

- Specialized leaf anatomy that reduces water loss
- A thick cuticle and a profusion of hairs over the leaf epidermis, which retard water loss
- Trichomes that diffract and diffuse sunlight, thereby decreasing the intensity of light impinging on the leaves and the risk of damage to the photosynthetic apparatus by excess light

In some xerophytes the stomata are strategically located in sunken cavities below the leaf surface (known as **stomatal crypts**), where they are sheltered from the drying effects of air currents (Figure 39.10). Trichomes surrounding the stomata slow air currents further. Cacti and similar plants have spines rather than typical leaves, and photosynthesis is confined to the fleshy stems. The spines may help the plants cope with desert conditions by reflecting solar radiation or by dissipating heat. The spines may also deter herbivores.

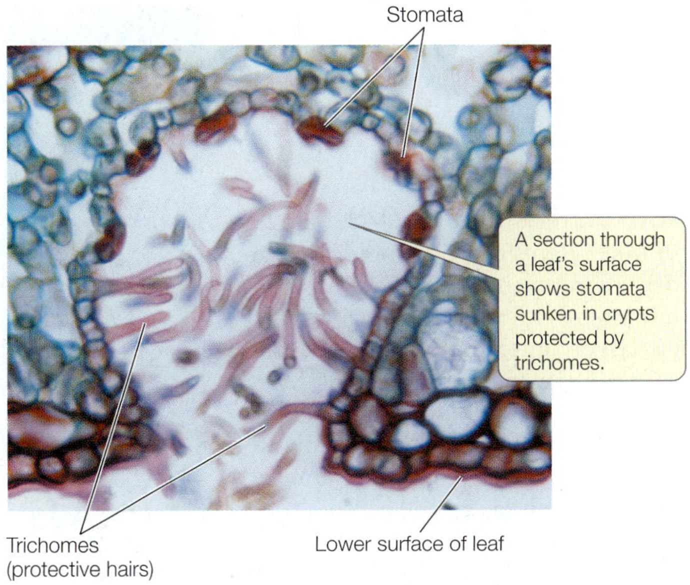
Stomata
A section through a leaf's surface shows stomata sunken in crypts protected by trichomes.
Trichomes (protective hairs)
Lower surface of leaf

39.10 Stomatal Crypts Stomata in the leaves of some xerophytes are located in sunken cavities called stomatal crypts. The trichomes (hairs) covering these crypts trap moist air.

WATER-STORING STRUCTURES **Succulence**—the possession of fleshy, water-storing leaves or stems—is another adaptation to dry environments (Figure 39.11). This adaptation allows plants to take up large amounts of water when it is available (such as after a brief thunderstorm) and then draw on the stored water during subsequent dry periods. Other adaptations of succulents include a reduced number of stomata and a variant form of photosynthesis (the CAM pathway; see Section 10.4), both of which reduce water loss.

39.11 Succulence The *Aloe* plant stores water in its fleshy leaves.

39.12 Mining Water with Deep Taproots In Death Valley, California, the root of this mesquite tree must reach far beneath the dunes for its water supply.

ROOT SYSTEMS THAT MAXIMIZE WATER UPTAKE Roots may also be adapted to dry environments. Cacti have shallow but extensive fibrous root systems that effectively intercept water at the soil surface following even light rains. Mesquite (*Prosopis*) (**Figure 39.12**) obtains water through taproots that grow to great depths, reaching water supplies far underground, as well as from condensation on its leaves. The Atacama Desert in northern Chile often goes for several years without measurable rainfall, but the landscape there has many surprisingly large mesquite shrubs.

SOLUTE ACCUMULATION Xerophytes and other plants that must cope with inadequate water supplies may accumulate high concentrations of the amino acid proline or of secondary metabolites in their vacuoles. This solute accumulation lowers the water potential in the plant's cells below that of the soil, which allows the plant to take up water via osmosis. Plants living in saline environments share this and several other adaptations with xerophytes, as we will see shortly.

Some plants grow in saturated soils

For some plants, the environmental challenge is too much water, the opposite of that faced by xerophytes. They live in environments so wet that the diffusion of oxygen to their roots is severely limited. These plants have shallow root systems that grow slowly; oxygen levels are likely to be highest near the surface of the soil, and slow growth decreases the roots' need for oxygen.

The root systems of some plants adapted to swampy environments, such as cypresses and some plants that grow in coastal mangrove habitats, have **pneumatophores**, which are extensions that grow out of the water and up into the air (**Figure 39.13A**). Pneumatophores contain lenticels (openings; see Figure 34.17) that allow oxygen to diffuse through them, aerating the submerged parts of the root system.

Many submerged or partly submerged aquatic plants have large air spaces in the leaf and stem parenchyma and in the petioles. Tissue containing such air spaces is called **aerenchyma** (**Figure 39.13B**). Aerenchyma stores oxygen produced by photosynthesis and permits its ready diffusion to parts of the plant where it is needed for cellular respiration. Aerenchyma also imparts buoyancy. Furthermore, because aerenchyma contains far fewer cells than most other plant tissues, metabolism in aerenchyma proceeds at a lower rate, so the need for oxygen is much reduced.

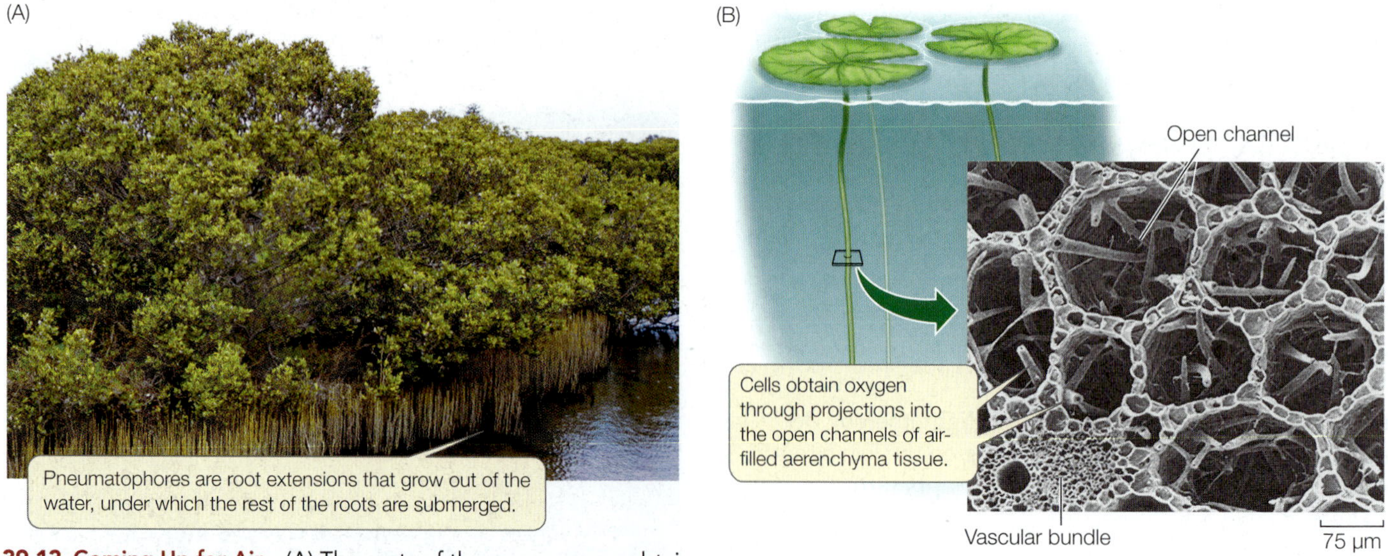
(A)
Pneumatophores are root extensions that grow out of the water, under which the rest of the roots are submerged.
(B)
Open channel
Cells obtain oxygen through projections into the open channels of air-filled aerenchyma tissue.
Vascular bundle
75 μm

39.13 Coming Up for Air (A) The roots of these mangroves obtain oxygen through pneumatophores. (B) This scanning electron micrograph of a cross section of a petiole of the yellow water lily shows the structure of the air-filled channels that make up aerenchyma tissue.

Plants can respond to drought stress

The adaptations of xerophytes for coping with dry environments are generally constitutive—they are always present—and under normal conditions they prevent the plants from experiencing drought stress. When conditions become so dry that even xerophytes are stressed, however, the plants turn to inducible responses. The same responses are found in many other plants, including those that are not adapted to grow in dry climates.

When the weather is abnormally dry, the water content of the soil is reduced and less water is available to plants. Water deficits in plant cells have two major biochemical effects:

- A reduction in membrane integrity as the polar–nonpolar forces that orient the lipid bilayer are reduced
- Changes in the three-dimensional structures of proteins

Plant growth is reduced when the structure of plant cells is compromised in these ways. Indeed, inadequate water supply is the single most important factor that limits production of our most important food crops.

When plants sense a water deficit in their roots, a signaling pathway is set in motion that initiates several measures to conserve water and maintain cellular integrity. This pathway begins with the production of the hormone abscisic acid in the roots. This hormone travels from the roots to the shoot, where it results in stomatal closure and initiates gene transcription that leads to other physiological events that conserve water and cellular integrity (**Figure 39.14**).

Many plant genes whose expression is altered by drought stress have been identified, largely through research using DNA microarrays, proteomics, and other molecular approaches (see Chapters 17 and 18). One group of proteins whose production is upregulated during drought stress is the *l*ate *e*mbryogenesis *a*bundant (LEA; pronounced "lee-yuh") group of proteins. These hydrophobic proteins also accumulate in maturing seeds as they dry out (hence their name). LEA proteins bind to membrane proteins and other cellular proteins to stabilize them, preventing their aggregation during desiccation. The importance of LEA proteins in coping with drought stress was shown by Ray Wu and colleagues at Cornell University, who found that transgenic rice plants expressing a high level of LEA protein in leaves and roots grew better than normal plants under drought conditions (**Figure 39.15**). Genes that encode LEA proteins occur in many plants and confer drought tolerance.

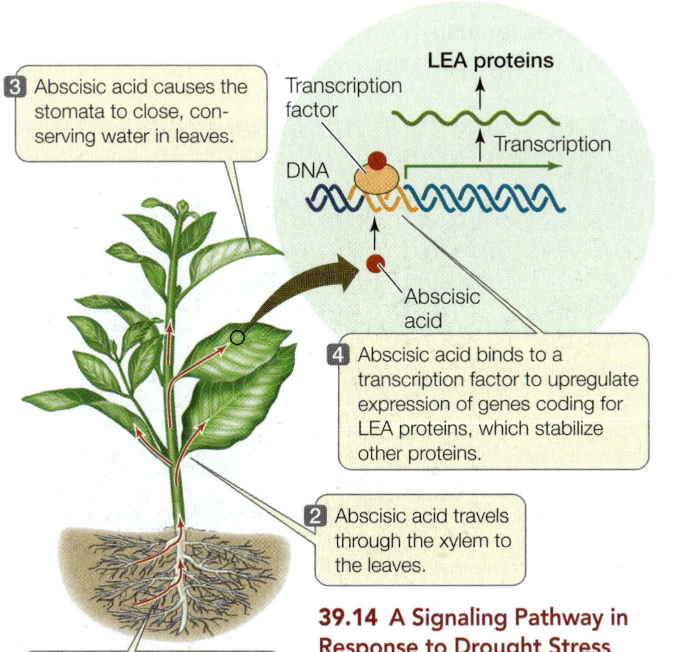

3 Abscisic acid causes the stomata to close, conserving water in leaves.
LEA proteins
Transcription factor
Transcription
DNA
Abscisic acid
4 Abscisic acid binds to a transcription factor to upregulate expression of genes coding for LEA proteins, which stabilize other proteins.
2 Abscisic acid travels through the xylem to the leaves.
1 Drought stress causes an increase in abscisic acid.

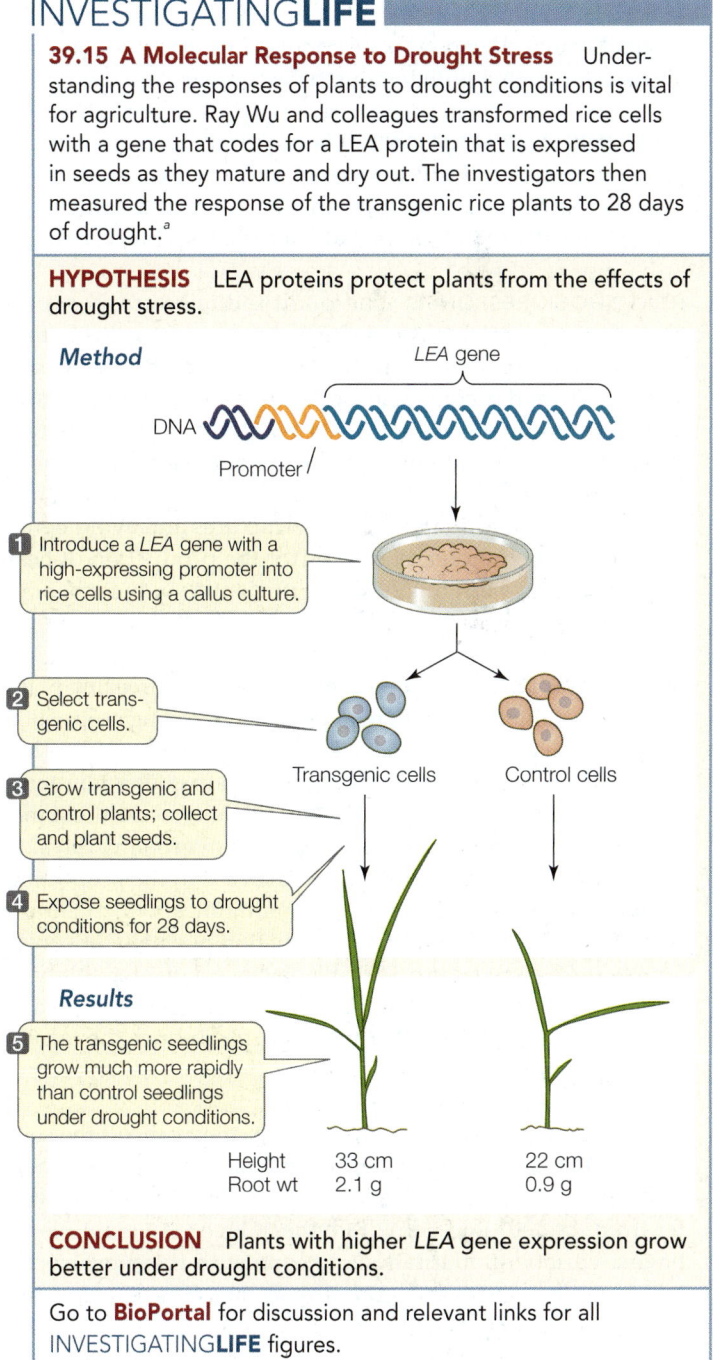

39.14 A Signaling Pathway in Response to Drought Stress Acclimation to drought stress begins in the root with the production of the hormone abscisic acid.

INVESTIGATING**LIFE**

39.15 A Molecular Response to Drought Stress Understanding the responses of plants to drought conditions is vital for agriculture. Ray Wu and colleagues transformed rice cells with a gene that codes for a LEA protein that is expressed in seeds as they mature and dry out. The investigators then measured the response of the transgenic rice plants to 28 days of drought.[a]

HYPOTHESIS LEA proteins protect plants from the effects of drought stress.

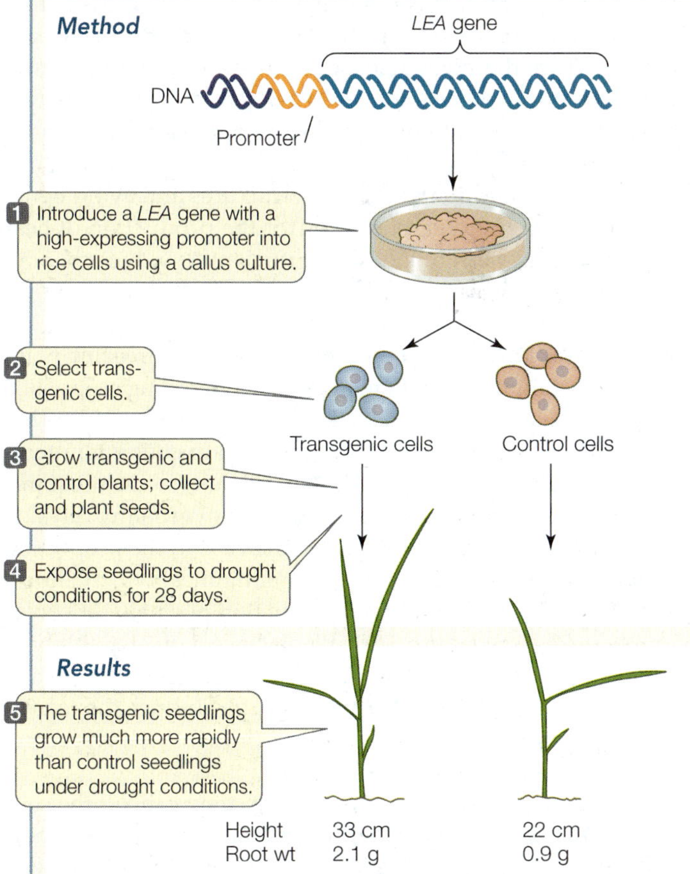

Method
LEA gene
DNA
Promoter
1 Introduce a *LEA* gene with a high-expressing promoter into rice cells using a callus culture.
2 Select transgenic cells.
Transgenic cells
Control cells
3 Grow transgenic and control plants; collect and plant seeds.
4 Expose seedlings to drought conditions for 28 days.
Results
5 The transgenic seedlings grow much more rapidly than control seedlings under drought conditions.
Height 33 cm 22 cm
Root wt 2.1 g 0.9 g

CONCLUSION Plants with higher *LEA* gene expression grow better under drought conditions.

Go to **BioPortal** for discussion and relevant links for all INVESTIGATING**LIFE** figures.

[a]Xu, D. et al. 1996. *Plant Physiology* 110:249–257.

Plants can cope with temperature extremes

Temperatures that are too high or too low can stress plants and even kill them. Plant species differ in their sensitivity to heat and cold, but all plants have their limits. Any temperature extreme can damage cellular membranes.

- High temperatures destabilize membranes and denature many proteins, especially some of the enzymes of photosynthesis.
- Low temperatures cause membranes to lose their fluidity and alter their permeabilities to solutes.
- Freezing temperatures may cause ice crystals to form, damaging membranes.

Plants have both constitutive adaptations and inducible responses for coping with temperature extremes.

ANATOMICAL ADAPTATIONS Many plants living in hot environments have constitutive adaptations similar to those of xerophytes. These adaptations include hairs and spines that dissipate heat and leaf forms that intercept less direct sunlight.

HEAT SHOCK RESPONSE The plant inducible response to heat stress is similar to the response to drought stress in that new proteins are made, often under the direction of an abscisic acid–mediated signaling pathway. Within minutes of experimental exposure to raised temperatures (typically a 5°C–10°C increase), plants synthesize several kinds of **heat shock proteins**. Among these proteins are chaperonins, which help other proteins maintain their structures and avoid denaturation. Threshold temperatures for the production of heat shock proteins vary, but 39°C is sufficient to induce them in most plants.

COLD-HARDENING Low temperatures above the freezing point can cause chilling injury in many plants, including crops such as rice, corn, and cotton as well as tropical plants such as bananas. Many plant species can acclimate to cooler temperatures through a process called **cold-hardening**, which requires repeated exposure to cool temperatures over many days. A key change during hardening is an increase in the proportion of unsaturated fatty acids in cell membranes, which allows them to retain their fluidity and function normally at cooler temperatures (see Figure 3.21). Plants have a greater ability to modify the degree of saturation of their membrane lipids than animals do. In addition, low temperatures induce the formation of proteins similar to heat shock proteins, which protect against chilling injury.

If ice crystals form within plant cells, they can kill the cells by puncturing organelles and plasma membranes. Furthermore, the growth of ice crystals outside the cells can draw water from the cells and dehydrate them. Freeze-tolerant plants have a variety of adaptations to cope with these problems, including the production of antifreeze proteins that slow the growth of ice crystals.

RECAP 39.3

Plants that live in continually dry or water-saturated environments have structural adaptations to cope with those conditions. Mechanisms that protect plants from drought stress are initiated by a signaling pathway involving abscisic acid. Heat shock proteins help plants acclimate to high and low temperatures.

- Describe two structural adaptations for growth in water-saturated soils. **See p. 808**
- What is the role of abscisic acid in acclimation to drought stress? **See p. 809 and Figure 39.14**
- What environmental conditions induce the formation of heat shock proteins, and what functions do those proteins serve? **See p. 810**

Just as climatic extremes can limit plant growth, the presence of certain substances, such as salt and heavy metals, can make an environment inhospitable to plant growth.

39.4 How Do Plants Deal with Salt and Heavy Metals?

A number of toxic solutes are found in soils, but worldwide, no toxic substance restricts angiosperm growth more than ordinary salt (sodium chloride). Saline—salty—habitats support, at best, limited types of vegetation. Saline habitats are found in diverse locales, from hot, dry deserts to moist, cool coastal marshes. Along the seashore, saline environments are created by ocean spray. The ocean itself is a saline environment, as are estuaries, where fresh and salt water meet and mingle. Salinization of agricultural land is an increasing global problem (**Figure 39.16**). Even where crops are irrigated with fresh water, sodium ions from the water accumulate in the soil to ever greater concentrations as the water evaporates.

Saline environments pose an osmotic challenge for plants. Because of its high salt concentration, a saline environment

39.16 Salty Soil Accumulation of salt from irrigation water with inadequate drainage has caused this soil in central California to become unsuitable for most plant growth.

has an unusually negative water potential (see Figure 35.2). To obtain water from such an environment, a plant must have an even more negative water potential; otherwise water will diffuse out of its cells, and the plant will wilt and die. Plants in saline environments are also challenged by the potential toxicity of sodium, which inhibits enzymes and protein synthesis.

Most halophytes accumulate salt

Halophytes—plants adapted to saline habitats—are found in a wide variety of flowering plant groups. Most halophytes share one adaptation: they take up sodium and, usually, chloride ions and transport those ions to their leaves. The accumulated ions are stored in the central vacuoles of leaf cells, away from more sensitive parts of the cells. Nonhalophytes accumulate relatively little sodium, even when placed in a saline environment; of the sodium that is absorbed by their roots, very little is transported to the shoot. The increased salt concentration in the tissues of halophytes lowers their water potential and allows them to take up water from their saline environment.

Some halophytes have other adaptations to life in saline environments. Some, for example, have **salt glands** in their leaves. These glands excrete salt, which collects on the leaf surface until it is removed by rain or wind (Figure 39.17). This adaptation, which reduces the danger of poisoning by accumulated salt, is found in some desert plants, such as *Frankenia palmeri*, and in some mangroves growing in seawater.

Salt glands can play multiple roles, as in the desert shrub *Atriplex halimus*. This shrub has glands that secrete salt into small bladders on the leaves. By lowering the water potential of the leaves, this salt not only helps them obtain water from the roots but also reduces their transpirational loss of water to the atmosphere.

The adaptations we have just discussed are specific to halophytes. Several other adaptations are shared by halophytes and xerophytes, including thick cuticles, succulence, and CAM photosynthesis.

39.17 Excreting Salt This saltwater mangrove plant has special salt glands that excrete salt, which appears here as crystals on the leaves.

Some plants can tolerate heavy metals

Salt is not the only toxic solute found in soils. High concentrations of some heavy metal ions, such as chromium, mercury, lead, and cadmium, are toxic to most plants; many of these ions are more toxic than sodium at equivalent concentrations.

Some geographic sites are naturally rich in heavy metals as a result of normal geological processes. In other places, acid rain leads to the release of toxic aluminum ions in the soil. Human activities, notably the mining of metallic ores, leave localized areas—known as tailings—with high concentrations of heavy metals and low concentrations of nutrients. Such sites are hostile to most plants, and seeds falling on them generally do not produce adult plants.

Most mine tailings rich in heavy metals, however, are not completely barren. They may support healthy plant populations that differ genetically from populations of the same species on the surrounding normal soils. How do these plants survive?

Initially, botanists believed that some plants were able to tolerate heavy metals by excluding them: that by not taking up the metal ions, the plants avoided being poisoned. Further investigations have shown, however, that tolerant plants growing on mine tailings do take up heavy metals, accumulating concentrations that would kill most plants. More than 200 plant species have been identified as **hyperaccumulators** that store large quantities of metals such as arsenic (As), cadmium (Cd), nickel (Ni), aluminum (Al), and zinc (Zn).

Perhaps the best-studied hyperaccumulator is alpine pennycress (*Thlaspi caerulescens*). Before the advent of chemical analysis, miners used to use the presence of this plant as an indicator of mineral-rich deposits. A *Thlaspi* plant may accumulate as much as 30,000 ppm Zn (most plants contain 100 ppm) and 1,500 ppm Cd (most plants contain 1 ppm). Studies of *Thlaspi* and other hyperaccumulators have revealed the presence of several common mechanisms:

- Increased ion transport into the roots
- Increased rates of translocation of ions to the leaves
- Accumulation of ions in vacuoles in the shoot
- Resistance to the ions' toxicity

Knowledge of these hyperaccumulation mechanisms and the genes underlying them has led to the emergence of **phytoremediation**, a form of bioremediation (see Section 18.6) that uses plants to clean up environmental pollution. Some phytoremediation projects use natural hyperaccumulators, whereas others use genes from hyperaccumulators to create transgenic plants that grow more rapidly and are better adapted to a particular polluted environment. In either case, the plants are grown in the contaminated soil, where they act as natural "vacuum cleaners" by taking up the contaminants (Figure 39.18). The plants are then harvested and disposed of to remove the contaminants. Perhaps the most dramatic use of phytoremediation occurred after an accident at the nuclear power plant at Chernobyl, Ukraine (then part of the Soviet Union), in 1986, when sunflower plants were used to remove uranium from the nearby soil. Phytoremediation is now widely used in cleaning up land after strip mining.

PHYTOREMEDIATION
Using Nature to Clean Itself Up
USGS

39.18 Phytoremediation Plants that accumulate heavy metals can be used to clean up contaminated soils. Here, poplars are being used to remove contaminants from an air force base.

After finding plants that accumulate valuable metals such as Ni, cobalt (Co), and silver (Ag), some scientists have proposed using those plants for *phytomining*. As in phytoremediation, the plants would be used to take up metals from the soil, but the metals would be extracted from the plants after they are harvested.

RECAP 39.4

Halophytes have several adaptations to saline habitats, most of which involve mechanisms that lower their water potential. Some plants can tolerate heavy-metal-rich soils that are toxic to most other plants.

- What are some of the roles of salt glands in halophyte leaves? See p. 811
- How are plants used for phytoremediation? See p. 811

?

What is the current status of artemisinin therapy for malaria?

ANSWER

Since 2000, artemisinin has become a mainstay of malaria treatment worldwide. Millions of people take it every day. Because the solubility of the natural molecule is quite limited, chemists have synthesized derivatives of it that are more soluble and easier to take. There is, however, a serious danger that the malaria parasite will develop resistance to artemisinin and its derivatives, so these drugs are used in combination with other synthetic drugs.

CHAPTER SUMMARY 39

39.1 How Do Plants Deal with Pathogens?

- Plants and pathogens have evolved together in a continuing "arms race": pathogens have evolved mechanisms for attacking plants, and plants have evolved mechanisms for defending themselves against those attacks.
- **Constitutive** defenses include plants' ability to strengthen their cell walls and block plasmodesmata when attacked, limiting the ability of viral pathogens to move from cell to cell.
- **Induced** defenses are triggered by a wide range of molecular **elicitors** and fall into two main categories: general immunity and specific immunity. **Review Figure 39.2, ANIMATED TUTORIAL 39.1**
- The **gene-for-gene concept** depends on a match between a plant's **resistance** (***R***) **genes** and a pathogen's ***Avirulence*** (***Avr***) **genes**. **Review Figure 39.3**
- In the **hypersensitive response** to infection by bacteria or fungi, cells produce two kinds of defensive molecules: **phytoalexins** and **pathogenesis-related** (**PR**) **proteins**. Some cells around the infected area die, sealing off the pathogens and the damage they have caused.
- The hypersensitive response is often followed by **systemic acquired resistance**, in which salicylic acid activates further synthesis of defensive compounds.
- Plants use RNA interference to develop specific immunity to invading RNA viruses.

39.2 How Do Plants Deal with Herbivores?

- Some plants produce **secondary metabolites** as defenses against herbivores. **Review Table 39.1, Figure 39.6**
- Hormones, including **jasmonates**, participate in signal transduction pathways leading to the production of defensive compounds. **Review Figure 39.7**
- Plants protect themselves against their own toxic defensive chemicals by isolating them in specialized compartments, by producing them only after the plant has already been damaged, or by having modified enzymes or receptors that are not affected by the toxic substance.

39.3 How Do Plants Deal with Environmental Stresses?

- Plants cope with environmental stresses by **adaptation** (genetically encoded resistance) or **acclimation** (increased tolerance) **Review Table 39.2**
- **Xerophytes** are plants that are adapted to dry environments.

continued

- Some xerophytic adaptations are structural, including thickened cuticles, specialized trichomes, **stomatal crypts**, **succulence**, and long taproots.
- Some plants accumulate solutes, making their water potential lower so they can tolerate drought.
- Adaptations to water-saturated habitats include **pneumatophores**, extensions of roots that allow oxygen uptake from the air, and **aerenchyma**, tissue in which oxygen can be stored and ready for diffusion throughout the plant.
- A signaling pathway involving abscisic acid initiates a plant's response to drought stress. **Review Figures 39.14, 39.15**
- Membranes and proteins can be damaged by extremely high or low temperatures. Plants respond to extreme temperatures by producing **heat shock proteins**.
- Some plants undergo **cold-hardening**, an acclimation process that includes changes in membrane lipids and production of heat shock proteins.
- Some plants resist freezing by producing antifreeze proteins.

39.4 How Do Plants Deal with Salt and Heavy Metals?

- Most **halophytes** accumulate salt. Some have **salt glands** that excrete salt to the leaf surface.
- Some plants living in soils that are rich in heavy metals are **hyperaccumulators** that take up large amounts of those metals into their tissues.
- **Phytoremediation** is the use of hyperaccumulating plants or their genes to clean up environmental pollution.

See ACTIVITY 39.1 for a concept review of this chapter.

Go to the Interactive Summary to review key figures, Animated Tutorials, and Activities Life10e.com/is39

CHAPTER**REVIEW**

REMEMBERING

1. Plants sometimes protect themselves from their own toxic secondary metabolites by
   a. producing special enzymes that destroy the toxin.
   b. storing precursors of the toxic substances in one compartment and the enzymes that convert those precursors to toxic products in another compartment.
   c. storing the toxic substances in mitochondria or chloroplasts.
   d. distributing the toxic substances to all cells of the plant.
   e. performing crassulacean acid metabolism.

2. Which statement about secondary metabolites is *not* true?
   a. They may be used in defense against fungi.
   b. Some are poisonous to herbivores.
   c. Some are amino acids that are normally part of proteins.
   d. Water-soluble molecules are stored in vacuoles.
   e. Some mimic the hormones of animals.

3. Which of the following is *not* an adaptation to dry environments?
   a. Increased solute concentration in the vacuoles
   b. Hairy leaves
   c. A heavier cuticle over the leaf epidermis
   d. Sunken stomata
   e. A root system that grows each rainy season and dies back when it is dry

4. Some plants adapted to swampy environments meet the oxygen needs of their roots by means of a specialized tissue called
   a. parenchyma.
   b. aerenchyma.
   c. collenchyma.
   d. sclerenchyma.
   e. chlorenchyma.

5. Halophytes
   a. may accumulate abscisic acid in their vacuoles.
   b. may have water potentials that are lower than those of other plants.
   c. only accumulate sodium.
   d. have low root-to-shoot ratios.
   e. rarely accumulate sodium.

6. Which of the following is *not* true of the general immunity response of a plant to a pathogen?
   a. It can result in changes in gene expression.
   b. It can result in the production of plant hormones.
   c. It is a weak response in comparison with specific immunity.
   d. It involves sealing off damaged tissues to contain the infection.
   e. It is produced when a plant receptor recognizes a molecular pattern in the pathogen.

UNDERSTANDING & APPLYING

7. How might the adaptations of herbivores affect plant evolution? How might plant adaptations affect the evolution of herbivores?

8. A tomato plant can be infected with the fungus *Cladosporium*. The host plant and pathogen can have various genes involved in the hypersensitive response. Fill in the table that describes the fungal strains and the results of infection of plants:

	Cladosporium genotype		
Tomato genotype	*R1R2*	*R3R4*	*R1R4*
Avr1Avr2	Healthy 	Diseased 	Healthy 
Avr2Avr3			
Avr1Avr4			

ANALYZING & EVALUATING

9. In the coming decades, climate change may have significant effects on the growth and productivity of plants, in particular the crops on which we depend for our food. Discuss the physiological effects, and possible genetic responses in terms of plant breeding, of the following:
   a. In Pakistan, reduced rainfall causes a reduction in wheat yields.
   b. In the Mekong Delta of Vietnam, rising sea level inundates rice fields, causing a drastic reduction in yields.
   c. Increased temperature and humidity in western Canada causes an increase in wheat rust.

10. The tobacco hornworm (*Manduca sexta*) is adapted to feeding on nicotine-producing plants. Using the genetically modified tobacco plants described in Figure 39.6, how might you test the hypothesis that dietary nicotine protects the tobacco hornworm against its parasite *Cotesia congregata*?

Go to BioPortal at **yourBioPortal.com** for Animated Tutorials, Activities, LearningCurve Quizzes, Flashcards, and many other study and review resources.

40 Physiology, Homeostasis, and Temperature Regulation

CHAPTER OUTLINE

OREGON TC
ING
KARA
ING
PAULA
ING NEW YORK CITY MARATHON

Limits to Performance Paula Radcliffe, photographed here during her winning performance at the 2008 New York City Marathon, collapsed from heat stress during the 2004 Olympic marathon. When the body is subjected to extreme heat, its homeostatic mechanisms may fail.

THE 2008 NEW YORK CITY MARATHON took place on a cold, clear, windy day in November. For the third time, the first-place woman in this 41-km race was world record holder Paula Radcliffe. Radcliffe had also been expected to win the women's marathon in the 2004 Olympics. But that race took place on an extremely hot (a high of 34°C), humid day in Athens. Overcome by heat stress, Radcliffe collapsed 6 km from the finish line. In contrast, the average temperature for the three New York marathons Radcliffe won was 7°C.

Based on a survey of many marathons, elite runners have their best times when temperatures are below 10°C; higher temperatures can mean serious problems. The 2012 Boston Marathon coincided with an unseasonable April heat wave, with temperatures exceeding 27°C. During the course of the race, 120 runners were rushed to hospitals with severe heat stress.

When a person's internal body temperature rises above 40°C, major organs begin to fail, a condition known as heat stroke. Every year some athletes suffer heat stroke, which leads to death in a high percentage of cases. Soldiers in desert environments are at extreme risk of heat stroke, as are workers in many occupations, including firefighting, agriculture, and construction.

Why is heat stroke a particular danger for those who must be active in the heat? The short answer is that working muscles generate heat. That heat leaves the muscles in the blood and is circulated around the body, raising the temperature of the body's internal tissues. Although some of the heated blood flows to the skin, where heat can be lost to the environment, humans are subject to the problems faced by all mammals in losing excess heat. First, their normal internal temperatures are not far from the environmental temperatures that cause heat stress, so they don't have much of a safety zone. Second, most mammalian skin surfaces are covered with an insulating layer of fur—great for conserving body heat in cold environments, but an impediment to heat loss in warm ones.

Evolutionary adaptation in mammals has resulted in the efficient heat-loss portals of non-furred areas such as the nose, tongue, and footpads. In these areas, specialized blood vessels can open up and act like radiators to disperse heat (conversely, these portals can close down to conserve heat). Humans are not furred, but our evolutionary ancestors were, and we retain these general mammalian blood vessel adaptations in our hands, feet, and face (which is why we blush).

?

Can we increase heat loss from our natural heat portals to protect against heat stress?

See answer on p. 831.

40.1 How Do Multicellular Animals Supply the Needs of Their Cells?

All animal cells must obtain nutrients and oxygen from the environment and must eliminate carbon dioxide and other waste products of metabolism to the environment. The cells of very small or very thin aquatic animals meet these needs by direct exchanges with the external environment. In such animals, no cell is far from direct contact with the water it lives in; the water contains nutrients, absorbs wastes, and provides a relatively unchanging physical environment. Most cells of larger animals do not have direct contact with the external environment, and their needs must be served by an environment that is wholly internal to the animal.

An internal environment makes complex multicellular animals possible

The cells of multicellular animals exist within an **internal environment** of extracellular fluid. A human, for example, is about 60 percent water. Two-thirds of that water is contained within our cells, while one-third is the extracellular fluid (ECF) that is our internal environment. About 20 percent of the ECF (about 3 liters) is the **plasma** that circulates in our blood vessels. The remaining 80 percent (about 11 liters) is the **interstitial fluid** that bathes every cell of the body (**Figure 40.1**). Individual cells get their nutrients from this interstitial fluid and dump their waste products into it. As long as conditions in the body's internal environment are held within certain limits, cells are protected from the changes and harsh conditions of the external environment. A stable internal environment makes it possible for an animal to occupy habitats that would kill its cells if they were directly exposed to the external conditions. How is the internal environment kept constant?

As multicellular organisms evolved, cells became specialized for maintaining specific aspects of the internal environment. In turn, the internal environment enabled these specializations, since each cell no longer had to provide for all of its own needs. Some cells evolved to be the body's interface between the internal and the external environments and to provide the necessary transport functions to get nutrients in and move wastes out. Other cells became specialized for internal functions such as circulation of the extracellular fluids, energy storage, movement, and information processing. The evolution of physiological systems to maintain the internal environment made it possible for multicellular animals to become larger, thicker, and more complex, and allowed them to occupy many different habitats.

The composition of the internal environment is constantly being challenged by the external environment and by the metabolic activity of the cells of the body. Organisms must maintain their internal environment in a state of **homeostasis**—a narrow range of stable physical and biochemical conditions under which the body functions optimally. If a physiological system fails to function properly, homeostasis is compromised, and cells are damaged and can die. To avoid the loss of homeostasis, physiological systems must be controlled and regulated in response to changes in both the external and internal environments. The maintenance of homeostasis is a central theme of physiology.

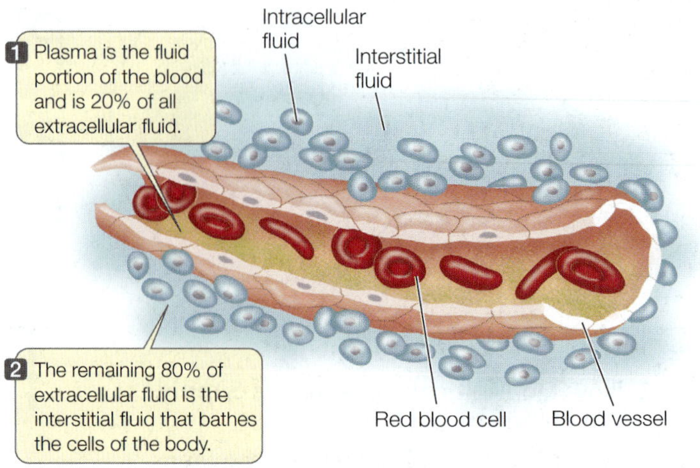
Intracellular fluid
Interstitial fluid
1 Plasma is the fluid portion of the blood and is 20% of all extracellular fluid.
2 The remaining 80% of extracellular fluid is the interstitial fluid that bathes the cells of the body.
Red blood cell
Blood vessel

40.1 The Internal Environment The "internal environment" is the extracellular fluid, or ECF, which accounts for about one-third of total body water. The ECF is made up of the blood plasma and the interstitial fluid. The physiological state of the ECF must remain stable within narrow limits, and maintaining that stability is the job of the body's organ systems.

Physiological systems are regulated to maintain homeostasis

The activities of physiological systems are *controlled*—speeded up or slowed down—by actions of the nervous and endocrine systems. But to *regulate* these systems and maintain homeostasis, information is required. As an analogy, think of the thermostat that controls the furnace or air conditioner to regulate the temperature of a house (**Figure 40.2**). The desired temperature is a **set point**, or reference point on the thermostat. The thermostat sensor responds to the temperature of the air, providing **feedback**—information that is compared with the set point. Any difference between the set point and the air temperature is an **error signal**. The error signal is converted into a corrective action—turning the air conditioner or furnace on or off.

Some components of physiological systems are called **effectors** because they *effect* changes in the internal environment. Muscles are a notable example of effectors. Another example are cells in the stomach that secrete digestive juices. Effectors are **controlled systems** because their activities are controlled by neural or hormonal signals from regulatory systems. **Regulatory systems** obtain, process, and integrate information, then issue commands to controlled systems. Important components of any regulatory system are the **sensors** (e.g., light-, temperature-, and pressure-sensitive cells) that provide feedback information to be compared with the internal set point.

Negative feedback is information used to counteract the influence that created an error signal. Whatever force is pushing the system away from its set point must be "negated." In our thermostat analogy, an air temperature below the set point causes the furnace to be turned on, which then reverses the direction of change in the air temperature.

Although not as common as negative feedback, **positive feedback** also exists in physiological systems. Rather than returning a system to a set point, positive feedback amplifies a response (i.e., it increases the deviation from the set point). An

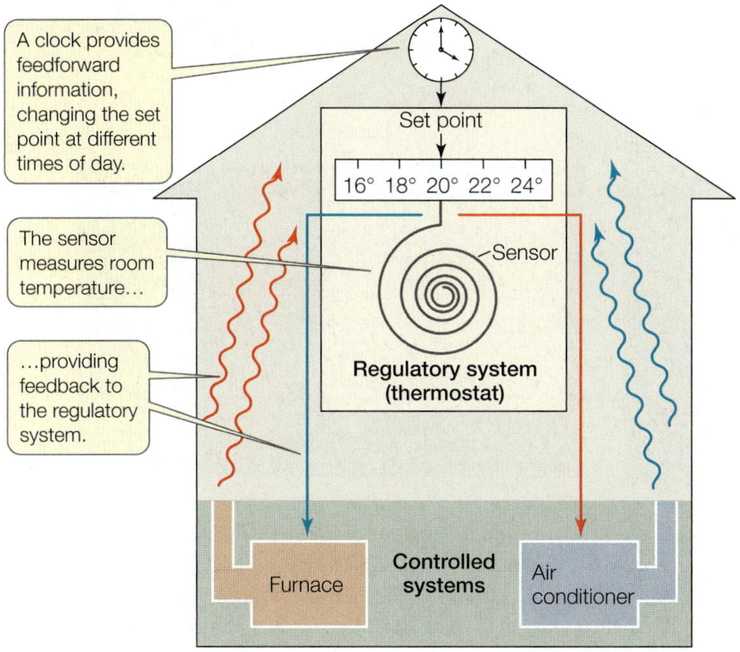
A clock provides feedforward information, changing the set point at different times of day.
The sensor measures room temperature...
...providing feedback to the regulatory system.
Set point
16° 18° 20° 22° 24°
Sensor
Regulatory system (thermostat)
Furnace
Controlled systems
Air conditioner

40.2 Thermostat Regulates Temperature A thermostat regulates the temperature of a room by turning the furnace or air conditioner on or off in response to the difference between feedback information (room temperature) and set points that are programmed into the thermostat.

example is sexual behavior, in which a little stimulation causes more behavior, which causes more stimulation, and so on. Positive feedback responses tend to reach a limit and terminate rapidly. The birth process is a good example. Contractions of the uterus stretch the birth canal, and that stretching stimulates more and stronger contractions until the baby is delivered, at which time contractions cease.

Feedforward information is another feature of regulatory systems. Its function is to change the set point in anticipation of a change in conditions. The timer on a thermostat provides feedforward information by changing the system's temperature set point, usually lowering it in the evening and raising it in the morning. Hearing the words "on your mark" before a race is feedforward information that raises your heart rate in anticipation of running. Feedforward information anticipates a change in the internal environment before that change occurs.

RECAP 40.1

The internal environment provides for the needs of all the cells that make up a complex multicellular animal. Organs and organ systems control the composition of the internal environment so as to maintain homeostasis.

- What are the three water-containing compartments in the human body? Why is the ECF crucial to survival in multicellular animals? See p. 816 and Figure 40.1
- Explain the difference between negative and positive feedback control mechanisms. See pp. 816–817
- Thinking about the thermoregulatory system in your own home, what is the set point, what is the feedback, and what would be feedforward information? See pp. 816–817 and Figure 40.2

Principles of control and regulation help organize our thinking about physiological systems. Once we understand how a system works, we can then ask how it is regulated. Part 9 of this book describes various physiological systems, how they function, and how they are regulated. But first we summarize some important structural features of the physiological systems that link the functioning of individual cells with the systemic functioning of multicellular animals.

40.2 What Are the Relationships between Cells, Tissues, and Organs?

Control, regulation, and the resulting maintenance of homeostasis is the work of a hierarchy of interacting physiological systems that make up a multicellular animal (see Figure 1.9A). Each system is composed of discrete **organs**, such as the stomach, heart, lungs, and kidneys. Organs are made up of assemblages of similar cells called **tissues**; tissues in turn are made up cells of different types. There are many specialized cell types, but there are only four kinds of tissues: epithelial, muscle, connective, and nervous tissues. The word "tissue" is often used in a general way to refer to a piece of an organ, such as "lung tissue" or "kidney tissue." An organ, however, always consists of more than one of the four tissue types. Here we will provide only brief descriptions of the four tissues types because each will be dealt with in more detail in the context of an organ or organ system in which it plays a major role.

Epithelial tissues are sheets of densely packed, tightly connected cells

Epithelial tissues are formed by sheets of cells, creating boundaries between the inside and the outside of the body and between body compartments. Epithelial tissues comprise the outer layer of the skin, line the blood vessels, and make up various ducts and tubules (Figure 40.3). They control which molecules and ions can move between different body compartments, such as the blood and the interstitial fluid. They selectively transport ions and molecules from one side of an epithelial membrane to the other; for example, the gut epithelium absorbs nutrients from the gut and secretes digestive juices into the gut.

The cells that make up epithelial tissues are of different types and configurations. The multilayered epithelial tissues of the skin are subject to a lot of wear and tear. Accordingly, epithelial cells in the deepest layer of the skin have a high rate of cell division, producing new cells that move progressively to the skin surface, die, and are shed (see Figure 40.3A). In contrast, gut epithelium consists of a single layer of tall, closely packed cells.

Epithelial cells have many specialized roles. Some secrete hormones, milk, mucus, digestive enzymes, or sweat. Others have cilia that move substances over surfaces or through tubes (see Figure 40.3B). Epithelial cells can also provide information to the nervous system. Smell- and taste-sensitive cells, for example, are epithelial cells that detect specific chemicals.

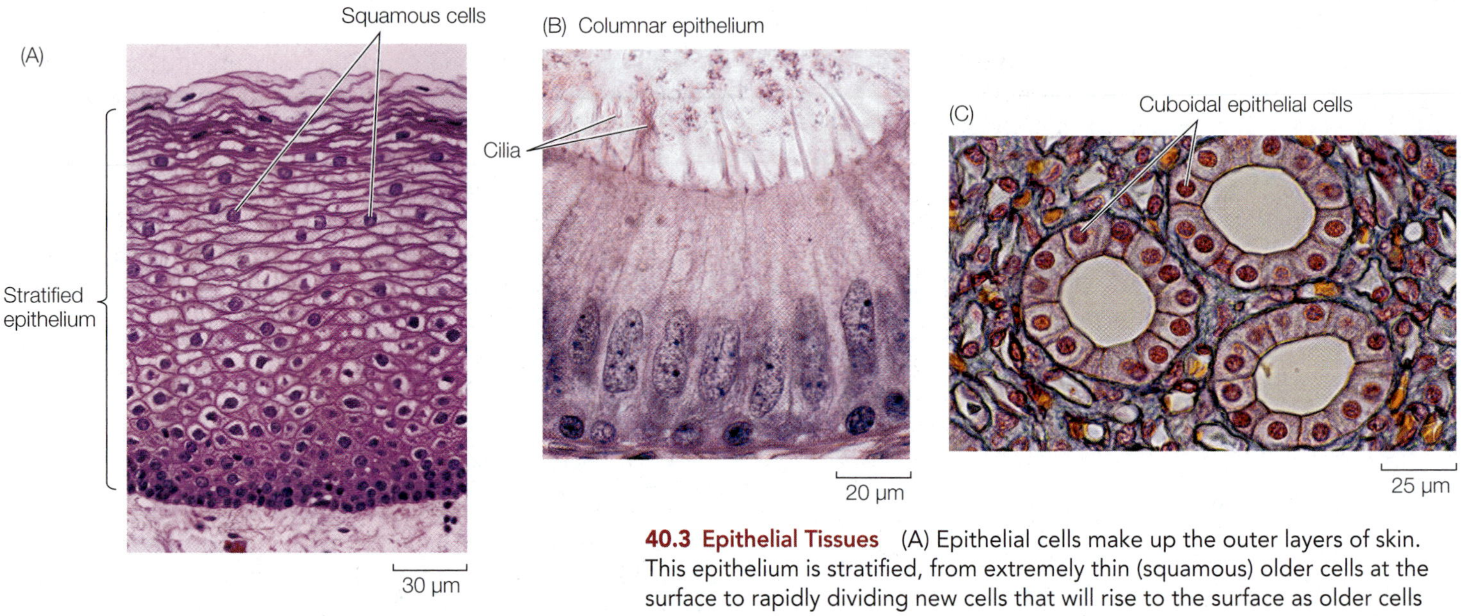
(A)
Squamous cells
Stratified epithelium
30 µm
(B) Columnar epithelium
Cilia
20 µm
(C)
Cuboidal epithelial cells
25 µm

40.3 Epithelial Tissues (A) Epithelial cells make up the outer layers of skin. This epithelium is stratified, from extremely thin (squamous) older cells at the surface to rapidly dividing new cells that will rise to the surface as older cells are shed. (B) Ciliated columnar epithelium from the male reproductive duct (the vas deferens). (C) A single layer of cuboidal epithelial cells forms a tubule in the kidney. These cells have many molecular transport functions.

Muscle tissues generate force and movement

Muscle tissues are the most abundant tissues in the animal body. They are the most prominent example of effectors (see p. 816).

The cells that make up muscle tissues contain long filaments of proteins called actin and myosin, which interact to cause muscle cells to contract and exert force. There are three types of muscle tissues (**Figure 40.4**):

- *Skeletal muscle* (so named because most of it is attached to bone) is responsible for locomotion and other body movements such as facial expressions, shivering, and breathing.
- *Cardiac muscle* makes up the heart and is responsible for the beating of the heart and the pumping of blood.
- *Smooth muscle* makes up the walls of many hollow internal organs such as the gut, bladder, and blood vessels.

Connective tissues include bone, blood, and fat

The cells that make up connective tissues are generally dispersed in an extracellular matrix that these cells themselves secrete. The composition and properties of the matrix differ among the different types of connective tissues, but protein fibers are always an important component of the matrix. The dominant protein in the extracellular matrix is collagen (see Figure 5.22), which makes up about 25 percent of total body protein. Collagen fibers are strong and resistant to stretch, giving strength to the skin and to the connections between bones and between bones and muscles. These fibers also provide a netlike framework for organs, giving them shape and structural strength. Elastin is another protein fiber found in the extracellular matrix of connective tissues. It recoils after

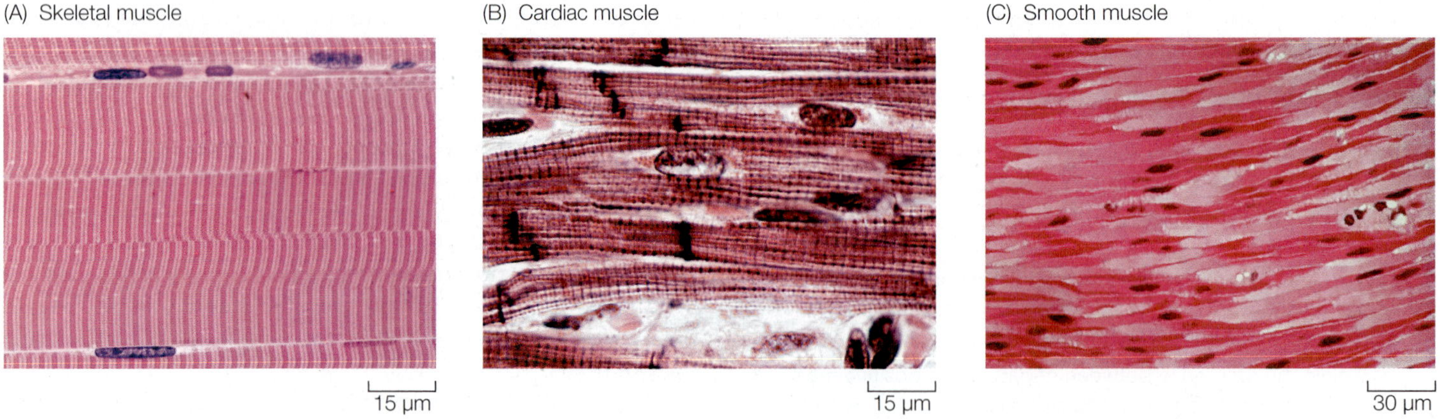
(A) Skeletal muscle
15 µm
(B) Cardiac muscle
15 µm
(C) Smooth muscle
30 µm

40.4 Muscle Cells Contain Protein Filaments Filaments of two specific proteins—actin and myosin—interact to cause contraction and generate force in muscle tissue. (A) The regular arrangement of actin and myosin filaments results in the striated (striped) appearance of skeletal muscle. (B) The individual cells of cardiac muscle are branched and form a strong structural meshwork. (C) The actin and myosin filaments of smooth muscle are not regularly arranged and thus it does not have a striated appearance.

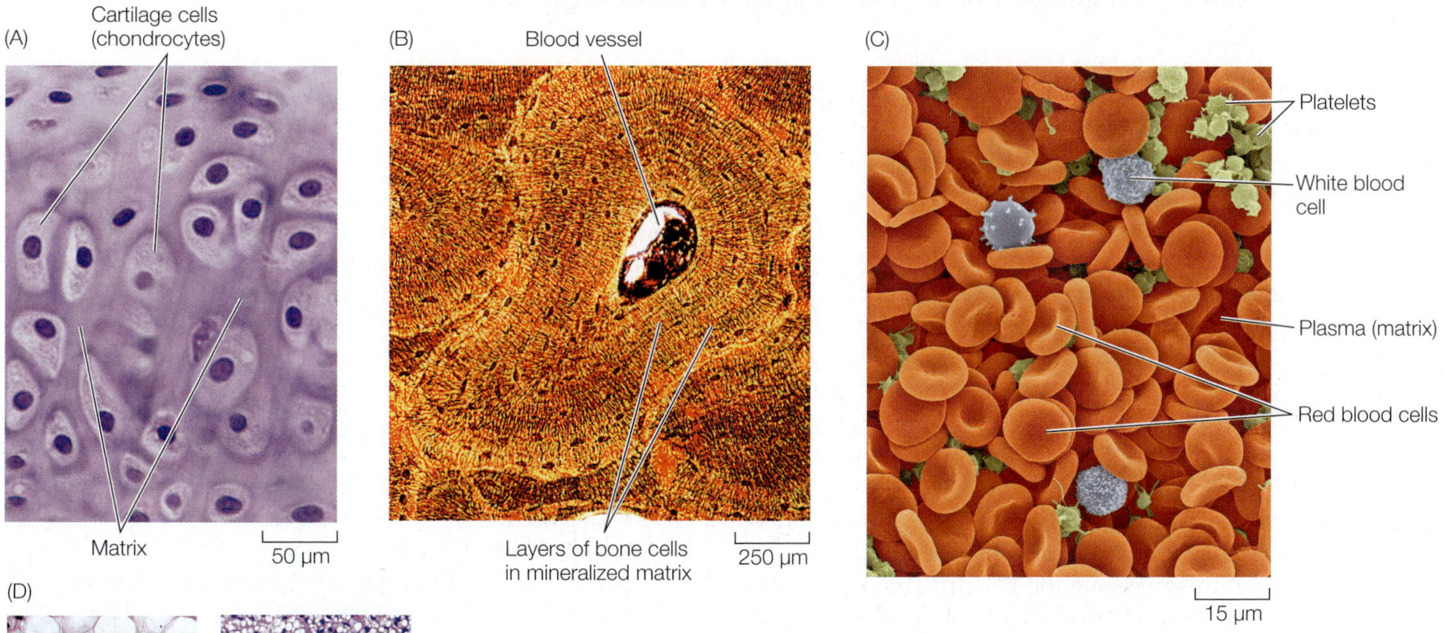
(A)
Cartilage cells (chondrocytes)
Matrix
50 μm
(B)
Blood vessel
Layers of bone cells in mineralized matrix
250 μm
(C)
Platelets
White blood cell
Plasma (matrix)
Red blood cells
15 μm

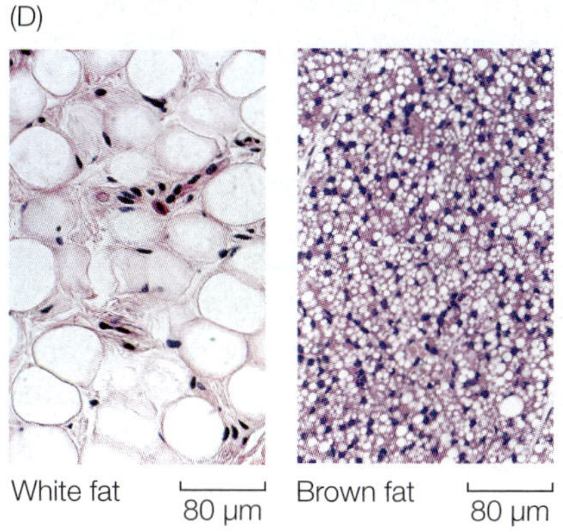
(D)
White fat
80 μm
Brown fat
80 μm

40.5 Connective Tissues (A) Cartilage cushions joints and makes structures such as the ear stiff but flexible. Cartilage cells, or chondrocytes, secrete an extracellular matrix rich in collagen and chondroitin sulfate. (B) Bone is the mineral-rich connective tissue of the vertebrate skeleton. (C) Blood is unique among the connective tissues, consisting of blood cells floating in an extracellular matrix of plasma. (D) "White" fat (left) is typical adipose tissue, with large droplets of energy-storing lipids and limited blood supply. In some mammals, specialized "brown" fat produces heat; this tissue is packed with mitochondria and blood vessels.

being stretched, so elastin fibers are abundant in tissues that are regularly stretched, such as the walls of the lungs and the large arteries.

- *Cartilage* and *bone* (**Figure 40.5A,B**) are connective tissues that provide firm structural support.
- *Blood* (**Figure 40.5C**) is a connective tissue consisting of cells dispersed in an extensive liquid extracellular matrix, the blood plasma.
- *Adipose* cells (**Figure 40.5D**) form loose connective tissue that stores lipids. Adipose tissue, or "fat," is a major source of stored energy. It also cushions organs, and layers of adipose tissue under the skin can provide a barrier to heat loss.

Neural tissues include neurons and glial cells

The many different cells of neural tissues are specialized for processing information. Nerve cells fall into one of two categories, neurons or glial cells.

- *Neurons* (**Figure 40.6A**) come in many shapes and sizes, but all neurons encode and conduct information as electrical signals. Most neurons release chemical signals that are received by target cells; the target cells can be other neurons,

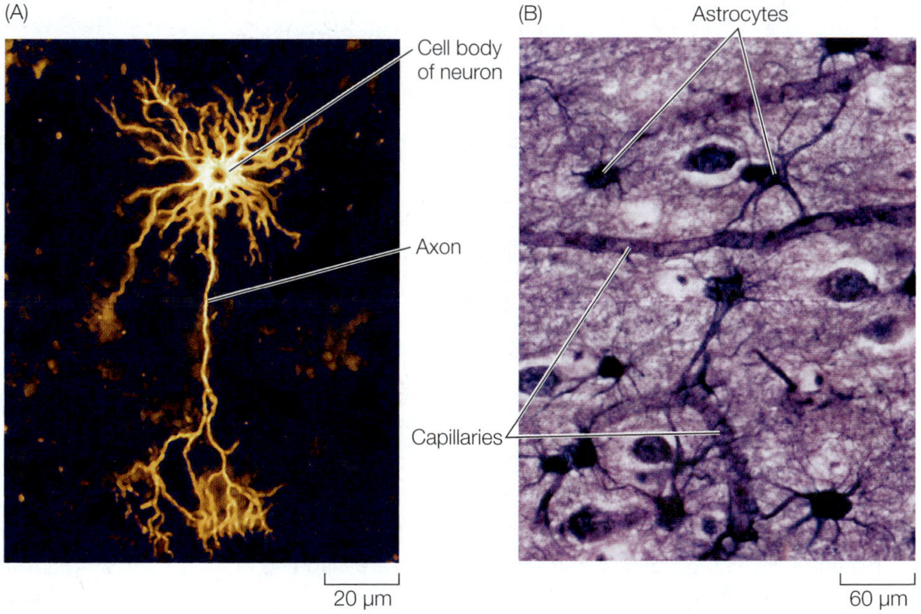
(A)
Cell body of neuron
Axon
20 μm
(B)
Astrocytes
Capillaries
60 μm

40.6 Neural Tissue Includes Neurons and Glial Cells (A) This human neuron consists of a cell body, several processes that receive input from other neurons, and one long axon that sends information to other cells. (B) A section through human brain tissue shows astrocytes, a type of glial cell.

muscle cells, or cells that secrete hormones and other molecules and substances, such as saliva.

- *Glial cells*, or *glia*, provide a variety of support functions for neurons (Figure 40.6B). One class of glia creates a barrier between the blood vessels and the neural tissue that protects the nervous system from potentially harmful chemicals circulating in the blood. Although glia do not generate electrical signals, they can communicate information through the release of chemical signals, and their roles in regulatory processes are becoming increasingly known.

Chapters 45, 46, and 47 detail the properties of nervous tissues.

Organs consist of multiple tissues

Organs are composed of an epithelium and one or more other kinds of tissue. Indeed, most organs include all four tissue types. The wall of the gut is a good example (Figure 40.7). Its inner surface is lined with a sheet of columnar epithelial cells. Different types of epithelial cells in this lining secrete hormones or digestive juices or absorb nutrients from the gut. Beneath the epithelial lining is a layer of connective tissue called the mucosa. Within this connective tissue are blood vessels, neurons, and glands (clusters of secretory epithelial cells). Concentric layers of smooth muscle tissue enable the gut to contract to mix food with digestive juices. A network of neurons between the muscle layers controls these movements.

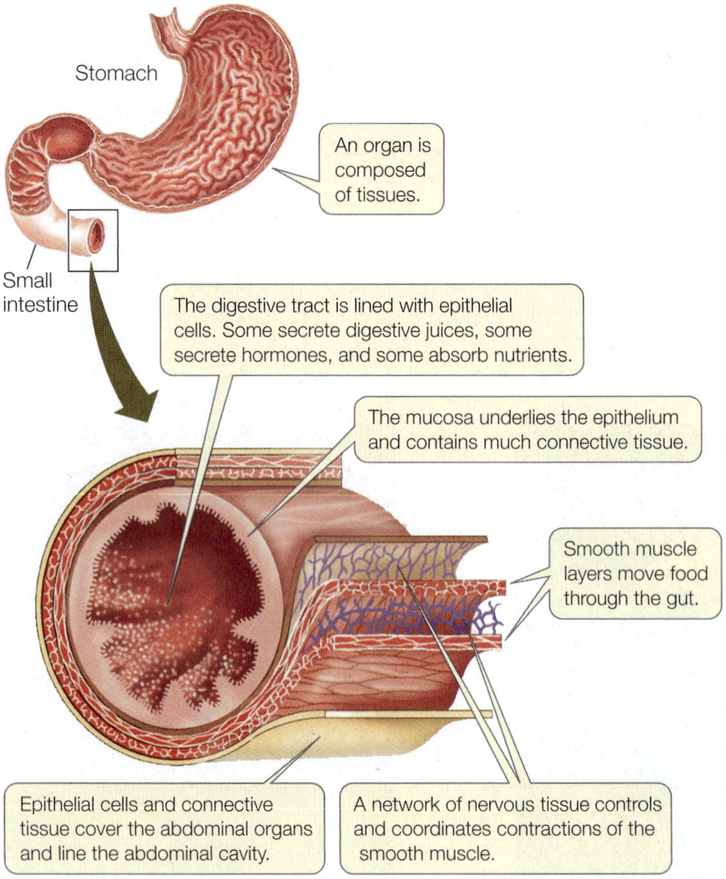

Stomach
An organ is composed of tissues.
Small intestine
The digestive tract is lined with epithelial cells. Some secrete digestive juices, some secrete hormones, and some absorb nutrients.
The mucosa underlies the epithelium and contains much connective tissue.
Smooth muscle layers move food through the gut.
Epithelial cells and connective tissue cover the abdominal organs and line the abdominal cavity.
A network of nervous tissue controls and coordinates contractions of the smooth muscle.

40.7 Tissues Form Organs The organs of the human digestive system, such as the stomach and small intestine, are made up of all four tissue types.

An individual organ is usually part of an **organ system,** a group of organs that work together to carry out certain functions. The stomach, small intestine, liver, and pancreas, for example, are parts of the digestive system. Thus we see an organizational hierarchy, with cells forming tissues that become part of organs, which in turn are organized in the functioning physiological systems of an individual organism.

RECAP 40.2

There are four tissue types in the animal body: epithelial, connective, nervous, and muscle. Individual organs include tissues of multiple types and are part of organ systems

- What is it crucial that the cells of epithelial tissue be densely packed into tight sheets? See p. 817 and Figure 40.3
- What is the key difference between neurons and glia, the two cell types that make up nervous tissue? See pp. 819–820
- What is meant by the hierarchical organization of physiological function? See p. 820 and Figure 40.7

Subsequent chapters will describe each of the organ systems mentioned above in much greater detail. The remainder of this chapter focuses on the mechanisms of homeostasis, using one important variable of the internal environment—temperature—as our example.

40.3 How Does Temperature Affect Living Systems?

Temperatures where organisms live vary enormously, from the boiling hot springs of Yellowstone National Park to the interior of Antarctica, where the temperature can fall below –80°C. Cells can function over only a narrow range of temperatures. If cells cool below 0°C, ice crystals form and damage cell structures. Some animals have adaptations, such as antifreeze molecules in their blood that help them resist freezing; others can survive freezing. Generally, however, cells must remain above 0°C to stay alive. The upper temperature limit for survival in most cells is about 45°C (although some specialized algae can grow in hot springs at 70°C, and some archaea live at near 100°C).

In general, proteins begin to denature and lose their function as temperatures rise above 40°C. Most cellular functions are limited to the range between 0°C and 40°C, which approximates the thermal limits for most organisms. Most species, however, have much narrower thermal limits. To stay within those limits in spite of environmental conditions, animals have evolved thermoregulatory adaptations. Those adaptations give them certain thermal tolerances, which determine their geographic ranges. When environments change rapidly, as may be happening as the global climate warms, animals may find themselves in situations that exceed their thermal tolerances.

Q_{10} is a measure of temperature sensitivity

Even between 0°C and 40°C, changes in tissue temperature create problems for animals. Most physiological processes, like the biochemical reactions that constitute them, are temperature-sensitive, going faster at higher temperatures (see Figure 8.20). The temperature sensitivity of a reaction or process is described in terms of $\mathbf{Q_{10}}$, a factor calculated by dividing the rate of a process or reaction at a certain temperature, R_T, by that rate at a temperature 10°C lower, R_{T-10}:

$$Q_{10} = \frac{R_T}{R_{T-10}}$$

Q_{10} can be measured for a simple enzymatic reaction or for a complex physiological process, such as rate of oxygen consumption. If a reaction or process is not temperature-sensitive, it has a Q_{10} of 1. Most biological Q_{10} values are between 2 and 3. A Q_{10} of 2 means that the reaction rate doubles as temperature increases by 10°C, and a Q_{10} of 3 indicates a tripling of the rate over a 10°C temperature range (**Figure 40.8**).

You will notice that the Q_{10} values (except for $Q_{10} = 1$) plotted in Figure 40.8 produce curves rather than straight lines. This is because the temperatures increase in additive intervals (10, 20, 30, etc.) but the reaction rates increase in a multiplicative fashion (2, 4, 8, 16, 32, etc.). Such curvilinear plots are common in biological data.

Changes in body temperature can disrupt an animal's physiology because not all of the biochemical reactions that constitute the metabolism of an animal have the same Q_{10}. These biochemical reactions are linked together in complex networks: the products of one reaction are the reactants for other reactions. Because different reactions have different Q_{10}'s, changes in tissue temperature will shift the rates of some reactions more than others, disrupting the overall network. Therefore, to maintain homeostasis, organisms must be able to compensate for or prevent changes in body temperature.

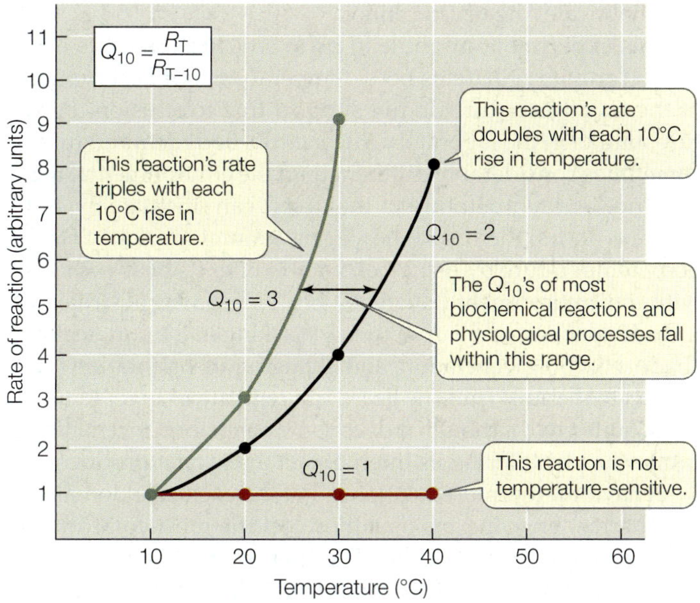

$Q_{10} = \frac{R_T}{R_{T-10}}$
This reaction's rate triples with each 10°C rise in temperature.
This reaction's rate doubles with each 10°C rise in temperature.
$Q_{10} = 2$
$Q_{10} = 3$
The Q_{10}'s of most biochemical reactions and physiological processes fall within this range.
$Q_{10} = 1$
This reaction is not temperature-sensitive.
Rate of reaction (arbitrary units)
Temperature (°C)

40.8 Q_{10} and Reaction Rate The larger the Q_{10} of a reaction or process, the faster its rate rises in response to an increase in temperature.

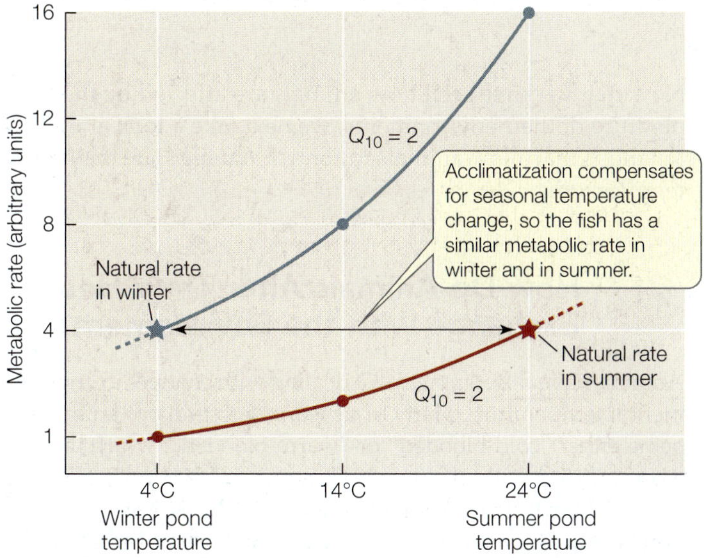

$Q_{10} = 2$
Acclimatization compensates for seasonal temperature change, so the fish has a similar metabolic rate in winter and in summer.
Natural rate in winter
Natural rate in summer
$Q_{10} = 2$
Metabolic rate (arbitrary units)
4°C
Winter pond temperature
14°C
24°C
Summer pond temperature

40.9 Metabolic Compensation In its natural environment, a fish's metabolic rate readjusts, or acclimatizes, to compensate for seasonal changes in temperature.

Animals acclimatize to seasonal temperatures

The body temperature of some animals (especially aquatic animals) is coupled to environmental temperature. The body temperature of a fish in a pond, for example, will be the same as the water temperature, which might range from 4°C in winter to 24°C in summer. If we bring that fish into the laboratory in the summer and measure the rates of any of its physiological or biochemical processes such as oxygen consumption, we will demonstrate a Q_{10} relationship. On the basis of that relationship, we can predict what the fish's metabolic rate will be in its pond in the winter. However, if we bring that fish back into the laboratory in the winter and measure its metabolic rate at winter pond temperature, we will find that rate to be higher than we predicted. The fish's biochemistry and physiology will have **acclimatized** to the seasonal change in water temperature so that it can remain active at winter temperatures (**Figure 40.9**). What could be the mechanism of this acclimatization? One possibility is that the fish may express isozymes with different temperature optima in summer and in winter. The ability of animals to acclimatize means that their metabolic functions are less sensitive to long-term changes in temperature than to short-term changes.

RECAP 40.3

Cells can survive only within a narrow range of temperatures, but even changes within that range can be disruptive because different physiological processes have different temperature sensitivities.

- Plot a $Q_{10} = 2.5$ curve for a physiological process. **See Figure 40.8**
- Explain how a change in body temperature can disrupt physiological processes. **See p. 821**

Now that we have seen how animals are affected by the temperature of their environments, we next take a look at the adaptations that allow animals to control and regulate their body temperatures.

40.4 How Do Animals Alter Their Heat Exchange with the Environment?

Animals have different ways of dealing with changes in environmental temperature. Many of us learned to think of animals as being either "cold-blooded" or "warm-blooded," which implies a comparison with our own body temperature and sets mammals and birds apart from other animals. This simple classification breaks down when we realize that mammals that hibernate become cold, and that many reptiles and insects can have body temperatures similar to ours when they are active. Physiologists sometimes classify animals according to whether they have a constant body temperature (homeotherms) or a variable body temperature (poikilotherms). But there are situations where applications of these descriptive terms also lead to strange conclusions. A deep-sea fish has a constant body temperature. Should it be classified with mammals? And the hibernator's body temperature in winter varies between a typical high mammalian level and a level that is close to environmental temperatures.

Another classification system is based on the source of heat that predominantly influences the body temperature of the animal. **Ectotherms** are animals whose body temperatures are determined primarily by external sources of heat. **Endotherms** can regulate their body temperatures by producing heat metabolically. Mammals and birds are endotherms—most of the time; other animals are ectotherms—most of the time. Therefore we have a third category: a **heterotherm** is an animal that sometimes behaves as an endotherm and at other times as an ectotherm. For example, a mammal that hibernates is an endotherm over the summer, but during its winter bouts of hibernation its internal heat production falls and it behaves much like an ectotherm. And some ectotherms can produce substantial amounts of metabolic heat, thus behaving like endotherms.

Endotherms produce substantial amounts of metabolic heat

Transfers of energy in biological systems are always inefficient, as Section 8.1 explained. With every transfer of energy—from food molecules to ATP, from ATP to biological work—some of the energy is lost as heat. This is true for both ectotherms and endotherms. In both cases, working muscles produce heat, as do the metabolic activities of all tissues. So why do endotherms produce more heat? The answer is that the cells of endotherms are less efficient at using energy than are the cells of ectotherms.

The cells of endotherms are more "leaky" to ions than are the cells of ectotherms. Therefore Na^+ ions are constantly diffusing into the cells, and K^+ ions are constantly diffusing out. Even an endotherm at rest must spend considerable amounts of energy to transport Na^+ out of the cells and transport K^+ back in. Because of their leaky membranes, endotherms expend more energy (and thus release more heat) than do ectotherms just to maintain the ion concentration gradients across their cell membranes. This situation is analogous to a leaky rowboat: the faster water comes in (i.e., the faster ions leak), the more metabolic energy has to be expended to bail the water out (i.e., pump ions) just to remain afloat.

We can speculate that a mutation resulting in seemingly faulty or leaky ion channels may underlie the evolution of endothermy. Such a mutation in a small ectotherm would have increased its energy expenditure and therefore its heat production. Increased heat production would allow the animal to be active for a longer time after sunset. Being active in the evening would open up a new world of ecological opportunities—a world in which there was less competition from similar-sized ectotherms.

Two major differences between endotherms and ectotherms are their resting metabolic rates—the sum total of all energy expenditures in their bodies when at rest—and their responses to changes in environmental temperature.

Ectotherms and endotherms respond differently to changes in environmental temperature

How do two similar-size animals, a lizard (an ectotherm) and a mouse (an endotherm), respond to changes in environmental temperature? We put each animal in a closed chamber and measure its body temperature and its resting metabolic rate as we change the temperature of the chamber from 37°C to 0°C. The body temperature of the lizard equilibrates with that of the chamber, whereas the body temperature of the mouse remains stable (**Figure 40.10A**). The metabolic rate of the lizard (already lower than that of the mouse) decreases as the temperature drops (**Figure 40.10B**). In contrast, the mouse's metabolic rate increases as the chamber temperature falls below 25°C. This increase in metabolism produces enough heat to prevent the mouse's body temperature from falling. In other words, the mouse can regulate its body temperature by increasing its metabolic rate; the lizard cannot.

This experiment might lead us to conclude that the ectotherm cannot regulate its body temperature, but observations of the lizard in nature do not support this conclusion. In nature, unlike in the laboratory, the lizard's body temperature is sometimes considerably different than the environmental temperature. Air temperature in the desert can fluctuate by 40°C in a few hours; the lizard, however, maintains a fairly stable body temperature by using behavior to alter its heat exchange with the environment (**Figure 40.11**). Its behavioral strategies include spending time in a burrow, basking in the sun, seeking shade, climbing vegetation, and changing its orientation with respect to the sun. Thus the lizard can regulate its body temperature quite well, although it does so by behavioral mechanisms rather than by altering its internal metabolic heat production.

Behavioral thermoregulation is not the exclusive domain of ectotherms. Endotherms usually select the most comfortable thermal environment possible. They may change posture, orient to the sun, move between sun and shade, and move between still air and moving air. Examples of more complex thermoregulatory behaviors include nest construction and social behaviors such as huddling. Humans put on or remove clothing and burn fossil fuels to generate the energy to heat or cool buildings.

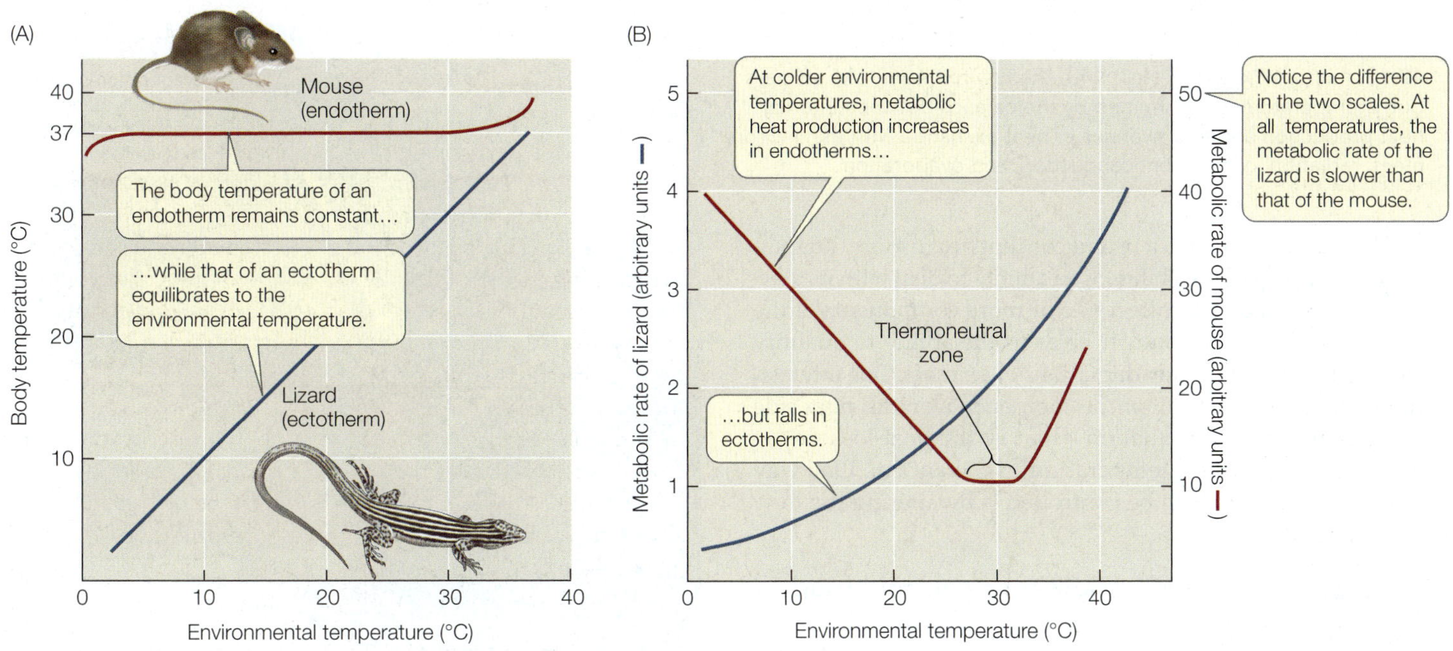
(A)
Mouse (endotherm)
The body temperature of an endotherm remains constant…
…while that of an ectotherm equilibrates to the environmental temperature.
Lizard (ectotherm)
Body temperature (°C)
Environmental temperature (°C)
(B)
At colder environmental temperatures, metabolic heat production increases in endotherms…
…but falls in ectotherms.
Thermoneutral zone
Notice the difference in the two scales. At all temperatures, the metabolic rate of the lizard is slower than that of the mouse.
Metabolic rate of lizard (arbitrary units —)
Metabolic rate of mouse (arbitrary units —)
Environmental temperature (°C)

40.10 Ectotherms and Endotherms React Differently to Environmental Temperatures (A) At the same environmental temperature, an ectotherm and an endotherm of approximately the same body size (here, a lizard and a mouse) have different body temperatures. (B) The metabolic rates of the lizard and mouse react in opposite ways to cooler temperatures. (The mouse's metabolic rate rises again at higher temperatures because, after a certain point, it takes metabolic energy to dissipate heat by sweating or panting.)

Energy budgets reflect adaptations for regulating body temperature

Both ectotherms and endotherms can influence their body temperatures by altering four avenues of heat exchange between their bodies and the environment (**Figure 40.12**):

- **Radiation** Heat moves from warmer objects to cooler ones via the exchange of infrared radiation (what you feel when you stand in front of a fire).
- **Convection** Heat transfers to a surrounding medium such as air or water as that medium flows over a surface (the wind-chill factor).
- **Conduction** Heat transfers directly between two objects at different temperatures when they come into contact (e.g., an icepack on a sprained ankle).
- **Evaporation** Heat transfers away from a surface when water evaporates on that surface (the effect of sweating).

The total balance of heat production and heat exchange can be expressed as an **energy budget**, based on the simple fact that if the body temperature of an animal is to remain constant, the heat entering the animal must equal the heat leaving it. The heat coming in is usually from metabolism and radiation (R_{abs}, for radiation absorbed). Heat leaves the body via the four mechanisms listed above—radiation emitted (R_{out}), convection, conduction, and evaporation. The energy budget takes the mathematical form

$$\underbrace{\text{metabolism} + R_{abs}}_{\text{heat}_{in}} = \underbrace{R_{out} + \text{convection} + \text{conduction} + \text{evaporation}}_{\text{heat}_{out}}$$

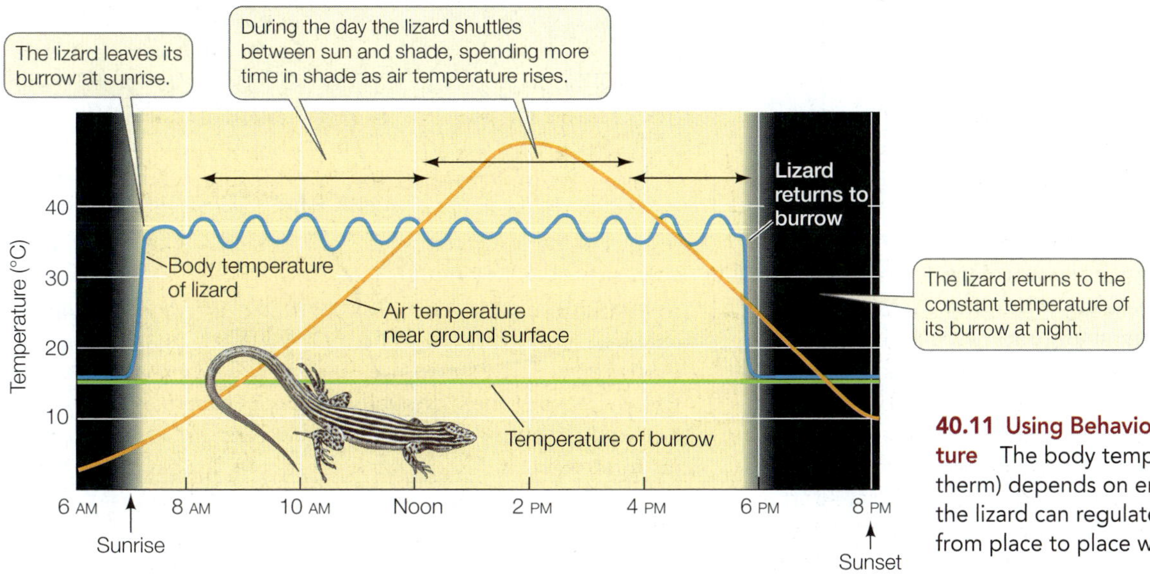
The lizard leaves its burrow at sunrise.
During the day the lizard shuttles between sun and shade, spending more time in shade as air temperature rises.
Lizard returns to burrow
Body temperature of lizard
Air temperature near ground surface
Temperature of burrow
The lizard returns to the constant temperature of its burrow at night.
Temperature (°C)
6 AM
8 AM
10 AM
Noon
2 PM
4 PM
6 PM
8 PM
Sunrise
Sunset

40.11 Using Behavior to Regulate Body Temperature The body temperature of a lizard (an ectotherm) depends on environmental temperature, but the lizard can regulate its temperature by moving from place to place within its environment.

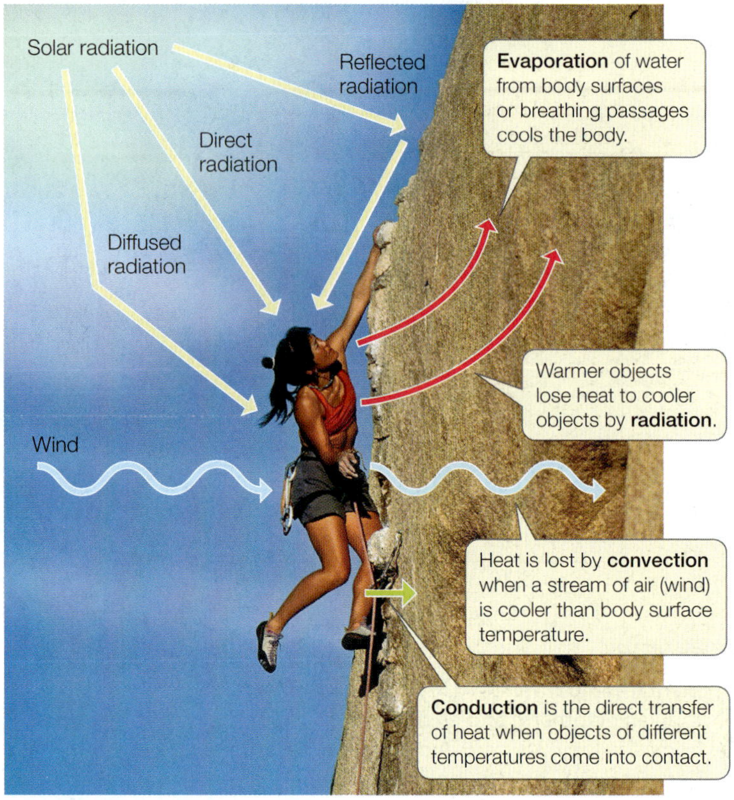
Solar radiation
Reflected radiation
Evaporation of water from body surfaces or breathing passages cools the body.
Direct radiation
Diffused radiation
Warmer objects lose heat to cooler objects by radiation.
Wind
Heat is lost by convection when a stream of air (wind) is cooler than body surface temperature.
Conduction is the direct transfer of heat when objects of different temperatures come into contact.

40.12 Animals Exchange Heat with the Environment An animal's body temperature is determined by the balance between internal heat production and four avenues of heat exchange with the environment: radiation, convection, conduction, and evaporation.

The energy budget is a useful concept because any adaptation that influences the ability of an animal to deal with its thermal environment must affect one or more components of the budget. So the energy budget gives us the ability to quantify and compare the thermal adaptations of animals. One interesting observation is that all of the components on the right side of the energy budget equation—that is, the heat-loss side—depend on the surface temperature of the animal. One way surface temperature can be controlled is by altering the flow of blood to the skin.

Both ectotherms and endotherms control blood flow to the skin

Heat exchange between the internal environment and the skin occurs largely through blood flow. As described at the beginning of this chapter, when body temperature rises because of exercise, blood flow to the skin increases, and the skin surface becomes warm. The heat that the blood brings from the body core to the skin is lost to the environment through the four avenues listed above, which helps bring the body temperature back to normal. In contrast, when body temperature is too low or the environment is too cold, the blood vessels supplying the skin constrict, reducing heat loss to the environment.

The ability to control blood flow to the skin can be an important adaptation for an ectotherm such as the marine iguana (a reptile) of the Galápagos archipelago (**Figure 40.13**). The Galápagos are volcanic islands that lie on the equator but are bathed by cold ocean currents. The iguanas bask on hot black lava rocks on the shore, but periodically they enter the cold ocean water to feed on seaweed. While the iguanas are feeding, they cool to the temperature of the sea. This cooling lowers their metabolism, making them slower, more vulnerable to predators, and incapable of efficient digestion. They therefore alternate between feeding in the cold seawater and basking on the hot rocks. It is advantageous for iguanas to retain body heat as long as possible while swimming and to warm up as fast as possible when basking. They can accomplish these results by changing their heart rate and thus the rate of blood flow to their skin.

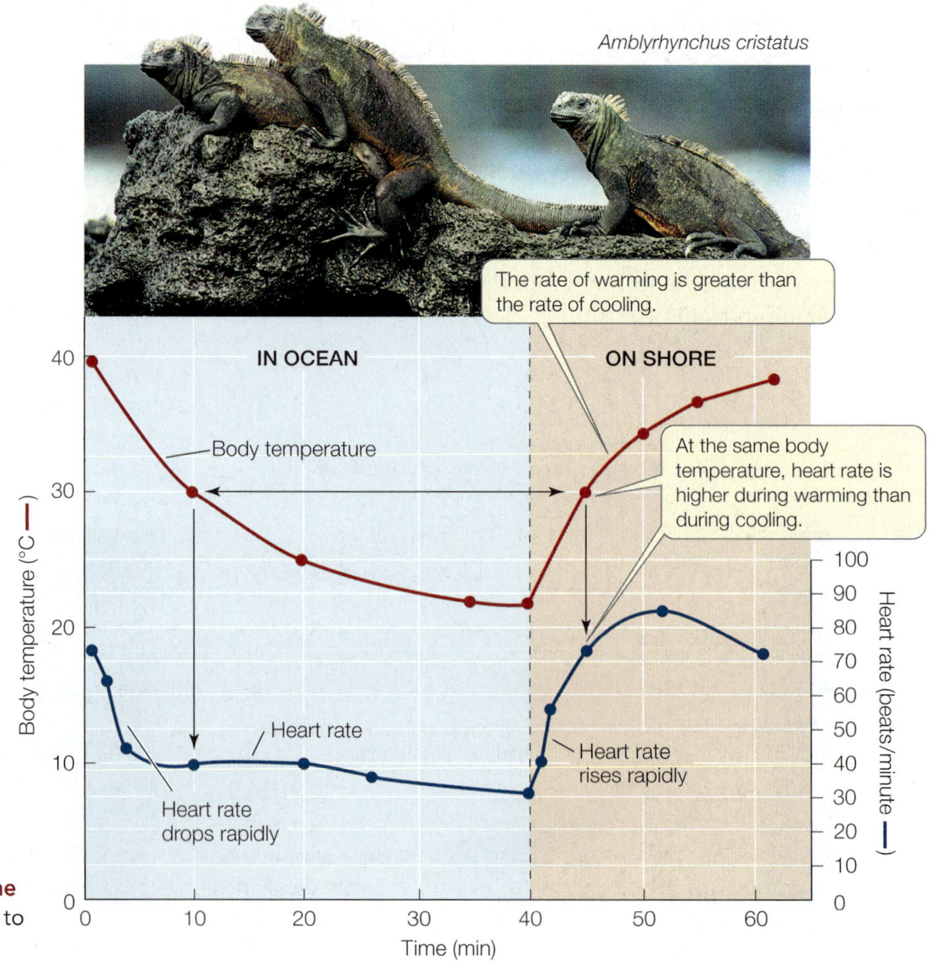
Amblyrhynchus cristatus
The rate of warming is greater than the rate of cooling.
IN OCEAN
ON SHORE
Body temperature
At the same body temperature, heart rate is higher during warming than during cooling.
Heart rate
Heart rate drops rapidly
Heart rate rises rapidly
Body temperature (°C —)
Heart rate (beats/minute —)
Time (min)
0
10
20
30
40
50
60
70
80
90
100

40.13 Some Ectotherms Regulate Blood Flow to the Skin Galápagos marine iguanas control blood flow to the skin to alter their heating and cooling rates.

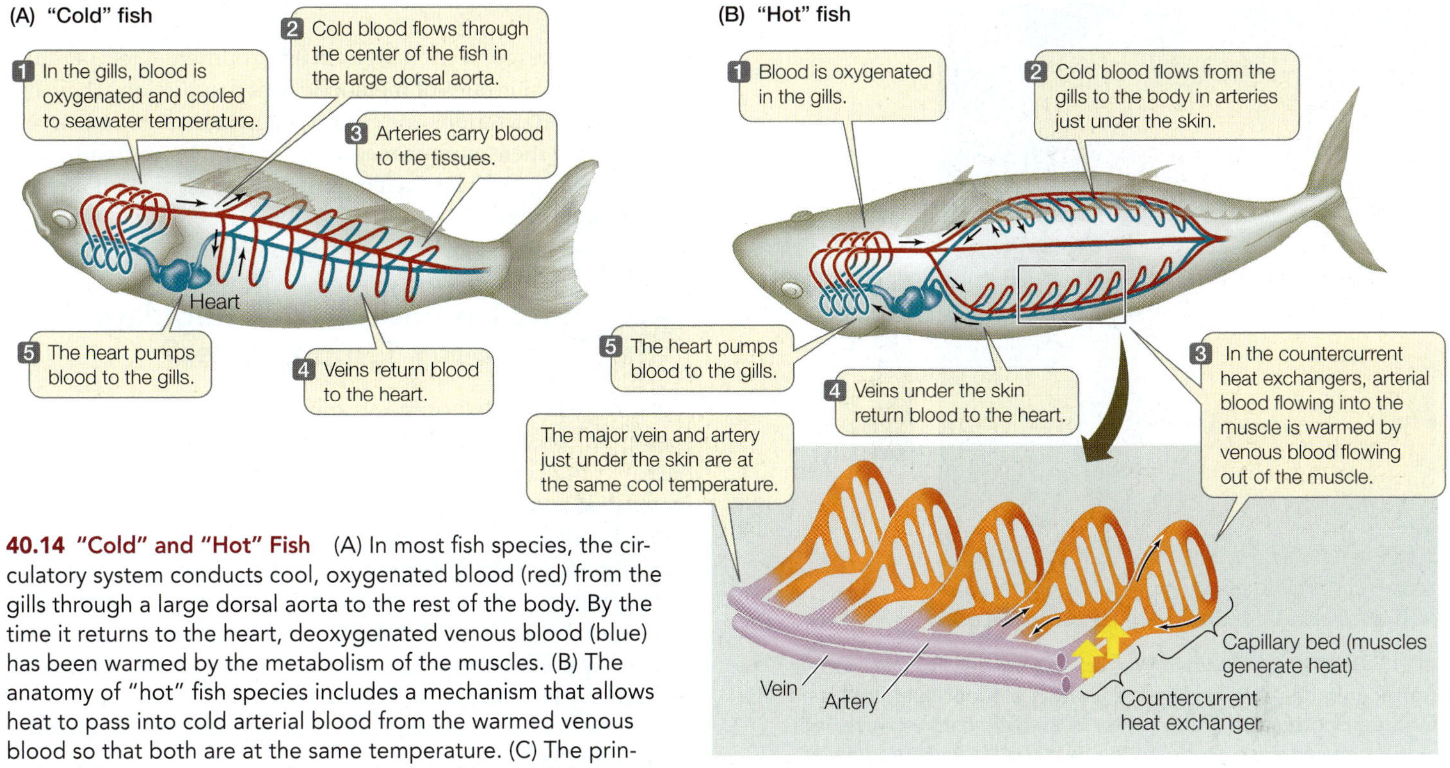

(A) "Cold" fish
1 In the gills, blood is oxygenated and cooled to seawater temperature.
2 Cold blood flows through the center of the fish in the large dorsal aorta.
3 Arteries carry blood to the tissues.
Heart
5 The heart pumps blood to the gills.
4 Veins return blood to the heart.
(B) "Hot" fish
1 Blood is oxygenated in the gills.
2 Cold blood flows from the gills to the body in arteries just under the skin.
5 The heart pumps blood to the gills.
4 Veins under the skin return blood to the heart.
3 In the countercurrent heat exchangers, arterial blood flowing into the muscle is warmed by venous blood flowing out of the muscle.
The major vein and artery just under the skin are at the same cool temperature.
Vein
Artery
Capillary bed (muscles generate heat)
Countercurrent heat exchanger

40.14 "Cold" and "Hot" Fish (A) In most fish species, the circulatory system conducts cool, oxygenated blood (red) from the gills through a large dorsal aorta to the rest of the body. By the time it returns to the heart, deoxygenated venous blood (blue) has been warmed by the metabolism of the muscles. (B) The anatomy of "hot" fish species includes a mechanism that allows heat to pass into cold arterial blood from the warmed venous blood so that both are at the same temperature. (C) The principle of this countercurrent heat exchange allows heat (orange) to be transferred with maximum efficiency.

What about furred mammals? Fur acts as insulation to keep body heat in, making it possible for mammals to function in cold environments. When they are active, however, mammals still must get rid of excess heat, and it does little good to transport that heat to the skin under the fur. Thus, as mentioned at the opening of this chapter, mammals have specialized blood vessels for transporting heat to their hairless skin surfaces. Heat loss from these areas is tightly controlled by the opening and closing of these blood vessels. When you are cold, the blood flow to your hands and feet decreases and they feel cold, but when you exercise, these same surfaces can get hot quickly.

Go to Media Clip 40.1
Thermoregulation in Animals
Life10e.com/mc40.1

Some fish conserve metabolic heat

The muscles of active fish produce substantial amounts of metabolic heat, but they have difficulty retaining any of that heat. Blood pumped from the fish heart goes directly to the gills, where it must come very close to the surrounding water to exchange respiratory gases. So any heat that the blood picks up from metabolically active muscles is lost to the water flowing across its gills. It is thus surprising that some large, rapidly swimming fish, such as giant bluefin tuna and great white sharks, can maintain temperature differences as great as 10°C to 15°C between their bodies and the surrounding water. The heat comes from their powerful swimming muscles, and the ability of these "hot" fish to conserve that heat is based on the remarkable arrangement of their blood vessels.

In the typical ("cold") fish circulatory system, oxygenated blood from the gills collects in a large dorsal vessel, the aorta, and travels through the center of the fish, distributing blood to all organs and muscles (Figure 40.14A). "Hot" fish have a smaller central dorsal aorta; most of their oxygenated blood is transported in large side vessels just under their skin (Figure 40.14B). The cold blood leaving the gills is thus kept close to the surface of the fish as the blood flows posteriorly to the swimming muscles. Smaller vessels transport this cold blood into the muscle mass, and these small vessels run parallel to the vessels carrying warm blood from the swimming muscles back toward the heart. Because the vessels carrying the cold blood into the muscles are in close contact with the vessels carrying warm blood away, heat flows from the warm to the cold blood by conduction and is therefore retained in the muscle mass.

Because heat is exchanged between blood vessels carrying blood in opposite directions, this adaptation is called **countercurrent heat exchange** (Figure 40.14C). It keeps the heat within the muscles, enabling these fish to have an internal body temperature considerably higher than the water temperature. Each 10°C rise in muscle temperature increases the fish's sustainable power output almost threefold, giving it a faster swimming capability.

Some ectotherms regulate metabolic heat production

Some ectotherms raise their body temperature by producing metabolic heat. For example, the powerful flight muscles of many insects must reach 35°C to 40°C before the insects can fly, and they must maintain these high temperatures during flight. Such insects warm up to fly by contracting their flight

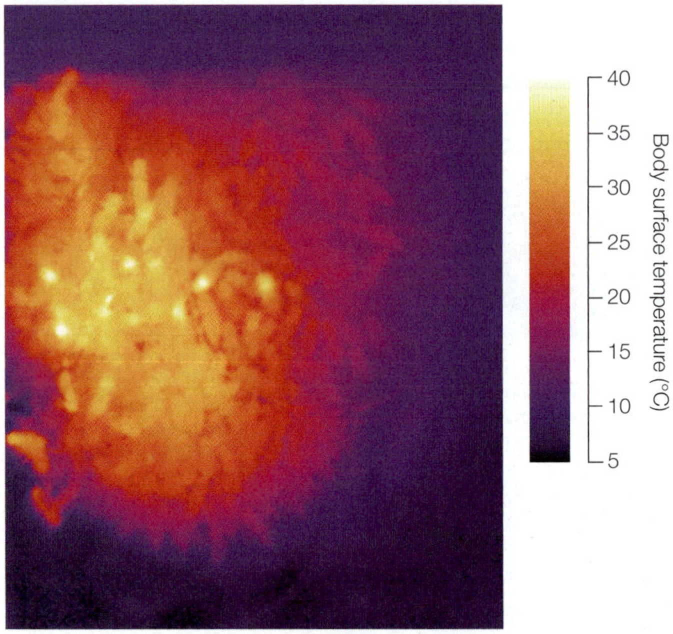
40
35
30
25
20
15
10
5
Body surface temperature (°C)

40.15 Bees Keep Warm in Winter Honey bee colonies survive winter cold because workers generate metabolic heat. In this infrared photograph of the center of an overwintering hive, individual bees are discernible by the heat their bodies produce as they cluster around their queen.

muscles in a manner analogous to shivering in mammals. The heat-producing ability of insects can be quite remarkable. Probably the most impressive case is a species of scarab beetle that lives mostly underground in mountains north of Los Angeles, California. These beetles come aboveground to mate, with the males flying in search of females. They undertake this mating ritual at night, in winter, and only during snowstorms.

Honey bees regulate temperature as a group. They live in large colonies consisting mostly of female worker bees that maintain the hive and rear the larval offspring of the single queen bee. During winter, worker bees cluster around the brood (eggs and larvae). They adjust their individual metabolic heat production and density of clustering so that the brood temperature remains remarkably constant, at about 34°C, even as the outside air temperature drops below freezing (**Figure 40.15**).

RECAP 40.4

Animals that can maintain constant high body temperatures because of their metabolic heat production are called endotherms. Those whose body temperatures are determined primarily by environmental sources of heat are called ectotherms. Heat exchange between an animal and its environment occurs via radiation, convection, conduction, and evaporation.

- Explain why the two curves in Figure 40.10A are different from the two curves in Figure 40.10B. **See p. 822**
- In terms of the energy budget, why is the control of blood flow to the skin so important for thermoregulation? **See pp. 823–824**
- Explain how countercurrent heat exchange makes it possible for some fish to have a body temperature higher than that of the surrounding water. **See p. 825 and Figure 40.14**

Endotherms respond to changes in environmental temperature by changing their rates of metabolic heat production. They also have other adaptations for controlling their rates of heat exchange with their environments. How do they regulate these various avenues of heat exchange to achieve a constant internal body temperature?

40.5 How Do Endotherms Regulate Their Body Temperatures?

Physiologists can determine an animal's metabolic rate by measuring its consumption of O_2 or production of CO_2. Within a narrow range of environmental temperatures, called the **thermoneutral zone** (see Figure 40.10B), the metabolic rates of endotherms (birds and mammals) are at low levels and independent of temperature. The metabolic rate of a resting animal at a temperature within the thermoneutral zone is known as the **basal metabolic rate**, or **BMR**. It is usually measured in animals that are quiet but awake and not using energy for digestion, reproduction, or growth. Thus the BMR is the rate at which a resting animal is consuming just enough energy to carry out its minimal body functions.

Basal metabolic rates correlate with body size

As you might expect, the BMR of an elephant is greater than that of a mouse. After all, the elephant is more than 100,000 times larger than the mouse. However, the BMR of the elephant is only about 7,000 times greater than that of the mouse. That means that a gram of mouse tissue uses energy at a rate 15 times greater than a gram of elephant tissue (**Figure 40.16**). Across all of the endotherms, BMR per gram of tissue increases as animals get smaller.

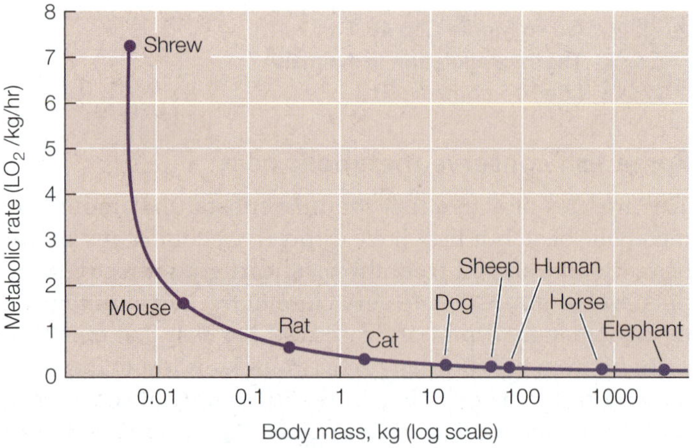
8
7
6
5
4
3
2
1
0
Metabolic rate (LO$_2$ /kg/hr)
Shrew
Mouse
Rat
Cat
Dog
Sheep
Human
Horse
Elephant
0.01
0.1
1
10
100
1000
Body mass, kg (log scale)

40.16 The Mouse-to-Elephant Curve On a weight-specific basis, the metabolic rate of small endotherms is much greater than that of larger endotherms. This classic illustration originally published in the 1930s, plots O_2 consumption per kilogram of body mass (a measure of metabolic rate) against a logarithmic plot of body mass.

WORKING WITH**DATA**:

A Mammal's BMR Is Proportional to Its Body Size

Original Paper

White, C. R. and R. S. Seymour. 2003. Mammalian basal metabolic rate is proportional to body mass. *Proceedings of the National Academy of Sciences USA* 100: 4046–4049.

Analyze the Data

Studies conducted in the late 1800s concluded that BMR varied as a function of body mass to the ⅔ power. This conclusion made sense because metabolism is a function of body *mass* (a cube function of a linear dimension), whereas heat loss is a function of body *surface area* (a square function of a linear dimension). Thus, because of this limitation on heat loss, BMR would have to decrease as body size increased.

Subsequent studies in the early 1900s, however, concluded that BMR varies as a function of body mass to the ¾ power. Recent theoretical considerations of fractal geometry led to support for this ¾ power function. The table at right gives a small subset of the data from ten species presented in the paper cited above.

QUESTION 1

Do you think these data support the ⅔ power function or the ¾ power function?

QUESTION 2

You know from the "mouse-to-elephant curve" in Figure 40.16 that these data relate to each other as an exponential function and therefore produce a curvilinear plot. You can linearize the presentation of these data by conducting a logarithmic transformation. Using a calculator, create a new data table that expresses the values below as $\log_{10}$ values. Plot these values with body size ($\log_{10}$ body mass) on the x axis and metabolic rate ($\log_{10}$ BMR) on the y axis. Then, using a linear regression program on your calculator or computer, determine the equation that describes the graph you created in Question 2.

QUESTION 3

Remember that you transformed your data into logarithmic values, so in your equation, $y = \log_{10}$ BMR and $x = \log_{10}$ body mass. Convert your equation to a non-logarithmic form. Does your answer support the conclusion expressed in the title of the paper listed above?

Species	Body size (g)	BMR (ml O_2/hr)
Bat	16	40
Mouse	30	63
Rat	400	146
Muskrat	1,000	640
Marmot	4,000	1,550
Beaver	10,000	4,500
Coyote	10,000	2,690
Baboon	17,000	5,150
Antelope	37,800	9,300
Moose	325,000	51,400

Go to **BioPortal** for all WORKING WITH**DATA** exercises

Why should this disproportionate difference exist? There are several possible reasons. As animals get bigger, they have a smaller ratio of surface area to volume (see Figure 5.2). Since heat production is related to the volume (i.e., mass) of the animal, but its capacity to dissipate heat is related to its surface area, it has been reasoned that larger animals evolved lower metabolic rates to avoid overheating. However, this explanation alone is insufficient because the relationship between body mass and metabolic rate holds for even very small organisms and for ectotherms, in which overheating is not a problem. Other hypotheses have also been proposed. For example, a larger animal has a greater proportion of support tissues (e.g., skin and bone), which are not as metabolically active as other tissue types. The real explanation is probably a mixture of different factors, but the relationship holds over a very broad range of species.

For an endotherm, a metabolic rate versus environmental temperature curve represents the integrated response of all the animal's thermoregulatory adaptations (**Figure 40.17**). The thermoneutral zone is bounded by a lower and an upper **critical temperature**. When the environmental temperature is within its thermoneutral zone, an endotherm's thermoregulatory responses do not require much energy and could be considered passive; such responses include changing posture, fluffing fur, and controlling blood flow to the skin. Outside its thermoneutral zone, however, an endotherm's thermoregulatory responses are active and require considerable metabolic energy.

Endotherms respond to cold by producing heat and adapt to cold by reducing heat loss

When environmental temperatures fall below the lower critical temperature, endotherms must produce heat to compensate for the heat they lose to the environment. Mammals can accomplish this by shivering and/or nonshivering heat production. Birds use only shivering heat production. Shivering uses the contractile machinery of skeletal muscles to convert ATP to ADP, with the energy from this process released as heat. Shivering muscles pull against each other so that little movement other than a tremor results. "Shivering heat production" is perhaps too narrow a term, however; increased muscle tone and increased body movements also contribute to increased heat production in cold environments.

Most nonshivering heat production occurs in specialized adipose tissue called **brown fat** (see Figure 40.5D). This tissue looks brown because of its abundant mitochondria and rich blood supply. In brown fat cells, a protein called thermogenin uncouples proton movement from ATP production,

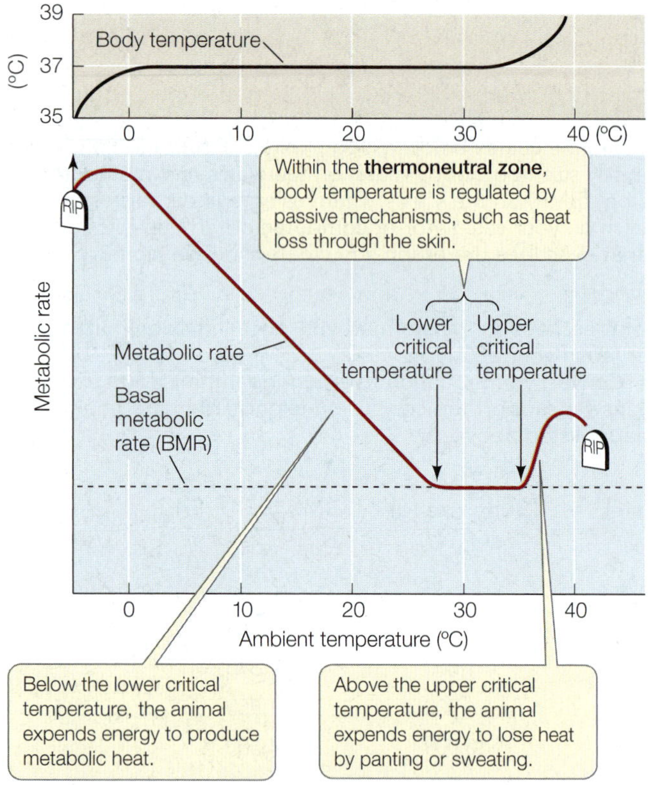

39
37
35
(°C)
Body temperature
0
10
20
30
40 (°C)
Within the **thermoneutral zone**, body temperature is regulated by passive mechanisms, such as heat loss through the skin.
Metabolic rate
Lower critical temperature
Upper critical temperature
Metabolic rate
Basal metabolic rate (BMR)
RIP
RIP
0
10
20
30
40
Ambient temperature (°C)
Below the lower critical temperature, the animal expends energy to produce metabolic heat.
Above the upper critical temperature, the animal expends energy to lose heat by panting or sweating.

40.17 Environmental Temperature and Mammalian Metabolic Rates A plot of an endotherm's metabolic rate versus environmental temperature represents the integrated response of all of its thermoregulatory mechanisms. Outside the thermoneutral zone, maintaining a constant body temperature requires expending energy. Outside extreme limits (0°C and 40°C in this instance), the animal cannot maintain its body temperature and dies.

allowing protons to leak across the inner mitochondrial membrane rather than having to pass through the ATP synthase and generate ATP (review the discussion of brown fat and the chemiosmotic mechanism in Chapter 9). As a result, metabolic fuels are consumed without producing ATP, but heat is still released. Pads of brown fat are found in newborns of many mammalian species (including humans), in some adult mammals that are small and acclimatized to cold, and in mammals that hibernate. Recently it has been discovered that adult humans have small amounts of brown fat distributed around the body and that its metabolic activity is stimulated by cold exposure. One study found less brown fat activity in obese than in lean individuals, leading to the hypothesis that individual differences in propensity for weight gain may be related to the amount of brown fat in an individual, as described in the opening stories of Chapters 9 and 41.

In addition to their ability to produce heat, endotherms that live in cold climates have evolved adaptations to reduce their heat loss. Heat is lost from the body surface, and cold-climate species have anatomical adaptations that give them smaller surface-to-volume ratios than their warm-climate relatives (**Figure 40.18**). These adaptations include rounder body shapes and shorter appendages.

The most common and important means of decreasing heat loss is to increase thermal insulation. Animals adapted to cold climates have much thicker layers of fur, feathers, or fat than do their warm-climate relatives. Fur and feathers are good insulators because they trap a layer of still, warm air close to the skin surface. If that air is displaced by water, insulation is drastically reduced. In many species, oil secretions spread through fur or feathers by grooming are critical for resisting wetting and maintaining a high level of insulation.

The ability to decrease blood flow to the skin is an important thermoregulatory adaptation for cold-climate endotherms. Constriction of blood vessels in the skin, and especially in the appendages, greatly improves an animal's ability to conserve heat. Countercurrent heat exchange like that we saw in "hot" fish is also an important adaptation in the appendages of endotherms. Blood flowing out to the paw of a wolf, the hoof of a caribou, or the foot of a bird parallels the flow of the blood returning to the body core. Heat is transferred from the outgoing to the returning blood, thus retaining heat in the animal's core.

Go to Activity 40.1 Thermoregulation in an Endotherm
Life10e.com/ac40.1

(A) *Lepus alleni*
(B) *Lepus arcticus*

40.18 Adaptations to Cold and Hot Climates (A) The antelope jackrabbit is found in the Sonoran Desert of Arizona. Its large ears serve as heat exchangers, passing heat from the animal's blood to the surrounding air. (B) The thick fur of the Arctic hare provides insulation and its rounded body shape lowers its surface-to-volume ratio. The ears and extremities are smaller than those of its warm-climate relatives so less heat is lost to the environment.

Evaporation of water can dissipate heat, but at a cost

As environmental temperature rises within an endotherm's thermoneutral zone, the animal dissipates more of its metabolic heat by increasing blood flow to the skin, seeking shade and cool breezes, and decreasing activity. When the temperature exceeds the upper critical temperature, however, overheating becomes a problem. For exercising animals (including athletes), overheating can occur even at low environmental temperatures. Large mammals, especially those in hot habitats such as elephants, rhinoceroses, and water buffaloes, have little or no insulating fur and seek out water to wallow in when the air temperature is high. Having water in contact with the skin greatly increases heat loss because the heat-absorbing capacity of water is much greater than that of air.

Evaporation from external or internal body surfaces through sweating or panting can also cool an endotherm. A gram of water absorbs about 580 calories of heat when it evaporates. If this evaporation occurs on the skin, most of the heat comes from the skin. However, sweat or saliva that falls off the body provides no cooling. Thus when the need for heat loss is greatest, water from the internal environment can be squandered with no cooling benefit. Water is heavy, so animals do not carry an excess supply of it, and many hot environments are also arid. In habitats that are both hot and dry, sweating and panting are cooling adaptations of last resort.

Sweating and panting are *active* processes that require expending metabolic energy. That is why the metabolic rate increases when the upper critical temperature is exceeded (see Figure 40.17). A sweating or panting animal is generating heat in the process of dissipating heat, which can be a losing battle.

The mammalian thermostat uses feedback information

The thermoregulatory mechanisms and adaptations of endotherms that we have just discussed are controlled by neural regulatory systems that integrate information from environmental and physiological sources and then issue commands to the effectors that alter the heat content of the body. These regulatory systems are similar in principle in birds and mammals but differ in many details. Here we focus on the nervous system thermostats of mammals.

The major thermoregulatory integrative center of mammals is at the base of the brain in a structure called the **hypothalamus**. As we will see in upcoming chapters, the hypothalamus is a key player in many animal regulatory systems. Experiments demonstrating its role have shown that slight cooling of the hypothalamus stimulates constriction of skin blood vessels and that stronger cooling increases metabolic heat production. As a result, cooling of the hypothalamus in an unchanging, thermoneutral environment will cause body temperature to rise. Conversely, hypothalamic heating causes the overall body temperature to fall (**Figure 40.19**).

In mammals, the temperature of the hypothalamus itself is the major feedback signal. The hypothalamus generates set points for thermoregulatory responses. When the temperature of the hypothalamus exceeds or drops below those set points, thermoregulatory responses are activated to reverse the direction of temperature change (**Figure 40.20**). The system integrates other sources of information in addition to hypothalamic temperature. For example, temperature sensors in the skin register environmental temperature. A change in skin temperature is feedforward information that shifts hypothalamic set points; the set point for metabolic heat production is higher when the skin is cold and lower when the skin is warm.

INVESTIGATING**LIFE**

40.19 The Hypothalamus Regulates Body Temperature

A mammal's hypothalamus was subjected directly to temperature manipulation. The body's responses to the manipulations were as expected if the hypothalamus is the mammalian "thermostat."[a]

HYPOTHESIS Heating or cooling the mammalian hypothalamus results in predictable changes in body temperature.

Method

1. Implant a probe into the hypothalamus of a living ground squirrel's brain. Use the probe to heat or cool the hypothalamus directly (i.e., without affecting the ambient temperature).
2. Manipulate the hypothalamic temperature T_H.

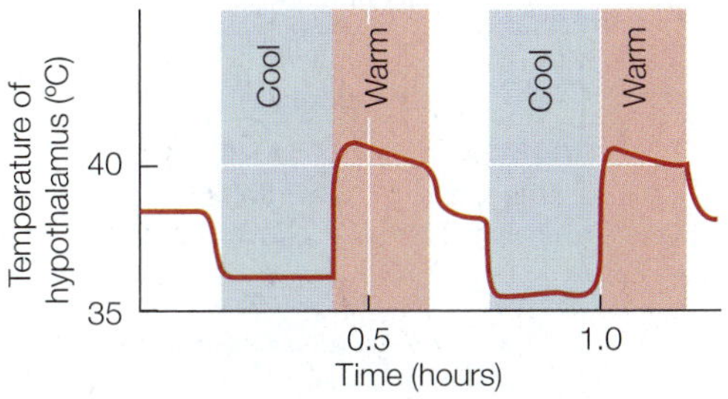

Temperature of hypothalamus (°C)
40
35
Cool
Warm
Cool
Warm
0.5
1.0
Time (hours)

3. Measure the animal's metabolic rate and body temperature throughout the period of hypothalamic manipulation.

Results

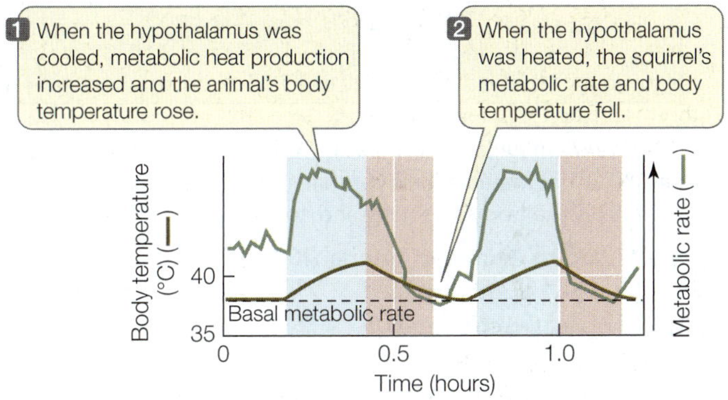

1 When the hypothalamus was cooled, metabolic heat production increased and the animal's body temperature rose.
2 When the hypothalamus was heated, the squirrel's metabolic rate and body temperature fell.
Body temperature (°C) (—)
Metabolic rate (—)
40
35
Basal metabolic rate
0
0.5
1.0
Time (hours)

CONCLUSION The ground squirrel's hypothalamus acts as a thermostat. When cooled it activates metabolic heat production; when warmed, it suppresses metabolic heat production and favors heat loss.

Go to **BioPortal** for discussion and relevant links for all INVESTIGATING**LIFE** figures.

[a]Heller, H. C. and G. W. Colliver. 1974. *American Journal of Physiology* 227: 583–589.

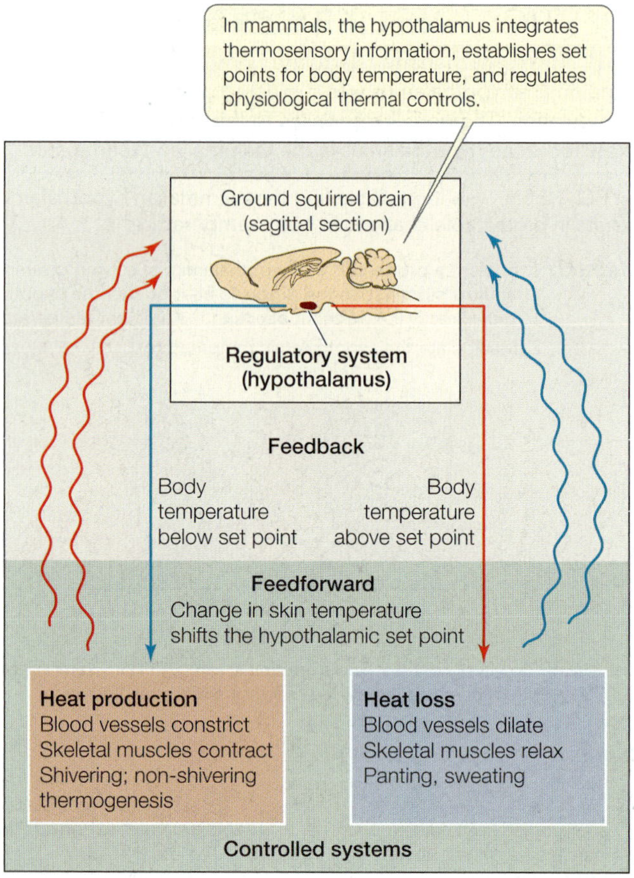
In mammals, the hypothalamus integrates thermosensory information, establishes set points for body temperature, and regulates physiological thermal controls.
Ground squirrel brain (sagittal section)
Regulatory system (hypothalamus)
Feedback
Body temperature below set point
Body temperature above set point
Feedforward
Change in skin temperature shifts the hypothalamic set point
Heat production
Blood vessels constrict
Skeletal muscles contract
Shivering; non-shivering thermogenesis
Heat loss
Blood vessels dilate
Skeletal muscles relax
Panting, sweating
Controlled systems

40.20 The Mammalian Thermostat Like the home thermostat in Figure 40.2, the mammalian hypothalamus is the regulatory system that controls the body's heating and cooling mechanisms.

Go to Animated Tutorial 40.1
The Hypothalamus
Life10e.com/at40.1

Hypothalamic set points are higher during wakefulness than during sleep, and they are higher during the active part of the daily cycle than the inactive part, even if the animal is awake at both times. Even when an endotherm is kept under constant environmental conditions, its body temperature displays a daily cycle of changes in set point. This kind of cycle is controlled by an internal biological clock, as we will discuss in Chapter 53.

Fever helps the body fight infections

Fever is an adaptive response that helps the body fight pathogens. A fever is a rise in body temperature in response to molecules called **pyrogens**. Pyrogens can be exogenous, such as foreign proteins produced by bacteria or viruses that invade the body, or endogenous, including substances produced by cells of the immune system in response to infection.

The presence of a pyrogen causes a rise in the body's hypothalamic set point for metabolic heat production. As a result, you shiver, put on a sweater, or crawl under a blanket, and your body temperature rises until it matches the new set point. At the higher body temperature you no longer feel cold, and you may not feel hot, but someone touching your forehead will say that you are "burning up." Taking aspirin lowers your set point to normal. Now you feel hot, take off clothes, and even sweat until your elevated body temperature returns to normal. Although modest fevers help the body fight infections, extreme fevers can be dangerous and must be controlled, usually with fever-reducing drugs.

Some animals conserve energy by turning down the thermostat

Hypothermia is a below-normal body temperature. It can result from starvation (lack of metabolic fuel), exposure to extreme cold, serious illness, or anesthesia. In each of these cases, the drop in body temperature is unregulated. However, many birds and mammals undergo *regulated* hypothermia to survive periods of cold and food scarcity.

Hummingbirds, for example, are very small endotherms with a high metabolic rate, and going even a single day without food could exhaust their metabolic reserves. Hummingbirds and other small endotherms can extend the period over which they can survive without food by dropping their body temperature 10°C to 20°C during the portion of day or night when they are normally inactive, thus lowering their metabolic rate and conserving energy. This adaptive hypothermia is called **daily torpor**.

Regulated hypothermia that lasts for days or even weeks, during which the body temperature falls close to the ambient temperature, is called **hibernation** (**Figure 40.21**). Many species of mammals, including bats, bears, marmots, and ground squirrels, hibernate, but only one species of bird (the poorwill) has been shown to hibernate. The metabolic rate needed to sustain a hibernating animal may be only one-fiftieth its basal metabolic rate, and many hibernating animals maintain body temperatures close to the freezing point. Arousal from hibernation occurs when the hypothalamic set point returns to the normal level. The ability of animals to enter daily torpor or deep hibernation to reduce their thermoregulatory set point so dramatically probably evolved as an extension of the set point decrease that accompanies sleep in all mammals and birds.

RECAP 40.5

Within the thermoneutral zone, an endotherm controls its body temperature principally by passive means. Above or below the thermoneutral zone, an endotherm must expend considerable metabolic energy to control its body temperature. Thermoregulatory responses in mammals are regulated by the hypothalamus.

- Describe how endotherms produce heat. **See pp. 827–828 and Figure 40.17**
- Why is dependence on evaporative water loss a dangerous strategy for dealing with hot environments? **See p. 829**
- What is the nature of the negative feedback information and feedforward information used by the mammalian thermostat? **See pp. 829–830 and Figure 40.20**

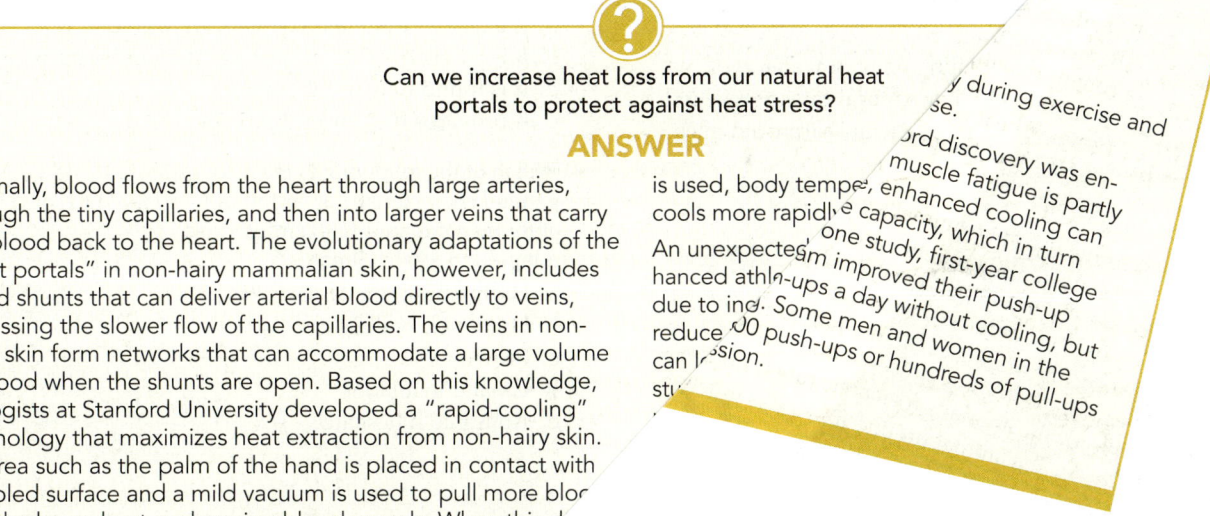

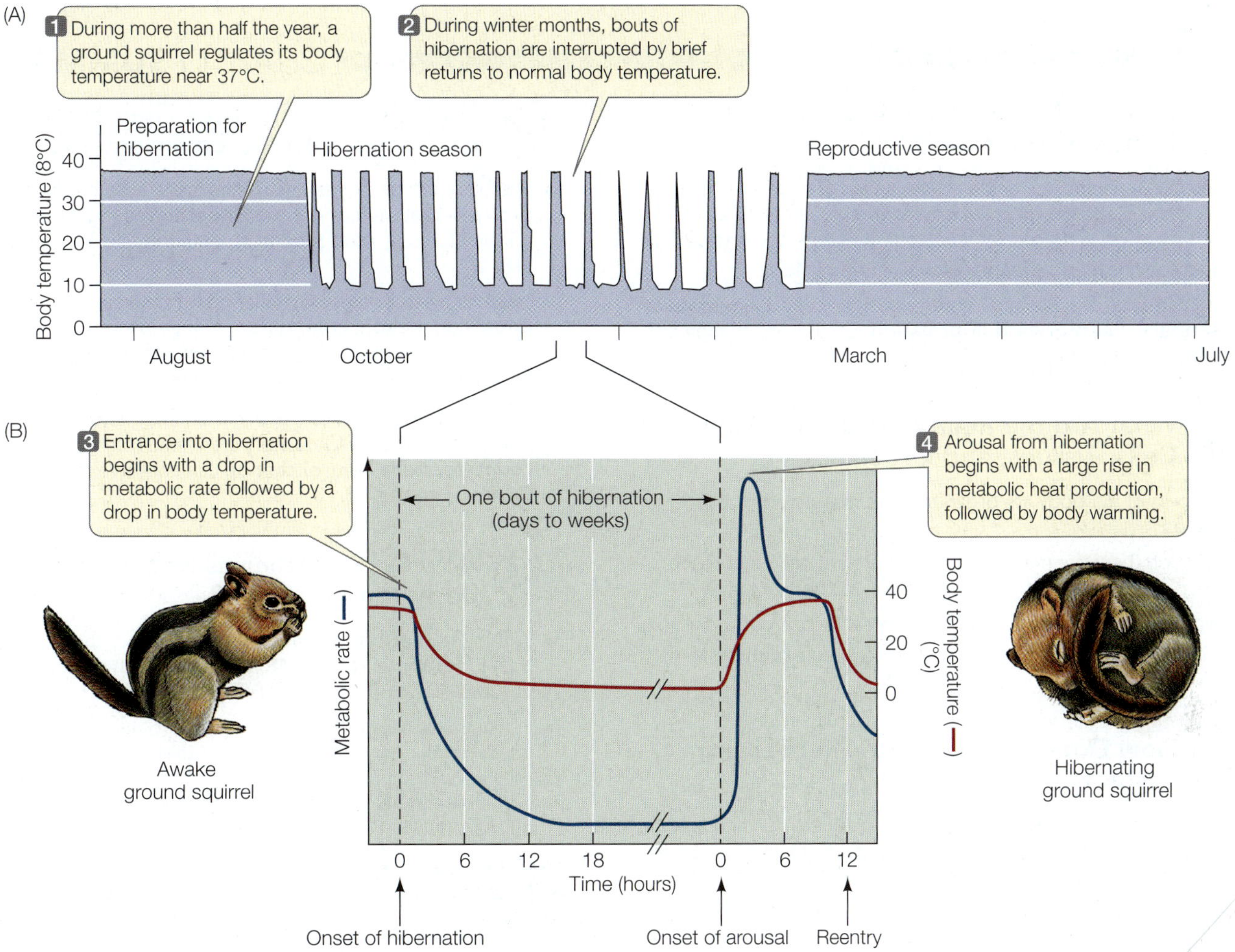
(A)
1 During more than half the year, a ground squirrel regulates its body temperature near 37°C.
2 During winter months, bouts of hibernation are interrupted by brief returns to normal body temperature.
Preparation for hibernation
Hibernation season
Reproductive season
Body temperature (8°C)
40
30
20
10
0
August
October
March
July
(B)
3 Entrance into hibernation begins with a drop in metabolic rate followed by a drop in body temperature.
4 Arousal from hibernation begins with a large rise in metabolic heat production, followed by body warming.
One bout of hibernation (days to weeks)
Metabolic rate (—)
Body temperature (—) (°C)
40
20
0
Awake ground squirrel
Hibernating ground squirrel
0
6
12
18
0
6
12
Time (hours)
Onset of hibernation
Onset of arousal
Reentry

40.21 Hibernation Patterns in a Ground Squirrel During most of the year, the ground squirrel regulates its body temperature around 37°C. During winter months, however, these animals hibernate in underground burrows, living off stored fuel (in the form of either fat or cached food). The metabolic demands for these stored fuels are decreased by bouts of torpor, during which body temperature drops close to that of the environment for long periods of time.

Can we increase heat loss from our natural heat portals to protect against heat stress?

ANSWER

Normally, blood flows from the heart through large arteries, through the tiny capillaries, and then into larger veins that carry the blood back to the heart. The evolutionary adaptations of the "heat portals" in non-hairy mammalian skin, however, includes gated shunts that can deliver arterial blood directly to veins, bypassing the slower flow of the capillaries. The veins in non-hairy skin form networks that can accommodate a large volume of blood when the shunts are open. Based on this knowledge, biologists at Stanford University developed a "rapid-cooling" technology that maximizes heat extraction from non-hairy skin. An area such as the palm of the hand is placed in contact with a cooled surface and a mild vacuum is used to pull more bloo into the large, heat-exchanging blood vessels. When this d

is used, body tempe
cools more rapidl
An unexpecte
hanced athl
due to inc
reduce
can l
st

CHAPTER SUMMARY 40

40.1 How Do Multicellular Animals Supply the Needs of Their Cells?

- Multicellular animals provide for the needs of all their cells by maintaining a stable **internal environment.** That environment consists of two extracellular fluid compartments: the **interstitial fluid** and the blood **plasma**. Review Figure 40.1
- Regulation of physiological systems is mostly through **negative feedback**. **Feedforward information** functions to change **set points**. Review Figure 40.2

40.2 What Are the Relationships between Cells, Tissues, and Organs?

- The cells of the body are organized into assemblages called **tissues**.
- Although there are many cell types, there are only four tissue types: **epithelial**, **muscle**, **connective**, and **neural** tissues.
- **Organs** are made up of tissues, and most organs contain all four types of tissues. Organs are grouped into **organ systems**. Review Figure 40.7

40.3 How Does Temperature Affect Living Systems?

- Life is possible only within a narrow range of environmental temperatures. Q_{10} is a measure of the sensitivity of a life process to temperature. A Q_{10} of 2 means that the reaction rate of that process doubles as temperature increases by 10°C. Review Figure 40.8
- Animals can acclimatize to seasonal changes in temperature ...ough biochemical and physiological adaptations. Review ...re 40.9

40.4 How Do Animals Alter Their Heat Exchange with the Environment?

- The body temperatures of **ectotherms** are determined primarily by external sources of heat. **Endotherms** regulate their body temperatures by producing heat metabolically. Review Figure 40.10
- The four avenues of heat exchange with the environment are **radiation**, **convection**, **conduction**, and **evaporation**. The balance between heat production and heat exchange can be expressed as an **energy budget**. Review Figure 40.12
- Control of blood flow to the skin is an important means of temperature regulation. Circulatory system adaptations such as **countercurrent heat exchange** can conserve metabolic heat. Review Figures 40.13, 40.14

40.5 How Do Endotherms Regulate Their Body Temperatures?

- Within the **thermoneutral zone**, resting endotherms have a **basal metabolic rate** (**BMR**) that scales with body size. Review Figures 40.16, 40.17, ACTIVITY 40.1
- In mammals, control of body temperature relies on commands from a regulatory center in the **hypothalamus**. This thermostat uses its own temperature as negative feedback information and skin temperature as feedforward information. Review Figure 40.20, ANIMATED TUTORIAL 40.1

Go to the Interactive Summary to review key figures, Animated Tutorials, and Activities Life10e.com/is40

C... REVIEW

...ERING

1. Wh... life c... ...wing best describes the limits to
   a. Th...
   b. The...
   c. The f... protei... and the boiling point of water
   d. The abi... dissipate... ture on biochemical rates of reaction ... ater and the temperature at which
   e. The effect... environme... to actively produce and
2. If the Q_{10} of an a... oxygen content of the ... is 2, then
   a. the animal is ... environment th...
   b. the animal is an ...
   c. the animal consu... per hour at 20°C a... ld
   d. the animal's metabo...
   e. the animal produces ... does at 30°C. ...it

3. Which statement about brown fat is true?
   a. It produces heat without producing much ATP.
   b. It insulates animals acclimatized to cold.
   c. It is a major source of heat production for birds.
   d. It is found only in hibernators.
   e. It provides fuel for muscle cells.
4. Which of the following is the most important and most general characteristic of endotherms adapted to cold climates compared with endotherms adapted to warm climates?
   a. Higher basal metabolic rates
   b. Higher Q_{10} values
   c. Brown fat
   d. Greater insulation
   e. Ability to hibernate

5. Which of the following would cause a decrease in the hypothalamic temperature set point for metabolic heat production?
   a. Entering a cold environment
   b. Taking an aspirin when you have a fever
   c. Arousing from hibernation
   d. Getting an infection that causes a fever
   e. Cooling the hypothalamus

UNDERSTANDING & APPLYING

6. What is the advantage of feedforward information for homeostasis? Can you suggest what some sources of feedforward information could be for regulation of breathing, blood pressure, and secretion of digestive juices?

7. Newton's law of cooling describes how a physical object comes into thermal equilibrium with its environment. This law can be expressed as

$$HL = K(T_o - T_a)$$

where HL is the rate of heat loss, K is the thermal conductance constant (how easily an object loses heat), T_o is the temperature of the object, and T_a is the ambient temperature. Apply this expression to the metabolic rate/environmental temperature curve for endotherms (see Figure 40.17). What would be the equivalent of HL? In Newton's law of cooling, K is a constant reflecting the properties of the object. What would K represent for an endotherm? What would 1/K represent? Using a version of Newton's law that replaces T_o with T_b (body temperature), explain why the curve projects to zero at an ambient temperature that equals body temperature.

8. What do you expect to be the effect of temperature on the ability of a heart to pump blood and the ability of skeletal muscle to contract? How does this create a physiological challenge for a great white shark or a giant bluefin tuna?

ANALYZING & EVALUATING

The following data from the White and Seymour paper cited in the Working with Data exercise (p. 827) should be used to answer Questions 9 and 10.

Species	Body mass (kg)	Heart size (g)	BMR (ml O_2/hr)
Mouse	0.03	0.2	63
Rat	0.4	1.5	146
Muskrat	1.0	8	640
Marmot	4	24	1,550
Coyote	10	60	2,690
Baboon	17	100	5,150

9. Plot the data for (1) BMR versus body mass and (2) heart mass versus body size. Describe the differences between these two curves.

10. The BMR is supported by the amount of blood pumped by the heart, and the amount of blood the heart can pump depends on its size (volume). What other factor would you propose to explain the differences between these two curves?

11. The observations on the Galápagos marine iguana in Figure 40.13 showed that this animal's body temperature rose faster in air than it fell in water. The inference was that the iguana was influencing its gain or loss of heat by altering the blood flow to its skin. However, the thermal properties of air and water are different, and the animal was breathing when in air, but not while swimming. In terms of the energy budget

$$\underbrace{\text{metabolism} + R_{abs}}_{\text{heat}_{in}} = \underbrace{R_{out} + \text{convection} + \text{conduction} + \text{evaporation}}_{\text{heat}_{out}}$$

what factors other than blood flow to the skin could influence these rates of heat exchange, and what experiment could you do to strengthen the argument that changes in blood flow to the skin were critical variables?

Go to BioPortal at **yourBioPortal.com** for Animated Tutorials, Activities, LearningCurve Quizzes, Flashcards, and many other study and review resources.

41 Animal Hormones

CHAPTER OUTLINE

Exercised Muscles Secrete a Hormone Research on mice has shown that irisin, a hormone produced by exercising muscles, travels to fat cells in the body, where it stimulates molecular changes that make the adipose tissue more metabolically active.

IF YOU ARE a person who eats a lot but never puts on weight, it is likely that you also exercise regularly. But do you stay thin simply because exercising burns the excess calories that would otherwise be stored as fat? We learned in earlier chapters that typical adipose tissue, or "white" fat, stores lipids, whereas "brown" fat metabolizes lipids to produce heat without producing ATP. Brown fat is known to be present in cold-acclimated rodents, hibernators, and newborn humans, but it was not thought to be present in adult humans. Recently, however, imaging techniques revealed brown fat activity in cold-exposed adults. What was interesting was that the amount of brown fat was inversely proportional to total body mass—lean people had more brown fat and obese people had less. It was suggested that the excess calories burned by the brown fat were responsible for low body mass.

A recently discovered signaling molecule may explain the difference in individual propensity to put on weight. The molecule has been classified as a hormone and given the name irisin, after the mythological character Iris. Irisin was discovered in a strain of mice bred for increased exercise endurance capacity, and it was shown that the molecule is released from muscles. It has been well documented that exercise training causes numerous structural and metabolic changes in muscle. But exercise training also improves many other aspects of health. How are these changes mediated? Analysis of fat tissue from the super athletic mice showed a remarkable finding: their white fat had properties of brown fat. It was then shown that this "browning" of the white fat was triggered by a blood-borne chemical signal produced by the exercising muscles. Thus the muscle is telling the white fat to change its properties to a tissue that is more metabolically active, burns more calories, and produces more heat. So exercise improves the condition of muscles, and it also causes the muscles to talk to adipose tissues, telling them to "shape up."

Why did the scientists call their signaling molecule a hormone and choose to name it after Iris? As we will discuss in this chapter, a hormone is a chemical message that circulates in the blood and activates distant target cells. In Greek mythology, Iris was a messenger of the gods, traveling the world with the speed of the wind.

> **?** How can we demonstrate that a molecule found in the blood is a hormone?
>
> See answer on p. 853.

41.1 What Are Hormones and How Do They Work?

In multicellular animals, physiological regulatory systems require information and cell-to-cell communication. Most intercellular communication is by means of chemical signals that bind to receptors, as described in Chapter 7. Examples of chemical signals include hormones, growth factors and morphogens, cytokines, and neurotransmitters. These signals operate in different contexts—the endocrine system, growth and development, the immune system, and the nervous system—but the principles of their function are the same: one cell releases a chemical signal that travels to and binds to a receptor on a second cell (the "target"), causing a response in the target cell.

Not all signaling is chemical; keep in mind that there are receptors in the nervous system that encode information from external physical sources, such as temperature, pressure, and light. And the nervous system uses electric signals called action potentials to get information from place to place in the body. Regardless of the system, the processing of information depends on which cells have receptors for the signals and how those target cells respond to the signals.

Some analogies might help distinguish how the immune, nervous, and endocrine informational systems operate. The immune system (the topic of Chapter 42) operates like an army of private security guards. The various cellular agents make their rounds of the body, and if they detect a security breach, they sound their alarms—cytokines—that activate the body's defenses. The nervous system (see Chapters 45–47) operates like a landline telephone system, with a central integration and command center that sends signals along specific wires to specific receivers. This chapter focuses on the endocrine system, which is more like a radio or television network, broadcasting signals that can be picked up by anyone who has an appropriate receiver that is turned on and tuned in.

Endocrine signaling can act locally or at a distance

The many and varied types of **endocrine cells** produce and release chemical signals directly into the extracellular fluid (ECF). These molecules may then diffuse through the ECF and enter the bloodstream. Endocrine signals that enter the blood are called **hormones**, and hormones activate target cells far from their site of release (Figure 41.1A). Irisin is an example of a hormone. You probably are familiar with several other hormones, such as testosterone, estrogen, adrenaline, and insulin.

Some endocrine signals are released in such tiny quantities, or are so rapidly inactivated by enzymes, or are taken up so efficiently by local cells that they never diffuse into the blood in sufficient amounts to act on distant cells (Figure 41.1B). Because these signals affect only target cells near their release site, they are called **paracrines** (*para*, "near"). An example of a paracrine signal is histamine, one of the mediators of inflammation. The most local action an endocrine signal can have is when it binds to receptors on or in the same cell that secreted it. When a chemical signal influences the cell that secreted it, it is called an **autocrine**. Hormones and paracrines can have

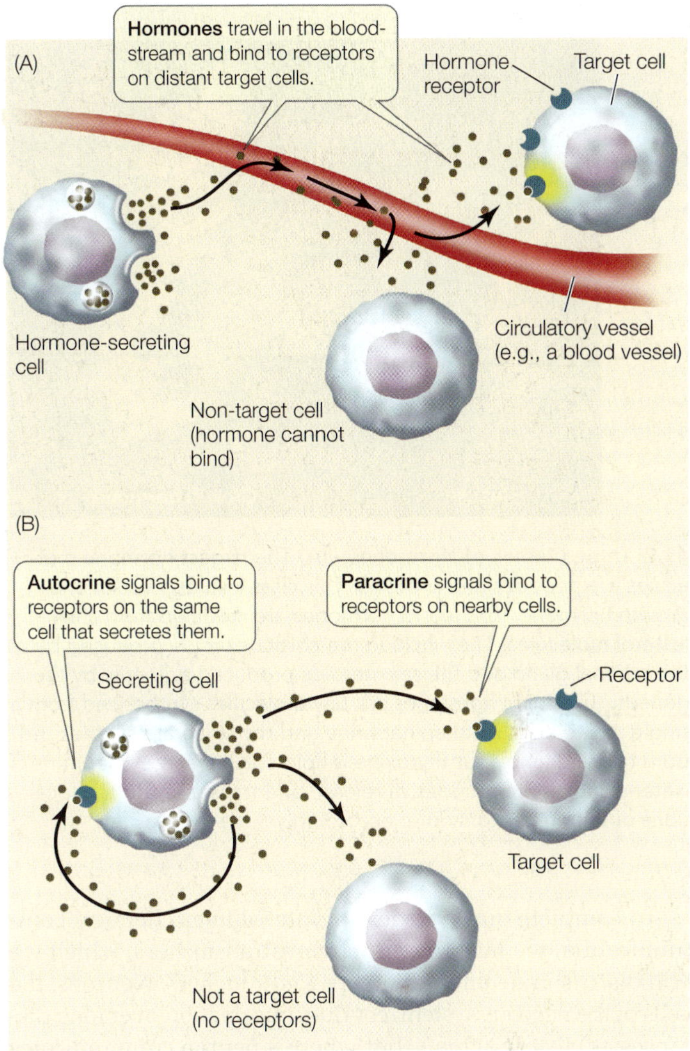
(A)
Hormones travel in the bloodstream and bind to receptors on distant target cells.
Hormone receptor
Target cell
Hormone-secreting cell
Circulatory vessel (e.g., a blood vessel)
Non-target cell (hormone cannot bind)
(B)
Autocrine signals bind to receptors on the same cell that secretes them.
Paracrine signals bind to receptors on nearby cells.
Secreting cell
Receptor
Target cell
Not a target cell (no receptors)

41.1 Chemical Signaling Systems (A) Hormones are distributed throughout the body in the bloodstream. (B) Paracrines and autocrines do not enter the blood. Paracrines simply diffuse to nearby cells, and autocrines influence the same cells that release them.

autocrine functions as a means of providing negative feedback to control their own rates of secretion.

Some endocrine cells exist as single cells within a tissue. Hormones of the digestive tract, for example, are secreted by isolated endocrine cells in the walls of the stomach and small intestine. Many hormones are secreted by aggregations of endocrine cells in secretory organs called **endocrine glands**. A single endocrine gland may secrete multiple hormones.

The name "endocrine" (Greek, "separated within") reflects the fact that the substances secreted by these cells are secreted directly into the "internal environment," the extracellular fluid (see Section 40.1). From the ECF they can diffuse locally or enter the bloodstream. **Exocrine glands**, in contrast, have ducts through which the products of exocrine cells are carried to an *external* environment such as the surface of the skin or a body cavity such as the gut (sweat and salivary glands are examples).

(A) Protein hormones

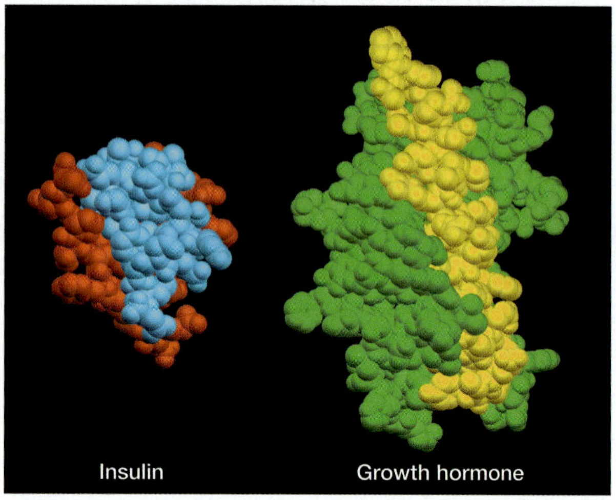
Insulin
Growth hormone

(B) Steroid hormones

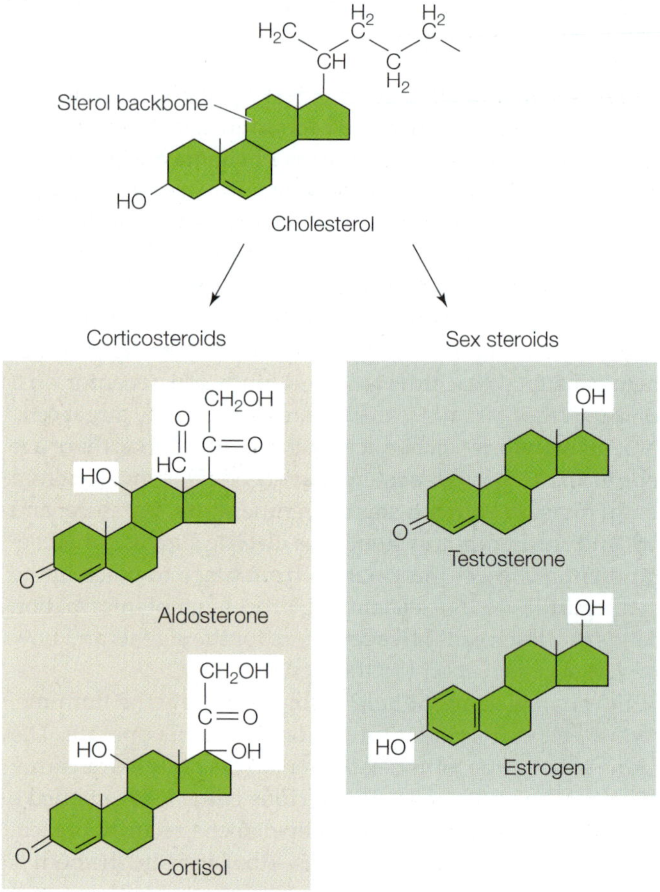
Sterol backbone
HO
Cholesterol
Corticosteroids
Sex steroids
CH2OH
O
C=O
HC
HO
O
Aldosterone
CH2OH
C=O
OH
HO
O
Cortisol
OH
O
Testosterone
OH
HO
Estrogen

(C) Amine hormones

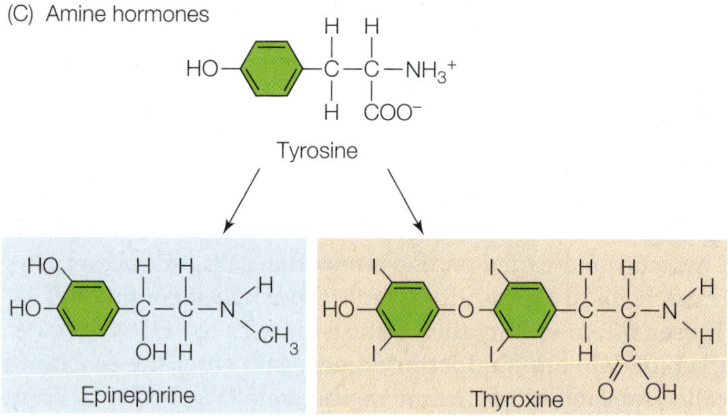
HO
H H
C C NH3+
H COO−
Tyrosine
HO
HO
OH H
CH3
Epinephrine
I
O
Thyroxine
OH

41.2 Three Classes of Hormones (A) The largest hormone molecules are peptides and proteins. This class includes insulin and growth hormone. (B) Steroid hormones are modified from cholesterol molecules. They include the corticosteroids produced by the adrenal gland and the sex steroids produced primarily by the gonads. (C) Amine hormones are tiny molecules synthesized from a single amino acid. Both epinephrine and thyroxine are synthesized from tyrosine units, but thyroxine is lipid-soluble and epinephrine is water-soluble. Their modes of release and transport and the locations of their receptors differ accordingly.

To complete our overview of intercellular chemical communication, we must mention neurotransmitters, which we will discuss in detail in Chapters 45 through 47. Neurons, the cells of the nervous system, conduct information over long distances as electric signals, but where a neuron communicates that information to another cell, it does so by releasing chemical signals called neurotransmitters. Most neurotransmitters act very locally (frequently on the same neuron that released them). Some, however, diffuse into the blood and act on distant targets; neurotransmitters that act in this manner are referred to as **neurohormones**.

Hormones can be divided into three chemical groups

There is enormous diversity in the chemical structure of hormones, but by and large they can be classified into three groups:

- The majority of hormones are peptides or proteins. These hormones (insulin is an example; Figure 41.2A) are water-soluble and thus easily transported in the blood without carrier molecules. Peptide and protein hormones can be packaged in vesicles within the cells that make them, and then released by exocytosis. Their receptors are on cell surfaces.
- **Steroid hormones** (such as estrogen and testosterone) are synthesized from the steroid cholesterol (Figure 41.2B), are lipid-soluble, and pass easily through cell membranes. Steroid hormones diffuse out of the cells that make them and are usually bound to carrier molecules in the blood. Their receptors are mostly intracellular.
- **Amine hormones** are mostly synthesized from the amino acid tyrosine (thyroxine is one example; Figure 41.2C). Some amine hormones are water-soluble and others are lipid-soluble; their modes of release differ accordingly.

Hormone action is mediated by receptors on or within their target cells

Water-soluble hormones cannot pass readily through plasma membranes, so their receptors must be located on the surfaces of target cells. These receptors are large transmembrane glycoprotein complexes with three domains: a binding domain that projects outside the plasma membrane, a transmembrane domain that anchors the receptor in the membrane, and a cytoplasmic domain that extends into the cytoplasm of the cell. When a hormone binds to the binding domain, the cytoplasmic domain initiates the target cell's response. Second messengers activate a cascade of intracellular events, eventually activating protein kinases or protein phosphatases (see Figures 7.6 and 7.7). In most cases these protein kinases and phosphatases activate or inactivate enzymes in the cytoplasm that lead to the cell's response, but the signaling cascade initiated by the receptor can also generate signals that enter the nucleus and alter gene expression (see Figure 7.10).

Lipid-soluble hormones can diffuse through plasma membranes, and therefore their receptors are usually inside cells, in either the cytoplasm or the nucleus (although some membrane-bound receptors for lipid-soluble hormones have recently been described). In most cases, the complex formed by the lipid-soluble hormone and its receptor acts by altering gene expression in the cell's nucleus (see Figure 7.8).

Hormone action depends on the nature of the target cell and its receptors

Wherever a hormone encounters a cell with an appropriate receptor, it can bind to that receptor and trigger a response. The nature of the response depends on the responding cell and its receptors. Thus the same hormone can cause different responses in different types of cells.

Consider the amine hormone adrenaline, or **epinephrine**. Suppose you are walking in the forest and almost step on a rattlesnake. You jump back, your heart starts to thump, and protective reactions are set in motion. The jump and the heart thumping are driven by your rapidly responding nervous system. Simultaneously with these muscular responses, your nervous system stimulates endocrine cells in the adrenal glands just above your kidneys to secrete epinephrine. Within seconds, epinephrine is diffusing into your blood and circulating around your body to activate the many components of the **fight-or-flight response** (Figure 41.3).

Epinephrine binds to receptors in your heart, causing a faster and stronger heartbeat. Your heart is now pumping more blood. Epinephrine also binds to receptors in certain blood vessels. By causing constriction of blood vessels supplying your skin, kidneys, and digestive tract (digesting lunch can wait!), the hormone diverts more blood to the muscles needed for your escape from danger.

Epinephrine binds to cells in the liver, stimulating them to break down glycogen and release its glucose units into the blood as a quick energy supply (see Figure 7.18). In fatty tissue, epinephrine stimulates the breakdown of fats to yield fatty acids—another source of energy. These are just some of the actions triggered by one hormone. In each case the cellular response depends on the cell's receptors and associated intracellular signaling cascade, but they all contribute to increasing your chances of surviving a dangerous situation.

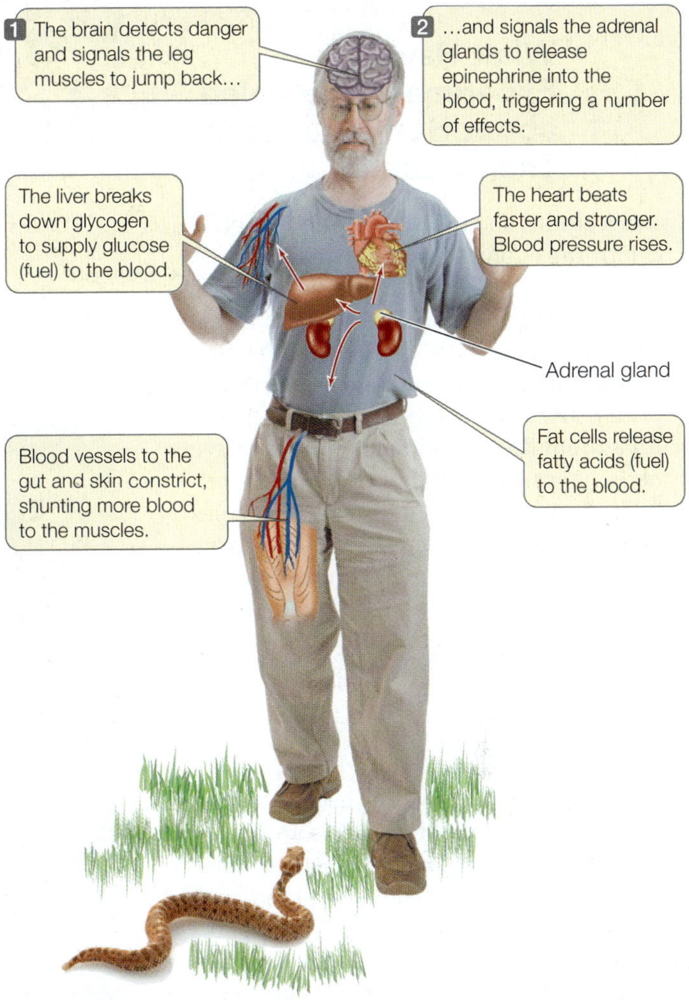
1 The brain detects danger and signals the leg muscles to jump back...
2 ...and signals the adrenal glands to release epinephrine into the blood, triggering a number of effects.
The liver breaks down glycogen to supply glucose (fuel) to the blood.
The heart beats faster and stronger. Blood pressure rises.
Adrenal gland
Fat cells release fatty acids (fuel) to the blood.
Blood vessels to the gut and skin constrict, shunting more blood to the muscles.

41.3 The Fight-or-Flight Response The brain of a person suddenly faced with a threatening situation sends a signal to the adrenal glands, which almost instantaneously release the hormone epinephrine. Epinephrine circulates around the body and induces the various components of the fight-or-flight response in different tissues.

RECAP 41.1

Hormones are chemical signals released by endocrine cells into the extracellular fluid, where they can diffuse into the blood and travel to distant target cells. The receptors for water-soluble hormones are on the surfaces of target cells; receptors for most lipid-soluble hormones are inside the target cells.

- What are the three general chemical categories of hormones? **See p. 836 and Figure 41.2**
- Describe the different methods by which water-soluble and lipid-soluble hormones reach their receptors. **See p. 837**
- How can a single hormone have diverse effects in the body? **See p. 837 and Figure 41.3**

Scientists discovered and elucidated the transformative actions of many hormones before their specific molecular structures were known. Ingenious experimental approaches have been used to show that a chemical released by one cell has effects on other distant cells. Subsequent studies have focused on the nature of the signal and the mechanisms of its actions.

41.2 What Have Experiments Revealed about Hormones and Their Action?

Intercellular chemical signaling was critical for the evolution of multicellularity. A protist, the slime mold *Dictyostelium*, uses a chemical signal (cAMP) to coordinate the aggregation of individual cells to form a multicellular fruiting structure (see Figure 27.18). The least complex of the multicellular animals—the sponges—do not have nervous systems, but they do have intercellular chemical communication. Molecules that diffuse between cells and communicate information have a long evolutionary history in the multicellular lineages. As we discussed in Chapter 37, plant growth is regulated by a variety of hormones, and hormones are identifiable throughout the animal kingdom.

Studying the evolution of hormonal signaling reveals an interesting generalization: the signal molecules themselves are highly conserved. We find the same chemical compounds over broad groups of organisms, although their functions may differ. As organisms have evolved to occupy different environments and have different lifestyles, the same hormone–receptor systems have diversified to serve different functions. A good example of this evolutionary diversification is seen in the hormone prolactin, described in **Figure 41.4**.

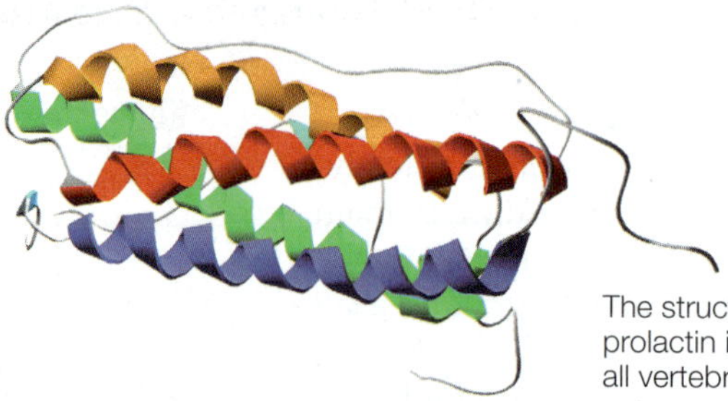

The structure of prolactin is similar in all vertebrate groups.

Fish
Required for osmoregulation in freshwater species. In saltwater species that return to fresh water to spawn (e.g., salmon), prolactin production in adults may play a role in generating the drive to return to natal streams.

Amphibians
In some species, creates a "water drive" that returns adults to breeding locations. Stimulates oviduct development and production of egg jelly in females. In some species, controls development of sexual characteristics.

Birds
In some species, stimulates nesting activity, incubation behavior, and parental care in both sexes. Stimulates the epithelial cells of the upper GI tract to proliferate and slough off to form "crop milk" to nourish the young.

Mammals
In females, stimulates growth of the mammary glands and milk production. In humans, it is responsible for the sensation of sexual gratification as well as the male refractory period following sexual intercourse.

41.4 Prolactin's Structure Is Conserved, but Its Functions Have Evolved The hormone prolactin is found in all vertebrate groups and has a long evolutionary history. Its probable function in early vertebrates was in regulating the body's salt and water balance (osmoregulation). It maintains this function in some species, and has evolved in others to control a number of physiological processes, most of which are associated with reproduction.

The existence and functions of hormones were experimentally demonstrated many years before specific molecules were isolated and identified chemically. That is not surprising when you consider their small size and the tiny amounts of certain hormones that exist in an organism.

The first hormone discovered was the gut hormone secretin

Secretin, a protein released from cells in the gut, stimulates the pancreas to secrete digestive fluids into the gut. At the start of the twentieth century, the prevailing view was that the secretion of digestive juices was controlled by the nervous system. This view was in part the result of the famous work of the Russian physiologist Ivan Pavlov, who discovered the neural control of salivary secretions of the mouth (see Figure 53.1). Pavlov failed, however, when he tried similar experiments with the secretions of the pancreas.

In 1902 William Bayliss and Ernest Starling, working on anesthetized dogs, surgically removed all of the nerves to the pancreas. They then showed that the denervated pancreas still produced secretions when stomach acid was injected into the gut. Then they took tissue from the gut, ground it up, and injected an extract from this preparation into the bloodstream; this injected fluid stimulated pancreatic secretions. Bayliss and Starling's work proved that a chemical extracted from one tissue could travel in the blood to cause a reaction in a different tissue. They called this type of chemical a "hormone," from the Greek word for "impetus."

INVESTIGATINGLIFE

41.5 Muscle Cells Can Produce a Hormone One effect of exercise is the "browning" of white fat (see Figure 40.5D), which increases the fat's metabolic activity and leads to other health benefits. Bruce Spiegelman and his colleagues investigated the possibility that exercised muscles secrete a hormone that changes the characteristics of white fat.[a]

HYPOTHESIS Exercised muscle cells produce a hormone that stimulates browning of fat cells.

Method

1. Two types of muscle cell cultures were prepared. One culture received a treatment that mimicked the effects of exercise on muscle cells.
2. After culture, the muscle cells were removed and their used media (the culture fluid) was added to cultures of developing fat cells.

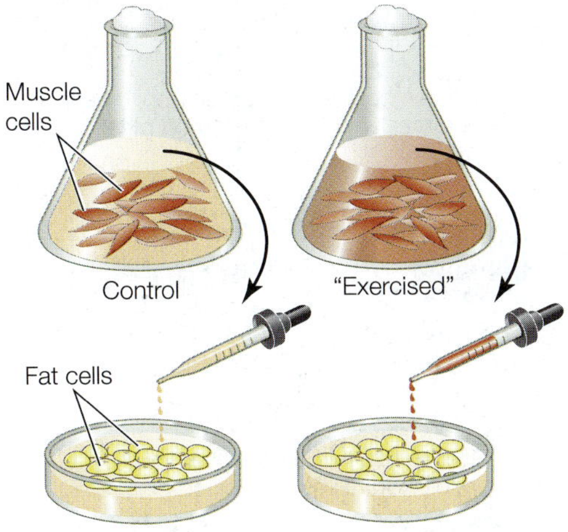
Muscle cells
Control
"Exercised"
Fat cells

Results

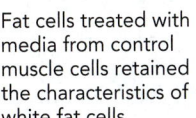

Fat cells treated with media from control muscle cells retained the characteristics of white fat cells.

Fat cells treated with media from "exercised" cells developed properties of brown fat.

CONCLUSION A substance secreted by exercised muscle cells stimulates "browning" of cultured fat.

Go to **BioPortal** for discussion and relevant links for all INVESTIGATING**LIFE** figures.

[a]Boström, P. et al. 2012. *Nature* 481: 463–469.

The discovery of irisin described at the start of this chapter was achieved using modern methods that did not rely on animal subjects (**Figure 41.5**). However, these experiments were similar in principle to those of Bayliss and Starling.

Early experiments on insects illuminated hormonal signaling systems

The British physiologist Sir Vincent Wigglesworth was a pioneer in the study of hormonal action and control in insects. Insects, like all arthropods, have rigid exoskeletons (see Chapter 32). Their growth is therefore episodic, punctuated with molts (shedding of the exoskeleton). Each growth stage between two molts is called an instar. In the 1930s and 1940s Wigglesworth studied the phenomenon of insect molting in a series of experiments on the bloodsucking bug *Rhodnius prolixus*. Newly hatched *Rhodnius* look like miniature adults but lack certain adult features. A juvenile bug molts five times before

WORKING WITH**DATA:**

Identifying a Hormone Secreted by Exercised Muscles

Original Paper

Boström, P. and 17 others. 2012. A PGC1-α-dependent myokine that drives brown-fat-like development of white fat and thermogenesis. *Nature* 481: 463–469.

Analyze the Data

Over the course of many experiments, Bruce Spiegelman and his colleagues demonstrated that exercising mammalian muscle cells secrete a substance (identified as irisin) that they hypothesized was a hormonal signal stimulating white fat cells to develop some of the characteristics of brown fat cells. They cultured muscle cells from wild-type mice (the equivalent of sedentary mice) and from transgenic mice that overexpress irisin (the equivalent of exercising mice). They then took the fluid media from the two culture types and added that media to cultures of white fat cells (see Figure 41.5). They compared the properties of the white fat cells exposed to "exercised" muscle cell media and to "sedentary" muscle cell media.

QUESTION 1

In Experiment 1 (left-hand bars in the figure below), the investigators measured the expression of uncoupling protein 1 (UCP1), a mitochondrial inner membrane protein abundant in brown fat that allows the energy in fat to be converted to heat (see Chapter 9). They then extracted mRNA from treated fat cells and quantified the amount of UCP1 mRNA. Looking at the results, what effect did the "exercised" muscle cell media have on the amount of UCP1 mRNA? Why does this support the idea that exercised muscle cells stimulate "browning" of white fat? In testing the hypothesis that a hormone was involved in stimulating UCP1 expression, why was it important to remove the muscle cells from the conditioned media before adding the media to the fat cells? (See Figure 41.5.)

QUESTION 2

In a second experiment (right-hand bars in the figure), the investigators repeated the procedure in Figure 41.5 but pretreated the culture media with an antibody that recognizes and *inactivates* the molecule of interest (i.e., irisin). What was the effect of pretreating the culture media with antibody? What additional information did this experiment provide?

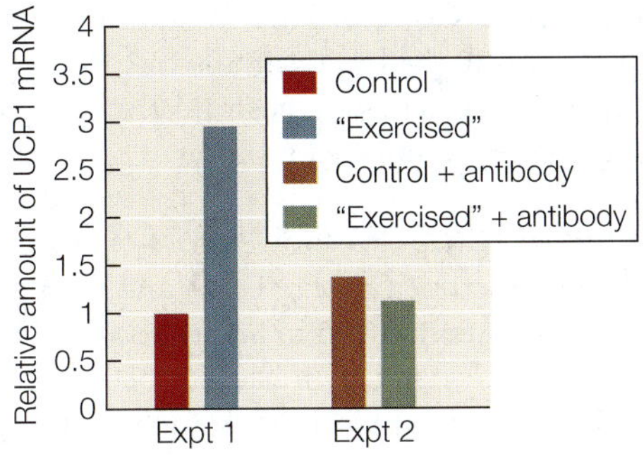
Relative amount of UCP1 mRNA
4
3.5
3
2.5
2
1.5
1
0.5
0
Expt 1
Expt 2
Control
"Exercised"
Control + antibody
"Exercised" + antibody

Go to **BioPortal** for all WORKING WITH**DATA** exercises

developing into a mature adult; a blood meal triggers each episode of molting and growth.

Rhodnius is an amazingly hardy experimental animal—it survives for quite a long time after its head is cut off. Wigglesworth's studies revealed that, if decapitated within an hour after a blood meal, *Rhodnius* can survive for up to a year, but it never molts. If decapitated a week after its blood meal, however, it does molt. Wigglesworth hypothesized that the time lag meant that the substance that triggers molting diffuses slowly from the head. He tested this hypothesis with the experiment described in **Figure 41.6**, which showed that molting was triggered by a substance that diffused throughout the body from a point of origin in the bug's head.

Wigglesworth's experiments yielded another curious result. Regardless of which instar was used, decapitated *Rhodnius* that molted always molted directly into adults, bypassing the usual juvenile instars. Additional experiments demonstrated that a different substance (distinct from that identified by the experiment in Figure 41.6) determines whether a bug molts into another juvenile instar or into an adult.

Because the head of *Rhodnius* is long, it is possible to remove just the front part of it, which contains the brain, while leaving the rear part intact. When fourth-instar bugs that had had a blood meal a week earlier were partly decapitated in this way, they molted into fifth-instar juveniles, not into adults.

This experiment was followed by more experiments in which Wigglesworth used glass tubes to connect individual bugs. When an unfed, completely decapitated fifth-instar bug was connected to a blood-fed, partly decapitated fourth-instar bug (i.e., with only the front part of its head removed), both bugs molted into juvenile forms. A substance from the rear part of the head of the fourth-instar bug prevented both bugs from molting into adults.

INVESTIGATING**LIFE**

41.6 A Diffusible Substance Triggers Molting The blood-sucking bug *Rhodnius prolixus* develops from hatchling to adult in a series of five molts (instars) that are triggered by ingesting blood. Sir Vincent Wigglesworth's experiments demonstrated that a blood meal stimulates production of some molt-inducing substance in the insect's head.[a]

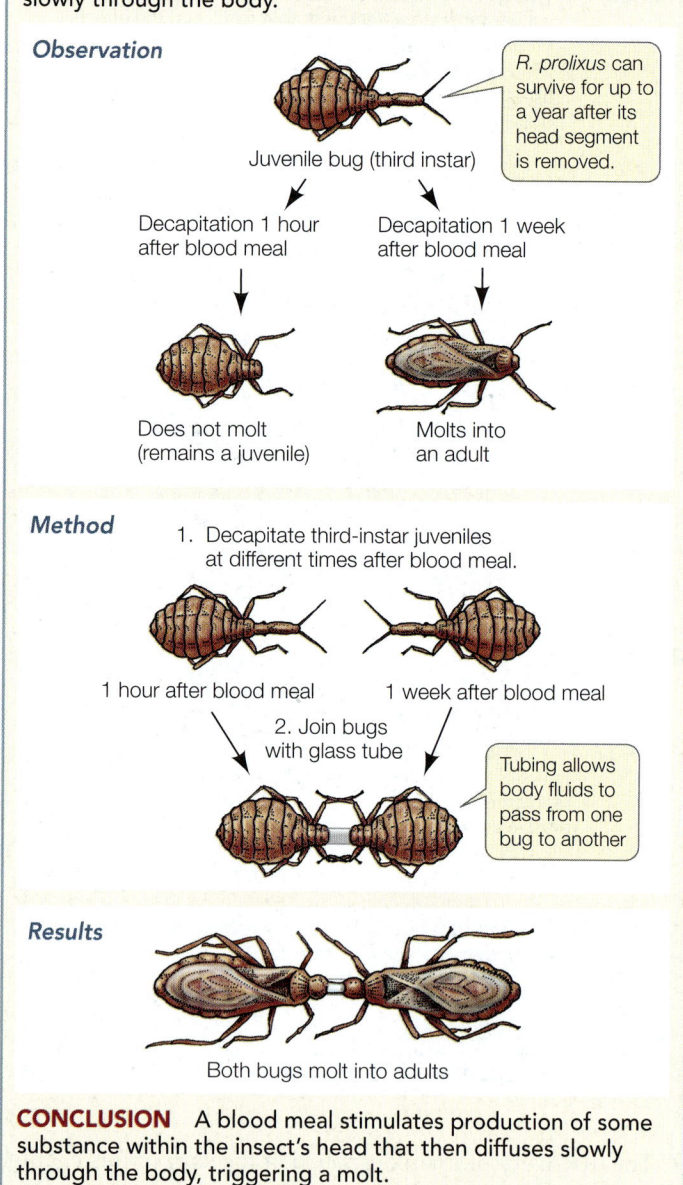

HYPOTHESIS The substance that controls molting in *R. prolixus* is produced in the head segment and diffuses slowly through the body.

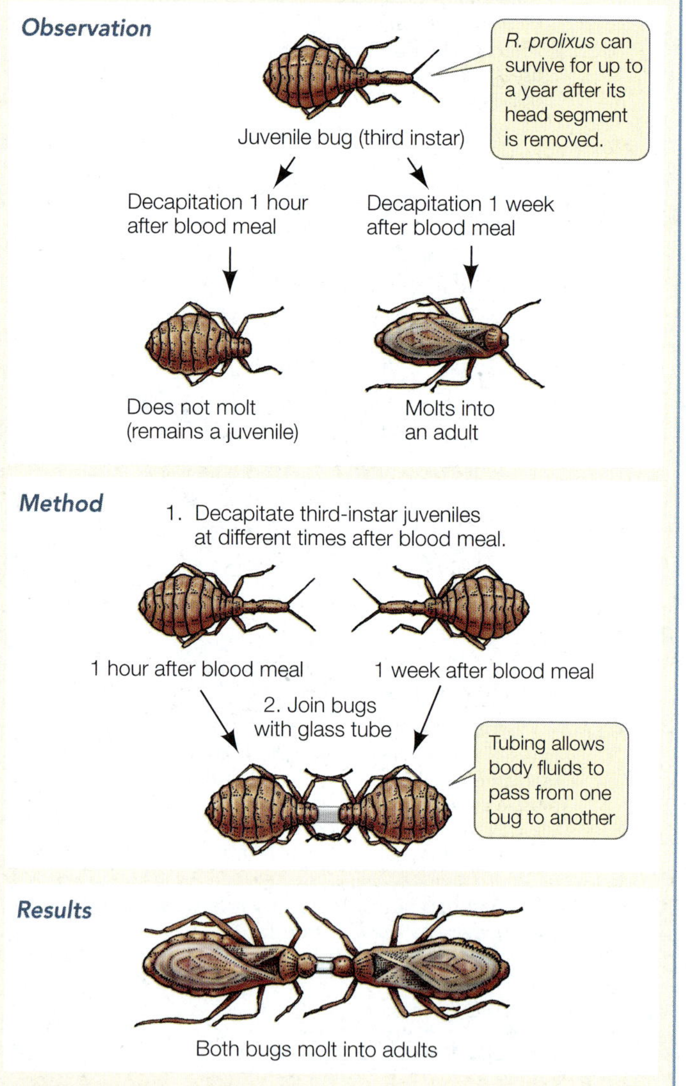
Observation
Juvenile bug (third instar)
R. prolixus can survive for up to a year after its head segment is removed.
Decapitation 1 hour after blood meal
Decapitation 1 week after blood meal
Does not molt (remains a juvenile)
Molts into an adult
Method
1. Decapitate third-instar juveniles at different times after blood meal.
1 hour after blood meal
1 week after blood meal
2. Join bugs with glass tube
Tubing allows body fluids to pass from one bug to another
Results
Both bugs molt into adults

CONCLUSION A blood meal stimulates production of some substance within the insect's head that then diffuses slowly through the body, triggering a molt.

Go to **BioPortal** for discussion and relevant links for all INVESTIGATING**LIFE** figures.

[a]Wigglesworth, V. B. 1934. *Quarterly Journal of Microscopical Science* 77: 191–223.

Three hormones regulate molting and maturation in arthropods

An arthropod's nervous system receives various types of information about the environment (e.g., day length, temperature, social cues, and nutrition) that help determine the optimal timing for the stages of growth and development. When conditions are right, the brain signals the prothoracic gland to produce the hormones that orchestrate physiological processes involved in development and molting.

PTTH AND ECDYSONE We now know that two hormones, prothoracicotropic hormone (PTTH) and ecdysone, work in sequence to regulate molting in arthropods. Cells in the brain produce PTTH, which is why it has also been called "brain hormone." PTTH is transported to and stored in paired structures called the corpora cardiaca attached to the brain (see Figure 41.7) After appropriate stimulation (which for *Rhodnius* is a blood meal), PTTH is released and diffuses through the extracellular fluid to an endocrine gland, the prothoracic gland. PTTH stimulates the prothoracic gland to secrete the hormone ecdysone. Ecdysone diffuses to target tissues and stimulates molting.

Ecdysone is a steroid hormone (see Figure 41.2B) and thus is related to the vertebrate hormones estrogen and testosterone (which also play roles in controlling growth and development). Ecdysone is lipid-soluble and readily passes through the plasma membrane of its target cells (mostly cells of the

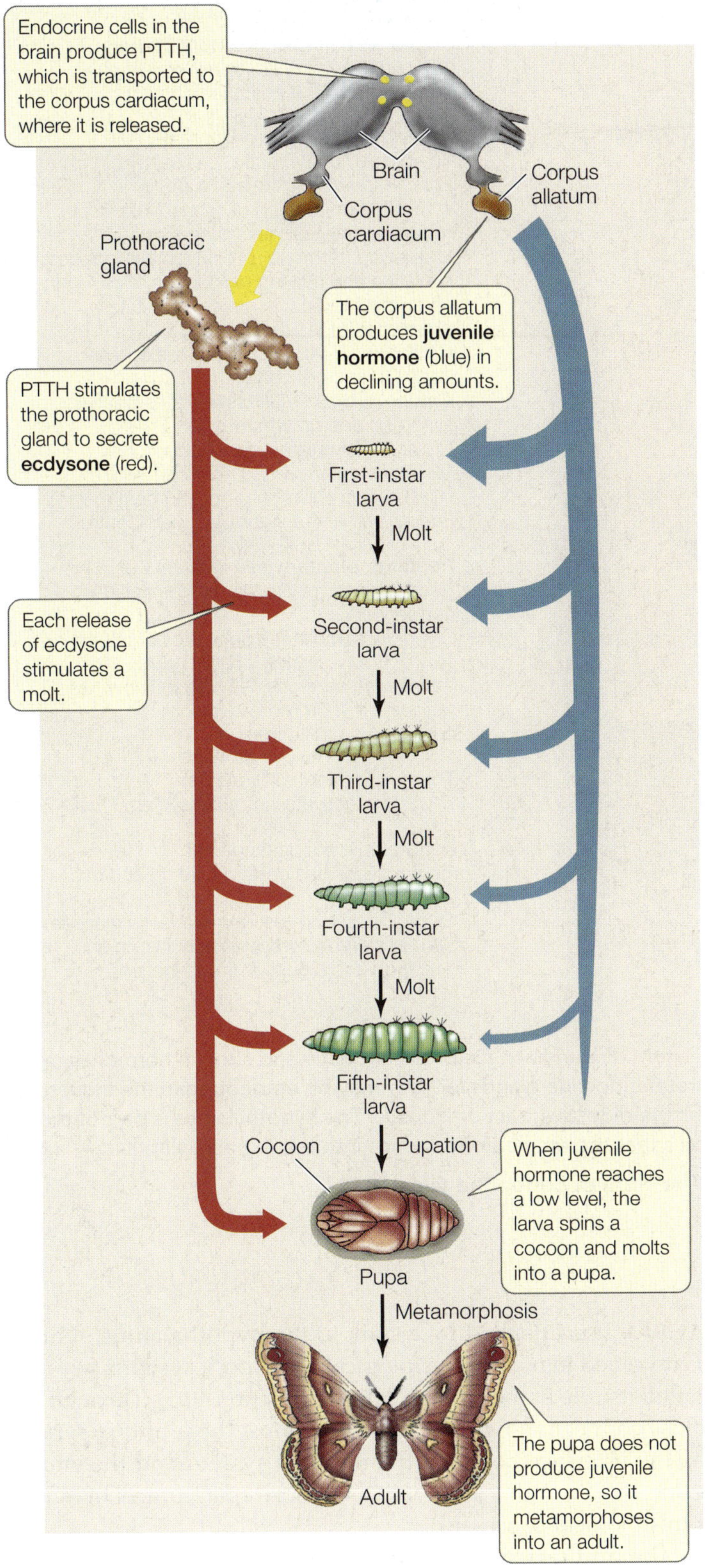
Endocrine cells in the brain produce PTTH, which is transported to the corpus cardiacum, where it is released.
Brain
Corpus cardiacum
Corpus allatum
Prothoracic gland
The corpus allatum produces **juvenile hormone** (blue) in declining amounts.
PTTH stimulates the prothoracic gland to secrete **ecdysone** (red).
First-instar larva
Molt
Each release of ecdysone stimulates a molt.
Second-instar larva
Molt
Third-instar larva
Molt
Fourth-instar larva
Molt
Fifth-instar larva
Cocoon
Pupation
When juvenile hormone reaches a low level, the larva spins a cocoon and molts into a pupa.
Pupa
Metamorphosis
Adult
The pupa does not produce juvenile hormone, so it metamorphoses into an adult.

41.7 Hormonal Control of Metamorphosis Three hormones control molting and metamorphosis in the silkworm moth *Hyalophora cecropia*.

Go to Animated Tutorial 41.1 Complete Metamorphosis
Life10e.com/at41.1

epidermis). In the target cells, ecdysone binds to a receptor that is probably related to the vertebrate testosterone receptor. The resulting hormone–receptor complex induces expression of the genes encoding enzymes involved in digesting the old cuticle and secreting a new one.

JUVENILE HORMONE What about the other substance noted in Wigglesworth's experiments—the one that produced sequential juvenile instars rather than adults? The substance responsible for preventing maturation is now known to be **juvenile hormone**, a molecule that is released continuously from the corpora allata (structures that are attached to the corpora cardiaca, which release PTTH). As long as juvenile hormone is present, *Rhodnius* molts into another juvenile instar. Normally *Rhodnius* stops producing juvenile hormone during the fifth instar, then molts into an adult, a life cycle called incomplete metamorphosis (see Section 32.4).

The role of juvenile hormone is more complex in insects that undergo complete metamorphosis (such as butterflies). These animals undergo dramatic developmental changes in their life cycles. The fertilized egg hatches into a larva, which feeds and molts several times, becoming bigger each time. After a fixed number of molts, it enters an inactive stage called pupation. The pupa undergoes major body reorganization and finally emerges as an adult.

An example of complete metamorphosis is provided by the silkworm moth *Hyalophora cecropia* (**Figure 41.7**). As long as juvenile hormone is present in high concentrations, larvae molt into larger larvae. When the level of juvenile hormone falls, larvae spin cocoons and molt into pupae. Because no juvenile hormone is produced in pupae, they molt into adults. Many modern pesticides use juvenile hormone analogs to prevent larvae developing into adults.

RECAP 41.2

The chemical structure of hormones is highly conserved across many organisms. Early experiments identifying secretin as a hormone defined the characteristics of hormonal signaling in mammals. Early studies on insects revealed much about the nature of hormone action, although the chemical structures of the signaling molecules involved were not established until much later.

- What is meant when we say that hormone molecules are highly conserved? **See p. 838 and Figure 41.4**
- Why did decapitation of *Rhodnius* prevent molting when done 1 day after feeding but not when done 1 week after feeding? **See pp. 840–841 and Figure 41.6**
- What is the role of juvenile hormone in metamorphosis? **See p. 841 and Figure 41.7**

The hormonal control of molting and maturation described above is a general arthropod hormonal control mechanism and exemplifies how the endocrine system works with the nervous system to integrate information and induce long-term effects. The next section will describe similar links between the nervous system and endocrine glands in mammals.

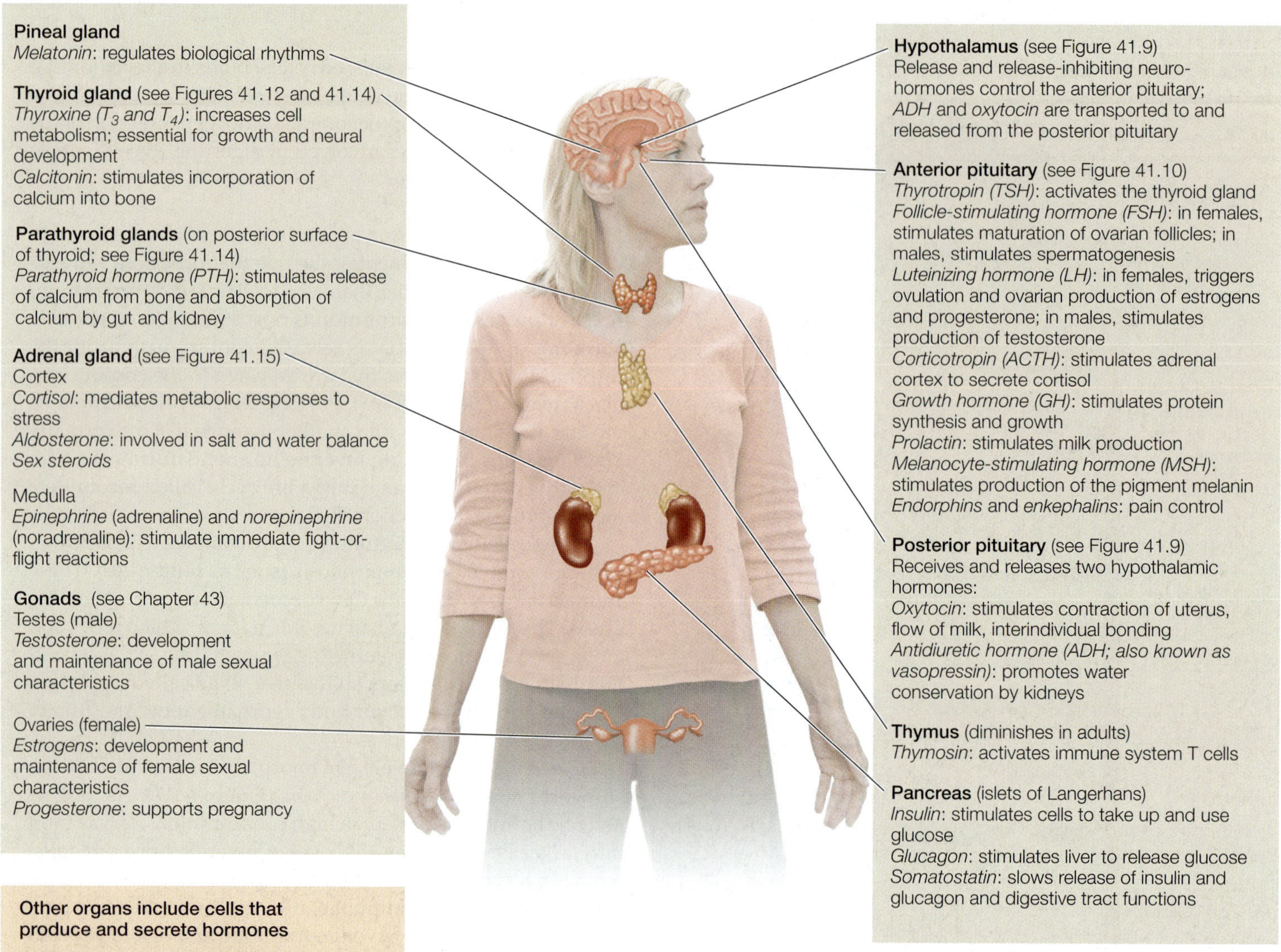

Pineal gland
Melatonin: regulates biological rhythms
Thyroid gland (see Figures 41.12 and 41.14)
Thyroxine (T_3 and T_4): increases cell metabolism; essential for growth and neural development
Calcitonin: stimulates incorporation of calcium into bone
Parathyroid glands (on posterior surface of thyroid; see Figure 41.14)
Parathyroid hormone (PTH): stimulates release of calcium from bone and absorption of calcium by gut and kidney
Adrenal gland (see Figure 41.15)
Cortex
Cortisol: mediates metabolic responses to stress
Aldosterone: involved in salt and water balance
Sex steroids
Medulla
Epinephrine (adrenaline) and *norepinephrine* (noradrenaline): stimulate immediate fight-or-flight reactions
Gonads (see Chapter 43)
Testes (male)
Testosterone: development and maintenance of male sexual characteristics
Ovaries (female)
Estrogens: development and maintenance of female sexual characteristics
Progesterone: supports pregnancy
Hypothalamus (see Figure 41.9)
Release and release-inhibiting neurohormones control the anterior pituitary; *ADH* and *oxytocin* are transported to and released from the posterior pituitary
Anterior pituitary (see Figure 41.10)
Thyrotropin (TSH): activates the thyroid gland
Follicle-stimulating hormone (FSH): in females, stimulates maturation of ovarian follicles; in males, stimulates spermatogenesis
Luteinizing hormone (LH): in females, triggers ovulation and ovarian production of estrogens and progesterone; in males, stimulates production of testosterone
Corticotropin (ACTH): stimulates adrenal cortex to secrete cortisol
Growth hormone (GH): stimulates protein synthesis and growth
Prolactin: stimulates milk production
Melanocyte-stimulating hormone (MSH): stimulates production of the pigment melanin
Endorphins and *enkephalins*: pain control
Posterior pituitary (see Figure 41.9)
Receives and releases two hypothalamic hormones:
Oxytocin: stimulates contraction of uterus, flow of milk, interindividual bonding
Antidiuretic hormone (ADH; also known as vasopressin): promotes water conservation by kidneys
Thymus (diminishes in adults)
Thymosin: activates immune system T cells
Pancreas (islets of Langerhans)
Insulin: stimulates cells to take up and use glucose
Glucagon: stimulates liver to release glucose
Somatostatin: slows release of insulin and glucagon and digestive tract functions

Other organs include cells that produce and secrete hormones

Organ	Hormone
Adipose tissue	Leptin
Heart	Atrial natriuretic peptide
Kidney	Erythropoietin
Stomach	Gastrin, ghrelin
Intestine	Secretin, cholecystokinin
Liver	Somatomedins, insulin-like growth factors

41.8 The Endocrine System of Humans Cells that produce and secrete hormones may be organized into discrete endocrine glands, or they may be embedded in the tissues of other organs, such as the digestive tract or kidneys. The hypothalamus is part of the brain, but it includes cells that secrete neurohormones into the extracellular fluid.

Go to Activity 41.1 The Human Endocrine Glands
Life10e.com/ac41.1

41.3 How Do the Nervous and Endocrine Systems Interact?

The list of hormones known to exist is long and growing longer (as shown by the recent identification of irisin). To make the subject manageable, we will focus primarily on information about the endocrine system of humans (**Figure 41.8**), much of which is applicable to other mammals. We will begin by considering the hormones involved in the integration of nervous system and endocrine system functions.

The pituitary is an interface between the nervous and endocrine systems

The **pituitary gland** sits in a depression at the bottom of the skull, just over the back of the roof of the mouth (**Figure 41.9A**). It is attached by a stalk to the hypothalamus, which is involved in many physiological regulatory systems (we detailed its role in thermoregulation in Section 40.5). Through its close connection with the hypothalamus, the pituitary serves as the interface between the nervous system and the endocrine system and is involved in the hormonal control of many physiological processes.

The pituitary has two parts with different developmental origins. The **anterior pituitary** originates as an outpocketing of the roof of the embryonic mouth cavity, whereas the **posterior pituitary** originates as an outpocketing of the floor of the developing brain. Thus the anterior pituitary originates from gut epithelial tissue and the posterior pituitary from neural tissue. Both parts interact with the nervous system but in different ways. The anterior pituitary contains endocrine cells controlled by neurohormones secreted by the hypothalamus. The posterior pituitary contains axons from hypothalamic neurons.

Go to Animated Tutorial 41.2
The Hypothalamic–Pituitary–Endocrine Axis
Life10e.com/at41.2

THE POSTERIOR PITUITARY Long axons extend into the posterior pituitary from neurons in the hypothalamus. The ends, or terminals, of those axons release two neurohormones, antidiuretic hormone and oxytocin (**Figure 41.9B**). These neurohormones are packaged in vesicles that are transported down the axons. The vesicles are stored in the nerve terminals until an action potential stimulates their release.

The main action of **antidiuretic hormone** (**ADH**) in mammals and birds is to increase the amount of water conserved by the kidneys. When ADH secretion is high, the kidneys produce only a small volume of highly concentrated urine. When ADH secretion is low, the kidneys produce a large volume of dilute urine. The posterior pituitary increases its release of ADH when blood pressure falls or the blood becomes too salty. ADH is also known as **vasopressin** because at high concentrations it causes the constriction of peripheral blood vessels as a means of elevating blood pressure.

When a woman is about to give birth, her posterior pituitary releases **oxytocin**, which stimulates the uterine contractions that deliver the baby. Oxytocin also brings about the flow of milk from the mother's breasts. The baby's suckling stimulates neurons in the mother's brain that cause the secretion of oxytocin. Even the sound of a baby can cause a nursing mother to secrete oxytocin and release breast milk—a good example of how the nervous system integrates information that regulates hormonally mediated processes.

Hormones, in turn, can influence the nervous system. Oxytocin, for example, promotes bonding (see the story that opens Chapter 7). If oxytocin release is experimentally blocked, mammalian mothers, from rats to sheep, will reject their newborn offspring, but if a virgin rat is given a dose of oxytocin, she will adopt strange pups as if they were her own. Oxytocin promotes pair bonding and trust in a variety of animals. In humans its secretion rises with intimate sexual contact, and it has been nicknamed the "cuddle hormone."

THE ANTERIOR PITUITARY The anterior pituitary produces and releases four peptide and protein hormones that act as **tropic hormones**, meaning they control the activities of other endocrine glands. These four tropic hormones are thyrotropin (thyroid-stimulating hormone), luteinizing hormone, follicle-stimulating hormone, and corticotropin. Each is produced by a different type of pituitary cell. We will say more about the tropic hormones when we describe their target glands—the thyroid, testes, ovaries, and adrenal cortex—later in this chapter and in Chapter 43. Other peptide and protein hormones produced by the anterior pituitary are prolactin (see Figure 41.4), growth hormone, enkephalins, and endorphins.

Growth hormone (**GH**) acts on a wide variety of tissues to promote growth. One of its important effects is to stimulate

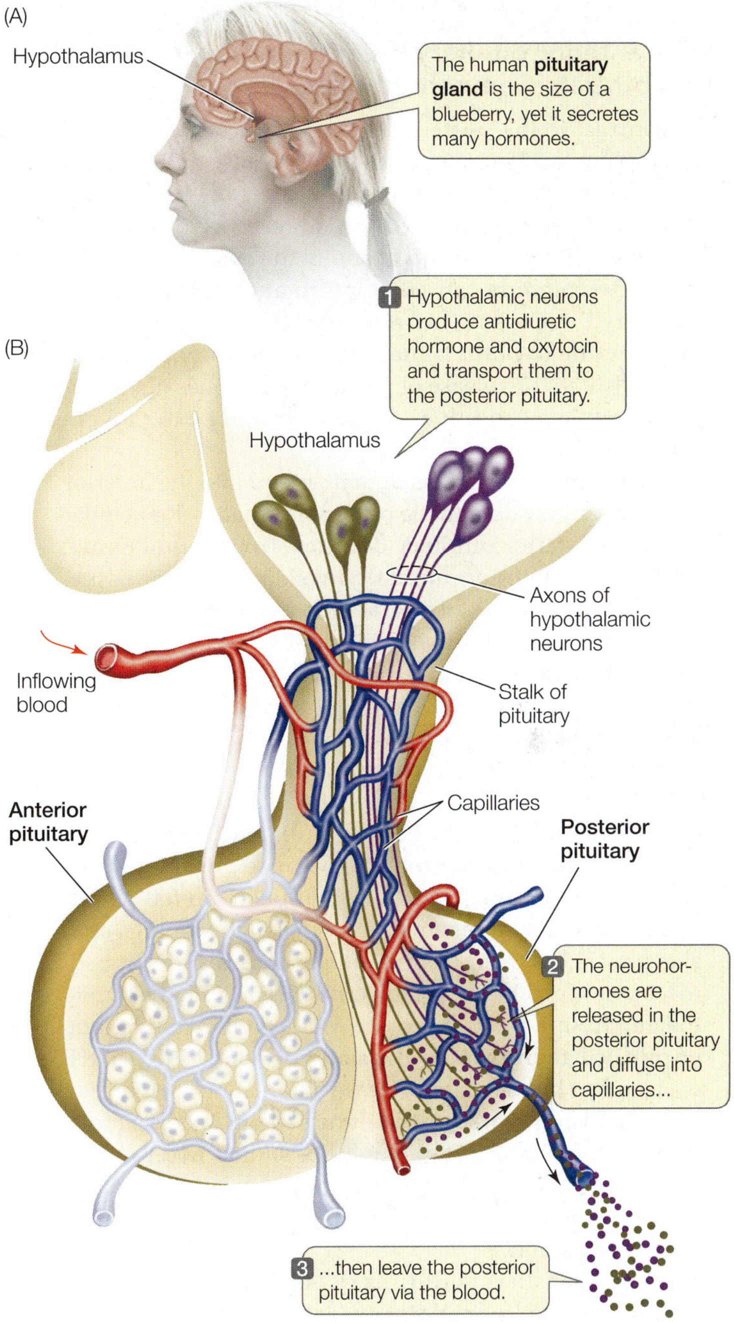

(A)
Hypothalamus
The human **pituitary gland** is the size of a blueberry, yet it secretes many hormones.
(B)
1 Hypothalamic neurons produce antidiuretic hormone and oxytocin and transport them to the posterior pituitary.
Hypothalamus
Axons of hypothalamic neurons
Inflowing blood
Stalk of pituitary
Capillaries
Anterior pituitary
Posterior pituitary
2 The neurohormones are released in the posterior pituitary and diffuse into capillaries...
3 ...then leave the posterior pituitary via the blood.

41.9 The Posterior Pituitary Releases Neurohormones Neurons in the hypothalamus produce two peptide neurohormones, which are stored and released by the posterior pituitary.

cells to take up amino acids. Growth hormone also promotes growth by stimulating the liver to produce chemical signals that stimulate the growth of bone and cartilage. Overproduction of growth hormone in children causes gigantism, in which affected individuals may grow to nearly 8 feet tall. Underproduction causes pituitary dwarfism, in which individuals fail to reach normal adult height.

Endorphins and **enkephalins** are the body's natural painkillers. In the brain, these molecules act as neurotransmitters in pathways that control pain. Their production in the anterior pituitary is normally quite small and probably has little significant effect.

The anterior pituitary is controlled by hypothalamic neurohormones

In contrast to the posterior pituitary, the anterior pituitary makes and secretes its own hormones, but its secretion of these hormones is controlled by the hypothalamus. The hypothalamus senses and receives information about conditions in the body and in the external environment and communicates that information to the anterior pituitary by releasing neurohormones. If the connection between the hypothalamus and the pituitary is experimentally cut, the release of pituitary hormones no longer changes when conditions in the internal or external environment change.

Hypothalamic neurons do not extend into the anterior pituitary as they do into the posterior pituitary. Remember that the posterior pituitary develops from neural tissue, whereas the anterior pituitary develops from gut tissue. Instead, a special set of **portal blood vessels** bridges the gap between the hypothalamus and the anterior pituitary (**Figure 41.10**). Secretions from neurons in the hypothalamus enter the blood and are conducted down the portal vessels to the anterior pituitary, where they stimulate the release of anterior pituitary hormones.

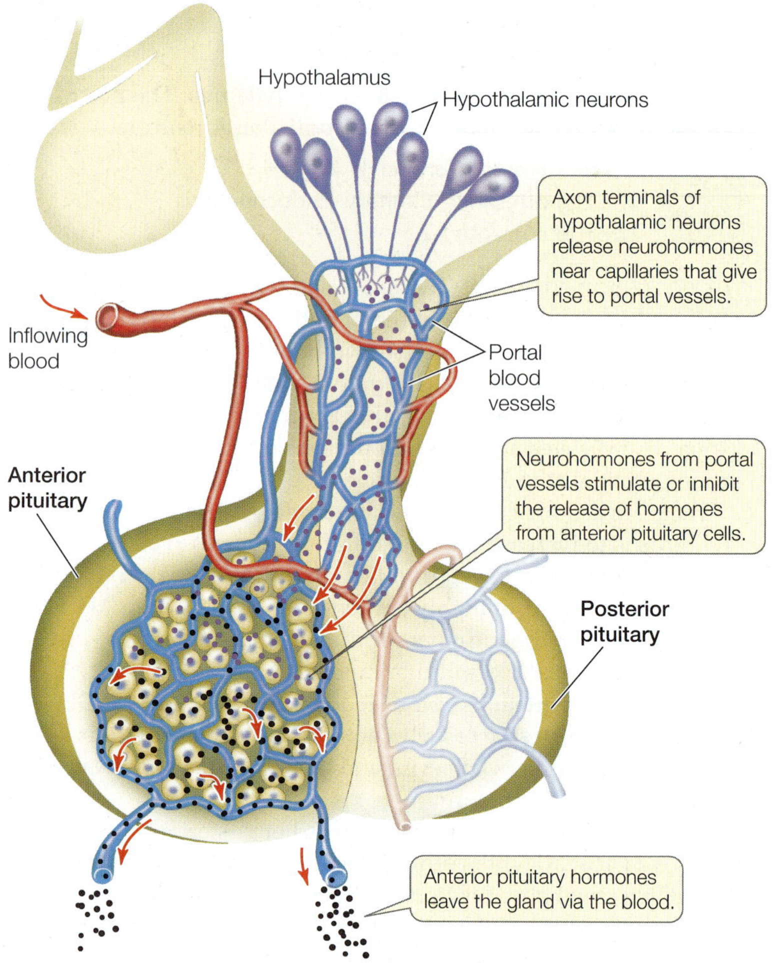
Hypothalamus
Hypothalamic neurons
Axon terminals of hypothalamic neurons release neurohormones near capillaries that give rise to portal vessels.
Inflowing blood
Portal blood vessels
Anterior pituitary
Neurohormones from portal vessels stimulate or inhibit the release of hormones from anterior pituitary cells.
Posterior pituitary
Anterior pituitary hormones leave the gland via the blood.

41.10 The Anterior Pituitary Is Controlled by the Hypothalamus Cells of the anterior pituitary produce four tropic hormones that control other endocrine glands, as well as several other peptide and protein hormones. These cells are controlled by neurohormones produced in the hypothalamus and delivered through portal blood vessels that run between the hypothalamus and the anterior pituitary through the pituitary stalk.

In the 1960s two large teams of scientists led by Roger Guillemin and Andrew Schally initiated the search for these hypothalamic secretions. Because the amounts of such neurohormones in any individual mammal would be tiny, massive numbers of hypothalami from pigs and sheep were collected from slaughterhouses and shipped to laboratories. One extraction effort began with the hypothalami from 270,000 sheep and yielded only 1 milligram of purified **thyrotropin-releasing hormone** (**TRH**). TRH was the first hypothalamic release-stimulating hormone to be isolated and characterized. It turned out to be a simple tripeptide consisting of glutamine, histidine, and proline. It causes certain anterior pituitary cells to release the tropic hormone thyrotropin, which in turn stimulates the activity of the thyroid gland.

Soon after discovering TRH, Guillemin's and Schally's teams identified **gonadotropin-releasing hormone** (**GnRH**), which stimulates certain anterior pituitary cells to release the tropic hormones that control the activity of the gonads (the ovaries and the testes). For these discoveries, Guillemin and Schally shared the 1977 Nobel Prize in Medicine with Rosalyn Yalow, who invented a technique that allows measurement of miniscule amounts of specific molecules (see Figure 41.19).

Many other hypothalamic neurohormones, including both releasing and release-inhibiting hormones, are now known. The major hypothalamic neurohormones that control anterior pituitary function are:

- Thyrotropin-releasing hormone
- Gonadotropin-releasing hormone
- Prolactin-releasing and release-inhibiting hormones
- Growth hormone-releasing hormone and growth hormone release-inhibiting hormone (somatostatin)
- Corticotropin-releasing hormone

Negative feedback loops regulate hormone secretion

In addition to being controlled by hypothalamic releasing and release-inhibiting hormones, the endocrine cells of the anterior pituitary are also under direct and indirect negative feedback control by the hormones of the target glands they stimulate (**Figure 41.11**). For example, cortisol, produced by the adrenal gland in response to corticotropin secreted by the anterior pituitary, reaches the pituitary in the circulating blood and inhibits further release of corticotropin. Cortisol also acts as a negative feedback signal to the hypothalamus, inhibiting the release of corticotropin-releasing hormone. In some cases a tropic hormone also exerts negative feedback control on the hypothalamic cells that produce the corresponding releasing hormone.

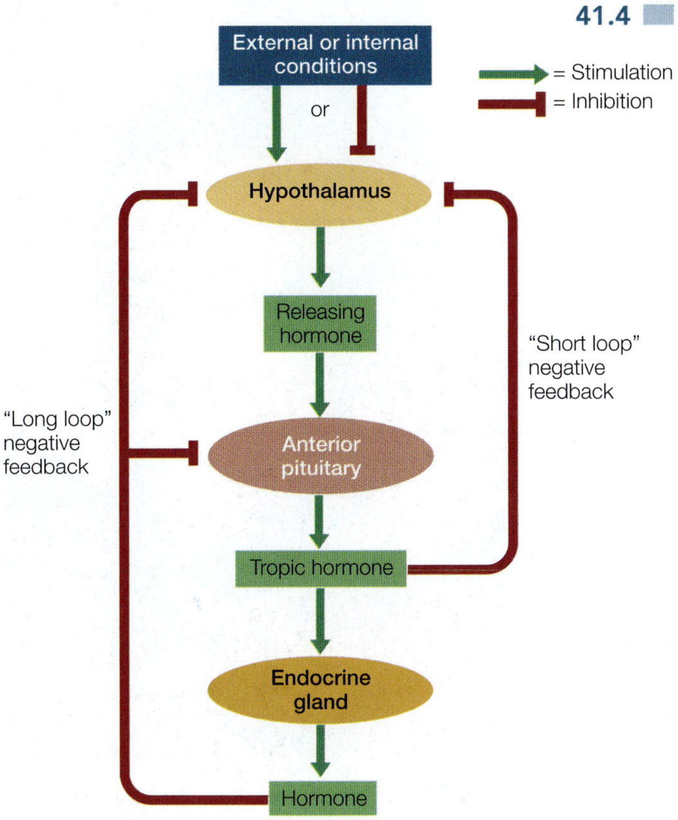
External or internal conditions
or
= Stimulation
= Inhibition
Hypothalamus
Releasing hormone
"Short loop" negative feedback
"Long loop" negative feedback
Anterior pituitary
Tropic hormone
Endocrine gland
Hormone

41.11 Multiple Feedback Loops Control Hormone Secretion Multiple negative feedback loops regulate the chain of command from hypothalamus to anterior pituitary to endocrine glands.

RECAP 41.3

The pituitary is the interface between the nervous system and the endocrine system. The posterior pituitary releases two neurohormones. The anterior pituitary, under the control of other neurohormones from the hypothalamus, releases hormones that control other endocrine glands.

- Describe the anatomical and functional relationships between the brain and the two parts of the pituitary. See pp. 842–843
- Describe the role of hypothalamic neurohormones and the portal blood vessels in the secretion of hormones from the anterior pituitary. See p. 844 and Figure 41.10

Now that we have described some of the mechanisms by which endocrine systems are controlled, we will take a more detailed look at the functions of the major endocrine glands of mammals, as exemplified by humans.

41.4 What Are the Major Endocrine Glands and Hormones?

Hormones help regulate functions in all mammalian physiological systems. In this section we will examine a few major examples of hormonal action in physiological processes. Although we will focus on humans, these systems are very similar in all mammals. Be aware that these are only a few examples; hormonal actions play an important role in virtually all the physiological systems that will be described in the chapters that follow.

The thyroid gland secretes thyroxine

The **thyroid gland** wraps around the front of the windpipe (trachea) and expands into a lobe on either side (see Figures 41.8 and 41.14). There are two cell types in the thyroid gland, each of which produces a specific hormone. **Thyroxine** is produced by epithelial cells that make up round, colloid-containing structures called follicles (**Figure 41.12A**). **Calcitonin** is produced by cells in the spaces between the follicles and is involved in blood calcium regulation (which we will describe shortly).

Thyroxine, a crucial signal in the regulation of cellular energy metabolism, begins as the glycoprotein thyroglobulin, which is synthesized by the follicle cells and packaged in secretory vesicles. The follicle cells actively take up iodide from the blood and move it into the lumen of the follicle. Each thyroglobulin molecule contains about 100 tyrosine units. When the secretory vesicles release thyroglobulin into the lumen of the follicle, they also release an enzyme that catalyzes the iodination of the tyrosine units in the thyroglobulin. When the thyroid gland is stimulated to release thyroxine, the follicle cells take up thyroglobulin from the follicle by endocytosis. These bits of thyroglobulin are then cleaved to form smaller molecules consisting of only two tyrosine units, and these molecules leave the follicle cells and enter the blood (**Figure 41.12B**). If these molecules are iodinated at the maximum of four sites on the tyrosine units, the hormone is tetraiodothyronine, or T_4:

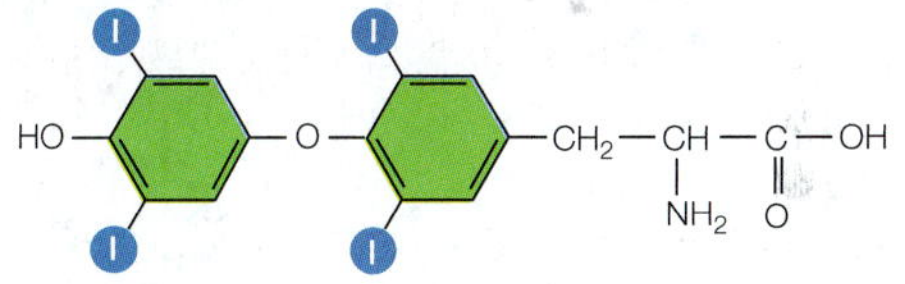

and if they are iodinated at only three sites, they are triiodothyronine, or T_3:

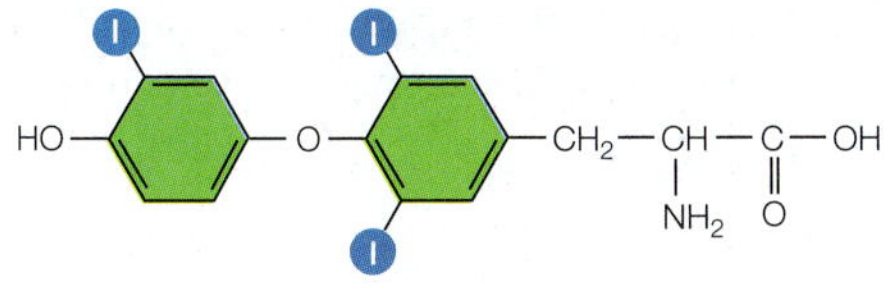

The thyroid usually releases about ten times as much T_4 as T_3. However, T_3 is a much more active hormone than T_4 is, so when you read about the effects of thyroxine, keep in mind that the actions discussed are primarily those of T_3. The difference in the activities of T_3 and T_4 makes it possible to control the effects of thyroxine in different tissues. Within target cells, T_4 can be converted to T_3 by an enzyme called a deiodinase. Another deiodinase can convert T_4 into an inactive hormone called reverse T_3. Deiodinase can also inactivate T_3 by converting it into T_2 or T_1. Thus each target cell can set a unique sensitivity to thyroid hormones using these enzymes to control the conversion of T_4 to T_3 or to reverse T_3.

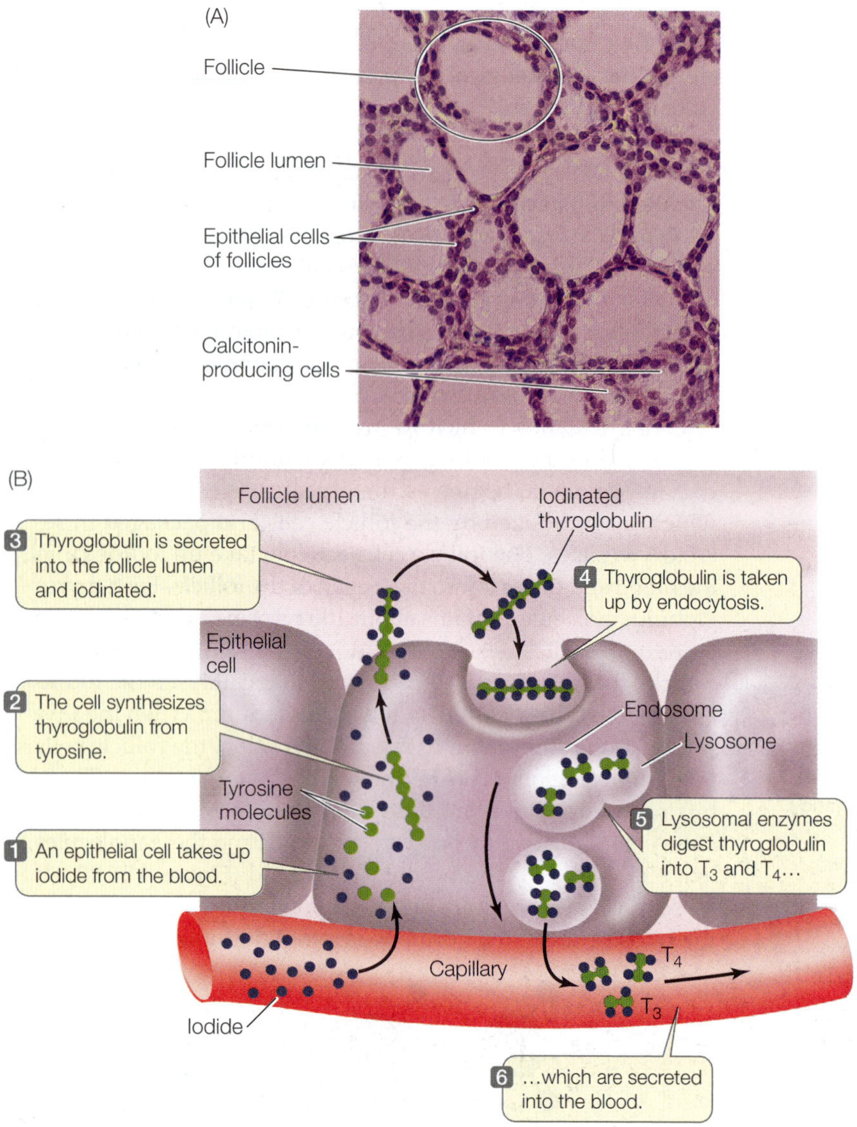
(A)
Follicle
Follicle lumen
Epithelial cells of follicles
Calcitonin-producing cells
(B)
Follicle lumen
Iodinated thyroglobulin
3 Thyroglobulin is secreted into the follicle lumen and iodinated.
4 Thyroglobulin is taken up by endocytosis.
Epithelial cell
2 The cell synthesizes thyroglobulin from tyrosine.
Endosome
Lysosome
Tyrosine molecules
5 Lysosomal enzymes digest thyroglobulin into T_3 and T_4...
1 An epithelial cell takes up iodide from the blood.
Capillary
T_4
T_3
Iodide
6 ...which are secreted into the blood.

41.12 The Thyroid Gland Consists of Many Follicles (A) Cross section through a thyroid gland, showing numerous follicles bounded by epithelial cells. Calcitonin-secreting cells are located in the spaces between the follicles. (B) The epithelial cells of the follicle synthesize thyroglobulin and secrete it into the lumen of the follicle, where it is iodinated and stored until it is processed by the epithelial cells to generate T_3 and T_4.

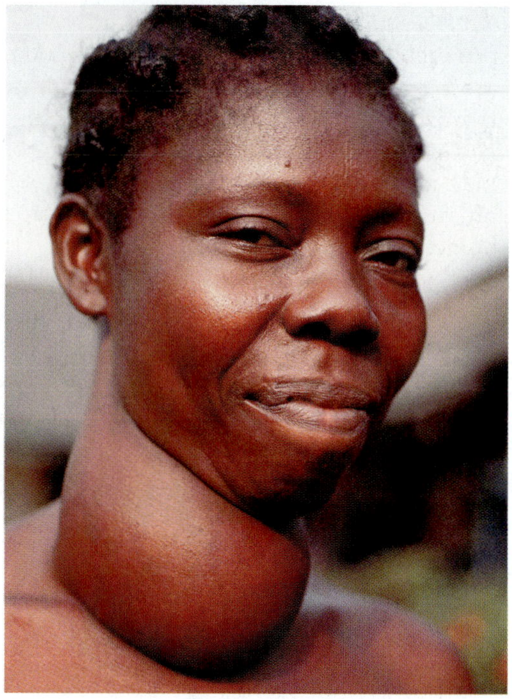

41.13 A Hypothyroid Goiter In this condition, dietary iodide deficiency leads to a lack of functional thyroxine, resulting in the oversynthesis of thyroglobulin and subsequent enlarged follicles.

TSH AND TRH REGULATE THYROXINE PRODUCTION The tropic hormone **thyroid-stimulating hormone** (**TSH**, also known as **thyrotropin**), produced by the anterior pituitary, activates the thyroxine-producing follicle cells in the thyroid. Thyrotropin-releasing hormone (TRH), produced in the hypothalamus and transported to the anterior pituitary through the portal blood vessels, activates the TSH-producing pituitary cells. The hypothalamus uses environmental information, such as temperature or day length, to determine whether to increase or decrease its secretion of TRH. This sequence of steps is regulated by a negative feedback loop (see Figure 41.11). Circulating thyroxine inhibits the response of pituitary cells to TRH, so less TSH is released when thyroxine levels are high, and more TSH is released when thyroxine levels are low. Circulating thyroxine also exerts negative feedback on the production and release of TRH by the hypothalamus.

Because thyroxine is lipid-soluble, it enters cells readily and binds to receptors in the nucleus. When combined with thyroxine, these receptors (which are found in most cells of the body) stimulate the transcription of numerous genes whose products are transport proteins, structural proteins, and enzymes involved in metabolic pathways; thus thyroxine elevates the metabolic rates of most cells and tissues. Exposure to cold for several days leads to an increased release of thyroxine, an increased conversion of T_4 to T_3, and therefore an increased basal metabolic rate (see Section 40.5).

During development and growth, thyroxine promotes amino acid uptake and protein synthesis. Insufficient thyroxine in a human fetus or growing child greatly retards physical and mental development, resulting in a condition known as cretinism.

GOITER A **goiter** is an enlarged thyroid gland (**Figure 41.13**) that can be associated with either hyperthyroidism (excess production of thyroxine) or hypothyroidism (thyroxine deficiency). The negative feedback loop whereby thyroxine controls TSH release helps explain how two seemingly opposite conditions can result in the same symptom.

- The most common cause of *hyperthyroid* goiter is Graves' disease, an autoimmune disease involving an antibody to the TSH receptor. This antibody binds to and activates the TSH receptors on the follicle cells, causing uncontrolled production and release of thyroxine. Blood levels of TSH are low due to negative feedback from high levels of thyroxine, but the thyroid remains maximally stimulated and grows bigger. People with hyperthyroidism have high metabolic rates, usually feel hot, and may develop a buildup of fat behind the eyeballs that causes their eyes to bulge.

- *Hypothyroid* goiter results when there is not enough circulating thyroxine to turn off TSH production. The most common cause is a deficiency of dietary iodide, without which the follicle cells cannot make thyroxine. Without sufficient thyroxine, TSH levels remain high and the thyroid continues to produce large amounts of thyroglobulin. Because sufficient iodine is not available, however, the thyroglobulin is poorly iodinated. When it is broken down by the follicle cells, it produces little functional T_3 or T_4. TSH levels remain high and stimulate more and more synthesis of thyroglobulin, and the thyroid gets bigger. The symptoms of hypothyroidism are low metabolism, intolerance of cold, and general physical and mental sluggishness.

Goiter affects about 5 percent of the world's population. The addition of iodide to table salt has greatly reduced the incidence of hypothyroid goiter in industrialized nations, but the condition is still common in the other parts of the world and is a leading cause of intellectual impairment.

Three hormones regulate blood calcium concentrations

The regulation of calcium concentration in the blood is crucial, and shifts in blood calcium concentration above or below a narrow range can cause serious problems. When blood calcium falls below this range, the nervous system becomes overly excited, resulting in muscle spasms and even seizures. When blood calcium rises above this range, the nervous system becomes depressed and muscles—including the heart—weaken. Regulation of blood calcium is difficult because only about 0.1 percent of the calcium in the body is located in the extracellular fluid. About 1 percent is in cells, and almost 99 percent is in the bones. Therefore the body must maintain a tiny pool of calcium in the blood at a precise concentration, and that tiny pool can be influenced greatly by relatively small shifts in the much larger pools of calcium in the cells and bones.

The body has multiple mechanisms for changing blood calcium levels, including:

- Deposition or absorption of bone
- Excretion or retention of calcium by the kidneys
- Absorption of calcium from the digestive tract

Go to Animated Tutorial 41.3
Hormonal Regulation of Calcium
Life10e.com/at41.3

These mechanisms are controlled by three hormones: calcitonin, parathyroid hormone, and calcitriol (synthesized from vitamin D).

CALCITONIN REDUCES BLOOD CALCIUM Calcitonin is released by the thyroid and lowers the concentration of calcium in the blood, mainly by regulating bone turnover (**Figure 41.14**). Bone is continuously remodeled through a dynamic process that involves both resorption of old bone and synthesis of new bone, as we will discuss in Section 48.3. Cells called osteoclasts break down bone and release calcium into the blood, and cells called osteoblasts take up calcium from the blood and deposit it in new bone. Calcitonin decreases the activity of osteoclasts and thereby favors removal of calcium from the blood and its deposition in bone by osteoblasts. The turnover of bone in adult humans is not very high, so calcitonin does not play a major role in calcium homeostasis in adults. It is probably more important in young individuals whose bones are actively growing.

PARATHYROID HORMONE INCREASES BLOOD CALCIUM The **parathyroid glands** are four tiny structures embedded in the

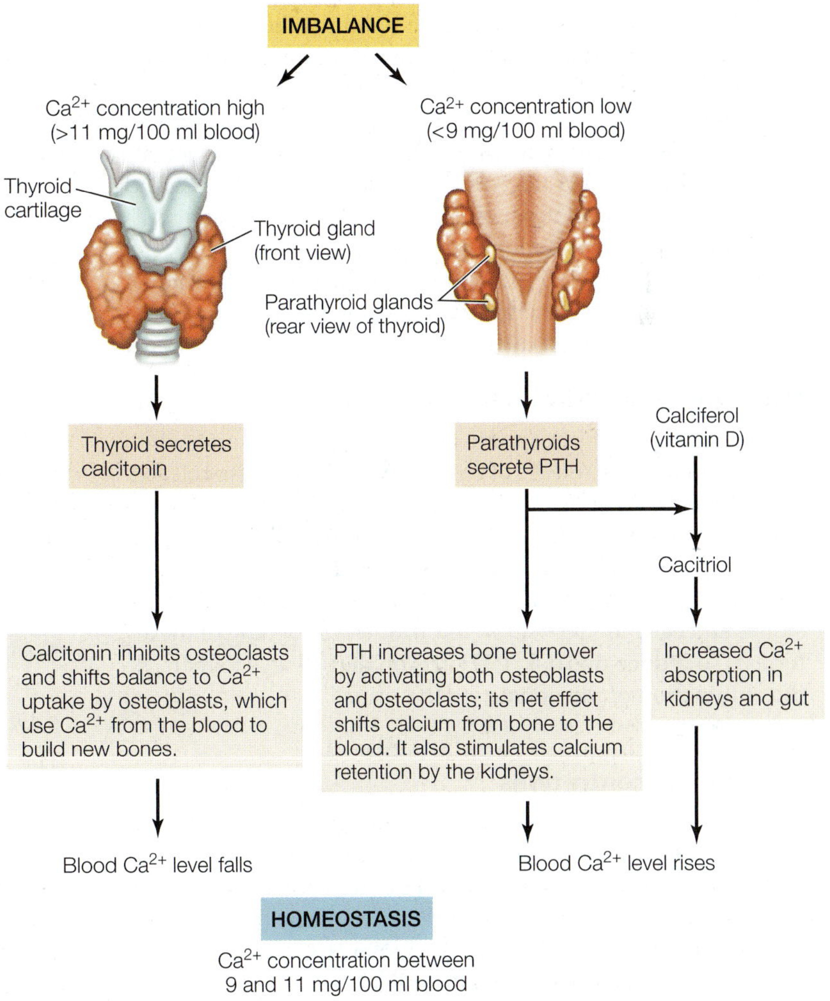
IMBALANCE
Ca^{2+} concentration high (>11 mg/100 ml blood)
Ca^{2+} concentration low (<9 mg/100 ml blood)
Thyroid cartilage
Thyroid gland (front view)
Parathyroid glands (rear view of thyroid)
Thyroid secretes calcitonin
Parathyroids secrete PTH
Calciferol (vitamin D)
Cacitriol
Calcitonin inhibits osteoclasts and shifts balance to Ca^{2+} uptake by osteoblasts, which use Ca^{2+} from the blood to build new bones.
PTH increases bone turnover by activating both osteoblasts and osteoclasts; its net effect shifts calcium from bone to the blood. It also stimulates calcium retention by the kidneys.
Increased Ca^{2+} absorption in kidneys and gut
Blood Ca^{2+} level falls
Blood Ca^{2+} level rises
HOMEOSTASIS
Ca^{2+} concentration between 9 and 11 mg/100 ml blood

41.14 Hormonal Regulation of Calcium Calcitonin, parathyroid hormone (PTH), and calcitriol (the active form of calciferol, or vitamin D) regulate Ca^{2+} levels in the blood.

posterior surface of the thyroid gland (see Figure 41.14). Their single hormone product, **parathyroid hormone** (**PTH**, also called parathormone), is the most important hormone in the regulation of blood calcium levels. Circulating calcium activates receptors in the plasma membrane of the parathyroid cells. When these receptors are active, they inhibit the synthesis and release of PTH. A fall in blood calcium removes this inhibition and triggers the synthesis and release of PTH. PTH stimulates bone turnover by actions on both osteoclasts and osteoblasts. The end result of these actions of PTH is a net increase of calcium in the blood. PTH also raises blood concentration of calcium by stimulating the kidneys to reabsorb it rather than excrete it in the urine.

CALCITRIOL INCREASES BLOOD CALCIUM It had long been known that fragile bones were common among people living at high latitudes, where winter days are short and the winter diet often lacks fish, dairy products, and fresh vegetables. Since the condition could be reversed by taking cod-liver oil, it was assumed that a vitamin deficiency was involved. That vitamin was vitamin D, but when its chemical identity was established, this molecule turned out not to be a vitamin at all.

A vitamin is a substance that the body requires in small quantities but cannot synthesize for itself and must therefore obtain from food (or from supplements such as vitamin pills). However, vitamin D—now more accurately named calciferol—is synthesized naturally from cholesterol when skin cells receive ultraviolet light. Calciferol is not an active hormone, but through actions of the liver and kidneys it is converted into the active form called **calcitriol**, which circulates in the blood and acts on distant cells (and therefore *is* a hormone). The conversion of calciferol to calcitriol is activated by PTH. Calcitriol promotes the absorption of calcium from food in the gut. Thus the combined actions of PTH and calcitriol raise blood calcium levels.

PTH lowers blood phosphate levels

Bones are made of phosphate as well as calcium, and when PTH stimulates the release of calcium from bone it also releases phosphate. Normal blood concentrations of calcium and phosphate are just below the levels at which they precipitate out of solution as calcium phosphate salts. Even a small rise may cause such precipitation, leading to maladies such as kidney stones and calcium deposits in the arteries (hardening of the arteries). To reduce this risk, PTH acts on the kidneys to increase the elimination of phosphate via the urine.

Insulin and glucagon regulate blood glucose concentrations

Before the 1920s, the disease diabetes mellitus was fatal. Characterized by weakness, lethargy, and a dramatic loss of body mass, this condition was known to be connected somehow with the pancreas—a large gland located just below the stomach (see Figure 41.8)—and with abnormal glucose metabolism. The exact links, however, were not clear.

Today we know that diabetes mellitus is caused by a lack of the protein hormone **insulin** (in type I diabetes) or by a lack of insulin responsiveness in target tissues (in type II diabetes). Glucose enters cells by diffusion, but cell membranes are not very permeable to glucose. Glucose transporter proteins in cell membranes facilitate the movement of glucose into cells, and the glucose transporters most common in muscle and adipose tissue are controlled by insulin. When insulin binds to its receptor on the cell membrane, it causes these glucose transporters to move from cytoplasmic vesicles to the cell membrane, thus making the cell more permeable to glucose. When insulin is not present, these transporters are returned to the cytoplasmic pool through endocytosis.

In the absence of insulin or insulin responsiveness, glucose entry into cells is impaired, resulting in so much glucose accumulating in the blood that it starts to spill over into the urine. A high concentration of glucose in the blood increases urine output by two mechanisms. First, it causes water to move from cells into the blood by osmosis, and this increase in blood volume results in increased urine production. Second, the increased glucose in the tubules of the kidneys pulls more water into the urine by osmosis. Diabetic individuals thus can become dehydrated, but more importantly they suffer from a lack of metabolic fuel. Because glucose uptake by muscle and adipose tissue is impaired in the absence of insulin, muscle cells must depend on fat and protein for fuel and adipose tissue cannot replenish its stores of triglycerides. If the condition is not treated, the body can waste away.

For centuries, the prospects for people suffering with diabetes were bleak. A change came almost overnight in 1921, when the physician Frederick Banting and a medical student, Charles Best, at the University of Toronto, discovered they could reduce the symptoms of diabetes by injecting an extract prepared from pancreatic tissue. The active component of this extract was found to be insulin, a small protein consisting of just 51 amino acids. In the United States today, insulin replacement therapy using manufactured insulin allows more than 1.5 million people with type I diabetes to lead almost normal lives.

ISLETS OF LANGERHANS Insulin is produced in clusters of endocrine cells in the pancreas. These clusters are called **islets of Langerhans** after the German medical student who discovered them. They contain three types of cells, each of which produces a specific hormone:

- Beta (β) cells produce and secrete insulin.
- Alpha (α) cells produce and secrete **glucagon**, a hormone that has effects mostly opposite from those of insulin.
- Delta (δ) cells produce the hormone **somatostatin**.

The rest of the pancreas is made up of exocrine tissue, which produces enzymes and other secretions that travel through ducts to the gut, where they participate in digestion.

After a meal, the concentration of glucose in the blood rises, stimulating the β cells of the islets to release insulin. Insulin causes target cells throughout the body to use circulating glucose as fuel and convert it into storage products such as glycogen and fat. When the gut is empty of food, blood glucose concentration falls and the islets stop releasing insulin. As a result, most cells shift to using glycogen and fat rather than glucose as fuel. If blood glucose concentration falls substantially below normal, the islet α cells release glucagon, which stimulates the liver to break down stored glycogen and release glucose into the blood. These actions will be discussed in greater detail in Section 51.4.

SOMATOSTATIN Somatostatin is released from the δ cells of the pancreas in response to rapid increases of glucose and amino acids in the blood. This hormone has paracrine functions within the islets, where it inhibits the release of both insulin and glucagon. Outside the pancreas it acts as a hormone, slowing the digestive activities of the gut and extending the period during which nutrients are absorbed. Somatostatin is also produced in very small amounts by cells in the hypothalamus. Hypothalamic somatostatin is transported in the portal blood vessels to the anterior pituitary, where it acts as a neurohormone to inhibit the release of growth hormone and thyrotropin.

The adrenal gland is two glands in one

An **adrenal gland** sits above each kidney, just below the middle of your back. Functionally and anatomically, each adrenal gland consists of a gland within a gland (**Figure 41.15**). The core, or **adrenal medulla**, produces: **epinephrine** (also known as adrenaline) and, to a lesser degree, **norepinephrine** (noradrenaline). The medulla develops from nervous tissue and is under the control of the nervous system. Surrounding the medulla is the **adrenal cortex**, which produces steroid hormones. The cortex is under hormonal control, largely by corticotropin produced by the anterior pituitary.

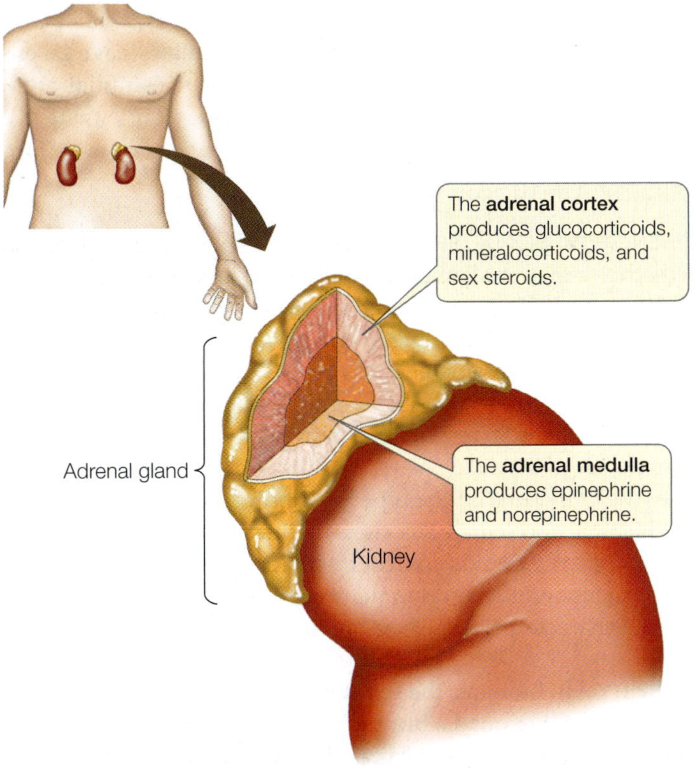

The adrenal cortex produces glucocorticoids, mineralocorticoids, and sex steroids.
The adrenal medulla produces epinephrine and norepinephrine.
Adrenal gland
Kidney

41.15 The Adrenal Is Really Two Glands An adrenal gland, consisting of an outer cortex and an inner medulla, sits above each kidney. The medulla and the cortex produce different hormones.

THE ADRENAL CORTEX The cells of the adrenal cortex use cholesterol to produce three classes of steroid hormones (see Figure 41.2B), collectively called **corticosteroids**:

- *Mineralocorticoids* influence the salt and water balance of the extracellular fluid.
- *Glucocorticoids* influence blood glucose concentrations as well as other aspects of fat, protein, and carbohydrate metabolism.
- *Sex steroids* play roles in sexual development, sexual behavior, and anabolism (tissue-building).

In adult humans, the adrenal cortex secretes only negligible amounts of sex steroids. The major producers of sex steroids are the gonads, as we will see in the following section.

Aldosterone, the primary mineralocorticoid, stimulates the kidneys to conserve sodium and excrete potassium, as we will discuss in Chapter 51. If the adrenal glands are removed from an animal, sodium must be added to its diet, or its sodium will be depleted and it will die.

The main glucocorticoid in humans is **cortisol**, which is critical for mediating the body's metabolic responses to stress. Within minutes of a stressful stimulus (one provoking fear or anger, for example), blood cortisol levels begin to rise. This response is much slower than the neurally mediated epinephrine and norepinephrine response to stress, but it lasts longer. Cells not critical for a sustained stress response are stimulated by cortisol to decrease their use of blood glucose and shift instead to using fats and proteins for energy. Cortisol also inhibits the immune system (because dealing with the immediate stressor is more important than feeling sick, having allergic reactions, or healing wounds). This explains why cortisol and drugs that mimic its action are useful for reducing inflammation and allergic responses.

Cortisol release is controlled from the anterior pituitary by **corticotropin** (also called **adrenocorticotropic hormone**, or **ACTH**), whose release is controlled in turn by **corticotropin-releasing hormone (CRH)** from the hypothalamus. The action of ACTH on the adrenal cortex is to stimulate the synthesis of cortisol. Like other steroid hormones, cortisol is not stored in vesicles and therefore is available for immediate release. As cortisol or other steroid hormones diffuse into the blood, they combine with carrier proteins, and their release from these proteins can have a long time course, thus stretching out their actions. Also, many of their actions stimulate gene expression in target cells, which also takes time but has a long-lasting effect.

Turning off the stress responses activated by cortisol is as important as turning them on. A study of stress in rats showed that old rats could turn on these stress responses as effectively as young rats, but they had lost the ability to turn them off as rapidly. As a result, they suffered from the well-known consequences of stress seen in humans: digestive system problems, cardiovascular problems, strokes, impaired immune system function, and increased susceptibility to cancers and other diseases. Acute stress responses are controlled by negative feedback from cortisol on both the ACTH-secreting cells of the anterior pituitary and CRH-secreting cells of the hypothalamus. With chronic or prolonged stress, these control mechanisms

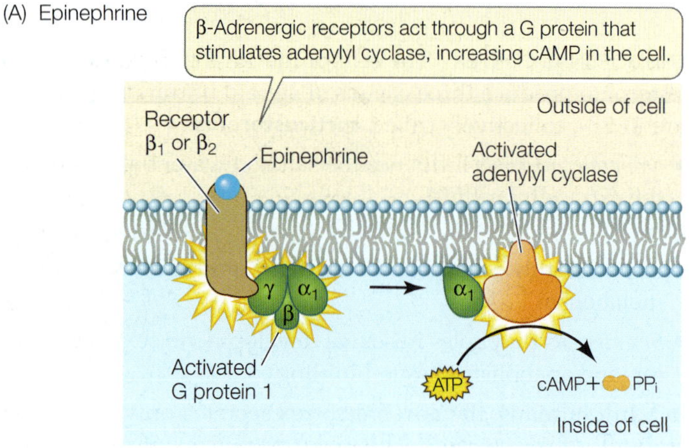
(A) Epinephrine
β-Adrenergic receptors act through a G protein that stimulates adenylyl cyclase, increasing cAMP in the cell.
Outside of cell
Receptor β1 or β2
Epinephrine
Activated adenylyl cyclase
γ
α1
β
Activated G protein 1
ATP
cAMP+ PPi
Inside of cell

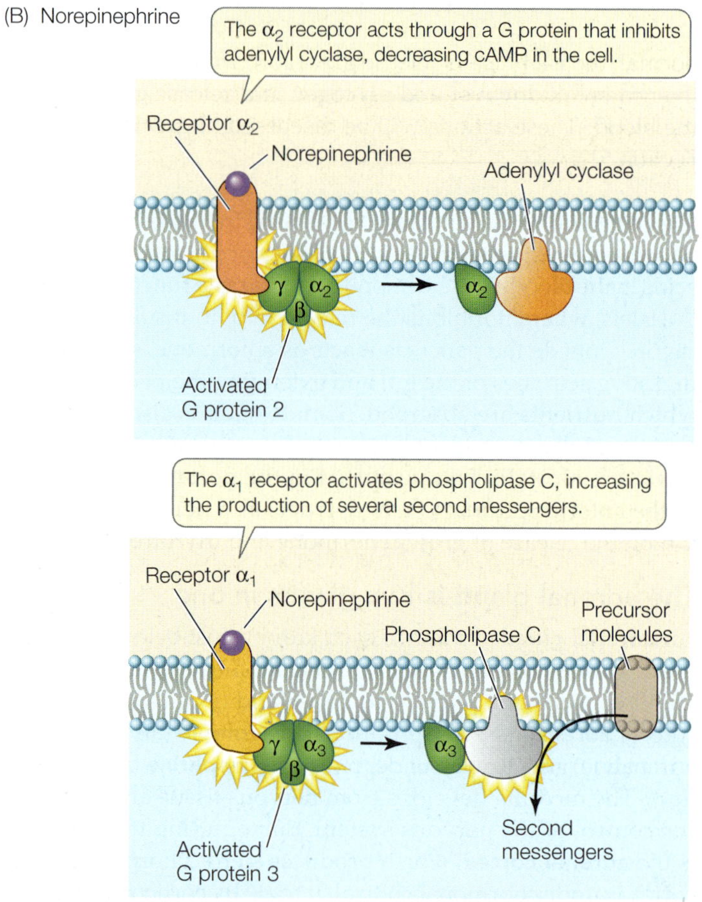
(B) Norepinephrine
The α2 receptor acts through a G protein that inhibits adenylyl cyclase, decreasing cAMP in the cell.
Receptor α2
Norepinephrine
Adenylyl cyclase
γ
α2
β
Activated G protein 2
The α1 receptor activates phospholipase C, increasing the production of several second messengers.
Receptor α1
Norepinephrine
Phospholipase C
Precursor molecules
α3
Activated G protein 3
Second messengers

41.16 Hormones Can Activate a Variety of Signal Transduction Pathways Epinephrine and norepinephrine bind to G protein-linked adrenergic receptors that act through different signal transduction pathways. Epinephrine (A) acts equally on both α- and β-adrenergic receptors; norepinephrine (B) acts mostly on α-adrenergic receptors.

become insufficient and cortisol must exert negative feedback through another brain region, the hippocampus. Prolonged exposure to cortisol, however, causes trauma to and loss of hippocampal cells, resulting in the decreased ability to turn off the stress response.

THE ADRENAL MEDULLA The adrenal medulla produces epinephrine and norepinephrine in response to stressful situations, arousing the body to action. As we saw earlier in this chapter, epinephrine increases heart rate and blood pressure and diverts blood flow to active muscles and away from the gut and skin. Norepinephrine has similar functions, but since it is also a neurotransmitter involved in many physiological regulatory processes, it has many ongoing functions in addition to its involvement in flight-or-fight reactions.

Epinephrine and norepinephrine are both water-soluble, and both bind to the same set of receptors on the surfaces of target cells. These **adrenergic receptors** are of two general types, α-adrenergic and β-adrenergic (Figure 41.16). The α-adrenergic receptors respond more strongly to norepinephrine than to epinephrine, whereas β-adrenergic receptors respond about equally to both epinephrine and norepinephrine. Because of this difference in receptor affinities, it is possible for drugs to blunt the flight-or-fight responses without disrupting physiological regulatory processes. Such drugs are called "beta blockers" because, by inhibiting β-adrenergic receptors, they can reduce the fight-or-flight response to epinephrine without disrupting the physiological regulatory functions of norepinephrine mediated through the α-adrenergic receptors. Beta blockers are commonly prescribed to reduce symptoms of anxiety such as dry mouth and elevated heart rate (palpitations).

Sex steroids are produced by the gonads

The **gonads**—the testes of the male and the ovaries of the female—produce hormones as well as sperm and ova. The male steroid hormones are collectively called **androgens**, and the dominant hormone is testosterone. The female steroids are **estrogens** and **progesterone**. The dominant estrogen is estradiol, which is synthesized from testosterone. Males and females both synthesize testosterone, but females have an enzyme (aromatase) that converts testosterone to estradiol.

Go to Media Clip 41.1
The Testosterone Factor
Life10e.com/mc41.1

PHENOTYPIC SEX DETERMINATION The sex steroids determine whether a mammalian embryo develops into a phenotypic female or male. In humans, the gonads of an early embryo are undifferentiated. Beginning in about the seventh week of development, the expression of genes on the Y chromosome of an XY individual normally causes the undifferentiated gonads to produce androgens. In response to androgens, the reproductive system develops the male phenotype. If no Y chromosome is present (i.e., the individual is genotype XX), androgens are not produced at this time and female structures develop (Figure 41.17). After birth, the sex steroids control the maturation of the reproductive organs and the development and maintenance of secondary sexual characteristics, such as breasts and facial hair.

PUBERTY In humans, the sex steroids are produced at low levels by juvenile gonads, but their production increases rapidly at puberty (around age of 12 or 13). Why does this sudden increase occur? In both juvenile and adult humans, the activities of the gonads are controlled by the tropic hormones luteinizing hormone (LH) and follicle-stimulating hormone (FSH), which

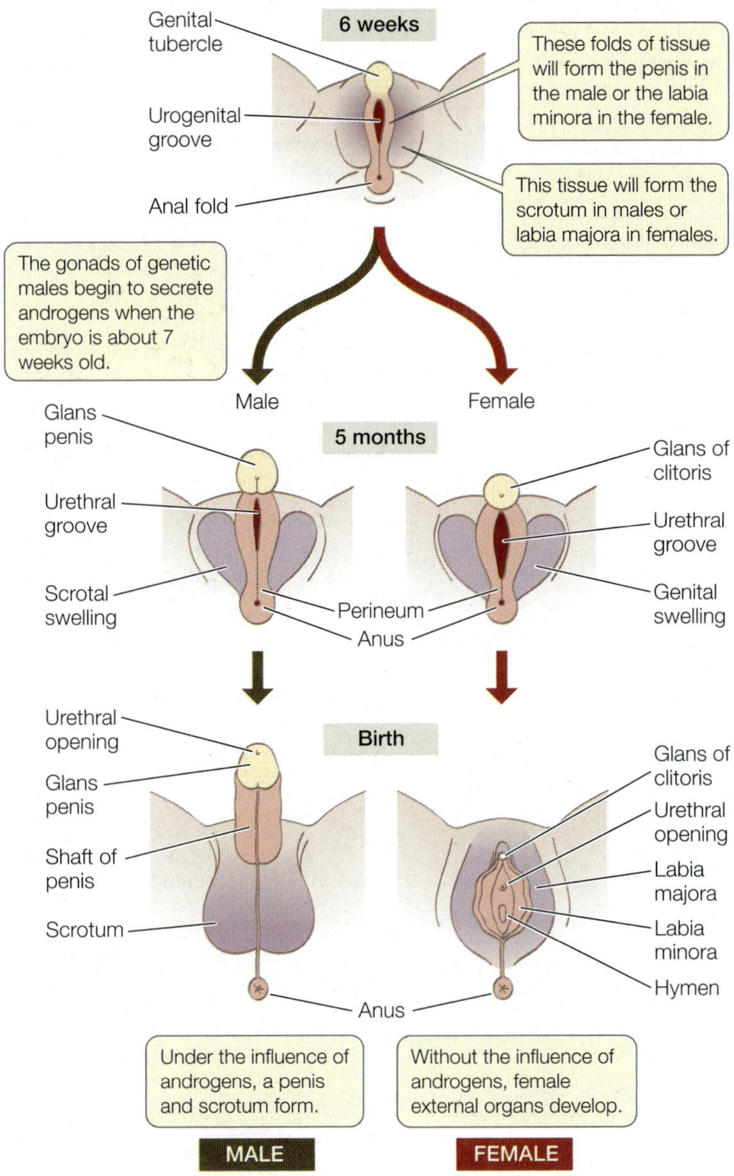

6 weeks
Genital tubercle
Urogenital groove
Anal fold
These folds of tissue will form the penis in the male or the labia minora in the female.
This tissue will form the scrotum in males or labia majora in females.
The gonads of genetic males begin to secrete androgens when the embryo is about 7 weeks old.
Male
Female
5 months
Glans penis
Urethral groove
Scrotal swelling
Perineum
Anus
Glans of clitoris
Urethral groove
Genital swelling
Birth
Urethral opening
Glans penis
Shaft of penis
Scrotum
Anus
Glans of clitoris
Urethral opening
Labia majora
Labia minora
Hymen
Under the influence of androgens, a penis and scrotum form.
Without the influence of androgens, female external organs develop.
MALE
FEMALE

41.17 Sex Steroids Direct the Development of Human Sex Organs The sex organs of early human embryos are undifferentiated. Androgens promote the development of male sex organs. In the absence of androgens, female sex organs form.

together are the **gonadotropins**. The production of gonadotropins by the anterior pituitary is under the control of gonadotropin-releasing hormone (GnRH) produced by the hypothalamus. Before puberty, the hypothalamus produces only low levels of GnRH. Puberty is initiated by a reduction in the sensitivity of hypothalamic GnRH-producing cells to negative feedback from sex steroids and from gonadotropins. As a result, GnRH production increases, stimulating increased production of gonadotropins and hence increased production of sex steroids.

In females, increasing levels of LH and FSH at puberty stimulate the ovaries to increase their production of the female sex hormones. The increased circulating levels of these hormones initiate the development of the traits of a sexually mature woman: enlarged breasts, vagina, and uterus; broadened hips; increased subcutaneous fat; pubic hair; and initiation of the menstrual and ovarian cycles (see Figure 43.13). In males, an increasing level of LH stimulates groups of cells in the testes to increase their synthesis of testosterone, which in turn initiates the physiological, anatomical, and psychological changes associated with adolescence. The voice deepens, hair begins to grow on the face and body, and the testes and penis grow larger. Testosterone also stimulates bone and skeletal muscle growth. FSH in males stimulates production of sperm.

The roles that sex steroids play in adult sexual behavior and reproduction will be described in Chapter 43.

Melatonin is involved in biological rhythms and photoperiodicity

The **pineal gland** is situated between the two hemispheres of the brain and is connected to the brain by a stalk. It synthesizes the amine hormone **melatonin** from the amino acid tryptophan. The pineal gland releases melatonin in the dark, and therefore melatonin levels indicate the length of the night. Exposure to light inhibits the release of melatonin.

In vertebrates, melatonin is involved in biological rhythms, including **photoperiodicity**—the phenomenon whereby seasonal changes in day length cause physiological changes. Many species, for example, come into reproductive condition when the days begin to lengthen (**Figure 41.18**). Humans are not strongly photoperiodic, but melatonin in humans may play a role in synchronizing daily biological rhythms to the daily cycle of light and dark.

Many chemicals may act as hormones

We have discussed the major mammalian endocrine glands and their hormones in this chapter, but many more hormones

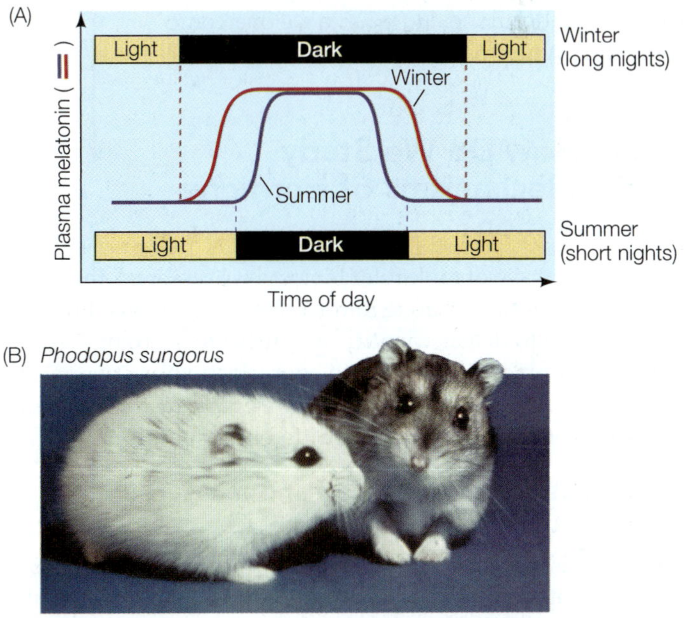

(A)
Light
Dark
Light
Winter (long nights)
Winter
Summer
Light
Dark
Light
Summer (short nights)
Plasma melatonin (=)
Time of day
(B) *Phodopus sungorus*

41.18 Melatonin Regulates Seasonal Changes (A) Melatonin release occurs in the dark and is inhibited by exposure to light. The duration of daily melatonin release thus changes as day length (photoperiod) changes, inducing dramatic seasonal physiological changes in some animals. (B) In winter, Siberian hamsters are white and do not reproduce. In summer, they are mottled brown and breed.

exist. As we discuss the organ systems of the body in the chapters that follow, we will frequently describe hormones that their tissues produce as well as hormones that control their functions.

RECAP 41.4

The major endocrine glands of mammals include the hypothalamus, pituitary gland, thyroid gland, parathyroid glands, pancreas, adrenal glands, gonads, and pineal gland. Each of these glands secretes and responds to hormones that play crucial roles in controlling physiology and development.

- Describe how thyroxine is produced and how its production and release are controlled. **See pp. 845–846 and Figure 41.12**
- How is the concentration of calcium in the blood regulated? **See pp. 847–848 and Figure 41.14**
- How does insulin control the rate of glucose uptake by cells? **See p. 848**
- What changes in the feedback control of sex steroids result in puberty? **See pp. 850–851**

Many hormones are released in very small quantities, and some disappear from the extracellular fluid rapidly. A hormone's receptors may be found on diverse cells around the body, and those cells can respond in different ways to the same hormone. How have we overcome these difficulties to learn how hormones work?

RESEARCH**TOOLS**

41.19 An Immunoassay Allows Measurement of Small Concentrations
An immunoassay uses labeled and unlabeled samples of a purified hormone (or other substance to be measured) and an antibody to that hormone to develop a standard curve against which a sample of unknown concentration can be measured.

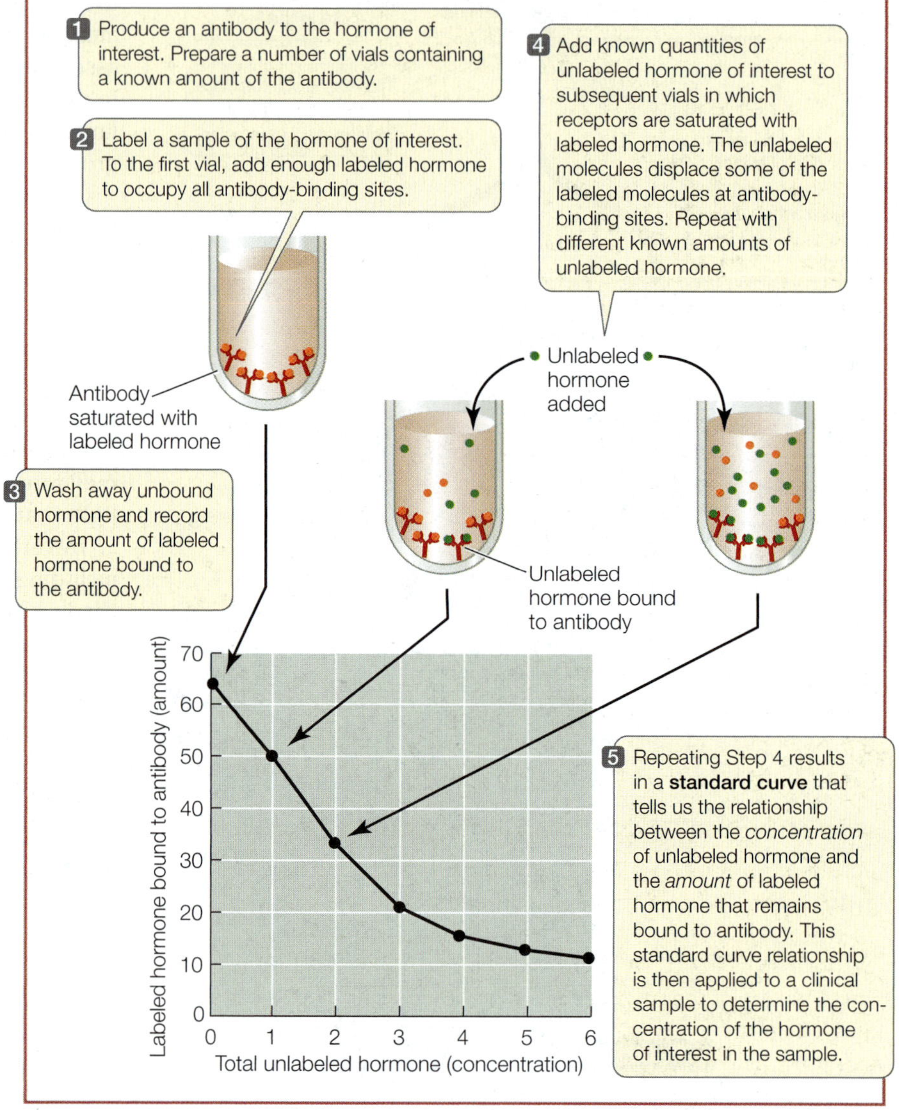

1 Produce an antibody to the hormone of interest. Prepare a number of vials containing a known amount of the antibody.
2 Label a sample of the hormone of interest. To the first vial, add enough labeled hormone to occupy all antibody-binding sites.
Antibody saturated with labeled hormone
3 Wash away unbound hormone and record the amount of labeled hormone bound to the antibody.
4 Add known quantities of unlabeled hormone of interest to subsequent vials in which receptors are saturated with labeled hormone. The unlabeled molecules displace some of the labeled molecules at antibody-binding sites. Repeat with different known amounts of unlabeled hormone.
Unlabeled hormone added
Unlabeled hormone bound to antibody
5 Repeating Step 4 results in a **standard curve** that tells us the relationship between the *concentration* of unlabeled hormone and the *amount* of labeled hormone that remains bound to antibody. This standard curve relationship is then applied to a clinical sample to determine the concentration of the hormone of interest in the sample.
Labeled hormone bound to antibody (amount)
70
60
50
40
30
20
10
0
0
1
2
3
4
5
6
Total unlabeled hormone (concentration)

41.5 How Do We Study Mechanisms of Hormone Action?

In the current age of molecular biology, we can break the study of hormone actions into different sets of problems. First we must be able to detect, identify, and measure hormones. Next we must be able to identify and characterize hormone receptors. Finally, we must understand the signal transduction pathways activated by hormones in different tissues.

Hormones can be detected and measured with immunoassays

As we have seen, testosterone has many dramatic and diverse effects, yet its concentration in the blood of adult human males is only about 30 to 100 *billionths* of a gram per milliliter. Measuring hypothalamic neurohormones requires calibrations in the range of *trillionths* of a gram per milliliter.

The ability to detect and measure minute quantities of hormones was an important breakthrough. Rosalyn Yalow developed a method she named radioimmunoassay because it used radioactive labels (she used radioactive isotopes of iodine) to track interactions between an antigen (the hormone or other substance of interest to be measured) and an antibody made to that antigen. Today we are more likely to use nonradioactive labels, so the technique is called simply **immunoassay** (**Figure 41.19**). Being able to measure hormones in the blood made it possible to study many important hormonal mechanisms.

An important characteristic of a hormone is the time course over which it acts. This time course can be measured by the hormone's half-life in the blood (defined as the length of time it takes for one-half of the hormone molecules to disappear). Soon after endocrine cells are stimulated to secrete their hormone, the hormone reaches its maximum concentration in the blood. By examining a series of blood samples using immunoassays, researchers can determine how long it takes for the circulating hormone to drop to half its maximum concentration. The fight-or-flight response to epinephrine is fast both in its onset and its termination; the half-life of epinephrine in the

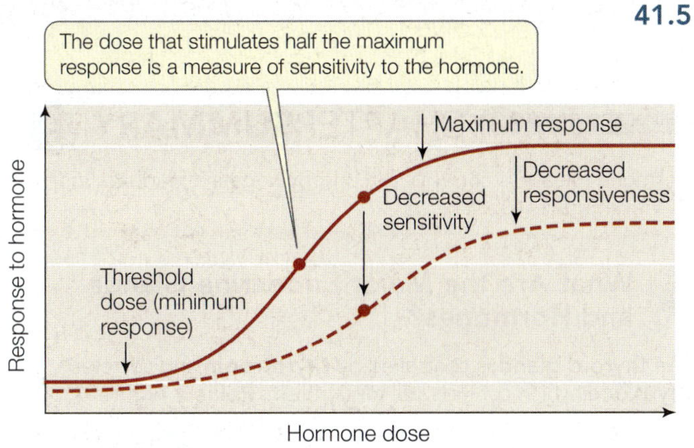
The dose that stimulates half the maximum response is a measure of sensitivity to the hormone.
Maximum response
Decreased responsiveness
Decreased sensitivity
Threshold dose (minimum response)
Response to hormone
Hormone dose

41.20 Dose–Response Curves Quantify the Body's Response to a Hormone Between the threshold and maximum doses, a dose–response curve frequently has an S shape. Anything that changes the responsiveness of a system—such as an increase or decrease in the number of receptors in target cells—affects the position of the curve.

blood is 1 to 3 minutes. The effects of other hormones, such as cortisol and thyroxine, are expressed over much longer periods, with half-lives measured in days.

Immunoassays have also facilitated the measurement of dose–response relationships. To evaluate a drug or a natural hormone for therapeutic use, it is critical to know the sensitivity of the body to that drug or hormone. Being able to measure the concentrations of drugs or hormones in the blood makes it possible to construct dose–response curves that help physicians adjust dosages appropriately (**Figure 41.20**).

A hormone can act through many receptors

Different receptors may be involved in mediating the actions of a single hormone. Because there are slight differences among the receptors for a particular hormone, it is possible to create drugs that are selective in blocking or stimulating specific responses. A number of receptors have been identified, isolated, and purified through biochemical separation techniques. For example, a hormone can be bound to a substrate such as resin beads packed into a glass column. When an extract of cells suspected of containing receptors to that hormone is added to the column, the receptors bind to the hormone molecules on the beads. The hormone–receptor complexes can then be washed off the beads and the receptors isolated. This technique is called **affinity chromatography**.

As more receptors are isolated and characterized, researchers discover that they frequently exist in families with common structural features. These common features result from common nucleotide sequences in the receptor genes. Genomic analyses have led to the discovery of many new receptors. Investigators "scan" the genome for sequences that bear homologies to known receptor gene sequences; when they get a "hit," they have found a candidate gene for a new receptor. They can then identify the molecule to which that receptor binds (its ligand), describe where the receptor occurs within the body, and characterize the receptor's physiological effects.

Knowing the molecular identity of a receptor and being able to measure its concentration makes it possible to study that receptor's regulation. We saw above that the release of hormones can be under negative feedback control. Similarly, the abundance of *receptors* for a hormone can be under feedback control. In some cases, continuous high concentrations of a hormone can decrease the number of its receptors, a process known as **downregulation**. An increased number of receptors can occur when hormone secretion is suppressed, resulting in **upregulation**. The regulation of receptor abundance is an important mechanism controlling the sensitivity of the body to hormonal signaling.

An example of downregulation occurs in type II diabetes, which is characterized by elevated insulin concentrations in the blood and a loss of insulin receptors. Although genetic factors are probably involved, a possible immediate cause of the disease is an overstimulation of pancreatic release of insulin by excessive carbohydrate intake, which leads to downregulation of the insulin receptors. An example of upregulation may be seen in people who have been on a regular dose of beta blockers (see p. 850). As the activity of the β-adrenergic receptors is blocked over time, more of these receptors are produced. If the person goes off the medication suddenly, the effects of the receptors are amplified, resulting in heightened anxiety. Changes in dosage in the long-term use of such medications thus are usually gradual and carefully supervised.

RECAP 41.5

Studying the mechanisms of hormone action requires the ability to measure hormone concentrations and to identify and characterize hormone receptors.

- Describe how an immunoassay is performed. **See p. 852 and Figure 41.19**
- How are receptors for a particular hormone identified? **See p. 853**

?

How can we demonstrate that a molecule found in the blood is a hormone?

ANSWER

In order for a molecule to be considered a hormone, we have to show that the molecule is released from cells in a certain tissue and that it has an effect on cells in another, distant tissue. We also have to show that the molecule is *necessary* to stimulate its presumed target cells and that it is *sufficient* to stimulate those target cells. In the experiment shown in Figure 41.6, when the bugs were decapitated soon after a blood meal, molting did not occur, but when they were decapitated a week after a blood meal, molting did occur. These results showed that some substance from the head, but not the head itself, was necessary for the molting response. Sufficiency was demonstrated by connecting two bugs with a glass tube allowing diffusion of substances between them. In this experiment, both bugs molted. In the irisin experiments (Figure 41.5), it was shown that the molecule came from muscle cells and acted on fat cells; that the molecule could stimulate "browning" of white fat cells in culture; (sufficiency); and that when the molecule was inactivated it blocked this "browning" of the white fat (necessity).

CHAPTER SUMMARY 41

41.1 What Are Hormones and How Do They Work?

- **Endocrine cells** secrete chemical signals that induce responses in other cells that have receptors for those molecules. In some cases endocrine cells are aggregated into **endocrine glands**.
- **Hormones** are endocrine signals that are secreted from a cell, circulate in the blood, and bind to target cells distant from the secreting cell. **Review Figure 41.1**
- Hormones fall into three general categories: proteins and peptides, steroids, and amines. Peptide and protein hormones and some amines are water-soluble; steroids and some amines are lipid-soluble. **Review Figure 41.2**
- Receptors for water-soluble hormones are located on the cell surface. Receptors for most lipid-soluble hormones are inside the cell.
- Hormones can cause different responses in different target cells. **Review Figure 41.3**

41.2 What Have Experiments Revealed about Hormones and Their Action?

- The chemical structures of hormones are highly conserved. Through evolution, however, hormones acquire different functions in different animal groups. **Review Figure 41.4**
- Early experiments identifying secretin as a hormone defined the characteristics of hormonal signaling. Modern experiments demonstrate these characteristics in order to identify hormone molecules. **Review Figure 41.5**
- Pioneering experiments illustrating hormonal action showed that two hormones, PTTH and ecdysone, control molting in arthropods. A third hormone, **juvenile hormone**, prevents maturation. **Review Figures 41.6, 41.7, ANIMATED TUTORIAL 41.1**

41.3 How Do the Nervous and Endocrine Systems Interact?

- In humans, the major endocrine glands are distributed around the body. **Review Figure 41.8, ACTIVITY 41.1**
- The **pituitary gland** is the interface between the nervous and endocrine systems. The **anterior pituitary** develops from embryonic mouth tissue; the **posterior pituitary** develops from the developing brain. **Review Figures 41.9, 41.10**
- The posterior pituitary secretes two **neurohormones**: **antidiuretic hormone** (**ADH**) and **oxytocin**. The anterior pituitary secretes **tropic hormones** (thyrotropin, corticotropin, luteinizing hormone, and follicle-stimulating hormone) as well as **growth hormone**, **prolactin**, **endorphins**, and **enkephalins**.
- The anterior pituitary is controlled by neurohormones produced by cells in the hypothalamus and transported through **portal blood vessels** to the anterior pituitary. **See ANIMATED TUTORIAL 41.2**
- Hormone release is controlled in part by negative feedback loops. **Review Figure 41.11**

41.4 What Are the Major Endocrine Glands and Hormones?

- The **thyroid gland** is controlled by **thyrotropin** and secretes **thyroxine**, which controls cell metabolism. **Review Figure 41.12**
- The level of calcium in the blood is regulated by three hormones. **Calcitonin** from the thyroid lowers blood calcium by promoting bone deposition. **Parathyroid hormone** (**PTH**) raises blood calcium by promoting bone turnover and decreasing calcium excretion. **Calcitriol** promotes calcium absorption from the digestive tract. **Review Figure 41.14, ANIMATED TUTORIAL 41.3**
- The pancreas secretes three hormones. **Insulin** stimulates glucose uptake by cells and lowers blood glucose, **glucagon** raises blood glucose, and **somatostatin** slows the rate of nutrient processing.
- The **adrenal gland** has two portions, one within the other. The inner portion, the **adrenal medulla**, releases **epinephrine** and **norepinephrine** in response to stress. The outer portion, the **adrenal cortex,** produces three classes of **corticosteroids**: glucocorticoids, mineralocorticoids, and small amounts of sex steroids. **Review Figure 41.15**
- **Aldosterone** is a mineralocorticoid that stimulates the kidneys to conserve sodium and excrete potassium. **Cortisol** is a glucocorticoid that is released in response to stressful stimuli but acts more slowly than the hormones of the adrenal medulla.
- Sex hormones (**androgens** in males, **estrogens** and **progesterone** in females) control sexual development, secondary sexual characteristics, and reproductive functions. **Review Figure 41.17**
- The **pineal gland** releases **melatonin**, a hormone involved in controlling biological rhythms. **Review Figure 41.18**

41.5 How Do We Study Mechanisms of Hormone Action?

- **Immunoassay** techniques are used to measure concentrations of hormones and other substances. **Review Figure 41.19**
- The body's sensitivity to a hormone is measured by a dose–response curve. **Review Figure 41.20**
- The sensitivity of a cell to a hormone can be altered by **down-regulation** or **upregulation** of the hormone's receptors in that cell.

See ACTIVITY 41.2 for a concept review of this chapter

Go to the Interactive Summary to review key figures, Animated Tutorials, and Activities
Life10e.com/is41

CHAPTER**REVIEW**

REMEMBERING

1. Prior to puberty
   a. the pituitary secretes luteinizing hormone and follicle-stimulating hormone, but the gonads are unresponsive.
   b. the hypothalamus does not secrete much gonadotropin-releasing hormone.
   c. males can stimulate massive muscle development through a vigorous training program.
   d. testosterone plays no role in development of the male sex organs.
   e. genetic females will develop male genitals unless estrogen is present.

2. Both epinephrine and cortisol are secreted in response to stress. Which of the following statements is also true for *both* of these hormones?
   a. They act to increase blood glucose availability.
   b. Their receptors are on the surfaces of target cells.
   c. They are secreted by the adrenal cortex.
   d. Their secretion is stimulated by corticotropin.
   e. They are secreted into the blood within seconds of the onset of stress.

3. The posterior pituitary
   a. synthesizes oxytocin.
   b. is under the control of hypothalamic releasing hormones.
   c. secretes tropic hormones.
   d. secretes neurohormones.
   e. is under feedback control by thyroxine.

4. PTH
   a. stimulates osteoblasts to lay down new bone.
   b. reduces blood calcium levels.
   c. stimulates calcitonin release.
   d. is produced by the thyroid gland.
   e. is released when blood calcium levels fall.

5. Steroid hormones
   a. are produced only by the adrenal cortex.
   b. have only cell-surface receptors.
   c. are water-soluble.
   d. act by altering the activity of proteins in the target cell.
   e. act by altering gene expression in the target cell.

UNDERSTANDING & APPLYING

6. Compare the characteristics you would expect of a hormone signaling system that controls a short-term process, such as digestion, with the characteristics you would expect of a hormone signaling system that controls a long-term process, such as embryonic development.

7. Some body builders, males and females, take high doses of synthetic male steroid hormones to enhance the growth of their muscles. Among other side effects, the ovarian cycles of the females stop and the males become sterile. Explain these consequences in terms of the hypothalamic/pituitary/gonadal hormonal axis.

8. Explain how both hyperthyroidism and hypothyroidism can both be associated with goiter.

ANALYZING & EVALUATING

9. Honey bees build honeycombs from beeswax that they produce. After extracting honey from the combs, beekeepers store the combs until they are needed again as the bees build up their honey reserves. Wax moths lay their eggs in honeycombs. The larvae eat the wax as they go through cycles of molting and growth until they metamorphose into adult moths. A beekeeper decided to store his honeycombs in a cold room. Months later when he went to get the spare combs out of the cold room, he found them full of large holes and gigantic moth larvae, but no adult moths. Hypothesize as to what aspect of the moth endocrine system was altered by the low temperature.

10. Neurons (the cells of the nervous system) do not require insulin. Why might this be so, and why is it important?

Go to BioPortal at **yourBioPortal.com** for Animated Tutorials, Activities, LearningCurve Quizzes, Flashcards, and many other study and review resources.

42 Immunology: Animal Defense Systems

CHAPTER**OUTLINE**

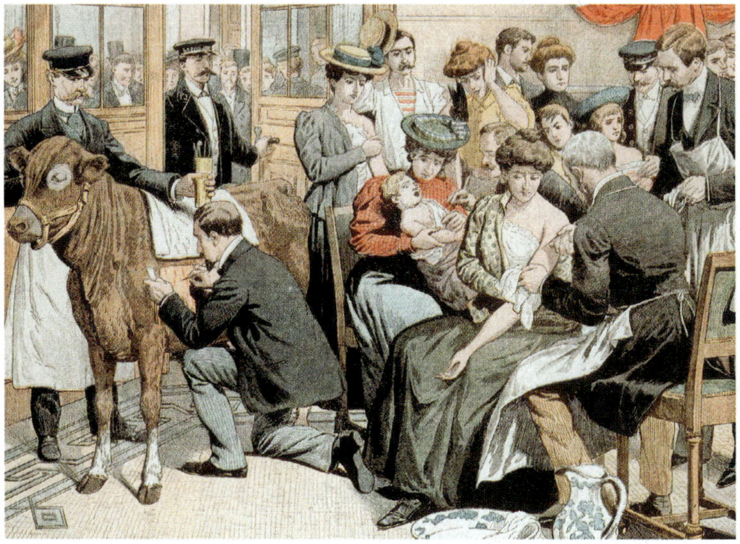

Fighting the Pox This 1905 illustration depicts Parisians being vaccinated against smallpox with serum from a cow infected with cowpox, a disease similar to smallpox that causes mild symptoms in humans.

TWO TINY VIALS IN DEEP FREEZERS, one in Atlanta and the other in Siberia, are all that is left of the smallpox virus, long a scourge of humanity. It last occurred as a human pathogen in 1978, after killing more than 300 million people in the twentieth century alone. Smallpox was eliminated by the human immune system with the help of vaccination. A vaccine is usually an inactive form of a pathogen or toxin that nevertheless provokes the immune system to produce antibodies: specific proteins directed against the target. The immune system destroys whatever is bound to the antibodies.

The eradication of smallpox was a spectacular international accomplishment. One might think that vaccination for other potentially lethal diseases such as the flu (influenza) would be widely accepted by the public. But in the U.S. and some other Western countries, this is not necessarily so. In fact, some surveys show that more than one-third of Americans refuse flu shots, and a significant number of parents refuse vaccination for their children.

Unfortunately, those who refuse vaccination may harm people other than themselves. A vaccination program can control or eradicate a disease only if a high percentage (typically above 80 percent) of people are vaccinated, thus disrupting the chain of infection from person to person. This level of vaccination results in "herd immunity," meaning that even those who cannot be vaccinated or who have weak immune systems are protected from infection. This protection is lost if the vaccination rate falls below the level needed for herd immunity. Those who are old or sick, and infants whose immune systems have not yet fully developed, are most at risk.

The best way to develop herd immunity in a population is compulsory vaccination. In many countries, vaccination is a prerequisite for school enrollment and military enlistment and is required during epidemics. For example, in the periodic smallpox epidemics during the twentieth century in the U.S., doctors accompanied by police would go into neighborhoods where the disease raged, vaccinating all those who were uninfected and removing infected people to quarantine. You can imagine the reaction of parents whose children were taken away, often to die. When Henning Jacobson, a Swedish immigrant in Massachusetts, refused vaccination during a smallpox epidemic, he was arrested and took his case to the U.S. Supreme Court. It ruled in 1905 that while personal freedom is important, each state was entitled to protect its citizens. This provided a legal framework for compulsory vaccination that continues to this day. But opposition and court challenges continue. These cases of "freedom v. immunity" are a political dilemma.

? Why do many people resist vaccination?

See answer on p. 877.

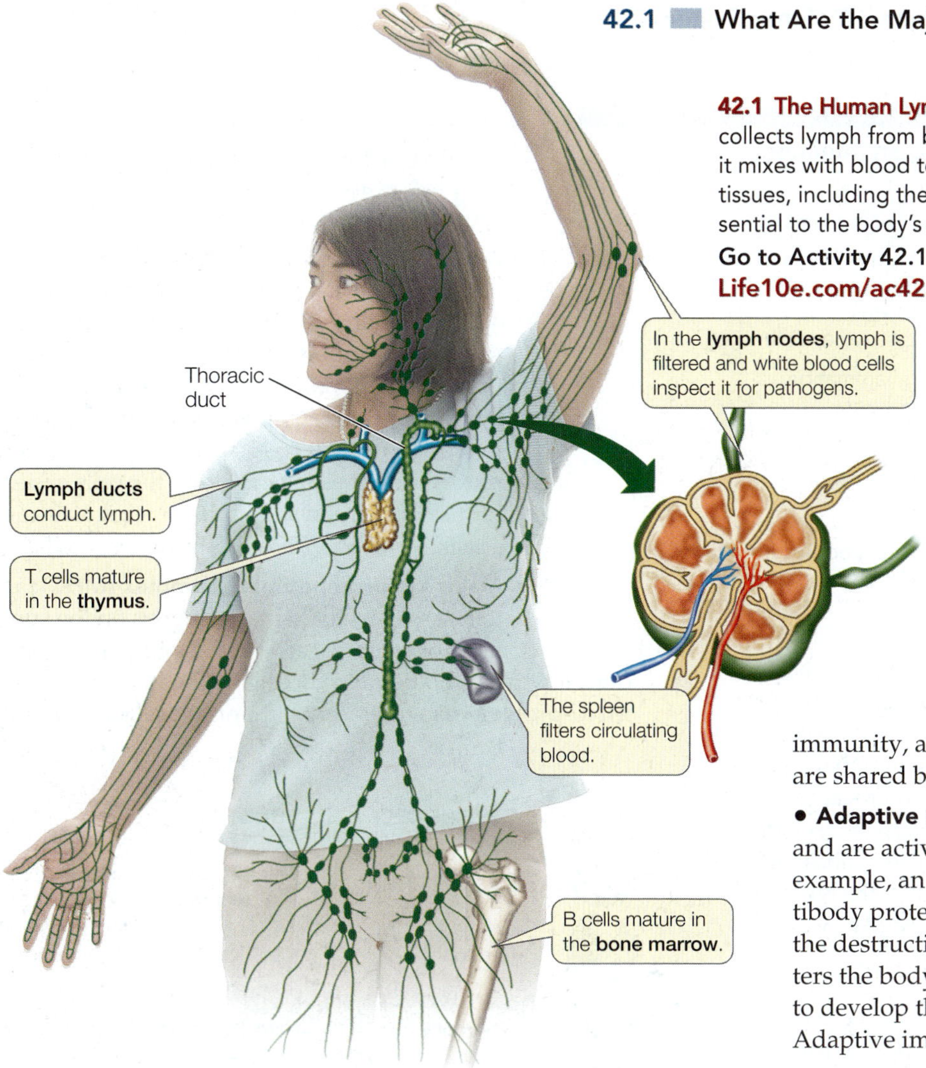

Thoracic duct
Lymph ducts conduct lymph.
T cells mature in the thymus.
The spleen filters circulating blood.
B cells mature in the bone marrow.
In the lymph nodes, lymph is filtered and white blood cells inspect it for pathogens.

42.1 The Human Lymphatic System A network of ducts and vessels collects lymph from body tissues and carries it toward the heart, where it mixes with blood to be pumped back to the tissues. Other lymphoid tissues, including the thymus, spleen, and bone marrow, are also essential to the body's defense system.

Go to Activity 42.1 The Human Defense System
Life10e.com/ac42.1

42.1 What Are the Major Defense Systems of Animals?

Animals have several ways of defending themselves against **pathogens**—harmful organisms and viruses that can cause disease. These defense systems are based on the distinction between *self*—the animal's own molecules—and *nonself*, or foreign, molecules. The defensive response involves three phases:

- *Recognition phase*. The organism must be able to discriminate between self and nonself.
- *Activation phase*. The recognition event leads to a mobilization of cells and molecules to fight the invader.
- *Effector phase*. The mobilized cells and molecules destroy the invader.

There are two general types of defense mechanisms:

- **Innate defenses**, or nonspecific defenses, provide the first line of defense against pathogens. They typically act very rapidly and include barriers such as the skin, molecules that are toxic to invaders, and phagocytic cells that ingest invaders. (Recall from Section 6.5 that phagocytosis is a form of endocytosis, in which a cell engulfs a large particle or another cell.) This system recognizes broad classes of organisms or molecules and responds quickly, within minutes or hours. All animals have some level of innate immunity, and certain innate immunity mechanisms are shared between animals and plants.
- **Adaptive defenses** are aimed at specific pathogens and are activated by the innate immune system. For example, an adaptive defense system can make an antibody protein that will recognize, bind to, and aid in the destruction of a specific virus, if that virus ever enters the body. Adaptive defenses are typically slower to develop than the innate response and longer-lasting. Adaptive immunity evolved in vertebrate animals.

Mammals have both kinds of defense mechanism and are the focus of this chapter. In mammals and other vertebrates, the innate and adaptive mechanisms operate together as a coordinated defense system. **Table 42.1** gives an overview of these processes during the course of an infection. Innate immunity is the body's first line of defense because the adaptive defenses often require days or even weeks to become effective.

Blood and lymph tissues play important roles in defense

The components of the mammalian defense system are dispersed throughout the body and interact with almost all of its other tissues and organs. The lymphoid tissues, which include the thymus, bone marrow, spleen, and lymph nodes, are essential parts of the defense system (**Figure 42.1**). The blood and lymph are complex systems with nondefensive functions that will be discussed in Chapter 50. They each have central roles in defense as well.

The blood and lymph both consist of liquids in which cells are suspended:

- **Blood plasma** is a yellowish solution containing ions, small molecule solutes, and soluble proteins. Suspended in the plasma are red blood cells, white blood cells, and platelets (cell fragments essential to blood clotting). Whereas red blood cells are normally confined to the closed circulatory system (the heart, arteries, capillaries, and veins), white blood cells and platelets are also found in the lymph.

TABLE **42.1**

Innate and Adaptive Immune Responses to an Infection

Response (Time after Infection by a Pathogen)	System	Mechanisms
Early (0–4 hr)	Innate, nonspecific (first line)	Barrier (skin and lining of organs) Dryness, low pH Mucus Lysozyme, defensins
Middle (>4–96 hr)	Innate, nonspecific (second line)	Inflammation Phagocytosis Natural killer cells Complement system Interferons
Late (>96 hr)	Adaptive, specific	Humoral immunity (antibodies from B cells) Cellular immunity (T cells)

- **Lymph** is a fluid that is derived from the blood (but lacking red blood cells) and other tissues and accumulates in intercellular spaces throughout the body. From these spaces, the lymph moves slowly into the vessels of the lymphatic system. Tiny lymph capillaries conduct this fluid to larger ducts that eventually join together, forming one large vessel, the thoracic duct, which joins a major vein (the left subclavian vein) near the heart. By this system of vessels, the lymph is eventually returned to the blood and the circulatory system.

At many sites along the lymph vessels are small, roundish structures called **lymph nodes**, which contain a type of white blood cell called a **lymphocyte**. As lymph passes through a lymph node, the lymphocytes encounter foreign cells and molecules that have entered the body, and if they are recognized as nonself, an immune response is initiated.

White blood cells play many defensive roles

One milliliter of human blood typically contains about *5 billion* red blood cells and *7 million* of the larger **white blood cells** (also called leukocytes). All of these cells originate from multipotent stem cells (constantly dividing undifferentiated cells that can form several different cell types; see Section 19.2) in the bone marrow. There are two major families of white blood cells: lymphocytes and phagocytes (**Figure 42.2**). Lymphocytes include the B cells and T cells; they are smaller than other white blood cells and are not phagocytic. **Phagocytes** include most of the other cells shown in Figure 42.2, and as their name suggests, they are phagocytic. Each kind of white blood cell has specialized functions. Some phagocytes are also referred to collectively as granulocytes because they contain numerous granules (vesicles containing defensive molecules). Defensive proteins and signals play fundamental roles in the interactions and functioning of these cells.

Immune system proteins bind pathogens or signal other cells

The cells that defend mammalian bodies work together, interacting with one another and with the cells of invading pathogens. These cell–cell interactions are accomplished by a variety of key proteins, including receptors, other cell surface proteins, and signaling molecules. Four of the major players are listed here, and will be discussed in more detail later in the chapter.

- **Antibodies** are proteins that bind specifically to certain substances identified by the immune system as nonself. They recognize and bind specific configurations of atoms. The molecules that bind antibodies are called **antigens**. This binding can directly inactivate viruses and toxins; on nonself cells, antibody–antigen complexes can act as tags, making the cells easier for the immune system cells to recognize and attack. Antibodies are produced by B cells.
- **Major histocompatibility complex** (**MHC**) proteins are used to display antigens on the surfaces of self cells, so that the antigens can be detected by antibodies and by cells of the immune system. MHC proteins also function as important self-identifying

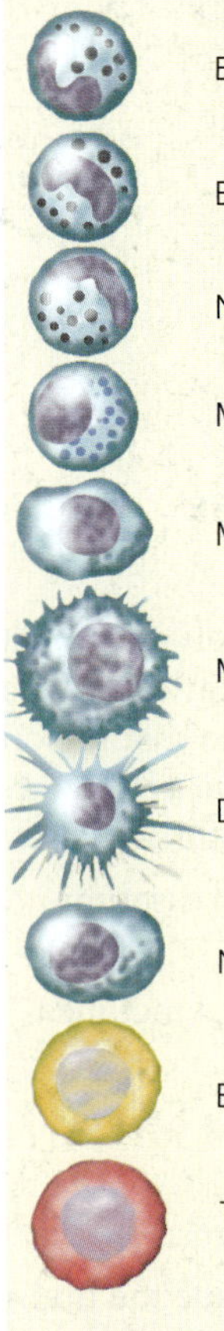

TYPE OF CELL	FUNCTION
Basophils (I, A)	Release histamine; may promote development of T cells
Eosinophils (A)	Kill antibody-coated parasites
Neutrophils (I)	Stimulate inflammation; engulf and digest microorganisms
Mast cells (I)	Release histamine when damaged
Monocytes (I, A)	Develop into macrophages and dendritic cells
Macrophages (I, A)	Engulf and digest microorganisms; activate T cells
Dendritic cells (A)	Present antigens to T cells
Natural killer cells (I)	Attack and lyse virus-infected or cancerous body cells
B lymphocytes (A)	Differentiate to form antibody-producing cells and memory cells
T lymphocytes (A)	Kill virus-infected cells or cancer cells; regulate activities of other white blood cells

42.2 White Blood Cells White blood cells have key roles in both innate (I) and adaptive (A) immunity. The lymphocytes are the B cells and T cells; the other cell types are phagocytes.

Go to Animated Tutorial 42.1
Cells of the Immune System
Life10e.com/at42.1

labels. There are two major classes of MHC proteins: MHC I proteins are found on the surfaces of most cells in the mammalian body, whereas MHC II proteins are found on immune system cells.

- **T cell receptors** are integral membrane proteins on the surfaces of T cells. They recognize and bind to antigens presented by the MHC proteins on the surfaces of other cells.
- **Cytokines** are soluble signaling proteins released by many cell types. They bind to cell surface receptors and alter the behavior of their target cells. Various cytokines activate or inactivate B cells, macrophages (see Figure 42.1), and T cells.

RECAP 42.1

All animals have innate defenses against pathogens, and vertebrates have innate and adaptive defenses. Both kinds of mechanisms are based on the ability to differentiate self from nonself. Innate defenses target a broad range of molecules and organisms, whereas adaptive defenses target specific pathogens.

- List the differences between innate and adaptive defenses. **See p. 857**
- What are the two classes of white blood cells, and how do they function in innate or adaptive immunity? **See p. 858 and Figure 42.2**

The outcome of a disease—the life or death of the host—often depends on the success of both rapid, innate responses and long-lasting, adaptive responses to invading pathogens. We will turn now to the innate defenses that protect vertebrates from disease.

42.2 What Are the Characteristics of the Innate Defenses?

Innate defenses are general protection mechanisms that attempt to stop pathogens from invading the body or to quickly eliminate those that do manage to invade. They are genetically programmed (innate) and "ready to go," in contrast to adaptive responses, which take time to develop after a pathogen or toxin has been recognized as nonself. In mammals, innate defenses include physical barriers as well as cellular and chemical defenses (**Figure 42.3**).

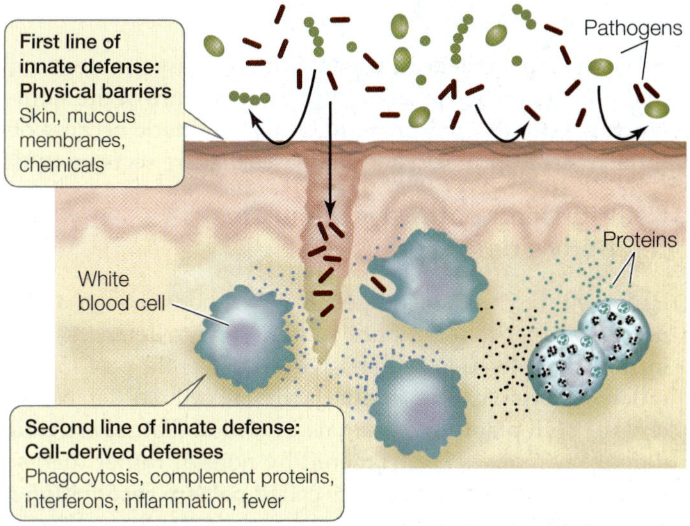
First line of innate defense: Physical barriers
Skin, mucous membranes, chemicals
Pathogens
Proteins
White blood cell
Second line of innate defense: Cell-derived defenses
Phagocytosis, complement proteins, interferons, inflammation, fever

42.3 Innate Immunity Physical barriers, cells, and proteins (complement and interferons) provide nonspecific defenses against invading pathogens.

Go to Media Clip 42.1
The Chase Is On: Phagocyte versus Bacteria
Life10e.com/mc42.1

Barriers and local agents defend the body against invaders

The first line of innate defense is encountered by a potential pathogen as soon as it lands on the surface of an animal. Consider a pathogenic bacterium that lands on human skin. The challenges faced by the bacterium just to reach its target are formidable:

- The *physical barrier of the skin*: Bacteria rarely penetrate intact skin; by the same token, broken skin increases the risk of infection.
- The *saltiness and dryness of skin*: This environment may not be hospitable to the growth of the bacterium.
- The *presence of normal flora*: Bacteria and fungi that normally live and sometimes reproduce in great numbers on our body surfaces without causing disease will compete with pathogens for space and nutrients.

If a pathogen lands inside the nose or another internal organ, it faces other innate defenses:

- **Mucus** is a slippery secretion produced by mucous membranes found at the inner surfaces of the nose (as well as the digestive, respiratory, and urogenital systems). Mucus traps microorganisms so they can be removed by the beating of cilia (see Figure 5.17), which continuously move the mucus and its trapped debris away.
- **Lysozyme** is an enzyme made by mucous membranes that attacks the cell walls of many bacteria, causing them to lyse (burst open).
- **Defensins**, also made by mucous membranes, are peptides of 18 to 45 amino acids that contain hydrophobic domains. They are toxic to a wide range of pathogens, including bacteria, microbial eukaryotes, and enveloped viruses. Defensins insert themselves into the plasma membranes of these organisms and make the membranes permeable, thus killing the invaders. Defensins are also produced in phagocytes, where they kill pathogens ingested by phagocytosis. Plants also produce defensins in response to pathogen exposure (see Chapter 39).

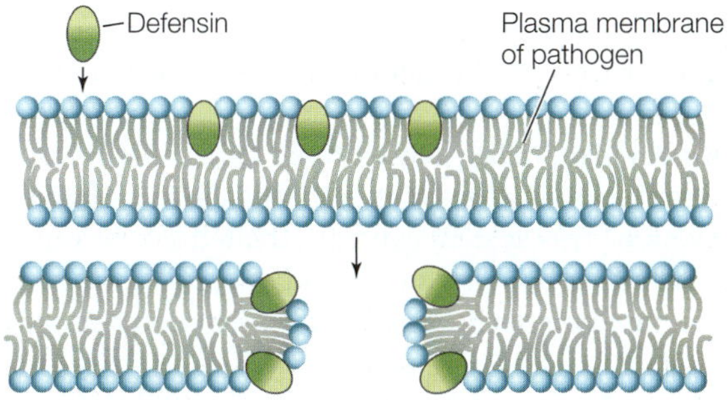
Defensin
Plasma membrane of pathogen

- Harsh conditions in an animal's internal environment can also kill pathogens. For example, gastric juice in the stomach is a deadly environment for many bacteria because of the hydrochloric acid and proteases that are secreted into it.

Cell signaling pathways stimulate the body's defenses

Pathogens that are able to penetrate the body's outer and inner surfaces encounter more complex innate defenses. These include the activation of defensive cells. Over the past 15 years, much has been learned about the innate cellular defense systems of both plants and animals. A critical feature is that the defense responses are triggered by nonself rather than self molecules. A class of receptors called **pattern recognition receptors** (**PRRs**) plays an important role in distinguishing self from nonself.

PRRs are present in cells that play roles in the innate immune systems of both plants and animals. In mammals these include macrophages, dendritic cells, and natural killer cells. The molecules recognized by PRRs are called **pathogen associated molecular patterns** (**PAMPs**). As we described in Chapter 39, these are molecules that are unique to large classes of microbes, such as bacterial lipopolysaccharides, which are found in bacterial cell membranes.

An invading pathogen can be regarded as a signal. In response to that signal, the body produces molecules (complement proteins, interferons, and other cytokines) that regulate phagocytosis and other defense processes. Not surprisingly, the link between signal and response is a signal transduction pathway, similar to the ones we considered in Section 7.3. A key group of PRRs in mammals is the **toll-like receptors**, which activate signal transduction pathways involved in both innate and adaptive defenses (Figure 42.4). The toll protein was first identified in insects, where it is involved in development and in sensing infection. Comparative genomics has revealed at least ten similar receptors in humans. Bruce Beutler, the scientist who first described these receptors, won the Nobel Prize in 2011. Binding of a PAMP to the receptor sets in motion a cascade of molecular changes, including the activation of the transcription factor NF-κB. (NF-κB stands for *n*uclear *f*actor *kappa* light chain enhancer of activated *B* cells.) The activated NF-κB enters the nucleus, where it activates the transcription of genes encoding defensive proteins.

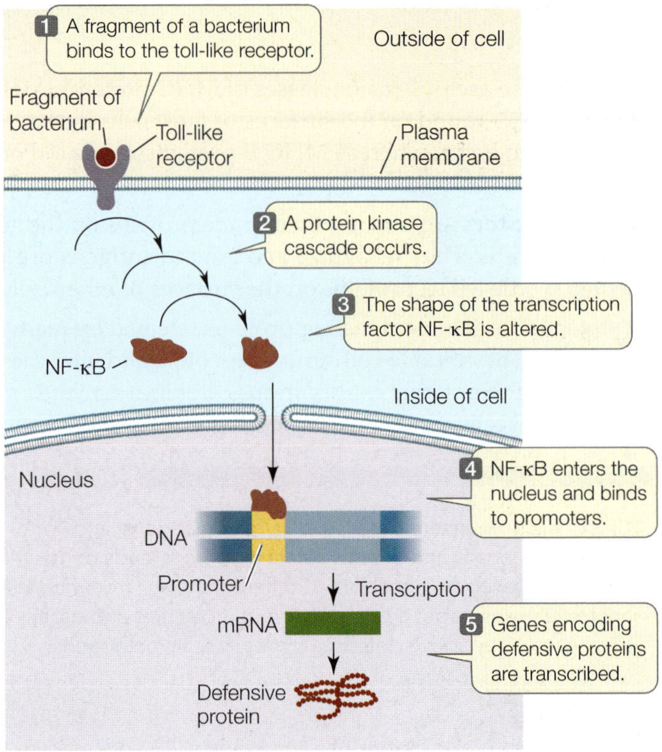
1 A fragment of a bacterium binds to the toll-like receptor.
Outside of cell
Fragment of bacterium
Toll-like receptor
Plasma membrane
2 A protein kinase cascade occurs.
3 The shape of the transcription factor NF-κB is altered.
NF-κB
Inside of cell
Nucleus
4 NF-κB enters the nucleus and binds to promoters.
DNA
Promoter
Transcription
mRNA
5 Genes encoding defensive proteins are transcribed.
Defensive protein

42.4 Cell Signaling and Defense Binding of a pathogenic molecule or fragment to the toll-like receptor initiates a signal transduction pathway that results in the transcription of genes whose products are involved in adaptive and innate defenses.

Specialized proteins and cells participate in innate immunity

Several proteins are produced by the body either before an infection occurs or in response to invasion by pathogens. Two important groups are the complement and interferon proteins.

COMPLEMENT PROTEINS Vertebrate blood contains more than 20 different proteins that make up the antimicrobial **complement system**. This system can be activated by various mechanisms, including both innate and adaptive defense responses. The proteins act in a characteristic sequence, or cascade, with each protein activating the next:

- First, the proteins attach to specific components on the surface of a microbe or to an antibody that has already bound to the microbe's surface. In either case, binding helps phagocytes recognize and destroy the microbe.
- Then, complement proteins activate the inflammatory response (see next page) and attract phagocytes to the site of infection.
- Finally, complement proteins lyse invading cells (such as bacteria).

INTERFERONS When a cell is infected by a pathogen, it produces small amounts of signaling proteins called **interferons** that increase the resistance of neighboring cells to infection. Interferons are a class of cytokines and have been found in many vertebrates. Various molecules, including double-stranded (viral) RNA, induce the production of interferons. Thus interferons are particularly important as a first line of defense against viruses. Interferons bind to receptors on the plasma membranes of uninfected cells, stimulating a signaling pathway that inhibits viral reproduction if the cells are subsequently infected. In addition, interferons stimulate the cells to hydrolyze bacterial or viral proteins to peptides, an initial step in adaptive immunity (see Section 42.3).

PHAGOCYTES Some phagocytes travel freely in the circulatory and lymphatic systems; others can move out of blood vessels and adhere to certain tissues. Pathogenic cells, viruses, or fragments of these invaders are recognized by phagocytes, which then ingest them by phagocytosis. Defensins, nitric oxide, and

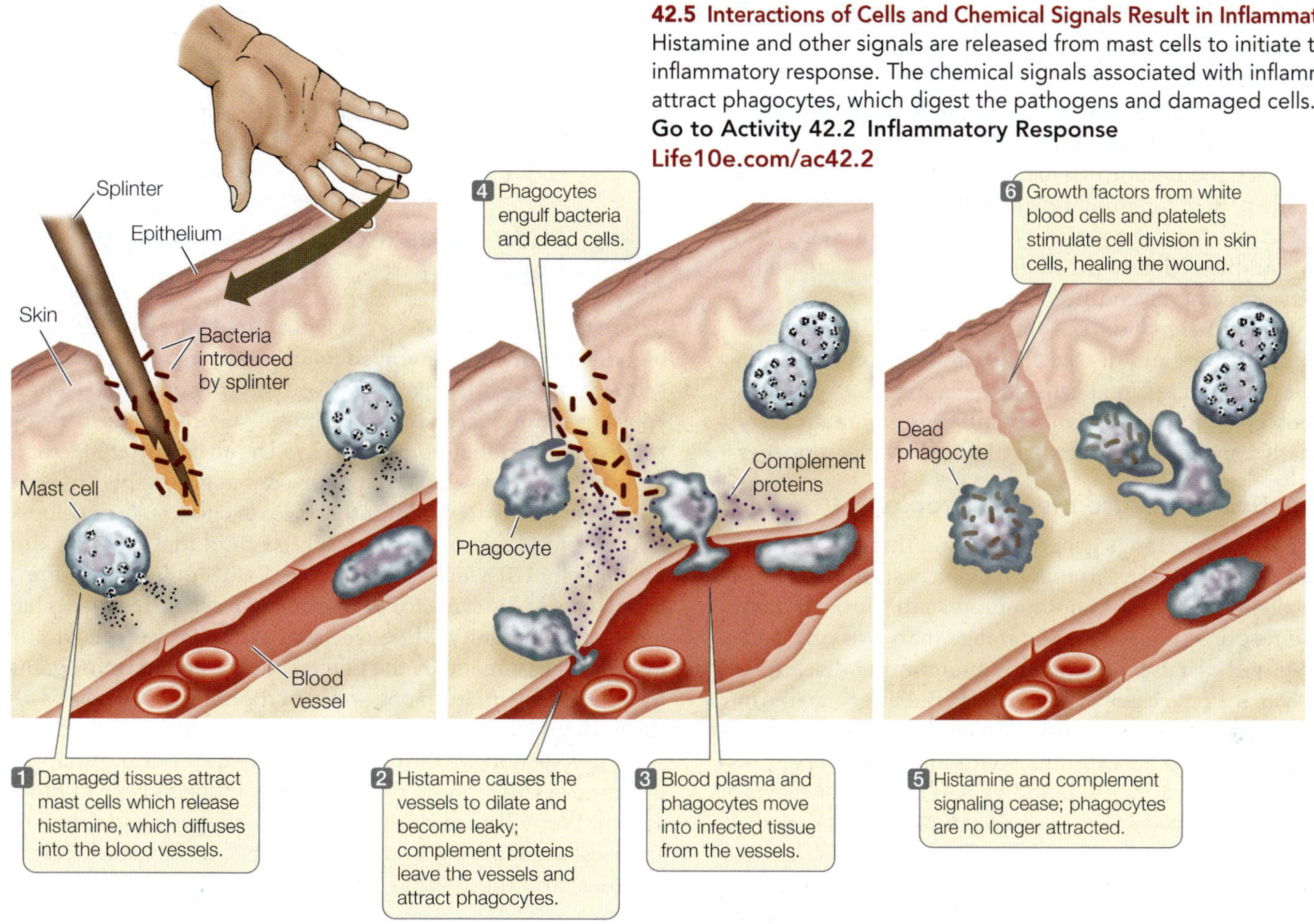
Splinter
Epithelium
Skin
Bacteria introduced by splinter
Mast cell
Blood vessel
4 Phagocytes engulf bacteria and dead cells.
Complement proteins
Phagocyte
6 Growth factors from white blood cells and platelets stimulate cell division in skin cells, healing the wound.
Dead phagocyte
1 Damaged tissues attract mast cells which release histamine, which diffuses into the blood vessels.
2 Histamine causes the vessels to dilate and become leaky; complement proteins leave the vessels and attract phagocytes.
3 Blood plasma and phagocytes move into infected tissue from the vessels.
5 Histamine and complement signaling cease; phagocytes are no longer attracted.

42.5 Interactions of Cells and Chemical Signals Result in Inflammation Histamine and other signals are released from mast cells to initiate the inflammatory response. The chemical signals associated with inflammation attract phagocytes, which digest the pathogens and damaged cells.
Go to Activity 42.2 Inflammatory Response
Life10e.com/ac42.2

reactive oxygen intermediates (see also Section 39.1) inside these phagocytes then kill the pathogens.

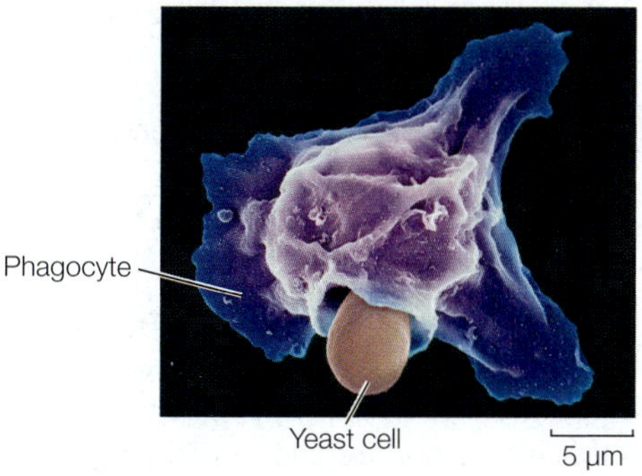
Phagocyte
Yeast cell
5 µm

NATURAL KILLER CELLS One class of lymphocytes, known as **natural killer cells**, can distinguish virus-infected cells and some tumor cells from their normal counterparts and initiate the apoptosis of these target cells. In addition to this innate defense action, natural killer cells interact with the adaptive defense mechanisms by lysing antibody-labeled target cells.

DENDRITIC CELLS These phagocytes act as messengers between the innate and adaptive systems. They can endocytose microbes, viruses, and even virus-infected host cells. Once inside a dendritic cell, these particles are digested to fragments, and if fragments have PAMPs, the dendritic cell "presents" an antigenic fragment on its surface, along with class II MHC proteins. In addition, the dendritic cell secretes signals that activate cells of the adaptive immune system.

Inflammation is a coordinated response to infection or injury

When mammalian tissue is damaged because of infection or injury, the body responds with **inflammation**. This response can happen almost anywhere in the body, internally as well as on the surface. Inflammation is an important phenomenon: it isolates the damaged area to stop the spread of the damage; it recruits cells and molecules to the damaged location to kill the invader; and it promotes healing. The first responders to tissue damage are **mast cells**, which adhere to the skin and the linings of organs and release numerous chemical signals, including:

- **Tumor necrosis factor**, a cytokine protein that kills target cells and activates immune cells.
- **Prostaglandins**, fatty acid derivatives involved in various responses, including the widening of blood vessels. Prostaglandins interact with nerve endings and are partly responsible for the pain caused by inflammation.
- **Histamine**, an amino acid derivative that leads to itchy, watery eyes and rashes seen with some types of allergic reactions.

The redness and heat of inflammation result from the dilation and leakiness of blood vessels in the infected or injured area (**Figure 42.5**). Phagocytes enter the inflamed area, where they engulf the invaders and dead tissue cells. Phagocytes are responsible for most of the healing associated with inflammation. They produce several cytokines, which (among other functions) can signal the brain to produce a fever. This rise in body temperature accelerates lymphocyte production and phagocytosis, thereby speeding the immune response. In some cases, pathogens are temperature-sensitive and their growth is inhibited. The pain of inflammation results from increased pressure due to swelling, the action of leaked enzymes on nerve endings, and the action of prostaglandins, which increase the sensitivity of the nerve endings to pain.

Following inflammation, pus may accumulate. Pus is a mixture of leaked fluid and dead cells: bacteria, neutrophils (the most abundant white blood cells—see Figure 42.2), and damaged body cells. Pus is a normal result of inflammation and is gradually consumed and further digested by macrophages.

Inflammation can cause medical problems

Although inflammation is generally a good thing, sometimes the inflammatory response is inappropriately strong, resulting in some allergies, cases of autoimmunity, and sepsis. In these cases the response causes more damage than was originally there. We will discuss allergy and autoimmune diseases in Section 42.6. In some cases of severe bacterial infection, the inflammatory response does not remain local. Instead it extends throughout the bloodstream in a condition called sepsis. As in a local infection or injury, blood vessels dilate, but they do so throughout the body. The lowering of blood pressure that results is a medical emergency and can be lethal.

The symptoms of swelling, pain, and fever caused by excessive inflammation can be bothersome to the point of incapacitation. Diseases such as rheumatoid arthritis and chronic obstructive pulmonary disease, and accidents such as athletic injuries, result in tissue damage and an inflammatory response. In order to manage excessive inflammation, drugs have been developed that act on the various cytokines and signal transduction pathways to reduce inflammation and its symptoms. For example, aspirin works by inhibiting an enzyme in the pathway for the synthesis of prostaglandins. Other anti-inflammatory drugs act on the prostaglandin pathway, on the actions of tumor necrosis factor, and on the actions of histamine.

RECAP 42.2

Innate immunity is the first line of defense against pathogens. Innate immunity includes physical barriers such as the skin, and cellular responses involving the recognition of self and nonself molecules. Recognition of nonself molecules by white blood cells leads to coordinated responses such as the production of defensive proteins and inflammation.

- How do complement proteins and interferons defend the body against microbes? See p. 860
- What are the roles of pathogen-associated patterns (PAMPs) and pattern recognition receptors (PRRs) in innate defenses? See p. 860 and Figure 42.4
- Describe the inflammatory response. See pp. 861–862 and Figure 42.5

Often the innate immune system, with its nonspecific defenses, is adequate to prevent or fight off a pathogenic infection. But in many cases this system works together with adaptive immunity, which detects and responds to specific pathogens. We will now turn to the development and functioning of adaptive immunity.

42.3 How Does Adaptive Immunity Develop?

Before the twentieth century, scientists had long suspected that blood was somehow involved in immunity against pathogens. More than a century ago, Emil von Behring and Shibasaburo Kitasato at the University of Marburg in Germany performed a key experiment that pointed to blood as an important factor in immunity (Figure 42.6). They showed that guinea pigs injected with a sublethal dose of diphtheria toxin or bacteria developed in their blood serum (the noncellular fluid that remains after blood is clotted) a factor that protected other guinea pigs from a lethal dose of the same toxin. In other words, the recipients had developed **immunity**. Moreover, the immunity was specific: the immune factor made by the guinea pigs protected only against the specific toxin, from one strain of diphtheria-causing bacteria, with which they had been injected.

In this section we outline the main features of the adaptive immune system, much of which does indeed occur in blood serum. We will consider the two major types of adaptive responses: the humoral immune response, which produces antibodies; and the cellular immune response, which destroys infected cells.

Adaptive immunity has four key features

Four important features of the adaptive immune system are:

- specificity
- the ability to distinguish self from nonself
- the ability to respond to an enormous diversity of nonself molecules
- immunological memory

SPECIFICITY Lymphocytes (B and T cells) are crucial components of adaptive immunity. T cell receptors and the antibodies produced by B cells recognize and bind to specific nonself substances (antigens), and this interaction initiates an adaptive immune response. The specific sites on antigens that the immune system recognizes are called **antigenic determinants**, or **epitopes**:

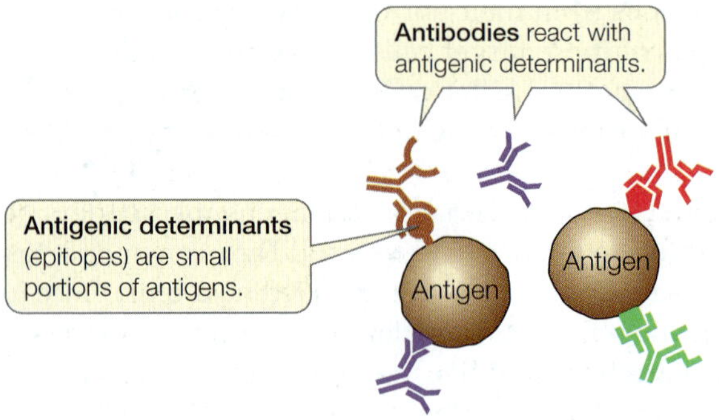
Antibodies react with antigenic determinants.
Antigenic determinants (epitopes) are small portions of antigens.
Antigen
Antigen

An antigenic determinant is a specific portion of a large molecule, such as a certain sequence of amino acids that may be present in a protein. Antigens are usually proteins or

INVESTIGATING**LIFE**

42.6 The Discovery of Adaptive Immunity Until the twentieth century, most people did not survive an attack of the bacterium that causes diphtheria, but a few did. Emil von Behring and Shibasaburo Kitasato performed a key experiment using an animal model, and demonstrated that the factor(s) responsible for immunity against diphtheria were in blood serum.[a]

HYPOTHESIS Serum from guinea pigs injected with a sublethal dose of diphtheria toxin protects other guinea pigs that are exposed to a lethal dose of the same toxin.

Method

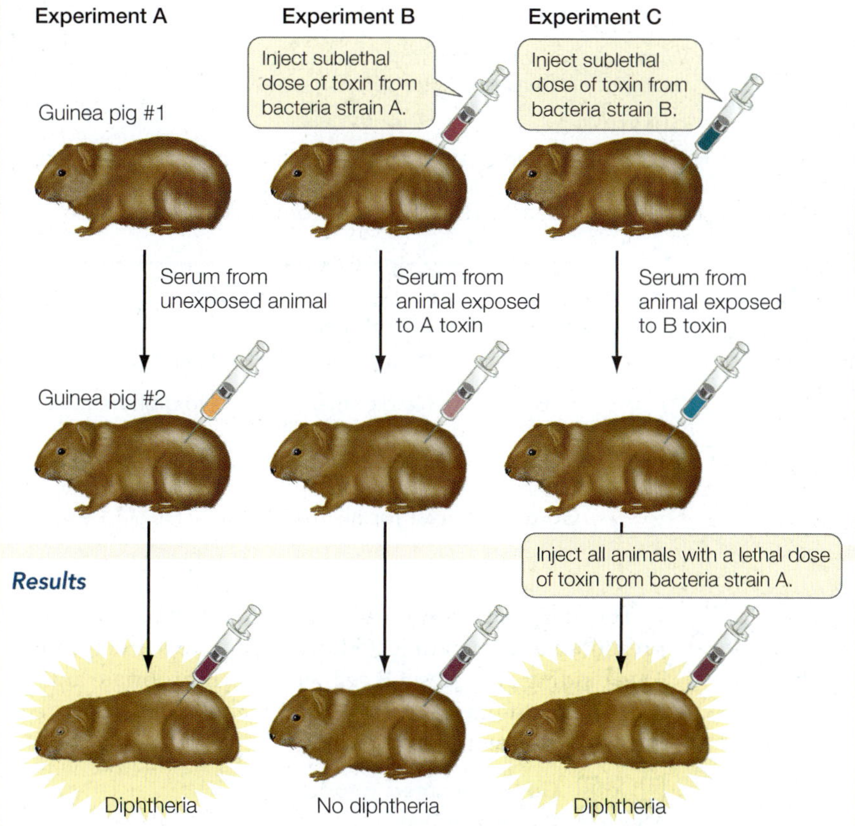
Experiment A
Experiment B
Experiment C
Inject sublethal dose of toxin from bacteria strain A.
Inject sublethal dose of toxin from bacteria strain B.
Guinea pig #1
Serum from unexposed animal
Serum from animal exposed to A toxin
Serum from animal exposed to B toxin
Guinea pig #2
Inject all animals with a lethal dose of toxin from bacteria strain A.
Results
Diphtheria
No diphtheria
Diphtheria

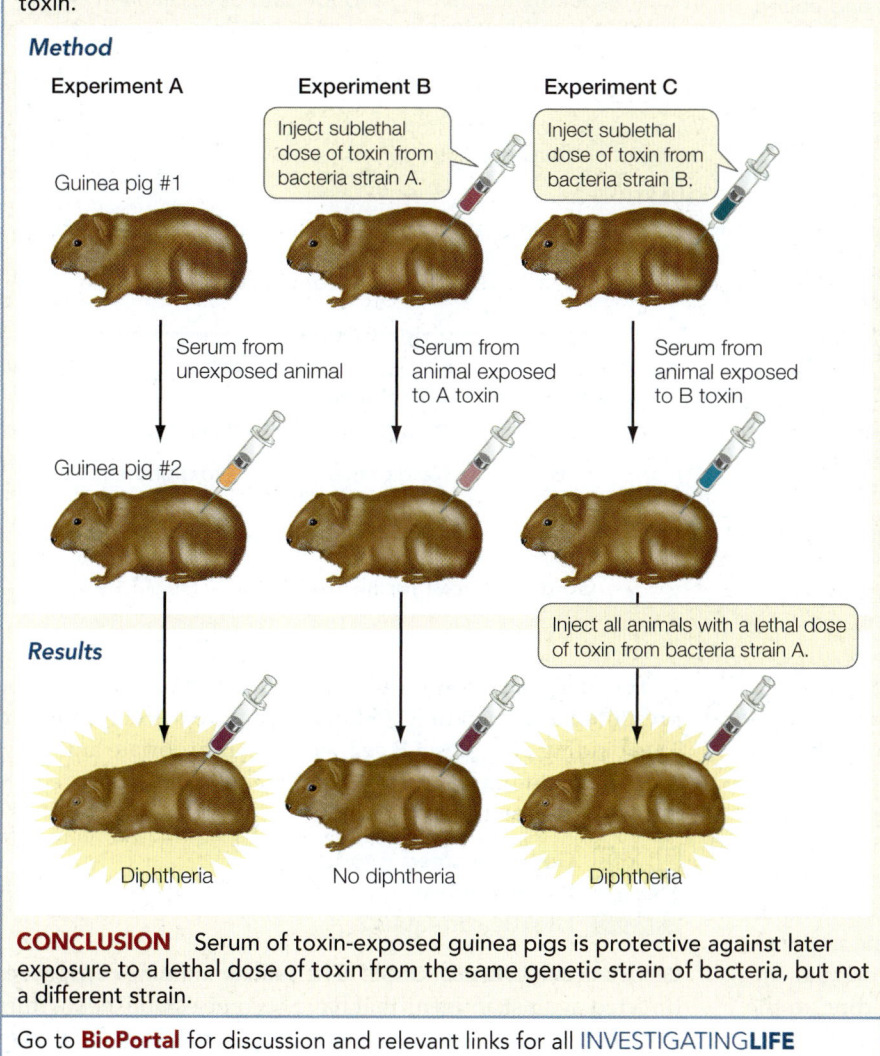

CONCLUSION Serum of toxin-exposed guinea pigs is protective against later exposure to a lethal dose of toxin from the same genetic strain of bacteria, but not a different strain.

Go to **BioPortal** for discussion and relevant links for all INVESTIGATING**LIFE** figures.

[a]Behring, E. and S. Kitasato. 1890. *Deustche medizinische Wochenschrift* 16: 1113–1114.

polysaccharides, and there can be multiple antigens on a single invading bacterium. A single antigenic molecule can have multiple, different antigenic determinants. The host animal responds to the presence of an antigen with highly specific defenses involving T cell receptors and antibodies. These receptors and soluble proteins bind to the antigenic determinants. Each T cell and each antibody is specific for a single antigenic determinant. For the remainder of the chapter, we will refer to antigenic determinants simply as "antigens."

DISTINGUISHING SELF FROM NONSELF We have seen how the innate immune system distinguishes between self and nonself molecules. The adaptive immune system has another set of mechanisms for distinguishing self from nonself. The human body contains tens of thousands of different proteins, each with a specific three-dimensional structure capable of generating immune responses. Thus every cell in the body bears a tremendous number of antigens. A crucial requirement of an individual's adaptive immune system is that it recognize the body's own antigens and not attack them. This is accomplished by clonal deletion, negative selection, and the action of Treg cells (we will discuss these mechanisms later in the chapter.)

DIVERSITY Challenges to the immune system are numerous. Pathogens take many forms: viruses, bacteria, protists, fungi, and multicellular parasites. Furthermore, each pathogenic species usually exists as many subtly different genetic strains, and each strain possesses multiple surface features. Estimates vary, but a reasonable guess is that humans can respond specifically to 10 million different antigens. Upon recognizing an antigen, the adaptive immune system responds by activating lymphocytes of the appropriate specificity. This capacity is accomplished by a special genetic recombination mechanism that we will describe in Section 42.4.

IMMUNOLOGICAL MEMORY After the innate immune system responds to a particular type of pathogen once, the adaptive immune system "remembers" that pathogen and can usually respond more rapidly and powerfully to the same threat in the future. This **immunological memory** usually saves us from repeats of childhood diseases such as chicken pox. Vaccination against specific diseases works because the adaptive immune system "remembers" the antigens that were introduced into the body.

All four of these features of adaptive immune defense characterize both the humoral immune response and the cellular immune response.

Two types of adaptive immune responses interact: an overview

The adaptive immune system mounts two types of responses against invaders: the humoral immune response and the cellular immune response. B cells that make antibodies are the workhorses of the humoral immune response, and cytotoxic (killer) T cells are the workhorses of the cellular immune response. These two responses operate simultaneously and cooperatively, sharing many mechanisms. A key event early in these two processes is the exposure of the nonself antigen's three-dimensional structure to the immune system. This occurs

WORKING WITH**DATA:**

The Discovery of Adaptive Immunity

Original Paper

Behring, E. and S. Kitasato. 1890. Uber das Zustandekommen der Diptherie-Immunitat und der Tetanus-Immunitat bel thieren. *Deustche medizinische Wochenschrift* 16: 1113–1114. In *Milestones in Microbiology: 1556–1940*, translated and edited by T. D. Brock, 1998, p. 138. ASM Press, Washington, D.C.

Analyze the Data

Until the early twentieth century, diphtheria, an infectious disease called the "strangling angel" for the way it disrupts breathing, exacted a terrible death toll, particularly on children. In the late 1800s a team of scientists led by Robert Koch, a leader in the then new field of microbiology, focused on diseases such as diphtheria and the equally lethal tetanus. Both diseases are caused by bacteria that kill their victims by secreting toxins into the bloodstream. A member of Koch's team from Japan, Shibasaburo Kitasato isolated the bacteria that cause diphtheria, and a recent medical school graduate, Emil Behring, set out to see how the disease could be treated. They discovered that animals injected with the disease-causing bacteria had substances in their blood that could prevent productive infections in naïve (previously unexposed) animals and even increase survival of animals already infected (see Figure 42.6). They called these substances anti-toxins (we know them as antibodies) and soon found that they were not species-specific: anti-toxin from a horse could be used to treat people. This and similar work on tetanus led quickly to the development of anti-diphtheria and anti-tetanus medicines. Behring shared the first Nobel Prize for Medicine in 1901.

In their experiments, Behring and Kitasato used different doses of toxin to test the immunity of guinea pig #2 in each experiment (see Figure 42.6). The results are shown in the table.

QUESTION 1

What can you conclude from these data in terms of the level of protection afforded by the serum?

QUESTION 2

These experiments could be performed with the same results with either intact bacteria that cause diphtheria or a bacteria-free filtrate of a 10-day-old culture of the bacteria. Explain.

	Symptoms	
Experiment	0.5 ml dose	10 ml dose
A	Diphtheria	Diphtheria
B	No diphtheria	No diphtheria
C	Diphtheria	Diphtheria

Go to **BioPortal** for all WORKING WITH**DATA** exercises

when an antigenic molecule, or a fragment of the molecule, is displayed on the surface of a cell and the unique epitope structure protrudes from the cell, where it is exposed to nearby T or B cells. Cells that can "present" the antigen to the immune system in this way are collectively referred to as **antigen-presenting cells**. Dendritic cells play a key role as antigen-presenting cells, although other cells can also perform this function. **Figure 42.7** provides a simplified overview of antigen presentation and the roles of T and B cells in the adaptive immune response.

The key player integrating the humoral and cellular immune responses is the T-helper (T_H) cell. By binding to the antigen on a presenting cell, the T_H cell stimulates events in both responses.

HUMORAL IMMUNE RESPONSE In the **humoral immune response** (from the Latin *humor*, "fluid"), antibodies react with antigens on pathogens in blood, lymph, and tissue fluids. An animal can produce a staggering diversity of antibodies capable of binding to almost any conceivable antigen the animal encounters. Antibodies are secreted by B cells and travel freely in the blood and lymph. A particular B cell also possesses receptors on its surface with the same specificity as the antibodies it produces.

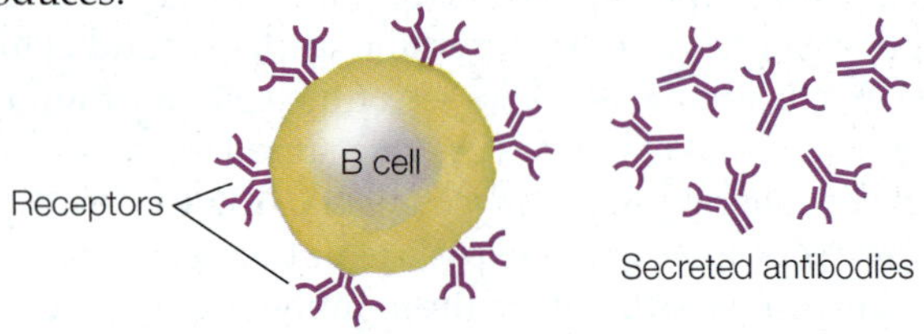
B cell
Receptors
Secreted antibodies

The first time a specific antigen invades the body, it may be presented and then detected by binding to a T cell receptor. This binding activates a B cell with the appropriate antibody; this cell proliferates, and its daughter cells make and secrete multiple copies of the antibody.

Go to Animated Tutorial 42.2
Pregnancy Test
Life10e.com/at42.2

CELLULAR IMMUNE RESPONSE The **cellular immune response** is directed against antigens that have become established within a cell of the host animal. It detects and destroys virus-infected or mutated cells, such as cancer cells expressing unique proteins caused by mutations.

T cells in the lymph nodes, bloodstream, and intercellular spaces carry out the cellular immune response. These T cells have integral membrane proteins—T cell receptors—that recognize and bind to antigens. T cell receptors are rather similar to antibodies in structure and function, each including specific molecular configurations that bind to specific antigens. Once a T cell is bound to an antigen, it initiates an immune response that typically results in the total destruction of the antigen-containing cell.

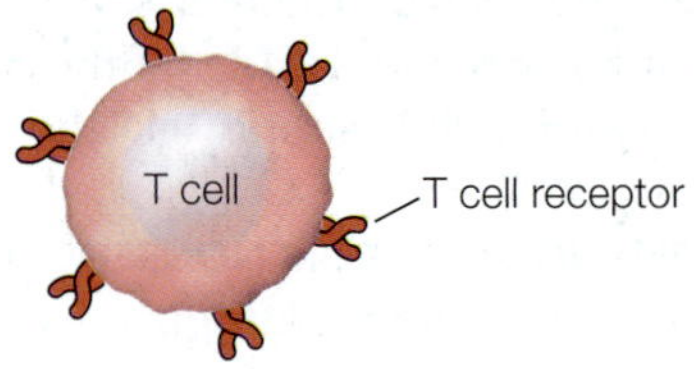
T cell
T cell receptor

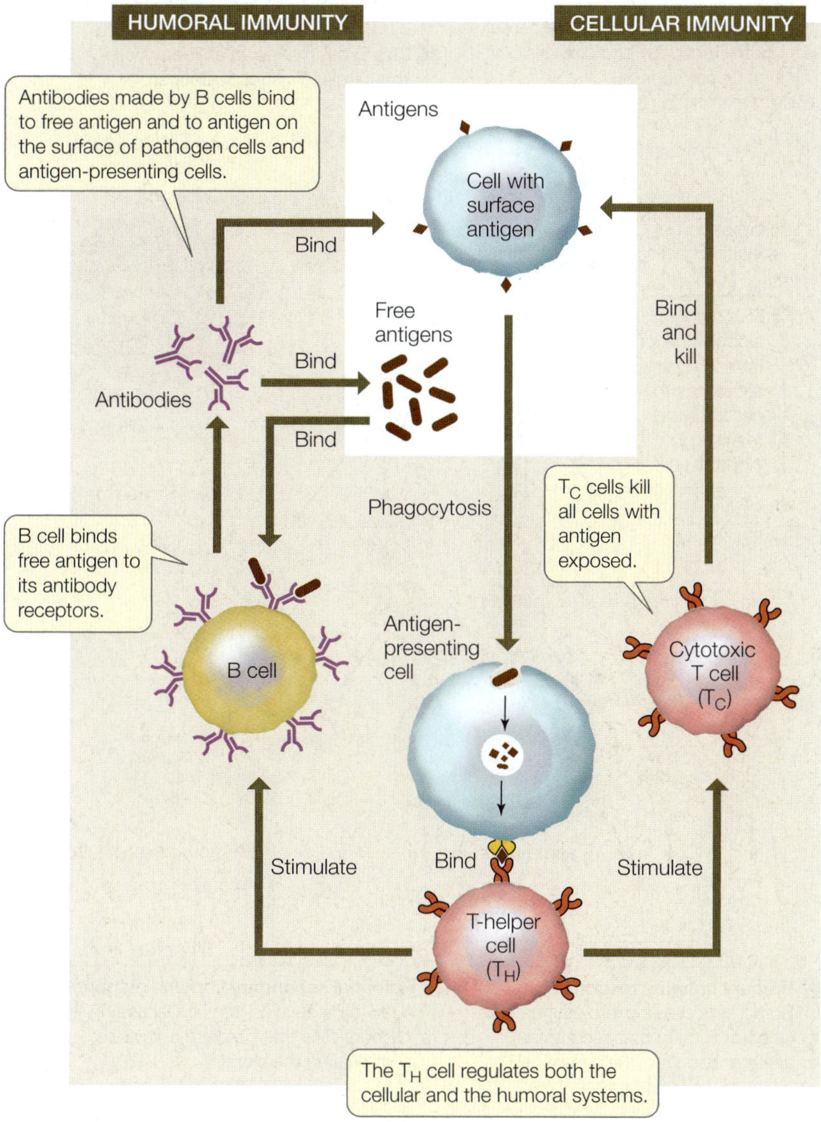
HUMORAL IMMUNITY
CELLULAR IMMUNITY
Antibodies made by B cells bind to free antigen and to antigen on the surface of pathogen cells and antigen-presenting cells.
Antigens
Cell with surface antigen
Bind
Free antigens
Bind
Bind
Bind and kill
Antibodies
Phagocytosis
T_C cells kill all cells with antigen exposed.
B cell binds free antigen to its antibody receptors.
B cell
Antigen-presenting cell
Cytotoxic T cell (T_C)
Stimulate
Bind
Stimulate
T-helper cell (T_H)
The T_H cell regulates both the cellular and the humoral systems.

42.7 The Adaptive Immune System Humoral immunity involves the production of antibodies by B cells. Cellular immunity involves the activation of cytotoxic T cells that bind to cells expressing the antigen. For further details, see Figure 42.16.

Adaptive immunity develops as a result of clonal selection

Before the reactions just described for the humoral and cellular immune responses can take place, the body needs to generate a vast diversity of lymphocytes that have the ability to bind different antigens. How does this tremendous diversity arise? As we will discuss in Section 42.4, this diversity is generated primarily by DNA changes—chromosomal rearrangements and other mutations—that occur just after the B and T cells are formed in the bone marrow. Millions of different B cells develop, each of which can produce only one kind of antibody. Similarly, there are millions of different T cells, each with one specific kind of T cell receptor. Thus the adaptive immune system is "predeveloped"—*all of the machinery available to respond to an immense diversity of antigens is already there, even before the antigens are ever encountered.*

As we have described, when a pathogen enters the vertebrate body it stimulates the innate immune system. In addition to triggering its own defensive responses, the innate immune system triggers adaptive defensive responses via specific antigens that are presented on the surfaces of antigen-presenting cells, particularly dendritic cells. This triggers the proliferation of lymphocytes that are specific for those particular antigens. How does this proliferation occur? The answer lies in the process of **clonal selection**: *antigen binding "selects" a particular B or T cell for proliferation*. When an antigen fits the surface receptor on a B or T cell and binds to it, that cell is activated. It divides to form a clone of cells (a genetically identical group derived from a single cell), all of which recognize and react to the same antigen. This process is illustrated for B cells in **Figure 42.8**. Binding and activation select a particular lymphocyte, while proliferation generates the clone, hence the term "clonal selection."

Clonal deletion helps the immune system distinguish self from nonself

Normally, the body is tolerant of its own molecules—the same molecules that would generate an immune response in another individual. One way that the immune system does this is through the process of **clonal deletion**. This occurs primarily in the thymus, during the early differentiation of T and B cells, when these cells encounter self antigens. Any immature B or T cell that shows the potential to mount an immune response against self antigens undergoes programmed cell death (apoptosis) within a short time.

Immunological memory results in a secondary immune response

The first time a vertebrate animal is exposed to a particular antigen there is a time lag (usually several days) before the B cell–produced antibody molecules and T cells specific to that antigen slowly increase. But for years afterward—sometimes for life—the immune system "remembers" that particular antigen, allowing the body to mount a faster response the next time it encounters the antigen. How does this happen?

The answer lies in the fact that activated lymphocytes divide and differentiate to produce *two types* of daughter cells: effector cells and memory cells.

- **Effector cells** carry out the attack on the antigen. Effector B cells, called **plasma cells**, secrete antibodies. Effector T cells release cytokines and other molecules that initiate reactions that destroy nonself or altered cells. Effector cells live only a few days.
- **Memory cells** (see Figure 42.8) are long-lived cells that retain the ability to start dividing on short notice to produce more effector and more memory cells. Memory B and T cells may survive in the body for decades, rarely dividing.

These two types of lymphocytes can respond to an antigen in two different ways:

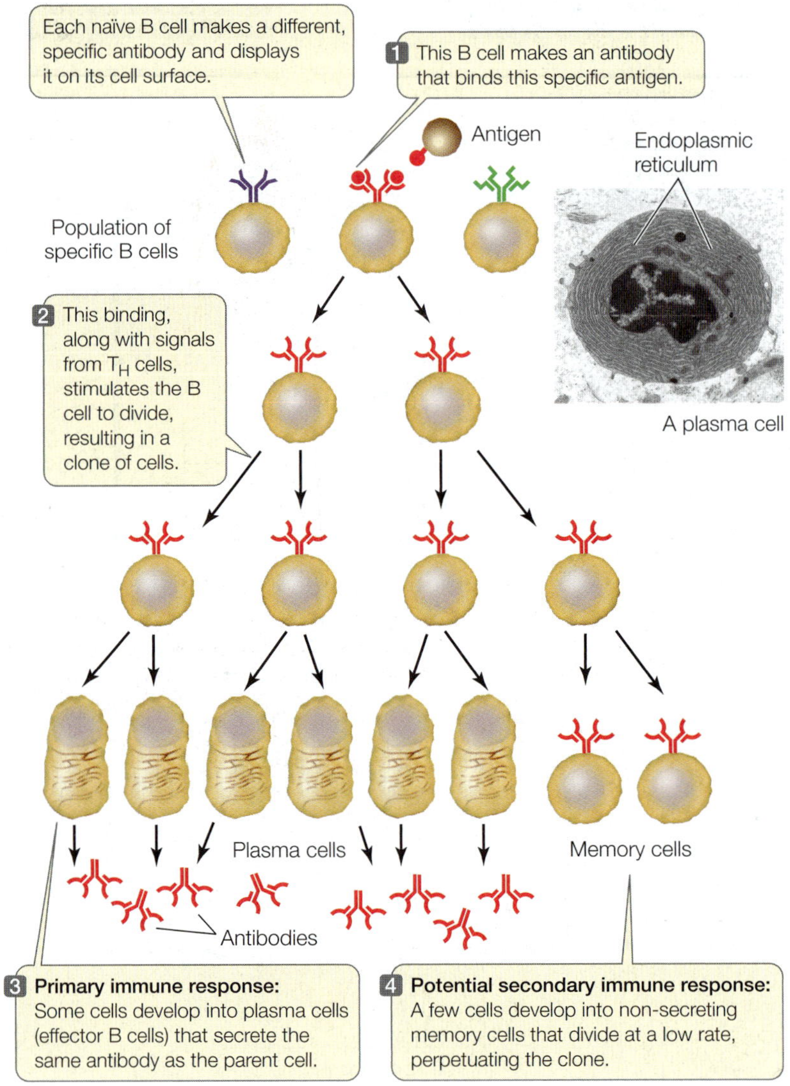
Each naïve B cell makes a different, specific antibody and displays it on its cell surface.
1 This B cell makes an antibody that binds this specific antigen.
Antigen
Endoplasmic reticulum
Population of specific B cells
2 This binding, along with signals from T_H cells, stimulates the B cell to divide, resulting in a clone of cells.
A plasma cell
Plasma cells
Memory cells
Antibodies
3 Primary immune response: Some cells develop into plasma cells (effector B cells) that secrete the same antibody as the parent cell.
4 Potential secondary immune response: A few cells develop into non-secreting memory cells that divide at a low rate, perpetuating the clone.

42.8 Clonal Selection in B Cells The binding of an antigen to a specific receptor on the surface of a B cell stimulates that cell to divide, producing a clone of genetically identical cells to fight that invader. Plasma cells have extensive endoplasmic reticulum for synthesizing antibodies.

- When the body first encounters a particular antigen, a **primary immune response** is activated, in which the "naïve" (previously unexposed) lymphocytes that recognize that antigen proliferate to produce clones of effector and memory cells.
- After a primary immune response to a particular antigen, subsequent encounters with the same antigen will trigger a much more rapid and powerful **secondary immune response**. The memory cells that bind with that antigen proliferate, launching a huge army of plasma cells and effector T cells.

Vaccines are an application of immunological memory

You will recall Behring's experiment on using the serum of diphtheria-exposed animals to protect other animals from the disease (see Figure 42.6). The animals that survived and donated serum had developed an adaptive immune response, including memory cells that lead to long-term protection.

Thanks to immunological memory, exposure to many diseases (including childhood diseases such as chicken pox) provides a natural immunity to those diseases. Furthermore, it is possible to provide artificial immunity against many life-threatening diseases by **vaccination**: the introduction of antigen into the body in a form that does not cause disease.

Vaccination initiates a primary immune response, generating memory cells without making the person ill. Later, if a pathogen carrying the same antigen attacks, specific memory cells already exist. They recognize the antigen and quickly overwhelm the invaders with a massive production of lymphocytes and antibodies (**Figure 42.9**).

Because the antigens used for immunization or vaccination are produced by pathogenic organisms, they must be altered so that they cannot cause disease but are still able to provoke an immune response. There are three principal ways to do this:

- *Inactivation* involves killing the pathogen with heat or chemicals.
- *Attenuation* involves reducing the virulence of a virus by repeatedly infecting cells with it in the laboratory; this results in mutations in the virus that render it nonpathogenic but still recognized as nonself.
- *Recombinant DNA technology* can be used to produce peptide fragments that bind to and activate lymphocytes but do not have the harmful part of a protein toxin.

For most of the 70 or so bacteria, viruses, fungi, and parasites known to cause serious human diseases, vaccines are already available or will be in the next few years. As you saw in the opening story, vaccination has completely or almost completely wiped out some deadly diseases, including smallpox, diphtheria, and polio, in industrialized countries.

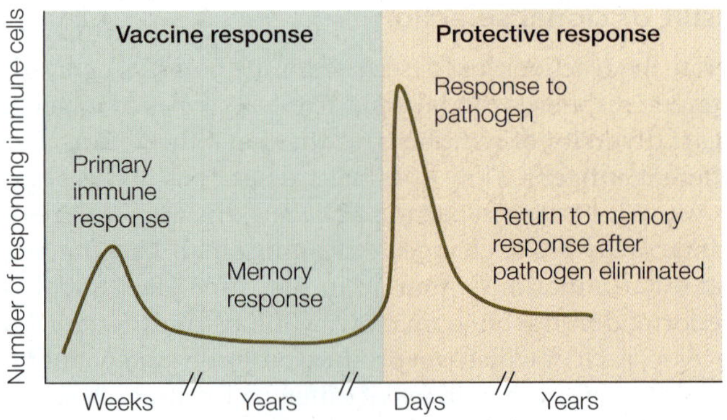
Number of responding immune cells
Vaccine response
Protective response
Response to pathogen
Primary immune response
Return to memory response after pathogen eliminated
Memory response
Weeks
Years
Days
Years

42.9 Vaccination Immunological memory from exposure to an antigen that does not cause disease can result in a massive response to the disease agent when it appears later.

RECAP 42.3

The adaptive immune system reacts against nonself or mutated self molecules called antigens. The system generates amazing diversity in both antibodies (produced by B cells) and in T cell receptors. In the primary immune response, B cells and T cells that recognize a particular pathogen proliferate by clonal selection. Immunological memory prepares the body for a much stronger secondary immune response.

- How does an antigen initiate an adaptive immune response? **See pp. 864–865 and Figure 42.7**
- Describe clonal selection. How does it contribute to immunological memory? **See pp. 865–866 and Figure 42.8**
- How do vaccines make use of immunological memory? **See p. 866 and Figure 42.9**

Now that we have discussed some general features of the adaptive immune system, let's focus in more detail on the B lymphocytes and the humoral immune response.

42.4 What Is the Humoral Immune Response?

Every day, billions of B cells survive the test of clonal deletion and are released from the bone marrow into the circulation. B cells are the basis for the humoral immune response.

Go to Animated Tutorial 42.3
Humoral Immune Response
Life10e.com/at42.3

Some B cells develop into plasma cells

A B cell begins by making a receptor protein on its cell surface. As we have seen, if a B cell is activated by antigen binding to this receptor, it gives rise to clones of plasma cells and memory cells. The plasma (effector B) cells secrete antibodies into the bloodstream (see Figure 42.8).

Usually, for a naïve B cell to develop into an antibody-secreting plasma cell, a T-helper (T_H) cell with the same specificity must also bind to the antigen (see Figure 42.7). The division and differentiation of the B cell is stimulated by chemical signals from the T_H cell.

As plasma cells develop, the number of ribosomes and the amount of endoplasmic reticulum in their cytoplasms increase greatly. These increases allow the cells to synthesize and secrete large amounts of antibody proteins—up to 2,000 molecules per second! *All the plasma cells arising from a given B cell produce antibodies that are specific for the antigen that originally bound to the parent B cell.* Thus antibody specificity is maintained as B cells proliferate.

Different antibodies share a common structure

Antibodies belong to a class of proteins called **immunoglobulins**. There are several types of immunoglobulins, but all

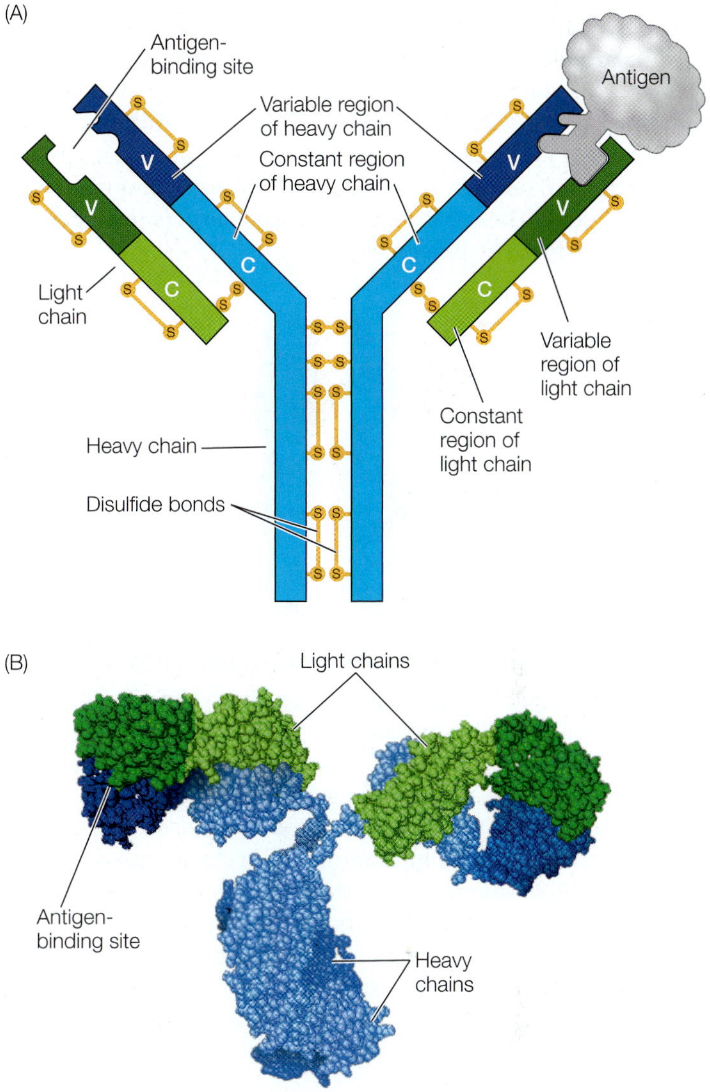

(A)
Antigen-binding site
Variable region of heavy chain
Constant region of heavy chain
Antigen
V
C
S
Light chain
Heavy chain
Disulfide bonds
Variable region of light chain
Constant region of light chain
(B)
Light chains
Antigen-binding site
Heavy chains

42.10 The Structure of an Immunoglobulin Four polypeptide chains (two light, two heavy) make up an immunoglobulin molecule. Both diagrammatic (A) and space-filling (B) representations of immunoglobulin are shown here.

Go to Activity 42.3 Immunoglobulin Structure
Life10e.com/ac42.3

contain a tetramer consisting of four polypeptide chains (**Figure 42.10**). In each immunoglobulin molecule, two of these polypeptides are identical light chains, and two are identical heavy chains. Disulfide bonds hold the chains together. Each polypeptide chain has a constant region and a variable region:

- The amino acid sequences of the **constant regions** are similar among the immunoglobulins. They determine the destination and function—the class—of each immunoglobulin.
- The amino acid sequences of the **variable regions** are different for each specific immunoglobulin. Their three-dimensional antigen-binding sites are determined by their secondary structures and are responsible for antibody specificity.

The two antigen-binding sites on each immunoglobulin molecule are identical, making the antibody bivalent (*bi*, "two"; *valent*, "binding"). This ability to bind two antigen molecules at once, along with the presence of multiple epitopes on the surfaces of many antigens (including large proteins, viruses, and bacteria) permits antibodies to form large complexes with the antigens. These complexes are easy targets for ingestion and breakdown by phagocytes.

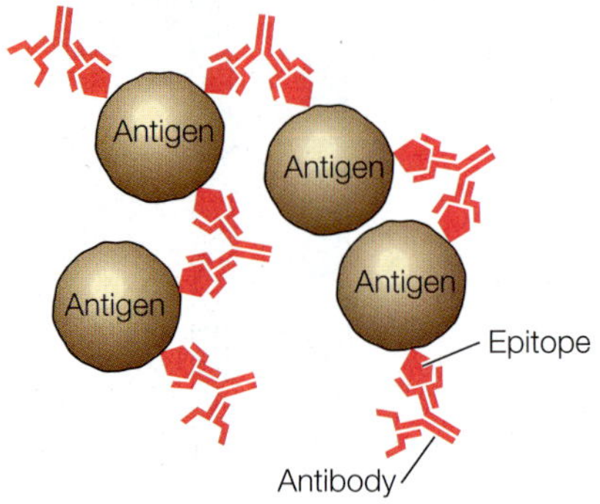
Antigen
Antigen
Antigen
Antigen
Epitope
Antibody

There are five classes of immunoglobulins

While the variable regions are responsible for the specificity of an immunoglobulin, the constant regions of the heavy chain determine the class of the immunoglobulin—for example, whether it will be an integral membrane receptor (e.g., on the surface of a B cell) or a soluble antibody that is secreted into the bloodstream. The five immunoglobulin classes are described in **Table 42.2**. The most abundant class is IgG; these soluble antibody proteins make up about 80 percent of the total immunoglobulin content of the bloodstream. They are made in greatest quantity during a secondary immune response. IgG molecules defend the body in several ways. For example, after some IgG molecules bind to antigens, they become attached by their heavy chains to macrophages. This attachment permits the macrophages to destroy the antigens by phagocytosis.

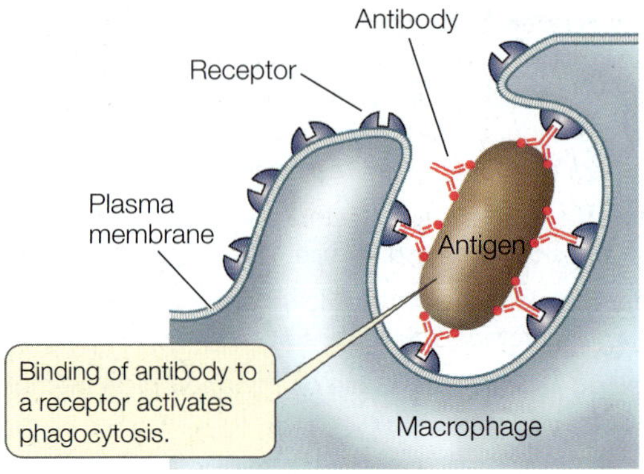
Antibody
Receptor
Plasma membrane
Antigen
Binding of antibody to a receptor activates phagocytosis.
Macrophage

Immunoglobulin diversity results from DNA rearrangements and other mutations

Each mature B cell makes one—and only one—specific antibody targeted to a single epitope. As we have seen, there are millions of possible epitopes to which a human is exposed or can be exposed. A simple calculation using approximate numbers shows that it would be impossible to have a unique gene for each of these epitopes:

One antibody gene ≈ 2,100 base pairs (bp) DNA

10 million different antibodies ≈ 21 billion bp DNA

This is seven times the size of the entire human genome! There must be another way to generate antibody diversity.

It turns out that instead of a single gene encoding each immunoglobulin, the genome of the differentiating B cell has a limited number of alleles for each of several regions (domains) of the protein, and that combinations of these alleles generate

TABLE 42.2
Antibody Classes

Class	General Structure	Location	Function
IgG	Monomer	Free in blood plasma; about 80 percent of circulating antibodies	Most abundant antibody in primary and secondary immune responses; crosses placenta and provides passive immunization to fetus
IgM	Pentamer	Surface of B cell; free in blood plasma	Antigen receptor on B cell membrane; first class of antibodies released by B cells during primary response
IgD	Monomer	Surface of B cell	Cell surface receptor of mature B cell; important in B cell activation
IgA	Dimer	Saliva, tears, milk, and other body secretions	Protects mucosal surfaces; prevents attachment of pathogens to epithelial cells
IgE	Monomer	Secreted by plasma cells in skin and tissues lining gastrointestinal and respiratory tracts	Binds to mast cells and basophils to sensitize them to subsequent binding of antigen, which triggers release of histamine that contributes to inflammation and some allergic responses

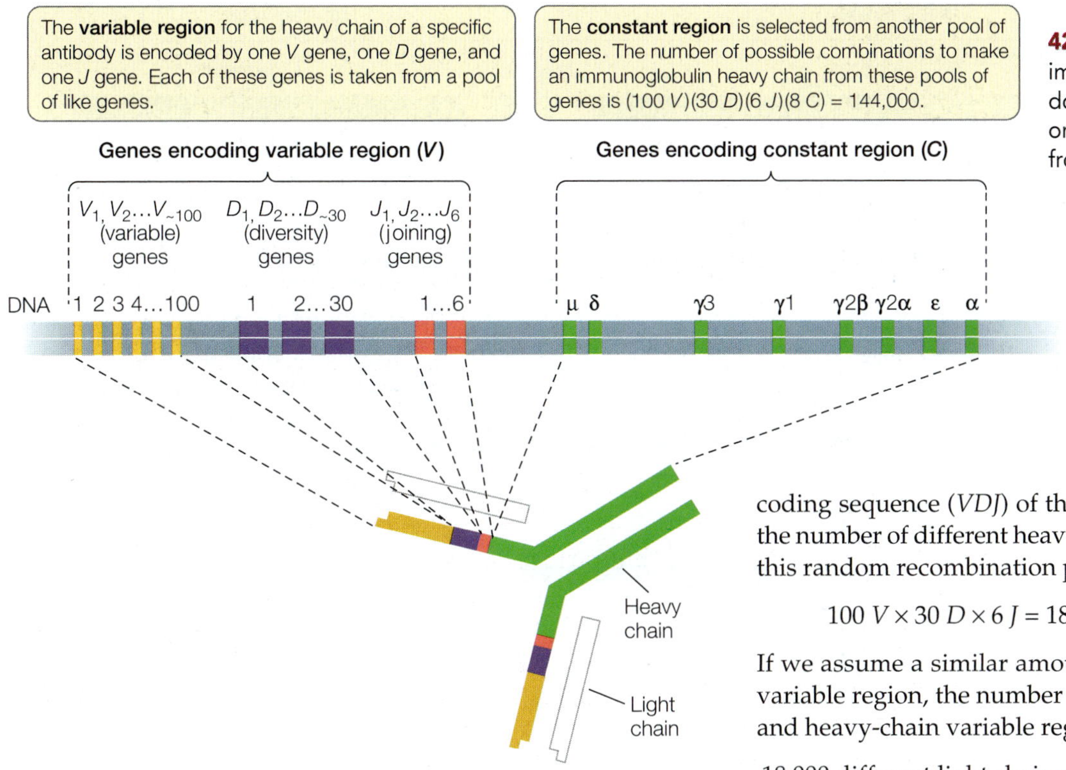

The **variable region** for the heavy chain of a specific antibody is encoded by one *V* gene, one *D* gene, and one *J* gene. Each of these genes is taken from a pool of like genes.
The **constant region** is selected from another pool of genes. The number of possible combinations to make an immunoglobulin heavy chain from these pools of genes is (100 *V*)(30 *D*)(6 *J*)(8 *C*) = 144,000.
Genes encoding variable region (*V*)
Genes encoding constant region (*C*)
$V_1, V_2...V_{\sim100}$ (variable) genes
$D_1, D_2...D_{\sim30}$ (diversity) genes
$J_1, J_2...J_6$ (joining) genes
DNA
1 2 3 4...100
1 2...30
1...6
μ δ
γ3
γ1
γ2β γ2α ε α
Heavy chain
Light chain

42.11 Heavy-Chain Genes Mouse immunoglobulin heavy chains have four domains, each of which is coded for by one of several possible genes selected from a cluster of similar genes.

diversity. First let's look at the unusual process of shuffling this genetic deck to generate the enormous immunological diversity that characterizes each individual mammal. This process affects the variable region, the part of the immunoglobulin that recognizes a particular epitope. Next we will see how similar events involving the constant region produce the five classes of immunoglobulins, which have different cellular locations or functions in the body.

Each gene encoding an immunoglobulin chain is in reality a "supergene" assembled by means of genetic recombination from several clusters of smaller genes scattered along part of a chromosome. Such a region in the mouse genome is shown in **Figure 42.11**. Every cell in the body has hundreds of immunoglobulin genes located in separate clusters that are potentially capable of participating in the synthesis of both the variable and constant regions of immunoglobulin chains. In most body cells and tissues, these genes remain intact and separated from one another. But during B cell development, these genes are cut out, rearranged, and joined together in DNA recombination events. One gene from each cluster is chosen randomly for joining, and the others are deleted.

In this manner, a unique immunoglobulin supergene is assembled from randomly selected "parts." Each B cell precursor assembles two supergenes, one for a specific heavy chain and the other, assembled independently, for a specific light chain. In humans and mice, two families of genes encode the variable region of the light chain, and three families encode the variable region of the heavy chain. For example, in mice the variable region of the heavy chain is assembled from 100 *V*, 30 *D*, and 6 *J* genes (see Figure 42.11). Each B cell randomly selects one gene from each of these clusters to make the final coding sequence (*VDJ*) of the heavy-chain variable region. So the number of different heavy chains that can be made through this random recombination process is quite large:

$$100\ V \times 30\ D \times 6\ J = 18{,}000 \text{ possible combinations}$$

If we assume a similar amount of diversity in the light chain variable region, the number of possible combinations of light- and heavy-chain variable regions is:

$$18{,}000 \text{ different light chains} \times 18{,}000 \text{ different heavy chains} = 324 \text{ million possibilities!}$$

Even more diversity is generated by various kinds of mutation that occur during the recombination events. These mutations can occur through imprecise recombination and the high spontaneous mutation rates in immunoglobulin genes.

These genetic events are irreversible—once the final coding sequence has been assembled for the variable regions of a B cell's light and heavy chains, that B cell's epitope specificity cannot change. This remarkable example of irreversible cell differentiation generates an enormous diversity of immunoglobulins from the same starting genome. A similar process results in the diversity of T cell receptors, which we will discuss in more detail in Section 42.5.

Once the pre-transcriptional processing is completed, each supergene is transcribed and then translated to produce an immunoglobulin light chain or heavy chain. These combine to form an active immunoglobulin protein, as shown in **Figure 42.12**. This figure shows the variable region joining with a μ constant region segment, which forms a gene for an immunoglobulin in the IgM class (see below). However, this genetic system is capable of still other kinds of changes. The B cell or plasma cell can switch the immunoglobulin class it produces while retaining its antigen specificity.

Go to Animated Tutorial 42.4
A B Cell Builds an Antibody
Life10e.com/at42.4

The constant region is involved in immunoglobulin class switching

Table 42.2 describes the different classes of immunoglobulins and their functions. Generally, a B cell makes only one class

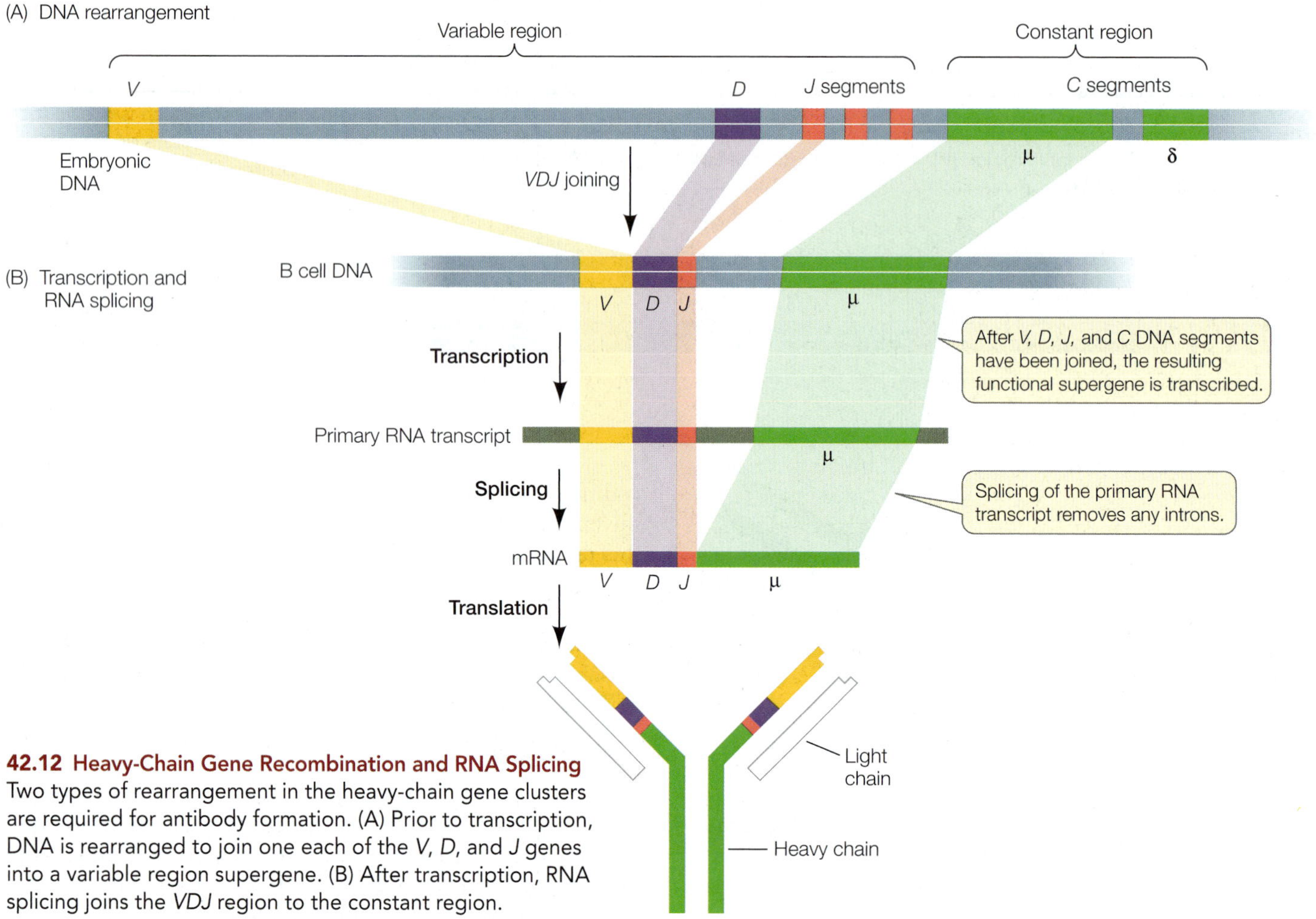

(A) DNA rearrangement
Variable region
Constant region
V
D
J segments
C segments
Embryonic DNA
μ
δ
VDJ joining
(B) Transcription and RNA splicing
B cell DNA
V D J
μ
After V, D, J, and C DNA segments have been joined, the resulting functional supergene is transcribed.
Transcription
Primary RNA transcript
μ
Splicing
Splicing of the primary RNA transcript removes any introns.
mRNA
V D J
μ
Translation
Light chain
Heavy chain

42.12 Heavy-Chain Gene Recombination and RNA Splicing Two types of rearrangement in the heavy-chain gene clusters are required for antibody formation. (A) Prior to transcription, DNA is rearranged to join one each of the *V*, *D*, and *J* genes into a variable region supergene. (B) After transcription, RNA splicing joins the *VDJ* region to the constant region.

at a time. But **class switching** can occur, in which a B cell changes the immunoglobulin class it synthesizes. For example, a B cell making IgM can switch to making IgG.

Early in its life, a B cell produces IgM molecules, which are the receptors responsible for its recognition of a specific antigen. At this time, the constant region of the heavy chain is encoded by the first constant region gene, the μ gene (see Figures 42.11 and 42.12). If the B cell later becomes a plasma cell during a humoral immune response, another deletion occurs in the cell's DNA, positioning the variable region genes (consisting of the same *V*, *D*, and *J* genes) next to a constant region gene farther away on the original DNA molecule (**Figure 42.13**). Such a DNA deletion results in the production of a new immunoglobulin with a different constant region of the heavy chain, and therefore a different function (see Table 42.2). However, this immunoglobulin has the same variable regions—and therefore the same antigen specificity—as the IgM produced by the parent B cell. The new immunoglobulin protein falls into one of the other four classes (IgA, IgD, IgE, or IgG), depending on which of the constant region genes is placed adjacent to the variable region genes.

What triggers class switching? T_H cells direct the course of an immune response and determine the nature of the attack on the antigen. These T cells induce class switching by sending cytokine signals. The cytokines bind to receptors

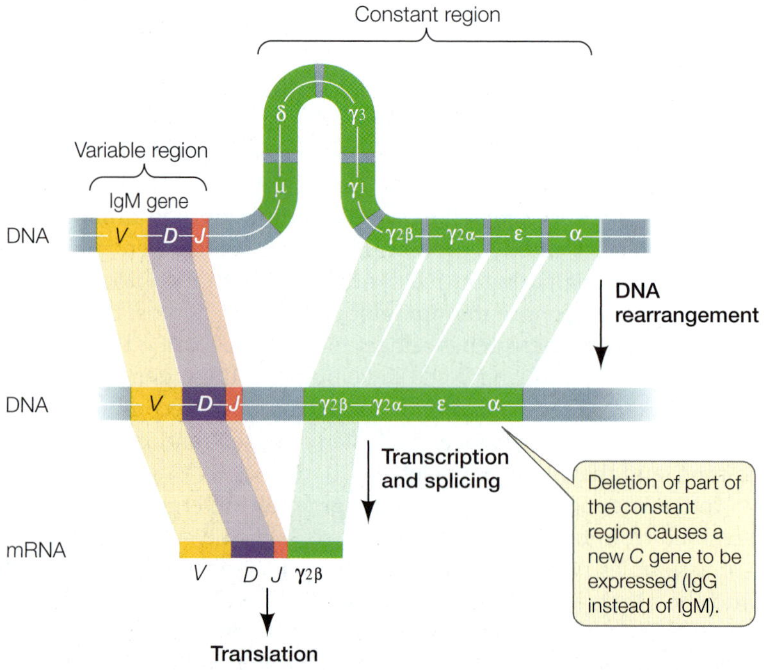

Constant region
Variable region
IgM gene
DNA
V D J
δ γ3
μ γ1
γ2β γ2α ε α
DNA rearrangement
DNA
V D J γ2β γ2α ε α
Transcription and splicing
Deletion of part of the constant region causes a new C gene to be expressed (IgG instead of IgM).
mRNA
V D J γ2β
Translation

42.13 Class Switching: Exchanging C Regions The supergene produced by joining *V*, *D*, *J*, and *C* genes (see Figure 42.12) may later be modified, causing a different *C* region to be transcribed. This modification, known as class switching, is accomplished by deletion of part of the constant region gene cluster. Shown here is class switching from IgM to IgG.

on the target B cells, generating signal transduction cascades that result in recombination and altered expression of the immunoglobulin genes.

Monoclonal antibodies have many uses

The specificity of antibodies suggested to scientists that they might be useful for detecting specific substances in the laboratory. However, the immune response to a complex antigen is polyclonal—that is, most antigens carry many different antigenic determinants and will produce a complex mixture of antibodies, each made by a different clone of B cells. Furthermore, as emphasized in our study of biochemistry, many biological molecules share regions of similar structure—all human steroid hormones, for example, have a similar multi-ring structure (see Figure 41.2). A polyclonal group of antibodies targeted to estrogen might be uninformative because some of the antibodies would bind to any steroid hormone present in the blood sample. More useful would be a clone of B cells that produce large amounts of an antibody that binds to only one specific epitope—a **monoclonal antibody**. Various methods can be used to produce monoclonal antibodies in the laboratory.

Monoclonal antibodies have many applications:

- *Immunoassays* involve the use of monoclonal antibodies to detect tiny amounts of molecules in tissues and fluids. For example, this technique is used in pregnancy tests to detect human chorionic gonadotropin, the hormone made by the developing embryo.
- *Immunotherapy* involves the use of monoclonal antibodies targeted against antigens on the surfaces of cancer cells. The coupling of a radioactive ligand or a toxin to the antibody makes it into a medical "smart bomb." In a related approach, binding of the antibody itself is enough to trigger a cellular immune response that destroys the cancer. This is the case with trastuzumab (Herceptin), a monoclonal antibody that binds to a growth factor receptor on some breast cancer cells.

RECAP 42.4

The humoral immune response is based on the synthesis by B cells of specific immunoglobulins. The specificity of an immunoglobulin derives from the amino acid sequence of its variable regions. B cells can make millions of immunoglobulins with different specificities by rearranging the genes that encode the variable regions of the heavy and light chains. Monoclonal antibodies are specific to one epitope and can be produced artificially for use in diagnostics and therapy.

- How does a B cell respond to an antigen? **See p. 867**
- How is the structure of an antibody molecule related to its function? **See pp. 867–868 and Figure 42.10**
- How can millions of antibodies with different specificities be generated from a relatively small number of genes? **See pp. 869–870 and Figures 42.11 and 42.12**
- What is the role of the constant region of the immunoglobulin in class switching? **See p. 870 and Figure 42.13**
- What are monoclonal antibodies, and how are they used? **See p. 871**

By making antibodies, B cells are the major players in the humoral immune response. We will now turn to the cellular immune response, where T cells are active at all stages.

42.5 What Is the Cellular Immune Response?

Two types of effector T cells (T-helper cells and cytotoxic T cells) are involved in the cellular immune response. They work along with proteins of the major histocompatibility complex (the MHC proteins), which present antigens on the surfaces of cells and contribute to the immune system's tolerance for the body's own cells.

Go to Animated Tutorial 42.5
Cellular Immune Response
Life10e.com/at42.5

T cell receptors bind to antigens on cell surfaces

Like B cells, T cells possess specific membrane receptors. The T cell receptor is not an immunoglobulin, however, but a glycoprotein with a molecular weight of about half that of an IgG. It is made up of two polypeptide chains, each encoded by a separate gene (**Figure 42.14**). The two chains have distinct regions with constant and variable amino acid sequences. As in the immunoglobulins, the variable regions provide the site for specific binding to antigens. But there is one major difference: whereas an antibody can bind to any antigen, whether it is present on the surface of a cell or not, a T cell receptor binds only to an antigen displayed by an MHC protein on the surface of an antigen-presenting or target cell.

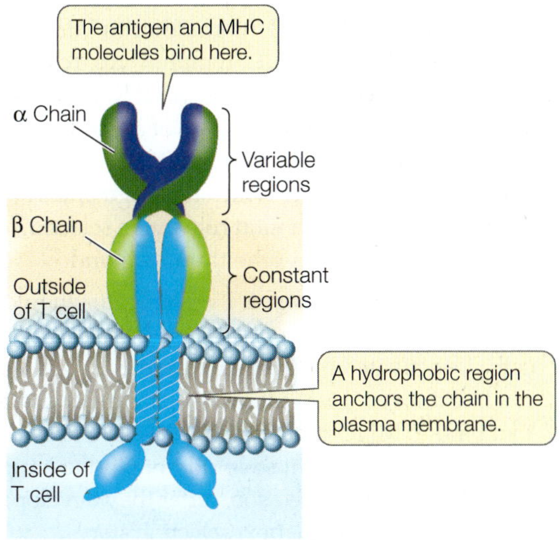

The antigen and MHC molecules bind here.
α Chain
Variable regions
β Chain
Outside of T cell
Constant regions
A hydrophobic region anchors the chain in the plasma membrane.
Inside of T cell

42.14 A T Cell Receptor The receptors on T lymphocytes are smaller than those on B lymphocytes, but their two polypeptides contain both variable and constant regions. As with the B cell receptors, the constant region fixes the receptor in the plasma membrane, while the variable regions establish the specificity for binding to antigen.

When a T cell is activated by contact with a specific antigen, it proliferates and forms a clone. Its descendants form clones of two types of effector T cells:

- **Cytotoxic T cells**, or **T_C cells**, recognize virus-infected or mutated cells and kill them by inducing lysis (see p. 874).
- **T-helper cells** (**T_H cells**, also called helper T cells) assist both the cellular and the humoral immune responses.

MHC proteins present antigen to T cells

T cell receptors do not bind directly to antigens. Instead, they bind to antigens that are bound to a cell surface glycoprotein, an MHC protein. With several gene families and hundreds of alleles, the MHC protein is a cell surface marker of genetic individuality. The diversity of MHC proteins also means that there are many possibilities for presenting an antigen to the T cell receptor.

There are two classes of MHC proteins. Both function to present antigens to the different T lymphocytes:

- **Class I MHC** proteins are present on the surface of every nucleated cell in the vertebrate body. They enable T_C cells to recognize virus-infected cells and kill them. Viral protein fragments that are antigenic are complexed with MHC I inside the cell and then the complex is carried to the plasma membrane. A T_C cell with the appropriate T cell receptor then binds to the MHC–antigen complex. To ensure binding, the T_C cell also has a cell surface protein called CD8 that recognizes and binds to MHC I.
- **Class II MHC** proteins are found mostly on the surfaces of B cells, macrophages, and other antigen-presenting cells, including dendritic cells (see Figure 42.2). When one of these cells ingests a pathogen such as a bacterium, the bacterial antigens are broken down in a phagosome. An MHC II molecule may bind to one of the fragments and carry it to the cell surface, where it is presented to a T_H cell (**Figure 42.15**). T_H cells have a surface protein called CD4 that recognizes and binds to MHC II.

Table 42.3 summarizes the information on MHC proteins, the cellular origins of antigens, and T lymphocytes. To accomplish its role in antigen presentation, each kind of MHC protein has an antigen-binding site that can hold a peptide of about 10 to 20 amino acids. The T cell receptor recognizes not just the antigenic fragment but the MHC I or II protein to which the fragment is bound.

MHC proteins play a vital role in the selection of T cells during their development in the thymus gland:

- *Binding to MHC proteins.* A T cell receptor should bind not to an antigen alone but to an antigen–MHC complex, because T cells are activated by antigen presented on the surface of cells, not free antigen. Here there is positive selection for T cells that bind to MHC proteins. Any T cells that do not recognize MHC proteins (and thus would not bind to antigen-presenting cells) are eliminated soon after they develop; the rest of the T cells go on to the next selection step.
- *Binding to self peptides bound to self MHC proteins.* In this case there is negative selection of T cells that bind to self antigens presented on MHC proteins. This eliminates the further production of T cells that react to self antigens. Negative selection through clonal deletion (see p. 874) is a mechanism that prevents the adaptive immune system from reacting to self molecules.

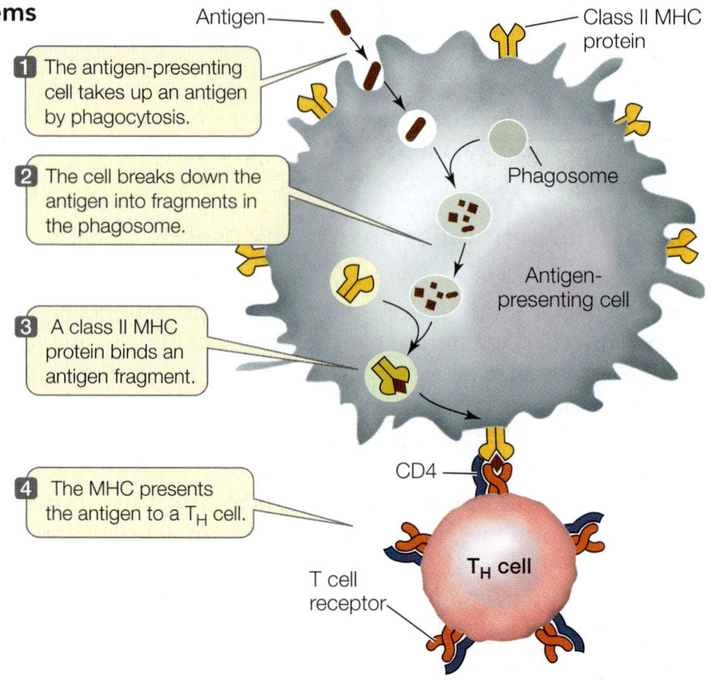

Antigen
Class II MHC protein
1 The antigen-presenting cell takes up an antigen by phagocytosis.
2 The cell breaks down the antigen into fragments in the phagosome.
Phagosome
Antigen-presenting cell
3 A class II MHC protein binds an antigen fragment.
CD4
4 The MHC presents the antigen to a T_H cell.
T cell receptor
T_H cell

42.15 Macrophages Are Antigen-Presenting Cells A fragment of an antigen is displayed by MHC II on the surface of a macrophage. T cell receptors on a specific T-helper cell can then bind to and interact further with the antigen–MHC II complex.

T-helper cells and MHC II proteins contribute to the humoral immune response

When a T_H cell survives the selection processes and binds to an antigen-presenting cell, it releases cytokines that activate the T_H cell to proliferate, producing a clone of T_H cells with the same specificity. The steps to this point constitute the activation phase of the humoral immune response, and they occur in the lymphoid tissues. Next comes the effector phase, in which the T_H cells activate naïve B cells with the same specificity to produce antibodies.

B cells are also antigen-presenting cells. B cells take up antigens bound to their surface immunoglobulin receptors by endocytosis, break them down, and display antigenic fragments on class II MHC proteins. When a T_H cell binds to the displayed antigen–MHC II complex, it releases cytokines that cause the B cell to produce a clone of plasma cells (**Figure 42.16A**). Finally, the plasma cells secrete antibodies, completing the effector phase of the humoral immune response.

TABLE 42.3 The Interaction between T Cells and Antigen-Presenting Cells

Presenting Cell Type	Antigen Presented	MHC Class	T Cell Type	T Cell Surface Protein
Any cell	Intracellular protein fragment	Class I	Cytotoxic T cell (T_C)	CD8
Macrophages, dendritic cells, and B cells	Fragments from extracellular proteins	Class II	Helper T cell (T_H)	CD4

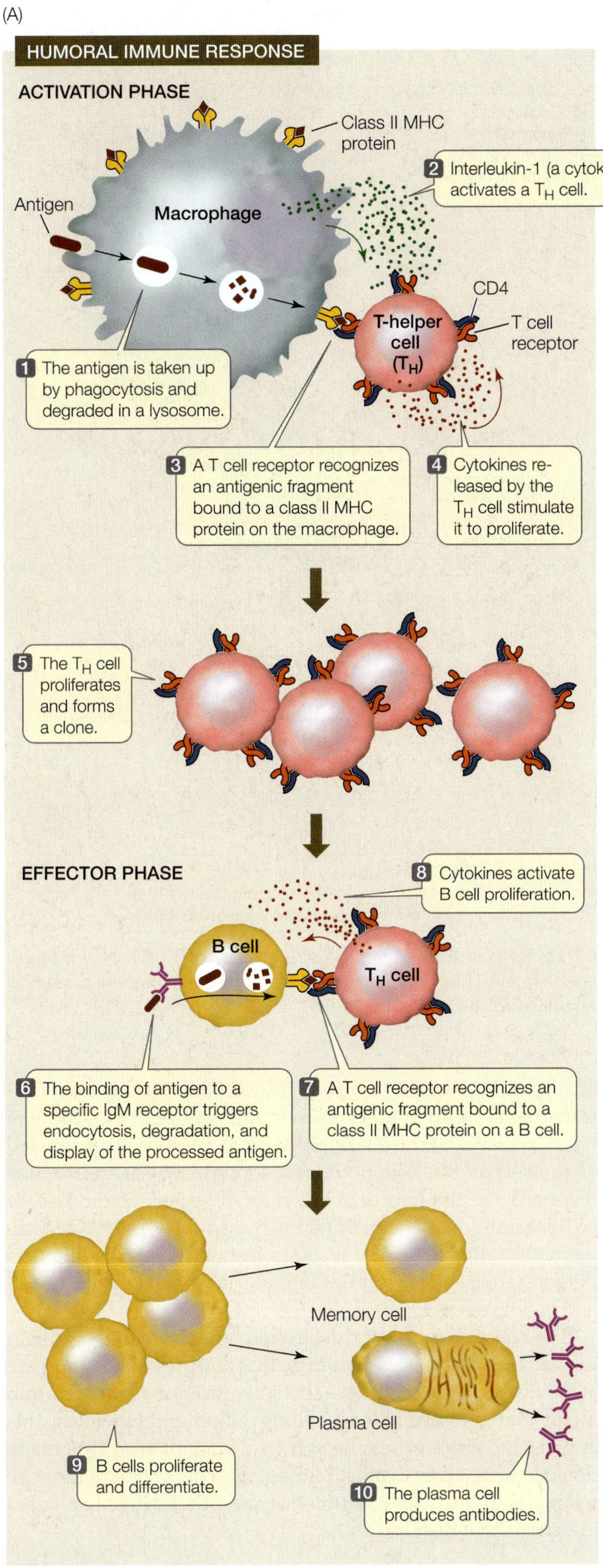
(A)
HUMORAL IMMUNE RESPONSE
ACTIVATION PHASE
Class II MHC protein
Antigen
Macrophage
2 Interleukin-1 (a cytokine) activates a T_H cell.
CD4
T-helper cell (T_H)
T cell receptor
1 The antigen is taken up by phagocytosis and degraded in a lysosome.
3 A T cell receptor recognizes an antigenic fragment bound to a class II MHC protein on the macrophage.
4 Cytokines released by the T_H cell stimulate it to proliferate.
5 The T_H cell proliferates and forms a clone.
EFFECTOR PHASE
8 Cytokines activate B cell proliferation.
B cell
T_H cell
6 The binding of antigen to a specific IgM receptor triggers endocytosis, degradation, and display of the processed antigen.
7 A T cell receptor recognizes an antigenic fragment bound to a class II MHC protein on a B cell.
Memory cell
Plasma cell
9 B cells proliferate and differentiate.
10 The plasma cell produces antibodies.

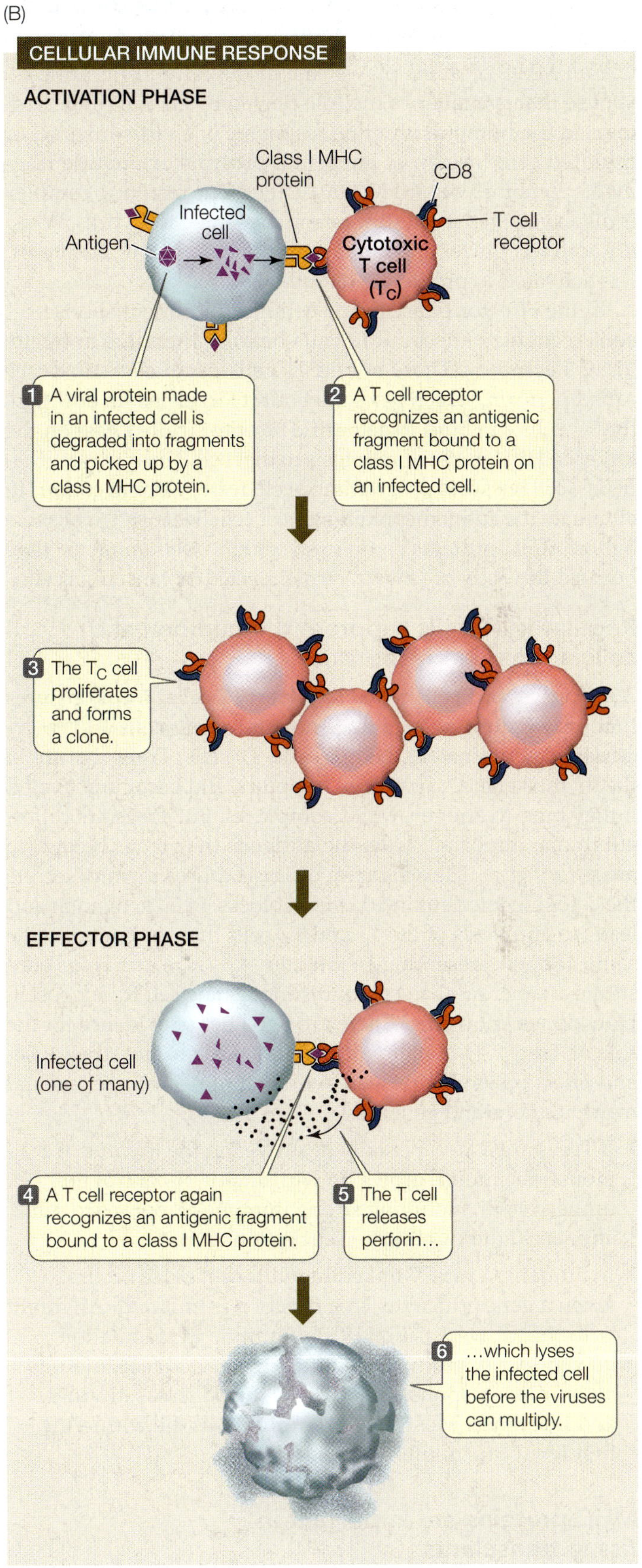
(B)
CELLULAR IMMUNE RESPONSE
ACTIVATION PHASE
Class I MHC protein
CD8
Infected cell
Antigen
Cytotoxic T cell (T_C)
T cell receptor
1 A viral protein made in an infected cell is degraded into fragments and picked up by a class I MHC protein.
2 A T cell receptor recognizes an antigenic fragment bound to a class I MHC protein on an infected cell.
3 The T_C cell proliferates and forms a clone.
EFFECTOR PHASE
Infected cell (one of many)
4 A T cell receptor again recognizes an antigenic fragment bound to a class I MHC protein.
5 The T cell releases perforin…
6 …which lyses the infected cell before the viruses can multiply.

42.16 Phases of the Humoral and Cellular Immune Responses Both the humoral and the cellular immune responses have activation and effector phases, all of which involve T cells.

Cytotoxic T cells and MHC I proteins contribute to the cellular immune response

Class I MHC proteins play a role in the cellular immune response that is similar to the role played by class II MHC proteins in the humoral immune response. In a virus-infected or mutated cell, foreign or abnormal proteins or peptide fragments combine with MHC I molecules. The resulting complex is displayed on the cell surface and presented to T_C cells. When a T_C cell recognizes and binds to this antigen–MHC I complex, it is activated to proliferate (**Figure 42.16B**).

In the effector phase of the cellular immune response, T_C cells recognize and bind to cells bearing the same antigen–MHC I complex. These bound T_C cells produce a substance called perforin, which lyses the bound target cell. In addition, the T_C cells can bind to a specific receptor (called Fas) on the target cell that initiates apoptosis in that cell. These two mechanisms, cell lysis and programmed cell death, work in concert to eliminate the antigen-containing host cell. Because T_C cells recognize MHC proteins complexed with nonself antigens, they help rid the body of its own virus-infected or cancerous cells.

Regulatory T cells suppress the humoral and cellular immune responses

A third class of T cells called **regulatory T cells** (**Tregs**) ensures that the immune system does not attack self cells and molecules indiscriminately. Like T_H and T_C cells, Tregs mature in the thymus gland, carry T cell receptors, and become activated if they bind to antigen–MHC complexes. But Tregs are different in one important way: the antigens that Tregs recognize are *self antigens*. The activation of Tregs causes them to secrete the cytokine interleukin-10, which blocks T cell activation and leads to apoptosis of the T_C and T_H cells that are bound to the same antigen-presenting cell (**Figure 42.17**). Thus Tregs constitute another mechanism for distinguishing self from nonself. How do we know this? There are two lines of evidence for the role of Tregs. As in many other biological studies, the cause-and-effect relationships were worked out using experimental manipulations and genetics:

- If Tregs are experimentally destroyed in the thymus of a mouse, the mouse grows up with an out-of-control immune system, mounting strong immune responses to self antigens (autoimmunity—see Section 42.6).
- In humans, a rare X-linked hereditary disease occurs when a gene critical to Treg function is mutated. An infant with this disease, called IPEX (*i*mmune dysregulation, *p*olyendocrinopathy and *e*nteropathy, *X*-linked), mounts an immune response that attacks the pancreas, thyroid, and intestines. Most affected individuals die within the first few years of life.

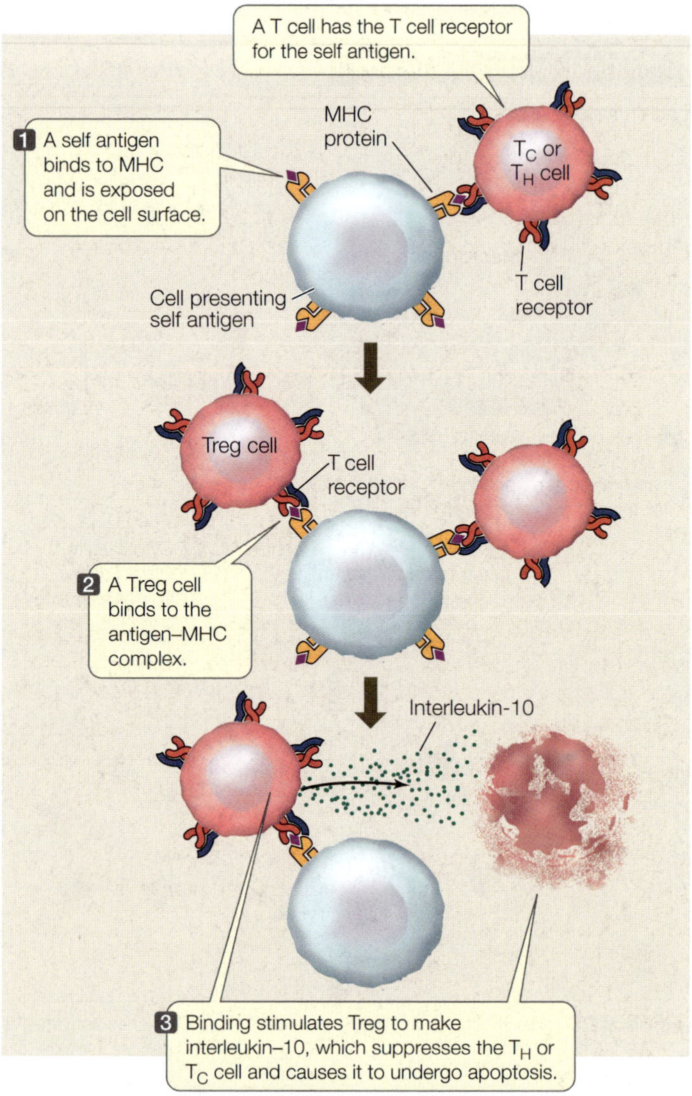
A T cell has the T cell receptor for the self antigen.
1 A self antigen binds to MHC and is exposed on the cell surface.
MHC protein
Tc or TH cell
T cell receptor
Cell presenting self antigen
Treg cell
T cell receptor
2 A Treg cell binds to the antigen–MHC complex.
Interleukin-10
3 Binding stimulates Treg to make interleukin–10, which suppresses the TH or TC cell and causes it to undergo apoptosis.

42.17 Tregs and Tolerance A special class of T cells called regulatory T cells (Tregs) inhibits the activation of the immune system in response to self antigens.

MHC proteins are important in tissue transplants

In humans, one consequence of the major histocompatibility complex became important with the development of organ transplant surgery. Because the proteins produced by the MHC are specific to each individual, they act as nonself antigens if transplanted into another individual. An organ or a piece of tissue transplanted from one person to another is recognized as nonself by the host body and soon provokes an immune response; the tissue is then killed, or "rejected," by the host's cellular immune system. But if the transplant is performed immediately after birth, or if it comes from a genetically identical person (an identical twin), the material is recognized as self and is not rejected.

The rejection problem can be overcome by treating a patient with a drug, such as cyclosporin, that suppresses the immune system. Cyclosporin blocks the activation of a transcription factor that is essential for T cell development. However, this approach compromises the ability of transplant recipients to defend themselves against pathogens. This problem must be managed by the use of antibiotics and other drugs to combat infections that develop.

RECAP 42.5

Specific receptors on T cells bind to antigen–MHC complexes displayed on cell surfaces. During development, T cells are selected that recognize MHC proteins. T-helper cells bind antigens on MHC II proteins and contribute to the humoral immune response. Cytotoxic T cells bind antigens on MHC I proteins and contribute to the cellular immune response. The cellular immune response acts against virus-infected or mutated body cells. Tregs suppress immune responses to self antigens.

- What are the roles of a T cell receptor in cellular immunity? **See p. 871 and Figure 42.14**
- What parts do MHC proteins play in the cellular immune response? **See p. 872 and Figure 42.15**
- What occurs during the cellular immune response to a virus-infected cell? **See pp. 873–874 and Figure 42.16**

Given the numerous and complex cellular interactions that activate the immune system and generate antibody diversity, you may have perceived many points at which the immune system could fail. We will now turn to several situations in which one or more components of this complex system malfunction.

42.6 What Happens When the Immune System Malfunctions?

Sometimes the immune system fails us in one way or another. It may overreact, as in an allergic reaction; it may attack self antigens, as in an autoimmune disease; or it may function weakly or not at all, as in an immune deficiency disease.

Allergic reactions result from hypersensitivity

An **allergic reaction** arises when the human immune system overreacts to (is hypersensitive to) a dose of antigen. Although the antigen itself may present no danger to the host, the inappropriate immune response may produce inflammation and other symptoms, which can cause serious illness or even death. Allergic reactions are the most familiar examples of this phenomenon. Allergic reactions may involve immediate hypersensitivity or delayed hypersensitivity.

IMMEDIATE HYPERSENSITIVITY **Immediate hypersensitivity** arises when an allergic individual is exposed to an antigen (in this case referred to as an allergen) from the environment, such as a food, pollen, or the venom of an insect. In response to the allergen, the individual makes large amounts of IgE. When this happens, mast cells in tissues and basophils in the blood bind the constant end of the IgE. If that individual is exposed to the same allergen again, binding of the allergen to the IgE causes the mast cells and basophils to rapidly release a large amount of histamine (**Figure 42.18**). This results in symptoms such as dilation of blood vessels, inflammation, and difficulty breathing. If not treated with antihistamines, a severe allergic reaction can lead to death. It is not known why some people produce excessive amounts of IgE in response to allergens. There is some evidence for genetic factors predisposing people to allergic responses.

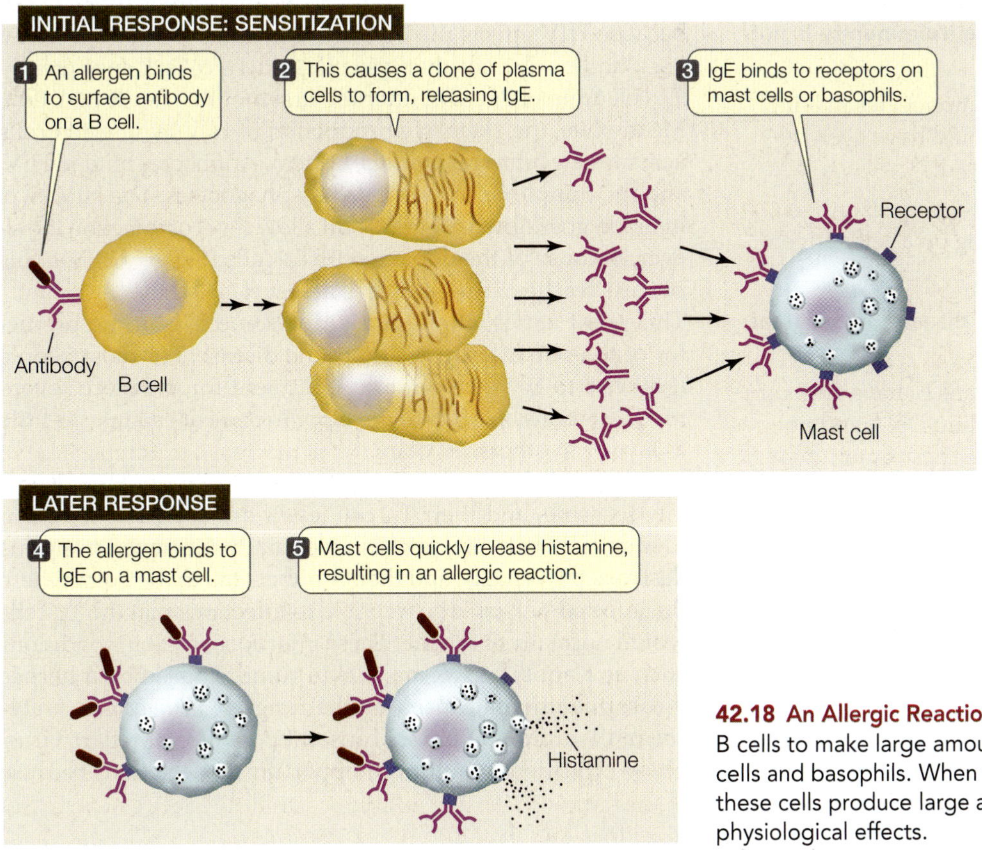

INITIAL RESPONSE: SENSITIZATION
1 An allergen binds to surface antibody on a B cell.
2 This causes a clone of plasma cells to form, releasing IgE.
3 IgE binds to receptors on mast cells or basophils.
Receptor
Antibody
B cell
Mast cell
LATER RESPONSE
4 The allergen binds to IgE on a mast cell.
5 Mast cells quickly release histamine, resulting in an allergic reaction.
Histamine

42.18 An Allergic Reaction An allergen is an antigen that stimulates B cells to make large amounts of IgE antibodies, which bind to mast cells and basophils. When the body encounters the allergen again, these cells produce large amounts of histamine, which has harmful physiological effects.

Allergy to pollen can be treated using a process called desensitization. The process involves injecting small amounts of the allergen (typically just an extract of the offending plant tissue) into the skin—enough to stimulate IgG production but not enough to stimulate IgE production. The next time the person is exposed to the allergen, IgG binds to it, tying it up before IgE can bind it and exert its harmful effects.

Desensitization does not work for food allergens because the IgE response to those substances is so strong that even a small amount of antigen provokes it. The best approach for those with food allergies—there are an estimated 3 million people in the U.S.—is to avoid foods containing the allergens. This can be difficult, but food labels listing all the ingredients are helpful. Molecular biologists are beginning to identify the antigens that act as allergens, with the hope of developing vaccines or genetically modified foods that lack the allergenic epitopes.

DELAYED HYPERSENSITIVITY **Delayed hypersensitivity** is an allergic reaction that does not begin until hours after exposure to an antigen. In this case the antigen is taken up by antigen-presenting cells and a T cell response is initiated. A T_H cell produces a clone of cells that secrete various cytokines, which cause such reactions as inflammation and rash. These events take time (hence the term "delayed"). An example is the rash that develops after exposure to poison ivy.

Autoimmune diseases are caused by reactions against self antigens

Errors in the selection of T cells in the thymus can result in T cells that bind to antigen–MHC complexes that carry self antigens. Although the precise origin of **autoimmunity** is not known, there are several hypotheses:

- *Failure of negative selection.* A clone of lymphocytes making antibodies against self antigens that should have been destroyed by clonal deletion is not destroyed.
- *Molecular mimicry.* T cells that recognize a nonself antigen, such as a virus, also recognize something on a self antigen that has a similar structure.

Autoimmunity does not always result in disease, but several autoimmune diseases are common:

- People with *systemic lupus erythematosis* (SLE) have antibodies to many cellular components, including DNA and nuclear proteins released from dying cells. These antinuclear antibodies can cause serious damage when they bind to normal tissue antigens and form large circulating antigen–antibody complexes, which become stuck in tissues and provoke inflammation.
- People with *rheumatoid arthritis* have difficulty in shutting down a T cell response to self antigens. These patients may have low activity of CTLA4, an inhibitory protein that blocks T cells from reacting to self antigens. This results in inflammation of the joints and other tissue damage.
- *Hashimoto's thyroiditis* is the most common autoimmune disease in women over 50. Immune cells attack thyroid tissue, resulting in fatigue, depression, weight gain, and other symptoms.
- *Insulin-dependent diabetes mellitus*, or type I diabetes, occurs most often in children. It is caused by an immune reaction against several proteins in the cells of the pancreas that manufacture the protein hormone insulin. This reaction kills the insulin-producing cells, so people with type I diabetes must take insulin daily in order to survive.

AIDS is an immune deficiency disorder

There are several inherited and acquired immune deficiency disorders. In some individuals, T or B cells never form; in others, B cells lose the ability to give rise to plasma cells. In either case, the affected individual is unable to mount an adaptive immune response and thus lacks a major line of defense against pathogens. The T_H cell is perhaps the most central component of the immune system because of its essential roles in both the humoral and cellular immune responses (see Figure 42.7). This cell is the target of **human immunodeficiency virus** (**HIV**), the retrovirus that results in **acquired immune deficiency syndrome** (**AIDS**).

HIV can be transmitted from person to person in body fluids containing the virus (such as blood, semen, vaginal fluid, or breast milk). The recipient tissue is either blood (by transfusion) or a mucous membrane lining an organ (the mucus contains a high concentration of lymphocytes). HIV initially infects macrophages, T_H cells, and antigen-presenting dendritic cells in the blood and tissues. At first there is an immune response to the viral infection, and T_H cells are activated. But because HIV infects the T_H cells, they are killed both by HIV itself and by T_C cells that lyse infected T_H cells. Consequently T_H cell numbers decline after the first month or so of infection. Meanwhile, the extensive production of HIV by infected cells activates the humoral immune system. Antibodies bind to HIV, and the complexes are removed by phagocytes. The HIV level in blood goes down. There is still a low level of infection, however, because of the depletion of T_H cells (**Figure 42.19**). This process reaches a low, steady-state level called the "set point." This point varies among individuals and is a strong predictor of the rate of progression of the disease. For most people it takes 8 to 10 years without treatment for the more severe manifestations of AIDS to develop. In some it can take as little as a year; in others, 20 years.

During this dormant period, people carrying HIV generally feel fine, and their T_H cell levels are adequate for them to mount immune responses. Eventually, however, the virus destroys the T_H cells, and their numbers fall to the point where the infected person is susceptible to infections that the T_H cells would normally eliminate. These infections result in conditions such as Kaposi's sarcoma, a skin tumor caused by a herpes virus; pneumonia caused by the fungus *Pneumocystis jirovecii*; and lymphoma tumors caused by the Epstein–Barr virus. These conditions result from opportunistic infections because

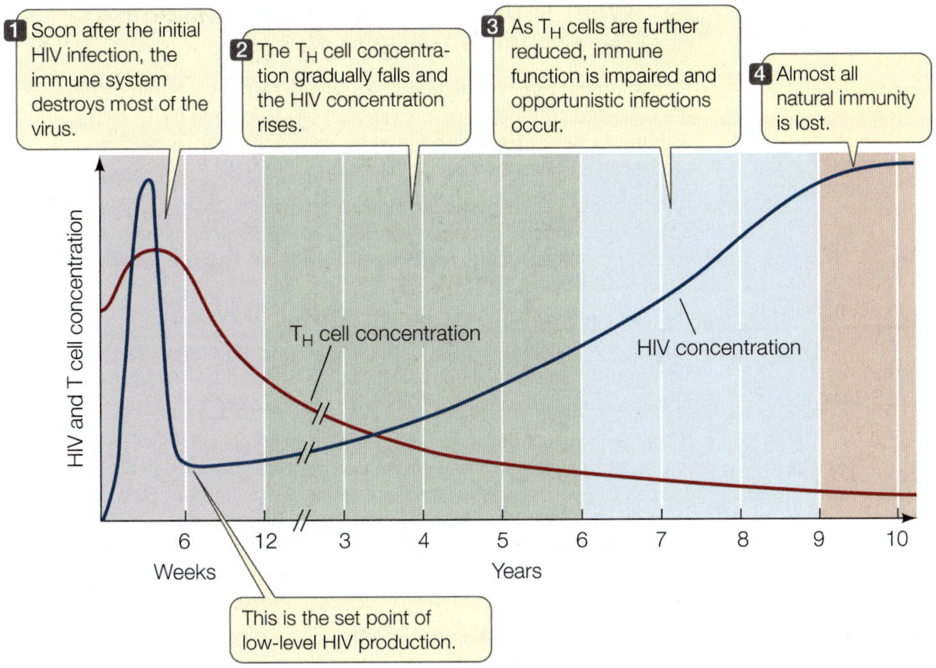

1 Soon after the initial HIV infection, the immune system destroys most of the virus.
2 The T_H cell concentration gradually falls and the HIV concentration rises.
3 As T_H cells are further reduced, immune function is impaired and opportunistic infections occur.
4 Almost all natural immunity is lost.
HIV and T cell concentration
T_H cell concentration
HIV concentration
6
12
3
4
5
6
7
8
9
10
Weeks
Years
This is the set point of low-level HIV production.

42.19 The Course of an HIV Infection An HIV infection may be carried, unsuspected, for many years before the onset of symptoms.

the pathogens take advantage of the crippled immune system of the host. They lead to death within a year or two.

The molecular biology of HIV and its life cycle have been intensively studied (see Figure 16.16). This has resulted in the development of drugs targeted to HIV proteins, such as the reverse transcriptase that makes cDNA from the viral RNA, and the viral protease that cuts the large precursor viral protein into its final active proteins. Treatment with combinations of such drugs has had spectacular success. Getting AIDS before the 1990s was a death sentence, with few sufferers surviving beyond a year or two. Today the survival of a treated, infected person is not much different from that of an uninfected person. Unfortunately, like many medical treatments, HIV drugs are not available to all who need them—particularly in poor regions of the world where AIDS is prevalent. As a result, there are about 1.7 million deaths per year worldwide from AIDS.

RECAP 42.6

Failures of the immune system include allergic reactions (caused by hypersensitivity to antigens), autoimmune diseases (caused by reactions against self antigens), and immune deficiency disorders.

- How does immediate hypersensitivity develop? **See p. 875 and Figure 42.18**
- What is an autoimmune disease? Give an example. **See p. 876**
- Describe the course of events in the human immune system during HIV infection. **See pp. 876–877 and Figure 42.19**

?

Why do many people resist vaccination?

ANSWER

There seem to be three main reasons why a person might refuse a vaccine. The first is complacency. Measles, which used to kill thousands of children every year in the U.S. and still does in poor countries, is no longer a highly visible threat to public health. Second, some people believe that vaccines, although exhaustively tested for safety, are actually unsafe and cause disease. The internet is full of such assertions. Third, people are wary of experts because of past false alarms. The discovery of the H1N1 flu virus (swine flu) in Mexico in 2009 led to a high alert and a mass vaccination program that turned out not to be necessary. Vaccination is a scientific success in terms of bolstering immunity and eradicating disease. But like any technology, its acceptance is a political issue.

42.1 What Are the Major Defense Systems of Animals?

- Animal defenses against **pathogens** are based on the body's ability to distinguish between self and nonself.
- **Innate** (nonspecific) **defenses** are inherited mechanisms that protect the body from many kinds of pathogens. They typically act rapidly.
- **Adaptive** (specific) **defenses** respond to specific pathogens. They develop more slowly than innate defenses but are long-lasting.
- Many defenses are implemented by cells and proteins carried in the blood plasma and **lymph**. **Review Figure 42.1, ACTIVITY 42.1**
- **White blood cells** fall into two broad groups. **Phagocytes** engulf pathogens by phagocytosis. **Lymphocytes**, which include B cells and T cells, participate in adaptive responses. **Review Figure 42.2, ANIMATED TUTORIAL 42.1**

42.2 What Are the Characteristics of the Innate Defenses?

- An animal's innate defenses include physical barriers such as the skin, and competing resident microorganisms known as normal flora. **Review Figure 42.3**
- The **complement system** consists of more than 20 different antimicrobial proteins that act to alter membrane permeability and kill targeted cells.
- Circulating defensive cells, such as phagocytes and **natural killer cells**, eliminate invaders.
- A cell signaling pathway involving the **toll-like receptor** stimulates the body's defenses. **Review Figure 42.4**
- **Inflammation** involves activation of several types of cells and proteins that act against invading pathogens. **Mast cells** release **histamines**, which cause blood vessels to dilate and become "leaky." **Review Figure 42.5, ACTIVITY 42.2**

42.3 How Does Adaptive Immunity Develop?

- The adaptive immune response recognizes specific **antigens**, responds to an enormous diversity of **antigenic determinants**, distinguishes self from nonself, and remembers the antigens it has encountered. **Review ANIMATED TUTORIAL 42.2**
- Each antibody and each T cell is specific for a single antigenic determinant. **T cell receptors** bind to antigens on the surfaces of virus-infected cells and abnormal cells.
- The **humoral immune response** is directed against pathogens in the blood, lymph, and tissue fluids. The **cellular immune response** is directed against an antigen established within a host cell. Both responses are mediated by antigenic fragments being presented on a cell surface along with the proteins of the **major histocompatibility complex** (**MHC**). **Review Figure 42.7**
- **Clonal selection** accounts for the specificity and diversity of the immune response and for **immunological memory**. **Review Figure 42.8**
- An activated B or T lymphocyte produces **effector cells** that attack the antigen, and **memory cells** that are long-lived and rarely divide. Effector B cells are called **plasma cells** and secrete specific **antibodies**.
- **Vaccination** is inoculation with modified pathogens or antigens that provoke an immune response but are not pathogenic. **Review Figure 42.9**

42.4 What Is the Humoral Immune Response? See ANIMATED TUTORIAL 42.3

- B cells are the basis of the humoral immune response. Naïve B cells are activated by binding of antigen and by stimulation by T_H cells with the same specificity, and then form plasma cells. These cells synthesize and secrete specific antibodies.
- An antibody is an **immunoglobulin**, a tetramer of four polypeptides: two identical light chains and two identical heavy chains, each consisting of a **constant region** and a **variable region**. **Review Figure 42.10, ACTIVITY 42.3**
- The variable regions determine the specificity of an immunoglobulin, and the constant regions of the heavy chain determine its class. There are five classes of immunoglobulins with different body locations and functions. **Review Table 42.2**
- B cell genomes undergo random recombination of genes coding for regions of the immunoglobulin polypeptide chains so that each cell can produce a specific antibody protein. The immunoglobulin chains derive from "supergenes" that are constructed from different combinations of *V*, *D*, *J*, and *C* genes. This DNA rearrangement and rejoining yields millions of different immunoglobulin chains. **Review Figures 42.11, 42.12, ANIMATED TUTORIAL 42.4**
- Once a B cell becomes a plasma cell, it may undergo **class switching**, in which a deletion of one or more constant region genes results in the production of an immunoglobulin with a different constant region and a different function. **Review Figure 42.13**
- A **monoclonal antibody** can be used in diagnosis and therapy.

42.5 What Is the Cellular Immune Response? See ANIMATED TUTORIAL 42.5

- T cells are the effectors of the cellular immune response. T cell receptors are somewhat similar in structure to the immunoglobulins, having variable and constant regions. **Review Figure 42.14**
- There are three types of T cells. **Cytotoxic T cells** (**T_C cells**) recognize and kill virus-infected cells or mutated cells. **T-helper cells** (**T_H cells**) direct both the cellular and humoral immune responses. **Regulatory T cells** (**Tregs**) inhibit the other T cells from mounting an immune response to self antigens.
- The genes of the major histocompatibility complex (MHC) encode membrane proteins that bind antigenic fragments and present them to T cells. **Review Figures 42.15, 42.16**
- Organ transplants are rejected when the host's immune system recognizes MHC proteins on transplanted tissue as nonself and initiates an immune defense attacking the foreign tissue.

42.6 What Happens When the Immune System Malfunctions?

- An **allergic reaction** is an inappropriate immune response caused by **immediate hypersensitivity** or **delayed hypersensitivity** to certain antigens. **Review Figure 42.18**
- Autoimmune diseases result when the immune system produces B and T cells that attack self antigens.
- Immune deficiency disorders result from failure of one or another part of the immune system. **Acquired immune deficiency syndrome** (**AIDS**) is a disorder that arises from depletion of the T_H cells as a result of infection with **human immunodeficiency virus** (**HIV**). **Review Figure 42.19**

Go to the Interactive Summary to review key figures, Animated Tutorials, and Activities
Life10e.com/is42

CHAPTER**REVIEW**

REMEMBERING

1. Phagocytes kill harmful bacteria by
   a. endocytosis.
   b. producing antibodies.
   c. complement proteins.
   d. T cell stimulation.
   e. inflammation.

2. Which statement about an antigenic determinant is *not* true?
   a. It is a specific chemical grouping.
   b. It may be part of many different molecules.
   c. It is the part of an antigen to which an antibody binds.
   d. It may be part of a cell.
   e. A single protein has only one on its surface.

3. T cell receptors
   a. are the primary receptors for the humoral immune system.
   b. are carbohydrates.
   c. cannot function unless the animal has previously encountered the antigen.
   d. are produced by plasma cells.
   e. are important in combating viral infections.

4. According to the clonal selection theory,
   a. an antibody changes its shape to match the antigen it meets.
   b. an individual animal contains only one type of B cell.
   c. an individual animal contains many types of B cells, each producing one kind of antibody.
   d. each B cell produces many types of antibodies.
   e. many clones of antiself lymphocytes appear in the bloodstream.

5. The extraordinary diversity of antibodies results in part from
   a. the action of monoclonal antibodies.
   b. the splicing of protein molecules.
   c. the action of cytotoxic T cells.
   d. the rearrangement of genes.
   e. their remarkable nonspecificity.

6. The major histocompatibility complex
   a. codes for specific proteins found on the surfaces of cells.
   b. plays no role in T cell immunity.
   c. plays no role in antibody responses.
   d. plays no role in skin graft rejection.
   e. is encoded by a single locus with multiple alleles.

UNDERSTANDING & APPLYING

7. Describe the part of an antibody molecule that interacts with an antigen. How is it similar to the active site of an enzyme? How does it differ from the active site of an enzyme?

8. Contrast immunoglobulins and T cell receptors with respect to their structures and functions.

9. The gene family (genes A, B, and D) determining MHC on the cell surface in humans is on a single chromosome. A father's MHC genotype is A1, A3, B5, B7, D9, D11. A mother's genotype is A2, A4, B6, B7, D11, D12. Their child's is A1, A4, B6, B7, D11, D12. What are the parents' haplotypes—that is, which alleles are linked on each of the two chromosomes of each parent? Assuming there is no recombination among the genes determining the MHC type, can these same two parents have a child who has the genotype A1, A2, B7, B8, D9, D11?

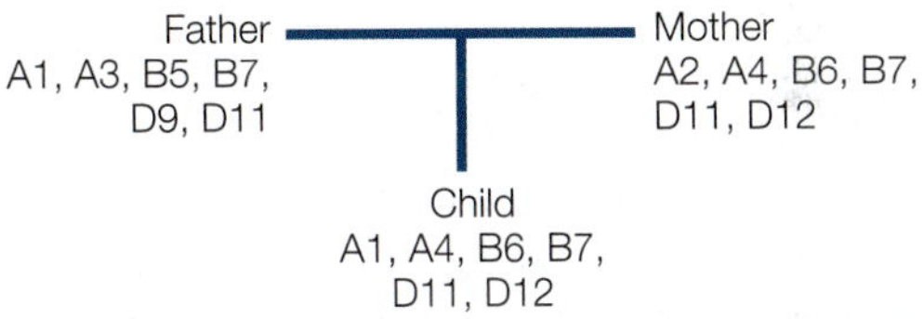
Father
A1, A3, B5, B7,
D9, D11
Mother
A2, A4, B6, B7,
D11, D12
Child
A1, A4, B6, B7,
D11, D12

ANALYZING & EVALUATING

10. Discuss the diversity of antibody specificities in an individual in relation to the diversity of enzymes. Does every cell in an animal contain genetic information for all the organism's enzymes? Does every cell contain genetic information for all the organism's immunoglobulins?

11. Development of an effective HIV vaccine requires that the person being vaccinated develop both cellular and humoral immunity against HIV. What experiments would you do to test whether both types of immunity developed in people given a potential new vaccine?

Go to BioPortal at **yourBioPortal.com** for Animated Tutorials, Activities, LearningCurve Quizzes, Flashcards, and many other study and review resources.

43 Animal Reproduction

CHAPTER**OUTLINE**

A Unique Reproductive Strategy Among honey bees and other hymenopteran insects, the only reproductive female is the queen, seen in the center. The female workers attending her are sterile.

THE HONEY BEE *Apis mellifera* has been studied throughout recorded history. Sex and reproduction in these social insects are intriguing. A honey bee hive contains a single reproductive female, the queen, and thousands of infertile female worker bees. There are few if any male bees in a hive. The queen lays eggs continuously while the worker bees build the honeycomb, forage, clean and defend the hive, and feed and care for the young. In honey bees, fertilized eggs develop into females and unfertilized eggs develop into males. The queen controls which of her eggs are fertilized. She carries sperm from males she has mated with and occasionally produces males by *not* releasing sperm onto her eggs. It is not in her best interest to produce many males because they don't contribute to raising young or maintaining the hive; they just hang around until they have a very rare chance to mate.

Eventually a hive needs a new queen. The old queen might die, or she and a retinue of workers might leave an overcrowded hive to start a new hive (a phenomenon known as swarming). Whether the queen dies or swarms, the remaining worker bees enlarge a few cells in the honeycomb that contain fertilized eggs laid by the old queen. The larvae from those eggs are fed special food that stimulates their growth and development into prospective queens.

The first queen to pupate and emerge from her royal chamber kills the other developing queens. She then leaves the hive for her mating flight. Males from all around sense a chemical message (a pheromone) that a virgin queen is available and congregate around her. While in flight, she mates with 15 to 20 males, and each coupling is an event. After a male manages to insert his penis into the queen's vagina, he literally explodes, leaving behind not only his sperm but also his sex organs (the latter will drop out later). The males die and the queen returns to the hive with a lifetime supply of sperm. The queen lives for about 2 years and lays as many as 3,000 eggs each day. Most of those eggs develop into female worker bees that devote their lives to feeding her and raising their sisters.

Natural selection has resulted in some amazing adaptations, none more so than those involved in reproduction. Sexual or asexual, bizarre or otherwise, the anatomy, physiology, biochemistry, and behavior surrounding the urge to propagate are fascinating.

?

How can a queen be so different from her worker sisters when they all share the same genome?

See answer on p. 899.

43.1 How Do Animals Reproduce without Sex?

Sexual reproduction is a nearly universal trait in animals, although many species can also reproduce asexually and some reproduce only asexually. Offspring produced asexually are genetically identical to one another and to their parents. Asexual reproduction is efficient because no time or energy is wasted on mating and every member of the population can convert resources into offspring. However, asexual reproduction does not generate genetic diversity in a population as sexual reproduction does, and this diversity is the raw material that enables natural selection to shape adaptations in response to environmental change. When environmental changes occur, lack of genetic diversity can be disadvantageous to a population.

A variety of animals, mostly invertebrates, reproduce asexually. They tend to be species that are attached to their substrates and cannot search for mates, or species that live in sparse populations and rarely encounter potential mates. Asexually reproducing species are likely to be found in relatively constant environments where genetic diversity is less important for species success. In fact, asexual reproduction is a good way to preserve a successful genotype in a particular environment—as long as that environment does not change.

Three common modes of asexual reproduction are budding, regeneration, and parthenogenesis.

Budding and regeneration produce new individuals by mitosis

Many simple multicellular animals produce offspring by **budding**. New individuals form as outgrowths or buds from the bodies of older animals. A bud grows by mitotic cell division, and the cells differentiate before the bud breaks away from the parent (Figure 43.1A). The bud is genetically identical to the parent, and it may grow as large as the parent before it becomes independent.

Regeneration is usually thought of as the replacement of damaged tissues or lost limbs, but in some cases pieces of an organism can regenerate complete individuals. Echinoderms, for example, have remarkable abilities to regenerate. If sea stars (starfish) are cut into pieces, each piece that includes an arm and a portion of the central disc can grow into a new animal (Figure 43.1B). In the early 1900s oyster fishermen in Narragansett Bay tried to eliminate the sea stars that were preying on their oysters. Whenever they encountered sea stars, they chopped them up with their knives and threw them back into the water. As a result, the sea star population increased explosively.

Regeneration can occur when an animal is broken by an outside force such as wave action in the intertidal zone. In some cases, breakage occurs in the absence of external forces. Some species of segmented marine worms develop segments with rudimentary heads bearing sensory organs. The segments then break apart and each one forms a new worm.

Parthenogenesis is the development of unfertilized eggs

Not all eggs must be fertilized to develop. A common mode of asexual reproduction in arthropods is the development of offspring from unfertilized eggs. This phenomenon, called **parthenogenesis**, also occurs in some species of fishes, amphibians, and reptiles. Most species that reproduce parthenogenetically also engage in sexual reproduction or at least sexual behavior at other times. In some species, parthenogenesis is part of the mechanism that determines sex. As we saw at the beginning of this chapter, in honey bees (as well as in most ants and wasps), males develop from unfertilized eggs and are haploid. Females develop from fertilized eggs and are diploid.

Parthenogenetic reproduction in some species requires sexual behavior even though sperm are not delivered to the female reproductive tract and eggs are not fertilized. David Crews and his students at the University of Texas extensively investigated one such case, that of parthenogenetic reproduction in a species of whiptail lizard. There are no males of this species. Females can act as males, engaging in all aspects of courtship display and mating, although no sperm are produced or transferred

(A) *Hydra* sp.

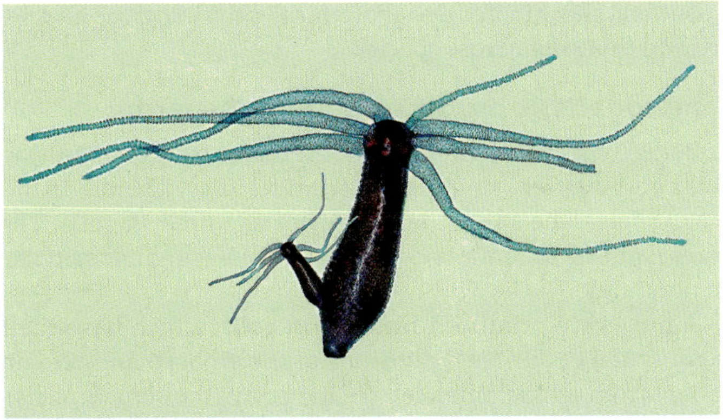

(B) *Fromia* sp.

43.1 Two Forms of Asexual Reproduction (A) Budding: A new individual forms as an outgrowth from an adult hydra. (The hydra in this photo was stained with a blue dye to make it easier to see.) (B) Regeneration: A single severed arm and a piece of the central disc of a mature sea star can regenerate into an entire animal.

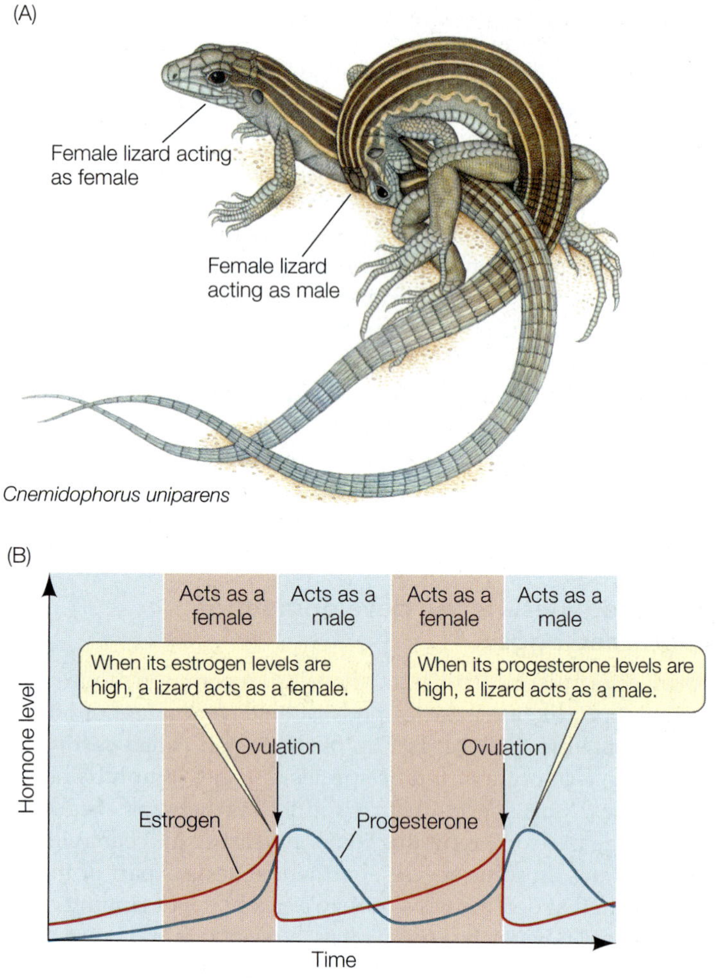

(A)
Female lizard acting as female
Female lizard acting as male
Cnemidophorus uniparens
(B)
Acts as a female
Acts as a male
Acts as a female
Acts as a male
When its estrogen levels are high, a lizard acts as a female.
When its progesterone levels are high, a lizard acts as a male.
Ovulation
Ovulation
Estrogen
Progesterone
Hormone level
Time

43.2 Asexual Reproduction May Require Sexual Behavior (A) Parthenogenetic whiptail lizards are all females, but they take turns acting the male role in reproductive behavior. The stimulation from sexual behavior is necessary for ovulation to occur. (B) The ovarian cycle determines the role an individual whiptail plays.

(**Figure 43.2**). Whether a specific female acts as a female or as a male depends on cyclical hormonal states. When estrogen levels are high, she acts as a female. When her progesterone levels peak, she acts as a male. The stimulation resulting from the sexual activity triggers the release of eggs from the ovaries of the acting female.

RECAP 43.1

Most animals reproduce sexually, but many can also or can only reproduce asexually, through budding, regeneration, or parthenogenesis.

- Explain why asexual reproduction might be disadvantageous for an animal living in a changing environment. **See p. 881**
- How is parthenogenesis related to sex determination in honey bees? **See p. 881**

Asexual reproduction is an efficient way to use resources. Since most animals reproduce sexually, however, the genetic diversity produced by sexual reproduction must confer a tremendous advantage.

43.2 How Do Animals Reproduce Sexually?

Given the efficiency of asexual reproduction in perpetuating an organism's genome, the prevalence of sexual reproduction is somewhat surprising. Because one gender (i.e., males) cannot produce offspring, it is a much less efficient strategy. And mating behaviors involve costs and risks. Costs include time and energy spent finding, attracting, and competing for a mate, as well as the "opportunity costs" of detracting from other activities such as feeding and caring for existing offspring. Risks include increased exposure to predation and the potential for physical damage. Despite these disadvantages, most eukaryotic organisms reproduce sexually. Thus it would seem that the production of genetic diversity is an evolutionary advantage that overwhelms the cost of sex (see Sections 21.2 and 21.4).

Sexual reproduction requires the joining of two haploid sex cells to form a diploid individual. These haploid cells, or **gametes**, are produced through gametogenesis, a process that involves meiotic cell divisions. Two events in meiosis contribute to genetic diversity: crossing over between homologous chromosomes and the independent assortment of chromosomes (see Sections 11.5 and 12.1). Sexual reproduction itself also contributes to genetic diversity. The genetic variation among the gametes of a single individual and the genetic variation between any two parents produce an enormous potential for genetic variation between any two offspring of a sexually reproducing pair of individuals.

Sexual reproduction in animals consists of three fundamental steps:

- **Gametogenesis**: making gametes
- **Spawning** or **mating**: bringing gametes together
- **Fertilization**: fusing gametes

The process of gametogenesis is similar across sexually reproducing animal species. Processes of fertilization are also quite similar in widely different species. Therefore, while our discussion of gametogenesis will focus generally on mammals, and our discussion of fertilization will feature sea urchins, the facts would not be dramatically different were we to consider many other animal groups. Adaptations for spawning and mating, in contrast, show incredible anatomical, physiological, and behavioral diversity across species.

Gametogenesis produces eggs and sperm

Gametogenesis occurs in the gonads: **testes** (singular *testis*) in males and **ovaries** (singular *ovary*) in females. The tiny gametes of males, the **sperm**, move by beating their flagella. The larger gametes of females—the eggs or **ova** (singular *ovum*)—are nonmotile.

Gametes are produced from **germ cells**, which have their origin in the earliest cell divisions of the embryo and remain distinct from all the other cells of the body (the somatic cells). Germ cells are sequestered in the body of the embryo until its gonads begin to form. The germ cells then migrate to the developing gonads, where they take up residence and proliferate by

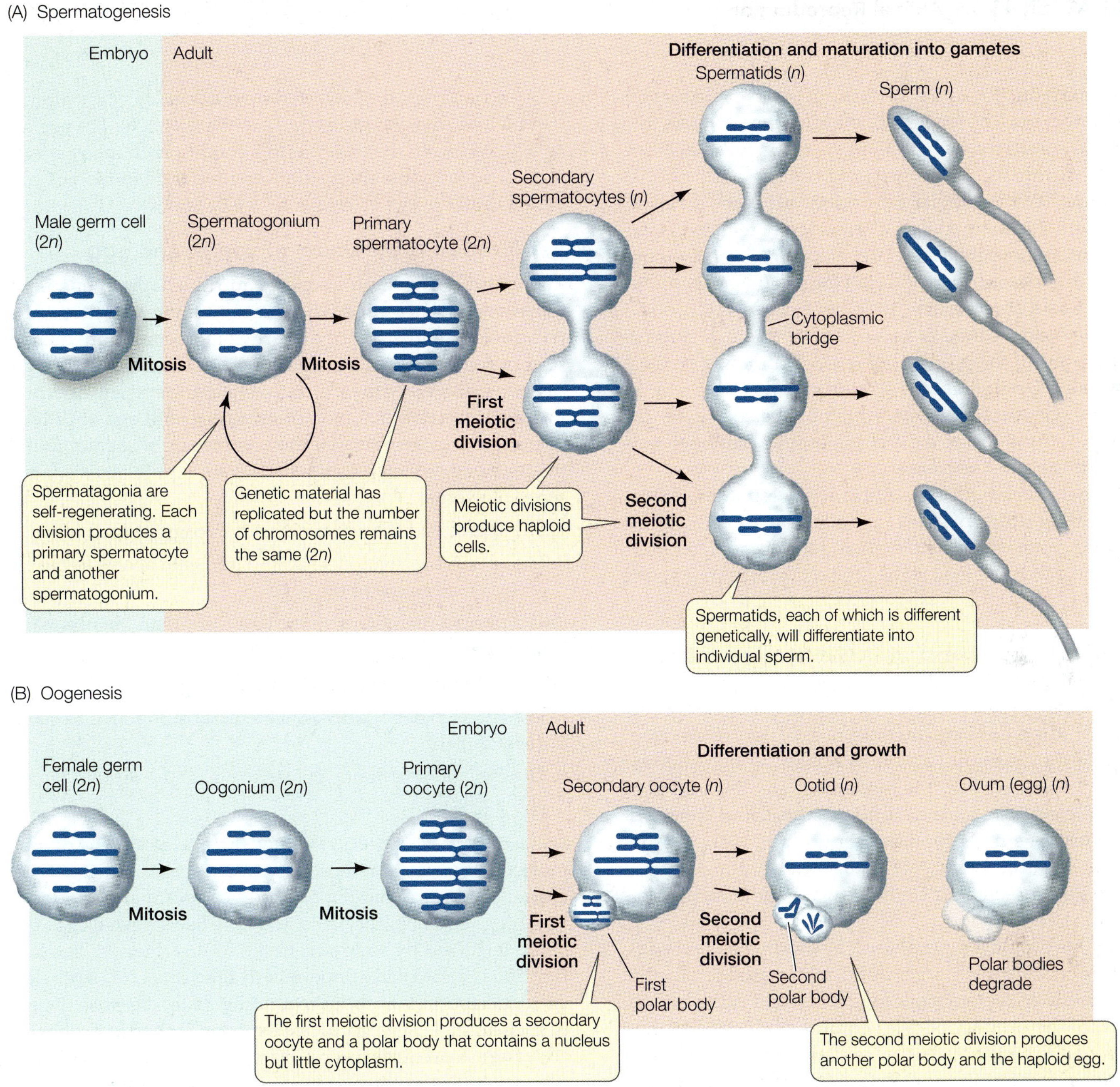

(A) Spermatogenesis
Embryo
Adult
Differentiation and maturation into gametes
Spermatids (n)
Sperm (n)
Secondary spermatocytes (n)
Male germ cell (2n)
Spermatogonium (2n)
Primary spermatocyte (2n)
Mitosis
Mitosis
Cytoplasmic bridge
First meiotic division
Second meiotic division
Spermatagonia are self-regenerating. Each division produces a primary spermatocyte and another spermatogonium.
Genetic material has replicated but the number of chromosomes remains the same (2n)
Meiotic divisions produce haploid cells.
Spermatids, each of which is different genetically, will differentiate into individual sperm.
(B) Oogenesis
Embryo
Adult
Differentiation and growth
Female germ cell (2n)
Oogonium (2n)
Primary oocyte (2n)
Secondary oocyte (n)
Ootid (n)
Ovum (egg) (n)
Mitosis
Mitosis
First meiotic division
Second meiotic division
First polar body
Second polar body
Polar bodies degrade
The first meiotic division produces a secondary oocyte and a polar body that contains a nucleus but little cytoplasm.
The second meiotic division produces another polar body and the haploid egg.

43.3 Gametogenesis Male and female germ cells proliferate by mitosis and produce diploid spermatogonia and oogonia that mature into primary spermatocytes and oocytes before entering meiosis. (A) Spermatogonia continue to divide by mitosis in adults, producing a steady supply of spermatocytes that divide meiotically to produce haploid spermatids, which differentiate into sperm. In many species, the progeny of spermatocytes remain in contact through cytoplasmic bridges until the sperm mature. (B) In mammals, oogonia cease division in the embryo, and primary oocytes remain arrested in prophase I of meiosis until they are ovulated and fertilized. Each oocyte will produce one haploid ootid, which matures into an ovum.

mitosis, producing **spermatogonia** (singular *spermatogonium*) in males and **oogonia** (singular *oogonium*) in females (**Figure 43.3**). Spermatogonia and oogonia are diploid, multiply by mitosis, and are stem cells. They are self-regenerating, and they also produce progeny—spermatocytes and oocytes—that will enter the next stage of gametogenesis.

In this next stage of gametogenesis, meiotic cell division reduces the chromosomes to the haploid number (see Section 11.5). The progeny of the spermatogonia and oogonia that enter meiosis are **primary spermatocytes** and **primary oocytes**. The steps of meiosis are similar in males and females, but there are important sex differences in gametogenesis.

SPERMATOGENESIS The initial proliferation of male germ cells into spermatogonia proceeds by mitosis in the embryo. But throughout the male life span, spermatogonia continue to divide by mitosis, with one daughter cell retaining the self-regenerating capacity of a spermatogonium while the other daughter cell becomes a primary spermatocyte. As illustrated in Figure 43.3A, primary spermatocytes then undergo the first meiotic division—the reduction division—to form two haploid **secondary spermatocytes**. The second meiotic division produces four haploid **spermatids** for each primary spermatocyte that enters meiosis. In mammals, the progeny of primary spermatocytes remain connected by cytoplasmic bridges after each division.

One reason that mammalian spermatocytes remain in cytoplasmic contact throughout their development is the asymmetry of sex chromosomes in males. Half the secondary spermatocytes receive an X chromosome, the other half a Y chromosome. The Y chromosome contains fewer genes than the X chromosome, and some of the products of genes found only on the X chromosome are essential for spermatocyte development. By remaining in cytoplasmic contact, all four spermatocytes can share the gene products of the X chromosomes, although only half of them have an X chromosome.

A spermatid bears little resemblance to a mature sperm. Through further differentiation (spermiogenesis), the spermatid becomes compact, streamlined, and grows a flagellum to become motile. We will look at the production of human sperm in Section 43.3.

OOGENESIS Oogonia, like spermatogonia, proliferate through mitosis (see Figure 43.3B). The resulting primary oocytes immediately enter prophase of the first meiotic division. In many species, including humans, the oocyte experiences developmental arrest at this point and may remain in that state for days, months, or years. In the human female, this period of arrest is at least 10 years (i.e., until puberty), and some primary oocytes remain in prophase I for up to 50 years (i.e., until menopause). In contrast, spermatogenesis continues, uninterrupted, to completion once the primary spermatocyte has differentiated.

During this prolonged prophase I, or shortly before it ends, the primary oocyte grows larger through increased production of ribosomes, RNA, cytoplasmic organelles, and energy stores. At this point the primary oocyte acquires all the energy, raw materials, and RNA that the ovum will need to survive its first cell divisions after fertilization. In fact, the nutrients in the egg must maintain the embryo until it is either nourished by the maternal circulatory system or can feed on its own.

When a primary oocyte resumes meiosis, its nucleus completes the first meiotic division near the surface of the cell. The daughter cells of this division receive grossly unequal shares of cytoplasm. This asymmetry represents another major difference from spermatogenesis, in which cytoplasm is apportioned equally. The daughter cell that receives almost all the cytoplasm becomes the **secondary oocyte**, and the one that receives almost none forms the **first polar body** (see Figure 43.3B).

The second meiotic division—that of the large, secondary oocyte—is also accompanied by an asymmetrical division of the cytoplasm. One daughter cell forms the large, haploid **ootid**, which eventually differentiates into a mature egg, and the other forms the **second polar body**. Polar bodies degenerate, so the end result of oogenesis is only one mature egg for each primary oocyte that entered meiosis. However, that egg is a large, well-provisioned cell.

A second period of arrested development occurs after the first meiotic division forms the secondary oocyte. The egg may be expelled from the ovary in this condition. In many species, including humans, the second meiotic division is not completed until the egg is fertilized by a sperm.

Fertilization is the union of sperm and egg

The union of the haploid sperm and the haploid egg in fertilization creates a single diploid cell, called a **zygote**, which will develop into an embryo. Fertilization does more than just restore the full genetic complement of the animal. The processes associated with fertilization help the egg and sperm get together, prevent the union of the sperm and egg of different species, and guarantee that only one sperm will enter and activate the egg metabolically. Fertilization involves a complex series of events:

1. The sperm and the egg chemically recognize each other.
2. The sperm is activated, enabling it to gain access to the plasma membrane of the egg.
3. The plasma membrane of the egg fuses with the plasma membrane of a single sperm.
4. The egg blocks entry of additional sperm.
5. The egg is metabolically activated and stimulated to start development.
6. The egg and sperm nuclei fuse to create the diploid nucleus of the zygote.

SPECIFICITY IN SPERM–EGG INTERACTIONS Specific recognition molecules mediate interactions between sperm and eggs. These molecules ensure that the activities of sperm are directed toward eggs and not other cells, and they help prevent eggs from being fertilized by sperm from the wrong species. The latter function is particularly important in aquatic species that release eggs and sperm into the surrounding water, because the eggs of such animals may readily be exposed to sperm of other species. The sea urchin is a good example of such a species, and its mechanisms of fertilization have been well studied.

Sea urchin eggs release chemical attractants that increase the motility of sperm and cause them to swim toward the egg. These chemical attractants are species-specific. For example, eggs of one species of sea urchin release a specific peptide consisting of 14 amino acids. This peptide binds to receptors present on sperm of the same species. The sperm respond by increasing their mitochondrial respiration and motility. Before exposure to the peptide, the sperm swim in tight little circles, but after binding to the peptide, they swim energetically up the concentration gradient of the peptide until they reach the egg that is releasing it.

When sperm reach an egg, they must get through two protective layers before they can fuse with the egg plasma membrane. The eggs of sea urchins are covered with a jelly coat, which surrounds a proteinaceous **vitelline envelope** (Figure 43.4A). The success of a sperm's assault on these protective layers depends on a membrane-enclosed structure at the front of the sperm head called an **acrosome**.

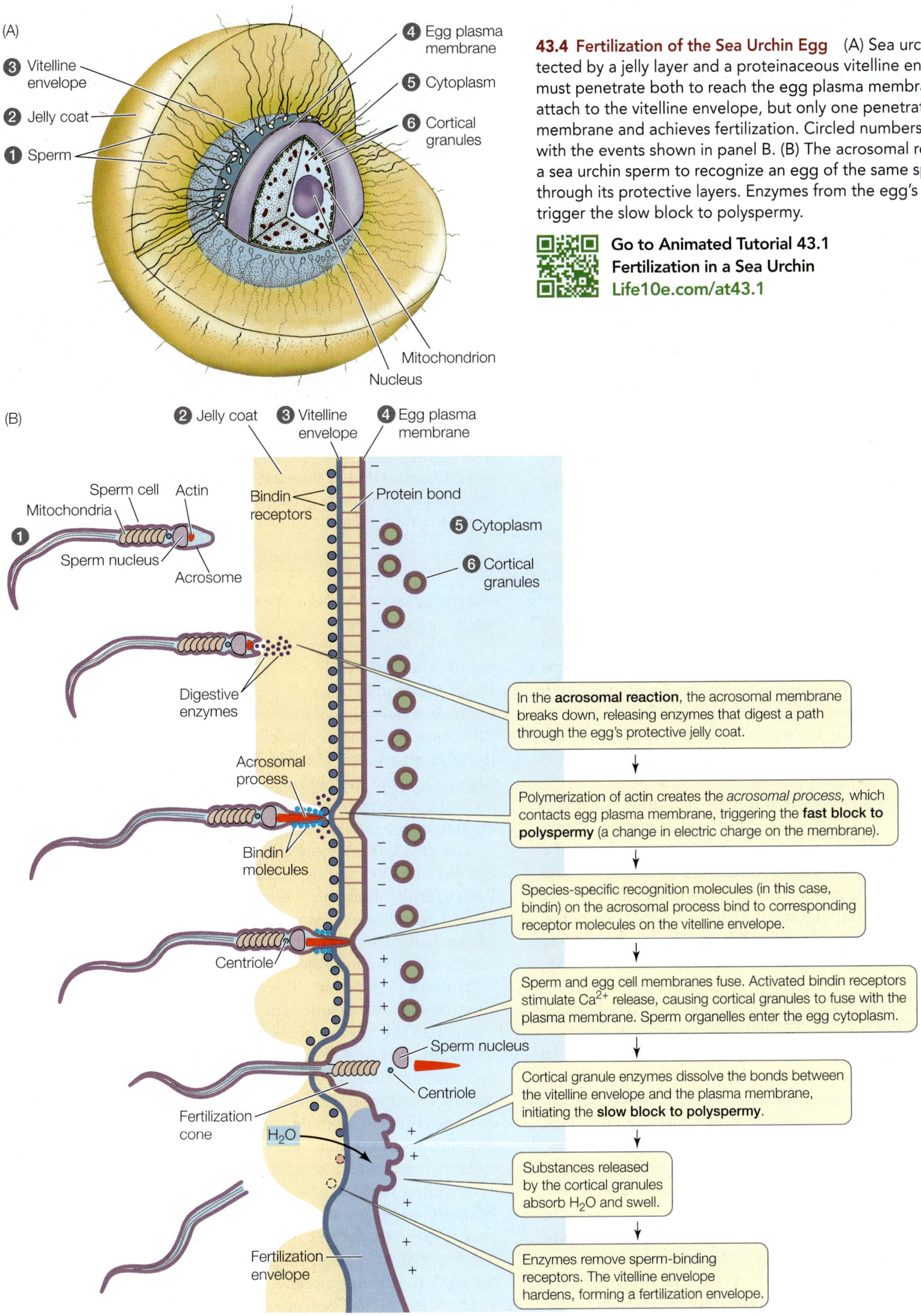
(A)
4 Egg plasma membrane
3 Vitelline envelope
5 Cytoplasm
2 Jelly coat
6 Cortical granules
1 Sperm
Mitochondrion
Nucleus
(B)
2 Jelly coat
3 Vitelline envelope
4 Egg plasma membrane
Sperm cell
Actin
Bindin receptors
Protein bond
Mitochondria
5 Cytoplasm
1
Sperm nucleus
6 Cortical granules
Acrosome
Digestive enzymes
In the **acrosomal reaction**, the acrosomal membrane breaks down, releasing enzymes that digest a path through the egg's protective jelly coat.
Acrosomal process
Polymerization of actin creates the *acrosomal process,* which contacts egg plasma membrane, triggering the **fast block to polyspermy** (a change in electric charge on the membrane).
Bindin molecules
Species-specific recognition molecules (in this case, bindin) on the acrosomal process bind to corresponding receptor molecules on the vitelline envelope.
Centriole
Sperm and egg cell membranes fuse. Activated bindin receptors stimulate Ca^{2+} release, causing cortical granules to fuse with the plasma membrane. Sperm organelles enter the egg cytoplasm.
Sperm nucleus
Centriole
Cortical granule enzymes dissolve the bonds between the vitelline envelope and the plasma membrane, initiating the **slow block to polyspermy**.
Fertilization cone
H_2O
Substances released by the cortical granules absorb H_2O and swell.
Fertilization envelope
Enzymes remove sperm-binding receptors. The vitelline envelope hardens, forming a fertilization envelope.

43.4 Fertilization of the Sea Urchin Egg (A) Sea urchin eggs are protected by a jelly layer and a proteinaceous vitelline envelope. Sperm must penetrate both to reach the egg plasma membrane. Many sperm attach to the vitelline envelope, but only one penetrates the egg cell membrane and achieves fertilization. Circled numbers match structures with the events shown in panel B. (B) The acrosomal reaction allows a sea urchin sperm to recognize an egg of the same species and pass through its protective layers. Enzymes from the egg's cortical granules trigger the slow block to polyspermy.

Go to Animated Tutorial 43.1
Fertilization in a Sea Urchin
Life10e.com/at43.1

The acrosome contains enzymes and other proteins. When the sperm makes contact with an egg of its own species, substances in the jelly coat trigger an acrosomal reaction, which begins with the breakdown of the plasma membrane covering the sperm head and the underlying acrosomal membrane (Figure 43.4B). The acrosomal enzymes are released and digest a hole through the jelly coat.

As a result of the polymerization of actin triggered by the acrosomal reaction, an acrosomal process extends out of the head of the sperm. The acrosomal process is coated with species-specific recognition molecules called bindin, and there are bindin receptors on the vitelline envelope of the egg. The interaction of bindin and bindin receptors enables the sperm to contact the egg plasma membrane. That contact results in fusion of the sperm and egg plasma membranes and the formation of a fertilization cone that engulfs the sperm head, bringing it into the egg cytoplasm. The sperm mitochondria, which largely constitute the midpiece of the sperm, are also drawn into the egg cytoplasm, but they degrade and disappear; this means that the mitochondria and mitochondrial genes of the new urchin are derived only from the egg.

In animals that practice internal fertilization, mating behaviors help guarantee species specificity, but egg–sperm recognition mechanisms still exist. The mammalian egg is surrounded by a thick layer called the **cumulus**, which consists of a loose assemblage of maternal cells in a gelatinous matrix (Figure 43.5). Beneath the cumulus is a glycoprotein envelope called the **zona pellucida**, which is functionally similar to the vitelline envelope of sea urchin eggs. When mammalian sperm are deposited in the female reproductive tract, they are metabolically activated and made capable of an acrosomal reaction should they encounter an egg. An activated sperm can penetrate the cumulus and interact with the zona pellucida.

Unlike the jelly coat of sea urchin eggs, the cumulus of mammalian eggs does not trigger the acrosomal reaction. When sperm make contact with the zona pellucida, a species-specific glycoprotein binds to recognition molecules on the head of the sperm. This binding triggers the acrosomal reaction, releasing acrosomal enzymes that digest a path through the zona pellucida. When the sperm head reaches the egg plasma membrane, other proteins facilitate its adhesion to and fusion with the egg plasma membrane.

The importance of the zona pellucida and its sperm-binding molecules as a species-specific recognition mechanism was revealed in experiments on mammalian eggs and sperm in culture dishes. When the zona pellucida was stripped from human eggs and the eggs were exposed to hamster sperm, fertilization took place, resulting in a hamster–human hybrid zygote. The hybrid zygote did not survive its first cell division, but the experiment demonstrated that a recognition mechanism in mammalian species resides in the zona pellucida.

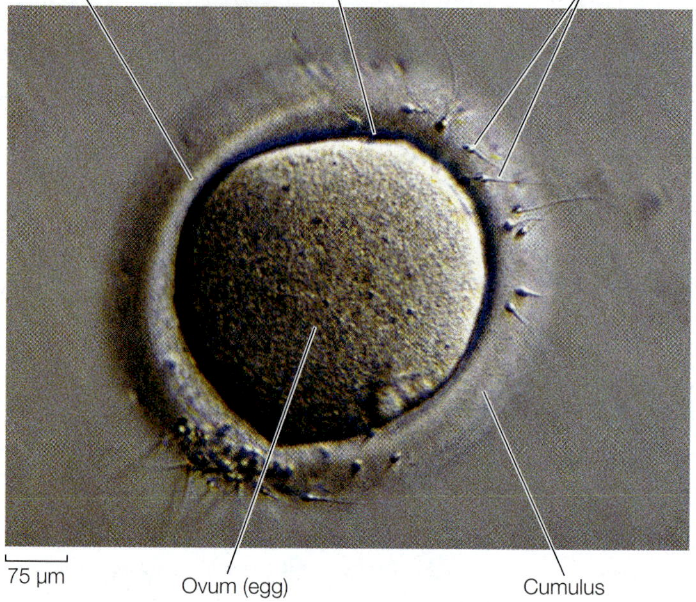
Zona pellucida
Egg plasma membrane
Sperm
75 μm
Ovum (egg)
Cumulus

43.5 Barriers to Sperm This human egg, like other mammalian eggs, is surrounded by the cumulus and zona pellucida. Sperm must penetrate both to fertilize the egg. Only one sperm will penetrate the zona pellucida and fuse with the plasma membrane.

BLOCKS TO POLYSPERMY The fusion of the sperm and egg plasma membranes and the entry of the sperm into the egg initiate a programmed sequence of events. The first responses to sperm entry are **blocks to polyspermy**. These blocks are mechanisms that prevent more than one sperm from entering the egg. If more than one sperm enters the egg, the embryo is unlikely to survive.

Blocks to polyspermy have been studied extensively in sea urchin eggs, which can be fertilized in a dish of seawater. Within seconds after the sperm membrane contacts the egg membrane, an influx of sodium ions changes the electric charge difference across the egg's plasma membrane. This fast block to polyspermy prevents the fusion of any other sperm with the egg plasma membrane, but it is transient. The change in membrane electric charge lasts only about a minute, but that is enough time to allow a slower block to sperm entry to develop.

The slow block to polyspermy involves converting the vitelline envelope to a physical barrier that sperm cannot penetrate. Before fertilization, the vitelline envelope is bonded to the egg plasma membrane. Just under the plasma membrane are vesicles called cortical granules (see Figure 43.4), which contain enzymes and other proteins.

The sea urchin egg, like all animal cells, sequesters calcium in its endoplasmic reticulum. Sperm entry into the egg stimulates the release of calcium ions from the endoplasmic reticulum and into the egg cytoplasm. This increase in cytosolic calcium causes the egg's cortical granules to fuse with the plasma membrane and release their contents. Cortical granule enzymes break the bonds between the vitelline envelope and the plasma membrane, and other proteins released from the cortical granules attract water into the space between them. As a result, the vitelline envelope rises to form a fertilization envelope. Cortical granule enzymes also degrade sperm-binding molecules on the surface of the fertilization envelope and cause it to harden, thus preventing additional sperm from contacting the egg's plasma membrane.

In mammalian eggs, sperm entry does not cause a rapid change in membrane potential, but it does trigger a release of calcium from the endoplasmic reticulum. As in the sea urchin egg, increased calcium causes the cortical granules to fuse egg with the egg plasma membrane. A fertilization envelope does not form around the mammalian egg, but the cortical granule enzymes destroy the sperm-binding molecules in the zona pellucida. The rise in cytosolic calcium also signals the egg to complete meiosis. The stage is set for the first cell division.

Getting eggs and sperm together

As we have just seen, sexual reproduction requires the production of haploid gametes (gametogenesis) and the joining together of those gametes to form a diploid zygote (fertilization). Spawning and mating behaviors get eggs and sperm close enough together that fertilization can occur. Fertilization can occur externally or internally.

EXTERNAL FERTILIZATION In an aquatic environment, animals can bring their gametes together by simply releasing them into the water. This practice, called **spawning**, results in **external fertilization**. Many aquatic animals are not very mobile, but they produce huge numbers of gametes that can travel far from the point of release. A female oyster, for example, will release millions of eggs when she spawns, and the number of sperm produced by a male oyster is astronomical.

Numbers alone, however, do not guarantee that gametes will meet. The reproductive activities of the males and females of a population must be synchronized, since released gametes have a limited life span. Seasonal breeders may use day length, changes in temperature, or changes in weather to time the production and release of their gametes. Mutual stimulation is also important. Release of gametes into the water by one individual can stimulate others to spawn.

Behavior can play an important role in bringing gametes together even when fertilization is external. Many species travel great distances to congregate with potential mates and release their gametes at the same time in a suitable environment. Salmon are an extreme example. They hatch and develop in freshwater streams and then migrate to the ocean, where they remain for years. When they are mature, they travel hundreds of miles back in the stream where they hatched to spawn. Males and females expend great amounts of energy to swim up the streams to the spawning grounds, where they pair up, prepare a depression in the streambed gravel, and together release their sperm and eggs. As the gametes drift down into the gravel, fertilization occurs.

INTERNAL FERTILIZATION Terrestrial animals cannot simply release their gametes into the environment. Sperm can move only through liquid, and delicate gametes released into air would dry out and die. Terrestrial animals avoid these problems by **internal fertilization**, the release of sperm into the female reproductive tract. Some aquatic animals also practice internal fertilization, but it is ubiquitous in terrestrial animals.

Animals have evolved an astonishing diversity of behavioral and anatomical adaptations for internal fertilization. As we saw above, gametogenesis occurs in the gonads, which are the **primary sex organs**. All additional anatomical components of an animal's reproductive system are called **accessory sex organs**. An obvious accessory sex organ in males of many species is the **penis**, which enables the male to deposit sperm in the female's **vagina**, the entry to her reproductive tract. Accessory sex organs include a variety of glands, tubules, ducts, and other structures.

Copulation is the physical joining of male and female accessory sex organs. Most male insects copulate and transfer sperm to the female's vagina through a penis. The **genitalia**—external sex organs—of insects often have species-specific shapes that ensure that the male and female genitalia match in a lock-and-key fashion. This mechanism ensures a tight, secure fit between the mating pair during the prolonged period of sperm transfer. In some insect species in which females mate with more than one male, the males have elaborate structures on their penises that can scoop sperm deposited by other males out of a female's reproductive tract, replacing it with their own.

Transfer of sperm in internal fertilization can also be indirect. Males of many invertebrate species (e.g., mites and scorpions) and a few vertebrates (e.g., salamanders) deposit spermatophores—packets of sperm protected from desiccation—in the environment. When a female mite encounters a spermatophore from a potential mate, she straddles it and opens a pair of plates in her abdomen so that the tip of the spermatophore enters her reproductive tract and allows the sperm to enter.

Male squids and spiders play a more active role in spermatophore transfer. The male spider secretes a drop containing sperm onto a bit of web, then uses a special structure on his foreleg to pick up the sperm-containing web and insert it through the female's genital opening. Male squids use one specialized tentacle to pick up a spermatophore and insert it into the female's genital opening.

Some individuals can function as both male and female

In most species, gametes are produced by individuals that are either male or female. Species that have separate male and female members are referred to as **dioecious** (Greek, "two houses"). In some species, however, a single individual may produce both sperm and eggs. Such species are **monoecious** ("one house") or **hermaphroditic**.

Almost all invertebrate groups contain some hermaphroditic species. An earthworm is an example of a *simultaneous* hermaphrodite, meaning an individual is both male and female at the same time. When two earthworms mate, they exchange sperm, and as a result, the eggs of each are fertilized (see Figure 32.12C). Some vertebrates are *sequential* hermaphrodites, meaning that an individual may function as a male or a female at different times in its life. An example is the anemone fish, or clown fish, a species that lives in small groups within large sea anemones (Figure 43.6). All anemone fish are born male. The largest fish in a group becomes a functional female. If that fish is removed from the group, the next-largest male becomes a female. The second-largest fish in the group is the only male in breeding condition.

What is the evolutionary advantage of hermaphroditism? Some simultaneous hermaphrodites, such as parasitic

Amphiprion sp.

43.6 When Size Determines Sex Anemone fish (also known as clown fish) live in groups of about a dozen centered on a single sea anemone. All anemone fish are born male; the largest fish in the group becomes a functional female. Thus a fish may function as a male and as a female at different times in its life.

tapeworms, have a low probability of meeting a potential mate—an individual tapeworm may be the only one in its host. Tapeworms can fertilize their own eggs, but most simultaneous hermaphrodites must mate with another individual. Because every simultaneous hermaphrodite is both male and female, however, the probability of encountering a possible mate doubles. In some sequential hermaphrodites, siblings are all male or all female at the same time, thus reducing the incidence of inbreeding.

The evolution of vertebrate reproductive systems parallels the move to land

The earliest vertebrates evolved in aquatic environments. The closest living relatives of those earliest vertebrates are modern-day fishes. They remain exclusively aquatic, and most practice external fertilization. The most primitive of the fishes, the lampreys and hagfishes, simply release their gametes into the environment. In most fishes, however, mating behaviors bring females and males into close proximity at the time of gamete release. In sharks and rays, fins have evolved into claspers that hold the male and female together and enable sperm to be transferred directly into the female reproductive tract.

Amphibians were the first vertebrates to live in terrestrial environments. They dealt with the challenge of a dry environment by returning to water to reproduce, as most amphibians still do today.

Reptiles were the first vertebrate group to solve the problem of reproduction in the terrestrial environment (**Figure 43.7**). Their solution was the amniote egg (see Figure 33.19A). A hard or leathery shell protects the embryo and impedes water loss while allowing the diffusion of oxygen into the egg and carbon dioxide out of the egg. The eggshell creates an obvious problem for fertilization. Sperm cannot penetrate the shell, so they must reach the egg before the shell forms. Hence internal fertilization and the evolution of accessory sex organs were necessary for the evolution of the amniote egg.

Male snakes and lizards have paired hemipenes, which can be filled with blood and thereby extruded from the male's body. Only one hemipenis is inserted into the female's reproductive tract at a time. It is usually rough or spiny at the end to achieve a secure hold while sperm are transferred down a groove on its surface. Retractor muscles pull the hemipenis back into the male's body when mating is completed. Some evolutionarily ancient bird species have erectile penises that channel sperm along a groove into the female's reproductive tract. Birds with more recent evolutionary origins, however, do not have erectile penises; instead, the male and female simply bring their genital openings close together to transfer sperm.

(A) *Chelonia mydas*

(B) *Merops apiaster*

43.7 The Shelled Egg The shelled egg was a major evolutionary step that allows reptiles and birds to reproduce in the terrestrial environment. (A) A female green sea turtle has deposited her eggs in the sand. (B) The shelled egg requires that sperm meet egg before the shell forms. Terrestrial animals thus must practice internal fertilization, as these European bee-eaters are doing.

Usually this involves the male standing on the female's back (see Figure 43.7B).

All mammals practice internal fertilization. With the exception of the prototherian mammals, the developing embryo is retained for some time in the female reproductive tract. Prototherian mammals (the monotremes; see Figure 33.24) lay eggs. The other mammals (the therians) vary enormously as to the developmental stage of their offspring at the time of birth.

Animals with internal fertilization are distinguished by where the embryo develops

Two patterns of care and nurture of the embryo have evolved in animals: oviparity (egg laying) and viviparity (live bearing).

Oviparous animals lay eggs in the environment, and their embryos develop outside the mother's body. Oviparous terrestrial animals such as insects, reptiles, and birds protect their eggs from desiccation with waterproof membranes or shells. Oviparity is possible because eggs are stocked with abundant nutrients to supply the needs of the embryo. Some oviparous animals engage in various forms of parental behavior to protect their eggs, but until the eggs hatch, the embryos depend entirely on the nutrients stored in the egg.

Viviparous animals retain the embryo within the mother's body during its early developmental stages. Although examples of viviparity exist in all vertebrate groups except the crocodiles, turtles, and birds (even some sharks retain fertilized eggs in their bodies and give birth to free-living offspring), there is a big difference between viviparity in mammals and in other species.

All mammals except the prototherians are viviparous and have a specialized portion of the female reproductive tract, the **uterus**, or womb, that holds the embryo and interacts with it to produce a **placenta**, which enables the exchange of nutrients and wastes between the blood of the mother and that of the embryo. Very few non-mammalian species have evolved such a connection between the embryo and the mother.

In most non-mammalian viviparous animals, such as garter snakes and the well-known aquarium fish the guppy, fertilized eggs are retained in the mother's body until they hatch. These embryos still receive nutrition from stores in the egg, so this reproductive adaptation is called **ovoviviparity**.

RECAP 43.2

Sexual reproduction involves gametogenesis, mating, and fertilization. Fertilization can be external or internal and involves mechanisms for ensuring that only one sperm from the right species enters the egg.

- Describe the steps by which a sea urchin sperm penetrates the egg. See Figure 43.4
- Explain how polyspermy is prevented and why it is crucial to do so. See pp. 886–887 and Figure 43.4
- What reproductive adaptations made life on land possible? See p. 888

Now that we have covered some of the general aspects of gametogenesis and fertilization and have briefly discussed the great diversity of mating systems, we will next consider the human reproductive systems in detail.

43.3 How Do the Human Male and Female Reproductive Systems Work?

In this section we describe the structures and functions of male and female reproductive systems in mammals, and we will use humans as our prime example. We will also discuss the hormonal regulation of both male and female systems. Our discussion covers:

- *The primary sex organs* (testes in males and ovaries in females) that produce gametes and serve endocrine functions.
- *The accessory sex organs*, which include the ducts through which the gametes pass, the various glands that empty into those ducts, and the external genitalia.
- *Secondary sexual characteristics*, which are not directly involved in reproduction but are responsible for the major differences in external appearance of men and women and are important in mating and in rearing offspring.

Male sex organs produce and deliver semen

Semen is the product of the male reproductive system. Semen contains sperm and a complex mixture of fluids and molecules that support the sperm and facilitate fertilization. Sperm make up less than 5 percent of the volume of the semen.

The male reproductive organs are diagrammed in Figure 43.8. Sperm are produced in the testes, the paired male gonads. The testes of most mammals are located outside the body cavity in a pouch of skin called the **scrotum**.

Why should the testes be located outside the body cavity? The optimal temperature for spermatogenesis in most mammals is slightly lower than the normal body temperature. The scrotum keeps the testes at this optimal temperature. Muscles in the scrotum contract in a cold environment, bringing the testes closer to the warmth of the body; in a hot environment they relax, cooling the testes by suspending them farther from the body.

Spermatogenesis takes place within the **seminiferous tubules** tightly coiled in each testis (Figure 43.9A). Between the seminiferous tubules are clusters of **Leydig cells** that produce testosterone (Figure 43.9B). Spermatogonia reside in the outer regions of the seminiferous tubules, just under the basement membrane. Moving inward toward the lumen of the tubule are germ cells in successive stages of spermatogenesis (Figure 43.9C). The germ cells are intimately associated with **Sertoli cells** that provide nutrients for the developing sperm.

When the second meiotic division is complete, each primary spermatocyte has produced four spermatids (see Figure 43.3A). The spermatids develop into spermatozoa as they migrate toward the lumen of the seminiferous tubule. The nucleus becomes compact, and the surrounding cytoplasm is lost. A flagellum—the sperm tail—develops. The mitochondria that provide the energy for sperm motility become condensed into a midpiece between the head and tail. An acrosome forms over the nucleus in the head of the sperm. Immature sperm are shed into the lumen of the seminiferous tubule.

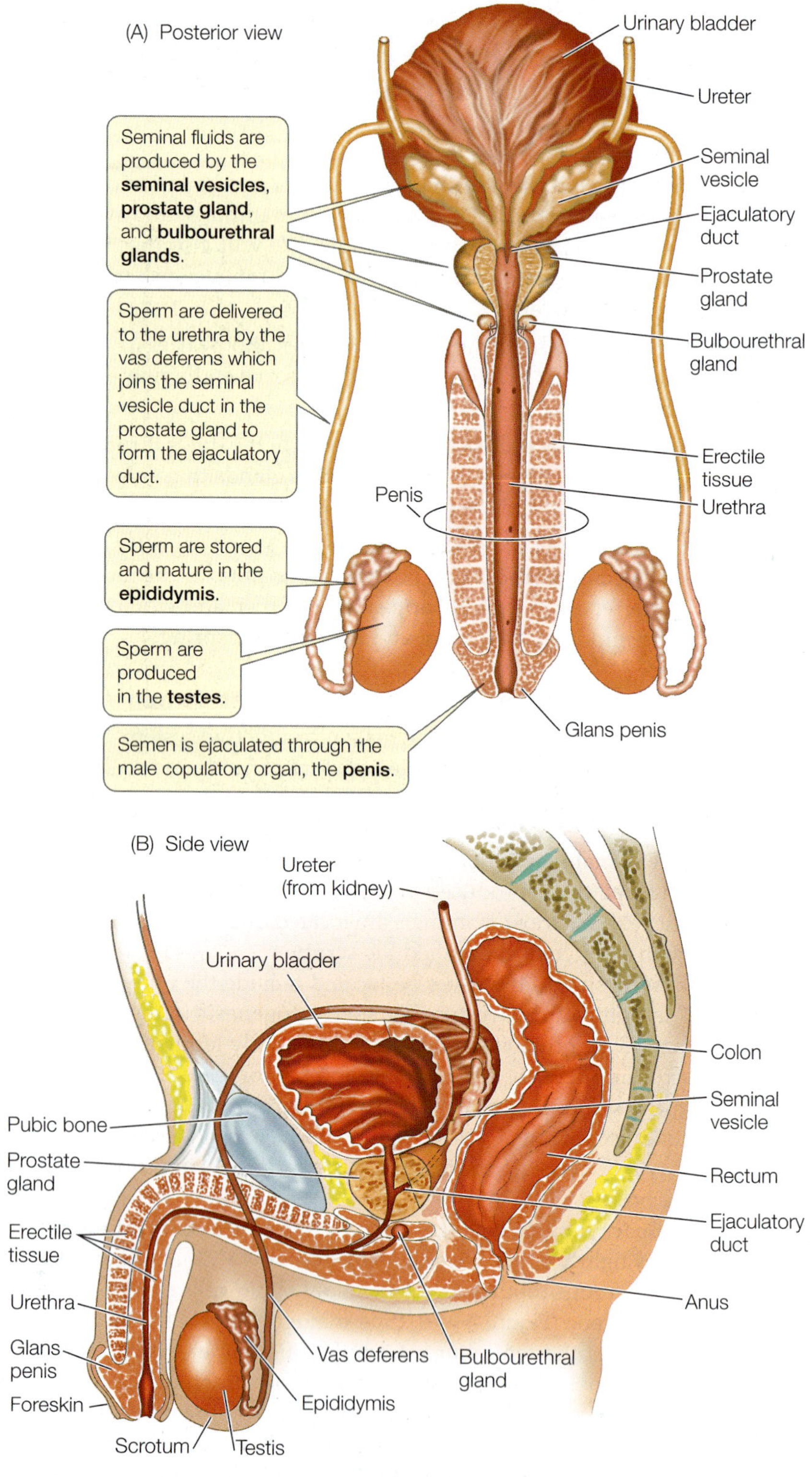

(A) Posterior view
Urinary bladder
Ureter
Seminal vesicle
Ejaculatory duct
Prostate gland
Bulbourethral gland
Erectile tissue
Urethra
Penis
Glans penis
Seminal fluids are produced by the **seminal vesicles**, **prostate gland**, and **bulbourethral glands**.
Sperm are delivered to the urethra by the vas deferens which joins the seminal vesicle duct in the prostate gland to form the ejaculatory duct.
Sperm are stored and mature in the **epididymis**.
Sperm are produced in the **testes**.
Semen is ejaculated through the male copulatory organ, the **penis**.
(B) Side view
Ureter (from kidney)
Urinary bladder
Pubic bone
Prostate gland
Erectile tissue
Urethra
Glans penis
Foreskin
Scrotum
Testis
Epididymis
Vas deferens
Bulbourethral gland
Colon
Seminal vesicle
Rectum
Ejaculatory duct
Anus

43.8 Reproductive Tract of the Human Male The male reproductive organs are shown (A) from the rear and (B) from the side.

Go to Activity 43.1 The Human Male Reproductive Tract
Life10e.com/ac43.1

From the seminiferous tubules, sperm move into the **epididymis** (see Figure 43.8), where they mature, become motile, and are stored. The epididymis connects to the **urethra** via the **vas deferens** (plural *vasa deferentia*) and the **ejaculatory duct**.

The urethra originates in the bladder, runs through the penis, and opens to the outside of the body at the tip of the penis. It serves as the common final duct for the urinary and reproductive systems.

The components of the semen other than sperm come from several accessory glands. About 60 percent of the volume of semen is secreted by the paired **seminal vesicles**, which empty into the vas deferens just before it joins the urethra. Seminal fluid is thick because it contains mucus and fibrinogen, a protein also found in the blood, where it can polymerize to form blood clots. Seminal fluid also contains the monosaccharide fructose, an energy source for the sperm.

The **prostate gland** contributes about 30 percent of the volume of the semen. Prostate fluid is alkaline, so it neutralizes the acidity in the male and female reproductive tracts and makes these environments more hospitable to sperm. The prostate also secretes a clotting enzyme that causes fibrinogen from the seminal vesicles to convert the semen into a clotted, gelatinous mass, facilitating the semen's propulsion into the upper regions of the female reproductive tract. Another prostate enzyme, fibrinolysin, then dissolves the clotted semen and liberates the sperm. Prostaglandins—hormones produced by the prostate—stimulate contractions of the female reproductive tract.

The **bulbourethral glands** produce a small volume of an alkaline, mucoid secretion that helps neutralize acidity in the urethra and lubricate it to facilitate the passage of semen immediately preceding the climax of sexual intercourse. These secretions can carry with them residual sperm from prior sexual activity. It is therefore possible for pregnancy to occur even if the penis is withdrawn just before climax (an ancient but ineffective birth control practice known as coitus interruptus).

The penis and the scrotum are the male genitalia. The shaft of the penis is covered with normal skin, but the highly sensitive tip, the **glans penis**, is covered with thinner, more sensitive skin that is especially responsive to sexual stimulation. A fold of skin called the foreskin covers the glans of the human penis. The procedure known as circumcision removes a portion of the foreskin.

Sexual stimulation triggers responses in the nervous system that result in penile **erection**. Nerve endings release a neurotransmitter that causes the endothelial cells lining the penile blood vessels to release a gaseous neurotransmitter, nitric oxide (NO). NO diffuses into the muscle cells that control the diameter of the penile arteries and stimulates them to produce the second messenger cGMP (see Figure 7.15). Increased cGMP in these muscle cells causes them to relax; the arteries dilate and carry more blood into the penis. Increased blood flow swells

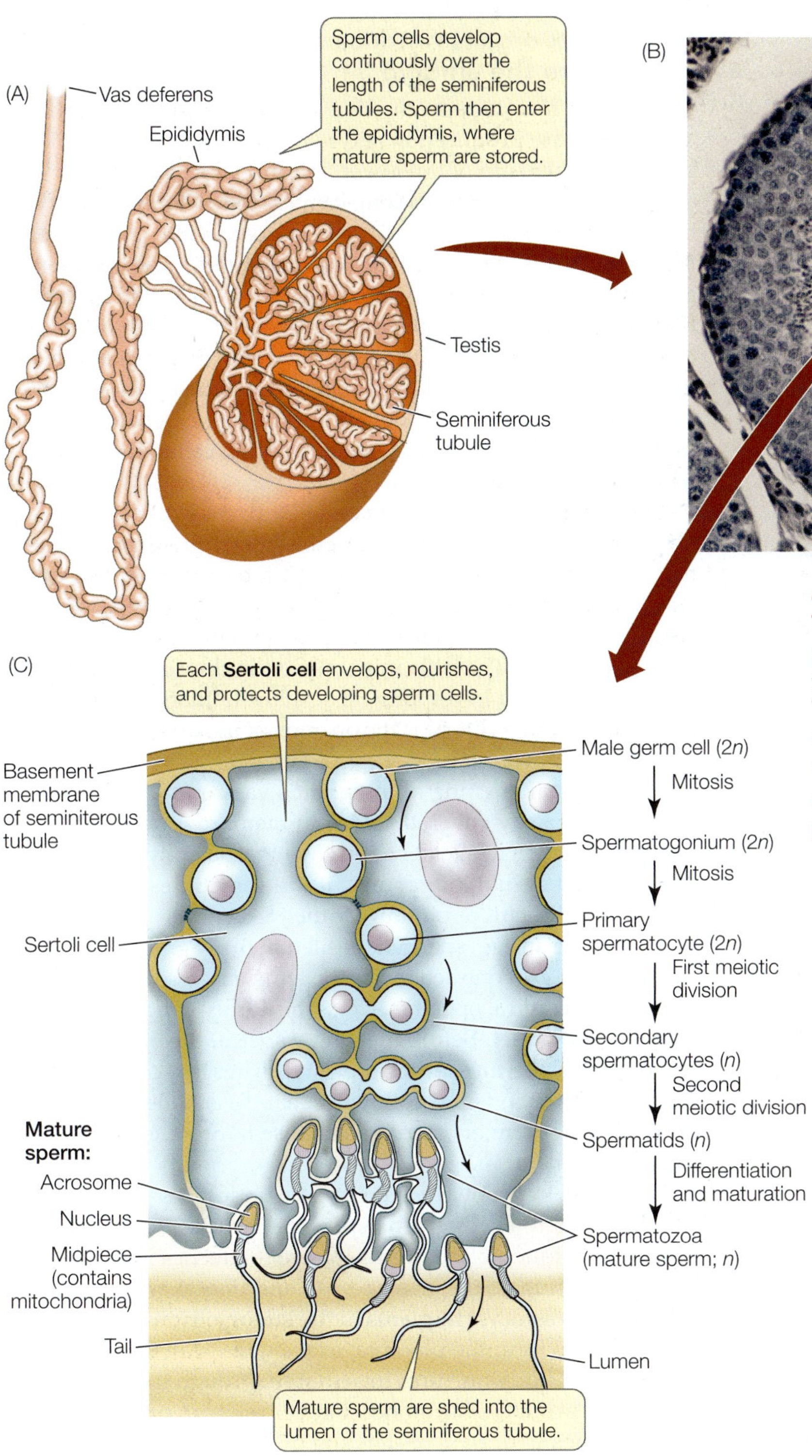
(A)
Vas deferens
Epididymis
Sperm cells develop continuously over the length of the seminiferous tubules. Sperm then enter the epididymis, where mature sperm are stored.
Testis
Seminiferous tubule
(B)
Leydig cells in the tissue between seminiferous tubules produce male sex hormones.
Lumen of seminiferous tubules
(C)
Each Sertoli cell envelops, nourishes, and protects developing sperm cells.
Basement membrane of seminiterous tubule
Male germ cell (2n)
Mitosis
Spermatogonium (2n)
Mitosis
Sertoli cell
Primary spermatocyte (2n)
First meiotic division
Secondary spermatocytes (n)
Second meiotic division
Mature sperm:
Acrosome
Nucleus
Midpiece (contains mitochondria)
Tail
Spermatids (n)
Differentiation and maturation
Spermatozoa (mature sperm; n)
Lumen
Mature sperm are shed into the lumen of the seminiferous tubule.

43.9 Spermatogenesis Takes Place in the Seminiferous Tubules (A) Seminiferous tubules fill the testes, continuously producing millions of sperm. (B) Cross section of seminiferous tubules and the Leydig cells in the spaces between them. (C) This longitudinal diagram shows how, as sperm mature, they move from the outer layer of the tubule toward the center, where they are shed into the lumen of the tubule.

Go to Activity 43.2 Spermatogenesis
Life10e.com/ac43.2

the shafts of spongy erectile tissue located along the length of the penis. The enlargement of these blood-filled cavities compresses the vessels that carry blood out of the penis, and the erectile tissue becomes engorged with blood. The penis becomes hard and erect, facilitating its insertion into the vagina. Many species of mammals (though not humans) have a bone in the penis, but these species still depend on erectile tissue for copulation.

At the climax of copulation, 2 to 6 milliliters of semen are propelled through the vasa deferentia and the urethra in two steps, emission and ejaculation. During emission, rhythmic contractions of smooth muscles in the vasa deferentia and accessory glands move the semen into the urethra at the base of the penis. Ejaculation is caused by contractions of other muscles at the base of the penis surrounding the urethra. These contractions force the coagulum of semen through the urethra and out of the penis. The muscle contractions of ejaculation are accompanied by feelings of intense pleasure known as orgasm. They are also accompanied by transient increases in heart rate, blood pressure, breathing, pupil dilation and skeletal muscle contractions throughout the body.

After ejaculation, NO release decreases and enzymes break down cGMP, causing the blood vessels flowing into the penis to constrict. The blood pressure in the erectile tissue decreases, relieving the compression of the blood vessels leaving the penis, and the erection declines.

Erectile dysfunction (ED), or impotence, is the inability to achieve or sustain an erection. ED may have different causes, including cardiovascular disease. Drugs used to treat ED act by inhibiting the breakdown of cGMP, thus enhancing the effect of NO released in the penis, which improves the ability to achieve and maintain an erection.

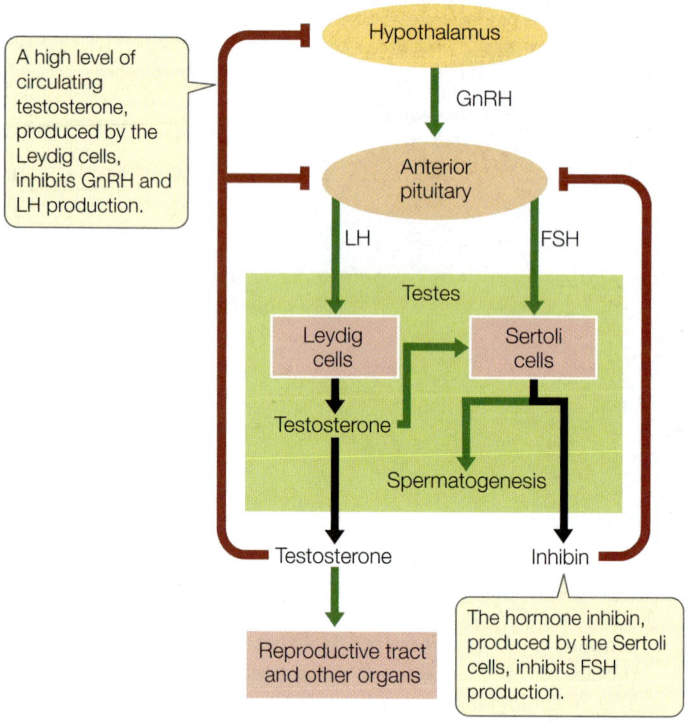
A high level of circulating testosterone, produced by the Leydig cells, inhibits GnRH and LH production.
Hypothalamus
GnRH
Anterior pituitary
LH
FSH
Testes
Leydig cells
Sertoli cells
Testosterone
Spermatogenesis
Testosterone
Inhibin
Reproductive tract and other organs
The hormone inhibin, produced by the Sertoli cells, inhibits FSH production.

43.10 Male Reproductive Hormones The male reproductive system is under hormonal control by the hypothalamus and the anterior pituitary. Red lines indicate inhibition; green lines indicate stimulation.

Male sexual function is controlled by hormones

Spermatogenesis and maintenance of male secondary sexual characteristics such as facial hair and a deep voice depend on testosterone produced by the Leydig cells of the testes. As described in Section 41.3, increased production of testosterone at puberty results from an increased release of gonadotropin-releasing hormone (GnRH) by the hypothalamus. GnRH stimulates anterior pituitary cells to secrete luteinizing hormone (LH) and follicle-stimulating hormone (FSH) (**Figure 43.10**). Higher levels of LH stimulate the Leydig cells to increase their production and release of testosterone. Testosterone exerts negative feedback on the anterior pituitary and the hypothalamus. At the time of puberty, the sensitivity of the hypothalamus to negative feedback from testosterone declines; as a result, the level of circulating testosterone increases.

Increased testosterone in pubertal boys causes the development of pubic and facial hair, a deeper voice, enlarged genitals, and an increased growth rate. Testosterone also promotes increased muscle mass and maturation of the testes. Continued production of testosterone after puberty is essential for the maintenance of secondary sexual characteristics and the production of sperm.

Spermatogenesis is controlled by the influence of FSH and testosterone on the Sertoli cells in the seminiferous tubules. The Sertoli cells also produce a hormone called inhibin that exerts negative feedback on the anterior pituitary cells producing and secreting FSH.

Female sex organs produce eggs, receive sperm, and nurture the embryo

The human female reproductive system is shown in **Figure 43.11**. Eggs are produced in and released from the ovaries located on either side of the lower abdominal cavity. When an egg leaves the ovary (**ovulation**), it enters the abdominal cavity, but it does not go far. The ovaries are close to the openings of the **oviducts** (also known as the Fallopian tubes). The openings are surrounded by undulating, fringed tissues called fimbria that sweep the egg into the oviduct. Cilia lining the oviduct propel the egg toward the **uterus**, a muscular, thick-walled cavity shaped in humans like an upside-down pear. The uterus is where the embryo will develop if the egg is fertilized. At its bottom, the uterus narrows into a region called the **cervix**, which opens into the vagina.

Externally, the vagina is enclosed by two sets of skin folds. The inner, more delicate folds are the labia minora; the outer, thicker folds are the labia majora. At the anterior tip of the labia minora is the clitoris, a small bulb of highly sensitive erectile tissue that has the same developmental origins as the male glans penis. The labia minora and the clitoris become engorged with blood in response to sexual stimulation. Between the vagina and the clitoris is the opening of the urethra through which urine flows.

The external opening of an infant's vagina is usually, but not always, partly covered by a thin membrane, the hymen. Eventually the hymen can be torn by vigorous physical activity or by first sexual intercourse; it can sometimes make first intercourse difficult or painful for the woman.

To fertilize an egg, sperm deposited in the vagina swim and are propelled by contractions of the female reproductive tract through the cervical opening, across the uterus, and most of the way up an oviduct. Before meeting a sperm, the egg is still a secondary oocyte, and fertilization stimulates it to complete its second meiotic division. Following that division, the haploid nuclei of the sperm and the egg can fuse to produce a diploid zygote nucleus.

While still in the oviduct, the zygote undergoes its first few cell divisions to become a **blastocyst**. The blastocyst moves down the oviduct to the uterus, where it attaches itself to the epithelial lining of the uterus—the **endometrium**. Once attached, the blastocyst implants in the endometrium and interacts with it to form the placenta, as we will see in Chapter 44. The mother's body nurtures the embryo through the placenta, which also produces hormones that help sustain pregnancy.

Just as the maturation of eggs and ovulation in the ovaries is a cyclical process, so are critical events in the uterus. In anticipation of receiving a blastocyst, the endometrium thickens and develops lots of blood vessels. If a blastocyst does not arrive within a certain window of time, the endometrium regresses. Thus female reproductive functions consist of two linked cycles: an **ovarian cycle** that produces eggs and hormones; and a **uterine**, or **menstrual**, **cycle** that prepares the endometrium for the arrival of a blastocyst. The two cycles must be synchronized so that a blastocyst arrives in the uterus at the optimal time to embed in the endometrium and continue its development.

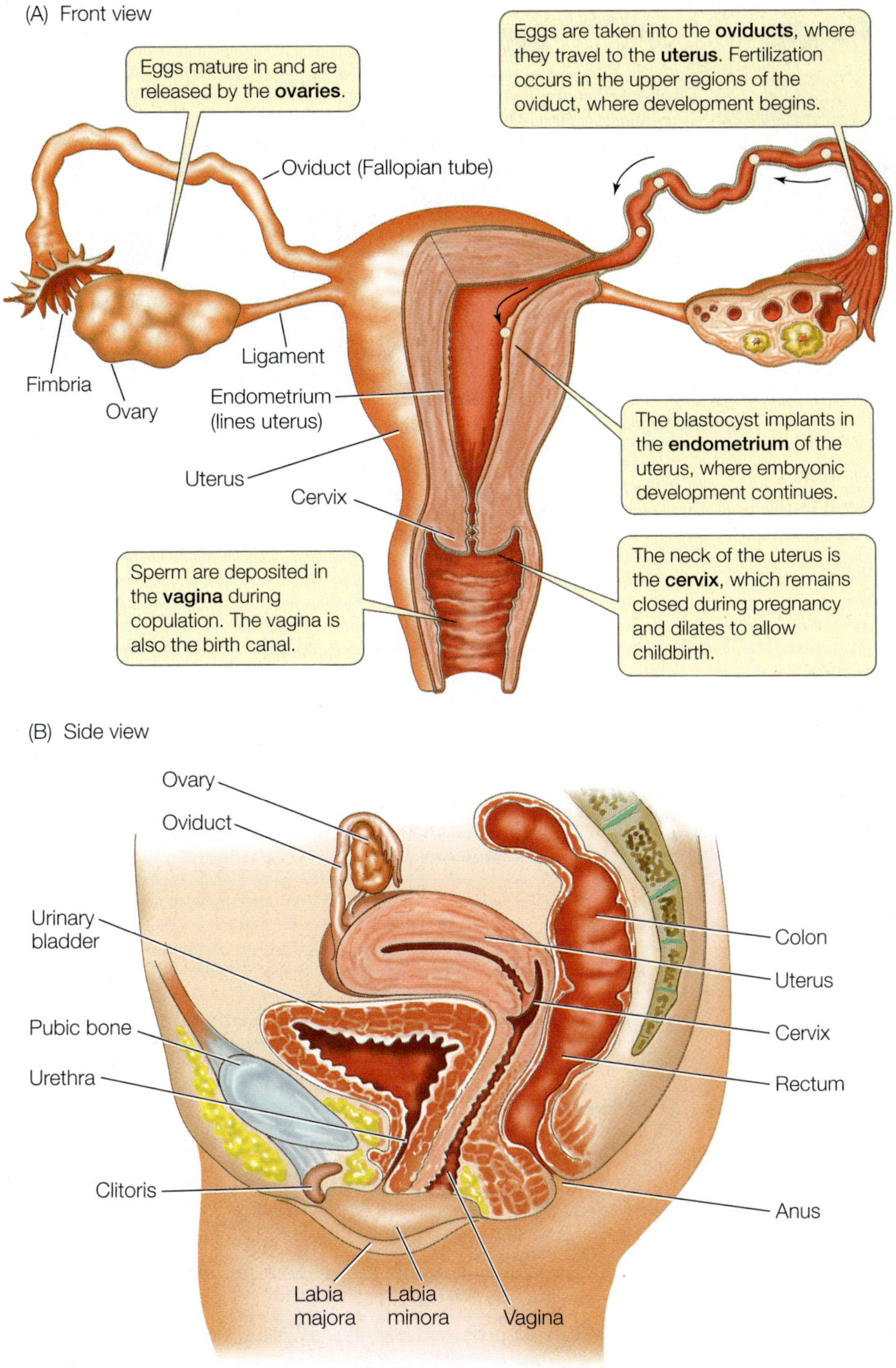
(A) Front view
Eggs mature in and are released by the **ovaries**.
Eggs are taken into the **oviducts**, where they travel to the **uterus**. Fertilization occurs in the upper regions of the oviduct, where development begins.
Oviduct (Fallopian tube)
Fimbria
Ovary
Ligament
Endometrium (lines uterus)
Uterus
Cervix
The blastocyst implants in the **endometrium** of the uterus, where embryonic development continues.
Sperm are deposited in the **vagina** during copulation. The vagina is also the birth canal.
The neck of the uterus is the **cervix**, which remains closed during pregnancy and dilates to allow childbirth.
(B) Side view
Ovary
Oviduct
Urinary bladder
Pubic bone
Urethra
Clitoris
Labia majora
Labia minora
Vagina
Colon
Uterus
Cervix
Rectum
Anus

43.11 Reproductive Tract of the Human Female The female reproductive organs are shown (A) from the front and (B) from the side.

Go to Activity 43.3 The Human Female Reproductive Tract
Life10e.com/ac43.3

The ovarian cycle produces a mature egg

A newborn baby girl has about a million primary oocytes in each ovary. By the time she reaches puberty, she has only about 200,000; the rest have degenerated. During a woman's fertile years, her ovaries go through about 450 ovarian cycles (**Figure 43.12A**). During each cycle, 10 to 20 primary oocytes begin to mature, but usually only one matures completely and is ovulated; the others degenerate. At around the age of 50, a woman reaches **menopause**—the end of fertility—and may have few if any oocytes left in each ovary.

A primary oocyte surrounded by a layer of ovarian cells is the functional unit of the ovary—the **follicle** (**Figure 43.12B**). The follicle cells surrounding the oocyte supply it with nutrients and growth factors. The follicle cells are also the main site of production of the female hormones estrogen and progesterone. Usually only one follicle matures completely and reaches the stage of ovulation, releasing its egg at midcycle.

The second half of the ovarian cycle is called the luteal phase because the follicle cells left in the ovary after the egg is ovulated develop into an endocrine gland—the **corpus luteum** (yellow body)—secreting estrogen and progesterone. If the egg is not fertilized and does not embed into the endometrium, the corpus luteum becomes inactive and degenerates in 12 to 14 days.

The uterine cycle prepares an environment for a fertilized egg

The uterine cycle parallels the ovarian cycle and consists of a buildup and then a breakdown of the endometrium (**Figure 43.13**). About 5 days into the ovarian cycle, the endometrium starts to thicken in preparation for receiving a blastocyst. The uterus attains its maximal state of preparedness about 5 days after ovulation and remains in that state for another 9 days. If a blastocyst has not arrived by that time, the endometrium breaks down and the sloughed-off tissue, including blood, flows from the body through the vagina—the process of **menstruation** (Latin *menses*, "months").

The uterine cycles of most mammals other than humans do not include menstruation; instead, the uterine lining typically is resorbed. In these species, the most obvious correlate of the ovarian cycle is a state of sexual receptivity called **estrus** ("heat") around the time of ovulation. You may be aware of the bloody discharge that occurs in dogs at the time of estrus. This discharge is not the same as menstruation—in fact it is exactly the opposite. Bleeding in dogs occurs during the *proliferation* of the uterine lining, which occurs just before ovulation. When the female mammal comes into estrus, she actively solicits male attention and may be aggressive to other females. Humans are unusual among mammals in that females are potentially sexually receptive throughout their ovarian cycles and at all seasons of the year.

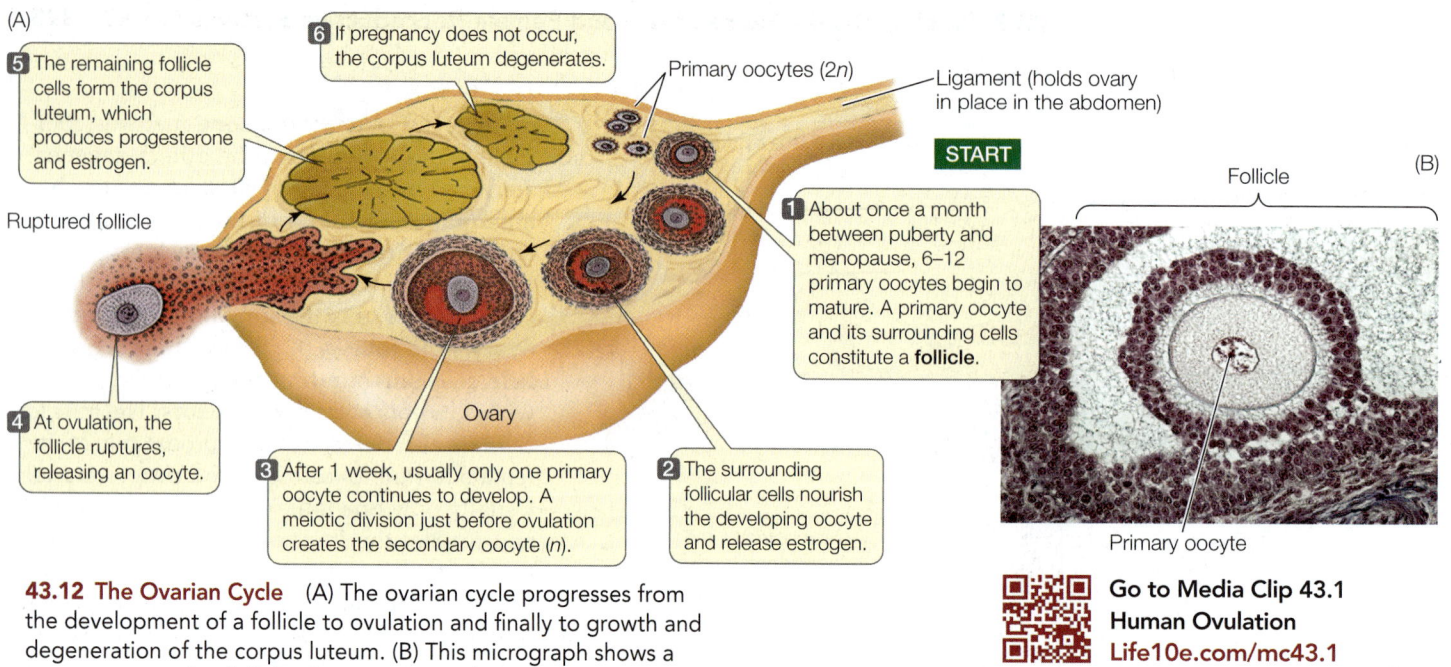

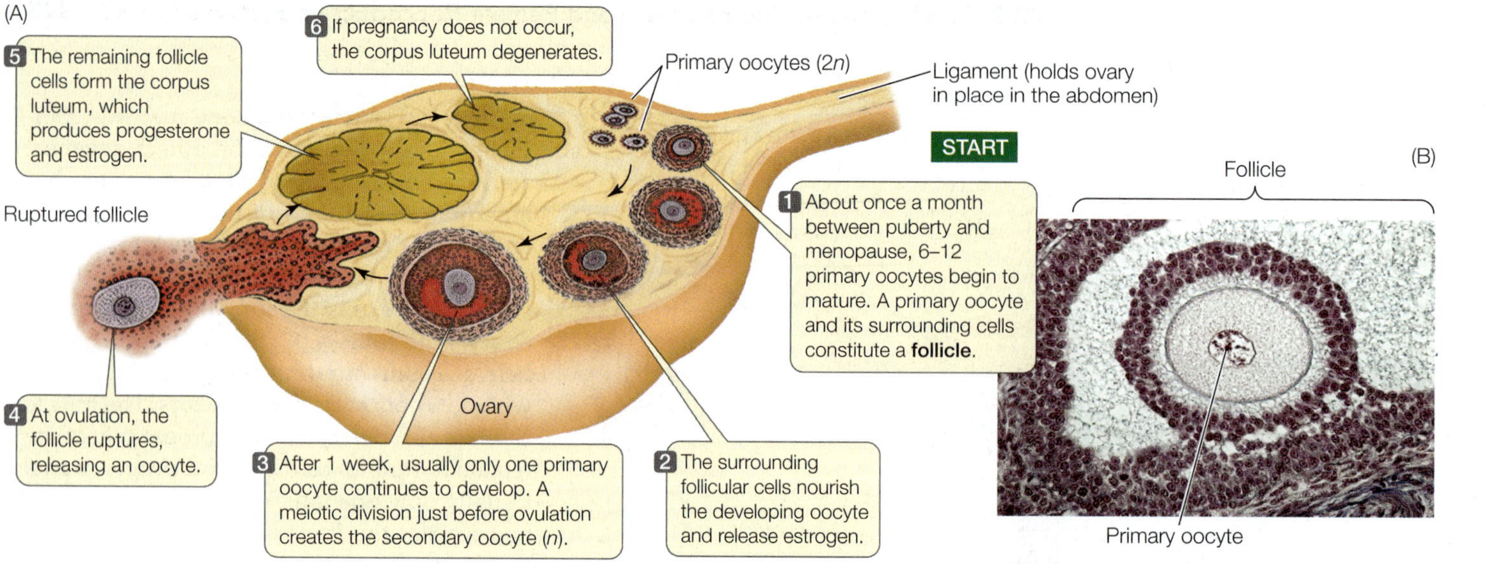
(A)
5 The remaining follicle cells form the corpus luteum, which produces progesterone and estrogen.
6 If pregnancy does not occur, the corpus luteum degenerates.
Primary oocytes (2n)
Ligament (holds ovary in place in the abdomen)
START
Ruptured follicle
1 About once a month between puberty and menopause, 6–12 primary oocytes begin to mature. A primary oocyte and its surrounding cells constitute a follicle.
Ovary
4 At ovulation, the follicle ruptures, releasing an oocyte.
3 After 1 week, usually only one primary oocyte continues to develop. A meiotic division just before ovulation creates the secondary oocyte (n).
2 The surrounding follicular cells nourish the developing oocyte and release estrogen.
Follicle
(B)
Primary oocyte

43.12 The Ovarian Cycle (A) The ovarian cycle progresses from the development of a follicle to ovulation and finally to growth and degeneration of the corpus luteum. (B) This micrograph shows a mature mammalian follicle; the oocyte is in the center.

Go to Media Clip 43.1
Human Ovulation
Life10e.com/mc43.1

Hormones control and coordinate the ovarian and uterine cycles

The ovarian and uterine cycles are coordinated and timed by the same hormones that initiate sexual maturation. Gonadotropins (FSH and LH) secreted by the anterior pituitary are the central elements of this control. Before puberty (that is, before about 11 years of age), the secretion of FSH and LH is low and the ovaries are inactive. At puberty the hypothalamus increases its release of GnRH, stimulating the anterior pituitary to secrete FSH and LH. In response to FSH and LH, ovarian tissue grows and produces estrogen. The rise in estrogen causes the maturation of the accessory sex organs and the development of female secondary sexual characteristics. Between puberty and menopause, interactions of GnRH, gonadotropins, and sex steroids control the ovarian and uterine cycles.

Menstruation marks the beginning of each uterine and ovarian cycle. A few days before menstruation begins, the anterior pituitary begins to increase its secretion of FSH and LH. In response, several follicles begin to mature in the ovaries, and these follicles steadily increase

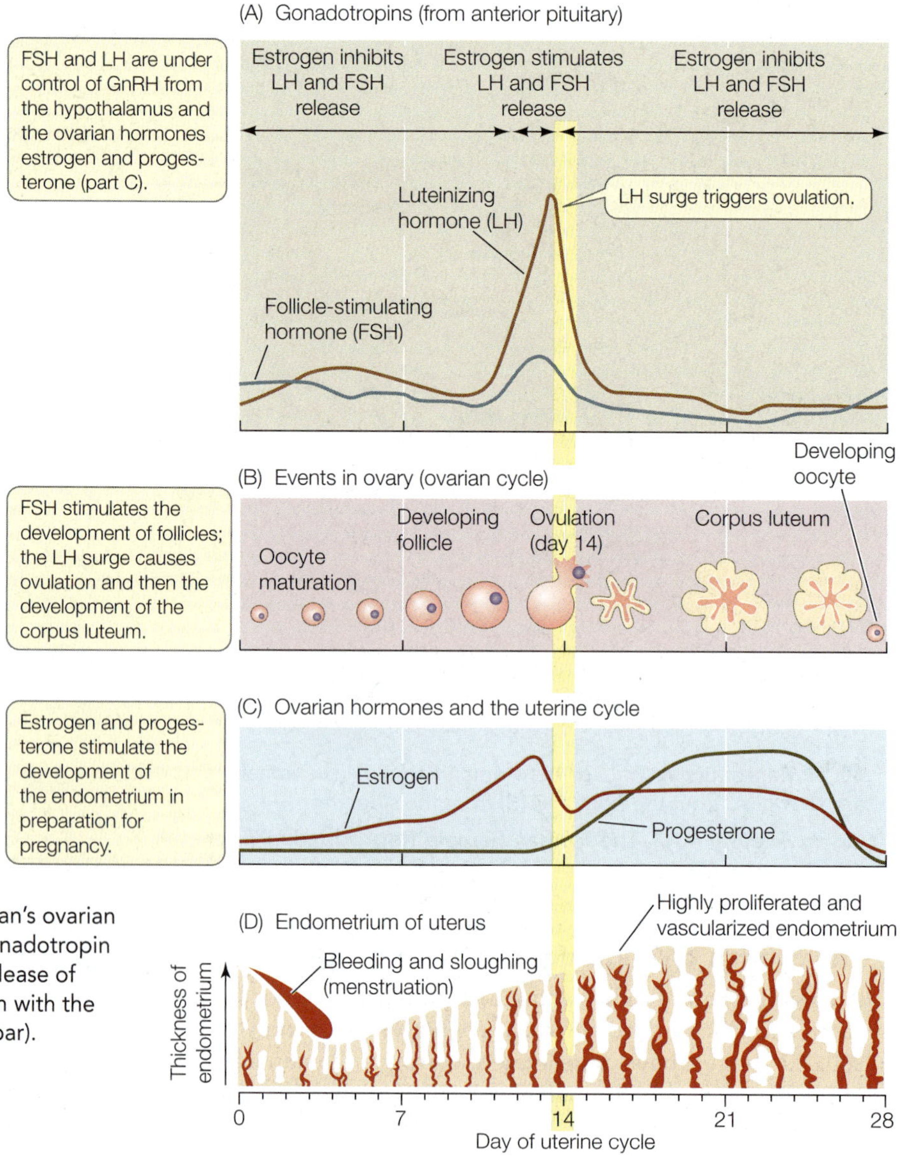
FSH and LH are under control of GnRH from the hypothalamus and the ovarian hormones estrogen and progesterone (part C).
(A) Gonadotropins (from anterior pituitary)
Estrogen inhibits LH and FSH release
Estrogen stimulates LH and FSH release
Estrogen inhibits LH and FSH release
Luteinizing hormone (LH)
LH surge triggers ovulation.
Follicle-stimulating hormone (FSH)
FSH stimulates the development of follicles; the LH surge causes ovulation and then the development of the corpus luteum.
(B) Events in ovary (ovarian cycle)
Developing oocyte
Oocyte maturation
Developing follicle
Ovulation (day 14)
Corpus luteum
Estrogen and progesterone stimulate the development of the endometrium in preparation for pregnancy.
(C) Ovarian hormones and the uterine cycle
Estrogen
Progesterone
(D) Endometrium of uterus
Highly proliferated and vascularized endometrium
Bleeding and sloughing (menstruation)
Thickness of endometrium
0
7
14
21
28
Day of uterine cycle

43.13 The Ovarian and Uterine Cycles During a woman's ovarian and uterine cycles, coordinated changes occur in (A) gonadotropin release by the anterior pituitary, (B) the ovary, (C) the release of female sex steroids, and (D) the uterus. The cycles begin with the onset of menstruation; ovulation is at midcycle (yellow bar).

Go to Animated Tutorial 43.2
The Ovarian and Uterine Cycles
Life10e.com/at43.2

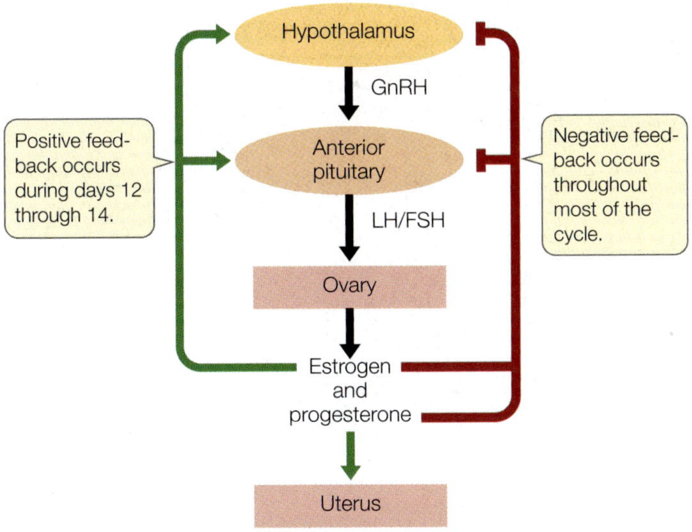
Hypothalamus
GnRH
Anterior pituitary
LH/FSH
Ovary
Estrogen and progesterone
Uterus
Positive feed-back occurs during days 12 through 14.
Negative feed-back occurs throughout most of the cycle.

43.14 Hormones Control the Ovarian and Uterine Cycles The ovarian and uterine cycles are under a complex series of positive and negative feedback controls involving several hormones.

their production of estrogen. After about a week, all but one of the follicles wither away.

Estrogen exerts negative feedback control on gonadotropin release by the anterior pituitary during the first 12 days of the ovarian cycle. Then, on about day 12, estrogen exerts positive rather than negative feedback control on the pituitary (**Figure 43.14**). As a result, a surge of LH and a lesser surge of FSH occur (see Figure 43.13A). The LH surge triggers the mature follicle to rupture and release its egg, and it stimulates the cells of the ruptured follicle to develop into a corpus luteum.

Estrogen and progesterone secreted by the corpus luteum following ovulation are crucial to growth and maintenance of the endometrium. These sex steroids also exert negative feedback control on the pituitary, inhibiting gonadotropin release and thus preventing new follicles from maturing.

If the egg is not fertilized, the corpus luteum degenerates on about day 26 of the cycle. Without production of progesterone by the corpus luteum, the endometrium sloughs off and menstruation occurs. The decrease in circulating steroids also releases the hypothalamus and pituitary from negative feedback control, so GnRH, FSH, and LH all begin to increase. The increase in these hormones induces the next round of follicle development, and the ovarian cycle begins again.

FSH receptors determine which follicle ovulates

Early in the ovarian cycle several follicles begin to develop, but only one reaches full maturity. What determines which one will survive? There are two types of follicle cells, arranged in two layers. The inner layer immediately surrounding the ovum is made up of granulosa cells, and the outer layer contains thecal cells. Thecal cells are stimulated by LH to produce testosterone

WORKING WITH**DATA:**

Circadian Timing, Hormone Release, and Labor

Original Paper

Olcese, J., S. Lozier, and C. Paradise. 2012. Melatonin and the circadian timing of human parturition. *Reproductive Sciences*, epub before print, May 3, 2012.

Analyze the Data

Pregnant women are more likely to go into labor during the night than in the daytime. Frequently, however, when women in labor are taken to a well-lit hospital, their contractions decrease. James Olcese and his colleagues thought that the timing of labor contractions might be controlled by a signal from the brain's circadian timing system—specifically, the release of the hormone melatonin from the pineal gland. Melatonin is always released at night, and its release is inhibited by light (see Section 41.5).

The researchers hypothesized that rising levels of melatonin potentiate contractility of the uterine muscles. To test this hypothesis, they collected two types of data from pregnant women close to term. They collected saliva samples from 5 women at 30-minute intervals overnight and measured the melatonin levels in the samples. They also recorded the frequency of contractions of two women who were having regular but premature contractions at night. In both situations the women were exposed to bright light between 11 P.M. and midnight. The results are shown in the table.

QUESTION 1

Plot the data in the tables. How would you interpret these data? Do they support the authors' hypothesis?

QUESTION 2

It is known that the posterior pituitary hormone oxytocin stimulates the uterine contractions leading to expulsion of the fetus from the uterus. If you cultured some uterine muscle tissue, how would you investigate the possible interaction of oxytocin and melatonin? What would your hypothesis be?

		Data Set 1	Data Set 2	
		Salivary melatonin (percent of maximum; mean for 5 subjects)	Contractions per hour (2 subjects)	
Time			Subject A	Subject B
7 P.M.		5		
8 P.M.		5	2	3
9 P.M.		12	4	5
10 P.M.		30	5	5
11 P.M.	Bright light	55	6	2
Midnight	Bright light	30	1	1
1 A.M.		50	2	0
2 A.M.		60	2	0
3 A.M.		65	2	11
4 A.M.		90	5	18
5 A.M.		85	5	15
6 A.M.		70	4	12
7 A.M.		60	2	4

Go to BioPortal for all WORKING WITHDATA exercises

(similar to Leydig cells in the male). Testosterone diffuses into the granulosa cells, where the enzyme aromatase converts the testosterone to estrogen. Estrogen, along with FSH, stimulates the growth and maturation of the granulosa cells (similar to the Sertoli cells in the male). The estrogen plays two important roles in the selection of the follicle that will ovulate: (1) estrogen stimulates the granulosa cells to express more FSH and LH receptors, and (2) estrogen entering the circulation feeds back on the pituitary to decrease the production of FSH. The granulosa cells also produce inhibin (similar to Sertoli cells in the male), and inhibin also decreases the production of FSH. As FSH and LH levels fall, the follicle with the most FSH and LH receptors survives and matures while the others regress.

In pregnancy, hormones from the extraembryonic membranes take over

If the egg is fertilized and a blastocyst arrives in the uterus and implants in the endometrium, a new hormone comes into play. A layer of cells covering the blastocyst begins to secrete **human chorionic gonadotropin**, or **hCG** (Figure 43.15A). This gonadotropin, a molecule similar to LH, stimulates the corpus luteum to continue to produce estrogen and progesterone to support the growth and maintenance of the endometrium and thereby prevent menstruation. Because it is present only in the blood of pregnant women, the presence of hCG is the basis for pregnancy testing. Pregnancy tests use an antibody to detect hCG in urine; they take only minutes and can be done at home.

Blastocyst and endometrial tissues form the placenta, which nourishes the embryo. The placenta also replaces the corpus luteum as the major producer of estrogen and progesterone. Continued high levels of estrogen and progesterone prevent the pituitary from secreting gonadotropins; thus the ovarian cycle ceases for the duration of pregnancy. This mechanism underlies the action of birth control pills, which contain synthetic hormones resembling estrogen and progesterone that exert negative feedback control on the hypothalamus and pituitary.

Childbirth is triggered by hormonal and mechanical stimuli

Throughout pregnancy, the muscles of the uterine wall periodically undergo slow, weak, rhythmic contractions called Braxton Hicks contractions. These contractions become stronger during the third trimester of pregnancy and are sometimes called false labor. The contractions of true labor mark the beginning of childbirth. Both hormonal and mechanical stimuli contribute to the onset of labor.

Progesterone inhibits and estrogen stimulates contractions of uterine muscle. Toward the end of the third trimester, the estrogen–progesterone ratio shifts in favor of estrogen. The onset of labor is marked by increased secretion of the hormone oxytocin by the posterior pituitaries of both mother and fetus. Oxytocin is a powerful stimulant of uterine muscle contraction. Manufactured oxytocin is used to induce labor when that is necessary.

Mechanical stimuli come from the stretching of the uterus by the fully grown fetus and the pressure of the fetal head on the cervix. These mechanical stimuli increase the release of oxytocin by the mother's posterior pituitary, which in turn

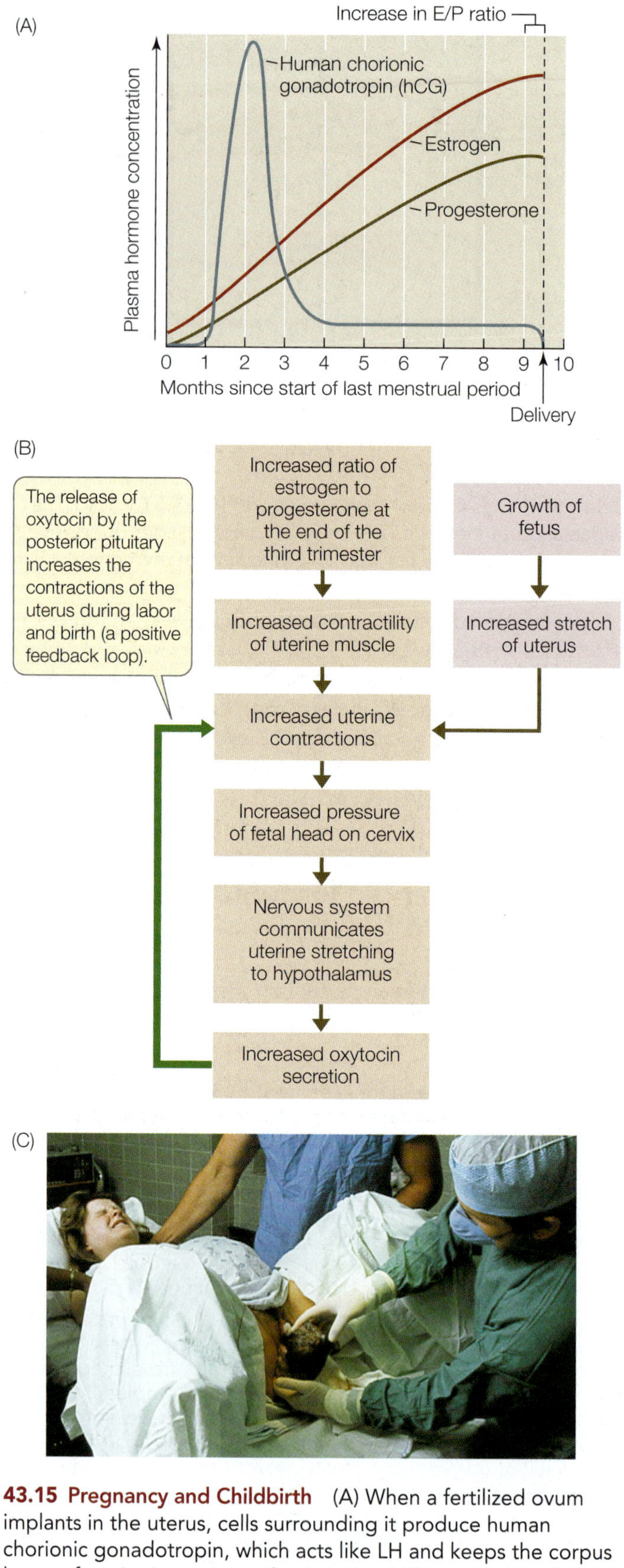
(A)
Increase in E/P ratio
Human chorionic gonadotropin (hCG)
Estrogen
Progesterone
Plasma hormone concentration
0 1 2 3 4 5 6 7 8 9 10
Months since start of last menstrual period
Delivery
(B)
The release of oxytocin by the posterior pituitary increases the contractions of the uterus during labor and birth (a positive feedback loop).
Increased ratio of estrogen to progesterone at the end of the third trimester
Growth of fetus
Increased contractility of uterine muscle
Increased stretch of uterus
Increased uterine contractions
Increased pressure of fetal head on cervix
Nervous system communicates uterine stretching to hypothalamus
Increased oxytocin secretion
(C)

43.15 Pregnancy and Childbirth (A) When a fertilized ovum implants in the uterus, cells surrounding it produce human chorionic gonadotropin, which acts like LH and keeps the corpus luteum functioning as an endocrine gland. The ovarian and uterine cycles are put on hold for the duration of pregnancy. (B) Both mechanical and hormonal signals are involved in stimulating the uterine contractions of labor and delivery. (C) A new person comes into the world head first.

increases the activity of uterine muscle, causing even more pressure on the cervix. This positive feedback loop converts the weak, slow, rhythmic Braxton Hicks contractions into stronger labor contractions (**Figure 43.15B**).

In the early stage of labor, hormonal changes and pressure created by the contractions cause the cervix to dilate (expand) until it is large enough to allow the baby to pass through. Gradually the contractions become more frequent and more intense. This stage of labor lasts an average of 12 to 15 hours in a first pregnancy, but is usually 8 hours or less in subsequent ones.

The second stage of labor begins when the cervix is fully dilated to a diameter of about 10 cm (**Figure 43.15C**). The baby's head can now move into the vagina. Passage of the fetus through the vagina is assisted by the mother's bearing down ("pushing") with her abdominal and other muscles. Once the head and shoulders of the baby clear the cervix, the rest of its body eases out rapidly, but it is still connected to the placenta by the umbilical cord. Once the baby clears the birth canal, it starts breathing and is independent of its mother's circulation. The umbilical cord may then be clamped and cut. The segment still attached to the baby dries up and sloughs off in a few days, leaving behind its distinctive signature, the belly button—more properly called the umbilicus. The third stage of labor, the detachment and expulsion of the placenta and fetal membranes, takes from a few minutes to an hour, and may be accompanied by uterine contractions. A baby's suckling at the breast immediately following birth stimulates additional secretion of oxytocin, which augments uterine contractions, reduces the size of the uterus, and helps stop bleeding.

RECAP 43.3

The reproductive systems of men and women produce gametes and hormones, and these functions are controlled by hypothalamic and anterior pituitary hormones. In women, the hormonal control of reproductive functions produces linked ovarian and uterine cycles.

- Describe the path the human sperm and ovum take in moving from their respective gonads to the point at which fertilization occurs. **See Figures 43.8 and 43.11**
- In males, increased production of GnRH at puberty stimulates the release of what two hormones of the anterior pituitary? What effect do these hormones have? **See p. 892**
- Explain the events in the ovarian cycle that result in release of a single ovum each month. What events prepare the uterus to receive the egg? **See Figures 43.12 and 43.13**

Understanding the physiology of human reproduction has led to numerous methods and technologies for controlling it, either to prevent unwanted pregnancies or to overcome infertility.

43.4 How Can Fertility Be Controlled?

Sexual issues and sexual behavior are dominant aspects of our society, and reproductive technologies have had huge impacts on our sexual and reproductive lives.

Humans use a variety of methods to control fertility

According to a recent study, almost half of the more than 6 million pregnancies that occur in the United States each year are unintended. For women of college age, a single act of unprotected intercourse in the two days prior to ovulation carries a chance of conception as high as 50 percent.

The only failure-proof methods of preventing pregnancy are either complete abstinence from sexual activity or surgical removal of the gonads. Those options are not acceptable to most people, and they turn to other methods to prevent pregnancy. Many of these methods prevent fertilization or implantation (conception) and are therefore referred to as **contraception**. **Table 43.1** lists some of the most commonly used contraceptive methods and their relative failure rates; note that these methods vary enormously in their effectiveness. Most methods are used by the woman, although some are used by the man.

Once a fertilized egg is successfully implanted in the uterus, any termination of the pregnancy is called an **abortion**. A spontaneous abortion is the medical term for what is commonly called a miscarriage. Spontaneous abortions frequently occur early in pregnancy and are usually the result of either a chromosomal abnormality in the fetus or a breakdown in the process of implantation. Many spontaneous abortions occur before the woman even realizes she is pregnant.

Abortions that result from medical intervention may be performed either for therapeutic purposes or for fertility control. A therapeutic abortion may be necessary to protect the health of the mother, or it may be performed because prenatal testing reveals that the fetus has a severe defect. In a medical abortion, the cervix is dilated and some of the endometrium, along with the implanted fetus, is removed. When performed in the first trimester of a pregnancy, a medical abortion carries less risk of death to the mother than a full-term pregnancy. The risk rises after the first 12 weeks of pregnancy but remains less than that of a full-term pregnancy through the second trimester.

Reproductive technologies help solve problems of infertility

About 15 percent of couples in the United States are infertile—that is, they can't have children. The reasons for infertility are many, and are about equally distributed between men and women. Several technologies have been developed to overcome barriers to both conceiving and bearing a child. The simplest of these is **artificial insemination**, in which the physician places sperm in the female's reproductive tract. This technique is useful if the male's sperm count is low, if his sperm lack motility, or if problems in the female's reproductive tract prevent the normal movement of sperm up to and through the oviducts. Artificial insemination is also widely used in the production of domesticated animals such as cattle.

Assisted reproductive technologies, or **ARTs**, involve procedures that remove unfertilized eggs from the ovary, combine them with sperm outside the body, and then place fertilized eggs or egg–sperm mixtures in the appropriate location in the female's reproductive tract for development to take place. The

TABLE 43.1 Methods of Contraception

Method	Mode of Action	Failure Rate[a]	Comments
Unprotected	No form of birth control.	85	High risk of pregnancy, especially for women 15–30.
Nontechnological methods			
Rhythm method	The couple abstains from intercourse between days 10 and 20 of the ovarian cycle (peak fertility).	15–35	High failure rate due to miscalculation and/or variation of individual cycles.
Coitus interruptus	The man withdraws his penis prior to ejaculation with the intention of not depositing sperm into the vagina.	20–40	Requires self-control, especially by the man. Very high failure rate.
Barrier methods[b]			
Condom	A sheath of impermeable material (often latex) is fitted over the erect penis. Semen is trapped in the condom, so no sperm are deposited in the vagina.	15	If fitted correctly, an intact condom can prevent pregnancy and provide protection against sexually transmitted diseases (STDs), including HIV (AIDS).
Spermicidal jellies	Applied inside the vagina, these chemical compounds kill or immobilize sperm.	25	Used alone, spermicidal compounds have a fairly high failure rate.
Diaphragms, cervical caps	Inserted by the woman prior to intercourse, these devices work by blocking the cervix so that sperm cannot pass into the uterus.	10–15	Approximately the same failure rate as condom use by men, but do not protect against STDs. Can be used in conjunction with spermicidal jelly for extra protection.
Hormone-based contraceptives			
Oral hormones ("the pill")	A daily pill for women containing a combination of synthetic estrogens and progesterone. These hormones mimic pregnancy to the extent that the ovarian cycle and ovulation are suspended. The uterine cycle is allowed to continue by including a week of non-hormone administration every 21–28 days.	0–3	Requires medical consultation and prescription. Taken correctly, oral contraceptives are extremely effective. In the U.S., more than 12 million women use them each year; they are sometimes prescribed to treat menstrual disorders.
Non-orally administered hormones	Making use of same hormonal actions as the pill, these methods include long-acting injections, patches that release hormones transdermally (through the skin), and a hormone-containing vaginal ring.	<1	Same as oral hormones. A slightly lower failure rate because the woman does not have to remember to take a daily pill.
Progestin-only pill (Plan B®)	An oral contraceptive meant to be taken within 72 hours after unprotected sex. A high dose of progestin in two pills prevents ovulation in the same manner birth control pills do.	5–40	Not an "abortion pill," this drug will not terminate an existing pregnancy. Currently available to women over 17 without a prescription. Failure rate varies widely depending on when taken.
Implantation blockers			
Intrauterine device (IUD)	A medical professional inserts a small plastic or metal device into the uterus. The resulting inflammation reaction (see Chapter 42) releases prostaglandins, which prevent implantation of the fertilized egg.	0.5–5	A highly effective contraceptive, it is the most widely used birth control device in China (and hence the world). With medical monitoring, can remain in place for several years.
Mifepristone (RU-486)	This drug blocks progesterone receptors necessary to maintain the endometrium during implantation and pregnancy.	0.5–6	Prevents implantation when taken up to several days after unprotected intercourse. Can terminate a pregnancy up to the time of the first missed menstrual period. In the U.S., available from specialized providers.
Sterilization			
Vasectomy	The vasa deferentia (see Figure 43.8) are cut and tied off so that sperm can no longer pass into the urethra. Sperm continue to be produced but are reabsorbed by the man's body. Male hormone levels and sexual responses are not affected.	0–0.15	A simple surgical procedure performed under local anesthetic in a doctor's office. Although theoretically it can be reversed, vasectomy should be considered permanent.
Tubal ligation	The oviducts (see Figure 43.11A) are tied off so that eggs cannot reach the uterus and sperm cannot reach the egg. As with vasectomy, hormone levels and sexual responses are not affected.	0–0.05	This surgical procedure is somewhat more complex than vasectomy. It is often performed in conjunction with childbirth when a woman has decided that her family is complete.

[a]Failure rate refers to the number of pregnancies per 100 women per year

[b]All of these barrier methods are routinely available without medical prescription.

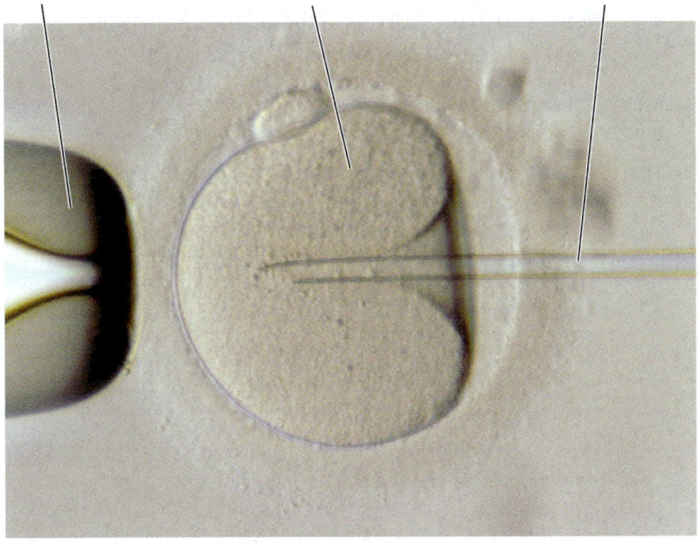
Pipette
holding egg
Egg
Pipette
injecting sperm

43.16 Intracytoplasmic Sperm Injection In this procedure, a sperm is injected directly into a mature egg cell. The fertilized egg is then placed in the female reproductive tract, where it can implant and develop into a fetus.

first successful ART was in vitro fertilization (IVF). In IVF the female is treated with hormones that stimulate many follicles in her ovaries to mature. Eggs are collected from these follicles, and sperm are collected from the male. Eggs and sperm are combined in a culture medium outside the body, where fertilization takes place. The resulting embryos can be injected into the mother's uterus in the blastocyst stage or kept frozen for implantation later. The first "test-tube baby" resulting from IVF was born in England in 1978. Since then, millions of babies have resulted from this ART.

A major cause of IVF failure is failure of sperm to gain access to the egg plasma membrane (see Figure 43.4). To solve this problem, methods have been developed to inject a sperm cell directly into the cytoplasm of an egg. In intracytoplasmic sperm injection (ICSI), an egg is held in place by suction applied to a polished glass pipette. A slender, sharp pipette is then used to penetrate the egg and inject a sperm (**Figure 43.16**). This ART was used successfully for the first time in 1992 by researchers in Belgium; now thousands of these procedures are performed in U.S. clinics each year, with a success rate of about 25 percent.

IVF, coupled with techniques of genetic analysis, can eliminate the risk that adults who are carriers of genetic diseases will produce affected children. It is now possible to take a cell from a human embryo at the 4- or 8-cell stage (see Figure 44.4) without damaging its developmental potential. The sampled cell can be subjected to molecular analysis to determine whether it carries the harmful gene. This procedure, called preimplantation genetic diagnosis (PGD), makes it possible to determine whether an embryo produced by IVF carries the genetic defect of concern.

RECAP 43.4

Controlling fertility is an important aspect of modern human life. Decreasing the probability of pregnancy is achieved through methods that prevent sperm and egg from meeting and from preventing implantation. Pregnancies can be facilitated through medical technology.

- Which method of contraception is the only one to offer protection against sexually transmitted diseases (STDs)? **See Table 43.1**
- Explain what a couple who are both carriers of a genetic disease could do to ensure that their offspring would not have the disease. **See p. 899**

The fertilized egg of a sexually reproducing organism is a single cell containing all the genetic information needed to create a new organism. Chapters 19 and 20 introduced some of the molecular aspects of the process of development in multicellular animals. Chapter 44 will describe the physiological and anatomical events of animal development.

?

How can a queen be so different from her worker sisters when they all share the same genome?

ANSWER

When a hive loses its queen, the workers create new queens from a few eggs that the old queen laid by feeding a substance called "royal jelly" to the larvae that hatch from the chosen eggs. This special food stimulates growth and has recently been discovered to have powerful epigenetic effects on honey bee development. As described in Chapter 16, the expression of a gene can be more or less permanently altered by the chemical modification of histones—proteins that are closely associated with the DNA—or by methylation or demethylation of the DNA itself (see Section 16.4). Because these changes do not alter the gene's nucleotide sequence, they are referred to as epigenetic ("outside the gene"). One component of royal jelly is phenyl butyrate, an inhibitor of histone deacetylation. Furthermore, the queen bee has well over 500 genes that are methylated differently than those same genes in workers. It is therefore likely that the royal jelly fed to a future queen dramatically alters the expression of her genome.

CHAPTER SUMMARY 43

43.1 How Do Animals Reproduce without Sex?

- Asexual reproduction produces offspring that are genetically identical to their parent and to one another; it produces no genetic diversity.
- Means of asexual reproduction include **budding**, **regeneration**, and **parthenogenesis**. Review Figures 43.1, 43.2

43.2 How Do Animals Reproduce Sexually?

- Sexual reproduction involves three basic steps: **gametogenesis**, **spawning** or **mating**, and **fertilization**.
- **Gametogenesis** and fertilization are similar in all animals, but spawning and mating include a great variety of anatomical, physiological, and behavioral adaptations.
- Gametogenesis occurs in **testes** and **ovaries**. In spermatogenesis (the production of **sperm**) and oogenesis (the production of eggs), the **germ cells** proliferate mitotically, undergo meiosis, and mature into gametes.
- Each **primary spermatocyte** can produce four haploid sperm through the two divisions of meiosis. Review Figure 43.3A
- **Primary oocytes** immediately enter prophase of the first meiotic division, and in many species, including humans, their development is arrested at this point. Each **oogonium** produces only one egg. Review Figure 43.3B
- Fertilization involves sperm activation, species-specific binding of sperm to egg, the acrosomal reaction, digestion of a path through the protective coverings of the egg, and fusion of sperm and egg plasma membranes. Fusion of these two membranes triggers **blocks to polyspermy**, which prevent additional sperm from entering the egg and, in mammals, signal the egg to complete meiosis and begin development. Review Figure 43.4, ANIMATED TUTORIAL 43.1
- **External fertilization** is common in aquatic species. **Internal fertilization** is necessary in terrestrial species and usually involves **copulation**.
- **Hermaphroditic**, or **monoecious**, species have both male and female reproductive systems in the same individual, either sequentially or simultaneously. **Dioecious** species have separate male and female individuals.
- Animals can be classified as **oviparous** or **viviparous**, depending on whether the early stages of development occur outside or inside the mother's body.

43.3 How Do the Human Male and Female Reproductive Systems Work?

- Men produce **semen** consisting of sperm suspended in seminal fluid (which nourishes the sperm and facilitates fertilization).
- Sperm are generated in the **seminiferous tubules** of the testes, mature in the **epididymis**, and are delivered to the **urethra** through the **vasa deferentia**. Other components of semen are produced in the **seminal vesicles**, **prostate gland**, and **bulbourethral gland**. Review Figures 43.8, 43.9, ACTIVITIES 43.1, 43.2
- All components of the semen join in the urethra at the base of the **penis** and are ejaculated through the erect penis by muscle contractions at the culmination of copulation.
- Spermatogenesis depends on testosterone secreted by the **Leydig cells** of the testes, which are under the control of hormones produced in the anterior pituitary and the hypothalamus. The production of these hormones is controlled by negative feedback from testosterone and from inhibin, a hormone produced by the **Sertoli cells** of the testes. Review Figure 43.10
- Eggs mature in the woman's ovaries and are released into the **oviducts**. Sperm deposited in the **vagina** during **copulation** move up through the **cervix** and **uterus** into the oviducts. Fertilization occurs in the upper regions of the oviducts. Review Figure 43.11, ACTIVITY 43.3
- The maturation and release of eggs constitute an **ovarian cycle**. The **uterine cycle** prepares the uterus for receipt of a blastocyst. If no blastocyst is implanted, the lining of the uterus sloughs off in the process of **menstruation**. Review Figure 43.13, ANIMATED TUTORIAL 43.2
- Both the ovarian and the uterine cycles are under the control of hypothalamic and pituitary hormones, which in turn are under the feedback control of estrogen and progesterone. Review Figure 43.14
- Childbirth is initiated by hormonal and mechanical stimuli that increase the contraction of uterine muscle. Review Figure 43.15

43.4 How Can Fertility Be Controlled?

- Methods of **contraception** include abstention from copulation and the use of technologies that decrease the probability of fertilization. Review Table 43.1
- **Assisted reproductive technologies** (**ARTs**) have been developed to increase fertility.

Go to the Interactive Summary to review key figures, Animated Tutorials, and Activities
Life10e.com/is43

CHAPTER**REVIEW**

REMEMBERING

1. Which statement about human oocytes is true?
   a. By birth, a human female infant has produced her lifetime supply of oocytes.
   b. At the onset of puberty, ovarian follicles produce new oocytes in response to hormonal stimulation.
   c. A woman stops producing oocytes at the onset of menopause.
   d. A woman produces oocytes throughout adolescence.
   e. Oocytes are stored in the oviducts.

2. Spermatogenesis and oogenesis differ in that
   a. spermatogenesis produces gametes with greater stores of raw materials than those produced by oogenesis.
   b. spermatocytes remain in prophase of the first meiotic division longer than oocytes.
   c. oogenesis produces four equally functional haploid cells per meiotic event, and spermatogenesis does not.
   d. spermatogenesis produces many gametes with meager energy reserves, whereas oogenesis produces relatively few, well-provisioned gametes.
   e. spermatogenesis begins before birth in humans, whereas oogenesis does not start until the onset of puberty.

3. Semen contains all of the following except
   a. fructose.
   b. mucus.
   c. clotting enzymes.
   d. testosterone.
   e. an active clot-dissolving enzyme.

4. Which of the following statements about the ovarian and uterine cycles is *not* true?
   a. Falling estrogen and progesterone levels induce menstruation.
   b. A sudden rise in LH induces ovulation.
   c. Estrogen levels reach highest levels in the follicular phase and progesterone reaches highest levels in the luteal phase of the ovarian cycle.
   d. If fertilization occurs, the corpus luteum secretes hCG.
   e. Estrogen is produced by follicle cells.

5. Contractions of muscles in the uterine wall are stimulated by
   a. progesterone.
   b. estrogen.
   c. prolactin.
   d. oxytocin.
   e. human chorionic gonadotropin.

6. Which of the following methods of contraception is *most likely* to fail?
   a. Rhythm method
   b. Birth control pills
   c. Diaphragm
   d. Vasectomy
   e. Condom

UNDERSTANDING & APPLYING

7. In terms of characteristics and functions, explain how you would pair Leydig cells, Sertoli cells, thecal cells, and granulosa cells, and why.

8. A drug called RU-486 blocks the actions of progesterone. It is called a contragestational drug because it can terminate a pregnancy after fertilization has occurred. How does RU-486 do this?

ANALYZING & EVALUATING

9. Females of the marine worm *Bonellia viridis* release fertilized eggs into the water. The eggs hatch into larvae that swim and then settle on a substrate. If the larvae land on sand, they burrow in and develop into females with large bodies and a proboscis that extends into the water above to collect food. If a larva lands on a female proboscis, it burrows into the female body and becomes a tiny male whose only function is to produce sperm to fertilize the female's eggs. What selection pressures could contribute to this extreme sexual dimorphism?

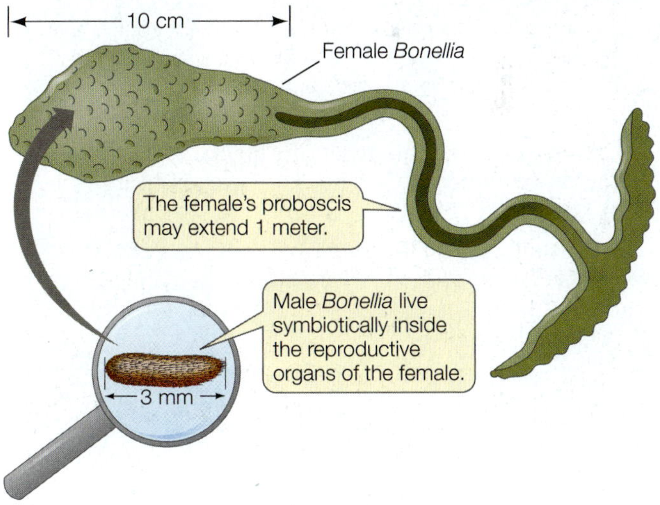
10 cm
Female *Bonellia*
The female's proboscis may extend 1 meter.
Male *Bonellia* live symbiotically inside the reproductive organs of the female.
3 mm

10. If a man carries a genetic mutation that prevents the maintenance of cytoplasmic bridges between his secondary spermatocytes and his spermatids, would you expect that he would father only sons or only daughters? Why?

Go to BioPortal at **yourBioPortal.com** for Animated Tutorials, Activities, LearningCurve Quizzes, Flashcards, and many other study and review resources.

44 Animal Development

CHAPTER OUTLINE

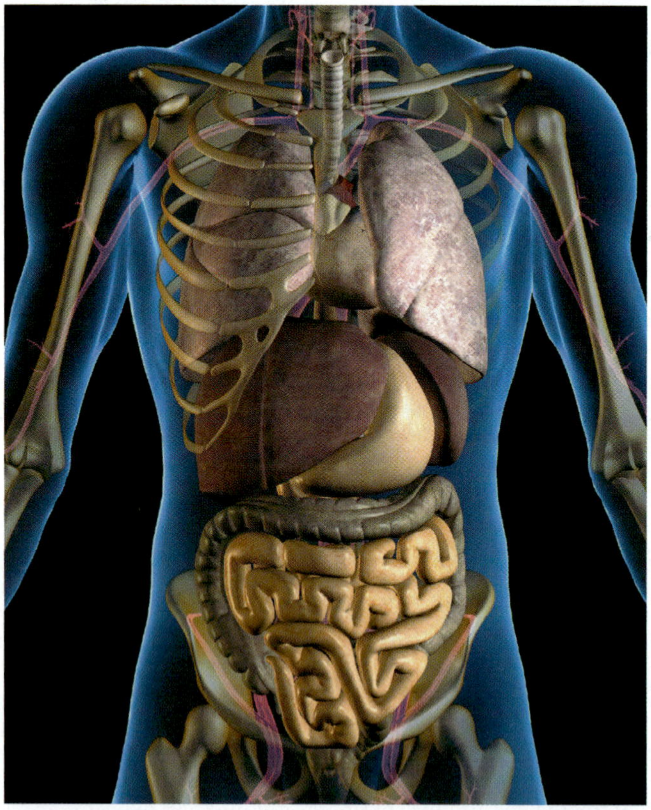

Go with the Flow The internal organs of humans are not all symmetrical, and some individuals are born with the mirror-image pattern of what is seen in most people—a condition called situs inversus. The left–right asymmetry of the internal organs is initiated by asymmetrical stimulation of primary cilia at a very early stage in development.

PLACE YOUR HAND over your heart. Next point to your appendix. Surely you first put your hand on the left side of your chest and then pointed to the right side of your lower abdomen. Like other vertebrates, humans are bilaterally symmetrical—but our symmetry is not absolute. Some of our internal organs are oriented differently with respect to the left and right sides of the body. In about 1 out of every 7,000 people, the arrangement of the internal organs is reversed, a condition known as situs inversus ("location inverted"). The difference arises from events very early in the development of the embryo, and most people with situs inversus lead normal lives.

As you will learn in this chapter, to get from an embryo with a single layer of cells to the next stage with two layers of cells, a midline slit forms as cells in one area of the embryo migrate inward. The place where the inward movement of cells starts is called the node. Cells at the node have motile cilia that beat in a clockwise motion and move extracellular fluid across the surface of the node. That fluid movement is always leftward—that is, from right to left. Why?

Imagine the movement of the cilia describes a clock face oriented so that 12 o'clock and 6 o'clock lie along the embryo's midline slit, with 12 o'clock closest to the future head. The cilia protrude from the cell surfaces at an angle. Thus when the cilia are rotating through the 6 o'clock position, they are closer to the cell surface and experience greater shear forces (resistance) than when they are rotating through the 12 o'clock position. In the bilaterally symmetrical early embryo, the circular beating of cilia at this critical spot—the node—creates left–right information.

How do we know the information generated by beating nodal cilia is critical for the left–right asymmetrical patterns of gene expression and developmental processes that follow? In mutant strains of mice that have no cilia or nonmotile cilia in the node, about half of the mice have reversed organ symmetry. Similarly, a spectrum of rare genetic disorders in humans, collectively called Kartagener's syndrome, are characterized by nonmotile cilia, and about 50 percent of these individuals also have situs inversus.

?

How does the directional flow of extracellular fluid across the node stimulate a left–right asymmetry in gene expression and development?

See answer on p. 921.

44.1 How Does Fertilization Activate Development?

Fertilization is the joining of sperm and egg to produce a diploid zygote. You might therefore think of it as the event that begins development. Keep two things in mind, however. First, in animals that reproduce asexually, development proceeds without fertilization. And second, in animals where fertilization does occur, it is preceded by critical events in the maturing egg that will influence subsequent development but are suspended in developmental arrest. Thus in studying fertilization we are asking how it activates or restarts multicellular development in sexually reproducing animals.

Fertilization does more than restore a full diploid complement of maternal and paternal genes. The fusion of sperm and egg plasma membranes accomplishes several things:

- It stimulates ion fluxes across the egg membrane.
- It sets up blocks to the entry of additional sperm into the egg.
- It changes the pH of the egg cytoplasm.
- It increases egg metabolism and stimulates protein synthesis.
- It initiates the rapid series of cell divisions that produce a multicellular embryo.

The mechanisms of fertilization were described in Section 43.2. Here we will take a closer look at the cellular and molecular interactions of sperm and egg that initiate the first steps of development.

The sperm and the egg make different contributions to the zygote

In most species, eggs are much larger than sperm. Egg cytoplasm is well stocked with organelles, nutrients, and a variety of molecules, including cytoplasmic determinants such as transcription factors and mRNAs (see Section 19.2). Nearly everything the embryo needs during its early stages of development comes from the mother, including its mitochondria (and therefore all of its mitochondrial DNA). In addition to its haploid nucleus, the sperm makes one other crucial contribution to the zygote in most species—the centriole. The centriole contributes to the zygote's centrosome, which organizes the mitotic spindles for subsequent cell divisions (see Figure 11.10). Centrioles are also the origin of the microtubules of the primary cilia, which are important in cell signaling, as we saw in the opening story about situs inversus.

Cytoplasmic determinants in the egg play important roles in setting up the signaling cascades that orchestrate the major events of development: determination, differentiation, morphogenesis, and growth.

Rearrangements of egg cytoplasm set the stage for determination

The unique attributes of amphibian eggs make them ideal models for illustrating how rearrangements of egg cytoplasm set the stage for determination. The molecules in the cytoplasm of the amphibian egg are not homogeneously distributed. The entry of the sperm into the egg stimulates rearrangements of the egg cytoplasm that introduce additional organization to the egg. Sperm entry establishes the polarity of the zygote, and informational molecules in the egg cytoplasm are organized with respect to that polarity. Therefore, when cell divisions begin, these informational molecules—which guide subsequent development—are not divided equally among daughter cells.

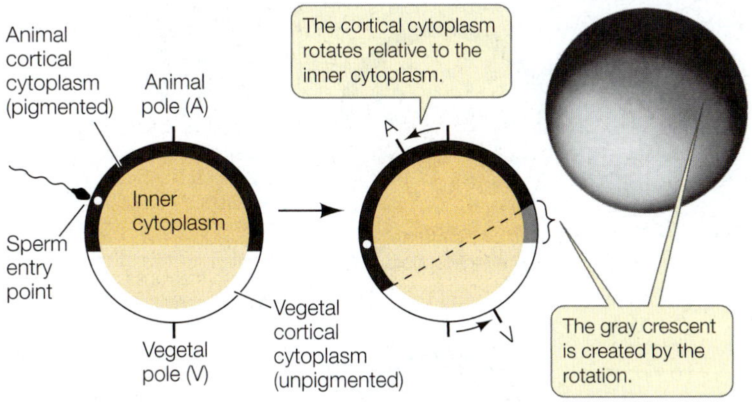
Animal cortical cytoplasm (pigmented)
Animal pole (A)
Inner cytoplasm
Sperm entry point
Vegetal pole (V)
Vegetal cortical cytoplasm (unpigmented)
The cortical cytoplasm rotates relative to the inner cytoplasm.
A
V
The gray crescent is created by the rotation.

44.1 The Gray Crescent In amphibian eggs, cortical rotation and rearrangement of the cytoplasm after fertilization create the gray crescent opposite the point of sperm entry. These events are important for specifying the body axes and other important events in later development.

Rearrangement of egg cytoplasm following fertilization is easily observed in some frog species because of pigments in the cytoplasm. The nutrients in an unfertilized frog egg are dense yolk granules that are concentrated by gravity in the lower half of the egg, called the **vegetal hemisphere**. The haploid nucleus of the egg is located at the opposite end, in the **animal hemisphere**. The outermost (cortical) cytoplasm of the animal hemisphere is heavily pigmented, and the underlying cytoplasm has more diffuse pigmentation. The vegetal hemisphere is not pigmented. Because of these differences, it is easy to observe how the cytoplasm rearranges when the egg is fertilized.

The frog egg is radially symmetrical. You can turn it on its vegetal–animal pole axis, and all sides are the same. Sperm-binding sites are localized on the surface of the animal hemisphere. When a sperm binds to and enters the egg, the egg's radial symmetry is converted to bilateral symmetry and an anterior–posterior axis is created. Cortical cytoplasm rotates toward the site of sperm entry (**Figure 44.1**). This rotation brings the animal and vegetal regions of cytoplasm into contact with each other, producing a band of pigmented cytoplasm on the side opposite the site of sperm entry. This band, called the **gray crescent**, marks the location of important developmental events in some species of amphibians.

The centriole that was the sperm's contribution to the egg initiates the cytoplasmic reorganization that coincides with the appearance of the gray crescent. The centriole organizes the microtubules in the vegetal hemisphere cytoplasm into a parallel array that guides the movement of the cortical cytoplasm. These

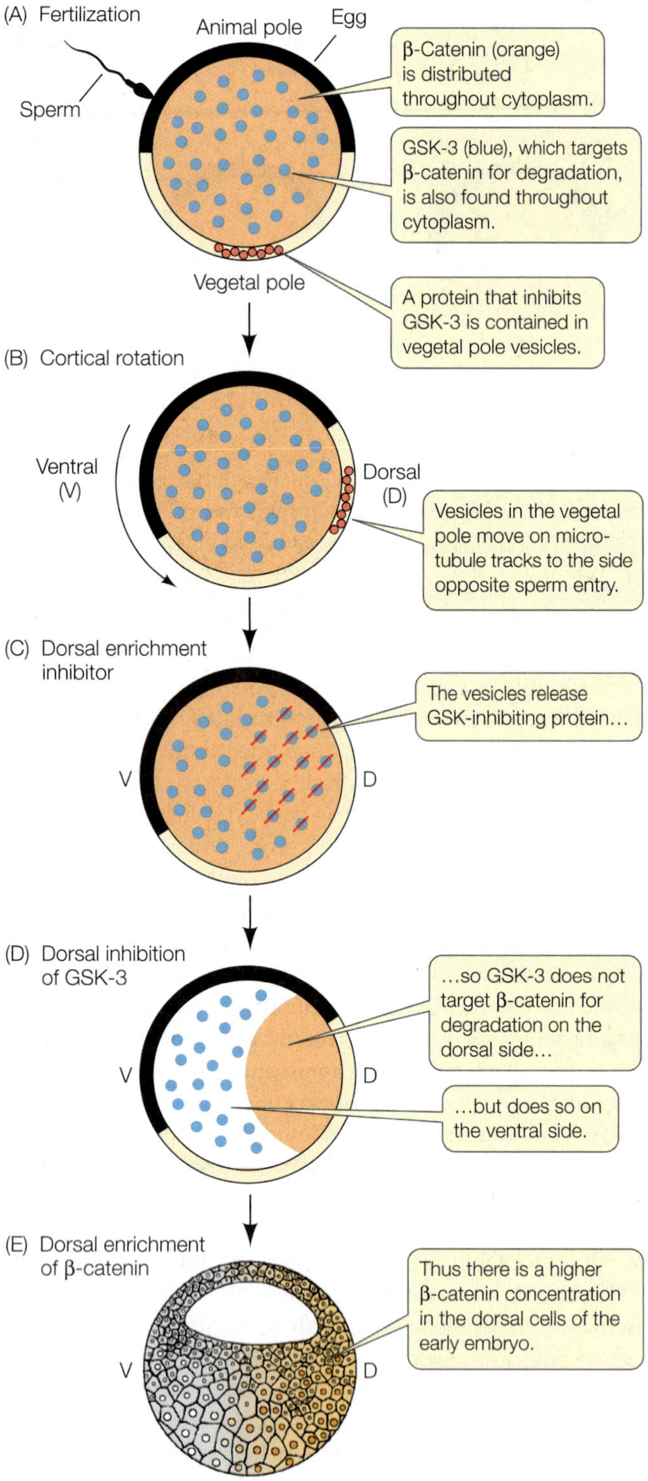
(A) Fertilization
Animal pole
Egg
Sperm
β-Catenin (orange) is distributed throughout cytoplasm.
GSK-3 (blue), which targets β-catenin for degradation, is also found throughout cytoplasm.
Vegetal pole
A protein that inhibits GSK-3 is contained in vegetal pole vesicles.
(B) Cortical rotation
Ventral (V)
Dorsal (D)
Vesicles in the vegetal pole move on microtubule tracks to the side opposite sperm entry.
(C) Dorsal enrichment inhibitor
V
D
The vesicles release GSK-inhibiting protein…
(D) Dorsal inhibition of GSK-3
V
D
…so GSK-3 does not target β-catenin for degradation on the dorsal side…
…but does so on the ventral side.
(E) Dorsal enrichment of β-catenin
V
D
Thus there is a higher β-catenin concentration in the dorsal cells of the early embryo.

44.2 Cytoplasmic Factors Set Up Signaling Cascades Cytoplasmic movement changes the distributions of critical developmental signals. In the frog zygote, the interaction of the protein kinase GSK-3, its inhibitor, and the protein β-catenin are crucial in specifying the dorsal–ventral axis of the embryo.

microtubules also appear to be directly responsible for movement of specific organelles and proteins, because these organelles and proteins move from the vegetal hemisphere to the gray crescent region even faster than the cortical cytoplasm rotates.

The movement of cytoplasm, proteins, and organelles changes the distribution of critical signals. A key transcription factor in early development is β-catenin, which is produced from maternal mRNA (mRNA produced and stored in the egg while it was maturing in the ovary). β-Catenin is found throughout the egg cytoplasm (**Figure 44.2**). Also present throughout the egg cytoplasm is a protein kinase, glycogen synthase kinase-3 (GSK-3), that phosphorylates β-catenin and thereby targets it for degradation. An inhibitor of GSK-3 is localized in the vegetal cortex of the egg. After sperm entry, this inhibitor moves along microtubules to the gray crescent, where it prevents the degradation of β-catenin. As a result, the concentration of β-catenin is higher on the dorsal than on the ventral side of the developing embryo.

RECAP 44.1

The egg is stocked with nutrients and informational molecules that power and direct the early stages of development. Fertilization activates the egg and stimulates rearrangement of the cytoplasm, setting up the body axes and organizing positional information that will control determination and differentiation.

- What are the contributions of the sperm and of the egg to the zygote? **See p. 903**
- Explain how β-catenin becomes more concentrated on the dorsal side of the embryo. **See p. 904 and Figure 44.2**

The uneven distribution of informational molecules in the cytoplasm of the fertilized egg is essential to later events in development. In the next section we will see how these informational molecules end up in different cells of the developing embryo.

44.2 How Does Mitosis Divide Up the Early Embryo?

β-Catenin plays a major role in the cell–cell signaling cascade that begins the process of cell determination and the formation of the embryo. But cell–cell signaling requires more than one cell. The single-celled zygote must become a multicellular embryo.

Cleavage repackages the cytoplasm

Cleavage is the sequence of early cell divisions that transforms the diploid zygote into a mass of undifferentiated cells that will develop as the embryo. Because the cytoplasm of the zygote is not homogeneous, these first cell divisions result in the differential distribution of nutrients and cytoplasmic determinants in the early embryo.

In most animals, cleavage proceeds with rapid DNA replication and mitosis but with no cell growth and little gene expression. The embryo becomes a solid ball of smaller and smaller cells. Eventually this ball forms a central fluid-filled cavity called a **blastocoel**, at which point the embryo is called a **blastula**. Its individual cells are called **blastomeres**. The pattern of cleavage in different species influences the form of their blastulas.

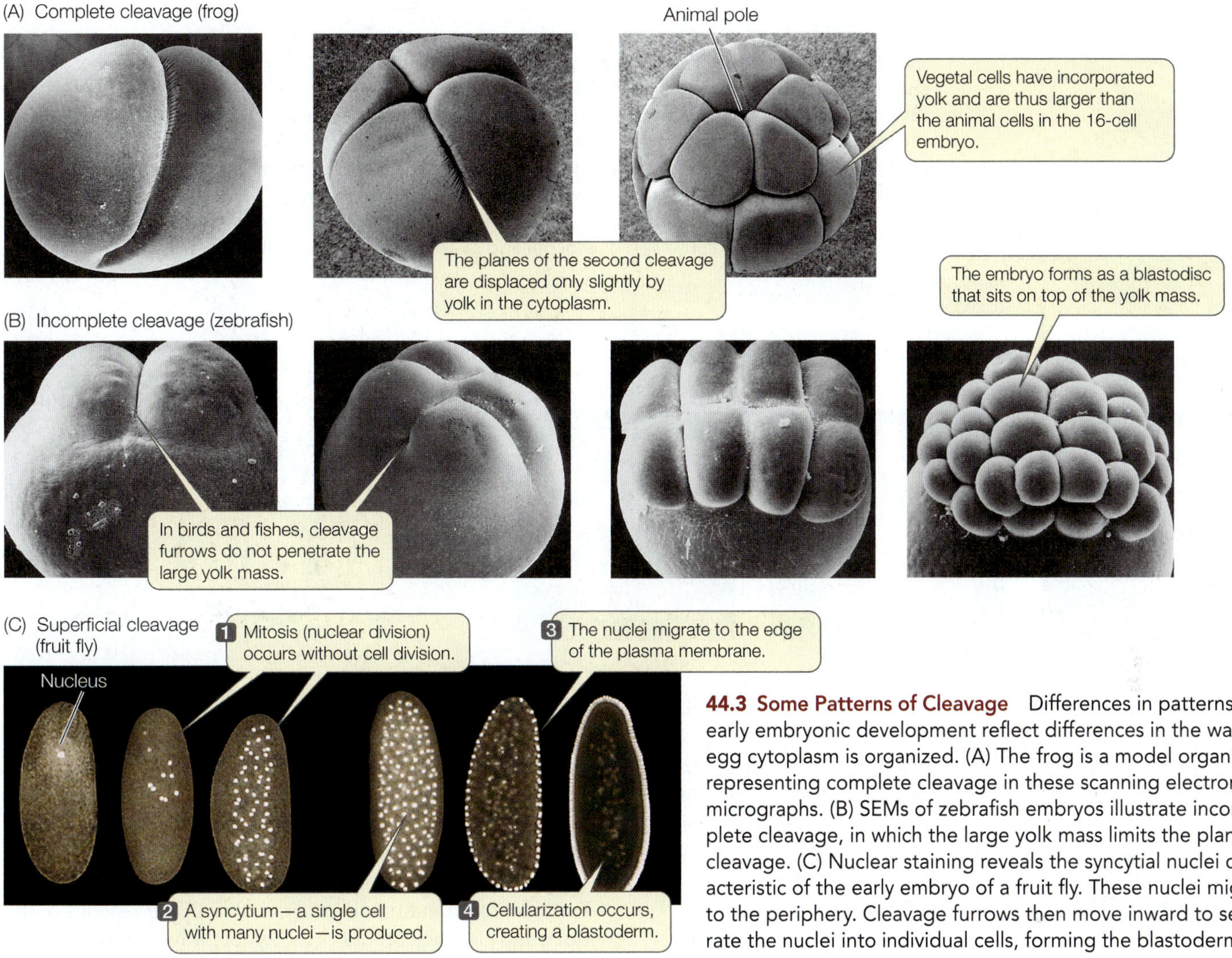
(A) Complete cleavage (frog)
Animal pole
Vegetal cells have incorporated yolk and are thus larger than the animal cells in the 16-cell embryo.
The planes of the second cleavage are displaced only slightly by yolk in the cytoplasm.
(B) Incomplete cleavage (zebrafish)
The embryo forms as a blastodisc that sits on top of the yolk mass.
In birds and fishes, cleavage furrows do not penetrate the large yolk mass.
(C) Superficial cleavage (fruit fly)
1 Mitosis (nuclear division) occurs without cell division.
3 The nuclei migrate to the edge of the plasma membrane.
Nucleus
2 A syncytium—a single cell with many nuclei—is produced.
4 Cellularization occurs, creating a blastoderm.

44.3 Some Patterns of Cleavage Differences in patterns of early embryonic development reflect differences in the way the egg cytoplasm is organized. (A) The frog is a model organism representing complete cleavage in these scanning electron micrographs. (B) SEMs of zebrafish embryos illustrate incomplete cleavage, in which the large yolk mass limits the planes of cleavage. (C) Nuclear staining reveals the syncytial nuclei characteristic of the early embryo of a fruit fly. These nuclei migrate to the periphery. Cleavage furrows then move inward to separate the nuclei into individual cells, forming the blastoderm.

- **Complete cleavage** occurs in most eggs that have little yolk. Early cleavage furrows divide the egg completely. The frog egg undergoes complete cleavage, but because its vegetal pole contains more yolk, the division of the cytoplasm is unequal and the blastomeres in the animal hemisphere are smaller than those in the vegetal hemisphere (Figure 44.3A).
- **Incomplete cleavage** occurs in species in which the egg contains a lot of yolk and the cleavage furrows do not penetrate it all. **Discoidal cleavage** is a type of incomplete cleavage common in fishes and in birds and other reptiles, in which the embryo forms as a disc of cells, or **blastodisc**, that sits on top of the dense yolk mass (Figure 44.3B).
- **Superficial cleavage** is a variation of incomplete cleavage that occurs in insects such as the fruit fly (*Drosophila*). Early in development, cycles of mitosis occur without cell division, producing a **syncytium**—a single cell with many nuclei (Figure 44.3C). The nuclei eventually migrate to the periphery of the egg, after which the plasma membrane of the egg grows inward, creating a **blastoderm** by partitioning the nuclei into individual cells surrounding a core of yolk.

The positions of the mitotic spindles during cleavage are not random but are defined by cytoplasmic determinants produced from the maternal genome and stored in the egg (see Section 19.2). The orientation of the mitotic spindles can determine the planes of cleavage and the arrangement of the blastomeres.

In complete cleavage, if the mitotic spindles of successive cell divisions form parallel or perpendicular to the animal–vegetal axis of the zygote, a pattern of **radial cleavage** occurs. The first two cell divisions are parallel to the animal–vegetal axis, and the third is perpendicular to it (see Figure 44.4A). **Spiral cleavage** results when the mitotic spindles are at oblique angles to the animal–vegetal axis. In spiral cleavage, each new cell layer is shifted to the left or right, depending on the orientation of the mitotic spindles. Most mollusks have spiral cleavage, reflected in some species by a coiling shell pattern (as seen in snails).

Early cell divisions in mammals are unique

Several features of early cell divisions in placental mammals (eutherians) are very different from those seen in other animal groups. First, this process in mammals is very slow. Cell divisions are 12 to 24 hours apart, compared with tens of minutes to a few hours in non-mammalian species. Also, the cell

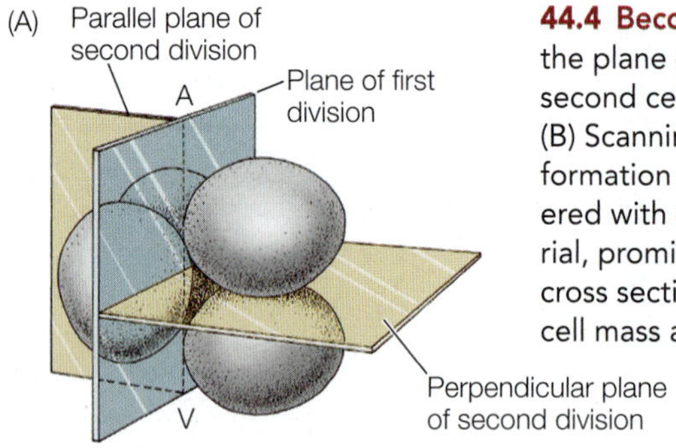
(A)
Parallel plane of second division
A
Plane of first division
Perpendicular plane of second division
V

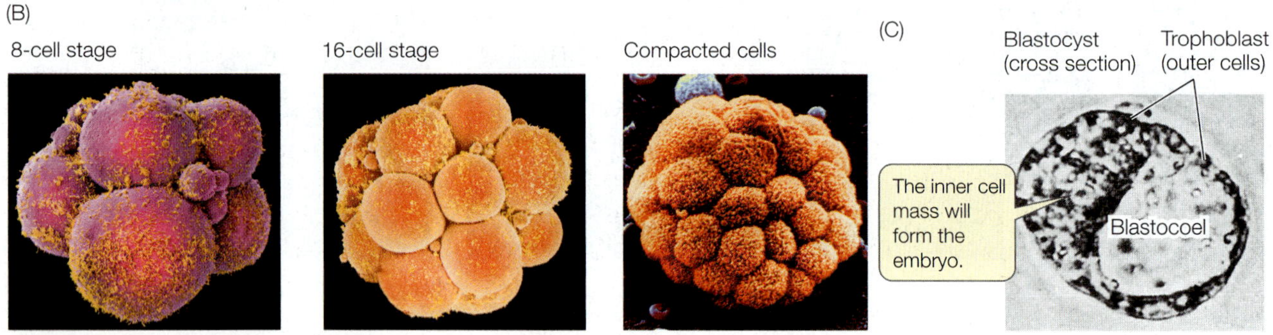
(B)
8-cell stage
16-cell stage
Compacted cells
(C)
Blastocyst (cross section)
Trophoblast (outer cells)
The inner cell mass will form the embryo.
Blastocoel

44.4 Becoming a Blastocyst (A) Mammals have rotational cleavage, in which the plane of the first cleavage is parallel to the animal–vegetal (A–V) axis, but the second cell division involves two planes (beige) at right angles to each other. (B) Scanning electron micrographs (color added) of early cleavage (leading to the formation of the blastocyst) in a human embryo. The cells' outer surfaces are covered with cilia (bright yellow). The small spheres, or "blebs," of cytoplasmic material, prominent at the 8-cell stage, disintegrate as cleavage progresses. (C) Seen in cross section under a light microscope, a mammalian blastocyst consists of an inner cell mass adjacent to a fluid-filled blastocoel and surrounded by trophoblast cells.

divisions of mammalian blastomeres are not in synchrony with each other. Because the blastomeres do not undergo mitosis at the same time, the number of cells in the embryo does not increase in the regular progression (2, 4, 8, 16, 32, etc.) typical of other species. This slow mammalian cleavage means that genes expressed during cleavage play roles in cleavage. In animals such as sea urchins and frogs, very little if any gene transcription occurs in the blastomeres, with cleavage being directed by molecules that were present in the egg before fertilization.

The pattern of mammalian cleavage is unique and is called rotational cleavage. The first cell division is parallel to the animal–vegetal axis as in radial cleavage, but in the second cell division, the two blastomeres divide at right angles to one other. One blastomere divides parallel to the animal–vegetal axis, while the other divides perpendicular to this axis (**Figure 44.4A**). As in other animals that have complete cleavage, the early cell divisions in a mammalian zygote produce a loosely associated ball of cells. After the 8-cell stage, however, the behavior of the mammalian blastomeres changes. They change shape to maximize their surface contact with one another, form tight junctions (see Figure 6.7), and become a compact mass of cells (**Figure 44.4B**).

Soon after the transition to the 32-cell stage, the cells separate into two groups. The **inner cell mass** will develop as the embryo, while the surrounding outer cells become an encompassing sac called the **trophoblast**. Trophoblast cells secrete fluid, creating a cavity—the blastocoel—with the inner cell mass at one end. At this stage the mammalian embryo is called a **blastocyst**, distinguishing it from the blastulas of other animal groups (**Figure 44.4C**). The pluripotent cells of the inner cell mass are known as **embryonic stem cells** and are the subject of much research because of their therapeutic potential (see Section 19.5).

Why is mammalian cleavage so different? A key factor is that mammalian eggs contain little or no yolk and must derive all nutrients from the mother. To support the developing embryo, a connection develops between the circulatory systems of the embryo and the mother. As we will see later in this chapter, the structures that provide this connection are the placenta and the umbilical cord. Thus the blastocyst of placental mammals must produce both the embryo (from the inner cell mass) and its support structures (from the trophoblast).

Fertilization in mammals occurs in the upper reaches of the oviduct, and cleavage occurs as the zygote travels down the oviduct to the uterus (**Figure 44.5**). When the blastocyst arrives in the uterus, the trophoblast adheres to the lining of the uterus (the **endometrium**), beginning the process of **implantation**. In humans, implantation begins about 6 days after fertilization and is aided by adhesion molecules and enzymes secreted by the trophoblast.

As the blastocyst moves down the oviduct to the uterus, it must not embed itself in the oviduct (Fallopian tube) wall, or the result will be an ectopic, or tubal, pregnancy—a very dangerous condition. Early implantation is prevented by the zona pellucida, which surrounded the egg (see Figure 43.5) and remains around the cleaving ball of cells. At about the time the blastocyst reaches the uterus, it hatches from the zona pellucida, and implantation can occur.

Specific blastomeres generate specific tissues and organs

Cleavage results in a repackaging of the egg cytoplasm into a large number of small cells surrounding the fluid-filled blastocoel. Except in mammals, there is little gene expression during cleavage. Nevertheless, cells in different regions of the blastula possess different complements of the nutrients and cytoplasmic determinants that were present in the egg. For example, Figure 44.2 illustrated the processes by which β-catenin becomes localized in the region of the zygote that will become the dorsal side of the embryo.

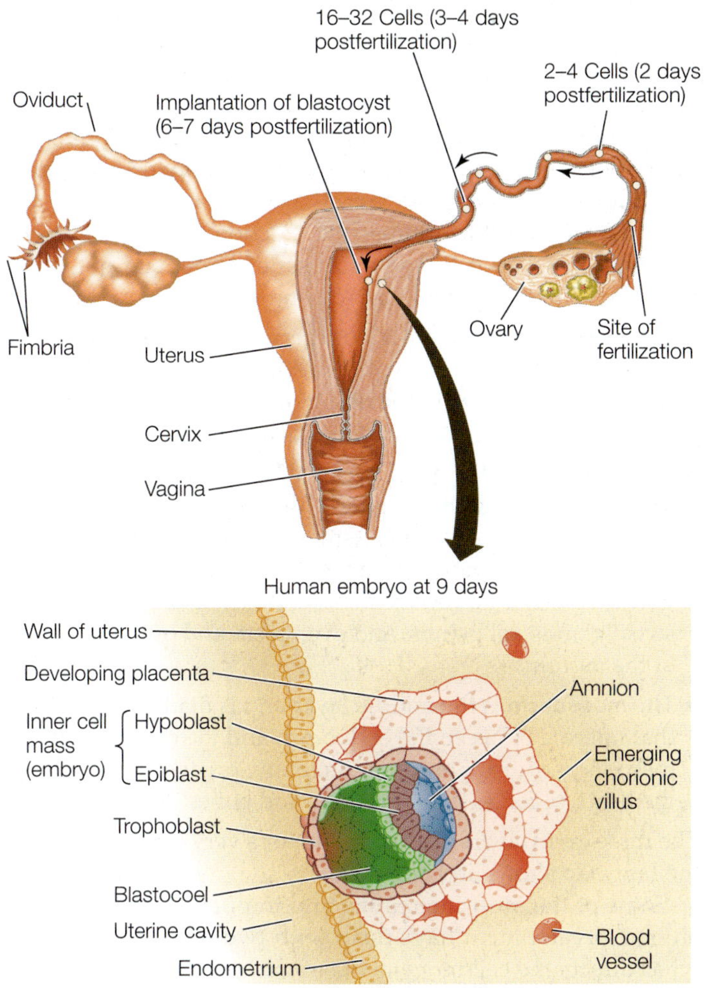
16–32 Cells (3–4 days postfertilization)
2–4 Cells (2 days postfertilization)
Oviduct
Implantation of blastocyst (6–7 days postfertilization)
Fimbria
Uterus
Ovary
Site of fertilization
Cervix
Vagina
Human embryo at 9 days
Wall of uterus
Developing placenta
Amnion
Inner cell mass (embryo)
Hypoblast
Epiblast
Emerging chorionic villus
Trophoblast
Blastocoel
Uterine cavity
Endometrium
Blood vessel

44.5 A Human Blastocyst at Implantation Adhesion molecules and proteolytic enzymes secreted by trophoblast cells allow the blastocyst to burrow into the endometrium. Once the blastocyst is implanted in the wall of the uterus, the trophoblast cells send out numerous projections—the chorionic villi—which increase the embryo's area of contact with the mother's bloodstream.

The blastocoel prevents cells from different regions of the blastula from coming into contact and interacting, but that will soon change. During the next stage of development, the cells of the blastula will move around and come into new associations with one another, communicate instructions to one another, and begin to differentiate. In many animals, these movements of the blastomeres are so regular and well orchestrated that it is possible to label each specific blastomere with a dye, thus producing **fate maps** that identify the tissues and organs formed from each blastomere's progeny (Figure 44.6).

Blastomeres become **determined**—committed to specific fates—at different times in different species. In some species, such as roundworms, the fates of blastomeres are restricted as early as the two-cell stage. If one of these blastomeres is experimentally removed, a particular portion of the embryo will not form. This type of development has been called **mosaic development** because each blastomere seems to contribute a specific set of "tiles" to the final "mosaic" that is the adult animal.

In contrast to mosaic development, the loss of some cells during cleavage in **regulative development** does not affect the developing embryo, because the remaining cells compensate for the loss. Regulative development is typical of many vertebrate species. Because development is regulative in humans, a single blastomere can be removed from early embryos without harming the remaining blastomeres or disrupting normal development. Cells removed from embryos produced by in vitro fertilization can be used for preimplantation genetic diagnosis to ensure that healthy embryos are selected for implantation in the mother.

If some blastomeres can change their fate to compensate for the loss of other cells during cleavage and blastula formation, can those cells form an entire embryo? To a certain extent, yes. During cleavage or early blastula formation in mammals, for example, if the blastomeres are physically separated into two groups, both groups can produce complete embryos. Since the two embryos come from the same zygote, they will be monozygotic twins—genetically identical.

Non-identical twins occur when two separate eggs are fertilized by two separate sperm. Thus, although identical twins are always of the same sex, non-identical twins have a 50 percent chance of being the same sex (that is, the same as two non-twin siblings). In about 1 out of 50,000 human pregnancies, genetic or environmental factors cause the inner cell mass to split partially. The result is twins who are conjoined at some point on their bodies and usually share some of their organs and limbs.

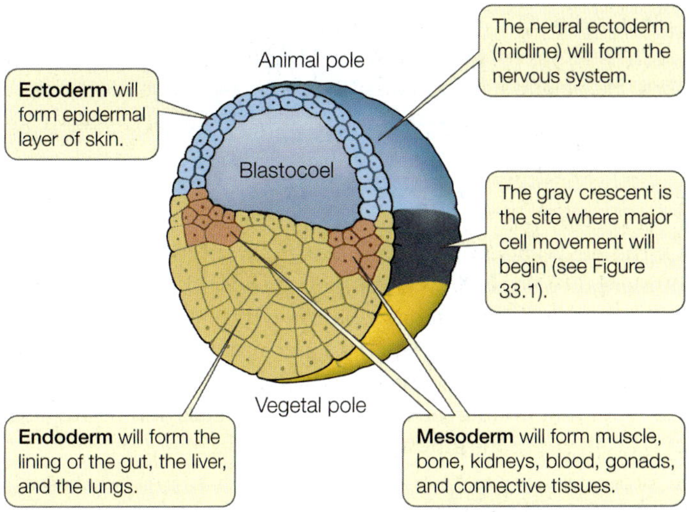
Animal pole
The neural ectoderm (midline) will form the nervous system.
Ectoderm will form epidermal layer of skin.
Blastocoel
The gray crescent is the site where major cell movement will begin (see Figure 33.1).
Vegetal pole
Endoderm will form the lining of the gut, the liver, and the lungs.
Mesoderm will form muscle, bone, kidneys, blood, gonads, and connective tissues.

44.6 Fate Map of a Frog Blastula Colors indicate the portions of the blastula that will form the three germ layers and subsequently the frog's tissues and organs. This cutaway view shows the inside of the blastula. The cells fated to become mesoderm are in the deeper layer of cells, whereas the superficial layers of cells will mostly form ectoderm and endoderm.

Germ cells are a unique lineage even in species with regulative development

Molecules present in the egg cytoplasm determine which lineage of cells will eventually populate the gonads and become the reproductive stem cells—oogonia and spermatogonia. In fruit flies, at the ninth nuclear division (recall that the egg is a multinucleate syncytium at this stage), a group of nuclei migrate to the posterior pole of the egg where they become surrounded by **pole plasm**—cytoplasm containing a complex mixture of fibrils, mitochondria, and specific proteins and mRNAs. As the cellularization of the blastoderm proceeds, the nuclei within the pole plasm give rise to the lineage of cells that will eventually migrate to the gonads (when they form) and produce germ cells (eggs and sperm).

As in fruit flies, the germ cell lineage in frogs starts with a special type of cytoplasm—the germ cell plasm—localized to one part of the egg. As a result of cleavage, the germ cell plasm becomes enclosed within some of the cells in the vegetal hemisphere; descendants of these cells will eventually migrate to the gonads once those structures form. The components of germ cell plasm have not been fully characterized, but one hypothesis is that they include inhibitors of transcription and translation that prevent these cells from differentiating into anything other than germ cells.

RECAP 44.2

Cleavage divides up the cytoplasm of the zygote such that different blastomeres contain different combinations of informational molecules. The amount of nutrients stored in the egg influences the pattern of cell cleavage that produces the blastula. Blastulation in mammals is different than in other vertebrates. The cells that will give rise to germ cell lineages are set aside very early in development.

- In general terms, describe the difference between complete and incomplete cleavage. **See p. 905 and Figure 44.3**
- What does a fate map tell us? How are fate maps constructed? **See p. 907 and Figure 44.6**
- Explain the statement that "the germ cell lineage exists outside of the processes of determination and differentiation." **See p. 908**

Of the next stage of development—gastrulation—the developmental biologist Lewis Wolpert once said, "It is not birth, marriage, or death, but gastrulation which is the most important time in your life." During gastrulation, cell movements create new cell-to-cell contacts, which in turn sets up signaling cascades that initiate the differentiation of cells and tissues and set the stage for the emergence of the body plan.

44.3 How Does Gastrulation Generate Multiple Tissue Layers?

The blastula is typically a fluid-filled ball of cells. How does this simple ball of cells become an embryo made up of multiple tissue layers with head and tail ends and dorsal and ventral sides? **Gastrulation** is the process whereby the blastula is transformed by massive movements of cells into an embryo with multiple tissue layers and distinct body axes. The resulting spatial relationships between tissues make possible the inductive interactions between cells that trigger differentiation and organ formation (see Figure 19.4).

In the triploblastic animals (see Section 33.1), three **germ layers** (also called cell layers or tissue layers, and not to be confused with germ cells) form during gastrulation:

- The **endoderm** is the innermost germ layer, created as some blastomeres move to the inside of the embryo. The endoderm gives rise to the lining of the digestive tract, respiratory tract, pancreas, thyroid, and liver.
- The **ectoderm** is the outer germ layer, formed from those cells remaining on the outside of the embryo. The ectoderm gives rise to the nervous system, including the eyes and ears; and to the epidermal layer of the skin and structures derived from skin, such as hair, feathers, nails or claws, sweat glands, oil glands, and even teeth and other tissues of the mouth.
- The **mesoderm** is the middle layer and is made up of cells that migrate between the endoderm and the ectoderm. The mesoderm contributes tissues to many organs, including the heart, blood vessels, muscles, and bones.

The three germ layers are illustrated for a very early embryo in the fate map shown in Figure 44.6.

Some of the most interesting and important challenges in animal development have dealt with two related questions: what directs the cell movements of gastrulation, and what is responsible for the resulting patterns of cell differentiation and organ formation? Scientists have made significant progress in answering both these questions at the molecular level. In the following discussion we will begin with sea urchin gastrulation because it is the simplest to conceptualize in spatial terms. We will then describe the more complex pattern of gastrulation in frogs, and then to the still more complex patterns in reptiles, birds, and mammals.

Go to Animated Tutorial 44.1
Gastrulation
Life10e.com/at44.1

Invagination of the vegetal pole characterizes gastrulation in the sea urchin

The sea urchin blastula is a hollow ball of cells only one cell layer thick. The end of blastulation is marked by slowing of the rate of mitosis, and the beginning of gastrulation is marked by a flattening of the vegetal hemisphere (**Figure 44.7**). Some cells at the vegetal pole break away from neighboring cells and migrate into the cavity. These cells become **mesenchyme**—cells of the middle germ layer, the mesoderm. Mesenchymal cells are not organized in tightly packed sheets or tubes like epithelial cells are; they act as independent units, migrating into and among the other tissue layers.

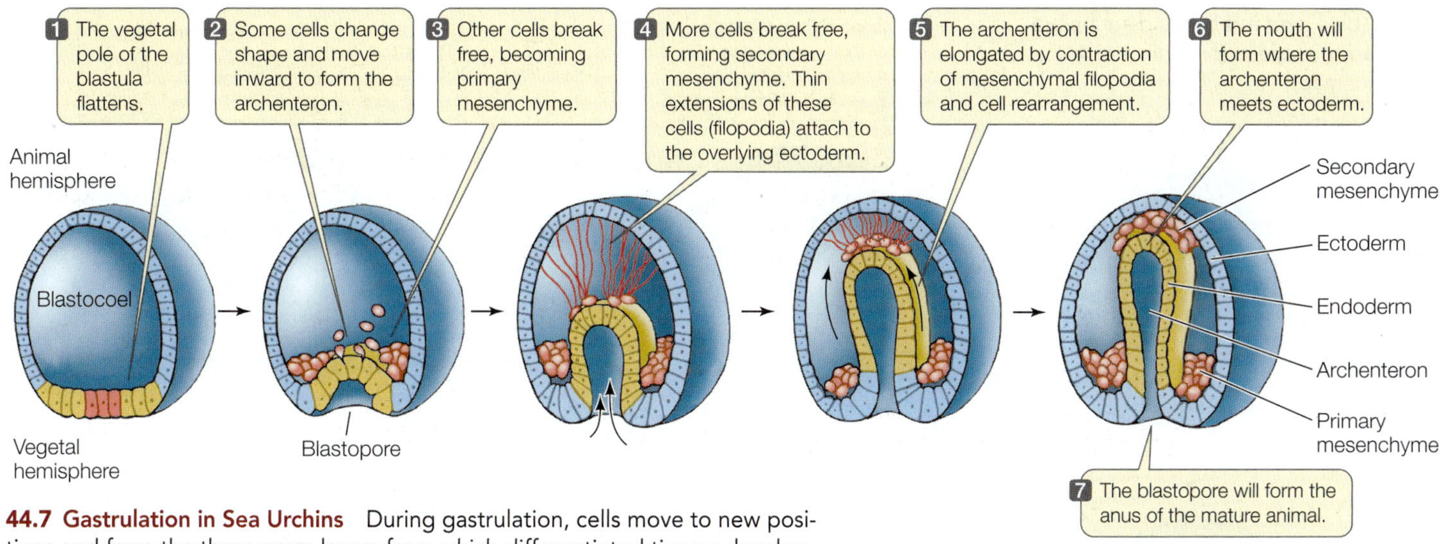
1 The vegetal pole of the blastula flattens.
2 Some cells change shape and move inward to form the archenteron.
3 Other cells break free, becoming primary mesenchyme.
4 More cells break free, forming secondary mesenchyme. Thin extensions of these cells (filopodia) attach to the overlying ectoderm.
5 The archenteron is elongated by contraction of mesenchymal filopodia and cell rearrangement.
6 The mouth will form where the archenteron meets ectoderm.
Animal hemisphere
Blastocoel
Vegetal hemisphere
Blastopore
Secondary mesenchyme
Ectoderm
Endoderm
Archenteron
Primary mesenchyme
7 The blastopore will form the anus of the mature animal.

44.7 Gastrulation in Sea Urchins During gastrulation, cells move to new positions and form the three germ layers from which differentiated tissues develop.

The flattening at the vegetal pole results from changes in the shape of individual blastomeres. These cells, which are originally rather cuboidal, become wedge-shaped, with smaller outer edges and larger inner edges. As a result, the vegetal pole bulges inward, or invaginates, as if someone were poking a finger into a hollow ball (see Figure 44.7). The invaginating cells become endoderm and form the primitive gut, called the **archenteron**. At the tip of the archenteron, more cells enter the blastocoel to form mesoderm.

Changes in cell shapes cause the initial invagination of the archenteron, but eventually the archenteron is pulled inward by the mesenchyme cells. These cells, attached to the tip of the archenteron, send out extensions called filopodia that adhere to the overlying ectoderm. When the filopodia contract, they pull the archenteron toward the ectoderm at the opposite end of the embryo from where the invagination began. The mouth of the animal forms where the archenteron makes contact with this overlying ectoderm. The opening created by the invagination of the vegetal pole is called the **blastopore**, and it will become the anus of the animal.

What mechanisms control the various cell movements of sea urchin gastrulation? The immediate answer is that specific properties of particular blastomeres change. For example, some vegetal cells change shape and bulge into the blastocoel, and these cells become mesenchyme. Once they lose contact with their neighboring cells on the surface of the blastula, they send out filopodia that then move along an extracellular matrix of proteins laid down by the cells lining the blastocoel.

A deeper understanding of gastrulation requires that we discover the molecular mechanisms whereby different blastomeres develop different properties. Cleavage systematically divides up the cytoplasm of the egg. The sea urchin blastula at the 64-cell stage is radially symmetrical, but it has polarity, as described in Section 19.2. It consists of tiers of cells. As in the frog blastula, the top is the animal pole and the bottom the vegetal pole.

If different tiers of blastula cells are separated experimentally, they show different developmental potentials; only cells from the vegetal pole are capable of initiating the development of a complete larva. It has been proposed that these differences are due to uneven distribution of various transcriptional regulatory proteins in the egg cytoplasm. As cleavage progresses, these proteins end up in different groups of cells. Therefore specific sets of genes are activated in different cells, determining their different developmental capacities.

Next we will turn to gastrulation in the frog and to the key signaling molecules involved.

Gastrulation in the frog begins at the gray crescent

Amphibian blastulas have considerable yolk and are more than one cell layer thick; gastrulation is therefore more complex in amphibians than in sea urchins. Variation is considerable across different species of amphibians, so this brief account describes results from studies done on different species to produce a generalized picture of amphibian development.

Amphibian gastrulation begins when certain cells in the gray crescent region (see Figure 44.1) change their shapes and cell-adhesion properties. These cells bulge inward toward the blastocoel while they remain attached to the outer surface of the blastula by slender necks; because of their shape, they are called bottle cells. Bottle cells mark the spot where the **dorsal lip** of the blastopore will form (**Figure 44.8**).

As the bottle cells move inward, the dorsal lip is created, and a sheet of cells moves over it into the blastocoel. This process is called **involution**. One group of involuting cells is the prospective endoderm; these cells form the primitive gut, or archenteron. Another group will move between the endoderm and the outermost cells to form the mesoderm. These rearrangements are due to changes in cell properties called **convergent extension**. The cells elongate in the direction of movement, but they also intercalate (move in between each other). If they just elongated, the migrating group of cells would become much

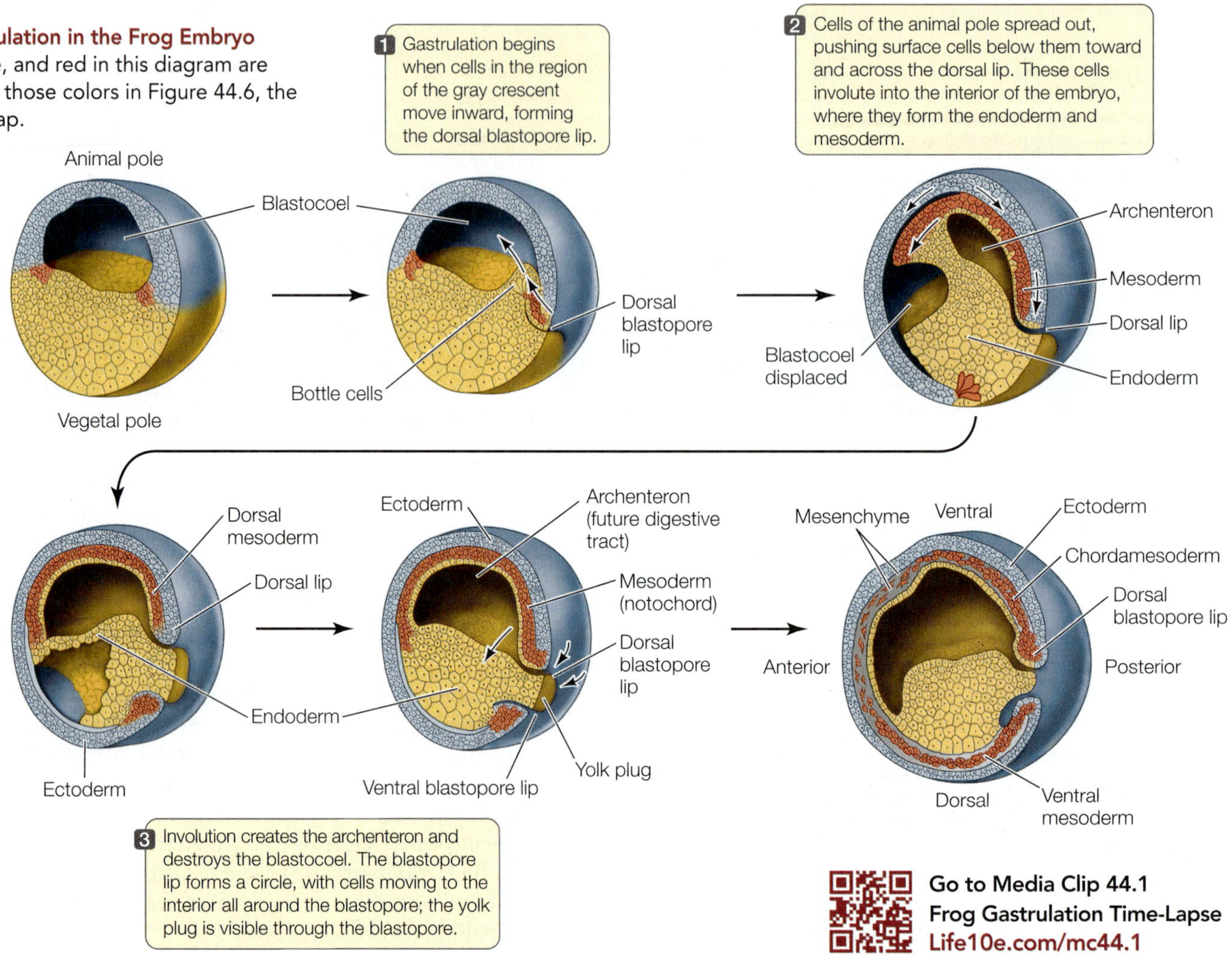
1 Gastrulation begins when cells in the region of the gray crescent move inward, forming the dorsal blastopore lip.
2 Cells of the animal pole spread out, pushing surface cells below them toward and across the dorsal lip. These cells involute into the interior of the embryo, where they form the endoderm and mesoderm.
Animal pole
Blastocoel
Vegetal pole
Bottle cells
Dorsal blastopore lip
Archenteron
Mesoderm
Dorsal lip
Blastocoel displaced
Endoderm
Dorsal mesoderm
Dorsal lip
Endoderm
Ectoderm
Ectoderm
Archenteron (future digestive tract)
Mesoderm (notochord)
Dorsal blastopore lip
Yolk plug
Ventral blastopore lip
Mesenchyme
Ventral
Ectoderm
Chordamesoderm
Dorsal blastopore lip
Anterior
Posterior
Dorsal
Ventral mesoderm
3 Involution creates the archenteron and destroys the blastocoel. The blastopore lip forms a circle, with cells moving to the interior all around the blastopore; the yolk plug is visible through the blastopore.

44.8 Gastrulation in the Frog Embryo
Yellow, blue, and red in this diagram are matched to those colors in Figure 44.6, the frog fate map.

Go to Media Clip 44.1
Frog Gastrulation Time-Lapse
Life10e.com/mc44.1

narrower; by intercalating, they maintain the width of the migrating cell group.

As gastrulation proceeds, cells from the animal hemisphere flatten and move toward the site of involution in a process called **epiboly**. The blastopore lip widens and eventually forms a complete circle surrounding a "plug" of yolk-rich cells. As cells continue to move inward through the blastopore, the archenteron grows, gradually displacing the blastocoel.

As gastrulation comes to an end, the amphibian embryo consists of three germ layers: ectoderm on the outside, endoderm on the inside, and mesoderm in between. The embryo also has a dorsal–ventral and anterior–posterior organization. Most importantly, the fates of specific regions of the endoderm, mesoderm, and ectoderm have been determined. The beautiful experiments revealing how determination takes place in the amphibian embryo are an old but exciting story.

The dorsal lip of the blastopore organizes embryo formation

In the early 1900s the German biologist Hans Spemann was studying the development of salamander eggs. He was interested in finding out whether the nuclei of blastomeres remain capable of directing the development of complete embryos or whether these nuclei lose some developmental potential. With great patience and dexterity, he formed loops from single hairs taken from a baby (in fact, his daughter) and tied them around fertilized eggs along the plane of the first cell division, effectively dividing the eggs in half, with the nucleus restricted to one side. That side went through cell divisions and developed into a salamander; the other half simply degenerated. Up until the 16-cell stage, if one nucleus escaped to the other side of the constriction, twin salamanders could develop. Thus each of the nuclei of the blastula (at least up to the 16-cell stage) was capable of directing and supporting development of the whole organism.

As often happens in science, Spemann's bisection experiments revealed a new phenomenon. Sometimes the half of the blastula receiving an escaped nucleus did not develop. When his loops bisected the gray crescent, both halves of the zygote developed into a complete embryo. When he tied the loops so the gray crescent was on only one side of the constriction, however, only that half of the zygote developed into a complete embryo (Figure 44.9). The half lacking gray crescent material underwent cell division, but even if it contained a nucleus, it became a clump of undifferentiated cells that Spemann called a "belly piece." Spemann hypothesized that cytoplasmic factors unequally distributed in the fertilized egg were necessary for gastrulation and the development of a normal salamander.

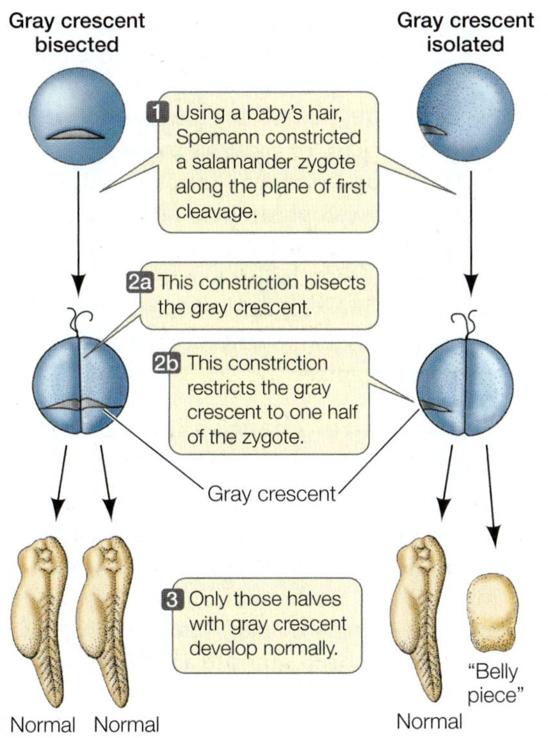
Gray crescent bisected
Gray crescent isolated
1 Using a baby's hair, Spemann constricted a salamander zygote along the plane of first cleavage.
2a This constriction bisects the gray crescent.
2b This constriction restricts the gray crescent to one half of the zygote.
Gray crescent
3 Only those halves with gray crescent develop normally.
Normal
Normal
Normal
"Belly piece"

44.9 Gastrulation and the Gray Crescent Spemann's research revealed that gastrulation and subsequent normal development in salamanders depend on cytoplasmic determinants localized in the gray crescent.

To further test the hypothesis that cells receiving different complements of cytoplasmic factors had different developmental fates, Spemann transplanted pieces of early gastrulas to various locations on other gastrulas. Guided by fate maps (see Figure 44.6), he was able to take a piece of ectoderm he knew would develop into the epidermis of the skin and transplant it to a region that normally becomes part of the nervous system, and vice versa.

When he performed these transplants in early gastrulas—when the blastopore was just beginning to form—the transplanted pieces always developed into tissues that were appropriate for the location where they were placed. Transplanted cells destined to become epidermis in their original location developed into nervous system tissue, and transplanted cells destined to become nervous system tissue in their original location developed into host epidermis. Thus Spemann learned that the fates of the transplanted cells had not been determined before the transplantation.

In late gastrulas, however, the same experiment yielded opposite results. Transplanted cells destined to become epidermis in their original location produced patches of skin cells in the host nervous system, and the transplanted cells from regions that would develop into nervous system tissue produced neural tissue in the skin of the recipient. At some point during gastrulation, the fates of the embryonic cells had become determined.

Spemann's next experiment, done with his student Hilde Mangold, produced momentous results: they transplanted the dorsal lip of the blastopore (Figure 44.10). When this small piece of tissue was transplanted into the presumptive belly area of another gastrula, it stimulated a second site of gastrulation—and a second complete embryo formed belly-to-belly with the original embryo. Because the dorsal lip of the blastopore was apparently capable of inducing the host tissue to form an entire embryo, Spemann and Mangold dubbed the dorsal lip tissue the **primary embryonic organizer**, or simply the **organizer**. For more than 90 years, the organizer has been an active area of research.

Transcription factors and growth factors underlie the organizer's actions

With the advent of modern molecular methods, the primary embryonic organizer has been studied intensively to discover the molecular mechanisms involved in its action. The distribution of the transcription factor β-catenin in the late blastula corresponds to the location of the organizer in the early gastrula, so β-catenin is a candidate for the initiator of organizer activity. To prove that a protein is an inductive signal, it has to be shown that it is both *necessary* and *sufficient* for the proposed effect. In other words, the effect should not occur if the candidate protein is not present (necessity), and the candidate protein should be capable of inducing the effect where it would otherwise not occur (sufficiency).

The criteria of necessity and sufficiency have been satisfied for β-catenin. If β-catenin mRNA transcripts are depleted by injections of antisense RNA into the egg (see Section 18.4), gastrulation does not occur. If β-catenin is experimentally overexpressed in another region of the blastula, it can induce a second axis of embryo formation, as the transplanted dorsal lip did in the Spemann–Mangold experiments. Thus β-catenin appears to be both necessary and sufficient for the formation of the primary embryonic organizer—but it is only one component of a complex signaling process.

How the presence of β-catenin creates the organizer, and how the organizer then induces the beginnings of the body plan, involves a complex series of interactions between transcription factors and growth factors that control gene expression. What follows is only a portion of this complex and still emerging story. What you should take from this description is not the names of the genes and gene products involved. Rather, we hope you will gain a basic appreciation for how signaling molecules interact to produce different combinations of signals that convey spatial and temporal information. This information guides cells into different paths of determination and differentiation.

Studies of early gastrulas revealed that primary embryonic organizer activity is generated by the interaction of β-catenin with signals coming from the vegetal cells. Together they activate the expression of the transcription factor Goosecoid. Expression of the *goosecoid* gene depends on two signaling pathways.

The first of these pathways involves a *goosecoid*-promoting transcription factor called Siamois. The *siamois* gene is normally

INVESTIGATING**LIFE**

44.10 The Dorsal Lip Induces Embryonic Organization In a classic experiment, Hans Spemann and Hilde Mangold transplanted the dorsal blastopore lip mesoderm of an early gastrula stage salamander embryo.[a] The results showed that the cells of this embryonic region, which they dubbed "the organizer," could direct the formation of an entire embryo.

HYPOTHESIS The early dorsal blastopore lip organizes cell differentiation in amphibian embryos.

Method

1. Excise a patch of mesoderm tissue from above the dorsal blastopore lip of an early gastrula stage salamander embryo (the donor).
2. Transplant the donor tissue onto a recipient embryo at the same stage. The donor tissue is transplanted onto a region of ectoderm that should become epidermis (skin).

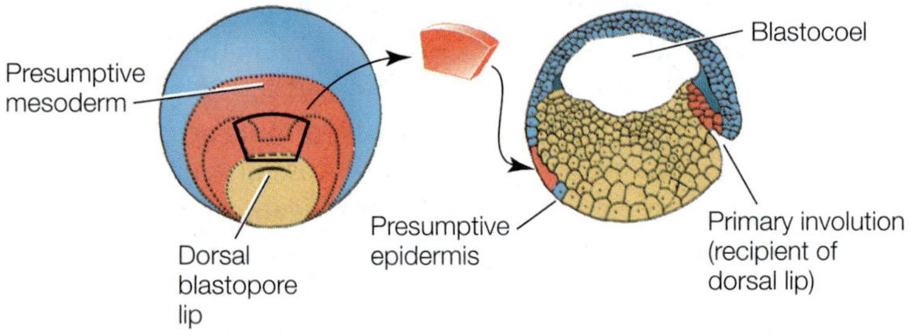
Presumptive
mesoderm
Dorsal
blastopore
lip
Presumptive
epidermis
Blastocoel
Primary involution
(recipient of
dorsal lip)

Results

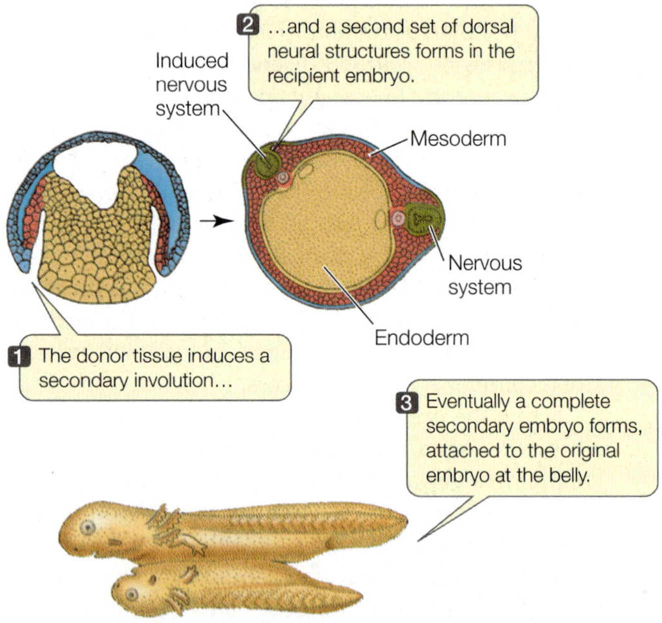
Induced
nervous
system
2 ...and a second set of dorsal neural structures forms in the recipient embryo.
Mesoderm
Nervous
system
Endoderm
1 The donor tissue induces a secondary involution...
3 Eventually a complete secondary embryo forms, attached to the original embryo at the belly.

CONCLUSION The cells of the dorsal blastopore lip can induce other cells to change their developmental fates.

Go to **BioPortal** for discussion and relevant links for all INVESTIGATING**LIFE** figures.

[a]Spemann, H. and H. Mangold. 1924. *Roux' Arch. Entw. Mech.* 100: 599–638. Viktor Hamburger's translation appeared in *Foundations of Experimental Embryology, 1964*, (B.H. Willier and J.M. Oppenheimer, eds.), pp. 146–184.

Go to Animated Tutorial 44.2
Tissue Transplants Reveal the Process of Determination
Life10e.com/at42.2

repressed by a ubiquitous transcription factor called Tcf-3, but in cells in which β-catenin is present, an interaction between Tcf-3 and β-catenin induces *siamois* expression (**Figure 44.11**). But Siamois protein alone is not sufficient for *goosecoid* expression.

The second pathway involves mRNAs from the original egg cytoplasm for a family of proteins called transforming growth factor-β (TGF-β). TGF-β interacts with the Siamois protein to turn on *goosecoid* transcription. Thus you can see it is a complex combination of factors that determines which cells become the primary organizer.

The organizer changes its activity as it migrates from the dorsal lip

Organizer cells begin the process of formation of the dorsal lip of the blastopore. Specifically, these cells are at the center of the dorsal lip and involute, moving forward on the midline (i.e., the middle of the anterior–posterior axis). The first organizer cells to enter the embryo migrate anteriorly to become the head endoderm and head mesoderm. Here they induce neighboring cells to participate in making structures of the head. Organizer cells that involute into the later embryo will induce structures of the trunk, and the last of the organizer cells to move inward from the dorsal lip will induce structures of the tail. How does the nature of the organizer cells change to enable them to induce head, trunk, or tail structures?

Inductive tissue interactions can suppress as well as activate. As we learned above, the early organizer cells express the transcription factor Goosecoid, which activates genes encoding soluble signals. As these cells move forward in the blastocoel, they come into contact with new populations of cells that produce several different growth factors. For head structures to form, certain of these growth factors have to be suppressed. The most anterior organizer cells, under the influence of Goosecoid, produce and release antagonists to those growth factors.

The induction of trunk structures requires suppression of a different set of growth factors. In organizer cells that involute later than the head organizers, Goosecoid is no longer the dominant transcription factor, and these cells express different growth factor antagonists. The induction of tail structures requires still different activities of the organizer cells that involute last. Thus the organizer cells express appropriate sets of growth factor antagonists at the right times to achieve different patterns of differentiation on the anterior–posterior axis.

The initiation of nervous system development also involves a suppressive tissue interaction. For a long time it was thought that the involuting organizer cells actively induced the overlying ectoderm to form neural tissue rather than becoming epidermis. We now

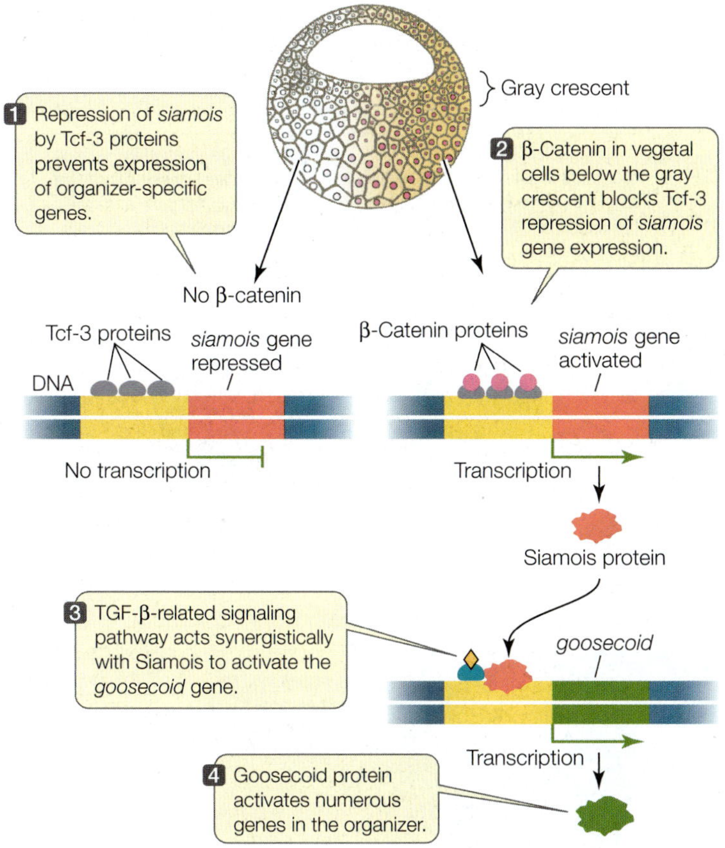
Gray crescent
1 Repression of *siamois* by Tcf-3 proteins prevents expression of organizer-specific genes.
2 β-Catenin in vegetal cells below the gray crescent blocks Tcf-3 repression of *siamois* gene expression.
No β-catenin
Tcf-3 proteins
siamois gene repressed
DNA
No transcription
β-Catenin proteins
siamois gene activated
Transcription
Siamois protein
3 TGF-β-related signaling pathway acts synergistically with Siamois to activate the *goosecoid* gene.
goosecoid
Transcription
4 Goosecoid protein activates numerous genes in the organizer.

44.11 Molecular Mechanisms of the Organizer In amphibians, the organizing potential of the gray crescent depends on the activity of the *goosecoid* gene, which in turn is activated by signaling pathways set up in the vegetal cells below the gray crescent.

know, however, that epidermis is not the default state of the dorsal ectoderm. Rather, the underlying mesoderm secretes factors called BMP proteins that induce the ectoderm to become epidermis. The role of the involuting organizer cells is to block that induction, allowing the overlying ectodermal cells to follow what is really their default pathway—differentiation into neural tissue (**Figure 44.12**).

Reptilian and avian gastrulation is an adaptation to yolky eggs

The eggs of reptiles and birds contain a mass of yolk, and the blastulas of these groups develop as a disc of cells on top of the yolk (see Figure 44.3B). We will use the chicken egg to show how gastrulation proceeds in a flat disc of cells rather than in a ball of cells.

Cleavage in the chick results in a flat, circular layer of cells called a blastodisc (**Figure 44.13**). Between the blastodisc and the yolk mass is a fluid-filled space. Some cells from the blastodisc break free and move into this space. These cells come together to form a continuous layer called the **hypoblast**, which will later contribute to extraembryonic membranes that will support and nourish the developing embryo. The overlying cells make up the **epiblast**, from which the embryo will form.

INVESTIGATING**LIFE**

44.12 Differentiation Can Be Due to Inhibition of Growth Factors When organizer cells involute to underlie dorsal ectoderm along the embryo midline, that overlying ectoderm becomes neural tissue rather than skin (epidermis). But do the organizer cells cause dorsal ectoderm to become neural tissue, or do they prevent this ectoderm from becoming skin?[a]

HYPOTHESIS The default state of amphibian dorsal ectoderm is neural; it is induced by underlying mesoderm to become epidermis.

Method

1. Excise the animal caps of late-stage frog blastulas and disperse the cells in culture medium so there is no cell-to-cell contact.

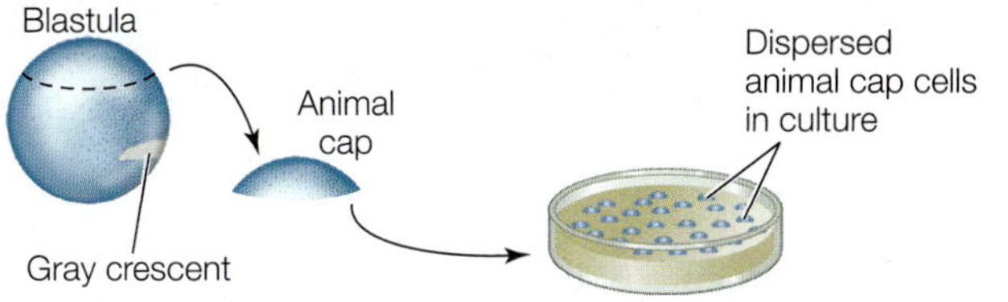
Blastula
Animal cap
Dispersed animal cap cells in culture
Gray crescent

2. Prepare four separate cultures of embryonic ectodermal cells. Incubate with no additions (control); with BMP4 (isolated from mesodermal cells) with BMP4 inhibitor (isolated from organizer cells) and with both molecules.

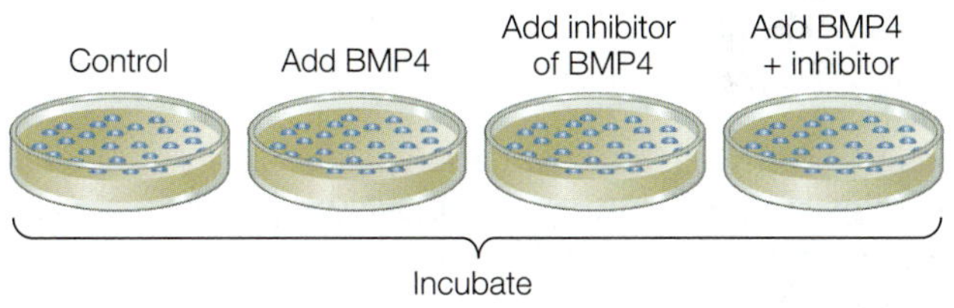
Control
Add BMP4
Add inhibitor of BMP4
Add BMP4 + inhibitor
Incubate

3. After incubation, extract mRNAs from the ectodermal cells and analyze for the presence of mRNAs for marker proteins NCAM (neural cell adhesion molecule, a neural protein) and/or keratin (an epidermal protein).

Results The control ectoderm (no inductive factors added) expresses the neural marker. In the presence of mesodermal BMP4, ectoderm expresses the epidermal marker. If BMP4 is inhibited, ectoderm expresses the neural marker.

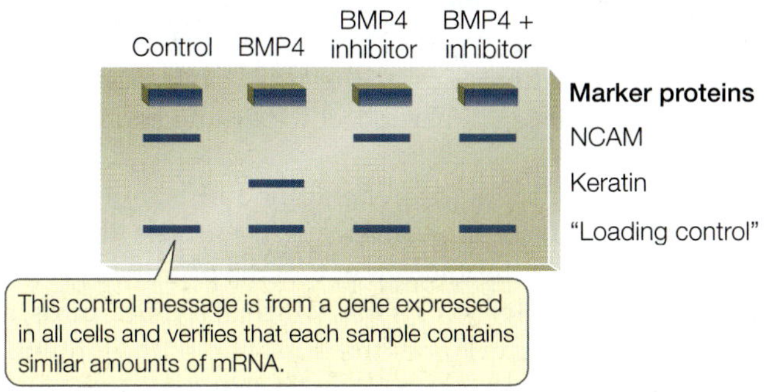
Control
BMP4
BMP4 inhibitor
BMP4 + inhibitor
Marker proteins
NCAM
Keratin
"Loading control"
This control message is from a gene expressed in all cells and verifies that each sample contains similar amounts of mRNA.

CONCLUSION The default state of amphibian dorsal ectoderm is neural. MP4 protein from mesoderm can induce ectoderm cells to differentiate into epidermis. Thus the organizer cells must secrete an inhibitor of BMP4.

Go to **BioPortal** for discussion and relevant links for all INVESTIGATING**LIFE** figures.

[a]Wilson, P. A. and A. Hemmati-Brivanlou. 1995. *Nature* 376: 331–333.

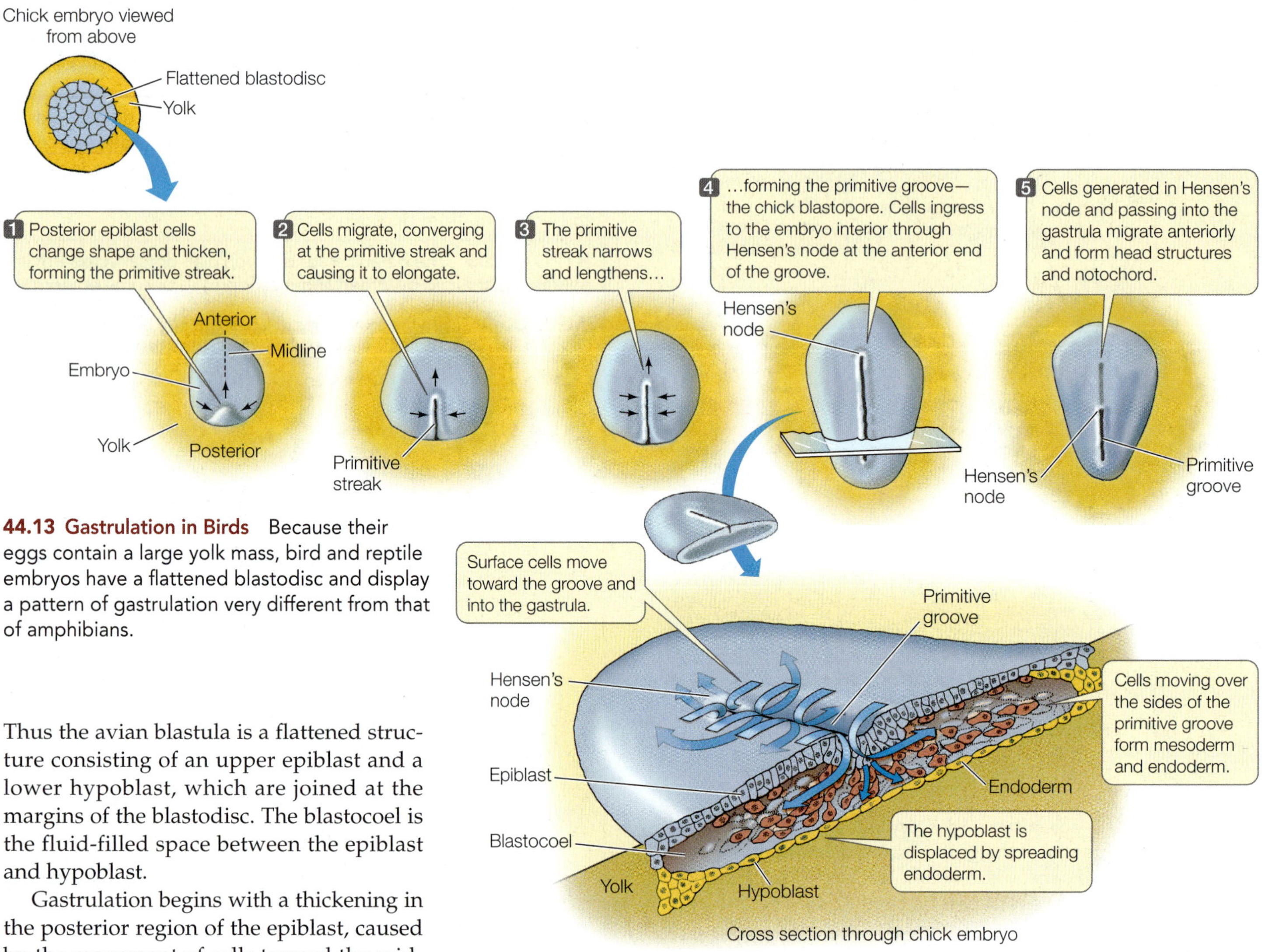

Chick embryo viewed from above
Flattened blastodisc
Yolk
1 Posterior epiblast cells change shape and thicken, forming the primitive streak.
2 Cells migrate, converging at the primitive streak and causing it to elongate.
3 The primitive streak narrows and lengthens...
4 ...forming the primitive groove—the chick blastopore. Cells ingress to the embryo interior through Hensen's node at the anterior end of the groove.
5 Cells generated in Hensen's node and passing into the gastrula migrate anteriorly and form head structures and notochord.
Anterior
Midline
Embryo
Yolk
Posterior
Primitive streak
Hensen's node
Hensen's node
Primitive groove
Surface cells move toward the groove and into the gastrula.
Primitive groove
Hensen's node
Epiblast
Blastocoel
Yolk
Hypoblast
Endoderm
Cells moving over the sides of the primitive groove form mesoderm and endoderm.
The hypoblast is displaced by spreading endoderm.
Cross section through chick embryo

44.13 Gastrulation in Birds Because their eggs contain a large yolk mass, bird and reptile embryos have a flattened blastodisc and display a pattern of gastrulation very different from that of amphibians.

Thus the avian blastula is a flattened structure consisting of an upper epiblast and a lower hypoblast, which are joined at the margins of the blastodisc. The blastocoel is the fluid-filled space between the epiblast and hypoblast.

Gastrulation begins with a thickening in the posterior region of the epiblast, caused by the movement of cells toward the midline and then forward along the midline (see Figure 44.13). The result is a midline ridge called the **primitive streak**. A depression called the primitive groove forms along the length of the primitive streak. The primitive groove functions as the blastopore, and cells migrate through it into the blastocoel to become endoderm and mesoderm.

In the chick embryo, no archenteron forms, but the endoderm and mesoderm migrate forward to form the gut and other structures. At the anterior end of the primitive groove is a thickening called **Hensen's node**, which in reptiles, birds, and mammals is the equivalent of the dorsal lip of the amphibian blastopore. Many signaling molecules that have been identified in the frog organizer are also expressed in Hensen's node. Cells moving into the blastocoel and moving anteriorly from Hensen's node become the notochord and organize the chick embryo in a manner similar to that of the frog embryo.

The embryos of placental mammals lack yolk

Mammalian embryos (with the exception of monotremes) derive their nourishment from the maternal circulation, and therefore mammalian eggs do not have large amounts of yolk constraining their cleavage and early development. Nevertheless, mammals evolved from reptilian ancestors, so it is not surprising that mammals, birds, and reptiles share certain patterns of early development. Earlier we described the development of the mammalian inner cell mass (the equivalent of the avian blastodisc) and the outer trophoblast.

As in avian development, in placental mammals the inner cell mass splits into an upper layer called the epiblast and a lower layer called the hypoblast. The embryo forms from the epiblast, while the hypoblast contributes to the extraembryonic membranes that will encase the developing embryo and help form the placenta (see Figure 44.5). The epiblast also contributes to the extraembryonic membranes; specifically, it splits off an upper layer of cells that will form the amnion. The amnion will grow to surround the developing embryo as a membranous sac filled with amniotic fluid. Gastrulation occurs in the mammalian epiblast just as it does in the avian epiblast. A primitive groove forms, and epiblast cells migrate through the groove to become layers of endoderm and mesoderm. At the top of the groove is the node, which, as we learned at the start of this chapter, is where the beating of nodal cell cilia creates an asymmetrical flow of extracellular fluid. The asymmetrical

WORKING WITH**DATA**:

Nodal Flow and Inverted Organs

Original Paper

Nonaka, S., H. Shiratori, and H. Hamata. 2002. Determination of left–right patterning of the mouse embryo by artificial nodal flow. *Nature* 418: 96–99.

Analyze the Data

The phenotype of the mutant mouse strain *inversus viscerum* (*iv*/*iv*) mimics human situs inversus. These mice have nonmotile primary cilia in the ventral (bellyward) node of the early embryo. Left alone, roughly half of all *iv*/*iv* mouse embryos will develop with normal left–right organ asymmetry while the other half show reversed asymmetry. Researchers in Japan used *iv*/*iv* embryos to test the hypothesis that the leftward flow of extracellular fluid created by the beating of nodal primary cilia is the stimulus for breaking bilateral symmetry in organ development.

The researchers anchored very early mouse embryos with their nodal ends pointed upward in a chamber filled with culture medium. Culture medium was then artificially pumped through the chamber either from the left (normal) or from the right (reversed) and at slow or fast flow rates. After 4 days the embryos were assessed as to the direction of looping (normal or reversed) of their developing heart tubes. The researchers compared the effects of speed and direction culture medium flow on wild-type and mutant embryos that were at the presomite stage when they were placed in the chamber.

A second set of experiments explored the the effects of fast, reversed flow on wild-type embryos that were at later stages of development (1, 2, or 3 somites) when placed in the chamber. The results are shown in the table below.

QUESTION 1

Do the data support the hypothesis that nodal flow is a stimulus that determines left–right organ asymmetry? Why or why not?

QUESTION 2

How would you explain the different results for the slow rightward flow in the presomite wild-type and *iv*/*iv* mice?

QUESTION 3

What do you conclude from the results on the 1-, 2-, and 3-somite wild-type embryos?

	Speed and direction of culture medium flow			
Genotype, stage	Fast left	Slow left	Slow right	Fast right
iv/*iv*, presomite	10N, 0R, 10T[a]	8N, 2R, 10T	3N, 25R, 34T	1N, 11R, 12T
Wild-type, presomite	9N, 0R, 10T	12N, 4R, 16T	13N, 3R, 16T	2N, 21R, 24T
Wild-type, 1-somite	—	—	—	9N, 0R, 12T
Wild-type, 2-somite	—	—	—	22N, 0R, 22T
Wild-type, 3-somite	—	—	—	14N, 0R, 14T

[a]The normal direction for fluid flow is left. Developmental outcomes indicated as follows: N, normal; R, reversed; T, total embryos in sample. Note that some embryos were neither normal nor reversed but were included in the total.

Go to **BioPortal** for all WORKING WITH**DATA** exercises

flow stimulates nonmotile cilia, generating signaling cascades that determine the left–right asymmetry of the internal organs.

RECAP 44.3

The cell movements of gastrulation convert the blastula into an embryo with three tissue layers. New contacts between cells set up inductive signaling interactions that determine cell fates. Dorsal lip tissue is the source of organizer cells that induce development of preliminary head, trunk, and tail structures.

- Describe and compare the cell movements that occur during gastrulation in a sea urchin, a frog, and a bird. **See Figures 44.7, 44.8, and 44.13**
- Explain the molecular basis for the inductive capabilities of the organizer. **See pp. 912–913 and Figures 44.11 and 44.12**

We have described how the fertilized egg develops into an embryo with three germ layers and how cellular signals trigger different patterns of differentiation. In the next section we will describe how organs and organ systems develop.

44.4 How Do Organs and Organ Systems Develop?

Gastrulation produces an embryo with three germ layers that are positioned to influence one another through inductive tissue interactions. During the next phase of development, called **organogenesis**, organs and organ systems develop simultaneously and in coordination with each other. In the chordates (see Section 33.1), an early process of organogenesis is **neurulation**, the initiation of the nervous system. We will examine neurulation in the amphibian embryo, but it occurs in a similar fashion in reptiles, birds, and mammals.

The stage is set by the dorsal lip of the blastopore

As we learned in the previous section, one group of cells that passes over the dorsal lip of the blastopore moves anteriorly and becomes the endodermal lining of the digestive tract. Another group of cells that involutes over the dorsal lip on the midline becomes chordamesoderm, so named because it forms a rod of mesoderm—the **notochord**—that extends down the

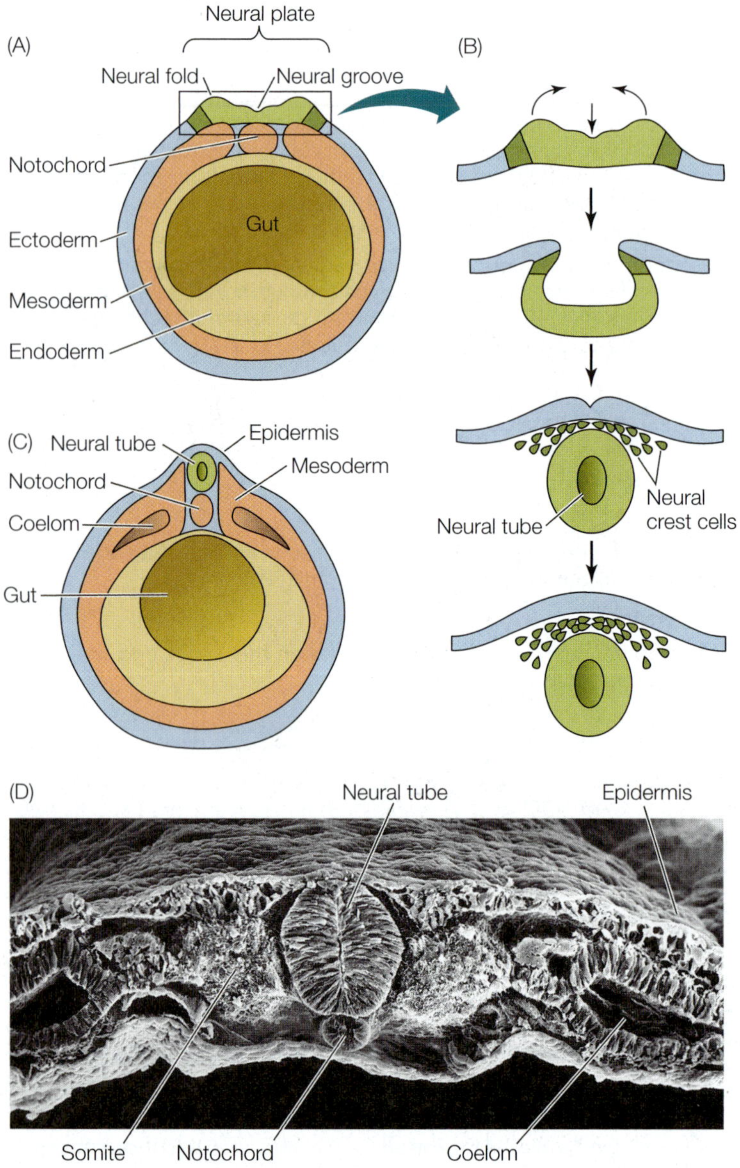

(A)
Neural plate
Neural fold
Neural groove
Notochord
Ectoderm
Mesoderm
Endoderm
Gut
(B)
Neural tube
Neural crest cells
(C)
Neural tube
Epidermis
Mesoderm
Notochord
Coelom
Gut
(D)
Neural tube
Epidermis
Somite
Notochord
Coelom

44.14 Neurulation in a Vertebrate (A) At the start of neurulation, the ectoderm of the neural plate (green) is flat. (B) The neural plate invaginates and folds, forming a tube. (C,D) The completely formed neural tube seen (C) in diagrammatic form and (D) in a scanning electron micrograph of a chick embryo.

center of the embryo. These cells also have important organizer functions (see Figure 44.8). The notochord gives structural support to the developing embryo and in vertebrates is replaced by the vertebral column. The organizing capacity of the chordamesoderm enables the overlying ectoderm to become neural ectoderm (see Figure 44.12). It does this by expressing signaling molecules (one appropriately called Noggin and another one called Chordin) that initiate differentiation of the different divisions of the nervous system.

Neurulation involves the formation of an internal neural tube from an external sheet of cells. The first signs of neurulation are flattening and thickening of the ectoderm overlying the notochord; this thickened area forms the neural plate (**Figure 44.14A**). The edges of the neural plate that run in an anterior–posterior direction continue to thicken to form ridges or folds. Between these neural folds, a groove forms and deepens as the folds roll over it to converge on the midline. The folds fuse, forming a cylinder, the **neural tube**, and a continuous overlying layer of epidermal ectoderm (**Figure 44.14B–D**).

Cells from the most lateral portions of the neural plate do not become part of the neural tube, but disassociate from it and come to lie between the neural tube and the overlying epidermis. These **neural crest cells** migrate outward to lead the development of the connections between the central nervous system (brain and spinal cord) and the rest of the body.

The neural tube develops bulges at the anterior end, which become the major divisions of the brain; the rest of the tube becomes the spinal cord. In humans, failure of the neural folds to fuse in this posterior region results in spina bifida, a birth defect in which the spinal cord is exposed because the vertebrae do not fuse. If the folds fail to fuse at the anterior end, an infant can develop without a forebrain (a condition called anencephaly). Although several genetic factors can cause these defects, other factors are environmental, including maternal diet. The incidence of neural tube defects in the United States in the early 1900s was as high as 1 in 300 live births; today it is less than 1 in 1,000. A major factor in this improvement has been the inclusion of folic acid (a B vitamin, also known as folate) in the mother's diet. It is essential for pregnant women to ingest sufficient folic acid.

Body segmentation develops during neurulation

The vertebrate body plan, like that of arthropods, consists of repeating segments that are modified during development. These segments are most evident as the repeating patterns of vertebrae, ribs, nerves, and muscles along the anterior–posterior axis.

As the neural tube forms, mesodermal tissues gather along the sides of the notochord to form separate, segmented blocks of cells called **somites** (**Figure 44.15**). Somites produce cells that will become the vertebrae, ribs, muscles of the trunk, and limbs.

Nerves that connect the brain and spinal cord with tissues and organs throughout the body are also arranged segmentally. The somites help guide the organization of these peripheral nerves, but the nerves are not of mesodermal origin. As we saw above, when the neural tube fuses, the neural crest cells break loose and migrate inward between the epidermis and the somites and through the somites. These neural crest cells have diverse fates, including the development of peripheral nerves.

As development progresses, the different segments of the body change. Regions of the spinal cord differ, regions of the vertebral column differ in that some vertebrae grow ribs of various sizes and others do not, forelegs arise in the anterior part of the embryo, and hind legs arise in the posterior region.

Hox genes control development along the anterior–posterior axis

How is mesoderm in the anterior part of a mouse embryo programmed to produce forelegs rather than hind legs? In Section 19.4, we saw how homeotic genes control body segmentation in *Drosophila*. We also learned that all homeotic genes contain a

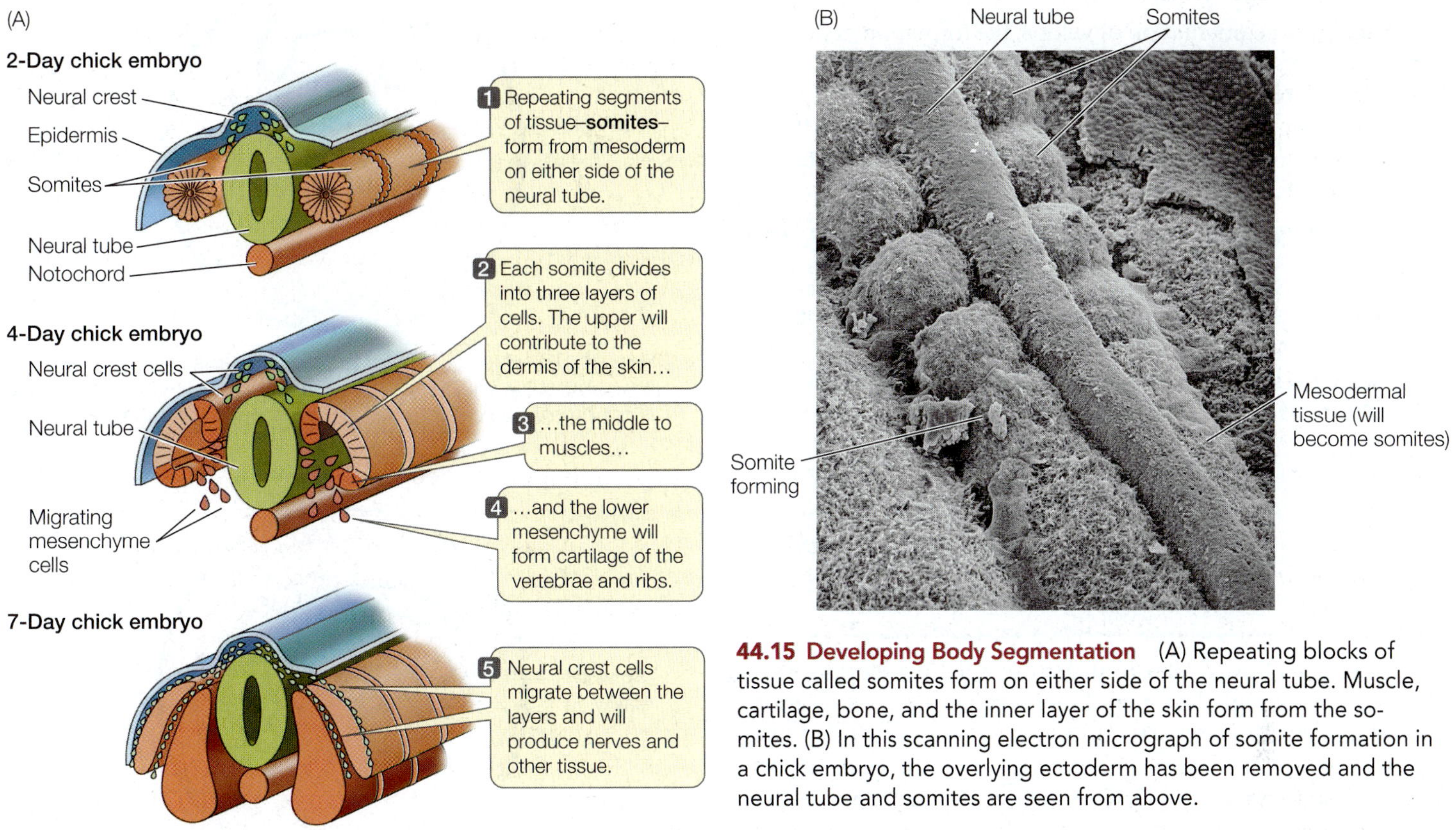

(A)
2-Day chick embryo
Neural crest
Epidermis
Somites
Neural tube
Notochord
1 Repeating segments of tissue–somites–form from mesoderm on either side of the neural tube.
4-Day chick embryo
Neural crest cells
Neural tube
Migrating mesenchyme cells
2 Each somite divides into three layers of cells. The upper will contribute to the dermis of the skin…
3 …the middle to muscles…
4 …and the lower mesenchyme will form cartilage of the vertebrae and ribs.
7-Day chick embryo
5 Neural crest cells migrate between the layers and will produce nerves and other tissue.
(B)
Neural tube
Somites
Mesodermal tissue (will become somites)
Somite forming

44.15 Developing Body Segmentation (A) Repeating blocks of tissue called somites form on either side of the neural tube. Muscle, cartilage, bone, and the inner layer of the skin form from the somites. (B) In this scanning electron micrograph of somite formation in a chick embryo, the overlying ectoderm has been removed and the neural tube and somites are seen from above.

DNA sequence called the **homeobox**. Some of the genes directing gastrulation in the frog are homeobox genes—for example, *goosecoid* and *siamois*. In vertebrates, the homeotic genes that control differentiation along the anterior–posterior body axis are called **Hox genes**.

In mammals, four Hox gene complexes reside on different chromosomes in clusters of about 10 genes each. Remarkably, the temporal and spatial expression of these genes follows the same pattern as their linear order on their chromosome. That is, the Hox genes closest to the 3′ end of each gene complex are expressed first and in the anterior of the embryo. The Hox genes at the 5′ end of the gene complex are expressed later and in a more posterior part of the embryo. As a result, different segments of the embryo receive different combinations of Hox gene products, which serve as transcription factors (**Figure 44.16**; see also Figure 20.2).

Whereas Hox genes give cells information about their position on the anterior–posterior body axis, other genes provide information about their dorsal–ventral position. Tissues in each segment of the body differentiate according to their dorsal–ventral location. The notochord provides many of these signals. One example of a dorsal–ventral difference is seen in the spinal cord; sensory nerve connections develop in the dorsal region, and motor nerve connections in the ventral region. The protein Sonic hedgehog (named for the video-game character), which is expressed in the mammalian notochord, induces cells in the overlying neural tube (i.e., the ventralmost cells of the tube) to become motor neurons.

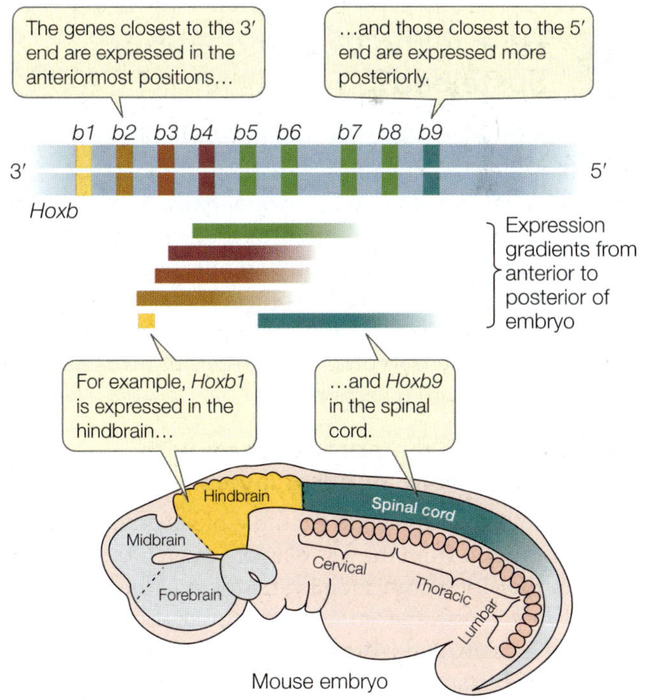

The genes closest to the 3′ end are expressed in the anteriormost positions…
…and those closest to the 5′ end are expressed more posteriorly.
b1 b2 b3 b4 b5 b6 b7 b8 b9
3′
5′
Hoxb
Expression gradients from anterior to posterior of embryo
For example, Hoxb1 is expressed in the hindbrain…
…and Hoxb9 in the spinal cord.
Hindbrain
Spinal cord
Midbrain
Forebrain
Cervical
Thoracic
Lumbar
Mouse embryo

44.16 Hox Genes Control Body Segmentation Hox genes are expressed along the anterior–posterior axis of the embryo in the same order as their arrangement between the 3′ and 5′ ends of the gene complex. As a result of gene duplication during evolution, vertebrates have four copies of the Hox gene complex shown.

After body segmentation develops, the formation of organs and organ systems progresses rapidly. The development of an organ involves extensive inductive interactions of the kind we saw in the example of the vertebrate eye (see Figure 19.4). These inductive interactions are a current focus of study for developmental biologists.

RECAP 44.4

Gastrulation sets up tissue interactions that initiate organogenesis. Neurulation is initiated by organizer mesoderm that forms the notochord.

- Describe the formation of the neural tube in vertebrates. See p. 916 and Figure 44.14
- How do somites relate to segmentation of the body axis? See p. 916 and Figure 44.15
- Explain what Hox genes are and how they instruct patterns of differentiation along the body axis. (You may want to refer back to Section 19.4 for more information on Hox genes.) See pp. 916–917 and Figure 44.16

We have seen how the basic structure of the developing embryo arises, through the establishment of the anterior–posterior and dorsal–ventral axes, the formation of the neural tube, and the emergence of a segmented body plan. In the next section we will examine the developmental events that result in the formation of structures that support the developing embryo: the extraembryonic membranes and the placenta.

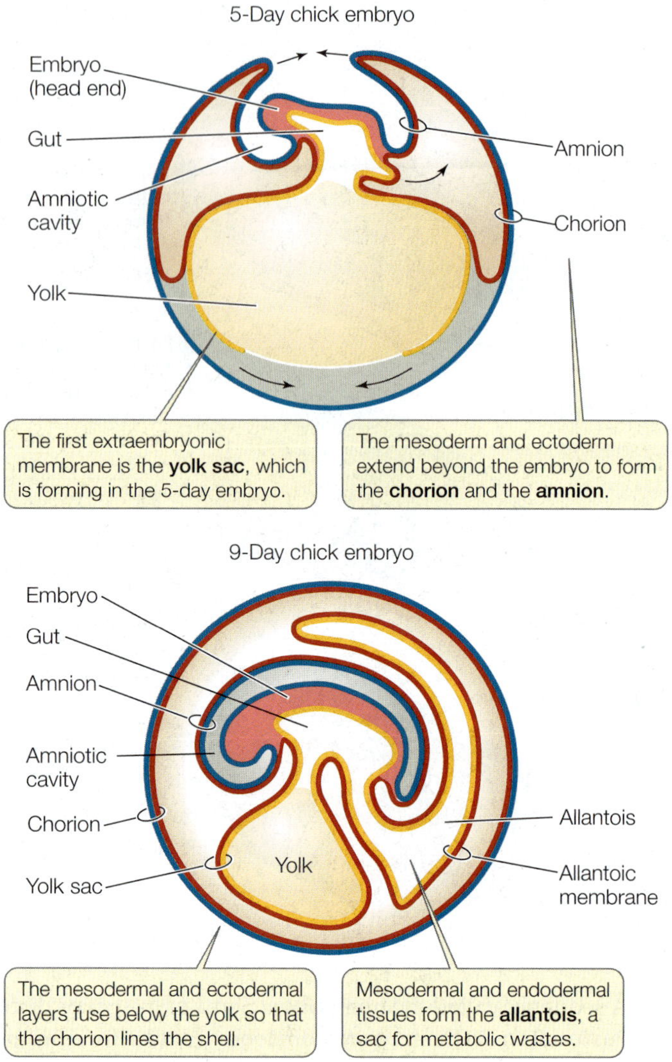
5-Day chick embryo
Embryo (head end)
Gut
Amniotic cavity
Yolk
Amnion
Chorion
The first extraembryonic membrane is the **yolk sac**, which is forming in the 5-day embryo.
The mesoderm and ectoderm extend beyond the embryo to form the **chorion** and the **amnion**.
9-Day chick embryo
Embryo
Gut
Amnion
Amniotic cavity
Chorion
Yolk sac
Yolk
Allantois
Allantoic membrane
The mesodermal and ectodermal layers fuse below the yolk so that the chorion lines the shell.
Mesodermal and endodermal tissues form the **allantois**, a sac for metabolic wastes.

44.17 The Extraembryonic Membranes In birds and mammals, the embryo constructs four extraembryonic membranes. In birds, the yolk sac encloses the yolk, and the amnion and chorion enclose the embryo. Fluids secreted by the amnion fill the amniotic cavity, providing an aqueous environment for the embryo. The chorion, along with the allantoic membrane, mediates gas exchange between the embryo and its environment. The allantois stores the embryo's waste products. (See also Figure 33.19.)

Go to Activity 44.1 Extraembryonic Membranes
Life10e.com/ac44.1

44.5 How Is the Growing Embryo Sustained?

There is more to a developing reptile, bird, or mammal than the embryo itself. As mentioned earlier, the embryos of these vertebrates are surrounded by several **extraembryonic membranes** that originate from the embryo but are not part of it. Extraembryonic membranes function in nutrition, gas exchange, and waste removal. In mammals they interact with tissues of the mother to form the placenta. The evolutionary relationships between the extraembryonic membranes of birds and mammals were discussed in Section 33.4.

Extraembryonic membranes form with contributions from all germ layers

The chicken provides a good example of how extraembryonic membranes form from the germ layers created during gastrulation. In the chick, four membranes form—the yolk sac, the allantoic membrane, the amnion, and the chorion. The **yolk sac** is the first to form, and it does so by extension of the hypoblast layer along with some adjacent mesoderm. The yolk sac grows to enclose the entire body of yolk in the egg (**Figure 44.17**). It constricts at the top to create a tube that is continuous with the gut of the embryo. However, yolk does not pass through this tube. Yolk is digested by the cells of the yolk sac, and the nutrients are transported to the embryo through blood vessels that form from mesoderm and line the outer surface of the yolk sac. The **allantoic membrane** is also an outgrowth of the extraembryonic endoderm plus adjacent mesoderm. It forms the **allantois**, a sac for storage of metabolic wastes.

Ectoderm and mesoderm combine and extend beyond the limits of the embryo to form the other extraembryonic membranes. Two layers of cells extend all along the inside of the eggshell, both over the embryo and below the yolk sac. Where they meet, they fuse, forming two membranes, the inner **amnion** and the outer **chorion**. The amnion surrounds the embryo, forming the amniotic cavity. The amnion secretes fluid into the cavity, providing a protective environment for the embryo. The outer membrane, the chorion, forms a continuous membrane just under the eggshell (see Figure 44.17). It limits water loss from the egg and also works with the enlarged allantoic membrane to exchange respiratory gases between the embryo and the outside world.

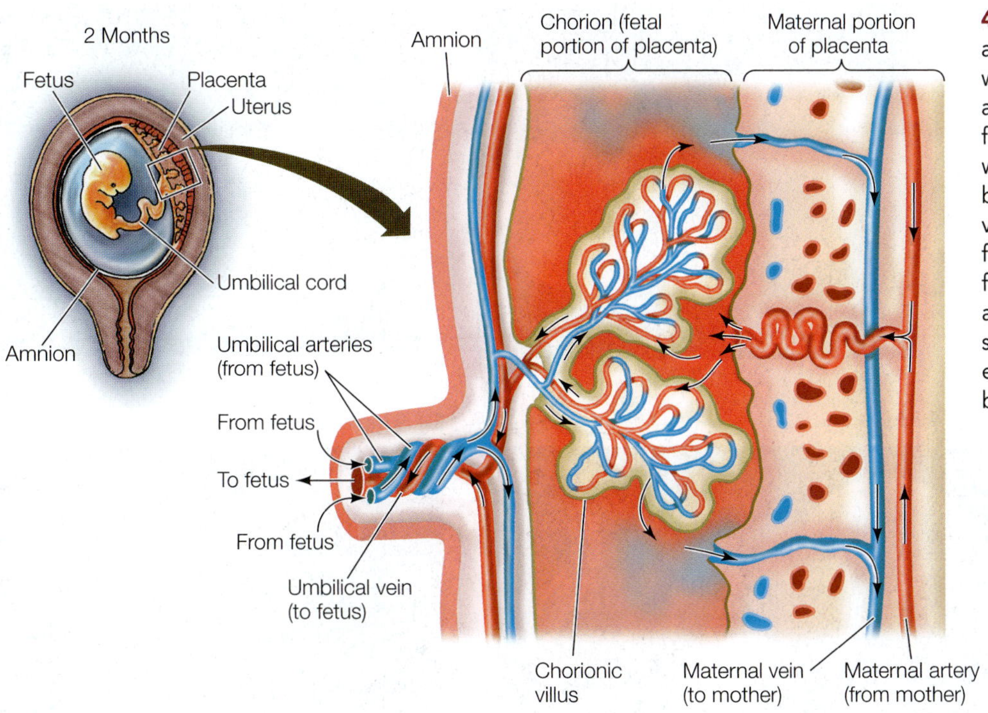

2 Months
Fetus
Placenta
Uterus
Umbilical cord
Amnion
Amnion
Chorion (fetal portion of placenta)
Maternal portion of placenta
Umbilical arteries (from fetus)
From fetus
To fetus
From fetus
Umbilical vein (to fetus)
Chorionic villus
Maternal vein (to mother)
Maternal artery (from mother)

44.18 The Mammalian Placenta In humans and most other mammals, nutrients and wastes are exchanged between maternal and fetal blood in the placenta, which forms from the chorion and tissues of the uterine wall. The embryo is attached to the placenta by the umbilical cord. Embryonic blood vessels invade the placental tissue to form fingerlike chorionic villi. Maternal blood flows into the spaces surrounding the villi, and placental blood flows through the villi so nutrients and respiratory gases can be exchanged between the maternal and fetal blood.

Extraembryonic membranes in mammals form the placenta

In placental mammals, the first extraembryonic membrane to form is the trophoblast (see Figures 44.4C and 44.5). When the blastocyst reaches the uterus and hatches from its encapsulating zona pellucida, trophoblast cells interact directly with the endometrium. Adhesion molecules expressed on the surfaces of these cells attach them to the uterine wall. By secreting proteolytic enzymes, the trophoblast burrows into the endometrium, beginning the process of implantation. Eventually the entire trophoblast is within the wall of the uterus. The trophoblast cells send out numerous projections, or villi, to increase the surface area of contact with maternal blood.

Meanwhile, the hypoblast cells proliferate to form what in the bird would be the yolk sac. But there is virtually no yolk in eggs of placental mammals, so the yolk sac contributes mesodermal tissues that interact with trophoblast tissues to form the chorion. The chorion, along with tissues of the uterine wall, produces the **placenta**, the organ that exchanges nutrients, respiratory gases, and metabolic wastes between the mother and the embryo (**Figure 44.18**).

At the same time the yolk sac is forming from the hypoblast, the epiblast produces the amnion, which grows to enclose the entire embryo in a fluid-filled amniotic cavity. The rupturing of the amnion and chorion and the loss of the **amniotic fluid** (the "water breaks") herald the onset of labor in humans.

An allantois also develops in mammals, but its importance depends on how well nitrogenous wastes can be transferred across the placenta. The human placenta deals effectively with the fetal nitrogenous wastes, so the human allantoic sac is small. In contrast, the pig placenta is not very good at clearing nitrogenous wastes from the fetus, so the pig's allantoic sac is large. In humans and other placental mammals, allantoic tissues contribute to the formation of the umbilical cord, by which the embryo is attached to the chorionic placenta (see Figure 33.19B). It is through the blood vessels of the umbilical cord that nutrients and oxygen from the mother reach the developing fetus, and wastes, including carbon dioxide and urea, are removed.

RECAP 44.5

The extraembryonic membranes of reptiles, birds, and mammals sustain the growing embryo. In reptiles and birds these membranes surround the embryo within the shelled egg. In mammals the extraembryonic membranes form the placenta, an organ that exchanges nutrients, respiratory gases, and metabolic wastes between the mother and the embryo.

- Describe each of the four extraembryonic membranes and their functions in the developing chick egg. **See p. 918 and Figure 44.17**
- Explain the role of the trophoblast in the early development of a mammalian embryo. **See p. 919 and Figure 44.5**

44.6 What Are the Stages of Human Development?

In humans, **gestation**, or pregnancy, lasts about 266 days, or 9 months. Gestation is shorter in smaller mammals—21 days in mice, for example—and in larger mammals it is longer—330 days in horses and 600 days in elephants. The events of human gestation can be divided into three **trimesters** of roughly 3 months each.

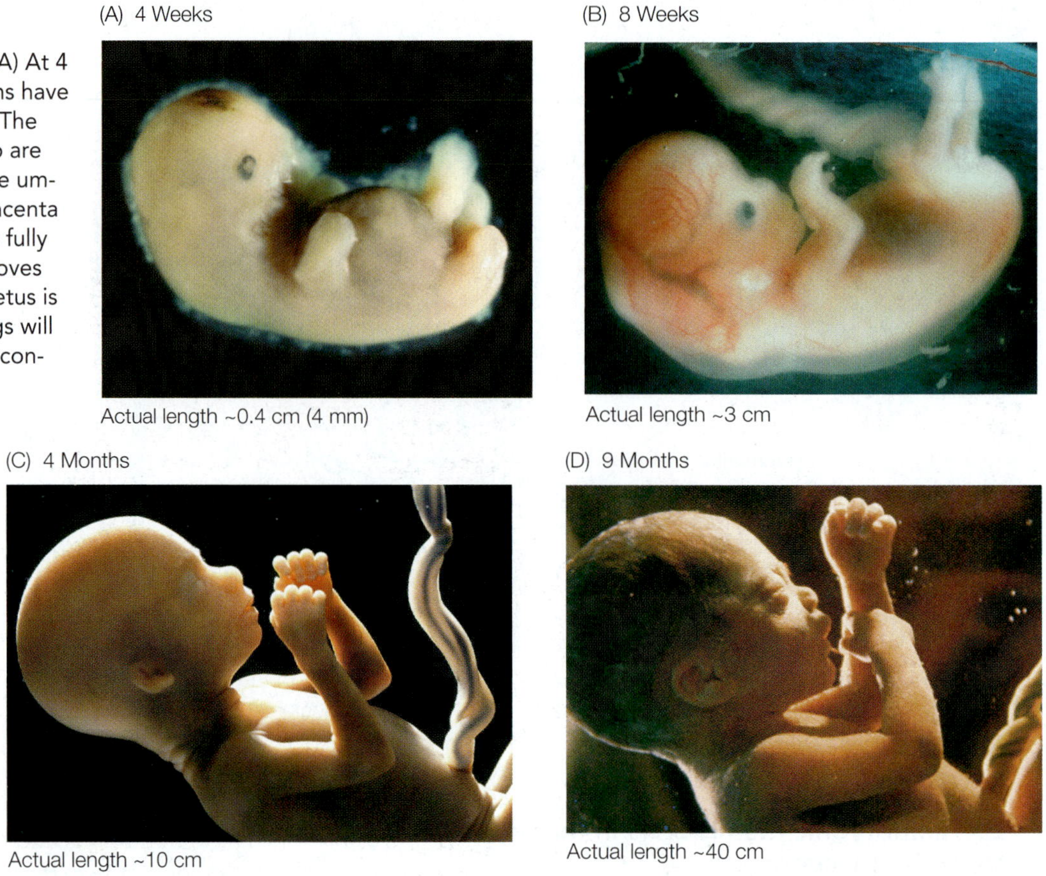
(A) 4 Weeks
Actual length ~0.4 cm (4 mm)
(B) 8 Weeks
Actual length ~3 cm
(C) 4 Months
Actual length ~10 cm
(D) 9 Months
Actual length ~40 cm

44.19 Stages of Human Development (A) At 4 weeks, most of the embryo's organ systems have been formed and the heart is beating. (B) The body structures of this 8-week-old embryo are forming rapidly, and it is visibly a male. The umbilical cord attaches the embryo to the placenta (upper left). (C) At 4 months, the fetus has fully formed limbs with fingers and toes and moves freely within the amniotic cavity. (D) This fetus is well along in its ninth month. Soon its lungs will be mature enough to trigger the onset of contractions and birth.

Organ development begins in the first trimester

Implantation of the human blastocyst begins about 6 days after fertilization. After implantation, gastrulation occurs, tissues differentiate, the placenta forms, and organs begin to develop. The heart begins to beat during week 4, and limbs are formed by week 8 (**Figure 44.19A,B**). By the end of the first trimester, most organs have started to form. The embryo is about 8 centimeters long and weighs about 40 grams (less than 2 ounces). At about this point in time, the human embryo is medically and legally referred to as a **fetus**. (The distinction between embryo and fetus is not made for other mammals; developing mice, for example, remain embryos until they are born.)

The first trimester is a time of rapid cell division and tissue differentiation. Signal transduction cascades and the resulting branching sequences of developmental processes are in their early stages. Therefore the first trimester is the period during which the embryo is most sensitive to damage from radiation, drugs, chemicals, and pathogens that can cause birth defects. An embryo can be damaged before the mother even realizes she is pregnant. A classic case is that of thalidomide, a drug widely prescribed in Europe in the late 1950s to treat nausea. Women who took this drug in the fourth and fifth weeks of pregnancy, when the embryo's limbs are beginning to form, gave birth to children with missing or severely malformed arms and legs.

Organ systems grow and mature during the second and third trimesters

During the second trimester the fetus grows rapidly to a weight of about 600 grams. The limbs of the fetus elongate, and the fingers, toes, and facial features become well formed (**Figure 44.19C**). Eyebrows and fingernails grow, and the fetus's nervous system develops rapidly. Fetal movements are first felt by the mother early in the second trimester and become progressively stronger and more coordinated.

The fetus grows rapidly during the third trimester (**Figure 44.19D**). As this final stage approaches its end, the internal organs mature. The digestive system begins to function, the liver stores glycogen, the kidneys produce urine, and the brain undergoes cycles of sleep and waking. A human infant is born when the last of its critical organs—the lungs—mature.

Although the first-trimester embryo is the most susceptible to adverse effects of drugs, chemicals, and diseases, the potential for serious effects from exposure to environmental factors exists throughout pregnancy. Protein malnutrition and exposure to alcohol and cigarette smoke are examples of factors that can result in low birth weight, mental retardation, and other developmental complications.

Developmental changes continue throughout life

Development does not end with birth. Growth continues until adult size is reached, and even when growth stops, organs of the body continue to repair and renew themselves through cycles of cell replacement by the progeny of undifferentiated stem cells. In humans especially, enormous developmental changes occur in the brain in the years between birth and adolescence. Especially in the early years, there is a great deal of plasticity in the organization of the nervous system as the connections between neurons develop.

For example, a child born with misaligned eyes (a condition known as strabismus) will use mostly one eye. The connections to the brain from one eye will become strong while connections from the other eye remain weak, and the child will develop with reduced visual acuity and depth perception. If eye alignment is corrected in the first 3 years of life, the connections between the eyes and the brain can improve and the child is likely to develop normal vision. After the age of 3, corrective measures are less likely to result in improvement and visual impairments may persist. Thus plasticity in human visual system development declines during early childhood. However, recent data indicate that it is not lost entirely and may be reactivated even in adulthood.

RECAP 44.6

Human gestation lasts 9 months and is divided into three trimesters. At the end of the first trimester, the fetus is very small but most of its organs have begun to form. In the second trimester, limbs elongate and the fetus moves. By the end of the third trimester, most organs have begun to function.

- Why is a first-trimester embryo particular sensitive to environmental risks? See p. 920

Having briefly outlined how multicellular tissues and organ systems emerge from a single cell (the fertilized egg), the remainder of this unit will discuss the physiological functioning of organ systems. The next three chapters will describe the workings of perhaps the most complicated of these, the vertebrate nervous system.

?

How does the directional flow of extracellular fluid across the node stimulate a left–right asymmetry in gene expression and development?

ANSWER

Surrounding the node are cells with nonmotile cilia that are believed to sense the direction of fluid flow across the node. There are two hypotheses about how these cilia function, one chemical and the other mechanical. Both sensory mechanisms are used by cilia elsewhere in the body. In the nose, cilia of the olfactory cells sense chemicals. In the ear, cilia are bent by sound waves, which opens ion channels. The leftward flow of fluid across the node certainly can create mechanical forces on the nonmotile cilia that differ on the two sides of the node. Other research, however, supports the possibility of chemical signaling. Research on the mouse node shows that the beating of the cilia causes proteins of a certain size range to form a concentration gradient across the node. More important, it was discovered that nodal cells secrete small vesicles that are swept to the left side of the node by the flow and burst when they contact the nonmotile cilia of surrounding cells. The contents of these vesicles could be the chemical signals that initiate left–right asymmetry in gene expression and development.

CHAPTER SUMMARY 44

44.1 How Does Fertilization Activate Development?

- The sperm and the egg contribute differentially to the zygote. The sperm contributes a haploid nucleus and, in most species, a centriole. The egg contributes a haploid nucleus, nutrients, ribosomes, mitochondria, mRNAs, and proteins.
- In amphibians, the cytoplasmic contents of the egg are not distributed homogeneously, and they are rearranged after fertilization to set up the major axes of the future embryo. The nutrient molecules are generally found in the **vegetal hemisphere**, whereas the nucleus is found in the **animal hemisphere**. **Review Figures 44.1, 44.2**

44.2 How Does Mitosis Divide Up the Early Embryo?

- **Cleavage** is a period of rapid cell division. Except in mammals, little if any gene expression occurs during cleavage. Cleavage can be complete or incomplete, and the pattern of cell divisions depends on the orientation of the mitotic spindles. The result of cleavage is a ball or mass of cells called a **blastula**. **Review Figure 44.3**
- Early cell divisions in mammals are unique in being slow and allowing for gene expression early in the process. These cell divisions produce a **blastocyst** composed of an **inner cell mass** that becomes the embryo and an outer cell mass that develops as the **trophoblast**. At the time of **implantation**, the trophoblast secretes molecules that help the blastocyst implant in the uterine wall. **Review Figures 44.4, 44.5**
- A **fate map** can be created by labeling specific **blastomeres** and observing what tissues and organs are formed by their progeny. **Review Figure 44.6**
- Some species undergo **mosaic development**, in which the fate of each cell is determined during early divisions. Other species, including vertebrates, undergo **regulative development**, in which remaining cells can compensate for cells lost in early cleavages.

44.3 How Does Gastrulation Generate Multiple Tissue Layers?

- **Gastrulation** involves massive cell movements that produce three **germ layers** and place cells from various regions of the blastula into new associations with one another. **Review Figure 44.7, ANIMATED TUTORIAL 44.1**
- The initial step of sea urchin and amphibian gastrulation is inward movement of certain blastomeres. The site of inward movement becomes the **blastopore**. Cells that move into the blastula become the **endoderm** and **mesoderm**; cells remaining on the outside become the **ectoderm**. Cytoplasmic factors in the vegetal pole cells are essential to initiate development. **Review Figures 44.7, 44.8**
- The **dorsal lip** of the amphibian blastopore is a critical site for cell determination. It has been called the **primary embryonic organizer** because it induces determination in cells that pass over it during gastrulation. **Review Figures 44.8, 44.9, 44.10, ANIMATED TUTORIAL 44.2**

continued

- The protein β-catenin activates a signaling cascade that induces the primary embryonic organizer and sets up the anterior–posterior body axis. **Review Figures 44.2, 44.11**
- Gastrulation in reptiles and birds differs from that in sea urchins and frogs because the large amount of yolk causes the blastula to form a flattened disc of cells. **Review Figure 44.13**
- Although their eggs have no yolk, placental mammals have a pattern of gastrulation similar to that of reptiles and birds.

44.4 How Do Organs and Organ Systems Develop?

- Gastrulation is followed by **organogenesis**, the process whereby tissues interact to form organs and organ systems.
- In the formation of the vertebrate nervous system, one group of cells that migrates over the blastopore lip is determined to become the **notochord**. The notochord organizes the overlying ectoderm to thicken, form parallel ridges, and fold in on itself to form a **neural tube** below the epidermal ectoderm. The nervous system develops from this neural tube. **Review Figure 44.14**
- The notochord and **neural crest cells** participate in the segmental organization of mesoderm into structures called **somites** along the body axis. Rudimentary organs and organ systems form during these stages. **Review Figure 44.15**
- In vertebrates, **Hox genes** determine the pattern of anterior–posterior differentiation along the body axis. Other genes, such as *Sonic hedgehog*, contribute to dorsal–ventral differentiation. **Review Figure 44.16**

44.5 How Is the Growing Embryo Sustained?

- The embryos of reptiles, birds, and mammals are protected and nurtured by four **extraembryonic membranes**. In birds and reptiles the **yolk sac** surrounds the yolk and provides nutrients to the embryo, the **chorion** lines the eggshell and participates in gas exchange, the **amnion** surrounds the embryo and encloses it in an aqueous environment, and the **allantois** stores metabolic wastes. **Review Figure 44.17, ACTIVITY 44.1**
- In mammals the chorion and the trophoblast cells interact with the maternal uterus to form a **placenta**, which provides the embryo with nutrients and gas exchange. The amnion encloses the embryo in an aqueous environment. **Review Figure 44.18**

44.6 What Are the Stages of Human Development?

- Human pregnancy, or **gestation**, can be divided into three trimesters. The embryo forms in the first trimester; during this time, it is most vulnerable to environmental factors that can lead to birth defects. During the second and third trimesters the **fetus** grows, the limbs elongate, and the organ systems mature.
- Development continues throughout childhood and throughout life.

Go to the Interactive Summary to review key figures, Animated Tutorials, and Activities Life10e.com/is44

CHAPTER**REVIEW**

REMEMBERING

1. How does cleavage in mammals differ from cleavage in frogs?
   a. Slower rate of cell division
   b. Formation of tight junctions
   c. Expression of the embryo's genome
   d. Early separation of cells that will not contribute to the embryo
   e. All of the above

2. Which statement about gastrulation is *true*?
   a. In frogs, gastrulation begins in the vegetal hemisphere.
   b. In sea urchins, gastrulation produces the notochord.
   c. In birds, cells from the surface of the blastodisc move down through the primitive groove to form the hypoblast.
   d. In mammals, gastrulation occurs in the hypoblast.
   e. In sea urchins, gastrulation produces only two germ layers.

3. Which of the following was a conclusion from the experiments of Spemann and Mangold?
   a. Cytoplasmic determinants of development are homogeneously distributed in the amphibian zygote.
   b. In the late blastula, certain regions of cells are determined to form skin or nervous tissue.
   c. The dorsal lip of the blastopore can be isolated and will form a complete embryo.
   d. The dorsal lip of the blastopore can initiate gastrulation.
   e. The dorsal lip of the blastopore gives rise to the neural tube.

4. Which of the following characterizes neurulation?
   a. The notochord forms a neural tube.
   b. The neural tube is formed from ectoderm.
   c. A neural tube forms around the notochord.
   d. The neural tube forms somites.
   e. In birds, the neural tube forms from the primitive groove.

5. Which statement about trophoblast cells is *true*?
   a. They are capable of producing monozygotic twins.
   b. They are derived from the hypoblast of the blastocyst.
   c. They are endodermal cells.
   d. They secrete proteolytic enzymes.
   e. They prevent the zona pellucida from attaching to the oviduct.

UNDERSTANDING & APPLYING

6. The glycogen synthase kinase-3 (GSK-3)-inhibiting protein in the amphibian egg is a product of the *Disheveled* gene. If you had both Disheveled protein and an inhibitor of Disheveled protein, what experiments might you do to test whether β-catenin was both necessary and sufficient to initiate gastrulation?

7. If you used a laser to kill a small number of cells on the midline of the dorsal blastopore lip of an amphibian embryo, what defects would you expect to see during subsequent development?

ANALYZING & EVALUATING

8. During gastrulation in birds, the *Sonic hedgehog* gene is expressed only on the left side of Hensen's node. What might be the cause of this expression pattern, and what is its significance? How could you test your hypotheses?

9. When oogonia or spermatogonia divide by mitosis, one daughter cell remains a germinal stem cell and one becomes a primary oocyte or spermatocyte. What mechanism could possibly account for these different cell fates?

10. There is controversy over therapeutic cloning as a way of obtaining embryonic stem cells to treat diseases. Given that human development is regulative, suggest a way to produce a source of isogenic (i.e., identically matching a person's own body) stem cells for an individual without resorting to therapeutic cloning.

Go to BioPortal at **yourBioPortal.com** for Animated Tutorials, Activities, LearningCurve Quizzes, Flashcards, and many other study and review resources.

45 Neurons, Glia, and Nervous Systems

CHAPTER OUTLINE

Trying and Learning To varying degrees, individuals with Down syndrome are unable to learn basic concepts and tasks despite intense effort on their own part and that of their teachers. Understanding the cellular basis of this learning disability may lead to new ways of alleviating it.

YOUR BRAIN ENABLES YOU to learn the material in this chapter. It enables you to read the words, understand the illustrations, and store that information so you can use it to answer the questions at the end of the chapter. You have to spend time studying in order to master this chapter; it may be hard work, but eventually you will have learned something about how the brain receives and processes information.

Imagine what it would be like if you could not learn any of this material—no matter how much you poured over the book, no matter how hard you tried. This is the situation faced by individuals with a learning disability that is part of the condition known as Down syndrome, which affects 1 out of every 700 children born in the United States.

Individuals with Down syndrome are born with three copies of chromosome 21. This smallest of the human somatic chromosomes contains only some 250 genes, but having an extra copy of these genes causes numerous developmental and functional problems, including a learning disability in which the brain does not properly take in and store new information. How can we understand the cause of this disability and perhaps find a way to remedy it?

One productive way of investigating the causes and possible treatments for a human disease or deficit is to develop an animal model. Through genetic engineering, researchers created a "Down syndrome mouse" that has most of the same genes triplicated as those in humans with Down syndrome. Using this mouse model, biologists found out that the learning disability of these mice was due to overinhibition in the brain, and that when this inhibition was reduced with drugs, the ability of these mice to learn was increased. But what do we mean by "overinhibition in the brain"?

We often think of the brain as a puppet master, pulling the strings that activate the muscles and organs of the body. In fact the brain is more like an orchestra conductor, making some sections louder, some softer, speeding up, slowing down. The brain must constantly maintain a delicate balance of excitation and inhibition, acting on some signals and ignoring others. In the brains of "Down syndrome mice," there is consistently too much inhibition. If inhibition is reduced using certain drugs, the mice appear able to learn. Research like this is a first step toward bringing potential therapies for humans into clinical trials in the hope of someday alleviating suffering.

> **?** What causes overinhibition in the nervous system, and how can it be reduced?
>
> See answer on p. 943.

45.1 What Cells Are Unique to the Nervous System?

Nervous systems are informational systems. They encode, process, and store a wide variety of information from the external and internal environments, and they use that information to control and regulate the physiological processes and behavioral actions of the organism. Nervous systems are able to carry out these functions because of the properties of two unique types of cells: nerve cells, or **neurons,** and glial cells, or **glia**. The many different types of neurons vary enormously in structure and appearance, but all neurons are **excitable**, meaning they can generate and transmit electric signals; the electric signals generated by neurons are known as **action potentials**. Generally glia do not generate action potentials, but they interact intimately with neurons, modulating their activity and supporting them in many ways.

The structure of neurons reflects their functions

All neurons have a basic structure that includes four regions (Figure 45.1).

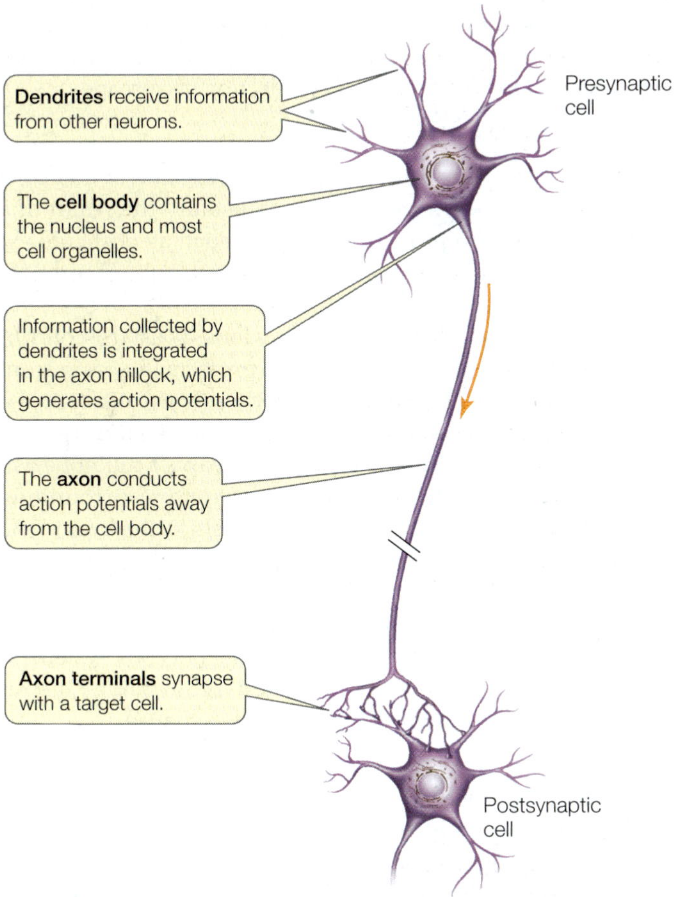
Dendrites receive information from other neurons.
Presynaptic cell
The cell body contains the nucleus and most cell organelles.
Information collected by dendrites is integrated in the axon hillock, which generates action potentials.
The axon conducts action potentials away from the cell body.
Axon terminals synapse with a target cell.
Postsynaptic cell

45.1 A Generalized Neuron The diagram shows the features typical of most neurons. The forms of these features, including the length of the single axon and the density and branching patterns of the dendrites, vary greatly across the many different types of neurons.

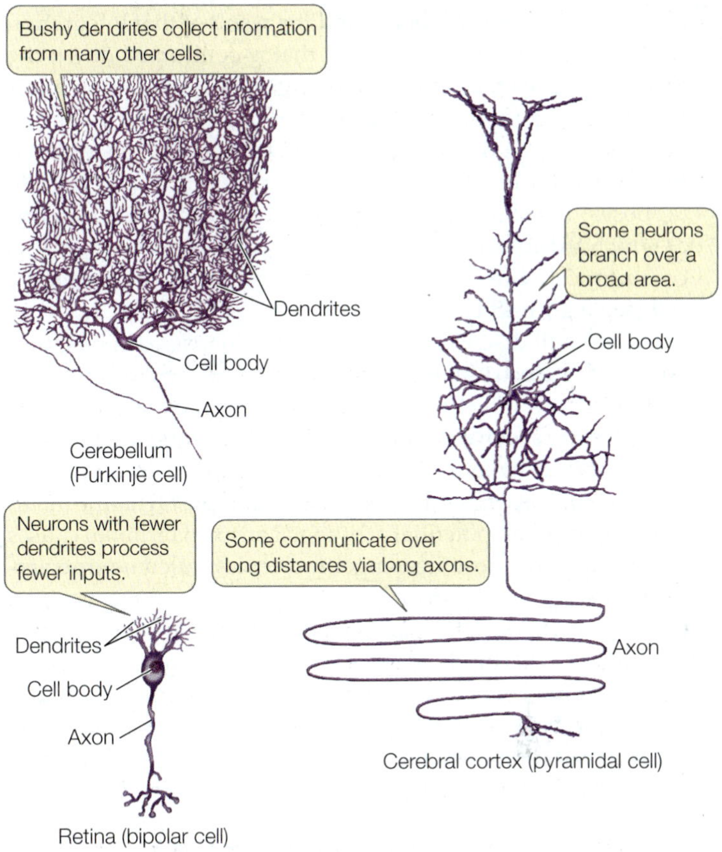
Bushy dendrites collect information from many other cells.
Dendrites
Cell body
Axon
Cerebellum (Purkinje cell)
Some neurons branch over a broad area.
Cell body
Neurons with fewer dendrites process fewer inputs.
Some communicate over long distances via long axons.
Dendrites
Cell body
Axon
Axon
Retina (bipolar cell)
Cerebral cortex (pyramidal cell)

45.2 Neurons Have Many and Varied Forms The morphological differences in neurons from different parts of the mammalian nervous system are related to their specific functional adaptations. The small sample here shows two neuronal types from the human brain (a Purkinje cell and a pyramidal cell) and a sensory neuron from the retina of the human eye.

- A **cell body** contains the nucleus and most of the cell's organelles.
- Shrublike projections called **dendrites** (Greek *dendron*, "tree") may extend from the cell body. Dendrites bring information from other neurons or sensory cells to the cell body. Neurons with few dendrites are receiving information from specific and limited sources, whereas neurons with large arrays of dendrites can collect and integrate information from a wider range of sources.
- In most neurons, one projection—the **axon**—is much longer than the others. Axons carry information in the form of action potentials away from the originating cell body (the **presynaptic cell**) to the receiving target cell (the **postsynaptic cell**).
- At the postsynaptic cell, the axon divides into a spray of fine nerve endings. At the tip of each of these tiny nerve endings is a swelling called the **axon terminal**.

A wide variety of forms can be seen among the many different types of neurons (Figure 45.2), but all neurons share the mechanisms whereby their plasma membranes generate and conduct action potentials. Information received by dendrites is integrated by the cell body, and the result of that integration can generate an action potential that is conducted down the axon toward its terminals at the target cell. Action potentials can travel at speeds up to 100 m/sec (360 km/hr), making it possible for an individual to sense, process, and act on information very quickly. The axon terminal comes extremely close to (apposes) the membrane of the target cell, forming a **synapse** at which the information conveyed by the action potential is communicated from the presynaptic cell to the postsynaptic cell (see Figure 45.1).

As we will discuss in Section 45.3, synapses can be either chemical or electrical. Electrical synapses allow the action potential to pass directly between two neurons. In vertebrates, however, most synapses are chemical. At chemical synapses a space about 25 nanometers wide (about 1/2000th the width of a human hair) separates the presynaptic and postsynaptic membranes. An action potential arriving at an axon terminal causes it to release chemical messenger molecules called **neurotransmitters**. The neurotransmitters diffuse across the space and bind to receptors on the plasma membrane of the postsynaptic (target) cell. This binding alters the activity of the postsynaptic neuron. Some neurotransmitter–receptor combinations inhibit activity of the postsynaptic neuron, and other neurotransmitter–receptor combinations excite it. Neurons integrate information by summing excitatory and inhibitory inputs.

Glia are the "silent partners" of neurons

The human brain has about 10 times more glial cells than neurons. A neurobiologist once said that "flashy neurons get all of the attention, but glial cells do most of the brain's work and are the cause of many of its diseases." It has been easier to study the functions of neurons because their action potentials can be observed. The mostly silent glia have been more difficult to study, and therefore we know much less about them. Our knowledge of glia will grow enormously in the years to come, and there are likely to be many surprises.

Like neurons, glia come in several forms and have diverse functions. In the brain and spinal cord, glia called **oligodendrocytes** wrap around the axons of neurons, covering them with concentric layers of insulating plasma membrane. Outside the brain and spinal cord, glia called **Schwann cells** wrap axons (Figure 45.3). **Myelin** is the covering produced by oligodendrocytes and Schwann cells, and it gives many parts of the nervous system a glistening white appearance. Not all axons are myelinated, but those that are can conduct action potentials more rapidly than can axons that are not myelinated, for reasons we will describe in Section 45.2.

Diseases that affect myelin can be devastating because they impair conduction of action potentials. The most common of these demyelinating diseases is multiple sclerosis—literally "multiple scars"—which occurs in about 1 in 700 people in the United States. Individuals with this autoimmune disease produce antibodies to proteins in the myelin in the brain and spinal cord. The symptoms and damage from the disease depend on where in the nervous system the antibody attacks occur. Motor impairment is common. An example of a demyelinating disease that attacks myelin outside the brain and spinal cord is Guillain–Barre syndrome, which is usually the result of a severe infection. Environmental factors such as pesticide exposure can also damage myelin. There are no known cures for demyelinating diseases.

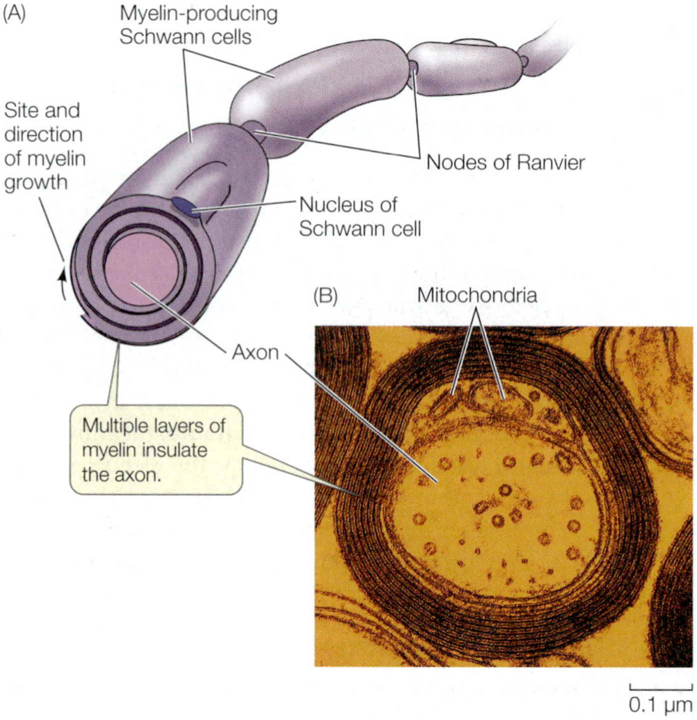
(A)
Myelin-producing Schwann cells
Site and direction of myelin growth
Nodes of Ranvier
Nucleus of Schwann cell
Axon
Multiple layers of myelin insulate the axon.
(B)
Mitochondria
0.1 μm

45.3 Wrapping Up an Axon (A) Schwann cells produce layers of myelin, a type of plasma membrane that provides electrical insulation to the axon. At the intervals between Schwann cells—the nodes of Ranvier—the axon is exposed. Action potentials travel along the axon by "jumping" from node to node. (B) A myelinated axon, seen in cross section through an electron microscope.

Glia called **astrocytes** (because they look like stars) contribute to the **blood–brain barrier**, which protects the brain from toxic chemicals in the blood. Blood vessels throughout the body are very permeable to many chemicals, including toxic ones, which would reach the brain if this barrier did not exist. Astrocytes help form the blood–brain barrier by surrounding the smallest, most permeable blood vessels in the brain. The barrier is not perfect, however. Because it consists of plasma membranes, it is permeable to fat-soluble substances such as anesthetics and alcohol (which explains why these substances have such rapid and marked effects on the nervous system).

In addition to their role in the blood–brain barrier, astrocytes have several known functions at the synapse:

- They can take up neurotransmitter that has been released into the synapse and thereby control communication between the pre- and postsynaptic cells.
- They can supply neurons with nutrients. Neurons have no energy reserves, but astrocytes store glycogen that they can break down to supply the neurons with fuel.
- They have signaling properties. Even though most astrocytes do not generate action potentials, they do release neurotransmitters that can alter the activities of neurons.

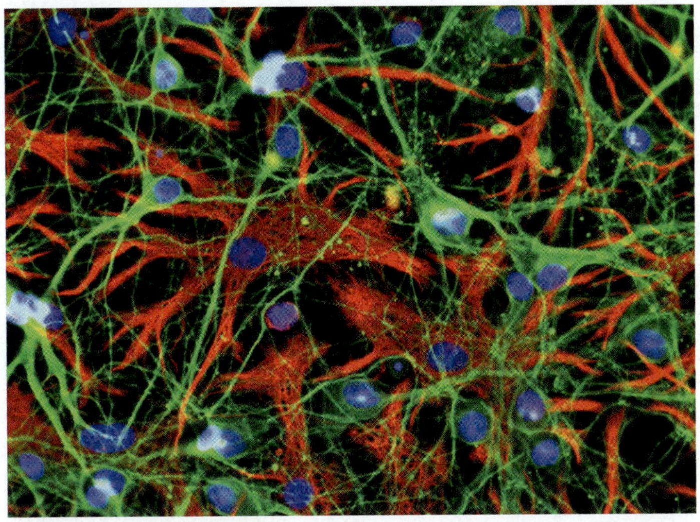

45.4 Astrocytes Communicate with Many Synapses Cell-type-specific antibodies have been used to label the astrocytes and their processes (red) and the neurons and their processes (green) in this fluorescence micrograph. The blue label visualizes the nuclei of both cell types.

- They aid in the repair and regeneration of neurons.
- They make contact with both blood vessels and neurons and can therefore signal changes in the composition of the blood.

Astrocytes play crucial yet poorly understood roles in modulating synapse activity. The projections of a single astrocyte may make contact with more than 100,000 synapses (**Figure 45.4**). The contact of the astrocyte with the neuronal components of the synapse is so intimate that it has inspired the concept of the **tripartite synapse**—the idea that a synapse includes not only the pre- and postsynaptic neurons but also connections from astrocytes.

Microglia are another type of glial cell. The blood–brain barrier typically prevents antibodies in the general circulation from entering the brain and spinal cord. Microglia that originate during development from stem cells in the bone marrow come to reside in the nervous system and act as macrophages and mediators of inflammatory responses, thus providing the nervous system with immune defenses.

RECAP 45.1

Nervous systems have two unique types of cells: neurons and glia. There are many types of neurons, but they all generate action potentials. A neuron has four regions: cell body, dendrites, axon, and axon terminals. Neurons communicate with target cells at synapses, which can be chemical or electrical. Although glia do not generate action potentials, they have a wide variety of functions that are only beginning to be elucidated.

- Describe the different parts of a neuron and their functions. **See pp. 925–926 and Figure 45.1**
- What are some types of glia, and what are their functions? **See pp. 926–927**
- What is meant by the "tripartite synapse"? **See p. 927**

The one feature common to all neurons is that they process information in the form of action potentials. In the next section we will focus on how action potentials are generated and transmitted.

45.2 How Do Neurons Generate and Transmit Electric Signals?

Sodium–potassium pumps create concentration gradients across all animal cell membranes, with Na^+ concentration higher outside the cell and K^+ concentration higher inside. Both inside and outside the cell, the positive charges of these ions are balanced by negatively charged ions. But *across the plasma membrane* there is an electric charge difference, with the inside of the cell being negative relative to the outside. This is because there are "leak channels" that allow only certain ions—usually K^+—to passively leak across the cell membrane.

Because there is a higher concentration of K^+ inside the cell than in the extracellular fluid outside, K^+ diffuses out of the cell down its concentration gradient. But when K^+ leaks out of the cell, it leaves behind an unbalanced negative electric charge that tends to pull K^+ back into the cell. An equilibrium is reached when the tendency for K^+ to diffuse out is countered by the electric charge pulling K^+ back in. The result is a charge difference—a **membrane potential**—across the membrane, with the inside of the cell negative relative to the outside.

Membrane potentials exist in all cells. In neurons the steady state membrane potential is called the **resting potential**. An action potential is a sudden, large, transient reversal in the resting potential that is generated by sudden openings and rapid closings of ion channels. Before describing the properties of ion channels and action potentials in detail, a review of some simple concepts of electricity may be useful.

Simple electrical concepts underlie neural function

Voltage (electric potential difference) is a force that causes electrically charged particles to move between two points. Voltage is to the flow of electrically charged particles as pressure is to the flow of water. If the negative and positive poles of a battery are connected by a wire, an electric current will flow through the wire because there is a voltage difference between the two poles. This flow of electric current can be used to do work, just as a current of water can be used to do work.

In wires, electric current is carried by electrons, but in solutions and across cell membranes, electric current is carried by ions. The major ions that carry electric charges across the plasma membranes of neurons are sodium (Na^+), potassium (K^+), calcium (Ca^{2+}), and chloride (Cl^-). Recall that ions with opposite charges attract one another and that those with like charges repel one another. How do these basic principles of bioelectricity establish the resting potential of the neural plasma membrane? And how is the flow of ions through membrane channels turned on and off to generate action potentials? We will address these questions next.

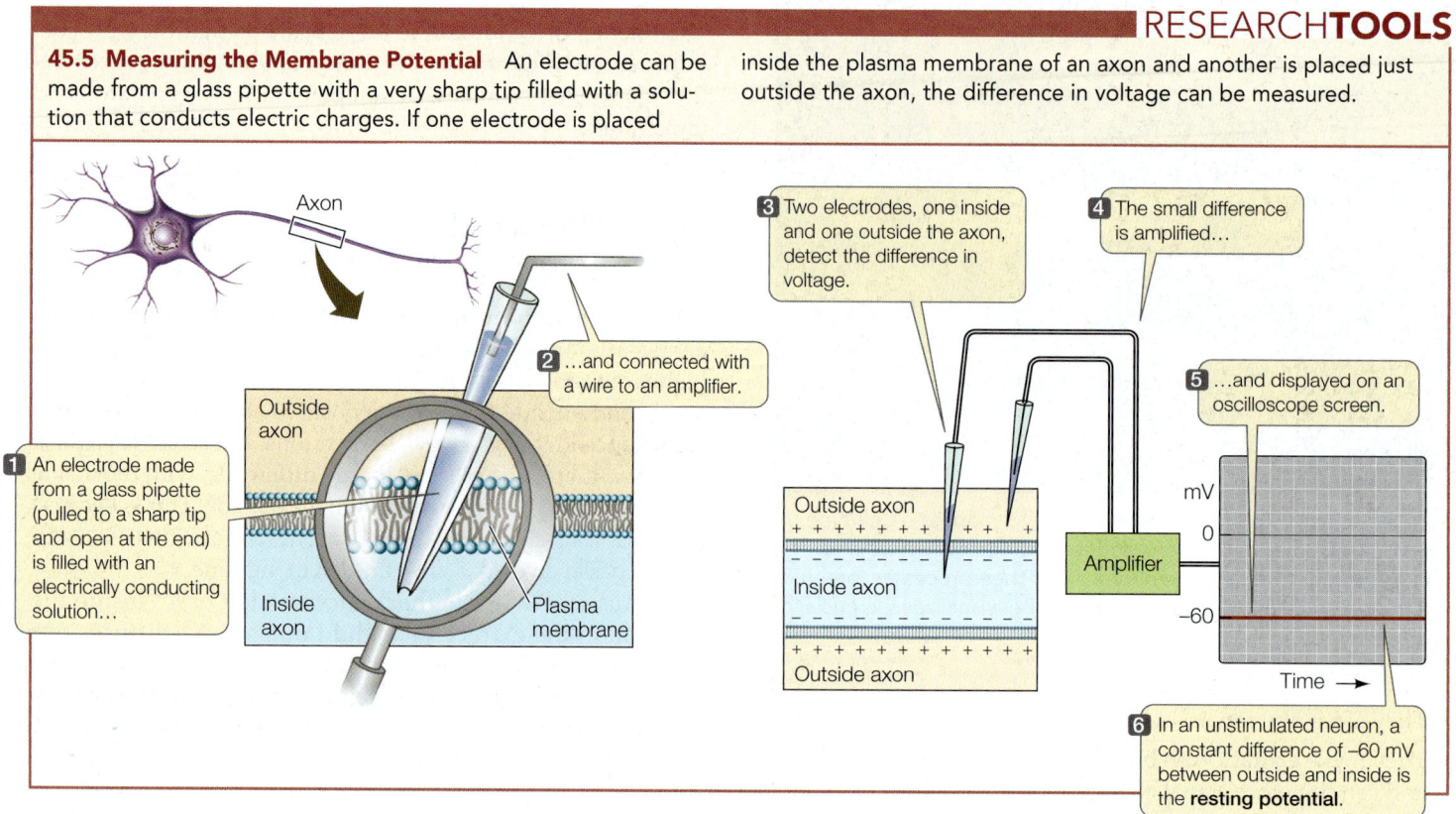

RESEARCH**TOOLS**

45.5 Measuring the Membrane Potential An electrode can be made from a glass pipette with a very sharp tip filled with a solution that conducts electric charges. If one electrode is placed inside the plasma membrane of an axon and another is placed just outside the axon, the difference in voltage can be measured.

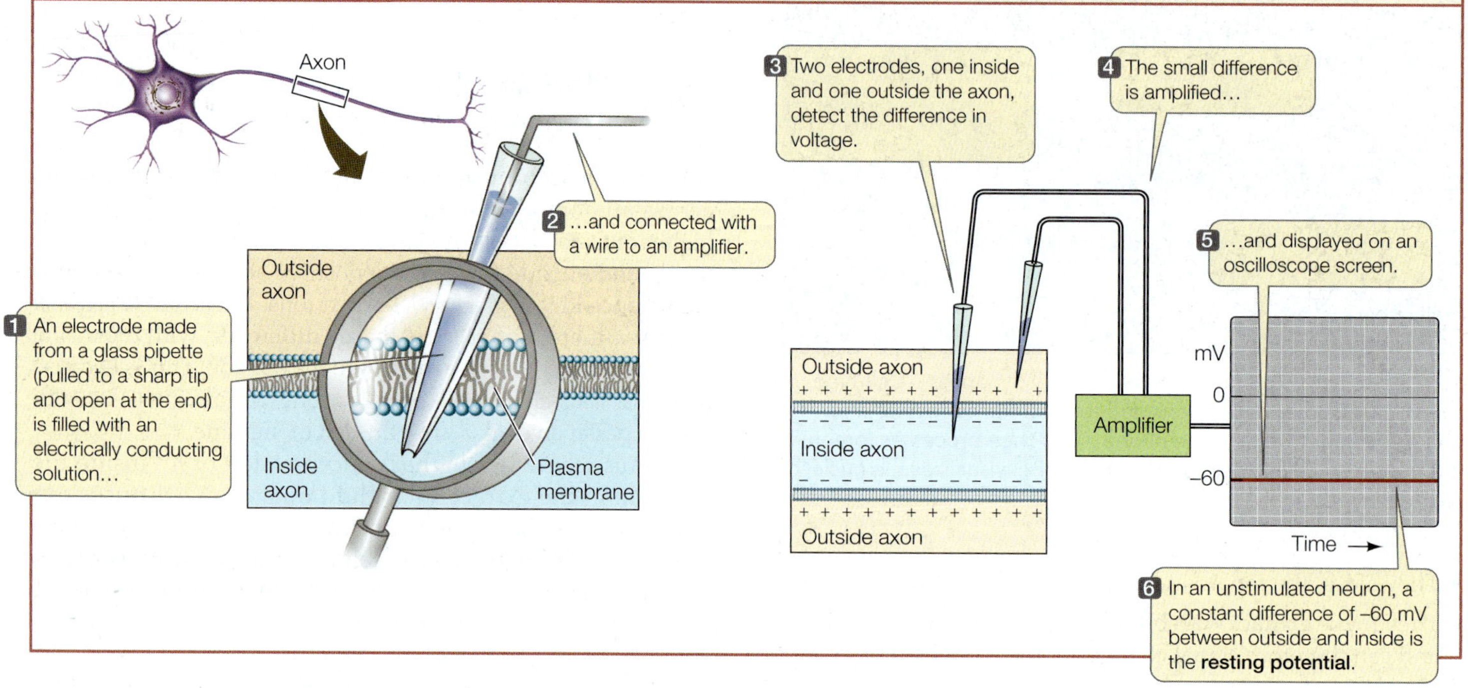

Axon
1 An electrode made from a glass pipette (pulled to a sharp tip and open at the end) is filled with an electrically conducting solution...
2 ...and connected with a wire to an amplifier.
Outside axon
Inside axon
Plasma membrane
3 Two electrodes, one inside and one outside the axon, detect the difference in voltage.
4 The small difference is amplified...
5 ...and displayed on an oscilloscope screen.
Outside axon
Inside axon
Outside axon
Amplifier
mV
0
–60
Time
6 In an unstimulated neuron, a constant difference of –60 mV between outside and inside is the **resting potential**.

Membrane potentials can be measured with electrodes

We can record electrical events in a cell using electrodes. **Figure 45.5** shows how this technique is applied across an unstimulated axon to measure the resting potential, which is usually between –60 and –70 millivolts (mV). The minus sign indicates that the inside of the cell is electrically negative compared with the outside.

The resting potential provides a means for neurons to respond to a stimulus. Because of the voltage difference across the membrane, and the different ion concentrations on either side of the membrane, ions would cross the membrane if they could. For example, Na^+ ions are more abundant outside the cell than inside, and the inside of the resting cell is negatively charged. Therefore if the membrane suddenly became permeable to Na^+, those positively charged ions would rush into the cell. Any chemical or physical stimulus that changes the permeability of the plasma membrane to ions will produce a change in the cell's membrane potential. The most extreme change in membrane potential is the action potential, a sudden and rapid reversal in the voltage across a portion of the plasma membrane. For 1 or 2 milliseconds, positively charged ions flow into the cell, making the inside of the cell *more positive* than the outside.

Go to Animated Tutorial 45.1
The Resting Membrane Potential
Life10e.com/at45.1

Ion transporters and channels generate membrane potentials

The plasma membranes of neurons, like those of all other cells, are lipid bilayers that are impermeable to ions but contain many protein molecules that serve as ion transporters and channels. Ion transporters and channels are responsible for the distribution of charges across the membrane that create resting and action potentials.

Ion transporters require energy to move ions against their concentration or electrical gradients and are therefore called ion pumps. A major ion transporter in the plasma membranes of neurons (and all other cells) is the **sodium–potassium pump**, so called because it actively expels Na^+ ions from inside the cell, exchanging them for K^+ ions from outside the cell (**Figure 45.6A**). The Na^+–K^+ pump is also known as sodium–potassium ATPase, a term emphasizing that it is an enzyme complex requiring ATP to do its work. The Na^+–K^+ pump keeps the concentration of K^+ inside the cell greater than the K^+ concentration of the extracellular fluid, and the concentration of Na^+ inside the cell less than that of the extracellular fluid. The concentration differences established by this active transporter means that K^+ would diffuse out of the cell and Na^+ would diffuse in if the ions could cross the lipid bilayer. How do these concentration gradients relate to the electrical gradients we discussed above?

Ion channels permit the diffusion of ions across membranes. These channels are water-filled pores formed by proteins that cross the lipid bilayer and are generally selective,

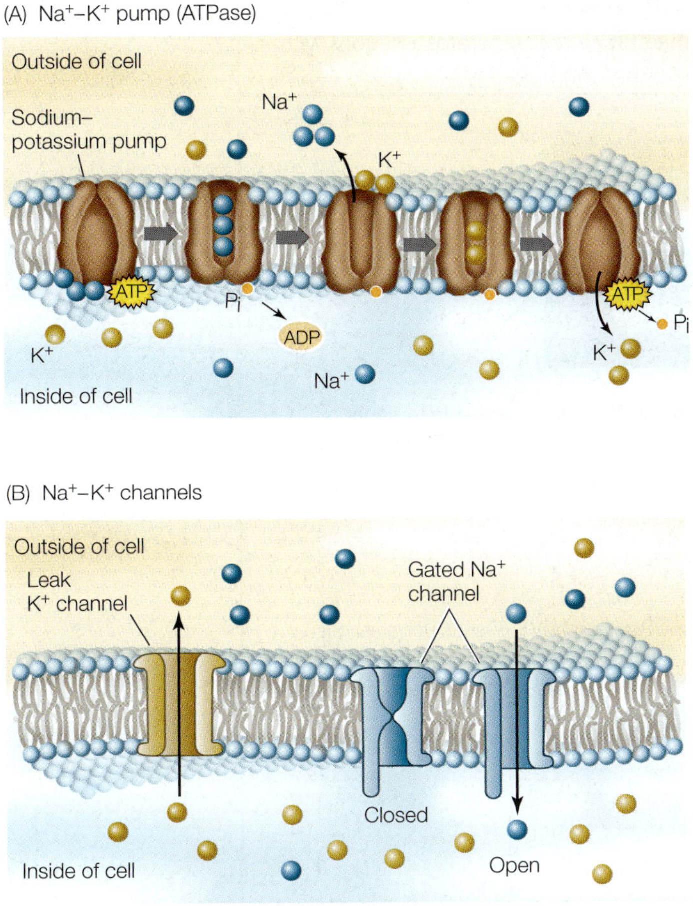
(A) Na+–K+ pump (ATPase)
Outside of cell
Sodium–potassium pump
Na+
K+
ATP
Pi
ADP
K+
Na+
Inside of cell
(B) Na+–K+ channels
Outside of cell
Leak K+ channel
Gated Na+ channel
Closed
Open
Inside of cell

45.6 Ion Transporters and Channels (A) The sodium–potassium pump is an active transporter that moves K^+ to the inside of a neuron and Na^+ to the outside. (B) Ion channels allow specific ions to diffuse down their concentration gradients; K^+ tends to leave neurons when potassium channels are open, and Na^+ tends to enter neurons when sodium channels are open. Leak channels like the K^+ channel shown are always open and create the resting membrane potential. Gated channels like the Na^+ channels shown are opened by chemical or electrical stimulation.

allowing some types of ions to pass through more easily than others (**Figure 45.6B**). Thus there are potassium channels, sodium channels, chloride channels, and calcium channels, and there are different kinds of channels for each ion. Ions can diffuse through these channels in either direction. The direction and magnitude of the net movement of ions through a channel depend on the concentration gradient of that ion type across the plasma membrane, as well as on the voltage difference across that membrane. These two motive forces acting on an ion are termed its **electrochemical gradient**. Although the electrochemical gradient drives the movement of ions through channels, that movement is modified by gates that open and close the channels.

Potassium channels are the most common open, or leak, channels in the plasma membranes of resting (nonstimulated) neurons. As a consequence, resting neurons are more permeable to K^+ than to any other ion. Thus open potassium channels are largely responsible for the resting membrane potential. Because the potassium channels make the plasma membrane permeable to K^+, and because the Na^+–K^+ pump keeps the concentration of K^+ inside the cell much higher than that outside the cell, K^+ tends to diffuse down its electrochemical gradient, out of the cell, through the channels. As these positively charged potassium ions diffuse out of the cell, they leave behind unbalanced negative charges, generating an electric potential across the membrane that tends to pull K^+ back into the cell.

The membrane potential at which the net diffusion of K^+ out of the cell ceases (that is, the point at which K^+ diffusion out due to the concentration gradient is balanced by its movement in due to the negative electric potential) is the **potassium equilibrium potential**, or $\boldsymbol{E_K}$. The value of E_K can be calculated from the concentrations of K^+ on the two sides of the membrane using the **Nernst equation** (**Figure 45.7**). This equation, developed in the late 1800s, shows that the existence of ion channels in neural membranes was hypothesized long before their specific structures and properties were described.

In the late 1940s, A. L. Hodgkin and A. F. Huxley at the University of Cambridge set out to study the electrical properties of axonal membranes. With the techniques available at that time, the necessary measurements could be made only if you had a very large axon to work with. Such an axon exists in nature, in the huge neuron that controls the escape response of squid. Hodgkin and Huxley used electrodes to measure the voltage across the plasma membrane of this large axon, as seen in Figure 45.7, and to pass electric current into it to change its membrane potential. They also changed the concentrations of Na^+ and K^+ both inside and outside the squid axon and measured the resulting changes in membrane potential. On the basis of their many careful experiments, Hodgkin and Huxley developed virtually all of our basic concepts about the electrical properties of neurons, and shared a Nobel Prize in 1963.

We now know that, in general, the resting potential is less negative than the E_K calculated from the Nernst equation. This means that the resting potential is not due solely to leak K^+ channels. The neuronal membrane is slightly permeable to other ions, especially Na^+ and Cl^-, and movements of these ions influence the resting potential. A different equation takes into account (1) all of the ions that can cross the membrane and (2) the relative permeability of the membrane to those ions. This equation, called the **Goldman equation**, predicts the membrane potential more accurately than does the Nernst equation (see p. 931).

Ion channels and their properties can now be studied directly

Because Hodgkin and Huxley were working long before there were laboratory techniques that could investigate ion channels, they could only hypothesize their properties. These hypotheses could not be tested until the late 1970s, when B. Sakmann and E. Neher developed a technique called **patch clamping**, for which they shared the Nobel Prize in 1991. Patch clamping, described in **Figure 45.8**, is widely used by neurobiologists, enabling them to record in real time the tiny electric currents caused by the openings and closings of single ion channels.

RESEARCH**TOOLS**

45.7 Using the Nernst Equation The Nernst equation calculates membrane potential when only one type of ion can cross a membrane that separates solutions with different concentrations of that ion.

1. Measure concentrations of ions inside and outside a neuron.

To measure the concentration of ions in a neuron, the neuron (and its axon) must be big. Squid have giant neurons that control their escape response (see Figure 45.17C). It is possible to sample the cytoplasm of these axons, which are about 1 mm in diameter.

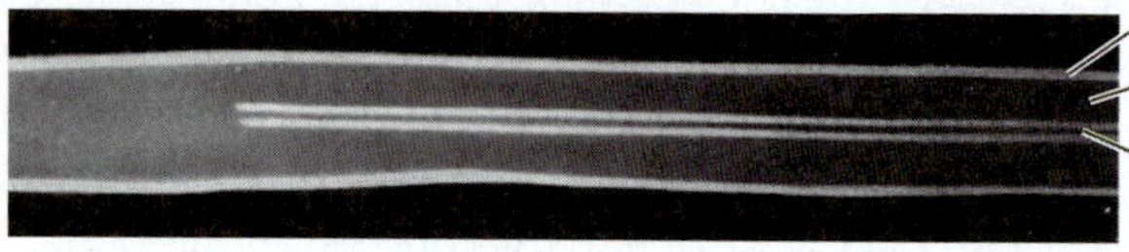
Squid axon
1 mm
Plasma membrane
Cytoplasm
Electrode

2. Use the Nernst equation to calculate what the membrane potential would be if it were permeable to each of the ions that are differently concentrated on the two sides of the membrane: Na^+, K^+, Ca^{2+}, and Cl^-.

The Nernst equation predicts the membrane potential resulting from membrane permeability to a single type of ion that differs in concentration on the two sides of the membrane. The equation is written

$$E_{ion} = 2.3\frac{RT}{zF}\log\frac{[ion]_o}{[ion]_i}$$

where E is the equilibrium (resting) membrane potential (the voltage across the membrane in mV), R is the universal gas constant, T is the absolute temperature, z is the charge on the ion (+1, +1, +2, or –1), and F is the Faraday constant. The subscripts o and i indicate the ion concentrations outside and inside the cell, respectively.

At this point you could just "plug and play," but do you understand this equation?

A concentration difference of ions across a membrane creates a *chemical* force that pushes the ions across the membrane; however, the resulting unbalanced *electric charges* will pull the ions back the other way. At *equilibrium*, the work done moving ions in each direction will be the same.

The *chemical* energy pushing the ions will equal $2.3\,RT\log\,[ion]_o/[ion]_i$
The *electric* energy pulling the ions will equal zEF. So, at equilibrium:

$$zEF = 2.3\,RT\log\frac{[ion]_o}{[ion]_i}$$

Rearranging the equation to solve for E, we get the Nernst equation:

$$E_{ion} = 2.3\frac{RT}{zF}\log\frac{[ion]_o}{[ion]_i}$$

We can simplify the equation by picking a temperature—let's use "room temperature," or 20°C—and solving for $2.3\,RT/F$. At 20°C, $2.3\,RT/F$ equals 58. Thus:

$$E_{ion} = 58/z\log\frac{[ion]_o}{[ion]_i}$$

3. Measuring ion concentrations in squid giant axon cytoplasm and in seawater, then solving the Nernst equation for each ion, we find:

Ion	Ion concentration (mM) in squid axon	Ion concentration (mM) in seawater	Predicted membrane potential (mV)
K^+	400	20	–75
Na^+	50	460	+56
Ca^{2+}	0.5	10	+38
Cl^-	50	560	–60

4. Since the measured membrane potential is –66 mV, it is clear that the resting potential of the axon is due to permeability of the membrane to more than just one type of ion.

Gated ion channels alter membrane potential

The ion channels called leak channels are always open, but other ion channels in the plasma membranes of neurons behave as if they contain "gates"; they are open under some conditions and closed under other conditions. **Voltage-gated channels** open or close in response to a change in the voltage across the plasma membrane. **Chemically gated channels** open or close depending on the presence or absence of a specific molecule that binds to the channel protein, or to a separate receptor that in turn alters the channel protein. **Mechanically gated channels** open or close in response to mechanical force applied to the plasma membrane. Gated channels play important roles in neural function.

Openings and closings of gated channels alter the membrane potential. Imagine what happens, for example, if sodium channels in the plasma membrane suddenly open. Na^+ diffuses into the neuron down its electrochemical gradient to approach

WORKING WITH**DATA:**

Equilibrium Membrane Potential: The Goldman Equation

Original Papers

Goldman, D. E. 1943. Potential, impedence and rectification in membranes. *Journal of General Physiology* 27: 37–60.

Hodgkin, A. L. and B. Katz. 1949. The effect of sodium ions on the electrical activity of the giant axon of the squid. *Journal of Physiology* 108: 37–77.

Analyze the Data

Figure 45.7 presented the Nernst equation by which the membrane potential for a single ion can be determined. But we also saw (at the end of Figure 45.7) that the equilibrium membrane potential is the product of more than one ion. The Goldman equation (sometimes called the Goldman, Hodgkin, Katz equation) calculates the equilibrium membrane potential by taking into account all of the ions that can diffuse across a given membrane and the relative permeabilities of the membrane to those ions. The ions involved in mammalian neurons here are K^+, Na^+, and Cl^-, and the Goldman equation is

$$V_m = \frac{RT}{F}\ln\left(\frac{p_K[K^+]_o + p_{Na}[Na^+]_o + p_{Cl}[Cl^-]_i}{p_K[K^+]_i + p_{Na}[Na^+]_i + p_{Cl}[Cl^-]_o}\right)$$

Relative permeabilities (p) are expressed as percentages. The membrane's permeability to potassium ions is the highest, so $p_K = 1.0$. Then $p_{Na} = 0.05$ and $p_{Cl} = 0.45$. Bracketed elements refer to the inside and outside ion concentrations, as in the Nernst equation.

QUESTION

The table gives the intra- and extracellular ion concentrations for a mammalian neuron. Use these values and the Goldman equation to calculate the membrane potential. Refer to Figure 45.7 for a comparison with calculations based on the Nernst equation. (Hint: In redrafting the equation, you can substitute "2.3*RT/F* log" for "*RT/F* ln".)

	Ion concentration (mM)	
	Intracellular	Extracellular
K^+	140	5
Na^+	10	145
Cl^-	20	110

Go to **BioPortal** for all WORKING WITH**DATA** exercises

RESEARCH**TOOLS**

45.8 Patch Clamping The patch clamp is a glass micropipette filled with an electrically conductive solution that has the same composition as extracellular fluids. When this pipette/electrode is positioned against the membrane of a cell and slight suction is applied, a seal forms. If a single ion channel (or a few ion channels) are within the patch of membrane bounded by the seal, the openings and closings of individual channels can be recorded by the electrode. If the pipette is retracted, it can tear the patched membrane away from the cell, and the activities of the ion channels in the patch can continue to be recorded.

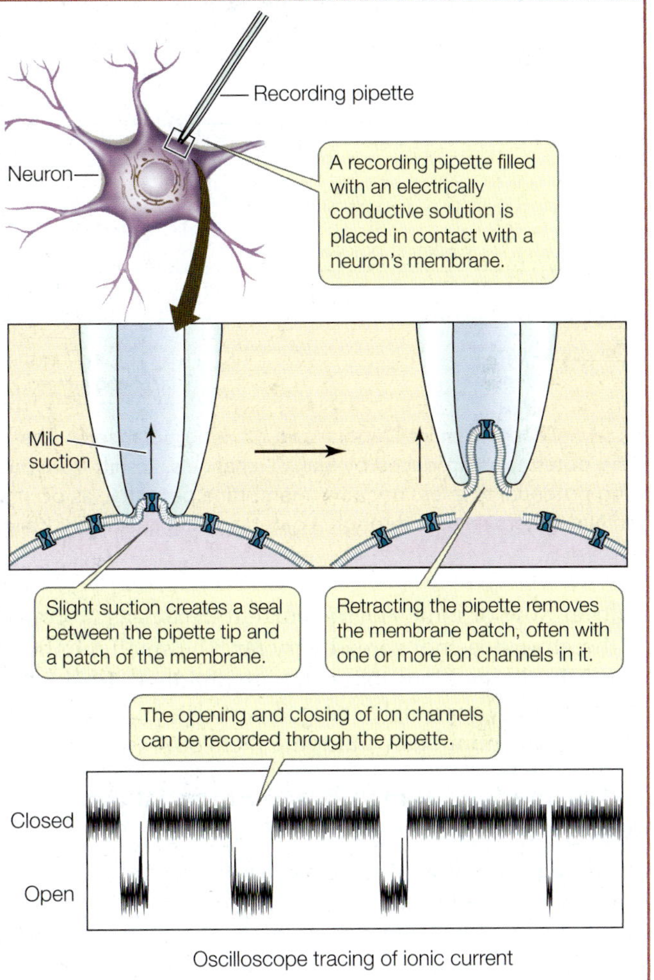
Recording pipette
Neuron
A recording pipette filled with an electrically conductive solution is placed in contact with a neuron's membrane.
Mild suction
Slight suction creates a seal between the pipette tip and a patch of the membrane.
Retracting the pipette removes the membrane patch, often with one or more ion channels in it.
The opening and closing of ion channels can be recorded through the pipette.
Closed
Open
Oscilloscope tracing of ionic current

the equilibrium potential for Na^+ (E_{Na}). Therefore the inside of the cell becomes less negative. When the inside of a neuron becomes less negative (or more positive) in comparison to its resting condition, its plasma membrane is **depolarized** (Figure 45.9).

An opposite change in the membrane potential occurs if gated K^+ channels open. When K^+ efflux from the neuron increases over the normal leak current (the movement of K^+ through the leak channels), the membrane potential becomes even more negative, and the plasma membrane is **hyperpolarized**.

The openings and closings of ion channels that result in changes in the voltage across the plasma membrane are the basic mechanisms by which neurons respond to stimuli, be they electrical, chemical, or mechanical. How do such local changes in membrane potential get communicated to other parts of the cell?

A local change in membrane potential causes a flow of ions that spreads the change in membrane potential to adjacent regions of the membrane. For example, when Na^+ enters a neuron through open sodium channels at one location, those positively charged ions are attracted to adjacent areas on the inside of the membrane that are more negative, and thus there is a rapid flow of ionic electric current (movement of charged ions) away from the site of the open Na^+ channels. However,

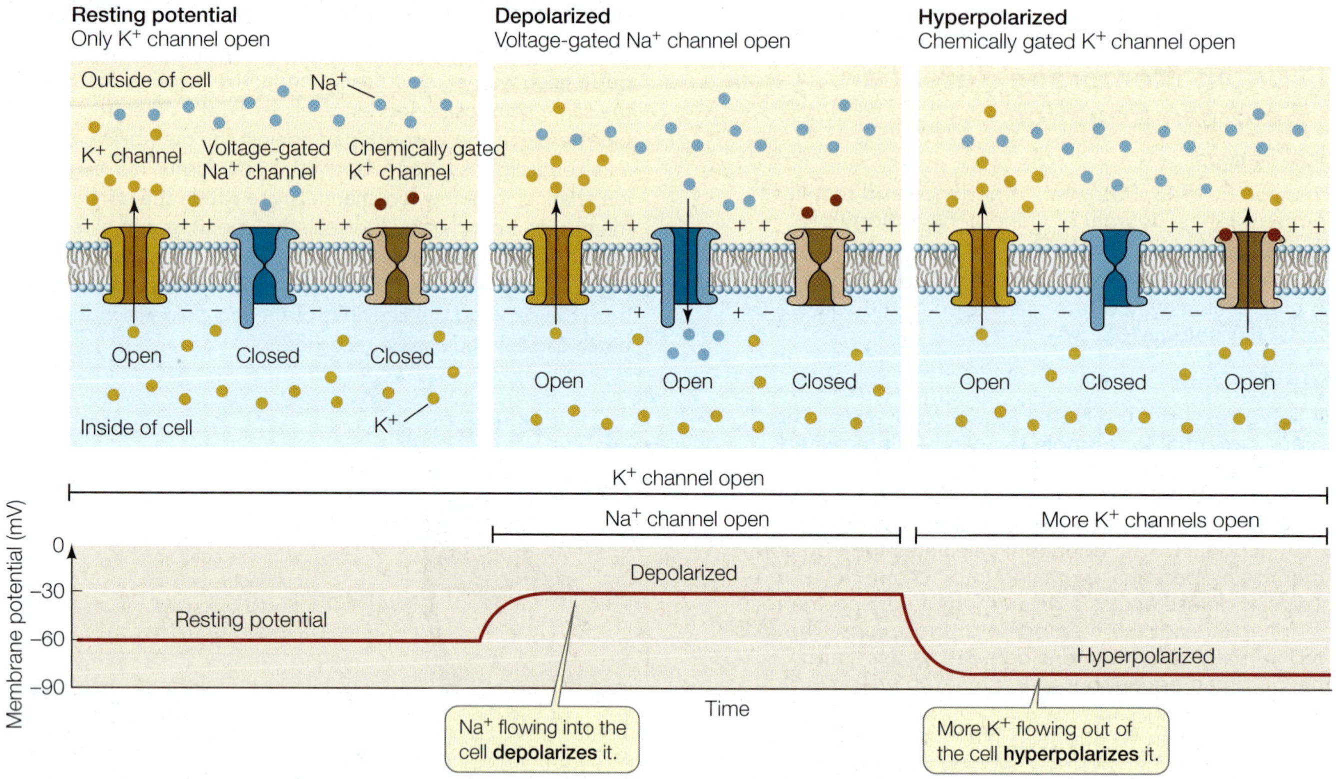

Resting potential
Only K+ channel open
Depolarized
Voltage-gated Na+ channel open
Hyperpolarized
Chemically gated K+ channel open
Outside of cell
Na+
K+ channel
Voltage-gated Na+ channel
Chemically gated K+ channel
Open
Closed
Closed
Inside of cell
K+
Open
Open
Closed
Open
Closed
Open
K+ channel open
Na+ channel open
More K+ channels open
Membrane potential (mV)
0
–30
–60
–90
Depolarized
Resting potential
Hyperpolarized
Time
Na+ flowing into the cell depolarizes it.
More K+ flowing out of the cell hyperpolarizes it.

45.9 Membranes Can Be Depolarized or Hyperpolarized The resting potential is produced by leak K^+ channels. A shift from the resting potential to a less negative membrane potential, as occurs when Na^+ enters the cell through a gated sodium channel, is called depolarization. Hyperpolarization occurs when the membrane potential becomes more negative, as when additional K^+ leaves the cell through gated K^+ channels, which occurs extensively in your brain when you fall asleep.

this local flow of ionic electric current decays as it spreads and therefore does not spread very far. The small number of electrically charged ions that enter are rapidly diluted by the large number of oppositely charged particles, and the Na^+/K^+ ATPases are continuously pumping them out.

Graded changes in membrane potential can integrate information

Even though the flow of ionic electric current along plasma membranes can only extend over short distances, it can cause graded changes in membrane potentials locally. A **graded membrane potential** is a change from the resting potential. Such changes can be due to chemical or mechanical influences on ion channels. Graded potentials are a means of integrating inputs to a cell because the membrane can respond to those inputs with proportional amounts of depolarization or hyperpolarization.

Graded potentials can transmit signals over very short distances and play an important role at the neuromuscular junction (see Section 45.3). In the next chapter we will learn how they play important roles in sensory systems. However, axons are too long to transmit information as a continuous flow of ionic electric current. Therefore axons code information as discrete action potentials that travel along their membranes. Graded potentials, however, play an important role in the generation of action potentials.

Sudden changes in Na^+ and K^+ channels generate action potentials

Action potentials are sudden, transient, large changes in membrane potential. In unmyelinated axons (those not wrapped in myelin by oligodendrocytes or Schwann cells), they can be conducted at speeds of up to 2 meters per second, but in myelinated axons the conduction velocity can be 100 meters per second. Think of running the 100-meter dash—the world record is slightly under 10 seconds.

If we place the tips of a pair of electrodes on either side of the plasma membrane of a resting axon and measure the voltage difference, the reading might be about –60 mV, as we saw in Figure 45.5. If these electrodes are in place when an action potential travels down the axon, they register a rapid change in membrane potential, from –60 mV to about +50 mV. The membrane potential then rapidly returns to its resting level of –60 mV as the action potential passes (**Figure 45.10**).

The action potential is generated by the actions of voltage-gated Na^+ and K^+ channels in the plasma membrane of the axon. At the resting potential, most of these channels are closed (balloon 1 in Figure 45.10). A slight depolarization of the membrane causes them to open. For example, if a neuron is stimulated sufficiently to cause the plasma membrane of its cell body to depolarize slightly, that graded potential can spread by local current flow to the **axon hillock**, the region of the cell body at the base of the axon (see Figure 45.1). Voltage-gated

45.10 The Course of an Action Potential Action potentials result from rapid changes in voltage-gated Na^+ and K^+ channels.

Go to Animated Tutorial 45.2
The Action Potential
Life10e.com/at45.2

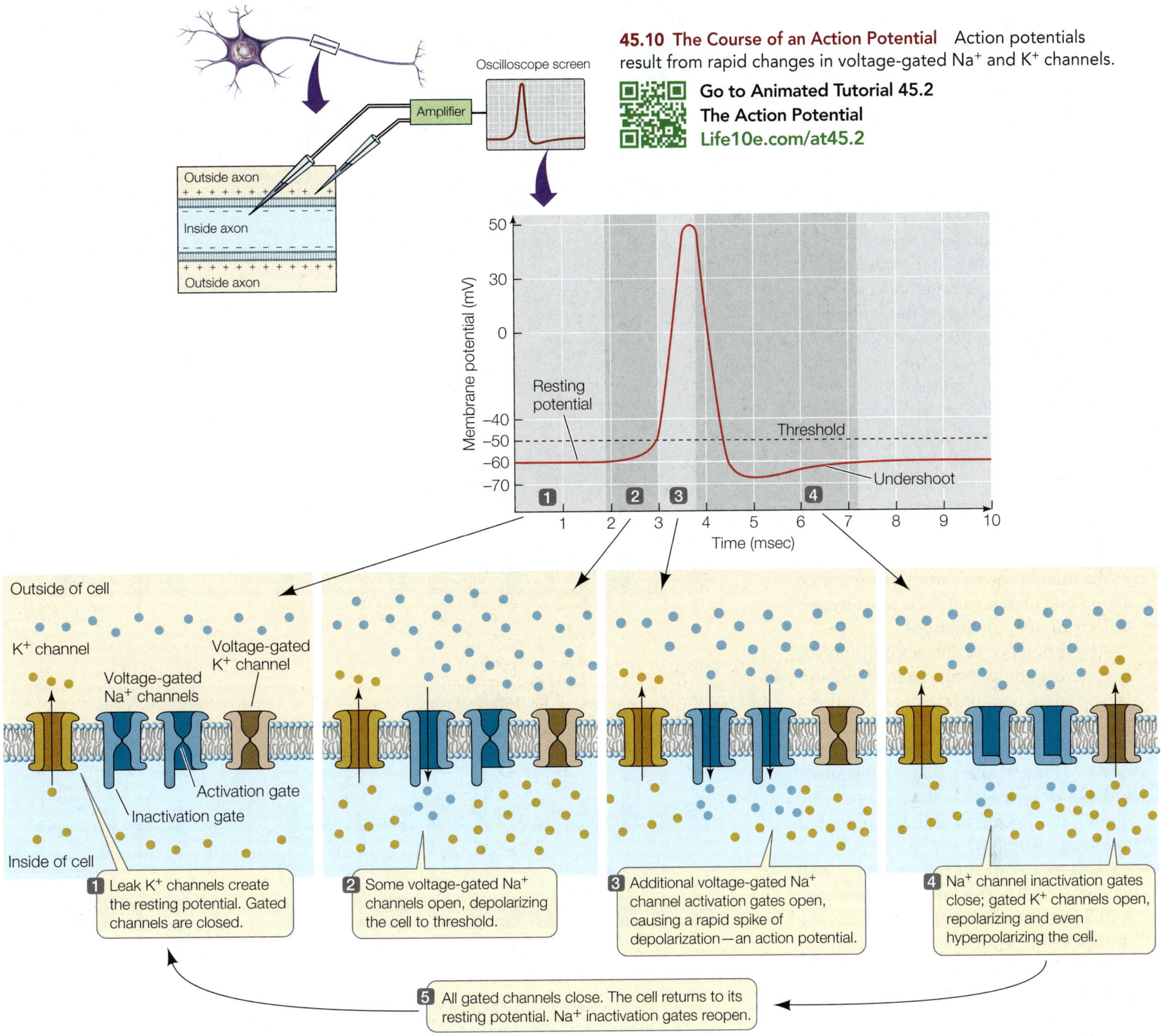

Oscilloscope screen
Amplifier
Outside axon
Inside axon
Outside axon
Membrane potential (mV)
50
30
0
–40
–50
–60
–70
Resting potential
Threshold
Undershoot
Time (msec)
1 2 3 4 5 6 7 8 9 10
Outside of cell
K+ channel
Voltage-gated Na+ channels
Voltage-gated K+ channel
Activation gate
Inactivation gate
Inside of cell
1 Leak K+ channels create the resting potential. Gated channels are closed.
2 Some voltage-gated Na+ channels open, depolarizing the cell to threshold.
3 Additional voltage-gated Na+ channel activation gates open, causing a rapid spike of depolarization—an action potential.
4 Na+ channel inactivation gates close; gated K+ channels open, repolarizing and even hyperpolarizing the cell.
5 All gated channels close. The cell returns to its resting potential. Na+ inactivation gates reopen.

Na^+ channels are concentrated in the axon hillock. A slight depolarization of the plasma membrane in this area causes some of these voltage-gated channels to open briefly—for less than a millisecond (balloon 2 in Figure 45.10). When these channels open, Na^+ rushes into the axon and depolarizes the membrane even more, causing more Na^+ channels to open—a positive feedback effect. When the membrane is depolarized about 5 to 10 mV above the resting potential, a **threshold** is reached; a large number of sodium channels open (balloon 3 in Figure 45.10), and the membrane potential becomes positive—an action potential. The rising phase of the action potential halts abruptly in 1 to 2 milliseconds, and the membrane potential rapidly becomes negative once again.

What causes the axon to return to resting potential? There are two contributing factors: the voltage-gated Na^+ channels close, and voltage-gated K^+ channels open (balloon 4 in Figure 45.10). Voltage-gated K^+ channels open more slowly than the Na^+ channels and stay open longer, allowing K^+ to carry excess positive charges out of the axon. As a result, the membrane potential returns to a negative value and usually becomes even more negative than the resting potential until the voltage-gated K^+ channels close (balloon 5 in Figure 45.10).

Another feature of the voltage-gated Na^+ channels is that once they open and close, they have a **refractory period** of 1 to 2 milliseconds during which they cannot open again. This property can be explained by the channels having two gates, an

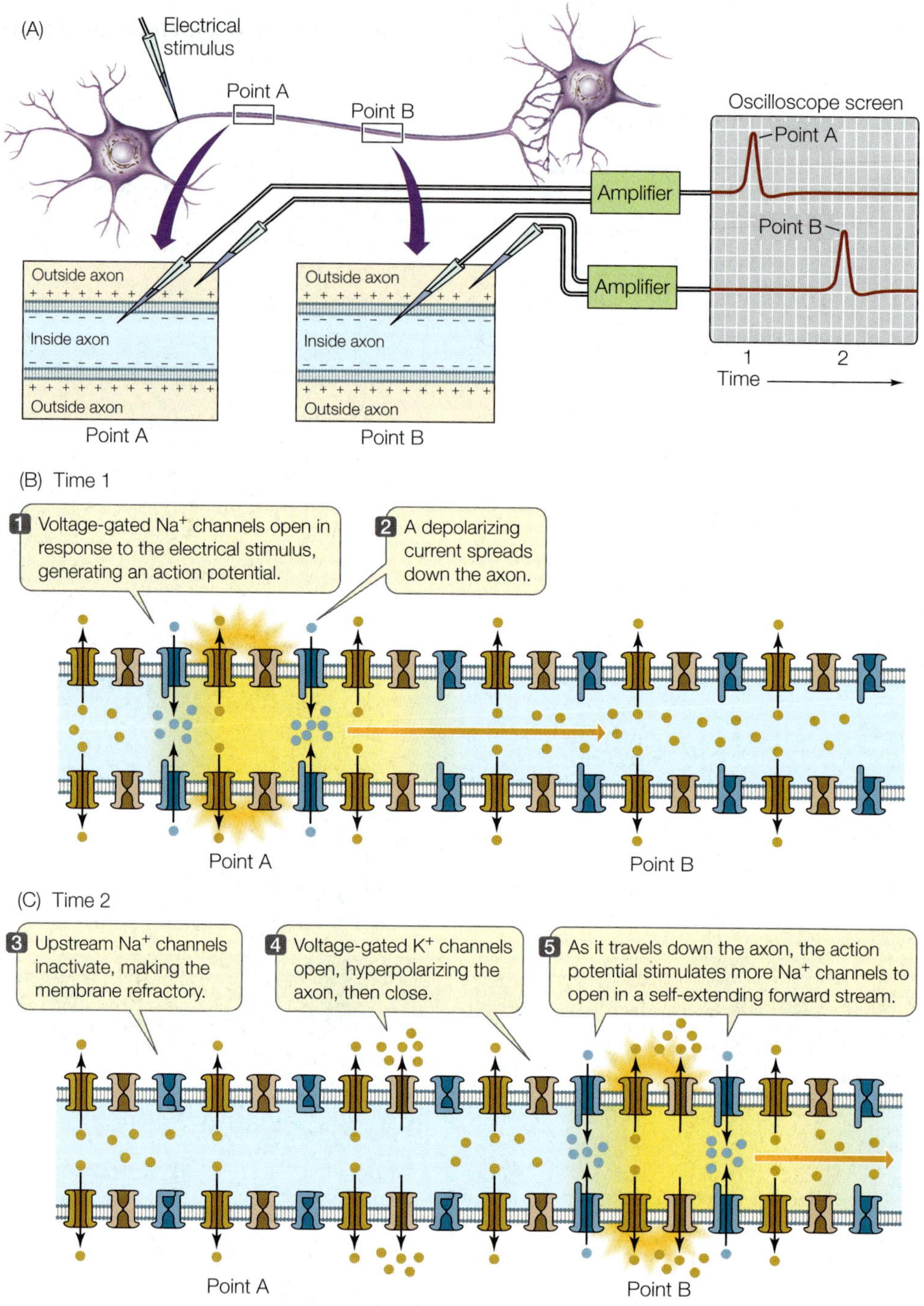

(A)
Electrical stimulus
Point A
Point B
Oscilloscope screen
Point A
Amplifier
Point B
Amplifier
Outside axon
Inside axon
Outside axon
Point A
Outside axon
Inside axon
Outside axon
Point B
1
2
Time
(B) Time 1
1 Voltage-gated Na+ channels open in response to the electrical stimulus, generating an action potential.
2 A depolarizing current spreads down the axon.
Point A
Point B
(C) Time 2
3 Upstream Na+ channels inactivate, making the membrane refractory.
4 Voltage-gated K+ channels open, hyperpolarizing the axon, then close.
5 As it travels down the axon, the action potential stimulates more Na+ channels to open in a self-extending forward stream.
Point A
Point B

45.11 Action Potentials Travel along Axons (A) There is no loss of signal as an action potential travels along an axon. (B) When an action potential is stimulated in one region of membrane, electric current flows to adjacent areas of membrane and depolarizes them. (C) The advancing wave of depolarization causes more Na^+ channels to open, and the action potential is generated anew in the next section of membrane. Meanwhile, in the region where the action potential has just fired, the Na^+ channels are inactivated and the voltage-gated K^+ channels are still open, rendering this section of the axon incapable of generating an action potential. Hence the action potential cannot "back up," but moves continuously forward, regenerating itself as it goes.

activation gate and an **inactivation gate** (see Figure 45.10). Under resting conditions, the activation gate is closed and the inactivation gate is open. Depolarization of the membrane to the threshold level causes both gates to change state, but the activation gate responds faster. As a result, the channel is open for a brief time between the opening of the activation gate and the closing of the inactivation gate. Inactivation gates remain closed for 1 to 2 milliseconds before they spontaneously open again, thus explaining why the membrane has a refractory period before it can fire another action potential. By the time the inactivation gate reopens, the activation gate is closed, and the membrane is poised to generate another action potential. Another contribution to the refractory period is the duration of the opening of the voltage-gated K^+ channels, as we saw above. The dip in the membrane potential following an action potential is called the **after-hyperpolarization** or **undershoot**.

The difference in the concentration of Na^+ across the plasma membrane and the negative resting potential constitute the "battery" that drives action potentials. How rapidly does the battery run down? It might seem that a substantial number of ions would have to cross the membrane for the membrane potential to change from –60 mV to +50 mV and back to –60 mV again. In fact, only a vanishingly small percent of the Na^+ concentrated just outside the plasma membrane moves through the channels during the passage of an action potential. Thus the effect of a single action potential on the concentration gradients of Na^+ and K^+ is very small, and it is possible in most cases for the sodium–potassium pump to keep the "battery" charged, even when the neuron is generating many action potentials every second.

Action potentials are conducted along axons without loss of signal

Action potentials can travel over long distances with no loss of signal. If we place two pairs of electrodes at two different locations along an axon, we can record an action potential at those two locations as it travels along the axon (**Figure 45.11A**). The magnitude of the action potential does not change between the two recording sites. This constancy is possible because an action potential is an all-or-none, self-regenerating event.

- An action potential is *all-or-none* because of the interaction between the voltage-gated Na^+ channels and the membrane potential. If the membrane is depolarized slightly, some voltage-gated Na^+ channels open. Some sodium ions cross the plasma membrane and depolarize it even more, opening more voltage-gated Na^+ channels, and so on, generating an action potential. This positive feedback mechanism ensures that action potentials always rise to their maximum value.

- An action potential is *self-regenerating* because it spreads by local current flow to adjacent regions of the plasma membrane. The resulting depolarization brings those neighboring areas of membrane to threshold. So when an action potential occurs at one location on an axon, it stimulates the adjacent region of axon to generate an action potential, and so on down the length of the axon.

We can use an electrode to stimulate an axon, causing it to depolarize and to fire an action potential that is then conducted along the axon. **Figure 45.11B** shows the changes in the ion channels in the membrane that are responsible for conducting the action potential along the axon without a reduction in amplitude. Normally an action potential is propagated in only one direction—away from the cell body. It cannot reverse itself because the voltage-gated Na^+ channels in the region of the membrane it came from are in their refractory period (**Figure 45.11C**).

Action potentials do not travel along all axons at the same speed. They travel faster in large-diameter axons than in small-diameter axons because the resistance to ionic current flow decreases as an axon's diameter gets bigger. They travel faster in myelinated than in nonmyelinated axons because they can move down the axon in short "jumps" (think of a kangaroo jumping down a path) (**Figure 45.12**). Invertebrates mostly depend on increased axon diameter for fast conduction, but vertebrates mostly depend on myelination of axons to increase conduction velocity.

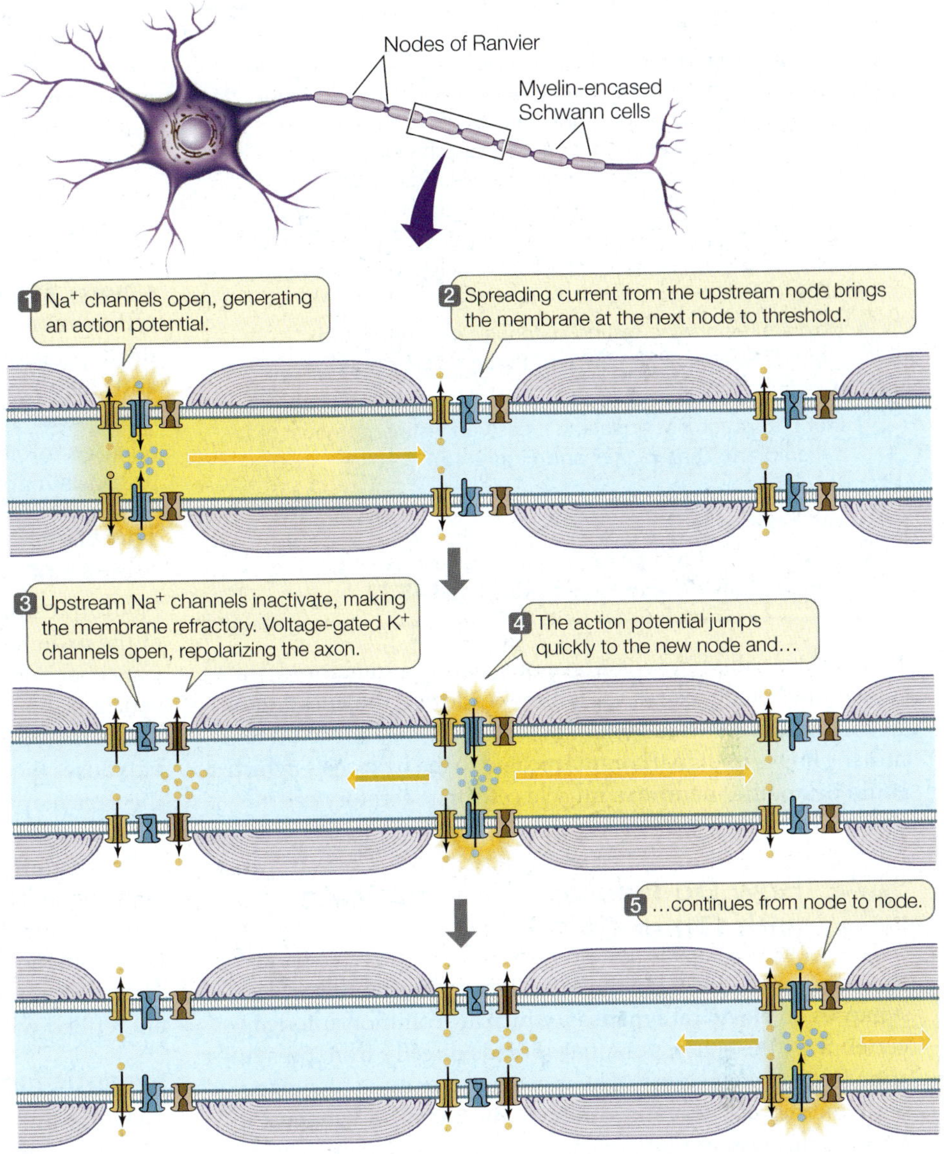
Nodes of Ranvier
Myelin-encased Schwann cells
1 Na^+ channels open, generating an action potential.
2 Spreading current from the upstream node brings the membrane at the next node to threshold.
3 Upstream Na^+ channels inactivate, making the membrane refractory. Voltage-gated K^+ channels open, repolarizing the axon.
4 The action potential jumps quickly to the new node and...
5 ...continues from node to node.

45.12 Saltatory Action Potentials Action potentials "jump" from node to node in myelinated axons, allowing faster transmission of information.

Action potentials jump along myelinated axons

In vertebrate nervous systems, increasing the speed of action potentials by increasing the diameter of axons is not feasible because of the huge number of axons involved. Each of our eyes, for example, has about a million axons connecting it to the brain. These axons conduct action potentials at about the same speed as does the squid giant axon—about 20 meters per second—yet the diameter of each is 200 times smaller than the squid axon's diameter. Imagine having optic tracts 200 times bigger. A different way of increasing conduction velocity of axons has evolved in vertebrates, and that adaptation is myelination.

When glia wrap around axons, they cover the axons with concentric layers of myelin (see Figure 45.3). However, they leave regularly spaced gaps called **nodes of Ranvier**, where the axon is not covered (see Figure 45.12). The leakage of ions across the regions of the plasma membrane that are wrapped in myelin is reduced, so ionic electric current can spread farther along the inside of a myelinated axon than it can along a nonmyelinated axon. Additionally, voltage-gated ion channels are clustered at the nodes of Ranvier. Thus an axon can fire action potentials only at nodes, and those action potentials cannot be propagated through the adjacent patch of membrane covered with myelin. The positive charges that flow into the axon at the node do, however, flow down the inside of the axon in the form of ionic electric current. When the current reaches the next node, the plasma membrane at that node is depolarized to threshold and fires another action potential. Action potentials therefore jump from node to node along the axon.

The speed of conduction is increased in these myelin-wrapped axons because ionic electric current flows much faster through the cytoplasm than ion channels can open and close. This form of rapid impulse propagation is called **saltatory conduction** (Latin *saltare*, "to jump").

RECAP 45.2

Neurons have membrane potentials due to ionic concentration differences across their membranes and because leak channels make the membrane differentially permeable to ions. Changes in ion channel permeabilities cause graded changes in membrane potentials. Sudden openings and closings of gated ion channels in the membrane produce action potentials. Action potentials are rapid, all-or-none changes in membrane potential that are conducted along axons from the cell body to the axon terminals.

- How are membrane resting potentials generated and altered? **See pp. 928–929 and Figures 45.6 and 45.9**
- What does the Nernst equation calculate, and why does that calculation not equal the measured resting potential of a membrane? **See p. 929 and Figure 45.7**
- How are action potentials generated? **See pp. 932–934 and Figure 45.10**
- How are action potentials transmitted along axons? **See pp. 934–936 and Figures 45.11 and 45.12**

Having described how action potentials are generated and transmitted along axons, we will next address the question of what happens when an action potential reaches the axon terminal. How is its signal communicated to the next cell—which could be another neuron, a muscle cell, or a secretory cell?

45.3 How Do Neurons Communicate with Other Cells?

Neurons communicate with each other and with other cells at synapses. In **electrical synapses**, which are common among invertebrates, the action potential spreads directly from presynaptic to postsynaptic cell. The most common type of synapse in the vertebrate nervous system is the **chemical synapse**, in which neurotransmitters released from a presynaptic cell induce changes in a postsynaptic cell. We will begin this section with a discussion of the synapses between neurons and muscle cells. We will then consider the diversity in synapses and how they integrate information.

The neuromuscular junction is a model chemical synapse

Neuromuscular junctions are synapses between neurons and skeletal muscle cells. They are excellent models for how chemical synaptic transmission works. Like other neurons, a motor neuron has only one axon, but that axon can branch into numerous axon terminals that form many synapses with muscle cells. At each axon terminal an enlarged knob or buttonlike structure contains vesicles filled with neurotransmitter molecules. The neurotransmitter used by all vertebrate neuromuscular synapses is **acetylcholine** (**ACh**). ACh is released by exocytosis when the membrane of a vesicle fuses with the presynaptic membrane of the axon terminal.

Where does the neurotransmitter come from? Some neurotransmitters, such as ACh, are synthesized in the axon terminal and packaged in vesicles. The enzymes required for ACh biosynthesis, however, are produced in the cell body of the motor neuron and are transported along microtubules down the axon to the terminals. In contrast, peptide neurotransmitters are produced in the cell body and packaged into membrane-bound vesicles by the Golgi apparatus. These vesicles are transported down the axon to the terminals.

The postsynaptic membrane of the neuromuscular junction is a modified part of the muscle cell plasma membrane called a **motor end plate**. It appears as a depression in the muscle cell membrane, and the terminals of the motor neuron sit in the depression. The space between the presynaptic membrane and the postsynaptic membrane is the **synaptic cleft**, which in chemical synapses is about 20 to 40 nanometers wide. ACh released into the cleft by the presynaptic cell diffuses across to the postsynaptic membrane (**Figure 45.13**).

The arrival of an action potential causes the release of neurotransmitter

Neurotransmitter is released when an action potential arrives at the axon terminal and causes the opening of voltage-gated Ca^{2+} channels in the presynaptic membrane. Because the Ca^{2+} concentration is greater outside the cell than inside, Ca^{2+} enters the axon terminal. This increase in Ca^{2+} inside the axon terminal causes the vesicles containing neurotransmitter to fuse with the presynaptic membrane and empty their contents into the synaptic cleft.

In neuromuscular synapses, vesicle fusion and emptying is all-or-none. The vesicle membrane is incorporated into the presynaptic membrane, which actually gets larger as a result—at least until the extra membrane is recycled through endocytosis. The membrane is reprocessed by the cell into new vesicles that are refilled with neurotransmitter.

Synaptic functions involve many proteins

The description above of the release of neurotransmitter from the presynaptic membrane may seem simple, but it involves hundreds of proteins that are responsible for various aspects of the process: vesicle formation, transport of neurotransmitter into vesicles, anchoring of vesicles to cytoskeletal elements, docking of the vesicles with the presynaptic membrane, fusion of the vesicular and cell membranes, and endocytosis of the vesicle membrane for recycling.

Some of these proteins are the targets of toxins. For example, botulinum and tetanus toxins from bacteria of the genus *Clostridium* act on several of the proteins necessary for the docking of vesicles to the presynaptic membrane, resulting in diseases that are frequently fatal. Botulinum toxin impairs muscle contraction, whereas tetanus toxin causes uncontrolled muscle contraction. Poisons can become medicines, however. Botulinum toxin (in the form of Botox®) is used therapeutically to subdue muscle spasms and cosmetically to subdue wrinkles.

The postsynaptic membrane responds to neurotransmitter

When acetylcholine is released at a synapse, some of it diffuses across the synaptic cleft and binds to ACh receptors on the postsynaptic membrane (**Figure 45.14**). The postsynaptic

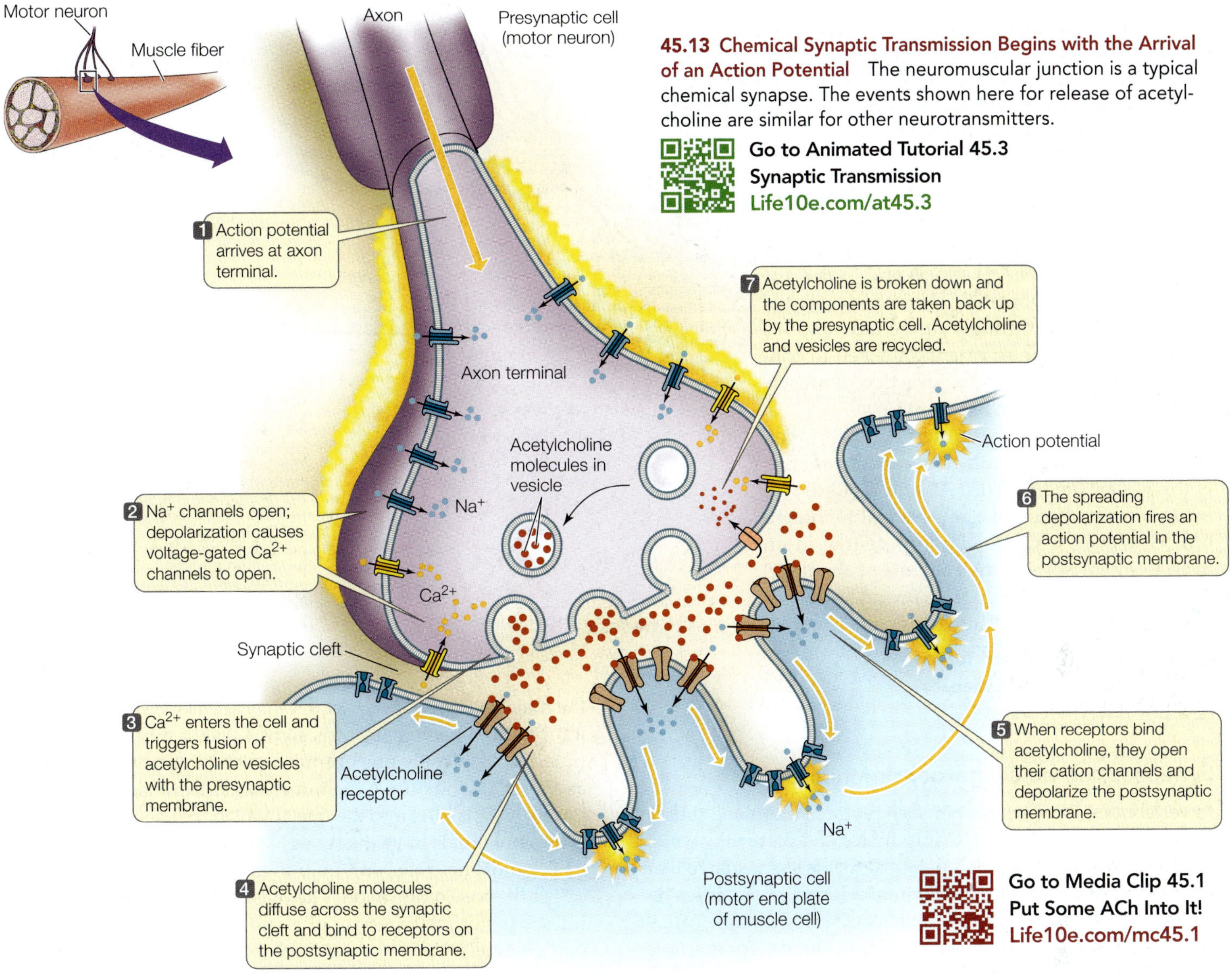

Motor neuron
Muscle fiber
Axon
Presynaptic cell (motor neuron)
1 Action potential arrives at axon terminal.
Axon terminal
Acetylcholine molecules in vesicle
7 Acetylcholine is broken down and the components are taken back up by the presynaptic cell. Acetylcholine and vesicles are recycled.
Action potential
2 Na^+ channels open; depolarization causes voltage-gated Ca^{2+} channels to open.
Na^+
Ca^{2+}
6 The spreading depolarization fires an action potential in the postsynaptic membrane.
Synaptic cleft
3 Ca^{2+} enters the cell and triggers fusion of acetylcholine vesicles with the presynaptic membrane.
Acetylcholine receptor
5 When receptors bind acetylcholine, they open their cation channels and depolarize the postsynaptic membrane.
Na^+
4 Acetylcholine molecules diffuse across the synaptic cleft and bind to receptors on the postsynaptic membrane.
Postsynaptic cell (motor end plate of muscle cell)

45.13 Chemical Synaptic Transmission Begins with the Arrival of an Action Potential The neuromuscular junction is a typical chemical synapse. The events shown here for release of acetylcholine are similar for other neurotransmitters.

Go to Animated Tutorial 45.3
Synaptic Transmission
Life10e.com/at45.3

Go to Media Clip 45.1
Put Some ACh Into It!
Life10e.com/mc45.1

membrane of the motor end plate is highly folded. ACh receptors are on the crests of the folds, and voltage-gated cation channels are at the bottoms of the folds and in the surrounding muscle cell membrane (see Figure 45.13). The ACh receptors are chemically gated channels that allow both Na^+ and K^+ to flow through, but since the electrochemical gradients favor a net influx of Na^+, the response of the motor end plate to ACh is to depolarize. That graded potential reflecting the number of receptors activated spreads to the depths of the folds of the motor end plate membrane and to surrounding muscle cell membrane, which contain voltage-gated Na^+ channels.

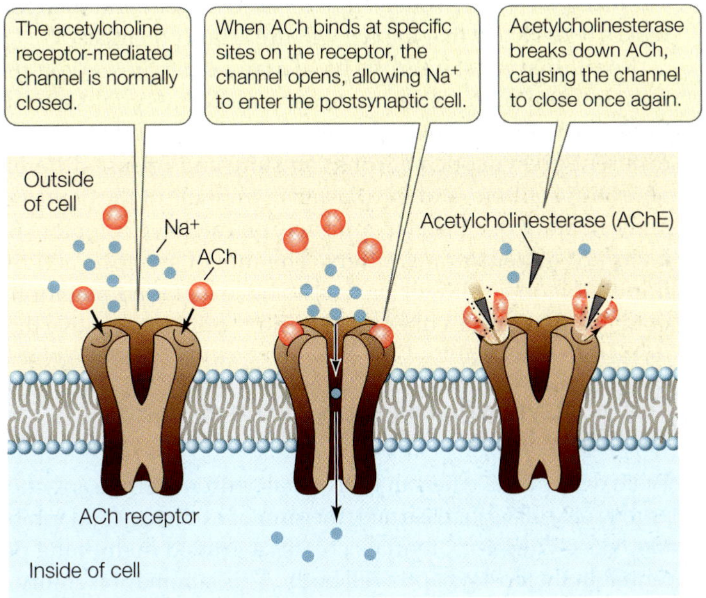

The acetylcholine receptor-mediated channel is normally closed.
When ACh binds at specific sites on the receptor, the channel opens, allowing Na^+ to enter the postsynaptic cell.
Acetylcholinesterase breaks down ACh, causing the channel to close once again.
Outside of cell
Na^+
ACh
Acetylcholinesterase (AChE)
ACh receptor
Inside of cell

45.14 Acetylcholine Receptors ACh receptors are chemically gated ion channels found in the motor end plate and other types of postsynaptic membranes. When one of these receptors binds ACh, its channel pore opens and Na^+ moves into the postsynaptic cell, depolarizing its membrane. The enzyme acetylcholinesterase (AChE) breaks down ACh in the synapse, closing the channel. The breakdown products (acetate and choline) are then taken up by the presynaptic membrane and resynthesized into more ACh.

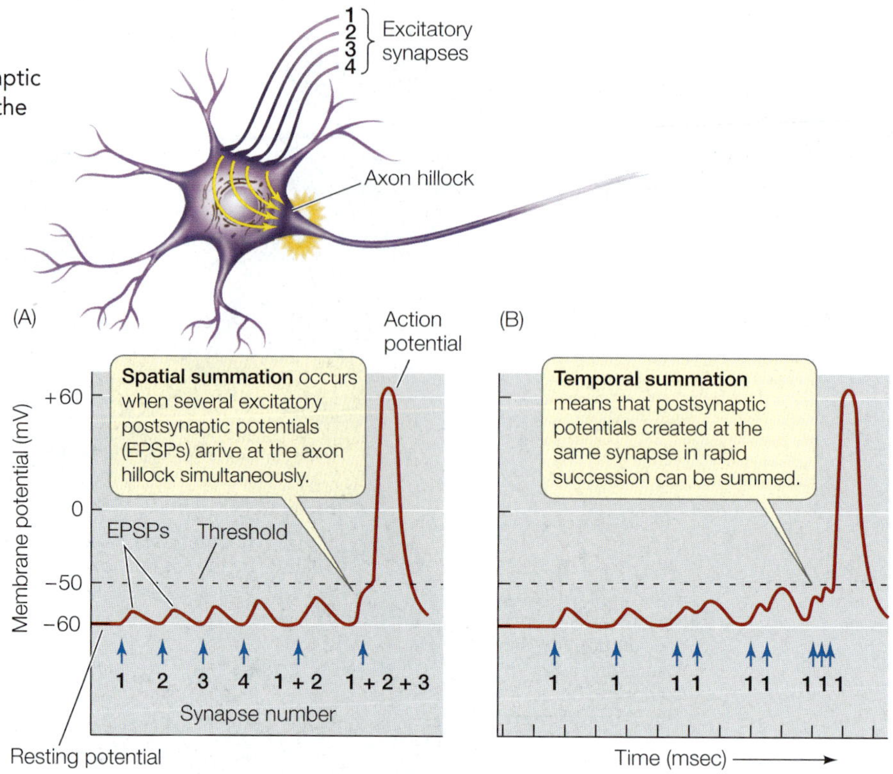

1
2
3
4
Excitatory synapses
Axon hillock
(A)
Action potential
(B)
Spatial summation occurs when several excitatory postsynaptic potentials (EPSPs) arrive at the axon hillock simultaneously.
Membrane potential (mV)
+60
0
–50
–60
EPSPs
Threshold
1 2 3 4 1 + 2 1 + 2 + 3
Synapse number
Resting potential
Temporal summation means that postsynaptic potentials created at the same synapse in rapid succession can be summed.
1 1 1 1 1 1 1 1 1
Time (msec)

45.15 The Postsynaptic Neuron Sums Information Individual neurons sum excitatory and inhibitory postsynaptic potentials over space (A) and time (B). When the sum of the potentials depolarizes the axon hillock to threshold, the neuron generates an action potential.

If the axon terminal of a motor neuron releases sufficient amounts of ACh to depolarize a motor end plate, that spreading depolarization activates the voltage-gated Na^+ channels and causes the firing of an action potential. This action potential is then conducted throughout the muscle cell's system of membranes, causing the cell to contract. We will discuss the contraction of muscle cells in greater detail in Section 48.1.

How much neurotransmitter is enough? Neither a single ACh molecule nor the contents of an entire vesicle (about 10,000 ACh molecules) will bring the membrane of a muscle cell to threshold. However, a single action potential in an axon terminal releases the contents of about 100 vesicles—more than enough to fire an action potential in the muscle cell and cause it to contract.

Go to Animated Tutorial 45.4
Neurons and Synapses
Life10e.com/at45.4

Synapses can be excitatory or inhibitory

In vertebrates, the synapses between motor neurons and muscle cells are always excitatory; that is, motor end plates always respond to ACh with a graded potential that is less negative than the resting potential (depolarization). However, synapses between neurons are frequently inhibitory; such a synapse causes hyperpolarization of the postsynaptic membrane. For example, recall that there are more chloride ions (Cl^-) outside the cell than inside it. If the receptors on the postsynaptic membrane open Cl^- channels, Cl^- ions enter the postsynaptic cell and hyperpolarize it. Hyperpolarization takes the membrane farther from the threshold potential for the voltage-gated Na^+ channels, and therefore makes it less likely that the cell will fire action potentials.

Recall that most neurons have many dendrites. Axon terminals from many other neurons can form synapses with those dendrites and with the cell body. The axon terminals of different presynaptic neurons can store and release different neurotransmitters, and the plasma membrane of the dendrites and cell body of a postsynaptic neuron can have receptors for a variety of neurotransmitters. The mix of synaptic activity impinging on a cell will cause it to have a graded membrane potential that can be either more positive or more negative than its resting potential.

The postsynaptic cell sums excitatory and inhibitory input

What determines when an individual neuron will fire an action potential? As we just learned, the sum of excitatory and inhibitory postsynaptic potentials creates a graded membrane potential in the postsynaptic cell body. This summation ability is the major mechanism by which the nervous system integrates information. Each neuron receives 1,000 or more synaptic inputs, but it has only one output: an action potential in a single axon. At any one time, the information from all of the active inputs is translated into the rate at which that neuron generates action potentials in its axon.

For most neurons, summation takes place in the axon hillock at the base of the axon. The plasma membrane of the axon hillock is not insulated by glia and has many voltage-gated Na^+ channels. Excitatory and inhibitory postsynaptic potentials from synapses anywhere on the dendrites or the cell body spread to the axon hillock by local current flow. If the resulting graded potential depolarizes the axon hillock to threshold, it fires an action potential. Because postsynaptic potentials decrease in strength as they spread from the site of the synapse, a synapse at the tip of a dendrite has less influence than a synapse on the cell body, near the axon hillock.

Excitatory and inhibitory postsynaptic potentials are summed over space and over time. **Spatial summation** adds up the simultaneous influences of synapses at different sites on the postsynaptic cell (**Figure 45.15A**). **Temporal summation** adds up postsynaptic potentials generated at the same site in a rapid sequence (**Figure 45.15B**).

Synapses can be fast or slow

Most neurotransmitter receptors induce changes in postsynaptic cells by opening or closing ion channels. How they do so is the basis for grouping receptors into two general categories:

- **Ionotropic receptors** are themselves ion channels. Neurotransmitter binding to an ionotropic receptor causes a direct change in ion movement across the plasma membrane of the postsynaptic cell. These proteins enable fast, short-lived responses.

The ACh receptor of the motor end plate is an example of an ionotropic receptor. Each of its five subunits extends through the plasma membrane. When assembled, the subunits create a central pore that allows ions to pass through (see Figure 45.14). Of several different kinds of subunit, only one kind has the ability to bind ACh. Each functional receptor has two ACh-binding subunits and three other subunits.

- **Metabotropic receptors** are not ion channels, but they induce signaling cascades in the postsynaptic cell that secondarily lead to changes in ion channels (see Figure 7.10A). Postsynaptic cell responses mediated by metabotropic receptors are generally slower and longer-lived than those induced by ionotropic receptors.

Metabotropic receptors are also transmembrane proteins, but instead of acting as ion channels, they initiate an intracellular signaling process that results in the opening or closing of an ion channel and other changes in the postsynaptic cell.

Electrical synapses are fast but do not integrate information well

Electrical synapses are different from chemical synapses because they couple neurons electrically. Electrical synapses contain numerous gap junctions. At these synapses, the presynaptic and postsynaptic cell membranes are separated by a space of only 2 or 3 nanometers, and membrane proteins called connexins link the two neurons by forming pores that connect the cytoplasm of the two cells (see Figure 7.19A). Ions and small molecules can pass directly from cell to cell through these pores. Transmission at electrical synapses is very fast and can proceed in either direction, whereas transmission at chemical synapses is slower and unidirectional.

Electrical synapses are less common in the nervous systems of vertebrates than are chemical synapses for several reasons. First, electrical continuity between neurons does not allow temporal summation of synaptic inputs. Second, an effective electrical synapse requires a large area of contact between the presynaptic and postsynaptic cells. This condition rules out the possibility of thousands of synaptic inputs to a single neuron—which is the norm in complex nervous systems. Third, electrical synapses cannot be inhibitory. Thus electrical synapses are useful for rapid communication, but they are less useful for processes of integration and learning.

The action of a neurotransmitter depends on the receptor to which it binds

More than 100 neurotransmitters are now recognized, and more will surely be discovered. ACh, as we have seen, is an important neurotransmitter because it is how the nervous system commands muscles to contract. ACh also plays roles in certain synapses between neurons in the brain, but it accounts for only a small percent of the total neurotransmitter content of the brain.

The workhorse neurotransmitters of the brain are three simple amino acids: glutamate, which is excitatory, and glycine and γ-aminobutyric acid (GABA), which are inhibitory. The integration of information at the cellular level is a balance between excitation and inhibition, so it is understandable that excessive inhibition such as we saw in the Down syndrome mice described at the start of this chapter could impair an animal's ability to process and integrate new information. Indeed, when GABA's inhibitory action in these mice was reduced by a drug that blocks the GABA receptor, the learning ability of the mice increased (**Figure 45.16**).

Another important group of neurotransmitters in the brain is the monoamines, which are derivatives of amino acids. They

INVESTIGATING**LIFE**

45.16 Reducing Neuronal Inhibition May Enhance Learning
Learning and memory in mice can be studied using the "novel object recognition task" described in the method below. Genetically engineered mice that are models for Down syndrome (i.e., the genes known to be present on human chromosome 21 have been triplicated in the mouse) cannot perform this task. Treatment of these model mice with a drug that partially blocks GABA receptors (GABA being an inhibitor of synaptic transmission) appears to increase their ability to learn.[a]

HYPOTHESIS Excessive inhibition of neurons by GABA impairs learning ability in a mouse model of Down syndrome.

Method
1. Place the mouse in arena with two objects and allow them to explore and investigate; remove mouse.
2. After 24 hours, change one of the two objects and return the mouse to the arena. Compare the amount of time the mouse spends with the old versus the novel object. If the mouse spends more time with the novel object, it remembers the old object (has "learned").
3. Repeat the experiment with normal and Down syndrome mice before and after the subject mice are treated with a drug that blocks GABA receptors (GABAr).

Results

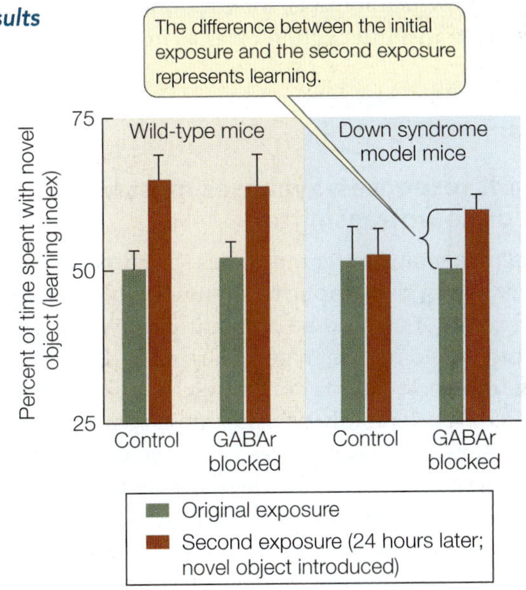

The difference between the initial exposure and the second exposure represents learning.
Wild-type mice
Down syndrome model mice
Percent of time spent with novel object (learning index)
75
50
25
Control
GABAr blocked
Control
GABAr blocked
Original exposure
Second exposure (24 hours later; novel object introduced)

CONCLUSION Blocking GABA receptors improved learning capacity in the Down syndrome model mice.

Go to **BioPortal** for discussion and relevant links for all INVESTIGATING**LIFE** figures.

[a]Fernandez, F. et al. 2007. *Nature Neuroscience* 10: 411–413.

include dopamine and norepinephrine (derivatives of tyrosine) and serotonin (a derivative of tryptophan). Peptides also function as neurotransmitters; for example, endorphins and enkephalins are the body's opiates and modulate the sensation of pain. Another peptide, substance P, transmits pain sensations. Even a gas, nitric oxide, is used by neurons as an intercellular messenger (see Figure 7.15).

Neurotransmission is complex in part because each neurotransmitter has multiple receptor types. ACh, for example, has two receptor types: nicotinic receptors, which are ionotropic; and muscarinic receptors, which are metabotropic. Both types of ACh receptors are found in the brain and spinal cord, where nicotinic receptors tend to be excitatory and muscarinic receptors tend to be inhibitory. ACh actions can differ outside the brain and spinal cord as well. ACh acting through nicotinic receptors causes the smooth muscle of the gut to increase its motility, but ACh acting through muscarinic receptors causes cardiac muscle to hyperpolarize and therefore to slow down. There are many more examples of neurotransmitters that have different effects in different tissues, but the important thing to remember is that the action of a neurotransmitter depends on the receptor to which it binds. In addition, keep in mind that turning neurotransmitter action off is as important as turning it on.

Go to Activity 45.1 Neurotransmitters
Life10e.com/ac45.1

To turn off responses, synapses must be cleared of neurotransmitter

If released neurotransmitter molecules simply remained in the synaptic cleft, the postsynaptic membrane would become saturated and receptors would be constantly activated. The postsynaptic cell would remain hyperpolarized or depolarized and would be unresponsive to short-term changes in the presynaptic cell. The more rapidly neurons can respond to input, the more information they can process in a given amount of time. Thus neurotransmitter must be cleared from the synaptic cleft shortly after it is released by the axon terminal.

Neurotransmitter action can be terminated in several ways. First, enzymes may destroy the neurotransmitter. Acetylcholine, for example, is rapidly destroyed by the enzyme acetylcholinesterase (AChE), which is present in the synaptic cleft in close association with ACh receptors on the postsynaptic membrane (see Figure 45.14). When AChE is inhibited, ACh lingers in the synaptic cleft, causing spastic (contracted) muscle paralysis and usually resulting in death. Some of the most deadly nerve gases developed for chemical warfare work by inhibiting AChE. Some agricultural insecticides, such as malathion, also inhibit AChE and can poison farm workers if used without safety precautions.

Neurotransmitter can also simply diffuse away from the cleft, or be taken up via active transport by nearby cell membranes, most notably glial cell membranes. The antidepressant drug commonly prescribed under the brand name Prozac slows the reuptake of the neurotransmitter serotonin, thus enhancing serotonin's activity at the synapse.

The diversity of receptors makes drug specificity possible

Many drugs used to treat the nervous system act by modulating specific synaptic interactions. Drugs that mimic or potentiate the effect of a neurotransmitter are called **agonists**; those that block the actions of neurotransmitters are called **antagonists**. For example, morphine is an agonist at the endorphin receptor and therefore blocks pain. Propranolol, a widely used β-blocker, is an antagonist of certain adrenergic receptors and therefore decreases panic attacks and anxiety. A major emphasis in neurobiology is to identify neurotransmitter receptor subtypes and design drugs that selectively bind to them to have highly specific effects on nervous system activity.

RECAP 45.3

Chemical synapses involve the release of neurotransmitter molecules stored in vesicles in the presynaptic terminal. Action potentials reaching that terminal cause the fusion of vesicles with the presynaptic membrane, releasing neurotransmitter that can then bind to receptors on the postsynaptic membrane and influence its membrane potential. There is a great diversity of neurotransmitters and their receptors.

- Describe the role of Ca^{2+} channels in synaptic events. **See p. 936 and Figure 45.13**
- How can some synapses be excitatory and others inhibitory? **See p. 938**
- How do neurons integrate the input from various synapses? **See p. 938 and Figure 45.15**

45.4 How Are Neurons and Glia Organized into Information-Processing Systems?

Nervous systems can process information because their neurons are organized into **neural networks**. These networks include three functional categories of neurons, which can be thought of as being involved with input, output, and integration:

- **Afferent neurons** carry sensory information into the nervous system. That information comes from specialized sensory cells that transduce (convert) various kinds of sensory stimuli (e.g., light, heat, pressure) into action potentials.
- **Efferent neurons** carry commands to physiological and behavioral effectors such as muscles and glands.
- **Interneurons** integrate and store information and communicate between afferent and efferent neurons.

Nervous systems range in complexity

Simple animals such as cnidarians (sea anemones, for example) process information with a limited number of simple neural networks that do little more than provide direct lines of communication from sensory cells to effectors; there is little or no integration or processing of signals (**Figure 45.17A**). The

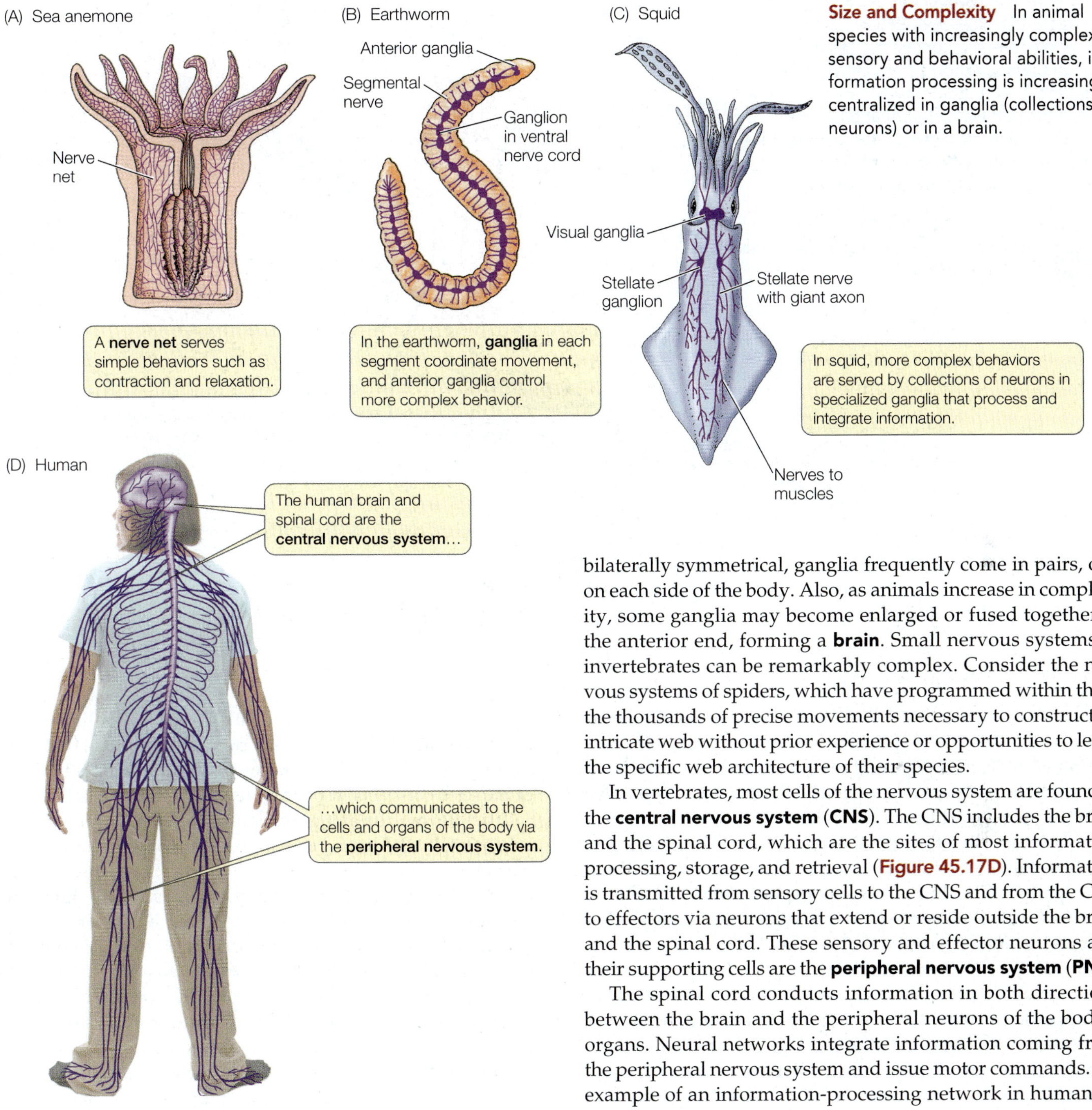
(A) Sea anemone
Nerve net
A **nerve net** serves simple behaviors such as contraction and relaxation.
(B) Earthworm
Anterior ganglia
Segmental nerve
Ganglion in ventral nerve cord
In the earthworm, **ganglia** in each segment coordinate movement, and anterior ganglia control more complex behavior.
(C) Squid
Visual ganglia
Stellate ganglion
Stellate nerve with giant axon
Nerves to muscles
In squid, more complex behaviors are served by collections of neurons in specialized ganglia that process and integrate information.
(D) Human
The human brain and spinal cord are the **central nervous system**...
...which communicates to the cells and organs of the body via the **peripheral nervous system**.

45.17 Nervous Systems Vary in Size and Complexity In animal species with increasingly complex sensory and behavioral abilities, information processing is increasingly centralized in ganglia (collections of neurons) or in a brain.

cnidarian's **nerve net** is most developed around the tentacles and the oral opening, where it facilitates detection of food or danger and causes tentacles to extend or retract.

Animals that are more complex and move about in search of food and mates must process and integrate larger amounts of information. Even earthworms fit this description, and their increased need for information processing is met by higher numbers of neurons organized into clusters of neuronal cell bodies called **ganglia** (singular *ganglion*). Ganglia serving different functions may be distributed around the body, as in earthworms and squid (**Figure 45.17B,C**). In animals that are bilaterally symmetrical, ganglia frequently come in pairs, one on each side of the body. Also, as animals increase in complexity, some ganglia may become enlarged or fused together at the anterior end, forming a **brain**. Small nervous systems of invertebrates can be remarkably complex. Consider the nervous systems of spiders, which have programmed within them the thousands of precise movements necessary to construct an intricate web without prior experience or opportunities to learn the specific web architecture of their species.

In vertebrates, most cells of the nervous system are found in the **central nervous system** (**CNS**). The CNS includes the brain and the spinal cord, which are the sites of most information processing, storage, and retrieval (**Figure 45.17D**). Information is transmitted from sensory cells to the CNS and from the CNS to effectors via neurons that extend or reside outside the brain and the spinal cord. These sensory and effector neurons and their supporting cells are the **peripheral nervous system** (**PNS**).

The spinal cord conducts information in both directions between the brain and the peripheral neurons of the body's organs. Neural networks integrate information coming from the peripheral nervous system and issue motor commands. An example of an information-processing network in humans is the spinal knee-jerk reflex.

The knee-jerk reflex is controlled by a simple neural network

A cross section of the human spinal cord reveals a central area of gray matter in the shape of a butterfly, surrounded by an area of white matter (see Figure 45.18). In the nervous system, **gray matter** is rich in neural cell bodies, and **white matter** contains myelinated axons. The gray matter of the spinal cord contains the cell bodies of the spinal neurons; the white matter contains myelinated axons that conduct information up and down the spinal cord.

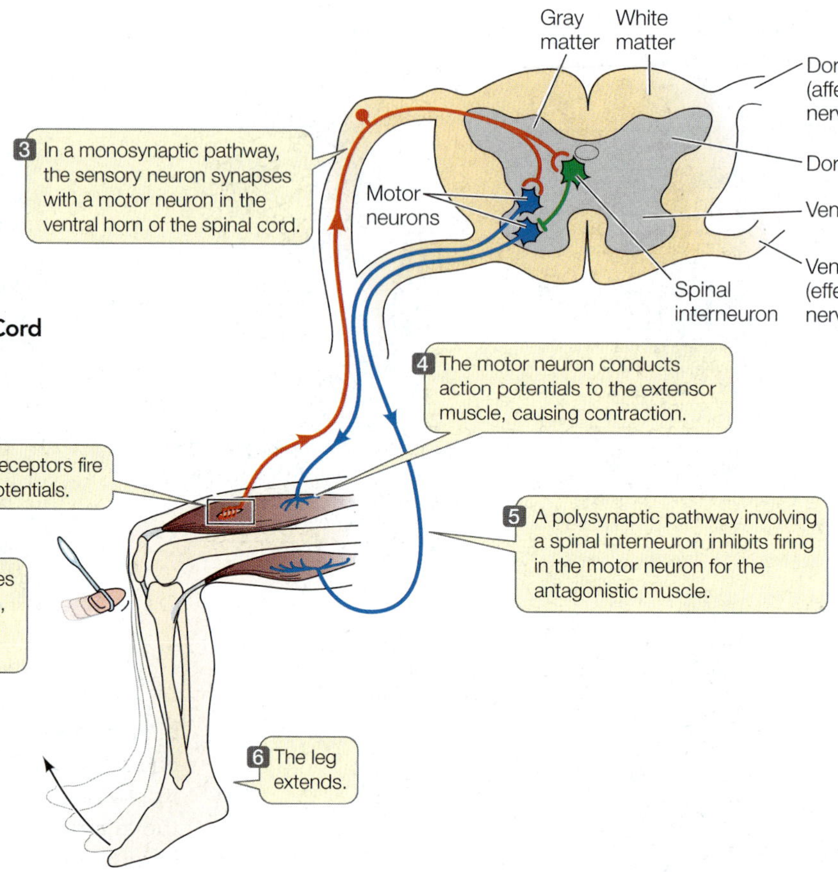

Gray matter
White matter
Dorsal root (afferent nerves)
Dorsal horn
Ventral horn
Ventral root (efferent nerves)
Spinal interneuron
Motor neurons
3 In a monosynaptic pathway, the sensory neuron synapses with a motor neuron in the ventral horn of the spinal cord.
4 The motor neuron conducts action potentials to the extensor muscle, causing contraction.
2 Stretch receptors fire action potentials.
5 A polysynaptic pathway involving a spinal interneuron inhibits firing in the motor neuron for the antagonistic muscle.
1 A hammer tap stretches the tendon in the knee, stretching receptors in the extensor muscle.
6 The leg extends.

45.18 A Neural Network in the Spinal Cord Generates the Knee-Jerk Reflex Sensory (afferent) information enters the spinal cord through the dorsal horns (red pathway), and motor (efferent) output leaves it via the ventral horns (blue pathways). Information travels to the brain in white matter tracts. Interneurons make connections within the spinal cord that result in a complex, coordinated behavior pattern.

Go to Animated Tutorial 45.5
Information Processing in the Spinal Cord
Life10e.com/at45.5

Spinal nerves extend from the spinal cord at regular intervals on each side. Each spinal nerve has two roots, one connecting with the dorsal horn of the gray matter, the other with the ventral horn. The afferent (sensory) axons in a spinal nerve enter the spinal cord through the dorsal root, and the efferent (motor) axons leave through the ventral root. The conversion of afferent to efferent information in the spinal cord without participation of the brain is called a **spinal reflex**. The neural network involved in one such spinal reflex—the "knee-jerk"—is diagrammed in Figure 45.18. The stimulus for this particular reflex is a tap on the knee with a small mallet. That tap stretches the tendon going over the knee, which attaches the muscle of the upper leg to bone in the lower leg. Stretching the tendon stretches muscle fibers in the upper leg, and stretch receptors in that muscle transduce the physical stimulus into action potentials. The action potentials are then conducted by a sensory neuron into the dorsal horn of the spinal cord. That sensory neuron goes all the way to the ventral horn and synapses onto a motor neuron, which fires action potentials. The axon of that motor neuron travels out through the ventral horn of the spinal cord and extends all the way to the same muscle that initially was stretched, causing that muscle to contract.

The function of this simple circuit is to sense an increased load on the muscle and to increase the strength of its contraction to compensate for that additional load. Because there is only one synapse between the afferent and efferent neurons in this simple network, it is called a monosynaptic reflex; most spinal networks are more complex. For example, limb movement is controlled by antagonistic sets of muscles that work against each other. When one member of an antagonistic set of muscles contracts, it bends (flexes) the limb; it is therefore called a flexor. The antagonist muscle, the extensor, straightens (extends) the limb. For a limb to move, one muscle of the pair must relax while the other contracts. Thus sensory input that activates the motor neuron of one muscle also inhibits its antagonist. This coordination is achieved by an interneuron, which makes an inhibitory synapse onto the motor neuron of the antagonistic muscle (see Figure 45.18). Thus the reciprocal inhibition of antagonistic muscles involves an interneuron between the sensory cell and the motor neuron of the inhibited muscle, and therefore at least two synapses.

The withdrawal reflex is an example of a polysynaptic spinal reflex that involves many interneurons. When you step on a tack, you immediately pull back your foot: the tack stimulates pain receptors in the foot, and the sensory neurons transmit action potentials into the dorsal horn of the spinal cord on the same side of the body. In the dorsal horn, these neurons synapse with interneurons that send information through their axons to the brain, resulting in the conscious sensation of pain. Before the brain is aware of the pain, however, synapses of the sensory neurons with other interneurons stimulate and inhibit a variety of different motor neurons in the spinal cord. Interneurons on the same side of the spinal cord coordinate the activity of the muscles that withdraw the foot and leg. To

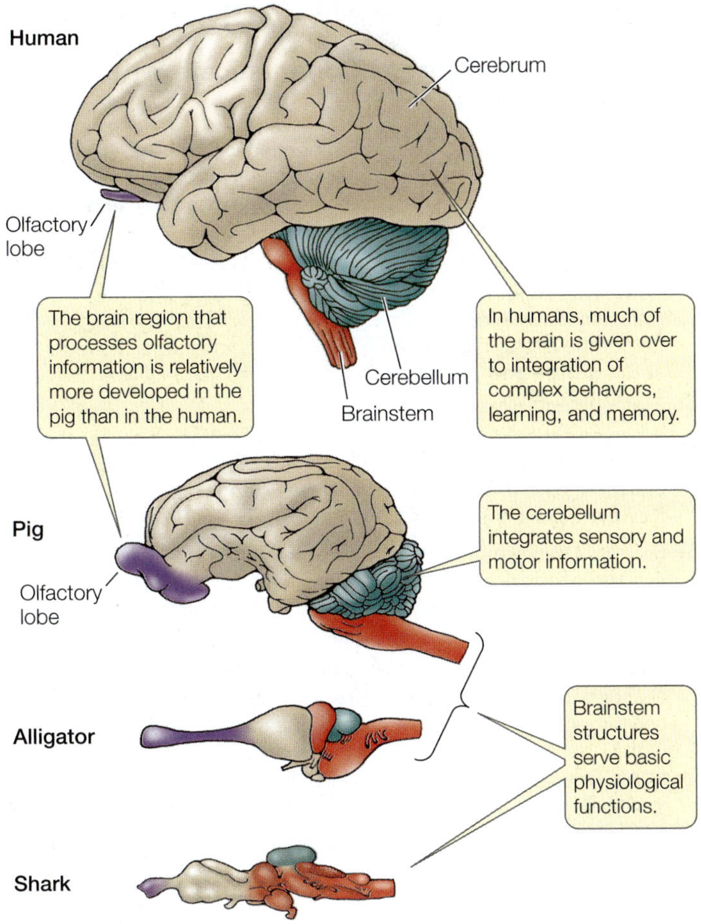
Human
Cerebrum
Olfactory lobe
The brain region that processes olfactory information is relatively more developed in the pig than in the human.
Cerebellum
Brainstem
In humans, much of the brain is given over to integration of complex behaviors, learning, and memory.
Pig
The cerebellum integrates sensory and motor information.
Olfactory lobe
Alligator
Brainstem structures serve basic physiological functions.
Shark

45.19 Brains Vary in Size and Complexity The brains of four vertebrate species—all of which may have a similar body mass—show immense differences. Note that the brainstem, which is involved in physiological regulation and stereotypic behavior, differs less among these species than does the cerebrum, which is responsible for complex behavior and learning.

pull away, however, the other leg has to extend and balance must be shifted. The coordination of these actions involves interneurons that make connections across the spinal cord to motor neurons on the opposite side. Thus a rather complex suite of movements is coordinated by a network in the spinal cord. By extension, you can appreciate how much more complex a neural network is required to enable you to execute complex movements in time with music and coordinated with another person—in other words, to dance.

The vertebrate brain is the seat of behavioral complexity

Vertebrates differ greatly in their behavioral complexity and in their physiological specializations, and their neural networks reflect this diversity. **Figure 45.19** shows the brains of four vertebrate species of similar body mass drawn to the same scale.

The human nervous system contains an estimated 10^{11} neurons. A given neuron in the brain can have 1,000 or more synapses. Thus the human brain can contain 10^{14} synapses (10^{11} neurons $\times$ 10^3 synapses per neuron). Then there are the glia. A single astrocyte might participate in 100,000 synapses while at the same time monitoring signals in the extracellular fluid and the blood. And beyond these numbers, synapses are not constant but can be highly plastic. They can increase or decrease in number and size, and they can become more or less sensitive.

This astronomical number of neurons and synapses is divided into thousands of distinct but interacting networks that function in parallel. The possible number of informational networks in the brain is almost infinite, and therein lies the incredible ability of the human brain to process information, to learn, to do complex tasks, to remember, and to have emotions.

RECAP 45.4

Nervous systems are composed of neural networks that include afferent neurons, interneurons, and efferent neurons. Nervous systems range in complexity from simple nerve nets to the human brain. The spinal knee-jerk reflex is an example of a simple neural network, but networks that control more complex behavior become increasingly complicated.

- What are ganglia, and why are they concentrated in the anterior region of many invertebrates? **See p. 941 and Figure 45.17**
- Explain the components of a spinal reflex network. How can the same stimulus cause contraction in one muscle and relaxation in another? **See p. 942 and Figure 45.18**

?

What causes overinhibition in the nervous system, and how can it be reduced?

ANSWER

Some neurotransmitter–receptor combinations excite activity in the postsynaptic neuron, depolarizing it and making it likely to fire an action potential. Other neurotransmitter–receptor combinations inhibit the responses of the postsynaptic cell. Neurons integrate information by summing excitatory and inhibitory synaptic inputs. Overinhibition in the nervous system can arise when too much inhibitory neurotransmitter (such as GABA) is released, or if postsynaptic cells have too many receptors for inhibitory neurotransmitters. Drugs that decrease the synthesis and release of the inhibitory neurotransmitter can reduce the inhibition, as can drugs that block the receptors for such neurotransmitters. In the case of the Down syndrome model mice, drugs that blocked the GABA receptors that are also Cl^- channels reduced the level of inhibition in the mice nervous systems, apparently increasing the ability of the mice to learn and form memories.

CHAPTER SUMMARY 45

45.1 What Cells Are Unique to the Nervous System?

- The cells of the nervous systems include many types of **neurons** and **glia**.
- All neurons are **excitable**, which means they can generate and conduct electric signals called **action potentials**. Glia support and modulate the activities of neurons but do not generate action potentials.
- A neuron generally receives information via its **dendrites**, of which there can be many, and transmits information via its single **axon**, which ends in **axon terminals**. **Review Figure 45.1**
- Where neurons and their target cells meet, information is transmitted across specialized junctions called **synapses**.
- Glia include **Schwann cells** and **oligodendrocytes,** both of which generate **myelin** sheets on axons. Glia also include **astrocytes**, which support neurons metabolically, modulate synaptic signaling, and contribute to the **blood–brain barrier**. **Review Figures 45.3, 45.4**

45.2 How Do Neurons Generate and Transmit Electric Signals?

- Neurons have an electric charge difference across their plasma membranes, the **membrane potential**. The membrane potential is created by ion transporters and channels. When a neuron is not firing action potentials, its membrane potential is referred to as the **resting potential**. **Review Figures 45.5, 45.6, ANIMATED TUTORIAL 45.1**
- The **sodium–potassium pump** concentrates K^+ on the inside of a neuron and Na^+ on the outside. Potassium leak channels allow K^+ to diffuse out of the neuron, leaving behind unbalanced negative charges. **Review Figures 45.6, 45.7**
- **Patch clamping** allows the study of single ion channels. **Review Figure 45.8**
- The resting potential is perturbed when ion channels open or close, changing the permeability of the plasma membrane to charged ions. Through this mechanism, the plasma membrane can become **depolarized** or **hyperpolarized** and therefore have a **graded membrane potential** response to input. **Review Figure 45.9**
- An action potential results from a rapid reversal in charge across a portion of the plasma membrane resulting from the sequential opening and closing of voltage-gated Na^+ and K^+ channels. These changes in voltage-gated channels occur when the plasma membrane depolarizes to a **threshold** level. **Review Figure 45.10, ANIMATED TUTORIAL 45.2**
- Action potentials are all-or-none, self-regenerating events. They are conducted down axons because local current flow depolarizes adjacent regions of membrane and brings them to threshold. **Review Figure 45.11**
- In myelinated axons, action potentials appear to jump between **nodes of Ranvier**, areas of axonal plasma membrane that are not covered by myelin. **Review Figure 45.12**

45.3 How Do Neurons Communicate with Other Cells?

- Neurons communicate with each other and with other cell types by transmitting information over **electrical synapses** or by the transmission of molecular signals called **neurotransmitters** over **chemical synapses**.
- The **neuromuscular junction** is a well-studied chemical synapse between a motor neuron and a skeletal muscle cell. Its neurotransmitter is **acetylcholine** (**ACh**), which causes depolarization of the postsynaptic membrane when it binds to its receptor at the **motor end plate**. **Review Figure 45.13, ANIMATED TUTORIAL 45.3**
- When an action potential reaches an axon terminal, it causes the release of neurotransmitters, which diffuse across the **synaptic cleft** and bind to receptors on the postsynaptic membrane. **Review Figures 45.13, 45.14, ANIMATED TUTORIAL 45.4**
- Synapses between neurons can be either excitatory or inhibitory. A postsynaptic neuron integrates information by summing excitatory and inhibitory postsynaptic potentials in both spatially and temporally. **Review Figure 45.15**
- **Ionotropic receptors** are ion channels or directly influence ion channels. **Metabotropic receptors** influence the postsynaptic cell through various signal transduction pathways and can result in the opening or closing of ion channels. The actions of ionotropic synapses are generally faster than those of metabotropic synapses.
- There are many different neurotransmitters and even more types of receptors. The action of a neurotransmitter depends on the type of receptor to which it binds. **See ACTIVITY 45.1**
- Neurotransmitter molecules cannot be allowed to accumulate in a synapse but must be cleared in order to turn off responses in the postsynaptic cell. This may be done by enzymatic degradation, simple diffusion, or reuptake of the neurotransmitter.

45.4 How Are Neurons and Glia Organized into Information-Processing Systems?

- In vertebrates, the brain and spinal cord form the **central nervous system** (**CNS**), which communicates with the rest of the body via the **peripheral nervous system** (**PNS**). The CNS increases in complexity from invertebrates to vertebrates and from fish to mammals. **Review Figures 45.17, 45.19**
- **Neural networks** include **afferent neurons** and **efferent neurons**, generally connected through **interneurons**.
- A **spinal reflex** is an example of a simple neural network that integrates information and controls a response. **Review Figure 45.18, ANIMATED TUTORIAL 45.5**

Go to the Interactive Summary to review key figures, Animated Tutorials, and Activities
Life10e.com/is45

CHAPTER**REVIEW**

REMEMBERING

1. The rising phase of an action potential is due to the
   a. closing of K^+ channels.
   b. opening of chemically gated Na^+ channels.
   c. closing of voltage-gated Ca^{2+} channels.
   d. opening of voltage-gated Na^+ channels.
   e. spread of positive current along the plasma membrane.

2. Which statement about synaptic transmission is *not* true?
   a. The synapses between neurons and skeletal muscle cells use ACh as their neurotransmitter.
   b. A single vesicle of neurotransmitter can cause a muscle cell to contract.
   c. The release of neurotransmitter at the neuromuscular junction causes the motor end plate to depolarize.
   d. In vertebrates, the synapses between motor neurons and muscle fibers are always excitatory.
   e. Inhibitory synapses cause the resting potential of the postsynaptic membrane to become more negative.

3. Which statement accurately describes an action potential?
   a. Its magnitude increases along the axon.
   b. Its magnitude decreases along the axon.
   c. All action potentials in a single neuron are of the same magnitude.
   d. During an action potential, the membrane potential of a neuron remains constant.
   e. An action potential permanently shifts a neuron's membrane potential away from its resting value.

4. Graded membrane potentials
   a. can be hyperpolarizing.
   b. can be depolarizing.
   c. integrate the many synaptic inputs to a cell.
   d. are important means of summing sensory inputs.
   e. All of the above

5. The binding of an inhibitory neurotransmitter to the postsynaptic receptors of a neuron at its resting potential results in
   a. depolarization of the membrane.
   b. generation of an action potential.
   c. hyperpolarization of the membrane.
   d. increased permeability of the membrane to sodium ions.
   e. increased permeability of the membrane to calcium ions.

UNDERSTANDING & APPLYING

6. If you stimulate an axon in the middle, action potentials are conducted in both directions. Yet when an action potential is generated at the axon hillock, it goes only toward the axon terminals and does not backtrack. Explain why action potentials are bidirectional in the first example and unidirectional in the second.

7. How does a neuron integrate excitatory and inhibitory synapses?

ANALYZING & EVALUATING

8. Benzodiazepines are drugs that potentiate the effects of GABA on its receptor. What effects might you expect to observe in a person taking these drugs?

9. The language of the nervous system consists of one "word," the action potential. How can this single word convey a diversity of information? How can that information be quantified, and how can it be integrated?

Go to BioPortal at **yourBioPortal.com** for Animated Tutorials, Activities, LearningCurve Quizzes, Flashcards, and many other study and review resources.

46 Sensory Systems

CHAPTER OUTLINE

Sensing Infrared Radiation The "hole" to the right of this diamondback rattlesnake's eye is one of its bilateral pit organs. Pit organs detect infrared radiation from the snake's preferred prey—small rodents—with unerring precision, even in total darkness. The forked tongue also provides positional information, picking up molecular signals that are transmitted to the brain by a specialized organ in the roof of the snake's mouth.

A RATTLESNAKE can see to strike a running rodent in complete darkness. How can this be, when "seeing" means using the eyes to detect light waves, and "complete darkness" means no light? It is possible because these definitions are based on human capabilities. What we call "light" is actually only a small portion (red, orange, yellow, green, blue, indigo, and violet) of the spectrum of electromagnetic radiation. Other animals see wavelengths humans cannot. In Chapter 27 we saw that insects perceive patterns on flowers that reflect ultraviolet wavelengths invisible to humans. Similarly, rattlesnakes "see" infrared wavelengths that we cannot (although at high enough levels of intensity, humans feel infrared wavelengths as heat).

It is not the snake's eyes that perceive infrared light. Rattlesnakes and their relatives have pit organs located between the nostril and the eye on each side of the skull that contain high densities of infrared-sensitive neurons. The two pits are positioned in such a way that sensory receptor cells in the pits receive directional information. The fields of "view" of the bilateral pits are overlapping and thus convey a three-dimensional perspective. Information from the pit organs goes to the same region of the brain as information from the eyes, so rattlesnakes actually do "see" the world in a range of electromagnetic radiation that is different from the human visual spectrum.

Our definition of silence is as arbitrary as our definition of darkness. "Sound" is actually pressure waves in the environment, and many animals are sensitive to pressure waves with frequencies, or pitches, we cannot hear. Elephants communicate using sound waves that are below human hearing range; such long waves travel great distances, an advantage to large animals that roam over extensive areas. Bats emit incredibly loud, brief sound pulses that are above our range of hearing. A flying bat hears echoes of these pulses bouncing off objects in the environment. The pulses are so loud and the echoes so weak that it is rather like a construction worker trying to overhear a whispered conversation while using a pneumatic drill.

"Reality" is what our eyes see, our ears hear, our noses smell, and what we touch and taste. But human beings sense only a limited range of the information available. Animals with different ranges of sensitivity process different sources of information and may perceive "reality" quite differently.

?

How can bats emit loud pulses of sound and not be deaf to the faint echoes that return within milliseconds?

See answer on p. 963.

46.1 How Do Sensory Receptor Cells Convert Stimuli into Action Potentials?

Sensory receptor cells, usually simply called **sensors** or **receptors**, transduce (convert) physical and chemical stimuli such as light and sound waves, pressure (touch), and odorant and taste molecules into neural signals. These signals are then transmitted to the central nervous system (CNS) for processing and interpretation. The first step in this process of **sensory transduction** is a change in the membrane potential of the receptor cell in response to a specific type of stimulus.

Sensory transduction involves changes in membrane potentials

Sensory transduction typically begins with a **receptor protein** that opens or closes ion channels in response to a specific stimulus such as heat, light, chemicals, mechanical force (including sound waves), or electric fields. The resulting change in ion flow alters the receptor cell's membrane potential. A change in the membrane potential of a receptor cell in response to a stimulus is called a **receptor potential**. Receptor potentials are *graded membrane potentials* that spread over only short distances. In order to signal over long distances in the nervous system, receptor potentials must generate action potentials, which they can do in two ways:

- The receptor potential may trigger action potentials in the receptor cell itself.
- The receptor potential may cause the receptor cell to release neurotransmitters that can induce a postsynaptic neuron to generate action potentials.

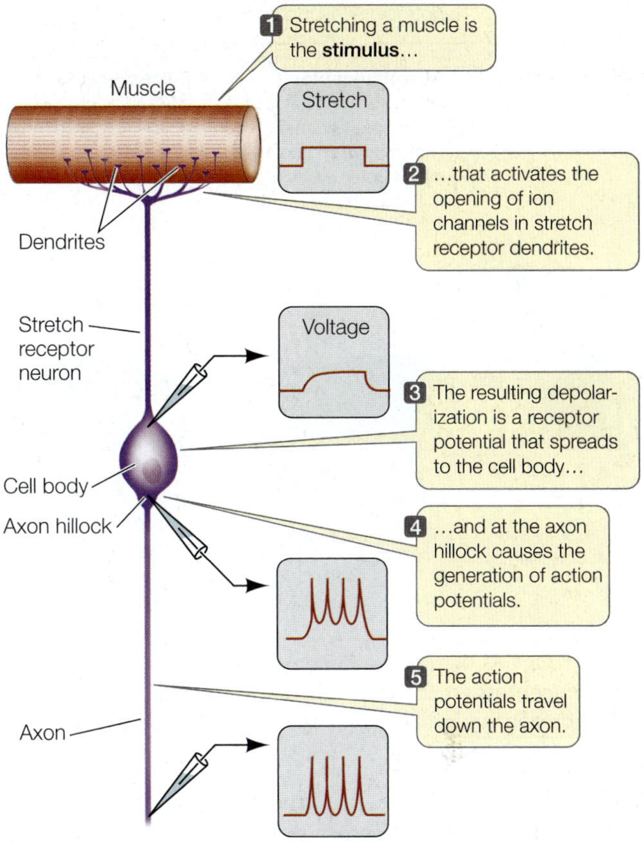

1 Stretching a muscle is the stimulus...
Muscle
Stretch
2 ...that activates the opening of ion channels in stretch receptor dendrites.
Dendrites
Stretch receptor neuron
Voltage
3 The resulting depolarization is a receptor potential that spreads to the cell body...
Cell body
Axon hillock
4 ...and at the axon hillock causes the generation of action potentials.
5 The action potentials travel down the axon.
Axon

46.1 Stimulating a Sensory Cell Produces a Receptor Potential Signal transduction in the stretch receptor of a crayfish can be investigated by measuring the membrane potential at different places on the stretch receptor neuron while stretching the muscle innervated by that sensory neuron.

A good model of how a receptor cell generates action potentials is the **stretch receptor** of a crayfish (**Figure 46.1**). Stretching the muscle to which the stretch receptor is attached causes receptor potentials. These receptor potentials spread to the base of the cell's axon, where they generate action potentials that travel down the axon to the CNS. The rate at which action potentials are fired depends on the magnitude of the receptor potential; that magnitude, in turn, depends on how much the muscle is stretched.

In a receptor cell that does not fire action potentials (such as the photoreceptors in the vertebrate eye), the spreading receptor potential reaches a presynaptic patch of plasma membrane and induces the release of a neurotransmitter. The intensity of the stimulus influences how much neurotransmitter is released. That neurotransmitter binds to receptor proteins on an associated sensory neuron, altering its membrane potential and causing it to increase or decrease its rate of firing action potentials. In a few cases, this second cell also does not generate action potentials, but simply changes the rate at which it releases neurotransmitter onto another neuron. Eventually, however, the stimulation of a sensory cell is always coded as a change in firing of action potentials in a sensory circuit.

Sensory receptor proteins act on ion channels

Sensory receptor proteins respond to stimuli by directly or indirectly opening or closing ion channels in the sensory cell (**Figure 46.2**), leading either to an action potential or to the release of neurotransmitter. Section 45.3 noted that synaptic receptor proteins are either ionotropic or metabotropic; the same distinction can be applied to sensory receptor proteins. Ionotropic sensory receptor proteins are either ion channels themselves or directly affect the opening of an ion channel. Examples are receptors that respond to physical force (mechanoreceptors) and those that respond to temperature (thermoreceptors). Electrosensors most likely do not have receptor proteins, but they are grouped with the ionotropic receptors; the plasma membrane of electrosensory cells is sensitive to the voltage across it and releases neurotransmitter in response to slight changes in membrane potential.

Metabotropic sensory receptor proteins influence ion channels indirectly, through G proteins and second messengers as described in Chapter 7. Examples are most chemoreceptors and photoreceptors.

Sensation depends on which neurons receive action potentials from sensory cells

All sensory systems process information in the form of action potentials. But the sensations we perceive—heat, pressure, light, smell, sound—differ because the messages from different kinds of sensory cells arrive at different places in the central nervous system. Action potentials arriving in the visual cortex

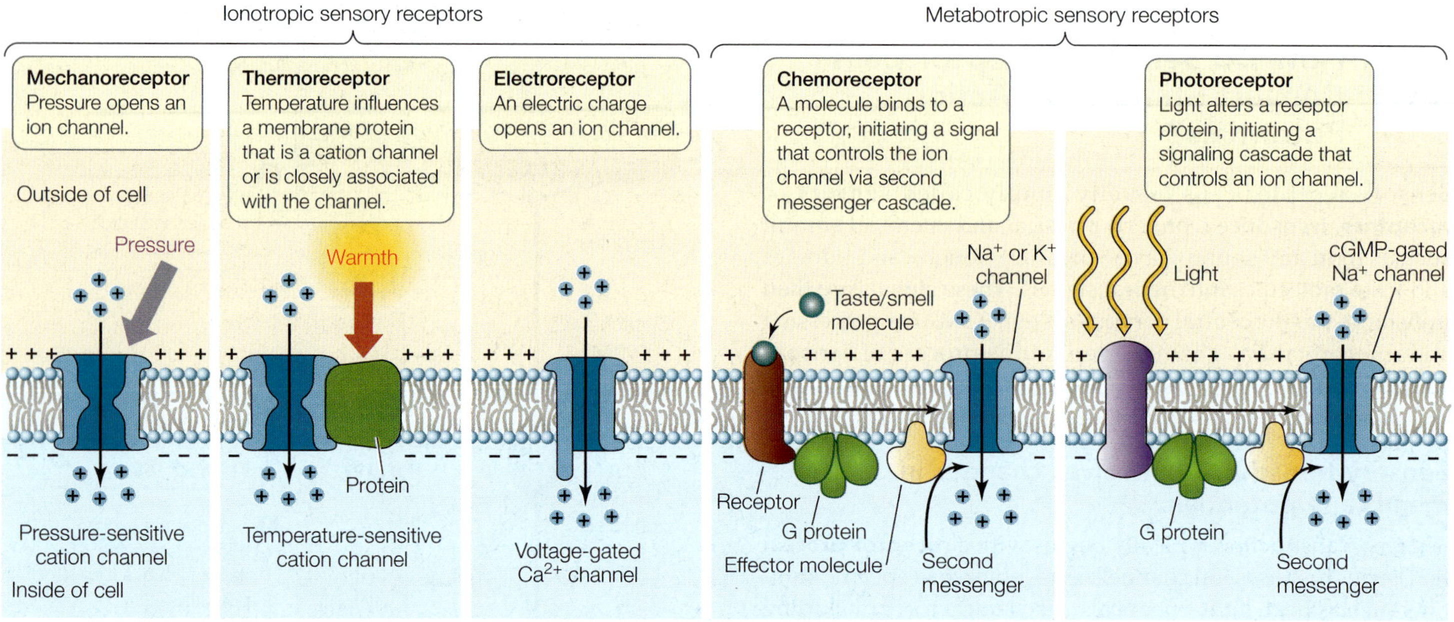

Ionotropic sensory receptors
Metabotropic sensory receptors
Mechanoreceptor
Pressure opens an ion channel.
Thermoreceptor
Temperature influences a membrane protein that is a cation channel or is closely associated with the channel.
Electroreceptor
An electric charge opens an ion channel.
Chemoreceptor
A molecule binds to a receptor, initiating a signal that controls the ion channel via second messenger cascade.
Photoreceptor
Light alters a receptor protein, initiating a signaling cascade that controls an ion channel.
Outside of cell
Pressure
Warmth
Protein
Pressure-sensitive cation channel
Inside of cell
Temperature-sensitive cation channel
Voltage-gated Ca2+ channel
Taste/smell molecule
Na+ or K+ channel
Receptor
G protein
Effector molecule
Second messenger
Light
cGMP-gated Na+ channel
G protein
Second messenger

46.2 Sensory Cell Membrane Receptor Proteins Respond to Stimuli The receptor proteins in mechanoreceptors are ion channels. The activated receptor proteins of metabotropic chemoreceptors and photoreceptors initiate signal transduction cascades that eventually open or close ion channels.

of the brain, for example, are interpreted as light, whereas those that arrive in the olfactory bulb are perceived as smells.

A small patch of skin on your arm contains some sensory receptor cells that increase their firing rates when the skin is warmed and others that increase their activity when the skin is cooled. Other sensory cells in the same patch of skin respond to touch, irritants such as insect bites, and painful stimuli. These receptor cells transmit their messages through axons that enter the CNS at the spinal cord. The synapses made by those axons in the spinal cord and the subsequent pathways of transmission determine whether the stimulation of the skin on your arm is perceived as warmth, cold, touch, itch, or pain; even though the action potentials carried by all of these sensory axons look the same, the connectivity of each axon is specific for a given sensory modality. The *intensity* of a given sensation is coded by the frequency of the action potentials.

Some sensory cells transmit information about the body's internal conditions of which we may not be consciously aware. The brain continuously receives information about body temperature, the concentrations of carbon dioxide and oxygen in the blood, arterial pressure, muscle tension, and the position of the limbs. All of this information is important for homeostasis but does not necessarily result in conscious sensation.

Some sensory receptor cells are assembled with other types of cells into **sensory organs**, such as eyes, ears, and noses, that enhance the ability of sensory cells to collect, filter, and amplify stimuli. **Sensory systems** include the sensory cells, their associated organs, and the neural networks that process the information.

Many receptors adapt to repeated stimulation

Some sensory cells give gradually diminishing responses to maintained or repeated stimulation. This phenomenon, known as **adaptation**, enables an animal to ignore background or unchanging conditions while remaining sensitive to changes and new information. (Note that this use of the term "adaptation" is different from its application in an evolutionary context.) When you get dressed, you feel each item of clothing touch your skin, but the sensation of clothes touching your skin is not constantly on your mind throughout the day. You are immediately aware of new sensations, however, such as your shoe coming untied or someone touching your back.

Some sensory cells adapt very little or very slowly; examples are some types of pain receptors and the mechanoreceptors that control balance. You do not want to ignore pain, which usually is signaling that something is wrong in your body, and to maintain equilibrium you must continuously know (albeit unconsciously) the tensions and forces on all of your joints and muscles.

RECAP 46.1

Sensory receptor cells have receptor proteins that respond to specific stimuli from the external or internal environment by opening or closing ion channels, which results in the generation of action potentials in sensory neurons.

- What is the difference between ionotropic and metabotropic sensory receptor proteins? **See p. 947 and Figure 46.2**
- How are we able to perceive action potentials—all of which are essentially the same—as different sensations? **See pp. 947–948**

Now that we have a general view of how sensory systems code and process information, we will next discuss how sensory systems gather and filter stimuli, transduce specific stimuli into action potentials, and transmit action potentials to the central nervous system, which perceives them in many different forms.

46.2 How Do Sensory Systems Detect Chemical Stimuli?

A colony of corals responds to a small amount of meat extract in seawater by extending their bodies and tentacles in search of food; a solution of a single amino acid can stimulate this response. Conversely, a small amount of seawater in which corals were crushed will stimulate a defensive retraction of the coral polyps. Humans also react strongly to certain chemical stimuli. When we smell freshly baked bread we salivate and feel hungry, whereas when we smell rotting meat we feel nauseated.

All animals receive information about chemical stimuli through **chemoreceptors**, which are receptor proteins that bind to specific molecules—their **ligands**—and are responsible for smell and taste. Chemoreceptors are also responsible for monitoring some aspects of the internal environment, such as the level of carbon dioxide in the blood.

Olfaction is the sense of smell

The sense of smell, **olfaction**, depends on chemoreceptors. In vertebrates the olfactory sensors are neurons embedded in a layer of epithelial tissue in the uppermost region of the nasal cavity. Axons from these neurons extend into the **olfactory bulb** (the olfactory integration area of the brain), whereas their dendrites end in olfactory cilia on the surface of the nasal epithelium. A protective layer of mucus covers the epithelium. Molecules from the environment must diffuse through this mucus to reach the receptor proteins on the olfactory cilia. When you have a cold, the amount of mucus in your nose increases, and the epithelium swells. With this in mind, study Figure 46.3, and you will easily understand why respiratory infections can cause you to lose your sense of smell.

An **odorant** is a molecule in the environment that binds to and activates an olfactory receptor protein on the cilia of **olfactory receptor neurons** (**ORNs**). Different olfactory receptor proteins bind specific subsets of odorant molecules. An odorant molecule binding to its receptor on an ORN activates a G protein. The G protein then activates an enzyme that causes an increase of a second messenger (cAMP in vertebrates) in the cytoplasm (see Figure 7.17). The second messenger binds to and opens cation channels in the ORN's plasma membrane, causing an influx of Na^+ into the ORN—which then depolarizes to threshold and fires action potentials. An interesting feature of ORNs is that they are continuously regenerated. Because they are embedded in nasal epithelium that, like other epithelial linings, is regularly shed, ORNs have to be constantly replaced.

The olfactory world has an enormous number of odors and a correspondingly large number of olfactory receptor proteins. In the 1990s Linda Buck and Richard Axel discovered a family of about 1,000 genes in mice (about 3 percent of the mouse genome) that code for olfactory receptor proteins. Each receptor protein that is expressed is found in a limited number of ORNs in the olfactory epithelium, and each ORN expresses just one receptor type. Using a combination of patch clamping (see Figure 45.8) and molecular techniques, the investigators were able to match specific gene products with the odorants they detect. For their discoveries of the molecular nature of the olfactory system, Buck and Axel received the Nobel Prize in 2004.

Olfactory sensitivity enables discrimination of many more odorants than there are olfactory receptors. An odorant molecule can be quite complex, and different regions of that molecule may bind to different receptor proteins. The next stage of processing olfactory information is in the olfactory bulb, where axons from ORNs expressing the same receptor protein cluster

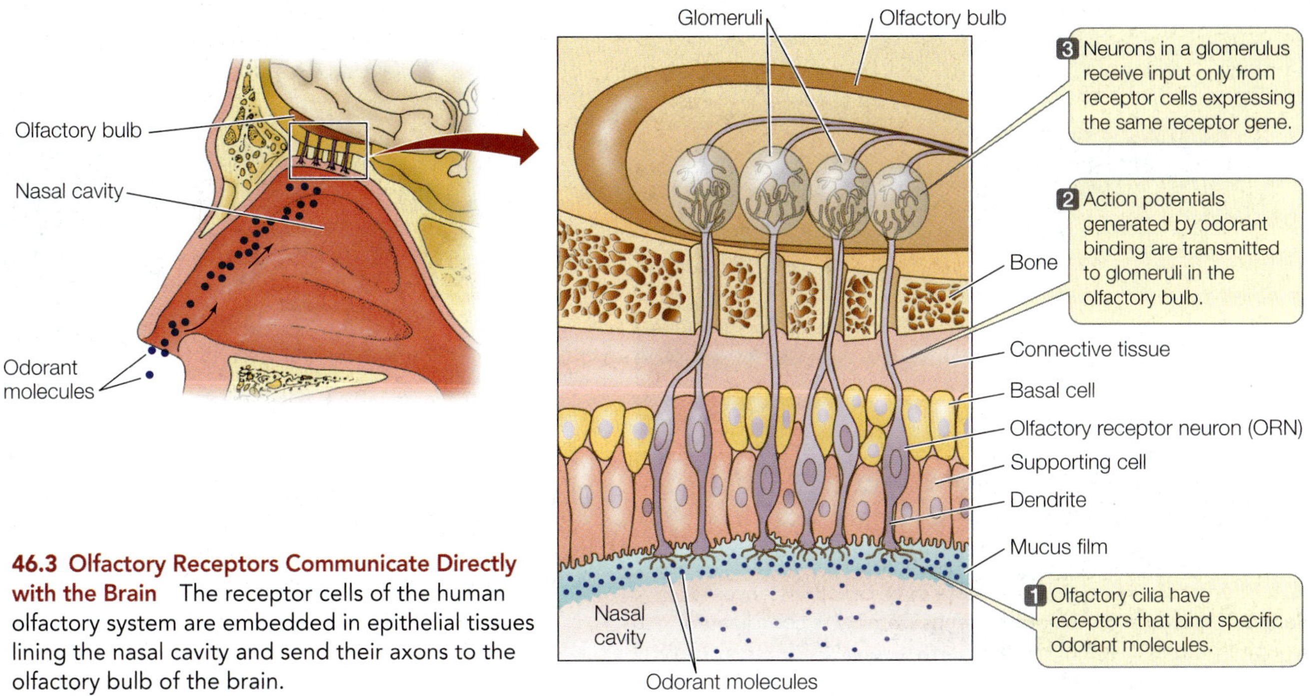
Olfactory bulb
Nasal cavity
Odorant molecules
Glomeruli
Olfactory bulb
3 Neurons in a glomerulus receive input only from receptor cells expressing the same receptor gene.
2 Action potentials generated by odorant binding are transmitted to glomeruli in the olfactory bulb.
Bone
Connective tissue
Basal cell
Olfactory receptor neuron (ORN)
Supporting cell
Dendrite
Mucus film
1 Olfactory cilia have receptors that bind specific odorant molecules.
Nasal cavity
Odorant molecules

46.3 Olfactory Receptors Communicate Directly with the Brain The receptor cells of the human olfactory system are embedded in epithelial tissues lining the nasal cavity and send their axons to the olfactory bulb of the brain.

together on olfactory bulb neurons, forming structures called glomeruli (see Figure 46.3). A complex odorant molecule can activate a unique combination of glomeruli in the olfactory bulb, so an olfactory system with hundreds of different receptor proteins can discriminate an astronomically large number of smells. The more odorant molecules that bind to ORNs, the greater the frequency of action potentials and thus the greater the intensity of the perceived smell.

Humans have a sensitive olfactory system, but in comparison with most mammals we depend far more on vision than on olfaction. The nasal epithelium of a typical dog is 15 to 20 times larger than a human's and has about *1 billion* odorant receptors, compared with about 20 million in the average human. For some scents, the threshold sensitivity of the dog is 100 million times lower; a dog's nose reveals a huge amount of information not available to people.

Some chemoreceptors detect pheromones

A specialized type of chemical signal used for communication among conspecifics (individuals of the same species) is called a **pheromone**. Individual animals secrete pheromone molecules into the environment, triggering behavioral responses in other individuals of the species. Pheromones may communicate alarm signals, mark food trails, or define territories, among many other uses. Their function in mating and mate attraction especially illustrates the remarkable sensitivity of chemosensory systems, and the silkworm moth *Bombyx mori* is an extensively studied example.

To attract a mate, the female silkworm moth releases a pheromone called bombykol from a gland at the tip of her abdomen (**Figure 46.4A**). The male silkworm moth has about 10,000 receptors for this molecule on each of his feathery antennae (**Figure 46.4B**). A single molecule of bombykol may be sufficient to generate action potentials in the antennal nerve, which transmits the signal to the male's CNS. The extreme sensitivity of the male bombykol receptors ensures that the sexual message sent by a female moth is likely to reach any male within a huge downwind area. When approximately 200 hairs per second are activated, a male orients upwind in search of the female. Because the rate of firing in the male's sensory nerves is proportional to the concentration of bombykol in the air, he can follow an airborne concentration gradient to "home in" on the signaling female.

The vomeronasal organ contains chemoreceptors

The **vomeronasal organ** (**VNO**) is a small, paired tubular structure embedded in the nasal epithelium of amphibians, reptiles, and many mammals (although not humans). In mammals the VNO is located on the septum dividing the two nostrils (see Figure 53.3).

The vomeronasal organ has a pore that opens into the nasal cavity. When the animal sniffs, the VNO pulsates and draws a sample of nasal fluid over the chemoreceptors embedded in its walls. The information from these chemoreceptors goes to an accessory olfactory bulb in the brain, and from there to brain regions involved in sexual and other instinctive behaviors.

In snakes the VNO opens into the roof of the mouth cavity. Each time the snake's forked tongue darts in and out, the forks fit into the VNO openings and present the chemoreceptors located there with a sample of molecules from the surrounding air (see the chapter-opening photo). Thus the snake uses its tongue to smell its environment, not to taste it. Why doesn't the snake simply use the flow of air to and from its lungs, as we do, to smell the environment? In reptiles, air flows to and from

(A)

The female moth releases
a pheromone from a gland
at the tip of her abdomen.
The pheromone can travel
thousands of meters
downwind.

(B)

A male moth detects
this pheromone in the
air passing over his
antennae, which are
covered with chemo-
sensitive hairs.

46.4 Pheromones Can Communicate over Great Distances Mating in silkworm moths of the genus *Bombyx* is coordinated by a pheromone called bombykol.

the lungs slowly (and can even stop entirely for long periods of time), but the tongue can dart in and out rapidly. It is a quick source of olfactory information.

Studies on mice have led to the hypothesis that the mammalian VNO is a specialized olfactory organ that detects pheromones. Lawrence Katz and his colleagues at Duke University recorded the activity of neurons in the mouse accessory olfactory bulb, which receives input from chemosensors in the VNO. These accessory olfactory neurons were activated when a mouse attached to recording electrodes sniffed another mouse placed in the same cage. However, the neurons fired differentially, depending on the gender and strain of the "intruder" mouse. Other studies on mouse behavior have supported a role for the VNO in gender identification and sexutal behaviors that are linked to pheromone perception (see Section 53.2).

Gustation is the sense of taste

In humans and other vertebrates, the sense of taste, **gustation**, depends on clusters of chemoreceptors called **taste buds**. The taste buds of terrestrial vertebrates are confined to the mouth cavity, but some fish have taste buds in the skin that enhance their ability to sense their environment. Some fish living in murky water are very sensitive to small amounts of amino acids in the water and can find food without the use of vision. The duck-billed platypus, a prototherian mammal (see Figure 33.26), has similar talents as a result of taste buds on the sensitive skin of its bill.

The human tongue has 5,000 to 10,000 taste buds embedded in the epithelium. Most of them are found on the sides of the papillae (**Figure 46.5**). (Look at your tongue in a mirror—the papillae make it look fuzzy.) The outer surface of a taste bud has a pore that exposes the tips of the sensory receptor cells. Microvilli (tiny hairlike projections) increase the surface area of these cells where their tips converge at the pore. These chemosensory cells generate action potentials and release neurotransmitter at their bases, where they form synapses with sensory neurons that convey the signals into the central nervous system.

The tongue does a lot of hard work, so its epithelium, along with cells of its taste buds, are shed and replaced at a rapid rate. Individual taste bud cells last about 10 days before they are replaced, but the sensory neurons associated with them live on, constantly forming new synapses as new taste buds form.

Humans perceive five taste classes: sweet, salty, sour, bitter, and umami. However, taste buds can distinguish among a variety of sweet-tasting molecules and a variety of bitter-tasting molecules, and small families of genes for receptor proteins responding to sweet and bitter tastes have been discovered. Umami is a savory, meaty taste that originates from receptors for amino acids, including monosodium glutamate (MSG), a commonly used flavor enhancer. In addition, spicy/hot tastes involve the activation of heat sensors, and minty tastes activate cold sensors. The full complexity of the chemosensitivity that enables us to enjoy the subtle flavors of food comes from the combined activation of gustatory and olfactory receptors,

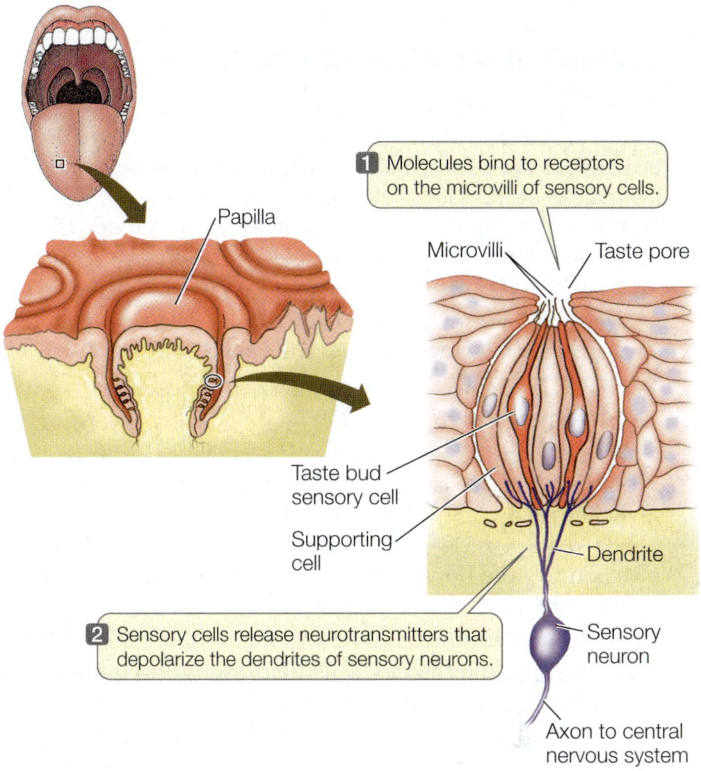
1 Molecules bind to receptors on the microvilli of sensory cells.
Papilla
Microvilli
Taste pore
Taste bud sensory cell
Supporting cell
Dendrite
2 Sensory cells release neurotransmitters that depolarize the dendrites of sensory neurons.
Sensory neuron
Axon to central nervous system

46.5 Taste Buds Are Clusters of Sensory Cells A human tongue has as many as 10,000 taste buds, most of which are found on the papillae.

which is why you may lose your sense of taste when you have a cold.

Gustation begins with receptor proteins in the membranes of the microvilli of the taste bud sensory cells (see Figure 46.5). The nature of these proteins and the mechanisms by which they depolarize sensory receptor cells differ for the different tastes. Saltiness receptor proteins are ionotropic and allow Na^+ to diffuse into taste bud sensory cells through open Na^+ channels, depolarizing the sensory cell. Sourness receptors probably are ionotropic as well, and their depolarization is also due to a direct effect of H^+ ions on Na^+ channels. In contrast, sweet, bitter, and umami taste reception is metabotropic, involving families of G protein-coupled receptor proteins. The bitter taste may have evolved as a protective mechanism, since it enables animals to detect toxic plant compounds such as quinine, caffeine, and nicotine. Because many such toxic compounds occur in plants, where they have evolved in response to herbivorous predators, a variety of receptors is essential. Similarly, a large number of molecules in food could indicate nutritional value, so a variety of receptors is of value. The diversity of sweet receptors helps explain why it has been possible to invent many chemically distinct artificial sweeteners.

Regardless of the mechanism of taste transduction by the sensory cells of the taste buds, all these cells release neurotransmitter onto sensory neurons. These neurons then generate action potentials that are conducted to the CNS, where they are interpreted as specific taste sensations.

RECAP 46.2

All animals receive information about chemical stimuli through chemoreceptors, which have diverse structures and bind to a tremendous variety of stimulus molecules. Chemoreceptors are the basis of the sensations of olfaction and gustation and the reception of pheromones. They also monitor some aspects of an animal's internal environment.

- Why are we able to distinguish so many different smells? Why do some animals experience more or different odors than others? **See pp. 949–950 and Figure 46.3**
- What is distinctive about pheromones as opposed to odorant molecules? **See p. 950**
- Describe how different substances in food are transduced into action potentials in taste buds. **See p. 951 and Figure 46.5**

We have now seen how chemoreceptors give rise to the sensations of smell and taste, and how some animals use chemoreception to communicate with others of their species. Next we will describe the sensory cells that respond to mechanical forces, including the vibrations we perceive as sound.

46.3 How Do Sensory Systems Detect Mechanical Forces?

Mechanoreceptors respond to mechanical (physical) forces. Physical distortion of a mechanoreceptor's plasma membrane causes ion channels to open, altering the membrane potential of the cell to create a graded receptor potential, which in turn leads to either the release of neurotransmitter or the generation of action potentials. The rate of action potentials tells the CNS the strength of the stimulus to the mechanoreceptor. A considerable diversity of mechanosensory cells involved in many sensory systems has evolved. The functions of these cells range from interpreting skin sensations to sensing blood pressure to hearing and maintaining balance.

Many different cells respond to touch and pressure

Human skin (and that of other mammals) is packed with diverse mechanoreceptors that generate varied sensations (**Figure 46.6**). The most important tactile receptors, found in both hairy and nonhairy skin, are **Merkel's discs**, which adapt rather slowly and provide continuous information about anything touching the skin. **Meissner's corpuscles**, found primarily in nonhairy skin, are very sensitive but adapt rapidly; they provide information about *changes* in things touching the skin. The rapid adaptation of Meissner's corpuscles is why you roll a small object between your fingers (rather than holding it still) to discern its shape and texture: as you roll it, the object continually stimulates Meissner's corpuscles.

Deeper in the skin, **Ruffini endings** adapt slowly and are good at providing information about vibrating stimuli of low frequencies, while **Pacinian corpuscles**, which adapt rapidly, provide information about vibrating stimuli of higher frequencies. Even deeper in the skin, the dendrites of sensory neurons wrap around hair follicles. When the surface hairs are displaced, those neurons are stimulated.

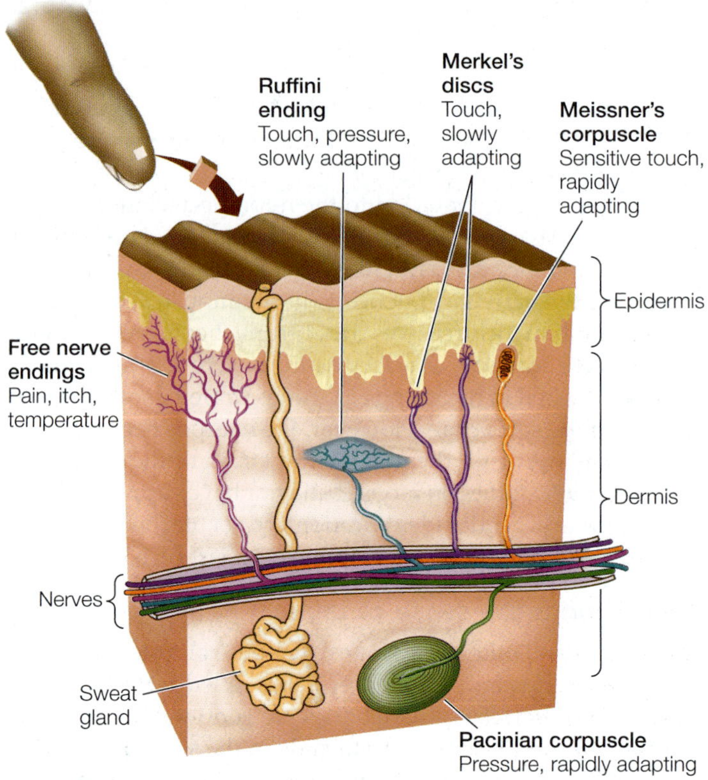

Ruffini ending
Touch, pressure, slowly adapting
Merkel's discs
Touch, slowly adapting
Meissner's corpuscle
Sensitive touch, rapidly adapting
Epidermis
Free nerve endings
Pain, itch, temperature
Dermis
Nerves
Sweat gland
Pacinian corpuscle
Pressure, rapidly adapting

46.6 The Skin Feels Many Sensations Even a very small patch of skin contains a variety of sensory cells, making the skin a multimodal receptor that can sense temperature, pressure, texture, pain, touch, and itch.

The density of tactile mechanoreceptor cells varies across the body's surface. By touching the skin with two toothpicks simultaneously, you can determine how far apart two stimuli have to be before a person can tell whether the sensations are produced by one toothpick or by two. On a person's back, for example, stimuli have to be relatively far apart before they are perceived as two discrete stimuli. But when this same "two-point spatial discrimination test" is applied to the lips or fingertips, a person can identify two stimuli as separate even when they are quite close together, meaning that receptor density is much greater in these regions.

Mechanoreceptors are also found in muscles, tendons, and ligaments

An animal receives information from mechanoreceptors about the position of its limbs and the stresses on its muscles and joints. These mechanoreceptors supply information continuously to the CNS, and this information is essential for postural control and the coordination of movements.

The mechanoreceptors in skeletal muscle are the **muscle spindles**. These stretch receptors are modified muscle cells embedded in connective tissue inside muscles and innervated by sensory neurons (**Figure 46.7A**). When a muscle is stretched, the muscle spindles are stretched as well, signaling the spindle neurons transmit action potentials to the central nervous system. Figure 46.1 showed how crayfish stretch receptors transduce physical force into action potentials; the actions of muscle spindles are similar. The CNS uses the information from

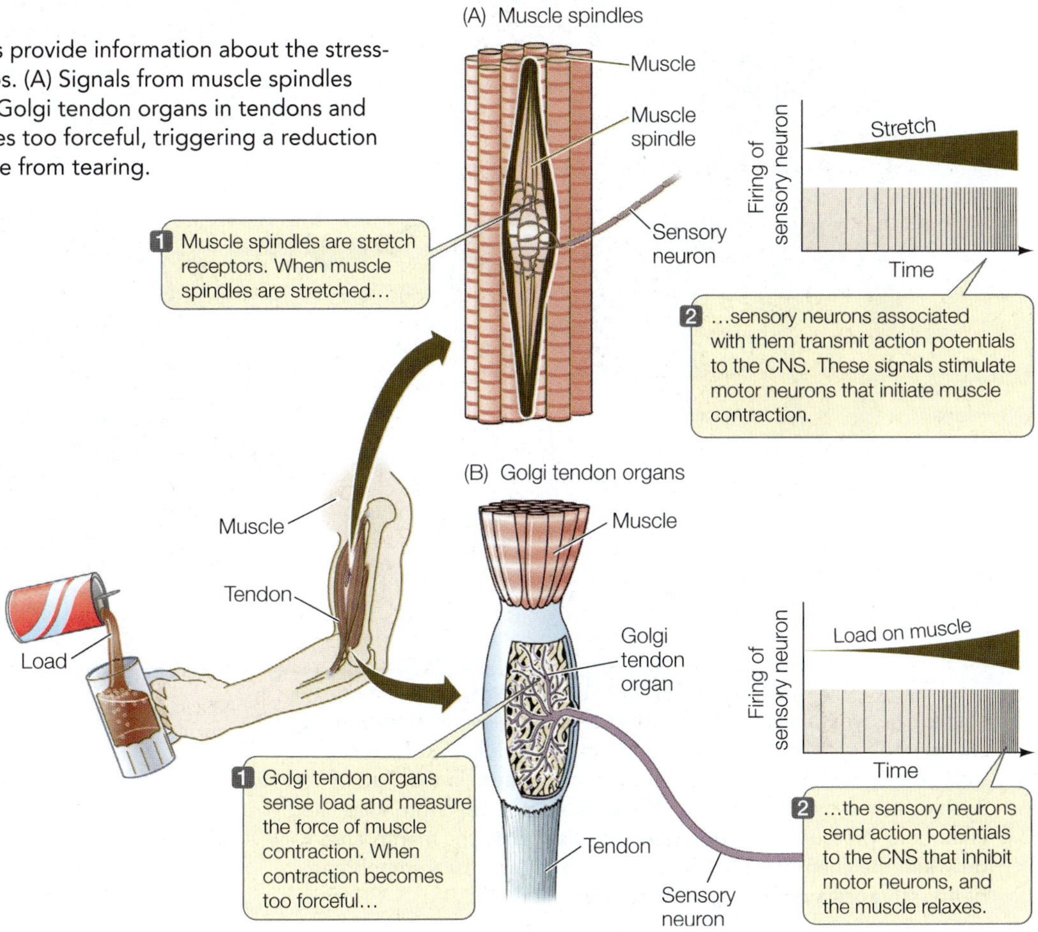
(A) Muscle spindles
Muscle
Muscle spindle
Sensory neuron
Firing of sensory neuron
Stretch
Time
1 Muscle spindles are stretch receptors. When muscle spindles are stretched…
2 …sensory neurons associated with them transmit action potentials to the CNS. These signals stimulate motor neurons that initiate muscle contraction.
Muscle
Tendon
Load
(B) Golgi tendon organs
Muscle
Golgi tendon organ
Tendon
Sensory neuron
Firing of sensory neuron
Load on muscle
Time
1 Golgi tendon organs sense load and measure the force of muscle contraction. When contraction becomes too forceful…
2 …the sensory neurons send action potentials to the CNS that inhibit motor neurons, and the muscle relaxes.

46.7 Stretch Receptors Stretch receptors provide information about the stresses on muscles and joints in an animal's limbs. (A) Signals from muscle spindles to the CNS initiate muscle contraction. (B) Golgi tendon organs in tendons and ligaments inhibit a contraction that becomes too forceful, triggering a reduction in muscle tension and protecting the muscle from tearing.

muscle spindles to adjust the strength of the muscle contraction so that it matches the load put on the muscle; thus a person can hold a mug steady while it is being filled.

Another type of mechanoreceptor, the **Golgi tendon organ**, is found in tendons and ligaments and provides information about the force generated by a contracting muscle (**Figure 46.7B**). When a contraction becomes too forceful, action potentials from the Golgi tendon organ inhibit the spinal cord motor neurons innervating that muscle, causing the muscle to relax and protecting it from tearing. (You may recall a cell organelle called the Golgi apparatus. What these two very different structures have in common is their discovery in the late nineteenth century by the Italian anatomist Camillo Golgi.)

Hair cells are mechanoreceptors of the auditory and vestibular systems

Hair cells are the mechanoreceptors for the vertebrate auditory system (sound-perceiving) and vestibular (equilibrium-maintaining) systems. Both of these systems are housed in the complex structures of the vertebrate ear. **Stereocilia**—fingerlike extensions of the cell membrane stiffened by cross-linked actin filaments—project from the surface of each hair cell like a set of organ pipes (**Figure 46.8A**). Stereocilia bend in response to waves of pressure; bending of the stereocilia in one direction depolarizes the hair cell, and bending in the other direction hyperpolarizes it (**Figure 46.8B**).

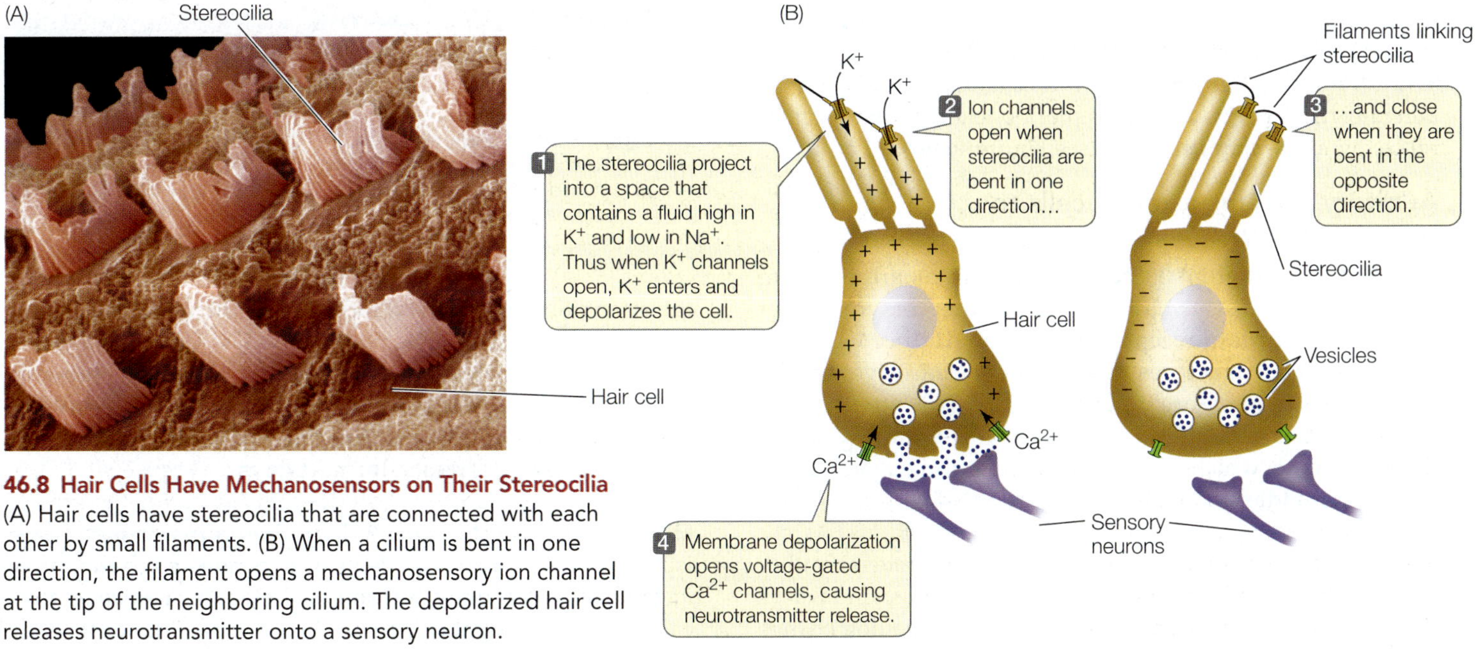
(A)
Stereocilia
Hair cell
(B)
K^+
K^+
1 The stereocilia project into a space that contains a fluid high in K^+ and low in Na^+. Thus when K^+ channels open, K^+ enters and depolarizes the cell.
2 Ion channels open when stereocilia are bent in one direction…
Filaments linking stereocilia
3 …and close when they are bent in the opposite direction.
Stereocilia
Hair cell
Vesicles
Ca^{2+}
Ca^{2+}
Sensory neurons
4 Membrane depolarization opens voltage-gated Ca^{2+} channels, causing neurotransmitter release.

46.8 Hair Cells Have Mechanosensors on Their Stereocilia (A) Hair cells have stereocilia that are connected with each other by small filaments. (B) When a cilium is bent in one direction, the filament opens a mechanosensory ion channel at the tip of the neighboring cilium. The depolarized hair cell releases neurotransmitter onto a sensory neuron.

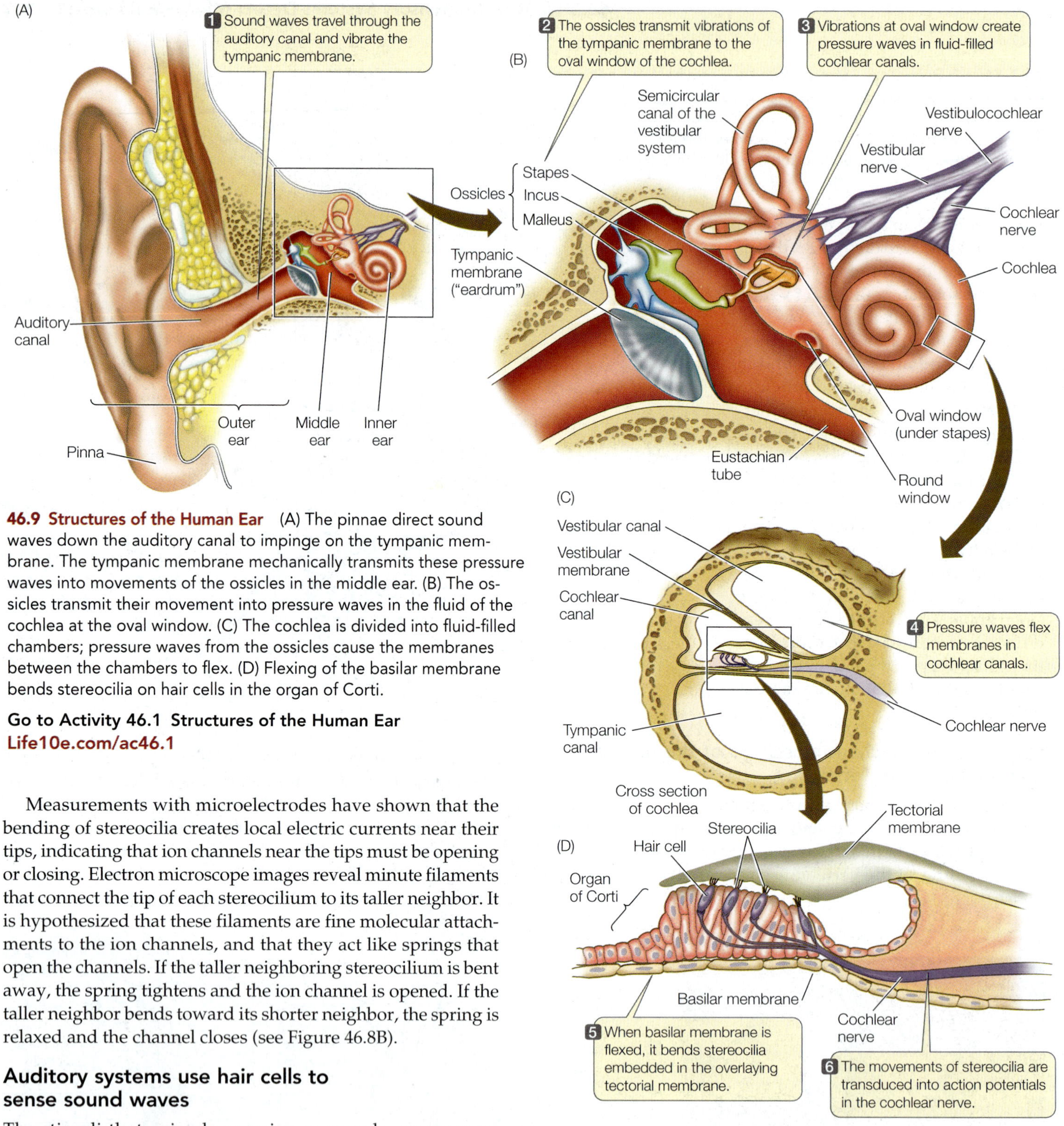
(A)
1 Sound waves travel through the auditory canal and vibrate the tympanic membrane.
Auditory canal
Pinna
Outer ear
Middle ear
Inner ear
(B)
2 The ossicles transmit vibrations of the tympanic membrane to the oval window of the cochlea.
3 Vibrations at oval window create pressure waves in fluid-filled cochlear canals.
Semicircular canal of the vestibular system
Vestibulocochlear nerve
Vestibular nerve
Ossicles
Stapes
Incus
Malleus
Cochlear nerve
Tympanic membrane ("eardrum")
Cochlea
Oval window (under stapes)
Eustachian tube
Round window
(C)
Vestibular canal
Vestibular membrane
Cochlear canal
4 Pressure waves flex membranes in cochlear canals.
Cochlear nerve
Tympanic canal
Cross section of cochlea
(D)
Hair cell
Stereocilia
Tectorial membrane
Organ of Corti
Basilar membrane
Cochlear nerve
5 When basilar membrane is flexed, it bends stereocilia embedded in the overlaying tectorial membrane.
6 The movements of stereocilia are transduced into action potentials in the cochlear nerve.

46.9 Structures of the Human Ear (A) The pinnae direct sound waves down the auditory canal to impinge on the tympanic membrane. The tympanic membrane mechanically transmits these pressure waves into movements of the ossicles in the middle ear. (B) The ossicles transmit their movement into pressure waves in the fluid of the cochlea at the oval window. (C) The cochlea is divided into fluid-filled chambers; pressure waves from the ossicles cause the membranes between the chambers to flex. (D) Flexing of the basilar membrane bends stereocilia on hair cells in the organ of Corti.

Go to Activity 46.1 Structures of the Human Ear
Life10e.com/ac46.1

Measurements with microelectrodes have shown that the bending of stereocilia creates local electric currents near their tips, indicating that ion channels near the tips must be opening or closing. Electron microscope images reveal minute filaments that connect the tip of each stereocilium to its taller neighbor. It is hypothesized that these filaments are fine molecular attachments to the ion channels, and that they act like springs that open the channels. If the taller neighboring stereocilium is bent away, the spring tightens and the ion channel is opened. If the taller neighbor bends toward its shorter neighbor, the spring is relaxed and the channel closes (see Figure 46.8B).

Auditory systems use hair cells to sense sound waves

The stimuli that animals perceive as sounds are pressure waves. Auditory systems use mechanoreceptors to convert pressure waves into receptor potentials. Auditory systems include special structures that gather sound waves, direct them to the sensory organ, and amplify their effect on the mechanoreceptors. A good example of an auditory system is the human ear, which can be divided into three major areas: the outer, middle, and inner ear (**Figure 46.9A**).

OUTER EAR The outer ear consists of the pinnae (singular, pinna) and the auditory canal. The pinnae collect sound waves and direct them into the auditory canals; watch a dog change the orientation of its ears to focus on a particular sound to get the idea of the role pinnae play in hearing. Covering the end of the auditory canal is the **tympanic membrane** (commonly called the eardrum), which vibrates in response to pressure waves traveling down the canal, thus converting the pressure waves to physical forces in the middle ear.

MIDDLE EAR On the other side of the tympanic membrane is the middle ear, an air-filled cavity connected to the throat at the back of the mouth through the **eustachian tube** (also called the pharyngotympanic tube). Because the eustachian tube is also filled with air, pressure equilibrates between the middle ear and the environment. When you have a cold, the tube can become blocked by mucus or tissue swelling and you have difficulty equilibrating the pressure in the middle ear with the outside air pressure (as you have to do when changing altitude in an airplane).

The middle ear contains the **ossicles**, three delicate bones individually named the malleus ("hammer"), incus ("anvil"), and stapes ("stirrup") (Figure 46.9B). The ossicles transmit the vibrations of the tympanic membrane to another flexible membrane, the **oval window**. The ossicles act as a lever (like a hammer pulling out a nail), translating a large movement of the tympanic membrane into a smaller movement of the oval window, but a movement of greater force. Because the oval window is much smaller than the tympanic membrane, the pressure the stapes transmits to the oval window is more than 20 times greater than the pressure exerted by a sound wave on the tympanic membrane.

Behind the oval window lies the fluid-filled inner ear. Movements of the oval window impart pressure changes to that enclosed fluid. These pressure waves are transduced into action potentials.

INNER EAR The inner ear is a bony structure consisting of two sets of canals. One is the organ of balance, the **vestibular system**, and the other is the organ of hearing, the **cochlea**. The cochlea ("snail" or "spiral shell") is a long, tapered, coiled structure. A cross section of the cochlea reveals that it is composed of three parallel canals separated by two membranes, the vestibular membrane and the basilar membrane (Figure 46.9C). Sitting on the basilar membrane is the **organ of Corti**, which transduces pressure waves into action potentials. The organ of Corti contains hair cells with stereocilia (see Figure 46.8). The tips of the hair cells are embedded in a gelatinous overhanging shelf called the tectorial membrane (Figure 46.9D).

Stereocilia are not motile, but because their tips are attached to the more rigid tectorial membrane, stereocilia bend when the basilar membrane flexes. The response of the hair cell is a graded membrane potential. The hair cells do not fire action potentials, but the changes in their membrane potential alter the rate at which the hair cells release neurotransmitter onto the sensory neurons whose axons make up the auditory nerve and transmit action potentials to the brain.

The vestibular and tympanic canals are separate until they reach the distal end of the cochlea (the end farthest from the oval window), where they join; thus they form one continuous canal that turns back on itself (see Figure 46.10). Just as the oval window is a flexible membrane at the beginning of the vestibular canal, the **round window** is a flexible membrane at the end of the tympanic canal. When the oval window vibrates, the waves of fluid pressure create traveling waves, or flexions, in the basilar membrane.

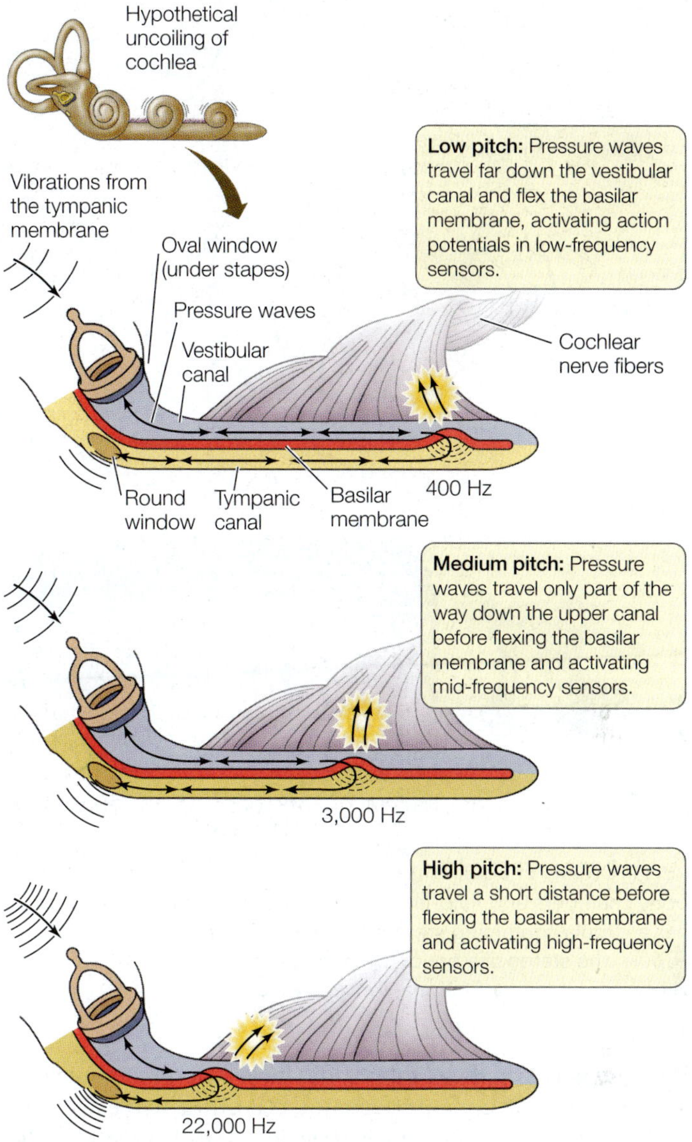
Hypothetical
uncoiling of
cochlea
Vibrations from
the tympanic
membrane
Oval window
(under stapes)
Pressure waves
Vestibular
canal
Cochlear
nerve fibers
Round
window
Tympanic
canal
Basilar
membrane
400 Hz
Low pitch: Pressure waves travel far down the vestibular canal and flex the basilar membrane, activating action potentials in low-frequency sensors.
Medium pitch: Pressure waves travel only part of the way down the upper canal before flexing the basilar membrane and activating mid-frequency sensors.
3,000 Hz
High pitch: Pressure waves travel a short distance before flexing the basilar membrane and activating high-frequency sensors.
22,000 Hz

46.10 Sensing Pressure Waves in the Inner Ear Pressure waves of different frequencies flex the basilar membrane at different locations. Information about sound frequency is specified by which hair cells are activated. For simplicity, this representation illustrates the cochlea as uncoiled, and leaves out the middle ear.

Go to Animated Tutorial 46.1
Sound Transduction in the Human Ear
Life10e.com/at46.1

Flexion of the basilar membrane is perceived as sound

What causes the basilar membrane to flex, and how does this mechanism distinguish sounds of different frequencies? Air is highly compressible but fluids are not; therefore a pressure wave can travel through air without displacing much air, whereas a pressure wave in fluid displaces that fluid. When the stapes pushes on the oval window, the fluid in the vestibular canal is displaced. If the movement of the oval window occurs slowly, the cochlear fluid pressure wave travels down the vestibular canal, around the bend, and back through the tympanic canal (Figure 46.10). At the end of the tympanic canal, the displacement pressure is dissipated by the outward bulging of the round window.

The basilar membrane is not uniform—it is thicker and stiffer at its base and wider and thinner at its apical end.

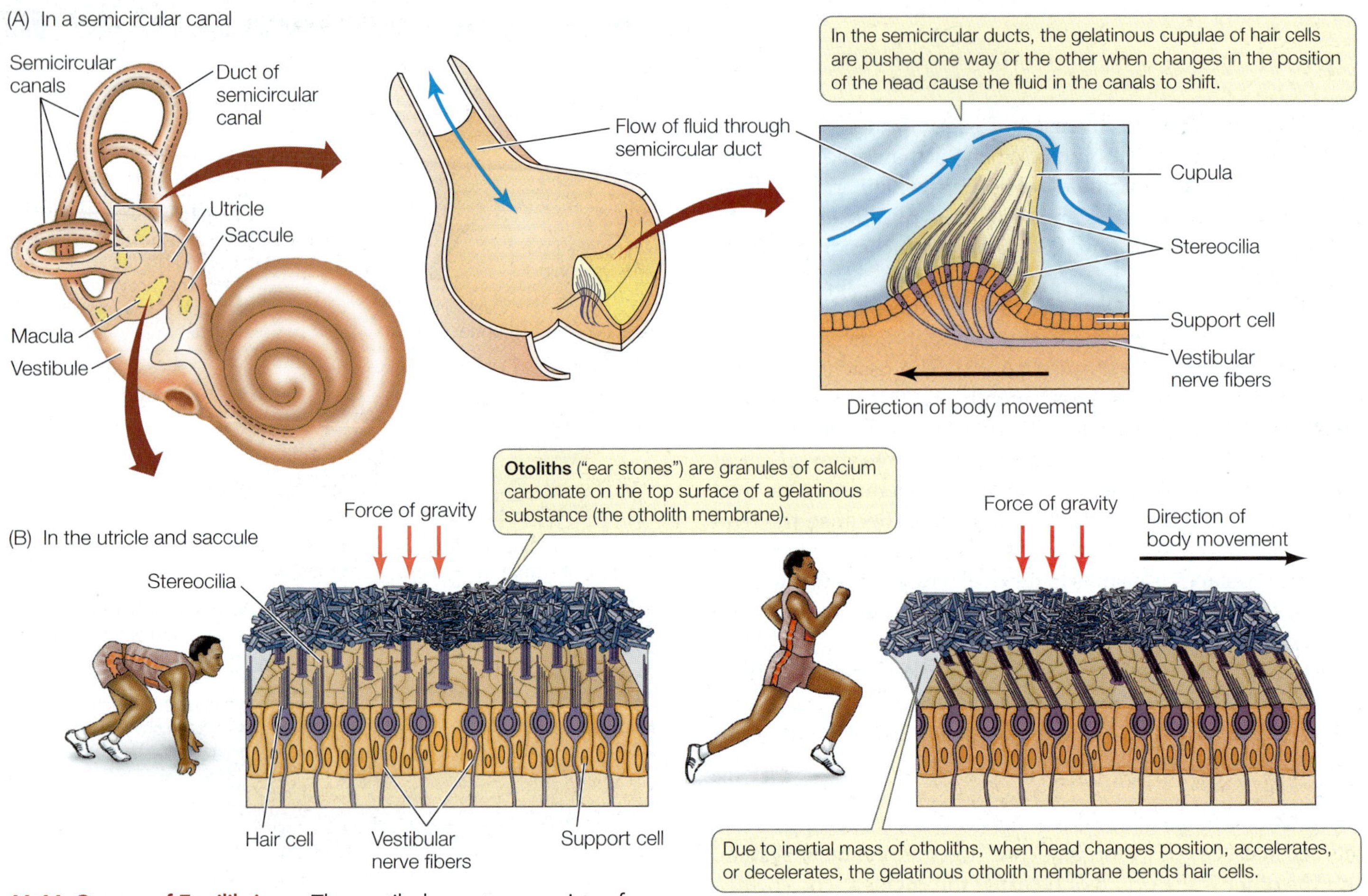

(A) In a semicircular canal
Semicircular canals
Duct of semicircular canal
Utricle
Saccule
Macula
Vestibule
Flow of fluid through semicircular duct
In the semicircular ducts, the gelatinous cupulae of hair cells are pushed one way or the other when changes in the position of the head cause the fluid in the canals to shift.
Cupula
Stereocilia
Support cell
Vestibular nerve fibers
Direction of body movement
(B) In the utricle and saccule
Force of gravity
Otoliths ("ear stones") are granules of calcium carbonate on the top surface of a gelatinous substance (the othlolith membrane).
Stereocilia
Hair cell
Vestibular nerve fibers
Support cell
Force of gravity
Direction of body movement
Due to inertial mass of otholiths, when head changes position, accelerates, or decelerates, the gelatinous otholith membrane bends hair cells.

46.11 Organs of Equilibrium The vestibular system consists of bony chambers and fluid-filled canals. (A) Each semicircular duct has a cupula containing stereocilia. When fluid moves against the cupula, the stereocilia bend. (B) In the saccule and utricle, stereocilia are bent by gravitational forces on the otoliths.

Pressure waves in the cochlear fluid have different frequencies and set up different patterns of traveling waves. High-frequency waves cause maximal flexion at the basal end of the basilar membrane, whereas low-frequency pressure waves result in maximal flexion at the apical end (see Figure 46.10). Thus different pitches of sound flex the basilar membrane at different locations and activate different sets of hair cells. Action potentials stimulated by the mechanoreceptors at different positions along the organ of Corti travel along the cochlear nerve and are transmitted to different regions of the brain's auditory cortex by the vestibulocochlear nerve.

Various types of damage can result in hearing loss

There are two general types of acquired hearing loss, or deafness. Conduction deafness is caused by the loss of function of the tympanic membrane and/or the ossicles of the middle ear. Repeated infections of the middle ear can cause scarring of the tympanic membrane and stiffening of the connections between the ossicles. The consequence is less efficient conduction of sound waves from the tympanic membrane to the oval window. With increasing age, the ossicles inevitably stiffen, resulting in a gradual loss of the ability to hear high-frequency sounds.

Nerve deafness is caused by damage to the inner ear or the auditory pathways. A common cause of nerve deafness is damage to the hair cells of the delicate organ of Corti by exposure to loud sounds such as jet engines, pneumatic drills, or highly amplified music. Consistent exposure to sounds above 85 decibels can damage hearing; this damage is cumulative and irreversible. Even using earphones can put you at risk for hearing loss because they generate high-pressure sound waves close to the tympanic membrane. Personal stereo earphones can reach 120 decibels, and people commonly use them at 100 decibels (equivalent to being at a rock concert).

The vestibular system uses hair cells to detect forces of gravity and momentum

Hair cells in the vestibular system of the inner ear detect the position and movement of the head—information that is essential for maintaining equilibrium (balance). Information from the vestibular system is also crucial for the control of eye movements. When you look at something, you can move your head while staying focused on the object because of your vestibulo-ocular reflex.

In the mammalian inner ear, the vestibular system consists of three bony **semicircular canals** and a bony chamber called the **vestibule**. Within each canal is a membranous semicircular duct, and within the vestibule are the membranous saccule and the utricle. The ducts and the saccule and utricle are filled with fluid. In the semicircular ducts, the fluid shifts when the head changes position (**Figure 46.11A**). Since the three semicircular canals have different orientations, the fluid in their ducts responds differentially to the direction of movement. Projecting into the base of each duct is a **cupula**, a gelatinous swelling enclosing a cluster of hair cell stereocilia. When the shifting fluid pushes on the cupula, it bends the stereocilia and causes a graded potential in their hair cell plasma membranes.

The stereocilia in the saccule and utricle are bent in a different way. These stereocilia are embedded gelatinous membranes that contain otoliths ("ear stones") that are crystals of calcium carbonate. When the head changes position or when it accelerates or decelerates, gravitational forces are exerted on the otoliths and the stereocilia bend (**Figure 46.11B**).

As in the cochlea, the hair cells of the vestibular system do not fire action potentials, but they release neurotransmitter at synapses with sensory neurons, which in turn fire action potentials.

Go to Animated Tutorial 46.2
Mechanoreceptors
Life10e.com/at46.2

RECAP 46.3

Sensations that derive from mechanoreceptors include touch, tickle, pressure, joint position, muscle load, hearing, and equilibrium.

- Describe some of the different mechanoreceptors in the skin and their properties. **See p. 952 and Figure 46.6**
- How do hair cells transduce force into action potentials? **See pp. 953–954 and Figure 46.8**
- How do different frequencies of sound result in action potentials being fired in different acoustic neurons? **See pp. 954–955 and Figures 46.9 and 46.10**

We turn next to another example of metabotropic sensory reception, one in which light is the stimulus. We will see how light energy is converted into action potentials that in higher vertebrates are perceived as vision, perhaps the most elaborate of the senses.

46.4 How Do Sensory Systems Detect Light?

Sensitivity to light—**photosensitivity**—confers on the simplest animals the ability to orient to the sun and sky, and it gives more complex animals rapid and extremely detailed information about objects in their environment. Photosensitivity is ubiquitous in the animal kingdom, and the molecular basis for that sensitivity is a family of visual that has been evolutionarily conserved.

In this section we will learn how a visual pigment molecule responds when stimulated by light energy and how that response is transduced into neural signals. We will also examine the structures of eyes, the organs that gather light energy and focus it onto **photoreceptor cells**, the metabotropic sensory receptors that transform light energy into action potentials, and the routes those impulses travel to the brain.

Rhodopsin is a vertebrate visual pigment

Photosensitivity depends on the ability of visual pigments to absorb photons of light and to undergo a change in conformation. One such pigment is **rhodopsin**, a well-studied vertebrate visual pigment. A rhodopsin molecule consists of the protein **opsin** (which by itself is not photosensitive) and 11-*cis*-retinal, a nonprotein, light-absorbing functional group cradled in the center of the opsin protein and covalently bound to it (**Figure 46.12**). The entire rhodopsin molecule sits within the plasma membrane of a photoreceptor cell, such as the rod cells of humans (see p. 960).

When 11-*cis*-retinal absorbs a photon of light energy, it changes into a different isomer of retinal, called all-*trans*-retinal. This change puts a strain on the bonds between retinal and opsin, changing the conformation of opsin and thus signaling the detection of light. In vertebrate eyes, the retinal and the opsin eventually separate from each other (a process called bleaching), which causes the molecule to lose its photosensitivity. A series of enzymatic reactions returns the all-*trans*-retinal to the 11-*cis* isomer, which then recombines with opsin so that it once again becomes the photosensitive pigment rhodopsin.

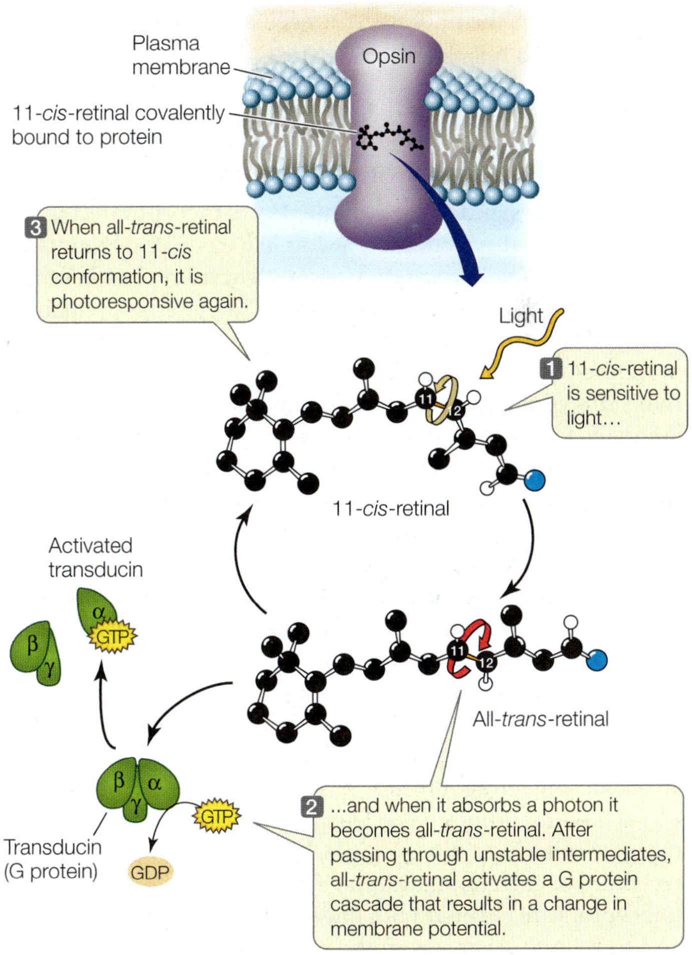

Plasma membrane
Opsin
11-*cis*-retinal covalently bound to protein
3 When all-*trans*-retinal returns to 11-*cis* conformation, it is photoresponsive again.
Light
1 11-*cis*-retinal is sensitive to light...
11-*cis*-retinal
Activated transducin
α
GTP
β
γ
All-*trans*-retinal
Transducin (G protein)
GDP
2 ...and when it absorbs a photon it becomes all-*trans*-retinal. After passing through unstable intermediates, all-*trans*-retinal activates a G protein cascade that results in a change in membrane potential.

46.12 Light Changes the Conformation of Rhodopsin The light-absorbing molecule 11-*cis*-retinal bonds with the protein opsin to form the vertibrate visual pigment rhodopsin.

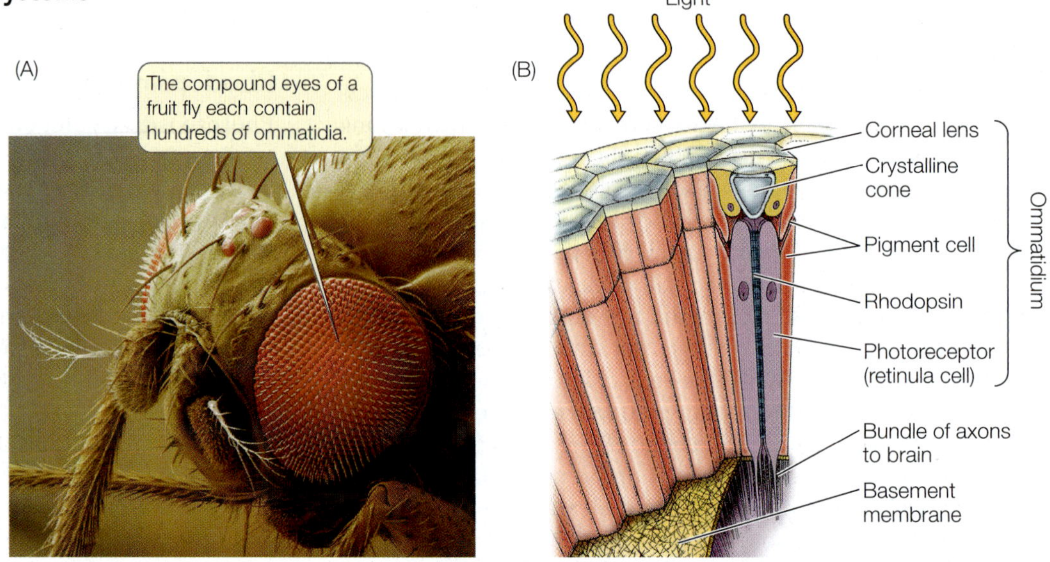
(A)
The compound eyes of a fruit fly each contain hundreds of ommatidia.
(B)
Light
Corneal lens
Crystalline cone
Pigment cell
Rhodopsin
Photoreceptor (retinula cell)
Ommatidium
Bundle of axons to brain
Basement membrane

46.13 Ommatidia: The Functional Units of Insect Eyes (A) The micrograph shows the compound eye of a fruit fly. (B) The rhodopsin-containing retinula cells are the photoreceptors in ommatidia.

How does the conformational change of rhodopsin transduce light into a cellular response? After retinal is converted from the 11-*cis* to the all-*trans* form, its interactions with opsin pass through several unstable intermediate stages. One of these stages triggers a cascade of reactions involving a G protein signaling mechanism that results in the alteration of membrane potential that is the photoreceptor cell's response to light.

Go to Animated Tutorial 46.3
Photosensitivity
Life10e.com/at46.3

Invertebrates have a variety of visual systems

Photoreceptors and visual pigments are incorporated into a variety of visual systems, from simple to complex. Flatworms obtain directional information about light from photoreceptor cells that are organized into **eye cups**. The eye cups are paired, bilateral structures, each partly shielded from light by a layer of pigmented cells lining the cup. The photoreceptors on the two sides of the animal are unequally stimulated unless the animal is facing directly toward or away from a light source. The flatworm generally uses directional information from the eye cups to move away from light.

Arthropods have **compound eyes** that provide them with information about patterns or images in the environment. These eyes are called compound because each eye consists of many optical units called **ommatidia** (singular *ommatidium*), each with its own narrow-angle lens (**Figure 46.13**). In contrast, a vertebrate eye consists of just one optical unit with a wide-angle lens. The number of ommatidia in a compound eye varies from only a few in some ants to 800 in fruit flies and to 30,000 in some dragonflies.

Each ommatidium has a lens that directs light onto photoreceptor cells. Flies, for example, have eight elongated photoreceptors in each ommatidium. The inner borders of the photoreceptors are covered with microvilli that contain rhodopsin and trap light. Axons from the photoreceptors send the light information to the nervous system. Since each ommatidium of a compound eye is directed at a slightly different part of the visual world, only a low-resolution (pixillated) image can be communicated from the compound eye to the CNS.

Image-forming eyes evolved independently in vertebrates and cephalopods

Both vertebrates and cephalopod mollusks have eyes with exceptional abilities to form detailed images of the visual world. Like cameras, both of these eye types focus inverted images on an internal surface that is sensitive to light. Considering that they evolved completely independently of each other, their degree of similarity is remarkable (**Figure 46.14**).

The vertebrate eye (see Figure 46.14A) is a spherical, fluid-filled structure bounded by a tough connective tissue layer called the **sclera**. At the front of the eye, the sclera forms the transparent **cornea**, through which light passes to enter the eye. Just inside the cornea is the pigmented **iris**, which gives the eye its color. The iris controls the amount of light that reaches the photoreceptor cells at the back of the eye, just as the diaphragm of a camera controls the amount of light reaching the film. The central opening of the iris is the **pupil**. The iris is under neural control. In bright light, the iris constricts and the pupil is very small. As light levels fall, the iris relaxes and the pupil enlarges.

Behind the iris is the crystalline protein **lens**, which makes fine adjustments in the focus of images falling on the photosensitive layer—the **retina**—at the back of the eye. The cornea and the fluids within the eye bend light rays passing through them so that they are focused on the retina. The lens makes fine adjustments to the focus and allows the eye to accommodate—that is, to focus on objects at various locations in the near visual field. To focus a camera on objects close at hand, you adjust the distance between the lens and the internal surface sensitive to light. Fishes, amphibians, and non-avian reptiles accommodate in a similar manner, moving the lenses of their eyes closer to or farther from their retinas. Mammals and birds use a different method; they alter the shape of the lens.

The mammalian lens is contained in a connective tissue sheath that tends to keep it in a spherical shape, but it is attached to suspensory ligaments that pull it into a flatter shape. Circular ciliary muscles counteract the pull of the suspensory ligaments, permitting the lens to round up. When the ciliary muscles are at rest, the flatter lens has the correct optical properties to focus distant images on the retina. Contracting the

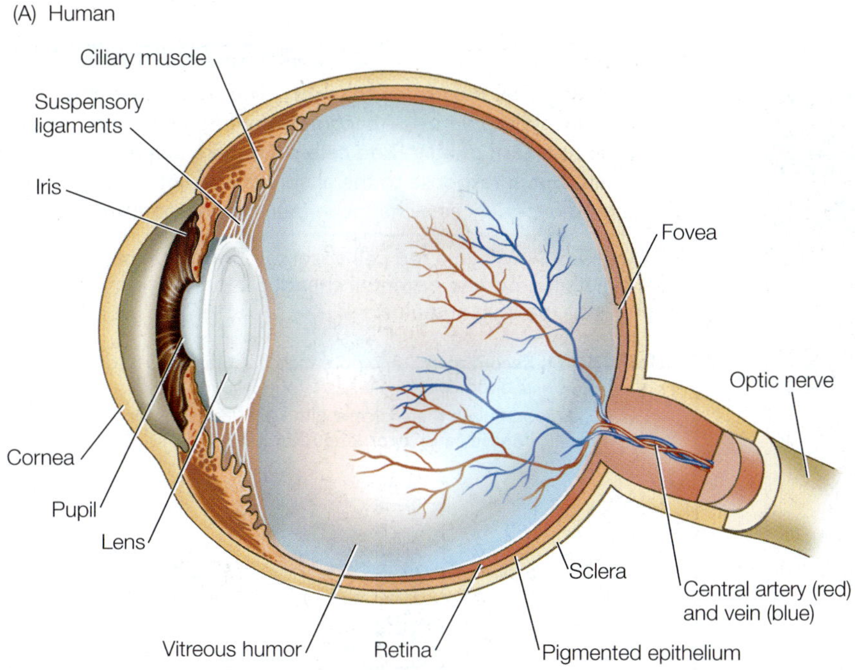
(A) Human
Ciliary muscle
Suspensory ligaments
Iris
Fovea
Optic nerve
Cornea
Pupil
Lens
Sclera
Central artery (red) and vein (blue)
Vitreous humor
Retina
Pigmented epithelium

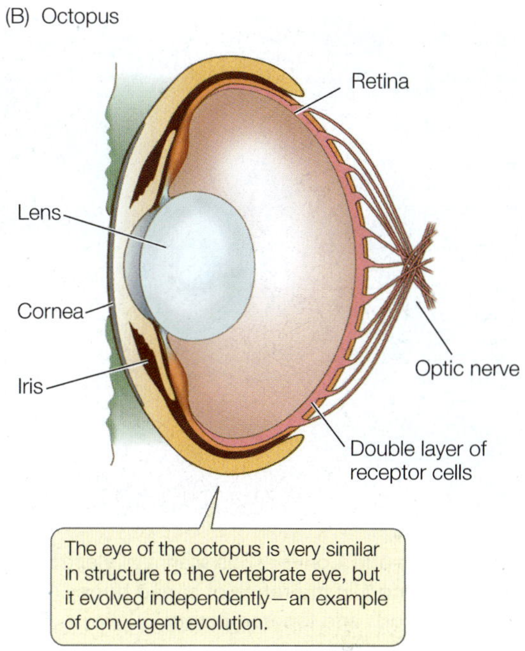
(B) Octopus
Retina
Lens
Cornea
Iris
Optic nerve
Double layer of receptor cells
The eye of the octopus is very similar in structure to the vertebrate eye, but it evolved independently—an example of convergent evolution.

46.14 Convergent Evolution of Eyes The lenses of vertebrate (A) and cephalopod (B) eyes focus images on layers of photoreceptor cells.

ciliary muscles rounds up the lens, changing its light-bending properties to bring close images into focus (**Figure 46.15**).

Lenses become less elastic with age, so we lose the ability to focus on objects close at hand without the help of corrective lenses. Most people over the age of 45 need the assistance of reading glasses or bifocal lenses.

Go to Activity 46.2 Structure of the Human Eye
Life10e.com/ac46.2

The vertebrate retina receives and processes visual information

During embryonic development, neural tissue grows out from the brain to form the retina. In addition to a layer of photoreceptor cells, the retina includes four additional types of cells that process visual information from the photoreceptors (see Figure 46.20). Light must pass through all the layers of retinal cells before being captured by photosensors. In humans and other day-active animals, the light that is not captured is absorbed by a black-pigmented epithelial tissue layer behind the retina. In contrast, nocturnal animals such as deer and raccoons have an iridescent reflective layer behind their retinas, which maximizes the capture of photons by reflecting them back onto the photoreceptors. This is why a deer in the headlights appears to have bright white eyes. Because humans do not have a white reflective layer in the retina, photographs taken indoors often have a "red-eye effect," caused by the flash of light from the camera being reflected by the abundant blood vessels in the retina.

The pigmented epithelium also plays a role in the renewal of the photoreceptors. The photoreceptor cells are always shedding discs from their distal ends as new ones are being generated by the inner segments of those cells. The pigmented epithelial cells phagocytose the shed discs. Each outer segment is totally renewed about every two weeks.

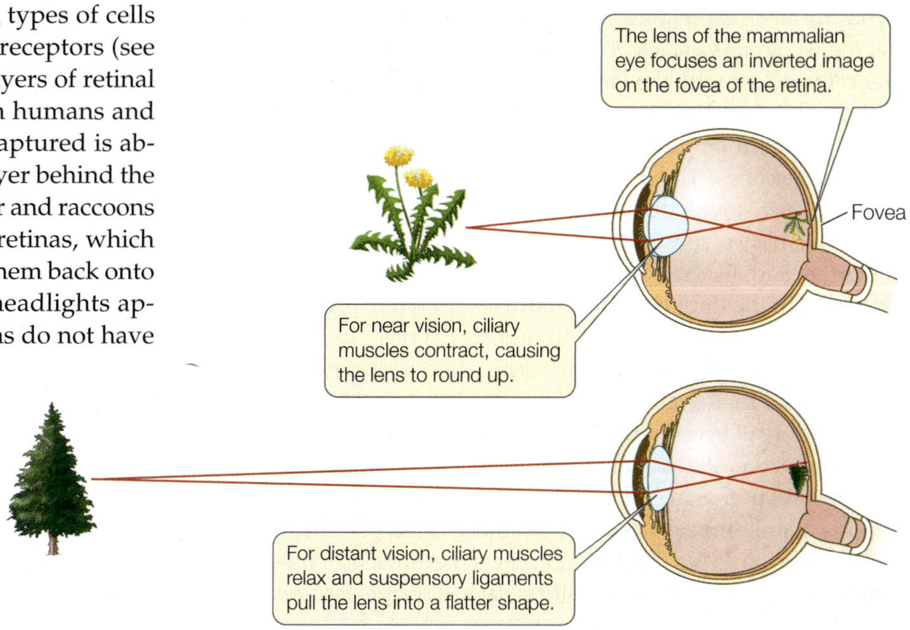
The lens of the mammalian eye focuses an inverted image on the fovea of the retina.
Fovea
For near vision, ciliary muscles contract, causing the lens to round up.
For distant vision, ciliary muscles relax and suspensory ligaments pull the lens into a flatter shape.

46.15 Staying in Focus Mammals and birds focus their eyes by changing the shape of the lens depending on the eye's distance from the object of focus.

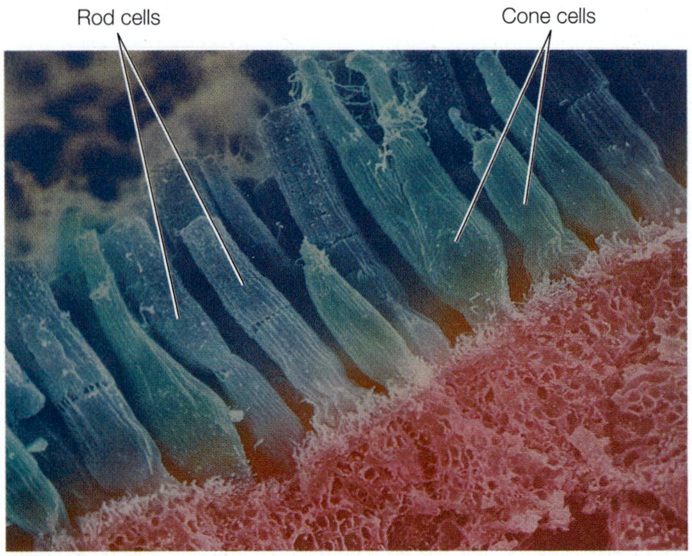
Rod cells
Cone cells

46.16 Rods and Cones This scanning electron micrograph of photoreceptors in the retina of a mud puppy (an amphibian) shows cylindrical rods and tapered cones.

Rod and cone cells are the photoreceptors of the vertebrate retina

The photoreceptors of the vertebrate retina are two types of modified neurons called **rod cells** and **cone cells** based on their shapes (**Figure 46.16**). Rod cells are highly light-sensitive and perceive shades of gray in dim light. Cone cells function at high light levels and are responsible for high-acuity color vision. The human retina has about 5 million cones and 100 million rods, but the density of each varies across the retina.

ROD CELLS Rod cells do not produce action potentials, but they release neurotransmitter from their bases, where they form synapses with the next neurons in the visual pathway. Each rod cell has an outer segment, an inner segment, and a synaptic terminal (**Figure 46.17**). The highly specialized outer segment contains a stack of discs made up of plasma membrane densely packed with the visual pigment rhodopsin (see p. 957). The function of these discs is to capture photons of light. The inner segment contains the cell nucleus, mitochondria, and other organelles. The synaptic terminal is where the rod cell communicates with other neurons.

To see how a rod cell responds to light, we can penetrate a single rod cell with an electrode and record its membrane potential in the dark and in the light, as shown in Figure 46.17. From what we have learned about other types of sensory receptors, we might expect that stimulation of the rod cell by light would make its membrane potential less negative, but the opposite is true—it becomes more negative.

When a rod cell is kept in the dark, it has a relatively depolarized resting potential compared with other neurons. In fact, the plasma membrane of the rod cell is almost as permeable to Na^+ as to K^+. In the dark, Na^+ continually enters the outer segment of the cell—the dark current. When light is flashed on the dark-adapted rod cell, its membrane potential becomes more negative—it hyperpolarizes (see Figure 46.17). The rate of neurotransmitter release changes as membrane potential changes. As the rod cell hyperpolarizes, its release of neurotransmitter decreases.

INVESTIGATING**LIFE**

46.17 A Rod Cell Responds to Light The plasma membrane of a rod cell hyperpolarizes—becomes more negative—in response to a flash of light. Rod cells do not fire action potentials, but in response to the absorption of light energy, the neuron experiences a change in membrane potential.[a]

HYPOTHESIS When a rod cell absorbs photons (light energy), its membrane potential changes in proportion to the strength of the light stimulus.

Method

1. Record membrane potentials from the inner segment of a rod cell.
2. Stimulate the rod cells with light flashes of varying intensity and record the results.

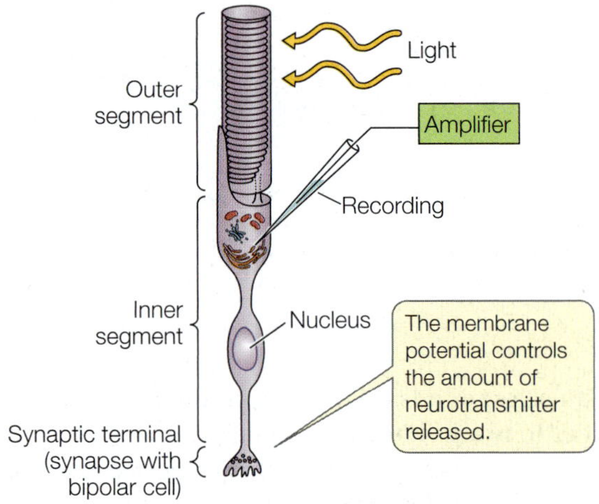
Light
Outer segment
Amplifier
Recording
Inner segment
Nucleus
The membrane potential controls the amount of neurotransmitter released.
Synaptic terminal (synapse with bipolar cell)

Results

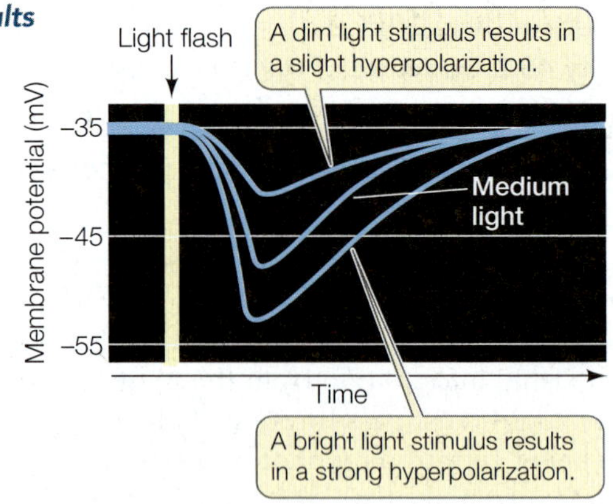
Light flash
A dim light stimulus results in a slight hyperpolarization.
Membrane potential (mV)
–35
–45
–55
Medium light
Time
A bright light stimulus results in a strong hyperpolarization.

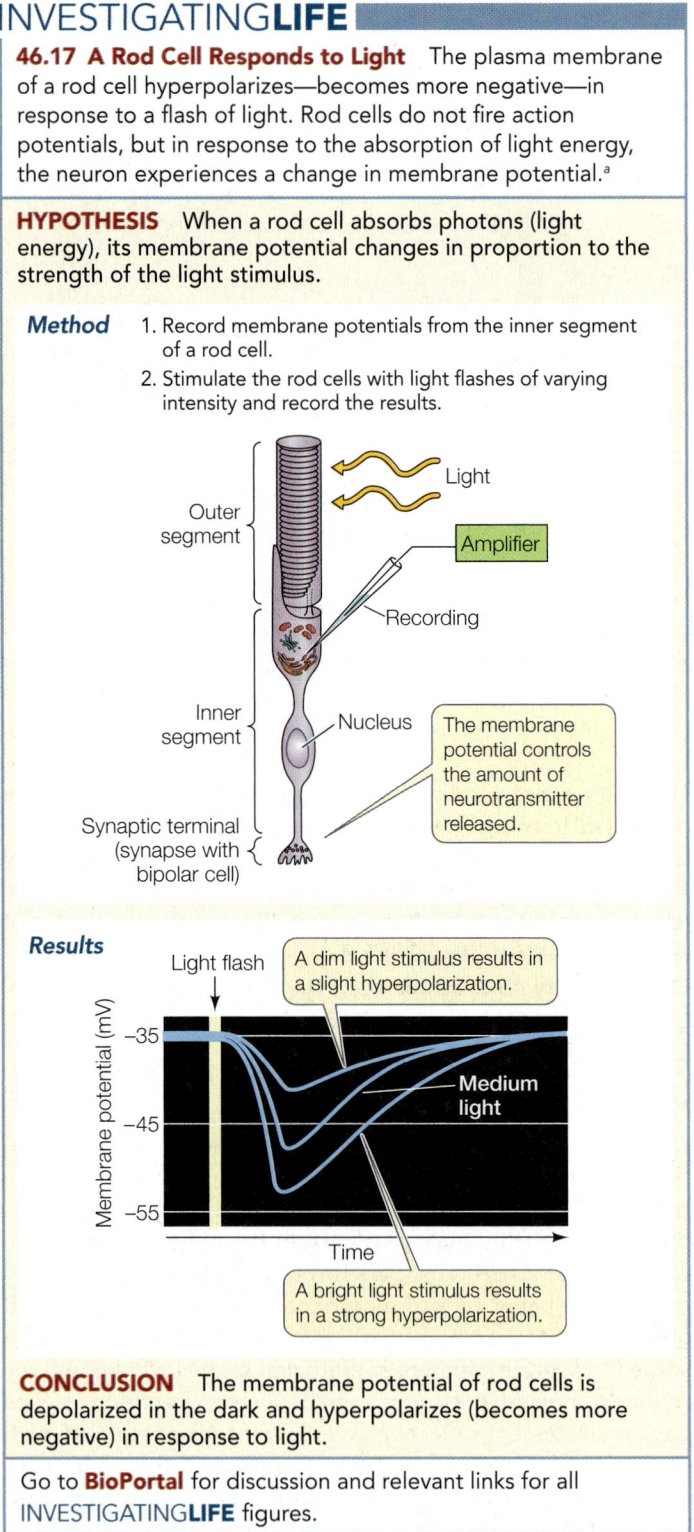

CONCLUSION The membrane potential of rod cells is depolarized in the dark and hyperpolarizes (becomes more negative) in response to light.

Go to **BioPortal** for discussion and relevant links for all INVESTIGATING**LIFE** figures.

[a]Baylor, D. A. et al. 1979. *Journal of Physiology* 288: 589–634.

WORKING WITH**DATA:**

Membrane Currents and Light Intensity in Rod Cells

Original Paper

Baylor, D. A., T. D. Lamb, and K.-W. Yau. 1979. The membrane current of single rod outer segments. *Journal of Physiology* 288: 589–611.

Analyze the Data

In a slightly different set of experiments than that described in Figure 46.17, researchers measured the effect of light on the current across the rod cell membrane. The figure at right shows a series of recordings of the membrane currents (i.e., inward currents of positive ions) generated when rod cells were illuminated by light flashes of varying intensities. The initial values for the membrane currents represent the condition of a rod cell in total darkness. The light flash was given at time 0, and the intensity of the flashes is indicated on the right side of the response curves.

QUESTION 1

Why is there no difference between the *maximum* currents induced by flashes of light at 7.8 and 16 photons per μm^2?

QUESTION 2

Why does a rod maintain its maximum current for longer in response to a flash of light at 16 photons per μm^2 than it does to one at 7.8 photons per μm^2?

QUESTION 3

If you measured membrane potential instead of an outward current, how would the resulting recordings differ from the one shown here?

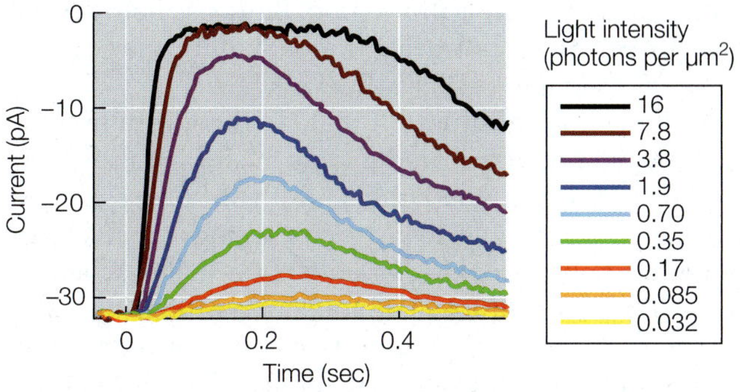

0
−10
−20
−30
Current (pA)
0
0.2
0.4
Time (sec)
Light intensity (photons per μm²)
16
7.8
3.8
1.9
0.70
0.35
0.17
0.085
0.032

Go to **BioPortal** for all WORKING WITH**DATA** exercises

How does the absorption of light by rhodopsin hyperpolarize the rod cell? When rhodopsin is excited by light, it initiates a cascade of events. The dark-adapted rod cell has open Na^+ channels, allowing a depolarizing dark current (**Figure 46.18**). Light photons excite rhodopsin, which activates a G protein called transducin. Activated transducin in turn activates a phosphodiesterase (PDE). Activated PDE converts cyclic GMP (cGMP) to GMP, which causes the Na^+ channels to close. Na^+ is pumped out and the cell hyperpolarizes.

This cascade may seem like a roundabout way of doing business, but its advantage is its enormous amplification ability. Each molecule of light-excited rhodopsin can activate several hundred transducin molecules, thus activating a large number of PDE molecules. The catalytic capacity of PDE is great—one PDE molecule can hydrolyze several hundred molecules of cGMP per second. The bottom line is that a single photon of light can result in the closure of a huge number of Na^+ channels.

CONE CELLS Cone cells are responsible for the high-acuity color vision of day-active vertebrates such as humans. It is therefore logical that the highest density of cone cells is in the area of the retina that receives light

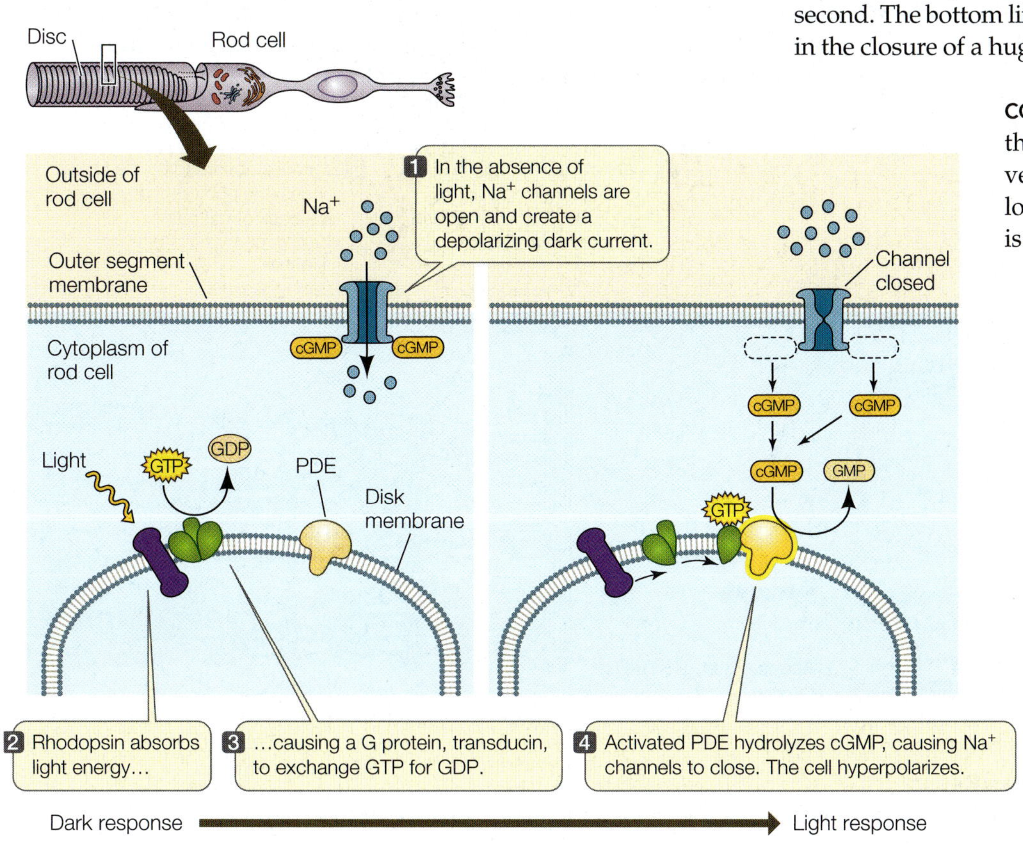

Disc
Rod cell
Outside of rod cell
Na+
1 In the absence of light, Na+ channels are open and create a depolarizing dark current.
Outer segment membrane
Channel closed
Cytoplasm of rod cell
cGMP
cGMP
GMP
Light
GTP
GDP
PDE
Disk membrane
2 Rhodopsin absorbs light energy…
3 …causing a G protein, transducin, to exchange GTP for GDP.
4 Activated PDE hydrolyzes cGMP, causing Na+ channels to close. The cell hyperpolarizes.
Dark response
Light response

46.18 Light Absorption Closes Sodium Channels The absorption of light by rhodopsin initiates a signaling cascade that hyperpolarizes the rod cell. (A) In the dark, Na^+ channels in the plasma membrane of the rod cell's outer segment are held open by cGMP, allowing positive charges to enter the cell (upper right of panel). When the rod cell is stimulated by light, it activates transducin (lower portion of panel). (B) Transducin activates a molecule of phosphodiesterase (PDE). (C) Activated PDE catalyzes the breakdown of cGMP to GMP. The depletion of cGMP results in closure of the Na^+ channels and hyperpolarization of the cell.

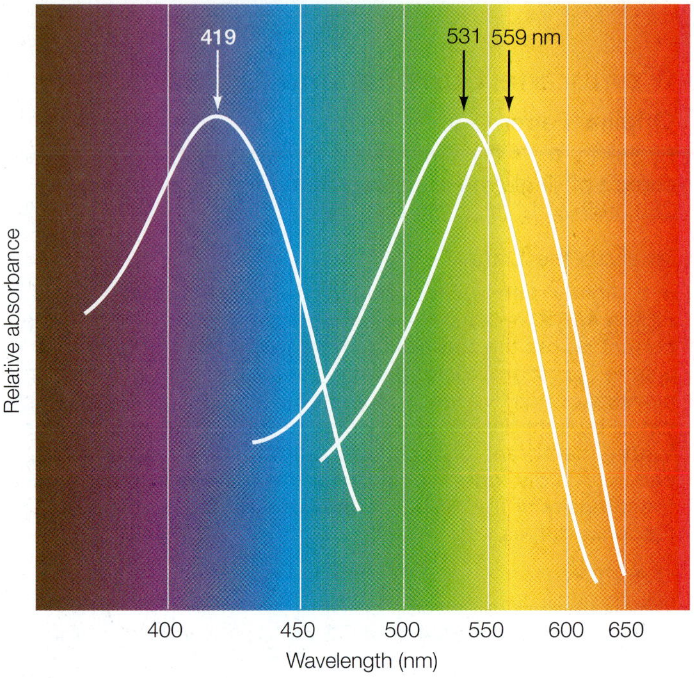

419
531
559 nm
Relative absorbance
400
450
500
550
600
650
Wavelength (nm)

46.19 Absorption Spectra of Cone Cells The three kinds of cone cells contain slightly different visual pigments that absorb different wavelengths of light.

from the center of the visual field, a region called the **fovea**. The human fovea has about 160,000 cones per square millimeter. But humans are not the champions of high-acuity vision; a hawk's fovea has almost twice that number of cones, making the hawk's vision much sharper than ours. Birds also have *two* foveae in each eye; one receives light from straight ahead, the other from a more lateral field of vision. The forward-looking foveae make binocular vision possible, while the lateral-looking foveae provide high-acuity vision. Birds use both sets of foveae by frequently turning their heads slightly; they cannot move their eyes in the sockets as humans can.

Cones have low sensitivity to light and contribute little to night vision. Night vision depends mostly on rod cells, which is why vision in dim light is mostly in shades of gray and acuity is low. You may have trouble seeing a small object at night when you are looking straight at it—that is, when its image is falling on your fovea. If you look a little to the side, so that the image falls on a rod-rich area of your retina, you see the object better. Astronomers looking for faint objects in the sky learned this trick a long time ago. The retinas of nocturnal animals, such as flying squirrels, contain a high percentage of rods. By contrast, some animals that are active only during the day (such as chipmunks) have mostly cones in their retinas.

The human retina has three kinds of cone cells, each containing slightly different opsin molecules that differ in the wavelengths of light they absorb best. Although the same 11-*cis*-retinal group is the light-absorber in all three kinds of cones (see Figure 46.12), its molecular interactions with opsin determine the spectral sensitivity of the cone cell as a whole (**Figure 46.19**). Because different wavelengths of light are differentially absorbed by the different cone cell visual pigments, the brain interprets the relative inputs from the different classes as a full range of color. Color blindness in humans results from the absence or dysfunction of one or more of the three classes of cone cells. Some mammals have only one or two classes of cone cells, whereas birds have four.

Information flows through layers of neurons in the retina

The human retina is organized into layers of neurons that receive visual information and process it before sending it to the brain (**Figure 46.20**). Closest to the lens (and thus to light

Go to Media Clip 46.1 Into the Eye
Life10e.com/mc46.1

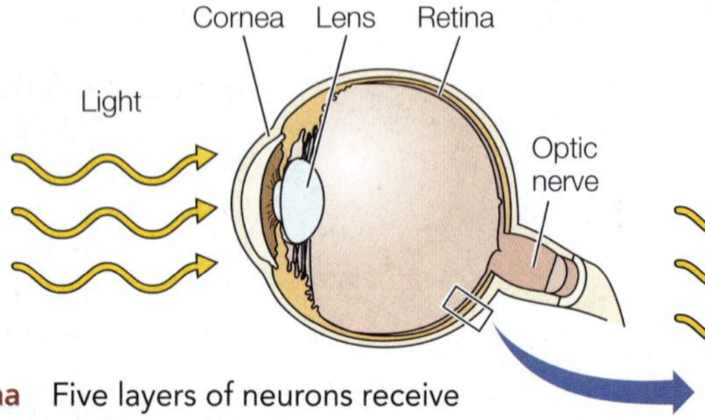

Cornea
Lens
Retina
Light
Optic nerve

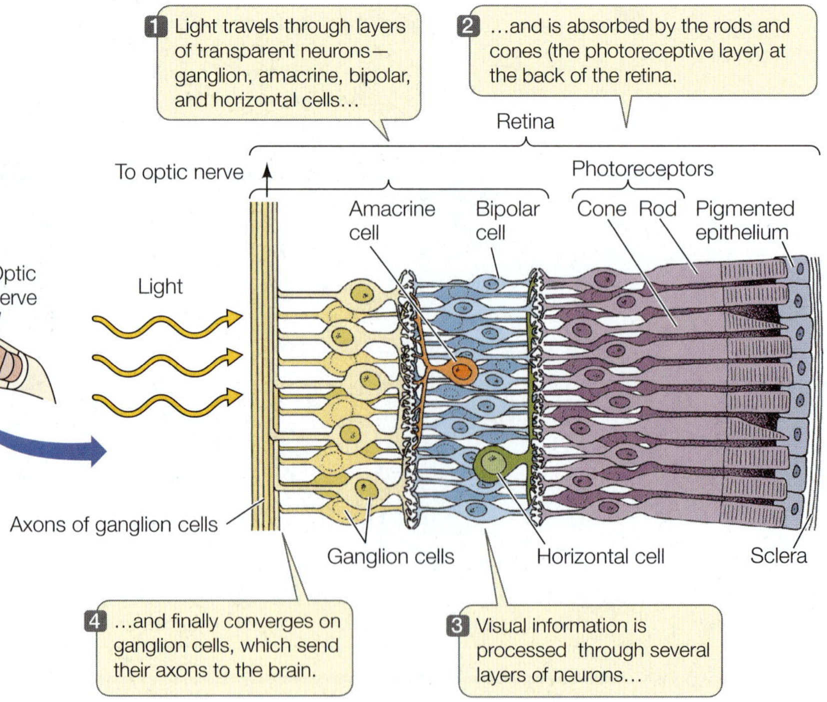

1 Light travels through layers of transparent neurons—ganglion, amacrine, bipolar, and horizontal cells…
2 …and is absorbed by the rods and cones (the photoreceptive layer) at the back of the retina.
Retina
To optic nerve
Photoreceptors
Amacrine cell
Bipolar cell
Cone
Rod
Pigmented epithelium
Light
Axons of ganglion cells
Ganglion cells
Horizontal cell
Sclera
4 …and finally converges on ganglion cells, which send their axons to the brain.
3 Visual information is processed through several layers of neurons…

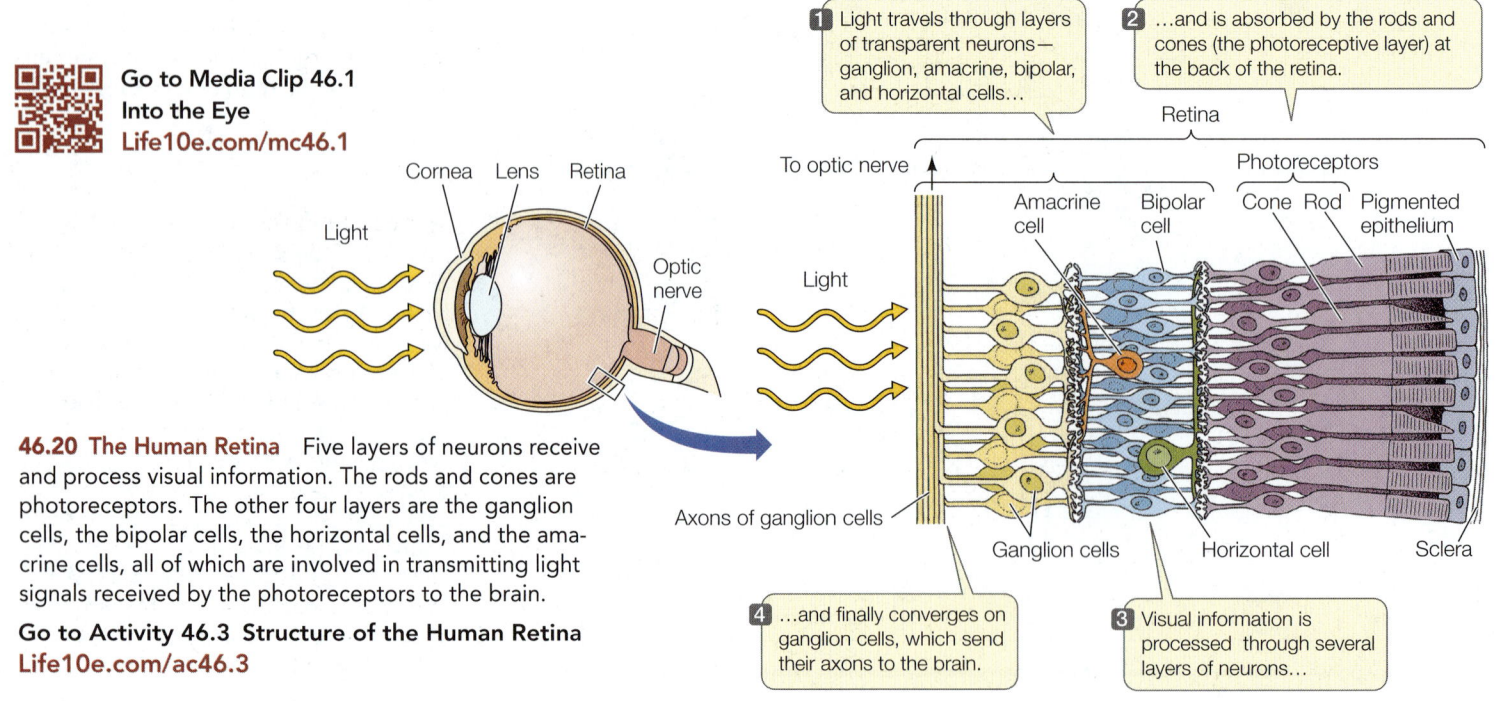

46.20 The Human Retina Five layers of neurons receive and process visual information. The rods and cones are photoreceptors. The other four layers are the ganglion cells, the bipolar cells, the horizontal cells, and the amacrine cells, all of which are involved in transmitting light signals received by the photoreceptors to the brain.

Go to Activity 46.3 Structure of the Human Retina
Life10e.com/ac46.3

input) is a layer of **ganglion cells**; a central layer contains three neuronal types, **bipolar cells**, **horizontal cells**, and **amacrine cells**; and at the "rear" of the retina lie the photoreceptors (rods and cones). The layers of cells between the photoreceptors and the ganglion cells process information about the visual field.

GANGLION AND BIPOLAR CELLS From our discussion of rod cells, we know that the photoreceptor cells at the back of the retina hyperpolarize in response to light and do not generate action potentials. The ganglion cells lie at the front of the retina, and they do fire action potentials. The axons of ganglion cells form the **optic nerve** that travels to the brain.

The ganglion cells are connected to the photoreceptors by bipolar cells. Changes in the membrane potential of rods and cones in response to light alter the rates at which the rods and cones release neurotransmitter at their synapses with the bipolar cells. In response to this neurotransmitter, the membrane potentials of the bipolar cells change, altering the rate at which they release neurotransmitter onto ganglion cells. The rate of neurotransmitter release from the bipolar cells determines the rate at which ganglion cells fire action potentials. Thus the direct flow of information in the retina is from photoreceptor to bipolar cell to ganglion cell. The ganglion cells send the information to the brain via the optic nerve.

Each human eye contains about 1.2 million ganglion cells but more than 100 million rods and cones. Therefore there must be convergence of information as it passes from the photoreceptors to the ganglion cells. A given bipolar cell can receive input from multiple rods or multiple cones, but not from both. The relationship between photoreceptors, bipolar cells, and ganglion cells depends on their location on the retina. In the fovea, a ganglion cell may receive input from as few as five photoreceptors, but in the periphery of the retina, a ganglion cell may receive input from thousands of photoreceptors. Visual acuity is a reflection of these quantitative relationships.

The patch of photoreceptors that communicates with a ganglion cell forms a circular receptive field. When light falls on a receptive field, its ganglion cell can be either excited or inhibited. As mentioned above, each ganglion cell sends an axon to the brain in the optic nerve. Thus the information coming from the retina to the brain is about the pattern of patches of light and dark falling on the retina.

HORIZONTAL AND AMACRINE CELLS The other two cell layers, the horizontal cells and the amacrine cells, consist of interneurons that communicate laterally across the retina. Horizontal cells form synapses with neighboring photoreceptors and bipolar cells. Thus light falling on one photoreceptor can influence the sensitivity of its neighbors to light. This lateral flow of information enables the retina to sharpen the perception of contrast between light and dark patterns. Amacrine cells form local interconnections between bipolar cells and ganglion cells. Some amacrine cells are highly sensitive to changing illumination or to motion. Others assist in adjusting the sensitivity of the eyes according to the overall level of light falling on the retina. When background light levels change, amacrine cell connections to the ganglion cells help the ganglion cells remain sensitive to temporal changes in stimulation. Thus even with large changes in background illumination, the eyes are sensitive to smaller, more rapid changes in the pattern of light falling on the retina.

RECAP 46.4

A family of photosensitive visual pigments are responsible for light sensitivity in all animals. Receptor cells, including rod and cone cells in humans, transduce the photosensitivity of visual pigments to light and use it to form images of the environment.

- Explain how a photon of light affects the membrane potential in a rod cell. **See pp. 960–961 and Figures 46.12, 46.17, and 46.18**
- What is the mechanism of color vision? **See pp. 961–962 and Figure 46.19**
- Describe the flow of signals that takes place in the eye in response to light. **See pp. 962–963 and Figure 46.20**

Knowing the path of information from sensory receptor cells to the central nervous system still does not tell us how that information is processed by the brain. What does the eye tell the brain, for example, in response to a pattern of light falling on the retina? In Chapter 47 we will describe how the mammalian brain reassembles sensory information into our perception of the world.

How can bats emit loud pulses of sound and not be deaf to the faint echoes that return within milliseconds?

ANSWER

Bat "echolocation" allows these mammals to navigate around objects and to find prey in total darkness. Bats emit sound waves at frequencies well above the range of human hearing; these sound waves bounce off objects, including potential prey. The bat perceives the echoes that bounce back almost immediately and uses these echoes to locate prey and other objects. The loud pulses of sound still being emitted don't "drown out" the weak echoes because small muscles in a bat's ears contract to dampen its hearing sensitivity while the sounds are being emitted, but relax in time for the bat to hear the echo—a truly remarkable ability, considering that the sound pulses are emitted at rates of 20 to 80 per second.

CHAPTER SUMMARY 46

46.1 How Do Sensory Receptor Cells Convert Stimuli into Action Potentials?

- Sensory receptor cells, also known as **sensors** or **receptors**, transduce information about an animal's external and internal environment into action potentials that the brain perceives as different forms of sensory information.
- **Receptor potentials** can spread to regions of the cell's plasma membrane that generate action potentials. Some sensors do not fire action potentials but release neurotransmitter onto sensory neurons that do fire action potentials. **Review Figure 46.1**
- Sensors have **receptor proteins** that cause ion channels to open or close, affecting the receptor cell's membrane potential. Metabotropic receptors act through signal transduction pathways to generate receptor potentials. Mechanoreceptors are ionotropic sensory receptors that open ion channels physically through forces such as pressure or stretch. **Review Figure 46.2**
- The interpretation of action potentials as particular sensations depends on which neurons in the central nervous system receive them.
- **Adaptation** enables the nervous system to ignore irrelevant or continuous stimuli while remaining responsive to relevant or new stimuli.

46.2 How Do Sensory Systems Detect Chemical Stimuli?

- **Chemoreceptors** are responsible for **olfaction**, **gustation**, and the sensing of **pheromones**.
- Mammalian **olfactory receptor neurons** (**ORNS**) project directly to the **olfactory bulb** of the brain. ORNs for the same **odorant** project to the same area of the olfactory bulb.
- Each ORN expresses one receptor protein that can bind a specific type of molecule or ion. Binding causes a second messenger to open ion channels, which creates an action potential. **Review Figure 46.3**
- In vertebrates, **taste buds** in the mouth cavity are responsible for gustation. The five basic tastes are sweet, salty, sour, bitter, and umami. **Review Figure 46.5**

46.3 How Do Sensory Systems Detect Mechanical Forces?

- The skin contains a variety of ionotropic **mechanoreceptors** that respond to touch and pressure. The density of mechanoreceptors in any skin area determines the sensitivity of that area. **Review Figure 46.6**
- Stretch receptors in **muscle spindles** and in the **Golgi tendon organ** inform the CNS of the positions of and loads on parts of the body. **Review Figure 46.7**
- **Hair cells** are mechanoreceptors of the auditory and vestibular systems. Physical bending of their **stereocilia** alters their receptor proteins and therefore their membrane potentials. **Review Figure 46.8**
- In mammalian auditory systems, ear pinnae collect and direct sound waves to the **tympanic membrane**, which vibrates in response to sound waves. The movements of the tympanic membrane are amplified through a chain of **ossicles** that conduct the vibrations to the **oval window**. Movements of the oval window create pressure waves in the fluid-filled **cochlea**. **Review Figure 46.9, ACTIVITY 46.1**
- The basilar membrane running down the center of the cochlea is distorted by pressure waves at specific locations that depend on the frequency of the wave. These distortions cause hair cells in the **organ of Corti** to bend and to release neurotransmitter, generating action potentials in the cochlear nerve that are transmitted to the auditory cortex of the brain. **Review Figure 46.10, ANIMATED TUTORIAL 46.1**
- Hair cells are also the mechanoreceptors of the organs of equilibrium in the mammalian vestibular system, which include the **semicircular ducts** and the **saccule** and **utricle**. **Review Figure 46.11, ANIMATED TUTORIAL 46.2**

46.4 How Do Sensory Systems Detect Light?

- **Photosensitivity** depends on the absorption of photons of light by visual pigment molecules that consist of a protein called **opsin** and a light-absorbing group. Absorption of light is the first step in a cascade of intracellular events leading to a change in the membrane potential of the **photoreceptor cell**. **Review Figure 46.12, ANIMATED TUTORIAL 46.3**
- Visual systems range from the simple **eye cups** of flatworms, which sense the direction of a light source, to the **compound eyes** of arthropods, which detect shapes and patterns, to the image-forming eyes of vertebrates and cephalopods. **Review Figures 46.13, 46.14**
- Vertebrate and cephalopod eyes focus detailed images of the visual field onto dense arrays of photoreceptors that transduce the visual image into neural signals. **Review Figures 46.14, 46.15, ACTIVITY 46.2**
- Vertebrates have two types of photoreceptors, **rod cells** and **cone cells**. Rod cells are more sensitive to light and are responsible for dim light vision. Cone cells are less sensitive to light but are responsible for high-acuity and color vision.
- Photoreceptors do not fire action potentials. When not stimulated by light they release neurotransmitter continuously. Light hyperpolarizes rod cells, and their release of neurotransmitter decreases. **Review Figures 46.17, 46.18**
- **Rhodopsin** is the visual pigment of rod cells. The visual pigments of cone cells have three different opsin components, which gives them different spectral sensitivities. **Review Figure 46.19**
- The vertebrate **retina** consists of layers of neurons lining the back of the eye. The light-absorbing photoreceptor cells are at the rear of the retina. The axons of the **ganglion cells** are bundled together in the **optic nerve**. Between the photoreceptors and the ganglion cells are neurons that process information from the photoreceptors. **Review Figure 46.20, ACTIVITY 46.3**

Go to the Interactive Summary to review key figures, Animated Tutorials, and Activities
Life10e.com/is46

CHAPTER**REVIEW**

REMEMBERING

1. Which statement about sensory systems is *not* true?
   a. Sensory transduction involves the conversion (direct or indirect) of a physical or chemical stimulus into changes in membrane potentials.
   b. In general, a stimulus causes a change in the flow of ions across the plasma membrane of a sensory receptor cell.
   c. The term "adaptation" refers to the process by which a sensory system becomes insensitive to a continuing source of stimulation.
   d. The more intense a stimulus, the greater the magnitude of each action potential fired by a sensory neuron.
   e. Sensory adaptation plays a role in the ability of organisms to discriminate between important and unimportant information.

2. Which statement about olfaction is *not* true?
   a. In general, mammals depend more on vision than on olfaction as their dominant sensory modality.
   b. Olfactory stimuli are recognized by the interaction between odorant molecules and receptor proteins on olfactory hairs.
   c. The more odorant molecules that bind to receptors, the more action potentials are generated.
   d. The greater the number of action potentials generated by an olfactory receptor, the greater the intensity of the perceived smell.
   e. The perception of different smells results from the activation of different combinations of olfactory receptors.

3. The membrane most directly responsible for the ability to discriminate different pitches of sound is the
   a. round window.
   b. oval window.
   c. tympanic membrane.
   d. tectorial membrane.
   e. basilar membrane.

4. Which statement is *not* true?
   a. The transmembrane potential of a rod cell becomes more negative when the rod cell is exposed to light.
   b. A photoreceptor releases the most neurotransmitter when in total darkness.
   c. Whereas in vision the intensity of a stimulus is encoded by the degree of hyperpolarization of photoreceptors, in hearing the intensity of a stimulus is encoded by changes in firing rates of sensory neurons.
   d. Stiffening of the ossicles in the middle ear can lead to deafness.
   e. The interaction among hammer (malleus), anvil (incus), and stirrup (stapes) conducts sound waves across the fluid-filled middle ear.

5. Which of the following statements about information flow in the vertebrate visual system is true?
   a. Action potentials in bipolar cells cause the release of neurotransmitter onto ganglion cells.
   b. Amacrine cells integrate the activity of neighboring rod and cone cells.
   c. When photons of light enter the eye, the first cells they encounter in the retina are ganglion cells.
   d. The highest density of rod cells in the human retina is centrally located in the fovea, resulting in high-acuity dim-light vision.
   e. Pigmented epithelial cells at the back of the retina provide information about the level of ambient light for contrast adjustments.

UNDERSTANDING & APPLYING

6. What are the similarities and differences in the functioning of olfactory receptors and taste receptors? How do these sensory cells enable the central nervous system to discriminate between an apple and an orange?

7. If you were blindfolded and sitting in a wheeled chair, how would you know if you were being pushed forward or backward?

8. To human ears, sounds are louder underwater than in air. Why is this so?

ANALYZING & EVALUATING

9. Fish have fluid-filled channels called lateral line canals running down the sides of their bodies. These canals contain hair cells, as illustrated below. Describe what you think the functions of these lateral-line hair cells are.

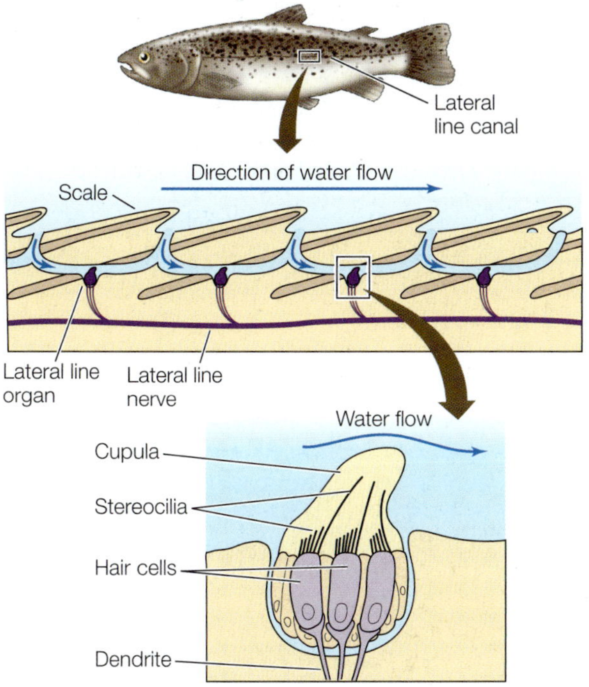

Lateral line canal
Direction of water flow
Scale
Lateral line organ
Lateral line nerve
Water flow
Cupula
Stereocilia
Hair cells
Dendrite

10. An owl can capture a mouse scurrying across a forest floor in total darkness. What sensory information do you think the owl uses, and how does it get directional information from that sensory information?

Go to BioPortal at **yourBioPortal.com** for Animated Tutorials, Activities, LearningCurve Quizzes, Flashcards, and many other study and review resources.

47 The Mammalian Nervous System: Structure and Higher Functions

CHAPTER OUTLINE

© Bluesky International Limited

A Mind-Expanding Maze The extraordinary ability of taxicab drivers in London to navigate its maze of streets and byways prompted a study that revealed London cabbies to have larger than normal posterior hippocampi—a brain region implicated in the memory of spatial relationships in the environment.

COMPARE the Google maps of London and New York City at the same scale. In which city do you think it is easier to drive a taxi? In the city with sequentially numbered avenues and streets laid out at right angles to each other, or the city with a maze of arbitrarily named streets going in every direction?

Eleanor Maguire at University College London was so impressed with the navigational abilities of London taxi drivers that she decided to see if there was anything "special" about their brains. Using the brain imaging technique known as MRI (magnetic resonance imaging), Maguire and her colleagues examined the brains of taxi drivers with varying numbers of years of experience and compared them with each other and with the brains of control subjects who were not taxi drivers. The studies revealed significant changes in the anatomy of the hippocampus among taxi drivers.

The hippocampus is an area of the brain involved in learning and memory. The posterior hippocampus in particular is implicated in the memory of spatial relationships among objects in the environment. Maguire found that the posterior hippocampi of taxi drivers was larger than that of the control subjects and that, among the cab drivers themselves, there was a positive correlation between the size of the posterior hippocampus and years of driving experience.

In M. A. Wilson's lab at the Massachusetts Institute of Technology, researchers recorded the activity of hippocampal neurons of rats as they navigated a maze. They located specific neurons, referred to as "place cells," that fire only when the rat is at a particular location in the maze. In a sense these researchers can see what the rat is thinking, because when the animal is not moving through the maze, the neurons will occasionally fire in the same pattern as when it is running the maze—or they will fire in the reverse sequence, representing where the rat has just been. But these replays of firing patterns are much faster; the rat *thinks* about its experience in the maze about 25 times faster than the actual experience. In a maze that had a choice point (go right or go left), the firing pattern seen when the animal was held at the start position predicted which direction it would turn. Similar firing patterns occur when the rats are sleeping. Are they dreaming about the maze? Are they transferring memory of today's experience in the maze into long-term memory?

> **?** Does the firing pattern of hippocampal place cells during sleep represent a memory of recent experience being transferred into long-term memory?
>
> See answer on p. 983.

47.1 How Is the Mammalian Nervous System Organized?

The organization of the mammalian nervous system can be described anatomically and functionally. In anatomical terms, all vertebrate nervous systems consist of three parts: a brain, a spinal cord, and a set of peripheral nerves that reach all parts of the body. As discussed in Section 45.4, the brain and spinal cord are the **central nervous system**, or **CNS**, and the nerves that connect the CNS to all the tissues and sensors of the body are the **peripheral nervous system**, or **PNS**. An additional division of the nervous system exists in the gut; we will discuss this **enteric nervous system** in Chapter 51.

Recall from Section 45.1 that a neuron is an electrically excitable cell that communicates via an axon. When used in the context of a nervous system, the term **nerve** refers to a bundle of axons that carries information about many things simultaneously. Some axons in a nerve may be carrying information to the CNS while other axons in the same nerve are carrying information from the CNS to the body's organs. A discussion of the functional organization of the nervous system refers to these paths of information flow. In this chapter we will divide the anatomy of the mammalian brain, spinal cord, and peripheral nervous system into smaller, discrete functional units.

Functional organization is based on flow and type of information

Figure 47.1 illustrates the major avenues of information flow through the human nervous system. The white boxes represent the four divisions of the peripheral nervous system; two of these bring information from the periphery to the CNS, and two transmit information from the CNS to the periphery.

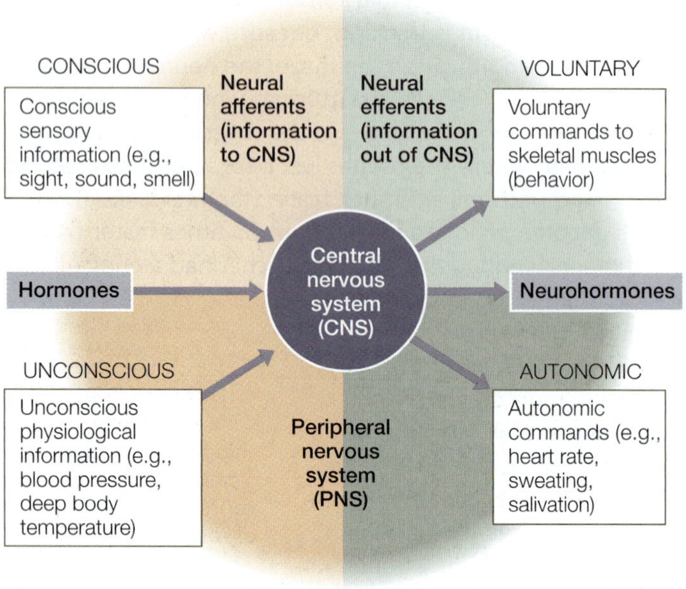
CONSCIOUS
Conscious sensory information (e.g., sight, sound, smell)
Neural afferents (information to CNS)
Neural efferents (information out of CNS)
VOLUNTARY
Voluntary commands to skeletal muscles (behavior)
Hormones
Central nervous system (CNS)
Neurohormones
UNCONSCIOUS
Unconscious physiological information (e.g., blood pressure, deep body temperature)
Peripheral nervous system (PNS)
AUTONOMIC
Autonomic commands (e.g., heart rate, sweating, salivation)

47.1 Organization of the Nervous System The peripheral nervous system carries information to (afferent) and from (efferent) the central nervous system (center circle in the diagram). The CNS also receives hormonal inputs and produces hormonal outputs.

- The **afferent** portion of the PNS carries information from sensory receptor cells to the CNS. We are *conscious* of much of this information (e.g., light, sound, skin temperature, limb position), but we are usually *unconscious* of the information involved in physiological regulation (e.g., blood pressure, deep body temperature, blood oxygen levels).
- The **efferent** portion of the PNS carries information from the CNS to the muscles and glands of the body. Efferent pathways are divided into a *voluntary* division that executes our conscious movements; and an *involuntary*, or *autonomic*, division that controls physiological functions.

In addition to the neural information it receives from the PNS, the central nervous system receives chemical information from hormones circulating in the blood. In turn, neurohormones released by neurons enter the circulation and affect neurons and other cells distant from the site of release (see Section 41.1).

The anatomical organization of the CNS emerges during development

Early in the development of a vertebrate embryo, a tube of neural tissue forms (see Section 44.4). At its anterior end, this neural tube forms three swellings that become the **hindbrain**, **midbrain**, and **forebrain**. The rest of the neural tube becomes the spinal cord. Peripheral nerves sprout from the midbrain and hindbrain (the cranial nerves) and from the spinal cord (the spinal nerves). From these early stages we see the linear axis of information flow in the nervous system. Although the developing brain will fold and become a complex structure, the information flow in the adult nervous system will follow paths that emerge from the simple, linear neural tube.

Each of these three regions of the embryonic brain develops into several structures in the adult brain (Figure 47.2). From the embryonic midbrain come structures that integrate information from the different senses and coordinate motor responses. From the hindbrain come the **medulla**, the **pons**, and the **cerebellum**. The medulla is continuous with the spinal cord, the pons is in front of the medulla, and the cerebellum is a dorsal outgrowth of the pons. The medulla and pons contain distinct groups of neurons involved in controlling physiological functions such as breathing, circulation, and basic motor patterns such as swallowing and vomiting. All information traveling between the spinal cord and higher brain areas must pass through the pons, the medulla, and the midbrain, which are collectively known as the **brainstem**.

The cerebellum is involved in coordinating muscle activity and maintaining balance. It is like the director of a movie; the cerebellum receives a "script" of the commands going to the muscles from higher brain areas, and it receives information about the actual performance coming up the spinal cord from the "actors"—the joints and muscles. The cerebellum compares the "script" with the performance and refines motor commands accordingly. Damage to the cerebellum results in loss of fine motor control and coordination.

The embryonic forebrain develops a central region called the **diencephalon** and a surrounding structure called the **telencephalon**. The diencephalon is the core of the forebrain and consists of an upper structure, the thalamus, and a lower

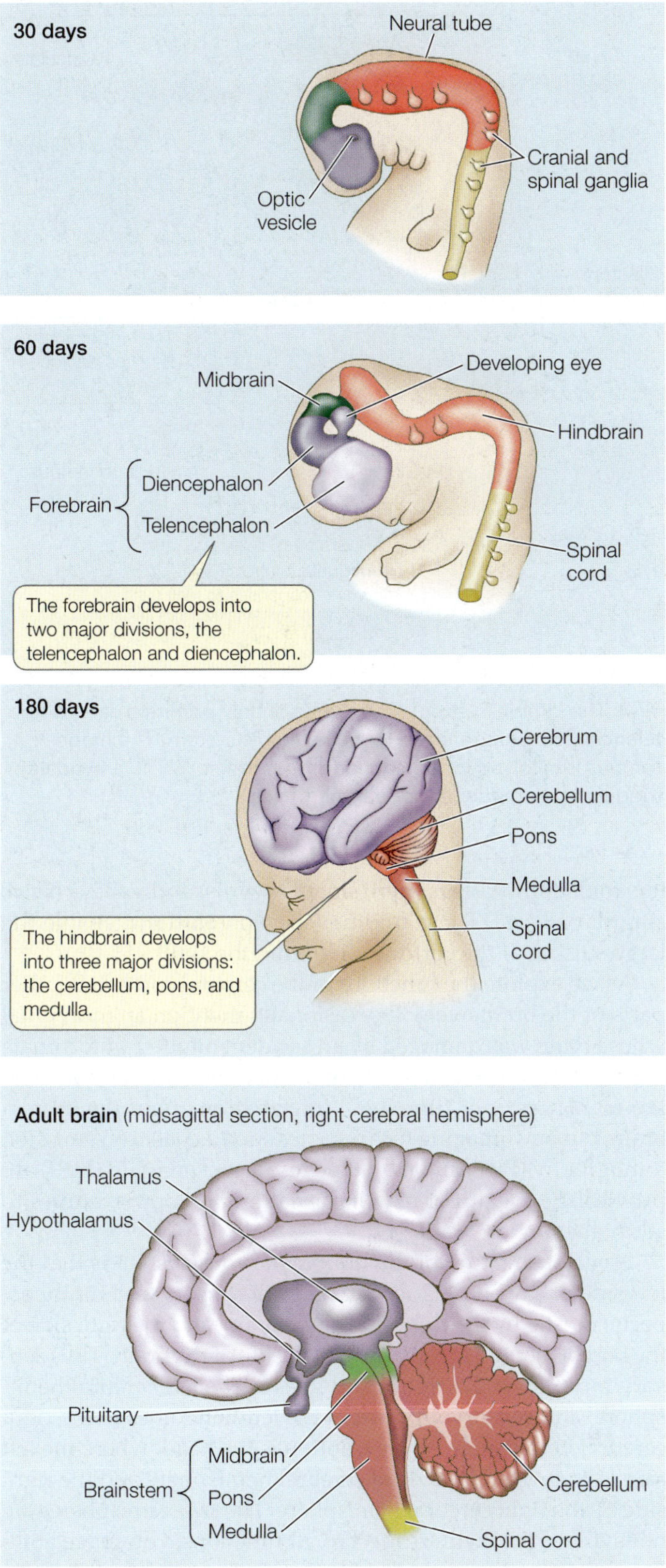
30 days
Neural tube
Cranial and spinal ganglia
Optic vesicle
60 days
Midbrain
Developing eye
Hindbrain
Forebrain
Diencephalon
Telencephalon
Spinal cord
The forebrain develops into two major divisions, the telencephalon and diencephalon.
180 days
Cerebrum
Cerebellum
Pons
Medulla
Spinal cord
The hindbrain develops into three major divisions: the cerebellum, pons, and medulla.
Adult brain (midsagittal section, right cerebral hemisphere)
Thalamus
Hypothalamus
Pituitary
Brainstem
Midbrain
Pons
Medulla
Cerebellum
Spinal cord

47.2 Development of the Central Nervous System In vertebrate embryos, the anterior end of the hollow neural tube differentiates into forebrain, midbrain, and hindbrain. Each of these regions develops into several structures of the adult brain. The remainder of the neural tube becomes the spinal cord.

structure, the hypothalamus. The thalamus is the final relay station for sensory information going to the telencephalon. The hypothalamus receives a lot of physiological information of which we are not conscious, and it uses that information to regulate many physiological functions and biological drives.

The telencephalon—also called the **cerebrum**—consists of the left and right **cerebral hemispheres**. The outer layer of the telencephalon is the **cerebral cortex**, a thin layer rich in cell bodies. If we compare vertebrate groups from fish through amphibians, reptiles, and mammals, the telencephalon increases in size, complexity, and importance—an evolutionary trend called telencephalization (see Figure 45.19). In humans, the telencephalon is by far the largest part of the brain and plays major roles in sensory perception, learning, memory, and conscious behavior.

The spinal cord transmits and processes information

The spinal cord conveys information to and from the brain. However, the spinal cord is more than an information pipe. As we saw in Section 45.4, the spinal cord carries out integrative functions as well. The knee-jerk reflex (see Figure 45.18) is an example of a circuit between the PNS and the spinal cord that controls a simple behavioral function. That simple circuit, however, can be built on to control more complex behaviors such as the withdrawal reflex, which involves readjusting tension in many muscles on both sides of the body to coordinate movement and maintain balance. Complex motor programs also exist in the spinal cords of many vertebrates. A shark can swim perfectly well after its spinal cord is separated from its brain, and there is the proverbial chicken running around with its head cut off. Central pattern generation in the spinal cord has been demonstrated in mammals by the fact that an experimental animal (usually a cat) with a spinal cord transection that has isolated its hind limbs from its brain can still coordinate its hind limb movements to walk on a treadmill.

The brainstem carries out many autonomic functions

Swallowing, salivating, breathing, eye movements, blood pressure regulation, and gut activity are only a few of the many autonomic functions that are localized in the medulla, the pons, and the midbrain. To carry out these functions, the brainstem has its own components of the PNS, the 12 paired **cranial nerves**. You encountered the olfactory nerve, the optic nerve, and the auditory nerve (cranial nerves I, II, and VIII) in Chapter 46. Another one, cranial nerve X, is called the vagus ("wandering") nerve because it travels into the body cavity and communicates with many of the organs, including the heart and the gut. We will encounter the vagus nerve in subsequent chapters.

Within the highly complex networks of axons and dendrites in the brainstem, there are many discrete groups of neurons that share a common characteristic such as the neurotransmitter they produce and release. Such an anatomically distinct group of neurons is called a **nucleus** (not to be confused with the nucleus of a single cell). Many of these brainstem nuclei send their axons to various regions of the brain to modulate their activity; for example, brainstem nuclei are involved in keeping the higher brain areas awake or allowing them to sleep. All of the sensory information coming up the neural axis from the spinal cord passes through the brainstem on its way to the forebrain, and many of these ascending neuronal tracts give off collateral branches to the awake-promoting nuclei in the brainstem. Because the neuronal circuitry in this part of the brain is so complicated and because activity in these ascending sensory pathways can promote wakefulness, the core of the brainstem has been termed the **reticular activating system** (reticular means net-like). Damage to the brain or spinal cord below the reticular activating system can result in paralysis but leave sleep–wake cycle behavior normal. Damage above the level of the reticular activating system can result in coma.

The core of the forebrain controls physiological drives, instincts, and emotions

As mentioned above, the diencephalon consists of the thalamus and the hypothalamus. The thalamus communicates sensory information to the cerebral cortex; the hypothalamus receives information about physiological conditions in the body and regulates many homeostatic functions. Section 40.5 described how the hypothalamus is involved in regulating body temperature, and Section 41.3 discussed the intimate association between the hypothalamus and the pituitary gland in the control of many homeostatic functions.

Surrounding the diencephalon of all vertebrates are phylogenetically older structures of the telencephalon. These structures comprise the **limbic system** (Figure 47.3), which is responsible for instincts, long-term memory formation, physiological drives such as hunger and thirst, and emotions such as fear. Within the limbic system are areas that, when stimulated with small electric currents, can cause intense sensations of pleasure, pain, or rage. A rat given the opportunity to stimulate its own pleasure centers by pressing a switch will ignore food and water, pushing the switch until it is exhausted.

Pleasure and pain centers in the limbic system are believed to play roles in learning and in physiological drives. One component of the limbic system—the **amygdala**—is involved in fear and fear memory. If a certain portion of the amygdala is damaged or chemically blocked, an animal cannot learn to be afraid of a stimulus or a situation that would normally induce a strong fear reaction. The amygdala is involved in post-traumatic stress disorder (PTSD). Another part of the limbic system, the **hippocampus**, is necessary in humans for the transfer of certain types of short-term memory to long-term memory, as we will discuss in Section 47.3.

Regions of the telencephalon interact to control behavior and produce consciousness

The cerebrum is the dominant structure in the mammalian brain. In humans it is so large that it covers all other parts of the brain except the cerebellum (Figure 47.4A). The cerebral cortex covering the cerebrum is only about 4 millimeters thick, but it covers a surface area larger than a square meter because

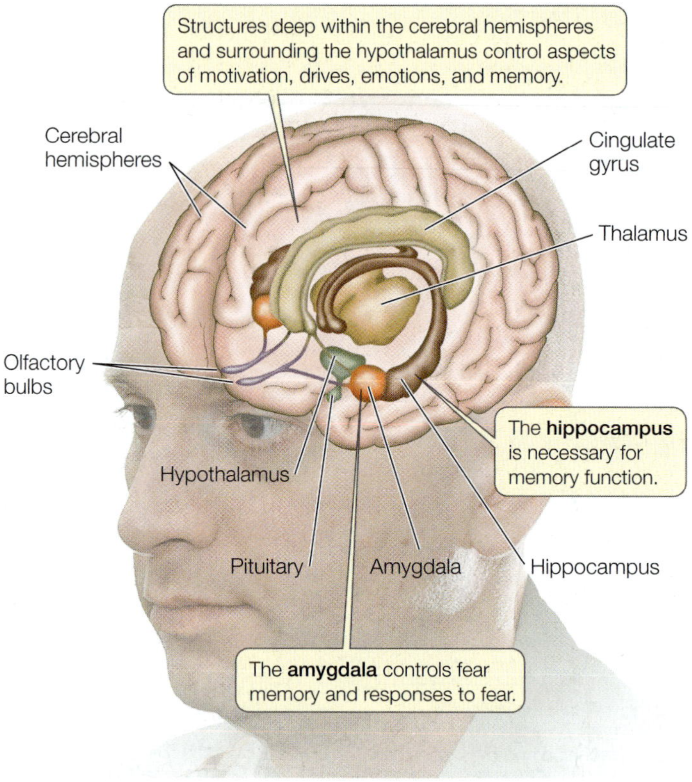
Structures deep within the cerebral hemispheres and surrounding the hypothalamus control aspects of motivation, drives, emotions, and memory.
Cerebral hemispheres
Cingulate gyrus
Thalamus
Olfactory bulbs
The **hippocampus** is necessary for memory function.
Hypothalamus
Pituitary
Amygdala
Hippocampus
The **amygdala** controls fear memory and responses to fear.

47.3 The Limbic System The evolutionarily primitive parts of the telencephalon are referred to as the limbic system. The hippocampus is involved in forming long-term memory. The amygdala triggers fear emotions and fear memories.

it is folded into ridges (**gyri**; singular *gyrus*) and valleys (**sulci**; singular *sulcus*). These foldings, or **convolutions**, enable the large surface of the cortex to fit within the skull.

As we explore the functions of the cerebral cortex and other parts of the brain, we will occasionally mention an individual whose brain was damaged by an accident or other unfortunate event. Until recently the study of such individuals has been the main source of functional information about the human brain, but new imaging technologies such as positron emission tomography (PET) and magnetic resonance imaging (MRI) are providing a wealth of new information and opportunities to study the human brain.

A curious feature of the human nervous system is that the left side of the body is served (in both sensory and motor aspects) mostly by the right side of the brain, and the right side of the body is served mostly by the left side of the brain. Thus sensory input from the right hand goes to the left cerebral hemisphere, and sensory input from the left hand goes to the right cerebral hemisphere. The exception is the head, where the left side is controlled by the left cerebral hemisphere and the right side by the right cerebral hemisphere. The two hemispheres are not symmetrical with respect to all functions. Language abilities, for example, reside predominantly in the left hemisphere.

Different regions of the cerebral cortex have specific functions (Figure 47.4B). Some of those functions are easily defined, such as receiving and processing sensory information or generating motor commands, but most of the cortex is involved in

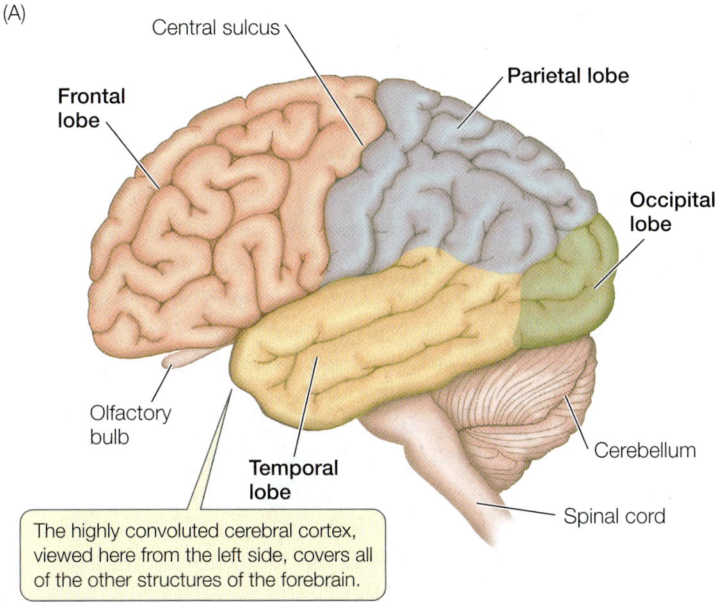
(A)
Central sulcus
Parietal lobe
Frontal lobe
Occipital lobe
Olfactory bulb
Temporal lobe
Cerebellum
Spinal cord
The highly convoluted cerebral cortex, viewed here from the left side, covers all of the other structures of the forebrain.

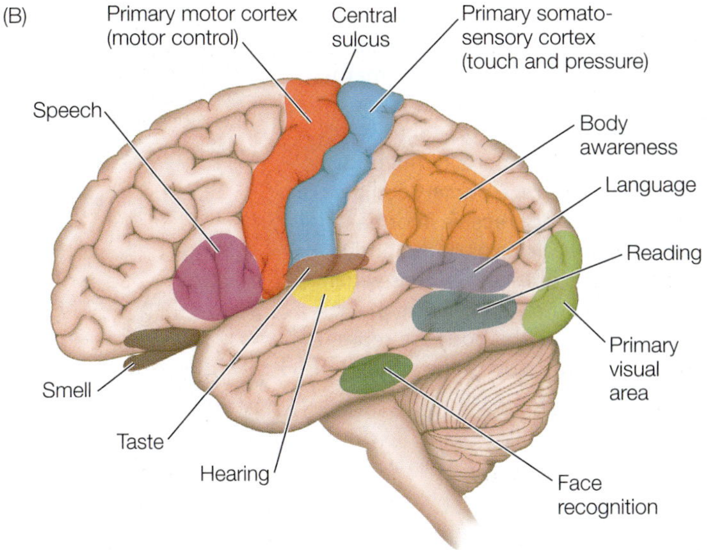
(B)
Primary motor cortex (motor control)
Central sulcus
Primary somato-sensory cortex (touch and pressure)
Speech
Body awareness
Language
Reading
Primary visual area
Smell
Taste
Hearing
Face recognition

47.4 The Human Cerebrum (A) Each cerebral hemisphere is divided into frontal, temporal, parietal, and occipital lobes. (B) Different functions are localized in particular areas of the four cerebral lobes.

Go to Activity 47.1 The Human Cerebrum
Life10e.com/ac47.1

higher-order information processing that is less easy to define. These latter areas are given the general name of **association cortex**, so named because they integrate, or *associate*, information from different sensory modalities and from memory.

To understand the cerebral cortex, it helps to have an anatomical road map. Viewed from the left side, the left cerebral hemisphere looks like a boxing glove for the right hand with the fingers pointing forward, the thumb pointing out, and the wrist at the rear. The "thumb" area is the **temporal lobe**, the fingers the **frontal lobe**, the back of the hand the **parietal lobe**, and the wrist the **occipital lobe** (see Figure 47.4A). The right cerebral hemisphere shows a mirror image of this arrangement. We will look at each lobe separately.

THE TEMPORAL LOBE The upper region of the temporal lobe receives and processes auditory information. The association areas of this lobe are involved in recognizing, identifying, and naming objects. Damage to the temporal lobe results in disorders called agnosias, in which the individual is aware of an object but cannot identify it.

Damage to a certain area of the temporal lobe results in the inability to recognize faces. Even old acquaintances cannot be identified by facial features, although they may be identified by other attributes such as voice, body features, and posture. Using monkeys, it has been possible to record the activity of neurons in this region that respond selectively to faces in general. These neurons do not respond to other stimuli in the visual field, and their responsiveness decreases if some of the facial features are missing or appear in inappropriate locations (**Figure 47.5**). Damage to other association areas of the temporal lobe causes deficits in understanding spoken language, although speaking, reading, and writing abilities may be intact.

THE FRONTAL LOBE The frontal and parietal lobes are separated by a deep valley called the central sulcus. A strip of the frontal lobe cortex just in front of the central sulcus is called the **primary motor cortex** (see Figure 47.4B). The neurons in this region control muscles in specific parts of the body; the parts of the body have been mapped onto the primary motor cortex, largely during neurosurgical procedures. As part of these procedures, electrodes were used to stimulate small areas of cortex. In the area just anterior to the central sulcus, stimulation causes specific muscles to contract. Parts of the body with fine motor control, such as the face and hands, have disproportionate representation (**Figure 47.6A**). Stimulation of neurons in the primary motor cortex causes twitches of muscles, not coordinated movements.

The association functions of the frontal lobe are diverse and are best described as having to do with feeling and planning. They contribute significantly to personality. People with

This neuron responds maximally to a complete face viewed from the front.
Firing rate of neuron

47.5 "Face Neurons" in One Region of the Temporal Lobe The electrode traces represent the firing rate of a neuron in the temporal lobe of a monkey in response to the pictures shown below them. Highest firing is stimulated by the appearance of a complete face.

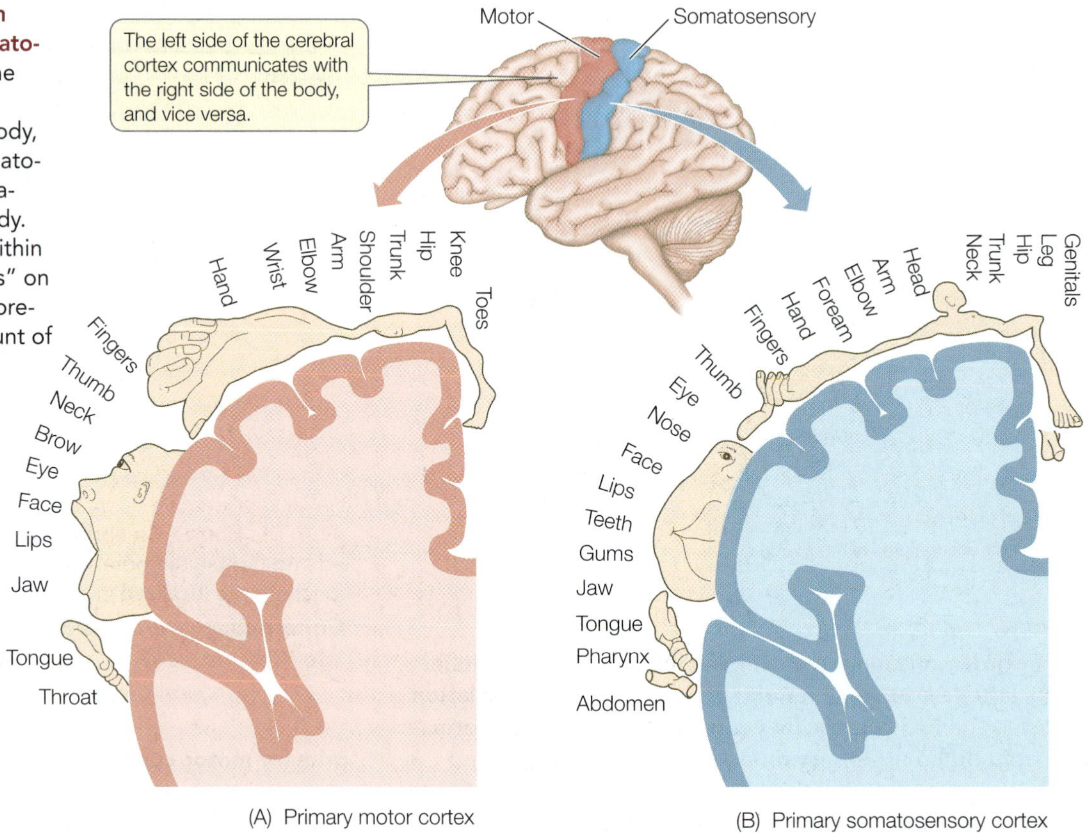
Motor
Somatosensory
The left side of the cerebral cortex communicates with the right side of the body, and vice versa.
Toes
Knee
Hip
Trunk
Shoulder
Arm
Elbow
Wrist
Hand
Fingers
Thumb
Neck
Brow
Eye
Face
Lips
Jaw
Tongue
Throat
(A) Primary motor cortex
Genitals
Leg
Hip
Trunk
Neck
Head
Arm
Elbow
Forearm
Hand
Fingers
Thumb
Eye
Nose
Face
Lips
Teeth
Gums
Jaw
Tongue
Pharynx
Abdomen
(B) Primary somatosensory cortex

47.6 The Body Is Represented in Primary Motor and Primary Somatosensory Cortexes Neurons in the primary motor cortex (A) control muscles in specific parts of the body, while neurons in the primary somatosensory cortex (B) receive information from specific parts of the body. The locations of these neurons within each cortex correspond to "maps" on which regions of the body are represented in proportion to the amount of innervation they receive.

frontal lobe damage have drastic alterations of personality and difficulty planning future events. A dramatic case of frontal lobe damage is the story of Phineas Gage, who in 1848 was an industrious and responsible young railroad construction foreman. Then a blasting accident shot a meter-long, 3-centimeter-wide iron tamping rod through his brain. The rod entered Gage's head below his left eye, passed through his frontal lobe, and exited the top of his head (**Figure 47.7**).

Remarkably, Gage survived, but he was a completely different person. He was quarrelsome, impatient, obstinate, and used profane language, which he had never done before. He lost his railroad job and spent his days as a drifter, earning money by telling his story and exhibiting his scars (and the tamping iron). He died of a seizure in 1860, at the age of 38. If you are in Boston, you can pay him a visit—his skull, death mask, and the tamping iron are on display in the Warren Anatomical Museum of Harvard Medical School.

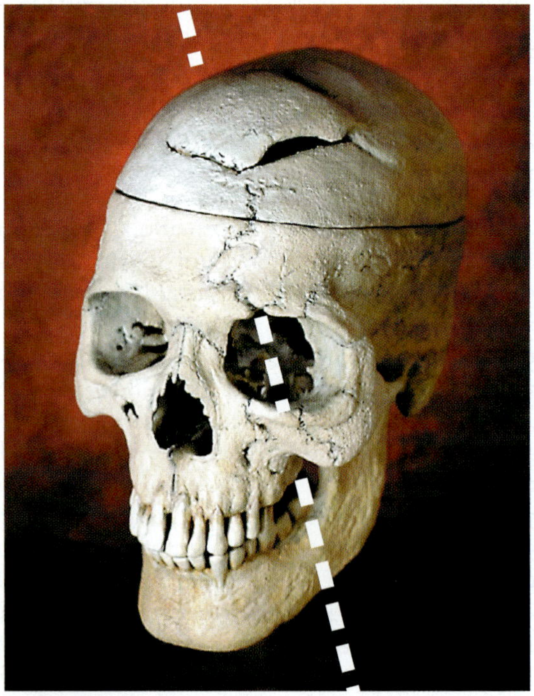

47.7 A Mind-Altering Experience Phineas Gage miraculously survived a nineteenth-century railroad construction accident in which an explosion blew an iron rod through his brain. His personality, however, was permanently altered from a responsible foreman to a quarrelsome drifter. The path of the iron rod through Gage's brain is superimposed on this reconstruction of his skull.

THE PARIETAL LOBE The strip of parietal lobe cortex just behind the central sulcus is the **primary somatosensory cortex** (see Figure 47.4B). This area receives touch and pressure information relayed from the body through the thalamus.

The entire body surface can be mapped onto the primary somatosensory cortex (**Figure 47.6B**). Areas of the body that have a high density of tactile mechanoreceptors and are capable of making fine discriminations in touch (such as the lips and fingers) have disproportionately large representation. If a very small area of the primary somatosensory cortex is stimulated

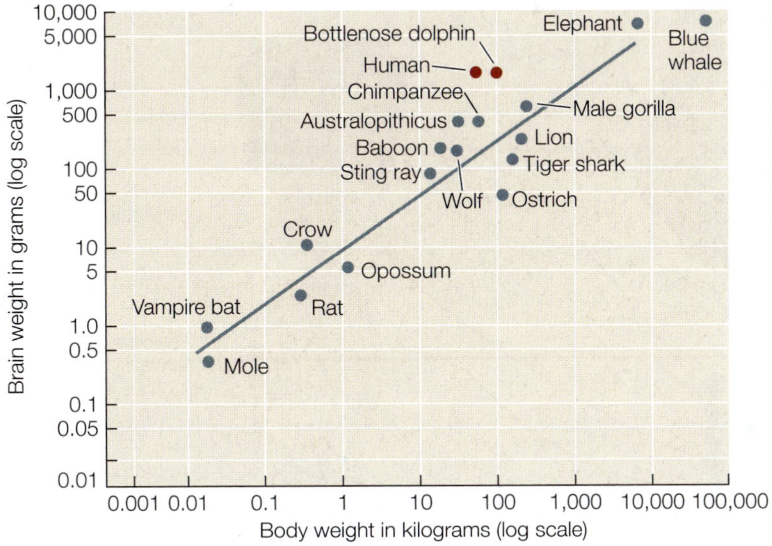

10,000
5,000
1,000
500
100
50
10
5
1.0
0.5
0.1
0.05
0.01
Brain weight in grams (log scale)
Bottlenose dolphin
Elephant
Blue whale
Human
Chimpanzee
Male gorilla
Australopithicus
Lion
Baboon
Tiger shark
Sting ray
Wolf
Ostrich
Crow
Opossum
Vampire bat
Rat
Mole
0.001 0.01 0.1 1 10 100 1,000 10,000 100,000
Body weight in kilograms (log scale)

47.8 Evolution of the Human Brain Brain size scales to body size across a wide range of vertebrates. The higher primates have larger brains than predicted by the correlation, and humans stand outside this relationship with much bigger brains. The increase in brain size in humans is mostly due to an increase in the cerebral cortex. The human brain is also highly convoluted, and more of it is devoted to associative functions.

electrically, the subject reports feeling specific sensations, such as touch, in a localized part of the body.

A major association function of the parietal lobe is attending to complex stimuli. Damage to the right parietal lobe causes a condition called contralateral neglect syndrome, in which the individual tends to ignore stimuli from the left side of the body or the left visual field. Such individuals have difficulty performing complex tasks, such as dressing the left side of the body; an afflicted man may not be able to shave the left side of his face. When asked to copy simple drawings, a person who exhibits this syndrome can do well with the right side of the drawing but not the left.

The parietal cortex is not symmetrical with respect to its role in attention. Damage to the left parietal cortex does not cause the same degree of neglect of the right side of the body. We will see similar asymmetries in cortical function later in the chapter when we discuss language.

THE OCCIPITAL LOBE The occipital lobe receives and processes visual information. The association areas of the occipital cortex are essential for making sense of the visual world and translating visual experience into language. Some deficits resulting from damage to these areas are specific. In one case, a woman with limited damage was unable to see motion. Her vision was intact, but she could see a waterfall only as a still image, and an approaching car only as a series of a stationary object at different distances.

The size of the human brain is off the curve

Humans are sometimes called "big-brain primates," and that is an accurate characterization. Across vertebrate species there is a correlation between body size and brain size (Figure 47.8). Higher primates such as chimpanzees, baboons, and gorillas all fall above this regression line, but humans stand out because they are so far above the regression line. Gorillas are much larger than humans, but they have smaller brains. Elephants and whales have large brains, but they fall closer to the regression line. Dolphins and humans stand out as having much larger brains than would be predicted by their body sizes.

The correlation of brain size to body size does not tell the whole story of human brain evolution, however. In Figure 45.19, which compares the brains of four vertebrates, we see that the forebrain is larger than other brain regions, and in mammals this is seen as an elaboration of the cerebral cortex. If we look just at mammals, another feature is the degree of convolution of the cortex. Since the cortex is a layered, two-dimensional array of neurons, the area of cortex is increased by convolutions, which are greatest in humans. And finally, the percent of the cortex that is association cortex (i.e., the brain regions devoted to the integration of information) is by far the greatest in humans. It is these evolutionary changes, primarily in the cortex, that provide the resources for the intellectual capacity of humans—a topic to which we will return at the end of the chapter.

RECAP 47.1

The central nervous system communicates with the rest of the body through the peripheral nervous system. We are conscious of some sensory information coming into the CNS, but we are not conscious of other afferent information used in physiological regulation. Different regions of the brain have specific functions. Evolution of the human brain has resulted in a greatly increased cerebral cortex devoted to integration of information.

- Explain how the major functional divisions of the nervous system relate to their origins in the embryonic neural tube. **See pp. 968–969 and Figure 47.2**
- Describe the spatial relations and functions of the major divisions of the telencephalon. **See pp. 970–971 and Figure 47.4**
- What features distinguish the human brain from the brains of other mammals? **See p. 973 and Figure 47.8**

Having briefly described the structure and function of different regions of the nervous system, we will now explore some examples of how information is processed by the neural circuitry in some specific brain regions.

47.2 How Is Information Processed by Neural Networks?

Specific functions are localized in specific parts of the nervous system and depend on the neural circuits, or networks, in those structures. A major focus of modern neuroscience is understanding how the various functions of the nervous

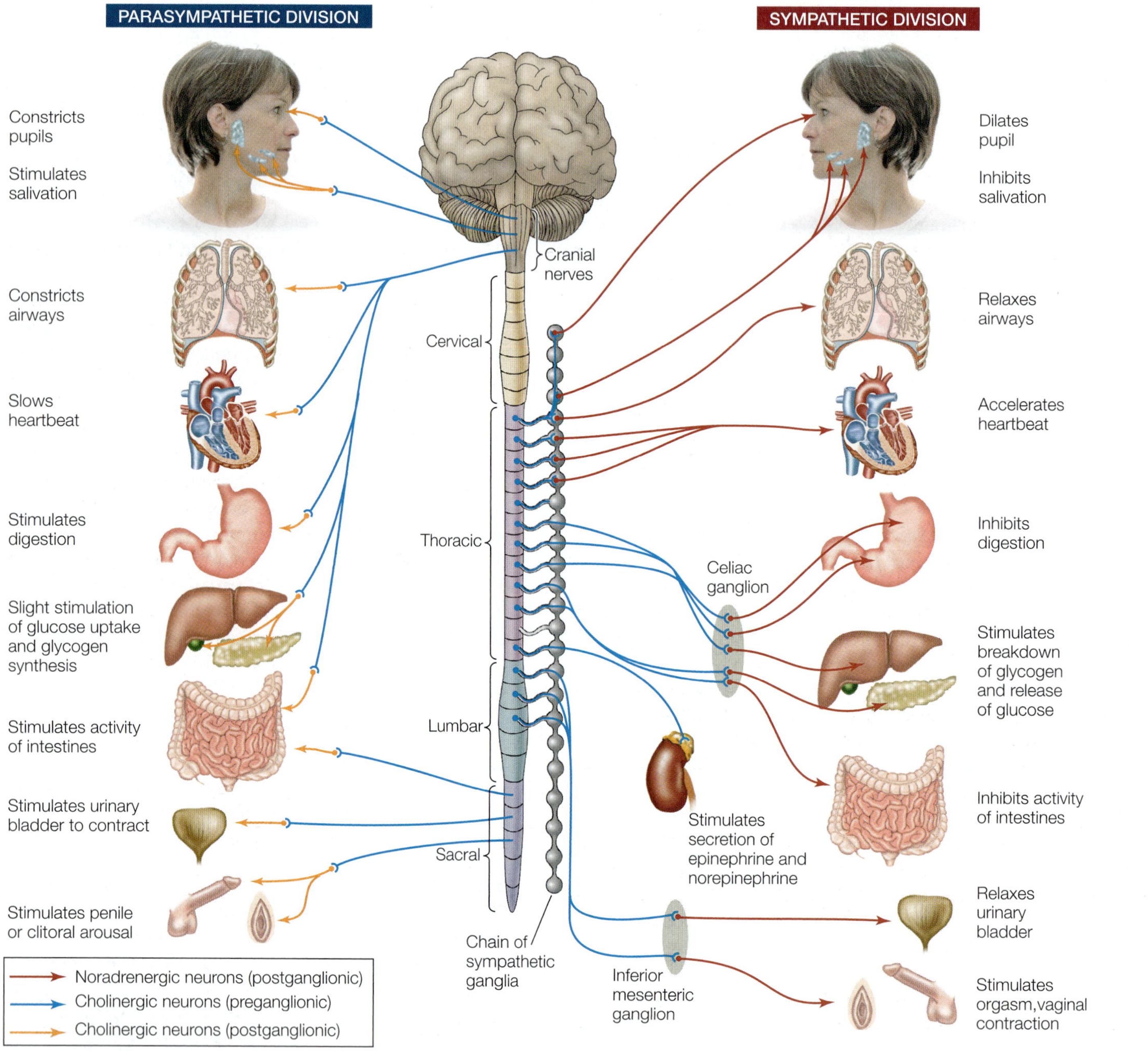
PARASYMPATHETIC DIVISION
SYMPATHETIC DIVISION
Constricts pupils
Stimulates salivation
Constricts airways
Slows heartbeat
Stimulates digestion
Slight stimulation of glucose uptake and glycogen synthesis
Stimulates activity of intestines
Stimulates urinary bladder to contract
Stimulates penile or clitoral arousal
Cranial nerves
Cervical
Thoracic
Lumbar
Sacral
Chain of sympathetic ganglia
Celiac ganglion
Inferior mesenteric ganglion
Dilates pupil
Inhibits salivation
Relaxes airways
Accelerates heartbeat
Inhibits digestion
Stimulates breakdown of glycogen and release of glucose
Stimulates secretion of epinephrine and norepinephrine
Inhibits activity of intestines
Relaxes urinary bladder
Stimulates orgasm,vaginal contraction
Noradrenergic neurons (postganglionic)
Cholinergic neurons (preganglionic)
Cholinergic neurons (postganglionic)

47.9 The Autonomic Nervous System The autonomic nervous system is divided into the sympathetic and parasympathetic divisions. The two divisions work in opposition to each other in their effects on most organs; one results in an increase and the other a decrease in activity.

system, ranging from simple reflexes to complex learning and memory, are accomplished by the interactions of neurons in circuits. Two extensively studied examples of how neural networks process information are the autonomic nervous system (an output pathway) and the visual system (an input pathway).

Pathways of the autonomic nervous system control involuntary physiological functions

The **autonomic nervous system**, or **ANS**, comprises the output pathways of the CNS that control involuntary functions, such as heart rate, blood flow, sweating, and digestive activities. Its control of diverse organs and tissues is crucial to homeostasis. The ANS has two divisions, **sympathetic** and **parasympathetic**, that work in opposition to each other in their effects on most organs: one division causes an increase in an activity and the other a decrease (Figure 47.9). The sympathetic and parasympathetic divisions are easily distinguished by their anatomy, neurotransmitters, and actions.

The best-known functions of the ANS are those of the sympathetic division that produce the fight-or-flight response: increasing heart rate, blood pressure, and cardiac output and preparing the body for emergencies (see Figure 41.3). In contrast, the parasympathetic division slows the heart and lowers blood pressure; its actions have been characterized as "rest and digest." It is tempting to think of the sympathetic division as speeding things up and the parasympathetic division as slowing things down, but it is not that simple; for example, the sympathetic division slows down the digestive system whereas the parasympathetic division accelerates it.

Whether sympathetic or parasympathetic, every autonomic efferent pathway begins with a cholinergic neuron (that is, a neuron that uses acetylcholine as its neurotransmitter) that has its cell body in the brainstem or spinal cord. These cells are called preganglionic neurons because the second neuron in the pathway with which they synapse resides in a collection of neurons outside the CNS called a **ganglion** (plural *ganglia*). The second neuron is called a postganglionic neuron because its axon extends out from the ganglion. The axon of the postganglionic neuron synapses with cells in the target organs (see Figure 47.9).

The postganglionic neurons of the sympathetic division are called noradrenergic because they use norepinephrine (also known as noradrenaline) as their neurotransmitter. In contrast, the postganglionic neurons of the parasympathetic division are cholinergic. In organs that receive both sympathetic and parasympathetic input, the target cells respond in an opposite manner to norepinephrine and to acetylcholine. This happens, for example, in a region of the heart called the pacemaker, which generates the heartbeat. Stimulating the sympathetic nerve to the heart or dripping norepinephrine onto pacemaker cells increases their firing rate and causes the heart to beat faster. In contrast, stimulating the parasympathetic nerve to the heart or dripping acetylcholine onto pacemaker cells decreases their firing rate and causes the heart to beat more slowly.

The sympathetic and parasympathetic divisions of the ANS can also be distinguished by anatomy. The preganglionic neurons of the parasympathetic division come from the cranial nerves of the brainstem and the sacral (lowest) region of the spinal cord; those of the sympathetic division come from the thoracic and lumbar regions of the spinal cord (see Figure 47.9). Most of the ganglia of the sympathetic division are lined up in two chains, one on either side of the spinal cord. The parasympathetic ganglia are close to the target organs.

The autonomic nervous system is an important link between the CNS and many physiological functions. Its control of diverse organs and tissues is crucial to homeostasis. Despite the complexity of the ANS, work by neurobiologists and physiologists over many decades has made it possible to understand its functions in terms of neuronal properties and circuits.

The visual system is an example of information integration by the cerebral cortex

The visual system is one of the most-studied input pathways to the central nervous system. In Section 46.4 we described how light falling on the retina produces signals that are transmitted through the cellular circuits of the retina, resulting in action potentials in the optic nerve. But how does the central nervous system use this information to reconstruct the visual world in the brain? The experiments that have investigated this question are some of the most famous experiments in neurobiology.

RETINAL RECEPTIVE FIELDS Section 46.4 described how a retinal ganglion cell collects information from a number of photoreceptors—an example of "convergence of information." Each ganglion cell is communicating to the brain something more than simply the presence or absence or intensity of light falling on a portion of the retina. What information does the retinal ganglion cell extract from the photoreceptors?

This question was addressed in classic experiments by Stephen Kuffler, then at Johns Hopkins University. He used electrodes to record the activity of single ganglion cells of cat eyes while stimulating their retinas with spots of light (**Figure 47.10**). These experiments were the starting point for understanding how the brain assembles information from single cells to create visual images—in other words, of how the brain sees.

Kuffler's experiments revealed that each ganglion cell has a well-defined **receptive field** composed of a group of photoreceptor cells that receive light from a small area of the entire visual field. Stimulating these photoreceptors with light activates the ganglion cell, which sends action potentials to the thalamus and on to the visual cortex (the area of the occipital lobe where visual information is processed; see Figure 47.4). Information from many photoreceptors is therefore communicated to the brain as a single message. Individual photoreceptors may contribute to the receptive fields of multiple ganglion cells, so that receptive fields overlap.

The receptive fields of most ganglion cells are circular, but whether a spot of light falling on a receptive field excites or inhibits its ganglion cell depends both on the nature of the receptive field and on where the spot of light falls on it. Receptive fields have a center and a concentric surround, and can be either "on-center" or "off-center." Light falling on the center of an on-center receptive field excites the ganglion cell, and light falling on the center of an off-center receptive field inhibits the ganglion cell. Light falling on the surround has the opposite effect: the surround for an on-center receptive field inhibits the ganglion cell, and the surround for an off-center field is excitatory. Thus the activity of the ganglion cell reflects how much of the light stimulus is on the center and how much is on the surround of its receptive field (see Figure 47.10).

Center effects are always stronger than surround effects. Thus a small dot of light directly on the center of a receptive field has the maximal effect, and a larger light stimulus illuminating the center and parts of the surround has a smaller effect. A uniform patch of light falling equally on the center and surround has very little effect on the firing rate of the ganglion cell for that receptive field.

Go to Animated Tutorial 47.1
Visual Receptive Fields
Life10e.com/at47.1

How are cells in the retina connected to each other to create receptive fields? Remember from Section 46.4 that photoreceptors synapse onto bipolar cells and bipolar cells onto ganglion cells. This pattern of connectivity describes the relationship between the photoreceptors in the center of a receptive field. The photoreceptors in the surround area modify communication between the center photoreceptors and their bipolar cells through the lateral connections of horizontal cells and amacrine cells. Thus the receptive field of a ganglion cell results

INVESTIGATINGLIFE

47.10 What Does the Eye Tell the Brain? Stephen Kuffler's experiments recorded the activity of single ganglion cells in the eyes of cats. These groundbreaking experiments revealed the existence of a circular receptive field for each of the retina's ganglion cells. Signals from photoreceptor cells in a receptive field are either excitatory or inhibitory to the ganglion cell, which sends action potentials via the optic nerve to the brain.[a]

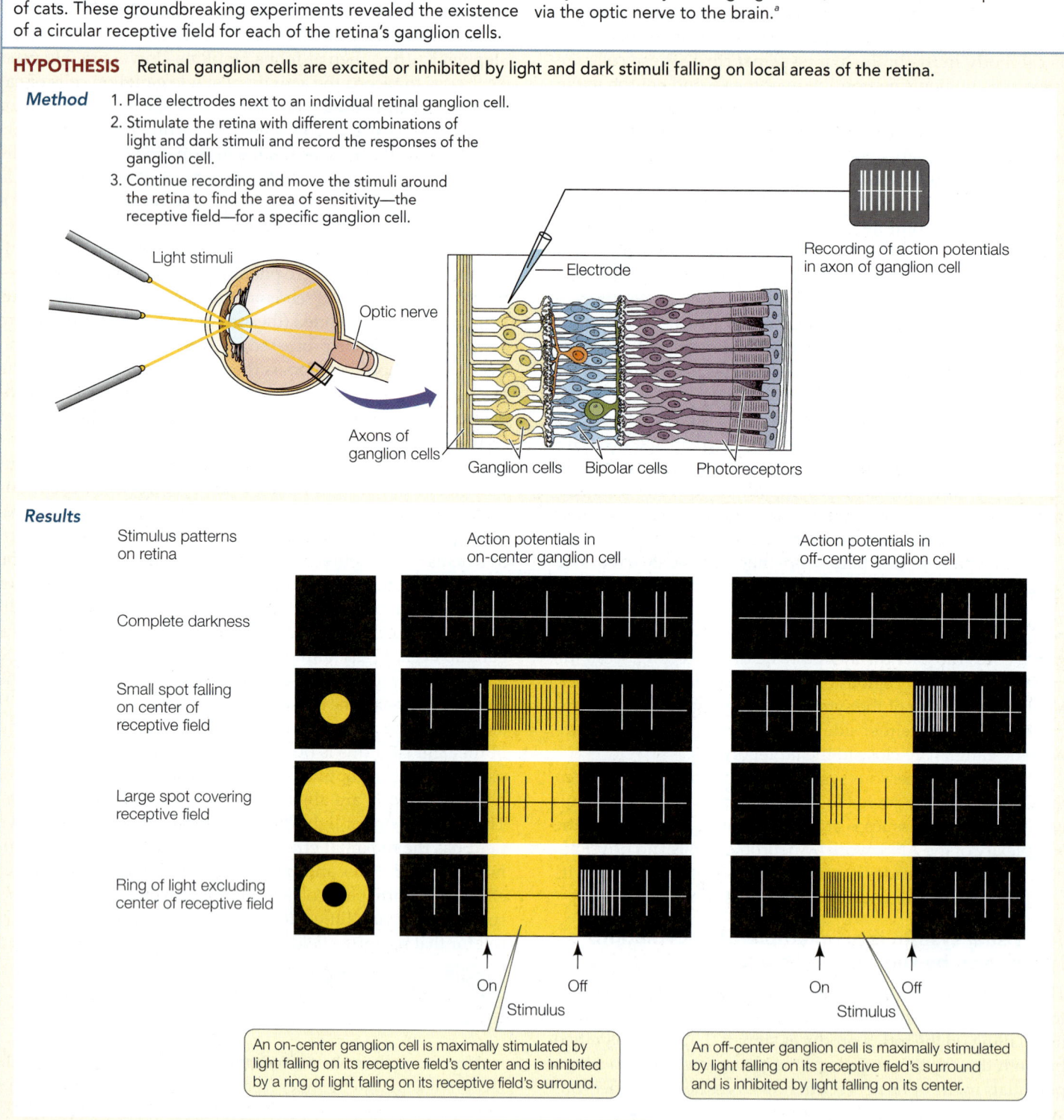

HYPOTHESIS Retinal ganglion cells are excited or inhibited by light and dark stimuli falling on local areas of the retina.

Method

1. Place electrodes next to an individual retinal ganglion cell.
2. Stimulate the retina with different combinations of light and dark stimuli and record the responses of the ganglion cell.
3. Continue recording and move the stimuli around the retina to find the area of sensitivity—the receptive field—for a specific ganglion cell.

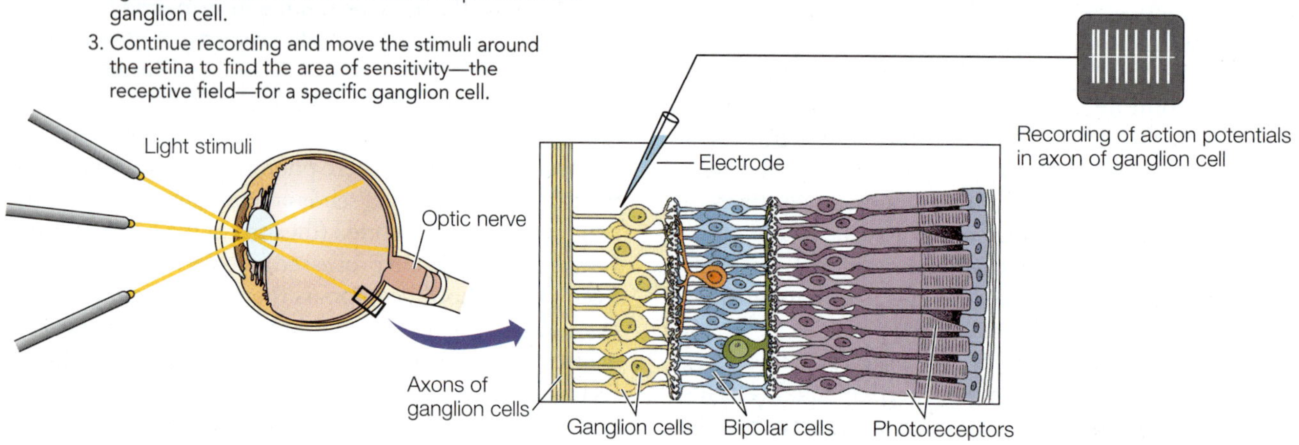

Light stimuli
Optic nerve
Axons of ganglion cells
Electrode
Recording of action potentials in axon of ganglion cell
Ganglion cells
Bipolar cells
Photoreceptors

Results

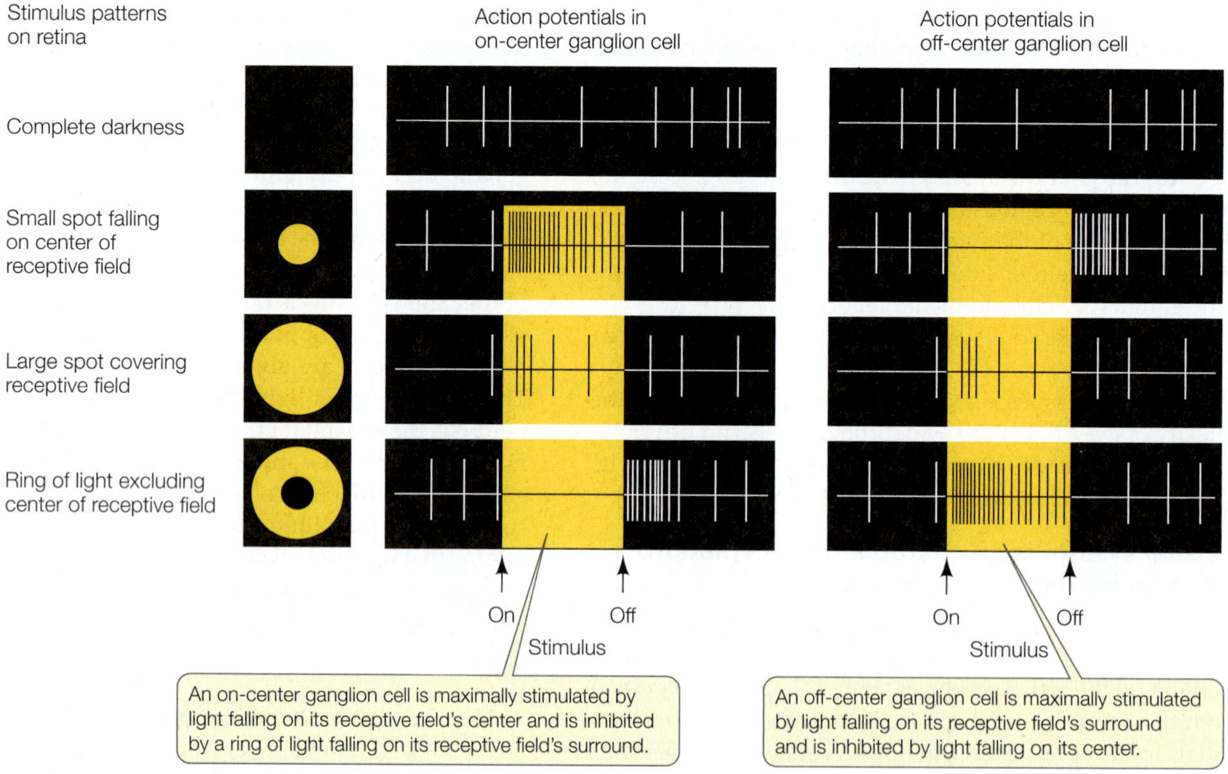

Stimulus patterns on retina
Action potentials in on-center ganglion cell
Action potentials in off-center ganglion cell
Complete darkness
Small spot falling on center of receptive field
Large spot covering receptive field
Ring of light excluding center of receptive field
On
Off
Stimulus
On
Off
Stimulus
An on-center ganglion cell is maximally stimulated by light falling on its receptive field's center and is inhibited by a ring of light falling on its receptive field's surround.
An off-center ganglion cell is maximally stimulated by light falling on its receptive field's surround and is inhibited by light falling on its center.

CONCLUSION Ganglion cells use a center-surround dichotomy to encode patterns of contrast between light and dark.

Go to **BioPortal** for discussion and relevant links for all INVESTIGATING**LIFE** figures.

[a]Kuffler, S. W. 1953. *Journal of Neurophysiology* 16: 37–68.

from a pattern of synapses between photoreceptors, horizontal cells, amacrine cells, and bipolar cells. A general lesson to learn from this seemingly confusing chain of events is that inhibition can be as important as excitation in neural circuits.

In summary, the neural circuitry of the retina results in the generation of signals in the axons of the optic nerve to the brain that communicate simple information about the contrasting patterns of light and dark falling on different parts of the retina. But once the action potentials in the optic nerve reach their destinations, how does the brain integrate them to construct visual images of the outside world?

Go to Animated Tutorial 47.2
Information Processing in the Retina
Life10e.com/at47.2

RECEPTIVE FIELDS OF CELLS IN THE VISUAL CORTEX The axons of the optic nerves terminate in a region of the thalamus that is a relay station receiving information from both the right and left eyes. From the thalamus, the information encoded in the activity of axons in the optic nerves is relayed to the visual cortex in the occipital lobes at the back of the brain. In the 1960s David Hubel and Torsten Wiesel of Harvard University studied the activity of neurons in the visual cortex by shining spots and bars of light on retinas while recording the activities of single cells in the cortex. They found that neurons in the visual cortex, like retinal ganglion cells, have receptive fields.

Neurons in the visual cortex respond selectively to bars of light of different orientations falling on the retina and in some cases to movement of those bars of light in different directions. The concept that emerges from these experiments is that the brain assembles a mental image of the visual world by analyzing edges in patterns of light falling on the retina. Each retina sends a million axons to the brain, but there are *hundreds of millions* of neurons in the visual cortex. The action potentials from one retinal ganglion cell are received by hundreds of cortical neurons, each responsive to a different combination of orientation, position, color, and movement of contrasting lines in the patterns of light and dark falling on the retina.

Three-dimensional vision results from cortical cells receiving input from both eyes

How do we perceive objects in three dimensions? The short answer is that a person's front-facing two eyes see overlapping, yet slightly different, visual fields—that is, humans have **binocular vision**. A person who is blind in one eye has difficulty discriminating distances. Animals whose eyes are on the sides of the head rather than facing front have minimal overlap in their fields of vision and, as a result, poor depth vision; however, they can see predators creeping up from all sides.

The story of how the brain integrates information from two eyes begins with the paths of the optic nerves. The two optic nerves run along the underside of the brain, join just under the hypothalamus, and then separate again (**Figure 47.11A**). The place where they join is called the **optic chiasm**. Axons from the half of each retina closest to the nose cross in the optic chiasm and go to the opposite side of the brain. The axons from the

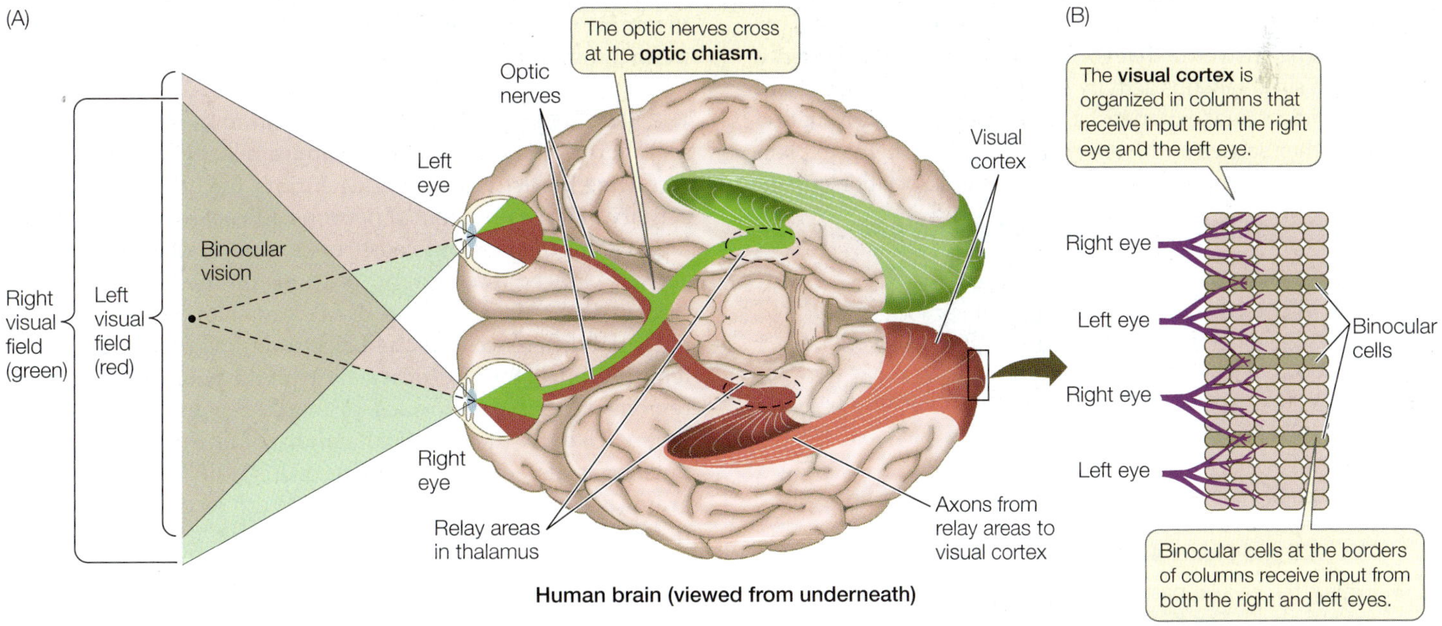
(A)
The optic nerves cross at the optic chiasm.
Optic nerves
Left eye
Visual cortex
Binocular vision
Right visual field (green)
Left visual field (red)
Right eye
Relay areas in thalamus
Axons from relay areas to visual cortex
Human brain (viewed from underneath)
(B)
The visual cortex is organized in columns that receive input from the right eye and the left eye.
Right eye
Left eye
Right eye
Left eye
Binocular cells
Binocular cells at the borders of columns receive input from both the right and left eyes.

47.11 Anatomy of Binocular Vision (A) Each eye transmits information to both sides of the brain; however, the right side of the brain processes all information from the left visual field (red), and the left side of the brain processes all information from the right visual field (green). (B) The visual cortex sorts visual field information according to whether it comes from the right or left eye.

outer half of each retina do not cross over at the optic chiasm; axons from the outer left retina go to the left side of the brain, and vice versa for axons from the outer right retina.

The functional consequence of the optic chiasm is that all of the visual information from the left side of your field of vision when you are looking straight ahead goes to the right side of your brain, and all of the visual information from the right side of your field of vision goes to the left side of your brain. These relationships are shown in red and green in Figure 47.11A.

Cells in the visual cortex are organized in stripes and columns. Stripes refer to the organization across the surface of the cortex, and columns to the organization through the depth of the cortex (**Figure 47.11B**). Stripes and columns alternate according to the source of their input: left eye, right eye, left eye, right eye, and so on. Cells closest to the border between two stripes or columns receive input from both eyes and thus are called binocular cells. Binocular cells interpret distance by measuring the *disparity* between the points at which the same stimulus falls on the two retinas.

What is disparity? Hold your finger out in front of you and look at it closing one eye and then the other. Your finger appears to jump back and forth because its image falls on a different position on each retina. Repeat the exercise with a distant object. It doesn't jump back and forth as much because there is less disparity in the positions of the image on the two retinas. Certain binocular cells respond optimally to a stimulus falling on both retinas with a particular disparity. Which set of binocular cells is stimulated depends on how far away the stimulus is.

When we look at something, we can detect its shape, color, depth, and movement. Where does all this information come together? Is there a single cell that fires only when a red sports car drives by? The answer to that is "no." Specific visual experience comes from simultaneous activity in a large collection of cells. In addition, most visual experiences are enhanced by information from the other senses and from memory, which helps explain why about 75 percent of the cerebral cortex is association cortex.

RECAP 47.2

Information in the nervous system is processed by cellular interactions in neural networks. The opposing actions of the sympathetic and parasympathetic divisions of the ANS can be understood in terms of neural pathways consisting of just two neurons. Vision involves a more complex interaction of neurons, organized into receptive fields, to process patterns of light and dark falling on the retina.

- Describe the anatomical and functional differences between the sympathetic and parasympathetic divisions. **See pp. 974–975 and Figure 47.9**
- Explain the cellular basis for the receptive fields of retinal ganglion cells. **See p. 975 and Figure 47.10**
- How do cells in the visual cortex interpret information about how far away an object is? **See pp. 977–978 and Figure 47.11**

By studying the neural circuitry of the visual system and the autonomic nervous system, you have gained some understanding of how information reaches the central nervous system and how the CNS controls various functions of the body. But what about the higher functions of the mammalian CNS—the complex functions between input and output, such as language, learning, memory, and dreams?

47.3 Can Higher Functions Be Understood in Cellular Terms?

The higher brain functions discussed in the remaining pages of this chapter are undeniably complex. Nevertheless, neuroscientists, using a wide range of techniques, are making considerable progress in understanding some of the cellular and molecular mechanisms involved in those processes. The following discussion will address several aspects of brain and behavior that present challenges to neuroscientists: sleep and dreaming, learning and memory, language use, and consciousness.

Sleep and dreaming are reflected in electrical patterns in the cerebral cortex

A dominant feature of behavior is the daily cycle of sleep and waking. All birds and mammals, probably all other vertebrates, and also many invertebrates, sleep. We humans spend one-third of our lives sleeping, yet we do not know why or how. We do know, however, that we need to sleep. Loss of sleep impairs alertness and performance. Many people in our society—certainly most college students—are chronically sleep-deprived. Accidents and serious mistakes that endanger lives can be attributed to impaired alertness caused by lack of sleep. Insomnia (difficulty in falling or staying asleep) is one of the most common medical complaints.

THE ELECTROENCEPHALOGRAM A common tool of sleep researchers is the **electroencephalogram**, or **EEG**. Rather than recording the activity of single neurons, the EEG characterizes activity in huge numbers of neurons. EEG electrodes are much larger than the very fine electrodes used to detect single cell activity. Placed at different locations on the head and scalp (**Figure 47.12A**), EEG electrodes record changes in the electric potential differences between electrodes over time. These differences reflect the electrical activity of the neurons in the brain regions under the electrodes, primarily regions of the cerebral cortex. Usually the electrical activity of one or more skeletal muscles is also recorded; this record is called an electromyogram (EMG). Movements of the eyes are recorded as an electroocculogram (EOG).

EEG, EMG, and EOG patterns reveal the transition from being awake to being asleep. They also reveal that there are different states of sleep. In mammals other than humans, two major sleep states are easily distinguished: **slow-wave sleep** and **rapid eye movement (REM) sleep**. Slow-wave sleep gets its name from the high-amplitude, slow-frequency waves in the EEG. REM sleep gets its name from jerky movements of the eyeballs that occur during this state. In humans, sleep states are characterized as non-REM sleep and REM sleep. Human

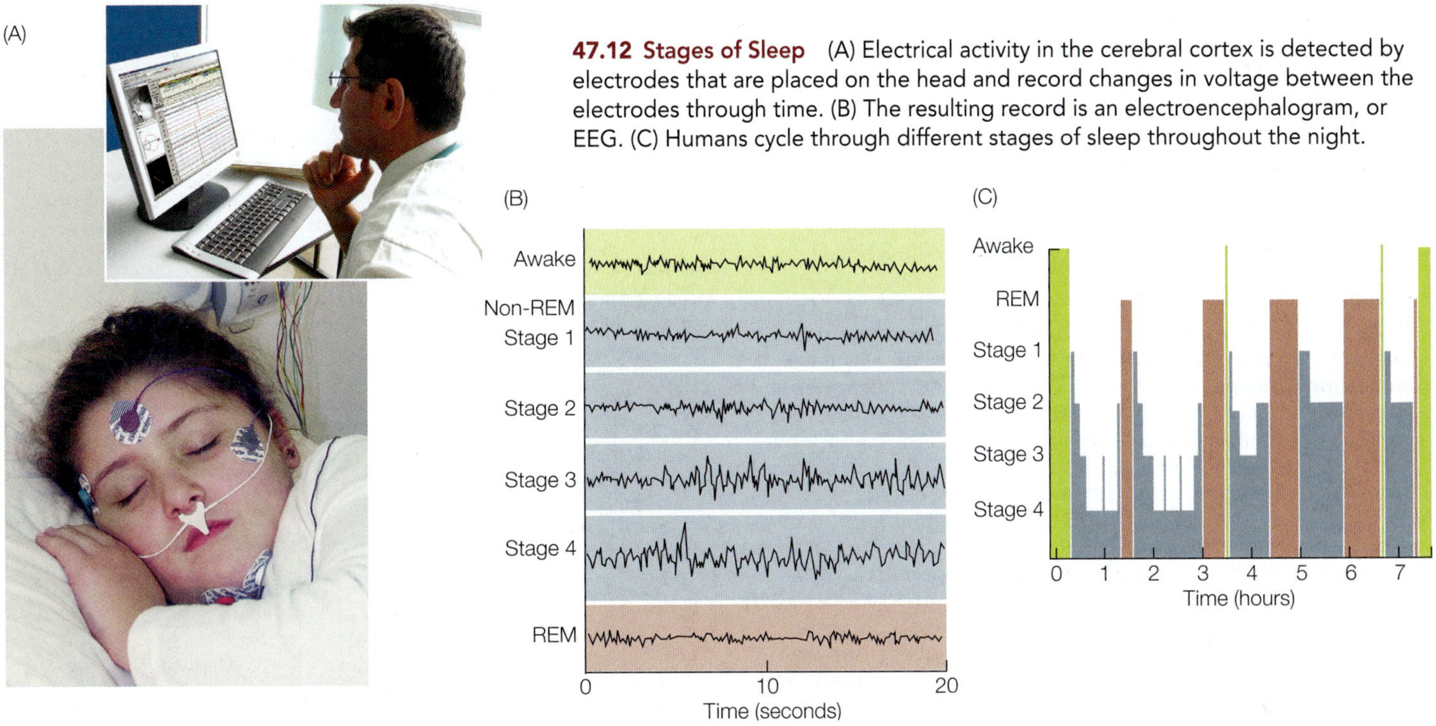
(A)
(B)
Awake
Non-REM
Stage 1
Stage 2
Stage 3
Stage 4
REM
0
10
20
Time (seconds)
(C)
Awake
REM
Stage 1
Stage 2
Stage 3
Stage 4
0
1
2
3
4
5
6
7
Time (hours)

47.12 Stages of Sleep (A) Electrical activity in the cerebral cortex is detected by electrodes that are placed on the head and record changes in voltage between the electrodes through time. (B) The resulting record is an electroencephalogram, or EEG. (C) Humans cycle through different stages of sleep throughout the night.

non-REM sleep is divided into four stages, of which the two deepest stages are considered true slow-wave sleep.

When you fall asleep, the first state entered is stage 1 non-REM sleep, which then progresses through stages 2, 3, and 4 (**Figure 47.12B**). Stages 3 and 4 are deep, restorative, slow-wave sleep. This first full cycle of non-REM sleep is followed by an episode of REM sleep. Throughout the night you experience four or five cycles of non-REM and REM sleep (**Figure 47.12C**). About 80 percent of your sleep is non-REM sleep. The most vivid dreams and nightmares occur during the 20 percent of sleep that is REM sleep.

CELLULAR CHANGES DURING SLEEP When we are awake, several nuclei in the brainstem reticular formation are continuously active. Axons from neurons in these nuclei extend to the thalamus and throughout the cerebral cortex, where they release depolarizing neurotransmitters (acetylcholine, norepinephrine, and serotonin). These broadly distributed neurotransmitters keep the resting potential of the neurons of the thalamus and cortex close to threshold and sensitive to synaptic inputs, thereby maintaining the responsiveness of the brain that characterizes being awake.

With the onset of sleep, activity in these brainstem nuclei decreases, and their axon terminals release less neurotransmitter. With the withdrawal of the depolarizing neurotransmitters, the resting potentials of the cells of the thalamus and cortex become more negative (hyperpolarized), and the cells are less sensitive to excitatory synaptic input. Their processing of information is inhibited, and consciousness is lost.

An interesting neural event happens as a result of the hyperpolarization with the onset of sleep: cells begin to fire action potentials in bursts. The synchronization of these bursts over broad areas of cerebral cortex results in the EEG slow-wave pattern that characterizes deep non-REM sleep. Studies of neurons of the thalamus and cortex have shown that their hyperpolarization during non-REM sleep is due to increased opening of K^+ channels, and that bursting is due to Ca^{2+} channels whose inactivation gates close rapidly and require hyperpolarization to be reopened. We can therefore explain the EEG pattern of non-REM sleep in terms of the properties of neurons and ion channels.

At the transition from non-REM to REM sleep, dramatic changes occur. Some of the brainstem nuclei that were inactive during non-REM sleep become active again, causing a general depolarization of cortical neurons. Thus in REM sleep the synchronized bursts of firing cease, and the EEG resembles that of the awake brain. Because the resting potentials of the neurons return to near threshold levels, the cortex can process information, and vivid dreams occur.

So why don't we act out our dreams? During REM sleep the brain inhibits both afferent (sensory) and efferent (motor) pathways; we are paralyzed during REM sleep. Limb twitches and the jerky eye movements are motor signals breaking through the inhibition. The bizarre nature of dreams may be due to the lack of sensory feedback to the cortex from the body and the outside world. In other words, a functioning cortex is out of touch with reality. The function of muscle paralysis during REM sleep may be to prevent the acting out of dreams.

Knowing the cellular mechanisms of sleep has not yet led to an understanding of its function. Many questions remain. Why do we have two sleep states with very different neurophysiological characteristics? Why does non-REM sleep always occur first? Why do the two states cycle during the rest period? We know sleep is essential for life, but we don't know why. One set of hypotheses is that sleep is necessary for the maintenance and repair of neural connections and for the neural changes

WORKING WITH**DATA:**

Sleep and Learning

Original Paper

Walker, M. P., T. Brakefield, A. Morgan, J. A. Hobson, and R. Stickgold. 2002. Practice with sleep makes perfect: Sleep-dependent motor skill learning. *Neuron* 35: 205–211.

Analyze the Data

Does sleep enhance learning and memory? To answer this question, we need to specify the kind of learning and have an accurate means of measuring it. Walker and colleagues at Harvard Medical School investigated procedural (motor skill) memory through a simple task. Subjects used their nondominant hand to type the number sequence 4-1-3-2-4 on a computer keyboard as fast and accurately as they could for 30 seconds. The computer provided scores for speed and accuracy. Training consisted of 12 trials of 30 seconds each, with 30-second rests between trials. Retesting consisted of two similar sessions. Subjects were trained either at 10 A.M. or at 10 P.M. with the first retest after 12 hours and the second retest after 24 hours. Thus, one group had a sleep phase between the two retest events, while the other group had a sleep phase between the training and the first retest event. The results are shown in the bar graphs (right).

Another group of subjects was similarly trained at 10 P.M. and tested the next morning at 10 A.M. However, during their overnight sleep session their EEGs were recorded. Overnight improvement in motor skill scores was then correlated with the quantitative analysis of their sleep stages. The results are given in the table below (remember that most people cycle through 4–5 complete sleep cycles over the course of a night).

QUESTION 1

What would you conclude from these data about the role of sleep or wake in the consolidation of a procedural memory?

QUESTION 2

Based on the tabular data and the text descriptions of sleep stages, what feature of sleep seems to be most important for procedural memory consolidation?

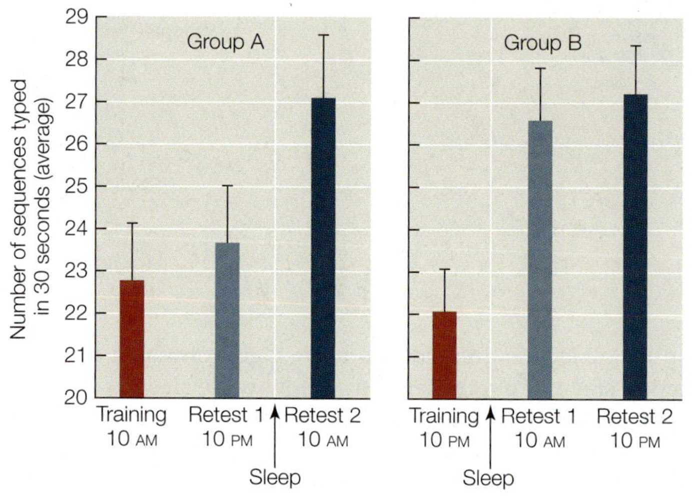
Number of sequences typed in 30 seconds (average)
29
28
27
26
25
24
23
22
21
20
Group A
Group B
Training 10 AM
Retest 1 10 PM
Retest 2 10 AM
Sleep
Training 10 PM
Retest 1 10 AM
Retest 2 10 PM
Sleep

Group C Stage of sleep cycle	Percent of total time asleep (mean ± standard error)	Correlation with motor skill improvement (correlation coefficient *r*)	Significance (*p* value)
Stage 1	3.1 ± 0.68	0.41	0.17
Stage 2	52.1 ± 1.94	0.66	0.01
Stage 3 (slow-wave)	10.2 ± 85	–0.26	0.40
Stage 4 (slow-wave)	11.4 ± 1.51	–0.17	0.59
REM sleep	18.4 ± 2.05	–0.32	0.30

Go to **BioPortal** for all WORKING WITH**DATA** exercises

involved in learning and memory—and possibly forgetting. These hypotheses are supported by many experiments showing that performance of a learned task or recall of declarative information on the day following training is impaired if sleep is prevented, and is best following a good night's sleep.

Language abilities are localized in the left cerebral hemisphere

No aspect of brain function is as integrally related to human consciousness and intellect as language. Therefore brain mechanisms that underlie the acquisition and use of language are extremely interesting to neuroscientists. A curious observation about language ability is that it resides in one cerebral hemisphere—which in 97 percent of people is the left hemisphere. This phenomenon is referred to as the **lateralization** of language functions.

Fascinating research on this subject was conducted by Roger Sperry and his colleagues at the California Institute of Technology. The two cerebral hemispheres are connected by a tract of white matter called the corpus callosum. In one severe form of epilepsy, bursts of action potentials causing seizures travel between hemispheres via the corpus callosum. Cutting the tract eliminates the problem, and patients function well following surgery. However, these "split-brain" subjects display interesting deficits in language ability.

After the surgery, if an object is shown in the right visual field and the left eye is closed (see Figure 47.11), the patient can describe it verbally and in writing. If the object is shown

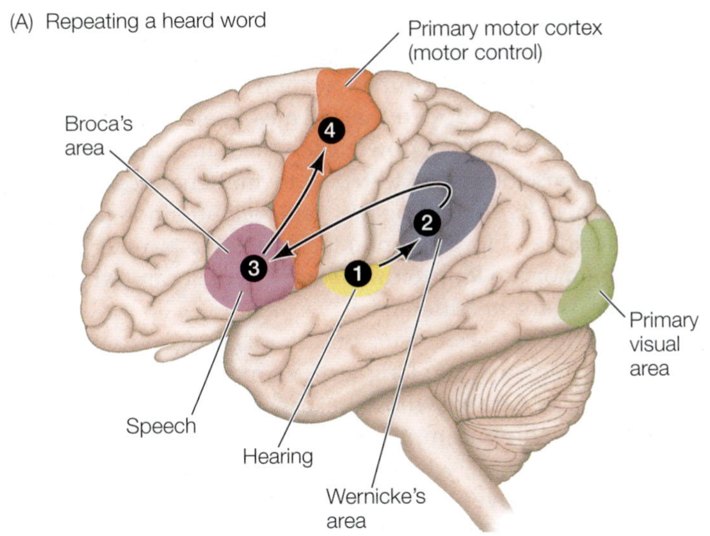
(A) Repeating a heard word
Primary motor cortex
(motor control)
Broca's area
Speech
Hearing
Wernicke's area
Primary visual area
(B) Speaking a written word
Angular gyrus

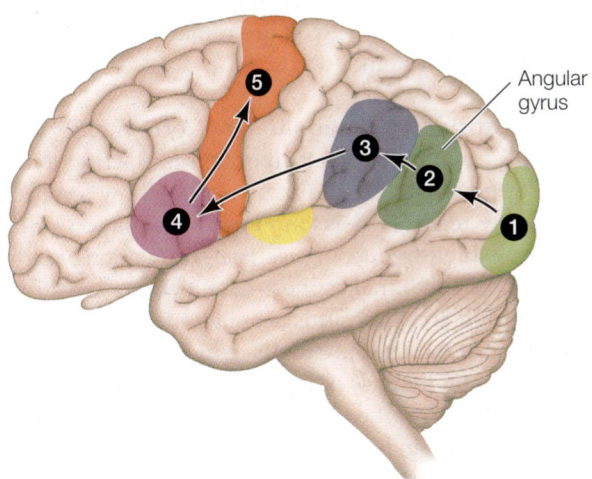

47.13 Language Areas of the Cortex Different regions of the left cerebral cortex participate in the processes of (A) repeating a word that is heard and (B) speaking a written word.

Go to Activity 47.2 Language Areas of the Cortex
Life10e.com/ac47.2

in the left visual field and the right eye is closed, the patient cannot describe it either verbally or in writing, but can use his or her left hand to point to a picture of the object. Without the connecting tissue between the two hemispheres, knowledge or experience of the right hemisphere can no longer be expressed in language.

Individuals who have suffered damage to the left hemisphere frequently suffer from some form of **aphasia**, a deficit in the ability to use or understand words. Studies of such individuals have identified several language areas in the left hemisphere.

- **Broca's area**, located in the frontal lobe just in front of the primary motor cortex, is essential for speech. Damage to Broca's area results in halting, slow, poorly articulated speech or even complete loss of speech, but the patient can still read and understand language.
- **Wernicke's area**, located in the temporal lobe close to its border with the occipital lobe, is more involved with sensory than with motor aspects of language. Damage to Wernicke's area can cause a person to lose the ability to speak sensibly while retaining the abilities to form the sounds of normal speech and to imitate its cadence. Such a patient cannot understand spoken or written language.
- The **angular gyrus**, located near Wernicke's area, is believed to be essential for integrating spoken and written language.

Normal language ability depends on the flow of information among various areas of the left cerebral cortex. Input from spoken language travels from the auditory cortex to Wernicke's area (**Figure 47.13A**). Input from written language travels from the visual cortex to the angular gyrus to Wernicke's area (**Figure 47.13B**). Commands to speak are formulated in Wernicke's area and travel to Broca's area and from there to the primary motor cortex. Damage to any one of those areas or the pathways between them can result in aphasia. Using modern methods of brain imaging, it is possible to see the metabolic activity in different brain areas when the brain is using language (**Figure 47.14**).

Some learning and memory can be localized to specific brain areas

Learning is the modification of behavior by experience. *Memory* is the ability of the nervous system to retain what is learned and experienced. Even very simple animals can learn and remember, but these two abilities are most highly developed in humans. Consider the amount of information associated with

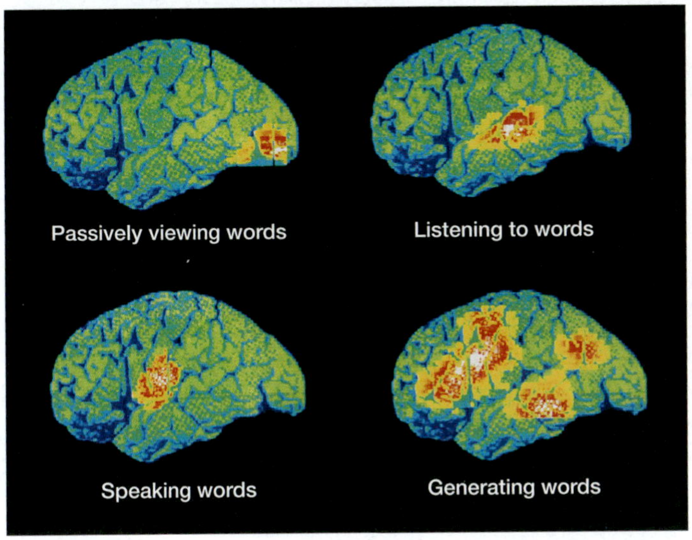
Passively viewing words
Listening to words
Speaking words
Generating words

47.14 Imaging Techniques Reveal Active Parts of the Brain Positron emission tomography (PET) scanning reveals the brain regions activated by different aspects of language use. Radioactively labeled glucose is given to the subject. Brain areas take up radioactivity in proportion to their metabolic use of glucose. The PET scan visualizes levels of radioactivity in specific brain regions when a particular activity is performed. The red and white areas are the most active.

learning a language. The capacity of memory and the rate at which memories can be retrieved are remarkable features of the human nervous system.

LEARNING Learning that leads to long-term memory and modification of behavior must involve long-lasting synaptic changes. A phenomenon that may explain how long-term synaptic changes might arise is **long-term potentiation**, or **LTP**. LTP results from high-frequency electrical stimulation of certain identifiable circuits that makes these circuits more sensitive to subsequent stimulation. In contrast, continuous, repetitive, low-level stimulation of these same circuits reduces their responsiveness, a phenomenon that has been called **long-term depression** (**LTD**). LTP and LTD may be fundamental cellular or molecular mechanisms involved in learning and memory.

Several kinds of learning exist. A form that is widespread among animal species is **associative learning**, in which two unrelated stimuli become linked to the same response. The simplest example of associative learning was described by the Russian physiologist Ivan Pavlov. Pavlov observed that a dog salivates at the sight or smell of food. He then showed that if a non-food stimulus (such as a sound) was paired with the food, after a number of paired experiences the dog would salivate at the sound even if no food was present (see Figure 53.1). Pavlov called this a **conditioned reflex**.

More complex forms of learning, often referred to as "observational learning," are the foundation of human intelligence. The general pattern of successful observational learning has three elements:

- We pay attention to another person's behavior.
- We retain a memory of what we have observed.
- We try to copy or use that information.

A key to this scheme of learning is the way in which we create and recall memories.

MEMORY Some of the first insights into memory processes came from clinical treatment of patients with severe epilepsy, a disorder characterized by uncontrollable, local increases in neural activity. The resulting seizures can endanger these individuals. Serious cases of epilepsy are sometimes treated by destroying the part of the brain from which the surge of activity originates. To find the right area, the surgery is done under local anesthesia, with the patient remaining conscious. As different regions of the brain are electrically stimulated with electrodes, the patient reports the resulting sensations. Stimulation of some regions of the association cortex elicits recall of vivid memories. Such observations provided the first evidence that specific areas in the brain are associated with specific memories and that memory can be attributed to networks of neurons. Destroying a small area of the brain does not completely erase a memory, however, so it is postulated that memory is a function distributed over many brain regions and can be stimulated via many different routes.

You experience several forms of memory everyday. You have **immediate memory** for events that are happening now. Immediate memory is almost perfectly photographic but lasts only seconds. **Short-term memory** contains less information but lasts longer—on the order of 10 to 15 minutes. When you are introduced to a group of several new people, you probably will have forgotten their names in an hour or so if you have not written them down, used them in a conversation, or made a conscious effort to repeat them. Repetition, use, or reinforcement by something that gets your attention (a title such as "President," for example) facilitates the transfer of short-term memory to **long-term memory**, which can last for days, months, years, or a lifetime.

Knowledge about neural mechanisms for the transfer of short-term memory to long-term memory has come from observations of persons who have lost parts of the limbic system, notably the hippocampus. A famous case is that of the man identified as H.M., whose hippocampus on both sides of the brain was removed in 1953 in an effort to control severe epilepsy. After the surgery, H.M. was unable to transfer information to long-term memory. If someone was introduced to him, had a conversation with him, and then left the room for several minutes, when that person returned, H.M did not recognize him—it was as if the conversation had never taken place. Up until his death 55 years later, H.M. remembered events that happened before his surgery but could not remember postsurgery events for more than 10 or 15 minutes.

Go to Media Clip 47.1
The Man with No Short-Term Memory
Life10e.com/mc47.1

Memory of people, places, events, and things is called **declarative memory** because you can consciously recall and describe them. **Procedural memory** cannot be consciously recalled and described; it is the memory of how to perform a motor task. When you learn to ride a bicycle, ski, or use a computer keyboard, you form procedural memories. Although H.M. was incapable of forming declarative memories, he could form procedural memories. When taught a motor task day after day, he could not recall the lessons of the previous day, yet his performance steadily improved. Thus procedural learning and memory must involve mechanisms different from those used in declarative learning and memory.

Memories can have considerable emotional content. As mentioned earlier, the limbic system plays a major role in controlling emotions. The amygdala, a component of the limbic system, is necessary for the emotion of fear and the formation of fear memories. Patients with a damaged amygdala do not associate fear reactions with their declarative memories. Memories can also have positive emotional content, and recalling those memories activates parts of the brain known to be associated with pleasurable sensations and reward, as revealed by brain imaging technologies.

We still cannot answer the question "What is consciousness?"

This chapter has only scratched the surface of the organization and functions of the human brain. Even with all of our knowledge of the human brain, and with all of the sophisticated new research tools, we still cannot answer the question "What is consciousness?"

The word "consciousness" is used in everyday language to refer to being awake in contrast to being asleep or in a coma. Here we are referring to the deeper meaning of being mentally aware of yourself, your environment, and events going on around you in such a way that you can plan for future events and make decisions based on experience, evidence, value systems, and predicted consequences. Speculations about consciousness have been the realm of philosophers, but we are getting closer to a neurobiological understanding.

The central requirement for conscious experience is a perception of self that can be integrated with information from the physical and social environment and information from past experience. The basis for a perception of self derives from the huge amount of somatosensory and visceral information that comes from all parts of the body. In the CNS of all vertebrates, this information is used for motor control and for homeostatic regulation. It enables animals to find food, seek mates, seek warmth, avoid cold, avoid danger, and so on. This afferent information goes to appropriate control and regulatory systems in the brainstem and forebrain.

In addition, some of this information goes to somatosensory areas of the cerebral cortex, so the animal is aware of certain information in the sense that it responds to it behaviorally. Visceral afferent information goes beyond its regulatory and control centers in the brainstem and hypothalamus to an area deep within the forebrain called the insular cortex, or **insula**. The insula appears to integrate physiological information from all over the body to create a sensation of how the body "feels." Thus when an animal's actions restore homeostasis, it "feels" better, and this is motivation to do the right thing for well-being.

In humans and the great apes, the insula is greatly expanded and has even acquired new types of spindle-shaped neurons not seen in other animals. The circuitry involving the insula has also evolved to communicate with parts of the brain that are involved in planning and decision making. In imaging studies, the insula is seen to be active in a great diversity of situations that involve strong feelings such as pleasure, disgust, humor, pain, lust, craving, humiliation, guilt, or empathy. Damage to the insula causes apathy, loss of ability to enjoy music, loss of sexual response, and even loss of the ability to distinguish good food from spoiled food. Humans and the few other species that have expanded insulas and the new spindle cells are the only species that can recognize themselves in a mirror. Could it be that this very discrete part of our brains and its circuitry are the neurobiological bases for self-awareness and conscious experience?

RECAP 47.3

Even complex functions of the nervous system are beginning to be understood in terms of the properties of neurons and neural networks.

- What events in the brain are associated with wakefulness and the stages of sleep? See p. 979
- Why do some neurobiologists think that the insula might be involved in conscious experience? See p. 983

?

Does the firing pattern of hippocampal place cells during sleep represent a memory of recent experience that is being transferred into long-term memory?

ANSWER

Unfortunately we cannot give a rat an exam to ask what it remembers. However, we can look in other brain areas for electrophysiological patterns of activity that correlate with the hippocampal place cell patterns. Wilson and colleagues described the fast replay of the hippocampal place cell patterns as "ripples." They observed that these ripples were tightly coupled to EEG brainwaves that spread to cortical areas of the brain. When they recorded in some of those other areas—the frontal cortex and the visual cortex—they found similar patterns of firing (i.e., ripples) in those areas that were synchronized to the hippocampal ripples. They hypothesized that the ripples did represent memory transcripts that were transferred to and stored in areas of the cortex.

CHAPTER SUMMARY 47

47.1 How Is the Mammalian Nervous System Organized?

- The brain and spinal cord make up the **central nervous system (CNS)**; the cranial and spinal nerves make up the **peripheral nervous system (PNS)**.
- The nervous system can be modeled conceptually in terms of the direction of information flow and whether we are conscious of the information. The afferent component carries information from the PNS to the CNS, and the efferent component directs information from the CNS to the peripheral parts of the body. **Review Figure 47.1**
- The vertebrate nervous system develops from a hollow dorsal neural tube. The brain forms from three swellings at the anterior end of the neural tube, which become the **hindbrain**, the **midbrain**, and the **forebrain**. The forebrain develops into the **cerebral hemispheres** (the **telencephalon**, or **cerebrum**) and the underlying thalamus and hypothalamus (which together compose the **diencephalon**). The midbrain and hindbrain develop into the **brainstem** and the **cerebellum**. **Review Figure 47.2**
- The spinal cord communicates information between the brain and the rest of the body.
- The **reticular activating system** is a complex network that directs incoming information to appropriate brainstem **nuclei** that control autonomic functions, and transmits the information to the forebrain that results in conscious sensation. The reticular activating system controls the level of arousal of the nervous system, including sleep and wakefulness.
- The **limbic system** is an evolutionarily primitive part of the telencephalon that is involved in emotions, physiological drives (such as hunger and thirst), instincts, and memory. **Review Figure 47.3**

continued

- The cerebral hemispheres are the dominant structures of the human brain. Their surfaces are layers of neurons called the **cerebral cortex**. The cerebral hemispheres can be divided into the **temporal**, **frontal**, **parietal**, and **occipital lobes**. Many motor functions are localized in parts of the frontal lobe. Information from many sensory receptors projects to a region of the parietal lobe. Visual information projects to the occipital lobe, and auditory information projects to a region of the temporal lobe. **Review Figures 47.4, 47.5, 47.6, ACTIVITY 47.1**

47.2 How Is Information Processed by Neural Networks?

- The **autonomic nervous system** (**ANS**) consists of efferent pathways that control the physiological function of organs and organ systems. Its **sympathetic** and **parasympathetic** divisions are characterized by their anatomy, neurotransmitters, and effects on target tissues. **Review Figure 47.9**
- The neural network of vision involves patterns of light falling on **receptive fields** in the retina. Receptive fields have a center and a surround, which have opposing effects on ganglion cell firing. **Review Figure 47.10, ANIMATED TUTORIALS 47.1, 47.2**
- Information from retinal ganglion cells is communicated via the optic nerve to the thalamus and then to the visual cortex. The visual cortex seems to assemble an image of the visual world by analyzing edges of patterns of light.
- **Binocular vision** is possible because information from both eyes is communicated to binocular cells in the visual cortex. These cells interpret distance by measuring the disparity between where the same stimulus falls on the two retinas. **Review Figure 47.11**

47.3 Can Higher Functions Be Understood in Cellular Terms?

- Humans have a daily cycle of sleep and waking. Sleep can be divided into **rapid eye movement** (**REM**) **sleep** and non-REM sleep. Deep non-REM sleep is known as **slow-wave sleep** because of its characteristic EEG patterns. **Review Figure 47.12**
- Language abilities are localized mostly in the left cerebral hemisphere, a phenomenon known as **lateralization**. Different areas of the left hemisphere—including **Broca's area**, **Wernicke's area**, and the **angular gyrus**—are responsible for different aspects of language. **Review Figures 47.13, 47.14, ACTIVITY 47.2**
- Some learning and memory processes have been localized to specific brain areas. Long-lasting changes in synaptic properties referred to as long-term potentiation (LTP) and **long-term depression** (**LDP**) may be involved in learning and memory.
- Complex memories can be elicited by stimulating small regions of association cortex. Damage to the hippocampus can destroy the ability to form long-term **declarative memory** but not **procedural memory**.
- A sense of the physiological state of the body may be created in the **insula** of the cortex from visceral afferent information. Evolution of this integrative function in higher primates and humans could be the basis for conscious experience.

See ACTIVITY 47.3 for a concept review of this chapter

Go to the Interactive Summary to review key figures, Animated Tutorials, and Activities
Life10e.com/is47

CHAPTER**REVIEW**

REMEMBERING

1. Which statement about the limbic system is *not* true?
   a. Damage to one structure in the limbic system makes it impossible to form a fear memory.
   b. The limbic system is involved in basic physiological drives, instincts, and emotions.
   c. The limbic system consists of primitive forebrain structures.
   d. The limbic system contains nuclei that maintain the cortex in an awake state.
   e. In humans, a part of the limbic system is necessary for the transfer of short-term memory to long-term memory.

2. Which of the following represents the largest portion of the human cerebral cortex?
   a. The frontal lobes
   b. The primary somatosensory cortex
   c. The temporal cortex
   d. The association cortex
   e. The occipital cortex

3. Which statement about the autonomic nervous system is true?
   a. The sympathetic division is afferent, and the parasympathetic division is efferent.
   b. The transmitter norepinephrine is always excitatory, and acetylcholine is always inhibitory.
   c. Each pathway in the autonomic nervous system includes two neurons, and the neurotransmitter of the first neuron is acetylcholine.
   d. The cell bodies of many sympathetic preganglionic neurons are in the brainstem.
   e. The cell bodies of most parasympathetic postganglionic neurons are in or near the thoracic and lumbar spinal cord.

4. Which statement is *not* true about some cells in the visual cortex?
   a. They receive input from only the left visual field.
   b. They respond most strongly to bars of light falling at specific locations on the retina.
   c. They receive inputs directly from single retinal ganglion cells.
   d. They receive input from both eyes.
   e. They respond most strongly to an object when it is a certain distance from the eyes.

5. Which of the following conclusions was supported by experiments on split-brain patients?
   a. Language abilities are localized mostly in the left cerebral hemisphere.
   b. Language abilities require both Wernicke's area and Broca's area.
   c. The ability to speak depends on Broca's area.
   d. The ability to read depends on Wernicke's area.
   e. The left hand is served by the left cerebral hemisphere.

UNDERSTANDING & APPLYING

6. A person receives a stab wound to the left side of his neck. Miraculously, blood vessels are spared. Afterward, however, the man's left pupil remains more constricted than his right pupil, and he drools out of the left side of his mouth. How can you explain these symptoms?

7. The eyes of some animals point in the same direction as do those of humans, but in other animals (such as birds) the eyes are positioned more laterally so they point in different directions. What would be the selective advantages of these different anatomical arrangements, and what kinds of animals would you expect to have one or the other?

8. Sleepwalking occurs in up to 15 percent of the population, although it is more common in children. While sleepwalking, individuals engage in mostly routine activities but are unaware of their actions when they awake. In what state of sleep do you think sleepwalking occurs, and why?

ANALYZING & EVALUATING

9. Patient X received a gunshot wound that destroyed the right side of his spinal cord at the midthoracic level. He has no conscious motor control over his right leg, but that leg is sensitive to painful stimuli, although not to touch. His right leg shows a reflex response when a painful stimulus is applied to his left foot, but he does not sense that stimulus as pain in his left foot. Considering that different types of information flow in different tracts in the spinal cord and that some types of information cross from one side of the cord to the other, what do the symptoms of this patient tell you about the routes of motor commands, pain sensory information, and mechanosensory information in the spinal cord?

10. High-density EEG recordings are made from arrays of 128 or 256 electrodes evenly spaced over the head. This allows measurements of local changes in EEG activity. When a person was trained in a task to use his left hand to do a difficult motor task, during subsequent sleep the slow-wave activity recorded over the posterior region of his right frontal cortex was greater than that recorded on the opposite side of his brain. Explain these results both in terms of neuroanatomy and possible functions of sleep.

Go to BioPortal at **yourBioPortal.com** for Animated Tutorials, Activities, LearningCurve Quizzes, Flashcards, and many other study and review resources.

48 Musculoskeletal Systems

CHAPTER**OUTLINE**

Champion Jumpers Relative to their size, many animals have more impressive jumping skills than humans. This leopard frog (*Rana pipiens*) can leap distances up to 20 times its body length.

THE OLYMPIC RECORD for the women's long jump is 7.4 meters, set in 1988 by Jackie Joyner-Kersee. Another world-record long jump that still stands was set two years earlier by Rosie the Ribeter, who jumped 6.5 meters. Rosie was a frog competing in the Calaveras County Jumping Frog Contest. In some ways Rosie's jump is more impressive. While Jackie's jump was about 5 times her body length (i.e., her height), Rosie's was about 20 times her body length.

Both jumps were powered by skeletal muscle. Muscle tissue responds to commands from the nervous system. The cellular mechanisms of muscle contraction are essentially the same in the frog and the human, so why is the frog's jump so much more impressive? The answer involves the concept of *leverage*, which depends on the muscles and skeletal elements working together.

Both frog and human jumping muscles pull on bones that are connected at joints to make levers. A lever makes it possible for the same force to move a large mass a small distance or a small mass a large distance. The ratio of a frog's leg length to its body mass is simply greater than that in a human. Thus the frog's legs are better at moving a small mass a long distance than are the human's legs.

Let's add a flea to our interspecies competition. The flea can jump more than *200 times* its body length. This incredible performance is not due to feats of leverage, because no muscle can contract fast enough to explain the take-off velocity of the flea. A different mechanism evolved in the flea—a kind of slingshot action. At the base of the flea's jumping legs is an elastic material that is compressed by muscles while the flea is resting. When a trigger mechanism is released, the elastic material recoils and "fires" the flea into the air.

In a contest of jumping efficiency, the uncontested champion would be the kangaroo. As a human runs faster, the number of strides and the energy expended per minute increase rapidly. Neither is true for the kangaroo. When moving at speeds from about 5 to 25 kilometers per hour, the kangaroo takes the same number of strides per minute and its metabolic rate *does not increase*.

The ability to move about in the environment is a distinguishing feature of most animals. Muscles and skeletons—musculoskeletal systems—enable animals to move.

How can the kangaroo increase its speed fivefold without increasing its metabolic expenditure?

See answer on p. 1003.

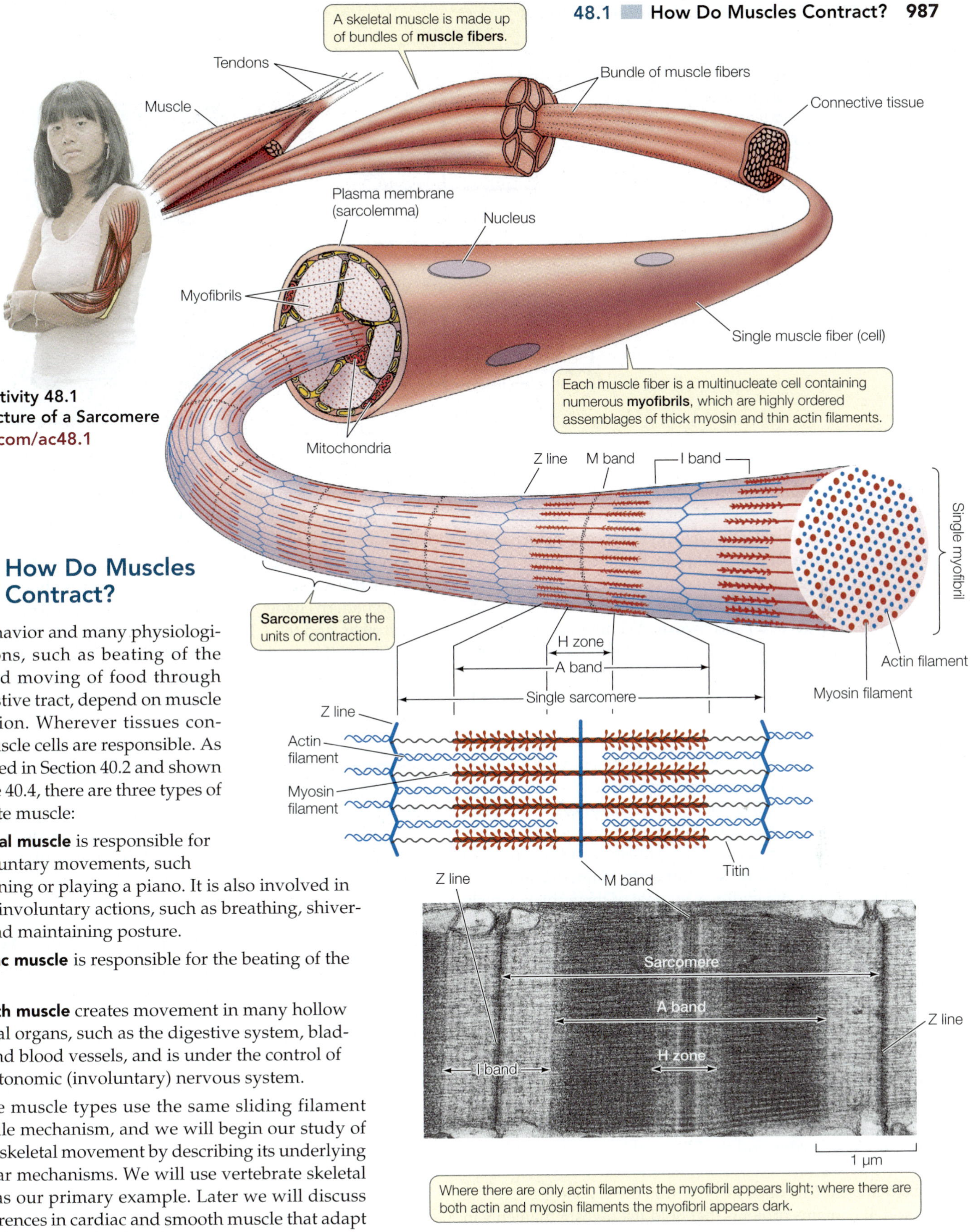

A skeletal muscle is made up of bundles of **muscle fibers**.
Tendons
Muscle
Bundle of muscle fibers
Connective tissue
Plasma membrane (sarcolemma)
Nucleus
Myofibrils
Single muscle fiber (cell)
Each muscle fiber is a multinucleate cell containing numerous **myofibrils**, which are highly ordered assemblages of thick myosin and thin actin filaments.
Mitochondria
Z line
M band
I band
Single myofibril
Sarcomeres are the units of contraction.
H zone
A band
Single sarcomere
Actin filament
Myosin filament
Z line
Actin filament
Myosin filament
Titin
M band
Z line
Sarcomere
A band
H zone
I band
Z line
1 μm
Where there are only actin filaments the myofibril appears light; where there are both actin and myosin filaments the myofibril appears dark.

48.1 The Structure of Skeletal Muscle A skeletal muscle is made up of bundles of muscle fibers. Each muscle fiber is a multinucleate cell containing numerous myofibrils, which are highly ordered assemblages of thick myosin and thin actin filaments. The arrangement of the actin and myosin filaments gives skeletal muscle fibers their characteristic striated appearance.

Go to Activity 48.1
The Structure of a Sarcomere
Life10e.com/ac48.1

48.1 How Do Muscles Contract?

Most behavior and many physiological actions, such as beating of the heart and moving of food through the digestive tract, depend on muscle contraction. Wherever tissues contract, muscle cells are responsible. As introduced in Section 40.2 and shown in Figure 40.4, there are three types of vertebrate muscle:

- **Skeletal muscle** is responsible for all voluntary movements, such as running or playing a piano. It is also involved in many involuntary actions, such as breathing, shivering, and maintaining posture.
- **Cardiac muscle** is responsible for the beating of the heart.
- **Smooth muscle** creates movement in many hollow internal organs, such as the digestive system, bladder, and blood vessels, and is under the control of the autonomic (involuntary) nervous system.

All three muscle types use the same sliding filament contractile mechanism, and we will begin our study of musculoskeletal movement by describing its underlying molecular mechanisms. We will use vertebrate skeletal muscle as our primary example. Later we will discuss the differences in cardiac and smooth muscle that adapt them to their particular functions.

Sliding filaments cause skeletal muscle to contract

Skeletal muscle is also called striated muscle because of its striped appearance (Figure 48.1; also see Figure 40.4A). Skeletal muscle cells, called **muscle fibers**, are large and have many

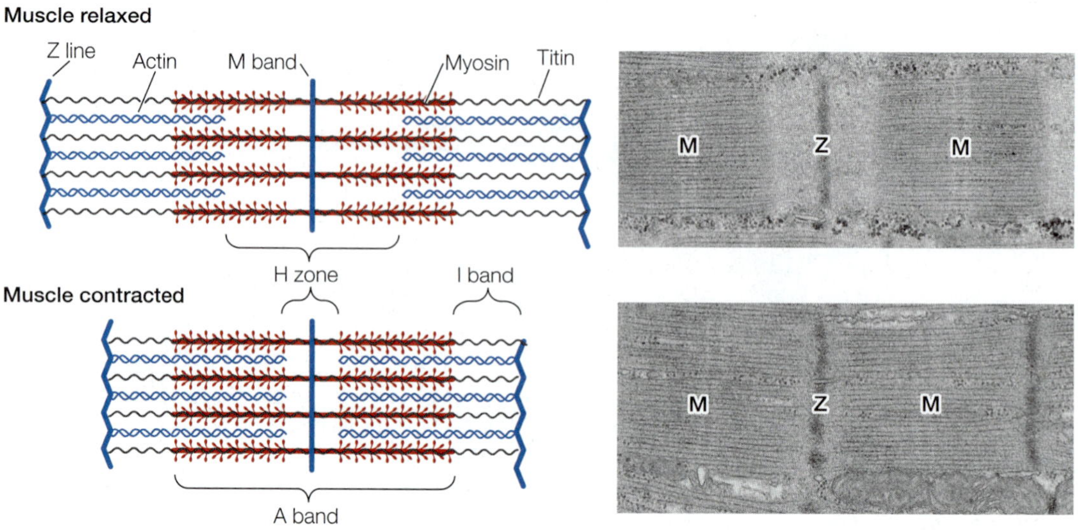
Muscle relaxed
Z line
Actin
M band
Myosin
Titin
H zone
I band
Muscle contracted
A band
M
Z
M
M
Z
M

48.2 Sliding Filaments The banding pattern of the sarcomere changes as it shortens. Observations of electron micrographs such as those on the right led to the sliding filament model of muscle contraction.

nuclei. These multinucleate cells form in development through the fusion of many individual embryonic muscle cells called myoblasts. A specific muscle such as your biceps (which bends your arm) is composed of hundreds or thousands of muscle fibers bundled together by connective tissue.

Muscle contraction is due to the interaction between the contractile proteins **actin** and **myosin**. Within muscle cells, actin and myosin molecules are organized into filaments. Actin filaments are also called thin filaments, and myosin filaments are thick filaments. The two kinds of filaments lie parallel to each other. When muscle contraction is triggered, the actin and myosin filaments slide past each other in a telescoping fashion.

What is the relationship between a skeletal muscle fiber and the actin and myosin filaments responsible for its contraction? Each muscle fiber (cell) is packed with **myofibrils**—bundles of thin actin and thick myosin filaments arranged in orderly fashion. In most regions of the myofibril, each thick myosin filament is surrounded by six thin actin filaments, and each thin actin filament sits within a triangle of three thick myosin filaments.

A longitudinal view of a myofibril reveals why skeletal muscle appears striated. The myofibril consists of repeating units called **sarcomeres**. Each sarcomere is made of overlapping filaments of actin and myosin, which create a distinct banding pattern (see Figure 48.1). Before the molecular nature of the muscle banding pattern was known, the bands were given names that are still used today. Each sarcomere is bounded by Z lines, which anchor the thin actin filaments. Centered in the sarcomere is the A band, which contains all the myosin filaments. The H zone and the I band, which appear light, are regions where actin and myosin filaments do not overlap in the relaxed muscle. The dark stripe within the H zone is called the M band; it contains proteins that hold the myosin filaments in their regular arrangement.

The bundles of myosin filaments are held in a centered position within the sarcomere by a protein called **titin**. Titin is the largest protein in the body; it runs the full length of the sarcomere from Z line to Z line. Each titin molecule runs right through a myosin bundle. Between the ends of the myosin bundles and the Z lines, titin molecules are very stretchable, like bungee cords. In a relaxed skeletal muscle, resistance to stretch is mostly due to the elasticity of the titin molecules.

As the muscle contracts, the sarcomeres shorten and the band pattern changes. The H zone and the I band become much narrower, and the Z lines move toward the A band as if the actin filaments were sliding into the H zone, the region occupied by the myosin filaments (**Figure 48.2**). In the mid 1950s this observation independently led two teams of British biologists to propose the **sliding filament model** of muscle contraction.

It is not uncommon in science for critical breakthroughs to be made simultaneously in different laboratories, but in this case the coincidences are remarkable. The leaders of the two teams were named Hugh Huxley and Andrew Huxley—but they were not related. Working in separate Cambridge University labs, the two groups proposed the sliding filament model at the same time, and both papers were published in the same issue of the journal *Nature*.

Actin–myosin interactions cause filaments to slide

To understand how the sliding filament model explains muscle contraction, we must first examine the structures of actin and myosin (**Figure 48.3**). A myosin molecule consists of two long polypeptide chains coiled together, each ending in a large globular head. A myosin filament is made up of many myosin molecules arranged in parallel, with their heads projecting sideways at each end of the filament.

An actin filament consists of actin monomers polymerized into a long molecule that looks like two strands of pearls twisted together. Twisting around the actin chains is another protein, **tropomyosin**, and attached to tropomyosin at intervals are molecules of **troponin**. We'll discuss the latter two proteins in more detail later in this section.

The myosin heads can bind specific sites on actin, to form cross-bridges between the myosin and the actin filaments. Moreover, when a myosin head binds to an actin filament, the head's conformation changes. As the head bends, it exerts a

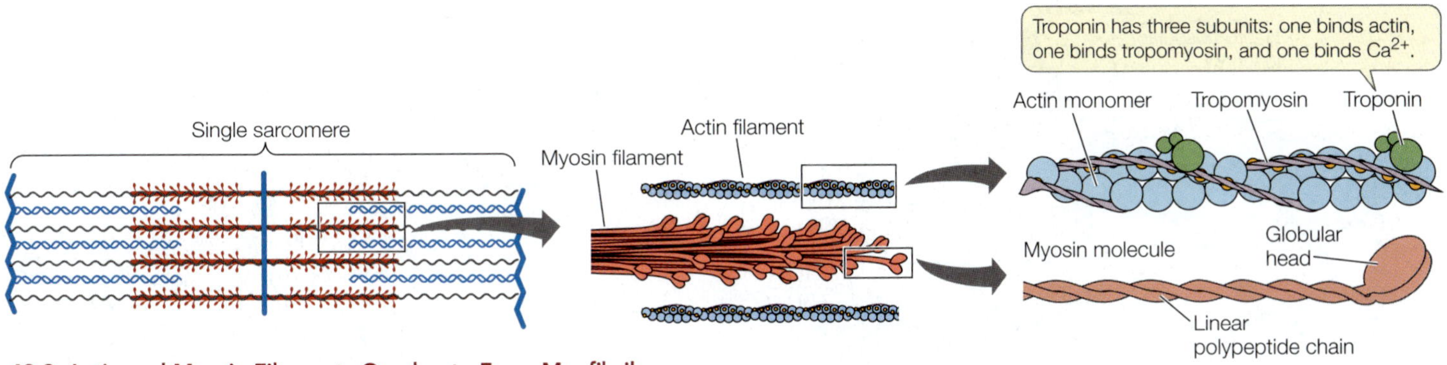
Troponin has three subunits: one binds actin, one binds tropomyosin, and one binds Ca^{2+}.
Single sarcomere
Actin filament
Myosin filament
Actin monomer
Tropomyosin
Troponin
Myosin molecule
Globular head
Linear polypeptide chain

48.3 Actin and Myosin Filaments Overlap to Form Myofibrils Myosin filaments are bundles of molecules with globular heads and polypeptide tails; the protein titin holds these filaments centered within the sarcomeres. Actin filaments consist of two chains of actin monomers twisted together. They are wrapped by chains of the polypeptide tropomyosin and are studded at intervals with another protein, troponin.

Go to Animated Tutorial 48.1
Molecular Mechanisms of Muscle Contraction
Life10e.com/at48.1

tiny force that causes the actin filament to move 5 to 10 nanometers relative to the myosin filament. When the myosin heads are bound to actin, they can bind and hydrolyze ATP. The energy released when this happens changes the conformation of the myosin head, causing it to release the actin and return to its extended position, from which it can bind to actin again.

Together these details help to explain the cycle of events that cause the actin and myosin filaments to slide past each other and shorten the sarcomere. They also explain rigor mortis—the stiffening of muscles soon after death. ATP binding causes myosin to release from actin, so when ATP production stops with death, myosin cannot release and the muscles stay contracted. Eventually, however, the proteins lose their integrity and the muscles soften. The timing of these events helps a medical examiner estimate the time of death.

We have been discussing the cycle of contraction in terms of a single myosin head. Remember that each myosin filament has many myosin heads at both ends and is surrounded by six actin filaments; thus the contraction of the sarcomere involves a great many cycles of interaction between actin and myosin molecules. That is why when a single myosin head breaks its contact with actin, the actin filaments do not slip backward.

Actin–myosin interactions are controlled by calcium ions

Like neurons, muscle cells are excitable—that is, their plasma membranes can generate and conduct action potentials. In skeletal muscle fibers, action potentials are initiated by motor neurons arriving at a **neuromuscular junction**. The axon terminals of motor neurons are generally highly branched and form synapses with hundreds of muscle fibers (Figure 48.4). A motor neuron and all of the fibers with which it forms synapses constitute a **motor unit**. The fibers contract simultaneously when the unit's motor neuron fires. A muscle can consist of many motor units. Thus there are two ways to increase a muscle's strength of contraction—increase the firing rate of an individual motor neuron, or recruit more motor neurons.

When an action potential arrives at a neuromuscular junction, the neurotransmitter acetylcholine is released from the motor neuron terminals, diffuses across the synaptic cleft, binds to receptors in the postsynaptic membrane, and causes ion channels in the motor end plate to open (see Figures 45.13 and 45.14). Most of the ions that flow through these channels are Na^+, and therefore the motor end plate is depolarized. The depolarization spreads to the surrounding plasma membrane of the muscle fiber, which contains voltage-gated sodium channels. When threshold is reached, the plasma membrane fires an action potential that is conducted rapidly to all points on the surface of the muscle fiber.

An action potential in a muscle fiber also travels deep within the cell. The plasma membrane is continuous with a system of tubules that descend into the muscle fiber cytoplasm (also called the **sarcoplasm**). The action potential that spreads over the plasma membrane also spreads through this system of transverse tubules, or **T tubules** (Figure 48.5).

The T tubules come very close to the endoplasmic reticulum (ER) of the muscle cell. In muscle cells the ER is called the **sarcoplasmic reticulum**, and it is a closed compartment surrounding every myofibril. Calcium pumps in the sarcoplasmic reticulum

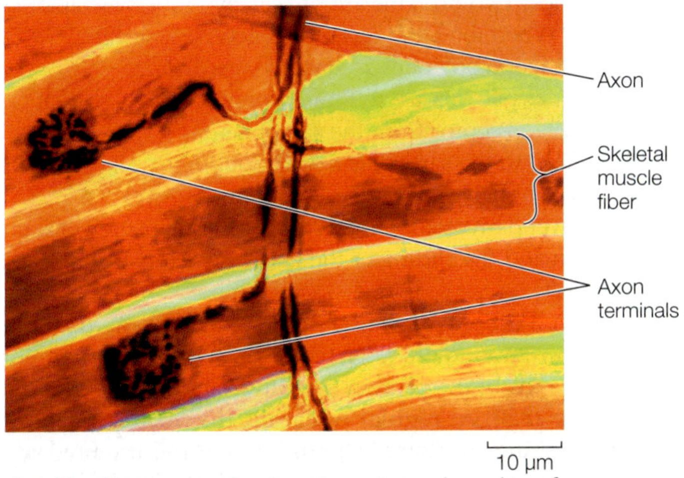
Axon
Skeletal muscle fiber
Axon terminals
10 μm

48.4 The Neuromuscular Junction Axons branching from a single motor neuron end in terminals that innervate multiple skeletal muscle fibers.

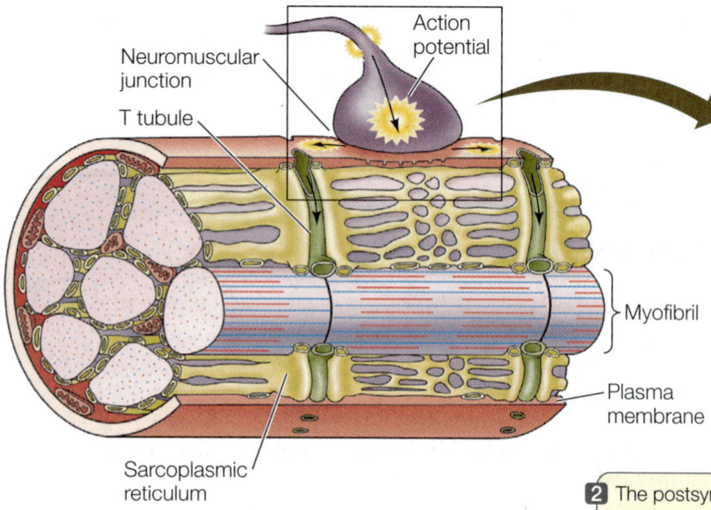

Neuromuscular junction
Action potential
T tubule
Myofibril
Plasma membrane
Sarcoplasmic reticulum

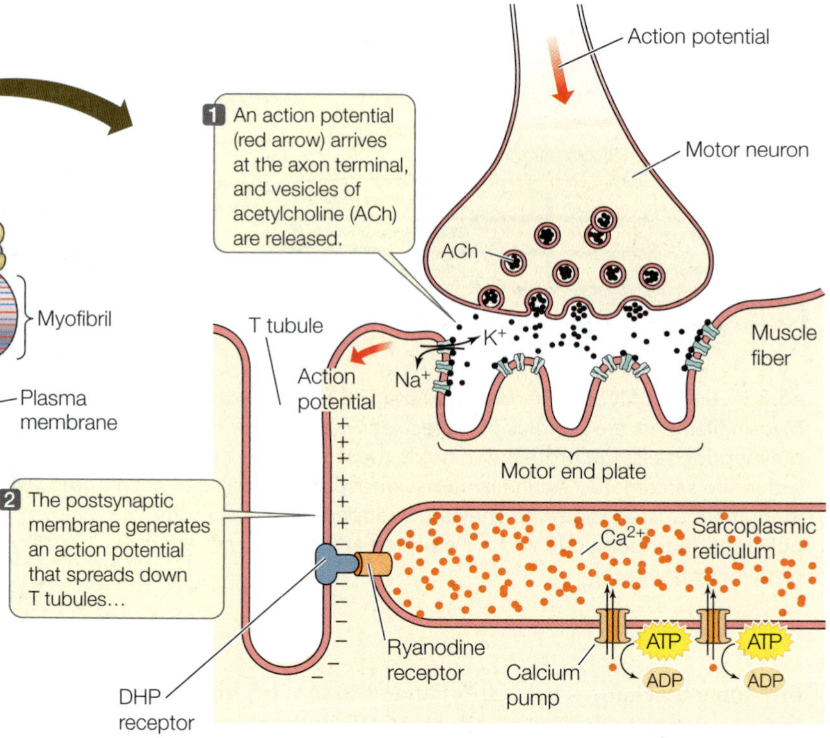

Action potential
Motor neuron
1 An action potential (red arrow) arrives at the axon terminal, and vesicles of acetylcholine (ACh) are released.
ACh
T tubule
K+
Na+
Action potential
Muscle fiber
Motor end plate
2 The postsynaptic membrane generates an action potential that spreads down T tubules…
Ca2+
Sarcoplasmic reticulum
Ryanodine receptor
Calcium pump
ATP
ADP
DHP receptor

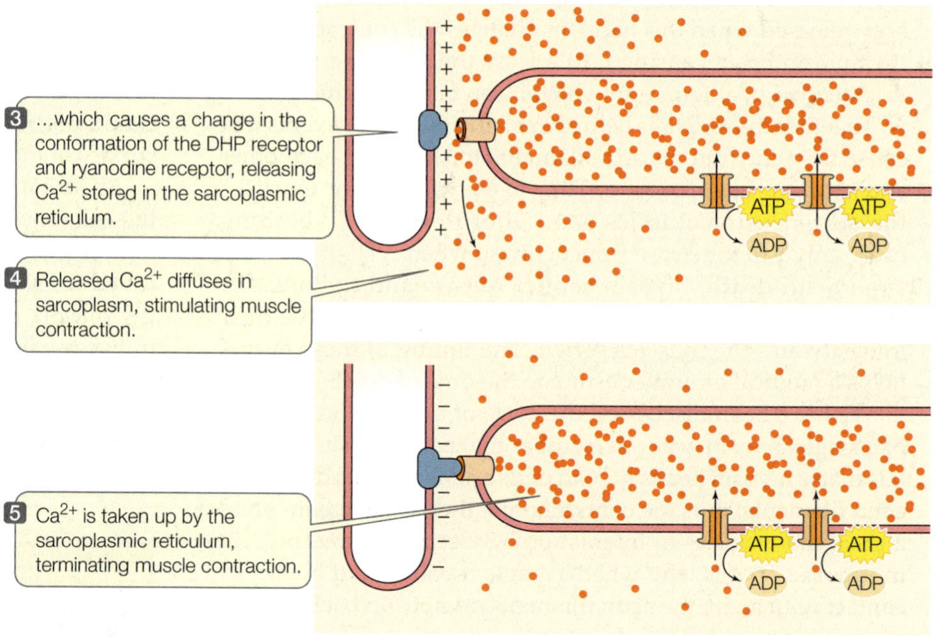

3 …which causes a change in the conformation of the DHP receptor and ryanodine receptor, releasing Ca2+ stored in the sarcoplasmic reticulum.
4 Released Ca2+ diffuses in sarcoplasm, stimulating muscle contraction.
5 Ca2+ is taken up by the sarcoplasmic reticulum, terminating muscle contraction.
ATP
ADP

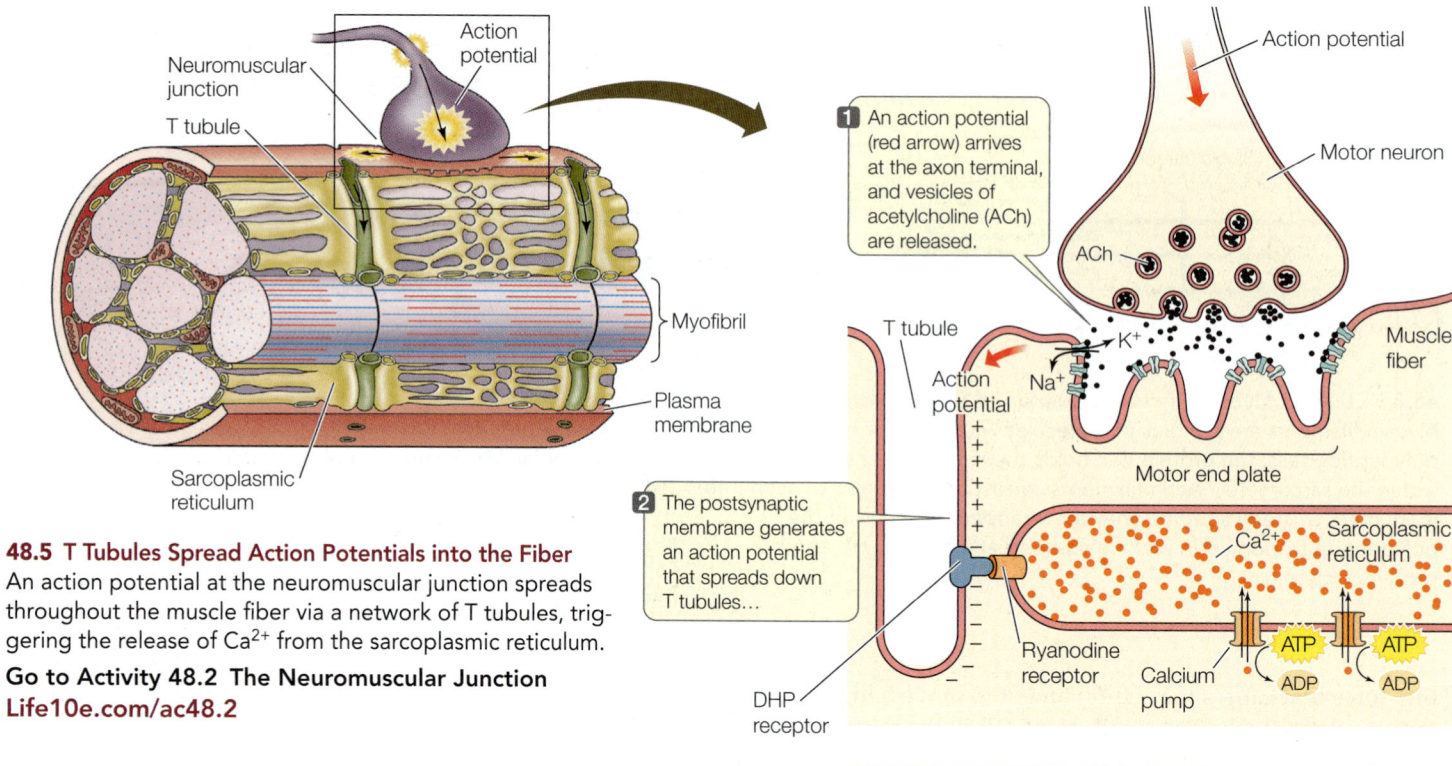

48.5 T Tubules Spread Action Potentials into the Fiber An action potential at the neuromuscular junction spreads throughout the muscle fiber via a network of T tubules, triggering the release of Ca^{2+} from the sarcoplasmic reticulum.

Go to Activity 48.2 The Neuromuscular Junction Life10e.com/ac48.2

take up Ca^{2+} from the sarcoplasm. Therefore when the muscle fiber is at rest, there is a high concentration of Ca^{2+} in the sarcoplasmic reticulum and a very low concentration of Ca^{2+} in the sarcoplasm.

Spanning the space between the membranes of the T tubules and the membranes of the sarcoplasmic reticulum are two proteins. One protein, the dihydropyridine (DHP) receptor, is located in the T tubule membrane; it is voltage-sensitive and changes its conformation in response to an action potential. The other protein, the ryanodine receptor, is located in the sarcoplasmic reticulum membrane and is a Ca^{2+} channel. These two proteins are physically connected. When the DHP receptor is activated by an action potential, it changes conformation; this allows Ca^{2+} to flow through the ryanodine receptor from the sarcoplasmic reticulum to the sarcoplasm. Ca^{2+} ions diffusing through the sarcoplasm surrounding the actin and myosin filaments trigger the interaction of actin and myosin and the sliding of the filaments. How do the Ca^{2+} ions do this?

An actin filament, as we have seen, is a helical arrangement of actin monomers. Twisted around the actin filament are two strands of the protein tropomyosin (**Figure 48.6**; see also Figure 48.3). At regular intervals, the filament also includes a globular protein, troponin. The troponin molecule has three subunits: one binds actin, one binds tropomyosin, and one binds Ca^{2+}.

When the muscle is at rest, the tropomyosin strands are positioned so that they block the sites on the actin filament where myosin heads can bind. When Ca^{2+} is released into the sarcoplasm, it binds to troponin, changing its conformation. Because the troponin is bound to the tropomyosin, this conformational change twists the tropomyosin enough to expose the actin–myosin binding sites. Thus the cycle of making and breaking actin–myosin bonds is initiated, the filaments are pulled past each other, and the muscle fiber contracts. When the calcium pumps in the sarcoplasmic reticulum membranes remove the Ca^{2+} ions from the sarcoplasm, the conformation of the tropomyosin returns to the state in which it blocks the binding of myosin heads to actin, and the muscle fiber returns to its resting condition. Figure 48.6 summarizes this cycle.

48.6 Release of Ca^{2+} from the Sarcoplasmic Reticulum Triggers Muscle Contraction When Ca^{2+} binds to troponin, it exposes myosin-binding sites on the actin. As long as binding sites and ATP are available, the cycle of actin and myosin interactions continues and the filaments slide past each other.

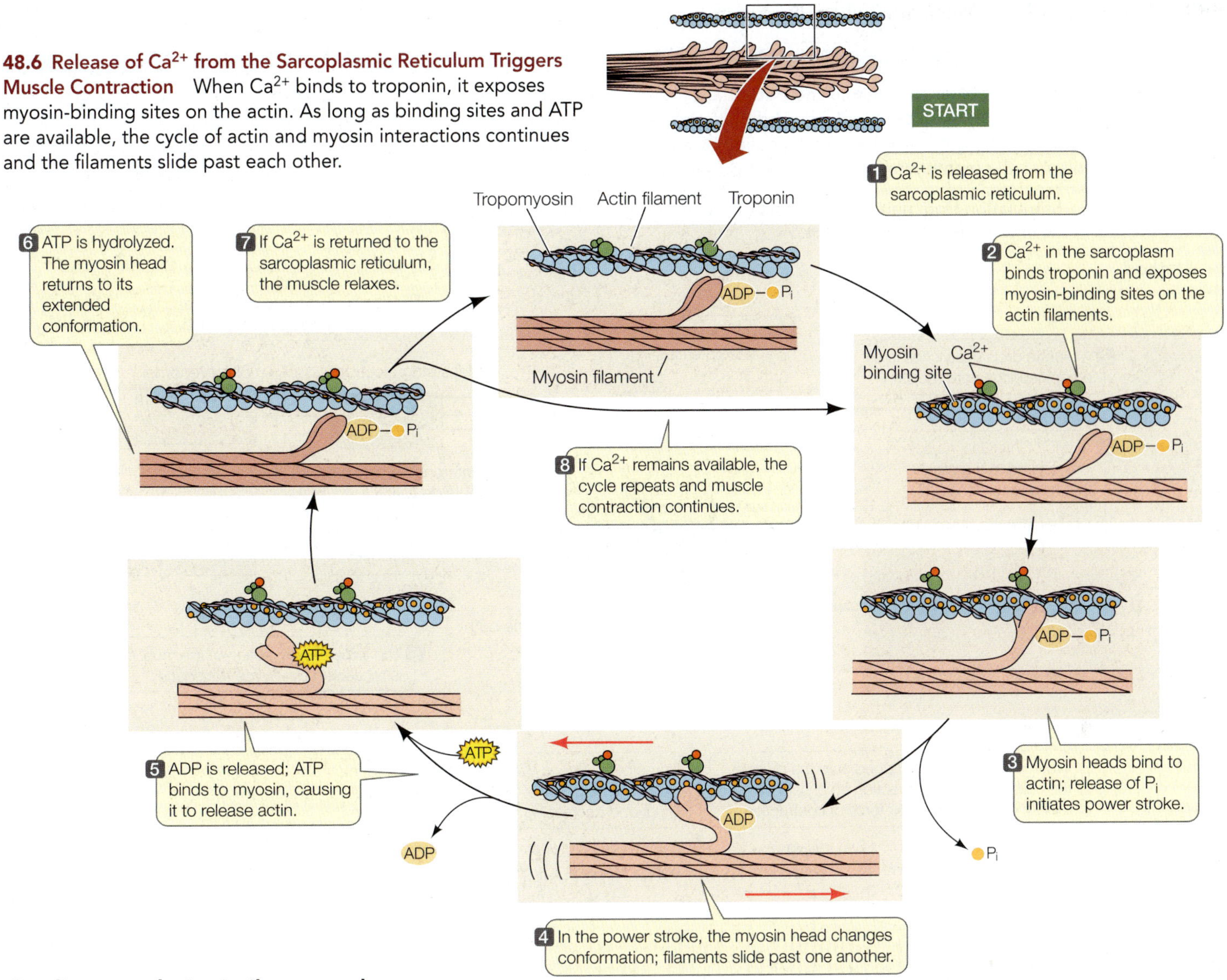

START
1 Ca2+ is released from the sarcoplasmic reticulum.
Tropomyosin
Actin filament
Troponin
Myosin filament
ADP — Pi
2 Ca2+ in the sarcoplasm binds troponin and exposes myosin-binding sites on the actin filaments.
Myosin binding site
Ca2+
3 Myosin heads bind to actin; release of Pi initiates power stroke.
Pi
ADP
4 In the power stroke, the myosin head changes conformation; filaments slide past one another.
ATP
5 ADP is released; ATP binds to myosin, causing it to release actin.
6 ATP is hydrolyzed. The myosin head returns to its extended conformation.
7 If Ca2+ is returned to the sarcoplasmic reticulum, the muscle relaxes.
8 If Ca2+ remains available, the cycle repeats and muscle contraction continues.

Cardiac muscle is similar to and different from skeletal muscle

Like skeletal muscle, cardiac muscle appears striated because of the regular arrangement of actin and myosin filaments into sarcomeres (**Figure 48.7**). The difference between cardiac and skeletal muscle is that cardiac muscle cells are much smaller and have only one nucleus each (uninucleate). Cardiac muscle cells branch, and the branches of adjoining cells interdigitate into a meshwork that is resistant to tearing. As a result, the heart walls can withstand high pressures while pumping blood, without the danger of developing leaks. Adding to the strength of cardiac muscle are intercalated discs that provide strong mechanical adhesions between adjacent cells. Gap junctions are an important feature of cardiac muscle. These structures in the intercalated discs allow cytoplasmic continuity between cells (see Figure 7.19A). Because of gap junctions, cardiac muscle cells are electrically coupled. An action potential initiated at one point in a sheet of cardiac muscle spreads rapidly, causing a large number of cardiac muscle cells to contract simultaneously.

Certain cardiac muscle cells are specialized for generating and conducting electric signals. These pacemaker and conducting cells have a low density of actin and myosin filaments, but they initiate and coordinate the rhythmic contractions of the

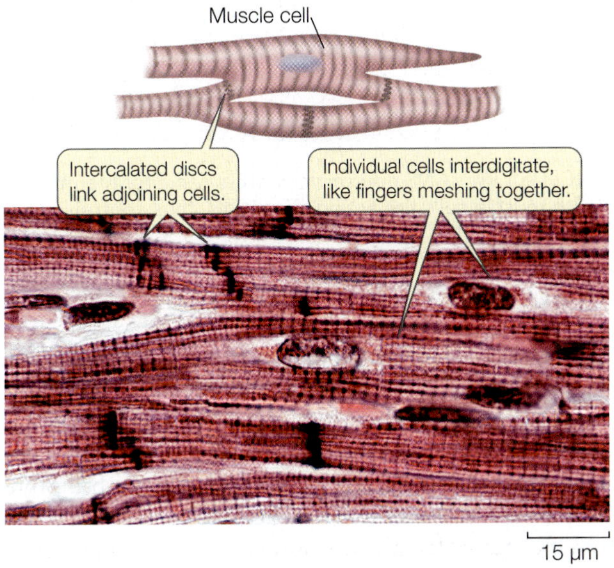

Muscle cell
Intercalated discs link adjoining cells.
Individual cells interdigitate, like fingers meshing together.
15 μm

48.7 Cardiac Muscle Cells Form a Strong Meshwork Cardiac muscle cells branch and interdigitate, forming a tear-resistant mesh that can withstand the pressure of blood pumping through the heart.

INVESTIGATING**LIFE**

48.8 Neurotransmitters Alter the Membrane Potential of Smooth Muscle Cells Earlier experiments showed that stretching the smooth muscles of the gut (as in the stretch applied by a full stomach) depolarizes the membranes, causing action potentials that activate the contractile mechanism. This follow-up experiment showed that the parasympathetic neurotransmitter acetylcholine and the sympathetic neurotransmitter norepinephrine act antagonistically to alter the membrane potential of smooth muscle.[a]

HYPOTHESIS Stimulation from neurotransmitters of the autonomic nervous system (ANS) induce contractions in the smooth muscles of the gut.

Method

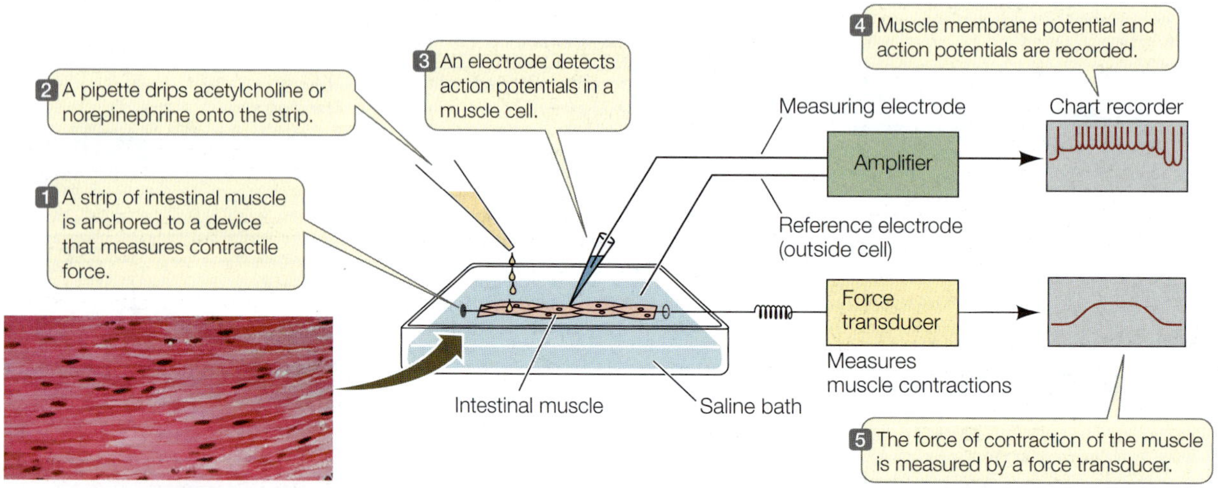

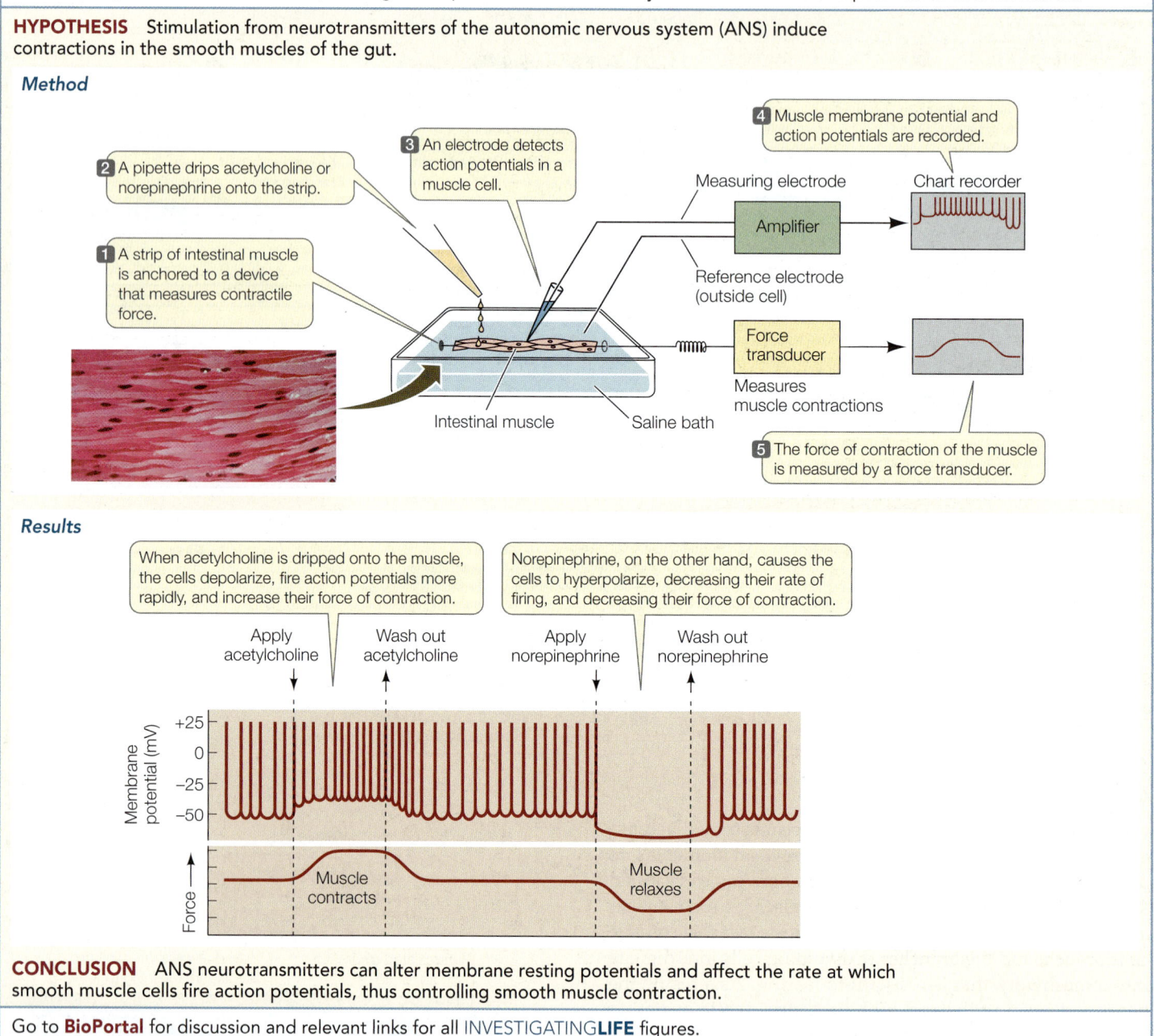

4 Muscle membrane potential and action potentials are recorded.
3 An electrode detects action potentials in a muscle cell.
2 A pipette drips acetylcholine or norepinephrine onto the strip.
1 A strip of intestinal muscle is anchored to a device that measures contractile force.
Measuring electrode
Chart recorder
Amplifier
Reference electrode (outside cell)
Force transducer
Measures muscle contractions
Intestinal muscle
Saline bath
5 The force of contraction of the muscle is measured by a force transducer.

Results

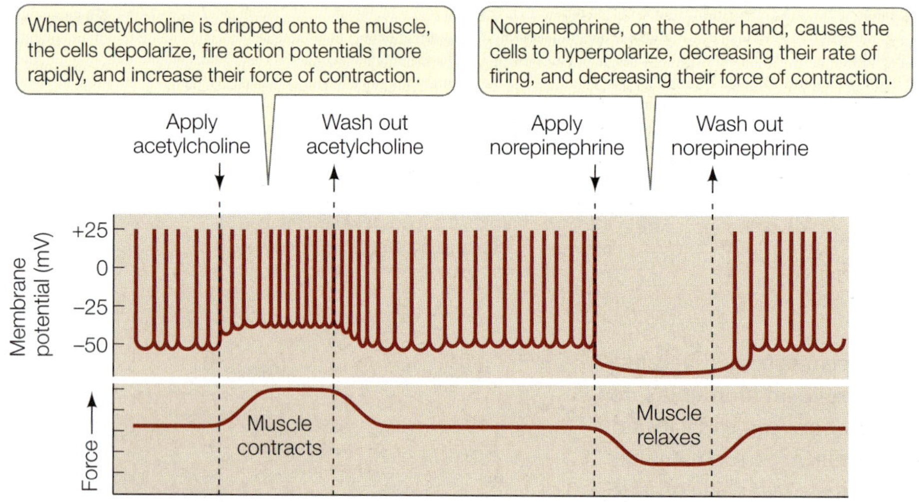

When acetylcholine is dripped onto the muscle, the cells depolarize, fire action potentials more rapidly, and increase their force of contraction.
Norepinephrine, on the other hand, causes the cells to hyperpolarize, decreasing their rate of firing, and decreasing their force of contraction.
Apply acetylcholine
Wash out acetylcholine
Apply norepinephrine
Wash out norepinephrine
Membrane potential (mV)
+25
0
–25
–50
Force
Muscle contracts
Muscle relaxes

CONCLUSION ANS neurotransmitters can alter membrane resting potentials and affect the rate at which smooth muscle cells fire action potentials, thus controlling smooth muscle contraction.

Go to **BioPortal** for discussion and relevant links for all INVESTIGATING**LIFE** figures.

[a]Bolton, T. B. et al. 1999. *Annual Review of Physiology* 61: 85–115.

heart. (The molecular basis for this pacemaking function will be covered in Section 50.3.) Pacemaker cells make the vertebrate heartbeat myogenic, meaning it is generated by the heart muscle itself. A heart removed from a vertebrate can continue to beat with no input from the nervous system. Although input from the autonomic nervous system modifies the *rate* of the pacemaker cells, it is not essential for their continued rhythmic function.

The mechanism of excitation–contraction coupling in cardiac muscle cells is different from that in skeletal muscle cells. The T tubules are larger, and the voltage-sensitive DHP receptor proteins in the T tubules are Ca^{2+} channels. These T tubule proteins are not physically connected with the ryanodine receptors in the sarcoplasmic reticulum. Instead, the ryanodine receptors are ion-gated Ca^{2+} channels. When an action potential spreads down the T tubules, it causes the voltage-gated channels to open, allowing extracellular Ca^{2+} to flow into the sarcoplasm. Increased sarcoplasmic Ca^{2+} concentration results in Ca^{2+} binding to the ryanodine receptors and opens calcium channels in the sarcoplasmic reticulum. The resulting huge rise in sarcoplasmic Ca^{2+} concentration stimulates fiber contraction. This mechanism is called Ca^{2+}-induced Ca^{2+} release.

Go to Media Clip 48.1
Be Still My Beating Stem Cell Heart
Life10e.com/mc48.1

Smooth muscle causes slow contractions of many internal organs

Smooth muscle provides the contractile force for most of our internal organs, which are under the control of the autonomic nervous system. Smooth muscle moves food through the digestive tract, controls the flow of blood through blood vessels, and empties the urinary bladder. Structurally, smooth muscle cells are the simplest muscle cells. They are smaller than skeletal muscle cells, usually long and spindle-shaped, and each has a single nucleus. They are "smooth" because the actin and myosin filaments are not as regularly arranged as they are in skeletal and cardiac muscle, and so do not produce the striated appearance.

Some smooth muscle tissue, such as that from the wall of the digestive tract, has interesting properties. The cells are arranged in sheets, and individual cells in a sheet are in electrical contact with one another through gap junctions, as they are in cardiac muscle. As a result, an action potential generated in the membrane of one smooth muscle cell can spread to all the cells in the sheet of tissue. Thus the cells in the sheet contract in a coordinated fashion.

The plasma membranes of smooth muscle cells are sensitive to stretch, with important consequences. If the wall of the digestive tract is stretched in one location (as by a mouthful of food passing down the esophagus to the stomach), the membranes of the stretched cells depolarize, reach threshold, and fire action potentials, which cause the cells to contract. Thus smooth muscle contracts after being stretched, and the harder it is stretched, the stronger it contracts. This behavior of smooth muscle is important for moving food through the digestive system.

The walls of blood vessels are mostly smooth muscle. This is especially true on the arterial side where the blood is under higher pressure. Changes in vascular smooth muscle tone are responsible for controlling the distribution of blood in the body.

The neural influences on smooth muscle come from the two divisions of the autonomic nervous system. The neurotransmitters of the sympathetic and parasympathetic postganglionic cells alter the membrane potential of smooth muscle cells. For example, in the digestive tract, acetylcholine causes smooth muscle cells to depolarize, making them more likely to fire action potentials and contract. Antagonistically, norepinephrine causes these muscle cells to hyperpolarize and thus be less likely to fire action potentials and contract (**Figure 48.8**). In contrast, norepinephrine acting through G protein-coupled receptors causes the smooth muscle in arteries serving the gut to contract. Remember that the action of the neurotransmitters depends on the receptors in the target tissues. Sympathetic activity is high in a fight-or-flight situation; in an emergency you don't need to digest your lunch, but you do need to send blood to the tissues critical for survival.

Although smooth muscle cell contraction is not controlled by the troponin–tropomyosin mechanism, calcium still plays a critical role. A Ca^{2+} influx into the sarcoplasm of a smooth muscle cell can be stimulated by action potentials, hormones, or stretching. The Ca^{2+} that enters the sarcoplasm combines with a protein called calmodulin. The calmodulin–Ca^{2+} complex activates an enzyme called myosin kinase that phosphorylates myosin heads. When the myosin heads in smooth muscle are phosphorylated, they undergo cycles of binding and releasing actin, causing muscle contraction. As Ca^{2+} is removed from the sarcoplasm, it dissociates from calmodulin, and the activity of myosin kinase falls. An additional enzyme, myosin phosphatase, dephosphorylates the myosin to help reduce actin–myosin interactions (**Figure 48.9**).

Go to Animated Tutorial 48.2
Smooth Muscle Action
Life10e.com/at48.2

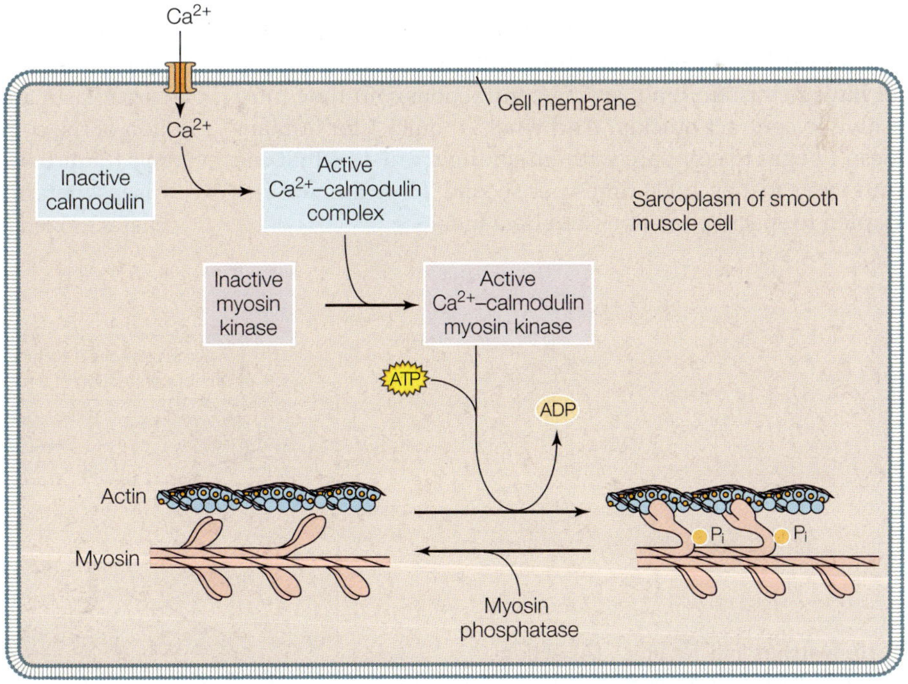

Ca2+
Cell membrane
Ca2+
Inactive calmodulin
Active Ca2+–calmodulin complex
Sarcoplasm of smooth muscle cell
Inactive myosin kinase
Active Ca2+–calmodulin myosin kinase
ATP
ADP
Actin
Myosin
Pi
Pi
Myosin phosphatase

48.9 The Role of Ca^{2+} in Smooth Muscle Contraction When a smooth muscle cell is stimulated by neurotransmitter, Ca^{2+} enters the sarcoplasm and binds to calmodulin, which in turn activates an enzyme that phosphorylates the myosin heads, causing them to bind to actin. As long as the myosin remains phosphorylated, actin and myosin go through cycles of binding and release. Thus in smooth muscle the Ca^{2+}-mediated change is on myosin, whereas in skeletal and cardiac muscle it is on the actin–tropomyosin filament.

RECAP 48.1

The contractile ability of muscle derives from interactions between actin and myosin filaments. The three types of muscle are skeletal, cardiac, and smooth. Contraction in all three depends on control by Ca^{2+} in the sarcoplasm. Tropomyosin and troponin are controlling elements in skeletal and cardiac muscle. Calmodulin is the controlling element in smooth muscle.

- Explain how the cellular and subcellular structures of skeletal muscle relate to the sliding filament theory of muscle contraction. **See pp. 987–989 and Figures 48.1, 48.2, and 48.3**
- What is the role of Ca^{2+} in the contractile mechanism of skeletal, cardiac, and smooth muscle? **See pp. 989–990 and Figures 48.5 and 48.6**
- What roles does ATP play in the actin and myosin interactions that produce contraction? **See Figure 48.6**

Now that we understand how muscles generate force, we can look at what determines the characteristics of a muscle, its performance, and how individual muscles can change their characteristics with regular use and conditioning.

48.2 What Determines Skeletal Muscle Performance?

The functions that different muscles perform place different demands on them. Some muscles, such as postural muscles, must sustain a load continuously over long periods of time. Muscles used in locomotion and heavy work must be able to vary their strength of contraction over a wide range. Other muscles, such as those that control your fingers, generally do not have to sustain long, strong contractions, but they must be able to contract quickly. And what is "quick" for humans doesn't begin to compare with insect flight muscles that can contract as fast as 1,000 times per second. How are muscles adapted to specific functions and demands?

The strength of a muscle contraction depends on how many fibers are contracting and at what rate

In skeletal muscle, the arrival of an action potential at a neuromuscular junction causes an action potential in a muscle fiber. The spread of that action potential through the muscle fiber's T tubule system causes a minimum unit of contraction, called a **twitch**. A twitch can be measured in terms of the tension, or force, it generates (**Figure 48.10A**). A single action potential stimulates a single twitch, but the ultimate force generated by an action potential varies enormously depending on how many muscle fibers it reaches. The level of tension an entire muscle generates depends on two factors: the number of motor units activated, and the frequency at which the motor units fire.

In muscles responsible for fine movements, such as those of the fingers, a motor neuron may innervate only one or a few muscle fibers, but in a muscle that produces large forces, such as the biceps, a motor neuron innervates a large number of muscle fibers.

At the level of a muscle fiber, a single action potential stimulates a single twitch. If action potentials reaching the muscle fiber are adequately separated in time, each twitch is a discrete, all-or-none phenomenon. If action potentials are fired more rapidly, however, new twitches are triggered before the myofibrils have a chance to return to their resting condition. As a result, the twitches sum, and the tension generated by the fiber increases and becomes more sustained. Thus an individual muscle fiber can show a graded response to increased levels of stimulation by its motor neuron.

Twitches sum at high levels of stimulation because the calcium pumps in the sarcoplasmic reticulum (see Figure 48.5) are not able to clear the Ca^{2+} ions from the sarcoplasm between action potentials. Eventually a stimulation frequency can be reached that results in the continuous presence of Ca^{2+} in the sarcoplasm at high enough levels to cause continuous activation of the contractile machinery—a condition known as **tetanus** (**Figure 48.10B**). (Do not confuse this condition with the disease tetanus, which is caused by a bacterial toxin and is characterized by spastic contractions of skeletal muscles.)

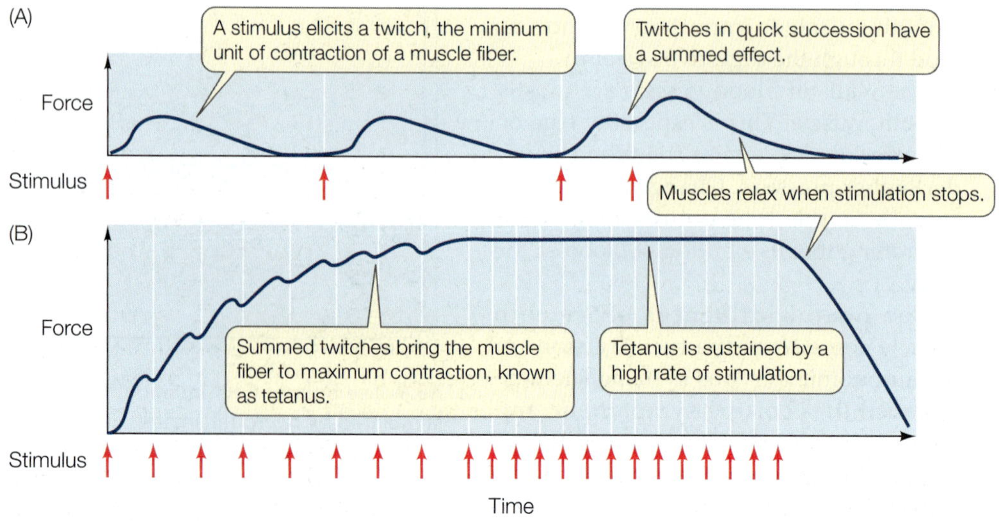
(A)
A stimulus elicits a twitch, the minimum unit of contraction of a muscle fiber.
Twitches in quick succession have a summed effect.
Force
Stimulus
Muscles relax when stimulation stops.
(B)
Force
Summed twitches bring the muscle fiber to maximum contraction, known as tetanus.
Tetanus is sustained by a high rate of stimulation.
Stimulus
Time

48.10 Twitches and Tetanus (A) Action potentials from a motor neuron cause a muscle fiber to twitch. Twitches in quick succession can be summed. (B) Summation of many twitches can bring the muscle fiber to the maximum level of contraction, known as tetanus.

How long a muscle fiber can maintain a tetanic contraction depends on its supply of ATP. Eventually the fiber will become fatigued and be unable to sustain the contraction. It may seem paradoxical that the *lack* of ATP causes fatigue, since the action of ATP is to break actin–myosin bonds. But remember that the energy released from the hydrolysis of ATP "re-cocks" the myosin heads, allowing them to cycle through another power stroke. When a muscle is contracting against a load, the cycle of making and breaking actin–myosin bonds must continue in order to prevent the load from stretching the muscle. The situation is like rowing a boat upstream. You cannot maintain your position relative to the stream bank by just holding the oars out against the current; you have to keep rowing. Likewise, actin–myosin bonds have to keep cycling to maintain tension in the muscle.

Many muscles of the body maintain a low level of tension even when the body is at rest. For example, the muscles of the neck, trunk, and limbs that maintain our posture against the pull of gravity are always working, even when we are standing or sitting still. Muscle tone comes from the activity of a small but changing number of motor units in a muscle; at any one time, some of the muscle's fibers are contracting and others are relaxed. The nervous system is constantly readjusting muscle tone.

Muscle fiber types determine endurance and strength

Not all skeletal muscle fibers are alike, and a single muscle often contains more than one type of fiber. The two major types of skeletal muscle fibers express different genes for their myosin molecules, and these myosin variants have different rates of ATPase activity. Those with high ATPase activity can recycle their actin–myosin cross-bridges rapidly and are therefore called fast-twitch fibers. Slow-twitch fibers have lower ATPase activity; they develop tension more slowly but can maintain it longer.

Slow-twitch fibers are also called oxidative or red muscle because they contain **myoglobin** (an oxygen-binding protein similar to the hemoglobin in red blood cells), have many mitochondria, and are well supplied with blood vessels. These characteristics both increase the fibers' capacity for oxidative metabolism and result in their red appearance. The maximum tension a slow-twitch fiber produces is low and develops slowly but is highly resistant to fatigue. Slow-twitch fibers have substantial reserves of fuel (glycogen and fat), so they can maintain steady, prolonged production of ATP as long as oxygen is available. Muscles with high proportions of slow-twitch fibers are good for long-term aerobic work (that is, work that requires oxygen). Long-distance runners, swimmers, cyclists, and other athletes whose activities require endurance have leg and arm muscles consisting mostly of slow-twitch fibers (**Figure 48.11**).

Some **fast-twitch fibers** are also called glycolytic or white muscle because, compared with slow-twitch fibers, they have few mitochondria, little or no myoglobin, and fewer blood vessels; thus they look pale. Fast-twitch glycolytic fibers can develop maximum tension more rapidly than slow-twitch fibers can, and that maximum tension is greater. However, fast-twitch fibers fatigue rapidly. The myosin of these fibers puts the energy of ATP to work very rapidly, but the fibers cannot replenish ATP quickly enough to sustain contraction for a long time. Fast-twitch fibers are especially good for short-term work that requires maximum strength. Weight lifters and sprinters have leg and arm muscles with high proportions of fast-twitch fibers.

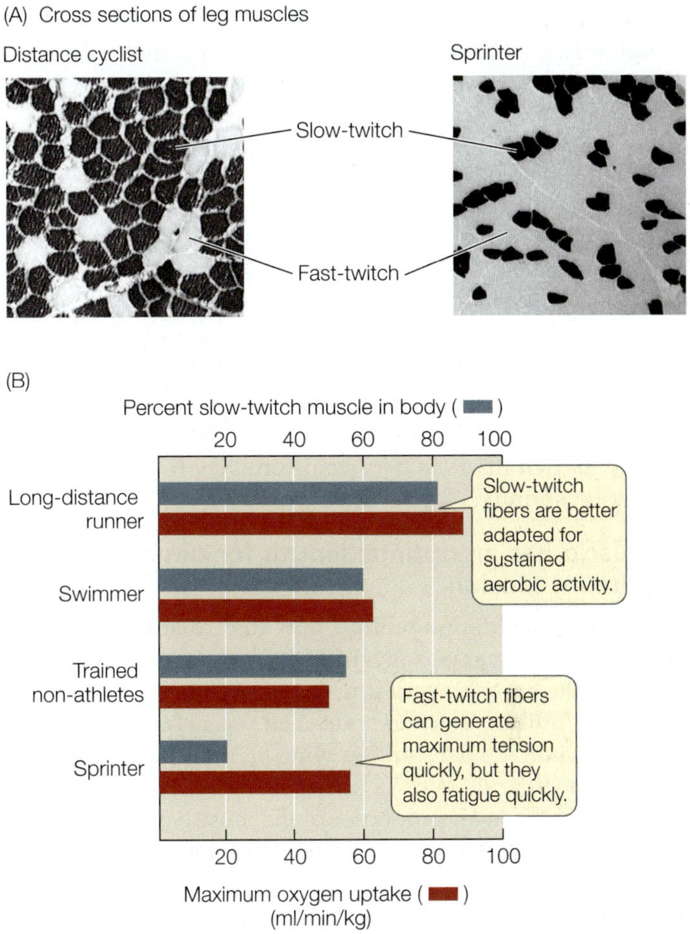
(A) Cross sections of leg muscles
Distance cyclist
Sprinter
Slow-twitch
Fast-twitch
(B)
Percent slow-twitch muscle in body
20
40
60
80
100
Long-distance runner
Swimmer
Trained non-athletes
Sprinter
Slow-twitch fibers are better adapted for sustained aerobic activity.
Fast-twitch fibers can generate maximum tension quickly, but they also fatigue quickly.
Maximum oxygen uptake (ml/min/kg)

48.11 Slow- and Fast-Twitch Muscle Fibers (A) The skeletal muscles in the micrographs were stained with a reagent that shows slow-twitch fibers as dark and fast-twitch muscle as light. (B) Athletes in different sports have different distributions of muscle fiber types.

The types of fibers that make up a muscle influence the performance properties of that muscle, and different muscles have different fiber compositions depending on their function. Postural muscles that maintain continuous contractions are mostly composed of slow-twitch fibers. An example is the soleus muscle that runs up the back of the leg from the heel. Its contraction extends the foot and is therefore used in walking, but its continuous contraction is required for standing. If the soleus muscle fatigued, we would fall forward. A person can walk or stand for a long period of time because the soleus muscle is resistant to fatigue. In contrast, a muscle that is used mostly for short-term work, such as the biceps, has a higher percentage of fast-twitch fibers than does the soleus. We can use our biceps to pick up a heavy weight, but we cannot hold that weight in a given position for a long period of time.

Can you change the fiber composition of your muscles to optimize your performance in a particular activity? To a limited extent, you can alter the properties of your muscle fibers through training. There are fast-twitch fibers that are somewhat oxidative and therefore intermediate in their properties between slow-twitch and fast glycolytic fibers. These intermediate fibers can become more oxidative with endurance training and more glycolytic with strength training. However, the most important determinant of your muscle fiber types is your genetic heritage. There is some truth to the statement that champions are born, not made. A person born with a high proportion of fast-twitch fibers in her legs is unlikely to become a champion marathon runner, and a person born with a high proportion of slow-twitch fibers in her legs is unlikely to become a champion sprinter.

A muscle has an optimal length for generating maximum tension

If you have ever done a pull-up, you know that two parts of this exercise are especially difficult. When you are hanging from the bar with your arms fully extended, it is hard to get the pull-up started; and when your chin has just about reached the bar, pulling yourself up the last small distance is difficult. Why is this? Part of the explanation comes from the lever properties of the muscle–joint interaction that we will discuss in Section 48.3, and part comes from the structure of the sarcomere.

You can see the relationship between the length of a muscle fiber and its ability to develop tension in **Figure 48.12**. When a muscle is stretched and the sarcomeres are lengthened, there is less overlap between the actin and myosin filaments; therefore fewer cross-bridges can form, and less force can be produced. In fact, if the sarcomeres are stretched too much, actin and myosin do not overlap and no force can be produced. How would a muscle recover from such a difficult situation? Like a bungee cord, titin molecules create enough elastic recoil to pull the actin and myosin fibrils back into an overlapping arrangement.

When the muscle is fully contracted, the actin and myosin filaments overlap so much that the myosin bundles are pressed up against the Z lines. Because they have no place to go, additional shortening is difficult.

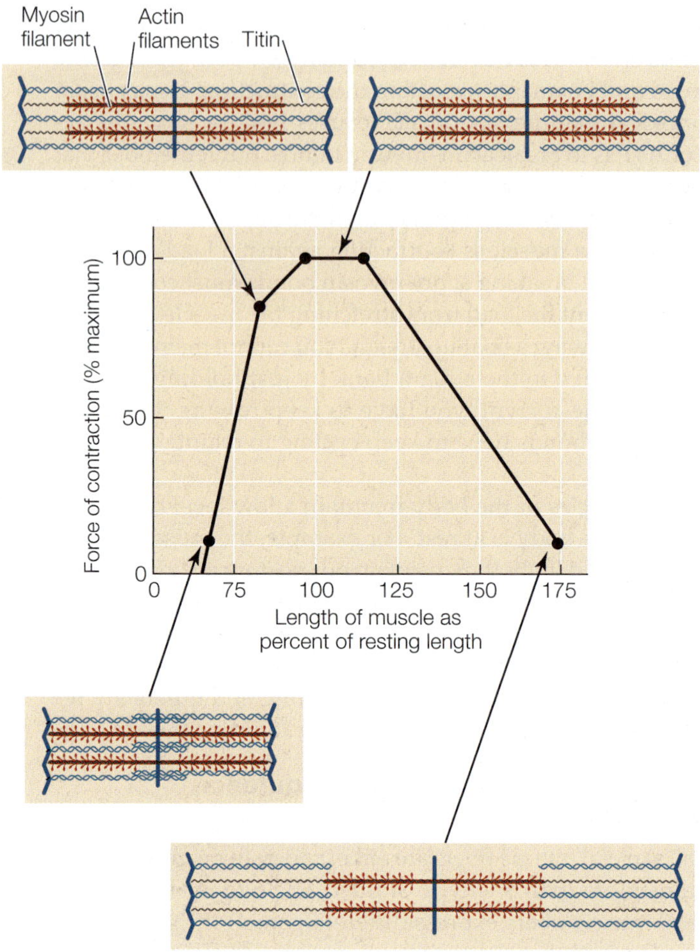
Myosin filament
Actin filaments
Titin
100
50
0
Force of contraction (% maximum)
0
75
100
125
150
175
Length of muscle as percent of resting length

48.12 Force and Length The amount of force a sarcomere can generate depends on its resting length. When a muscle is stretched, the sarcomeres lengthen, there is less overlap between the actin and myosin filaments, and less force is produced. Overstretched sarcomeres produce no force because there is no overlap between the actin and myosin.

Exercise increases muscle strength and endurance

Different types of exercise produce different physical conditioning responses. In general, anaerobic activities, such as weight lifting, increase strength, and aerobic activities, such as jogging, increase endurance. Strength is the maximum force a muscle can exert, and endurance is work capacity or how long a given workload can be sustained. What are the physiological bases for these differences?

Strength is a function of the cross-sectional area of muscles: the more actin and myosin filaments in a muscle fiber, and the more muscle fibers in a muscle, the more tension it can produce. When athletes undertake strength training, they use weights or exercises such as pull-ups to repeatedly contract specific muscles under heavy loads. Repetitions are usually done until the muscle is completely fatigued. Such stress on a muscle does minor tissue damage—hence the soreness the day after a hard workout—but it also induces the formation of new actin and myosin filaments in existing muscle fibers. The muscle fibers, and hence the muscles, get bigger and stronger. In extreme cases, and after serious muscle damage, new muscle fibers can also be produced from stem cells called satellite cells in the muscle. In general, however, the major effect of strength training is to produce bigger, rather than more, muscle fibers.

Aerobic exercise has a completely different effect on muscles: it enhances their oxidative capacity. This effect comes from increases in the number of mitochondria, in enzymes involved in energy use, and in the density of capillaries that deliver oxygen to the muscle. Myoglobin also increases in skeletal muscle cells. Although it is similar to hemoglobin, myoglobin has a higher affinity for oxygen. Therefore myoglobin accepts oxygen from the blood, facilitates the diffusion of oxygen throughout the muscle, and provides a store of oxygen for use when oxygen delivery by the blood is insufficient. By increasing the capacity of muscle to use oxygen to produce ATP, aerobic training increases the length of time that a given workload can be sustained.

Muscle ATP supply limits performance

Muscles have three systems for supplying the ATP they need for contraction:

1. The *immediate system* uses preformed ATP and creatine phosphate.
2. The *glycolytic system* metabolizes carbohydrates to lactate and pyruvate.
3. The *oxidative system* metabolizes carbohydrates or fats all the way to H_2O and CO_2.

The capacity of these three systems and the rates at which they can produce ATP determine both work capacity and endurance (**Figure 48.13**).

ATP is present in muscles in very small amounts. However, muscle fibers also contain a storage compound called creatine phosphate (CP). This molecule stores energy in a phosphate bond, which it can transfer to ADP. The total energy available in all the muscles of your body in the form of ATP and CP—the immediate energy system—is only about 10 kilocalories. When at rest, you metabolize a kilocalorie of energy in less than a minute. Even though the energy available from ATP and CP is limited, it is available immediately, and it enables fast-twitch fibers to generate a lot of force quickly. During burst activity, the immediate system is exhausted in seconds.

The glycolytic system activates within a few seconds to replace the ATP depleted at the onset of muscle activity. The glycolytic enzymes are located in the cytoplasm of the muscle fiber, and therefore the ATP they generate is rapidly available to the myosin filaments. However, as noted in Chapter 9, glycolysis alone is an inefficient way to produce ATP, and it leads to the accumulation of lactic acid, which slows the process. Thus the glycolytic system and the immediate system together can provide most of the energy for active muscles for less than a minute (see Figure 48.13).

Oxidative metabolism becomes fully active in about a minute, producing relatively huge amounts of ATP because it can completely metabolize carbohydrates and fats. However, it requires many reactions (see Section 9.3), and it takes place in the mitochondria, so O_2 and substrate must diffuse into the mitochondria, and the formed ATP must diffuse from the mitochondria to the myosin filaments in the muscle. These processes are not instantaneous, so the rate at which oxidative metabolism can make ATP available to do work is slower than the rate at which the other two systems can supply ATP.

The fuel supply available to the muscles influences how long someone can sustain a high level of aerobic exercise.

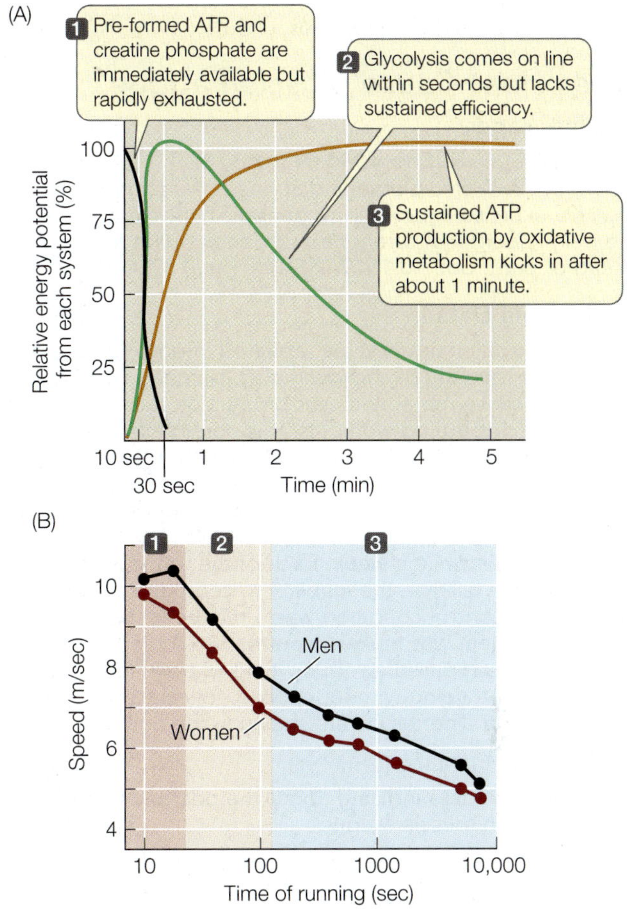
(A)
1 Pre-formed ATP and creatine phosphate are immediately available but rapidly exhausted.
2 Glycolysis comes on line within seconds but lacks sustained efficiency.
3 Sustained ATP production by oxidative metabolism kicks in after about 1 minute.
Relative energy potential from each system (%)
100
75
50
25
10 sec
30 sec
1
2
3
4
5
Time (min)
(B)
1
2
3
Speed (m/sec)
10
8
6
4
Men
Women
10
100
1000
10,000
Time of running (sec)

48.13 Supplying Fuel for High Performance (A) Muscles have three systems for obtaining the ATP they need for contraction during exertion such as running. (B) Looking at a plot of world-record times for running events of different durations, you can see that the performance of the athletes corresponds to the time courses of the three energy systems.

From the circulating blood, muscle receives glucose and free fatty acids, which it can metabolize to generate ATP. At high levels of aerobic exercise, however, most of the fuel used by muscles to produce ATP comes from the reserve of glycogen stored in the muscle itself. Depletion of muscle glycogen results in fatigue.

The rate at which muscle glycogen is replenished depends on diet: it is high with a high-carbohydrate diet, low with a high-fat diet, and intermediate with a mixed diet. This fact is the basis for a practice called "carbo-loading." For 3 to 5 days, athletes exercise at a level that depletes muscle glycogen. Then, 2 or 3 days before the event, they taper down their level of training and eat a diet rich in complex carbohydrates. The result can be glycogen supercompensation, in which the restoration of muscle glycogen stores "overshoots" and reaches above-normal levels.

Insect muscle has the greatest rate of cycling

Insect flight muscle can produce a wingbeat frequency of up to 1,000 cycles per second. Since neural action potentials last 1 to 3

WORKING WITHDATA:

Does Heat Cause Muscle Fatigue?

Original Paper

Grahn, D. A., V. H. Cao, C. M. Nguyen, M. T. Lieu, and H. C. Heller. 2012. Work volume and strength training responses to resistive exercise improve with periodic heat extraction from the palm. *Journal of Strength and Conditioning Research*. Epub ahead of print. doi: 10.1519/JSC.0b013e31823f8c1a

Analyze the Data

Physical conditioning requires repeated intense physical activity, and the capacity of such workouts is limited by muscle fatigue. Because metabolic heat production raises the temperature of muscles during workouts, it is possible that the rise in temperature contributes to muscle fatigue and limits the capacity of workouts. To test this idea, investigators used the rapid-cooling technology described in the opening of Chapter 40 to extract heat from subjects during 3-minute rests between ten sets of pull-ups twice a week. Each of the ten sets of pull-ups was to muscle failure—the inability to complete an additional pull-up. The control condition was 3-minute rests without cooling. Each subject was his own control; the subjects were randomly assigned to begin with 6 weeks of training with cooling or 6 weeks of training without cooling, followed by a reversal of the treatments. The results are shown in the figure.

QUESTION 1

What do these data indicate about the possible role of muscle temperature in muscle fatigue?

QUESTION 2

What do these data indicate about the relationship between workout capacity and physical conditioning effects?

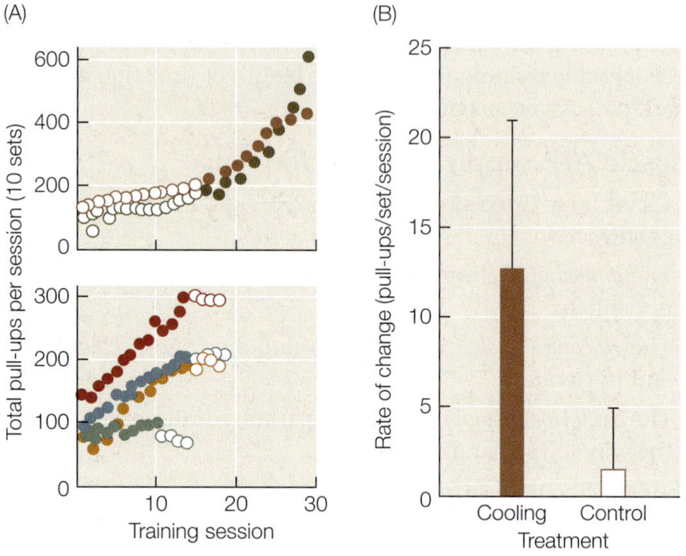
(A)
(B)
600
400
200
0
300
200
100
0
10
20
30
Training session
Total pull-ups per session (10 sets)
25
20
15
10
5
0
Rate of change (pull-ups/set/session)
Cooling
Control
Treatment

THE EFFECT OF COOLING ON WORKOUT CAPACITY (A) The results for individual subjects. Open symbols indicate control treatment, closed symbols indicate cooling treatment; different colors indicate different subjects. (B) Regression analysis of the pull-up data, which shows the rate of increase in workout capacity over the course of the experiment. Mean ± standard deviation of rate of change in pull-ups per set during the two treatment phases ($n = 6$, $P < 0.001$ paired t-tests).

Go to **BioPortal** for all WORKING WITH**DATA** exercises

milliseconds, that number of cycles per second would push the capacity of motor neurons, let alone the mechanism of cycling of striated muscle contraction/relaxation. The extremely fast wingbeat of a hummingbird may only be about 50 cycles per second. How do insects do it?

The mechanism of excitation/contraction coupling is different in insect flight muscle. Vertebrate striated muscle and much of invertebrate striated muscle is called "synchronous" because the cycling of the contractile mechanism is linked to the firing of the motor neurons. This is not true of insect flight muscle, which is therefore called "asynchronous" muscle. The firing of action potentials in the insect flight motor neurons is not particularly fast, but it does cause depolarization of the muscle cell membrane, the spreading of an action potential throughout the membrane, and the release of Ca^{2+} from the sarcoplasmic reticulum. However, once the asynchronous muscle fiber is stimulated, its cycle of contraction/relaxation proceeds at its own characteristic frequency as long as the Ca^{2+} is available to bind to the troponin. The contraction of the muscle fiber deactivates the actin–myosin binding, which in turn permits a stretching of the muscle, which in turn activates the actin–myosin binding. Thus contractile cycling and the resulting wingbeat frequency are not tied to the firing rate of the flight motor neurons.

RECAP 48.2

Depending on the function a muscle serves, it may need to generate maximum force rapidly, sustain activity for a long period, or contract and relax at a very rapid rate. Properties of muscles can facilitate these types of activities.

- How can a single motor unit exert varying levels of force? **See p. 994 and Figure 48.10**
- Describe the differences between slow-twitch and fast-twitch fibers. **See p. 995 and Figure 48.11**
- How does exercise influence muscle strength and endurance? **See p. 996**
- How do the different sources of ATP influence performance in different types of exercise? **See p. 997 and Figure 48.13**
- Explain how insects can beat their wings ten or more times faster than hummingbirds can. **See p. 998**

Regardless of how much force a muscle can generate, how long it can sustain a workload, or how fast it can contract and relax, a muscle needs something to pull on; otherwise it would just be a lump of pulsating, quivering tissue. Let's look now at how muscle and skeletal elements work together to produce movement.

48.3 How Do Skeletal Systems and Muscles Work Together?

Muscles can contract and exert force, or they can relax. To create significant movement, they must have something to pull on and something that stretches the muscle back to a longer position. In some cases muscles pull on each other, as in the trunk of an elephant or the arms of an octopus. In most cases, however, **skeletal systems** are the rigid supports against which muscles pull to create directed movement. In this section we will examine the three types of skeletal systems: hydrostatic skeletons, exoskeletons, and endoskeletons.

A hydrostatic skeleton consists of fluid in a muscular cavity

Cnidarians, annelids, and other soft-bodied invertebrates have **hydrostatic skeletons** consisting of a volume of fluid enclosed in a body cavity surrounded by muscle (see Section 31.2). When muscles oriented in one direction contract, the fluid-filled body cavity bulges out in a perpendicular direction.

An earthworm uses its hydrostatic skeleton to crawl (Figure 48.14). The earthworm's body cavity is divided into many separate segments filled with extracellular fluid. The body wall surrounding each segment has two muscle layers: a circular layer and a longitudinal layer. If the circular muscles in a segment contract, the compartment in that segment narrows and elongates. If the longitudinal muscles in a segment contract, the compartment shortens and bulges outward. Alternating contractions of the earthworm's circular and longitudinal muscles create waves of narrowing and widening, lengthening and shortening, that travel down the body. Bulging, shortened segments serve as anchors as long, narrow segments project forward and longitudinal contractions pull other segments forward. Bristles help the widest parts of the body to hold firm against the substrate, so the body moves forward.

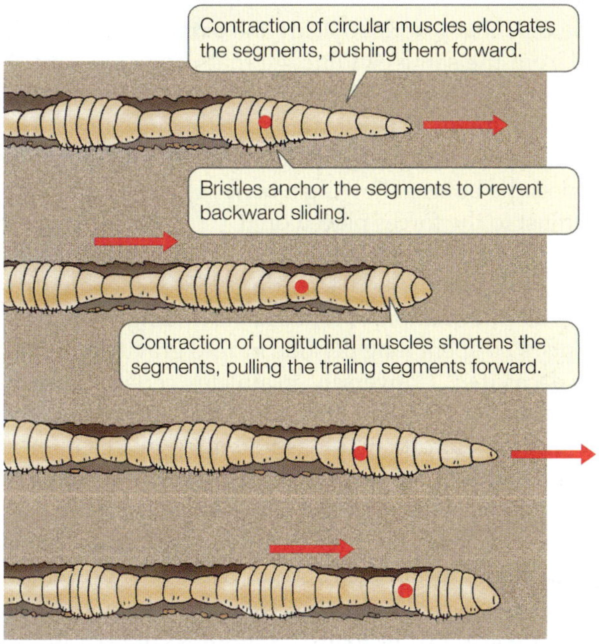
Contraction of circular muscles elongates the segments, pushing them forward.
Bristles anchor the segments to prevent backward sliding.
Contraction of longitudinal muscles shortens the segments, pulling the trailing segments forward.

48.14 A Hydrostatic Skeleton Alternating waves of muscle contraction move the earthworm through the soil. The red dot enables you to follow the changes in one segment as the worm moves forward.

Exoskeletons are rigid outer structures

An **exoskeleton** is a hardened, rigid outer surface to which muscles can be attached. Contractions of the muscles cause jointed segments of the exoskeleton to move relative to each other. The simplest example of an exoskeleton is the shell of a mollusk. Some marine mollusks, such as clams, have shells composed of protein strengthened by crystals of calcium carbonate (a rock-hard material). These shells can be massive, affording significant protection against predators. The shells of land mollusks (snails) generally lack the hard mineral component and are much lighter.

The most complex exoskeletons are found among the arthropods. A type of exoskeleton called a **cuticle** covers the outer surfaces of the arthropod body and all its appendages. It is made up of **chitin** secreted by a layer of cells just below the exoskeleton. Chitin stiffens and hardens the cuticle everywhere except at the joints, where flexibility must be retained. Muscles attached to the inner surfaces of the arthropod cuticle move its parts around the joints (see Figure 32.4).

A drawback of the rigid arthropod exoskeleton is that it cannot expand. Therefore, if the animal is to become larger, it must **molt**, shedding its exoskeleton and forming a new, larger one. A molting animal is vulnerable because the new exoskeleton takes time to harden. The animal's body is temporarily unprotected, and without a firm exoskeleton against which its muscles can exert maximum tension, it is unable to move rapidly. Soft-shelled crabs, a gourmet delicacy, are crabs caught while they are molting.

Vertebrate endoskeletons consist of cartilage and bone

The **endoskeleton** of vertebrates is an internal scaffolding. Muscles are attached to it and pull against it. Endoskeletons are composed of rodlike, platelike, and tubelike bones connected to one another at a variety of joints that allow a wide range of movements. An advantage of endoskeletons over the exoskeletons of arthropods is that bones in the body can grow without the animal shedding its skeleton.

The human skeleton consists of 206 bones, some of which are shown in Figure 48.15. It can be divided into an axial skeleton, which includes the skull, vertebral column, sternum, and ribs; and an appendicular skeleton, which includes the pectoral girdle, pelvic girdle, and bones of the arms, legs, hands, and feet.

The vertebrate endoskeleton consists of two kinds of connective tissue, cartilage and bone, which are produced by two kinds of connective tissue cells. Cartilage cells produce an extracellular matrix that is a tough, rubbery mixture of polysaccharides and proteins—mainly fibrous collagen. Collagen fibers run in all directions like reinforcing cords through the gel-like matrix and give it the well-known strength and resiliency of

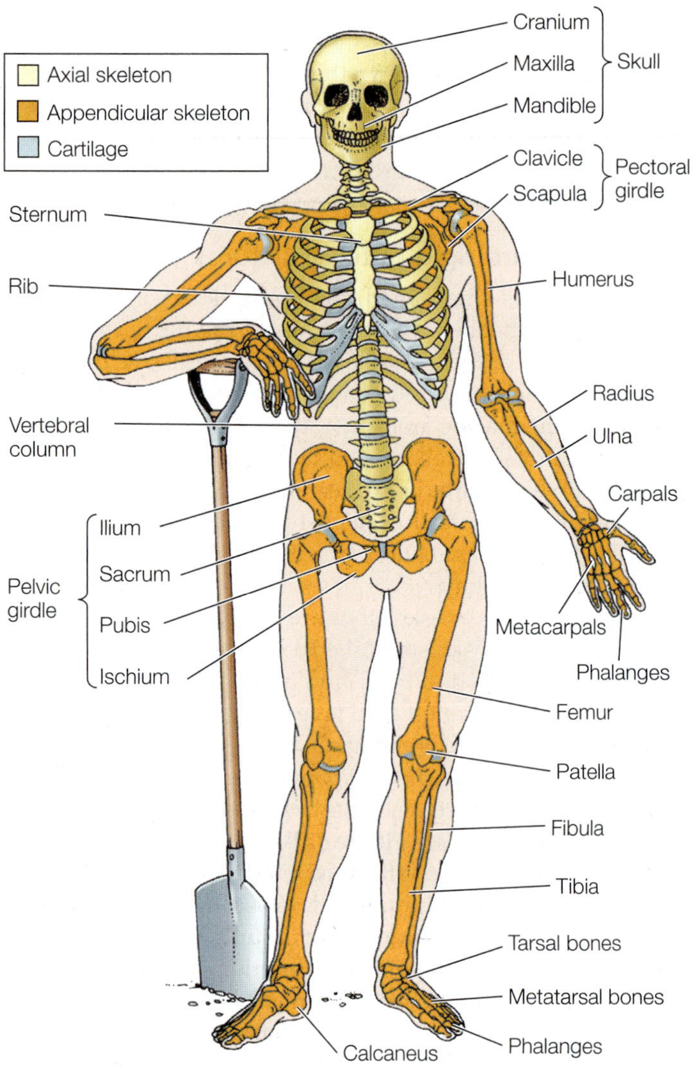
Axial skeleton
Appendicular skeleton
Cartilage
Cranium
Maxilla
Mandible
Skull
Clavicle
Scapula
Pectoral girdle
Sternum
Rib
Humerus
Radius
Ulna
Vertebral column
Carpals
Ilium
Sacrum
Pubis
Ischium
Pelvic girdle
Metacarpals
Phalanges
Femur
Patella
Fibula
Tibia
Tarsal bones
Metatarsal bones
Calcaneus
Phalanges

48.15 The Human Endoskeleton Cartilage and bone make up the internal skeleton of a human being.

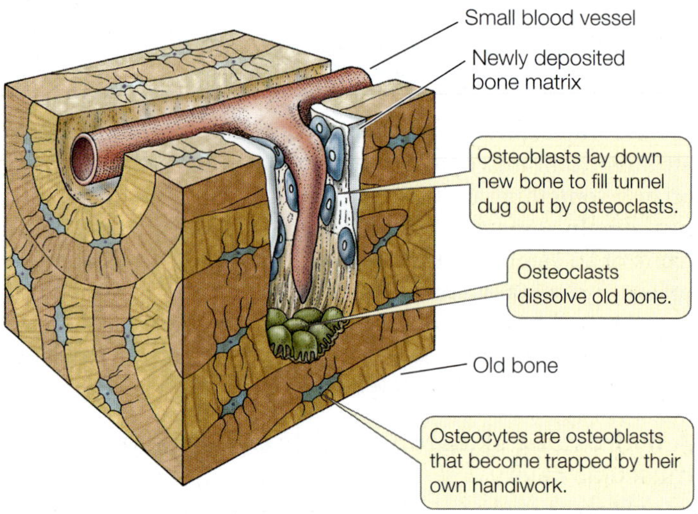
Small blood vessel
Newly deposited bone matrix
Osteoblasts lay down new bone to fill tunnel dug out by osteoclasts.
Osteoclasts dissolve old bone.
Old bone
Osteocytes are osteoblasts that become trapped by their own handiwork.

48.16 Bone Is Living Tissue Bones are constantly being remodeled by osteoblasts, which lay down bone, and osteoclasts, which resorb bone.

"gristle." This matrix, called **cartilage**, is found in parts of the endoskeleton where both stiffness and resiliency are required, such as on the surfaces of joints where bones move against one another. Cartilage is also the supportive tissue in stiff but flexible structures such as the larynx (voice box), nose, and ear pinnae. Sharks and rays are called cartilaginous fishes because their skeletons are composed entirely of cartilage. In most other vertebrates, cartilage is the principal component of the embryonic skeleton, but during development most of it is gradually replaced by bone.

Bone also contains collagen fibers, but it gets its rigidity and hardness from an extracellular matrix of insoluble calcium phosphate crystals. Bone serves as a reservoir of calcium for the rest of the body and is in dynamic equilibrium with soluble calcium in the extracellular fluids of the body. This equilibrium is under the control of calcitonin and parathyroid hormone (see Figure 41.14). If too much calcium is taken from the skeleton, the bones are seriously weakened.

The living cells of bone—osteoblasts, osteocytes, and osteoclasts—are responsible for the constant dynamic remodeling of bone (Figure 48.16). **Osteoblasts** lay down new matrix material on bone surfaces. These cells gradually become surrounded by matrix and eventually become enclosed within the bone, at which point they cease laying down matrix but continue to exist within small lacunae (cavities) in the bone. In this state they are called **osteocytes**. Despite the vast amounts of matrix between them, osteocytes remain in contact with one another through long cellular extensions that run through tiny channels in the bone. Communication between osteocytes is important in controlling the activities of the cells that are laying down or removing bone.

The cells that resorb bone are the **osteoclasts**. They are derived from the same cell lineage that produces white blood cells. Osteoclasts erode bone, forming cavities and tunnels. Osteoblasts follow osteoclasts, depositing new bone. Thus the interplay of osteoblasts and osteoclasts constantly replaces and remodels the bones, allowing a bone to recover from damage and adjust to the forces placed on it.

How the activities of the bone cells are coordinated is not understood, but stress placed on bones somehow provides them with information. A remarkable finding in studies of astronauts who spent long periods in zero gravity was that their bones decalcified. Conversely, in athletes, certain bones thicken during training. Both thickening and thinning of bones are experienced by anyone who has had a leg in a cast for a long time: the bones of the uninjured leg carry the person's weight and thicken while the bones of the inactive leg in the cast thin.

Because of the positive effects of physical stress on bone deposition, weight-bearing exercise is effective in preventing and treating osteoporosis, which is the loss of bone density (and hence strength). More than 25 million people in the United States suffer from this debilitating condition. Although osteoporosis is most commonly a problem for postmenopausal women, it can occur in younger people as a result of malnutrition. For example, the condition known as female athlete triad includes eating disorders, cessation of menstrual cycling, and osteoporosis. These are interactive conditions in which the eating disorder and excessive training lead to malnutrition that can result in endocrine disruption and osteoporosis. Excessive training and malnutrition can lead to bone loss in males as well.

Bones develop from connective tissues

Bones are divided into two types on the basis of how they develop. **Membranous bone** forms on a scaffold of connective tissue membrane. **Cartilage bone** forms first as a cartilaginous structure resembling the future mature bone, then gradually hardens, or ossifies, to become bone. The outer bones of the skull are membranous bones; the bones of the limbs are cartilage bones.

Cartilage bones can grow throughout the ossification process. The long bones of the legs and arms, for example, ossify first at the centers and later at each end (**Figure 48.17**). Growth can continue until these areas of ossification join. The membranous bones forming the skull cap grow until their edges meet.

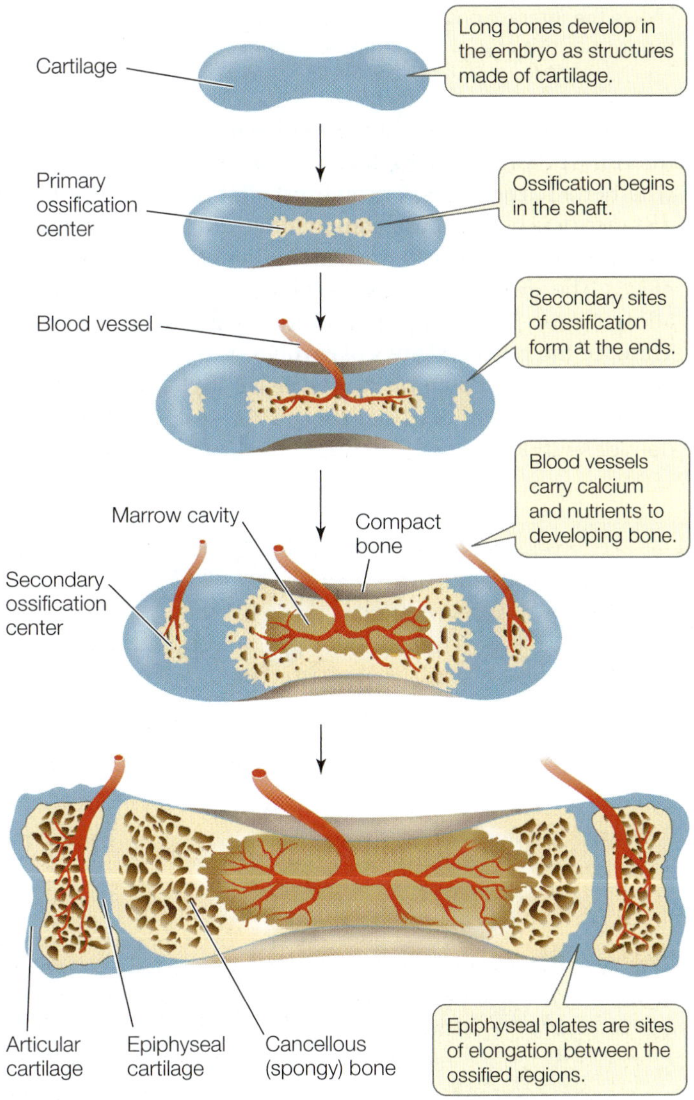
Cartilage
Long bones develop in the embryo as structures made of cartilage.
Primary ossification center
Ossification begins in the shaft.
Blood vessel
Secondary sites of ossification form at the ends.
Marrow cavity
Compact bone
Blood vessels carry calcium and nutrients to developing bone.
Secondary ossification center
Articular cartilage
Epiphyseal cartilage
Cancellous (spongy) bone
Epiphyseal plates are sites of elongation between the ossified regions.

48.17 The Growth of Long Bones In the long bones of human limbs, ossification occurs first at the centers and later at each end.

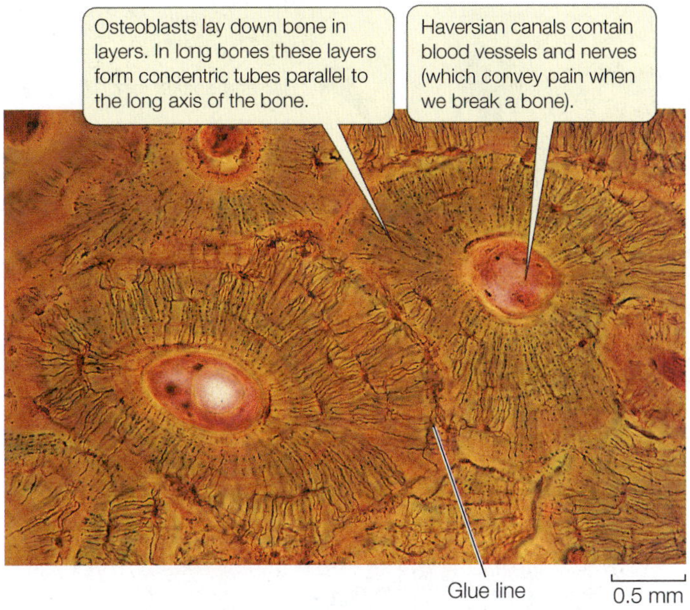
Osteoblasts lay down bone in layers. In long bones these layers form concentric tubes parallel to the long axis of the bone.
Haversian canals contain blood vessels and nerves (which convey pain when we break a bone).
Glue line
0.5 mm

48.18 Most Compact Bone Is Composed of Haversian Systems A micrograph of a section of a long bone shows Haversian systems with their central canals. Glue lines separate Haversian systems.

The soft spot on the top of a baby's head (the fontanelle) is the point at which the skull bones have not yet joined.

The structure of bone may be **compact** (solid and hard) or **cancellous** (having numerous internal cavities that make it appear spongy, although it is rigid). The architecture of a specific bone depends on its position and function, but most bones have both compact and cancellous regions. The shafts of the long bones of the limbs, for example, are cylinders of compact bone surrounding central cavities that contain the bone marrow, where the cellular elements of the blood are made. The ends of the long bones are cancellous (see Figure 48.17). Cancellous bone is lightweight because of its numerous cavities, but it is also strong because its internal meshwork constitutes a support system. It can withstand considerable forces of compression. The rigid, tubelike shaft of compact bone can withstand compression and bending forces. Architects and nature alike use hollow tubes as lightweight structural elements.

Most of the compact bone in mammals is called Haversian bone because it is composed of structural units called **Haversian systems** (**Figure 48.18**). Each Haversian system is a set of thin, concentric bony cylinders, between which are the osteocytes in their lacunae. Through the center of each Haversian system runs a narrow canal containing blood vessels and nerves. Adjacent Haversian systems are separated by boundaries called glue lines. Haversian bone is resistant to fracturing because cracks tend to stop at glue lines.

Bones that have a common joint can work as a lever

Muscles and bones work together around **joints**, where two or more bones come together. Different kinds of joints allow motion in different directions (**Figure 48.19**), but muscles can exert force in only one direction. Therefore muscles create

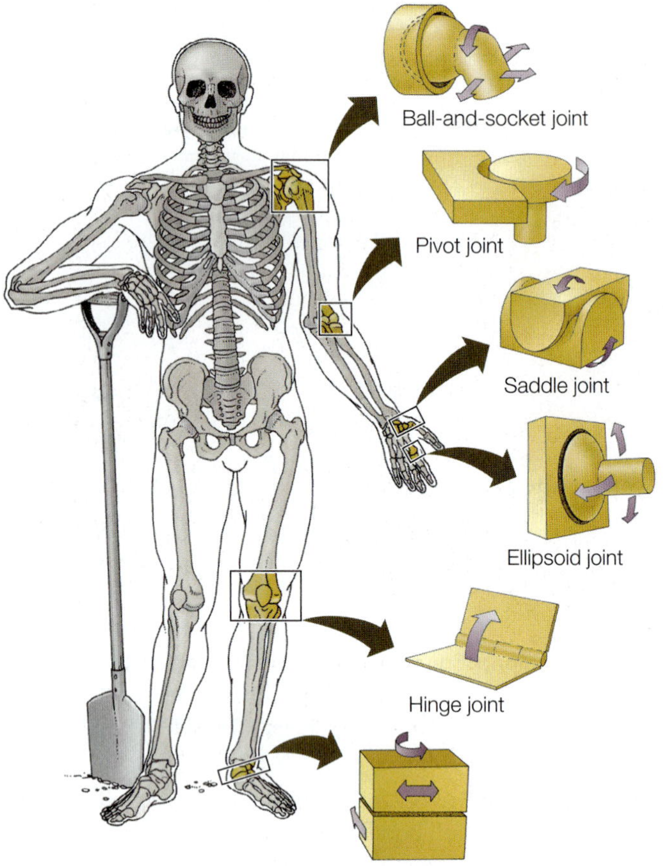

Ball-and-socket joint
Pivot joint
Saddle joint
Ellipsoid joint
Hinge joint
Plane joint

48.19 Types of Joints The designs of joints are similar to mechanical counterparts and enable a variety of movements.

Go to Activity 48.3 Joints
Life10e.com/ac48.3

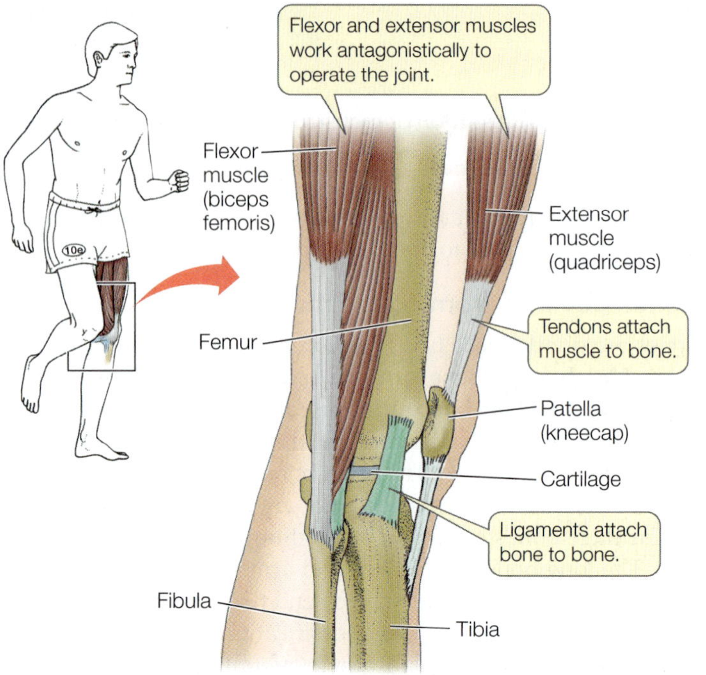

Flexor and extensor muscles work antagonistically to operate the joint.
Flexor muscle (biceps femoris)
Extensor muscle (quadriceps)
Tendons attach muscle to bone.
Femur
Patella (kneecap)
Cartilage
Ligaments attach bone to bone.
Fibula
Tibia

48.20 Joints, Ligaments, and Tendons A side view of the knee shows the interactions of muscle, bone, cartilage, ligaments, and tendons at this crucial and vulnerable human joint.

movement around joints by working in antagonistic pairs: when one muscle contracts, the other relaxes. When both contract, the joint becomes rigid (which is important for maintaining posture, for example).

With respect to a particular joint, such as the knee, we refer to the muscle that bends, or flexes, the joint as the **flexor**, and the muscle that straightens, or extends, the joint as the **extensor**. The bones that meet at the joint are held together by **ligaments**, which are flexible bands of connective tissue. Other straps of connective tissue, called **tendons**, attach the muscles to the bones (Figure 48.20). In many kinds of joints, only the tendon spans the joint, sometimes moving over the surfaces of the bones like a rope over a pulley. The tendon of the quadriceps muscle traveling over the knee joint is what is tapped to elicit the knee-jerk reflex (see Figure 45.18).

Bones constitute a system of levers that are moved around joints by the muscles. A lever has an effort arm and a load arm that work around a fulcrum (pivot). The length ratio of the two arms determines whether a particular lever can exert a lot of force over a short distance or is better at translating force into large or fast movements. Compare the jaw and knee joints, for example (Figure 48.21). The effort arm of the jaw is long relative to the load arm, allowing the jaw to apply great force over a small distance. Think of the powerful jaws of carnivores that can easily crack bones. The effort arm of the lower leg, by contrast, is short relative to the load arm, so you can run fast, jump high, and deliver swift kicks.

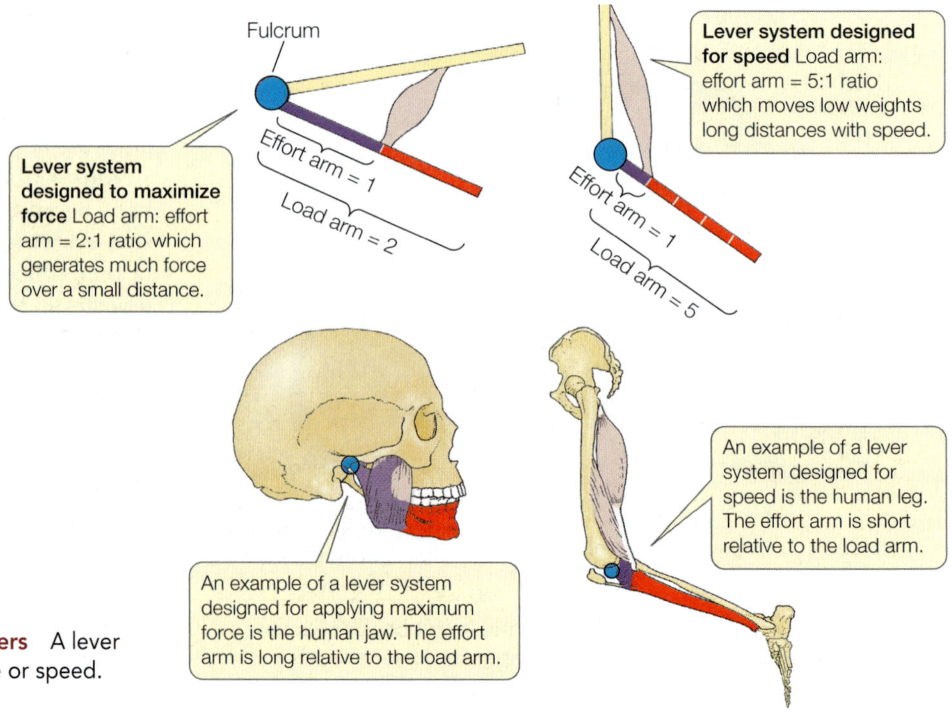

Fulcrum
Effort arm = 1
Load arm = 2
Lever system designed to maximize force Load arm: effort arm = 2:1 ratio which generates much force over a small distance.
Lever system designed for speed Load arm: effort arm = 5:1 ratio which moves low weights long distances with speed.
Effort arm = 1
Load arm = 5
An example of a lever system designed for applying maximum force is the human jaw. The effort arm is long relative to the load arm.
An example of a lever system designed for speed is the human leg. The effort arm is short relative to the load arm.

48.21 Bones and Joints Work Like Systems of Levers A lever system can be designed for maximizing either force or speed.

RECAP 48.3

Muscles can only contract and relax; to achieve organized movement, they must pull against rigid structures—other muscles, hydrostatic skeletons, exoskeletons, or endoskeletons.

- How do the muscles and fluid-filled body cavity of an earthworm interact to enable the animal to move? **See p. 999 and Figure 48.14**
- In terms of levers, explain how specific joints can produce maximum force versus maximum speed. **See p. 1002 and Figure 48.21**

How can the kangaroo increase its speed fivefold without increasing its metabolic expenditure?

ANSWER

In kangaroos, as in other vertebrates, the muscles used to jump are attached to bones by tendons. Tendons are elastic. The kangaroo's tendons stretch when it lands, and their recoil helps power the next jump—similar to the action of a pogo stick. To move faster, the kangaroo simply increases the length of its stride, thereby increasing both the stretch on its tendons each time it lands and the magnitude of the recoil at the initiation of each jump.

CHAPTER SUMMARY 48

48.1 How Do Muscles Contract?

- **Skeletal muscle** consists of bundles of **muscle fibers**. Each skeletal muscle fiber is a large cell containing multiple nuclei.
- Skeletal muscles contain numerous **myofibrils**, which are bundles of **actin** and **myosin** filaments. The regular, overlapping arrangement of the actin and myosin filaments into **sarcomeres** gives skeletal muscle its striated appearance. **Review Figure 48.1, ACTIVITY 48.1**
- The changes in the banding patterns of sarcomeres led to the **sliding filament model** of muscle contraction. **Review Figure 48.2**
- The molecular mechanism of muscle contraction involves the binding of the globular heads of myosin molecules to actin. **Review Figures 48.3, 48.6, ANIMATED TUTORIAL 48.1**
- All the fibers activated by a single motor neuron constitute a **motor unit**. Each nerve ending of the motor neuron forms a synapse with the muscle cell membrane. Action potentials spread across the muscle cell membrane and through the **T tubules**, causing Ca^{2+} to be released from the **sarcoplasmic reticulum**. **Review Figure 48.5, ACTIVITY 48.2**
- Ca^{2+} binds to **troponin** and changes its conformation, pulling the **tropomyosin** strands away from the myosin-binding sites on the actin filament. The muscle fiber continues to contract until the Ca^{2+} is returned to the sarcoplasmic reticulum. **Review Figure 48.6**
- **Cardiac muscle** cells are striated, uninucleate, branching, and electrically connected by gap junctions, so that action potentials spread rapidly throughout sheets of cardiac muscle and cause coordinated contractions.
- **Smooth muscle** provides contractile force for internal organs. Smooth muscle cells respond to stretch and to neurotransmitters from the autonomic nervous system. **Review Figure 48.8, ANIMATED TUTORIAL 48.2**

48.2 What Determines Muscle Performance?

- In skeletal muscle, a single action potential causes a minimum unit of contraction called a **twitch**. Twitches occurring in rapid succession can be summed to achieve sustained tension, known as **tetanus**. **Review Figure 48.10**
- **Slow-twitch fibers** facilitate extended, aerobic work; **fast-twitch fibers** generate maximum forces for short periods of time. The ratio of slow-twitch to fast-twitch fibers in the muscles of an individual is largely genetically determined. **Review Figure 48.11**
- The force that a muscle fiber can produce depends on its initial state of extension or contraction. **Review Figure 48.12**
- Anaerobic exercise stimulates the enlargement of muscle fibers through production of new microfilaments. Aerobic exercise stimulates greater oxidative capacity of muscle fibers.
- Muscle performance depends on a supply of ATP. **Review Figure 48.13**

48.3 How Do Skeletal Systems and Muscles Work Together?

- **Skeletal systems** provide supports against which muscles can pull.
- **Hydrostatic skeletons** are fluid-filled body cavities that can be squeezed by muscles. **Review Figure 48.14**
- **Exoskeletons** are hardened outer surfaces to which internal muscles are attached.
- **Endoskeletons** are internal systems of rigid rodlike, platelike, and tubelike supports, consisting of **bone** and **cartilage** to which muscles are attached. **Review Figure 48.15**
- Bone is continually remodeled by **osteoblasts**, which lay down new bone, and **osteoclasts**, which erode bone. **Review Figure 48.16**
- Bones develop from connective tissue membranes (**membranous bone**) or from cartilage (**cartilage bone**) through ossification. **Review Figure 48.17**
- Bone can be **compact** (solid and hard) or **cancellous** (containing numerous internal spaces). Most of the compact bone of mammals is composed of **Haversian systems**. **Review Figure 48.18**
- **Joints** enable muscles to power movements in different directions. Muscles and bones work together around joints as systems of levers. **Review Figures 48.19, 48.21, ACTIVITY 48.3**
- **Tendons** connect muscles to bones; **ligaments** connect bones to one another. **Review Figure 48.20**

Go to the Interactive Summary to review key figures, Animated Tutorials, and Activities
Life10e.com/is48

CHAPTER**REVIEW**

REMEMBERING

1. The role of Ca^{2+} in the control of muscle contraction is to
   a. cause depolarization of the T tubule system.
   b. change the conformation of troponin, thus exposing myosin-binding sites.
   c. change the conformation of myosin heads, thus causing microfilaments to slide past each other.
   d. bind to tropomyosin and break actin–myosin cross-bridges.
   e. block the ATP-binding site on myosin heads, enabling muscles to relax.

2. Fifteen minutes into a 10-kilometer run, what is the major energy source of the leg muscles?
   a. Preformed ATP
   b. Glycolysis
   c. Oxidative metabolism
   d. Pyruvate and lactate
   e. High-protein drink consumed right before the race

3. Which statement about skeletal muscle contraction is *not* true?
   a. A single action potential at the neuromuscular junction is sufficient to cause a muscle to twitch.
   b. Once maximum muscle tension is achieved, no ATP is required to maintain that level of tension.
   c. An action potential in the muscle cell activates contraction by releasing Ca^{2+} into the sarcoplasm.
   d. Summation of twitches leads to a graded increase in the tension that can be generated by a single muscle fiber.
   e. The tension generated by a muscle can be varied by controlling how many of its motor units are active.

4. Which statement about the structure of skeletal muscle is true?
   a. The light bands of the sarcomere are the regions where actin and myosin filaments overlap.
   b. When a muscle contracts, the A bands of the sarcomere lengthen.
   c. The myosin filaments are anchored in the Z lines.
   d. When a muscle contracts, the H zone of the sarcomere shortens.
   e. The sarcoplasm of the muscle cell is contained within the sarcoplasmic reticulum.

5. Insects can beat their wings at exceptionally high frequencies because
   a. their wing muscles have mostly fast-twitch fibers.
   b. their motor neurons can fire action potentials at a very high frequency.
   c. their wings have exoskeletal supports.
   d. their wing muscles have extensive sarcoplasmic reticulum that cycles Ca^{2+} very fast.
   e. their wing muscles can generate a rapid oscillation of contraction that is asynchronous with motor neuron firing.

UNDERSTANDING & APPLYING

6. Sarcomeres can shorten only by less than 50 percent of their resting length. Yet muscles cause movements of a wide range of magnitudes—compare the range of movement of a toe and a leg. What are two design features of muscles and skeletons that can maximize a muscle's ability to produce a wide range of movements of an appendage?

7. If an adolescent breaks a leg bone close to the ankle joint, after the break heals, that leg may not grow as long as the other one. Why?

8. Stand with your arms held out at right angles to your body for as long as you can. Which will fatigue first: your shoulders or your legs? Why? What does this tell you about the muscles of your shoulders and legs?

ANALYZING & EVALUATING

9. A single action potential in a muscle cell lasts about 2 milliseconds, but the single muscle twitch it generates reaches its peak after about 30 msec, and does not return to resting level until about 150 msec after the action potential. Explain three factors that explain the time difference between the action potential and the twitch.

10. Malignant hyperthermia is a rare but frequently lethal condition stimulated by certain anesthetics. Suddenly the individual's muscles become rigid, heart rate shoots up, and body temperature rises rapidly. Those at risk for malignant hyperthermia have a mutation in the gene that codes for the ryanodine receptor. The mutation results in excessive opening of the Ca^{2+} channels in the sarcoplasmic reticulum. Knowing this cause of the condition, how can you explain all of its manifestations—temperature, heart rate, and muscle tension?

Go to BioPortal at **yourBioPortal.com** for Animated Tutorials, Activities, LearningCurve Quizzes, Flashcards, and many other study and review resources.

49 Gas Exchange

CHAPTER OUTLINE

Proboscis as Snorkel Elephants can cross deep bodies of water either swimming or by walking on the bottom—whichever allows it to keep its trunk above water.

ELEPHANT TRUNKS HAVE MANY USES. They can pluck leaves, pull up plants, and pick up peanuts and tree trunks; they suck in water to squirt into the mouth for drinking or to spray over the body for cooling; they smell the air around them; they communicate by touch; and mothers can wallop their calves when they misbehave.

Of course, elephants also use their trunks to breathe, and it is a useful appendage indeed when they go into deep water—for example, to cross rivers. Whether the elephant swims or walks on the bottom, it uses its trunk as a snorkel. As long as the tip of the trunk is above water, the elephant can breathe.

You may be familiar with diving using a snorkel about a foot long. When you want to dive deeper, you hold your breath. Water is heavy, and as you swim deeper, water pressure increases and presses on your body. The air in your lungs is compressible, so as you swim deeper, the volume of your lungs decreases.

Think what would happen if you tried to suck air through a long snorkel in deep water. Because the snorkel would be open to the air above, the pressure in your lungs would be the same as at the water surface, but the surrounding water would be pressing on your body. To expand your lungs, you would have to exert enough force to push against the surrounding water. Not so easy! We solve this problem with scuba equipment: by breathing air from a pressurized tank with a regulator valve, we keep the air in our lungs at the same pressure as the surrounding water.

The chest of a snorkeling elephant can be 3 meters below the surface—and for every 3 meters, water pressure increases by 0.3 atmospheres (about 5 pounds per square inch, PSI). Even though the elephant's respiratory muscles are strong enough to work against this pressure, there is still a serious problem for the blood vessels lining the cavity in which the lungs are suspended. The pressure difference across the walls of those blood vessels is the difference between the pressure in the vessels (blood pressure plus the compressive water pressure) and the air pressure at the water surface (where the trunk opens to the air). This difference is at least 5 PSI—a pressure that in other mammals would burst small blood vessels and fill the chest cavity with blood.

?

How do elephants avoid damage to the blood vessels of their thoracic cavities when they snorkel?

See answer on p. 1022.

49.1 What Physical Factors Govern Respiratory Gas Exchange?

The **respiratory gases** that animals must exchange are oxygen (O_2) and carbon dioxide (CO_2). Cells need to obtain O_2 from the environment to produce an adequate supply of ATP by cellular respiration (see Chapter 9). CO_2 is an end product of cellular respiration, and it must be removed from the body to prevent toxic effects. Gas exchange systems of animals consist of (1) specialized body surface areas where these gases can move between the body and the environment, and (2) mechanisms that ventilate the environmental side of those surfaces with air or water and perfuse the internal side of those surfaces with extracellular fluids.

Diffusion is the only means by which respiratory gases are exchanged between an animal's internal body fluids and the outside medium (air or water). There are no active transport mechanisms to move respiratory gases across biological membranes. Because diffusion is a physical process, knowing what physical factors influence rates of diffusion helps us understand the diverse adaptations of gas exchange systems. (You may want to review the discussion about the physical nature of diffusion in Section 6.3.)

Diffusion of gases is driven by partial pressure differences

Diffusion results from the random motion of molecules, and the net movement of molecules within a medium such as air or water is down their concentration gradient. Concentrations in solutions are simply the amount of solute per volume of solution. Concentrations of gases, however, vary with pressure because gases are compressible. For example, there are twice as many gas molecules in a liter of gas at 2 atmospheres of pressure as there are in a liter of gas at 1 atmosphere of pressure.

Biologists express the concentrations of different gases in a mixture as the **partial pressures** of those gases. To calculate the partial pressure of a gas such as oxygen in a mixture of gases such as air, we have to know the total pressure. In most cases, for an air-breathing animal the total pressure is atmospheric pressure. At sea level, atmospheric pressure is about 760 millimeters of mercury (mm Hg), depending on the weather. Because dry air is 20.9 percent O_2, the partial pressure of oxygen (P_{O_2}) at sea level is 20.9 percent of 760 mm Hg, or about 159 mm Hg. If two gas mixtures are separated by a membrane permeable to O_2, O_2 will diffuse from the mixture where its partial pressure is higher to the mixture where its partial pressure is lower.

To calculate the concentration of a gas in a solution, we have to know the solubility of that gas in that particular solvent. The amount of a gas in a liquid depends both on its partial pressure in the gas phase in contact with the liquid *and* on its solubility in that liquid. The bottom line is that diffusion of a gas between the gas phase and the liquid phase is a function of its partial pressures in those two phases; the gas diffuses from the phase with the higher partial pressure to the phase with the lower partial pressure, and at equilibrium the partial pressures in the two phases are equal. However, the *amount* of the gas that can be contained by the liquid depends on the solubility of that gas in that liquid. Furthermore, the solubility of a gas in a particular liquid can vary widely depending on conditions. What follows is a practical illustration of these facts.

Solubility of a gas in a liquid, such as oxygen in water, is a function of temperature—solubility is higher at low temperatures. So if we have similar containers of water in equilibrium with air, but at different temperatures, the concentrations of oxygen in these water containers will be different (less O_2 in the warmer ones), but the partial pressures of oxygen will be the same. Thus for water-breathing animals, the warmer the water is, the less O_2 there is per liter of water. The important point is that for a gas in solution, its concentration is not the same as its partial pressure, but partial pressures are what drive diffusion. Thus in our continuing discussions of respiratory gas exchange, we will always use partial pressure rather than concentration when referring to the amount of gas in solution.

Fick's law applies to all systems of gas exchange

Whether in air or water, the diffusion rates of respiratory gases depend on their partial pressure gradients and on other factors that can be described quantitatively with a simple equation called **Fick's law of diffusion**. All environmental variables that limit respiratory gas exchange and all adaptations that maximize respiratory gas exchange are reflected in one or more components of this equation. Fick's law is written as

$$Q = DA\frac{P_1 - P_2}{L}$$

where

- Q is the rate at which a gas such as O_2 diffuses between two locations
- D is the diffusion coefficient, which is a characteristic of the diffusing substance, the medium, and the temperature. For example, perfume has a higher D than motor oil vapor, and all substances diffuse faster at higher temperatures and faster in air than in water. Temperature is not expressed explicitly in Fick's law because the diffusion coefficient is usually determined at room temperature (about 20°C)
- A is the area across which the gas is diffusing
- P_1 and P_2 are the partial pressures of the gas at the two locations
- L is the path length, or distance, between the two locations
- $(P_1 - P_2)/L$ is a partial pressure gradient.

The strict dependence of animals on diffusion for gas exchange with their environments has selected for various adaptations that maximize Q, many of which we will describe in this chapter. Animals can maximize D for respiratory gases by using air rather than water as their gas exchange medium whenever possible. All other adaptations for maximizing respiratory gas exchange must influence the surface area (A) for gas exchange or the partial pressure gradient across that surface area.

Air is a better respiratory medium than water

The slow diffusion of O_2 molecules in water affects both air- and water-breathing animals. Eukaryotic cells carry out cellular respiration in their mitochondria, which are located in the cytoplasm—an aqueous medium. Cells are bathed in extracellular fluid—also an aqueous medium. In addition, all respiratory surfaces must be protected from drying out by a thin film of fluid through which O_2 must diffuse. Even in air-breathing animals, the slow rate of O_2 diffusion in water limits the efficiency of O_2 distribution from gas exchange surfaces to the sites of cellular respiration.

Diffusion of O_2 in water is so slow that even animal cells with low rates of metabolism cannot function more than a few millimeters away from a good source of environmental O_2. Therefore there are severe size and shape limits on the many species of invertebrates that lack internal systems for distributing O_2. Most of these species are very small, but some have grown larger by evolving a flat, thin body with a large external surface area (**Figure 49.1A**). Others have a very thin body built around a central cavity through which water can circulate (**Figure 49.1B**). A critical factor enabling larger, more complex animal bodies has been the evolution of specialized respiratory systems with large surface areas that are highly permeable to respiratory gases (**Figure 49.1C**).

(A) *Pseudobiceros* sp.
Gills
Central cavity
Channels
(C) *Ambystoma mexicanum*
(B) *Callyspongia plicifera*

49.1 Keeping in Touch with the Medium (A) No cell in the leaf-like body of this marine flatworm is more than a millimeter away from seawater. (B) Sponges have body walls perforated by many channels, which allow water to flow between the outside world and a central cavity. No cell in the sponge is more than a millimeter away from seawater. (C) A feathery fringe of gills on this aquatic salamander provides a large surface area for gas exchange. Blood circulating through the gills comes into close contact with the respiratory medium.

O_2 can be obtained more easily from air than from water for several reasons:

- The O_2 content of air is much higher than the O_2 content of an equal volume of water. The maximum O_2 content of a bubbling stream in equilibrium with air is less than 10 milliliters of O_2 per liter of water. The O_2 content of the air over the stream is about 200 ml of O_2 per liter of air.
- O_2 diffuses about 8,000 times more rapidly in air than in water. That is why the O_2 content of a stagnant pond can be zero only a few millimeters below the surface.
- An animal has to work (expend energy) to ventilate its gas exchange surfaces with water or air. More energy is required to move water than air because water is 800 times denser than air and about 50 times more viscous.

You can appreciate how important these facts were for the evolutionary transition of life to the terrestrial environment, because they meant that there were fewer constraints on the evolution of higher metabolic rates.

High temperatures create respiratory problems for aquatic animals

Animals that use water for their respiratory exchange medium are in a double bind when environmental temperatures rise. Most water-breathing animals are ectotherms—their body temperatures are closely tied to the temperature of the water around them. As the water temperature rises, an ectotherm's body temperature and metabolic rate rise (see Figure 40.8). Thus water breathers need more O_2 as the water gets warmer (**Figure 49.2**). But as mentioned above, warm water holds less dissolved gas than cold water does (think of the gases that escape when you open a warm bottle of soda). In addition, since a water-breathing animal performs work to move water across its gas exchange surfaces, it must expend more energy to breathe as water temperature rises. Therefore, as water temperature goes up, a water-breathing animal must extract more and more O_2 from an environment that is increasingly O_2 deficient, and a lower percentage of that O_2 is available to support activities other than breathing.

O_2 availability decreases with altitude

Just as a rise in water temperature reduces the supply of O_2 available to water-breathing animals, an increase in altitude reduces the O_2 supply for air breathers. At all altitudes, O_2 makes up 20.9 percent of the dry air; however, as you go up in altitude, the total amount of gas per unit of volume decreases, as reflected in the atmospheric pressure. For example, at 5,800 meters, atmospheric pressure is only half what it is at sea level,

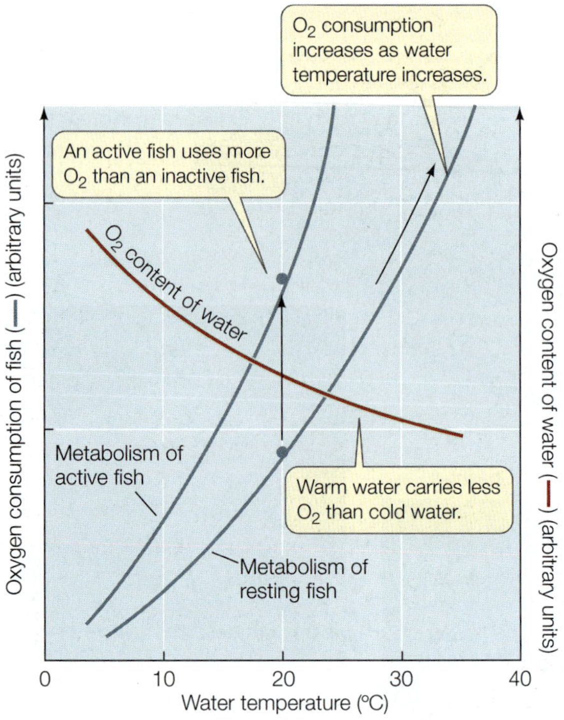
O2 consumption increases as water temperature increases.
An active fish uses more O2 than an inactive fish.
O2 content of water
Metabolism of active fish
Warm water carries less O2 than cold water.
Metabolism of resting fish
Oxygen consumption of fish (—) (arbitrary units)
Oxygen content of water (—) (arbitrary units)
0
10
20
30
40
Water temperature (°C)

49.2 The Double Bind of Water Breathers Fishes need more O_2 when the water is warmer, but warm water carries less O_2 than cold water.

so the P_{O_2} at that altitude is only about 80 mm Hg. At the summit of Mount Everest (8,850 m), P_{O_2} is only about 50 mm Hg—roughly one-third what it is at sea level.

Because the movement of O_2 across respiratory gas exchange surfaces and into the body depends on diffusion, its rate of movement depends on the P_{O_2} difference between the air and the body fluids. Therefore the drastically reduced P_{O_2} in the air at high altitudes constrains O_2 uptake. Because of this, mountain climbers attempting the highest peaks usually breathe O_2 from pressurized bottles.

CO_2 is lost by diffusion

Respiratory gas exchange is a two-way process: CO_2 diffuses out of the body as O_2 diffuses in. The direction and rate of diffusion of the respiratory gases across the exchange surfaces depend on the partial pressure gradients of the gases. The partial pressure gradients of O_2 and CO_2 across these exchange surfaces are quite different. The amount of CO_2 in the atmosphere is extremely low (0.03%), so for air-breathing animals there is always a large concentration gradient for diffusion of CO_2 from the body to the environment. Whereas the partial pressure gradient for O_2 decreases with increasing altitude, the gradient driving CO_2 out of the body hardly changes. The partial pressure of carbon dioxide (P_{CO_2}) in the atmosphere is close to zero both at sea level and atop Mount Everest.

In general, getting rid of CO_2 is not a problem for water-breathing animals because CO_2 is much more soluble in water than is O_2. Even in stagnant water, where the P_{CO_2} is higher than in moving water, the lack of O_2 becomes a problem for an animal long before CO_2 exchange difficulties arise.

RECAP 49.1

Respiratory gases are exchanged only by diffusion. Air is a better respiratory medium than water because a given volume of air has more O_2 than the same volume of water. O_2 diffuses faster in air than in water, and less work is required to move air over respiratory exchange surfaces.

- Describe how the variables in Fick's law of diffusion relate to respiratory systems. **See p. 1006**
- Why does a rise in water temperature create a double-bind situation for water-breathing animals? **See p. 1007 and Figure 49.2**
- Explain the concept of partial pressures of gases and how it relates to diffusion rates of O_2 and CO_2 at different altitudes. **See pp. 1006 and 1007–1008**

Now that we have discussed the physical factors that influence diffusion rates of respiratory gases between animals and their environments, let's look at some of the adaptations that have evolved for maximizing respiratory gas exchange.

49.2 What Adaptations Maximize Respiratory Gas Exchange?

As you might expect from the components of Fick's law of diffusion, adaptations to maximize respiratory gas exchange can be categorized as those that:

- Increase the surface area for gas exchange (*A*)
- Maximize the partial pressure difference driving diffusion ($P_1 - P_2$)
- Minimize the diffusion path length (*L*)
- Minimize the diffusion that takes place in an aqueous medium (maximize *D*)

Respiratory organs have large surface areas

A variety of anatomical adaptations maximize the specialized body surface areas (*A*) for respiratory gas exchange. Water-breathing animals generally have gills, and air-breathing animals have tracheae or lungs. **External gills** are highly branched and folded extensions of the body surface that provide a large surface area for gas exchange (**Figure 49.3A**; see also Figure 49.1C). External gills are found in larval amphibians and in the larvae of many insects. Because they consist of thin, delicate tissues, external gills minimize the path length (*L*) traversed by diffusing molecules of O_2 and CO_2. External gills are vulnerable to damage, however, and are tempting morsels for predators, so in many animals protective body cavities for gills have evolved. Such **internal gills** are found in most mollusks and arthropods and in all fishes (**Figure 49.3B**).

Lungs are internal cavities for respiratory gas exchange with air (**Figure 49.3C**). Their structure is quite different from that of gills. Lungs have a large surface area because they are highly divided; and because they are elastic, they can be inflated with air and deflated.

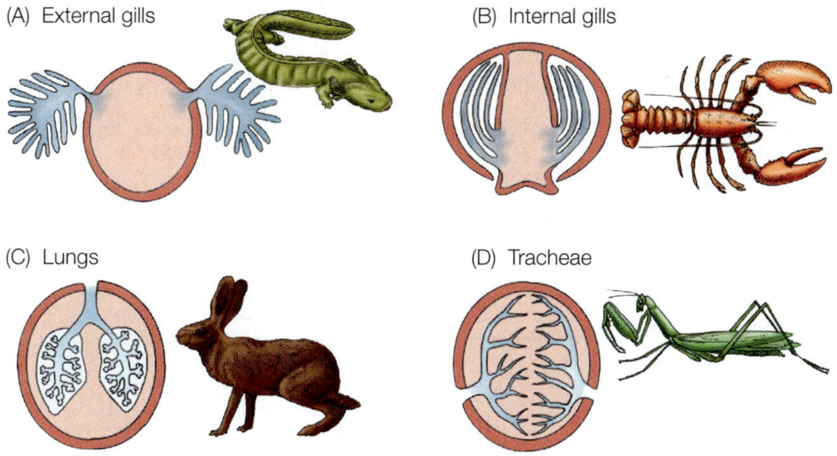
(A) External gills
(B) Internal gills
(C) Lungs
(D) Tracheae

49.3 Gas Exchange Systems Large surface areas (blue in these diagrams) for the diffusion of respiratory gases are common features of animals. External (A) and internal (B) gills are adaptations for gas exchange with water. Lungs (C) and tracheae (D) are organs for gas exchange with air.

Insects have a respiratory gas exchange system consisting of a network of air-filled tubes called **tracheae** that branch through all tissues of the insect's body (Figure 49.3D). The terminal branches of these tubes are so numerous that they have an enormous surface area compared with the external surface area of the insect's body.

Ventilation and perfusion of gas exchange surfaces maximize partial pressure gradients

Partial pressure gradients $[(P_1 - P_2)/L]$ drive diffusion across gas exchange surfaces; the larger the gradient, the greater the rate of gas exchange. These gradients can be maximized in several ways:

- *Minimization of path length*: Very thin tissues in gills and lungs reduce the diffusion path length (L).
- *Ventilation*: Actively moving the external medium over the gas exchange surfaces (i.e., breathing) regularly exposes those surfaces to fresh respiratory medium containing maximum O_2 and minimum CO_2 concentrations. This maximizes the partial pressure gradients.
- *Perfusion*: Actively moving the internal medium (e.g., blood) over the internal side of the exchange surfaces transports CO_2 to those surfaces and O_2 away from them, thus maximizing the partial pressure gradients driving diffusion.

This chapter describes four gas exchange systems. First we will look at the unique gas exchange system of insects. Then we will describe two highly efficient gas exchange systems, fish gills and bird lungs. Lastly we will examine the structure and function of human lungs.

Insects have airways throughout their bodies

The tracheal system that enables insects to exchange respiratory gases extends to all tissues in the insect body. Thus respiratory gases diffuse through air most of the way to and from every cell. The insect respiratory system communicates with the outside environment through gated openings called spiracles in the sides of the abdomen and thorax (Figure 49.4A,B). The spiracles open to allow gas exchange and then close to decrease water loss. They open into tubes called tracheae that branch into even finer tubes, or tracheoles, which end in tiny air capillaries that are the actual gas exchange surfaces (Figure 49.4C). In the insect's flight muscles and other highly active tissues, every mitochondrion is close to an air capillary.

Fish gills use countercurrent flow to maximize gas exchange

The internal gills of fishes are supported by gill arches that lie between the mouth cavity and the protective opercular flaps on

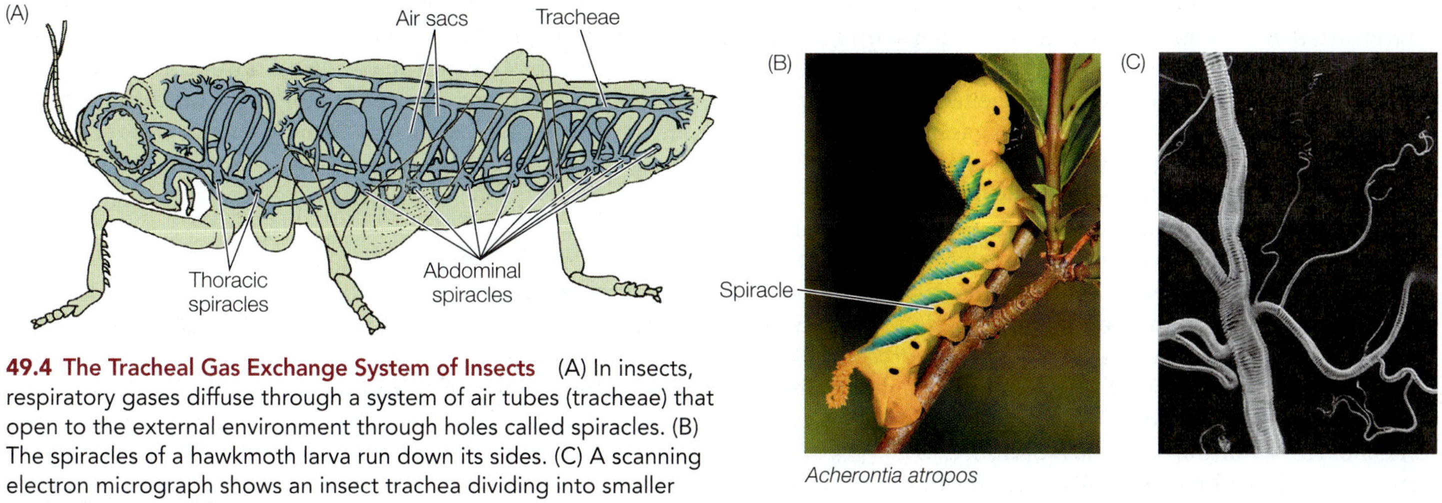
(A)
Air sacs
Tracheae
Thoracic spiracles
Abdominal spiracles
(B)
Spiracle
(C)
Acherontia atropos

49.4 The Tracheal Gas Exchange System of Insects (A) In insects, respiratory gases diffuse through a system of air tubes (tracheae) that open to the external environment through holes called spiracles. (B) The spiracles of a hawkmoth larva run down its sides. (C) A scanning electron micrograph shows an insect trachea dividing into smaller tracheoles and still finer air capillaries.

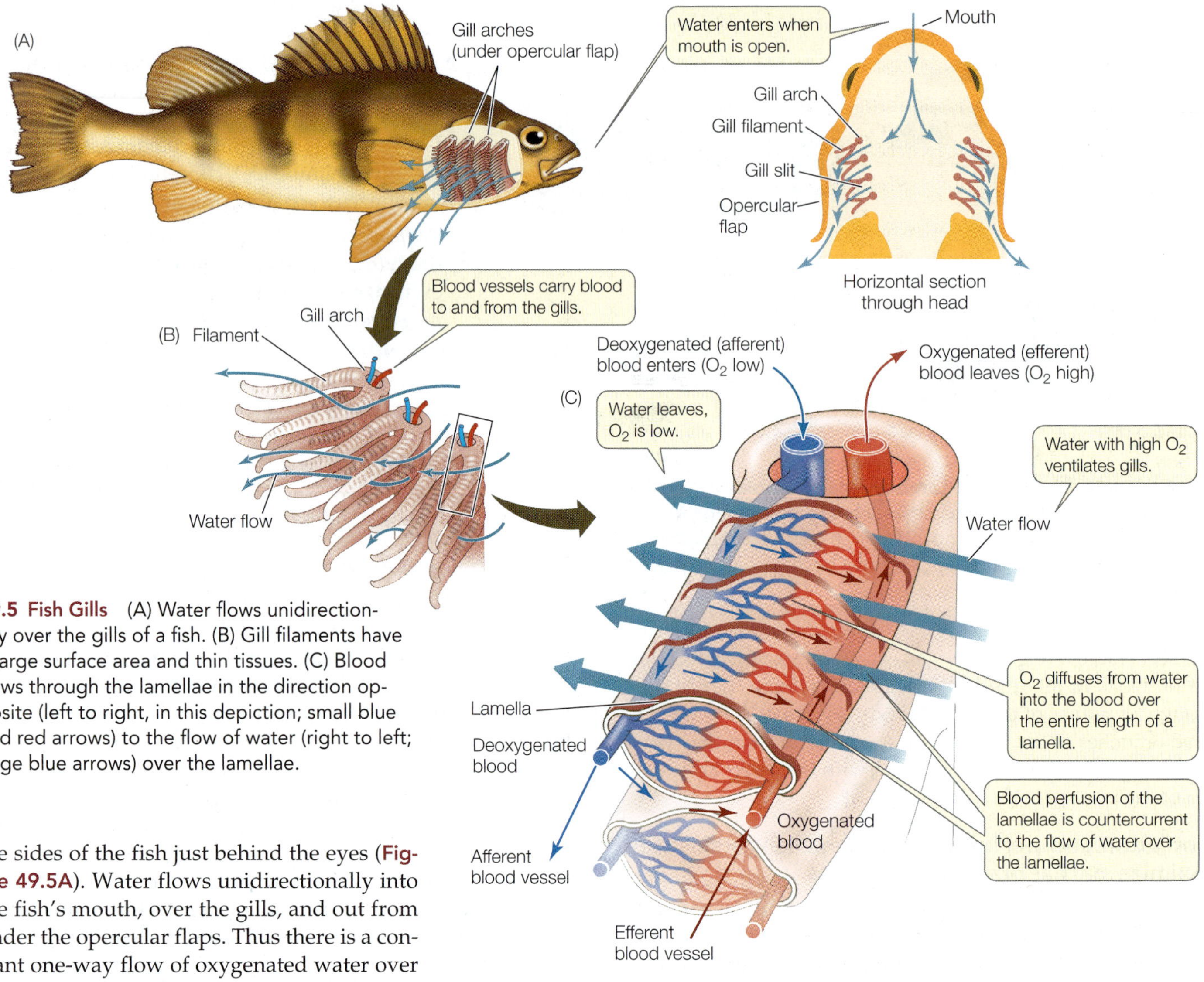
(A)
Gill arches
(under opercular flap)
Water enters when mouth is open.
Mouth
Gill arch
Gill filament
Gill slit
Opercular flap
Horizontal section through head
(B) Filament
Gill arch
Blood vessels carry blood to and from the gills.
Water flow
(C)
Deoxygenated (afferent) blood enters (O_2 low)
Oxygenated (efferent) blood leaves (O_2 high)
Water leaves, O_2 is low.
Water with high O_2 ventilates gills.
Water flow
O_2 diffuses from water into the blood over the entire length of a lamella.
Blood perfusion of the lamellae is countercurrent to the flow of water over the lamellae.
Lamella
Deoxygenated blood
Oxygenated blood
Afferent blood vessel
Efferent blood vessel

49.5 Fish Gills (A) Water flows unidirectionally over the gills of a fish. (B) Gill filaments have a large surface area and thin tissues. (C) Blood flows through the lamellae in the direction opposite (left to right, in this depiction; small blue and red arrows) to the flow of water (right to left; large blue arrows) over the lamellae.

the sides of the fish just behind the eyes (Figure 49.5A). Water flows unidirectionally into the fish's mouth, over the gills, and out from under the opercular flaps. Thus there is a constant one-way flow of oxygenated water over the gills, maximizing the P_{O_2} on the external gill surfaces. On the internal side of the gill membranes, the circulation of blood minimizes the P_{O_2} by sweeping O_2 away as rapidly as it diffuses across.

Gills have an enormous surface area for gas exchange because they are so highly divided. Each gill consists of hundreds of ribbonlike gill filaments (Figure 49.5B). The upper and lower flat surfaces of each gill filament are covered with rows of evenly spaced folds, or lamellae. The lamellae are the actual gas exchange surfaces. Because the lamellae are exceedingly thin, the path length (*L*) for diffusion of gases between blood and water is minimized. The surfaces of the lamellae consist of highly flattened epithelial cells, so the water and the fish's red blood cells are separated by little more than 1 to 2 micrometers.

The flow of blood perfusing the inner surfaces of the lamellae, like the flow of water over the gills, is unidirectional. **Afferent** blood vessels bring deoxygenated blood to the gills, while **efferent** blood vessels take oxygenated blood away from the gills (Figure 49.5C). Blood flows through the lamellae in the direction opposite to the flow of water over the lamellae. This **countercurrent flow** optimizes the P_{O_2} gradient between water and blood, making gas exchange more efficient than it would be in a system using concurrent (parallel) flow (Figure 49.6).

Some fishes, including anchovies, tuna, and certain sharks, ventilate their gills by swimming almost constantly with their mouths open. Most fishes, however, ventilate their gills by means of a two-pump mechanism. The closing and contracting of the mouth cavity pushes water over the gills, and the expansion of the opercular cavity prior to opening of the opercular flaps pulls water over the gills.

These adaptations for maximizing the surface area (*A*) for diffusion, minimizing the path length (*L*) for diffusion, and maximizing the P_{O_2} gradient allow fishes to extract an adequate supply of O_2 from meager environmental sources.

Birds use unidirectional ventilation to maximize gas exchange

Birds are remarkable for their ability to sustain high levels of activity for a long time—for example, on long-distance flights—even at high altitudes where mammals cannot even survive. The first team to climb Mount Everest (8,850 m) was surprised to see birds flying over the mountain when they themselves could barely move without supplemental O_2. Bar-headed geese regularly migrate over Mt. Everest and surrounding peaks, but the highest recorded flight of a bird is from a Ruppell's griffon, a vulture, that was sucked into a jet engine at 11,278 m.

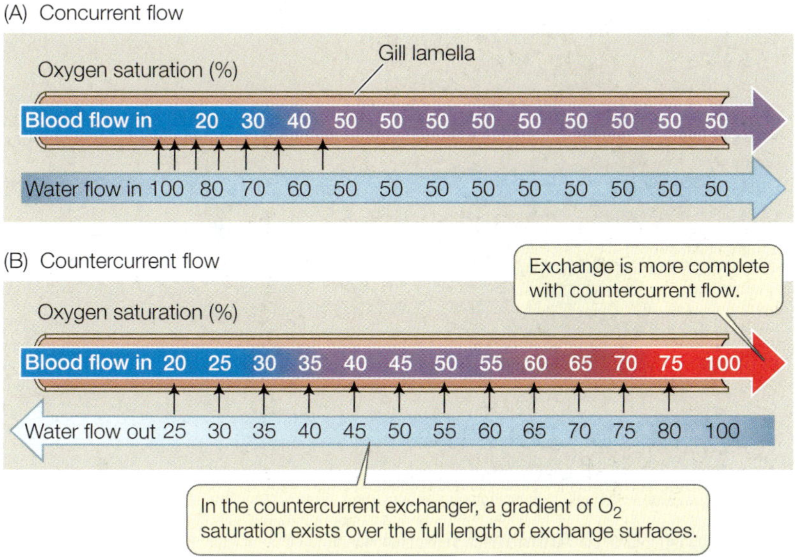
(A) Concurrent flow
Gill lamella
Oxygen saturation (%)
Blood flow in 20 30 40 50 50 50 50 50 50 50 50 50
Water flow in 100 80 70 60 50 50 50 50 50 50 50 50 50
(B) Countercurrent flow
Exchange is more complete with countercurrent flow.
Oxygen saturation (%)
Blood flow in 20 25 30 35 40 45 50 55 60 65 70 75 100
Water flow out 25 30 35 40 45 50 55 60 65 70 75 80 100
In the countercurrent exchanger, a gradient of O_2 saturation exists over the full length of exchange surfaces.

49.6 Countercurrent Flow Enables More Complete Exchange In these models of concurrent and countercurrent gas exchange, the numbers represent the O_2 saturation percentages of blood and water. (A) In a concurrent exchanger, the saturation percentages of blood and water reach equilibrium halfway across the exchange surface. (B) A countercurrent exchanger allows more complete gas exchange because the water is always more O_2-saturated than the blood; thus a gradient of O_2 saturation is maintained.

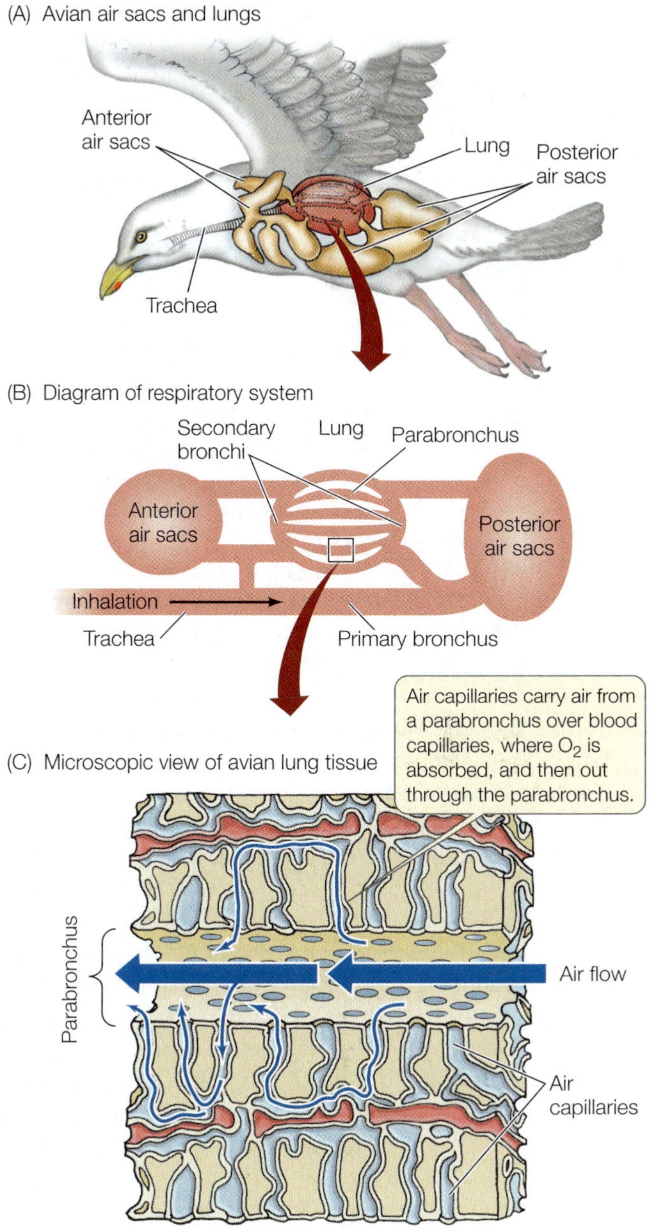
(A) Avian air sacs and lungs
Anterior air sacs
Lung
Posterior air sacs
Trachea
(B) Diagram of respiratory system
Secondary bronchi
Lung
Parabronchus
Anterior air sacs
Posterior air sacs
Inhalation
Trachea
Primary bronchus
(C) Microscopic view of avian lung tissue
Air capillaries carry air from a parabronchus over blood capillaries, where O_2 is absorbed, and then out through the parabronchus.
Parabronchus
Air flow
Air capillaries

49.7 The Respiratory System of a Bird (A) Air sacs and air spaces in the bones are unique to birds. (B) The lung is divided into numerous parabronchi. Primary and secondary bronchi connect the lung to the air sacs. (C) Air flows through bird lungs unidirectionally in parabronchi. Air capillaries, the site of gas exchange, branch off the parabronchi.

Humans cannot survive at such altitudes without supplemental O_2. Yet the lungs of a bird are smaller than the lungs of a similar-sized mammal, and bird lungs expand and contract less during a breathing cycle than do mammalian lungs. Furthermore, bird lungs *are compressed* during inhalation and *expand* during exhalation. How do birds accomplish such remarkable feats of respiratory gas exchange?

The structure of bird lungs allows air to flow unidirectionally through the lungs, rather than bidirectionally through all the same airways, as it does in mammals. Because mammalian lungs are never completely emptied of air during exhalation, there is always some lung volume that is not ventilated with fresh air. The air remaining in lungs and airways after exhalation is called **dead space**. Bird lungs, by contrast, have very little dead space, and the fresh incoming air is not mixed with stale air. In this way, a high P_{O_2} gradient is maintained.

An important and unique feature of the avian respiratory system is its **air sacs**, which occupy much of the body cavity of the bird (Figure 49.7A). The air sacs are interconnected with each other, with the lungs, and with air spaces in some of the bones. The air sacs receive inhaled air, but they are not gas exchange surfaces; they are integrally involved, however, in the flow of air through the lungs. As in other air-breathing vertebrates, air enters and leaves a bird's gas exchange system through the **trachea** (commonly known as the windpipe, and not to be confused with the air-conducting tracheae of insects) (Figure 49.7B). The trachea divides into two smaller airways, the **primary bronchi** (singular *bronchus*). The primary bronchi extend all the way to the posterior air sacs and branch into **secondary bronchi**. The posterior air sacs also have connections to the secondary bronchi. Secondary bronchi divide into tubelike **parabronchi** that run parallel to one another through the lungs.

Branching off the parabronchi are numerous tiny air capillaries (Figure 49.7C). Air flows through the parabronchi and diffuses into the air capillaries, which are the gas exchange surfaces. They are so numerous that they provide an enormous surface area for gas exchange. The parabronchi coalesce into larger bronchi that take the air out of the lungs and into anterior air sacs and back to the trachea. Thus the anatomy of a bird's airways allows air to flow unidirectionally through the lungs: trachea, bronchi, posterior air sacs, parabronchi, anterior air sacs, trachea.

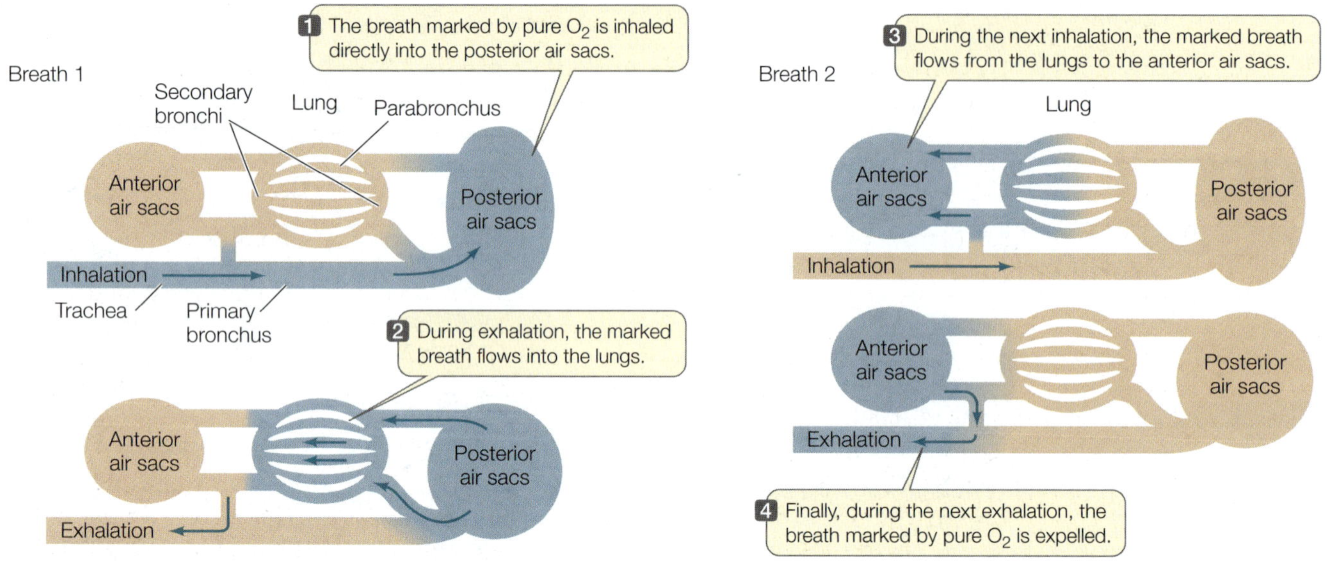
Breath 1
1 The breath marked by pure O_2 is inhaled directly into the posterior air sacs.
Secondary bronchi
Lung
Parabronchus
Anterior air sacs
Posterior air sacs
Inhalation
Trachea
Primary bronchus
2 During exhalation, the marked breath flows into the lungs.
Anterior air sacs
Posterior air sacs
Exhalation
Breath 2
3 During the next inhalation, the marked breath flows from the lungs to the anterior air sacs.
Lung
Anterior air sacs
Posterior air sacs
Inhalation
Anterior air sacs
Posterior air sacs
Exhalation
4 Finally, during the next exhalation, the breath marked by pure O_2 is expelled.

49.8 The Path of Air Flow through Bird Lungs The air a bird takes in by breathing (blue) travels through the lungs in one direction, from the posterior to the anterior air sacs. Each breath of air remains in the system for two breathing cycles.

Go to Animated Tutorial 49.1
Airflow in Birds
Life10e.com/at49.1

The puzzle of how birds breathe was solved by placing small O_2 electronic sensors at different locations in birds' air sacs and airways. The birds were then exposed to pure O_2 for just a single breath, which made it possible to track that particular inhalation. The experiment demonstrated that a single breath remains in a bird's gas exchange system for two cycles of inhalation and exhalation, and that the air sacs work like bellows; inhalation expands the sacs, and exhalation compresses them to maintain a continuous, unidirectional flow of fresh air through the lungs (**Figure 49.8**).

The advantages of the bird gas exchange system are similar to those of fish gills. The air sacs keep fresh air flowing unidirectionally over the gas exchange surfaces without interruption. Thus a bird can supply its gas exchange surfaces with a continuous flow of fresh air that has a P_{O_2} very close to that of the ambient air. Even when the P_{O_2} of the ambient air is only slightly above that of the blood, O_2 can diffuse from air to blood.

Tidal ventilation produces dead space that limits gas exchange efficiency

Lungs evolved in early lungfish as outpocketings of the digestive tract. Although their structure has evolved considerably, lungs remain dead-end sacs in all air-breathing vertebrates except birds (and likely their extinct reptilian ancestors). Because of this, ventilation cannot be constant and unidirectional but must be **tidal**: air flows in and exhaled gases flow out by the same route. Since the lungs and airways can never be completely emptied of air, they always contain dead space. We can easily measure the volumes of air exchanged during breathing, but we have to use an indirect method to measure the dead space contained in the lungs and airways. Measures of dead space are important in assessing lung health and disease.

A spirometer is a device that measures the volume of air breathed in and out (**Figure 49.9**). Using a human as an example, the amount of air that moves in and out per breath when at rest is called the **tidal volume** (**TV**) (about 500 ml for an average human adult). When we breathe in as much as possible,

RESEARCH**TOOLS**

49.9 Measuring Lung Ventilation A spirometer is a device that measures the volume of air a person breathes through a mouthpiece. The combined tidal volume, inspiratory reserve volume, and expiratory reserve volume are the lungs' vital capacity.

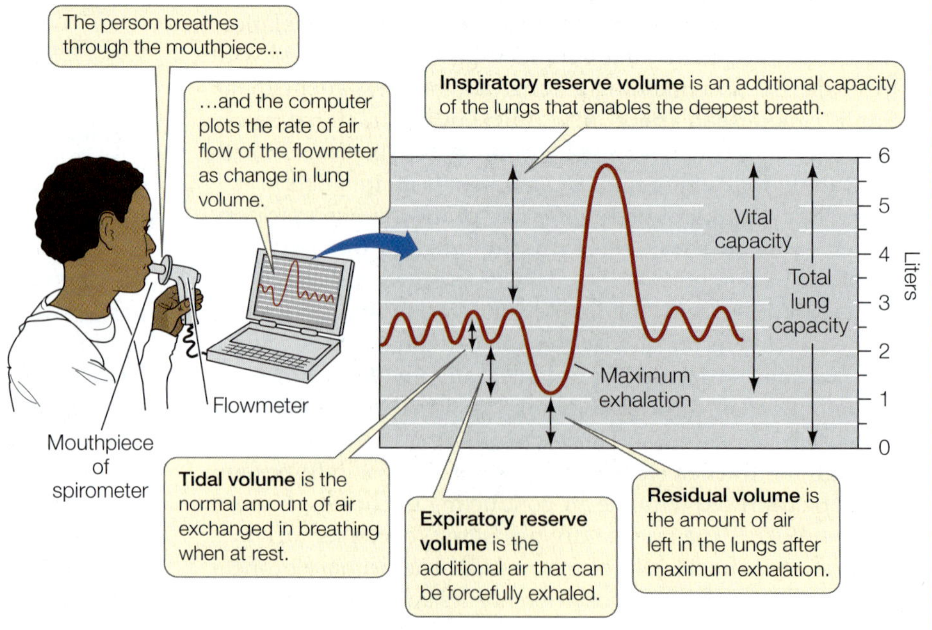
The person breathes through the mouthpiece...
...and the computer plots the rate of air flow of the flowmeter as change in lung volume.
Inspiratory reserve volume is an additional capacity of the lungs that enables the deepest breath.
6
5
4
3
2
1
0
Liters
Vital capacity
Total lung capacity
Maximum exhalation
Flowmeter
Mouthpiece of spirometer
Tidal volume is the normal amount of air exchanged in breathing when at rest.
Expiratory reserve volume is the additional air that can be forcefully exhaled.
Residual volume is the amount of air left in the lungs after maximum exhalation.

the additional volume is the **inspiratory reserve volume** (**IRV**). Conversely, if we forcefully exhale as much air as possible, the additional amount of air expelled is the **expiratory reserve volume** (**ERV**). The maximum capacity for air exchange in one breath, or the **vital capacity** (**VC**), is the sum of TV + IRV + ERV. The vital capacity of an athlete is generally greater than that of a nonathlete, and vital capacity decreases with age because of stiffening of the lung tissue.

Vital capacity is not the entire lung capacity because of the dead space, also called the **residual volume** (**RV**). We can't measure RV directly with the spirometer, but we can measure it indirectly using the helium dilution method. Briefly, a person breathes from a closed reservoir with a known volume of air containing a known amount of helium (He). The helium is not absorbed from the lungs, so it becomes evenly distributed between the lungs and the reservoir as the subject inhales and exhales. Because the fixed amount of He becomes dispersed in a larger volume of air, its concentration decreases. That decrease in He concentration enables us to calculate the subject's **functional residual volume** (**FRV**), which is the ERV + RV. Since we *can* measure the ERV with the spirometer, we can subtract the ERV from the FRV to obtain the RV.

Why is the RV important? Referring to Figure 49.9, you will see that for a normal person the ERV is about 1,000 ml and the RV is 1,000 ml. Thus the FRV is 2,000 ml, but the tidal volume is only 500 ml; this means that the air that reaches the alveoli (the actual gas exchange surfaces) with each breath consists of only 500 ml of fresh air diluted by 2,000 ml of stale air. The maximum P_{O_2} in this mixed air is much below the P_{O_2} of the outside air, and because of the tidal ventilation pattern, the P_{O_2} in the alveoli is steadily dropping during the breathing cycle. The RV is important because it contributes to the FRC and to the dilution of the O_2 in the inhaled air. Any disease or condition that increases the RV (such as emphysema or pulmonary fibrosis) compromises a patient's respiratory ability. Similarly, considering the mixing of fresh air with the FRV, you can understand why reductions in tidal volume can be a problem—and therefore why patients recovering from surgery are encouraged to breathe deeply, even if it hurts.

Offsetting the inefficiencies of tidal breathing, mammalian lungs have some design features to maximize the rate of gas exchange: an enormous surface area and a very short path length for diffusion.

RECAP 49.2

The major adaptations that increase animals' efficiency of respiratory gas exchange are a large surface area for exchange and a maximized partial pressure gradient across that surface.

- Describe three different ways that the partial pressure gradient for O_2 exchange is maximized across fish gills. **See pp. 1009–1010 and Figures 49.5 and 49.6**
- What respiratory adaptations enable birds to fly at extremely high altitudes? **See pp. 1010–1011 and Figures 49.7 and 49.8**
- Explain why residual volume limits the efficiency of tidal breathing. **See pp. 1012–1013 and Figure 49.9**

Despite their limitations, mammalian lungs serve the respiratory needs of mammals, including humans, well. Next we will look at the human respiratory system as an example.

49.3 How Do Human Lungs Work?

Air enters the lungs through the oral cavity or through the nasal passage, which join together in the **pharynx** (**Figure 49.10A**). Below the pharynx, the esophagus conducts food to the stomach, and the trachea conducts air to the lungs. At the beginning of this airway is the **larynx**, or voice box, which houses the vocal cords. The larynx is the lump that you can see or feel on the front of your neck. The trachea is about 2 centimeters in diameter. C-shaped bands of cartilage prevent the thin walls of the trachea from collapsing as air pressure changes during the breathing cycle. If you run your fingers down the front of your neck just below your larynx, you can feel a few of these bands of cartilage.

The trachea branches into two bronchi, one leading to each lung. The bronchi branch repeatedly to generate a treelike structure of progressively smaller airways extending to all regions of the lungs. After four branchings, the cartilage supports disappear, marking the transition to **bronchioles**. After about 16 branchings, the bronchioles are less than a millimeter in diameter, and tiny, thin-walled air sacs called **alveoli** begin to appear. Alveoli are the sites of gas exchange. After the first alveoli there are about six more branchings of the airways that end in clusters of alveoli (**Figure 49.10B**). Because the airways conduct air only to and from the alveoli and do not themselves participate in gas exchange, their volume is dead space.

Human lungs have about 300 million alveoli. Although each alveolus is very small, their combined surface area for diffusion of respiratory gases is about 70 square meters—about one-fourth the size of a basketball court. Each alveolus is made of very thin cells. Between and surrounding the alveoli are networks of capillaries whose walls are also very thin. Thus where capillary meets alveolus, the length of the diffusion path between air and blood is less than 2 micrometers (**Figure 49.10C**).

Diseases of the bronchioles and alveoli are the third leading cause of death in the United States as of 2010. Among these diseases, the most lethal is emphysema, a condition in which inflammation damages and eventually destroys the walls of the alveoli. As a result, the lungs have fewer but larger alveoli, the RV increases, and the lungs lose elasticity. Although genetic factors can contribute to emphysema, the principal cause of the disease is smoking.

Respiratory tract secretions aid ventilation

Mammalian lungs produce two secretions that do not directly influence their gas exchange but do affect the process of ventilation: mucus and surfactant.

Many cells lining the airways produce sticky mucus that captures bits of dirt and microorganisms that are inhaled. Other cells lining the airways have cilia whose beating continually sweeps the mucus, with its trapped debris, up toward the

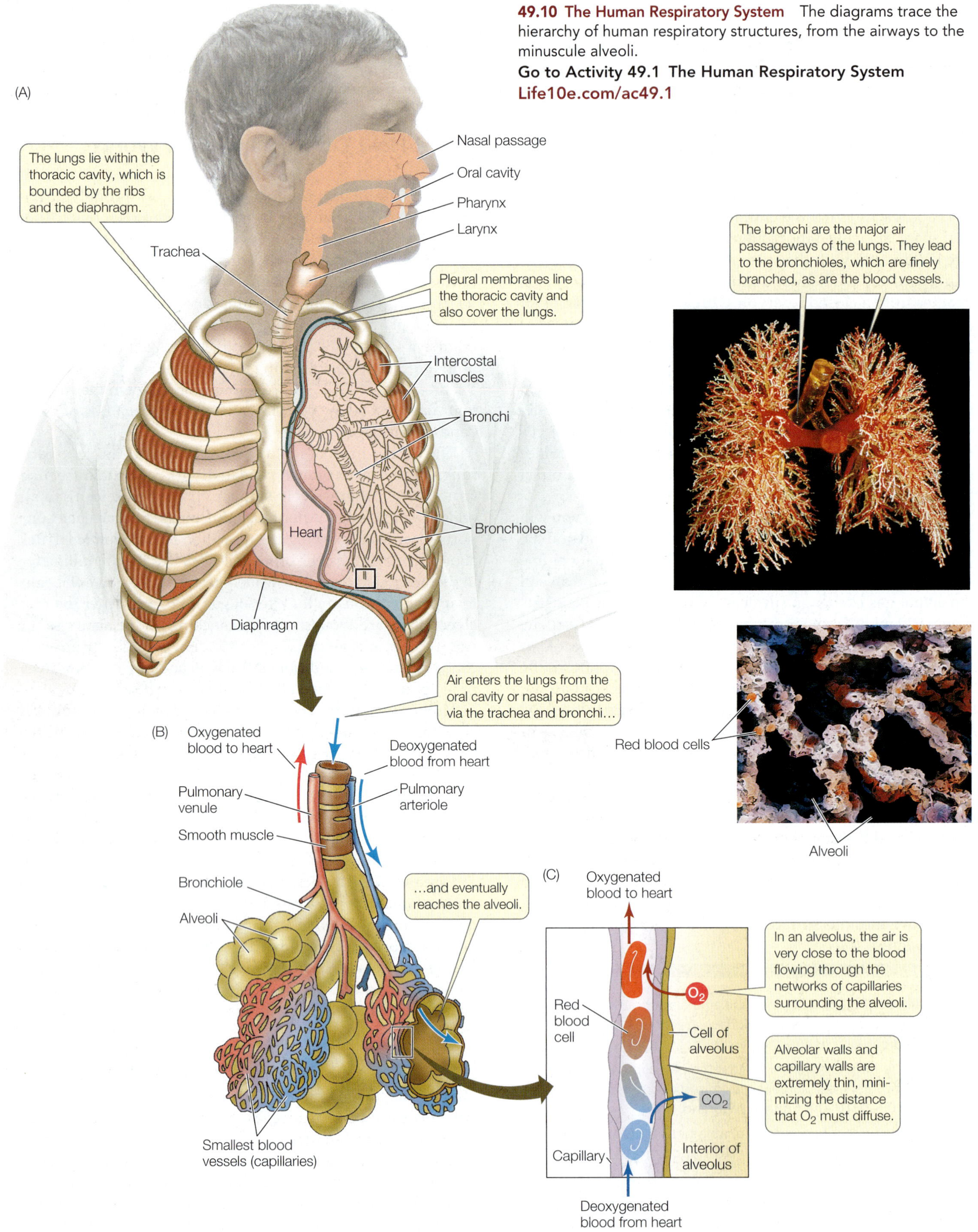
(A)
The lungs lie within the thoracic cavity, which is bounded by the ribs and the diaphragm.
Nasal passage
Oral cavity
Pharynx
Larynx
Trachea
Pleural membranes line the thoracic cavity and also cover the lungs.
Intercostal muscles
Bronchi
Bronchioles
Heart
Diaphragm
The bronchi are the major air passageways of the lungs. They lead to the bronchioles, which are finely branched, as are the blood vessels.
Red blood cells
Alveoli
Air enters the lungs from the oral cavity or nasal passages via the trachea and bronchi…
(B)
Oxygenated blood to heart
Deoxygenated blood from heart
Pulmonary venule
Pulmonary arteriole
Smooth muscle
Bronchiole
Alveoli
…and eventually reaches the alveoli.
Smallest blood vessels (capillaries)
(C)
Oxygenated blood to heart
Red blood cell
O_2
Cell of alveolus
CO_2
Capillary
Interior of alveolus
Deoxygenated blood from heart
In an alveolus, the air is very close to the blood flowing through the networks of capillaries surrounding the alveoli.
Alveolar walls and capillary walls are extremely thin, minimizing the distance that O_2 must diffuse.

49.10 The Human Respiratory System The diagrams trace the hierarchy of human respiratory structures, from the airways to the minuscule alveoli.

Go to Activity 49.1 The Human Respiratory System
Life10e.com/ac49.1

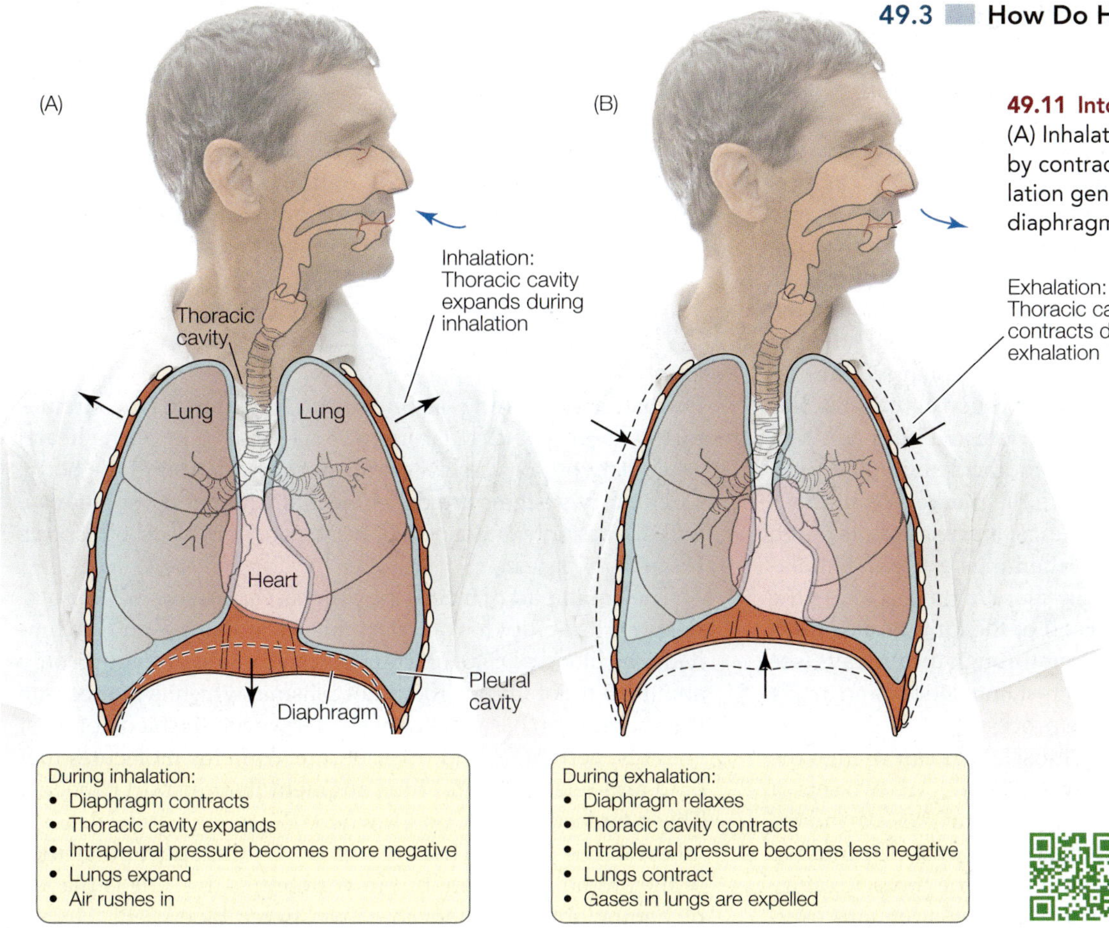
(A)
Inhalation:
Thoracic cavity
expands during
inhalation
Thoracic
cavity
Lung
Lung
Heart
Diaphragm
Pleural
cavity
During inhalation:
• Diaphragm contracts
• Thoracic cavity expands
• Intrapleural pressure becomes more negative
• Lungs expand
• Air rushes in
(B)
Exhalation:
Thoracic cavity
contracts during
exhalation
During exhalation:
• Diaphragm relaxes
• Thoracic cavity contracts
• Intrapleural pressure becomes less negative
• Lungs contract
• Gases in lungs are expelled

49.11 Into the Lungs and Out Again (A) Inhalation is an active process spurred by contraction of the diaphragm. (B) Exhalation generally is a passive process as the diaphragm relaxes.

Go to Animated Tutorial 49.2 Airflow in Mammals
Life10e.com/at49.2

pharynx, where it can be swallowed or spit out. This phenomenon, called the mucus escalator, can be adversely affected by inhaled pollutants. Smoking one cigarette can immobilize the cilia of the airways for hours. A smoker's cough results from the need to clear the obstructing mucus from the airways when the mucus escalator is out of order.

A **surfactant** is a substance that reduces the surface tension of a liquid. **Surface tension** gives the surface of a liquid the properties of an elastic membrane, and it is why certain insects, such as water-striders, can walk on water. As discussed in Section 2.4, surface tension is the result of chemical forces of attraction between water molecules. The attractive forces working on the water molecules at the surface pull from below and from the sides but not from above. This imbalance of forces creates surface tension. The thin film of fluid covering the air-facing surfaces of the alveoli has surface tension that contributes to the lungs' elasticity. To inflate the lungs, enough force has to be generated to overcome both the elasticity of the lung tissue and the surface tension in the alveoli.

Lung surfactant is a fatty, detergent-like substance that is critical for reducing the work necessary to inflate the lungs. Certain cells in the alveoli release surfactant molecules when they are stretched. If a baby is born more than a month prematurely, these cells may not have developed the ability to produce surfactant. A baby with this condition, known as respiratory distress syndrome, will have great difficulty breathing and may die from exhaustion and lack of O_2. Common treatments for premature babies have been to put them on respirators to assist their breathing and to give them hormones to speed lung development. A newer approach is to apply surfactant to the lungs via an aerosol.

Lungs are ventilated by pressure changes in the thoracic cavity

Human lungs are suspended in the **thoracic cavity**, a closed compartment bounded on the bottom by a sheet of muscle called the **diaphragm** (see Figure 49.10A). Each lung is covered by a continuous sheet of tissue called the **pleural membrane** that also lines the thoracic cavity adjacent to that lung. There is no real space between the pleural membranes of the lung and the thoracic cavity, but there is a thin film of fluid. This fluid lubricates the inner surfaces of the pleural membranes so they can slip and slide against each other. Just as we mentioned above in the explanation of surface tension, there are forces of attraction between the molecules of fluid in the pleural membranes. As a result, it is difficult to pull the pleural membranes apart. Think of two wet panes of glass or two wet microscope slides; you can slide them past each other, but it is difficult to separate them. While the inner surfaces of the pleural membranes are "stuck" to each other by surface tension, they can move relative to each other during breathing movements.

Inhalation and exhalation involve changes in the volume of the thoracic cavity (**Figure 49.11**). Because the pleural membranes covering the cavity wall and the lung surface are stuck to each other by surface tension, any attempt to increase the volume of the thoracic cavity increases the tension between the pleural membranes. Even between breaths, there is tension between the pleural membranes because the rib cage is pulling outward and the elasticity of the lung tissue is pulling inward. This slight negative pressure keeps the alveoli partly inflated even at the end of an exhalation. If the thoracic cavity is punctured—by a knife wound, for example—air can leak into the space between the pleural membranes and cause the lung to deflate. If the wound is not sealed, breathing movements pull air in between the pleural membranes rather than into the lung (a "sucking chest wound"), and there is no ventilation of the alveoli in that lung—a condition called "collapsed lung."

At rest, inhalation is initiated by contraction of the muscular diaphragm (see Figure 49.11A). As the domed diaphragm contracts, it pulls down, expanding the thoracic cavity and pulling on the pleural membranes. Since the pleural membranes cannot separate, they pull on the lungs; air rushes in through the trachea from the outside and the lungs expand. Exhalation begins when contraction of the diaphragm ceases. As the diaphragm relaxes, the elastic recoil of the lung tissues pulls the diaphragm up and pushes air out through the airways (see Figure 49.11B). When a person is at rest, inhalation is an active process and exhalation is a passive process.

The diaphragm is not the only muscle that can change the volume of the thoracic cavity. Between the ribs are two sets of **intercostal muscles**. The external intercostal muscles expand the thoracic cavity by lifting the ribs up and outward. The internal intercostal muscles decrease the volume of the thoracic cavity by pulling the ribs down and inward. During strenuous exercise, the external intercostal muscles increase the volume of air inhaled, making use of the inspiratory reserve volume, and the internal intercostal muscles increase the amount of air exhaled, making use of the expiratory reserve volume. The abdominal muscles can also aid in breathing. When they contract, they cause the abdominal contents to push up on the diaphragm and thereby contribute to the expiratory reserve volume.

Remember that ventilation and perfusion work together to maximize the partial pressure gradients across the alveolar membranes. Ventilation delivers O_2 to the environmental side of the exchange surface, where it diffuses across and is swept away by the perfusing blood, which carries it to the tissues. The reverse is true for CO_2. Perfusion delivers CO_2 to the exchange surface, where it diffuses and is swept away by ventilation.

RECAP 49.3

The mammalian respiratory system consists of a highly branching system of airways that lead to alveoli—the gas exchange surfaces. Respiratory muscles ventilate the alveoli by creating pressure differences between the lungs and the outside air. CO_2 and O_2 are exchanged across thin capillary and alveoli walls by diffusion.

- Describe the path that a breath of air takes from the nose to the gas exchange surfaces. **See p. 1013 and Figure 49.10**
- What roles do mucus and surfactant play in maintaining the function of the mammalian respiratory system? **See pp. 1013 and 1015**
- Explain the anatomical and functional relationships between the thoracic cavity, the pleural membranes, and the lungs. **See pp. 1015–1016 and Figure 49.11**

Having discussed how respiratory gases get to and from the environmental side of the gas exchange membranes through ventilation, we will now look at how these gases get to and from the internal side of those membranes through perfusion.

49.4 How Does Blood Transport Respiratory Gases?

Perfusion of the lungs is one of the functions of the circulatory system. The circulatory system uses a pump (the heart) and a network of vessels to transport blood around the body. Circulatory systems are the subject of Chapter 50, so here we will discuss only one aspect of perfusion: how blood transports respiratory gases.

The liquid part of blood, the plasma, carries some O_2 in solution, but its ability to transport this nonpolar molecule is limited. The blood plasma of a human can contain in solution only about 0.3 ml of O_2 per 100 ml of plasma, which is inadequate to support even basal metabolism. However, the blood of most animals, vertebrate and invertebrate, contains molecules that bind and release O_2 and thus augment its transport capacity. These molecules pick up O_2 where P_{O_2} is high and release it where P_{O_2} is lower. There are many O_2 transport molecules in the animal kingdom, but in vertebrates this role is played by hemoglobin, a protein contained in red blood cells. Hemoglobin increases the capacity of blood to carry 60 times more oxygen than it could carry in solution, making high rates of metabolism possible.

Hemoglobin combines reversibly with O_2

Red blood cells contain enormous numbers of hemoglobin molecules. **Hemoglobin** is a protein consisting of four polypeptide subunits (see Figure 3.11), each of which surrounds a heme group—an iron-containing ring structure that can reversibly bind a molecule of O_2. Thus each hemoglobin molecule can bind and release up to four O_2 molecules, enabling the blood to carry a large amount of O_2 to the body's tissues.

Hemoglobin's ability to pick up or release O_2 depends on the P_{O_2} in its environment. When the P_{O_2} of the blood plasma is high, as it usually is in the lung capillaries, each hemoglobin molecule can carry its maximum load of four O_2 molecules. As the blood circulates through the rest of the body, it releases some of the O_2 it is carrying when it encounters lower P_{O_2} values in the body's tissues.

The relationship between P_{O_2} and the amount of O_2 bound to hemoglobin is not linear but S-shaped (sigmoidal). The hemoglobin–oxygen binding curve reflects interactions between the four subunits of the hemoglobin molecule. At low P_{O_2} values, only one subunit will bind an O_2 molecule (**Figure 49.12A**). When it does so, the shape of that subunit changes, altering the quaternary structure of the entire hemoglobin molecule. That structural change makes it easier for the other subunits to bind an O_2 molecule; that is, their O_2 *affinity* is increased. Therefore a smaller increase in P_{O_2} is necessary to get the hemoglobin molecules to bind a second O_2 molecule (that is, to become

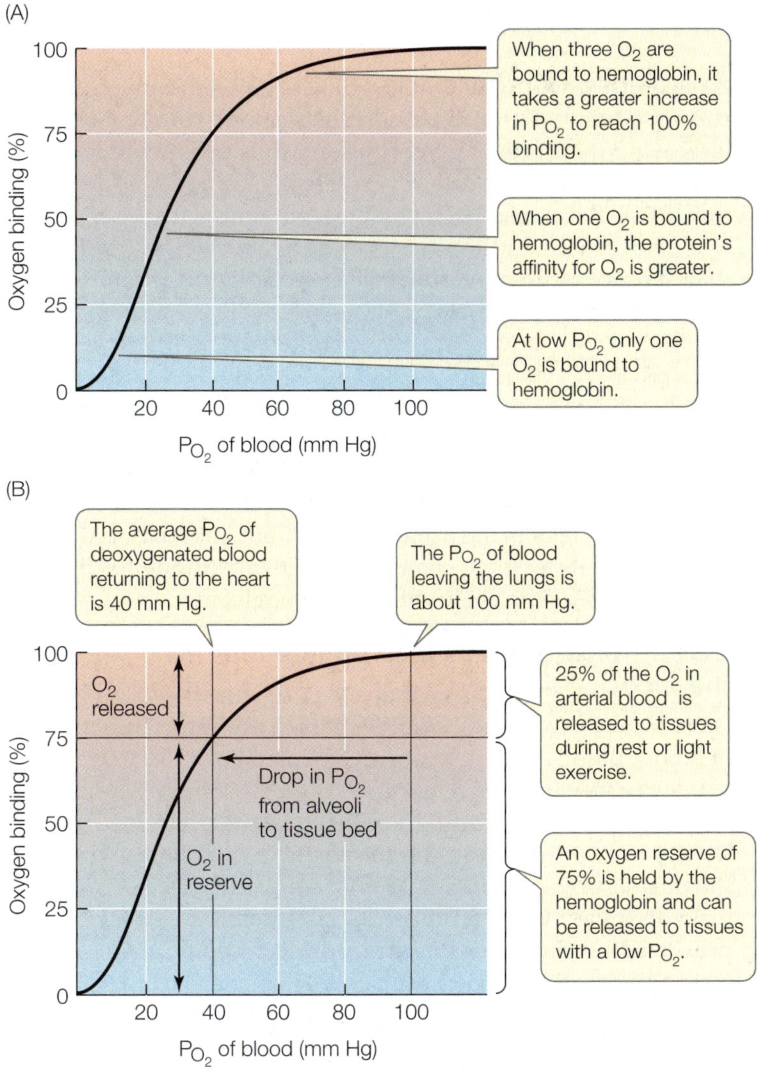
(A)
When three O_2 are bound to hemoglobin, it takes a greater increase in P_{O_2} to reach 100% binding.
When one O_2 is bound to hemoglobin, the protein's affinity for O_2 is greater.
At low P_{O_2} only one O_2 is bound to hemoglobin.
Oxygen binding (%)
100
75
50
25
0
20
40
60
80
100
P_{O_2} of blood (mm Hg)
(B)
The average P_{O_2} of deoxygenated blood returning to the heart is 40 mm Hg.
The P_{O_2} of blood leaving the lungs is about 100 mm Hg.
O_2 released
Drop in P_{O_2} from alveoli to tissue bed
O_2 in reserve
25% of the O_2 in arterial blood is released to tissues during rest or light exercise.
An oxygen reserve of 75% is held by the hemoglobin and can be released to tissues with a low P_{O_2}.
Oxygen binding (%)
P_{O_2} of blood (mm Hg)

49.12 Binding of O_2 to Hemoglobin Depends on P_{O_2} (A) The sigmoidal shape of hemoglobin's oxygen binding curve reflects positive cooperativity among hemoglobin's subunits. (B) Hemoglobin in blood leaving the lungs is 100 percent saturated (four O_2 molecules are bound to each hemoglobin molecule). Most hemoglobin molecules drop only one of their four O_2 molecules as they circulate through the body and are still 75 percent saturated when the blood returns to the lungs. The steep portion of this O_2-binding curve comes into play when tissue P_{O_2} falls below the normal 40 mm Hg, at which point hemoglobin "unloads" its O_2 reserves.

50% saturated) than was necessary to get them to bind one O_2 molecule (to become 25% saturated). This change in affinity is reflected in the increased steepness of the O_2 binding curve. The influence of O_2 binding by one subunit on the O_2 affinity of the other subunits is called **positive cooperativity**.

Once the third O_2 molecule is bound, the relationship seems to change, as a larger increase in P_{O_2} is required for the hemoglobin to reach 100 percent saturation. This upper bend of the sigmoid curve is due to a probability phenomenon. The closer we get to having all subunits occupied, the less likely it is that any particular O_2 molecule will find a place to bind. Therefore it takes a relatively greater P_{O_2} to achieve 100 percent saturation.

The O_2-binding/dissociation properties of hemoglobin help get O_2 to the tissues that need it most (**Figure 49.12B**). In the lungs, where the P_{O_2} is about 100 mm Hg, hemoglobin is 100 percent saturated. The P_{O_2} in blood returning to the heart from the body (at rest) is usually about 40 mm Hg. You can see that at this P_{O_2} the hemoglobin is still about 75 percent saturated. This means that as the blood circulates around the body, it releases only about one in four of the O_2 molecules it carries. This system seems inefficient, but it is really quite adaptive, because the hemoglobin keeps 75 percent of its O_2 in reserve to meet peak demands of highly active tissues.

When a tissue becomes starved of O_2 and its local P_{O_2} falls below 40 mm Hg, the hemoglobin flowing through that tissue is on the steep portion of its binding/dissociation curve. That means relatively small decreases in P_{O_2} below 40 mm Hg will result in the release of lots of O_2 to the tissue. Thus hemoglobin is very effective in making O_2 available to tissues precisely when and where it is needed most.

The O_2 transport function of hemoglobin can rapidly and tragically be disrupted by a common by-product of incomplete combustion: carbon monoxide (CO). If CO from a faulty heating system, engine exhaust, or burning charcoal accumulates in a closed space, the results can be deadly. Because CO binds to hemoglobin with a 240-fold higher affinity than O_2, it prevents hemoglobin from transporting O_2. In the United States, more than 5,000 people die each year from CO poisoning.

Go to Animated Tutorial 49.3
Hemoglobin: Loading and Unloading
Life10e.com/at49.3

Myoglobin holds an O_2 reserve

Muscle cells have their own O_2-binding molecule, **myoglobin**. Myoglobin consists of just one polypeptide chain associated with an iron-containing ring structure that can bind one O_2 molecule (see Figure 24.11). Myoglobin has a higher affinity for O_2 than hemoglobin does, so it picks up and holds O_2 at P_{O_2} values at which hemoglobin is releasing its bound O_2 (**Figure 49.13**).

Myoglobin facilitates the diffusion of O_2 in muscle cells and provides an O_2 reserve for times when metabolic demands are high and blood flow is interrupted. Interruption of blood flow in muscles is common because contracting muscles squeeze blood vessels. When tissue P_{O_2} values are low and hemoglobin can no longer supply more O_2, myoglobin releases its bound O_2. Diving mammals such as seals have high concentrations of myoglobin in their muscles, which is one reason they can stay under water for so long. (We will discuss more adaptations for diving in Chapter 50.) Even in non-diving animals, muscles called on for extended periods of work frequently have more myoglobin than muscles that are used for short, intermittent periods, as noted in Section 48.2.

Hemoglobin's affinity for O_2 is variable

Various factors influence the O_2-binding/dissociation properties of hemoglobin, thereby influencing O_2 delivery to tissues. Three of these factors are the chemical composition of the hemoglobin, the blood pH, and the presence of 2,3-bisphosphoglyceric acid (BPG) in red blood cells.

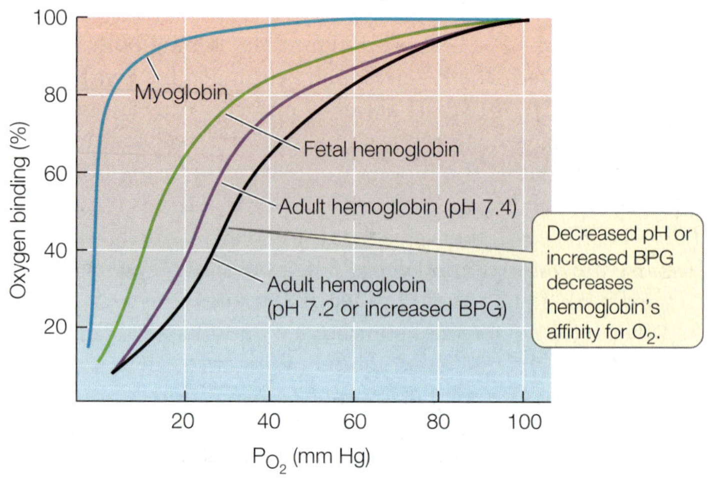
100
80
60
40
20
Oxygen binding (%)
Myoglobin
Fetal hemoglobin
Adult hemoglobin (pH 7.4)
Adult hemoglobin
(pH 7.2 or increased BPG)
Decreased pH or increased BPG decreases hemoglobin's affinity for O_2.
20
40
60
80
100
P_{O_2} (mm Hg)

49.13 Oxygen-Binding Adaptations Myoglobin and the different hemoglobins have different O_2-binding properties adapted to different circumstances. Fetal hemoglobin, for example, has a higher affinity for O_2 than does adult hemoglobin, facilitating O_2 transfer in the placenta. When high metabolism lowers the pH of the blood, or low O_2 increases bisphosphoglyceric acid (BPG), hemoglobin releases more of its O_2.

Go to Activity 49.2 Oxygen-Binding Curves
Life10e.com/ac49.2

HEMOGLOBIN COMPOSITION There is more than one type of hemoglobin, because the chemical composition of the polypeptide chains that form the hemoglobin molecule varies. The normal hemoglobin of adult humans has two each of two kinds of polypeptide chains—two α-globin chains and two β-globin chains. This normal adult hemoglobin has the O_2-binding characteristics shown in **Figure 49.13**.

Before birth, the human fetus has a different form of hemoglobin, consisting of two α-globin and two γ-globin chains. The functional difference between fetal and adult hemoglobin is that fetal hemoglobin has a higher affinity for O_2. Therefore the fetal hemoglobin–oxygen binding/dissociation curve is shifted to the left compared with the adult curve (see Figure 49.13). You can see from these curves that if both types of hemoglobin are at the same P_{O_2} (as they are in the placenta), fetal hemoglobin will pick up O_2 that the adult hemoglobin releases. This difference in O_2 affinities enables the efficient transfer of O_2 from the mother's blood to the fetus's blood.

HEMOGLOBIN AND PH The O_2-binding properties of hemoglobin are also influenced by physiological conditions. The influence of pH (hydrogen ion concentration) on the function of hemoglobin is known as the **Bohr effect**. As blood passes through metabolically active tissue such as exercising muscle, it picks up acidic metabolites such as lactic acid, fatty acids, and CO_2. As a result, blood pH falls. The excess H^+ binds preferentially to deoxygenated hemoglobin and decreases its affinity for O_2 and the O_2-binding/dissociation curve of hemoglobin shifts to the right (see Figure 49.13). This shift means the hemoglobin will release more O_2 in tissues where pH is low—another way that O_2 is supplied where and when it is most needed.

2,3-BISPHOSPHOGLYCERIC ACID BPG is a metabolite of glycolysis. Mammalian red blood cells respond to low P_{O_2} by increasing their rate of glycolysis and thus producing more BPG, which is an important regulator of hemoglobin function. BPG, like excess H^+, reversibly combines with *deoxygenated* hemoglobin and lowers its affinity for O_2. The result is that at any P_{O_2}, hemoglobin releases more of its bound O_2 than it otherwise would. In other words, BPG shifts the O_2-binding/dissociation curve of mammalian hemoglobin to the right.

When humans go to high altitudes, or when they cease being sedentary and begin to exercise, their red blood cells are exposed to a lower P_{O_2} and their level of BPG goes up, making it easier for hemoglobin to deliver more O_2 to tissues. The reason fetal hemoglobin has a left-shifted O_2-binding/dissociation curve is that its γ-globin chains have a lower affinity for BPG than do the β-globin chains of adult hemoglobin.

CO_2 is transported as bicarbonate ions in the blood

Delivering O_2 to tissues is only half the respiratory function of blood. Blood also must take CO_2, a metabolic waste product, away from tissues (**Figure 49.14**). CO_2 is highly soluble and readily diffuses through cell membranes, moving from its site of production in the tissues into the blood, where the partial pressure of CO_2 (P_{CO2}) is lower. However, very little dissolved CO_2 is transported by the blood. Most CO_2 produced by the tissues is transported to the lungs in the form of bicarbonate ions, HCO_3^-. CO_2 is converted to HCO_3^-, transported to the lungs, and then converted back to CO_2 in several steps.

When CO_2 dissolves in water, some of it slowly reacts with the water molecules to form carbonic acid (H_2CO_3), some of which then dissociates into a proton (H^+) and a bicarbonate ion (HCO_3^-). This reversible reaction is expressed as follows:

$$CO_2 + H_2O \rightleftharpoons H_2CO_3 \rightleftharpoons H^+ + HCO_3^-$$

In the extracellular fluid, the reaction between CO_2 and H_2O proceeds slowly. But it is a different story in the endothelial cells of the capillaries and in the red blood cells, where the enzyme carbonic anhydrase speeds up the conversion of CO_2 to H_2CO_3. The newly formed H_2CO_3 dissociates, and the resulting bicarbonate ions enter the plasma in exchange for Cl^- (see Figure 49.14). By converting CO_2 to H_2CO_3, carbonic anhydrase reduces the P_{CO_2} in these cells and in the plasma, facilitating the diffusion of CO_2 from tissue cells to endothelial cells, plasma, and red blood cells. Some CO_2 is also carried in chemical combination with hemoglobin.

In the lungs, the reactions involving CO_2 and bicarbonate ions are reversed. Remember that an enzyme such as carbonic anhydrase only speeds up a reversible reaction; it does not determine its direction. The direction is determined by concentrations of reactants and products. Ventilation keeps the P_{CO_2} in the alveoli low, so CO_2 diffuses from the blood plasma into the alveoli, lowering the P_{CO_2} in the blood, which favors the conversion of HCO_3^- into CO_2.

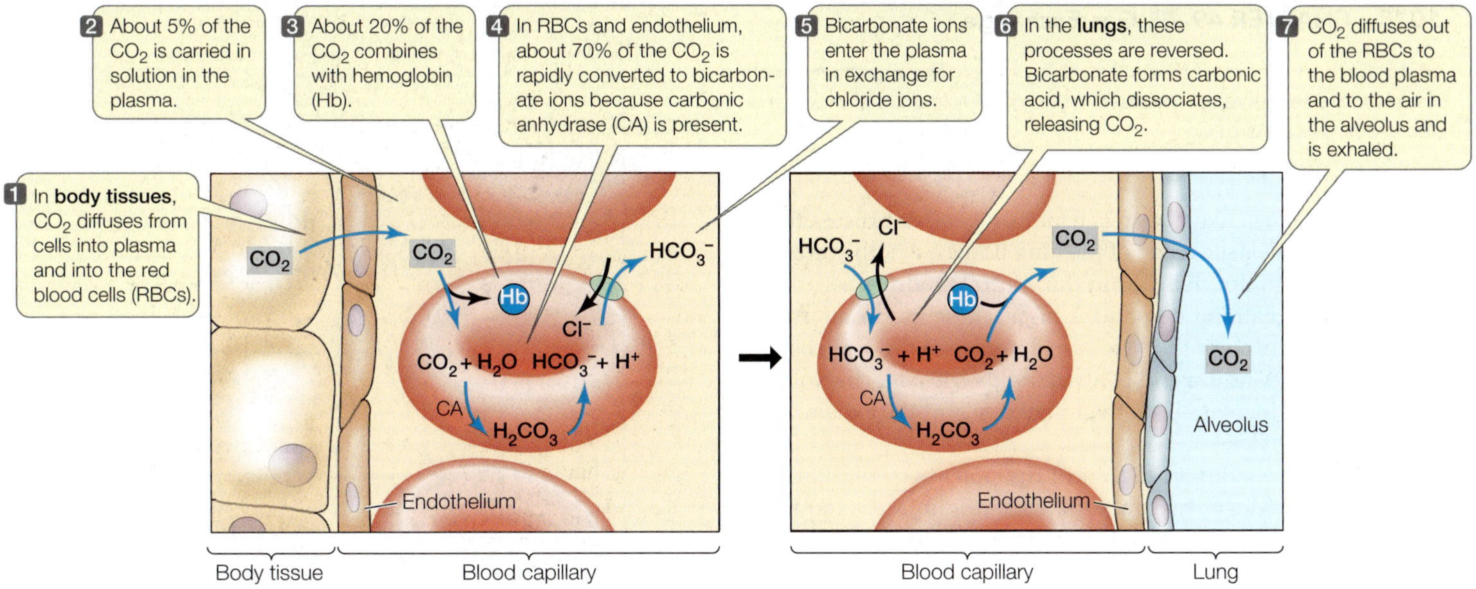

2 About 5% of the CO_2 is carried in solution in the plasma.
3 About 20% of the CO_2 combines with hemoglobin (Hb).
4 In RBCs and endothelium, about 70% of the CO_2 is rapidly converted to bicarbonate ions because carbonic anhydrase (CA) is present.
5 Bicarbonate ions enter the plasma in exchange for chloride ions.
6 In the lungs, these processes are reversed. Bicarbonate forms carbonic acid, which dissociates, releasing CO_2.
7 CO_2 diffuses out of the RBCs to the blood plasma and to the air in the alveolus and is exhaled.
1 In body tissues, CO_2 diffuses from cells into plasma and into the red blood cells (RBCs).
CO_2
CO_2
Hb
HCO_3^-
Cl^-
$CO_2 + H_2O$ $HCO_3^- + H^+$
CA
H_2CO_3
Endothelium
HCO_3^-
Cl^-
CO_2
Hb
$HCO_3^- + H^+$ $CO_2 + H_2O$
CA
H_2CO_3
CO_2
Alveolus
Endothelium
Body tissue
Blood capillary
Blood capillary
Lung

49.14 Carbon Dioxide Is Transported as Bicarbonate Ions Carbonic anhydrase (CA) in capillary endothelial cells and in red blood cells facilitates conversion of CO_2 produced by tissues into bicarbonate ions carried by the plasma. In the lungs, the process is reversed as CO_2 is exhaled.

RECAP 49.4

O_2 is transported from the lungs to the body's tissues in reversible combination with hemoglobin. Each hemoglobin molecule can reversibly combine with four O_2 molecules; the percent saturation of the binding sites is a function of the P_{O_2} in the hemoglobin's environment.

- Explain the advantage of having hemoglobin hold on to three O_2 molecules at the usual P_{O_2} of mixed venous blood. **See pp. 1016–1017 and Figure 49.12**
- How is the O_2-binding/dissociation curve of hemoglobin influenced by pH? By BPG? By development from fetus to newborn infant? **See p. 1018 and Figure 49.13**
- How is CO_2 transported in the blood? **See p. 1018 and Figure 49.14**

We must breathe every minute of our lives, but most of us usually don't worry about it, or even think about it very often. In the next section we will examine how the regular breathing cycle is generated and controlled by the central nervous system.

49.5 How Is Breathing Regulated?

Breathing is an involuntary function of the central nervous system. The breathing pattern easily adjusts itself around other activities (such as speech and eating), and breathing rates change to match the metabolic demands of our bodies. How is this accomplished?

Breathing is controlled in the brainstem

The basic breathing rhythm is an involuntary function driven by a rhythm-generating neural circuit in the brainstem. Breathing is also under voluntary control that enables activities such as talking, singing, sighing, and breath-hold diving. Voluntary control of breathing is eventually and inevitably overridden by feedback information that conveys the body's need for an adequate O_2 supply and CO_2 elimination.

Breathing ceases if the spinal cord is severed in the neck region, showing that breathing is generated in the brain. If the brainstem is cut just above the medulla (the segment of the brainstem just above the spinal cord), an irregular breathing pattern remains (**Figure 49.15**). A group of respiratory motor neurons in the dorsal medulla increase their firing rates just before an inhalation begins. The axons of these neurons leave the CNS in the neck region to form the phrenic nerve, which innervates the diaphragm. As more and more of these neurons fire—and fire faster and faster—the diaphragm contracts. All of a sudden the neurons stop firing, the diaphragm relaxes, and

Go to Media Clip 49.1
Sea Bed Hunting on One Breath
Life10e.com/mc49.1

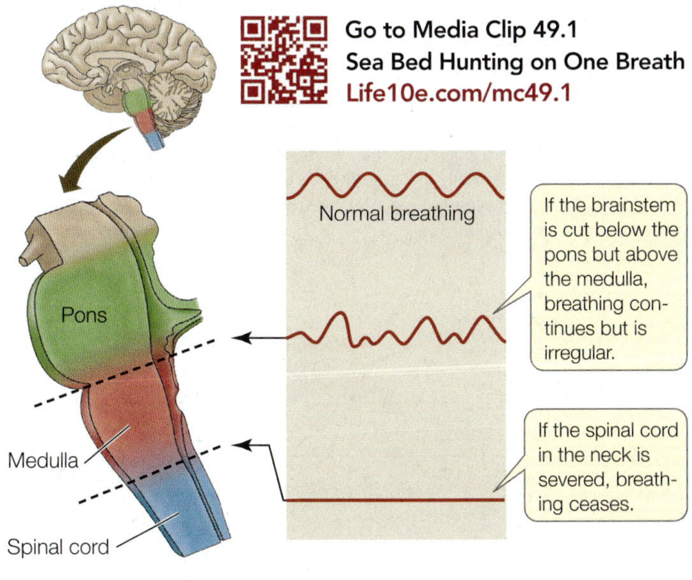

Normal breathing
If the brainstem is cut below the pons but above the medulla, breathing continues but is irregular.
Pons
Medulla
If the spinal cord in the neck is severed, breathing ceases.
Spinal cord

49.15 Breathing Is Controlled in the Brainstem Basic breathing rhythm is generated in the medulla and is modified by neurons in or above the pons.

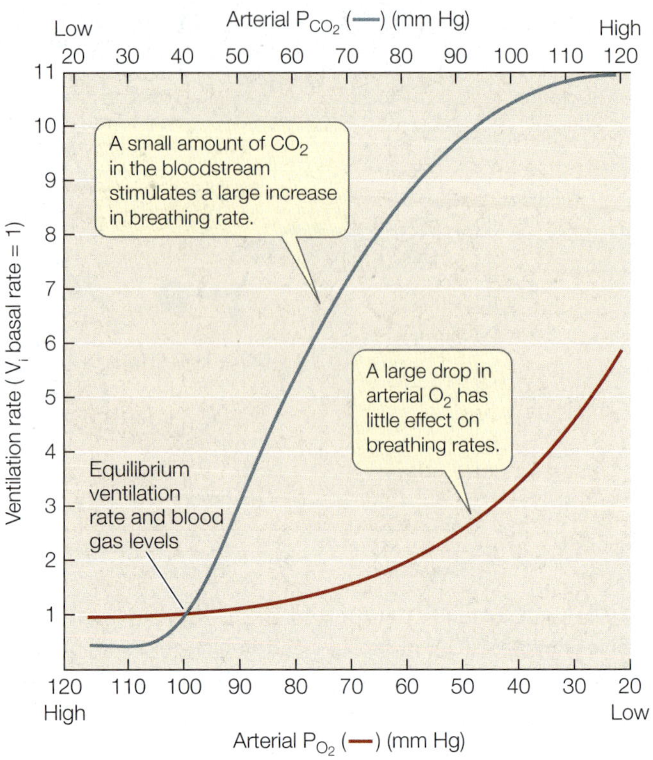
Low
Arterial P_{CO_2} (—) (mm Hg)
High
20 30 40 50 60 70 80 90 100 110 120
11 10 9 8 7 6 5 4 3 2 1
Ventilation rate (V_i basal rate = 1)
A small amount of CO_2 in the bloodstream stimulates a large increase in breathing rate.
A large drop in arterial O_2 has little effect on breathing rates.
Equilibrium ventilation rate and blood gas levels
120 110 100 90 80 70 60 50 40 30 20
High
Low
Arterial P_{O_2} (—) (mm Hg)

49.16 Carbon Dioxide Affects Breathing Rate The breathing mechanism is more sensitive to increased levels of CO_2 in arterial blood than to decreased amounts of O_2.

exhalation begins. Exhalation is usually a passive process that depends on the elastic recoil of the lung tissues. Another group of respiratory motor neurons in the ventral medulla becomes active when breathing demand is high. These motor neurons communicate through thoracic spinal nerves to the intercostal muscles. By expanding and contracting the rib cage, the intercostal muscles increase both the inhalation and the exhalation volumes.

Neurons in the lower region of the pons help regularize the basic respiratory rhythm. Still higher brain areas modify breathing to accommodate speech, ingestion of food, coughing, and emotional states.

Regulating breathing requires feedback

When breathing or metabolism changes, it alters the P_{O_2} and P_{CO_2} in the blood. We should therefore expect the blood levels of one or both of these gases to provide feedback information to the breathing rhythm generator in the medulla. Humans and other mammals are remarkably insensitive to falling levels of O_2 in arterial blood but are extremely sensitive to increases in CO_2. That is, arterial P_{O_2} can deviate considerably from normal without causing much of an increase in ventilation rate, but even a small rise in arterial P_{CO_2} causes a large increase in ventilation (**Figure 49.16**). This relationship is reversed for water-breathing animals, in which O_2 is the primary feedback stimulus for gill ventilation.

We might ask whether it is an increase in the P_{CO_2} of the blood that stimulates increased breathing when we exercise. To answer this question, C. R. Bainton observed dogs running on treadmills at different inclines (**Figure 49.17**). As the incline of the treadmill gradually increased, the dogs ran at the same speed but were working harder because they were running uphill. The P_{CO_2} of their blood increased as the incline of the

INVESTIGATING**LIFE**

49.17 The Respiratory Control System Is Sensitive to P_{CO_2}
What is the metabolic feedback signal that controls ventilation rate during exercise? To find out, Cedric Bainton conducted a series of experiments with dogs trained to run on treadmills.[a]

HYPOTHESIS Rising levels of blood CO_2 during exercise is the feedback signal that stimulates an increase in respiratory rate.

Method

1. Dogs are trained to run on a treadmill.
2. The dogs are equipped with instruments that measure respiratory rate and with arterial catheters that enable sampling of blood.
3. As a dog runs, the incline of the treadmill is changed to gradually increase the metabolic workload.
4. Ventilation rate (V; L/min) is plotted as a function of arterial P_{CO_2} (mm Hg).

Results When the workload is altered by slowly changing the incline of the treadmill (no change in speed), the ventilation rate is a function of P_{CO_2}.

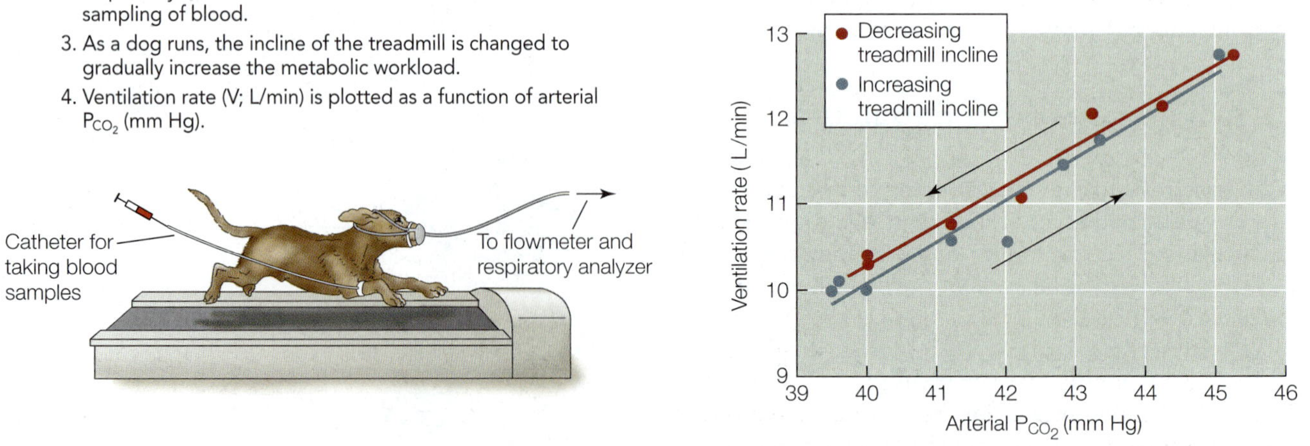
Catheter for taking blood samples
To flowmeter and respiratory analyzer
13 12 11 10 9
Ventilation rate (L/min)
Decreasing treadmill incline
Increasing treadmill incline
39 40 41 42 43 44 45 46
Arterial P_{CO_2} (mm Hg)

CONCLUSION CO_2 level in arterial blood is the metabolic feedback signal that regulates respiration in response to workload.

Go to **BioPortal** for discussion and relevant links for all INVESTIGATING**LIFE** figures.

[a]Bainton, C. R. 1972. *Journal of Applied Physiology* 33: 778–787.

WORKING WITH**DATA:**

The Respiratory Control System Is Not Always Regulated by P_{CO_2}

Original Paper

Bainton, C. R. 1972. Effect of speed vs. grade and shivering on ventilation in dogs during active exercise. *Journal of Applied Physiology* 33: 778–787.

Analyze the Data

Experiments in which dogs run at different speeds while their P_{CO_2} and V are measured produce results different from those shown in Figure 49.17. When subjects run at different speeds, their P_{CO_2}s are the same and yet their V changes. To resolve the differences between these types of experiments, Bainton recorded P_{CO_2} and V in dogs, breath-by-breath immediately after a change in treadmill speed from 3 to 6 mph. The results for one experiment are shown in the table. When this dog was running at 3 mph, its average P_{CO_2} was 39 mm Hg and its V was 9 L/min. After the dog had run at 6 mph for a few minutes its average P_{CO_2} was 41 mm Hg and its V was 15.

QUESTION 1

Draw a plot with V on the y-axis and P_{CO_2} on the x-axis that shows the average values at 3 mph and 6 mph and also the breath-by-breath values (in the table) for the transition. Do the data support the hypothesis that P_{CO_2} is the feedback signal controlling ventilation rate? Why or why not?

QUESTION 2

How do you explain the average values after the dog had been running at the higher speed for a few minutes?

QUESTION 3

Relate these results to those obtained in the experiment in which the incline of the treadmill was gradually raised and lowered. Explain the differences between the results in terms of the P_{CO_2} sensing mechanism (Hint: use the concept of "set point" in your explanation).

Breath	P_{CO_2} (mm Hg)	V (L/min)
1	38.0	13.0
2	37.0	14.0
3	36.5	11.2
4	37.2	11.2
5	37.2	12.0
6	36.8	13.0
7	37.2	13.0

Go to **BioPortal** for all WORKING WITH**DATA** exercises

treadmill and respiratory gas exchange rate increased. Bainton observed a similar relationship between blood P_{CO_2} and respiratory rate as he lowered the treadmill back to its original position. These results indicated that blood P_{CO_2} could act as a signal to regulate breathing rate.

Before concluding that blood P_{CO_2} is the *only* signal that controls breathing rate, Bainton changed the experiment. Instead of increasing the incline of the treadmill, he increased its speed (see the Working with Data exercise above). In this experiment the P_{CO_2} of the blood remained constant as treadmill speed increased and as the respiratory gas exchange rate increased. Bainton concluded that blood P_{CO_2} is the primary metabolic feedback information for breathing. However, when an animal starts to run or changes its running speed, additional feedback information from receptors in muscles and joints changes its sensitivity to CO_2—an example of feedforward information. As noted in Section 40.1, feedforward information can change the sensitivity or the set point of a regulatory system.

Where are partial pressures of gases in the blood sensed? The major site of P_{CO_2} sensitivity is an area on the ventral surface of the medulla. The primary sensitivity of these chemosensitive cells is not to CO_2, however. Rather, they are stimulated by H^+ ions. The H^+ ion concentration, or pH, in the environment of these cells is a direct reflection of the P_{CO_2} of the blood. When the P_{CO_2} of the blood is higher than that of the extracellular fluid in this area, CO_2 diffuses out of the blood. That CO_2 interacts with H_2O to form carbonic acid (H_2CO_3), which dissociates into H^+ ions and HCO_3^- ions (see Figure 49.14). The H^+ ions that are produced stimulate the chemosensitive cells that increase respiratory gas exchange. Thus even though we measure blood P_{CO_2} as the stimulus for breathing, the real stimulus is pH.

Sensitivity to blood P_{O_2} resides in nodes of neural tissue on the large blood vessels leaving the heart: the aorta and the carotid arteries (**Figure 49.18**). These **carotid bodies** and

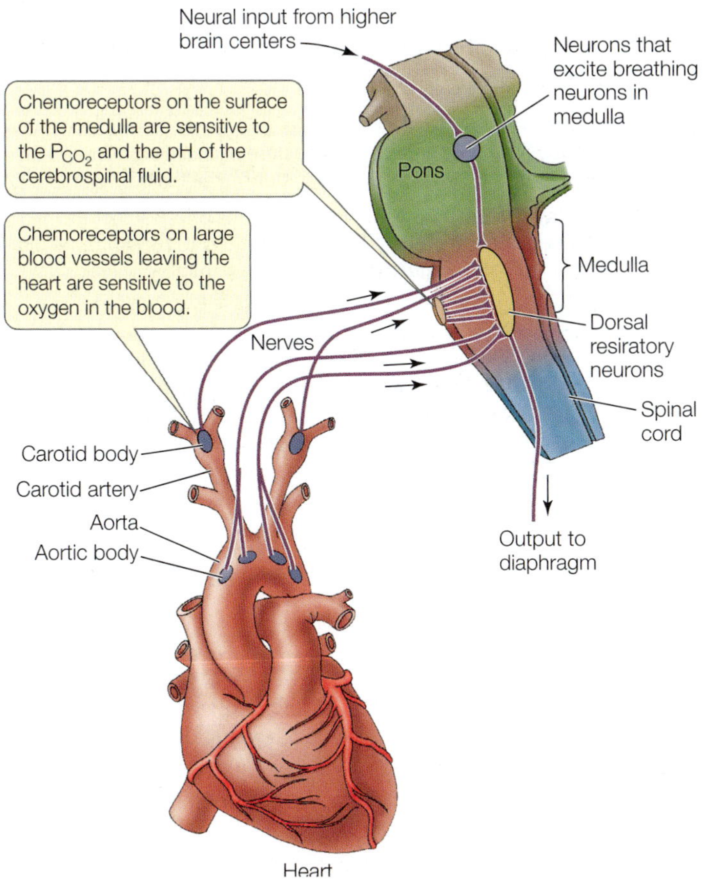

Neural input from higher brain centers
Neurons that excite breathing neurons in medulla
Chemoreceptors on the surface of the medulla are sensitive to the P_{CO_2} and the pH of the cerebrospinal fluid.
Pons
Chemoreceptors on large blood vessels leaving the heart are sensitive to the oxygen in the blood.
Medulla
Nerves
Dorsal resiratory neurons
Spinal cord
Carotid body
Carotid artery
Aorta
Aortic body
Output to diaphragm
Heart

49.18 Feedback Information Controls Breathing The body uses feedback information from chemosensors in the heart and brain to match breathing rate to metabolic demand.

aortic bodies are chemosensors. If the blood supply to these structures decreases, or if the blood P_{O_2} falls dramatically, the chemosensors are activated and send nerve impulses to the breathing control center. Although we are not very sensitive to changes in blood P_{O_2}, the carotid and aortic bodies can stimulate increases in breathing during exposure to very high altitudes or when blood volume or blood pressure is very low. Also, there is a synergism between CO_2 and O_2 sensing. When blood P_{CO_2} increases, there is an increased sensitivity to low O_2, and vice versa.

RECAP 49.5

The rhythmic contractions of the respiratory muscles that drive breathing are generated by neurons in the brainstem.

- What is the primary chemical stimulus for controlling the respiratory rate, and where is it sensed? **See pp. 1019–1020 and Figures 49.16 and 49.18**
- Explain what feedforward information is and how exercise provides feedforward information in the respiratory control system. **See pp.1020–1021 Figure 49.17**
- What are the functions of the carotid and aortic bodies? **See pp. 1021–1022 and Figure 49.18**

?

How do elephants avoid damage to the blood vessels of their thoracic cavities when they snorkel?

ANSWER

In all mammals except elephants, the surfaces of the lungs and the chest cavity move easily against each other. As the lungs inflate and deflate, their surface slips and slides against that of the chest cavity. In elephants, however, dense connective tissue attaches the surface of the lungs to the surface of the chest cavity. This connective tissue acts as reinforcement for the fragile blood vessels on the surface of the chest cavity and prevents them from rupturing and filling the cavity with blood, which would destroy the function of the lungs—exchange of oxygen and carbon dioxide. The elephant is the only mammal that does not have a functional pleural space.

CHAPTER SUMMARY 49

49.1 What Physical Factors Govern Respiratory Gas Exchange?

- Most cells require a constant supply of O_2 and continuous removal of CO_2. These **respiratory gases** are exchanged between an animal's body fluids and its environment by diffusion.
- **Fick's law of diffusion** shows how various physical factors influence the diffusion rate of gases. Adaptations to maximize respiratory gas exchange influence one or more variables of Fick's law.
- In water-breathing animals, gas exchange is limited by the low diffusion rate and low amount of O_2 in water. If water temperature rises, water-breathing animals face a double bind in that the amount of O_2 in water decreases, but their metabolism and the amount of work required to move water over the gas exchange surfaces increase. **Review Figure 49.2**
- In air, the **partial pressure** of oxygen (P_{O_2}) decreases with altitude.

49.2 What Adaptations Maximize Respiratory Gas Exchange?

- Adaptations to maximize gas exchange include increasing the surface area for gas exchange and maximizing partial pressure gradients across those exchange surfaces by ventilating the outer surface with the respiratory medium, and perfusing the inner surface with blood. **Review Figure 49.3**
- Insects distribute air throughout their bodies in a system of **tracheae**, tracheoles, and air capillaries. **Review Figure 49.4**
- The **gills** of fishes have large gas exchange surface areas that are ventilated continuously and unidirectionally with water. The **countercurrent flow** of blood helps increase the efficiency of gas exchange. **Review Figures 49.5, 49.6**
- The gas exchange system of birds includes **air sacs** that communicate with the lungs but are not used for gas exchange. Air flows unidirectionally through bird lungs; gases are exchanged in air capillaries that run between **parabronchi**. **Review Figure 49.7**
- Each breath of air remains in a bird's respiratory system for two breathing cycles. The air sacs work as bellows to supply the air capillaries with a continuous unidirectional flow of fresh air. **Review Figure 49.8, ANIMATED TUTORIAL 49.1**
- In all air-breathing vertebrates except birds, breathing is **tidal**. This is a less efficient form of gas exchange than that of fishes and birds. Although the volume of air exchanged with each breath can vary considerably in tidal breathing, the inhaled air is always mixed with stale air. **Review Figure 49.9**

49.3 How Do Human Lungs Work?

- In mammalian lungs, the gas exchange surface area provided by the millions of **alveoli** is enormous, and the diffusion path length between the air and perfusing blood is short. **Surface tension** in the alveoli would make inflation of the lungs difficult if the alveoli did not produce **surfactant**. **Review Figure 49.10, ACTIVITY 49.1**
- Inhalation occurs when contractions of the **diaphragm** increase volume and reduce pressure in the **thoracic cavity**, thereby pulling on the **pleural membranes**. Relaxation of the diaphragm increases pressure in the thoracic cavity and results in exhalation. **Review Figure 49.11, ANIMATED TUTORIAL 49.2**
- During periods of heavy metabolic demands such as strenuous exercise, the **intercostal muscles**, located between the ribs, increase the volume of air inhaled and exhaled.

continued

49.4 How Does Blood Transport Respiratory Gases?

- O_2 is reversibly bound to **hemoglobin** in red blood cells. Each hemoglobin molecule can carry a maximum of four O_2 molecules. Because of **positive cooperativity**, hemoglobin's affinity for O_2 depends on the P_{O_2} to which the hemoglobin is exposed. Therefore hemoglobin picks up O_2 as it flows through respiratory exchange structures and gives up O_2 in metabolically active tissues. **Review Figure 49.12, ANIMATED TUTORIAL 49.3**
- **Myoglobin** serves as an O_2 reserve in muscle.
- There is more than one type of hemoglobin. Fetal hemoglobin has a higher affinity for O_2 than does adult hemoglobin, allowing fetal blood to pick up O_2 from the maternal blood in the placenta. **Review Figure 49.13, ACTIVITY 49.2**
- CO_2 is transported in the blood principally as bicarbonate ions (HCO_3^-). **Review Figure 49.14**

49.5 How Is Breathing Regulated?

- The basic breathing rhythm is an involuntary function generated by neurons in the medulla and modulated by higher brain centers. The most important feedback stimulus for breathing is the level of CO_2 in the blood. **Review Figures 49.16, 49.17**
- The breathing rhythm is sensitive to feedback from chemoreceptors on the ventral surface of the medulla and in the **carotid** and **aortic bodies** on the large vessels leaving the heart. **Review Figure 49.18**

See ACTIVITY 49.3 for a concept review of this chapter

Go to the Interactive Summary to review key figures, Animated Tutorials, and Activities Life10e.com/is49

CHAPTER**REVIEW**

REMEMBERING

1. Which statement about the gas exchange system of birds is *not* true?
   a. Respiratory gases are not exchanged in the air sacs.
   b. A bird can achieve more complete exchange of O_2 from air to blood than humans can.
   c. Air passes through birds' lungs in only one direction.
   d. The gas exchange surfaces in bird lungs are the parabronchi.
   e. A breath of air remains in the system for two breathing cycles.

2. In the human gas exchange system,
   a. the lungs and airways are completely collapsed after a forceful exhalation.
   b. exhalation is driven by contraction of the diaphragm.
   c. the P_{O_2} of the blood leaving the lungs is greater than that of the exhaled air.
   d. the amount of air that is moved per breath during normal, at-rest breathing is termed the total lung capacity.
   e. P_{CO_2} in the air reaching the alveoli during inhalation is close to zero, as it is in the outside air.

3. The hemoglobin of a human fetus
   a. is the same as that of an adult.
   b. has a higher affinity for O_2 than adult hemoglobin has.
   c. has only two protein subunits instead of four.
   d. is supplied by the mother's red blood cells.
   e. has a higher affinity for BPG than adult hemoglobin has.

4. Most CO_2 in the blood is carried
   a. in the cytoplasm of red blood cells.
   b. as CO_2 dissolved in the plasma.
   c. in the plasma as bicarbonate ions.
   d. bound to plasma proteins.
   e. in red blood cells bound to hemoglobin.

5. Myoglobin
   a. binds O_2 at P_{O_2} values at which hemoglobin is releasing its bound O_2.
   b. has a lower affinity for O_2 than hemoglobin does.
   c. consists of four polypeptide chains, just as hemoglobin does.
   d. provides an immediate source of O_2 for muscle cells at the onset of activity.
   e. can bind four O_2 molecules at once.

UNDERSTANDING & APPLYING

6. The blood of a certain species of fish that lives in Antarctica has no hemoglobin. What anatomical and behavioral characteristics would you expect to find in this fish, and why is its distribution limited to the waters of Antarctica?

7. In patients with emphysema, the fine structures of alveoli break down, resulting in the formation of larger air cavities in the lungs. Also, the lung tissue becomes less elastic. Give at least two explanations for why such patients have a low tolerance for exercise.

8. At what point in the breathing cycle would the pleural cavity pressure be going down while the alveolar pressure is going up? At what point in the breathing cycle would the alveolar pressure be most positive in relation to atmospheric pressure?

ANALYZING & EVALUATING

9. Blood banks store whole blood for a much shorter period than they store blood plasma. This is because when blood that has been stored for too long is infused into a patient, it can actually decrease the O_2 available to the patient's tissues. Refer to Figure 9.5 on glycolysis and explain why the affinity of hemoglobin for O_2 increases with time in storage. Hint: Remember that whole blood is living tissue and requires energy, and that a bag of blood is a closed system. Then think about what fuel can supply energy in a closed system.

10. When you suddenly travel to high altitude, you notice an unusual breathing pattern when you are resting. For a while you stop breathing completely; then suddenly you start breathing rapidly for a short time; then you stop breathing again. This can go on and on in a cyclical pattern called Cheyne–Stokes breathing. Think in terms of the changes in partial pressure gradients when you go to high altitudes and explain how the breathing control system could produce this breathing pattern.

11. South American camelids—llamas, alpacas, guanacos, and vicuñas—are native to the Andes Mountains. In the natural habitat of these mammals, more than 5,000 m above sea level, the P_{O_2} is below 85 mm Hg, and the P_{O_2} in the camelids' lungs is about 50 mm Hg. The hemoglobins of these camelids are different from that of humans; the figure shows the O_2-binding/dissociation curves for human and llama hemoglobins.
    a. Compared with human hemoglobin, does llama hemoglobin have a higher or a lower affinity for O_2?
    b. Why would llama hemoglobin be advantageous at high altitudes?
    c. How would the characteristics of llama hemoglobin affect the transfer of O_2 from the blood to the tissues of the llama?

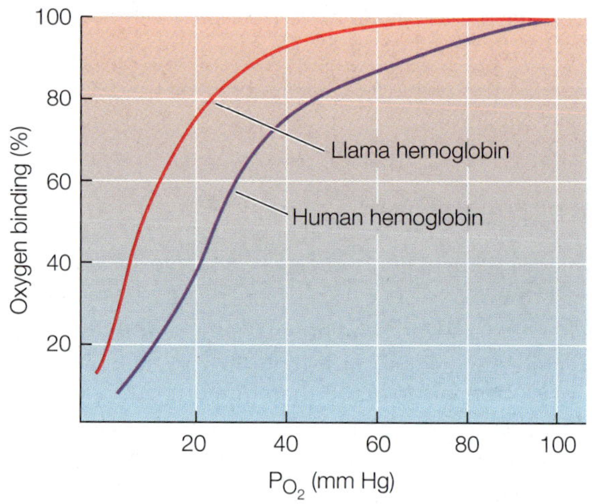
Oxygen binding (%)
100
80
60
40
20
Llama hemoglobin
Human hemoglobin
20
40
60
80
100
P_{O_2} (mm Hg)

Go to BioPortal at **yourBioPortal.com** for Animated Tutorials, Activities, LearningCurve Quizzes, Flashcards, and many other study and review resources.

50 Circulatory Systems

CHAPTER**OUTLINE**

An Athlete's Heart Basketball star Reggie Lewis died of heart failure at the age of 27. Although the exact medical situation underlying Lewis's heart problems remains clouded in controversy, great strides have been made in diagnosing a genetic heart condition that can be particularly dangerous for athletes.

ON APRIL 29, 1993, the Boston Celtics met the Charlotte Hornets in an NBA playoff game. The Celtics' star, 27-year-old Reggie Lewis, had just scored 10 points in 3 minutes when he slumped forward and fell to the floor. The team doctor examined him and allowed him to return to the game. But Lewis's legs were "wobbly," and he played only briefly.

Lewis had been experiencing dizzy spells for about a month before the playoff incident. After the playoffs he underwent rigorous testing by cardiologists, who diagnosed him as having a dangerous arrhythmia (irregular heartbeat) indicative of diseased cardiac muscle. Accepting this diagnosis would mean the end of his professional athletic career, and Lewis sought a second opinion. The second medical team concluded that Lewis had undergone a transient irregular heartbeat attributable to normal athletic stresses. The condition was thought to be due to an enlarged heart—a condition common in high-performing athletes. In July 1993, after an hour spent shooting baskets in a pick-up game, Reggie Lewis collapsed and died of heart failure.

Your heart is a muscular pump that, at rest, beats an average of 60 to 70 times per minute. With each beat, it circulates about 70 milliliters of blood through the body. Without taking work or exercise into account, that is 300 liters per hour, 7,200 liters per day, 2.6 million liters per year—no time-outs.

Heart failure is the leading cause of death in the United States, accounting for some one-fourth (about 600,000) of the deaths each year. Heart failure most commonly results from blockage of the vessels that supply the heart muscle with blood, and its risk increases with age. But heart failure is also the leading cause of death among young athletes.

In athletes, the most common cause of heart failure is not blocked vessels but a gene mutation that affects the contractile proteins of the heart. This mutation can lead to a thickening of the walls of the heart. About 0.5 percent of the population has this mutation; most are unaware of it, and most live their entire lives without symptoms.

Diagnosis is improving. In 2008, 33-year-old Cuttino Mobley announced his retirement from basketball after his physical exam revealed arrhythmia and an MRI showed the symptomatic thickening of the heart walls. "Getting the MRI saved my life," Mobley said.

?

How can the same mutation-based heart condition be fatal to an athlete but innocuous in most other people?

See answer on p. 1046.

50.1 Why Do Animals Need a Circulatory System?

A **circulatory system** consists of a muscular pump (the heart), a fluid (blood), and a series of conduits (blood vessels) through which the fluid can be pumped around the body. Heart, blood, and vessels are also known collectively as a **cardiovascular system** (Greek *kardia*, "heart"; Latin *vasculum*, "vessel"). The function of a circulatory system is to transport things around the body. Preceding chapters discussed how circulatory systems transport heat, hormones, respiratory gases, blood cells, platelets, and cells and molecules of the immune system. Succeeding chapters will add nutrients and waste products to that list. In this section we will describe the general types of circulatory systems found in animals.

Some animals do not have a circulatory system

Single-celled organisms serve all of their needs through direct exchanges with the environment. Such organisms are found mostly in aquatic or very moist terrestrial environments. Similarly, many multicellular aquatic organisms are small or thin enough that all of their cells are close to the external environment. Such species may not have a circulatory system because nutrients, respiratory gases, and wastes can diffuse directly between the cells of their bodies and the environment.

The cells of some larger aquatic multicellular animals without a circulatory system are served by highly branched central cavities called gastrovascular systems that bring the external environment into the animal. All of the cells of a sponge are in contact with, or very close to, the water that surrounds the animal and circulates through its central cavity (see Figure 49.1B). Very small animals without a circulatory system can maintain high levels of metabolism and activity, but larger animals without a circulatory system such as sponges, jellyfishes, and flatworms tend to be inactive, slow, or even sedentary. Large, active animals, however, require a circulatory system.

Circulatory systems can be open or closed

The cells of large, mobile animals are supported by the extracellular fluid. All nutrients—oxygen, fuel, essential molecules—come from that fluid, and the waste products of cell metabolism go into it. Circulatory systems have muscular chambers, or **hearts**, that move the extracellular fluid around the body. In open circulatory systems, extracellular fluid is the same as the fluid in the circulatory system and is called hemolymph. This fluid leaves the vessels of the circulatory system, percolates between cells and through tissues, and then flows back into the heart or vessels of the circulatory system to be pumped out again. In contrast, closed circulatory systems completely contain the circulating fluid (blood) in a continuous system of vessels. Blood cells and large molecules stay within the system, but water and low-molecular-weight solutes leak out of the smallest vessels, the capillaries, which are highly permeable.

In animals with a closed circulatory system, **extracellular fluid** refers to both the fluid in the circulatory system and the fluid outside it. The fluid in the circulatory system is the blood plasma; the fluid around the cells is the **interstitial fluid** (see Figure 40.1). A 70-kilogram person has a total extracellular fluid volume of about 14 liters. Less than a quarter of it—about 3 liters—is the blood plasma.

Open circulatory systems move extracellular fluid

Open circulatory systems are found in arthropods, mollusks, and some other invertebrate groups. In these systems a muscular pump, or heart, helps move the hemolymph through vessels leading to different regions of the body. The fluid leaves the vessels to filter through the tissues before returning to the heart. In the generalized arthropod shown in **Figure 50.1A**, the fluid returns directly to the heart through openings called ostia. Ostia have valves that allow hemolymph to enter the relaxed heart but prevent it from flowing in the reverse direction when the heart contracts. In mollusks, open vessels collect hemolymph from different regions of the body and return it to the heart (**Figure 50.1B**).

Lest you think that open circulatory systems are inefficient and can support only sluggish lifestyles such as those of mollusks, remember that crabs scuttling along the beach, yellow jackets buzzing around your picnic, and scorpions dashing across the desert all have open circulatory systems.

Closed circulatory systems circulate blood through a system of blood vessels

In **closed circulatory systems**, a system of vessels keeps circulating blood separate from the interstitial fluid. Blood is pumped through this vascular system by one or more muscular hearts, and some components of the blood never leave the vessels. Closed circulatory systems characterize vertebrates and some invertebrate groups, among them annelids.

A simple example of a closed circulatory system is that of the earthworm (**Figure 50.1C**). One large ventral blood vessel carries blood from the worm's anterior end to its posterior end. Smaller vessels branch off and transport the blood to even smaller vessels serving the tissues in each body segment. In the smallest vessels, respiratory gases, nutrients, and metabolic wastes diffuse between the blood and interstitial fluid. The blood then flows from these vessels into larger vessels that lead into one large dorsal vessel, which carries the blood from the posterior to the anterior end of the body. Five pairs of muscular vessels connect the large dorsal and ventral vessels in the anterior end, thus completing the circuit. The dorsal vessel and the five connecting vessels serve as hearts for the earthworm; their contractions keep the blood circulating. The direction of circulation is determined by one-way valves in the dorsal and connecting vessels.

Closed circulatory systems have several advantages compared with open systems:

- Fluid can flow more rapidly through vessels than through intercellular spaces and can therefore transport nutrients and wastes to and from tissues more rapidly.

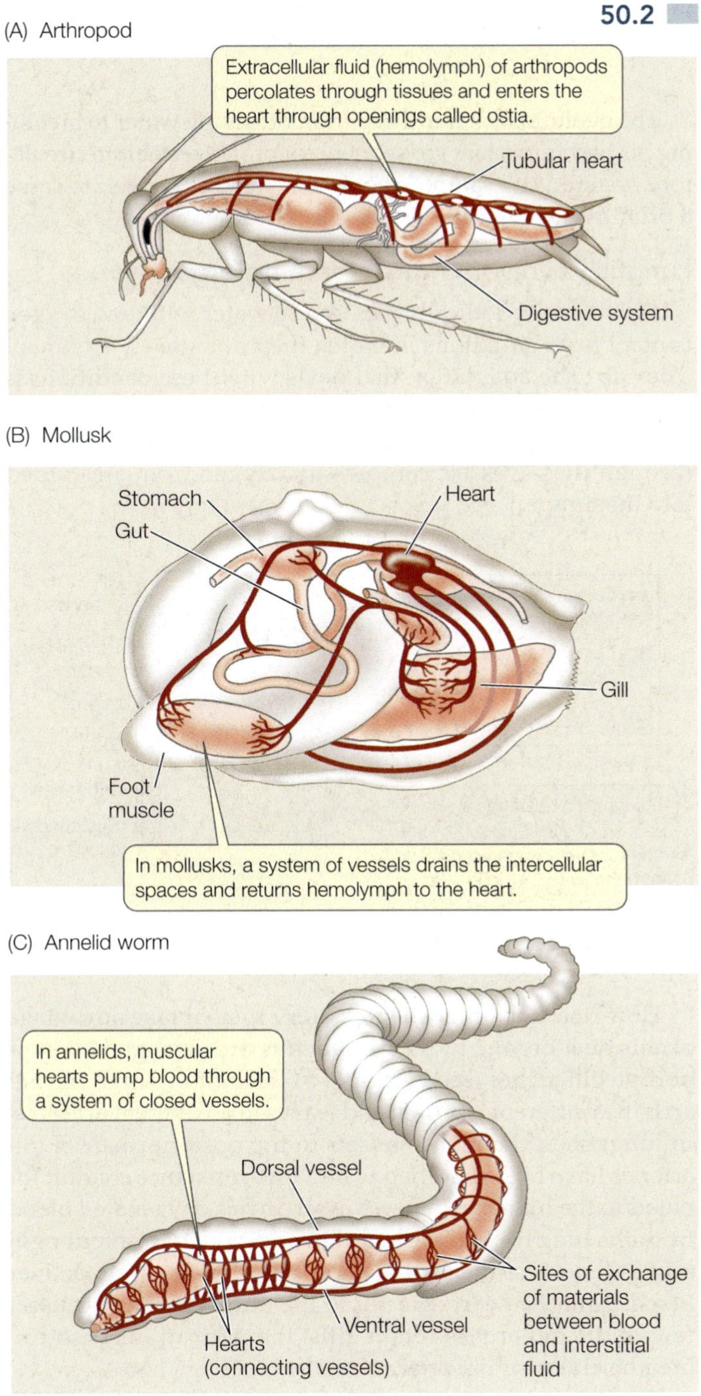
(A) Arthropod
Extracellular fluid (hemolymph) of arthropods percolates through tissues and enters the heart through openings called ostia.
Tubular heart
Digestive system
(B) Mollusk
Stomach
Gut
Heart
Gill
Foot muscle
In mollusks, a system of vessels drains the intercellular spaces and returns hemolymph to the heart.
(C) Annelid worm
In annelids, muscular hearts pump blood through a system of closed vessels.
Dorsal vessel
Sites of exchange of materials between blood and interstitial fluid
Ventral vessel
Hearts (connecting vessels)

50.1 Circulatory Systems Arthropods, illustrated here by an insect (A), and mollusks such as clams (B) have an open circulatory system. Hemolymph is pumped by a tubular heart and directed to different regions of the body through vessels that open into intercellular spaces. (C) Annelids such as earthworms have a closed circulatory system, in which the cellular and macromolecular elements of the blood are confined in a system of vessels and the blood is pumped through those vessels by one or more muscular hearts.

- By changing the diameter (and hence the resistance) of specific vessels, closed systems can control the flow of blood to selective tissues and organs to match their needs.
- Specialized cells and large molecules that aid in transporting hormones and nutrients can be kept in the vessels but can drop their cargo in the tissues where it is needed.

With all of these "advantages" of closed circulatory systems, you might wonder how some species with open circulatory systems can sustain such high levels of activity. In the case of insects, the answer is clear: insects do not depend on their circulatory systems for respiratory gas exchange (see Chapter 49). The open systems of some other species, such as the crab referred to above, involve elaborate systems of vessels that can direct the movement of tissue fluids from one part of the animal's body to another.

RECAP 50.1

Circulatory systems consist of a pump and an open or closed set of vessels through which a fluid transports oxygen, nutrients, wastes, and a variety of other substances.

- Circulatory systems are transport systems. What do they transport? **See p. 1026**
- Why are many species with open circulatory systems rather limited with respect to metabolic activity, and why does this limitation not apply to insects? **See p. 1026 and Figure 50.1**
- What are some advantages of a closed circulatory system? **See pp. 1026–1027**

This overview of the open and closed systems found among invertebrates introduced some basic concepts about circulatory systems. Next we will turn to describing the closed circulatory systems of vertebrates.

50.2 How Have Vertebrate Circulatory Systems Evolved?

Vertebrates have a closed circulatory system and a heart with two or more chambers. When a heart chamber contracts, it squeezes the blood, putting it under pressure. Blood then flows out of the heart and into vessels, where pressure is lower. Resistance to flow in the vessels dissipates the pressure imparted to the blood by the heart. One-way valves prevent the backflow of blood as the heart cycles between contraction and relaxation.

As we explore the features of the circulatory systems of different classes of vertebrates, a general evolutionary theme will emerge: as circulatory systems become more complex, the blood that flows to the gas exchange organs (gills or lungs; see Chapter 49) becomes increasingly separated from the blood that flows to the rest of the body.

In fishes, the phylogenetically oldest vertebrates, blood is pumped from the heart to the gills and then to the tissues of the body and back to the heart—a single circuit. In birds and mammals, blood is pumped from the heart to the lungs and back to the heart in a **pulmonary circuit**, and then from the heart to the rest of the body and back to the heart in a **systemic circuit**. In all other vertebrates we see various adaptations for separating the blood flow into pulmonary and systemic circuits.

Both pulmonary and systemic circuits begin with vessels called **arteries** that carry blood away from the heart. Arteries branch into smaller vessels called **arterioles** that feed blood into capillary beds. **Capillaries** are the tiny, thin-walled vessels where materials are exchanged between the blood and the tissue fluid. Small vessels called **venules** drain capillary beds. The venules join to form larger vessels called **veins** that ultimately deliver blood back to the heart.

We can trace the evolutionary history of vertebrate circulatory systems by comparing the circulatory systems of ray-finned fish, lungfish, amphibians, non-avian reptiles, birds, and mammals.

Circulation in fish is a single circuit

The fish heart has four chambers. Blood returning from all parts of the body collects in a **sinus venosus**, which feeds into the muscular **atrium**. When the atrium contracts, it pumps blood into the more muscular chamber, the **ventricle**. Contraction of the ventricle pushes blood into the last part of the fish heart, the **bulbus arteriosus**, which is a highly elastic chamber. The pressure imparted to the blood by the ventricle stretches the bulbus arteriosus, and its elastic recoil dampens the blood pressure oscillations generated by the beating of the heart. The arterial blood leaving the bulbus arteriosus under pressure flows through the gills, where respiratory gases are exchanged. Blood leaving the gills collects in a large dorsal artery, the **aorta**, which distributes blood to smaller arteries and arterioles leading to all the organs and tissues of the body. In the tissues, blood flows through beds of tiny capillaries, collects in venules and veins, and eventually returns to the sinus venosus of the heart. The unidirectional flow of blood in this circuit is enabled by one-way valves between the sinus venosus and the atrium, between the atrium and the ventricle, and between the ventricle and the bulbus arteriosus.

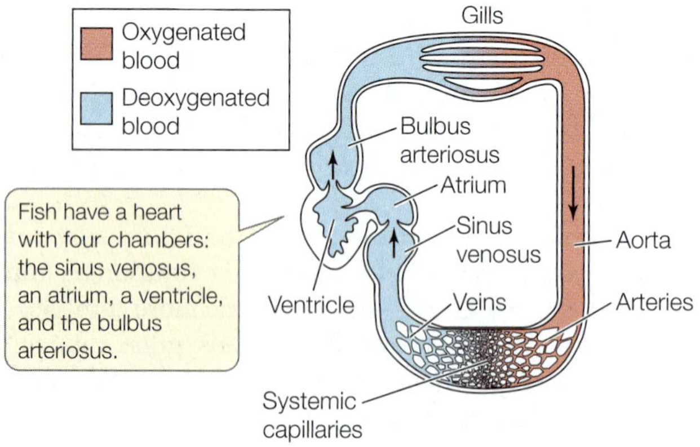
Oxygenated blood
Deoxygenated blood
Fish have a heart with four chambers: the sinus venosus, an atrium, a ventricle, and the bulbus arteriosus.
Gills
Bulbus arteriosus
Atrium
Sinus venosus
Aorta
Ventricle
Veins
Arteries
Systemic capillaries

Most of the pressure imparted to the blood by the contraction of the ventricle is dissipated as a result of resistance to flow in the many narrow spaces in the gill lamellae. Therefore blood leaving the gills and entering the aorta is under low pressure, limiting the capacity of the fish circulatory system to supply the tissues with oxygen and nutrients. Yet this limitation on arterial blood pressure does not seem to limit swimming performance. Some species, such as tuna and marlin, can swim at remarkably high rates of speed for long distances.

The evolutionary transition from breathing water to breathing air had important consequences for the vertebrate circulatory system. An example of how the system changed to serve a primitive lung can be seen in the African lungfish.

Lungfish evolved a gas-breathing organ

Lungfish are periodically exposed to water with low oxygen content or to situations in which their aquatic environment dries up. The adaptation that deals with these conditions is an outpocketing of the gut that serves as a lung. The lung contains many thin-walled blood vessels, so blood flowing through those vessels can pick up oxygen from air gulped into the lung.

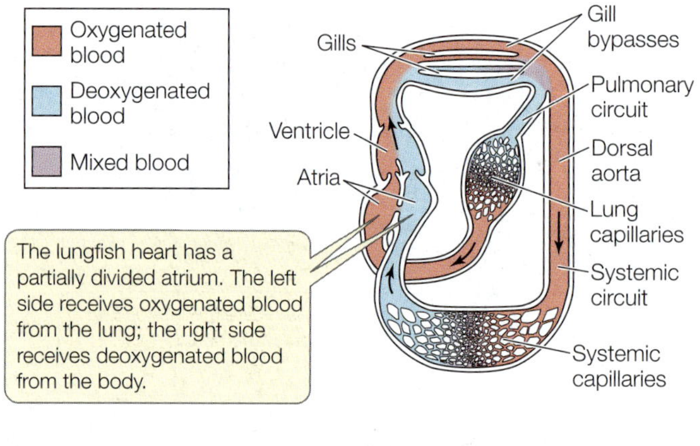
Oxygenated blood
Deoxygenated blood
Mixed blood
Gills
Gill bypasses
Pulmonary circuit
Ventricle
Dorsal aorta
Atria
Lung capillaries
The lungfish heart has a partially divided atrium. The left side receives oxygenated blood from the lung; the right side receives deoxygenated blood from the body.
Systemic circuit
Systemic capillaries

How does the lungfish circulatory system take advantage of this new organ? In fishes, the gills are arranged on supportive gill arches (see Figure 49.5). Blood flows into the gill arch in an afferent arteriole and leaves in an efferent arteriole. In lungfishes, the blood vessels in the posterior pair of gill arteries have been modified to be a low-resistance conduit for blood to the lung, and a new vessel carries oxygenated blood from the lung back to the heart. In addition, two anterior gill arches have lost their gills, and their blood vessels deliver blood from the heart directly to the dorsal aorta. Because a few of the gill arches retain gills, the African lungfish can breathe either air or water.

The lungfish heart partially separates its flow of blood into pulmonary and systemic circuits; it has a partially divided atrium. The left side receives oxygenated blood from the lung and the right side receives deoxygenated blood from the sinus venosus. These two bloodstreams stay mostly separate as they flow through the ventricle and the bulbus arteriosus. As a result, oxygenated blood goes mostly to the anterior gill arteries leading to the dorsal aorta, and deoxygenated blood goes mostly to the other gill arches that have functional gills as well as to the gill arteries that serve the lung.

We can conclude that the lungfish lung evolved as a means of supplementing oxygen uptake from the gills. When the water is oxygenated, the lungfish can obtain oxygen through its gills; but in oxygen-depleted water, it can depend on getting oxygen from its lung. Associated modifications of the lungfish vascular system set the stage for the evolution of separate pulmonary and systemic circulations in higher vertebrates.

Amphibians have partial separation of systemic and pulmonary circulation

In adult amphibians, a single ventricle pumps blood to the lungs and the rest of the body, but two atria receive blood returning to the heart. The left atrium receives oxygenated blood from the lungs, and the right atrium receives deoxygenated blood from the body.

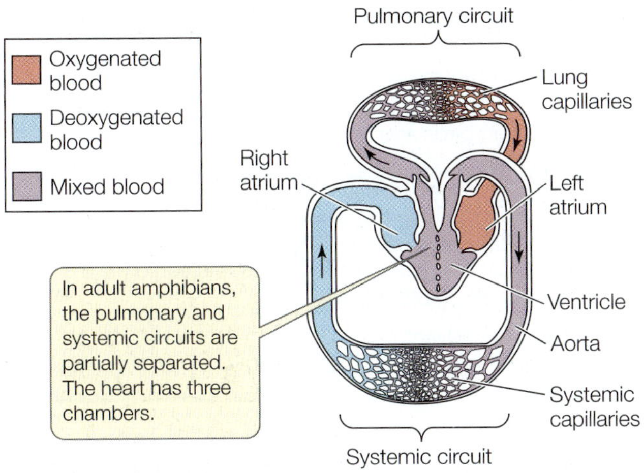
Oxygenated blood
Deoxygenated blood
Mixed blood
Pulmonary circuit
Lung capillaries
Right atrium
Left atrium
In adult amphibians, the pulmonary and systemic circuits are partially separated. The heart has three chambers.
Ventricle
Aorta
Systemic capillaries
Systemic circuit

Because both atria deliver blood to the same ventricle, the oxygenated and deoxygenated blood could mix, in which case blood going to the tissues would not carry a full load of oxygen. Mixing is limited, however, because anatomical features of the ventricle direct the flow of deoxygenated blood from the right atrium to the pulmonary circuit and the flow of oxygenated blood from the left atrium to the aorta. Partial separation of pulmonary and systemic circulation has the advantage of allowing blood destined for the tissues to sidestep the large pressure drop that occurs in the gas exchange organ. Blood leaving the amphibian heart for the tissues moves directly to the aorta, and hence to the body, at a higher pressure than if it had first flowed through the lungs.

Amphibians have another adaptation for oxygenating their blood: they can pick up a considerable amount of oxygen in blood flowing through small blood vessels in their skin.

Reptiles have exquisite control of pulmonary and systemic circulation

As described in Chapter 33, the reptiles include turtles, snakes, lizards, crocodilians, and birds (see Figure 33.20). Crocodilians and birds have cardiovascular systems with two completely separated ventricles, creating a four-chambered heart. All other reptiles have ventricles that are not completely separated into left and right chambers.

Consider the behavior, ecology, and physiology of ectothermic reptiles (i.e., excluding birds). Many are active, powerful, fast animals, but their activity comes in bursts that are interspersed with long periods of inactivity. At these times the animals' metabolic rates are much lower than the resting metabolic rates of the endothermic birds and mammals. So enormous is the range of metabolic demand in ectothermic reptiles that they do not need to breathe continuously. Some species are accomplished divers and spend long periods under water, where they cannot breathe air.

When these animals are not breathing, it would be a waste of energy for them to pump blood through their lungs. Thus they have evolved the capability to send blood to the lungs and the rest of the body when they are breathing, but when they are not breathing, they can bypass the pulmonary circuit and pump all the blood to the body. How do they do this?

In ectothermic reptiles with a three-chambered heart—that is, the turtles, snakes, and lizards—the ventricle is partially divided into left and right halves by a septum. Oxygenated blood from the lungs enters the left side of the ventricle through the left atrium. Deoxygenated blood from the body enters the right side of the ventricle through the right atrium. These species have two aortas, left and right. The left aorta is positioned so that it receives oxygenated blood from the left side of the ventricle. The right aorta, however, is positioned so that it can receive blood from either the right or left side of the ventricle.

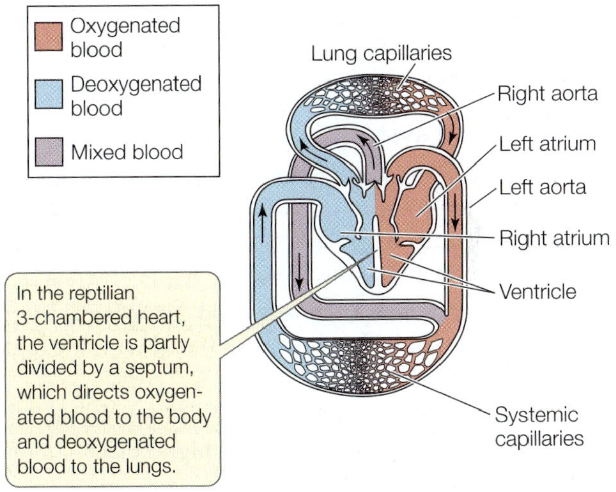
Oxygenated blood
Deoxygenated blood
Mixed blood
Lung capillaries
Right aorta
Left atrium
Left aorta
Right atrium
Ventricle
In the reptilian 3-chambered heart, the ventricle is partly divided by a septum, which directs oxygenated blood to the body and deoxygenated blood to the lungs.
Systemic capillaries

When the animal is breathing air, the resistance in the pulmonary circuit is lower than the resistance in the systemic circuit, so blood from the right side of the ventricle tends to flow into the pulmonary artery rather than the right aorta. When the animal is not breathing, pulmonary vessels constrict, resistance in the pulmonary circuit goes up, and blood from the right side of the ventricle tends to flow into the right aorta. As a result, blood from both sides of the ventricle flows through both aortas to the systemic circuit.

Crocodilians, like birds, have two completely separated ventricles. Unlike birds, they have two aortas, one originating in each ventricle. But there is a connection between the two aortas just as they leave the heart, and this connection enables them to alter the proportions of blood going to their pulmonary and systemic circuits. When a crocodile or alligator is breathing and resistance in the pulmonary circuit is low, backpressure from the stronger left ventricle closes the valve between the right ventricle and the right aorta, forcing all of the blood from the right ventricle to flow into the pulmonary circuit. When the animal stops breathing, pulmonary vessels constrict, resistance in the pulmonary circuit rises, and blood from the right ventricle flows into the right aorta. This ability of all ectothermic reptiles to direct blood to their pulmonary or systemic circuits is highly adaptive for their lifestyle of intermittent breathing.

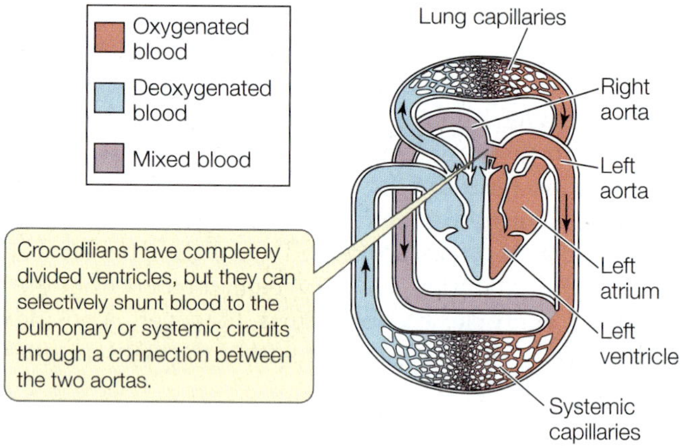

Oxygenated blood
Deoxygenated blood
Mixed blood
Lung capillaries
Right aorta
Left aorta
Left atrium
Left ventricle
Systemic capillaries
Crocodilians have completely divided ventricles, but they can selectively shunt blood to the pulmonary or systemic circuits through a connection between the two aortas.

Birds and mammals have fully separated pulmonary and systemic circuits

The four-chambered hearts of birds and mammals have completely separate pulmonary and systemic circuits. Separate circuits have several advantages for these active animals with continuously high metabolic rates:

- Oxygenated and deoxygenated blood cannot mix; therefore the systemic circuit always receives blood with the highest oxygen content.
- Respiratory gas exchange is maximized because the blood with the lowest oxygen content and highest CO_2 content is sent to the lungs.
- Separate systemic and pulmonary circuits can operate at different pressures.

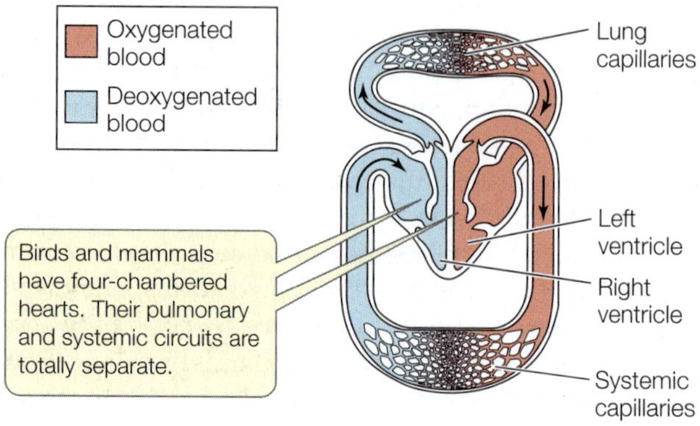

Oxygenated blood
Deoxygenated blood
Lung capillaries
Left ventricle
Right ventricle
Systemic capillaries
Birds and mammals have four-chambered hearts. Their pulmonary and systemic circuits are totally separate.

The tissues of birds and mammals have high nutrient demands and thus a very high density of blood vessels, requiring the heart to generate a high blood pressure to perfuse all the vessels of the systemic circuit. The pulmonary circuit of these animals receives a blood flow equal to that of the systemic circuit, but the lungs have far fewer blood vessels. Thus the pulmonary circuit of birds and mammals can function at lower pressures, and the four-chambered heart makes that possible.

Go to Activity 50.1 Vertebrate Circulatory Systems
Life10e.com/ac50.1

RECAP 50.2

The closed circulatory system of vertebrates has evolved from a single circuit system in fishes to separate pulmonary and systemic circuits in birds and mammals.

- Explain why fish cannot supply blood to their tissues at high pressure. **See p. 1028**
- By comparing lungfish and amphibian circulatory systems, explain how a heart with two separate atria could have evolved. **See pp. 1028–1029**
- What are some advantages of separate pulmonary and systemic circuits? **See pp. 1029–1030**

50.3 How Does the Mammalian Heart Function?

Mammals have hearts consisting of a right and a left atrium and a right and a left ventricle. We can use the human heart as our example (**Figure 50.2**). The right ventricle pumps blood through the pulmonary circuit, and the left ventricle pumps blood through the systemic circuit.

One-way valves between the atria and ventricles, the **atrioventricular (AV) valves**, prevent backflow of blood into the atria when the ventricles contract. The **pulmonary valve** and **aortic valve**, one-way valves between the ventricles and the major arteries, prevent backflow of blood into the ventricles when they relax.

In this section we will describe the flow of blood through the heart and body and examine the unique electrical properties of cardiac muscle that result in the rhythmic contractions of the heart.

Blood flows from right heart to lungs to left heart to body

The heart's right atrium receives deoxygenated blood from the **superior** (upper) **vena cava** and the **inferior** (lower) **vena cava** (see Figure 50.2)—the large veins that collect blood returning to the heart from the upper and lower body, respectively. The veins of the heart itself also drain into the right atrium. From the right atrium, the blood flows through the right AV valve into the right ventricle. Most of the filling of the ventricle results from passive flow while the heart is relaxed between beats. Just at the end of this period of passive ventricular filling, the atrium contracts and adds a little more blood to the

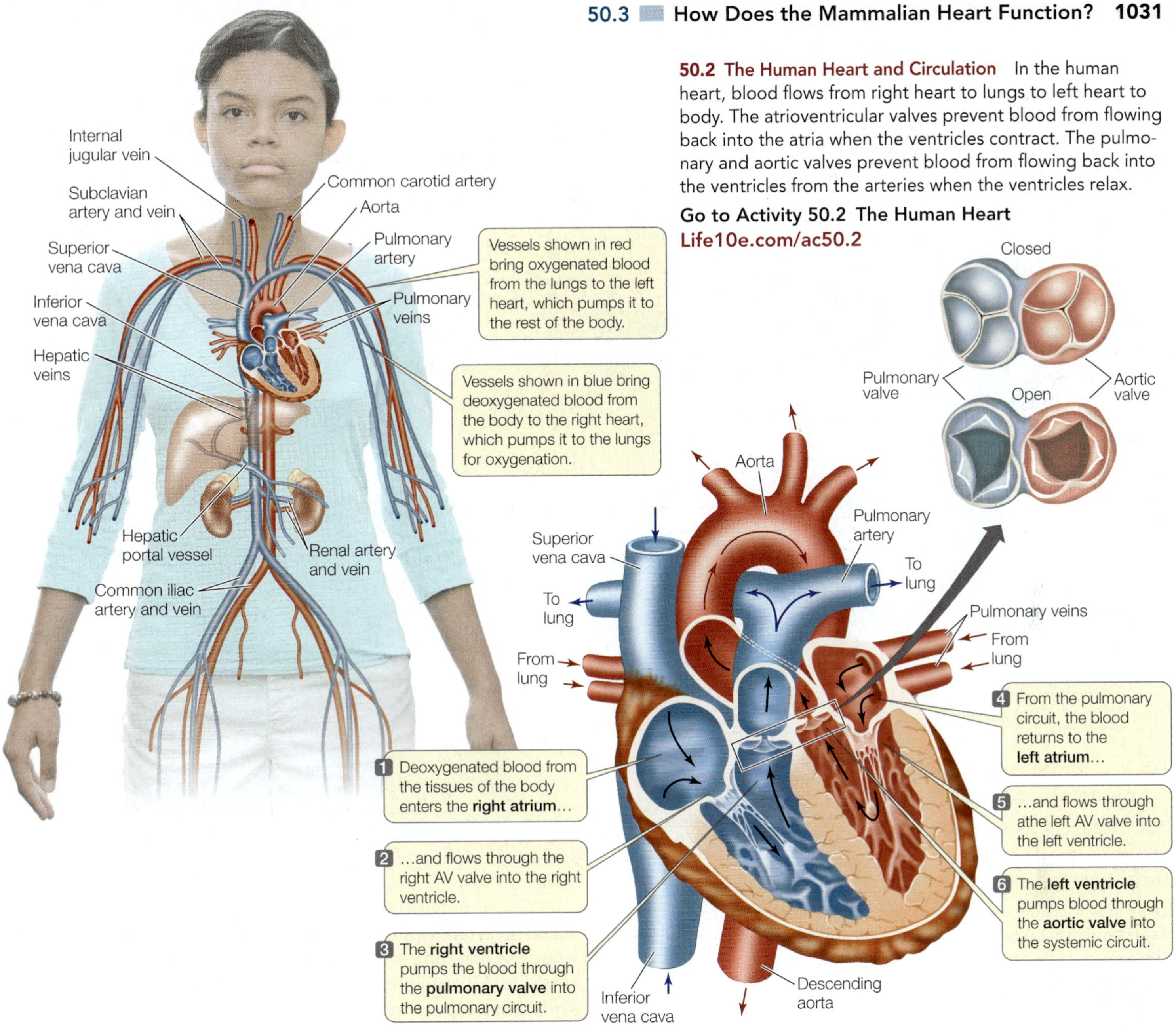

Internal jugular vein
Subclavian artery and vein
Superior vena cava
Inferior vena cava
Hepatic veins
Common carotid artery
Aorta
Pulmonary artery
Pulmonary veins
Vessels shown in red bring oxygenated blood from the lungs to the left heart, which pumps it to the rest of the body.
Vessels shown in blue bring deoxygenated blood from the body to the right heart, which pumps it to the lungs for oxygenation.
Hepatic portal vessel
Renal artery and vein
Common iliac artery and vein
Closed
Open
Pulmonary valve
Aortic valve
Aorta
Superior vena cava
Pulmonary artery
To lung
To lung
From lung
Pulmonary veins
From lung
1 Deoxygenated blood from the tissues of the body enters the right atrium...
2 ...and flows through the right AV valve into the right ventricle.
3 The right ventricle pumps the blood through the pulmonary valve into the pulmonary circuit.
4 From the pulmonary circuit, the blood returns to the left atrium...
5 ...and flows through athe left AV valve into the left ventricle.
6 The left ventricle pumps blood through the aortic valve into the systemic circuit.
Inferior vena cava
Descending aorta

50.2 The Human Heart and Circulation In the human heart, blood flows from right heart to lungs to left heart to body. The atrioventricular valves prevent blood from flowing back into the atria when the ventricles contract. The pulmonary and aortic valves prevent blood from flowing back into the ventricles from the arteries when the ventricles relax.

Go to Activity 50.2 The Human Heart
Life10e.com/ac50.2

ventricular volume. The right ventricle then contracts, causing the AV valve to close and pumping the blood into the **pulmonary artery** leading to the lungs.

Pulmonary veins return the oxygenated blood from the lungs to the left atrium, from which the blood enters the left ventricle through the left AV valve. As on the right side of the heart, most left ventricular filling is passive, but ventricular filling reaches a maximum when the atria contract.

The walls of the left ventricle are powerful muscles that contract around the blood with a wringing motion starting from the bottom. When pressure in the left ventricle is high enough to push open the aortic valve, blood rushes into the aorta to begin its circulation throughout the body. In Figure 50.2, observe that the walls of the left ventricle are thicker than those of the right ventricle. The left ventricle has to propel blood through many more kilometers of blood vessels than does the right ventricle, and must therefore push against more resistance, even though both ventricles pump the same volume of blood.

Both sides of the heart contract at the same time. Contraction of the two atria, followed by contraction of the two ventricles and then relaxation, is the **cardiac cycle**. The cardiac cycle is divided into two phases: **systole** (pronounced sís-toll-ee), when the ventricles contract, and **diastole** (die-ás-toll-ee), when the ventricles relax (**Figure 50.3**). At the very end of diastole (step 1 in Figure 50.3), just before the ventricles contract, the atria contract and top off the volume of blood in the ventricles.

The sounds of the cardiac cycle—the "lub-dup" heard through a stethoscope—are created by the heart valves slamming shut. The closing and opening of these valves are simple mechanical events resulting from pressure differences on the two sides of the valves. As the ventricles begin to contract (step 2 in Figure 50.3), the pressure in them rises above the pressure in the atria, so the

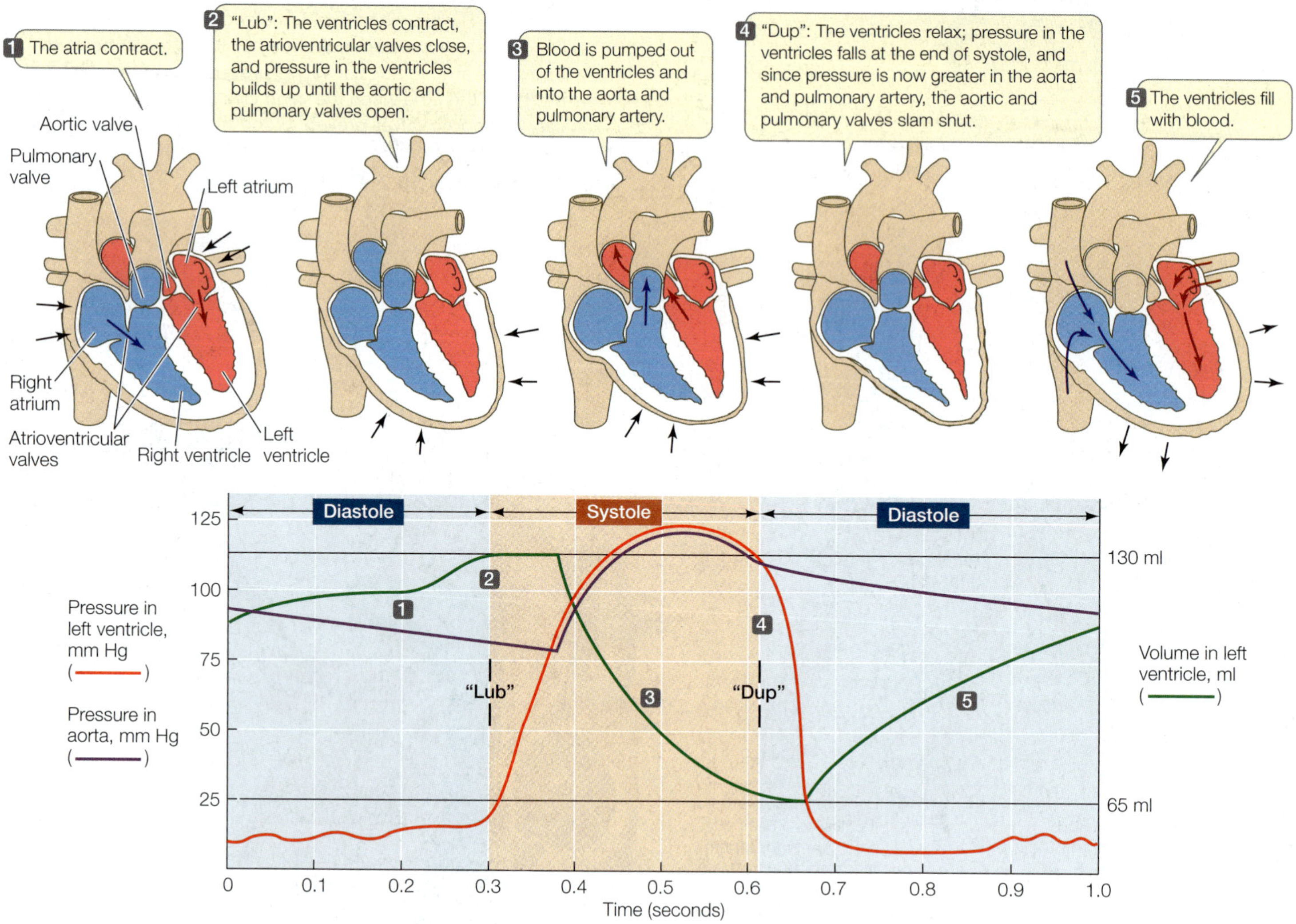
1 The atria contract.
2 "Lub": The ventricles contract, the atrioventricular valves close, and pressure in the ventricles builds up until the aortic and pulmonary valves open.
3 Blood is pumped out of the ventricles and into the aorta and pulmonary artery.
4 "Dup": The ventricles relax; pressure in the ventricles falls at the end of systole, and since pressure is now greater in the aorta and pulmonary artery, the aortic and pulmonary valves slam shut.
5 The ventricles fill with blood.
Aortic valve
Pulmonary valve
Left atrium
Right atrium
Atrioventricular valves
Right ventricle
Left ventricle
Diastole
Systole
Diastole
130 ml
65 ml
Pressure in left ventricle, mm Hg
Pressure in aorta, mm Hg
Volume in left ventricle, ml
"Lub"
"Dup"
125
100
75
50
25
0
0.1
0.2
0.3
0.4
0.5
0.6
0.7
0.8
0.9
1.0
Time (seconds)

50.3 The Cardiac Cycle The rhythmic contraction (systole) and relaxation (diastole) of the ventricles is called the cardiac cycle. The representation below shows pressure and volume changes during the cardiac cycle for the left ventricle only.

Go to Animated Tutorial 50.1
The Cardiac Cycle
Life10e.com/at50.1

AV valves close ("lub"). When the ventricles begin to relax (step 4 in Figure 50.3), the high pressure in the aorta and pulmonary artery closes the aortic and pulmonary valves ("dup").

Defective valves that do not close completely produce turbulent blood flow and the sounds known as heart murmurs. For example, if an AV valve does not close completely, blood will flow back into the atrium with a "whoosh" at the beginning of systole.

Blood pressure changes associated with the cardiac cycle can be measured in the large artery in your arm by using an inflatable pressure cuff and a pressure gauge, together called a sphygmomanometer, and a stethoscope (**Figure 50.4**). This method measures the minimum pressure necessary to compress an artery so blood does not flow through it at all (the systolic value) and the minimum pressure that causes intermittent flow through the artery (the diastolic value). A conventional blood pressure reading is expressed as the systolic value placed over the diastolic value. Healthy values for a young adult might be 120 millimeters of mercury (mm Hg) during systole and 70 mm Hg during diastole, or 120/70.

The heartbeat originates in the cardiac muscle

Cardiac muscle has unique adaptations that enable it to function as a pump. First, cardiac muscle cells are in electrical contact with one another through gap junctions, which enable action potentials (see Section 45.2) to spread rapidly from cell to cell. Because a spreading action potential stimulates contraction, large groups of cardiac muscle cells contract in unison. This coordinated contraction is essential for pumping blood effectively.

Second, some cardiac muscle cells are **pacemaker cells** that can initiate action potentials without stimulation from the nervous system. When they fire action potentials, they stimulate neighboring cells to contract. The primary pacemaker of the heart is a group of modified cardiac muscle cells, the **sinoatrial node**, located at the junction of the superior vena cava and right atrium (see Figure 50.7). The resting membrane potentials of these cells are less negative than those of other cardiac muscle cells and are not stable; instead they gradually become even less negative until they reach threshold for initiating an

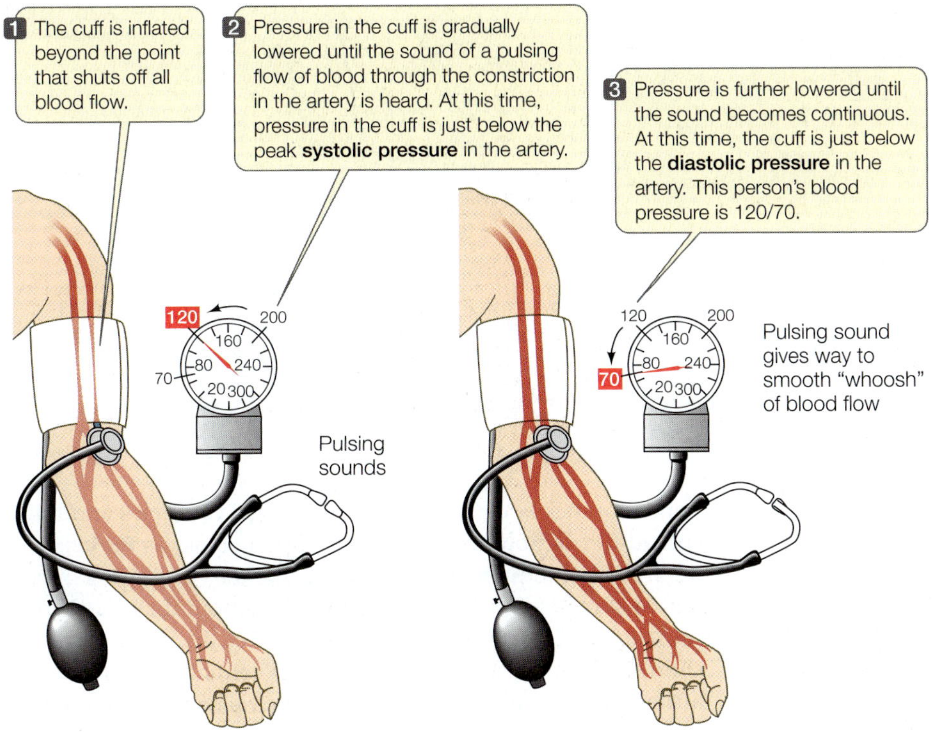
1 The cuff is inflated beyond the point that shuts off all blood flow.
2 Pressure in the cuff is gradually lowered until the sound of a pulsing flow of blood through the constriction in the artery is heard. At this time, pressure in the cuff is just below the peak **systolic pressure** in the artery.
3 Pressure is further lowered until the sound becomes continuous. At this time, the cuff is just below the **diastolic pressure** in the artery. This person's blood pressure is 120/70.
120
200
160
80
240
70
20
300
Pulsing sounds
120
200
160
80
240
70
20
300
Pulsing sound gives way to smooth "whoosh" of blood flow

50.4 Measuring Blood Pressure Blood pressure in the major artery of the arm can be measured with a device called a sphygmomanometer, which combines an inflatable cuff and a pressure gauge. A stethoscope is also used to detect sounds created by the blood vessels in response to changes in pressure during the cardiac cycle.

action potential. The action potentials of pacemaker cells are very different from those of neurons and other muscle cells (see Figure 45.10). They are slower to rise; they are broader; and they are slower to return to resting potential (**Figure 50.5A**). These properties of pacemaker cells are due to the ion channels in their membranes.

Pacemaker potentials involve Na^+, Ca^{2+}, and K^+ channels (**Figure 50.5B**). As we discussed in Section 45.2, when Na^+ or Ca^{2+} channels open, positive charges flow into the cell and the membrane potential becomes less negative. When K^+ channels open, positive charges flow out of the cell and the membrane potential becomes more negative. Because the Na^+ channels of pacemaker cells are open more of the time than are those of other cardiac muscle cells, the pacemaker resting potential is less negative. The action potential of pacemaker cells is due to voltage-gated Ca^{2+} channels rather than voltage-gated Na^+ channels as in neurons, skeletal muscle, and other cardiac muscle cells. These Ca^{2+} channels open and close more slowly than voltage-gated Na^+ channels, explaining the shape of pacemaker action potentials.

The unstable resting potential of pacemaker cells is due to the behavior of cation channels. As in neurons and skeletal muscle cells, there are voltage-gated K^+ channels that slowly open on the rising phase of the action potential. The opening of these channels allows K^+ ions to leave the cell and thereby restore the negative charge on the cell membrane. That restoration of a negative membrane potential causes the opening of a unique class of voltage-gated cation channels that mostly conduct Na^+. At the same time, the voltage-gated K^+ channels that opened during the action potential slowly close. The result is that there are more Na^+ ions coming into the cell than there are K^+ ions leaving, and the cell membrane potential gradually becomes less negative (see Figure 50.5).

The gradual rise in membrane potential closes the channels that allow Na^+ to move into the cell, but as the membrane

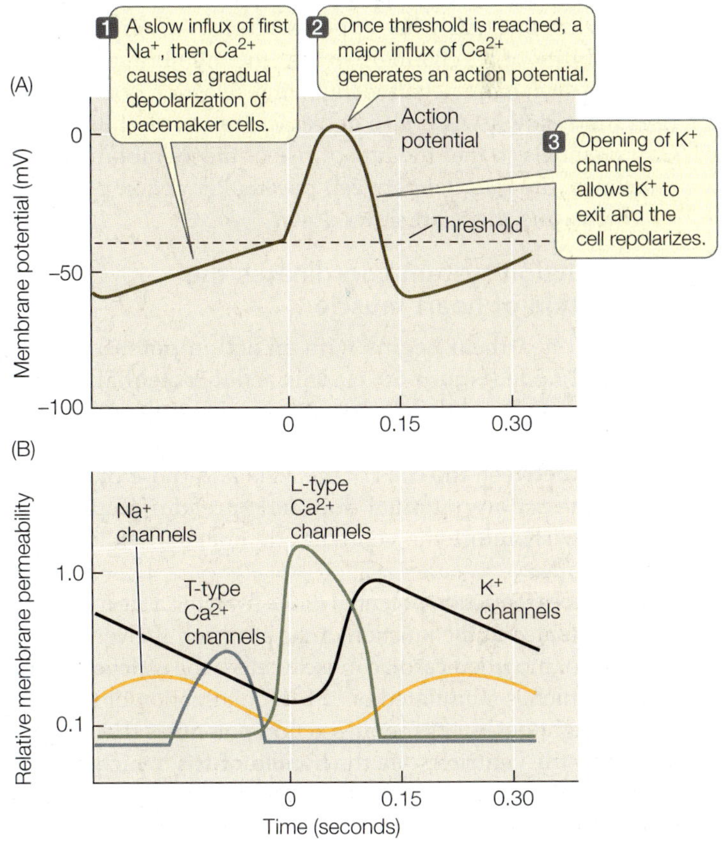
(A)
1 A slow influx of first Na+, then Ca2+ causes a gradual depolarization of pacemaker cells.
2 Once threshold is reached, a major influx of Ca2+ generates an action potential.
Action potential
3 Opening of K+ channels allows K+ to exit and the cell repolarizes.
Threshold
Membrane potential (mV)
0
–50
–100
0
0.15
0.30
(B)
Relative membrane permeability
1.0
0.1
Na+ channels
L-type Ca2+ channels
T-type Ca2+ channels
K+ channels
0
0.15
0.30
Time (seconds)

50.5 The Pacemaker Potential (A) The resting potential of the pacemaker cells of the sinoatrial node is less negative and gradually drifts upward between action potentials. Action potentials are slow to rise and are broad. These characteristics are due to properties of Na^+, K^+, and two types of Ca^{2+} channels. (B) At rest, sinoatrial pacemaker cells are more permeable to Na^+ than are neurons or other muscle cells. During rest, the cation channels gradually close, but some voltage-gated Ca^{2+} channels open; these are designated T-type channels because their opening is transient. When a threshold is reached, other voltage-gated Ca^{2+} channels open (designated as L-type for long-lasting), generating the action potential.

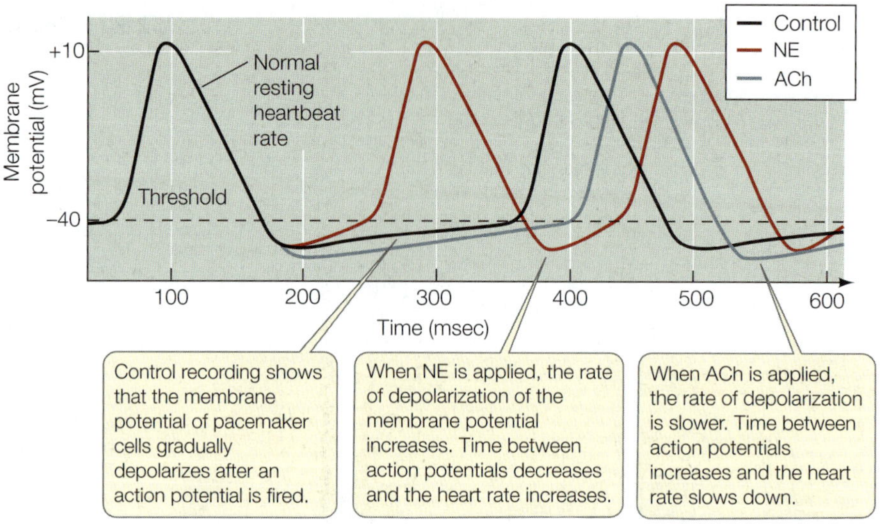

Control
NE
ACh
+10
–40
Membrane potential (mV)
Normal resting heartbeat rate
Threshold
100
200
300
400
500
600
Time (msec)
Control recording shows that the membrane potential of pacemaker cells gradually depolarizes after an action potential is fired.
When NE is applied, the rate of depolarization of the membrane potential increases. Time between action potentials decreases and the heart rate increases.
When ACh is applied, the rate of depolarization is slower. Time between action potentials increases and the heart rate slows down.

50.6 The Autonomic Nervous System Controls Heart Rate Isolated pacemaker cells continue firing action potentials in a culture dish. Neurotransmitter signals from the two divisions of the autonomic nervous system speed up and slow down the rate at which the pacemaker membrane potential drifts upward, thereby controlling the rate at which pacemaker cells fire action potentials.

becomes less negative, some Ca^{2+} channels open, causing the membrane potential to continue its gradual rise. Eventually this rising membrane potential reaches threshold for the major voltage-gated Ca^{2+} channels, and another action potential is generated. The intricate interaction of these ion channels through their effects on membrane potential causes the rhythmic generation of action potentials that characterizes pacemaker cells.

The autonomic nervous system controls the heartbeat (speeds it up or slows it down) by influencing the rate at which the resting potentials of pacemaker cells drift upward (**Figure 50.6**). Norepinephrine (NE) released onto pacemaker cells by sympathetic nerves increases the permeability of the Na^+ channels and the Ca^{2+} channels. The result is that the resting potential of the pacemaker cells drifts up more rapidly, the interval between action potentials is decreased, and the heart beats faster. Conversely, the parasympathetic neurotransmitter acetylcholine (ACh) has opposite effects. ACh increases the permeability of K^+ channels so that the membrane potential becomes even more negative following an action potential and rises more slowly. ACh also decreases the permeability of the Ca^{2+} channels so that the rate of rise of the membrane potential slows, the interval between pacemaker action potentials lengthens, and the heart slows down.

A conduction system coordinates the contraction of heart muscle

A normal heartbeat begins with an action potential in the sinoatrial node (**Figure 50.7**). This action potential spreads rapidly throughout the electrically coupled cells of the atria, causing them to contract in unison. Because there are no gap junctions between the cells of the atria and those of the ventricles, the action potential does not spread directly to the ventricles. Therefore the ventricles do not contract in unison with the atria.

How does the action potential move from the atria to the ventricles? Situated at the junction of the atria and the ventricles is a nodule of modified cardiac muscle cells—the **atrioventricular node**—which is stimulated by the depolarization of the atria. With a slight delay, it generates action potentials that are conducted to the ventricles via the **bundle of His**, which consists of modified cardiac muscle fibers that do not contract but do conduct action potentials. These fibers divide into right and left bundle branches that run to the tips of the ventricles and then spread throughout the ventricular muscle mass as **Purkinje fibers**. These conducting fibers ensure that the cardiac action potential spreads rapidly and evenly throughout the ventricular muscle mass, starting at the very bottom of the ventricles. The short delay in the spread of the action potential imposed by the atrioventricular node ensures that the atria contract before the ventricles do, so that the blood passes progressively from the atria to the ventricles to the arteries.

Electrical properties of ventricular muscles sustain heart contraction

Electrical properties of ventricular muscle fibers allow them to contract for about 300 milliseconds—much longer than those of skeletal muscle fibers. As in neuronal and skeletal muscle action potentials, the rising phase of the ventricular muscle cell action potentials is due to the opening of voltage-gated Na^+ channels. Unlike neurons and skeletal muscle fibers, however, ventricular muscle cells remain depolarized for a long time. This extended plateau of the action potential is due to sustained opening of voltage-gated Ca^{2+} channels (**Figure 50.8**). Like other muscle, cardiac muscle is stimulated to contract when Ca^{2+} is available to bind with troponin (see Figure 48.6). As long as Ca^{2+} remains in the sarcoplasm, the ventricular muscle cells continue to contract.

To terminate systole and allow the ventricles to fill again, Ca^{2+} must be rapidly cleared from the sarcoplasm of the ventricular cells. Ca^{2+} pumps in the sarcoplasmic reticulum membrane actively transport Ca^{2+} ions out of the sarcoplasm and into the sarcoplasmic reticulum; there, the ions are sequestered until the next action potential triggers another round of Ca^{2+} release and muscle contraction. Thus the rate of cycling of Ca^{2+} into and out of the sarcoplasmic reticulum puts limits on the heart rate and strength of contraction of the ventricle (**Figure 50.9**).

The drug digitalis has been used since the late 1700s to treat weakened hearts or hearts with irregular patterns of contraction. Digitalis strengthens and slows the heartbeat by slowing the reuptake of Ca^{2+} by the sarcoplasmic reticulum and thereby increasing the concentration of Ca^{2+} in the sarcoplasm. Before being introduced into the practice of medicine, digitalis prepared from the purple foxglove plant (*Digitalis purpurea*) was a folk remedy for heart problems.

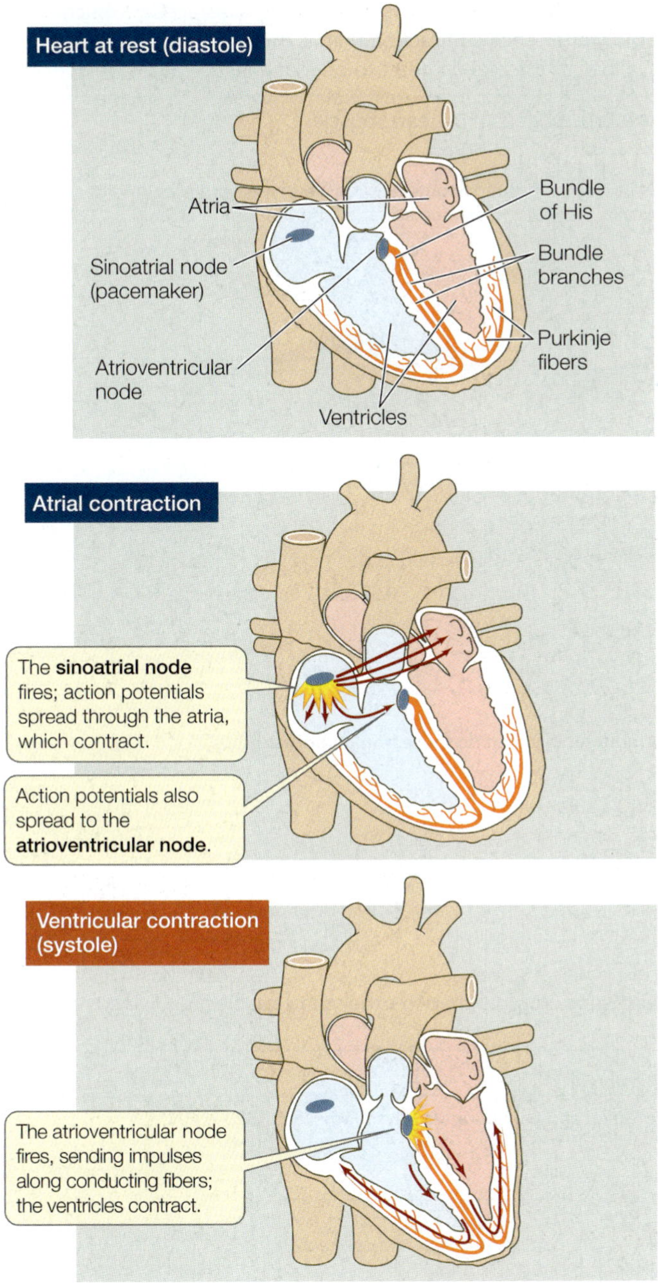

Heart at rest (diastole)
Atria
Sinoatrial node (pacemaker)
Atrioventricular node
Ventricles
Bundle of His
Bundle branches
Purkinje fibers
Atrial contraction
The **sinoatrial node** fires; action potentials spread through the atria, which contract.
Action potentials also spread to the **atrioventricular node**.
Ventricular contraction (systole)
The atrioventricular node fires, sending impulses along conducting fibers; the ventricles contract.

50.7 The Heartbeat Pacemaker cells in the sinoatrial node initiate the heartbeat by firing action potentials that spread through the electrically coupled atrial muscle. The atrial action potential eventually spreads to the atrioventricular node which, with a delay, conducts it through the bundle of His and Purkinje fibers to the cells of the ventricles.

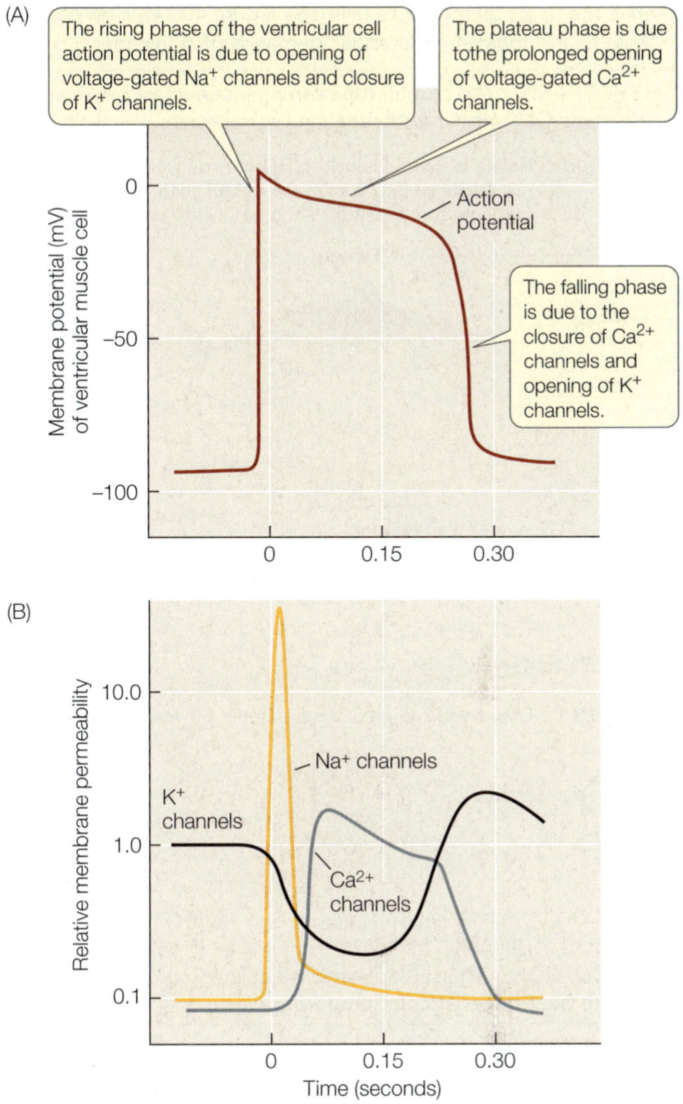

(A)
The rising phase of the ventricular cell action potential is due to opening of voltage-gated Na+ channels and closure of K+ channels.
The plateau phase is due tothe prolonged opening of voltage-gated Ca2+ channels.
Action potential
The falling phase is due to the closure of Ca2+ channels and opening of K+ channels.
Membrane potential (mV) of ventricular muscle cell
0
−50
−100
0
0.15
0.30
(B)
Relative membrane permeability
10.0
1.0
0.1
Na+ channels
K+ channels
Ca2+ channels
0
0.15
0.30
Time (seconds)

50.8 The Action Potential of Ventricular Muscle Fibers (A) The three phases of the action potential of ventricular muscle fibers are due to the opening and closing of voltage-gated channels. (B) At the initiation of the action potential, voltage-gated Na+ channels open rapidly but briefly. At the same time, but more slowly, K+ channels are closing and Ca^{2+} channels are opening. The open Ca^{2+} channels sustain the depolarization. Repolarization occurs when the Ca^{2+} channels close, and the slow opening of the K^+ channels also contribute to the repolarization.

The ECG records the electrical activity of the heart

Electrical events in the cardiac muscle during the cardiac cycle can be recorded by electrodes placed on the surface of the body. Such a recording is an **electrocardiogram**, or **ECG**. **EKG** is also used because German physicians who invented the method used the Greek spelling (*kardia*) and called it the electrokardiogramm. The ECG is an important tool for diagnosing heart problems (Figure 50.10A).

The action potentials that sweep through the muscles of the atria and ventricles before they contract are such massive, localized electrical events that they cause electric currents to flow throughout the body. Electrodes placed at different locations on the skin detect those currents at different times and register a voltage difference between them. The appearance of the ECG depends on the placement of the electrodes. Electrodes placed on the right wrist and left ankle produced the normal ECG shown in Figure 50.10B. The wave patterns of the ECG are designated P, Q, R, S, and T, each letter representing a particular event in the cardiac muscle, as shown in the figure.

INVESTIGATINGLIFE

50.9 Hot Fish, Cold Heart In "hot" fish like the bluefin tuna, highly active tissues such as the swimming muscles remain warm while the heart temperature falls (see Figure 40.14). How can a "cold heart" meet the metabolic demands of active muscles? Barbara Block and her colleagues demonstrated that the sarcoplasmic reticulum in the heart muscle cells of a bluefin takes up Ca^{2+} much more rapidly after each contraction than is the case for "cold" fish species, allowing the bluefin heart to beat at a faster rate.[a]

HYPOTHESIS The bluefin tuna can maintain a fast heart rate at low temperatures because its heart muscle cells cycle Ca^{2+} more rapidly than in typical fishes.

Method Isolate membrane-enclosed vesicles derived from sarcoplasmic reticulum (SR) from hearts of bluefin (a "hot" fish) and from albacore and yellowfin tuna ("cold" fish). Incubate the SR vesicles in a solution containing Ca^{2+} and measure their rate of Ca^{2+} uptake. Repeat the experiment at a range of temperatures.

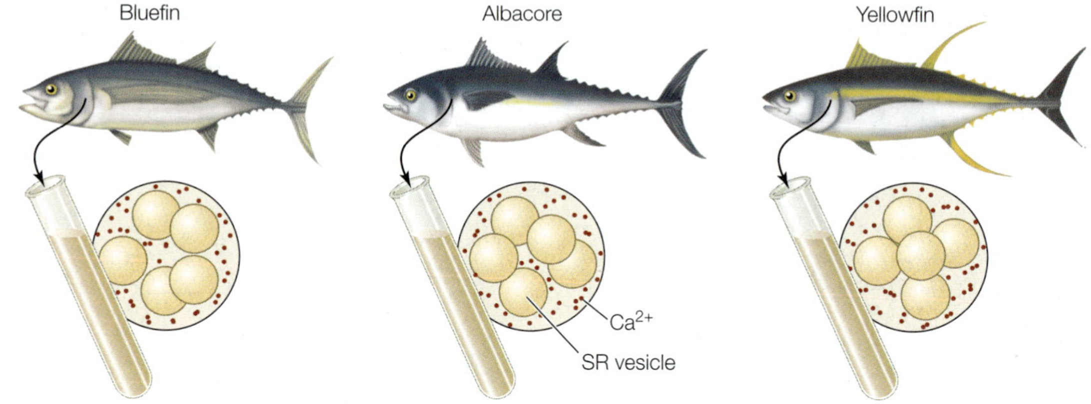
Bluefin
Albacore
Yellowfin
Ca2+
SR vesicle

Results Over a wide range of temperatures, SR vesicles from bluefin take up Ca^{2+} more rapidly than do those from the other fishes.

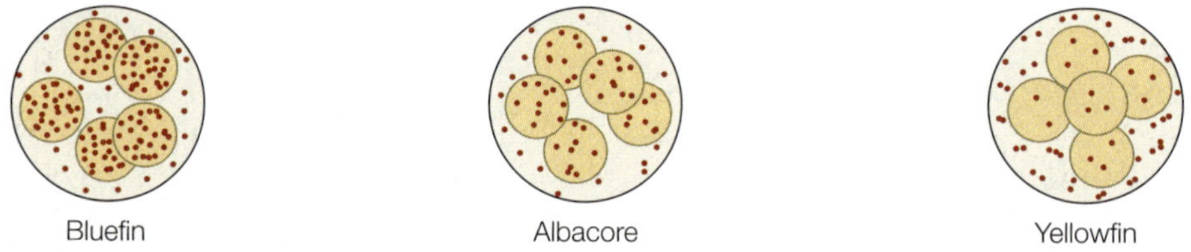
Bluefin
Albacore
Yellowfin

CONCLUSION The sarcoplasmic reticulum in bluefin heart muscle cycles Ca^{2+} more rapidly than in other tuna species, allowing the heart to beat faster at cold temperatures.

Go to **BioPortal** for discussion and relevant links for all INVESTIGATING**LIFE** figures.

[a]Landeira-Fernandez, A. et al. 2004. *American Journal of Physiology: Regulatory, Integrative, and Comparative Physiology* 286: R398–R404.

(A)

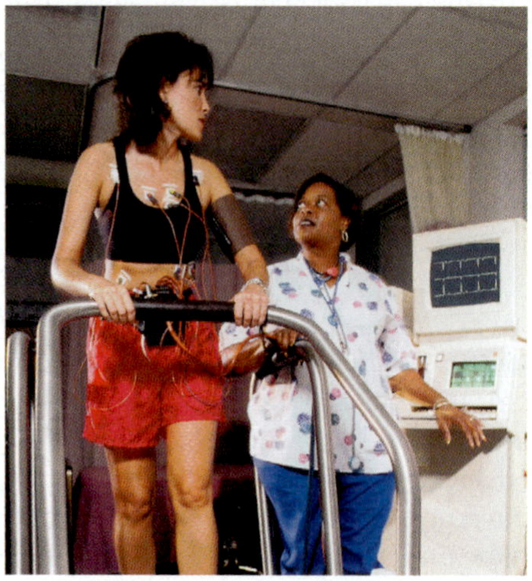

(B)

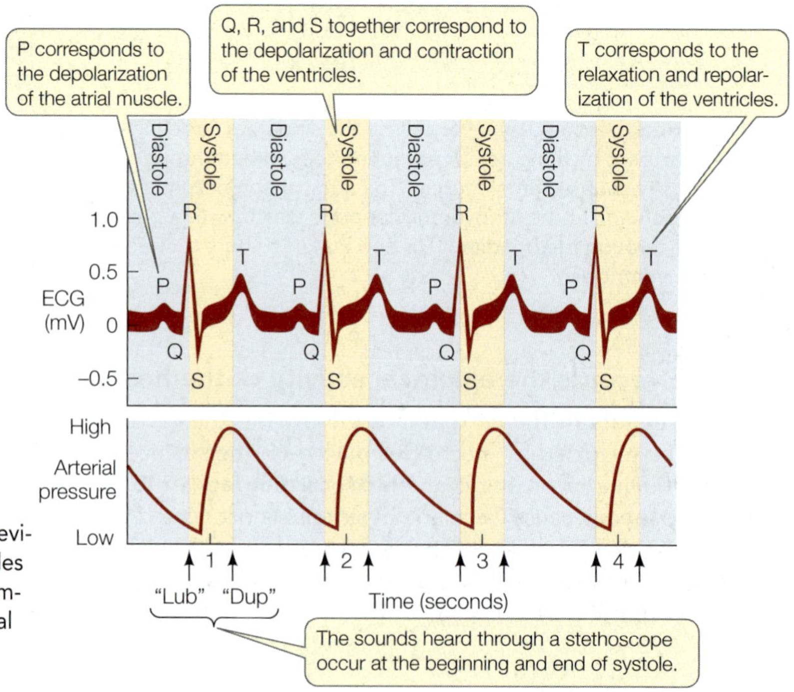
P corresponds to the depolarization of the atrial muscle.
Q, R, and S together correspond to the depolarization and contraction of the ventricles.
T corresponds to the relaxation and repolarization of the ventricles.
Diastole
Systole
1.0
0.5
0
–0.5
ECG (mV)
P
Q
R
S
T
High
Arterial pressure
Low
1
2
3
4
"Lub"
"Dup"
Time (seconds)
The sounds heard through a stethoscope occur at the beginning and end of systole.

50.10 The Electrocardiogram (A) An electrocardiogram (abbreviated as ECG or EKG) is used to monitor heart function. Electrodes attached to the person on the treadmill record an ECG that is amplified and displayed on a monitor. (B) Variations from the normal pattern shown here can be used to diagnose heart problems.

WORKING WITH**DATA:**

Warm Fish with Cold Hearts

Original paper

Landeira-Fernandez, A., M. M. Morrissette, J. M. Blanc, and B. A. Block. 2004. Temperature dependence of the Ca^{2+}-ATPase (SERCA2) in the ventricles of tuna and mackerel. *American Journal of Physiology: Regulatory, Integrative, and Comparative Physiology* 286: R398–R404.

Analyze the Data

"Hot fish" like the bluefin tuna (see Figure 40.14) face a dilemma when swimming in cold waters. As in all fish, the rate at which their hearts can beat goes down as the heart temperature goes down, which lowers the supply of oxygenated blood to the tissues of the body. But, unlike in "cold fish," the swimming muscles in bluefin tuna remain warm even in cold waters, so their demand for oxygen remains high. How does a bluefin tuna maintain an adequate blood supply to the swimming muscles at cold temperatures? The heart muscle of the bluefin tuna is adapted to cycle Ca^{2+} more rapidly than the heart muscle of "cold fish," allowing the heart to contract more frequently. Barbara Block and her colleagues hypothesized that this adaptation could be due to an increased ability to sequester Ca^{2+} back into the sarcoplasmic reticulum after each contraction (see Figure 50.9). The researchers isolated vesicles derived from the sarcoplasmic reticulum of heart muscle from three different fish: one "hot fish" (bluefin tuna) and two closely related "cold fish" (albacore and yellowfin tuna). The vesicles were incubated in a Ca^{2+} solution and the rate of Ca^{2+} uptake into the vesicles measured at a range of temperatures. The results are shown in the top panel of the figure at the right.

QUESTION 1

How do these data support the hypothesis that bluefin tuna have an increased ability to sequester Ca^{2+} in the sarcoplasmic reticulum? How would these data look if the hypothesis weren't true?

QUESTION 2

Based on these data, which fish would you expect to be able to achieve the highest cardiac output at 25°C, albacore or yellowfin? Albacore at 25°C or bluefin at 15°C?

QUESTION 3

The high rate of Ca^{2+} uptake observed in the bluefin vesicles means that the Ca^{2+}/ATPase pumps in the sarcoplasmic reticulum of the bluefin must be either different from those of other fish *or* more numerous. To distinguish between these two possibilities, the researchers plotted their data in a different way; this time they expressed the uptake rates as a percentage of the maximum rate (the rate at 30°C). This plot is shown in the bottom panel of the figure. What does this plot tell you about the properties of the Ca^{2+}/ATPase pumps in the different fish? What is the explanation for the higher rate of Ca^{2+} uptake in bluefin tuna?

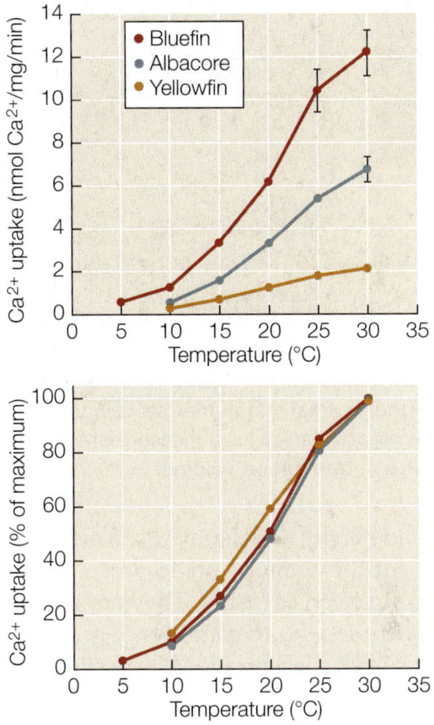
Bluefin
Albacore
Yellowfin
Ca2+ uptake (nmol Ca2+/mg/min)
14
12
10
8
6
4
2
0
0 5 10 15 20 25 30 35
Temperature (°C)
Ca2+ uptake (% of maximum)
100
80
60
40
20
0
0 5 10 15 20 25 30 35
Temperature (°C)

Go to **BioPortal** for all WORKING WITH**DATA** exercises

RECAP 50.3

The mammalian heart has two atria and two ventricles. Modified cardiac muscle tissue in the right atrium functions to spontaneously generate pacemaker action potentials. Other modified cardiac muscle tissue between the atria and ventricles and throughout the ventricles conducts those signals and coordinates the heart contraction. Broad action potentials in ventricular muscle reflect Ca^{2+} cycling in the ventricular muscle cells and make sustained contractions possible.

- Trace the path of blood through both sides of the heart, naming the major blood vessels and heart valves. **See pp. 1030–1031 and Figure 50.2**
- Differentiate systole and diastole and describe the events of the cardiac cycle. **See p. 1031 and Figure 50.3**
- How do cells of the sinoatrial node generate the heartbeat? **See pp. 1032–1034 and Figures 50.5 and 50.7**
- What determines the duration of the contraction of the ventricles during systole? **See p. 1034 and Figure 50.8 and 50.9**

Next we will consider the composition of the blood and the characteristics of the vessels through which blood circulates around the body, illustrating once again how structure serves function. We will also consider the role of the lymphatic vessels that return interstitial fluid to the blood.

50.4 What Are the Properties of Blood and Blood Vessels?

Blood is a connective tissue. It consists of cells suspended in an extracellular matrix of complex, yet specific, composition. The unusual feature of blood is that the extracellular matrix is a liquid, so blood is a fluid tissue.

The cells of the blood can be separated from the fluid matrix, called **plasma**, by centrifugation (**Figure 50.11**). If a sample of blood is spun in a centrifuge, all the cells move to the bottom of the tube, leaving the clear, straw-colored plasma on top. The packed-cell volume, or hematocrit, is the percentage of the

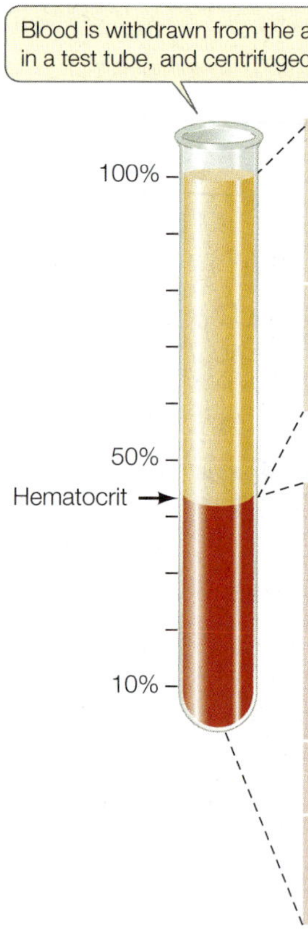
Blood is withdrawn from the arm, placed in a test tube, and centrifuged.
100%
50%
Hematocrit
10%

Plasma portion

		Salts	Plasma proteins
Components	Water	Sodium, potassium, calcium, magnesium, chloride, bicarbonate	Albumin Fibrinogen Immunoglobulins
Functions	Solvent	Osmotic balance, pH buffering, regulation of membrane potentials	Osmotic balance, pH buffering, clotting, immune responses

Transported by blood:
- Nutrients (e.g., glucose, vitamins)
- Waste products of metabolism
- Respiratory gases (O_2 and CO_2)
- Hormones
- Heat

Cellular portion

	Erythrocytes (red blood cells)	Leukocytes (white blood cells; see Chapter 42)	Platelets (cell fragments)
Components	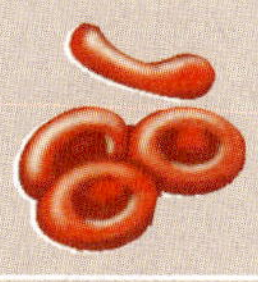	Basophil, Eosinophil, Neutrophil, Lymphocyte, Monocyte 	
Number per µl of blood	4–6 million	5,000–10,000	250,000–400,000
Functions	Transport oxygen and carbon dioxide	Destroy foreign cells, produce antibodies; roles in allergic responses	Blood clotting

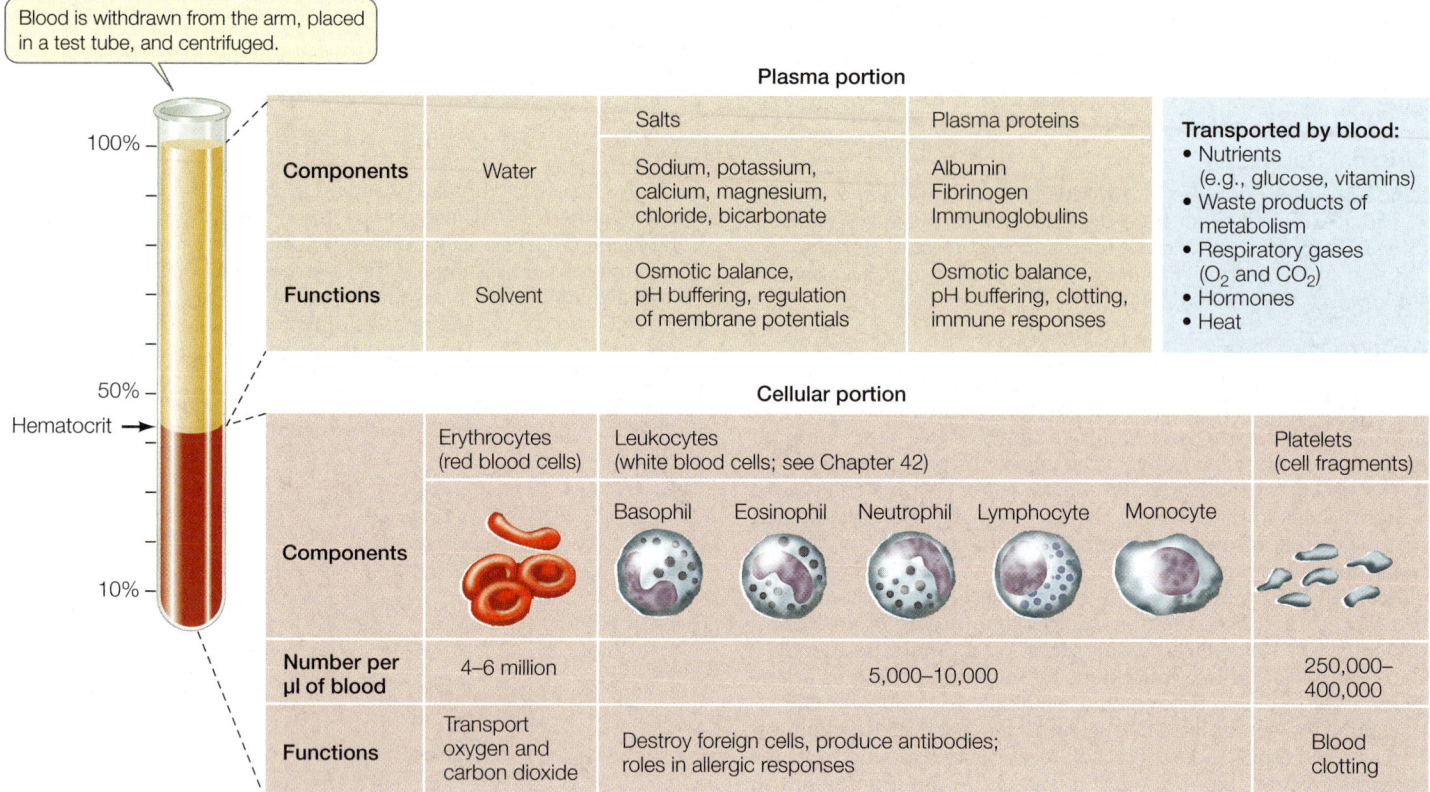

50.11 The Composition of Blood Blood consists of a complex aqueous solution (the plasma) and numerous cell types and cell fragments. The hematocrit (arrow) is a measure of the cellular portion as a percentage of total blood volume.

blood volume made up by red blood cells. Normal hematocrit is about 42 percent for women and 46 percent for men, but these values can vary considerably. They are usually higher, for example, in people who live and work at high altitudes, because the low oxygen concentrations there stimulate the production of more red blood cells.

Here we will consider two elements in blood: the red blood cells and the platelets, which are pinched-off fragments of cells. White blood cells, or leukocytes, are the cells of the immune system, discussed in Chapter 42.

Red blood cells transport respiratory gases

Most of the cells in the blood are **erythrocytes**, or red blood cells. Mature red blood cells are biconcave, flexible discs packed with hemoglobin. Their function is to transport respiratory gases. Their shape gives them a large surface area for gas exchange, and their flexibility enables them to squeeze through narrow capillaries. Men have 4.5 to 6.0 million red blood cells per microliter of blood, and women have 3.5 to 5.0 million.

Red blood cells, as well as all the other cellular components of blood, are generated by stem cells in the bone marrow, particularly in the ribs, breastbone, pelvis, vertebrae, and the long bones of the limbs. Red blood cell production is controlled by a hormone, **erythropoietin**, which is released by cells in the kidneys in response to insufficient oxygen—**hypoxia**. Many tissues respond to hypoxia by expressing a transcription factor called hypoxia-inducible factor 1 (HIF-1). When the kidneys become hypoxic and express HIF-1, one of the actions of the transcription factor is to activate the gene encoding erythropoietin. Increased circulating erythropoietin extends the lives of mature red blood cells and stimulates production of new red blood cells in the bone marrow.

Under normal conditions, your bone marrow produces about 2 million red blood cells every second. Developing red blood cells divide many times while still in the bone marrow, and during this time they produce hemoglobin. When hemoglobin makes up about 30 percent of the immature red blood cell, the nucleus, endoplasmic reticulum, Golgi apparatus, and mitochondria begin to break down. This process is almost complete when the newly mature red blood cell squeezes between the endothelial cells of blood vessels in the bone marrow and enters the circulation. Loss of nuclei from the red blood cells occurs in most mammalian species, but the red cells of a few mammals and of all other vertebrates are nucleated.

Each red blood cell circulates for about 120 days. As it gets older, its membrane becomes less flexible and more fragile, so older red blood cells are more likely to rupture as they bend to fit through narrow capillaries. Red blood cells are particularly squeezed in the **spleen**, an organ that sits near the stomach in the upper left side of the abdominal cavity. The spleen has many sinuses (cavities) that serve as reservoirs for red blood cells. To get into the sinuses, however, the red blood cells must squeeze between spleen cells. When old red blood cells are ruptured by this squeezing, their remnants are taken up and degraded by macrophages (a class of white blood cells that ingest debris and foreign materials).

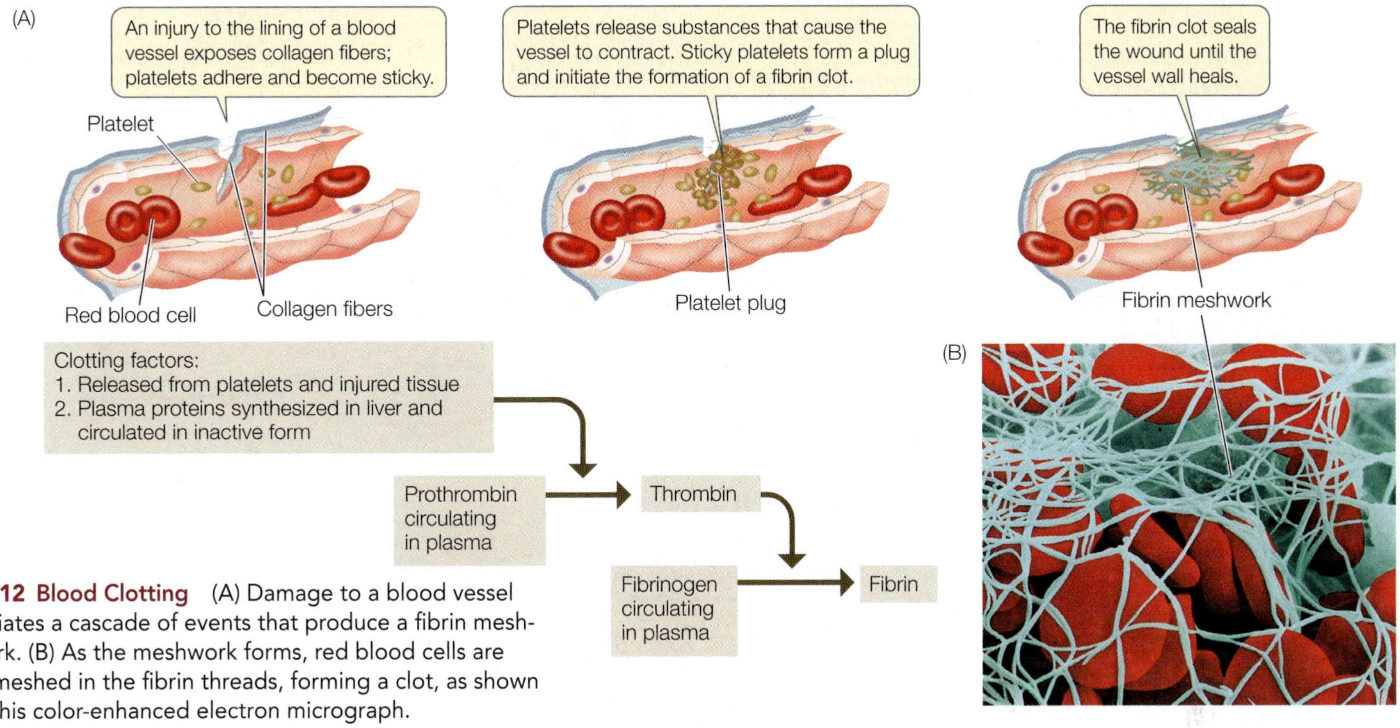
(A)
An injury to the lining of a blood vessel exposes collagen fibers; platelets adhere and become sticky.
Platelets release substances that cause the vessel to contract. Sticky platelets form a plug and initiate the formation of a fibrin clot.
The fibrin clot seals the wound until the vessel wall heals.
Platelet
Red blood cell
Collagen fibers
Platelet plug
Fibrin meshwork
Clotting factors:
1. Released from platelets and injured tissue
2. Plasma proteins synthesized in liver and circulated in inactive form
Prothrombin circulating in plasma
Thrombin
Fibrinogen circulating in plasma
Fibrin
(B)

50.12 Blood Clotting (A) Damage to a blood vessel initiates a cascade of events that produce a fibrin meshwork. (B) As the meshwork forms, red blood cells are enmeshed in the fibrin threads, forming a clot, as shown in this color-enhanced electron micrograph.

Platelets are essential for blood clotting

Besides producing erythrocytes and leukocytes, the bone marrow stem cells described in Section 42.1 also produce cells called megakaryocytes. Megakaryocytes are large cells that remain in the bone marrow and release cell fragments called **platelets** into the circulation. A platelet is just a tiny fragment of a cell without cell organelles, but it is packed with enzymes and chemicals necessary for its function: sealing leaks in blood vessels and initiating **blood clotting** (Figure 50.12).

Damage to a blood vessel exposes collagen fibers. An encounter with collagen fibers activates a platelet. The platelet swells, becomes irregularly shaped and sticky, and releases chemicals that activate other platelets and initiate the clotting of blood. The sticky platelets also form a plug at the damaged site.

Blood clotting requires many steps and many clotting factors, most of which are circulating in the blood in an inactive form. The absence of any one of these proteins can impair clotting and cause excessive bleeding. Because the liver produces most of the clotting factors, liver diseases such as hepatitis and cirrhosis can result in excessive bleeding. People with hemophilia experience uncontrolled bleeding because of a genetic inability to produce one of the clotting factors.

Blood clotting factors participate in a cascade of chemical activations of other substances circulating in the blood. The cascade begins with blood vessel and other tissue damage that exposes the blood to proteins such as collagen that are normally separated from the blood by endothelial cells lining the blood vessels. This exposure activates platelets and begins the clotting factor cascade. The end result of this cascade is to convert an inactive circulating enzyme, **prothrombin**, to its active form, **thrombin**. Thrombin cleaves molecules of **fibrinogen**, a plasma protein, forming insoluble threads of **fibrin**. The fibrin threads form the meshwork that binds platelets, seals the vessel, and provides a scaffold for the formation of scar tissue (see Figure 50.12).

Arteries withstand high pressure, arterioles control blood flow

Blood circulates through the vertebrate body in a system of closed vessels, and the properties of those different classes of vessels reflect their functions. The walls of the large arteries have many extracellular collagen and elastin fibers, which enable them to withstand the high blood pressures generated by the heart (**Figure 50.13A**). These elastic tissues have another important function: as we saw with the bulbus arteriosus in fish, they are stretched during systole, and thereby store some of the energy imparted to the blood by the heart. Elastic recoil during diastole returns this energy to the blood by squeezing it and pushing it forward. As a result, even though pressure in the arteries pulsates with the beating of the heart, the flow of blood is smoother than it would be through a system of rigid pipes.

Smooth muscle cells in the walls of the arteries and arterioles constrict or dilate those vessels. When the diameter of the vessels changes, their resistance to blood flow also changes, and the amount of blood flowing through them changes as a result. Neural and hormonal mechanisms act on smooth muscle cells in the walls of the arteries and arterioles, controlling the flow of blood through these vessels. The arterioles are referred to as resistance vessels because their resistance can vary to control the blood flow to specific tissues.

Materials are exchanged in capillary beds by filtration, osmosis, and diffusion

Beds of capillaries lie between arterioles and venules (**Figure 50.13B**). Few cells are more than a few cell diameters away from a capillary. (Notable exceptions include developing

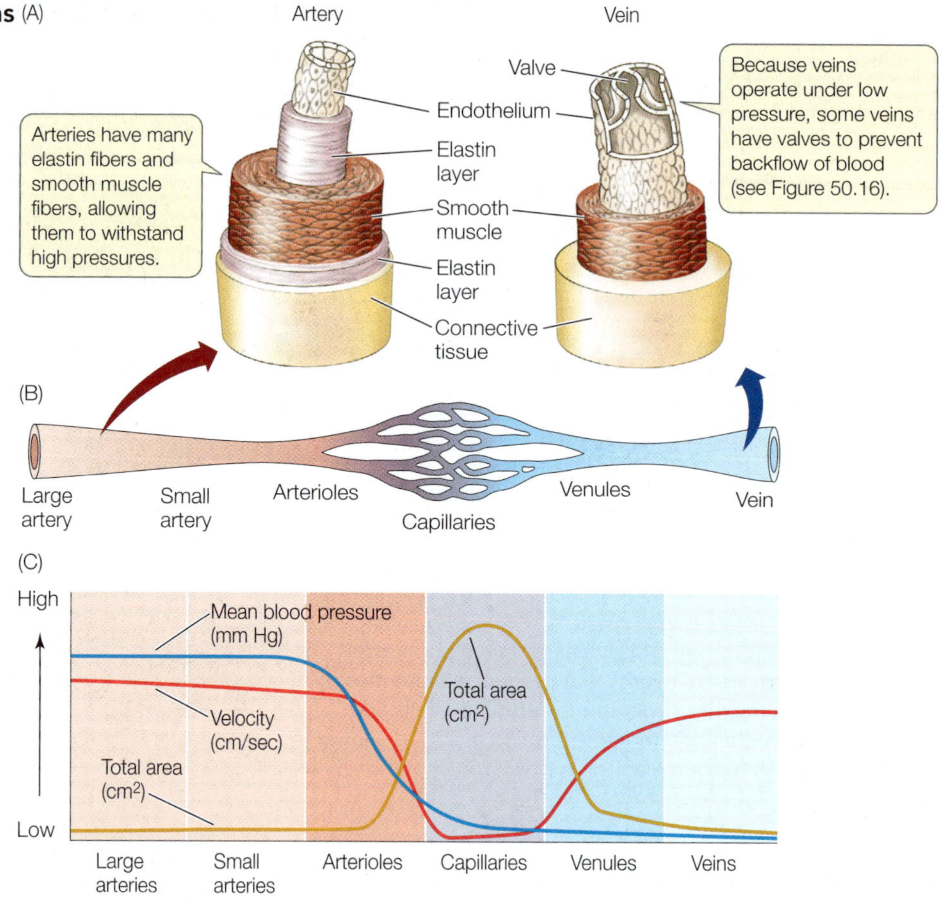
(A)
Artery
Vein
Valve
Endothelium
Elastin layer
Smooth muscle
Elastin layer
Connective tissue
Arteries have many elastin fibers and smooth muscle fibers, allowing them to withstand high pressures.
Because veins operate under low pressure, some veins have valves to prevent backflow of blood (see Figure 50.16).
(B)
Large artery
Small artery
Arterioles
Capillaries
Venules
Vein
(C)
High
Low
Mean blood pressure (mm Hg)
Velocity (cm/sec)
Total area (cm2)
Total area (cm2)
Large arteries
Small arteries
Arterioles
Capillaries
Venules
Veins

50.13 Anatomy of Blood Vessels (A) The different anatomical characteristics of arteries and veins match their functions. (B) Blood from the arterial system feeds into capillary beds, where exchanges with the interstitial fluid occur. The venous system returns the blood to the heart. (C) The area encompassed by each vessel type is graphed along with the pressure and velocity of the blood within them.

Go to Activity 50.3 Structure of a Blood Vessel
Life10e.com/ac50.3

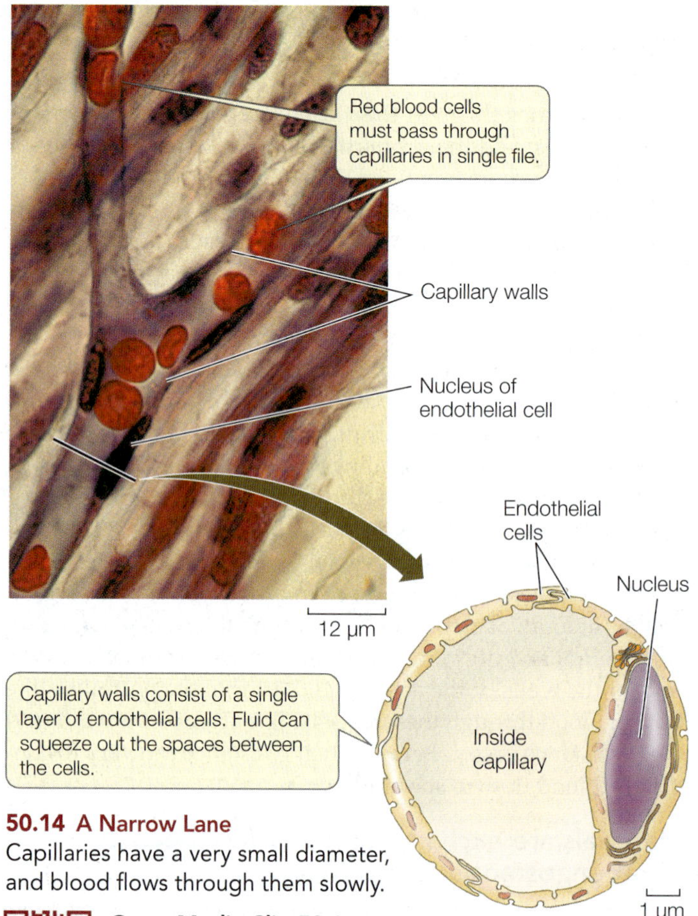
Red blood cells must pass through capillaries in single file.
Capillary walls
Nucleus of endothelial cell
12 µm
Endothelial cells
Nucleus
Inside capillary
Capillary walls consist of a single layer of endothelial cells. Fluid can squeeze out the spaces between the cells.
1 µm

50.14 A Narrow Lane
Capillaries have a very small diameter, and blood flows through them slowly.

Go to Media Clip 50.1
Capillary Flow: A Tight Squeeze
Life10e.com/mc50.1

oocytes and the cells of the lens and cornea.) The cells' needs are served by the exchange of materials between blood and interstitial fluid across the capillary walls. Capillary walls are thin and permeable to water and many solutes. Also, blood flows slowly through capillaries, allowing time for exchange.

It may seem strange that blood flows through the large arteries rapidly at high pressures, but when it reaches the small capillaries the pressure and rate of flow decrease (**Figure 50.13C**). When you place your thumb over the opening of a garden hose, the pressure in the hose increases, which in turn increases the velocity of the water spraying out. But keep in mind that the arteries branch into many arterioles, which give rise to a huge number of capillaries. Even though each capillary has a diameter so small that red blood cells must pass through in single file, there are so many capillaries that their total cross-sectional area is much greater than that of any other class of vessels. As a result, all the capillaries together have a much greater capacity for blood than do the arterioles. Returning to our garden hose analogy, if we connected the hose to large number of lawn sprinklers, the pressure and flow in each sprinkler would be low.

Capillary walls consist of a single layer of endothelial cells (**Figure 50.14**). Capillaries are permeable to water, some ions, and some small molecules but not to large molecules such as proteins. At the arterial (high pressure) end, blood pressure squeezes water and small solutes out through spaces between the cells of the capillary walls into the surrounding intercellular space.

Why don't water and small-molecular-weight solutes collect in the intercellular spaces? How is the blood volume maintained if fluid is continuously leaking out of the capillaries?

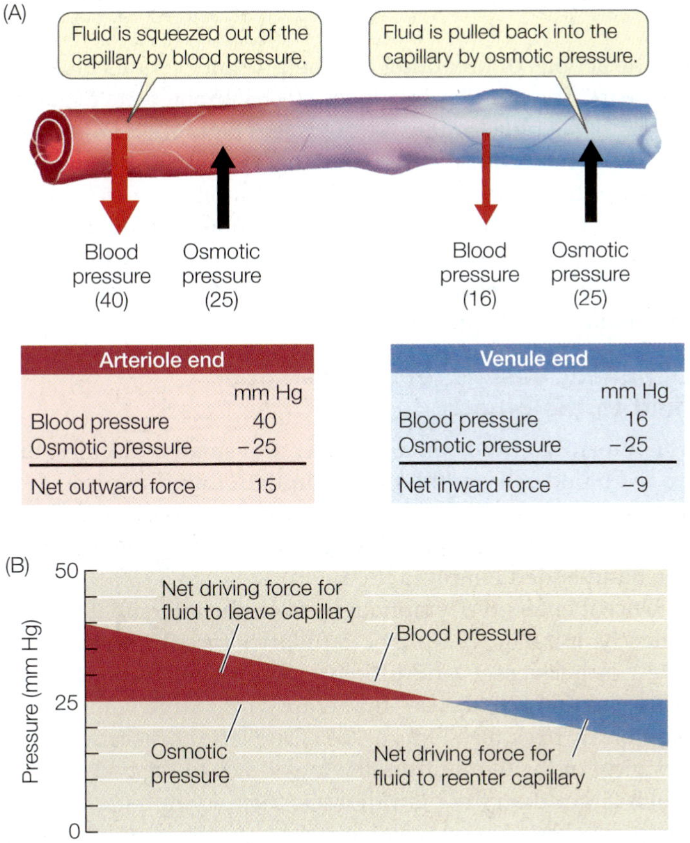
(A)
Fluid is squeezed out of the capillary by blood pressure.
Fluid is pulled back into the capillary by osmotic pressure.
Blood pressure (40)
Osmotic pressure (25)
Blood pressure (16)
Osmotic pressure (25)
Arteriole end
mm Hg
Blood pressure 40
Osmotic pressure −25
Net outward force 15
Venule end
mm Hg
Blood pressure 16
Osmotic pressure −25
Net inward force −9
(B)
Pressure (mm Hg)
50
25
0
Net driving force for fluid to leave capillary
Blood pressure
Osmotic pressure
Net driving force for fluid to reenter capillary

50.15 Starling's Forces Starling's model explains how blood volume is maintained in the capillary beds. (A) When blood pressure is greater than osmotic pressure, fluid leaves the capillaries; when blood pressure falls below osmotic pressure, fluid returns to the capillaries. (B) The balance of these two forces changes over the capillary bed as blood pressure falls.

An answer to this question was put forth more than 100 years ago by the physiologist E. H. Starling. Starling suggested that water movement across capillary walls is a result of two opposing forces, which are now known as **Starling's forces**:

- Blood pressure squeezes water and small solutes out of the capillaries.
- Osmotic pressure pulls water back into the capillaries.

Blood pressure is high at the arterial end of a capillary bed and steadily drops as blood moves toward the venous end (**Figure 50.15**). The osmotic pressure is due to the large protein molecules that cannot leave the capillaries, and it is relatively constant along the capillaries. As long as the blood pressure is above the osmotic pressure, fluid leaves the capillaries. At the venule end of most capillaries, blood pressure falls below the osmotic pressure, so fluid returns to the capillaries. The actual numbers for a normal capillary bed in a resting person suggest that there would be a *slight* net loss of fluid to the intercellular spaces. This loss, about 4 liters per day, percolates between cells as the interstitial fluid before it returns to the venous blood via the lymphatic system, which we will discuss later in this chapter.

Several observations support Starling's model. In people with severe liver disease or protein starvation, a fall in blood protein concentration leads to an accumulation of fluid in the extracellular spaces, which results in tissue swelling, or **edema**. Edema is also characteristic of the inflammation response accompanying tissue damage or allergic responses (see Figure 42.5). **Histamine**, a mediator of inflammation released by certain white blood cells, increases capillary permeability and relaxes the smooth muscles of the arterioles, raising blood pressure in the capillaries and leading to fluid leakage into tissues.

A few situations are not explained by Starling's hypothesis. During strenuous exercise, the blood pressure in the arterioles serving the muscles rises substantially but does not result in edema. In birds, the blood pressure in arterioles is much higher than in mammals, and the osmotic pressure is lower. If edema is not a chronic problem in exercising muscles and in birds, what is missing from Starling's model?

Recent research suggests that bicarbonate ions (HCO_3^-) in the blood plasma contribute significantly to the osmotic attraction that draws water back into the capillaries. The CO_2 produced by cellular metabolism diffuses into the endothelial cells lining the capillaries, where it is converted into HCO_3^- and released into the plasma. When an individual is at rest, the increasing HCO_3^- concentration can cause the osmotic pressure of the blood at the venous end to be 30 mm Hg higher than at the arterial end, and during strenuous exercise this difference can be much higher. Thus it appears that CO_2 and HCO_3^- are major factors that pull water back into the capillaries.

All capillaries are permeable to O_2, CO_2, and small ions. Lipid-soluble substances readily pass through the capillary walls. Water and small solutes pass through spaces in the capillary wall, and in some cases through holes called **fenestrations**. These less selective capillaries are found in the digestive tract, where nutrients are absorbed, and in the kidneys, where wastes are filtered. The capillary walls also contain transporters that can facilitate the passage of specific molecules, such as glucose and lactate. Overall, permeability varies widely in different capillary beds—an important consideration in the design and delivery of drugs.

The capillaries of the brain are a special case, being rather impermeable and wrapped by glia. Not much can pass through them other than lipid-soluble substances (including alcohol and anesthetics). This high selectivity of brain capillaries is known as the **blood–brain barrier**. Even in the brain, however, there are specific regions where the capillaries are more permeable, enabling the brain to detect non-lipid-soluble hormones.

Blood flows back to the heart through veins

The pressure of the blood flowing from capillaries to venules is extremely low and is insufficient to propel blood back to the heart. The walls of veins are more expandable than the walls of arteries, and blood tends to accumulate in veins. As much as 60 percent of your total blood volume may be in your veins when you are resting. Because of their high capacity to stretch and store blood, veins are called capacitance vessels.

Blood flow through veins that are above the level of the heart is assisted by gravity. Below the level of the heart, however, venous return is against gravity. The most important force propelling blood from these regions is the squeezing of the veins by the contractions of surrounding skeletal muscles.

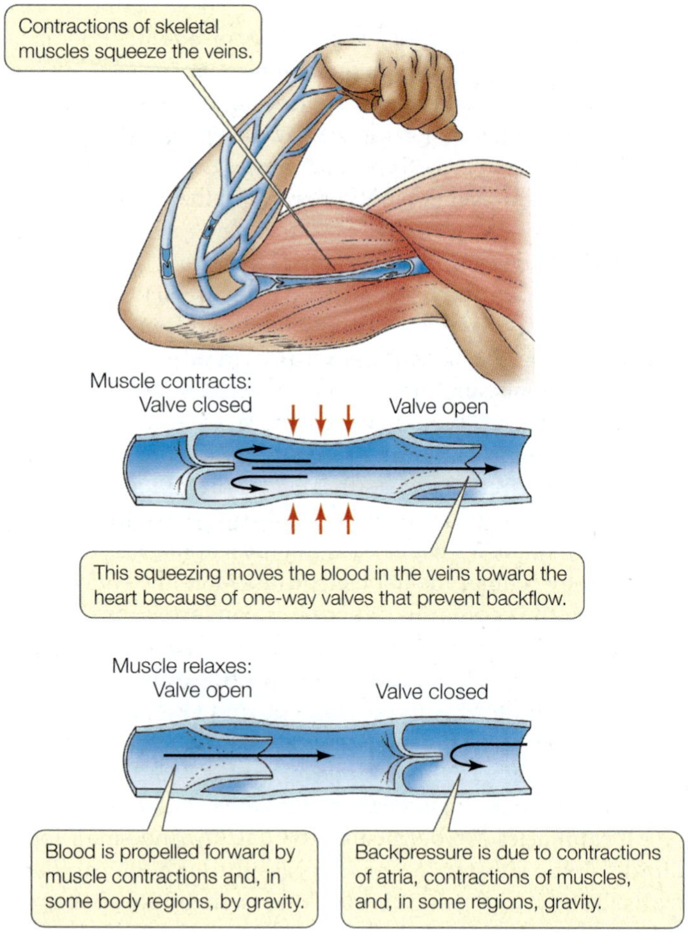
Contractions of skeletal muscles squeeze the veins.
Muscle contracts:
Valve closed
Valve open
This squeezing moves the blood in the veins toward the heart because of one-way valves that prevent backflow.
Muscle relaxes:
Valve open
Valve closed
Blood is propelled forward by muscle contractions and, in some body regions, by gravity.
Backpressure is due to contractions of atria, contractions of muscles, and, in some regions, gravity.

50.16 One-Way Flow Veins have valves that prevent blood from flowing backward, and contractions of skeletal muscle help move blood toward the heart.

As muscles contract, the vessels are compressed and blood is squeezed through them. Blood flow may be temporarily obstructed during a prolonged muscle contraction, but when muscles relax, blood is free to move again. One-way valves in the veins of the extremities prevent backflow of blood. Thus whenever a vein is squeezed, blood is propelled forward toward the heart (**Figure 50.16**).

In a resting person, gravity causes blood accumulation in the veins of the lower body and exerts backpressure on the capillary beds. This backpressure shifts the balance between blood pressure and osmotic pressure, causing increased loss of fluid to the intercellular spaces. That is why your feet swell during a long airline flight.

Because of the one-way valves in the veins of the legs, the contractions of leg muscles act as auxiliary vascular pumps when an animal walks or runs and facilitate the return of blood to the heart from the lower body. As a greater volume of blood is returned to the heart, the heart contracts more forcefully and its pumping action is enhanced. The heartbeat gets stronger because of a property of cardiac muscle cells described by the **Frank–Starling law**: if the cardiac muscle cells are stretched, as they are when the volume of returning blood increases, they contract more forcefully.

The actions of breathing also help return venous blood to the heart. The muscles involved in inhalation create negative pressure that pulls air into the lungs (see Figure 49.11), and this negative pressure also pulls blood toward the chest, increasing venous return to the right atrium. In addition, some of the largest veins closest to the heart contain smooth muscle that contracts at the onset of exercise. Contraction of veins can rapidly increase venous return and stimulate the heart in accord with the Frank–Starling law, increasing cardiac output.

Lymphatic vessels return interstitial fluid to the blood

The interstitial fluid contains water and small molecules, but no red blood cells, and less protein than found in plasma. A separate system of vessels—the **lymphatic system**—returns interstitial fluid to the blood. Each capillary bed contains at least one blind-ended lymph capillary.

Once it enters the lymphatic vessels, the interstitial fluid is called **lymph**. Fine lymphatic capillaries merge into progressively larger vessels and ultimately into two lymphatic vessels—the **thoracic ducts**—that empty into large veins at the base of the neck (see Figure 42.1). The left thoracic duct carries most of the lymph from the lower part of the body and is much larger than the right thoracic duct. Lymphatic vessels, like veins, have one-way valves that keep the lymph flowing toward the thoracic duct. Therefore lymph, like blood, is propelled toward the heart by skeletal muscle contractions and breathing movements. Lymph nodes along the major lymphatic vessels are a major site of lymphocyte production and of the phagocytic action that removes microorganisms and other foreign materials from the circulation (see Section 42.1).

Vascular disease is a killer

As mentioned at the start of this chapter, cardiovascular disease is responsible for about one-fourth of all deaths each year in the United States, and the same is true for Europe. The immediate cause of most of these deaths is not a defect in heart muscle as in athletes, but is heart attack or stroke—both of which are usually the end result of a disease called **atherosclerosis** ("hardening of the arteries") that begins many years before symptoms are detected.

Healthy arteries have a smooth internal lining of endothelial cells (**Figure 50.17A**) that can be damaged by chronic high blood pressure, smoking, a high-fat diet, or microorganisms. Deposits called **plaque** begin to form at sites of endothelial damage. First, the damaged endothelial cells attract certain white blood cells to the site. These cells are then joined by smooth muscle cells migrating from the deeper layers of the arterial wall. Lipids, especially cholesterol, are deposited in these cells, so that the developing plaque becomes fatty. Fibrous connective tissue made by the invading smooth muscle cells in the plaque, along with deposits of calcium, make the artery wall less elastic—hence "hardening of the arteries." The growing plaque deposit narrows the artery and causes turbulence in the blood flow. Blood platelets stick to the plaque (see Figure 50.12) and initiate formation of an intravascular blood clot, a **thrombus**, which can block the artery (**Figure 50.17B**).

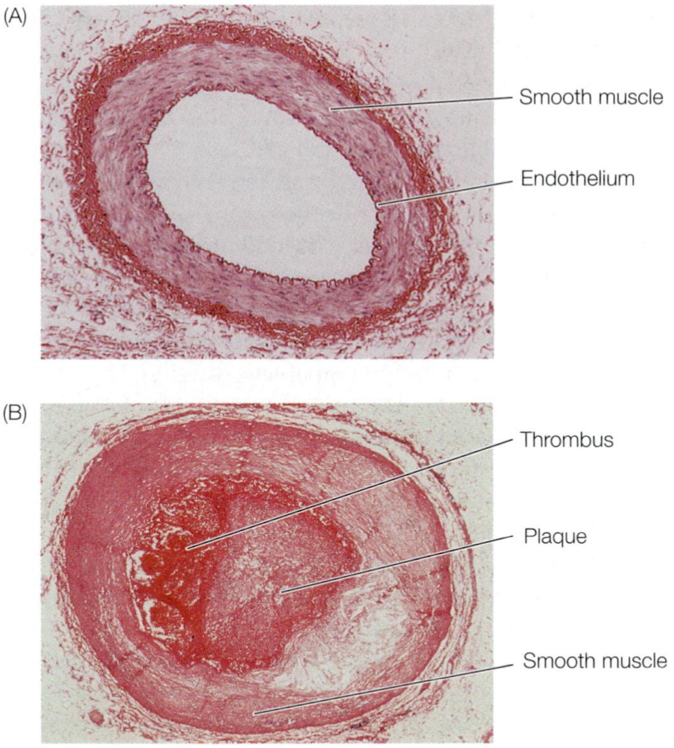
(A)
Smooth muscle
Endothelium
(B)
Thrombus
Plaque
Smooth muscle

50.17 Atherosclerotic Plaque (A) A healthy, clear artery. (B) An atherosclerotic artery, clogged with plaque and a thrombus.

The blood supply to the heart muscle flows through the **coronary arteries**, which are highly susceptible to atherosclerosis. As these arteries narrow, blood flow to the heart muscle decreases, causing the symptoms of chest pain and shortness of breath during mild exertion. A person with atherosclerosis is at high risk of forming a thrombus in a coronary artery. This condition, called **coronary thrombosis**, can totally block the vessel, causing a **myocardial infarction** (heart attack).

A piece of a thrombus that breaks loose, called an **embolus**, is likely to travel to and become lodged in a vessel of smaller diameter, blocking its flow (an **embolism**). Arteries already narrowed by plaque formation are likely places for an embolism. An embolism in an artery in the brain causes the cells fed by that artery to die. This event is a **stroke**. The specific damage resulting from a stroke, such as memory loss, speech impairment, or paralysis, depends on the location of the blocked artery.

Important risk factors for developing atherosclerosis are your genetic predisposition and your age. Environmental risk factors also play a large role, however. These include high-fat and high-cholesterol diets, smoking, and a sedentary lifestyle. Certain untreated medical conditions such as hypertension (high blood pressure), obesity, and diabetes are also risk factors for atherosclerosis. For those who have a genetic predisposition to atherosclerosis, it is even more important to minimize environmental risk factors. Changes in diet and behavior and treatment of predisposing medical conditions can prevent and reverse early atherosclerosis and help fend off this silent killer.

RECAP 50.4

Blood is a fluid tissue with cellular components that play roles in transport of respiratory gases, immune system function, and blood clotting. The properties of the arteries, arterioles, capillaries, venules, and veins reflect their functions. Exchanges between the blood and interstitial fluids occur in the smallest of those vessels, the capillaries.

- How are the structural differences among the various classes of vessels related to their functions? **See pp. 1039–1040 and Figure 50.13**
- Why are arterioles called resistance vessels and veins called capacitance vessels? **See p. 1039 and p. 1041**
- What factors control the movement of fluids between the vascular and extravascular spaces? **See pp. 1040–1041 and Figures 50.14 and 50.15**
- What propels blood from the lower part of the body back to the heart? **See pp. 1041–1042 and Figure 50.16**

Every tissue in the body requires an adequate flow of oxygen-saturated blood. Blood flow depends on the maintenance of an appropriate blood pressure, and the distribution of blood flow throughout the body depends on control of the resistance in the blood vessels supplying different tissues.

50.5 How Is the Circulatory System Controlled and Regulated?

When we investigate how a physiological process is regulated, we start by identifying the critical components of that process, how they can be controlled, and the information used to govern that control. Because blood flow depends on pressure, we can identify the pressure in the aorta as a critical variable of the circulatory system. The pressure in the aorta oscillates between systole and diastole, so we define our variable as the mean arterial pressure (MAP). MAP is determined by the cardiac output (CO) and the resistance to flow in the blood vessels, or total peripheral resistance (TPR):

$$\text{MAP} = \text{CO} \times \text{TPR}$$

Since CO is equal to the heart rate (HR) times how much blood the heart pumps with each beat (stroke volume [SV]), the critical relationships can be expressed as:

$$\text{MAP} = \text{HR} \times \text{SV} \times \text{TPR}$$

HR, SV, and TPR are controlled by neural and hormonal mechanisms at both the local and systemic levels. At the local level, each tissue controls its own blood flow through autoregulatory mechanisms that cause the arterioles supplying that tissue to constrict or dilate.

The collective autoregulatory actions in the capillary beds in all tissues of the body determine TPR and therefore MAP. If many arterioles suddenly dilate, TPR goes down and MAP falls. If many arterioles constrict, TPR goes up and MAP goes up. Changes in MAP provide information about changing needs of the body. In addition, as blood flows through capillary beds, its composition changes—its CO_2 content goes up and its O_2

content goes down. Thus blood composition also provides information the body uses to regulate the circulatory system.

The nervous and endocrine systems respond to changes in MAP and blood composition by changing breathing rate, heart rate, stroke volume, and peripheral resistance to match the metabolic needs of the body.

Go to Animated Tutorial 50.2
Blood Pressure and Heart Rate Regulation
Life10e.com/at50.2

Autoregulation matches local blood flow to local need

The amount of blood that flows through a capillary bed is controlled by the smooth muscle of the arteries and arterioles feeding that bed. **Figure 50.18** illustrates the flow of blood in a typical capillary bed. Blood flows into the bed from an arteriole. Smooth muscle "cuffs," or **precapillary sphincters**, on the arteriole can shut off the supply of blood to the capillary bed. When the precapillary sphincters are relaxed and the arteriole is open, the arterial blood pressure pushes blood into the capillaries.

Autoregulation depends on the sensitivity of the smooth muscle to its local chemical environment. Low O_2 concentrations and high CO_2 concentrations cause the smooth muscle to relax, thus increasing the supply of blood, which brings in more O_2 and carries away CO_2—a response known as hyperemia, which means "excess blood." Increases in other by-products of metabolism, such as lactic acid, hydrogen ions, potassium, and adenosine (all of which increase in exercising muscle), also promote hyperemia. Hence activities that increase the metabolism of a tissue also induce hyperemia in that tissue.

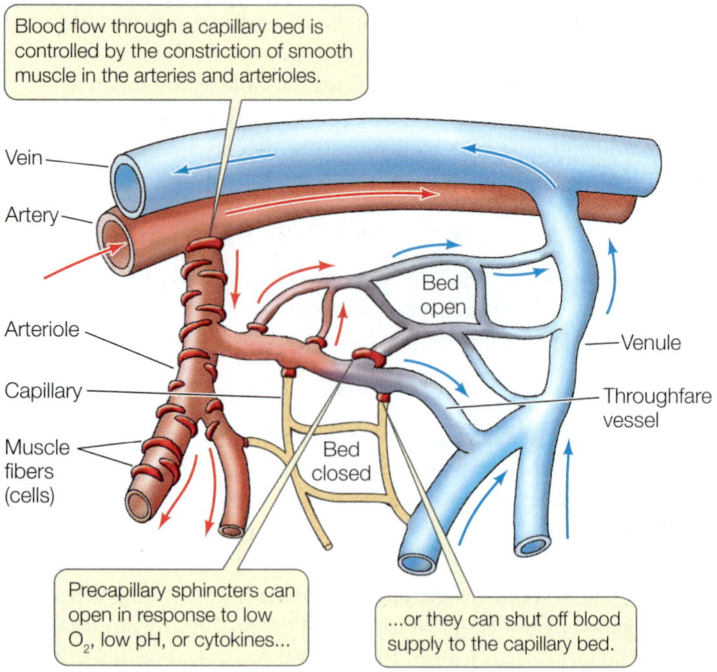
Blood flow through a capillary bed is controlled by the constriction of smooth muscle in the arteries and arterioles.
Vein
Artery
Arteriole
Capillary
Muscle fibers (cells)
Bed open
Bed closed
Venule
Throughfare vessel
Precapillary sphincters can open in response to low O_2, low pH, or cytokines...
...or they can shut off blood supply to the capillary bed.

50.18 Local Control of Blood Flow Low O_2 concentrations or high levels of metabolic by-products cause the smooth muscle of the arteries and arterioles to relax, thus increasing the supply of blood to the capillary bed.

Arterial pressure is regulated by hormonal and neural mechanisms

Control and regulation of the circulatory system begins with the local autoregulatory mechanisms that alter the resistance of arteries and arterioles feeding capillary beds. The demands of the capillary beds influence MAP and blood composition. Both of these provide information for the control of endocrine and neural responses that act to return blood pressure and composition to normal. Thus circulatory functions are matched to the regional and overall needs of the body.

Arteries and arterioles are innervated by the autonomic nervous system, particularly the sympathetic division. The sympathetic postganglionic neurotransmitter norepinephrine binds to receptors in smooth muscle in blood vessels in the gut and other tissues not essential for "fight or flight" and causes these vessels to constrict, resulting in reduced blood flow through them and an elevation in MAP. As we discussed earlier in this chapter, increased sympathetic activity increases heart rate, and by increasing the strength of the cardiac muscle contraction, it also increases stroke volume.

Hormones also play a role in regulating arterial pressure. Epinephrine has actions similar to those of norepinephrine and is released from the adrenal medulla during massive sympathetic activation stimulated by a fall in arterial pressure or by activation of the fight-or-flight response to a dangerous threat. Another hormone, **angiotensin**, is produced when blood pressure in the kidneys falls (**Figure 50.19**). These hormones influence arterioles located in peripheral tissues (extremities) or in tissues whose functions need not be maintained continuously (such as the digestive system). By reducing blood flow in those arterioles, the hormones increase central blood pressure and blood flow to essential organs such as the heart, brain, and kidneys.

The autonomic nervous system activity that controls heart rate and constriction of blood vessels originates in a cardiovascular control center in the medulla. Many inputs converge on this central integrative network and influence the commands it issues via parasympathetic and sympathetic nerves (**Figure 50.20**). Of special importance is incoming information about changes in blood pressure and composition from both **baroreceptors** (stretch receptors) and chemoreceptors in the walls of the large arteries leading to the brain—the aorta and the carotid arteries.

Increased activity in baroreceptors of the large arteries signals rising blood pressure and inhibits sympathetic nervous system signaling to arteries and arterioles while increasing parasympathetic signaling to the heart's pacemaker. As a result, the heart slows and arterioles in peripheral tissues dilate, reducing blood pressure. If pressure in the large arteries falls, the activity of the baroreceptors decreases, stimulating sympathetic output to the arteries and arterioles while reducing parasympathetic output to the heart's pacemaker. As a result, the heart beats faster and the arterioles in peripheral tissues constrict, increasing blood pressure.

Another hormone that helps stabilize blood pressure is antidiuretic hormone (ADH, also called vasopressin), which is secreted by the posterior pituitary in response to a fall in the activity of the baroreceptors, signaling a fall in arterial pressure. ADH causes the kidneys to resorb more water and thereby

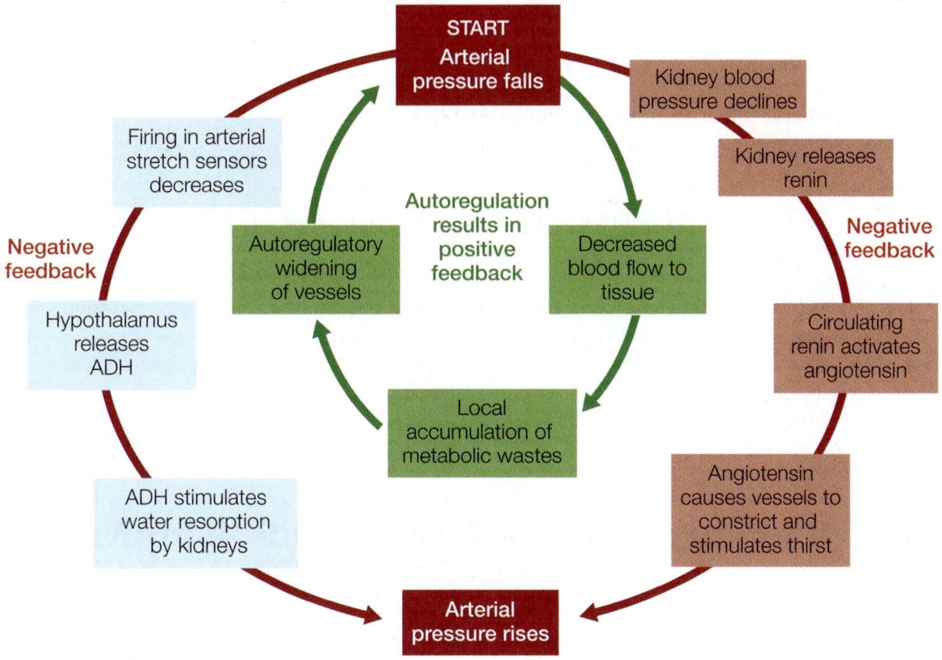

START
Arterial pressure falls
Kidney blood pressure declines
Kidney releases renin
Negative feedback
Circulating renin activates angiotensin
Angiotensin causes vessels to constrict and stimulates thirst
Arterial pressure rises
ADH stimulates water resorption by kidneys
Hypothalamus releases ADH
Negative feedback
Firing in arterial stretch sensors decreases
Autoregulation results in positive feedback
Autoregulatory widening of vessels
Decreased blood flow to tissue
Local accumulation of metabolic wastes

50.19 Control of Blood Pressure through Local and Systemic Mechanisms A drop in arterial pressure reduces blood flow to tissues, resulting in local accumulation of metabolic wastes. This change in the extracellular environment stimulates autoregulatory opening of the arteries. A fall in central blood pressure is prevented by negative feedback mechanisms (including the release of antidiuretic hormone, ADH) that constrict arteries in less essential tissues and stimulate maintenance of blood volume and blood pressure.

maintain blood volume and increase blood pressure (see Section 52.6). Increased activity of the baroreceptors inhibits the release of ADH, and as a result the kidneys excrete more water, reducing blood volume and contributing to a fall in arterial pressure (see Figure 50.19).

Other information that causes the cardiovascular control center to increase heart rate and blood pressure comes from chemoreceptors in the medulla, aorta, and the carotid arteries. As we discussed in Section 49.5, the medullary chemosensors are activated by increases in arterial CO_2 levels, and the carotid and aortic bodies are activated by falls in arterial O_2 levels. Chemosensors send signals to the cardiovascular regulatory center as well as to the respiratory regulatory center.

RECAP 50.5

The delivery of blood to tissues is controlled locally by autoregulatory mechanisms that dilate or constrict arterioles. These local actions are translated into alterations in central blood pressure and composition that are detected by neural and hormonal mechanisms, which then mediate corrective cardiovascular adjustments.

- How do autoregulatory changes in blood flow to capillary beds result in adjustments to MAP? **See p. 1044 and Figure 50.18**
- What are the roles of hormones in regulating blood pressure? **See p. 1044 and Figure 50.19**
- Describe the role of baroreceptors and chemoreceptors in regulating blood pressure. **See pp. 1044–1045 and Figure 50.20**

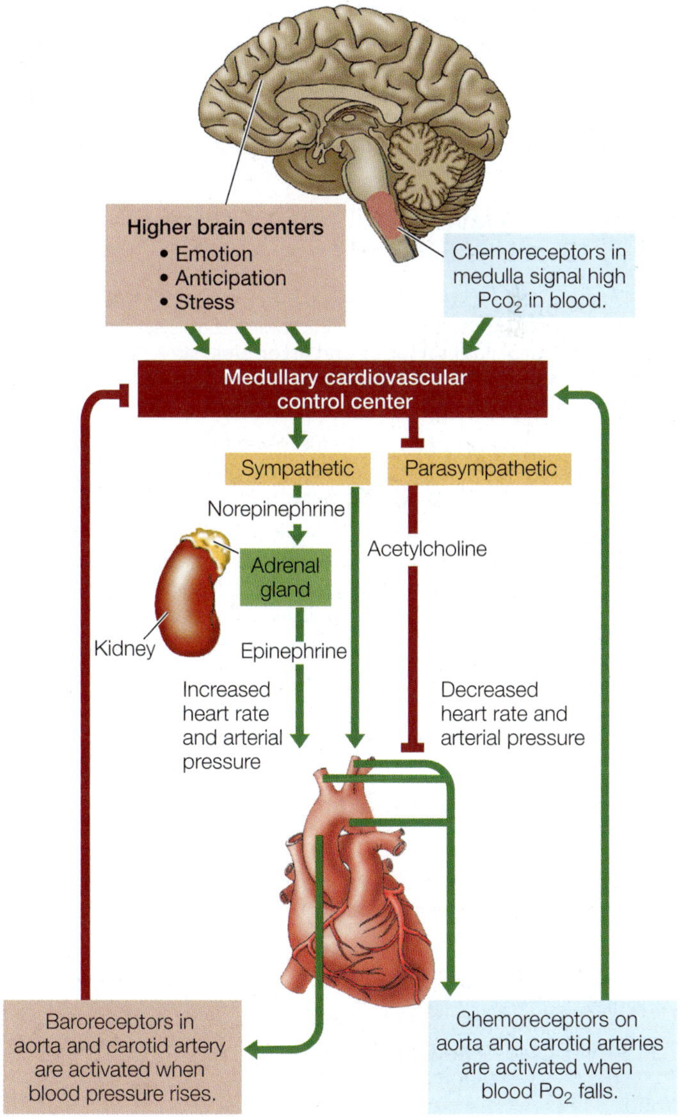

Higher brain centers
• Emotion
• Anticipation
• Stress
Chemoreceptors in medulla signal high P_{CO_2} in blood.
Medullary cardiovascular control center
Sympathetic
Parasympathetic
Norepinephrine
Adrenal gland
Acetylcholine
Kidney
Epinephrine
Increased heart rate and arterial pressure
Decreased heart rate and arterial pressure
Baroreceptors in aorta and carotid artery are activated when blood pressure rises.
Chemoreceptors on aorta and carotid arteries are activated when blood P_{O_2} falls.

50.20 Regulating Cardiac Output The autonomic nervous system controls heart rate in response to information about blood pressure and blood composition originating in baroreceptors and chemoreceptors shown at the bottom of the figure. Information from these sensors goes to the cardiovascular control center in the medulla, where it is integrated with other information. The medullary center generates responses in the sympathetic and parasympathetic nervous systems that control cardiac output.

How can the same mutation-based heart condition be fatal to an athlete but innocuous in most other people?

ANSWER

The mutation responsible for the condition described in the opening story encodes a contractile protein necessary for the pumping action of cardiac muscle. The mutant form of this protein renders the heart less efficient at pumping blood, and the heart compensates by getting larger. In most people, the heart enlarges enough to meet their needs without causing other problems. High-performing athletes, however, regularly engage in heavy exercise that places huge demands on their hearts, and in athletes with this mutation heart enlargement can be excessive. Eventually thickening of the ventricular walls, especially those of the left ventricle, can disrupt the electrical impulses that coordinate contractions of the heart muscle. When heavy demand is placed on such an enlarged heart, muscle fiber contractions can suddenly become uncoordinated or blocked, rendering the heart incapable of pumping blood.

CHAPTER SUMMARY 50

50.1 Why Do Animals Need a Circulatory System?

- The metabolic needs of the cells of many small animals are met by direct exchange of materials with the external medium. The metabolic needs of the cells of larger animals are met by a **circulatory system** that transports nutrients, respiratory gases, and metabolic wastes throughout the body.
- In **open circulatory systems**, extracellular fluid leaves vessels and percolates through tissues. In **closed circulatory systems**, the blood is contained in a system of vessels. Closed circulatory systems have the ability to selectively direct blood, hormones, and nutrients to specific tissues. **Review Figure 50.1**

50.2 How Have Vertebrate Circulatory Systems Evolved?

- The circulatory system of vertebrates consists of a heart and a closed system of vessels containing blood that is separate from the interstitial fluid. **Arteries** and **arterioles** carry blood from the heart; **capillaries** are the site of exchange between blood and interstitial fluid; **venules** and **veins** carry blood back to the heart.
- The vertebrate circulatory system evolved from a single circuit in fishes to partially or completely separate pulmonary and systemic circuits in amphibians, reptiles, and mammals.
- In the single-circuit system of fishes, blood flow is unidirectional and is propelled by one-way valves between the **sinus venosus** and the **atrium**, between the atrium and the **ventricle**, and between the ventricle and the **bulbus arteriosus**.
- In birds and mammals, blood circulates through two completely separate circuits. The **pulmonary circuit** transports blood between the heart and lungs, and the **systemic circuit** transports oxygen-rich blood between the heart and tissues. **See ACTIVITY 50.1**

50.3 How Does the Mammalian Heart Function?

- The mammalian heart has four chambers. Valves in the heart prevent the backflow of blood. **Review Figure 50.2, ACTIVITY 50.2**
- The **cardiac cycle** has two phases: **systole**, in which the ventricles contract, and **diastole**, in which the ventricles relax. The sequential heart sounds ("lub-dup") are made by the closing of the heart valves. **Review Figure 50.3, ANIMATED TUTORIAL 50.1**
- Blood pressure can be measured using a sphygmomanometer and a stethoscope. **Review Figure 50.4**
- **Pacemaker cells** of the sinoatrial node set the heart rate as a result of the properties of their ion channels. The autonomic nervous system controls heart rate: sympathetic activity increases heart rate, and parasympathetic activity decreases it by altering the rate of depolarization of the pacemaker cell resting membrane potentials following the termination of systole. **Review Figures 50.5, 50.6**
- The **sinoatrial node** controls the cardiac cycle by initiating a wave of depolarization in the atria, which is conducted to the ventricles through a system consisting of the **atrioventricular node**, **bundle of His**, and **Purkinje fibers**. **Review Figure 50.7**
- Sustained contraction of ventricular muscle cells is due to long-duration action potentials that are generated by voltage-gated Na^+ and Ca^{2+} channels. **Review Figures 50.8, Figure 50.9**
- An **electrocardiogram** (**ECG** or **EKG**) records electrical events associated with the contraction and relaxation of the cardiac muscles. **Review Figure 50.10**

50.4 What Are the Properties of Blood and Blood Vessels?

- Blood consists of a **plasma** portion (water, salts, and proteins) and a cellular portion (**erythrocytes** or red blood cells, platelets, and white blood cells). All of the cellular components are produced from stem cells in the bone marrow. **Review Figure 50.11**
- Erythrocytes transport oxygen. Their production in the bone marrow is stimulated by **erythropoietin**, which is produced in response to **hypoxia** (low oxygen levels) in the tissues.
- **Platelets**, along with circulating proteins, are involved in **blood clotting**, which results in a meshwork of **fibrin** threads that help seal damaged vessels. **Review Figure 50.12**
- Abundant smooth muscle cells allow vessels to change their diameter, altering their resistance and thus blood flow. Arteries have elastic fibers that enable them to withstand high pressures. **Review Figure 50.13, ACTIVITY 50.3**
- Capillary beds are the site of exchange of materials between blood and tissue fluid.
- **Starling's forces** suggest that blood volume is maintained in the capillary beds by an exchange of fluids driven by both blood pressure and osmotic pressure. **Review Figure 50.15**

continued

- An accumulation of fluid in the extracellular spaces leads to **edema**. Bicarbonate ions in the blood plasma contribute to the osmotic forces that draw water back into capillaries.
- The ability of a specific molecule to cross a capillary wall depends on the architecture of the capillary, the type of substance, and the concentration gradient between the blood and the tissue fluid.
- Veins have a high capacity for storing blood. Aided by gravity, by contractions of skeletal muscle, and by the actions of breathing, they return blood to the heart. **Review Figure 50.16**
- The **Frank–Starling law** describes forces that increase cardiac output, such as stretch of the cardiac muscles cells caused by increased venous return.
- The **lymphatic system** returns the interstitial fluid to the blood.

50.5 How Is the Circulatory System Controlled and Regulated?

- Blood flow through capillary beds is controlled by local **autoregulatory mechanisms**, hormones, and the autonomic nervous system. **Review Figure 50.18, ANIMATED TUTORIAL 50.2**
- Blood pressure is controlled in part by the hormones ADH and **angiotensin**, which stimulate contraction of blood vessels. **Review Figure 50.19**
- Heart rate is controlled by the autonomic nervous system, which responds to information about blood pressure and blood composition that is integrated by regulatory centers in the medulla. **Review Figure 50.20**

Go to the Interactive Summary to review key figures, Animated Tutorials, and Activities Life10e.com/is50

CHAPTER REVIEW

REMEMBERING

1. Which statement about vertebrate circulatory systems is *not* true?
   a. In fish, oxygenated blood from the gills returns to the heart through the left atrium.
   b. In mammals, deoxygenated blood leaves the heart through the pulmonary artery.
   c. In amphibians, deoxygenated blood enters the heart through the right atrium.
   d. In reptiles, the blood in the pulmonary artery has a lower oxygen content than the blood in the aorta.
   e. In birds, the pressure in the aorta is higher than the pressure in the pulmonary artery.

2. Which statement about the human heart is true?
   a. The walls of the right ventricle are thicker than the walls of the left ventricle.
   b. Blood flowing through atrioventricular valves is always deoxygenated blood.
   c. The second heart sound is due to the closing of the aortic valve.
   d. Blood returns to the heart from the lungs in the vena cava.
   e. During systole, the aortic valve is open and the pulmonary valve is closed.

3. The pacemaker action potentials in the heart
   a. are due to opposing actions of norepinephrine and acetylcholine.
   b. are generated by the bundle of His.
   c. depend on the gap junctions between the cells that make up the atria and those that make up the ventricles.
   d. are due to spontaneous depolarization of the plasma membranes of modified cardiac muscle cells.
   e. reflect large depolarizations of membrane potential due to opening of voltage-gated Na^+ channels.

4. Blood velocity through capillaries is slow because
   a. much blood volume is lost from the capillaries.
   b. the pressure in venules is high.
   c. the total cross-sectional area of capillaries is larger than that of arterioles.
   d. the osmotic pressure in capillaries is very high.
   e. erythrocytes must pass through in single file.

5. Autoregulation of blood flow to a tissue is due to
   a. sympathetic innervation.
   b. the release of ADH by the hypothalamus.
   c. increased activity of baroreceptors.
   d. chemoreceptors in the aorta and the carotid arteries.
   e. the effect of the local chemical environment on arterioles.

UNDERSTANDING & APPLYING

6. At the beginning of a race, cardiac output increases immediately before there is any change in blood O_2 or CO_2 concentrations. Explain two factors that contribute to this effect. Include the Frank–Starling law in your answer.

7. Is there a time in the mammalian cardiac cycle when all four heart valves are open? Explain.

8. A sudden and massive loss of blood results in a decrease in blood pressure. Describe several mechanisms that help return blood pressure to normal and how they do so.

ANALYZING & EVALUATING

9. The hearts of most mammals stop beating when their temperature falls more than a few degrees below 20°C, but the hearts of hibernating mammals can beat at 0°C. What adaptations of cardiac muscle might explain this capacity of the hearts of hibernators?

10. You can describe the cycle of events in a ventricle of the heart by a graph that plots the pressure in the ventricle on the y axis and the volume of blood in the ventricle on the x axis. Draw such a graph for a left ventricle using maximum and minimum pressures of 125 and 25 mm Hg and maximum and minimum ventricular volumes of 130 and 60 ml. Where would the heart sounds be on this graph? How would the graph differ for the left and the right ventricles?

Go to BioPortal at **yourBioPortal.com** for Animated Tutorials, Activities, LearningCurve Quizzes, Flashcards, and many other study and review resources.

51 Nutrition, Digestion, and Absorption

CHAPTER**OUTLINE**

Efficiency Genes The Pima are an example of a human population that repeatedly experienced periods of severe food deprivation. These historic occurrences may have imposed selection for genes that improve the efficiency of managing the energy obtained from food. With modern diets and lifestyles, these "efficiency genes" can contribute to obesity.

FOR THOUSANDS of years the Pima of southwestern North America were hunters and gatherers who supplemented their diet with subsistence agriculture. Their environment was arid, so they developed sophisticated irrigation systems; even so, they frequently encountered drought and subsequent starvation. Today most individuals of the ethnic Pima population in North America are clinically obese. In fact, as a population they are one of the heaviest in the world.

With obesity come related health problems such as diabetes, high blood pressure, and heart disease. Diabetes incidence in the Pima is seven times the national average; two-thirds of adults over the age of 40 are diabetic. Moreover, diabetes is occurring in younger individuals than ever before. What has caused such a radical health change in an entire population? Two interacting factors are involved: genetics and lifestyle.

Geneticists hypothesize that recurring episodes of starvation produce strong selective pressure for "thrifty genes"—particular alleles of the genes involved in digestion, absorption, and energy storage that result in greater-than-average efficiency in converting food into energy reserves, such as fat. Thrifty genes would carry a strong selective advantage when food is scarce. The Pima display a "thrifty" phenotype. They have low resting metabolic rates and convert food into fat readily. For many Pima, consuming a standard amount of glucose causes their insulin levels to rise three times higher than it does in Americans of European ancestry. Insulin is the hormone that facilitates conversion of dietary sugar into fat.

The other factor in the Pima obesity epidemic is an abrupt change in their traditional lifestyle. Instead of eating their traditional diet, the Pima now eat a high-calorie, high-fat Western diet, and they engage in less physical activity than their ancestors did.

Another population of Pima lives in the Sierra Madre of northern Mexico. Genetically they are the same as the Arizona population. However, they eat traditional foods and live a traditional lifestyle that involves much physical activity. Whereas the Arizona Pima engage in an average of only 2 hours of physical work per week, the Mexico Pima average 23 hours per week. Obesity and diabetes are not prevalent among the Mexico Pima.

A high-calorie diet and sedentary lifestyle affect not just the Pima but contribute to the overall increase in obesity throughout the U.S. population. Researchers are studying the Pima to learn more about the genetics of obesity and related pathologies.

?

Are there genes that predispose a person to obesity?

See answer on p. 1068.

(A)

(B)

51.1 Heterotrophs Get Energy from Autotrophs (A) Herbivores get their energy directly from autotrophs. The large herbivores of the African grasslands must consume huge amounts of plant matter to fulfill their nutritional needs. (B) A carnivore's energy is indirectly obtained from autotrophs, since the energy stored in a prey animal was originally obtained from autotrophs.

51.1 What Do Animals Require from Food?

Animals are **heterotrophs**—they derive their nutrition from eating other organisms. In contrast, **autotrophs** (most plants, some bacteria, some archaea, and some protists) can use solar energy or inorganic chemical energy to synthesize all of their components. Directly and indirectly, heterotrophs take advantage of—indeed, depend on—the organic synthesis carried out by autotrophs and have evolved an enormous diversity of adaptations to exploit this resource (**Figure 51.1**). In this section we will describe how animals use food, be it plants or other animals, to obtain energy and building blocks of complex molecules. We will also consider the need for special mineral nutrients and organic molecules and the diseases that result when they are lacking in the diet.

Energy needs and expenditures can be measured

Energy—the capacity to do work—comes in different forms, including electric, heat, chemical, and nuclear energy. As discussed in Chapter 8, a calorie (note the small *c*) is a unit of heat energy; specifically, it is the amount of heat necessary to raise the temperature of 1 gram of water 1°C. Because this is such a tiny amount of energy, physiologists commonly use the **kilocalorie** (**kcal**) as a unit of measure (1 kcal = 1,000 calories). Nutritionists also use the kilocalorie as a standard unit of energy, but they traditionally refer to it as the **Calorie** (**Cal**), which is capitalized to distinguish it from the single calorie.

Just about any food container you pick up in the U.S. carries the label "Nutrition Facts" that includes the item "Calories." How do Calories (or kcal) relate to the discussion in Chapter 9 about how energy in the chemical bonds of food molecules is transferred to the high-energy phosphate bonds of adenosine triphosphate (ATP) and is then used to do cellular work? And why do we use heat energy as a measure of nutrition?

The reason is found in the laws of thermodynamics, which tell us that energy cannot be created or destroyed, but can be converted from one form to another (see Section 8.1). However, every energy conversion is inefficient and a large portion of the original energy always ends up as heat. Whether we are using the energy in glucose to make ATP or are using that ATP to power muscle contraction or ion transport, most of the available chemical energy is lost as heat. The bottom line is that if an animal is not growing, not doing any external work, and not changing its body temperature, the heat it loses to the environment is a measure of its total energy expenditure, or metabolism.

An animal's energy needs must be met by the ingestion, digestion, and assimilation of food. The basal energy expenditure of a human is 1,300 to 1,500 Cal/day for an adult female and 1,600 to 1,800 Cal/day for an adult male. Physical activity adds to this basal energy requirement. For a person doing sedentary work, about 30 percent of the Calories expended are used for skeletal muscle activity; for a person doing heavy physical labor, more than 95 percent of caloric expenditure is for skeletal muscle activity.

The components of food that provide energy are fats, carbohydrates, and proteins. Fats yield 9.5 Cal/gram, carbohydrates 4.2 Cal/gram, and proteins about 4.1 Cal/gram. **Figure 51.2** shows some equivalencies of food, energy, and energy consumption.

Even though the units calorie, kilocalorie, and Calorie remain in popular use, most scientists now use the International System of Units (ISU). In this system the basic unit of energy is the joule: 1 joule = 0.239 calories, and the measure of energy use is 1 joule/second = 1 watt. You are familiar with light bulb ratings, so think about that when you convert kcal/day into watts. The 1,700 Cal/day energy expenditure of the average man converts to 82 watts (note that a watt includes the time dimension, so it is a rate of energy use).

Thus it is possible to quantify the caloric value of any food an animal eats. It is also possible to quantify the caloric

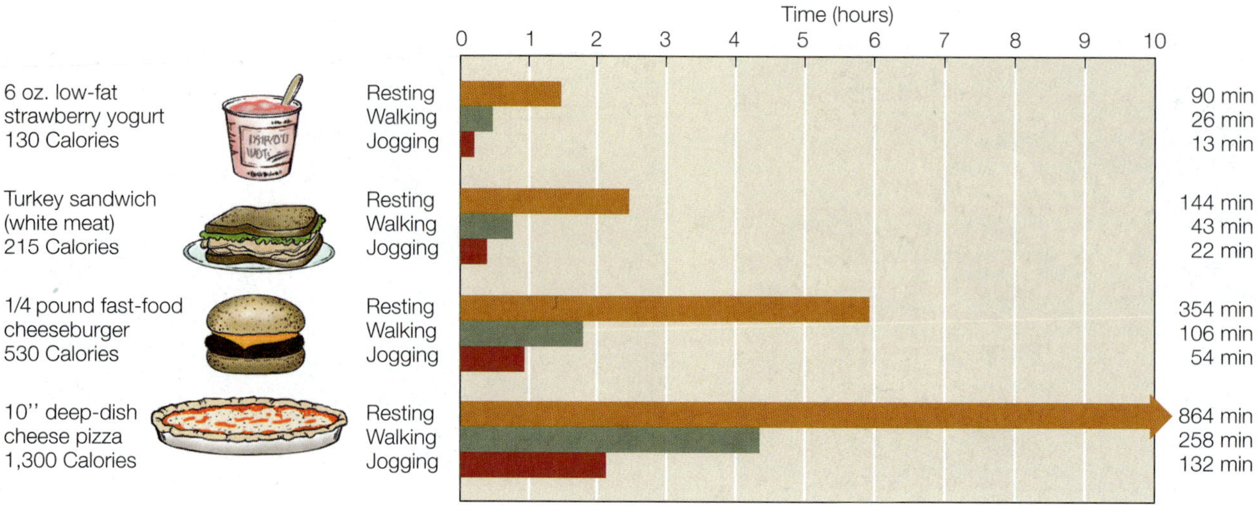
Time (hours)
0 1 2 3 4 5 6 7 8 9 10
6 oz. low-fat strawberry yogurt 130 Calories
Resting Walking Jogging
90 min 26 min 13 min
Turkey sandwich (white meat) 215 Calories
Resting Walking Jogging
144 min 43 min 22 min
1/4 pound fast-food cheeseburger 530 Calories
Resting Walking Jogging
354 min 106 min 54 min
10" deep-dish cheese pizza 1,300 Calories
Resting Walking Jogging
864 min 258 min 132 min

51.2 Food Energy and How We Use It The energy contained in several common food items is shown at the left. The graphs indicate about how long it would take a person with a basal energy requirement of about 1,800 Cal/day to use the equivalent amount of energy while resting, walking, or jogging.

expenditure of any activity or behavior an animal performs. By comparing calories consumed with calories expended, we can construct **energy budgets** that allow ecologists and evolutionary biologists to apply a cost–benefit analysis to feeding behavior, as we will explain in Section 53.4.

Sources of energy can be stored in the body

Although the cells of the body use energy continuously, most animals do not eat continuously and so must store fuel molecules that can be released as needed between meals. Carbohydrates are stored in liver and muscle cells as glycogen, but the total glycogen stored represents only about a day's basal energy requirement. Fat is the most important form of stored energy in the bodies of animals. Not only does fat have more energy per gram than glycogen, but it can be stored with little associated water, making it more compact. Migrating birds store energy as fat to fuel their long flights; if they had to store the same amount of energy as glycogen, they would be too heavy to fly. Proteins are not used as energy storage compounds, although the body's structural protein can be metabolized as an energy source of last resort.

If an animal takes in too little food to meet its energy requirements, it must start metabolizing some of the molecules of its own body. This "self-consumption" begins with the energy storage compounds glycogen and fat. Once fat reserves are seriously depleted, the body increases its metabolism of proteins for energy (**Figure 51.3A**). The first proteins to be sacrificed are those of the blood plasma. The loss of plasma proteins decreases the osmotic concentration of the plasma, resulting in increased loss of fluid from the blood to the interstitial spaces (edema; see Section 50.4). Accumulation of fluid in the extremities and abdomen is the classic sign of kwashiorkor, a disease caused by chronic protein deficiency (**Figure 51.3B**). Continued protein loss damages the body's organs, leading eventually to death.

When an animal consistently takes in more food than it needs to meet its energy requirements, the excess nutrients are stored as increased body mass. First glycogen reserves build

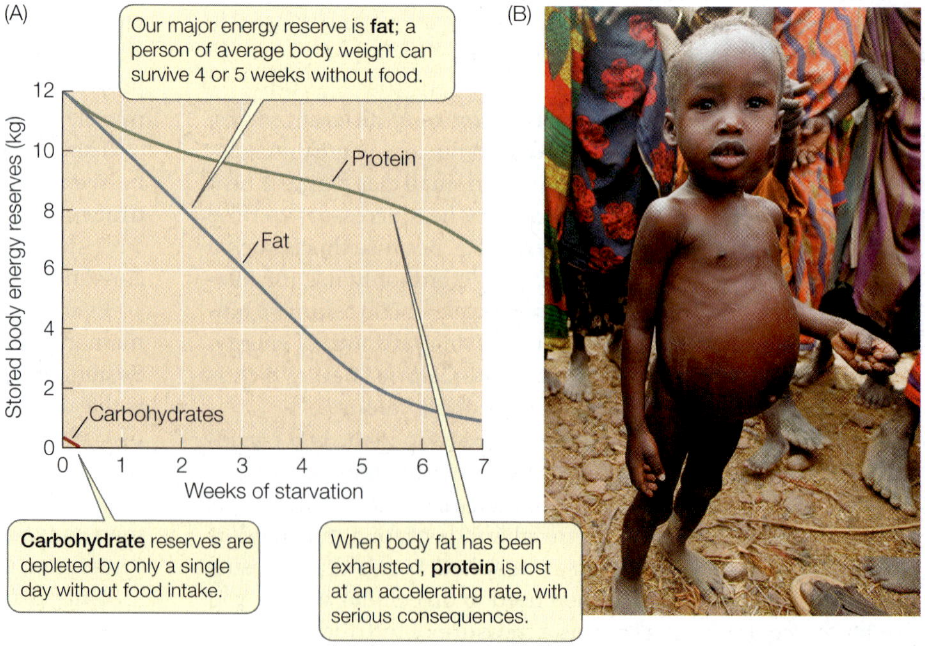
(A)
Our major energy reserve is **fat**; a person of average body weight can survive 4 or 5 weeks without food.
Stored body energy reserves (kg)
12
10
8
6
4
2
0
Protein
Fat
Carbohydrates
0 1 2 3 4 5 6 7
Weeks of starvation
Carbohydrate reserves are depleted by only a single day without food intake.
When body fat has been exhausted, **protein** is lost at an accelerating rate, with serious consequences.
(B)

51.3 The Course of Starvation (A) In a person subjected to undernutrition, the body's energy reserves are eventually depleted. (B) The swollen abdomen, face, hands, and feet of this girl are due to edema. Along with her spindly limbs, these are symptoms of kwashiorkor, a syndrome resulting from the body breaking down blood proteins and muscle tissue to fuel metabolism.

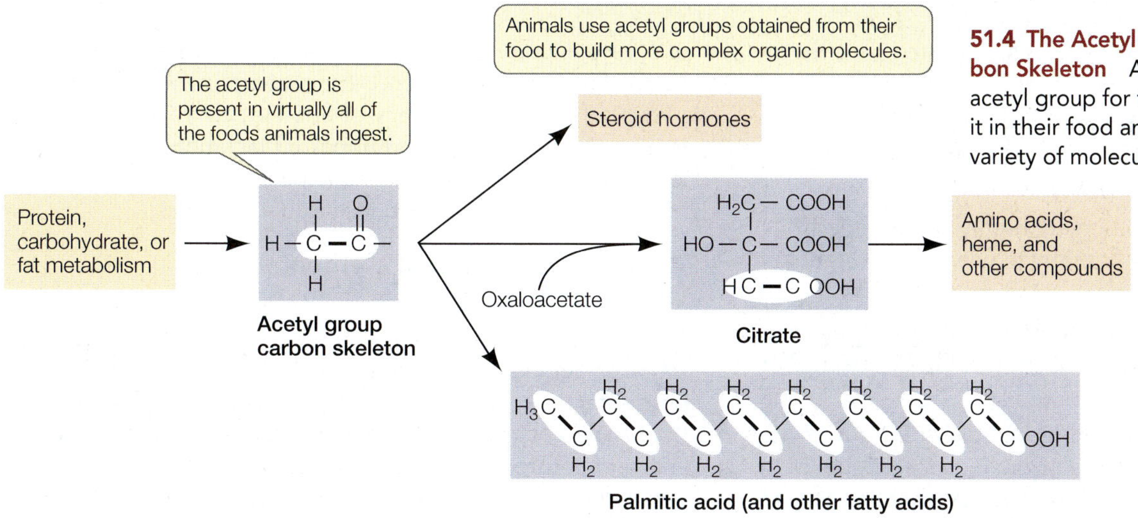
Animals use acetyl groups obtained from their food to build more complex organic molecules.
The acetyl group is present in virtually all of the foods animals ingest.
Steroid hormones
Protein, carbohydrate, or fat metabolism
Acetyl group carbon skeleton
Oxaloacetate
Citrate
Amino acids, heme, and other compounds
Palmitic acid (and other fatty acids)

51.4 The Acetyl Group Is an Acquired Carbon Skeleton Animals cannot synthesize the acetyl group for themselves, but they ingest it in their food and use it to synthesize a wide variety of molecules.

up; then additional dietary carbohydrates, fats, and proteins are converted to body fat. In some species, such as hibernators, seasonal overnutrition is an important adaptation for surviving periods when food is not available. In humans, however, overnutrition can be a serious health hazard, increasing the risk of high blood pressure, heart attack, diabetes, and other disorders, as seen in the Pima discussed in the chapter opener.

Food provides carbon skeletons for biosynthesis

Every animal must take in certain organic molecules that it cannot synthesize for itself but needs to form the building blocks of its own complex organic molecules. The acetyl group (CH_3CO—) is one such required building block, supplying the **carbon skeleton** of larger organic molecules (Figure 51.4). Animals cannot synthesize acetyl groups from carbon, oxygen, and hydrogen molecules but must obtain them from food. Acetyl groups can be derived from the metabolism of almost any food, but they originate in plants.

Acetyl groups are never in short supply for an adequately nourished animal. However, some groups supplying carbon skeletons can be deficient in an animal's diet even if caloric intake is adequate. One such group includes certain amino acids, the building blocks of proteins. Animals can synthesize some of their own amino acids using carbon skeletons from acetyl or other groups and transferring to them amino groups (—NH_2) derived from other amino acids. However, most animals cannot synthesize all the amino acids they need and thus must obtain certain **essential amino acids** from food. If an animal does not take in enough of even one of its essential amino acids, its protein synthesis is impaired and its capacity to maintain enzymatic and transport functions is challenged.

Essential amino acids vary by species. Most researchers agree that adult humans must obtain eight essential amino acids from their food: isoleucine, leucine, lysine, methionine, phenylalanine, threonine, tryptophan, and valine. All eight are available in milk, eggs, meat, and soybean products, but most plant foods do not contain adequate quantities of all eight, so a strict vegetarian diet carries a risk of protein malnutrition.

A **complementary diet** of plant foods, however, supplies all eight essential amino acids (Figure 51.5). In general, grains (such as rice, wheat, and corn) are complemented by legumes (such as beans and peas). Long before the chemical basis for complementarity was understood, societies with little access to meat developed complementary diets. Many Central and South American peoples traditionally eat beans with corn, and the native peoples of North America complemented their beans with squash.

Human infants are thought to require four additional amino acids in their diets: histidine, tyrosine, cysteine, and arginine. Also, some amino acids are required by individuals with certain metabolic disorders who cannot synthesize them adequately. For example, individuals with the genetic disease phenylketonuria lack the enzyme for converting phenylalanine to tyrosine (see Section 15.2) and must obtain tyrosine from their diets. They must keep their dietary intake of phenylalanine low to prevent its accumulation to toxic levels.

Why are dietary proteins completely digested to their constituent amino acids before being used by the body? Wouldn't it be more energy-efficient to reuse some dietary proteins

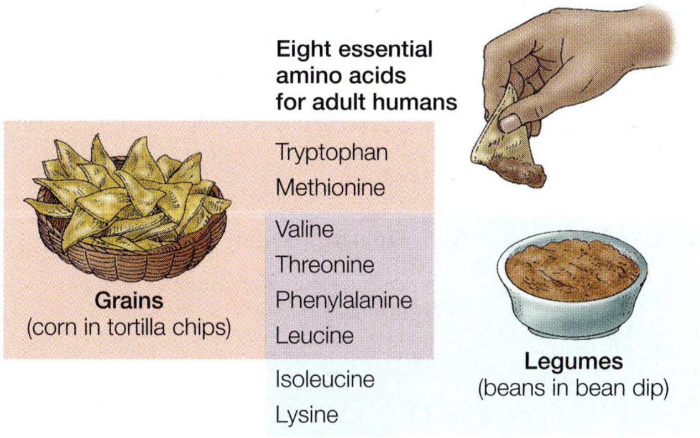
Eight essential amino acids for adult humans
Tryptophan
Methionine
Valine
Threonine
Phenylalanine
Leucine
Isoleucine
Lysine
Grains
(corn in tortilla chips)
Legumes
(beans in bean dip)

51.5 A Strategy for Vegetarians By combining cereal grains with legumes, an adult vegetarian can obtain all eight essential amino acids.

directly? There are several reasons why ingested proteins are not used "as is":

- Macromolecules such as proteins are not readily absorbed by the cells of the gut, but their constituent monomers (such as amino acids) are readily absorbed.
- Protein structure and function are highly species-specific. A protein that functions optimally in one species might not function well in another.
- Foreign proteins entering the body directly from the gut would be recognized as invaders and would be attacked by the immune system.

Humans can synthesize almost all the lipids required by the body using acetyl groups obtained from food (see Figure 51.4), but we must have a dietary source of certain **essential fatty acids**—notably, linoleic acid—that we cannot synthesize. Linoleic acid is needed by mammals to synthesize other unsaturated fatty acids, such as arachidonic acid, which is a component of several signaling molecules, including prostaglandins. Essential fatty acids are also necessary components of membrane phospholipids. A deficiency of linoleic acid can lead to problems such as infertility and impaired lactation, but because it is commonly present in vegetable oils, a deficiency is unlikely in an adequately nourished individual.

Animals need mineral elements for a variety of functions

Table 51.1 lists the principal mineral elements that animals require. Elements required in large amounts are called **macronutrients**; those required in only tiny amounts (generally less than 100 mg/day) are called **micronutrients**. Some micronutrients are required in such minute amounts that deficiencies are never observed, but they are nevertheless essential elements.

Calcium is an example of a macronutrient. It is the fifth most abundant element in the body; a 70-kg person contains about 1.2 kg of calcium. Calcium phosphate is the principal structural

TABLE 51.1 Mineral Elements Required by Animals

Element	Source in Human Diet	Major Functions
MACRONUTRIENTS		
Calcium (Ca)	Dairy foods, eggs, green leafy vegetables, whole grains, legumes, nuts, meat	Found in bones and teeth; blood clotting; nerve and muscle action; enzyme activation
Chlorine (Cl)	Table salt (NaCl), meat, eggs, vegetables, dairy foods	Water balance; digestion (as HCl); principal negative ion in extracellular fluid
Magnesium (Mg)	Green vegetables, meat, whole grains, nuts, milk, legumes	Required by many enzymes; found in bones and teeth
Phosphorus (P)	Dairy, eggs, meat, whole grains, legumes, nuts	Component of nucleic acids, ATP, and phospholipids; bone formation; buffers; metabolism of sugars
Potassium (K)	Meat, whole grains, fruits, vegetables	Nerve and muscle action; protein synthesis; principal positive ion in cells
Sodium (Na)	Table salt, dairy foods, meat, eggs	Nerve and muscle action; water balance; principal positive ion in extracellular fluid
Sulfur (S)	Meat, eggs, dairy foods, nuts, legumes	Found in proteins and coenzymes; detoxification of harmful substances
MICRONUTRIENTS		
Chromium (Cr)	Meat, dairy, whole grains, legumes, yeast	Glucose metabolism
Cobalt (Co)	Meat, tap water	Found in vitamin B_{12}; formation of red blood cells
Copper (Cu)	Liver, meat, fish, shellfish, legumes, whole grains, nuts	Found in active site of many redox enzymes and electron carriers; production of hemoglobin; bone formation
Fluorine (F)	Most water supplies	Found in teeth; helps prevent tooth decay
Iodine (I)	Fish, shellfish, iodized salt	Found in thyroid hormones
Iron (Fe)	Liver, meat, green vegetables, eggs, whole grains, legumes, nuts	Found in active sites of many redox enzymes and electron carriers, hemoglobin, and myoglobin
Manganese (Mn)	Organ meats, whole grains, legumes, nuts, tea, coffee	Activates many enzymes
Molybdenum (Mo)	Organ meats, dairy, whole grains, green vegetables, legumes	Found in some enzymes
Selenium (Se)	Meat, seafood, whole grains, eggs, milk, garlic	Fat metabolism
Zinc (Zn)	Liver, fish, shellfish, and many other foods	Found in some enzymes and some transcription factors; insulin physiology

material in bones and teeth. Muscle contraction, neural function, and many other intracellular functions in animals require calcium ions (Ca^{2+}). The turnover of calcium in the extracellular fluid is high, as bones are constantly being remodeled and calcium is constantly entering and leaving cells. Calcium is lost from the body in urine, sweat, and feces, so it must be replaced regularly. Adult humans require 800 to 1,000 mg of calcium per day in their diet.

Iron is an example of a micronutrient. It is found throughout the body because it is the oxygen-binding atom in hemoglobin and myoglobin and is a component of enzymes in the electron transport chain. Nevertheless, the total amount of iron in a 70-kg person is only about 4 grams, and since iron is recycled efficiently in the body and is not lost in the urine, we require only about 15 mg per day in our food. Despite the small amount required, insufficient iron is the most common mineral nutrient deficiency in the world today. Iron deficiency leads to anemia, a condition that renders individuals weak and tired all the time.

Go to Activity 51.1 Mineral Elements Required by Animals
Life10e.com/ac51.1

Animals must obtain vitamins from food

Like essential amino acids and fatty acids, **vitamins** are carbon compounds that an animal requires for growth and metabolism but cannot synthesize for itself. They are required in very small amounts compared with the essential amino acids and fatty acids that are incorporated into large body structures. Most vitamins function as coenzymes or parts of coenzymes.

The list of vitamins varies from species to species. Most mammals, for example, can make their own ascorbic acid. Primates (including humans) cannot, so for primates, ascorbic acid is a vitamin—vitamin C. If we do not get vitamin C in our food, we develop scurvy, a disease characterized by bleeding gums, loss of teeth, subcutaneous hemorrhages, and slow wound healing. Scurvy was a frequently fatal problem for sailors on long voyages until late in the eighteenth century, when a Scottish physician, James Lind, discovered that the disease could be prevented if the sailors ate fresh greens and citrus fruit. The British Admiralty made limes standard provisions for its ships (and British sailors have been called "limeys" ever since). When the active ingredient in limes was isolated, it was named ascorbic ("without scurvy") acid.

Humans require 13 vitamins; these are divided into two groups, water-soluble and fat-soluble (Table 51.2). When water-soluble vitamins are ingested in excess of bodily needs, they are simply eliminated in the urine. (This is the fate of much of the large doses of vitamin C that people take.) Fat-soluble vitamins, however, can accumulate in body fat and may build up to toxic levels in the liver if taken in excess.

TABLE **51.2**
Vitamins in the Human Diet

Vitamin	Source	Function	Deficiency Symptoms
WATER-SOLUBLE			
B_1 (thiamin)	Liver, legumes, whole grains	Coenzyme in cellular respiration	Beriberi, loss of appetite, fatigue
B_2 (riboflavin)	Dairy, meat, eggs, green leafy vegetables	Coenzyme in FAD	Lesions in corners of mouth, eye irritation, skin disorders
Niacin	Meat, fowl, liver, yeast	Coenzyme in NAD and NADP	Pellagra, skin disorders, diarrhea, mental disorders
B_6 (pyridoxine)	Liver, whole grains, dairy foods	Coenzyme in amino acid metabolism	Anemia, slow growth, skin problems, convulsions
Pantothenic acid	Liver, eggs, yeast	Found in acetyl CoA	Adrenal problems, reproductive problems
Biotin	Liver, yeast, bacteria in gut	Found in coenzymes	Skin problems, loss of hair
B_{12} (cobalamin)	Liver, meat, dairy foods, eggs	Formation of nucleic acids, proteins, red blood cells	Pernicious anemia
Folic acid	Vegetables, eggs, liver, whole grains	Coenzyme in formation of heme and nucleotides	Anemia
C (ascorbic acid)	Citrus fruits, tomatoes, potatoes	Formation of connective tissues; antioxidant	Scurvy, slow healing, poor bone growth
FAT-SOLUBLE			
A (retinol)	Fruits, vegetables, liver, dairy	Found in visual pigments	Night blindness
D (calciferol)	Fortified milk, fish oils, sunshine	Absorption of calcium and phosphate	Rickets
E (tocopherol)	Meat, dairy foods, whole grains	Muscle maintenance, antioxidant	Anemia
K (menadione)	Intestinal bacteria, liver	Blood clotting	Blood clotting problems

The fat-soluble vitamin D (calciferol), which is essential for absorbing and metabolizing calcium, is a special case because the body can synthesize it. (As noted in Section 41.4, vitamin D is by definition a hormone.) Certain lipids present in the human body can be converted into vitamin D by the action of ultraviolet light on the skin. Thus vitamin D must be obtained in the diet by individuals with inadequate exposure to the sun.

The need for vitamin D may have been an important factor in the evolution of skin color. For humans living in equatorial and low latitudes, dark skin pigmentation is adaptive, as it is a protection against the damaging effects of ultraviolet radiation. These peoples generally expose extensive areas of skin to the sun on a regular basis, so their skin synthesizes adequate amounts of vitamin D. Most races that adapted to life in the higher latitudes lost this dark skin pigmentation, probably because lighter skin facilitates vitamin D production in the relatively small areas of skin exposed to sunlight during the short days of winter. The dark-skinned Inuit peoples of the Arctic are an exception to the correlation between latitude and skin pigmentation, but the Inuit obtain ample vitamin D from the large amounts of animal fat (especially whale blubber) and fish oils in their diet.

Go to Activity 51.2 Vitamins in the Human Diet
Life10e.com/ac51.2

Nutrient deficiencies result in diseases

The lack of any essential nutrient in the diet produces a state of **malnutrition**, and chronic malnutrition leads to a characteristic **deficiency disease** (see Table 51.2). We have discussed kwashiorkor (protein deficiency) and scurvy (vitamin C deficiency). Another deficiency disease, beriberi, was directly involved in the discovery of vitamins.

Beriberi, which means "extreme weakness," became prevalent in Asia in the nineteenth century when it became standard practice to mill rice to a white polish and discard the hulls present in brown rice. A critical observation was that chickens and pigeons developed beriberi-like symptoms when they were fed only polished rice. In 1912 Casimir Funk, a Polish scientist working in England, cured pigeons of beriberi by feeding them discarded rice hulls.

At the time of Funk's discovery, all diseases were thought either to be caused by microorganisms or to be inherited. Funk suggested that beriberi and some other diseases are dietary in origin and result from deficiencies in specific substances. Funk coined the term "vitamines" from "vital amines" because he mistakenly thought that all these substances vital for life were compounds with amino groups. In 1926 thiamin (vitamin B_1)—the substance lost in the rice milling process—was the first vitamin to be isolated in pure form.

Deficiency diseases can also result from an inability to absorb or process an essential nutrient even if it is present in the diet. Vitamin B_{12} (cobalamin), for example, is present in all foods of animal origin. Since plants neither use nor produce vitamin B_{12}, a strictly vegetarian diet (not supplemented with dairy products or vitamin pills) can lead to a B_{12} deficiency disease called pernicious anemia, characterized by a failure of red blood cells to mature. The most common cause of pernicious anemia, however, is not a lack of vitamin B_{12} in the diet but an inability to absorb it. Normally cells in the stomach lining secrete a peptide called intrinsic factor, which binds to vitamin B_{12} and makes it absorbable by the small intestine. Conditions that damage the stomach lining, such as alcoholism or gastritis, can thus lead to pernicious anemia.

Inadequate mineral nutrition can also lead to deficiency diseases. Examples are hypothyroidism and goiter resulting from iodine deficiency (see Section 41.4), and anemia resulting from iron deficiency. Iodine deficiency is almost unheard of in the developed world because we add iodide to salt. However, it is still a major health problem in large segments of the human population.

RECAP 51.1

As heterotrophs, animals must obtain the energy and molecular building blocks for biosynthesis from their food. Energy can come from the metabolism of carbohydrates, fats, and proteins. Molecular building blocks include carbon skeletons, vitamins, and minerals.

- Explain the different roles of dietary proteins, carbohydrates, and fats in producing metabolic energy. **See p. 1050 and Figure 51.3**
- Explain the nutritional roles of an essential carbon skeleton, a micronutrient, and a macronutrient. **See pp. 1051–1052, Figure 51.4, and Table 51.1**
- Why should fat-soluble vitamins not be taken in excess? **See p. 1053**

We have surveyed the essential elements of nutrition in animals. Next we will look at various methods and adaptations by which animals obtain the food they need, and the mechanisms they use to extract nutrients from their food.

51.2 How Do Animals Ingest and Digest Food?

Heterotrophic organisms can be classified by how they acquire their nutrition. **Saprobes** (also called saprotrophs) are organisms—most of which are either protists or fungi—that absorb nutrients from dead organic matter. **Detritivores** or **decomposers**, such as earthworms and crabs, actively feed on dead organic material. Animals that feed on living organisms are **predators**: **herbivores** prey on plants, **carnivores** prey on animals, and **omnivores** prey on both. **Filter feeders**, such as clams and blue whales, prey on small organisms by filtering them from the aquatic environment. **Fluid feeders** include mosquitoes, aphids, leeches, and hummingbirds. The anatomical adaptations that enable a species to exploit a particular source of nutrition are usually obvious, but physiological and biochemical adaptations are also important, although less obvious.

The food of herbivores is often low in energy and hard to digest

Most vegetation is coarse, difficult to break down, and has low energy content. Therefore herbivores spend a great deal of time feeding and processing their food. Many have striking adaptations for feeding, such as the trunk (a flexible, gripping nose) of the elephant or the huge bill of the fruit-eating toucan, which can be half as long as its body. Many types of grinding, rasping, cutting, and shredding mouthparts have evolved in invertebrates for ingesting plant material, and the teeth of herbivorous vertebrates have been shaped by selection to tear, crush, and grind coarse plant matter.

The digestive processes of herbivores can also be quite specialized. An example is the koala, which almost exclusively eats leaves of eucalyptus trees. These leaves are very fibrous, low in usable energy and protein, and high in toxic chemicals. The koala has strong jaws for grinding the leaves, a very long gut for fermenting them, enzymes in its liver for detoxifying chemicals in the leaves, and a low metabolic rate (i.e., it expends little energy) to compensate for low energy intake.

Carnivores must find, capture, and kill prey

The predatory behaviors of many carnivores are legendary—the hunting skills of hawks, wolves, and tigers, for example. Carnivores have evolved stealth, speed, power, large jaws, sharp teeth, and strong gripping appendages. They also have evolved remarkable means of detecting prey. Bats use echolocation, pit vipers sense infrared radiation from the warm bodies of their prey, and certain fish detect electric fields created in the water by their prey. There are many fascinating examples of adaptations for capturing prey, such as the immobilizing venom of many snakes, the long sticky tongues of chameleons, and the webs of spiders.

Some predators digest their prey externally. For example, a spider injects its insect prey with digestive enzymes and then sucks out the liquefied contents, leaving behind the empty exoskeletons frequently seen in old spider webs. The majority of animals, however, digest their food internally. For many, the process of digestion begins with the physical breaking down of the food items by the teeth.

Vertebrate species have distinctive teeth

Teeth are adapted for the acquisition and initial processing of specific types of foods. Because they are among the hardest structures of the body, an animal's teeth remain in the environment long after it dies. Paleontologists use teeth to identify animals that lived in the distant past and to deduce their feeding behavior.

All mammalian teeth have the same general, three-layered structure (**Figure 51.6A**). An extremely hard material called **enamel**, composed principally of calcium phosphate, covers the crown of the tooth. Both the crown and root contain a layer of bony material called **dentine**, inside of which is a **pulp cavity** containing blood vessels, nerves, and the cells that produce the dentine.

There is a great deal of homology in the dentition of mammals, but the shapes and organization of mammalian teeth are adaptations to different diets (**Figure 51.6B**). In general, incisors are used for cutting, chopping, and gnawing; canines are used for stabbing, gripping, and ripping; and molars and premolars (the cheek teeth) are used for shearing, crushing, and grinding. The highly varied diet of humans is reflected in our multipurpose set of teeth, as is common among omnivores.

Current methods make it possible to analyze residues derived from food in and on tooth enamel of prehistoric animals.

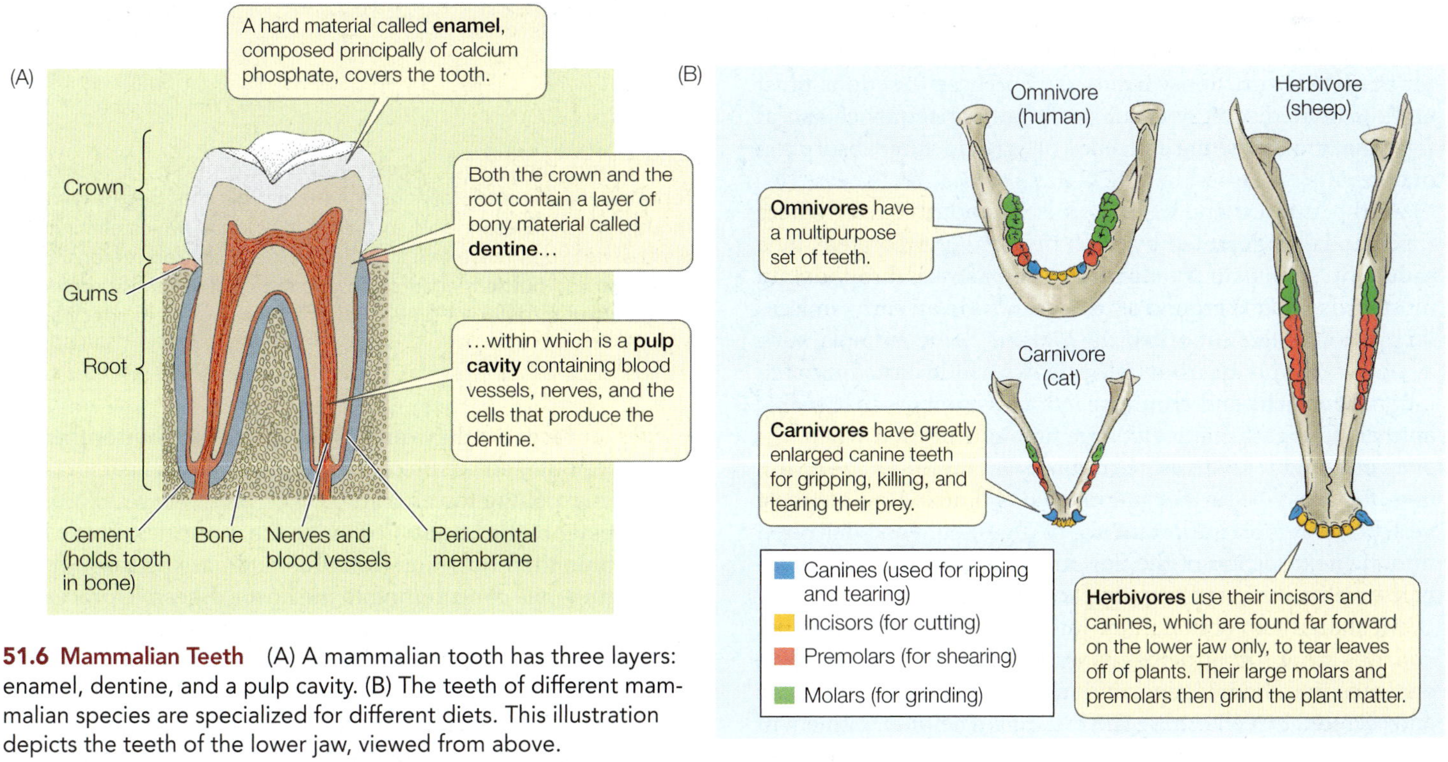
(A)
A hard material called **enamel**, composed principally of calcium phosphate, covers the tooth.
Crown
Gums
Root
Both the crown and the root contain a layer of bony material called **dentine**...
...within which is a **pulp cavity** containing blood vessels, nerves, and the cells that produce the dentine.
Cement (holds tooth in bone)
Bone
Nerves and blood vessels
Periodontal membrane
(B)
Omnivore (human)
Herbivore (sheep)
Omnivores have a multipurpose set of teeth.
Carnivore (cat)
Carnivores have greatly enlarged canine teeth for gripping, killing, and tearing their prey.
Canines (used for ripping and tearing)
Incisors (for cutting)
Premolars (for shearing)
Molars (for grinding)
Herbivores use their incisors and canines, which are found far forward on the lower jaw only, to tear leaves off of plants. Their large molars and premolars then grind the plant matter.

51.6 Mammalian Teeth (A) A mammalian tooth has three layers: enamel, dentine, and a pulp cavity. (B) The teeth of different mammalian species are specialized for different diets. This illustration depicts the teeth of the lower jaw, viewed from above.

Go to Activity 51.3 Mammalian Teeth Life10e.com/ac51.3

Such analysis of some hominin fossils (*Paranthropus boisei*; see Figure 33.34) from South Africa has revealed that they were predominantly grass eaters. What was the evidence for this conclusion? Tropical grasses are C_4 plants (see Section 10.4), whereas leafy trees that produce fruits and nuts are C_3 plants. The carbon residues on the *Paranthropus* teeth were those produced by C_4 plants, presumably grasses and sedges.

Digestion usually begins in a body cavity

Animals take food into a body cavity that is continuous with the outside environment. They secrete digestive enzymes into that cavity, and the enzymes break down the food into nutrient molecules that can be absorbed by the cells lining the cavity.

The simplest digestive system is found in the simplest animals, the sponges. Water flows from the environment through the body of the sponge in water channels, and individual cells capture food particles from the water (see Figure 31.2B). A slightly more specialized digestive system is a **gastrovascular cavity**, which connects to the outside world through a single opening. Cnidarians, such as jellyfish, capture prey using stinging nematocysts and use their tentacles to cram the prey into their gastrovascular cavity. Enzymes in the gastrovascular cavity partially digest the prey. Cells lining the cavity take in small food particles by endocytosis. The vesicles created by endocytosis then fuse with lysosomes containing digestive enzymes, and intracellular digestion completes the breakdown of the food. Nutrients are released to the cytoplasm as the vesicles break down.

Tubular guts have an opening at each end

The guts of most animals are tubular: a **mouth** takes in food; molecules are digested and absorbed throughout the length of the gut; and solid digestive wastes are eliminated through an **anus**. Different regions in the tubular gut are specialized for particular functions (**Figure 51.7**). These functions must be coordinated so they occur in the proper sequence and at rates that maximize the efficiency of digestion and absorption of nutrients.

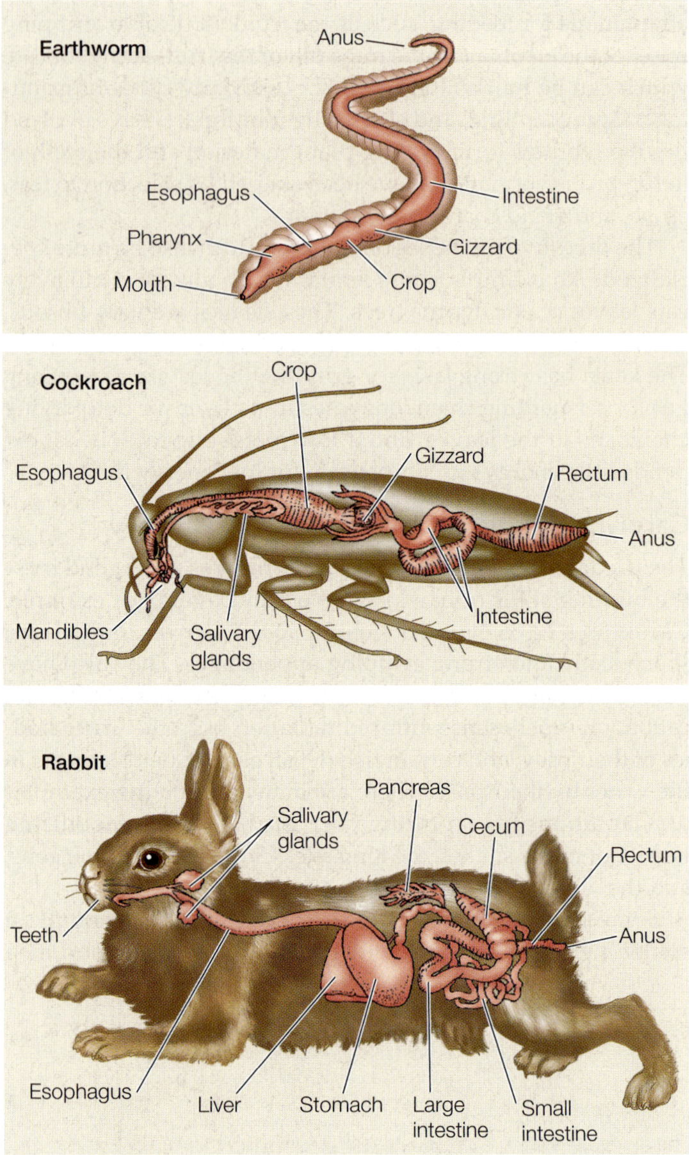

Earthworm
Anus
Esophagus
Intestine
Pharynx
Gizzard
Mouth
Crop
Cockroach
Crop
Gizzard
Esophagus
Rectum
Anus
Intestine
Mandibles
Salivary glands
Rabbit
Pancreas
Salivary glands
Cecum
Rectum
Teeth
Anus
Esophagus
Liver
Stomach
Large intestine
Small intestine

51.7 Compartments for Digestion and Absorption Most invertebrates and all vertebrates have a tubular gut that begins with a mouth, which takes in food, and ends in an anus, which eliminates wastes. Between these two structures are specialized regions for digestion and nutrient absorption; the structures in these regions are adapted to different diets and vary from species to species.

At the anterior end of the gut is the mouth cavity where food can be fragmented by teeth (in many vertebrates), by a **radula** (in snails), or by **mandibles** (in many arthropods). In most birds, food is ground by small stones in an early, muscular portion of the gut called the **gizzard**. Some animals, such as snakes, simply ingest whole prey with little or no fragmentation. **Stomachs** and **crops** are storage chambers that enable animals to ingest relatively large amounts of food when it is available, and then digest it gradually. In these storage chambers, food may be further fragmented and mixed, and in most vertebrates it is an important site of digestion. Food delivered into the next section of the gut, the **intestine**, is in small particles, well mixed, and usually partially digested.

Most digestion occurs in the intestine, and nutrients, water, and ions are absorbed across its walls. Glands secrete digestive enzymes into the intestine, and other enzymes are produced and secreted by cells lining the intestine. The final segment of the intestine recovers water and ions and stores undigested wastes, or **feces**, so they can be released to the environment at an appropriate time or place. A muscular **rectum** near the anus assists in expelling feces.

Endosymbiotic bacteria colonize the intestines. These bacteria obtain their nutrition from the food passing through the host's gut while contributing to the host's digestive processes. Members of the leech genus *Hirudo*, for example, produce no enzymes that can digest the proteins in the blood they suck from vertebrates; instead they depend on bacteria to perform this service. The resulting amino acids are subsequently used by both the leech and the bacteria. The microorganisms in

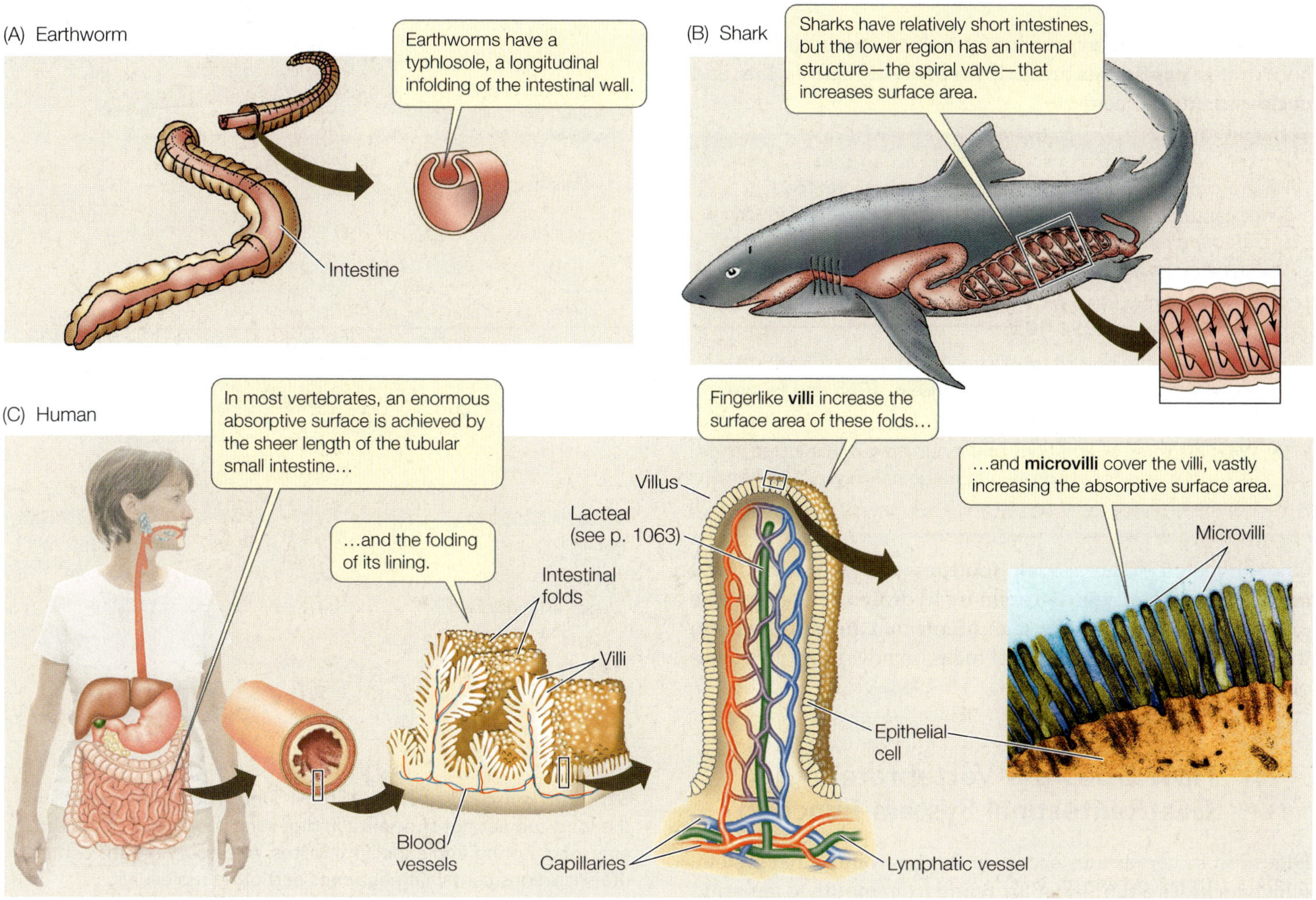

(A) Earthworm
Earthworms have a typhlosole, a longitudinal infolding of the intestinal wall.
Intestine
(B) Shark
Sharks have relatively short intestines, but the lower region has an internal structure—the spiral valve—that increases surface area.
(C) Human
In most vertebrates, an enormous absorptive surface is achieved by the sheer length of the tubular small intestine...
...and the folding of its lining.
Fingerlike villi increase the surface area of these folds...
...and microvilli cover the villi, vastly increasing the absorptive surface area.
Villus
Lacteal (see p. 1063)
Intestinal folds
Villi
Microvilli
Epithelial cell
Blood vessels
Capillaries
Lymphatic vessel

51.8 Intestinal Surface Area and Nutrient Absorption Maximizing the surface area of the gut increases an animal's ability to absorb nutrients.

the human gut are called the "forgotten organ" because they provide important services in digestion, prevent the establishment of harmful microorganisms, and even produce some vitamins (vitamin K and biotin). This forgotten organ is huge. It is estimated that the human body consists of 10^{13} cells, but our guts contain probably ten times that number of unicellular organisms representing at least 500 different species (see Figure 26.21).

In many animals, the parts of the gut that absorb nutrients have greater surface areas than would be expected of a simple tube. The simplest way to increase the surface area of a tube without changing its diameter is to produce an infolding. Such an infolding of the gut, called a typhlosole, is seen in earthworms (**Figure 51.8A**). Sharks, which have large stomachs but short intestines, have a unique adaptation called a spiral valve (**Figure 51.8B**). The lower region of the intestine is enlarged and has an internal structure that forces the food to pass through it in a spiral fashion (like going down a spiral staircase). The walls of the spiral present a large surface area for absorption of nutrients. This spiral passage, however, will not accommodate large chunks of food or undigestible matter, so food remains in the stomach for a long time to be broken down, and some larger, undigestible items (such as pieces of surfboard) do not leave the stomach except by regurgitation.

In humans, as in most other vertebrates, the wall of the intestine is highly folded, with the individual folds bearing legions of tiny fingerlike projections called **villi** (**Figure 51.8C**). The cells that line the surfaces of the villi, in turn, have microscopic projections called **microvilli**. The microvilli give the intestine an enormous internal surface area for absorbing nutrients.

Digestive enzymes break down complex food molecules

Protein, carbohydrate, and fat macromolecules are broken down into their simplest monomeric units by hydrolytic enzymes produced at different locations in the digestive tract. Many are secreted into the lumen of the gut, and others remain associated with the membranes of the microvilli. All of these enzymes cleave the chemical bonds of macromolecules through hydrolysis, a reaction that adds a water molecule (see Figure 3.4B). Digestive enzymes are classified according to the substances they hydrolyze: **proteases** break the bonds between adjacent amino acids in proteins; **carbohydrases** hydrolyze carbohydrates; **peptidases** break down peptides; **lipases**, fats; and **nucleases**, nucleic acids.

RECAP 51.2

Heterotrophs have diverse adaptations for acquiring food. Once captured or ingested, food is digested extracellularly by secreted enzymes to release nutrients, which are absorbed into the animal's body, usually via a tubular gut.

- Why do herbivores typically spend a great deal of their time feeding? **See p. 1054–1055**
- What characteristics of tooth structure distinguish herbivores, carnivores, and omnivores? **See p. 1055 and Figure 51.6.**
- What are three adaptations of the human gut that increase its surface area for absorption of nutrients? **See p. 1057 and Figure 51.8**

Once ingested by an animal, food may be fragmented and moved into the gut for digestion by hydrolytic enzymes. The processes of digestion release nutrients that are absorbed into the animal's body. Next we will focus on how those processes occur in vertebrates.

51.3 How Does the Vertebrate Gastrointestinal System Function?

Digestion in vertebrates occurs in the gastrointestinal system, which includes a tubular gut running from mouth to anus and several accessory structures that produce secretions that play important roles in digestion (**Figure 51.9**). In this section we will consider three important processes of this system: the movement of food through it, the sequential steps of digestion, and the absorption of nutrients. We will use as our primary example a typical vertebrate, the human.

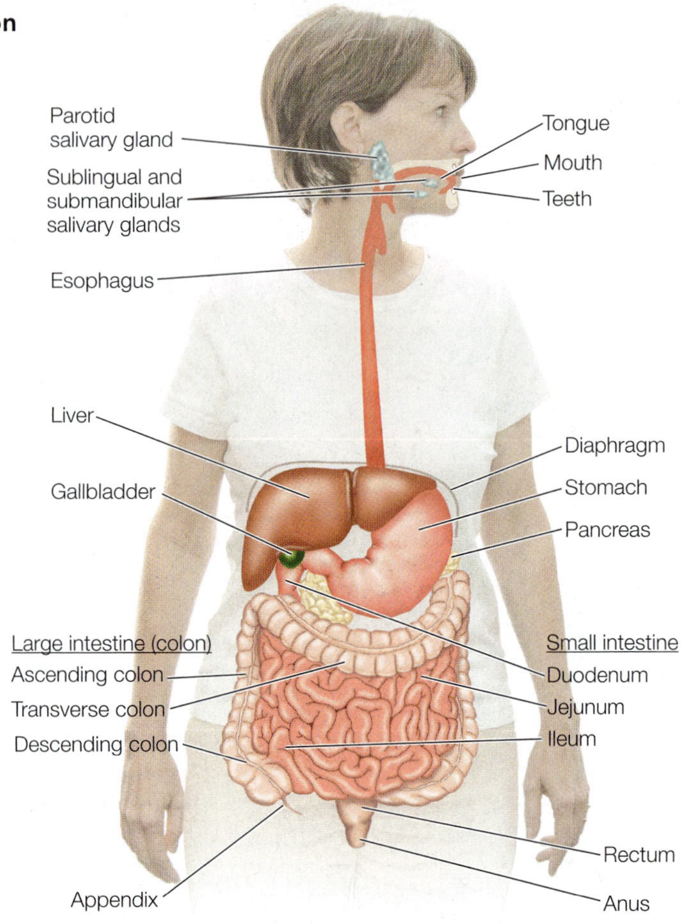

Parotid salivary gland
Sublingual and submandibular salivary glands
Esophagus
Liver
Gallbladder
Large intestine (colon)
Ascending colon
Transverse colon
Descending colon
Appendix
Tongue
Mouth
Teeth
Diaphragm
Stomach
Pancreas
Small intestine
Duodenum
Jejunum
Ileum
Rectum
Anus

51.9 The Human Digestive System Different compartments in the long tubular gut specialize in digesting food, absorbing nutrients, and storing and expelling wastes. Accessory organs contribute secretions containing enzymes and other molecules.

Go to Activity 51.4 The Human Digestive System
Life10e.com/ac51.4

The vertebrate gut consists of concentric tissue layers

The tissues of the vertebrate gut are arranged in concentric layers that have a similar organization throughout its length (**Figure 51.10**). Starting in the internal cavity, or **lumen**, the first layer is the **mucosa**, which consists of delicate epithelial cells with underlying connective tissue. Some cells of this mucosal epithelium secrete mucus to lubricate and protect the walls of the gut; some secrete digestive enzymes; and some secrete hormones. Mucosal epithelial cells in the stomach secrete hydrochloric acid and, as we noted in Section 51.1, some secrete intrinsic factor to aid the absorption of vitamin B_{12}. In some regions of the gut, nutrients are absorbed by mucosal epithelial cells. The apical plasma membranes of these absorptive cells have microvilli that increase the surface area over which absorption can take place (see Figure 51.8C).

At the base of the mucosa are smooth muscle cells that move the mucosa to improve contact with gut contents, and just under the mucosa is the submucosal tissue layer. Here we find the blood and lymph vessels that carry absorbed nutrients to the rest of the body. The **submucosa** also contains a network of nerves; the neurons in this network have sensory functions (responsible for stomach aches) and also control various secretory functions of the gut.

External to the submucosa are two layers of smooth muscle responsible for the large movements of the gut. Innermost is the circular muscle layer, with its cells oriented around the gut. Outermost is the longitudinal muscle layer, with its cells oriented along the length of the gut (see Figure 51.10). The circular muscles constrict the gut, and the longitudinal muscles shorten it. Between the two layers of smooth muscle is another nerve network that controls and coordinates the movements of the gut. The coordinated activity of the two smooth muscle layers mixes the content of the gut and moves it continuously toward the rectum. The stomach has a third layer of smooth muscle that is closest to the lumen. The orientation of its fibers is oblique to the longitudinal and circular layers. This third layer is important in generating churning motions of the stomach that mix the food and digestive juices.

Go to Media Clip 51.1
Following Food from Mouth to Gut
Life10e.com/mc51.1

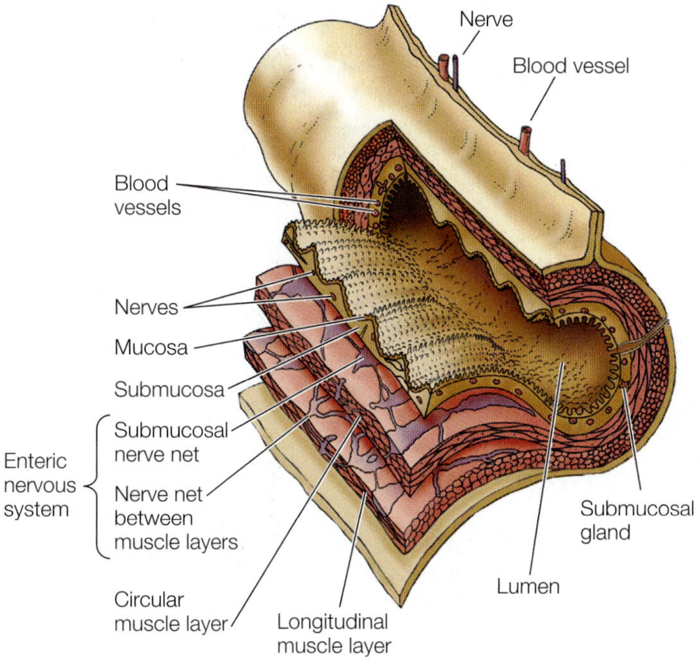
Nerve
Blood vessel
Blood vessels
Nerves
Mucosa
Submucosa
Enteric nervous system
Submucosal nerve net
Nerve net between muscle layers
Circular muscle layer
Longitudinal muscle layer
Lumen
Submucosal gland

51.10 Tissue Layers of the Vertebrate Gut The organization of tissue layers is the same in all compartments of the gut, but specialized adaptations of specific tissues characterize different regions.

The nerve nets in the submucosa and between the smooth muscle layers are called the **enteric nervous system**, and they are unusual. Whereas most neurons of the peripheral nervous system either receive synapses from neurons in the central nervous system (CNS) or contribute synapses to neurons in the CNS, most of the neurons in the enteric nervous system form synapses only with other neurons in their network. Thus they are responsible for communication within the gut. The CNS can influence activity in the enteric nervous system and receive information from it, but the gut truly has "a mind of its own."

A tissue membrane called the **peritoneum** surrounds the gut, as it does all of the organs of the abdominal cavity as well as lining the wall of the cavity. The peritoneum includes connective and epithelial tissues that secrete a fluid that lubricates the organs so they can easily slide against each other in the body cavity.

Mechanical activity moves food through the gut and aids digestion

In humans and most other mammals, food is chewed in the mouth and mixed with saliva. Periodically the tongue pushes a bolus (mass) of the chewed food toward the throat. By making contact with the soft palate at the back of the mouth cavity, the food bolus initiates swallowing, which is a complex series of reflexes. Swallowing propels the food through the pharynx (where the mouth cavity and nasal passages join) and into the **esophagus** (food tube). To prevent food from entering the trachea (windpipe), the larynx (voice box) closes, and a flap of tissue called the **epiglottis** covers the entrance to the larynx (Figure 51.11A).

Once a bolus of food enters the esophagus, it is moved toward the stomach both by the force of gravity and by waves of muscle contraction called **peristalsis** (Figure 51.11B). The muscle of the upper region of the esophagus is striated (i.e., skeletal muscle) and is controlled by the central nervous system reflexes of swallowing. The muscles of the rest of the esophagus are smooth muscle controlled by the autonomic and enteric nervous systems.

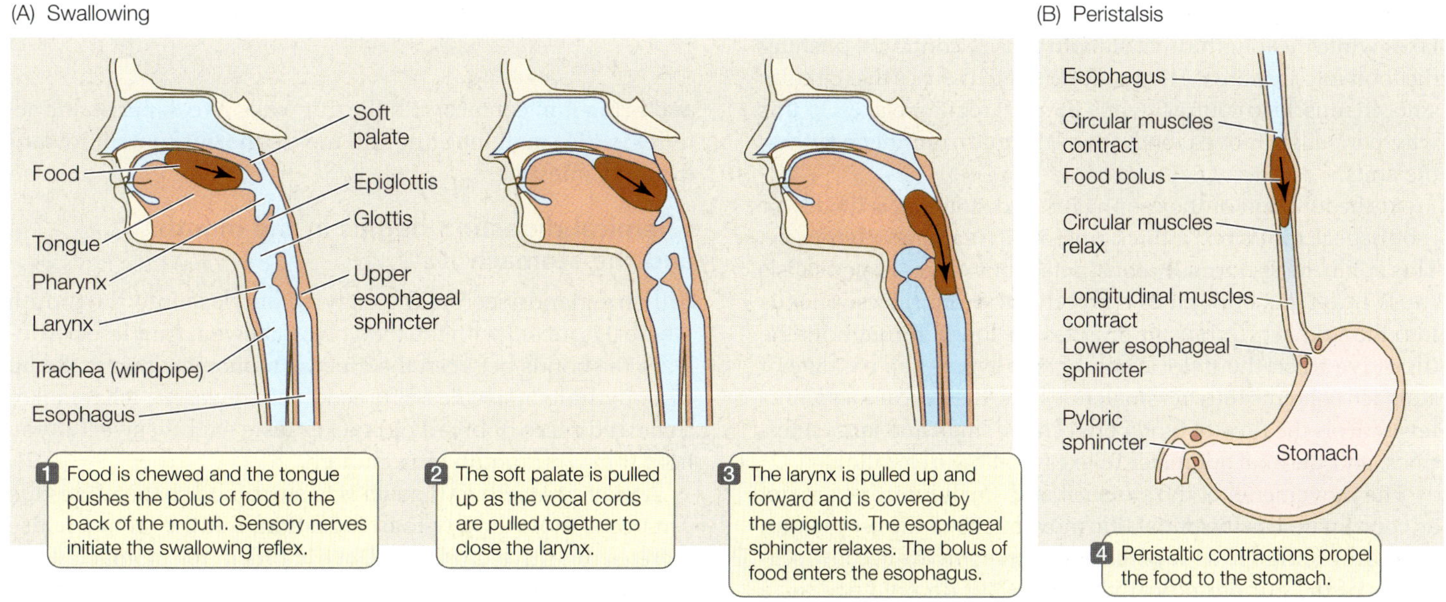
(A) Swallowing
Food
Tongue
Pharynx
Larynx
Trachea (windpipe)
Esophagus
Soft palate
Epiglottis
Glottis
Upper esophageal sphincter
1 Food is chewed and the tongue pushes the bolus of food to the back of the mouth. Sensory nerves initiate the swallowing reflex.
2 The soft palate is pulled up as the vocal cords are pulled together to close the larynx.
3 The larynx is pulled up and forward and is covered by the epiglottis. The esophageal sphincter relaxes. The bolus of food enters the esophagus.
(B) Peristalsis
Esophagus
Circular muscles contract
Food bolus
Circular muscles relax
Longitudinal muscles contract
Lower esophageal sphincter
Pyloric sphincter
Stomach
4 Peristaltic contractions propel the food to the stomach.

51.11 Swallowing and Peristalsis (A) Food pushed to the back of the mouth triggers the swallowing reflex. (B) Once a food bolus enters the esophagus, peristalsis propels it from mouth to anus by coordinated actions of the circular and longitudinal muscle layers of the gut.

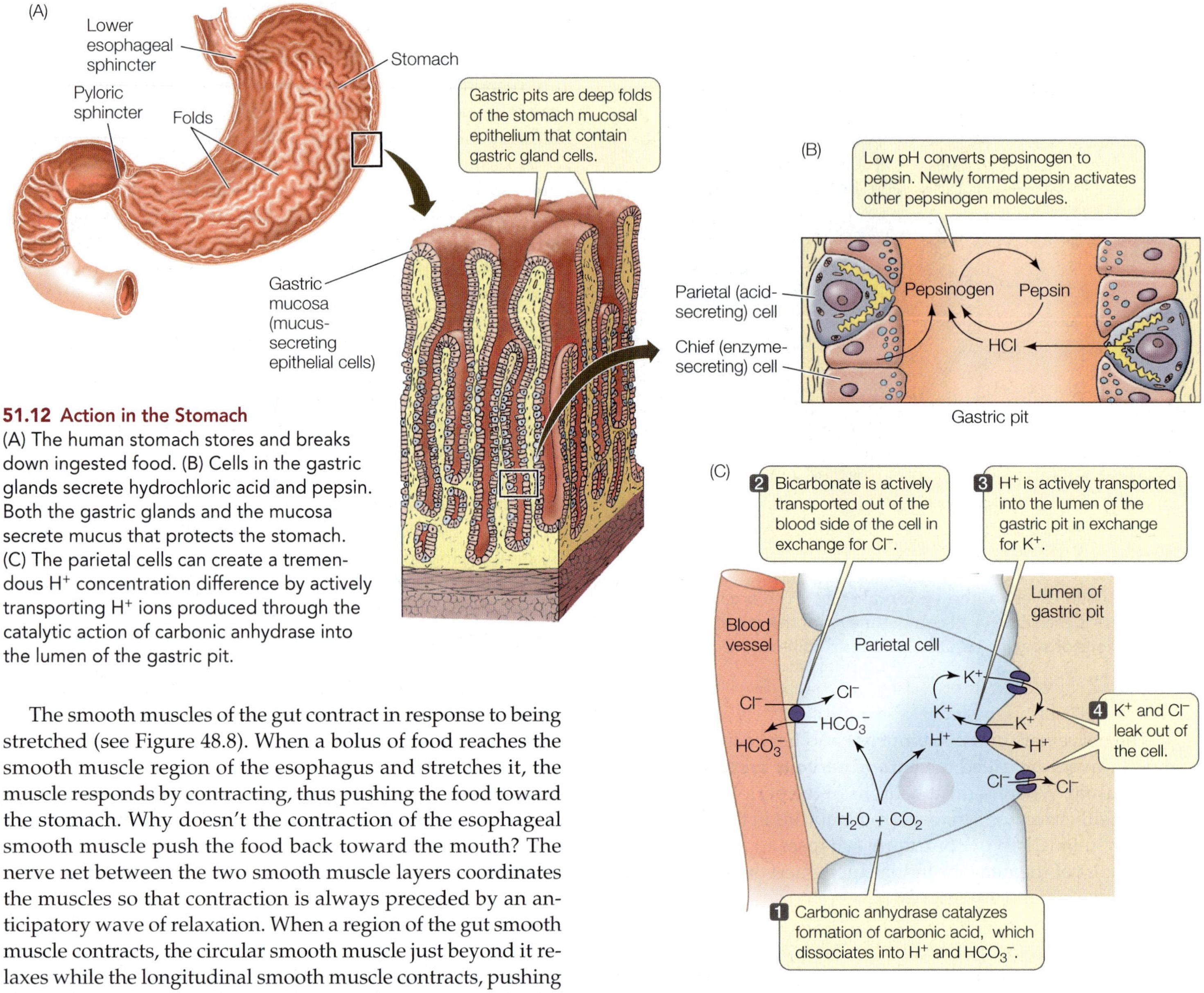

(A)
Lower esophageal sphincter
Stomach
Pyloric sphincter
Folds
Gastric pits are deep folds of the stomach mucosal epithelium that contain gastric gland cells.
Gastric mucosa (mucus-secreting epithelial cells)
(B)
Low pH converts pepsinogen to pepsin. Newly formed pepsin activates other pepsinogen molecules.
Parietal (acid-secreting) cell
Chief (enzyme-secreting) cell
Pepsinogen
Pepsin
HCl
Gastric pit
(C)
2 Bicarbonate is actively transported out of the blood side of the cell in exchange for Cl⁻.
3 H⁺ is actively transported into the lumen of the gastric pit in exchange for K⁺.
Lumen of gastric pit
Blood vessel
Parietal cell
Cl⁻
HCO_3^-
K⁺
H⁺
4 K⁺ and Cl⁻ leak out of the cell.
$H_2O + CO_2$
1 Carbonic anhydrase catalyzes formation of carbonic acid, which dissociates into H⁺ and HCO_3^-.

51.12 Action in the Stomach
(A) The human stomach stores and breaks down ingested food. (B) Cells in the gastric glands secrete hydrochloric acid and pepsin. Both the gastric glands and the mucosa secrete mucus that protects the stomach. (C) The parietal cells can create a tremendous H^+ concentration difference by actively transporting H^+ ions produced through the catalytic action of carbonic anhydrase into the lumen of the gastric pit.

The smooth muscles of the gut contract in response to being stretched (see Figure 48.8). When a bolus of food reaches the smooth muscle region of the esophagus and stretches it, the muscle responds by contracting, thus pushing the food toward the stomach. Why doesn't the contraction of the esophageal smooth muscle push the food back toward the mouth? The nerve net between the two smooth muscle layers coordinates the muscles so that contraction is always preceded by an anticipatory wave of relaxation. When a region of the gut smooth muscle contracts, the circular smooth muscle just beyond it relaxes while the longitudinal smooth muscle contracts, pushing the food into that area. The resulting stretch causes that circular smooth muscle to contract while the next region relaxes. In this way peristalsis moves food down the gut from the mouth to the anus.

At the junction of the esophagus and stomach is the lower esophageal **sphincter**, a thick ring of circular smooth muscle. This sphincter is normally constricted, but waves of peristalsis cause it to relax enough to let food pass from the esophagus into the stomach. Sphincter muscles are found throughout the digestive tract: the pyloric sphincter governs the passage of stomach contents into the small intestine; the ileocaecal sphincter controls the flow of food between the small and large intestines; and the anal sphincter relaxes to allow defecation.

The movements of the stomach and small intestine are not as coordinated as the peristaltic movements of the esophagus. They are referred to as **segmentation movements** because segments of the gut periodically contract but do not generate a peristaltic wave of contraction that moves the food in one direction. Segmentation movements move the content of the gut in either direction and thereby mix it with the digestive juices and bring it in contact with the gut walls. Weak peristaltic activity in the small intestine still moves the gut content toward the large intestine.

Chemical digestion begins in the mouth and the stomach

Salivary glands secrete the enzyme amylase into the mouth where it is mixed with the food being chewed. Amylase hydrolyzes the bonds between the glucose monomers that make up carbohydrate molecules. The action of amylase is what makes a chewed piece of bread or cracker taste slightly sweet if you hold it in your mouth long enough.

A main role of the stomach is to store food so that digestion can occur more slowly than ingestion, but the stomach also secretes digestive enzymes. **Gastric pits** in the stomach walls are lined with three types of secretory cells (**Figure 51.12A**). **Chief cells** secrete a proteolytic enzyme, pepsin, that begins the digestion of protein. **Parietal cells** pump out H^+ ions, and the lowered pH kills most ingested microorganisms. This mix of substances could damage the stomach walls, but the epithelial cells of the gastric mucosa secrete mucus that provides a protective coating for the walls of the gastric pits and stomach.

Chief cells actually secrete an inactive digestive enzyme, or **zymogen**, called pepsinogen. The extremely low pH of the stomach juices initiates the conversion of pepsinogen to pepsin by cleaving away a sequence of amino acids that masks the active site of the enzyme. Newly activated pepsin activates other pepsinogen molecules, creating a positive feedback process called **autocatalysis** (**Figure 51.12B**).

The parietal cells of the gastric pits can produce about 2 liters of hydrochloric acid (HCl) per day—enough to bring the pH of the stomach contents below 1, which is the same as battery acid and ten times more acidic than pure lemon juice. This means that across their plasma membranes, gastric pits can create a H^+ ion concentration difference of 3 million-fold. Such a feat of transport is not seen anywhere else in the body. How do the gastric pits do it? Enzymes and transporters are involved.

The enzyme carbonic anhydrase in parietal cells catalyzes the hydration of CO_2 to H_2CO_3, which dissociates into H^+ and bicarbonate ion (HCO_3^-). An antiporter transport protein (see Figure 6.13) exchanges HCO_3^- for Cl^- on the blood side of the gastric pits, and an antiporter on the gastric pit side exchanges H^+ for K^+ (**Figure 51.12C**). However, this K^+ can leak out again down its concentration gradient. Thus the inward transport of K^+ acts like an endless conveyer belt moving H^+ out into the stomach lumen. Cl^- also passively leaks out of the gastric lumen side of the parietal cells to maintain electrical neutrality.

The stomach gradually releases its contents to the small intestine

Contractions of the smooth muscles in the walls of the stomach churn its contents, thoroughly mixing them with the stomach secretions. The acidic, fluid mixture of gastric juice and partly digested food in the stomach is called **chyme**. A few substances can be absorbed across the stomach wall, including alcohol (hence its rapid effects), aspirin, and caffeine, but even these substances are absorbed in rather small quantities from the stomach.

Contractions of the stomach walls push the chyme toward the bottom of the stomach. These waves of contractions cause the pyloric sphincter to relax briefly so that little squirts of the chyme can enter the small intestine. In this manner the human stomach empties itself gradually over a period of approximately 4 hours. This slow introduction of food into the small intestine enables it to work on a little material at a time.

Most chemical digestion occurs in the small intestine

In the **small intestine**, the digestion of carbohydrates and proteins continues, and the digestion of fats and absorption of nutrients begin. The small intestine takes its name from its diameter; it is in fact a very large organ, about 6 meters long in an adult human. Given its length and the folds, villi, and microvilli of its lining, its inner surface area is roughly the size of a tennis court. Across this surface the small intestine absorbs all the nutrient molecules derived from food.

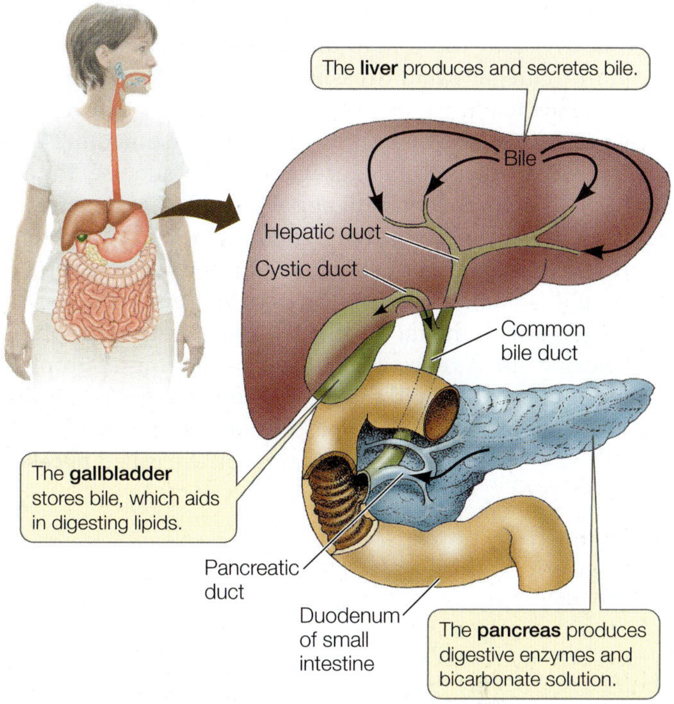
The **liver** produces and secretes bile.
Bile
Hepatic duct
Cystic duct
Common bile duct
The **gallbladder** stores bile, which aids in digesting lipids.
Pancreatic duct
Duodenum of small intestine
The **pancreas** produces digestive enzymes and bicarbonate solution.

51.13 Ducts of the Gallbladder and Pancreas Bile produced in the liver leaves the liver via the hepatic duct. Branching off this duct is the gallbladder, which stores bile. Below the gallbladder, the hepatic duct is called the common bile duct and is joined by the pancreatic duct before entering the duodenum.

The small intestine of humans has three sections. The initial section (about 25 cm long) is called the **duodenum** and is the site of most digestion; the **jejunum** and the **ileum** (together about 600 cm) carry out 90 percent of the absorption of nutrients (see Figure 51.9).

Digestion in the small intestine requires many specialized enzymes, as well as several other secretions. Two accessory organs that are not part of the digestive tract—the liver and the pancreas—produce many of these secretions and deliver them to the lumen of the intestine through ducts.

LIVER The liver synthesizes **bile salts** from cholesterol and secretes them as **bile**. Bile also includes other substances, such as phospholipids and bilirubin (the breakdown product of hemoglobin). Bile flows from the liver through the hepatic duct. A side branch off the hepatic duct called the cystic duct goes to the **gallbladder**, where bile is stored. Below this junction, the hepatic duct is called the common bile duct. Before it reaches the duodenum, the common bile duct is joined by the pancreatic duct (**Figure 51.13**).

Fat entering the duodenum stimulates cells of the duodenal epithelium to release the hormone **cholecystokinin** (**CCK**), which stimulates the walls of the gallbladder to contract rhythmically. As a result, bile is squeezed out of the gallbladder and through the cystic duct to the common bile duct. A small sphincter at the junction of the common bile duct with the duodenum relaxes in response to waves of peristalsis and allows squirts of bile to enter the duodenal lumen.

To understand the role of bile in fat digestion, think of an oil-and-vinegar salad dressing. The oil, which is hydrophobic, tends to aggregate in large globules. For that reason, many salad dressings include an emulsifier—something that prevents oil droplets from aggregating. Mayonnaise, for example, is oil and vinegar with egg yolk added as an emulsifier. Bile salts emulsify fats in the chyme. One end of each bile salt molecule is lipophilic (soluble in fat), and the other end is hydrophilic (soluble in water). The lipophilic ends of bile molecules merge with the fat droplets, leaving their hydrophilic ends sticking out. As a result, bile salts prevent the fat droplets from sticking together and thereby greatly enlarge the surface area of the fats exposed to the lipases—the enzymes that digest fats. The very small fat particles that result are called **micelles** (Figure 51.14A).

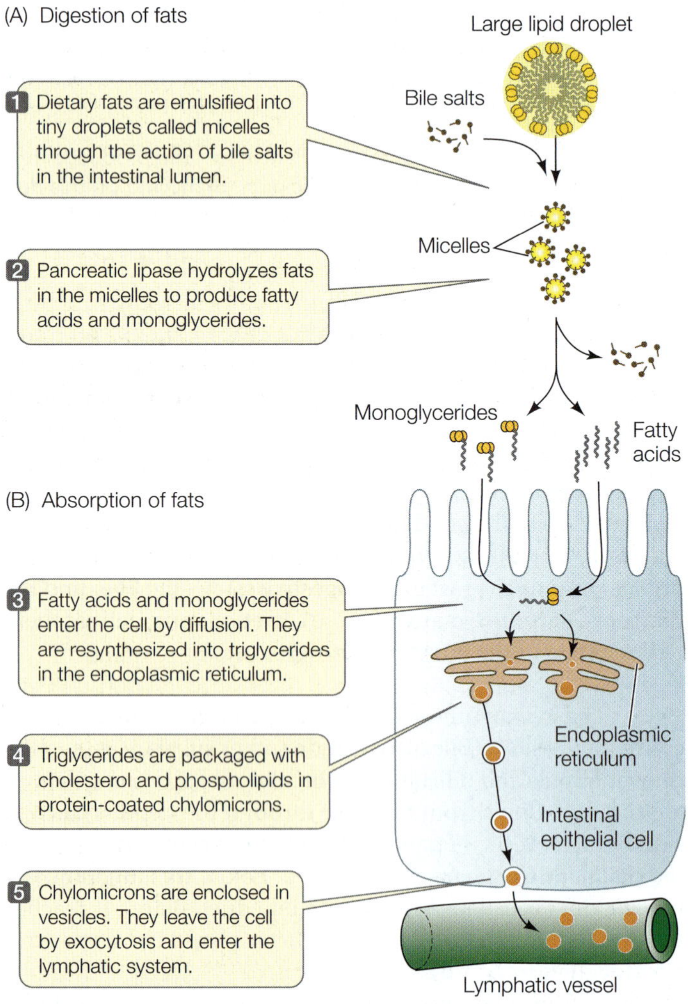
(A) Digestion of fats
Large lipid droplet
Bile salts
1 Dietary fats are emulsified into tiny droplets called micelles through the action of bile salts in the intestinal lumen.
Micelles
2 Pancreatic lipase hydrolyzes fats in the micelles to produce fatty acids and monoglycerides.
Monoglycerides
Fatty acids
(B) Absorption of fats
3 Fatty acids and monoglycerides enter the cell by diffusion. They are resynthesized into triglycerides in the endoplasmic reticulum.
4 Triglycerides are packaged with cholesterol and phospholipids in protein-coated chylomicrons.
Endoplasmic reticulum
Intestinal epithelial cell
5 Chylomicrons are enclosed in vesicles. They leave the cell by exocytosis and enter the lymphatic system.
Lymphatic vessel

51.14 Digesting Fats (A) Dietary fats are broken up by bile into small micelles that present a large surface area to lipases. (B) The products of fat digestion are absorbed by intestinal mucosal cells, where they are resynthesized into triglycerides and exported to lymphatic vessels.

Go to Animated Tutorial 51.1
The Digestion and Absorption of Fats
Life10e.com/at51.1

PANCREAS The **pancreas** is a large gland that lies just behind and below the stomach (see Figures 51.9 and 51.13). It is both an endocrine gland (secreting hormones into the extracellular fluid; see Section 41.1) and an exocrine gland (secreting digestive juices through the pancreatic duct to the gut lumen). The exocrine tissues of the pancreas produce a host of digestive enzymes, including lipases, amylases, proteases, and nucleases (Table 51.3). As in the stomach, the protease enzymes are released as zymogens; if proteases were not in this inactive state, they would digest the pancreas and its ducts before ever reaching the duodenum. Once in the duodenum, the zymogen trypsinogen is activated by the enzyme enterokinase (secreted by cells lining the duodenum) to produce the active protease **trypsin**. Trypsin cleaves other zymogens, releasing other proteases as well as more active trypsin.

The mixture of zymogens produced by the pancreas can be dangerous if the pancreatic duct is blocked or if the pancreas is injured by infection or physical trauma such as a blow to the abdomen. A few activated trypsin molecules can initiate a chain reaction of enzyme activity that digests the tissues of the pancreas (a condition called pancreatitis), destroying both its endocrine and exocrine functions.

The pancreas also produces a secretion rich in bicarbonate ions (HCO_3^-). Bicarbonate ions are alkaline (basic) and neutralize the acidic pH of the chyme that enters the duodenum from the stomach. Intestinal enzymes function best at a neutral or slightly alkaline pH.

TABLE **51.3** Major Digestive Enzymes of Humans

Source/Enzyme	Action
SALIVARY GLANDS	
Salivary amylase	Starch → Maltose
STOMACH	
Pepsin	Proteins → Peptides; autocatalysis
PANCREAS	
Pancreatic amylase	Starch → Maltose
Lipase	Fats → Fatty acids and glycerol
Nuclease	Nucleic acids → Nucleotides
Trypsin	Proteins → Peptides; zymogen activation
Chymotrypsin	Proteins → Peptides
Carboxypeptidase	Peptides → Shorter peptides and amino acids
SMALL INTESTINE	
Aminopeptidase	Peptides → Shorter peptides and amino acids
Dipeptidase	Dipeptides → Amino acids
Enterokinase	Trypsinogen → Trypsin
Nuclease	Nucleic acids → Nucleotides
Maltase	Maltose → Glucose
Lactase	Lactose → Galactose and glucose
Sucrase	Sucrose → Fructose and glucose

Nutrients are absorbed in the small intestine

The final step in digesting proteins and carbohydrates and absorbing their components occurs among the microvilli. Mucosal epithelial cells produce peptidases that cleave small peptides into absorbable amino acids. These epithelial cells also produce the enzymes maltase, lactase, and sucrase that cleave the common disaccharides into absorbable monosaccharides—glucose, galactose, and fructose. There is also some lipase activity for fat digestion.

Many humans stop producing the enzyme lactase in childhood and thereafter have difficulty digesting lactose (the sugar in milk). Lactose is a disaccharide and cannot be absorbed without being cleaved into its constituents, glucose and galactose. Unabsorbed lactose is metabolized by bacteria in the large intestine, causing gas, diarrhea, and abdominal cramps.

The mechanisms by which cells of the intestinal epithelium absorb nutrients and inorganic ions are diverse and include diffusion, facilitated diffusion, osmosis, active transport, and co-transport. Many inorganic ions such as sodium, calcium, and iron are actively transported by these cells. For example, active Na^+ transporters exist on the basal and lateral sides of the epithelial cells. They maintain a low concentration of Na^+ in those cells so that Na^+ can diffuse in from the chyme in the intestinal lumen. About 30 grams of Na^+ are transported this way every day, and Cl^- follows.

The transport of Na^+ and other ions is also important for water absorption because it creates an osmotic concentration gradient. At least 7 to 8 liters of water per day move through the spaces between the epithelial cells in response to this osmotic gradient. Because the water moves through *spaces* between the cells and not through the cells themselves, it can carry with it nutrients that are in solution—a transport mechanism called "solvent drag."

Many different kinds of transport proteins exist in the epithelial cell membranes. Some, such as the transport protein for fructose, only facilitate diffusion, and that requires a concentration gradient. This mechanism works for fructose because once fructose enters the cell it is converted to glucose. Thus the concentration of fructose in the cell is always low and the concentration gradient is maintained. Transport proteins known as symporters (see Figure 6.13) exploit the concentration gradient of Na^+ between the inside and outside of the cell that is maintained by the Na^+/K^+ ATPase common to all cells. Symporters combine the transport of Na^+ and another molecule, such as glucose, galactose, or an amino acid. As Na^+ moves down its concentration gradient into the cell, the "hitchhiking" molecules are carried along with it.

The absorption of the products of fat digestion is relatively simple. Triglycerides are hydrolyzed to diglycerides, monoglycerides, and fatty acids, all of which are lipid-soluble and thus able to pass through the plasma membranes of the microvilli. In the intestinal epithelial cells, these molecules are resynthesized into triglycerides, combined with cholesterol and phospholipids, and coated with protein to form water-soluble **chylomicrons** (Figure 51.14B). Rather than enter the blood directly, chylomicrons pass into blind-ended lymph vessels called **lacteals** that are inside each villus (see Figure 51.8C). They then flow through the lymphatic system, entering the bloodstream through the thoracic ducts at the base of the neck. After a meal rich in fats, chylomicrons can be so abundant in the blood that they give the plasma a milky appearance. Chylomicrons deliver their triglyceride and cholesterol cargo as they circulate through tissues.

The bile salts that emulsify fats are not absorbed along with the monoglycerides, diglycerides, and the fatty acids, but are shuttled back and forth between the gut contents and the microvilli. In the ileum, bile salts are actively reabsorbed and returned to the liver via the bloodstream.

Absorbed nutrients go to the liver

Blood leaving the digestive tract flows to the liver in the **hepatic portal vein**. This large vein delivers the blood to small spaces called sinusoids between groups of liver cells. These cells absorb the nutrients coming from the digestive tract and either store them or convert them to molecules the body needs. Glucose, sucrose, and fructose are used to synthesize glycogen. Amino acids are used to build proteins. Lipids from the chylomicrons are either stored as triglycerides or used to make lipoproteins, which are released by the liver and carry the triglycerides and cholesterol to other tissues (see Section 51.4).

Water and ions are absorbed in the large intestine

The motility of the small intestine gradually pushes its contents into the large intestine, or **colon**. Most of the available nutrients have been removed from the chyme that enters the colon, but it contains a lot of water and inorganic ions. Segmentation movements bring the colon's contents into contact with its walls and promote reabsorption of ions and water, producing feces, a semisolid mass of waste products. Absorption of too much water from the colon can cause constipation. The opposite condition, diarrhea, results if too little water is absorbed; in this case, water in the colon is excreted with the feces. Excessive diarrhea caused by diseases such as cholera can produce such rapid loss of water and electrolytes that death can occur in hours.

Fecal matter is stored in the descending colon and in the rectum until defecation occurs. This is usually once a day and is preceded by strong peristaltic activity. The distension of the walls of the rectum by fecal matter initiates a parasympathetic reflex, causing the rectal muscle to contract and the internal anal sphincter to relax. In addition, there is an external anal sphincter that is under conscious control so that defecation is not entirely an involuntary act.

Herbivores rely on microorganisms to digest cellulose

As the primary component of plant cell walls, cellulose is the principal component of the food of herbivores. Most herbivores, however, cannot produce cellulases, the enzymes that break down cellulose. (Exceptions include earthworms, shipworms, and the silverfish that eat books and stored papers.) From termites to cattle, herbivores rely on microorganisms in their digestive tracts to digest cellulose.

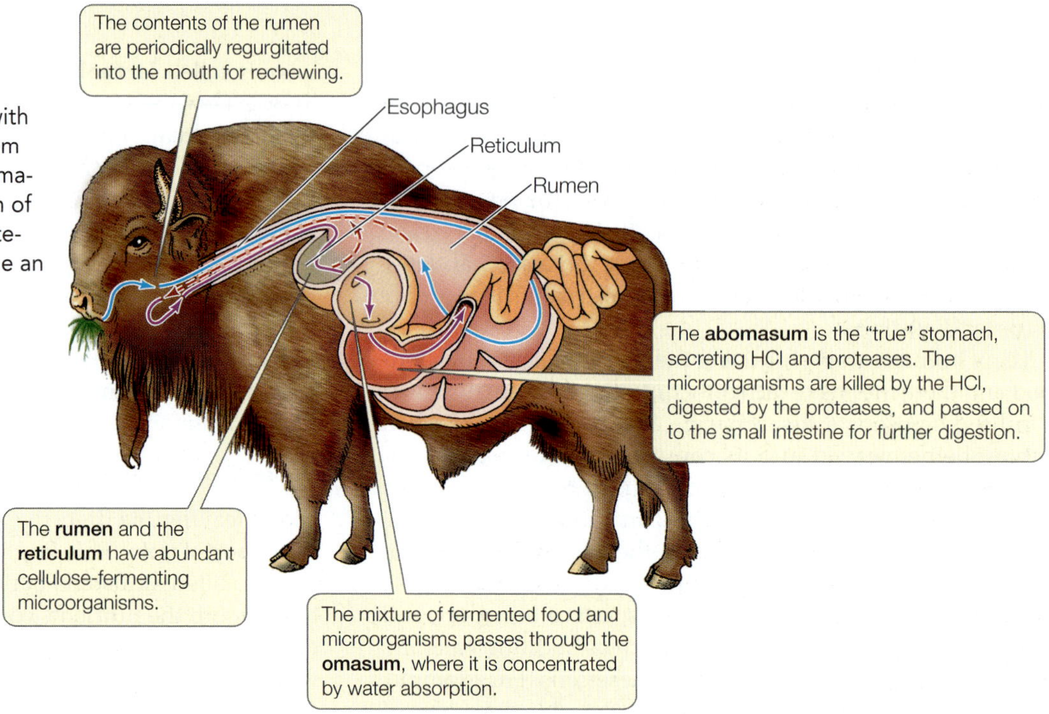
The contents of the rumen are periodically regurgitated into the mouth for rechewing.
Esophagus
Reticulum
Rumen
The **abomasum** is the "true" stomach, secreting HCl and proteases. The microorganisms are killed by the HCl, digested by the proteases, and passed on to the small intestine for further digestion.
The **rumen** and the **reticulum** have abundant cellulose-fermenting microorganisms.
The mixture of fermented food and microorganisms passes through the **omasum**, where it is concentrated by water absorption.

51.15 A Ruminant's Stomach Bison, like their relatives domestic cattle, have a specialized stomach with four compartments that enables them to obtain energy from coarse plant material through bacterial fermentation of the otherwise indigestible plant material. The bacteria themselves become an important source of nutrition.

The stomachs of **ruminants** (cud chewers) such as cattle are large, four-chambered organs that take advantage of their endosymbiotic microorganisms (**Figure 51.15**). The first two chambers, the **rumen** and the **reticulum**, are packed with microorganisms that break down cellulose by fermentation. The ruminant periodically regurgitates the contents of the rumen (the cud) into the mouth for more chewing. When swallowed again, the vegetal fibers present more surface area to the microorganisms. The microorganisms metabolize cellulose and other nutrients to simple fatty acids, which are nutrients for their host.

Enormous numbers of microorganisms leave the rumen along with the partially digested food. This mass is concentrated by water absorption in the **omasum** before it enters the true stomach, the **abomasum**, where the microorganisms are killed by secreted hydrochloric acid, digested by proteases, and passed on to the small intestine for further digestion and absorption. A cow derives more than 100 grams of protein per day from digestion of its endosymbiotic microorganisms. The rate of multiplication of microorganisms in the rumen offsets their loss, so a well-balanced, mutually beneficial relationship is maintained.

Some mammalian herbivores have a microbial fermentation chamber called a **cecum** extending from the large intestine. An example is the rabbit (see Figure 51.7). Since the cecum empties into the large intestine, absorption of some nutrients produced by the microorganisms is inefficient, because of the large intestine's limited surface area. Such species frequently produce two kinds of feces—ones that are pure waste and ones that contain cecal material. In a behavior known as **coprophagy**, these species reingest the cecal feces directly from the anus so they can digest and absorb the nutrients that would otherwise be lost. In humans the cecum is small and ends in the vestigial **appendix**, which serves no digestive function.

RECAP 51.3

The vertebrate gastrointestinal system is a tubular gut that is adapted to ingest food, fragment it, digest it, and absorb nutrients. Peristalsis moves food through the gut. Segmentation movements mix the gut contents. Digestion and absorption of nutrients occur mostly in the small intestine; water and ions are absorbed in the large intestine.

- What digestive functions occur in the mouth and stomach? **See p. 1060–1061 and Figure 51.12**
- How do bile salts assist in the digestion of fats? **See p. 1062 and Figure 51.14**
- Describe how symporters drive the absorption of nutrients. **See p. 1063**

The steps included in ingestion and digestion of food—from fragmentation in the mouth to the digestive processes in the gastrointestinal tract—make the nutrients in food available for absorption and ultimately for metabolism. We look next at how the processes of digestion are controlled and how nutrients are handled by the body once food has been digested.

51.4 How Is the Flow of Nutrients Controlled and Regulated?

The vertebrate gut is an assembly line in reverse—a *dis*assembly line. As with a standard assembly line, the control and coordination of the sequential processes of digestion are critical. Both neuronal and hormonal controls govern these processes. Once the products of digestion are absorbed, their availability to the cells of the body must also be controlled.

You have certainly experienced salivation at the sight or smell of food. That response is an unconscious reflex, as is swallowing. Many such autonomic reflexes coordinate activity in different regions of the digestive tract. For example, the introduction of food into the stomach stimulates increased activity in the colon that may lead to defecation.

Neuronal messages travel from one region of the digestive tract to another in the enteric nervous system. One function of the gut's nervous system is coordinating the movement of food through the gut. Of course, this intrinsic nervous system communicates information to the CNS and receives input from the CNS, but its most important role is to coordinate actions throughout the digestive tract. In spite of this marvelous intrinsic nervous system, however, much of the control and regulation of the digestive system and nutrient management involves hormonal mechanisms.

Hormones control many digestive functions

Several hormones control the activities of the digestive tract and its accessory organs (**Figure 51.16**). The first hormone ever discovered came from the duodenum; it was called **secretin** because it causes the pancreas to secrete digestive juices. We now know that secretin is only one of several hormones that control pancreatic secretion; specifically, secretin stimulates the pancreas to secrete a solution rich in bicarbonate ions.

The stimulus that causes the duodenum to release secretin is low pH caused by the arrival of acidic chyme from the stomach. Similarly, the presence of fats and proteins in the chyme stimulates the release of cholecystokinin (CCK), the hormone that stimulates the gallbladder to release bile. CCK also stimulates the pancreas to release digestive enzymes. Both CCK and secretin slow the movements of the stomach, thus slowing the delivery of chyme into the small intestine and allowing more complete digestion in the duodenum.

The presence of food in the stomach stimulates cells in the lower region of the stomach to secrete a hormone called **gastrin**. Gastrin returns to the stomach in the blood and stimulates the secretion of digestive juices and also the movements of the stomach. Gastrin release begins to be inhibited when the pH of the stomach contents falls below 3—an example of negative feedback.

Most animals do not eat continuously, so they can be either in an **absorptive state** (food in the gut) or in a **postabsorptive state** (no food in the gut). Nutrient requirements for energy metabolism and biosynthesis are continuous, however. Thus nutrient traffic must be controlled so that reserves accumulate in the liver, muscle, and adipose (fat) tissue while the animal is in the absorptive state and are then used efficiently during the postabsorptive state.

The liver directs the traffic of the molecules that fuel metabolism

When fuel molecules are abundant in the blood, the liver stores them in the form of glycogen and fats. The liver also synthesizes blood plasma proteins from circulating amino acids. When levels of fuel molecules in the blood decline, the liver taps its reserves and delivers nutrients into the blood.

The liver has an enormous capacity to interconvert fuel molecules. Liver cells can convert monosaccharides into either

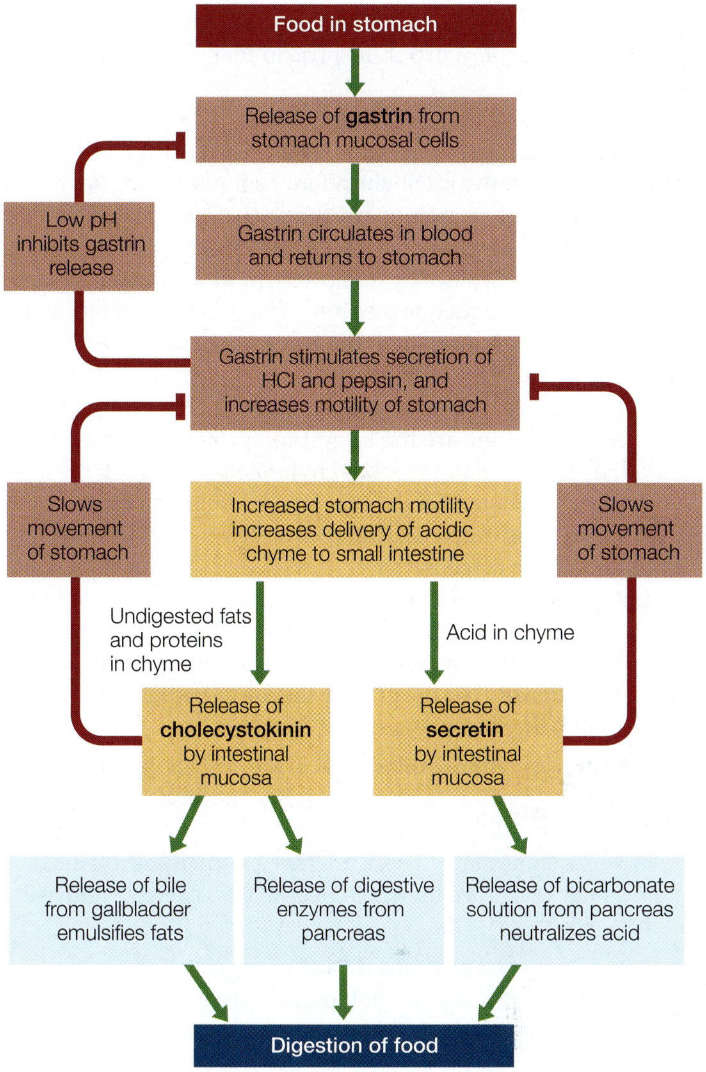

Food in stomach
Release of gastrin from stomach mucosal cells
Low pH inhibits gastrin release
Gastrin circulates in blood and returns to stomach
Gastrin stimulates secretion of HCl and pepsin, and increases motility of stomach
Slows movement of stomach
Increased stomach motility increases delivery of acidic chyme to small intestine
Slows movement of stomach
Undigested fats and proteins in chyme
Acid in chyme
Release of cholecystokinin by intestinal mucosa
Release of secretin by intestinal mucosa
Release of bile from gallbladder emulsifies fats
Release of digestive enzymes from pancreas
Release of bicarbonate solution from pancreas neutralizes acid
Digestion of food

51.16 Hormones Control Digestion The hormones gastrin, cholecystokinin, and secretin are involved in feedback loops that control the sequential processing of food in the digestive tract. Red lines indicate inhibitory actions; green lines indicate stimulatory actions.

glycogen or fats, and vice versa. The liver can also convert certain amino acids and some other molecules, such as pyruvate and lactate, into glucose—the process of gluconeogenesis (see Section 9.5). Gluconeogenesis provides an indirect pathway for exercising muscle to contribute to blood glucose levels. At high levels of aerobic activity, muscle cells break down their stores of glycogen to provide metabolic fuel (see Section 48.2). The glucose released from muscle glycogen cannot leave muscle cells as it can leave liver cells. However, when the activity of the muscle becomes anaerobic, pyruvate and lactate build up, leave the muscle cells, and enter the circulation. Circulating pyruvate and lactate are taken up by the liver and converted to glucose that can then move out of the liver cells and into the blood.

The liver is also the major controller of fat metabolism through its production of lipoproteins. A **lipoprotein** is a particle made up of a core of hydrophobic fat and cholesterol with a covering of hydrophilic protein that allows it to be suspended in water.

LIPOPROTEINS: THE GOOD, THE BAD, AND THE UGLY Lipoproteins move fats, the most abundant fuel reserve in the body, from sites of absorption or synthesis to sites of storage, and from sites of storage to sites of use. We saw in Section 51.3 how, in the intestine, bile solves the problem of processing hydrophobic fats in an aqueous medium. The transport of fats in the circulatory system presents the same problem, and lipoproteins provide the solution.

The chylomicrons (see p. 1063) produced by the mucosal cells of the intestine are the largest lipoprotein particles in the blood. As the circulation carries chylomicrons through the liver and adipose tissue, lipoprotein lipases begin to break them down, and their triglyceride and cholesterol cargo is absorbed into the liver or fat cells.

Lipoproteins other than chylomicrons are synthesized in the liver. These lipoproteins can be classified according to their density. Fat has a low density (it floats on water) and protein has a high density, so the greater the fat-to-protein ratio in the lipoprotein, the lower its density.

- **High-density lipoproteins** (**HDLs**) remove cholesterol from tissues and carry it to the liver, where it can be used to synthesize bile. HDL consists of about 50% protein, 35% lipids, and 15% cholesterol. These are the "good" lipoproteins, and their levels are higher in people who exercise and are fit.
- **Low-density lipoproteins** (**LDLs**) transport cholesterol around the body for use in biosynthesis and for storage. LDL consists of about 25% protein, 25% lipids, and 50% cholesterol. These are the "bad" lipoproteins associated with a high risk for cardiovascular disease.
- **Very low-density lipoproteins** (**VLDLs**) contain mostly triglyceride fats, which they transport to fat cells in adipose tissues around the body. VLDL consists of about 2% protein, 94% lipids, and 3% cholesterol. These are the "ugly" lipoproteins, as they are associated with excessive fat deposition as well as a high risk for cardiovascular disease.

INSULIN AND GLUCAGON CONTROL FUEL METABOLISM During the absorptive state, blood glucose levels rise as carbohydrates are digested and absorbed (**Figure 51.17**). During this time, beta cells of the pancreas release the hormone **insulin**, which plays a major role in directing glucose to where it will be used or stored. The actions of insulin vary in different tissues, but they are all aimed at promoting the use of glucose for metabolic fuel and directing the excess glucose into storage as either glycogen or fat.

Glucose enters cells by diffusion. This diffusion is facilitated by transporters, but they are not active transporters—they require a concentration gradient, which is why it is important to regulate blood glucose levels so there is always an adequate glucose concentration gradient across the cell membranes.

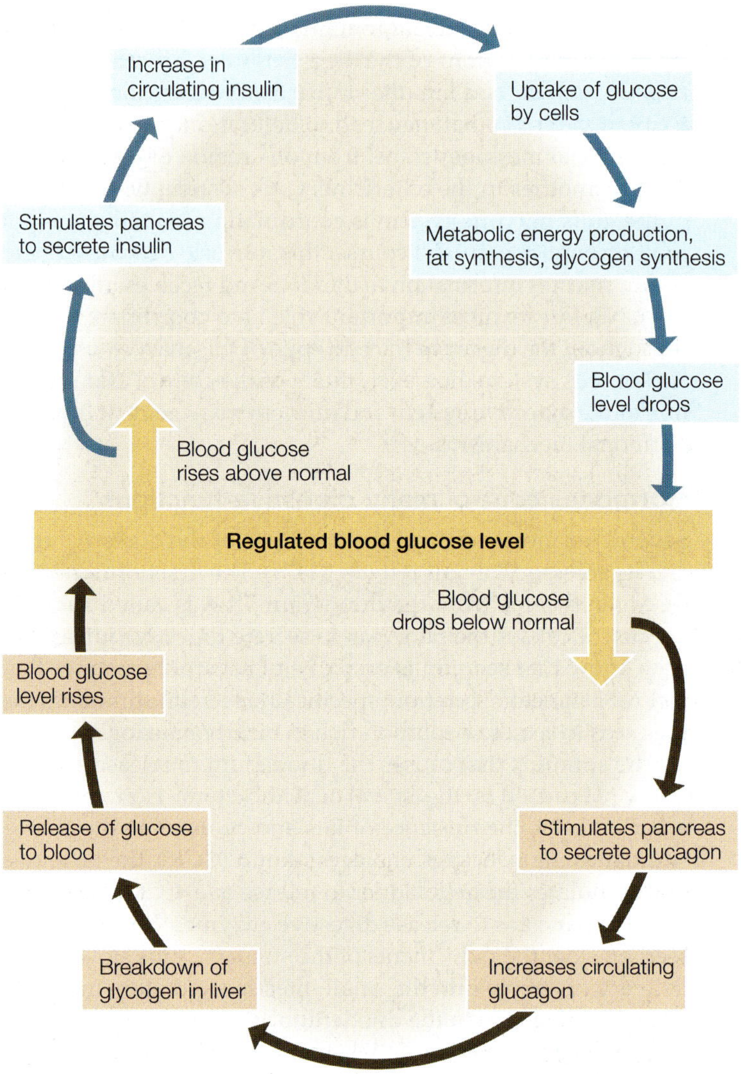

Increase in circulating insulin
Uptake of glucose by cells
Stimulates pancreas to secrete insulin
Metabolic energy production, fat synthesis, glycogen synthesis
Blood glucose level drops
Blood glucose rises above normal
Regulated blood glucose level
Blood glucose drops below normal
Blood glucose level rises
Release of glucose to blood
Stimulates pancreas to secrete glucagon
Breakdown of glycogen in liver
Increases circulating glucagon

51.17 Regulating Glucose Levels in the Blood Insulin (blue) and glucagon (brown) interactions maintain the homeostasis of circulating glucose. It is important for blood glucose to remain stable because it is the essential source of fuel for the nervous system.

Go to Animated Tutorial 51.2
Insulin and Glucose Regulation
Life10e.com/at51.2

There are several kinds of glucose transporters, and those in skeletal muscle and adipose tissues are normally sequestered in cytoplasmic vesicles until insulin binds its receptors on the cell surface and triggers the insertion of transporters into the plasma membrane.

Insulin plays many roles in controlling how cells use the glucose they take up from the circulation. In adipose cells, insulin inhibits lipase and promotes fat synthesis from glucose. In the liver, insulin activates an enzyme that phosphorylates glucose as it enters the liver cell so it cannot diffuse back out again, enhancing the overall diffusion of glucose into the cells. Insulin also activates the enzymes in liver cells that catalyze the synthesis of glycogen.

During the postabsorptive state, a fall in blood glucose decreases the release of insulin, and the uptake of glucose by most cells is curtailed (see Figure 51.17). To maintain blood glucose levels, liver cells break down their stored glycogen, releasing glucose into the blood. The liver and adipose tissues break down fats and supply fatty acids to the blood, and most cells preferentially use those fatty acids as their metabolic fuel. *The most important control of fuel metabolism in the postabsorptive state is the lack of insulin.*

One tissue that does not switch fuel sources when an animal is postabsorptive is the nervous system. The cells of the nervous system require a constant supply of glucose and can use other fuels only to a very limited extent. Most neurons do not require insulin to absorb glucose from the blood, but they do need an adequate glucose concentration gradient to drive the facilitated diffusion of glucose across their plasma membranes. Therefore it is critical that blood glucose levels are maintained when an animal is postabsorptive. The overall dependence of neural tissues on glucose, and their requirement for constant blood glucose levels, are the reasons it is so important for other cells of the body to shift to fat metabolism during the postabsorptive state.

The metabolism of fuel molecules during the postabsorptive state is mostly controlled by the lack of insulin, but if blood glucose falls below a certain level, another pancreatic hormone, **glucagon**, is called into play. Glucagon's effect is opposite that of insulin: it stimulates liver cells to break down glycogen and to carry out gluconeogenesis. Thus, under the influence of glucagon, the liver produces glucose and releases it into the blood.

The brain plays a major role in regulating food intake

Multiple brain areas and signals are involved in the regulation of food intake. Long ago it was discovered that damaging a region in the center of the rat hypothalamus resulted in the rats increasing their food intake and becoming obese. Damage to the lateral hypothalamus, however, led to decreased food intake and the rats became thin. In both cases the rats eventually reached a new equilibrium body weight; thus it appeared that a capacity for regulation remained, but the set point was altered.

We now know that another region of the hypothalamus, the **arcuate nucleus**, plays an important role in integrating a variety of feedback signals that influence food intake and body mass. Cells within the arcuate nucleus send axons to the ventromedial and dorsal hypothalamus, as well as to other brain areas that influence food intake and metabolism. One group of arcuate neurons projects to brain areas that inhibit food intake while the other projects to brain areas that stimulate food intake. But what stimulates or inhibits the activity of the arcuate neurons?

Several factors have been identified that reflect the body's energy balance. Three of these are the proteins insulin, leptin, and ghrelin. Insulin, as detailed earlier, is released when blood glucose levels are high. **Leptin** (Greek *leptos*, "thin") is released by fat cells in proportion to how much lipid they contain. Evidence that leptin is a satiety signal is described in **Figure 51.18**. **Ghrelin** is released by the stomach when it is empty; its levels rise before meals and fall after meals. In the arcuate nucleus, insulin and leptin activate the neurons that inhibit feeding and inhibit the neurons that stimulate feeding. Ghrelin has the opposite effect on these two groups of neurons.

INVESTIGATING**LIFE**

51.18 A Single-Gene Mutation Leads to Obesity in Mice

In mice the *Ob* gene codes for the protein leptin, a satiety factor that signals the brain when enough food has been consumed. The recessive *ob* allele is a loss-of-function allele, so *ob/ob* mice do not produce leptin; they do not experience satiety and become obese. The *Db* gene encodes the leptin receptor, so mice homozygous for the recessive loss-of-function allele *db*, even if they produce leptin, cannot use it and so become obese.[a]

HYPOTHESIS Mice who cannot produce the satiety signal protein leptin will not become obese if they are able to obtain leptin from an outside source.

Method

1. Create two strains of genetically obese laboratory mice, one of which lacks functional leptin (genotype *ob/ob*) and one which lacks the receptor for leptin (genotype *db/db*).
2. Create parabiotic pairs by surgically joining the circulatory systems of a non-obese (wild-type) mouse with a partner from one of the obese strains.
3. Allow mice to feed at will.

Parabiotic pair

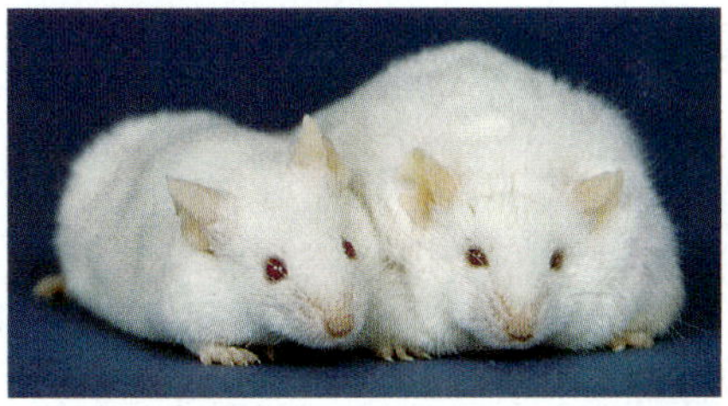

Wild-type mouse (*Ob*/– and *Db*/–)

Genetically obese mouse (either *ob/ob* or *db/db*)

Results Parabiotic *ob/ob* mice obtain leptin from the wild-type partner and lose fat. Parabiotic *db/db* mice remain obese because they lack the leptin receptor and thus the leptin they obtain from their partner has no effect.

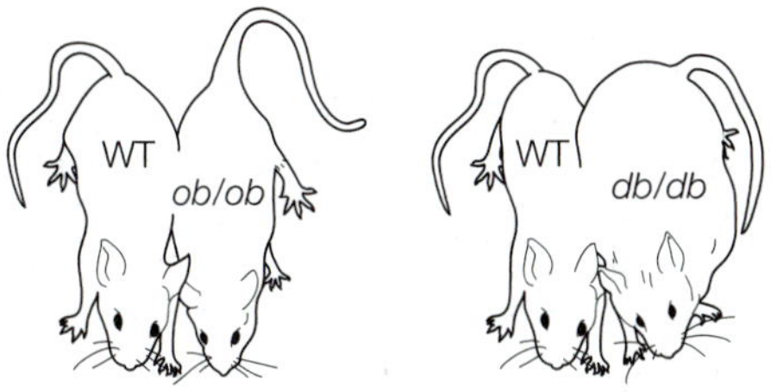
WT
ob/ob
WT
db/db

CONCLUSION The protein leptin is a satiety signal that acts to prevent overeating and resultant obesity.

Go to **BioPortal** for discussion and relevant links for all INVESTIGATING**LIFE** figures.

[a]Coleman, D. L. and K. P. Hummel. 1969. *American Journal of Physiology* 217: 1298–1304.

Go to Animated Tutorial 51.3
Parabiotic Mice
Life10e.com/at51.3

WORKING WITH**DATA:**

Is Leptin a Satiety Signal?

Original Papers

Coleman, D. L., and K. P. Hummel. 1969. Effects of parabiosis of normal with genetically diabetic mice. *American Journal of Physiology* 217: 1298–1304.

Coleman, D. L. 1973. Effects of parabiosis of obese with diabetes and normal mice. *Diabetologia* 9: 294–297.

Analyze the Data

The study summarized in Figure 51.18 supported the hypothesis that there are two separate genes that signal the brain when enough food has been obtained—one that encodes a satiety signal and another that encodes a receptor protein for the signal. Subsequent studies revealed that the genes in question encode the hormone leptin and its receptor.

In addition to the experiment in Figure 51.18, the researchers purified leptin and injected it into wild-type mice (*Ob/Ob*) and into genetically obese mice that lacked the leptin gene (*ob/ob*). The data in the table below were collected before the injections began (baseline) and 10 days later.

QUESTION 1

Do the data support the hypothesis that leptin is a satiety signal? Explain your answer.

QUESTION 2

What factors might explain the loss of body mass in the leptin-injected *ob/ob* mice?

	Baseline		Day 10	
Parameter	*ob/ob*	*Ob/Ob*	*ob/ob*	*Ob/Ob*
Food intake (g/day)	12.0	5.5	5.0	6.0
Body mass (g)	64	35	50	38
Metabolic rate (ml O_2/kg/hr)	900	1,150	1,100	1,150
Body temperature (°C)	34.8	37.0	37.0	37.0

Go to **BioPortal** for all WORKING WITH**DATA** exercises

An integrative signal that could be playing a central role in regulation of feeding is the enzyme AMP-activated protein kinase (AMPK). When most cells are nutrient deprived, they produce AMPK, which stimulates the oxidation of substrates to replenish ATP. Fasting increases AMPK levels in the hypothalamus, and feeding reduces them. Insulin and leptin decrease AMPK activity in the hypothalamus, and ghrelin increases it. Thus AMPK could be a final common pathway for various signals controlling food intake.

RECAP 51.4

The major controlling factors of gut function are an intrinsic nervous system and the hormones gastrin, secretin, and cholecystokinin. Insulin is the major hormonal controller of fuel metabolism. The hypothalamus controls food intake by generating sensations of hunger and satiety influenced by feedback from blood glucose and hormones, including insulin, leptin, and ghrelin.

- What are the roles of the three different classes of lipoproteins? **See p. 1066**
- By what actions does insulin promote uptake and storage of energy during the absorptive state? **See pp. 1066–1067 and Figure 51.17**
- What evidence supports the hypothesis that leptin influences satiety? **See pp. 1067–1068 and Figure 51.18**

?

Are there genes that predispose a person to obesity?

ANSWER

Many studies have demonstrated that a propensity for obesity is genetic and therefore heritable. About 30 studies comparing monozygotic and dizygotic twins led to the conclusion that the heritability of body mass is between 64 and 84 percent. Studies of monozygotic twins raised apart concluded that heritability of body mass was around 68 percent.

Single-gene causes of obesity have been observed but are rare. An example is a mutation of the leptin gene which has been found in a very small number of severely obese individuals, and these responded rapidly to leptin replacement therapy. To date the most commonly documented single-gene association with obesity is a mutation of the *MC4R* gene that encodes the melanocortin-4 receptor found on many hypothalamic neurons that inhibit feeding behavior. In a study population of Danish men, 2.5 percent of those who qualified as obese had mutations of the *MC4R* gene.

The Pima community of Arizona was among the first to be involved in a genome-wide survey. Many candidate genes were shown to be associated with obesity, but only weakly so. The conclusion must be that obesity is heritable, but except in a few very rare cases, it is a highly multigenic trait. Body mass is a function of Calories in minus Calories out, and genes influencing either side of this equation may be involved in obesity.

CHAPTER SUMMARY 51

51.1 What Do Animals Require from Food?

- Animals are **heterotrophs** that derive their energy and molecular building blocks, directly or indirectly, from **autotrophs**.
- Carbohydrates, fats, and proteins in food supply animals with energy. A measure of the energy content of food is the **kilocalorie** (**kcal**). Excess caloric intake is stored as glycogen and fat. **Review Figure 51.2**
- For many animals, food provides essential **carbon skeletons** that they cannot synthesize themselves. **Review Figure 51.4**
- Most researchers consider 8 amino acids to be essential for adult humans; some believe that infants require as many as 12 **essential amino acids** in their diet. **Macronutrients** are mineral elements needed in large quantities; **micronutrients** are needed in small amounts. **Review Figure 51.5, Table 51.1, ACTIVITY 51.1**
- **Vitamins** are organic molecules that must be obtained in food. **Review Table 51.2, ACTIVITY 51.2**
- **Malnutrition** results when any essential nutrient is lacking from the diet. Chronic malnutrition causes **deficiency disease**.

51.2 How Do Animals Ingest and Digest Food?

- Animals can be characterized by how they acquire nutrients: **saprobes** and **detritivores**, or **decomposers**, depend on dead organic matter, **filter feeders** strain the aquatic environment for small food items, **herbivores** eat plants, and **carnivores** eat other animals. Behavioral and anatomical adaptations reflect these feeding strategies. **See ACTIVITY 51.3**
- Digestion involves the breakdown of complex food molecules into monomers that can be absorbed and used by cells. In most animals, digestion takes place in a tubular gut. **Review Figure 51.7**
- Absorptive areas of the vertebrate gut are characterized by a large surface area produced by extensive folding and numerous **villi** and **microvilli**. **Review Figure 51.8**
- Hydrolytic enzymes break down proteins, carbohydrates, and fats into their monomeric units.

51.3 How Does the Vertebrate Gastrointestinal System Function?

- The vertebrate gut can be divided into several compartments with different functions. **Review Figure 51.9, ACTIVITY 51.4**
- The cells and tissues of the vertebrate gut are organized in the same way throughout its length. The innermost tissue layer, the **mucosa**, is the secretory and absorptive surface. The **submucosa** contains blood and lymph vessels and a nerve network that is sensory and also controls gut secretions. External to the submucosa are two smooth muscle layers. Between the two muscle layers is another nerve network that controls the movements of the gut. **Review Figure 51.10**
- Swallowing is a reflex that pushes a bolus of food into the **esophagus**. **Peristalsis** and **segmentation movements** of the gut move the bolus down the esophagus and through the entire length of the gut. **Sphincters** block the gut at certain locations, but they relax as a wave of peristalsis approaches. **Review Figure 51.11**
- Digestion begins in the **mouth**, where amylase is secreted with the saliva. Digestion of protein begins in the **stomach**, where parietal cells secrete HCl and chief cells secrete pepsinogen, a zymogen that becomes pepsin when activated by low pH and **autocatalysis**. The mucosa also secretes mucus, which protects the tissues of the gut. **Review Figure 51.12**
- In the **duodenum**, pancreatic enzymes carry out most of the digestion of food. **Bile** from the liver and **gallbladder** emulsifies fats into **micelles**. Bicarbonate ions from the **pancreas** neutralize the pH of the **chyme** entering from the stomach to produce an environment conducive to the actions of pancreatic enzymes such as **trypsin**. **Review Figure 51.13, Table 51.3**
- Final enzymatic cleavage of polypeptides and disaccharides occurs among the microvilli of the intestinal mucosa. Amino acids, monosaccharides, and inorganic ions are absorbed by the microvilli. Specific transporter proteins are sometimes involved. Symporters often power the absorption of nutrients.
- Fats broken down by **lipases** are absorbed mostly as monoglycerides and fatty acids and are resynthesized into triglycerides within the gut epithelium. The triglycerides are combined with cholesterol and phospholipids and coated with protein to form **chylomicrons**, which pass out of the mucosal cells and into lymphatic vessels in the submucosa. **Review Figure 51.14, ANIMATED TUTORIAL 51.1**
- Water and ions are absorbed in the large intestine as waste matter and consolidated into **feces**, which are periodically eliminated.
- Microorganisms in some compartments of the gut digest materials that their host cannot. **Review Figure 51.15**

51.4 How Is the Flow of Nutrients Controlled and Regulated?

- Autonomic reflexes coordinate activity of the digestive tract, which has an intrinsic nervous system that can act independently of the CNS.
- The actions of the stomach and small intestine are largely controlled by the hormones **gastrin**, **secretin**, and **cholecystokinin**. **Review Figure 51.16**
- The liver plays a central role in directing the traffic of fuel molecules. In the **absorptive state**, the liver takes up and stores fats and carbohydrates, converting monosaccharides to glycogen or fats. The liver also takes up amino acids and uses them to produce blood plasma proteins, and can engage in gluconeogenesis.
- Fat and cholesterol are shipped out of the liver as **low-density lipoproteins**. **High-density lipoproteins** act as acceptors of cholesterol and bring fat and cholesterol back to the liver.
- **Insulin** largely controls fuel metabolism during the absorptive state and promotes glucose uptake as well as glycogen and fat synthesis. In the **postabsorptive state**, lack of insulin blocks the uptake and use of glucose by most cells of the body except neurons. If blood glucose levels fall, **glucagon** secretion increases, stimulating the liver to break down glycogen and release glucose to the blood. **Review Figure 51.17, ANIMATED TUTORIAL 51.2**
- Food intake is governed by sensations of hunger and satiety, which are determined by brain mechanisms responding to feedback signals such as insulin, **leptin**, and **ghrelin**. **Review Figure 51.18, ANIMATED TUTORIAL 51.3**

Go to the Interactive Summary to review key figures, Animated Tutorials, and Activities
Life10e.com/is51

CHAPTERREVIEW

REMEMBERING

1. Which statement about essential amino acids is true?
   a. They are not found in vegetarian diets.
   b. They are stored by the body until they are needed.
   c. Without them, the body is undernourished.
   d. All animals require the same ones.
   e. Humans can acquire all of theirs by eating milk, eggs, and meat.

2. The digestive enzymes of the small intestine
   a. do not function well at a low pH.
   b. are produced and released in response to circulating secretin.
   c. are produced and released under neural control.
   d. are all secreted by the pancreas.
   e. are all activated by an acidic environment.

3. Which statement about nutrient absorption by the intestinal mucosal cells is true?
   a. Carbohydrates are absorbed as disaccharides.
   b. Fats are absorbed as fatty acids and monoglycerides.
   c. Amino acids move across the plasma membrane only by diffusion.
   d. Bile transports fats across the plasma membrane.
   e. Most nutrients are absorbed in the duodenum.

4. Chylomicrons are like the tiny micelles of dietary fat in the lumen of the small intestine in that both
   a. are coated with bile.
   b. are lipid-soluble.
   c. travel through the lymphatic system.
   d. contain triglycerides.
   e. are coated with lipoproteins

5. Which of the following is stimulated by cholecystokinin?
   a. Stomach motility
   b. Release of bile
   c. Secretion of hydrochloric acid
   d. Secretion of bicarbonate ions
   e. Secretion of mucus

UNDERSTANDING & APPLYING

6. Several popular diet books recommend high fat and protein intake and low carbohydrate intake as a means of losing body mass. What is the rationale for a high-fat and high-protein diet?

7. Explain how insulin is involved in at least five differences between the management of fuel molecules in the absorptive and the postabsorptive state.

8. Trace the history of a fatty acid molecule from a slice of cheese pizza to a plaque on a coronary artery. Into what possible forms and structures might it have been converted as it passed through the body? Describe a direct and an indirect route it could have taken.

EVALUATING & ANALYZING

9. The pancreatic duct cells and the parietal cells of the stomach both express high levels of carbonic anhydrase. Explain how the actions of this enzyme are similar in both cases, but how the consequences in terms of influences on the pH of the blood are different.

10. Body mass is a function of Calories in minus Calories out. If genome scans of obese individuals show that most of the associated genes play roles in the hypothalamic regulatory pathways, what does that tell us about which side of the energy balance equation is the most important in contributing to obesity?

Go to BioPortal at **yourBioPortal.com** for Animated Tutorials, Activities, LearningCurve Quizzes, Flashcards, and many other study and review resources.

52 Salt and Water Balance and Nitrogen Excretion

CHAPTER OUTLINE

Blood as Fast Food The vampire bat *Desmodus rotundus* is able to adjust its excretory physiology rapidly from water-excreting to water-conserving, depending on whether it is ingesting or digesting its blood meal.

BLOOD, SWEAT, AND TEARS taste salty because they have the ionic composition of the extracellular fluid that bathes the cells of the body. The volume and composition of the extracellular fluid must be regulated and kept relatively free of wastes. Maintaining homeostasis of the extracellular fluid is the job of the excretory system. It is a challenging job that sometimes requires getting rid of excess fluids and conserving ions, and at other times requires conserving fluids and excreting excess ions.

The nature of the challenge depends on an animal's environment and lifestyle. Some desert animals rarely if ever encounter open water, so drinking is not an option. Animals that live in fresh water have the opposite challenge: water continuously enters their bodies by osmosis and in their food. Animals that live in salt water face a challenge similar to that of the desert dwellers—they need to conserve water and excrete ions. The physiological mechanisms all animals have to maintain salt and water balance are similar, but they are used in different ways to solve the unique problems of each species. Consider, for example, vampire bats.

Vampire bats feed on the blood of animals such as goats and cattle, using sharp incisor teeth to make a small incision (usually on the legs and ankles of a sleeping victim) and then lapping up the blood. The bat's saliva has an anticoagulant that keeps the blood flowing. Blood contains nutritious protein, but it consists mostly of water. Blood meals may be few and far between, so the bat quickly consumes as much as it can—up to half its body mass. To maximize its protein intake and keep its weight low enough to fly, it rapidly eliminates the water from its meal. Within minutes of starting to feed, the bat is producing copious dilute urine.

Once feeding ends, this high rate of water loss must stop—now the bat is metabolizing protein and must excrete large amounts of nitrogenous wastes while conserving water. Within minutes, the bat's excretory system switches from producing abundant, dilute urine to producing a tiny amount of highly concentrated urine. To conserve water, vampire bats (and some desert rodents) can produce urine that is 15 times more concentrated than their own blood. In one feeding cycle, the vampire bat rapidly transitions from an excretory physiology typical of a mammal living in an environment with abundant fresh water to that of a desert mammal that never sees water.

> ?
>
> How do desert rodents and vampire bats make highly concentrated urine?
>
> See answer on p. 1090.

52.1 How Do Excretory Systems Maintain Homeostasis?

Homeostasis of the extracellular fluid (the blood plasma and interstitial fluid; see Section 40.1) is critical for several reasons:

- The solute concentration of the extracellular fluid determines the water balance of the cells of the body.
- The specific ionic composition of the extracellular fluid influences many functions of various types of cells. Consider, for example, the importance of ion concentration gradients between the extracellular fluid and the cytoplasm of nerve and muscle cells (see Sections 45.2 and 48.1).
- The health of cells requires the elimination of nitrogenous wastes.

The problems that have to be solved to maintain homeostasis of the extracellular fluid depend on the environment in which a species lives (salt water, fresh water, or terrestrial) and its lifestyle, as we saw in the case of the vampire bat in the opening story. Animals depend on **excretory systems** to maintain the volume, concentration, and composition of their extracellular fluids, and to excrete wastes.

Water enters or leaves cells by osmosis

The volume of cells depends on whether they take up water from or lose water to the extracellular fluid. The movement of water across cell plasma membranes depends on differences in solute concentration on the two sides of the membrane and on the permeability of the membrane. This is the process of osmosis, which we discussed in Sections 6.3 and 35.1. If the solute concentration of the extracellular fluid is less than that of the cytoplasm, water moves into the cells, causing them to swell and possibly burst (see Figure 6.9). If the solute concentration of the extracellular fluid is greater than that of the cytoplasm, the cells lose water and shrink. Thus the solute concentration of the extracellular fluid affects both the volume and the solute concentration of the cells.

Animal physiologists use the term **osmolarity** in discussing osmosis. The osmolarity of a solution is the number of moles of osmotically active solutes per liter of solvent. Thus a 1 molar solution of glucose is also a 1 osmolar (1 osmole per liter) solution, but a 1 molar solution of sodium chloride (NaCl) is a 2 osmolar solution, because each NaCl molecule dissociates into two osmotically active ions.

Excretory systems control extracellular fluid osmolarity and composition

Excretory systems control the osmolarity and composition of the extracellular fluids by excreting solutes that are present in excess (such as NaCl when we eat lots of salty food) and conserving solutes that are valuable or in short supply (such as glucose and amino acids). Excretory systems also eliminate the toxic waste products of protein metabolism. The output of the excretory system is called **urine**.

Three basic processes are common to a wide variety of animal excretory systems: filtration, secretion, and reabsorption. Filtered extracellular fluid contains no cells or large molecules, such as proteins. In animals with closed circulatory systems, the blood plasma is usually filtered from capillaries into associated tubules. The walls of the capillaries and of the tubules are the filter, and the filtration is driven by blood pressure. As the filtrate flows through the tubules, its composition and concentration are modified through processes of secretion and reabsorption to form the urine that leaves the body.

In all of the discussions that follow about the movement of water across membranes, it is important to remember that there are no mechanisms for the active transport of water. The movement of water is due either to a pressure difference (filtration) or to a difference in solute concentration (osmosis). Water always flows down a pressure gradient or up a solute concentration gradient.

Aquatic invertebrates can conform to or regulate their osmotic and ionic environments

Most invertebrates that live in seawater conform to the osmotic concentration of their environment over a fairly wide range of salinities and are therefore called **osmoconformers** (Figure 52.1). The osmolarity of seawater in the open ocean is about 1,000 milliosmoles/liter (mosm/l), but it can vary quite a bit in estuaries where it is diluted by an influx of fresh water or in evaporating tide pools as the salt gets concentrated. Osmoconformity can result in considerable energetic savings, as it costs metabolic energy to move ions across membranes to achieve

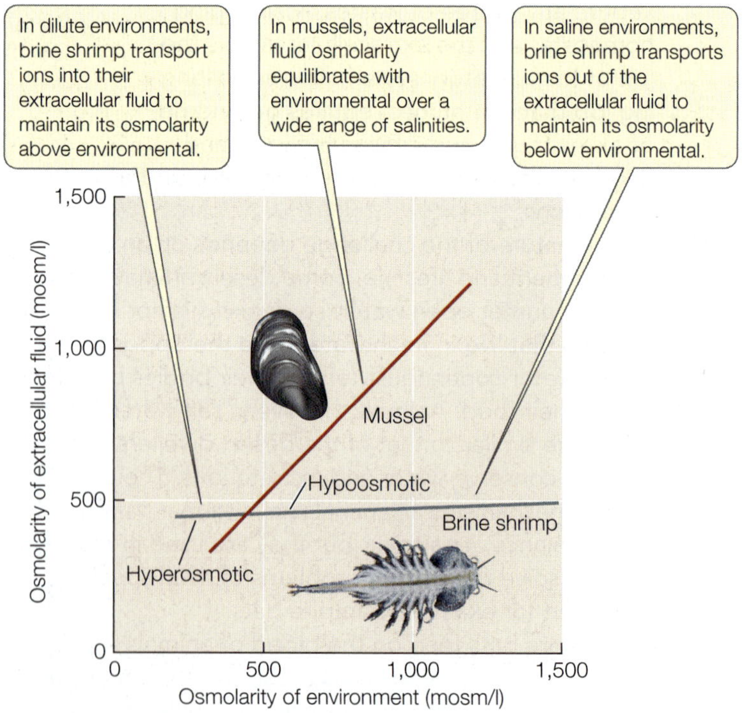

In dilute environments, brine shrimp transport ions into their extracellular fluid to maintain its osmolarity above environmental.
In mussels, extracellular fluid osmolarity equilibrates with environmental over a wide range of salinities.
In saline environments, brine shrimp transports ions out of the extracellular fluid to maintain its osmolarity below environmental.
Osmolarity of extracellular fluid (mosm/l)
1,500
1,000
500
0
Mussel
Hypoosmotic
Brine shrimp
Hyperosmotic
0
500
1,000
1,500
Osmolarity of environment (mosm/l)

52.1 Some Marine Invertebrates Osmoregulate Some aquatic animals, such as mussels, experience an extreme range of salt concentrations in their environment and are osmoconformers over much of that range. Other aquatic animals, such as brine shrimp, are osmoregulators in that they maintain a relatively constant osmolarity of their extracellular fluids as environmental salinity varies.

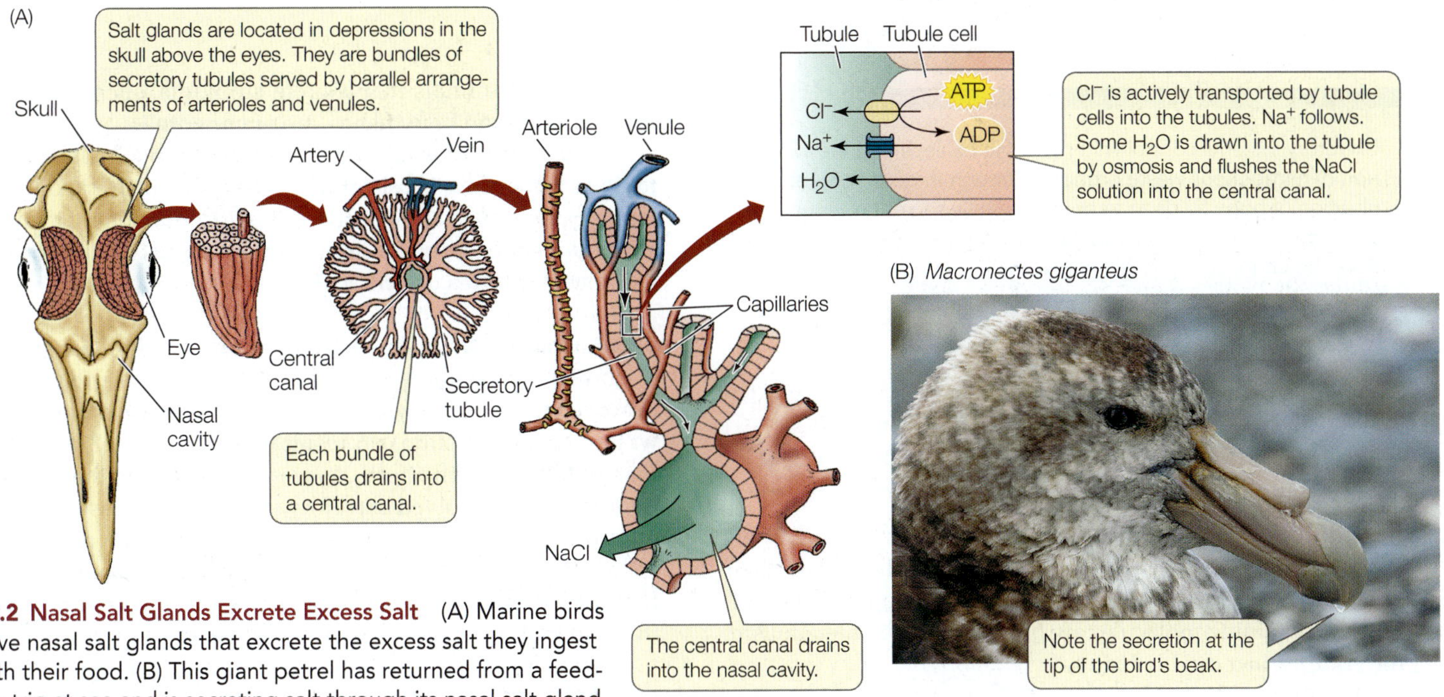
(A)
Salt glands are located in depressions in the skull above the eyes. They are bundles of secretory tubules served by parallel arrangements of arterioles and venules.
Skull
Eye
Nasal cavity
Artery
Vein
Central canal
Each bundle of tubules drains into a central canal.
Arteriole
Venule
Secretory tubule
Capillaries
NaCl
The central canal drains into the nasal cavity.
Tubule
Tubule cell
Cl⁻
Na⁺
H₂O
ATP
ADP
Cl⁻ is actively transported by tubule cells into the tubules. Na⁺ follows. Some H₂O is drawn into the tubule by osmosis and flushes the NaCl solution into the central canal.
(B) Macronectes giganteus
Note the secretion at the tip of the bird's beak.

52.2 Nasal Salt Glands Excrete Excess Salt (A) Marine birds have nasal salt glands that excrete the excess salt they ingest with their food. (B) This giant petrel has returned from a feeding trip at sea and is secreting salt through its nasal salt gland.

osmotic regulation. However, osmoconformity is not an option for freshwater animals. No cell can do without ions or nutrients, so freshwater animals always have to expend energy to conserve salts and excrete water.

Some marine invertebrates maintain a rather constant osmolarity of their extracellular fluids as the osmolarity of the environment changes, and these animals are therefore called **osmoregulators** (see Figure 52.1). Osmoregulators in the marine environment mostly maintain their osmotic concentration considerably below that of the environment and are therefore engaged in **hypoosmotic regulation**. Occasionally, however, seawater is diluted by an influx of fresh water, as in estuaries, and animals must maintain their osmotic concentrations above that of the environment—a process called **hyperosmotic regulation**.

The brine shrimp *Artemia* illustrated in Figure 52.1 is an osmoregulator with an enormous range of tolerances. *Artemia* are found in huge numbers in the most salty environments known, such as Utah's Great Salt Lake and in coastal evaporation ponds where salt is concentrated for commercial purposes (see Figure 26.17) and can reach an osmolarity of 2,500 mosm/l. No animal could survive with internal osmolarities that high; such a solute concentration would cause proteins to denature. *Artemia* are able to exploit these environments because of their ability to regulate hypoosmotically by actively transporting NaCl from their extracellular fluid out across their gill membranes to the environment. *Artemia* cannot survive in fresh water, but they can live in dilute seawater by reversing the direction of transport of NaCl across their gill membranes to maintain the osmolarity of their extracellular fluids above that of the environment, thus becoming hyperosmotic regulators.

Virtually all marine invertebrates are ionic conformers with respect to certain ions such as Na^+ and Cl^- in their extracellular fluids—that is, the extracellular concentrations of these ions are the same as in seawater. Thus these animals avoid the considerable metabolic costs of moving ions across membranes to maintain concentration differences between their extracellular fluids and the seawater. However, most marine animals (with the exception of echinoderms) are ionic regulators with respect to certain ions; they employ active transport mechanisms to maintain these ions in their extracellular fluid at optimal concentrations.

Vertebrates are osmoregulators and ionic regulators

All aquatic vertebrates, with two exceptions, regulate the osmolarity of their extracellular fluids at 300 mosm/l. In doing so, they are selective in which ions they conserve and which ions they excrete; thus they are ionic regulators as well as osmoregulators. One exception is the hagfish, a primitive jawless fish and a very ancient vertebrate group (see Figure 33.10). Hagfish are osmoconformers as well as ionic conformers for many of the ions found in seawater. The other exception is the chondrichthyans (cartilaginous fishes, the sharks and rays). Chondrichthyans retain in their extracellular fluid two organic solutes, urea and trimethylamine oxide (TMAO), that are products of protein metabolism. As a result, their extracellular fluid is slightly hyperosmotic to the seawater and they gain water by osmosis.

Terrestrial vertebrates obtain their salts mostly from food and regulate the ionic composition of their extracellular fluids by conserving some ions and excreting others. For example, herbivores have to conserve Na^+ because the plants they eat have low concentrations of Na^+. By contrast, birds that feed on marine animals must excrete the excess sodium they ingest with their food. Such birds, which include penguins and gulls, excrete excess salt through their nasal salt glands. These glands use active transport of Cl^- ions (with Na^+ and some water following passively) into a series of canals and ducts to produce a concentrated solution of NaCl that empties into the nasal cavity (**Figure 52.2**). These birds can be seen frequently sneezing or shaking their heads to get rid of the salty droplets excreted from their nasal salt glands.

RECAP 52.1

Excretory systems control water and salt balance and the excretion of nitrogenous waste products through three mechanisms: filtration of body fluids to form urine, active secretion of substances into the urine, and active reabsorption of substances from the urine.

- Describe the two mechanisms used to move water across membranes. See p. 1072
- What different salt and water balance problems might animals encounter in marine, freshwater, and terrestrial environments? What are some of the ways they meet those challenges? See pp. 1072–1073 and Figures 52.1 and 52.2

In addition to maintaining salt and water balance, animals must eliminate the waste products of metabolism from their extracellular fluids. The major problem is nitrogen. When nitrogen-containing molecules are broken down by metabolism, the end product can be toxic.

52.2 How Do Animals Excrete Nitrogen?

The end products of the metabolism of carbohydrates and fats are water and carbon dioxide, which are not difficult to eliminate. Proteins and nucleic acids, however, contain nitrogen, so their metabolism produces nitrogenous wastes in addition to water and carbon dioxide.

Animals excrete nitrogen in a number of forms

The most common nitrogenous waste is **ammonia** (NH_3). Because it is highly toxic, ammonia is either excreted continuously to prevent its accumulation or is detoxified by conversion into **urea** or **uric acid** (Figure 52.3).

AMMONIA Ammonia is highly soluble in water and diffuses rapidly, so its continuous excretion is relatively simple for many aquatic animals that continuously lose ammonia from their blood to the environment by diffusion across their gill membranes. Animals that excrete ammonia, such as aquatic invertebrates and bony fishes, are **ammonotelic**.

If ammonia builds up in the extracellular fluids, it becomes toxic at rather low levels and is a dangerous metabolite for terrestrial animals and for those aquatic animals that cannot continuously excrete ammonia. These animals must convert ammonia into urea or uric acid.

UREA **Ureotelic** animals, such as mammals, amphibians, chondrichthans (sharks and rays), and hagfish excrete urea as their principal nitrogenous waste product. Urea is quite soluble in water, but its excretion can result in a large loss of water that many animals can ill afford. As we will see later in this chapter, mammals have evolved excretory systems that conserve water by producing urine that has a high concentration of urea. As we mentioned in Section 52.1, sharks, rays, and hagfish retain high concentrations of urea and TMAO in their extracellular fluid, so that it is hyperosmotic to seawater.

URIC ACID Animals that conserve water by excreting nitrogenous wastes mostly as uric acid are **uricotelic**. Insects, reptiles (including birds), and some amphibians are uricotelic. Uric acid is not very soluble in water, so it forms a colloidal suspension in the urine and is excreted as a semisolid (for example, the whitish material in bird droppings). A uricotelic animal loses very little water as it disposes of its nitrogenous wastes.

Most species produce more than one nitrogenous waste

Humans are ureotelic, but we also excrete uric acid. The uric acid in human urine comes largely from the metabolism of nucleic acids and caffeine. If uric acid levels in the extracellular fluid rise too high, uric acid crystals can precipitate in joints and cause the age-old malady called gout. Because solubility goes down with temperature, uric acid crystals usually precipitate first in the extremities, especially the big toe. Pain in the big toe is a telltale symptom of gout.

Humans can also excrete ammonia, which is an important mechanism for regulating the pH of the extracellular fluids. As

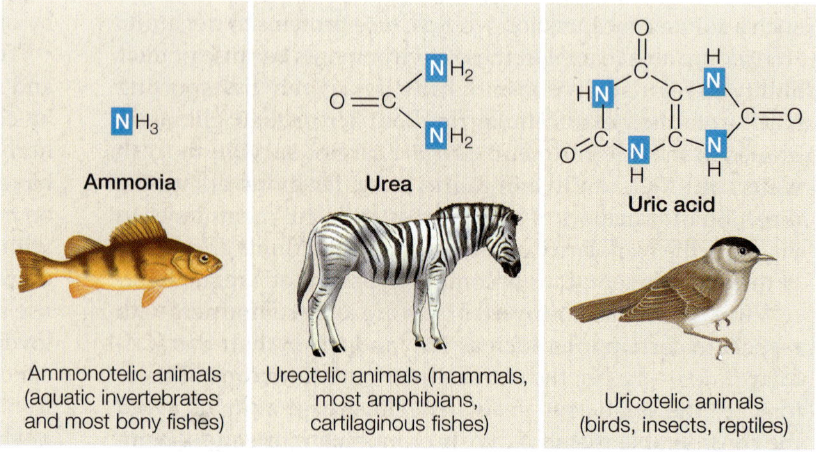

NH3
Ammonia
Urea
Uric acid
Ammonotelic animals (aquatic invertebrates and most bony fishes)
Ureotelic animals (mammals, most amphibians, cartilaginous fishes)
Uricotelic animals (birds, insects, reptiles)

52.3 Waste Products of Metabolism The metabolism of proteins and nucleic acids produces nitrogenous wastes. Many aquatic animals, including most fishes, excrete nitrogenous wastes as ammonia, which is highly diffusible and soluble in an aqueous environment. Most terrestrial animals and some aquatic animals excrete either urea or uric acid. Urea is more soluble in water and is the major nitrogenous excretory product for mammals, amphibians, and some fishes. Uric acid is not very soluble in water and is the major nitrogenous excretory product for reptiles, birds, insects, and some amphibians.

we will see later in this chapter, excreted ammonia buffers the urine and enables the excretion of excess hydrogen ions.

Species that live in different habitats at different developmental stages may use more than one mechanism of nitrogen excretion. The tadpoles of frogs and toads, for example, excrete ammonia across their gill membranes, but adult frogs and toads generally excrete urea. Some adult amphibians that live in arid habitats excrete uric acid.

RECAP 52.2

Ammonia is a common metabolic waste product of nitrogen-containing molecules. Most aquatic animals excrete ammonia by diffusion into the water. Terrestrial animals and some aquatic animals detoxify ammonia by conversion to urea or uric acid.

- Explain the significance of high concentrations of urea in the blood of sharks. See p. 1074
- Why might you expect a species from an arid habitat to use uric acid as its primary nitrogenous waste product? See p. 1074 and Figure 52.3

Animals exhibit a variety of adaptations for dealing with the challenges of salt and water balance in different environments. All of these adaptations, however, are based on two basic mechanisms—namely, filtration and tubular processing of the filtrate to conserve some solutes and excrete others.

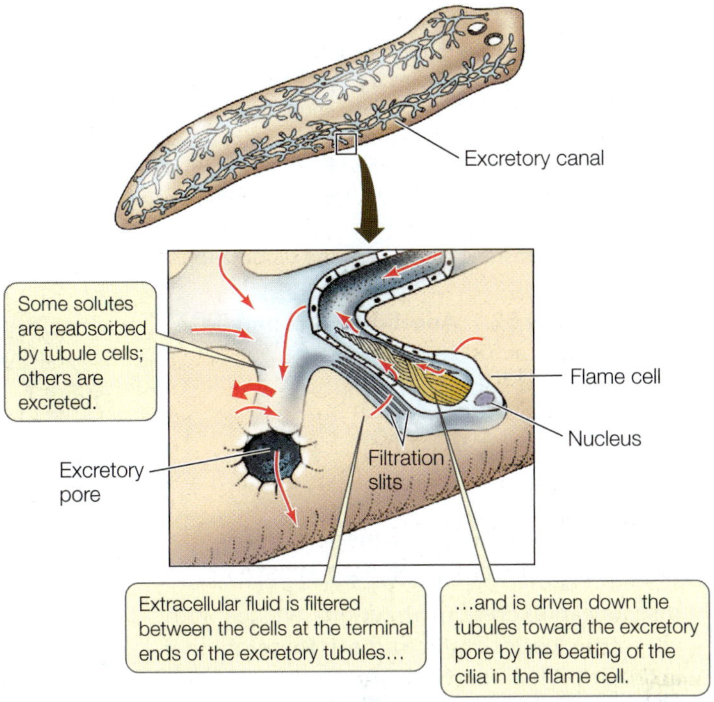
Excretory canal
Some solutes are reabsorbed by tubule cells; others are excreted.
Flame cell
Nucleus
Excretory pore
Filtration slits
Extracellular fluid is filtered between the cells at the terminal ends of the excretory tubules...
...and is driven down the tubules toward the excretory pore by the beating of the cilia in the flame cell.

52.4 Protonephridia in Flatworms The protonephridia of the flatworm *Planaria* consist of tubules ending in flame cells. In the region of the flame cells, body fluid is filtered between the tubule cells. The composition of the filtrate is modified as it flows down the tubule.

52.3 How Do Invertebrate Excretory Systems Work?

Freshwater and terrestrial invertebrates have a wide variety of adaptations for maintaining salt and water balance and excreting nitrogen. In this section we will explore three examples of invertebrate excretory systems: protonephridia, metanephridia, and Malpighian tubules. Each of these systems produces an extract of interstitial fluid lacking large molecules. They then change the solute composition (ions and small molecules) of that fluid to form an excretory product.

The protonephridia of flatworms excrete water and conserve salts

Many free-living flatworms, such as *Planaria*, live in fresh water. These animals excrete water through an elaborate network of tubules running throughout their bodies. The tubules end in flame cells, so called because each cell has a tuft of cilia projecting into the tubule (Figure 52.4). The beating of the cilia gives the appearance of a flickering flame. A flame cell and a tubule together form a **protonephridium** (plural *protonephridia*; Greek *proto*, "before"; *nephros*, "kidney").

Extracellular fluid enters the tubules by filtration. The beating of the cilia causes a slight negative pressure in the tubule, and movements of the animal create positive pressure in the extracellular fluid. This pressure difference causes extracellular fluid to be filtered through tiny spaces between tubule cells. The filtrate flows toward the animal's excretory pore, and along the way the cells of the tubules modify the composition of the fluid by reabsorption and secretion of specific ions and molecules. Because more ions are reabsorbed than are secreted, the urine that leaves the flatworm's body is less concentrated than the extracellular fluid. Thus the protonephridium conserves ions and excretes water and wastes.

The metanephridia of annelids process coelomic fluid

Filtration of body fluids and modification of urine by tubules are highly developed processes in annelid worms such as the earthworm. Annelids are segmented, and in each segment they have a fluid-filled body cavity called a coelom (see Figure 32.11). Annelids have a closed circulatory system through which blood is pumped under pressure. The pressure causes the blood to be filtered across the thin, permeable capillary walls into the coelom. Some waste products, such as ammonia, diffuse directly from the tissues into the coelom. Where does this coelomic fluid go?

Each segment of the earthworm contains a pair of **metanephridia** (singular *metanephridium*; Greek *meta*, "akin to"). Each metanephridium begins as a ciliated, funnel-like opening called a nephrostome. The nephrostome resides in one segment and continues as a tubule in the next segment. The tubule ends in a pore, called a nephridiopore, that opens to the outside of the animal (Figure 52.5). Coelomic fluid is swept into the metanephridia through the ciliated nephrostomes. As the fluid passes through the tubules, their cells actively reabsorb certain molecules from it and actively secrete other molecules into it. What leaves the animal through the nephridiopores is a dilute urine containing nitrogenous wastes and other solutes.

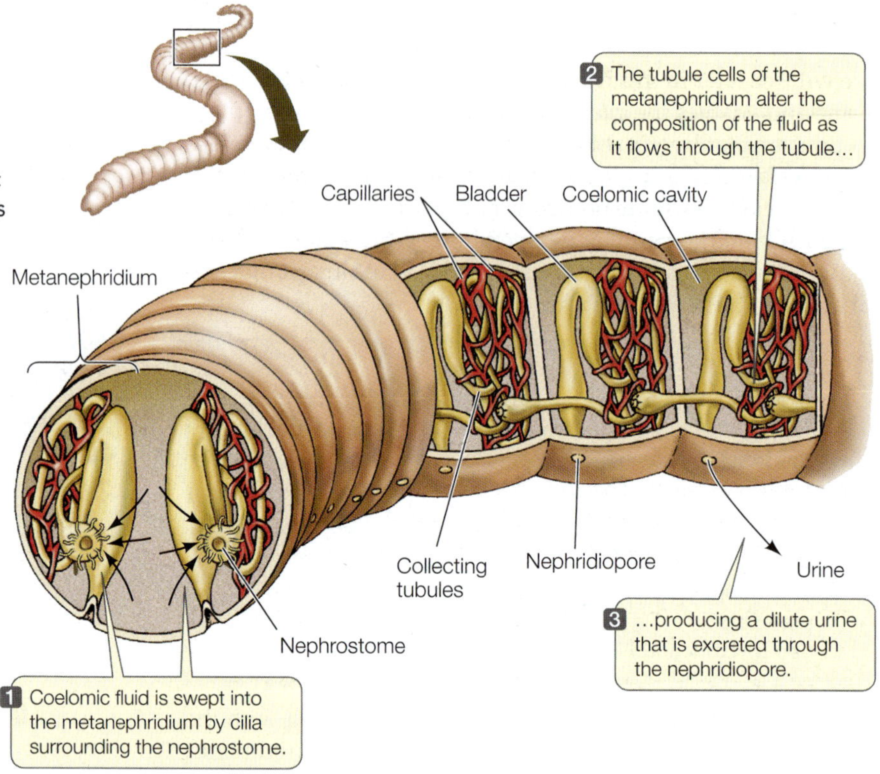
2 The tubule cells of the metanephridium alter the composition of the fluid as it flows through the tubule…
Capillaries
Bladder
Coelomic cavity
Metanephridium
Collecting tubules
Nephridiopore
Urine
3 …producing a dilute urine that is excreted through the nephridiopore.
Nephrostome
1 Coelomic fluid is swept into the metanephridium by cilia surrounding the nephrostome.

52.5 Metanephridia in Earthworms The metanephridia of annelids are arranged segmentally. The cross section at the left end shows a pair of metanephridia. Three longitudinal sections (right) show only one metanephridium of the two in each segment. Coelomic fluid enters the nephrostome and flows through tubules leading to the nephridiopore. A close association of the tubules and blood capillaries facilitates the active exchange of substances between the blood and the tubular fluid.

Go to Activity 52.1 Annelid Metanephridia
Life10e.com/ac52.1

Malpighian tubules of insects use active transport to excrete wastes

Insects can excrete nitrogenous wastes with very little loss of water and can therefore live in the driest habitats on Earth. The insect excretory system consists of **Malpighian tubules**. An individual insect has from 2 to more than 100 of these blind-ended tubules that open into the gut between the midgut and hindgut (**Figure 52.6**).

Insects have an open circulatory system and therefore cannot use a pressure difference to filter extracellular fluids into the Malpighian tubules. Instead, the cells of the tubules actively transport uric acid, potassium ions, and sodium ions from the extracellular fluid into the tubules. The high concentration of solutes in the tubules causes water to follow osmotically, which flushes the tubule contents toward the gut.

The epithelial cells of the hindgut and rectum actively transport sodium and potassium ions from the gut contents back into the extracellular fluid. This local transport of salts

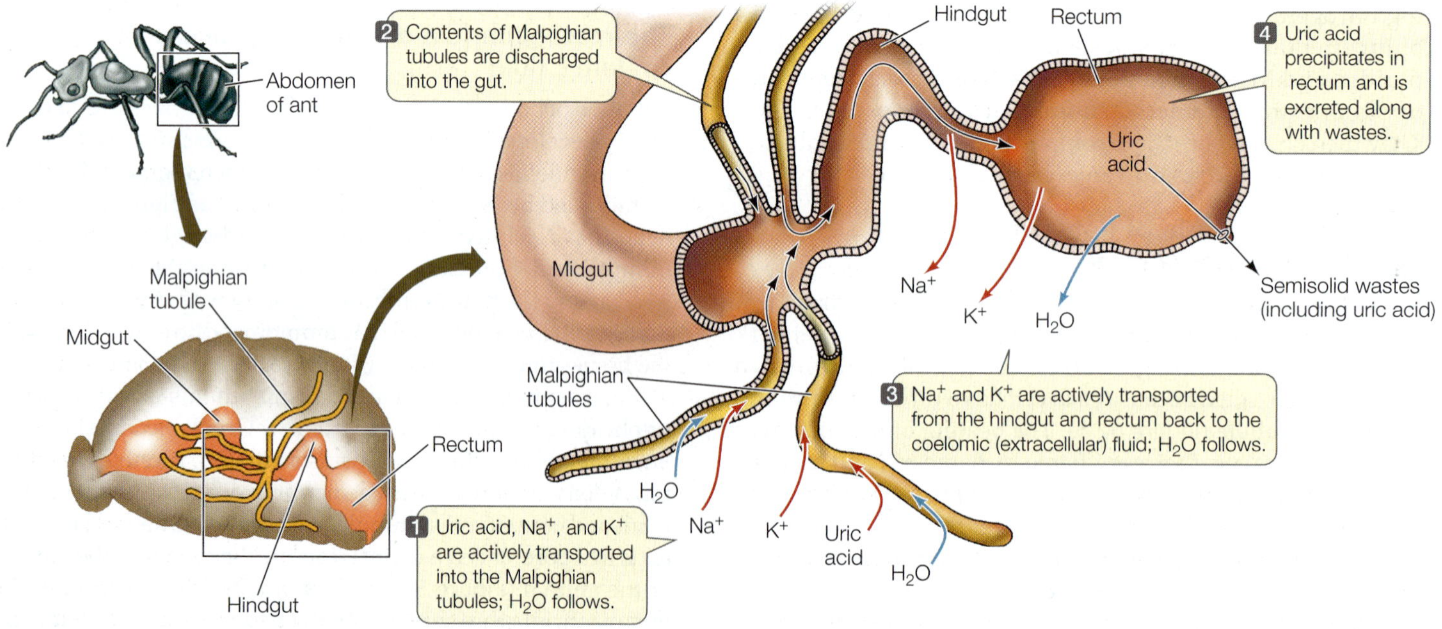
Abdomen of ant
Malpighian tubule
Midgut
Rectum
Hindgut
2 Contents of Malpighian tubules are discharged into the gut.
Hindgut
Rectum
4 Uric acid precipitates in rectum and is excreted along with wastes.
Uric acid
Midgut
Na^+
K^+
H_2O
Semisolid wastes (including uric acid)
Malpighian tubules
3 Na^+ and K^+ are actively transported from the hindgut and rectum back to the coelomic (extracellular) fluid; H_2O follows.
H_2O
Na^+
K^+
Uric acid
H_2O
1 Uric acid, Na^+, and K^+ are actively transported into the Malpighian tubules; H_2O follows.

52.6 Malpighian Tubules in Insects The blind, thin-walled Malpighian tubules are attached to the junction of the insect's midgut and hindgut and project into the spaces containing extracellular fluid. This system makes it possible to excrete wastes with very little loss of water.

creates an osmotic gradient that pulls water out of the rectal contents. As its concentration increases, the uric acid forms a colloidal suspension, freeing even more water to be reabsorbed. Remaining in the rectum is the uric acid mixed with other wastes; this semisolid matter is what the insect excretes. If you ever park your car under a tree being visited by bees, you will find the little excretory droplets from the bees all over the car. The Malpighian tubule system is a highly effective mechanism for excreting nitrogenous wastes and some salts without giving up much water.

RECAP 52.3

Protonephridia and metanephridia work by creating a filtrate of the body fluids that is modified by the secretion and reabsorption of specific substances before being excreted. Insect Malpighian tubules actively secrete uric acid and other solutes into closed tubules.

- Describe how an earthworm filters its blood and produces urine. **See p. 1075 and Figure 52.5**
- How do Malpighian tubules make it possible for some insect species to survive where there is no free water? **See pp. 1076–1077 and Figure 52.6**

Having described how several invertebrate groups handle nitrogen excretion, we will next consider the nephron—the basic unit of the vertebrate excretory system—and how it evolved to be able to respond to a variety of salt and water balance challenges and maintain a relatively constant internal environment.

52.4 How Do Vertebrates Maintain Salt and Water Balance?

The main excretory organ of vertebrates is the **kidney**, and the functional unit of the kidney is the **nephron**, which has a blood vessel component and a tubule component. The vascular component begins with a knot of capillaries that are highly permeable and filter the blood into the tubule component. The blood vessels also carry substances to the tubules for secretion and carry away substances that the tubule cells reabsorb. Nephrons can filter large volumes of blood and achieve bulk reabsorption of salts and other valuable molecules such as glucose, making the vertebrate kidney well adapted for the excretion of excess water.

The evolution of vertebrates is thought to have begun with a marine predecessor that moved into a freshwater habitat. The excretory system of this vertebrate ancestor would have evolved to excrete large quantities of water, enabling the osmoregulatory capacities seen in present-day vertebrates. But if the early vertebrates evolved to excrete water, how did subsequent vertebrate lineages adapt to environments where water must be conserved and salts excreted? The answer to this question differs among vertebrate groups. Even among the marine fishes, the excretory adaptations of the bony fishes differ from those of the chondrichthyans. Reptiles, birds, and mammals have excretory systems that conserve water. Reptiles and birds achieve this mainly by being uricotelic and producing a semisolid excretory product that contains little water. Mammals, in contrast, are ammonotelic; they excrete a liquid waste product but have evolved the ability to produce a highly concentrated urine.

Marine fishes must conserve water

Marine bony fishes osmoregulate their extracellular fluids to maintain them at one-third to one-half the osmolarity of seawater. Their only source of water is the sea around them, so they must conserve water and excrete excess solutes. Marine bony fishes cannot produce urine that is more concentrated than their extracellular fluids, so they minimize water loss by producing very little urine. In contrast, freshwater fishes produce lots of dilute urine.

If marine bony fishes cannot excrete excess solutes in their urine, how do they deal with the large salt loads they ingest with food? Marine bony fishes do not absorb from their guts some of the ions they take in, especially divalent ions such as Mg^{2+} or SO_4^{2-}. NaCl, the major salt ingested, is actively excreted across the gill membranes. As we mentioned earlier, bony fishes can lose their nitrogenous waste, ammonia, by diffusion across their gill membranes.

Chondrichthyans (sharks and rays) are osmoconformers but not ionic conformers. As we have discussed, they raise the osmolarity of their body fluids in a unique way. Unlike bony fishes, they retain urea and trimethylamine oxide in their extracellular fluids so that it is hyperosmotic to seawater. These species have adapted to a concentration of urea in the body fluids that would be toxic to other vertebrates. Sharks and rays still have the problem of excreting the large amounts of salts they take in with their food. They solve this problem by having a gland in the rectum that actively secretes NaCl by a mechanism similar to that of the nasal salt glands of seabirds.

Terrestrial amphibians and reptiles must avoid desiccation

Most amphibians live in or near fresh water, and they stay in humid habitats when they do venture from the water. Like freshwater fishes, most amphibians produce large amounts of dilute urine and conserve salts. Some amphibians, however, have adapted to habitats that require water conservation.

Amphibians living in dry terrestrial environments have skin with a reduced permeability to water. Some secrete a waxy substance over the skin to waterproof it. Several species of frogs that live in arid regions of Australia burrow deep into the ground and remain there during long dry periods. They enter **estivation**, a state of very low metabolic activity and therefore low water turnover. When it rains, the frogs come out of estivation, feed, and reproduce. Their most interesting adaptation is an enormous urinary bladder. Before entering estivation, they fill the bladder with dilute urine, which can amount to one-third of their body weight. This dilute urine serves as a water reservoir that is gradually reabsorbed into the blood during the long period of estivation. Aboriginal peoples have learned to locate these buried frogs and use them as an emergency source of water.

Reptiles occupy habitats ranging from aquatic to extremely hot and dry. In fact, snakes, lizards, and birds are among the most prominent members of many desert faunas. Three major adaptations have freed reptiles from the close association with water that is necessary for most amphibians (see Section 33.4):

- Reptiles are amniotes that do not need fresh water to reproduce because they employ internal fertilization and lay eggs with shells that retard evaporative water loss.
- Reptiles have a dry epidermis (skin) that retards evaporative water loss.
- Reptiles excrete nitrogenous wastes as uric acid semisolids, losing little water in the process.

Mammals can produce highly concentrated urine

Mammals occupy diverse habitats, many of which present special excretory system challenges. The most challenging environments are those in which water is severely limited. Mammals have a variety of adaptations to conserve water, but chief among them is the ability to produce urine that is more concentrated than their extracellular fluids. They are able to concentrate their urine because of adaptations of their kidneys that we will explore in detail in Section 52.5. To understand how these adaptations work, however, we must first describe the structure and function of the vertebrate nephron.

The nephron is the functional unit of the vertebrate kidney

Urine formation in vertebrate nephrons involves three main processes (Figure 52.7):

- *Filtration*. Each nephron has a dense bed of capillaries called a **glomerulus** (plural *glomeruli*). The glomerulus is highly permeable to water, ions, and small molecules but impermeable to large molecules. Blood pressure drives the movement of water and small-molecular-weight solutes out of the glomerular capillaries.
- *Tubular reabsorption*. The filtrate from the glomerulus flows into the **renal tubule**. Cells in the renal tubule modify the filtrate by reabsorbing specific ions, nutrients, and water, returning these to the blood, and leaving behind and concentrating excess ions and waste products such as urea.
- *Tubular secretion*. The filtrate in the renal tubule is further modified by tubule cells transporting substances into the tubule. These are substances that the body needs to excrete.

Blood enters the glomerular capillaries via an **afferent arteriole** and leaves the glomerulus in an **efferent arteriole**. This short vessel is called an arteriole because it feeds another capillary bed, called the **peritubular capillaries**, that intimately surrounds the renal tubules. Peritubular capillaries deliver substances to the renal tubule cells that these cells secrete into the urine. The peritubular capillaries also carry away substances that the tubule cells reabsorb from the urine.

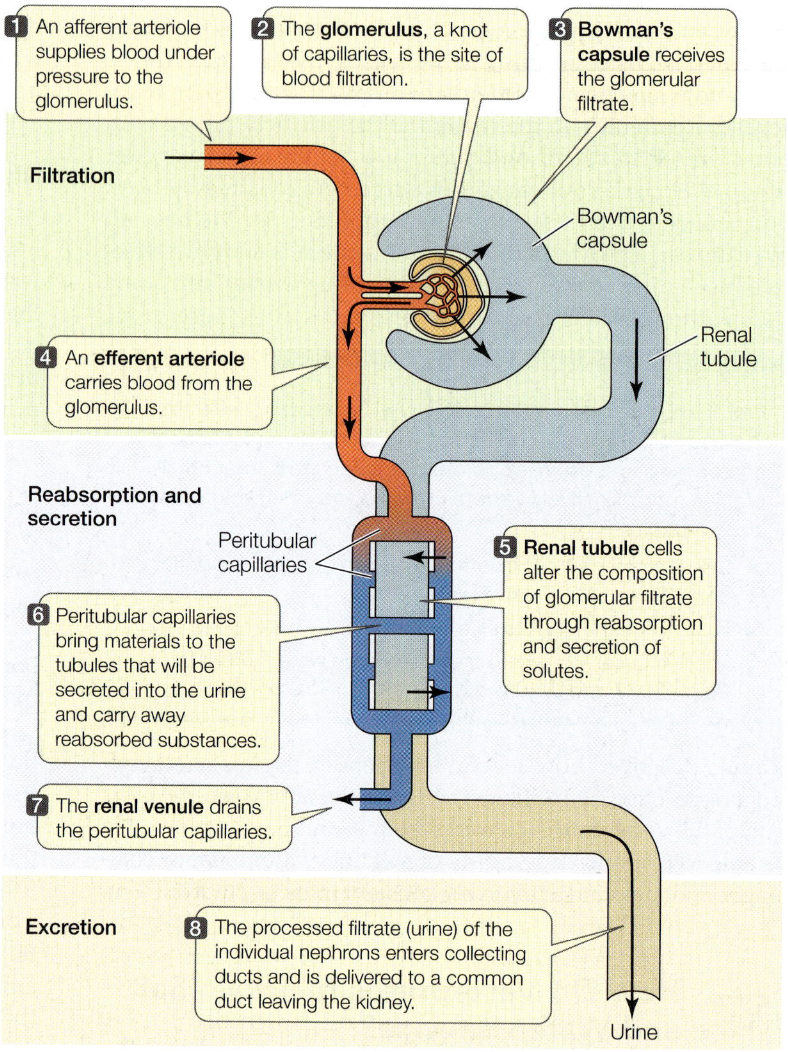

1 An afferent arteriole supplies blood under pressure to the glomerulus.
2 The glomerulus, a knot of capillaries, is the site of blood filtration.
3 Bowman's capsule receives the glomerular filtrate.
Filtration
Bowman's capsule
Renal tubule
4 An efferent arteriole carries blood from the glomerulus.
Reabsorption and secretion
Peritubular capillaries
5 Renal tubule cells alter the composition of glomerular filtrate through reabsorption and secretion of solutes.
6 Peritubular capillaries bring materials to the tubules that will be secreted into the urine and carry away reabsorbed substances.
7 The renal venule drains the peritubular capillaries.
Excretion
8 The processed filtrate (urine) of the individual nephrons enters collecting ducts and is delivered to a common duct leaving the kidney.
Urine

52.7 The Vertebrate Nephron The vertebrate nephron consists of a renal tubule closely associated with two capillary beds, the glomerulus and the peritubular capillaries.

Go to Activity 52.2 The Vertebrate Nephron
Life10e.com/ac52.2

Blood is filtered into Bowman's capsule

The renal tubule begins with **Bowman's capsule** (see Figure 52.7), which encloses the glomerulus (Figure 52.8A and B). The glomerulus appears to be pushed into Bowman's capsule much like a fist pushed into an inflated balloon. The cells of the capsule that are in direct contact with the glomerular capillaries are called **podocytes** (Figure 52.8C). These highly specialized cells have numerous armlike extensions, each with hundreds of fine, fingerlike processes. The podocytes wrap around the capillaries so that their processes interdigitate and intimately cover the capillaries.

The glomerulus filters the blood to produce a fluid (the renal filtrate) that lacks cells and large molecules. The walls of the capillaries, the basal lamina of the capillary endothelium, and the

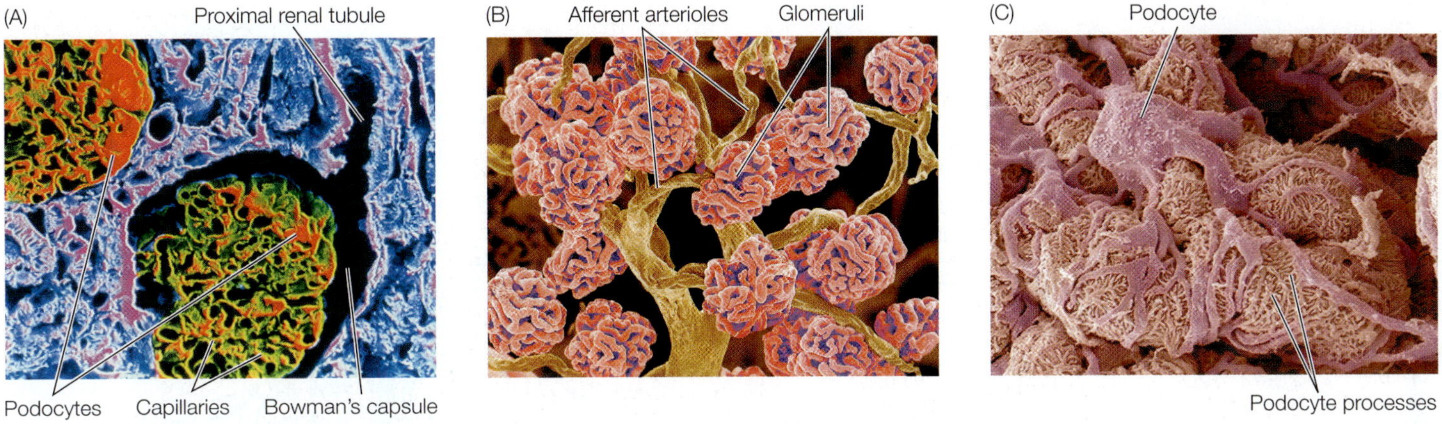
(A)
Proximal renal tubule
Podocytes
Capillaries
Bowman's capsule
(B)
Afferent arterioles
Glomeruli
(C)
Podocyte
Podocyte processes

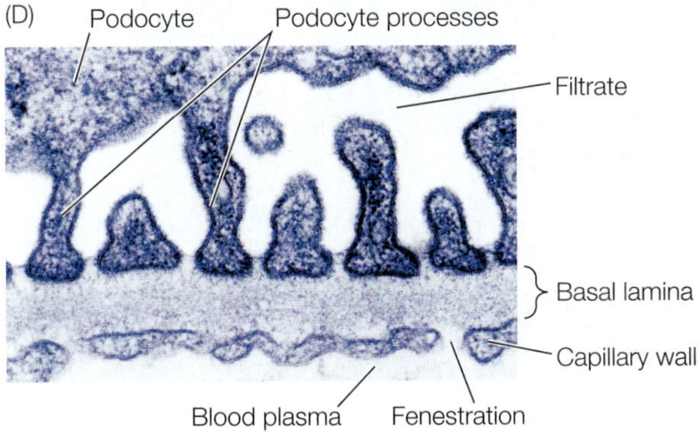
(D)
Podocyte
Podocyte processes
Filtrate
Basal lamina
Capillary wall
Blood plasma
Fenestration

52.8 A Tour of the Nephron Scanning electron micrographs illustrate the anatomical basis for blood filtration by the kidneys. (A) This cross section of an intact glomerulus shows the tubule cells that form Bowman's capsule. (B) In a preparation showing only the blood vessels (tubular tissue has been digested away), the glomeruli appear as balls of capillaries served by arterioles. (C) Higher magnification of a glomerulus with the tubule cells intact shows the podocytes that wrap around the glomerular capillaries. (D) The glomerular filter has three layers: the fenestrated endothelial cells of the capillaries, the meshwork of collagen fibers making up the basement membrane, and the filtration slits between the podocyte processes. All three layers have negative charges, which contribute to their ability to prevent the passage of protein molecules.

podocytes of Bowman's capsule all participate in filtration. Fenestrations in the walls of the capillaries (see Section 50.4) allow water and many solute molecules, but not red blood cells, to pass through. The meshwork of the basal lamina and the spaces between the processes of the podocytes are even finer and prevent large molecules from leaving the capillaries (**Figure 52.8D**). The arterial pressure of the blood entering the permeable capillaries causes the filtration of water and small molecules out of the glomerular capillaries and into the Bowman's capsule. The glomerular filtration rate is high because blood pressure in the glomerular capillaries is unusually high, and because the capillaries of the glomerulus, along with their covering of podocytes, are more permeable to water than other capillary beds in the body are.

The renal tubules convert glomerular filtrate to urine

The composition of the filtrate that enters the renal tubule is similar to that of the blood plasma, with the exception of high-molecular-weight solutes such as proteins. Reabsorption and secretion cause the composition of this fluid to change as it passes down the renal tubule. Cells of the tubule actively reabsorb certain molecules from the tubule fluid (which are returned to the blood flowing through the peritubular capillaries). For example, glucose and amino acids are reabsorbed. Most NaCl is reabsorbed. Other substances in the blood of the peritubular capillaries are actively secreted into the tubule fluid. An example is paraminohippuric acid (PAH), which is produced in the liver from benzoic acid, a common food preservative. Because of the actions of the renal tubules, the excreted urine is very different from the original filtrate.

RECAP 52.4

The kidney is the major excretory organ of vertebrates. Its functional unit is the nephron, which includes a glomerulus that filters blood and a renal tubule that secretes and reabsorbs solutes, modifying the filtrate to produce urine. The nephron is a mechanism for excreting excess water while conserving valuable solutes.

- Explain the differences in the osmoregulatory adaptations of freshwater fish, marine bony fish, and chondrichthyans. **See p. 1077**
- What are the functional relationships between the glomerular and peritubular capillaries? **See p. 1078 and Figure 52.7**
- Describe how blood is filtered by the glomerulus. **See pp. 1078–1079**
- How is the composition of the urine made different from the composition of the blood? **See p. 1079**

The adaptations that enable the mammalian kidney to produce urine more concentrated than extracellular fluids were important steps in vertebrate evolution, and they were largely achieved through changes in the structure and regional functions of the renal tubules. These changes converted a kidney that evolved to excrete water into an organ that conserves water.

52.5 How Does the Mammalian Kidney Produce Concentrated Urine?

Mammals have high body temperatures and high metabolic rates, and therefore have the potential for a high rate of water

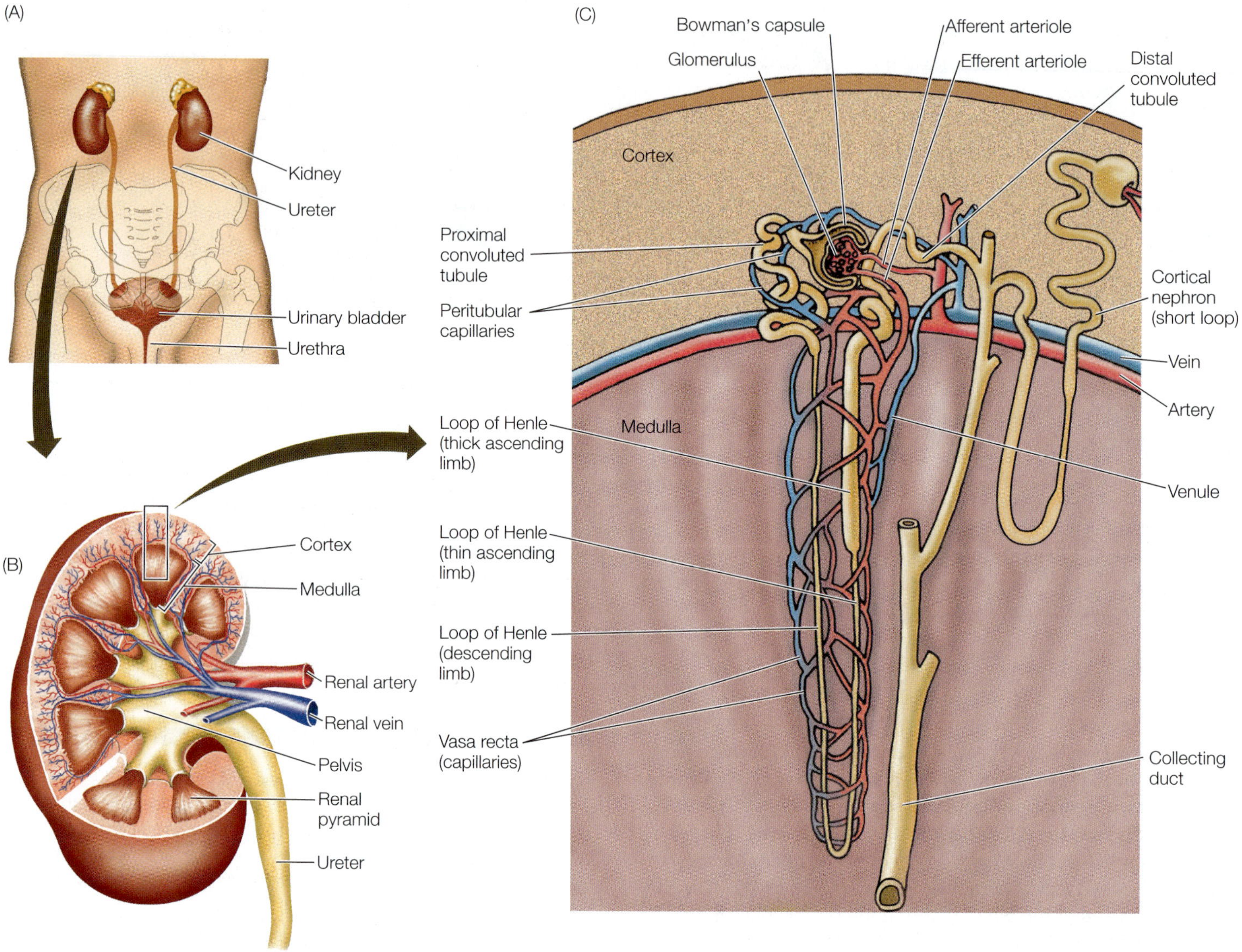

(A)
Kidney
Ureter
Urinary bladder
Urethra
(B)
Cortex
Medulla
Renal artery
Renal vein
Pelvis
Renal pyramid
Ureter
(C)
Bowman's capsule
Glomerulus
Afferent arteriole
Efferent arteriole
Distal convoluted tubule
Cortex
Proximal convoluted tubule
Peritubular capillaries
Cortical nephron (short loop)
Vein
Artery
Medulla
Loop of Henle (thick ascending limb)
Venule
Loop of Henle (thin ascending limb)
Loop of Henle (descending limb)
Vasa recta (capillaries)
Collecting duct

52.9 The Human Excretory System (A) The human kidneys lie against the back wall of the abdominal cavity, in the region of the middle back. (B) A highly organized internal structure is the basis for kidney function. Certain parts of the nephrons are in the organ's outer region, called the cortex; other parts are in the internal region, called the medulla. (C) The glomeruli and the proximal and distal convoluted tubules are located in the cortex of the kidney. The loops of Henle run in parallel as straight sections down into the renal medulla and back up to the cortex. Collecting ducts run from the cortex to the inner surface of the medulla, where they open into the ureter. The vasa recta are peritubular capillaries that parallel the loops of Henle.

Go to Activity 52.3 The Human Excretory System
Life10e.com/ac52.3

loss. Having an excretory system that minimizes water loss made it possible for these highly active species to occupy arid habitats.

Go to Media Clip 52.1
Inside the Bladder
Life10e.com/mc52.1

Kidneys produce urine and the bladder stores it

Mammalian excretory systems are similar, so we will use that of humans as our example. Humans have two kidneys at the back of the upper region of the abdominal cavity (Figure 52.9A). Each kidney filters blood, processes the filtrate into urine, and releases that urine into a duct called the **ureter**. The ureter of each kidney leads to the **urinary bladder**, where the urine is stored until it is excreted through the **urethra**, a short tube that opens to the outside of the body.

Two sphincter muscles surrounding the base of the urethra control urination. One of these sphincters is a smooth muscle and is controlled by the autonomic nervous system. As the bladder fills, stretch receptors in the walls of the bladder trigger a spinal reflex that relaxes this sphincter. This reflex is the only control of urination in infants, hence their frequent "accidents." The other sphincter is a skeletal muscle and is controlled by the voluntary nervous system. When the bladder is *very* full, only deliberate conscious effort prevents urination. Toilet training of children teaches them to control this sphincter.

WORKING WITH**DATA:**

What Kidney Characteristics Determine Urine Concentrating Ability?

Original Paper

Schmidt-Nielsen, B. and R. O'Dell. 1961. Structure and concentrating mechanism in the mammalian kidney. *American Journal of Physiology* 200: 1119–1124

Analyze the Data

In the 1950s it was hypothesized that the loops of Henle in mammalian kidneys are a countercurrent multiplier mechanism responsible for the ability of mammals to concentrate their urine. Bodil Schmidt-Nielsen and Roberta O'Dell extended this hypothesis to make a prediction: the longer the loops of Henle, the greater the concentration gradient a mammal could establish. There was a problem with their prediction, however. Since kidney size varies with body size, it presumes that a very small mammal could not have loops of Henle longer than those of a large mammal—yet many mammals that inhabit arid environments (and thus need to conserve water by producing highly concentrated urine) are extremely small. Schmidt-Nielsen and O'Dell refined their prediction to say that the *relative* lengths of the loops of Henle should correlate with ability to concentrate urine. Since the longest loops of Henle can only be as long as the renal medulla is thick, they conceived of a measure they called "relative medullary thickness," or RMT (see the table).

Another possibility existed, however. As can be seen in Figure 52.9C, not all loops of Henle extend all the way to the tip of the medulla—there are short and long loops of Henle. Could the concentrating ability of the kidney be a function of the *proportion* of the loops that are long? Schmidt-Nielsen and O'Dell measured aspects of kidney size and function in several animals from habitats of different aridity and determined the maximum concentration of urine the animals could produce when water-deprived. Some of their data are summarized in the table below.

QUESTION 1

Is the length of the loops of Henle relative to overall kidney size predictive of ability to concentrate urine? (To answer this, plot RMT versus FPD.)

QUESTION 2

Are animals with a higher percentage of long loops of Henle more capable of producing concentrated urine? (Plot percent long loops of Henle versus FPD.)

QUESTION 3

Is percent long loops of Henle or RMT the better predictor of concentration ability?

Animal	Kidney size (mm)[a]	Relative medullary thickness, RMT[b]	Percent long loops of Henle	Freezing point depression, FPD (°C)[c]
Human	64.0	3.0	14	2.6
Pig	66.0	1.6	3	2.0
Domestic cat	24.0	4.8	100	5.80
Beaver	36.0	1.3	0	0.96
Lab rat	14.0	5.8	28	4.85
Sand rat	13.0	10.7	100	9.2
Kangaroo rat	5.9	8.5	27	10.4
Jerboa	4.5	9.3	33	12.0

[a] Kidney size = cube root of overall length × width × thickness
[b] RMT = thickness of medulla in mm × 10 divided by kidney size.
[c] FPD = the number of degrees below 0°C at which urine freezes. FPD is a measure of solute concentration: the more concentrated the urine, the higher the FPD.

Go to **BioPortal** for all WORKING WITH**DATA** exercises

Nephrons have a regular arrangement in the kidney

The kidney is shaped like a kidney bean; when sliced along its long axis, its key anatomical features are revealed (**Figure 52.9B**). The ureter and the **renal artery** and **renal vein** enter the kidney on its concave (punched-in) side. The ureter extends into the kidney in several branches, the ends of which envelop kidney tissues called **renal pyramids**. The renal pyramids make up the internal core, or **medulla**, of the kidney. The medulla is covered by an outer layer, or **cortex**, that has a granular appearance. Between the cortex and the medulla, the renal artery divides into the many arterioles that serve the nephrons. In this same region, the renal vein collects blood from the many venules that drain the peritubular capillaries.

The organization of the nephrons within the kidney is very regular. All of the glomeruli with their Bowman's capsules are located in the cortex. The initial segment of a renal tubule is called the **proximal convoluted tubule**—"proximal" because it is closest to the glomerulus, and "convoluted" because it is twisted (**Figure 52.9C**). All of the proximal convoluted tubules are also located in the cortex.

At the point at which the renal tubule descends into the medulla, it becomes thin, straightens, and descends directly down into the medulla. In the medulla the tubule makes a hairpin turn and ascends back to the cortex, forming what is called the **loop of Henle**. Some nephrons have longer loops of Henle than others. Some 20 to 30 percent of human nephrons that have glomeruli deep in the cortex (i.e., near the border with the medulla) have long loops of Henle that go deep into the medulla. Nephrons that have glomeruli farther up in the cortex generally have short loops of Henle that descend only a short distance into the medulla. As we will see, the long loops are the critical adaptation of the mammalian nephron that enables the kidney to concentrate the urine.

The ascending limb of the loop of Henle becomes the **distal convoluted tubule** when it reaches the cortex—"distal" because it is farther from the glomerulus. The distal convoluted tubules of many nephrons join a common **collecting duct** in the cortex. The collecting ducts descend back down through the renal pyramid, parallel to and past the tips of the loops of Henle, and empty into a funnel-shaped structure called the pelvis. Divisions of the pelvis that surround each renal pyramid join together to leave the kidney as the ureter (see Figure 52.9B).

The organization of the blood vessels of the kidney closely parallels the organization of the nephrons (see Figure 52.9C). Smaller arteries branch from the renal artery and radiate into the cortex, forming the afferent arterioles that carry blood to each glomerulus. Each glomerulus is drained by an efferent arteriole that gives rise to the peritubular capillaries, most of which surround the proximal and distal convoluted tubules. As we have seen, the intimate associations of the glomerular and peritubular capillaries with the renal tubules permit exchanges between the blood and the specialized regions of the tubules.

Some of the peritubular capillaries run into the medulla in parallel with the loops of Henle and the collecting ducts, forming a vascular network called the **vasa recta**. All of the peritubular capillaries from a nephron join back together into a venule that joins with venules from other nephrons and eventually leads to the renal vein. As we will see, the concentrating ability of the mammalian kidney depends on water reabsorption in the renal medulla, and the vasa recta are the avenue by which that water gets out of the renal medulla and back into the circulation.

Most of the glomerular filtrate is reabsorbed by the proximal convoluted tubule

Most of the water and solutes filtered by the glomerulus are reabsorbed and do not appear in the urine. We can reach this conclusion by comparing the rate of filtration by the glomeruli with the rate of urine production. The kidneys receive about 1 liter of blood per minute, or about 1,500 liters of blood per day. How much of this huge volume is filtered out of the glomeruli? The answer is about 12 percent. This is still a large volume—180 liters per day! We normally urinate less than 2 liters per day, so about 99 percent of the fluid volume that is filtered out of the glomerulus is returned to the blood. Where and how is this enormous fluid volume reabsorbed?

The proximal convoluted tubule (PCT) is responsible for most of the reabsorption of water and solutes from the glomerular filtrate. The cells of this section of the renal tubule have many microvilli that increase their apical (facing into the tubule) surface area for reabsorption, and they have many mitochondria—an indication that they are metabolically active. PCT cells actively transport Na^+ (with Cl^- following) and other solutes, such as glucose and amino acids, out of the tubule fluid.

Almost all glucose and amino acid molecules that are filtered from the blood are actively reabsorbed by PCT cells and transported into the extracellular fluid. The active transport of solutes from the proximal tubule into the interstitial fluid causes water to follow osmotically. The water and solutes moved into the interstitial fluid are taken up by the peritubular capillaries and returned to the venous blood. These processes accomplish the reabsorption of more than 75 percent of the fluid that initially enters the nephron.

Despite the bulk reabsorption of water and solutes by the proximal convoluted tubule, the overall osmolarity of the fluid flowing through the PCT does not change. Thus the process that is occurring in the PCT is called isosmotic reabsorption. The fluid that enters the loop of Henle has the same osmolarity as the blood plasma, although its composition is different. How then does the kidney produce urine that is more concentrated than the blood plasma?

The loop of Henle creates a concentration gradient in the renal medulla

Humans can produce urine that is four times more concentrated than their blood plasma. The vampire bat we encountered at the beginning of this chapter can produce urine that is 15 times more concentrated than its blood plasma. The concentrating ability of the mammalian kidney arises from a **countercurrent multiplier** mechanism made possible by the anatomical arrangement of the loops of Henle. The term "countercurrent" refers to the opposing directions in which the tubule fluid in the descending and ascending limbs flows. The term "multiplier" refers to the ability of this system to create a solute concentration gradient in the renal medulla.

The loops of Henle do not themselves produce concentrated urine; rather, they increase the osmolarity of the extracellular fluid in the medulla in a graduated way. In humans, for example, the extracellular fluid at the top of the medulla bordering the cortex will be about 300 mosm/L (the concentration of blood plasma). But at the bottom of the medulla, where the loops of Henle make their hairpin turns, the extracellular fluid can be 1,200 mosm/L (see Figure 52.10). How do the loops produce this effect?

The cells that make up the different segments of the loop of Henle differ anatomically and functionally. Cells of the descending limb and the initial cells of the ascending limb are thin, with no microvilli and few mitochondria. They are not specialized for transport. Partway up the ascending limb, the cells become specialized for active transport. These cells are thick and have many mitochondria. Accordingly, the segments of the loop of Henle are named the thin descending limb, the thin ascending limb, and the thick ascending limb (**Figure 52.10**).

The countercurrent multiplier mechanism may be more easily understood by first considering events occurring in the thick ascending limb (Figure 52.10, note 1). The cells of the thick ascending limb reabsorb Na^+ and Cl^- from the tubule fluid and move it into the interstitial fluid. (In the following discussion, we will distinguish between the two components of extracellular fluid—the blood plasma and the interstitial fluid.) The thick ascending limb is not permeable to water, so the reabsorption of Na^+ and Cl^- from the tubular fluid raises the concentration of those solutes in the surrounding interstitial fluid

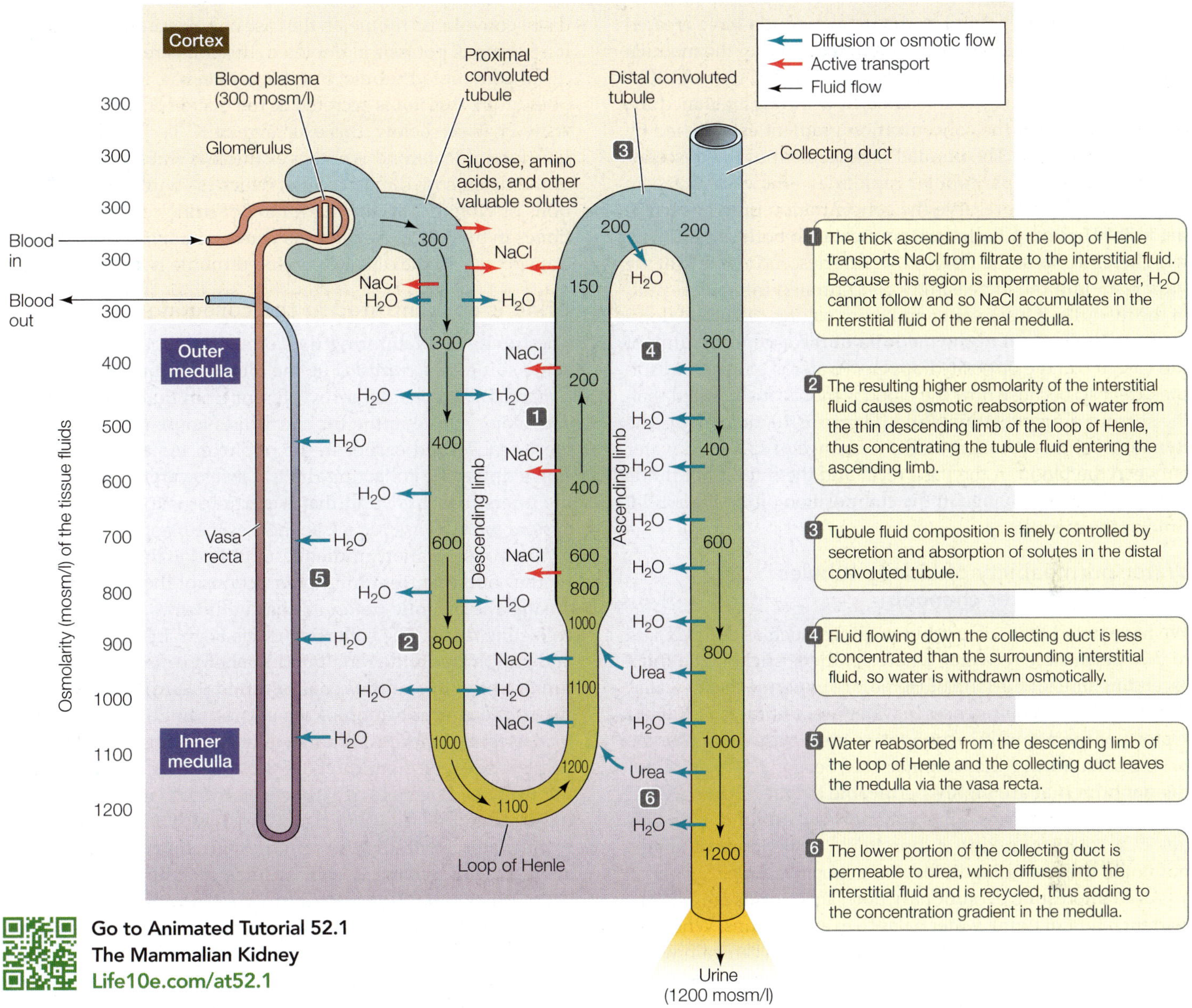
Cortex
Outer medulla
Inner medulla
Osmolarity (mosm/l) of the tissue fluids
300
300
300
300
300
400
500
600
700
800
900
1000
1100
1200
Blood in
Blood out
Blood plasma (300 mosm/l)
Glomerulus
Proximal convoluted tubule
Glucose, amino acids, and other valuable solutes
Distal convoluted tubule
Collecting duct
Diffusion or osmotic flow
Active transport
Fluid flow
Vasa recta
Descending limb
Ascending limb
Loop of Henle
NaCl
H2O
Urea
Urine (1200 mosm/l)
1 The thick ascending limb of the loop of Henle transports NaCl from filtrate to the interstitial fluid. Because this region is impermeable to water, H2O cannot follow and so NaCl accumulates in the interstitial fluid of the renal medulla.
2 The resulting higher osmolarity of the interstitial fluid causes osmotic reabsorption of water from the thin descending limb of the loop of Henle, thus concentrating the tubule fluid entering the ascending limb.
3 Tubule fluid composition is finely controlled by secretion and absorption of solutes in the distal convoluted tubule.
4 Fluid flowing down the collecting duct is less concentrated than the surrounding interstitial fluid, so water is withdrawn osmotically.
5 Water reabsorbed from the descending limb of the loop of Henle and the collecting duct leaves the medulla via the vasa recta.
6 The lower portion of the collecting duct is permeable to urea, which diffuses into the interstitial fluid and is recycled, thus adding to the concentration gradient in the medulla.

52.10 Concentrating the Urine A countercurrent multiplier mechanism enables the mammalian kidney to produce urine that is far more concentrated than the blood plasma. The composition—but not the concentration—of the filtrate is changed by the proximal convoluted tubule, which reabsorbs valuable molecules (including NaCl). Bulk reabsorption of water follows osmotically. The urine concentration process begins in the thick ascending limb of the loop of Henle, which reabsorbs NaCl but is impermeable to H_2O. Some of the reabsorbed NaCl enters the descending limb and is thereby trapped in the renal medulla, creating a concentration gradient in the interstitial fluid. As urine in the collecting duct passes through this concentration gradient, it can lose water osmotically and become highly concentrated.

Go to Animated Tutorial 52.1
The Mammalian Kidney
Life10e.com/at52.1

and decreases the concentration of the tubular fluid entering the distal convoluted tubule.

The thin descending limb, in contrast, is highly permeable to water but not very permeable to Na^+ and Cl^-. Since the local interstitial fluid has been made more concentrated by the Na^+ and Cl^- reabsorbed from the neighboring thick ascending limb, water is withdrawn osmotically from the fluid in the descending limb. Therefore the fluid in the descending limb becomes more concentrated as it flows toward the hairpin turn at the bottom of the renal medulla (Figure 52.10, note 2).

The thin ascending limb, like the thick ascending limb, is not permeable to water. It is, however, permeable to Na^+ and Cl^-. As the concentrated tubule fluid flows up the thin ascending limb, it is more concentrated than the surrounding interstitial fluid, so Na^+ and Cl^- diffuse out. When the tubule fluid reaches the thick ascending limb, active transport continues to move Na^+ and Cl^- from the tubule fluid to the interstitial fluid.

As a result of the processes described above, the tubule fluid reaching the distal convoluted tubule is less concentrated than the blood plasma (Figure 52.10, note 3), and the solutes that have been left behind in the renal medulla have created a concentration gradient in the interstitial fluid of the medulla (indicated by the background color gradient in Figure 52.10).

You may wonder why the blood flow through the medulla does not wash out the concentration gradient established by the loops of Henle. The parallel arrangement of the descending and ascending peritubular capillaries—the vasa recta—in the medulla helps preserve the concentration gradient in the medulla. These capillaries are permeable to both salt and water. Therefore, as blood flows down the descending limb of the vasa recta into the increasingly concentrated interstitial fluid of the medulla, it loses water and gains solutes. As blood flows up from the bottom of the medulla in the ascending limb of the vasa recta, the opposite happens (water is gained and solutes are lost) because now the blood is more concentrated than the surrounding interstitial fluid (Figure 52.10, notes 4–6). The dynamics of this countercurrent exchange of salts and water between the blood in the vasa recta and the interstitial fluids result in little net change in the composition of the interstitial fluid in the medulla.

Water permeability of kidney tubules depends on water channels

We have noted that some tubule regions, such as the PCT, are highly permeable to water whereas others, such as the thick ascending limb of the loop of Henle, are impermeable to water. What causes these differences in water permeability in different regions of the nephron? Aquaporins are a class of membrane proteins that form water channels (see Section 6.3). Regions of the nephron that are highly permeable to water have greater numbers of aquaporins. Thus aquaporins are abundant in kidney PCT cells and in descending limbs of the loops of Henle, but not in the ascending limbs of the loop of Henle.

As an interesting evolutionary note, aquaporins are also important in maintaining water balance in amphibians. When not in an aqueous environment, many amphibians can gain water from a moist substrate because they have aquaporins in the epithelial cells of their belly skin. Thus water can cross their skin into the interstitial fluid by osmosis.

The distal convoluted tubule fine-tunes the composition of the urine

The first portion of the distal convoluted tubule is similar to the thick ascending limb of the loop of Henle. Na^+ and Cl^- are transported out of the tubule fluid, and water cannot follow. As a result, the tubule fluid becomes even more dilute. The later sections of the distal convoluted tubule, however, can be permeable to water, and water can be osmotically drawn from the tubule into the interstitial fluid. As the tubule fluid flows from the distal tubule to the collecting duct, it can be below or equal to the osmolarity of the blood plasma.

An important function of the distal tubule is the fine-tuning of the ionic composition of the urine. Even though bulk reabsorption of substances such as calcium, phosphate, bicarbonate, and potassium occurs in the proximal convoluted tubule, changes in the concentrations of these substances occur in the distal convoluted tubule. In the case of potassium, for example, if a person is potassium depleted, this ion is reabsorbed in the distal convoluted tubule, but if a person has an abundance of potassium, this ion is secreted in the distal convoluted tubule. As we will see below, this exchange of K^+ is controlled by the hormone aldosterone. Another example is reabsorption of Ca^{2+} in the distal convoluted tubule, which is controlled by the actions of vitamin D. The fine-tuning of urine composition continues in the collecting duct. As you can imagine, the list of ion transporters in the distal convoluted tubule is large.

Urine is concentrated in the collecting duct

The tubule fluid entering the collecting duct is at about the same solute *concentration* as the blood plasma, but its solute *composition* is considerably different from that of the plasma. The major solute in the tubular fluid is now urea, since salts were reabsorbed earlier in the nephron. As the tubule fluid flows down the collecting duct, it loses water osmotically to the interstitial fluid, and that water returns to the circulatory system via the vasa recta (see Figure 52.10, note 4).

The concentration gradient established in the renal medulla by the countercurrent multiplier actions of the loops of Henle creates the osmotic potential that withdraws water from the collecting ducts. The collecting ducts begin in the renal cortex and run through the renal medulla before emptying into the ureter at the tips of the renal pyramids. During this journey, the solute concentration of the surrounding interstitial fluid increases, and more and more water can be absorbed from the urine in the collecting duct. By the time it reaches the ureter, the urine can become greatly concentrated, with urea as the major solute.

As water is withdrawn from the collecting duct, some urea also leaks out into the medullary interstitial fluid, adding to its osmotic potential. This urea diffuses back into the loop of Henle and is returned to the collecting duct. The recycling of urea in the renal medulla contributes significantly to the concentration gradient and therefore the ability of the kidney to concentrate the urine in the collecting duct. The ability of a mammal to concentrate its urine is determined by the maximum concentration gradient it can establish in its renal medulla.

The kidneys help regulate acid–base balance

Besides regulating salt and water balance and excreting nitrogenous wastes, the kidneys have another important role: they regulate the hydrogen ion concentration (the pH) of the extracellular fluids. pH is a critical variable because it influences the structure and function of proteins.

One way to minimize pH changes in a chemical solution is to add a buffer—a substance that can either absorb or release hydrogen ions (see Section 2.4). The major buffer in the blood is bicarbonate ions (HCO_3^-; see Figure 49.14) that are formed from the dissociation of carbonic acid, which in turn is formed by the hydration of CO_2 according to the following equilibrium reaction:

$$CO_2 + H_2O \rightleftharpoons H_2CO_3 \rightleftharpoons H^+ + HCO_3^-$$

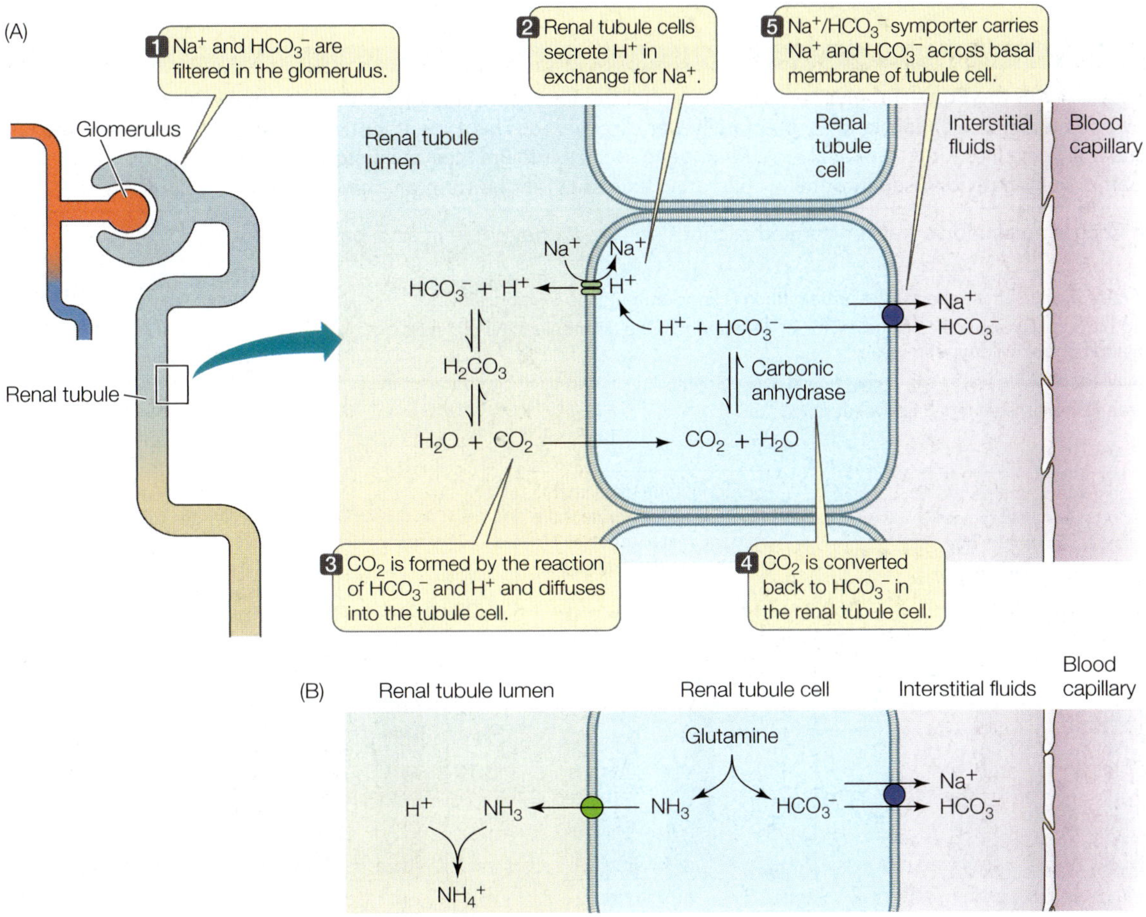
(A)
1 Na^+ and HCO_3^- are filtered in the glomerulus.
2 Renal tubule cells secrete H^+ in exchange for Na^+.
5 Na^+/HCO_3^- symporter carries Na^+ and HCO_3^- across basal membrane of tubule cell.
Glomerulus
Renal tubule
Renal tubule lumen
Renal tubule cell
Interstitial fluids
Blood capillary
$HCO_3^- + H^+$
H_2CO_3
$H_2O + CO_2$
$H^+ + HCO_3^-$
Carbonic anhydrase
$CO_2 + H_2O$
3 CO_2 is formed by the reaction of HCO_3^- and H^+ and diffuses into the tubule cell.
4 CO_2 is converted back to HCO_3^- in the renal tubule cell.
(B)
Renal tubule lumen
Renal tubule cell
Interstitial fluids
Blood capillary
Glutamine
H^+
NH_3
NH_4^+
HCO_3^-
Na^+

52.11 The Kidney Excretes Acids and Conserves Bases (A) Bicarbonate ions are filtered out of the blood at the glomerulus, and renal tubule cells secrete hydrogen ions into the tubule fluid. In the renal tubule, the filtered bicarbonate buffers the secreted hydrogen ions and keeps the urine from becoming too acidic. The CO_2 formed by the reaction of bicarbonate and hydrogen ions is converted back to bicarbonate by the renal tubule cells and transported back into the interstitial fluid. (B) Excretion of ammonia (NH_3) by renal tube cells is also important for acid–base balance.

From this equation, you can see that if excess hydrogen ions are added to this reaction mixture, the reaction will move to the left and absorb the excess H^+. If hydrogen ions are removed from the reaction mixture, however, the reaction will move to the right and supply more H^+.

The HCO_3^- buffer system is important for controlling the pH of the blood, and therefore of the interstitial fluids as well, because the reaction can be pushed to the right and pulled to the left physiologically. The lungs control the levels of CO_2 in the blood, thus altering the acid portion of the reaction. CO_2 is considered the acid portion of the reaction because if you add additional CO_2, the reaction shifts to the right, producing more H^+ ions. The kidneys control the base portion of the reaction by removing H^+ from the blood and returning HCO_3^- to the blood. How does this occur?

HCO_3^- is filtered in the glomerulus and is therefore present in the tubular fluid. Tubule cells transport H^+ into the tubule fluid in exchange for Na^+. In the tubule, the excreted H^+ combines with the filtered HCO_3^- to produce H_2CO_3 that then dissociates into H_2O and CO_2. The CO_2 diffuses into the tubule cells, where in the presence of the enzyme carbonic anhydrase it produces HCO_3^- that is transported out of the basal end of the cell into the interstitial fluid and thence to the blood (Figure 52.11A). Thus for each H^+ secreted into the tubule fluid, a HCO_3^- ion is released into the blood.

Another mechanism for H^+ secretion and HCO_3^- reabsorption involves ammonia (NH_3). The metabolism of glutamine in tubule cells produces NH_3 and HCO_3^- (Figure 52.11B). The HCO_3^- is reabsorbed into the interstitial fluid. The NH_3 is transported into the tubule fluid and combines with H^+ to form ammonium (NH_4^+), which is excreted in the urine. This process results in the net excretion of H^+ from the body. The NH_3 is transported into tubules by means of an NH_3 transporter that has been characterized recently in an effort to identify novel proteins coming from the sequencing of the human and other genomes (Figure 52.12).

Kidney failure is treated with dialysis

Loss of kidney function (renal failure) results in the retention of salts and water (hence high blood pressure), retention of urea (uremic poisoning), and a decreasing pH (acidosis). A person who suffers complete renal failure will die within 2 weeks if not treated. A drastic but highly successful treatment is kidney transplant, but it is usually necessary to sustain a patient for

INVESTIGATING**LIFE**

52.12 An Ammonium Transporter in the Renal Tubules? An important way the kidney excretes hydrogen ions and buffers the blood is to secrete ammonia (NH_3) into the renal tubules. It was thought that ammonia simply diffused into the tubules until the function of Rhcg, a protein in the Rh blood antigen family was discovered. This experiment demonstrated that loss of the *Rhcg* gene in mice impairs their ability to buffer their blood pH by excreting excess H^+ (in the form of ammonium)[a]

HYPOTHESIS The protein Rhcg is an ammonia transporter and is critical for the kidney's role in acid–base balance.

Method

1. Create a line of mice in which the gene for the protein Rhcg is knocked out (see Section 18.4).
2. Measure starting blood pH, plasma bicarbonate (HCO_3^-) levels, and urine ammonium (NH_4^+) levels in experimental and control (wild-type) mice.
3. For 6 days, administer drinking water containing a mild acid to control mice and to Rhcg knockout mice.
4. Measure the three variables (see item 2 above) at days 2 and 6.

Results

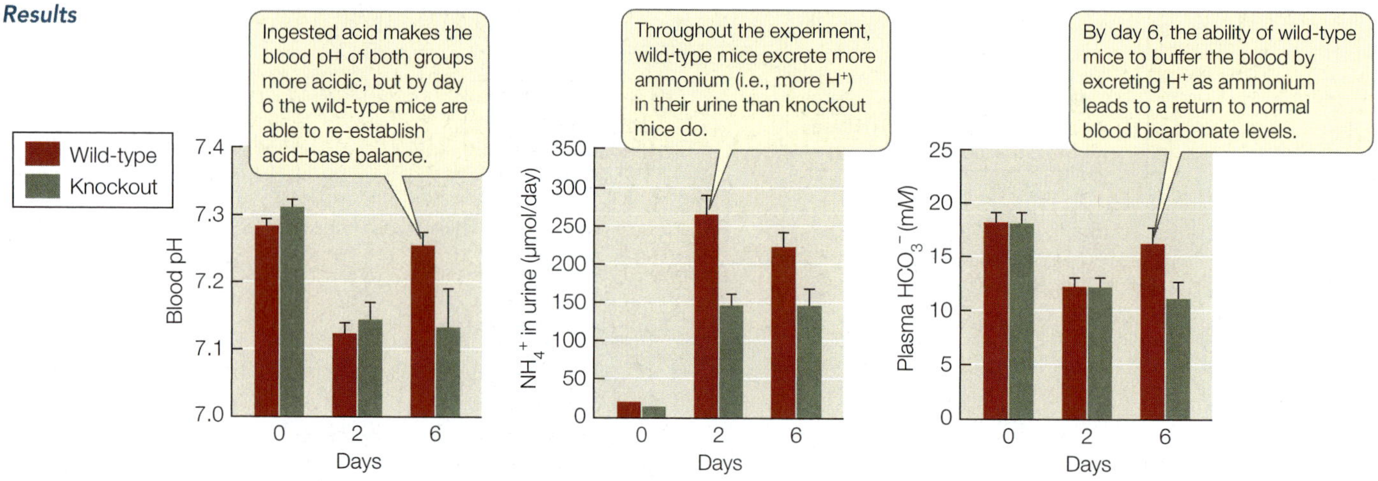
Wild-type
Knockout
Blood pH
7.4
7.3
7.2
7.1
7.0
0
2
6
Days
Ingested acid makes the blood pH of both groups more acidic, but by day 6 the wild-type mice are able to re-establish acid–base balance.
NH_4^+ in urine (μmol/day)
350
300
250
200
150
100
50
0
Throughout the experiment, wild-type mice excrete more ammonium (i.e., more H+) in their urine than knockout mice do.
Plasma HCO_3^- (mM)
25
20
15
10
5
By day 6, the ability of wild-type mice to buffer the blood by excreting H+ as ammonium leads to a return to normal blood bicarbonate levels.

CONCLUSION Lack of functional Rhcg protein impairs a mouse's ability to secrete ammonium ions in its urine, thus limiting the capacity to regulate acid–base balance. This protein is probably an ammonia transporter in the renal tubules.

Go to **BioPortal** for discussion and relevant links for all INVESTIGATING**LIFE** figures.

[a] Biver, S. et al. 2008. *Nature* 456: 339–343.

considerable time while waiting for a kidney to become available. Therefore artificial kidneys, or renal dialysis machines, are essential modes of treatment.

In a dialysis machine, the patient's blood flows through many small channels made of semipermeable membranes (Figure 52.13). A dialysis solution flows on the other side of these membranes, through which small molecules can diffuse. Molecules and ions diffuse from an area of higher concentration to an area of lower concentration, so the composition of the dialysis fluid is crucial. The concentrations of the molecules or ions that need to be conserved must be at the same concentration in the dialysis fluid as they are in the blood. The concentrations of molecules and ions that need to be removed from the blood are zero in the dialysis fluid. The total osmotic potential of the dialysis fluid must equal that of the plasma.

About 500 ml of the patient's blood is in the dialysis machine at any one time, and the unit processes several hundred milliliters of blood per minute. A patient with no kidney function must be on the dialysis machine for 4 to 6 hours three times a week.

RECAP 52.5

The anatomical organization of nephrons makes it possible for the mammalian kidney to produce a urine more concentrated than the blood, thereby conserving water to maintain extracellular fluid volume. Bulk reabsorption of salts, other valuable solutes, and water takes place in the proximal convoluted tubule. The loops of Henle act as a countercurrent multiplier, creating a concentration gradient of the interstitial fluids in the renal medulla. Collecting ducts run through the renal medulla and lose water osmotically to the surrounding interstitial fluids, concentrating the urine.

- Explain how the countercurrent multiplier mechanism of the nephron makes it possible for the kidney to form a concentrated urine. **See pp. 1082–1083 and Figure 52.10**
- Why doesn't the blood flow through the kidney wash out the concentration gradient in the medulla? **See p. 1082 and Figure 52.10**
- How does the kidney contribute to acid–base balance? **See pp. 1085–1086 and Figures 52.11 and 52.12**

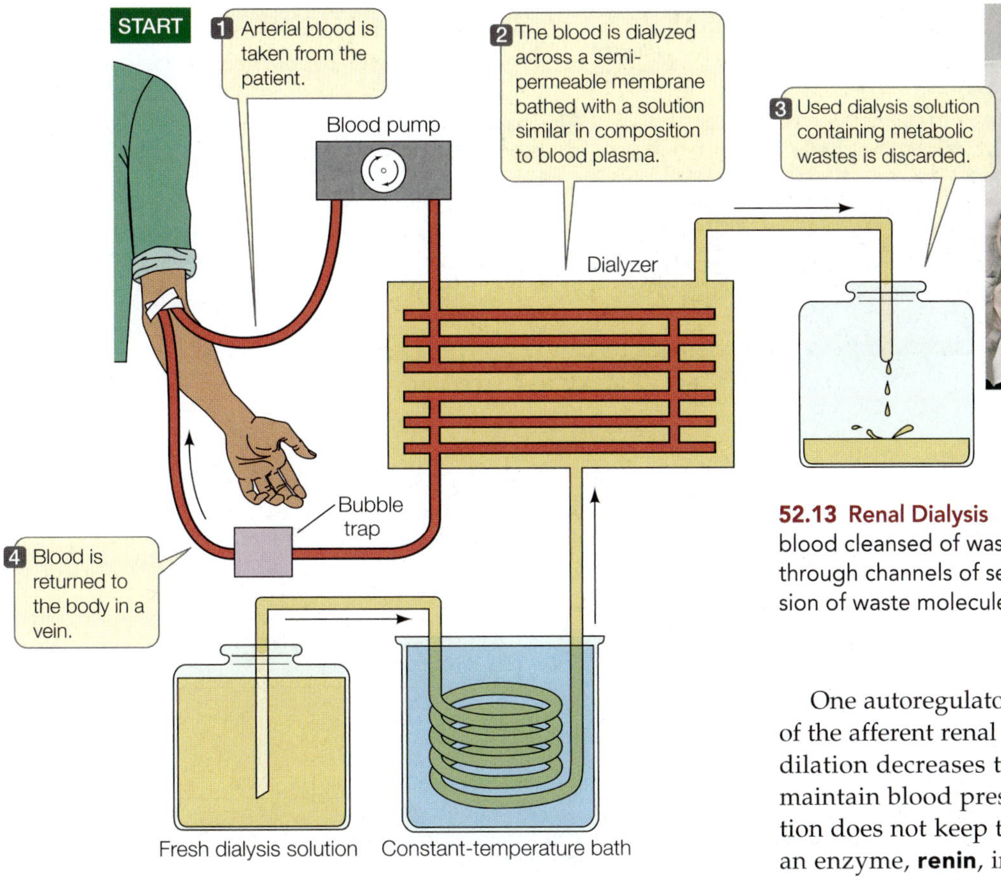
START
1 Arterial blood is taken from the patient.
Blood pump
2 The blood is dialyzed across a semi-permeable membrane bathed with a solution similar in composition to blood plasma.
Dialyzer
3 Used dialysis solution containing metabolic wastes is discarded.
Bubble trap
4 Blood is returned to the body in a vein.
Fresh dialysis solution
Constant-temperature bath

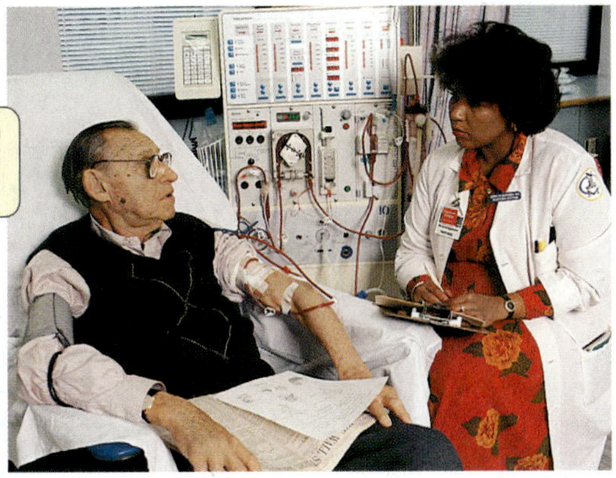

52.13 Renal Dialysis Patients with kidney failure can have their blood cleansed of wastes by renal dialysis machines. Blood flows through channels of semipermeable membranes that allow diffusion of waste molecules from the blood to a dialysis fluid.

The kidney contributes to homeostasis in several ways, including regulating extracellular fluid volume, maintaining the osmotic concentration and ionic composition of the extracellular fluid, and regulating pH. As we will see next, the kidneys also play a major role in regulating blood pressure.

52.6 How Are Kidney Functions Regulated?

Several regulatory mechanisms act on the kidneys to maintain blood pressure, blood osmolarity, and blood composition. We will discuss these mechanisms separately, but keep in mind that they are always working together.

Glomerular filtration rate is regulated

If the kidneys stop filtering blood, they cannot accomplish any of their functions. The maintenance of a constant **glomerular filtration rate** (**GFR**) depends on an adequate blood supply to the kidneys at an adequate blood pressure. Renal arteries usually deliver blood to the kidneys at high pressure because they are early branches off the aorta. In addition, autoregulatory mechanisms ensure adequate blood supply and blood pressure for kidney function regardless of what is happening elsewhere in the body. The kidney's autoregulatory adjustments compensate for decreases in cardiac output or decreases in blood pressure so that the GFR remains constant.

One autoregulatory mechanism is the dilation (expansion) of the afferent renal arterioles when blood pressure falls. This dilation decreases the resistance in the arterioles and helps maintain blood pressure in the glomerulus. If arteriole dilation does not keep the GFR from falling, the kidney releases an enzyme, **renin**, into the blood. Renin converts a circulating protein, angiotensinogen, into angiotensin I, which is then acted on by angiotensin-converting enzyme (ACE) to form the active hormone angiotensin II, or simply **angiotensin** (Figure 52.14). Angiotensin has several effects that help restore the GFR to normal:

- It constricts the efferent renal arterioles, raising the resistance for blood leaving the glomerulus. Like putting a finger over the end of a garden hose, this restriction of drainage elevates blood pressure in the glomerular capillaries.
- It constricts peripheral blood vessels all over the body, an action that elevates blood pressure.
- It stimulates the adrenal cortex to release the hormone **aldosterone**. Aldosterone stimulates sodium reabsorption by the kidney, making its reabsorption of water more effective. Enhanced water reabsorption helps maintain blood volume and therefore blood pressure.
- It acts on the brain to stimulate thirst. Increased water intake in response to thirst increases blood volume and blood pressure.

Thus the renin-angiotensin-aldosterone system, or RAAS, coordinates many responses to maintain blood pressure and kidney function.

Regulation of GFR uses feedback information from the distal tubule

A remarkable anatomical feature of the nephron is that where its renal tubule returns to the cortex and becomes the distal

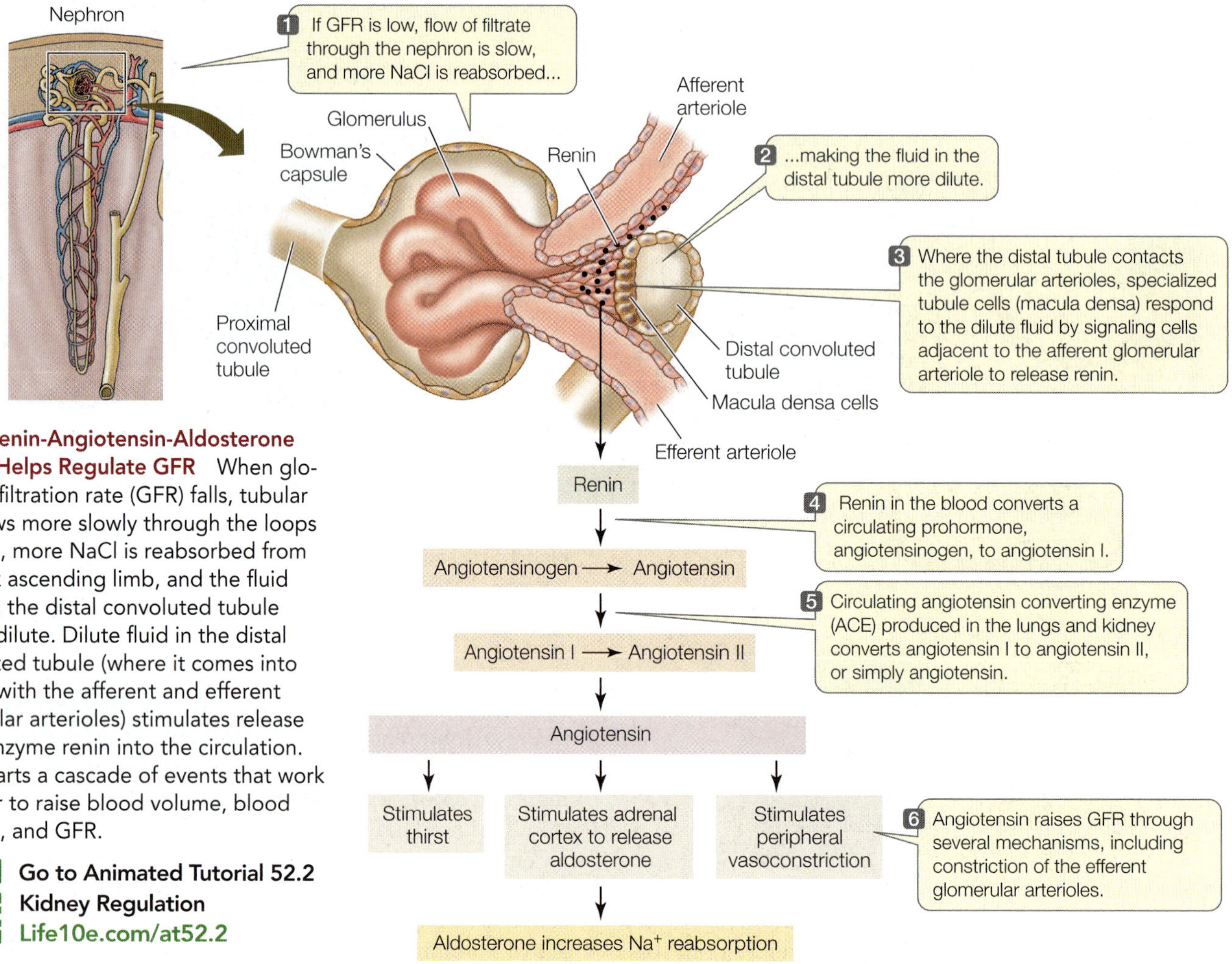
Nephron
1 If GFR is low, flow of filtrate through the nephron is slow, and more NaCl is reabsorbed...
Glomerulus
Bowman's capsule
Renin
Afferent arteriole
2 ...making the fluid in the distal tubule more dilute.
3 Where the distal tubule contacts the glomerular arterioles, specialized tubule cells (macula densa) respond to the dilute fluid by signaling cells adjacent to the afferent glomerular arteriole to release renin.
Proximal convoluted tubule
Distal convoluted tubule
Macula densa cells
Efferent arteriole
Renin
4 Renin in the blood converts a circulating prohormone, angiotensinogen, to angiotensin I.
Angiotensinogen → Angiotensin
5 Circulating angiotensin converting enzyme (ACE) produced in the lungs and kidney converts angiotensin I to angiotensin II, or simply angiotensin.
Angiotensin I → Angiotensin II
Angiotensin
Stimulates thirst
Stimulates adrenal cortex to release aldosterone
Stimulates peripheral vasoconstriction
6 Angiotensin raises GFR through several mechanisms, including constriction of the efferent glomerular arterioles.
Aldosterone increases Na+ reabsorption

52.14 Renin-Angiotensin-Aldosterone System Helps Regulate GFR When glomerular filtration rate (GFR) falls, tubular fluid flows more slowly through the loops of Henle, more NaCl is reabsorbed from the thick ascending limb, and the fluid reaching the distal convoluted tubule is more dilute. Dilute fluid in the distal convoluted tubule (where it comes into contact with the afferent and efferent glomerular arterioles) stimulates release of the enzyme renin into the circulation. Renin starts a cascade of events that work together to raise blood volume, blood pressure, and GFR.

Go to Animated Tutorial 52.2
Kidney Regulation
Life10e.com/at52.2

convoluted tubule, it makes contact with the afferent and efferent arterioles of its glomerulus. At this location the cells of the tubule are modified to form a structure called the macula densa (see Figure 52.14), and the arteriole cells are also modified and are called juxtaglomerular cells. The macula densa cells sense the concentration of NaCl in the fluid entering the distal convoluted tubule. If glomerular filtration rate is high, flow through the loop of Henle is high and the cells of the thick ascending limb cannot reabsorb all of the NaCl passing through it. If GFR is low, flow through the loop of Henle is slower and the cells of the thick ascending limb can reabsorb more of the NaCl. If the NaCl level in the distal convoluted tubule drops too low, the macula densa cells signal the juxtaglomerular cells to release renin and trigger the RAAS response. Thus the concentration of NaCl in the fluid passing over the macula densa is a function of the GFR and is information that enables fine control of the RAAS system.

Blood osmolarity and blood pressure are regulated by ADH

Cells in the hypothalamus can stimulate the release of a hormone called **antidiuretic hormone** (**ADH**, also called vasopressin) from the posterior pituitary. ADH can act on cells of the collecting duct to insert aquaporins (water channels) into their plasma membranes. The aquaporins increase the permeability of these membranes to water, and therefore more water is reabsorbed from the collecting duct fluid into the interstitial spaces of the renal medulla. The higher the circulating levels of ADH, the greater the number of aquaporins. Various factors can stimulate or inhibit the release of ADH. Of key importance to kidney function are osmoreceptors that monitor blood osmolarity and stretch receptors that monitor blood pressure (**Figure 52.15**).

Osmoreceptor neurons in the hypothalamus are activated by a rise in blood osmolarity, and they increase the release of ADH. ADH helps regulate blood osmolarity by controlling water reabsorption. The osmoreceptors also stimulate thirst. The resulting water retention and water intake dilute the blood as they expand blood volume.

Stretch receptors in the walls of the aorta and the carotid arteries (see Figure 50.19) that detect an increase in blood pressure will *inhibit* the release of ADH. With less circulating ADH, less water is reabsorbed, which decreases blood volume and hence acts to lower blood pressure.

If blood pressure falls, as when you lose blood volume through hemorrhage or excessive evaporative water loss,

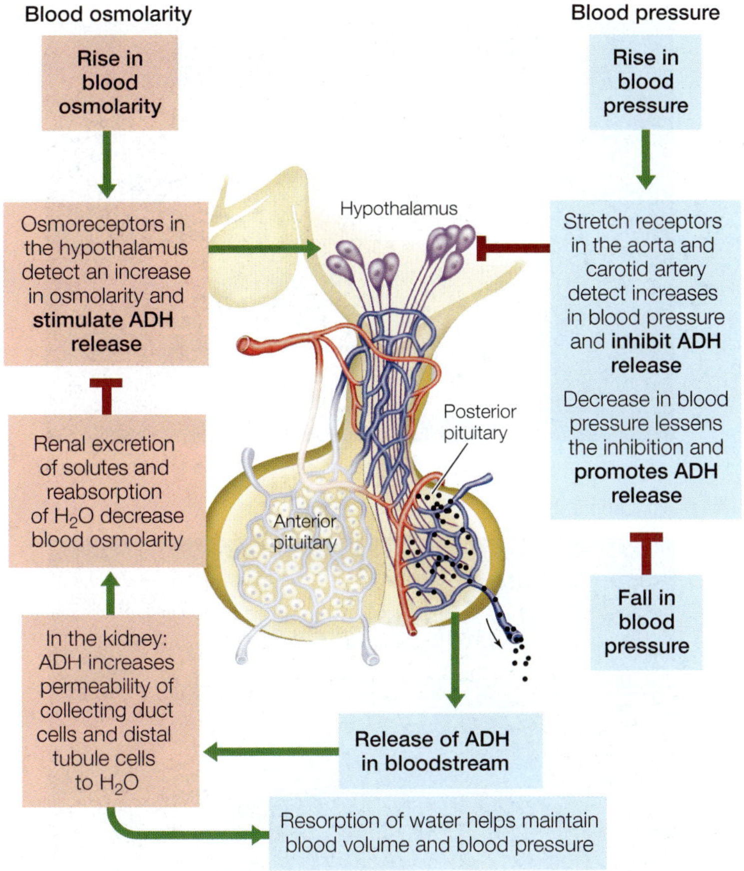
Blood osmolarity
Blood pressure
Rise in blood osmolarity
Rise in blood pressure
Hypothalamus
Osmoreceptors in the hypothalamus detect an increase in osmolarity and stimulate ADH release
Stretch receptors in the aorta and carotid artery detect increases in blood pressure and inhibit ADH release
Decrease in blood pressure lessens the inhibition and promotes ADH release
Renal excretion of solutes and reabsorption of H_2O decrease blood osmolarity
Posterior pituitary
Anterior pituitary
Fall in blood pressure
In the kidney: ADH increases permeability of collecting duct cells and distal tubule cells to H_2O
Release of ADH in bloodstream
Resorption of water helps maintain blood volume and blood pressure

52.15 Antidiuretic Hormone Increases Blood Pressure and Promotes Water Reabsorption ADH is produced by neurons in the hypothalamus and released from nerve endings in the posterior pituitary. The release of ADH is stimulated by hypothalamic osmoreceptors and inhibited by stretch receptors in the great arteries. Red lines indicate inhibitory actions; green lines show stimulatory actions.

activity of the stretch receptors in the aorta and carotid arteries decreases. Input via cranial nerves to the hypothalamus from these receptors inhibits the release of ADH, so when the firing rates of these stretch receptors fall, ADH release increases. More ADH results in more efficient water reabsorption and therefore a protection of blood volume and blood pressure.

Alcohol inhibits ADH release, explaining why excessive beer drinking leads to excessive urination and dehydration, which contributes to the symptoms of a hangover.

As mentioned earlier, the presence of aquaporins in plasma membranes determines their water permeability. Aquaporins play an important and unique role in the collecting duct. Several members of the aquaporin family of water channels are found in the plasma membranes of cells in the distal tubules. At least two aquaporins (AQP-3 and AQP-4) are localized in the basolateral membranes (facing the blood vessels). A different aquaporin, called AQP-2, is found in the apical plasma membranes (facing into the tubule). The presence of AQP-2 in these membranes is controlled by ADH, as described in Figure 52.16.

INVESTIGATING**LIFE**

52.16 ADH Induces Insertion of Aquaporins into Plasma Membranes Aquaporin proteins make some regions of renal tubules permeable to water. One aquaporin, AQP-2, is responsible for the permeability of the collecting duct cells. M. A. Knepper and colleagues did an experiment to find out how antidiuretic hormone acts on these proteins to control the level of permeability in renal cells.[a]

HYPOTHESIS Antidiuretic hormone (ADH) controls the location of aquaporin proteins.

Method

1. Isolate collecting ducts from rat kidney.
2. Use immunochemical staining to localize the AQP-2 aquaporins in collecting duct cells both with and without the presence of ADH. Also localize the aquaporins after ADH is applied and then washed away.

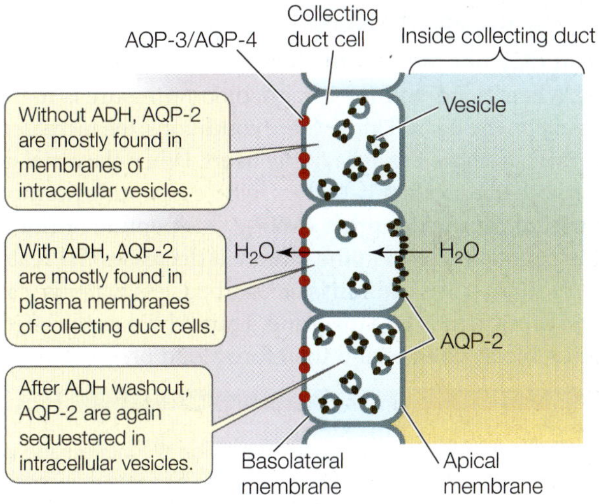
AQP-3/AQP-4
Collecting duct cell
Inside collecting duct
Vesicle
Without ADH, AQP-2 are mostly found in membranes of intracellular vesicles.
With ADH, AQP-2 are mostly found in plasma membranes of collecting duct cells.
H_2O
H_2O
AQP-2
After ADH washout, AQP-2 are again sequestered in intracellular vesicles.
Basolateral membrane
Apical membrane

3. Measure the water permeability of the collecting duct cells under the same three conditions.

Results

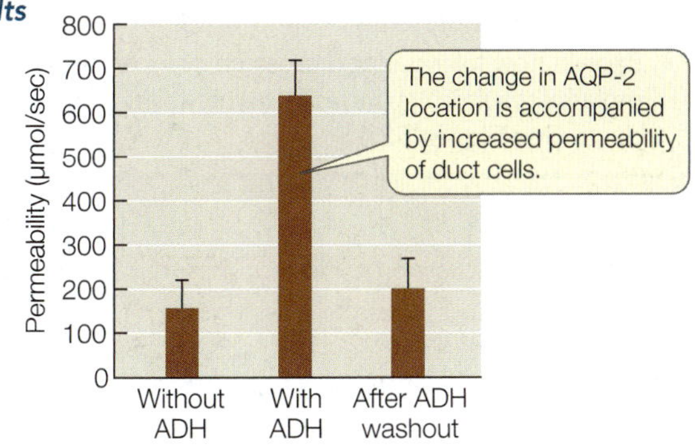
800
700
600
500
400
300
200
100
0
Permeability (μmol/sec)
The change in AQP-2 location is accompanied by increased permeability of duct cells.
Without ADH
With ADH
After ADH washout

CONCLUSION In the absence of ADH, AQP-2 is sequestered intracellularly. When ADH is present, AQP-2 channels are inserted into the plasma membranes, making the cells more permeable to water.

Go to **BioPortal** for discussion and relevant links for all INVESTIGATING**LIFE** figures.

[a]Nielsen, S. et al. 1995. *Proceedings of the National Academy of Sciences USA* 92: 1013–1017.

When ADH levels are low, such as when a person is well hydrated, most of the AQP-2 protein is sequestered in the membranes of intracellular vesicles, and the collecting duct permeability is low. A rise in ADH levels stimulates the insertion of these vesicles along with their AQP-2 channels into the apical plasma membranes. As a result, the membranes become more permeable to water. Water that enters the collecting duct cells passes into the interstitial fluid through the aquaporins in the basolateral membranes. ADH also stimulates the synthesis of new AQP-2 proteins. Thus circulating ADH controls the number of AQP-2 water channels in the plasma membranes of the collecting duct cells, and therefore the permeability of the collecting duct to water.

The heart produces a hormone that helps lower blood pressure

You may not think of the heart as an endocrine organ, but it is. When blood volume is high, blood pressure is high, putting strain on the heart. Under these conditions, the increased venous return stretches the atria of the heart. When the atrial muscle fibers are overly stretched, they release a peptide hormone called **atrial natriuretic peptide** (**ANP**). This peptide hormone enters the circulation, and in the kidney it decreases the reabsorption of sodium. If less sodium is reabsorbed, less water is reabsorbed, and more passes into the urine. Thus ANP has the effect of lowering blood volume and therefore blood pressure.

RECAP 52.6

Glomerular filtration is essential for kidney function and is sustained by autoregulatory mechanisms. Sensors that monitor blood pressure and blood osmolarity may stimulate or inhibit the release of hormones that regulate kidney function.

- Explain how falling GFR results in an increase in circulating angiotensin and how angiotensin restores GFR. **See pp. 1087–1088 and Figure 52.14**
- Explain how falling blood pressure or increasing blood osmolarity results in changes in permeability of the collecting ducts. **See Figure 52.15**

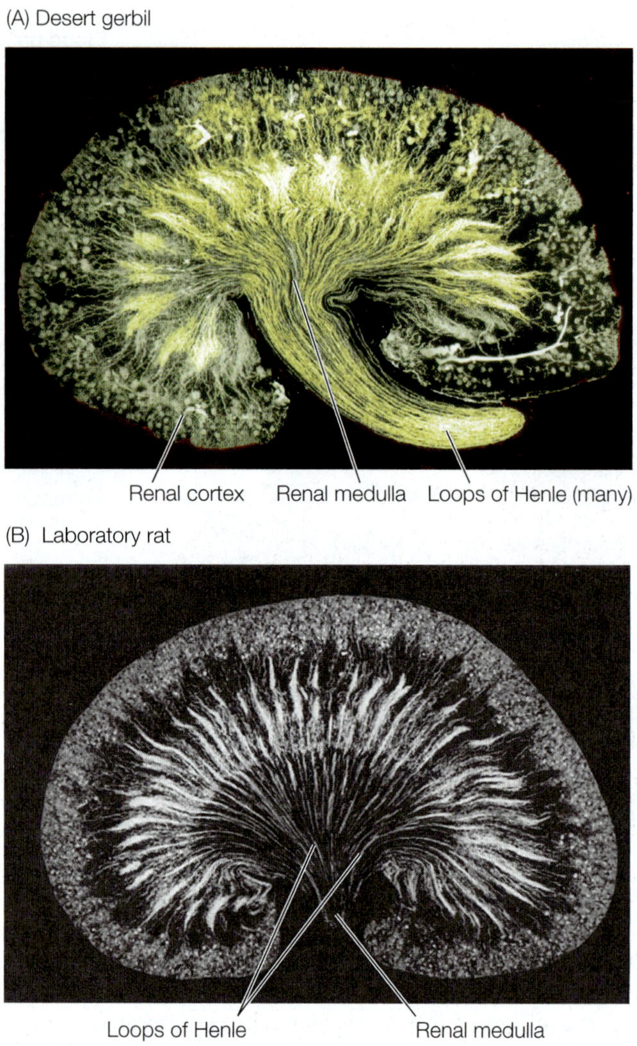
(A) Desert gerbil
Renal cortex
Renal medulla
Loops of Henle (many)
(B) Laboratory rat
Loops of Henle
Renal medulla

52.17 The Ability to Concentrate The ability of the mammalian kidney to concentrate urine depends on the lengths of its loops of Henle relative to the overall size of the kidney. (A) The kidney of a desert gerbil has a single renal pyramid with loops of Henle so long that the pyramid extends far into the ureter (ureter not shown). (B) An ordinary laboratory rat has much shorter loops of Henle.

How do desert rodents and vampire bats make highly concentrated urine?

ANSWER

As we described in the opening story, some desert rodents and vampire bats conserve water by producing urine that can be as much as 15 times more concentrated than their blood. In fact, the urine is so concentrated that it contains crystals of solute. For comparison, the maximum urine concentration of humans is about four times that of the blood.

Recall from Section 52.5 that the mammalian body's ability to produce concentrated urine depends on the concentration gradient set up in the renal medulla by the loops of Henle. An important adaptation for increasing the concentration gradient is to increase the lengths of the loops of Henle relative to overall kidney size. Some tiny desert gerbils, for example, have such extremely long loops of Henle that the renal pyramid (each of the rodent kidneys has only one, in contrast to humans) extends far out of the concave surface of the kidney and into the ureter (**Figure 52.17**). The large concentration gradient that results draws most of the water out of the urine as it passes down the collecting duct. Desert gerbils are so effective in conserving water that they can survive on the water released by the metabolism of their food. The vampire bat kidney is particularly good at concentrating urea. Within minutes after feeding, the urea concentration in its urine can increase by more than 500-fold.

52.1 How Do Excretory Systems Maintain Homeostasis?

- **Excretory systems** maintain the osmolarity and volume of the extracellular fluids and eliminate the waste products of nitrogen metabolism through the processes of filtration, reabsorption, and secretion. **Urine** is the output of excretory systems.
- There is no active transport of water, so water must be moved across membranes by a difference in either **osmolarity** or pressure.
- Water enters and leaves cells by osmosis. To achieve cellular water balance, animals must maintain the osmolarity of their extracellular fluids within an acceptable range.
- Marine animals can be **osmoconformers** or **osmoregulators**. Freshwater animals must be osmoregulators and must continually excrete water and conserve salts. Terrestrial animals are osmoregulators, but the nature of their regulation depends on environment and lifestyle.
- Apart from regulating osmolarity of cells and extracellular fluids, animals must also regulate their ionic composition by conserving some ions and secreting others. Salt glands are adaptations for secretion of NaCl. **Review Figure 52.2**

52.2 How Do Animals Excrete Nitrogen?

- Aquatic animals that breathe water can eliminate nitrogenous wastes such as **ammonia** by diffusion across their gill membranes. Terrestrial animals and some aquatic animals must detoxify ammonia by converting it to **urea** or **uric acid** before excretion. **Review Figure 52.3**
- Depending on the form in which they excrete their nitrogenous wastes, animals are classified as **ammonotelic**, **ureotelic**, or **uricotelic**.

52.3 How Do Invertebrate Excretory Systems Work?

- The **protonephridia** of flatworms consist of flame cells and excretory tubules. Extracellular fluid is filtered into the tubules, which process the filtrate to produce a dilute urine. **Review Figure 52.4**
- In annelid worms, blood pressure causes filtration of the blood across capillary walls. The filtrate enters the coelomic cavity, where it is taken up by **metanephridia**, which alter the composition of the filtrate by active transport mechanisms. **Review Figure 52.5, ACTIVITY 52.1**
- The **Malpighian tubules** of insects receive ions and nitrogenous wastes by active transport across the tubule cells. Water follows by osmosis. Ions and water are reabsorbed from the rectum, so the insect excretes semisolid wastes. **Review Figure 52.6**

52.4 How Do Vertebrates Maintain Salt and Water Balance?

- Marine bony fishes produce little urine. Chondrichthyans retain urea and TMAO, so the osmolarity of their body fluids remains close to that of seawater.
- Reptiles and birds have skin with low water permeability and excrete nitrogenous wastes as uric acid in a semisolid form.
- Mammals produce urine that is more concentrated than their extracellular fluids.
- The **nephron**, the functional unit of the vertebrate **kidney**, consists of a **glomerulus**, in which blood is filtered, a **renal tubule**, which use processes of active secretion and reabsorption to convert the glomerular filtrate into urine to be excreted, and a system of **peritubular capillaries**, which surround the tubule and support its functions of secretion and reabsorption. **Review Figure 52.7, ACTIVITY 52.2**

52.5 How Does the Mammalian Kidney Produce Concentrated Urine?

- The concentrating ability of the mammalian kidney is a function of its anatomy, which enables countercurrent exchange. **Review Figure 52.9**
- The glomeruli and the **proximal** and **distal convoluted tubules** are located in the **cortex** of the kidney. Certain molecules are actively reabsorbed from the glomerular filtrate by the tubule cells, and other molecules are actively secreted. Straight sections of renal tubules called **loops of Henle** and **collecting ducts** are arranged in parallel in the **medulla** of the kidney. **Review ACTIVITY 52.3**
- Salts, water, and valuable molecules such as glucose and amino acids are reabsorbed in the proximal convoluted tubule without the renal filtrate becoming more concentrated, although its composition changes.
- The loops of Henle create a concentration gradient in the interstitial fluid of the renal medulla by a **countercurrent multiplier** mechanism. Urine flowing down the collecting ducts to the **ureter** is concentrated by the osmotic reabsorption of water caused by the concentration gradient in the surrounding interstitial fluid. **Review Figure 52.10, ANIMATED TUTORIAL 52.1**
- Hydrogen ions secreted by the renal tubules are buffered in the urine by bicarbonate and other chemical buffering systems. **Review Figures 52.11, 52.12**

52.6 How Are Kidney Functions Regulated?

- Kidney function in mammals is controlled by autoregulatory mechanisms that maintain a constant high **glomerular filtration rate** (**GFR**) even if blood pressure varies.
- An important autoregulatory mechanism is the release of **renin** by the kidney when blood pressure falls. Renin activates **angiotensin**, which causes the constriction of efferent glomerular arterioles and peripheral blood vessels, causes the release of **aldosterone** (which enhances water reabsorption), and stimulates thirst. **Review Figure 52.14, ANIMATED TUTORIAL 52.2**
- Changes in blood pressure and osmolarity influence the release of antidiuretic hormone (ADH), which controls the permeability of the collecting duct to water and therefore the amount of water that is reabsorbed from the urine. ADH stimulates the expression of and controls the intracellular location of aquaporins, which serve as water channels in the membranes of collecting duct cells. **Review Figures 52.15, 52.16**
- When the volume of blood returning to the heart increases and stretches the atrial walls, **atrial natriuretic peptide** (**ANP**) is released, which causes increased excretion of salt and water.

See Activity 52.4 for a review of the major human organ systems.

Go to the Interactive Summary to review key figures, Animated Tutorials, and Activities Life10e.com/is52

CHAPTERREVIEW

REMEMBERING

1. Which statement is *true*?
   a. Most marine invertebrates are osmoregulators.
   b. All freshwater invertebrates are osmoconformers.
   c. Marine bony fish and chondrichthyans have similar extracellular osmolarities.
   d. Freshwater fishes are ionic regulators.
   e. Marine mammals gain water osmotically.

2. The excretion of nitrogenous wastes
   a. by humans can be in the form of urea and uric acid.
   b. by mammals is never in the form of uric acid.
   c. by marine fishes is mostly in the form of urea.
   d. does not contribute to the osmolarity of the urine.
   e. requires more water if the waste product is the rather insoluble uric acid.

3. What is the role of renal podocytes?
   a. They prevent red blood cells and large molecules from entering the renal tubules.
   b. They reabsorb most of the glucose that is filtered from the plasma.
   c. They control the glomerular filtration rate by changing the resistance of renal arterioles.
   d. They provide a large surface area for tubular secretion and reabsorption.
   e. They release renin when the glomerular filtration rate falls.

4. Which part of the nephron is responsible for most of the difference in mammals between the glomerular filtration rate and the urine production rate?
   a. The glomerulus
   b. The proximal convoluted tubule
   c. The loops of Henle
   d. The distal convoluted tubule
   e. The collecting duct

5. Which of the following would *not* be a response stimulated by a large drop in blood pressure?
   a. Constriction of afferent renal arterioles
   b. Increased release of renin
   c. Increased release of antidiuretic hormone
   d. Increased thirst
   e. Constriction of efferent renal arterioles

UNDERSTANDING & APPLYING

6. Persons with uncontrolled diabetes mellitus can have very high levels of glucose in their blood. Why do such individuals have a high level of urine production?

7. Patients with high blood pressure are sometimes treated with ACE inhibitors. Explain the rationale of this treatment.

8. Inulin is a molecule that is filtered out of the glomerulus but is not secreted or reabsorbed by the renal tubules. If you injected inulin into an animal and after a brief time measured the concentration of inulin in the animal's blood and urine, how could you determine the animal's glomerular filtration rate? Assume that the rate of urine production is 1 milliliter per minute.

ANALYZING & EVALUATING

9. After you did the inulin experiment to measure glomerular filtration rate, how could you use that information to determine whether another substance is secreted or reabsorbed by the renal tubules? Assume you can measure the concentration of that substance in the blood and urine. Urine production is still 1 milliliter per minute.

10. Acetazolamide is a drug used by mountain climbers for short periods of time to help them acclimate to high altitude. The drug is an inhibitor of carbonic anhydrase. Referring back to Chapter 49, how do you think this drug helps in the acclimation process? In the context of this chapter, explain why a side effect of this drug is frequent urination.

Go to BioPortal at **yourBioPortal.com** for Animated Tutorials, Activities, LearningCurve Quizzes, Flashcards, and many other study and review resources.

53 Animal Behavior

CHAPTER**OUTLINE**

Big Babies Chicks hatched from eggs laid by parasitic cowbirds (*Molothrus ater*) are frequently found in the nests of much smaller species such as the yellow warbler (*Dendroica petechia*). The cowbird chick grows rapidly, exceeding the size of the host parent. The cowbird chick's demand for food is so great that the host's own offspring may be ignored and die of neglect.

BROWN-HEADED COWBIRDS (*Molothrus ater*) are abundant and ubiquitous in North America, ranging over an enormous diversity of habitats. They are brood, or nest, parasites, laying their eggs in the nests of other species. A female cowbird may lay 40 eggs a season in as many different nests, but she never incubates her eggs or feeds her young. Cowbird eggs have been found in the nests of at least 220 other species. The host incubates the egg along with its own and feeds the young cowbird—often to the detriment of its own chicks because cowbird chicks are usually larger and more demanding.

Cowbirds have enormous reproductive success and it is not difficult to understand the evolution of brood parasitism from the standpoint of the parasite. But what about the selective pressures on the individuals that are parasitized? Cowbird eggs usually look very different from the host's own eggs, so why haven't host species evolved the ability to recognize an intruder egg and remove it from the nest? Indeed, the European common cuckoo (*Cuculus canorus*) is also a nest parasite, but its eggs closely mimic the eggs of its host species, presumably an adaptive response to an improved ability among host birds to discriminate and remove cuckoo eggs (an example of an evolutionary "arms race").

But in some cases eliminating the parasite's eggs can have negative consequences for the host. Researchers have reported observing what they hypothesize to be a kind of "mafia behavior" by cowbirds, in which the female cowbird periodically returns to inspect the nests she has parasitized. If she finds her egg missing, she destroys the nest along with the host's eggs. Female cowbirds have also been observed to destroy the unparasitized nest and eggs of a potential host. Researchers suggest this "farming behavior" may force the owner of the destroyed nest to rebuild and lay a new clutch of eggs—at which time the cowbird can lay her egg in the new nest.

Students of animal behavior seek to understand how behaviors such as nest parasitism evolve, how they develop, and what their underlying mechanisms are. That quest, as in this cowbird example, frequently requires research that includes the environment and interacting species. Behavioral biology is a highly integrative field that incorporates approaches from virtually all the biological subdisciplines. The study of animal behavior can give us insights into our own behavior.

?

Could cowbird behaviors create selective pressure for host species *not* to develop egg discrimination behavior?

See answer on p. 1117.

53.1 What Are the Origins of Behavioral Biology?

Humans have studied animal behavior since prehistoric times. Understanding the habits of potential prey, as well as those of their predators, was of great value to hunters. Appreciation of behavioral traits led to the domestication of animal species. Accounts of animal behaviors such as seasonal appearances and disappearances, mating displays, aggression, prey capture, parental care, and communication are found throughout recorded history. Yet the scientific study of animal behavior did not truly get under way until the early 1900s.

Conditioned reflexes are a simple behavioral mechanism

In the late 1800s, the Russian physiologist Ivan Pavlov was studying the neural control of digestive juice secretion when he observed that not only did the dog he was experimenting on salivate when it smelled food, the animal also salivated whenever the technician who routinely fed the dog entered the room—even when no food was present. Following up this observation, Pavlov substituted a sound stimulus for the technician; a metronome ticked while the dog was fed. After several trials, the dog salivated when it heard the metronome, even if no food was offered.

Salivation in response to the sight, smell, or taste of food is a natural reflex response to a stimulus, but salivation in response to a sound was a learned response. The pairing of a sound with the experience of receiving food conditioned the dog's nervous system to generate a response, which Pavlov dubbed the **conditioned reflex** (Figure 53.1). Pavlov received a Nobel prize in 1904 for his work showing that a simple behavior controlled by the nervous system could be modified through experience. This work stimulated much new research because Pavlov had developed an experimental model of learning.

Going beyond the conditioning of autonomic reflexes, the psychologist B. F. Skinner showed that any random action of an animal could become a conditioned response to a stimulus if a reward was temporally associated with the action and the stimulus. A rat, for example, could be conditioned to press a lever in response to a stimulus if it got a reward when it behaved as the experimenter desired. Because the animal was conditioned to perform an *operation* on its environment, this experimental protocol was known as **operant conditioning** and was viewed as another model of learning.

The experimental approaches to behavior initiated by Pavlov and Skinner had powerful effects on the nature of research on animal behavior, and the focus of scientists using their approaches was quite specific:

- They focused on laboratory environments rather than natural environments because in the laboratory, the variables in their experiments could be precisely controlled.
- They focused on only a few species (predominantly the albino rat) as model systems rather than studying diverse species from nature.

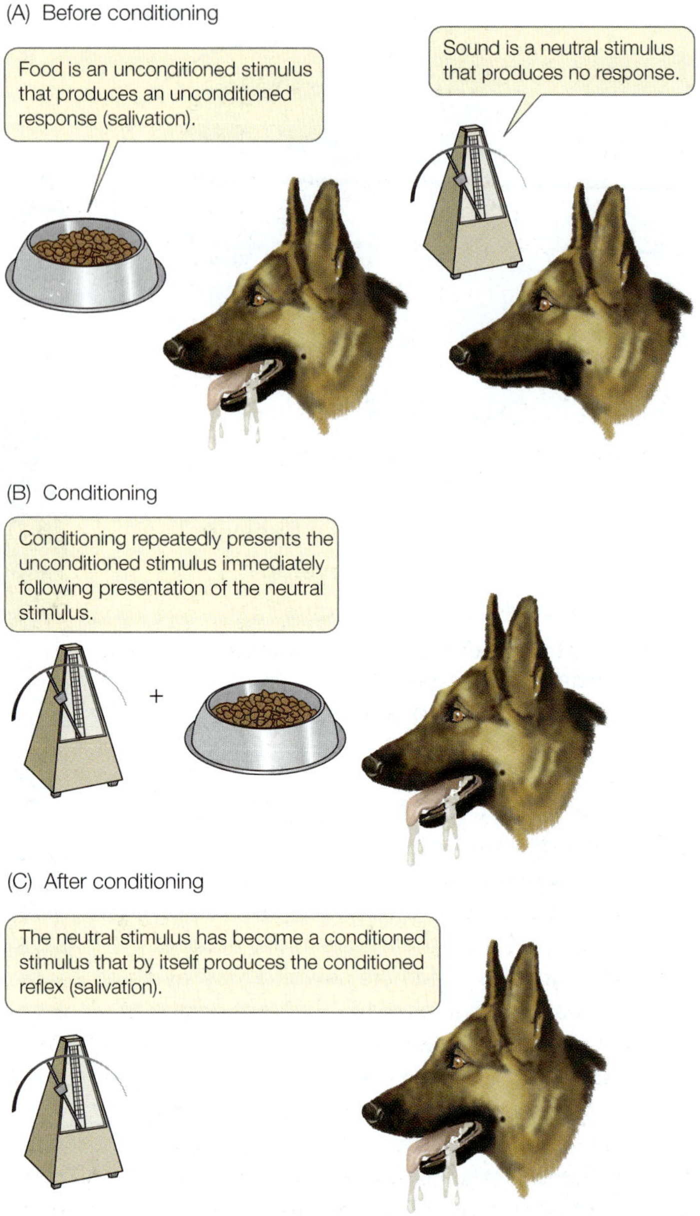
(A) Before conditioning
Food is an unconditioned stimulus that produces an unconditioned response (salivation).
Sound is a neutral stimulus that produces no response.
(B) Conditioning
Conditioning repeatedly presents the unconditioned stimulus immediately following presentation of the neutral stimulus.
+
(C) After conditioning
The neutral stimulus has become a conditioned stimulus that by itself produces the conditioned reflex (salivation).

53.1 The Conditioned Reflex Ivan Pavlov discovered that when a normal response is paired with an artificial or neutral stimulus, an animal learns to produce the response even when only the artificial stimulus is presented.

- They focused on questions of learning and memory, largely to the exclusion of other types of behavior (e.g., mating, feeding, communication).

Thus defined, the field of animal behavior research became known as behaviorism and was largely the domain of psychologists.

Ethologists focused on the behavior of animals in their natural environment

An alternative approach to the study of animal behavior arose at the same time as behaviorism, but largely in Europe. Scientists there focused on describing the characteristics of animals in their natural environment, an approach that became known as **ethology** (Greek *ethos*, "character"; *logos*, "study"). In contrast to the behaviorists, the ethologists were interested in a wide variety of species, their evolutionary relationships, and

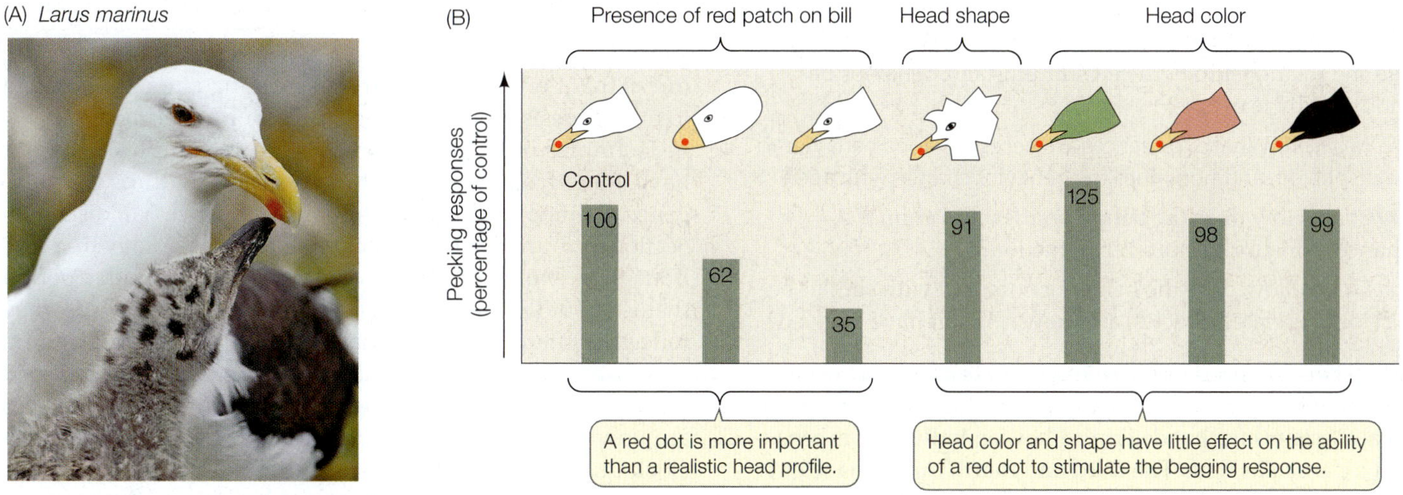
(A) *Larus marinus*
(B)
Presence of red patch on bill
Head shape
Head color
Pecking responses (percentage of control)
Control
100
62
35
91
125
98
99
A red dot is more important than a realistic head profile.
Head color and shape have little effect on the ability of a red dot to stimulate the begging response.

53.2 Releasing a Fixed Action Pattern (A) In many gull species, chicks instinctively peck at the red dot on the parent's lower bill, a behavior that induces the parent to regurgitate food into the chick's mouth. (B) Tinbergen's work showed that the red dot on the parent's lower bill is the critical component that releases the pecking response.

the ways in which their behaviors were adapted to their environments. The leaders of the ethology movement were Karl von Frisch, who discovered the dance language of honey bees; Konrad Lorenz, who discovered that the strong bond between parent and offspring develops during a "critical period" following birth; and Niko Tinbergen, who studied inborn patterns of behavior commonly known as instincts. These three scientists shared the Nobel Prize in 1973 for "their discoveries concerning organization and elicitation of individual and social behavior patterns."

The ethologists were mainly interested in species-specific or instinctive behaviors that therefore had to have genetic components. Behaviors were thought to be genetically determined if they:

- are performed without learning
- are stereotypic (that is, they are performed the same way each time)
- cannot be modified by learning

The ethologists called such behaviors **fixed action patterns**.

To demonstrate that a behavior was genetically determined, ethologists performed **deprivation experiments** in which an animal was raised in an environment devoid of opportunities to learn its species-specific behavior. An example of a natural deprivation experiment is the web spinning of a spider. The parents of a young spider die before it hatches, and in a seasonal environment it has no model webs to copy when it spins its first web, which requires thousands of stereotyped sequential movements. Yet a young spider creates a perfect, species-specific web the first and every time it spins a web. Thus the information for the web spinning behavior has to be genetically programmed into its nervous system.

Fixed action patterns are usually responses to specific stimuli. The ethologists carefully characterized such stimuli, which they called **releasers**. In general, releasers are simple subsets of the information available in the environment. For example, Tinbergen studied the begging behavior of gull chicks. Adult gulls have a red dot on their lower bill. When a parent returns to the nest to feed its chicks, the chicks peck on the red dot, which stimulates the parent to regurgitate food (**Figure 53.2A**). Experimenters investigated what stimulated the chicks to peck their parents' bills. Models of gull heads of different shapes and colors were tested (**Figure 53.2B**), as were models of a beak without a head. The results showed that the red dot was necessary for the release of chick pecking behavior. In fact, a pencil with a red eraser elicited a more robust pecking response than an accurate model of a gull head without a red dot.

Ethologists probed the causes of behavior

The ethologists demonstrated the genetic basis for fixed action patterns by interbreeding closely related species. Konrad Lorenz studied the courtship behaviors of different species of dabbling ducks. Some of these species, such as mallards, teals, pintails, and gadwalls, are closely related and can interbreed, but they rarely do so in nature. Each male duck performs a courtship display consisting of a precise series of movements that is typical of his species. A female is not likely to accept him unless the entire display is successfully and correctly completed.

When Lorenz crossbred these duck species, the hybrid offspring expressed some elements of each parent's courtship display, but in novel combinations. Furthermore, Lorenz observed that hybrids sometimes exhibited display elements that were not in the repertoire of either parent species but were seen in other dabbling duck species. Lorenz's interbreeding studies demonstrated that the stereotypic motor patterns of the courtship displays are inherited. The observation that females were not interested in males performing hybrid displays was evidence that sexual selection had shaped these genetically determined behaviors (see Section 21.2).

The ethologists recognized the importance of development and motivation in behavior, and they laid the foundation for the application of modern biological methods to the study of animal behavior. Tinbergen outlined the challenges for investigators as four questions:

- *Causation:* What is the stimulus for the behavior, and how can the relationship between stimulus and behavior be modified by learning?
- *Development:* What experiences are necessary for a behavior to be displayed, and how does the behavior change with age?
- *Function:* How does the behavior affect the animal's chances for survival and reproduction?
- *Evolution:* How does the behavior compare with similar behaviors in related species, and how might it have evolved?

The first two questions refer to the **proximate causes** of behavior: the immediate genetic, physiological, neurological, and developmental mechanisms that determine how an individual is behaving at a particular time. The third and fourth questions refer to the **ultimate causes** of behavior: the evolutionary processes that produced the animal's capacity and tendency to behave in particular ways. In the sections that follow, we will describe many experiments on animal behavior. For each one, ask yourself which of Tinbergen's four questions it addresses and whether it focuses on proximate or ultimate causes of behavior.

RECAP 53.1

Early scientific studies of animal behavior took two approaches. Behaviorists focused on the study of conditioned behavior in a few species of laboratory animals and asked questions about learning. Ethologists studied genetically determined behavior in many species in their natural environments and asked evolutionary questions.

- Describe the difference between conditioned reflexes and operant conditioning. See p. 1094 and Figure 53.1
- What is the relationship between a releaser and a fixed action pattern? See p. 1095 and Figure 53.2
- Explain the difference between proximate and ultimate causes of behavior and how Tinbergen's questions address these two types of causes. See p. 1096

The work of the ethologists left no doubt that behavior can be genetically determined, but how? Genes code for proteins, whereas behaviors are highly complex traits involving sensory input and intricate patterns of control over responses to that input. Is it reasonable to think that a single gene can have a specific effect on a behavior?

53.2 How Do Genes Influence Behavior?

Most behaviors are complex traits that depend on many genes. Nevertheless, evidence from multiple approaches shows that alterations in single genes can result in discrete behavioral phenotypes on which natural selection can operate.

Breeding experiments can produce behavioral phenotypes

Behavioral geneticists identify individuals with unusual behavioral phenotypes and conduct breeding experiments to see if those traits are inherited and to identify the genes involved. The origin of the unusual behavioral phenotypes may come from diversity in natural populations such as the dabbling ducks that Lorenz crossed; from screening animals subjected to mutagenesis; or from artificial selection experiments.

Examples of single genes involved in complex behavior have come from studies of the circadian rhythms of mutagenized animals. Circadian rhythms, which we will discuss in Section 53.5, are rhythms of behavior and physiology that continue to be expressed under constant conditions with a period that is about (but not exactly) 24 hours—in other words, they reflect an internal clock mechanism. The *per* gene (for period) was discovered in fruit flies, and variants of this gene caused the circadian period to be long, short, or eliminated. *Per* genes are now known to be an important element of the circadian clock in a wide variety of species, including humans. Other genes involved in circadian behavior have been discovered in subsequent mutagenesis and screening experiments.

Very few behaviors can be linked to a single gene, however. Most behaviors are complex traits and are influenced by many genes. A technique called quantitative trait analysis (or QTL, for "quantitative trait loci," referring to locations on chromosomes) has been developed to identify multiple genes that influence a given trait. QTL analysis requires animals, such as mice, with completely sequenced and well-mapped genomes that have many identifiable genetic markers (i.e., unique DNA sequences). When individuals from two strains of a species that differ with respect to the trait of interest are crossed, the trait can be quantified in the offspring and the offspring can be genotyped to show which of the genetic markers inherited from each parent correlate with the trait of interest in the offspring. Those markers are assumed to be close to the genes that influence the trait. QTL analysis thus points to regions on chromosomes where there are genes that influence the trait of interest.

Knockout experiments can reveal the roles of specific genes

Some modern molecular genetic approaches start with identified genes and eliminate or silence them to see what effects their elimination has on a behavioral phenotype (see Section 18.4). As you might expect, knocking out genes involved in sensory pathways can have pronounced effects on behavior. One example is a gene for a specific olfactory receptor in mice.

As we saw in Section 46.2, mice have two olfactory organs: the nasal olfactory epithelium common to all mammals, and a small organ in the nasal passages called the vomeronasal organ, or VNO (**Figure 53.3**). Catherine Dulac at Harvard University discovered that a large number of pheromone receptors were expressed in that organ. (As described in Section 46.2, pheromones are signaling molecules released into the environment.) Dulac hypothesized that when the receptors in the male's VNO bound to sex pheromones produced by female mice, they stimulated mating behavior. To test this hypothesis, Dulac created a genetically engineered male mouse in which a gene for VNO receptor signaling was knocked out. Contrary to the prediction of the hypothesis, the knockout males in fact did pursue and mate with females placed in their cages. However, they also pursued and tried to mate with *males* placed in their cages. Normally a male mouse reacts aggressively to

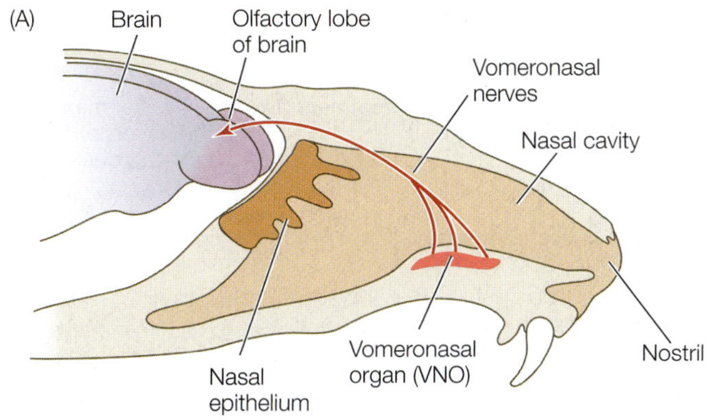
(A)
Brain
Olfactory lobe of brain
Vomeronasal nerves
Nasal cavity
Nostril
Vomeronasal organ (VNO)
Nasal epithelium

(B)

53.3 The Mouse Vomeronasal Organ Identifies Gender (A) The mouse VNO is located adjacent to the nasal passages. It contains pheromone receptors whose input travels to a specific region of the olfactory bulb (the accessory olfactory bulb). (B) In male mice, information from the VNO appears to be crucial in identifying gender and thus a potential sexual partner.

a strange male, but the knockout male could not discriminate between males and females placed in his cage. Thus properly functioning VNO receptors appear to be essential not for sexual attraction, but for gender identification. It is possible to imagine how selection working on this one gene could modify the intensity of male–male aggression and lead to changes in social behavior.

Behaviors are controlled by gene cascades

Male courtship behavior in *Drosophila melanogaster*, the laboratory fruit fly, is stereotypic, species-specific, and requires no learning—a classic fixed action pattern. When a male encounters a potential mate, he follows her, taps her body with his foreleg, extends and vibrates one wing, and licks her genitals (**Figure 53.4A**). The development of this complex male behavior is under the control of a single gene, called *fruitless* (*fru*), just as the development of male anatomy is under the control of another gene, called *doublesex* (*dsx*). In both males and females, these two genes are part of a gene

53.4 The *fruitless* Gene (A) Male fruit flies display stereotypic, species-specific courtship behavior. (B) Sexual differentiation in *Drosophila* is controlled by a cascade of genes, including the *fru* gene, whose expression results in male sexual behavior.

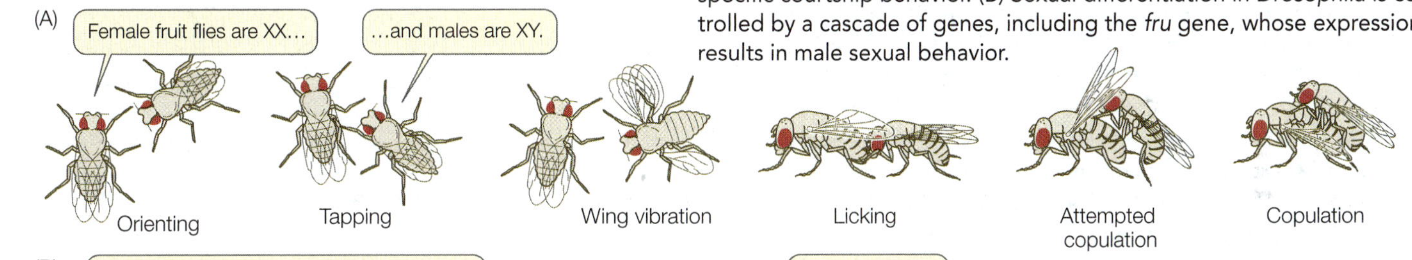
(A)
Female fruit flies are XX…
…and males are XY.
Orienting
Tapping
Wing vibration
Licking
Attempted copulation
Copulation

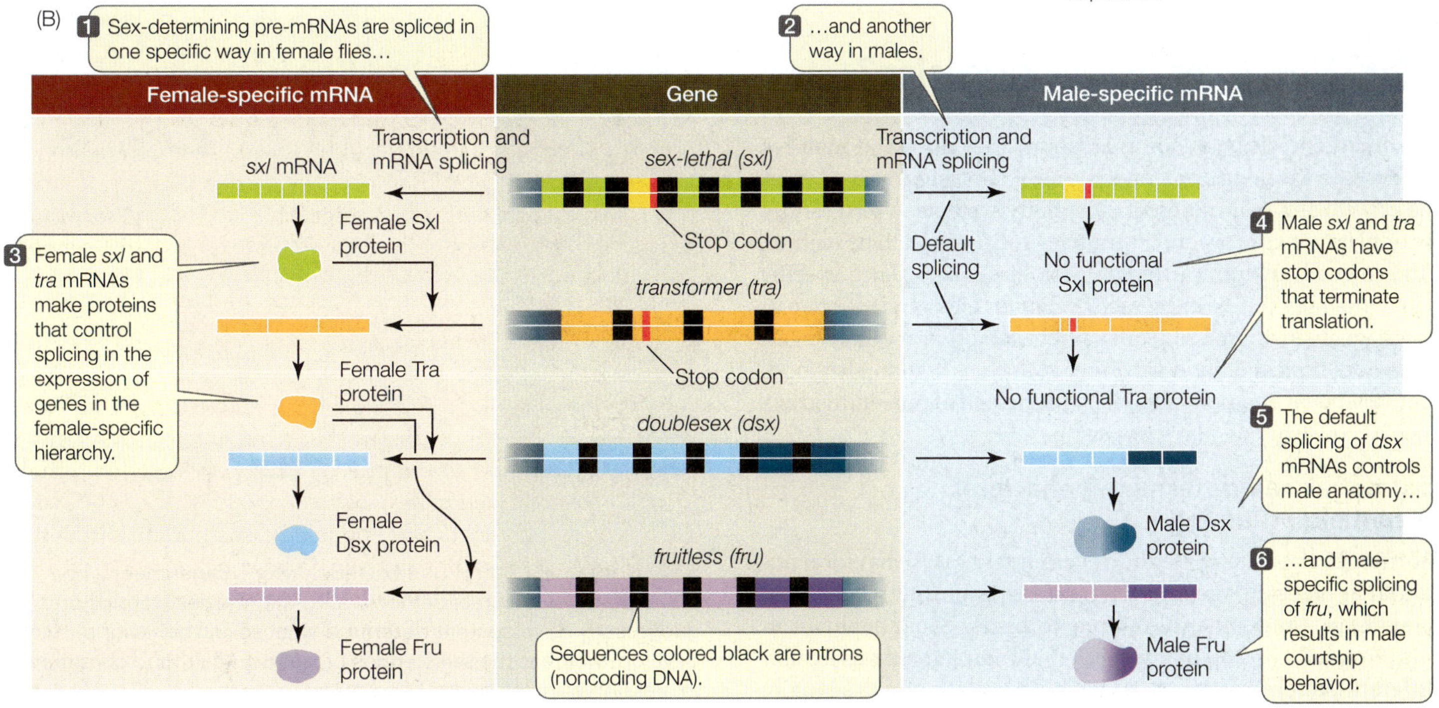
(B)
1 Sex-determining pre-mRNAs are spliced in one specific way in female flies…
2 …and another way in males.
Female-specific mRNA
Gene
Male-specific mRNA
Transcription and mRNA splicing
sxl mRNA
sex-lethal (sxl)
Transcription and mRNA splicing
Stop codon
Female Sxl protein
Default splicing
No functional Sxl protein
3 Female sxl and tra mRNAs make proteins that control splicing in the expression of genes in the female-specific hierarchy.
transformer (tra)
4 Male sxl and tra mRNAs have stop codons that terminate translation.
Stop codon
Female Tra protein
No functional Tra protein
doublesex (dsx)
5 The default splicing of dsx mRNAs controls male anatomy…
Female Dsx protein
Male Dsx protein
fruitless (fru)
6 …and male-specific splicing of fru, which results in male courtship behavior.
Female Fru protein
Sequences colored black are introns (noncoding DNA).
Male Fru protein

expression cascade that results in different *dsx* and *fru* gene products in males and females (**Figure 53.4B**). The female version of the Dsx protein controls the development of female anatomy, and the expression of *fru* in the male nervous system results in the organization of the neural circuitry controlling male sexual behavior.

There are two take-home lessons from this example. First, genes that control aspects of behavior, like other genes, are generally embedded in gene cascades that offer multiple opportunities for simple genetic changes that will alter the phenotype of even complex behaviors. Second, certain genes, such as *dsx* and *fru*, influence a range of other genes that contribute to complex behaviors. Modifications in any one of those genes or its expression can alter behavior. Thus even though no behavior is coded for by a single gene, alterations in single genes can influence behavior in ways that affect an animal's fitness.

RECAP 53.2

Breeding experiments show that behavioral phenotypes can be inherited and modified by natural selection. Although most behaviors are controlled by complex cascades of genes, molecular genetic methods have shown that a single, identifiable gene in the cascade can influence a complex behavior.

- How can knockout experiments reveal genes controlling behavior? **See p. 1096 and Figure 53.3**
- What is the evidence that the vomeronasal organ in mice is responsible for gender identification? **See pp. 1096–1097**
- Describe some of the ways a gene can control expression of a complex behavior. **See pp. 1097–1098 and Figure 53.4**

How can the genetic cascades that underlie complex behaviors be programmed to respond selectively to specific sets of stimuli? How can their expression be limited to appropriate times in an animal's life? The answers to these questions can by found by studying how behaviors develop over the life span.

53.3 How Does Behavior Develop?

The emergence of behavior as an animal develops and matures depends on the development of the nervous system as well as on the growth and maturation of other body systems. A bird cannot fly until its wings grow and its muscles and flight feathers mature. But even with anatomical and physiological competence, specific behaviors may not be expressed. Behaviors that are adaptive at one stage in an animal's life may not be adaptive at other stages. Behaviors typical of juvenile animals, such as begging for food, may disappear and new behavior patterns of a mature individual, such as courtship displays, appear.

Hormones can determine behavioral potential and timing

Hormones can determine the development of a behavioral potential at an early age and the expression of that behavior at a later age. An excellent example of this is sexual behavior in rats (**Figure 53.5**). Normally, adult male and female rats exhibit different patterns of sexual behavior: females adopt a sexually

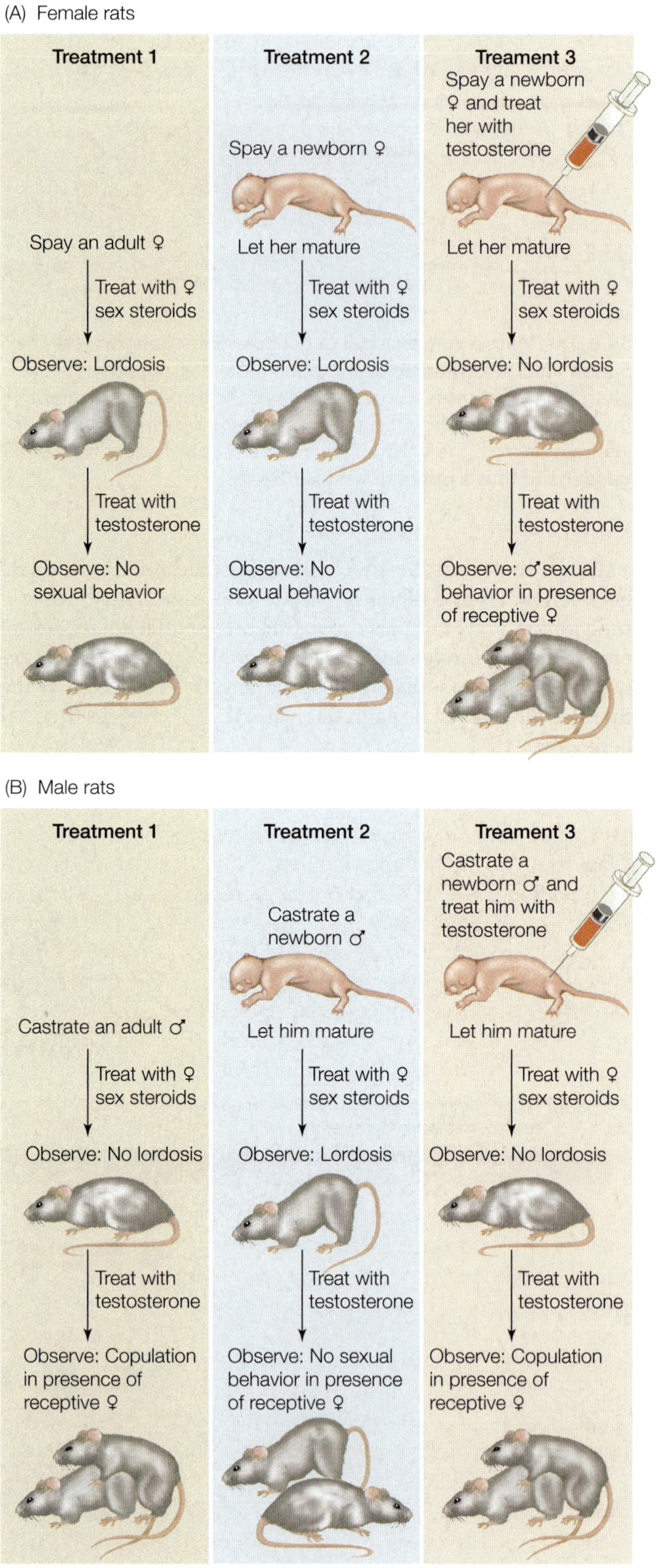
(A) Female rats
Treatment 1
Spay an adult ♀
Treat with ♀ sex steroids
Observe: Lordosis
Treat with testosterone
Observe: No sexual behavior
Treatment 2
Spay a newborn ♀
Let her mature
Treat with ♀ sex steroids
Observe: Lordosis
Treat with testosterone
Observe: No sexual behavior
Treament 3
Spay a newborn ♀ and treat her with testosterone
Let her mature
Treat with ♀ sex steroids
Observe: No lordosis
Treat with testosterone
Observe: ♂ sexual behavior in presence of receptive ♀
(B) Male rats
Treatment 1
Castrate an adult ♂
Treat with ♀ sex steroids
Observe: No lordosis
Treat with testosterone
Observe: Copulation in presence of receptive ♀
Treatment 2
Castrate a newborn ♂
Let him mature
Treat with ♀ sex steroids
Observe: Lordosis
Treat with testosterone
Observe: No sexual behavior in presence of receptive ♀
Treament 3
Castrate a newborn ♂ and treat him with testosterone
Let him mature
Treat with ♀ sex steroids
Observe: No lordosis
Treat with testosterone
Observe: Copulation in presence of receptive ♀

53.5 Hormonal Control of Sexual Behavior Experimental hormone treatments of rats demonstrated that the sex steroids present during early development determine what sexual behavior patterns develop, whereas the sex steroids present in adulthood control the expression of those behavior patterns.

receptive posture, called lordosis, in the presence of males, and males copulate with receptive females. Neither sex, however, expresses these behaviors until the animals have reached adulthood. Experiments in which newborn and adult rats were neutered (to remove the influence of sex steroids naturally produced by their gonads) and artificially treated with hormones led to the following conclusions:

- Development of male sexual behavior requires that the brain of the newborn rat be exposed to testosterone, but development of female sexual behavior does not require exposure to estrogen.
- Testosterone masculinizes the nervous systems of both genetic males and genetic females.
- Exposure to sex steroids in adulthood is necessary for the expression of sexual behavior, but testosterone produces male sexual behavior only in adult rats whose brains were masculinized when they were newborns, and estrogen produces female sexual behavior only in adult rats whose brains were not masculinized when they were newborns.

Thus the sex steroids that are present at birth determine which pattern of behavior develops, and the sex steroids that are present in adulthood determine when that pattern is expressed.

Some behaviors can be acquired only at certain times

Responsiveness to simple releasers is sufficient for certain behaviors such as begging behavior in gull chicks, but more complex information that cannot be genetically programmed is required for other behaviors. An example is parent–offspring recognition. When animals live in close proximity to other individuals, as in a herd or a nesting colony, it is important for parent and offspring to learn each other's identity soon after birth so they will be able to find each other in a crowded situation. In many such cases, a parent–offspring bond is formed by **imprinting**. What characterizes imprinting is that an animal learns a specific set of stimuli during a limited time called a **critical period**, also known as a **sensitive period**.

Konrad Lorenz demonstrated that young graylag geese (*Anser anser*) imprint on their parents between 12 and 16 hours after hatching. By positioning himself to be present during this critical period, Lorenz succeeded in imprinting goslings on himself. The imprinted goslings followed him around as if he were their parent (Figure 53.6A). In a subsequent experiment he had his assistants wear boots with different patterns on them. The goslings imprinted on the boots, and even in a situation that mixed different groups of goslings, they always sorted themselves out by following their "parental" boots.

Imprinting requires only a brief exposure, but its effects are strong and long-lasting. Emperor penguins (*Aptenodytes forsteri*) reproduce during the coldest, darkest time of year in Antarctica. The parents walk up to 150 kilometers inland to form a dense colony, where the female lays her egg. She then walks back to the ocean to feed while her mate incubates the egg. By the time she returns, the chick has hatched. She then takes over its care and feeding, and the father walks back to the ocean to feed. Generally he is away so long that the mother must leave to find food and to avoid starvation. Thus after being away for weeks, the father must find his chick in a crowded, milling colony of chicks, all calling for their parents (Figure 53.6B). Yet he can unerringly locate his own offspring by recognizing its call, which he imprinted on before he left to feed.

(A) *Anser anser*

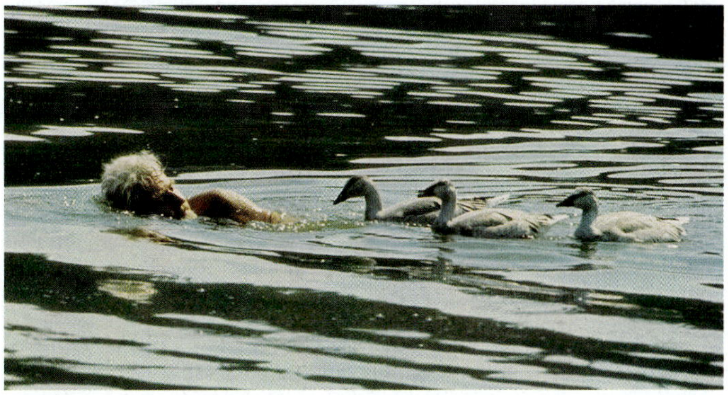

(B) *Aptenodytes forsteri*

53.6 Imprinting Helps Parents and Offspring Recognize Each Other (A) Greylag geese that imprinted on Konrad Lorenz as hatchlings followed him everywhere he went. (B) Imprinting allows a male emperor penguin to find his own chick among many others.

The critical or sensitive period for imprinting may be determined by a brief developmental or hormonal state. For example, if a mother goat does not nuzzle and lick her newborn within 10 minutes after its birth, she will not recognize it as her own offspring later. For goats, the sensitive period is associated with peaking levels of the hormone oxytocin in the mother's circulatory system at the time she gives birth and is sensing the olfactory cues emanating from her newborn kid. A female goat rendered incapable of smelling before giving birth is unable to differentiate between her own kid and other kids after giving birth.

Birdsong learning involves genetics, imprinting, and hormonal timing

Male songbirds use species-specific song to claim and advertise a breeding territory, compete with other males, and declare dominance. They also use song to attract females, which recognize the song of their species even though they do not

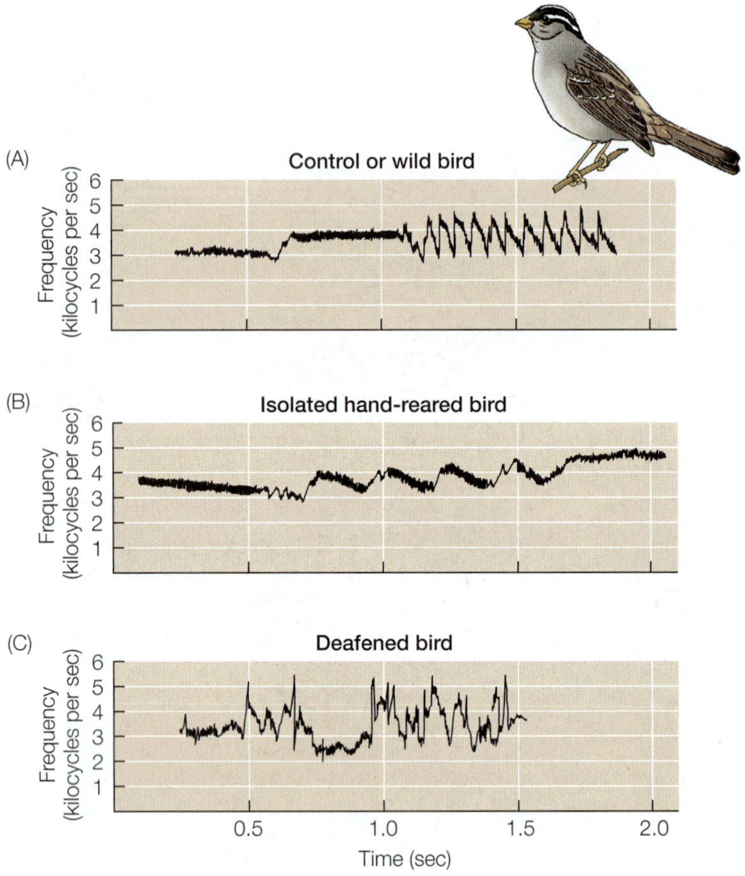
(A)
Control or wild bird
(B)
Isolated hand-reared bird
(C)
Deafened bird
Frequency (kilocycles per sec)
6
5
4
3
2
1
0.5
1.0
1.5
2.0
Time (sec)

53.7 Sensitive Periods for Song Learning (A) Sonogram showing the species-specific song of an adult male white-crowned sparrow (*Zonotrichia leucophrys*). (B) Song of an adult male raised in isolation (never having heard the song as a nestling). (C) Song of an adult male who heard the song as a nestling but was deafened prior to ever singing himself. Marler's experiments showed that the bird must first acquire a song memory by hearing the song as a nestling, and must then be able to hear himself as he attempts to match his singing to that song memory.

sing it. For males of many species, such as the white-crowned sparrow (*Zonotrichia leucophrys*), learning is an essential step in the acquisition of song, but *what* they can learn seems to be influenced by genes, and there is a limited developmental time frame for learning. A hatchling in nature hears his father and other white-crowned sparrows singing. He also hears the songs of many other bird species. But he does not sing until he approaches sexual maturity almost a year later, and when he does, he sings his father's type of song.

Studies of song learning in this species were initiated in the 1960s by Peter Marler at the University of California at Berkeley. Marler incubated eggs of white-crowned sparrows and hand-reared the hatchlings in the laboratory. He was able to expose them to different recorded songs at different times in their development, and discovered that adult male sparrows cannot produce their species-specific song unless they heard it as nestlings in the first 2 months of their lives (**Figure 53.7**). If nestlings hear recordings of white-crowned sparrow song during those first 2 months, they begin to sing (poorly) as they approach sexual maturity. Through trial and error, the maturing birds match their singing to their stored song memory, and from then on they sing their species-specific song. To reach this point, the young bird must be able to hear himself sing. If he is deafened before he begins to sing, he will not be able to match his stored song memory correctly. If he is deafened *after* he has sung his correct species-specific song, however, he will continue to sing normally. We say that at this point the behavior pattern is "crystallized." Thus there are two sensitive periods for song learning, the first in the nestling stage when song memory is imprinted, and the second as the bird approaches sexual maturity, when he learns to match that song memory.

In nature, nestling male white-crowned sparrows hear the songs of many species, so why do they learn only the song of their own species? Marler investigated this question in his isolation experiments by playing recordings of other songs to the hatchlings. The young male sparrows did not learn the songs of other species, even if they heard them many times, but hearing songs of their own species just a few times was sufficient for imprinting. Thus although male sparrows must learn their song, they seem to have a genetic predisposition to learn their own song and not the songs of other species. Marler called this phenomenon "an instinct to learn." There is an important reason why male white-crowned sparrows learn only their species-specific song: female white-crowned sparrows also listen to their fathers' songs while they are nestlings, and when they mature, they choose mates that sing like their fathers did.

More recent investigations have revealed additional complexity in white-crowned sparrow song learning capacity. First, it was demonstrated that each population of white-crowns has its own dialect, and that males from one population can learn the dialect of another population. Second, it was observed that in nature, white-crowns are occasionally heard singing the songs of other species. Could Marler's laboratory experiments have missed a critical natural variable?

Luis Baptista, a curator of birds at the California Academy of Sciences, took up the study of white-crowned sparrows to explore the effects of social interactions on song learning. He discovered that when a hand-reared white-crowned sparrow nestling was exposed to the sight and sound of a related species in an adjacent cage while recordings of his own species' song played in the background, the white-crowned sparrow sang the song of the other species when he matured. Thus social experience has a powerful effect on what the young bird can and will learn.

To ask the question of what the adaptive significance of plasticity in the male singing behavior is, it is necessary to take female choice into consideration. Sarah Woolley and Allison Doupe at the University of California, San Francisco have done that with zebra finches (*Taeniopygia guttata*), a species that has become a valuable model system for studying the neurobiology of birdsong. When male zebra finches sing in isolation, their song is variable ("undirected"), almost as if they are improvising. When they sing in the presence of females, however, their song is very stereotyped ("directed"). Experiments by Woolley and Doupe

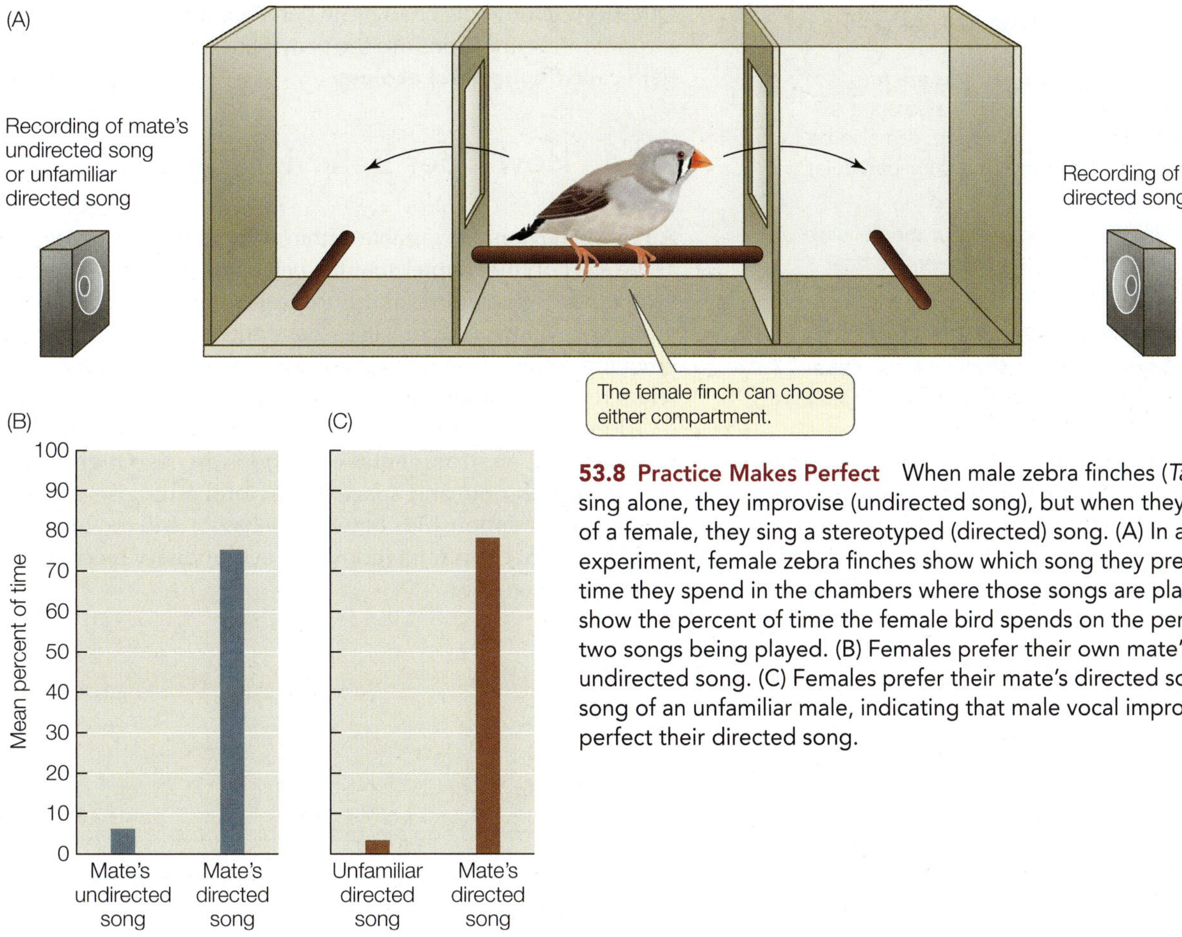
(A)
Recording of mate's undirected song or unfamiliar directed song
Recording of mate's directed song
The female finch can choose either compartment.
(B)
(C)
Mean percent of time
100
90
80
70
60
50
40
30
20
10
0
Mate's undirected song
Mate's directed song
Unfamiliar directed song
Mate's directed song

53.8 Practice Makes Perfect When male zebra finches (*Taeniopygia guttata*) sing alone, they improvise (undirected song), but when they sing in the presence of a female, they sing a stereotyped (directed) song. (A) In a behavioral choice experiment, female zebra finches show which song they prefer by the amount of time they spend in the chambers where those songs are played. (B, C) The graphs show the percent of time the female bird spends on the perch nearest each of the two songs being played. (B) Females prefer their own mate's directed song to his undirected song. (C) Females prefer their mate's directed song to the directed song of an unfamiliar male, indicating that male vocal improvisations may serve to perfect their directed song.

showed that female zebra finches preferred directed song and that the directed song resulted in greater activation of auditory areas in their brains. The strongest effects were seen in mated females hearing the directed song of their mates (**Figure 53.8**). Clearly females are differentially responsive to small variations in the male's song. It appears that the males practice variations in their song but then consolidate the song into a "performance" version that attracts females.

The timing and expression of birdsong are under hormonal control

As we have seen, both male and female songbirds hear their species-specific song as nestlings, but only the males of most species sing as adults, and most do so only in spring. Hormones underlie both the difference in song expression between male and female songbirds and the timing of song expression.

To determine whether the absence of testosterone is the main reason female songbirds don't sing, investigators injected adult female songbirds with testosterone in spring. In response to these injections, females sang their species-specific song just as males did. Apparently females form a memory of their species-specific song when they are nestlings, and have the physical capacity to sing, but under normal circumstances they simply lack the hormonal stimulation.

How does testosterone cause a songbird to sing? A study by Fernando Nottebohm at Rockefeller University revealed that each spring an increase in circulating testosterone levels causes certain parts of the male's brain necessary for learning and developing song to grow larger. Individual neurons in those regions of the brain increase in size and grow longer extensions, and the number of neurons in those regions increases. Thus hormones can control behavior by changing brain structure as well as brain function, both developmentally and in response to environmental cues.

What triggers the release of testosterone in response to the onset of spring? Takashi Yoshimura conducted a DNA microarray analysis of brain samples from another bird species, the Japanese quail (*Coturnix japonica*), which is a model species for avian genetic analyses. Knowing that photoperiod was the environmental cue, he examined the responses of 38,000 genes to determine which genes were activated by changes in day length. Fourteen hours after dawn on the first day of the critical photoperiod that induces singing behavior, genes in the brain that code for thyroid-stimulating hormone were switched on. This hormone acts on the pituitary gland, which in turn regulates the production of other hormones that stimulate the growth of the testes and the production of testosterone (see Figure 43.10).

RECAP 53.3

Hormones can control what behavior patterns are programmed in the brain long before those behaviors are expressed. Learning and expression of behaviors can also be controlled by hormones, and therefore timed for particular stages in the life cycle.

- What hormonal conditions are necessary for the development of adult sexual behavior in male and female rats? **See pp. 1098–1099 and Figure 53.5**
- What is the adaptive value of imprinting? **See p. 1099**
- Describe the series of events necessary for a male white-crowned sparrow to sing its species-specific song in spring. **See pp. 1100–1101 and Figure 53.7**

Complex behaviors are the product of interactions of genetic, physiological, and environmental factors. Many genes are involved in shaping behavior, and therefore there are multiple opportunities for selection to favor behavioral modifications. Questions about how changes in behavior adapt animals to environmental conditions are the province of an evolution-based field called **behavioral ecology**.

53.4 How Does Behavior Evolve?

Animal behaviors are variable within and among species. We saw at the start of this chapter that the behaviors of North American brown-headed cowbirds and European common cuckoos—two groups of avian brood parasites—have taken subtly different evolutionary pathways, and that these differences have resulted in different responses on the part of the host birds these parasites affect.

Along with the forces of intra- and interspecific interactions on evolution, environmental conditions are also highly variable over both time and space. Behavioral ecologists strive to discover the relationships between behavior and environment with the intent of understanding the evolutionary mechanisms underlying behavior.

WORKING WITH**DATA**:

Why Put Up with a Parasite?

Original Paper

Hoover, J. P. and S. K. Robinson. 2007. Retaliatory mafia behavior by a parasitic cowbird favors host acceptance of parasitic eggs. *Proceedings of the National Academy of Sciences USA* 104: 4479–4483.

Analyze the Data

A nest or brood parasite lays her eggs in the nest of a different species, who then raises the parasite's chicks at the expense of the host's own reproductive success. Brood parasitism has evolved in several bird species, including cuckoos and cowbirds. Why do the host birds accept this behavior, especially if the parasite's eggs and/or hatchlings are clearly distinct from their own? Why don't the host parents just push the parasite's eggs or offspring out of the nest?

In several cases researchers have observed what they call "mafia behavior," in which the parasitic female returns to the host nest and, if her eggs have been removed, destroys the nest completely. Jeffrey Hoover and Scott Robinson studied the interactions of the brown-headed cowbird (*Molothrus ater*), a North American brood parasite, and one of its common host species, the prothonotary warbler (*Protonotaria citrea*). These experiments placed warbler nests in five treatment categories, two of which protected the warbler nests from any approaching cowbird ("cowbird access denied") and three of which allowed the cowbirds access to the nests (see the table). Hoover and Robinson then tracked the incidence of nest destruction by cowbirds and the average number of warbler offspring fledged per warbler nest in each treatment category, as summarized in the two histograms below.

Treatment	Description
1	Cowbird egg ejected, cowbird access allowed
2	Nonparasitized nest, cowbird access allowed
3	Cowbird egg accepted, cowbird access allowed
4	Cowbird egg ejected, cowbird access denied
5	Nonparasitized nest, cowbird access denied

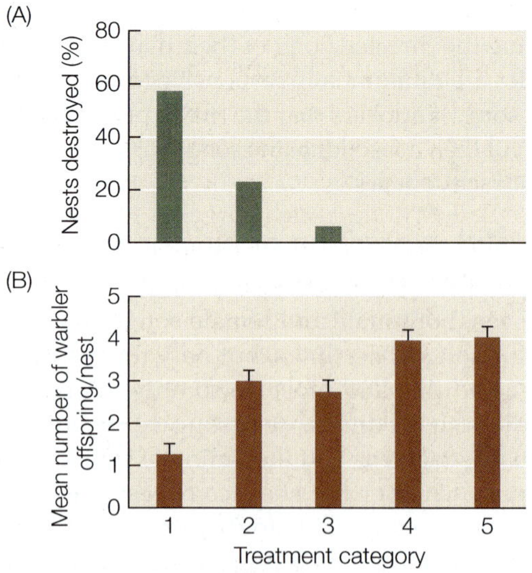
(A)
Nests destroyed (%)
80
60
40
20
0
(B)
Mean number of warbler offspring/nest
5
4
3
2
1
0
1
2
3
4
5
Treatment category

QUESTION 1

Under which conditions were warblers most successful? Under which conditions were warbler nests most likely to be destroyed?

QUESTION 2

Do these results demonstrate a cost–-benefit approach to the evolution of behavior? Explain your answer.

Go to **BioPortal** for all WORKING WITH**DATA** exercises

Animals are faced with many choices

Over an animal's lifetime, its behavior is largely a sequence of choices: where and when to move, where to build a nest, what to eat, when to fight and when to flee, with whom to associate, with whom to mate. Making wrong choices reduces fitness. Behavioral ecologists seek to discover what information animals use to make behavioral choices and how that information relates to aspects of the environment that influence their fitness.

Where an animal lives is referred to as its **habitat**. In most cases the habitat provides not only a protected nest site, but also food and access to mates. The environmental cues animals use to select their habitat may be quite simple. For example, seabirds select cliffs or offshore rocks for nesting, and both of those sites offer protection from predators. Animals with very specialized food requirements select habitats where those foods are abundant. The general hypothesis that guides behavioral ecologists is that the cues animals use to select habitats are reliable predictors of conditions suitable for future survival and reproduction.

For many species, the presence of **conspecifics**—other members of the same species—can be a valuable cue. Observing conspecifics can provide animals with information about the quality of a habitat. After all, you can't argue with success. In Europe, collared flycatchers (*Ficedula albicollis*) in the spring breeding season are nosy neighbors, regularly visiting the nests of conspecifics. Researchers hypothesized that this behavior allows the flycatchers to assess the quality of the habitat by seeing how well their neighbors are faring. To test this hypothesis, they created some areas with supersized broods—normally an indication of abundant food—by taking young birds from some nests and adding them to nests in another area. The next year, flycatchers preferentially settled in the areas where broods had been artificially enlarged.

Behaviors have costs and benefits

Behavioral ecologists often use a cost–benefit approach to investigate the relationship between behavior, environment, and fitness. A **cost–benefit approach** assumes that an animal has only a limited amount of time and energy, and therefore cannot afford to engage in behaviors that cost more to perform than they bring it in benefits. A cost–benefit approach provides a framework that behavioral ecologists can use to make observations, construct hypotheses, and design experiments to investigate why behavior patterns evolve as they do.

The benefits of a behavior can be measured in terms of the enhancement in fitness an animal accrues by performing the behavior. The cost of a behavior typically has three components:

- **Energetic cost** is the difference between the energy the animal expends performing the behavior and not performing it.
- **Risk cost** is the increased chance of being injured or killed as a result of performing the behavior.
- **Opportunity cost** is the benefit the animal forgoes by not being able to perform other behaviors during the same time interval.

Territorial behavior carries significant costs

Cost–benefit analysis has been used extensively in the study of **territorial behavior**, which is aggressive behavior that actively denies other animals access to a habitat or resource. Optimal habitats and resources are frequently in short supply, so conspecifics must compete for them. Many animals—usually males—defend all-purpose territories that provide a nest site, food, and access to mates. The territory holder stakes out his boundaries by engaging in aggressive interactions with neighbors, and must then patrol those boundaries constantly and respond to trespassers. These aggressive interactions usually consist of highly stereotypic, species-specific displays such as birdsong. Through territorial behavior, the male obtains the resources he needs for reproductive success, but he also pays a price.

Territorial displays require considerable expenditure of energy, they make a male more vulnerable to predation, and they detract from the time he has for feeding or engaging in parental behavior. Michael Moore and Catherine Marler at Arizona State University performed an experiment to estimate the costs incurred by male Yarrow's spiny lizards (*Sceloporus jarrovii*) when defending a territory. These lizards defend territories that include the home ranges of several females. Their territorial behavior is normally most intense during September and October, when the circulating testosterone levels of the males are high and the females are most receptive to mating. The researchers varied the intensity of the lizards' territorial behavior by implanting testosterone capsules in some males in summer, when they are not normally highly territorial (**Figure 53.9**).

Testosterone-treated males spent more time patrolling their territories, performed more displays, and expended about one-third more energy than control males (an energetic cost). As a result, they had less time to feed (an opportunity cost), captured fewer insects, stored less energy, and had a higher death rate (a risk cost). In summer, when females are not normally receptive, these high costs of vigorous territorial defense outweigh the reproductive benefits of territoriality. Thus natural selection has favored seasonal variation in the level of the hormone controlling territorial behavior in this species.

Go to Animated Tutorial 53.1
The Costs of Defending a Territory
Life10e.com/at53.1

The cost–benefit approach explains the diversity of territorial behaviors seen in different species. Even if a resource is absolutely essential to an animal, if it cannot be defended economically, the animal will not engage in territorial behavior. Food is essential for all animals, but if the food is widely distributed in space or fluctuating in availability, there is no benefit to balance the high costs of trying to defend it. For example,

INVESTIGATING**LIFE**

53.9 The Costs of Defending a Territory By using testosterone implants to increase territorial behavior, Michael Moore and Catherine Marler measured the costs to male Yarrow's spiny lizards (*Sceloporus jarrovii*) of defending a territory during the summer, when they do not normally do so.[a]

HYPOTHESIS Yarrow's spiny lizards do not defend a territory during summer because the energetic costs of territorial behavior in that season outweigh the benefits.

Method

1. During the summer, when female lizards are not sexually receptive, insert testosterone capsules under the skin of some males; leave other males untreated as controls.
2. Observe the patterns of territorial behavior and the survival rate of the two groups of males.

Results

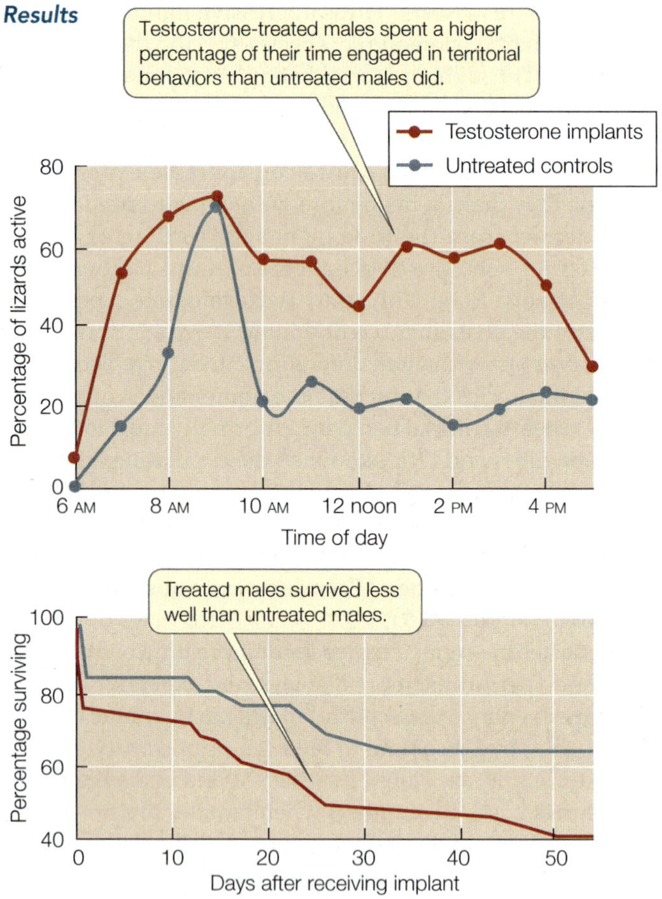
Testosterone-treated males spent a higher percentage of their time engaged in territorial behaviors than untreated males did.
Testosterone implants
Untreated controls
Percentage of lizards active
0
20
40
60
80
6 AM
8 AM
10 AM
12 noon
2 PM
4 PM
Time of day
Treated males survived less well than untreated males.
Percentage surviving
40
60
80
100
0
10
20
30
40
50
Days after receiving implant

CONCLUSION For these lizards, the cost of defending territories during summer significantly reduces their survival rate without increasing their reproductive success.

Go to **BioPortal** for discussion and relevant links for all INVESTIGATING**LIFE** figures.

[a]Marler, C. A. and M. C. Moore. 1988. *Behavioral Ecology and Sociobiology* 23: 21–26.

the open ocean where seabirds feed cannot be defended. But safe nest sites on islands or rocky cliffs are in short supply, and they can be defended. Thus the territories of seabirds may be no larger than the distance the birds can reach while sitting on their nests (**Figure 53.10A**).

In some cases the resource that is defended is the female herself. Elephant seals spend most of their lives at sea, but females come to land at traditional beach sites to give birth to their pups. Male elephant seals arrive at these sites ahead of time and stake out territories through vigorous fighting (**Figure 53.10B**). When the females arrive on the beaches, they enter the territories of the males. As long as the male territory holder can fend off challengers, he will be able to mate with all the females using his piece of the beach.

One unusual form of male territorial behavior arises in situations in which neither food, nest sites, nor females are defended. A **lek** is an area where males gather for the purpose of engaging in intense displays of their territorial prowess aimed at impressing females and winning the opportunity to mate. Even though space is not limited, each male defends a small piece of real estate on which he performs a display (**Figure 53.10C**). Those territories closest to the center of the lek are the prime sites, and males compete intensely for those locations. The females visit into the lek, observe the males, and generally mate with the males holding the prime sites. The benefit of this system to the female is that she is inseminated by a successful competitor, and therefore her offspring will carry the genes that contributed to his success. This is another example of sexual selection (see Section 21.2). The costs of lekking to males are high, as they engage in continuous, intense territorial behavior that precludes eating, drinking, and sleeping until they are displaced. The benefit is the chance to maximize their fitness by mating with many females.

Cost–benefit analysis can be applied to foraging behavior

When an animal forages (searches for food), among the decisions it must make are how much time to spend in each location before giving up and moving on, what resources at each location are actually edible, and which of the different types of potential food should be eaten and which should be left alone. By applying cost–benefit approaches to feeding behavior, scientists have produced a body of knowledge known as **optimal foraging theory**, which helps them identify the fitness value of feeding choices. The primary benefit of foraging is the nutritional value of the food obtained: the energy, minerals, and vitamins it contains (see Section 51.1). The costs of foraging are similar to those of other behaviors: energy expended, time lost from other activities that could enhance fitness, and the risk of increased exposure to predators.

Go to Animated Tutorial 53.2
Foraging Behavior
Life10e.com/at53.2

Animals frequently have to make choices among food items that may differ not only in terms of energy content, but also abundance or ease of acquisition and processing. Optimal foraging theory predicts that in such situations, animals will make choices that will maximize the rate at which they obtain energy. The more rapidly a foraging animal satisfies its

(A) *Diomedea melanophris*

(B) *Mirounga angustirostris*

(C) *Centrocercus urophasianus*

53.10 Animals Defend Territories of Different Sizes (A) The nesting territories of many seabirds consist of only as much space as they can defend without leaving the nest. (B) Male elephant seals fight vigorously to defend areas of beach where females haul out of the water to give birth to their pups. (C) Male greater sage-grouse gather at a lek in Colorado to perform displays aimed at impressing females and winning the opportunity to mate.

energetic requirements, the lower the opportunity costs and risk costs of foraging.

Earl Werner and Donald Hall of Michigan State University performed laboratory experiments with bluegill sunfish (*Lepomis macrochirus*) to test this energy maximization hypothesis. In preparation for their experiments, they measured the energy content of water fleas (*Daphnia*) of different sizes (the different food types), how much time bluegill sunfish (the foragers) needed to capture and eat those different food types, the energy they spent pursuing and capturing the different food types, and the rates at which they encountered the different food types under different food densities. Werner and Hall then stocked experimental environments with different densities and proportions of large, medium, and small water fleas. They made two predictions from the energy maximization hypothesis: first, that in an environment with abundant large water fleas, the fish would ignore smaller water fleas; and second, that in an environment stocked with low densities of all three sizes of water fleas, the fish would eat every water flea they encountered. The proportions of large, medium, and small water fleas eaten by the fish under different conditions were close to those predicted by the hypothesis (**Figure 53.11**).

The energy maximization hypothesis considers food items in terms of the energy they provide, but animals have nutrient requirements in addition to energy that can play a role in shaping their foraging behavior. Essential minerals, for example, are in short supply in some animals' diets, and those animals may incur large energetic costs and risks to obtain them (**Figure 53.12**). Foods may also have medicinal value. Chimpanzees, for example, have been observed eating the pith of the plant *Vernonia amygdalina*. The pith contains small quantities of a secondary metabolite (vernonioside B1) that is toxic to chimps at high concentrations, but at low concentrations can kill their intestinal parasites. Chimps that consume this pith have fewer parasites.

RECAP 53.4

Behavioral ecologists seek to explain relationships between variation in behavior and variation in environmental conditions. They seek to discover what information animals use to make behavioral choices and how that information relates to aspects of the environment that influence their fitness. Cost–benefit analysis has been applied to territorial and foraging behavior.

- What might the presence of conspecifics in a habitat tell an animal about that habitat? **See p. 1103**
- Describe three types of territorial behavior and the costs and benefits of each. **See pp. 1103–1104 and Figure 53.9**
- How can cost–benefit analysis be applied to a behavior? **See pp. 1104–1105 and Figures 53.9 and 53.11**

Whereas behavioral ecologists are interested in understanding how the natural environment influences the fitness value of behavioral choices—the ultimate causes of those behaviors, in Tinbergen's terms—other behavioral biologists focus on the physiological mechanisms and principles that underlie behavior—the proximate causes.

INVESTIGATING**LIFE**

53.11 Bluegill Sunfish Are Energy Maximizers Based on energy maximization calculations, Earl Werner and Donald Hall predicted (1) that in an environment with abundant large food items (i.e., the water flea *Daphnia*), bluegill sunfish (*Lepomis macrochirus*) would ignore smaller food items and feed preferentially on larger water fleas; and (2) that in an environment where all sizes of *Daphnia* were scarce, the fish would eat every one they encountered. Such a strategy is in keeping with the cost–benefit hypothesis of foraging behavior.[a]

HYPOTHESIS Bluegill food item selection will match the energy maximization predictions of cost–benefit analysis.

Method

1. Measure the respective energy content of large, medium, and small water fleas (*Daphnia*).
2. Use cost–benefit energy maximization calculations to predict the rate at which bluegill sunfish will consume the different sized water fleas under different levels of food abundance (i.e., density of water fleas).
3. Provide bluegills with *Daphnia* of different sizes in varying proportions (represented by the different colored bars) and at different densities.

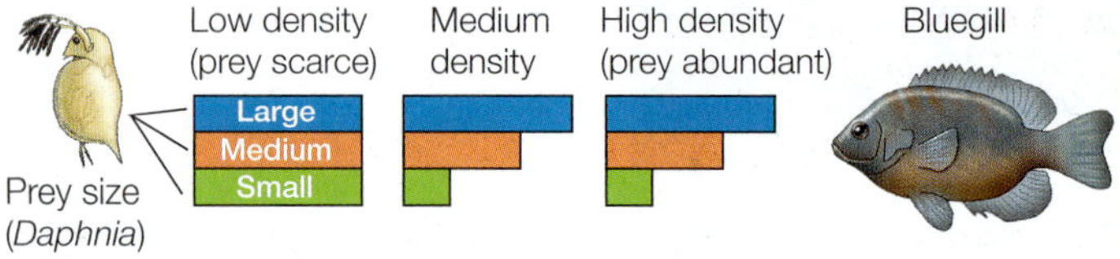

Low density (prey scarce)
Medium density
High density (prey abundant)
Bluegill
Large
Medium
Small
Prey size (Daphnia)

4. Note the proportions of small, medium, and large *Daphnia* actually eaten by the fish under the different conditions and compare these proportions to the predictions of energy maximization.

Results

The food choices made by bluegills match the predictions of the energy maximization hypothesis.

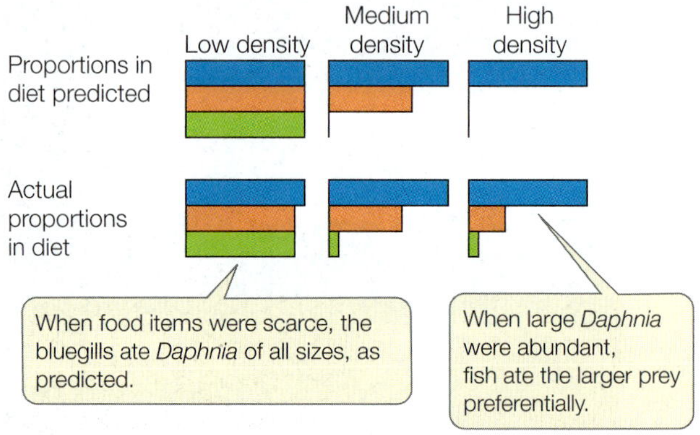

Low density
Medium density
High density
Proportions in diet predicted
Actual proportions in diet
When food items were scarce, the bluegills ate Daphnia of all sizes, as predicted.
When large Daphnia were abundant, fish ate the larger prey preferentially.

CONCLUSION Bluegills select food items in accordance with the predictions of energy maximization calculations.

Go to **BioPortal** for discussion and relevant links for all INVESTIGATING**LIFE** figures.

[a]Werner, E. E. and D. J. Hall. 1974. *Ecology* 55: 1042–1052.

(A) *Ara chloropterus*

(B)

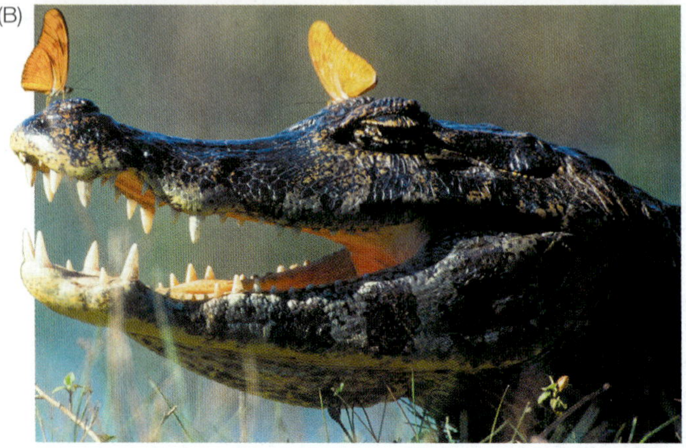

53.12 Herbivores Seek Out Unusual Sources of Minerals (A) Red-and-green macaws of the Amazon jungle obtain essential minerals by eating dried clay. (B) Pierid butterflies obtain needed salts by drinking secretions from the skin and nostrils of a caiman.

53.5 What Physiological Mechanisms Underlie Behavior?

Control of behavior involves the nervous and endocrine systems. Execution of behavior involves the musculoskeletal system as well as other effector mechanisms, such as those that produce secretions, color changes, electrical impulses, sound, and even light. We have already considered many of the physiological systems that are involved in these processes, including hormones, reproductive systems, nervous systems, sensory systems, and feeding mechanisms. The field of behavioral physiology, which encompasses aspects of all of these systems, is enormous, so here we will dig deeper into just three different phenomena studied by behavioral physiologists: the timing of behavior, navigation, and communication.

Biological rhythms coordinate behavior with environmental cycles

Earth turns on its axis once every 24 hours, generating daily cycles of light and dark, temperature, humidity, and tides. In addition, Earth is tilted on its axis, so the light–dark cycle

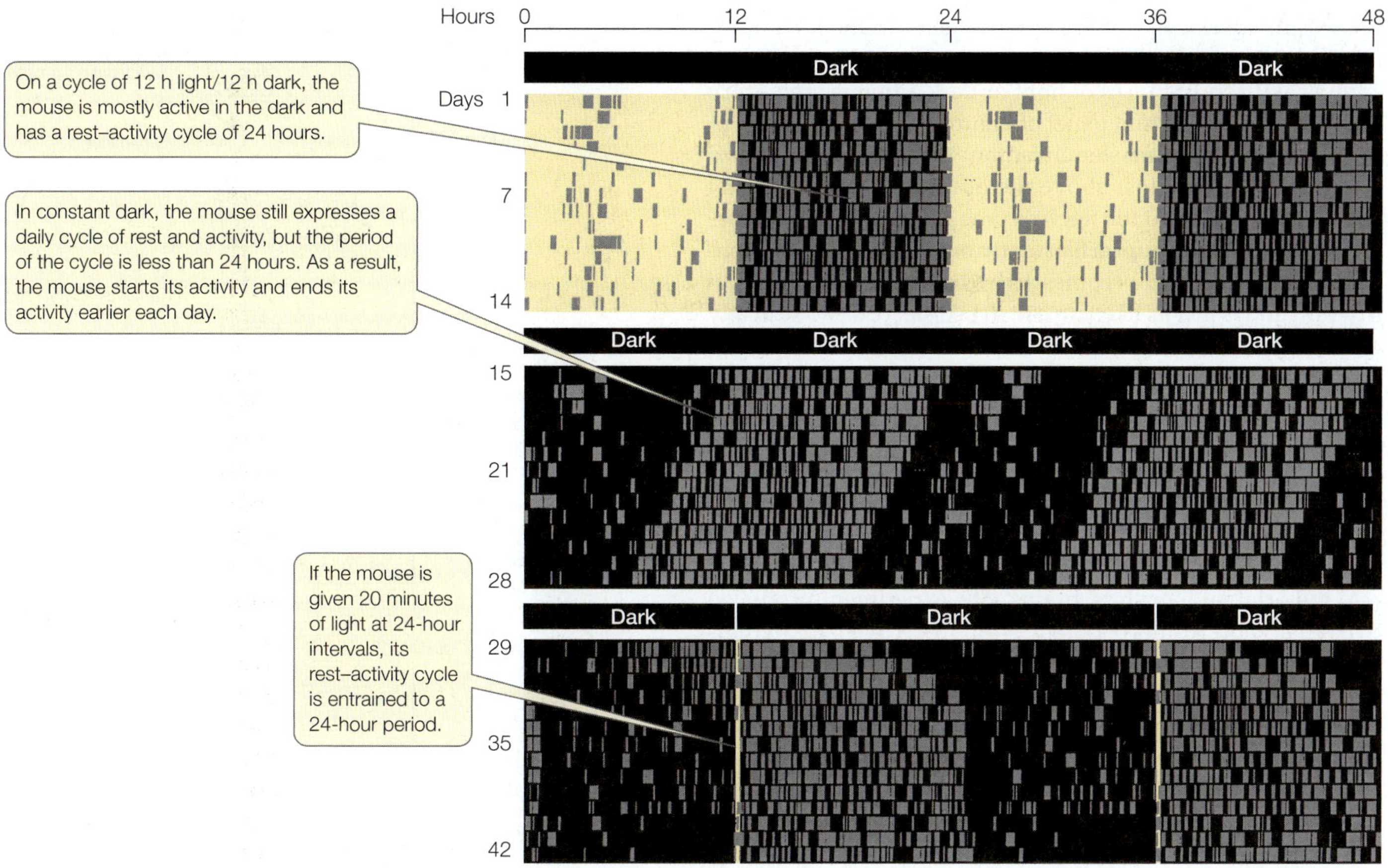
Hours
0
12
24
36
48
Dark
Dark
Days 1
7
14
15
21
28
29
35
42
On a cycle of 12 h light/12 h dark, the mouse is mostly active in the dark and has a rest–activity cycle of 24 hours.
In constant dark, the mouse still expresses a daily cycle of rest and activity, but the period of the cycle is less than 24 hours. As a result, the mouse starts its activity and ends its activity earlier each day.
If the mouse is given 20 minutes of light at 24-hour intervals, its rest–activity cycle is entrained to a 24-hour period.

53.13 Circadian Rhythms Are Entrained by Environmental Cues The activity–rest cycle of a laboratory mouse (a nocturnal animal) responds to the light–dark cycle under which it is kept. The gray bars indicate times when the mouse is running on an activity wheel. Two days of activity are recorded on each horizontal line; the data for each day are plotted twice—once on the *right* half of each line (hours 24–48) and again on the *left* half of the line below it (hours 0–24). This double plotting is merely to make the pattern easier to see.

Go to Animated Tutorial 53.3
Circadian Rhythms
Life10e.com/at53.3

changes as Earth revolves around the sun. These daily and seasonal cycles profoundly influence the physiology and behavior of animals. Animals tend to be active either during the day (diurnal) or at night (nocturnal) and have sensory capabilities appropriate to this distinction. Therefore it is adaptive to organize behavior on a cycle that corresponds with the environmental cycle of light and dark. Similarly, a behavior that is adaptive at one time of year (such as midsummer) may not be adaptive at another times (midwinter). Thus it is important for animals to organize their behavior with respect to time of the day or year and to be able to anticipate those times.

CIRCADIAN RHYTHMS Experimental animals kept in constant darkness and at a constant temperature, with food and water available all the time, still demonstrate daily cycles of activities such as locomotor activity, sleeping, eating, drinking, learning, and just about anything else that can be measured. The persistence of these daily cycles in the absence of environmental time cues suggests that animals have an internal clock. Because these daily cycles are not exactly 24 hours long, they are known as **circadian rhythms** (*circa*, "about"; *dies*, "day").

As described in Section 37.4, any biological rhythm can be viewed as a series of cycles, and the length of one of those cycles is the period of the rhythm. Any point in the cycle is a phase of that cycle. When two rhythms completely match, they are in phase, and if a rhythm is shifted (as in the resetting of a clock), it is phase-advanced or phase-delayed. Because the period of a circadian rhythm is not exactly 24 hours, it must be phase-advanced or phase-delayed each day to remain in phase with the daily cycle of the environment. In other words, the rhythm has to be **entrained** to the cycle of light and dark in the animal's environment.

An animal kept under constant conditions will not be entrained to the light–dark cycle of the environment, and its circadian clock will run according to its natural period—it will be **free-running**. If the period is less than 24 hours, the animal will begin its activity a little earlier each day (Figure 53.13). The period of the free-running circadian rhythm is under genetic control. Different species may have different average periods, and within a species, mutations can lead to different period lengths.

Under natural conditions, environmental time cues, such as the onset of light or dark, entrain the free-running rhythm to the light–dark cycle of the environment. In the laboratory it is possible to entrain the circadian rhythms of free-running animals with short pulses of light or dark administered every 24 hours (see the bottom panel of Figure 53.13).

In mammals, the master circadian "clock" consists of two clusters of neurons just above the optic chiasm (the area of the brain where the optic nerves come together). These structures are called the **suprachiasmatic nuclei** (**SCN**). If they are destroyed, the animal becomes arrhythmic (loses its circadian rhythm) and is just as likely to eat, drink, sleep, or wake at any time of day.

In 1990, a notable study by Martin Ralph and his colleagues, then at the University of Virginia, showed that the SCN is the source of circadian rhythms. When the SCNs of adult hamsters with typical 24-hour rhythms were destroyed, the animals became arrhythmic. After several weeks of this arrhythmic behavior, the researchers transplanted SCN tissue from hamster fetuses bred for an atypical (mutant) short-day rhythmicity into the original hamsters' brains. The experiment produced two remarkable results (**Figure 53.14**). First, circadian rhythms were restored by the transplanted SCN tissue, demonstrating that the SCN is sufficient to generate circadian rhythms—a unique case of a behavior being restored by a neural transplant. Second, the restored circadian rhythms had the period length of the *donor* strain, demonstrating that the specific phenotype of the behavior was a property of the donor neural tissue, and thus wholly generated by the SCN.

The molecular mechanism of the circadian clock involves negative feedback loops. Although there are a number of genes involved, including the *per* genes discussed in Section 53.2, we can generalize about the mechanism by saying that when certain "clock genes" are expressed in SCN cells, the mRNA enters the cytoplasm, where it is translated. The resulting proteins combine, and the dimer returns to the nucleus as a transcription factor that shuts off the expression of the clock genes. The period of this cycle is about a day. These findings show that it is possible to understand circadian rhythms of behavior at all levels, from the molecular rhythm generators to the environmental stimuli that entrain them to the daily cycle of light and dark.

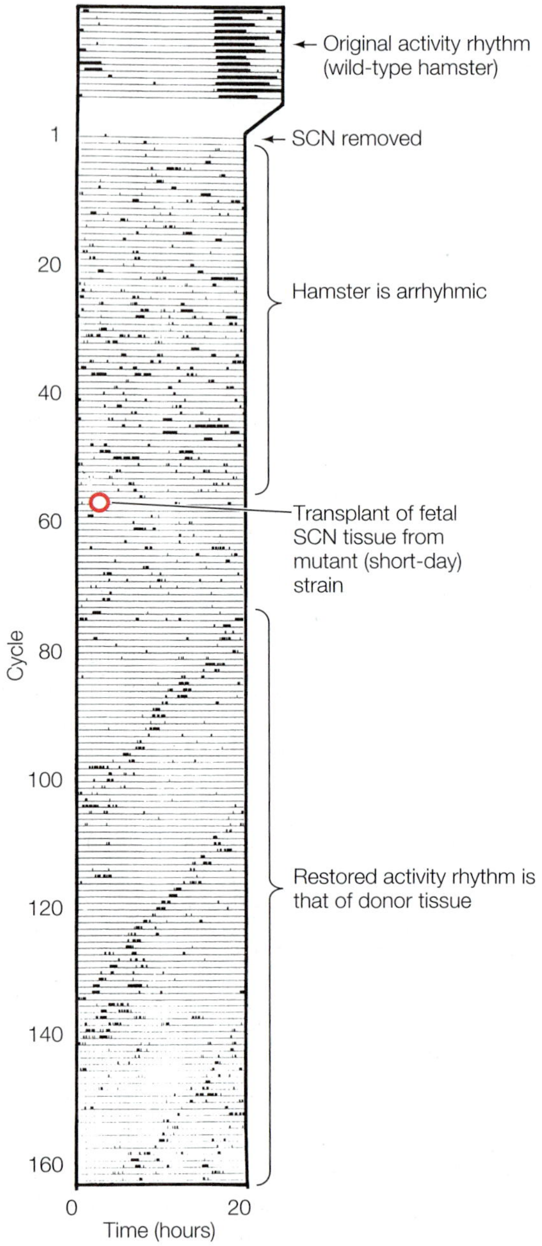
Original activity rhythm (wild-type hamster)
SCN removed
Hamster is arrhyhmic
Transplant of fetal SCN tissue from mutant (short-day) strain
Restored activity rhythm is that of donor tissue
Cycle
1
20
40
60
80
100
120
140
160
0
20
Time (hours)

53.14 The Brain Clock Can Be Transplanted In this experiment, the activity rhythm of a wild-type (i.e., genetically typical) hamster was measured; this animal had a circadian period of 24.5 hours (top of bar). After its SCN was removed (lesioned), the hamster became arrhythmic. SCN tissue from a fetal "short-day" hamster (a mutant strain with a 19-hour circadian period) was then transplanted into the region where the lesion was made in the wild-type hamster. The transplanted tissues restored circadian rhythm in the lesioned hamster, but the restored rhythm had the period of the donor animal.

CIRCANNUAL RHYTHMS Seasonal changes in the environment present challenges to many species. Most animals reproduce most successfully if they time their reproductive behavior to coincide with the most favorable time of year for the survival of their offspring. Many species require considerable advance preparation for reproduction. Migratory animals must arrive on their breeding grounds at the right time, and animals that have specialized structures used in mating displays, such as the antlers of deer, moose, and caribou, must grow these structures before the breeding season arrives.

For many species, a change in day length—the **photoperiod**—is a reliable indicator of seasonal changes to come. For others, however, change in day length is not a reliable seasonal cue. Hibernators, for example, spend long months in dark burrows underground but must be physiologically prepared to breed almost as soon as they emerge in the spring. A bird overwintering near the equator cannot use changes in photoperiod as a cue to time its migration to its temperate-zone breeding grounds. When held under constant laboratory conditions, such animals show endogenous circannual rhythms that keep track of the time of year. Unlike circadian rhythms, the neural basis for circannual rhythms is unknown.

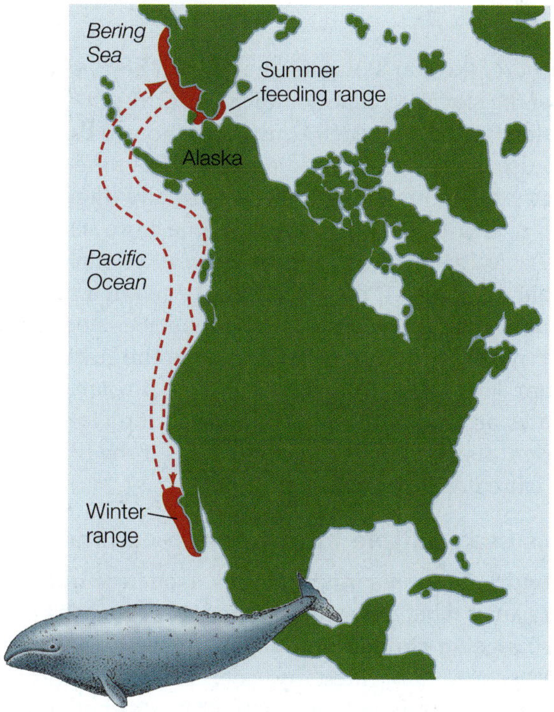
Bering Sea
Summer feeding range
Alaska
Pacific Ocean
Winter range

53.15 Piloting Gray whales (*Eschrichtius robustus*) migrate south in winter from the Bering Sea to the coast of Baja California by piloting, in part by following the western coast of North America.

Animals must find their way around their environment

To locate suitable habitats, find food and mates, and avoid predators and bad weather, an animal needs to be able to find its way around its environment. Within its local habitat, an animal can orient to landmarks. But what if its destination is a considerable distance away?

PILOTING: ORIENTATION BY LANDMARKS Most animals find their way by knowing and remembering the structure of their environment. This form of navigation is called **piloting**. Gray whales, for example, migrate seasonally between the Bering Sea and the coastal lagoons of Mexico (**Figure 53.15**). They find their way in part by following the west coast of North America. Coastlines, mountain chains, rivers, water currents, and wind patterns can all serve as piloting cues for the whales. But some remarkable cases of long-distance orientation and movement cannot be explained by piloting.

HOMING: RETURN TO A SPECIFIC LOCATION The ability to return to a nest site, burrow, or other specific location is called **homing**. Homing can be accomplished by piloting in a known environment, but some animals that travel long distances through unfamiliar territory perform much more sophisticated homing. The ability of pigeons to return to their home loft even after being transported to remote sites is well known. How do they find their way home? Experiments have shown that pigeons use the sun as a compass, but they can still find their way home when the sun is not visible. Other experiments have shown that pigeons equipped with frosted contact lenses can find their way home, suggesting that visual cues are not essential. Most amazing has been the demonstration that pigeons can detect Earth's magnetic field and orient to it much as a human orients with a compass. Taken together, the studies of homing by pigeons suggest that they can use multiple, redundant sources of directional information and can switch among those sources depending on the circumstances.

Go to Animated Tutorial 53.4
Homing Simulation
Life10e.com/at53.4

MIGRATION: NAVIGATION OVER GREAT DISTANCES For as long as humans have inhabited temperate latitudes, they have been aware that entire populations of animals, especially birds, disappear and reappear seasonally. Not until the early nineteenth century, however, were patterns of migration traced by marking individual birds with identification bands around their legs. Only when individuals could be unmistakably identified was it possible to show that the same birds and their offspring returned to the same breeding grounds year after year, and that these same birds could be found during the nonbreeding season at locations hundreds or even thousands of kilometers from their breeding grounds.

Many homing and migrating species take direct routes to their destinations through environments they have never experienced because they use mechanisms of navigation other than piloting. Humans use two major forms of navigation:

- *Distance–direction navigation* requires knowing in what direction and how far away the destination is. With a compass to determine direction and a means of measuring distance, humans can navigate.
- *Bicoordinate navigation*, also known as true navigation, requires knowing the latitude and longitude (the map coordinates) of both the current position and the destination, as well as a compass to determine direction.

Many non-human animals seem to have a "compass sense" that allows them to use environmental cues to determine direction, and some seem to have a "map sense" that allows them to determine their position.

The behavior of many animals suggests that they are capable of bicoordinate navigation. Gray-headed albatrosses (*Thalassarche chrysostoma*), for example, breed on oceanic islands in the Southern Hemisphere. When a young albatross leaves its parents' nest, it flies widely over the southern oceans for 8 or 9 years (appropriately, the genus name *Thalassarche* is Greek for "first on the sea"). At that time, it reaches reproductive maturity and flies back to the island where it was raised, where it mates and builds a nest (**Figure 53.16**). How can the bird find a tiny island in an enormous ocean after years of wandering? A circadian clock probably gives the albatross information about the time of day, and additional information from the position of the sun might allow it to determine its map coordinates—much as sailors did in the days before global positioning satellites.

The ability to locate a position by calculating the angles between celestial objects such as the sun and stars and the horizon

(A)

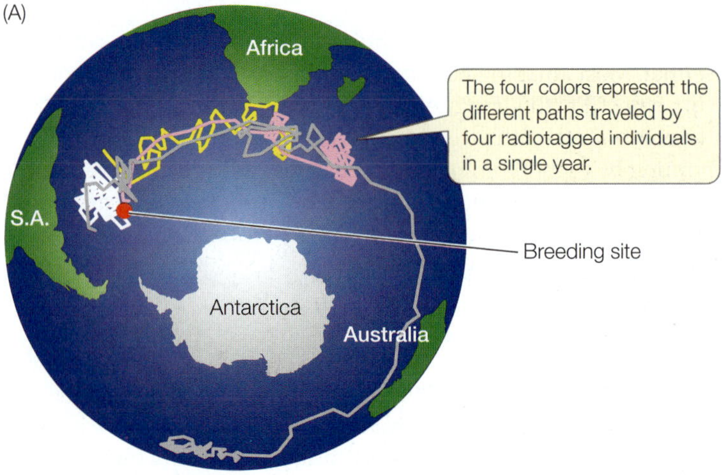
Africa
The four colors represent the different paths traveled by four radiotagged individuals in a single year.
S.A.
Breeding site
Antarctica
Australia

(B) *Thalassarche chrysostoma*

53.16 Coming Home (A) Gray-headed albatrosses are born on islands in the subantarctic oceans. Young birds roam widely over the southern oceans for 8 or 9 years. (B) Once they reach maturity, the birds return to the island where they were hatched to mate and raise their own young. A courting couple is shown here.

at specific times of day is called celestial navigation. During the day, the sun can serve as a compass if you know what time it is, and animals can use their circadian clocks for that information. This capacity has been demonstrated by "clock-shifting" experiments such as the one shown in **Figure 53.17**. Similar experiments have shown that many animal species can orient by means of a time-compensated solar compass.

Go to Animated Tutorial 53.5
Time-Compensated Solar Compass
Life10e.com/at53.5

Many animals are normally nocturnal; in addition, many diurnal bird species migrate at night and thus cannot use the sun to determine direction. The stars offer two sources of information about direction: moving constellations and a fixed point. The positions of constellations (like that of the sun) change because Earth is rotating. With a star map and a clock, direction can be determined using any constellation. But one point that does not change position during the night is the point directly over the axis on which Earth turns. In the Northern Hemisphere, the star Polaris—the "North Star"—lies in that region of the sky and reliably indicates north.

Stephen Emlen at Cornell University showed that birds can learn to use the stars for orientation. As the time of year approaches when young birds would normally migrate to their winter range, young captive birds become more active and orient their activity in the direction they would fly. How do they know that direction? If these birds are raised in a planetarium with a natural star pattern, but one that does not rotate, the birds do not learn to orient, and their premigratory activity is random. However, if the planetarium sky rotates, and even if it rotates around a different point than the North Star, the birds orient their premigratory activity as if the fixed point in the sky were north.

Animals use multiple modalities to communicate

As individual animals interact, they exchange information; therefore animal behavior can evolve into systems of information exchange, or **communication**. The behaviors of individuals may become elaborated into communication signals, but only if the transmission of information benefits both the sender and the receiver. To understand why these conditions must be met, consider male courtship displays. These displays will be favored if they increase the male's probability of mating and passing on his genes, and sexual selection will occur if the display conveys information to a female (the receiver) about his qualities as a potential father for her offspring.

Animals communicate using a variety of sensory modalities that vary in the nature of the signal produced, the specificity of the information conveyed, the speed and persistence of the signal, and its suitability in different environments. Behavioral physiologists interested in communication must take into consideration the sensory and motor characteristics of their study animals, the physics of the communication modalities they use, and the environment in which the communication takes place.

CHEMICAL SIGNALS Because of the diversity of their molecular structures, pheromones can communicate very specific, information-rich messages (see Section 46.2). Pheromones are effective day and night, and they can cover a broad range of transmission distances. Pheromones used in different types of communication vary in their volatility (ease of vaporization) and diffusibility; these chemical properties are functions of the nature and size of the pheromone molecule. Pheromones that act as alarm signals, for example, are highly volatile and diffusible, so their message spreads rapidly but disappears rapidly. Territory-marking and trail-marking pheromones have low volatility and diffusibility and stay effective for a long time, so they can convey directional information. Sex pheromones, such as that of the gypsy moth (see Figure 46.4), are intermediate in these properties, so they can spread a long distance but do not disappear rapidly.

Pheromones are an effective way to exchange species-specific information, and because the recipient must have the proper receptor molecule to detect the pheromone, it is not a signal that is easily intercepted by predators. Pheromonal signals cannot

INVESTIGATING**LIFE**

53.17 A Time-Compensated Solar Compass Experiments show that pigeons use the sun to establish directions for navigation and finding food. "Clock-shifting" experiments demonstrate that the birds' circadian clocks factor into their ability to judge direction correctly based on the sun's position.[a]

HYPOTHESIS Pigeons determine compass direction from the position of the sun with respect to their internal circadian clocks.

Method 1. Place a pigeon in a circular cage from which it can see the sun and sky, but not the horizon or any other visual cue.

2. Surround the cage with multiple food bins but place food only in the southernmost bin, thus training the bird to look for food in the south. (Rotating the cage but always placing the food in the southernmost bin confirms that the bird is navigating to find south.)

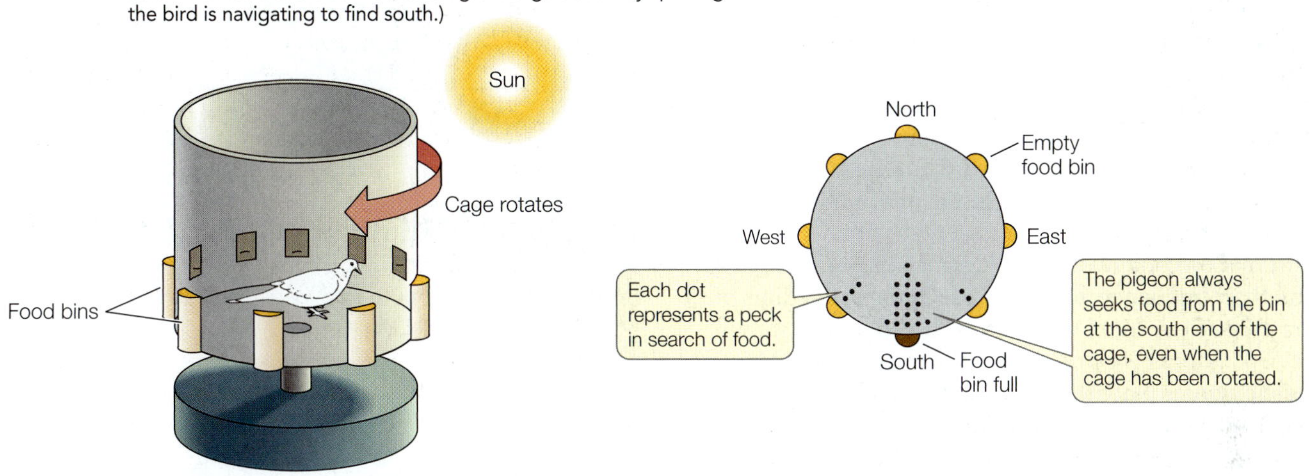
Sun
Cage rotates
Food bins
North
Empty food bin
West
East
Each dot represents a peck in search of food.
South
Food bin full
The pigeon always seeks food from the bin at the south end of the cage, even when the cage has been rotated.

3. Place the trained pigeon in a room with a controlled light cycle for 2 weeks. Turn the lights on at midnight and off at noon to phase-advance its circadian rhythm by 6 hours (i.e., 6 A.M. feels like noon to the bird).

4. Return the pigeon to the circular cage under natural light and observe its food-seeking behavior.

Results A 6-hour shift in the circadian clock results in a 90-degree error in the pigeon's orientation.

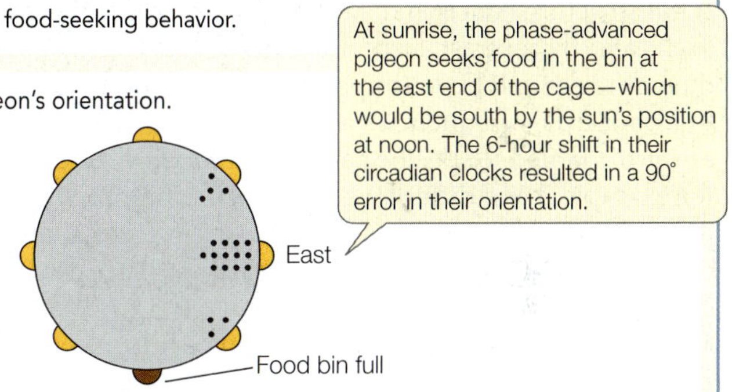
At sunrise, the phase-advanced pigeon seeks food in the bin at the east end of the cage—which would be south by the sun's position at noon. The 6-hour shift in their circadian clocks resulted in a 90° error in their orientation.
East
Food bin full

CONCLUSION Pigeons have the ability to determine direction using the position of the sun as a compass.

Go to **BioPortal** for discussion and relevant links for all INVESTIGATING**LIFE** figures.

[a]Kramer, G. 1952. *Ibis* 94: 265-285.

be changed rapidly, but they can convey static, complex information. Mammals that mark their territories with pheromones reveal a great deal of information about themselves: species, individual identity, reproductive status, size (indicated by the height of the marking), and how recently the animal has been in the area (indicated by the strength of the scent).

VISUAL SIGNALS Visual signals offer the advantage of rapid delivery of information over considerable distances (depending on the environment and the visual acuity of the receiver); they also convey the exact position of the signaler. Signal content can be enhanced by movements (as in a courtship display) or by different postures. Effective visual signals, however, require sufficient light, and the receiver must be looking at the signaler. Thus visual communication is not particularly useful at night or in environments that lack light, such as caves and ocean depths. Some species have overcome this constraint with light-emitting mechanisms. Fireflies, for example, use an enzymatic mechanism to create flashes of light. By emitting flashes in species-specific patterns, fireflies advertise for mates at night.

Another drawback of visual signals is that they can be intercepted by other species. There are predatory firefly species, for example, that mimic the flash pattern of females of other species. A male that approaches the mimicking "female" becomes a meal rather than a mate. Thus deception can be part of animal communication systems, just as it is part of human communication.

ACOUSTIC SIGNALS Sound cannot convey complex information as rapidly as visual signals can. But acoustic signals, unlike visual signals, can be used at night and in dark environments. They are not hindered by objects that would interfere with visual signals, so they can be transmitted in complex environments such as forests. They are often better than visual signals at getting the attention of a receiver because the receiver does not have to be looking at the signaler for the message to be received. Sounds are also useful for communicating over long distances. Even though the intensity of a sound decreases with distance from the source, loud sounds can transmit information over much longer distances than visual signals can. The complex songs of humpback whales, when produced at ocean depths of about 1,000 meters, can be heard hundreds of kilometers away, allowing these whales to locate one another across vast expanses of ocean.

The information content of acoustic signals can be increased by varying their frequency, as we can see in the sonograms of the species-specific song of white-crowned sparrows shown in Figure 53.7, and as we practice in our own speech. However, acoustic signals place the signaler at risk for detection by predators. This danger can be minimized by adjustments of frequency and signal structure that decrease the directional information the receiver can extract from the signal. Alarm calls tend to be pure tones (a single frequency) without much temporal structure (starts and stops). It is very difficult to localize such calls. By contrast, territorial calls tend to cover a broad frequency range and have temporal structure. These calls are easy to localize. The frequencies and structures of acoustic signals are also adapted to specific habitats. Different vegetation types, for example, have different sound-absorbing properties: pure tones at lower frequencies carry better in forests, and more complex calls at higher frequencies carry well in open habitats.

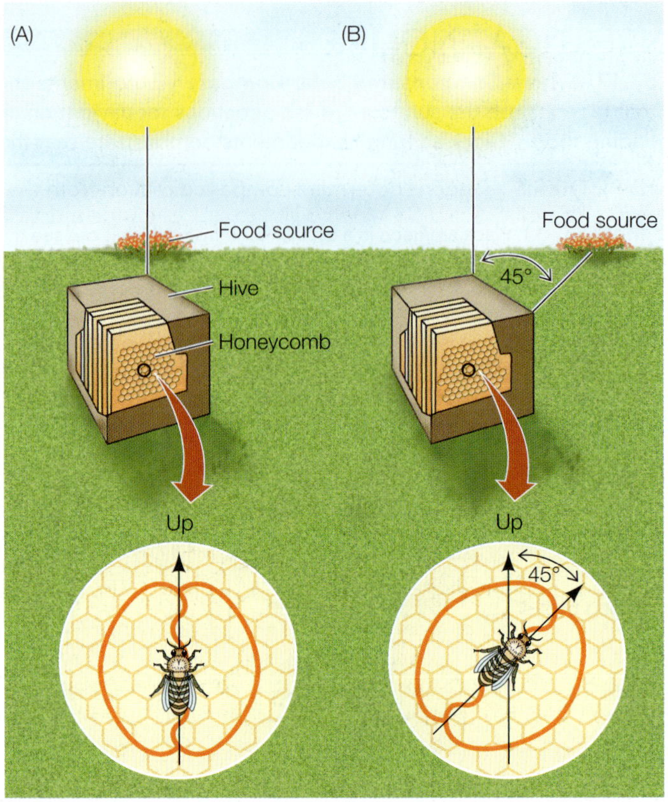
(A)
(B)
Food source
Hive
Honeycomb
Food source
45°
Up
Up
45°

53.18 The Honey Bee Waggle Dance (A) A honey bee (*Apis mellifera*) runs straight up on the vertical surface of the honeycomb in the dark hive while wagging her abdomen to tell her hivemates that there is a food source in the direction of the sun. (B) When her waggle runs are at an angle from the vertical, the other bees know that the same angle separates the direction of the food source from the direction of the sun.

Go to Activity 53.1 Honey Bee Dance Communication
Life10e.com/ac53.1

MECHANOSENSORY SIGNALS Animals in close contact with one another can communicate by touch. A classic case of mechanosensory communication is the dance of honey bees (*Apis* spp.), first described by Karl von Frisch. Honey bees have a spectacular ability to navigate and can accurately communicate the location of food sources as far away as 10 kilometers. When a forager bee finds food, she returns to the hive and communicates her discovery to her hivemates by performing a vigorous **waggle dance** in the dark hive on the vertical surface of the honeycomb. Other bees follow the dancer and receive her message.

The waggle dance conveys information about both the distance and the direction of the food source. The dancing bee repeatedly traces out a figure-eight pattern as she runs on the honeycomb. She alternates half-circles to the left and right with vigorous wagging of her abdomen in the short, straight run between turns (**Figure 53.18**). Bees use the sun as their compass, and the angle of the straight run indicates the direction of the food source relative to the position of the sun projected down to the horizon. Even under cloudy conditions, the forager can provide directions to the food source because she can see polarized light. The bee's circadian clock allows the dancer to adjust her dance to take into account the sun's movement during her return flight. The clocks of the recruits enable them to adjust their flight direction to accommodate the sun's movement.

The distance to the food source is communicated by the duration of the waggle portion of the dance. The farther away the food source is, the longer the duration of each waggle run. When food is close to the hive, the waggle portion of the dance becomes so short that it appeared to von Frisch that it was a different dance, which he called a round dance. Recent research, however, has shown that the round and waggle dances are really on a continuum that grades one into the other. Thus the honey bee has one dance language that communicates both the distance and the direction to a food source.

When challenged to prove that the bees were not simply using an odor trail to find the indicated food source, von Frisch responded with a very common sense observation. Bees returning from a new food source fly around barriers such as buildings, but the recruits going out to the food source fly over the barriers in a "beeline," which would be impossible if they were following an odor trail. Careful observation is still one of the best tools for studying behavior.

COMMUNICATION IN MULTIPLE SENSORY MODALITIES Avoiding ambiguity is a high priority in any signaling system. Signal specificity is enhanced if multiple sensory modalities are used. Courtship behavior in fruit flies, for example, involves visual, tactile, chemical, and acoustic signals (see Figure 53.4A). The male fruit fly orients toward the female's line of vision (visual signal) and taps her body with his foreleg (tactile signal). Upon detecting pheromones in her cuticle (chemical signal), the male begins to vibrate one wing, producing a species-specific courtship song (acoustic signal). The male then extends his mouthparts to taste the female's genitalia (chemical and tactile signals); if she is receptive, he initiates copulation. If at any point sensory feedback indicates to either the male or the female that their pairing is inappropriate, the courtship abruptly ends.

RECAP 53.5

Biological rhythms allow an animal to anticipate changes in its environment. In mammals, a circadian clock located in the suprachiasmatic nuclei generates a rhythm that is entrainable by environmental information. The navigational abilities of animals range from simple piloting by landmarks to distance–direction and bicoordinate navigation. Behaviors may evolve into communication signals if the transmission of information benefits both the sender and the receiver.

- What is meant by a free-running rhythm? Describe how it can be entrained to the 24-hour day. **See pp. 1107–1108 and Figure 53.13**
- Explain the difference between piloting, distance–direction navigation, and bicoordinate navigation. **See pp. 1109–1110**
- Explain how a time-compensated solar compass works. **See pp. 1110 and Figure 53.17**
- Describe an advantage and a disadvantage for each of three modalities of communication. **See pp. 1110–1112**

When behaviors involve multiple individuals—or different species, as is the case in the chapter-opening story—we have to consider how natural selection operates on all of the interacting parties, whether they are sending signals or receiving them. This concern becomes particularly relevant in the case of social behaviors.

53.6 How Does Social Behavior Evolve?

The evolution of social behavior became a field of study in its own right in 1975, with the publication of E. O. Wilson's landmark book *Sociobiology*. We begin our consideration of this enormous field with simple interactions that involve a single male and a single female, but already we see diversity. Species differ in their mating systems, which vary from monogamous to promiscuous; in the amount of parental care they give their young; and in the degree to which the male contributes to raising the young. Beyond these relatively simple mating systems, there are associations of larger numbers of reproductive individuals in **polygynous** mating systems, in which a male has more than one mate, or **polyandrous** mating systems, in which a female has more than one mate. Even more complex interactions exist in which extended families participate in raising young, and finally there are societies such as honey bee colonies in which large numbers of nonreproductive individuals assist a single reproductive individual. Sociobiology seeks to understand the evolution of the diversity of social behaviors by asking how the behaviors contribute to the fitness of *all* of the individuals involved.

Mating systems maximize the fitness of both partners

At the start of Chapter 7 we learned about the mating behavior of two species of voles. Prairie voles (*Microtus ochrogaster*) are monogamous, forming strong pair bonds that can last for life, and both parents participate in rearing the young. In contrast, montane voles (*M. montanus*) are promiscuous: males mate with many females, and the young are raised by the females alone. Behavioral physiologists have explained the proximate mechanisms behind these stark behavioral differences in terms of the release of neurohormones and the distribution of the receptors for those hormones in the brains of the two species. The ultimate question—and the one asked here—is why two such different mating systems evolved in two species that are so closely related.

We begin with the premise that there is an asymmetry in the contributions of male and female animals to their offspring at the time of fertilization. Females produce a limited number of eggs, and each egg is generously stocked with resources. Males produce an almost infinite number of sperm, which contain next to no resources. So the energetic and opportunity costs of reproduction are greater for the female than for the male. In mammals this asymmetry increases throughout gestation as the female bears most of the costs. By the time of birth or hatching, the female's investment in the young is much greater than the male's investment, and the main way for the female to maximize her fitness is to make sure her young are healthy and survive to pass on her genes.

The male has different options for maximizing his fitness. He can simply move on after inseminating the female and seek additional mates as a means of maximizing his reproductive success—as in the case of the meadow vole. Or he can stay with the female he inseminated, protect her, and help care for their young—as in the case of the prairie vole. Which strategy maximizes his fitness depends on a number of factors that are influenced by the species' environment, such as the likelihood that a female and her offspring will survive without a male's help, and a male's likelihood of finding another fertile female. Thus sociobiologists seek to quantify these factors in nature as a means of explaining observed differences in mating systems.

POLYGYNY Systems in which a male has more than one mate involve a different asymmetry. In situations in which a male can sequester a group of females from other males, he can increase his fitness by increasing the number of females in his group.

As we saw in Section 53.4, male elephant seals accomplish this by protecting an area of beach where females give birth. Male baboons do so by herding females. Male red-winged blackbirds may acquire more than one mate by defending high-quality nesting territories where females prefer to build their nests. Since sex ratios in all these species are close to 50:50, a large differential in male fitness is established, with some males having high reproductive success while many males have none. Thus selection favors males that are successful in competing with other males to obtain and protect access to many females. In general, bigger, stronger males are the winners, and sexual dimorphism in body size evolves. The elephant seal is an extreme example: males may weigh more than three times as much as females. When species with polygynous mating systems are compared, there is a strong correlation between the number of females a male controls and the degree of sexual dimorphism.

Why do females participate in these polygynous mating systems? Why doesn't a female seek out a nice, kind, noncompetitive male? In some cases she has no choice. If a female elephant seal wants to have her pup on a safe beach, she must enter the territory of a male. If a female red-winged blackbird wants to nest in an optimal territory, she will have to share the attentions of the territory's owner with other females. However, even if the female has a choice of mates, she is likely to maximize her fitness by mating with a male that is strong and dominant enough to control a number of females. Why? If her mate is a dominant male, her male offspring are likely to have their father's traits, become dominant males, and give her more grandchildren. The ultimate result of females selecting males for their prowess and dominance in male–male competition is the lek mating system (see Figure 53.10C), in which the *only* thing a male offers a female is the display of his dominance over other males.

POLYANDRY Mating systems in which one female mates with multiple males are relatively rare, but it is seen in some birds and a few mammal species in which paternal care for the young can have a large effect on fitness. An example of a polyandrous species is the golden lion tamarin (*Leontopithecus rosalia*), a primate native to Brazil's tropical rainforests (**Figure 53.19**). Tamarins are very small—adults weigh less than 1 kilogram—and so face high predation pressure. Females usually give birth to twins, and thus newborns constitute a higher percentage of maternal weight than is typical of other primates. The young also grow more rapidly than other primates, so nursing costs are high. For all these reasons, young tamarins cared for by their mother alone are unlikely to survive.

What can a male tamarin do to help guarantee his reproductive success? Watching out for predators is one obvious contribution; gathering food for the female and her young is another. Like other primate parents, tamarins carry their young most of the time, but most other primates have single offspring. When tamarin mothers are carrying twins, they spend 92 percent of the time resting, compared with 58 percent of the time when they are not carrying young. Resting is not compatible with foraging and filling the mother's high energy requirements. When a male is present, however, he carries the young about

Leontopithecus rosalia

53.19 Polyandry in a Small Primate The endangered golden lion tamarins of Brazil are small primates whose unique life history has given rise to polyandry in some groups, with males playing a major role in rearing the young.

one-third of the time, so the mother has much more time for foraging and feeding.

If one male tamarin is helpful in protecting and raising young, then two should be even more helpful. Some females can attract a second mate by being sexually receptive to him. Neither male can be sure that any eventual offspring are his, so it is in the best interest of both to help in their rearing. Of the social groups observed in field studies, only 22 percent had one male and one female, whereas 61 percent had multiple males and one female.

Fitness can include more than your own offspring

As humans, we readily understand the concept of extended family—brothers, sisters, aunts, uncles, nieces, nephews. Extended families are a common form of social organization in other species as well, and members of these families may cooperate in territory defense, predator avoidance, foraging, and rearing of young. If behavior is favored when it increases the fitness of the individual performing it, then how can we explain the evolution of social behaviors that do not lead to the performer having more offspring and that may even appear to be **altruistic**—benefiting another individual at a cost to the performer?

An individual's fitness is increased by having offspring because those offspring carry the parent's genes into the next generation. Fitness gained by producing offspring is referred to as **direct fitness**. However, an individual's genes are carried into the next generation by more than his or her own offspring. In diploid organisms, two offspring of the same parents share, on average, 50 percent of the same alleles, and an individual is likely to share 25 percent of its alleles with its siblings' offspring (nieces or nephews). Therefore, by helping parents and other relatives raise their offspring, an individual

Aphelocoma coerulescens

53.20 Helpers at the Nest Young Florida scrub-jays often forego reproduction in their first few years of adulthood to help their parents raise their siblings. These young birds help their parents feed the nestlings, defend the territory, and protect the nest from predators.

increases the transmission of those shared alleles to the next generation. **Inclusive fitness** is the individual's direct fitness plus its **indirect fitness**: the reproductive success of the individual's relatives, to the extent that those relatives share the individual's alleles.

The maximization of inclusive fitness is the mechanism driving **kin selection**, selection for behaviors that increase the reproductive success of relatives even when they come at a cost to the performer. One example is "helping at the nest" behavior, which was studied extensively in Florida scrub-jays (*Aphelocoma coerulescens*) by Glen Woolfenden and John Fitzpatrick. Scrub-jay pairs mate for life and establish large territories, which they defend aggressively. The mating pair may be assisted in rearing their young by three to five helpers (**Figure 53.20**). The helpers guard against predators, feed the young, clean the nest, and fly with fledglings. Why are these birds helping others rather than rearing their own young? Through a long-term study, Woolfenden and Fitzpatrick were able to establish a number of important facts:

- The helpers are prior offspring of the mating pair and are usually 1 to 3 years old.
- Young birds that attempt to breed have almost zero reproductive success.
- Mating pairs with helpers have approximately three times the reproductive success of those without helpers.

These results support the conclusion that helper scrub-jays are maximizing their inclusive fitness by helping their parents raise siblings until they are mature enough to have a reasonable probability of successfully raising their own offspring.

The concept of kin selection was formalized by W. D. Hamilton in what has become known as **Hamilton's rule**. He argued that, for an apparent altruistic behavior to be adaptive, the fitness benefit of that act to the recipient times the degree of relatedness between the performer and the recipient has to be greater than the cost to the performer. This relationship was clearly stated years before by the eminent geneticist J. B. S. Haldane, who said during an argument about altruism that he would not be willing to risk his life to save his brother, but for two brothers or eight cousins, he would consider it.

Eusociality is the extreme result of kin selection

Hamilton's rule can be applied to explain **eusociality**: social groups that include nonreproductive members (that is, members who as individuals cannot reproduce). The most obvious examples of eusociality occur among the Hymenoptera, an insect group that includes wasps, bees, and ants. In a honey bee colony, for example, the thousands of individuals in the colony are sterile females.

Go to Media Clip 53.1
Social Shrimps
Life10e.com/mc53.1

The key to understanding the evolution of eusociality in hymenopterans is their sex determination mechanism, **haplodiploidy**, in which diploid individuals are female and haploid individuals are male. The queen carries a lifetime supply of sperm obtained during her single mating flight, and she controls whether her eggs are fertilized or not. An unfertilized egg develops into a haploid male; a fertilized egg develops into a diploid female. The queen's daughters share all of their father's genes and, on average, half of their mother's genes. As a result, the sterile female workers in the hive—all sisters—share, on average, 75 percent of their alleles (**Figure 53.21**). Were these females to reproduce, they would share only 50 percent of their alleles with their own female offspring. Thus they potentially increase their inclusive fitness more by raising sisters than by producing and caring for their own offspring.

Eusociality may also arise if it is costly or dangerous to establish new colonies. Nearly all eusocial animals construct elaborate nests or burrow systems within which their offspring are reared. Such a structure represents an enormous investment of resources. Naked mole-rats are eusocial mammals that live in elaborate underground tunnel systems (**Figure 53.22**). A colony includes 70 to 80 individuals but only 1 reproductive female and a few reproductive males. The other colony members are sterile workers that dig and maintain the tunnels, guard against intruders, harvest food (tubers), and use their feces to feed the queen and her offspring. Individuals attempting to found new colonies have a high risk of failing or being captured by predators. When chances of individual reproductive success are practically zero,

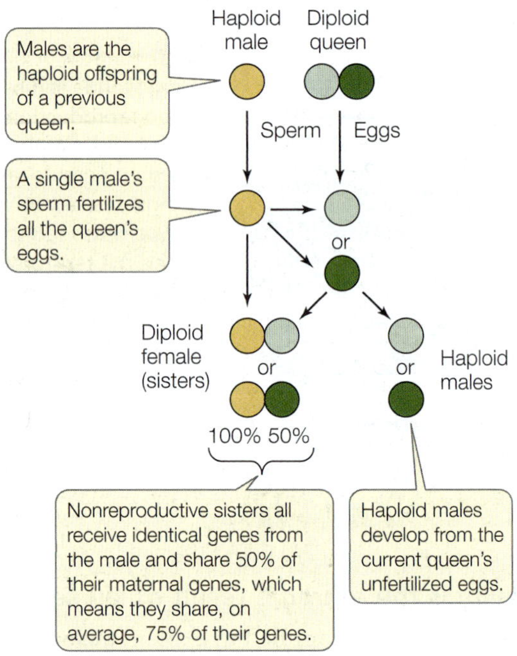
Haploid male
Diploid queen
Males are the haploid offspring of a previous queen.
Sperm
Eggs
A single male's sperm fertilizes all the queen's eggs.
or
Diploid female (sisters)
or
or
Haploid males
100% 50%
Nonreproductive sisters all receive identical genes from the male and share 50% of their maternal genes, which means they share, on average, 75% of their genes.
Haploid males develop from the current queen's unfertilized eggs.

53.21 Favoring Sisters over Daughters Female honey bees are diploid and males are haploid. Thus if a female worker bee were to reproduce, she would share approximately 50 percent of her genes with her daughters. However, she shares an average of 75 percent of her genes with her sisters. In terms of inclusive fitness, then, a sister is more valuable than a daughter for this species.

an individual can best maximize its inclusive fitness by staying with and helping maintain the colony.

Group living has benefits and costs

Apart from their direct influences on reproductive success, social systems can contribute to survival in many ways, but they can also involve costs. Thus the cost–benefit approach of behavioral ecology is relevant to understanding the evolution of social behavior.

Heterocephalus glaber

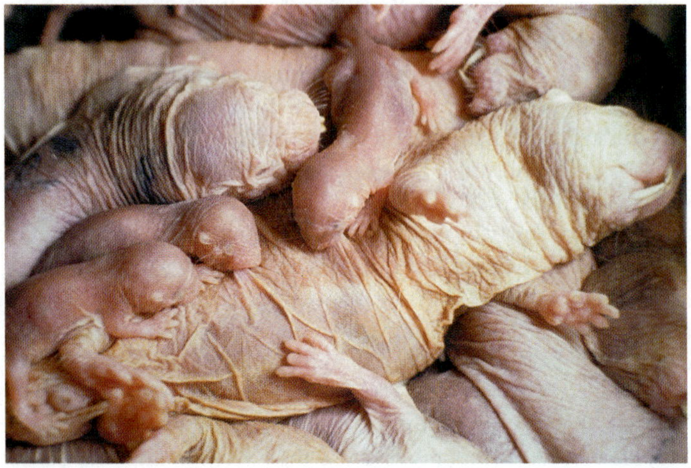

53.22 A Eusocial Mammal Naked mole-rats live in a large colony with one reproductive female and a few reproductive males. Their home is an elaborate tunnel system excavated by the colony over time.

An obvious example of a benefit of group living is improved foraging efficiency. By hunting in packs, African wild dogs (see Figure 59.17) employ cooperative strategies that enable them to bring down larger prey than could a single dog. The larger the pack, the greater the hunting success rate. Once the prey is killed, the presence of conspecifics also reduces the risk that the wild dogs will lose their prey to larger scavengers, such as hyenas.

Living in a group can also reduce the risk of individuals becoming prey themselves. Many small birds forage in flocks. To test the hypothesis that flocking provides protection against predators, R. E. Kenward released a trained goshawk (*Accipiter gentilis*) near wild common wood-pigeons (*Columba palumbus*) in England. The hawk was most successful when it attacked solitary pigeons. Its success in capturing a pigeon in a flock decreased as the number of pigeons in the flock increased (**Figure 53.23A**). The larger the flock, the sooner some individual in the flock spotted the hawk and flew away. This escape behavior stimulated other individuals in the flock to take flight as well.

Alarm calling is another means of reducing predation risk, but the caller incurs a risk cost by calling attention to itself. Belding's ground squirrels live in large colonies in open meadows. When one squirrel announces the presence of a predator with loud, sharp barks, all the nearby squirrels dive into their burrows (**Figure 53.23B**). Paul Sherman showed that callers double their risk of being preyed on—so why do they do it? Research by Sherman and by others has shown that this altruistic behavior is a product of kin selection. In this polygynous species, males establish large territories in the spring that include the territories of several females, whom they inseminate. The females then drive off the males. Female offspring settle near their mothers, so neighboring females in a colony tend to be sisters, and they defend each other's young. Sherman showed that males are less likely to give alarm calls than females, and that females are more likely to give alarm calls when related individuals are nearby.

Social behavior has many costs as well as benefits. Foraging in a group may reduce the amount of food available to each individual, and the foraging individuals may interfere with one another's foraging activities. Individuals living in groups may face more competition for mates, as well as for food, than solitary individuals would. A large group may actually attract the attention of predators. And living at high population densities can increase the risk of disease transmission. The study of disease transmission in wild animal populations is a relatively new field, but such studies have made it apparent that species living in social groups are more prone to outbreaks of disease than are solitary species.

Can the concepts of sociobiology be applied to humans?

Ever since the publication of E. O. Wilson's *Sociobiology*, applications of the concepts of evolutionary genetics to human behavior have been hugely controversial. The intensity of the debate may stem from impressions that sociobiological approaches resemble previously discredited pseudoscientific movements such as social Darwinism, biological determinism, and eugenics, all of which have been used as rationales

(A)
Goshawk
Wood pigeon
The more pigeons in the flock, the sooner the hawk is spotted…
Hawk's distance (meters) when spotted by pigeons (●)
Hawk's attack success (%) (●)
80
60
40
20
1
2–10
11–50
>50
Number of pigeons in flock
…and the lower the hawk's attack success.
(B) *Spermophilus beldingi*

53.23 Group Living Provides Protection from Predators Animals that live in groups can spread the cost of looking out for predators. (A) The larger the number of common wood-pigeons in a flock, the greater the chances that one of the pigeons will spot a predatory goshawk before it attacks, and the lower the chances that the hawk will capture one of the pigeons. (B) A male Belding's ground squirrel gives an alarm call upon spotting a predator. Although this behavior increases his individual risk of becoming prey, he increases the survival chances of many of his close relatives.

for racism and discrimination. The proponents of sociobiology maintain that theirs is an objective science and that questions of "what is" should not be conflated with questions of "what ought to be." For example, we can demonstrate the genetic basis of sexual dimorphism in body size and muscle mass in humans, and we can compare this dimorphism with that in other mammals. However, any attempt to use such data as a political or legal defense for polygamy would not be science and should not be confused with science.

The fact that our biochemistry, our cell biology, physiology, and anatomy are shaped by our genes is beyond argument. However, it is also clear that these genetically shaped characteristics are also influenced by factors such as environment, nutrition, social interactions, and culture. Why should it be different for behavior? Studies of identical twins reared apart have produced evidence for inheritance of uncanny similarities in behavioral propensities. Studies of isolated human cultures around the world have also revealed remarkable similarities in social organization. None of these studies, however, would even begin to challenge the dominant role of learning and culture in the shaping of human behavior.

RECAP 53.6

Social behavior can be understood by asking how it contributes to the fitness of the individuals involved. Asymmetry between the sexes in parental investment is a key factor in the evolution of mating systems. According to the theory of kin selection, an individual can increase its fitness by helping related individuals with whom it shares alleles. In extreme cases, kin selection has given rise to eusociality.

- What environmental conditions can lead to monogamy, polygamy, or polyandry? **See pp. 1113–1114**
- Explain how an individual can increase its fitness by helping its relatives. **See pp. 1114–1115**
- Why is eusociality so common among hymenopterans? **See p. 1115 and Figure 53.21**
- What are some of the costs and benefits of group living? **See p. 1116 and Figure 53.23**

Knowledge of the behavior of particular species—how they use the environment, how they obtain food, how they organize their activities spatially and temporally—is essential for understanding how species interact in nature. These interactions are one focus of the science of ecology, the subject of Part Ten of this book.

?

Could cowbird behaviors create selective pressure for host species *not* to develop egg discrimination behavior?

ANSWER

Results of studies by Jeffrey Hoover and Scott Robinson, some of which were described in the Working with Data exercise on p. 1102, showed that because of nest destruction and other behaviors by cowbirds, hosts that tolerated the cowbird nest parasitism may successfully rear more of their own offspring than those that discriminate against cowbird eggs—a selection pressure on host birds not to discriminate. Of course, when the experimenters prevented cowbird access to the host nests, the hosts had even greater reproductive success. Thus the host birds were in effect paying "protection money" in the form of nurturing cowbird eggs (and thus reducing but not totally eliminating their own reproductive success), leading the researchers to dub the cowbirds' actions "mafia behavior."

CHAPTER SUMMARY 53

53.1 What Are the Origins of Behavioral Biology?

- Ivan Pavlov's discovery of **conditioned reflexes** and B. F. Skinner's research on **operant conditioning** as a model for learning led to an approach called behaviorism that mainly carried out laboratory experiments on rats and a few other animal models. **Review Figure 53.1**
- **Ethology** focuses on both the **proximate causes** of behavior (the immediate cause of the behavior, and how the behavior develops) and on the **ultimate causes** (how the behavior affects the animal's evolutionary fitness).
- A major focus of the ethologists was **fixed action patterns** and their **releasers**. They performed **deprivation experiments** as well as breeding experiments to demonstrate that certain behaviors are genetically determined. **Review Figure 53.2**.

53.2 How Do Genes Influence Behavior?

- Breeding experiments can reveal whether a behavioral phenotype is inherited. Quantitative trait analysis can reveal candidate genes that influence a behavior. Gene knockout experiments can reveal the roles of specific genes underlying a behavioral phenotype. **Review Figure 53.3**
- Most behaviors are complex traits involving many genes that function in cascades and offer many points for a change in a single gene to influence behavior. **Review Figure 53.4**

53.3 How Does Behavior Develop?

- Hormones can determine the pattern of behavior that develops and the timing of its expression. **Review Figure 53.5**
- **Imprinting** is a process by which an animal learns a specific set of stimuli during a limited **critical** or **sensitive period**. That critical period may be determined by hormones.
- The development and expression of song in white-crowned sparrows involves a genetic predisposition to learn the species-specific song, a critical period for imprinting of a song memory, and hormonally controlled timing of song expression. Social interactions may also play a role. **Review Figures 53.7, 53.8**

53.4 How Does Behavior Evolve?

- An animal's behavior involves a series of choices that influence its fitness. To make these choices, animals use environmental cues that are reliable predictors of the potential effects of their choice on their fitness.
- The **cost–benefit approach** can be used to investigate the fitness value of specific behaviors. The cost of a behavior typically has three components: **energetic cost**, **risk cost**, and **opportunity cost**. **Review Figure 53.9, ANIMATED TUTORIAL 53.1**
- According to **optimal foraging theory**, animals should practice feeding behaviors that maximize their energetic gain at the least cost. **Review Figure 53.11, ANIMATED TUTORIAL 53.2**

53.5 What Physiological Mechanisms Underlie Behavior?

- **Circadian rhythms** control the daily cycle of behavior. Without environmental time cues, circadian rhythms free-run with a period that is genetically programmed. They are normally **entrained** to the light–dark cycle by environmental cues. **Review Figure 53.13, ANIMATED TUTORIAL 53.3**
- Forms of navigation used by animals to find their way in the environment include **piloting** (orienting to landmarks), distance–direction navigation, and bicoordinate navigation. Navigation mechanisms include celestial navigation and a time-compensated solar compass. **Review Figures 53.15–53.17, ANIMATED TUTORIALS 53.4, 53.5**
- The behaviors of individuals may become **communication signals** if the transmission of information benefits both the sender and the receiver. **Review Figure 53.18, ACTIVITY 53.1**
- Chemical communication signals (pheromones) can be highly specific and have different time courses. Visual signals can convey complex messages rapidly, but only if the recipient can see the sender. Acoustic signals can travel over long distances, do not require a focused recipient, and can be modified to reveal or conceal directional information. Tactile signals can convey complex messages when animals are in close proximity.

53.6 How Does Social Behavior Evolve?

- **Polygynous** mating systems, in which one male controls and mates with many females, can result in great variation in male reproductive success. **Polyandry**—a female mating with multiple males—can evolve in circumstances in which a male can make a substantial contribution to the survival of his offspring.
- The fitness an individual gains by producing offspring (**direct fitness**) plus the fitness it gains by increasing the reproductive success of relatives with whom it shares alleles (**indirect fitness**) is called **inclusive fitness**. **Kin selection** may favor **altruistic** behavior toward relatives, despite its cost to the performer, if it increases the performer's inclusive fitness.
- As a result of **haplodiploidy**, the sex determination mechanism of hymenopteran insects, nonreproductive female workers (sisters) share more alleles with one another than reproductive females share with their own offspring. **Review Figure 53.21**
- Haplodiploidy has probably facilitated the evolution of **eusocial** behavior in this group through kin selection. Eusociality has also arisen in diploid species in which chances of individual reproductive success are extremely low.
- Group living confers benefits such as greater foraging efficiency and protection from predators, but it also has costs, such as increased competition for food and ease of transmission of diseases.

See ACTIVITY 53.2 for a concept review of this chapter

Go to the Interactive Summary to review key figures, Animated Tutorials, and Activities Life10e.com/is53

CHAPTER**REVIEW**

REMEMBERING

1. Which of the following is *not* true of a fixed action pattern?
   a. Its expression may depend on hormonal conditions.
   b. It is induced by complex, species-specific stimuli.
   c. It is highly stereotypic and species-specific.
   d. It can be expressed even if the animal has never seen it performed.
   e. Its genetic basis can be demonstrated by breeding experiments.

2. Which of the following is *not* a component of the cost of performing a behavior?
   a. Its energetic cost
   b. The risk of being injured
   c. Its opportunity cost
   d. The risk of being attacked by a predator
   e. Its information cost

3. Birds that migrate at night
   a. inherit a star map.
   b. determine direction by knowing the time and the position of a constellation in the rotating night sky.
   c. orient to a fixed point in the rotating night sky.
   d. imprint on one or more key constellations.
   e. determine distance, but not direction, from the stars.

4. If a bird is trained to seek food on the western side of a cage open to the sky, and is then placed in a chamber with a controlled light cycle so that its circadian rhythm becomes phase-delayed by 6 hours (i.e., its circadian rhythm is 6 hours behind real time), when it is returned to the open cage at noon in real time, it will seek food in the
   a. north.
   b. south.
   c. east.
   d. west.

5. Which of the following statements about communication is true?
   a. Complex information cannot be conveyed by pheromones.
   b. Visual signaling is advantageous in complex environments.
   c. Acoustic communication always reveals the location of the signaler.
   d. One advantage of pheromones is that the message can persist over time.
   e. The dance of honey bees is an example of visual signaling.

6. A cost commonly associated with group living is
   a. increased risk of predation.
   b. interference with foraging.
   c. increased exposure to diseases and parasites.
   d. increased competition for mates.
   e. All of the above

7. Altruistic behavior
   a. can increase an individual's inclusive fitness.
   b. depends on haplodiploid sex determination.
   c. is most common among unrelated individuals.
   d. always causes a net decrease in the performer's fitness.
   e. characterizes a monogamous mating system.

8. A group is said to be eusocial if
   a. the group's members interact intensively.
   b. some members produce many more offspring than others do.
   c. a dominance hierarchy exists among group members.
   d. young individuals remain in the group to help their parents rear other offspring.
   e. the group contains nonreproductive helper individuals.

UNDERSTANDING & APPLYING

9. Adult male dogs lift a hind leg when they urinate, whereas young puppies and adult female dogs squat. If a newborn male puppy receives an injection of estrogen, it will never lift its leg to urinate; for the rest of its life, it will always squat. How might this result be explained?

10. In most vertebrate species with helpers, the helpers are individuals capable of reproducing, and often breed later. Among eusocial insects, sterile castes have evolved repeatedly. What differences between vertebrates and insects might explain the failure of sterile castes to evolve in vertebrates?

11. Studies of birdsong in zebra finches showed that males appear to have "practice" (undirected) songs and "performance" (directed) songs (see Figure 53.8). Field studies show that white-crowned sparrow song has local dialects and that the songs of male birds living only a few kilometers apart are distinguishable. Assuming white-crowned sparrows, like zebra finches, have directed and undirected songs, what could be the adaptive significance of this "practice" versus "performance" behavior?

ANALYZING & EVALUATING

12. Cowbirds are nest parasites, as seen in the opening story of this chapter. What do you think would characterize the acquisition of song in cowbirds? In a given geographical region, cowbirds tend to parasitize the nests of particular bird species. How do you think female cowbirds learn this behavior? How would you test your hypothesis?

13. Some honey bee hives show hygienic behavior: if pupae die in their cells, some workers uncap the cells while other workers remove the carcasses. Some hives do not show this behavior. In a classical behavior genetics study, crosses were made between hygienic and nonhygienic hives. The results were that about 25% of the resulting hives were hygienic and about 75% were not. Of those that were not, about one-third showed uncapping behavior and about one-third showed carcass removal behavior if cells were uncapped. But there was considerable variability in these data. A recent gene-mapping study (using QTL analysis) revealed seven significant correlations between specific allele frequencies and hygienic behavior. What do the classical data suggest about the genetic basis for hygienic behavior? What do the gene-mapping data suggest? How can you resolve the differences?

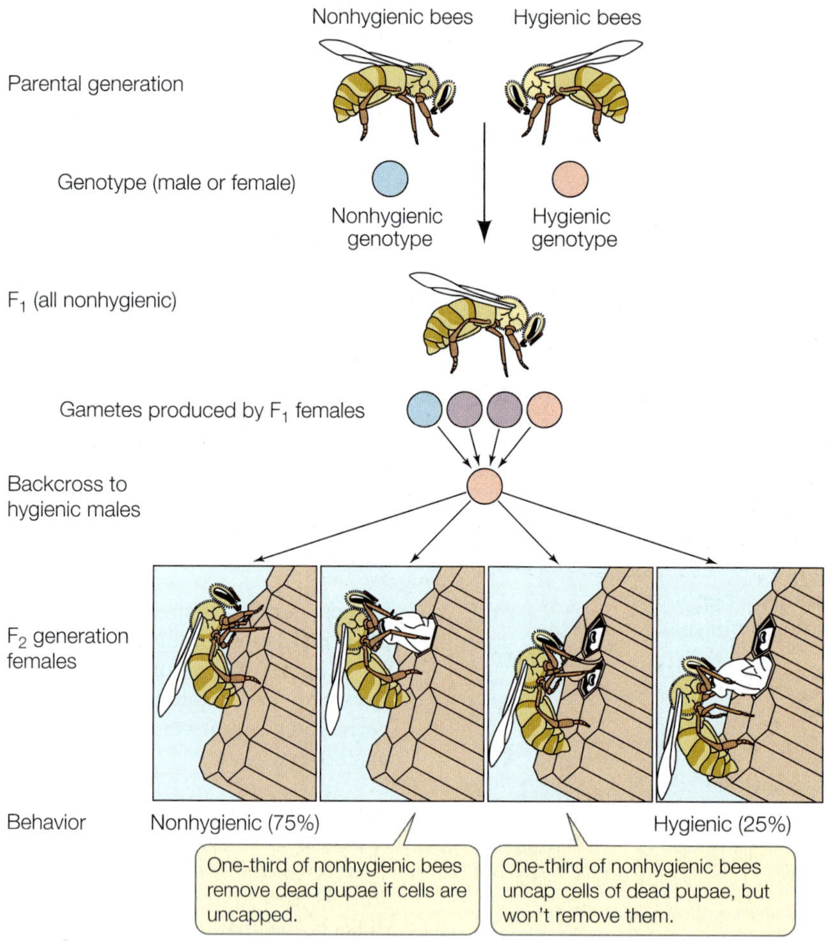
Nonhygienic bees
Hygienic bees
Parental generation
Genotype (male or female)
Nonhygienic genotype
Hygienic genotype
F_1 (all nonhygienic)
Gametes produced by F_1 females
Backcross to hygienic males
F_2 generation females
Behavior
Nonhygienic (75%)
Hygienic (25%)
One-third of nonhygienic bees remove dead pupae if cells are uncapped.
One-third of nonhygienic bees uncap cells of dead pupae, but won't remove them.

Go to BioPortal at **yourBioPortal.com** for Animated Tutorials, Activities, LearningCurve Quizzes, Flashcards, and many other study and review resources.

54 Ecology and the Distribution of Life

CHAPTER**OUTLINE**

Fynbos in the Spring This unique vegetation community is restricted to a small region on the western coast of the South African Cape.

FYNBOS IS AN AFRIKAANS WORD derived from the Dutch *fijnbosch*, meaning "fine bush." It is the name given to an unusual assemblage of plant species found in the western Cape region of South Africa. The Dutch colonists who settled there were probably disappointed by the absence of large, sturdy trees; most of the region's shrubby vegetation lacks the heft desirable for construction. Most of the plants also have small, slender leaves that are too leathery and tough for any purpose useful to humans.

Fynbos plants may not be useful for construction, but they include some of the world's most beautiful and popular garden flowers, including geraniums, gladioli, and proteas. Although it covers less than 90,000 square kilometers (smaller than the state of Maine), the fynbos is home to 6,000 endemic species—that is, species found nowhere else in the world.

The fynbos is prone to sweeping wildfires every 15 to 30 years. These fires kill mature plants and trees, which are replaced by vigorous new plants that germinate from seeds buried in the soil or stored in fire-safe cones. If fires are too infrequent, forest trees become established and displace the unique fynbos species. About one-third of these plants depend on ants for dispersal and germination, and their seeds are equipped with a fleshy, lipid-rich structure (an elaiosome). Ants pick a seed up, nibble off the elaiosome, and then store the seed for later consumption, burying it deep enough to avoid injury by fire.

Today more than 1,700 fynbos species are threatened with extinction. For a long time, conservation efforts were stymied by the inability of fynbos seeds to germinate in the absence of extensive wildfires. In 1990 two South African botanists discovered that seeds of the rare fynbos plant *Audounia capitata* germinated in response to smoke from burning wood. Even the "liquid smoke" produced by the food-flavoring industry turned out to be enough to stimulate germination. Investigators in Australia and California then found that the chemicals in smoke stimulate germination in plants in other, entirely different families that also depend on intermittent fires for maintenance.

Fynbos is found only in a small area on the western coast of the South African Cape. Yet this *type* of vegetation—tough, shrubby, and fire-adapted—is characteristic of many areas around the world that experience hot, dry summers and cool, wet winters. This weather pattern is known as a Mediterranean climate, and similar plant communities are found not only along the Mediterranean coasts of Europe and the Near East but also along the western coasts of Australia, North America, and South America.

> ?
>
> What is it about the western edges of continents that promotes tough, shrubby plant communities such as fynbos?
>
> See answer on p. 1146.

54.1 What Is Ecology?

Ecology is the scientific investigation of interactions among organisms and between organisms and their physical environment. Ecology is a relatively new branch of the biological sciences; in fact, it did not even have a formal name until 1866. In 1859 Charles Darwin described the focus of *On the Origin of Species* as being "the coadaptation of organic beings to each other and to their physical conditions of life." Ernst Haeckel, a German biologist who was profoundly influenced by Darwin, constructed a new word for this new enterprise: "ecology," from the Greek root *oikos*, "household," where "household" embraces all of an organism's environment. As Haeckel put it, "Ecology is the study of all those complex interrelations referred to by Darwin as 'the conditions of the struggle for existence.'"

Ecology provides explanations of the perceptible, palpable world. Although the *consequences* of enzyme–substrate interactions may be visible, the interactions themselves are not readily visible to the casual observer. In contrast, interactions among organisms, and between organisms and their environment, can often be observed. Analysis of those observations, however, often requires persistence, ingenuity, and additional investigation. That butterflies visit flowers has been observed and admired for centuries; that butterflies perform an essential service to the flowers they visit by transporting pollen is an ecological insight less than 250 years old.

The need for sound science in making decisions about our own species' interactions with the environment is one key reason for studying ecology. Humans are part of the biotic environment, and our activities have profound effects on a tremendous variety of other organisms, as well as having incalculable effects on abiotic energy flow and nutrient cycling through the physical environment. An understanding of ecology greatly improves our ability to grow food for ourselves reliably and sustainably, to manage pests and diseases safely and effectively, and to deal with natural disasters such as floods and fires. The greater our understanding of ecological interactions, the greater the likelihood that we can accomplish these things without causing a cascade of unanticipated consequences for ourselves and other organisms.

Ecology is not the same as environmentalism

In defining what ecology is, it is important to emphasize what ecology is *not*. "Ecology" is sometimes equated with "environmentalism," but the two terms are not equivalent. Ecology is a science that generates knowledge about interactions in the natural world; as a field of inquiry, it is not inherently focused on human concerns. **Environmentalism** is the use of ecological knowledge, along with economics, ethics, and many other considerations, to inform both personal decisions and public policy relating to stewardship of natural resources and ecosystems.

Ecologists study biotic and abiotic components of ecosystems

From its beginnings, ecology has encompassed both the living, or **biotic**, components and the physical and chemical, or **abiotic**, components of ecosystems. The abiotic, physical characteristics of Earth's atmosphere, for example, determine surface temperatures and precipitation patterns, which in turn limit where organisms can live. The biotic components of an organism's environment are other organisms, so ecology includes the study of interactions within species and between species. Ecology also encompasses the study of the movement of energy and nutrients through **ecosystems**—the networks of interacting organisms in an area and the physical environment they occupy.

In Haeckel's time, a main concern of ecology was understanding the distribution and abundance of organisms. That remains true today, but the field has advanced and diversified considerably. A continuous influx of new tools—mathematical models, molecular techniques, and satellite imaging, to name just a few—as well as new or enhanced connections to other fields (particularly the physical sciences) have dramatically changed ecological research. The ultimate goal of ecology, however, remains the same: to provide objective data on the interactions of the different components of the biotic and abiotic environments and, through analysis of the data, to understand these interactions and their various results.

RECAP 54.1

Ecology is the scientific investigation of interactions among organisms and between organisms and their physical environment.

- What are some reasons to consider ecology a useful scientific enterprise? See p. 1122
- How does ecology differ from environmentalism? See p. 1122

We will begin our study of ecology as the discipline began in Haeckel's time, focusing on factors that determine the distribution and abundance of organisms. First we will look at the physical forces that result in climate variation.

54.2 Why Do Climates Vary Geographically?

The terms "weather" and "climate" both refer to atmospheric conditions—temperature, humidity, precipitation, and wind direction and velocity—but they refer to different time scales. **Weather** is the short-term state of atmospheric conditions at a particular place and time, whereas **climate** refers to the average atmospheric conditions, and the extent of their variation, at a particular place over a longer time. In other words, climate is what you expect; weather is what you get. The responses of organisms to weather are usually short-term—seeking shelter from a sudden rainstorm, for example, or shivering to keep warm when the temperature drops. Responses to climate, on the other hand, tend to be evolutionary adaptations that arise within populations over time and affect physiology, morphology, and behavior. These adaptations are among the forces driving speciation. If organisms cannot adapt to the climate of a particular place, they will not be found there.

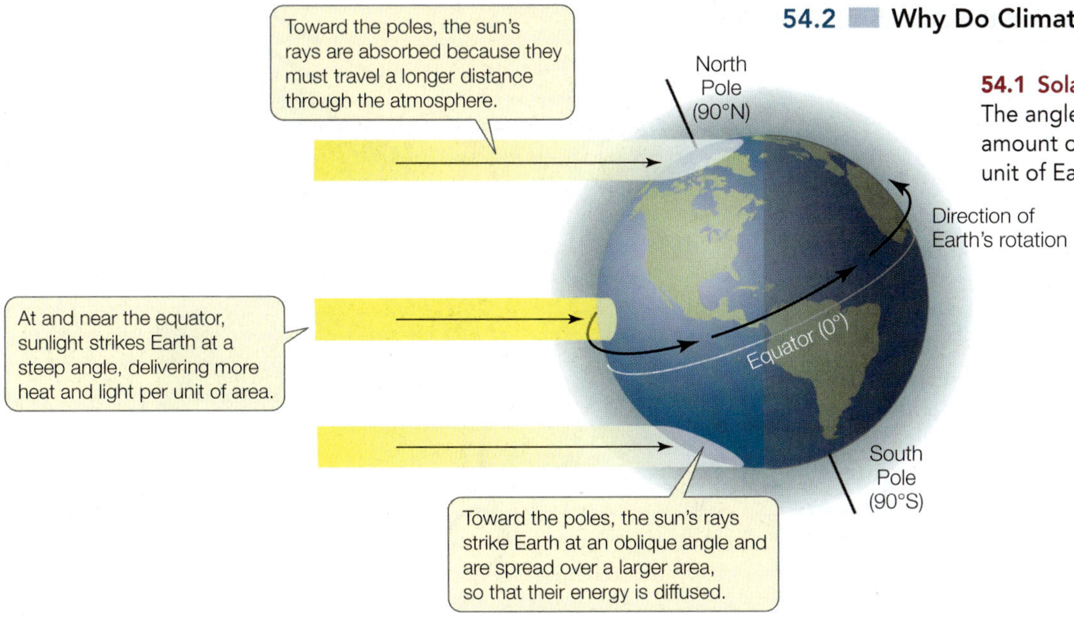
Toward the poles, the sun's rays are absorbed because they must travel a longer distance through the atmosphere.
North Pole (90°N)
Direction of Earth's rotation
At and near the equator, sunlight strikes Earth at a steep angle, delivering more heat and light per unit of area.
Equator (0°)
South Pole (90°S)
Toward the poles, the sun's rays strike Earth at an oblique angle and are spread over a larger area, so that their energy is diffused.

54.1 Solar Energy Input Varies with Latitude The angle of incoming sunlight affects the amount of solar energy that reaches a given unit of Earth's surface.

Solar radiation varies over Earth's surface

Solar energy input determines the air's temperature and is the major determinant of climate. The intensity of solar radiation varies over the course of a year and from place to place due to the shape of Earth, its orbit around the sun, and the tilt of its axis. The amount of solar energy reaching a given point on Earth's surface depends primarily on the angle of the sun's rays. At high latitudes (i.e., areas toward the North and South poles), sunlight strikes Earth's surface at an angle, so the incoming solar energy is distributed over a larger area (and thus is less intense) than at the equator, where sunlight strikes the surface perpendicularly (**Figure 54.1**). Moreover, when coming in at an angle, the sun's radiation must pass through more of Earth's atmosphere, so more of its energy is absorbed or reflected before reaching the surface. Because of this difference in solar energy input, air at the poles is colder than air at the equator. The average air temperature over the course of a year decreases about 0.76°C for every degree of latitude (about 110 km) at sea level. Air temperatures also decrease with elevation, so temperatures at sea level are warmer than temperatures on mountaintops at the same latitude.

Because Earth's axis is tilted at an angle of 23.5 degrees, the amount of sunlight a particular region of Earth receives varies over the course of a year as Earth orbits the sun (**Figure 54.2**). This tilt causes seasonal variation in temperature and day length. Higher latitudes experience greater seasonal variation than lower latitudes do. Around the equator, day length and seasonal temperatures change only slightly over the course of the year, although there are seasonal shifts in precipitation patterns.

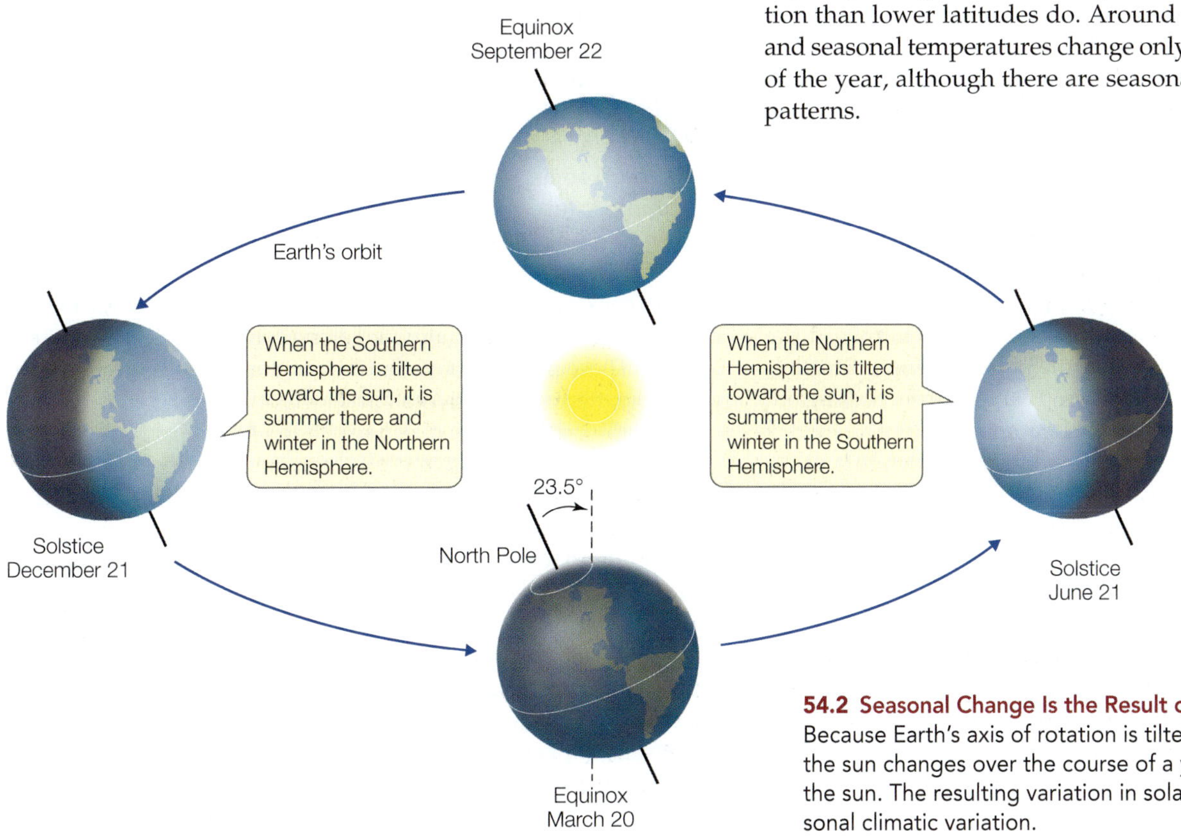
Equinox September 22
Earth's orbit
When the Southern Hemisphere is tilted toward the sun, it is summer there and winter in the Northern Hemisphere.
When the Northern Hemisphere is tilted toward the sun, it is summer there and winter in the Southern Hemisphere.
Solstice December 21
23.5°
North Pole
Solstice June 21
Equinox March 20

54.2 Seasonal Change Is the Result of the Tilt of Earth's Axis Because Earth's axis of rotation is tilted, orientation relative to the sun changes over the course of a year as the planet orbits the sun. The resulting variation in solar radiation creates seasonal climatic variation.

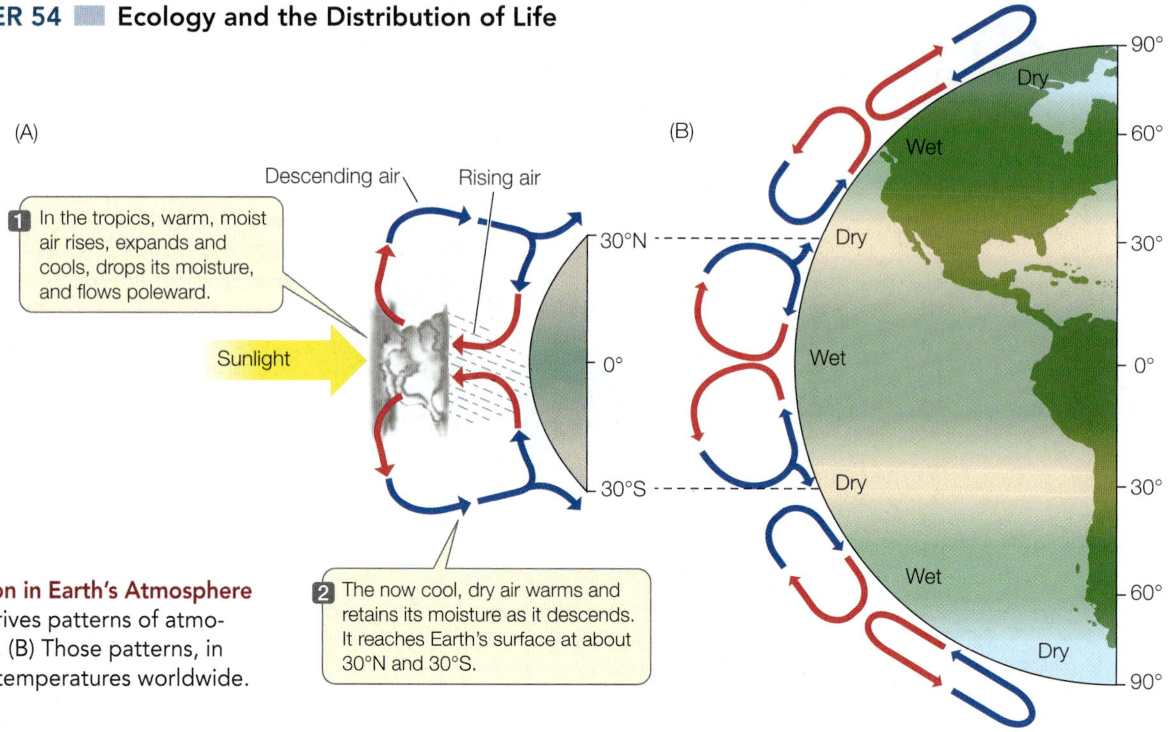
(A)
Descending air
Rising air
1 In the tropics, warm, moist air rises, expands and cools, drops its moisture, and flows poleward.
Sunlight
30°N
0°
30°S
2 The now cool, dry air warms and retains its moisture as it descends. It reaches Earth's surface at about 30°N and 30°S.
(B)
90°
Dry
60°
Wet
30°
Dry
Wet
0°
Dry
30°
Wet
60°
Dry
90°

54.3 Air Circulation in Earth's Atmosphere (A) Solar energy drives patterns of atmospheric circulation. (B) Those patterns, in turn, influence air temperatures worldwide.

Solar energy input determines atmospheric circulation patterns

Air in the region surrounding the equator receives the greatest input of solar energy. When a parcel of air is warmed, it expands, becomes less dense, and rises. As it rises, however, it cools. Cool air cannot hold as much water vapor as warm air, so the expanding, cooling air releases moisture in the form of precipitation. Thus as the sun warms air in the tropics, that air rises into the atmosphere, cools, and releases large amounts of rainfall. As it rises, it is replaced by surface air flowing in from the north and south (**Figure 54.3A**).

High in the atmosphere, the tropical air is pushed to the north and south as newly warmed air rises to replace it. As it reaches latitudes around 30°N and 30°S, it cools and sinks. This cool, dry air, which lost its moisture as it rose over the equator, now *takes up* moisture from the ground rather than releasing it. Earth's great deserts—including the Sahara of Africa, the Gobi of China, and the deserts of Australia and the American Southwest—are located at these latitudes.

While some of the descending air flows back toward the equator, some of it flows toward the poles, setting up further cyclic movements of air. At about 60° latitude, air rises again. At the poles, where there is little solar energy input, cold, dry air descends (**Figure 54.3B**). These cyclic movements of air masses are largely responsible for determining air temperatures and precipitation patterns across Earth's surface.

Atmospheric circulation and Earth's rotation result in prevailing winds

The velocity of Earth's rotation around its axis is fastest at the equator, where Earth's diameter is greatest, and slowest close to the poles. An air mass that is not moving either to the north or the south has the same rotational velocity as Earth does at the same latitude. However, as an air mass moves toward the equator (driven by the circulation patterns described above and in Figure 54.3), its rotational movement becomes slower than that of the planet beneath it and it is deflected to the west. Conversely, the rotational movement of an air mass moving toward either pole is faster than that of the surface beneath it and is deflected to the east. This interaction of Earth's rotation and north–south air mass movement sets up a pattern of circulating surface air referred to as the **prevailing winds** (**Figure 54.4**). Prevailing winds blow from east to west in the tropics (the **trade winds**); from west to east in mid-latitudes (the **westerlies**); and from east to west again above 60°N or 60°S latitude (the **easterlies**).

Prevailing winds drive ocean currents

Wind moves the water it blows over by means of frictional drag. Thus global air circulation patterns drive the circulation patterns of surface ocean waters, known as **currents**. The trade winds, for example, cause currents to converge at the equator and move westward until they encounter a continental land mass. At that point, the strong Equatorial Countercurrent brings some of the water back eastward. The remaining water divides, some moving northward and some southward along continental shores (**Figure 54.5**). These patterns of water movement set up rotating circulation patterns called **gyres** (Greek *gyros*, "spiral"). These great circular currents rotate clockwise in the Northern Hemisphere and counterclockwise in the Southern Hemisphere.

Because the ocean currents transport heat, they have a tremendous effect on Earth's climates. The poleward movement of warm water from the tropics transfers large amounts of heat to high latitudes. The Gulf Stream, for example, carries warm water from the tropical Atlantic Ocean (including the Gulf of Mexico) north across the Atlantic to northern Europe, making the European climate considerably milder than that of corresponding latitudes in North America. Similarly, currents flowing toward the equator from high latitudes bring cool, wet winters to some western coastal regions that are otherwise warm and dry.

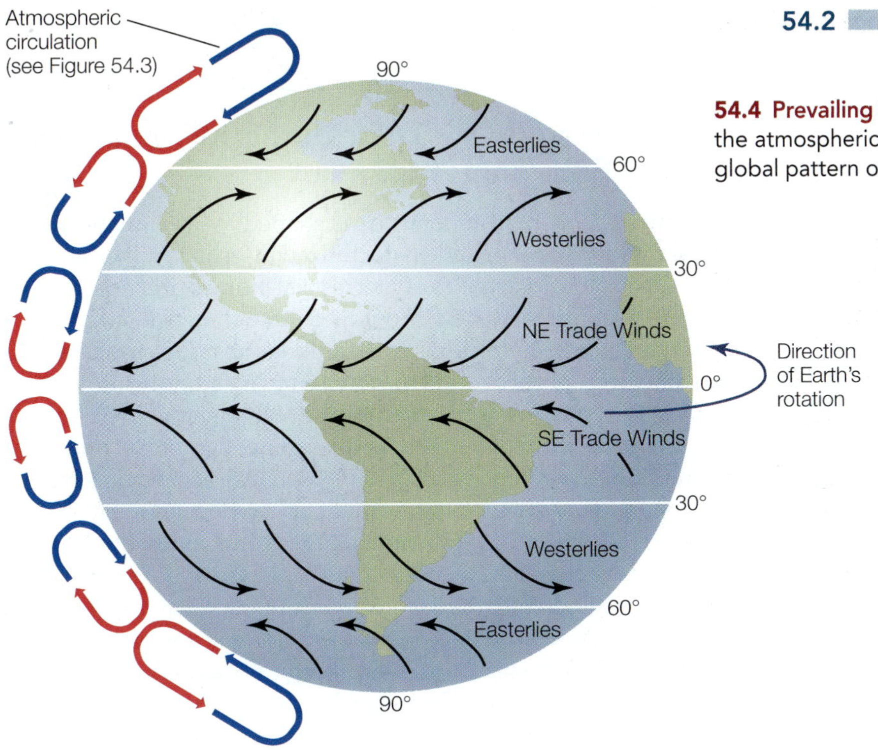
Atmospheric circulation (see Figure 54.3)
90°
Easterlies
60°
Westerlies
30°
NE Trade Winds
Direction of Earth's rotation
0°
SE Trade Winds
30°
Westerlies
60°
Easterlies
90°

54.4 Prevailing Winds The speed of Earth's rotation combines with the atmospheric circulation of air masses (see Figure 54.3) to create a global pattern of prevailing surface winds across the planet.

Organisms adapt to climatic challenges

The patterns we have just described give rise to a mosaic of climatic conditions across Earth. The climatic conditions in a region—especially temperature and precipitation—act as selective agents on the organisms that live there. As a consequence, many organisms display adaptations to climatic conditions, which can involve physiological, morphological, or behavioral specializations.

Metabolic specializations help many organisms cope with an environment that periodically becomes too hot, too cold, too dry, or short of food. Organisms that must deal with extremely low temperatures, for example, are capable of surviving by a variety of strategies, chief among which is producing antifreezes. Microbes, sponges, Arctic fishes, and many temperate-zone insects are among the organisms that produce antifreezes to lower the freezing point of their cell contents or body fluids (see Section 40.3). One of only two flowering plant species native to Antarctica, the freeze-tolerant hairgrass (*Deschampsia antarctica*), survives by producing proteins that inhibit the formation of damaging large ice crystals. In Alaska the wood frog (*Rana sylvatica*), the only amphibian found north of the Arctic Circle, can survive temperatures of –6°C for more than a month with up to 65 percent of its body fluid frozen solid (**Figure 54.6**). The frog avoids damage to its cells by allowing fluids to freeze in extracellular spaces.

Adaptation to climatic conditions is often reflected in differences in morphology. For example, some endotherms living in cold climates have proportionally rounder shapes and shorter appendages than their relatives adapted to warmer climates, which gives them a smaller surface area relative to their volume and allows them to conserve heat more easily (see Figure 40.18).

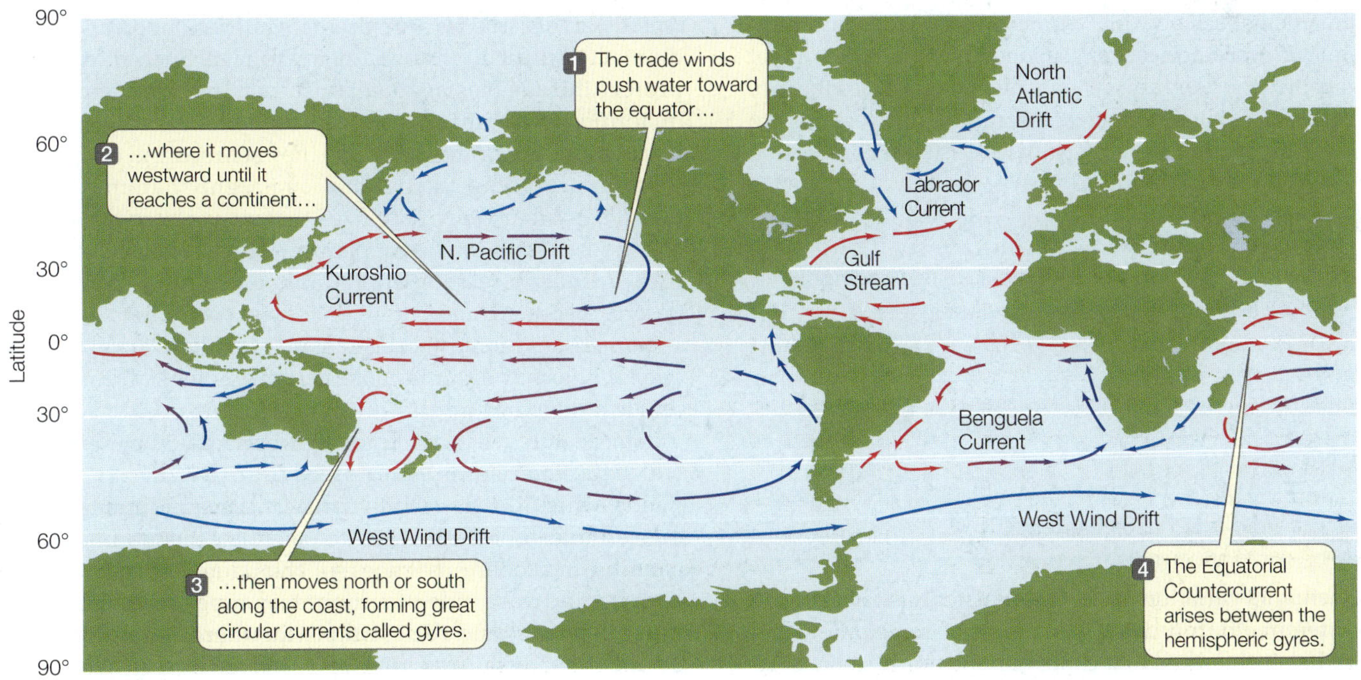
90°
60°
30°
0°
30°
60°
90°
Latitude
1 The trade winds push water toward the equator…
2 …where it moves westward until it reaches a continent…
3 …then moves north or south along the coast, forming great circular currents called gyres.
4 The Equatorial Countercurrent arises between the hemispheric gyres.
North Atlantic Drift
Labrador Current
N. Pacific Drift
Kuroshio Current
Gulf Stream
Benguela Current
West Wind Drift
West Wind Drift

54.5 Oceanic Circulation The surface currents of the ocean are driven primarily by the prevailing winds shown in Figure 54.4.

Rana sylvatica

54.6 An Evolutionary Adaptation to Climate Wood frogs survive frigid Alaskan winters by allowing up to 65 percent of their body fluids to freeze.

Behavioral mechanisms for temperature regulation often complement physiological and morphological adaptations, particularly among ectotherms. An important behavioral adaptation to changing climatic conditions is changing one's location to find a more suitable microclimate. A **microclimate** is a subset of climatic conditions in a small specific area that generally differ from those in the environment at large. For example, some desert lizards maintain their body temperature by spending time in an underground burrow at night (see Figure 40.11). While surface temperatures may fluctuate wildly, the microclimate of the burrow is buffered against such changes.

In addition to such localized movements, long-distance movements can be key adaptations to climatic challenges. Many organisms seek new places to live when local conditions deteriorate. If repeated seasonal changes alter an environment in predictable ways, organisms may evolve life cycles that appear to anticipate those changes. Migration, one response to such cyclic environmental changes, was discussed in Section 53.5.

RECAP 54.2

Latitudinal differences in solar energy input create patterns of atmospheric circulation, which in turn drive oceanic circulation. These air and oceanic circulation patterns determine Earth's climates.

- How does latitudinal variation in solar energy input drive global air circulation patterns? See p. 1124 and Figure 54.3
- How do global air circulation patterns drive ocean currents? See pp. 1124–1125 and Figure 54.5

The tremendous variation in Earth's climates has given rise to many different environments, all home to assemblages of organisms that are adapted to the local abiotic conditions. Ecologists have found it useful to classify and name these environments based on their ecological similarities.

54.3 How Is Life Distributed in Terrestrial Environments?

A **biome** is an environment that is shaped by its climatic and geographic attributes and characterized by ecologically similar organisms. Ecologists classify terrestrial biomes principally by their dominant plants. By providing three-dimensional structure and by modifying physical conditions near the ground, the dominant plants of a terrestrial environment strongly influence the existence of the other organisms living there.

The distribution of biomes is determined largely by annual patterns of temperature and precipitation. The same biome may be found in several widely separated places, depending in large part on the presence of suitable climatic conditions (**Figure 54.7**). Different assemblages of species may be found in geographically separate regions, but due to convergent evolution in response to similar selective forces (see Section 22.1), organisms in the same biome are likely to share many physiological, morphological, and behavioral adaptations. The shrubby stature and tough leaves of fynbos vegetation described at the opening of this chapter, for example, are features shared by the dominant plants of other regions that have a Mediterranean climate.

In some biomes, such as temperate deciduous forest, precipitation is relatively constant throughout the year, but temperature varies strikingly between summer and winter. In other biomes, both temperature and precipitation change seasonally. In the tropics, seasonal temperature fluctuations are small and annual cycles are dominated by wet and dry seasons. Tropical biome types are determined primarily by the length of the dry season.

Other abiotic factors—particularly soil characteristics and wildfires—also influence the structure and life cycles of the dominant vegetation in an area and, consequently, the ecological attributes of the other organisms living there. For example, Australian desert soils are extremely nutrient-poor, and plants there have difficulty growing new foliage. Such plants often protect their leaves against consumers by producing large quantities of chemicals that reduce the leaves' digestibility. Since these leaves are not eaten, they senesce and drop to the ground, providing an abundance of highly flammable litter that feeds intense periodic fires that sweep across the landscape. As a result, succulent plants—which are easily killed by fire—are not found in Australia, although they are common in deserts on other continents.

Sometimes biomes occur in close geographic proximity to one another but differ because certain geological features alter local temperature and precipitation patterns. Major topographic features such as mountains or large lakes have regional effects on temperature and precipitation. When prevailing winds bring air masses into contact with a mountain range, for example, the air must rise to pass over the mountains, expanding and cooling as it does so. Thus clouds frequently form on the windward side of a mountain range (the side facing into the winds) and release moisture there as rain or snow. On the leeward side (that is, opposite from the direction of the winds), the now-dry air descends, warms, and once again picks up moisture. This pattern often results in a dry area called a

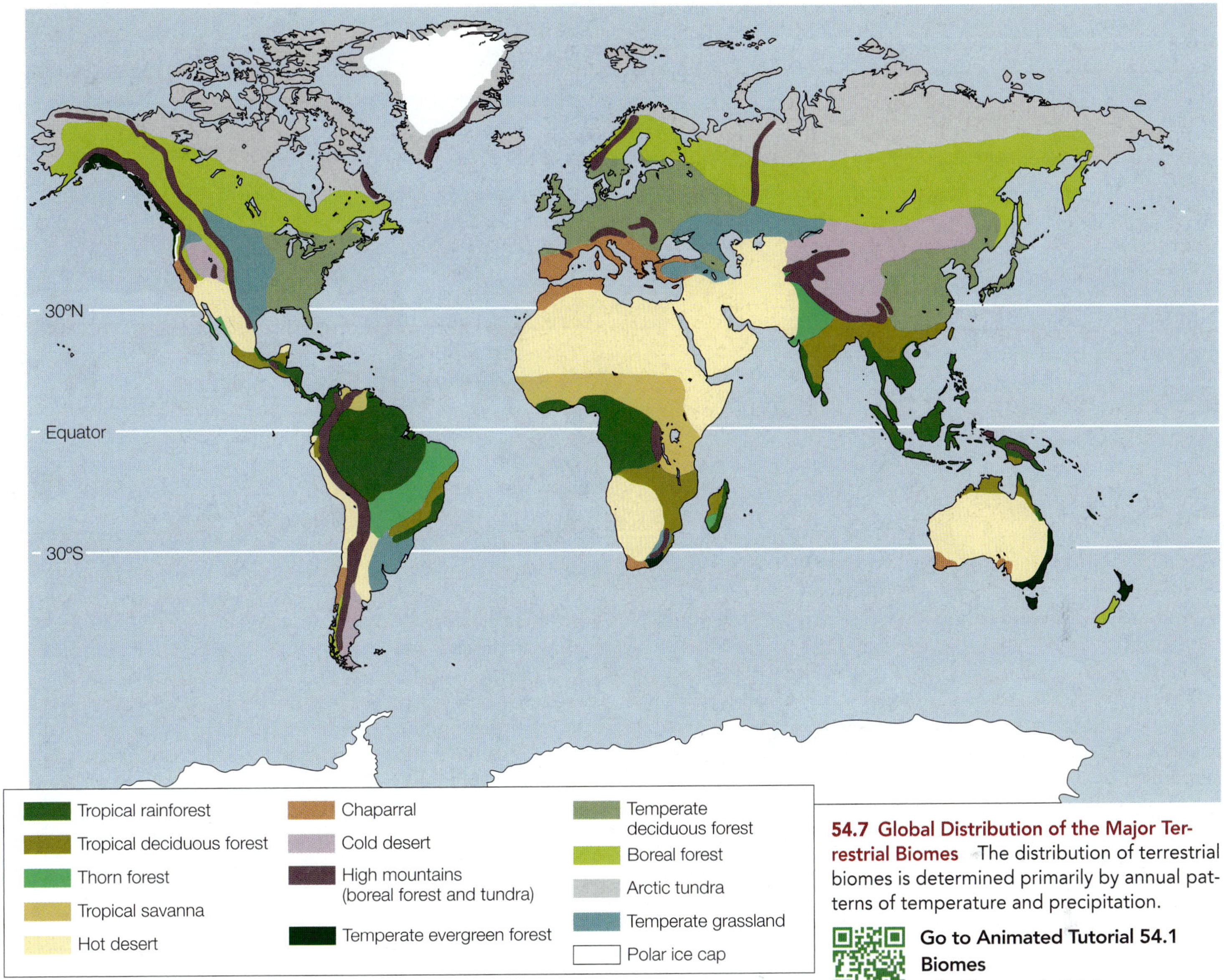
30°N
Equator
30°S
Tropical rainforest
Tropical deciduous forest
Thorn forest
Tropical savanna
Hot desert
Chaparral
Cold desert
High mountains (boreal forest and tundra)
Temperate evergreen forest
Temperate deciduous forest
Boreal forest
Arctic tundra
Temperate grassland
Polar ice cap

54.7 Global Distribution of the Major Terrestrial Biomes The distribution of terrestrial biomes is determined primarily by annual patterns of temperature and precipitation.

Go to Animated Tutorial 54.1 Biomes
Life10e.com/at54.1

rain shadow on the leeward side of the mountain range (**Figure 54.8**). The Atacama Desert, on the leeward side of the Chilean Coast Range, is one such area.

The descriptions of biomes presented here are very general and fall far short of encompassing the variation that can be found in each biome. Moreover, the boundaries ecologists draw between biomes tend to be arbitrary. Although sometimes an abrupt change is apparent in a landscape, more often one biome gradually merges into another. Despite these uncertainties, recognizing the major biomes of the world is useful because these environments share certain ecological attributes irrespective of their locations.

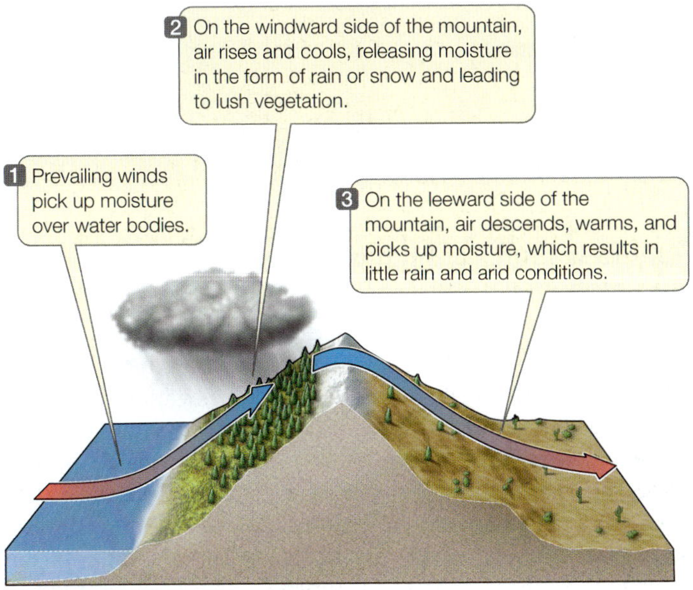
2 On the windward side of the mountain, air rises and cools, releasing moisture in the form of rain or snow and leading to lush vegetation.
1 Prevailing winds pick up moisture over water bodies.
3 On the leeward side of the mountain, air descends, warms, and picks up moisture, which results in little rain and arid conditions.

54.8 Rain Shadow Mean annual precipitation tends to be lower on the leeward side of a mountain range than on the windward side.

Go to Animated Tutorial 54.2 Rain Shadow
Life10e.com/at54.2

TUNDRA

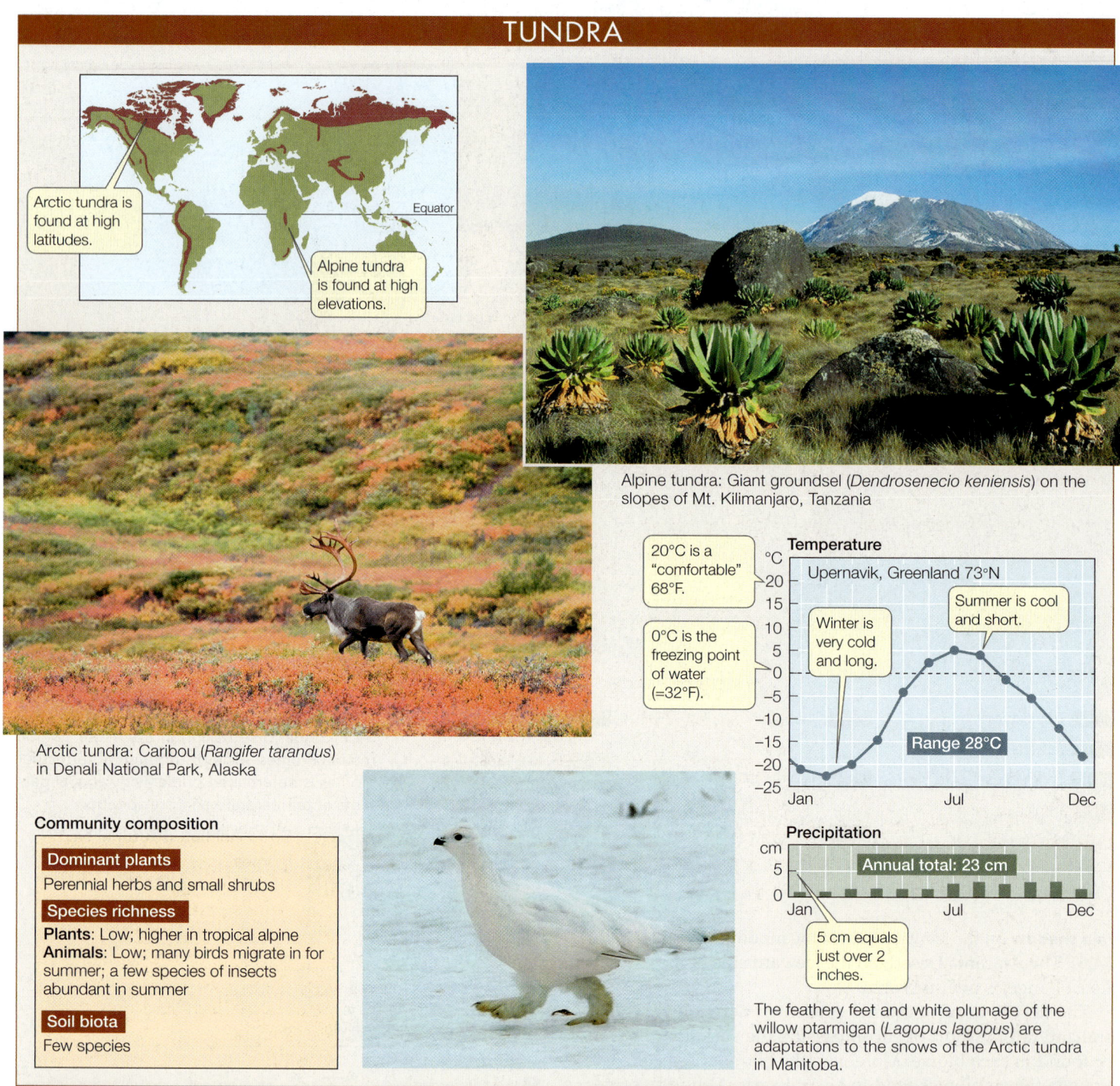

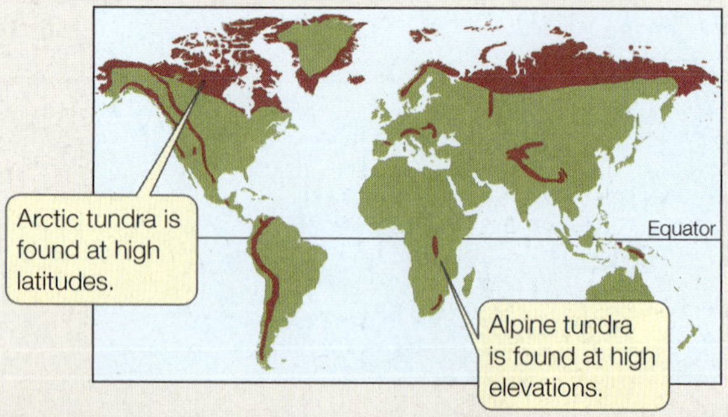
Arctic tundra is found at high latitudes.
Equator
Alpine tundra is found at high elevations.

Alpine tundra: Giant groundsel (*Dendrosenecio keniensis*) on the slopes of Mt. Kilimanjaro, Tanzania

Arctic tundra: Caribou (*Rangifer tarandus*) in Denali National Park, Alaska

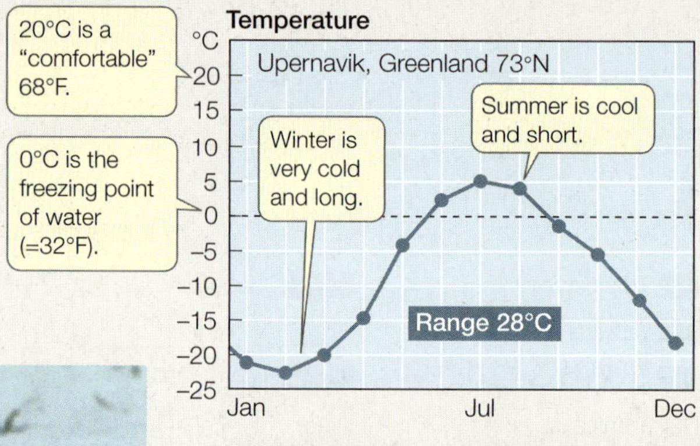
Temperature
Upernavik, Greenland 73°N
°C
20
15
10
5
0
−5
−10
−15
−20
−25
Jan
Jul
Dec
20°C is a "comfortable" 68°F.
0°C is the freezing point of water (=32°F).
Winter is very cold and long.
Summer is cool and short.
Range 28°C

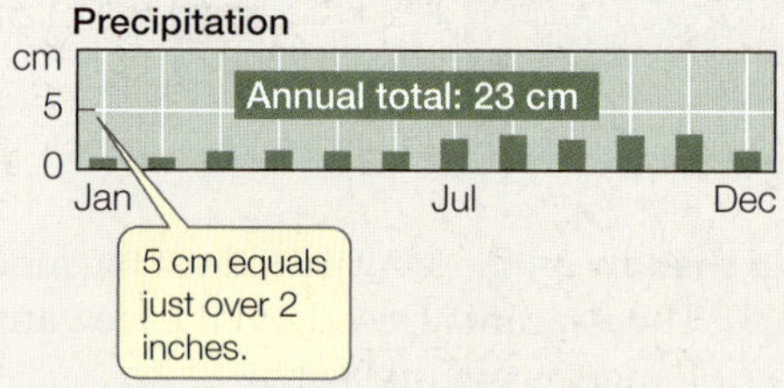
Precipitation
cm
5
0
Jan
Jul
Dec
Annual total: 23 cm
5 cm equals just over 2 inches.

Community composition

Dominant plants

Perennial herbs and small shrubs

Species richness

Plants: Low; higher in tropical alpine
Animals: Low; many birds migrate in for summer; a few species of insects abundant in summer

Soil biota

Few species

The feathery feet and white plumage of the willow ptarmigan (*Lagopus lagopus*) are adaptations to the snows of the Arctic tundra in Manitoba.

Tundra is found at high latitudes and high elevations

The tundra biome is characterized by low temperatures and a short growing season. These conditions prevail not only at the high latitudes of the Arctic but also at high elevations in mountains at all latitudes. In Arctic tundra, the vegetation consists of low-growing perennial plants and is underlain by permafrost—soil permeated with permanently frozen water. The top few centimeters of the soil thaw during the short summers, when the sun may be above the horizon 24 hours a day. Thus even though there is little precipitation near the poles, the soil in lowland Arctic tundra is wet because water cannot drain through the permafrost.

Tundra found at high elevations outside polar regions is called alpine tundra. Tropical alpine tundra is not underlain by permafrost, so photosynthesis and most other biological activities continue (albeit slowly) throughout the year. A variety of plant growth forms are present in tropical alpine tundra, including low-growing shrubs, perennials, and grasses.

Many tundra plants have hairy leaves that trap heat. The flowers of some species move over the course of the day, tracking the sun's warmth. Most animals are either summer

BOREAL and TEMPERATE EVERGREEN FOREST

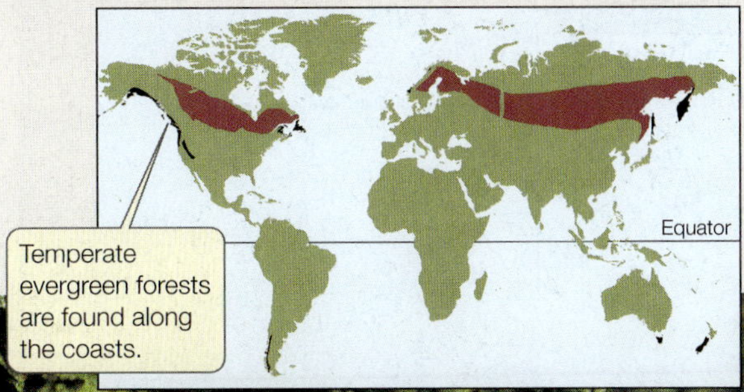
Equator
Temperate evergreen forests are found along the coasts.

Bull moose (*Alces alces*) in boreal forest, Alberta, Canada

Southern beech (*Nothofagus*) in the temperate evergreen forest of Fiordland National Park, New Zealand

Community composition

Dominant plants
Trees, shrubs, and perennial herbs

Species richness
Plants: Low in trees, higher in understory
Animals: Low, but with summer peaks in migratory birds

Soil biota
Very rich in deep litter layer

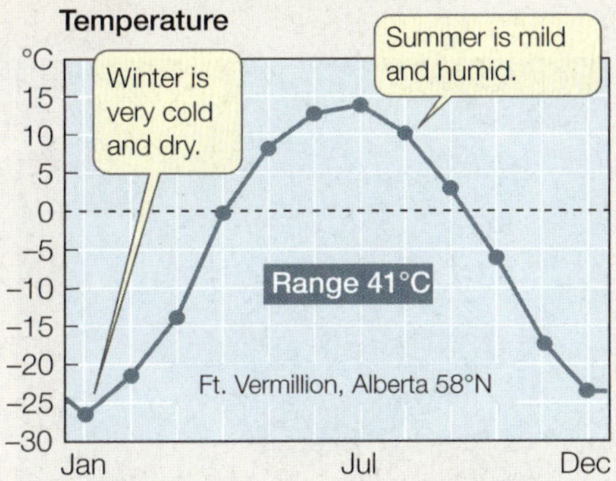
Temperature
°C
15
10
5
0
−5
−10
−15
−20
−25
−30
Winter is very cold and dry.
Summer is mild and humid.
Range 41°C
Ft. Vermillion, Alberta 58°N
Jan
Jul
Dec

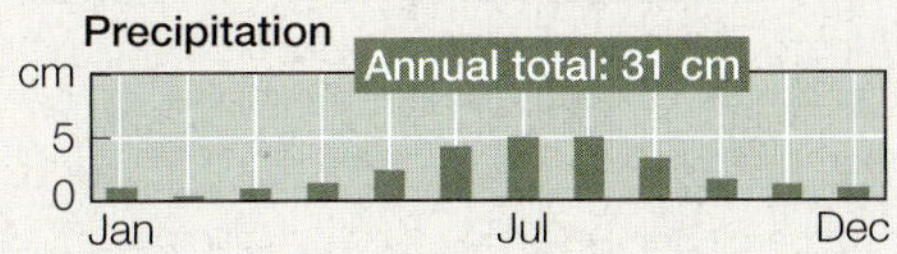
Precipitation
cm
5
0
Annual total: 31 cm
Jan
Jul
Dec

The boreal owl (*Aegolius funereus*) feeds on insects and small birds and mammals of the coniferous evergreen forests of Europe and North America. The owl lives in the forests year-round, dispersing occasionally when food becomes extremely scarce.

migrants or are dormant for much of the year. Resident birds and mammals, such as the willow ptarmigan (*Lagopus lagopus*) and Arctic fox (*Vulpes lagopus*), have thick fur or feathers that may change color with the seasons, from brown in summer to white in winter.

Evergreen trees dominate boreal and temperate evergreen forests

The boreal forest biome (also known as taiga) occurs at latitudes below Arctic tundra and at elevations below alpine tundra on temperate-zone mountains. Winters in the boreal forest are long and very cold; summers are short, although often relatively warm. The boreal forests of the Northern Hemisphere are dominated by coniferous gymnosperm species such as spruces and firs. The evergreen leaves of conifers are needlelike rather than flat; their reduced surface area cuts down on evaporative water loss. The short summers favor evergreen leaves, which are ready to photosynthesize as soon as temperatures warm. In winter, downward-drooping limbs allow the trees to shed snow easily.

The dominant mammals of the boreal forest, such as moose and hares, eat leaves, but the seeds in conifer cones support a variety of rodents, birds, and insects. Many small mammals

TEMPERATE DECIDUOUS FOREST

Equator

Fallen oak leaves on the winter forest floor, New York State

Black bears (*Ursus americanus*) are common in North American deciduous forests.

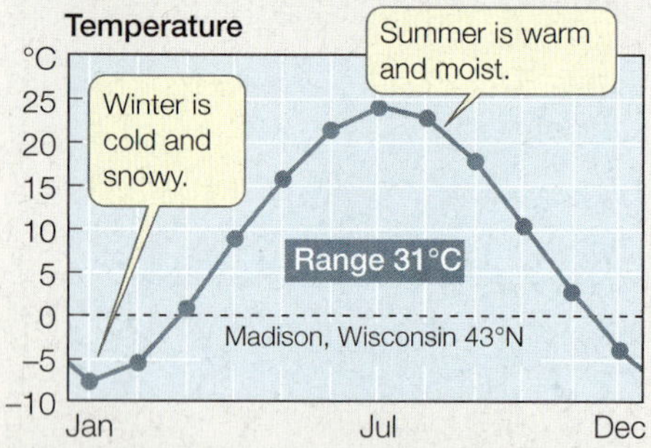
Temperature
°C
25
20
15
10
5
0
−5
−10
Winter is cold and snowy.
Summer is warm and moist.
Range 31°C
Madison, Wisconsin 43°N
Jan
Jul
Dec

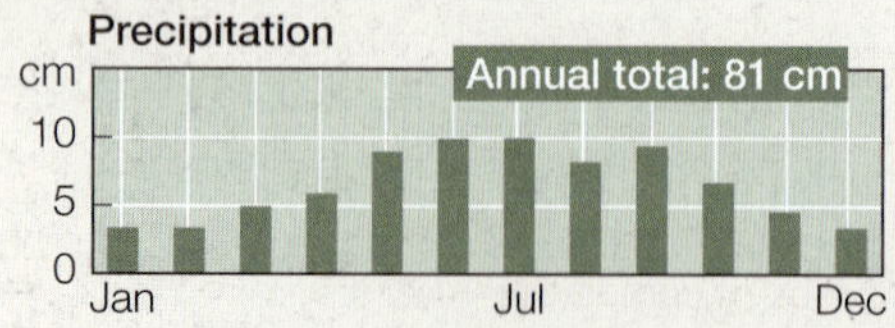
Precipitation
cm
10
5
0
Annual total: 81 cm
Jan
Jul
Dec

Mourning cloak butterflies (*Nymphalis antiopa*) overwinter as adults, settled in crevices and under tree bark in the deciduous forests of eastern North America. Their early appearance is a harbinger of spring.

Community composition

Dominant plants
Trees and shrubs

Species richness
Plants: Many tree species in southeastern U.S. and eastern Asia, rich shrub layer
Animals: Rich; many migrant birds, richest amphibian communities on Earth, rich summer insect fauna

Soil biota
Rich

hibernate in winter, but voles, lemmings, and mice remain active under the snowpack, serving as food for predators such as foxes and owls.

Temperate evergreen forests grow along the coasts of continents in both hemispheres at middle to high latitudes, where winters are mild and wet and summers are cool and dry. In the Northern Hemisphere, the dominant trees in temperate evergreen forests are conifers, some of which are the world's most massive tree species (including the giant sequoia and coast redwood). In the Southern Hemisphere, the dominant trees are southern beeches (*Nothofagus*), some of which are evergreen.

Temperate deciduous forests change with the seasons

The temperate deciduous forest biome is found in eastern North America, eastern Asia, and Europe. Temperatures in these regions fluctuate dramatically between summer and winter, although precipitation is fairly evenly distributed throughout the year. Deciduous trees, which dominate these forests, lose their leaves during the cold winters and produce new leaves that photosynthesize rapidly during the warm, moist summers.

Many more tree species live here than in boreal forests. The temperate forests richest in species are those of the southern

TEMPERATE GRASSLANDS

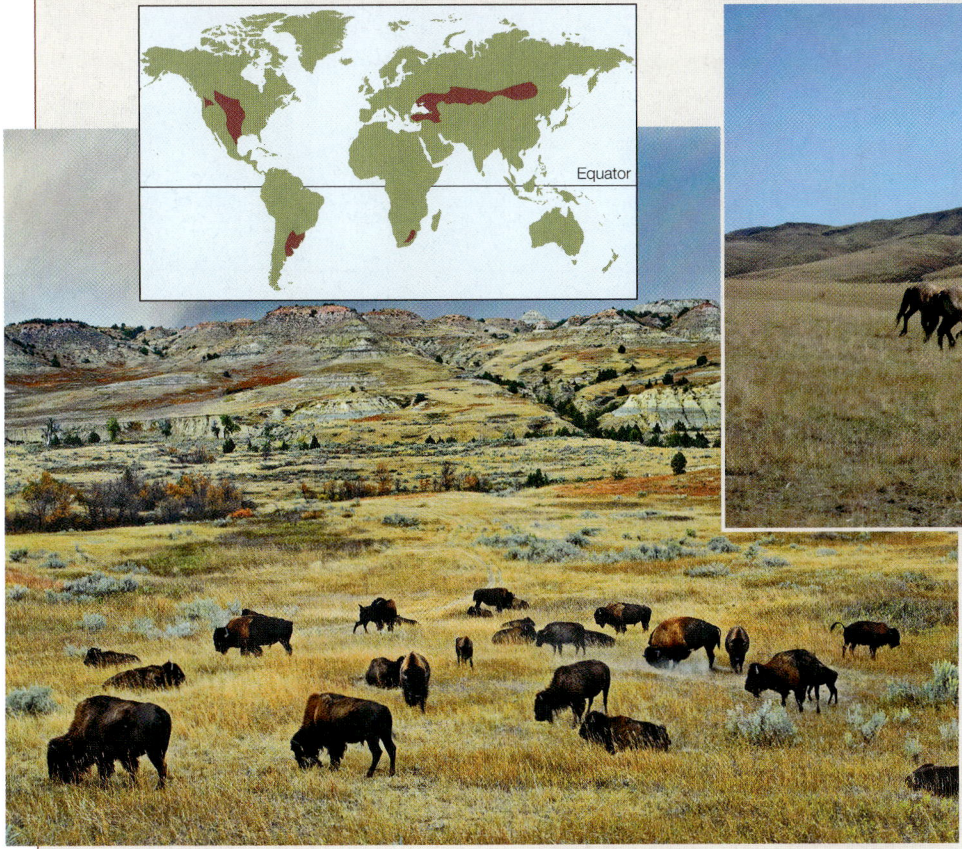
Equator

American bison herd grazing shortgrass prairie, North Dakota

Przewalski's horse in Khustain Nuruu National Park, Mongolia

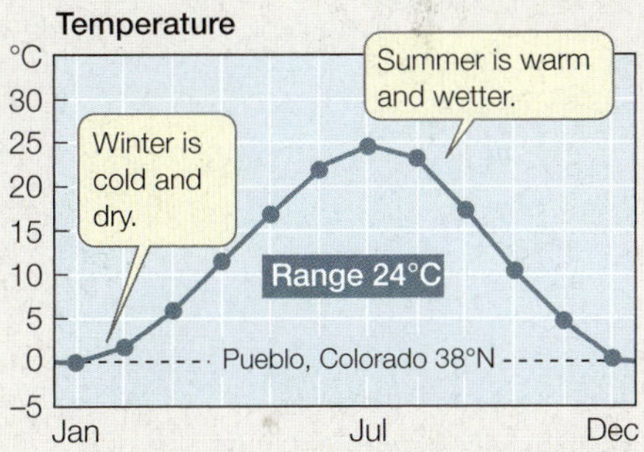
Temperature
°C
30
25
20
15
10
5
0
–5
Summer is warm and wetter.
Winter is cold and dry.
Range 24°C
Pueblo, Colorado 38°N
Jan
Jul
Dec

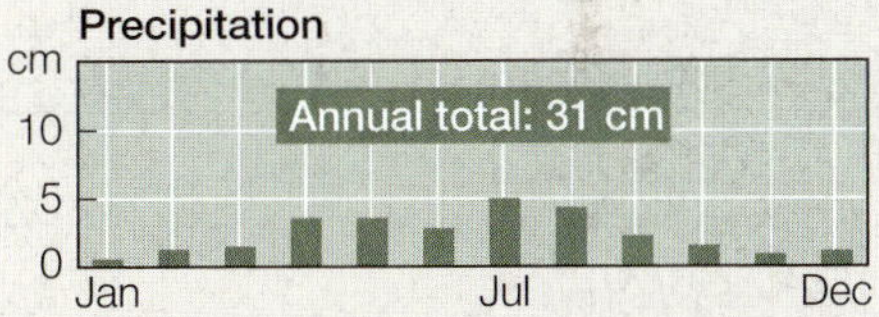
Precipitation
cm
10
5
0
Annual total: 31 cm
Jan
Jul
Dec

Community composition

Dominant plants
Perennial grasses and forbs

Species richness
Plants: Fairly high
Animals: Relatively few birds because of simple structure; mammals fairly rich

Soil biota
Rich

The flightless Darwin's rhea (*Rhea pennata*) grazes the grasslands of Patagonia.

Appalachian Mountains of the United States and those found in eastern China and Japan—areas that were not covered by glaciers during the Pleistocene. Many plant genera are shared among the three geographically separate regions where this biome is found.

Although many animals are permanent residents of deciduous forests, some (including many birds) migrate to escape the winter cold. Others that remain through the winter acquire massive fat stores in fall and hibernate (see Section 40.5), often in underground burrows. Many insects pass the winter in a state of diapause (suspended development), the onset of which is triggered by the decreasing hours of daylight—a reliable predictor of winter.

Temperate grasslands are widespread

Temperate grasslands are found in many parts of the world, all of which are relatively dry for much of the year. Most grasslands, such as the pampas of Argentina, the veldt of South Africa, and the Great Plains of North America, have hot summers and relatively cold winters. In some grasslands, most of the precipitation falls in winter (as in California grasslands); in others, the majority falls in summer (as in the Great Plains and the Russian steppe).

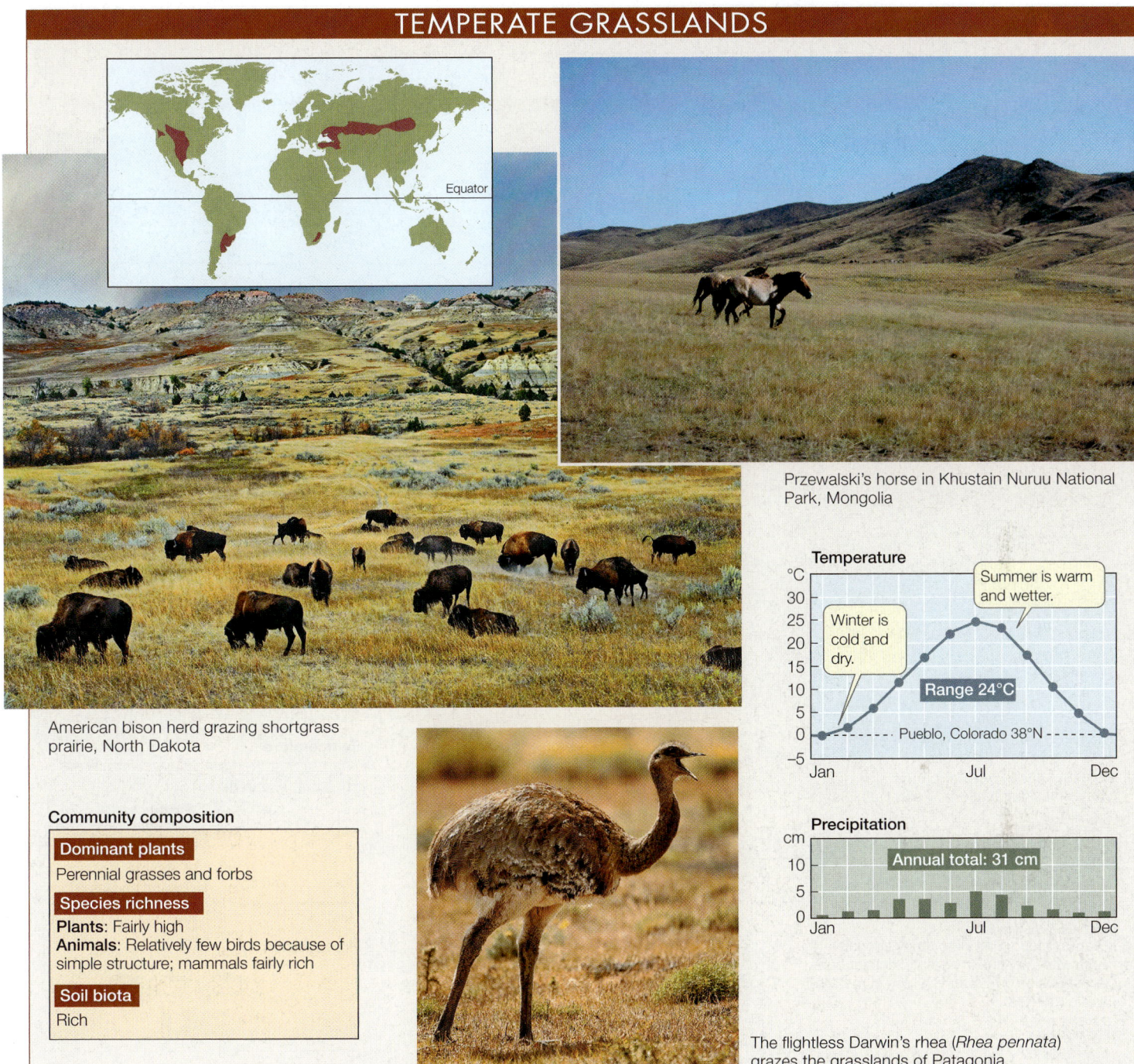

Grassland vegetation is structurally simple but rich in species of perennial grasses and forbs (herbaceous plants other than grasses). This abundant plant biomass supports herds

HOT DESERT

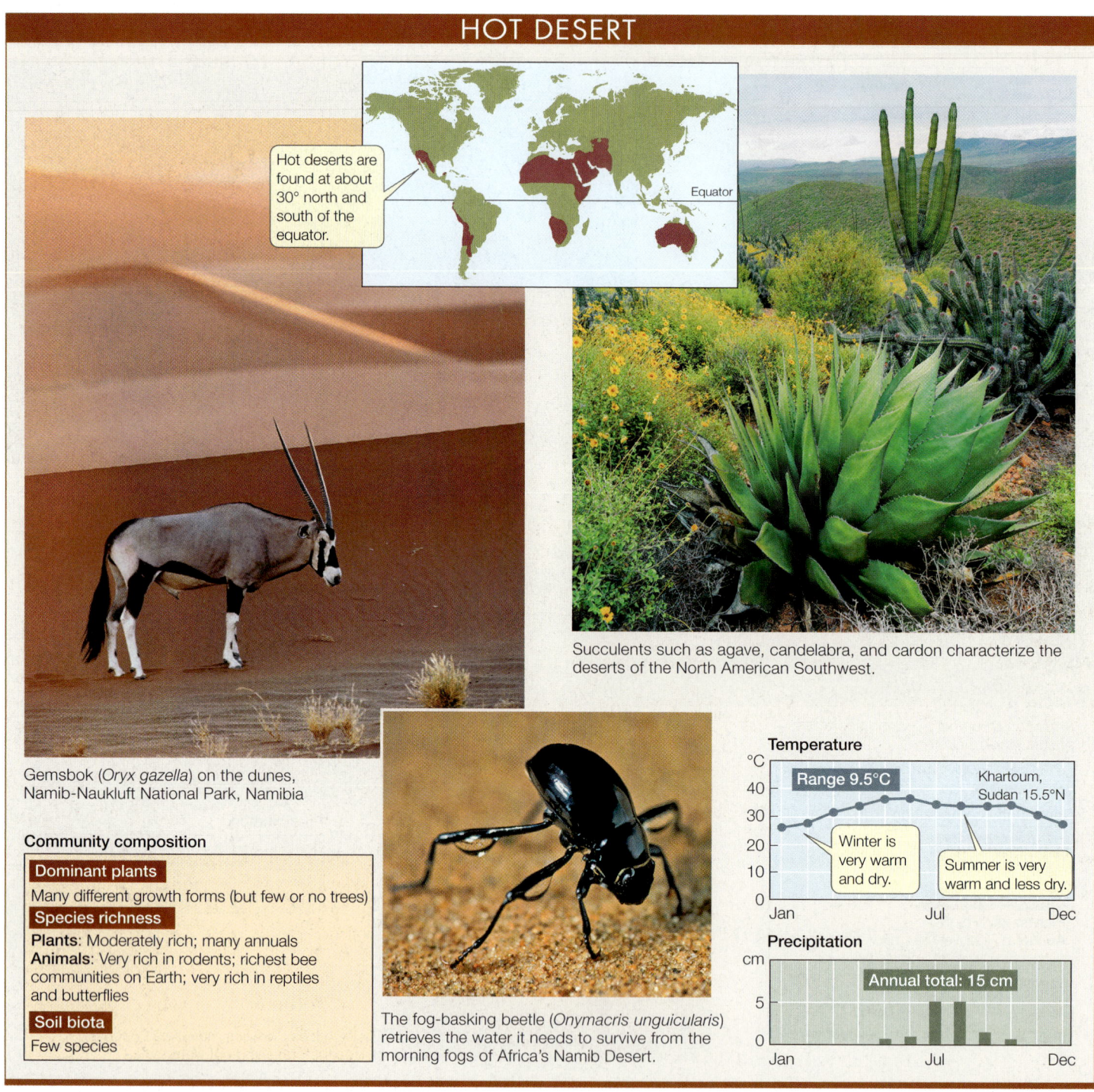

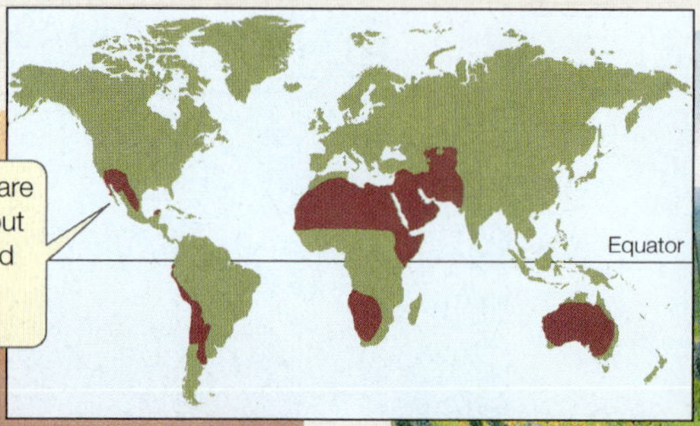
Hot deserts are found at about 30° north and south of the equator.
Equator

Gemsbok (*Oryx gazella*) on the dunes, Namib-Naukluft National Park, Namibia

Succulents such as agave, candelabra, and cardon characterize the deserts of the North American Southwest.

Community composition

Dominant plants

Many different growth forms (but few or no trees)

Species richness

Plants: Moderately rich; many annuals
Animals: Very rich in rodents; richest bee communities on Earth; very rich in reptiles and butterflies

Soil biota

Few species

The fog-basking beetle (*Onymacris unguicularis*) retrieves the water it needs to survive from the morning fogs of Africa's Namib Desert.

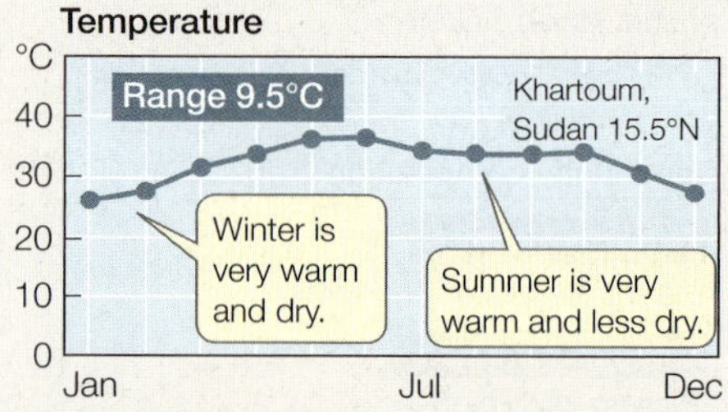
Temperature
°C
Range 9.5°C
Khartoum, Sudan 15.5°N
40
30
20
10
0
Winter is very warm and dry.
Summer is very warm and less dry.
Jan
Jul
Dec

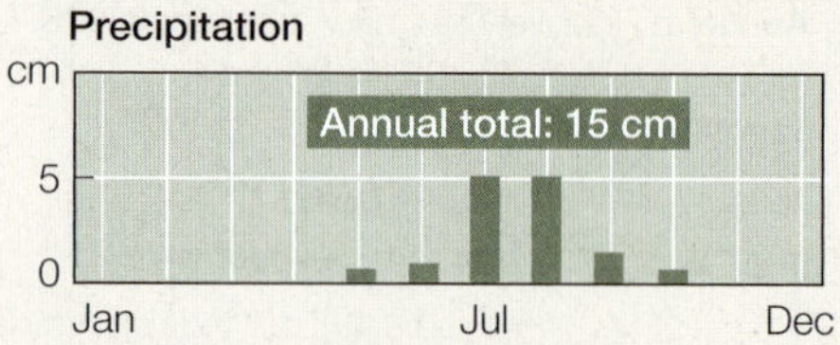
Precipitation
cm
Annual total: 15 cm
5
0
Jan
Jul
Dec

of large grazing mammals. Grassland plants are adapted to grazing and to fire. They store much of their energy underground and resprout quickly after being burned or grazed. There are comparatively few trees in temperate grasslands because trees cannot survive the periodic fires. If grasslands do not burn periodically, many of the species that typify this biome will be replaced by fire-intolerant species that are superior competitors.

The topsoil of grasslands is usually rich and deep, and thus exceptionally well suited to growing crops such as corn and wheat. As a consequence, most of the world's temperate grasslands have been turned over to agriculture and no longer exist in their natural state.

Hot deserts form around 30° latitude

The hot desert biome is concentrated in two belts, centered around 30°N and 30°S latitude (where warm, dry air descends and picks up moisture; see Figure 54.3). The driest of these regions, where rains rarely penetrate, are far from the oceans, as in the center of Australia and the middle of the Sahara in Africa.

Desert plants have several structural and physiological adaptations that help them conserve water, as described in

COLD DESERT

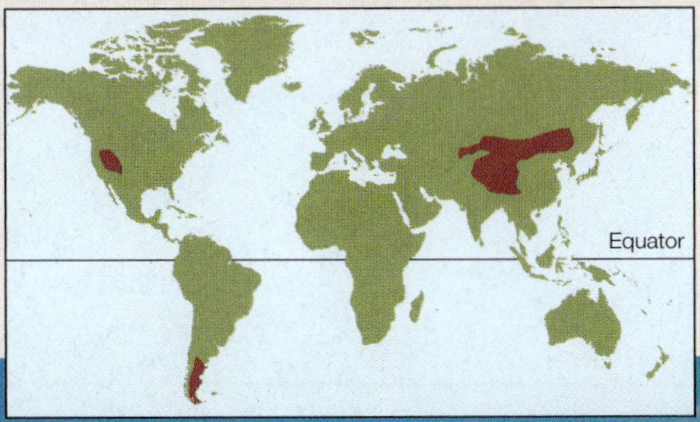
Equator

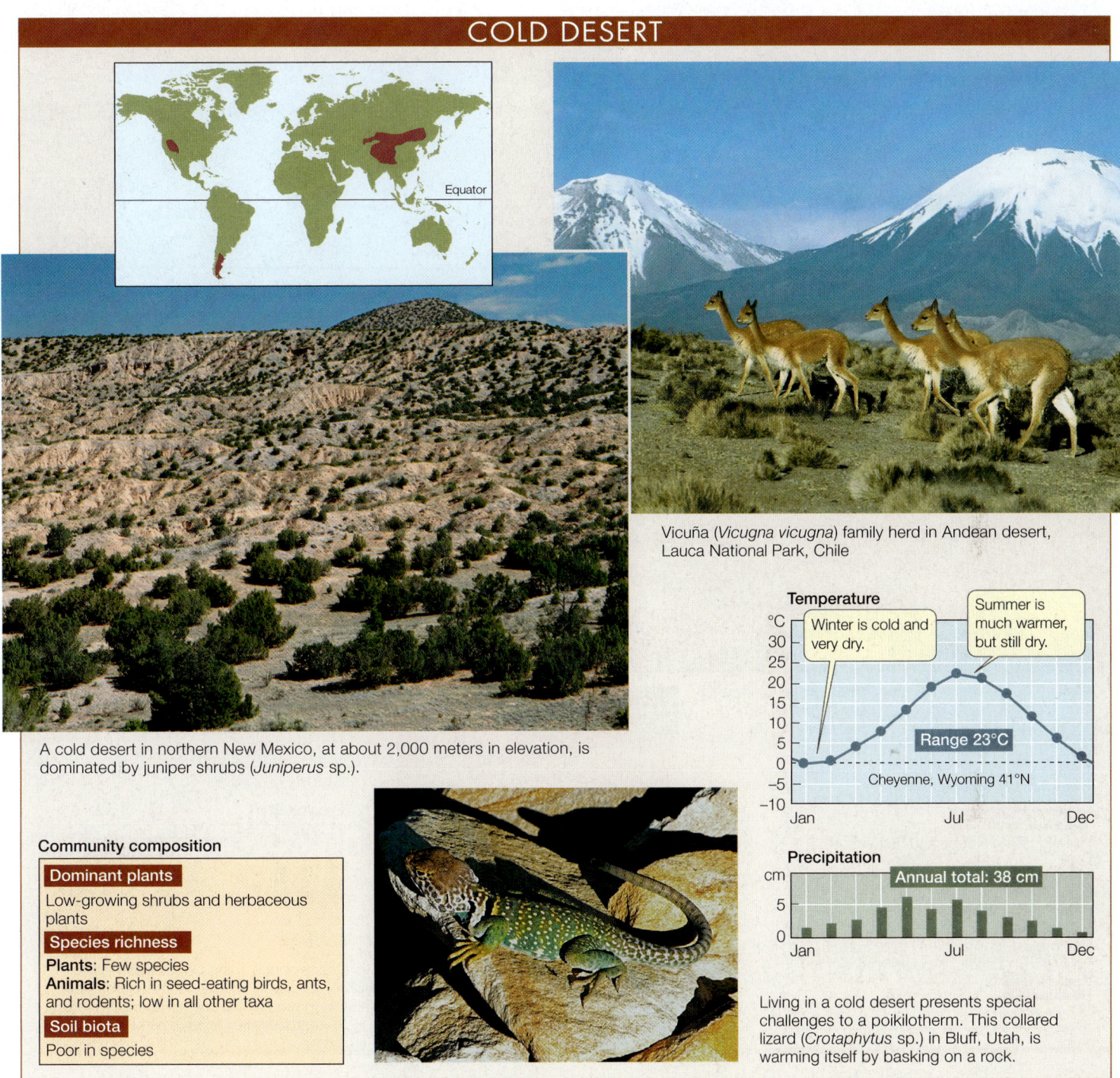

A cold desert in northern New Mexico, at about 2,000 meters in elevation, is dominated by juniper shrubs (*Juniperus* sp.).

Vicuña (*Vicugna vicugna*) family herd in Andean desert, Lauca National Park, Chile

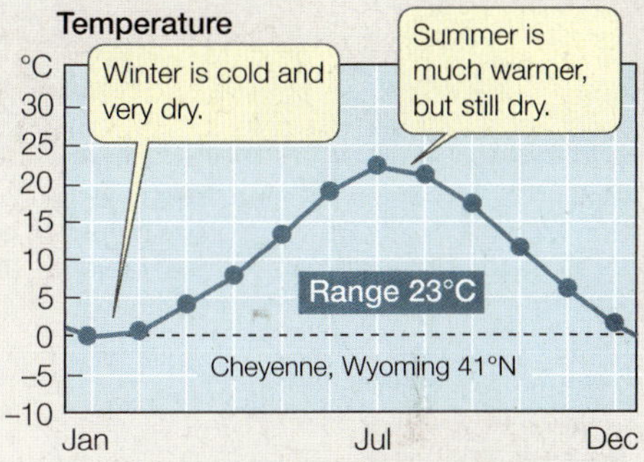
Temperature
°C
30
25
20
15
10
5
0
−5
−10
Winter is cold and very dry.
Summer is much warmer, but still dry.
Range 23°C
Cheyenne, Wyoming 41°N
Jan
Jul
Dec

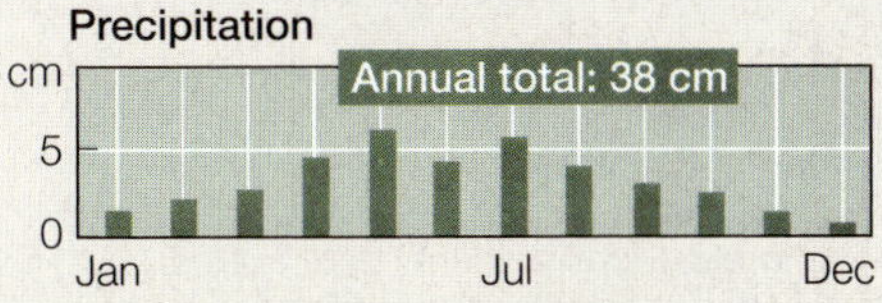
Precipitation
cm
5
0
Annual total: 38 cm
Jan
Jul
Dec

Community composition

Dominant plants

Low-growing shrubs and herbaceous plants

Species richness

Plants: Few species
Animals: Rich in seed-eating birds, ants, and rodents; low in all other taxa

Soil biota

Poor in species

Living in a cold desert presents special challenges to a poikilotherm. This collared lizard (*Crotaphytus* sp.) in Bluff, Utah, is warming itself by basking on a rock.

Section 39.3. Many desert plants are xerophytes: species with adaptations for reducing water loss or storing water. The aboveground parts of many desert plants are covered with a waxy cuticle to prevent water loss; leaves may be reduced to spines to minimize surface area, as in the Cactaceae of the Western Hemisphere and the Euphorbiaceae in much of the rest of the world. Other desert plants are succulents that store water in fleshy leaves or stems. Most perennial plants go dormant during dry seasons, and then grow rapidly as soon as rains return. Their seeds tend to be heat- and drought-resistant and accumulate in a dormant state in the soil.

Small desert animals are inactive during the hottest part of the day, remaining in underground burrows. Desert mammals have physiological adaptations for conserving water, including a reduced number of sweat glands and kidneys that produce highly concentrated urine. Many desert animals require no water beyond what they can extract from the carbohydrates in their food.

Cold deserts are high and dry

The cold desert biome is found in dry regions at mid- to high latitudes, especially in the interiors of continents where mountain ranges block moisture-rich air (see Figure 54.8). Blocked by

CHAPARRAL

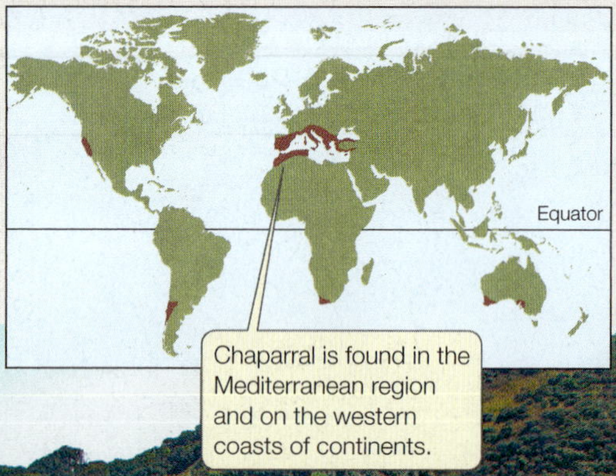
Equator
Chaparral is found in the Mediterranean region and on the western coasts of continents.

A Spanish ibex (*Capra pyrenaica*) among maquis, a chaparral vegetation found in Spain and Corsica.

Maritime chaparral on Montara Mountain in California's San Mateo County. The light green shrubs in the foreground are *Arctostaphylos montaraensis*, a species that is endemic to this mountain.

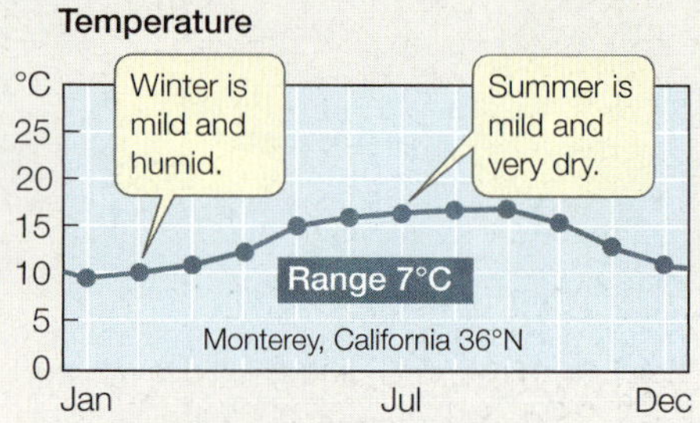
Temperature
°C
25
20
15
10
5
0
Winter is mild and humid.
Summer is mild and very dry.
Range 7°C
Monterey, California 36°N
Jan
Jul
Dec

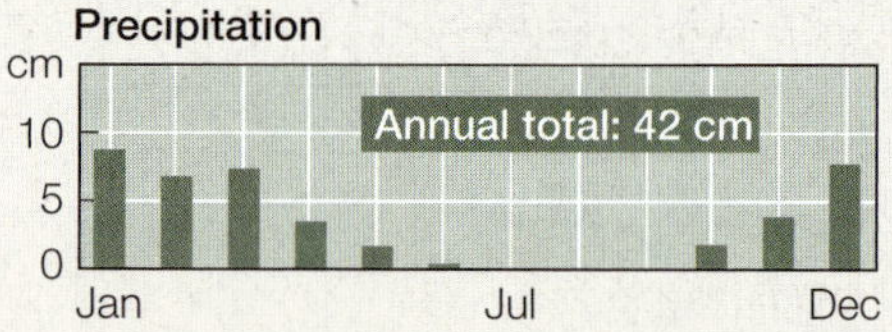
Precipitation
cm
10
5
0
Annual total: 42 cm
Jan
Jul
Dec

Community composition

Dominant plants
Low-growing shrubs and herbaceous plants

Species richness
Plants: Extremely high in South Africa and Australia
Animals: Rich in rodents and reptiles; very rich in insects, especially bees

Soil biota
Moderately rich

Two fynbos species, the Cape sugarbird (*Promerops cafer*) and its nectar source, the pincushion protea (*Leucospermum cordifolium*) in Helderberg Nature Reserve, South Africa.

two mountain ranges (the Andes and the Chilean Coast range), the Atacama Desert is the driest place on Earth; average yearly rainfall is less than 1 millimeter.

Cold deserts are dominated by a few species of low-growing shrubs. The surface layers of the soil are recharged with moisture in winter, and plant growth is concentrated in spring. Cold deserts are relatively species-poor, but the plants that do grow there tend to produce large numbers of seeds, supporting many species of seed-eating birds, ants, and rodents. Burrowing behavior is widespread among cold desert dwellers but—in contrast to hot desert animals—they burrow to escape cold temperatures, not excessive heat.

Chaparral has hot, dry summers and wet, cool winters

The chaparral biome is found on the western coasts of continents at mid-latitudes (around 40°). Winters in this biome are cool and wet; summers are warm and dry. Such climates are found in the Mediterranean region of Europe (for which the Mediterranean climate is named), coastal California, central Chile, extreme southern Africa, and southwestern Australia. The fynbos of the Cape region of South Africa, described at the opening of this chapter, is part of the chaparral biome.

The dominant plants of chaparral vegetation are low-growing shrubs and trees with tough evergreen leaves that conserve water.

THORN FOREST and TROPICAL SAVANNA

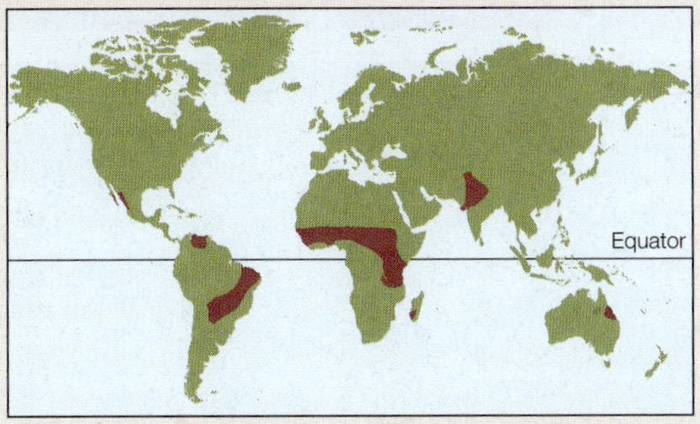
Equator

Family groups of African elephants (*Loxodonta africana*) in Kenya's Ol Malo Wildlife Sanctuary converge into a large herd for migration.

Madagascar ocotillo (*Alluadia procera*) dominate this thorn forest.

Community composition

Dominant plants

Shrubs and small trees; grasses

Species richness

Plants: Moderate in thorn forest; low in savanna

Animals: Rich mammal faunas; moderately rich in birds, reptiles, and insects

Soil biota

Rich

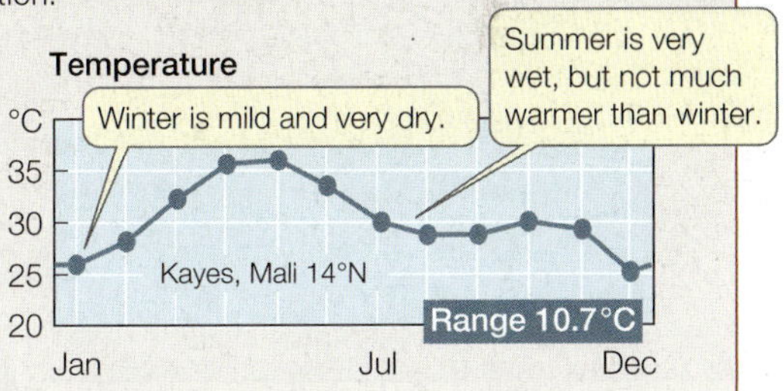
Temperature
°C
Winter is mild and very dry.
Summer is very wet, but not much warmer than winter.
35
30
25
20
Kayes, Mali 14°N
Range 10.7°C
Jan
Jul
Dec

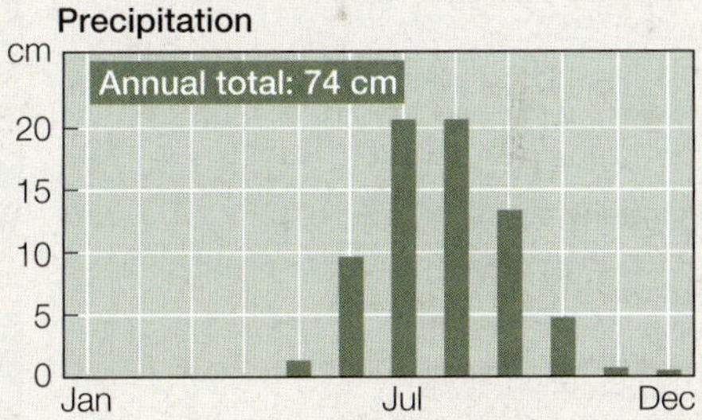
Precipitation
cm
Annual total: 74 cm
20
15
10
5
0
Jan
Jul
Dec

Termite colonies build huge mounds on the African savanna, providing a food source for mammals such as this baboon (*Papio* sp.).

The shrubs carry out most of their growth and photosynthesis in early spring, when insects are active and birds breed. Many chaparral species produce strong-smelling defensive chemicals to reduce losses of their hard-to-replace foliage to herbivores. Annual plants are abundant and produce large quantities of seeds that fall onto the soil, supporting many small rodents, most of which store seeds in underground burrows. Burrowing to avoid midday heat and nocturnal foraging are strategies used by many chaparral animals. Chaparral vegetation is adapted to periodic fires; the seeds of some species do not germinate until after they have survived a fire. Many shrubs of Northern Hemisphere chaparral produce bird-dispersed fruits that ripen in late fall, when large numbers of migrant birds arrive from the north. In the fynbos of South Africa, seeds equipped with elaiosomes are transported by ants, which bury them deep enough to survive the periodic fires.

Thorn forests and tropical savannas have similar climates

The thorn forest and tropical savanna biomes are found primarily at latitudes below the hot deserts of Africa, South America, and Australia. Little or no rain falls in these biomes in winter, but rainfall may be heavy during summer. Thorn forests contain many plants similar to those found in hot deserts,

TROPICAL DECIDUOUS FOREST

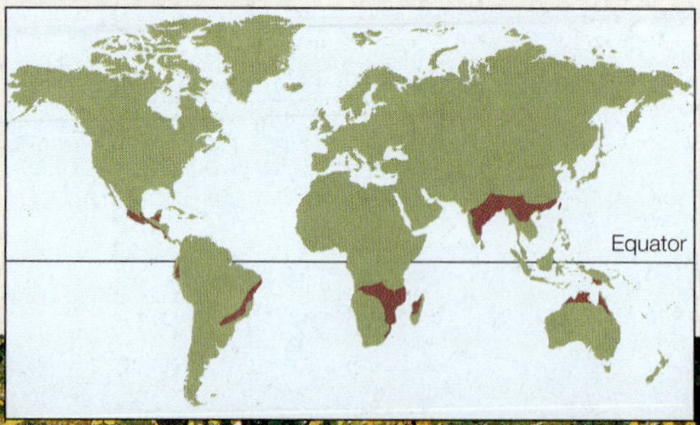
Equator

A forest canopy of pijio trees (*Cavanillesia platanifolia*) during the dry season in Cerro Blanco on the Ecuadoran coast

Bengal tiger (*Panthera tigris tigris*) in the forests of Bandhavgarh, India.

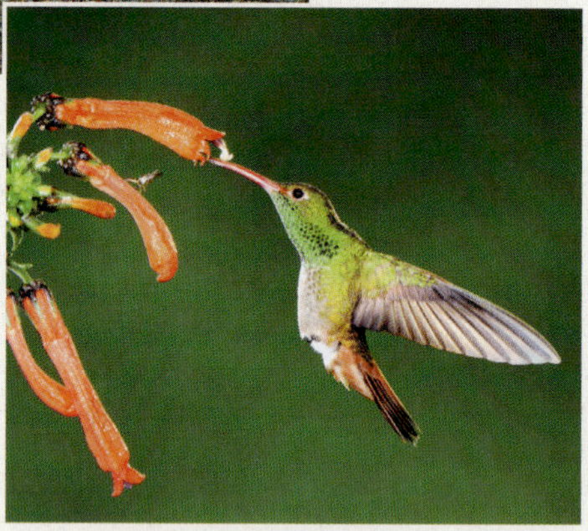

Hummingbirds such as *Amazilia tzacatl* are major pollinators in this biome.

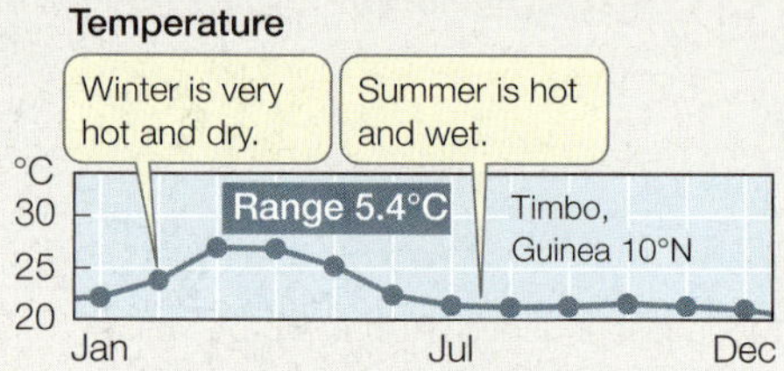
Temperature
Winter is very hot and dry.
Summer is hot and wet.
°C
30
25
20
Range 5.4°C
Timbo, Guinea 10°N
Jan
Jul
Dec

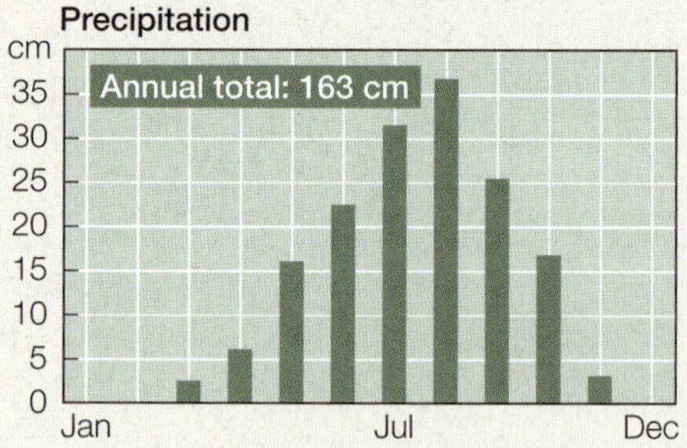
Precipitation
cm
35
30
25
20
15
10
5
0
Annual total: 163 cm
Jan
Jul
Dec

Community composition

Dominant plants
Deciduous trees

Species richness
Plants: Moderately rich in tree species
Animals: Rich mammal, bird, reptile, and amphibian communities; rich in insects

Soil biota
Rich, but poorly known

including succulents. The dominant plants are spiny shrubs and small trees, many of which drop their leaves during the long, dry winter. Trees of the genus *Acacia* are common in thorn forests and savannas worldwide. In Africa, *Andansonia* (baobab) trees are also a hallmark of these biomes.

Savanna is characterized by expanses of grasses and grass-like plants surrounding scattered individual trees. The largest tropical savannas are found in central and eastern Africa, where they are populated by herds of grazing and browsing mammals and the large carnivores that prey on them. The migration of vast herds of herbivores in search of "greener pastures" during the dry season is another characteristic of this impressive region. Grazers and browsers maintain the savannas by disproportionately damaging shrubs and trees, which cannot withstand as much tissue loss as can the grasses. If savanna vegetation is not grazed, browsed, or burned, it typically reverts to dense thorn forest.

Tropical deciduous forests occur in hot lowlands

As the temperature and the length of the rainy season increase toward the equator, the tropical deciduous forest biome replaces thorn forest. Tropical deciduous forests have taller trees

TROPICAL RAINFOREST
Equator
Tropical rainforests are located at low latitudes.
Community composition
Dominant plants
Trees and vines
Species richness
Plants: Extremely high
Animals: Extremely high in mammals, birds, amphibians, and arthropods
Soil biota
Very rich but poorly known
Temperature
°C
25
20
15
10
The weather is warm and rainy all year.
Range 2.2°C
Iquitos, Peru 3°S
Jan
Jul
Dec
Precipitation
cm
30
25
20
15
10
5
0
Annual total: 262 cm
Jan
Jul
Dec

The canopy of an Ecuadoran tropical rainforest, seen from above

A golden lion tamarin (*Leontopithecus rosalia*) near Rio de Janeiro, Brazil

The three-toed sloth (*Bradypus variegatus*) is almost totally arboreal, spending its life in the rainforest canopies of Central and South America.

and fewer succulent plants than thorn forests or savannas, and they support a much greater number of plant and animal species. Most of the trees, except for those growing along rivers, lose their leaves during the long, hot dry season. Growth increases in the rainy season; many plants flower while they are still leafless.

Most plant species in this biome are pollinated by animals. In the Sierra Madre Occidental, a mountain range in the extreme southwestern United States extending into western Mexico, tropical deciduous forests are part of a "nectar corridor," a series of patches of flowering plants that are used as refueling stops by long-distance migrants traveling north from overwintering sites to their breeding sites in the Rocky Mountains.

The soils of this biome are among the best in the tropics for agriculture because they contain more nutrients than the soils of wetter areas. As a result, most tropical deciduous forests worldwide have been cleared for agriculture and grazing.

Tropical rainforests are rich in species

The tropical rainforest biome, or simply the rainforest, is found in equatorial regions where total rainfall exceeds 250 centimeters

WORKING WITH**DATA:**

Walter Climate Diagrams

Original Source

Devised by the German biogeographer Heinrich Walter in 1979, this graphic technique plots temperature and precipitation data in a simple way that visualizes a "growing season"—those months when average temperatures are above 0°C and the average precipitation trace falls above the temperature trace.

Analyze the Data

Walter climate diagrams are predicated on the "rule of thumb" that plant growth requires temperatures above 0°C and at least 20 mm of precipitation for each 10°C above 0°. They have two *y*-axis scales, one for temperature and one for precipitation; these axes align 0 mm of precipitation with a temperature of 0°C. The *x* axis shows the 12 months, with the summer solstice placed in the center of the *x* axis.

The Walter diagram shown at the right is for London, England. Average yearly temperature and precipitation data for three other cities are given in the table. Using these data, create Walter diagrams for each city. Use your diagrams along with the information in the preceding sections of this chapter to answer the questions.

QUESTION 1

Based solely on your diagrams, which biome do you think is represented by each location? What physical attribute other than temperature and precipitation might significantly affect the biomes of these locations?

QUESTION 2

How do you explain the temperature disparity between London and Moscow, both of which lie in a similar latitude of Europe?

QUESTION 3

Perth lies on the western coast of Australia, in the Southern Hemisphere. How does this affect the configuration of your Walter diagram? (Hint: Where did you place the summer solstice?) Without considering this climate data, what biome would you expect to find based solely on Perth's latitude and coastal location?

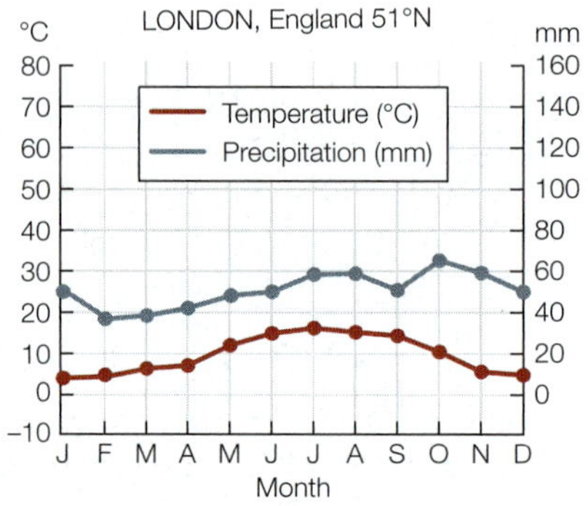
LONDON, England 51°N
°C
mm
Temperature (°C)
Precipitation (mm)
80
70
60
50
40
30
20
10
0
–10
160
140
120
100
80
60
40
20
0
J F M A M J J A S O N D
Month

Location and latitude	Jan	Feb	Mar	Apr	May	June	July	Aug	Sept	Oct	Nov	Dec
MOSCOW, Russia 56°N												
Temp. (°C)	–10.5	–9.0	–4.0	4.5	12.0	16.5	18.5	16.5	11.0	4.0	–2.0	–7.5
Precip. (mm)	34.5	29.0	32.5	38.0	51.0	65.5	81.5	72.0	58.0	50.5	44.0	42.5
DENVER, U.S.A. 38°N												
Temp. (°C)	–1.0	0.5	4.0	9.0	14.0	19.5	23.0	22.0	17.0	11.0	4.0	–0.5
Precip. (mm)	14.0	15.5	33.5	44.5	62.5	43.0	47.0	37.5	28.5	26.0	23.0	15.0
PERTH, Australia 32°S												
Temp. (°C)	30.0	30.0	28.0	24.5	21.0	18.5	17.5	18.0	19.5	21.5	24.5	27.5
Precip. (mm)	8.5	12.5	19.0	45.0	121.5	182.0	174.0	135.5	80.0	53.5	21.0	13.5

Go to BioPortal for all WORKING WITHDATA exercises

annually and the dry season lasts no longer than 2 or 3 months. With no seasons unsuitable for growth, it is the most species-rich of all biomes, with up to 500 species of trees per square kilometer. Although these forests cover less than 2 percent of Earth's surface, they are home to more than half of all known species.

Along with the immense number of species they support, rainforests have the highest overall productivity of all terrestrial ecological communities. However, most mineral nutrients are tied up in the vegetation. The soils usually cannot support agriculture without massive applications of fertilizers. These forests are home to many epiphytes—plants that grow on other plants, deriving their nutrients and moisture from air and water rather than soil.

The rainforests provide humans with a dazzling range of products, including fruits, nuts, medicines, fuel, pulp, and furniture wood. Many more useful species undoubtedly await discovery, as only a small proportion of this biome's species have been inventoried. The rainforests, however, are currently being cut down or converted to agriculture at a rate of almost 20 million hectares per year.

RECAP 54.3

Ecologists recognize several terrestrial environment types called biomes. The geographic distribution of biomes is determined primarily by temperature and precipitation, but is also influenced by soil characteristics and fire.

- How do temperate grasslands differ from tropical savannas? In what ways are they similar? **See pp. 1131 and 1135**
- What primary factor distinguishes a tropical biome? **See pp. 1135–1137**

About 70 percent of Earth's surface is covered by saltwater oceans and seas that support abundant life. The small percentage of the aquatic world that consists of fresh water also hosts a significant proportion of Earth's aquatic organisms.

54.4 How Is Life Distributed in Aquatic Environments?

Aquatic biomes do not depend on plants for their structure in the way that terrestrial biomes do. Salinity is the primary factor that distinguishes the aquatic biomes. The marine biome is characterized by salt water, freshwater biomes by low salinity, and estuaries by the mixing of fresh water and salt water.

The marine biome can be divided into several life zones

Earth's oceans form one large, interconnected water mass on which the atmospheric factors that distinguish terrestrial biomes have little influence. However, light penetration, water temperature, water pressure, water movement (i.e., waves and tides), and salinity all vary spatially, so most marine organisms have restricted ranges and display adaptations to particular physical conditions.

The physical and biological discontinuities within the marine biome divide it into several distinct **life zones** (Figure 54.9). These zones are identified principally by their distance from shore and from the water's surface. The depth of an ocean basin varies from the shoreline to the relatively shallow continental shelf and to the deepest part of the ocean, sometimes known as the abyssal plain.

Water depth affects how much light is available to sustain the photosynthetic organisms that form the base of the marine food chain. In both marine and freshwater environments, the layer of water reached by enough sunlight to support photosynthesis is called the **photic zone**. Approximately 90 percent of all aquatic life is found in the photic zone.

The **coastal zone** extends from the shoreline to the edge of the continental shelf; it is characterized by relatively shallow, well-oxygenated water and relatively stable temperatures and salinities. These conditions support high densities of phytoplankton (photosynthetic floating protists), which in turn support some of the world's most important fisheries. Structure in coastal-zone communities may be provided by a variety of organisms. In warm coastal waters, corals generate complex reef structures that support ecosystems rivaling the rainforest in diversity. "Forests" of multicellular algal species (seaweeds and giant kelps) grow along many coasts at higher latitudes.

The area of the coastal zone that is affected by wave action is called the **littoral zone**. The principal autotrophs in this zone—sea grasses and algae—are consumed by a variety of invertebrates as well as small fishes. The portion of the coastal zone lying between the high and low tide levels is the **intertidal zone**, where tidal movements create conditions of highly variable temperature and salinity. Intertidal organisms, including clams, barnacles, copepods, and burrowing worms, are alternately exposed to air and submerged under water.

Throughout most of the oceans, the dominant autotrophs are phytoplankton. In the open ocean, or **pelagic zone**, the principal consumers of phytoplankton are zooplankton—mainly small crustaceans and larval stages of marine animals—which

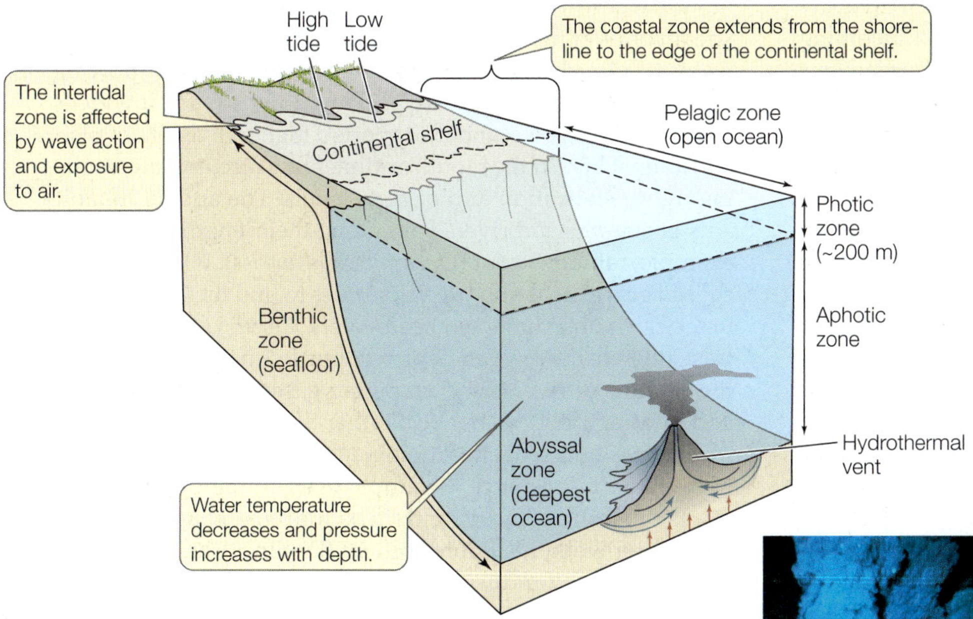
High tide
Low tide
The coastal zone extends from the shoreline to the edge of the continental shelf.
The intertidal zone is affected by wave action and exposure to air.
Continental shelf
Pelagic zone (open ocean)
Photic zone (~200 m)
Aphotic zone
Benthic zone (seafloor)
Abyssal zone (deepest ocean)
Hydrothermal vent
Water temperature decreases and pressure increases with depth.

Forests of giant kelp (*Macrocystis* sp.) dominate many coastal communities.

A large sailfish (*Istiophorus albicans*) feeds on sardines (*Sardinella aurita*) in pelagic waters of the Gulf of Mexico.

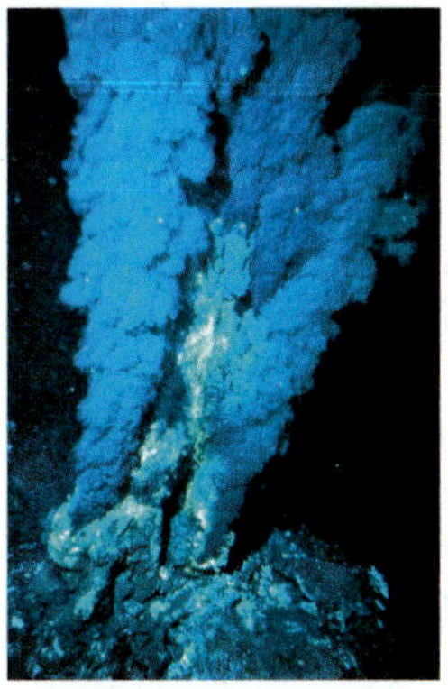

Heat and minerals from hydrothermal vents nourish unique deep-ocean communities.

54.9 Life Zones of the Marine Biome The ocean's life zones are primarily determined by light penetration. More than 90 percent of ocean-dwelling species live in the sunlit photic zone, which comprises less than 2 percent of the volume of open water. Wave action and exposure to air affect those life zones where the ocean meets the shore.

54.10 Life Zones in a Freshwater Lake Like oceans, standing bodies of fresh water can be divided into life zones based on water depth and distance from shore.

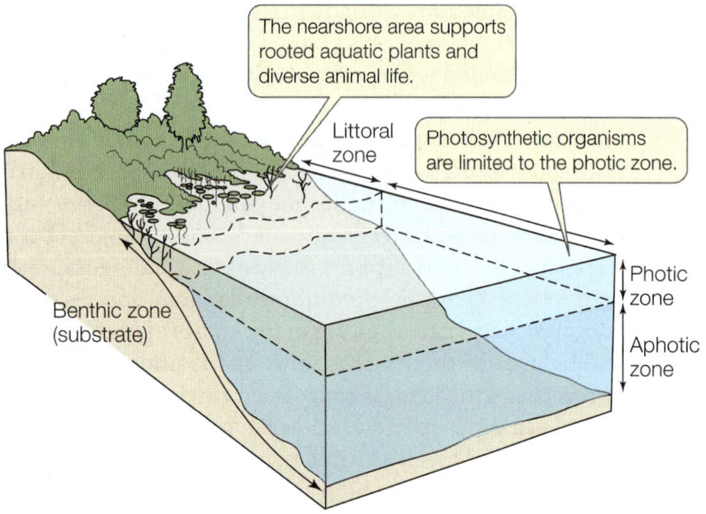
The nearshore area supports rooted aquatic plants and diverse animal life.
Littoral zone
Photosynthetic organisms are limited to the photic zone.
Photic zone
Aphotic zone
Benthic zone (substrate)

A school of bleak (*Alburnus scoranza*) among giant reeds along a lakeshore in Macedonia.
An osprey (*Pandion haliaetus*) snatches a large fish from the open waters of a Canadian lake.

in turn support many larger free-swimming vertebrate and invertebrate species.

The ocean bottom—the sediment surface—is referred to as the **benthic zone**. Many benthic organisms are adapted to life on the seafloor substrate. They include sessile animals such as sponges, bryozoans, ribbon worms, and brachiopods as well as motile bottom feeders such as crabs and sea slugs.

Where the water is too deep for light to penetrate, little photosynthesis can take place, and both plant and animal diversity are low. Depths reached by less than 1 percent of incoming sunlight constitute the **aphotic zone**. Many of the organisms inhabiting these regions subsist on decaying organic matter that sinks down from the photic zone. Some produce their own light by means of bioluminescent organs (see Figure 33.14D). Even deep-ocean trenches and rift valleys support hydrothermal vent ecosystems sustained by chemoautotrophic prokaryotes that can metabolize the nutrients in seawater without the aid of sunlight (see Section 26.3).

Freshwater biomes may be rich in species

In contrast to the vast oceans, bodies of fresh water cover less than 3 percent of Earth's surface, but they are home to about 10 percent of all aquatic species. Freshwater biomes, as the name implies, are characterized by low levels of salinity—generally below 1 percent.

Freshwater biomes are distinguished by the degree and direction of water movement. The water in streams and rivers flows (generally) in one direction, from the source to the mouth. Lakes and ponds are bodies of standing water. Wetlands constitute an intermediate biome, with water levels that fluctuate.

Lakes and ponds are found on every continent. They vary in size and persistence: some small ponds may exist for only a single season, whereas lakes thousands of square kilometers in size can persist for centuries or longer. Like the oceans, bodies of standing water can be divided into life zones based on depth and distance from shore (**Figure 54.10**). The zone along the shoreline is characterized by warm temperatures and high species diversity. The photic upper layers of the open-water zone of open water teem with phytoplankton and the fish that feed on them; below that lies an aphotic zone where little light penetrates, oxygen levels are low, and there is little biotic diversity. As in the marine biome, the benthic zone comprises the sediments and other substrates at the lake bottom.

The physical features of a stream or river change along its length as water flows from the point of origin (the source) to its mouth, where it empties into a lake or an ocean. The source of a stream or river may be snowmelt, a spring, or a lake. The headwaters (those close to the source) tend to be cool, fast-flowing, and well oxygenated. As a river flows downstream, it widens, its rate of flow slows, and it supports a higher diversity of plant and animal life. At the mouth, sediment can accumulate, reducing light penetration and oxygen levels. The animal inhabitants of streams and rivers vary along their length as well. For example, rainbow trout (*Oncorhynchus mykiss*), which lay their eggs in gravel beds and use visual cues to find their prey, thrive in the clear, unsedimented headwaters. Certain catfish, by contrast, tolerate the oxygen-depleted, murky shallow waters near the mouths of rivers by exchanging gases through their skin and locating prey by chemical rather than visual cues.

The freshwater wetland biome is highly variable in terms of size and persistence. Swamps, marshes, and bogs are all forms of freshwater wetlands. The unifying characteristic of freshwater wetlands is intermittent flooding. The fluctuations in water level are due to inputs in the form of groundwater, surface water, and rainwater and outputs in the form of evapotranspiration, water flow below the surface, and surface runoff. Plants found in freshwater wetlands include duckweed and other floating water plants with tiny roots, emergent water plants with roots that are completely submerged, such as cattails, and trees and shrubs that grow on the margins. Although many kinds of animals are found in wetlands, frogs and other amphibians, which have a life cycle with both aquatic and terrestrial phases, fare especially well in these water-saturated terrestrial environments.

Estuaries have characteristics of both freshwater and marine environments

Estuaries form where rivers meet the ocean and salt water mixes with fresh water. Depending on local conditions, estuarine environments vary in size and species composition. In the upper part of the intertidal zone, estuaries can support salt marshes, with salt-tolerant rushes, grasses, and low-growing shrubs. Mangrove forests can be found along shorelines and in river deltas in tropical and subtropical latitudes. Dominating these forests are mangroves (*Rhizophora*). These trees display many remarkable adaptations—including aerial roots that are impervious to salt—that make them highly tolerant of high salinity, periodic anoxia, and occasional inundation. Sea grass beds can form in the subtidal zone, dominated by flowering plants such as eelgrass (*Zostera*) that can survive entirely underwater.

Diversity in estuaries tends to be very high. Having characteristics of both freshwater and marine systems, estuaries are home to many unique species and play an important role for other species as a conduit between marine and freshwater environments. Some salmon species that hatch in rivers, for example, spend many months in estuaries adjusting to higher salinities before swimming out to sea to grow into adults. The importance of estuaries as nurseries of marine life cannot be overstated.

Estuarine environments have long been a source of benefits for humans, not the least of which is their role in purifying terrestrial runoff and groundwater. In many places around the world, however, overfishing, habitat destruction, and pollution threaten the viability of estuarine ecosystems.

RECAP 54.4

The marine biome can be divided into several life zones determined by distance from the surface, which influences how much light is available to sustain photosynthesis, and distance from shore. Lakes and ponds, which are freshwater biomes, are also divided into life zones according to water depth and light penetration. Salt and fresh water mix in bodies of water known as estuaries.

- How does light penetration affect diversity in different life zones of the oceans? **See p. 1139 and Figure 54.9**
- How do estuaries link freshwater and marine systems, and why is this biome so important? **See p. 1141**

Biomes and life zones have similar physical characteristics in different parts of the world, and biome-adapted organisms share similar characteristics in widely separated regions. Yet biomes in different regions rarely have particular species in common, so climate alone cannot explain why species live where they do.

54.5 What Factors Determine the Boundaries of Biogeographic Regions?

Climate interacts with local abiotic features to influence where and how organisms live, but these are not the only factors that determine where organisms can be found. Evolutionary history—where and when groups of organisms originated and diverged—is key to determining the distributions of organisms. Evolutionary history, in turn, is greatly influenced by geological history, which has had a profound influence on the dispersal of species.

Geological history influences the distribution of organisms

Until European naturalists traveled the globe in the nineteenth century, they had no way of knowing how organisms were distributed in other parts of the world. Alfred Russel Wallace, who along with Charles Darwin advanced the idea that natural selection could account for the evolution of life on Earth (see Section 21.1), was one of those global travelers. Wallace spent seven years in the Malay Archipelago, where he noticed some remarkable patterns in the distributions of organisms. For example, he described the dramatically different birds that inhabited the adjacent islands Bali and Lombok:

> *In Bali we have barbets, fruit-thrushes and woodpeckers; on passing over to Lombock these are seen no more, but we have an abundance of cockatoos, honeysuckers, and brush-turkeys, which are equally unknown in Bali, or any island further west. The strait here is fifteen miles wide, so that we may pass in two hours from one great division of the earth to another, differing as essentially in their animal life as Europe does from America.*

Wallace pointed out that these differences could not be explained by climate or by soil characteristics, because in those respects Bali and Lombok are essentially identical.

Wallace saw that, based on the distributions of plant and animal species, he could draw a line that divided the Malay Archipelago into two distinct halves (**Figure 54.11**). He correctly deduced that the dramatic differences in flora and fauna were related to the depth of the channel separating Bali and Lombok. This channel is so deep that it would have remained full of water, and thus would have been a barrier to the movement of terrestrial animals, even during the glaciations of the Pleistocene epoch, when sea level dropped more than 100 meters and Bali and the islands to the west were connected to the Asian mainland.

With these insights, Wallace established the conceptual foundations of **biogeography**, the scientific study of the patterns of distribution of populations, species, and ecological communities across Earth. In *The Geographical Distribution of Animals*, published in 1876, he detailed the factors known at the time that influence the distributions of animals, including past glaciation, land bridges, deep ocean channels, and mountain ranges. He earned some measure of scientific immortality in that the Malay discontinuity that first piqued his curiosity is known to this day as "Wallace's line."

The biotas of different parts of the world differ enough to allow us to divide Earth into several continental-scale areas called **biogeographic regions** (**Figure 54.12**), each containing characteristic assemblages of species occupying many different biomes. The boundaries of these biogeographic regions are drawn where assemblages of species change dramatically, often over short distances. The biotas of biogeographic regions

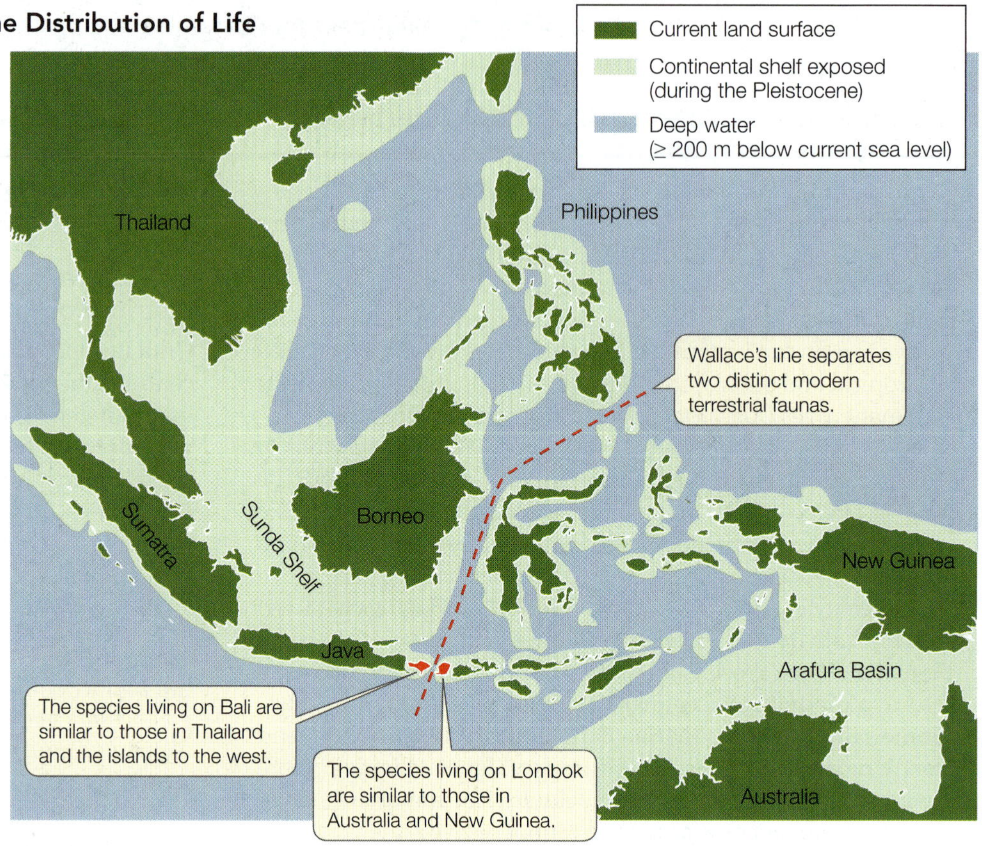

Current land surface
Continental shelf exposed (during the Pleistocene)
Deep water (≥ 200 m below current sea level)
Thailand
Philippines
Wallace's line separates two distinct modern terrestrial faunas.
Sumatra
Sunda Shelf
Borneo
New Guinea
Java
Arafura Basin
The species living on Bali are similar to those in Thailand and the islands to the west.
The species living on Lombok are similar to those in Australia and New Guinea.
Australia

54.11 Wallace's Line Wallace's line corresponds to a deep-water channel between the islands of Bali and Lombok. This channel would have blocked the movement of terrestrial organisms even during the Pleistocene glaciations, when sea level was 100 meters lower than it is today.

differ because oceans, mountains, deserts, and other barriers restrict the dispersal of organisms from one region to another. Although organisms do disperse between adjacent biogeographic regions, such interchanges have not been frequent or massive enough to eliminate the striking differences between them.

Go to Activity 54.1
Major Biogeographic Regions
Life10e.com/ac54.1

Two scientific advances changed the field of biogeography

For many decades after observing that the biotas of the major biogeographic regions are strikingly different, biogeographers speculated about the causes of these differences. The field remained primarily descriptive, however, until the second half of the twentieth century when two scientific advances transformed biogeography into a dynamic, multidisciplinary field. These advances were (1) the acceptance of the theory of continental drift and (2) the development of phylogenetic taxonomy.

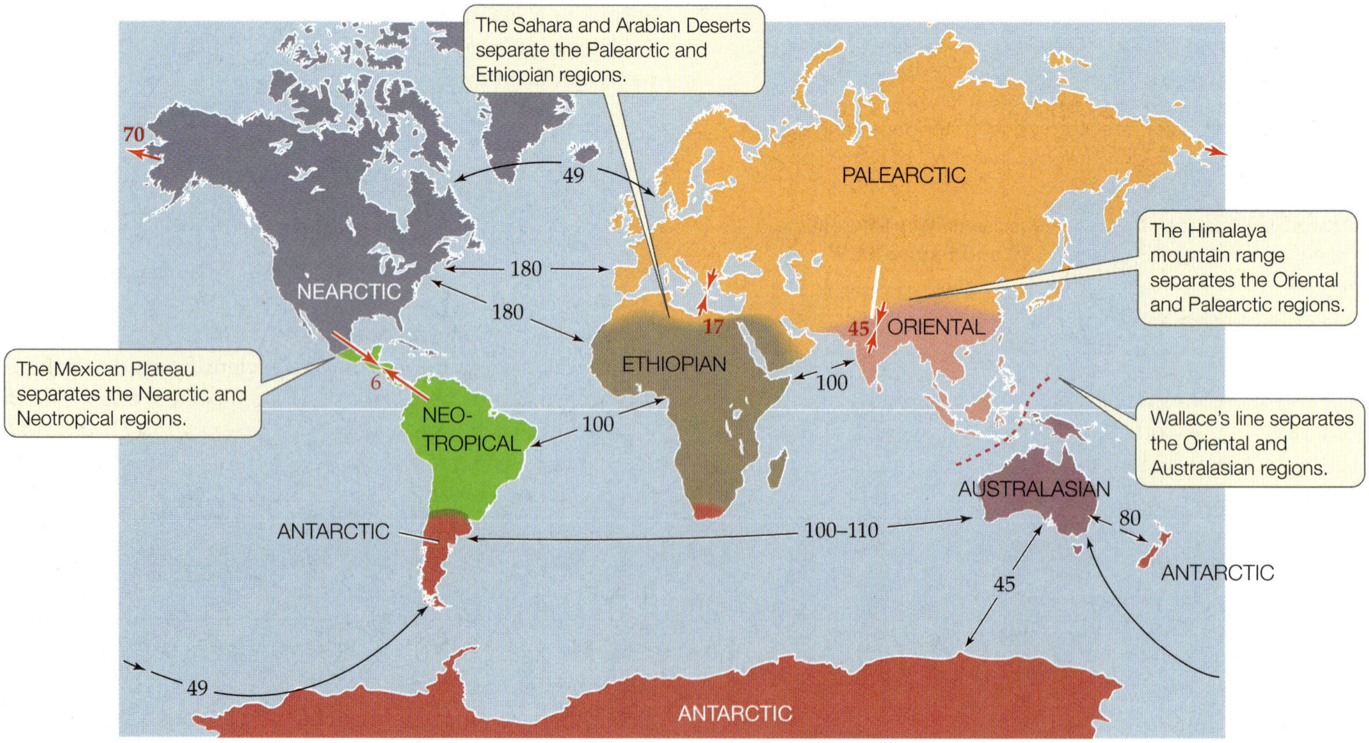

The Sahara and Arabian Deserts separate the Palearctic and Ethiopian regions.
70
49
PALEARCTIC
The Himalaya mountain range separates the Oriental and Palearctic regions.
180
NEARCTIC
180
17
45
ORIENTAL
ETHIOPIAN
The Mexican Plateau separates the Nearctic and Neotropical regions.
6
100
NEO-TROPICAL
100
Wallace's line separates the Oriental and Australasian regions.
AUSTRALASIAN
ANTARCTIC
100–110
80
ANTARCTIC
45
49
ANTARCTIC

54.12 Earth's Biogeographic Regions The major biogeographic regions are separated by climatic, topographic, and/or aquatic barriers to dispersal that cause their biotas to differ strikingly from one another. The red arrows on the map show the time (in millions of years) since land masses came together. Black arrows show the time since land masses separated.

(A)

(B)

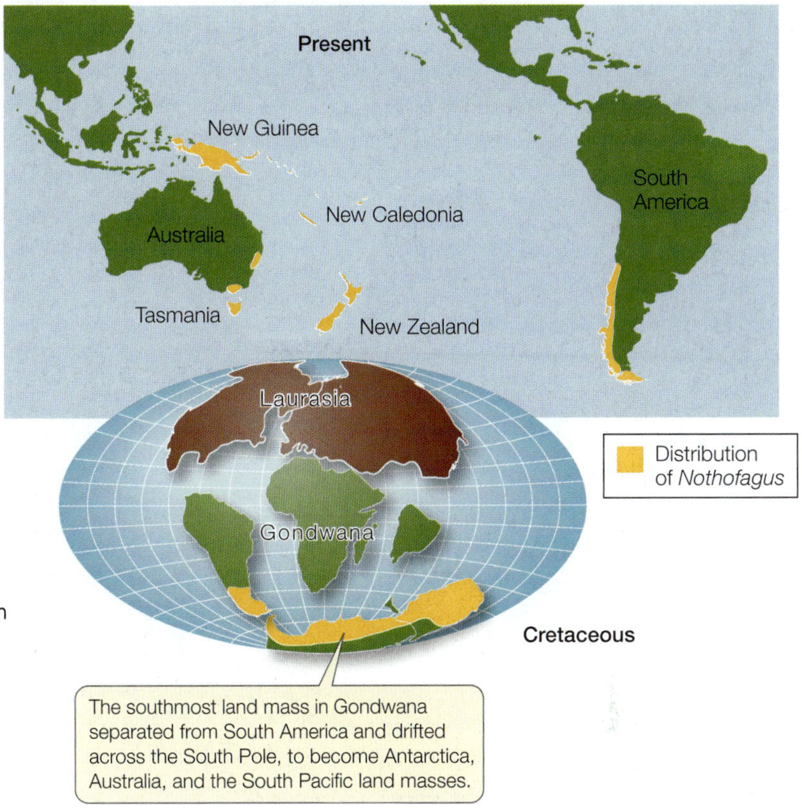
Present
New Guinea
New Caledonia
Australia
Tasmania
New Zealand
South America
Laurasia
Gondwana
Cretaceous
Distribution of *Nothofagus*
The southmost land mass in Gondwana separated from South America and drifted across the South Pole, to become Antarctica, Australia, and the South Pacific land masses.

54.13 *Nothofagus* Has a Gondwanan Distribution The modern range of southern beeches (A) is best explained by their origin in Gondwana during the Cretaceous. (B) When Gondwana broke apart, *Nothofagus* remained in South America, Australia, New Zealand, and the islands of the South Pacific.

CONTINENTAL DRIFT By the 1960s scientists knew that continents can and do move (see Section 25.2). We now know that over the course of the Triassic and Jurassic periods, the supercontinent Pangaea divided into two great land masses, Laurasia and Gondwana (see Figure 25.12), which subsequently separated into the continents we know today. Those groups of organisms that are represented on two or more continents are believed to be ancient groups whose ancestors were widely distributed over these great land masses before they broke apart. After the breakup, however, their descendants evolved independently, so groups that did not originate until after the continents separated have more discrete distributions. Thus continental drift is at least partly responsible for the existence of the biogeographic regions shown in Figure 54.12.

Continental drift explains certain biogeographic distributions that would otherwise be difficult to understand. For example, the southern beeches—trees of the genus *Nothofagus*—are found in both the Neotropical and the Australasian biogeographic regions. Their distribution suggests that the genus originated in Gondwana during the Cretaceous period and was geographically separated by the breakup of that land mass (**Figure 54.13**).

What evidence do we have that *Nothofagus* did not simply leapfrog from one biogeographic region to another? Fossilized *Nothofagus* pollen from 55 to 34 million years ago has been found in Australia, New Zealand, western Antarctica, and South America, suggesting that *Nothofagus* was once continuously distributed across a single land mass (see Figure 54.13B). Moreover, the modern distribution of aphid genera that feed exclusively on *Nothofagus* parallels the distribution of the trees. There are no air or water currents between Chile and New Zealand that would be likely to disperse insects, indicating that the aphids arose at a time when their host plants grew on a common land mass.

PHYLOGEOGRAPHY As we saw in Chapter 22, taxonomists have developed powerful methods of reconstructing the phylogenetic relationships among organisms. Biogeographers have adapted these methods to help them understand how organisms came to occupy their present-day distributions. Biogeographers can transform phylogenetic trees into **area phylogenies** by replacing the names of the taxa on a tree with the names of the places where those taxa now live or once lived.

Suppose, for example, we wonder why zebras, which are members of the horse family (Equidae), live in Africa when the fossil record indicates that the Equidae originated in North America. An area phylogeny of living equid species suggests that the ancestors of today's horses (represented by the oldest fossils) dispersed from North America to Asia, and then from Asia to Africa, and that the subsequent speciation of zebras took place entirely in Africa (**Figure 54.14**).

Go to Media Clip 54.1
Rafting to Madagascar
Life10e.com/mc54.1

Discontinuous distributions may result from vicariant or dispersal events

The appearance of a physical barrier to dispersal that splits the range of a species is called a **vicariant event**. A vicariant event

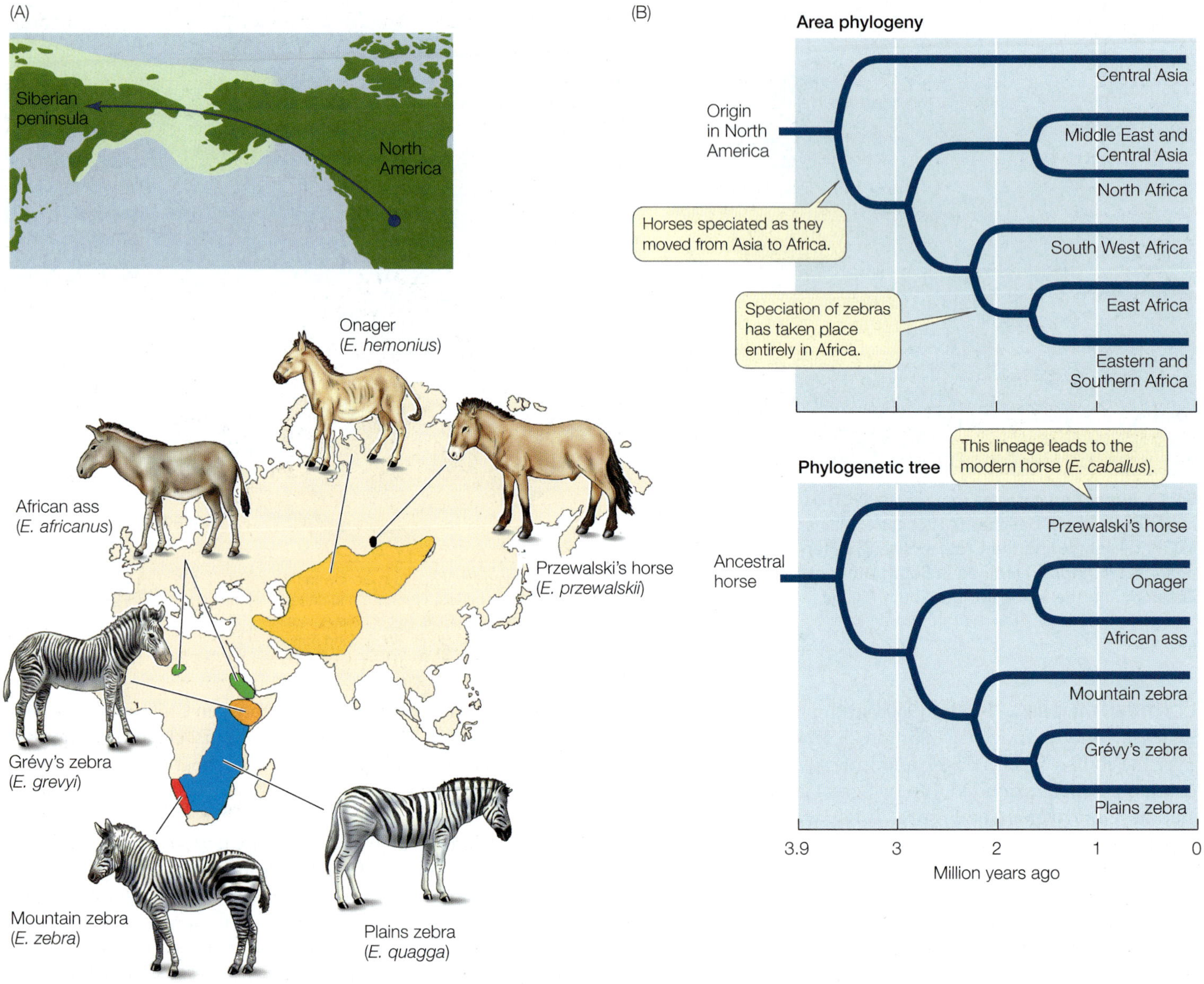
(A)
Siberian peninsula
North America
Onager
(E. hemonius)
African ass
(E. africanus)
Przewalski's horse
(E. przewalskii)
Grévy's zebra
(E. grevyi)
Mountain zebra
(E. zebra)
Plains zebra
(E. quagga)
(B)
Area phylogeny
Origin in North America
Central Asia
Middle East and Central Asia
North Africa
Horses speciated as they moved from Asia to Africa.
South West Africa
East Africa
Speciation of zebras has taken place entirely in Africa.
Eastern and Southern Africa
Phylogenetic tree
This lineage leads to the modern horse (E. caballus).
Ancestral horse
Przewalski's horse
Onager
African ass
Mountain zebra
Grévy's zebra
Plains zebra
3.9
3
2
1
0
Million years ago

54.14 Phylogenetic Tree to Area Phylogeny The conversion of a phylogenetic tree into an area phylogeny helps biogeographers explain how the current distribution of a taxon came about. (A) The ancestor of all Asian and African equids (genus *Equus*; horses and their relatives) migrated across the Bering Strait land bridge (light green) some 10 million years ago. (B) Organismal and area phylogenies explain the Asian and African distribution of the descendants of these ancestral horses.

divides the species into two or more discontinuous populations, even though no individuals have dispersed to new areas. If, however, members of a species cross an existing barrier and establish a new population, the discontinuous range of the species is considered to be the result of dispersal.

Given that the processes of vicariance and dispersal both influence distribution patterns, how can biogeographers determine the role of each process when reconstructing the evolutionary history of a particular species? By studying area phylogenies, a biogeographer may discover evidence suggesting that the distribution of an ancestral species was influenced by a vicariant event, such as continental drift or a change in sea level. If that inference is correct, then it is reasonable to assume that ancestral species in other lineages would have been affected by the same event and that similar distribution patterns should therefore be seen in other taxonomic groups. Differences in distribution patterns among taxonomic groups may indicate that they responded differently to the same vicariant events, that they diverged at different times, or that they had very different dispersal histories. By analyzing such similarities and differences, biogeographers seek to discover the relative roles of vicariant events and dispersal in determining today's distribution patterns.

The parsimony principle used in the reconstruction of phylogenies (see Section 22.2) can also be helpful in biogeographic studies. For example, the New Zealand flightless

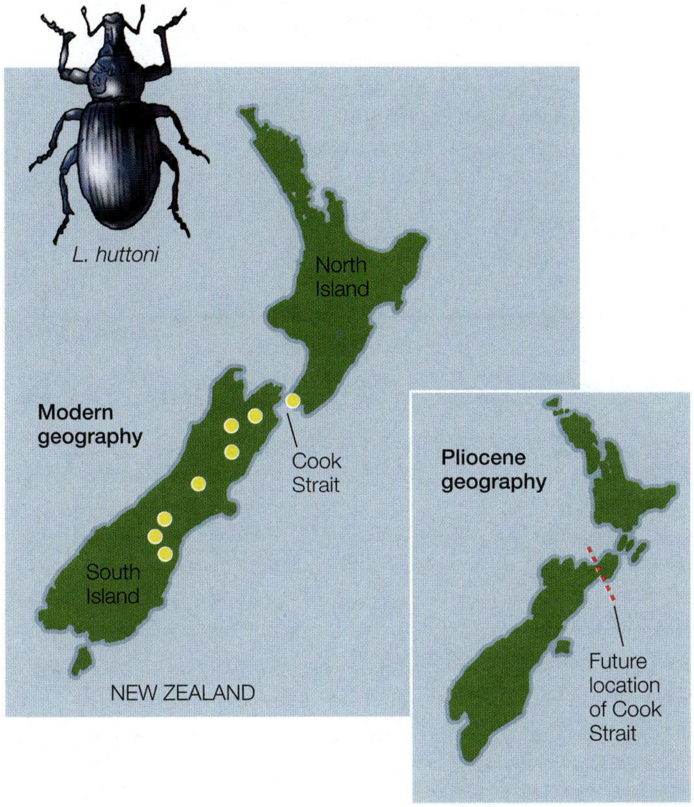
L. huttoni
North Island
Modern geography
Cook Strait
South Island
NEW ZEALAND
Pliocene geography
Future location of Cook Strait

54.15 A Vicariant Distribution Yellow circles indicate the current distribution of the flightless weevil *Lyperobius huttoni*. A comparison of New Zealand's present-day geography with its geography in the Pliocene, when the southern part of today's North Island was part of South Island, suggests that a vicariant event—a physical split separating populations—explains this distribution.

weevil *Lyperobius huttoni* is found in the mountains of South Island and on sea cliffs at the extreme southwestern corner of North Island (**Figure 54.15**). At first glance, its distribution might suggest that, even though this weevil cannot fly, some individuals in the distant past managed to cross Cook Strait, the 25-km wide body of water that separates the two islands. However, more than 60 other animal and plant species, including other flightless insects, are found on both sides of Cook Strait. Irrespective of their ability to fly, wade, or swim, it is unlikely that all 60 of these species made the same ocean crossing independently at different times over the course of their evolutionary history. In fact, geological evidence indicates that the present-day southwestern tip of North Island was once united with South Island. Thus a single vicariant event—the separation of the northern tip of South Island from the remainder of the island by the newly formed Cook Strait—could have produced the distribution pattern shared by all 60 species today.

In other cases, the evidence points to dispersal. There are, for example, more than 135 species of long-horned beetles endemic to the Hawaiian Islands. The islands have never been attached to any continent (they arose by volcanism from the deep ocean floor), so no separation of continuous populations ever took place. Therefore this distribution must have been the result of long-distance dispersal. The beetles, which are most closely related to a genus found in North and Central America, probably colonized the islands and subsequently speciated as a consequence of specializing on different host plants on different islands (see Section 23.3).

Humans exert a powerful influence on biogeographic patterns

One more force capable of generating distribution patterns that span multiple biogeographic regions is human activity. Of the insects found both in Europe and in North America, it has been estimated that more than half have been transported between the continents by humans, either by accident or deliberately. Many of these transported species have unintended consequences for other species in their new regions. Where the fynbos has been invaded by the Argentine ant, for example, seedlings often fail to appear after fires. As we saw at the beginning of this chapter, the seeds of some fynbos plants survive fire only with the help of ants. The ants must be attracted to a seed, pick it up off the ground, nibble off the lipid-rich elaiosome, and then bury the seed deep enough in the soil to avoid injury by fire. Argentine ants, which humans accidentally transported to South Africa from South America, are attracted to the seeds and eat the elaiosomes, but these tiny ants are too small to carry off large seeds and they cannot bury seeds deep enough in the ground to survive fires. In places where Argentine ants have displaced native ant species, replacement of large-seeded plants by seedlings after fires can drop by tenfold compared with areas that have not been invaded. The effects of such accidental species introductions by humans will be discussed at length in Chapter 59.

RECAP 54.5

Earth can be divided into seven biogeographic regions, each with unique assemblages of species. Vicariant events generate distribution patterns by splitting the ranges of species; distribution patterns may also change when species disperse across barriers.

- What determines the boundaries of Earth's major biogeographic regions? How are these boundaries different from those of the biomes described in Section 54.3? **See p. 1126, p. 1141, and Figures 54.7 and 54.12**
- Explain how the concepts of continental drift and phylogeography transformed the field of biogeography. **See pp. 1142–1143**
- How do vicariance and dispersal interact to generate species distribution? **See pp. 1143–1145**

Earth's physical environment and geological history are major factors influencing the distribution of organisms. Next we will turn to the influence of organisms and populations of organisms on one another. Abiotic, intraspecific, and interspecific forces all interact in the complex processes of population dynamics.

What is it about the western edges of continents that promotes tough, shrubby plant communities such as fynbos?

ANSWER

The fynbos is an example of the chaparral biome found in the Mediterranean regions and on the western coasts of continents at mid-latitudes (around 40°). The Mediterranean climate promotes plant communities that are nurtured by cool, damp winters and can survive the dry summer conditions and periodic fire. Continental locations of this biome are related to the proximity of the cold ocean currents that flow toward the equator offshore (see Figure 54.5). The prevailing winds set up rotating gyres of ocean water. Because of the direction of their rotation, warm water tends to move toward the poles along the east coasts of continents, whereas cool water moves toward the equator from higher latitudes along west coasts. These cool offshore currents bring cool, wet winters similar to those experienced by the Mediterranean region to large areas of continental western coastal regions.

CHAPTER SUMMARY 54

54.1 What Is Ecology?

- **Ecology** is the scientific investigation of interactions among organisms, between organisms and their physical environment, and the patterns of distribution and abundance resulting from these interactions.
- **Environmentalism** is the use of ecological knowledge to inform our decisions about the stewardship of natural resources.
- An organism's environment encompasses both **abiotic** (physical and chemical) components and **biotic** components (other living organisms).

54.2 Why Do Climates Vary Geographically?

- **Weather** refers to atmospheric conditions at a particular place and time. **Climate** is the average of atmospheric conditions, and the variation in those conditions, found in a particular place over an extended period of time.
- The solar energy that reaches a given unit of Earth's surface depends primarily on the angle of the sun's radiation, which in turn is a function of latitude. The tilt of Earth's axis results in seasonal variation in temperature and day length. **Review Figures 54.1, 54.2**
- Latitudinal variation in solar energy input drives atmospheric circulation patterns. **Review 54.3**
- Global surface wind patterns are driven by atmospheric circulation and Earth's rotation; these **prevailing winds** in turn drive ocean surface **currents**. **Review Figure 54.4, 54.5**
- Organisms respond to climatic challenges with physiological, morphological, and behavioral adaptations.

54.3 How Is Life Distributed in Terrestrial Environments?

- A **biome** is an environment that is shaped by its climatic and geographic attributes and characterized by ecologically similar organisms. **Review Figure 54.7**
- The distribution of terrestrial biomes is determined primarily by climate, but other factors, such as soil characteristics and fire, also influence vegetation.
- Biomes include Arctic and alpine tundra, boreal forest, temperate evergreen and temperate deciduous forests, temperate grasslands, hot and cold deserts, chaparral, thorn forest and savanna, tropical deciduous forest, and tropical rainforest. **See ANIMATED TUTORIAL 54.1, 54.2**

54.4 How Is Life Distributed in Aquatic Environments?

- Aquatic biomes do not depend on plants for their structure in the way terrestrial biomes do. Salinity is the primary factor that distinguishes aquatic biomes.
- The marine biome is characterized by high salinity. Marine **life zones** are determined by distance from the surface, which influences how much light is available to sustain photosynthetic organisms, and by distance from the shore. **Review Figure 54.9**
- Freshwater biomes are distinguished by their water movement (standing versus flowing water). Standing water (lakes and ponds), like ocean basins, can be divided into life zones distinguished by depth and distance from shore. **Review Figure 54.10**
- The physical conditions in streams and rivers change along their length as water flows from the source to the mouth.
- In freshwater wetlands, water levels fluctuate because of variation in water input and output.
- **Estuaries** are bodies of water where salt and fresh water mix. This biome supports many unique species.

54.5 What Factors Determine the Boundaries of Biogeographic Regions?

- **Biogeography** is the scientific study of the patterns of distribution of populations, species, and ecological communities.
- The boundaries of the **biogeographic regions** are drawn where assemblages of species change dramatically over short distances. These boundaries are generally continental in scale and correspond to present or past barriers to dispersal. **Review Figures 54.11, 54.12, ACTIVITY 54.1**
- Continental drift explains some discontinuous distributions that include more than one biogeographic region. **Review Figure 54.13**
- Biogeographers can transform phylogenetic trees into **area phylogenies** to understand how organisms came to occupy their present-day distributions. **Review Figure 54.14**
- Both **vicariant events** and dispersal across barriers generate discontinuous species distributions. **Review Figure 54.15**.

Go to the Interactive Summary to review key figures, Animated Tutorials, and Activities Life10e.com/is54

CHAPTER**REVIEW**

REMEMBERING

1. Ecology and environmentalism are
   a. synonymous; the terms can be used interchangeably.
   b. differentiated by the emphasis placed by ecology on the biotic rather than the abiotic world.
   c. differentiated by the lack of utility of ecology in solving world problems.
   d. differentiated by the inherent focus of environmentalism on human concerns.
   e. both scientific fields of inquiry that generate knowledge about the natural world but use completely different tools.

2. Energy from the sun determines
   a. air temperature.
   b. air and wind circulation patterns.
   c. ocean surface currents.
   d. All of the above
   e. None of the above

3. The amount of solar energy that reaches a given unit of Earth's surface depends primarily on
   a. the angle of the sun's rays.
   b. the moisture content of the air.
   c. the amount of cloud cover.
   d. the strength of the winds.
   e. day length.

4. The marine biome can be divided into life zones because
   a. the rate of photosynthesis in the oceans is low.
   b. ocean currents keep organisms close to where they were born.
   c. water temperature, salinity, and food supply all vary within the ocean.
   d. trade winds keep warm and cold waters separate.
   e. continents provide barriers to movement of planktonic life forms.

5. You are choosing a location on which to grow corn (*Zea mays*) and want to minimize the amount of land you need to cultivate for maximum yield. In which biome would you locate your farm?
   a. Tropical rainforest because this biome is home to tremendous plant diversity.
   b. Temperate evergreen forest because some of the world's largest plants are found there.
   c. Temperate grassland because the topsoil is deep and rich.
   d. Arctic tundra because summer days are long and soil moisture is abundant.
   e. All of these biomes are equally well suited for efficient cultivation of corn.

UNDERSTANDING & APPLYING

6. The Chilean matorral is located on the west coast of South America between 32°S and 37°S. The region experiences wet, cool winters and hot, dry summers. Describe the appearance and life cycles of the seed plants you would expect to find there. How would they most likely disperse their seeds? Describe some specific adaptations you might expect to encounter in at least some of the plant species there.

7. In 2008 spider expert Rudy Jocqué discovered two new species of spiders in the genus *Australutica* in South Africa. Prior to this find, the only species known in the genus were all found in Australia. One of the new species was found among the oldest rock formations in southern Africa (formed about 150 million years ago). Based on what you know about continental drift, do you think undiscovered *Australutica* species might exist anywhere else in the world? Where besides Australia and southern Africa would you look for such species?

ANALYZING & EVALUATING

8. Refer to the summary chart of biome characteristics below. If mean average global temperatures increase 5°C by 2100, as predicted by some climate models, which biomes would be expected to decrease in geographic extent?

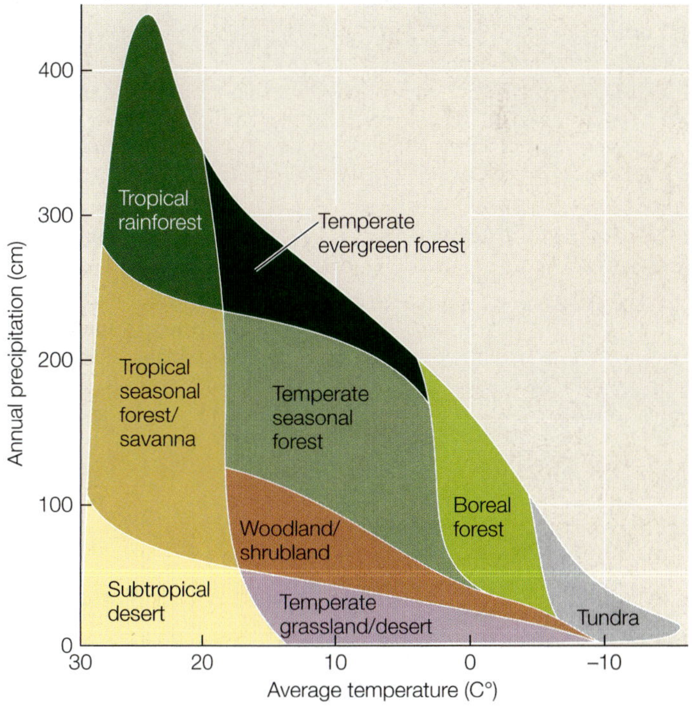

Annual precipitation (cm)
400
300
200
100
0
Tropical rainforest
Temperate evergreen forest
Tropical seasonal forest/ savanna
Temperate seasonal forest
Boreal forest
Woodland/ shrubland
Subtropical desert
Temperate grassland/desert
Tundra
30
20
10
0
–10
Average temperature (C°)

9. Today by far the greatest number of species of fruit flies (genus *Drosophila*) is found in the Hawaiian Islands. Would you conclude that the genus originated in Hawaii and spread to other regions? Under what circumstances do you think it might be accurate to conclude that a group of organisms originated in the same region where the greatest number of species live today? (Hint: Review the discussion of equid phylogeny and Figure 54.14.)

10. The map below is from a 2008 paper by H. I. McCallum and colleagues (*Ecology and Society* 13: 41–57). The paper's title is "Will Wallace's Line save Australia from avian influenza?" Based on the map and what you have learned about biogeography, how would you answer the title's question? What factor(s) do you think might be influencing the geographic spread of this disease?

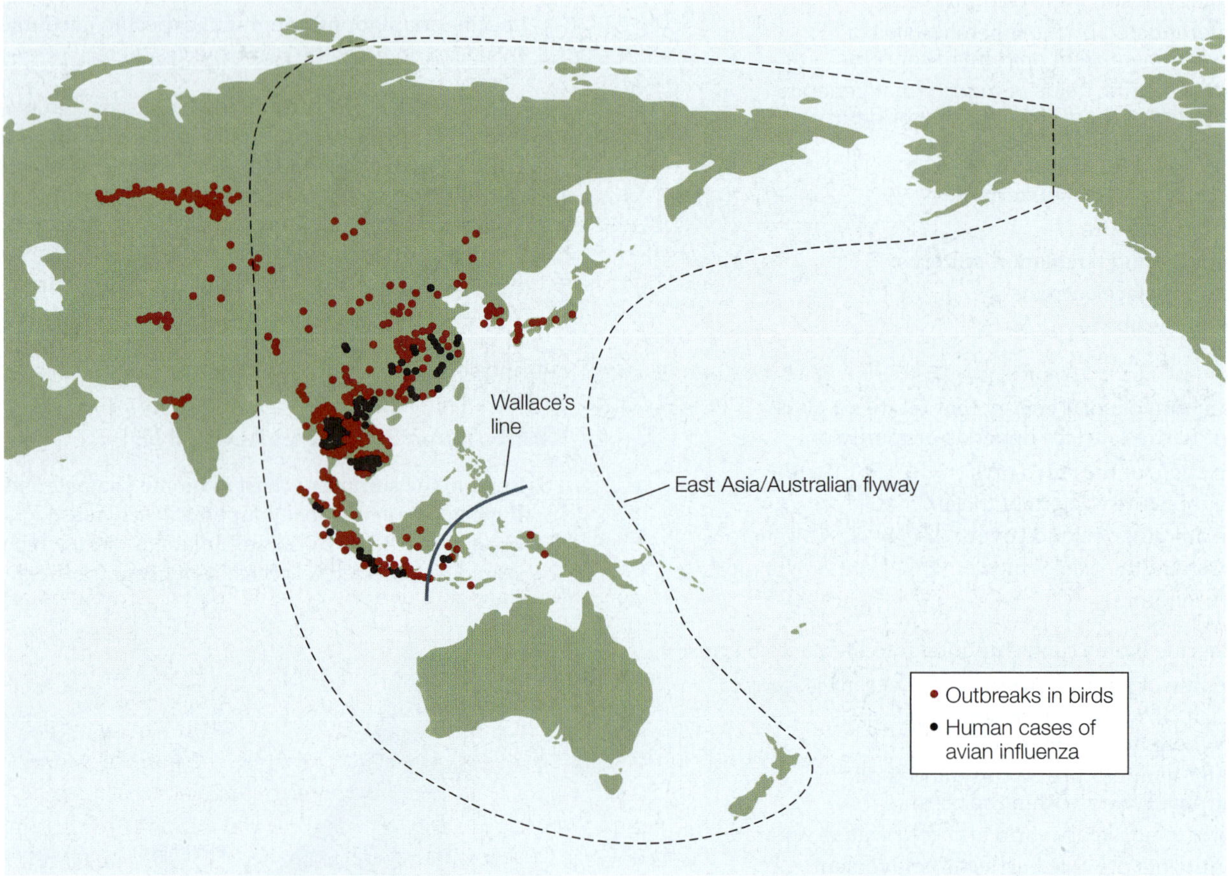
Wallace's line
East Asia/Australian flyway
Outbreaks in birds
Human cases of avian influenza

Go to BioPortal at **yourBioPortal.com** for Animated Tutorials, Activities, LearningCurve Quizzes, Flashcards, and many other study and review resources.

55 Population Ecology

CHAPTER**OUTLINE**

55.1 **How Do Ecologists Measure Populations?**

55.2 **How Do Ecologists Study Population Dynamics?**

55.3 **How Do Environmental Conditions Affect Life Histories?**

55.4 **What Factors Limit Population Densities?**

55.5 **How Does Habitat Variation Affect Population Dynamics?**

55.6 **How Can We Use Ecological Principles to Manage Populations?**

Reindeer Games Part of the St. Matthew reindeer herd is seen in this photograph taken in 1963, shortly before a particularly severe winter destroyed most of this isolated population. The herd had grown exponentially since being introduced to the island as a food source for a Coast Guard station during World War II.

DURING World War II the U.S. Coast Guard established a LORAN (Long-Range Aids to Navigation) tracking station on the tiny island of St. Matthew in Alaska, an isolated and unoccupied patch of tundra more than 300 kilometers from the nearest village. As an emergency food supply for the 19 men assigned to the island in 1944, the Coast Guard brought in 29 reindeer (*Rangifer tarandus*) by barge and released them.

The reindeer thrived on the thick, lush mat of lichens that covered the island. Other than the men, the island had no reindeer predators; the only other terrestrial vertebrate occupants were Arctic foxes, one species of vole, and a few ground-nesting birds. Then the war ended and the men left the island, leaving the reindeer behind in an environment with plentiful food and no natural predators.

In 1957 David Klein, a U.S. Fish and Wildlife biologist, visited St. Matthew. He and an assistant counted more than 1,350 reindeer, most of which appeared healthy. However, they also noticed areas of overgrazed lichen. In 1963 Klein and three colleagues hitched a ride to the island on a Coast Guard cutter. This time they counted more than 6,000 reindeer, packed in at a density of 47 per square mile. The island was covered with reindeer droppings, and the animals were smaller than the ones sighted 6 years earlier.

The winter of 1963–1964 brought punishing storms, record low temperatures, and tremendous snowfalls to St. Matthew. When Klein returned in summer 1966, the island was littered with reindeer skeletons. Klein found only 42 living reindeer, 41 of which were adult females; the lone male appeared to have deformed antlers. Lichens had disappeared, replaced almost entirely by sedges and grasses, on which reindeer cannot subsist. By 1980, the reindeer too had disappeared from the island.

Introducing large mammals to small islands is inherently risky, as the experience of reindeer on St. Matthew illustrates. But such introductions do not always end in disaster. Reindeer herds introduced to the subantarctic island of South Georgia almost a century ago have persisted, and their populations appear to be stable.

Why do populations of a particular species in one place explode and crash, but in another, seemingly similar, place remain stable? That knowledge is critical for understanding why some species become pests in some places and not in others, for managing sustainable harvests of economically important species, and for designing plans for conserving endangered species.

> **?** Why did introduced reindeer populations persist on the island of South Georgia but not on the island of St. Matthew?
>
> See answer on p. 1166.

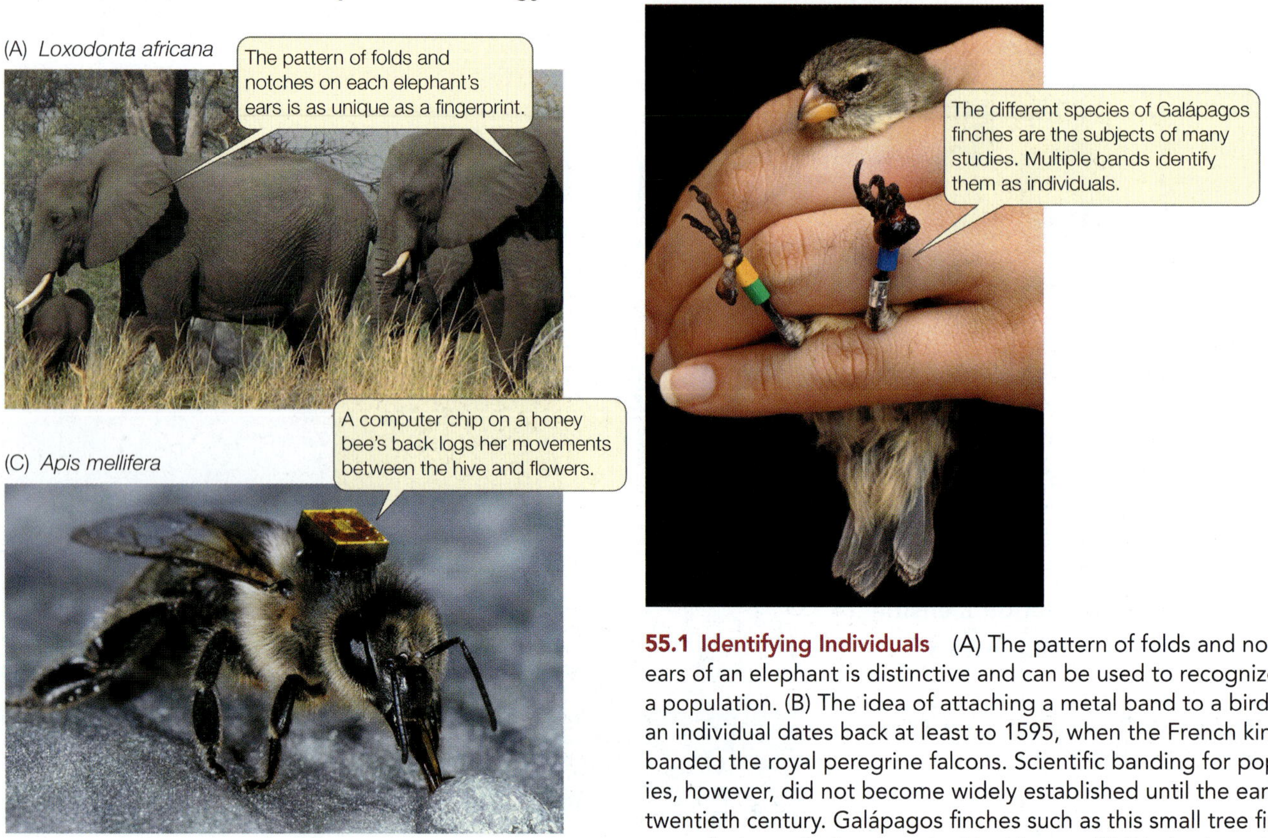
(A) Loxodonta africana
The pattern of folds and notches on each elephant's ears is as unique as a fingerprint.
(B) Camarhyncus parvulus
The different species of Galápagos finches are the subjects of many studies. Multiple bands identify them as individuals.
(C) Apis mellifera
A computer chip on a honey bee's back logs her movements between the hive and flowers.

55.1 Identifying Individuals (A) The pattern of folds and notches on the ears of an elephant is distinctive and can be used to recognize individuals in a population. (B) The idea of attaching a metal band to a bird's leg to identify an individual dates back at least to 1595, when the French king Henry IV banded the royal peregrine falcons. Scientific banding for population studies, however, did not become widely established until the early years of the twentieth century. Galápagos finches such as this small tree finch have been extensively studied and marked in this way. (C) Worker bees in a hive are individually indistinguishable to ecologists, who have come up with ingenious methods of marking. This female honey bee sports a computer chip on her back, which not only identifies her but also logs her movements between the hive and flowers.

55.1 How Do Ecologists Measure Populations?

Well before ecology became a distinct biological discipline, people engaged in population management. Whenever we grow crops or raise livestock, we are explicitly increasing the size of populations of domesticated plants and animals. Pest control strategies aim to reduce population sizes of organisms whose presence we consider undesirable. Game wardens, park managers, and conservation biologists aim to maintain stable populations of fish, wildlife, and threatened or endangered species. All of these activities require an understanding of **population dynamics**: the patterns and processes of change in populations. The study of population dynamics also allows us to understand the changes in populations we make inadvertently in the course of other human activities—as when the Coast Guard introduced reindeer to St. Matthew Island.

A **population** consists of the individuals of a species that interact with one another within a given area at a particular time. Populations are important units for study because groups of individuals that interact in time and space have ecological characteristics that individuals do not. At any given moment, an individual organism occupies only one point in space and is a particular age and size. The members of a population, however, are distributed over space, and they vary in age and size.

Population **density** is the number of individuals per unit of area or volume. Density is a function of processes that add individuals to the population (births and immigration) and processes that reduce the number of individuals in the population (deaths and emigration). Populations also have a characteristic **age structure**, or distribution of individuals across age categories, and a characteristic **dispersion pattern**, or spatial distribution of individuals in the environment.

These properties of populations, which are constantly changing because of births, deaths, and movements of individuals, influence the stability of populations and affect the ways in which populations of one species interact with populations of other species. Thus, to study populations, ecologists need to count the individuals in a given area and determine their ages.

Ecologists use a variety of approaches to count and track individuals

How the individuals in a population are counted depends on the nature of the organism under study. Populations of animals are usually more challenging to count than populations of trees. Most animals can move, so to avoid double counting, individuals must be identified. Nevertheless, counting every tree in a forest can be logistically difficult, even though the trees are standing still.

In some species, individuals are large and distinct enough, and populations small enough, that investigators can identify all the individuals and count them. Biologists performed this type of count, called a **full census**, on the African elephant population of Samburu and Buffalo Springs National Reserves in Kenya. By monitoring the elephants for 21 months, the biologists learned to recognize each of the 760 individuals in the population, primarily by their unique and distinctive ear markings (**Figure 55.1A**).

For most species, however, recognizing individuals is impossible or impractical. If biologists are to identify individuals of such species, they must be marked in some way. No single form of artificial marking works for all species. Plants can be marked with tags tied to their branches or by stakes in the ground nearby. Birds can be marked by colored bands on their legs (**Figure 55.1B**), and butterflies and beetles with small dabs of colored paint in different patterns. Honey bees can be monitored with fully automatic radio frequency identification (RFID) technology—the same technology used for tracking supermarket purchases (**Figure 55.1C**). Individual bees are marked with a chip, and a reader is placed at the hive entrance to register movements of the marked bees. Small mammals can be marked by bleaching or dyeing their fur in strategic places.

In most species, populations are too large, and their individual members too small, too similar in appearance, or too mobile, for a full census to be conducted. Thus population sizes are often estimated from representative samples using statistical methods.

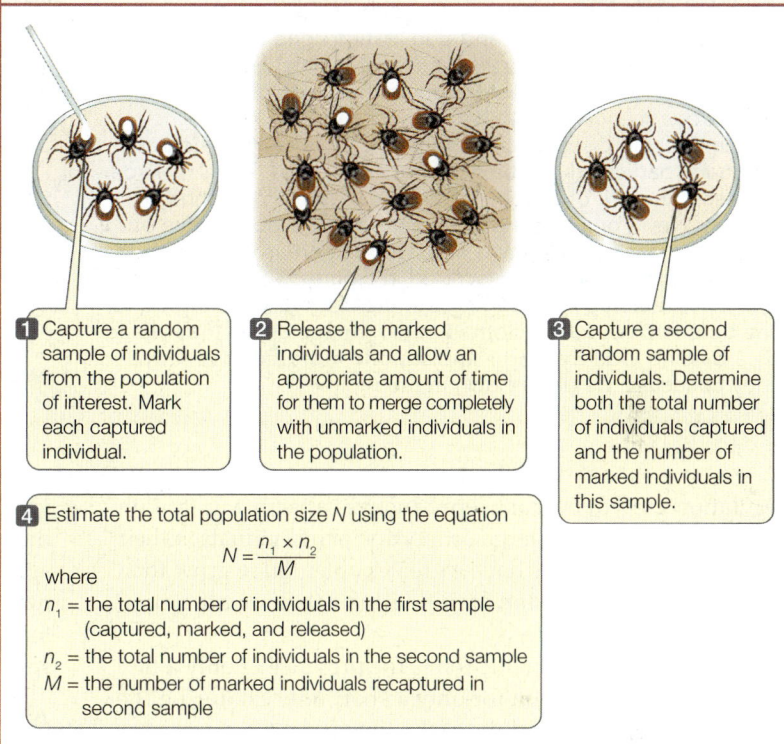

RESEARCH**TOOLS**

55.2 The Mark–Recapture Method The method described here is used to estimate population sizes for animal populations in which the individuals are highly mobile (such as *Ixodes scapularis*, the black-legged tick). Once a sampling area has been determined, investigators capture, mark, and then release individuals of the organism of interest. The proportion of these marked individuals that appears in a second sample is assumed to be the same as the proportion of the total individuals in that sample to the area's total population.

1 Capture a random sample of individuals from the population of interest. Mark each captured individual.

2 Release the marked individuals and allow an appropriate amount of time for them to merge completely with unmarked individuals in the population.

3 Capture a second random sample of individuals. Determine both the total number of individuals captured and the number of marked individuals in this sample.

4 Estimate the total population size N using the equation

$$N = \frac{n_1 \times n_2}{M}$$

where

n_1 = the total number of individuals in the first sample (captured, marked, and released)

n_2 = the total number of individuals in the second sample

M = the number of marked individuals recaptured in second sample

Ecologists can estimate population densities from samples

Ecologists usually measure the densities of terrestrial organisms as the number of individuals per unit of area. They may measure the densities of organisms living in soil, air, or water as the number or mass per unit of volume. Ecologists obtain these measurements from sample units, then extrapolate from these samples to estimate the total population density.

Estimating population densities is easiest for sessile organisms. Investigators need only count the individuals in a sample of representative locations and extrapolate the counts to the entire geographic range of the population. Individuals may be counted within marked and measured areas called **quadrats**. Plants are often counted along a **transect**—a line drawn across an area within the population's range (often designated by a string marked at regular intervals). Any individual that touches the line is counted. By making repeated counts with either of these methods, investigators can make reasonably good estimates of the size of a population.

Counting mobile organisms is more difficult because individuals move into and out of sampling areas. In such cases, investigators may use the **mark–recapture method** (**Figure 55.2**). They begin by capturing, marking, and then releasing a number of individuals. Later, after the marked individuals have had time to mix with unmarked individuals in the population (but before enough time has elapsed for births, deaths, and individual movement to affect the population size significantly), another sample of individuals is captured. This sample is then used to obtain an estimate of the total size of the population in the sampling area. This is done by applying the equation described in Figure 55.2, which assumes that the *proportion* of marked individuals in the second sample (i.e., individuals that were captured and marked in the first sample) is about the same as the proportion of individuals in the sampling area that were captured in the first sample.

For some species, however, this assumption does not apply. Some captured animals learn to avoid traps or leave the sampling area and are thus less likely to be recaptured than are unmarked individuals. Others become "trap-happy" (some mice, for example, reenter live traps repeatedly in order to snack on the peanut butter bait). In some cases the act of marking may reduce an individual's chances of survival due to the stress of handling or inadvertent alterations of appearance that make marked individuals more conspicuous to predators. Ecologists use statistical techniques to correct for these errors and improve the accuracy of population estimates.

Determining the size and density of populations is important, but these numbers are only a starting point for understanding population dynamics because not all individuals contribute equally to population growth.

A population's age structure influences its capacity to grow

The **age structure** of a population—the distribution of individuals across age categories—has a profound effect on

WORKING WITH**DATA**:

Monitoring Tick Populations

Original Paper

Falco, R. C. and O. Fish. 1988. Prevalence of *Ixodes dammini* near the homes of Lyme disease patients in Westchester County, New York. *American Journal of Epidemiolology* 127: 826–830.

Analyze the Data

Lyme disease is a chronic and debilitating condition caused by spirochete bacteria of the genus *Borrelia*, which infect humans by way of the bite of an intermediate host, the black-legged tick *Ixodes scapularis* (also known as the deer tick). The incidence of Lyme disease has increased dramatically in the past 20 years, particularly in the northeastern United States. In order to assess the risk of exposure to this disease in Westchester County, New York, investigators measured the abundance of deer ticks in suburban lawns near wooded areas using the mark–recapture method described in Figure 55.2. (Ticks are typically collected by dragging a white cloth along the ground; the ticks latch onto the cloth in much the same way they would to a passing leg.) By drag-sampling one representative lawn, they collected the data shown in the table.

QUESTION 1

Refer to Figure 55.2. Using the equation and other information described in that figure, estimate the total number of ticks in the sampled lawn from the data below.

QUESTION 2

The lawn was approximately 700 m^2 in size. What is the approximate density of ticks per square meter?

QUESTION 3

What do you think might be the implications of this study for residents of this neighborhood?

	Original capture event	Second capture event (3 weeks later)
Adult ticks captured	180	33
No. of marked ticks	180[a]	8

[a]All ticks captured in the first event were marked with acrylic paint and released.

Go to **BioPortal** for all WORKING WITH**DATA** exercises

population growth because reproductive capacity varies with age. Populations with a large proportion of individuals in their peak reproductive years have a greater potential to grow than do populations dominated by individuals that are too young or too old to reproduce.

In some species, reproduction is the province of only a tiny fraction of the population for only a short interval during the life cycle. For example, adults of the tiny insect *Clunio marinus* (the "one-hour midge") mate, lay eggs, and die within about an hour after completing their larval development. In contrast, some vertebrates, such as elephants, are capable of reproducing for years.

Consider the results of a long-term study of the age structure of the elephant population of Kidepo Valley National Park, Uganda, and how it changed over time. Relative to the population in 1970, the 2000 population had more elephants over 25 years of age (**Figure 55.3**). The change was the result of years of differential mortality among young elephants due to drought, and among adult males due to ivory poaching. The age structure as of 2000 portends an increase in the population's growth rate, given that female African elephants become fertile around age 10 to 15 and can continue producing offspring through their fifties.

A population's dispersion pattern reflects how individuals are distributed in space

Dispersion refers to the distribution of individuals in space. Dispersion affects how individuals in a population interact with one another and thus can have important effects on population growth. In addition, ecologists must understand the dispersion patterns of a species to choose appropriate sampling areas and statistical methods for estimating population sizes.

Ecologists recognize three basic dispersion patterns:

Go to Animated Tutorial 55.1
Age Structure and Survivorship
Life10e.com/at55.1

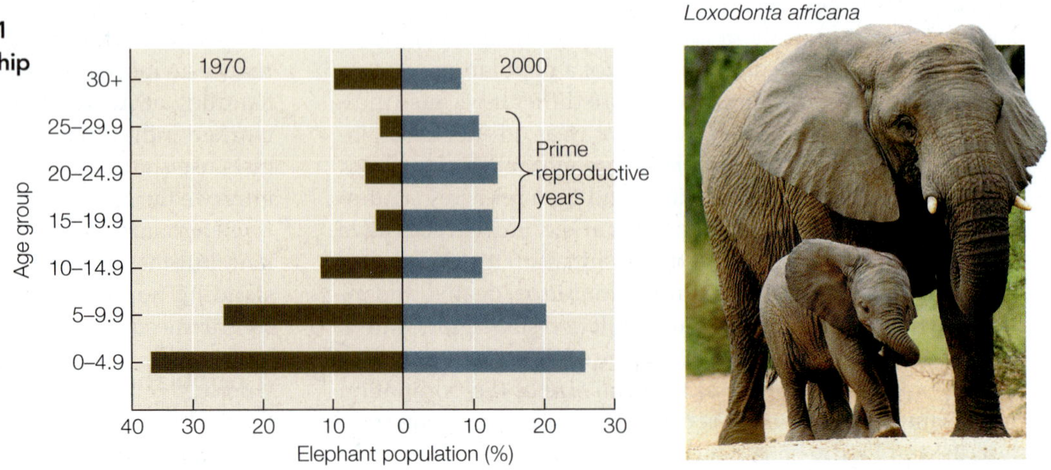
1970
2000
30+
25–29.9
20–24.9
15–19.9
10–14.9
5–9.9
0–4.9
Age group
Prime reproductive years
40 30 20 10 0 10 20 30
Elephant population (%)
Loxodonta africana

55.3 Changes in Age Structure Influence Population Growth
The elephant population in Kidepo Valley National Park, Uganda, was monitored between 1970 and 2000. During this time, the proportion of the population in the prime reproductive age range (15–30 years) grew considerably. Such an age structure in a population is likely to result in a high rate of growth.

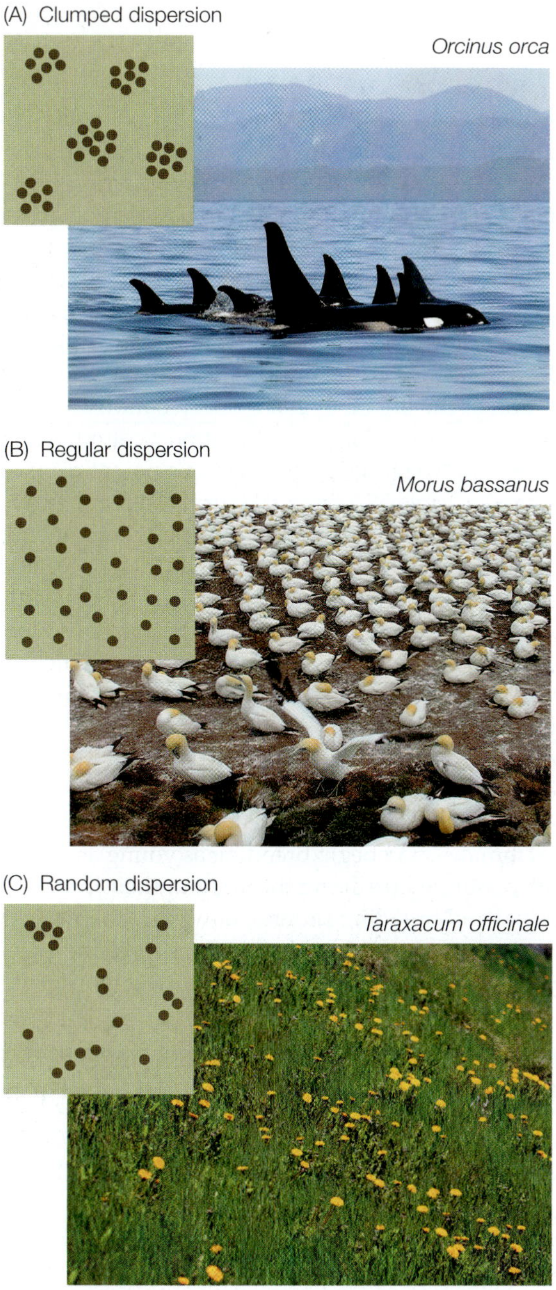
(A) Clumped dispersion
Orcinus orca
(B) Regular dispersion
Morus bassanus
(C) Random dispersion
Taraxacum officinale

55.4 Dispersion Patterns (A) Orcas hunting together in pods display a clumped dispersion pattern. (B) Nesting seabirds often stake out territories with a radius defined by their wingspans—an amount of space they can defend without leaving the nest. This behavior results in a regular dispersion pattern. (C) Dandelion seeds are dispersed by the wind in random fashion, so the plants that grow from those seeds show a random dispersion pattern.

- A **clumped dispersion pattern** occurs when the presence of one individual at any point in space increases the probability of others being near that point (**Figure 55.4A**).
- A **regular dispersion pattern** occurs when the presence of one individual at any point in space reduces the probability of others being near that point (**Figure 55.4B**).
- A **random dispersion pattern** occurs when the presence of one individual at any point in space does not affect the probability of other individuals being near that point (**Figure 55.4C**).

Dispersion patterns can vary among species, or among populations within species. Spatial variation in environmental conditions strongly influences dispersion patterns. Small-scale differences in temperature, humidity, or wind speed can make particular places more or less suitable for certain organisms. Aphids, for example, cluster along the protruding veins of leaves, where they are sheltered from wind. Interactions among individuals may also bring about characteristic dispersion patterns. Social life, with the cooperation it involves, tends to promote clumped dispersion patterns. By contrast, intraspecific competition for food, space, or mates tends to space individuals apart in regular dispersion patterns.

RECAP 55.1

To understand the dynamics of populations, ecologists measure their density, age structure, and dispersion patterns.

- What are some of the ways in which population density can be measured? **See p. 1151**
- How can the age structure of a population influence its growth? **See pp. 1151–1152 and Figure 55.3**
- How do environmental factors influence dispersion patterns? **See p. 1153 and Figure 55.4**

Once the sizes, densities, and other important traits of populations have been measured, these data can be used to describe various "survival strategies" found among populations and to understand and predict changes within and among populations.

55.2 How Do Ecologists Study Population Dynamics?

Quantifying population density and age structure provides useful information about populations, but those traits alone cannot explain how, when, and why populations change in size. In order to understand population growth, ecologists must measure population *processes* as well as population traits. The study of population processes is known as **demography**.

Demographic events determine the size of a population

The size of a population changes over time because of **demographic events**: births, deaths, immigration, and emigration. Over any given interval of time, the size of a population increases by the number of individuals added to the population by births and by immigration (the movement of individuals into the population from elsewhere) and decreases by the number of individuals lost from the population by deaths and by emigration (individuals leaving the population to go elsewhere). This relationship is expressed mathematically as

$$N_1 = N_0 + (B - D) + (I - E) \quad (55.1)$$

where

N_1 = the number of individuals at time 1

N_0 = the number of individuals at time 0

B = the number of individuals born between time 0 and time 1

D = the number that died between time 0 and time 1

I = the number that immigrated between time 0 and time 1

E = the number that emigrated between time 0 and time 1

Using Equation 55.1 to estimate N_1 over multiple time intervals helps researchers estimate the *rate of change in population size over time*—that is, the growth rate of the population.

Life tables track demographic events

The study of population dynamics requires keeping track of demographic events in populations and determining the rate (number per unit of time) at which they occur. An individual is born only once and dies only once; birth rates and death rates are properties of populations. A **life table** is a tool that ecologists use for these purposes. Life insurance companies use similar tables (called actuarial tables) to determine how much to charge people of different ages for insurance policies. Data from life tables can be used to identify the principal mortality factors, or causes of death, at particular life stages, to predict future population trends, and to develop strategies for managing populations of species of commercial or ecological value.

COHORT LIFE TABLES Life tables can be constructed by a number of methods. To construct a cohort life table, investigators start with a **cohort**—a group of individuals born within the same time frame, or age class—and record their deaths until no individuals from the cohort remain alive. This type of life table is sometimes called a horizontal life table because it is based on data collected *across* the entire life span.

The age classes used in a cohort life table depend on the life cycle of the organism of interest. Age-dependent cohort life tables track demographic events as a function of calendar age. Stage-dependent cohort life tables track demographic events at various stages of the life cycle (e.g., eggs, larvae, pupae, and adults in insects). They are commonly used when survival and reproduction depend more on developmental stage than on calendar age, as is the case, for example, with insects and other animals that undergo metamorphosis.

Using the data in a cohort life table, investigators can calculate **mortality**: the proportion of individuals of each age class that die before reaching the next age class. By following a cohort, investigators can also calculate the average individual's chance of dying during a particular time interval, a value known as the per capita death rate, or d. By the same token, they can calculate **survivorship** (represented by the term l_x), which is the likelihood of an individual member of the cohort surviving to reach age x (**Table 55.1**).

ESTIMATING REPRODUCTIVE CAPACITY A cohort life table can also be used to track the degree to which individuals in different age categories contribute to reproduction (and hence population growth). Investigators can use the data in the table to calculate the number of offspring the average individual produces, or the per capita birth rate, b. Because only females produce offspring, life tables generally track the number of female offspring produced by each female during each time period—a factor called **fecundity** (indicated by the term m_x). The portion of the life table that tracks fecundity is called a fecundity schedule (see Table 55.1, rightmost column). Such data allow scientists to estimate a population's potential for growth.

The data in Table 55.1 track the survivorship (l_x) and fecundity (m_x), respectively, of a cohort of the cactus finch species *Geospiza scandens* on Isla Daphne in the Galápagos. Peter and Rosemary Grant followed a cohort of 210 birds from the time they hatched in 1978 until 1991, at which time only 3 individuals—all males—remained alive. All of the cactus finches on the island were banded so that the Grants could recognize them as individuals (see Figure 55.1B).

The *G. scandens* life table shows that mortality was high during the first year of life, then dropped dramatically for several years before increasing in later years. The fecundity data indicate that females may begin breeding as young as 1 year of age and may continue breedomg throughout their lives.

Survival and breeding success, however, are not correlated exclusively with age. Other observations of conditions on Isla Daphne revealed a correlation of fecundity with rainfall, which

TABLE **55.1**
Life Table for the 1978 Cohort of *Geospiza scandens* on Isla Daphne

Age Class (years)	Number Alive	Survivorship[a]	Mortality[b]	Fecundity[c]
0–1	210	—	0.57	0.00
1–2	91	0.43	0.14	0.05
2–3	78	0.37	0.10	0.67
3–4	70	0.33	0.07	1.50
4–5 Increased rain	65	0.31	0.05	0.66
5–6	62	0.30	0.32	5.50
6–7 Drought	42	0.20	0.45	0.69
7–8	23	0.11	0.35	0.00
8–9	15	0.07	0.07	0.00
9–10	14	0.07	0.21	2.20
10–11	11	0.05	0.09	0.00
11–12	10	0.05	0.60	0.00
12–13	4	0.02	0.25	—
13	3	0.01	—	—

[a] Survivorship (l_x) = the proportion of the original cohort (here, of 210 individuals) who survive to age x.
[b] Mortality (d) = the proportion of individuals of age x who die before reaching age $x + 1$.
[c] Fecundity (m_x) = number of fledgling females per female per breeding season. Of the 210 birds in this cohort, 90 were females.

55.5 Survivorship Curves Ecologists recognize three general types of survivorship curves. Notice that the number of survivors has been plotted on a logarithmic scale. Three species provide real-world examples of the three types of life histories.

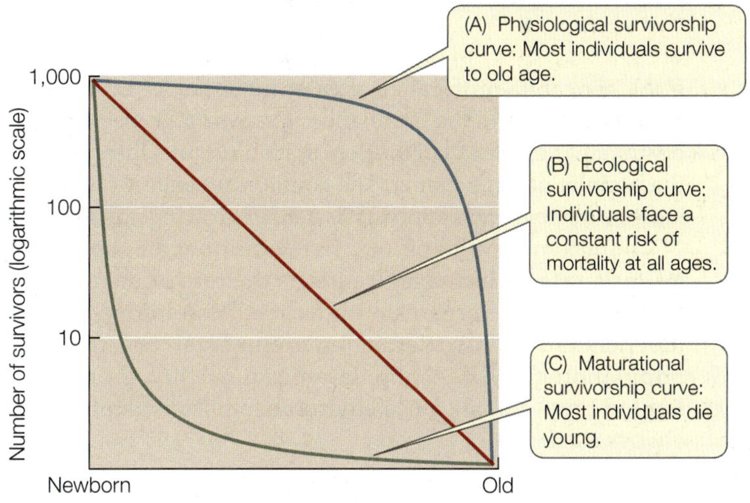
(A) Physiological survivorship curve: Most individuals survive to old age.
(B) Ecological survivorship curve: Individuals face a constant risk of mortality at all ages.
(C) Maturational survivorship curve: Most individuals die young.
1,000
100
10
Number of survivors (logarithmic scale)
Newborn
Old

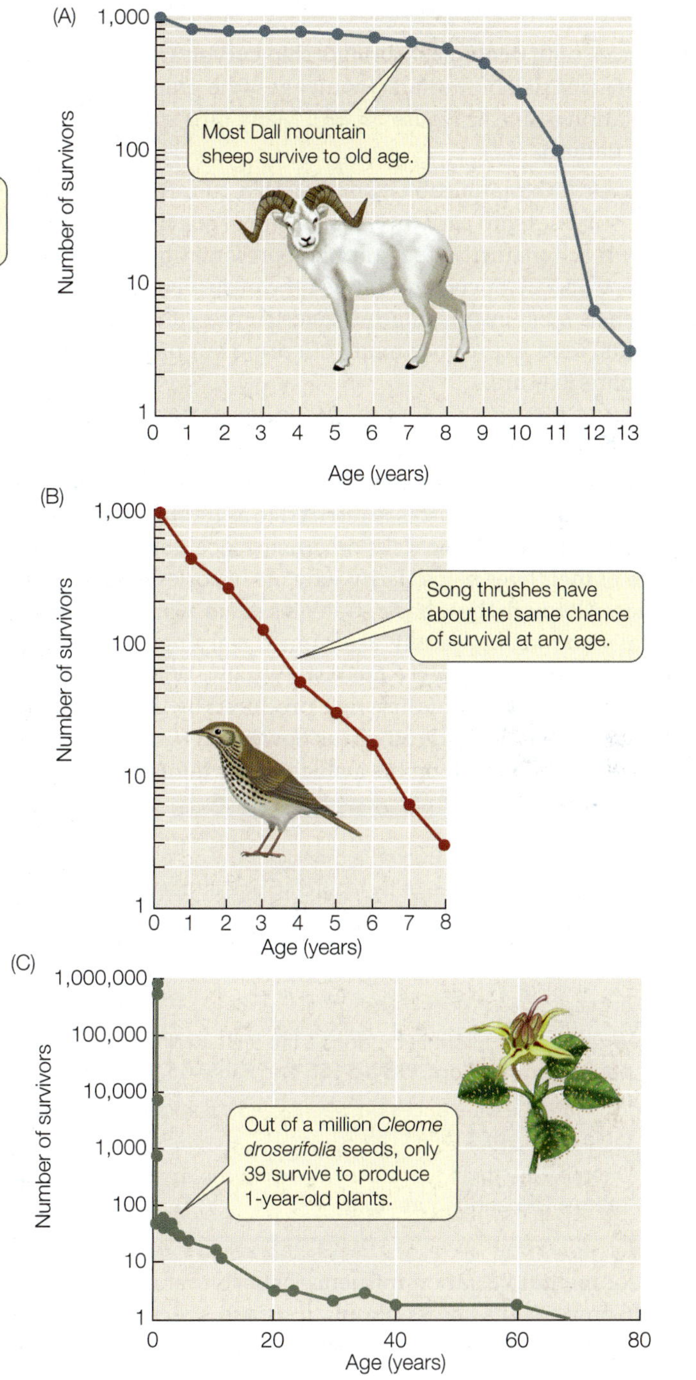
(A)
1,000
100
10
1
Number of survivors
Most Dall mountain sheep survive to old age.
0 1 2 3 4 5 6 7 8 9 10 11 12 13
Age (years)
(B)
1,000
100
10
1
Number of survivors
Song thrushes have about the same chance of survival at any age.
0 1 2 3 4 5 6 7 8
Age (years)
(C)
1,000,000
100,000
10,000
1,000
100
10
1
Number of survivors
Out of a million *Cleome droserifolia* seeds, only 39 survive to produce 1-year-old plants.
0 20 40 60 80
Age (years)

in the Galápagos varies dramatically from year to year. This life table and other ecological data, taken together, suggest that the survival of adult birds and the number of offspring they are able to fledge depend on food availability—that is, on cactus flower and fruit production, which is strongly correlated with rainfall (see Table 55.1). In short, life table data can be useful in identifying the ecological factors that affect population dynamics, at least over the short term.

Fecundity schedules vary greatly among species not only because organisms differ in the number of offspring they can produce, but also because they vary in the timing of reproduction. Whereas female *G. scandens* can begin breeding at the age of 1 year and in favorable conditions may fledge multiple broods each season, African elephant females do not produce offspring until they are at least 15 years old and usually produce only one calf every 5 years or so.

VERTICAL LIFE TABLES Because not all species can be easily followed over time, some life tables are constructed by sampling a population at a single time. These life tables cut across all age categories and thus are known as **vertical life tables**. One way to construct a vertical life table is to record information from a death assemblage, a collection of bodies or fossils of individuals that lived together in a particular place at a given time. Similarly, the birth and death dates on tombstones in a cemetery, for example, can be used to construct a vertical life table for a human population and to estimate its probability of reaching different ages.

Survivorship curves reflect life history strategies

The construction of life tables has allowed ecologists to observe common life history patterns, reflecting common solutions to ecological challenges, across a tremendous diversity of organisms. For example, the proportions of individuals surviving through each life stage (survivorship, l_x) can be taken from a life table and plotted graphically to construct a **survivorship curve**. Typically, a survivorship curve is constructed for a hypothetical cohort, usually of 1,000 individuals, by plotting the numbers of individuals expected to survive to reach each age category on a logarithmic scale.

Ecologists have noticed that survivorship curves tend to take one of three general shapes:

- Species with **physiological survivorship curves** experience high overall survivorship through adulthood but steep declines late in life (the graphic representation is concave; Figure 55.5A). Species with this type of survivorship curve (such as humans, elephants, and many other large mammals) typically have low fecundity but provide parental care to their offspring, which reduces the risk of death in early stages of development.

- Species with **ecological survivorship curves** are faced with a constant risk of mortality at all ages (the graphic representation is linear; Figure 55.5B). Many bird species display this pattern.
- Species with **maturational survivorship curves** experience low survivorship early in life and higher survivorship once they reach maturity (the graphic representation is convex; Figure 55.5C). Species with this type of survivorship curve (such as most insects, marine invertebrates, and annual plants) tend to produce many offspring but provide little or no parental care.

These different survivorship curves reflect differences in the ways in which organisms partition their time and energy among growth, maintenance, and reproduction; the way a species partitions its energetic resources is referred to as its **life history strategy**. Understanding the risks organisms face during different stages of their lives helps clarify why life histories differ among species. Although these basic strategies are to a large degree genetically and taxonomically determined, varying environmental conditions can influence life history traits. Witness how, for example, the number of offspring produced by cactus ground finches in a year depends on cactus flower and fruit availability, which in turn depends on the availability of rainfall.

RECAP 55.2

Life tables can be constructed either by following a cohort of individuals through time or by recording age at death in a vertical life table. Survivorship curves can be constructed by plotting the likelihood of survival to different ages. Differences in the shape of these curves can shed light on differences in life history strategies.

- What kinds of information does a life table provide about a population? See pp. 1154–1155 and Table 55.1
- What are the differences between a vertical and a cohort life table? See pp. 1154–1155
- Describe the three types of survivorship curves. See pp. 1155–1156 and Figure 55.5

Environmental variation influences survivorship and fecundity. Comparisons across populations and species reveal different patterns in life history traits, which allow organisms to cope with different environmental challenges.

55.3 How Do Environmental Conditions Affect Life Histories?

Because resources and mortality factors vary greatly among environments, life history strategies also vary dramatically. Those variations, in turn, determine how fast populations can grow.

Survivorship and fecundity determine a population's growth rate

To see how a population is likely to grow, ecologists can use life table data to calculate the population's **per capita growth rate**, symbolized ***r***. A population's growth rate is the difference between the per capita birth rate (b) and the per capita death rate (d) (leaving aside, for the moment, immigration and emigration). In other words, it is the average rate of change in population size per individual per unit of time. It is expressed by the equation

$$r = b - d \qquad (55.2)$$

If the per capita birth rate is greater than the per capita death rate, then $r > 0$ and the population is growing. If the per capita death rate is greater than the per capita birth rate, then $r < 0$ and the population is declining. The equilibrium state, $r = 0$, would indicate a stable population that is neither growing nor declining.

The maximum value of r (r_{max}) is referred to as the population's **intrinsic rate of increase**. It reflects the rate of increase that is inherent in the organism under ideal conditions—that is, independent of any external (environmental) constraints on population growth. A population can reach r_{max} only for a limited time, if at all, since environmental constraints almost always exist.

Life history traits vary with environmental conditions

Birth rates and death rates are both influenced by environmental factors, so r changes as the environment changes. The life history traits most influenced by environmental conditions include:

- age at first reproduction (generation time)
- number of broods per female (the number of times a female produces offspring)
- number of offspring per brood (the number of offspring produced each time a female reproduces)

These traits vary not only between species, but also between populations of the same species.

Opportunities for reproduction for some species are limited to certain locations or certain times of year, whereas other species and populations can breed continuously over their life span. Many desert wildflowers grow and flower only during the spring rainy season, and they may not be able to reproduce at all in years when rainfall is inadequate. In contrast, some tropical vines flower continuously in their warm, moist rainforest environment.

Species that can reproduce multiple times over the course of their adult lives are **iteroparous** (*itero*, "repeat"; *pario*, "beget"). **Semelparous** species (*semel*, "once") reproduce only once (Figure 55.6). Generally speaking, semelparous species produce many more offspring in a single brood than iteroparous species do over their entire lifetimes; semelparity is thus sometimes referred to as "big bang" reproduction.

Semelparity is typical of organisms that experience no great survival advantage upon reaching adulthood; it includes some fishes, many insects, and all annual plants. In contrast, iteroparity is typical of organisms whose survival chances increase once they reach maturity. For example, because environmental conditions within the nests of social insects such as honey bees and ants are remarkably stable, iteroparity is the rule among these species; some queens live 10 years or longer and reproduce over their entire adult lives.

Orgyia antiqua

55.6 Big Bang Reproduction Semelparous species reproduce only once and invest a great deal of energy in producing the maximum number of offspring. Female rusty tussock moths do not fly but remain with their empty cocoons, which are attached to the plants that are the caterpillar-stage food source. The stationary female lays a large number of eggs and then dies. When the eggs hatch the following spring, the larvae are surrounded by foliage they can eat.

Life history traits are influenced by interspecific interactions

Predation and other interactions among species can influence life history strategies in many ways. Some populations of guppies (*Poecilia reticulata*) in Trinidad, for example, live in streams where they are attacked and eaten by larger fish. But some streams have waterfalls that predatory fishes are unable to negotiate. Guppies that live in the predator-free areas upstream from those waterfalls have lower death rates than guppies below the falls. To see whether the risk of being eaten by a predator influenced the life history strategies of these guppies, David Reznick and his colleagues collected guppies from high-predation and low-predation sites and raised them in the laboratory. Some guppies from each group were provided with plentiful food, and others with limited food, to simulate the variation the fish would typically encounter in their home streams. In the laboratory, where no predators were present, guppies from high-predation sites matured earlier, reproduced more frequently, and produced more offspring in each brood than did guppies from low-predation sites, no matter how much food they received. The investigators concluded that predation had selected for early and frequent reproduction.

RECAP 55.3

The difference between birth rate and death rate provides an estimate of a population's per capita growth rate, or *r*. That rate is strongly influenced by the population's life history strategy, which in turn is highly dependent on environmental conditions.

- Give some examples of life history traits that vary with environmental conditions. **See p. 1156**
- Explain the difference between iteroparity and semelparity. **See p. 1156**
- How can predation affect the evolution of life history strategies? **See p. 1157**

In any given species, environmental factors may influence the growth of populations differently in different places and at different times.

55.4 What Factors Limit Population Densities?

What would happen if all the offspring produced by a population survived to reproduce themselves? The prospects are alarming. In 1911, L. O. Howard, then chief entomologist of the U.S. Department of Agriculture, estimated that, if all their offspring were to survive, a pair of flies beginning to reproduce on April 15 would produce a population of 5,598,720,000,000 adults by September 10 of the same year. Other entomologists took issue with Howard's calculation—they pegged the number much *higher*. Given such amazing reproductive capacities, it is clear there are forces at work that limit the growth of fly populations (and populations of every other organism).

All populations have the potential for exponential growth

As the number of individuals in a population increases, the number of reproducing individuals also increases, so the number of new individuals added per unit of time accelerates, even though the per capita rate of increase remains constant. If births and deaths occur continuously at constant rates, a graph of the population size over time forms a continuous upward curve (**Figure 55.7**). This pattern is known as **exponential growth**. Mathematically, the change in the number of individuals N over an interval of time T can be expressed as $\Delta N/\Delta T$ where Δ is the mathematical representation for "change in." The ratio between N and T can be expressed as the average contribution of each individual to population growth r (remember from Equation 55.2 that $r = b - d$) multiplied by the number of individuals in the population. In mathematical terms,

$$\frac{\Delta N}{\Delta T} = rN$$

Using the notation of differential calculus, which in this case simply indicates that the time interval represented by Δ is short, this equation can be expressed as

$$\frac{dN}{dT} = rN \qquad (55.3)$$

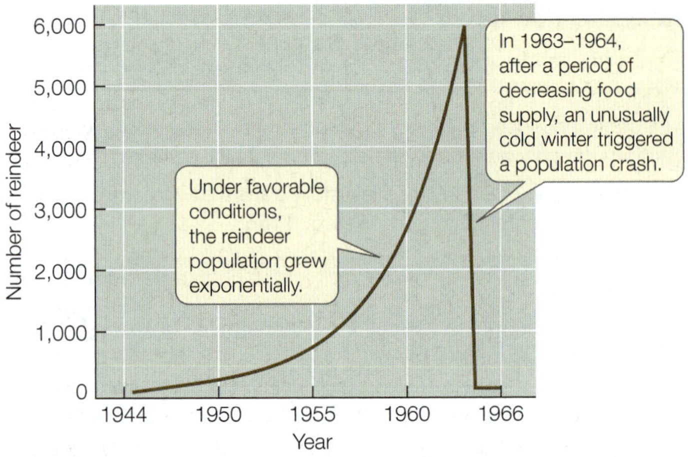
6,000
5,000
4,000
3,000
2,000
1,000
0
Number of reindeer
1944
1950
1955
1960
1966
Year
Under favorable conditions, the reindeer population grew exponentially.
In 1963–1964, after a period of decreasing food supply, an unusually cold winter triggered a population crash.

55.7 Exponential Population Growth Can Lead to a Population Crash The reindeer herd introduced on St. Matthew Island experienced favorable conditions and grew exponentially for many years. A single catastrophically cold winter triggered a population crash that eventually resulted in the death of the entire population.

Go to Animated Tutorial 55.2
Exponential Population Growth
Life10e.com/at55.2

The term dN/dT is the rate of change in the size of the population over time, and the expression rN is sometimes called the **biotic potential** of the population.

Some populations may grow at rates close to their biotic potential, but only for short periods. During the 20 years following their introduction, the reindeer population described at the opening of this chapter grew exponentially, as seen in Figure 55.7. When the herd was first introduced, it had ample habitat, abundant food, and no predators, so there was nothing to limit the population's growth. Favorable climate conditions also allowed the population to grow exponentially. A sudden change in climate conditions—deep snow that made foraging difficult—was a major factor leading to the population's crash. Although the crash was precipitated by unusually harsh weather conditions, the relatively poor physical condition of the reindeer in the herd, caused by overcrowding and overgrazing of the lichens that were their principal food source, contributed to the massive mortality.

Go to Media Clip 55.1
The Biotic Potential of a Population
Life10e.com/mc55.1

Logistic growth occurs as a population approaches its carrying capacity

No real population can maintain exponential growth for very long. As a population increases in density, the resources it requires—such as food, nest sites, and shelter—become depleted. In the absence of adequate resources to sustain more individuals, birth rates drop and death rates rise.

Any given environment has only enough resources to support a finite number of individuals of a species indefinitely.

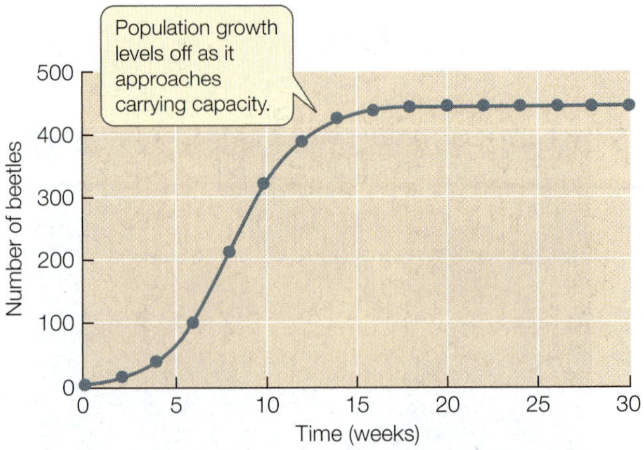
Population growth levels off as it approaches carrying capacity.
500
400
300
200
100
0
Number of beetles
0
5
10
15
20
25
30
Time (weeks)

55.8 Logistic Population Growth Levels Off In an environment with limited resources, a population typically stops growing exponentially before it reaches the environmental carrying capacity (*K*). The data here were recorded from a laboratory population of sawtoothed grain beetles maintained on a constant food supply; it is a typical logistic growth pattern, which forms an S-shaped curve.

Go to Animated Tutorial 55.3
Logistic Population Growth
Life10e.com/at55.3

That number of individuals, referred to as the environment's **carrying capacity (*K*)**, is a function of its resources. The growth of a population typically slows down as its density approaches the environmental carrying capacity. A population that exhibits decreasing growth as resources become more scarce displays a pattern called **logistic growth**, in which a graph of population size over time forms an S-shaped curve. **Figure 55.8** shows this growth pattern in a laboratory population of beetles maintained on a constant food supply.

To generate the S-shaped logistic growth curve from the equation for exponential growth we add a term,

$$\frac{K-N}{K}$$

This quantity represents the reduction in population growth caused by preemption of available resources and is referred to as **environmental resistance**. As long as the population size is less than the carrying capacity (i.e., $N < K$), only a fraction of the available resources are being used. As the population size approaches the carrying capacity, however, the fraction of resources available for any new individual becomes smaller. The implication is that each individual added to the population depresses population growth by an equal amount. Thus

rate of change in population size over time =
biotic potential × environmental resistance

or, in mathematical terms,

$$\frac{dN}{dT} = rN \times \frac{K-N}{K} \qquad (55.4)$$

Population growth should stop when $N = K$ because at that point, $K - N = 0$, so $(K - N)/K = 0$, and thus $dN/dT = 0$ and the population remains at a constant size.

Go to Activity 55.1 Logistic Population Growth
Life10e.com/ac55.1

r-strategists	*K*-strategists
HABITAT Can inhabit a broad range of habitats. High tolerance for both environmental instability and low-quality resources.	HABITAT Specific habitat requirements, including environmental stability. Efficient users of specific and usually high-quality resources.
PHYSIOLOGY Rapid embryonic development, rapid maturation to reproductive age, small body size.	PHYSIOLOGY Extended embryonic development, long maturation to reproductive age, large body size.
REPRODUCTIVE STRATEGY Random mating. Reproduce once (semelparity) resulting in a large number of offspring. Little or no parental investment in each offspring.	REPRODUCTIVE STRATEGY Mate choice, pair bonds. Reproduce many times (iteroparity), each event producing few offspring. Large parental investment in each offspring.
SURVIVORSHIP Short life span, density-independent mortality, typically a maturational survivorship curve (see Figure 55.4).	SURVIVORSHIP Long life span, density-dependent mortality, typically physiological or ecological survivorship curve (see Figure 55.4).
POPULATION FLUCTUATION Short periods of exponential population growth followed by periodic or seasonal population crashes.	POPULATION FLUCTUATION Slowly rising population growth that stabilizes and levels off at carrying capacity (*K*).
EXAMPLES Dandelions, house flies, rabbits	EXAMPLES Oak trees, bluebirds, polar bears

55.9 Two Life History Strategies Species whose life histories are geared to achieve the maximum possible rate of population increase are referred to as *r*-strategists; those whose population dynamics are bounded by carrying capacity are *K*-strategists. The life histories of most species combine elements of both strategies.

Population growth can be limited by density-dependent or density-independent factors

When resources are limited, adding more individuals to a population risks making things worse for everyone. Factors with an effect on population size that increases in proportion to population density are called **density-dependent** regulation factors. These factors include the following:

- *Food supply.* As a population increases, it may deplete its food supply, reducing the amount of food available to each individual. Poor nutrition may then increase the death rate or decrease the birth rate.
- *Predators* may be attracted to areas with high densities of their prey. If predators capture a larger proportion of the prey population than they did when that population was small, the death rate of the prey population rises.
- *Pathogens* may spread more easily in dense populations than in populations with fewer individuals per unit of area, resulting in a rise in the death rate.

Not all factors that change population size act in a density-dependent manner, however. A period of extreme cold, or an exceptionally strong hurricane, may kill a large proportion of the individuals in a population regardless of the population's density; such an event is **density-independent**. Abiotic factors tend to act in a density-independent manner, whereas biotic factors (such as food supply) tend to be density-dependent. For an ecological process to regulate population size (i.e., to maintain the population at a certain level), it must exhibit density dependence such that some sort of negative feedback is applied when populations increase.

Different population regulation factors lead to different life history strategies

Species vary in their capacity to reproduce, as well as in the extent to which they are vulnerable to density-dependent and density-independent mortality factors. Some of this variation in life history traits appears to result from adaptation to different habitat conditions. Generally, unpredictable habitats are associated with high fecundity and correspondingly high intrinsic rates of increase as organisms make the most of rare opportunities to reproduce. Conversely, predictable habitats, where organisms have a high probability of reproductive success, are associated with low fecundity and low *r*.

Species whose life history strategies allow for high intrinsic rates of increase are called ***r*-strategists**, and species whose life history strategies allow them to persist at or near the carrying capacity (*K*) of their environment are called ***K*-strategists** (Figure 55.9). Keep in mind, however, that these categories are not absolute; many species fall along a continuum between these two strategies.

For *r*-strategists, life is uncertain. Individuals tend to reproduce only once and to produce large numbers of offspring. They can generally use a wide variety of resources and tolerate a wide range of conditions. *K*-strategists are adapted to predictable environments, are long-lived, and reproduce several times; their smaller numbers of offspring have a high probability of surviving to adulthood. *K*-strategists tend to be more specialized in their resource use and less tolerant of variation in resource quality.

That life history strategies can evolve is suggested by genetic correlations among suites of life history traits. Such genetic correlations imply either simultaneous selection on two or more life history traits or linkages among the genes that code for those traits. Across *Drosophila melanogaster* strains, for example, a high intrinsic rate of increase is correlated with the ability to reproduce under starvation conditions and with the ability to develop on a variety of media in the laboratory—both of which are consistent with the *r* strategy of tolerating a wide range of resources and conditions.

Several ecological factors explain species' characteristic population densities

Density-dependent and density-independent factors can explain how populations grow or decline, but they do not explain why

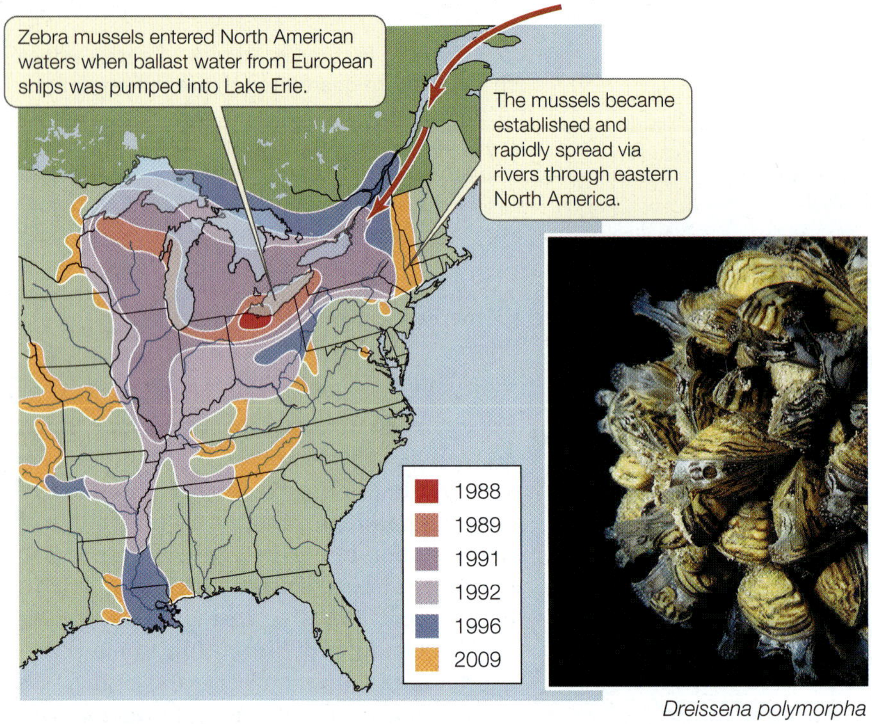
Zebra mussels entered North American waters when ballast water from European ships was pumped into Lake Erie.
The mussels became established and rapidly spread via rivers through eastern North America.
1988
1989
1991
1992
1996
2009
Dreissena polymorpha

55.10 Newly Introduced Populations Can Grow Rapidly Between 1988 and 2009, the range of zebra mussels in eastern North America increased exponentially, through rapid population growth as well as by inadvertent transport on barges moving among waterways. Female mussels can lay more than 1 million eggs in a single season, and in North America the species has few natural predators. Humans can unwittingly transport zebra mussel larvae from one body of water to another on their fishing boats and other watercraft, and in recent years the invasive pest has begun to appear in lakes and streams of the American West.

some species are common whereas others are rare—that is, why the characteristic densities of species differ. Many factors explain why typical population densities vary so greatly among species, but three of these factors are especially influential:

- *Species that use abundant resources generally reach higher population densities than species that use scarce resources.* Thus, on average, the fruit fly *Drosophila melanogaster*, which feeds on yeasts and other microbes found on just about any kind of rotten fruit, reaches substantially higher population densities than do other fruit fly species that feed on the microbes found on specific fruits.
- *Species with small body sizes generally reach higher population densities than species with large body sizes.* In general, population density decreases as body size increases because, on a per capita basis, small individuals require less energy to survive than large individuals.
- *Complex social organization may facilitate high population densities.* Highly social species, including ants, termites, and humans, can achieve remarkably high population densities.

Some newly introduced species reach high population densities

Species that are introduced into a new region, where their normal predators and pathogens are absent, sometimes reach population densities much higher than those in their native ranges. Sometimes these high population densities are only temporary; these densities decline if and when new mortality factors exert an influence. In the absence of such factors, however, populations in the newly colonized habitat may remain so dense that the introduced species becomes a major problem for native species.

The population of zebra mussels (*Dreissena polymorpha*) in North America demonstrates the speed with which newly introduced populations can grow. Zebra mussels first appeared in Lake St. Clair, between Lake Erie and Lake Huron, in 1988. They were probably transported there from Europe in the ballast water of transoceanic cargo ships. They spread rapidly and today occupy most of the Great Lakes and the Mississippi River drainage (**Figure 55.10**), reaching densities as high as 400,000 individuals per square meter in some places. Because they attach to any stable underwater substrate, zebra mussels can cover the bottoms of boats and clog municipal water supply intakes and power plant pipelines. They even settle on other aquatic organisms, causing problems for native mussels. Such high densities are never found in their native Europe, where more than 100 species of predators and parasites keep zebra mussel population densities under control.

Evolutionary history may explain species abundances

The three key factors that explain variation in population densities cannot explain all differences in species abundance. For example, although Douglas firs (*Pseudotsuga menziesii*) and giant sequoias (*Sequoiadendron giganteum*) are both large trees that use the same sources of energy (sunlight) and nutrients (soil), Douglas firs are abundant throughout western North America, whereas giant sequoias are restricted to a few groves in the Sierra Nevada. Similarly, each of several species of desert pupfish (genus *Cyprinodon*) is restricted to a single spring in Death Valley, California, whereas smallmouth bass (*Micropterus dolomieu*) can be found in many of the rivers and lakes of eastern North America. To explain these differences, it is important to know not just the contemporary ecology of these organisms, but also their evolutionary history.

As Chapter 23 described, a new species can originate in several ways. Species that arise by polyploidy or by founder events inevitably begin with a very small, local population. Desert pupfish species appear to have evolved in isolation as increasing aridity in Death Valley over the past 50,000 years cut once continuous populations off from one another. Conversely, when a species is declining toward extinction (as may be happening to the giant sequoia), its range shrinks until it vanishes when the last individual dies.

RECAP 55.4

Population sizes are limited by the carrying capacity of the environment, which is determined by the availability of resources. Species associated with unpredictable habitats tend to be *r*-strategists, whereas species associated with predictable habitats tend to be *K*-strategists.

- Why can populations grow exponentially only for short periods? **See pp. 1157–1158 and Figures 55.7 and 55.8**
- What is the difference between density-dependent and density-independent factors that influence populations size? **See p. 1159**
- Describe the characteristics of *r*-strategists and *K*-strategists. **See p. 1159 and Figure 55.9**

A species is rarely found in all of the habitats that seem suitable for it. Geological history and the evolutionary histories of species supply one type of explanation for this observation (see Section 54.5). This chapter next explores another area of explanation: spatial variation in habitat suitability.

55.5 How Does Habitat Variation Affect Population Dynamics?

Most natural history field guides have maps showing the geographic range over which a species can be found. But not even the most abundant species is found everywhere within its mapped range. Every species has particular habitat requirements that determine where within its potential range it will occur.

Many populations live in separated habitat patches

Most organisms live in distinct **habitat patches**, areas of a particular habitat type surrounded by areas of less suitable habitat. For example, caterpillars of the Bay checkerspot butterfly (*Euphydryas editha bayensis*) feed on only two species of annual plants (California plantain and purple owl's clover) that grow only on outcrops of serpentine rock on hills south of San Francisco, California. The butterflies are restricted to patches of these plants and cannot establish populations in the surrounding habitats that lack them.

Some populations living in separated habitat patches are effectively divided into discrete **subpopulations** that are linked together by regular movement of individuals between patches. The larger population that includes all such subpopulations is known as a **metapopulation**. Each subpopulation has a probability of "birth" (colonization of its habitat patch) and "death" (extinction in that patch). Each subpopulation grows in the ways we have described, but because the subpopulations are much smaller than the metapopulation, local disturbances and random fluctuations in numbers of individuals are more likely to cause the extinction of a subpopulation than of the entire metapopulation, as we will explain in Chapter 59. However, if individuals move frequently between subpopulations, immigration may prevent declining subpopulations from becoming extinct, a process called the **rescue effect**.

The Bay checkerspot butterfly provides a dramatic illustration of the dynamics of metapopulations. In 1960 Paul Ehrlich and his colleagues at Stanford University began studying a population of this butterfly in the Jasper Ridge Biological Preserve near the Stanford campus. They determined that the Jasper Ridge population was actually one of several subpopulations within a large, very fragmented metapopulation. They followed the Jasper Ridge subpopulation, as well as several other subpopulations within this metapopulation, over a number of years and found that the subpopulations varied enormously and asynchronously in size. Larval survival depended on climate factors (particularly temperature), the timing of rainfall, and host plant survival.

During drought years, most host plants died early in spring, before the caterpillars had developed enough to enter their summer resting stage. A severe drought in 1975–1977 led to extinctions of some of the subpopulations. One of the empty patches was repopulated a few years later, most likely by individuals from the largest single subpopulation, Morgan Hill, which as late as 1989 contained several hundred thousand butterflies (**Figure 55.11**). In 1998, however, the Morgan Hill subpopulation, which had historically been the largest in the metapopulation, went extinct. Ehrlich and his colleagues examined 70 years of climate data for the region and concluded that increasing climate variation accounted for the extinction.

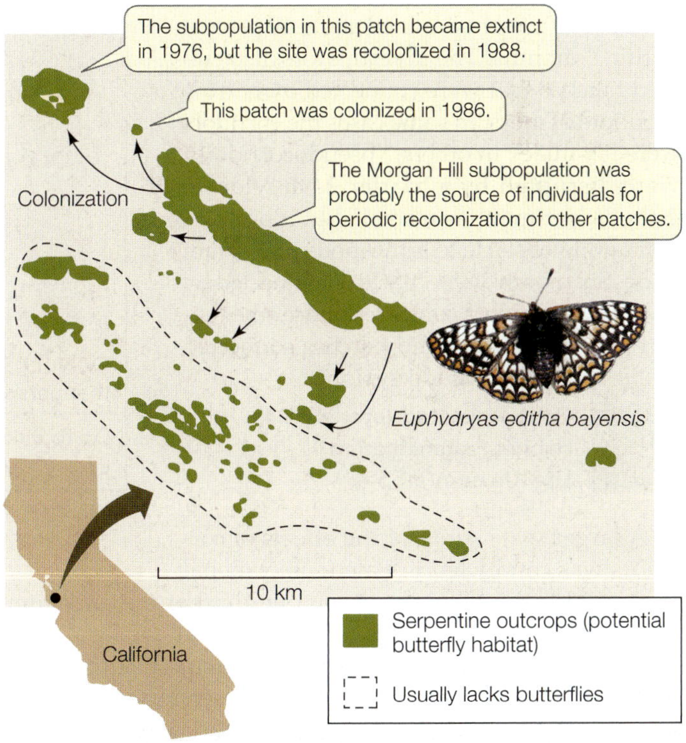

The subpopulation in this patch became extinct in 1976, but the site was recolonized in 1988.
This patch was colonized in 1986.
Colonization
The Morgan Hill subpopulation was probably the source of individuals for periodic recolonization of other patches.
Euphydryas editha bayensis
10 km
California
Serpentine outcrops (potential butterfly habitat)
Usually lacks butterflies

55.11 A Checkerboard of Checkerspots The Bay checkerspot butterfly metapopulation is divided into several subpopulations confined to patches of habitat (serpentine rock outcrops) that contain the species' food plants.

Without a large, stable source subpopulation to provide emigrants for recolonization, as the Morgan Hill subpopulation did during the 1970s drought, it is unlikely that any of the other subpopulations will persist without human intervention.

Corridors may allow subpopulations to persist

In any metapopulation, habitat between patches through which organisms can move, known as **corridors**, plays a critical role in facilitating dispersal to maintain subpopulations. What constitutes a corridor depends on the dispersal ability of the organism. Studying corridors is experimentally daunting because long distances may separate patches; moreover, movements of animals, particularly those that fly, can be difficult to monitor. Therefore one of the first tests of the importance of corridors was a small-scale manipulative experiment using mosses growing on rocks, which provide habitat for several small arthropod species, including springtails (minute wingless hexapods) and mites.

The investigators created patches of habitat by clearing away the mosses surrounding the patches (**Figure 55.12**). In small, completely isolated patches, the number of small arthropod species present declined about 40 percent within 6 months. The investigators also created patches that were connected by narrow corridors of moss. In some cases the corridors were left intact; in others, "pseudocorridors" were disrupted by a barrier 2 cm wide. A 2-cm barrier may seem small, but it presents a daunting obstacle to arthropods only 2 mm wide. Six months later, patches connected by unbroken corridors contained more small arthropod species than did patches connected by the disrupted pseudocorridors.

Go to Animated Tutorial 55.4
Habitat Fragmentation
Life10e.com/at55.4

INVESTIGATING**LIFE**

55.12 Corridors Can Rescue Some Populations Data from the experiments by Andrew Gonzales and Enrique Chaneton summarized here suggest that corridors between patches of habitat increase the chances of recolonization, and thus of subpopulation persistence.[a]

HYPOTHESIS Subpopulations of a fragmented metapopulation are more likely to persist if there is no barrier to recolonization.

Method

1. On replicate moss-covered boulders, scrape off the continuous cover of moss to create a "landscape" of moss "mainland" with patches surrounded by bare rock. A central 50 cm × 50 cm moss "mainland" (M) is surrounded by 12 circular patches of moss, each 10 cm^2 (subpopulations). In the "insular" treatment (I), the patches are surrounded by bare rock (which is inhospitable to moss-dwelling small arthropods, and thus a barrier to recolonization). In the "corridor" treatment (C), the patches are connected to the mainland by a 7 × 2 cm strip of live moss. In the "broken-corridor" treatment (B), the configuration is the same as the "corridor" treatment, except that the moss strip is cut by a 2-cm strip of bare rock.

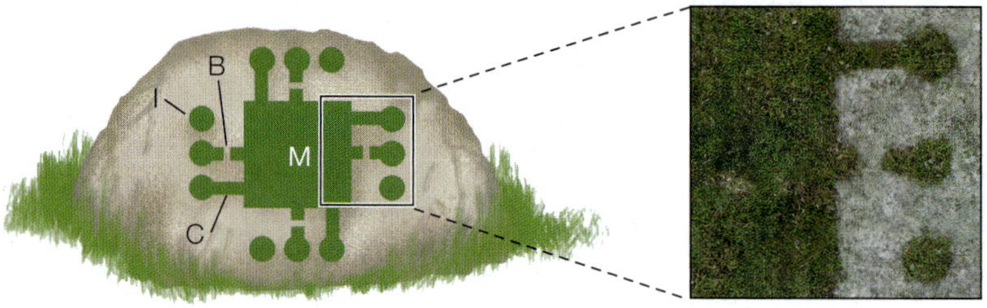
B
I
M
C

2. After 6 months, determine the number of small arthropod species present in each of the mainlands and small patches.

Results Patches connected to the mainland by corridors retained as many species as did the mainland to which they were connected. Fewer species remained in the broken-corridor and insular treatments.

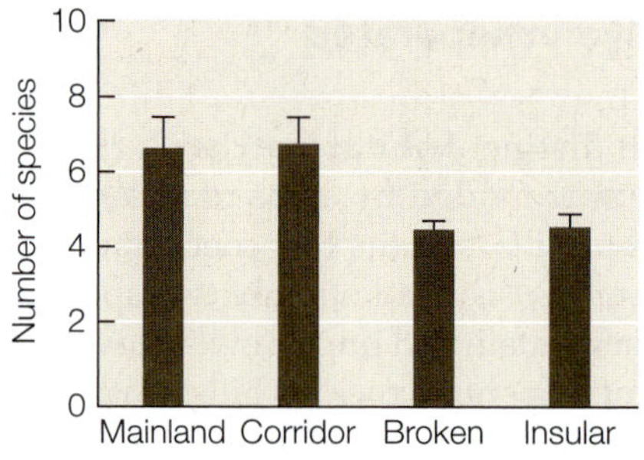
Number of species
0
2
4
6
8
10
Mainland
Corridor
Broken
Insular

CONCLUSION Barriers to recolonization reduce the number of subpopulations that persist in a metapopulation.

Go to **BioPortal** for discussion and relevant links for all INVESTIGATING**LIFE** figures.

[a]Gonzalez, A. and E. J. Chaneton. 2002. *Journal of Animal Ecology* 71: 594–602.

A larger scale study of the effects of corridors was conducted in Palenque National Park in Mexico. Although about one-third of the park comprises tropical rainforest, that forest is surrounded by a patchwork of cattle pasture, river habitat, and forest fragments. Investigators moved individual birds, representing a wide range of species, from one patch of forest in the park to another. Some individuals were moved between forest patches that were in close proximity but were not connected by forest corridors (i.e., they were completely surrounded by cattle pastures). The rest were moved between patches that were in close proximity and surrounded by cattle pastures, but were physically connected by narrow corridors of forest habitat. The investigators then monitored the return of the captured birds to their home forest patches. Across all species, birds were more than six times more likely to be recaptured in home forest patches connected by corridors to the patches where they were released than in home forest patches that were unconnected to the patches where they were released. Even narrow forest corridors may be beneficial to tropical forest birds, which experience an increased risk of predation and greater physiological stress when they have to fly across open areas.

RECAP 55.5

A metapopulation consists of separate subpopulations living in distinct habitat patches. Corridors that facilitate movement between patches increase the chances of subpopulation persistence.

- What effects do patches of unsuitable habitat have on population structure? **See p. 1161 and Figure 55.11**
- What effects do corridors between habitat patches have on subpopulations? **See p. 1162 and Figure 55.12**

For many centuries, people have tried to reduce populations of species they consider undesirable and maintain or increase populations of desirable or useful species. Such efforts to manage populations are more likely to be successful if they are based on knowledge of how those populations grow and what determines their densities.

55.6 How Can We Use Ecological Principles to Manage Populations?

If we wish to manage other species—that is, to increase or decrease their populations—we need to understand their life histories and the dynamics of their populations. The principles of population dynamics can also help us understand the effects our own population and its activities are having on other species.

Management plans must take life history strategies into account

Knowing the life history strategy of a species can be helpful in managing populations of commercial value. The black rockfish (*Sebastes melanops*), an important game fish that lives off the Pacific coast of North America, provides one such example. Rockfish have an indeterminate growth pattern—they continue to grow throughout their lives. As in many other animals, the number of eggs a female rockfish produces is proportional to her size, so larger females produce more eggs than smaller females. In addition, older, larger females are better able to provision the eggs they produce with oil droplets, which provide energy to the newly hatched larvae, giving them a head start in life (**Figure 55.13**). Larvae from eggs with larger oil droplets, produced by larger females, grow faster and survive better than do larvae from eggs with smaller oil droplets.

These life history traits have important implications for the management of rockfish populations. Because fishermen prefer to catch big fish, intensive fishing off the Oregon coast from 1996 to 1999 reduced the average age of female rockfish from 9.5 to 6.5 years. Thus the females reproducing in 1999 were, on average, smaller than the females reproducing in 1996. This change decreased the average number of eggs produced by females and reduced the average growth rate of larvae by about 50 percent. This reduction in reproductive ability was linked to a decrease in the ability of the rockfish population to recover from intensive fishing. Maintaining productive populations of rockfish may require setting aside no-fishing zones where some females can be protected from fishing and allowed to grow to large sizes.

(A) *Sebastes melanops*

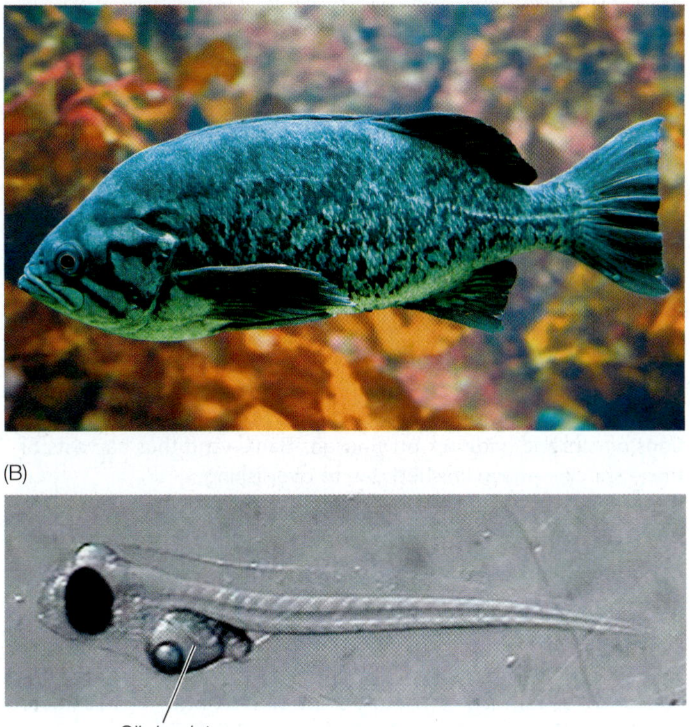
(B)
Oil droplet

55.13 Energy Stocks Give Rockfish a Head Start (A) Among rockfish, older, larger females are more reproductively successful, producing both more eggs and eggs with larger nutritive oil droplets. (B) The oil droplet attached to this rockfish larva provides it with nutrition to fuel its growth until it can feed on its own.

Management plans must be guided by the principles of population dynamics

If we look at a logistic growth curve (see Figure 55.8), we can see that the number of births tends to be highest when a population is well below its carrying capacity. Therefore if we wish to maximize the number of individuals that can be harvested from a population, we should manage the population so that it is far enough below the carrying capacity to have a high birth rate. Hunting and fishing regulations are established with this objective in mind.

Populations that have high intrinsic rates of increase can persist even if harvest rates are high. In such populations (which include many fish species), each female may produce thousands or millions of eggs. In many of these fast-reproducing populations, the growth rates of individuals are density-dependent. Therefore if prereproductive individuals are harvested at a high rate, the remaining individuals may grow faster. Some fish populations can be harvested heavily on a sustained basis because a relatively small number of females can produce sufficient numbers of eggs to maintain the population.

Fish can, however, be overharvested, as illustrated by the story of the black rockfish. Many fish populations have been greatly reduced because so many individuals were harvested that the few surviving reproductive adults could not maintain the population. Georges Bank, off the northeast coast of North America—a source of cod, haddock, and other prime food

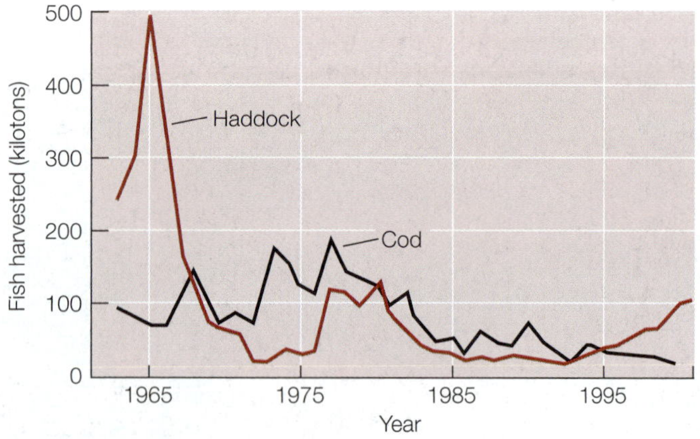
Fish harvested (kilotons)
500
400
300
200
100
0
1965
1975
1985
1995
Year
Haddock
Cod

55.14 Overharvesting Can Reduce Fish Populations Populations of cod and haddock on Georges Bank—and thus harvests of these species—have crashed due to overfishing.

Bufo marinus

55.15 Biological Control Gone Awry The Central American cane toad not only failed to control destructive beetles in Australia's sugarcane fields, but increased dramatically in abundance and now threatens many native Australian species (including the native frog this individual is eating).

fishes—was exploited so heavily during the twentieth century that many fish stocks have been reduced to levels insufficient to support a commercial fishery (**Figure 55.14**). The haddock population has rebounded enough to support a fishery because commercial fishing of that species ceased and was restarted only after the population had recovered. In contrast, managers reduced fishing pressure on cod only slowly, and the cod population has failed to increase.

Many rapidly reproducing species can recover if overharvesting is stopped, but recovery is more difficult for slowly reproducing species. Twentieth-century whalers hunted the blue whale (*Balaenoptera musculus*), Earth's largest animal, nearly to extinction. These whales reproduce very slowly: they live up to 10 years before becoming reproductively mature, produce only one offspring at a time, and have long intervals between births. Not surprisingly, the population has failed to recover.

Whether we want to manage the sizes of populations of desirable species for sustainable harvesting or of undesirable species for control purposes, the same principles apply. If the dynamics of a pest population are influenced primarily by density-dependent regulation factors, then killing part of that population will only reduce it to a density at which it will grow faster. A more effective approach to reducing such a population is to remove its resources, thereby lowering the carrying capacity of its environment. For example, we can rid our cities of rats more easily by making garbage unavailable (reducing the carrying capacity of the rats' environment) than by poisoning rats (which only increases their reproductive rate).

Biological control is the use of natural enemies (predators, parasites, or pathogens) to reduce the population density of an economically damaging species. In many cases the target species is a pest only because it has been introduced to a new area. Natural enemies used for biological control are often obtained from the native region of the pest species. Biological control became popular in the nineteenth century after an outbreak of cottony-cushion scale, an Australian insect that attacks citrus, appeared in the citrus groves in California. A predaceous ladybeetle and a parasitic fly were then introduced from Australia. Within a year of their release, these insects brought the scales under control.

Sometimes, however, introduced natural enemies not only fail to have any effect on the pest they were imported to control but also, freed of their own enemies, become pests themselves. This fact underlies the horror story of the cane toad (*Bufo marinus*) in Australia. This Central American toad (**Figure 55.15**) was introduced to control cane beetles attacking Australian sugarcane fields. But Australian cane beetles stay high on the upper stalks of the plants; the toads could not reach that high, and thus had no effect on the beetle population. Unfortunately, they had massive effects on other species.

All stages of the *B. marinus* life cycle are poisonous, and Australian reptiles (including snakes and lizards) and mammals that eat them usually die. With no enemies to limit their population growth, cane toads grow fast and outcompete native amphibian species for resources. The toads have spread from northern Australia down the east coast, where they threaten native frog species by preying on them as well as by competing with them. The Australian government is forced to spend millions of dollars in attempting to reduce their numbers.

Human population growth has been exponential

In 1798 Thomas Robert Malthus, in his *Essay on the Principle of Population*, pointed out that the human population was growing exponentially but its food supply was not, and argued that at some point, famine and death would be the ultimate fate of the human race. Malthus could not have anticipated the technological innovations over the next 200 years that would greatly enhance the capacity of humans to produce

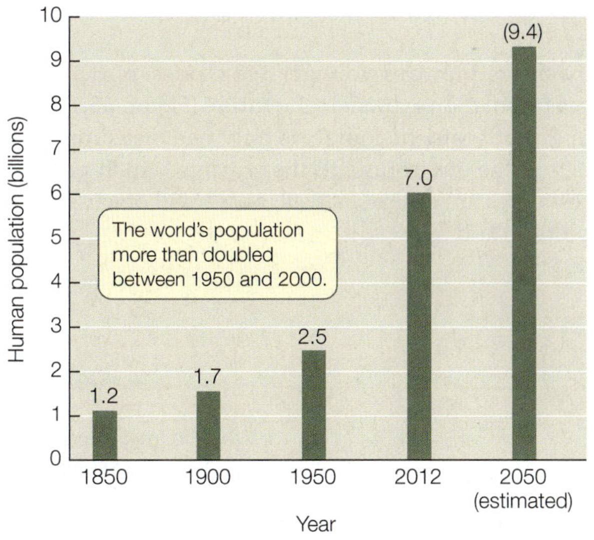

Human population (billions)
The world's population more than doubled between 1950 and 2000.
1.2
1.7
2.5
7.0
(9.4)
1850
1900
1950
2012
2050 (estimated)
Year

55.16 The Human Population Is Growing Exponentially The growth rate of the human population has slowed somewhat, but its large size means that millions more people are still being added every year.

food. Today, however, the size of the human population is once again a serious concern as we confront the effects of our contributions to pollution, habitat destruction, and extinctions of other species.

For thousands of years, Earth's carrying capacity for humans was low because of the relative inefficiency with which we could obtain food and water. The development of social systems and communication, the domestication of plants and animals, ever-increasing crop and livestock yields due to ongoing technological advances, and our increasing proficiency at managing diseases all contributed to unprecedented growth of the human population. It took more than 10,000 years for the population to reach 1 billion, which happened in the early nineteenth century. Today, a mere 200 years later, the planet is home to more than 7 billion human beings (**Figure 55.16**). Population growth has slowed somewhat from its post-World War II highs—estimates place the current worldwide rate of increase to be about 1.1 percent per year—but with a base of 7 billion, even a minimal growth rate means millions more individuals each year.

Human populations are not growing at the same pace across the world. As we saw for elephants in Figure 55.2, in long-lived species the timing of births and deaths affects a population's age distribution for many years. Between 1946 and 1964, the United States experienced a "baby boom." During those years almost 75 million babies were born, and the average number of children per family grew from 2.5 to 3.8. U.S. birth rates declined during the 1960s, but in the 1970s and 1980s the baby boomers became parents, generating another demographic bulge—a "baby boom echo" (**Figure 55.17A**). Today this "echo generation" is on the threshold of becoming the dominant age class.

The age structure of the U.S. population is typical of that of many industrialized nations, but a few highly developed nations (particularly in Europe) are experiencing population declines. In the developing world, however, many countries are experiencing exponential growth rates and have populations

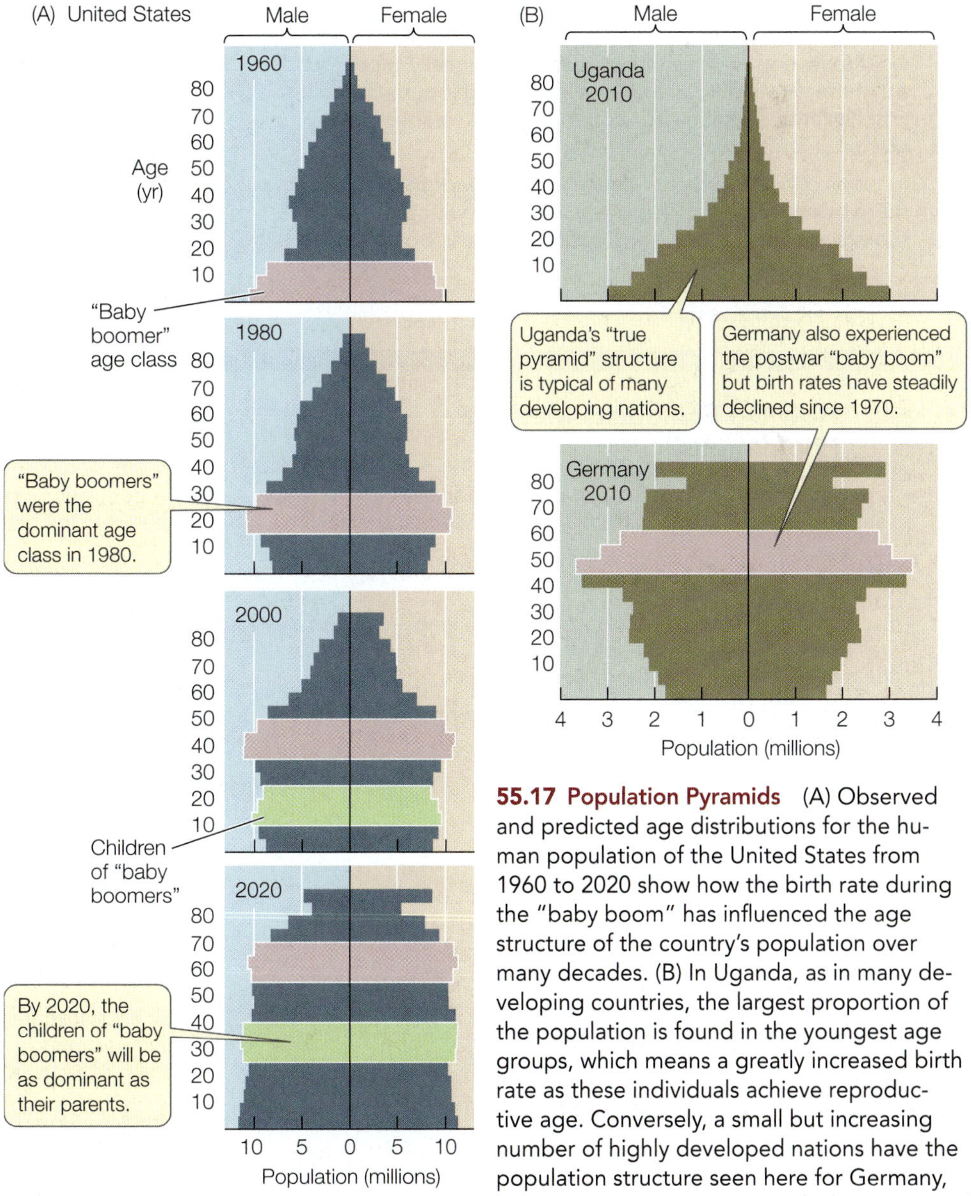

(A) United States
Male
Female
1960
Age (yr)
"Baby boomer" age class
1980
"Baby boomers" were the dominant age class in 1980.
2000
Children of "baby boomers"
2020
By 2020, the children of "baby boomers" will be as dominant as their parents.
Population (millions)
(B)
Uganda 2010
Uganda's "true pyramid" structure is typical of many developing nations.
Germany also experienced the postwar "baby boom" but birth rates have steadily declined since 1970.
Germany 2010
Population (millions)

55.17 Population Pyramids (A) Observed and predicted age distributions for the human population of the United States from 1960 to 2020 show how the birth rate during the "baby boom" has influenced the age structure of the country's population over many decades. (B) In Uganda, as in many developing countries, the largest proportion of the population is found in the youngest age groups, which means a greatly increased birth rate as these individuals achieve reproductive age. Conversely, a small but increasing number of highly developed nations have the population structure seen here for Germany, which presages a population decline.

highly skewed toward younger age classes, portending high rates of population growth in the future (Figure 55.17B). Population growth rate, of course, is only one measure of human impact on the environment; the ways in which populations use resources are critical as well, as we will discuss further in Chapters 58 and 59.

RECAP 55.6

Efforts to manage populations are more likely to be successful if they are based on an understanding of life histories and population dynamics.

- Describe an effective strategy for reducing a pest population and explain why it is effective. See p. 1164
- How have humans changed the carrying capacity of Earth for our own population? See p. 1165

?

Why did introduced reindeer populations persist on the island of South Georgia but not on the island of St. Matthew?

ANSWER

The different fates of the reindeer populations on these two islands reflect differences not only in the physical conditions on the islands but also in the history and purpose of the reindeer introductions. In physical terms, average climate conditions on St. Matthew are harsher than those on South Georgia; catastrophic weather events such as the winter that essentially wiped out the St. Matthew reindeer are far less frequent on South Georgia. Stability in population size is often related to stability in environmental conditions.

In terms of history, the reindeer on South Georgia were brought there by men involved in the whaling trade with the goal of establishing a food supply for ships traveling through the area. As a consequence, the population experienced regular harvesting (initially by whalers, who shot the reindeer for food, and later by scientists who shot them for research purposes).

That the reindeer population on South Georgia has not crashed, however, does not mean it is at a desirable size. Reindeer densities range from 40 to 85 animals per square kilometer—almost 10 times higher than densities in areas where reindeer are native. At these high densities, the reindeer are having negative effects on South Georgia's native plants and animals, such as the burrow-nesting white-chinned petrel. In February 2011, plans were made to eradicate the reindeer on South Georgia in the hope of preserving the native species.

CHAPTER SUMMARY 55

55.1 How Do Ecologists Measure Populations?

- A **population** consists of the individuals of a species that interact with one another within a particular area at a particular time.
- The **density** of a population is the number of individuals per unit of area or volume.
- Ecologists have developed many ways of counting individuals as well as ways of estimating population sizes from a sample, such as the **mark–recapture method**. Review Figure 55.2
- Populations have a characteristic **age structure** and pattern of **dispersion**. Review Figures 55.3, 55.4, ANIMATED TUTORIAL 55.1

55.2 How Do Ecologists Study Population Dynamics?

- **Demographic events**—births, deaths, immigration, and emigration—determine the size of a population.
- **Life tables** provide summaries of demographic events in a population. A cohort life table tracks a **cohort** of individuals born at the same time and records the **survivorship** and **fecundity** of those individuals over time. Review Table 55.1
- Life table data can be used to construct a **survivorship curve**. Ecologists describe three general types of survivorship curves, which reflect different life history patterns. Review Figure 55.5
- The **life history strategy** of an organism describes how it partitions its time and energy among growth, maintenance, and reproduction.

55.3 How Do Environmental Conditions Affect Life Histories?

- A population's **per capita growth rate** (r) is the difference between the per capita birth rate (b) and the per capita death rate (d).
- Life history traits within a species may vary with habitat.
- Interactions with other species and the abiotic environment can influence the evolution of a species' life history traits.

55.4 What Factors Limit Population Densities?

- Populations can exhibit **exponential growth** for short periods, but eventually their resources become depleted, causing birth rates to drop and death rates to rise. Review Figure 55.7, ANIMATED TUTORIAL 55.2
- **Logistic growth** is the pattern seen when the growth of a population slows as its density approaches the environmental **carrying capacity** (K). Review Figure 55.8, ANIMATED TUTORIAL 55.3, ACTIVITY 55.1
- Species that are ***r*-strategists** have life histories that allow for high intrinsic rates of increase. ***K*-strategists** persist at or near the carrying capacity (K) of their environment. Many species' life history strategies fall along a continuum between these two extremes. Review Figure 55.9
- Population densities are determined by both **density-dependent** and **density-independent** factors. Several factors—including resource abundance, body size, and social organization—influence population densities.

continued

55.5 How Does Habitat Variation Affect Population Dynamics?

- No species is found everywhere within its range. Members of most species live in distinct **habitat patches**.
- A **metapopulation** consists of separate **subpopulations** among which some individuals move on a regular basis. **Review Figure 55.11**
- Extinction of a subpopulation may be prevented by immigration of individuals from another subpopulation, a process known as the **rescue effect**. **Corridors** between patches may facilitate such movement. **Review Figure 55.12, ANIMATED TUTORIAL 55.4**

55.6 How Can We Use Ecological Principles to Manage Populations?

- To manage populations, it is important to understand their life histories and population dynamics. To maximize the number of individuals that can be harvested from a population, the population should be kept well below carrying capacity.
- Reducing the carrying capacity of the environment for a pest species is a more effective way to reduce its population than killing its members.
- Earth's carrying capacity for humans depends on our use of resources and the effects of our activities on the environment. Human populations grow at different rates in different parts of the world. **Review Figures 55.16, 55.17**

Go to the Interactive Summary to review key figures, Animated Tutorials, and Activities Life10e.com/is55

CHAPTER**REVIEW**

REMEMBERING

1. A group of individuals of the same species born at the same time is known as a
   a. deme.
   b. subpopulation.
   c. Mendelian population.
   d. cohort.
   e. taxon.

2. A population whose size remains constant at its carrying capacity is exhibiting
   a. exponential growth.
   b. geometric growth.
   c. logistic growth.
   d. J-shaped growth.
   e. negative growth.

3. The process by which immigrants prevent a subpopulation from becoming extinct is called the
   a. colonization effect.
   b. rescue effect.
   c. metapopulation effect.
   d. genetic drift effect.
   e. salvage effect.

4. Which of the following mortality factors is *least* likely to act in a density-dependent manner?
   a. Predation
   b. Disease
   c. Food supply
   d. Fire
   e. All of these factors act in a density-dependent manner.

5. The best way to reduce the population of an undesirable species in the long term is to
   a. reduce the carrying capacity of the environment for the species.
   b. selectively kill reproducing adults.
   c. selectively kill prereproductive individuals.
   d. attempt to kill individuals of all ages.
   e. sterilize individuals.

6. Populations that are most readily overharvested are characterized by having
   a. very long-lived adults.
   b. short prereproductive periods and many offspring.
   c. short prereproductive periods and few offspring.
   d. long prereproductive periods and few offspring.
   e. long prereproductive periods and many offspring.

UNDERSTANDING & APPLYING

7. Most organisms that humans manage for higher densities are long-lived and have low reproductive rates, whereas most organisms that humans want to reduce in numbers are short-lived but have high reproductive rates. What is the significance of these differences for management strategies and the effectiveness of management practices?

8. In the mid-nineteenth century, the human population of Ireland was largely dependent on a single food crop, the potato. When a disease caused the potato crop to fail, the Irish population declined drastically for three reasons: (1) a large percent of the population emigrated to the United States and other countries; (2) the average age of a woman at marriage increased from about 20 to about 30 years; and (3) many people starved to death. None of these social changes were planned at the national level, yet they all contributed to adjusting the population size to the new carrying capacity. Discuss the ecological principles involved, using examples from other species.

ANALYZING & EVALUATING

9. One method of controlling introduced pest species is to introduce a natural enemy (a predator, parasite, or pathogen) from the pest's native habitat to reduce its population density. However, some species introduced to control a pest have become pests themselves. Some scientists argue that biological controls should not be used under any circumstances for pest management. Others argue that, provided they are properly studied and thoroughly vetted, we should continue to use biological control organisms as part of our set of tools for managing pests. Which view do you support, and why?

10. Section 55.5 described two studies of the effects of corridors on metapopulation dynamics—one on tiny arthropods with limited dispersal abilities (see Figure 55.12) and another on birds of tropical forests in Mexico (see p. 1162). Given the differences in size and mobility between tiny, wingless arthropods and forest birds, is it possible to come up with a general definition of a corridor? How could an investigator conduct a single experiment to determine the effects of corridors on multiple organisms that differ widely in size and mobility? Is it important to consider more than one group of organisms in trying to understand the effects of corridors in fragmented habitats?

Go to BioPortal at **yourBioPortal.com** for Animated Tutorials, Activities, LearningCurve Quizzes, Flashcards, and many other study and review resources.

56 Species Interactions and Coevolution

CHAPTER OUTLINE

Fungus Farmers *Atta cephalotes* is one Central American species of leafcutter ant. Leafcutter ants harvest and transport leaf fragments to their nests where the vegetation will nourish a thriving crop of fungus, which the ants consume.

NOT MANY INSECTS can claim to have been the subject of a Hollywood filmmaking feud, but ants are an exception. In 1998 two animated films from competing studios, *Antz* and *A Bug's Life*, were released within a month of each other. Whether as animated entertainment or the subject of scientific study, the behaviors of these social insects have long fascinated humans.

The 50 or so species of leafcutter or "parasol" ants owe their name to their habit of clipping bits of leaves and holding the pieces above their heads like parasols as they cart them off to their nests. The ants don't eat the leaf matter they collect, however. The cut leaves will serve as a substrate for growing the fungi on which leafcutter ants feed.

When a new queen ant leaves her mother's nest, she takes with her a portion of the fungal mass on which she was raised. After mating, she digs into the soil to form a tunnel ending in a chamber, in which she places the pellet of fungus and lays eggs. Her offspring eat the fungus and develop into pint-size workers, which then collect leaf material to "feed" the fungus. As the fungus garden expands, the ants construct more nest chambers. In 3 years the number of workers in a nest can reach 8 million and the nest can measure more than 30 meters across. More than 2 kilograms of leaves each day are needed to maintain an average colony's fungus garden, so these ants can easily strip an area of vegetation. Fungus production on a large scale necessitates a division of labor; different individuals harvest leaf pieces, care for the fungus, defend the nest, clear trails for the leaf collectors, and guard the leaf fragments being carried back to the nest.

The fungi in leafcutter nests cannot exist without the ants, which supply the fungi with leaves to grow on and add fertilizer in the form of their fecal droplets. Leafcutter ants even evaluate leaves, avoiding those that contain fungus-killing chemicals. The fungal gardens, however, are vulnerable to invasion by undesirable microbes. To fend off one such invader, the green mold *Escovopsis,* the ants bring in another partner. They have special structures for carrying *Pseudonocardia* bacteria, which manufacture powerful antibiotics that suppress the unwelcome mold but do no harm to the cultivated fungus. Also present are *Klebsiella* bacteria, which fix atmospheric nitrogen to help fertilize the fungus garden and satisfy ant nutritional requirements. No doubt other organisms lurk in the fungus gardens awaiting discovery by curious ecologists or future filmmakers.

?

The fungi in leafcutter nests cannot survive without the ants, but can leafcutter ants survive without the fungus?

See answer on p. 1185.

56.1 What Types of Interactions Do Ecologists Study?

One of life's certainties is that, at some point between birth and death, every individual will encounter and interact with individuals of other species. These interactions have consequences that can affect each individual's fitness. Thus they can influence the densities of populations and the distributions of species, and, over the long term, they can lead to evolutionary change in one or more of the interacting species.

Interactions among species can be grouped into several categories

Although the kinds of interactions that take place among living things on Earth are essentially limitless, ecologists group interactions among species into a few basic categories. These categories reflect whether the outcome of the interactions is positive (+), negative (–), or neutral (0) for each of the species involved (**Figure 56.1**). We will introduce five broad categories of species interactions in this chapter.

ANTAGONISTIC INTERACTIONS **Antagonistic interactions** are those in which one species benefits and the other is harmed. Antagonistic interactions include three basic types:

- **Predation**, in which an individual of one species kills and consumes multiple individuals of other species (its **prey**).
- **Herbivory**, in which an individual of another species consumes part or (rarely) all of a plant.
- **Parasitism**, in which one species consumes only certain tissues in one or a few individuals of another species (its **host**) without necessarily killing them. Some parasites are pathogens that cause symptoms of disease in their hosts.

MUTUALISM **Mutualism** is a type of interaction between species that benefits both species. The interaction between leaf-cutter ants and fungi described at the opening of this chapter is an example of mutualism: the ants feed and cultivate the fungi, and the fungi, in turn, serve as food for the ants. Mutualisms exist between widely varied pairs of partners, including not only animals and fungi but also fungi and plants, animals and plants, animals and animals, and microbes and all other kinds of organisms.

COMPETITION **Competition** between species refers to interactions in which two or more species use the same resource. The outcomes of these interactions depend on resource availability. In some cases competitors can coexist by using the resource in different ways; if the resource is in extremely short supply,

(A)

Categories of Species Interactions

	Type of interaction	Effect on species 1	Effect on species 2
Antagonistic interactions	Predation (predator-prey)	+	–
	Herbivory (plant-herbivore)	+	–
	Parasitism (parasite/pathogen host)	+	–
	Mutualism	+	+
	Competition	–	–
	Commensalism (commensal-host)	+	0
	Amensalism	0	–

(B)

Parasitism, Predation, Mutualism
The buffalo's hide is infested with parasitic ticks. Oxpecker birds eat the ticks, to the mutual benefit of the birds and the buffalo.
Herbivory
The African buffalo feeds on the grasses of the savanna.
Amensalism, Commensalism
The large mammal unwittingly destroys insects and their nests. The white cattle egrets feed on insects disturbed by the buffalo's passage.
Predation
Carnivores such as timber wolves hunt and kill herbivorous mammals.
Competition
The grizzly bear is attempting to take over the wolves' kill.

56.1 Types of Species Interactions (A) Interactions among species can be grouped into categories based on whether their influence on each of the interacting species is positive (+), negative (–), or neutral (0). (B) Even small scenes can encompass many different species interactions.

Go to Activity 56.1 Ecological Interactions
Life10e.com/ac56.1

however, the outcome can be negative for all competing species. At some point a resource may be in such short supply that a population can no longer sustain itself; when a resource becomes limiting in this way, competition becomes intense. Competition can occur along with almost any other kind of interaction: between predators that depend on the same prey species, between herbivores that feed on the same host plant, or between pathogenic microbes attacking the same host. The limiting resource need not be food; species may compete for water, for space, for nesting sites, or even (in the case of plants) for sunlight.

COMMENSALISM AND AMENSALISM Antagonistic interactions, mutualism, and competition all affect the fitness of both participants, but there are two other types of interactions that affect only one participant. **Commensalism** is a type of interaction in which one participant benefits but the other is unaffected. Most examples of commensalism (Latin, "at the same table") involve one species feeding in, on, or around another species. For instance, one species may associate with another species that, by virtue of its own feeding behavior, makes food more accessible. Cattle egrets, for example, feed on insects disturbed by large grazing animals, but their activities have no effect on the grazers (see Figure 56.1B).

Another form of commensalism involves association for the purpose of transport, often to reach food resources that are rare and short-lived. Piles of mammal dung, for example, are a valuable resource for some detritivores, but they can be hard to find and never last long. Many kinds of detritivores that cannot fly—mites, nematodes, and even fungi—attach themselves to the bodies of dung beetles, which not only can fly but are also very good at locating fresh dung. These hitchhikers have no known effect on the dung beetles' fitness.

Amensalism is a type of interaction in which one participant is unaffected while the other is harmed. A herd of elephants moving through a forest crushes insects and plants with each step, but the elephants are unaffected by this carnage. Amensal interactions tend to be more random, and thus less predictable, than other types of interactions.

Interaction types are not always clear-cut

Although ecologists find it useful to group interactions among species into a few basic categories, the boundaries between categories are not always clear. For example, sea anemones in the Pacific Ocean sting and eat small fish, but a select few fish species (mostly in the genus *Amphiprion*) live inside sea anemones and are unaffected by their stings. Safe from their predators, the anemonefish move freely among the stinging tentacles to scavenge the cnidarians' leavings (**Figure 56.2**).

Amphiprion ocellaris

56.2 Interactions between Species Are Not Always Clear-Cut Ecologists long believed that the relationship between sea anemones and anemonefish was a commensalism: that the fish, by living among the anemone's stinging tentacles, gained protection from its predators. But could it also be considered a mutualism, if the fish's feces provide the anemone with beneficial nutrients?

Anemonefish must acclimate to the anemone's venom, and the anemone, in turn, must acclimate to the fish. The acclimation process appears to involve a change in the mucus coat of the fish; wiping off the mucus of an acclimated fish results in immediate stinging, whereas anemones do not sting fish with intact mucus. The benefits of this relationship to the anemonefish are clear: it escapes its own predators by hiding behind the anemone's stinging tentacles, and it has no need to forage widely for food. But does the anemone benefit from the association? By defecating while in residence, the anemonefish may provide nitrogen-rich nutrients to the anemone. On the other hand, the fish may occasionally steal the anemone's prey, which has a negative effect on the anemone's fitness.

The interaction types described in this section are in reality part of a continuum, and over evolutionary time they may shift from one type to another. Their outcomes depend on both ecological and evolutionary circumstances, including the presence and influence of other species.

Some types of interactions result in coevolution

All types of interactions have the potential to influence the population densities of the interacting species. By contributing to the differential survival or reproduction of individuals with different traits, they can also alter genotype frequencies within the interacting populations over time. Thus these interactions have both ecological consequences, as when they affect the distribution and abundance of a species, and evolutionary consequences, as when they lead to adaptations. In some cases an adaptation in one species may lead to the evolution of a reciprocal adaptation in a species it interacts with, a process known as **coevolution**.

Darwin observed that evolutionary change occurs not only in response to physical conditions, as described in Chapter 54, but also in response to interactions among species. In his introduction to *On the Origin of Species*, Darwin pointed out that woodpeckers have feet, tails, beaks, and tongues "admirably adapted to catch insects under the bark of trees" as a result of their long-standing interactions with their insect prey.

While abiotic factors also act as agents of selection, they differ in a fundamental way from biotic agents of selection in that they do not themselves undergo change as a result of the interaction. Snow and ice do not become more deadly as a result of encountering cold-resistant organisms, but predators can, over evolutionary time, become swifter, more powerful, or more efficient at capturing their prey. In response, prey species may become swifter, tougher, less conspicuous, or more poisonous, all of which decrease the likelihood of being consumed.

A series of reciprocal adaptations can lead to what has been dubbed a coevolutionary **arms race**. The arms race analogy, first used in the context of interactions between herbivores and plants, can be applied to most antagonistic interactions. The evolution of traits that increase the fitness of a predator, herbivore, or parasite species exerts selection pressure on its prey or host species to counter the consumer's adaptation. The prey or host adaptation, in turn, exerts selection pressure on the consumer to improve its fitness even more, resulting in an escalating series of reciprocal adaptations.

The types of interactions most likely to lead to coevolution are those that occur predictably and with high frequency over time and that have a strong effect on the interacting species. Thus most amensal and commensal interactions are less likely to coevolve than are many antagonistic and mutualistic interactions.

Go to Animated Tutorial 56.1
Coevolution: Strategies for Survival
Life10e.com/at56.1

RECAP 56.1

Species interactions can be grouped into categories based on whether they benefit or harm each of the species involved. Some species interactions can lead to reciprocal adaptations and coevolution.

- Describe the categories of interspecific interactions. See p. 1170 and Figure 56.1
- What is meant by a coevolutionary arms race? See p. 1172

Sections 31.3 and 51.1 looked at a number of heterotrophic feeding strategies from the consumer's point of view. In the next section we will see how the antagonistic interactions—predation, herbivory, and parasitism—influence both consumer and resource species.

56.2 How Do Antagonistic Interactions Evolve?

Every species serves as a food resource, in one way or another, for at least one other species. Consumers can increase their fitness by acquiring food, whereas resource species can increase their fitness by avoiding being consumed. Thus the interests of consumer and resource species set up an antagonistic relationship that can lead to a coevolutionary arms race. These consumptive relationships need not, however, be fatal; organisms make meals of one another in many different ways.

Predator–prey interactions result in a range of adaptations

Predator–prey interactions are probably the most familiar, and the most dramatic, type of antagonistic interaction. Predators invariably kill the prey individuals they consume, and over its lifetime, a predator kills and consumes many prey individuals. Predators tend to be less specialized than other types of consumers.

The fitness of predators depends on balancing the cost of pursuing, subduing, and handling prey against the energetic benefit of consuming it, as we saw in Section 53.4. Thus many predators are larger than their prey, and many of them use strength or swiftness to capture prey. This is true of predators of all sizes: tigers pursuing deer and tiger beetles pursuing smaller insects are both fast, powerful predators and both are equipped with strong jaws (**Figure 56.3**). Predators that are smaller than their prey rely on other strategies that increase their efficiency. Many spiders, for example, capture their prey in webs. The tiny short-tailed shrew, among the smallest mammalian predators, produces venomous saliva that paralyzes not

(A) *Panthera tigris* (B) *Cicindela campestris*

56.3 Predators Use Many Weapons (A) Tigers embody most people's image of a predator—a large animal that uses stealth, speed, strength, teeth, and claws to capture prey. (B) The 1.3-centimeter green tiger beetle is also formidable to its prey, including caterpillars. The beetle's huge jaws account for much of its body length, and it is one of the speediest runners among the insects.

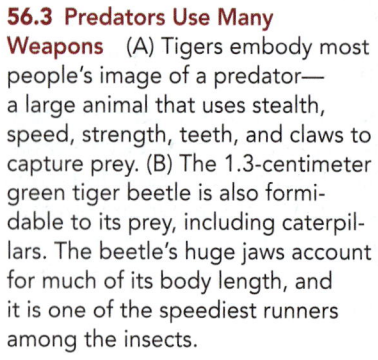

(A) *Hyla versicolor*
(B) *Mimetica* sp.

56.4 Avoiding Consumption by Avoiding Detection (A) The gray tree frog can change its coloration to blend in with its substrate. (B) Resemblance to an inedible object can be an effective defense against visually hunting predators. Birds searching for insect prey are likely to bypass a katydid that looks like a partially eaten leaf.

only earthworms and snails but also prey much larger than itself, including mice and small birds.

Prey species have many different kinds of defenses against predators. Many animals can escape from predators simply by flying or running away. Others have morphological defenses. Tough skin, shells, spines, or hair can foil even a determined predator. In turn, however, adaptations evolve in predators that may overcome these defenses.

AVOIDING DETECTION Prey species can often escape predators by hiding. One form of hiding is camouflage, or background matching, also called **crypsis**. Some animals can even change their coloration to match the substrate they find themselves on (Figure 56.4A). The camouflage of some species allows them to resemble objects their predators consider inedible. The katydid in Figure 56.4B, for example, looks very much like a dead leaf, even down to the likeness of a spot of fungal decay.

Because the vision of many types of predators is adapted to spot moving prey, many prey species simply stop moving if they are being pursued. "Playing possum," a term that is sometimes applied to this strategy, refers to the ability of the opossum (*Didelphis virginiana*) to simulate death.

CHEMICAL DEFENSES Many animals use chemical defenses to escape or repel their predators. Chemical defenses are generally the province of animal prey that are small, weak, sessile, or otherwise unprotected. Among the mollusks, for example, the weaker a species' shell, the more likely it is to use chemical defenses; for example, the sea slug in Figure 56.5B has no shell but is highly toxic. Some vertebrates also rely on chemicals to repel their predators.

Many insects produce sprays, oozes, or froths when attacked. Bombardier beetles, for example, possess a pair of glands near the anal opening. Each gland has two compartments lined with a protective cuticle. The inner compartment contains a mix of relatively nontoxic chemicals, along with hydrogen peroxide. The outer compartment contains enzymes. When the beetle is disturbed, it discharges the contents of the inner compartment into the outer compartment, which leads to an instant, energy-releasing chemical reaction. Oxygen is one of the end products generated by this reaction, and the resulting pressure discharges the mixture with an audible "pop." Because of the energy released by the reaction, the temperature of the spray is approximately 100°C. The reaction of predators—including humans—to this hot, explosive secretion is predictable, and bombardier beetles have very few enemies.

Go to Media Clip 56.1
Bombardier Beetle Sprays Its Enemies
Life10e.com/mc56.1

But adaptions may evolve in predators that overcome their prey's chemical defenses, as we saw in the case of the rough-skinned newt and the garter snakes that have become insensitive to its protective toxin (see Figure 21.20). Some predators are not only undeterred by their prey's defensive chemicals, but ingest them and sequester them in their bodies as defenses against their own predators. Sea slugs are able to feed on a variety of well-defended prey with impunity and are masters at acquiring defenses from their food. Some species that feed on sponges concentrate toxic chemicals expropriated from their prey, whereas others, which feed on hydrozoans, incorporate the stinging cells of their prey, still active, into their own bodies.

APOSEMATISM Some prey species that defend themselves with toxic chemicals advertise that fact. This form of advertisement is called **aposematism**, or **warning coloration**. Aposematic prey species exploit the fact that predators can learn to avoid certain warning signals. Their warning signals may be visual (many toxic species are brightly colored) or acoustical (the rattlesnake's warning rattle, for example), depending on what sensory cues their predators use to find prey.

Many toxic prey sport bright colors or striking patterns to protect themselves against visually orienting predators. Such warning coloration increases the probability that a predator will learn to recognize and avoid the toxic species (Figure 56.5). Some vertebrate predators that rely on visual cues can learn quickly to associate certain color patterns with an unpleasant dining experience. Thus aposematic species are characteristically tough enough to survive a brief encounter with a predator. Any encounter that results in the death of the aposematic individual is unlikely to result in selection for its aposematic pattern. Sometimes field researchers find aposematic butterflies with damage inflicted by a bird beak—an indication of having survived being tasted by an uneducated avian predator.

MIMICRY SYSTEMS Even some nontoxic species benefit from warning coloration. We have seen that some prey species avoid consumption by mimicking inedible objects (see Figure 56.4B).

(A) Danaus plexippus larva
(B) Chromodoris sp.
(C) Dendrobates reticulatus

56.5 Some Prey Come with Warning Labels Some toxic prey warn potential predators with aposematic coloration. (A) Milkweed plants are toxic, and many of the insects that feed on them, such as this monarch butterfly larva, incorporate the plant's toxic chemicals into their systems. (B) Nudibranchs (sea slugs) are mollusks without protective shells; however, they may possess stinging nematocysts (acquired from their hydrozoan prey). (C) Poison dart frogs of Central and South American sequester highly toxic chemicals in their brightly colored skin.

Others do so by mimicking aposematic species. This strategy has led to the evolution of mimicry systems of two types. In **Batesian mimicry**, a benign, edible species (the mimic) closely resembles a dangerous, toxic species (the model) and benefits from the avoidance behavior learned by the model species' predators (**Figure 56.6A**). Mimicry may extend beyond physical appearance; many mimics also simulate distinctive behaviors of their models. In the Kalahari Desert of southern Africa, adult *Eremias lugubris* lizards are cryptically colored to blend in with the sand, but juvenile lizards of this species are conspicuously black and white, resembling the dangerous oogpister beetles native to the same region. Oogpisters (Afrikaans for "piss in your eye"), like the bombardier beetles described earlier, can emit a noxious spray over a considerable distance. Young lizards will press their tails to the ground and arch their backs, thus enhancing their resemblance to an oogpister about to "fire."

(A) Batesian mimics
This harmless blenny...
...closely resembles a venomous related species.
Petroscirtes breviceps
Meiacanthus grammistes

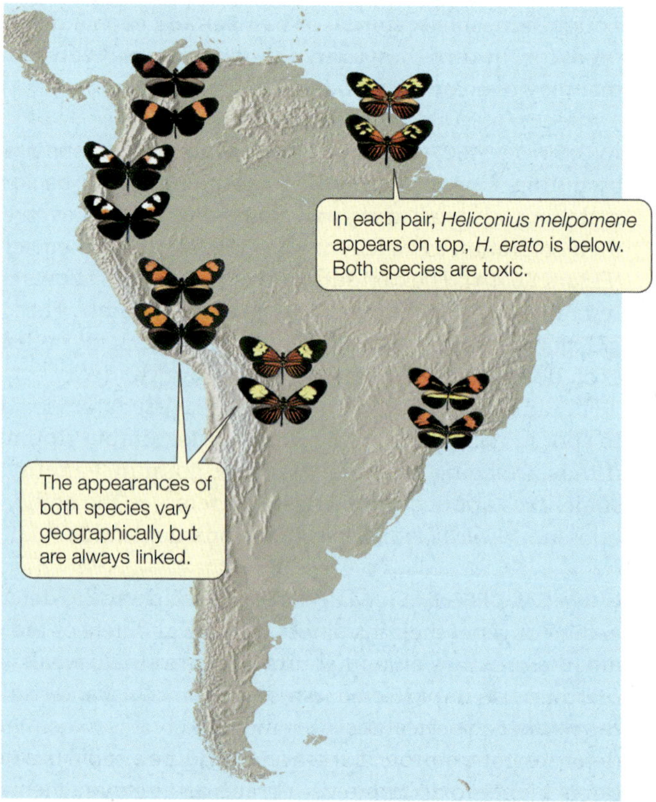
(B) Müllerian mimics
In each pair, Heliconius melpomene appears on top, H. erato is below. Both species are toxic.
The appearances of both species vary geographically but are always linked.

56.6 Truth in Labeling? (A) Batesian mimics are vulnerable species that gain protection by mimicking the aposematic signals of dangerous species. The appearance of the harmless blenny species *Petroscirtes breviceps* closely resembles that of the fanged striped blenny, which possesses a pair of grooved fangs with associated venom glands. (The male fish seen here are guarding eggs that females laid inside discarded bottles on the seafloor.) (B) The shared aposematic coloration of Müllerian mimics is an honest advertisement of their toxicity. As caterpillars, all of the longwing butterflies (genus *Heliconius*) of South America feed on toxic passionflower plants and incorporate the toxins into their adult bodies. The *Heliconius* species living together in a particular region have similar warning coloration.

In **Müllerian mimicry**, a number of aposematic species converge on a common color pattern; all benefit from providing a stronger recognition signal to predators. Many of the Neotropical longwing butterflies (*Heliconius*), which as caterpillars feed on toxic passionflower plants and incorporate the plant toxins into their bodies, are Müllerian mimics, and *Heliconius* species living together in a particular geographic region are likely to have similar coloration and share a common warning patern (**Figure 56.6B**). Genome sequencing of *Heliconius* Müllerian mimics has identified one gene, *optix*, that codes for a transcription factor that can, by changing gene expression patterns, create the same color patterns in *Heliconius* species that are not very closely related, thus leading to the evolution of mimetic color patterns within a geographic region.

Herbivory is a widespread interaction

The most common interaction among Earth's multicellular organisms is that between plants and the herbivores that eat them. Herbivores have a relatively easy time acquiring food, since plants are sessile and cannot claw, bite, or run away. Every major class of vertebrates includes at least a few herbivores. In marine systems, organisms that feed on plants and algae include mollusks, crustaceans, echinoderms, and annelids. But in terms of numbers of individuals as well as numbers of species, the vast majority of the world's herbivores are insects.

More than 90 percent of herbivorous insects are **oligophagous**, or specialists that dine on just one or a few, often taxonomically related, plant species. **Polyphagous** species, in contrast, feed on as many as hundreds of unrelated plant species. Vertebrate herbivores are generally polyphagous; a cow grazing in a pasture, for example, can consume many different plant species in a single afternoon. There are exceptions to this pattern, however. Australian koalas famously feed exclusively on the foliage of eucalyptus trees, and the diet of giant pandas is made up almost entirely of bamboo.

Herbivores, particularly insects, generally consume only parts of their food plants and usually do not kill them. In most natural ecosystems, insects rarely remove more than 20 percent of plant biomass. For that reason, some ecologists question the ability of insects to exert selection pressure on plant traits. Mortality is not, however, the only form of selection that leads to evolutionary change; herbivores can reduce plant fitness if the plants they attack produce fewer offspring.

PLANT DEFENSES AGAINST HERBIVORES The defenses of plants against their diverse consumers are necessarily highly diverse. For most plant species, chemistry is the principal defense mechanism. As we saw at the opening of this chapter, the leaves of some plants contain chemicals that prevent them from being consumed by fungi—and thus, incidentally, from being harvested by leafcutter ants. The amazing variety of secondary metabolites produced by plants to defend themselves against herbivores is the topic of Section 39.2. Many plants, however, have additional defenses.

Some plants protect themselves by being physically difficult to ingest. Thorns and spines are effective deterrents to browsing vertebrate herbivores. Smaller herbivores, including many insects, can be deterred by small hooked hairs on leaf surfaces. The soft bodies of leafhoppers can be pierced by these hairs, which fix the insect in place until it eventually dies from starvation or loss of blood. The plant's cuticle may also act as a physical barrier. Most grasses contain silica, which wears down sharp edges of herbivore teeth. Insects that feed only on grasses tend to have chisel-like mandibles that slice through leaf tissue, and their heads are enlarged to accommodate the larger jaw muscles needed to process their food.

RECIPROCAL ADAPTATIONS IN HERBIVORES AND PLANTS The concept of coevolution was first described in the context of interactions between herbivores and plants. In 1959 the entomologist Gottfried Fraenkel reached the conclusion after many years of study that all green plants are essentially nutritionally equivalent for insects. Why, then, are so many insects such picky eaters? Fraenkel proposed the novel hypothesis that ecological factors underlie the diversity of secondary metabolites that deter insect herbivores. A few years later, the entomologist Paul Ehrlich and the botanist Peter Raven proposed the following evolutionary scenario to account for patterns of host plant use among herbivorous insects (specifically, in their case, butterfly families):

- Certain plants, by mutation or recombination, evolve a novel secondary metabolite.
- If the chemical reduces the plant's appeal to herbivores, then plant genotypes producing the chemical are favored by natural selection.
- Freed from mortality associated with herbivory, plants possessing the novel chemical undergo an adaptive radiation.
- Certain herbivores, by mutation or recombination, evolve resistance to the chemical, and these resistant herbivores undergo their own adaptive radiation.
- With sufficient selection pressure, a resistant herbivore can evolve to use the chemical as a defense against its own predators.

This stepwise coevolutionary process explains not only the biochemical diversity of flowering plants but also the tremendous diversity of herbivorous insects. The ecological scenario outlined by Ehrlich and Raven is another example of a coevolutionary arms race.

A spectacular variety of adaptations to plant defenses has evolved in herbivores. Many herbivores circumvent plant defenses by behavioral means. For example, the secondary metabolites produced by a plant called St. Johnswort (*Hypericum perforatum*) require exposure to sunlight for optimal toxicity, so some insects that feed on this plant roll its leaves into a light-impervious cylinder and feed in comfort in the dark. The laticifer-cutting beetles described in Section 39.2 have a different method of detoxifying their food plant. Many large polyphagous herbivores, such as deer, horses, and the like, graze on a wide variety of plant species, minimizing their exposure to any particular defensive chemical. Long-lived and with relatively good memories, they can learn to avoid plants with an unpleasant taste.

Unlike large mammalian herbivores, caterpillars and many other insect herbivores may spend their entire lives feeding on a single individual plant. Such oligophagous diets are associated with highly specialized detoxification systems. The diamondback moth caterpillar eats plants in the cabbage family, which are rich in toxic mustard oil glycosides. In its gut is an enzyme that breaks down the glycosides into harmless byproducts, allowing it to eat these plants with impunity.

Some herbivores take resistance a step further by storing, or sequestering, plant toxins in specialized organs or tissues that are insensitive to those toxins. In this way they can accumulate large quantities of toxins in their bodies with no ill effects. This strategy also makes the expropriated chemicals available for defense against the herbivores' own enemies. The caterpillar of the monarch butterfly, for example, is insensitive to the neurotoxic glycosides in its milkweed host plants, but most of its enemies, including insect-eating birds, cannot tolerate these compounds (as the caterpillar's aposematic coloration suggests; see Figure 56.5A).

Yet the plants continue their side of the coevolutionary arms race. As we have seen, longwing butterflies principally consume passionflower plants. These oligophagous butterflies lay eggs only on passionflower plants, and their larvae sequester host plant toxins in their bodies as they feed on the leaves. Some passionflower species, however, have modified leaf structures that resemble the eggs of butterflies. Some longwing butterfly species will not lay eggs on plants already containing eggs, so the egg mimics reduce the plant's probability of being consumed (**Figure 56.7**).

Spots on the leaves of *Passiflora* mimic butterfly eggs.

56.7 Using Mimicry to Avoid Herbivory The leaves of some passionflower species develop structures that resemble the eggs of their principal herbivores, longwing butterflies (*Heliconius* spp.; see Figure 56.6B). Females of many longwing butterfly species will not lay eggs on plants already containing eggs, so the egg mimics deter these females, thus protecting the plant from being eaten by hatchling caterpillars.

Parasite–host interactions may be pathogenic

Parasitism is an interaction in which one species consumes only certain tissues in one or a few host individuals of another species without necessarily killing them. Keeping the host alive is important for parasites that are highly specialized; killing the host would leave the parasite with no way to make a living.

MICROPARASITES **Microparasites** are many orders of magnitude smaller than their hosts and generally live and reproduce inside their hosts. Microparasites include in their ranks viruses, bacteria, and protists. Multiple generations may reside within a single host individual, and a host may harbor thousands or millions of them. Many microparasites, in the process of acquiring nutrients at the expense of their host, cause symptoms of disease—that is, they are pathogens. Section 39.1 describes the array of secondary metabolites that plants produce to defend themselves against pathogens, and Chapter 42 describes the immune system defenses of animals.

Infection by pathogens may in some cases result in the death of the host, but death is by no means the inevitable outcome of these interactions. If a pathogen strain is to persist in a host population, the pathogens must continually infect new host individuals. A less deadly strain that kills a smaller proportion of host individuals may be able to infect a larger number of new hosts. Thus pathogen and host may reach a state of coexistence as increased host resistance (ability to withstand the effects of a pathogen) and decreased pathogen virulence (ability to cause disease) evolve. Yet new virulent strains may also arise, reminding us that the arms race goes on.

The pathogens' hosts fall into three classes: susceptible (capable of being infected), infected, or recovered (and thus, in many cases, immune). A pathogen can readily invade a host population dominated by susceptible individuals, but as the infection spreads, fewer susceptible individuals remain to be infected. Eventually a point is reached at which most infected individuals no longer transmit the infection to susceptible individuals. Thus rates of infection typically rise, then fall, and do not rise again until a sufficiently large population of susceptible host individuals has reappeared.

MACROPARASITES While microparasites generally live and reproduce inside their hosts, larger **macroparasites** are associated with their hosts in a slightly less intimate way. Although macroparasites rarely cause the same kinds of disease symptoms that pathogenic microparasites cause, they may nevertheless affect host survival and reproduction and can thereby act as selective agents on their hosts. **Ectoparasites** are macroparasites that live outside the bodies of their hosts. **Endoparasites**, such as the tapeworms described in Section 31.3, are macroparasites that spend at least part of their life cycle inside the bodies of their hosts.

Some ectoparasites—leeches, mosquitoes, and the like—are only casually associated with their hosts, interacting with them just long enough to eat their fill and then moving on. Others spend their entire lives on their hosts; these sedentary ectoparasites have a number of attributes that keep them attached

(A) *Pthirus pubis*

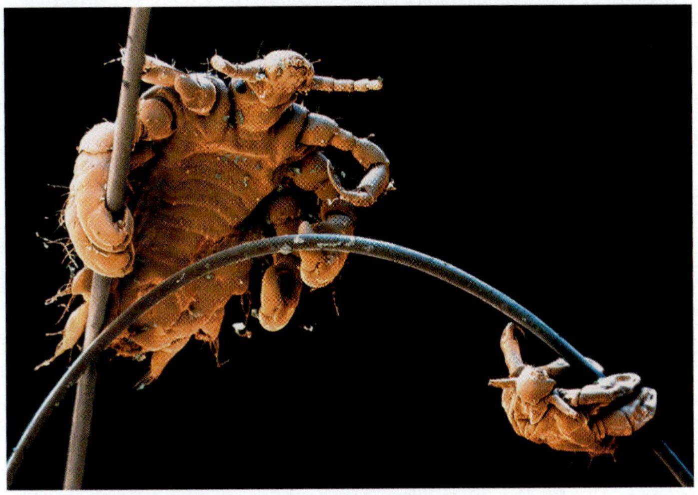
0.5 μm

(B) *Macaca fuscata*

56.8 Ectoparasites and Primates (A) Ectoparasites such as crab lice tend to be tiny, wingless, flattened, and equipped with strong claws for gripping. Humans are the only known host of this species, which infests the pubic hair. (B) Reciprocal grooming behaviors among primates is believed to have evolved in response to ectoparasites. Japanese macaques form social groups in which this behavior plays a significant role.

to their hosts. Crab lice, which are generally found in the pubic region of their human hosts, have claws on the tips of their legs that clamp around pubic hairs with great precision (**Figure 56.8A**). Pulling off a crab louse will often leave the legs behind, still firmly attached to the hair. Other adaptations that reduce the ability of irritated hosts to remove an ectoparasite include flattened bodies and a thick, tough cuticle. Most sedentary ectoparasitic insects are highly specialized, sometimes feeding on only a single host species.

Most hosts actively work to rid themselves of their ectoparasites. Grooming behavior—an important component of the social interactions of many primates—may have evolved in response to ectoparasites. The Japanese macaque (*Macaca fuscata*), for example, is prone to infestation by two species of lice, which tend to lay their multitudinous eggs on the outer surfaces of the host's back, arms, and legs. To keep louse populations in check, macaques form and maintain social bonds that ensure the consistent presence of grooming partners (**Figure 56.8B**). Some biologists believe that humans' hairlessness and bipedal posture (which freed the hands for manipulating small objects), as well as the opposable thumb, were evolutionary responses to ectoparasites.

RECAP 56.2

Predator–prey, herbivore–plant, and parasite–host interactions are all antagonistic. Consumers have adaptations for finding and using their resource species efficiently. Their resource species in turn have adaptations that reduce their probability of being discovered, captured, or eaten.

- What are some of the adaptations that help prey species avoid consumption by predators? **See pp. 1172–1175**
- How are aposematism and mimicry related? **See pp. 1173–1175 and Figures 56.4–56.6**
- Explain the scenario for coevolution between insect herbivores and their host plants proposed by Ehrlich and Raven. **See p. 1175**

Like antagonistic interactions, mutually beneficial interactions between species can result in coevolution. A mutually beneficial exchange of goods or services can ensure the predictability and frequency of such interactions over evolutionary time; thus many mutualistic interactions are tightly coevolved.

56.3 How Do Mutualistic Interactions Evolve?

Mutualisms are interactions between two species that benefit both partners. There are few taxonomic limits on mutualistic interactions: many organisms have mutualistic partners from other domains and distant branches on the tree of life. Mutualistic interactions often arise in environments where resources are in short supply. Consequently, many mutualisms involve an exchange of food for housing or defense. Corals and their photosynthetic endosymbionts (see Figure 27.21) and lichens formed from fungi and photosynthetic algae (see Section 30.2) are examples of mutualistic interactions in which food is exchanged for housing. In another common type of mutualism, sessile organisms, particularly flowering plants, rely on more mobile species for mating or dispersal. In this chapter we will focus on mutualisms that involve animals, which can form mutualistic associations with other animals, with plants, and with a wide range of microorganisms.

Many mutualisms are asymmetrical—in other words, one party benefits more than the other. One or both partners may evolve adaptations that ensure that the exchange benefits both of them. Reciprocal adaptations are most likely to arise in mutualistic interactions if an increase in dependence on a partner provides an increase in the benefits realized from the interaction. If increased dependence provides no selective advantage, mutualists (particularly species in asymmetrical mutualisms) may evolve into parasites, lose their partners and live independently, or even go extinct.

Some mutualistic partners exchange food for care or transport

Some organisms, such as the leafcutter ants described at the opening of this chapter, get their food by "farming" fungi. Fungus farming has been documented in a wide variety of species, including beetles, termites, and even a snail. In most cases the farmers provide housing, nutrition, and care for the fungal partner. The fungus provides food for the host, producing enzymes that degrade plant proteins and cellulose and thus converting plant materials that the insects could not have digested by themselves into an edible form.

Over the past 50 years the fungus-farming southern pine beetle (*Dendroctonus frontalis*) has destroyed huge tracts of valuable pine forests in the southeastern United States (**Figure 56.9**). The beetle owes much of its efficiency to its mutualistic partners. Masses of adult beetles attack a pine tree at once, overwhelming the tree's ability to defend itself (the tree's defense is to release large quantities of resin under pressure to force out the beetles). The beetles then excavate a series of galleries through the vascular tissue underneath the bark in which to lay their eggs. Female beetles also carry spores of their partner fungus into the galleries. The fungus grows on and breaks down the gallery walls, and the beetles feed directly on the fungus and the partially digested wood. The beetles also introduce a bacterium that produces an antibiotic to keep harmful bacteria from attacking the fungus (such as the leafcutter ants at the opening of this chapter). This insect–fungus–bacteria consortium overcomes the tree's antiherbivore defenses, to the partners' mutual benefit but to the great detriment of pine forests.

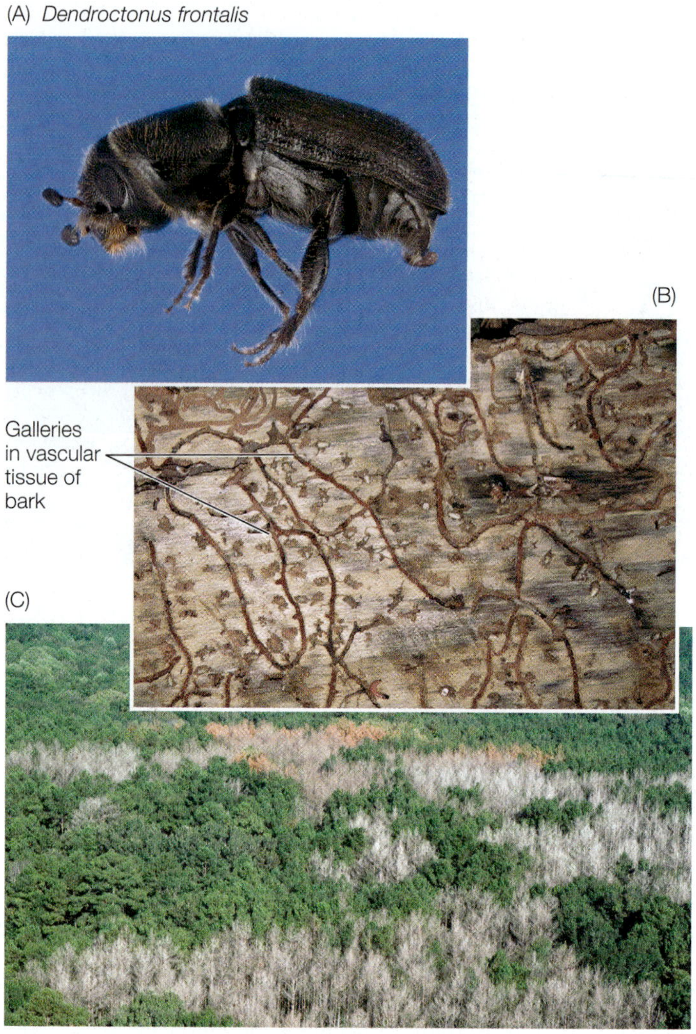
(A) Dendroctonus frontalis
(B)
Galleries in vascular tissue of bark
(C)

56.9 A Mutualistic Interaction Brings Death to Pine Trees
(A) The southern pine beetle has a mutualistic relationship with a fungus, which it "farms" within the vascular tissue of pine trees. (B) The beetles excavate galleries inside the trees' vascular tissue. Here they lay eggs and farm fungus; the fungus digests wood and provides nutrition for the larvae. (C) Masses of pine beetles have overwhelmed this forest, resulting in widespread death of pine trees.

Some mutualistic partners exchange food or housing for defense

Some plants are not only food resources for insects, they are also mutualistic partners. The best known of these interactions is that between ants and acacia trees in Central America. In 1874, in Nicaragua, the naturalist Thomas Belt observed a peculiar interaction between bullhorn acacia trees (*Acacia cornigera*) and *Pseudomyrmex* ants, known as acacia ants because they are found only in association with acacias. Bullhorn acacias get their common name from the enlarged, hollow thorns, in which the ants build nests. The trees also produce rewards for the ants, both in nectar-producing extrafloral structures and modified leaflet tips that are rich in oil and protein. These structures have no apparent purpose other than providing food for ants.

Belt suggested that the notoriously aggressive acacia ants defend the plants against herbivores in exchange for food and shelter. But his idea was not tested until Daniel Janzen conducted experiments in 1966. By removing ants from some acacias with insecticide, Janzen demonstrated that trees without ants suffered a reduction in growth and an increase in mortality (**Figure 56.10**). Although this experimental design was imperfect—the insecticides removed non-ants as well as ants from the experimental acacias and may have also influenced the ability of the trees to grow—it was the first experimental demonstration that plants may benefit from an association with ants, which is now a widely accepted concept. In fact, since Janzen conducted his experiment, additional work on ants and acacias has revealed that the ants do more than simply defend the plant against herbivorous enemies; they also clip weeds from around the base of the plants, presumably reducing competition for nutrients.

Go to Animated Tutorial 56.2
Mutualism
Life10e.com/at56.2

INVESTIGATING**LIFE**

56.10 Are Ants and Acacias Mutualists? Bullhorn acacia trees (*Acacia cornigera*) have numerous structures that provide food and shelter for ants of the genus *Pseudomyrmex* (acacia ants). Daniel Janzen's experiments demonstrated that the trees benefit greatly from their association with these ants, and that the energy expended in growing ant-attractive structures is repaid with increased growth and survival.[a]

HYPOTHESIS *Acacia cornigera* trees deprived of their *Pseudomyrmex* ant populations will survive and grow less well than trees populated by ant colonies.

Method

1. Define a population of *A. cornigera* trees; randomly designate some of them as untreated controls and the rest as experiment subjects.
2. Fumigate the experimental trees with insecticide to eliminate all *Pseudomyrmex* ants.
3. Apply Tanglefoot® (a sticky material) to the base of the experimental trees to prevent ants from recolonizing them.
4. Record the survival and growth rates of the trees in both groups over a 10-month period.

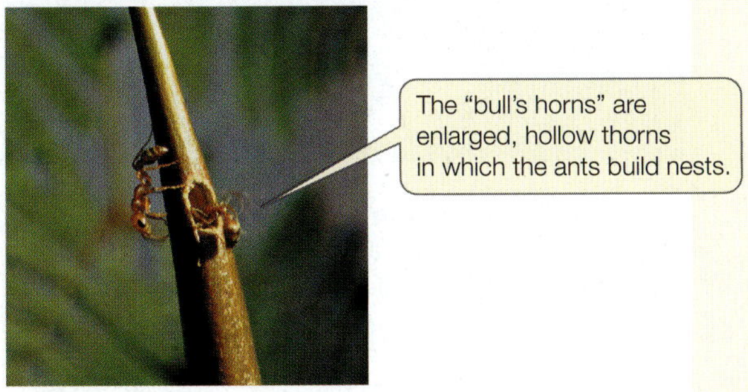
The "bull's horns" are enlarged, hollow thorns in which the ants build nests.

Results

After 10 months, control trees (with ants) had considerably higher survival and growth rates than did trees without ant populations.

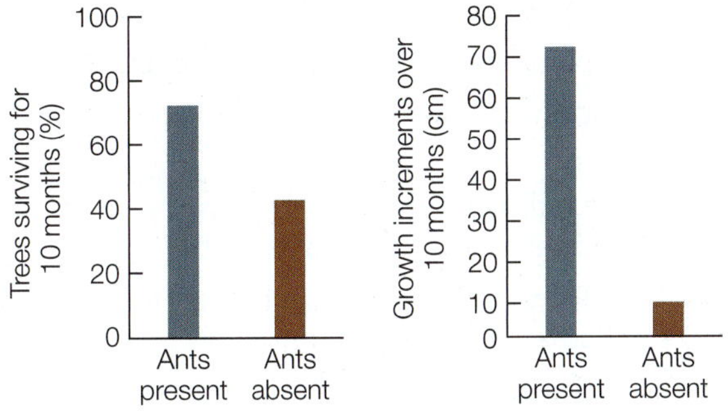
Trees surviving for 10 months (%)
100
80
60
40
20
0
Ants present
Ants absent
Growth increments over 10 months (cm)
80
70
60
50
40
30
20
10
0
Ants present
Ants absent

CONCLUSION *Pseudomyrmex* ants provide substantial fitness benefits to *Acacia cornigera* trees.

Go to **BioPortal** for discussion and relevant links for all INVESTIGATING**LIFE** figures.

[a]Janzen, D. H. 1966. *Evolution* 20: 249–275.

WORKING WITH**DATA:**

A Complex Species Interaction

Original Paper

Ness, J. H. 2006. A mutualism's indirect costs: The most aggressive plant bodyguards also deter pollinators. *Oikos* 113: 506–514.

Analyze the Data

As mentioned in the text, bullhorn acacias produce sugary substances on extrafloral nectaries that attract ants. Certain ant species act as pugnacious bodyguards for the plants, attacking other insects that may threaten the plants. In this way the ants and plants act as mutualists (see Figure 56.10). However, not all insects that are discouraged by ant bodyguards necessarily threaten the plant. In another such mutualism, the fishhook barrel cactus (*Ferocactus wislizeni*) of the western United States produces extrafloral nectaries that attract several species of ants. In 2003 and 2004, Joshua Ness collected data on visitation by bee pollinators from *F. wislizeni* colonized by each of four ant species. Use the data in the table to answer the questions, explaining your reasoning in each case.

QUESTION 1

Which ant species is the best defender against herbivores?

QUESTION 2

In the presence of which ant species are bees most likely to forage?

QUESTION 3

Why do you think that the reproductive success of these plants (as measured by seed mass and number of seeds produced) varies according to which ants are guarding them?

	Ant species present[a]			
Trait	*Crematogaster opuntiae*	*Forelius* sp.	*Solenopsis aurea*	*Solenopsis xyloni*
No. ants/flower	1.3 ± 0.2	2.5 ± 0.3	4.0 ± 0.5	3.8 ± 0.3
Flowers occupied by ants (%)	10.8 ± 3.1	52.0 ± 6.8	30.9 ± 4.1	44.2 ± 4.3
Flowers occupied by pollinators (%)	23.3 ± 4.5	14.2 ± 4.9	13.6 ± 3.1	4.2 ± 1.2
Bee foraging time/flower (sec)	43 ± 12.0	44 ± 18	76 ± 19.0	15 ± 3.0
Individual seed mass (mg)	237 ± 6.0	253 ± 15	232 ± 8.0	203 ± 8
No. seeds/fruit	1017 ± 88.0	1037 ± 206	1239 ± 107	871 ± 10.7
Seed mass/fruit (g)	2.32 ± 0.16	2.43 ± 0.38	2.82 ± 0.21	1.77 ± 0.20

[a]Only one ant species was present on each cactus plant studied. Values are means ± standard error.

Go to **BioPortal** for all WORKING WITH**DATA** exercises

Plants and pollinators exchange food for pollen transport

For about three-fourths of the planet's 250,000 flowering plant species, reproduction requires the transport of pollen by an animal partner. The benefit from the plant's perspective is clear: the animal partner moves pollen from one sessile individual to another and thereby promotes sexual reproduction and thus genetic diversity. In this section we will focus on the benefits accruing to the animal pollinator.

A mutualistic pollination system requires several features:

- An attractant or reward that entices a pollinator to visit the plant
- Behavior by the pollinator that ensures it will visit more than one individual of the same plant species
- Anatomical features that allow the pollinator to transport the plant's pollen

Floral characteristics influence the type of pollinator that is attracted to a flower. Ultraviolet color patterns, for example, are highly attractive to bees (see Figure 29.15) but are invisible to most other pollinators. The depth and width of a flower can restrict the size and shape of the pollinator mouthparts that can gain access to its nectar (see Figure 23.13). The timing of a plant's flowering can also restrict the number of potential pollinator species and encourage pollinator fidelity.

Flowers entice pollinators in many ways. The most direct reward for pollinators is the pollen itself, which sometimes serves as food. Pollen was probably the original attractant in the evolutionary history of plant–pollinator interactions. Plant reproduction would not be served, however, if pollinators were to eat *all* of a plant's pollen; thus plants have evolved various adaptations to ensure that they benefit from the exchange. For example, some plants have two types of anthers: feeding anthers to produce pollen for pollinators, and fertilization anthers to produce pollen for reproduction. These two types of anthers are shaped and positioned differently, so that as the pollinator dines on pollen from the feeding anthers, the fertilization anthers deposit pollen on a part of its body that will transfer it to the stigma of another flower of the same species.

Compared with pollen, nectar—a sugar-rich solution produced by some angiosperms—is a relatively new evolutionary development. Of the floral rewards, nectar has the greatest appeal and is consumed by the widest range of animal pollinators, including birds (such as hummingbirds) and mammals (such as bats) as well as insects. While nectar is particularly effective for attracting potential pollinators, it is also prone to removal by "nectar thieves": animals such as ants that consume the nectar without transporting pollen. Nectar thieves lower plant fitness by depleting nectar that would otherwise attract actual pollinators.

Plants may also take advantage of their pollinators. Some orchid species have evolved flowers that resemble the females of particular wasp species (sometimes even producing the same chemical substance the female wasp uses to attract mates). The plants are pollinated by male wasps that attempt to copulate with the flower (**Figure 56.11**).

Argogorytes mystaceus (male wasp)
The flower of Ophrys insectifera closely resembles a female wasp.

56.11 Taking Advantage of a Pollinator Flowers of the orchid *Ophrys insectifera* look and smell like female *Argogorytes mystaceus* wasps. A male wasp of this species expends energy in a futile attempt to mate with the orchid's flower, getting pollen on his body in the process. The wasp then carries the pollen to the next flower he visits in his quest for a genuine mate.

Plants not only need to attract pollinators, but must also ensure that those pollinators carry their pollen to other members of the same species. Repeat visits by a pollinator to different individuals of a particular plant species increase the likelihood that the pollen will end up on the appropriate stigma; thus some plants have adaptations to limit the diversity of their animal visitors. Botanists have long wondered why certain plants produce small amounts of toxic substances in their nectar. The nectar of tobacco flowers, for example, contains trace amounts of nicotine, an insecticidal neurotoxin. Many flower visitors, including hummingbirds, can ingest only tiny amounts of nicotine-laced nectar before moving on to other flowers. To other pollinators, however, nicotine may actually be addictive. Honey bees, for example, overwhelmingly prefer artificial nectar spiked with nicotine in laboratory tests. Putting small amounts of a potentially addictive substance in nectar may be one way tobacco plants improve their odds of a repeat visit by the right pollinator species.

Most flowers can be successfully pollinated by several different animal species. The evolution of broad suites of floral characteristics that attract certain groups of pollinators is an example of **diffuse coevolution**: the evolution of similar suites of traits in species experiencing similar selection pressures (**Table 56.1**). Scarlet gilia (*Ipomopsis aggregata*), a common wildflower in the Rocky Mountains, has successfully combined two different pollinator attraction strategies. Early in its growing season, it produces red flowers that attract hummingbirds; later in the season, the gilia shifts to producing white flowers because by then the most abundant pollinators are hawk moths, which cannot see red but are attracted to white.

A few plant–pollinator relationships are much more exclusive; these relationships lead to highly specific, rather than diffuse, coevolution. Yucca plants, for example, are pollinated only by a group of moths collectively known as yucca moths, whose

TABLE 56.1 Pollination Syndromes Resulting from Diffuse Coevolution

	Suite of anatomical traits			
Preferred pollinator	Flower shape	Flower color	Reward	Odor
Bees	Irregular	Many	Nectar, pollen	Sweet
Flesh flies	Irregular	Purplish	None	Carrion
Beetles	Bowl	White or pale	Pollen	Faint
Butterflies	Tubular	Many	Nectar	Faint
Moths	Often pendant	White or pale	Nectar	Heavy
Hummingbirds	Tubular	Red	Nectar	Imperceptible
Bats	Cuplike	White or pale	Copious nectar	Musty

larvae feed exclusively on yucca seeds. The stigma of the yucca flower is located deep within the pistil, and fertilization will not occur unless pollen is physically placed there. The specialized mouthparts of female yucca moths have distinctive long tentacles, which the moths use to pack masses of pollen from one yucca flower into transportable balls that they then carry to another flower. The moth pushes the pollen ball deep into the recess in which the flower's stigma is tucked, then turns around and deposits her eggs inside the flower's ovule (**Figure 56.12**). When the eggs hatch, the moth caterpillars will consume some—but not all—of the flower's developing seeds. Neither of these species can reproduce in the absence of the other.

Plants and frugivores exchange food for seed transport

Many animals that eat fruits (called **frugivores**) provide a valuable service to the plants that produce those fruits by dispersing seeds. Seed dispersal by animals not only offers plants the advantages of delivery to potential germination sites away from the parent plant (described in Section 38.1), but comes with the bonus of organic fertilizer for the seeds. Interactions between plants and frugivores, however, are not always reciprocal; in many cases, one party benefits more than the other. Whereas the frugivore is paid "in advance" for its transportation services, the seeds may never reach an appropriate destination for germination (your windshield, for example, will not do). From the plant's perspective, its partnership with frugivores requires a delicate balance between discouraging them from eating fruits before the seeds are capable of germinating and attracting them when the seeds are ready. In addition, the plant must protect the seeds from destruction in the frugivore's digestive tract and defend them against inappropriate consumers that would damage the seeds or fail to disperse them at all.

The chemical process of fruit ripening ensures that fruits are most attractive to frugivores when the seeds are mature and ready for dispersal. In many fruits, ripening is accompanied by a decrease in organic acids, which make many unripe fruits sour. Color changes, which result from loss of chlorophyll and the accumulation of other pigments (the conversion of peppers from green to red during ripening is an example), have enormous signal value to many frugivores. Green, unripe fruits are generally difficult for vertebrate frugivores to see against green foliage; red and bicolored red and black fruits contrast with foliage. Fruit softens as it ripens, allowing for gentle processing by the frugivore

Yucca filamentosa
Tegeticula yuccasella
The female moth collects and stores pollen grains in specialized mouthparts...
...and pushes the pollen deep into the pistil.
This female is laying eggs in the yucca's ovule.

56.12 Pistil-Packing Mama Yucca flowers are pollinated only by yucca moths, and the larvae of yucca moths feed only on the seeds of yucca plants. The moth *Tegeticula yuccasella* is the exclusive pollinator of *Yucca filamentosa*.

Dicaeum hirundinaceum

56.13 A Frugivore Plants and Fertilizes a Seed at the Same Time After a mistletoebird eats the fruit of the parasitic mistletoe plant, the seeds inside the fruit pass through the bird's digestive tract intact. As the seeds are voided, their sticky outer coat makes them stick to the bird's feathers. As the bird wipes itself clean on a branch, the seed sticks to the branch, where it germinates.

and rapid passage through its gut. Another conspicuous change in ripening fruits is an increase in sugar content—the "reward" most sought by frugivores. Seed coats, fruit pulp, and epidermis may all contain secondary chemicals designed to discourage inappropriate frugivores from consuming the fruit.

Because of the often asymmetrical nature of the mutualism between frugivores and plants, relatively few highly specialized frugivores exist. One apparently reciprocal interaction is between mistletoes—parasitic plants that grow on trees—and the mistletoebird that serves as the plants' primary dispersal agent in Asia and Australia (**Figure 56.13**). This bird dines largely on the fleshy berries of mistletoe. The seeds, covered with a gluelike outer coat, experience little enzymatic or mechanical damage as they pass through the thin-walled guts of the birds that swallow them. When the seeds are voided with a bird's droppings, the sticky outer coat causes the seeds to adhere to the bird's feathers, prompting it to wipe its bottom across the tree branch on which it is perched. Once the seed is wiped on the branch, the gluey coat keeps it there—in an ideal location for a mistletoe seed to germinate.

RECAP 56.3

Mutualistic interactions involve an exchange of benefits. Most plants rely on mutualisms with animals for fertilization and seed dispersal.

- Give examples of benefits that are exchanged in at least two mutualisms between plants and animals.
- What three features are required by a plant–pollinator mutualism? **See p. 1180**
- How are plant–pollinator interactions different from plant–frugivore interactions? **See pp. 1180–1181**
- Describe some adaptations in plants that help maintain the balance in their relationships with frugivores. **See p. 1181–1182**

From our discussion of interactions between two species that benefit both, we will now move to a type of interaction that benefits neither: competition. This type of interaction is widespread because it can arise wherever two or more species require the same resources.

56.4 What Are the Outcomes of Competition?

Antagonistic interactions can be quite attention-getting; the scene of a lion stalking a gazelle is almost emblematic of the African savanna. But at the same time a predator is interacting with its prey, it may also be interacting with other predators that hunt the same prey species. Lions are not the only predators of gazelles; cheetahs, hyenas, and even crocodiles hunt and kill gazelles, potentially reducing the supply of food available for lions.

Whenever any resource is not sufficiently abundant to meet the needs of all the organisms with an interest in that resource, organisms must compete with one another to gain enough of that resource to survive. Competition not only influences the evolution of species but also plays an important role in determining the structure and composition of communities, as we will see in the next chapter.

Competition is widespread because all species share resources

Virtually no species enjoys exclusive access to any given resource. The vast majority of species must compete for at least some resources with other species. As we saw in Section 55.4, limited resources are the main reason why populations do not grow indefinitely. When resources are limited, individuals in the population compete for those resources. Such **intraspecific competition**—competition among individuals of the same species—may result in reduced growth and reproductive rates for some individuals, may exclude some individuals from better habitats, and may cause the deaths of others. **Interspecific competition**—competition among individuals of different species—affects individuals in much the same way. At some point an essential resource may be in such short supply that a population is in danger of becoming unable to sustain itself; when a resource becomes limiting in this way, competition becomes intense and can influence the persistence and evolution of species.

The principle of competitive exclusion holds that no two species can share the same limiting resource indefinitely. If one species can prevent all members of another species from using the resource, the inferior competitor may go locally extinct, a result called **competitive exclusion**. In other cases, selection pressures resulting from interspecific competition cause changes in the ways in which the competing species use the limiting resource. If those changes allow the species to coexist, the result is called **resource partitioning**.

Whether it is interspecific or intraspecific, competition occurs by two major mechanisms. **Interference competition** occurs when a competitor interferes with another competitor's access to a limiting resource. **Exploitation competition** occurs when a limiting resource is available to all competitors and the

Geospiza fuliginosa

Nectar Use and Wingspan of *G. fuliginosa*		
Island	Time spent feeding on nectar (%)	Mean wingspan (mm)
Bees absent		
Pinta	10	59.8
Marchena	28	58.2
Bees present		
Fernandina	1	64.8
Santa Cruz	14	64.0
San Salvador	0	63.8
Española	0	64.7
Isabela	7	64.5

Xylocopa darwinii on *Opuntia* flower

56.14 Competition with Bees Influences Finch Morphology On islands in the Galápagos archipelago where *Geospiza fuliginosa* is the sole pollinators of cactus flowers, a short wingspan increases these birds' ability to negotiate the flowers. On islands where carpenter bees compete with the birds for cactus nectar, *G. fuliginosa* individuals have a longer wingspan and feed more heavily on other foods.

outcome of competition depends on the relative efficiency with which the competitors use the resource.

Interference competition may restrict habitat use

Interference competition can take many forms. A graphic example involves the desert ant *Conomyrma bicolor* and the honeypot ant *Myrmecocystus mexicanus*. These two ant species occupy the same type of habitat—arid areas containing little vegetation—and they feed on similar foods—the sugary excretions of aphids and other sap-feeding insects as well as occasional arthropods, none of which is in great supply. When *C. bicolor* workers find the entrance of a honeypot ant nest, they pick up small stones in their mandibles, carry them to the rim of the nest opening, and drop them down the hole—up to 200 stones in a 5-minute interval. This activity is enough to stop the honeypot ants from going out foraging. Some honeypot ant colonies, under constant stone-dropping attack for several weeks, may be almost entirely deprived of food.

Even microorganisms interfere with one another's use of resources. In the highly structured environment of the rhizosphere, or "root-world," of the soil, competitive interactions can be locally intense. Many soil bacteria produce substances that subdue their microbial competitors. Actinobacteria, for example, produce chemicals that interfere with essentially every life process in other kinds of bacteria. Many of the chemicals that these remarkably well defended microbes produce to defeat their competitors are used as antibiotics by mutualistic partners, such as the bark beetles described in Section 56.3, as well as in human pharmacology.

Exploitation competition may lead to coexistence

Exploitation competition may lead to coexistence, provided that the species relying on the same resource evolve adaptations to divide up, or partition, that resource. For example, in many Rocky Mountain communities, at least three species of bees consume the nectar of *Agave schottii*, the shindagger agave. The three bee species differ in where and when they collect shindagger nectar. Honey bees tend to forage in places with the greatest numbers of shindagger flowers, bumblebees in places with intermediate numbers of flowers, and carpenter bees where flowers are few and far between. Honey bees also tend to be most active when nectar output is greatest. With their larger nests and greater numbers of offspring to support, honey bees require greater foraging efficiency and greater energy intake. Foraging sites that are not worth their while are left to the other bees.

In some cases individuals within a species display different behaviors or morphologies depending on whether they are competing for resources with another species. Darwin remarked in *On the Origin of Species* that "Natural Selection leads to divergence of character; for more living beings can be supported on the same area the more they diverge in structure, habits, and constitutions." This "divergence of character" is referred to today as **character displacement**. On some of the islands of the Galápagos archipelago, for example, certain cactus species are pollinated exclusively by the small ground finch *Geospiza fuliginosa*, for which cactus nectar is an important food source (see Figure 23.8). On other islands, a carpenter bee (*Xylocopa darwinii*) competes with the finches for cactus nectar; the birds consequently feed more heavily on seeds and insects. On the islands where bees are absent, the birds feed on nectar more often and have smaller wingspans than on islands where they share cacti with bees (**Figure 56.14**).

Sometimes organisms respond to competition by changing their location to avoid confrontations. The African wild dog (*Lycaon pictus*) is a carnivore that lives and forages in packs (groups of related individuals). Frequent vocalizations, called twitters, function to keep the pack together, but these acoustical signals also can alert their chief competitors for prey, African lions (*Panthera leo*). Lions hearing the dogs' twitters can use them to locate dog packs and steal their kills. The dogs avoid competing with lions by selecting areas for their dens where the likelihood of being overheard by lions is low; thus wild dog densities are inversely correlated with lion densities. The wild dog is considered a **fugitive species**—a species that leaves an otherwise suitable habitat in order to avoid competition with another species.

Species may compete indirectly for a resource

Species may compete indirectly for a resource even when they are not present in the same habitat at the same time. Sometimes a species so alters the quality of a resource that it is rendered less usable by other species that may encounter it afterward. For example, feeding by sap-sucking leafhoppers on potato plants early in the growing season can cause curled leaves and chlorosis (loss of chlorophyll). Potato beetles feed on potato plants later in the growing season. Beetles that consume leaves damaged by leafhoppers suffer reduced growth and survival rates. Even though these two herbivores do not feed at the same time, one species influences the use of the shared food resource by its competitor.

Indirect competition can also result when two species share a common predator. For example, the parasitoid wasp *Venturia canescens* is a consumer of two different species of caterpillars that infest stored food products such as flour, the Indianmeal moth caterpillar (*Plodia interpunctella*) and the Mediterranean flour moth caterpillar (*Ephestia kuehniella*). The two caterpillar species can coexist in a flour bin, but when the wasp is present, it preferentially attacks and kills the flour moth caterpillars. Thus in the presence of the wasp, the competitive balance between the two caterpillar species is altered in the meal moth's favor. This type of competition is indirect because the outcome of competition depends not on how the two competitors use the shared resource, but on how the two competitors interact with a shared predator.

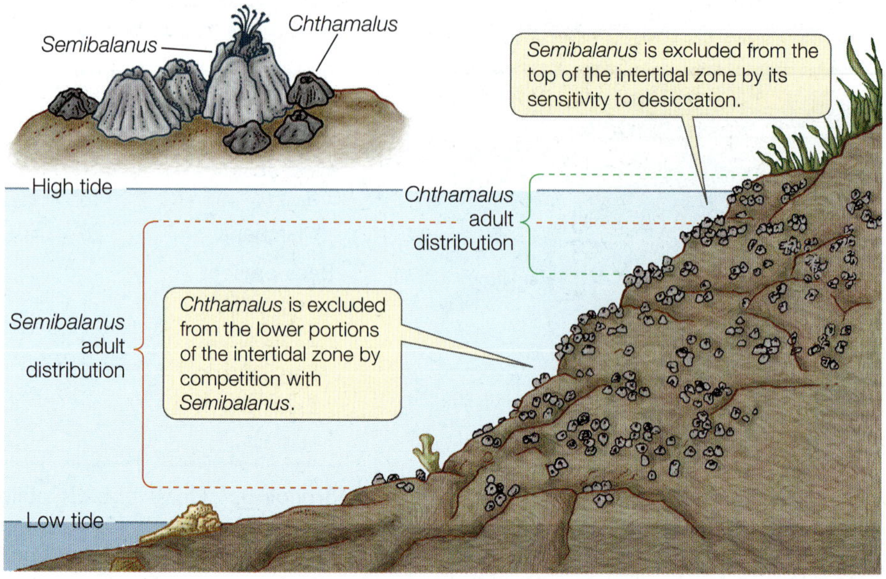
Semibalanus
Chthamalus
Semibalanus is excluded from the top of the intertidal zone by its sensitivity to desiccation.
High tide
Chthamalus adult distribution
Semibalanus adult distribution
Chthamalus is excluded from the lower portions of the intertidal zone by competition with Semibalanus.
Low tide

56.15 Interspecific Competition Can Restrict a Species' Range Interspecific competition with rock barnacles (*Semibalanus*) restricts stellate barnacles (*Chthamalus*) to a smaller portion of the intertidal zone than they could otherwise occupy. Larvae of both species settle throughout the intertidal, but at lower levels, rock barnacles grow much faster and eliminate the stellate barnacle larvae. In the upper reaches of the intertidal, however, the greater susceptibility of rock barnacles to desiccation (drying out) allows stellate barnacles to outcompete them. The two species can coexist in a small portion of the intertidal zone.

Competition may determine a species' niche

Competition is important in determining where a species can be found. A species' **niche** is the set of physical and biological conditions it requires to survive, grow, and reproduce. Thus a species' niche is partly defined by the resources available in the environment. Although a species might be physiologically able to live under a wide range of conditions, competitors may restrict its use of resources in a particular location. Thus every species has a **fundamental niche**, defined by its physiological capabilities, and a **realized niche**, defined by its interactions with other species.

Two species of barnacles, the rock barnacle (*Semibalanus balanoides*) and Poll's stellate barnacle (*Chthamalus stellatus*), compete for space on the rocky shorelines of the North Atlantic Ocean. The planktonic larvae of both species settle in the intertidal zone and metamorphose into sessile adults. The smaller stellate barnacles generally live at higher levels in the intertidal zone, where they face longer periods of exposure and desiccation (drying out) than do rock barnacles, which live at a lower level. There is little overlap between the areas occupied by adults of the two species (**Figure 56.15**). What explains their distinct distributions in the intertidal zone?

In a famous study conducted more than 50 years ago, Joseph Connell experimentally removed one or the other species from its characteristic zone and observed the response of the remaining species. Stellate barnacle larvae normally settle in large numbers throughout much of the intertidal zone, including the lower levels where rock barnacles are found (their fundamental niche), but they thrive at those lower levels only when rock barnacles are not present (their realized niche). Connell found that the rock barnacles grew so fast they smothered, crushed, or undercut the stellate barnacle larvae. However, removing stellate barnacles from their spots higher in the intertidal zone did not lead to their replacement by rock barnacles; the rock barnacles are less tolerant of desiccation and failed to thrive there even when stellate barnacles were absent. The result of the competitive interaction between the two species is a distinctive pattern of intertidal zonation, with stellate barnacles restricted in their distribution by competition and rock barnacles restricted in their distribution by their physiological limitations.

RECAP 56.4

Competition occurs when two or more species require a resource that is in limited supply. No two species can share the same limiting resource indefinitely. The outcome of competition may be competitive exclusion, in the form of local extinction, or coexistence, in the form of resource partitioning.

- How does exploitation competition differ from interference competition? **See pp. 1182–1183**
- How can competition lead to character displacement? **See p. 1183 and Figure 56.14**
- Explain the difference between an organism's fundamental niche and its realized niche. **See p. 1184 and Figure 56.15**

56.16 A Fungal Garden Cutaway view of a South American leafcutter ant nest chamber filled with fungus. Several winged ants (*Atta colombica*) can be seen in the crevices of the fungal mass.

The study of interactions among the species in a community is a large part of community ecology—the topic of the next chapter. Every kind of interaction we have studied in this chapter influences the nature and structure of communities. Competition helps determine which species persist and which go extinct, as well as dictating how many different species can be supported by a particular resource. Similarly, antagonistic interactions have important effects on the distribution and abundance of consumer and resource species, and the presence of mutualistic partners may dictate whether a particular species can exist in a particular community.

?

The fungi in leafcutter nests cannot survive without the ants, but can leafcutter ants survive without the fungus?

ANSWER

The relationship between the leafcutter ants and their fungi is a coevolved mutualism. The ants are so specialized behaviorally that they would starve without their fungus gardens. Recent sequencing of the genome of one leafcutter species, *Atta cephalotes*, revealed that this species has lost several genes possessed by other ants. Those genes include the ones encoding the entire pathway for biosynthesizing the amino acid arginine, which the leafcutter obtains from its fungal food, as well as several genes that break down plant toxins (which the fungus does for the leafcutter as it metabolizes the leaf substrate). In the case of leafcutter ants and fungi, we can assume that the increasing dependence of each species on its partner provided it with an increase in benefits from the partnership, resulting in reciprocal adaptations.

CHAPTER**SUMMARY** 56

56.1 What Types of Interactions Do Ecologists Study?

- Species interactions can be grouped into categories. **Antagonistic interactions** include **predation**, **herbivory**, and **parasitism**, all of which benefit a consumer while harming the species that is consumed. **Mutualism** benefits both participants, whereas **competition** harms both. **Commensalism** benefits one participant with no effect on the other; **amensalism** has no effect on one participant but harms the other. **Review Figure 56.1, ACTIVITY 56.1**
- The evolution of an adaptation in one species may lead to the evolution of an adaptation in a species with which it interacts, a process known as **coevolution**. A series of reciprocal adaptations among consumers and their resource species can lead to a coevolutionary **arms race**. **See ANIMATED TUTORIAL 56.1**

56.2 How Do Antagonistic Interactions Evolve?

- Predators kill the individuals they consume (their prey). Over its lifetime, a predator kills and consumes many prey individuals.
- Some prey species avoid detection by means such as **crypsis**. Others defend themselves by physical or chemical means. Chemically defended animals often advertise their toxicity with **aposematism**, or warning coloration. **Review Figure 56.4, 56.5**
- In **Batesian mimicry**, a nontoxic species mimics a toxic species. In **Müllerian mimicry**, two or more toxic species converge to resemble one another. **Review Figure 56.6**
- Herbivores generally consume only parts of their food plants and usually do not kill them.
- Many herbivores have evolved resistance to the defensive secondary metabolites produced by plants, and some have incorporated them into their own defenses against predators.
- Parasites consume certain tissues in one or a few host individuals of another species without necessarily killing them. **Microparasites** include viruses, bacteria, and protists; large numbers of these organisms can live and reproduce within the body of the host and are often pathogenic. **Macroparasites** are less intimately associated with their hosts but can nonetheless affect host fitness.

56.3 How Do Mutualistic Interactions Evolve?

- Mutualistic interactions involve an exchange of benefits. Many mutualisms arise in environments where resources are in short supply.
- Reciprocal adaptations are most likely to arise when an increase in dependency on a partner provides an increase in the benefits realized from the interaction.
- Some animals "farm" fungal species, which provide them with food. Other mutualisms involve an exchange of food or housing for defense. **Review Figures 56.9, 56.10, ANIMATED TUTORIAL 56.2**

continued

- Many mutualisms between plants and animals involve an exchange of food for transport. In plant–pollinator interactions, animals that collect and transport pollen are rewarded with pollen or nectar.
- Broad suites of floral characteristics that are attractive to certain types of pollinators exemplify **diffuse coevolution**. Some plant–pollinator mutualisms, however, are much more specific and exclusive. **Review Figure 56.11, Table 56.1**
- Plants that depend on **frugivores** for seed dispersal must balance the need to discourage frugivores from eating fruits before the seeds are mature, attract frugivores when the seeds are mature, and protect the seeds from destruction in a frugivore's digestive tract.

56.4 What Are the Outcomes of Competition?

- Competition occurs whenever a resource is not sufficient to meet the needs of all organisms with an interest in that resource.
- Competition may result in local extinction of the inferior competitor, an outcome called **competitive exclusion**. Alternatively, selection pressures resulting from competition may change the ways in which the competing species use a limiting resource, an outcome called **resource partitioning**. **Interference competition** occurs when an individual interferes with a competitor's access to a limiting resource. **Exploitation competition** occurs when a limiting resource is available to all competitors and the outcome of competition depends on the relative efficiency with which competitors use the resource.
- Exploitation competition may lead to **character displacement**, in which attributes of a species vary depending on whether a competitor is present or absent. **Review Figure 56.14**
- Species may compete indirectly even when they are not present in the same place at the same time, as, for example, when they share a common predator.
- A species' **niche** is the set of physical and biological conditions it requires to persist. Although a species may be able to persist under a wide range of resource conditions (its **fundamental niche**), competitors may restrict its use of resources in a particular location (its **realized niche**). **Review Figure 56.15**

Go to the Interactive Summary to review key figures, Animated Tutorials, and Activities Life10e.com/is56

CHAPTER**REVIEW**

REMEMBERING

1. Predation, herbivory, and parasitism are all examples of
   a. antagonistic interactions.
   b. mutualistic interactions.
   c. commensal interactions.
   d. amensal interactions.
   e. competitive interactions.

2. In a coevolutionary arms race, after a plant evolves a novel chemical defense against an herbivore,
   a. the herbivore can be expected to go extinct.
   b. the plant can be expected to undergo a range restriction because of the cost of producing the novel chemical.
   c. the herbivore can be expected to evolve resistance to the plant's defense.
   d. the plant can be expected to experience reduced fitness because of the cost of producing the novel chemical.
   e. the plant can be expected to stop producing other types of defenses.

3. Damage caused to shrubs by branches falling from overhead trees is an example of
   a. interference competition.
   b. predation.
   c. amensalism.
   d. commensalism.
   e. diffuse coevolution.

4. A hummingbird sips nectar from the flowers of a plant, pollinating those flowers in the process. This interaction is best classified as
   a. parasitism, because the hummingbird consumes the flower's nectar.
   b. predation, because the hummingbird eats the plant's seeds.
   c. commensalism, because the hummingbird benefits from consuming nectar and the plant is unaffected.
   d. mutualism, because the plant provides nectar for the hummingbird and the hummingbird transports pollen for the plant.
   e. Not enough information is provided to classify this interaction.

5. One factor that can constrain the realized niche occupied by an organism is
   a. crypsis
   b. aposematism.
   c. mimicry.
   d. commensalism.
   e. competition.

UNDERSTANDING & APPLYING

6. The different types of interspecific interactions are part of a continuum, and their outcomes often depend on circumstances. Refer to the Working with Data exercise on p. 1179. How does this example exemplify the various types of interspecific interactions (mutualism, competition, predation, parasitism, commensalism)? Do you think a continuum is represented here? What aspects of the situation described by the data could change the interactions?

7. Like the southern pine beetle in Figure 56.9, the mountain pine beetle (*Dendroctonus ponderosae*) attacks pine trees with the help of a symbiotic fungus that infects the host tree. In 2009 these beetles infested almost 4 million acres of pine forest across Montana, Wyoming, Colorado, Idaho, Utah, Oregon, and Washington. How would you determine which pine trees are susceptible to mountain pine beetle attack? How could the fact that this beetle has a symbiotic partner affect approaches for managing the outbreak?

8. *Salmonella serovar typhimurium* is a bacterium that lives in the intestines of a wide variety of animals—including humans, in which it can cause gastroenteritis. In the United States, raw chicken and other poultry are frequent dietary sources of this bacterium and thus present a significant health hazard. Although warning labels now appear on packaged raw poultry, the food industry is testing new ways to reduce the likelihood of contamination by reducing *Salmonella* populations in chickens before they are slaughtered. In one such test, broiler chicks were given a culture of three species of bacteria: *Lactobacillus plantarum*, *L. acidophilus*, and *Lactococcus lactis*. These birds, along with control birds that had not ingested the cultures, were exposed to *Salmonella* gut colonization and then tested to see if they maintained populations of *Salmonella* in their guts. Chicks that had been given bacterial cultures consistently had significantly lower populations of *Salmonella* than the control group.
   a. What ecological principle is being applied by the poultry industry?
   b. What other ecological outcomes might this experiment have produced?
   c. What other problems might this ecological principle be useful in tackling?

ANALYZING & EVALUATING

9. Many ectoparasites feed on only a narrow range of host species. Until recently, investigators used close genetic relationships among bird lice as evidence of close genetic relationships among their host bird species. Some ornithologists thought that flamingoes were closely related to ducks and geese, citing as evidence the observation that three of the four genera of bird lice found on flamingoes also parasitize ducks and geese. DNA analysis, however, showed that flamingoes are not close relatives of ducks, but are more closely related to grebes, another group of waterfowl. If this analysis is accurate, what would you predict about the lice parasitizing grebes? How could you use modern methods of molecular analysis to determine relationships among bird lice and their hosts? Given that grebes, ducks, and flamingoes are all water birds, what other factors might contribute to host shifts in ectoparasitic lice?

10. Even though nectar serves no function in the life of a plant other than to attract and reward pollinators, some plants produce toxic compounds in their nectar. As we've seen, in some cases these substances are addictive and encourage pollinators to revisit the same plant species; honey bees, for example, may visit tobacco flowers repeatedly because they become "addicted" to nicotine (see Section 56.3). But another study had a different outcome. Some researchers created genetically modified tobacco plants that produced different levels of nicotine in their nectar and found that higher concentrations led to shorter visits by pollinating hummingbirds and hawk moths—and more successful pollination than in plants whose flowers hosted longer visits. Why might shorter visits increase pollination success? What other factors might influence how much nicotine a tobacco plant should produce in its nectar?

57 Community Ecology

CHAPTER OUTLINE

Dead Reckoning Forensic entomologists use pig carcasses like this one as models to measure successional changes in corpse communities. They are able to apply these measurements to human corpses because pigs are in the same weight range as humans and, like humans, pigs are mostly hairless.

ONE DAY IN MARCH OF 1996, the body of an apparent suicide victim was discovered under the bushes near the railroad tracks in Cologne, Germany. The badly decomposed corpse contained masses of maggots—fly larvae—and the dried outer skin was peppered with pale yellow insect eggs. Examining the body, Mark Benecke, a forensic entomologist, recovered a single adult fly, which he identified as *Piophila casei*, the cheese skipper.

A diverse community of insect species colonizes human corpses, and its composition varies predictably as decomposition progresses. There are three major stages of decomposition: autolysis (degradation of proteins and lipids), putrefaction, and finally, decomposition. Each stage is characterized by a distinctive faunal community of species that use decomposing corpses in many ways. Some species, such as cheese skippers, consume dead flesh; others, such as hide beetles, eat hair and nails; still others prey on the flesh-eating insects or consume their excrement. Among these species, the cheese skippers are latecomers, arriving at corpses only when autolysis is well advanced and the body's proteins start breaking down—typically after 1 to 3 months. The abundance of cheese skipper eggs suggested to Benecke that these insects had undergone at least two generations in the corpse. Knowing that completion of the insect's life cycle required from 11 to 19 days under local weather conditions, he calculated that the first adults probably laid eggs about 90 days after death, and that 22 to 38 more days were needed to complete two generations. Thus he calculated that death must have occurred between 112 and 128 days earlier. As it happened, a 38-year-old woman had been reported missing about 4 months before the body was found; the estimated postmortem interval helped investigators identify her as the suicide victim found by the tracks.

What species occur where and when is the concern of community ecologists. The species present in a community can change in predictable ways, at spatial scales ranging from a dead body in a patch of shrubbery to an Amazonian rainforest, and at time scales ranging from days to millennia. But the ecological processes affecting communities are similar whatever the scale. The study of the seemingly esoteric changes in the composition of carrion communities has not only allowed ecologists to help find evidence that can identify a missing person or convict a murderer, but has also added to our understanding of ecological communities.

?

How do the insect species in a corpse community influence one another's ability to survive?

See answer on p. 1203.

(A) *Sarracenia purpurea*

(B)

57.1 Ecological Communities Exist at Different Scales (A) The microorganisms and tiny invertebrates (such as insect larvae) existing within a single pitcher plant constitute an ecological community. (B) Lake Superior in central North America is Earth's largest freshwater lake. It encompasses a large biological community with boundaries defined by its shoreline. Despite having such a defined boundary, the lake community is subject to effects from species and activities far beyond its contained waters.

57.1 What Are Ecological Communities?

In ecological terms, a **community** is a group of species that coexist and interact within a defined area. Although each species has unique interactions with the other species in its community (as we saw in the previous chapter), ecologists often find it useful to study the properties of the community as a whole.

Communities vary greatly in size and scope. The organisms colonizing a dead body constitute one type of definable community. Another type of easily identifiable community is that living within the purple pitcher plant (*Sarracenia purpurea*), a plant common in North American wetlands (**Figure 57.1A**). Each plant has several leaves that form rainwater-collecting "pitchers," and the plant derives nitrogen from insect prey that get trapped in the pitchers and drown (see Section 36.5). However, the pitchers are also occupied by thriving communities of living microorganisms and tiny invertebrates, including bacteria, protists, rotifers, and mosquito larvae.

The boundaries defining a community, particularly a large one, are not always so easy to recognize. The community of organisms inhabiting Lake Superior, for example, is for the most part bounded by the shores of the lake (**Figure 57.1B**). However, even though Lake Superior may appear to be self-contained, many of its components originate far away. For example, mallards and other ducks may consume seeds in one location, and in the 5 to 11 hours it takes for the seeds to move from one end of the duck digestive tract to the other, the birds may fly up to 1,400 kilometers before depositing the seeds in their excrement.

When borders become unclear, ecologists may designate boundaries somewhat arbitrarily, based on their ability to study the community. By the same token, because a community can contain thousands of different species from microscopic bacteria to towering trees, it is usually impractical or impossible to study all species within a community. Ecologists often define a community of interest taxonomically—for example, the fish community of Lake Superior.

Communities are characterized (1) by their species composition—the number and kinds of species they contain; and (2) by the relative abundances of those species. Species composition is determined by the same factors that determine the distributions of species because, as we saw in Chapter 54, a species can occur in a particular place only if it can colonize that place and if environmental conditions allow it to persist there. Thus even the same type of community may contain different numbers of species in different places. The insect community of human corpses, for example, varies with local climate and burial conditions (e.g., whether the body is buried, deposited on the soil surface, or submerged in a river or pond).

Although communities vary in size and complexity, ecologists have devised methods for quantifying basic properties of community structure and organization irrespective of their scale. These methods have revealed patterns that reflect underlying community assembly rules and general principles.

What determines how many species constitute a community in any particular place? One important factor is the amount of energy available to sustain organisms.

Energy enters communities through primary producers

Sunlight is the ultimate source of energy for most of Earth's communities. Sunlight makes photosynthesis possible, and photosynthesis, in the vast majority of communities, makes energy available to other organisms in an edible form. All nonphotosynthetic organisms (heterotrophs) consume, either directly or indirectly, the energy-rich organic molecules produced by the plants and other photosynthetic organisms that get their energy directly from sunlight (autotrophs). Photosynthetic autotrophs, along with a handful of chemoautotrophs (organisms that obtain chemical energy from inorganic molecules in their environment), are known as **primary producers**.

Gross primary productivity (GPP) is the rate at which all the primary producers in a particular community turn solar energy into stored chemical energy via photosynthesis. The energy that is accumulated by primary producers is called gross

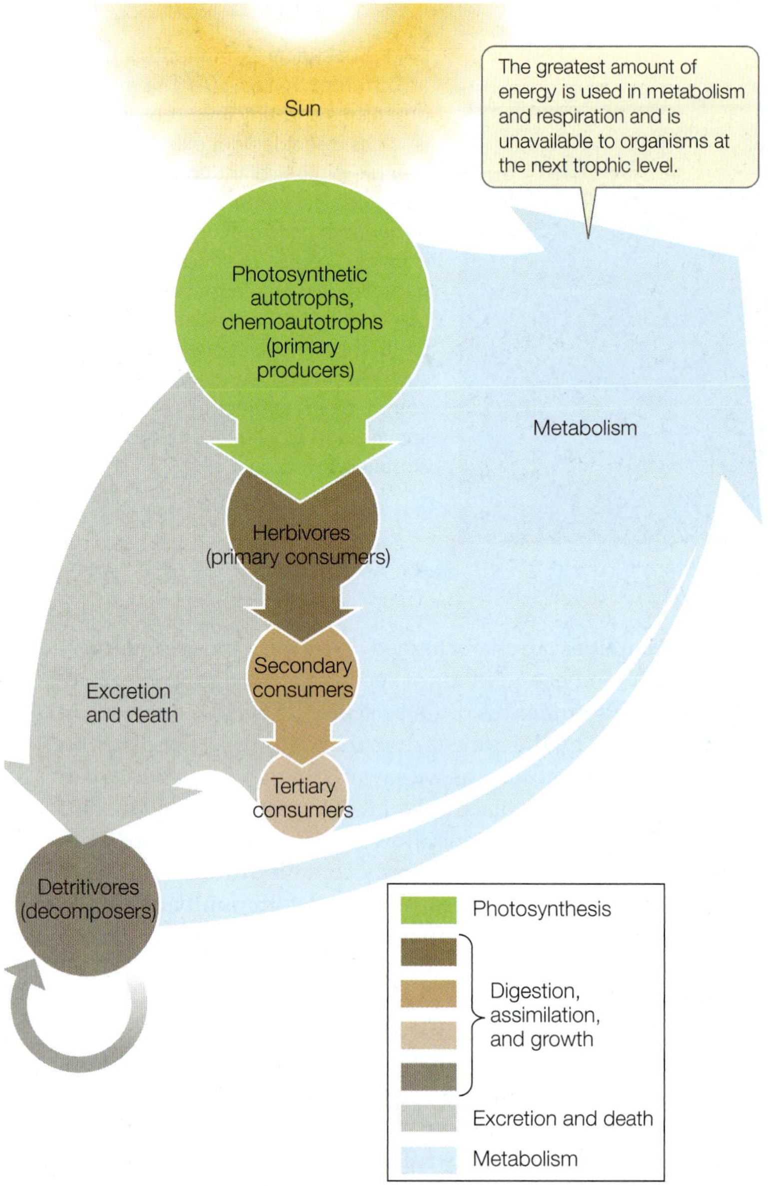

Sun
The greatest amount of energy is used in metabolism and respiration and is unavailable to organisms at the next trophic level.
Photosynthetic autotrophs, chemoautotrophs (primary producers)
Metabolism
Herbivores (primary consumers)
Secondary consumers
Excretion and death
Tertiary consumers
Detritivores (decomposers)
Photosynthesis
Digestion, assimilation, and growth
Excretion and death
Metabolism

57.2 Energy Flow through Trophic Levels Much of the energy accumulated at each trophic level is lost (often as heat) to metabolism and respiration by the organisms at that level. In this diagram, the width of each arrow is roughly proportional to the amount of energy flowing through that channel. Arrows indicate directions of energy flow.

Go to Activity 57.1
Energy Flow through an Ecological Community
Life10e.com/ac57.1

primary production. (The terms "productivity" and "production" are often used interchangeably: "productivity" is the rate of energy accumulation; "production" is a measure of accumulated energy as a product.)

Not all GPP accumulated by primary producers becomes available to heterotrophs because primary producers use some of that energy for their own respiration and other metabolic processes. **Net primary productivity (NPP)** is the rate at which energy is incorporated into primary producers' bodies through growth and reproduction. Thus net primary production can be measured as the amount of primary producer biomass (the weight of organic matter) that is available for consumption by heterotrophs. This relationship is described mathematically as

$$NPP = GPP - R$$

where R is the energy lost through respiration. These relationships are represented in a highly simplified form in Figure 57.2.

Consumers use diverse sources of energy

A **food chain** is a diagram that depicts the linear sequence of who eats whom in a given community. Food chains can be interwoven into a more realistic depiction of community feeding relationships, called a **food web** (Figure 57.3). Most communities contain so many species interacting in so many different ways that it is impossible to enumerate (or even identify) all of the links. Nevertheless, simplified food webs are useful in envisioning the sequence of energy flow through a community. An organism's **trophic level** indicates where in that sequence it obtains its energy (Table 57.1). Primary producers start the chain of trophic levels. At the next level are **primary consumers**—the herbivores that dine on the primary producers. Organisms that eat herbivores, called **secondary consumers**, are the next trophic level. Those that eat secondary consumers are tertiary consumers, and so on. The waste products and dead bodies of organisms (known as **detritus**) provide another source of energy, as we saw at the opening of this chapter. Organisms that consume such materials are called **detritivores** or **decomposers** (see Figure 57.2).

Go to Media Clip 57.1
A Food Chain in Africa
Life10e.com/mc57.1

Most species in a community eat and are eaten by more than one other species, so food webs and trophic levels are necessarily oversimplified. Some organisms, called **omnivores**, feed on multiple trophic levels. Opossums, for example, are famously omnivorous. Investigators in Portland, Oregon, dissected the stomachs of road-killed opossums and found remains of mammals, birds, insects, earthworms, snails, fruits, bulbs, seeds, leaves, grass, pet food, and garbage, along with some items they couldn't identify—a diet that would be difficult to depict in a food web.

Fewer individuals and less biomass can be supported at higher trophic levels

One important way to characterize a community is by the distribution of energy and biomass within it. The flow of energy through a community is governed by the physical laws that regulate energy transformations, foremost among which are the first and second laws of thermodynamics. Recall from Section 8.1 that the amount of energy in the universe is constant, and that when energy is converted from one form to another, part of it becomes unavailable to do work. As Figure 57.2 showed, energy is lost to metabolism and respiration at each trophic level.

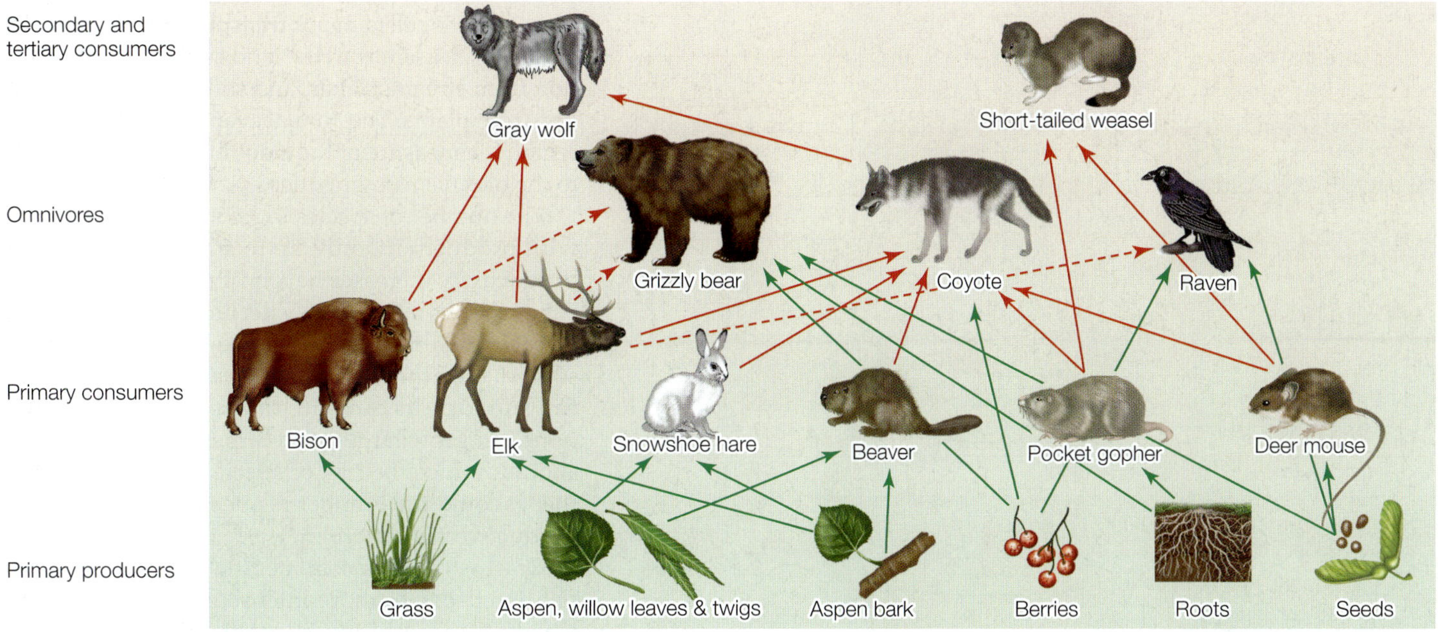
Secondary and tertiary consumers
Omnivores
Primary consumers
Primary producers
Gray wolf
Short-tailed weasel
Grizzly bear
Coyote
Raven
Bison
Elk
Snowshoe hare
Beaver
Pocket gopher
Deer mouse
Grass
Aspen, willow leaves & twigs
Aspen bark
Berries
Roots
Seeds

57.3 Food Webs Show Trophic Interactions in a Community
This simplified food web for the grasslands of Yellowstone National Park includes only large vertebrates and the plants on which they depend. The arrows show who eats whom. Species whose sole source of food is plants (green arrows) are primary consumers. Carnivores that kill and eat animals (red arrows) are secondary and tertiary consumers. Omnivores such as grizzly bears, coyotes, and ravens eat both plant and animal tissues; ravens and grizzlies also eat carrion (dashed red arrows), so these species are also detritivores.

Go to Activity 57.2 The Major Trophic Levels
Life10e.com/ac57.2

On average, only about 10 percent of the energy of one trophic level is transferred to the next, for a number of reasons:

- *Heat loss.* Organisms incorporate much of the energy they accumulate into biomass, but they use much more of it for respiration and other metabolic processes. That energy is dissipated as heat and is lost to the community.
- *Biomass availability.* Not all of the biomass in a community can be ingested. Grazers routinely miss blades of grass; effective plant defenses prevent herbivory; prey can escape predators or leave the community.
- *Indigestibility.* Not all of the biomass ingested can be assimilated by consumers. Tree bark, for example, contains lignin and cellulose, which cannot be digested by most herbivores.

The overall transfer of energy from one trophic level to the next (which can be expressed as the ratio of consumer production to producer production) is called **ecological efficiency**.

Pyramid diagrams such as those in **Figure 57.4A** can be used to illustrate the proportions of energy transferred from each trophic level to the next and to compare those proportions among different communities. Pyramid diagrams can also be used to illustrate the amount of biomass or numbers of individuals found at each trophic level (**Figure 57.4B**). As these diagrams show, a given environment typically supports fewer individuals, less biomass, and fewer species at higher trophic levels than at lower trophic levels. Progressive energy loss through the inefficiencies of energy transfer also limits the number of trophic levels in a food chain or food web; largely for this reason, most communities support only three to five trophic levels. One conspicuous exception to this general pattern occurs in the open oceans. The phytoplankton that are the primary producers

TABLE 57.1
The Major Trophic Levels

Trophic Level	Source of Energy	Examples
Photosynthesizers (primary producers)	Solar energy	Green plants, photosynthetic bacteria, diatoms
Herbivores (primary consumers)	Tissues of primary producers	Copepods, grasshoppers, bark beetles, deer, geese, white-footed mice
Primary carnivores (secondary consumers)	Tissues of herbivores	Spiders, warblers, wolves, anchovies
Secondary carnivores (tertiary consumers)	Tissues of primary carnivores	Tuna, falcons, killer whales
Omnivores	Several trophic levels	Humans, opossums, crabs, robins
Decomposers (detritivores)	Dead tissues and waste products of other organisms	Many fungi, many bacteria, vultures, earthworms, termites

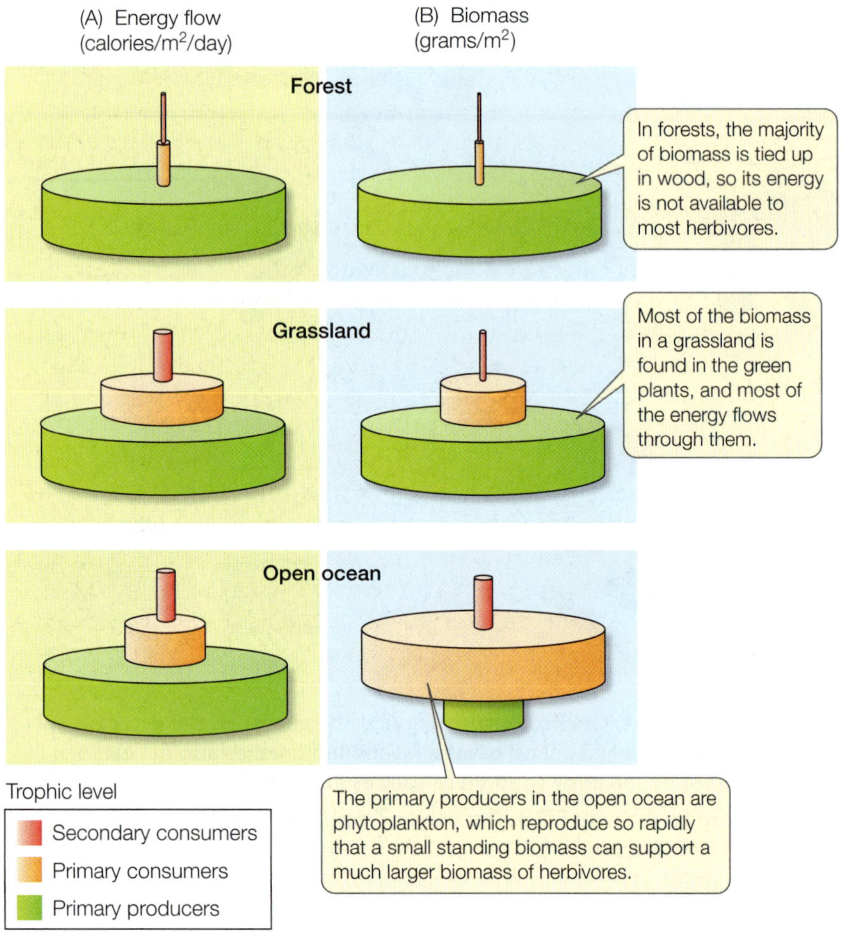
(A) Energy flow
(calories/m²/day)
(B) Biomass
(grams/m²)
Forest
In forests, the majority of biomass is tied up in wood, so its energy is not available to most herbivores.
Grassland
Most of the biomass in a grassland is found in the green plants, and most of the energy flows through them.
Open ocean
Trophic level
Secondary consumers
Primary consumers
Primary producers
The primary producers in the open ocean are phytoplankton, which reproduce so rapidly that a small standing biomass can support a much larger biomass of herbivores.

57.4 Energy and Biomass Distributions Pyramid diagrams allow ecologists to compare patterns of energy flow through trophic levels (A) and the amount of biomass present at different trophic levels in different communities (B). Biomass distribution in the open ocean differs ftrom that in most other communities because most of the biomass is not at the primary producer level.

in this community grow and reproduce so much faster than the zooplankton and small fish that consume them that their smaller biomass, with its rapid rate of primary production, can actually support a larger biomass of primary consumers.

Productivity and species diversity are linked

Just as the diversity of a trophic level tends to be positively correlated with the amount of energy available to it, the species diversity of a community tends to be positively correlated with its productivity, up to a point (**Figure 57.5**). A number of factors that influence productivity vary among communities. The most obvious of these factors is energy input. The amount of solar energy reaching Earth's surface varies by latitude, as we saw in Figure 54.1. The ability of plants to photosynthesize, however, depends not only on the supply of energy from the sun but also on the supply of water and nutrients. The number of tree species across different regions of North America, for example, can be best predicted not by measuring incoming solar radiation but by measuring an ecological attribute of those regions called **evapotranspiration**: the amount of water released from the land surface by evaporation from streams, lakes, and soil and by transpiration from plants. The annual evapotranspiration of a region is a measure of the amount of water available to the organisms living there.

The number of species in a community increases with productivity only up to a point, however. If productivity increases beyond that point, the number of species may actually decline (see Figure 57.5). Why should that be? As local productivity increases, so does the number of individuals the local habitat can support (its carrying capacity). Thus populations can grow larger, and larger population sizes should reduce the risk of species extinction. Why, then, should the number of species decrease when productivity is very high?

One hypothesis postulates that interspecific competition becomes more intense when productivity is very high, resulting in competitive exclusion of some species (see Section 56.4). This hypothesis is supported by the results of a long-term experiment at the Rothamsted Experiment Station in England, begun in 1857 and still going on. Fertilizer has been added regularly to selected plots of land to increase their productivity, and fertilized and unfertilized plots have been monitored continuously. Over 150 years, the number of plant species in the unfertilized plots has remained roughly constant, whereas the number of species in the fertilized plots has declined, supporting the premise that species diversity can decline when productivity rises.

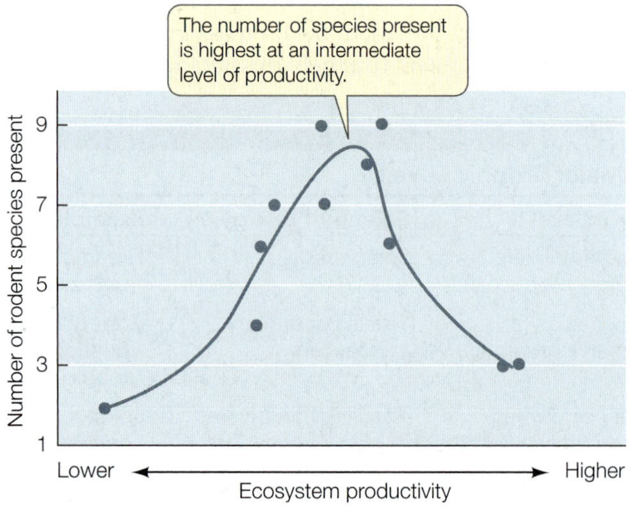
The number of species present is highest at an intermediate level of productivity.
Number of rodent species present
9
7
5
3
1
Lower
Higher
Ecosystem productivity

57.5 Species Diversity Peaks at Intermediate Productivity The number of rodent species living in ecosystems of varying productivity in the Gobi Desert exemplifies a pattern in which species richness increases only up to a certain point. Beyond that point, it can actually decline with productivity.

RECAP 57.1

An ecological community is a group of species that coexist and interact within a defined area. Energy enters an ecological community through the primary producers and moves through the food web by means of trophic interactions.

- Explain the difference between gross primary productivity and net primary productivity. **See pp. 1189–1190 and Figure 57.2**
- Describe three trophic levels that might appear in a food web. **See pp. 1190–1191, Table 57.1, and Figure 57.3**
- What is a typical distribution of energy among the trophic levels of a community? **See pp. 1190–1191 and Figures 57.2 and 57.4**

Energy and biomass distributions primarily reflect the abiotic factors that influence the community. In the next section we will see how the interactions among species described in Chapter 56 shape community structure.

57.2 How Do Interactions among Species Influence Communities?

The antagonistic interactions described in Section 56.2, which transfer energy between trophic levels, have a strong influence on ecological communities. But species are not identical bags of biomass through which energy flows. Species in one part of a food web can affect many other species without necessarily eating them.

Species interactions can cause trophic cascades

The interactions of a single consumer species with other species in its community can cause a progression of successive effects throughout an entire food web, a pattern called a **trophic cascade**. The reintroduction of wolves into Lamar Valley in Yellowstone National Park initiated just such a pattern.

The food web in the grasslands of Yellowstone National Park is far more complex than the "streamlined" rendition in Figure 57.3 indicates. Gray wolves in the park feed on elk, bison, and coyotes. Although they share some of these prey species with coyotes and grizzly bears, wolves exert particularly strong effects on the park community's structure and dynamics, as demonstrated by the effects of their absence during most of the twentieth century.

By 1926, unrestricted hunting had eliminated wolves from the Yellowstone community. To prevent elk from exceeding the park's carrying capacity, the park service culled elk herds (that is, they selectively killed some members of each herd) until 1968, when, in response to public pressure, the culls were stopped and the elk population rapidly increased (**Figure 57.6A**). The elk browsed aspen trees so intensely that the number of young trees recruited (added to the population) declined precipitously, and by 1960 no new trees were recruited at all (**Figure 57.6B**). The elk also severely browsed streamside willows, with the result that beavers, which depend on willows for food, were nearly exterminated from Lamar Valley. In regions of the park where elk were absent, however, aspen and willow trees flourished. This observation suggested that the decline of the trees in Lamar Valley was

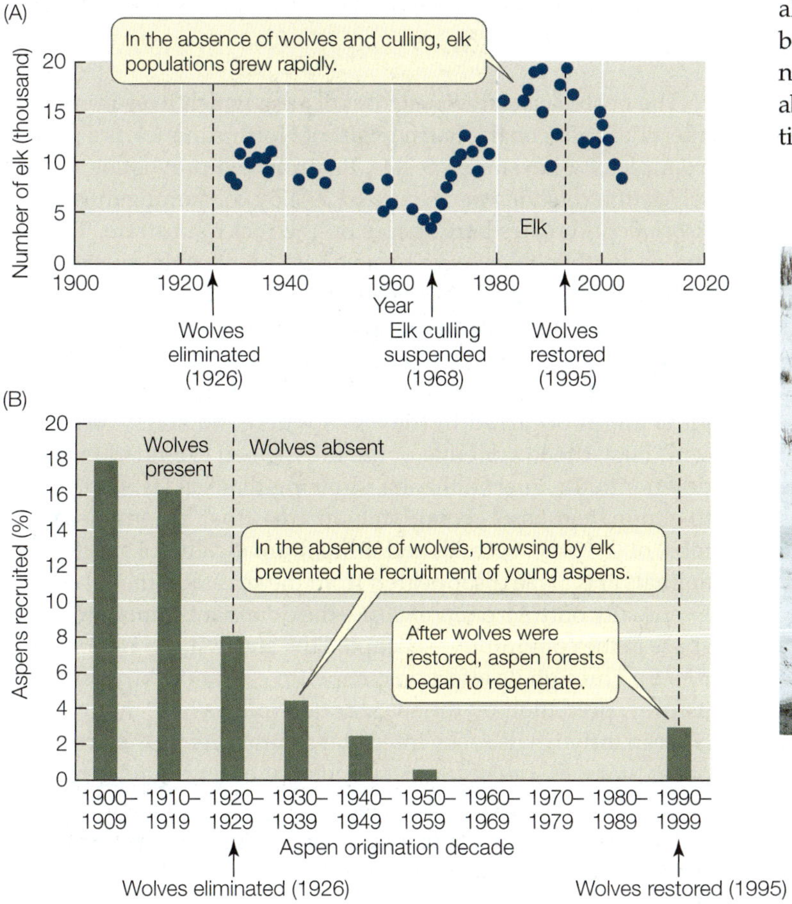

(A)
In the absence of wolves and culling, elk populations grew rapidly.
Number of elk (thousand)
0
5
10
15
20
Elk
1900
1920
1940
1960
1980
2000
2020
Year
Wolves eliminated (1926)
Elk culling suspended (1968)
Wolves restored (1995)
(B)
Wolves present
Wolves absent
In the absence of wolves, browsing by elk prevented the recruitment of young aspens.
After wolves were restored, aspen forests began to regenerate.
Aspens recruited (%)
0
2
4
6
8
10
12
14
16
18
20
1900–1909
1910–1919
1920–1929
1930–1939
1940–1949
1950–1959
1960–1969
1970–1979
1980–1989
1990–1999
Aspen origination decade
Wolves eliminated (1926)
Wolves restored (1995)

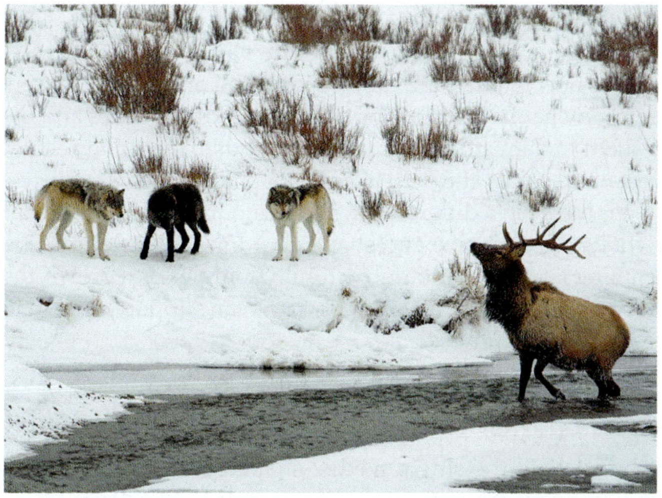

57.6 Wolves Initiated a Trophic Cascade in Yellowstone National Park (A) Number of elk in Wyoming's Yellowstone National Park. (B) Aspen recruitment (that is, the number of new trees becoming established) in the presence and absence of wolves.

indeed due to elk browsing rather than to climate conditions or some other factor.

In 1995, after wolves had been absent for 70 years, park managers reintroduced them to Yellowstone, and their population grew rapidly. The wolves preyed primarily on elk. The elk population of Lamar Valley dropped, and elk avoided the aspen groves, where they were especially vulnerable to wolf predation. Young aspen began to grow, willows regrew along streams, and the number of beaver colonies increased from one in 1996 to seven in 2003. Thus the presence or absence of a single predator influenced not only populations of its prey but also populations of its prey's food resource and of other species that depended on that resource.

Herbivores, too, can have indirect effects on other trophic levels. The savannas of central Kenya are dominated by large grazing mammals such as zebras, eland, elephants, Grant's gazelles, giraffes, buffaloes, and hartebeests. A team of investigators used exclosures (areas protected by a barrier, such as a fence, designed to keep organisms out) to investigate the influence of these grazers on the savanna community. They created six exclosures and compared community structure within the exclosures with that in control sites where large mammals could graze freely. Over 19 months of monitoring, they found that trees, beetles, and an insectivorous lizard species were all more abundant within the exclosures. The elimination of grazing increased the abundance of trees, which in turn increased the number of beetles by providing more food and habitat, which in turn increased the number of lizards, which feed preferentially on beetles.

Trophic cascades can be initiated by the behaviors of species as well as by their diets. Beavers preferentially cut down some species of trees to build their dams; by so doing, they alter the composition of the vegetation. In addition, the beavers' activities create wetlands, meadows, and ponds that provide habitat for species that would otherwise not be able to live in the area.

Organisms that build structures that alter existing habitats or create new habitats are called **ecosystem engineers**. Beavers create new habitats by cutting down (and killing) trees and using them to dam streams and create ponds. Trees in terrestrial forests, and corals and kelp in aquatic communities, are ecosystem engineers in a different way: they modify the environment by changing their size and shape over time. For example, by becoming larger and more structurally complex as it grows, the stilt palm (*Socratea exorrhiza*) of Panama can support more than 60 species of epiphytes.

Keystone species have disproportionate effects on their communities

Architects call the single wedge-shaped stone in the center of an archway the "keystone" because it holds all the other stones in place. Ecologists thus refer to any species that exerts an influence on the other members of its community that is disproportionate to its abundance as a **keystone species**. Keystone species can influence both the number of species and the number of trophic levels in a community.

Pisaster ochraceus
Mytilus californianus
Eliminating *P. ochraceus* undermined the community's diversity, leaving only *M. californianus* to occupy the substrate.
Number of species
25
20
15
10
5
Pisaster not removed
Pisaster removed
1962
1964
1966
1968
1970
1972
1974
Year

57.7 Ochre Sea Stars Are a Keystone Species (A) Along the Pacific coast of northern North America, ochre sea stars (*Pisaster ochraceus*) consume large quantities of the mussel *Mytilus californianus*, creating bare substrate that is then inhabited by a variety of intertidal species. (B) Experiments by Robert Paine and his colleagues demonstrated that when sea stars were excluded from this intertidal community, *M. californianus* outcompeted all other species for space on the substrate and eliminated the community's diversity.

The ochre sea star (*Pisaster ochraceus*), which lives in rocky intertidal zones on the Pacific coast of North America, is a good example of a keystone species. Its preferred prey is the mussel *Mytilus californianus* (**Figure 57.7A**) By consuming mussels, *P. ochraceus* creates bare spaces on the rocky substrate. These spaces are then taken over by a variety of intertidal species.

In a classic experiment, Robert Paine of the University of Washington demonstrated the disproportionate influence of *P. ochraceus* on species diversity by removing the sea stars from selected sites repeatedly over a 5-year period. Two major changes occurred in the areas where sea stars were absent. First, the lower edge of the mussel bed extended farther down into the intertidal zone, showing that sea stars are able to eliminate mussels completely in areas that are submerged most of the time. Second, and more dramatically, 18 species of animals and algae disappeared from the sea star removal sites. Eventually only *M. californianus*—the dominant competitor for space in the community—occupied the entire substrate (**Figure 57.7B**). Through its effect on competitive relationships among its prey, predation by the sea star determines how many species can thrive in its rocky intertidal community. In the absence of sea stars, *M. californianus* crowds out other organisms in a broad belt of the intertidal zone.

Species other than consumers can be keystone species. Fig trees in tropical forests, for example, produce fruits several times every year, so their fruits are abundant at times when few, if any, other trees are fruiting. Dozens of frugivores depend on figs when no other fruits are present. Fig-eating animals include fruit bats, parrots, toucans, pigeons, flycatchers, trogons, orioles, rodents, howler monkeys, and even fish, which eat figs that fall into nearby streams. All of these animals provide prey for a diverse community of predators. Moreover, the trunks of fig trees provide habitat for several thousand species of insects, reptiles, rodents, and birds. Without fig trees, rainforest communities around the world would be profoundly diminished in terms of the number of species they contain.

RECAP 57.2

Some interactions between consumers and their resource species result in a trophic cascade of indirect effects on species at other trophic levels. A keystone species has effects on its community that are disproportionate to its abundance.

- Describe an example of a consumer whose interactions with other species cause indirect effects across trophic levels. See p. 1194
- What are some of the ways in which keystone species can affect other species in their communities? See p. 1194

We have seen how certain keystone species influence the species composition of their communities. But a simple count of species is only one measure of community diversity. The next section will look more closely at the different ways in which ecologists measure species diversity.

Community A

Community B

57.8 Species Richness and Species Evenness Both Contribute to Diversity These two hypothetical mushroom communities are the same size (12 individuals) and have the same species richness (four species), but community A has a more even distribution of species and is thus more diverse than community B.

57.3 What Patterns of Species Diversity Have Ecologists Observed?

Communities clearly vary in their diversity, both geographically, on scales ranging from local to global, and over time, on scales ranging from a day to centuries. Comparing the diversity of two or more communities can be challenging because diversity has many different components depending on the scale at which it is measured. In some cases it is important to measure species diversity within a single community or habitat, but it can also be useful to measure diversity across an entire region, encompassing a range of communities or habitats.

Diversity comprises both the number and the relative abundance of species

The most straightforward way to quantify the diversity of a community is to count the number of species present in a sample. This number is the **species richness** of the community. The larger the area that is sampled, the greater the likelihood that rare species in the community will be found, and the more accurate the resulting assessment of its species richness will be.

To say that two communities of the same size have the same species richness tells only part of the story, however. How abundant each species is in the community also affects diversity. This aspect of diversity—the distribution of abundances of individuals across species—is called **species evenness**. Imagine that we take samples of 12 individuals in each of two different communities. Our sample from community A contains 3 individuals of each of four species (an even distribution of individuals). Our sample from community B, however, contains 9 individuals of one species and only 1 individual of each of the other three species (an uneven distribution). Even though the species richness of the two communities is the same (four), community B is less diverse because the less abundant species are encountered infrequently compared with the single most abundant species (**Figure 57.8**). Thus estimates of diversity should account for both species richness and species evenness.

A study of diversity patterns in an agricultural region of southern England demonstrates how measures of diversity can be applied at different spatial scales. Investigators sampled plants and macroinvertebrates in three types of freshwater communities—rivers, ponds, and ditches—in this area of English countryside. Rivers had high species richness, but most rivers in the area contained the same assortment of species. In contrast, the ponds displayed a wide range of variation in species richness. Thus the ponds contributed more to the overall diversity of the area than the rivers did because they contained

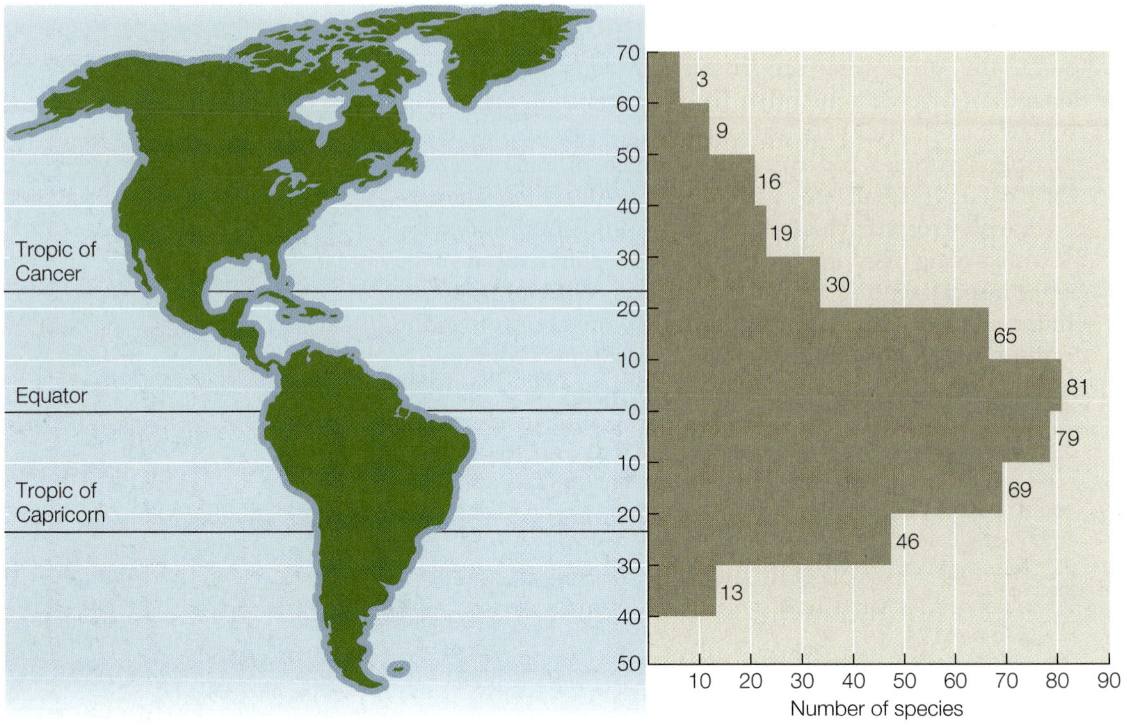
Tropic of Cancer
Equator
Tropic of Capricorn
70
60
50
40
30
20
10
0
10
20
30
40
50
3
9
16
19
30
65
81
79
69
46
13
10 20 30 40 50 60 70 80 90
Number of species

57.9 Latitudinal Gradients in Diversity Among swallowtail butterflies (Papilionidae), species diversity decreases with latitude both north and south of the equator. Similar latitudinal gradients of diversity have been observed in many other taxa.

more unique or rare species. Surprisingly, the ditches, many of which contained water for only a short time, had the lowest species richness of the three community types, but many of the species in the ditches were found nowhere else in the area, including insects that live only in temporary bodies of water (including a very rare water beetle). Ponds and ditches, despite being relatively species-poor, contributed disproportionately to regional diversity because the few species they did support were found nowhere else in the region. Partitioning diversity within a community, between communities, and across an entire region in this way can provide insights into the ecological characteristics of the species making up those communities as well as the processes by which those communities were assembled.

Ecologists have observed latitudinal gradients in diversity

Diversity can be measured at a wide range of scales, but the broader the scale at which it is measured, the more difficult it can be to understand the ecological process underlying the differences observed. Understanding diversity at a global scale has proved to be a challenge.

About 200 years ago, the German explorer and naturalist Alexander von Humboldt spent 5 years traveling around Latin America. He remarked in the account of his voyages that "the nearer we approach the tropics, the greater the increase in the variety of structure, grace of form, and mixture of colors, as also in perpetual youth and vigour of organic life." Humboldt would not have been surprised to learn that, if he had sailed toward the poles, the diversity he observed would have decreased. These latitudinal gradients in diversity have been observed repeatedly in both hemispheres and in a wide variety of taxa, including birds, mammals, flowering plants, and insects (**Figure 57.9**).

Although most ecologists agree that latitudinal gradients in diversity exist, there is less consensus as to why they exist. At least four hypotheses have been advanced to account for latitudinal gradients in diversity:

- The *time hypothesis* argues that over evolutionary time, organisms in tropical regions have had more time to diversify under relatively stable climate conditions than have those in more temperate regions.
- The *spatial heterogeneity hypothesis* suggests that tropical regions have higher spatial heterogeneity—more different types of microclimates, vegetation, soils, and so forth—and thus contain more distinct habitats and many more species.
- The *specialization hypothesis* attributes latitudinal gradients to greater interspecific competition in the tropics, which leads to narrower realized niches (see Figure 56.15).
- The *predation hypothesis* proposes that predation intensity is greater in the tropics. Where predation is high, it argues, prey populations are held to levels so low that interspecific competition never comes into play, and rare species can persist.

Corroborative evidence can be found for each of these hypotheses, varying with taxon, locality, and scale, and not all of the hypotheses are mutually exclusive. It may be that multiple factors are responsible for this widespread ecological pattern.

The theory of island biogeography suggests that immigration and extinction rates determine diversity on islands

While latitudinal diversity gradients prevail on a global scale, other factors influence species diversity at smaller spatial scales. For example, small islands tend to have fewer species

WORKING WITH**DATA:**

Latitudinal Gradients in Pitcher Plant Communities

Original Paper

Buckley, H. L., T. E. Miller, A. Ellison, and N. J. Gotelli. 2003. Reverse latitudinal trend in the species richness of an entire community at two spatial scales. *Ecology Letters* 6: 825–829.

Analyze the Data

In 2003 Hannah Buckley and several colleagues examined the diversity of species present in the community of water-filled leaves—pitchers—of *Sarracenia purpurea* (see Figure 57.1A). The researchers determined the relative abundances of all species of invertebrates (including insect larvae), heterotrophic protists (protozoa), and bacteria present in each of 20 pitcher plants collected at each of 39 sites that spanned the plant's entire north–south range. They summarized the data at two scales: as the average number of species per site (the site-wide scale) and as the average number of species per pitcher at each site (the pitcher scale). Their data are plotted in the graphs at the right.

QUESTION 1

Overall, does this community display a typical diversity gradient with latitude? Of the individual taxa, which ones depart from the typical diversity gradient?

QUESTION 2

The top predator in this system is the larva of the mosquito *Wyeomyia smithii*, which filter-feeds on bacteria and protozoans. This species is significantly more abundant at low latitudes (i.e., farther south in the Northern Hemisphere) than at high latitudes. How might the abundance of a top predator explain the pattern observed by these investigators?

QUESTION 3

Do you think *Wyeomyia smithii* is a keystone species? Why or why not?

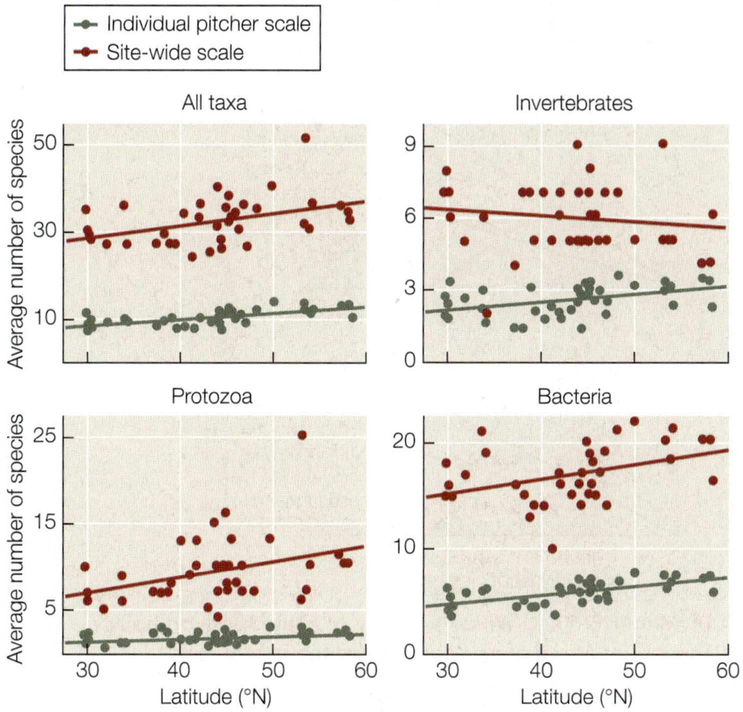
Individual pitcher scale
Site-wide scale
All taxa
Invertebrates
Protozoa
Bacteria
Average number of species
Average number of species
50
30
10
9
6
3
0
25
15
5
20
10
0
30
40
50
60
30
40
50
60
Latitude (°N)
Latitude (°N)

Go to **BioPortal** for all WORKING WITH**DATA** exercises

than large islands, irrespective of latitude (**Figure 57.10**). The biologist Edward O. Wilson was struck by this **species–area relationship**, which he encountered through his exhaustive collection of ant species from all over the world.

With Robert MacArthur, Wilson developed the theory of **island biogeography**. They based their theory on just two processes: the immigration of new species to an island and the extinction of species already present on that island (**Figure 57.11A**). The premise of island biogeography is that the number of species on an island represents a balance, or equilibrium, between the rate at which species immigrate to and colonize the island and the rate at which resident species go locally extinct.

The rate of immigration is determined in part by the number of species in the source area providing the immigrants, known as the **species pool**. In the case of oceanic islands, the species pool comprises all the species on the nearest land mass (usually a continent). Not all species that reach the island will persist there, however. The more species there are on an island, the greater the likelihood that there will be competition for limited resources, and the higher the likelihood that any of the species on the island will go extinct. Therefore at some point the number of species arriving from

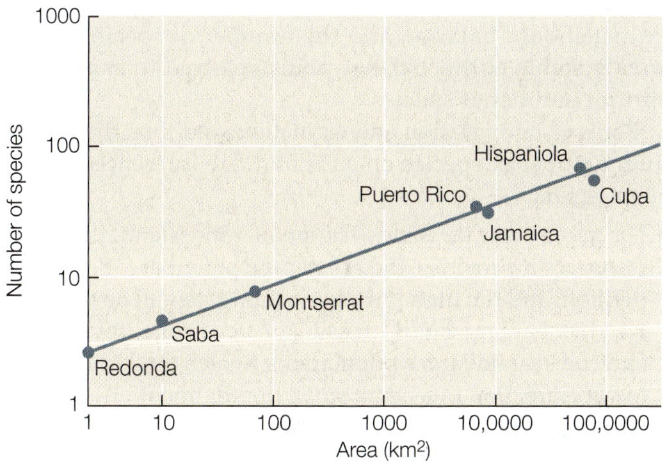
1000
100
10
1
Number of species
Hispaniola
Puerto Rico
Cuba
Jamaica
Montserrat
Saba
Redonda
1
10
100
1000
10,0000
100,0000
Area (km2)

57.10 The Species–Area Relationship E. O. Wilson and Robert MacArthur plotted the number of species of reptiles and amphibians against the size of several islands in the Antilles of the Caribbean. Larger islands consistently contained more species, regardless of latitude, a fact these scientists incorporated in their theory of island biogeography.

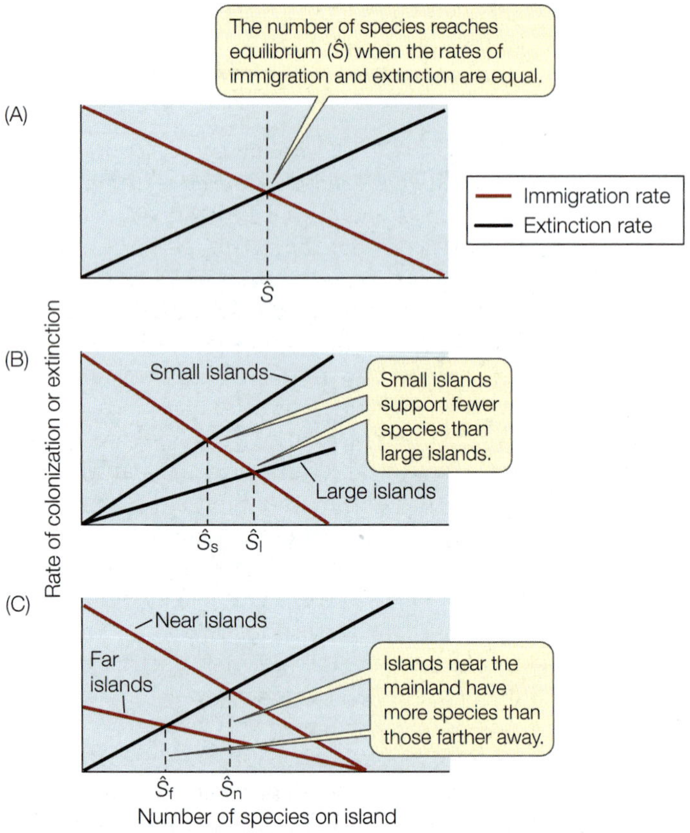

The number of species reaches equilibrium (Ŝ) when the rates of immigration and extinction are equal.
(A)
Immigration rate
Extinction rate
Ŝ
Rate of colonization or extinction
(B)
Small islands
Small islands support fewer species than large islands.
Large islands
Ŝs
Ŝl
(C)
Near islands
Far islands
Islands near the mainland have more species than those farther away.
Ŝf
Ŝn
Number of species on island

57.11 MacArthur and Wilson's Theory of Island Biogeography (A) The rate of arrival of new species and the rate of extinction of species already present determine the equilibrium number of species on an island. These rates and the eventual equilibrium numbers are affected by the size of the island (B) and by the island's distance from the mainland (C). For simplicity these rates are depicted as linear, but in reality immigration and extinction rates tend to be curvilinear.

the source area and the number of resident species going extinct should balance, and the number of species should remain stable at this balance point, referred to as the equilibrium number of species.

Rates of immigration and extinction, and thus the equilibrium number of species on an island, are influenced by two other factors:

- *The size (area) of the island.* The smaller the island, the fewer resources it provides, the greater the potential for competition, and the higher the extinction rate will be (**Figure 57.11B**). Larger islands provide greater habitat diversity and can sustain larger populations (which tend to have lower extinction rates than small populations).
- *Distance of the island from the species pool.* The farther the island is from the source of immigrants, the lower the immigration rate—the rate at which new species arrive—will be (**Figure 57.11C**).

Between 1966 and 1969, Wilson and his student Daniel Simberloff conducted an ingenious experiment to test the theory of island biogeography (although given concerns over environmental impacts, this experiment might not have been approved today). Simberloff and Wilson identified four small, isolated clumps of red mangrove (*Rhizophora mangle*), all approximately the same size (11–12 meters in diameter), in the Florida Keys. These mangrove islands were small enough to allow an accurate count of the arthropod species on each one. They were also small enough for the research team to enclose each island in a tent and gas it with methyl bromide to kill all the arthropods. After this defaunation, Simberloff and Wilson monitored and tracked recolonization of the islands by arthropods (**Figure 57.12**). After 2 years, species richness on all but the farthest island was close to what it was before defaunation. This observation is consistent with the idea that the number of species on the islands prior to defaunation represented an equilibrium number of species.

The theory of island biogeography can be applied equally well to **habitat islands**—isolated patches of suitable habitat surrounded by extensive areas of unsuitable habitat. Thus a pond in the English countryside or a forest surrounded by housing subdivisions may acquire an equilibrium number of species in much the same way an oceanic island does. The theory of island biogeography also has important applications for the conservation of endangered species. As habitat islands decrease in size because of human encroachment, more and more species become vulnerable to population declines, especially those that require large areas in order to live and breed successfully. Even more broadly, the processes of immigration and extinction contribute to determining community composition at continental scales.

Go to Animated Tutorial 57.1
Biogeography Simulation
Life10e.com/at57.1

RECAP 57.3

Measures of species diversity encompass both species richness and species evenness. Species diversity is highest in the tropics, decreasing at higher latitudes. Island biogeography theory states that species diversity on an island or other isolated habitat represents a balance between immigration and extinction rates.

- What is the difference between species richness and species evenness? **See p. 1195 and Figure 57.8**
- Explain how scale can influence estimates of diversity. **See pp. 1195–1196**
- What factors influence the equilibrium number of species on an island, according to the theory of island biogeography? **See pp. 1197–1198 and Figure 57.11**

The composition of a community is dynamic: as we have seen, it can change over time and space. Processes of change such as immigration to and extinction on islands are generally predictable and consistent. But community composition can change dramatically in response to less predictable forces as well.

INVESTIGATING**LIFE**

57.12 The Theory of Island Biogeography Can Be Tested

By experimentally removing all the arthropods on four small mangrove islands of equal size but different distance from the mainland, Simberloff and Wilson were able to observe the process of recolonization and compare the results with the predictions of island biogeography theory.

HYPOTHESIS If mangrove islands are populated by an equilibrium number of species, then the rate at which they will accumulate species after defaunation will decrease with distance from a mainland source of colonists, as will their eventual species richness.

Method

1. Census the terrestrial arthropods on 4 small mangrove islands of equal size (11–12 m diameter) but different distance from a mainland source of colonists.
2. Erect scaffolding and tent the islands. Fumigate with methyl bromide (a chemical that kills arthropods but does not harm plants).

3. Remove tenting. Monitor recolonization for the following 2 years, periodically censusing terrestrial arthropod species.

Results Recolonization was fastest on the closer islands, slowest on the one farthest from the mainland. Two years after defaunation, each island had about the same number of species it had before the experiment.

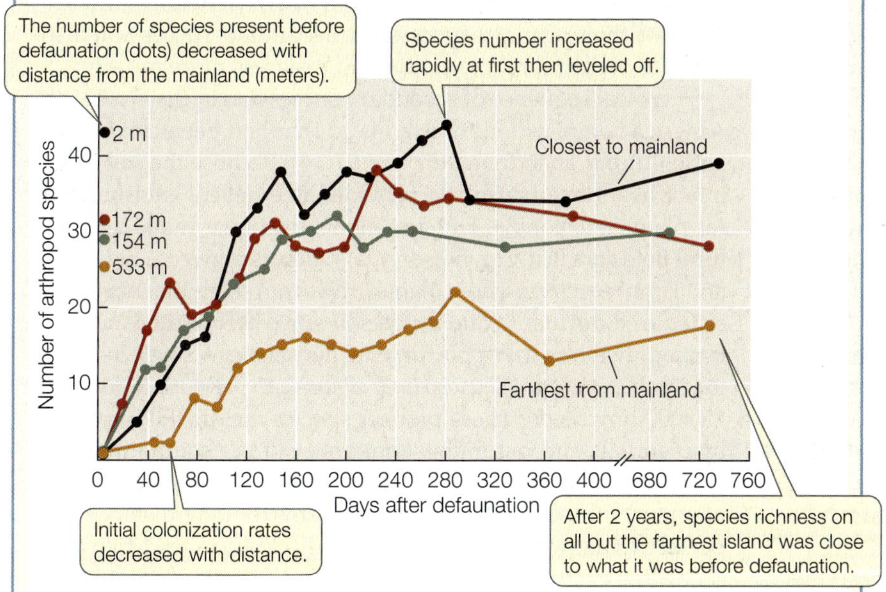
The number of species present before defaunation (dots) decreased with distance from the mainland (meters).
Species number increased rapidly at first then leveled off.
2 m
Closest to mainland
172 m
154 m
533 m
Number of arthropod species
40
30
20
10
Farthest from mainland
0
40
80
120
160
200
240
280
320
360
400
680
720
760
Days after defaunation
Initial colonization rates decreased with distance.
After 2 years, species richness on all but the farthest island was close to what it was before defaunation.

CONCLUSION The data support the theory that species richness on islands represents a dynamic balance between colonization and extinction rates.

Go to **BioPortal** for discussion and relevant links for all INVESTIGATING**LIFE** figures.

Simberloff, D. S. and E. O. Wilson 1970. *Ecology* 51: 934–937

57.4 How Do Disturbances Affect Ecological Communities?

An ecological **disturbance** is a disruption in a community caused by a discrete external force, often abiotic in nature. Disturbances may remove some species from a community but may open up space and resources for other species.

The magnitude of a disturbance's effects varies enormously. Some disturbances are limited to small areas—for example, a log carried by waves may crush algae and animals attached to rocks in an intertidal community. In contrast, hurricanes, forest fires, and volcanic eruptions affect communities over hundreds or thousands of hectares. Although small-scale disturbances are far more frequent, a few large-scale events may be responsible for most of the changes in a community. A single hurricane, for example, may fell more trees than several years of "normal" storms, and the movements of a glacier change community composition across millennia.

A community's history of disturbance may explain patterns of species diversity that would otherwise be puzzling. The Province Islands, located just above the Antarctic Circle in the South Indian Ocean, provide an example. The climate of these islands is cool, with average temperatures above freezing for only 6 months of the year; precipitation is high, and gale force winds are common. The vegetation is primarily tundra. South African entomologist S. L. Chown, an expert on life in the Antarctic, compared the insect faunas on the four largest of these islands and found that the two largest, Marion and Kerguelen, housed fewer arthropod species (16 and 22 species, respectively) than the significantly smaller Cochons and Possessions (26 and 38 species). Why do the species diversity patterns on these islands fail to conform to the theory of island biogeography? One possible explanation lies in their different disturbance histories. The two largest islands were once covered by glaciers and thus experienced considerably more disturbance than did the two smaller islands.

Succession is the predictable pattern of change in a community after a disturbance

How does a community recover from a disturbance, particularly one as massive as a glacier? The pattern of change in community composition following a disturbance is known as **succession**. The most common type of succession is **directional succession**, which is characterized by an orderly (or at least predictable) progression

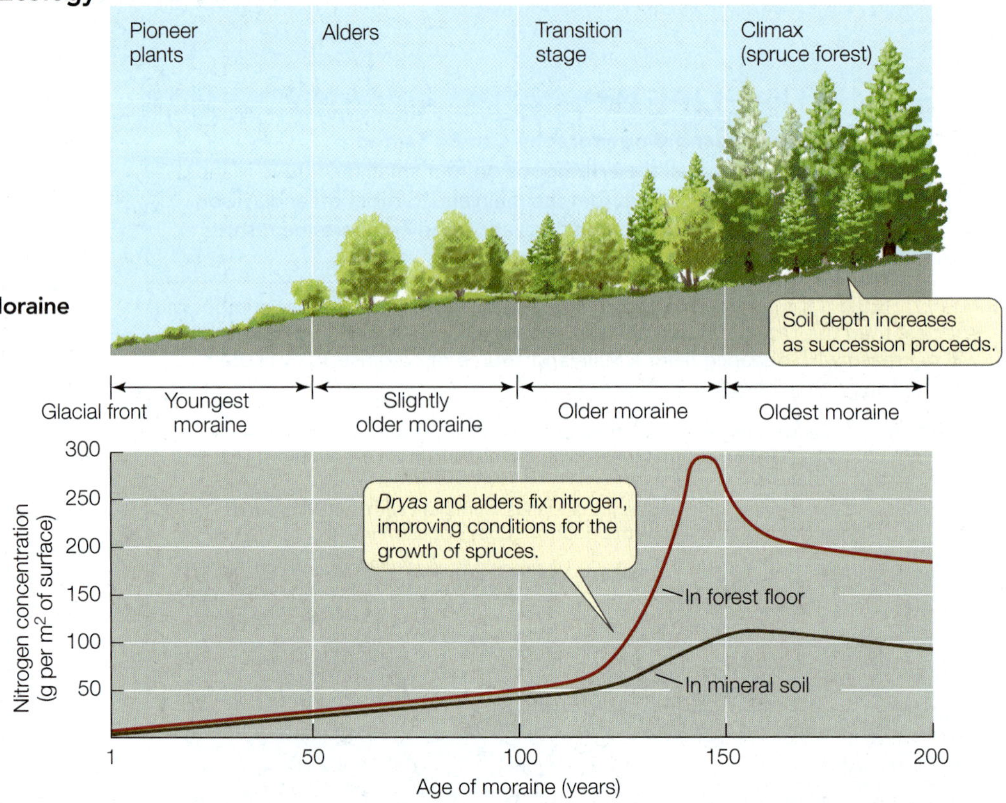
Pioneer plants
Alders
Transition stage
Climax (spruce forest)
Soil depth increases as succession proceeds.
Glacial front
Youngest moraine
Slightly older moraine
Older moraine
Oldest moraine
300
250
200
150
100
50
Nitrogen concentration (g per m2 of surface)
Dryas and alders fix nitrogen, improving conditions for the growth of spruces.
In forest floor
In mineral soil
1
50
100
150
200
Age of moraine (years)

57.13 Primary Succession As the community occupying a glacial moraine at Glacier Bay, Alaska, changes from an assemblage of pioneer plants such as *Dryas* to a spruce forest, soil depth increases and nitrogen accumulates in the soil.

Go to Animated Tutorial 57.2
Primary Succession on a Glacial Moraine
Life10e.com/at57.2

of community assemblages. Species come and go until a particular community—one that is capable of perpetuating itself under the local climate and soil conditions—persists for a relatively long time. This persistent stage is called the **climax community**.

Directional succession is easiest to observe after a disturbance strips away all preexisting living organisms and exposes a bare substrate. The type of directional change that occurs under these circumstances is known as **primary succession**. Glaciers (see Figure 1.16), volcanic activity, and in some cases, floods can initiate primary succession.

Primary succession can be seen in the successive changes in plant growth form and community composition in the wake of the retreat of the glacier in Glacier Bay, Alaska, over the last 200 years (**Figure 57.13**). The glacier scraped the landscape down to bare rock and left a series of moraines—gravel deposits dropped where the glacial front was stationary for a number of years. No human observer was present to measure changes over the entire 200-year period, but ecologists have inferred the temporal pattern of succession by studying the vegetation on moraines of different ages. The youngest moraines, closest to the current glacial front, are populated with bacteria, fungi, and photosynthetic microorganisms that can support themselves on bare rock. Slightly older moraines farther from the glacial front are home to lichens, which break down rocks and, when they die, decompose and contribute to the buildup of soil. Mosses and a few species of shallow-rooted herbs such as mountain avens (*Dryas octopetala*) become established and contribute to soil-building as they die and decompose. Still farther from the glacial front, successively older moraines have deeper soil layers that support shrubby willows, alders, and spruces.

Nitrogen is virtually absent from glacial moraines, so the plants that grow best on recently formed moraines at Glacier Bay are *Dryas* and alders (*Alnus*), both of which have nitrogen-fixing bacteria in nodules on their roots (see Figure 36.7B). Nitrogen fixation by these plants improves the soil so that spruces can grow. Spruces then outcompete and displace the early colonists. If the local climate does not change dramatically, a climax community dominated by spruce trees is likely to persist for many centuries on old moraines at Glacier Bay.

Directional succession following a disturbance that some organisms, particularly those in the soil, survive is called **secondary succession**. Secondary succession is often initiated by human activities (such as the clearing of a forest) as well as by natural disasters (such as storms and fires). Generally easier to monitor than primary succession, secondary succession also tends to occur more frequently and progress more rapidly.

A typical sequence of secondary succession in eastern North America begins when forested land that had been cleared for agriculture is abandoned (**Figure 57.14**). Because the soil is left intact, abandoned farmland provides an excellent environment for plants to colonize. The first plants to appear in old agricultural fields are fast-growing annuals such as pigweed, ragweed, and lamb's-quarter. These plants grow from seeds that have persisted in a dormant state in the soil since before the land was farmed, awaiting the opportunity to germinate. A single individual lamb's-quarter (*Chenopodium album*) can produce more than 150,000 tiny seeds. These pioneer species are quickly replaced by biennials and perennials that are stronger competitors for resources, such as milkweed, goldenrod, and thistles. The seeds of many of these plants have mechanisms for long-distance dispersal that allow them to colonize newly cleared sites. Milkweed seeds, for example, come equipped with a silky "parachute" that catches the wind and allows them to travel long distances. Eventually shrubby plants such as dogwood, eastern red cedar, and sumac become established, followed by tree species such as cottonwood, cherry, and red maple. Ultimately, shade-tolerant tree species whose seedlings can survive the shady conditions under the established trees, including beech and sugar maple, dominate the landscape. The beech–maple forest is the climax community for much of the region.

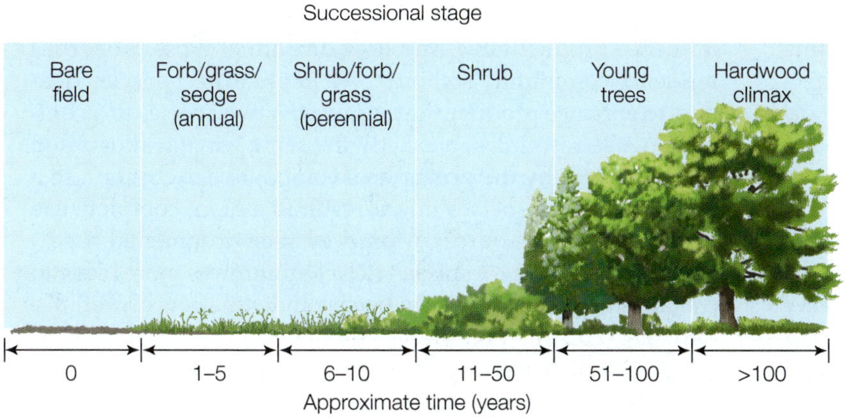
Successional stage
Bare field
Forb/grass/ sedge (annual)
Shrub/forb/ grass (perennial)
Shrub
Young trees
Hardwood climax
0
1–5
6–10
11–50
51–100
>100
Approximate time (years)

57.14 Secondary Succession Land that was once an agricultural field ultimately supports a long-lasting climax community characterized by shade-tolerant trees.

Directional succession, irrespective of where it takes place, is characterized by certain trends. In general, the pioneer species of early successional stages tend to be good dispersers with high rates of increase (*r*-strategists; see Section 55.4). Early stages of succession are characterized by high productivity and simple food webs; most nutrients are present in detritus or in the soil. As succession proceeds, nutrients accumulate in living biomass, food webs become more complex, and abiotic sources of nutrients become less important. Species typical of late successional stages tend to be good competitors with relatively low intrinsic rates of increase (*K*-strategists).

Both facilitation and inhibition influence succession

To some extent, the progress of succession depends on the activity of successive colonists, each of which modifies the environment in such a way as to facilitate colonization by other species. Predators are unlikely to colonize a habitat with no prey species, nor can primary consumers exist before plants are established. The fixation of nitrogen by *Dryas* and alders, which allows spruces to become established in Glacier Bay (see Figure 57.13), is an example of such **facilitation**.

Although secondary succession is often described in terms of changes in plant species composition, colonization by plants is actually facilitated by heterotrophs. Many of the first organisms to arrive on bare soil after a disturbance are detritivores, which process dead organic matter and release nutrients (especially nitrogen), thus facilitating the establishment of plants. In a study of intensively burned 20-year-old pine plantations in northern Germany, the first organisms to colonize these burned forests were algae, slime molds, liverworts, mosses, and mushrooms. These were followed by algae-feeding flies, fungus-feeding beetles, and moss-feeding springtails. Flowering plants such as fireweed moved in, at which point leaf-feeding insects appeared, soon followed by predaceous ground beetles and wolf spiders. Even in the corpse communities described at the opening of this chapter, early colonists can make resources available for subsequent colonizers: some spider beetles, for example, cannot feed on rotting flesh directly, but rather feed on the excrement of the early flesh-eating colonizers.

In other cases the effect of early colonists does not facilitate but rather inhibits colonization by other species. The roots of some old-field species such as goldenrod and thistle exude chemicals that inhibit the germination and growth of potential competitors. Eventually, when these plants grow old and die, other plant species can become established. Similarly, in rocky intertidal communities, when wave action turns over boulders and clears rock surfaces, the green alga *Ulva* colonizes the cleared spaces quickly and efficiently, preventing colonization by the slower-growing perennial red algae. Certain crab species then selectively graze on *Ulva*, helping to undermine this inhibition and thus promoting the establishment of the red algae that dominate the community in later stages.

Cyclical succession requires adaptation to periodic disturbances

Some forms of disturbance recur with regularity, even if their recurrence is not always predictable. Such recurrent disturbances are associated with a pattern called **cyclical succession** because the climax community depends on the periodic disturbances in order to persist. The lodgepole pine (*Pinus contorta*) forests of southern Oregon, for example, are maintained by periodic forest fires (**Figure 57.15**). In this fire-adapted community, fires return nutrients to the soil in the form of burned organic matter and thus provide favorable conditions for seed germination. The cones of lodgepole pines are sealed shut by resins; only when they are subjected to high temperatures that melt the resins do they open and release their seeds.

57.15 Some Communities Are Adapted to Disturbance Forest communities dominated by lodgepole pine (*Pinus contorta*) are adapted to periodic fires. Fire removes mature trees weakened by pest infestation, revitalizes the soil, and provides favorable conditions for seed germination and new growth.

Lodgepole pine trees are attacked by the mountain pine beetle (*Dendroctonus ponderosae*) and are also prone to infection by the fungus *Phaeolus schweinitzii*, which causes the roots and heartwood of the trees to rot. Trees that have lived long enough to experience and be scarred by a fire are much more likely to become infected by the fungus than are trees that have not been scarred. Fungus-infected, weakened trees are preferentially attacked by beetles. After a beetle outbreak, in which many mature, fire-scarred trees are killed, the dead trees serve as potential fuel for a fire that will free up their nutrients for use by the remaining trees as well as new seedlings.

Heterotrophic succession generates distinctive communities

Plants play a vital role in most patterns of succession because, as autotrophs, they are the source of energy for the other organisms in the community. Successional changes, however, can take place without the participation of plants. Detritus-based communities—found in dung, dead plants, and carrion—undergo a series of changes known as **heterotrophic succession**. In these communities, energy resources are greatest when the habitat first becomes available to colonists and are depleted as succession takes place. There is no mechanism, such as photosynthesis, for generating more energy. Thus in contrast to most other forms of succession, biomass and species diversity decrease over time because the resource base declines. In addition, these temporary habitats are not really self-contained, so predators, which do not have to confine themselves to one dead body to live out their lives, can outnumber primary consumers (detritivores, in these communities), in apparent violation of the laws of thermodynamics.

RECAP 57.4

Ecological disturbances may remove some species from a community but may open up space and resources for other species. Ecological succession—a predictable pattern of change in community composition—typically follows a disturbance.

- How do primary succession and secondary succession differ? See p. 1200 and Figures 57.13 and 57.14
- Describe some ways in which early colonists facilitate or inhibit colonization by the species that follow them in a pattern of succession. See p. 1201

Now that we have seen how communities change in composition over time, let's look at what allows certain communities (such as climax communities) to persist over time and withstand disturbance with little change. In this context we will return to diversity as an important arbiter of community stability.

57.5 How Does Species Richness Influence Community Stability?

Up to a point, higher productivity favors higher species richness, as we saw in Section 57.1. Does species richness, in turn, influence productivity? And how do both of those properties influence community stability? We might expect species richness to enhance productivity because no two species in a community use resources in exactly the same way (as illustrated in Section 56.4 by the principle of competitive exclusion), so a mixture of more species might result in a more complete use of the available resources. Moreover, if environmental conditions should change, a species-rich community is more likely to contain some species that can persist under the new conditions. Thus a species-rich community should be more stable—that is, it should change less over time in either productivity or species composition—than a species-poor community.

Species richness is associated with productivity and stability

To test the hypothesis that species-rich communities are more stable than species-poor communities, David Tilman and his colleagues at the University of Minnesota cleared 120 outdoor plots, in which they planted grasses in mixtures ranging from 2 to 22 grass species. At the end of each growing season, they measured total plant cover (a measure of grass biomass, and thus of net primary production) and the population densities of all the grasses in each plot. Over a period of 11 years, which included a serious drought, the plots with more species were more productive (**Figure 57.16A**), and their productivity was less variable from year to year. These findings were consistent with the hypothesis that species richness promotes productivity and keeps productivity stable. Moreover, in the plots with greater species richness, soil nitrogen was used more efficiently (**Figure 57.16B**). However, the population densities of *individual* species in the plots were not stable over the years (regardless of a plot's species richness) because different species performed better during drought years and wet years. In other words, higher species richness increased the stability of productivity in the plots, but not the stability of their species composition.

Researchers continue to debate whether species diversity is responsible for maintaining stability or is simply correlated with stability. This question is important because many of the alterations humans have made in the structure of natural communities have reduced their species richness, and many of these human-altered communities—notably agricultural communities—are notoriously unstable.

Diversity, productivity, and stability differ between natural and managed communities

Although ecologists have been debating the relationships among species richness, productivity, and stability for only a few decades, humans have been experimenting with those relationships, albeit inadvertently, for millennia—since plants were domesticated and agriculture was invented. Since the dawn of agriculture, crops have been susceptible to diseases and insect outbreaks: massive (often sudden) increases in populations of species that destroy or damage crops.

The practice of growing crops as **monocultures**—plantings of a single crop species—is one reason why managed

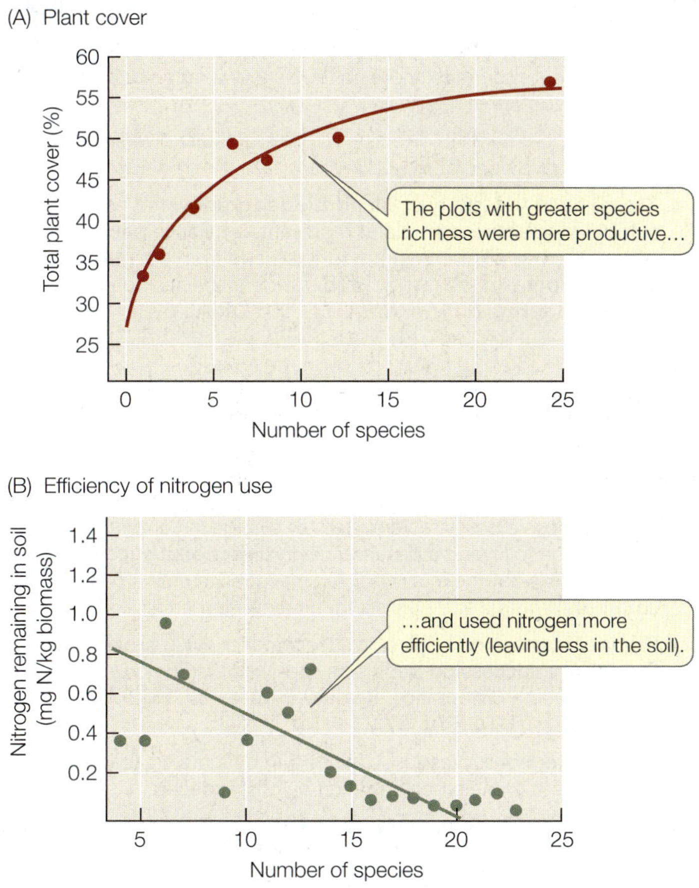
(A) Plant cover
60
55
50
45
40
35
30
25
Total plant cover (%)
0
5
10
15
20
25
Number of species
The plots with greater species richness were more productive...
(B) Efficiency of nitrogen use
1.4
1.2
1.0
0.8
0.6
0.4
0.2
Nitrogen remaining in soil (mg N/kg biomass)
5
10
15
20
25
Number of species
...and used nitrogen more efficiently (leaving less in the soil).

57.16 Species Richness Enhances Community Productivity Tilman and colleagues cultivated a total of 120 grassland plots, containing from 2 to 22 grass species, for 11 years. (A) Total plant cover (a measure of grass biomass, and thus of net primary production). (B) The amount of nitrogen remaining in the soil is a measure of resource use efficiency.

agricultural communities are particularly unstable. Most farmers have little tolerance for the presence of any potential competitors for their crops and actively eliminate weeds (and the herbivore species that live with them) from their fields. Thus a typical agricultural community has very low species diversity. So the answer to the question of whether diversity causes or is merely correlated with stability may be sought in modern farming practices. The predisposition of agricultural communities to play host to outbreaks may well result from human influences on community structure.

For the last 20 years, ecologists have been using traditional subsistence agricultural plots as experimental models for testing the relationships between diversity and stability. Throughout the world, many farmers with small land holdings grow multiple crops on the same plot. In Costa Rica, for example, farmers often grow corn together with sweet potato. Such corn–sweet potato dicultures contain fewer sweet potato pests and many more parasitoid wasps (which feed on those pests) than sweet potato monocultures do. The wasps feed on the corn pollen, and the tall corn plants act as a structural barrier, shade plant, and source of disruptive chemical signals that interfere with the ability of the sweet potato pests to find their host plants.

In recent years such applications of community ecology have been paying dividends. Although monoculture is overwhelmingly the dominant agricultural practice, polycultures are under development for agricultural production systems as varied as carp and shrimp farming, vermicomposting (raising worms for compost), and biofuel feedstock production.

RECAP 57.5

Communities with higher species diversity tend to be more productive and more stable than less diverse communities because they use resources more efficiently. The instability of modern agricultural monocultures suggests that diversity results in stability.

- What relationships have ecologists observed between species diversity, community productivity, and community stability? **See p. 1202 and Figure 57.16**
- Describe some agricultural practices that might result in more stable ecological communities. **See pp. 1202–1203**

?

How do the insect species in a corpse community influence one another's ability to survive?

ANSWER

The animal communities in decomposing corpses consist primarily of insects, but their exact composition varies with the factors that influence the rate and nature of decomposition: climate, season, and the condition of the body, including whether it is immersed or buried, wrapped or exposed. Typically, immediately after death blow flies, bluebottle flies, and house flies arrive to lay eggs. As a detectable odor develops, other flies, including greenbottles and flesh flies, arrive. Fat breakdown, with its accompanying release of volatile fatty acids, attracts a range of carrion-feeding beetles. As proteins decompose, cheese skippers can colonize. Other species have less interest in the corpse than in the corpse-eaters: rove beetles prey on the maggots that develop from the flies' eggs. During decay, skin beetles, hide beetles, and clothes moths (which can feed on the keratin in mammalian hair) dominate.
In the final stages of decomposition, spider beetles and other scavengers arrive to feed on the excrement and shed exoskeletons of the insects that have been consuming the corpse. This succession varies tremendously with climate and geography, but within any particular region it is sufficiently predictable that it is admissible in court as evidence of the time of death.

CHAPTER SUMMARY 57

57.1 What Are Ecological Communities?

- A **community** is a group of species that coexist and interact within a defined area.
- **Gross primary productivity** (GPP) is the rate at which the **primary producers** in a community turn solar energy into chemical energy via photosynthesis. **Net primary production** represents the energy incorporated into primary producer **biomass**. Review Figure 57.2, ACTIVITY 57.1
- A **food web** is a diagram of the feeding relationships in a community. Review Figure 57.3
- The organisms in a community can be divided into **trophic levels** based on the energy sources they use. Primary producers get their energy from sunlight; **primary consumers** get their energy by eating primary producers; **secondary consumers** get their energy by eating primary consumers; and so on. Review Table 57.1, ACTIVITY 57.2
- Organisms that consume the dead bodies of other organisms or their waste products are called **detritivores** or **decomposers**. **Omnivores** are organisms that feed at multiple trophic levels.
- **Ecological efficiency** is the overall transfer of energy from one trophic level to the next. Pyramid diagrams illustrate the proportions of energy or biomass that flow to each successive trophic level. Review Figure 57.4
- Species diversity tends to increase with productivity up to a point; however; if productivity increases beyond that point, species diversity may decline. Review Figure 57.5

57.2 How Do Interactions among Species Influence Communities?

- The interactions of a consumer with other species can result in a **trophic cascade**: a series of indirect effects across successive trophic levels. Review Figure 57.6
- Organisms that build structures that create habitat for other species are known as **ecosystem engineers**.
- **Keystone species** have an influence on their community that is disproportionate to their abundance. Review Figure 57.7

57.3 What Patterns of Species Diversity Have Ecologists Observed?

- Species diversity encompasses **species evenness** as well as **species richness**. Review Figure 57.8
- Species diversity can be measured at multiple spatial scales: within a single community or habitat, or over a range of communities in a geographic region.
- Latitudinal gradients in diversity, with the greatest diversity at low latitudes, have been observed in many taxa. Review Figure 57.9
- According to the theory of **island biogeography**, the equilibrium number of species on an island represents a balance between the rate at which species immigrate to the island from the mainland **species pool** and the rate at which resident species go extinct. Review Figure 57.11, ANIMATED TUTORIAL 57.1

57.4 How Do Disturbances Affect Ecological Communities?

- A **disturbance** is a disruption in a community caused by a discrete external force, often abiotic in nature.
- **Succession** is a predictable pattern of change in community composition following a disturbance. In **directional succession**, species come and go in a predictable sequence until a **climax community** forms and persists for an extended time.
- **Primary succession** begins on sites that lack living organisms. **Secondary succession** begins on sites where some organisms have survived a disturbance. Review Figures 57.13, 57.14, ANIMATED TUTORIAL 57.2
- In any pattern of succession, species that become established may facilitate or inhibit colonization by other species.
- In **cyclical succession**, the climax community is maintained by periodic disturbances.
- **Heterotrophic succession** in detritus-based communities does not rely on photosynthesis and therefore differs in a number of ways from other types of succession.

57.5 How Does Species Richness Influence Community Stability?

- Species-rich communities use resources more efficiently, and thus tend to vary less in productivity, than do less diverse communities. Review Figure 57.16
- **Monocultures** are subject to pest outbreaks, whereas agricultural communities containing greater species diversity tend to be more stable.

Go to the Interactive Summary to review key figures, Animated Tutorials, and Activities
Life10e.com/is57

CHAPTER**REVIEW**

REMEMBERING

1. An ecological community is
   a. a group of species that coexist and interact within a defined area.
   b. a group of species that coexist and interact in an area together with the abiotic environment.
   c. all the species in an area that belong to a particular trophic level.
   d. all the species that are members of a local food web.
   e. All of the above

2. A trophic level consists of those organisms
   a. whose energy has passed through the same number of steps to reach them.
   b. that use similar foraging methods to obtain food.
   c. that are eaten by a similar set of predators.
   d. that eat both plants and other animals.
   e. that compete with one another for food.

3. Net primary production is
   a. the total amount of photosynthesis in a community.
   b. the total amount of primary producer biomass available for consumption by heterotrophs.
   c. the total amount of biomass produced by all autotrophs and heterotrophs in a community.
   d. the total amount of biomass consumed by heterotrophs.
   e. gross primary productivity minus secondary productivity.

4. Pyramid diagrams of energy and biomass distribution for forests and for grasslands differ because
   a. forests are more productive than grasslands.
   b. forests are less productive than grasslands.
   c. large mammals avoid living in forests.
   d. wood presents more nutritional challenges to herbivores than grasses do.
   e. grasses grow faster than trees.

5. The theory of island biogeography
   a. predicts that the equilibrium number of species on an island is a balance between the rate of immigration of new species and the rate of extinction of resident species.
   b. predicts that the rate of immigration of new species will decline with island distance from the mainland species pool.
   c. predicts that the rate of extinction of resident species will decrease as island size increases.
   d. applies to isolated habitat patches as well as to oceanic islands.
   e. All of the above

6. Which of the following events is *not* followed by primary succession?
   a. A glacier recedes.
   b. A volcano erupts.
   c. A fire destroys a forest.
   d. A hurricane creates a bare-sand beach.
   e. All of these disturbances are followed by primary succession.

7. Early stages of succession are characterized by
   a. species that are good dispersers.
   b. species with high rates of reproduction.
   c. simple food webs.
   d. nutrients that are available primarily from detritus and abiotic sources.
   e. All of the above

UNDERSTANDING & APPLYING

8. Recent analyses of human gut flora using genomic methods have revealed tremendous microbial diversity, including many previously unknown species (see Figure 26.21). This microbial diversity has effects on human health. Describe the ecological methods you could use to investigate (a) how microbial diversity varies between individuals or across populations; and (b) which particular microbes might be keystone species or ecosystem engineers.

9. Jan Beck and Ian Kitching examined patterns of hawk moth diversity in the 113 islands of Thailand and mainland Malaysia. How do the findings of their study as summarized in the figure below relate to the theory of island biogeography?

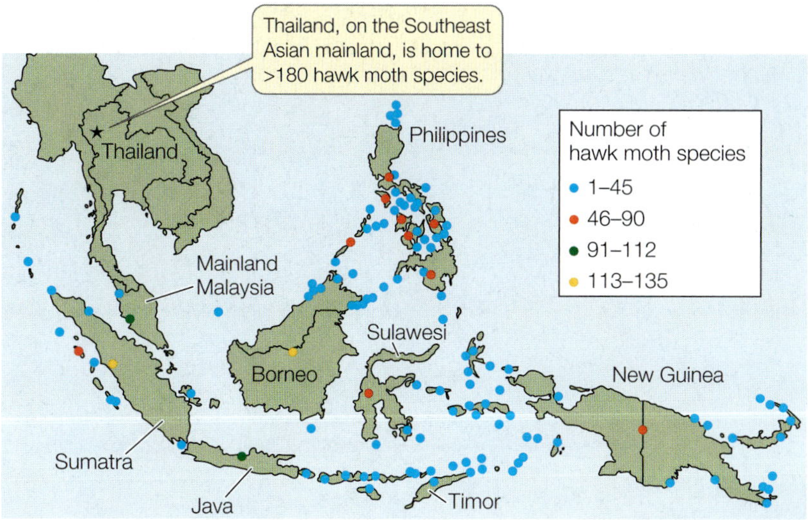

Thailand, on the Southeast Asian mainland, is home to >180 hawk moth species.
Thailand
Philippines
Number of hawk moth species
1–45
46–90
91–112
113–135
Mainland Malaysia
Sulawesi
Borneo
New Guinea
Sumatra
Java
Timor

ANALYZING & EVALUATING

10. Marek Sammul, Lauri Oksanen, and Merike Mägi investigated the effect of productivity on species richness in 16 different plant communities in western Estonia and northern Norway. When they removed one perennial species (the goldenrod *Solidago virgaurea*) from these communities, they found that its competitors, particularly the grass *Anthoxanthum odoratum*, increased in biomass, most noticeably in communities with high productivity (where living plant biomass was greater than 200 g/m^2). In less productive communities, such increases could not be detected. How might interspecific competition lead to a decrease in species richness at high levels of productivity? What other hypotheses might explain this puzzling relationship, and how would you test them?

11. Sea lampreys (*Petromyzon marinus*) are parasitic fish that fasten onto the bodies of host fish with their disc-shaped mouths and remain attached for long periods, feeding on the host's blood and other body fluids (see Figure 33.11B). This invasive species was responsible for reducing populations of sport fish in the Great Lakes, and consequently has been the target of many extermination campaigns. Recent studies, however, have revealed that *P. marinus* spawns in fast-moving freshwater streams, where these fish build elaborate nests by burrowing and by moving stones around, to create nesting mounds. While they are nesting, sea lampreys do not parasitize other fish. Also, when adult lampreys die, their decomposing remains help restore nutrients to freshwater habitats. Later in the season, salmon and brook trout move from the ocean to freshwater streams; they spawn in the same habitats as the lampreys and are known to use abandoned lamprey nests with great success. Given all this, should lampreys be eliminated as damaging parasites of game fish, or should they be encouraged as ecosystem engineers? What other kinds of information might you need to decide on an ecologically sound lamprey management strategy?

Go to BioPortal at **yourBioPortal.com** for Animated Tutorials, Activities, LearningCurve Quizzes, Flashcards, and many other study and review resources.

58 Ecosystems and Global Ecology

CHAPTER OUTLINE

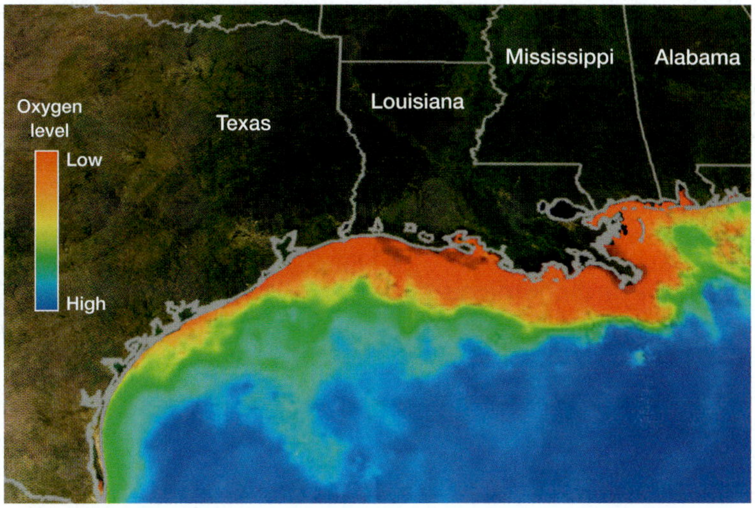
Mississippi
Alabama
Louisiana
Texas
Oxygen level
Low
High

The Gulf Dead Zone High concentrations of nitrogen and phosphorus in the runoff from agricultural lands in the U.S. interior are carried by the Mississippi River to the Gulf of Mexico. This nutrient enrichment of Gulf waters creates a "dead zone" in which many aquatic organisms cannot survive.

HOW CAN A CORNFIELD in Illinois, 1,500 kilometers from the nearest ocean, affect the price of sushi? When farmers in the Midwest apply chemical fertilizers to their fields, nitrogen and phosphorus from those fertilizers are dissolved in rainwater and washed into streams, which carry them into the Mississippi River. The river water eventually reaches the Gulf of Mexico, where the enormous inputs of dissolved nitrogen and phosphorus nourish explosive blooms of floating photosynthetic organisms (phytoplankton), including algae.

During the day, phytoplankton photosynthesize and produce oxygen. At night, however, they take up oxygen from the water to carry out cellular respiration. When these short-lived organisms die, their bodies sink to the bottom, where bacterial decomposition further depletes oxygen. This process results in a state of hypoxia (the reduction of dissolved oxygen to levels below 2 milligrams per liter of water), suffocating other organisms and creating a "dead zone." There are more than 140 coastal dead zones around the world. The one in the northern Gulf of Mexico, which has been mapped since 1985, is among the largest, spanning an area the size of New Jersey.

The Atlantic croaker (*Micropogonias undulatus*) lives along the coasts of eastern North America and the Gulf of Mexico. These bottom-feeding fish, which consume invertebrates, can reach 30 centimeters in length when mature. Their white, firm flesh is used in imitation crabmeat, or surimi, a popular ingredient in sushi. But recently croakers in the dead zone have experienced strange symptoms that may forecast a decrease in their populations. The croaker sex ratio, which is 50:50 among fish caught east of the Delta, is skewed toward males in the dead zone, and about one-fifth of the females in dead zone samples were found to have male germ cells in their ovaries. Laboratory experiments confirmed that only 10 weeks of exposure to hypoxia could induce these sexual defects. Depriving fish brains of oxygen apparently inhibits production of the neurohormones that promote normal ovary development. The croaker owes its name to its ability to make a drumming sound, but if the dead zone continues to expand and resulting sexual defects reduce their population growth rates, these fish might end up "croaking," both literally and figuratively.

Human activities such as farming change the movement patterns of mineral nutrients, and these changes affect not just croakers but many organisms that live far from where the human activities take place.

> ?
>
> How can we determine to what extent dead zones result from human actions and to what extent they are the result of natural processes?
>
> See answer on p. 1225.

58.1 How Does Energy Flow through the Global Ecosystem?

An **ecosystem** includes all of the organisms in an ecological community as well as the physical and chemical factors that influence those organisms. In other words, ecosystems have both biotic and abiotic components. Ecosystems can occupy a wide range of spatial scales, from the entire planet to a watershed, a specific forest, a lake or pond, or even a patch of lichen on a rock. To some degree, ecologists must define the boundaries of ecosystems arbitrarily. In this chapter we will focus on the global ecosystem, noting that all smaller-scale ecosystems are linked by the global flows of energy and chemical elements that are considered in this chapter.

Energy flows and chemicals cycle through ecosystems

Earth is essentially a closed system with respect to chemical elements, but it is an open system with respect to energy. The sun delivers a nearly constant amount of energy to Earth every day and has done so for billions of years. When captured by primary producers such as plants and photosynthetic bacteria, that energy flows through the trophic levels of food webs in one direction (see Figure 57.3). Much of the energy that enters each trophic level is used to power the metabolism of producers and consumers; that energy is eventually dissipated as heat and is lost from the ecosystem (see Figure 57.2). Chemical elements, by contrast, are not altered when they are transferred between organisms. Furthermore, they are not lost from the global ecosystem, although they may become unavailable to organisms for long periods; instead chemical elements cycle continually between living organisms and the abiotic components of ecosystems (**Figure 58.1**).

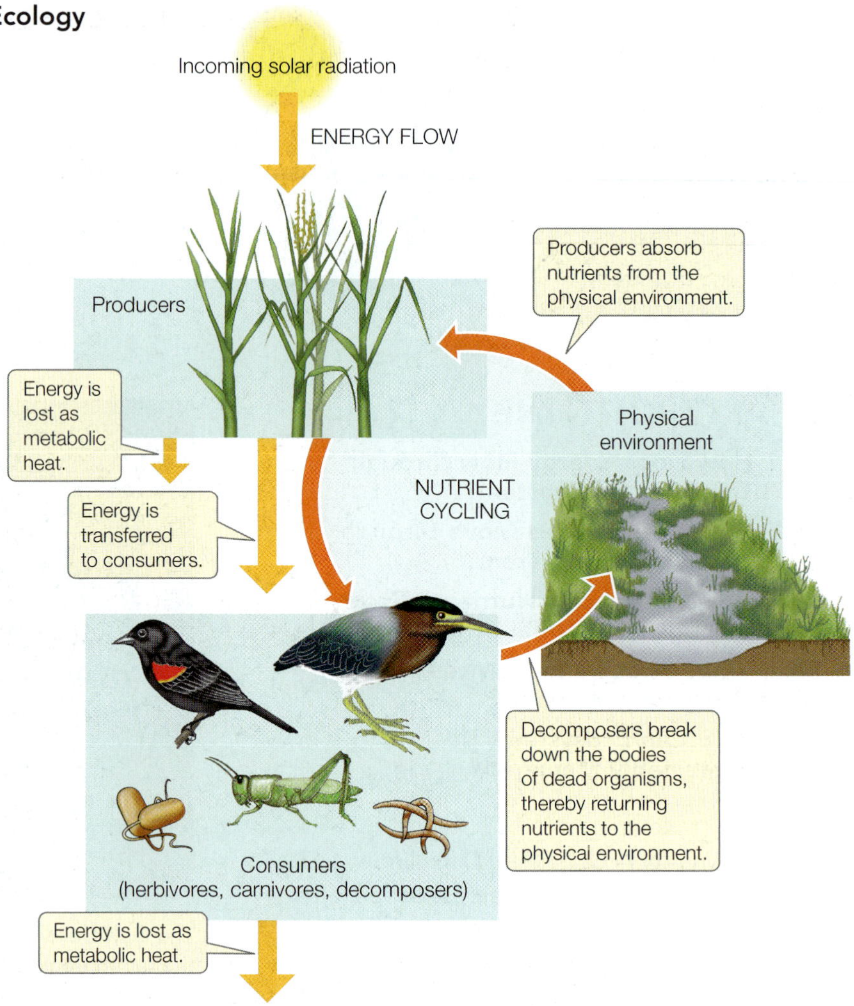
Incoming solar radiation
ENERGY FLOW
Producers
Producers absorb nutrients from the physical environment.
Energy is lost as metabolic heat.
Energy is transferred to consumers.
NUTRIENT CYCLING
Physical environment
Decomposers break down the bodies of dead organisms, thereby returning nutrients to the physical environment.
Consumers (herbivores, carnivores, decomposers)
Energy is lost as metabolic heat.

58.1 Energy Flows and Chemical Nutrients Cycle through Ecosystems Each time one organism eats another, a portion of the solar energy originally captured by a primary producer is lost as heat (gold arrows). As a result, energy flows through the ecosystem in a single direction. Chemical elements that organisms use as nutrients, however, cycle repeatedly between organisms and the physical environment (orange arrows).

In this first section we examine how the geographic distribution of incoming solar radiation influences the amount of energy assimilated by primary producers and how human activities are modifying energy flow through the global ecosystem.

The geographic distribution of energy flow is uneven

Nearly all energy used by organisms comes (or once came) from the sun. The only exceptions are found in those few ecosystems in which solar energy is not the main energy source (such as some caves and deep-sea hydrothermal vent ecosystems). Even the fossil fuels—coal, oil, and natural gas—on which the economy of modern human civilization is based are reserves of captured solar energy locked in the remains of organisms that lived (and died) millions of years ago (see Section 25.2).

As described in Section 57.1, solar energy enters ecosystems by way of plants and other photosynthetic organisms. These primary producers use some of the energy they assimilate for their own metabolism; the rest—net primary production (NPP)—is stored in their bodies or used for their growth and reproduction (see Figure 57.2). Because only the energy of NPP is potentially available to other organisms, which obtain it by consuming primary producers, NPP can be used as a rough measure of energy influx into an ecosystem. NPP varies among ecosystem types, but because ecosystem types also vary greatly in their geographic extent, the most productive ecosystem types do not necessarily contribute the most to Earth's net primary production (**Figure 58.2**).

The geographic distribution of NPP reflects the geographic variation in incoming solar radiation described in Section 54.2 and the climate patterns that result from it. In other words, the distribution of temperatures and moisture makes some ecosystems more productive than others (**Figure 58.3**). Close to the equator at sea level, temperatures are high throughout the year and the water supply is adequate for plant growth much of the time. In these climates, productive forests thrive. In deserts, plant growth is limited by lack of moisture and NPP is low. At higher latitudes, even though moisture is generally available, NPP is low because

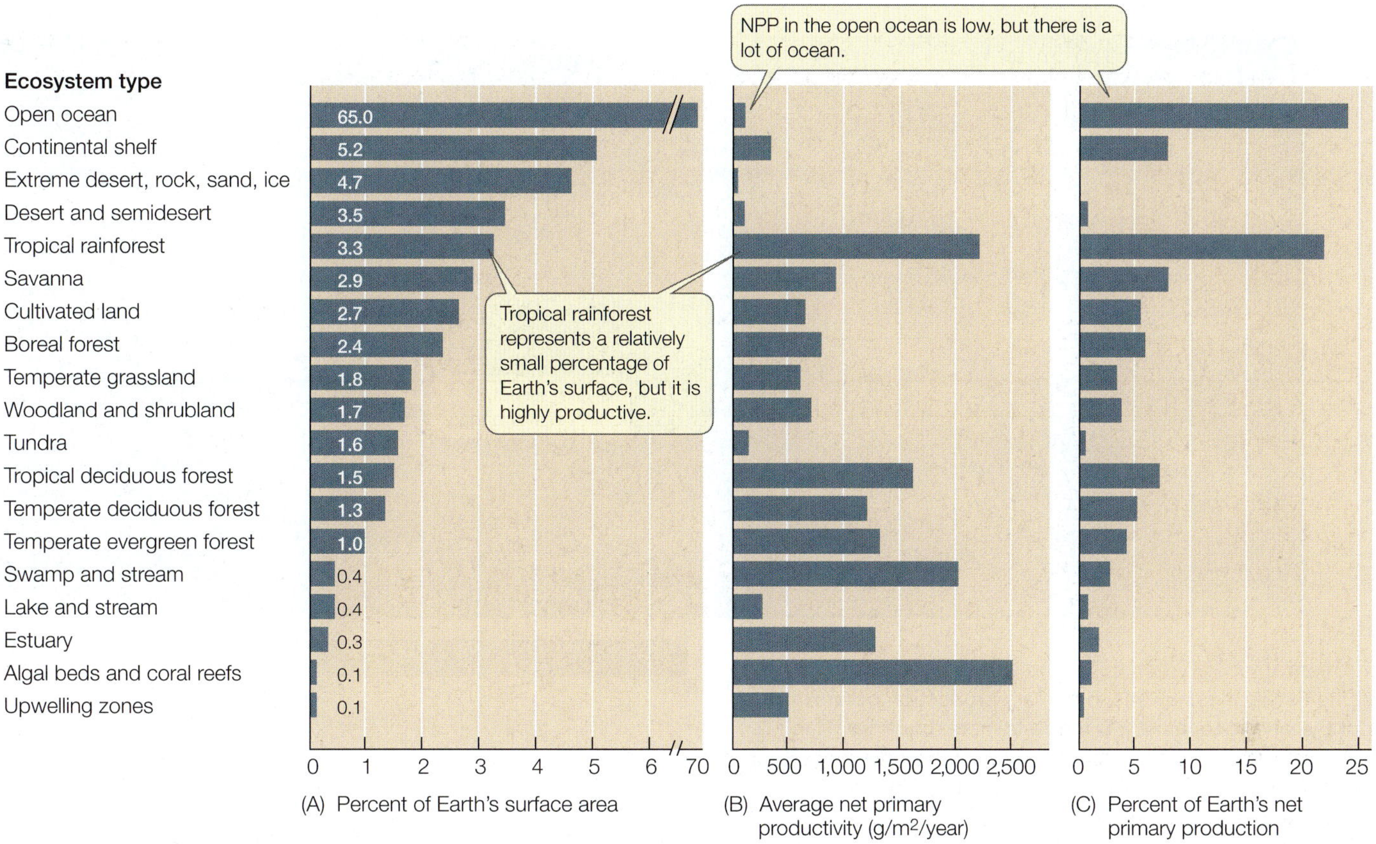

NPP in the open ocean is low, but there is a lot of ocean.
Ecosystem type
Open ocean
Continental shelf
Extreme desert, rock, sand, ice
Desert and semidesert
Tropical rainforest
Savanna
Cultivated land
Boreal forest
Temperate grassland
Woodland and shrubland
Tundra
Tropical deciduous forest
Temperate deciduous forest
Temperate evergreen forest
Swamp and stream
Lake and stream
Estuary
Algal beds and coral reefs
Upwelling zones
65.0
5.2
4.7
3.5
3.3
2.9
2.7
2.4
1.8
1.7
1.6
1.5
1.3
1.0
0.4
0.4
0.3
0.1
0.1
Tropical rainforest represents a relatively small percentage of Earth's surface, but it is highly productive.
0 1 2 3 4 5 6 70
0 500 1,000 1,500 2,000 2,500
0 5 10 15 20 25
(A) Percent of Earth's surface area
(B) Average net primary productivity (g/m2/year)
(C) Percent of Earth's net primary production

58.2 Energy Flow Contributions by Ecosystem Type The contributions of different ecosystem types to global energy flow can be measured by (A) their geographic extent and (B) their average net primary productivity. (C) Combining these two measures gives us a proportional contribution of each ecosystem type to Earth's total net primary production.

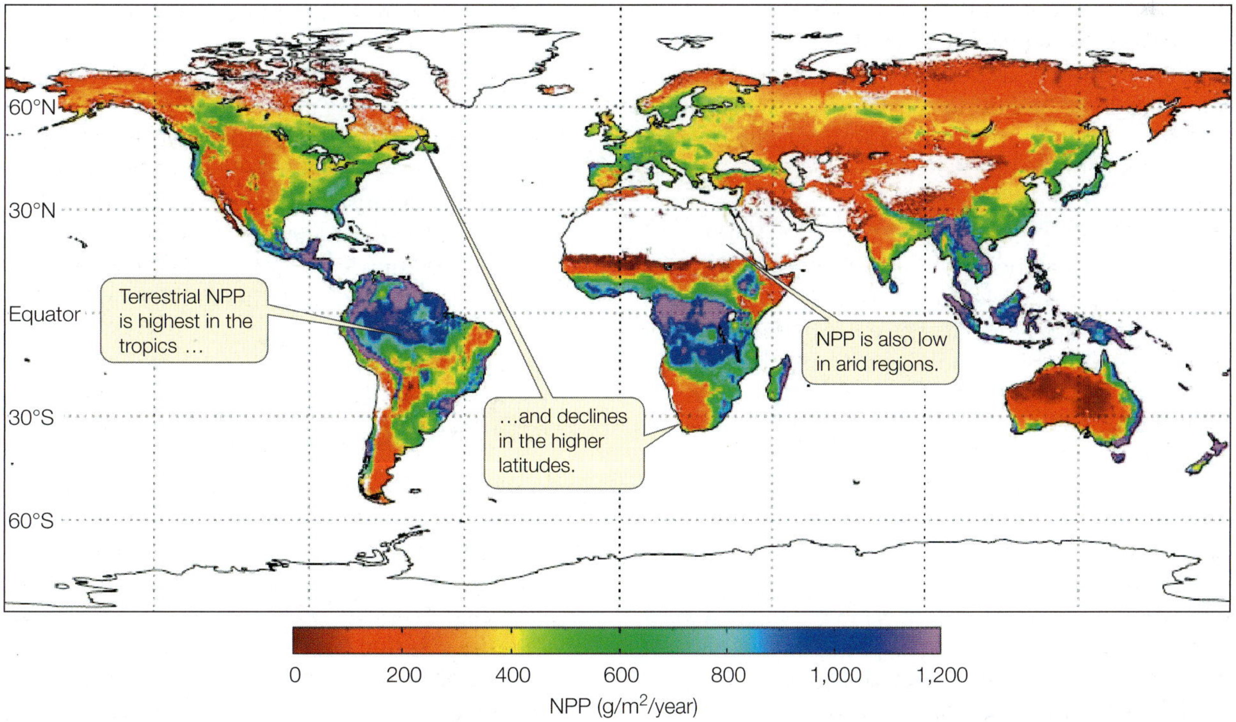

60°N
30°N
Equator
30°S
60°S
Terrestrial NPP is highest in the tropics ...
...and declines in the higher latitudes.
NPP is also low in arid regions.
0 200 400 600 800 1,000 1,200
NPP (g/m2/year)

58.3 Geographic Variation in Terrestrial NPP This map of estimated terrestrial net primary production is based on satellite sensor data accumulated from 2000 through 2005. White spaces represent unvegetated areas, including deserts and ice caps.

...and in
coastal zones.
NPP is highest in
zones of upwelling...

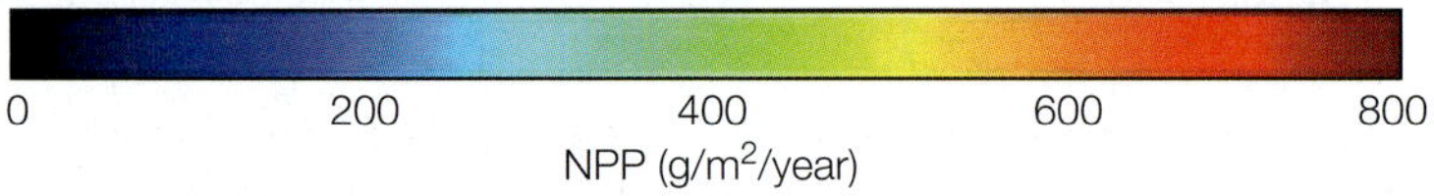
0
200
400
600
800
NPP (g/m2/year)

58.4 Geographic Variation in Marine NPP The availability of nutrients determines how much primary production occurs in any part of the photic zone. NPP is highest where runoff from land brings nutrients into shallow coastal waters and where upwellings bring nutrients from the seafloor to the surface.

it is relatively cold much of the year and the growing season is short (see Working with Data, p. 1138).

Production in aquatic ecosystems is limited by light, which decreases rapidly with depth (see Figure 54.9), and by temperature, which also decreases with depth, except in areas on the seafloor near hydrothermal vents. Production in aquatic ecosystems is also strongly limited by nutrient availability, as we'll see in the following section. Net primary production in the oceans tends to be highest in coastal zones, where runoff from land and upwellings from deeper waters bring nutrients into shallow waters (**Figure 58.4**).

Human activities modify the flow of energy

The effects of human activities on energy flow through the global ecosystem have accelerated markedly in the last 150 years. Some human activities decrease net primary production, as when forests are cut down and replaced by cities; some increase it, as when prairies are converted to extensive agricultural fields.

Humans consume about one-quarter of Earth's average annual net primary production. More than 50 percent of this consumption results from the croplands and rangelands that occupy more than one-third of Earth's ice-free surface; another 40 percent results from productivity changes brought about by alterations in land use; and 7 percent represents biomass consumed in fires caused by humans. The percentage varies strikingly among regions, however; urban areas consume as much as 300 times the NPP they generate, but people in sparsely inhabited parts of the Amazon Basin consume vanishingly small amounts of the NPP generated there.

RECAP 58.1

Nearly all energy used by living organisms comes from the sun. Energy flow through ecosystems, as measured by net primary production, varies with geographic location.

- How does the distribution of temperature and moisture influence the geographic distribution of net primary production in terrestrial systems? **See p. 1208 and Figure 58.3**
- What percentage of Earth's average annual net primary production is appropriated by humans, and how does that percentage vary regionally? **See p. 1210**

Ecosystem productivity is influenced not only by energy flow but also by the availability of the nutrients and other materials required by organisms to build their bodies and to power their metabolism. The next section will describe how biochemical materials and nutrients move around the abiotic environment and become available to living organisms.

58.2 How Do Materials Move through the Global Ecosystem?

In contrast to the energy that powers biological processes, which comes from the sun, the chemical elements that make up the bodies of organisms come from within the Earth system itself. As we have seen, these elements cycle continually through the global ecosystem. But they are not always available in the right place, or in the right form, to be useful to organisms. Because nutrient availability, in addition to energy input, influences productivity, ecologists are interested in knowing where on Earth nutrients are located and how they move from one location to another.

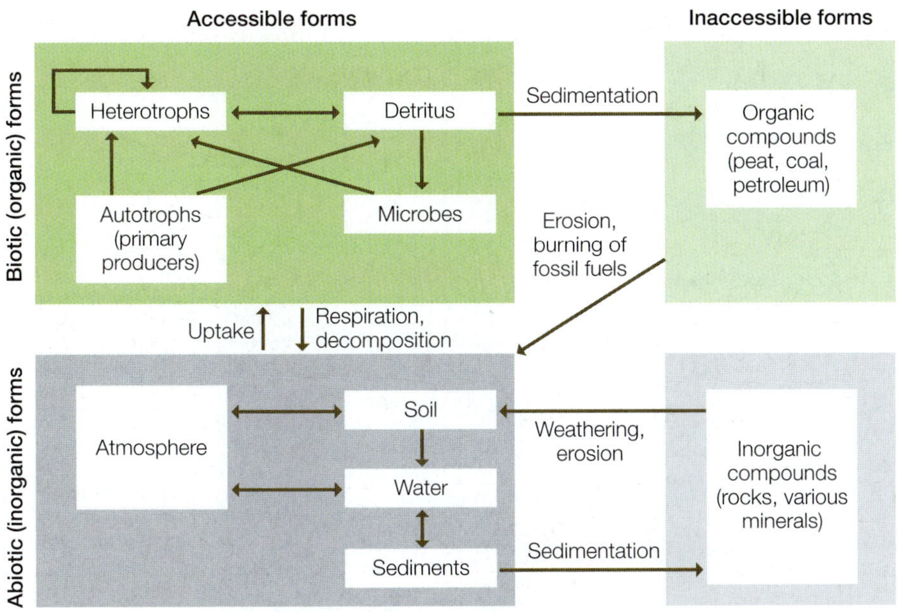
Accessible forms
Inaccessible forms
Biotic (organic) forms
Heterotrophs
Detritus
Autotrophs (primary producers)
Microbes
Sedimentation
Organic compounds (peat, coal, petroleum)
Erosion, burning of fossil fuels
Uptake
Respiration, decomposition
Abiotic (inorganic) forms
Atmosphere
Soil
Water
Sediments
Weathering, erosion
Inorganic compounds (rocks, various minerals)
Sedimentation

58.5 Chemical Elements Cycle through the Biosphere The different forms and locations of the chemical elements determine whether or not they are accessible to living organisms. Biological, geological, and chemical processes cycle matter among these biotic and abiotic components of the global ecosystem.

Energy from the sun, combined with energy from the radioactive decay in Earth's interior, drives the biological, geological, and chemical processes that transform chemical elements and move them around the planet (Figure 58.5). In this section we'll examine the properties of some of the abiotic and biotic components of the global ecosystem—referred to as compartments—through which elements move, as well as the processes that move them. The rate at which an element moves through a compartment is called its **flux**; the term "flux" is also applied to the movements of energy. Elements may accumulate, or "pool," in some compartments.

Elements move between biotic and abiotic compartments of ecosystems

All materials in the bodies of organisms ultimately originate from abiotic sources, but organisms acquire these materials in many different ways. Autotrophs such as plants take up certain elements directly from soil, water, and the atmosphere and incorporate them into organic molecules to build biomass. Heterotrophs generally acquire elements by consuming the biomass produced by other organisms, then reassemble those elements, via chemical reactions, in different ways. Some heterotrophs, however, acquire some elements by housing mutualistic microbes that convert those elements into forms that are usable by their hosts.

Respiration by living organisms returns certain elements to the atmosphere as gases. After organisms die, the materials in their bodies become detritus and are broken down by decomposers into simpler biochemical components. In this way the elements are freed to be taken up again by autotrophs. Elements that do not get taken up by autotrophs can accumulate in soil, water, or sediments.

At times in the remote past, great quantities of organic material were removed from active cycling when organisms died in large numbers and were buried in sediments that lacked oxygen. In such anaerobic environments, decomposers could not efficiently break down organic molecules to their inorganic forms. Instead organic molecules accumulated and were eventually transformed into deposits of oil, natural gas, coal, or peat—the **fossil fuels** that modern humans use as a combustible source of energy.

The movement of elements through food webs from uptake to decomposition—that is, through the biotic compartments of ecosystems—occurs primarily on a local scale. In contrast, abiotic processes can move elements far beyond the boundaries of the local ecosystem. The various abiotic compartments of the global ecosystem differ in fundamental ways, and the quantities of different elements in each compartment (e.g., atmosphere, ocean, soil), how long those elements remain there (their **residence time**), the forms they take, and the rates at which they enter and leave also differ. Moreover, the cycling of nutrients among compartments is influenced by the ways in which energy flows through them.

The atmosphere contains large pools of the gases required by living organisms

The outermost compartment of the global ecosystem is the atmosphere, a thin layer of gases surrounding Earth. The atmosphere is 78.08 percent nitrogen gas (N_2), 20.95 percent oxygen gas (O_2), 1 percent water vapor, 0.93 percent argon, and 0.03 percent carbon dioxide (CO_2). It also contains traces of hydrogen gas, neon, helium, krypton, xenon, ozone, and methane. It contains Earth's biggest pool of nitrogen as well as a large proportion of its oxygen.

About 80 percent of the mass of the atmosphere lies in its lowest layer, the **troposphere**. This layer extends upward from Earth's surface about 17 km in the tropics and subtropics, but only about 10 km at high latitudes. Most global air circulation takes place within the troposphere, and virtually all of the atmospheric water vapor is found there (Figure 58.6).

The **stratosphere**, which extends from the top of the troposphere up to about 50 km above Earth's surface, contains very little water vapor. Most materials enter the stratosphere from the region of the troposphere that encircles the equator, where air heated by the sun's energy rises to high altitudes (see Section 54.2).

The stratosphere contains a layer of ozone (O_3) that absorbs most of the biologically damaging ultraviolet radiation that enters the atmosphere. Over the last several decades, this **ozone layer** has been seriously damaged by human activities, particularly the widespread release of chlorinated fluorocarbons (CFCs), such as the refrigerant freon. CFCs remain stable as they ascend to the stratosphere, where they interact with and break

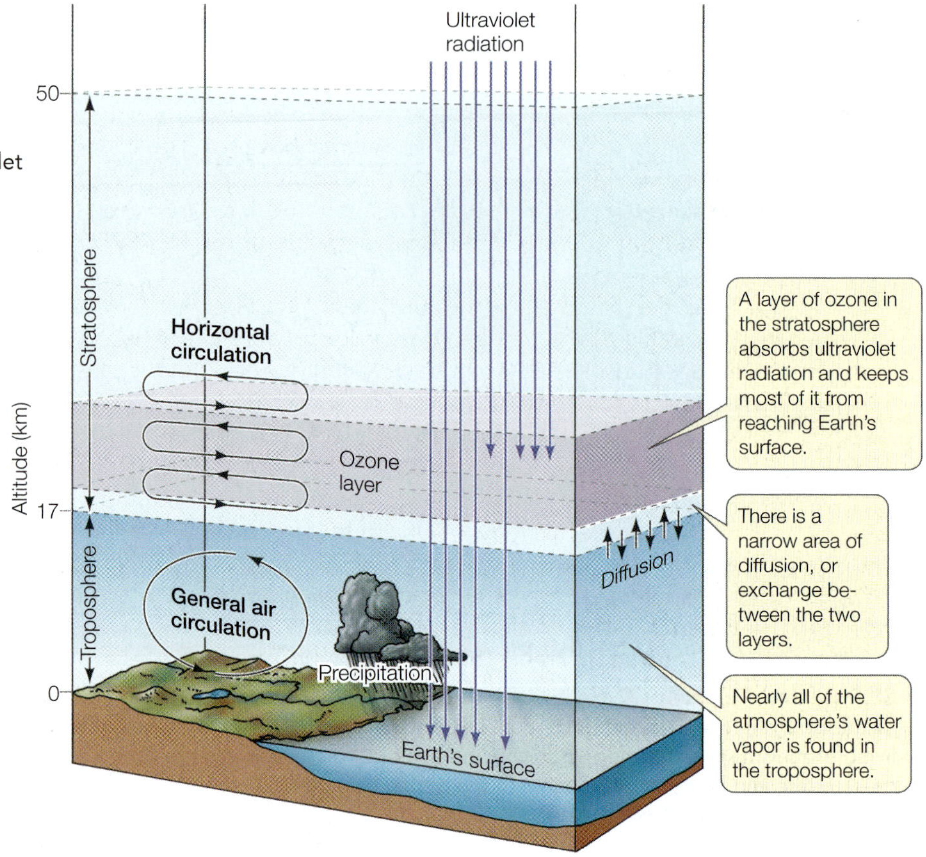
Ultraviolet radiation
50
17
0
Altitude (km)
Stratosphere
Troposphere
Horizontal circulation
Ozone layer
General air circulation
Precipitation
Diffusion
Earth's surface
A layer of ozone in the stratosphere absorbs ultraviolet radiation and keeps most of it from reaching Earth's surface.
There is a narrow area of diffusion, or exchange between the two layers.
Nearly all of the atmosphere's water vapor is found in the troposphere.

58.6 Earth's Atmosphere Has Two Layers The troposphere and the stratosphere differ in their circulation patterns, the amount of water vapor they contain, and the amount of ultraviolet radiation they receive.

down ozone molecules in the presence of ultraviolet light. The more ozone that is lost, the more ultraviolet light can reach Earth's surface. Increases in ultraviolet radiation are associated with increased rates of skin cancer, cataract formation, and crop damage. Since 1989, when a global treaty was enacted to phase out production of CFCs, atmospheric levels of CFCs have for the most part stabilized or declined, and there is encouraging evidence from both satellite and ground station measurements that stratospheric ozone levels are slowly returning to their normal state.

The atmosphere moderates temperatures at and near Earth's surface by trapping heat energy. If Earth had no atmosphere, its average surface temperature would be about –18°C, rather than its actual +17°C. Carbon dioxide, methane (CH_4), nitrous oxide (N_2O), water vapor, and certain other gases in the atmosphere are known as **greenhouse gases** because they are transparent to sunlight but trap heat radiating from Earth's surface back toward space (**Figure 58.7**). Ozone, when it is present in the troposphere rather than the stratosphere, also acts as a greenhouse gas. As we'll see later in this chapter, human activities are increasing the concentrations of greenhouse gases in the atmosphere, and those increases are altering the climate.

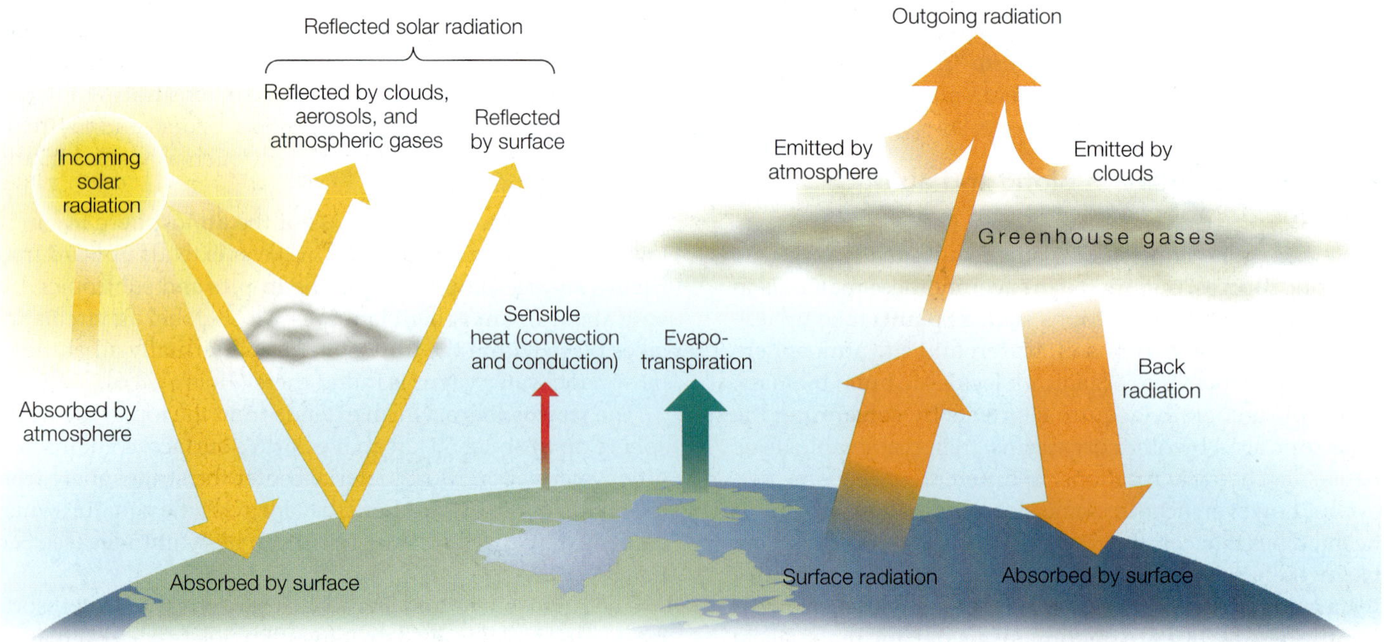
Reflected solar radiation
Reflected by clouds, aerosols, and atmospheric gases
Reflected by surface
Incoming solar radiation
Absorbed by atmosphere
Absorbed by surface
Sensible heat (convection and conduction)
Evapo-transpiration
Outgoing radiation
Emitted by atmosphere
Emitted by clouds
Greenhouse gases
Back radiation
Surface radiation
Absorbed by surface

58.7 Radiant Energy Warms the Planet Solar energy input (yellow arrows) is absorbed by Earth's atmosphere and surface. Much of this energy is radiated from Earth's surface in the form of heat (orange arrows). Much of this radiation is prevented from escaping back into space by greenhouse gases in the atmosphere. The widths of the arrows here are roughly proportional to the sizes of the energy fluxes.

Go to Animated Tutorial 58.1
Earth's Radiation Balance
Life10e.com/at58.1

The terrestrial surface is influenced by slow geological processes

About one-fourth of Earth's surface consists of land above sea level. Because the geological processes that move elements through minerals and soils are so slow, regional and local variations in the supply of particular elements greatly influence terrestrial ecosystem processes.

Nearly all the rocks that underlie the continents have been transformed at least once through a complex cycle of plate tectonic processes (see Figure 25.3). The physical and chemical processes of weathering break down surface rocks into soil. The type of soil in an area and the chemical nutrients that soil contains are determined in large part by the underlying rock from which the soil forms, although climate, topography, the local biota, and the length of time that soil-forming processes have been acting also influence the nature of soil (see Section 36.3). Chemical elements in rocks are released by weathering and by certain biological processes; these elements are then carried in solution into streams and groundwater, which transport them ultimately into the oceans. Structural features of the land surface affect how rapidly and in what direction wind and water currents can transport elements.

Water transports elements among compartments

The high heat capacity of water and its ability to change states from gas to liquid to solid at temperatures found on Earth (see Section 2.4) mean that it can move freely among all compartments, including the biosphere. Water is a powerful solvent that can carry materials in solution within and among all the compartments of the global ecosystem.

FRESH WATERS The liquid fresh waters of the global ecosystem consist of streams, lakes, and **groundwater** (water occupying pore spaces in rock, sand, and soil). Only a small fraction of Earth's water resides in lakes and streams at any given time, but because water moves so rapidly through the freshwater compartment, most of Earth's water spends some time there. Some mineral nutrients enter fresh waters from the atmosphere in rainfall, but most are released from rocks by weathering. They are dissolved and carried into streams by surface runoff or by movements of groundwater in a process called erosion. After entering streams, mineral nutrients are usually carried rapidly to lakes or to the oceans.

TURNOVER IN LAKES The nutrients in lakes are taken up by and incorporated into the bodies of the organisms living there. Those organisms eventually die and sink to the lake bottom, where decomposition of their tissues by microbes releases the nutrients but consumes most of the oxygen in the bottom water. The surface waters of deep lakes thus quickly become depleted of nutrients while deeper waters become depleted of oxygen. The waters of most deep lakes in temperate climates have an annual cycle of **turnover**, vertical movements of the water column that bring nutrients to the surface and oxygen to deeper water (**Figure 58.8**).

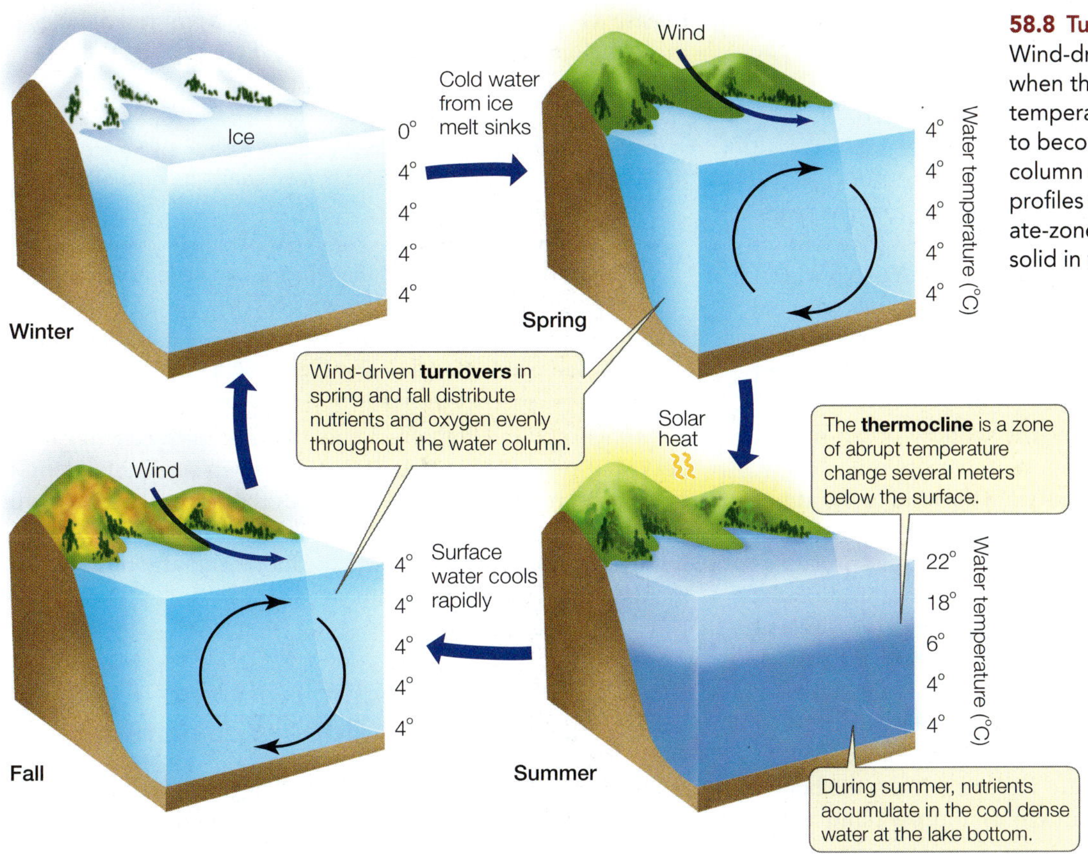
Ice
0°
4°
4°
4°
4°
Winter
Cold water from ice melt sinks
Wind
4°
4°
4°
4°
4°
Water temperature (°C)
Spring
Wind-driven **turnovers** in spring and fall distribute nutrients and oxygen evenly throughout the water column.
Solar heat
The **thermocline** is a zone of abrupt temperature change several meters below the surface.
Wind
4°
4°
4°
4°
4°
Surface water cools rapidly
Fall
22°
18°
6°
4°
4°
Water temperature (°C)
Summer
During summer, nutrients accumulate in the cool dense water at the lake bottom.

58.8 Turnover in a Temperate-Zone Lake Wind-driven turnovers in spring and fall, when the lake water is a relatively uniform temperature, allow nutrients and oxygen to become evenly distributed in the water column of a lake. The vertical temperature profiles shown here are typical of temperate-zone lakes whose surface water freezes solid in winter.

Turnover depends on the unique physical properties of water. Liquid water is densest at 4°C, a few degrees above its freezing point of 0°C (at which point it becomes solid and floats). Above 4°C, water expands. Thus in winter the coldest liquid water in a lake is at the surface, often just beneath a layer of ice, and the dense waters of the depth remain at 4°C. In spring, when the sun melts the ice and warms the surface water to 4°C, there is a time at which water density is uniform throughout the lake, and even modest winds will readily mix the entire water column. As spring and summer progress, the surface water becomes warmer still, and the depth of the warm water layer gradually increases. However, there is a well-defined depth, called the **thermocline**, at which the temperature changes abruptly. Only if the lake is shallow enough so that water warms up all the way to the bottom does the temperature of the deepest water rise above 4°C.

Another turnover occurs in fall as the process reverses itself. The surface of the lake cools until the water there is denser than the warmer water below it, at which point it sinks and is replaced by warmer water from below. Once again, water density becomes uniform, and winds can mix the entire water column.

A similar process contributes to the formation of the Gulf of Mexico dead zone described at the opening of this chapter, but in that case the density differences are caused by differences in salinity rather than in temperature. Fresh water is less dense than salt water, so when the nutrient-rich fresh water of the Mississippi River flows into the Gulf, it does not mix with the salt water but floats on top of it. Dying algae sink to the bottom, and bacterial decomposition depletes the available oxygen there. The difference in water density prevents oxygen from the surface water from mixing with the hypoxic salt water below.

Go to Media Clip 58.1
Tracking Dead Zones from Space
Life10e.com/mc58.1

OCEANS Over time scales of hundreds to thousands of years, most materials that cycle through the global ecosystem end up in the oceans, which hold almost 97 percent of Earth's water. The oceans are enormous, but they exchange materials with the atmosphere only at their surfaces, so they respond very slowly to inputs from that compartment. They receive materials from land primarily in runoff from rivers.

Except on continental shelves—the shallow ocean waters surrounding large land masses—ocean waters mix slowly. Most materials that enter the oceans from the land or the atmosphere gradually sink to the seafloor, where they may remain for millions of years, until intermittent plate tectonic processes lift seafloor sediments above sea level. Thus concentrations of mineral nutrients in most ocean waters are very low (except where human activities have released materials into the water).

Near the coasts of continents, however, offshore winds may push the warmer surface waters away from shore, causing cold water from the bottom to rise and bringing nutrients back to the surface waters. An area where water from depths below 50 meters rises in this way is known as an **upwelling zone**. Upwelling zones support high rates of primary production by phytoplankton, which in turn support dense consumer populations. Most of the world's great fisheries are concentrated in upwelling zones. For example, the upwelling zone off the coast of Peru is the source of much of the world's supply of anchovies. This rich fish population supports vast seabird communities, which in turn produce enormous quantities of guano, or excrement, that have provided Peru with an important raw material to support a major fertilizer industry.

Fire is a major mover of elements

Every year 200 to 400 million hectares of savannas, 5 to 15 million hectares of boreal forests, and smaller expanses of other biomes catch fire and burn. Lightning ignites some of these fires, but humans start most of them to manage vegetation (as when they cut down and burn forests to clear land for growing crops). Fires rapidly consume the energy stored in, and release the chemical elements from, the vegetation they burn. Some nutrients, such as nitrogen, are readily vaporized by fire. Nitrogen enters the atmosphere in smoke or is carried into groundwater by rain falling on burned ground.

Fires also release large amounts of carbon into the atmosphere. The global annual flux of carbon to the atmosphere from savanna and forest fires is estimated at 1.7 to 4.1 petagrams (1 pg = 10^{15} g or 10^{12} kg). Biomass burning (which includes combustion of wood and alcohol, wildfires, and land clearing, but not the burning of fossil fuels) is responsible for about 40 percent of Earth's annual flux of CO_2 into the atmosphere and contributes to the production of other greenhouse gases as well. Large-scale wildfires in the western and southeastern United States can release as much CO_2 into the atmosphere as motor vehicles in those regions release over an entire year.

RECAP 58.2

Biological, geological, and chemical processes move materials within and among biotic and abiotic compartments of the global ecosystem.

- How does the atmosphere keep temperatures at and close to Earth's surface warmer than they would be in its absence? **See pp. 1211–1212 and Figure 58.7**
- Describe the process of turnover in a temperate-zone lake in fall. **See pp. 1213–1214 and Figure 58.8**

As we learned in Chapters 3 and 10, most of the chemical energy that primary producers convert from sunlight is stored in carbon-containing compounds. In the next section we will consider how carbon and other chemical elements required by living organisms cycle through the biotic and abiotic compartments of the global ecosystem.

58.3 How Do Specific Nutrients Cycle through the Global Ecosystem?

Each of the chemical elements that organisms use in large quantities cycles in a distinctive way through the biotic and

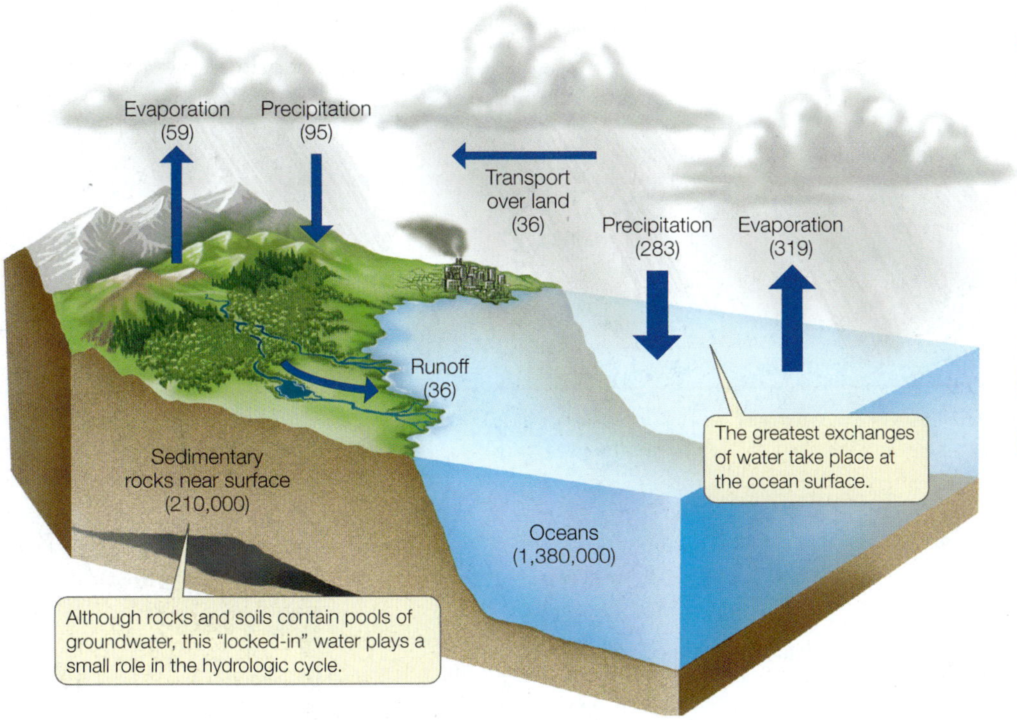
Evaporation (59)
Precipitation (95)
Transport over land (36)
Precipitation (283)
Evaporation (319)
Runoff (36)
The greatest exchanges of water take place at the ocean surface.
Sedimentary rocks near surface (210,000)
Oceans (1,380,000)
Although rocks and soils contain pools of groundwater, this "locked-in" water plays a small role in the hydrologic cycle.

58.9 The Global Hydrologic Cycle
As described in Section 2.4, the unique properties of water are essential to life as we know it on Earth. The numbers in parentheses show the estimated amounts of water (expressed as exagrams, one exagram equalling 10^{18} g or 10^{15} kg) held in or exchanged annually by fluxes (arrows) among compartments of the global ecosystem. The widths of the arrows are proportional to the sizes of the fluxes.

Go to Animated Tutorial 58.2
The Global Hydrologic Cycle
Life10e.com/at58.2

abiotic compartments of the global ecosystem. Because geological, chemical, and biological processes are all important in moving materials around the planet, the pattern of movement of an element is called its **biogeochemical cycle**. The nature of each biogeochemical cycle depends on the physical and chemical properties of the element and on the ways in which it is used by organisms.

Water cycles rapidly through the ecosystem

In addition to being a compartment of the global ecosystem where nutrients are found and a medium that transports those nutrients between other compartments, water is itself a material. Water cycles through the ecosystem in the global **hydrologic cycle** (Figure 58.9).

Energy from the sun drives the hydrologic cycle, taking up water by evaporation from the vast surfaces of the oceans. The cycle operates because more water evaporates from the ocean surfaces than is returned to the oceans in the form of precipitation. On land, water evaporates from soils, lakes, and rivers and is taken up from the leaves of plants by transpiration. However, the total amount evaporated and transpired from terrestrial surfaces is less than the amount that falls on them as precipitation. Excess terrestrial precipitation eventually returns to the oceans via streams, coastal runoff, and groundwater flows. More than half of this volume of water is carried back to the oceans by Earth's four largest rivers: the Amazon in South America, the Nile in Africa, the Mississippi in North America, and the Yangtze in Asia.

Despite their relatively small volume, rivers play a disproportionate role in the hydrologic cycle because the average residence time of a water molecule in rivers is only a few years. By comparison, the average residence time of a water molecule in lakes ranges from a few years to centuries. The larger the lake, the longer the residence time; the residence time for water in the top portion of Lake Superior, for example, is 1,500 to 2,000 years, and the water at the bottom of this massive lake never cycles. In the oceans the average residence time of a water molecule is about 3,000 years. Other pools of water include glaciers (with residence times of 20–100 years), seasonal snow cover (a few months), and soil moisture (1–2 months). The average residence time of water in the bodies of organisms is particularly brief, averaging just under a week.

Although large amounts of groundwater are present in underground pools called **aquifers**, this water has a long residence time underground and plays only a small role in the hydrologic cycle. In some places, however, aquifers are being depleted because humans are using groundwater more rapidly than it can be replaced, primarily by pumping it for irrigation. Much of the groundwater being used today in the Northern Hemisphere was deposited during the most recent ice age, when regional precipitation was much greater than it is now. Using this groundwater for irrigation and other purposes has increased flows of water to the oceans and has contributed to the sea level rise of the past century.

The effects of groundwater depletion are already being felt. On the North China Plain, depletion of shallow aquifers is forcing people to sink wells more than 1,000 meters deep to reach groundwater. Worldwide, more than 1 billion people have no access to safe drinking water. If current water consumption patterns continue, by 2025 at least 48 percent of the current world population will live in areas with inadequate water supplies. However, per capita water consumption in the United States and Europe is declining as a result of increasing use of water-efficient home appliances as well as implementation of new regulations that restrict water use. If such trends continue, global water use in 2025 could be lower than it is today, despite continued population growth.

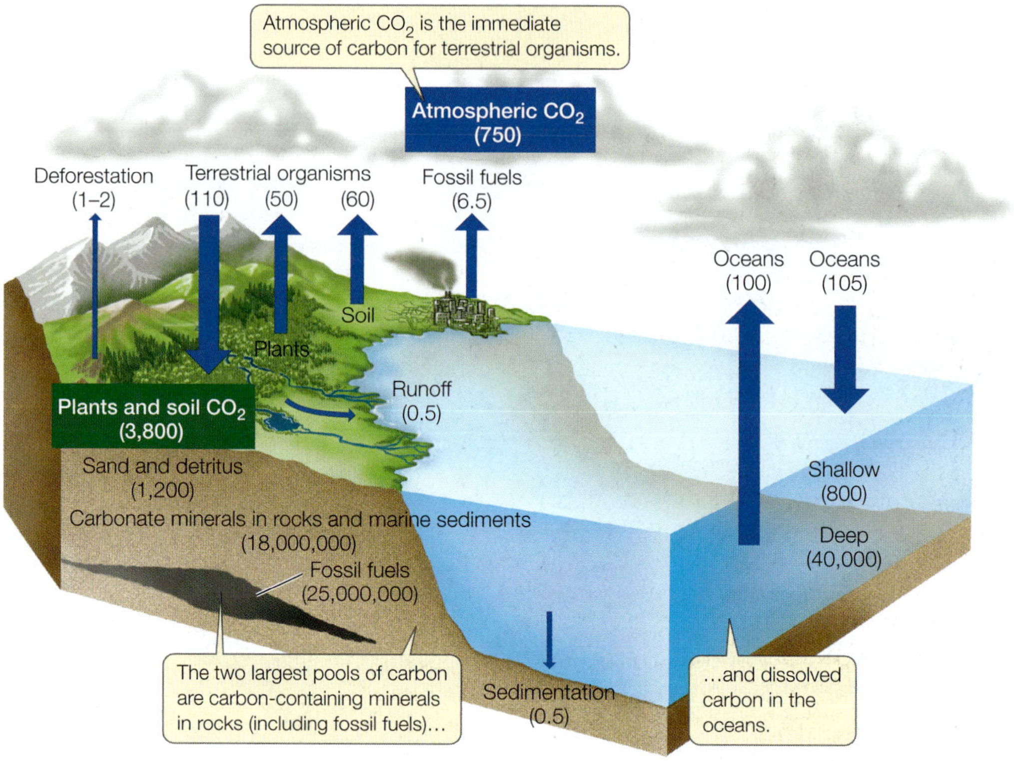
Atmospheric CO_2 is the immediate source of carbon for terrestrial organisms.
Atmospheric CO_2 (750)
Deforestation (1–2)
Terrestrial organisms (110) (50) (60)
Fossil fuels (6.5)
Oceans (100)
Oceans (105)
Soil
Plants
Plants and soil CO_2 (3,800)
Runoff (0.5)
Sand and detritus (1,200)
Shallow (800)
Carbonate minerals in rocks and marine sediments (18,000,000)
Deep (40,000)
Fossil fuels (25,000,000)
The two largest pools of carbon are carbon-containing minerals in rocks (including fossil fuels)...
Sedimentation (0.5)
...and dissolved carbon in the oceans.

58.10 The Global Carbon Cycle Carbon is the basis of the organic molecules essential to life. The numbers in parentheses show the quantities of carbon in petagrams (1 pg = 10^{15} g or 10^{12} kg) held in or exchanged annually by fluxes (arrows) among compartments of the global ecosystem. The widths of the arrows are proportional to the sizes of the fluxes.

Go to Animated Tutorial 58.3 The Global Carbon Cycle Life10e.com/at58.3

The carbon cycle has been altered by human activities

As described in Part One of this book, all of the important macromolecules that make up living organisms contain carbon, and much of the energy that organisms use to fuel their metabolic activities is stored in carbon-containing (organic) compounds. Carbon in the atmosphere, in the form of CO_2, is taken up by autotrophs and incorporated into organic molecules by photosynthesis. All heterotrophic organisms obtain carbon by consuming autotrophs or other heterotrophs, their remains, or their waste products.

On land, biological processes move carbon directly between organisms and the atmosphere as terrestrial organisms take up carbon during photosynthesis and return it to the atmosphere through respiration and metabolism. In contrast, carbon dioxide moves into ocean waters from the atmosphere primarily by simple diffusion at the ocean surface; this dissolved CO_2 is the source of the carbon used by marine primary producers (**Figure 58.10**). Even taken together, however, the amounts of carbon in the atmosphere, in soils, and in living and dead organisms are dwarfed by the vast quantities of carbon stored in terrestrial rocks, in fossil fuels, in marine sediments, and in seawater in the form of carbonate ions (CO_3^{-2}) or bicarbonate ions (HCO_3^-).

At times in the remote past, quantities of carbon were removed from active cycling when organisms died in large numbers and were buried in sediments lacking oxygen. In such anaerobic environments, with no detritivores to reduce organic carbon to CO_2, organic molecules accumulate and are eventually transformed into deposits of oil, natural gas, coal, or peat. Humans have discovered and used these fossil fuels at ever-increasing rates during the past 150 years. As a result, CO_2, one of the final products of burning fossil fuels, is being released into today's atmosphere faster than it is dissolving in the oceans or being incorporated into terrestrial biomass (**Figure 58.11**). Based on a variety of calculations, atmospheric scientists estimate that before the Industrial Revolution, the concentration of CO_2 in Earth's atmosphere was probably about 265 parts per million. Today it is 392 parts per million, representing a rate of increase more than 10 times faster than at any other time for millions of years.

WHERE HAS ALL THE CARBON GONE? Less than half of the CO_2 released into the atmosphere by human activities remains in

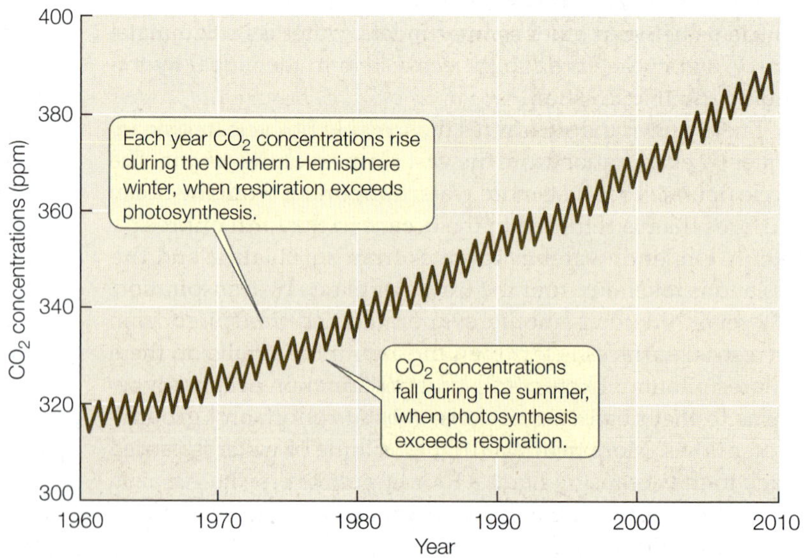
400
380
360
340
320
300
CO_2 concentrations (ppm)
Each year CO_2 concentrations rise during the Northern Hemisphere winter, when respiration exceeds photosynthesis.
CO_2 concentrations fall during the summer, when photosynthesis exceeds respiration.
1960
1970
1980
1990
2000
2010
Year

58.11 Atmospheric CO_2 Concentrations Are Increasing Carbon dioxide concentrations, expressed as parts per million by volume of dry air, have been recorded since 1960 on top of Mauna Loa, Hawaii, far from most sources of human-generated CO_2 emissions. Although concentrations vary seasonally, the trend has been consistently upward.

the atmosphere. Where does the rest of the CO_2 wind up? Much of it is dissolved in the oceans in inorganic forms. Over decades to centuries, the oceans, which contain 50 times more carbon than the atmosphere, determine atmospheric CO_2 concentrations. The rate at which CO_2 diffuses from the atmosphere into the oceans depends in part on photosynthesis by phytoplankton in the surface waters. These organisms remove dissolved CO_2 from water, thereby increasing the rate at which atmospheric CO_2 is absorbed by surface waters. In addition, many marine organisms (including clams, oysters, corals, and planktonic foraminiferans) incorporate carbon in their shells and other structures in the form of calcium carbonate ($CaCO_3$), which is synthesized by combining bicarbonate ions (HCO_3^-) and calcium ions (Ca^{2+}) dissolved in seawater. When these organisms die, those shells and their embedded carbon sink to the ocean floor.

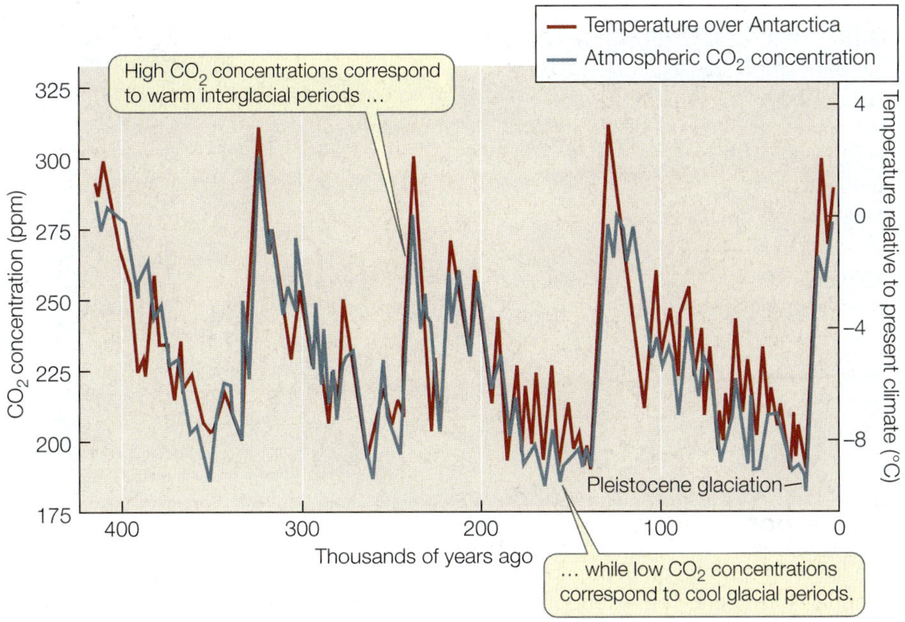
Temperature over Antarctica
Atmospheric CO_2 concentration
High CO_2 concentrations correspond to warm interglacial periods ...
CO_2 concentration (ppm)
325
300
275
250
225
200
175
Temperature relative to present climate (°C)
4
0
–4
–8
Pleistocene glaciation
400
300
200
100
0
Thousands of years ago
... while low CO_2 concentrations correspond to cool glacial periods.

58.12 Higher Atmospheric CO_2 Concentrations Correlate with Warmer Temperatures Atmospheric concentrations of CO_2 (measured in air bubbles trapped in Antarctic ice) have varied with temperatures over Antarctica (estimated by a technique known as oxygen isotope analysis).

Today's oceans absorb 20 to 25 million tons of CO_2 from the atmosphere each day—more than at any time during the past 20 million years. As a result, water near the ocean surface is becoming more acidic. As CO_2 concentrations in the atmosphere rise, more of the gas diffuses into the water at the ocean surface, where it reacts with water to form carbonic acid (H_2CO_3). As levels of carbonic acid rise, the pH of seawater drops. This increase in acidity can have negative effects on many marine organisms, particularly corals. The combination of decreasing pH and increasing water temperature to which corals are being exposed kills their symbiotic algae, "bleaching" the corals and killing them as well (see Figure 27.21). Because so many other reef species depend on corals and the structure they provide, the entire reef community can collapse if the corals fail to thrive.

Photosynthesis by terrestrial vegetation, principally in forests and savannas, typically absorbs about the same amount of carbon that is released by terrestrial metabolism—about half of it released by plants and half by microbes in the soil that break down plant detritus. The photosynthetic consumption of CO_2 currently exceeds the metabolic production of CO_2, which means Earth's terrestrial vegetation is storing carbon that would otherwise be increasing atmospheric CO_2 concentrations—but we cannot count on terrestrial vegetation to store the vast amounts of excess CO_2 that human activities produce. Furthermore, climate warming (another result of increasing atmospheric CO_2 concentrations, as we will see next) increases plant metabolism and is thus likely to increase the flux of CO_2 from vegetation into the atmosphere.

ATMOSPHERIC CO_2 AND GLOBAL CLIMATE CHANGE Carbon dioxide is a greenhouse gas, so we would expect increasing atmospheric CO_2 concentrations to raise temperatures at Earth's surface. What evidence do we have that this is occurring? Measurements of gases in air trapped in the Antarctic and Greenland ice caps show that atmospheric CO_2 concentrations have been higher when Earth has been warmer and lower when it has been cooler (**Figure 58.12**). For example, the atmospheric CO_2 concentration was very low during the most recent glaciation, between 30,000 and 15,000 years ago, when temperatures were presumably much colder than they are today. In contrast, during a warm interval 5,000 years ago, atmospheric CO_2 concentration may have been slightly higher than it is today.

How global climates and ecosystems will change in response to this rapid warming is a subject of intense investigation. Complex computer models of the global ecosystem indicate that a doubling of today's atmospheric CO_2 concentration would increase mean annual temperatures worldwide and would probably result in droughts in the central regions of continents, but would increase precipitation in coastal areas. Global warming has already resulted in the shrinking of Arctic sea ice (**Figure 58.13**). The five smallest expanses of Arctic sea ice ever measured were recorded in the past six years; overall Arctic sea ice extent is declining by 3.5 to 12 percent per decade. Based on scientific climate models, if glacial ice continues to melt and temperatures continue to rise, sea level will rise (because of both thermal expansion of ocean waters and the addition of glacial meltwater), increasing the probability of flooding of coastal cities and agricultural lands.

Global climate warming is having profound effects on the distributions and abundances of species and, consequently, on species interactions. One clear example is an increase in the severity of insect infestations in certain temperate forest communities. Pine trees in some temperate forests are attacked by pine bark beetles, which carry with them a symbiotic fungus that infects the trees and helps the beetles overcome the trees' defenses (see Figure 56.9). In Colorado, cold winters have

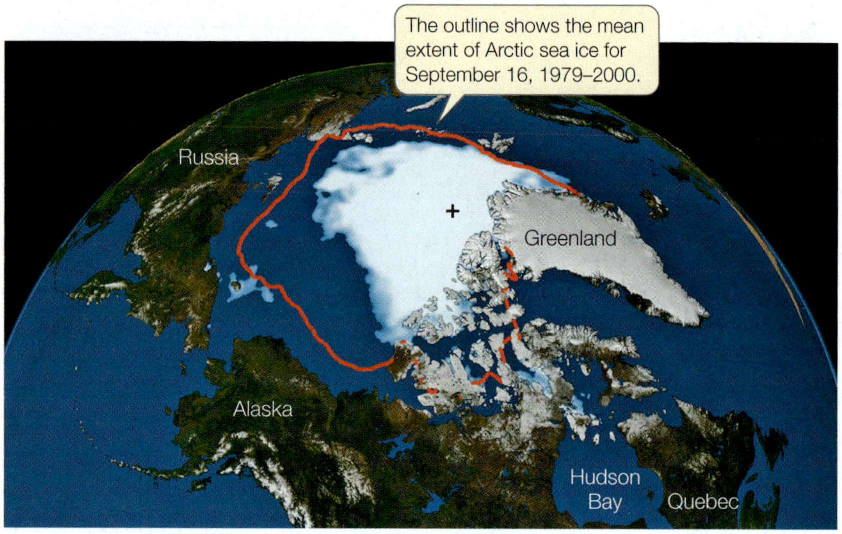
The outline shows the mean extent of Arctic sea ice for September 16, 1979–2000.
Russia
Greenland
Alaska
Hudson Bay
Quebec

58.13 Shrinking Ice Caps A NASA satellite image shows the extent of Arctic sea ice (bright white) on September 16, 2012. The plus symbol marks the geographic North Pole. The extent of the ice cap in 2012 was about one-half that of the September 16 mean for the years 1979–2000.

historically limited the ability of these beetles to kill many trees. In 2008, however, Colorado experienced an outbreak of mountain pine beetles that destroyed more than 400,000 hectares (1 million acres) of trees. The lack of an extended period of below-zero temperatures in the previous winter had allowed large numbers of overwintering beetles to survive.

The nitrogen cycle depends on both biotic and abiotic processes

Nitrogen gas is the most abundant gas in Earth's atmosphere, but most organisms cannot use nitrogen in its gaseous form. Only a few species of microorganisms can convert atmospheric N_2 into forms such as ammonia and nitrate that are usable by plants, a process called nitrogen fixation (see Figure 36.10). Other microorganisms carry out denitrification, the principal process that removes nitrogen from the biosphere and returns it to the atmosphere as N_2. Collectively this microbial processing of nitrogen accounts for about 95 percent of all natural nitrogen flux on Earth (**Figure 58.14**).

Abiotic weathering is an important source of nitrogen in some terrestrial ecosystems. In temperate forests growing on land underlain by nitrogen-rich sedimentary rocks, for example, the soils and foliage have 50 percent higher levels of nitrogen than in temperate forests growing on land underlain by nitrogen-poor igneous rocks.

All living organisms require nitrogen, and the inability of the vast majority of organisms to use N_2 means that usable nitrogen is often in short supply. Populations of nitrogen-fixing organisms rarely increase in abundance to the extent that nitrogen is no longer limiting because the end products of their nitrogen fixation are rapidly lost from ecosystems (ammonia by vaporization and denitrification; and nitrate, which is highly water-soluble, by leaching).

Human activities that fix nitrogen, such as the manufacture of artificial fertilizers, have had some unanticipated effects on the nitrogen cycle. The extensive use of artificial fertilizers on agricultural crops, coupled with the burning of fossil fuels (which generates nitric oxide and nitrogen dioxide), has resulted in total nitrogen fixation by humans being nearly equal

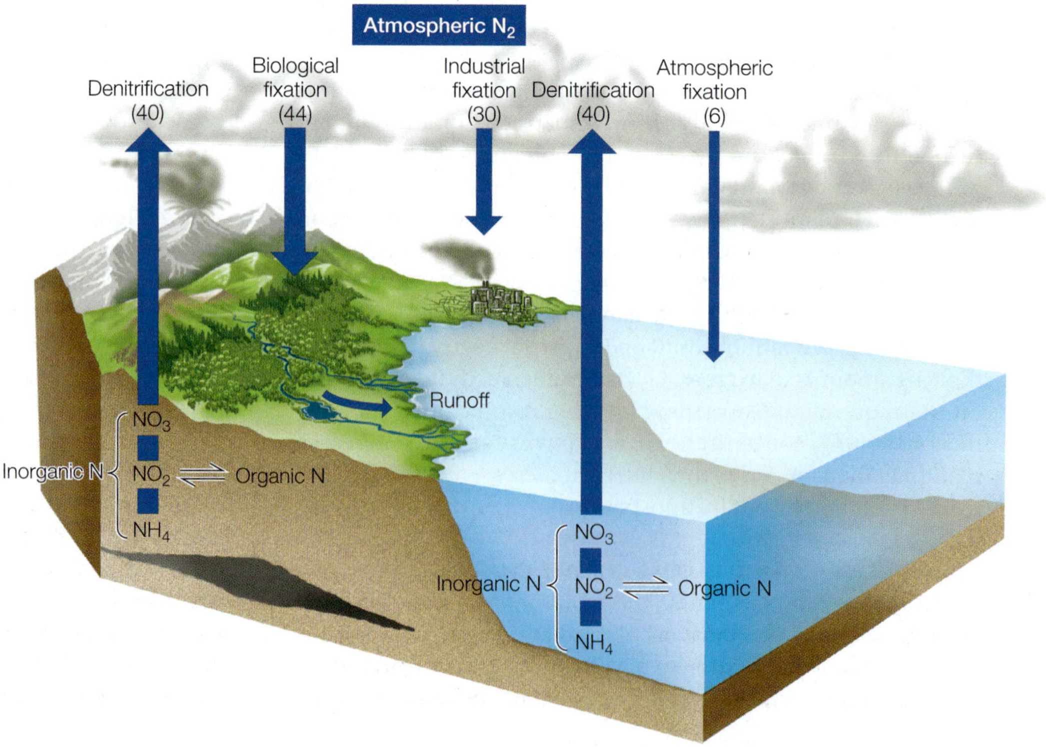
Atmospheric N2
Denitrification (40)
Biological fixation (44)
Industrial fixation (30)
Denitrification (40)
Atmospheric fixation (6)
Runoff
NO3
Inorganic N
NO2
Organic N
NH4
NO3
Inorganic N
NO2
Organic N
NH4

58.14 The Global Nitrogen Cycle The largest pool of nitrogen is held in the atmosphere in the form of nitrogen gas, N_2. Nitrogen cycles through the biosphere primarily via the processes of nitrogen fixation, which converts inorganic nitrogen to an organic form usable by plants, and denitrification, which returns N_2 to the atmosphere. The numbers in parentheses show the quantities of nitrogen in teragrams (1 tg = 10^{12} g or 10^9 kg) exchanged annually by compartments of the global ecosystem. The widths of the arrows are proportional to the sizes of the fluxes.

Go to Animated Tutorial 58.4
The Global Nitrogen Cycle
Life10e.com/at58.4

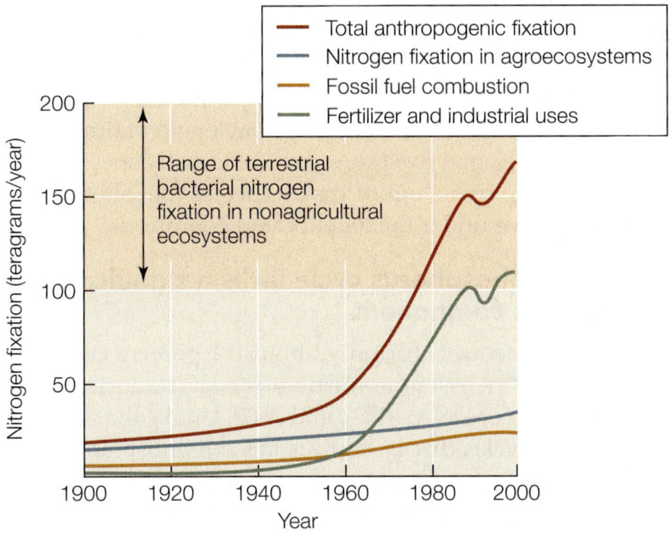
Total anthropogenic fixation
Nitrogen fixation in agroecosystems
Fossil fuel combustion
Fertilizer and industrial uses
Range of terrestrial bacterial nitrogen fixation in nonagricultural ecosystems
Nitrogen fixation (teragrams/year)
200
150
100
50
1900
1920
1940
1960
1980
2000
Year

58.15 Human Activities Have Increased Nitrogen Fixation Most of the nitrogen fixed by industrial processes is used in agricultural fertilizers. Some fixation is a by-product of fossil fuel combustion. Fixation by natural processes in managed agroecosystems (e.g., by legumes grown as crops) also contributes to anthropogenic (human-caused) effects on nitrogen flux.

to global natural nitrogen fixation (**Figure 58.15**). This human-generated nitrogen flux has been increasing over the past half-century and is expected to continue to increase.

Eutrophication is an increase in biomass production in a body of water due to inputs of nutrients. Eutrophication occurs naturally as part of the aging process in lakes. As lakes become more shallow with the accumulation of sediments brought in by streams, their water warms more rapidly with the onset of summer, and exploding populations of photosynthetic cyanobacteria and single-celled algae, called blooms, can deplete oxygen levels (see Figure 26.9C). Human nutrient inputs greatly increase the likelihood and frequency of these blooms. When more nitrogen fertilizer is applied to croplands than can be taken up by the crops, the excess nitrogen moves out of the system in surface runoff, or downward into groundwater, and ultimately ends up in rivers, lakes, and oceans. The dead zone that has formed near the mouth of the Mississippi River in the Gulf of Mexico, described at the opening of this chapter, is a result of water flows from agricultural fields in the U.S. interior carrying high concentrations of nitrogen from fertilizer.

The human increase in nitrogen fixation has also increased atmospheric concentrations of the greenhouse gas nitrous oxide (N_2O), resulting in the production of tropospheric ozone—also a greenhouse gas—and smog. Some of the nitrogen that enters the atmosphere falls back to land in precipitation or as dry particles. This deposition of nitrogen from the atmosphere has increased dramatically during recent decades. Nitrogen deposition affects the composition of terrestrial vegetation by favoring those plant species that are best adapted to take advantage of high nutrient levels, which then outcompete other species. Spatial variation in nitrogen deposition rates has allowed ecologists to determine that plant species richness in grasslands declines as the rate of nitrogen deposition increases. Rates of nitrogen deposition are high enough over much of Europe and eastern North America to cause substantial reductions in species richness in grasslands on both continents.

The burning of fossil fuels affects the sulfur cycle

As a component of proteins, sulfur is required by all organisms. Most of Earth's sulfur supply is locked up in rocks on land and as sulfate salts in deep-sea sediments, but some sulfur moves between the atmosphere and land. Emissions of the gases sulfur dioxide (SO_2) and hydrogen sulfide (H_2S) from volcanoes account for between10 and 20 percent of the total natural abiotic flux of sulfur to the atmosphere, but they occur only intermittently (although volcanic eruptions spew great quantities of sulfur over broad areas, they are rare events). In the atmosphere, H_2S can combine with oxygen to form SO_2, which dissolves in atmospheric water and reaches the ground as sulfuric acid in precipitation and fog.

When sulfur in the soil comes in contact with atmospheric oxygen, it is converted to sulfate salts, which can be taken up by plants and incorporated into proteins. This sulfur ultimately is returned to the atmosphere via microbial decomposition. In marine systems, too, microbial decomposition is important in returning sulfur to the atmosphere. Many marine phytoplankton and seaweeds manufacture large quantities of a sulfur-containing compound (dimethylsulfoniopropionate, or DMSP) to maintain their salt and water balance. When broken down, DMSP releases dimethyl sulfide (CH_3SCH_3), the principal odorant of rotting seaweed stench. Because the quantities of phytoplankton in the oceans are enormous, dimethyl sulfide production accounts for about half of the biotic component of the global sulfur cycle.

Atmospheric sulfur plays an important role in global climate. Even if air is moist, clouds do not form readily unless there are small particles in the atmosphere around which water can condense. Dimethyl sulfide is the major component of such particles, so increases in atmospheric sulfur concentrations increase cloud cover and reduce the amount of incoming solar radiation that ultimately reaches Earth's surface.

Human use of fossil fuels alters the sulfur cycle as well as the carbon and nitrogen cycles. The combustion of fossil fuels releases sulfur in the form of SO_2, as well as nitrogen in the form of nitrogen dioxide (NO_2), into the atmosphere. Both compounds react with water molecules in the atmosphere to form sulfuric acid (H_2SO_4) and nitric acid (HNO_3), respectively. These acids can travel hundreds of kilometers in the atmosphere before they settle to Earth as dry particles or in precipitation. Rain or snow that contains enough nitric and sulfuric acid to lower its pH is called **acid precipitation**.

Acid precipitation now falls in all major industrialized countries and is particularly widespread in eastern North America and Europe. The normal pH of unpolluted precipitation is about 5.6, but precipitation in New England now averages about pH 4.5, and there have been occasional rainfalls

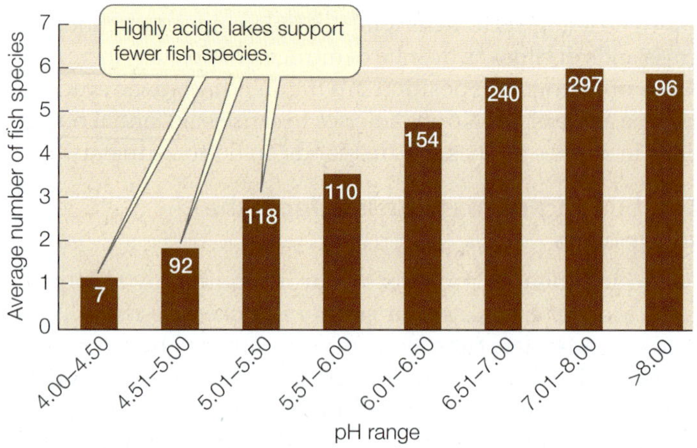
Highly acidic lakes support fewer fish species.
Average number of fish species
pH range
7
92
118
110
154
240
297
96
4.00–4.50
4.51–5.00
5.01–5.50
5.51–6.00
6.01–6.50
6.51–7.00
7.01–8.00
>8.00

58.16 Acidification Reduces Fish Species Richness The average number of fish species found in lakes sampled in the Adirondack region of New York is directly correlated with pH. Numbers in the bars indicate the number of lakes in each pH range.

and snowfalls with a pH as low as 3.0. Precipitation with a pH of about 3.5 or lower damages the leaves of plants and reduces rates of photosynthesis. Acidification of lakes in the Adirondack region of New York State has reduced fish species richness by causing the extinction of acid-sensitive species (**Figure 58.16**). Many invertebrates that are primary consumers in aquatic communities are sensitive to pH; in particular, multiple species of mayflies and caddisflies (important food resources for fish) have experienced local population reductions in acidified streams around the world. Even when its effects are not lethal, acidification can have subtle effects on behavior that reduce the viability of aquatic organisms. Diving beetles, for example, lose their ability to regulate their underwater oxygen supply when pH levels drop significantly.

Regulations instituted by the U.S. Clean Air Act in 1990 have raised the pH of precipitation in much of the eastern United States, primarily by lowering sulfur emissions. There are indications that once emissions have been reduced, acidified aquatic systems can recover quickly. David Schindler at the University of Alberta studied the effects of acid precipitation by adding enough H_2SO_4 to two small Canadian lakes to reduce their pH from about 6.6 to a moderately acidic level of 5.2. In both lakes, nitrifying bacteria failed to survive, and nitrogen cycling within the lake was blocked. When Schindler stopped adding acid to one of the lakes, its pH returned to its original value in about a year and nitrification resumed. That said, the larger organisms in acidified aquatic systems may be slower to bounce back. In Wales, investigators conducted a 25-year study of 14 rivers to determine the impacts of acid rain reduction. They were disappointed to find that, even after that length of time, only 4 insect species—2 mayfly species and 2 caddisfly species—had recolonized the rivers out of the 29 species that should have been able to live under the ameliorated conditions.

The global phosphorus cycle lacks a significant atmospheric component

Phosphorus accounts for only about 0.1 percent of Earth's crust, but it is an essential nutrient for all life forms. It is a key component of DNA, RNA, and ATP. Unlike the other biogeochemical cycles discussed thus far, the phosphorus cycle lacks a significant gaseous component (**Figure 58.17**). Some phosphorus is transported on dust particles, but very little of the phosphorus cycle takes place in the atmosphere. Most of Earth's phosphorus is present in the form of phosphate salts in rocks and deep-sea sediments. Abiotic cycling of phosphorus takes millions of years because the processes of sedimentary rock formation, uplift, and weathering all take a long time. In contrast, phosphorus often cycles rapidly among organisms, and it is often a limiting factor for their growth, particularly for plants. Artificial fertilizers routinely include phosphorus as well as nitrogen.

Human activity has radically accelerated some parts of the phosphorus cycle. One consequence of the massive use of artificial fertilizers described above is that between 10.5 and 15.5 teragrams (1 tg = 10^{12} g or 10^9 kg) of phosphorus accumulate in

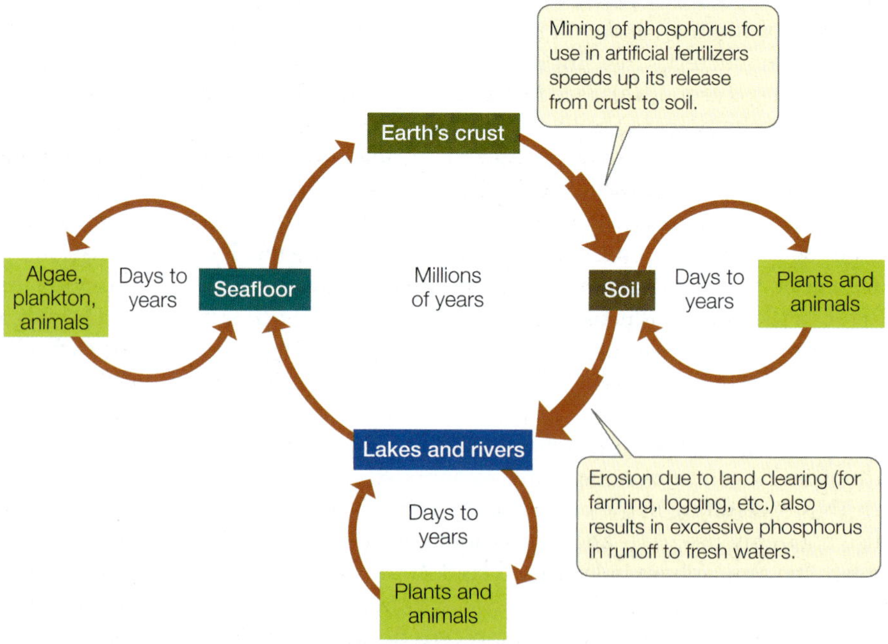
Mining of phosphorus for use in artificial fertilizers speeds up its release from crust to soil.
Earth's crust
Algae, plankton, animals
Days to years
Seafloor
Millions of years
Soil
Days to years
Plants and animals
Lakes and rivers
Days to years
Plants and animals
Erosion due to land clearing (for farming, logging, etc.) also results in excessive phosphorus in runoff to fresh waters.

58.17 The Phosphorus Cycle The widths of the arrows are proportional to the sizes of the fluxes. Two large increases in phosphorus flux are the result of human activities.

soils each year, primarily in agricultural fields. When the concentration of phosphorus in soils exceeds the capacity of plants to take it up, the excess moves into streams and lakes. Soil erosion due to the clearing of land for purposes such as agriculture and logging also increases the amount of phosphorus and other nutrients in runoff. Phosphorus is a limiting nutrient in many lakes, so when it enters those lakes through runoff, eutrophication results, in much the same way that nitrate enrichment can cause eutrophication.

The capacity of phosphorus-charged runoff to cause extensive eutrophication was graphically illustrated almost 50 years ago. A technological innovation in the formulation of laundry detergents in the early 1960s was the use of sodium tripolyphosphate (STPP) ($Na_5P_3O_{10}$) as a water softener and dirt-breaking agent to enhance cleaning efficiency. Widespread adoption of these new detergents led to massive phosphorus enrichment of lakes and streams and resulting eutrophication and blooms of phytoplankton. Water quality declined so much across the United States that phosphate-based detergents were banned in many states. In today's detergents, phosphonates—forms of phosphorus that do not appear to promote algal growth—and aluminum silicates perform the functions STPP used to serve.

Detergents, however, are only one agent contributing to eutrophication. Human waste is rich in phosphorus, as are manure from domesticated animals and industrial wastes of various types. Two hundred years ago, Lake Erie, one of the Great Lakes on the border between the United States and Canada, had only moderate phytoplankton populations and clear, oxygenated water. With increasing industrialization in the early part of the twentieth century, nutrient concentrations in the lake increased greatly, and algae proliferated. At the water filtration plant in Cleveland, Ohio, algae increased from 81 individuals per milliliter in 1929 to 2,423 per milliliter in 1962. Populations of bacteria also increased; *Escherichia coli* levels rose so high that many of the lake's beaches were declared health hazards. Since 1972 the United States and Canada have invested more than U.S. $9 billion to improve municipal waste treatment facilities and reduce discharges of pollutants. As a result, the amount of phosphorus added to Lake Erie has decreased more than 80 percent from its highest level, and phosphorus concentrations in the lake have declined substantially. The deeper waters of Lake Erie are still oxygen-poor during the summer months, but the rate of oxygen depletion is declining.

We could greatly reduce phosphorus pollution by recovering and recycling phosphorus. The phosphorus discarded in sewage and animal wastes could supply much of the needs of the detergent and fertilizer industries. More careful application of fertilizers on agricultural lands could reduce the rate of phosphorus accumulation in soils without reducing crop yields. However, reduction of phosphorus concentrations in soils will take many decades after remedial actions are initiated, and eutrophication of lakes and streams may persist even after these actions are taken.

Other biogeochemical cycles are also important

Other elements are important to the global ecosystem because they are essential nutrients for organisms, even though they are needed only in very small amounts. One such element is iron (Fe), an essential micronutrient for almost all organisms. Iron is a key component of the enzymes involved in chlorophyll synthesis as well as an essential component of many animal enzymes. Iron confers oxygen-binding ability on hemoglobin in vertebrate blood. Members of the cytochrome P450 family of enzymes, which in most aerobic organisms play a central role in detoxifying environmental poisons, rely on iron for their catalytic activity.

Iron is readily available on land in rocks and minerals. It moves into coastal waters in streams and into the open oceans in atmospheric dust. Because iron is insoluble in oxygenated water, it rapidly sinks to the ocean floor. Therefore in most marine communities the rate of photosynthesis is limited by iron. In 1996 investigators launched an ecosystem-scale experiment in which surface waters of the equatorial Pacific Ocean were seeded with dissolved iron. The response was a tremendous phytoplankton bloom, accompanied by massive uptake of nitrate and carbon dioxide, which had apparently been underused because of iron limitation.

Iodine is an example of an element that is globally rare but is an essential micronutrient for living organisms. Endothermic vertebrates in particular require iodine in concentrations that exceed the supply in many environments. It is an essential component of the hormone thyroxine, which governs many metabolic processes (see Section 41.4). Iodine is found on land in mineral deposits and in seawater as an inorganic salt.

Biogeochemical cycles interact

The biogeochemical cycles of different elements interact with one another in complex ways, and perturbations of one cycle can have profound effects on other cycles. In recent years, human-induced perturbations have made these interactions glaringly apparent. For example, nitrate released by human activities can have profound effects on the biogeochemical cycle of arsenic. The bottom sediments of some urban lakes contain arsenic levels in excess of 2,000 parts per million. Nitrate is a powerful oxidant, so nitrate pollution can increase the oxidation of arsenic in lake sediments, releasing it into the water in a form that is carcinogenic and has negative effects on embryonic development. Every year scientists discover interactions of which they were previously unaware, and studies exploring biogeochemical interactions are increasing in number.

Changes in atmospheric CO_2 concentrations have been a particular focus of investigation in recent years because of their potential for interacting with other biogeochemical cycles through photosynthesis. A case in point is a study of the effect of elevated atmospheric CO_2 concentrations on rates of nitrogen fixation by microorganisms associated with plant roots. Bruce Hungate and his colleagues grew a nitrogen-fixing vine (i.e., a legume) called Elliott's milkpea (*Galactia elliottii*) under

INVESTIGATING**LIFE**

58.18 Effects of Atmospheric CO_2 Concentration on Nitrogen Fixation Some scientists have hypothesized that rising concentrations of CO_2 in the atmosphere could lead to increased rates of photosynthesis, increased rates of nitrogen fixation, and eventually to the fixation of large amounts of carbon in the soil—which could potentially reduce global warming. In a 7-year-long experiment, Bruce Hungate and his colleagues at Northern Arizona University expected to find that nitrogen fixation in a leguminous vine would be enhanced by increased atmospheric CO_2.[a]

HYPOTHESIS Exposure of legumes to elevated CO_2 concentrations will enhance nitrogen fixation by their symbiotic bacteria.

Method Grow plots of the leguminous vine *Galactia elliottii* under baseline (typical) and artificially elevated concentrations of CO_2. Measure nitrogen fixation over 7 years.

Results

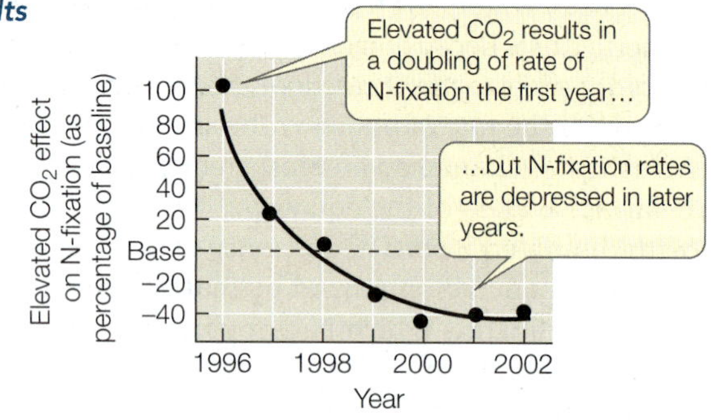
Elevated CO2 results in a doubling of rate of N-fixation the first year...
...but N-fixation rates are depressed in later years.
Elevated CO2 effect on N-fixation (as percentage of baseline)
100
80
60
40
20
Base
–20
–40
1996
1998
2000
2002
Year

In an attempt to explain these results, (which did not support the hypothesis), the investigators measured the concentrations of iron and molybdenum—two micronutrients that are essential for nitrogen fixation—in the leaves of the 7-year-old plants.

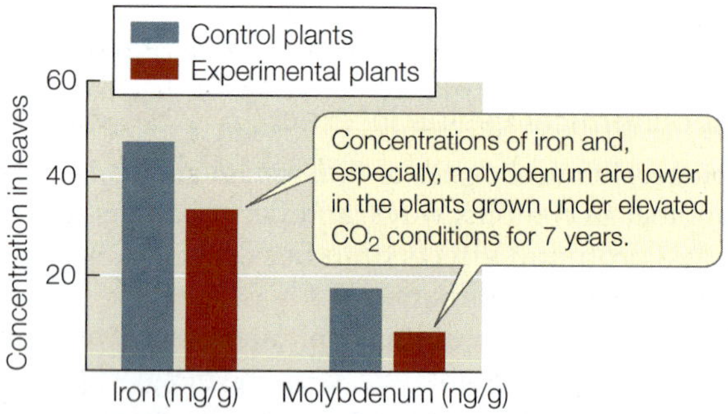
Control plants
Experimental plants
Concentration in leaves
60
40
20
Concentrations of iron and, especially, molybdenum are lower in the plants grown under elevated CO2 conditions for 7 years.
Iron (mg/g)
Molybdenum (ng/g)

CONCLUSION Although enhanced CO_2 levels initially increase nitrogen fixation, lowered levels of essential micronutrients in plants growing under these conditions soon leads to decreased rates of nitrogen fixation.

Go to **BioPortal** for discussion and relevant links for all INVESTIGATING**LIFE** figures.

[a]Hungate, B. A. et al. 2004. *Science* 304: 1291.

WORKING WITH**DATA:**

How Does Molybdenum Concentration Affect Nitrogen Fixation?

Original Paper

Hungate, B. A. et al. 2004. CO_2 elicits long-term decline in nitrogen fixation. *Science* 304: 1291.

Analyze the Data

The experiments in Figure 58.18 were conducted in an oak woodland where *G. elliottii* grew naturally. The investigators used open-top chambers to produce a 350-ppm increase in the concentration of CO_2 in the air around the plants. The study site had a sandy, acidic soil known to have low concentrations of molybdenum.

Because nitrogen-fixing plants are sensitive to light availability, an alternative explanation for the results in Figure 15.18 is that increased shading resulting from greater leaf area of the CO_2-stimulated plants could have caused the subsequent decline in fixation. To test this possibility, the investigators computed the leaf-area index (LAI), a measure of the amount of leaf-surface area per unit of ground area. They found no correlation between LAI and nitrogen fixation, but they did find a positive correlation between concentrations of molybdenum in *G. elliottii* leaves and the rate of nitrogen fixation.

QUESTION 1

The authors claim that this regression analysis provides strong evidence in favor of low availability of molybdenum rather than low light availability as the reason for the decline in rate of nitrogen fixation. Do you agree? Why or why not?

QUESTION 2

What does this experiment suggest about the kinds of future studies and the range of ecosystem types and nutrient elements that should be investigated to determine the likely overall response of Earth's ecosystems to increasing atmospheric concentrations of CO_2?

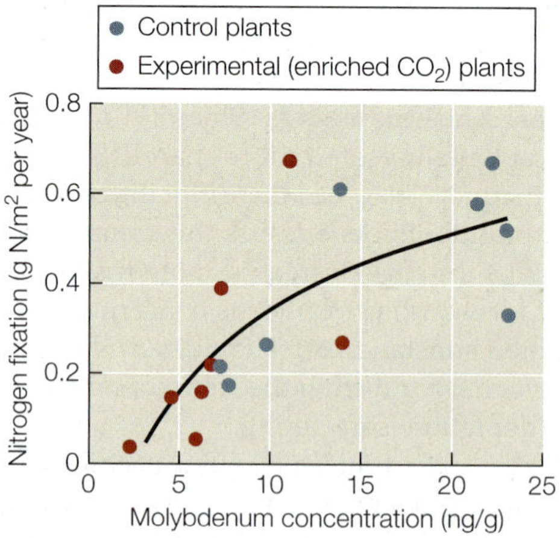
Control plants
Experimental (enriched CO2) plants
Nitrogen fixation (g N/m2 per year)
0.8
0.6
0.4
0.2
0
0
5
10
15
20
25
Molybdenum concentration (ng/g)

Go to **BioPortal** for all WORKING WITH**DATA** exercises

artificially increased CO_2 concentrations. Higher CO_2 concentrations led to an enhancement of nitrogen fixation during the first year of the experiment, but surprisingly, the positive effect disappeared by the third year, and elevated CO_2 concentrations actually *reduced* nitrogen fixation below baseline levels during the fourth, fifth, sixth, and seventh years of the experiment (**Figure 58.18**).

The investigators suspected that a lack of micronutrients such as iron and molybdenum had caused the reduction in nitrogen fixation, so they measured concentrations of those elements in the leaves of the *G. elliottii* grown under high CO_2 concentrations. They found that concentrations of molybdenum in those plants were particularly low. Hungate and his colleagues proposed a mechanism by which this could occur: elevated CO_2 concentrations could increase the acidity of water in soil by enhancing the rate of carbonic acid formation. By enhancing photosynthesis, elevated CO_2 concentrations could increase the accumulation of organic matter in the soil. Both types of change would increase the tendency of iron and molybdenum to bind to soil particles, which would reduce their availability to nitrogen-fixing bacteria and cause a decrease in nitrogen fixation rates.

RECAP 58.3

The pattern of movement of a chemical element through the biotic and abiotic compartments of the global ecosystem is its biogeochemical cycle. Human activities have affected many biogeochemical cycles, especially those of water, carbon, nitrogen, sulfur, and phosphorus. Increasing concentrations of carbon dioxide and other greenhouse gases in the atmosphere are implicated in global climate change.

- Describe the global hydrologic cycle and explain what drives it. **See p. 1215 and Figure 58.9**
- How do biological processes move carbon from the atmosphere to land and then return it to the atmosphere? **See p. 1216 and Figure 58.10**
- What are some results of human-induced alterations of the sulfur and nitrogen cycles? **See pp. 1219–1221**
- Name two elements that can cause eutrophication in aquatic ecosystems and describe their effects on those systems. **See pp. 1219 and 1221**

The biogeochemical cycles of chemical elements are intimately involved in ecosystem function. Just as human alterations of those cycles are having many effects on ecosystems worldwide, the resulting changes in those ecosystems are having profound effects on human lives.

58.4 What Goods and Services Do Ecosystems Provide?

Although it seems obvious today that humans depend on natural ecosystems for survival, explicit recognition of the value of those ecosystems is rather recent. Environmental writers introduced the idea of "natural capital" in the 1940s; it was in 1970 that ecosystems were first said to provide people with a variety of "goods and services." The goods include food, clean water, clean air, fiber, building materials, and fuel; the services include flood control and water quality, soil stabilization, pollination, and climate regulation. Most of these benefits either are irreplaceable, or the technology necessary to replace them is prohibitively expensive. For example, fresh drinking water can be provided by desalinating seawater, but only at great cost. The aesthetic, psychological, spiritual, and recreational benefits of ecosystems are less tangible, but no less important, and no more easily replaced.

Humans have increasingly altered Earth's ecosystems in ways that increase the systems' capacity to provide us with necessities such as food, fresh water, timber, fiber, and fuel. The benefits of these ecosystem alterations have not been equally distributed, and some human populations have been harmed by manipulations of natural ecosystems. Moreover, short-term increases in some ecosystem goods and services often comes at the cost of the long-term degradation of others.

Although humans have been altering natural ecosystems for millennia, the pace and scope of the shift to intense human use have increased considerably in the past century. More land was converted to cropland between 1950 and 1980 than in the 150 years between 1700 and 1850. Ecosystem conversions have been particularly rapid in tropical and subtropical biomes. Aquatic ecosystems have suffered losses at an increasing pace as well. In freshwater ecosystems, so much water is now impounded behind dams that artificial reservoirs hold about six times as much water today as do natural rivers. These freshwater systems are being rapidly depleted: the amount of water withdrawn from rivers—most of it for agriculture—has doubled since 1960. In terms of nutrient cycling, more than half of all the artificial nitrogen fertilizer ever used on Earth has been used since 1985.

Human alteration of ecosystems has had many positive effects on human health and prosperity, but it necessarily involves trade-offs. Agriculture, for example, feeds and employs huge numbers of people. But the spread of agriculture into marginal lands may degrade soils and compromise the ability of ecosystems to provide clean water, as when overuse of artificial fertilizers results in eutrophication. Extensive use of pesticides controls insect pests, but also reduces populations of pollinators and the services they provide to both crops and native plants.

Similarly, the loss of wetlands and other natural buffers has reduced the ability of ecosystems to regulate flooding and other natural hazards. The damage from the tsunami that hit Indonesia and other Southeast Asian countries in December 2004 was greater in many places than it would have been had the mangrove forests that protect the coast not been cut down and converted to cropland. Hurricane Katrina, which struck the U.S. Gulf Coast less than a year later, would not have caused as much flooding in New Orleans had the wetlands surrounding the city been intact. Katrina's devastating effects were due in part to a situation that had been developing for decades.

New Orleans is located on the Mississippi River delta. Much of the city lies below sea level, buffered by dams and levees constructed by the Army Corps of Engineers. The upstream dams that protect New Orleans from flooding also prevent the river from depositing the sediments that have sustained the surrounding delta wetlands for centuries. Oil and natural gas producers have cut thousands of small canals through those wetlands in order to lay pipelines and install drilling rigs, and the extraction of oil and gas from beneath the land has caused it to sink. Increased dredging of shipping lanes and rising sea levels due to global warming have contributed to a rise in salinity, killing off many of the great cypress tree swamps. These extensive alterations resulted in the loss of more than 80 percent (more than 50,000 hectares, or 1.2 million acres) of the delta wetlands between 1930 and 2005. By the time Katrina made landfall, those wetlands could no longer protect New Orleans from flooding. Storm surges raced along the paths carved by canals and shipping lanes to breach the levees, inundating much of the city.

As it flooded New Orleans, Hurricane Katrina raised awareness and appreciation of an ecosystem hitherto taken for granted by most people. Crucial not only for their flood control services, the delta wetlands provide winter habitat for some 70 percent of the migrating birds in the huge Mississippi Valley. They are also the spawning grounds for marine organisms, some of which are commercially valuable. The delta's famous shrimping industry contributes about 30 percent by weight of the total commercial fish harvest in the continental United States. The importance of coastal wetlands—and, indeed, of a wide range of other ecosystems—to human well-being mandates careful ecosystem management to guarantee a sustained flow of ecosystem goods and services.

RECAP 58.4

Ecosystems provide human society with indispensable goods and services. Altering ecosystems can compromise their ability to provide these goods and services.

- What are some of the essential goods and services that ecosystems provide to humans? See p. 1223
- Give an example of a human effort to increase the provision of some ecosystem goods or services that caused the degradation of others. See p. 1223

How can we meet the challenge of obtaining goods and services from ecosystems without compromising their ability to provide those goods and services over the long term? What options exist for sustainable management of ecosystems?

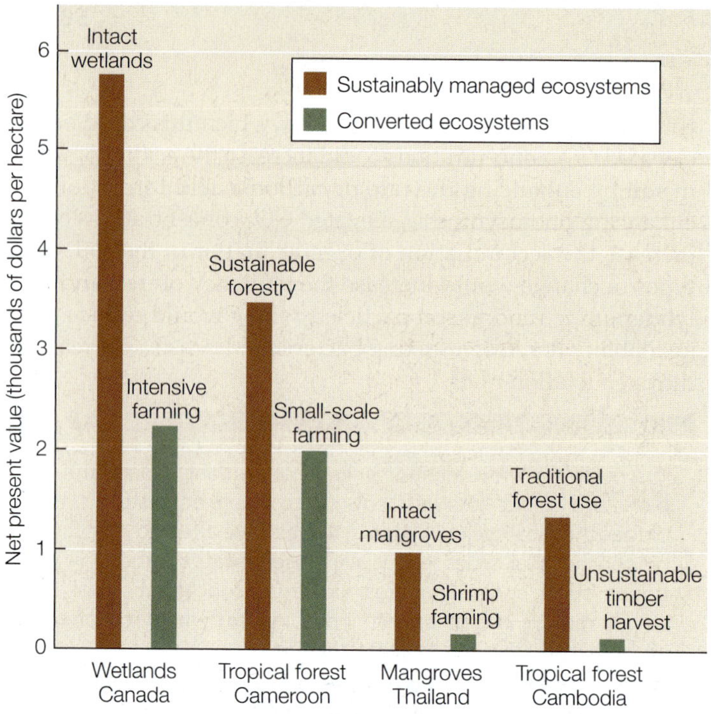
Net present value (thousands of dollars per hectare)
0
1
2
3
4
5
6
Sustainably managed ecosystems
Converted ecosystems
Intact wetlands
Intensive farming
Sustainable forestry
Small-scale farming
Intact mangroves
Shrimp farming
Traditional forest use
Unsustainable timber harvest
Wetlands Canada
Tropical forest Cameroon
Mangroves Thailand
Tropical forest Cambodia

58.19 The Economic Value of Sustainably Managed Ecosystems Many types of ecosystems are able provide more goods and services when they are sustainably managed than when they are completely converted to human use and intensively exploited.

58.5 How Can Ecosystems Be Sustainably Managed?

Practices that allow us to conserve or enhance ecosystems so as to benefit from specific ecosystem goods and services over the long term without compromising others are referred to as **sustainable**. In many cases the total economic value of a sustainably managed ecosystem is higher than that of a converted or intensively exploited ecosystem (**Figure 58.19**). Furthermore, the long-term economic benefits of preventing overexploitation of ecosystems are enormous. For example, the collapse of the cod fishery on Georges Bank due to overfishing (see Figure 55.14) resulted in the loss of tens of thousands of jobs.

Impeding the establishment of policies that encourage sustainable practices is the impression that ecosystem services are "public goods" with no market value. People who do not stand to profit from the services provided by natural ecosystems have no incentive to pay for them, whereas individuals who own converted ecosystems can reap great economic benefits. Government action may be needed to create incentives encouraging sustainable ecosystem management. Examples of such action might include:

- Elimination of subsidies that promote damaging exploitation of ecosystems. For example, the billions of dollars paid by governments in developed nations to subsidize domestic agriculture (with the aim of insulating farmers from economic risk) have led to greater food production than the global market warrants, promoted excessive use of fertilizers, and reduced the profitability of agriculture in developing countries.
- More sustainable use of fresh water from rivers and aquifers could be achieved by charging users the full cost of providing water, by developing methods to use water more efficiently, and by altering the allocation of water rights so that the incentives favor conservation rather than wasteful consumption.
- More sustainable use of marine fisheries could be achieved by establishing more protected marine reserves and "no-take" zones where fish can grow to reproductive age. Recent discussions of marine fisheries management center on what is colloquially known as the BOFFF ("Big, Old, Fat, Female Fish") hypothesis, which proposes that it is most important to protect the largest and oldest females in a population because they outreproduce younger fish by an enormous margin (see Section 55.6).

Raising public awareness is essential to the implementation of sustainable management programs. Most people do not realize the long-term value of ecosystem goods and services or understand how human activities affect the functioning of ecosystems. Maintaining and enhancing ecosystem goods and services that have no established market value is especially difficult. Perhaps the most difficult aspect of ecosystem function to maintain in the face of increasingly intensive human use of the global ecosystem will be biological diversity. The final chapter of this book is devoted to this important topic.

How can we determine to what extent dead zones result from human actions and to what extent they are the result of natural biogeochemical processes?

ANSWER

Hypoxia occurs naturally in aquatic systems, but throughout history it has been restricted primarily to deep-water ecosystems such as deep ocean basins, fjords, and the bottoms of the largest lakes. The appearance of dead zones in shallow coastal waters and estuaries is a twentieth-century phenomenon (the Gulf of Mexico dead zone was first identified in 1972). This timing, as well as the observation that dead zones are increasing in number and size, suggests human involvement.

The White House Office of Science and Technology Policy's Committee on Environment and Natural Resources initiated a scientific assessment of the causes and consequences of Gulf hypoxia in 1997. As part of this effort, sediment cores were taken to determine historic levels of algal deposition in sediment. These cores revealed a clear pattern of increases in the second half of the twentieth century. Sophisticated computer models demonstrated a significant association between river loads of dissolved inorganic nitrogen and rates of oxygen depletion, and the most dramatic increases coincided with historic records of changes in human activities that increased nitrate loads in the river system. Given that the Gulf of Mexico provides almost three-fourths of the shrimp and two-thirds of the oysters harvested in the U.S., as well as recreational fishing resources and ecologically vital forage fishes, further expansion of the dead zone could have devastating economic consequences.

CHAPTER SUMMARY 58

58.1 How Does Energy Flow through the Global Ecosystem?

- An **ecosystem** includes all of the organisms in an ecological community as well as the physical and chemical factors that influence those organisms.
- Energy flows and chemical elements cycle through ecosystems. **Review Figure 58.1**
- Terrestrial net primary production varies across the globe, reflecting differences in solar energy input and the climate patterns that result from them. **Review Figures 58.2, 58.3**
- Productivity in aquatic ecosystems is limited by light, temperature, and nutrient availability. **Review Figure 58.4**
- Humans appropriate about one-quarter of Earth's average annual net primary production, although this amount varies regionally.

58.2 How Do Materials Move through the Global Ecosystem?

- Chemical elements cycle through biotic and abiotic compartments of the global ecosystem. **Review Figure 58.5**
- The movement of elements through the biotic compartment of ecosystems, from uptake by autotrophs to decomposition, generally occurs on a local scale.
- Most global air circulation takes place in the lowest layer of the atmosphere, the **troposphere**. An **ozone layer** in the **stratosphere** absorbs ultraviolet radiation. **Review Figure 58.6**
- Carbon dioxide, water vapor, and other **greenhouse gases** in the atmosphere are transparent to sunlight but trap heat, thus warming Earth's surface. **Review Figure 58.7, ANIMATED TUTORIAL 58.1**
- Because the geological processes that move elements on land are so slow (on the scale of millions of years), there are large regional and local variations in the supply of particular elements within the terrestrial compartments.
- Some nutrients enter fresh waters from the atmosphere in rainfall, but most are released from rocks by weathering. They are usually carried rapidly to lakes or to the oceans.
- **Turnover** occurs regularly in temperate-zone lakes in both spring and fall, bringing nutrients to the surface and oxygen to the deeper waters. **Review Figure 58.8**
- Most materials that cycle through biotic and abiotic compartments end up in the oceans, where they eventually sink to the bottom.
- Fires release the chemical elements from the vegetation they burn. Those vaporized elements enter the atmosphere, where they can be carried into groundwater by rain.

58.3 How Do Specific Nutrients Cycle through the Global Ecosystem?

- The pattern of movement of a chemical element through the biotic and abiotic compartments of the global ecosystem is its **biogeochemical cycle**.
- The **hydrologic cycle** is driven by the sun, which evaporates more water from the ocean surface than it returns by precipitation. The excess precipitation that falls on land eventually returns to the oceans, primarily in rivers. **Review Figure 58.9, ANIMATED TUTORIAL 58.2**
- Groundwater plays a minor role in the hydrologic cycle, but underground **aquifers** are being seriously depleted by human activities.
- Carbon is removed from the atmosphere by photosynthesis and returned to the atmosphere by metabolism and burning. **Review Figure 58.10, ANIMATED TUTORIAL 58.3**
- The concentration of CO_2 in the atmosphere has increased greatly in the last 150 years, largely because of the burning of fossil fuels. This buildup of CO_2 is warming the global climate. **Review Figures 58.11–58.13**
- As a result of agricultural use of fertilizers and the burning of fossil fuels, total nitrogen fixation by humans is nearly equal to natural nitrogen fixation. **Review Figures 58.14. 58.15, ANIMATED TUTORIAL 58.4**
- Human alteration of the nitrogen cycle has resulted in excesses of nitrogen compounds in bodies of water, leading to **eutrophication** and dead zones.
- The burning of fossil fuels releases sulfur and nitrogen into the atmosphere, leading to **acid precipitation**. **Review Figure 58.16**
- Agricultural use of fertilizers and clearing of land have dramatically increased the input of phosphorus into soils and fresh waters. **Review Figure 58.17**

continued

58.4 What What Goods and Services Do Ecosystems Provide?

- The goods and services provided by ecosystems include food, clean water, flood control, pollination, pest control, climate regulation, spiritual fulfillment, and aesthetic enjoyment. Most ecosystem services either are irreplaceable or the technology necessary to replace them is prohibitively expensive.
- Efforts to enhance the capacity of an ecosystem to provide some goods and services often come at the cost of the system's ability to provide others.

58.5 How Can Ecosystems Be Sustainably Managed?

- The total economic value of an ecosystem managed in a **sustainable** manner often is higher than that of a converted or intensively exploited ecosystem. **Review Figure 58.19**
- Recognition of the value of ecosystem goods and services that are now perceived as "public goods" may induce government action to protect them. Public education is needed to make people aware of how much they benefit from ecosystem goods and services.

See ACTIVITY 58.1 for a concept review of this chapter

Go to the Interactive Summary to review key figures, Animated Tutorials, and Activities Life10e.com/is58

CHAPTERREVIEW

REMEMBERING

1. What features of Earth influence its ecosystem dynamics?
   a. Lithospheric plates that move continuously
   b. Atmospheric gases that moderate surface temperatures
   c. Large amounts of water in liquid form
   d. A diversity of living organisms
   e. All of the above

2. Marine upwelling zones are important because
   a. they help scientists measure the chemistry of deep ocean waters.
   b. they bring to the surface organisms that are difficult to observe elsewhere.
   c. ships can sail faster in these zones.
   d. they increase marine productivity by bringing nutrients back to surface ocean waters.
   e. they bring oxygenated water to the surface.

3. The hydrologic cycle operates as it does because
   a. water flows into the oceans via rivers.
   b. water evaporates from the leaves of plants.
   c. more water evaporates from the surface of the oceans than is returned to the oceans as precipitation.
   d. precipitation falls on land.
   e. more water falls on the oceans as precipitation than evaporates from its surface.

4. Carbon dioxide is called a greenhouse gas because
   a. it is used in greenhouses to increase plant growth.
   b. it is transparent to heat but traps sunlight.
   c. it is transparent to sunlight but traps heat.
   d. it is transparent to both sunlight and heat.
   e. it traps both sunlight and heat.

5. The biogeochemical cycle of phosphorus differs from the cycles of carbon and nitrogen in that
   a. phosphorus lacks an atmospheric component.
   b. phosphorus lacks a liquid phase.
   c. only phosphorus is cycled through marine organisms.
   d. living organisms do not need phosphorus.
   e. The phosphorus cycle does not differ importantly from the carbon and nitrogen cycles.

6. Maintaining the capacity of ecosystems to provide goods and services is important because
   a. most ecosystem services cannot be replicated by any other means.
   b. replacing them with technological substitutes is prohibitively expensive.
   c. technological substitutes take up valuable land.
   d. governments cannot function without taxing ecosystem services.
   e. It is not important. Humans could survive quite well even if ecosystem services declined greatly.

UNDERSTANDING & APPLYING

7. The waters of Lake Washington, the second largest lake in the state of Washington and lying adjacent to the city of Seattle, returned to their preindustrial condition within 10 years after sewage was diverted from the lake to Puget Sound, an arm of the Pacific Ocean. Would all lakes being polluted with sewage clean themselves up as quickly as Lake Washington if sewage inputs were stopped? What characteristics of a lake are most important to its rate of recovery following reduction of nutrient inputs?

8. A government official authorizes construction of a large coal-burning power plant in a former wilderness area. Its smokestacks discharge great quantities of combustion wastes. List and describe all the likely effects of this action on ecosystems at local, regional, and global levels. If the wastes were thoroughly scrubbed from the stack gases, which of the effects you have just outlined would still happen?

ANALYZING & EVALUATING

9. What types of experiments would you conduct to assess the likely consequences of fertilization of the oceans with iron to increase rates of photosynthesis? At what spatial and temporal scales should these experiments be conducted?

10. One mechanism proposed for reducing the anthropogenic (human-caused) flux of carbon into the atmosphere is called "cap and trade," whereby a government sets a "cap," or limit, on carbon emissions by polluters, but allows facilities that emit less than their emission allowance to sell their excess credits to polluters who would otherwise exceed their emission allowance. What benefits and drawbacks can you see to such an approach to reducing carbon emissions?

11. A string of powerful hurricanes struck the east coast of the United States over the course of a single year's hurricane season. Some people claim that this disaster was due to warming of the oceans caused by greenhouse gases in the atmosphere. Others assert that global warming is not responsible because hurricanes have occurred for many centuries. How would you evaluate these conflicting claims?

59 Biodiversity and Conservation Biology

CHAPTER**OUTLINE**

A Natural Christmas Tree During the 1980s, large numbers of migrating bald eagles (*Haliaeetus leucocephalus*) turned Montana's Glacier National Park into a tourist attraction each fall as the birds stopped to feed on spawning kokanee salmon. When the salmon population fell victim to introduced lake trout, the eagles also disappeared and tourism declined.

MANY HUMAN ACTIVITIES have factored in the extinction of animal species; even good intentions can pose a threat. Flathead Lake in northwestern Montana originally had fewer than a dozen native fish species, including bull trout (*Salvelinus confluentus*) and westslope cutthroat trout (*Oncorhynchus clarkii lewisi*). To encourage sport fishing, non-native sport fish species were introduced. Most of these introductions were unsuccessful, but the kokanee salmon (*Oncorhynchus nerka*), introduced from western Canada, eventually prospered and by the mid 1980s was the dominant sport fish.

Because kokanee were popular with anglers, efforts were made to establish them in nearby lakes. Fisheries managers introduced opossum shrimp (*Mysis diluviana*) into neighboring lakes to provide a food source for juvenile kokanee, which feed on zooplankton. But kokanee are daytime feeders that use vision to find their prey; opossum shrimp remain on lake bottoms during the day, thus escaping predation by the young kokanee.

Somehow the shrimp made their way to Flathead Lake, where they proved to be a bonanza for lake trout (*Salvelinus namaycush*), another introduced species but one that had never become abundant because of limited food supply on the lake bottom, where the trout feed as juveniles. With the new food source, the lake trout population exploded. Adult lake trout are voracious consumers of other fish, and kokanee numbers plummeted as they fell prey to the lake trout. By 1992 kokanee were gone from Flathead Lake. The native bull trout may be next; this species was officially designated "vulnerable to extinction" in 1999 and its future is uncertain.

These changes have had economic impacts well beyond sport fishing. In the 1980s, flocks of migrating bald eagles gorged on the abundant kokanee spawning upstream of Flathead Lake in Glacier National Park. The sight was a tremendous tourist draw every fall. Without the salmon, fewer eagles visit the area, and without the eagles, there are fewer tourists. This example of unanticipated effects of human activities is far from unique. Given what we know about species interactions, it should not be surprising that a species introduced in the wrong place can endanger other species.

> ?
>
> How can adverse impacts of species introductions be anticipated before lasting damage occurs?
>
> See answer on p. 1245.

59.1 What Is Conservation Biology?

Virtually all natural ecosystems on Earth have been altered by human activities. Many habitats have disappeared completely, and many others have been greatly modified. Even Earth's climate and its global biogeochemical cycles have been altered, as we saw in Chapter 58. One consequence of these changes has been a rapid increase in the rate at which species go extinct.

Conservation biology is a scientific discipline devoted to protecting and managing Earth's biodiversity. The discipline draws heavily on the principles of ecology, ethology, evolutionary biology, and wildlife management particularly when elucidating the factors that determine whether a given population will persist.

Early conservation efforts were characterized by tensions between people whose principal goal was to conserve natural resources for their economic benefits and people who believed that nature has intrinsic value independent of human economic interests. Today conservation biologists study the full array of goods and services that humans derive from species and ecosystems, including aesthetic and psychological benefits. We now know that understanding the global ecosystem and the effects of human activities on that system is essential to the long-term well-being of *Homo sapiens*.

Conservation biology is an applied discipline, which is to say that it involves the practical application of scientific knowledge to solve problems. Workers in conservation biology are guided by three basic principles:

- *The processes of evolution unite all forms of life.* To effectively protect and manage biodiversity, we must understand the evolutionary processes that generate and maintain it.
- *The ecological world is dynamic.* Because populations and communities change continuously over time, there is no static "balance of nature" that can serve as a goal of conservation activities.
- *Humans are a part of ecosystems.* Human interests and activities must be incorporated into conservation goals and practices.

59.1 Extinct Megafauna The extinction of some large North American mammals during the Pleistocene may have been driven by the arrival of *Homo sapiens*. This portion of a museum re-creation shows Columbian mammoths (*Mammuthus columbi*), ancient bison (*Bison antiquus*, an extinct ancestor of the American bison) and western horses (*Equus occidentalis*).

Conservation biology aims to protect and manage biodiversity

The term **biodiversity**, a contraction of "biological diversity," has multiple definitions. We may speak of biodiversity as the degree of genetic variation within a species. Genetic variation can be measured as the number of alleles at a locus, the number of polymorphic loci in a genome, or the number of individuals in a population that are polymorphic at given loci. As we have seen throughout this book, genetic variation allows organisms to adapt to environmental change. Biodiversity can also be defined in terms of species richness in a particular community. At a larger scale, biodiversity also embraces ecosystem diversity—particularly the complex interactions within and between ecosystems. While we may study these components of biodiversity separately, in life they are intimately interconnected.

One conspicuous manifestation of biodiversity loss is species extinction. Extinction is a constant theme in the history of life; most of the species that have lived on Earth over the ages are extinct today. Consider, for example, the anaerobic organisms that were lost as early photosynthetic prokaryotes and eukaryotes added oxygen to Earth's atmosphere, as described in Section 25.2. Extinctions have occurred throughout Earth's history at what is referred to as a "background" rate as changes in environmental conditions have favored some species and negatively affected others. But the rate of extinctions taking place today rivals those of the five great mass extinctions (see Table 25.1 and Figure 25.2). The past mass extinction episodes were the result of cataclysmic natural disturbances, whereas the majority of modern extinctions can be attributed to effects of human activities.

Humans have a tremendous capacity to alter ecosystems and, accordingly, to cause extinctions. When humans first arrived in North America from Siberia about 14,000 years ago, they encountered a diverse and spectacular fauna of large mammals, including saber-toothed cats, dire wolves, mammoths, mastodons, giant ground sloths, and giant beavers (**Figure 59.1**). Most of this megafauna went extinct within a few thousand years after humans arrived. Although several hypotheses have been advanced to account for the geologically rapid and simultaneous disappearance of so many large animals, overhunting by humans is the most likely explanation. Losses of megafauna coinciding with the arrival of humans have also been documented in Australia and Hawaii.

Over the past 400 years, increasing industrialization and urbanization have accelerated the rate of species extinctions astronomically. The renowned evolutionary biologist Edward O. Wilson estimates that Earth is losing some 30,000 species per year, putting us in the midst of a sixth mass extinction. The mass extinction events in Earth's past occurred relatively far apart in time, and each one provided ecological opportunities for other groups, which subsequently underwent adaptive radiations. Protecting Earth's biodiversity today requires maintaining the processes that generate new species as well as bringing extinction rates closer to background levels.

Biodiversity has great value to human society

Conservation biologists are concerned about the escalating loss of Earth's biodiversity for many reasons:

- *Humans depend on thousands of other species for food, fiber, and medicine.* Humans have domesticated countless plants as sources of food, and more than 2,000 plant species are used for fiber worldwide. In India alone, more than 7,000 species of plants are used in traditional medicine, and in the United States more than one-fourth of all medical prescriptions contain or are based on plant products. Hundreds of animal species also supply us with food, clothing, and medicine.
- *Losing species can threaten ecosystem functioning.* Throughout Part 10 we have described many complex interactions among species. When species are lost, entire communities and ecosystems may change or be lost completely and humans may lose the goods and services those ecosystems provide.
- *Humans derive enormous psychological benefits, including aesthetic pleasure, from interacting with other organisms.* These aesthetic benefits give biodiversity economic value. Trees growing on a residential lot, for example, can increase the lot's property value by an amount that is greater than the value of the lumber that could be made from the trees.
- *Extinctions deprive the scientific community of opportunities to study and understand ecological relationships among organisms.* The more species that are lost, the more difficult it will be to understand the structure and functioning of ecological communities and ecosystems.
- *Living in ways that cause the extinction of other species raises ethical issues.* Losses of biodiversity increasingly concern philosophers, ethicists, and religious leaders, who believe species to have intrinsic value.

All of these concerns, to varying degrees, may be integrated by conservation biologists into strategies for protecting biodiversity.

RECAP 59.1

Conservation biology is an applied scientific discipline aimed at protecting and managing biodiversity, which is rapidly decreasing due to extinctions that are the result of human activities.

- Explain the multiple meanings of the term "biodiversity." **See p. 1229**
- What are some of the ways in which biodiversity is valuable to humans? **See p. 1230**

Conservation biologists must understand biodiversity as it exists today as well as how and why it is changing. An important goal of conservation biologists is to predict which species are most likely to go extinct and how soon extinction is likely to happen.

59.2 How Do Conservation Biologists Predict Changes in Biodiversity?

How many, and which, species will go extinct will depend both on human activities and on natural events. Conservation biologists attempt to track the extinctions that are occurring and to predict the ones that are likely to occur during the coming century.

Our knowledge of biodiversity is incomplete

Tracking and predicting extinctions is difficult for several reasons. First, we do not know how many species live on Earth today. Many species that are likely to go extinct in the near future have not even been named and described by scientists. Insects provide a case in point: although more than 1 million species have been described (see Section 32.4), estimates of the number of species yet to be discovered range from 2 million to more than 50 million. Even in the case of larger organisms, our understanding of biodiversity is far from complete. For example, in an 18-month period in 2005–2006, more than 50 species of animals and plants previously unknown to science were discovered in the rainforests of Borneo. Worldwide, an annual inventory of newly described species counted 19,232 species discovered in 2009 alone; this list included 9,738 insects, 2,184 plants, 1,360 fungi, 71 mammals, and 7 birds.

Second, we do not know where species live. The ranges of most described species, particularly those that are small, reclusive, and rare to start with, are poorly known. One tiny North American true bug, *Corixidea major* (so rare that it has no common name), had been found in only one location near Clarksville, Tennessee, until entomologists collecting insects attracted to lights at night discovered it in Virginia and Florida, extending its known range by more than 1,000 kilometers.

Third, it is difficult to determine whether a species is truly extinct. Rarely is the death of the last surviving member of a species recorded with certainty, as it was in the case of the last passenger pigeon (*Ectopistes migratorius*), a female named Martha, who died in the Cincinnati Zoo on September 1, 1914. The status of rare, reclusive species with poorly known life histories is much more difficult to determine, as has been the case with the ivory-billed woodpecker (*Campephilus principalis*) in the southeastern United States, which is thought to be extinct despite reported sightings (**Figure 59.2A**). Pygmy tarsiers (*Tarsius pumilus*), tiny primates weighing less than 60 grams, were thought to have gone extinct from their native cloud forests on the island of Sulawesi in Indonesia. In 2008—85 years after the last reported sighting of a living *T. pumilus*—a research team from Texas A&M University discovered individuals of this species living in one of the island's national parks (**Figure 59.2B**).

Fourth, we rarely know all of the connections among species. At the opening of this chapter, we saw how the introduction

(A) *Campephilus principalis*

(B) *Tarsius pumilus*

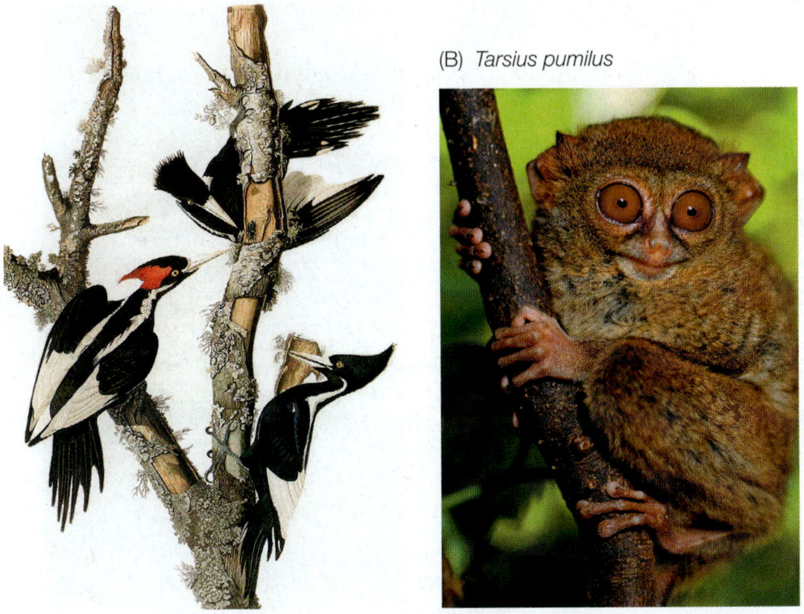

59.2 Is It Really Extinct? (A) The ivory-billed woodpecker (shown here in a nineteenth-century Audubon print) was presumed to be extinct until reports of sightings in 2004. Clear proof of the living bird has so far eluded ornithologists. (B) In 2008, pygmy tarsiers were discovered in an Indonesian national park after having been presumed extinct for 85 years.

of one non-native species altered Flathead Lake's entire food web and placed other species at risk. How many species are at risk is unknown, however, because the ecological interactions among all of the lake inhabitants have never been thoroughly characterized.

Go to Media Clip 59.1
New Species Found in the Twenty-first Century
Life10e.com/mc59.1

We can predict the effects of human activities on biodiversity

Despite these gaps in our understanding of biodiversity, methods exist for estimating probable rates of extinction resulting from human activities. To estimate the risk that a particular population will become extinct, conservation biologists develop statistical models that incorporate information about a population's size, its genetic variation, its life history traits, and the physiology and behavior of its members. The International Union for the Conservation of Nature (IUCN) has published categories that define a species' danger of extinction. Species in imminent danger of extinction in all or most of their range are classified as "endangered" or "critically endangered"; those believed to be susceptible to extinction in the near future are classified as "vulnerable." Biologists consider species in any one of these three categories to be "threatened" (**Figure 59.3**).

Rarity in and of itself is not always a cause for concern. Some species that are specialized for living exclusively in rare and unusual habitats have probably never been especially abundant and are well adapted to being rare. The Cayman crab fly (*Drosophila endobranchia*), for example, has probably never been abundant. It is found only in the Cayman Islands, where it parasitizes only two species of terrestrial crabs. Small population sizes are of concern, however, especially for species whose populations shrink suddenly. These "newly rare" species are usually at high risk of extinction, as large and rapid reductions in population size can lead to genetic drift and loss of genetic variation (see Chapter 21).

Certain aspects of species' life histories are particularly important in predicting the ability of populations to recover from reductions in size (see Chapter 55). In fishes and mammals, for example, one of the best predictors of extinction risk is age at maturity, a life history trait that has a profound influence on rates of population growth. Ecological niche requirements can also influence the ability of populations to recover from rapid declines. Species with specialized habitat or dietary requirements, for

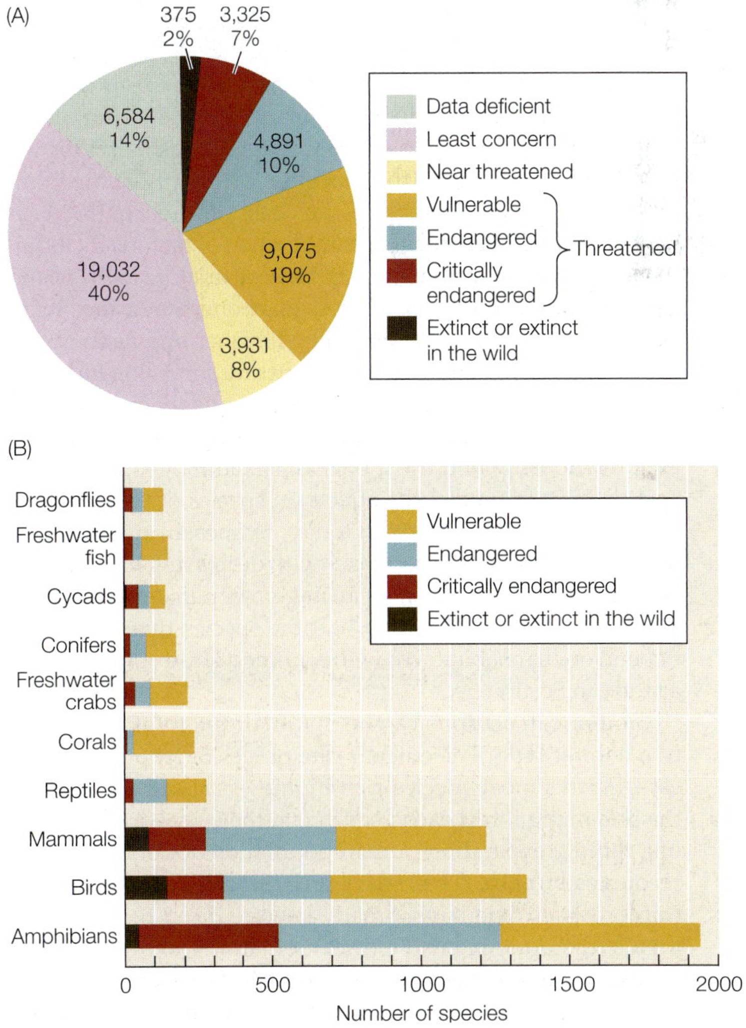
(A)
375
2%
3,325
7%
6,584
14%
4,891
10%
9,075
19%
19,032
40%
3,931
8%
Data deficient
Least concern
Near threatened
Vulnerable
Endangered
Critically endangered
Threatened
Extinct or extinct in the wild
(B)
Dragonflies
Freshwater fish
Cycads
Conifers
Freshwater crabs
Corals
Reptiles
Mammals
Birds
Amphibians
Vulnerable
Endangered
Critically endangered
Extinct or extinct in the wild
0
500
1000
1500
2000
Number of species

59.3 Species at Risk of Extinction (A) The breakdown by extinction risk category of all 47,677 species assessed by th IUCN. (B) The bars show the numbers and proportions of species in the various extinction risk categories (together termed "threatened") in several taxonomic groups that have been comprehensively assessed.

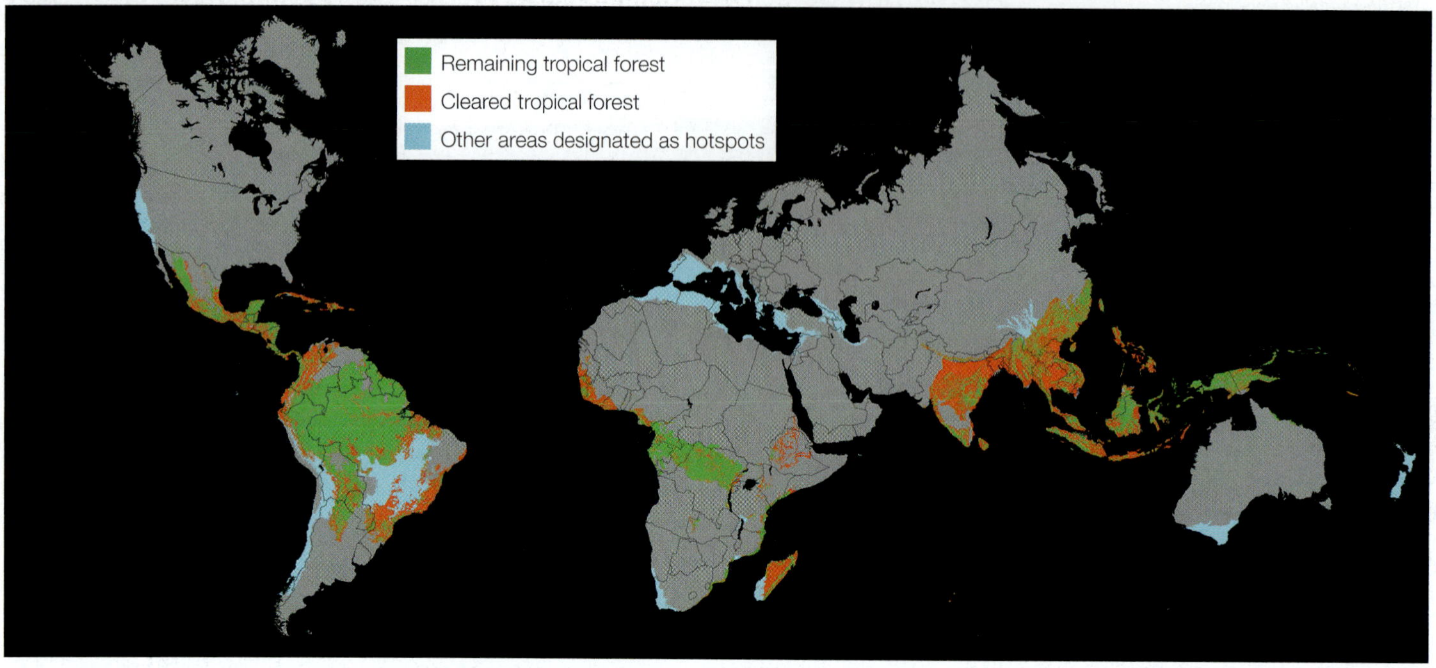
Remaining tropical forest
Cleared tropical forest
Other areas designated as hotspots

59.4 The Disappearing Rainforest Since around 1950, tropical forests have been destroyed at tremendous rates as land is cleared for agriculture, highways, timber resources, and other needs of an exploding human population. Rainforests have long been recognized as centers of biodiversity, or "hotspots," that harbor vast numbers of species (see Section 59.4 and Figure 59.10).

example, are more likely to become extinct than species with more generalized requirements.

In addition, populations reduced to a small size or confined to a small range can easily be eliminated by local disturbances. For example, populations of the Cozumel thrasher (*Toxostoma guttatum*), a member of the mockingbird family known only on the island of Cozumel off the coast of Mexico, had been declining since 1970 due to a combination of factors, including the unintentional introduction of boa constrictors to the island. Then, beginning in 1988, a series of strong hurricanes had a catastrophic effect on the remaining thrasher populations. Surveys done in 2006 failed to document any surviving individuals, and today *Toxostoma guttatum* is most likely extinct.

Conservation biologists apply the principles of the species–area relationship and the theory of island biogeography (see Section 57.3) to predict the effects on species of habitat loss—the major cause of extinction today. By measuring the rate at which species richness decreases with decreasing habitat patch size, they have found that, on average, a 90 percent loss of habitat area results in the loss of half the species that live in and depend on that habitat. We will examine a key example of such a study in Section 59.3.

Similar calculations can be made for the total global area of a habitat type. The current rate of loss of tropical rainforest—Earth's most species-rich biome—is about 2 percent of the remaining forest each year due to the increasing demands of a rapidly expanding human population not only for forest resources but also for cleared agricultural land. Most of the rainforests of Asia have already been reduced to small fragments, the only extensive tracts remaining being found on the islands of New Guinea and, to a much lesser extent, Borneo (**Figure 59.4**). Between 2000 and 2010, the highest rate of tropical deforestation took place in Central America. If the current rate of loss continues, close to 1 million rainforest species (a conservative estimate) could become extinct before the end of this century.

RECAP 59.2

Predicting changes in biodiversity is difficult because our knowledge of biodiversity is incomplete. The species–area relationship can be used to predict rates of extinction in areas that are subject to habitat loss.

- What are some of the gaps in our current knowledge of biodiversity? **See pp. 1230–1231**
- What are some of the factors that render a species especially vulnerable to extinction? **See pp. 1231–1232**

Many factors can place species at risk of extinction, but human activities have had a disproportionate impact on the mass extinction that Earth is currently experiencing. Understanding how particular human activities present challenges to species survival is essential for developing ways to mitigate biodiversity losses.

59.3 What Human Activities Threaten Species Persistence?

Human activities that threaten the persistence of species include habitat alteration and destruction, introductions of nonnative species, overexploitation, and climate alteration. Conservation biologists determine how these activities affect species and use that information to devise strategies to protect species that are endangered or threatened.

Habitat losses endanger species

Global Biodiversity Outlook 3, a report published by the United Nations in 2010, identified five principal pressures on biodiversity. Topping the list is habitat loss and degradation, including fragmentation or outright destruction of habitat by human activities.

Many habitats—particularly freshwater habitats—are being degraded by pollution. Many toxic substances released into natural habitats by human activities have negative effects on the reproduction, development, and behavior of species, reducing both their survivorship and their competitive ability. Among the most troublesome toxic pollutants today are heavy metal waste products of mining and manufacturing, polycyclic aromatic hydrocarbons arising from fossil fuel combustion, and synthetic organic chemicals released into the environment to control pests.

Pollutants do not necessarily have to be toxic to cause problems. Nondegradable plastic trash dumped in the ocean poses a choking hazard to marine birds and mammals, which can mistake floating bits of plastic for prey that they then try to eat. Fish, corals, and other sea life can become entangled in discarded plastic, often resulting in their death.

Habitat loss can also occur through outright habitat elimination. As we saw in Section 58.4, natural ecosystems are being converted to human use at an increasing rate. Physical destruction of a particular habitat, as when tropical rainforest is cut down and the land converted to agricultural use, eliminates species that cannot survive anywhere else. Habitat loss also affects nearby habitats that are not destroyed. As portions of a habitat are lost to human activities, the remaining habitat becomes **fragmented** into habitat patches that become ever smaller and more isolated.

Small habitat patches are qualitatively different from larger patches of the same habitat in ways that affect species persistence. Small patches cannot maintain populations of species that require large areas, and they can support only small populations of those species that can survive in small patches. In addition, the fraction of a patch influenced by external factors increases disproportionately as patch size decreases (**Figure 59.5**). Close to the edges of a forest patch, winds are stronger, temperatures are higher, humidity is lower, and light levels are higher than they are farther inside the forest. Species from surrounding habitats often colonize the edges of a patch, where they compete with or prey on the species living in the patch. These influences are known as **edge effects**. A proliferation of edges can benefit some species, such as those that depend on resources in multiple habitats and must travel among them. For many other species, however, edge effects render habitats unsuitable or promote the establishment of competitors, predators, or parasites.

One effect of forest fragmentation in midwestern North America has been an increase in the abundance of the brown-headed cowbird (*Molothrus ater*). This bird is a brood parasite—that is, it lays its eggs in the nests of other bird species and its hatchlings are raised by the host parents, to the detriment of their own young (see the opening of Chapter 53). Historically,

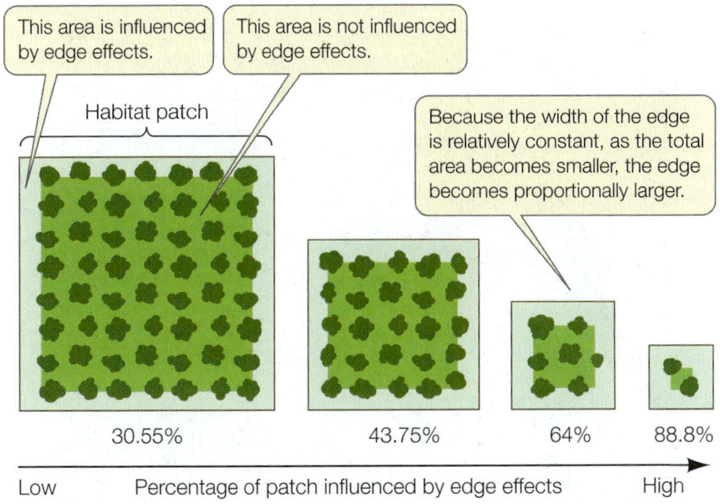

This area is influenced by edge effects.
This area is not influenced by edge effects.
Habitat patch
Because the width of the edge is relatively constant, as the total area becomes smaller, the edge becomes proportionally larger.
30.55%
43.75%
64%
88.8%
Low
Percentage of patch influenced by edge effects
High

59.5 Edge Effects The smaller a patch of habitat, the greater the proportion of that patch that is influenced by conditions in the surrounding environment.

Go to Animated Tutorial 59.1
Edge Effects
Life10e.com/at59.1

cowbirds followed bison and other grazing mammals, feeding on insects kicked up by the herds; thus their eggs were laid primarily in nests of grassland host species. Forest fragmentation, however, opened up new opportunities for the cowbirds, which can now lay their eggs in the nests of forest birds in forest edges. Fragmented forests, with relatively more edge than intact forests, thus favor the proliferation of cowbirds at the expense of forest species.

Because so many habitats have already undergone fragmentation by the time they are first investigated by ecologists, determining the effects of fragmentation on the original communities can be difficult. Timely surveys can provide some of this information. For example, a major research project was launched in 1979 in a tropical rainforest near Manaus, Brazil, that was slated for conversion to pasture. The landowners agreed to preserve forest plots of certain sizes and configurations laid out by biologists (**Figure 59.6A**). The biologists counted the species in the future "fragments" while they were still part of the continuous forest, then monitored these plots after the surrounding forest was cut (**Figure 59.6B**). Species soon began to disappear from isolated plots. The first species to be eliminated were monkeys that travel over large areas. Army ants and the birds that follow army ant swarms also disappeared quickly.

Species that are lost from small habitat fragments are unlikely to become reestablished there because dispersing individuals are unlikely to find the isolated fragments. As Section 55.5 pointed out, however, a species may persist in a small patch if it is connected to other patches by habitat corridors through which individuals can disperse. Among the experimental forest plots in Brazil, those that were completely isolated

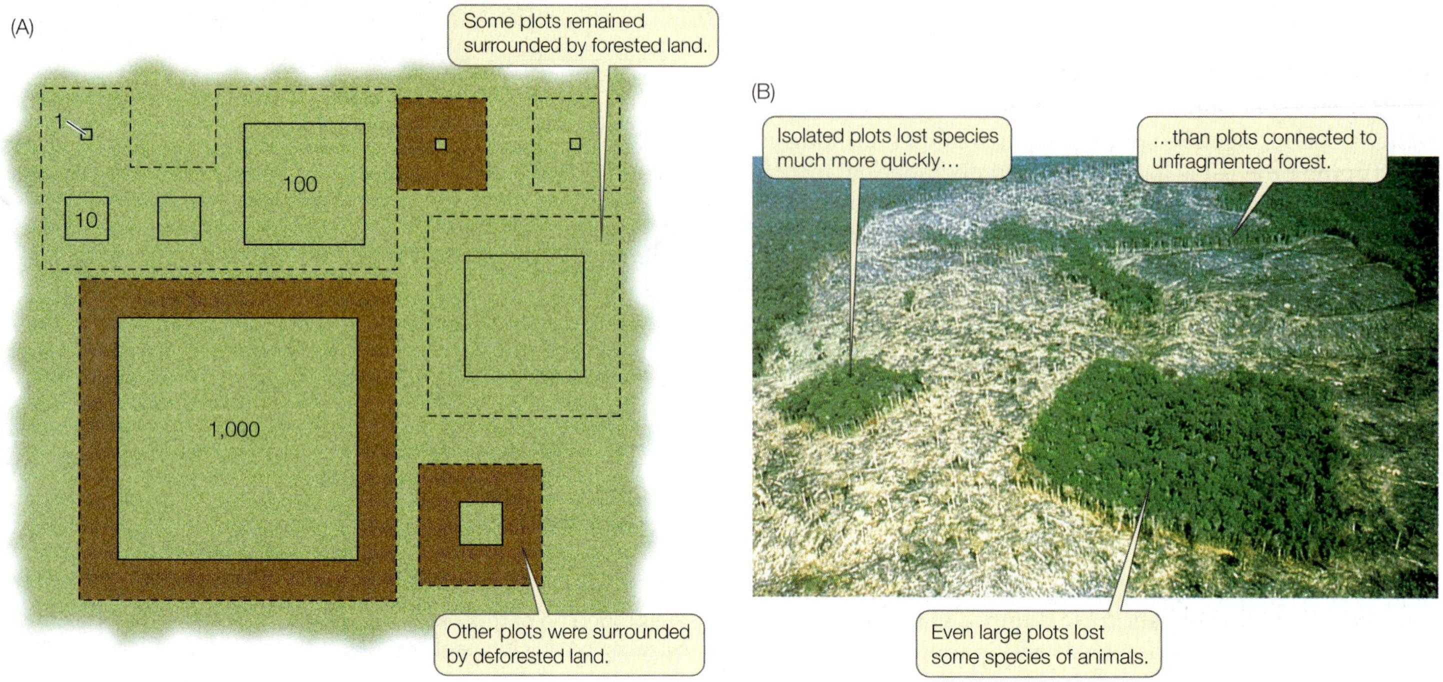
(A)
Some plots remained surrounded by forested land.
1
10
100
1,000
Other plots were surrounded by deforested land.
(B)
Isolated plots lost species much more quickly…
…than plots connected to unfragmented forest.
Even large plots lost some species of animals.

59.6 Species Losses in Fragmented Brazilian Forest Biologists studied plots of tropical rainforest near Manaus, Brazil, before and after they were isolated by forest clearing. (A) The landowners agreed to preserve forest plots of certain sizes and configurations according to a plan laid out by the biologists. (B) Some of the plots after clearing. The results of the study demonstrated that small, isolated habitat plots lost species more quickly than did larger plots of the same habitat.

lost species more rapidly than did those that were connected to unfragmented forest by corridors. Since the experiment began, some of the pastures that surrounded the experimental plots have been abandoned, and young forests now grow in them. Within 9 years of abandonment, army ants and some of the birds that follow them recolonized forest fragments connected to larger forest fragments by young forests that served as dispersal corridors. Other birds that forage in the forest canopy also reestablished themselves. Young forest is not a suitable permanent habitat for most of these species, but they can disperse through it to find more appropriate habitat.

Insight into the importance of corridors has led to new regional conservation initiatives, among the most notable of which is the Yellowstone to Yukon Conservation Initiative. This joint Canada–United States nonprofit organization has as its goal the sustainable preservation of the mountain ecosystem extending from Yellowstone National Park in the United States to Yukon, Canada. This stretch of land, the largest intact ecosystem of its kind on the planet, contains high-quality habitat for many of North America's most imperiled animals, including grizzly bears, gray wolves, lynx, and native fishes. The initiative works with landowners to find sustainable ways of preserving high-quality, well-connected wildlife habitat in the region. Managing the entire region in this way will not only provide habitat for these species, but will also provide room for their populations to shift in response to global climate change.

Overexploitation has driven many species to extinction

Overexploitation was once the most important cause of species extinctions. Although habitat loss now presents a greater threat to more species, many species are still threatened by overexploitation. Particularly at risk are species with life history traits that are linked to slow population growth (see Section 55.3), which make recovery from losses less likely. Elephants and rhinoceroses, for example, are slow to reach reproductive maturity and produce relatively few offspring over the course of their lives; they are at risk in much of Africa and Asia because poachers kill elephants for their valuable ivory tusks and rhinoceroses for their horns (primarily based on a long-prevalent but false belief that imbibing drinks made with powdered rhinoceros horn boosts a man's sexual potency).

The principal threat to the continued survival of tigers, whose numbers have declined by more 90 percent since 1900 (**Figure 59.7A**), is the use of their body parts in traditional medicine—bones to treat rheumatism, eyes to cure epilepsy, and penises to enhance virility. In 2009 a bowl of tiger penis soup could be obtained for $300 in Taiwan. There is some hope that the availability of inexpensive drugs for treating erectile dysfunction will reduce the incidence of poaching of these and other endangered species, but hopes are dim for eliminating poaching altogether in Asia and Africa. Demand for traditional animal-based medicines remains high, and animal aphrodisiacs provide an economic boon for impoverished hunters and a status symbol for the rich.

Massive international trade in exotic pets and aquarium fishes, ornamental plants, and tropical forest hardwoods has decimated many species. The Banggai cardinalfish (*Pterapogon kauderni*; **Figure 59.7B**) is on the brink of extinction entirely because of the pet trade; almost a million of these critically endangered fish are hauled out of the waters annually near Sulawesi, Indonesia, to satisfy the demand from saltwater aquarium enthusiasts.

(A)

(B) *Pterapogon kauderni*

59.7 Endangered by Exploitation (A) Skins confiscated at the China–Myanmar border illustrate the extent of poaching of endangered tigers (*Panthera tigris*). Beyond the value of their pelts, tiger bones and other body parts are highly prized in Asian traditional medicine. (B) The international pet trade has brought the Banggai cardinalfish to the brink of extinction. Each year almost a million of these critically endangered fish are hauled out of Indonesian coral reef waters to satisfy the demand from saltwater aquarium enthusiasts.

Boiga irregularis

59.8 An Agent of Extinction Since it was accidentally introduced onto the tiny Pacific island of Guam, the brown tree snake (*Boiga irregularis*) has eaten 15 species of land birds to extinction.

Burgeoning human populations in need of food are also placing unprecedented pressure on species harvested from the wild. Humans have captured wild fish for food for at least 40,000 years, but in recent centuries innovations in technology and increasing demand have led to removal of fish from wild populations at rates that far exceed the capacity of the remaining individuals to reproduce. An estimated 25 percent of the world's wild fisheries are currently at risk of overexploitation and collapse. Deep-sea fish that are slow to mature and produce relatively few offspring, such as the orange roughy (*Hoplostethus atlanticus*), are especially sensitive to overexploitation.

Invasive predators, competitors, and pathogens threaten many species

As people travel, they deliberately or inadvertently move species to regions outside their original ranges. Some of these non-native species become **invasive**—that is, they reproduce rapidly, spread widely, and have negative effects on the native species of the region. As we saw in Section 55.4, species that are introduced into a region where their natural enemies are absent may reach very high population densities. Moreover, the native species in an invader's new range may not have evolved defenses against these new antagonists and competitors.

Invasive species are spread in several ways. Marine organisms have been spread throughout the oceans by ballast water, taken on by ships at the port of departure and discharged at the destination port along with its content of surviving animals and plants. The notorious zebra mussel (*Dreissena polymorpha*; see Figure 55.10) is thought to have arrived in North America in this way. The brown tree snake (*Boiga irregularis*; **Figure 59.8**) arrived on Guam in air cargo shortly after World War II. Until then, the only snake on Guam was a tiny insect-eating species. The number of *B. irregularis* on Guam remained low for some 20 years, but in the 1960s the species began to multiply and today can be found at densities up to 5,000 individuals per square kilometer. The snake has exterminated 15 species of land birds, including 3 found only on Guam.

Over the past 400 years, Europeans colonizing new continents have deliberately introduced plants and animals to their new homes in an effort to reconstruct their familiar surroundings. Many of these introductions have had disastrous effects on native flora and fauna. In Australia the introduction of European rabbits and foxes for sport hunting and of dogs and cats as pets has led to the extermination of nearly half the small- to medium-sized native marsupials over the last 100 years. Some species have been introduced deliberately to control other invasive species, and then have themselves caused even greater problems. One such example is the cane toad (*Bufo marinus*), introduced into Australia to control sugarcane pests (see Figure 55.15).

It can be difficult for people to imagine that plants that are desirable and attractive in their place of origin can "go rogue" when they escape from cultivation in a new region. Some of today's most noxious weeds were deliberately transported and

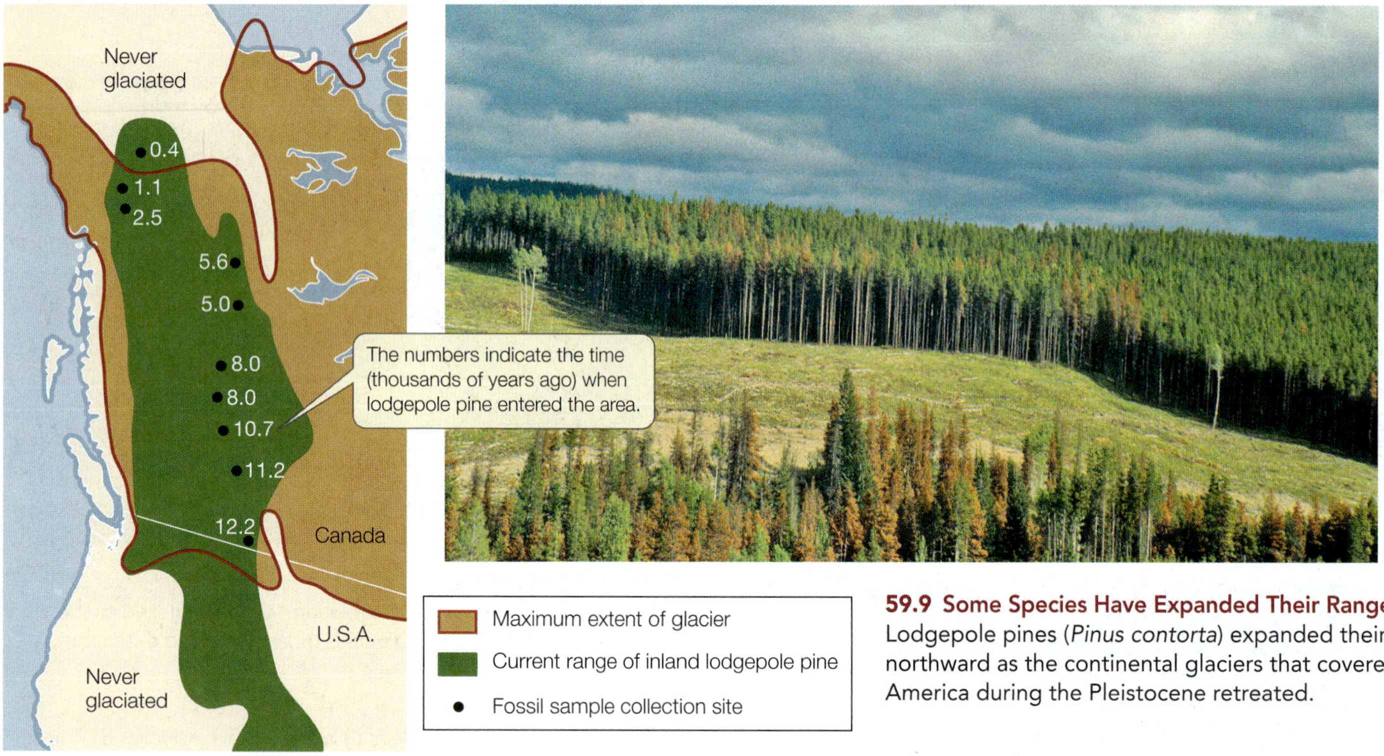
Never glaciated
0.4
1.1
2.5
5.6
5.0
8.0
8.0
10.7
11.2
12.2
The numbers indicate the time (thousands of years ago) when lodgepole pine entered the area.
Canada
U.S.A.
Never glaciated
Maximum extent of glacier
Current range of inland lodgepole pine
Fossil sample collection site

59.9 Some Species Have Expanded Their Range Lodgepole pines (*Pinus contorta*) expanded their range northward as the continental glaciers that covered North America during the Pleistocene retreated.

planted in new places for their beauty, fragrance, or culinary value. Once established in their new environments, however, these invasive plants have had profoundly negative effects. While native plants must devote considerable energy and resources to defending themselves against native herbivores, invasive plants are less prone to attack, in part because their natural enemies have been left behind in their original range. Therefore invasive plants can devote more resources to growth and reproduction and fewer to producing defensive secondary compounds. The majority of plants considered by U.S. farmers to be weeds are non-native, and controlling them, primarily with chemical herbicides, costs billions of dollars every year.

Introduced pathogens have also wreaked havoc among native species, as exemplified by the effects of avian malaria in the Hawaiian Islands. Before the arrival of Europeans, no mosquitoes existed in the islands. The first mosquito species was found there in 1827, and over the next century several others followed. At the start of the twentieth century, the microbial pathogen that causes avian malaria arrived, most likely carried by imported caged birds. Not having been exposed to malaria over the course of their evolutionary history, Hawaii's many endemic bird species were exceptionally vulnerable to infection. Today nearly all species living below 1,500 meters elevation (the current upper limit of the range of the mosquito vectors) have been eliminated, mostly by avian malaria. Species living at higher elevations have fared better, but the range of the mosquitoes appears to be expanding upward as the climate warms, placing the surviving endemic species at risk.

Rapid climate change can cause species extinctions

As we saw in Section 58.3, human-generated emissions of greenhouse gases are hastening global climate warming, and that warming is likely to become an increasingly important cause of extinctions. Across North America, for example, average annual temperatures are predicted to increase anywhere from 2°C to 5°C by the end of the twenty-first century. If the climate warms to that extent, the average temperature found at any given location in North America today will shift 500 to 800 kilometers to the north. Those species that cannot adapt to the warmer climate will have to shift their geographic ranges by that distance within less than a century if they are to persist. Some biomes, such as alpine tundra, could disappear entirely as temperate forests expand up mountain slopes.

Efforts to control or reverse global warming present a challenge to people worldwide. Conservation biologists can contribute to discussions about how to respond to climate change by predicting how it may affect organisms and looking for ways to mitigate those effects. Their research activities include analyses of past climate changes and studies of sites currently undergoing rapid climate change. It would be helpful to know, for example, how rapidly species responded to the end of the most recent ice age, about 10,000 years ago. Which species did and did not keep pace with the warming climate? How much, and in what ways, do past ecological communities differ from those of today as a result of differences in the rates at which species' ranges shifted?

Species that can disperse easily, such as birds and insects that can fly considerable distances, may be able to shift their ranges as rapidly as the climate changes, provided they can find appropriate habitats. However, the ranges of other species are likely to shift more slowly. For example, after the glaciers retreated in North America about 8,000 years ago, the ranges of some pine trees, which have lightweight seeds that can be carried great distances by wind, expanded northward, so that today they grow as far north as the current climate permits (**Figure 59.9**). Native earthworms, on the other hand, fared less well—the glaciers may well have eliminated all earthworm species in Canada, and they have not been replaced by other North American species, which have moved their ranges northward only slowly. (Many of the earthworms found in the United States today are non-native species accidentally introduced from elsewhere.)

If Earth's surface warms as predicted, entirely new climates will develop, especially at low elevations in the tropics, where a warming of even 2°C would result in conditions warmer than those found anywhere in the humid tropics today. Adaptation to those climates may prove difficult even for many tropical organisms. Since the mid-1980s, the average minimum nightly temperature at La Selva Biological Station, in the Caribbean lowlands of Costa Rica, has increased from about 20°C to 22°C. On warmer nights, trees use more of their energy reserves to maintain themselves. As a result, even this small rise in temperature has reduced the average growth rates of six different tree species by about 20 percent.

RECAP 59.3

Several human activities threaten the persistence of species, including habitat degradation, fragmentation, and destruction; overexploitation; introductions of invasive species; and activities that cause rapid climate change.

- Describe three ways in which habitat loss is occurring today. **See p. 1233**
- Why are rates of species loss high in small habitat patches? **See pp. 1233–1234 and Figures 59.5 and 59.6**
- How can dispersal ability and climate change interact to affect the probability of extinction? **See pp. 1236–1237**

Demonstrating that species are endangered is an empty exercise if we cannot implement a plan of action to save them. In the next section we will consider some of the positive steps that are being taken to protect biodiversity.

59.4 What Strategies Are Used to Protect Biodiversity?

Conservation biologists use scientific theory, empirical data, and tools from a variety of disciplines to help protect endangered and threatened species and ecosystems. They identify the factors that present risks to species and use that information to devise action plans. Implementing those plans, however, often requires the cooperation of many different groups of people, so conservation biologists also work with landowners, politicians, lawyers, environmental activists, and the general public. It is thus very useful to examine the actions that conservation biologists and policy makers take to protect biodiversity in order to determine which approaches have been most successful and to understand what aspects have contributed to their success.

Protected areas preserve habitat and prevent overexploitation

The establishment of **protected areas**, in which habitat alteration and exploitation are restricted or prohibited, is an important component of efforts to conserve biological diversity. Protected areas allow populations to maintain themselves in the preserved habitat and may also serve as nurseries from which individuals can disperse into exploited areas, replenishing populations that might otherwise become extinct.

Deciding which areas to protect is a challenging enterprise. Two robust criteria are species richness (the total number of species living in an area; see Section 57.3) and **endemism** (the number of species in an area that are found nowhere else—a measure of its uniqueness). Using these two criteria, biologists have identified regions of unusual richness and endemism, which they have labeled **biodiversity hotspots** (Figure 59.10). These hotspots occupy slightly less than 16 percent of Earth's land surface, but they are home to approximately 77 percent of its terrestrial vertebrate species. Most of these hotspots are also areas of high human population density, which means habitat loss is ongoing and often rapid. Developing a conservation strategy for any of these regions requires not only a detailed analysis of the distributions of species and the locations of special habitat resources (such as caves, freshwater springs, or migratory stopover areas for birds), but also an analysis of factors that threaten and factors that support biodiversity in the region.

In 2010, in an effort to pinpoint sites with threatened species that are found nowhere else, conservation biologists identified 587 "centers of imminent extinction." These sites are concentrated in tropical forests, on islands, and in mountainous regions (Figure 59.11). Only about half the sites are even partially protected by law, and most of them are surrounded by land that is undergoing rapid development. Unless protective actions are taken soon, species extinctions at these sites are inevitable. Identifying biodiversity hotspots and centers of imminent extinction has been helpful in prioritizing conservation efforts worldwide, encouraging international cooperation, and raising public awareness of critical threats to species persistence.

Degraded ecosystems can be restored

When a species is endangered as a consequence of habitat degradation rather than outright habitat loss, protecting the species may require restoring the habitat to a more natural state. Many degraded ecosystems recover only slowly, if at all, without human assistance. Practitioners of **restoration ecology** are developing methods aimed at just such habitat reconstitution.

Because the soil that supports them is so rich, grasslands all over the world have been converted to agriculture. By the middle of the twentieth century, for example, most North American prairies had been converted to cropland or were heavily grazed by domestic livestock. The herds of large mammals that roamed the prairies when European settlers arrived have been reduced to tiny remnant populations confined to small areas. Most of these populations are too small to maintain their genetic diversity or to function in their original ecological roles. The species *have* survived, however, so opportunities exist to reintroduce them if their habitat can be restored.

A major prairie restoration project is underway in northeastern Montana. When Lewis and Clark mapped this region 200

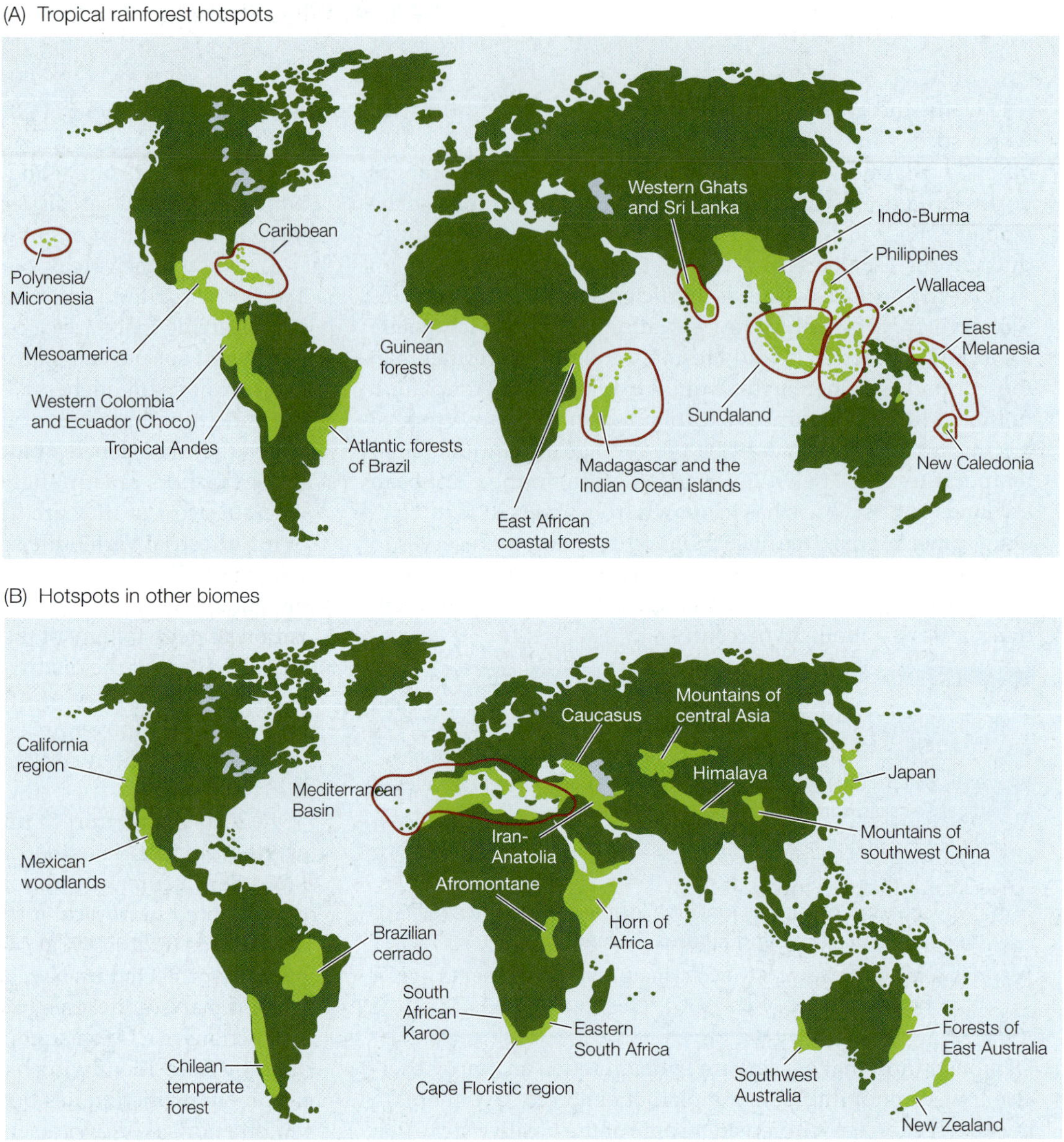
(A) Tropical rainforest hotspots
Polynesia/
Micronesia
Caribbean
Mesoamerica
Western Colombia
and Ecuador (Choco)
Tropical Andes
Atlantic forests
of Brazil
Guinean
forests
East African
coastal forests
Madagascar and the
Indian Ocean islands
Western Ghats
and Sri Lanka
Indo-Burma
Philippines
Wallacea
Sundaland
East
Melanesia
New Caledonia
(B) Hotspots in other biomes
California
region
Mexican
woodlands
Brazilian
cerrado
Chilean
temperate
forest
Mediterranean
Basin
Caucasus
Mountains of
central Asia
Himalaya
Japan
Mountains of
southwest China
Iran-
Anatolia
Afromontane
Horn of
Africa
South
African
Karoo
Eastern
South Africa
Cape Floristic region
Southwest
Australia
Forests of
East Australia
New Zealand

59.10 Hotspots of Biodiversity (A) Almost half of the world's terrestrial biodiversity hotspots are regions of tropical rainforest habitat. There are only three remaining areas of extensive unbroken rainforest (Amazonia, the Congo Basin of Africa, and the island of New Guinea; see Figure 59.4). (B) Eighteen additional hotspots represent non-rainforest ecosystems.

years ago, they saw large herds of bison, elk, deer, and pronghorn as well as abundant populations of their predators. The goal of the restoration project, which is run by the World Wildlife Fund and the American Prairie Foundation in cooperation with public land managers and several other private conservation organizations, is to restore the native prairie and its fauna in a 15,000-km^2 area near the Missouri River (**Figure 59.12**).

This ambitious project is feasible for three reasons. First, the private land in the area is owned by a small number of ranchers, each of whom owns extensive grazing leases on public lands administered by either U.S. federal agencies or the State of Montana. Second, most of the land has never been plowed,

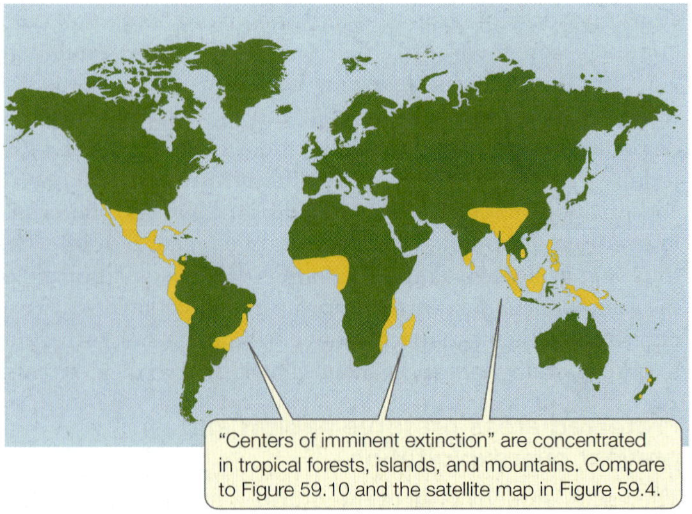
"Centers of imminent extinction" are concentrated in tropical forests, islands, and mountains. Compare to Figure 59.10 and the satellite map in Figure 59.4.

59.11 Centers of Imminent Extinction Areas shown in yellow include many of the world's 587 "centers of imminent extinction" (as designated in 2010 by the Alliance for Zero Extinction, a coalition of more than 80 conservation groups). Although there are scattered centers in other regions, the areas highlighted here harbor an estimated 1,000 endemic species (species found nowhere else) known to be at high risk of extinction.

(A)

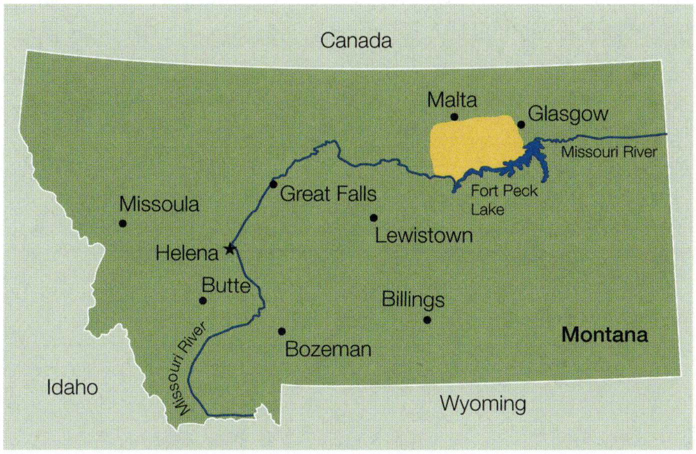
Canada
Malta
Glasgow
Missouri River
Fort Peck Lake
Great Falls
Missoula
Lewistown
Helena
Butte
Billings
Montana
Bozeman
Missouri River
Idaho
Wyoming

(B) *Cynomys ludovicianus*

(C) *Bison bison*

59.12 Restoring a North American Prairie (A) A major prairie restoration project (yellow area) is under way north of the Missouri River in the state of Montana. (B) Native prairie dogs maintain the vegetation by digging extensive burrows and clipping plants. (C) The first bison were reintroduced to the area in 2005.

so native vegetation may recover rapidly when grazing pressures are reduced. Third, the area's human population is decreasing. Ranchers are aging, and their children are abandoning the hard work and uncertain profits of ranching for careers in urban settings. Once a free-ranging herd of several thousand bison and large numbers of elk—along with their predators (wolves)—has been established, nature-minded tourists are expected to flock to the area to view the wildlife spectacle. Over the long term, the restored ecosystem should deliver major economic benefits to the region.

In the United States, the sense that humans are capable of creating functioning ecosystems to replace those lost to development underlies policies that allow developers to destroy habitats. Destruction of wetlands, in particular, is often permitted because developers assert that those ecosystems can be replaced. However, creating new wetlands requires detailed ecological knowledge that generally surpasses what is currently available.

In southern California, where 90 percent of the coastal wetlands have been destroyed, wetland restoration is a high priority. Species have been lost from degraded coastal wetlands, so restoration requires species introductions. Early attempts at restoration, in which one or two common wetland species were introduced, did not succeed; other wetland-associated species failed to recolonize the "rehabilitated" wetlands. To understand why, conservation biologists established a large field experiment at the Tijuana Estuary near San Diego (Figure 59.13). Here they found that experimental plots planted with species-rich mixtures were covered with vegetation faster, developed a complex vegetation structure (which is important to insects and birds) more rapidly, and accumulated nitrogen (required for plant growth) faster than did species-poor plots (Figure 59.14). This outcome represents a practical demonstration of the relationship between community stability and species richness (see Section 57.5).

Disturbance patterns sometimes need to be restored

Many species depend on particular patterns of disturbance—such as fires or windstorms—to maintain their populations (see Section 57.4). Recognition of the need for periodic disturbance to maintain healthy ecosystems is a relatively new dimension

59.13 A Wetlands Laboratory The Tijuana Estuary near San Diego is a shallow-water wetland habitat. Experiments at this natural research reserve have advanced efforts to restore this valuable ecosystem.

INVESTIGATING**LIFE**

59.14 Species Richness Can Enhance Wetland Restoration In a large-scale experiment in the Tijuana Estuary, John Callaway and other ecologists from the Southern California Wetlands Recovery Project compared different methods for restoring shallow-water wetlands. They found that several measures of ecosystem function improved more rapidly in species-rich than in species-poor plantings.[a]

HYPOTHESIS Faster progress toward restoring a shallow-water wetland community to its original condition will be made by planting mixtures of species than by planting a single species.

Method

1. In an area of wetland denuded of vegetation, mark off replicate experimental plots, all of the same size.
2. Choose 8 native species typical of the region. Plant some plots with 1 of the 8 species by itself, others with different subsets of 3 species, and others with different subsets of 6 species. Plant the same total number of seedlings in each plot. Leave control plots unplanted.
3. Measure the percent ground cover, number of canopy layers, and soil nitrogen levels at 6-month intervals over the next 18 months.

Results In the plots with higher species richness, more of the ground was covered by plants, the vegetation structure was more complex, and more nitrogen accumulated in the soil.

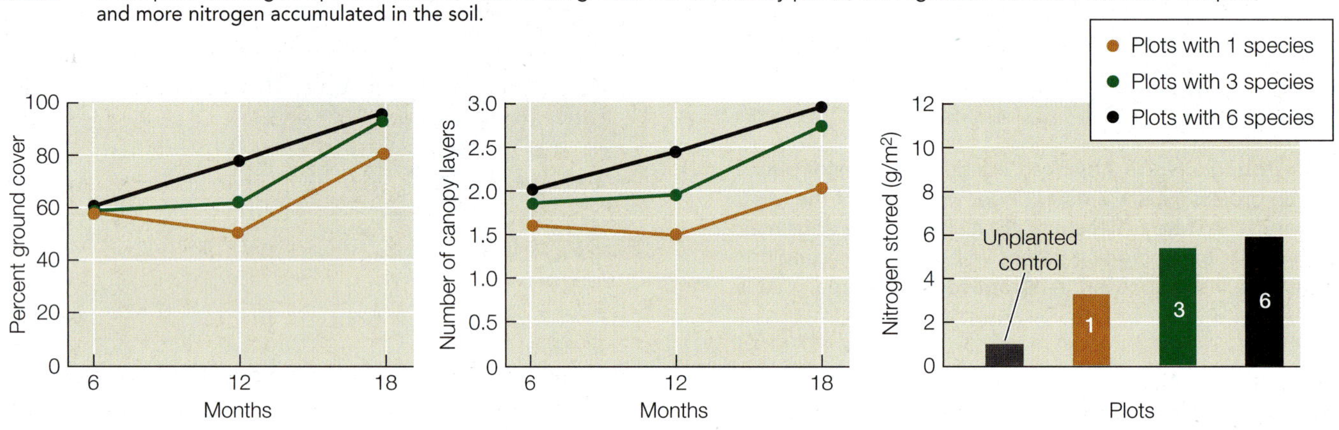
Plots with 1 species
Plots with 3 species
Plots with 6 species
Percent ground cover
100
80
60
40
20
0
6
12
18
Months
Number of canopy layers
3.0
2.5
2.0
1.5
1.0
0.5
0
Nitrogen stored (g/m²)
12
10
8
6
4
2
Unplanted control
1
3
6
Plots

CONCLUSION Planting a rich mixture of species leads to more rapid restoration of shallow-water wetlands.

Go to **BioPortal** for discussion and relevant links for all INVESTIGATING**LIFE** figures.

[a] Callaway, J. C., G. Sullivan, and J. B. Zedler. 2003. *Ecological Applications* 13: 1626–1639.

of conservation biology. For example, although many plant species require periodic fires for successful establishment and survival, for many years the official policy of the U.S. Forest Service, symbolized by the iconic mascot Smokey Bear, was to suppress all forest fires. Today, however, controlled burning is common, particularly in western North America. In order to use fire as an ecosystem management tool, it is important to know the historical pattern of fires in an area, which can be determined in part by studies of the annual growth rings and fire scars of trees (**Figure 59.15**). A schedule of controlled burning that recreates the historical pattern can reduce forest floor litter, avoiding a buildup of fuel that can lead to intense, tree-killing canopy fires.

Ending trade is crucial to saving some species

Most endangered species cannot survive any further reductions in their breeding populations, so it is important to prevent their exploitation. The legal mechanism for prohibiting trade in these species or their products is an international agreement called the Convention on International Trade in Endangered Species (CITES). Most nations of the world are members of CITES. CITES rules currently prohibit international trade in items such as whale meat, rhinoceros horns, and many species of parrots, orchids, and others.

The recent history of elephant poaching and trade in ivory illustrates how complex preventing exploitation of endangered species can be. CITES instituted a ban on international trade in African elephant ivory in 1989, but demand for ivory remains strong, especially in Asia. As a result, poaching of elephants continues in the forests of central and East Africa, where the animals are threatened. However, some countries, including Malawi and Zambia, have so many elephants that government officials kill them to control populations and prevent the animals from damaging crops. These countries would like to sell the ivory from culled elephants to fund conservation efforts, but other countries are worried that if restrictions are relaxed, poaching will escalate everywhere.

Control of ivory trade might be possible if scientists could determine where the ivory comes from. Conservation biologist Samuel Wasser and his colleagues identified 16 DNA markers from elephant feces collected by park rangers in Malawi and Zambia. The source of an elephant tusk could then be determined by matching DNA extracted from the ivory with the geographically based frequencies of the 16 DNA markers in the dung samples. Such safeguards were partially responsible for the controversial decision to sanction sales of ivory from Namibia, Botswana, Zimbabwe, and South Africa in 2008, the first such sales in close to a decade. More than 100 tons of

1994
1896
1900···
1885
1869
1864
1851
1844
1834
1828
1823
1820

59.15 A Natural Disturbance Pattern As revealed by scars (arrows) in the growth rings of this ponderosa pine (*Pinus ponderosa*), low-intensity, nonlethal ground fires were frequent in the pine forests of the southwestern United States prior to fire suppression.

elephant tusks—the equivalent of 20,000 dead elephants—were auctioned off to authorized buyers from China and Japan, for use primarily in folk medicine. This legal sale generated some $15 million for elephant conservation efforts.

Although the 2008 sales were monitored by CITES, concerns remain that the flood of legal ivory will be intermingled with poached ivory. One promising development in curbing illegal sales was the decision by eBay, the international internet marketplace, to ban sales of ivory on its platform as of January 2009. An independent investigation by the International Fund for Animal Welfare stated that two-thirds of online sales of protected wildlife products take place on eBay, so conservationists hope eBay's actions will be effective in drying up markets. Notwithstanding such efforts, illegal poaching, smuggling, and trafficking of ivory have all increased, and representatives from 175 countries attending the 2010 meeting of CITES in Qatar voted to ban sales of stockpiled elephant ivory for at least 3 years.

Species invasions must be controlled or prevented

The best way to reduce the damage caused by invasive species is to prevent their introduction in the first place. Given the tremendous volume of global trade, it might seem impossible to curtail their spread, but some promising strategies do exist. For example, transoceanic transport of invasive species in ballast water (responsible for the devastation caused by the invasive zebra mussel; see Figure 55.10) could be largely eliminated by the simple procedure of deoxygenating ballast water before it is pumped out. This practice not only kills most organisms in the water but also extends the life of ballast tanks—an economic benefit to shippers.

In 1996 the U.S. Congress responded to concerns about ballast water with legislative action. After years of wrangling, in 2012 the U.S. Coast Guard amended its regulations on managing ballast water to set standards for "the allowable concentration of living organisms in ballast water discharged from ships in waters of the United States." The Coast Guard relied on scientific reports issued by the National Academy of Sciences and the U.S. Environmental Protection Agency Science Advisory Board to specify the most stringent discharge standards achievable with current technology. Despite the adoption of these strict standards for protecting U.S. waterways, the challenge of achieving global uniformity remains. The transport of invasive aquatic organisms in ballast water is an international problem whose potential solutions continue to run up against political and economic barriers.

Regulating the importation and sale of non-native plant species has been more successful in reducing deliberate introductions. In 2002, members of the American horticulture industry crafted a voluntary code of conduct for their profession, stating that the invasive potential of a plant should be assessed prior to its introduction and marketing. Horticulturists were encouraged to work with biologists to determine which species are currently invasive, or are likely to become so, and to identify suitable alternative species.

Conservation biologists have developed a "decision tree" based on the traits that characterize plant species that have become invasive (**Figure 59.16**). The tree is used to help horticulturists and regulators determine whether a non-native plant species should be allowed into North America. Although the protocols stipulated by this decision tree cannot eliminate all detrimental introductions, if followed conscientiously they can greatly reduce the risk of such events.

Biodiversity has economic value

Many studies have demonstrated the market value of protecting biodiversity. Markets already exist for many products of biodiversity; to cite just one example, sales of pharmaceuticals derived from plants worldwide amount to more than $30 billion annually. Thus the argument for conservation is compelling not only from an ecological or ethical perspective but also from an economic perspective. The interdisciplinary field of **ecological economics** provides tools for assessing the economic value of biodiversity

Crucial to ecological economics is recognizing and determining the value of services that thriving ecosystems provide to human society (see Section 58.4). These services depend on biodiversity, but it is difficult to assess their value in monetary terms. Ecosystem services that depend on the maintenance of biodiversity include:

- *Provisioning services*, including the availability of food and water for human consumption.

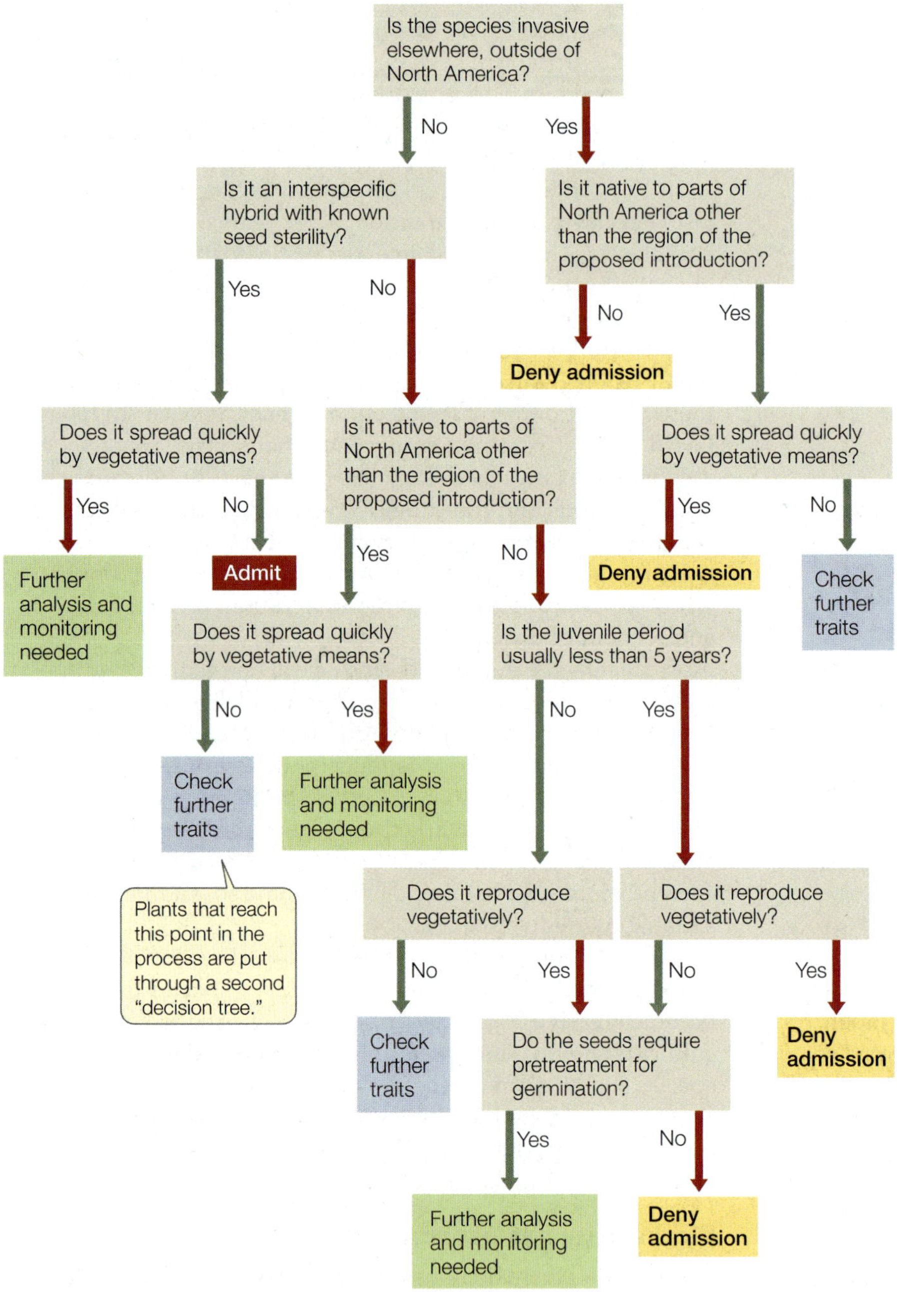
Is the species invasive elsewhere, outside of North America?
No
Yes
Is it an interspecific hybrid with known seed sterility?
Is it native to parts of North America other than the region of the proposed introduction?
Yes
No
No
Yes
Deny admission
Does it spread quickly by vegetative means?
Is it native to parts of North America other than the region of the proposed introduction?
Does it spread quickly by vegetative means?
Yes
No
Yes
No
Further analysis and monitoring needed
Admit
Yes
No
Deny admission
Check further traits
Does it spread quickly by vegetative means?
Is the juvenile period usually less than 5 years?
No
Yes
No
Yes
Check further traits
Further analysis and monitoring needed
Plants that reach this point in the process are put through a second "decision tree."
Does it reproduce vegetatively?
Does it reproduce vegetatively?
No
Yes
No
Yes
Check further traits
Do the seeds require pretreatment for germination?
Deny admission
Yes
No
Further analysis and monitoring needed
Deny admission

59.16 A "Decision Tree" This diagram sets criteria for evaluating proposed introductions of non-native plant species and helps regulators identify those species with the potential to become invasive.

- *Regulating services*, including the contributions of ecosystems to purification of water, flood control, pollination, and regulation of disease or pest outbreaks.
- *Supporting services*, including formation of soils and maintenance of nutrient cycles.
- *Cultural services*, including the provision of nonmaterial benefits such as recreational activities, psychological well-being, and spiritual enrichment (one example being the opportunities to fish for salmon and to watch eagles at Flathead Lake, described at the opening of this chapter).

The following three examples illustrate how biodiversity can offer a range of benefits to human populations that more than justify investing in conservation.

WILD DOGS AND ECOTOURISM **Ecotourism**—environmentally responsible travel to natural areas, the proceeds of which support conservation efforts and the economic well-being of the local communities—is a major source of income for many developing nations. For example, tourists visiting Africa often express interest in seeing wild dogs (*Lycaon pictus*; **Figure 59.17**). However, diseases such as rabies and canine distemper, along with habitat loss, roadkills, deliberate extermination due to a perceived threat to livestock, and many other factors have decimated wild dog populations, making this the second most endangered carnivore in Africa. (Another canid, the Ethiopian wolf *Canis simensis* is first.) South Africa is home to about 400 of Africa's remaining 5,000 dogs, most of which live in Kruger National Park. Their endangered status has piqued tourist interest in these charismatic animals; a survey of visitors to South Africa revealed that nearly three-fourths of them would be willing to pay an extra U.S. $12 for the opportunity to see wild dogs. Conservation biologists are working with lodge owners and ranchers elsewhere in South Africa and in Kenya to encourage them to reestablish wild dogs in areas from which they have disappeared.

FYNBOS Studies by a group of economists, ecologists, and land managers have attempted to calculate the value of the economic benefits provided by the spectacularly species-rich fynbos community (described at the opening of Chapter 54) to the Western Cape Province of South Africa. More than two-thirds of the 8,500 plant species in the fynbos community are endemic, thriving despite summer droughts, nutrient-poor soils, and periodic fires. Some of the endemic plants, including proteas, are harvested for cut and dried flowers. An international market has developed for rooibos, a fynbos shrub used for herbal tea. Income also comes from hundreds of thousands of ecotourists who visit the region. The fynbos also provides recreational opportunities for local residents in urban areas nearby. Perhaps most importantly, however, the highland watershed in which fynbos thrive provides about two-thirds of the Western Cape's water supply.

In recent years several trees and shrubs from other continents have invaded the fynbos. Taller and faster-growing than the endemics, they displace the native vegetation, increasing the intensity and severity of fires. Moreover, because the

59.17 A Sight for Ecotourists' Eyes Tourists visiting Africa to experience its wildlife often express a desire to see the endangered African wild dog (*Lycaon pictus*). Protecting such species (as seen here in South Africa's Mala Mala Game Reserve) can be in the economic interest of a region.

invaders transpire more water than the endemics, they could decrease stream flows to less than half of those from areas covered with native plants, drastically reducing the water supply for people in the region (**Figure 59.18A**).

We can get an idea of the economic value of fynbos biodiversity to the Western Cape Province by estimating the cost of maintaining or replacing the water supply services the fynbos provides. The South African government has launched an effort to maintain fynbos biodiversity by felling and digging out invasive trees and shrubs and by controlled fires (**Figure 59.18B**). This effort costs between $140 and $830 per hectare, depending on the densities of the invaders. Alternatively, the services provided by fynbos vegetation could be replaced, but at a much higher cost. A sewage purification plant that would deliver the same amount of water to the Western Cape Province as a well-managed watershed of 10,000 hectares would cost $135 million to build and $2.6 million per year to operate. Desalination of seawater would cost four times as much. Thus the available alternatives would deliver water at a cost somewhere between 1.8 and 6.7 times higher than the cost of maintaining natural vegetation in the watershed. Maintaining the fynbos is less expensive and more labor-intensive—thus generating more employment—than the technologically sophisticated methods that could substitute for the services it provides.

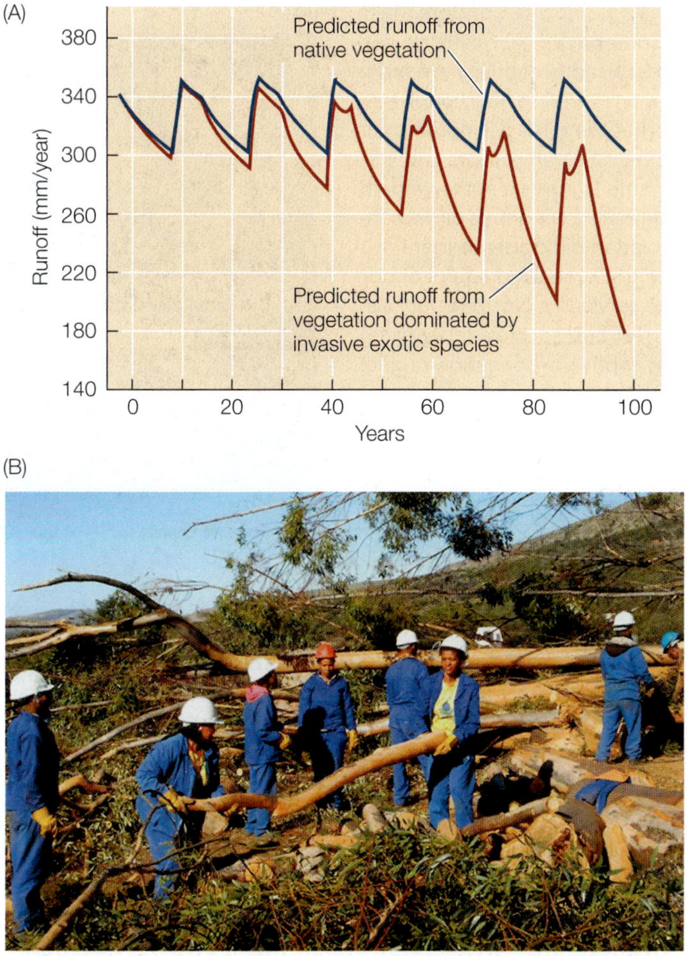
(A)
Predicted runoff from native vegetation
Predicted runoff from vegetation dominated by invasive exotic species
Runoff (mm/year)
380
340
300
260
220
180
140
0
20
40
60
80
100
Years
(B)

59.18 Biodiversity Maintains Ecosystem Functioning (A) A computer simulation of change in stream flows over time from fynbos watersheds that have and have not been invaded by nonnative trees and shrubs. (B) The Working for Water Programme has launched an effort to preserve the Western Cape Province's water supply by physically removing invasive species.

COFFEE AND POLLINATORS Taylor Ricketts and colleagues at Stanford University assessed the economic value of the pollination services provided by the bees that live in, and depend on, tropical forest patches adjacent to a coffee plantation in Costa Rica. They found that coffee production was highest at the sites that were closest to forest patches. They also hand-pollinated some coffee plants to show that the difference in production was a result of pollination services rather than other environmental conditions. The investigators calculated that the value of pollination services to the plantation on which the experiments were carried out was about $60,000 per year, more than the current conservation incentive payments offered to landowners to preserve forest patches.

Changes in human-dominated landscapes can help protect biodiversity

Establishing protected areas is an essential component of efforts to maintain biodiversity, but this action alone is insufficient to stem global biodiversity loss. The extensive landscapes in which people live and extract resources must also play important roles in biodiversity conservation. The good news is that, when carefully used, these lands can contribute much more to conservation than they currently do. The practice of using ecosystems for residences, resources, or recreation in ways that sustain their biodiversity is known as **reconciliation ecology**.

(A)
(B) *Gymnogyps californianus*

59.19 California Condors Make a Comeback (A) California condors raised in captivity are fed by humans wearing hand puppets so that the birds will not imprint on their human captors and will be able to survive in the wild. (B) Numbered wing tags allow conservation biologists to identify and track released adult condors. The survival of North America's largest bird species depends on this captive propagation project.

Reconciliation ecology is based on the principle that most ecosystem services are provided locally, and that people are more motivated to work to protect their local interests than they are to work on national or global issues. The National Wildlife Federation has established a successful program in which people petition to have their backyards certified as wildlife-friendly. Criteria for certification include planting shrubs that provide food for birds and refraining from applying pesticides.

Even some industrial sites can support biodiversity. The Turkey Point power plant in southern Florida uses large amounts of water to cool its generating units. To cool the heated water before discharging it, the Florida Power & Light Company dug a system of 38 canals that covers 6,000 acres. These cooling canals are separated by low-lying berms that support a variety of native and non-native plants. Red mangroves grow along the edges of the canals. Today they support a thriving population of American crocodiles, a highly endangered species. Crocodiles living in the canals yield about 10 percent of all young crocodiles born in the United States. Having discovered the biodiversity value of its cooling system, the company employs biologists to monitor the crocodiles and works actively to ensure their continued reproductive success.

Captive breeding programs can maintain a few species

A few of the world's endangered species can be maintained in captivity while the external threats to their persistence are reduced or removed. However, captive propagation is only a temporary measure that buys time to deal with those threats. Zoos, aquariums, and botanical gardens do not have enough space to maintain adequate populations of more than a small fraction of Earth's endangered and threatened species. Nonetheless, captive propagation can play an important role by maintaining species during critical periods, providing a source of individuals for reintroduction into the wild, and raising public awareness of threatened and endangered species.

The California condor, North America's largest bird, survives today only because of captive propagation (**Figure 59.19**). Two centuries ago, condors ranged from British Columbia to northern Mexico, but by 1978 the wild population was plunging toward extinction. Many of the birds, which are scavengers, had died from ingesting animal carcasses containing lead shot or bullets. To save the condor from certain extinction, biologists captured all the remaining condors—only 22 individuals—and initiated a captive breeding program in 1983.

The first captive-bred birds were released in the mountains north of Los Angeles in 1992. Since that time, there have also been releases in northern Arizona and Baja California. Today captive-bred birds use the same roosting sites, bathing pools, and mountain ridges that their wild-born predecessors did. In 2003 a wild-born chick fledged in the wild for the first time in more than two decades. By 2012 the number of condors living in the wild had reached 226, with another 179 living in captivity.

Most of the major threats to condor survival, including power lines, pesticides, and museum collectors, have been mitigated. Lead poisoning is still a problem, but as of July 1, 2008, under the Ridley–Tree Condor Preservation Act, California hunters are required to use non-lead bullets when hunting within the condor's range. Passage of this legislation marks a change in public attitudes from the days when cattle ranchers, in the mistaken belief that the condors killed livestock, vociferously opposed their reintroduction into the wild.

Earth is not a ship, a spaceship, or an airplane

Biologists often use metaphors to convey complex concepts, and ecologists in particular are fond of them. To convey the notion that human activities may be overwhelming the capacity of Earth's ecosystems to accommodate them, Herman Daly in 1966 introduced the metaphor of the Plimsoll line.

Since the nineteenth century, the Plimsoll line, or load line, has been painted on the side of ships that travel in international waters. If the Plimsoll line of a fully loaded ship anchored in port sinks below the water, there is a significant risk that, once at sea, the ship will capsize and the crew and cargo will be lost. Another ecological metaphor (also introduced in 1966, by Kenneth Boulding) is the concept of Spaceship Earth, which views humans as passengers on a planetary spaceship with limited supplies who must learn how to reuse and recycle in a "cyclical ecological system." And finally, in their 1985 book *Extinction*, Paul and Anne Ehrlich used an airplane metaphor to frame their "rivet hypothesis." According to this metaphor, ecosystems, like well-constructed airplanes, have redundant design features that allow certain parts to fail without causing a catastrophic breakdown: "A dozen rivets, or a dozen species, might never be missed. On the other hand, a thirteenth rivet popped from a wing flap, or the extinction of a key species ... could lead to a serious accident."

Even though they are remarkable human inventions, ships, spaceships, and airplanes are hopelessly inadequate to convey the enormous complexity of life. Moreover, relying on these metaphors creates a sense that we can rely on a few individuals—the captain or pilot and the "ship's" crew—to ensure that the planet doesn't capsize, run out of supplies, or crash. In fact, *every individual has a responsibility to understand how Earth provides essential resources and services and to recognize the effects that using those resources has on other living beings.*

Humans share Earth with a staggering diversity of organisms, and our continued existence depends on our interactions with them. We may be unique as a species, however, in that we have the intellectual capacity to recognize, quantify, and if need be, mitigate the effects of those interactions. Learning all we can about life on Earth is perhaps our best tool for improving the quality of our own existence on the planet for the long haul.

RECAP 59.4

To conserve biodiversity, it is necessary to set aside protected areas, restore ecosystems and natural disturbance patterns, restrict trade in endangered species and transport of invasive species, increase populations of endangered species, and otherwise recognize the benefits of maintaining functioning ecosystems.

- What are some of the priorities that conservation biologists consider when establishing protected areas? **See p. 1237 and Figures 59.10 and 59.11**
- What are some of the material and non-material ecosystem services that depend on maintaining biodiversity? **See pp. 1241 and 1241–1243**
- Explain the difference between restoration ecology and reconciliation ecology. **See pp. 1237 and 1243–1244**

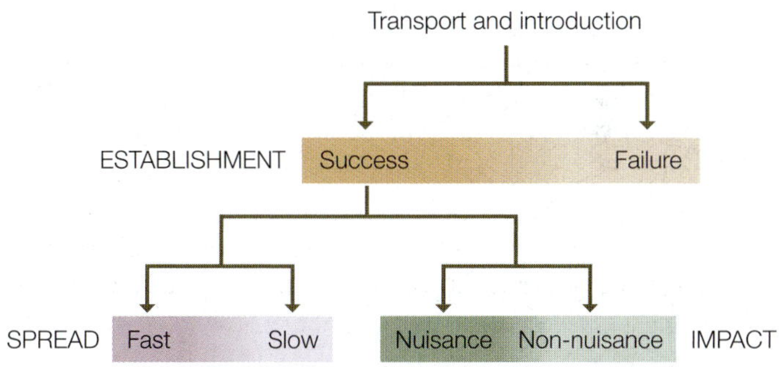
Transport and introduction
ESTABLISHMENT
Success
Failure
SPREAD
Fast
Slow
Nuisance
Non-nuisance
IMPACT

59.20 Three Steps to Invasion Invasion by a non-native species can be modeled as a process consisting of three steps, each with an independent probability of success or failure. To be considered invasive, the species must become established in a new environment, it must spread rapidly, and it must have an impact on its new environment.

How can adverse impacts of species introductions be anticipated before lasting damage occurs?

ANSWER

Conservation biologists use a range of tools to assess the risks of deliberate species introductions. In some cases, sufficient information is available from past introductions to allow biologists to identify the specific life history traits that make a species likely to become invasive. These traits can be used with a decision tree (see Figure 59.16), to guide decision-making. Mathematical models can also be constructed and used to assess the risks of both accidental and deliberate introductions. In one such model designed to identify potential fish invaders of the Great Lakes, investigators recognized three distinct stages of invasion: establishment, spread, and impact (**Figure 59.20**). They systematically compared the traits of fish species that had succeeded at each stage with those of species that had failed at that stage. The characteristics they used were minimum temperature threshold, breadth of diet, and relative growth. By identifying the characteristics of the successful versus failed invaders, investigators were then able to recreate the invasion process and predict with 94 percent accuracy the fish species that succeeded or failed at each stage.

CHAPTERSUMMARY 59

59.1 What Is Conservation Biology?

- **Conservation biology** is an applied scientific discipline devoted to protecting and managing biodiversity.
- Conservation biologists recognize that an understanding of the evolutionary processes that generate biodiversity is essential to protecting it. They also understand that ecosystems are dynamic, and that humans are part of those ecosystems.
- Species extinctions have always occurred, but they are currently occurring at a rate that rivals those of the five great mass extinctions in Earth's history.
- There are many compelling reasons for protecting biodiversity, including the maintenance of the species and ecosystems that provide humans with goods and services.

59.2 How Do Conservation Biologists Predict Changes in Biodiversity?

- Although our understanding of biodiversity is incomplete, biologists have identified many species that are threatened with extinction and have developed a classification system designed to aid in establishing policies for their protection. See Figure 59.3
- Biologists use the species–area relationship and the theory of island biogeography to estimate rates of extinction likely to be caused by habitat loss.
- To estimate a species' risk of extinction, statistical models take into account data on population sizes, demographic traits, genetic variation, physiology, and behavior.
- Rarity is not always a cause for concern, but species whose populations are shrinking rapidly are usually at risk of extinction.

59.3 What Human Activities Threaten Species Persistence?

- Habitat loss is the most important cause of species endangerment worldwide. As habitats become increasingly **fragmented**, more species are lost from those habitats. Small habitat patches can support only small populations and are adversely influenced by **edge effects**. Review Figures 59.4, 59.5, 59.6, ANIMATED TUTORIAL 59.1
- Overexploitation has historically been the most important cause of species extinctions, and it is still a major threat to biodiversity today.
- Some species introduced to regions outside their original range become **invasive**, causing extinctions of native species that have not evolved defenses against these new antagonists and competitors.
- Climate change is likely to become an increasingly important cause of extinctions for those species that cannot shift their ranges as rapidly as the climate warms. Review Figure 59.9

59.4 What Strategies Are Used to Protect Biodiversity?

- Establishing **protected areas** is crucial to conserving biodiversity. Protected areas are selected by taking into account species richness, **endemism**, and imminence of threats. Review Figures 59.10, 59.11
- **Restoration ecology** is an important conservation strategy because many degraded ecosystems will recover very slowly, if at all, without human assistance. Review Figure 59.14
- International trade in endangered species is controlled by regulations that most countries endorse.
- Conservation biologists work to determine which species are likely to become invasive and prevent their introduction to new areas. Review Figure 59.16
- Recognition of the economic value of biodiversity can help justify conservation efforts. Review Figure 59.18
- Even within landscapes where people live and extract resources, steps may be taken to protect biodiversity. This approach is known as **reconciliation ecology**.
- Captive breeding programs can maintain some endangered species for the short term while threats to their persistence in other natural habitats are reduced or removed.

See ACTIVITY 59.1 for a concept review of this chapter

Go to the Interactive Summary to review key figures, Animated Tutorials, and Activities
Life10e.com/is59

CHAPTERREVIEW

REMEMBERING

1. Which of the following is not currently a major cause of species extinctions?
   a. Habitat destruction
   b. Meteorite impacts
   c. Overexploitation
   d. Introductions of non-native predators
   e. Introductions of non-native pathogens

2. Species extinctions matter to human society because
   a. many important medications contain or are based on a plant product.
   b. people derive aesthetic pleasure from interacting with other organisms.
   c. causing species extinctions raises serious ethical issues.
   d. biodiversity helps maintain valuable ecosystem services.
   e. All of the above

3. As a habitat patch gets smaller, it
   a. cannot support populations of species that require large areas.
   b. supports only small populations of many species.
   c. is influenced to an increasing degree by edge effects.
   d. is invaded by species from surrounding habitats.
   e. All of the above

4. Global warming is a concern because
   a. the rate of change in climate is projected to be faster than the rate at which many species can shift their ranges.
   b. it is already too hot in the tropics.
   c. climates have been so stable for thousands of years that many species lack the ability to tolerate variable temperatures.
   d. climate change will be especially harmful to rare species.
   e. None of the above

5. Scientists can determine the historical frequency of fires in an area by
   a. examining charcoal in sites of ancient villages.
   b. measuring carbon in soils.
   c. radioactively dating fallen tree trunks.
   d. examining fire scars in growth rings of trees.
   e. determining the age structure of forests.

6. Captive propagation is a useful conservation tool provided that
   a. there is space in zoos, aquariums, and botanical gardens to preserve the species indefinitely.
   b. genetic uniformity of captive populations can be maintained.
   c. the threats that endangered the species are being alleviated so that captive-reared individuals can later be released back into the wild.
   d. there are sufficient caretakers.
   e. None of the above; captive propagation should never be used because it directs attention away from the need to protect species in their natural habitats.

7. A plant species is most likely to become invasive when introduced to a new area if it
   a. grows tall.
   b. has become invasive in other places where it has been introduced.
   c. is closely related to species living in the area where it has been introduced.
   d. has specialized dispersers of its seeds.
   e. has a long life span.

UNDERSTANDING & APPLYING

8. Conservation biologists have debated extensively which is better: many small protected areas (which may contain more species) or a few large protected areas (which may be the only ones that can support populations of species that require large areas). What ecological processes should be evaluated in making judgments about the sizes and locations of protected areas? How can principles of island biogeography be applied in this context?

9. The desert bighorn sheep of the southwestern United States is endangered. Its major predator, the puma, is also of conservation concern in the region. Under what conditions, if any, would it be appropriate to suppress the population of one rare species to assist another rare species?

ANALYZING & EVALUATING

10. During World War I, doctors adopted a "triage" system for dealing with wounded soldiers. The wounded were divided into three categories: those almost certain to die no matter what was done to help them, those likely to recover even if not assisted, immediately, and those whose probability of survival would be greatly increased if they were given immediate medical attention. Limited medical resources were directed primarily at the third category. What are some implications of adopting a similar approach toward species preservation?

11. Utilitarian arguments dominate discussions about the importance of preserving biological richness. In your opinion, what role should ethical and moral arguments play?

Go to BioPortal at **yourBioPortal.com** for Animated Tutorials, Activities, LearningCurve Quizzes, Flashcards, and many other study and review resources.

Appendix A The Tree of Life

Phylogeny is the organizing principle of modern biological taxonomy. A guiding principle of modern phylogeny is monophyly. A monophyletic group is considered to be one that contains an ancestral lineage and all of its descendants. Any such group can be extracted from a phylogenetic tree with a single cut.

The tree shown here provides a guide to the relationships among the major groups of extant (living) organisms in the tree of life as we have presented them throughout this book. The position of the branching "splits" indicates the relative branching order of the lineages of life, but the time scale is not meant to be uniform. In addition, the groups appearing at the branch tips do not necessarily carry equal phylogenetic "weight." For example, the ginkgo [75] is indeed at the apex of its lineage; this gymnosperm group consists of a single living species. In contrast, a phylogeny of the eudicots [83] could continue on from this point to fill many more trees the size of this one.

The glossary entries that follow are informal descriptions of some major features of the organisms described in Part Seven of this book. Each entry gives the group's common name, followed by the formal scientific name of the group (in parentheses). Numbers in square brackets reference the location of the respective groups on the tree.

It is sometimes convenient to use an informal name to refer to a collection of organisms that are not monophyletic but nonetheless all share (or all lack) some common attribute. We call these "convenience terms"; such groups are indicated in these entries by quotation marks, and we do not give them formal scientific names. Examples include "prokaryotes," "protists," and "algae." Note that these groups cannot be removed with a single cut; they represent a collection of distantly related groups that appear in different parts of the tree. We also use quotation marks here to designate two groups of fungi that are not believed to be monophyletic.

Go to BioPortal at **yourBioPortal.com** for an interactive version of this tree, with links to photos, distribution maps, species lists, and identification keys.

– A –

acorn worms (*Enteropneusta*) Benthic marine hemichordates [119] with an acorn-shaped proboscis, a short collar (neck), and a long trunk.

"algae" Convenience term encompassing various distantly related groups of aquatic, photosynthetic eukaryotes [4].

alveolates (*Alveolata*) [5] Unicellular eukaryotes with a layer of flattened vesicles (alveoli) supporting the plasma membrane. Major groups include the dinoflagellates [51], apicomplexans [50], and ciliates [49].

amborella (*Amborella*) [78] An understory shrub or small tree found only on the South Pacific island of New Caledonia. Thought to be the sister group of the remaining living angiosperms [15].

ambulacrarians (*Ambulacraria*) [29] The echinoderms [118] and hemichordates [119].

amniotes (*Amniota*) [36] Mammals, reptiles, and their extinct close relatives. Characterized by many adaptations to terrestrial life, including an amniotic egg (with a unique set of membranes—the amnion, chorion, and allantois), a water-repellant epidermis (with epidermal scales, hair, or feathers), and, in males, a penis that allows internal fertilization.

amoebozoans (*Amoebozoa*) [84] A group of eukaryotes [4] that use lobe-shaped pseudopods for locomotion and to engulf food. Major amoebozoan groups include the loboseans, plasmodial slime molds, and cellular slime molds.

amphibians (*Amphibia*) [128] Tetrapods [35] with glandular skin that lacks epidermal scales, feathers, or hair. Many amphibian species undergo a complete metamorphosis from an aquatic larval form to a terrestrial adult form, although direct development is also common. Major amphibian groups include frogs and toads (anurans), salamanders, and caecilians.

amphipods (*Amphipoda*) Small crustaceans [116] that are abundant in many marine and freshwater habitats. They are important herbivores, scavengers, and micropredators, and are an important food source for many aquatic organisms.

angiosperms (*Anthophyta* or *Magnoliophyta*) [15] The flowering plants. Major angiosperm groups include the monocots [82], eudicots [83], and magnoliids [81].

animals (*Animalia* or *Metazoa*) [19] Multicellular heterotrophic eukaryotes. The majority of animals are bilaterians [22]. Other groups of animals include the sponges [20], ctenophores [95], placozoans [96], and cnidarians [97]. The closest living relatives of the animals are the choanoflagellates [91].

annelids (*Annelida*) [105] Segmented worms, including earthworms, leeches, and polychaetes. One of the major groups of lophotrochozoans [24].

anthozoans (*Anthozoa*) One of the major groups of cnidarians [97]. Includes the sea anemones, sea pens, and corals.

anurans (*Anura*) Comprising the frogs and toads, this is the largest group of living amphibians [128]. They are tail-less, with a shortened vertebral column and elongate hind legs modified for jumping. Many species have an aquatic larval form known as a tadpole.

apicomplexans (*Apicomplexa*) [50] Parasitic alveolates [5] characterized by the possession of an apical complex at some stage in the life cycle.

arachnids (*Arachnida*) Chelicerates [114] with a body divided into two parts: a cephalothorax that bears six pairs of appendages (four pairs of which are usually used as legs) and an abdomen that bears the genital opening. Familiar arachnids include spiders, scorpions, mites and ticks, and harvestmen.

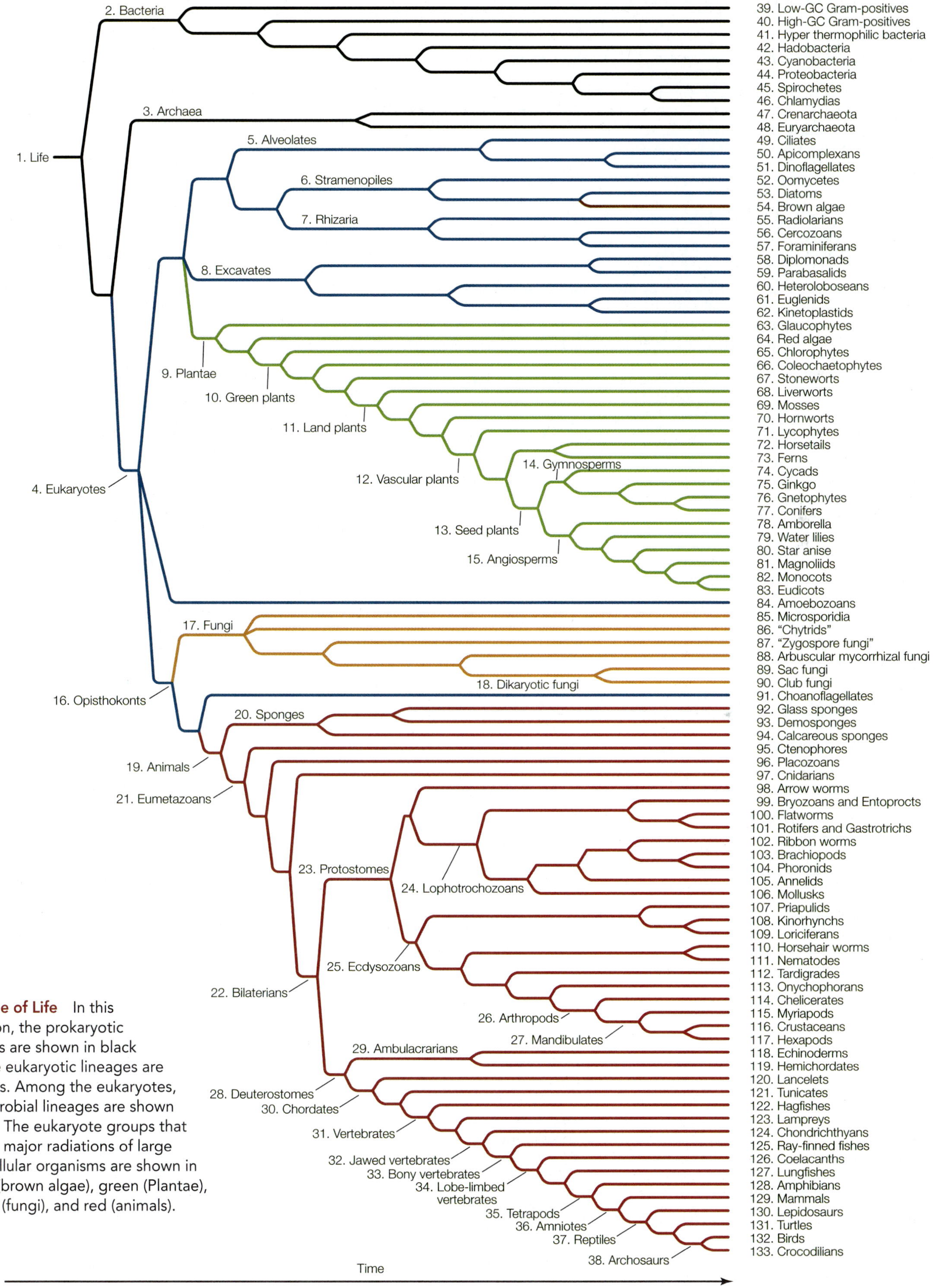
1. Life
2. Bacteria
3. Archaea
4. Eukaryotes
5. Alveolates
6. Stramenopiles
7. Rhizaria
8. Excavates
9. Plantae
10. Green plants
11. Land plants
12. Vascular plants
13. Seed plants
14. Gymnosperms
15. Angiosperms
16. Opisthokonts
17. Fungi
18. Dikaryotic fungi
19. Animals
20. Sponges
21. Eumetazoans
22. Bilaterians
23. Protostomes
24. Lophotrochozoans
25. Ecdysozoans
26. Arthropods
27. Mandibulates
28. Deuterostomes
29. Ambulacrarians
30. Chordates
31. Vertebrates
32. Jawed vertebrates
33. Bony vertebrates
34. Lobe-limbed vertebrates
35. Tetrapods
36. Amniotes
37. Reptiles
38. Archosaurs
39. Low-GC Gram-positives
40. High-GC Gram-positives
41. Hyper thermophilic bacteria
42. Hadobacteria
43. Cyanobacteria
44. Proteobacteria
45. Spirochetes
46. Chlamydias
47. Crenarchaeota
48. Euryarchaeota
49. Ciliates
50. Apicomplexans
51. Dinoflagellates
52. Oomycetes
53. Diatoms
54. Brown algae
55. Radiolarians
56. Cercozoans
57. Foraminiferans
58. Diplomonads
59. Parabasalids
60. Heteroloboseans
61. Euglenids
62. Kinetoplastids
63. Glaucophytes
64. Red algae
65. Chlorophytes
66. Coleochaetophytes
67. Stoneworts
68. Liverworts
69. Mosses
70. Hornworts
71. Lycophytes
72. Horsetails
73. Ferns
74. Cycads
75. Ginkgo
76. Gnetophytes
77. Conifers
78. Amborella
79. Water lilies
80. Star anise
81. Magnoliids
82. Monocots
83. Eudicots
84. Amoebozoans
85. Microsporidia
86. “Chytrids”
87. “Zygospore fungi”
88. Arbuscular mycorrhizal fungi
89. Sac fungi
90. Club fungi
91. Choanoflagellates
92. Glass sponges
93. Demosponges
94. Calcareous sponges
95. Ctenophores
96. Placozoans
97. Cnidarians
98. Arrow worms
99. Bryozoans and Entoprocts
100. Flatworms
101. Rotifers and Gastrotrichs
102. Ribbon worms
103. Brachiopods
104. Phoronids
105. Annelids
106. Mollusks
107. Priapulids
108. Kinorhynchs
109. Loriciferans
110. Horsehair worms
111. Nematodes
112. Tardigrades
113. Onychophorans
114. Chelicerates
115. Myriapods
116. Crustaceans
117. Hexapods
118. Echinoderms
119. Hemichordates
120. Lancelets
121. Tunicates
122. Hagfishes
123. Lampreys
124. Chondrichthyans
125. Ray-finned fishes
126. Coelacanths
127. Lungfishes
128. Amphibians
129. Mammals
130. Lepidosaurs
131. Turtles
132. Birds
133. Crocodilians
Time

The Tree of Life In this rendition, the prokaryotic lineages are shown in black and the eukaryotic lineages are in colors. Among the eukaryotes, the microbial lineages are shown in blue. The eukaryote groups that contain major radiations of large multicellular organisms are shown in brown (brown algae), green (Plantae), orange (fungi), and red (animals).

arbuscular mycorrhizal fungi (*Glomeromycota*) [88] A group of fungi [17] that associate with plant roots in a close symbiotic relationship.

archaeans (*Archaea*) [3] Unicellular organisms lacking a nucleus and lacking peptidoglycan in the cell wall. Once grouped with the bacteria, archaeans possess distinctive membrane lipids.

archosaurs (*Archosauria*) [38] A group of reptiles [37] that includes dinosaurs and crocodilians [133]. Most dinosaur groups became extinct at the end of the Cretaceous; birds [132] are the only surviving dinosaurs.

arrow worms (*Chaetognatha*) [98] Small planktonic or benthic predatory marine worms with fins and a pair of hooked, prey-grasping spines on each side of the head.

arthropods (*Arthropoda*) The largest group of ecdysozoans [25]. Arthropods are characterized by a stiff exoskeleton, segmented bodies, and jointed appendages. Includes the chelicerates [114], myriapods [115], crustaceans [116], and hexapods (insects and their relatives) [117].

ascidians (*Ascidiacea*) "Sea squirts"; the largest group of tunicates [121]. They are sessile (as adults), marine, saclike filter feeders.

– B –

bacteria (*Eubacteria*) [2] Unicellular organisms lacking a nucleus, possessing distinctive ribosomes and initiator tRNA, and generally containing peptidoglycan in the cell wall. Different bacterial groups are distinguished primarily on nucleotide sequence data.

barnacles (*Cirripedia*) Crustaceans [116] that undergo two metamorphoses—first from a feeding planktonic larva to a nonfeeding swimming larva, and then to a sessile adult that forms a "shell" composed of four to eight plates cemented to a hard substrate.

bilaterians (*Bilateria*) [22] Those animal groups characterized by bilateral symmetry and three distinct tissue types (endoderm, ectoderm, and mesoderm). Includes the protostomes [23] and deuterostomes [28].

birds (*Aves*) [132] Feathered, flying (or secondarily flightless) tetrapods [35].

bivalves (*Bivalvia*) Major mollusk [106] group; clams and mussels. Bivalves typically have two similar hinged shells that are each asymmetrical across the midline.

bony vertebrates (*Osteichthyes*) [33] Vertebrates [31] in which the skeleton is usually ossified to form bone. Includes the ray-finned fishes [125], coelacanths [126], lungfishes [127], and tetrapods [35].

brachiopods (*Brachiopoda*) [103] Lophotrochozoans [24] with two similar hinged shells that are each symmetrical across the midline. Superficially resemble bivalve mollusks, except for the shell symmetry.

brittle stars (*Ophiuroidea*) Echinoderms [118] with five long, whip-like arms radiating from a distinct central disk that contains the reproductive and digestive organs.

brown algae (*Phaeophyta*) [54] Multicellular, almost exclusively marine stramenopiles [6] generally containing the pigment fucoxanthin as well as chlorophylls *a* and *c* in their chloroplasts.

bryozoans (*Ectoprocta* or *Bryozoa*) [99] A group of marine and freshwater lophotrochozoans [24] that live in colonies attached to substrates; also known as ectoprocts or moss animals. They are the sister group of entoprocts.

– C –

caecilians (*Gymnophiona*) A group of burrowing or aquatic amphibians [128]. They are elongate, legless, with a short tail (or none at all), reduced eyes covered with skin or bone, and a pair of sensory tentacles on the head.

calcareous sponges (*Calcarea*) [94] Filter-feeding marine sponges with spicules composed of calcium carbonate.

cellular slime molds (*Dictyostelida*) Amoebozoans [84] in which individual amoebas aggregate under stress to form a multicellular pseudoplasmodium.

cephalochordates (*Cephalochordata*) [120] *See* lancelets.

cephalopods (*Cephalopoda*) Active, predatory mollusks [106] in which the molluscan foot has been modified into muscular hydrostatic arms or tentacles. Includes octopuses, squids, and nautiluses.

cercozoans (*Cercozoa*) [56] Unicellular eukaryotes [4] that feed by means of threadlike pseudopods. Group together with foraminiferans [57] and radiolarians [55] to comprise the rhizaria [7].

charophytes (*Charales*) [67] *See* stoneworts.

chelicerates (*Chelicerata*) [114] A major group of arthropods [26] with pointed appendages (chelicerae) used to grasp food (as opposed to the chewing mandibles of most other arthropods). Includes the arachnids, horseshoe crabs, pycnogonids, and extinct sea scorpions.

chimaeras (*Holocephali*) A group of bottom-dwelling, marine, scaleless chondrichthyan fishes [124] with large, permanent, grinding tooth plates (rather than the replaceable teeth found in other chondrichthyans).

chitons (*Polyplacophora*) Flattened, slow-moving mollusks [106] with a dorsal protective calcareous covering made up of eight articulating plates.

chlamydias (*Chlamydiae*) [46] A group of very small Gram-negative bacteria; they live as intracellular parasites of other organisms.

chlorophytes (*Chlorophyta*) [65] The most abundant and diverse group of green algae, including freshwater, marine, and terrestrial forms; some are unicellular, others colonial, and still others multicellular. Chlorophytes use chlorophylls *a* and *c* in their photosynthesis.

choanoflagellates (*Choanozoa*) [91] Unicellular eukaryotes [4] with a single flagellum surrounded by a collar. Most are sessile, some are colonial. The closest living relatives of the animals [19].

chondrichthyans (*Chondrichthyes*) [124] One of the two main groups of jawed vertebrates [32]; includes sharks, rays, and chimaeras. They have cartilaginous skeletons and paired fins.

chordates (*Chordata*) [30] One of the two major groups of deuterostomes [28], characterized by the presence (at some point in development) of a notochord, a hollow dorsal nerve cord, and a post-anal tail. Includes the lancelets [120], tunicates [121], and vertebrates [31].

"chytrids" [90] Convenience term used for a paraphyletic group of mostly aquatic, microscopic fungi [17] with flagellated gametes. Some exhibit alternation of generations.

ciliates (*Ciliophora*) [49] Alveolates [5] with numerous cilia and two types of nuclei (micronuclei and macronuclei).

clitellates (*Clitellata*) Annelids [105] with gonads contained in a swelling (called a clitellum) toward the head of the animal. Includes earthworms (oligochaetes) and leeches.

club fungi (*Basidiomycota*) [90] Fungi [17] that, if multicellular, bear the products of meiosis on club-shaped basidia and possess a long-lasting dikaryotic stage. Some are unicellular.

club mosses (*Lycopodiophyta*) [71] Vascular plants [12] characterized by microphylls. *See* lycophytes.

cnidarians (*Cnidaria*) [97] Aquatic, mostly marine eumetazoans [21] with specialized stinging organelles (nematocysts) used for prey capture and defense, and a blind gastrovascular cavity. The sister group of the bilaterians [22].

coelacanths (*Actinista*) [126] A group of marine lobe-limbed vertebrates [34] that was diverse from the Middle Devonian to the Cretaceous, but is now known from just two living species. The pectoral and anal fins are on fleshy stalks supported by skeletal elements, so they are also called lobe-finned fishes.

coleochaetophytes (*Coleochaetales*) [66] Multicellular green algae characterized by flattened growth form composed of thin-walled cells. Thought to be the sister-group to the stoneworts [67] plus land plants [11].

conifers (*Pinophyta* or *Coniferophyta*) [77] Cone-bearing, woody seed plants [13].

copepods (*Copepoda*) Small, abundant crustaceans [116] found in marine, freshwater, or wet terrestrial habitats. They have a single eye, long antennae, and a body shaped like a teardrop.

craniates (*Craniata*) Some biologist exclude the hagfishes [122] from the vertebrates [31], and use the term craniates to refer to the two groups combined.

crenarchaeotes (*Crenarchaeota*) [47] A major and diverse group of archaeans [3], defined on the basis of rRNA base sequences. Many are extremophiles (inhabit extreme environments), but the group may also be the most abundant archaeans in the marine environment.

crinoids (*Crinoidea*) Echinoderms [118] with a mouth surrounded by feeding arms, and a U-shaped gut with the mouth next to the anus. They attach to the substratum by a stalk or are free-swimming. Crinoids were abundant in the middle and late Paleozoic, but only a few hundred species have survived to the present. Includes the sea lilies and feather stars.

crocodilians (*Crocodylia*) [133] A group of large, predatory, aquatic archosaurs [38]. The closest living relatives of birds [132]. Includes alligators, caimans, crocodiles, and gharials.

crustaceans (*Crustacea*) [116] Major group of marine, freshwater, and terrestrial arthropods [26] with a head, thorax, and abdomen (although the head and thorax may be fused), covered with a thick exoskeleton, and with two-part appendages. Crustaceans undergo metamorphosis from a nauplius larva. Includes decapods, isopods, krill, barnacles, amphipods, copepods, and ostracods.

ctenophores (*Ctenophora*) [95] Radially symmetrical, diploblastic marine animals [19], with a complete gut and eight rows of fused plates of cilia (called ctenes).

cyanobacteria (*Cyanobacteria*) [43] A group of unicellular, colonial, or filamentous bacteria that conduct photosynthesis using chlorophyll *a*.

cycads (*Cycadophyta*) [74] Palmlike gymnosperms with large, compound leaves.

cyclostomes (*Cyclostomata*) This term refers to the possibly monophyletic group of lampreys [123] and hagfishes [122]. Molecular data support this group, but morphological data suggest that lampreys are more closely related to jawed vertebrates [32] than to hagfishes.

– D –

decapods (*Decapoda*) A group of marine, freshwater, and semiterrestrial crustaceans [116] in which five of the eight pairs of thoracic appendages function as legs (the other three pairs, called maxillipeds, function as mouthparts). Includes crabs, lobsters, crayfishes, and shrimps.

demosponges (*Demospongiae*) [93] The largest of the three groups of sponges [20], accounting for 90 percent of all sponge species. Demosponges have spicules made of silica, spongin fiber (a protein), or both.

deuterostomes (*Deuterostomia*) [28] One of the two major groups of bilaterians [22], in which the mouth forms at the opposite end of the embryo from the blastopore in early development (contrast with protostomes). Includes the ambulacrarians [29] and chordates [30].

diatoms (*Bacillariophyta*) [53] Unicellular, photosynthetic stramenopiles [6] with glassy cell walls in two parts.

dikaryotic fungi (*Dikarya*) [18] A group of fungi [17] in which two genetically different haploid nuclei coexist and divide within the same hypha; includes club fungi [90] and sac fungi [89].

dinoflagellates (*Dinoflagellata*) [51] A group of alveolates [5] usually possessing two flagella, one in an equatorial groove and the other in a longitudinal groove; many are photosynthetic.

diplomonads (*Diplomonadida*) [58] A group of eukaryotes [4] lacking mitochondria; most have two nuclei, each with four associated flagella.

– E –

ecdysozoans (*Ecdysozoa*) [25] One of the two major groups of protostomes [23], characterized by periodic molting of their exoskeletons. Nematodes [111] and arthropods [26] are the largest ecdysozoan groups.

echinoderms (*Echinodermata*) [118] A major group of marine deuterostomes [28] with five-fold radial symmetry (at some stage of life) and an endoskeleton made of calcified plates and spines. Includes sea stars, crinoids, sea urchins, sea cucumbers, and brittle stars.

elasmobranchs (*Elasmobranchii*) The largest group of chondrichthyan fishes [124]. Includes sharks, skates, and rays. In contrast to the other group of living chondrichthyans (the chimaeras), they have replaceable teeth.

embryophytes *See* land plants [11].

entoprocts (*Entoprocta*) [99] A group of marine and freshwater lophotrochozoans [24] that live as single individuals or in colonies attached to substrates. They are the sister group of bryozoans, from which they differ in having both their mouth and anus inside the lophophore (the anus is outside the lophophore in bryozoans).

eudicots (*Eudicotyledones*)[83] A group of angiosperms [15] with pollen grains possessing three openings. Typically with two cotyledons, net-veined leaves, taproots, and floral organs typically in multiples of four or five.

euglenids (*Euglenida*) [61] Flagellate excavates characterized by a pellicle composed of spiraling strips of protein under the plasma membrane; the mitochondria have disk-shaped cristae. Some are photosynthetic.

eukaryotes (*Eukarya*) [4] Organisms made up of one or more complex cells in which the genetic material is contained in nuclei. Contrast with archaeans [3] and bacteria [2].

eumetazoans (*Eumetazoa*) [21] Those animals [19] characterized by body symmetry, a gut, a nervous system, specialized types of cell junctions, and well-organized tissues in distinct cell layers (although there have been secondary losses of some or most of these characteristics in a few eumetazoan lineages).

euphyllophytes (*Euphyllophyta*) The group of vascular plants [12] that is sister to the lycophytes [71] and which includes all plants with megaphylls.

euryarchaeotes (*Euryachaeota*) [48] A major group of archaeans [3], diagnosed on the basis of rRNA sequences. Includes many methanogens, extreme halophiles, and thermophiles.

eutherians (*Eutheria*) A group of viviparous mammals [129], eutherians are well developed at birth (contrast to prototherians and marsupials, the other two groups of mammals). Most familiar mammals outside the Australian and South American regions are eutherians (see Table 33.1).

excavates (*Excavata*) [8] Diverse group of unicellular, flagellate eukaryotes, many of which possess a feeding groove; some lack mitochondria.

– F –

ferns Vascular plants [12] usually possessing large, frondlike leaves that unfold from a "fiddlehead." Not a monophyletic group, although most fern species are encompassed in a monophyletic clade, the leptosporangiate ferns [73].

flatworms (*Platyhelminthes*) [100] A group of dorsoventrally flattened and generally elongate soft-bodied lophotrochozoans [24]. May be free-living or parasitic, found in marine, freshwater, or damp terrestrial environments. Major flatworm groups include the tapeworms, flukes, monogeneans, and turbellarians.

flowering plants *See* angiosperms [15].

flukes (*Trematoda*) A group of wormlike parasitic flatworms [100] with complex life cycles that involve several different host species. May be paraphyletic with respect to tapeworms.

foraminiferans (*Foraminifera*) [57] Amoeboid organisms with fine, branched pseudopods that form a food-trapping net. Most produce external shells of calcium carbonate.

fungi (*Fungi*) [17] Eukaryotic heterotrophs with absorptive nutrition based on extracellular digestion; cell walls contain chitin. Major fungal groups include the microsporidia [85], "chytrids" [86], "zygospore fungi" [87], arbuscular mycorrhizal fungi [88], sac fungi [89], and club fungi [90].

– G –

gastropods (*Gastropoda*) The largest group of mollusks [106]. Gastropods possess a well-defined head with two or four sensory tentacles (often terminating in eyes) and a ventral foot. Most species have a single coiled or spiraled shell. Common in marine, freshwater, and terrestrial environments.

gastrotrichs (*Gastrotricha*) [101] Tiny (0.06–3.0 mm), elongate acoelomate lophotrochozoans [24] that are covered in cilia. They live in marine, freshwater, and wet terrestrial habitats. They are simultaneous hermaphrodites.

ginkgo (*Ginkgophyta*) [75] A gymnosperm [14] group with only one living species. The ginkgo seed is surrounded by a fleshy tissue not derived from an ovary wall and hence not a fruit.

glass sponges (*Hexactinellida*) [92] Sponges [20] with a skeleton composed of four- and/or six-pointed spicules made of silica.

glaucophytes (*Glaucophyta*) [63] Unicellular freshwater algae with chloroplasts containing traces of peptidoglycan, the characteristic cell wall material of bacteria.

gnathostomes (*Gnathostomata*) *See* jawed vertebrates [32].

gnetophytes (*Gnetophyta*) [76] A gymnosperm [14] group with three very different lineages; all have wood with vessels, unlike other gymnosperms.

green plants (*Viridiplantae*) [10] Organisms with chlorophylls *a* and *b*, cellulose-containing cell walls, starch as a carbohydrate storage product, and chloroplasts surrounded by two membranes.

gymnosperms (*Gymnospermae*) [14] Seed plants [13] with seeds "naked" (i.e., not enclosed in carpels). Probably monophyletic, but status still in doubt. Includes the conifers [77], gnetophytes [76], ginkgo [75], and cycads [74].

– H –

hadobacteria (*Hadobacteria*)[42] A group of extremophilic bacteria [2] that includes the genera *Deinococus* and *Thermus*.

hagfishes (*Myxini*) [122] Elongate, slimy-skinned vertebrates [31] with three small accessory hearts, a partial cranium, and no stomach or paired fins. *See also* craniata; cyclostomes.

hemichordates (*Hemichordata*) [119] One of the two primary groups of ambulacrarians [29];

marine wormlike organisms with a three-part body plan.

heteroloboseans (*Heterolobosea*) [60] Colorless excavates [8] that can transform among amoeboid, flagellate, and encysted stages.

hexapods (*Hexapoda*) [117] Major group of arthropods [26] characterized by a reduction (from the ancestral arthropod condition) to six walking appendages, and the consolidation of three body segments to form a thorax. Includes insects and their relatives (see Table 23.2).

high-GC Gram-positives (*Actinobacteria*) [40] Gram-positive bacteria with a relatively high (G+C)/(A+T) ratio of their DNA, with a filamentous growth habit.

hornworts (*Anthocerophyta*) [70] Nonvascular plants with sporophytes that grow from the base. Cells contain a single large, platelike chloroplast.

horsehair worms (*Nematomorpha*) [110] A group of very thin, elongate, wormlike freshwater ecdysozoans [25]. Largely nonfeeding as adults, they are parasites of insects and crayfish as larvae.

horseshoe crabs (*Xiphosura*) Marine chelicerates [114] with a large outer shell in three parts: a carapace, an abdomen, and a tail-like telson. There are only five living species, but many additional species are known from fossils.

horsetails (*Sphenophyta* or *Equisetophyta*) [72] Vascular plants [12] with reduced megaphylls in whorls.

hydrozoans (*Hydrozoa*) A group of cnidarians [97]. Most species go through both polyp and mesuda stages, although one stage or the other is eliminated in some species.

hyperthermophilic bacteria [41] A group of thermophilic bacteria [2] that live in volcanic vents, hot springs, and in underground oil reservoirs; includes the genera *Aquifex* and *Thermotoga*.

– I –

insects (*Insecta*) The largest group within the hexapods [117]. Insects are characterized by exposed mouthparts and one pair of antennae containing a sensory receptor called a Johnston's organ. Most have two pairs of wings as adults. There are more described species of insects than all other groups of life [1] combined, and many species remain to be discovered. The major insect groups are described in Table 23.2.

"invertebrates" Convenience term encompassing any animal [19] that is not a vertebrate [31].

isopods (*Isopoda*) Crustaceans [116] characterized by a compact head, unstalked compound eyes, and mouthparts consisting of four pairs of appendages. Isopods are abundant and widespread in salt, fresh, and brackish water, although some species (the sow bugs) are terrestrial.

– J –

jawed vertebrates (*Gnathostomata*) [32] A major group of vertebrates [31] with jawed mouths. Includes chondrichthyans [124], ray-finned fishes [125], and lobe-limbed vertebrates [34].

– K –

kinetoplastids (*Kinetoplastida*) [62] Unicellular, flagellate organisms characterized by the presence in their single mitochondrion of a kinetoplast (a structure containing multiple, circular DNA molecules).

kinorhynchs (*Kinorhyncha*) [108] Small (< 1 mm) marine ecdysozoans [25] with bodies in 13 segments and a retractable proboscis.

korarchaeotes (*Korarchaeota*) A group of archaeans [3] known only by evidence from nucleic acids derived from hot springs. Its phylogenetic relationships within the Archaea are unknown.

krill (*Euphausiacea*) A group of shrimplike marine crustaceans [116] that are important components of the zooplankton.

– L –

lampreys (*Petromyzontiformes*) [123] Elongate, eel-like vertebrates [31] that often have rasping and sucking disks for mouths.

lancelets (*Cephalochordata*) [120] A group of weakly swimming, eel-like benthic marine chordates [30].

land plants (*Embryophyta*) [11] Plants with embryos that develop within protective structures; also called embryophytes. Sporophytes and gametophytes are multicellular. Land plants possess a cuticle. Major groups are the liverworts [68], mosses [69], hornworts [70], and vascular plants [12].

larvaceans (*Larvacea*) Solitary, planktonic tunicates [121] that retain both notochords and nerve cords throughout their lives.

lepidosaurs (*Lepidosauria*) [130] Reptiles [37] with overlapping scales. Includes tuataras and squamates (lizards, snakes, and amphisbaenians).

leptosporangiate ferns (*Pteridopsida* or *Polypodiopsida*) [73] Vascular plants [12] usually possessing large, frondlike leaves that unfold from a "fiddlehead," and possessing thin-walled sporangia.

life (*Life*) [1] The monophyletic group that includes all known living organisms. Characterized by a nucleic-acid based genetic system (DNA or RNA), metabolism, and cellular structure. Some parasitic forms, such as viruses, have secondarily lost some of these features and rely on the cellular environment of their host.

liverworts (*Hepatophyta*) [68] Nonvascular plants lacking stomata; stalk of sporophyte elongates along its entire length.

lobe-limbed vertebrates (*Sarcopterygii*) [34] One of the two major groups of bony vertebrates [33], characterized by jointed appendages (paired fins or limbs).

loboseans (*Lobosea*) A group of unicellular amoebozoans [84]; includes the most familiar amoebas (e.g., *Amoeba proteus*).

"lophophorates" Convenience term used to describe several groups of lophotrochozoans [24] that have a feeding structure called a lophophore (a circular or U-shaped ridge around the mouth that bears one or two rows of ciliated, hollow tentacles). Not a monophyletic group.

lophotrochozoans (*Lophotrochozoa*) [24] One of the two main groups of protostomes [23]. This group is morphologically diverse, and is supported primarily on information from gene sequences. Includes bryozoans and entoprocts [99], flatworms [100], rotifers and gastrotrichs [101], ribbon worms [102], brachiopods [103], phoronids [104], annelids [105], and mollusks [106].

loriciferans (*Loricifera*) [109] Small (< 1 mm) ecdysozoans [25] with bodies in four parts, covered with six plates.

low-GC Gram-positives (*Firmicutes*) [39] A diverse group of bacteria [2] with a relatively low (G+C)/(A+T) ratio of their DNA, often but not always Gram-positive, some producing endospores.

lungfishes (*Dipnoi*) [127] A group of aquatic lobe-limbed vertebrates [34] that are the closest living relatives of the tetrapods [35]. They have a modified swim bladder used to absorb oxygen from air, so some species can survive the temporary drying of their habitat.

lycophytes (*Lycopodiophyta*) [71] Vascular plants [12] characterized by microphylls; includes club mosses, spike mosses, and quillworts.

– M –

magnoliids (*Magnoliidae*) [81] A major group of angiosperms [15] possessing two cotyledons and pollen grains with a single opening. The group is defined primarily by nucleotide sequence data; it is more closely related to the eudicots and monocots than to three other small angiosperm groups.

mammals (*Mammalia*) [129] A group of tetrapods [35] with hair covering all or part of their skin; -females produce milk to feed their developing young. Includes the prototherians, marsupials, and -eutherians.

mandibulates (*Mandibulata*) [27] Arthropods [26] that include mandibles as mouth parts. Includes myriapods [115], crustaceans [116], and hexapods [117].

marsupials (*Marsupialia*) Mammals [129] in which the female typically has a marsupium (a pouch for rearing young, which are born at an extremely early stage in development). Includes such familiar mammals as opossums, koalas, and kangaroos.

metazoans (*Metazoa*) *See* animals [19].

microbial eukaryotes *See* "protists."

microsporidia (*Microsporidia*) [85] A group of parasitic unicellular fungi [17] that lack mitochondria and have walls that contain chitin.

mollusks (*Mollusca*) [106] One of the major groups of lophotrochozoans [24], mollusks have bodies composed of a foot, a mantle (which often secretes a hard, calcareous shell), and a visceral mass. Includes monoplacophorans, chitons, bivalves, gastropods, and cephalopods.

monilophytes (*Monilophyta*) A group of vascular plants [12], sister to the seed plants [13], characterized by overtopping and possession of megaphylls; includes the horsetails [72] and ferns [73].

monocots (*Monocotyledones*) [82] Angiosperms [15] characterized by possession of a single cotyledon, usually parallel leaf veins, a fibrous root system, pollen grains with a single opening, and floral organs usually in multiples of three.

monogeneans (*Monogenea*) A group of ectoparasitic flatworms [100].

monoplacophorans (*Monoplacophora*) Mollusks [106] with segmented body parts and a single, thin, flat, rounded, bilateral shell.

mosses (*Bryophyta*) [69] Nonvascular plants with true stomata and erect, "leafy" gametophytes; sporophytes elongate by apical cell division.

moss animals *See* bryozoans [99].

myriapods (*Myriapoda*) [115] Arthropods [26] characterized by an elongate, segmented trunk with many legs. Includes centipedes and millipedes.

– N –

nanoarchaeotes (*Nanoarchaeota*) A group of extremely small, thermophilic archaeans [3] with a much-reduced genome. The only described example can survive only when attached to a host organism.

nematodes (*Nematoda*) [111] A very large group of elongate, unsegmented ecdysozoans [25] with thick, multilayer cuticles. They are among the most abundant and diverse animals, although most species have not yet been described. Include free-living predators and scavengers, as well as parasites of most species of land plants [11] and animals [19].

neognaths (*Neognathae*) The main group of birds [132], including all living species except the ratites (ostrich, emu, rheas, kiwis, cassowaries) and tinamous. *See* palaeognaths.

– O –

oligochaetes (*Oligochaeta*) Annelid [105] group whose members lack parapodia, eyes, and anterior tentacles, and have few setae. Earthworms are the most familiar oligochaetes.

onychophorans (*Onychophora*) [113] Elongate, segmented ecdysozoans [25] with many pairs of soft, unjointed, claw-bearing legs. Also known as velvet worms.

oomycetes (*Oomycota*) [52] Water molds and relatives; absorptive heterotrophs with nutrient-absorbing, filamentous hyphae.

opisthokonts (*Opisthokonta*) [16] A group of eukaryotes [4] in which the flagellum on motile cells, if present, is posterior. The opisthokonts include the fungi [17], animals [19], and choanoflagellates [91].

ostracods (*Ostracoda*) Marine and freshwater crustaceans [116] that are laterally compressed and protected by two clamlike calcareous or chitinous shells.

– P –

palaeognaths (*Palaeognathae*) A group of secondarily flightless or weakly flying birds [132]. Includes the flightless ratites (ostrich, emu, rheas, kiwis, cassowaries) and the weakly flying tinamous.

parabasalids (*Parabasalia*) [59] A group of unicellular eukaryotes [4] that lack mitochondria; they possess flagella in clusters near the anterior of the cell.

phoronids (*Phoronida*) [104] A small group of sessile, wormlike marine lophotrochozoans [24] that secrete chitinous tubes and feed using a lophophore.

placoderms (*Placodermi*) An extinct group of jawed vertebrates [32] that lacked teeth. Placoderms were the dominant predators in Devonian oceans.

placozoans (*Placozoa*) [96] A poorly known group of structurally simple, asymmetrical, flattened, transparent animals found in coastal marine tropical and subtropical seas. Most evidence suggests that placozoans are secondarily simplified eumetazoans [21].

Plantae [9] The most broadly defined plant group. In most parts of this book, we use the word "plant" as synonymous with "land plant" [11], a more restrictive definition.

plasmodial slime molds (*Myxogastrida*) Amoebozoans [84] that in their feeding stage consist of a coenocyte called a plasmodium.

pogonophorans (*Pogonophora*) Deep-sea annelids [105] that lack a mouth or digestive tract; they feed by taking up dissolved organic matter, facilitated by endosymbiotic bacteria in a specialized organ (the trophosome).

polychaetes (*Polychaeta*) A group of mostly marine annelids [105] with one or more pairs of eyes and one or more pairs of feeding tentacles; parapodia and setae extend from most body segments. May be paraphyletic with respect to the clitellates.

priapulids (*Priapulida*) [107] A small group of cylindrical, unsegmented, wormlike marine ecdysozoans [25] that takes its name from its phallic appearance.

"prokaryotes" Not a monophyletic group; as commonly used, includes the bacteria [2] and archaeans [3]. A term of convenience encompassing all cellular organisms that are not eukaryotes.

proteobacteria (*Proteobacteria*) [44] A large and extremely diverse group of Gram-negative bacteria that includes many pathogens, nitrogen fixers, and photosynthesizers. Includes the alpha, beta, gamma, delta, and epsilon proteobacteria.

"protists" This term of convenience is used to encompass a large number of distinct and distantly related groups of eukaryotes, many but far from all of which are microbial and unicellular. Essentially a "catch-all" term for any eukaryote group not contained within the land plants [11], fungi [17], or animals [19].

protostomes (*Protostomia*) [23] One of the two major groups of bilaterians [22]. In protostomes, the mouth typically forms from the blastopore (if present) in early development (contrast with deuterostomes). The major protostome groups are the lophotrochozoans [24] and ecdysozoans [25].

prototherians (*Prototheria*) A mostly extinct group of mammals [129], common during the Cretaceous and early Cenozoic. The five living species—four echidnas and the duck-billed platypus—are the only extant egg-laying mammals.

pterobranchs (*Pterobranchia*) A small group of sedentary marine hemichordates [119] that live in tubes secreted by the proboscis. They have one to nine pairs of arms, each bearing long tentacles that capture prey and function in gas exchange.

pycnogonids (*Pycnogonida*) Treated in this book as a group of chelicerates [114], but sometimes considered an independent group of arthropods [26]. Pycnogonids have reduced bodies and very long, slender legs. Also called sea spiders.

– R –

radiolarians (*Radiolaria*) [55] Amoeboid organisms with needlelike pseudopods supported by microtubules. Most have glassy internal skeletons.

ray-finned fishes (*Actinopterygii*) [125] A highly diverse group of freshwater and marine bony vertebrates [33]. They have reduced swim bladders that often function as hydrostatic organs and fins supported by soft rays (lepidotrichia). Includes most familiar fishes.

red algae (*Rhodophyta*) [64] Mostly multicellular, marine and freshwater algae characterized by the presence of phycoerythrin in their chloroplasts.

reptiles (*Reptilia*) [37] One of the two major groups of extant amniotes [36], supported on the basis of similar skull structure and gene sequences. The term "reptiles" traditionally excluded the birds [132], but the resulting group is then clearly paraphyletic. As used in this book, the reptiles include turtles [131], lepidosaurs [130], birds [132], and crocodilians [133].

rhizaria (*Rhizaria*) [7] Mostly amoeboid unicellular eukaryotes with pseudopods, many with external or internal shells. Includes the foraminiferans [57], cercozoans [56], and radiolarians [55].

rhyniophytes (*Rhyniophyta*) A group of early vascular plants [12] that appeared in the Silurian and became extinct in the Devonian. Possessed dichotomously branching stems with terminal sporangia but no true leaves or roots.

ribbon worms (*Nemertea*) [102] A group of unsegmented lophotrochozoans [24] with an eversible proboscis used to capture prey. Mostly marine, but some species live in fresh water or on land.

rotifers (*Rotifera*) [101] Tiny (< 0.5 mm) lophotrochozoans [24] with a pseudocoelomic body cavity that functions as a hydrostatic organ, and a ciliated feeding organ called the corona that surrounds the head. Rotifers live in freshwater and wet terrestrial habitats.

roundworms (*Nematoda*) [111] *See* nematodes.

– S –

sac fungi (*Ascomycota*) [89] Fungi that bear the products of meiosis within sacs (asci) if the organism is multicellular. Some are unicellular.

salamanders (*Caudata*) A group of amphibians [128] with distinct tails in both larvae and adults and limbs set at right angles to the body.

salps *See* thaliaceans.

sarcopterygians (*Sarcopterygii*) [34] *See* lobe-limbed vertebrates.

scyphozoans (*Scyphozoa*) Marine cnidarians [97] in which the medusa stage dominates the life cycle. Commonly known as jellyfish.

sea cucumbers (*Holothuroidea*) Echinoderms [118] with an elongate, cucumber-shaped body and leathery skin. They are scavengers on the ocean floor.

sea spiders *See* pycnogonids.

sea squirts *See* ascidians.

sea stars (*Asteroidea*) Echinoderms [118] with five (or more) fleshy "arms" radiating from an indistinct central disk. Also called starfishes.

sea urchins (*Echinoidea*) Echinoderms [118] with a test (shell) that is covered in spines. Most are globular in shape, although some groups (such as the sand dollars) are flattened.

"seed ferns" A paraphyletic group of loosely related, extinct seed plants that flourished in the Devonian and Carboniferous. Characterized by large, frondlike leaves that bore seeds.

seed plants (*Spermatophyta*) [13] Heterosporous vascular plants [12] that produce seeds; most produce wood; branching is axillary (not dichotomous). The major seed plant groups are gymnosperms [14] and angiosperms [15].

sow bugs *See* isopods.

spirochetes (*Spirochaetes*) [45] Motile, Gram-negative bacteria with a helically coiled structure and characterized by axial filaments.

sponges (*Porifera*) [20] A group of relatively asymmetric, filter-feeding animals that lack a gut or nervous system and generally lack differentiated tissues. Includes glass sponges [92], demosponges [93], and calcareous sponges [94].

springtails (*Collembola*) Wingless hexapods [117] with springing structures on the third and fourth segments of their bodies. Springtails are extremely abundant in some environments (especially in soil, leaf litter, and vegetation).

squamates (*Squamata*) The major group of lepidosaurs [130], characterized by the possession of movable quadrate bones (which allow the upper jaw to move independently of the rest of the skull) and hemipenes (a paired set of eversible penises, or penes) in males. Includes the lizards (a paraphyletic group), snakes, and amphisbaenians.

star anise (*Austrobaileyales*) [80] A group of woody angiosperms [15] thought to be the sister-group of the clade of flowering plants that includes eudicots [83], monocots [82], and magnoliids [81].

starfish (*Asteroidea*) *See* sea stars.

stoneworts (*Charales*) [67] Multicellular green algae with branching, apical growth and plasmodesmata between adjacent cells. The closest living relatives of the land plants [11], they retain the egg in the parent organism.

stramenopiles (*Heterokonta* or *Stramenopila*) [6] Organisms having, at some stage in their life cycle, two unequal flagella, the longer possessing rows of tubular hairs. Chloroplasts, when present, surrounded by four membranes. Major stramenopile groups include the brown algae [54], diatoms [53], and oomycetes [52].

– T –

tapeworms (*Cestoda*) Parasitic flatworms [100] that live in the digestive tracts of vertebrates as adults, and usually in various other species of animals as juveniles.

tardigrades (*Tardigrada*) [112] Small (< 0.5 mm) ecdysozoans [25] with fleshy, unjointed legs and no circulatory or gas exchange organs. They live in marine sands, in temporary freshwater pools, and on the water films of plants. Also called water bears.

tetrapods (*Tetrapoda*) [35] The major group of lobe-limbed vertebrates [34]; includes the amphibians [128] and the amniotes [36]. Named for the presence of four jointed limbs (although limbs have been secondarily reduced or lost completely in several tetrapod groups).

thaliaceans (*Thaliacea*) A group of solitary or colonial planktonic marine tunicates [121]. Also called salps.

therians (*Theria*) Mammals [129] characterized by viviparity (live birth). Includes eutherians and marsupials.

theropods (*Theropoda*) Archosaurs [38] with bipedal stance, hollow bones, a furcula ("wishbone"), elongated metatarsals with three-fingered feet, and a pelvis that points backwards. Includes many well-known extinct dinosaurs (such as *Tyrannosaurus rex*), as well as the living birds [132].

tracheophytes *See* vascular plants [12].

trilobites (*Trilobita*) An extinct group of arthropods [26] related to the chelicerates [114]. Trilobites flourished from the Cambrian through the Permian.

tuataras (*Rhyncocephalia*) A group of lepidosaurs [130] known mostly from fossils; there are only two living tuatara species. The quadrate bone of the upper jaw is fixed firmly to the skull. Sister group of the squamates.

tunicates (*Tunicata*) [121] A group of chordates [30] that are mostly saclike filter feeders as adults, with motile larval stages that resemble tadpoles.

turbellarians (*Turbellaria*) A group of free-living, generally carnivorous flatworms [100]. Their monophyly is questionable.

turtles (*Testudines*) [131] A group of reptiles [37] with a bony carapace (upper shell) and plastron (lower shell) that encase the body in a fashion unique among the vertebrates.

– U –

urochordates (*Tunicata*) [121] *See* tunicates.

– V –

vascular plants (*Tracheophyta*) [12] Plants with xylem and phloem. Major groups include the lycophytes [71] and euphyllophytes.

vertebrates (*Vertebrata*) [31] The largest group of chordates [30], characterized by a rigid endoskeleton supported by the vertebral column and an anterior skull encasing a brain. Includes hagfishes [122], lampreys [123], and the jawed vertebrates [32], although some biologists exclude the hagfishes from this group. *See also* craniates.

– W –

water bears *See* tardigrades.

water lilies (*Nymphaeaceae*) [79] A group of aquatic, freshwater angiosperms [15] that are rooted in soil in shallow water, with round floating leaves and flowers that extend above the water's surface. They are the sister-group to most of the remaining flowering plants, with the exception of the genus *Amborella* [78].

– Y –

"yeasts" Convenience term for several distantly related groups of unicellular fungi [17].

– Z –

"zygospore fungi" (*Zygomycota*, if monophyletic) [87] A convenience term for a probably paraphyletic group of fungi [17] in which hyphae of differing mating types conjugate to form a zygosporangium.

Appendix B Statistics Primer

This appendix is designed to help you conduct simple statistical analyses and understand their application and importance. This introduction will help you complete the Apply the Concept and Analyze the Data problems throughout this book. The formulas for a number of statistical tests are presented here, but the presentation is designed primarily to help you understand the purpose and reasoning of the various tests. Once you understand the basis of the analysis, you may wish to use one of many free, online web sites for conducting the tests and calculating relevant test statistics (such as http://faculty.vassar.edu/lowry/VassarStats.html).

Why Do We Do Statistics?

ALMOST EVERYTHING VARIES We live in a variable world, but within the variation we see among biological organisms there are predictable patterns. We use statistics to find and analyze these patterns. Consider any group of common things in nature—all women aged 22, all the cells in your liver, or all the blades of grass in your yard. Although they will have many similar characteristics, they will also have important differences. Men aged 22 tend to be taller than women aged 22, but, of course, not every man will be taller than every woman in this age group.

Natural variation can make it difficult to find general patterns. For example, scientists have determined that smoking increases the risk of getting lung cancer. But we know that not all smokers will develop lung cancer and not all nonsmokers will remain cancer-free. If we compare just one smoker to just one nonsmoker, we may end up drawing the wrong conclusion. So how did scientists discover this general pattern? How many smokers and nonsmokers did they examine before they felt confident about the risk of smoking?

Statistics helps us to find general patterns, even when nature does not always follow those patterns.

AVOIDING FALSE POSITIVES AND FALSE NEGATIVES When a woman takes a pregnancy test, there is some chance that it will be positive even if she is not pregnant, and there is some chance that it will be negative even if she is pregnant. We call these kinds of mistakes *false positives* and *false negatives*.

Doing science is a bit like taking a medical test. We observe patterns in the world, and we try to draw conclusions about how the world works from those observations. Sometimes our observations lead us to draw the wrong conclusions. We might conclude that a phenomenon occurs, when it actually does not; or we might conclude that a phenomenon does not occur, when it actually does.

For example, the planet Earth has been warming over the past century (see Concept 46.4). Ecologists are interested in whether plant and animal populations have been affected by global warming. If we have long-term information about the locations of species and temperatures in certain areas, we can determine whether species movements coincide with temperature changes. Such information can, however, be very complicated. Without proper statistical methods, one may not be able to detect the true impact of temperature or, instead, may think a pattern exists when it does not.

Statistics helps us to avoid drawing the wrong conclusions.

How Does Statistics Help Us Understand the Natural World?

Statistics is essential to scientific discovery. Most biological studies involve five basic steps, each of which requires statistics:

- **Step 1: Experimental Design**
  Clearly define the scientific question and the methods necessary to tackle the question.
- **Step 2: Data Collection**
  Gather information about the natural world through experiments and field studies.
- **Step 3: Organize and Visualize the Data**
  Use tables, graphs, and other useful representations to gain intuition about the data.
- **Step 4: Summarize the Data**
  Summarize the data with a few key statistical calculations.
- **Step 5: Inferential Statistics**
  Use statistical methods to draw general conclusions from the data about the way the world works.

Step 1: Experimental Design

We conduct experiments to gain knowledge about the world. Scientists come up with scientific ideas based on prior research and their own observations. These ideas may take the form of a question like "Does smoking cause cancer?," a hypothesis like "Smoking increases the risk of cancer," or a prediction like "If a person smokes, he/she will increase his/her chances of developing cancer." Experiments allow us to test such scientific ideas, but designing a good experiment can be quite challenging.

We use statistics to guide us in designing experiments so that we end up with the right kinds of data. Before embarking on an experiment, we use statistics to determine how much data will be required to test our idea, and to prevent extraneous factors from misleading us. For example, suppose we want to conduct an experiment on fertilizers to test the hypothesis that nitrogen increases plant growth. If we include too few plants, we will not be able to determine whether or not nitrogen has an effect on growth, and the experiment will be for naught. If we include too many plants, we will waste valuable time and resources. Furthermore, we should design the experiment so that we can detect differences that are actually caused by nitrogen fertilization rather than by variation, for example, in sunlight or precipitation experienced by the plants.

Step 2: Data Collection

TAKING SAMPLES When biologists gather information about the natural world, they typically collect a few representative pieces of information. For example, when evaluating the efficacy of a candidate drug for medulloblastoma brain cancer, scientists may test the drug on tens or hundreds of patients, and then draw conclusions about its efficacy for all patients with these tumors. Similarly, scientists studying the relationship between body weight and clutch size (number of eggs) for female spiders of a particular species may examine tens to hundreds of spiders to make their conclusions.

We use the expression "sampling from a population" to describe this general method of taking representative pieces of information from the system under investigation (**Figure B1**). The pieces of information in a **sample** are called **observations**. In the cancer therapy example, each observation was the change in a patient's tumor size six months after initiating treatment, and the population of interest was all individuals with medulloblastoma tumors. In the spider example, each observation was a pair of measurements—body size and clutch size—for a single female spider, and the population of interest was all female spiders of this species.

Sampling is a matter of necessity, not laziness. We cannot hope (and would not want) to collect *all* of the female spiders of the species of interest on Earth! Instead, we use statistics to determine how many spiders we must collect in order to confidently infer something about the general population and then use statistics again to make such inferences.

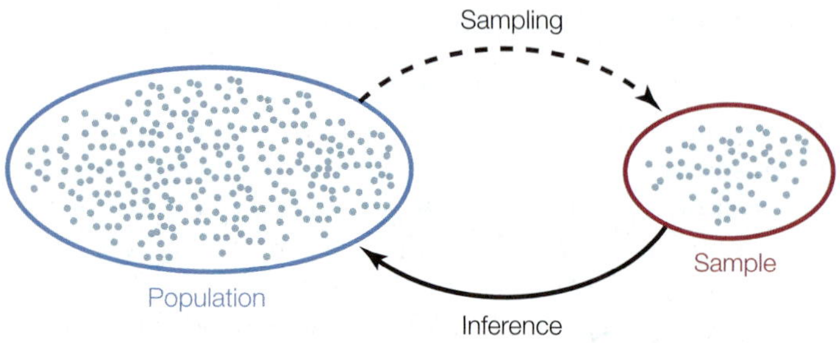
Sampling
Population
Sample
Inference

FIGURE B1 Sampling From a Population Biologists take representative samples from a population, use descriptive statistics to characterize their samples, and then use inferential statistics to draw conclusions about the original population.

TABLE B1
Poinsettia Colors

Color	Frequency	Proportion
Red	108	0.59
Pink	34	0.19
White	40	0.22
Total	182	1.00

DATA COME IN ALL SHAPES AND SIZES In statistics, we use the word *variable* to mean a measurable characteristic of an individual or a system. Some variables are on a numerical scale, like the daily high temperature (a numerical value constrained by the precision of our thermometer), or the clutch size of a spider (a whole number: 0, 1, 2, 3,...). We call these **quantitative variables**. Quantitative variables that only take on whole number values are called **discrete variables**, whereas variables that can also take on any fractional value are called **continuous variables**.

Other variables take categories as values, like a human blood type (A, B, AB, or O) or an ant caste (queen, worker, or male). We call these **categorical variables**. Categorical variables with a natural ordering, like a final grade in Introductory Biology (A, B, C, D, or F), are called **ordinal variables**.

Each class of variables comes with its own set of statistical methods. We will introduce a few common methods in this Appendix that will help you work on the problems presented in this book, but you should consult a biostatistics textbook for more advanced tests and analyses for other data sets and problems.

Step 3: Organize and Visualize the Data

Tables and graphs can help you gain intuition about your data, design appropriate statistical tests, and anticipate the outcome of your analysis. A **frequency distribution** lists all possible values and the number of occurrences of each value in the sample.

TABLE B2
Fish Weights of *Abramis brama* from Lake Laengelmavesi

Weight (grams)	Frequency	Relative Frequency
201–300	2	0.06
301–400	3	0.09
401–500	8	0.24
501–600	3	0.09
601–700	8	0.24
701–800	3	0.09
801–900	1	0.03
901–1000	6	0.18
Total	34	1.00

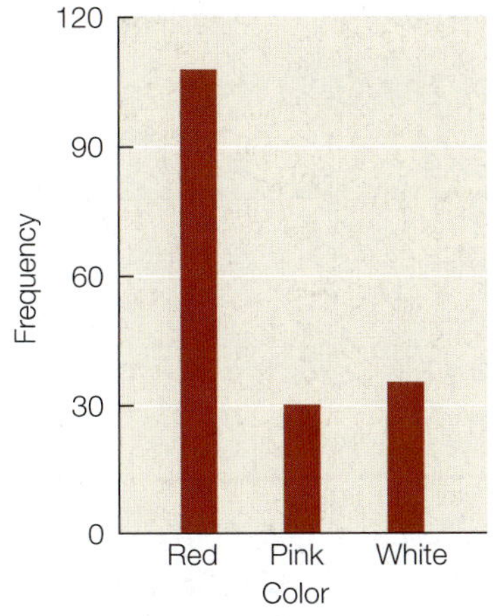
Frequency
120
90
60
30
0
Red
Pink
White
Color

B2 Bar Charts Compare Categorical Data This bar chart shows the frequency of three poinsettia colors that result from an experimental cross.

Table B1 shows a frequency distribution of the colors of 182 poinsettia plants (red, pink, or white) resulting from an experimental cross between two parent plants. For categorical data like this, we can visualize the frequency distribution by constructing a **bar chart**. The heights of the bars indicate the number of observations in each category (**Figure B2**). Another way to display the same data is in a **pie chart**, which shows the proportion of each category represented like pieces of a pie (**Figure B3**).

For quantitative data, it is often useful to condense your data by grouping (or binning) it into **classes**. In **Table B2**, we see a grouped frequency distribution of fish weights for a sample of 34 fish (*Abramis brama*) caught in Lake Laengelmavesi in Finland. The second column (*Frequency*) gives the number of observations in each class and the third column (*Relative Frequency*) gives the overall proportion of observations falling into each class.

Histograms depict frequency distributions for quantitative data. The histogram in **Figure B4** shows the relative frequencies of each weight class in this study. When grouping quantitative data, it is necessary to decide how many classes to include. It is often useful to look at multiple histograms before deciding which grouping offers the best representation of the data.

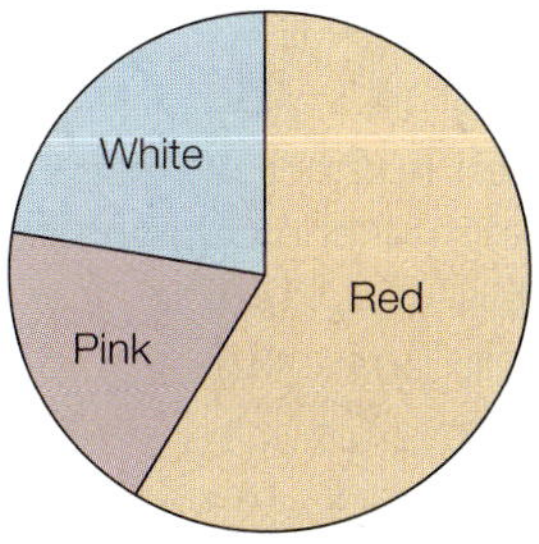
White
Pink
Red

B3 Pie Charts Show Proportions of Categories This pie chart shows the proportions of the three poinsettia colors presented in Table B1.

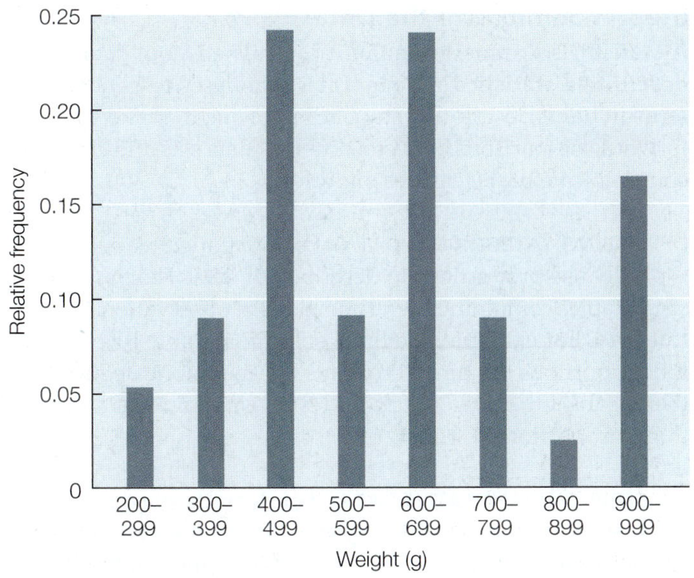
Relative frequency
0.25
0.20
0.15
0.10
0.05
0
200–299
300–399
400–499
500–599
600–699
700–799
800–899
900–999
Weight (g)

B4 Histograms Depict Frequency Distributions of Quantitative Data This histogram shows the relative frequency of different weight-classes of fish (*Abramis brama*).

Sometimes we wish to compare two quantitative variables. For example, the researchers at Lake Laengelmavesi investigated the relationship between fish weight and length and thus also measured the length of each fish. We can visualize this relationship using a **scatter plot** in which the weight and length of each fish is represented as a single point (**Figure B5**). We say that these two variables have a **linear relationship** since the points in their scatter plot fall roughly on a straight line.

Tables and graphs are critical to interpreting and communicating data, and thus should be as self-contained and comprehensible as possible. Their content should be easily understood simply by looking at them. Axes, captions, and units should be clearly labeled, statistical terms should be defined, and appropriate groupings should be used when tabulating or graphing quantitative data.

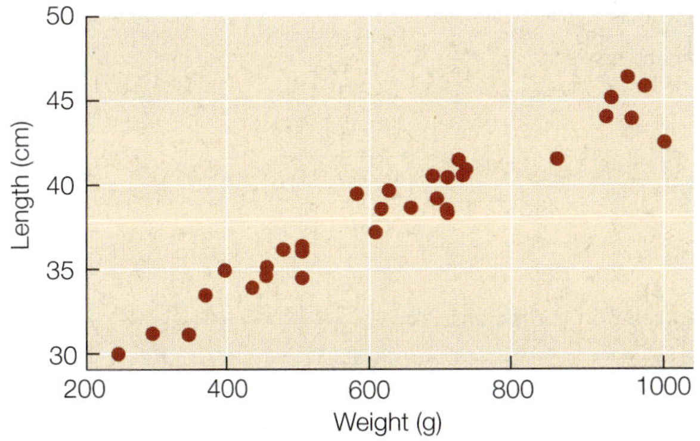
Length (cm)
50
45
40
35
30
200
400
600
800
1000
Weight (g)

B5 Scatter Plots Contrast Two Variables Scatter plot of *Abramis brama* weights and lengths (measured from nose to end of tail). These two variables have a linear relationship since the data points lie close to a straight line.

Step 4: Summarize the Data

A **statistic** is a numerical quantity calculated from data, while **descriptive statistics** are quantities that describe general patterns in data. Descriptive statistics allow us to make straightforward comparisons between different data sets and concisely communicate basic features of our data.

DESCRIBING CATEGORICAL DATA For categorical variables, we typically use proportions to describe our data. That is, we construct tables containing the proportions of observations in each category. For example, the third column in Table B1 provides the proportions of poinsettia plants in each color category, and the pie chart in Figure B3 provides a visual representation of those proportions.

DESCRIBING QUANTITATIVE DATA For quantitative data, we often start by calculating the average value or **mean** of our sample. This familiar quantity is simply the sum of all the values in the sample divided by the number of observations in our sample (**Figure B6**). The mean is only one of several quantities that roughly tell us where the *center* of our data lies. We call these quantities **measures of center**. Other commonly used measures of center are the **median**—the value that literally lies in the middle of the sample—and the **mode**—the most frequent value in the sample.

RESEARCH**TOOLS**

B6 Descriptive Statistics for Quantitative Data

Below are the equations used to calculate the descriptive statistic we discuss in this appendix. You can calculate these statistics yourself, or use free internet resources to help you make your calculations.

Notation:

$x_1, x_2, x_3, \ldots x_n$ are the n observations of variable X in the sample.

$\sum_{i=1}^{n} x_i = x_1 + x_2 + x_3, \ldots + x_n$ is the sum of all of the observations. (The Greek letter sigma, Σ, is used to denote "sum of.")

In regression, the independent variable is X, and the dependent variable is Y. b_0 is the vertical intercept of a regression line. b_1 is the slope of a regression line.

Equations

1. Mean: $$\bar{x} = \frac{\sum_{i=1}^{n} x_i}{n}$$

2. Standard deviation: $$s = \sqrt{\frac{\Sigma(x_i - \bar{x})^2}{n-1}}$$

3. Correlation coefficient: $$r = \frac{\Sigma(x_i - \bar{x})(y_i - \bar{y})}{\sqrt{\Sigma(x_i - \bar{x})^2 (y_i - \bar{y})^2}}$$

4. Least-squares regression line: $Y = b_0 + b_1 X$

where $$b_1 = \frac{\Sigma(x_i - \bar{x})(y_i - \bar{y})}{\Sigma(x_i - \bar{x})^2}$$ and $$b_0 = \bar{y} - b_1 \bar{x}$$

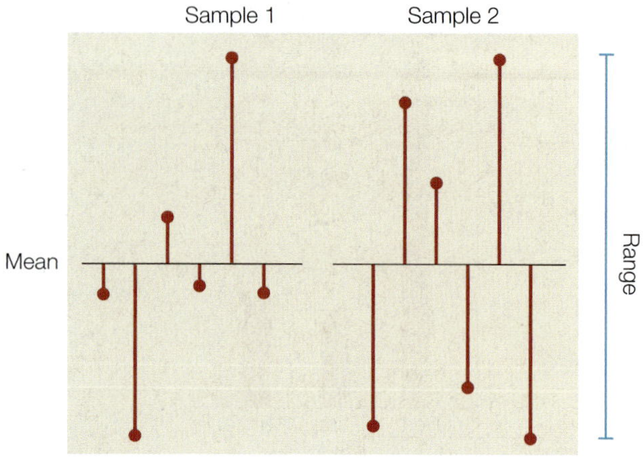
Sample 1
Sample 2
Mean
Range

B7 Measures of Dispersion Two samples with the same mean (black horizontal lines) and range (blue vertical line). Red lines show the deviations of each observation from the mean. Samples with large deviations have large standard deviations. The left sample has a smaller standard deviation than the right sample.

It is often just as important to quantify the variation in the data as it is to calculate its center. There are several statistics that tell us how much the values differ from one another. We call these **measures of dispersion**. The easiest to one understand and calculate is the **range**, which is simply the largest value in the sample minus the smallest value. The most commonly used measure of dispersion is the **standard deviation**, which calculates the extent to which the data are spread out from the mean. A deviation is the difference between an observation and the mean of the sample, and the standard deviation is a number that summarizes all of the deviations. Two samples can have the same range, but very different standard deviations if one is clustered closer to the mean than the other. In **Figure B7**, for example, the left sample has a lower standard deviation (s = 2.6) than the right sample (s = 3.6), even though the two samples have the same means and ranges.

To demonstrate these descriptive statistics, we return to the Lake Laengelmavesi study. The researchers also caught and recorded the weights of six fish in the species *Leusiscus idus*: 270, 270, 306, 540, 800, and 1,000 grams. The mean weight in this sample (equation 1 in Figure B6) is:

$$\bar{x}\frac{\Delta N}{\Delta T} = \frac{(270+270+306+540+800+1000)}{6} = 531$$

Since there is an even number of observations in the sample, then the median weight is the value halfway between the two middle values:

$$\frac{306+540}{2} = 423$$

The mode of the sample is 270, the only value that appears more than once. The standard deviation (equation 2 in Figure B6) is:

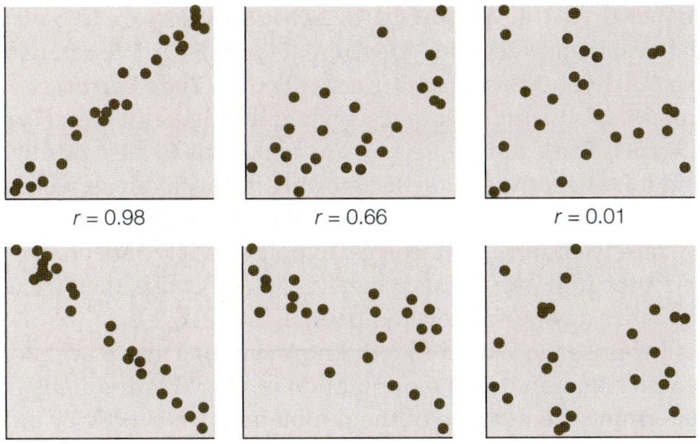
r = 0.98
r = 0.66
r = 0.01
r = –0.98
r = –0.66
r = –0.01

B8 Correlation Coefficients The correlation coefficient (*r*) indicates both the strength and the direction of the relationship.

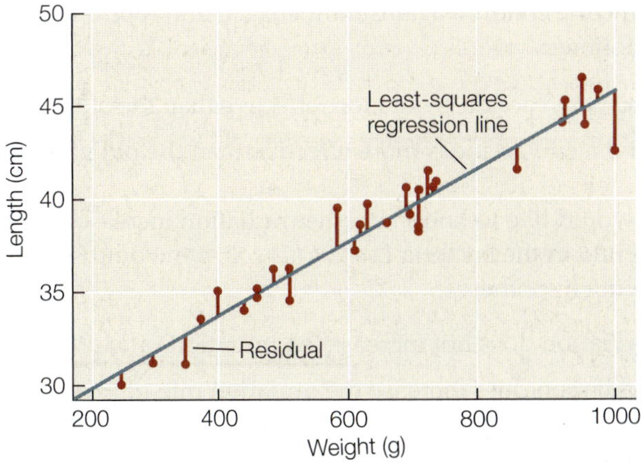
50
45
40
35
30
Length (cm)
200
400
600
800
1000
Weight (g)
Least-squares regression line
Residual

B9 Linear Regression Estimates the Typical Relationship Between Two Variables Linear-least squares regression line for *Abramis brama* weights and lengths (measured from nose to end of tail). The regression line (blue line) is given by the equation $Y = 26.1 + 0.02X$. It is the line that minimizes the sum of the squares of the residuals (red lines).

$$s = \sqrt{\frac{(270-531)^2 + (270-531)^2 + (306-531)^2 + (540-531)^2 + (800-531)^2 + (1000-531)^2}{5}} = 309.6$$

and the range is 1000 – 270 = 730.

DESCRIBING THE RELATIONSHIP BETWEEN TWO QUANTITATIVE VARIABLES Biologists are often interested in understanding the relationship between two different quantitative variables: How does the height of an organism relate to its weight? How does air pollution relate to the prevalence of asthma? How does lichen abundance relate to levels of air pollution? Recall that scatter plots visually represent such relationships.

We can quantify the strength of the relationship between two quantitative variables using a single value called the Pearson product–moment **correlation coefficient** (equation 3 in Figure B6). This statistic ranges between –1 and 1, and tells us how closely the points in a scatter plot conform to a straight line. A negative correlation coefficient indicates that one variable decreases as the other increases; a positive correlation coefficient indicates that the two variables increase together, and a correlation coefficient of zero indicates that there is no linear relationship between the two variables (**Figure B8**).

One must always keep in mind that *correlation does not mean causation*. Two variables can be closely related without one causing the other. For example, the number of cavities in a child's mouth correlates positively with the size of their feet. Clearly cavities do not enhance foot growth; nor does foot growth cause tooth decay. Instead the correlation exists because both quantities tend to increase with age.

Intuitively, the straight line that tracks the cluster of points on a scatter plot tells us something about the *typical* relationship between the two variables. Statisticians do not, however, simply eyeball the data and draw a line by hand. They often use a method called least-squares **linear regression** to fit a straight line to the data (equation 4 in Figure B6). This method calculates the line that minimizes the overall vertical distances between the points in the scatter plot and the line itself. These distances are called **residuals** (**Figure B9**). Two parameters describe the regression line: b_0 (the vertical intercept of the line, or the expected value of variable Y when $X = 0$), and b_1 (the slope of the line, or how much values of Y are expected to change with changes in values of X).

Step 5: Inferential Statistics

Data analysis often culminates with statistical inference—an attempt to draw general conclusions about the system under investigation. As depicted in Figure B1, the primary reason we collect data is to gain insight into the larger system from which the data are collected. When we test a new medulloblastoma brain cancer drug on ten patients, we do not simply want to know the fate of those ten individuals; rather, we hope to predict its efficacy on the much larger group of all medulloblastoma patients.

STATISTICAL HYPOTHESES When it comes to inferring something about the real world from our data, we often have a *"Whether or not"* question in mind. For example, we would like to know whether or not global warming impacts biodiversity; whether or not the clutch size of a spider increases with body size; or whether or not soil nitrogen increases the growth of a particular plant species.

Before making statistical inferences from data, we must formalize our *"Whether or not"* question into a pair of opposing hypotheses—a **null hypothesis** (denoted H_0) and an **alternative hypothesis** (denoted H_A). The alternative hypothesis is the *"Whether"*—it is formulated to describe the effect that we expect our data to support; the null hypothesis is the *"or not"*—it is formulated to represent the absence of the effect. In other words, we typically conduct our experiment seeking to demonstrate something new (the alternative hypothesis) and thereby reject the idea that it does not occur (the null hypothesis).

Suppose, for example, we would like to know *whether or not* a new vaccine is more effective than an existing vaccine at immunizing children against influenza. Our hypotheses would be as follows:

H_0: The new vaccine is not more effective than the old vaccine.

H_A: The new vaccine is more effective than the old vaccine.

If we would like to know whether radiation increases the mutation rate in the bacteria *Escherichia coli*, we would set up the following hypotheses:

H_0: Radiation does not increase the mutation rate of *E. coli*.

H_A: Radiation does increase the mutation rate of *E. coli*.

STATISTICAL BURDEN OF PROOF In the U.S. justice system, people are innocent until proven guilty. In statistics, the world is *null until proven alternative*. Statistics requires overwhelming proof in favor of the alternative hypothesis before rejecting the null hypothesis. In other words, scientists favor existing ideas and resist adopting new ideas until compelling evidence suggests otherwise. This is based on a philosophy that it is worse to accept new claims when they are false than to miss out on discovering some true facts about world.

When testing a new influenza vaccine, the burden of proof is on the new vaccine. Suppose we were to vaccinate three children with the new vaccine (Group A), three with the old vaccine (Group B) and leave three children unvaccinated (Group C). If no children from Group A, one child from Group B, and one child from Group C became infected, would we have enough evidence to conclude that the new vaccine is superior to the old vaccine? No, we would not. If the study were enlarged, and two out of 100 children in group A, seven out of 100 children in group B, and 22 out of 100 children in group C become infected, would we then have sufficient evidence to choose the new vaccine? Perhaps, but we need to use statistics to be sure.

This is the traditional burden of proof in biology and science in general. As a consequence, scientists are more likely to miss out on discovering something new (and true) about the world than they are to make a false discovery. In recent years, scientists have begun to question this approach and develop an alternative statistical approach, called **Bayesian inference**, which makes it easier to favor new hypotheses. In this primer, we discuss only traditional statistical methods, often called **frequentist statistics**.

Our conclusion	The real world: Null hypothesis true (*not more females*)	The real world: Null hypothesis false (*more females*)
Null hypothesis true (*not more females*)	√	Type 2 error (*false negative*)
Null hypothesis false (*more females*)	Type 1 error (*false positive*)	√

B10 Two Types of Error Possible outcomes of a statistical test. Statistical inference can result in correct and incorrect conclusions about the population of interest.

JUMPING TO THE WRONG CONCLUSIONS There are two ways that a statistical test can go wrong (**Figure B10**). We can reject the null hypothesis when it is actually true (**Type I error**) or we can accept the null hypothesis when it is actually false (**Type II error**). These kinds of errors are analogous to false positives and false negatives in medical testing, respectively. If we mistakenly reject the null hypothesis when it is actually true, then we falsely endorse the incorrect hypothesis. If we are unable to reject the null hypothesis when it is actually false, then we fail to realize a yet undiscovered truth.

Suppose we would like to know whether there are more females than males in a population of 10,000 individuals. To determine the makeup of the population, we choose 20 individuals randomly and record their sex. Our null hypothesis is that there are *not* more females than males; and our alternative hypothesis is that there are. The following scenarios illustrate the possible mistakes we might make:

- *Scenario 1*: The population actually has 40% females and 60% males. Although our random sample of 20 people is likely to be dominated by males, it is certainly possible that, by chance, we will end up choosing more females than males. If this occurs, and we mistakenly reject the null hypothesis (that there are *not* more females than males), then we make a Type I error.
- *Scenario 2*: The population actually has 60% females and 40% males. If, by chance, we end up with a majority of males in our sample and thus fail reject the null hypothesis, then we make a Type II error.

Fortunately, statistics has been developed precisely to avoid these kinds of errors and inform us about the reliability of our conclusions. The methods are based on calculating the **probabilities** of different possible outcomes. Although you may have heard or even used the word "probability" on multiple occasions, it is important that you understand its mathematical meaning. A probability is a numerical quantity that expresses the likelihood of some event. It ranges between zero and one; zero means that there is no chance the event will occur and one means that the event is guaranteed to occur. This only makes sense if there is an element of chance, that is, if it is possible the event will occur and possible that it will not occur. For example, when we flip a fair coin, it will land on heads with probability 0.5 and land on tails with probability 0.5. When we select individuals randomly from a population with 60% females and 40% males, we will encounter a female with probability 0.6 and a male with probability 0.4.

Probability plays a very important role in statistics. To draw conclusions about the real world (the population) from our sample, we first calculate the probability of obtaining our sample if the null hypothesis is true. Specifically, statistical inference is based on answering the following question:

Suppose the null hypothesis is true. What is the probability that a random sample would, by chance, differ from the null hypothesis as much as our sample differs from the null hypothesis?

If our sample is highly improbable under the null hypothesis, then we rule it out in favor of our alternative hypothesis. If, instead, our sample has a reasonable probability of occurring under the null hypothesis, then we conclude that our data are consistent with the null hypothesis and we do not reject it.

Returning to the sex ratio example, we consider two new scenarios:

- *Scenario 3*: Suppose we want to infer whether or not females constitute the majority of the population (our alternative hypothesis) based on a random sample containing 12 females and eight males. We would calculate the probability that a random sample of 20 people includes at least 12 females assuming that the population, in fact, has a 50:50 sex ratio (our null hypothesis). This probability is 0.13, which is too high to rule out the null hypothesis.
- *Scenario 4*: Suppose now that our sample contains 17 females and three males. If our population is truly evenly divided, then this sample is much less likely than the sample in scenario 3. The probability of such an extreme sample is 0.0002, and would lead us to rule out the null hypothesis and conclude that there are more females than males.

This agrees with our intuition. When choosing 20 people randomly from an evenly divided population, we would be surprised if almost all of them were female, but would not be surprised at all if we ended up with a few more females than males (or a few more males than females). Exactly how many females do we need in our sample before we can confidently infer that they make up the majority of the population? And how confident are we when we reach that conclusion? Statistics allows us to answer these questions precisely.

STATISTICAL SIGNIFICANCE: AVOIDING FALSE POSITIVES Whenever we test hypotheses, we calculate the probability just discussed, and refer to this value as the **P-value** of our test. Specifically, the *P*-value is the probability of getting data as extreme as our data (just by chance) if the null hypothesis is, in fact, true. In other words, it is the likelihood that chance alone would produce data that differ from the null hypothesis as much as our data differ from the null hypothesis. How we measure the difference between our data and the null hypothesis depends on the kind of data in our sample (categorical or quantitative) and the nature of the null hypothesis (assertions about proportions, single variables, multiple variables, differences between variables, correlations between variables, etc.).

For many statistical tests, *P*-values can be calculated mathematically. One option is to quantify the extent to which the data depart from the null hypothesis and then use look-up tables (available in most statistics textbooks, or on the internet) to find the probability that chance alone would produce a difference of that magnitude. Most scientists, however, find *P*-values primarily by using statistical software rather than hand calculations combined with look-up tables. Regardless of the technology, the most important steps of the statistical analysis are still left to the researcher: constructing appropriate null and alternative hypotheses, choosing the correct statistical test, and drawing correct conclusions.

After we calculate a *P*-value from our data, we have to decide whether it is small enough to conclude that our data are inconsistent with the null hypothesis. This is decided by comparing the *P*-value to a threshold called the **significance level**, which is often chosen even before making any calculations. We reject the null hypothesis only when the *P*-value is less than or equal to the significance level, denoted α. This ensures that, if the null hypothesis is true, we have at most a probability α of accidentally rejecting it. Therefore, the lower the value of α, the less likely you are to make a Type I error (lower left cell of Figure B10). The most commonly used significance level is $\alpha = 0.05$, which limits the probability of a Type I error to 5%.

If our statistical test yields a *P*-value that is less than our significance level α, then we conclude that the effect described by our alternative hypothesis is statistically significant at the level α and we reject the null hypothesis. If our *P*-value is greater than α, then we conclude that we are unable to reject the null hypothesis. In this case, we do not actually reject the alternative hypothesis, rather we conclude that we do not yet have enough evidence to support it.

POWER: AVOIDING FALSE NEGATIVES The **power** of a statistical test is the probability that we will correctly reject the null hypothesis when it is false (lower right cell of Figure B10). Therefore, the higher the power of the test, the less likely we are to make a Type II error (upper right cell of Figure B10). The power of a test can be calculated, and such calculations can be used to improve your methodology. Generally, there are several steps that can be taken to increase power and thereby avoid false negatives:

- **Decrease the significance level**, α. The higher the value of α, the harder it is to reject the null hypothesis, even if it is actually false.
- **Increase the sample size**. The more data one has, the more likely one is to find evidence against the null hypothesis, if it is actually false.
- **Decrease variability in the sample**. The more variation there is in the sample, the harder it is to discern a clear effect (the alternative hypothesis) when it actually exists.

It is always a good idea to design your experiment to reduce any variability that may obscure the pattern you seek to detect. For example, it is possible that the chance of a child contracting influenza varies depending on whether he or she lives in a crowded (e.g., urban) environment or one that is less so (e.g., rural). To reduce variability, a scientist might choose to test a new influenza vaccine only on children from one environment or the other. After you have minimized such extraneous variation, you can use power calculations to choose the right combination of α and sample size to reduce the risks of Type I and Type II errors to desirable levels.

There is a trade-off between Type I and Type II errors: As α increases, the risk of a Type I decreases but the risk of a Type II error increases. As discussed above, scientists tend to be more concerned about Type I errors than Type II errors. That is, they believe that it is worse to mistakenly believe a false hypothesis than it is to fail to make a new discovery. Thus, they prefer to use low values of α. However, there are many real-world scenarios in which it would be worse to make a Type II error than a Type I error. For example, suppose a new cold medication is being tested for dangerous (life-threatening) side effects. The null hypothesis is that there are no such side effects. A Type II error might lead regulatory agencies to approve a harmful medication that could cost human lives. In contrast, a Type I error would simply mean one less cold medication among the many that already line pharmacy shelves. In such cases, policymakers take steps to avoid a Type II error, even if, in doing so, they increase the risk of a Type I error.

STATISTICAL INFERENCE WITH QUANTITATIVE DATA There are many forms of statistical inference for quantitative data. When measuring a single quantitative variable, like birth weight in lambs, calcium concentration in the blood of pregnant women, or migration rate of birds, we often wish to infer the mean value of the population from which we drew the sample. However, the mean of a randomly chosen sample will not necessarily be the same or even close to the population mean. Suppose we wanted to know the average weight of newborn lambs on a particular farm. By chance, we may end up with a random sample that includes an excess of lightweight lambs and therefore a sample mean that is less than the overall mean in the population.

To infer the population mean from the sample data, we can calculate a **confidence interval for the mean**. This is a statistically derived range of values that is centered on the sample mean and is likely to include the population mean. For example, based on the sample of 34 *Abramis brama* weights from Lake Laengelmavesi (see Table B2; Figure B4), the 95% confidence interval for the mean weight ranges from 554 grams to 698 grams. The true average weight for this species of fish is likely, but not guaranteed, to fall within this range.

RESEARCH**TOOLS**

B11 The t-Test

What is the *t*-test? It is a standard method for assessing whether the means of two groups are statistically different from each another.

Step 1: State the null and alternative *hypotheses*:

H_0: The two populations have the same mean.

H_A: The two populations have different means.

Step 2: Choose a significance level, α, to limit the risk of a Type 1 error.

Step 3: Calculate the *test statistic*: $t_s = \dfrac{\bar{y}_1 - \bar{y}_2}{\sqrt{\dfrac{s_1^2}{n_1} + \dfrac{s_2^2}{n_2}}}$

Notation: $\bar{y}_1$ and $\bar{y}_2$ are the sample means; s_1 and s_2 are the sample standard deviations; and n_1 and n_2 are the sample sizes.

Step 4: Use the test statistic to assess whether the data are consistent with the null hypothesis:

Calculate the *P-value* (*P*) using statistical software or by hand using statistical tables.

Step 5: Draw conclusions from the test:

If $P \leq \alpha$, then reject H_0, and conclude that the population distribution is significantly different.

If $P > \alpha$, then we do not have sufficient evidence to conclude that the means differ.

Biologists frequently wish to compare the mean values in two or more groups; for example, newborn lamb weights on several different farms, calcium concentration in women in early and late stages of pregnancy, or migration rates in birds of different species. Based on the means and standard deviations calculated for each of the samples, they infer whether or not the means in the different populations are statistically different from one another. There are several statistical methods for this, and the correct method depends on the number of groups, the experimental design, and the nature of the data.

Figure B11 describes the steps of a *t*-test, a simple method for comparing the means in two different groups. To illustrate, we can apply a *t*-test to the Lake Laengelmavesi data to assess whether the two fish species *Abramis brama* and *Leusiscus idus* have significantly different mean weights. We begin by stating our hypotheses and choosing a significance level:

H_0: *Abramis brama* and *Leusiscus idus* have the same mean weight.

H_A: *Abramis brama* and *Leusiscus idus* have different mean weights.

$\alpha = 0.05$

The test statistic is calculated using the means, standard deviations, and sizes of the two samples:

$$t_s = \frac{626 - 531}{\sqrt{\dfrac{207^2}{34} + \dfrac{310^2}{6}}} = 0.724$$

We can use statistical software or one of the free statistical sites on the internet to find the *P*-value for this result to be $P = 0.497$. Since *P* is considerably greater than α, we fail to reject the null hypothesis and conclude that our study does not provide evidence that the two species have different mean weights.

You may want to consult an introductory statistics textbook to learn more about confidence intervals, *t*-tests, and other basic statistical tests.

STATISTICAL INFERENCE WITH CATEGORICAL DATA With categorical data, we often wish to infer the distribution of the different categories within the populations from which our samples are drawn. In the simplest case, we have a single categorical variable with two or more categories. If there are just two categories, we can construct a **confidence interval for the proportion** of the population that belongs to one of the two categories. This is a statistically derived range of values that is centered on the sample proportion and is likely to include the population proportion. If there are three or more categories, we can use a **chi-square goodness-of-fit** test to determine whether the distribution of the different categories in the population is consistent with a specific distribution.

Figure B12 outlines the steps of a chi-square goodness-of-fit-test. As an example, consider the data described in Table B1. Many plant species have simple Mendelian genetic systems in which parent plants produce progeny with three different colors of flowers in a ratio of 2:1:1. However, a botanist believes that these particular poinsettia plants have a different genetic system that does not produce a 2:1:1 ratio of red, pink, and

RESEARCH**TOOLS**

B12 The Chi-Square Goodness-of-Fit Test

What is the chi-square goodness-of-fit test? It is a standard method for assessing whether a sample came from a population with a specific distribution.

Step 1: State the null and alternative *hypotheses*:

H_0: The population has the specified distribution.

H_A: The population does not have the specified distribution.

Step 2: Choose a significance level, α, to limit the risk of a Type 1 error.

Step 3: Determine the *observed frequency* and *expected frequency* for each category:

The observed frequency of a category is simply the number of observations in the sample of that type.

The expected frequency of a category is the probability of the category specified in H_0 multiplied by the overall sample size.

Step 4: Calculate the *test statistic*: $\chi_s^2 = \sum_{i=1}^{c} \frac{(O_i - E_i)^2}{E_i}$

Notation: C is the total number of categories, O_i is the observed frequency of category i, and E_1 is the expected frequency of category i.

Step 5: Use the test statistic to assess whether the data are consistent with the null hypothesis:

Calculate the *P-value* (*P*) using statistical software or by hand using statistical tables.

Step 6: Draw conclusions from the test:

If $P \leq \alpha$, then reject H_0, and conclude that the population distribution is significantly different than the distribution specified by H_0.

If $P > \alpha$, then we do not have sufficient evidence to conclude that population has a different distribution.

white plants. A chi-square goodness-of-fit can be used to assess whether or not the data are consistent with this ratio, and thus whether or not this simple genetic explanation is valid. We start by stating our hypotheses and significance level:

H_0: The progeny of this type of cross have the following probabilities of each flower color:

$\Pr\{\text{Red}\} = .50$, $\Pr\{\text{Pink}\} = .25$, $\Pr\{\text{White}\} = .25$

H_A: At least one of the probabilities of H_0 is incorrect.

$\alpha = 0.05$

We next use the probabilities in H_0 and the sample size to calculate the expected frequencies:

	Red	Pink	White
Observed	108	34	40
Expected	(.50)(182) = 91	(.25)(182) = 45.5	(.25)(182) = 45.5

Based on these quantities, we calculate the chi-square test statistic:

$$\chi_s^2 = \sum_{i=1}^{C} \frac{(O_i - E_i)^2}{E_i} = \frac{(108-91)^2}{91} + \frac{(34-45.5)^2}{45.5} + \frac{(40-45.5)^2}{45.5} = 6.747$$

We find the P-value for this result to be $P = 0.0343$ using statistical software. Since P is less than α, we reject the null hypothesis and conclude that the botanist is correct: The plant color patterns cannot be explained by the simple Mendelian genetic model under consideration.

This introduction is only meant to provide a brief introduction to the concepts of statistical analysis, with a few example tests. **Figure B13** provides a summary of some of the commonly used statistical tests that you may encounter in biological studies.

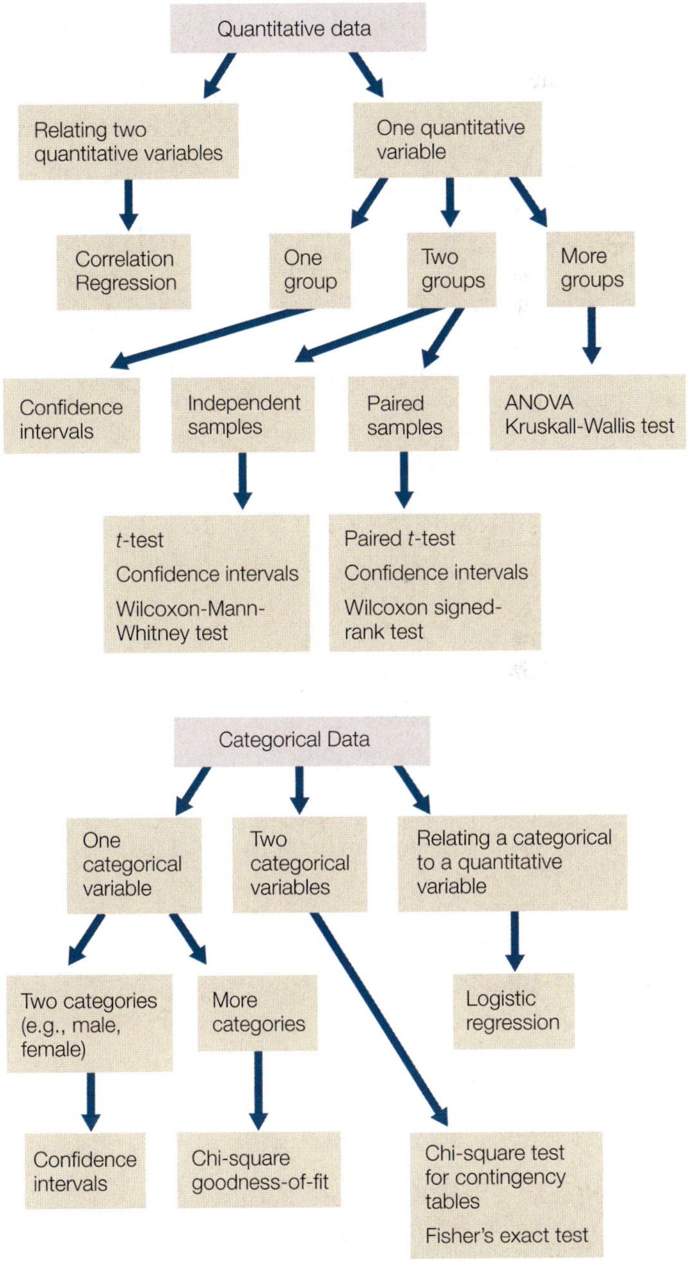

Quantitative data
Relating two quantitative variables
One quantitative variable
Correlation Regression
One group
Two groups
More groups
Confidence intervals
Independent samples
Paired samples
ANOVA Kruskall-Wallis test
t-test
Confidence intervals
Wilcoxon-Mann-Whitney test
Paired t-test
Confidence intervals
Wilcoxon signed-rank test
Categorical Data
One categorical variable
Two categorical variables
Relating a categorical to a quantitative variable
Two categories (e.g., male, female)
More categories
Logistic regression
Confidence intervals
Chi-square goodness-of-fit
Chi-square test for contingency tables
Fisher's exact test

B13 Some Common Methods of Statistical Inference
This flow chart shows some of the commonly used methods of statistical inference for different combinations of data. Detailed descriptions of these methods can be found in most introductory biostatistics textbooks.

Appendix C Some Measurements Used in Biology

MEASURES OF	UNIT	EQUIVALENTS	METRIC → ENGLISH CONVERSION
Length	meter (m)	base unit	1 m = 39.37 inches = 3.28 feet = 1.196 yards
	kilometer (km)	1 km = 1000 (10^3) m	1 km = 0.62 miles
	centimeter (cm)	1 cm = 0.01 (10^{-2}) m	1 cm = 0.39 inches
	millimeter (mm)	1 mm = 0.1 cm = 10^{-3} m	1 mm = 0.039 inches
	micrometer (µm)	1 µm = 0.001 mm = 10^{-6} m	
	nanometer (nm)	1 nm = 0.001 µm = 10^{-9} m	
Area	square meter (m^2)	base unit	1 m^2 = 1.196 square yards
	hectare (ha)	1 ha = 10,000 m^2	1 ha = 2.47 acres
Volume	liter (L)	base unit	1 L = 1.06 quarts
	milliliter (mL)	1 mL = 0.001 L = 10^{-3} L	1 mL = 0.034 fluid ounces
	microliter (µL)	1 µL = 0.001 mL = 10^{-6} L	
Mass	gram (g)	base unit	1 g = 0.035 ounces
	kilogram (kg)	1 kg = 1000 g	1 kg = 2.20 pounds
	metric ton (mt)	1 mt = 1000 kg	1 mt = 2,200 pounds = 1.10 ton
	milligram (mg)	1 mg = 0.001 g = 10^{-3} g	
	microgram (µg)	1 µg = 0.001 mg = 10^{-6} g	
Temperature	degree Celsius (°C)	base unit	°C = (°F – 32)/1.8
			0°C = 32°F (water freezes)
			100°C = 212°F (water boils)
			20°C = 68°F ("room temperature")
			37°C = 98.6°F (human internal body temperature)
	Kelvin (K)*	K = °C – 273	0 K = –460°F
Energy	joule (J)		1 J ≈ 0.24 calorie = 0.00024 kilocalorie†

*0 K (–273°C) is "absolute zero," a temperature at which molecular oscillations approach 0—that is, the point at which motion all but stops.

†A *calorie* is the amount of heat necessary to raise the temperature of 1 gram of water 1°C. The *kilocalorie*, or nutritionist's calorie, is what we commonly think of as a calorie in terms of food.

Answers to Chapter Review Questions

CHAPTER 1

1. e   2. b   3. e   4. d   5. e

6. In science, we formulate hypotheses about how the world works, then try to reject those hypotheses with experiments. The experiments must be designed so that we would expect them to uncover problems with our hypothesis. If the experiments are incapable of rejecting a hypothesis, then the experiments are not a rigorous test of the hypothesis.
7. The independent DNA found in mitochondria and chloroplasts is evidence of the origin of these eukaryotic organelles from ancient bacteria that became incorporated in the eukaryotic cell. Since the ancestors of these organelles once existed as independent organisms, they have their own genomes.
8. Controlled experiments, by definition, are able to control many variables in carefully maintained experiments, often in laboratory conditions. Comparative experiments, in contrast, often contain many additional variables that cannot be controlled by the investigator. Comparative experiments often incorporate realistic variation from uncontrolled factors, which accounts for their higher overall variability.
9. If two species share particular changes in the gene we compare, and those changes are not shared by other species we examine, we would expect the two species with the common changes to be more closely related to one another. By comparing many such changes in many genes, we can group species based on their relative evolutionary divergence from one another. For example, we share more changes in our genes with chimpanzees than we do with gorillas. From this, we can deduce that humans and chimpanzees shared a more recent common ancestor than they shared with gorillas.
10. Mitochondrial DNA is often used to follow the history of maternal lineages in a population or species. Nuclear DNA is not used in such cases because it is typically inherited from both parents. This difference can be useful in many circumstances. For example, we might examine a hybrid individual between two species. Equal portions of nuclear DNA from both species could confirm that the individual is a direct hybrid between the two species. If we examine the mitochondrial DNA, however, we can learn which of the two parental species was the female in the cross—and therefore learn by default which was the male.

CHAPTER 2

1. b   2. d   3. c   4. c   5. a   6. d

7.

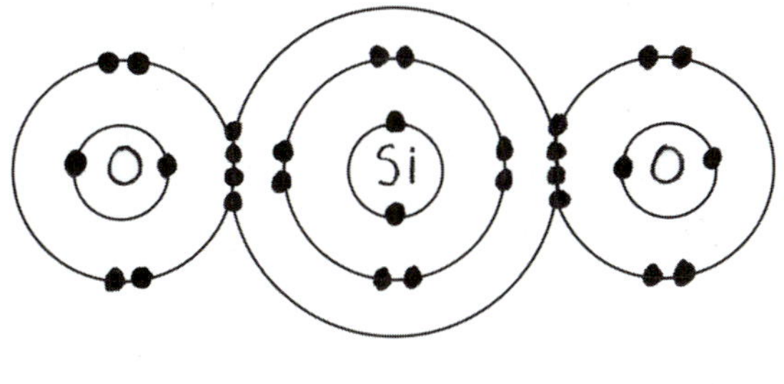
O
Si
O

8. An easy way to answer this question is to make a simple table:

	Covalent H—H	Hydrogen H····O
Electrons	Shared	Remain with H and O
Polarity	Nonpolar	Polar; + at H end
Strength	Stronger	Weaker

9. C—H: nonpolar; hydrophobic
C═O: polar; δ– at O; hydrophilic
O—P: polar; δ– at O; hydrophilic
C—C: nonpolar; hydrophobic

10. This is an example of Van der Waals forces, which act over a short distance and do not involve polarity.
11. The human body has the same elements as Earth's crust but in very different proportions.

CHAPTER 3

1. e   2. e   3. c   4. a   5. c   6. b

7. The observations support explanation "a." Glycine is small and nonpolar. Glutamic acid and arginine are larger and polar (charged). Serine and alanine are small: the protein retains its shape. But serine is polar (it has –OH as its R group), and that does not affect the structure. Valine is larger and nonpolar, and this affects shape. So the issue is size.
8. Mannose and galactose have the same atomic formula, $C_6H_{12}O_6$, but the arrangement of atoms is different: compare carbons 2 and 4. These sugars have the hydroxyl (–OH) functional group. Its polarity helps the sugars dissolve in water. The –OH group also can participate in bonding the sugar to other molecules through condensation reactions (see Figures 3.4 and 3.17).
9. High temperature disrupts weak interactions such as hydrogen bonds. Heat shock proteins might work by stabilizing the protein so that the weak interactions are not necessary to preserve its structure.
10. A change from lysine is a change in primary structure. The change could affect tertiary structure if the protein folds as a result of electrostatic attractions between charged amino acids (+ to –). In this case, the presence of a negatively charged amino acid (aspartic acid) where there should be a positively charged one (lysine) might prevent correct folding if a negatively charged amino acid elsewhere in the polypeptide chain is involved in folding (it is attracted to a + amino acid). The same forces might be at work in the interaction of separate chains for quaternary structure.
11. See Figure 3.10. Heat breaks hydrogen bonds and other weak interactions that maintain protein shape. Disulfide bonds also required for normal protein shape. Styling and perms partially denature keratin, then renature the protein in a new shape. Your investigation might involve measuring keratin protein structure of hair before and after disrupting hydrogen bonds and disulfide bonds.

CHAPTER 4

1. c   2. c   3. c   4. c   5. b   6. b

7. The presence of O_2 in the atmosphere produces an oxidizing condition that prevents the reduction reactions noted by the Miller–Urey experiment.
8. Oligonucleotides of RNA can fold because of hydrogen bonds forming between bases on the single chain and to a lesser extent because of weak interactions of base stacking when bases come near one another. Short strands of about 20 oligonucleotides are enough to produce uniquely folded RNA.
9. Cells provided concentration and compartmentation chemicals for the reactions needed for life, as well as differential permeability to distinguish life's chemical composition from that of the environment.
10. If microbes survived heat, the initial part of Pasteur's experiment might begin with microbes already present. They would grow in both the open and closed flasks. To get the results he did, Pasteur's flasks must not have contained such microbes. An answer for the proposed experiment on heat-stable microbes might be to inactivate them using reagents, such as mercaptoethanol, that destroy proteins.
11. A suggested experiment might be to dry the samples after the Miller–Urey experiment (allowing condensation reactions—polymerization) and then apply energy in the form of heat. This condition might have existed in volcanic rock in early Earth.

CHAPTER 5

1. b   2. d   3. e   4. a   5. d   6. b   7. a

8. Four membranes: two in the chloroplast and two in the mitochondrion
Two membranes: the lysosomal membrane and the plasma membrane (via vesicle; the molecules do not themselves cross any membranes)

No membranes: ribosomes do not have membranes. However, if the ribosomes were associated with the endoplasmic reticulum (ER), the answer would be two membranes: into the ER and out of the ER.

9.

	Animal Cell ECM	Plant Cell Wall
Composition	Collagen fibers in proteoglycan matrix	Cellulose fibers in polysaccharide and protein matrix
Rigidity	Less rigid	More rigid (especially secondary cell walls)
Connections	Some specialized proteins and junctions	Plasmodesmata

10. Microtubules line the long axons of nerve cells, where they act as tracks for vesicles that carry substances down the neuron. Without microtubules, the contents of these vesicles cannot be delivered to their destination, which can result in nerve problems.
    Microtubules are a key part of the mitotic spindle, which is used to move chromosomes during cell division. Depolymerization of microtubules can thus result in loss of dividing cells.
11. For a lysosomal enzyme, the pathway would be ribosome → interior of ER → Golgi → Golgi vesicles → lysosome.
    For an extracellular protein (animal cells), the pathway would be ribosome → interior of ER → Golgi → Golgi vesicles → plasma membrane → extracellular region.

CHAPTER 6

1. c 2. a 3. d 4. c 5. b 6. e 7. c

8. The pumping of Ca^{2+} requires a lipid bilayer membrane to separate compartments, a protein pump in the membrane, and ATP to provide energy for pumping.
9. Diatom wall components move from the Golgi apparatus to the cell wall by exocytosis.
10. Living in a hypotonic environment (cells hypertonic) results in a tendency for water to enter the organism by osmosis, which can cause swelling and dilute cell contents. Some organisms get around this by using reverse pinocytosis (exocytosis) to remove fluid.
11. Experiments might involve the following:
    To measure membrane fluidity, label a small amount of a lipid or protein with a dye and allow it to incorporate into a cell's membrane. This may make a localized labeled spot on the cell. The localized region will be seen to diffuse over the cell over time. In the cancer cells, this rate of diffusion may be faster.
    To measure cell adhesion, dissociate cancer and normal tissue cells. Incubate for a period of time and determine the rate at which cancer cells and normal lung cells bind to cells from the other tissues besides lung. The cancer cells may bind to a greater extent than normal cells.

CHAPTER 7

1. d 2. c 3. d 4. a 5. d 6. a 7. d 8. c

9. Different cells can have different target molecules to which cAMP binds, and these target molecules can have different activities and functions. Binding of cAMP changes the structure (e.g., tertiary structure of a protein) and therefore the function of a target molecule. So cAMP can have many effects.
10. Characteristics of direct communication: the size of signal molecules is limited by the size of openings between cells, it is not specific, it is fast, and there can be cytoplasmic connection between cells.
    Characteristics of receptor-mediated communication: the signal molecules can be larger, it is specific, it is slower, and there is no direct cytoplasmic connection.
    Direct communication is useful for a rapid, coordinated response of many cells.
11. See Figure 7.10. A mutation of the *Raf* gene that activates cell division might involve a protein product that does not need binding of Ras to be active. Cell division would occur without activated Ras, thereby eliminating the need for growth factor binding.
    A mutation of the *MAP kinase* gene would stimulate cell division if the resulting MAP kinase protein did not need to be phosphorylated by MEK to be active. No signaling cascade would be needed for the mutant protein to enter the nucleus and stimulate cell division.
12. Experiments might involve applying a solution containing the antibody to the upper part of the *Hydra* body. The antibody would block diffusion of the signal molecule from the apex to the upper body and—if the hypothesis is correct—would allow a bud to form in the upper body. A sham experiment, in which the solution without antibody is applied, would be a control. In this case, a bud would not form in the upper body.

CHAPTER 8

1. c 2. e 3. c 4. c 5. d 6. d

7. Endergonic reactions are coupled in time and space with exergonic reactions, which release the energy needed for the endergonic reactions.
8. A cytoplasmic enzyme generally has a globular structure with a hydrophilic exterior and an active site for substrate binding. An ion channel generally has a more linear structure with a hydrophobic membrane-spanning region and no active site.
9.

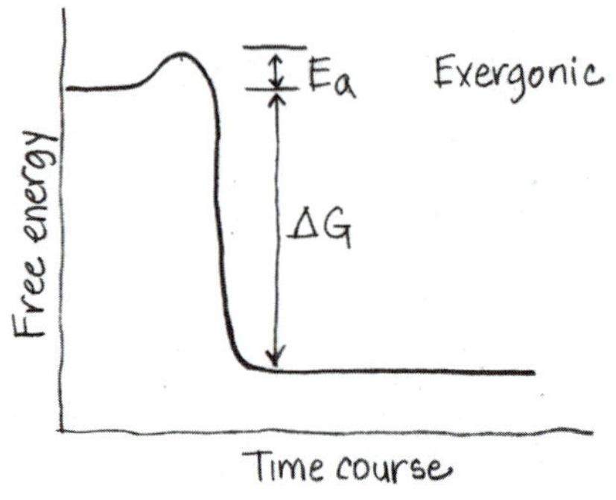

Ea
Exergonic
ΔG
Free energy
Time course

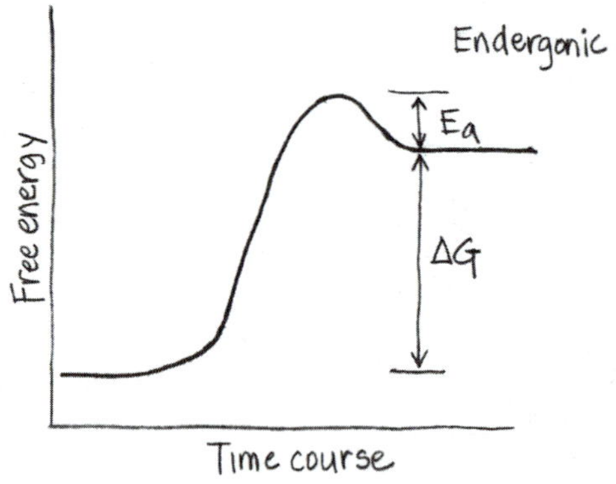

Endergonic
Ea
ΔG
Free energy
Time course

10. (a) The presence of water may prevent O_2 from reaching the enzyme. (b) Boiling denatures proteins, so polyphenol oxidase is irreversibly altered by boiling and its active site is destroyed. (c) Proteins have an optimal pH at which ionized R groups are appropriately charged to give the protein its tertiary structure. A pH 3 of may not be that optimal pH for polyphenol oxidase, so the enzyme is denatured and inactive.
11. See Figure 8.17. A competitive inhibitor binds to the active site of the enzyme and shifts the equilibrium to enzyme molecules in the active form.
12. To determine whether catalase has an allosteric or nonallosteric mechanism, perform an experiment with varying amounts of substrate and plot rate of catalase versus substrate concentration. An S-shaped curve will indicate an allosteric mechanism. A hyperbolic curve indicates a nonallosteric enzyme.
    To determine if a pollutant is a competitive or noncompetitive inhibitor, add the pollutant to the catalase to lower the rate of reaction, then add increasing amounts of substrate. A competitive inhibitor will be removed from the active site and the rate of reaction will increase. A noncompetitive inhibitor will not allow the rate to increase as more substrate is added. (There are more sophisticated kinetic experiments that you will learn in a biochemistry course).

CHAPTER 9

1. d 2. d 3. e 4. c 5. d 6. a

7. If cytochrome *c* remains reduced and cannot accept electrons, the electron transport (respiratory) chain stays reduced and NADH and $FADH_2$ remain reduced. This prevents oxidation reactions in the citric acid cycle and pyruvate oxidation, so pyruvate cannot be converted to acetyl CoA. Instead, pyruvate is converted to lactic acid, regenerating some NAD that can be used so that glycolysis can continue. Because the electron transport chain is not working, there is no proton gradient set up in the mitochondria, and ATP is not made by oxidative phosphorylation.
8. See Figure 9.13. Some amino acids are converted to intermediates of glycolysis. Once they enter glycolysis these intermediates are further metabolized to a

glycolytic intermediate that can be converted to glycerol, which is incorporated into triglycerides. Glycolysis and pyruvate oxidation produce acetyl CoA, which is converted to fatty acids and incorporated into lipids.

Glucose is converted in glycolysis to acetyl CoA, which is then converted to fatty acids as above.

9. (a) Oxidation (removal of H from C2 and C3 of succinate)
   (b) Exergonic (because it is an oxidation)
   (c) It requires the redox coenzyme NAD or FAD.
   (d) The fumarate is converted to other intermediates that regenerate oxaloacetate, the acceptor for the citric acid cycle.
   (e) The reduced coenzyme (NADH or FADH2) is reoxidized in the electron transport chain.
10. Anaerobes use alternate electron acceptors to generate energy, such as sulfur, sulfate, and nitrate. Also, they use substrate-level phosphorylation (direct transfer of phosphate to ADP) to make ATP.
11. The proton gradient in the experiment described in Figure 9.9 was generated artificially from the solution and did not require electron transport (a respiratory chain). The presence of antimycin A thus would have no effect on the experiment.

CHAPTER 10

1. e 2. b 3. d 4. d 5. d 6. d

7. In the dark, photosynthetic electron transport stops at photosystem II → reduced PQ (plastoquinone). Initially, the chlorophylls in light-harvesting complexes remain reduced, so reaction-center chlorophylls remain reduced and thus photosystem II remains reduced.

   In the dark, the Calvin cycle stops at the reduction phase, which requires NADH. No RuBP is regenerated, so there is no rubisco activity. The initial reactions are no oxidation of photosystem I, and no reduction of NADP to NADPH.
8. These processes can be compared using a table:

	Cyclic Electron Transport	Noncyclic Electron Transport
Products	ATP	ATP, NADPH, O_2
Source of electrons	Electron transport	Electron transport (photosystem I) or water (photosystem II)

9. See Figure 10.18. CO_2 carbons end up in 3PG, which is converted to pyruvate. Pyruvate goes to the citric acid cycle, where some of the intermediates are converted to amino acids, which are incorporated into protein.

   In the Calvin cycle, some 3PG is converted to G3P, which can enter glycolysis. Some of the intermediates of glycolysis are converted to amino acids, which are incorporated into protein.
10. (a) O_2
    (b) NADPH
    (c) 3PG
11. (a) Here is the pathway followed by ^{14}C: $^{14}CO_2$ → cells → photosynthesis → carbohydrate → combustion → $^{14}CO_2$. Release of $^{14}CO_2$ upon combustion would be evidence of photosynthesis (and life).
    (b) In this case: $^{14}CO_2$ → heat denatured cells, no photosynthesis. If living things were present, $^{14}CO_2$ would be released in experiment (a), but not in experiment (b).

CHAPTER 11

1. d 2. b 3. d 4. d 5. e 6. d 7. d 8. c

9. See Figure 11.19. In mitotic prophase, there is no pairing of homologous chromosomes, and crossing over is rare. In meiotic prophase I, homologous pairs of chromosomes align, and crossing over is common.

   In mitotic anaphase, sister chromatids separate, with one going to each pole. In meiotic anaphase I, sister chromatids do not separate; homologous pairs of chromosomes separate, with one pair going to each pole.
10. Normally, p53 induces expression of p21, which binds to the G1/S Cdk and prevents cyclin from activating it. Without active Cdk, the cell cycle ceases. If p53 is mutated such that it is nonfunctional, p21 is not induced and the cyclin–Cdk complex can form and stimulate the cell cycle at S phase.
11. Cancers often have multiple mutations in different cells of the tumor. If some of these mutations affect different parts of the cell cycle, targeting the different phases may be a useful therapy.
12. Your proposed experiments should involve isolating the synchronous meiotic cells from the lily anthers and establishing them in the lab. As the cells proceed through the meiotic cell cycle they can be analyzed at different stages for the presence and biochemical activity of various cyclins and Cdks.

CHAPTER 12

1. e 2. a 3. d 4. d 5. d 6. b 7. b 8. d 9. b

10. $BB \times bb$; $bb \times bb$; $Bb \times bb$; $Bb \times Bb$
11. 1/32
12. (a) Autosomal dominant
    (b) 1/4
13. (a) Males (XY) contain only one allele and will show only one color, black (X^BY) or yellow (X^bY). Females can be heterozygous (X^BX^b).
    (b) X^bY, yellow
14. The body color (G/g) and wing size (A/a) genes are linked; eye color (R/r) is unlinked to the other two genes. The distance between the linked genes is 18.5 units.
15. Yellow, blue, and white in a 1:2:1 ratio.
16. F_1 will all be wild type, *PpSwsw*. F_2 will have phenotypes in the ratio 9:3:3:1; see Figure 12.6 (p. 238) for analogous genotypes.
17. (a) F_1 will all be *PpByby* and will have wild-type eye color and wings. The ratio of phenotypes in F_2 will be 3:1, *PPByBy* (wild-type eyes and wings) to *ppbyby* (pink eyes and blistery wings.
    (b) F_1 will all be *PpbyBy* with wild-type eye color and wings; they will produce just two kinds of gametes (*Pby* and *pBy*). Combine them carefully and see the 1:2:1 phenotypic ratio fall out in the F_2: 1 wild-type eyes/blistery wings : 2 wild-type eyes/wild-type wings : 1 pink eyes/wild-type wings.
    (c) Pink–blistery
    (d) See Figures 11.16 and 11.18 (pp. 220–222). Crossing over took place in F_1.
18. *Rraa* and *RrAa*
19. (a) $w^+ > w^e > w$
    (b) The parents are w^ew and w^+Y. The progeny are w^+w^e, w^+w, w^eY, and wY.
20. (a) BX^a, BY, bX^a, bY
    (b) The mother is bbX^AX^a, the father BbX^aY, the son BbX^aY, and the daughter bbX^aX^a.
21. 75 percent
22. Because the gene is carried on mitochondrial DNA, it is passed through the mother only. Thus if the women does not have the disease but her husband does, their child will not be affected. However, if the woman has the disease but her husband does not, their child will have the disease.
23. The cross $RRYY \times rryy$ produces *RrYy* (round, yellow) F_1 offspring. If the seed shape and seed color genes were linked with no recombination between them, the F_2 would also be all *RrYy*. A distance of 10 map units between two genes means that on average 10% of the F_2 offspring will have recombinant phenotypes, in this case round green (5%) and wrinkled yellow (5%).

    The cross in Figure 12.19 is *BbVgvg* (gray, normal) × *bbvgvg* (black, vestigial). If there were no linkage between the genes, then the gray, normal parent would produce four types of gametes: *BVg*, *bVg*, *Bvg*, and *bvg*. When these combine with the *bvg* gametes produced by the other parent, four types of offspring in a 1:1:1:1 ratio will result: *BbVgvg* (gray, normal), *bbVgvg* (black, normal), *Bbvgvg* (gray, vestigial), and *bbvgvg* (black, vestigial).

CHAPTER 13

1. a 2. c 3. b 4. b 5. d 6. c 7. d

8. At 3,000 bp per minute in two directions, each origin grows at 6,000 bp per minute. There are 300 minutes in S phase, so the total bp possible for one origin is ($300 \times 6{,}000$) = 1,800,000. If there are 120 million bp to replicate, then the total number of origins is 120 million/1.8 million = 66 origins. If there are 3 μm of DNA, this means there are about 22 origins per micrometer of DNA.
9. DNA replication adds new nucleotides to the 3′ end of DNA, where there is an —OH group on the sugar at the 3′ position. If there is no —OH group, there cannot be a condensation reaction and formation of a bond to the next nucleotide, so replication stops.
10. After ten rounds there would still have been some DNA (about 1/512th) as hybrid because the original heavy DNA template strands would still have been there. This tiny amount might not have been detectable in the centrifuge, however.
11. The proposed experiments might use S strain pneumococcus and transform R strain as in Figure 13.1. Incubate separate batches of S strain bacteria in ^{32}P or ^{35}S. Make cell-free extracts of the S strains. Incubate with R cells and look for their transformation to the S phenotype. Then check to see if there is ^{32}P or ^{35}S label in the newly transformed cells. It would be expected that only ^{32}P label (DNA) would enter the cells.

CHAPTER 14

1. b 2. a 3. d 4. b 5. d 6. d 7. d 8. e

9. For 192 amino acids, the triplet genetic code mandates 576 bp of coding sequence. Add the start and stop codons and the total is 582. This is shorter than the actual DNA gene because of promoter and terminator of transcription sequences; introns; and ribosome binding sequences. All except the transcription signals are transcribed into the pre-mRNA. The mature mRNA has the introns removed.

10. Errors in transcription can be tolerated because many copies of each RNA are made; if a few have errors, there are enough perfect ones to overcome any problem. Errors in DNA replication are harmful because DNA is replicated only once in the life of the cell.

11. In the poly CA experiment, threonine is ACA or CAC and histidine is ACA or CAC. In the poly CAA experiment, threonine is CAA, ACA, or AAC. Therefore in the first experiment threonine must be ACA and histidine CAC.

12. Enzymes: $4 \rightarrow 2 \rightarrow 3 \rightarrow 1 \rightarrow 5$

    Compounds: $C \rightarrow F \rightarrow E \rightarrow D \rightarrow G \rightarrow T$

CHAPTER 15

1. a 2. c 3. b 4. b 5. d 6. b

7. (a) In a loss of function mutation, a phenotype is not present; for example, there may be a loss of enzyme activity. In a gain of function mutation, a new phenotype is present; for example, a new signaling protein may be active.

   (b) In a missense mutation, a single base pair change results in a codon change and thus an amino acid change in a protein. In a nonsense mutation, a single base pair change results in a codon change to a stop codon and thus premature termination of a protein.

   (c) In a spontaneous mutation, DNA changes as a result of unprovoked chemical changes or replication errors. In an induced mutation, DNA changes as a result of outside physical or chemical agents.

8. (a) The mutation that leads to PKU is rare in the human population; most people do not have the harmful allele and the highest probability is that the father is homozygous normal. Because the mother has PKU (she is homozygous mutant), the developing fetus is heterozygous.

   (b) High levels of phenylalanine cause brain damage. If the mother's phenylalanine levels were too high, the baby would be born with brain problems.

   (c) The woman should be on a phenylalanine-restricted diet.

9. Testing for the cystic fibrosis (CF) allele could be done by allele-specific oligonucleotide hybridization with probes for the normal and CF alleles; see Figure 15.18. Or direct DNA sequencing of the CF gene could be done. A person who is a carrier will test positive for both the normal and the mutant alleles.

   To do gene therapy, the normal allele for CF could be inserted into a viral vector that can infect cells in the lung and airway tissues. Then the virus could be sprayed onto these tissues.

10. Early identification of people with multifactorial diseases, even before symptoms appear, could lead to therapeutic interventions to prevent disease development. Ethical issues might include insurability, hiring eligibility, and social stigma.

11. An enzyme test for HEXA would reveal intermediate levels in people who are carriers. This could be done on accessible cells (e.g., blood) if the gene is expressed there. A DNA test could involve sequencing the gene by allele-specific oligonucleotide hybridization (see the answer to Question 9). The advantage of DNA testing is that it can be done on any cells from the body (not just cells that express the enzyme).

    Investigation of the stop codon hypothesis would involve isolating the HEXA protein from patients with Tay-Sachs disease and showing that it is shorter in primary structure than the protein encoded by the normal allele.

12. (a) The amino acid sequence would be Leu-Ile-Ser-Ile-Ala. This is a missense mutation.

    (b) The mutation replaces proline with serine. Proline is a nonpolar amino acid that is usually part of bends or loops in a protein; serine is a polar amino acid with a smaller side chain. The mutation is likely to affect enzyme activity because it is likely to affect protein structure.

    (c) See p. 317. This region of the gene could be amplified by PCR and then digested with *Eco*RV. The mutant DNA will be cut, but the wild-type DNA won't be.

CHAPTER 16

1. b 2. a 3. e 4. b 5. c 6. d

7. The easiest way to answer this question is to construct a simple table:

	Lysogenic Bacteriophage	HIV
(a) Viral entry to host cell	Attachment of viral protein to host cell membrane	Membrane fusion of virus to host cell membrane
(b) Virus release from host	Host cell lysis	Budding and exocytotic release
(c) Viral genome replication	Host DNA polymerase	Virus reverse transcriptase followed by host RNA polymerase
(d) New virus production	Host transcription of virus genes and host-mediated translation of virus proteins	Same as in lysogenic bacteria

8. In a prokaryotic gene, the promoter is a DNA sequence, there are few transcription factors, and there is one RNA polymerase. In a eukaryotic gene, the promoter is a DNA sequence, there are many transcription factors, and there are several RNA polymerases.

9. Here is the structure of the gene:

   E1 – I1 – E2 – I2 – E3 – I3 – E4

   (E = exon; I = intron). Assuming that initiation of transcription begins at E1, the possible proteins are composed of exons 1234; 134; 124; and 14.

10. To keep a constant, low-level expression of repressor protein, the regulatory gene would have an inefficient promoter, and synthesis of the repressor would be constitutive.

11. You could sequence the relevant genes of colon cancer cells and look for mutations that lead to aberrant function, then isolate the proteins involved and determine that their functions are indeed abnormal. To show epigenetic silencing, you might sequence the promoters of the genes and look for epigenetic changes (e.g., cytosine methylation, which would be increased if there is transcriptional silencing). Then you could examine the tumor cells to see if the active proteins are there but in small amounts.

CHAPTER 17

1. c 2. b 3. e 4. e 5. b 6. b 7. c 8. c 9. a

10. One gene can produce several proteins by alternative splicing, which makes the proteome highly complex. In addition, many proteins are modified after translation, and this contributes to even more protein diversity. The metabolome is highly variable from cell to cell and from one time to another. It is determined not only genetically but also by responses to environmental conditions.

11. While all of these plants have the same basic genes for "life" as well as "plant" functions (e.g., photosynthesis, cell-wall formation, flowering), there are some genes (and proteins) that are specialized for each plant (e.g., rice genes for growing under water, genes for timing flowering, genes for seed-storage proteins).

12. (a) Extract genomic DNA from the patient's cells and analyze it for SNP polymorphisms. If the SNP that correlates with kidney cancer is present, he has an increased susceptibility.

    (b) Isolate both normal and cancerous kidney cells. Do a metabolomic profile on the kidney cancer cells and the normal kidney cells using chemical analyses for small molecules. By comparing the profiles, generate a metabolomic "signature" for the kidney cancer cells. Next, examine the metabolomic profile of kidney tissue from the patient and compare it with the metabolomic signature for kidney cancer cells.

    (c) For possible drugs involved in kidney cancer treatment, isolate many cancers (or examine stored tissues) and do a SNP analysis, correlating tumor response to the drug with the SNP polymorphism. Then isolate some of the patient's tumor cells and examine the DNA for SNPs that relate to drug response. Use the drug that the patient's genome indicates will be effective.

CHAPTER 18

1. b 2. c 3. e 4. a 5. e 6. d 7. b 8. c

9. Both PCR and cloning begin with a gene sequence. In PCR, the sequence is amplified in the test tube. In cloning, the sequence is amplified by an organism (typically bacteria). In PCR, amplification is achieved by synthesizing primers that bind to either end of a target DNA sequence and adding nucleotides and DNA polymerase. The doubled DNA is them denatured, and the process is repeated 20 to 40 times.

In cloning, the target DNA is inserted by restriction and ligation into a vector, which has an origin of replication that will function in the organism where amplification will occur. The vector is added to the host cells, which are cultured and divide many times, amplifying the target DNA along with the host chromosome. The vector is then removed from the host cells and cut with a restriction enzyme, releasing amplified, cloned target DNA.

PCR is much simpler and faster but has artifacts where inappropriate fragments of DNA are amplified or sequence errors are introduced by DNA polymerase. Cloning yields the correct DNA without mutations but involves host cell culture and time-consuming DNA purification steps. See Figure 18.12. A simple table can answer this question.

	Conventional	Recombinant DNA
(a) Sources of new genes	Other plants of same species	Any organism or synthetic DNA
(b) Number of genes transferred	Often many	One
(c) How long it takes	At least one growing season, usually many	Weeks

10. (a) The target gene would be inserted into an expression vector with a promoter such that the gene would be expressed in the developing seed. The vector could be added to cultured wheat cells, and those cells carrying the vector selected (the vector could carry a reporter gene for resistance to an antibiotic). The cells could be induced to form a wheat plantlet, which would be transferred to the field and the seeds examined for the new protein.
    (b) The target gene could be inserted into a sheep expression vector containing the lactoglobulin promoter so that the gene would be expressed in milk glands. The recombinant vector would be inserted into sheep egg cells. After the female offspring grew up, their milk could be tested for the presence of the human enzyme.
11. Public concerns include the artificiality of unnatural interference with nature, the safety of these foods for human consumption, and environmental dangers if non-host plants receive recombinant genes.

CHAPTER 19

1. c 2. b 3. a 4. e 5. a 6. b 7. c

8. (a) All neuronal precursors might undergo apoptosis and no neurons would form.
   (b) The *p21* gene would be activated and the cell cycle would be blocked; in the presence of other factors, muscle cells would form.
   (c) There might be no gradient of the protein in the developing limb and therefore no differential development of digits—all the digits would be fingers.
   (d) The hunchback protein gradient would not form properly and the embryo would not establish its anterior–posterior axis.
9. (a) No apoptosis would lead to too many cells in developing organs, and the organs would not form properly.
   (b) No gradient of hunchback protein would form, and there would be no posterior end determination in the developing fruit fly.
10. A mutation that caused expression of class A genes instead of class C genes. This would lead to an AB combination instead of AC, and petals would develop instead of stamens.
11. Mechanisms might include cell-cycle inhibition as a result of Cdk blocker; induction of transcription of certain genes; and cytoplasmic segregation, so that when a cell divides only one daughter cell gets a factor important in determination.
12. One could analyze mRNA in egg cells, in the parent differentiated cells, and in the reprogrammed cells. This could be done by reverse transcriptase PCR or by gene expression arrays.

CHAPTER 20

1. c 2. a 3. a 4. c 5. c

6. If the expression of Gremlin were blocked, this protein could not inhibit BMP4 signaling. The cells in the webbing of the feet would undergo apoptosis, and the duck would be born with unwebbed feet.
7. All of the hatchlings at any temperature would be expected to develop into males. Aromatase is required to convert testosterone into estrogen, which is required for female development.
8. The coexpression of *Hoxc6* and *Hoxc8* appears to be important in the development of thoracic vertebrae (the vertebrae with ribs). This is a short region in mice, and mice have only a small number of thoracic vertebrae, and therefore a short body. In snakes, the coexpression of *Hoxc6* and *Hoxc8* along a much greater length of the embryo results in a much larger number of thoracic vertebrae, and therefore a much-elongated body.
9. The results support the conclusion that higher levels of BMP4 expression result in greater cartilage diameter on the beaks of developing chickens.
10. The observations are consistent with the hypothesis that there has been selection in some human populations for mutations on the enhancer that controls expression of the glycoprotein in red blood cells. This genetic change would be expected to have a selective advantage in human populations that are exposed to malaria at high levels, because the mutation confers greater resistance to malaria in humans that carry it.

CHAPTER 21

1. d 2. d 3. d 4. e 5. b

6. Humans select traits in domestic plant and animal populations based on our interest in the trait, rather than on how it affects the natural reproductive rate or survivorship of the organisms. Many of the traits artificially selected by humans would not be advantageous in wild populations. For example, humans have selected many cattle breeds for high body fat and high body weight. These traits result in large calves, which in turn result in calving difficulties for cows. Ranchers often have to assist in the birth of such calves, because the calf (and likely its mother) would often die without such assistance. In a natural population, there would be selection for smaller calf size and birth weight, which would increase the successful reproductive rate and survivorship.
7. Behaviors can respond to environmental cues that are predictive of future conditions, and these behaviors can be selected for if they are under genetic control. For example, day length becomes shorter as we move closer to winter, so individual mammals have a survival advantage if they respond to shortening days by going into hibernation. In this case, the environmental cue (day length) is predictive of future environmental conditions (the cold of winter).
8. Natural selection cannot act when there is no effect on the effective reproductive rate of the organism. Diseases such as Alzheimer's usually occur long after the reproductive years have passed. As long as the disease does not affect the relative likelihood of the survival of the affected person's offspring (as a result of reduced parental care, for example), we would not expect natural selection to lead to any reduction in Alzheimer's disease in human populations.
9. (a) Frequency of allele *a*: 0.60; of allele *A*: 0.40
   (b) Frequency of genotype *aa*: 0.40; of genotype *Aa*: 0.40; of genotype *AA*: 0.20
   (c) Expected frequency of genotype *aa*: 0.36; of genotype *Aa*: 0.48; of genotype *AA*: 0.16
   (d) We would expect some level of deviation because the assumptions of Hardy–Weinberg equilibrium are so restrictive. For example, the finite population size, the presence of mutation, any migration of individuals into or out of the population, gene flow from mating with adjacent populations, nonrandom mating within the population, or selection in the population could all lead to deviations from Hardy–Weinberg expectations.
10. The black mice and white mice are highly unlikely to be mating randomly with each other. The combined population is far from Hardy–Weinberg expectations, with far too few heterozygous (*Aa*) individuals. The much higher frequency of the *a* allele among the black mice, and of the *A* allele among the white mice, suggests that the black and white mice are mostly mating within color types, with few between-color-type matings. Another possibility, though, is that the population consists of two subpopulations (one of mostly black mice, the other of mostly white mice) that have only recently come together in the same location. These two hypotheses could be distinguished by following the mice through another generation. If mating is now occurring at random, we would expect the genotype frequencies to be similar to Hardy–Weinberg expectations after one generation of random mating.

CHAPTER 22

1. e 2. a 3. e 4. a 5. e 6. d

7. The classification is not currently monophyletic. Both genera could be monophyletic if Species 4 were moved from Genus B to Genus A; monophyly could also be achieved if all the species were included in the same genus.
8. Fossils can give us direct evidence of the character states of extinct lineages. For example, all modern birds lack teeth. Is the lack of teeth an ancestral or a derived condition? If we examine extinct species of theropods (the larger group of dinosaurs that includes the living birds), we see that they had teeth. Therefore we know that the lack of teeth is a derived condition in modern birds.
9. The estimated average rate of change is 0.9 amino acid change/500 million years, or 0.0018 amino acid change/million years. If we express this as a percentage rather than as a proportion, we would say there is (on average) 0.18 percent change in amino acid sequences per million years.

10. The West Nile virus in the United States appears to be most closely related to a strain of the virus isolated in Israel. A reasonable hypothesis is that the virus emerged in Africa in the 1930s and subsequently moved into Asia and Europe, probably multiple times. Then in the late 1990s, a strain of the virus from Israel appears to have been transported to New York, perhaps carried by mosquitoes on an airplane or in a cargo shipment. Once in the United States, the virus spread quickly in native bird populations across North America.

CHAPTER 23

1. e 2. c 3. a 4. e

5. If the only difference between the diverging lineages is at a single locus, then both of the new alleles must be functional when they interact with the products of other gene loci (in both lineages). Any interlocus genetic incompatibility produced by these new alleles would be expected to affect the parental lineages as well. In addition, there are many greater numbers of possible incompatibilities across different gene loci than there are within a single locus. Rather than two deleterious changes at the same locus (one in each lineage), the Dobzhansky–Muller model allows neutral changes at any pair of loci whose products interact. It is the negative interaction of these products in the hybrid between the two lineages that results in genetic compatibility.
6. If two different fusions of chromosomes occur in two different lineages, then the resulting chromosomes cannot pair normally in meiosis in the hybrids. If you attempt to diagram meiosis in the hybrid that would result from a cross of the divergent lineages in Figure 23.4, you will see that homologous pairings require parts of different chromosomes to align with one another. These chromosomes will then be pulled in two different directions as the cell divides in meiosis I, resulting either in a likely failure of the cell to divide, or an uneven distribution of the chromosome arms in the two daughter cells. Production of normal cells with an even distribution of the various chromosomes arms will be limited, so the hybrids will produce few, if any, normal gametes.
7. A likely possibility is that the incompatible alleles have not yet become fixed in the various strains, so only some combinations of crosses result in genetic incompatibility.
8. Species that arise in allopatry initially occur in separate, but usually adjacent, ranges (see Figure 23.6). Therefore we would expect many closely related species to exhibit this same pattern. The ranges of highly mobile species are more likely to change over time, so the pattern should be strongest among relatively sedentary species.
9. There are many possible designs of experiments that might prove informative. Here is an example of one that would examine the effect of flower position on pollinator attraction: Take one species of flower and divide the flowers into two groups. Position each flower to be either upright or pendant, then record the number and type of pollinators that are attracted to flowers in each group. Test to see if the differences between the two groups are statistically significant.
10. (a)

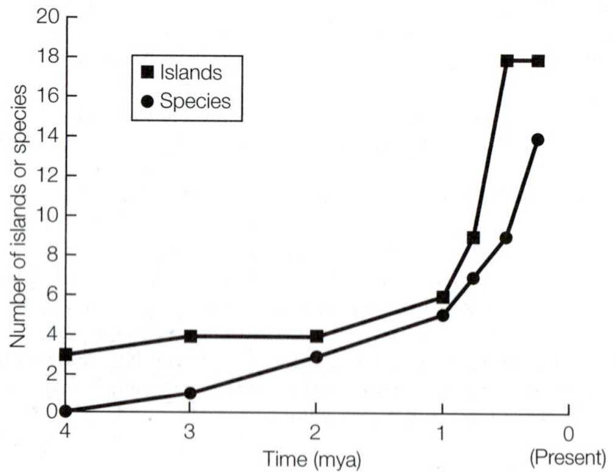

20
18
16
14
12
10
8
6
4
2
0
Number of islands or species
Islands
Species
4
3
2
1
0
Time (mya)
(Present)

(b) Yes, because the curve for the number of species lags behind the curve for the number of islands, but the two curves exhibit very similar changes in slope through time. As new islands arise, new opportunities for speciation also arise. The number of species at any one time is always just below the number of distinct islands.

(c) There are currently 18 islands in the archipelago and only 14 finch species. This suggests that there are still opportunities for additional speciation by geographic isolation. Based on our graph from Question a, we expect populations of species that occur on two or more islands to diverge into distinct species over time. To test this hypothesis, we could collect samples of each population and examine genetic divergence among the samples. Significant genetic divergence among the populations on different islands suggests that the distance between the islands is a significant barrier to gene flow, so the populations are expected to diverge into distinct species over time.

CHAPTER 24

1. a 2. a 3. a 4. e 5. b 6. e

7. Molecular clocks work best when they are applied within a group of organisms with similar generation times and populations sizes. Population size makes little difference if all or most changes are neutral, but rates of change among deleterious and beneficial changes are affected by population size. In addition, it is important to make comparisons among homologous genes and proteins, since rates of evolution across different genes are likely to vary widely. When molecular clocks are used to make comparisons across species with very different generation times, it is necessary to account for the different generation times.
8. New mutations are introduced into the experiment shown in Figure 24.14 through the errors made in the PCR amplification step. In other words, the mutation rate is a function of the error rate of the DNA polymerase. Using a different DNA polymerase with a higher error rate would increase the overall mutation rate of the experiment, and that would increase the variation in the population of molecules. Another possible answer is to add a mutagen to the PCR amplification step, which would also increase the mutation rate of the experiment. Any process that increases the mutation rate would be expected to increase the genetic variation present in the pool of molecules prior to the next round of selection.
9. This problem can be investigated by sequencing and comparing the genes for opsins in surface-dwelling (eyed) and cave-dwelling (eyeless) crayfishes. If the genes of the eyeless species are no longer under any selection, we would expect to observe a similar rate of synonymous and nonsynonymous substitutions in the genes. If there has been strong selection for a new function (something other than vision), we would expect a higher rate of nonsynonymous substitutions compared with synonymous substitutions (indicating positive selection). We would compare these rates to the rates seen in the surface-dwelling (eyed) species. In the surface-dwelling species, we would expect to see a higher rate of synonymous compared with nonsynonymous substitutions, which is expected under purifying selection.
10. (a) Codon numbers 12, 15, and 61 are likely to be evolving under positive selection for change because these three codons have each experienced a higher rate of nonsynonymous substitutions (which give rise to amino acid replacements) relative to the rate of synonymous substitutions.

(b) Codon numbers 80, 137, 156, and 226 are likely evolving under purifying selection, as the vast majority of changes at these codons are synonymous substitutions, which do not result in amino acid replacements. Substitutions that result in amino acid changes (nonsynonymous substitutions) undoubtedly occur, but they are usually selected against in the population. Codon number 165 has experienced similar numbers of synonymous and nonsynonymous substitutions. However, since there are approximately three times as many possible nonsynonymous substitutions as there are synonymous substitutions, the number of synonymous substitutions is slightly higher than expected if the rates of each type of substitution are equal. Codon 165 may be evolving under weak purifying selection; it is the codon that is closest to neutral among the codons shown in the table.

CHAPTER 25

1. b 2. c 3. a 4. c 5. b 6. c

7. There are many possible answers, but four familiar examples include the study of Earth's past atmosphere by examining the chemical composition of rocks; the study of past climates by examining the growth rings of trees; the study of continental drift by examining the geological record; and the study of the origins of the universe (the "Big Bang") by examining the speed at which galaxies are moving apart.
8. Relative dating provides us an order for events; we learn that Event 1 happened before Event 2. But absolute dating provides us with an estimate of the timing of those events. It is important to know not just that Event 1 occurred before Event 2, but also how much time separated the two events.
9. Multicellular organisms require higher concentrations of oxygen, and the levels of oxygen increased throughout the Precambrian. By the end of the Precambrian, atmospheric oxygen levels were sufficiently high to support a variety of multicellular organisms. In addition, the end of widespread glaciation (the "snowball Earth" period) near the end of the Precambrian probably allowed multicellular organisms to flourish.
10. There are many possible experiments that could be devised. For example, the effects of changing oxygen concentrations on other species (besides flying insects, such as the *Drosophila* used in the described experiment) could be tested. An ideal study organism would have a short generation time (so that many generations could be followed in the course of the experiment) and would be easy to raise in the laboratory. For example, guppies could be raised in elevated and reduced oxygen concentrations, and evolution in the size of the swim bladder (a site of oxygen uptake) could be evaluated as a response.

CHAPTER 26

1. e 2. c 3. e 4. b 5. b 6. d

7. Ribosomal RNA genes are universally present across organisms. They evolve slowly, so they can be compared among the most distantly related species. They are present in multiple copies, so they were relatively easy to isolate and sequence in the earliest days of gene sequencing. Also, since they are required for protein synthesis, and already present in all cellular species, the possibility of lateral gene transfer is greatly reduced. In contrast, different types of metabolism have arisen repeatedly in the history of prokaryotes, so species with similar types of metabolism may not be closely related. Cell structure is useful for identifying some major groups of prokaryotes (Gram-positive versus Gram-negative groups, for example), but the differences are too few to be of great use in classifying most species.
8. A laterally transferred gene does not represent descent from a common ancestor and thus does not reflect a true evolutionary relationship.

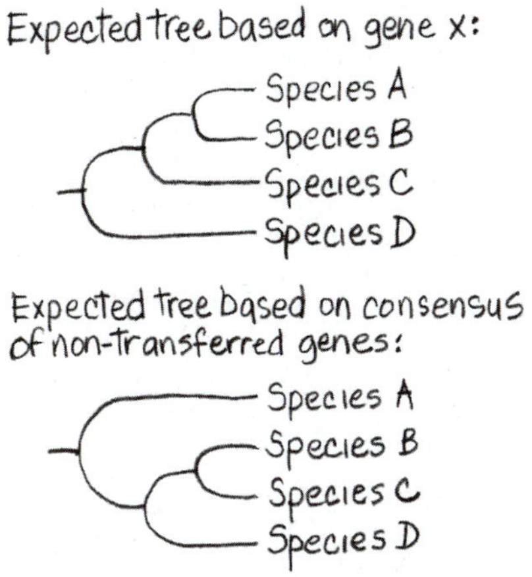
Expected tree based on gene x:
Species A
Species B
Species C
Species D
Expected tree based on consensus of non-transferred genes:
Species A
Species B
Species C
Species D

9. Most, but not all, biologists consider viruses to be living organisms. Viruses have their own genomes, and they are composed of proteins much like cellular organisms. They evolved from other living species, and they are clearly a part of life. However, viruses are not composed of cells, and they depend on cellular hosts to carry out many of their biological processes. For these reasons, some biologists consider them to be nonliving components of their cellular hosts rather than distinct living organisms.
10. There are many possible answers, but one widely used approach for detecting new life forms (in any environment, including high-temperature environments) is to directly isolate and amplify conserved gene sequences. The ribosomal RNA genes are often used for such detection because they evolve very slowly and are required for protein synthesis. DNA could be extracted from high-temperature environments, and any ribosomal DNA genes that were present could be amplified and sequenced. The sequences would then be compared with the ribosomal RNA genes of other known species of prokaryotes to classify the organisms living in the extreme environment.

CHAPTER 27

1. e 2. c 3. e

4. (a) Foraminiferans have external shells of calcium carbonate, whereas radiolarians have long, stiff pseudopods and radial symmetry. The external shells of foraminiferans and the internal skeletons of radiolarians are both important components of ocean sediments and sedimentary rocks.
   (b) Ciliates are covered with numerous hairlike cilia, whereas dinoflagellates generally have two flagella (one in an equatorial groove, and the other in longitudinal groove). Both ciliates and dinoflagellates have sacs, called alveoli, just beneath their plasma membranes, which identify them as alveolates.
   (c) Diatoms are unicellular and are typically composed of two nested plates (like a petri dish). Brown algae are large, multicellular organisms composed of branched elements or leaflike growths. Both diatoms and brown algae are photosynthetic.
   (d) The vegetative unit of a plasmodial slime mold is a plasmodium: a wall-less mass of cytoplasm containing numerous diploid nuclei. The vegetative unit of cellular slime molds consists of separate, single amoeboid cells. In both groups, when environmental conditions become unfavorable, the vegetative units form fruiting structures.
5. The independence of sex and reproduction in ciliates suggests that sex has functions apart from reproduction. Sex is important for recombining genes, which is important for several reasons. Sex allows populations of organisms to avoid the accumulation of deleterious alleles, and it allows the formation of new combinations of beneficial alleles. Thus even organisms that reproduce asexually generally have some other means of achieving sexual recombination of their genomes.
6.

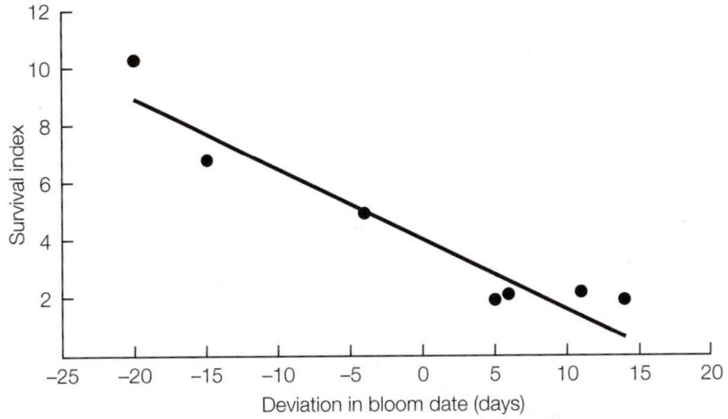
Survival index
12
10
8
6
4
2
–25 –20 –15 –10 –5 0 5 10 15 20
Deviation in bloom date (days)

Using the formula for a correlation coefficient shown in Appendix B, $r = -0.948$.

7. The results show that earlier bloom dates are associated with higher survival indices. The relationship between these two measures is very strong and nearly linear, resulting in a correlation coefficient of $r = -0.948$. As noted in the question, larval haddock depend on these blooms for both cover from predation and as a food source. A reasonable hypothesis for this is that earlier blooms provide better cover and more food for the larval haddock, so survivorship of the larval fish is higher in years when the phytoplankton blooms occur earlier. Another (not mutually exclusive) possibility is that the earlier blooms benefit other species that the haddock consume as food, or that the potential predators of haddock target the phytoplankton instead of the haddock.
8. The three rRNA genes of corn are not one another's closest relatives because the nuclear, mitochondrial, and chloroplast genomes have different origins, and the relationships shown in the tree reconstruct the endosymbiotic events that gave rise to mitochondria and chloroplasts.
9. The mitochondrial rRNA gene of corn is more closely related to the rRNA gene of *E. coli* than it is to the nuclear rRNA genes of other eukaryotes because the mitochondria were derived from an endosymbiosis with a proteobacterium. Likewise, the chloroplast rRNA gene of corn is more closely related to the rRNA gene of *Chlorobium* than it is to the nuclear rRNA gene of corn because the chloroplasts were derived from an endosymbiosis with a cyanobacterium.
10. The human and yeast mitochondrial rRNA genes would be expected to cluster on the tree closest to the corn mitochondrial rRNA gene because all of these genes are descended from the same endosymbiotic event (the origin of mitochondria). The human and yeast mitochondrial rRNA genes would be more closely related to each other than either is to the corn mitochondrial rRNA gene because fungi and animals are more closely related to each other than either is to plants (as can be seen in the relationships of the nuclear rRNA genes).

CHAPTER 28

1. c 2. e 3. b 4. b 5. d

6. Microphylls are usually small and typically have a single vascular strand. In contrast, megaphylls are larger and typically contain branched veins. Microphylls may have originated as sterile sporangia. Megaphylls may have originated from flattening of branching stems, between which photosynthetic tissues developed. Among modern plants, microphylls are found in lycophytes, whereas megaphylls are characteristic of the euphyllophytes (such as ferns and seed plants).
7. One advantage of heterospory is that it allows a greater degree of outcrossing, since there are separate male and female gametophytes.
8. Both mosses and ferns are homosporous, and both alternate between a diploid sporophyte and a haploid gametophyte generation. However, the gametophyte generation is the large, dominant portion of the moss life cycle, whereas the sporophyte generation in the large, dominant portion of the fern life cycle. The sporophyte of a moss is completely dependent on the gametophyte, whereas the sporophyte of a fern becomes independent of the gametophyte.
9. Yes. Heterospory is an example of a trait that appears to have evolved multiple times among different groups of vascular plants.
10. One possibility is to examine leaf size as a function of thermal environment among close relatives of living plant species. We might predict that large leaf size is restricted in hot, dry climates but favored in cooler, wetter climates.

CHAPTER 29

1. d 2. a 3. d 4. a 5. a

6. To be functional as a reproductive organ, a flower would need to have at least a carpel or a stamen. The petals function largely for pollinator attraction, and so can easily be lost in wind-pollinated species. The sepals function mostly to protect the flower in bud, and so may be lost in species with simplified flowers.

7. The fossil record is not a complete record of life on Earth. Early angiosperms may have been limited in distribution, or may have lived in environments that were not compatible with easy fossil formation. It is likely that the early angiosperms were not very abundant or widespread. They apparently underwent a rapid radiation in the Cretaceous, where they become common in the fossil record.

8. .

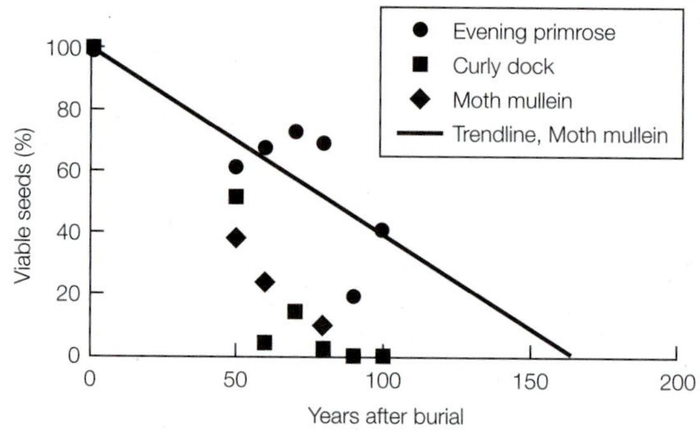
Evening primrose
Curly dock
Moth mullein
Trendline, Moth mullein
Viable seeds (%)
100
80
60
40
20
0
0
50
100
150
200
Years after burial

9. One approach to this problem is to calculate a linear trend line for survivorship of moth mullein (*Verbascum blattaria*) seeds by calculating a linear regression line (see Appendix B) and then projecting it forward in time to the point where it intersects zero percent survival. The resulting regression equation is $y = 102.09 - 0.62x$. The graph above shows this approach, which predicts that the last *Verbascum blattaria* seeds would germinate in about Year 165 of the experiment (set y to 0, and solve for x using the regression equation; the result is $x = 164.7$ years). This approach assumes a linear decline in viability of the seeds. It may be more reasonable to assume an exponential decay in seed viability (similar to radioactive decay; see Figure 25.1). If seeds decay exponentially, then we would expect some low level of survivorship of *Verbascum blattaria* seeds well beyond Year 165.

10. At least four factors are related to seed survivorship:
    1. Size of the seed: larger seeds have more food reserves (endosperm).
    2. Density of the seed coat: tougher seed coats provide better protection of the seed.
    3. Level of dormancy of the embryos: deeper dormancy results in longer survivorship.

CHAPTER 30

1. d 2. b 3. c 4. e 5. c

6. If it is a fungus, we would expect to detect chitin in the cells walls, whereas chitin would be absent if it is a plant. We could also examine the specimen for the presence of chloroplasts, which would be expected to produce the green coloration in a plant but not in a fungus. If it is a vascular plant, we would expect to observe vascular tissues in the sample, which would be absent if it is a fungus. We could also sequence a conserved gene from the sample, such as a ribosomal RNA gene, and compare the sequence phylogenetically with other fungi and plants.

7. It is because the nuclei remain separate in dikaryons, even though the two nuclei are contained within a single cell.

8. Fungi play a critical role in decomposition of plants and parts of animals. If there were no more fungi, there would be enormous accumulation of the remains of dead organisms, especially the cellulose and lignin of plants.

9. Site 5 shows the highest diversity and density of lichens and so is probably farthest from the city center. Site 4 is next, followed by Site 1, then Site 3, and finally Site 2. In addition to distance from the city center and prevailing wind direction, other predictive factors could include distance to point-pollution sources (such as factories or power plants) and distance to major highways (a source of pollution from automobile exhaust). Other answers are also possible; it is important for such studies to control for factors such as the species of tree examined and the exposure of the branches to similar light and humidity conditions.

10. One common source of fungal contaminants in plant samples is symbiotic fungi, including endophytic fungi and mycorrhizal fungi. In either case, it is usually possible to collect specific tissues from the plant that exclude these symbionts. If our hypothesis about the source of the fungal genes is correct, then the fungal sequences should be absent from these symbiont-free tissues.

CHAPTER 31

1. d 2. b 3. c 4. d 5. d

6. (a) In radial symmetry, body parts are symmetrical across multiple planes that run through a single axis at the body's center. Animals with radial symmetry have no front or rear ends, and they are often sessile or drift freely with currents. If they move under their own power, they can typically move slowly equally well in any direction. In contrast, bilaterally symmetrical animals have mirror-image right and left halves divided by a single plane that runs along an anterior–posterior midline. They have a front end that usually contains a concentration of sensory systems and nervous tissues in a distinct head. Bilateral animals usually move forward in the direction of the head, so the head encounters new environments first.

   (b) Among the bilaterian animals, there are two distinct forms of gastrulation, or the initial indentation of a hollow sphere of cells early in development to form the blastopore. In protostomes, the blastopore eventually develops into the mouth of the animal. In deuterostomes, the blastopore becomes the anus.

   (c) Diploblastic animals have embryos with two cell layers (an outer ectoderm and an inner endoderm). The embryos of triploblastic animals have an additional cell layer between the ectoderm and the endoderm, known as mesoderm.

   (d) Acoelomate animals lack a body cavity enclosed by mesoderm. Pseudocoelomate animals have a body cavity enclosed in mesoderm; this body cavity contains the gut and internal organs composed of endoderm, but these latter organs are not lined with mesoderm. Coelomate animals have a body cavity that is enclosed with mesoderm, and the internal organs are also lined with mesoderm.

7. The answer to this question will depend on the reader's opinion. However, most biologists would answer that the phylogenetic analyses that result from analysis of animal genomes provide the most definitive evidence of animal monophyly.

8. Bilateral organisms have an anterior and a posterior end. As the animal moves through the environment, the anterior end encounters potential food or predators first. It is therefore advantageous for the sensory organs and central nervous system to be concentrated at the anterior end.

9. A slow metabolic rate requires a low energy budget, and hence a low intake of food.

10. Placement of glass microscope slides (or other smooth substrates for placozoan attachment) in warm tropical waters often results in colonization by placozoans. The glass slides can be suspended in water in survey areas, then later retrieved and examined for the presence of placozoans.

CHAPTER 32

1. e 2. d 3. b 4. d 5. d 6. e

7. Segmentation allows an animal to move different parts of its body independently, which allows for much greater control of movement. However, segmentation tends to constrain the body shape of an organism. Loss of segmentation is often favored in some parasitic and burrowing organisms that live in confined spaces.

8. Several answers are possible, but some examples of key innovations that appear to be associated with major episodes of diversification in protostomes include the evolution of the cuticle in ecdysozoans, the evolution of shells in mollusks, the evolution of jointed limbs in arthropods, and the evolution of wings for flight in insects.

9. Insects have been highly successful in terrestrial environments, in part because flight gives insects greater access to plants. Many insect species are specialists on one or a few plant species, and plant diversity is far greater on land and in freshwater environments than in the oceans. Although some insects live in fresh water for part or all of their life cycles, these freshwater environments are closely associated with surrounding terrestrial environments. Crustaceans have been much more successful in the oceans than have insects, and crustaceans may simply outcompete insects in marine environments.

10. All entomologists agree that many more species of insects remain to be discovered, but many entomologists think that Erwin's estimates were high. Each estimate is highly dependent on how representative *Luehea seemannii* is as a tropical forest tree. If the average tropical forest tree has many fewer host-specific beetle species than does *Luehea seemannii*, then these estimates would be inflated. Likewise, overestimating the number of tropical forest trees, or the percentage of ground-dwelling beetles, or the percentage of all insects that are not beetles, would lead to further inflation of the estimates. In addition, species diversity of beetles may be higher in Panama than in other areas of the tropics. However, any of these estimates could be underestimates as well. Each of Erwin's assumptions is now being tested; these tests require extensive work on additional species of trees, additional groups of insects, and in additional areas of the world.

CHAPTER 33

1. d 2. a 3. d 4. a 5. e 6. b

7. The four appendages common to most vertebrates are the two pectoral appendages and the two pelvic appendages. In most swimming vertebrates, these appendages function as fins. They are commonly used for propulsion (especially the pectoral fins) but are also used for steering, stabilization, and manipulation of the body position in water. Among tetrapods, the appendages are often modified into limbs used for walking, running, jumping, burrowing, climbing, grasping, and manipulating objects. There have been several reversals to fin-like limbs used by aquatic tetrapods (several times among amphibians, turtles, birds, and mammals, for example). There have also been at least three origins of the pectoral limbs of tetrapods into wings for powered flight (among birds, bats, and the extinct pterosaurs). There have also been several other modifications of the limbs for gliding (in fishes, amphibians, reptiles, and mammals). One or both pairs of appendages have been lost (or greatly reduced) in many groups of fishes, amphibians, reptiles (including birds), and mammals. Some well-known examples of limb reduction or loss include the completely legless caecilians and snakes, the loss of external hind limbs in whales and manatees, and the greatly reduced forelimbs of flightless birds.
8. Amphibians exchange gases and fluids through their permeable skins. This makes them highly vulnerable to many environmental toxins. Many species of amphibians have a biphasic life cycle, so they are vulnerable to habitat degradation and loss of both aquatic and terrestrial environments. Most amphibians do not move long distances, so they do not easily move into new habitats when their local environment is destroyed. For these reasons, they are also sensitive to rapid climate changes. Many species of amphibians have highly specialized habitat requirements and live in very restrictive ranges. Habitat loss or changes within these restricted ranges often result in extinction.
9. Fossil remains of extinct theropod dinosaurs shows that many features once thought to be restricted to birds, such as feathers, actually evolved much earlier among the theropods. Other typical "bird" morphological features, such as air-filled bones and a furcula (wishbone), are also typical of the larger group of theropods. Among living reptiles, DNA sequence analyses clearly unite birds with the crocodilians (the other living archosaurs). The combined evidence from many sources that birds are a surviving group of theropod dinosaurs is now overwhelming.
10. Hair evolved in the ancestor of mammals; feathers evolved among theropod dinosaurs (seen today among the birds). Among the living tetrapods, birds and mammals are endothermic. Hair and feathers provide body insulation for mammals and birds, respectively. Without these forms of insulation, the maintenance of metabolic body heat would be difficult. Fossil evidence shows that many extinct theropod dinosaurs also had feathers, so many paleobiologists predict that they were endothermic as well. Endothermy would also be expected in large, active predators—a description that fits our current view of many theropod dinosaurs.

CHAPTER 34

1. b 2. e 3. a 4. b 5. b 6. d 7. c 8. c

9. The cell types can be compared using a table.

Structure/Function	Sclerenchyma	Collenchyma
Cell walls	Secondary, thickened	Primary, thicker at corners
Flexibility	Less flexible	More flexible
Cell conditions	Some dead (apoptosis)	Alive
Presence	Wood, bark	Petioles, growing areas

10. Primary growth involves cell division and cell enlargement, and typically results in growth of an organ in length. Secondary growth involves growth of an organ in thickness, by the addition of more cell layers. Only some angiosperms undergo secondary growth. Herbaceous plants such a peonies have only primary growth. Woody plants such as trees have both primary and secondary growth.
11. The initials are still 1.5 meters above the ground today because the plant grows in height at its apex.
12. Some examples might include a larger root apical meristem to produce thicker carrots and reduced internode growth to produce compact heads of cabbage.

CHAPTER 35

1. c 2. d 3. b 4. b 5. d 6. e

7. Epidermal cells have external walls with a waxy cuticle, which makes them repel water. The epidermal cells of roots might have a thinner (or absent) cuticle, as they take up water; and leaves and stems a thicker cuticle, to conserve water. In addition, the epidermis of leaves, and to a lesser extent stems, has stomata, which regulate gas exchange (including loss of water vapor from the leaf interior).
8. A source is an organ such as a leaf that produces more sugars than it uses. A sink is an organ such as a root that produces less sugars than it needs and so imports sugars from a source. In a deciduous tree, a leaf might be a source in the summer, and roots a sink. But then in spring, the roots might be a source for the buds (newly emerging leaves).
9. To cross the fewest membranes and still get from the soil solution to the atmosphere by way of the stele a water molecule would follow this route: soil to root symplast to stele symplast to stele apoplast to xylem to leaf apoplast to leaf interior air space to stoma to atmosphere. A water molecule could follow this path by crossing as few as two membranes: 1) from soil into a root hair or root cortical cell across a root cell membrane; 2) out of a stele cell into the stele apoplast across a stele cell membrane. Getting from the soil solution to a mesophyll cell in a leaf would require crossing at least three plasma membranes: the two membranes listed for the previous route plus the mesophyll cell membrane (to get from the leaf apoplast into the mesophyll cell).
10. The mutation in the *HARDY* gene might cause increased expression of a gene that inhibits cation accumulation in stomata, thereby keeping them more closed and conserving water. To test this hypothesis, you could look at the response of stomata to light in the leaves of mutant versus wild-type plants. The stomata of wild-type plants should open rapidly in response to light (see Figure 35.9); the stomata of HARDY mutant plants might open more slowly or less wide.
11. (a) Yes. The difference in water potential between the soil and the leaf (1.7 MPa) is enough to overcome gravity and draw water to the top of the tree.
    (b) No. If the soil water potential decreased to –1.0 MPa, it would be more negative than inside the root cells and water would leave the roots (and enter the soil).
    (c) If all the stomata closed, the leaf water potential would not be as negative. This in turn would make the xylem water potential less negative, and so on down to the roots. This would make the difference between the leaf water potential and root water potential insufficient for water to flow from the roots to the leaves (toward a more negative water potential).

CHAPTER 36

1. d 2. d 3. c 4. a 5. c 6. d

7. The ability of chemists to detect low concentrations of elements is fairly recent. Before then, nutrient solutions thought to be pure often were not.
8. Heavy irrigation after a prolonged dry period may produce runoff of topsoil (the A horizon) and leaching of ions (especially anions) into the subsoil, making fewer nutrients available to plant roots. Converting land use from virgin deciduous forest to crops will change the composition of living organisms in the soil, as many organisms that live in association with tree roots will disappear. The soil structure and texture will also change, because roots will no longer be present to hold the soil together and make air spaces. The soil chemistry will change, because crops take up nutrients from the soils and the nutrients are removed from the system when the crops are harvested.
9. See Figure 36.10, the nitrogen cycle. There are numerous species that fix nitrogen. Loss of one species might allow others to expand and replace it. Loss of all the species would mean that only abiotic methods could be used for nitrogen fixation. This might reduce overall nitrogen in the soil, meaning less would be available for plant growth.
10. The experiment with mutant *Arabidopsis* suggests that *Arabidopsis* uses either its own or exogenous strigolactones for growth regulation and has the appropriate receptor and response mechanisms. This reinforces the idea that an ancient mechanism to attract beneficial microbes also is used for modern plant growth regulation. Or the reverse might be true: the original function of strigolactone might have been as a plant hormone and its role in plant–microbe interactions might have evolved later.
11. Because holoparasitic plants can gain reduced carbon through association with hosts, the genes encoding photosynthesis functions are not under selection pressure, because having them would not confer any survival and reproductive advantage for the parasites. So any mutation that renders such a photosynthesis gene nonfunctional will not be deleterious.

CHAPTER 37

1. a 2. d 3. b 4. b 5. a 6. b

7. Fire produces ash, which enriches the soil with plant nutrients. A seed that germinated as a result of fire could have an advantage in such a nutrient-rich soil.
8. If a single species has two mechanisms for breaking seed dormancy, then if environmental conditions for Mechanism A are not present, environmental conditions for Mechanism B might be. This enables the plant to respond to a wider array of environmental conditions. In addition, if the cue for, say, Mechanism A turns out to be misleading (not predictive of favorable conditions) and the

seedling dies, there is still a second seed that can germinate at a different time (by Mechanism B), when conditions might be more favorable.

9. The charcoal in the bag absorbs ethylene gas, which is released by ripening fruits. The lack of ethylene prevents over-ripening and decay.
10. To test for the relationship between corn stunt spiroplasma disease and gibberellins, you could measure gibberellins in plants infected with the bacterium and in normal plants; you might expect the spiroplasma-infected plants to exhibit a reduction in gibberellins. Another approach would be to infect normal plants with the spiroplasma and then spray gibberellins on them; you might expect this to reverse the stunt phenotype.
11. (a) See Figure 37.2 Add a mutagen to hundreds of corn seeds and plant them. In a screen, look for plants that are shorter, and propagate these.
    (b) See Figure 37.11. If the transcription factor in the gibberellin signal transduction pathway is inactivated, the plants will be insensitive to gibberellin and be stunted. A mutation that inactivates the gibberellin receptor would have the same effect.
    (c) Other potential effects might include reduced seed germination and reduced seedling growth due to lack of mobilization of stored reserves in the seed (see Figure 37.5). If the mutant is completely gibberellin-insensitive, these effects will *not* be overcome by adding gibberellin to the seeds as they germinate. If, however, the mutant is a dwarf because of reduced amounts of gibberellin in the plant (because of a mutation that affects gibberellin biosynthesis, for example), the germination effects could be reversed with exogenous gibberellin.

CHAPTER 38

1. b 2. e 3. b 4. e 5. a 6. c
7. In triploid cells undergoing meiosis, there cannot be pairing of homologous chromosomes in meiosis I. So meiosis I is abnormal and functional gametes do not form.
   A fruit is formed from the ovary wall of the flower.
   Seedless grapes are probably propagated by cuttings (vegetative reproduction).
8. Poinsettias are short-day plants; they bloom at a time of year when days are getting shorter (in the Northern Hemisphere).
9. No, it isn't necessary. Just a flash of light during a long night is enough to convert P_r to P_{fr} and to change the photoperiod.
10. (a) The mutation stabilized the CO protein.
    (b) The mutation caused nonfunction of the FD protein.
    (c) The mutation increased expression of the FLC protein.
    (d) The mutation caused constitutive expression of the CO protein.
11. Several approaches might be taken, such as a genetic screen for meiotic cells that do not separate chromosomes at anaphase I, or a search for proteins (and then their genes) that bind to SWII protein.

CHAPTER 39

1. b 2. c 3. a 4. b 5. c 6. c
7. A plant might make a secondary metabolite that kills an insect pest. Plants making this metabolite would be selected for in evolution. However, the insect might develop resistance to the metabolite. Then the insect population would increase while the plant population decreased—until another defense mechanism evolves. This is coevolution. For more examples, see Chapter 56.
8.

Avr2Avr3	Healthy	Healthy	Diseased
Avr1Avr4	Healthy	Healthy	Healthy

9. (a) The effects of reduced rainfall could include dehydration and osmotic stress. Genetic responses might include alterations in leaf anatomy, with a thicker cuticle to reduce evaporation; a more extensive root system to obtain water; and accumulation of solutes in the roots, which would reduce root water potential and result in more water uptake in dry soils.
   (b) Flooding reduces the amount of O_2 available to the plants and results in reduced respiration. Adaptations might include increased production of pneumatophores or aerenchyma to supply air to submerged plant tissues.
   (c) Wheat rust is a fungal pathogen. Plants can adapt by increasing the ability to seal off infected areas and reduce the spread of the fungus within the plant, by developing specific immunity, and by increasing production of phytoalexin and pathogenesis-related proteins that kill the fungus.
10. You could feed one group of hornworms on normal plants and another group on genetically modified plants. The two groups could then be exposed to the parasite. If nicotine is protective, the hornworms that fed on normal plants should have fewer parasites.

CHAPTER 40

1. c 2. c 3. a 4. d 5. b
6. Feedforward information makes it possible to anticipate a physiological challenge to homeostasis and to take preemptive action by changing a set point or the sensitivity of a regulatory system. Feedforward information for the regulation of breathing could be the onset of exercise; for blood pressure it could be the fight-or-flight response to a threat; and for secretion of digestive juices it could be the sight, smell, or expectation of food.
7. In the metabolic rate/environmental temperature curve in Figure 40.17, the equivalent of *HL* would be metabolic rate, as long as the animal's temperature is not rising or falling and the animal is not doing external work. *K* would represent the animal's thermal conductance, or how easily it loses heat; $1/K$ would be a measure of the animal's insulation. The curve projects to 0 at an ambient temperature equal to body temperature because this portion of the curve represents the extra metabolic effort necessary to compensate for heat loss to the environment. If body temperature and environmental temperature were the same, there would be no heat loss to the environment.
8. Biological processes proceed more slowly at lower temperatures. Thus the lower the temperature of the heart or skeletal muscle, the slower will be its ability to generate a contractile force. This could pose a physiological challenge for highly active fish such as great white sharks or giant bluefin tuna that depend on fast swimming and endurance to catch prey. An evolutionary adaptation to this challenge can be seen in these fishes' vascular anatomy: blood from their hearts goes to the gills, where it exchanges respiratory gases but also comes into thermal equilibrium with the cold ocean water. Thus these fishes are sending cold blood to their body tissues.
9. Basal metabolic rate (ml O_2/hr) vs. body mass (kg):

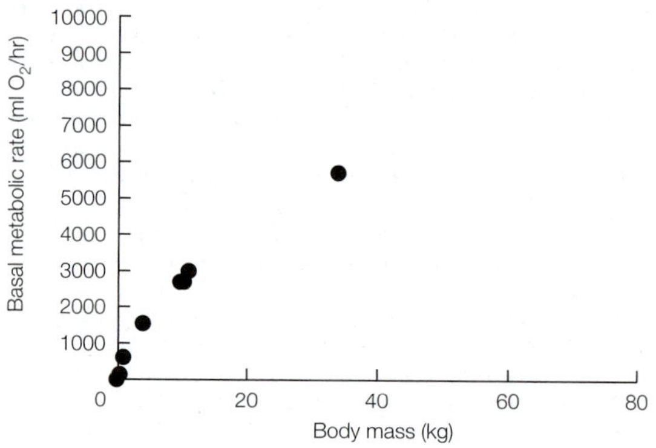

Basal metabolic rate (ml O2/hr)
10000
9000
8000
7000
6000
5000
4000
3000
2000
1000
0
20
40
60
80
Body mass (kg)

Heart size (g) vs. body mass (kg):

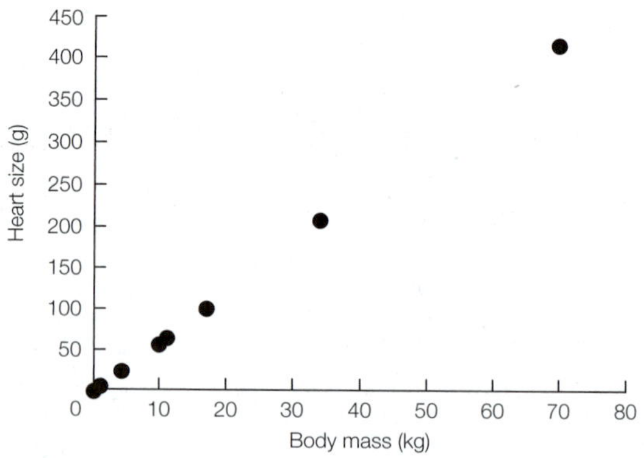

Heart size (g)
450
400
350
300
250
200
150
100
50
0
10
20
30
40
50
60
70
80
Body mass (kg)

10. The heart has to pump blood against the resistance of the vascular system. Since blood vessels go to all tissues of the body, the total length of blood vessels will be directly proportional to body size, and therefore total peripheral resistance will be directly proportional to body size. The other factor that determines how much blood the heart can pump is the heart rate. The total cardiac output is a function of the size of the heart and the number of times it beats per unit of time.
11. Several factors that determine the heat transfer rate between the iguana and its environment are different in these two conditions. When basking on the lava rocks, the radiation absorbed will be higher and the heat conducted across the skin will be different in comparison to being in the water. Also, when the animal breathes it will be losing heat through evaporation of water in its airways.
    An experiment in which the animal heats up and cools down in the same environment would offer stronger support for the conclusion that blood flow

to the skin is a critical factor. The iguana could be placed in water and in air at two different temperatures (e.g., 20°C and 40°C) to compare the rates of heating and cooling.

CHAPTER 41

1. b 2. a 3. d 4. e 5. e

6. The time course of a hormone signaling system depends on several factors, including the rate of release of the hormone, its half-life in the blood, and the interactions it has with its receptors. A hormone signaling system that controls a short-term process such as digestion would be expected to have a rapid (e.g., vesicular) release, a short half-life, and rapid action; these are attributes of a peptide hormone. A hormone signaling system that controls a long-term process such as embryonic development would be expected to have continuous release, a long half-life, and to be slow acting, attributes of a steroid hormone.
7. The high levels of synthetic male steroid hormones exert actions through the testosterone receptors that exist in both males and females. The usual effects of promoting muscle hypertrophy and male secondary sexual characteristics (e.g., hair, a deeper voice) will occur. In addition, the high level of negative feedback on the pituitary gonadotropes and on the production of GnRH by the hypothalamus will reduce testicular and ovarian functions in the males and females. Decrease of circulating estrogens in the females will cause reduction of breast tissue.
8. The most common cause of hypothyroidism is lack of iodine in the diet. Thyroglobulin continues to be produced, but there is a lack of functional T_3 or T_4. As a result, the levels of TRH and TSH rise because of the lack of negative feedback. The elevated TSH induces continued production of thyroglobulin, resulting in goiter. The most common cause of hyperthyroidism is an autoimmune disease called Graves disease. An antibody to the TSH receptor is produced, and the binding of that antibody to the receptor causes activation of the signaling pathway that increases the production of thyroglobulin and the development of goiter. If there is adequate iodine in the diet, however, this condition results in increased secretion of T_3 and T_4, which produces the symptoms of hyperthyroidism.
9. The large size of the larvae together with the lack of adult moths indicates that cycles of growth and molt of the larvae continued without the induction of pupation. The low temperature probably prevented the usual decline in the production and secretion of juvenile hormone by the corpora allata.
10. Insulin controls the entry of glucose into most cells of the body, but not neurons. When insulin levels fall, as happens during the postabsorptive state (i.e., after ingested nutrients have been fully digested), glucose entry into cells slows down and the cells convert to using other sources of energy. Neurons, however, always require glucose; their lack of insulin means that their access to an adequate glucose supply is protected during the postabsorptive state because there is no decrease in their ability to take up glucose from the blood. Also, the decrease in glucose use by other cells of the body preserves the glucose in the blood for use by the nervous tissue.

CHAPTER 42

1. a 2. e 3. e 4. c 5. d 6. a

7. See Figure 42.10. The antigen-binding site of an antibody has heavy and light chains in a unique three-dimensional configuration that binds a particular antigenic determinant. This is similar to an enzyme active site that binds a substrate. In both cases, binding is noncovalent. A major difference is in the result of binding: an antigen does not change its covalent structure when it binds to an antibody, whereas a substrate does change covalently when it binds to an active site.
8. Both immunoglobulins and T cell receptors have constant and variable protein regions, bind antigens, and have great variability in primary structure. T cell receptors are only membrane proteins of T cells. Immunoglobulins can be either membrane proteins of B cells or secreted proteins in the blood.
9. The father's haplotypes are A1B7D11 and A3B5D9; the mother's are A4B6D12 and A2B7D11. These two parents could not have the child with the genotype indicated.
10. There are thousands of different enzymes in an individual but potentially millions of different specific antibodies. Every cell in an animal has the genetic information for all enzymes. Each immunoglobulin, however, is derived from a unique gene (produced by DNA rearrangements) in a B cell or a clone.
11. Experiments might involve testing vaccinated people for neutralizing antibodies against HIV (humoral immunity) and looking for T cell activity against HIV-infected cells (cellular immunity).

CHAPTER 43

1. a 2. d 3. d 4. d 5. d 6. a

7. Leydig and thecal cells have similar functions and characteristics. They are both removed from direct contact with the developing gametes, and they both produce testosterone. Sertoli cells and granulosa cells are both in direct contact with the developing gametes, and they support their development by providing nutrients.
8. Progesterone actions are required to maintain the endometrium in a condition that can support implantation and not degenerate, as occurs during menstruation. By blocking progesterone receptors, RU-486 prevents implantation and the maintenance of the endometrium.
9. Conditions that could favor the evolution of this sexual dimorphism are a sessile existence, dispersed populations, and availability of suitable habitats. If a larva lands on a suitable substrate, it will have high reproductive success if it is a female and can produce lots of eggs; eggs typically have resources that enable them to travel considerable distances if released into the water, and their probability of being fertilized is therefore high. A larva would have a lower probability of success if it developed into a solitary male and produced sperm; sperm have less ability to survive travel over long distances in the water to encounter eggs. However, if a larva lands on a female, it is guaranteed high reproductive success if it can fertilize all of the eggs that female produces. Therefore, by attaching itself to the female and minimizing all of its own physiological processes other than sperm production, a larva can achieve high reproductive success at very little cost.
10. It is likely that the man's offspring would all be daughters. The Y chromosome lacks some essential genes that are on the X chromosome; in the absence of the cytoplasmic bridges, the developing sperm that contain a Y chromosome would lack those gene products. Thus all viable sperm the man produces would contain an X chromosome. This would result in female offspring, since the mother would also contribute an X chromosome.

CHAPTER 44

1. e 2. a 3. d 4. b 5. b

6. You could inject Disheveled protein into the side of the fertilized egg opposite the gray crescent and see if a secondary organizer formed in that region of the resulting blastula. You could also inject the inhibitor of the Disheveled protein into the region of the gray crescent and see if that prevented the formation of the organizer.
7. You would be destroying the cells that normally migrate from the dorsal blastopore lip to form the notochord, and that also produce the signals that determine the anterior–posterior differentiation of the embryo. Thus you might see defects in the development of the nervous system or abnormal segmental development of the body.
8. The flow of fluid over Henson's node may be asymmetrical and thereby create different physical forces on the primary cilia of cells on either side of the node. These differential forces could influence the expression of *Sonic hedgehog*. Experiments to test this hypothesis could include using gene knockouts that eliminate the motile cilia around Henson's node, or experiments in which early embryos are cultured under conditions in which the flow across Henson's node is opposite that of the normal pattern. The prediction for the first experiments would be that the right–left asymmetry of organs in the embryos would be randomized. The prediction for the second experiment would be that the normal asymmetry of organ development would be reversed.
9. A possible mechanism would be cytoplasmic factors that are not distributed randomly or evenly throughout the cytoplasm of the oogonia or the spermatogonia. Thus when they divide by mitosis, factors that control the fates of the daughter cells could be received by one daughter cell but not the other.
10. At an early stage of blastulation (e.g., the 16- or 32-cell stage,) a few blastomeres could be removed from the embryo and cultured separately to produce a population of stem cells. The embryo could go on to develop normally, and the stem cells could be frozen for later use.

CHAPTER 45

1. d 2. b 3. c 4. e 5. c

6. When the stimulus occurs at some point along an axon and an action potential is stimulated, depolarizing current will flow in both directions, bringing adjacent areas of the axon to threshold. However, once an action potential is fired, the Na^+ channel inactivation gates close and make that section of the axon refractory to further stimulation until they open again. Thus the action potential cannot reverse its direction of propagation, and if the action potential begins at the axon hillock, it cannot reverse its direction of propagation and is unidirectional.
7. Excitatory synapses cause a depolarization of the neuronal membrane, and inhibitory synapses hyperpolarize it. These two influences are summed by virtue of the resulting membrane potential. If it depolarizes enough to reach threshold, an action potential will be fired at the axon hillock.
8. Because the GABA receptor is inhibitory, benzodiazepines would be expected to slow cognitive processes and make a person more likely to fall asleep.

9. The type of information that an action potential transmits depends on the nature of the sensory cell that generated the action potential and on the nature of the cell that receives input as a result of that action potential. Thus photoreceptors transduce light into action potentials, and those action potentials are interpreted as light in the visual circuits that receive those action potentials. Intensity of the stimulus is coded as the frequency of action potentials. Integration is achieved by the summation of excitatory and inhibitory influences on the target cells.

CHAPTER 46

1. d 2. a 3. e 4. e 5. c

6. Olfactory and taste receptors are both chemosensors that respond to specific molecules in their environment. Olfactory receptors, however, are neurons, whereas taste receptors are epithelial cells that communicate with neurons that are associated with them. Olfactory receptors express a family of genes for olfactory receptor proteins that are then localized on cilia that project out of the olfactory epithelium. All of these olfactory receptor proteins are G protein-linked and are metabotropic. Taste receptors are also located on cilia of the epithelial taste sensor cells. Bitter, sweet, and umami receptors are G protein-linked metabotropic receptors, but salt and sour receptors are ionotropic. The discrimination between an apple and an orange depends on integration of information from both the olfactory and the taste receptors.
7. The sensation of directional motion arises from the vestibular system, which includes the semicircular canals and the vestibule, which contains membranous structures containing a fluid—endolymph. At the base of each semicircular canal is a gelatinous projection, a cupula, that encases a cluster of stereocilia. Movement of the head causes movement of the endolymph, which then exerts force on the cupula and bends the stereocilia, generating action potentials in the vestibular nerves. The vestibule includes two membranous structures called the saccule and the utricle. In these structures, stereocilia tips are in contact with otoliths, which are membranous structures containing crystals of calcium carbonate. When the head is accelerated forward or backward, the momentum of the otoliths causes the stereocilia to bend in a direction that indicates the direction of movement.
8. Underwater, the external ear canals are filled with water. Unlike air, water is not compressible and therefore sound waves are transmitted through water as vibrational movements of the water. These movements exert greater forces on the tympanic membrane than air pressure waves do.
9. As happens in humans in the vestibular system, movements of the fish cause movements of the water in the lateral line canals. The resulting forces are transduced into action potentials by the hair cells and provide information about the movement of the fish through the water. Additionally, vibrations in the water generated by other organisms or physical events will cause movement of the water in the lateral line canal and be transduced into action potentials, providing information to the fish about its environment.
10. The owl depends on auditory stimuli to locate the mouse in total darkness. Directional information comes from the bilateral placement of the ears, which are equally stimulated when the owl is directly facing the source of the sound. The face of the owl is disc-shaped, which helps collect sound waves and direct them to the ears.

CHAPTER 47

1. d 2. d 3. c 4. c 5. a

6. The stab wound must have severed the sympathetic nerves on the left side of the man's neck. Activity in these nerves causes dilation of the pupil. Severing the sympathetic nerves on the left side would remove all sympathetic activity reaching the pupil on the left side, and therefore it would be more constricted. Similarly, sympathetic activity decreases activity in the salivary glands, whereas parasympathetic activity increases salivation. Thus withdrawal of sympathetic input would release the salivary glands from any inhibition that would counteract even low levels of parasympathetic input.
7. Eyes positioned on the sides of the head enable a wider field of vision. Eyes pointing in the same direction create a narrow field of vision but make depth perception possible. You would expect prey species to benefit from wider fields of vision. You would expect predator species to benefit from depth perception, which would facilitate pursuit and capture of prey.
8. Sleepwalking is more likely to occur in non-REM sleep for two reasons: there is motor inhibition in REM sleep, which renders the individual paralyzed; and the nature of sleepwalking activities does not match with the vivid, bizarre content of REM-sleep dreams.
9. We can break this question down into the different observations. First, the loss of motor control of the right leg indicates that motor commands are ipsilateral—they descend on the same side as the limb that is being controlled. The ability to sense painful stimuli applied to the right leg but not to the left foot indicates that the pain pathways cross over to the opposite, or contralateral, side of the spinal cord before they ascend to the brain. The reflex movement of the right leg to a stimulus applied to the left foot indicates that reflex information is processed at the local level and does not require processing at higher levels of the central nervous system. The conclusion is that motor commands in the spinal cord are ipsilateral to the muscles being controlled but that the pain information ascending to the brain is contralateral. Finally, the different responses to pain and to touch indicate that these two modalities of somatosensory information travel in different tracks: touch ipsilaterally and pain contralaterally.
10. The fact that more slow-wave activity was seen over the right frontal cortex than over the left in response to exercising and training the left hand shows that the right side of the brain controls the left side of the body, and visa versa. This increase in slow-wave activity during sleep following the exercise/training suggests that this sleep slow-wave activity reflects either a restorative process or a learning process.

CHAPTER 48

1. b 2. c 3. b 4. d 5. c

6. One feature is the length of the muscle; half the length of a long muscle is more than half the length of a short muscle. Another feature is the location of insertion of the muscle on the bone; this determines the relative lengths of the effort arm and the load arm of the lever system created by the muscle, bone, and joint. If the ratio of load to effort arm is small, a large movement that can generate only relatively small forces is possible. If the ratio of load to effort arm is large, only small movements that can exert large force are possible.
7. If the break and healing have damaged the epiphyseal plate, and the primary and secondary areas of ossification fuse, the bone can no longer grow at that end.
8. The shoulders will fatigue first because they are not normally responsible for maintaining posture; they are adapted for rapid movements and sudden applications of large force. Thus shoulder muscles have a higher proportion of fast-twitch fibers. The leg muscles are postural muscles and have a higher proportion of slow-twitch fibers.
9. The action potential is conducted throughout the muscle cell by the system of T tubules. In the T tubules, the action potential causes conformational change of the DHP–ryanodine receptor complex. That change opens Ca^{2+} channels in the sarcoplasmic reticulum, and Ca^{2+} diffuses into and throughout the sarcoplasm. Ca^{2+} binds with the troponin units, causing the tropomyosin to expose the actin–myosin binding sites, cross-bridges to form, and the muscle to contract. When the Ca^{2+} concentration in the sarcoplasm falls as a result of being pumped back into the sarcoplasmic reticulum, the process reverses and actin–myosin binding sites are no longer available. The difference in time course of the contraction versus the action potential is due to the time that it takes for the Ca^{2+} to be released, diffuse throughout the sarcoplasm, and then be sequestered back into the sarcoplasmic reticulum.
10. The increased amount and duration of Ca^{2+} in the sarcoplasm causes increased contraction of the muscles and therefore an increase in muscle tension. The increase in muscle tension requires additional expenditure of ATP, raising metabolism and producing more heat. The increased metabolism causes elevated heart rate. This is in addition to the effect of the increased Ca^{2+} in the cardiac muscle itself.

CHAPTER 49

1. d 2. e 3. b 4. c 5. a

6. This fish would not be very active. It would move slowly. It would have a larger heart and larger blood vessels than fishes with hemoglobin, to accommodate a high flow of blood at low pressure. Its gill membranes would be well developed. It would have a high blood volume. Its jaws would show adaptations for sit-and-wait capture rather than pursuit. This fish occurs only in Antarctic waters because Antarctic waters are very cold and therefore the solubility of O_2 in those waters is high.
7. The total surface area for gas exchange is much smaller in a large air cavity than it is in many smaller cavities (alveoli) that add up to the same total volume. If the lung tissue is less elastic, the vital capacity of the lungs will go down, meaning less air can be exchanged during the breathing cycle. The less elastic lung tissue is also less permeable to respiratory gases.
8. Close to the end of inhalation, the pleural cavity pressure is reaching its maximum negative value. At the same time, the alveolar pressure is rising back up to being the same as the atmospheric pressure. Alveolar pressure would be most positive relative to atmospheric pressure at the midpoint of the exhalation phase.
9. Blood cells require energy. When in storage, their initial energy source is the glucose in the blood plasma, but as that supply gets depleted, blood cells also metabolize the intermediates in the glycolytic pathway. That includes 2,3-BPG. As the 2,3-BPG gets metabolized, there is less of it to bind to deoxygenated hemoglobin and therefore the affinity of the hemoglobin for O_2 increases. When

the affinity of the hemoglobin for O_2 gets too high, it can lower the P_{O_2} in the plasma to levels that are below the P_{O_2} in the plasma of a patient.

10. When you go up in altitude, the P_{O_2} in the air you breath goes down, but the P_{CO_2} was already low at low altitude and remains low at high altitude. Thus at higher altitude there is less of a concentration gradient driving diffusion of O_2 into the blood, but no decrease in the concentration gradient driving CO_2 out of the blood. Thus the main stimulus for breathing goes down as the need to increase breathing goes up. As a result, the blood becomes hypoxic and triggers breathing by activating the carotid and aortic chemosensors. The increase in breathing blows off even more CO_2, and breathing slows, which causes another bout of hypoxia and even a rise in blood CO_2. That triggers another bout of rapid breathing, and this cycle repeats.
11. (a) The llama hemoglobin has a higher affinity for O_2.
    (b) Llama hemoglobin would be advantageous at high altitudes because it can become 100 percent saturated at the low P_{O_2} of the high-altitude environment. Therefore the hemoglobin can carry a full load of O_2 to the tissues.
    (c) Llama hemoglobin allows the transfer of O_2 to occur at lower tissue P_{O_2}s.

CHAPTER 50

1. a 2. c 3. d 4. c 5. e

6. One factor is that at the beginning of a race there is a feedforward signal from the sympathetic nervous system that increases heart rate. Another factor is that the increased heart rate, together with the increased venous return to the heart from the exercising muscles, stretches the ventricles, which then contract with more force as described by the Frank–Starling law. This is due to the fact that a slight stretching of the sarcomeres optimizes the overlap of the actin and myosin fibrils for a maximum contraction. Increased breathing also increases the venous return and induces the Frank–Starling law.
7. There is no time when all four heart valves are open at the same time; if there were, the heart could not pump efficiently. Throughout diastole, the aortic and pulmonary valves are closed and the atrioventricular valves are open. At the beginning of systole, the atrioventricular valves close. There is a brief moment when all four valves are closed, until the aortic and pulmonary valves open, and stay open until the end of systole.
8. There are rapid responses and longer-term responses. A fall in blood pressure lowers the firing rate of baroreceptors in the great arteries. The decrease in baroreceptor input to areas of the brainstem that regulate cardiac function results in increased sympathetic and decreased parasympathetic output to the heart. This increases the heart rate and the force of contraction of the cardiac muscle (the fight-or-flight response). A slower response to blood loss is mediated by the kidney, which responds to the decreased blood pressure by increasing the release of renin, which in turn increases the activation of angiotensin circulating in the blood. Active angiotensin increases blood pressure by constricting peripheral blood vessels and stimulating thirst. Another slow response, stimulated by the fall in baroreceptor activity, is the release of ADH from the posterior pituitary. ADH increases the reabsorption of water by the kidney.
9. The cardiac muscle must be capable of generating and conducting action potentials. This may involve different variants of the ion channels involved in action potential generation and conduction. The cardiac muscle must be able to convert the action potential into the opening of Ca^{2+} channels in the sarcoplasmic reticulum, so there could be adaptive changes in the DHP and ryanodine receptors. Once Ca^{2+} is released into the sarcoplasm, it has to be resequestered into the sarcoplasmic reticulum, and that Ca^{2+} pump is likely to be adapted to operate at lower temperatures in the hibernator.
10. Your graph should look like this:

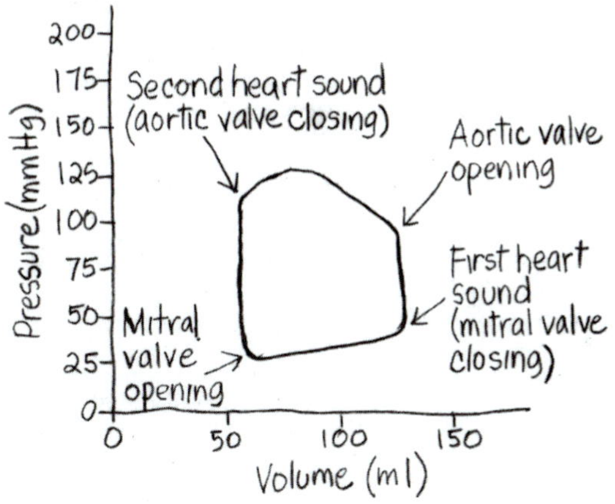
200
175
150
125
100
75
50
25
0
Pressure (mmHg)
Second heart sound
(aortic valve closing)
Aortic valve
opening
First heart
sound
(mitral valve
closing)
Mitral
valve
opening
0
50
100
150
Volume (ml)

A similar graph for the right ventricle would have the same volumes but lower pressures.

CHAPTER 51

1. e 2. a 3. b 4. d 5. b

6. The rationale for a high-fat and high-protein diet is that it minimizes the secretion of insulin. Insulin promotes the uptake, metabolism, and storage of glucose by various cells of the body; it also inhibits the action of lipase in the adipose tissue. Thus with low insulin, tissues are more likely to metabolize fats and less likely to store them.
7. The following points might be included in your answer.
    - Insulin stimulates most cells of the body to take up glucose from the blood by stimulating the insertion of glucose transporters into the plasma membranes of those cells in the absorptive state.
    - Insulin inhibits lipase in the adipose tissue, so breakdown of stored lipids is decreased in the absorptive state.
    - Insulin stimulates the synthesis of triglycerides in the adipose tissue.
    - Insulin activates the enzyme that phosphorylates glucose as it enters cells, thereby preventing it from diffusing back out of the cells. This maximizes the uptake of glucose by cells.
    - Insulin activates the liver enzymes that synthesize glycogen.
    - The lack of insulin in the postabsorptive state decreases the uptake of glucose by most cells of the body and activates the enzymes of lipolysis and glycogenolysis.
8. Triglyceride in food is emulsified by bile in the duodenum to form micelles, which are broken down by pancreatic lipase into free fatty acids and monoglycerides. Both are absorbed across the plasma membranes of intestinal epithelial cells. In the intestinal epithelial cells, the free fatty acids and monoglycerides are resynthesized into triglycerides and packaged into chylomicrons, which also contain cholesterol and are coated with lipoproteins. The chylomicrons are secreted from the basal ends of the epithelial cells into the center of the intestinal villi, where they enter the lymphatic vessels and circulate through the lymphatic vessels to the thoracic duct, where they enter the blood.
   A *direct route* the fatty acid could take would be for the chylomicrons circulating in the blood to come into contact with the damaged endothelium of the coronary arteries and be absorbed into the plaque.
   In an *indirect route*, the chylomicrons could be taken up by liver or adipose tissue cells and the triglyceride stored. In the liver, the triglyceride can be repackaged to form low-density lipoproteins or very low-density lipoprotein particles, depending on the amount of cholesterol in the particle. These lipoproteins leave the liver and circulate in the blood. When they come into contact with the damaged endothelial cells of the coronary arteries, they bind to lipoprotein receptors, and their triglyceride and cholesterol are absorbed into the plaque.
9. Carbonic anhydrase catalyzes the hydration of CO_2 to produce carbonic acid that dissociates into H^+ ions and HCO_3^+ ions. In the parietal cells of the stomach, H^+ is secreted into gastric pits and then flows into the stomach lumen. The bicarbonate ions are transported out of the basal side of the cells, where they are absorbed into the blood, raising blood pH. In the ducts of the pancreas, bicarbonate is transported into the lumen of the ducts; H^+ ions are transported out of the basal sides of the cells, where they are absorbed into the blood, lowering its pH.
10. The hypothalamic regulatory pathways control hunger and satiety, not energy use. This tells us that the side of the energy balance equation that is most important in contributing to obesity is Calories in. Of course, obesity can also result in decreased physical activity, which means that decreased Calories out can be a secondary factor in the causation of obesity.

CHAPTER 52

1. d 2. a 3. a 4. b 5. a

6. The glucose contributes to the osmotic concentration of the glomerular filtrate and therefore of the tubular fluid. This results in a greater volume of urine flowing through the collecting ducts and being excreted.
7. ACE inhibitors decrease the production of angiotensin II, the active form, from angiontensin I. Decreasing the level of angiotensin in the blood increases the glomerular filtration rate and therefore the production of urine. Losing more water in the urine lowers the blood volume and therefore the blood pressure. Blocking angiotensin also results in dilation of peripheral blood vessels, which lowers blood pressure, and decreases thirst, which helps maintain a lower vascular volume. Angiotensin stimulates the release of aldosterone, which promotes Na^+ reabsorption and therefore water retention.
8. The rate at which the inulin is filtered is equal to the concentration of inulin in the blood ($[I_b]$) times the glomerular filtration rate (GFR). The rate at which inulin is excreted is equal to the concentration of inulin in the urine ($[I_u]$) times

the urine flow rate (V = 1 ml/min). Since all inulin that is filtered leaves the body in the urine, the rate at which it is filtered must equal the rate at which it is excreted. Therefore $[I_b] \times GFR = [I_u] \times V$, and $GFR = [I_u] \times V/[I_b]$.

9. Any difference in the rate at which the substance (S) is filtered versus the rate at which it is excreted will be due to tubular reabsorption of tubular secretion. If more of S is excreted than filtered, then S must be secreted by the renal tubules. If less of S is excreted than filtered, then S must be reabsorbed by the tubules. Therefore $[S_b] \times GFR - [S_u] \times V$ = the rate of reabsorption (if negative) or secretion (if positive) of S.

10. At high altitude, the concentration difference driving O_2 across the alveolar membranes and into the blood deceases, but the concentration difference driving CO_2 across the alveolar membranes and into the expired air does not change. The increased breathing rate driven by hypoxia will therefore blow off too much CO_2 and the pH of the blood will increase, which will suppress breathing. Bicarbonate is filtered into the glomerular fluid; as that fluid passes through the renal tubules, H^+ ions are excreted, resulting in the formation of H_2CO_3, which dissociates into CO_2 and H_2O. The CO_2 is reabsorbed into the tubule cells, where carbonic anhydrase catalyses its hydration to H^+ ions that are secreted back into the tubular fluid and bicarbonate ions that are secreted back into the extracellular fluid. Blocking carbonic anhydrase results in fewer H^+ ions being secreted into the tubular fluid and more bicarbonate ions remaining in the tubular fluid to be excreted. The retention of H^+ and excretion of bicarbonate lower the blood pH, and that stimulates increased breathing. However, the increased bicarbonate in the tubular fluid raises the osmotic concentration of that fluid, causing more water to be excreted.

CHAPTER 53

1. b 2. e 3. c 4. a 5. d 6. e 7. a 8. e

9. The development of brain circuitry controlling the sexual dimorphism in urination behavior is influenced by the levels of estrogen or testosterone that are circulating in the blood during the early postnatal period. Estrogen must prevent the development of the neuronal patterns of connectivity that are responsible for the male behavior.

10. The major difference between the eusocial insects and vertebrates is the haplodiploid mechanism of sex determination in the insects. This means that a female worker that is a daughter of the queen shares more genes with sisters—the queen's offspring—than she would with her own offspring, which would have a different father. Thus raising a sister contributes more to the female worker's inclusive fitness than raising a daughter would. Haplodiploidy is not found among vertebrates, so this powerful selective force does not operate among them.

11. The variability in the undirected song allows the male to adapt his song to the local variant. The directed song is probably more effective in attracting a female because it more accurately identifies the male as a member of the local population, and is also more effective in competition with neighboring males.

12. Since male cowbirds will not hear the song of his father, his song pattern should be genetically determined. Females probably learn the song of host species and thereby learn to identify and locate potential host nests. You could test the hypothesis about males by raising them in isolation to see what songs develop as they mature. You could test the hypothesis about the females by giving them a choice test such as that done with the zebra finches in testing directed and undirected song preferences (see Figure 53.8). Give each female cowbird a three-chambered cage, and play the songs of host species and non-host species in the opposite end chambers. Place the test bird in the middle chamber. Record the amount of time the female spends in each end chamber as a measure of preference. Another experiment would be to bring a host nest with cowbird eggs into the laboratory and raise the cowbirds in the presence of recordings of another species song. Then repeat the choice experiment with the host's and the other species' songs being played.

13. The classical data suggest that the hygienic behavior is controlled by two genes to give the typical Mendelian ratio of 3 to 1 with 50% hybrid, 25% homozygous dominant, and 25% homozygous recessive. The QTL analysis, however, indicates that more than two genes are involved. The difference in these results could be due to there being two major-effect genes and several modifying genes.

CHAPTER 54

1. d 2. d 3. a 4. c 5. c

6. Based on its location and its weather conditions, the mattoral should have vegetation typical of that found in other areas with Mediterranean climates. Vegetation should be (and is) tough, shrubby and fire-adapted, with slender, leathery leaves and seeds either stored in fire-safe cones or equipped with elaiosomes and dispersed by ants.

7. The fact that species in the genus are known in Australia and in southern Africa, it is likely that the genus originated before the breakup of Gondwanaland, the supercontinent comprising Antarctica, South America, Africa and Australia. Other species in the genus thus might be found in South America (contemporary conditions are too cold to maintain spider life in Antarctica today but the possibility exists that fossil species in the genus might be found).

8. Examining the *x* axis of the figure suggests that, if average low temperatures shift upward by five degrees C, the tundra biome would experience a substantial decrease in geographic extent (and may cease to exist altogether); boreal forest might also experience a reduction in geographic extent.

9. The extensive radiation of *Drosophila* species in Hawaii suggests that the genus originated here. Hawaii is an isolated island system and its distance from any continents means that dispersal to such remote islands of ancestral species would likely be very rare. Other lines of evidence to suggest that the genus originated in Hawaii would be whether fossil ancestral taxa are found only in Hawaii or anywhere else in the world. If ancestral taxa are not known from any other continent, it's very likely that the genus originated and subsequently diversified in Hawaii (and then dispersed elsewhere).

CHAPTER 55

1. d 2. c 3. b 4. d 5. a 6. d

7. In both of these cases, human management strategies will be working against the organisms' intrinsic rates of increase. Populations of long-lived organisms with low reproductive rates grow slowly. Such organisms can be categorized as *K*-strategists, which tend to persist at or near the carrying capacity of the environment. They are adapted to predictable environments, and they tend to be more specialized in their resource use, and less tolerant of variation in resource quality, than other organisms. They produce few offspring, but each offspring has a high probability of surviving to adulthood. Recall that the number of births in a population tends to be highest when that population is well below its carrying capacity. For large, long-lived species that we wish to harvest, we should manage the population so that it is far enough below the carrying capacity to have a high birth rate. But some species (such as whales) reproduce so slowly that they cannot sustain any kind of substantial harvest rate. Short-lived organisms with high reproductive rates can be categorized as *r*-strategists. These organisms can generally use a wide variety of resources and tolerate a wide range of conditions, and they can produce large numbers of offspring when conditions are suitable. If we wish to decrease the numbers of a short-lived, rapidly reproducing pest species (such as rats), killing individuals will only increase the birth rate. A better approach is to reduce the species' resources (e.g., clean up garbage) in order to decrease the carrying capacity for the species. In each case, of course, managers need to understand the specific life history and population dynamics of the species they wish to manage.

8. Humans are subject to the same population dynamics that other species are. Resource abundance is a density-dependent population-regulating factor. Like the reindeer population on St. Matthew, the Irish population crashed when its food supply was diminished. Three kinds of changes in the rates of demographic events contributed to the decrease in population size. Recall that $N_1 = N_0 + (B - D) + (I - E)$. First, the emigration rate (E) increased. Second, the age at first reproduction, and thus generation time, increased, so the birth rate (B) decreased. In other words, the population's life history traits changed with changing environmental conditions. Finally, as a direct effect of the food shortage, the death rate (D) increased. A look at the social history of the Irish potato famine might tell us more about the roles of the Irish population's own growth before the famine and whether intraspecific competition (discussed further in Chapter 56) was involved in regulating the food supply—whether other populations monopolized resources and limited the access of the Irish people to those resources.

9. Those who support the view that biological controls should not be used under any circumstances might cite the example of the cane toad in Australia, which not only failed to control the cane beetles it was introduced to control but became a serious pest in its own right. They might note that many species introduced into new regions, where their normal predators and pathogens are absent, reach population densities much higher than those in their native ranges, and they might argue that there is no reason to think this generality would not apply to species introduced as biological control agents. Those who support the view that biological controls can be used safely and effectively might cite the example of the successful control of the cottony-cushion scale in California, which was brought under control within a year by the introduction of a predaceous ladybeetle and a parasitic fly. They might also argue that horror stories like that of the cane toad could be avoided by proper study of the ecology of the proposed biological control agents before they are introduced. Studies in test plots prior to release almost certainly would have revealed that Australian cane beetles stay high on the upper stalks of cane plants, out of reach of the toads, and studies of the toads' life histories might have revealed their generalized and voracious appetites. Strict requirements for extensive testing for specificity and efficacy prior to release can greatly reduce the risk that biological control agents themselves will become pests after they are introduced. But opponents of biological control might respond that, because natural systems are so complex, even careful study might fail to reveal the real risks of introducing a particular species into a new environment.

10. Corridors have to be defined in terms of specific organisms and their dispersal abilities. Corridors consist of habitat between patches through which the organisms of interest can move. An area that serves as a corridor for birds might not be effective as a corridor for small arthropods. On the other hand, the small arthropods would need less habitat area to maintain a viable population. Thus, designing a single study to determine the effects of corridors would be very difficult. A single experiment might be able to determine effects of corridors on animals that are similar in size and mobility but not if organisms differ widely in those attributes. To understand the effects of corridors in fragmented habitats, it is very important to consider multiple organisms because organisms interact within these habitats. Investigators in the Palenque National Park in Mexico discovered that birds are more likely to be recaptured in home forest patches connected by corridors to the patches where they were released than in home forest patches unconnected to the patches where they were released. However, their ability to navigate these corridors successfully depends on the presence of other species, including predators, and their ability to survive in home patches depends on the presence of other species, including prey species, as well. Designing a single study to determine the effects of corridors would be very difficult.

CHAPTER 56

1. a 2. c 3. c 4. d 5. e

6. The interactions among ants, cacti, and pollinators described in the Working With Data exercise represent a diversity of types of interactions. By fending off potential herbivores, the pugnacious ant bodyguards act as mutualists of the cactus, as do the bee species that visit the flowers and serve as pollinators. The five ant species all appear to use the extrafloral nectaries on this plant in similar ways and thus may be competitors for extrafloral nectar. Each type of interaction depends on the relative abundance and activities of the interacting species. Cactus plants that grow where there are no herbivorous insects may have no need of pugnacious bodyguards; under those circumstances, the ants might be considered parasites for removing extrafloral nectar without providing any defensive services. The mutualism between ants and plants could also break down if bee pollinators are scarce and the most aggressive ant bodyguard prevents any bees at all from pollinating the flowers.

7. Which pine trees are susceptible to mountain pine beetle attack could be determined by direct observation and by experimentation. Within infested stands, investigators may be able to identify individual pine trees that do not harbor beetle populations and characterize properties that may make them resistant to the beetles (e.g., ability to produce large quantities of resin). Conversely, trees with especially high beetle populations might have properties that make them particularly susceptible (having a history of surviving fire or lightning strike). Experiments can also be conducted under controlled conditions, in which investigators test various species to determine if beetles display a preference for particular species. The fact that this beetle has a symbiotic partner upon which it depends in order to feed on pine trees suggests that a novel strategy for managing the outbreak could be to identify a fungicide that kills the symbiont, thereby rendering the beetle incapable of colonizing and killing the trees. Although there are no such programs currently in use today, many researchers are exploring this dimension of interaction ecology to devise novel methods for pest management.

8. (a) By establishing a microbial population that excludes undesirable species, the poultry industry is applying the principle of competitive exclusion. The principle states that two or more species utilizing a limited resource in similar ways cannot coexist. In the broiler chicks given a culture of three species of bacteria, a microbial community was established in which introduced *Salmonella* could not compete.

   (b) Other ecological outcomes this experiment might have produced include ultimate domination of the gut flora by only one species of bacterium or coexistence of all four bacterial species. Whether these species coexist or whether one or more species goes extinct depends on the availability of resources in the chicken intestines.

   (c) The principle of competitive exclusion might be useful in tackling other problems involving a community of organisms growing under confined conditions. This ecological principle provides part of the rationale for the use of probiotics by humans to improve a variety of conditions. Probiotics are live organisms that are consumed in food for health benefits, which are thought to accrue by altering the microbial balance, inhibiting the growth of deleterious species. Probiotics are being investigation for treatment of intestinal inflammatory diseases, pathogen-related diarrhea, and infections of the urogenital tract.

9. If parasites and their hosts coevolve, then the phylogenetic relationships among parasites should reflect the phylogenetic relationships among the hosts. DNA analysis revealed that flamingoes are actually more closely related to grebes than they are to ducks and geese. This relationship leads to the prediction that lice on flamingoes should be more closely related to the lice on grebes than they are to the lice on ducks and geese. Modern methods of molecular analysis that can be used to determine relationships among bird lice and their hosts include constructing a DNA-based phylogeny of multiple species of waterbirds and their lice and then comparing the phylogenies of the hosts and parasites to see if they are congruent (see Chapter 22). In addition to acquiring parasites by shared ancestry between hosts and parasites, bird species may also acquire parasites by virtue of the fact that they share habitats and come in contact with other bird species, each of which has its own parasite fauna. Because flamingoes, ducks, grebes, and geese are all waterbirds, the possibility exists that flamingoes may have acquired some of its louse parasites by this process of host switching.

10. Among the requirements for a mutualistic pollination system is behavior by the pollinator that ensures it will visit more than one individual of the same plant species. Visiting more than one individual provides a pollinator with the opportunity to carry pollen from one plant individual to the receptive stigmatic surface of another individual of the same species. A pollinator that encounters a feeding deterrent that limits the amount of nectar it can imbibe in a single visit is more likely to continue foraging for nectar on another plant individual. The process of taking a larger number of smaller meals by seeking nectar from flowers of different plant individuals increases the likelihood that the pollinator will carry pollen from one individual to another. Too much nicotine in nectar, however, may reduce the likelihood of pollination if it deters future visits to conspecifics altogether or if it impairs the behavior of its pollinator (nicotine is a neurotoxin). Another factor limiting the amount of nicotine is the cost to the plant of biosynthesizing the compound; investing in increased amounts of nicotine may leave fewer resources to invest in producing flowers, seeds and fruits.

CHAPTER 57

1. a 2. a 3. b 4. d 5. a 6. c 7. e

8. The diversity of microbes in the human gut can be compared between individuals or across populations by using the same methods employed for comparing diversity of macroscopic communities. Diversity encompasses both the number of different species present, and richness, or abundances of individuals across species. To determine which microbes might be keystone species, selective antibiotics can be used to eliminate particular species and the effects of that elimination on community composition then monitored. Applying the methods for assessing diversity that were developed for macroscopic communities to assessing diversity for microbial communities is limited, however, by our ability to isolate, identify, and quantify all of the microbial species present. Although molecular methods of identifying microbial species has vastly expanded this capacity, there remain challenges in recognizing and categorizing the full expanse of microbial diversity.

9. According to the theory of island biogeography, the number of species on an island represents a balance between the rate at which species immigrate to and colonize the island and the rate at which resident species go locally extinct. With increasing distance from a source pool, the equilibrium number of species on an island decreases; with increasing size of an island, the species number increases. The pattern of hawk moth diversity documented by Beck and Hitching in the 113 islands of Thailand and mainland Malaysia conforms to several of the predictions of island biogeography. The continental source of colonists includes over 180 species. Borneo, a large island close to Thailand, has a larger number of species (between 113 and 135) than does New Guinea (with 46 to 90 species), which is roughly comparable in size and farther away from Thailand. Generally speaking, too, the prediction that larger islands support higher numbers of species is also upheld; Borneo has a larger number of species than the much smaller Philippines (between 46 and 112 species), even though the two places are about equidistant from Thailand.

10. The pattern documented by Marek Sammul, Lauri Oksanen, and M. Magi—that removal of one perennial species from plant communities resulted in an increase in biomass of its competitors in highly productive communities—has been documented in other communities. One hypothesis postulates that interspecific competition becomes more intense when productivity is very high. Goldenrod (*Solidago virgaurea*) is apparently a superior competitor that, when present in a community, can suppress other species. In less productive communities, competition is less intense, so release from competition with goldenrod does not result in increased growth of any remaining species. The results of this study parallel those of a long-term experiment at the Rothamsted Experiment Station in England, in which fertilizer added regularly to selected plots of land to increase their productivity resulted in a decline in the number of plant species compared with the other plots in the study that were unfertilized; in these less productive, unfertilized plots, species diversity remained essentially the same. An alternative hypothesis could be that goldenrod inhibits the growth of co-occurring plant species (as some colonizing species do in early stages of succession). This hypothesis could be tested directly by extracting root exudates of goldenrod and testing their ability to inhibit germination and growth of other species in the community.

11. Whether lampreys should be eliminated as damaging parasites of game fish, or encouraged as ecosystem engineers that create nesting sites that might increase the reproductive success of game fish depends on many factors. Some of the factors are ecological; it is important to quantify the impact of existing

sea lamprey populations on survivorship of game fish as well as to estimate the lamprey population size that does not influence survivorship. As well, the beneficial impact of nutrient enrichment and provisioning of nesting habitat should be measured. In addition, designing an ecologically sound lamprey management strategy will also require consideration of local cultural values; assessments of the economic value of the sport fishing industry to the local community and the aesthetic and cultural value placed by the local community on maintaining a more natural assemblage of fish species should be made and factored into management plans.

CHAPTER 58

1. e 2. d 3. c 4. c 5. a 6. b

7. The rate of turnover would be important to the recovery rate of a lake, and that would depend on its location. In Lake Washington, which is located in a temperate climate, turnover would occur every spring and fall. When sewage was flowing into the lake, the nutrients it contained would have led to eutrophication and thus to oxygen depletion in the bottom water. Once the flow of sewage stopped, however, biomass production would have decreased. There would have been fewer dead organisms to sink to the lake bottom, less accumulation of nutrients on the lake bottom, and less oxygen-consuming decomposition there. Fall and spring turnover would have brought the accumulated nutrients to the lake surface and oxygen to the bottom, improving conditions for organisms that were typical of the lake's preindustrial community. If the lake had been located in a climate where seasonal temperature changes were not great enough to cause turnover, the excess nutrients that had accumulated would have remained on the lake bottom, and eutrophic conditions would have persisted much longer. It's also possible that other conditions in the area might affect the lake's recovery time. Acid precipitation, for example, can affect the viability of freshwater organisms, and, if it were a problem in the region, it might slow the lake community's recovery.

8. A local effect might be nitrogen deposition. Coal—an organic fossil fuel—contains nitrogen, and its combustion would release nitrogen compounds (such as nitrogen dioxide, NO_2, and nitrous oxide, N_2O) through the smokestacks into the atmosphere. Some of this nitrogen would fall back to land in precipitation or as dry particles. The resulting increase of nitrogen in the soil would favor those plant species that are best adapted to take advantage of high nutrient levels, which would then outcompete other species. Thus the composition of the plant community would change and species diversity would be likely to decrease. Nitrogen deposition might also contribute to eutrophication in lakes, and emission of nitrogen into the atmosphere would contribute to smog.

   A regional effect might be acid precipitation. The combustion of fossil fuels releases NO_2 and sulfur dioxide (SO_2) into the atmosphere; both compounds react with water molecules in the atmosphere to form nitric acid (HNO_3) and sulfuric acid (H_2SO_4), respectively. These acids can travel hundreds of kilometers in the atmosphere, so their emission would affect ecosystems far from the smokestacks. Acid precipitation can damage the leaves of plants and reduce their rate of photosynthesis, and it can reduce fish and invertebrate species richness in freshwater lakes.

   A global effect would be climate change. The combustion of fossil fuels releases large amounts of CO_2, as well as lesser amounts of N_2O—both greenhouse gases. The presence of N_2O in the atmosphere also results in the production of trophospheric ozone; it, too, acts as a greenhouse gas as well as contributing to smog. The increasing concentrations of greenhouse gases in the atmosphere are already resulting in global climate warming. This climate change is having a number of worrisome effects on the global ecosystem, such as the shrinking of Arctic sea ice, rising sea level and potential coastal flooding, and profound changes in the abundances and distributions of species.

   The SO_2 released when coal is burned not only produces acid rain, but also contributes to global warming. Scrubbers remove the SO_2 but the scrubbing process contributes to pollution in another form—the process generates solid waste byproducts that contain sulfur, which must be deposited in a landfill, along with other solid waste products generated by burning coal.

9. Iron (Fe) is needed by organisms in only small amounts, but it is nevertheless an essential micronutrient. It is scarce in ocean waters because it is insoluble in oxygenated water, so that iron that enters the oceans sinks rapidly to the seafloor. The experiment described in the text demonstrated that iron is a limiting nutrient in the oceans: when the investigators added dissolved iron to surface waters in the equatorial Pacific Ocean, the large phytoplankton bloom that resulted was accompanied by an increase in the uptake of nitrate and carbon dioxide, showing that these nutrients had been available but underused. This experiment showed that adding iron to ocean waters increased photosynthesis, but to better understand the effects of iron fertilization, more such experiments would have to be carried out, still on an ecosystem scale, but over a longer time span. Investigators would have to observe the effects of the iron increase on the entire food web. The fertilized ecosystem would have to be compared with an unfertilized control ecosystem far enough away from the experimental one that the added iron, and its effects, would not reach it.

10. If the "cap and trade" system worked as intended, it could put the brakes on the ongoing increases quickly, holding emissions to the level that prevailed when the law went into effect. It might also be more acceptable to polluters than an outright ban on or regulation of emissions. And it might encourage some companies to invest more in cleaner technology, since doing so might give them credits to sell, or at least spare them having to buy credits. The drawbacks might include the likelihood that the government would have to set up a system to administer and enforce the law. They also include the fact that it is not easy to pass such a law (the United States, for example, has not succeeded in doing so). The biggest polluters might be reluctant to increase their costs of doing business by paying for credits and might thus be likely to lobby against such a law. From a different viewpoint, environmentalists might argue that a cap and trade system is an inadequate response to global warming—that we must not only stop increases in emissions, but decrease them dramatically. They might also argue that there is a moral hazard in allowing anyone to "pay to pollute"—that it might legitimize pollution.

11. No one hurricane—or even several—can be ascribed to global warming. Remember the difference between weather and climate, described in Chapter 54: "Weather is the short-term state of atmospheric conditions at a particular place and time, whereas climate refers to the average atmospheric conditions, and the extent of their variation, at a particular place over a longer time. In other words, climate is what you expect; weather is what you get." But neither does the observation that hurricanes have occurred for many centuries show that the climate has *not* changed. To address this question, we would have to compile temperature and hurricane data over long periods. First, we would have to show that the average temperatures of ocean waters are increasing over time—which has been done. Second, we would have to show that warmer water is correlated with more or stronger hurricanes—that there have been more and stronger hurricanes during years, or longer periods, when the water was warmer than in periods when it was cooler. Such a correlation would supply evidence that warming of the oceans is increasing hurricane frequency and intensity. It is more difficult to demonstrate that the warming of ocean waters is caused by increasing concentrations of greenhouse gases in the atmosphere, but most scientists believe that the evidence supports that claim.

CHAPTER 59

1. b 2. e 3. e 4. a 5. d 6. c 7. b

8. Conservation biologists and others who wish to preserve biodiversity are usually working with limited resources, so they often face hard choices. They might choose to focus their efforts in biodiversity hotspots and centers of imminent extinction, but within those areas, they would face many other choices. When we discussed the principles of island biogeography, we described the species–area relationship: large islands can support a larger equilibrium number of species than small islands. The same is true of "habitat islands." Protected areas often act as habitat islands, as many are surrounded by habitat that has been made unsuitable for many species by human activities. We must also consider edge effects, keeping in mind that not all of the area we protect will actually remain suitable habitat for communities and species of interest. Therefore, if we wish to preserve natural communities with their full diversity, the larger the preserved area, the better. However, if our concern is focused on one or a few endangered species, we may wish to preserve several separated areas of habitat; that way, if a disturbance or disease should wipe out the population in one area, the entire species will not become extinct. The area or areas we choose, however, must be large enough for the species to maintain a viable population in order to avoid loss of genetic variation. We might favor areas where corridors could be maintained to allow individuals to disperse from the protected area or areas and maintain other populations. But it is rare that a protected area can be designed based on these criteria alone; the plans are also dependent on the willingness of landowners, governments, and area residents to support the preservation of the area.

9. Some might argue that if the sheep constitute only a single population, but there are other populations of pumas, that the puma should be removed from the sheep's range. However, if the puma is a keystone predator in the region, removing it might have unforeseen negative consequences for other species in the community—recall the example of wolves in Yellowstone National Park, described in Chapter 57. If neither the sheep nor the puma is an introduced species—that is, if predator and prey had survived together for a long time before becoming threatened—it might be worth taking a look at what has changed. Have the sheep experienced a loss of habitat or resources, so that their populations are now too small to withstand the rate of predation that they once did? Those observations might suggest an alternative to suppressing the puma population: Could former sheep habitat be restored so that a larger sheep population could be supported, or as a last resort, could the sheep be bred in captivity and then introduced to a new, puma-free area?

10. To some extent, international organizations already have a triage system of sorts in in place. The International Union for the Conservation of Nature (IUCN), e.g., divides species in imminent danger of extinction in all or most of

their range as "endangered" or "critically endangered", differentiating them from those who are less likely to go extinct in the near future (and thus are classified as "vulnerable"). To some degree, this classification system can result in prioritization of rescue efforts. One problem with applying the triage system of World War I to species conservation, however, is that medical science is far more successful at predicting the certainty of death than ecological science is at predicting the certainty of extinction. After all, medical science is focused on one species, which has been the subject of intense scrutiny beginning in the earliest days of scientific research. The cost of erring in assigning certainty to extinction is the loss of an entire species—a unique combination of genes that, at least with current technology, can never be reconstructed.

11. Opinion as to the extent to which ethical and moral arguments should enter into discussions of protecting biodiversity varies widely. Your answer might take into consideration a wide range of cultural, historical, and economic factors, as well as the considerations brought to bear by modern scientific knowledge.

Glossary

A

A horizon *See* topsoil.

abiotic (a′ bye ah tick) [Gk. *a*: not + *bios*: life] Nonliving. (Contrast with biotic.)

abomasum The true stomach of a ruminant.

abortion Any termination of pregnancy, whether induced or natural (in which case it is called a spontaneous abortion), that occurs after a fertilized egg is successfully implanted in the uterus.

abscisic acid (ABA) (ab sighs′ ik) A plant growth substance with growth-inhibiting action. Causes stomata to close; involved in a plant's response to salt and drought stress.

abscission (ab sizh′ un) [L. *abscissio*: break off] The process by which leaves, petals, and fruits separate from a plant.

absorption (1) Of light: complete retention, without reflection or transmission. (2) Of water or other molecules: soaking up (taking in through pores or by diffusion).

absorption spectrum A graph of light absorption versus wavelength of light; shows how much light is absorbed at each wavelength.

absorptive heterotrophs Organisms (primarily fungi) that feed by **absorptive heterotrophy**, i.e., by secreting digestive enzymes into the environment to break down large food molecules, then absorbing the breakdown products.

absorptive state State in which food is in the gut and nutrients are being absorbed. (Contrast with postabsorptive state.)

abyssal zone (uh biss′ ul) [Gk. *abyssos*: bottomless] The deepest parts of the ocean.

accessory pigments Pigments that absorb light and transfer energy to chlorophylls for photosynthesis.

accessory sex organs Anatomical structures that allow transfer of sperm from male to female for internal fertilization. (contrast with primary sex organs.)

acclimation, acclimatization Acclimation refers to increased tolerance for environmental extremes (e.g., extreme cold) after prior exposure to them. Acclimatization refers to intrinsic seasonal adjustments in the "set points" of an animal's physiological functioning (e.g., metabolic rate).

acetyl coenzyme A (acetyl CoA) A compound that reacts with oxaloacetate to produce citrate at the beginning of the citric acid cycle; a key metabolic intermediate in the formation of many compounds.

acetylcholine (ACh) A neurotransmitter that carries information across vertebrate neuromuscular junctions and some other synapses. It is then broken down by the enzyme acetylcholinesterase (AChE).

acid [L. *acidus*: sharp, sour] A substance that can release a proton in solution. (Contrast with base.)

acid growth hypothesis The hypothesis that auxin increases proton pumping, thereby lowering the pH of the cell wall and activating enzymes that loosen polysaccharides. Proposed to explain auxin-induced cell expansion in plants.

acid precipitation Precipitation that has a lower pH than normal as a result of acid-forming precursor molecules introduced into the atmosphere by human activities.

acidic Having a pH below 7.0 (i.e., a hydrogen ion concentration greater than 10^{-7} molar). (Contrast with basic.)

acoelomate An animal that does not have a coelom.

acrosome (a′ krow soam) [Gk. *akros*: highest + *soma*: body] The structure at the forward tip of an animal sperm which is the first to fuse with the egg membrane and enter the egg cell.

ACTH *See* corticotropin.

actin [Gk. *aktis*: ray] A protein that makes up the cytoskeletal microfilaments in eukaryotic cells and is one of the two contractile proteins in muscle. See also myosin.

action potentials Generated by neurons, these are electrical signals that transmit information via waves of depolarization or hyperpolarization of the cell membrane.

action spectrum A graph of a biological process versus light wavelength; shows which wavelengths are involved in the process.

activation energy (E_a) The energy barrier that blocks the tendency for a chemical reaction to occur.

activator A transcription factor that stimulates transcription when it binds to a gene's promoter. (Contrast with repressor.)

active site The region on the surface of an enzyme or ribozyme where the substrate binds, and where catalysis occurs.

active transport The energy-dependent transport of a substance across a biological membrane against a concentration gradient—that is, from a region of low concentration (of that substance) to one of high concentration. (*See also* primary active transport, secondary active transport; contrast with facilitated diffusion, passive transport.)

adaptation (a dap tay′ shun) (1) In evolutionary biology, a particular structure, physiological process, or behavior that makes an organism better able to survive and reproduce. Also, the evolutionary process that leads to the development or persistence of such a trait. (2) In sensory neurophysiology, a sensory cell's loss of sensitivity as a result of repeated stimulation.

adaptive defenses One of the two general types of defenses against pathogens. Involves antibody proteins and other proteins that recognize, bind to, and aid in the destruction of specific viruses and bacteria. Present only in vertebrate animals. (Contrast with innate defenses.)

adaptive radiation A series of evolutionary events that results in an array (radiation) of related species that live in a variety of environments, differing in the characteristics each uses to exploit those environments.

additive growth Population growth in which a constant number of individuals is added to the population during successive time intervals. (Contrast with multiplicative growth.)

adenine (A) (a′ den een) A nitrogen-containing base found in nucleic acids, ATP, NAD, and other compounds.

adenosine triphosphate *See* ATP.

adrenal gland (a dree′ nal) [L. *ad*: toward + *renes*: kidneys] An endocrine gland located near the kidneys of vertebrates, consisting of two parts, the **adrenal cortex** and **adrenal medulla.**

adrenaline *See* epinephrine.

adrenergic receptors G protein-linked receptor proteins that bind to the hormones epinephrine and norepinephrine, triggering specific responses in the target cells.

adrenocorticotropic hormone (ACTH) *See* corticotropin.

adsorption Binding of a gas or a solute to the surface of a solid.

adventitious roots (ad ven ti′ shus) [L. *adventitius*: arriving from outside] Roots originating from the stem at ground level or below; typical of the fibrous root system of monocots.

aerenchyma In plants, parenchymal tissue containing air spaces.

aerobic (air oh′ bic) [Gk. *aer*: air + *bios*: life] In the presence of oxygen; requiring or using oxygen (as in **aerobic metabolism**). (Contrast with anaerobic.)

afferent (af′ ur unt) [L. *ad*: toward + *ferre*: to carry] Carrying to, as in neurons that carries impulses to the central nervous system (**afferent neurons**), or a blood vessel that carries blood to a structure. (Contrast with efferent.)

age structure The distribution of the individuals in a population across all age groups.

agonist A chemical substance (e.g., a neurotransmitter) that elicits a specific response in a cell or tissue. (Contrast with antagonist.)

air sacs Structures in the respiratory system of birds that receive inhaled air; they keep fresh air flowing unidirectionally through the lungs, but are not themselves gas exchange surfaces.

alcoholic fermentation *See* fermentation.

aldosterone (al dohs′ ter own) A steroid hormone produced in the adrenal cortex of mammals. Promotes secretion of potassium and reabsorption of sodium in the kidney.

aleurone layer In some seeds, a tissue that lies beneath the seed coat and surrounds the endosperm. Secretes digestive enzymes that break down macromolecules stored in the endosperm.

allantoic membrane In animal development, an outgrowth of extraembryonic endoderm plus adjacent mesoderm that forms the allantois, a saclike structure that stores metabolic wastes produced by the embryo.

allantois (al′ lun toh is) [Gk. *allant*: sausage] An extraembryonic membrane enclosing a sausage-shaped sac that stores the embryo's nitrogenous wastes.

allele (a leel′) [Gk. *allos*: other] The alternate form of a genetic character found at a given locus on a chromosome.

allele frequency The relative proportion of a particular allele in a specific population.

allergic reaction [Ger. *allergie*: altered] An overreaction of the immune system to amounts of an antigen that do not affect most people; often involves IgE antibodies.

allopatric speciation (al′ lo pat′ rick) [Gk. *allos*: other + *patria*: homeland] The formation of two species from one when reproductive isolation occurs because of the interposition of (or crossing of) a physical geographic barrier such as a river. Also called geographic speciation. (Contrast with sympatric speciation.)

allopolyploidy The possession of more than two chromosome sets that are derived from more than one species.

allosteric regulation (al lo steer′ ik) [Gk. *allos*: other + *stereos*: structure] Regulation of the activity of a protein (usually an enzyme) by the binding of an effector molecule to a site other than the active site.

α (alpha) helix A prevalent type of secondary protein structure; a right-handed spiral.

alternation of generations The succession of multicellular haploid and diploid phases in some sexually reproducing organisms, notably plants.

alternative splicing A process for generating different mature mRNAs from a single gene by splicing together different sets of exons during RNA processing.

altruistic Pertaining to behavior that benefits other individuals at a cost to the individual who performs it.

alveolus (al ve′ o lus) (plural: alveoli) [L. *alveus*: cavity] A small, baglike cavity, especially the blind sacs of the lung.

amensalism (a men′ sul ism) Interaction in which one animal is harmed and the other is unaffected. (Contrast with commensalism, mutualism.)

amine An organic compound containing an amino group (NH_2).

amine hormones Small hormone molecules synthesized from single amino acids (e.g., thyroxine and epinephrine).

amino acid An organic compound containing both NH_2 and COOH groups. Proteins are polymers of amino acids.

amino acid replacement A change in the nucleotide sequence that results in one amino acid being replaced by another.

ammonia NH_3, the most common nitrogenous waste.

ammonotelic (am moan′ o teel′ ic) [Gk. *telos*: end] Pertaining to an organism in which the final product of breakdown of nitrogen-containing compounds (primarily proteins) is **ammonia**. (Contrast with ureotelic, uricotelic.)

amnion (am′ nee on) The fluid-filled sac within which the embryos of reptiles (including birds) and mammals develop.

amniote egg A shelled egg surrounding four extraembryonic membranes and embryo-nourishing yolk. This evolutionary adaptation permitted mammals and reptiles to live and reproduce in drier environments than can most amphibians.

amphipathic (am′ fi path′ ic) [Gk. *amphi*: both + *pathos*: emotion] Of a molecule, having both hydrophilic and hydrophobic regions.

amplitude The magnitude of change over the course of a regular cycle.

amygdala A component of the limbic system that is involved in fear and the memory of fearful experiences.

amylase (am′ ill ase) An enzyme that catalyzes the hydrolysis of starch, usually to maltose or glucose.

anabolic reaction (an uh bah′ lik) [Gk. *ana*: upward + *ballein*: to throw] A synthetic reaction in which simple molecules are linked to form more complex ones; requires an input of energy and captures it in the chemical bonds that are formed. (Contrast with catabolic reaction.)

anaerobic (an ur row′ bic) [Gk. *an*: not + *aer*: air + *bios*: life] Occurring without the use of molecular oxygen, O_2. (Contrast with aerobic.)

anaphase (an′ a phase) [Gk. *ana*: upward] The stage in cell nuclear division at which the first separation of sister chromatids (or, in the first meiotic division, of paired homologs) occurs.

ancestral trait The trait originally present in the ancestor of a given group; may be retained or changed in the descendants of that ancestor.

androgen (an′ dro jen) Any of the several male sex steroids (most notably testosterone).

aneuploidy (an′ you ploy dee) A condition in which one or more chromosomes or pieces of chromosomes are either lacking or present in excess.

angiosperms Flowering plants; one of the two major groups of living seed plants. (*See also* gymnosperms.)

angiotensin (an′ jee oh ten′ sin) A peptide hormone that raises blood pressure by causing peripheral vessels to constrict. Also maintains glomerular filtration by constricting efferent vessels and stimulates thirst and the release of aldosterone.

angular gyrus A part of the human brain believed to be essential for integrating spoken and written language.

animal hemisphere The metabolically active upper portion of some animal eggs, zygotes, and embryos; does not contain the dense nutrient yolk. (Contrast with vegetal hemisphere.)

anion (an′ eye on) [Gk. *ana*: upward] A negatively charged ion. (Contrast with cation.)

annual A plant whose life cycle is completed in one growing season. (Contrast with biennial, perennial.)

antagonist A biochemical (e.g., a drug) that blocks the normal action of another biochemical substance.

antagonist interactions Interactions between two species in which one species benefits and the other is harmed. Includes predation, herbivory, and parasitism.

antenna system *See* light-harvesting complex.

anterior Toward or pertaining to the tip or headward region of the body axis. (Contrast with posterior.)

anterior pituitary The portion of the vertebrate pituitary gland that derives from gut epithelium. Produces trophic hormones.

anther (an′ thur) [Gk. *anthos*: flower] A pollen-bearing portion of the stamen of a flower.

antheridium (an′ thur id′ ee um) [Gk. *antheros*: blooming] The multicellular structure that produces the sperm in nonvascular land plants and ferns.

antibody One of the myriad proteins produced by the immune system that specifically binds to a foreign substance in blood or other tissue fluids and initiates its removal from the body.

anticodon The three nucleotides in transfer RNA that pair with a complementary triplet (a codon) in messenger RNA.

antidiuretic hormone (ADH) *See* vasopressin

antigen (an′ ti jun) Any substance that stimulates the production of an antibody or antibodies in the body of a vertebrate.

antigen-presenting cell In cellular immunity, a cell that ingests and digests an antigen, and then exposes fragments of that antigen to the outside of the cell, bound to proteins in the cell's plasma membrane.

antigenic determinant The specific region of an antigen that is recognized and bound by a specific antibody. Also called an epitope.

antiparallel Pertaining to molecular orientation in which a molecule or parts of a molecule have opposing directions.

antiporter A membrane transport protein that moves one substance in one direction and another in the opposite direction. (Contrast with symporter, uniporter.)

antisense RNA A single-stranded RNA molecule complementary to, and thus targeted against, an mRNA of interest to block its translation.

anus (a′ nus) An opening through which solid digestive wastes are expelled, located at the posterior end of a tubular gut.

aorta (a or′ tah) [Gk. *aorte:* aorta] The main trunk of the arteries leading to the systemic (as opposed to the pulmonary) circulation.

aortic body A chemosensor in the aorta that senses a decrease in blood supply or a dramatic decrease in partial pressure of oxygen in the blood.

aortic valve A one-way valve between the left ventricle of the heart and the aorta that prevents backflow of blood into the ventricle when it relaxes.

apex (a′ pecks) The tip or highest point of a structure, as of a growing stem or root.

aphasia a deficit in the ability to use or understand words.

aphotic zone In bodies of water (lakes and oceans), the region below the reach of light.

apical dominance In plants, inhibition by the apical bud of the growth of axillary buds.

apical hook A form taken by the stems of many eudicot seedlings that protects the delicate shoot apex while the stem grows through the soil.

apical meristem The meristem at the tip of a shoot or root; responsible for a plant's primary growth.

apomixis (ap oh mix′ is) [Gk. *apo*: away from + *mixis*: sexual intercourse] The asexual production of seeds.

apoplast (ap′ oh plast) In plants, the continuous meshwork of cell walls and extracellular spaces through which material can pass without crossing a plasma membrane. (Contrast with symplast.)

apoptosis (ap uh toh′ sis) A series of genetically programmed events leading to cell death.

aposematism Warning coloration; bright colors or striking patterns of toxic or toxix-mimic species that act as a warning to predators.

appendix In the human digestive system, the vestigial equivalent of the cecum (blind pouc), which serves no digestive function.

aquaporin A transport protein in plant and animal cell membranes through which water passes in osmosis.

aquatic (a kwa′ tic) [L. *aqua*: water] Pertaining to or living in water. (Contrast with marine, terrestrial.)

aqueous (a′ kwee us) Pertaining to water or a watery solution.

aquifer A large pool of groundwater.

archegonium (ar′ ke go′ nee um) The multicellular structure that produces eggs in nonvascular land plants, ferns, and gymnosperms.

archenteron (ark en′ ter on) [Gk. *archos*: first + *enteron*: bowel] The earliest primordial animal digestive tract.

area phylogeny Phylogenetic tree n which the names of the taxa are replaced with the names of the places where those taxa live or lived.

arms race A series of reciprocal adaptations between species involved in antagonistic interactions, in which adaptations that increase the fitness of a consumer species exert selection pressure on its resource species to counter the consumer's adaptation, and vice versa.

arteriole A small blood vessel arising from an artery that feeds blood into a capillary bed.

artery A muscular blood vessel carrying oxygenated blood away from the heart to other parts of the body. (Contrast with vein.)

artificial insemination An infertility treatment that involves the artificial introduction of sperm into the woman's reproductive tract.

artificial selection The selection by human plant and animal breeders of individuals with certain desirable traits.

ascus (ass′ cus) (plural: asci) [Gk. *askos*: bladder] In sac fungi, the club-shaped sporangium within which spores (ascospores) are produced by meiosis.

asexual reproduction Reproduction without sex.

assisted reproductive technologies (ARTs) Any of several procedures that remove unfertilized eggs from the ovary, combine them with sperm outside the body, and then place fertilized eggs or egg–sperm mixtures in the appropriate location in a female's reproductive tract for development.

association cortex In the vertebrate brain, the portion of the cortex involved in higher-order information processing, so named because it integrates, or associates, information from different sensory modalities and from memory.

associative learning A form of learning in which two unrelated stimuli become linked to the same response.

asthenosphere (ass thenn′ o sphere) [Gk. *asthenes*: weak] The viscous, malleable (changeable) layer of Earth's mantle. It is overlain by the solid lithospheric plates.

astrocyte [Gk. *astron*: star] A type of glial cell that contributes to the blood–brain barrier by surrounding the smallest, most permeable blood vessels in the brain.

atherosclerosis (ath′ er oh sklair oh′ sis) [Gk. *athero*: gruel, porridge + *skleros*: hard] A disease of the lining of the arteries characterized by fatty, cholesterol-rich deposits in the walls of the arteries. When fibroblasts infiltrate these deposits and calcium precipitates in them, the disease become arteriosclerosis, or "hardening of the arteries."

atom [Gk. *atomos*: indivisible] The smallest unit of a chemical element. Consists of a nucleus and one or more electrons.

atomic mass *See* atomic weight.

atomic number The number of protons in the nucleus of an atom; also equals the number of electrons around the neutral atom. Determines the chemical properties of the atom.

atomic weight The average of the mass numbers of a representative sample of atoms of an element, with all the isotopes in their normally occurring proportions. Also called atomic mass.

ATP (adenosine triphosphate) An energy-storage compound containing adenine, ribose, and three phosphate groups. When it is formed from ADP, useful energy is stored; when it is broken down (to ADP or AMP), energy is released to drive endergonic reactions.

ATP synthase An integral membrane protein that couples the transport of protons with the formation of ATP.

atrial natriuretic peptide A hormone released by the atrial muscle fibers of the heart when they are overly stretched, which decreases reabsorption of sodium by the kidney and thus blood volume.

atrioventricular node A modified node of cardiac muscle that organizes the action potentials that control contraction of the ventricles.

atrium (a′ tree um) [L. *atrium*: central hall] An internal chamber. In the hearts of vertebrates, the thin-walled chamber(s) entered by blood on its way to the ventricle(s). Also, the outer ear.

auditory system A sensory system that uses mechanoreceptors to convert pressure waves into receptor potentials; includes structures that gather sound waves, direct them to a sensory organ, and amplify their effect on the mechanoreceptors.

autocatalysis [Gk. *autos*: self + *kata*: to break down] A positive feedback process in which an activated enzyme acts on other inactive molecules of the same enzyme to activate them.

autocrine A chemical signal that binds to and affects the cell that makes it. (Contrast with paracrine.)

autoimmune diseases Diseases (e.g., rheumatoid arthritis) that result from failure of the immune system to distinguish between self and nonself, causing it to attack tissues in the organism's own body.

autoimmunity An immune response by an organism to its own molecules or cells.

autonomic nervous system (ANS) The portion of the peripheral nervous system that controls such involuntary functions as those of guts and glands. Also called the involuntary nervous system.

autophagy The programmed destruction of a cell's components.

autopolyploidy The possession of more than two entire chromosomes sets that are derived from a single species.

autoregulatory mechanisms In mammalian circulatory systems, local control of blood flow through capillary beds by constriction or dilation of incoming arterioles in response to local metabolite concentrations.

autosome Any chromosome (in a eukaryote) other than a sex chromosome.

autotroph (au' tow trowf') [Gk. *autos:* self + *trophe*: food] An organism that is capable of living exclusively on inorganic materials, water, and some energy source such as sunlight (photoautotrophs) or chemically reduced matter (see chemoautotrophs). (Contrast with heterotroph.)

auxin (awk' sin) [Gk. *auxein*: to grow] In plants, a substance (the most common being indoleacetic acid) that regulates growth and various aspects of development.

avirulence (*Avr*) genes Genes in a pathogen that may trigger defenses in plants. *See* gene-for-gene resistance.

Avogadro's number The number of atoms or molecules in a mole (weighed out in grams) of a substance, calculated to be 6.023×10^{23}.

axillary bud A bud that forms in the angle (axil) where a leaf meets a stem.

axon [Gk. axle] The process (branching structure) of a neuron that conducts action potentials away from the cell body. *See also* dendrites.

axon hillock The junction between an axon and the neuron's cell body; where action potentials are generated.

axon terminal The end portion of an axon, which passes action potentials to another cell. Axon terminals can form synapses and release neurotransmitter.

B

B cell A type of lymphocyte involved in the humoral immune response of vertebrates. Upon recognizing an antigenic determinant, a B cell develops into a plasma cell, which secretes an antibody. (Contrast with T cell.)

B horizon *See* subsoil.

bacillus (bah sil' us) [L: little rod] Any of various rod-shaped bacteria.

bacterial conjugation *See* conjugation.

bacteriophage (bak teer' ee o fayj) [Gk. *bakterion*: little rod + *phagein*: to eat] Any of a group of viruses that infect bacteria. Also called phage.

bacteroids Nitrogen-fixing organelles that develop from endosymbiotic bacteria.

bark All tissues external to the vascular cambium of a plant.

barometric pressure Atmospheric pressure; the total pressure of the gas mixture in air.

baroreceptor [Gk. *baros*: weight] A pressure-sensing cell or organ. Sometimes called a stress receptor.

basal metabolic rate (BMR) The minimum rate of energy turnover in an awake (but resting) bird or mammal that is not expending energy for thermoregulation.

base (1) A substance that can accept a hydrogen ion in solution. (Contrast with acid.) (2) In nucleic acids, the purine or pyrimidine that is attached to each sugar in the sugar–phosphate backbone.

base pair (bp) In double-stranded DNA, a pair of nucleotides formed by the complementary base pairing of a purine on one strand and a pyrimidine on the other. (*See* complementary base pairing.)

basic Having a pH greater than 7.0 (i.e., having a hydrogen ion concentration lower than 10^{-7} molar). (Contrast with acidic.)

basidioma (plural: basiomata) A fruiting structure produced by club fungi.

basidium (bass id' ee yum) In club fungi, the characteristic sporangium in which four **basidiospores** are formed by meiosis and then borne externally before being shed.

basilar membrane A membrane in the human inner ear whose flexion in response to sound waves activates hair cells; flexes at different locations in response to different pitch.

Batesian mimicry The convergence in appearance of an edible species (mimic) with an unpalatable species (model).

behavioral ecology An evolutionary approach to the study of animal behavior that studies how behaviors are adaptive in different environmental conditions.

behaviorism One of two classical approaches to the study of proximate causes of animal behavior, derived from the discoveries of Ivan Pavlov and focused on laboratory studies. (Compare with ethology.)

benefit Improvement in survival and reproductive success resulting from performing a behavior or having a trait.

benthic zone [Gk. *benthos*: bottom] The bottom of the ocean.

β (beta) pleated sheet A type of protein secondary structure; results from hydrogen bonding between polypeptide regions running antiparallel to each other.

biased gene conversion A mechanism of concerted evolution in which a DNA repair system appears biased in favor of using particular nucleotide sequences as templates for repair, resulting in the rapid spread of the favored sequence across all copies of the gene. (*See* concerted evolution.)

bicarbonate ion Ion (HCO_3^-) resulting from dissociation of carbonic acid in water; important in pH regulation and carbon dioxide (CO_2) transport.

biennial A plant whose life cycle includes vegetative growth in the first year and flowering and senescence in the second year. (Contrast with annual, perennial.)

bilateral symmetry The condition in which only the right and left sides of an organism, divided by a single plane through the midline, are mirror images of each other.

bilayer A structure that is two layers in thickness. In biology, most often refers to the phospholipid bilayer of membranes. (*See* phospholipid bilayer.)

bile A secretion of the liver made up of bile salts synthesized from cholesterol, various phospholipids, and bilirubin (the breakdown product of hemoglobin). Emulsifies fats in the small intestine.

binary fission Reproduction of a prokaryote by division of a cell into two comparable progeny cells.

binocular vision Overlapping visual fields of an animal's two eyes; allows the animal to see in three dimensions.

binomial nomenclature A taxonomic naming system in which each species is given a binomial (Gk.: two names), a genus name followed by a species name.

biodiversity hotspots Regions identified by conservation biologists as being particularly in need of protection because they harbor great species richness and endemism (i.e., large numbers of species, many of which are found nowhere else).

biofilm A community of microorganisms embedded in a polysaccharide matrix, forming a highly resistant coating on almost any moist surface.

biogeochemical cycle Movement of inorganic elements such as nitrogen, phosphorus, and carbon through living organisms and the physical environment.

biogeographic region One of several defined, continental-scale regions of Earth, each of which has a biota distinct from that of the others. (Contrast with biome.)

biogeography The scientific study of the patterns of distribution of populations, species, and ecological communities across Earth.

bioinformatics The use of computers and/or mathematics to analyze complex biological information, such as DNA sequences.

biological control The use of natural enemies (predators, parasites, or pathogens) to reduce the population density of an economically damaging (pest) species.

biological species concept The definition of a species as a group of actually or potentially interbreeding natural populations that are reproductively isolated from other such groups. (Contrast with lineage species concept; morphological species concept.)

biology [Gk. *bios*: life + *logos*: study] The scientific study of living things.

bioluminescence The production of light by biochemical processes in an organism.

biome (bye' ome) A major division of the ecological communities of Earth, characterized primarily by distinctive vegetation. A given biogeographic region contains many different biomes.

bioremediation The use by humans of other organisms to remove contaminants from the environment.

biosphere (bye' oh sphere) All regions of Earth (terrestrial and aquatic) and Earth's atmosphere in which organisms can live.

biota (bye oh' tah) All of the organisms—animals, plants, fungi, and microorganisms—found in a given area. (Contrast with flora, fauna.)

biotechnology The use of living cells or organisms to produce materials useful to humans.

biotic (bye ah' tick) [Gk. *bios*: life] Alive. (Contrast with abiotic.)

biotic interchange The mixing of biotas previously separated by physical, climatic, or other barriers, for example when two formerly separated land masses fuse.

blastocoel (blass' toe seal) [Gk. *blastos*: sprout + *koilos*: hollow] The central, hollow cavity of a blastula.

blastocyst (blass' toe cist) An early embryo formed by the first divisions of the fertilized egg (zygote). In mammals, a hollow ball of cells.

blastodisc (blass' toe disk) An embryo that forms as a disk of cells on the surface of a large yolk mass; comparable to a blastula, but occurring in animals such as birds and reptiles, in which the massive yolk restricts complete cleavage.

blastomere Any of the cells produced by the early divisions of a fertilized animal egg.

blastopore The opening created by the invagination of the vegetal pole during gastrulation of animal embryos.

blastula (blass' chu luh) An early stage of the animal embryo; in many species, a hollow sphere of cells surrounding a central cavity, the blastocoel. (Contrast with blastodisc.)

block to polyspermy Any of several responses to entry of a sperm into an egg that prevent more than one sperm from entering the egg.

blood A fluid connective tissue that is pumped throughout the body. A component of the circulatory system, blood transports gases such as oxygen and carbon dioxide as well as other essential elements.

blood clotting A cascade of events involving platelets and circulating proteins (clotting factors) that seals damaged blood vessels.

blood–brain barrier The selective impermeability of blood vessels in the brain that prevents most chemicals from diffusing from the blood into the brain.

blue-light receptors Pigments in plants that absorb blue light (400–500 nm). These pigments mediate many plant responses including photo-tropism, stomatal movements, and expression of some genes.

body plan The general structure of an animal, the arrangement of its organ systems, and the integrated functioning of its parts.

Bohr effect A shift in the O_2 binding curve of hemoglobin in response to excess H^+ ions such that the hemoglobin releases more O_2 in tissues where pH is low.

Bohr model A model for atomic structure that depicts the atom as largely empty space, with a central nucleus surrounded by electrons in orbits, or electron shells, at various distances from the nucleus.

bond *See* chemical bond.

bone A rigid component of vertebrate skeletal systems that contains an extracellular matrix of insoluble calcium phosphate crystals as well as collagen fibers.

Bowman's capsule An elaboration of the renal tubule, composed of podocytes, that surrounds and collects the filtrate from the glomerulus.

brain The centralized integrative center of a nervous system.

brainstem The portion of the vertebrate brain between the spinal cord and the forebrain, made up of the medulla, pons, and midbrain.

brassinosteroids Plant steroid hormones that mediate light effects promoting the elongation of stems and pollen tubes.

Broca's area A portion of the human brain essential for speech. Located in the frontal lobe just in front of the primary motor cortex.

bronchioles The smallest airways in a vertebrate lung, branching off the bronchi.

bronchus (plural: bronchi) The major airway(s) branching off the trachea into the vertebrate lung.

brown fat In mammals, fat tissue that is specialized to produce heat. It has many mitochondria and capillaries, and a protein that uncouples oxidative phosphorylation.

budding Asexual reproduction in which a more or less complete new organism grows from the body of the parent organism, eventually detaching itself.

buffer A substance that can transiently accept or release hydrogen ions and thereby resist changes in pH.

bulbourethral glands Secretory structures of the human male reproductive system that produce a small volume of an alkaline, mucoid secretion that helps neutralize acidity in the urethra and lubricate it to facilitate the passage of semen.

bulk flow The movement of a solution from a region of higher pressure potential to a region of lower pressure potential.

bundle of His Fibers of modified cardiac muscle that conduct action potentials from the atria to the ventricular muscle mass.

bundle sheath cell Part of a tissue that surrounds the veins of plants.

C

C horizon *See* parent rock.

C_3 plants Plants that produce 3PG as the first stable product of carbon fixation in photosynthesis and use ribulose bisphosphate as a CO_2 receptor.

C_4 plants Plants that produce oxaloacetate as the first stable product of carbon fixation in photosynthesis and use phosphoenolpyruvate as CO_2 acceptor. C_4 plants also perform the reactions of C_3 photosynthesis.

calcitonin Hormone produced by the thyroid gland; lowers blood calcium and promotes bone formation. (Contrast with parathyroid hormone.)

calcitriol A hormone derived from vitamin D whose actions include stimulating the cells of the digestive tract to absorb calcium from ingested food.

calorie [L. *calor*: heat] The amount of heat required to raise the temperature of 1 gram of water by 1°C. Physiologists commonly use the kilocalorie (kcal) as a unit of measure (1 kcal = 1,000 calories). Nutritionists also use the kilocalorie, but refer to it as the **Calorie** (capital C).

Calvin cycle The stage of photosynthesis in which CO_2 reacts with RuBP to form 3PG, 3PG is reduced to a sugar, and RuBP is regenerated, while other products are released to the rest of the plant. Also known as the Calvin–Benson cycle.

calyx (kay' licks) [Gk. *kalyx*: cup] All of the sepals of a flower, collectively.

CAM *See* crassulacean acid metabolism.

Cambrian explosion The rapid diversification of multicellular life that took place during the Cambrian period.

cAMP (cyclic AMP) A compound formed from ATP that acts as a second messenger.

cancellous bone A type of bone with numerous internal cavities that make it appear spongy, although it is rigid. (Contrast with compact bone.)

canopy The leaf-bearing part of a tree. Collectively, the aggregate of the leaves and branches of the larger woody plants of an ecological community.

capillaries [L. *capillaris*: hair] Very small tubes, especially the smallest blood-carrying vessels of animals between the termination of the arteries and the beginnings of the veins. **Capillary beds** are networks of capillaries where materials are exchanged between the blood and the interstitial fluid.

capsid The outer shell of a virus that encloses its nucleic acid.

carbohydrates Organic compounds containing carbon, hydrogen, and oxygen in the ratio 1:2:1 (i.e., with the general formula $C_nH_{2n}O_n$). Common examples are sugars, starch, and cellulose.

carbon skeleton The chains or rings of carbon atoms that form the structural basis of organic molecules. Other atoms or functional groups are attached to the carbon atoms.

carbon-fixation reactions The phase of photosynthesis in which chemical energy captured in the light reactions is used to drive the reduction of CO_2 to form carbohydrates.

carboxylase An enzyme that catalyzes the addition of carboxyl functional groups (O=C—OH) to a substrate.

cardiac cycle Contraction of the two atria of the heart, followed by contraction of the two ventricles and then relaxation.

cardiac muscle A type of muscle tissue that makes up, and is responsible for the beating of, the heart. Characterized by branching cells with single nuclei and a striated (striped) appearance. (Contrast with smooth muscle, skeletal muscle.)

cardiovascular system [Gk. *kardia*: heart + L. *vasculum*: small vessel] The heart, blood, and vessels are of a circulatory system.

carnivore [L. *carn*: flesh + *vorare*: to devour] An organism that eats animal tissues. (Contrast with detritivore, frugivore, herbivore, omnivore.)

carotenoid (ka rah′ tuh noid) A yellow, orange, or red lipid pigment commonly found as an accessory pigment in photosynthesis; also found in fungi.

carotid body A chemosensor in the carotid artery that senses a decrease in blood supply or a dramatic decrease in partial pressure of oxygen in the blood.

carpel (kar′ pel) [Gk. *karpos*: fruit] The organ of the flower that contains one or more ovules.

carrier (1) In facilitated diffusion, a membrane protein that binds a specific molecule and transports it through the membrane. (2) In respiratory and photosynthetic electron transport, a participating substance such as NAD that exists in both oxidized and reduced forms. (3) In genetics, a person heterozygous for a recessive trait.

carrying capacity (*K*) The maximum number of individuals in a population (i.e., maximum population size) that can be supported by the resources present in a given environment.

cartilage In vertebrates, a tough connective tissue found in joints, the outer ear, and elsewhere. Forms the entire skeleton in some animal groups.

cartilage bone A type of bone that begins its development as a cartilaginous structure resembling the future mature bone, then gradually hardens into mature bone. (Contrast with membranous bone.)

Casparian strip A band of cell wall containing suberin and lignin, found in the endodermis. Restricts the movement of water across the endodermis.

caspase One of a group of proteases that catalyze cleavage of target proteins and are active in apoptosis.

catabolic reaction (kat uh bah′ lik) [Gk. *kata*: to break down + *ballein*: to throw] A synthetic reaction in which complex molecules are broken down into simpler ones and energy is released. (Contrast with anabolic reaction.)

catabolite repression In the presence of abundant glucose, the diminished synthesis of catabolic enzymes for other energy sources.

catalyst (kat′ a list) [Gk. *kata*: to break down] A chemical substance that accelerates a reaction without itself being consumed in the overall course of the reaction. Catalysts lower the activation energy of a reaction. Enzymes are biological catalysts.

cation (cat′ eye on) An ion with one or more positive charges. (Contrast with anion.)

caudal [L. *cauda*: tail] Pertaining to the tail, or to the posterior part of the body.

cDNA *See* complementary DNA.

cDNA library A collection of complementary DNAs derived from mRNAs of a particular tissue at a particular time in the life cycle of an organism.

cecum (see′ cum) [L. blind] A blind branch off the large intestine. In many nonruminant mammals, the cecum contains a colony of microorganisms that contribute to the digestion of food.

cell The simplest structural unit of a living organism. In multicellular organisms, many individual cells serve as the building blocks of tissues and organs.

cell adhesion molecules (CAMs) Molecules on animal cell surfaces that affect the selective association of cells into tissues during development of the embryo. Also a component of desmosomes.

cell cycle The stages through which a cell passes between one mitotic division and the next. Includes all stages of interphase and mitosis. (*See* mitosis.)

cell cycle checkpoints Points of transition between different phases of the cell cycle, which are regulated by cyclins and cyclin-dependent kinases (Cdks).

cell division The reproduction of a cell to produce two new cells. In eukaryotes, this process involves nuclear division (mitosis) and cytoplasmic division (cytokinesis).

cell fate The type of cell that an undifferentiated cell in an embryo will become in the adult.

cell junctions Specialized structures associated with the plasma membranes of epithelial cells. Some contribute to cell adhesion, others to intercellular communication.

cell potency In multicellular organisms, an undifferentiated cell's potential to become a cell of a specific type. (*See* multipotent; pluripotent; totipotent.)

cell recognition Binding of cells to one another mediated by membrane proteins or carbohydrates.

cell theory States that cells are the basic structural and physiological units of all living organisms, and that all cells come from preexisting cells.

cell wall A relatively rigid structure that encloses cells of plants, fungi, many protists, and most prokaryotes, and which gives these cells their shape and limits their expansion in hypotonic media.

cellular immune response Immune system response mediated by T cells and directed against parasites, fungi, intracellular viruses, and foreign tissues (grafts). (Contrast with humoral immune response.)

cellular respiration The catabolic pathways by which electrons are removed from various molecules and passed through intermediate electron carriers to O_2, generating H_2O and releasing energy.

cellular specialization In multicellular organisms, the division of labor such that different cell types become responsible for different functions (e.g., reproduction or digestion) within the organism.

cellulose (sell′ you lowss) A straight-chain polymer of glucose molecules, used by plants as a structural supporting material.

central dogma The premise that information flows from DNA to RNA to polypeptide (protein).

central nervous system (CNS) That portion of the nervous system that is the site of most information processing, storage, and retrieval; in vertebrates, the brain and spinal cord. (Contrast with peripheral nervous system.)

central vacuole In plant cells, a large organelle that stores the waste products of metabolism and maintains turgor.

centrifuge [L. *centrum*: center + *fugere*: to flee] A laboratory device in which a sample is spun around a central axis at high speed. Used to separate suspended materials of different densities.

centriole (sen′ tree ole) A paired organelle that helps organize the microtubules in animal and protist cells during nuclear division.

centromere (sen′ tro meer) [Gk. *centron*: center + *meros*: part] The region where sister chromatids join.

centrosome (sen′ tro soam) The major microtubule organizing center of an animal cell.

cephalization (sef ah luh zay′ shun) [Gk. *kephale*: head] The evolutionary trend toward increasing concentration of brain and sensory organs at the anterior end of the animal.

cerebellum (sair uh bell′ um) [L. diminutive of *cerebrum*, brain] The brain region that controls muscular coordination; located at the anterior end of the hindbrain.

cerebral cortex The thin layer of gray matter (neuronal cell bodies) that overlies the cerebrum.

cerebrum (su ree′ brum) [L. brain] The dorsal anterior portion of the forebrain, making up the largest part of the brain of mammals; the chief coordination center of the nervous system and the major information-processing areas of the vertebrate brain consists of two **cerebral hemispheres**.

cervix (sir′ vix) [L. neck] The opening of the uterus into the vagina.

cGMP (cyclic guanosine monophosphate) An intracellular messenger that is part of signal transmission pathways involving G proteins. (*See* G protein.)

channel protein An integral membrane protein that forms an aqueous passageway across the membrane in which it is inserted and through which specific solutes may pass.

chaperone A protein that guards other proteins by counteracting molecular interactions that threaten their three-dimensional structure.

character In genetics, an observable feature, such as eye color. (Contrast with trait.)

character displacement An evolutionary phenomenon in which species that compete for the same resources within the same territory tend to diverge in morphology and/or behavior.

chemical bond An attractive force stably linking two atoms.

chemical equilibrium *See* equilibrium

chemical evolution The theory that life originated through the chemical transformation of inanimate substances.

chemical reaction The change in the composition or distribution of atoms of a substance with consequent alterations in properties.

chemical synapse Neural junction at which neurotransmitter molecules released from a presynaptic cell induce changes in a postsynaptic cell. (Contrast with electrical synapse.)

chemically gated channel A type of membrane channel that opens or closes depending on the presence or absence of a specific molecule that binds either to the channel protein itself or to a separate receptor that alters the three-dimensional shape of the channel protein.

chemiosmosis Formation of ATP in mitochondria and chloroplasts, resulting from a pumping of protons across a membrane (against a gradient of electrical charge and of pH), followed by the return of the protons through a protein channel with ATP synthase activity.

chemoautotroph Organisms that obtain energy by oxidizing inorganic substances, using some of that energy to fix carbon. Also known as chemolithotrophs.

chemoheterotroph An organism that must obtain both carbon and energy from organic substances. (Contrast with chemoautotroph, photoautotroph, photoheterotroph.)

chemoreceptor A sensory receptor cell that senses specific molecules (such as odorant molecules or pheromones) in the environment.

chiasma (kie az′ muh) (plural: chiasmata) [Gk. cross] An X-shaped connection between paired homologous chromosomes in prophase I of meiosis. A chiasma is the visible manifestation of crossing over between homologous chromosomes.

chief cells One of three types of secretory cell found in the gastric pits of the stomach wall. Chief cells secrete the protein-digesting enzyme pepsin. (*See* mucosal epithelium; parietal cells.)

chitin (kye′ tin) [Gk. *kiton*: tunic] The characteristic tough but flexible organic component of the exoskeleton of arthropods, consisting of a complex, nitrogen-containing polysaccharide. Also found in cell walls of fungi.

chlorophyll (klor′ o fill) [Gk. *kloros*: green + *phyllon*: leaf] Any of several green pigments associated with chloroplasts or with certain bacterial membranes; responsible for trapping light energy for photosynthesis.

chloroplast [Gk. *kloros*: green + *plast*: a particle] An organelle bounded by a double membrane containing the enzymes and pigments that perform photosynthesis. Chloroplasts occur only in eukaryotes.

choanocyte (ko′ an uh site) The collared, flagellated feeding cells of sponges.

cholecystokinin (CCK) (ko′ luh sis tuh kai′ nin) A hormone produced and released by the lining of the duodenum when it is stimulated by undigested fats and proteins. It stimulates the gallbladder to release bile and slows stomach activity.

chorion (kor′ ee on) [Gk. *khorion*: afterbirth] The outermost of the membranes protecting mammal, bird, and reptile embryos; in mammals it forms part of the placenta.

chromatid (kro′ ma tid) A newly replicated chromosome, from the time molecular duplication occurs until the time the centromeres separate (during anaphase of mitosis or of meiosis II).

chromatin The nucleic acid–protein complex that makes up eukaryotic chromosomes.

chromatin remodeling A mechanism for epigenetic gene regulation by the alteration of chromatin structure.

chromosomal mutation Loss of or changes in position/direction of a DNA segment on a chromosome.

chromosome (krome′ o sowm) [Gk. *kroma*: color + *soma*: body] In bacteria and viruses, the DNA molecule that contains most or all of the genetic information of the cell or virus. In eukaryotes, a structure composed of DNA and proteins that bears part of the genetic information of the cell.

chylomicron (ky low my′ cron) Particles of lipid coated with protein, produced in the gut from dietary fats and secreted into the extracellular fluids.

chyme (kime) [Gk. *kymus*: juice] Created in the stomach; a mixture of ingested food with the digestive juices secreted by the salivary glands and the stomach lining.

cilia (sil′ ee ah) (singular: cilium) [L. eyelashes] Hairlike organelle used for locomotion by many unicellular organisms and for moving water and mucus by many multicellular organisms. Generally shorter than flagella.

circadian rhythm (sir kade′ ee an) [L. *circa*: approximately + *dies*: day] A rhythm of growth or activity that recurs about every 24 hours.

circannual rhythm [L. *circa*: + *annus*: year) A rhythm of growth or activity that recurs on a yearly basis.

circulatory system A physiological system consisting of a muscular pump (heart), a fluid (blood or hemolymph), and a series of conduits (blood vessels) that transports materials around the body.

11-*cis*-retinal The nonprotein, light-absorbing component of the visual pigment rhodopsin. (*See* rhodopsin.)

***cis-trans* isomers** In molecules with a double bond (typically between two carbon items), identifies on which side of the double bond similar atoms or functional groups are found. If they are on the same side, the molecule is a *cis* isomer; in a *trans* isomer, similar atoms are on opposite sides of the double bond. (*See* isomer.)

citric acid cycle In cellular respiration, a set of chemical reactions whereby acetyl CoA is oxidized to carbon dioxide and hydrogen atoms are stored as NADH and $FADH_2$. Also called the Krebs cycle.

clade [Gk. *klados*: branch] A monophyletic group made up of an ancestor and all of its descendants.

class I MHC molecules Cell surface proteins that participate in the cellular immune response directed against virus-infected cells.

class II MHC molecules Cell surface proteins that participate in the cell–cell interactions (of T-helper cells, macrophages, and B cells) of the humoral immune response.

class switching Occurs when a B cell changes the immunoglobulin class it synthesizes (e.g., a B cell making IgM switches to making IgG).

cleavage The first few cell divisions of an animal zygote. *See also* complete cleavage, incomplete cleavage.

climate The long-term average atmospheric conditions (temperature, precipitation, humidity, wind direction and velocity) found in a region. (Contrast with weather.)

climax community The final stage of succession; a community that is capable of perpetuating itself under local climatic and soil conditions and persists for a relatively long time.

clinal variation [Gk. *klinein*: to lean] Gradual change in the phenotype of a species over a geographic gradient.

cloaca The opening through which both urinary wastes and digestive wastes are expelled in most amphibians and in reptiles (including birds).

clonal deletion Inactivation or destruction of lymphocyte clones that would produce immune reactions against the animal's own body.

clonal lineages Asexually reproduced groups of nearly identical organisms.

clonal selection Mechanism by which exposure to antigen results in the activation of selected T- or B-cell clones, resulting in an immune response.

clone [Gk. *klon*: twig, shoot] (1) Genetically identical cells or organisms produced from a common ancestor by asexual means. (2) To produce many identical copies of a DNA sequence by its introduction into, and subsequent asexual reproduction of, a cell or organism.

closed circulatory system Circulatory system in which the circulating fluid is contained within a continuous system of vessels. (Contrast with open circulatory system.)

clumped dispersion pattern *See* dispersion

CO (CONSTANS) Gene coding for a transcription factor that activates the synthesis of florigen (FT); involved in the induction of flowering.

co-repressor In the regulation of bacterial operons, a molecule that binds to the repressor, causing it to change shape and bind to the operator, thereby inhibiting transcription.

coastal zone The marine life zone that extends from the shoreline to the edge of the continental shelf. Characterized by relatively shallow, well-oxygenated water and relatively stable temperatures and salinities.

coccus (kock' us) (plural: cocci) [Gk. *kokkos*: berry, pit] Any of various spherical or spheroidal bacteria.

cochlea (kock' lee uh) [Gk. *kokhlos*: snail] A spiral tube in the inner ear of vertebrates; it contains the sensory cells involved in hearing.

codominance A condition in which two alleles at a locus produce different phenotypic effects and both effects appear in heterozygotes.

codon Three nucleotides in messenger RNA that direct the placement of a particular amino acid into a polypeptide chain. (Contrast with anticodon.)

coelom (see' loam) [Gk. *koiloma*: cavity] An animal body cavity, enclosed by muscular mesoderm and lined with a mesodermal layer called peritoneum that also surrounds the internal organs.

coelomate Possessing a coelom.

coenocytic (seen' a sit ik) [Gk. *koinos*: common + *kytos*: container] Referring to the condition, found in some fungal hyphae, of "cells" containing many nuclei but enclosed by a single plasma membrane. Results from nuclear division without cytokinesis.

coenzyme A nonprotein organic molecule that plays a role in catalysis by an enzyme.

coenzyme A (CoA) A coenzyme used in various biochemical reactions as a carrier of acyl groups.

coevolution Evolutionary processes in which an adaptation in one species leads to the evolution of an adaptation in a species with which it interacts; also known as reciprocal adaptation.

cofactor An inorganic ion that is weakly bound to an enzyme and required for its activity.

cohesin A protein involved in binding chromatids together.

cohesion The tendency of molecules (or any substances) to stick together.

cohort (co' hort) [L. *cohors*: company of soldiers] A group of similar-aged organisms.

cold-hardening A process by which plants can acclimate to cooler temperatures; requires repeated exposure to cool temperatures over many days.

coleoptile A sheath that surrounds and protects the shoot apical meristem and young primary leaves of a grass seedling as they move through the soil.

collagen [Gk. *kolla*: glue] A fibrous protein found extensively in bone and connective tissue.

collecting duct In vertebrates, a tubule that receives urine produced in the nephrons of the kidney and delivers that fluid to the ureter for excretion.

collenchyma (cull eng' kyma) [Gk. *kolla*: glue + *enchyma*: infusion] A type of plant cell, living at functional maturity, which lends flexible support by virtue of primary cell walls thickened at the corners. (Contrast with parenchyma, sclerenchyma.)

colon [Gk. *kolon*] The portion of the gut between the small intestine and the anus. Also called the large intestine.

commensalism [L. *com*: together + *mensa*: table] A type of interaction between species in which one participant benefits while the other is unaffected.

communication A signal from one organism (or cell) that alters the functioning or behavior of another organism (or cell).

community Any ecologically integrated group of species of microorganisms, plants, and animals inhabiting a given area.

compact bone A type of bone with a solid, hard structure. (Contrast with cancellous bone.)

companion cell In angiosperms, a specialized cell found adjacent to a sieve tube element.

comparative experiment Experimental design in which data from various unmanipulated samples or populations are compared, but in which variables are not controlled or even necessarily identified. (Contrast with controlled experiment.)

comparative genomics Computer-aided comparison of DNA sequences between different organisms to reveal genes with related functions.

competition In ecology, use of the same resource by two or more species when the resource is present in insufficient supply for the combined needs of the species.

competitive exclusion A result of competition between species for resources, in which one species completely eliminates the other from a given habitat.

competitive inhibitor A nonsubstrate that binds to the active site of an enzyme and thereby inhibits binding of its substrate. (Contrast with noncompetitive inhibitor.)

complement system A group of eleven proteins that play a role in some reactions of the immune system. The complement proteins are not immunoglobulins.

complementary base pairing The AT (or AU), TA (or UA), CG, and GC pairing of bases in double-stranded DNA, in transcription, and between tRNA and mRNA.

complementary DNA (cDNA) DNA formed by reverse transcriptase acting with an RNA template; essential intermediate in the reproduction of retroviruses; used as a tool in recombinant DNA technology; lacks introns.

complete cleavage Pattern of cleavage that occurs in eggs that have little yolk. Early cleavage furrows divide the egg completely and the blastomeres are of similar size. (Contrast with incomplete cleavage.)

complete metamorphosis A change of state during the life cycle of an organism in which the body is almost completely rebuilt to produce an individual with a very different body form. Characteristic of insects such as butterflies, moths, beetles, ants, wasps, and flies.

complex ions Groups of covalently bonded atoms that carry an electric charge (e.g., NH_4^+, the ammonium ion).

complex life cycle In reference to parasitic species, a life cycle that requires more than one host to complete.

composite transposon Two transposable elements located near one another that transpose together and carry the intervening DNA sequence with them. (*See* transposable element.)

compound (1) A substance made up of atoms of more than one element. (2) Made up of many units, as in the **compound eyes** of arthropods.

concentration gradient A difference in concentration of an ion or other chemical substance from one location to another, often across a membrane. (*See* active transport; facilitated diffusion.)

concerted evolution The common evolution of a family of repeated genes, such that changes in one copy of the gene family are replicated in other copies of the gene family, and thus evolve "in concert." (*See* biased gene conversion; unequal crossing over.)

condensation reaction A chemical reaction in which two molecules become connected by a covalent bond and a molecule of water is released ($AH + BOH \rightarrow AB + H_2O$.) (Contrast with hydrolysis reaction.)

conditional mutation A mutation that results in a characteristic phenotype only under certain environmental conditions.

conditioned reflex A form of associative learning first described by Ivan Pavlov, in which a natural response (such as salivation in response to food) becomes associated with a normally unrelated stimulus (such as the sound of a bell).

conduction The transfer of heat from one object to another through direct contact.

cone In conifers, a reproductive structure consisting of spore-bearing scales extending from a central axis. (Contrast with strobilus.)

cone cell In the vertebrate retina, a type of photoreceptor cell responsible for color vision.

conidium (ko nid' ee um) (plural: conidia) [Gk. *konis*: dust] A type of haploid fungal spore borne at the tips of hyphae, not enclosed in sporangia.

conjugation (kon ju gay' shun) [L. *conjugare*: yoke together] (1) A process by which DNA is passed from one cell to another through a conjugation tube, as in bacteria. (2) A nonreproductive sexual process by which *Paramecium* and other ciliates exchange genetic material.

connective tissue A type of tissue that connects or surrounds other tissues; its cells are embedded in a collagen-containing matrix. One of the four major tissue types in multicellular animals, including cartilage, bone, blood, and fat.

connexon In a gap junction, a protein channel linking adjacent animal cells.

conservation biology An applied science that carries out investigations with the aim of maintaining the diversity of life on Earth.

conserved Pertaining to a gene or trait that has evolved very slowly and is similar or even identical in individuals of highly divergent groups.

conspecifics Individuals of the same species.

constant region The portion of an immunoglobulin molecule whose amino acid composition determines its class and does not vary among immunoglobulins in that class. (Contrast with variable region.)

constitutive Always present; produced continually at a constant rate. (Contrast with inducible.)

constitutive genes Genes that are expressed all the time. (Contrast with inducible genes.)

constitutive proteins Proteins that an organism produces all the time, and at a relatively constant rate.

consumer An organism that eats the tissues of some other organism.

consumer–resource interactions Interactions in which organisms gain their nutrition by eating other living organisms or are eaten themselves.

continental drift The gradual movements of the world's continents that have occurred over billions of years.

contraception Birth control methods that prevent fertilization or implantation (conception).

contractile vacuole (kon trak' tul) A specialized vacuole that collects excess water taken in by osmosis, then contracts to expel the water from the cell.

controlled experiment An experiment in which a sample is divided into groups whereby experimental groups are exposed to manipulations of an independent variable while one group serves as an untreated control. The data from the various groups are then compared to see if there are changes in a dependent variable as a result of the experimental manipulation. (Contrast with comparative experiment.)

controlled system A set of components in a physiological system that is controlled by commands from a regulatory system. (Contrast with regulatory system.)

convection The transfer of heat to or from a surface via a moving stream of air or fluid.

convergent evolution Independent evolution of similar features from different ancestral traits.

convolutions Foldings of the vertebrate brain's cerebral cortex into ridges called gyri (sing. **gyrus**) and valleys called sulci (sing. **sulcus**). The level of cortical convolution increases taxonomically and is especially extensive in humans.

copulation Reproductive behavior that results in a male depositing sperm in the reproductive tract of a female.

cork cambium [L. *cambiare*: to exchange] In plants, a lateral meristem that produces secondary growth, mainly in the form of waxy-walled protective cells, including some of the cells that become bark.

cork In plants, a protective outermost tissue layer composed of cells with thick walls waterproofed with suberin.

cornea The clear, transparent tissue that covers the eye and allows light to pass through to the retina.

corolla (ko role' lah) [L. *corolla*: a small crown] All of the petals of a flower, collectively.

corpus luteum (kor' pus loo' tee um) (plural: corpora lutea) [L. yellow body] A structure formed from a follicle after ovulation; produces hormones important to the maintenance of pregnancy.

corridor A connection between habitat patches through which organisms can disperse; plays a critical role in maintaining subpopulations.

cortex [L. *cortex*: covering, rind] (1) In plants, the tissue between the epidermis and the vascular tissue of a stem or root. (2) In animals, the outer tissue of certain organs, such as the adrenal gland (adrenal cortex) and the brain (cerebral cortex).

corticosteroids Steroid hormones produced and released by the cortex of the adrenal gland.

corticotropin A tropic hormone produced by the anterior pituitary hormone that stimulates cortisol release from the adrenal cortex. Also called adrenocorticotropic hormone (ACTH).

corticotropin-releasing hormone A hormone produced by the hypothalamus that controls the release of cortisol from the anterior pituitary.

cortisol A corticosteroid that mediates stress responses.

cost A decrease in fitness resulting from performing a behavior or having a trait.

cost–benefit approach An approach to evolutionary studies that assumes an animal has a limited amount of time and energy to devote to each of its activities, and that each activity has fitness costs as well as benefits. (*See also* trade-off.)

cotyledon (kot' ul lee' dun) [Gk. *kotyledon*: hollow space] A "seed leaf." An embryonic organ that stores and digests reserve materials; may expand when seed germinates.

countercurrent flow An arrangement that promotes the maximum exchange of heat, or of a diffusible substance, between two fluids by having the fluids flow in opposite directions through parallel vessels close together.

countercurrent heat exchanger In "hot" fish, an adaptation of the circulatory system such that arterial blood flowing into the muscles is warmed by venous blood flowing out of the muscles, thereby conserving body heat by countercurrent exchange.

countercurrent multiplier The mechanism that increases the concentration of the interstitial fluid in the mammalian kidney through countercurrent flow in the loops of Henle and selective permeability and active transport of ions by segments of the loops of Henle.

covalent bond Chemical bond based on the sharing of electrons between two atoms.

CpG islands DNA regions rich in C residues adjacent to G residues. Especially abundant in promoters, these regions are where methylation of cytosine usually occurs.

crassulacean acid metabolism (CAM) A metabolic pathway enabling the plants that possess it to store carbon dioxide at night and then perform photosynthesis during the day with stomata closed.

critical night length In the photoperiodic flowering response of short-day plants, the length of night above which flowering occurs and below which the plant remains vegetative. (The reverse applies in the case of long-day plants.)

critical period *See* sensitive period.

crop A simple food storage sac, the first of two stomachlike organs in many animals (including reptiles, earthworms, and various insects). (*See also* gizzard.)

cross section A section taken perpendicular to the longest axis of a structure. Also called a transverse section.

crossing over The mechanism by which linked genes undergo recombination. In general, the term refers to the reciprocal exchange of corresponding segments between two homologous chromatids.

crosstalk Interactions between different signal transduction pathways.

crypsis [Gk. *kryptos*: hidden] The resemblance of an organism to some part of its environment, which helps it to escape detection by enemies.

cryptochromes [Gk. *kryptos*: hidden + *kroma*: color] Photoreceptors mediating some blue-light effects in plants and animals.

ctene (teen) [Gk. *cteis*: comb] In ctenophores, a comblike row of cilia-bearing plates. Ctenophores move by beating the cilia on their eight ctenes.

culture (1) A laboratory association of organisms under controlled conditions. (2) The collection of knowledge, tools, values, and rules that characterize a human society.

cumulus A thick gelatinous layer that protects a mammalian ovum.

cupula Gelatinous swelling in the semicircular canals of the vestibular system. A cupula encloses hair cell stereocilia that react to shifting fluid in the canal ducts.

currents Circulation patterns in the surface waters of oceans driven by the prevailing winds.

cuticle (1) In plants, a waxy layer on the outer body surface that retards water loss. (2) In ecdysozoans, an outer body covering that provides protection and support and is periodically molted.

cyclic AMP *See* cAMP.

cyclic electron transport In photosynthetic light reactions, the flow of electrons that produces ATP but no NADPH or O_2.

cyclical succession Pattern of change in community composition (succession) in which the climax community depends on periodic disturbances (e.g., fire) in order to persist. (Contrast with directional succession.)

cyclin A protein that activates a cyclin-dependent kinase, bringing about transitions in the cell cycle.

cyclin-dependent kinase (Cdk) A protein kinase whose target proteins are involved in transitions in the cell cycle and which is active only when complexed with additional protein subunits, called cyclins.

cytokine A regulatory protein made by immune system cells that affects other target cells in the immune system.

cytokinesis (sy' toe kine ee' sis) [Gk. *kytos*: container + *kinein*: to move] The division of the cytoplasm of a dividing cell. (Contrast with mitosis.)

cytokinin (sy' toe kine' in) A member of a class of plant growth substances that plays roles in senescence, cell division, and other phenomena.

cytoplasm The contents of the cell, excluding the nucleus.

cytoplasmic determinants In animal development, gene products whose spatial distribution may determine such things as embryonic axes.

cytoplasmic segregation The asymmetrical distribution of cytoplasmic determinants in a developing animal embryo.

cytosine (C) (site' oh seen) A nitrogen-containing base found in DNA and RNA.

cytoskeleton The network of microtubules and microfilaments that gives a eukaryotic cell its shape and its capacity to arrange its organelles and to move.

cytosol The fluid portion of the cytoplasm, excluding organelles and other solids.

cytotoxic T cells (T_C) Cells of the cellular immune system that recognize and directly eliminate virus-infected cells. (Contrast with T-helper cells.)

D

DAG *See* diacylglycerol.

data Quantified observations about a system under study.

daughter chromosomes During mitosis, the separated chromatids from the beginning of anaphase onward.

dead space The lung volume that fails to be ventilated with fresh air (because the lungs are never completely emptied during exhalation).

dead zones Regions in aquatic ecosystems that are devoid of aquatic life because eutrophication has resulted in severe oxygen depletion.

deciduous [L. *deciduus*: falling off] Pertaining to a woody plant that sheds its leaves but does not die.

declarative memory Memory of people, places, events, and things that can be consciously recalled and described. (Contrast with procedural memory.)

decomposer An organism that metabolizes organic compounds in debris and dead organisms, releasing inorganic material; found among the bacteria, protists, and fungi. *See also* detritivore, saprobe.

deductive logic Logical thought process that starts with a premise believed to be true then predicts what facts would also have to be true to be compatible with that premise. (Contrast with inductive logic.)

defensin A type of protein made by phagocytes that kills bacteria and enveloped viruses by insertion into their plasma membranes.

deficiency disease A condition (e.g., scurvy and beriberi) caused by chronic lack of any essential nutrient.

degeneracy The situation in which a single amino acid may be represented by any of two or more different codons in messenger RNA. Most of the amino acids can be represented by more than one codon.

deletion A mutation resulting from the loss of a continuous segment of a gene or chromosome. Such mutations almost never revert to wild type. (Contrast with duplication, point mutation.)

demethylase An enzyme that catalyzes the removal of the methyl group from cytosine, reversing DNA methylation.

demography The study of population structure and of the processes (**demographic events**, including births and deaths) by which it changes.

denaturation Loss of activity of an enzyme or nucleic acid molecule as a result of structural changes induced by heat or other means.

dendrites [Gk. *dendron*: tree] Branching fibers (processes) of a neuron. Dendrites are usually relatively short compared with the axon, and commonly carry information to the neuronal cell body.

denitrification Metabolic activity by which nitrate and nitrite ions are reduced to form nitrogen gas; carried out by certain soil bacteria.

denitrifiers Bacteria that release nitrogen to the atmosphere as nitrogen gas (N_2).

density *See* population density.

density-dependent Pertaining to an effect on population size that increases in proportion to population density.

density-independent Pertaining an effect on population size that acts independently of population density.

deoxyribonucleic acid *See* DNA.

deoxyribonucleoside triphosphates (dNTPs) The raw materials for DNA synthesis: deoxyadenosine triphosphate (dATP), deoxythymidine triphosphate (dTTP), deoxycytidine triphosphate (dCTP), and deoxyguanosine triphosphate (dGTP). Also called deoxyribonucleotides.

deoxyribose A five-carbon sugar found in nucleotides and DNA.

dependent variable In a scientific experiment, the response that is measured and analyzed as the independent variable is manipulated (See independent variable.)

depolarization A change in the resting potential across a membrane so that the inside of the cell becomes less negative, or even positive, compared with the outside of the cell. (Contrast with hyperpolarization.)

derived trait A trait that differs from the ancestral trait. (Contrast with synapomorphy.)

dermal tissue system The outer covering of a plant, consisting of epidermis in the young plant and periderm in a plant with extensive secondary growth. (Contrast with ground tissue system and vascular tissue system.)

descent with modification Darwin's premise that all species share a common ancestor and have diverged from one another gradually over time.

desmosome (dez' mo sowm) [Gk. *desmos*: bond + *soma*: body] An adhering junction between animal cells.

desmotubule A membrane extension connecting the endoplasmic retituclum of two plant cells that traverses the plasmodesma.

determinate growth A growth pattern in which the growth of an organism or organ ceases when an adult state is reached; characteristic of most animals and some plant organs. (Contrast with indeterminate growth.)

determination In development, the process whereby the fate of an embryonic cell or group of cells (e.g., to become epidermal cells or neurons) is set (becomes **determined**).

detritivore (di try' ti vore) [L. *detritus*: worn away + *vorare*: to devour] An organism that obtains its energy from the dead bodies or waste products (**detritus**) of other organisms.

development The process by which a multicellular organism, beginning with a single cell, goes through a series of changes, taking on the successive forms that characterize its life cycle.

developmental plasticity The capacity of an organism to alter its pattern of development in response to environmental conditions.

diacylglycerol (DAG) In hormone action, the second messenger produced by hydrolytic removal of the head group of certain phospholipids.

diapause A period of developmental or reproductive arrest, entered in response to day length, that enables an organism to better survive.

diaphragm (dye' uh fram) [Gk. *diaphrassein*: barricade] (1) A sheet of muscle that separates the thoracic and abdominal cavities in mammals; responsible for breathing. (2) A method of birth control in which a sheet of rubber is fitted over the woman's cervix, blocking the entry of sperm.

diastole (dye ass' toll ee) [Gk. dilation] The portion of the cardiac cycle when the heart muscle relaxes. (Contrast with systole.)

dichotomous (dye cot' oh mus) [Gk. *dichot*: split in two; *tomia*: removed) A branching pattern in which the shoot divides at the apex producing two equivalent branches that subsequently never overlap.

diencephalon The portion of the vertebrate forebrain that develops into the thalamus and hypothalamus.

differential gene expression The hypothesis that, given that all cells contain all genes, what makes one cell type different from another is the difference in transcription and translation of those genes.

differentiation The process whereby originally similar cells follow different developmental pathways; the actual expression of determination.

diffuse coevolution The evolution of similar traits in suites of species experiencing similar selection pressures imposed by other suites of species with which they interact.

diffusion Random movement of molecules or other particles, resulting in even distribution of the particles when no barriers are present.

digestive vacuole In protists, an organelle specialized for digesting food ingested by endocytosis.

dihybrid cross A mating in which the parents differ with respect to the alleles of two loci of interest.

dikaryon (di care' ee ahn) [Gk. *di*: two + *karyon*: kernel] A cell or organism carrying two genetically distinguishable nuclei. Common in fungi.

dioecious (die eesh' us) [Gk. *di*: two + *oikos*: house] Pertaining to organisms in which the two sexes are "housed" in two different individuals, so that eggs and sperm are not produced in the same individuals. Examples: humans, fruit flies, date palms. (Contrast with monoecious.)

diploblastic Having two cell layers. (Contrast with triploblastic.)

diploid (dip' loid) [Gk. *diplos*: double] Having a chromosome complement consisting of two copies (homologs) of each chromosome. Designated $2n$. (Contrast with haploid.)

direct development Pattern of development (notably among insects) in which hatchlings look like miniature versions of adults. (Contrast with metamorphosis.)

direct fitness That component of fitness resulting from an organism producing its own offspring. (Contrast with inclusive fitness, kin selection.)

directional selection Selection in which phenotypes at one extreme of the population distribution are favored. (Contrast with disruptive selection, stabilizing selection.)

directional succession Change in community composition after a disturbance (succession) that is characterized by an orderly progression culminating in a persistent state (the climax community). (Contrast with cyclical succession.)

disaccharide A carbohydrate made up of two monosaccharides (simple sugars).

discoidal cleavage In animal development, a type of incomplete cleavage that is common in fishes, reptiles, and birds, the eggs of which contain a dense yolk mass.

dispersal Movement of organisms away from a parent organism or from an existing population.

dispersion The distribution of individuals in space within a population. **Clumped dispersion** occurs when individuals tend to occupy the same space; **regular dispersion** is when the presence of one individual decreases the probability of another individual occupying the same space; and **random dispersion** assumes there is equal probability of any individual occupying any given space.

disruptive selection Selection in which phenotypes at both extremes of the population distribution are favored. (Contrast with directional selection; stabilizing selection.)

distal Away from the point of attachment or other reference point. (Contrast with proximal.)

distal convoluted tubule The portion of a renal tubule from where it reaches the renal cortex, just past the loop of Henle to where it joins a collecting duct. (Compare with proximal convoluted tubule.)

disturbance A short-term event that disrupts populations, communities, or ecosystems by changing the environment.

disulfide bridge The covalent bond between two sulfur atoms (–S—S–) linking two molecules or remote parts of the same molecule.

DNA (deoxyribonucleic acid) The fundamental hereditary material of all living organisms. In eukaryotes, stored primarily in the cell nucleus. A nucleic acid using deoxyribose rather than ribose.

DNA fingerprint An individual's unique pattern of allele sequences, commonly short tandem repeats and single nucleotide polymorphisms.

DNA helicase An enzyme that unwinds the double helix.

DNA ligase Enzyme that unites broken DNA strands during replication and recombination.

DNA methylation The addition of methyl groups to bases in DNA, usually cytosine or guanine.

DNA methyltransferase An enzyme that catalyzes the methylation of DNA.

DNA microarray A small glass or plastic square onto which thousands of single-stranded DNA sequences are fixed so that hybridization of cell-derived RNA or DNA to the target sequences can be performed.

DNA polymerase Any of a group of enzymes that catalyze the formation of DNA strands from a DNA template.

DNA replication The creation of a new strand of DNA in which DNA polymerase catalyzes the exact reproduction of an existing (template) strand of DNA.

DNA transposons Mobile genetic elements that move without making an RNA intermediate. (Contrast with retrotransposons.)

domain (1) An independent structural element within a protein. Encoded by recognizable nucleotide sequences, a domain often folds separately from the rest of the protein. Similar domains can appear in a variety of different proteins across phylogenetic groups (e.g., "homeobox domain"; "calcium-binding domain"). (2) In phylogenetics, the three monophyletic branches of life (Bacteria, Archaea, and Eukarya).

dominance In genetics, the ability of one allelic form of a gene to determine the phenotype of a heterozygous individual in which the homologous chromosomes carry both it and a different (recessive) allele. (Contrast with recessive.)

dormancy A condition in which normal activity is suspended, as in some spores, seeds, and buds.

dorsal [L. *dorsum*: back] Toward or pertaining to the back or upper surface. (Contrast with ventral.)

dorsal lip In amphibian embryos, the dorsal segment of the blastopore. Also called the "organizer," this region directs the development of nearby embryonic regions.

double fertilization In angiosperms, a process in which the nuclei of two sperm fertilize one egg. One sperm's nucleus combines with the egg nucleus to produce a zygote, while the other combines with the same egg's two polar nuclei to produce the first cell of the triploid endosperm (the tissue that will nourish the growing plant embryo).

double helix Refers to DNA and the (usually right-handed) coil configuration of two complementary, antiparallel strands.

downregulation A negative feedback process in which continuous high concentrations of a hormone can decrease the number of its receptors. (Contrast with upregulation.)

duodenum (do' uh dee' num) The beginning portion of the vertebrate small intestine. (Contrast with ileum, jejunum.)

duplication A mutation in which a segment of a chromosome is duplicated, often by the attachment of a segment lost from its homolog. (Contrast with deletion.)

E

ecdysone (eck die' sone) [Gk. *ek*: out of + *dyo*: to clothe] In insects, a hormone that induces molting.

ecological economics Interdisciplinary field that works to assess the economic value of biodiversity.

ecological efficiency The overall transfer of energy from one trophic level to the next, expressed as the ratio of consumer production to producer production.

ecological survivorship curve *See* survivorship curves

ecological system One or more organisms plus the external environment with which they interact.

ecology [Gk. *oikos*: house] The scientific study of the interaction of organisms with their living (biotic) and nonliving (abiotic) environments.

ecosystem (eek' oh sis tum) The organisms of a particular habitat, such as a pond or forest, together with the physical environment in which they live.

ecosystem engineer An organism that builds structures that alter existing habitats or create new habitats.

ecosystem services Processes by which ecosystems maintain resources that benefit human society.

ecotourism Ecologically responsible travel to natural places.

ectoderm [Gk. *ektos*: outside + *derma*: skin] The outermost of the three embryonic germ layers first delineated during gastrulation. Gives rise to the skin, sense organs, and nervous system.

ectotherm [Gk. *ektos*: outside + *thermos*: heat] An animal that is dependent on external heat sources for regulating its body temperature (Contrast with endotherm.)

edema (i dee' mah) [Gk. *oidema*: swelling] Tissue swelling caused by the accumulation of fluid.

edge effects Changes in ecological processes in a community caused by physical and biological factors originating in an adjacent community.

effector A component of a physiological system that responds to information by *effecting* changes (making change happen) in the internal environment; examples include muscles and the secretory cells of the digestive tract.

effector cells In cellular immunity, B cells and T cells that attack an antigen, either by secreting antibodies that bind to the antigen or by releasing molecules that destroy any cell bearing the antigen.

effector protein In cell signaling, a protein responsible for the cellular reponse to a signal transduction pathway.

efferent (ef' ur unt) [L. *ex*: out + *ferre*: to bear] Carrying outward or away from, as in neurons that carry impulses outward from the central to the peripheral nervous system (**efferent neurons**), or a blood vessel that carries blood away from a structure. (Contrast with afferent.)

egg In all sexually reproducing organisms, the female gamete; in birds, reptiles, and some other vertebrates, a structure within which early embryonic development occurs. *See also* amniote egg, ovum.

electrical synapse A type of synapse at which action potentials spread directly from presynaptic cell to postsynaptic cell. (Contrast with chemical synapse.)

electrocardiogram (ECG or EKG) A graphic recording of electrical potentials from the heart.

electrochemical gradient The concentration gradient of an ion across a membrane plus the voltage difference across that membrane.

electroencephalogram (EEG) A graphic recording of electrical potentials from the brain.

electromagnetic radiation A self-propagating wave that travels though space and has both electrical and magnetic properties.

electron A subatomic particle outside the nucleus carrying a negative charge and very little mass.

electron shell The region surrounding the atomic nucleus at a fixed energy level in which electrons orbit.

electron transport The passage of electrons through a series of proteins with a release of energy which may be captured in a concentration gradient or in chemical form such as NADH or ATP.

electronegativity The tendency of an atom to attract electrons when it occurs as part of a compound.

electrophoresis *See* gel electrophoresis.

element A substance that cannot be converted to a simpler substance by ordinary chemical means.

elongation (1) In molecular biology, the addition of monomers to make a longer RNA or protein during transcription or translation. (2) Growth of a plant axis or cell primarily in the longitudinal direction.

embolus (em' buh lus) [Gk. *embolos*: stopper] A circulating blood clot. Blockage of a blood vessel by an embolus or a bubble of gas is called an **embolism**. (Contrast with thrombus.)

embryo [Gk. *en*: within + *bryein*: to grow] A young animal, or young plant sporophyte, while it is still contained within a protective structure such as a seed, egg, or uterus.

embryo sac In angiosperms, the female gametophyte. Found within the ovule, it consists of eight or fewer cells, membrane bounded, but without cellulose walls between them.

embryonic stem cell (ESC) A pluripotent cell in the blastocyst.

emergent property A property of a complex system that is not exhibited by its individual component parts.

emigration The deliberate and usually oriented departure of an organism from the habitat in which it has been living.

endemic (en dem′ ik) [Gk. *endemos*: native] Confined to a particular region, thus often having a comparatively restricted distribution.

endergonic A chemical reaction in which the products have higher free energy than the reactants, thereby requiring free energy input to occur. (Contrast with exergonic.)

endocrine cells Cells that secrete substances into the extracellular fluid. (*See also* endocrine gland.)

endocrine gland (en′ doh krin) [Gk. *endo*: within + *krinein*: to separate] An aggregation of secretory cells that secretes hormones into the blood. The endocrine system consists of all endocrine cells and endocrine glands in the body that produce and release hormones. (Contrast with exocrine gland.)

endocytosis A process by which liquids or solid particles are taken up by a cell through invagination of the plasma membrane. (Contrast with exocytosis.)

endoderm [Gk. *endo*: within + *derma*: skin] The innermost of the three embryonic germ layers delineated during gastrulation. Gives rise to the digestive and respiratory tracts and structures associated with them.

endodermis In plants, a specialized cell layer marking the inside of the cortex in roots and some stems. Frequently a barrier to free diffusion of solutes.

endomembrane system A system of intracellular membranes that exchange material with one another, consisting of the Golgi apparatus, endoplasmic reticulum, and lysosomes when present.

endometrium The epithelial lining of the uterus.

endoplasmic reticulum (ER) [Gk. *endo*: within + L. *reticulum*: net] A system of membranous tubes and flattened sacs found in the cytoplasm of eukaryotes. Exists in two forms: rough ER, studded with ribosomes; and smooth ER, lacking ribosomes.

endorphins Molecules in the mammalian brain that act as neurotransmitters in pathways that control pain.

endoskeleton [Gk. *endo*: within + *skleros*: hard] An internal skeleton covered by other, soft body tissues. (Contrast with exoskeleton.)

endosperm [Gk. *endo*: within + *sperma*: seed] A specialized triploid seed tissue found only in angiosperms; contains stored nutrients for the developing embryo.

endospore [Gk. *endo*: within + *spora*: to sow] In some bacteria, a resting structure that can survive harsh environmental conditions.

endosymbiosis theory [Gk. *endo*: within + *sym*: together + *bios*: life] The theory that the eukaryotic cell evolved via the engulfing of one prokaryotic cell by another.

endothelium The single layer of epithelial cells lining the interior of a blood vessel.

endotherm [Gk. *endo*: within + *thermos*: heat] An animal that can control its body temperature by the expenditure of its own metabolic energy. (Contrast with ectotherm.)

endotoxin A lipopolysaccharide that forms part of the outer membrane of certain Gram-negative bacteria that is released when the bacteria grow or lyse. (Contrast with exotoxin.)

energetic cost The difference between the energy an animal expends in performing a behavior and the energy it would have expended had it rested.

energy The capacity to do work or move matter against an opposing force. The capacity to accomplish change in physical and chemical systems.

energy budget A quantitative description of all paths of energy exchange between an animal and its environment.

enhancers Regulatory DNA sequences that bind transcription factors that either activate or increase the rate of transcription.

enkephalins Molecules in the mammalian brain that act as neurotransmitters in pathways that control pain.

enteric nervous system The nerve nets in the submucosa and between the smooth muscle layers of the vertebrate gut.

enthalpy (*H*) The total energy of a system.

entrain To advance or delay an organism's circadian clock each day so that it is in phase with the light-dark cycle of the organism's environment.

entropy (*S*) (en′ tro pee) [Gk. *tropein*: to change] A measure of the degree of disorder in any system. Spontaneous reactions in a closed system are always accompanied by an increase in entropy.

enveloped virus A virus enclosed within a phospholipid membrane derived from its host cell.

environment Whatever surrounds and interacts with or otherwise affects a population, organism, or cell. May be external or internal.

environmental genomics Sequencing technique used when biologists are unable to work with the whole genome of a prokaryote species but instead examine individual genes collected from a random sample of the organism's environment.

environmental resistance Reduction in a population's growth rate caused by preemption of available resources by other individuals in the population.

environmentalism The use of ecological knowledge, along with economics, ethics, and many other considerations, to inform both personal decisions and public policy relating to stewardship of natural resources and ecosystems.

enzyme (en′ zime) [Gk. *zyme*: to leaven (as in yeast bread)] A catalytic protein that speeds up a biochemical reaction.

enzyme–substrate complex (ES) An intermediate in an enzyme-catalyzed reaction; consists of the enzyme bound to its substrate(s).

epi- [Gk. upon, over] A prefix used to designate a structure located on top of another; for example, epidermis, epiphyte.

epiblast The upper or overlying portion of the avian blastula which is joined to the hypoblast at the margins of the blastodisc.

epiboly The movement of cells over the surface of the blastula toward the forming blastopore.

epidermis [Gk. *epi*: over + *derma*: skin] In plants and animals, the outermost cell layers. (Only one cell layer thick in plants.)

epididymis (epuh did′ uh mus) [Gk. *epi*: over + *didymos*: testicle] Coiled tubules in the testes that store sperm and conduct sperm from the seminiferous tubules to the vas deferens.

epigenetics The scientific study of changes in the expression of a gene or set of genes that occur without change in the DNA sequence.

epinephrine (ep i nef′ rin) [Gk. *epi*: over + *nephros*: kidney] The "fight or flight" hormone produced by the medulla of the adrenal gland; it also functions as a neurotransmitter. (Also known as adrenaline.)

epistasis Interaction between genes in which the presence of a particular allele of one gene determines whether another gene will be expressed.

epithelial tissue A type of animal tissue made up of sheets of cells that lines or covers organs, makes up tubules, and covers the surface of the body; one of the four major tissue types in multicellular animals.

epitope *See* antigenic determinant.

equilibrium Any state of balanced opposing forces and no net change.

equilibrium potential The membrane potential at which an ion is at electrochemical equilibrium, i.e., there is no net flux of the ion across the membrane.

ER *See* endoplasmic reticulum.

error signal In regulatory systems, any difference between the set point of the system and its current condition.

erythrocyte (ur rith′ row site) [Gk. *erythros*: red + *kytos*: container] A red blood cell.

erythropoietin A hormone produced by the kidney in response to lack of oxygen that stimulates the production of red blood cells.

esophagus (i soff′ i gus) [Gk. *oisophagos*: gullet] That part of the gut between the pharynx and the stomach.

essential amino acids Amino acids that an animal cannot synthesize for itself and must obtain from its food.

essential element A mineral nutrient required for normal growth and reproduction in plants and animals.

essential fatty acids Fatty acids that an animal cannot synthesize for itself and must obtain from its food.

ester linkage A condensation (water-releasing) reaction in which the carboxyl group of a fatty acid reacts with the hydroxyl group of an alcohol. Lipids, including most membrane lipids, are formed in this way. (Contrast with ether linkage.)

estivation (ess tuh vay' shun) [L. *aestivalis*: summer] A state of dormancy and hypometabolism that occurs during the summer; usually a means of surviving drought and/or intense heat. (Contrast with hibernation.)

estrogen Any of several steroid sex hormones; produced chiefly by the ovaries in mammals.

estrus (es' trus) [L. *oestrus*: frenzy] The period of heat, or maximum sexual receptivity, in some female mammals. Ordinarily, the estrus is also the time of release of eggs in the female.

estuary Aquatic biome in which salt water and fresh water mix, as when a river meets the ocean. Includes such ecosystems as salt marshes and mangrove forests.

ether linkage The linkage of two hydrocarbons by an oxygen atom (HC—O—CH). Ether linkages are characteristic of the membrane lipids of the Archaea. (Contrast with ester linkage.)

ethology [Gk. *ethos*: character + *logos*: study] An approach to the study of animal behavior that focuses on studying many species in natural environments and addresses questions about the evolution of behavior. (Compare with behaviorism.)

ethylene One of the plant growth hormones, the gas $H_2C{=}CH_2$. Involved in fruit ripening and other growth and developmental responses.

euchromatin Diffuse, uncondensed chromatin. Contains active genes that will be transcribed into mRNA. (Contrast with heterochromatin.)

eudicots Angiosperms with two embryonic cotyledons. (*See also* monocots.)

Eukarya One of the three domains of life; organisms made up of one or more eukaryotic cells. (*See also* eukaryotes.)

eukaryotes (yew car' ree oats) [Gk. *eu*: true + *karyon*: kernel or nucleus] Organisms whose cells contain their genetic material inside a nucleus. Includes all life other than the viruses, archaea, and bacteria. (Contrast with prokaryotes.)

eusocial Pertaining to a social group that includes nonreproductive individuals, as in honey bees.

eustachian tube A connection between the middle ear and the throat that allows air pressure to equilibrate between the middle ear and the outside world.

eutrophication (yoo trofe' ik ay' shun) [Gk. *eu*: truly + *trephein*: to flourish] The addition of nutrient materials to a body of water, resulting in changes in ecological processes and species composition therein.

evaporation The transition of water from the liquid to the gaseous phase.

evolution Any gradual change. Most often refers to organic or Darwinian evolution, which is the genetic and resulting phenotypic change in populations of organisms from generation to generation. (*See* macroevolution, microevolution; contrast with speciation.)

evolutionary developmental biology (evo-devo) The study of the interplay between evolutionary and developmental processes, with a focus on the genetic changes that give rise to novel morphology. Key concepts of evo-devo include modularity, genetic toolkits, genetic switches, and heterochrony.

evolutionary radiation The proliferation of many species within a single evolutionary lineage.

evolutionary reversal The reappearance of an ancestral trait in a group that had previously acquired a derived trait.

evolutionary theory The understanding and application of the mechanisms of evolutionary change to biological problems.

excision repair DNA repair mechanism that removes damaged DNA and replaces it with the appro-priate nucleotide.

excitable Capable of generating an action potential.

excitatory Input from a neuron that causes depolarization of the recipient cell.

excited state The state of an atom or molecule when, after absorbing energy, it has more energy than in its normal, ground state.

excretion Release of metabolic wastes by an organism.

excretory systems In animals, organs that maintain the volume, solute concentration, and composition of the extracellular fluid by excreting water, solutes, and nitrogenous wastes in the form of urine.

exergonic A chemical reaction in which the products of the reaction have lower free energy than the reactants, resulting in a release of free energy. (Contrast with endergonic.)

exocrine gland (eks' oh krin) [Gk. *exo*: outside + *krinein*: to separate] Any gland, such as a salivary gland, that secretes to the outside of the body or into the gut. (Contrast with endocrine gland.)

exocytosis A process by which a vesicle within a cell fuses with the plasma membrane and releases its contents to the outside. (Contrast with endocytosis.)

exon A portion of a DNA molecule, in eukaryotes, that codes for part of a polypeptide. (Contrast with intron.)

exoskeleton (eks' oh skel' e ton) [Gk. *exos*: outside + *skleros*: hard] A hard covering on the outside of the body to which muscles are attached. (Contrast with endoskeleton.)

exotoxin A highly toxic, usually soluble protein released by living, multiplying bacteria. (Contrast with endotoxin.)

expanding triplet repeat A three-base-pair sequence in a human gene that is unstable and can be repeated a few to hundreds of times. Often, the more the repeats, the less the activity of the gene involved. Expanding triplet repeats occur in some human diseases such as Huntington's disease and fragile-X syndrome.

experiment A testing process to support or disprove hypotheses and to answer questions. The basis of the scientific method. *See* comparative experiment, controlled experiment.

expiratory reserve volume The amount of air that can be forcefully exhaled beyond the normal tidal expiration. (Contrast with inspiratory reserve volume, tidal volume, vital capacity.)

exploitation competition Competition in which individuals reduce the quantities of their shared resources. (Contrast with interference competition.)

exponential growth Growth, especially in the number of organisms in a population, which is a geometric function of the size of the growing entity: the larger the entity, the faster it grows. (Contrast with logistic growth.)

expression vector A DNA vector, such as a plasmid, that carries a DNA sequence for the expression of an inserted gene into mRNA and protein in a host cell.

expressivity The degree to which a genotype is expressed in the phenotype; may be affected by the environment.

extensor A muscle that extends an appendage. (Contrast with flexor.)

external fertilization The release of gametes into the environment; typical of aquatic animals. Also called spawning. (Contrast with internal fertilization.)

external gills Highly branched and folded extensions of the body surface that provide a large surface area for gas exchange with water; typical of larval amphibians and many larval insects.

extinction The termination of a lineage of organisms.

extracellular matrix A material of heterogeneous composition surrounding cells and performing many functions including adhesion of cells.

extraembryonic membranes Four membranes that support but are not part of the developing embryos of reptiles, birds, and mammals, defining these groups phylogenetically as amniotes. (*See* amnion, allantois, chorion, yolk sac.)

extreme halophiles A group of euryarchaeotes that live exclusively in very salty environments.

extremophiles Archaea and bacteria that live and thrive under conditions (e.g., extremely high temperatures) that would kill most organisms.

eye cups Photosensory organs in flatworms; components of one of the simplest visual systems in animals.

F

5′ end (5 prime) The end of a DNA or RNA strand that has a free phosphate group at the 5′ carbon of the sugar (deoxyribose or ribose).

F_1 The first filial generation; the immediate progeny of a parental (P) mating.

F_2 The second filial generation; the immediate progeny of a mating between members of the F1 generation.

facilitated diffusion Passive movement through a membrane involving a specific carrier protein; does not proceed against a concentration gradient. (Contrast with active transport, diffusion.)

facilitation In succession, modification of the environment by a colonizing species in a way that allows colonization by other species. (Contrast with inhibition.)

facultative anaerobe A prokaryote that can shift its metabolism between anaerobic and aerobic modes depending on the presence or absence of O_2. (Alternatively, facultative aerobe.)

fast-twitch fibers Skeletal muscle fibers that can generate high tension rapidly, but fatigue rapidly ("sprinter" fibers). Characterized by an abundance of enzymes of glycolysis. (Compare to slow-twitch fibers.)

fat (1) A triglyceride that is solid at room temperature. (Contrast with oil.) (2) Adipose tissue, one type of connective tissue. (See brown fat, white fat.)

fate map A diagram of the blastula showing which cells (blastomeres) are "fated " to contribute to specific tissues and organs in the mature body.

fatty acid A molecule made up of a long nonpolar hydrocarbon chain and a polar carboxyl group. Found in many lipids.

fauna (faw′ nah) All the animals found in a given area. (Contrast with flora.)

***FD* (*FLOWERING LOCUS D*)** Gene coding for a transcription factor in the shoot apical meristem that binds to florigen; involved in the induction of flowering.

feces [L. *faeces*: dregs] Waste excreted from the digestive system.

fecundity The average number of offspring produced by each female.

feedback In regulatory systems, information about the relationship between the set point of the system and its current state (Contrast with feedforward information).

feedback inhibition A mechanism for regulating a metabolic pathway in which the end product of the pathway can bind to and inhibit the enzyme that catalyzes the first committed step in the pathway. Also called end-product inhibition.

feedforward information In regulatory systems, information that changes the set point of the system. (Contrast with feedback.)

fermentation (fur men tay′ shun) [L. *fermentum*: yeast] The anaerobic degradation of a substance such as glucose to smaller molecules such as lactic acid or alcohol with the extraction of energy.

fertilization Union of gametes. Also known as syngamy.

fertilizer Any of a number of substances added to soil to improve the soil's capacity to support plant growth. May be organic or inorganic.

fetus Medical and legal term for the stages of a developing human embryo from about the eighth week of pregnancy (the point at which all major organ systems have formed) to the moment of birth.

fiber In angiosperms, an elongated, tapering sclerenchyma cell, usually with a thick cell wall, that serves as a support function in xylem. (*See also* muscle fiber.)

fibrin A protein that polymerizes to form long threads that provide structure to a blood clot.

fibrinogen A circulating protein that can be stimulated to fall out of solution and provide the structure for a blood clot.

fibrous root system A root system typical of monocots composed of numerous thin adventitious roots that are all roughly equal in diameter. (Contrast with taproot system.)

Fick's law of diffusion An equation that describes the factors that determine the rate of diffusion of a molecule from an area of higher concentration to an area of lower concentration.

fight-or-flight response A rapid physiological response to a sudden threat mediated by the hormone epinephrine.

filament In flowers, the part of a stamen that supports the anther.

filter feeder An organism that feeds on organisms much smaller than itself that are suspended in water or air by means of a straining device.

first filial generation *See* F_1.

first law of thermodynamics The principle that energy can be neither created nor destroyed.

fission *See* binary fission.

fitness The contribution of a genotype or phenotype to the genetic composition of subsequent generations, relative to the contribution of other genotypes or phenotypes. (*See* also inclusive fitness.)

fixed action pattern In ethology, a genetically determined behavior that is performed without learning, stereotypic (performed the same way each time), and not modifiable by learning.

flagellum (fla jell′ um) (plural: flagella) [L. *flagellum*: whip] Long, whiplike appendage that propels cells. Prokaryotic flagella differ sharply from those found in eukaryotes.

flexor A muscle that flexes an appendage. (Contrast with extensor.)

flora (flore′ ah) All of the plants found in a given area. (Contrast with fauna.)

floral meristem In angiosperms, a meristem that forms the floral organs (sepals, petals, stamens, and carpels).

floral organ identity genes In angiosperms, genes that determine the fates of floral meristem cells; their expression is triggered by the products of meristem identity genes.

florigen A plant hormone involved in the conversion of a vegetative shoot apex to a flower.

flower The sexual structure of an angiosperm.

fluid feeder An animal that feeds on fluids it extracts from the bodies of other organisms; examples include nectar-feeding birds and blood-sucking insects.

fluid mosaic model A molecular model for the structure of biological membranes consisting of a fluid phospholipid bilayer in which suspended proteins are free to move in the plane of the bilayer.

flux [L.: flow] In ecology, the flow of an element into or out of a compartment of the biosphere.

follicle [L. *folliculus*: little bag] In female mammals, an immature egg surrounded by nutritive cells.

follicle-stimulating hormone (FSH) A gonadotropin produced by the anterior pituitary.

food chain A portion of a food web, most commonly a simple sequence of prey species and the predators that consume them.

food web The complete set of food links between species in a community; a diagram indicating which ones are the eaters and which are eaten.

forebrain The region of the vertebrate brain that comprises the cerebrum, thalamus, and hypothalamus.

fossil Any recognizable structure originating from an organism, or any impression from such a structure, that has been preserved over geological time.

fossil fuels Fuels, including oil, natural gas, coal, and peat, formed over geologic time from organic material buried in anaerobic sediments.

founder effect Random changes in allele frequencies resulting from establishment of a population by a very small number of individuals.

fovea [L. *fovea*: a small pit] In the vertebrate retina, the area of most distinct vision.

frame-shift mutation The addition or deletion of a single or two adjacent nucleotides in a gene's sequence. Results in the misreading of mRNA during translation and the production of a nonfunctional protein. (Contrast with missense mutation, nonsense mutation, silent mutation.)

Frank–Starling law The stroke volume of the heart increases with increased return of blood to the heart.

free energy (*G*) Energy that is available for doing useful work, after allowance has been made for the increase or decrease of disorder.

frequency-dependent selection Selection that changes in intensity with the proportion of individuals in a population having the trait.

frontal lobe The largest of the brain lobes in humans; involved with feeling and planning functions; includes the primary motor cortex.

frugivore [L. *frugis*; fruit + *vorare*: to devour] An animal that eats fruit.

fruit In angiosperms, a ripened and mature ovary (or group of ovaries) containing the seeds. Sometimes applied to reproductive structures of other groups of plants.

***FT* (*FLOWERING LOCUS T*)** Gene that codes for florigen, a small, diffusible protein involved in the induction of flowering.

fugitive species A species that leave an otherwise suitable habitat in order to avoid competition with another species.

full census A count of every individual in a population. Can only be achieve if individuals are large and distinct enough to be identifiable by the census taker; population sizes are more usually estimated using sampling methods.

functional genomics The assignment of functional roles to the proteins encoded by genes identified by sequencing entire genomes.

functional group A characteristic combination of atoms that contributes specific properties (such as charge or polarity) when attached to larger molecules (e.g., carboxyl group; amino group).

fundamental niche A species' niche as defined by its physiological capabilities. (Contrast with realized niche.)

G

G protein A membrane protein involved in signal transduction; characterized by binding GDP or GTP.

G protein–linked receptors A class of receptors that change configuration upon ligand binding such that a G protein binding site is exposed on the cytoplasmic domain of the receptor, initiating a signal transduction pathway.

G1 In the cell cycle, the gap between the end of mitosis and the onset of the S phase.

G1-to-S transition In the cell cycle, the point at which G1 ends and the S phase begins.

G2 In the cell cycle, the gap between the S (synthesis) phase and the onset of mitosis.

gain of function mutation A mutation that results in a protein with a new function. (Contrast with loss of function mutation.)

gallbladder In the human digestive system, an organ in which bile is stored.

gametangium (gam uh tan′ gee um) (plural: gametangia) [Gk. *gamos*: marriage + *angeion*: vessel] Any plant or fungal structure within which a gamete is formed.

gamete (gam′ eet) [Gk. *gamete*/ *gametes*: wife, husband] The mature sexual reproductive cell: the egg or the sperm.

gametogenesis (ga meet′ oh jen′ e sis) The specialized series of cellular divisions that leads to the production of gametes. (*See also* oogenesis, spermatogenesis.)

gametophyte (ga meet′ oh fyte) In plants and photosynthetic protists with alternation of generations, the multicellular haploid phase that produces the gametes. (Contrast with sporophyte.)

ganglion (gang′ glee un) (plural: ganglia) [Gk. lump] A cluster of neurons that have similar characteristics or function.

ganglion cells Cells at the front of the human retina that transmit information from the bipolar cells to the brain.

gap genes In *Drosophila* development, segmentation genes that define broad areas along the anterior–posterior axis of the early embryo. Part of a developmental cascade that includes maternal effect genes, pair rule genes, segment polarity genes, and Hox genes.

gap junction A 2.7-nanometer gap between plasma membranes of two animal cells, spanned by protein channels. Gap junctions allow chemical substances or electrical signals to pass from cell to cell.

gastric pits Deep infoldings in the walls of the stomach lined with secretory cells.

gastrin A hormone secreted by cells in the lower region of the stomach that stimulates the secretion of digestive juices as well as movements of the stomach.

gastrovascular cavity Serving for both digestion (gastro) and circulation (vascular); in particular, the central cavity of the body of jellyfish and other cnidarians.

gastrulation Development of a blastula into a gastrula. In embryonic development, the process by which a blastula is transformed by massive movements of cells into a *gastrula*, an embryo with three germ layers and distinct body axes.

gated channel A membrane protein that changes its three-dimensional shape, and therefore its ion conductance, in response to a stimulus. When open, it allows specific ions to move across the membrane.

gel electrophoresis (e lek′ tro fo ree′ sis) [L. *electrum*: amber + Gk. *phorein*: to bear] A technique for separating molecules (such as DNA fragments) from one another on the basis of their electric charges and molecular weights by applying an electric field to a gel.

gene [Gk. *genes*: to produce] A unit of heredity. Used here as the unit of genetic function which carries the information for a polypeptide or RNA.

gene duplication The generation of extra copies of a gene in a genome over evolutionary time. A mechanism by which genomes can acquire new functions.

gene expression The transcription and translation into a protein of the information (nucleotide sequence) contained in a gene.

gene family A set of similar genes derived from a single parent gene; need not be on the same chromosomes. The vertebrate globin genes constitute a classic example of a gene family.

gene flow Exchange of genes between populations through migration of individuals or movements of gametes.

gene pool All of the different alleles of all of the genes existing in all individuals of a population.

gene therapy Treatment of a genetic disease by providing patients with cells containing functioning alleles of the genes that are nonfunctional in their bodies.

gene tree A graphic representation of the evolutionary relationships of a single gene in different species or of the members of a gene family.

gene-for-gene concept In plants, a mechanism of resistance to pathogens in which resistance is triggered by the specific interaction of the products of a pathogen's *Avr* genes and a plant's *R* genes.

general transcription factors In eukaryotes, transcription factors that bind to the promoters of most protein-coding genes and are required for their expression. Distinct from transcription factors that have specific regulatory effects only at certain promoters or classes of promoters.

genetic code The set of instructions, in the form of nucleotide triplets, that translate a linear sequence of nucleotides in mRNA into a linear sequence of amino acids in a protein.

genetic drift Changes in gene frequencies from generation to generation as a result of random (chance) processes.

genetic linkage Association between genes on the same chromosome such that they do not show random assortment and seldom recombine; the closer the genes, the lower the frequency of recombination.

genetic map The positions of genes along a chromosome as revealed by recombination frequencies.

genetic marker (1) In gene cloning, a gene of identifiable phenotype that indicates the presence of another gene, DNA segment, or chromosome fragment. (2) In general, a DNA sequence such as a single nucleotide polymorphism whose presence is correlated with the presence of other linked genes on that chromosome.

genetic screen A technique for identifying genes involved in a biological process of interest. Involves creating a large collection of randomly mutated organisms and identifying those individuals that are likely to have a defect in the pathway of interest. The mutated gene(s) in those individuals can then be isolated for further study.

genetic structure The frequencies of different alleles at each locus and the frequencies of different genotypes in a Mendelian population.

genetic switches Mechanisms that control how the genetic toolkit is used, such as promoters and the transcription factors that bind them. The signal cascades that converge on and operate these switches determine when and where genes will be turned on and off.

genetic toolkit A set of developmental genes and proteins that is common to most animals and is hypothesized to be responsible for the evolution of their differing developmental pathways.

genetics The scientific study of the structure, functioning, and inheritance of genes, the units of hereditary information.

genome (jee' nome) The complete DNA sequence for a particular organism or individual.

genome sequencing Determination of the nucleotide base sequence of the entire genome of an organism.

genomic imprinting The form of a gene's expression is determined by parental source (i.e., whether the gene is inherited from the male or female parent).

genomic library All of the cloned DNA fragments generated by the breakdown of genomic DNA into smaller segments.

genomics The scientific study of entire sets of genes and their interactions.

genotype (jean' oh type) [Gk. *gen*: to produce + *typos*: impression] An exact description of the genetic constitution of an individual, either with respect to a single trait or with respect to a larger set of traits. (Contrast with phenotype.)

genotype frequency The proportion of a genotype among individuals in a population.

genus (jean' us) (plural: genera) [Gk. *genos*: stock, kind] A group of related, similar species recognized by taxonomists with a distinct name used in binomial nomenclature.

geographic range The region within which a species occurs.

germ cell [L. *germen*: to beget] A reproductive cell or gamete of a multicellular organism. (Contrast with somatic cell.)

germ layers The three embryonic layers formed during gastrulation (ectoderm, mesoderm, and endoderm). Also called cell layers or tissue layers.

germ line mutation Mutation in a cell that produces gametes (i.e., a germ line cell). (Contrast with somatic mutation.)

germination Sprouting of a seed or spore.

gestation (jes tay' shun) [L. *gestare*: to bear] The period during which the embryo of a mammal develops within the uterus. Also known as pregnancy.

ghrelin A hormone produced and secreted by cells in the stomach that stimulates appetite.

gibberellin (jib er el' lin) A class of plant growth hormones playing roles in stem elongation, seed germination, flowering of certain plants, etc.

gill An organ specialized for gas exchange with water.

gizzard (giz' erd) [L. *gigeria*: cooked chicken parts] The second of two stomachlike organs in birds, other reptiles, earthworms, and various insects, that grinds up food, sometimes with the aid of fragments of stone. (*See also* crop.)

glia (glee' uh) [Gk. *glia*: glue] One of the two classes of neural cells (along with neurons, with which glia interact); glia do not typically conduct action potentials. Types of glia include astrocytes, oligodendrocytes, and Schwann cells.

global nitrogen cycle The movement of nitrogen through the biosphere. Steps in the cycle include the fixation of nitrogen gas (N_2) to ammonia; nitrification of the fixed nitrogen to nitrate by bacteria; nitrate reduction by plants; and denitrification back to N_2 by bacteria.

glomerular filtration rate (GFR) The rate at which the blood is filtered in the glomeruli of the kidney.

glomerulus (glo mare' yew lus) [L. *glomus*: ball] Sites in the kidney where blood filtration takes place. Each glomerulus consists of a knot of capillaries served by afferent and efferent arterioles.

glucagon Hormone produced by alpha cells of the pancreatic islets of Langerhans. Glucagon stimulates the liver to break down glycogen and release glucose into the circulation.

gluconeogenesis The biochemical synthesis of glucose from other substances, such as amino acids, lactate, and glycerol.

glucose [Gk. *gleukos*: sugar, sweet wine] The most common monosaccharide; the monomer of the polysaccharides starch, glycogen, and cellulose.

glyceraldehyde 3-phosphate (G3P) A phosphorylated three-carbon sugar; an intermediate in glycolysis and photosynthetic carbon fixation.

glycerol (gliss' er ole) A three-carbon alcohol with three hydroxyl groups; a component of phospholipids and triglycerides.

glycogen (gly' ko jen) [Gk. *glyk*: sweet] An energy storage polysaccharide found in animals and fungi; a branched-chain polymer of glucose, similar to starch.

glycolipid A lipid to which sugars are attached.

glycolysis (gly kol' li sis) [Gk. *gleukos*: sugar + *lysis*: break apart] The enzymatic breakdown of glucose to pyruvic acid.

glycoprotein A protein to which sugars are attached.

glycosidic linkage Bond between carbohydrate (sugar) molecules through an intervening oxygen atom (–O–).

glycosylation The addition of carbohydrates to another type of molecule, such as a protein.

glyoxysome (gly ox' ee soam) An organelle found in plants, in which stored lipids are converted to carbohydrates.

Golgi apparatus (goal' jee) A system of concentrically folded membranes found in the cytoplasm of eukaryotic cells; functions in secretion from the cell by exocytosis.

Golgi tendon organ A mechanoreceptor found in tendons and ligaments; provides information about the force generated by a contracting muscle.

gonad (go' nad) [Gk. *gone*: seed] An organ that produces gametes in animals: either an ovary (female gonad) or testis (male gonad).

gonadotropin A trophic hormone that stimulates the gonads.

gonadotropin-releasing hormone (GnRH) Hormone produced by the hypothalamus that stimulates the anterior pituitary to secrete gonadotropins.

Gondwana The large southern land mass that existed from the Cambrian (540 mya) to the Jurassic (138 mya). Present-day remnants are South America, Africa, India, Australia, and Antarctica.

graded membrane potential Small local change in membrane potential caused by opening or closing of ion channels.

grafting Artificial transplantation of tissue from one organism to another. In horticulture, the transfer of a bud or stem segment from one plant onto the root of another as a form of asexual reproduction.

gram stain A differential purple stain useful in characterizing bacteria. The peptidoglycan-rich cell walls of gram-positive bacteria stain purple; cell walls of gram-negative bacteria generally stain orange.

gravitropism [L. *gravitas*: weight, force; Gk. *tropos*: to turn] A directed plant growth response to gravity.

gray crescent In frog development, a band of diffusely pigmented cytoplasm on the side of the egg opposite the site of sperm entry. Arises as a result of cytoplasmic rearrangements that establish the anterior–posterior axis of the zygote.

gray matter In the nervous system, tissue that is rich in neuronal cell bodies. (Contrast with white matter.)

greenhouse gases Gases in the atmosphere, such as carbon dioxide and methane, that are transparent to sunlight, but trap heat radiating from Earth's surface, causing heat to build up at Earth's surface.

gross primary production The amount of energy captured by the primary producers in a community.

gross primary productivity (GPP) The rate at which the primary producers in a community turn solar energy into stored chemical energy via photosynthesis.

ground meristem That part of an apical meristem that gives rise to the ground tissue system of the primary plant body.

ground tissue system Those parts of the plant body not included in the dermal or vascular tissue systems. Ground tissues function in storage, photosynthesis, and support.

growth An increase in the size of the body and its organs by cell division and cell expansion.

growth factor A chemical signal that stimulates cells to divide.

growth hormone A peptide hormone released by the anterior pituitary that stimulates many anabolic processes.

guanine (G) (gwan' een) A nitrogen-containing base found in DNA, RNA, and GTP.

guard cells In plants, specialized, paired epidermal cells that surround and control the opening of a stoma (pore). *See* stoma.

gustation The sense of taste.

gut An animal's digestive tract.

gymnosperms Seed plants that do not produce flowers or fruits; one of the two major groups of living seed plants. (*See also* angiosperms.)

gyrus (ji' rus) [Gk. *gyros*: spiral) *See* convolutions.

H

habitat The particular environment in which an organism lives.

habitat patches Also called **habitat islands**; areas of suitable habitat for a species that are separated by substantial areas of unsuitable habitat.

Hadley cells Patterns of vertical atmospheric circulation that influence surface winds and precipitation patterns according to latitude.

hair cell A type of mechanoreceptor in animals. Detects sound waves and other forms of motion in air or water.

half-life The time required for half of a sample of a radioactive isotope to decay to its stable, nonradioactive form, or for a drug or other substance to reach half its initial dosage.

halophyte (hal' oh fyte) [Gk. *halos*: salt + *phyton*: plant] A plant that grows in a saline (salty) environment.

Hamilton's rule The principle that, for an apparent altruistic behavior to be adaptive, the fitness benefit of that act to the recipient times the degree of relatedness of the performer and the recipient must be greater than the cost to the performer.

haplodiploidy A sex determination mechanism in which diploid individuals (which develop from fertilized eggs) are female and haploid individuals (which develop from unfertilized eggs) are male; typical of hymenopterans.

haploid (hap' loid) [Gk. *haploeides*: single] Having a chromosome complement consisting of just one copy of each chromosome; designated 1*n* or *n*. (Contrast with diploid.)

haplotype Linked nucleotide sequences that are usually inherited as a unit (as a "sentence" rather than as individual "words").

Hardy–Weinberg equililbrium In a sexually reproducing population, the allele frequency at a given locus that is not being acted on by agents of evolution; the conditions that would result in no evolution in a population.

haustorium (haw stor' ee um) (plural: haustoria)[L. *haustus*: draw up] A specialized hypha or other structure by which fungi and some parasitic plants draw nutrients from a host plant.

Haversian systems Units of organization in compact bone that reflect the action of intercommunicating osteoblasts.

heart In circulatory systems, a muscular pump that moves extracellular fluid around the body.

heat of vaporization The energy that must be supplied to convert a molecule from a liquid to a gas at its boiling point.

heat shock proteins Chaperone proteins expressed in cells exposed to high or low temperatures or other forms of environmental stress.

helical Shaped like a screw or spring (helix); this shape occurs in DNA and proteins.

helper T cells *See* T-helper cells.

hemiparasite A parasitic plant that can photosynthesize, but derives water and mineral nutrients from the living body of another plant. (Contrast with holoparasite.)

hemizygous (hem' ee zie' gus) [Gk. *hemi*: half + *zygotos*: joined] In a diploid organism, having only one allele for a given trait, typically the case for X-linked genes in male mammals and Z-linked genes in female birds. (Contrast with homozygous, heterozygous.)

hemoglobin (hee' mo glow bin) [Gk. *heaema*: blood + L. *globus*: globe] Oxygen-transporting protein found in the red blood cells of vertebrates (and found in some invertebrates).

Hensen's node In avian embryos, a structure at the anterior end of the primitive groove; determines the fates of cells passing over it during gastrulation.

hepatic (heh pat' ik) [Gk. *hepar*: liver] Pertaining to the liver.

herbivore (ur' bi vore) [L. *herba*: plant + *vorare*: to devour] An animal that eats plant tissues. (Contrast with carnivore, detritivore, omnivore.)

heritable trait A trait that is at least partly determined by genes.

hermaphroditism (her maf' row dite ism) The coexistence of both female and male sex organs in the same organism.

hetero- [Gk.: *heteros*: other, different] A prefix indicating two or more different conditions, structures, or processes. (Contrast with homo-.)

heterochromatin Densely packed, dark-staining chromatin; any genes it contains are usually not transcribed.

heterochrony [Gk: different time] Alteration in the timing of developmental events, contributing to the evolution different phenotypes in the adult.

heterocyst A large, thick-walled cell type in the filaments of certain cyanobacteria that performs nitrogen fixation.

heterometry [Gk: different measure] Alteration in the level of gene expression, and thus in the amount of protein produced, during development, contributing to the evolution of different phenotypes in the adult.

heteromorphic (het' er oh more' fik) [Gk.: different form] Having a different form or appearance, as two heteromorphic life stages of a plant. (Contrast with isomorphic.)

heterosis The superior fitness of heterozygous offspring as compared with that of their dissimilar homozygous parents. Also called hybrid vigor.

heterosporous (het' er os' por us) Producing two types of spores, one of which gives rise to a female megaspore and the other to a male microspore. (Contrast with homosporous.)

heterotherm An animal that regulates its body temperature at a constant level at some times but not others, such as a hibernator.

heterotopy [Gk: different place] Spatial differences in gene expression during development, controlled by developmental regulatory genes and contributing to the evolution of distinctive adult phenotypes.

heterotroph (het' er oh trof) [Gk. *heteros*: different + *trophe*: feed] An organism that requires preformed organic molecules as food. (Contrast with autotroph.)

heterotrophic succession Succession in detritus-based communities, which differs from other types of succession in taking place without the participation of plants.

heterotypy [Gk.: different kind] Alteration in a developmental regulatory gene itself rather than the expression of the genes it controls. (Contraste with heterochrony; heterometry; heterotopy.)

heterozygous (het' er oh zie' gus) [Gk. *heteros*: different + *zygotos*: joined] In diploid organisms, having different alleles of a given gene on the pair of homologs carrying that gene. (Contrast with homozygous.)

heterozygous carrier An individual that carries a recessive allele for a phenotype of interest (e.g., a genetic disease); the individual does not show the phenotype, but may have progeny with the phenotype if the other parent also carries the recessive allele.

hexose [Gk. *hex*: six] A sugar containing six carbon atoms.

hibernation [L. *hibernum*: winter] The state of inactivity of some animals during winter; marked by a drop in body temperature and metabolic rate. (Contrast with estivation)

high-density lipoproteins (HDLs) Lipoproteins that remove cholesterol from tissues and carry it to the liver; HDLs are the "good" lipoproteins associated with good cardiovascular health.

high-throughput sequencing Rapid DNA sequencing on a micro scale in which many fragments of DNA are sequenced in parallel.

highly repetitive sequences Short (less than 100 bp), nontranscribed DNA sequences, repeated thousands of times in tandem arrangements.

hindbrain The region of the developing vertebrate brain that gives rise to the medulla, pons, and cerebellum.

hippocampus [Gk. sea horse] A part of the forebrain that takes part in long-term memory formation.

histamine (hiss' tah meen) A substance released by damaged tissue, or by mast cells in response to allergens. Histamine increases vascular permeability, leading to edema (swelling). (Contrast with histone deacetylase.)

histone Any one of a group of proteins forming the core of a nucleosome, the structural unit of a eukaryotic chromosome.

histone acetyltransferases Enzymes involved in chromatin remodeling. Add acetyl groups to the tail regions of histone proteins.

histone deacetylase In chromatin remodeling, an enzyme that removes acetyl groups from the tails of histone proteins. (Contrast with histone acetyltransferases.)

HIV Human immunodeficiency virus, the retrovirus that causes acquired immune deficiency syndrome (AIDS).

holoparasite A fully parasitic plant (i.e., one that does not perform photosynthesis).

homeobox 180-base-pair segment of DNA found in certain homeotic genes. A specific sequence within the homeobox—the **homeodomain**—regulates the expression of other genes and through this regulation controls large-scale developmental processes. (*See* homeotic genes.)

homeostasis (home' ee o sta' sis) [Gk. *homos*: same + *stasis*: position] The maintenance of a steady state, such as a constant temperature, by means of physiological or behavioral feedback responses.

homeotic genes Genes that act during development to determine the formation of an organ from a region of the embryo. (Compare with Hox genes.)

homeotic mutation Mutation in a homeotic gene that results in the formation of a different organ than that normally made by a region of the embryo.

homing In animal navigation, the ability to return to a nest site, burrow, or other specific location.

hominid Lineage that includes all modern and extinct Great Apes (i.e., humans, gorillas, chimpanzees, orangutans, and their ancestors.)

hominin Lineages that includes modern humans (*Homo sapiens*) and their extinct ancestors (e.g., Australopithecines; *Homo erectus*.)

homo- [Gk. *homos*: same] A prefix indicating two or more similar conditions, structures, or processes. (Contrast with hetero-.)

homolog (1) In cytogenetics, one of a pair (or larger set) of chromosomes having the same overall genetic composition and sequence. In diploid organisms, each chromosome inherited from one parent is matched by an identical (except for mutational changes) chromosome—its homolog—from the other parent. (2) In evolutionary biology, one of two or more features in different species that are similar by reason of descent from a common ancestor.

homologous pair A pair of matching chromosomes made up of a chromosome from each of the two sets of chromosomes in a diploid organism.

homologous recombination Exchange of segments between two DNA molecules based on sequence similarity between the two molecules. The similar sequences align and crossover. Used to create knockout mutants in mice and other organisms.

homology (ho mol' o jee) [Gk. *homologia*: of one mind; agreement] A similarity between two or more features that is due to inheritance from a common ancestor. The structures are said to be *homologous*, and each is a *homolog* of the others.

homoplasy (home' uh play zee) [Gk. *homos*: same + *plastikos*: shape, mold] The presence in multiple groups of a trait that is not inherited from the common ancestor of those groups. Can result from convergent evolution, evolutionary reversal, or parallel evolution.

homosporous Producing a single type of spore that gives rise to a single type of gametophyte, bearing both female and male reproductive organs. (Contrast with heterosporous.)

homotypic Pertaining to adhesion of cells of the same type. (Contrast with heterotypic.)

homozygous (home' oh zie' gus) [Gk. *homos*: same + *zygotos*: joined] In diploid organisms, having identical alleles of a given gene on both homologous chromosomes. An individual may be a homozygote with respect to one gene and a heterozygote with respect to another. (Contrast with heterozygous.)

horizons The horizontal layers of a soil profile, including the topsoil (A horizon), subsoil (B horizon) and parent rock or bedrock (C horizon).

hormone (hore' mone) [Gk. *hormon*: to excite, stimulate] A chemical signal produced in minute amounts at one site in a multicellular organism and transported to another site where it acts on target cells.

host An organism that harbors a parasite or symbiont and provides it with nourishment.

Hox genes Conserved homeotic genes found in vertebrates, *Drosophila*, and other animal groups. Hox genes contain the homeobox and specify pattern and axis formation in these animals.

human chorionic gonadotropin (hCG) A hormone secreted by the placenta which sustains the corpus luteum and helps maintain pregnancy.

Human Genome Project A publicly and privately funded research effort, successfully completed in 2003, to produce a complete DNA sequence for the entire human genome.

humoral immune response The response of the immune system mediated by B cells that produces circulating antibodies active against extracellular bacterial and viral infections. (Contrast with cellular immune response.)

humus (hew' mus) The partly decomposed remains of plants and animals on the surface of a soil.

hybrid (high' brid) [L. *hybrida*: mongrel] (1) The offspring of genetically dissimilar parents. (2) In molecular biology, a double helix formed of nucleic acids from different sources.

hybrid vigor *See* heterosis.

hybrid zone A region of overlap in the ranges of two closely related species where the species may hybridize.

hybridize (1) In genetics, to combine the genetic material of two distinct species or of two distinguishable populations within a species. (2) In molecular biology, to form a double-stranded nucleic acid in which the two strands originate from different sources.

hydrocarbon A compound containing only carbon and hydrogen atoms.

hydrogen bond A weak electrostatic bond which arises from the attraction between the slight positive charge on a hydrogen atom and a slight negative charge on a nearby oxygen or nitrogen atom.

hydrologic cycle The movement of water from the oceans to the atmosphere, to the soil, and back to the oceans.

hydrolysis reaction (high drol' uh sis) [Gk. *hydro*: water + *lysis*: break apart] A chemical reaction that breaks a bond by inserting the components of water ($AB + H_2O \rightarrow AH + BOH$). (Contrast with condensation reaction.)

hydrophilic (high dro fill′ ik) [Gk. *hydro*: water + *philia*: love] Having an affinity for water. (Contrast with hydrophobic.)

hydrophobic (high dro foe′ bik) [Gk. *hydro*: water + *phobia*: fear] Having no affinity for water. Uncharged and nonpolar groups of atoms are hydrophobic. (Contrast with hydrophilic.)

hydroponic Pertaining to a method of growing plants with their roots suspended in nutrient solutions instead of soil.

hydrostatic pressure Pressure generated by compression of liquid in a confined space. Generated in plants, fungi, and some protists with cell walls by the osmotic uptake of water. Generated in animals with closed circulatory systems by the beating of a heart.

hydrostatic skeleton A fluid-filled body cavity that transfers forces from one part of the body to another when acted on by surrounding muscles.

hydroxyl group The —OH group found on alcohols and sugars.

hyper- [Gk. *hyper*: above, over] Prefix indicating above, higher, more. (Contrast with hypo-.)

hyperaccumulators Plant species that store large quantities of heavy metals such as arsenic, cadmium, nickel, aluminum, and zinc.

hyperpolarization A change in the resting potential across a membrane so that the inside of a cell becomes more negative compared with the outside of the cell. (Contrast with depolarization.)

hypersensitive response A defensive response of plants to microbial infection in which phytoalexins and pathogenesis-related proteins are produced and the infected tissue undergoes apoptosis to isolate the pathogen from the rest of the plant.

hypertonic Having a greater solute concentration. Said of one solution compared with another. (Contrast with hypotonic, isotonic.)

hypha (high′ fuh) (plural: hyphae) [Gk. *hyphe*: web] In the fungi and oomycetes, any single filament.

hypo- [Gk. *hypo*: beneath, under] Prefix indicating underneath, below, less. (Contrast with hyper-.)

hypoblast The lower tissue portion of the avian blastula which is joined to the epiblast at the margins of the blastodisc.

hypothalamus The part of the brain lying below the thalamus; it coordinates water balance, reproduction, temperature regulation, and metabolism.

hypothermia Below-normal body temperature.

hypothesis A tentative answer to a question, from which testable predictions can be generated. (Contrast with theory.)

hypotonic Having a lesser solute concentration. Said of one solution in comparing it to another. (Contrast with hypertonic, isotonic.)

hypoxia A deficiency of oxygen.

I

ileum The final segment of the small intestine. (*See also* duodenum, jejunum.)

imbibition Water uptake by a seed; first step in germination.

immediate hypersensitivity A rapid, extensive overreaction of the immune system against an allergen, resuting in the release of large amounts of histamine. (Contrast with delayed hypersensitivity.)

immediate memory A form of memory for events happening in the present that is almost perfectly photographic, but lasts only seconds.

immunity [L. *immunis*: exempt from] In animals, the ability to avoid disease when invaded by a pathogen by deploying various defense mechanisms.

immunoassay The use of antibodies to measure the concentration of an antigen in a sample.

immunoglobulins A class of proteins containing a tetramer consisting of four polypeptide chains—two identical light chains and two identical heavy chains—held together by disulfide bonds; active as receptors and effectors in the immune system.

immunological memory The capacity to more rapidly and massively respond to a second exposure to an antigen than occurred on first exposure.

imperfect flower A flower lacking either functional stamens or functional carpels. (Contrast with perfect flower.)

implantation The process by which the early mammalian embryo becomes attached to and embedded in the lining of the uterus.

imprinting In animal behavior, a rapid form of learning in which an animal learns, during a brief critical period, to make a particular response (which is then maintained for life) to some object or other organism. *See also* genomic imprinting.

in vitro [L.: in glass] A biological process occurring outside of the organism, in the laboratory. (Contrast with in vivo.)

in vitro evolution A method based on natural molecular evolution that uses artificial selection in the laboratory to rapidly produce molecules with novel enzymatic and binding functions.

in vivo [L.: in life] A biological process occurring within a living organism or cell. (Contrast with in vitro.)

inclusive fitness The sum of an individual's genetic contribution to subsequent generations both via production of its own offspring and via its influence on the survival of relatives who are not direct descendants. (Contrast with direct fitness)

incomplete cleavage A pattern of cleavage that occurs in many eggs that have a lot of yolk, in which the cleavage furrows do not penetrate all of it. (*See also* discoidal cleavage, superficial cleavage; contrast with complete cleavage.)

incomplete dominance Condition in which the heterozygous phenotype is intermediate between the two homozygous phenotypes.

incomplete metamorphosis Insect development in which changes between instars are gradual. (Contrast with direct development; complete metamorphosis.)

independent assortment During meiosis, the random separation of genes carried on nonhomologous chromosomes into gametes so that inheritance of these genes is random. This principle was articulated by Mendel as his second law.

independent variable In a scientific experiment, a critical factor that is manipulated while all other factors are held constant. (Contrast with dependent variable.)

indeterminate growth A open-ended growth pattern in which an organism or organ continues to grow as long as it lives; characteristic of some animals and of plant shoots and roots. (Contrast with determinate growth.)

individual fitness See direct fitness.

induced fit A change in the shape of an enzyme caused by binding to its substrate that exposes the active site of the enzyme.

induced mutation A mutation resulting from exposure to a mutagen from outside the cell. (Contrast with spontaneous mutation.)

induced pluripotent stem cells (iPS cells) Multipotent or pluripotent animal stem cells produced from differentiated cells in vitro by the addition of several genes that are expressed.

induced responses Defensive responses that a plant produces only in the presence of a pathogen, in contrast to constitutive defenses, which are always present.

inducer (1) A compound that stimulates the synthesis of a protein. (2) In embryonic development, a substance that causes a group of target cells to differentiate in a particular way.

inducible genes Genes that are expressed only when their products—**inducible proteins**—are needed. (Contrast with constitutive genes.)

inducible Produced only in the presence of a particular compound or under particular circumstances. (Contrast with constitutive.)

induction In embryonic development, the process by which a factor produced and secreted by certain cells determines the fates other cells.

inductive logic Involves making observations and then formulating one or more possible scenarios—hypotheses—that might explain those observations. (Contrast with deductive logic.)

inflammation A nonspecific defense against pathogens; characterized by redness, swelling, pain, and increased temperature.

inflorescence A structure composed of several to many flowers.

inflorescence meristem A meristem that produces floral meristems as well as other small leafy structures (bracts).

ingroup In a phylogenetic study, the group of organisms of primary interest. (Contrast with outgroup.)

inhibitor A substance that blocks a biological process.

inhibitory Input from a neuron that causes hyperpolarization of the recipient cell.

initials Cells that perpetuate plant meristems, comparable to animal stem cells. When an initial divides, one daughter cell develops into another initial, while the other differentiates into a more specialized cell.

initiation complex In protein translation, a combination of a small ribosomal subunit, an mRNA molecule, and the tRNA charged with the first amino acid coded for by the mRNA; formed at the onset of translation.

initiation site The place within a promoter where transcription begins.

innate defenses In animals, one of two general types of defenses against pathogens. Nonspecific and present in most animals. (Contrast with adaptive immunity.)

inner cell mass Derived from the mammalian blastula (bastocyst), the inner cell mass that will give rise to the yolk sac (via hypoblast) and embryo (via epiblast).

inorganic fertilizer A chemical or combination of chemicals applied to soil or plants to make up for a plant nutrient deficiency. Often contains the macronutrients nitrogen, phosphorus, and potassium (N-P-K).

inositol trisphosphate (IP_3) An intracellular second messenger derived from membrane phospholipids.

inspiratory reserve volume The amount of air that can be inhaled above the normal tidal inspiration. (Contrast with expiratory reserve volume, tidal volume, vital capacity.)

instar (in′ star) An immature stage of an insect between molts.

insula (in′ su lah) [L. *insula*: island] An area deep within the forebrain that appears to integrate physiological information from all over the body to create a sensation of how the body "feels" and may be involved in human consciousness. Also called the insular cortex.

insulin (in′ su lin) [L. *insula*: island] A hormone synthesized in islet cells of the pancreas that promotes the conversion of glucose into the storage material, glycogen.

integral membrane proteins Proteins that are at least partially embedded in the plasma membrane. (Contrast with peripheral membrane proteins.)

integrin In animals, a transmembrane protein that mediates the attachment of epithelial cells to the extracellular matrix.

integument [L. *integumentum*: covering] A protective surface structure. In gymnosperms and angiosperms, a layer of tissue around the ovule which will become the seed coat.

intercostal muscles Muscles between the ribs that can augment breathing movements by elevating and suppressing the rib cage.

interference competition Competition in which individuals actively interfere with one another's access to resources. (Contrast with exploitation competition.)

interference RNA (RNAi) *See* RNA interference.

interferons Glycoproteins produced by virus-infected animal cells; interferons increase the resistance of neighboring cells to the virus.

internal environment In multicelluar organisms, includes blood plasma and interstitial fluid, i.e., the extracellular fluids that surround the cells.

internal fertilization The release of sperm into the female reproductive tract; typical of most terrestrial animals. (Contrast with external fertilization.)

internal gills Gills enclosed in protective body cavities; typical of mollusks, arthropods, and fishes.**interneuron** A neuron that communicates information between two other neurons.

interneuron A neuron that communicates information between two other neurons.

internode The region between two nodes of a plant stem.

interphase In the cell cycle, the period between successive nuclear divisions during which the chromosomes are diffuse and the nuclear envelope is intact. During interphase the cell is most active in transcribing and translating genetic information.

interspecific competition Competition between members of two or more species. (Contrast with intraspecific competition; see also exploitation competition, interference competition.)

interstitial fluid Extracellular fluid that is not contained in the vessels of a circulatory system.

intertidal zone A nearshore region of oceans that is periodically exposed to the air as the tides rise and fall.

intestine The portion of the gut following the stomach, in which most digestion and absorption occurs.

intraspecific competition Competition among members of the same species. (Contrast with interspecific competition.)

intrinsic rate of increase The rate at which a population is capable of growing when its density is low and environmental conditions are highly favorable.

intron Portion of a of a gene within the coding region that is transcribed into pre-mRNA but is spliced out prior to translation. (Contrast with exon.)

invasive species An exotic species that reproduces rapidly, spreads widely, and has negative effects on the native species of the region to which it has been introduced.

invasiveness The ability of a pathogen to multiply in a host's body. (Contrast with toxigenicity).

inversion A rare 180° reversal of the order of genes within a segment of a chromosome.

involution Cell movements that occur during gastrulation of frog embryos, giving rise to the archenteron.

ion (eye′ on) [Gk. *ion*: wanderer] An electrically charged particle that forms when an atom gains or loses one or more electrons.

ion channel An integral membrane protein that allows ions to diffuse across the membrane in which it is embedded.

ion exchange In plants**, a** process by which protons produced by the plant's root displace mineral cations from clay particles in the surrounding soil.

ionic attraction An electrostatic attraction between positively and negatively charged ions.

ionotropic receptors A receptor that directly alters membrane permeability to a type of ion when it combines with its ligand.

iris (eye′ ris) [Gk. *iris*: rainbow] The round, pigmented membrane that surrounds the pupil of the eye and adjusts its aperture to regulate the amount of light entering the eye.

island biogeography A theory proposing that the number of species on an island (or in another geographically defined and isolated area) represents a balance, or equilibrium, between the rate at which species immigrate to the island and the rate at which resident species go extinct.

islets of Langerhans Clusters of hormone-producing cells in the pancreas.

iso- [Gk. *iso*: equal] Prefix used for two separate entities that share some element of identity.

isomers Molecules consisting of the same numbers and kinds of atoms, but differing in the bonding patterns by which the atoms are held together.

isomorphic (eye so more′ fik) [Gk. *isos*: equal + *morphe*: form] Having the same form or appearance, as when the haploid and diploid life stages of an organism appear identical. (Contrast with heteromorphic.)

isotonic Having the same solute concentration; said of two solutions. (Contrast with hypertonic, hypotonic.)

isotope (eye′ so tope) [Gk. *isos*: equal + *topos*: place] Isotopes of a given chemical element have the same number of protons in their nuclei (and thus are in the same position on the periodic table), but differ in the number of neutrons.

isozymes Enzymes of an organism that have somewhat different amino acid sequences but catalyze the same reaction.

iteroparous [L. itero, to repeat + pario, to beget] Reproducing multiple times in a lifetime. (Contrast with semelparous.)

J

jasmonate Also called jasmonic acid, a plant hormone involved in triggering responses to pathogen attack as well as other processes.

jejunum (jih jew′ num) The middle division of the small intestine, where most absorption of nutrients occurs. (*See also* duodenum, ileum.)

joint In skeletal systems, a junction between two or more bones.

juvenile hormone In insects, a hormone maintaining larval growth and preventing maturation or pupation.

K

***K*-strategist** A species whose life history strategy allows it to persist at or near the carrying capacity (*K*) of its environment. (Contrast with *r*-strategist.)

karyogamy The fusion of nuclei of two cells. (Contrast with plasmogamy.)

karyotype The number, forms, and types of chromosomes in a cell.

keystone species Species that have a dominant influence on the composition of a community.

kidneys A pair of excretory organs in vertebrates.

kilocalorie (kcal) *See* Calorie.

kin selection That component of inclusive fitness resulting from helping the survival of relatives containing the same alleles by descent from a common ancestor. (Contrast with direct fitness.)

kinase *See* protein kinase.

kinetic energy (kuh-net′ ik) [Gk. *kinetos*: moving] The energy associated with movement. (Contrast with potential energy.)

kinetochore (kuh net′ oh core) Specialized structure on a centromere to which microtubules attach.

knockout A molecular genetic method in which a single gene of an organism is permanently inactivated.

Koch's postulates A set of rules for establishing that a particular microorganism causes a particular disease.

Krebs cycle See citric acid cycle.

L

lagging strand In DNA replication, the daughter strand that is synthesized in discontinuous stretches. (*See* Okazaki fragments.)

large intestine *See* colon.

larva (plural: larvae) [L. *lares*: guiding spirits] An immature stage of any animal that differs dramatically in appearance from the adult.

lateral [L. *latus*: side] Pertaining to the side.

lateral gene transfer The transfer of genes from one species to another, common among bacteria and archaea.

lateral meristem Either of the two meristems, the vascular cambium and the cork cambium, that give rise to a plant's secondary growth.

lateral root A root extending outward from the taproot in a taproot system; typical of eudicots.

lateralization A phenomenon in humans in which language functions come to reside in one cerebral hemisphere, usually the left.

laticifers (luh tiss′ uh furs) In some plants, elongated cells containing secondary plant products such as latex.

Laurasia The northernmost of the two large continents produced by the breakup of Pangaea.

law of independent assortment *See* independent assortment.

law of segregation *See* segregation.

laws of thermodynamics [Gk. *thermos*: heat + *dynamis*: power] Laws derived from studies of the physical properties of energy and the ways energy interacts with matter. (*See also* first law of thermodynamics, second law of thermodynamics.)

leaching In soils, a process by which mineral nutrients in upper soil horizons are dissolved in water and carried to deeper horizons, where they are unavailable to plant roots.

leading strand In DNA replication, the daughter strand that is synthesized continuously. (Contrast with lagging strand.)

leaf (plural: leaves) In plants, the chief organ of photosynthesis.

leaf primordium (plural: primordia) An outgrowth on the side of the shoot apical meristem that will eventually develop into a leaf.

leghemoglobin In nitrogen-fixing plants, an oxygen-carrying protein in the cytoplasm of nodule cells that transports enough oxygen to the nitrogen-fixing bacteria to support their respiration, while keeping free oxygen concentrations low enough to protect nitrogenase.

lek A display ground within which male animals compete for and defend small display areas as a means of demonstrating their territorial prowess and winning opportunities to mate.

lens In the vertebrate eye, a crystalline protein structure that makes fine adjustments in the focus of images falling on the retina.

leptin A hormone produced by fat cells that is believed to provide feedback information to the brain about the status of the body's fat reserves.

leukocyte *See* white blood cell.

lichen (lie′ kun) An organism resulting from the symbiotic association of a fungus and either a cyanobacterium or a unicellular alga.

life cycle The entire span of the life of an organism from the moment of fertilization (or asexual generation) to the time it reproduces in turn.

life history strategy The way in which an organism partitions its time and energy among growth, maintenance, and reproduction.

life history The time course of growth and development, reproduction, and death during an average individual organism's life.

life table A summary of information about the progression of individuals in a population through the various stages of their life cycles.

life zones In the aquatic (marine and freshwater) biomes, the regions defined by light penetration and water movement such as wave action. Life zones include, e.g., the intertidal, pelagic (open water) and bethic (bottom) zones.

ligament A band of connective tissue linking two bones in a joint.

ligand (lig′ and) Any molecule that binds to a receptor site of another (usually larger) molecule.

light reactions The initial phase of photosynthesis, in which light energy is converted into chemical energy. Followed by the **light-independent reactions** in which the energy captured in the light reactions is used to drive the reduction of CO_2 to form carbohydrates.

light-harvesting complex In photosynthesis, a group of different molecules that cooperate to absorb light energy and transfer it to a reaction center. Also called *antenna system*.

lignin A complex, hydrophobic polyphenolic polymer in plant cell walls that crosslinks other wall polymers, strengthening the walls, especially in wood.

limbic system A group of evolutionarily primitive structures in the vertebrate telencephalon that are involved in emotions, drives, instinctive behaviors, learning, and memory.

limiting resource The required resource whose supply (or lack thereof) most strongly influences the size of a population.

limnetic zone The open-water life zone of a lake

lineage A series of populations, species, or genes descended from a single ancestor over evolutionary time.

lineage species concept The definition of a species as a branch on the tree of life, which has a history that starts at a speciation event and ends either at extinction or at another speciation event. (Contrast with biological species concept; morphological species concept.)

linkage *See* genetic linkage.

lipase (lip′ ase; lye′ pase) An enzyme that digests fats.

lipid (lip′ id) [Gk. *lipos*: fat] Nonpolar, hydrophobic molecules that include fats, oils, waxes, steroids, and the phospholipids that make up biological membranes.

lipid bilayer *See* phospholipid bilayer.

lipoproteins Lipids packaged inside a covering of protein so that they can be circulated in the blood.

lithoosphere (lith′ o sphere) [Gk. *lithos*: strong] The crust of sold rock plates that overlays the viscous mantle of Earth. The movements of the lithosphere are the source of plate tectonics. (Constrast with asthenosphere.)

littoral zone The nearshore life zone of a lake that is shallow and is affected by wave action and fluctuations in water level.

liver A large digestive gland. In vertebrates, it secretes bile and is involved in the formation of blood.

loam A type of soil consisting of a mixture of sand, silt, clay, and organic matter. One of the best soil types for agriculture.

locus (low′ kus) (plural: loci, low′ sigh) In genetics, a specific location on a chromosome. May be considered synonymous with *gene*.

logistic growth Growth, especially in the size of an organism or in the number of organisms in a population, that slows steadily as the entity approaches its maximum size. (Contrast with multiplicative growth.)

long-day plant (LDP) A plant that requires long days (actually, short nights) in order to flower. (Compare to short-day plant.)

long-term potentiation (LTP) A long-lasting increase in the responsiveness of a neuron resulting from a period of intense stimulation.

loop of Henle (hen′ lee) Long, hairpin loop of the mammalian renal tubule that runs from the cortex down into the medulla and back to the cortex; creates a concentration gradient in the interstitial fluids in the medulla.

lophophore A U-shaped fold of the body wall with hollow, ciliated tentacles that encircles the mouth of animals in several different groups. Used for filtering prey from the surrounding water.

loss of function mutation A mutation that results in the loss of a functional protein. (Contrast with gain of function mutation.)

low-density lipoproteins (LDLs) Lipoproteins that transport cholesterol around the body for use in biosynthesis and for storage; LDLs are the "bad" lipoproteins associated with a high risk of cardiovascular disease.

lumen (loo′ men) [L. *lumen*: light] The open cavity inside any tubular organ or structure, such as the gut or a renal tubule.

lung An internal organ specialized for respiratory gas exchange with air.

luteinizing hormone (LH) A gonadotropin produced by the anterior pituitary that stimulates the gonads to produce sex hormones.

lymph [L. *lympha*: liquid] A fluid derived from blood and other tissues that accumulates in intercellular spaces throughout the body and is returned to the blood by the lymphatic system.

lymph node A specialized structure in the vessels of the lymphatic system. Lymph nodes contain lymphocytes, which encounter and respond to foreign cells and molecules in the lymph as it passes through the vessels.

lymphatic system A system of vessels that returns interstitial fluid to the blood.

lymphocyte One of the two major classes of white blood cells; includes T cells, B cells, and other cell types important in the immune system.

lysis (lie′ sis) [Gk. *lysis*: break apart] Bursting of a cell.

lysogeny A form of viral replication in which the virus becomes incorporated into the host chromosome and remains inactive. Also called a lysogenic cycle. (Contrast with lytic cycle.)

lysosome (lie′ so soam) [Gk. *lysis*: break away + *soma*: body] A membrane-enclosed organelle originating from the Golgi apparatus and containing hydrolytic enzymes. (Contrast with secondary lysosome.)

lysozyme (lie′ so zyme) An enzyme in saliva, tears, and nasal secretions that hydrolyzes bacterial cell walls.

lytic cycle A viral reproductive cycle in which the virus takes over a host cell's synthetic machinery to replicate itself, then bursts (lyses) the host cell, releasing the new viruses. (Contrast with lysogeny.)

M

M phase The portion of the cell cycle in which mitosis takes place.

macroevolution [Gk. *makros*: large] Evolutionary changes occurring over long time spans and usually involving changes in many traits. (Contrast with microevolution.)

macromolecule A giant (molecular weight > 1,000) polymeric molecule. The macromolecules are the proteins, polysaccharides, and nucleic acids.

macronutrient In plants, a mineral element required in concentrations of at least 1 milligram per gram of plant dry matter; in animals, a mineral element required in large amounts. (Contrast with micronutrient.)

macrophage (mac′ roh faj) Phagocyte that engulfs pathogens by endocytosis.

MADS box DNA-binding domain in many plant transcription factors that is active in development.

maintenance methylase An enzyme that catalyzes the methylation of the new DNA strand when DNA is replicated.

major histocompatibility complex (MHC) A complex of linked genes, with multiple alleles, that control a number of cell surface antigens that identify self and can lead to graft rejection.

malignant Pertaining to a tumor that can grow indefinitely and/or spread from the original site of growth to other locations in the body. (Contrast with benign.)

malnutrition A condition caused by lack of any essential nutrient.

Malpighian tubule (mal pee′ gy un) A type of protonephridium found in insects.

mantle (1) In mollusks, a fold of tissue that covers the organs of the visceral mass and secretes the hard shell that is typical of many mollusks. (2) In geology, the Earth's crust below the solid lithospheric plates.

map unit The distance between two genes as calculated from genetic crosses; a recombination frequency.

marine [L. *mare*: sea, ocean] Pertaining to or living in the ocean. (Contrast with aquatic, terrestrial.)

mark–recapture method A method of estimating population sizes of mobile organisms by capturing, marking, and releasing a sample of individuals, then capturing another sample at a later time.

mass extinction A period of evolutionary history during which rates of extinction are much higher than during intervening times.

mass number The sum of the number of protons and neutrons in an atom's nucleus.

mast cells Cells, typically found in connective tissue, that release histamine in response to tissue damage.

maternal effect genes Genes coding for morphogens that determine the polarity of the egg and larva in fruit flies. Part of a developmental cascade that includes gap genes, pair rule genes, segment polarity genes, and Hox genes.

mating type A particular strain of a species that is incapable of sexual reproduction with another member of the same strain but capable of sexual reproduction with members of other strains of the same species.

maturational survivorship curves *See* survivorship curves

maximum likelihood A statistical method of determining which of two or more hypotheses (such as phylogenetic trees) best fit the observed data, given an explicit model of how the data were generated.

mechanically gated channel A molecular channel that opens or closes in response to mechanical force applied to the plasma membrane in which it is inserted.

mechanoreceptor A cell that is sensitive to physical movement and generates action potentials in response.

medulla (meh dull′ luh) (1) The inner, core region of an organ, as in the adrenal medulla (adrenal gland) or the renal medulla (kidneys). (2) The portion of the brainstem that connects to the spinal cord.

medusa (plural: medusae) In cnidarians, a free-swimming, sexual life cycle stage shaped like a bell or an umbrella.

megagametophyte In heterosporous plants, the female gametophyte; produces eggs. (Contrast with microgametophyte.)

megaphyll The generally large leaf of a fern, horsetail, or seed plant, with several to many veins. (Contrast with microphyll.)

megaspore [Gk. *megas*: large + *spora*: to sow] In plants, a haploid spore that produces a female gametophyte.

megastrobilus In conifers, the female (seed-bearing) cone. (Contrast with microstrobilus.)

meiosis (my oh' sis) [Gk. *meiosis*: diminution] Division of a diploid nucleus to produce four haploid daughter cells. The process consists of two successive nuclear divisions with only one cycle of chromosome replication. In *meiosis I*, homologous chromosomes separate but retain their chromatids. The second division *meiosis II*, is similar to mitosis, in which chromatids separate.

melatonin A hormone released by the pineal gland. Involved in photoperiodicity and circadian rhythms.

membrane A phospholipid bylayer forming a barrier that separates the internal contents of a cell from the nonbiological environment, or enclosing the organelles within a cell. The membrane regulates the molecular substances entering or leaving a cell or organelle.

membrane potential The difference in electrical charge between the inside and the outside of a cell, caused by a difference in the distribution of ions.

membranous bone A type of bone that develops by forming on a scaffold of connective tissue. (Contrast with cartilage bone.)

memory cells Long-lived lymphocytes produced after exposure to antigen. They persist in the body and are able to mount a rapid response to subsequent exposures to the antigen.

Mendel's laws *See* independent assortment; segregation.

menopause In human females, the end of fertility and menstrual cycling.

menstruation The process by which the endometrium breaks down, and the sloughed-off tissue, including blood, flows from the body.

meristem [Gk. *meristos*: divided] Plant tissue made up of undifferentiated actively dividing cells.

meristem culture A method for the asexual propagation of plants, in which pieces of shoot apical meristem are cultured to produce plantlets.

meristem identity genes In angiosperms, a group of genes whose expression initiates flower formation, probably by switching meristem cells from a vegetative to a reproductive fate.

mesenchyme (mez' en kyme) [Gk. *mesos*: middle + *enchyma*: infusion] Embryonic or unspecialized cells derived from the mesoderm.

mesoderm [Gk. *mesos*: middle + *derma*: skin] The middle of the three embryonic germ layers first delineated during gastrulation. Gives rise to the skeleton, circulatory system, muscles, excretory system, and most of the reproductive system.

mesoglea (mez' uh glee uh) [Gk. *mesos*: middle + *gloia*, glue] A thick, gelatinous noncellular layer that separates the two cellular tissue layers of ctenophores, cnidarians, and scyphozoans.

mesophyll (mez' uh fill) [Gk. *mesos*: middle + *phyllon*: leaf] Chloroplast-containing, photosynthetic cells in the interior of leaves.

messenger RNA (mRNA) Transcript of a region of one of the strands of DNA; carries information (as a sequence of codons) for the synthesis of one or more proteins.

meta- [Gk.: between, along with, beyond] Prefix denoting a change or a shift to a new form or level; for example, as used in metamorphosis.

metabolic pathway A series of enzyme-catalyzed reactions so arranged that the product of one reaction is the substrate of the next.

metabolism (meh tab' a lizm) [Gk. *metabole*: change] The sum total of the chemical reactions that occur in an organism, or some subset of that total (as in respiratory metabolism).

metabolome The quantitative description of all the small molecules in a cell or organism.

metabotropic receptor A receptor that that indirectly alters membrane permeability to a type of ion when it combines with its ligand.

metagenomics The practice of analyzing DNA from environmental samples without isolating intact organisms.

metamorphosis (met' a mor' fo sis) [Gk. *meta*: between + *morphe*: form, shape] A change occurring between one developmental stage and another, as for example from a tadpole to a frog. (*See* complete metamorphosis, incomplete metamorphosis.)

metanephridia The paired excretory organs of annelids.

metaphase (met' a phase) The stage in nuclear division at which the centromeres of the highly supercoiled chromosomes are all lying on a plane (the metaphase plane or plate) perpendicular to a line connecting the division poles.

metapopulation A population divided into subpopulations, among which there are occasional exchanges of individuals.

methylation The addition of a methyl group (—CH_3) to a molecule.

MHC *See* major histocompatibility complex.

micelle A particle of lipid covered with bile salts that is produced in the duodenum and facilitates digestion and absorption of lipids.

microbiomes The diverse communities of bacteria that live on or within the body and are essential to bodily function.

microclimate A subset of climatic conditions in a small specific area, which generally differ from those in the environment at large, as in an animal's underground burrow.

microevolution Evolutionary changes below the species level, affecting allele frequencies. (Contrast with macroevolution.)

microfibril Crosslinked cellulose polymers, forming strong aggregates in the plant cell wall.

microfilament In eukaryotic cells, a fibrous structure made up of actin monomers. Microfilaments play roles in the cytoskeleton, in cell movement, and in muscle contraction.

microgametophyte In heterosporous plants, the male gametophyte; produces sperm. (Contrast with megagametophyte.)

microglia Glial cells that act as macrophages and mediators of inflammatory responses in the central nervous system.

micronutrient In plants, a mineral element required in concentrations of less than 100 micrograms per gram of plant dry matter; in animals, a mineral element required in concentrations of less than 100 micrograms per day. (Contrast with macronutrient.)

microphyll A small leaf with a single vein, found in club mosses and their relatives. (Contrast with megaphyll.)

micropyle (mike' roh pile) [Gk. *mikros*: small + *pylon*: gate] Opening in the integument(s) of a seed plant ovule through which pollen grows to reach the female gametophyte within.

microRNA A small, noncoding RNA molecule, typically about 21 bases long, that binds to mRNA to inhibit its translation.

microspore [Gk. *mikros*: small + *spora*: to sow] In plants, a haploid spore that produces a male gametophyte.

microstrobilus In conifers, male pollen-bearing cone. (Contrast with megastrobilus.)

microtubules Tubular structures found in centrioles, spindle apparatus, cilia, flagella, and cytoskeleton of eukaryotic cells. These tubules play roles in the motion and maintenance of shape of eukaryotic cells.

microvilli (sing.: microvillus) Projections of epithelial cells, such as the cells lining the small intestine, that increase their surface area.

midbrain One of the three regions of the vertebrate brain. Part of the brainstem, it serves as a relay station for sensory signals sent to the cerebral hemispheres.

middle lamella (la mell' ah) [L. *lamina*: thin sheet] A layer of polysaccharides that separates plant cells; a shared middle lamella lies outside the primary walls of the two cells.

mineral nutrients Inorganic ions required by organisms for normal growth and reproduction.

mismatch repair A mechanism that scans DNA after it has been replicated and corrects any base-pairing mismatches.

missense mutation A change in a gene's sequence that changes the amino acid at that site in the encoded protein. (Contrast with frame-shift mutation, nonsense mutation, silent mutation.)

mitochondria (my′ toe kon′ dree uh) (singular: mitochondrion) [Gk. *mitos*: thread + *chondros*: grain] Energy-generating organelles in eukaryotic cells that contain the enzymes of the citric acid cycle, the respiratory chain, and oxidative phosphorylation.

mitochondrial matrix The fluid interior of a mitochondrion, enclosed by the inner mitochondrial membrane.

mitosis (my toe′ sis) [Gk. *mitos*: thread] Nuclear division in eukaryotes leading to the formation of two daughter nuclei, each with a chromosome complement identical to that of the original nucleus.

mitosomes Reduced structures derived from mitochondria found in some organisms.

model systems Also known as **model organisms**, these include the small group of species that are the subject of extensive research. They are organisms that adapt well to laboratory situations and findings from experiments on them can apply across a broad range of species. Classic examples include white mice and the fruit fly *Drosophila*.

moderately repetitive sequences DNA sequences repeated 10–1,000 times in the eukaryotic genome. They include the genes that code for rRNAs and tRNAs, as well as the DNA in telomeres.

Modern Synthesis An understanding of evolutionary biology that emerged in the early twentieth century as the principles of evolution were integrated with the principles of modern genetics.

modularity In evolutionary developmental biology, the principle that the molecular pathways that determine different developmental processes operate independently from one another. *See also* developmental module.

mole A quantity of a compound whose weight in grams is numerically equal to its molecular weight expressed in atomic mass units. Avogadro's number of molecules: 6.023×10^{23} molecules.

molecular clock The approximately constant rate of divergence of macromolecules from one another over evolutionary time; used to date past events in evolutionary history.

molecular evolution The scientific study of the mechanisms and consequences of the evolution of macromolecules.

molecular toolkit *See* genetic toolkit.

molecular weight The sum of the atomic weights of the atoms in a molecule.

molecule A chemical substance made up of two or more atoms joined by covalent bonds or ionic attractions.

molting The process of shedding part or all of an outer covering, as the shedding of feathers by birds or of the entire exoskeleton by arthropods.

monoclonal antibody Antibody produced in the laboratory from a clone of hybridoma cells, each of which produces the same specific antibody.

monocots Angiosperms with a single embryonic cotyledon; one of the two largest clades of angiosperms. (*See also* eudicots.)

monoculture In agriculture, a large-scale planting of a single species of domesticated crop plant.

monoecious (mo nee′ shus) [Gk. *mono*: one + *oikos*: house] Pertaining to organisms in which both sexes are "housed" in a single individual that produces both eggs and sperm. (In some plants, these are found in different flowers within the same plant.) Examples include corn, peas, earthworms, hydras. (Contrast with dioecious.)

monohybrid cross A mating in which the parents differ with respect to the alleles of only one locus of interest.

monomer [Gk. *mono*: one + *meros:* unit] A small molecule, two or more of which can be combined to form oligomers (consisting of a few monomers) or polymers (consisting of many monomers).

monophyletic (mon′ oh fih leht′ ik) [Gk. *mono*: one + *phylon*: tribe] Pertaining to a group that consists of an ancestor and all of its descendants. (Contrast with paraphyletic, polyphyletic.)

monosaccharide A simple sugar. Oligosaccharides and polysaccharides are made up of monosaccharides.

monosomic Pertaining to an organism with one less than the normal diploid number of chromosomes.

monosynaptic reflex A neural reflex that begins in a sensory neuron and makes a single synapse before activating a motor neuron.

morphogen A diffusible substance whose concentration gradient determines a developmental pattern in embryonic animals and plants.

morphogenesis (more′ fo jen′ e sis) [Gk. *morphe*: form + *genesis*: origin] The development of form; the overall consequence of determination, differentiation, and growth.

morphological species concept The definition of a species as a group of individuals that look alike. (Contrast with biological species concept; lineage species concept.)

morphology (more fol′ o jee) [Gk. *morphe*: form + *logos*: study, discourse] The scientific study of organic form, including both its development and function.

mortality Death, or the death rate of a population.

mosaic development Pattern of animal embryonic development in which each blastomere contributes a specific part of the adult body. (Contrast with regulative development.)

motif *See* structural motif.

motile (mo′ tul) Able to move from one place to another. (Contrast with sessile.)

motor cortex The region of the cerebral cortex that contains motor neurons that directly stimulate specific muscle fibers to contract.

motor end plate The depression in the postsynaptic membrane of the neuromuscular junction where the terminals of the motor neuron sit.

motor neuron A neuron carrying information from the central nervous system to a cell that produces movement.

motor proteins Specialized proteins that use energy to change shape and move cells or structures within cells.

motor unit A motor neuron and the muscle fibers it controls.

mouth An opening through which food is taken in, located at the anterior end of a tubular gut.

mRNA *See* messenger RNA.

mucosal epithelium An epithelial cell layer containing cells that secrete mucus; found in the digestive and respiratory tracts. Also called mucosa.

mucus A viscous substance secreted by mucous membranes (e.g., mucosal epithelium). A barrier defense against pathogens in innate immunity in animals and a protective coating in many animal organ systems.

Muller's ratchet The accumulation—"ratcheting up"—of deleterious mutations in the nonrecombining genomes of asexual species.

Müllerian mimicry Convergence in appearance of two or more unpalatable species.

multifactorial The interaction of many genes and proteins with one or more factors in the environment. For example, cancer is a disease with multifactorial causes.

multipotent Having the ability to differentiate into a limited number of cell types. (Contrast with pluripotent, totipotent.)

muscle fiber A single muscle cell. In the case of skeletal muscle, a syncitial, multinucleate cell.

muscle tissue Excitable tissue that can contract through the interactions of actin and myosin; one of the four major tissue types in multicellular animals. There are three types of muscle tissue: skeletal, smooth, and cardiac.

mutagen (mute′ ah jen) [L. *mutare*: change + Gk. *genesis*: source] Any agent (e.g., a chemical, radiation) that increases the mutation rate.

mutation A change in the genetic material not caused by recombination.

mutualism A type of interaction between species that benefits both species.

mycelium (my seel′ ee yum) [Gk. *mykes*: fungus] In the fungi, a mass of hyphae.

mycologists Scientists who study fungi.

mycorrhiza (my′ ko rye′ za) (plural: mycorrhizae) [Gk. *mykes*: fungus + *rhiza*: root] An association of the root of a plant with the mycelium of a fungus.

myelin (my′ a lin) Concentric layers of plasma membrane that form a sheath around some axons; myelin provides the axon with electrical insulation and increases the rate of transmission of action potentials.

myocardial infarction (MI) Blockage of an artery that carries blood to the heart muscle; a "heart attack."

MyoD The protein encoded by the *myo*blast *d*eterming gene. A transcription factor involved in the differentiation of myoblasts (muscle precursor cells).

myofibril (my′ oh fy′ bril) [Gk. *mys*: muscle + L. *fibrilla*: small fiber] A polymeric unit of actin or myosin in a muscle.

myoglobin (my′ oh globe′ in) [Gk. *mys*: muscle + L. *globus*: sphere] An oxygen-binding molecule found in muscle. Consists of a heme unit and a single globin chain; carries less oxygen than hemoglobin.

myosin One of the two contractile proteins of muscle. See also actin.

N

natural history The characteristics of a group of organisms, such as how the organisms get their food, reproduce, behave, regulate their internal environments (their cells, tissues, and organs), and interact with other organisms.

natural killer cell A type of lymphocyte that attacks virus-infected cells and some tumor cells as well as antibody-labeled target cells.

natural selection The differential contribution of offspring to the next generation by various genetic types belonging to the same population. The mechanism of evolution proposed by Charles Darwin.

nauplius (naw′ plee us) [Gk. *nauplios*: shellfish] A bilaterally symmetrical larval form typical of crustaceans.

necrosis (nec roh′ sis) [Gk. *nekros*: death] Premature cell death caused by external agents such as toxins.

negative feedback In regulatory systems, information that decreases a regulatory response, returning the system to the set point. (Contrast with positive feedback.)

negative regulation A type of gene regulation in which a gene is normally transcribed, and the binding of a repressor protein to the promoter prevents transcription. (Contrast with positive regulation.)

nematocyst (ne mat′ o sist) [Gk. *nema*: thread + *kystis*: cell] An elaborate, threadlike structure produced by cells of jellyfishes and other cnidarians, used chiefly to paralyze and capture prey.

neoteny (knee ot′ enny) [Gk. *neo*: new, recent; *tenein*, to extend] The retention of juvenile or larval traits by the fully developed adult organism.

nephron (nef′ ron) [Gk. *nephros*: kidney] The functional unit of the kidney, consisting of a structure for receiving a filtrate of blood and a tubule that reabsorbs selected parts of the filtrate.

Nernst equation A mathematical statement that calculates the potential across a membrane permeable to a single type of ion that differs in concentration on the two sides of the membrane.

nerve A structure consisting of many neuronal axons and connective tissue.

nerve nets Diffuse, loosely connected aggregations of nervous tissues in certain non-bilatarian animals such as cnidarians.

nervous tissue Tissue specialized for processing and communicating information; one of the four major tissue types in multicellular animals.

net primary productivity (NPP) The rate at which energy captured by photosynthesis is incorporated into the bodies of primary producers through growth and reproduction.

neural crest cells During vertebrate neurulation, cells that migrate outward from the neural plate and give rise to connections between the central nervous system and the rest of the body.

neural network An organized group of neurons that contains three functional categories of neurons—afferent neurons, interneurons, and efferent neurons—and is capable of processing information.

neural tube An early stage in the development of the vertebrate nervous system consisting of a hollow tube created by two opposing folds of the dorsal ectoderm along the anterior–posterior body axis.

neurohormone A chemical signal produced and released by neurons that subsequently acts as a hormone.

neuromuscular junction Synapse (point of contact) where a motor neuron axon stimulates a muscle fiber cell.

neuron (noor′ on) [Gk. *neuron*: nerve] A nervous system cell that can generate and conduct action potentials along an axon to a synapse with another cell.

neurotransmitter A substance produced in and released by a neuron (the presynaptic cell) that diffuses across a synapse and excites or inhibits another cell (the postsynaptic cell).

neurulation Stage in vertebrate development during which the nervous system begins to form.

neutral allele An allele that does not alter the functioning of the proteins for which it codes.

neutral theory A view of molecular evolution that postulates that most mutations do not affect the amino acid being coded for, and that such mutations accumulate in a population at rates driven by genetic drift and mutation rates.

neutron (new′ tron) One of the three fundamental particles of matter (along with protons and electrons), with mass slightly larger than that of a proton and no electrical charge.

niche (nitch) [L. *nidus*: nest] The set of physical and biological conditions a species requires to survive, grow, and reproduce.

nitrate reduction The process by which nitrate (NO_3^-) is reduced to ammonia (NH_3).

nitric oxide (NO) An unstable molecule (a gas) that serves as a second messenger causing smooth muscle to relax. In the nervous system it operates as a neurotransmitter.

nitrification The oxidation of ammonia (NH_3) to nitrate (NO_3^-) in soil and seawater, carried out by chemoautotrophic bacteria (nitrifiers).

nitrogen fixation Conversion of atmospheric nitrogen gas (N_2) into a more reactive and biologically useful form (ammonia), which makes nitrogen available to living things. Carried out by **nitrogen fixers**—bacteria, some of them free-living and others living within plant roots.

nitrogenase An enzyme complex found in nitrogen-fixing bacteria that mediates the stepwise reduction of atmospheric N_2 to ammonia and which is strongly inhibited by oxygen.

nitrogenous wastes The potentially toxic nitrogen-containing end products—ammonia, urea, or uric acid—of protein and nucleic acid catabolism in animals. Eliminated from the body by excretion.

node [L. *nodus*: knob, knot] In plants, a (sometimes enlarged) point on a stem where a leaf is or was attached.

node of Ranvier A gap in the myelin sheath covering an axon; the point where the axonal membrane can fire action potentials.

nodule A specialized structure in the roots of nitrogen-fixing plants that houses nitrogen-fixing bacteria, in which oxygen is maintained at a low level by leghemoglobin.

non-REM sleep A state of deep, restorative sleep characterized by high-amplitude slow waves in the EEG. (Contrast with REM sleep.)

noncompetitive inhibitor A nonsubstrate that inhibits the activity of an enzyme by binding to a site other than its active site. (Contrast with competitive inhibitor.)

noncyclic electron transport In photosynthesis, the flow of electrons that forms ATP, NADPH, and O_2.

nondisjunction Failure of sister chromatids to separate in meiosis II or mitosis, or failure of homologous chromosomes to separate in meiosis I. Results in aneuploidy.

nonpolar Having electric charges that are evenly balanced from one end to the other. (Contrast with polar.)

nonrandom mating Selection of mates on the basis of a particular trait or group of traits.

nonsense mutation Change in a gene's sequence that prematurely terminates translation by changing one of its codons to a stop codon.

nonsynonymous substitution A change in a gene from one nucleotide to another that changes the amino acid specified by the corresponding codon (i.e., AGC → AGA, or serine → arginine). (Contrast with synonymous substitution.)

norepinephrine A neurotransmitter found in the central nervous system and also at the postganglionic nerve endings of the sympathetic nervous system. Also called noradrenaline.

normal flora Microorganisms that normally live and reproduce on or in the body without causing disease, and which form a nonspecific defense against pathogens by competing with them for space and nutrients. See also microbiota.

notochord (no' tow kord) [Gk. *notos*: back + *chorde*: string] A flexible rod of gelatinous material serving as a support in the embryos of all chordates and in the adults of tunicates and lancelets.

nucleic acid (new klay' ik) A polymer made up of nucleotides, specialized for the storage, transmission, and expression of genetic information. DNA and RNA are nucleic acids.

nucleic acid hybridization A technique in which a single-stranded nucleic acid probe is made that is complementary to, and binds to, a target sequence, either DNA or RNA. The resulting double-stranded molecule is a hybrid.

nucleoid (new' klee oid) The region that harbors the chromosomes of a prokaryotic cell. Unlike the eukaryotic nucleus, it is not bounded by a membrane.

nucleolus (new klee' oh lus) A small, generally spherical body found within the nucleus of eukaryotic cells. The site of synthesis of ribosomal RNA.

nucleoside A nucleotide without the phosphate group; a nitrogenous base attached to a sugar.

nucleosome A portion of a eukaryotic chromosome, consisting of part of the DNA molecule wrapped around a group of histone molecules, and held together by another type of histone molecule. The chromosome is made up of many nucleosomes.

nucleotide The basic chemical unit in nucleic acids, consisting of a pentose sugar, a phosphate group, and a nitrogen-containing base.

nucleotide substitution A change of one base pair to another in a DNA sequence.

nucleus (new' klee us) [L. *nux*: kernel or nut] (1) In cells, the centrally located compartment of eukaryotic cells that is bounded by a double membrane and contains the chromosomes. (2) In the brain, an identifiable group of neurons that share common characteristics or functions.

null hypothesis In statistics, the premise that any differences observed in an experiment are simply the result of random differences that arise from drawing two finite samples from the same population.

nutrient A food substance; or, in the case of mineral nutrients, an inorganic element required for completion of the life cycle of an organism.

O

obligate anaerobe An anaerobic prokaryote that cannot survive exposure to O_2.

occipital lobe One of the four lobes of the brain's cerebral hemisphere; processes visual information.

odorant A molecule that can bind to an olfactory receptor.

oil A triglyceride that is liquid at room temperature. (Contrast with fat.)

Okazaki fragments Newly formed DNA making up the lagging strand in DNA replication. DNA ligase links Okazaki fragments together to give a continuous strand.

olfaction (ole fak' shun) [L. *olfacere*: to smell] The sense of smell.

olfactory bulb Structure in the vertebrate forebrain that receives and processes input from olfactory receptor neurons.

olfactory receptor neurons (ORNs) Neurons with receptors for different odorants.

oligodendrocyte A type of glial cell that myelinates axons in the central nervous system.

oligophagous [Gk. *oligo*: few; *phagein*, eat] An animal that feeds on a limited number of foods; generally used of insects that feed on only one or a few plant species.

oligosaccharide A polymer containing a small number of monosaccharides.

omasum One of the four chambers of the stomach in ruminants; concentrates food by water absorption before it enters the true stomach (abomasum).

ommatidia [Gk. *omma*: eye] The units that make up the compound eye of some arthropods.

omnivore [L. *omnis*: everything + *vorare*: to devour] An organism that eats both animal and plant material. (Contrast with carnivore, detritivore, herbivore.)

oncogene [Gk. *onkos*: mass, tumor + *genes*: born] A gene that codes for a protein product that stimulates cell proliferation. Mutations in oncogenes that result in excessive cell proliferation can give rise to cancer.

one gene–one polypeptide The idea, now known to be an oversimplification, that each gene in the genome encodes only a single polypeptide—that there is a one-to-one correspondence between genes and polypeptides.

oocyte *See* primary oocyte, secondary oocyte.

oogenesis (oh' eh jen e sis) [Gk. *oon*: egg + *genesis*: source] Gametogenesis leading to production of an ovum.

oogonium (oh' eh go' nee um) (plural: oogonia) (1) In some algae and fungi, a cell in which an egg is produced. (2) In animals, the diploid progeny of a germ cell in females.

ootid In oogenesis, the daughter cell of the second meiotic division that differentiates into the mature ovum.

open circulatory system Circulatory system in which extracellular fluid leaves the vessels of the circulatory system, percolates between cells and through tissues, and then flows back into the circulatory system to be pumped out again. (Contrast with closed circulatory system.)

operator The region of an operon that acts as the binding site for the repressor.

operon A genetic unit of transcription, typically consisting of several structural genes that are transcribed together; the operon contains at least two control regions: the promoter and the operator.

opportunity cost The sum of the benefits an animal forfeits by not being able to perform some other behavior during the time when it is performing a given behavior.

opsin (op' sin) [Gk. *opsis*: sight] The protein portion of vertebrate visual pigments; associated with the pigment molecule 11-*cis*-retinal. See also rhodopsin.

optic chiasm [Gk. *chiasma*: cross] Structure on the lower surface of the vertebrate brain where the two optic nerves come together.

optic nerve The nerve that carries information from the retina of the eye to the brain.

optical isomers Two molecular isomers that are mirror images of each other.

optimal foraging theory The application of a cost–benefit approach to feeding behavior to identify the fitness value of feeding choices.

oral [L. *os*: mouth] Pertaining to the mouth, or that part of the body that contains the mouth.

orbital A region in space surrounding the atomic nucleus in which an electron is most likely to be found.

organ [Gk. *organon*: tool] A body part, such as the heart, liver, brain, root, or leaf. Organs are composed of different tissues integrated to perform a distinct function. Organs, in turn, are integrated into organ systems.

organ identity genes In angiosperms, genes that specify the different organs of the flower. (Compare with homeotic genes.)

organ of Corti Structure in the inner ear that transforms mechanical forces produced from pressure waves ("sound waves") into action potentials that are sensed as sound.

organ system An interrelated and integrated group of tissues and organs that work together in a physiological function.

organelle (or gan el′) Any of the membrane-enclosed structures within a eukaryotic cell. Examples include the nucleus, endoplasmic reticulum, and mitochondria.

organic (1) Pertaining to any chemical compound that contains carbon. (2) Pertaining to any aspect of living matter, e.g., to its evolution, structure, or chemistry.

organic fertilizers Substances added to soil to improve the soil's fertility; derived from partially decomposed plant material (compost) or animal waste (manure).

organism Any living entity.

organizer Region of the early amphibian embryo that directs early embryonic development. Also known as the primary embryonic organizer.

organogenesis The formation of organs and organ systems during development.

origin of replication (*ori*) DNA sequence at which helicase unwinds the DNA double helix and DNA polymerase binds to initiate DNA replication.

orthologs [Gk. *ortho*: true, direct] Homologous genes whose divergence can be traced to speciation events.

osmoconformer An aquatic animal that equilibrates the osmolarity of its extracellular fluid to be the same as that of the external environment. (Contrast with osmoregulator.)

osmolarity The concentration of osmotically active particles in a solution.

osmoregulation Regulation of the chemical composition of the body fluids of an organism.

osmoregulator An aquatic animal that actively regulates the osmolarity of its extracellular fluid. (Contrast with osmoconformer.)

osmosis (oz mo′ sis) [Gk. *osmos*: to push] Movement of water across a differentially permeable membrane, from one region to another region where the water potential is more negative.

ossicle (oss′ ick ul) [L. *os*: bone] The calcified construction unit of echinoderm skeletons.

osteoblast (oss′ tee oh blast) [Gk. *osteon*: bone + *blastos*: sprout] A cell that lays down the protein matrix of bone.

osteoclast (oss′ tee oh clast) [Gk. *osteon*: bone + *klastos*: broken] A cell that dissolves bone.

osteocyte An osteoblast that has become enclosed in lacunae within the bone it has built.

outgroup In phylogenetics, a group of organisms used as a point of reference for comparison with the groups of primary interest (the ingroup).

oval window The flexible membrane that, when moved by the bones of the middle ear, produces pressure waves in the inner ear.

ovarian cycle In human females, the monthly cycle of events by which eggs and hormones are produced. (Contrast with uterine cycle).

ovary (oh′ var ee) [L. *ovum*: egg] Any female organ, in plants or animals, that produces an egg.

overtopping Plant growth pattern in which one branch differentiates from and grows beyond the others.

oviduct In mammals, the tube serving to transport eggs to the uterus or to the outside of the body.

oviparity Reproduction in which eggs are released by the female and development is external to the mother's body. (Contrast with viviparity.)

ovoviviparity Pertaining to reproduction in which fertilized eggs develop and hatch within the mother's body but are not attached to the mother by means of a placenta.

ovulation Release of an egg from an ovary.

ovule (oh′ vule) In plants, a structure comprising the megasporangium and the integument, which develops into a seed after fertilization.

ovum (oh′ vum) (plural: ova) [L. egg] The female gamete.

oxidation (ox i day′ shun) Relative loss of electrons in a chemical reaction; either outright removal to form an ion, or the sharing of electrons with substances having a greater affinity for them, such as oxygen. Most oxidations, including biological ones, are associated with the liberation of energy. (Contrast with reduction.)

oxidation–reduction (redox) reaction A reaction in which one substance transfers one or more electrons to another substance. (*See* oxidation; reduction.)

oxidative phosphorylation ATP formation in the mitochondrion, associated with flow of electrons through the respiratory chain.

oxygenase An enzyme that catalyzes the addition of oxygen to a substrate from O_2.

oxytocin A hormone released by the posterior pituitary that promotes social bonding.

ozone layer A layer of ozone (O_3, a greenhouse gas) in the atmosphere that absorbs a high portion of the sun's potentially mutagenic ultraviology radiation.

P

pacemaker cells Cardiac cells that can initiate action potentials without stimulation from the nervous system, allowing the heart to initiate its own contractions.

pair rule genes In *Drosophila* (fruit fly) development, segmentation genes that divide the early embryo into units of two segments each. Part of a developmental cascade that includes maternal effect genes, gap genes, segment polarity genes, and Hox genes.

paleomagnetic dating A method for determining the age of rocks based on properties relating to changes in the patterns of Earth's magnetism over time.

pancreas (pan′ cree us) A gland located near the stomach of vertebrates that secretes digestive enzymes into the small intestine and releases insulin into the bloodstream.

Pangaea (pan jee′ uh) [Gk. *pan*: all, every] The single land mass formed when all the continents came together in the Permian period.

para- [Gk. *para*: akin to, beside] Prefix indicating association in being along side or accessory to.

parabronchi Passages in the lungs of birds through which air flows.

paracrine [Gk. *para*: near] Pertaining to a chemical signal, such as a hormone, that acts locally, near the site of its secretion. (Contrast with autocrine.)

parallel evolution The repeated evolution of similar traits, especially among closely related species; facilitated by conserved developmental genes.

paralogs Homologous genes whose divergence can be traced to gene duplication events. (Contrast with orthologs.)

paraphyletic (par′ a fih leht′ ik) [Gk. *para*: beside + *phylon*: tribe] Pertaining to a group that consists of an ancestor and some, but not all, of its descendants. (Contrast with monophyletic, polyphyletic.)

parasite An organism that consumes parts of an organism much larger than itself (known as its host). Parasites sometimes, but not always, kill their host.

parasympathetic nervous system The division of the autonomic nervous system that works in opposition to the sympathetic nervous system. (Contrast with sympathetic nervous system.)

parathyroid glands Four glands on the posterior surface of the thyroid gland that produce and release parathyroid hormone.

parathyroid hormone (PTH) A hormone secreted by the parathyroid glands that stimulates osteoclast activity and raises blood calcium levels. Also called parathormone.

parenchyma (pair eng′ kyma) A plant tissue composed of relatively unspecialized cells without secondary walls.

parent rock The soil horizon consisting of the rock that is breaking down to form the soil. Also called bedrock, or the C horizon.

parental (P) generation The individuals that mate in a genetic cross. Their offspring are the first filial (F_1) generation.

parietal cells One of three types of secretory cell found in the gastric pits of the stomach wall. Parietal cells produce hydrochloric acid (HCl), creating an acidic environment that destroys many of the harmful microorganisms ingested with food. (See chief cells; mucosal epithelium.)

parietal lobe One of four lobes of the cerebral hemisphere; processes complex stimuli and includes the primary somatosensory cortex.

parsimony Preferring the simplest among a set of plausible explanations of any phenomenon.

parthenocarpy Formation of fruit from a flower without fertilization.

parthenogenesis [Gk. *parthenos*: virgin] Production of an organism from an unfertilized egg.

particulate theory In genetics, the theory that genes are physical entities that retain their identities after fertilization.

passive transport Diffusion across a membrane; may or may not require a channel or carrier protein. (Contrast with active transport.)

patch clamping Technique for isolating a tiny patch of membrane to allow the study of ion movement through a particular channel.

pathogen (path' o jen) [Gk. *pathos*: suffering + *genesis*: source] An organism that causes disease.

pattern formation In animal embryonic development, the organization of differentiated tissues into specific structures such as wings.

pedigree The pattern of transmission of a genetic trait within a family.

pelagic zone [Gk. *pelagos*: deep sea] The open ocean; a marine life zone.

penetrance The proportion of individuals with a particular genotype that show the expected phenotype.

penis An accessory sex organ of male animals that enables the male to deposit sperm in the female's reproductive tract.

pentaradial symmetry Symmetry in five or multiples of five; a feature of adult echinoderms.

pentose [Gk. *penta:* five] A sugar containing five carbon atoms.

PEP carboxylase The enzyme that combines carbon dioxide with PEP to form a 4-carbon dicarboxylic acid at the start of C_4 photosynthesis or of crassulacean acid metabolism (CAM).

pepsin [Gk. *pepsis*: digestion] An enzyme in gastric juice that digests protein.

pepsinogen Inactive secretory product that is converted into pepsin by low pH or by enzymatic action.

peptide hormones Relatively large hormone molecules made up of amino acids; encoded by genes and produced by translation.

peptide linkage The bond between amino acids in a protein; formed between a carboxyl group and amino group (—CO—NH—) with the loss of water molecules.

peptidoglycan The cell wall material of many bacteria, consisting of a single enormous molecule that surrounds the entire cell.

peptidyl transferase A catalytic function of the large ribosomal subunit that consists of two reactions: breaking the bond between an amino acid and its tRNA in the P site, and forming a peptide bond between that amino acid and the amino acid attached to the tRNA in the A site.

per capita birth rate (*b*) In population growth models, the number of offspring that an average individual produces in some time interval.

per capita death rate (*d*) In population growth models, the average individual's chance of dying in some time interval.

per capita growth rate (*r*) In population models, the average individual's contribution to total population growth rate.

perennial (per ren' ee al) [L. *per*: throughout + *annus*: year] A plant that survives from year to year. (Contrast with annual, biennial.)

perfect flower A flower with both stamens and carpels; a hermaphroditic flower. (Contrast with imperfect flower.)

pericycle [Gk. *peri*: around + *kyklos*: ring or circle] In plant roots, tissue just within the endodermis, but outside of the root vascular tissue. Meristematic activity of pericycle cells produces lateral root primordia.

periderm The outer tissue of the secondary plant body, consisting primarily of cork.

period (1) A category in the geological time scale. (2) The duration of a single cycle in a cyclical event, such as a circadian rhythm.

peripheral membrane proteins Proteins associated with but not embedded within the plasma membrane. (Contrast with integral membrane proteins.)

peripheral nervous system (PNS) The portion of the nervous system that transmits information to and from the central nervous system, consisting of neurons that extend or reside outside the brain or spinal cord and their supporting cells. (Contrast with central nervous system.)

peristalsis (pair' i stall' sis) Wavelike muscular contractions proceeding along a tubular organ, propelling the contents along the tube.

peritoneum The mesodermal lining of the body cavity in coelomate animals.

peroxisome An organelle that houses reactions in which toxic peroxides are formed and then converted to water.

petal [Gk. *petalon*: spread out] In an angiosperm flower, a sterile modified leaf, nonphotosynthetic, frequently brightly colored, and often serving to attract pollinating insects.

petiole (pet' ee ole) [L. *petiolus*: small foot] The stalk of a leaf.

P_{fr} *See* phytochrome.

pH The negative logarithm of the hydrogen ion concentration; a measure of the acidity of a solution. A solution with pH = 7 is said to be neutral; pH values higher than 7 characterize basic solutions, while acidic solutions have pH values less than 7.

phage (fayj) *See* bacteriophage.

phagocyte [Gk. *phagein*: to eat + *kystos*: sac] One of two major classes of white blood cells; one of the nonspecific defenses of animals; ingests invading microorganisms by **phagocytosis**.

pharmacogenomics The study of how an individual's genetic makeup affects his or her response to drugs or other agents, with the goal of predicting the effectiveness of different treatment options.

pharming The use of genetically modified animals to produce medically useful products in their milk.

pharynx [Gk. throat] The part of the gut between the mouth and the esophagus.

phenotype (fee' no type) [Gk. *phanein*: to show] The observable properties of an individual resulting from both genetic and environmental factors. (Contrast with genotype.)

phenotypic plasticity *See* developmental plasticity.

pheromone (feer' o mone) [Gk. *pheros*: carry + *hormon*: excite, arouse] A chemical substance used in communication between organisms of the same species.

phloem (flo' um) [Gk. *phloos*: bark] In vascular plants, the vascular tissue that transports sugars and other solutes from sources to sinks.

phosphate group The functional group $—OPO_3H_2$.

phosphodiester linkage The connection in a nucleic acid strand, formed by linking two nucleotides.

phospholipid A lipid containing a phosphate group; an important constituent of cellular membranes. (*See* lipid.)

phospholipid bilayer The basic structural unit of biological membranes; a sheet of phospholipids two molecules thick in which the phospholipids are lined up with their hydrophobic "tails" packed tightly together and their hydrophilic, phosphate-containing "heads" facing outward. Also called lipid bilayer.

phosphorylation Addition of a phosphate group.

photic zone The life zone in lakes and oceans that is penetrated by light and therefore supports photosynthetic organisms.

photoautotroph An organism that obtains energy from light and carbon from carbon dioxide. (Contrast with chemolithotroph, chemoheterotroph, photoheterotroph.)

photoheterotroph An organism that obtains energy from light but must obtain its carbon from organic compounds. (Contrast with chemoautotroph, chemoheterotroph, photoautotroph.)

photomorphogenesis In plants, a process by which physiological and developmental events are controlled by light.

photon (foe' ton) [Gk. *photos*: light] A quantum of visible radiation; a "packet" of light energy.

photoperiodism Control of an organism's physiological or behavioral responses by the length of the day or night (the **photoperiod**).

photophosphorylation Mechanism for ATP formation in chloroplasts in which electron transport is coupled to the transport of hydrogen ions (protons, H^+) across the thylakoid membrane. Compare with chemiosmosis.

photoreceptors (1) In plants, pigments that trigger a physiological response when they absorb a photon. (2) In animals, the sensory receptor cells that sense and respond to light energy. (See cone cells; rod cells.)

photorespiration Light-driven uptake of oxygen and release of carbon dioxide, the carbon being derived from the early reactions of photosynthesis.

photosynthesis (foe tow sin' the sis) [Gk.: creating from light] Metabolic processes carried out by green plants and some microorganisms by which visible light is trapped and the energy used to synthesize compounds such as ATP and glucose.

photosystem [Gk. *phos:* light + *systema:* assembly] A light-harvesting complex in the chloroplast thylakoid composed of pigments and proteins. **Photosystem I** absorbs light at 700 nm, passing electrons to ferrodoxin and from there to NADPH. **Photosystem II** absorbs light at 680 nm and passes electrons to the electron transport chain in the chloroplast.

phototropism [Gk. *photos*: light + *trope*: turning] A directed plant growth response to light.

phycobilin Photosynthetic pigment that absorbs red, yellow, orange, and green light and is found in cyanobacteria and some red algae.

phylogeny (fy loj' e nee) [Gk. *phylon*: tribe, race + *genesis*: source] The evolutionary history of a particular group of organisms or their genes. A **phylogenetic tree** is a graphic representation of these lines of evolutionary descent.

physiological survivorship curves *See* survivorship curves.

physiology (fiz' ee ol' o jee) [Gk. *physis*: natural form] The scientific study of the functions of living organisms and the individual organs, tissues, and cells of which they are composed.

phytoalexins Substances toxic to pathogens, produced by plants in response to fungal or bacterial infection.

phytochrome (fy' tow krome) [Gk. *phyton*: plant + *chroma*: color] A plant pigment regulating a large number of developmental and other phenomena in plants. It has two isomers: P_r, which absorbs red light, and P_{fr}, which absorbs far red light. P_{fr} is the active form.

phytomers In plants, the repeating modules that compose a shoot, each consisting of one or more leaves, attached to the stem at a node; an internode; and one or more axillary buds.

phytoplankton Photosynthetic floating organisms. (*See* plankton.)

phytoremediation A form of bioremediation that uses plants to clean up environmental pollution.

pigment A substance that absorbs visible light.

piloting A form of navigation in which an animal finds its way by remembering landmarks in its environment.

pineal gland Gland located between the cerebral hemispheres that secretes melatonin.

pinocytosis Endocytosis by a cell of liquid containing dissolved substances.

pistil [L. *pistillum*: pestle] The structure of an angiosperm flower within which the ovules are borne. May consist of a single carpel, or of several carpels fused into a single structure. Usually differentiated into ovary, style, and stigma.

pith In plants, relatively unspecialized tissue found within a cylinder of vascular tissue.

pituitary gland A small gland attached to the base of the brain in vertebrates. Its hormones control the activities of other glands. Also known as the hypophysis.

placenta (pla sen' ta) The organ in female mammals that provides for the nourishment of the fetus and elimination of the fetal waste products.

plankton Aquatic organisms that float in the water column, dependent on currents and wind for movement. Plankton include many protists, some algae, and larval animals. (See also phytoplankton.)

planula (plan' yew la) [L. *planum*: flat] A free-swimming, ciliated larval form typical of the cnidarians.

plaque (plack) [Fr.: a metal plate or coin] (1) A circular clearing in a layer (lawn) of bacteria growing on the surface of a nutrient agar gel. (2) An accumulation of prokaryotic organisms on tooth enamel. Acids produced by these microorganisms cause tooth decay. (3) A region of arterial wall invaded by fibroblasts and fatty deposits.

plasma (plaz' muh) The liquid portion of blood, in which blood cells and other particulates are suspended.

plasma cell An antibody-secreting cell that develops from a B cell; the effector cell of the humoral immune system.

plasma membrane The membrane that surrounds the cell, regulating the entry and exit of molecules and ions. Every cell has a plasma membrane, and it is often called the cell membrane.

plasmid A DNA molecule distinct from the chromosome(s); that is, an extrachromosomal element; found in many bacteria. May replicate independently of the chromosome.

plasmodesmata (singular: plasmodesma) [Gk. *plassein*: to mold + *desmos*: band] Cytoplasmic strands connecting two adjacent plant cells.

plasmogamy The fusion of the cytoplasm of two cells. (Contrast with karyogamy.)

plastid A class of plant cell organelles that includes the chloroplast, which houses biochemical pathways for photosynthesis.

plate tectonics [Gk. *tekton*: builder] The scientific study of the structure and movements of Earth's lithospheric plates, which are the cause of continental drift.

platelet A membrane-bounded body without a nucleus, arising as a fragment of a cell in the bone marrow of mammals. Important to blood-clotting action.

pleiotropy (plee' a tro pee) [Gk. *pleion*: more] The determination of more than one character by a single gene.

pleural membrane [Gk. *pleuras*: rib, side] The membrane lining the outside of the lungs and the walls of the thoracic cavity. Inflammation of these membranes is a condition known as pleurisy.

pluripotent [L. *pluri*: many + *potens*: powerful] Having the ability to form all of the cells in the body. (Contrast with multipotent, totipotent.)

podocytes Cells of Bowman's capsule of the nephron that cover the capillaries of the glomerulus, forming filtration slits.

point mutation A mutation that results from the gain, loss, or substitution of a single nucleotide.

polar A molecule with separate and opposite electric charges at two ends, or poles; the water molecule (H_2O) is the most prevalent example. (Contrast with nonpolar.)

polar body A nonfunctional nucleus produced by meiosis during oogenesis.

polar covalent bond A covalent bond in which the electrons are drawn to one nucleus more than the other, resulting in an unequal distribution of charge.

polar nuclei In angiosperms, the two nuclei in the central cell of the megagametophyte; following fertilization they give rise to the endosperm.

polarity (1) In chemistry, the property of unequal electron sharing in a covalent bond that defines a polar molecule. (2) In development, the difference between one end of an organism or structure and the other.

pollen [L. *pollin*: fine flour] In seed plants, microscopic grains that contain the male gametophyte (microgametophyte) and gamete (microspore).

pollen tube A structure that develops from a pollen grain through which sperm are released into the megagametophyte.

pollination The process of transferring pollen from an anther to the stigma of a pistil in an angiosperm or from a strobilus to an ovule in a gymnosperm.

poly- [Gk. *poly*: many] A prefix denoting multiple entities.

poly A tail A long sequence of adenine nucleotides (50–250) added after transcription to the 3′ end of most eukaryotic mRNAs.

polyandry Mating system in which one female mates with multiple males.

polygyny Mating system in which one male mates with multiple females.

polymer [Gk. *poly*: many + *meros*: unit] A large molecule made up of similar or identical subunits called monomers. (Contrast with monomer.)

polymerase chain reaction (PCR) An enzymatic technique for the rapid production of millions of copies of a particular stretch of DNA where only a small amount of the parent molecule is available.

polymorphic (pol′ lee mor′ fik) [Gk. *poly*: many + *morphe*: form, shape] Coexistence in a population of two or more distinct traits.

polyp (pah′ lip) [Gk. *poly*: many + *pous*: foot] In cnidarians, a sessile, asexual life cycle stage.

polypeptide A large molecule made up of many amino acids joined by peptide linkages. Large polypeptides are called proteins.

polyphyletic (pol′ lee fih leht′ ik) [Gk. *poly*: many + *phylon*: tribe] Pertaining to a group that consists of multiple distantly related organisms, and does not include the common ancestor of the group. (Contrast with monophyletic, paraphyletic.)

polyploid (pol′ lee ploid ee) Possessing more than two entire sets of chromosomes.

polyribosome (polysome) A complex consisting of a threadlike molecule of messenger RNA and several (or many) ribosomes. The ribosomes move along the mRNA, synthesizing polypeptide chains as they proceed.

polysaccharide A macromolecule composed of many monosaccharides (simple sugars). Common examples are cellulose and starch.

pons [L. *pons*: bridge] Region of the brainstem anterior to the medulla.

pool The total amount of an element in a given compartment of the biosphere.

population Any group of organisms coexisting at the same time and in the same place and capable of interbreeding with one another.

population bottleneck A period during which only a few individuals of a normally large population survive.

population density The number of individuals in a population per unit of area or volume.

population dynamics The patterns and processes of change in populations.

population genetics The study of genetic variation and its causes within populations.

population size The total number of individuals in a population.

positional information In development, the basis of the spatial sense that induces cells to differentiate as appropriate for their location within the developing organism; often comes in the form of a morphogen gradient.

positive cooperativity Occurs when a molecule can bind several ligands and each one that binds alters the conformation of the molecule so that it can bind the next ligand more easily. The binding of four molecules of O_2 by hemoglobin is an example of positive cooperativity.

positive feedback In regulatory systems, information that amplifies a regulatory response, increasing the deviation of the system from the set point. (Contrast with negative feedback.)

positive regulation A form of gene regulation in which a regulatory macromolecule is needed to turn on the transcription of a structural gene; in its absence, transcription will not occur. (Contrast with negative regulation.)

positive selection Natural selection that acts to establish a trait that enhances survival in a population. (Contrast with purifying selection.)

post- [L. *postere*: behind, following after] Prefix denoting something that comes after.

postabsorptive state State in which no food remains in the gut and thus no nutrients are being absorbed. (Contrast with absorptive state.)

posterior Toward or pertaining to the rear. (Contrast with anterior.)

posterior pituitary A portion of the pituitary gland derived from neural tissue; involved in the storage and release of antidiuretic hormone and oxytocin.

postsynaptic cell The cell that receives information from a neuron at a synapse. (Contrast with presynaptic cell.)

postzygotic isolating mechanisms Barriers to the reproductive process that occur after the union of the nuclei of two gametes. (Contrast with prezygotic isolating mechanisms.)

potential energy Energy not doing work but with the potential to do so, such as the energy stored in chemical bonds. (Contrast with kinetic energy.)

P_r *See* phytochrome.

pre-mRNA (precursor mRNA) Initial gene transcript before it is modified to produce functional mRNA. Also known as the primary transcript.

Precambrian The first and longest period of geological time, during which life originated.

precapillary sphincter A cuff of smooth muscle that can shut off the blood flow to a capillary bed.

predator An organism that kills and eats other organisms.

pressure flow model An effective model for phloem transport in angiosperms. It holds that sieve element transport is driven by an osmotically generated pressure gradient between source and sink.

pressure potential (Ψ_p) The hydrostatic pressure of an enclosed solution in excess of the surrounding atmospheric pressure. (Contrast with solute potential, water potential.)

presynaptic cell The neuron that transmits information to another cell at a synapse. (Contrast with postsynaptic cell.)

prey [L. *praeda*: booty] An organism consumed by a predator as an energy source.

prezygotic isolating mechanisms Barriers to the reproductive process that occur before the union of the nuclei of two gametes (Contrast with postzygotic isolating mechanisms.)

primary active transport Active transport in which ATP is hydrolyzed, yielding the energy required to transport an ion or molecule against its concentration gradient. (Contrast with secondary active transport.)

primary cell wall In plant cells, a structure that forms at the middle lamella after cytokinesis, made up of cellulose microfibrils, hemicelluloses, and pectins. (Contrast with secondary cell wall.)

primary consumer An organism (herbivore) that eats plant tissues.

primary endosymbiosis The engulfment of a cyanobacterium by a larger eukaryotic cell that gave rise to the first photosynthetic eukaryotes with chloroplasts.

primary growth In plants, growth that is characterized by the lengthening of roots and shoots and by the proliferation of new roots and shoots through branching. (Contrast with secondary growth.)

primary immune response The first response of the immune system to an antigen, involving recognition by lymphocytes and the production of effector cells and memory cells. (Contrast with secondary immune response.)

primary meristem Meristem that produces the tissues of the primary plant body.

primary motor cortex *See* motor cortex

primary oocyte (oh′ eh site) [Gk. *oon*: egg + *kytos*: container] The diploid progeny of an oogonium. In many species, a primary oocyte enters prophase of the first meiotic division, then remains in developmental arrest for a long time before resuming meiosis to form a secondary oocyte and a polar body.

primary plant body That part of a plant produced by primary growth. Consists of all the *nonwoody* parts of a plant; many herbaceous plants consist entirely of a primary plant body. (Contrast with secondary plant body.)

primary producer A photosynthetic or chemosynthetic organism that synthesizes complex organic molecules from simple inorganic ones.

primary sex determination Genetic determination of gametic sex, male or female. (Contrast with secondary sex determination.)

primary somatosensory cortex *See* somatosensory cortex.

primary spermatocyte The diploid progeny of a spermatogonium; undergoes the first meiotic division to form secondary spermatocytes.

primary structure The specific sequence of amino acids in a protein.

primary succession Succession of ecological communities that begins in an area devoid of life, such as on recently exposed glacial till or lava flows. (Contrast with secondary succession.)

primase An enzyme that catalyzes the synthesis of a primer for DNA replication.

primer Strand of nucleic acid, usually RNA, that is the necessary starting material for the synthesis of a new DNA strand, which is synthesized from the 3′ end of the primer.

primordium (plural: primordia) [L. origin] The most rudimentary stage of an organ or other part.

pro- [L.: first, before, favoring] A prefix often used in biology to denote a developmental stage that comes first or an evolutionary form that appeared earlier than another. For example, prokaryote, prophase.

probe A segment of single-stranded nucleic acid used to identify DNA molecules containing the complementary sequence.

procambium Primary meristem that produces the vascular tissue.

procedural memory Memory of motor tasks.; these memories cannot be consciously recalled or described. (Contrast with declarative memory.)

processive Pertaining to an enzyme that catalyzes many reactions each time it binds to a substrate, as DNA polymerase does during DNA replication.

products The molecules that result from the completion of a chemical reation.

progesterone [L. *pro*: favoring + *gestare*: to bear] A female sex hormone that maintains pregnancy.

prokaryotes Unicellular organisms that do not have nuclei or other membrane-enclosed organelles. Includes Bacteria and Archaea. (Contrast with eukaryotes.)

prolactin A hormone released by the anterior pituitary, one of whose functions is the stimulation of milk production in female mammals.

prometaphase The phase of nuclear division that begins with the disintegration of the nuclear envelope.

promoter A DNA sequence to which RNA polymerase binds to initiate transcription.

prop roots Adventitious roots in some monocots that function as supports for the shoot.

prophage (pro′ fayj) The noninfectious units that are linked with the chromosomes of the host bacteria and multiply with them but do not cause dissolution of the cell. Prophage can later enter into the lytic phase to complete the virus life cycle.

prophase (pro′ phase) The first stage of nuclear division, during which chromosomes condense from diffuse, threadlike material to discrete, compact bodies.

prostaglandin Any one of a group of specialized lipids with hormone-like functions. It is not clear that they act at any considerable distance from the site of their production.

prostate gland In male humans, surrounds the urethra at its junction with the vas deferens; supplies an acid-neutralizing fluid to the semen.

prosthetic group Any nonprotein portion of an enzyme.

proteases Digestive enzymes that digest proteins.

proteasome In the eukaryotic cytoplasm, a huge protein structure that binds to and digests cellular proteins that have been tagged by ubiquitin.

protein (pro′ teen) [Gk. *protos*: first] Long-chain polymer of amino acids with twenty different common side chains. Occurs with its polymer chain extended in fibrous proteins, or coiled into a compact macromolecule in enzymes and other globular proteins. The component amino acids are encoded in the triplets of messenger RNA, and proteins are the products of genes.

protein kinase (kye′ nase) An enzyme that catalyzes the addition of a phosphate group from ATP to a target protein.

protein kinase cascade A series of reactions in response to a molecular signal, in which a series of protein kinases activate one another in sequence, amplifying the signal at each step.

proteoglycan A glycoprotein containing a protein core with attached long, linear carbohydrate chains.

proteolysis [protein + Gk. *lysis*: break apart] An enzymatic digestion of a protein or polypeptide.

proteome The complete set of proteins that can be made by an organism. Because of alternative splicing of pre-mRNA, the number of proteins that can be made is usually much larger than the number of protein-coding genes present in the organism's genome.

prothrombin The inactive form of thrombin, an enzyme involved in blood clotting.

protoderm Primary meristem that gives rise to the plant epidermis.

proton (pro′ ton) [Gk. *protos*: first, before] (1) A subatomic particle with a single positive charge. The number of protons in the nucleus of an atom determine its element. (2) A hydrogen ion, H^+.

proton pump An active transport system that uses ATP energy to move hydrogen ions across a membrane, generating an electric potential.

proton-motive force Force generated across a membrane having two components: a chemical potential (difference in proton concentration) plus an electrical potential due to the electrostatic charge on the proton.

protonephridium The excretory organ of flatworms, made up of a tubule and a flame cell.

protoplast The living contents of a plant cell; the plasma membrane and everything contained within it.

provirus Double-stranded DNA made by a virus that is integrated into the host's chromosome and contains promoters that are recognized by the host cell's transcription apparatus.

proximal convoluted tubule The initial segment of a renal tubule, closest to the glomerulus. (Compare with distal convoluted tubule.)

proximal Near the point of attachment or other reference point. (Contrast with distal.)

proximate causes The immediate genetic, physiological, neurological, and developmental mechanisms responsible for a behavior or morphology. (Contrast with ultimate cause.)

pseudocoelomate (soo′ do see′ low mate) [Gk. *pseudes*: false + *koiloma*: cavity] Having a body cavity, called a pseudocoel, consisting of a fluid-filled space in which many of the internal organs are suspended, but which is enclosed by mesoderm only on its outside.

pseudogene [Gk. *pseudes*: false] A DNA segment that is homologous to a functional gene but is not expressed because of changes to its sequence or changes to its location in the genome.

pseudopod (soo′ do pod) [Gk. *pseudes*: false + *podos*: foot] A temporary, soft extension of the cell body that is used in location, attachment to surfaces, or engulfing particles.

pulmonary [L. *pulmo*: lung] Pertaining to the lungs.

pulmonary circuit The portion of the circulatory system by which blood is pumped from the heart to the lungs or gills for oxygenation and back to the heart for distribution. (Contrast with systemic circuit.)

pulmonary valve A one-way valve between the right ventricle of the heart and the pulmonary artery that prevents backflow of blood into the ventricle when it relaxes.

Punnett square Method of predicting the results of a genetic cross by arranging the gametes of each parent at the edges of a square.

pupa (pew′ pa) [L. *pupa*: doll, puppet] In certain insects (the Holometabola), the encased developmental stage between the larva and the adult.

pupil The opening in the vertebrate eye through which light passes.

purifying selection The elimination by natural selection of detrimental characters from a population. (Contrast with positive selection.)

purine (pure′ een) One of the two types of nitrogenous bases in nucleic acids. Each of the purines—adenine and guanine—pairs with a specific pyrimidine.

Purkinje fibers Specialized heart muscle cells that conduct excitation throughout the ventricular muscle.

pyrimidine (pe rim′ a deen) One of the two types of nitrogenous bases in nucleic acids. Each of the pyrimidines—cytosine, thymine, and uracil—pairs with a specific purine.

pyrogen [Gk.: *pry, fire*;] Molecule that produces a rise in body temperature (fever); may be produced by an invading pathogen or by cells of the immune system in response to infection.

pyruvate The ionized form of pyruvic acid, a three-carbon acid; the end product of glycolysis and the raw material for the citric acid cycle.

pyruvate oxidation Conversion of pyruvate to acetyl CoA and CO_2 that occurs in the mitochondrial matrix in the presence of O_2.

Q

Q_{10} A value that compares the rate of a biochemical process or reaction over 10°C temperature ranges. A process that is not temperature-sensitive has a Q_{10} of 1; values of 2 or 3 mean the reaction speeds up as temperature increases.

qualitative Based on observation of an unmeasured quality of a trait, as in brown vs. blue.

quantitative Based on numerical values obtained by measurement, as in quantitative data.

quantitative trait loci A set of genes determining a complex character (trait) that exhibits quantitative variation (variation in amount rather than in kind).

quaternary structure The specific three-dimensional arrangement of protein subunits. Contrast with primary, secondary, tertiary structure.

quorum sensing The use of chemical communication signals to trigger density-linked activities such as biofilm formation in prokaryotes.

R

R group The distinguishing group of atoms of a particular amino acid; also known as a side chain.

***r*-strategist** A species whose life history strategy allows for a high intrinsic rate of population increase (*r*). (Contrast with *K*-strategist.)

radial symmetry The condition in which any two halves of a body are mirror images of each other, providing the cut passes through the center; a cylinder cut lengthwise down its center displays this form of symmetry.

radiation The transfer of heat from warmer objects to cooler ones via the exchange of infrared radiation. *See also* electromagnetic radiation; evolutionary radiation.

radicle An embryonic root.

radioisotope A radioactive isotope of an element. Examples are carbon-14 (^{14}C) and hydrogen-3, or tritium (^3H).

radiometric dating A method for determining the age of objects such as fossils and rocks based on the decay rates of radioactive isotopes.

rapid eye movement sleep *See* REM sleep.

reactant A chemical substance that enters into a chemical reaction with another substance.

reaction center A group of electron transfer proteins that receive energy from light-absorbing pigments and convert it to chemical energy by redox reactions.

realized niche A species' niche as defined by its interactions with other species. (Contrast with fundamental niche.)

receptive field The area of visual space that activates a particular cell in the visual system.

receptor *See* receptor protein, sensory receptor cell.

receptor potential The change in the resting potential of a sensory cell when it is stimulated.

receptor protein A protein that can bind to a specific molecule, or detect a specific stimulus, within the cell or in the cell's external environment.

receptor-mediated endocytosis Endocytosis initiated by macromolecular binding to a specific membrane receptor.

recessive In genetics, an allele that does not determine phenotype in the presence of a dominant allele. (Contrast with dominance.)

reciprocal crosses A pair of matings in one of which a female of genotype A mates with a male of genotype B and in the other of which a female of genotype B mates with a male of genotype A.

recognition sequence *See* restriction site.

recombinant Pertaining to an individual, meiotic product, or chromosome in which genetic materials originally present in two individuals end up in the same haploid complement of genes.

recombinant DNA A DNA molecule made in the laboratory that is derived from two or more genetic sources.

recombinant frequency The proportion of offspring of a genetic cross that have phenotypes different from the parental phenotypes due to crossing over between linked genes during gamete formation.

recombination frequency The proportion of offspring of a genetic cross that have phenotypes different from the parental phenotypes due to crossing over between linked genes during gamete formation.

reconciliation ecology The practice of making exploited lands more biodiversity-friendly. Compare with restoration ecology.

rectum The terminal portion of the gut, ending at the anus.

redox reaction A chemical reaction in which one reactant is oxidized (loses electrons) and the other is reduced (gains electrons). Short for reduction–oxidation reaction.

reduction Gain of electrons by a chemical reactant. (Contrast with oxidation.)

refractory period The time interval after an action potential during which another action potential cannot be elicited from an excitable membrane.

regeneration The development of a complete individual from a fragment of an organism.

regulative development A pattern of animal embryonic development in which the fates of the first blastomeres are not absolutely fixed. (Contrast with mosaic development.)

regulatory gene A gene that codes for a protein (or RNA) that in turn controls the expression of another gene.

regulatory sequence A DNA sequence to which the protein product of a regulatory gene binds.

regulatory system A system that uses feedback information to maintain a physiological function or parameter at an optimal level. (Contrast with controlled system.)

regulatory T cells (Treg) The class of T cells that mediates tolerance to self antigens.

reinforcement The evolution of enhanced reproductive isolation between populations due to natural selection for greater isolation.

Reissner's membrane *See* tectonic membrane.

releaser Sensory stimulus that triggers performance of a stereotyped behavior pattern.

REM (rapid-eye-movement) sleep A sleep state characterized by vivid dreams, skeletal muscle relaxation, and rapid eye movements. (Contrast with non-REM sleep.)

renal [L. *renes*: kidneys] Relating to the kidneys.

renal tubule A structural unit of the kidney that collects filtrate from the blood, reabsorbs specific ions, nutrients, and water and returns them to the blood, and concentrates excess ions and waste products such as urea for excretion from the body.

renin An enzyme released from the kidneys in response to a drop in the glomerular filtration rate. Together with angiotensin converting enzyme, converts an inactive protein in the blood into angiotensin.

replication The duplication of genetic material.

replication complex The close association of several proteins operating in the replication of DNA.

replication fork A point at which a DNA molecule is replicating. The fork forms by the unwinding of the parent molecule.

replicon A region of DNA replicated from a single origin of replication.

reporter gene A genetic marker included in recombinant DNA to indicate the presence of the recombinant DNA in a host cell.

repressor A protein encoded by a regulatory gene that can bind to a promoter and prevent transcription of the associated gene. (Contrast with activator.)

reproductive isolation Condition in which two divergent populations are no longer exchanging genes. Can lead to speciation.

rescue effect The process by which individuals moving between subpopulations of a metapopulation may prevent declining subpopulations from becoming extinct.

residence time The length of time a chemical element (e.g., carbon or nitrogen) remains in a given compartment of the ecosystem (e.g., in an organic body, in soil, in the atmosphere).

residual volume (RV) In tidal ventilation, the dead space that remains in the lungs at the end of exhalation.

resistance (*R*) genes Plant genes that confer resistance to specific strains of pathogens.

resource Something in the environment required by an organism for its maintenance and growth that is consumed in the process of being used.

resource partitioning A situation in which selection pressures resulting from interspecific competition cause changes in the ways in which the competing species use the limiting resource, thereby allowing them to coexist.

respiration (res pi ra' shun) [L. *spirare*: to breathe] (1) Cellular respiration. (2) Breathing.

respiratory chain The terminal reactions of cellular respiration, in which electrons are passed from NAD or FAD, through a series of intermediate carriers, to molecular oxygen, with the concomitant production of ATP.

respiratory gases Oxygen (O_2) and carbon dioxide (CO_2); the gases that an animal must exchange between its internal body fluids and the outside medium (air or water).

resting potential The membrane potential of a living cell at rest. In cells at rest, the interior is negative to the exterior. (Contrast with action potential.)

restoration ecology The science and practice of restoring damaged or degraded ecosystems.

restriction enzyme Any of a type of enzyme that cleaves double-stranded DNA at specific sites; extensively used in recombinant DNA technology. Also called a restriction endonuclease.

restriction fragment length polymorphism *See* RFLP.

restriction point (R) The specific time during G1 of the cell cycle at which the cell becomes committed to undergo the rest of the cell cycle.

restriction site A specific DNA base sequence that is recognized and acted on by a restriction endonuclease.

reticular activating system A central region of the vertebrate brainstem that includes complex fiber tracts conveying neural signals between the forebrain and the spinal cord, with collateral fibers to a variety of nuclei that are involved in autonomic functions, including arousal from sleep.

reticulum One of the four chambers of the ruminant stomach. Along with the rumen, where food is partially digested with the assistance of gut bacteria.

retina (rett' in uh) [L. *rete*: net] The light-sensitive layer of cells in the vertebrate or cephalopod eye.

retrotransposons Mobile genetic elements that are reverse transcribed into RNA as part of their transfer mechanism. (Contrast with DNA transposons.)

retrovirus An RNA virus that contains reverse transcriptase. Its RNA serves as a template for cDNA production, and the cDNA is integrated into a chromosome of the host cell.

reverse genetics Method of genetic analysis in which a phenotype is first related to a DNA variation, then the protein involved is identified.

reverse transcriptase An enzyme that catalyzes the production of DNA (cDNA), using RNA as a template; essential to the reproduction of retroviruses.

reversion mutation A second- or third-round mutation that reverts the DNA to its original sequence or to a new sequence that results in a non-mutant phenotype.

RFLP Restriction fragment length polymorphism, the coexistence of two or more patterns of restriction fragments resulting from underlying differences in DNA sequence.

rhizoids (rye' zoids) [Gk. root] Hairlike extensions of cells in mosses, liverworts, and a few vascular plants that serve the same function as roots and root hairs in vascular plants. The term is also applied to branched, rootlike extensions of some fungi and algae.

rhizome (rye' zome) An underground stem (as opposed to a root) that runs horizontally beneath the ground.

rhodopsin A vertebrate visual pigment involved in transducing photons of light into changes in the membrane potential of certain photoreceptor cells.

ribonucleic acid *See* RNA.

ribose A five-carbon sugar in nucleotides and RNA.

ribosomal RNA (rRNA) Several species of RNA that are incorporated into the ribosome. Involved in peptide bond formation.

ribosome A small particle in the cell that is the site of protein synthesis.

ribozyme An RNA molecule with catalytic activity.

ribulose bisphosphate carboxylase/oxygenase *See* rubisco.

risk cost The increased chance of being injured or killed as a result of performing a behavior, compared to resting.

RNA (ribonucleic acid) An often single-stranded nucleic acid whose nucleotides use ribose rather than deoxyribose and in which the base uracil replaces thymine found in DNA. Serves as genome from some viruses. (*See* ribosomal RNA, transfer RNA, messenger RNA, and ribozyme.)

RNA interference (RNAi) A mechanism for reducing mRNA translation whereby a double-stranded RNA, made by the cell or synthetically, is processed into a small, single-stranded RNA, whose binding to a target mRNA results in the latter's breakdown.

RNA polymerase An enzyme that catalyzes the formation of RNA from a DNA template.

RNA splicing The last stage of RNA processing in eukaryotes, in which the transcripts of introns are excised through the action of small nuclear ribonucleoprotein particles (snRNP).

rod cells Light-sensitive cells in the vertebrate retina; these sensory receptor cells are sensitive in extremely dim light and are responsible for dim light, black and white vision.

root The organ responsible for anchoring the plant in the soil, absorbing water and minerals, and producing certain hormones. Some roots are storage organs.

root apical meristem Undifferentiated tissue at the apex of the root that gives rise to the organs of the root.

root cap A thimble-shaped mass of cells, produced by the root apical meristem, that protects the meristem; the organ that perceives the gravitational stimulus in root gravitropism.

root hair A long, thin process from a root epidermal cell that absorbs water and minerals from the soil solution.

root system The organ system that anchors a plant in place, absorbs water and dissolved minerals, and may store products of photosynthesis from the shoot system.

rough endoplasmic reticulum (RER) The portion of the endoplasmic reticulum whose outer surface has attached ribosomes. (Contrast with smooth endoplasmic reticulum.)

round window A flexible membrane at the end of the lower canal of the cochlea in the human ear. (*See also* oval window.)

rRNA *See* ribosomal RNA.

rubisco Contraction of ribulose bisphosphate carboxylase/oxygenase, the enzyme that combines carbon dioxide or oxygen with ribulose bisphosphate to catalyze the first step of photosynthetic carbon fixation or photorespiration, respectively.

rumen One of the four chambers of the ruminant stomach. Along with the reticulum, where food is partially digested with the assistance of gut bacteria.

ruminant Herbivorous, cud-chewing mammals such as cows or sheep, characterized by a stomach that consists of four compartments: the rumen, reticulum, omasum, and abomasum.

S

S phase In the cell cycle, the stage of interphase during which DNA is replicated. (Contrast with G1 phase, G2 phase, M phase.)

salt glands Glands on the leaves of some halophytic plants that secrete salt, thereby ridding the plants of excess salt.

saltatory conduction [L. *saltare*: to jump] The rapid conduction of action potentials in myelinated axons; so called because action potentials appear to "jump" between nodes of Ranvier along the axon.

saprobe [Gk. *sapros*: rotten] An organism (usually a bacterium or fungus) that obtains its carbon and energy by absorbing nutrients from dead organic matter.

sarcomere (sark' o meer) [Gk. *sark*: flesh + *meros*: unit] The contractile unit of a skeletal muscle.

sarcoplasm The cytoplasm of a muscle cell.

sarcoplasmic reticulum The endoplasmic reticulum of a muscle cell.

saturated fatty acid A fatty acid in which all the bonds between carbon atoms in the hydrocarbon chain are single bonds—that is, all the bonds are saturated with hydrogen atoms. (Contrast with unsaturated fatty acid.)

Schwann cell A type of glial cell that myelinates axons in the peripheral nervous system.

scientific method A means of gaining knowledge about the natural world by making observations, posing hypotheses, and conducting experiments to test those hypotheses.

scion In horticulture, the bud or stem from one plant that is grafted to a root or root-bearing stem of another plant (the stock).

sclereid One of the principle types of cells in sclerenchyma.

sclerenchyma (skler eng' kyma) [Gk. *skleros*: hard + *kymus*: juice] A plant tissue composed of cells with heavily thickened cell walls. The cells are dead at functional maturity. The principal types of sclerenchyma cells are fibers and sclereids.

scrotum In most mammals, a pouch outside the body cavity that contains the testes.

second filial generation See F_2.

second law of thermodynamics The principle that when energy is converted from one form to another, some of that energy becomes unavailable for doing work.

second messenger A compound, such as cAMP, that is released within a target cell after a hormone (the first messenger) has bound to a surface receptor on a cell; the second messenger triggers further reactions within the cell.

second polar body In oogenesis, the daughter cell of the second meiotic division that subsequently degenerates. (*See also* ootid.)

secondary active transport A form of active transport that does not use ATP as an energy source; rather, transport is coupled to ion diffusion down a concentration gradient established by primary active transport.

secondary cell wall A thick, cellulosic structure internal to the primary cell wall formed in some plant cells after cell expansion stops (Contrast with primary cell wall.)

secondary consumer An organism that eats primary consumers. (Contrast with primary consumer.)

secondary endosymbiosis The engulfment of a photosynthetic eukaryote by another eukaryotic cell that gave rise to certain groups of photosynthetic eukaryotes (e.g., euglenids).

secondary growth In plants, growth that contributes to an increase in girth. (Contrast with primary growth.)

secondary immune response A rapid and intense response to a second or subsequent exposure to an antigen, initiated by memory cells. (Contrast with primary immune response.)

secondary lysosome Membrane-enclosed organelle formed by the fusion of a primary lysosome with a phagosome, in which macromolecules taken up by phagocytosis are hydrolyzed into their monomers. (Contrast with lysosome.)

secondary metabolite A compound synthesized by a plant that is not needed for basic cellular metabolism. Typically has an antiherbivore or antiparasite function.

secondary oocyte In oogenesis, the daughter cell of the first meiotic division that receives almost all the cytoplasm. (*See also* first polar body.)

secondary plant body That part of a plant produced by secondary growth; consists of woody tissues. (Contrast with primary plant body.)

secondary sex determination Formation of secondary sexual characteristics (i.e., those other than gonads), such as external sex organs and body hair. (Contrast with primary sex determination.)

secondary spermatocyte One of the products of the first meiotic division of a primary spermatocyte.

secondary structure Of a protein, localized regularities of structure, such as the α helix and the β pleated sheet. (Contrast with primary, tertiary, quarternary structure.)

secondary succession Succession of ecological communities after a disturbance that did not eliminate all the organisms originally living on the site. (Contrast with primary succession.)

secretin (si kreet' in) A peptide hormone secreted by the upper region of the small intestine when acidic chyme is present. Stimulates the pancreatic duct to secrete bicarbonate ions.

sedimentary rock Rock formed by the accumulation of sediment grains on the bottom of a body of water. Often contain stratified fossils that allow geologists and biologist to date evolutionary events relative to each other.

seed A fertilized, ripened ovule of a gymnosperm or angiosperm. Consists of the embryo, nutritive tissue, and a seed coat.

seedling A plant that has just completed the process of germination.

segment polarity genes In *Drosophila* (fruit fly) development, segmentation genes that determine the boundaries and anterior–posterior organization of individual segments. Part of a developmental cascade that includes maternal effect genes, gap genes, pair rule genes, and Hox genes.

segmentation Division of an animal body into segments.

segmentation genes Genes that determine the number and polarity of body segments.

segregation In genetics, the separation of alleles, or of homologous chromosomes, from each other during meiosis so that each of the haploid daughter nuclei produced contains one or the other member of the pair found in the diploid parent cell, but never both. This principle was articulated by Mendel as his first law.

selectable marker A gene, such as one encoding resistance to an antibiotic, that can be used to identify (select) cells that contain recombinant DNA from among a large population of untransformed cells.

selective permeability Allowing certain substances to pass through while other substances are excluded; a characteristic of membranes.

self-incompatability In plants, the possession of mechanisms that prevent self-fertilization.

semelparous [L. *semel*: once + *pario*: to beget] Reproducing only once in a lifetime. (Contrast with iteroparous.)

semen (see' men) [L. *semin*: seed] The thick, whitish liquid produced by the male reproductive system in mammals, containing the sperm.

semicircular canals Three canals in the human inner ear that form part of the vestibular system.

semiconservative replication The way in which DNA is synthesized. Each of the two partner strands in a double helix acts as a template for a new partner strand. Hence, after replication, each double helix consists of one old and one new strand.

seminiferous tubules The tubules within the testes within which sperm production occurs.

senescence [L. *senescere*: to grow old] Aging; deteriorative changes with aging; the increased probability of dying with increasing age.

sensitive period The life stage during which some particular type of learning must take place, or during which it occurs much more easily than at other times. Typical of song learning among birds. Also known as the critical period.

sensory receptor cell Cell that is responsive to a particular type of physical or chemical stimulation. Sometimes referred to as a sensor.

sensory system A set of organs and tissues for detecting a stimulus; consists of sensory cells, the associated structures, and the neural networks that process the information.

sensory transduction The transformation of environmental stimuli or information into neural signals.

sepal (see' pul) [L. *sepalum*: covering] One of the outermost structures of the flower, usually protective in function and enclosing the rest of the flower in the bud stage.

septate [L. wall] Divided, as by walls or partitions.

septum (plural: septa) (1) A partition or cross-wall appearing in the hyphae of some fungi. (2) The bony structure dividing the nasal passages.

sequence alignment A method of identifying homologous positions in DNA or amino acid sequences by pinpointing the locations of deletions and insertions that have occurred since two (or more) organisms diverged from a common ancestor.

Sertoli cells Cells in the seminiferous tubules of the testes that nurture the developing sperm.

sessile (sess' ul) [L. *sedere*: to sit] Permanently attached; not able to move from one place to another. (Contrast with motile.)

set point In a regulatory system, the threshold sensitivity to the feedback stimulus.

sex chromosome In organisms with a chromosomal mechanism of sex determination, one of the chromosomes involved in sex determination (in humans and many other animals, these are the X and Y chromosomes).

sex-linked inheritance Pattern of inheritance characteristic of genes located on the sex chromosomes of organisms having a chromosomal mechanism for sex determination.

sex pilus A thin connection between two bacteria through which genetic material passes during conjugation.

sexual reproduction Reproduction involving the union of gametes.

sexual selection Selection by one sex of characteristics in individuals of the opposite sex. Also, the favoring of characteristics in one sex as a result of competition among individuals of that sex for mates.

shared derived trait *See* synapomorphy.

shoot apical meristem Undifferentiated tissue at the apex of the shoot that gives rise to the organs of the shoot.

shoot system In plants, the organ system consisting of the leaves, stem(s), and flowers.

short-day plant (SDP) A plant that flowers when nights are longer than a critical length specific for that plant's species. (Compare to long-day plant.)

short tandem repeats (STRs) Short (1–5 base pairs), moderately repetitive sequences of DNA. The number of copies of an STR at a particular location varies between individuals and is inherited.

side chain *See* R group.

sieve tube element The characteristic cell of the phloem in angiosperms, which contains cytoplasm but relatively few organelles, and whose end walls (**sieve plates**) contain pores that form connections with neighboring cells.

sigma factor In prokaryotes, a protein that binds to RNA polymerase, allowing the complex to bind to and stimulate the transcription of a specific class of genes (e.g., those involved in sporulation).

signal sequence The sequence within a protein that directs the protein to a particular organelle.

signal transduction pathway The series of biochemical steps whereby a stimulus to a cell (such as a hormone or neurotransmitter binding to a receptor) is translated into a response of the cell.

silencer A gene sequence binding transcription factors that repress transcription. (Contrast with promoter.)

silent mutation A change in a gene's sequence that has no effect on the amino acid sequence of a protein either because it occurs in noncoding DNA or because it does not change the amino acid specified by the corresponding codon . (Contrast with frame-shift mutation, missense mutation, nonsense mutation.)

silent substitution *See* synonymous substitution.

similarity matrix A matrix used to compare the degree of divergence among pairs of objects. For molecular sequences, constructed by summing the number or percentage of nucleotides or amino acids that are identical in each pair of sequences.

simple diffusion Diffusion that does not involve a direct input of energy or assistance by carrier proteins.

single nucleotide polymorphisms (SNPs) Inherited variations in a single nucleotide base in DNA that differ between individuals.

single-strand binding protein In DNA replication, a protein that binds to single strands of DNA after they have been separated from each other, keeping the two strands separate for replication.

sink In plants, any organ that imports the products of photosynthesis, such as roots, developing fruits, and immature leaves. (Contrast with source.)

sinoatrial node (sigh' no ay' tree al) [L. *sinus*: curve + *atrium*: chamber] The pacemaker of the mammalian heart.

siRNAs (small interfering RNAs) Short, double-stranded RNA molecules used in RNA interference.

sister chromatid Each of a pair of newly replicated chromatids.

sister species Two species that are each other's closest relatives.

skeletal muscle A type of muscle tissue characterized by multinucleated cells containing highly ordered arrangements of actin and myosin microfilaments. Also called striated muscle. (Contrast with cardiac muscle, smooth muscle.)

skeletal systems Organ systems that provide rigid supports—**skeletons**—against which muscles can pull to create directed movements. See also endoskeleton, exoskeleton.

sliding DNA clamp Protein complex that keeps DNA polymerase bound to DNA during replication.

sliding filament model Mechanism of muscle contraction based on the formation and breaking of crossbridges between actin and myosin filaments, causing the filaments to slide together.

slow-twitch fibers Skeletal muscle fibers specialized for sustained aerobic work; contain myoglobin and abundant mitochondria, and are well-supplied with blood vessels. Also called oxidative or red muscle fibers. (Compare to fast-twitch fibers.)

slow-wave sleep *See* non-REM sleep.

small interfering RNAs *See* siRNAs.

small intestine The portion of the gut between the stomach and the colon; consists of the duodenum, the jejunum, and the ileum.

small nuclear ribonucleoprotein particle (snRNP) A complex of an enzyme and a small nuclear RNA molecule, functioning in RNA splicing.

smooth endoplasmic reticulum (SER) Portion of the endoplasmic reticulum that lacks ribosomes and has a tubular appearance. (Contrast with rough endoplasmic reticulum.)

smooth muscle Muscle tissue consisting of sheets of mononucleated cells innervated by the autonomic nervous system. (Contrast with cardiac muscle, skeletal muscle.)

sodium–potassium (Na^+–K^+) pump Antiporter responsible for primary active transport; it pumps sodium ions out of the cell and potassium ions into the cell, both against their concentration gradients. Also called a sodium–potassium ATPase.

soil horizon *See* horizons.

soil solution The aqueous portion of soil, from which plants take up dissolved mineral nutrients.

solute A substance that is dissolved in a liquid (solvent) to form a solution.

solute potential (Ψ_s) A property of any solution, resulting from its solute contents; it may be zero or have a negative value. The more negative the solute potential, the greater the tendency of the solution to take up water through a differentially permeable membrane. (Contrast with pressure potential, water potential.)

solution A liquid (the solvent) and its dissolved solutes.

solvent Liquid in which a substance (solute) is dissolved to form a solution.

somatic cell [Gk. *soma*: body] All the cells of the body that are not specialized for reproduction. (Contrast with germ cell.)

somatic mutation Permament genetic change in a somatic cell (as opposed to a germ cell, the egg or sperm). These mutations affect the individual only; they are not passed on to offspring. (Contrast with germ line mutation.)

somatosensory cortex An area of the parietal lobe that receives touch and pressure information from mechanoreceptors throughout the body; neurons in this area are arranged according to the parts of the body with which they communicate.

somatostatin Peptide hormone made in the hypothalamus that inhibits the release of other hormones from the pituitary and intestine.

somite (so' might) One of the segments into which an embryo becomes divided longitudinally, leading to the eventual segmentation of the animal as illustrated by the spinal column, ribs, and associated muscles.

source In plants, any organ that exports the products of photosynthesis in excess of its own needs, such as a mature leaf or storage organ. (Contrast with sink.)

spatial summation In the production or inhibition of action potentials in a postsynaptic cell, the interaction of depolarizations and hyperpolarizations produced at different sites on the postsynaptic cell. (Contrast with temporal summation.)

spawning *See* external fertilization.

speciation (spee' see ay' shun) The process of splitting one population into two populations that are reproductively isolated from one another.

species (spee' sees) [L. kind] The base unit of taxonomic classification, consisting of an ancestor–descendant group of populations of evolutionarily closely related, similar organisms. The more narrowly defined "biological species" consists of individuals capable of interbreeding with each other but not with members of other species.

species composition The particular mix of species a community contains and the abundances of those species.

species evenness A measure of species diversity that reflects the distribution of the species' abundances in a community.

species richness The total number of species living in a region.

species–area relationship The relationship between the size of an area and the numbers of species it supports.

specific defenses Defensive reactions of the vertebrate immune system that are based on the reaction of an antibody to a specific antigen. (Contrast with nonspecific defenses.)

specific heat The amount of energy that must be absorbed by a gram of a substance to raise its temperature by one degree centigrade. By convention, water is assigned a specific heat of one.

sperm [Gk. *sperma*: seed] The male gamete.

spermatid One of the products of the second meiotic division of a primary spermatocyte; four haploid spermatids, which remain connected by cytoplasmic bridges, are produced for each primary spermatocyte that enters meiosis.

spermatogenesis (spur mat' oh jen' e sis) [Gk. *sperma*: seed + *genesis*: source] Gametogenesis leading to the production of sperm.

spermatogonia In animals, the diploid progeny of a germ cell in males.

spherical symmetry The simplest form of symmetry, in which body parts radiate out from a central point such that an infinite number of planes passing through that central point can divide the organism into similar halves.

sphincter (sfink' ter) [Gk. *sphinkter*: something that binds tightly] A ring of muscle that can close an orifice, for example, at the anus.

spicule [L. arrowhead] A hard, calcareous skeletal element typical of sponges.

spinal cord Along with the brain, part of the central nervous system; transmits information between the body and the brain and mediates simple reflexes.

spinal reflex The conversion of afferent to efferent information in the spinal cord without participation of the brain.

spindle apparatus [O.E. *spindle*, a short stick with tapered ends] Array of microtubules emanating from both poles of a dividing cell during mitosis and playing a role in the movement of chromosomes at nuclear division.

spleen Organ that serves as a reservoir for venous blood and eliminates old, damaged red blood cells from the circulation.

spliceosome RNA–protein complex that splices out introns from eukaryotic pre-mRNAs.

splicing *See* RNA splicing.

spontaneous mutation A genetic change caused by internal cellular mechanisms, such as an error in DNA replication. (Contrast with induced mutation.)

sporangiophore A stalked reproductive structure produced by zygospore fungi that extends from a hypha and bears one or many sporangia.

sporangium (spor an' gee um) (plural: sporangia) [Gk. *spora*: seed + *angeion*: vessel or reservoir] In plants and fungi, any specialized stucture within which one or more spores are formed.

spore [Gk. *spora*: seed] (1) Any asexual reproductive cell capable of developing into an adult organism without gametic fusion. In plants, haploid spores develop into gametophytes, diploid spores into sporophytes. (2) In prokaryotes, a resistant cell capable of surviving unfavorable periods.

sporocyte Specialized cells of the diploid sporophyte that will divide by meiosis to produce four haploid spores. Germination of these spores produces the haploid gametophyte.

sporophyte (spor' o fyte) [Gk. *spora*: seed + *phyton*: plant] In plants and protists with alternation of generations, the diploid phase that produces the spores. (Contrast with gametophyte.)

stabilizing selection Selection against the extreme phenotypes in a population, so that the intermediate types are favored. (Contrast with disruptive selection.)

stamen (stay' men) [L. *stamen*: thread] A male (pollen-producing) unit of a flower, usually composed of an anther, which bears the pollen, and a filament, which is a stalk supporting the anther.

starch [O.E. *stearc*: stiff] A polymer of glucose; used by plants to store energy.

Starling's forces The two opposing forces responsible for water movement across capillary walls: blood pressure, which squeezes water and small solutes out of the capillaries, and osmotic pressure, which pulls water back into the capillaries.

start codon The mRNA triplet (AUG) that acts as a signal for the beginning of translation at the ribosome. (Contrast with stop codon.)

stele (steel) [Gk.: pillar] The central cylinder of vascular tissue in a plant stem.

stem cell In animals, an undifferentiated cell that is capable of continuous proliferation. A stem cell generates more stem cells and a large clone of differentiated progeny cells. (*See also* embryonic stem cell.)

stem In plants, the organ that holds leaves and/or flowers and transports and distributes materials among the other organs of the plant.

stereocilia Fingerlike extensions of hair cell membranes whose bending initiates sound perception. (*See* hair cell.)

steroid Any of a family of lipids whose multiple rings share carbons. The steroid cholesterol is an important constituent of membranes and is the base of steroid hormones such as testosterone.

sticky ends On a piece of two-stranded DNA, short, complementary, one-stranded regions produced by the action of a restriction endonuclease. Sticky ends facilitate the joining of segments of DNA from different sources.

stigma [L. *stigma*: mark, brand] The part of the pistil at the apex of the style that is receptive to pollen, and on which pollen germinates.

stimulus [L. *stimulare*: to goad] Something causing a response; something in the environment detected by a receptor.

stock In horticulture, the root or root-bearing stem to which a bud or piece of stem from another plant (the scion) is grafted.

stoma (plural: stomata) [Gk. *stoma*: mouth, opening] Small opening in the plant epidermis that permits gas exchange; bounded by a pair of guard cells whose osmotic status regulates the size of the opening.

stomach An organ that physically (and sometimes enzymatically) breaks down food, preparing it for digestion in the midgut.

stomatal crypt In plants, a sunken cavity below the leaf surface in which a stoma is sheltered from the drying effects of air currents.

stop codon Any of the three mRNA codons that signal the end of protein translation at the ribosome: UAG, UGA, UAA.

stratosphere The upper part of Earth's atmosphere, above the troposphere; extends from approximately 18 kilometers upward to approximately 50 kilometers above Earth's surface.

stratum (plural strata) [L. *stratos*: layer] A layer of sedimentary rock laid down at a particular time in the past.

stretch receptor A modified muscle cell embedded in the connective tissue of a muscle that acts as a mechanoreceptor in response to stretching of that muscle.

striated muscle *See* skeletal muscle.

strigolactones Signaling molecules produced by plant roots that attract the hyphae of mycorrhizal fungi.

strobilus (plural: strobili) One of several conelike structures in various groups of plants (including club mosses, horsetails, and conifers) associated with the production and dispersal of reproductive products.

stroke An embolism in an artery in the brain that causes the cells fed by that artery to die. The specific damage, such as memory loss, speech impairment, or paralysis, depends on the location of the blocked artery.

stroma The fluid contents of an organelle such as a chloroplast or mitochondrion.

structural gene A gene that encodes the primary structure of a protein not involved in the regulation of gene expression.

structural isomers Molecules made up of the same kinds and numbers of atoms, in which the atoms are bonded differently.

structural motif A three-dimensional structural element that is part of a larger molecule. For example, there are four common motifs in DNA-binding proteins: helix-turn-helix, zinc finger, leucine zipper, and helix-loop-helix.

style [Gk. *stele*: pillar or column] In the angiosperm flower, a column of tissue extending from the tip of the ovary, and bearing the stigma or receptive surface for pollen at its apex.

sub- [L. under] A prefix used to designate a structure that lies beneath another or is less than another. For example, subcutaneous (beneath the skin); subspecies

subduction In plate tectonics, the movement of one lithospheric plate under another.

suberin A waxlike lipid that is a barrier to water and solute movement across the Casparian strip of the endodermis.

submucosa (sub mew koe' sah) The tissue layer just under the epithelial lining of the lumen of the digestive tract.

subsoil The soil horizon lying below the topsoil and above the parent rock (bedrock); the zone of infiltration and accumulation of materials leached from the topsoil. Also called the B horizon.

substrate (sub' strayte) (1) The molecule or molecules on which an enzyme exerts catalytic action. (2) The base material on which a sessile organism lives.

succession The gradual, sequential series of changes in the species composition of an ecological community following a disturbance. See also cyclical succession, directional selection, heterotrophic succession.

succulence In plants, possession of fleshy, water-storing leaves or stems; an adaptation to dry environments.

sulcus (sul' kus; plural sulci) [Gk.: plowed furrow] *See* convolutions.

superficial cleavage A variation of incomplete cleavage in which cycles of mitosis occur without cell division, producing a syncytium (a single cell with many nuclei).

suprachiasmatic nuclei (SCN) In mammals, two clusters of neurons just above the optic chiasm that act as the master circadian clock.

surface area-to-volume ratio For any cell, organism, or geometrical solid, the ratio of surface area to volume; this is an important factor in setting an upper limit on the size a cell or organism can attain.

surface tension The attractive intermolecular forces at the surface of liquid; an especially important property of water.

surfactant A substance that decreases the surface tension of a liquid. Lung surfactant, secreted by cells of the alveoli, is mostly phospholipid and decreases the amount of work necessary to inflate the lungs.

survivorship The fraction of individuals that survive from birth to a given life stage or age.

survivorship curves Graphic plot of ages at death of a hypothetical cohort, usually of 1,000 individuals, by plotting the numbers of individuals expected to survive to reach each age category. There are three general shapes. Ecological survivorship is linear: individuals face a constant risk of mortality regardless of their age. Physiological survivorship curves are concave: high survivorship through adulthood with steep declines late in life. Maturational survivorship curves are convex, with high mortality early in life but higher survivorship once individuals reach maturity.

suspensor In the embryos of seed plants, the stalk of cells that pushes the embryo into the endosperm and is a source of nutrient transport to the embryo.

sustainable Pertaining to the use and management of ecosystems in such a way that humans benefit over the long term from specific ecosystem goods and services without compromising others.

symbiosis (sim' bee oh' sis) [Gk. *sym*: together + *bios*: living] The living together of two or more species in a prolonged and intimate relationship.

symmetry Pertaining to an attribute of an animal body in which at least one plane can divide the body into similar, mirror-image halves. (*See* bilateral symmetry, radial symmetry.)

sympathetic nervous system The division of the autonomic nervous system that works in opposition to the parasympathetic nervous system. (Contrast with parasympathetic nervous system.)

sympatric speciation (sim pat' rik) [Gk. *sym*: same + *patria*: homeland] Speciation due to reproductive isolation without any physical separation of the subpopulation. (Contrast with allopatric speciation.)

symplast The continuous meshwork of the interiors of living cells in the plant body, resulting from the presence of plasmodesmata. (Contrast with apoplast.)

symporter A membrane transport protein that carries two substances in the same direction. (Contrast with antiporter, uniporter.)

synapomorphy A trait that arose in the ancestor of a phylogenetic group and is present (sometimes in modified form) in all of its members, thus helping to delimit and identify that group. Also called a shared derived trait.

synapse (sin' aps) [Gk. *syn*: together + *haptein*: to fasten] A specialized type of junction where a neuron meets its target cell (which can be another neuron or some other type of cell) and information in the form of neurotransmitter molecules is exchanged across a synaptic cleft.

synapsis (sin ap′ sis) The highly specific parallel alignment (pairing) of homologous chromosomes during the first division of meiosis.

synaptic cleft The space between the presynaptic cell and the postsynaptic cell in a chemical synapse.

synergids [Gk. *syn*: together + *ergos*: work] In angiosperms, the two cells accompanying the egg cell at one end of the megagametophyte.

syngamy *See* fertilization.

synonymous (silent) substitution A change of one nucleotide in a sequence to another when that change does not affect the amino acid specified (i.e., UUA → UUG, both specifying leucine). (Contrast with nonsynonymous substitution, missense mutation, nonsense mutation.)

systematics The scientific study of the diversity and relationships among organisms.

systemic acquired resistance A general resistance to many plant pathogens following infection by a single agent.

systemic circuit Portion of the circulatory system by which oxygenated blood from the lungs or gills is distributed throughout the rest of the body and returned to the heart. (Contrast with pulmonary circuit.)

systems biology The scientific study of an organism as an integrated and interacting system of genes, proteins, and biochemical reactions.

systole (sis′ tuh lee) [Gk.: contraction] Contraction of a chamber of the heart, driving blood forward in the circulatory system. (Contrast with diastole.)

T

3′ end (3 prime) The end of a DNA or RNA strand that has a free hydroxyl group at the 3′ carbon of the sugar (deoxyribose or ribose).

T cell A type of lymphocyte involved in the cellular immune response. The final stages of its development occur in the thymus gland. (Contrast with B cell; *see also* cytotoxic T cell, T-helper cell.)

T cell receptor A protein on the surface of a T cell that recognizes the antigenic determinant for which the cell is specific.

T tubules A system of tubules that runs throughout the cytoplasm of a muscle fiber, through which action potentials spread.

T-helper (T_H) cell Type of T cell that stimulates events in both the cellular and humoral immune responses by binding to the antigen on an antigen-presenting cell; target of the HIV-I virus, the agent of AIDS. (Contrast with cytotoxic T cells.)

taproot system A root system typical of eudicots consisting of a primary root (*taproot*) that extends downward by tip growth and outward by initiating lateral roots. (Contrast with fibrous root system.)

target cell A cell with the appropriate receptors to bind and respond to a particular hormone or other chemical mediator.

taste bud A structure in the epithelium of the tongue that includes a cluster of chemoreceptors innervated by sensory neurons.

TATA box An eight-base-pair sequence, found about 25 base pairs before the starting point for transcription in many eukaryotic promoters, that binds a transcription factor and thus helps initiate transcription.

taxon (plural: taxa) [Gk. *taxis*: put in order] A biological group (typically a species or a clade) that is given a name.

T_C cells *See* cytotoxic T cells.

tectonic membrane One of two membranes (the other is the basilar membrane) that extend along the length of the cochlea in the human ear. Also known as Reissner's membrane.

telencephalon The outer, surrounding structure of the embryonic vertebrate forebrain, which develops into the cerebrum.

telomerase An enzyme that catalyzes the addition of telomeric sequences lost from chromosomes during DNA replication.

telomeres (tee′ lo merz) [Gk. *telos*: end + *meros*: units, segments] Repeated DNA sequences at the ends of eukaryotic chromosomes.

telophase (tee′ lo phase) [Gk. *telos*: end] The final phase of mitosis or meiosis during which chromosomes become diffuse, nuclear envelopes re-form, and nucleoli begin to reappear in the daughter nuclei.

template A molecule or surface on which another molecule is synthesized in complementary fashion, as in the replication of DNA.

template strand In double-stranded DNA, the strand that is transcribed to create an RNA transcript that will be processed into a protein. Also refers to a strand of RNA that is used to create a complementary RNA.

temporal lobe One of the four lobes of the cerebral hemisphere; receives and processes auditory and visual information; involved in recognizing, identifying, and naming objects.

temporal summation In the production or inhibition of action potentials in a postsynaptic cell, the interaction of depolarizations or hyperpolarizations produced by rapidly repeated stimulation of a single point on the postsynaptic cell. (Contrast with spatial summation.)

tendon A collagen-containing band of tissue that connects a muscle with a bone.

tepal A sterile, modified, nonphotosynthetic leaf of an angiosperm flower that cannot be distinguished as a petal or a sepal.

termination In molecular biology, the end of transcription or translation.

terminator A sequence at the 3′ end of mRNA that causes the RNA strand to be released from the transcription complex.

terrestrial (ter res′ tree al) [L. *terra*: earth] Pertaining to or living on land. (Contrast with aquatic, marine.)

territorial behavior Aggressive actions engaged in to defend a habitat or resource such that other animals are denied access.

tertiary consumers Carnivores that consume primary carnivores (secondary consumers).

tertiary endosymbiosis The mechanism by which some eukaryotes acquired the capacity for photosynthesis; for example, a dinoflagellate that apparently lost its chloroplast became photosynthetic by engulfing another protist that had acquired a chloroplast through secondary endosymbiosis.

tertiary structure In reference to a protein, the relative locations in three-dimensional space of all the atoms in the molecule. The overall shape of a protein. (Contrast with primary, secondary, and quaternary structures.)

test cross Mating of a dominant-phenotype individual (who may be either heterozygous or homozygous) with a homozygous-recessive individual.

testis (tes′ tis) (plural: testes) [L. *testis*: witness] The male gonad; the organ that produces the male gametes.

tetanus [Gk. *tetanos*: stretched] (1) A state of sustained maximal muscular contraction caused by rapidly repeated stimulation. (2) In medicine, an often fatal disease ("lockjaw") caused by the bacterium *Clostridium tetani*.

tetrad [Gk. *tettares*: four] During prophase I of meiosis, the association of a pair of homologous chromosomes or four chromatids

thalamus [Gk. *thalamos*: chamber] A region of the vertebrate forebrain; involved in integration of sensory input.

theory [Gk. *theoria*: analysis of facts] A far-reaching explanation of observed facts that is supported by such a wide body of evidence, with no significant contradictory evidence, that it is scientifically accepted as a factual framework. Examples are Newton's theory of gravity and Darwin's theory of evolution. (Contrast with hypothesis.)

thermoneutral zone (TNZ) [Gk. *thermos*: temperature] The range of temperatures over which an endotherm does not have to expend extra energy to thermoregulate.

thermophile (ther′ muh fyle)[Gk. *thermos*: temperature + *philos*: loving] An organism that lives exclusively in hot environments.

thoracic cavity [Gk. *thorax*: breastplate] The portion of the mammalian body cavity bounded by the ribs, shoulders, and diaphragm. Contains the heart and the lungs.

thoracic duct The connection between the lymphatic system and the circulatory system.

threshold The level of depolarization that causes an electrically excitable membrane to fire an action potential.

thrombin An enzyme involved in blood clotting; cleaves fibrinogen to form fibrin.

thrombus (throm′ bus) [Gk. *thrombos*: clot] A blood clot that forms within a blood vessel and remains attached to the wall of the vessel.

thylakoid (thigh′ la koid) [Gk. *thylakos*: sack or pouch] A flattened sac within a chloroplast. Thylakoid membranes contain all of the chlorophyll in a plant, in addition to the electron carriers of photophosphorylation. Thylakoids stack to form grana.

thymine (T) Nitrogen-containing base found in DNA.

thymus [Gk. *thymos*: warty] A ductless, glandular lymphoid tissue, involved in development of the immune system of vertebrates. In humans, the thymus degenerates during puberty.

thyroid gland [Gk. *thyreos*: door-shaped] A two-lobed gland in vertebrates. Produces the hormone thyroxine.

thyrotropin Hormone produced by the anterior pituitary that stimulates the thyroid gland to produce and release thyroxine. Also called thyroid-stimulating hormone (TSH).

thyrotropin-releasing hormone (TRH) Hormone produced by the hypothalamus that stimulates the anterior pituitary to release thyrotropin.

thyroxine Hormone produced by the thyroid gland; controls many metabolic processes.

tidal The bidirectional form of ventilation used by all vertebrates except birds; air enters and leaves the lungs by the same route.

tight junction A junction between epithelial cells in which there is no gap between adjacent cells.

tissue A group of similar cells organized into a functional unit; usually integrated with other tissues to form part of an organ.

tissue system In plants, any of three organized groups of tissues—dermal tissue, vascular tissue, and ground tissue—that are established during embryogenesis and have distinct functions.

titin A protein that holds bundles of myosin filaments in a centered position within the sarcomeres of muscle cells. The largest protein in the human body.

tonoplast The membrane of the plant central vacuole.

topsoil The uppermost soil horizon; contains most of the organic matter of soil, but may be depleted of most mineral nutrients by leaching. Also called the A horizon.

totipotent [L. *toto*: whole, entire + *potens*: powerful] Possessing all the genetic information and other capacities necessary to form an entire individual. (Contrast with multipotent, pluripotent.)

toxigenicity The ability of some pathogenic bacteria to produce chemical substances that harm the host.

trachea (tray′ kee ah) [Gk. *trakhoia*: tube] A tube that carries air to the bronchi of the lungs of vertebrates. When plural (*tracheae*), refers to the major airways of insects.

tracheary element Either of two types of xylem cells—tracheids and vessel elements—that undergo apoptosis before assuming their transport function.

tracheid (tray′ kee id) A type of tracheary element found in the xylem of nearly all vascular plants, characterized by tapering ends and walls that are pitted but not perforated. (Contrast with vessel element.)

trade-off The relationship between the fitness benefits conferred by an adaptation and the fitness costs it imposes. For an adaptation to be favored by natural selection, the benefits must exceed the costs.

trait In genetics, a specific form of a character: eye color is a character; brown eyes and blue eyes are traits. (Contrast with character.)

transcription The synthesis of RNA using one strand of DNA as a template.

transcription factors Proteins that assemble on a eukaryotic chromosome, allowing RNA polymerase II to perform transcription.

transcription initiation site The part of a gene's promoter where synthesis of the gene's RNA transcript begins.

transduction (1) Transfer of genes from one bacterium to another by a bacteriophage. (2) In sensory cells, the transformation of a stimulus (e.g., light energy, sound pressure waves, chemical or electrical stimulants) into action potentials.

transfection Insertion of recombinant DNA into animal cells.

transfer RNA (tRNA) A family of folded RNA molecules. Each tRNA carries a specific amino acid and anticodon that will pair with the complementary codon in mRNA during translation.

transformation (1) A mechanism for transfer of genetic information in bacteria in which pure DNA from a bacterium of one genotype is taken in through the cell surface of a bacterium of a different genotype and incorporated into the chromosome of the recipient cell. (2) Insertion of recombinant DNA into a host cell.

transgenic Containing recombinant DNA incorporated into the genetic material.

transition state In an enzyme-catalyzed reaction, the reactive condition of the substrate after there has been sufficient input of energy (activation energy) to initiate the reaction.

translation The synthesis of a protein (polypeptide). Takes place on ribosomes, using the information encoded in messenger RNA.

translational repressor A protein that blocks translation by binding to mRNAs and preventing their attachment to the ribosome. In mammals, the production of ferritin protein is regulated by a translational repressor.

translocation (1) In genetics, a rare mutational event that moves a portion of a chromosome to a new location, generally on a nonhomologous chromosome. (2) In vascular plants, movement of solutes in the phloem.

transmembrane protein An integral membrane protein that spans the phospholipid bilayer.

transpiration [L. *spirare*: to breathe] The evaporation of water from plant leaves and stem, driven by heat from the sun, and providing the motive force to raise water (plus mineral nutrients) from the roots.

transpiration–cohesion–tension mechanism Theoretical basis for water movement in plants: evaporation of water from cells within leaves (transpiration) causes an increase in surface tension, pulling water up through the xylem. Cohesion of water occurs because of hydrogen bonding.

transposable element (transposon) A segment of DNA that can move to, or give rise to copies at, another locus on the same or a different chromosome.

transversion A mutation that changes a purine to a pyrimidine or vice versa.

tree of life A term that encompasses the evolutionary history of all life, or a graphic representation of that history.

triglyceride A simple lipid in which three fatty acids are combined with one molecule of glycerol.

trimesters The three stages of human pregnancy, approximately 3 months each in length.

tripartite synapse The idea that a synapse includes not only the pre- and postsynaptic neurons involved but also encompasses many connections with glial cells called astrocytes.

triploblastic Having three cell layers.

trisomic Containing three rather than two members of a chromosome pair.

tRNA *See* transfer RNA.

trochophore (troke′ o fore) [Gk. *trochos*: wheel + *phoreus*: bearer] A radially symmetrical larval form typical of annelids and mollusks, distinguished by a wheel-like band of cilia around the middle.

trophic cascade The progression over successively lower trophic levels of the indirect effects of a predator.

trophic interactions The consumer–resource relationships among species in a community.

trophic level [Gk *trophes*: nourishment] A group of organisms united by obtaining their energy from the same part of the food web of a biological community.

trophoblast [Gk *trophes*: nourishment + *blastos*: sprout] At the 32-cell stage of mammalian development, the outer group of cells that will become part of the placenta and thus nourish the growing embryo. (Contrast with inner cell mass.)

tropic hormones Hormones produced by the anterior pituitary that control the secretion of hormones by other endocrine glands.

tropomyosin [troe poe my' oh sin] One of the three protein components of an actin filament; controls the interactions of actin and myosin necessary for muscle contraction.

troponin One of the three components of an actin filament; binds to actin, tropomyosin, and Ca^{2+}.

true-breeding A genetic cross in which the same result occurs every time with respect to the trait(s) under consideration, due to homozygous parents.

trypsin A protein-digesting enzyme. Secreted by the pancreas in its inactive form (trypsinogen), it becomes active in the duodenum of the small intestine.

tube feet A unique feature of echinoderms; extensions of the water vascular system, which functions in gas exchange, locomotion, and feeding.

tubulin A protein that polymerizes to form microtubules.

tumor [L. *tumor*: a swollen mass] A disorganized mass of cells. Malignant tumors spread to other parts of the body.

tumor necrosis factor A family of cytokines (growth factors) that causes cell death and is involved in inflammation.

tumor suppressor A gene that codes for a protein product that inhibits cell proliferation; inactive in cancer cells. (Contrast with oncogene.)

turgor pressure [L. *turgidus*: swollen] *See* pressure potential.

turnover In freshwater ecosystems, vertical movements of water that bring nutrients and dissolved CO_2 to the surface and O_2 to deeper water.

twitch A muscle fiber's minimum unit of contraction, stimulated by a single action potential.

tympanic membrane [Gk. *tympanum*: drum] The eardrum.

U

ubiquitin A small protein that is covalently linked to other cellular proteins identified for breakdown by the proteosome.

ultimate causes In ethology, the evolutionary processes that produce an animal's capacity and tendency to behave in particular ways. (Contrast with proximate causes.)

unequal crossing over When a highly repeated gene sequence becomes displaced in alignment during meiotic crossing over, so that one chromosome receives many copies of the sequence while the second chromosome receives fewer copies. One of the mechanisms of concerted evolution. (See also biased gene conversion.)

uniporter [L. *unus*: one + *portal*: doorway] A membrane transport protein that carries a single substance in one direction. (Contrast with antiporter, symporter.)

unipotent An undifferentiated cell that is capable of becoming only one type of mature cell. (Contrast with totipotent, multipotent, pluripotent.)

unsaturated fatty acid A fatty acid whose hydrocarbon chain contains one or more double bonds. (Contrast with saturated fatty acid.)

upregulation A process by which the abundance of receptors for a hormone increases when hormone secretion is suppressed. (Contrast with downregulation.)

upwelling zones Areas of the ocean where cool, nutrient-rich water from deeper layers rises to the surface.

uracil (U) A pyrimidine base found in nucleotides of RNA.

urea A compound that is the main form of nitrogen excreted by many animals, including mammals.

ureotelic Pertaining to an organism in which the final product of the breakdown of nitrogen-containing compounds (primarily proteins) is urea. (Contrast with ammonotelic, uricotelic.)

ureter (your' uh tur) Long duct leading from the vertebrate kidney to the urinary bladder or the cloaca.

urethra (you ree' thra) In most mammals, the canal through which urine is discharged from the bladder and which serves as the genital duct in males.

uric acid A compound that serves as the main excreted form of nitrogen in some animals, particularly those which must conserve water, such as birds, insects, and reptiles.

uricotelic Pertaining to an organism in which the final product of the breakdown of nitrogen-containing compounds (primarily proteins) is uric acid. (Contrast with ammonotelic, ureotelic.)

urinary bladder A structure in which urine is stored until it can be excreted to the outside of the body.

urine (you' rin) In vertebrates, the fluid waste product containing the toxic nitrogenous by-products of protein and nucleic acid metabolism.

uterine cycle In human females, the monthly cycle of events by which the endometrium is prepared for the arrival of a blastocyst. (Contrast with ovarian cycle).

uterus (yoo' ter us) [L. *utero*: womb] A specialized portion of the female reproductive tract in mammals that receives the fertilized egg and nurtures the embryo in its early development. Also called the womb.

V

vaccination Injection of virus or bacteria or their proteins into the body, to induce immunity. The injected material is usually attenuated (weakened) before injection and is called a *vaccine*.

vacuole (vac' yew ole) Membrane-enclosed organelle in plant cells that can function for storage, water concentration for turgor, or hydrolysis of stored macromolecules.

vagina (vuh jine' uh) [L. sheath] In female animals, the entry to the reproductive tract.

van der Waals forces Weak attractions between atoms resulting from the interaction of the electrons of one atom with the nucleus of another. This type of attraction is about one-fourth as strong as a hydrogen bond.

variable In a controlled experiment, a factor that is manipulated to test its effect on a phenomenon.

variable region The portion of an immunoglobulin molecule or T cell receptor that includes the antigen-binding site and is responsible for its specificity. (Contrast with constant region.)

vas deferens (plural: vasa deferentia) Duct that transfers sperm from the epididymis to the urethra.

vasa recta Blood vessels that parallel the loops of Henle and the collecting ducts in the renal medulla of the kidney.

vascular (vas' kew lar) [L. *vasculum*: a small vessel] Pertaining to organs and tissues that conduct fluid, such as blood vessels in animals and xylem and phloem in plants.

vascular bundle In vascular plants, a strand of vascular tissue, including xylem and phloem as well as thick-walled fibers.

vascular cambium (kam' bee um) [L. *cambiare*: to exchange] In plants, a lateral meristem that gives rise to secondary xylem and phloem.

vascular tissue system The transport system of a vascular plant, consisting primarily of xylem and phloem.

vasopressin A hormone that promotes water reabsorption by the kidney. Produced by neurons in the hypothalamus and released from nerve terminals in the posterior pituitary. Also called antidiuretic hormone or ADH.

vector (1) An agent, such as an insect, that carries a pathogen affecting another species. (2) A plasmid or virus that carries an inserted piece of DNA into a bacterium for cloning purposes in recombinant DNA technology.

vegetal hemisphere The lower portion of some animal eggs, zygotes, and embryos, in which the dense nutrient yolk settles. The *vegetal pole* is to the very bottom of the egg or embryo. (Contrast with animal hemisphere.)

vegetative Nonreproductive, nonflowering, or asexual.

vegetative meristem An apical meristem that produces leaves.

vegetative reproduction Asexual reproduction through the modification of stems, leaves, or roots.

vein [L. *vena*: channel] A blood vessel that returns blood to the heart. (Contrast with artery.)

vena cavae In the circulatory systems of crocodilians, birds, and mammals, large veins that empty into the right atrium of the heart.

ventral [L. *venter*: belly, womb] Toward or pertaining to the belly or lower side. (Contrast with dorsal.)

ventricle A muscular heart chamber that pumps blood through the lungs or through the body.

venule A small blood vessel draining a capillary bed that joins others of its kind to form a vein. (Contrast with arteriole.)

vernalization [L. *vernalis*: spring] Events occurring during a required chilling period, leading eventually to flowering.

vertebral column [L. *vertere*: to turn] The jointed, dorsal column that is the primary support structure of vertebrates.

very low-density lipoproteins (VLDLs) Lipoproteins that consist mainly of triglyceride fats, which they transport to fat cells in adipose tissues throughout the body; associated with excessive fat deposition and high risk for cardiovascular disease.

vesicle Within the cytoplasm, a membrane-enclosed compartment that is associated with other organelles; the Golgi complex is one example.

vessel element A type of tracheary element with perforated end walls; found only in angiosperms. (Contrast with tracheid.)

vestibular system (ves tib' yew lar) [L. *vestibulum*: an enclosed passage] Structures within the inner ear that sense changes in position or momentum of the head, affecting balance and motor skills.

vicariant event (vye care' ee unt) [L. *vicus*: change] The splitting of a taxon's range by the imposition of some barrier to dispersal.

villus (vil' lus) (plural: villi) [L. *villus*: shaggy hair or beard] A hairlike projection from a membrane; for example, from many gut walls.

virion (veer' e on) The virus particle, the minimum unit capable of infecting a cell.

virulence [L. *virus*: poison, slimy liquid] The ability of a pathogen to cause disease and death.

virus Any of a group of ultramicroscopic particles constructed of nucleic acid and protein (and, sometimes, lipid) that require living cells in order to reproduce. Viruses evolved multiple times from different cellular species.

vital capacity (VC) The maximum capacity for air exchange in one breath; the sum of the tidal volume and the inspiratory and expiratory reserve volumes.

vitamin [L. *vita*: life] An organic compound that an organism cannot synthesize, but nevertheless requires in small quantities for normal growth and metabolism.

vitelline envelope The inner, proteinaceous protective layer of a sea urchin egg.

viviparity (vye vi par' uh tee) Reproduction in which fertilization of the egg and development of the embryo occur inside the mother's body. (Contrast with oviparity.)

vivipary Premature germination in plants.

voltage A measure of the difference in electrical charge between two points.

voltage-gated channel A type of gated channel that opens or closes when a certain voltage exists across the membrane in which it is inserted.

vomeronasal organ (VNO) Chemosensory structure embedded in the nasal epithelium of amphibians, reptiles, and many mammals. Often specialized for detecting pheromones.

W

warning coloration *See* aposematism

water potential (psi, Ψ) In osmosis, the tendency for a system (a cell or solution) to take up water from pure water through a differentially permeable membrane. Water flows toward the system with a more negative water potential. (Contrast with solute potential, pressure potential.)

water vascular system In echinoderms, a network of water-filled canals that functions in gas exchange, locomotion, and feeding.

wavelength The distance between successive peaks of a wave train, such as electromagnetic radiation.

weather The state of atmospheric conditions in a particular place at a particular time. (Contrast with climate.)

weathering The mechanical and chemical processes by which rocks are broken down into soil particles.

Wernicke's area A region in the temporal lobe of the human brain that is involved with the sensory aspects of language.

white blood cells Cells in the blood plasma that play defensive roles in the immune system. Also called leukocytes.

white matter In the central nervous system, tissue that is rich in axons. (Contrast with gray matter.)

wild type Geneticists' term for standard or reference type. Deviants from this standard, even if the deviants are found in the wild, are usually referred to as mutant. (Note that this terminology is not usually applied to human genes.)

wood Secondary xylem tissue.

X-Y-Z

xerophyte (zee' row fyte) [Gk. *xerox*: dry + *phyton*: plant] A plant adapted to an environment with limited water supply.

xylem (zy' lum) [Gk. *xylon*: wood] In vascular plants, the tissue that conducts water and minerals; xylem consists, in various plants, of tracheids, vessel elements, fibers, and other highly specialized cells.

yolk [M.E. *yolke*: yellow] The stored food material in animal eggs, rich in protein and lipids.

yolk sac In reptiles, birds, and mammals, the extraembryonic membrane that forms from the endoderm of the hypoblast; it encloses and digests the yolk.

zeaxanthin A blue-light receptor involved in the opening of plant stomata.

zona pellucida A jellylike substance that surrounds the mammalian ovum when it is released from the ovary.

zone of cell division The apical and primary meristems of a plant root; the source of all cells of the root's primary tissues.

zone of cell elongation The part of a plant root, generally above the zone of cell division, where cells are expanding (growing), primarily in the longitudinal direction.

zone of maturation The part of a plant root, generally above the zone of cell elongation, where cells are differentiating.

zoospore (zoe' o spore) [Gk. *zoon*: animal + *spora*: seed] In algae and fungi, any swimming spore. May be diploid or haploid.

zygospore Multinucleate, diploid cell that is a resting stage in the life cycle of zygospore fungi.

zygote (zye' gote) [Gk. *zygotos*: yoked] The cell created by the union of two gametes, in which the gamete nuclei are also fused. The earliest stage of the diploid generation.

zymogen The inactive precursor of a digestive enzyme; secreted into the lumen of the gut, where a protease cleaves it to form the active enzyme.

Illustration Credits

Frontispiece © Steve Bloom Images/Alamy.

Table of Contents Page XXI: © FLPA/Alamy. Page XXII: © Biophoto Associates/Science Source/Photo Researchers, Inc. Page XXIII: Protein data from Sobolevsky et al. 2009. *Nature* 462: 745. Membrane data from Heller et al. 1993. *J. Phys. Chem.* 97: 8343. Page XXIV: © Manabu Kagami/amanaimages/Corbis. Page XXV: © Steve Gschmeissner/SPL/Photo Researchers, Inc. Page XXVI: © zhaoyan/Shutterstock. Page XXVII: © Power and Syred/Science Source/Photo Researchers, Inc. Page XXVIII: © Dr. Fred Hossler/Visuals Unlimited, Inc. Page XXIX: Drawings by Elizabeth Gould, c. 1845. Page XXXI: © SciMAT/Science Source/Photo Researchers, Inc. Page XXXII: David McIntyre. Page XXXIII: © Joe Belanger/Shutterstock. Page XXXIV: © Nigel Cattlin/Alamy. Page XXXV: Courtesy of Andrew D. Sinauer. Page XXXVI: © Martin Harvey/Corbis. Page XXXVIII: © Anthony Bannister/Gallo Images/Corbis. Page XXXIX: © Thomas Deerinck, NCMIR/Science Source/Photo Researchers, Inc. Page XL: David McIntyre. Page XLI: © Cathy Keifer/Shutterstock. Page XLII: © Angelo Gandolfi/Naturepl.com. Page XLIII: © Carol Buchanan/AGE Fotostock. Page XLIV: © Wild Wonders of Europe/O. Haarberg/Naturepl.com.

Chapter 1 *Opener*: © Pamela S. Turner. 1.1A: © Eye of Science/SPL/Photo Researchers, Inc. 1.1B: © Science Photo Library RF/Photolibrary.com. 1.1C: © Steve Gschmeissner/Photo Researchers, Inc. 1.1D: David McIntyre. 1.1E: © Glen Threlfo/Auscape/Minden Pictures. 1.1F: © Piotr Naskrecki/Minden Pictures. 1.1G: © Tui De Roy/Minden Pictures. 1.3A: © Kwangshin Kim/Photo Researchers, Inc. 1.3B: © Dr. Gopal Murti/Visuals Unlimited, Inc. 1.4A: © Walter Geiersperger/Corbis. 1.4B: © Roger Garwood & Trish Ainslie/Corbis. 1.6A: © Arco Images GmbH/Alamy. 1.6B: © Heather Angel/Natural Visions/Alamy. 1.6C: © Juniors Bildarchiv/Alamy. 1.6D: © Stephen Dalton/Naturepl.com. 1.9A: © A & J Visage/Alamy. 1.9B: © Stefan Huwiler/Rolfnp/Alamy. 1.11: From T. Hayes et al., 2003. *Environ. Health Perspect.* 111: 568. 1.13: Courtesy of Scott Bauer/USDA ARS. 1.14: © Kim Kulish/Corbis. 1.15A: Courtesy of Wayne Whippen. 1.16: Courtesy of the U.S. Geological Survey. 1.17: © Mark Moffett/Minden Pictures/Corbis.

Chapter 2 *Opener*: © Phil Degginger/Alamy. 2.3: Used with permission of Mayo Foundation for Medical Education and Research, mayoclinic.com. 2.14: © Pablo H Caridad/Shutterstock. 2.15A: © Michael Cole/Corbis. 2.15B: © kawhia/Shutterstock. 2.15C: David McIntyre. Page 38: © Jean Claude Carton/Bruce Coleman USA/AGE Fotostock.

Chapter 3 *Opener*: © Dennis Kunkel Microscopy, Inc. 3.9: Data from PDB 1IVM. T. Obita, T. Ueda, & T. Imoto, 2003. *Cell. Mol. Life Sci.* 60: 176. 3.11: Data from PDB 2HHB. G. Fermi et al., 1984. *J. Mol. Biol.* 175: 159. 3.18C *left*: © Biophoto Associates/Photo Researchers, Inc. 3.18C *middle*: © Dennis Kunkel Microscopy, Inc. 3.18C *right*: © Don W. Fawcett/Photo Researchers, Inc. 3.19 *Ear*: David McIntyre. 3.19 *Beetle*: © Pan Xunbin/Shutterstock. Page 57: David McIntyre.

Chapter 4 *Opener*: © Anup Shah/Naturepl.com. 4.7: Courtesy of the Argonne National Laboratory. 4.11B: Courtesy of Janet Iwasa, Szostak group, MGH/Harvard. 4.12: © Stanley M. Awramik/Biological Photo Service. 4.12 *inset*: © Dennis Kunkel Microscopy, Inc.

Chapter 5 *Opener*: © Roger J. Bick & Brian J. Poindexter/UT-Houston Medical School/Photo Researchers, Inc. 5.1: After N. Campbell, 1990. *Biology*, 2nd Ed., Benjamin Cummings. 5.1 *Protein*: Data from PDB 1IVM. T. Obita, T. Ueda, & T. Imoto, 2003. *Cell. Mol. Life Sci.* 60: 176. 5.1 *T4*: © Dept. of Microbiology, Biozentrum/SPL/Photo Researchers, Inc. 5.1 *Bacterium*: © Jim Biddle/Centers for Disease Control. 5.1 *Plant cells*: © Michael Eichelberger/Visuals Unlimited, Inc. 5.1 *Frog egg*: David McIntyre. 5.1 *Bird*: © Steve Byland/Shutterstock. 5.1 *Baby*: Courtesy of Sebastian Grey Miller. 5.3 *Light microscope*: © Radu Razvan/Shutterstock. 5.3 *Bright-field*: Courtesy of the IST Cell Bank, Genoa. 5.3 *Phase-contrast*: © Michael W. Davidson, Florida State U. 5.3 *DIC*: © Michael W. Davidson, Florida State U. 5.3 *Stained*: © Richard J. Green/SPL/Photo Researchers, Inc. 5.3 *Fluorescence*: © Michael W. Davidson, Florida State U. 5.3 *Confocal*: © Dr. Gopal Murti/SPL/Photo Researchers, Inc. 5.3 *Electron microscope*: © Sinclair Stammers/Photo Researchers, Inc. 5.3 *TEM*: © Dr. Gopal Murti/Visuals Unlimited, Inc. 5.3 *SEM*: © K. R. Porter/SPL/Photo Researchers, Inc. 5.3 *Freeze-fracture*: © D. W. Fawcett/Photo Researchers, Inc. 5.4: © J. J. Cardamone Jr. & B. K. Pugashetti/Biological Photo Service. 5.5A: © Dennis Kunkel Microscopy, Inc. 5.5B: Courtesy of David DeRosier, Brandeis U. 5.6 *Nuclear*: From Y. Mizutani et al., 2001. *J. Cell Sci.* 114: 3727. 5.6 *Mitochondrial*: From L. Argaud et al., 2004. *Cardiovasc Res.* 61: 115. 5.6 *ER*: From Y. Mizutani et al., 2001. *J. Cell Sci.* 114: 3727. 5.7 *Mitochondrion*: © K. Porter, D. Fawcett/Visuals Unlimited, Inc. 5.7 *Cytoskeleton*: © Don Fawcett, John Heuser/Photo Researchers, Inc. 5.7 *Nucleolus*: © Richard Rodewald/Biological Photo Service. 5.7 *Peroxisome*: © E. H. Newcomb & S. E. Frederick/Biological Photo Service. 5.7 *Cell wall*: © Biophoto Associates/Photo Researchers, Inc. 5.7 *Ribosome*: From M. Boublik et al., 1990. *The Ribosome*, p. 177. Courtesy of American Society for Microbiology. 5.7 *Centrioles*: © Barry F. King/Biological Photo Service. 5.7 *Plasma membrane*: Courtesy of J. David Robertson, Duke U. Medical Center. 5.7 *Rough ER*: © Don Fawcett/Science Source/Photo Researchers, Inc. 5.7 *Smooth ER*: © Don Fawcett, D. Friend/Science Source/Photo Researchers, Inc. 5.7 *Chloroplast*: © W. P. Wergin, E. H. Newcomb/Biological Photo Service. 5.7 *Golgi apparatus*: Courtesy of L. Andrew Staehelin, U. Colorado. 5.8A: © Barry King, U. California, Davis/Biological Photo Service. 5.8B: © Biophoto Associates/Science Source/Photo Researchers, Inc. 5.9: © B. Bowers/Photo Researchers, Inc. 5.10: © Sanders/Biological Photo Service. 5.11: © K. Porter, D. Fawcett/Visuals Unlimited, Inc. 5.12: © W. P. Wergin, E. H. Newcomb/Biological Photo Service. 5.13: © Biophoto Associates/Photo Researchers, Inc. 5.14: Courtesy of Vic Small, Austrian Academy of Sciences, Salzburg, Austria. 5.16: Courtesy of N. Hirokawa. 5.17A *upper*: © SPL/Photo Researchers, Inc. 5.17A *lower*, 5.17B: © W. L. Dentler/Biological Photo Service. 5.19: From N. Pollock et al., 1999. *J. Cell Biol.* 147: 493. Courtesy of R. D. Vale. 5.20: © Michael Abbey/Visuals Unlimited, Inc. 5.21: © Biophoto Associates/Photo Researchers, Inc. 5.22 *left*: Courtesy of David Sadava. 5.22 *upper right*: From J. A. Buckwalter & L. Rosenberg, 1983. *Coll. Rel. Res.* 3: 489. Courtesy of L. Rosenberg. 5.22 *lower right*: © J. Gross, Biozentrum/SPL/Photo Researchers, Inc. 5.24: Courtesy of Noriko Okamoto and Isao Inouye. Page 79: Courtesy of Dr. Siobhan Marie O'Connor. Page 92 *Poppy*: David McIntyre. Page 92 *Chromoplast*: © Richard Green/Photo Researchers, Inc. Page 93 *Potatoes*: David McIntyre. Page 93 *Leucoplast*: Courtesy of R. R. Dute.

Chapter 6 *Opener*: © Mike Franklin/FilmMagic/Getty Images. 6.2: After L. Stryer, 1981. *Biochemistry*, 2nd Ed., W. H. Freeman. 6.4: © D. W. Fawcett/Photo Researchers, Inc. 6.7A: Courtesy of D. S. Friend, U. California, San Francisco. 6.7B: Courtesy of Darcy E.

Kelly, U. Washington. 6.7C: Courtesy of C. Peracchia. 6.9A *top*: © Stanley Flegler/Visuals Unlimited, Inc. 6.9A *bottom*: © Ed Reschke/Getty Images. 6.9B *top*: © David M. Phillips/Photo Researchers, Inc. 6.9B *bottom*: © Ed Reschke/Getty Images. 6.9C *top*: © David M. Phillips/Photo Researchers, Inc. 6.9C *bottom*: © Ed Reschke/Getty Images. 6.11: From G. M. Preston et al., 1992. *Science* 256: 385. 6.17: From M. M. Perry, 1979. *J. Cell Sci.* 39: 26. Page 124: © blickwinkel/Alamy.

Chapter 7 *Opener*: Courtesy of Todd Ahern. 7.3A: Data from PDB 3EML. V. P. Jaakola et al., 2008. *Science* 322: 1211. 7.3B: © Georgii Dolgykh/istock. 7.14: © Stephen A. Stricker, courtesy of Molecular Probes, Inc. 7.20: Courtesy of David Kirk. Page 143: © Biophoto Associates/Photo Researchers, Inc.

Chapter 8 *Opener*: © Sinauer Associates. 8.1: Courtesy of Violet Bedell-McIntyre. 8.5B: © Alamy. 8.9: Data from PDB 148L. Kuroki et al., 1993. *Science* 262: 2030. 8.11A: Data from PDB 1AL6. B. Schwartz et al., 1997. 8.11B: Data from PDB 1BB6. V. B. Vollan et al., 1999. *Acta Crystallogr. D. Biol. Crystallogr.* 55: 60. 8.11C: Data from PDB 1AB9. N. H. Yennawar, H. P. Yennawar, & G. K. Farber, 1994. *Biochemistry* 33: 7326. 8.12: Data from PDB 1IG8 (P. R. Kuser et al., 2000. *J. Biol. Chem.* 275: 20814) and 1BDG (A. M. Mulichak et al., 1998 *Nat. Struct. Biol.* 5: 555).

Chapter 9 *Opener*: © Poulsons Photography/Shutterstock. 9.8: From Y. H. Ko et al., 2003. *J. Biol. Chem.* 278: 12305. Courtesy of P. Pedersen. 9.14: © Ana Abejon/istock.

Chapter 10 *Opener*: Courtesy of David F. Karnosky. 10.1: © Andrew Syred/SPL/Photo Researchers, Inc. 10.11: Courtesy of Lawrence Berkeley National Laboratory. 10.15A, 10.17A: © E. H. Newcomb & S. E. Frederick/Biological Photo Service. 10.19: © Aflo Foto Agency/Alamy. Table 10.1 *Rice*: © Alan49/Shutterstock. Table 10.1 *Maize*: © piyagoon/Shutterstock. Table 10.1 *Cactus*: © Dan Eckert/istock.

Chapter 11 *Opener*: © Obstetrics and Gynaecology/Photo Researchers, Inc. 11.1A: © SPL/Photo Researchers, Inc. 11.1B: © Biodisc/Visuals Unlimited, Inc. 11.1C: © Robert Valentic/Naturepl.com. 11.2B: © John J. Cardamone Jr./Biological Photo Service. 11.8 *Chromosome*: © Biophoto Associates/Photo Researchers, Inc. 11.8 *Nucleus*: © D. W. Fawcett/Photo Researchers, Inc. 11.9 *inset*: © Biophoto Associates/Science Source/Photo Researchers, Inc. 11.10: © Nasser Rusan. 11.11B: © Conly L. Rieder/Biological Photo Service. 11.13A: © Robert Brons/Biological Photo Service. 11.13B: © B. A. Palevitz, E. H. Newcomb/Biological Photo Service. 11.14: © Robert E. Ford/Biological Photo Service. 11.15 *left*: © Andrew Syred/SPL/Photo Researchers, Inc. 11.15 *center*: David McIntyre. 11.15 *right*: Courtesy of Andrew D. Sinauer. 11.16: © C. A. Hasenkampf/Biological Photo Service. 11.17: Courtesy of J. Kezer. 11.21: Courtesy of Dr. Thomas Ried and Dr. Evelin Schröck, NIH. 11.22: © Sergey Skleznev/Shutterstock. 11.23A: © Gopal Murti/Photo Researchers, Inc. 11.24: © Dennis Kunkel Microscopy, Inc. Page 231: Courtesy of Paul Schulte.

Chapter 12 *Opener*: © Gerry Pearce/Alamy. 12.1: © the Mendelianum. 12.10 *Dark*: © Marina Golskaya/istock. 12.10 *Chinchilla*: © purelook/istock. 12.10 *Point*: © Carolyn A. McKeone/Photo Researchers, Inc. 12.10 *Albino*: © ZTS/Shutterstock. 12.13: Courtesy of Madison, Hannah, and Walnut. 12.14: Courtesy of the Plant and Soil Sciences eLibrary (http://plantandsoil.unl.edu); used with permission from the Institute of Agriculture and Natural Resources at the University of Nebraska. 12.15: © Mark Taylor/Naturepl.com. 12.16: © Peter Morenus/U. of Connecticut. 12.23A: © David Scharf/Getty Images. Page 258 *Rose*: © Margo Harrison/Shutterstock. Page 258 *Pea, Walnut, and Single*: David McIntyre.

Chapter 13 *Opener*: Portrait by Albert Edelfelt, courtesy of the National Library of Medicine. 13.3: © Lee D. Simon/Photo Researchers, Inc. 13.6B: © Science Source/Photo Researchers, Inc. 13.7A: © A. Barrington Brown/Photo Researchers, Inc. 13.7B: Data from S. Arnott & D. W. Hukins, 1972. *Biochem. Biophys. Res. Commun.* 47(6): 1504. 13.14A: Data from PDB 1SKW. Y. Li et al., 2001. *Nat. Struct. Mol. Biol.* 11: 784. 13.19B: © Dr. Peter Lansdorp/Visuals Unlimited, Inc.

Chapter 14 *Opener*: © CDC/Janice Carr/AGE Fotostock. 14.3: Data from PDB 1MSW. Y. W. Yin & T. A. Steitz, 2002. *Science* 298: 1387. 14.7: From D. C. Tiemeier et al., 1978. *Cell* 14: 237. 14.11: Data from PDB 1EHZ. H. Shi & P. B. Moore, 2000. *RNA* 6: 1091. 14.13: Data from PDB 1GIX and 1G1Y. M. M. Yusupov et al., 2001. *Science* 292: 883. 14.17B: Courtesy of J. E. Edström and *EMBO J.*

Chapter 15 *Opener*: © Steve Lipofsky/Corbis. 15.3: © Stanley Flegler/Visuals Unlimited, Inc. 15.9: From C. Harrison et al., 1983. *J. Med. Genet.* 20: 280. 15.11B: © David M. Martin, M.D./SPL/Photo Researchers, Inc. 15.13: © Philippe Plailly/Photo Researchers, Inc. 15.14B: U.S. Army photo. 15.16 *Butterfly*: © Bershadsky Yuri/Shutterstock. 15.16 *Bacteria*: Courtesy of Janice Haney Carr/CDC. 15.16 *Fungus*: © Warwick Lister-Kaye/istock. 15.17: © Simon Fraser/Photo Researchers, Inc.

Chapter 16 *Opener*: © Beyond Fotomedia GmbH/Alamy. 16.12A: © Dennis Kunkel Microscopy, Inc. 16.12B: © Lee D. Simon/Photo Researchers, Inc. 16.21: Courtesy of Irina Solovei, University of Munich (LMU), Germany. Page 336: Data from PDB 2PE5. R. Daber et al., 2007. *J. Mol. Biol.* 370: 609.

Chapter 17 *Opener*: © moodboard RF/Photolibrary.com. 17.7: Courtesy of Tom Deerinck and Mark Ellisman of the National Center for Microscopy and Imaging Research at the University of California at San Diego. 17.11B: Courtesy of O. L. Miller, Jr. 17.13: Courtesy of Christoph P. E. Zollikofer, Marcia S. Ponce de León, and Elisabeth Daynès. 17.16: From P. H. O'Farrell, 1975. *J. Biol. Chem.* 250: 4007. Courtesy of Patrick H. O'Farrell. 17.18 *left*: © kostudio/Shutterstock. 17.18 *right*: © Bruce Stotesbury/PostMedia News/Zuma Press.

Chapter 18 *Opener*: U.S. Coast Guard photo by Petty Officer 2nd Class Etta Smith. 18.3: © Dr. Jack Bostrack/Visuals Unlimited, Inc. 18.4: © Stephen Sewell/istock. 18.13: Courtesy of the Golden Rice Humanitarian Board, www.goldenrice.org. 18.14: Courtesy of Eduardo Blumwald. Page 387: © Dr. George Chapman/Visuals Unlimited, Inc.

Chapter 19 *Opener*: © Frank Franklin II/AP/Corbis. 19.5A: From J. E. Sulston & H. R. Horvitz, 1977. *Dev. Bio.* 56: 100. 19.9A: David McIntyre. 19.12A: From A. Ephrussi and D. St. Johnston, 2004. *Cell* 116: 143. 19.12B: Courtesy of Ruth Lehmann. 19.12C *left*: From E. A. Wimmer, 2012. *Science* 287: 2476. 19.12C *right*: From D. Tautz, 1988. *Nature* 332: 284. 19.13B: Courtesy of C. Rushlow and M. Levine. 19.13C: Courtesy of T. Karr. 19.13D: Courtesy of S. Carroll and S. Paddock. 19.15: Courtesy of F. R. Turner, Indiana U. 19.17: From I. Wilmut et al., 1997. *Nature* 385: 810. Page 402: Courtesy of D. Daily and W. Sullivan.

Chapter 20 *Opener*: © Theo Allofs/Corbis. 20.1 *Mouse*: © orionmystery@flickr/Shutterstock. 20.1 *Fly*: David McIntyre. 20.1 *Shark*: © Kristian Sekulic/Shutterstock. 20.1 *Squid*: © Gergo Orban/Shutterstock. 20.3: © David M. Phillips/Photo Researchers, Inc. 20.5: © Bone Clones, www.boneclones.com. 20.6: Courtesy of J. Hurle and E. Laufer. 20.7: Courtesy of J. Hurle. 20.8: From M. Kmita and D. Duboule, 2003. *Science* 301: 331. 20.9 *Cladogram*: After R. Galant & S. Carroll, 2002. *Nature* 415: 910. 20.9 *Insect*: © Stockbyte/PictureQuest. 20.9 *Centipede*: © Burke/Triolo/Brand X Pictures/PictureQuest. 20.10: From Wang et al., 2005. *Nature* 436: 714. Courtesy of John Doebley. 20.11: © Neil Hardwick/Alamy. 20.12: © Rob Valentic/ANTPhoto.com. 20.13 *Caterpillars*: © Erick Greene. 20.13 *Adult*: Courtesy of John Gruber. 20.14: © Nigel Cattlin, Holt Studios International/Photo Researchers, Inc. 20.16: Courtesy of Mike Shapiro and David Kingsley.

Chapter 21 *Opener*: © Pasieka/Photo Researchers, Inc. 21.1 *H.M.S. Beagle*: Painting by Ronald Dean, reproduced by permission of the artist and Richard Johnson, Esquire. 21.1 *Darwin*: © The Art Gallery Collection/Alamy. 21.5A: © Luis César Tejo/Shutterstock. 21.5B: © Duncan Usher/Alamy. 21.5C: © PetStockBoys/Alamy. 21.5D: © Arco Images GmbH/Alamy. 21.8: © Simon G/Shutterstock. 21.14: Courtesy of David Hillis. 21.19A: © Reinhard Dirscherl/Alamy. 21.19B: © Marevision/AGE Fotostock. 21.20A: Courtesy of Edmund D. Brodie, Jr.

Chapter 22 *Opener*: Courtesy of Misha Matz. 22.6 *Sea squirt larva*: Courtesy of William Jeffery. 22.6 *Sea squirt adult*: © WaterFrame/Alamy. 22.6 *Frog larva*: David McIntyre. 22.6 *Frog adult*: © Mark Kostich/Shutterstock. 22.10: © Alexandra Basolo. 22.11 *L. bicolor*: Courtesy of Steve Matson. 22.11 *L. liniflorus*: Courtesy of

Anthony Valois/National Park Service. 22.14A, B: © Krieger, C./AGE Fotostock. 22.14C: © Photos by Andy/Shutterstock. 22.16: Courtesy of Misha Matz.

Chapter 23 *Opener*: David McIntyre, courtesy of Exotic Fish and Pet World, Southampton, MA. 23.1A *left*: © Roger K. Burnard/Biological Photo Service. 23.1A *right*: © Richard Codington/Alamy. 23.1B: © Stubblefield Photography/Shutterstock. 23.2A: David McIntyre. 23.2B: © Gerry Bishop/Visuals Unlimited, Inc. 23.4: © Barry Mansell/Naturepl.com. 23.10: © Tim Gainey/Alamy. 23.12 *G. olivacea*: © Phil A. Dotson/Photo Researchers, Inc. 23.12 *G. carolinensis*: © Suzanne L. Collins/Photo Researchers, Inc. 23.13A: © Gustav Verderber/Visuals Unlimited, Inc. 23.13B: © Nathan Derieg. 23.13C: © Bob Gibbons/Alamy. 23.13D: © Daniel L. Geiger/SNAP/Alamy. 23.14: Courtesy of Donald A. Levin. 23.15 *upper*: © Gerhard Schulz/AGE Fotostock. 23.15 *lower*: © Christophe Courteau/Naturepl.com. 23.17A: © blickwinkel/Alamy. 23.17B: © W. Peckover/VIREO. 23.18 *Madia*: © Peter K. Ziminsky/Visuals Unlimited, Inc. 23.18 *Argyroxiphium*: © Ron Dahlquist/Getty Images. 23.18 *Wilkesia*: © Photo Resource Hawaii/Alamy. 23.18 *Dubautia*: © Noble Proctor/The National Audubon Society Collection/Photo Researchers, Inc. 23.19: Courtesy of William R. Rice. Page 484 *left*: © Gustav Verderber/Visuals Unlimited, Inc. Page 484 *right*: © Nathan Derieg.

Chapter 24 *Opener*: © Jane Burton/Naturepl.com. 24.3 *Rice*: data from PDB 1CCR. H. Ochi et al., 1983. *J. Mol. Biol.* 166: 407. *Tuna*: data from PDB 5CYT. T. Takano, 1984. 24.4: From P. B. Rainey & M. Travisano, 1998. *Nature* 394: 69. © Macmillan Publishers Ltd. 24.7A *Langur*: © blickwinkel/Alamy. 24.7A *Longhorn*: Courtesy of David Hillis.

Chapter 25 *Opener*: © Graham Cripps/NHMPL. *Opener inset*: © Natasha Litova/istock. 25.3B: © Ric Ergenbright/Corbis. 25.5: Courtesy of Dave Harlow, U.S. Geological Survey. 25.6: © Martin Bond/SPL/Photo Researchers, Inc. 25.7: David McIntyre. 25.9A: © Ted Kinsman/Photo Researchers, Inc. 25.9B: © Georgette Douwma/Photo Researchers, Inc. 25.11: © PjrStudio/Alamy. 25.13A: From S. Xiao et al., 1998. *Nature* 391: 553. © Macmillan Publishers Ltd. 25.13B *top*: Courtesy of Martin Smith. 25.13B *bottom*: © Sinclair Stammers/Photo Researchers, Inc. 25.13C *top*: Courtesy of Chip Clark, National Museum of Natural History. 25.13C *bottom*: © Albert J. Copley/Visuals Unlimited, Inc. 25.14 *Cambrian*: © John Sibbick/NHMPL. 25.14 *Marella*: Courtesy of the Amherst College Museum of Natural History, The Trustees of Amherst College. 25.14 *Ottoia*: © Alan Sirulnikoff/Photo Researchers, Inc. 25.14 *Anomalocaris*: © Kevin Schafer/Alamy. 25.14 *Devonian*: © The Field Museum, #GEO86500_125d. 25.14 *Archaeopteris*: © John Cancalosi/Getty Images. 25.14 *Eusthenopteron*: © Wolfgang Kaehler/Alamy. 25.14 *Permian*: © Karen Carr Studio Inc. 25.14 *Dragonfly*: Image by Roy J. Beckemeyer. 25.14 *Walchia*: © The Natural History Museum, London. 25.14 *Triassic*: © The Natural History Museum (WAC)/Naturepl.com. 25.14 *Ferns*: © Ken Lucas/Visuals Unlimited, Inc. 25.14 *Plateosaurus*: © Koichi Kamoshida/J Press/Zuma Press. 25.14 *Cretaceous*: © Anness Publishing/NHMPL. 25.14 *Chasmosaurus*: © Oleksiy Maksymenko/Alamy. 25.14 *Sapindopsis*: © Barbara J. Miller/Biological Photo Service. 25.14 *Tertiary*: © Publiphoto/Photo Researchers, Inc. 25.14 *Hyracotherium*: Courtesy of the Amherst College Museum of Natural History, The Trustees of Amherst College. 25.14 *Plesiadapis*: © The Natural History Museum, London. 25.15: Courtesy of Conrad C. Labandeira, Department of Paleobiology, National Museum of Natural History, Smithsonian Institution.

Chapter 26 *Opener*: © Steven Haddock and Steven Miller. 26.2: Courtesy of the Centers for Disease Control. 26.3: © Dennis Kunkel Microscopy, Inc. 26.5: © Dr. Kari Lounatmaa/Photo Researchers, Inc. 26.6: © Dr. Gary Gaugler/Visuals Unlimited, Inc. 26.7: © Don W. Fawcett/Photo Researchers, Inc. 26.8: © David Phillips/Visuals Unlimited, Inc. 26.9A: © Paul W. Johnson/Biological Photo Service. 26.9B: © Dr. Terry Beveridge/Visuals Unlimited, Inc. 26.9C: © RJH Catalog/Alamy. 26.10A: © J. A. Breznak & H. S. Pankratz/Biological Photo Service. 26.10B: © James Cavallini/Photo Researchers, Inc. 26.11: Courtesy of Randall C. Cutlip. 26.12: © Kwangshin Kim/Photo Researchers, Inc. 26.13: © Dean A. Glawe/Biological Photo Service. 26.14: From K. Kashefi & D. R. Lovley, 2003. *Science* 301: 934. Courtesy of Kazem Kashefi. 26.16: © Arco Images GmbH/Alamy. 26.17: © Nancy Nehring/istock. 26.18: From H. Huber et al., 2002. *Nature* 417: 63. © Macmillan Publishers Ltd. Courtesy of Karl O. Stetter. 26.20B: © Science Photo Library RF. 26.22: © Juergen Berger/Photo Researchers, Inc. 26.23A: © Science Photo Library RF/Photolibrary.com. 26.23B, C: © Russell Kightley/Photo Researchers, Inc. 26.23D: © Science Photo Library RF. 26.23E: © animate4.com ltd./Photo Researchers, Inc. 26.23F: © Russell Kightley/Photo Researchers, Inc. 26.24: © Nigel Cattlin/Alamy. 26.26: Courtesy of Margaret McFall-Ngai. Page 548: Courtesy of the Centers for Disease Control.

Chapter 27 *Opener*: © Don Paulson/PureStock/AGE Fotostock. 27.4: © Dennis Kunkel Microscopy, Inc. 27.5A: © SPL/Photo Researchers, Inc. 27.5B: © Aaron Bell/Visuals Unlimited, Inc. 27.5C: © Steve Gschmeissner/Photo Researchers, Inc. 27.8: © Scenics & Science/Alamy. 27.9A: © Marevision/AGE Fotostock. 27.9B: © Carl W. May/Biological Photo Service. 27.10: David McIntyre. 27.11, 27.12A: © Robert Brons/Biological Photo Service. 27.12B: © Manfred Kage/Photo Researchers, Inc. 27.13A: © J. Paulin/Visuals Unlimited, Inc. 27.13B: © Dr. David M. Phillips/Visuals Unlimited, Inc. 27.15: © Michael Abbey/Photo Researchers, Inc. 27.16: © Wim van Egmond/Visuals Unlimited, Inc. 27.17A: © Matt Meadows/Getty Images. 27.17B: © Ed Reschke/Getty Images. 27.18: Courtesy of R. Blanton and M. Grimson. 27.20B: © London School of Hygiene/SPL/Photo Researchers, Inc. 27.21: Courtesy of M.A. Coffroth and Cindy Lewis, University at Buffalo.

Chapter 28 *Opener*: Courtesy of the U.S. Coast Guard. 28.2: © Dr. Peter Siver/Visuals Unlimited, Inc. 28.3A: © Wim van Egmond/Visuals Unlimited, Inc. 28.3B: © David Wrobel/Visuals Unlimited, Inc. 28.4A: © Carolina Biological/Visuals Unlimited, Inc. 28.4B: © Marevision/AGE Fotostock. 28.5A: © Larry Mellichamp/Visuals Unlimited, Inc. 28.5B: © Bob Gibbons/Alamy. 28.7 *top*: © Biodisc/Visuals Unlimited, Inc. 28.7 *bottom*: © J. Robert Waaland/Biological Photo Service. 28.8A: David McIntyre. 28.8B: © Dr. Brad Mogen/Visuals Unlimited, Inc. 28.8C: © Verbiesen Henk/AGE Fotostock. 28.9A: © mediacolor's/Alamy. 28.9B: David McIntyre. 28.10A: © Dr. John D. Cunningham/Visuals Unlimited, Inc. 28.10B: © Danilo Donadoni/AGE Fotostock. 28.11: © Daniel Vega/AGE Fotostock. 28.12: © Publiphoto/Photo Researchers, Inc. 28.14A: © Ed Reschke/Getty Images. 28.14B: © Stanislav Sokolov/istock. 28.14C: Courtesy of the Talcott Greenhouse, Mount Holyoke College. 28.14D: © Ted Mead/Getty Images. 28.15 *inset*: © John N. A. Lott/Biological Photo Service. 28.16A: Courtesy of the Biology Department Greenhouses, U. Massachusetts, Amherst. 28.16B: David McIntyre. 28.19: © Patrick Pleul/epa/Corbis.

Chapter 29 *Opener*: © Mitsuhiko Imamori/Minden Pictures. 29.1 *Cycad*: David McIntyre. 29.1 *Ginkgo*: © hypnotype/Shutterstock. 29.1 *Conifer*: © Irina Tischenko/istock. 29.1 *Magnolia*: © Dole/Shutterstock. 29.3: © Wildlife GmbH/Alamy. 29.4: © Susumu Nishinaga/Photo Researchers, Inc. 29.6A: Courtesy of Jane Sinauer. 29.6B: David McIntyre. 29.6C: © Juan Carlos Muñoz/AGE Fotostock. 29.6D: © Pepbaix/Alamy. 29.7A *left*: © Fritz Poelking/Blickwinkel/AGE Photostock. 29.7A *right*: © Scenics & Science/Alamy. 29.7B *left*: © Gunter Marx/Alamy. 29.7B *right*: © Scenics & Science/Alamy. 29.9: Courtesy of Jim Peaco/National Park Service. 29.10A, B: David McIntyre. 29.10C: © Nigel Cattlin/Alamy. 29.11A: © Phiseksit/Shutterstock. 29.11B: David McIntyre. 29.15: © Ted Kinsman/Photo Researchers, Inc. 29.17A: © Anne Power/Shutterstock. 29.17B: © Klaus Hackenberg/Corbis. 29.17C: © Brian A Jackson/Shutterstock. 29.17D: © Paul Thompson/Corbis. 29.17E: © Arco Images GmbH/Alamy. 29.17F: Courtesy of Keith Weller/USDA ARS. 29.19A: Photo by David McIntyre, courtesy of the U. Massachusetts Biology Department Greenhouses. 29.19B: © Cerealphoto/AGE Fotostock. 29.19C: © Holmes Garden Photos/Alamy. 29.19D: © dora modly-paris/Shutterstock. 29.19E: © Florapix/Alamy. 29.20A: © Floris Slooff/istock. 29.20B: © George Clerk/istock. 29.20C: © Jose B. Ruiz/Naturepl.com. 29.20D: © rotofrank/istock. 29.21A: © BonkersAboutTravel/Alamy. 29.21B: David McIntyre. 29.21C: Courtesy of David Hillis. 29.21D: © A & J Visage/Alamy. 29.22: © Rob Walls/Alamy. 29.23: © Janos Csernoch/Alamy.

Chapter 30 *Opener*: © Biophoto Associates/Photo Researchers, Inc. 30.2: © Steve Gschmeissner/Science Photo Library/Corbis. 30.3A: © Dr. Jeremy Burgess/Photo Researchers, Inc. 30.4: © Arco Images GmbH/Alamy. 30.5A: © Biophoto Associates/Photo Researchers, Inc. 30.6: © N. Allin & G. L. Barron/Biological Photo Service. 30.7A, C: Courtesy of David Hillis. 30.7B: David McIntyre. 30.9A: © R. L. Peterson/Biological Photo Service. 30.9B: © M. F. Brown/Biological Photo Service. 30.12: © Eye of Science/Photo Researchers, Inc. 30.13: © John Taylor/Visuals Unlimited, Inc. 30.14A: © J. Robert Waaland/Biological Photo Service. 30.14B: © Dr. Jeremy Burgess/Photo Researchers, Inc. 30.15: Photo by David McIntyre; manure courtesy of Myrtle Jackson. 30.16A: © Dr. Cecil H. Fox/Photo Researchers, Inc. 30.16B: © Biophoto Associates/Photo Researchers, Inc. 30.17A: © blickwinkel/Alamy. 30.17B: © Matt Meadows/Getty Images. 30.18: © Dennis Kunkel Microscopy, Inc. 30.19A: David McIntyre. 30.19B: © Mike Norton/Shutterstock. 30.20, 30.21: Courtesy of David Hillis. 30.22: © Dr. Gary Gaugler/Visuals Unlimited, Inc. 30.23: © Biophoto Associates/Photo Researchers, Inc.

Chapter 31 *Opener*: © Ana Yuri Signorovitch. 31.3: Courtesy of J. B. Morrill. 31.5A: © Ed Robinson/Getty Images. 31.5B: © Steve Gschmeissner/Photo Researchers, Inc. 31.5C: © Konrad Wothe/Minden Pictures/Corbis. 31.6A: © Jurgen Freund/Naturepl.com. 31.6B: © John Bell/istock. 31.6C: © Stockphoto4u/istock. 31.7A: © John A. Anderson/istock. 31.7B: © Kevin Schafer/DigitalVision/Photolibrary.com. 31.7B *inset*: © Mike Rogal/Shutterstock. 31.8A: © Doug Lindstrand/Alaska Stock Images/AGE Fotostock. 31.8B: © blickwinkel/Alamy. 31.9A: © Cathy Keifer/Shutterstock. 31.9B: © Don Johnston/AGE Fotostock. 31.9C: David McIntyre. 31.11 *inset*: © Scott Camazine/Phototake. 31.12: © Gerd Guenther/Photo Researchers, Inc. 31.13A: © First Light/Alamy. 31.13B: © Accent Alaska.com/Alamy. 31.14A: © Helmut Heintges/Corbis. 31.14B: © F1online digitale Bildagentur GmbH/Alamy. 31.15A: © Jurgen Freund/Naturepl.com. 31.15B: David McIntyre. 31.15C: © Robert Brons/Biological Photo Service. 31.16B: © Larry Jon Friesen. 31.17A: Courtesy of Wim van Egmond. 31.18, 31.19: Adapted from F. M. Bayerand & H. B. Owre, 1968. *The Free-Living Lower Invertebrates*, Macmillan Publishing Co. 31.20A: © Charles Wyttenbach/Biological Photo Service. 31.20B: © Georgette Douwma/Naturepl.com. 31.20C, D: © Larry Jon Friesen. 31.21A: © Jurgen Freund/Naturepl.com. 31.21B: © Stephan Kerkhofs/Shutterstock. 31.22: Adapted from F. M. Bayerand & H. B. Owre, 1968. *The Free-Living Lower Invertebrates*, Macmillan Publishing Co.

Chapter 32 *Opener*: © Mark Moffett/Minden Pictures. 32.2: © blickwinkel/Alamy. 32.3A: From D. C. García-Bellido & D. H. Collins, 2004. *Nature* 429: 40. Courtesy of Diego García-Bellido Capdevila. 32.3B: © Nature's Images/Photo Researchers, Inc. 32.6A: © Larry Jon Friesen. 32.7B: © Roland Birke/Getty Images. 32.7C: Courtesy of David Walter and Heather Proctor. 32.7D: © Michael Abbey/Photo Researchers, Inc. 32.8B: © Larry Jon Friesen. 32.9: © David Wrobel/Visuals Unlimited, Inc. 32.10A: © Fred Bavendam/Minden Pictures. 32.12A: © WaterFrame/Alamy. 32.12B: Courtesy of Cindy Lee Van Dover. 32.12C: © Pakhnyushcha/Shutterstock. 32.12D: © Larry Jon Friesen. 32.13B: © Marevision/AGE Fotostock. 32.13C: © Francesco Tomasinelli/Photo Researchers, Inc. 32.13D: © H. Wes Pratt/Biological Photo Service. 32.13E: © moodboard/Photolibrary.com. 32.14A: © Larry Jon Friesen. 32.14B: © Laura Romin & Larry Dalton/Alamy. 32.14C: © Jeff Rotman/Naturepl.com. 32.15A: Courtesy of Jen Grenier and Sean Carroll, U. Wisconsin. 32.15B: Courtesy of Graham Budd. 32.15C: Courtesy of Reinhardt Møbjerg Kristensen. 32.16B: © Grave/Photo Researchers, Inc. 32.16C: © Steve Gschmeissner/Photo Researchers, Inc. 32.17: © Pascal Goetgheluck/Photo Researchers, Inc. 32.18A: © Steve Gschmeissner/Photo Researchers, Inc. 32.18B: © George Grall/National Geographic Society/Corbis. 32.19: © Gerald & Buff Corsi/Visuals Unlimited, Inc. 32.20A: © David Shale/Naturepl.com. 32.20B: © Joe McDonald/Corbis. 32.21A: © Kelly Swift, www.swiftinverts.com. 32.21B: © Larry Jon Friesen. 32.21C: © Nigel Cattlin/Alamy. 32.21D: SEM by Eric Erbe; colorization by Chris Pooley/USDA ARS. 32.22A: © Rod Williams/Naturepl.com. 32.22B: © John R. MacGregor/Getty Images. 32.23A, B: © Larry Jon Friesen. 32.23C: © Solvin Zankl/Naturepl.com. 32.23D: © Larry Jon Friesen. 32.23E: © Norbert Wu/Minden Pictures. 32.25: © Scenics & Science/Alamy. 32.27A: © Cisca Castelijns/Foto Natura/Minden Pictures/Corbis. 32.27B: © Piotr Naskrecki/Minden Pictures/Corbis. 32.27C: © Pete Oxford/Naturepl.com. 32.27D: David McIntyre. 32.27E: © Papilio/Alamy. 32.27F: © CorbisRF/Photolibrary.com. 32.27G: © Rafael Campillo/AGE Fotostock. 32.27H: © Jean Claude Carton/Bruce Coleman USA/AGE Fotostock.

Chapter 33 *Opener*: © Michael Tyler/ANTPhoto.com. 33.2: From S. Bengtson, 2000. Teasing fossils out of shales with cameras and computers. *Palaeontologia Electronica* 3(1). 33.3A: © Triarch/Visuals Unlimited, Inc. 33.4: Courtesy of Samuel Chow (CybersamX)/Flickr. 33.5A: © Hal Beral/Visuals Unlimited, Inc. 33.5B, C: © WaterFrame/Alamy. 33.5D: © Marevision/AGE Fotostock. 33.5E: © Robert L. Dunne/Photo Researchers, Inc. 33.6A: © C. R. Wyttenbach/Biological Photo Service. 33.7A: © Stan Elems/Visuals Unlimited, Inc. 33.7B: © Larry Jon Friesen. 33.8A: © Marevision/AGE Fotostock. 33.8B: © Gavin Newman/Alamy. 33.11A: © Ken Lucas/Biological Photo Service. 33.11B *left*: © Marevision/AGE Fotostock. 33.11B *right*: © anne de Haas/istock. 33.12B: © Roger Klocek/Visuals Unlimited, Inc. 33.13A: © Wayne Lynch/AGE Fotostock. 33.13B: © Kelvin Aitken/AGE Fotostock. 33.13C: © Norbert Wu/Minden Pictures. 33.14A: © David Fleetham/Alamy. 33.14B, C: © Larry Jon Friesen. 33.14D: © Norbert Wu/Minden Pictures. 33.15A: © Hoberman Collection/Corbis. 33.15B: © Tom McHugh/Photo Researchers, Inc. 33.15C: © Ted Daeschler/Academy of Natural Sciences/VIREO. 33.18A: © Morley Read/Naturepl.com. 33.18B: © Michael & Patricia Fogden/Minden Pictures. 33.18C: © Jack Goldfarb/Design Pics, Inc./Photolibrary.com. 33.18D: Courtesy of David Hillis. 33.21A: © C. Alan Morgan/Getty Images. 33.21B: © Cathy Keifer/Shutterstock. 33.21C: © Larry Jon Friesen. 33.21D: © Gordon Chambers/Alamy. 33.22A: © Susan Flashman/istock. 33.22B: © Gerry Ellis, DigitalVision/PictureQuest. 33.23A: From X. Xu et al., 2003. *Nature* 421: 335. © Macmillan Publishers Ltd. 33.23B: © Tom & Therisa Stack/Painet, Inc. 33.24: © Melinda Fawver/istock. 33.25A: © Tim Zurowski/All Canada Photos/Getty Images. 33.25B: © Salvador III Manaois/Alamy. 33.25C: © Tom Vezo/Minden Pictures. 33.25D: © Marco Kopp/istock. 33.26A: © John N. A. Lott/Biological Photo Service. 33.26B: © Dave Watts/Visuals Unlimited, Inc. 33.27A: © Ingo Arndt/Naturepl.com. 33.27B: © Greg Harold/Auscape/Minden Pictures/Corbis. 33.27C: © R. Wittek/Arco Images/AGE Fotostock. 33.29A: © Robert McGouey/All Canada Photos/Corbis. 33.29B: © ANT Photo Library/Photo Researchers, Inc. 33.29C: © John E Marriott/All Canada Photos/AGE Fotostock. 33.29D: © Michael S. Nolan/AGE Fotostock. 33.31: © John Warburton-Lee Photography/Alamy. 33.32A: © mike lane/Alamy. 33.32B: © De Agostini Editore/AGE Fotostock. 33.33A: © Steve Bloom Images/Alamy. 33.33B: © Anup Shah/AGE Fotostock. 33.33C: © Lars Christensen/istock. 33.33D: © Anup Shah/Minden Pictures. 33.35A: © Cyril Ruoso/Minden Pictures. 33.35B: Courtesy of Andrew D. Sinauer. 33.35C: © Arco Images GmbH/Alamy. 33.35D: David McIntyre.

Chapter 34 *Opener*: © Picture Contact BV/Alamy. 34.6A: © Dr. Ken Wagner/Visuals Unlimited, Inc. 34.6B: © Phil Gates/Biological Photo Service. 34.6C: © Biophoto Associates/Photo Researchers, Inc. 34.6D: © Jack M. Bostrack/Visuals Unlimited, Inc. 34.7A: © John D. Cunningham/Visuals Unlimited, Inc. 34.7B: © J. Robert Waaland/Biological Photo Service. 34.7C: © Herve Conge/ISM/Phototake. 34.8 *upper*: © Biodisc/Visuals Unlimited, Inc. 34.8 *lower*: © M. I. Walker/Photo Researchers, Inc. 34.9B: © John N. A. Lott/Biological Photo Service. 34.10A: © Ed Reschke/Getty Images. 34.10B: © Dr. James W. Richardson/Visuals Unlimited, Inc. 34.11: © Larry Jon Friesen. 34.12A: © modesigns58/istock. 34.12B: © Adrian Sherratt/Alamy. 34.12C: © Science Photo Library/Alamy. 34.13A *left*: David McIntyre. 34.13A *right*: © Andrew Syred/Photo Researchers, Inc. 34.13B *left*: © Garry DeLong/Photo Researchers, Inc. 34.13B *right*: © Steve Gschmeissner/Photo Researchers, Inc. 34.14: David McIntyre. 34.15B: Courtesy of Thomas Eisner, Cornell U. 34.15C: © Susumu Nishinaga/Photo Researchers, Inc. 34.18: © Biodisc/Visuals Unlimited, Inc. 34.19: © Phil Gates/Biological Photo Service.

Chapter 35 *Opener*: © John Carr/Eye Ubiquitous/Corbis. 35.3: © Nigel Cattlin/Alamy. 35.8A: © Susumu Nishinaga/Photo Researchers, Inc. 35.10: © R. Kessel & G. Shih/

Visuals Unlimited, Inc. Page 735 *Aphid and stylet*: © M. H. Zimmermann.

Chapter 36 *Opener*: © Russ Munn/AgStock Images/Corbis. 36.1: David McIntyre. 36.5: David McIntyre. 36.11A: © J. H. Robinson/The National Audubon Society Collection/Photo Researchers, Inc. 36.11B: © Kim Taylor/Naturepl.com. 36.12: Courtesy of Susan and Edwin McGlew.

Chapter 37 *Opener*: © Micheline Pelletier/Sygma/Corbis. 37.2: From J. M. Alonso and J. R. Ecker, 2006. *Nature Reviews Genetics* 7: 524. 37.3A: Courtesy of J. A. D. Zeevaart, Michigan State U. 37.3B: From W. M. Gray, 2004. *PLoS Biol.* 2(9): e311. 37.4: © Sylvan Wittwer/Visuals Unlimited, Inc. 37.9: © Ed Reschke/Getty Images. 37.13, 37.16: David McIntyre. Page 757: © Gerald & Buff Corsi/Visuals Unlimited, Inc. Page 768: David McIntyre. Page 769: Courtesy of Adel A. Kader. Page 771: Courtesy of Eugenia Russinova, VIB Department of Plant Systems Biology, Ghent University, Belgium. Page 777: Clemson University - USDA Cooperative Extension Slide Series, Bugwood.org.

Chapter 38 *Opener*: © Remco Zwinkels/Minden Pictures. 38.1A: © kukuruxa/Shutterstock. 38.1B *left*: © Tish1/Shutterstock. 38.1B *right*: © Pierre BRYE/Alamy. 38.1C: © Bill Beatty/Visuals Unlimited, Inc. 38.3: © Rolf Nussbaumer Photography/Alamy. 38.4: © Christian Guatier/Biosphoto. 38.7A: David McIntyre. 38.7B: © Michael Moreno/istock. 38.7C: © Scenics & Science/Alamy. 38.9: Courtesy of Richard Amasino. 38.15: Courtesy of Richard Amasino and Colleen Bizzell. 38.17A: © ooyoo/istock. 38.17B: © Nigel Cattlin/Alamy. 38.17C: © Jerome Wexler/Visuals Unlimited, Inc. 38.18: David McIntyre. Page 784 *Thistle*: © John N. A. Lott/Biological Photo Service. Page 784 *Burrs*: © Scott Camazine/Alamy. Page 795: © yykkaa/Shutterstock.

Chapter 39 *Opener*: © Birgit Betzelt/actionmedeor/hand/dpa/Corbis. 39.4: © Holt Studios International Ltd/Alamy. 39.5A: © Kim Taylor/Naturepl.com. 39.5B: David McIntyre. 39.8: Courtesy of Thomas Eisner, Cornell U. 39.9: © Jon Mark Stewart/Biological Photo Service. 39.10: © Dr. Jack Bostrack/Visuals Unlimited, Inc. 39.11: © TH Foto/Alamy. 39.12: © Simon Fraser/SPL/Photo Researchers, Inc. 39.13: © John N. A. Lott/Biological Photo Service. 39.16: Courtesy of Scott Bauer/USDA. 39.17: © Jurgen Freund/Naturepl.com. 39.18: Courtesy of Ryan Somma.

Chapter 40 *Opener*: © PCN Black/Alamy. 40.3A: © Gladden Willis/Visuals Unlimited, Inc. 40.3B: From Ross, Pawlina, and Barnash, 2009. *Atlas of Descriptive Histology*. Sinauer Associates: Sunderland, MA. 40.3C: © Ed Reschke/Getty Images. 40.4A: From Ross, Pawlina, and Barnash, 2009. *Atlas of Descriptive Histology*. Sinauer Associates: Sunderland, MA. 40.4B: © Manfred Kage/Photo Researchers, Inc. 40.4C: © SPL/Photo Researchers, Inc. 40.5A: © Chuck Brown/Photo Researchers, Inc. 40.5B: From Ross, Pawlina, and Barnash, 2009. *Atlas of Descriptive Histology*. Sinauer Associates: Sunderland, MA. 40.5C: © Dennis Kunkel Microscopy, Inc. 40.5D: From Ross, Pawlina, and Barnash, 2009. *Atlas of Descriptive Histology*. Sinauer Associates: Sunderland, MA. 40.6A: © James Cavallini/Photo Researchers, Inc. 40.6B: © Innerspace Imaging/SPL/Photo Researchers, Inc. 40.12: © Greg Epperson/istock. 40.13: © Gerry Ellis/DigitalVision. 40.15: Courtesy of Anton Stabentheiner. 40.18A: © Robert Shantz/Alamy. 40.18B: © Jim Brandenburg/Minden Pictures.

Chapter 41 *Opener*: © Christian Liepe/Corbis. 41.2A *Insulin*: Data from PDB 2HIU. Q. X. Hua et al., 1995. *Nat. Struct. Biol.* 2: 129. 41.2A *HGH*: Data from PDB 1HGU. L. Chantalat et al., 1995. *Protein Pept. Lett.* 2: 333. 41.3 *Snake*: © Ameng Wu/istock. 41.4 *Prolactin*: Data from PDB 1RW5. K. Teilum et al., 2005. *J. Mol. Biol.* 351: 810. 41.4 *Fish*: © Alaska Stock LLC/Alamy. 41.4 *Amphibian*: © Gustav W. Verderber/Visuals Unlimited, Inc. 41.4 *Birds*: © Dave Cole/Alamy. 41.4 *Mammals*: © Ale Ventura/PhotoAlto/Photolibrary.com. 41.12A: © Ed Reschke/Getty Images. 41.13: © Scott Camazine/Photo Researchers, Inc. 41.18: Courtesy of Gerhard Heldmaier, Philipps U.

Chapter 42 *Opener*: © Heritage Images/Corbis. 42.8: © Steve Gschmeissner/Photo Researchers, Inc. Page 861: © Science Photo Library/Photo Researchers, Inc.

Chapter 43 *Opener*: David McIntyre. 43.1A: © P&R Photos/AGE Fotostock. 43.1B: © Constantinos Petrinos/Naturepl.com. 43.2A: © Patricia J. Wynne. 43.5: © david gregs/Alamy. 43.6: © Jane Gould/Alamy. 43.7A: © Morales/AGE Fotostock. 43.7B: © Dave Watts/Naturepl.com. 43.9B: © Michael Webb/Visuals Unlimited, Inc. 43.12B: © P. Bagavandoss/Photo Researchers, Inc. 43.15C: © S. I. U. School of Med./Photo Researchers, Inc. 43.17: Courtesy of The Institute for Reproductive Medicine and Science of Saint Barnabas, New Jersey.

Chapter 44 *Opener*: © Mads Abildgaard/istock. 44.1: Courtesy of Richard Elinson, U. Toronto. 44.3A *left*: From H. W. Beams and R. G. Kessel, 1976. *American Scientist* 64: 279. 44.3A *center*, *right*: © Dr. Lloyd M. Beidler/Photo Researchers, Inc. 44.3B: From H. W. Beams and R. G. Kessel, 1976. *American Scientist* 64: 279. 44.3C: Courtesy of D. Daily and W. Sullivan. 44.4B *left*, *center*: © Dr. Yorgos Nikas/Science Source/Photo Researchers, Inc. 44.4B *right*: © Petit Format/Science Source/Photo Researchers, Inc. 44.4C: From J. G. Mulnard, 1967. *Arch. Biol.* (Liege) 78: 107. Courtesy of J. G. Mulnard. 44.14D: Courtesy of K. W. Tosney and G. Schoenwolf. 44.15B: Courtesy of K. W. Tosney. 44.19A: © CNRI/SPL/Photo Researchers, Inc. 44.19B: © Dr. G. Moscoso/SPL/Photo Researchers, Inc. 44.19C: © Tissuepix/SPL/Photo Researchers, Inc. 44.19D: © Petit Format/Photo Researchers, Inc.

Chapter 45 *Opener*: © John Birdsall/AGE Fotostock. 45.3B: © C. Raines/Visuals Unlimited, Inc. 45.4: Courtesy of Philip Haydon. 45.7: From A. L. Hodgkin & R. D. Keynes, 1956. *J. Physiol.* 148: 127.

Chapter 46 *Opener*: © Casey K. Bishop/Shutterstock. 46.4: David McIntyre. 46.8A: © Dr. Fred Hossler/Visuals Unlimited, Inc. 46.13A: © Cheryl Power/Photo Researchers, Inc. 46.16: © Omikron/Science Source/Photo Researchers, Inc.

Chapter 47 *Opener*: © Bluesky International Limited. 47.7: Photo from "Brain: The World Inside Your Head," © Evergreen Exhibitions. 47.12A: Courtesy of Compumedics. 47.14: © Wellcome Dept. of Cognitive Neurology/SPL/Photo Researchers, Inc.

Chapter 48 *Opener*: © Oxford Scientific/Getty Images. 48.1 *Micrograph*: © Frank A. Pepe/Biological Photo Service. 48.2: © Tom Deerinck/Visuals Unlimited, Inc. 48.4: © Kent Wood/Getty Images. 48.7: © Manfred Kage/Photo Researchers, Inc. 48.8: © SPL/Photo Researchers, Inc. 48.11: Courtesy of Jesper L. Andersen. 48.18: © Robert Brons/Biological Photo Service.

Chapter 49 *Opener*: © Steve Bloom, stevebloom.com. 49.1A: © Ross Armstrong/AGE Fotostock. 49.1B: © WaterFrame/Alamy. 49.1C: © Photoshot Holdings Ltd/Alamy. 49.4B: © Andrew Darrington/Alamy. 49.4C: Courtesy of Thomas Eisner, Cornell U. 49.10 *Bronchi*: © SPL/Photo Researchers, Inc. 49.10 *Alveoli*: © P. Motta/Photo Researchers, Inc. 49.17: After C. R. Bainton, 1972. *J. Appl. Physiol.* 33: 775.

Chapter 50 *Opener*: © NBAE/Getty Images. 50.10A: © Brand X Pictures/Alamy. 50.11: After N. Campbell, 1990. *Biology*, 2nd Ed., Benjamin Cummings. 50.12B: © CNRI/Photo Researchers, Inc. 50.14: © Science Source/Photo Researchers, Inc. 50.17A: © Chuck Brown/Science Source/Photo Researchers, Inc. 50.17B: © Biophoto Associates/Science Source/Photo Researchers, Inc.

Chapter 51 *Opener*: © Marilyn "Angel" Wynn/Nativestock.com. 51.1: Courtesy of Andrew D. Sinauer. 51.3: © Dai Kurokawa/epa/Corbis. 51.8C *Microvilli*: © Biophoto Associates/Photo Researchers, Inc. 51.18: © Science VU/Jackson/Visuals Unlimited, Inc.

Chapter 52 *Opener*: © Michael & Patricia Fogden/Corbis. 52.2B: © Morales/AGE Fotostock. 52.8A: © CNRI/SPL/Photo Researchers, Inc. 52.8B: © Susumu Nishinaga/Photo Researchers, Inc. 52.8C: © Science Photo Library RF. 52.8D: © Dr. Donald Fawcett & D. Friend/Visuals Unlimited, Inc. 52.13: © Hank Morgan/Photo Researchers, Inc. 52.17: From L. Bankir & C. de Rouffignac, 1985. *Am. J. Physiol.* 249: R643-R666. Courtesy of Lise Bankir, INSERM Unit, Hôpital Necker, Paris.

Chapter 53 *Opener*: © E. R. Degginger/Photo Researchers, Inc. 53.2A: © FLPA/Alamy. 53.3B: © Maximilian Weinzierl/Alamy. 53.6A: © Nina Leen/Time Life Pictures/Getty Images. 53.6B: © Wayne Lynch/All Canada Photos/Corbis. 53.10A: © Interfoto/Alamy. 53.10B: ©

Shaun Cunningham/Alamy. 53.10C: © Shattil & Rozinski/Naturepl.com. 53.12A: © Jeremy Woodhouse/Photolibrary.com. 53.12B: © Elvele Images Ltd/Alamy. 53.16B: © Momatiuk - Eastcott/Corbis. 53.19: © Tui De Roy/Minden Pictures. 53.22: © J. Jarvis/Visuals Unlimited, Inc. 53.23B: © Richard R. Hansen/Photo Researchers, Inc.

Chapter 54 *Opener*: © Bob Gibbons/OSF/Getty Images. 54.6: David McIntyre. *Tundra, left*: © John Schwieder/Alamy. *Tundra, right*: © Imagestate Media Partners Limited - Impact Photos/Alamy. *Ptarmigan*: © Rolf Hicker Photography/Alamy. *Boreal, left*: © John E Marriott/All Canada Photos/AGE Fotostock. *Boreal, right*: © Robert Harding Picture Library Ltd/Alamy. *Boreal owl*: © John Cancalosi/Alamy. *Temperate deciduous, left*: © Carr Clifton/Minden Pictures. *Temperate deciduous, right*: © Jack Milchanowski/AGE Fotostock. *Butterfly*: © Rick & Nora Bowers/Alamy. *Temperate grasslands, left*: © Tim Fitzharris/Minden Pictures. *Temperate grasslands, right*: © Mikhail Yurenkov/Alamy. *Rhea*: © Joe McDonald/Visuals Unlimited, Inc. *Hot desert, left*: © Peter Lilja/AGE Fotostock. *Hot desert, right*: © David M. Schrader/Shutterstock. *Beetle*: © Michal P Fogden/Bruce Coleman USA/AGE Fotostock. *Cold desert, left*: David McIntyre. *Cold desert, right*: © Tui De Roy/Minden Pictures. *Lizard*: Courtesy of Andrew D. Sinauer. *Chaparral*: © Carl W. May/Biological Photo Service. *Maquis*: © José Antonio Jiménez/AGE Fotostock. *Sugarbird*: © Nigel Dennis/AGE Fotostock. *Thorn forest*: © Cyril Ruoso/JH Editorial/Minden Pictures/Corbis. *Savanna*: © Tui De Roy/Minden Pictures. *Termite mound*: Courtesy of Andrew D. Sinauer. *Tropical deciduous, left*: © Staffan Widstrand/Naturepl.com. *Tropical deciduous, right*: © Pete Oxford/Naturepl.com. *Hummingbird*: © Rolf Nussbaumer/Naturepl.com. *Tropical evergreen, left*: © Luciano Candisani/Minden Pictures. *Tropical evergreen, right*: Courtesy of Andrew D. Sinauer. *Sloth*: © Kevin Schafer/AGE Fotostock. 54.9 *Kelp*: © Mauricio Handler/National Geographic Society/Corbis. 54.9 *Fish*: © Bluegreen Pictures/Alamy. 54.9 *Vent*: Courtesy of P. Rona/OAR/NOAA National Undersea Research Program. 54.10 *Reeds*: © Wild Wonders of Europe/Maitland/Naturepl.com. 54.10 *Osprey*: © Robert Bannister/Alamy. 54.13A: © Chris Mattison/Alamy.

Chapter 55 *Opener*: Courtesy of Dave Klein. 55.1A: Courtesy of Andrew D. Sinauer. 55.1B: © Pete Oxford/Minden Pictures/Corbis. 55.1C: © Heidi & Hans-Juergen Koch/Minden Pictures. 55.3 *inset*: © Kitch Bain/Shutterstock. 55.4A: © Rolf Hicker/All Canada Photos/Corbis. 55.4B: © Ilmarin/Shutterstock. 55.4C: © Ilya Bushuev/istock. 55.6: © M. & C. Photography/Getty Images. 55.10: © Wolfgang Pölzer/Alamy. 55.11 *inset*: © T. W. Davies/California Academy of Sciences. 55.12: David McIntyre. 55.13A: © Rick & Nora Bowers/Alamy. 55.13B: Courtesy of Colin Chapman. 55.15: © Kathie Atkinson.

Chapter 56 *Opener*: © Mark Moffett/Minden Pictures. 56.1B *upper*: Courtesy of Andrew D. Sinauer. 56.1B *lower*: © Michael Mauro/Minden Pictures. 56.2: © Jodi Jacobson/Shutterstock. 56.3A: © E. A. Kuttapan/Naturepl.com. 56.3B: © Stephen Dalton/Minden Pictures. 56.4A: David McIntyre. 56.4B: © Jason Bazzano/AGE Fotostock. 56.5A: © Cathy Keifer/Shutterstock. 56.5B: © Kelpfish/Shutterstock. 56.5C: © Nicola Vernizzi/Shutterstock. 56.6A *left*: © Georgette Douwma/Naturepl.com. 56.6A *right*: © WaterFrame/Alamy. 56.6B: Courtesy of Andrew Brower. 56.7: David McIntyre. 56.8A: © Eye of Science/Photo Researchers, Inc. 56.8B: © Wolfgang Kaehler/Corbis. 56.9A, C: Courtesy of Erich G. Vallery, USDA Forest Service/Bugwood.org. 56.9B: Courtesy of James Denny Ward, USDA Forest Service/Bugwood.org. 56.10: © Neil Lucas/Naturepl.com. 56.11: © blickwinkel/Alamy. 56.12 *left*: © Trevor Sims/AGE Fotostock. 56.12 *center, right*: Courtesy of Olle Pellmyr. 56.13: © ANT Photo Library/Photo Researchers, Inc. 56.14 *left*: © Krystyna Szulecka Photography/Alamy. 56.14 *right*: © Pete Oxford/Minden Pictures. 56.16: © Mark Moffett/Minden Pictures.

Chapter 57 *Opener*: © Dr. James L. Castner/Visuals Unlimited, Inc. 57.1A: © Bill Beatty/Visuals Unlimited, Inc. 57.1B: © aaron peterson.net/Alamy. 57.6 *inset*: © Donald M. Jones/Minden Pictures. 57.7: © Chris Howes/Wild Places Photography/Alamy. 57.12: Courtesy of E. O. Wilson. 57.15: Courtesy of the National Park Service.

Chapter 58 *Opener*: Courtesy of Goddard SVS/NASA. 58.3: From Zhao et al. 2006. *J. Geophys. Res.—Biogeosciences* 111: G01002. 58.4: Image by Robert Simmon, NASA GSFC Earth Observatory, based on data provided by Watson Gregg, NASA GSFC. 58.13: Courtesy of the Scientific Visualization Studio at NASA Goddard Space Flight Center.

Chapter 59 *Opener*: © Thomas Mangelsen/Minden Pictures. 59.1: © Mark Hallett Paleoart/Photo Researchers, Inc. 59.2B: © Aurora Photos/Alamy. 59.6B: Richard Bierregaard, Courtesy of the Smithsonian Institution, Office of Environmental Awareness. 59.7A: © Xi Zhinong/Minden Pictures/Corbis. 59.7B: © Georgette Douwma/Naturepl.com. 59.8: © John Mitchell/Photo Researchers, Inc. 59.9: © John Worrall. 59.12B: © Craig Hosterman/Shutterstock. 59.12C: © Lee Prince/Shutterstock. 59.13: © Raymond Gehman/Corbis. 59.15: Courtesy of Christopher Baisan and the Laboratory of Tree-Ring Research, U. Arizona, Tucson. 59.16: After S. H. Reichard & C. W. Hamilton, 1997. *Conservation Biology* 11: 193. 59.17: © Peter Johnson/Corbis. 59.18B: Courtesy of the Working for Water Programme. 59.19A: Courtesy of the U.S. Fish and Wildlife Service. 59.19B: © Tom Vezo/Naturepl.com.

Index

Numbers in *italic* indicate the information will be found in a figure, caption, or table.

B

C

D

E

F

H

J

K

N

O

P

T

W

ONLINE RESOURCES

The following online resources are referenced throughout the textbook, on the pages indicated. Each is available via an instant access code (QR code), a direct Web address, and in BioPortal at **yourBioPortal.com**. (Many additional resources are available within BioPortal—see the inside front cover for more information.)

MEDIA CLIPS

ANIMATED TUTORIALS

ONLINE RESOURCES continued

The following online resources are referenced throughout the textbook, on the pages indicated. Each is available via an instant access code (QR code), a direct Web address, and in BioPortal at **yourBioPortal.com**. (Many additional resources are available within BioPortal—see the inside front cover for more information.)

ANIMATED TUTORIALS (continued)